SIXTH EDITION

Manual of
Clinical Microbiology

SIXTH EDITION

Manual of
Clinical Microbiology

EDITOR IN CHIEF

PATRICK R. MURRAY

Departments of Pathology and Medicine, Washington University
School of Medicine, St. Louis, Missouri

EDITORS

ELLEN JO BARON

Department of Medicine, University of California,
Los Angeles, California

MICHAEL A. PFALLER

Department of Pathology, University of Iowa Hospitals and Clinics,
Iowa City, Iowa

FRED C. TENOVER

Hospital Infections Program, Centers for Disease Control and Prevention,
Atlanta, Georgia

ROBERT H. YOLKEN

Department of Pediatric Infectious Diseases, Johns Hopkins Hospital,
Baltimore, Maryland

ASM PRESS
Washington, D.C.

Copyright © 1970, 1974, 1980, 1985, 1991, 1995
American Society for Microbiology
1325 Massachusetts Avenue, N.W.
Washington, DC 20005

Library of Congress Cataloging-in-Publication Data

Manual of clinical microbiology.—6th ed./editor in chief, Patrick R. Murray:
  editors, Ellen Jo Baron ... [et al.]
    p. cm.
  Includes bibliographical references and index.
  ISBN 1-55581-086-1 (hard cover)
  1. Diagnostic microbiology—Handbooks, manuals, etc.   2. Medical microbiology—Handbooks,
manuals, etc.   I. Murray, Patrick R.   II. American Society for Microbiology.
  [DNLM: 1. Microbiology.   QW 4 M294 1995]
QR67.M36 1995
616'.01—dc20
DNLM/DLC
for Library of Congress

94-44246
CIP

10 9 8 7 6 5 4 3 2 1

Cover figure: *Nocardia asteroides* complex in an expectorated sputum specimen stained with fluorescein.
The photograph was taken with a Zeiss fluorescent microscope at a magnification of ×1,000.
(*Courtesy of Patrick R. Murray.*)

Contents

Emeritus Editors

ALBERT BALOWS
EDITOR IN CHIEF
Emory University School of Medicine, Atlanta, Georgia

WILLIAM J. HAUSLER, JR.
University of Iowa Hygienic Laboratory, Iowa City, Iowa

KENNETH L. HERRMANN
Centers for Disease Control and Prevention (retired),
Atlanta, Georgia

HENRY D. ISENBERG
Long Island Jewish Medical Center, Long Island Campus
for Albert Einstein College of Medicine, New Hyde Park,
New York

H. JEAN SHADOMY
Departments of Pathology and Dermatology, Emory University
School of Medicine, Atlanta, Georgia

Editorial Board

Contributors

SHARON L. ABBOTT
Microbial Diseases Laboratory, California Department of Health Services, Berkeley, CA 94704

STEPHEN D. ALLEN
Department of Pathology, Indiana University Hospital, Indianapolis, IN 46202

ROBERT D. ARBEIT
Research Service, Veterans Affairs Medical Center, and Departments of Medicine and Microbiology, Boston University School of Medicine, Boston, MA 02130

ANN M. ARVIN
Department of Pediatrics, Stanford University School of Medicine, Stanford, CA 94305

LAWRENCE R. ASH
Department of Epidemiology, School of Public Health, University of California at Los Angeles, Los Angeles, CA 90024

DAVID M. ASHER
Laboratory of Central Nervous System Studies, National Institute of Neurological Disorders and Stroke, Bethesda, MD 20892

TAMMY L. BANNERMAN
Centers for Disease Control and Prevention, Atlanta, GA 30333

EDWARD BARKER
Cancer Research Institute, University of California, San Francisco, CA 94143

SUSAN W. BARNETT
Department of Viral Immunobiology, Chiron Corporation, 4560 Horton Street, Emeryville, CA 94608

ELLEN JO BARON
Department of Medicine, University of California, Los Angeles, CA 90024

BLAINE L. BEAMAN
Department of Medical Microbiology and Immunology, University of California School of Medicine, Davis, CA 95616

JACQUES BILLE
Laboratoire de Bactériologie Médicale, CHUV-BH 19, CH-1011, Lausanne, Switzerland

KAREN K. BIRON
Division of Virology, Burroughs Wellcome Co., Research Triangle Park, NC 27709

LORRAINE V. JONES BRANDO
Johns Hopkins University School of Medicine, Baltimore, MD 21205

DON J. BRENNER
Emerging Bacterial and Mycotic Diseases Branch, Centers for Disease Control and Prevention, Atlanta, GA 30333

DAVID A. BRUCKNER
Clinical Microbiology, Department of Pathology and Laboratory Medicine, UCLA Medical Center, Los Angeles, CA 90024

FRANK L. BRYAN
Food Safety Consultation and Training, Lithonia, GA 30058

SANDRA BULLOCK-IACULLO
Centers for Disease Control and Prevention, Atlanta, GA 30333

JOHN D. BURCZAK
Abbott Laboratories, Abbott Diagnostic Division, One Abbott Park Road, Abbott Park, IL 60064

WILLY BURGDORFER
Rocky Mountain Laboratories, National Institute of Allergy and Infectious Diseases, Hamilton, MT 59840

JOSEPH M. CAMPOS
Department of Laboratory Medicine, Children's National Medical Center, Washington, DC 20010

AMY M. CARNAHAN
Department of Medical and Research Technology, University of Maryland School of Medicine, Baltimore, MD 21201

JEAN B. CEDERNA
Division of Infectious Diseases, Department of Internal Medicine, University of Iowa College of Medicine, Iowa City, IA 52242

KIMBERLE CHAPIN
Department of Pathology, University of South Alabama
Medical Center, Mobile, AL 36617

MAX A. CHERNESKY
McMaster University Regional Virology and Chlamydiology
Laboratory, St. Joseph's Hospital, and Faculty of Health
Sciences, McMaster University, Hamilton, Ontario L8N
4A6, Canada

MARY L. CHRISTENSEN
Department of Pathology, Northwestern University Medical
School, Chicago, IL 60611, and Virology Laboratory,
Department of Pathology, Children's Memorial Hospital,
Chicago, IL 60614

JILL E. CLARRIDGE
Department of Pathology and Microbiology and
Immunology, Baylor College of Medicine, and Pathology and
Laboratory Medicine Service, Veterans Affairs Medical
Center, Houston, TX 77030

PATRICE COURVALIN
Unité des Agents Antibactériens, Institut Pasteur, 75724
Paris Cedex 15, France

NANCY J. COX
Influenza Branch, Centers for Disease Control and
Prevention, Atlanta, GA 30333

RICHARD F. D'AMATO
Division of Clinical Microbiology and Infection Control
and Environmental Health, Catholic Medical Center of
Brooklyn and Queens, Inc., Jamaica, NY 11432

GREGORY A. DASCH
Infectious Diseases Department, Naval Medical Research
Institute, Bethesda, MD 20889

JACQUELINE E. DAWSON
Viral and Rickettsial Zoonoses Branch, Division of Viral and
Rickettsial Diseases, National Center for Infectious Diseases,
Centers for Disease Control and Prevention, Atlanta,
GA 30333

GERALD A. DENYS
Department of Pathology and Laboratory Medicine,
Methodist Hospital, Indianapolis, IN 46206

JULES L. DIENSTAG
Harvard Medical School and Massachusetts General
Hospital, Boston, MA 02114

DENNIS M. DIXON
Division of Microbiology and Infectious Diseases, National
Institute of Allergy and Infectious Diseases, Bethesda,
MD 20892

GARY V. DOERN
Clinical Microbiology Laboratories, University of
Massachusetts Medical Center, Worcester, MA 01655

ANA ESPINEL-INGROFF
Division of Infectious Diseases, Medical College of Virginia,
Richmond, VA 23298

JOSEPH J. ESPOSITO
Division of Viral and Rickettsial Diseases, National Center
for Infectious Diseases, Centers for Disease Control and
Prevention, Atlanta, GA 30333

RICHARD R. FACKLAM
Streptococcus Laboratory, Childhood and Vaccine
Preventable Diseases Branch, Division of Bacterial and
Mycotic Diseases, Centers for Disease Control and
Prevention, Atlanta, GA 30333

J. J. FARMER III
Emerging Bacterial and Mycotic Diseases Branch, Division
of Bacterial and Mycotic Diseases, National Center for
Infectious Diseases, Centers for Disease Control and
Prevention, Atlanta, GA 30333

MARY JANE FERRARO
Clinical Microbiology Laboratory, Massachusetts General
Hospital, and Departments of Pathology and Medicine,
Harvard Medical School, Boston, MA 02114

SYDNEY M. FINEGOLD
Medical Service, VA Medical Center West Los Angeles,
UCLA School of Medicine, Los Angeles, CA 90073

BETTY A. FORBES
Department of Clinical Pathology, State University of New
York Health Science Center, Syracuse, NY 13210

HARVEY M. FRIEDMAN
Department of Medicine, University of Pennsylvania School
of Medicine, and Clinical Virology Laboratory, Children's
Hospital of Philadelphia, Philadelphia, PA 19104

THOMAS R. FRITSCHE
Departments of Laboratory Medicine and Microbiology,
University of Washington School of Medicine, Seattle,
WA 98195

ROBERT A. FROMTLING
Merck Research Laboratories, Rahway, NJ 07065

PAMELA A. FUSELIER
Microbiology Specialists Inc., 8911 Interchange Drive,
Houston, TX 77054

LYNNE S. GARCIA
Clinical Microbiology, Department of Pathology and
Laboratory Medicine, UCLA Medical Center, Los Angeles,
CA 90024

ANNE A. GERSHON
Department of Pediatrics, Columbia University College of
Physicians and Surgeons, New York, NY 10032

MARY J. R. GILCHRIST
Clinical Microbiology, Veterans Affairs Medical Center, and
Pathology and Laboratory Medicine, University of
Cincinnati College of Medicine, Cincinnati, OH 45220

PETER H. GILLIGAN
Clinical Microbiology-Immunology Laboratories, University
of North Carolina Hospitals, Departments of Microbiology
Immunology and Pathology, University of North Carolina
School of Medicine, Chapel Hill, NC 27514

KATHARINE K. GRADY
Centers for Disease Control and Prevention, Atlanta,
GA 30333

PAUL A. GRANATO
Department of Microbiology and Immunology, State
University of New York Health Science Center, Syracuse,
NY 13210

LARRY D. GRAY
Department of Pathology and Laboratory Services, Bethesda Hospitals, Cincinnati, OH 45242, and Department of Pathology and Laboratory Medicine, University of Cincinnati College of Medicine, Cincinnati, OH 45267

DIETER H. M. GRÖSCHEL
Department of Pathology, University of Virginia Health Sciences Center, Charlottesville, VA 22908

W. KEITH HADLEY
Department of Laboratory Medicine, University of California and San Francisco General Hospital, San Francisco, CA 94110

KEVIN C. HAZEN
Department of Pathology, University of Virginia Health Sciences Center, Charlottesville, VA 22908

GEORGE R. HEALY
905 Vistavia Circle, Decatur, GA 30033
(Centers for Disease Control and Prevention [retired])

JOHN E. HERRMANN
Division of Infectious Diseases and Immunology, University of Massachusetts Medical School, Worcester, MA 01655

LOREEN A. HERWALDT
Department of Internal Medicine, University of Iowa Hospitals and Clinics, Iowa City, IA 52242

JOHN C. HIERHOLZER
Respiratory and Enteric Viruses Branch, Division of Viral and Rickettsial Diseases, Center for Infectious Diseases, Centers for Disease Control and Prevention, Atlanta, GA 30333

SHARON L. HILLIER
Department of Obstetrics, Gynecology, and Reproductive Sciences, University of Pittsburgh/Magee-Womens Hospital, Pittsburgh, PA 15213

JANET A. HINDLER
Clinical Microbiology Laboratory, UCLA Medical Center, Los Angeles, CA 90024

RICHARD L. HODINKA
Departments of Pathology and Pediatrics, Clinical Virology Laboratory, Children's Hospital of Philadelphia and University of Pennsylvania School of Medicine, Philadelphia, PA 19104

LARRY A. HOLCOMB
University Hygienic Laboratory, University of Iowa, Iowa City, IA 52242

F. BLAINE HOLLINGER
Baylor College of Medicine, Houston, TX 77030

DANNIE G. HOLLIS
Division of Bacterial and Mycotic Diseases, Centers for Disease Control and Prevention, Atlanta, GA 30333

BARRY HOLMES
National Collection of Type Cultures, Central Public Health Laboratory, London NW9 5HT, England

HARVEY T. HOLMES
Microbiology Laboratories, St. Joseph Mercy Hospital, Pontiac, MI 48053

CLARK B. INDERLIED
Clinical and Molecular Microbiology Laboratory, Department of Pathology and Laboratory Medicine, Childrens Hospital Los Angeles, and Department of Pathology, University of Southern California, Los Angeles, CA 90027

HENRY D. ISENBERG
Long Island Jewish Medical Center, Long Island Campus for Albert Einstein College of Medicine, New Hyde Park, NY 11042

PETER B. JAHRLING
U.S. Army Medical Research Institute of Infectious Diseases, Fort Detrick, Frederick, MD 21702

J. MICHAEL JANDA
Microbial Diseases Laboratory, California Department of Health Services, Berkeley, CA 94704

ROBERT C. JERRIS
Clinical Microbiology, DeKalb Medical Center, Decatur, GA 30033

JAMES H. JORGENSEN
Department of Pathology, University of Texas Health Science Center, San Antonio, TX 78284

HANNELE R. JOUSIMIES-SOMER
Anaerobe Reference Laboratory, National Public Health Institute, Helsinki, Finland

JULIUS KANE
Daniel Medical Laboratories, 1111 Flint Road, Unit 1, Downsview, Ontario M3J 3C7, and Department of Microbiology, St. Joseph Health Centre, Toronto, Ontario M6R 1B5, Canada

ARNOLD F. KAUFMANN
Emerging Bacterial and Mycotic Diseases Branch, Division of Bacterial and Mycotic Diseases, National Center for Infectious Diseases, Centers for Disease Control and Prevention, Atlanta, GA 30333

MICHAEL J. KENNEDY
Animal Health Discovery Research, Veterinary Infectious Diseases, The Upjohn Company, Kalamazoo, MI 49001

NANCY B. KIVIAT
Departments of Pathology and Epidemiology, University of Washington, Seattle, WA 98104

WESLEY E. KLOOS
Department of Genetics, North Carolina State University, Raleigh, NC 27695

JOAN S. KNAPP
Division of Sexually Transmitted Diseases Laboratory Research, Centers for Disease Control and Prevention, Atlanta, GA 30333

LAURA A. KOUTSKY
Departments of Pathology and Epidemiology, University of Washington, Seattle, WA 98104

JOHN M. KRAMER
Food Hygiene Laboratory, Central Public Health Laboratory, London NW9 5HT, United Kingdom

DAVID LAKATUA
Department of Pathology, St. Paul-Ramsey Medical Center, St. Paul, MN 55101

MARIE L. LANDRY
Clinical Virology Laboratory, Yale New Haven Hospital, New Haven, CT 06504, and Virology Reference Laboratory, Veterans Affairs Medical Center, West Haven, CT 06516

DAVISE H. LARONE
Department of Pathology, Lenox Hill Hospital, New York, NY 10021

SANDRA A. LARSEN
Division of Sexually Transmitted Diseases Research, Centers for Disease Control and Prevention, Atlanta, GA 30333

PHILIP LaRUSSA
Department of Pediatrics, Columbia University College of Physicians and Surgeons, New York, NY 10032

HELEN LEE
Abbott Laboratories, Abbott Diagnostic Division, 1401 Sheridan Road, North Chicago, IL 60064

DAVID A. LENNETTE
Virolab, Inc., 1204 Tenth Street, Berkeley, CA 94710

EVELYNE T. LENNETTE
Virolab, Inc., 1204 Tenth Street, Berkeley, CA 94710

JAMES D. MacLOWRY
Department of Laboratories, Group Health Association, 1709 New York Avenue, N.W., Washington, DC 20006

JAMES B. MAHONY
McMaster University Regional Virology and Chlamydiology Laboratory, St. Joseph's Hospital, and Faculty of Health Sciences, McMaster University, Hamilton, Ontario L8N 4A6, Canada

EUGENE O. MAJOR
Laboratory of Molecular Medicine and Neuroscience, National Institute of Neurological Disorders and Stroke, Bethesda, MD 20892

MARIO J. MARCON
Department of Laboratory Medicine, Children's Hospital, and Departments of Pathology and Pediatrics, Ohio State University College of Medicine, Columbus, OH 43205

FREDERIC JOHN MARSIK
Becton Dickinson Microbiology Systems, 250 Schilling Circle, Cockeysville, MD 21030

ROBERT F. MASSUNG
Division of Viral and Rickettsial Diseases, National Center for Infectious Diseases, Centers for Disease Control and Prevention, Atlanta, GA 30333

JOSEPH E. McDADE
National Center for Infectious Diseases, Centers for Disease Control and Prevention, Atlanta, GA 30333

MICHAEL R. McGINNIS
Center for Tropical Diseases, Department of Pathology, University of Texas Medical Branch at Galveston, Galveston, TX 77555

JOHN E. McGOWAN, JR.
Clinical Microbiology Laboratory, Grady Memorial Hospital, Atlanta, GA 30335

JAMES C. McLAUGHLIN
Departments of Pathology and Microbiology, University of New Mexico School of Medicine, Albuquerque, NM 87106

JOSEPH L. MELNICK
Division of Molecular Virology, Baylor College of Medicine, Houston, TX 77030

WILLIAM G. MERZ
Microbiology Division, Department of Pathology, Johns Hopkins University, Baltimore, MD 21287

BEVERLY METCHOCK
Clinical Microbiology Laboratory, Grady Memorial Hospital, Atlanta, GA 30335

J. MICHAEL MILLER
Hospital Infections Program, National Center for Infectious Diseases, Centers for Disease Control and Prevention, Atlanta, GA 30333

THOMAS G. MITCHELL
Department of Microbiology and Immunology, Duke University Medical Center, Durham, NC 27710

ROBERT C. MOELLERING, JR.
Department of Medicine, New England Deaconess Hospital, Harvard Medical School, Boston, MA 02215

BERNARD J. MONCLA
Department of Microbiology and Biochemistry, School of Dental Medicine, University of Pittsburgh, Pittsburgh, PA 15261

NELSON P. MOYER
University Hygienic Laboratory, University of Iowa, Iowa City, IA 52242

PATRICK R. MURRAY
Departments of Pathology and Medicine, Washington University School of Medicine, St. Louis, MO 63110

IRVING NACHAMKIN
Department of Pathology and Laboratory Medicine, University of Pennsylvania School of Medicine, Philadelphia, PA 19104

VALERIE L. NG
Department of Laboratory Medicine, University of California and San Francisco General Hospital, San Francisco, CA 94110

FREDERICK S. NOLTE
Emory University Hospital and Department of Pathology and Laboratory Medicine, Emory University School of Medicine, Atlanta, GA 30322

STEVEN J. NORRIS
Departments of Pathology and Laboratory Medicine and Microbiology and Molecular Genetics, University of Texas Medical School at Houston, Houston, TX 77225

CAROLINE M. O'HARA
Hospital Infections Program, National Center for Infectious Diseases, Centers for Disease Control and Prevention, Atlanta, GA 30333

JAMES G. OLSON
Viral and Rickettsial Zoonoses Branch, Division of Viral and Rickettsial Diseases, National Center for Infectious Diseases, Centers for Disease Control and Prevention, Atlanta, GA 30333

ØRJAN OLSVIK
Hospital Infections Program, Centers for Disease Control and Prevention, Atlanta, GA 30333

ANDREW B. ONDERDONK
Brigham and Women's Hospital, Department of Pathology, Harvard Medical School, Boston, MA 02115

THOMAS C. ORIHEL
Department of Tropical Medicine, School of Public Health and Tropical Medicine, Tulane University, New Orleans, LA 70112

BETH E. OSTERGAARD
Section of Clinical Pharmacy, St. Paul-Ramsey Medical Center, St. Paul, MN 55101, and College of Pharmacy, University of Minnesota, Minneapolis, MN 55454

ARVIND A. PADHYE
Emerging Bacterial and Mycotic Diseases Branch, Division of Bacterial and Mycotic Diseases, National Center for Infectious Diseases, Centers for Disease Control and Prevention, Atlanta, GA 30333

JOSEPHINE PALMER
MDS Physicians Services Division, 100 International Boulevard, Etobicoke, Ontario M9W 6J6, Canada

LESTER PASARELL
Department of Pathology, University of Texas Medical Branch, Galveston, TX 77555

J. R. PATTISON
University College London Medical School, London WC1E 6JJ, United Kingdom

JOANNE L. PATTON
Division of Viral and Rickettsial Diseases, National Center for Infectious Diseases, Centers for Disease Control and Prevention, Atlanta, GA 30333

DAVID H. PERSING
Divisions of Clinical Microbiology and Experimental Pathology, Department of Laboratory Medicine and Pathology, Mayo Clinic Foundation, Rochester, MN 55905

LANCE R. PETERSON
Clinical Microbiology Laboratory, Northwestern Memorial Hospital, Chicago, IL 60611

MARTIN PETRIC
Department of Microbiology, Hospital for Sick Children, Toronto, Ontario M5G 1X8, Canada

MICHAEL A. PFALLER
Department of Pathology, University of Iowa Hospitals and Clinics, Iowa City, IA 52242

M. JOHN PICKETT
Department of Microbiology and Molecular Genetics, University of California, Los Angeles, CA 90024

NORMAN PIENIAZEK
Division of Parasitic Diseases, Center for Infectious Diseases, Centers for Disease Control and Prevention, Atlanta, GA 30333

RAYMOND P. PODZORSKI
Department of Pathology, Wayne State University School of Medicine, Detroit, MI 48201

TANJA POPOVIC
Division of Bacterial and Mycotic Diseases, Centers for Disease Control and Prevention, Atlanta, GA 30333

CHARLES G. PROBER
Department of Pediatrics, Stanford University School of Medicine, Stanford, CA 94305

RICHARD QUINTILIANI, JR.
Unité des Agents Antibactériens, Institut Pasteur, 75724 Paris Cedex 15, France

ROSELYN J. RICE
Office of the Director, National Center for Infectious Diseases, Centers for Disease Control and Prevention, Atlanta, GA 30333

MALCOLM D. RICHARDSON
Regional Mycology Reference Laboratory, Department of Dermatology, University of Glasgow and Western Infirmary, Glasgow G66 4UE, United Kingdom

GLENN D. ROBERTS
Section of Clinical Microbiology, Mayo Clinic and Mayo Foundation, Rochester, MN 55905

JOCELYN ROCOURT
Centre National de Référence pour la Lysotypie et le Typage Moléculaire de Listeria, Centre Collaborateur OMS pour la Listériose d'Origine Alimentaire, Institut Pasteur, 75724 Paris Cedex 15, France

PATRICIA A. ROSA
Rocky Mountain Laboratories, National Institute of Allergy and Infectious Diseases, Hamilton, MT 59840

HARLEY A. ROTBART
Departments of Pediatrics and Microbiology, University of Colorado School of Medicine, Denver, CO 80262

JOHN C. ROTSCHAFER
Section of Clinical Pharmacy, St. Paul-Ramsey Medical Center, St. Paul, MN 55101, and College of Pharmacy, University of Minnesota, Minneapolis, MN 55454

KATHRYN L. RUOFF
Microbiology Laboratories, Massachusetts General Hospital, Boston, MA 02114

WILLIAM A. RUTALA
Division of Infectious Diseases, University of North Carolina School of Medicine, Chapel Hill, NC 27599

DANIEL F. SAHM
Department of Pathology, Washington University School of Medicine, St. Louis, MO 63110

MAX SALFINGER
Clinical Mycobacteriology Laboratory, Wadsworth Center for Laboratories and Research, New York State Department of Health, Albany, NY 12201

IRA F. SALKIN
Wadsworth Center for Laboratories and Research, New York State Department of Health, Albany, NY 12201

AIMO A. SALMI
Department of Virology, University of Turku, SF-20520 Turku, Finland

MYRON SASSER
Microbial ID, Inc., Newark, DE 19711

MICHAEL A. SAUBOLLE
Department of Pathology, Good Samaritan Regional Medical Center, Phoenix, AZ 85006

JULIUS SCHACHTER
Chlamydia Research Laboratory, Department of Laboratory Medicine, University of California, San Francisco, San Francisco General Hospital, San Francisco, CA 94110

PETER SCHANTZ
Division of Parasitic Diseases, Center for Infectious Diseases, Centers for Disease Control and Prevention, Atlanta, GA 30333

WILEY A. SCHELL
Medical Mycology Research Center, Department of Medicine, Division of Infectious Diseases and International Health, Duke University Medical Center, Durham, NC 27710, and Mycology Reference Laboratory, Durham Veterans Affairs Medical Center, Durham, NC 27705

RON B. SCHIFMAN
Department of Pathology, University of Arizona, Pathology and Laboratory Medicine Service, Tucson Veterans Affairs Medical Center, Tucson, AZ 85723

TOM G. SCHWAN
Rocky Mountain Laboratories, National Institute of Allergy and Infectious Diseases, Hamilton, MT 59840

DAVID L. SEWELL
Pathology and Laboratory Medicine Service, Veterans Affairs Medical Center, Department of Pathology, Oregon Health Sciences University, Portland, OR 97201

GILLIAN S. SHANKLAND
Regional Mycology Reference Laboratory, Department of Dermatology, University of Glasgow and Western Infirmary, Glasgow G66 4UE, United Kingdom

JESSIE SHIH
Abbott Laboratories, Abbott Diagnostic Division, One Abbott Park Road, Abbott Park, IL 60064

ROBYN Y. SHIMIZU
Clinical Microbiology, Department of Pathology and Laboratory Medicine, UCLA Medical Center, Los Angeles, CA 90024

LYNNE SIGLER
University of Alberta Microfungus Collection and Herbarium, Devonian Botanic Garden and Medical Microbiology and Infectious Diseases, Edmonton, Alberta, Canada T6G 2E1

LEONARD N. SLATER
Department of Medicine, University of Oklahoma Health Sciences Center, Oklahoma City, OK 73104

JAMES W. SMITH
Department of Pathology, Indiana University School of Medicine, Indianapolis, IN 46202

JEAN S. SMITH
Rabies Laboratory, Viral and Rickettsial Zoonoses Branch, Division of Viral and Rickettsial Diseases, National Center for Infectious Diseases, Centers for Disease Control and Prevention, Atlanta, GA 30333

CAROL A. SPIEGEL
Department of Pathology and Laboratory Medicine, University of Wisconsin, and Clinical Microbiology, University of Wisconsin Hospitals and Clinics, Madison, WI 53792

WALTER E. STAMM
Division of Infectious Diseases, Department of Medicine, University of Washington, Harborview Medical Center, Seattle, WA 98104

SAMUEL L. STANLEY, JR.
Departments of Medicine and Molecular Microbiology, Washington University School of Medicine, St. Louis, MO 63110

JACK T. STAPLETON
Division of Infectious Diseases, Department of Internal Medicine, University of Iowa College of Medicine and Iowa City Veterans Affairs Medical Center, Iowa City, IA 52242

SHARON P. STEINBERG
Department of Pediatrics, Columbia University College of Physicians and Surgeons, New York, NY 10032

JOHN A. STEWART
Division of Viral and Rickettsial Diseases, National Center for Infectious Diseases, Centers for Disease Control and Prevention, Atlanta, GA 30333

SCOTT J. STEWART
Laboratory of Intracellular Parasites, Rocky Mountain Laboratories, National Institute of Allergy and Infectious Diseases, Hamilton, MT 59840

BARBARA A. STRAIN
Department of Pathology, University of Virginia Health Sciences Center, Charlottesville, VA 22908

ALEXANDER J. SULZER
Centers for Disease Control and Prevention, Atlanta, GA 30333

PAULA H. SUMMANEN
Research Service, VA Medical Center West Los Angeles, Los Angeles, CA 90073

RICHARD C. SUMMERBELL
Mycology Laboratory, Laboratory Services Branch, Box 9000, Terminal A, Toronto, Ontario M5W 1R5, Canada

BALA SWAMINATHAN
Division of Bacterial and Mycotic Diseases, National Center for Infectious Diseases, Centers for Disease Control and Prevention, Atlanta, GA 30333

JANA M. SWENSON
Hospital Infections Program, Centers for Disease Control and Prevention, Atlanta, GA 30333

ELLA M. SWIERKOSZ
Departments of Pediatrics and Pathology, St. Louis University School of Medicine, St. Louis, MO 63104

DAVID TAYLOR-ROBINSON
MRC Sexually Transmitted Disease Research Group, St. Mary's Hospital, London W2 1NY, United Kingdom

FRED C. TENOVER
Hospital Infections Program, Centers for Disease Control and Prevention, Atlanta, GA 30333

JOHN TICEHURST
Department of Pathology, Johns Hopkins University School of Medicine, Baltimore, MD 21287

DEBRA A. TRISTRAM
Department of Pediatrics, Division of Infectious Diseases, State University of New York at Buffalo School of Medicine and Biomedical Sciences and Children's Hospital of Buffalo, Buffalo, NY 14222

THEODORE F. TSAI
Division of Vector-Borne Infectious Diseases, National Center for Infectious Diseases, Centers for Disease Control and Prevention, Fort Collins, CO 80522

PETER C. B. TURNBULL
Centre for Applied Microbiology and Research, Porton Down, Salisbury, Wiltshire SP4 0JG, United Kingdom

GOVINDA S. VISVESVARA
Division of Parasitic Diseases, Center for Infectious Diseases, Centers for Disease Control and Prevention, Atlanta, GA 30333

ALEXANDER VON GRAEVENITZ
Department of Medical Microbiology, University of Zurich, CH-8028 Zurich, Switzerland

DAVID H. WALKER
Department of Pathology, University of Texas Medical Branch, Galveston, TX 77555

RICHARD J. WALLACE
Department of Microbiology, University of Texas Health Sciences Center, Tyler, TX 75710

THOMAS J. WALSH
Infectious Diseases Section, National Cancer Institute, Bethesda, MD 20892

JOSEPH L. WANER
Pediatric Infectious Diseases, University of Oklahoma Health Sciences Center, Oklahoma City, OK 73190

NANCY G. WARREN
Association of State and Territorial Public Health Laboratory Directors, Washington, DC 20036

JOHN A. WASHINGTON
Department of Clinical Pathology, Cleveland Clinic Foundation, Cleveland, OH 44195

ALICE S. WEISSFELD
Microbiology Specialists Inc., 8911 Interchange Drive, Houston, TX 77054

IRENE WEITZMAN
Clinical Microbiology Service, Columbia Presbyterian Medical Center, New York, NY 10032

DAVID F. WELCH
Department of Pediatrics, University of Oklahoma Health Sciences Center, Oklahoma City, OK 73104

ROBERT C. WELLIVER
Department of Pediatrics, Division of Infectious Diseases, State University of New York at Buffalo School of Medicine and Biomedical Sciences and Children's Hospital of Buffalo, Buffalo, NY 14222

RICHARD P. WENZEL
Department of Internal Medicine, University of Iowa College of Medicine, Iowa City, IA 52242

HANNAH M. WEXLER
Veterans Affairs Medical Center, West Los Angeles, West Los Angeles, CA 90073

ROBBIN S. WEYANT
Emerging Bacterial and Mycotic Diseases Branch, Division of Bacterial and Mycotic Diseases, National Center for Infectious Diseases, Centers for Disease Control and Prevention, Atlanta, GA 30333

JUDITH C. WILBER
Nucleic Acid Systems, Chiron Corporation, 4560 Horton Street, Emeryville, CA 94608, and Laboratory Medicine, University of California, San Francisco, San Francisco, CA 94143

MARIANNA WILSON
Division of Parasitic Diseases, Center for Infectious Diseases, Centers for Disease Control and Prevention, Atlanta, GA 30333

WASHINGTON C. WINN, JR.
University of Vermont College of Medicine, Medical Center Hospital of Vermont, Burlington, VT 05401

GAIL L. WOODS
Department of Pathology, University of Texas Medical Branch, Galveston, TX 77555

JOSEPH D. C. YAO
Infectious Diseases and Clinical Microbiology, The Vancouver Clinic Inc., 700 N.E. 87th Avenue, Vancouver, WA 98664

THEDI ZIEGLER
Influenza Branch, Centers for Disease Control and Prevention, Atlanta, GA 30333, and Department of Virology, University of Turku, SF-20520 Turku, Finland

Preface

From the first edition of the *Manual of Clinical Microbiology* (MCM), published in 1970, and with each subsequent edition, MCM has been universally recognized as an invaluable reference text for clinical microbiologists, infectious disease specialists, pathologists, medical technologists, clinicians, and students and their teachers. The source of this success is the leadership and expertise of the volume editors, editorial board, and authors. It was with some trepidation that the new volume editors assumed their mantle of responsibility and appointed many new editors and authors for MCM6. This task was made more manageable by the guidance offered by the emeritus editors and in particular the former editor in chief, Albert Balows. The emeritus editors provided an important frame of reference for the development of MCM6. For their help and support we thank them.

We also recognize that the most difficult tasks are frequently unrecognized, so we want to acknowledge the tremendous amount of work performed by the editorial board of MCM6. They had the responsibility of merging the efforts of numerous individual authors into a coordinated, coherent text. The task was particularly formidable because 11 of the 15 editors were newly appointed. No amount of praise can adequately recognize the energy, enthusiasm, and expertise with which all the editors accomplished their responsibilities.

The essence of MCM6 is the information contained in the 123 chapters. We thank the many authors, more than half of whom are new to MCM, for their contributions. As authors ourselves, we recognize the difficulties of working on a project of this nature. No author wants an editor to change his or her message or journalistic style or to insist on reducing the length of a presentation. The fact that all authors were willing to sacrifice their autonomy for a consistently uniform text made our editorial efforts easier. We believe that this coordinated effort between editors and authors produced a book that will serve as an essential clinical microbiology reference text for the present and the future.

The success of any book depends on its timely publication. We thank the many secretaries who helped prepare the texts within the guidelines of the Manual and the time constraints of a publication of this nature. Likewise, we thank the professionals at ASM Press for their help from the planning of MCM6 through the final stages of production. Although many individuals were instrumental in the development of this book, two deserve special recognition: Patrick Fitzgerald and Susan Birch, who worked tirelessly to ensure the successful publication of this manual.

Finally, we hope the users of MCM6 will find that the tradition of excellence of the *Manual of Clinical Microbiology* has been maintained in this edition. That accomplishment will be the most satisfying thank you for all who have worked so hard creating MCM6.

PATRICK R. MURRAY
ELLEN JO BARON
MICHAEL A. PFALLER
FRED C. TENOVER
ROBERT H. YOLKEN

GENERAL ISSUES IN CLINICAL MICROBIOLOGY

I

VOLUME EDITOR
PATRICK R. MURRAY

SECTION EDITOR
SUSAN E. SHARP

Introduction to the Sixth Edition of the *Manual of Clinical Microbiology*

PATRICK R. MURRAY

1

Many changes have occurred since the fifth edition of the *Manual of Clinical Microbiology* (MCM5) was published in 1991. A new editorial board was appointed for MCM6, and through the guidance and wisdom of the emeritus editors, we selected many new section editors and authors for this edition. It might be expected that the new editorial board would seize this opportunity to chart a completely new course for MCM6, seek new challenges and explore new philosophical boundaries, revamp the format of the Manual, and create a new presentation of image and substance. However, I believe we were wise to recognize the strengths of MCM that developed through the preceding five editions. Certainly, the proliferation of microbial discoveries, new diagnostic methods, and new therapeutic challenges have been incorporated into MCM6, but the basic format that provided the foundation of success for the previous five editions has been preserved. We have also retained the fundamental philosophy of the Manual. Like its predecessors, MCM6 is intended to guide clinical microbiologists in the selection, performance, and interpretation of laboratory procedures for diagnostic and therapeutic applications. It is a reference source defining not only what is done in clinical microbiology laboratories but also what these activities signify. It should form a bridge between taxonomy and procedure manuals, focusing on the "when" and "why" of testing as well as the "how."

The American Society for Microbiology has published two new textbooks since MCM5 was published: the *Clinical Microbiology Procedures Handbook* (CMPH), which is a step-by-step handbook of laboratory procedures for the clinical microbiologist, and *Diagnostic Molecular Microbiology*, which introduces the microbiologist to new genetic methods for the detection and identification of organisms. CMPH tells clinical microbiologists what should be done in the diagnostic laboratory, while *Diagnostic Molecular Microbiology* describes emerging technology. We view both of these books as complements to MCM6.

As stated above, MCM6 retains many of the organizational characteristics of MCM5. This edition is subdivided into 10 sections. The first four sections examine general issues in clinical microbiology, laboratory management and regulatory issues, specific diagnostic techniques, and infection control principles and procedures. The next five sections deal with the microbiology of various organisms: bacteriology, rickettsiology, mycology, virology, and parasitology. The final section consists of 14 chapters that describe antimicrobial agents and the methods used to test them. Significantly, the section on quality control, media, reagents, and stains was not retained in MCM6. In part, the limitations of space dictated this decision, but it was also believed that information about quality control should be moved to the laboratory management section of the text; information about culture media, reagents, and stains is now found in numerous other texts.

Although many features were adopted from MCM5, some innovative aspects are noteworthy. We believe the introduction of identification flow charts and the expanded use of summary tables will prove to be beneficial for the reader, particularly in the chapters on specimen handling and processing (chapters 3, 21, 60, and 70), microscopy (chapter 4), and microbial classification and identification (chapters 20, 54, 59, and 69). A chapter on regulatory issues in clinical microbiology (chapter 6) has been added, reflecting the expanding role of federal and local regulations in the clinical laboratories. Previous chapters on laboratory safety and waste management were combined (chapter 7), and the topic of sterilization, disinfection, and antisepsis was subdivided into procedures relevant to the laboratory (chapter 8) and the hospital environment (chapter 19). Seven chapters on immunodiagnostic techniques in MCM5 were consolidated into one chapter (chapter 11), with the realization that much of the information is available in the *Manual of Clinical Laboratory Immunology*. A chapter describing procedures for investigating food-borne diseases was added (chapter 18). The coverage of streptococci and other catalase-negative cocci was expanded into three chapters (chapters 23 to 25), as was the discussion of members of the family *Enterobacteriaceae* (chapters 32 to 34). A chapter reviewing *Bartonella* spp. and the organisms previously classified as *Rochalimaea* spp. has been added, and discussion of *Pneumocystis* spp. has been placed in a chapter in the mycology section, reflecting current thought regarding the taxonomic classification of this organism. Information on the hepatitis viruses has been expanded into four chapters (chapters 90 to 93), and a chapter on free-living amebae has been added to the parasitology section. Descriptions of antimicrobial agents and the relevant susceptibility test procedures have been expanded to include antibacterial agents (chapter 113), agents for mycobacterial infections (chapter 119), antifungal agents (chapter 120), antiviral agents (chapter 121), and antiparasitic agents (chapter 122). A new chapter on genetic methods for detecting resistance (chapter 117) has also been added.

Despite the many changes in editorial board, authorship, and subject matter, we believe that MCM6 faithfully reflects its heritage by providing an important reference source for clinical microbiologists, infectious disease specialists, pathologists, medical technologists, clinicians, and students and their teachers.

Indigenous and Pathogenic Microorganisms of Humans

HENRY D. ISENBERG AND RICHARD F. D'AMATO

2

CHANGING CONCEPTS OF INFECTIOUS DISEASE

Traditionally, clinical microbiologists have been responsible for identifying microorganisms in a clinical specimen as accurately and quickly as possible. There is also a tacit agreement between clinicians and microbiologists that the microorganisms of interest are the so-called pathogens. This attitude suggests that a very small number of microorganisms incite infectious disease processes regardless of their quantity, their portal of entry, or the presence of other microorganisms. Most significant, this view neglects the determinative roles of the host and the environment in the clinically overt manifestations of infectious disease, placing the onus for the disease squarely on the microorganism.

Dubos (7) and Burnet (2) have expanded the appraisal by Theobald Smith (20) of the numerous host and parasite interactions that culminate in clinically overt infectious disease. Studies by these investigators showed that the host's general health, previous contact with particular microorganisms, and medical history and a variety of toxic, traumatic, or iatrogenic insults are significant determinants of infectious disease. Implied is the understanding that, given the opportunity by an unrelated lowering of host resistance, indigenous microflora may become involved in infectious disease. The numerous factors and conditions of microbial and host origins that must be considered in infection have been reviewed recently (11, 12).

The dilemma of the clinical microbiologist is deciding which of the microorganisms isolated from a clinical specimen are involved in disease. There are very few microorganisms to which the term pathogenic can be applied invariably if pathogenic is defined as causing infectious disease at all times (16), yet the clinical microbiologist is expected to decide the causal relationship between a microorganism and a disease even though most of the organisms from clinical material are at best only sometimes pathogenic. All that can be presented here is a very limited discussion and outline of some of the factors that affect the categorizing of microorganisms as harmless, potentially hazardous, or actively involved in the observed pathology.

Microorganisms, i.e., bacteria, yeasts, fungi, protozoa, and viruses, are ubiquitous in and on the human body (17). From birth, people live in a microbial biosphere composed of innumerable microorganisms representing types, variants, strains, species, genera, etc. The composition of this microbial environment is dynamic. Numerous additions and deletions, both qualitative and quantitative, constantly take place. Many populated and sterile areas are found in and on the human body, as are areas that are sparsely populated or that harbor transient microbiota. These temporary habitats of microorganisms include the larynx, trachea, bronchi, accessory nasal sinuses, esophagus, stomach and upper portions of the small intestine, upper urinary tract (including the posterior urethra), and the corresponding distal areas of the male and female genital organs. The persistent finding of numerous microorganisms in these temporarily inhabited areas or in blood or other sterile body sites provides, according to Rosebury (16), as reliable a marker as can be found for the imaginary line that divides health from disease. Conclusions concerning the significance of a microorganism isolated from usually sterile areas must be based on properly obtained specimens that have been properly handled and transported, that have been examined promptly, and that yield a large number of microorganisms unusual in a particular locale or present there occasionally in very small numbers.

One other aspect of microbiology cannot be ignored. Not all of the microorganisms that may be present in a given specimen can be cultivated; others have not yet been cultivated or are very difficult to cultivate. In most instances, the clinical microbiologist insists on examining stained preparations from clinical material, especially when the specimen is obtained directly from a pathological lesion or from suspect areas usually populated by a mixed microbiota. Innovative techniques of molecular biology and immunochemistry are now being developed for direct application to specimens, considerably augmenting the diagnostic capabilities of clinical microbiologists.

No hard and fast rule divides the human microbiota into clear-cut categories of harmless commensal organisms and pathogenic species. The microbial species encountered are listed in the various tables of this chapter along with their general anatomical locales and involvement in disease. The frequency of their presence and their roles in infectious processes are graded in the tables as a guide. The clinical presentation of each patient must be considered when findings based on these tables, which are not exhaustive, are interpreted. For example, etiological agents of disease, such as *Echinococcus* spp., that are usually confined to tissues or organs that require biopsy for a definitive diagnosis are not listed in the tables.

BODY AREAS

Lists of the indigenous and pathogenic microorganisms of various body areas have been compiled (4), but the lists are constantly growing. All of the organisms listed in reference 4 were isolated from lesions or other appropriate specimens, although many times the causal relationship between the microbe recovered and the disease was tenuous at best. Despite the dearth of basic information and the cautious presentation of these listings, they have been accepted as authoritative. Organisms not mentioned as either indigenous or pathogenic are treated with an extreme degree of suspicion or, worse, ignored entirely. This practice still prevails despite considerable improvement in cultural technology and a changed approach to the basic tenets of infectious disease. Therefore, it must be stated categorically that the following compilations are not exhaustive; the omission or inclusion of a microbial species in any category does not imply that it cannot be isolated from a particular or any other body area or that it cannot incite disease, complicate underlying disease, or colonize anatomical abnormalities of congenital, traumatic, or iatrogenic origin. In the communities and especially in the hospitals of developed countries, antimicrobial drugs exert selective pressures that permit the entry of drug-resistant microorganisms into the intimate human biosphere from an inexhaustible pool in nature.

RESPIRATORY TRACT

See also references 4, 9, 15, 16, 21, 22, and 24.

Areas That Usually Harbor Microorganisms

Mouth

The mouth consists of the buccal cavity, teeth, tongue, gingivae, palates, and saliva. The following organisms are commonly found in this region. Various pigmented micrococci, *Staphylococcus epidermidis*, *Staphylococcus aureus*, and *Staphylococcus*-like anaerobic varieties are especially numerous in the saliva and on tooth surfaces, but they are not usually encountered in gingival crevices of healthy individuals. *S. aureus* and the anaerobic cocci are rare in the predentulous mouth. The viridans streptococci, including both the *Streptococcus mitis* and the *Streptococcus salivarius* groups, are ubiquitously distributed on all surfaces of the mouth, with certain species favoring specific sites (10). Enterococci may be present. *Streptococcus pyogenes* is present in a small percentage (5 to 10%) of healthy mouths, appearing usually in throat cultures of asymptomatic individuals. The finding of group A streptococci in healthy individuals is restricted to the saliva or tooth surfaces of adults; peptostreptococci are also commonly found. *Streptococcus pneumoniae* may be present in the predentulous mouth and has been recovered from saliva and tooth surfaces of as many as 25% of healthy adults. Pigmented *Neisseria* spp., *Moraxella* (*Branhamella*) *catarrhalis*, *Veillonella* spp., and aerobic corynebacteria are very common in saliva and in gingival crevices. Several *Actinomyces* spp. can be isolated from the mouth. The lactobacilli are found in saliva, whereas the leptotrichiae are more frequently encountered on tooth surfaces. The family *Enterobacteriaceae* is well represented; *Escherichia coli* and the *Klebsiella-Entero-*

bacter group are among the most common enteric organisms isolated, especially in saliva and on teeth after the initiation of broad-spectrum antimicrobial therapy. At times, these enteric bacteria may be joined by oxidase-producing or nonfermenting gram-negative rods. *Haemophilus influenzae* and *Haemophilus parainfluenzae* are often recovered from healthy mouths, as are *Prevotella*, *Porphyromonas*, *Bacteroides*, and *Fusobacterium* spp.; *Campylobacter sputorum*; and a variety of spirochetes, especially *Treponema denticola* and *Treponema refringens*. *Capnocytophaga* spp. are common colonizers. Several mycoplasmata have been isolated from the saliva of healthy individuals. *Candida albicans* and occasionally other *Candida* spp. exist in the oral cavity without disease production. The protozoa *Entamoeba gingivalis* and *Trichomonas tenax* are examples of ameba and flagellates found in the gingival crevices of some healthy adults.

The large array of microorganisms found in healthy mouths makes it inevitable that an even greater number of microorganisms will be found when infection is present; undoubtedly, some appear in lesions accidentally. Findings from gum lesions, root canals, caries, etc., reflect this state. Debilitated patients or those on prolonged chemotherapy often have lesions of the tongue from which fungi are isolated; other patients with nutritional deficiencies or possible hygienic neglect may have membranous lesions involving the entire oral cavity. Numerous bacteria can be demonstrated in this disease picture, which is often referred to as Vincent's angina. Only very few of the component bacteria have been cultivated, and the causal relationship between the microbes and the disease is not clear. *Candida* spp., especially *C. albicans*, may take advantage of debilitation, various degrees of immunosuppression, chemotherapy, and malnutrition to produce lesions in the oropharynx commonly designated thrush. Viruses involved in respiratory and other disease may be found in the mouth and in saliva (Table 1). In immunocompromised patients, filamentous fungi, particularly the zygomycetes, may cause disease.

Throat, Including the Nasopharynx, Oropharynx, and Tonsils

Micrococci are among the organisms usually found in the throat, nasopharynx, oropharynx, and tonsils. Coagulase-negative staphylococci may be found in the nasopharynx and in the tonsillar areas of children older than infants. *S. aureus* is frequently present in the nasopharynx, the oropharynx, and the tonsils. The tonsils may harbor anaerobic cocci, and viridans streptococci are almost always present in the throat. Various hemolytic streptococci, among them *S. pyogenes*, can be found in the nasopharynx and especially in the tonsillar regions of healthy individuals but in small numbers. Enterococci may be present on tonsils. *S. pneumoniae* and the neisseriae that tolerate 20 to 25°C are present in the healthy throat, sometimes accompanied in the nasopharynx by *Neisseria meningitidis*, usually without any evidence of a disease process or contact with individuals ill with meningococcemia or meningococcal meningitis. *Veillonella* spp., corynebacteria, and *Actinomyces* spp. are present, especially on the tonsils. *E. coli*, the *Klebsiella-Enterobacter* group, and *Proteus* spp. can be found with various frequencies in different areas of the throat. A large number of healthy throats harbor *H. influenzae* or *H. parainfluenzae*, and the tonsils especially contain various *Bacteroides*, *Prevotella*, *Porphyromonas* and *Fusobacterium* spp.; vibrios; and spirochetes. In addition, throat cultures from healthy individuals may indicate the presence of *Bacteroides*

TABLE 1 Microorganisms encountered in respiratory tract specimens

Organism	Frequency of isolation[a]	Disease involvement[b]	Organism	Frequency of isolation[a]	Disease involvement[b]
Absidia spp.	C	2	*Lactobacillus* spp.	B	1
Acinetobacter spp.	B	2	*Legionella* spp.	B	2
Actinomyces spp.	A	2	*Leptotrichia buccalis*	B	1
Adenovirus	B	3			
Aerococcus viridans	B	1	Measles virus	B	3
Agrobacterium tumefaciens	C	2	*Micrococcus* spp.	A	1
Arachnia propionica	B	1	*Moraxella* (*Branhamella*)	B	2
Arcanobacterium haemolyticum	B	3	catarrhalis		
Ascaris lumbricoides larvae	C	3	*Moraxella* spp.	B	1
Aspergillus spp.	B	2	*Mucor* spp.	C	2
			Mumps virus	B	3
Bacillus anthracis	C	3	*Mycobacterium* spp.	B	2
Bacillus spp.	B	1	*Mycobacterium tuberculosis* group	B	3
Bacteroides spp.	A	2			
Bifidobacterium spp.	B	1	*Necator americanus*	C	3
Bipolaris spp.	C	2	*Neisseria gonorrhoeae*	C	3
Blastomyces dermatitidis	C	3	*Neisseria meningitidis*	B	2
Bordetella pertussis	B	3	*Neisseria* spp.	A	1
Borrelia spp.	B	1	*Nocardia* spp.	B	3
Brucella spp.	C	3			
			Papillomaviruses	C	2
Campylobacter spp.	B	1	*Paracoccidioides brasiliensis*	C	3
Candida spp.	B	2	*Paragonimus* ova	C	3
Capnocytophaga spp.	B	2	Parainfluenza viruses	B	2
Cardiobacterium hominis	B	1	*Pasteurella* spp.	C	2
Chlamydia spp.	B	3	*Penicillium* spp.	C	1
Clostridium spp.	B	2	*Peptostreptococcus* spp.	A	2
Coccidioides immitis	B	3	*Pneumocystis carinii*	B	2
Coronavirus	B	2	*Porphyromonas* spp.	A	2
Corynebacterium diphtheriae (toxigenic)	C	3	*Prevotella* spp.	A	2
Corynebacterium spp.	B	1	*Pseudoallescheria boydii*	C	2
Coxiella burnetii	C	3			
Cryptococcus neoformans	C	3	Respiratory syncytial virus	B	3
Cryptosporidium spp.	C	3	Rhinoviruses	B	2
Cytomegalovirus	B	2	*Rhizomucor* spp.	C	2
			Rhizopus spp.	C	2
Echinococcus protoscolices or hooklets	C	3	*Rhodococcus* spp.	C	2
			Rothia dentocariosa	B	1
Eikenella corrodens	B	2	Rubella virus	B	3
Entamoeba gingivalis	B	1			
Entamoeba histolytica trophozoites	C	3	*Sarcinosporon inkin*	B	2
			Selenomonas spp.	C	1
Enterobacteriaceae	B	2	*Sporothrix schenckii*	C	3
Enterococcus spp.	C	1	*Staphylococcus* spp.	A	2
Enterovirus	B	2	*Stomatococcus mucilaginosus*	B	1
Epstein-Barr virus	C	2	*Streptococcus pneumoniae*	B	2
			Streptococcus pyogenes	B	2
Flavobacterium spp.	C	2	*Streptococcus* spp.	A	2
Fonsecaea spp.	C	2	*Strongyloides stercoralis* larvae	C	3
Francisella tularensis	C	3			
Fusobacterium spp.	A	2	*Treponema* spp.	B	2
			Trichomonas tenax	B	1
Gemella spp.	B	1			
			Varicella-zoster virus	B	3
Haemophilus spp.	A	2	*Veillonella* spp.	A	1
Herpes simplex virus	B	3	*Vibrio* spp.	B	1
Histoplasma capsulatum	B	3			
Human immunodeficiency virus	C	3	*Wolinella recta*	B	1
Influenza viruses	B	3	*Xanthomonas maltophilia*	B	2
Kingella spp.	C	2			

[a]A, commonly encountered in clinical specimens; B, occasionally encountered in clinical specimens; C, rarely encountered in clinical specimens.
[b]1, When present, rarely if ever involved in disease production; 2, when present, occasionally involved in disease production; 3, when present, commonly involved in disease production.

pneumosintes (very minute gram-negative rods capable of passing most bacterial filters), mycoplasmas, *C. albicans*, *Candida* spp., and the same protozoa described for the mouth.

Infectious disease of the throat can be caused by several microorganisms. Some lesions may be initiated by viruses that are followed quickly by bacteria and fungi. In most geographical areas, *S. pyogenes* remains the major bacterial pathogenic bacterium involved in throat disease. When it is detected, it is imperative to initiate therapy and prevent sequelae. Pneumococci do not contribute to pathological processes in the throat and indeed constitute part of the usual flora of many healthy adults. Although some hemophiline bacteria are normally present in the throat, *H. influenzae* serogroup B has been involved in epiglottitis, a disease of young children and, rarely, young adults that has a grave prognosis. *S. aureus* may be involved in a variety of disease processes in the throat. Its presence is probably the most difficult to interpret. Repeated isolation of this bacterium in large numbers from a carefully cultured lesion diminishes doubts of its significance. As a rule, *S. aureus* is present in small numbers in the healthy throat. Thrush, already described for the mouth, may also involve the throat and esophagus. Candidal lesions in the throat may accompany a variety of drug regimens and may also occur in patients with debilitating and neoplastic diseases and immunosuppression and in neonates. The throat, and especially the tonsils, may be a site for primary lesions of syphilis. *Neisseria gonorrhoeae* may also be detected in the throat and may give rise to a local disease in this area that can rapidly disseminate (8). Viral pharyngitis may be caused by adenovirus, type A coxsackieviruses, other enteroviruses, herpes simplex viruses, Epstein-Barr virus, rhinoviruses, and parainfluenza and influenza viruses (3, 6).

Nose

The nares are the usual habitat of the staphylococci. *S. epidermidis*, *S. aureus*, and other species are recovered with great frequency from this site. Although both organisms may not be present at all times in any one individual, their recovery from the nares is to be expected. On rare occasions, viridans streptococci can be isolated from the nasal passages. In children, and especially in infants, enterococci and other bacteria reflecting the fecal microbiota are not uncommon. On occasion, the nares of healthy persons have also yielded *S. pyogenes* and *S. pneumoniae*. The nonpathogenic neisseriae are transients in this location as well. Healthy contacts of patients with meningococcal disease may harbor *N. meningitidis*. Additional residents of the nose are various corynebacteria and occasionally *Moraxella lacunata*. Upper respiratory tract viruses may be found in this site (Table 1).

Actual disease of the nasal passages must be differentiated carefully from disease of adjacent areas, including the skin, nasopharynx, oropharynx, tonsils, and sinuses. In premature and newborn infants, lesions due to hemolytic *E. coli*, *Pseudomonas aeruginosa*, and *C. albicans* may be encountered, but these usually represent more generalized disease, as is indicated by the isolation of *Acinetobacter lwoffii*, *Moraxella* spp., *Acinetobacter baumannii*, and various flavobacteria. Ozaena, a disease characterized by atrophy of the nasal mucosa (15), is caused by gram-negative rods. An infective granuloma, rhinoscleroma, most often involves the nose, although the pharynx and the remaining upper respiratory tract may be involved as well. *Klebsiella pneumoniae* subsp. *rhinoscleromatis* is regarded by most investiga-

tors as the etiological agent of this disease, which is uncommon in the United States. Filamentous fungi such as the zygomycetes may be involved in serious disease processes of the nasal sinuses.

Usually Sterile Areas

The larynx, trachea, bronchi, bronchioles, alveoli, and accessory nasal sinuses are usually sterile. Contamination by occasional microorganisms is usual, but the various defense mechanisms of these organs remove such offenders quickly and efficiently. The usual specimen submitted to the clinical microbiology laboratory for establishing infectious disease of the lower respiratory tract is the sputum specimen. Bronchial washings, bronchoalveolar lavage and bronchoscopy specimens, thoracentesis fluid, and aspirates from tracheostomies or lung lesion biopsies are now frequently submitted for microbiological analysis. Sputum and some aspirates are invariably contaminated by the microbiota of the throat, nose, and mouth. Rapid processing is imperative to prevent these contaminating microorganisms from obscuring etiological agents. The most commonly encountered bacteria in the sputum are various viridans streptococci, *Neisseria* spp., staphylococci, corynebacteria, the Enterobacteriaceae, *M. catarrhalis*, *P. aeruginosa*, and *C. albicans*. These organisms may be involved in infectious processes, they may be contaminants, or they may reflect superinfection after therapy.

Acute infectious bronchitis, most frequently seen during winter in children and the aged and associated with adenoviruses, influenza viruses, and parainfluenza viruses, may be complicated by *S. aureus*, *S. pneumoniae*, *S. pyogenes*, *H. influenzae*, and *Mycoplasma pneumoniae* (and *Bordetella pertussis* when whooping cough is present). Chronic bronchitis, of unknown etiology and associated frequently with pulmonary emphysema, is assuming increasing importance. Sputum from such patients displays a variety of microorganisms, not necessarily the same species on repeat examination. The presence of bacteria associated with pathology of the respiratory tract, such as *M. catarrhalis*, pneumococci, group A streptococci, *H. influenzae*, *M. pneumoniae*, and staphylococci, may be significant.

Acute mediastinitis is usually the result of perforation of the esophagus in conjunction with instrumentation, obstruction, external wounds, downward propagation of deep cellulitis of the neck, forceful vomiting, and occasionally the extension of infectious disease of the lungs, pleural cavity, or pericardium. The bacteria that contribute most frequently to the disease picture are the peptostreptococci; *Bacteroides*, *Prevotella*, and *Porphyromonas* spp.; fusobacteria; and occasionally clostridia. Pneumonia may be caused by *S. pneumoniae*, *S. pyogenes*, *H. influenzae*, *S. aureus*, Enterobacteriaceae, *Francisella tularensis*, *C. albicans*, *P. aeruginosa*, *Proteus* spp., *Legionella* spp., *Pneumocystis carinii*, *M. pneumoniae*, and *Chlamydia pneumoniae* (14). In newborn or very young infants, chlamydiae and respiratory syncytial virus should be ruled out as etiological agents.

Pulmonary abscesses may result from staphylococcal or *K. pneumoniae* pneumonias or the aspiration of particulate matter contaminated with oropharyngeal flora. The organisms most commonly cultured from such lesions are *Bacteroides*, *Prevotella*, *Porphyromonas*, and *Fusobacterium* spp.; peptostreptococci; staphylococci; clostridia; klebsiellae; escherichiae; and pseudomonads. The examination of such abscesses should include a search for acid-fast organisms, aerobic and anaerobic actinomycetes, and the fungi associated with pulmonary lesions. Empyema is usually a secondary disease, i.e., an extension of the primary disease. Special attention must be accorded certain clinical variants of empyema, especially empyema of infants, usually caused by *S. aureus*, and empyema caused by *Entamoeba histolytica*, *Actinomyces* spp., or *Nocardia* spp.

GASTROINTESTINAL TRACT
See also references 3, 4, 6, 9, 15, 21, 22, and 24.

Areas That Usually Harbor Microorganisms
The part of the gastrointestinal tract that invariably harbors microorganisms is the large intestine, although fecal organisms are also recovered from aspirates of the lower ileum of healthy individuals (16). Past the ileocecal valve, the contents of the large intestine reflect the fecal flora. It is surprising that information on normal fecal microflora is comparatively meager (5, 16). It seems almost superfluous to list the organisms that can be encountered; most are ignored in the search for enteric pathogens. The presence of viruses in the intestinal tracts of healthy individuals has not been explored sufficiently to allow one to make definitive statements. Rosebury (16) lists the following microorganisms as indigenous: *S. epidermidis*, *S. aureus*, viridans streptococci, enterococci, occasionally *S. pyogenes* and related serogroups, peptostreptococci, lactobacilli (especially in infants), corynebacteria, mycobacteria, clostridia, actinomycetes, the *Enterobacteriaceae*, *P. aeruginosa*, *Alcaligenes faecalis*, *Flavobacterium* spp., *Bacteroides* spp., *Fusobacterium* spp., *Eubacterium* spp., *Propionibacterium* spp., *Bifidobacterium* spp., yeasts, filamentous fungi, and a large variety of protozoa such as *Entamoeba coli*, *Endolimax nana*, *Iodamoeba bütschlii*, *Trichomonas hominis*, and *Chilomastix mesnili*. Not all microbial groups are present in each individual at all times. Undoubtedly, many other genera can be found with some frequency. The dominance of the anaerobic bacteria in the healthy adult colon must be emphasized, even though these bacteria are not sought during routine stool cultures. Geographical distribution and dietary and sanitary habits are also influential in the selection of a resident microflora. A role for the healthy intestinal microbiota in the production of quantities of carcinogens has been advocated. These microbial activities may bring new dimensions to clinical microbiology and our understanding of neoplasia (5).

Acute diarrheal diseases are a major cause of morbidity throughout the world. While some diarrheal disease is caused by bacterial and parasitic agents, a significant proportion, particularly in children, is caused by or associated with a number of fastidious or noncultivatable viral agents. The most important of these viral gastroenteritis agents are the rotaviruses, the Norwalk-like viruses, and the "enteric" adenoviruses (types 40 and 41). In addition, the astroviruses and caliciviruses are important causes of diarrhea in infants and children under age 2 years.

The most readily recognized bacterial agents of gastroenteritis are salmonellae, shigellae, campylobacters, and *Helicobacter pylori*. All members of the genus *Salmonella* are capable of evoking the various clinical symptoms of salmonellosis and its complications. The various shigellae can cause diarrhea or the syndrome known as bacillary dysentery. *Campylobacter jejuni* is a major cause of diarrhea; *Campylobacter laridis* and *Campylobacter coli* have also been involved in gastroenteritis. *H. pylori* may be involved in gastritis and ulcer production in the stomach or duodenum (18). *Vibrio* spp. and their attendant disease are encountered increasingly in many parts of the world. Enteropathogenic, enterotoxigenic, enteroinvasive, enterohemorrhagic, and enteroaggregative or enteroadherent *E. coli* are known to cause pathology in the human intestine. *Bacillus cereus* and staphylococcal food poisonings also affect the gastrointestinal tract, as do food poisoning due to clostridial toxins, botulism, and pseudomembranous enterocolitis caused by *Clostridium difficile*. The toxins rather than the microorganisms cause the symptoms and should be the object of analytical efforts; the bacteria need not be demonstrated in the feces or vomitus. Acute diarrhea may at times be associated with *P. aeruginosa*, *Proteus* spp., *Aeromonas* spp., *Plesiomonas shigelloides*, or *C. albicans*. *Entamoeba histolytica* is the etiological agent of amebic dysentery and its complications. Large numbers of *Giardia lamblia* may cause an infectious colitis. Localized pathology of the large intestine may involve *Aeromonas* spp., *P. shigelloides*, or *Yersinia enterocolitica*, the last causing not only occasional gastroenteritis but also the symptoms of appendicitis. Anal carriage of *S. pyogenes* in the nosocomial spread of the disease and of the gonococcus in clinical or cryptic disease must not be overlooked in specific investigations of these infections. A large number and variety of protozoa and helminths cause pathology in the gastrointestinal tract (Table 2).

Usually Sterile Areas
The esophagus and the stomach are contaminated with bacteria whenever food is ingested. The microbial population does not survive well in these two sections of the gastrointestinal tract. Similarly, the small intestine (except the distal ileum), the liver, and the gallbladder usually are free from microbial contamination or harbor transient microbial populations. This is true also of the peritoneum. When microorganisms are present in these areas, they are secondary to underlying diseases such as carcinoma or they reach these sites because the large intestine has been punctured or ruptured. Peritonitis, regardless of its initial cause, can be incited by any of the fecal organisms listed in Table 2. A mixture of aerobic and anaerobic, gram-positive and gram-negative microorganisms participates. Peritonitis may give rise to intra-abdominal abscesses, especially pelvic, paracolic, intermesenteric, subphrenic, and retroperitoneal. All may show a mixed microbial population or single causative agents such as *S. aureus*, *Bacteroides* spp., enterococci, *P. aeruginosa*, or *E. coli*. Bacteria have been isolated in several cholecystitis cases, possible as secondary opportunists. *E. coli*, enterococci, peptostreptococci, and clostridia are the most frequently encountered bacteria, but many other common and uncommon representatives of the fecal microflora in health and disease may be found. Bacterial cholangitis usually is secondary to intra- or extrahepatic obstruction in the area of the bile duct. Cholangitis caused by salmonellae, staphylococci, and streptococci may be accompanied by septicemia caused by these bacteria (3). Again, the organisms usually recovered reflect the microbiota of the intestinal tract. Amebic abscesses of the liver constitute a complication of primary amebic dysentery. Pyogenic hepatic abscesses in the antibiotic era yield *E. coli* most frequently. Pancreatic abscess formation is a secondary complication of pancreatitis. The microorganisms most frequently found are *S. aureus* and *E. coli*. Clostridial diseases are uncommon but not rare (3). They include acute gaseous cholecystitis caused by *Clostridium perfringens*, but other

TABLE 2 Microorganisms encountered in gastrointestinal tract specimens

Organism[a]	Frequency of isolation[b]	Disease involvement[c]	Organism[a]	Frequency of isolation[b]	Disease involvement[c]
Absidia spp.	C	1	*Iodamoeba bütschlii*	B	1
Acidaminococcus fermentans	C	1	*Isospora belli*	C	2
Acinetobacter spp.	B	1			
Adenoviruses	B	2	*Lactobacillus* spp.	B	1
Aeromonas spp.	B	2			
Alcaligenes spp.	B	2	*Metagonimus yokogawai*	C	3
Anaerobiospirillum succiniciproducens	C	2	Microsporidian genera (several)	C	3
Ancylostoma duodenale	B	3	*Mucor* spp.	C	2
Angiostrongylus costaricensis	C	3	*Mycobacterium* spp.	B	2
Anisakis larvae	C	3	*Mycoplasma* spp.	C	1
Ascaris lumbricoides	B	3			
Astroviruses	C	2	*Necator americanus*	B	3
			Neisseria gonorrhoeae	C	3
Bacillus spp.	B	2	*Neisseria* spp.	B	1
Bacteroides spp.	A	2	Norwalk and Norwalk-like viruses	B	3
Balantidium coli	C	3			
Bifidobacterium spp.	B	1			
Blastocystis hominis	B	2	*Opisthorchis viverrini*	C	3
Caliciviruses	C	2	*Paracoccidioides brasiliensis*	C	3
Campylobacter spp.	B	3	*Paragonimus* spp.	C	3
Candida spp.	B	2	*Pediococcus* spp.	C	1
Capillaria spp.	C	3	*Peptostreptococcus* spp.	B	2
Chilomastix mesnili	B	1	*Plesiomonas shigelloides*	C	2
Clonorchis sinensis	C	3	*Pseudomonas* spp.	B	2
Clostridium spp.	A	2			
Coronaviruses	C	2	*Retortamonas intestinalis*	C	1
Corynebacterium spp.	B	1	*Rhizomucor* spp.	C	1
Cryptosporidium spp.	C	3	*Rhizopus* spp.	C	1
			Rhodotorula spp.	C	1
Dicrocoelium dentriticum ova	C	1	Rotavirus	B	3
Dientamoeba fragilis	B	2	*Ruminococcus bromii*	C	1
Diphyllobothrium latum	B	3			
Dipylidium caninum	C	3	*Saccharomyces* spp.	B	1
			Sapporo agent	B	2
Endolimax nana	B	1	*Sarcocystis* spp.	C	2
Entamoeba coli	B	1	*Schistosoma* spp.	B	3
Entamoeba hartmanni	B	1	*Selenomonas* spp.	B	1
Entamoeba histolytica	B	3	*Staphylococcus* spp.	B	2
Enterobacteriaceae	A	2	*Streptococcus* spp.	B	2
Enterobius vermicularis	B	3	*Strongyloides stercoralis*	B	3
Enterococcus spp.	A	2	*Succinimonas* spp.	B	1
Enteromonas hominis	C	1	*Succinivibrio* spp.	B	1
Enteroviruses	A	1			
Eubacterium spp.	A	1	*Taenia* spp.	C	3
			Trichinella spiralis larvae	C	3
Fasciola hepatica	C	3	*Trichomonas hominis*	B	1
Fasciolopsis buski	C	3	*Trichosporon beigelii*	C	1
Fusobacterium spp.	A	2	*Trichostrongylus* spp.	C	2
			Trichuris trichiura	B	2
Gastrodiscoides hominis	C	3			
Geotrichum candidum	C	1	*Veillonella* spp.	B	2
Giardia lamblia	B	3	*Vibrio* spp.	B	2
Helicobacter pylori	B	3			
Heterophyes heterophyes	C	3			
Histoplasma capsulatum	C	3			
Hymenolepis spp.	C	3			

[a]Diseases other than gastroenteritis may be caused by organisms listed. Several species of the genera listed may be found; not all of the species are involved in disease processes.

[b]A, commonly encountered in clinical specimens; B, occasionally encountered in clinical specimens; C, rarely encountered in clinical specimens.

[c]1, When present, rarely if ever involved in disease production; 2, when present, occasionally involved in disease production; 3, when present, commonly involved in disease production.

TABLE 3 Microorganisms encountered in genitourinary tract specimens

Organism	Frequency of isolation[a]	Disease involvement[b]	Organism[a]	Frequency of isolation[b]	Disease involvement[c]
Acinetobacter spp.	B	2	*Lactobacillus* spp.	A	1
Actinomyces spp.	B	2	*Leptospira interrogans*	C	3
Adenoviruses	B	1	*Listeria* spp.	C	2
Alcaligenes spp.	C	2			
Aspergillus spp.	C	1	*Mobiluncus* spp.	B	2
			Molluscum contagiosum virus	C	3
Bacillus spp.	C	1	*Moraxella* (*Branhamella*) *catarrhalis*	B	1
Bacteroides spp.	A	2	*Moraxella* spp.	B	1
Bifidobacterium spp.	C	1	Mumps virus	C	3
Bilophila wadsworthia	C	2	*Mycobacterium* spp.	B	2
Blastomyces dermatitidis	C	3	*Mycoplasma hominis*	B	2
Brugia malayi	C	3			
			Neisseria gonorrhoeae	B	3
Calymmatobacterium granulomatis	C	3	*Neisseria meningitidis*	C	3
Campylobacter spp.	C	2	*Neisseria* spp.	B	1
Candida spp.	B	2			
Chlamydia spp.	B	2	Papillomaviruses	C	3
Clostridium spp.	A	2	*Penicillium* spp.	C	1
Corynebacterium spp.	B	2	*Peptostreptococcus* spp.	B	1
Cryptococcus spp.	C	2	*Porphyromonas* spp.	A	2
Cytomegalovirus	B	2	*Prevotella* spp.	A	2
			Propionibacterium spp.	B	1
Dioctophyma renale	C	3	*Pseudomonas* spp.	B	2
			Rhodotorula spp.	C	1
Entamoeba histolytica	C	3			
Enterobacteriaceae	A	2	*Saccharomyces* spp.	C	1
Enterobius vermicularis	B	2	*Schistosoma haematobium*	C	3
Enterococcus spp.	B	2	*Staphylococcus* spp.	A	2
			Streptococcus spp.	B	2
Flavobacterium spp.	C	2			
Fusobacterium spp.	B	1	*Torulopsis glabrata*	B	2
			Treponema pallidum	B	3
Gardnerella vaginalis	B	2	*Trichomonas vaginalis*	B	3
Geotrichum candidum	C	1			
			Ureaplasma urealyticum	B	1
Haemophilus ducreyi	B	3			
Hepatitis A and B viruses	B	2	*Wuchereria bancroftii*	C	3
Herpes simplex virus	B	3			
Histoplasma capsulatum	C	3	Zygomycetes	C	1
Human immunodeficiency virus	B	3			

[a]A, commonly encountered in clinical specimens; B, occasionally encountered in clinical specimens; C, rarely encountered in clinical specimens.
[b]1, When present, rarely if ever involved in disease production; 2, when present, occasionally involved in disease production; 3, when present, commonly involved in disease production.

clostridia, *Enterobacteriaceae*, and aerobic and anaerobic streptococci have been associated with this entity. Invasion of the bile ducts from the intestinal tract results in clostridial choledochitis; *C. perfringens* exclusively has been isolated from patients with this disease. Enteritis necroticans is caused by heat-resistant *C. perfringens* type C. Clostridial cellulitis of the abdominal wall is a complication of perforation by intestinal neoplasms with local peritonitis or of colon or biliary tract surgery (Table 2).

GENITOURINARY TRACT

See Table 3. See also references 3, 4, 6, 9, 15, 21, 22, and 24.

Areas That Usually Harbor Microorganisms

External Genitalia

Rosebury (16) and Skinner and Carr (17) have listed the following organisms as present on the surface of the genitalia: *S. epidermidis*, viridans streptococci, enterococci, peptostreptococci, corynebacteria, mycobacteria, various *Enterobacteriaceae*, *Bacteroides* spp., *Prevotella* spp., *Fusobacterium* spp., mycoplasmas, *C. albicans*, and other yeasts.

The external genitalia are subject to the same infectious diseases as other skin areas. The special lesions of the external genitalia are venereal in nature and include the lesions of syphilis, chancroid caused by *Haemophilus ducreyi*, granuloma inguinale (which is primarily seen in the tropics and subtropics and is caused by *Calymmatobacterium granulomatis*), lymphogranuloma venereum (resulting from infection with certain serogroups of *Chlamydia trachomatis*), herpes genitalis, and condylomata acuminata (venereal warts) caused by papillomaviruses.

Anterior Urethra

An appreciable number and variety of microorganisms can usually be recovered from the anterior urethra in healthy individuals of both sexes: coagulase-negative and occasionally coagulase-positive staphylococci, enterococci, various nonpathogenic neisseriae, corynebacteria, rarely certain mycobacteria, the various enteric gram-negative rods, chlamydiae, *A. baumannii*, *Gardnerella vaginalis*, mycoplasmas, and yeasts, including *C. albicans*. *Trichomonas vaginalis* may on occasion gain access to this part of the urethra without overt disease.

It is difficult to delineate exactly where the anterior portion of the urethra ends, especially when disease is present. Urethritis may be caused specifically by the gonococcus or nonspecifically by a variety of bacteria, including the staphylococci, chlamydiae, mycoplasmas, fecal gram-negative rods, and *Listeria* spp. In the female, contamination of the anterior portion of the urethra by vaginolabial microbiota is unavoidable. It is appropriate to emphasize that *N. gonorrhoeae* can be recovered from the anterior urethra of the female and from males with cryptic disease. In young women, bacteriuria caused by *Staphylococcus saprophyticus* is reported with increasing frequency.

Vagina

The usual microbiota of the vagina from menarche to menopause is dominated by lactobacilli, designated Döderlein's bacilli and comprising glycogen-fermenting *Lactobacillus acidophilus* and related species. Prepubescent females from shortly after birth and postmenopausal women harbor skin microbiota in this region. Despite the control over the vaginal environment exerted by the lactobacilli, many other microorganisms can be cultivated from vaginal samples of healthy women: *S. aureus*, *S. epidermidis*, viridans streptococci, enterococci, peptostreptococci, group B streptococci, the low-temperature-tolerant neisseriae, corynebacteria, some actinomycetes, *Enterobacteriaceae*, acinetobacters, chlamydiae, *G. vaginalis*, occasionally clostridia and other anaerobic rods, mycoplasmas, *C. albicans*, other yeasts, and *T. vaginalis*.

The major infectious diseases of the female pudenda are the sexually transmitted diseases: chlamydiosis, syphilis, gonorrhea, and herpes simplex virus infections as well as candidiasis, trichomoniasis, and vaginosis caused by a variety of organisms that may act at times as opportunistic secondary invaders reflecting disturbances in the microbiological balance as a complication of disease elsewhere and its treatment. *G. vaginalis* and *Mobiluncus* spp. rank high among the so-called nonspecific vaginitides, but fecal and skin organisms, including staphylococci, enterococci, and *Listeria* spp., may contribute to clinically overt disease (16). *Campylobacter fetus* and related unclassified organisms have been isolated from the vaginas of women who have aborted repeatedly. The vaginal microflora, especially the anaerobic bacteria, may contribute to the complications of septic abortions. Postpartum sepsis and salpingitis, frequently caused by facultatively and obligately anaerobic gram-negative rods and to a lesser degree by clostridia and staphylococci, probably reflect the microbiota of the vagina. Similarly, colonization and infectious disease of the newborn reflect the microbial population of the vagina. Among the viruses, herpes simplex virus and papillomaviruses rank as significant causes of pathology.

Usually Sterile Areas

As a rule, the remaining structures of the genitourinary tract are without permanent microbiota. Infectious disease of the kidneys is still not completely understood, with two major theories advanced to explain the seeding of this organ by infectious organisms entering either by the hematogenous or the ascending route. There are many predisposing factors for pyelonephritis, especially host factors and underlying diseases, which appear to be mandatory for the establishment of infection. The major bacterial offenders are the gram-negative rods, especially *E. coli*, *Proteus* spp., *P. aeruginosa*, and the *Klebsiella-Enterobacter* group; enterococci are not uncommon and can be demonstrated in adequate numbers repeatedly in patients with untreated acute pyelonephritis. *Mycobacterium tuberculosis* may infect the kidney and be demonstrated in urine. Perinephric abscesses usually caused by *S. aureus* are not detected in urine unless the abscess ruptures into the kidney. The prostate in the male may become a secondary focus of gonorrheal disease. However, other microorganisms, including staphylococci, enterococci, listeriae, group B streptococci, *E. coli*, mycoplasmas, pseudomonads, achromobacters, and rarely trichomonads, may lodge in this organ, especially in middle-aged and older males. Urine samples from persons acutely or chronically infected with these agents commonly contain some viruses, notably cytomegalovirus. Similarly, some viruses, including human immunodeficiency virus, hepatitis B virus, and cytomegalovirus, may be found in the semen of chronically infected persons.

SKIN, WOUNDS, AND BURNS

See also references 3, 4, 6, 9, 15, 16, 21, 22, and 24.

The outermost border of the body is constantly populated with microorganisms that reflect the contacts, habits, profession, and environment of the individual. Table 4 lists numerous microorganisms, but those usually found consist mostly of *Staphylococcus* spp., *Corynebacterium* spp., *Propionibacterium* spp., *Mycobacterium* spp., and various yeasts. Certain areas, especially those adjoining the various body openings, reflect the microbiota of the adjacent sites.

The most common and most obvious skin diseases are caused by the staphylococci. *S. aureus* is involved most frequently in boils and furuncles, and *S. epidermidis*, often accompanied by *Propionibacterium acnes*, is found in pimples and acne. Pustulosis, especially of the newborn, is also caused by *S. aureus*. The agents of impetigo are *S. pyogenes* and *S. aureus*. Coagulase-positive staphylococci are also involved in superficial folliculitis and sycosis barbae. Although streptococci and other oral bacteria may also be found, *S. aureus* is the major etiological agent of acute suppurative parotitis. Actinomycosis is usually caused by aerobic and anaerobic members of the order *Actinomycetales* and represents a chronic, suppurative or granulomatous disease culminating in abscesses. Deep cellulitis of the neck is most frequently caused by streptococci, both aerobic and anaerobic, and staphylococci. *S. aureus* is responsible for hydradenitis suppurativa, a disease more common in women and usually involving the axillae, and it is the most frequent disease-producing agent in acute mastitis and breast abscesses. Paronychia most often involve staphylococci, streptococci and, less commonly, *C. albicans*. Staph-

TABLE 4 Microorganisms encountered in skin, wound, and burn specimens

Organism	Frequency of isolation[a]	Disease involvement[b]	Organism[a]	Frequency of isolation[b]	Disease involvement[c]
Absidia spp.	A	1	Madurella spp.	C	2
Acinetobacter spp.	B	2	Malassezia spp.	B	2
Actinomadura madurae	C	2	Mansonella streptocerca	C	3
Actinomyces spp.	B	2	Micrococcus spp.	A	1
Aeromonas spp.	B	2	Microsporum spp.	B	3
Agrobacterium tumefaciens	C	2	Moraxella spp.	B	2
Ancylostoma spp.	C	3	Mucor spp.	A	1
Arcanobacterium haemolyticum	C	2	Mycobacterium spp.	B	2
Aspergillus spp.	B	1			
			Neisseria spp.	B	2
Bacillus anthracis	C	3	Neotestudina rosatii	C	2
Bacillus spp.	B	1	Nocardia spp.	C	2
Bacteroides spp.	B	2			
Bartonella bacilliformis	C	3	Oerskovia spp.	C	2
Basidiobolus ranarum	C	2	Onchocerca volvulus	C	3
Bifidobacterium spp.	B	2			
Bipolaris spp.	C	2	Papillomaviruses	B	3
Blastomyces dermatitidis	B	3	Paracoccidioides brasiliensis	C	3
Brugia spp.	C	3	Pasteurella spp.	C	2
			Penicillium spp.	B	1
Candida spp.	A	2	Peptostreptococcus spp.	B	2
Cephalosporium spp.	B	2	Plesiomonas shigelloides	C	2
Cladosporium carrionii	C	2	Poxviruses	B	3
Clostridium spp.	B	2	Propionibacterium spp.	A	2
Coccidioides immitis	B	3	Pseudallescheria boydii	C	2
Conidiobolus coronatus	C	2	Pseudomonas spp.	B	2
Corynebacterium diphtheriae	C	3			
Corynebacterium spp.	B	2	Rhinocladiella aquaspersa	C	2
Cryptococcus neoformans	C	3	Rhizomucor spp.	B	1
Cryptococcus spp.	C	1	Rhizopus spp.	B	1
			Rhodococcus spp.	C	2
Dematiaceous fungi	B	2	Rhodotorula spp.	B	1
Dirofilaria spp.	C	3	Rickettsia spp.	B	3
Dracunculus medinensis	C	3			
			Sarcinosporon inkin	B	2
Ehrlichia spp.	C	3	Scopulariopsis brevicaulis	B	2
Entamoeba histolytica trophozoites	C	3	Spirillum minus	C	3
			Sporothrix schenckii	B	2
Enterobacteriaceae	B	2	Staphylococcus spp.	A	2
Enterococcus spp.	B	2	Streptobacillus moniliformis	C	2
Epidermophyton floccosum	B	3	Streptococcus spp.	B	2
Erysipelothrix spp.	C	3	Streptomyces somaliensis	C	2
Eubacterium spp.	B	2	Strongyloides spp.	C	3
Flavobacterium spp.	B	2			
Fonsecaea spp.	C	2	Taenia solium cysticerci	C	3
Fusarium spp.	B	2	Treponema spp.	B	3
Fusobacterium spp.	B	2	Trichophyton spp.	B	3
Geotrichum candidum	B	1	Varicella-zoster virus	B	3
			Veillonella spp.	B	2
Herpes simplex virus	A	3	Vibrio spp.	C	2
Histoplasma capsulatum	B	3			
			Wangiella dermatitidis	C	2
Leishmania spp.	C	3			
Leptospira spp.	C	3	Xanthomonas maltophilia	B	2
Listeria spp.	C	2			
Loboa loboi	C	2			

[a] A, commonly encountered in clinical specimens; B, occasionally encountered in clinical specimens; C, rarely encountered in clinical specimens.
[b] 1, When present, rarely if ever involved in disease production; 2, when present, occasionally involved in disease production; 3, when present, commonly involved in disease production.

ylococci and streptococci are also frequently recovered from patients with tenosynovitis. Clostridia are usually responsible for anaerobic cellulitis, mixtures of bacteria are responsible for crepitant cellulitis, and S. pyogenes is involved in necrotizing fasciitis or hemolytic streptococcal gangrene. Erysipelas is a subcutaneous form of streptococcal cellulitis. Erysipeloid, an acute, cutaneous, bluish red inflammation of the hands and wrists that is usually seen in males in certain vocations, is caused by Erysipelothrix spp. Other microorganisms occasionally or rarely involved in skin and wound infections are also listed in Table 4. Many viral diseases are accompanied by exanthems and a variety of other skin lesions, some of which are typical for a specific etiological agent. Some of these lesions may yield the etiological agent. The dermatophytic fungi cause skin lesions, most of which yield the causative agent from properly obtained specimens; laboratorians must be aware of the special lipid requirements of certain fungi such as Malassezia furfur.

The microbiota of wounds reflects their anatomical site, the mode of infliction (i.e., traumatic or surgical), the environment, and the degree of microbial contamination of adjacent areas that were perforated in the process. These considerations supplement rather than substitute for the general considerations that maintain an adequate host-parasite equilibrium. Traumatic wounds are usually complicated by aerobic indigenous microorganisms, especially S. aureus, group A streptococci, enterococci, P. aeruginosa, E. coli, Proteus spp., flavobacteria, and Acinetobacter spp. Among the anaerobic bacteria associated with traumatic wounds, the histotoxic and neurotoxic clostridia are most prominent, causing, under proper conditions, gas gangrene or tetanus. The most common clostridia of gas gangrene are C. perfringens type A, Clostridium septicum, and Clostridium novyii, but these organisms may also be harmless contaminants of wounds. The diagnosis of gas gangrene is strictly clinical; the isolation of clostridia may alert a clinician but does not constitute the diagnosis of clostridial cellulitis or clostridial myonecrosis. Clostridium tetani should not present a problem in properly immunized individuals but should not be ignored in deep puncture wounds when the immunization history is not known. Criminally inflicted stab wounds, often intentionally contaminated with feces, require not only expert surgical but also exquisite microbiological attention.

Infections complicating surgery may be of two kinds. The first, the so-called wound infection, usually denotes a complication of a clean surgical procedure, i.e., a procedure performed under the best available aseptic conditions on tissues usually sterile and not found during surgery to be grossly contaminated. These wounds may yield S. aureus, coagulase-negative staphylococci, enterococci, or gram-negative rods and rarely microorganisms such as S. pyogenes, corynebacteria, pneumococci, clostridia, and Bacillus subtilis. The second kind of infection, which occurs when the surgeon performs all or part of the surgical procedures in a contaminated area of the body, results in infectious disease complications in a certain percentage of cases and often reflects the state of health of the patient. The microorganisms mirror the microbiota of the particular anatomic site, but minority members of the microflora frequently attain a majority in the usually sterile tissue subjected to surgery and may gravely complicate the recovery of the patient. Among the bacteria involved in this complication are E. coli, P. aeruginosa, Proteus spp., Providencia spp., the Klebsiella-Enterobacter-Serratia group, flavobacteria, Acinetobacter spp., and Bacteroides spp.

Microbial contamination of severe burns can be a life-threatening complication. The organism most commonly encountered and most difficult to control is P. aeruginosa, often found in conjunction with flavobacteria and other gram-negative rods, enterococci, and staphylococci, which abound in the institutional environment or on the uninvolved skin of the patient. Early microbiological analysis of burns usually shows a large number of a great variety of microorganisms, many of which are eventually supplemented by the bacteria cited above.

EYE

See also references 3, 4, 6, 9, 15, 16, 21, 22, and 24.

The healthy conjunctiva of the eye may harbor a number of skin organisms. The staphylococci are among those most frequently recovered from the healthy eye; viridans streptococci, S. pyogenes, pneumococci, neisseriae, and corynebacteria are recovered rarely. The fecal gram-negative rods have been isolated even less frequently from this site, but H. influenzae has an incidence second only to that of the coagulase-negative staphylococci. Conjunctival cultures have also yielded C. albicans and other yeasts. Eye disease has been caused by S. aureus and P. aeruginosa (very often iatrogenically) and occasionally by S. pneumoniae. More often, infectious diseases peculiar to the eye have yielded M. lacunata H. influenzae, Haemophilus aegyptius (H. influenzae biotype III), and, on rare occasions, Bacillus spp. Still rarer etiological agents of eye infections and occasional commensal organisms are listed in Table 5.

EAR

See also references 3, 4, 6, 9, 15, 16, 21, 22, and 24.

The external auditory canal usually reflects the microbiota of the skin. Perhaps S. pneumoniae and the gram-negative rods, including P. aeruginosa, have been recovered with greater frequency from this site than from other skin areas. The middle and inner ears are usually sterile. Diseases of the outer ear reflect the disorders of the skin. Other parts of the auditory organ may be involved during generalized disease or may be affected only locally by S. aureus, S. pneumoniae, S. pyogenes, P. aeruginosa, H. influenzae, and occasionally other microorganisms (Table 6).

BLOOD, CENTRAL NERVOUS SYSTEM, EXUDATES, AND TRANSUDATES

See also references 3, 4, 6, 9, 15, 16, 21, 22, and 24.

The blood and spinal fluid of a healthy individual are usually sterile. Occasionally, microorganisms may be isolated from blood cultures of healthy asymptomatic individuals. Often these microorganisms represent low-level contamination of the circulation from inapparent skin lesions or other sources. Such findings complicate the interpretation of laboratory results but are usually of no consequence. However, during several infectious diseases or during infectious complications of a primary disease, microorganisms may be recovered from blood. They may be present transiently or persistently. Positive blood cultures may result from traumatic and surgical wounds as well as from burns, injury to bones and joints, brain abscesses, furunculosis, cellulitis, meningitis, pneumonia, lung abscesses, empyema, mucoviscidosis, mycotic aneurysm, cardiac anomalies, peritonitis, intestinal or biliary obstruction, cholangitis, carci-

TABLE 5 Microorganisms encountered in eye specimens

Organism	Frequency of isolation[a]	Disease involvement[b]	Organism[a]	Frequency of isolation[b]	Disease involvement[c]
Absidia spp.	C	2	*Lactobacillus* spp.	C	2
Acanthamoeba spp.	C	3	*Leptospira* spp.	C	3
Achromobacter xylosoxidans	C	1	*Loa loa*	C	3
Acinetobacter spp.	B	2			
Acremonium spp.	B	2	Measles virus	B	3
Actinobacillus actinomycetemcomitans	C	2	*Moraxella* (*Branhamella*) *catarrhalis*	B	1
Actinomyces spp.	C	2	*Moraxella* spp.	B	2
Adenoviruses	B	2	*Mucor* spp.	B	2
Alternaria spp.	B	2	Mumps virus	B	3
Arachnia spp.	B	2	*Mycobacterium* spp.	C	2
Aspergillus spp.	B	2	*Mycobacterium tuberculosis* group	C	3
Aureobasidium spp.	C	1			
Azotobacter spp.	C	1	*Neisseria gonorrhoeae*	B	3
			Neisseria meningitidis	B	3
Bacillus spp.	B	2	*Neisseria* spp.	B	1
Bacteroides spp.	C	2	*Nocardia* spp.	C	3
Bifidobacterium spp.	C	2	*Nosema* spp.	C	3
Blastomyces dermatitidis	B	3			
Brucella spp.	C	3	*Onchocerca volvulus*	C	3
Candida spp.	B	2	*Paecilomyces* spp.	B	1
Chlamydia trachomatis	B	3	*Peptostreptococcus* spp.	C	2
Cladosporium spp.	B	1	Poxvirus	C	3
Clostridium spp.	C	2	*Propionibacterium* spp.	B	2
Coccidioides immitis	C	3	*Pseudoallescheria boydii*	C	3
Corynebacterium diphtheriae	C	3	*Pseudomonas* spp.	B	2
Corynebacterium spp.	B	2			
Cryptococcus neoformans	C	3	*Rhizomucor* spp.	B	2
Curvularia spp.	B	1	*Rhizopus* spp.	B	2
Cytomegalovirus	B	3	Rubella virus	B	3
Drechslera spp.	B	1	*Sporothrix schenckii*	C	3
			Staphylococcus spp.	A	2
Enterobacteriaceae	B	2	*Streptococcus* spp.	B	2
Epstein-Barr virus	C	3			
Exophiala jeanselmei	C	1	*Taenia solium* larvae	C	3
			Toxocara spp.	C	3
Francisella tularensis	C	3	*Toxoplasma gondii*	B	3
Fusarium spp.	B	2	*Treponema pallidum*	B	3
Fusobacterium spp.	C	2			
			Varicella-zoster virus	B	3
Haemophilus influenzae biotype III (*H. aegyptius*)	B	3	*Veillonella* spp.	C	2
Herpes simplex virus	B	3			

[a]A, commonly encountered in clinical specimens; B, occasionally encountered in clinical specimens; C, rarely encountered in clinical specimens.
[b]1, When present, rarely if ever involved in disease production; 2, when present, occasionally involved in disease production; 3, when present, commonly involved in disease production.

noma, urinary obstruction, nephropathies, postpartum endometritis, and septic abortion. Immunosuppressive, cytotoxic, or radiation therapies and conditions such as arteriosclerosis, chronic debility, diabetic acidosis, hematological diseases, hepatic insufficiency, and malignancies of all types may lead to positive blood cultures. No rule concerning the types of microorganisms of significance can be stated. Certainly, staphylococci (not only *S. aureus* but also coagulase-negative staphylococci) have been recovered repeatedly. The latter can be involved in bacterial endocarditis and have complicated cardiac catheterization and

prosthesis procedures. The gram-negative rods, especially *E. coli*, the obligately anaerobic bacteria of the colon, *Klebsiella-Enterobacter* organisms, *P. aeruginosa*, *Proteus* spp., and *Providencia* spp., have been recovered after trauma or surgery on contaminated body areas or as acquired microorganisms complicating protracted hospitalization of patients with a variety of diseases. Recovery of salmonellae from the blood of individuals with systemic salmonellosis is not uncommon. Other fevers of unknown origin may yield brucellae, pasteurellae, pneumococci, *N. meningitidis*, *Listeria monocytogenes*, fungi, and parasites. Bacterial endocardi-

TABLE 6 Microorganisms encountered in ear specimens

Organism	Frequency of isolation[a]	Disease involvement[b]
Achromobacter xylosoxidans	C	2
Adenoviruses	B	2
Aspergillus spp.	B	2
Bacteroides spp.	C	2
Bifidobacterium spp.	C	2
Branhamella (*Moraxella*) *catarrhalis*	B	2
Candida spp.	B	2
Chlamydia trachomatis	C	2
Clostridium spp.	C	2
Corynebacterium diphtheriae	C	3
Corynebacterium spp.	C	1
Coxsackievirus	B	2
Enterobacteriaceae	B	2
Enteroviruses	B	2
Eubacterium spp.	C	1
Fusobacterium spp.	C	2
Haemophilus influenzae	B	3
Herpes simplex virus	C	2
Influenza viruses	B	2
Mycobacterium spp.	C	2
Parainfluenza viruses	B	2
Peptococcus spp.	C	2
Peptostreptococcus spp.	C	2
Propionibacterium spp.	B	2
Pseudomonas aeruginosa	B	2
Respiratory syncytial virus	B	2
Staphylococcus spp.	B	2
Streptococcus spp.	B	2
Streptococcus pneumoniae	B	3
Varicella-zoster virus	B	2
Veillonella spp.	C	2
Vibrio spp.	C	2

[a]A, commonly encountered in clinical specimens; B, occasionally encountered in clinical specimens; C, rarely encountered in clinical specimens.

[b]1, When present, rarely if ever involved in disease production; 2, when present, occasionally involved in disease production; 3, when present, commonly involved in disease production.

tis patients may harbor viridans streptococci, staphylococci, enterococci, corynebacteria, *Bacteroides* spp., Hacek organisms, and yeasts. Fungi have been found in blood cultures from patients with lung abscesses, mycotic aneurysms, or hematological disorders and from immunocompromised individuals. Clostridia have been recovered from the blood of accident victims and from patients with cholangitis and postpartum complications. In the newborn with bacteremia or septicemia during the first week of life, the organisms most often responsible are *Streptococcus agalactiae* (23) and *S. aureus*.

Some viruses, including human immunodeficiency virus, hepatitis B virus, hepatitis C virus (formerly called hepatitis non-A non-B virus), and cytomegalovirus, persist in circulating blood and blood cells of individuals in the absence of clinical disease. The potential risk of transmitting these agents via transfusion of or exposure to blood products from chronically infected persons is well recognized. Blood is the major diagnostic specimen for all plasmodia, *Babesia* spp., *Brugia* spp., *Loa loa*, *Mansonella* spp., *Wuchereria* spp., some microsporidia, and hemoflagellates.

Positive cultures obtained from cerebrospinal fluids also reflect a variety of conditions. Several conditions besides meningitis, including trauma, infectious complications of surgery, cranial and spinal epidural abscesses, subdural abscess, septic thrombophlebitis of the venous sinuses, and brain abscesses, contribute to positive findings. There are systemic infectious diseases that may afflict the meninges, including severe pneumococcal pneumonia, salmonellosis, *H. influenzae* pneumonia and bacteremia, tuberculosis, listeriosis, *Acinetobacter* septicemia of the newborn, generalized candidiasis, and advanced sepsis with every type of gram-negative rod, especially *E. coli* and *P. aeruginosa*. Neonatal meningitis is most commonly caused by *S. agalactiae*, gram-negative rods, and listeriae. In children, meningitis is caused primarily by *H. influenzae* serotype B in individuals not immunized against this bacterium; in most adults, the causative agents are *S. pneumoniae*, meningococci, and pneumococci; the aged are most frequently infected with the pneumococci. *S. aureus*, Enterobacteriaceae, *P. aeruginosa*, and *L. monocytogenes* are sometimes recovered. *Flavobacterium meningosepticum* has been isolated occasionally. Viruses regularly associated with aseptic meningitis and easily recovered from cerebrospinal fluid include members of the enteroviruses (coxsackieviruses and echoviruses) and mumps virus. Viruses associated with true encephalitis (herpes simplex virus, the arboviruses, and rabies virus) are rarely recovered. Traumatic or surgical injury may be complicated by any one of many microorganisms, but the staphylococci, enterococci, and gram-negative rods predominate. In intracranial abscesses, anaerobic microorganisms are most common. These include peptostreptococci; *Bacteroides*, *Porphyromonas*, *Prevotella*, *Fusobacterium*, and *Actinomyces* spp.; *P. acnes*; and, less frequently, *Veillonella* spp. Cranial epidural abscesses have yielded staphylococci, streptococci, pneumococci, and occasionally gram-negative rods. Spinal epidural abscesses have involved *S. aureus* and *P. aeruginosa*. Other gram-positive cocci have been isolated, as have *Salmonella* spp., which have caused epidural abscesses while infecting vertebrae (3). Subdural abscess is a comparatively rare complication of *H. influenzae* meningitis in children. *M. tuberculosis* infection and leptospiral disease of the meninges must also be considered. *Cryptococcus neoformans* and *Toxoplasma gondii* have assumed important infectious roles in AIDS patients. Free-living amebae such as *Naegleria fowleri* may be encountered in cerebrospinal fluid on rare occasions.

Transudates and exudates accompany a large variety of clinical conditions. They may be sterile or contain a varied microflora. Any microorganisms that involve the afflicted organs may be recovered. The organisms found reflect the anatomical site of the disease nidus or adjacent areas. The organisms outlined in the foregoing sections may also be encountered. These include the anaerobic bacteria, fungi, *M. tuberculosis*, and legionellae.

SURGICAL SPECIMENS

See also references 3, 4, 6, 9, 15, 16, 21, 22, and 24.

The microbiology of surgical specimens depends on the anatomic site and the underlying disease, the care exercised at the time of excision and during subsequent handling, and the available information that will help narrow the number of possible etiological agents. The usual pyogenic organisms may be recovered, often as single components. Systemic fungi, nocardiae, actinomycetes, and various mycobacteria, including, of course, M. *tuberculosis*, as well as mixtures of anaerobic bacteria have been found. In addition to yielding S. *aureus*, primary osteomyelitis patients have yielded peptostreptococci, *Salmonella* spp. (especially children with sickle cell disease), gonococci, *Veillonella* spp., representatives of the fecal gram-negative rods, and rarely actinomycetes and *Blastomyces dermatitidis*. Foreign bodies, including prostheses, old sutures, and indwelling catheters, have been sources of many species of organisms, among them staphylococci, P. *aeruginosa*, and *Enterobacteriaceae*.

AUTOPSY SPECIMENS

Autopsy specimens can yield invaluable information when the examination is performed promptly and aseptically and when cultures are plated immediately and accompanied by smears. Routinely, the kidney, spleen, liver, and lung are negative unless distinct lesions are cultured or an appreciable portion of the organ is diseased. If more than 6 h has elapsed between death and the examination, heart blood cultures and the tissues will reflect postmortem or perimortem contamination by organisms from the large bowel. An experienced and interested prosector will culture the areas of tissues possibly involved in an infectious disease process. A large variety of suspected and totally unexpected microorganisms can be recovered and demonstrated subsequently in tissue sections. The microorganisms range from the usual bacteria to lesions populated by brucellae, nocardiae, penicillia, aspergilli, zygomycetes, pneumococci, group A streptococci, clostridia, salmonellae, and shigellae, to name but a few.

MICROBIAL NUMBERS AND CLINICAL DISEASE

The preceding considerations underline the lack of understanding, in quantitative terms, of the many factors that contribute to overt symptoms of infectious disease in any particular person. The notion that disease production in the host resides solely in the pathogenic and virulent properties of the microorganisms can no longer be accepted. The application of molecular biology has initiated a better understanding of the mechanisms of microbial attachment to host cells; the advantages of certain plasmids and episomes in the expression of filaments, fimbriae, and fibrillae; the utility of some lipoteichoic acids; and the protection of colonizing microorganisms by exopolysaccharides (1). However, the specific and nonspecific defense mechanisms of a host, the general health of the individual, and the various stresses to which the host has been subjected have considerable influence, if not the determining role, in the initiation and progression of an infectious disease. These roles are as multifarious as the genera and species of microorganisms that constitute the microbial ecosystem of the person. An understanding of microbial ecological relationships among the constituents of this large pool of microbes

is lacking, especially with regard to the health of the host. We know very little about the selectivity exercised by the tissues and organs of the host in the establishment of a local, autochthonous microbiota. There are but a few indications that age, physical environment, and nutrition may affect the minimal infective dose of some microorganisms with more pronounced pathogenic proclivities. As with other areas of human biology, it is the abnormal condition that permits an intimation of understanding. Microorganisms of nosocomial significance are widely distributed in nature but are usually involved in disease complications only in medical facilities. They may be minority members of an individual's microbiota, or they may reside in the hospital environment, where in some instances they are concentrated as a result of antimicrobial agent-associated selective pressures (13).

Although it is obvious that much must be learned before a clear understanding of infectious disease is established (19), it is equally clear that the concern and responsibility of the clinical microbiologist must transcend the narrow borders of microbial identification and encompass aspects of the history, condition, treatment, and present and past environments and community of the patient. This broader view is required to allow the proper distinction of significant microorganisms isolated from pathological specimens. Without such information, the unclear lines of separation between the indigenous and pathogenic microbiota of humans will be confused still more, depriving clinicians of information and the ability to intercede effectively against microorganisms for the benefit of their patients (12).

REFERENCES

1. **Beachey, E. H.** 1980. *Bacterial Adherence*. Chapman & Hall, Ltd., London.
2. **Burnet, F. M.** 1962. *Natural History of Infectious Disease*, 3rd ed. Cambridge University Press, Cambridge.
3. **Dalton, H. P., and H. C. Nottebart, Jr.** 1986. *Interpretive Medical Microbiology*. Churchill Livingstone, New York.
4. **D'Amato, R. F., M. F. Sierra, and M. R. McGinnis.** 1977. Infectious diseases and etiologic or associated bacterial/fungal agents, p. 11–38. *In* D. Seligson (ed.), *CRC Handbook Series in Clinical Laboratory Science*. CRC Press, Inc., Cleveland.
5. **Drasar, B. S., and M. J. Hill.** 1974. *Human Intestinal Flora*. Academic Press, Inc. (London), Ltd., London.
6. **Drew, W. L.** 1986. Laboratory methods in basic virology, p. 632–677. *In* S. M. Finegold and E. J. Baron (ed.), *Bailey and Scott's Diagnostic Microbiology*. The C. V. Mosby Co., St. Louis.
7. **Dubos, R. J.** 1958. The evolution and the ecology of microbial diseases, p. 4–27. *In* R. J. Dubos (ed.), *Bacterial and Mycotic Infections of Man*, 3rd ed. J. B. Lippincott Co., Philadelphia.
8. **Fiumara, N. J., H. M. Wise, Jr., and M. Many.** 1967. Gonorrheal pharyngitis. *N. Engl. J. Med.* **276:**1248–1250.
9. **Garcia, L. S., and D. A. Bruckner (ed.).** 1993. *Diagnostic Medical Parasitology*, 2nd ed. American Society for Microbiology, Washington, D.C.
10. **Gibbons, R. J.** 1975. Attachment of oral streptococci to mucosal surfaces, p. 127–131. *In* D. Schlessinger (ed.), *Microbiology—1975*. American Society for Microbiology, Washington, D.C.
11. **Isenberg, H. D.** 1988. Pathogenicity and virulence: another view. *Clin. Microbiol. Rev.* **1:**40–53.
12. **Isenberg, H. D., and A. Balows.** 1981. Bacterial pathogenicity in man and animals, p. 8–122. *In* M. P. Starr, H. Stolp, H. G. Truper, A. Balows, and H. G. Schlegel (ed.), *The Prokaryotes*. Springer-Verlag, New York.

13. **Isenberg, H. D., and J. I. Berkman.** 1971. The role of drug-resistant and drug-selected bacteria in nosocomial disease. *Ann. N.Y. Acad. Sci.* **182:**52–58.

14. **Jones, G. L., and G. A. Hebert (ed.).** 1979. *"Legionnaires' ": the Disease, the Bacterium and Methodology.* Center for Disease Control, Atlanta.

15. **Krieg, N. R., and J. G. Holt (ed.).** 1984. *Bergey's Manual of Systematic Bacteriology,* vol. 1. The Williams & Wilkins Co., Baltimore.

16. **Rosebury, T.** 1961. *Microorganisms Indigenous to Man.* McGraw-Hill Book Co., Inc., New York.

17. **Skinner, F. A., and J. G. Carr.** 1974. *The Normal Microflora of Man.* Academic Press, Inc. (London), Ltd., London.

18. **Smibert, R. M.** 1978. The genus *Campylobacter. Annu. Rev. Microbiol.* **32:**673–709.

19. **Smith, H.** 1972. The little-known determinants of microbial pathogenicity, p. 1–24. *In* H. Smith and J. H. Pearce (ed.), *Microbial Pathogenicity in Man and Animals.* Cambridge University Press, Cambridge.

20. **Smith, T.** 1934. *Parasitism and Disease.* Princeton University Press, Princeton, N.J.

21. **Sneath, P. H. A., N. S. Mair, M. E. Sharpe, and J. G. Holt (ed.).** 1986. *Bergey's Manual of Systematic Bacteriology,* vol. 2. The Williams & Wilkins Co., Baltimore.

22. **Staley, J. T., M. P. Bryant, N. Pfennig, and J. G. Holt (ed.).** 1989. *Bergey's Manual of Systematic Bacteriology,* vol. 3. The Williams & Wilkins Co., Baltimore.

23. **Wilkinson, H. W.** 1978. Group B streptococcal infections. *Annu. Rev. Microbiol.* **32:**41–57.

24. **Williams, S. T., M. E. Sharpe, and J. G. Holt (ed.).** 1978. *Bergey's Manual of Systematic Bacteriology,* vol. 4. The Williams & Wilkins Co., Baltimore.

Specimen Collection, Transport, and Storage

J. MICHAEL MILLER AND HARVEY T. HOLMES

3

In terms of the effectiveness of the laboratory, nothing is more important than the appropriate selection, collection, and transportation of a specimen. When specimen collection and management are not priorities, the laboratory can contribute little to patient care. Consequently, all members of the medical staff involved in these processes must understand the critical nature of maintaining specimen quality throughout the testing process. It is the responsibility of the laboratory to provide the necessary information in a form that can be easily incorporated into each department's nursing manual, which should include specific criteria for safety, selection, collection, transportation, specimen acceptability, and labeling. This chapter provides practical guidelines for the proper collection and handling of specimens destined for analysis in the clinical microbiology laboratory.

SAFETY

1. All specimen collection procedures must be performed while wearing gloves, laboratory coat, and, where appropriate, masks and/or goggles (36).
2. All specimen containers should be leakproof and transported within a sealable, leakproof plastic bag having a separate compartment for paperwork (37).
3. Never transport syringes with needles to the laboratory. Instead transfer the contents to a sterile tube, or remove the needle (with a protective device), recap the syringe, and place the syringe in a sealable, leakproof plastic bag (43).
4. Do not transport leaking specimen containers to the laboratory or process them. They should be autoclaved, and another specimen must be submitted (34).

SELECTION AND COLLECTION

1. Avoid commensal contamination from indigenous flora to ensure representative sampling of the infectious process (Table 1) (5).
2. Select the correct anatomic site from which to obtain the specimen, and use the proper technique and supplies, as shown in the tables, to collect specimens.
3. For selecting appropriate sites for anaerobic cultures, refer to Table 2; biopsy samples or needle aspirates are the specimens of choice, while anaerobic swabs are the least

desirable (23, 34). Never refrigerate specimens for anaerobic culture; instead, maintain them at room temperature (19).
4. Collect adequate volumes. Insufficient material may yield false-negative results.
5. Label each specimen with patient's name, identification number, source (i.e., specific site), date, time of collection, and initials of collector (12).
6. Place the specimen in a container designed to promote survival of suspected agents and to eliminate leakage and potential safety hazards.

TRANSPORTATION

1. All specimens must be *promptly* transported to the laboratory, preferably within 2 h (24). If processing is delayed, store specimens for bacterial agents under the specific conditions shown in Table 3.
2. In general, do not store bacterial agents for more than 24 h. Viruses, however, usually remain stable for 2 to 3 days at 4°C (24, 25).
3. Optimal transport of clinical specimens, including anaerobic cultures, depends primarily on the volume of material obtained. Submit small amounts within 15 to 30 min; biopsy tissue may be maintained for up to 20 to 24 h if it is stored at 25°C in an anaerobic transport system (23).
4. Environmentally sensitive organisms include *Shigella* spp. (process immediately), *Neisseria gonorrhoeae*, *Neisseria meningitidis*, and *Haemophilus influenzae* (sensitive to cold

TABLE 1 Sources of contamination with commensal microbiota

Site of infection	Source of contamination
Bladder	Urethra and perineum
Blood	Venipuncture site
Endometrium	Vagina
Fistula	Gastrointestinal tract
Middle ear	External ear canal
Nasal sinus	Nasopharynx
Subcutaneous infections and superficial wounds	Skin and mucous membranes

TABLE 2 Suitability of various clinical materials for anaerobic culture

Acceptable material	Unacceptable material
Aspirate (by needle or syringe)	Bronchoalveolar washing not protected
Bartholin's gland	Cervical swab
Bile	Endotracheal aspirate
Blood	Endocervical swab, contaminated
Bone marrow	Lochia
Bronchoscopic, protected brushing	Nasopharyngeal swab
Culdocentesis specimen	Perineal swab
Fallopian tube	Prostatic or seminal fluid
IUD,[a] for *Actinomyces* spp.	Sputum, expectorated
Ovary	Sputum, induced
Placenta, via cesarean section	Stool[b] or rectal samples
Sinus aspirate	Throat swab
Stool, for *Clostridium* spp.	Tracheostomy aspirate
Surgery, swab	Urethral swab
Surgery, tissue	Urine, from bladder or catheter
Transtracheal aspirate	Urine, voided
Uterus, endometrial aspirate	Vaginal or vulval swab
Urine, suprapubic aspirate	

[a]IUD, intrauterine device.
[b] There are a few exceptions, e.g., for botulism (especially infant botulism), *Clostridium perfringens*-caused food-borne disease, and *Clostridium difficile*-caused antibiotic-associated pseudomembranous colitis. Some malabsorption syndromes may require detection of overcolonization of the upper intestine.

TABLE 4 Specimens to be discouraged because of questionable microbial information

Specimen type	Alternative or comment
Burn wound, swab	Submit tissue or aspirate.
Colostomy, discharge	Do not process.
Decubitus, swab	Submit tissue or aspirate.
Foley catheter tip	Do not process.
Gastric aspirate of newborn	Do not process.
Lochia	Do not process.
Perirectal abscess, swab	Submit tissue or aspirate.
Gangrenous lesion, swab	Submit tissue or aspirate.
Periodontal lesion, swab	Submit tissue or aspirate.
Varicose ulcer, swab	Submit tissue or aspirate.
Vomitus	Do not process.

TABLE 3 Storage conditions for bacterial transport systems[a]

Transport system	Specimen to hold at storage temp of:		Transport system	Specimen to hold at storage temp of:	
	4°C	25°C		4°C	25°C
No preservative	Autopsy tissue	CSF, bacterial agents	Directly inoculated medium		Corneal scraping
	Bronchial wash	Synovial fluid			Blood culture
	Catheter, i.v.				*Bordetella* spp. (RL, BG)
	CSF, viral agents				Gonorrheal specimen (JEMBEC)
	Lung biopsy sample				Vitreous humor
	Pericardial fluid				
	Sputum				
	Urine, CATH, CC, MS				
			Transport medium[b]	Burn wound biopsy sample	Bone marrow
Anaerobic transport		Abdominal fluid		*Campylobacter* spp.	*Bordetella* spp.
		Actinomyces spp.		Ear, external, sample	Cervical swab
		Amniotic fluid		*Shigella* spp.	Conjunctival swab
		Anaerobic culture		*Vibrio* spp.	*Corynebacterium* spp.
		Bile		*Yersinia* spp.	Ear, internal, swab
		Cul-de-sac			Genital culture
		Deep lesion			Nasopharyngeal swab
		Lung aspirate			*Neisseria* spp.
		Sinus aspirate			*Salmonella* spp.
		Tissue, surgery			Upper respiratory culture
		Transtracheal aspirate			
		Urine, suprapubic			

[a]BG, Bordet-Gengou; CATH, catheter; CC, clean catch; CSF, cerebrospinal fluid; i.v., intravenous; MS, midstream; RL, Regan-Lowe medium.
[b]Stuart's medium, charcoal-impregnated swabs originally formulated for *N. gonorrhoeae* transport; Amies medium, modified Stuart's medium but incorporates charcoal in medium instead of swab; Cary-Blair medium, similar to Stuart's medium but modified for fecal specimens, with pH increased from 7.4 to 8.4.

TABLE 5 Bacteriology and mycology specimen collection guidelines[a]

Specimen type (reference)	Collection		Time and temp		Replica limits	Comments
	Guidelines	Device and/or minimum vol	Transport[b]	Storage		
Abscess (44)	Remove surface exudate by wiping with sterile saline or 70% EtOH.					Tissue or fluid is always superior to swab specimen. If swabs must be used, collect 2, 1 for culture and 1 for Gram stain. Preserve them with Stuart's or Amies medium.
Open	Aspirate if possible, or pass swab deep into lesion and firmly sample lesion's advancing edge.	Swab transport system	≤2 h, RT	≤24 h, RT	1/day from same source	
Closed	Aspirate abscess wall material with needle and syringe. Aseptically transfer *all* material into anaerobic transport device or vial.	Anaerobic transport system, ≥1 ml	≤2 h, RT	≤24 h, RT	1/day from same source	Sampling of surface area can introduce colonizing bacteria not involved in infectious process.
Bite wound (13, 17, 44)	See Abscess.					Do not culture animal bite wounds ≤12 h old (agents are usually not recovered) unless they are on face or hand or unless signs of infection are present.
Blood culture (14, 30, 32, 40, 41)	Disinfection of culture bottle: apply 70% isopropyl alcohol to rubber stoppers and wait 1 min. Disinfection of venipuncture site 1. Cleanse site with 70% alcohol. 2. Swab concentrically, starting at center, with iodophor. 3. Allow iodine to dry. 4. *Do not palpate vein at this point.* 5. Collect blood. 6. After venipuncture, remove iodine from skin with alcohol.	Bacteria: blood culture vials Adult, 10–20 ml/set Infant, 1–2 ml/set Fungi 1. Biphasic culture 2. Lysis centrifuge	≤2 h, RT	≤24 h, RT, or per instructions	3 sets in 24 h	Acute sepsis: 2–3 sets from separate sites, all within 10 min Endocarditis, acute: 3 sets from 3 sites over 1–2 h Endocarditis, subacute: 3 sets from 3 sites taken ≥15 min apart; if negative at 24 h, obtain 3 more sets. Fever of unknown origin: 2–3 sets from separate sites ≥1 h apart; if negative at 24 h, obtain 2–3 more sets.
Catheter i.v. (16, 29, 31)	1. Cleanse skin around catheter site with alcohol. 2. Aseptically remove and clip 5-cm distal tip of catheter directly into sterile tube.	Sterile screw-cap tube or cup	≤15 min, RT	≤24 h, 4°C	None	Acceptable i.v. catheters for semiquantitative culture (Maki method): central, CVP, Hickman, Broviac, peripheral, arterial, umbilical, hyperalimentation, Swan-Ganz

(Continued on next page)

TABLE 5 Bacteriology and mycology specimen collection guidelines[a] *(Continued)*

Specimen type (reference)	Collection		Time and temp		Replica limits	Comments
	Guidelines	Device and/or minimum vol	Transport[b]	Storage		
	3. Transport directly to microbiology laboratory to prevent drying.					
Foley (34)	Do not culture, since growth represents distal urethral flora.					Not acceptable for culture.
Cellulitis (44, 46)	1. Cleanse site by wiping with sterile saline or 70% alcohol. 2. Aspirate area of maximum inflammation (commonly center rather than leading edge) with fine needle and syringe. 3. Draw small amt of sterile saline into syringe. 4. Remove needle (with protective device), and cap.	Capped syringe or sterile tube	≤15 min, RT	≤24 h, RT	None	Yield of potential pathogens is only 25–35%.
CSF (18, 39, 47)	1. Disinfect site with 2% iodine tincture. 2. Insert needle with stylet at L3-L4, L4-L5, or L5-S1 interspace. 3. Upon reaching subarachnoid space, remove stylet, and collect 1–2 ml of fluid into each of 3 leakproof tubes.	Sterile screw-cap tube Bacteria, ≥1 ml Fungi, ≥2 ml AFB, ≥2 ml Virus, ≥1 ml	Bacteria: never refrigerate ≤15 min, RT; Virus: send on ice ≤15 min, 4°C	≤24 h, RT ≤72 h, 4°C	None	Obtain blood cultures also. If only 1 tube of CSF is collected, submit it to microbiology laboratory first; otherwise, generally submit tube 2.
Decubitus ulcer (44)	1. Cleanse surface with sterile saline. 2. If biopsy sample is not available, *vigorously* swab base of lesion. 3. Place swab in appropriate transport system.	Swab transport or anaerobic system	≤2 h, RT	≤24 h, RT	1/day from same source	Decubitus swab provides little clinical information; discourage collection of it. Tissue biopsy sample or needle aspirate is specimen of choice.

Specimen	Collection procedure	Transport device	Transport time/temp	Transport time/temp	Replica limits	Comments
Dental culture: gingival, periodontal, periapical, Vincent's stomatitis	1. Carefully cleanse gingival margin and supragingival tooth surface to remove saliva, debris, and plaque. 2. Using periodontal scaler, carefully remove subgingival lesion material and transfer it to anaerobic transport system. 3. Prepare smears collected in same fashion.	Anaerobic transport system	≤2 h, RT	≤24 h, RT	1/day	Periodontal lesions should be processed only by laboratories equipped to provide specialized techniques for detection and enumeration of specific agents.
Ear Inner (2)	Tympanocentesis is reserved for complicated, recurrent, or chronic persistent otitis media. 1. For intact ear drum, clean ear canal with soap solution, and collect fluid via syringe aspiration technique. 2. For ruptured ear drum, collect fluid on flexible-shaft swab via auditory speculum.	Sterile tube, swab transport medium, or anaerobic system.	≤2 h, RT	≤24 h, RT	1/day from same source	If aspirate or biopsy, use anaerobic transport system, and transport for ≤2 h at RT for both aerobic and anaerobic cultures. Throat or nasopharyngeal cultures are not predictive of agents responsible for otitis media.
Outer (2)	1. Use moistened swab to remove any debris or crust from ear canal. 2. Obtain sample by firmly rotating swab in outer canal.	Swab transport	≤2 h, RT	≤24 h, 4°C	1/day from same source	For otitis externa, *vigorous* swabbing is required because surface swabbing may miss streptococcal cellulitis.
Eye Conjunctiva (1, 26)	1. Sample both eyes with separate swabs (premoistened with sterile saline) by rolling swab over each conjunctiva. 2. Inoculate medium at time of collection. 3. Smear swabs onto 2 slides for staining.	Direct culture inoculation: BAP and CHOC or swab transport medium	Plates: ≤15 min, RT; Swabs: ≤2 h, RT	≤24 h, RT	None	Sample both conjunctiva to determine indigenous microflora. Uninfected eye serves as a control.
Corneal scrapings (1, 26)	1. Obtain conjunctival swab specimens as described above. 2. Instill 2 drops of local anesthetic.	Direct culture inoculation: BHI+Bld, CHOC, and inhibitory mold agar	≤15 min, RT	≤24 h, RT	None	Take swabs for culture prior to anesthetic application; corneal scrapings can be obtained after.

(Continued on next page)

TABLE 5 Bacteriology and mycology specimen collection guidelines^a (*Continued*)

Specimen type (reference)	Collection		Time and temp		Replica limits	Comments
	Guidelines	Device and/or minimum vol	Transport^b	Storage		
	3. Using sterile spatula, scrape ulcers or lesions, and inoculate scraping directly onto medium. 4. Apply remaining material to 2 clean glass slides for staining.					
Feces						
Routine culture (42)	Pass directly into clean, dry container. Transport to microbiology laboratory within 1 h of collection, or transfer to enteric transport system.	Sterile, leakproof, wide-mouth container or enteric transport system, ≥2 g	Unpreserved: ≤1 h, RT	≤24 h, 4°C Enteric transport system: ≤48 h, RT	1/day	Do not routinely perform stool cultures for patients whose length of stay was >3 days and admitting diagnosis was not gastroenteritis. However, consider *Clostridium difficile.*
Clostridium difficile (3, 27, 42)	Pass liquid or soft stool directly into clean, dry container. Soft stool is defined as stool assuming shape of its container.	Sterile, leakproof, wide-mouth container, ≥5 ml	≤1 h, RT 1–24 h, 4°C >24 h, −20°C		1/2 days	Patients should be passing ≥5 stools with liquid or soft consistency per 24 h.
Escherichia coli O157: H7 (42)	Pass liquid and/or bloody stool into clean, dry container.	Sterile, leakproof, wide-mouth container or enteric transport system, >2 ml	Unpreserved: ≤1 h, RT	≤24 h, 4°C Enteric transport system: ≤48 h, RT	1/day	Bloody or liquid stools collected within 6 days of onset among patients with abdominal cramps have highest yield.
Leukocytes (21)	Pass directly into clean, dry container. Transport to microbiology laboratory within 1 h of collection, or transfer to ova and parasite transport system (10% formalin and/or PVA).	Sterile, leakproof, wide-mouth container or 10% formalin and/or PVA, >2 ml	Unpreserved: ≤1 h, RT	≤24 h, 4°C In formalin or PVA: indefinite, RT	1/day	
Rectal swab	1. Carefully insert swab ≈1 in. (2.54 cm) beyond anal sphincter. 2. Gently rotate swab to sample anal crypts.	Swab transport	≤2 h, RT	≤24 h, RT	1/day	Reserved for detecting gonorrhea, *Shigella* and *Campylobacter* spp., HSV, and anal carriage of *Streptococcus pyogenes* or for patients unable to pass specimen. Feces should be evident on the swab.
Fistulas	See Abscess.					

Specimen	Collection	Transport device	Transport time	Storage	No.	Comments
Fluids: abdominal, ascites, bile, joint, pericardial, peritoneal, pleural, synovial	1. Disinfect overlying skin with 2% iodine tincture. 2. Obtain specimen via percutaneous needle aspiration or surgery. 3. Transport immediately to laboratory. 4. Always submit as much fluid as possible; *never submit swab dipped in fluid.*	Sterile screw-cap tube or anaerobic transport system, ≥1 ml	≤15 min, RT	≤24 h, RT, except pericardial fluid and fungal cultures	None	Hold pericardial fluid at 4°C for ≤24 h if processing is delayed. Store fluids for fungal cultures at 4°C.
Gangrenous tissue	See Abscess.					Discourage sampling of superficial tissue; tissue biopsy samples or aspirates are preferred.
Gastric: wash or lavage fluid (6)	Collect in early morning before patients eat and while they are still in bed. 1. Introduce nasogastric tube orally or nasally to stomach. 2. Perform lavage with 25–50 ml of chilled, sterile, distilled water. 3. Recover sample and place in leakproof sterile container. 4. Before removing tube, release suction and clamp it.	Sterile leakproof container	≤15 min, RT, or neutralize within 1 h of collection	≤24 h, 4°C	1/day	Specimen must be processed promptly, because mycobacteria die rapidly in gastric washings. Neutralize each 35–50 ml of gastric washings with 1.5 ml of 40% anhydrous Na_2HPO_4.
Genital: female Amniotic (48)	1. Aspirate via amniocentesis, cesarean section, or intrauterine catheter. 2. Transfer fluid to anaerobic transport system.	Anaerobic transport system, ≥1 ml	≤15 min, RT	≤24 h, RT	None	Swabbing or aspiration of vaginal membrane is *not* acceptable because of vaginal contamination.
Bartholin	1. Disinfect skin with 2% iodine tincture. 2. Aspirate fluid from ducts.	Anaerobic transport system, ≥1 ml	≤2 h, RT	≤24 h, RT	1/day	
Cervix (15)	1. Visualize cervix with speculum without lubricant. 2. Remove mucus and/or secretions from cervix with swab, and discard swab.	Swab transport	≤2 h, RT	≤24 h, RT	1/day	Viral and chlamydial tests require separate collection and transport kits.

(Continued on next page)

TABLE 5 Bacteriology and mycology specimen collection guidelines[a] (Continued)

Specimen type (reference)	Collection		Time and temp		Replica limits	Comments
	Guidelines	Device and/or minimum vol	Transport[b]	Storage		
	3. Firmly yet gently, sample endocervical canal with sterile swab.					
Cul-de-sac	Submit aspirate or fluid.	Anaerobic transport system, >1 ml	≤2 h, RT	≤24 h, RT	1/day	
Endometrium	1. Collect transcervical aspirate via telescoping catheter. 2. Transfer entire amount to anaerobic transport system.	Anaerobic transport system, ≥1 ml	≤2 h, RT	≤24 h, RT	1/day	
Products of conception	1. Submit portion of tissue in sterile container. 2. If obtained by cesarean section, immediately transfer to anaerobic transport system.	Sterile tube or anaerobic transport system	≤2 h, RT	≤24 h, RT	1/day	Do not process lochia. See Table 2.
Urethra (15)	1. Remove exudate from urethral orifice. 2. Collect discharge material on swab by massaging urethra against pubic symphysis through vagina.	Swab transport	≤2 h, RT	≤24 h, RT	1/day	If no discharge can be obtained, wash external urethra with betadine soap, and rinse with water. Then insert urethrogenital swab 2–4 cm into urethra, and rotate swab for 2 s.
Vagina (15)	1. Wipe away excessive amt of secretion or discharge. 2. Obtain secretions from mucosal membrane of vaginal vault with sterile swab. 3. If smear is also requested, obtain it with second swab.	Swab transport	≤2 h, RT	≤24 h, RT	1/day	For intrauterine devices, place entire device into sterile container, and submit at room temperature.
Genital: female or male Lesion	1. Clean lesion with sterile saline, and remove lesion's surface with sterile scalpel blade. 2. Allow transudate to accumulate. 3. While pressing base of lesion, firmly sample exudate with sterile swab.	Swab transport	≤2 h, RT	≤24 h, RT	1/day	To rule out syphilis, touch glass slide to transudate, add coverslip, and transport directly to laboratory in humidified chamber (petri dish with moist gauze).

Specimen	Collection procedure	Container and minimum volume	Transport time and temperature	Replicate limits	Comments
Genital: male Prostate (43)	1. Clean glans with soap and water. 2. Massage prostate through rectum. 3. Collect fluid on sterile swab or in sterile tube.	Swab transport or sterile tube	≤2 h, RT / ≤24 h, RT	1/day	More relevant results may be obtained by also using urine specimens obtained immediately before and after massage.
Urethra	Insert urethrogenital swab 2–4 cm into urethral lumen, rotate swab, and leave it in place for at least 2 s.	Swab transport	≤2 h, RT / ≤24 h, RT	1/day	
Hair: dermatophytosis (20)	1. With forceps, collect at least 10–12 affected hairs with bases of shafts intact. 2. Place in clean tube or container.	Clean container, 10 hairs	≤24 h, RT	1/day from same site	Collect scalp scales, if present, along with scrapings of active borders of lesions. Note any antifungal therapy taken recently.
Nail: dermatophytosis (20)	1. Wipe nail with 70% alcohol. Use gauze (not cotton). 2. Clip away generous portion of affected area, and collect material or debris from under nail. 3. Place material in clean container.	Clean container, enough scrapings to cover head of thumbtack	≤24 h, RT	1/day	
Pilonidal cyst	See Abscess.				
Respiratory tract, lower BAL, BBW, tracheal aspirate	1. Place aspirate or washing into sputum trap. 2. Place brush in sterile container with saline.	Sterile container, >1 ml	≤2 h, RT / ≤24 h, 4°C	1/day	
Sputum, expectorate (4)	1. Collect specimen under *direct* supervision of nurse or physician. 2. Have patient rinse or gargle with water.	Sterile container, >1 ml	≤2 h, RT / ≤24 h, 4°C	1/day	For pediatric patients unable to produce specimen, respiratory therapist should collect via suction. Best specimen should have ≤10 squamous cells/100 fields.

(Continued on next page)

TABLE 5 Bacteriology and mycology specimen collection guidelines[a] (Continued)

| Specimen type (reference) | Collection | | Time and temp | | Replica limits | Comments |
	Guidelines	Device and/or minimum vol	Transport[b]	Storage		
	3. Instruct patient to cough *deeply* to produce lower respiratory specimen (not postnasal fluid). Collect into sterile container.					
Sputum, induced (4)	1. Have patient rinse mouth with water after brushing gums and tongue. 2. With aid of nebulizer, have patient inhale ≈25 ml of 3–10% sterile saline. 3. Collect induced sputum into sterile container.	Sterile container	≤2 h, RT	≤24 h, RT	1/day	*Histoplasma capsulatum* and *Blastomyces dermatitidis* survive for only short periods once specimen is obtained.
Respiratory tract, upper Oral	1. Remove oral secretions or debris from surface of lesion with swab, and discard swab. 2. Using second swab, vigorously sample lesion, avoiding any areas of normal tissue.	Swab transport	≤2 h, RT	≤24 h, RT	1/day	Discourage sampling of superficial tissue for bacterial evaluation. Tissue biopsy samples or needle aspirates are specimens of choice.
Nasal	1. Use swab premoistened with sterile saline. Insert ≈2 cm into nares. 2. Rotate swab against nasal mucosa.	Swab transport	≤2 h, RT	≤24 h, RT	1/day	Anterior nose cultures are reserved for detecting staphylococcal and streptococcal carriers or for nasal lesions.
Nasopharynx (2)	1. Gently insert calcium alginate swab into posterior nasopharynx via nose. 2. Rotate swab slowly for 5 s to absorb secretions. Remove swab; inoculate medium at bedside, or place swab in transport medium.	Direct medium inoculation or swab transport	Plates: ≤15 min, RT Swabs: ≤2 h, RT	≤24 h, RT	1/day	Dacron, rayon, or Calgiswab may be necessary for specific organism isolation.
Throat	1. Depress tongue with tongue depressor. 2. Sample posterior pharynx, tonsils, and inflamed areas with sterile swab.	Swab transport	≤2 h, RT	≤24 h, RT	1/day	Throat cultures are contraindicated for patient with inflamed epiglottis.

Specimen	Collection procedure	Container	Transport time and temp	Replicate limits	Comments
Skin: dermatophytosis (20)	1. Cleanse affected area with 70% alcohol. 2. Gently scrape surface of skin at *active margin of lesion. Do not draw blood.* 3. Place sample in clean container.	Clean container, enough scrapings to cover head of thumbtack	≤24 h, RT	1/day from same site	If specimen is submitted between glass slides, tape slides together and submit in envelope.
Tissue	1. Submit in sterile container. 2. For small samples, add several drops of sterile saline to keep moist. 3. *Do not allow tissue to dry out.* 4. Place in anaerobic transport system.	Anaerobic transport system	≤15 min, RT	None	Always submit as much tissue as possible. *Never* submit swab that has simply been rubbed over surface.
Urine Female, midstream (11)	1. Thoroughly clean urethral area with soap and water. 2. Rinse area with wet gauze pads. 3. While holding labia apart, begin voiding. 4. After several ml have passed, collect midstream portion without stopping flow of urine.	Sterile wide-mouth container, ≥1 ml, or urine transport kit	Unpreserved: ≤2 h, RT ≤24 h, 4°C Preserved: ≤24 h, RT	1/day	
Male, midstream (11)	1. Clean the glans with soap and water. 2. Rinse area with wet gauze pads. 3. While holding foreskin retracted, begin voiding. 4. After several ml have passed, collect midstream portion without stopping flow of urine.	Sterile wide-mouth container, ≥1 ml, or urine transport kit	Unpreserved: ≤2 h, RT ≤24 h, 4°C Preserved: ≤24 h, RT	1/day	
Straight catheter (11)	1. Thoroughly clean urethral area with soap and water. 2. Rinse area with wet gauze pads. 3. Aseptically insert catheter into bladder. 4. Allow ≈15 ml to pass; then collect urine to be submitted in sterile container.	Sterile leakproof container	Unpreserved: ≤2 h, RT ≤24 h, 4°C Preserved: ≤24 h, RT	1/day	Not recommended for routine urine culture because of potential contamination problems. Procedure may introduce urethral flora into bladder.

(Continued on next page)

TABLE 5 Bacteriology and mycology specimen collection guidelines[a] (Continued)

Specimen type (reference)	Collection		Time and temp[b]		Replica limits	Comments
	Guidelines	Device and/or minimum vol	Transport[b]	Storage		
Indwelling catheter (43)	1. Disinfect catheter collection port with 70% alcohol. 2. Use needle and syringe to aseptically collect 5–10 ml of urine. 3. Transfer sample to sterile tube or container.	Sterile leakproof container	Unpreserved: ≤2 h, RT	≤24 h, 4°C Preserved: ≤24 h, RT	1/day	
Wound	See Abscess.					

[a]Abbreviations: AFB, acid-fast bacilli; BAL, bronchoalveolar lavage; BAP, blood agar plate; BBW, bronchial brushing or washing; BHI+Bld, brain heart infusion plus 10% sheep blood; CHOC, chocolate agar; CSF, cerebrospinal fluid; CVP, central venous pressure; EtOH, ethanol; HSV, herpes simplex virus; i.v., intravenous; PVA, polyvinyl alcohol fixative; RT, room temperature.
[b]Transport all specimens in leakproof plastic bags with a separate compartment for the requisition.

temperatures). Never refrigerate spinal fluid, genital, eye, or internal ear specimens (34).

SPECIMEN ACCEPTABILITY AND REJECTION CRITERIA

At times, specimens arriving in the laboratory may have been improperly selected, collected, or transported. Processing and reporting results from these specimens may provide misleading information to the physician that can lead to misdiagnosis and inappropriate therapy. Therefore, the laboratory must adhere to a strict policy of specimen acceptance and rejection.

Listed below are several examples of situations for which specific laboratory policies must be formulated and enforced to ensure specimen quality.

1. No label. Do not process. Contact physician or nurse. For noninvasive specimens (urine, sputum, or throat), have the floor resubmit it. For specimens from invasive procedures (needle aspirates, body fluids, or tissues), process only after consulting with the physician who obtained the specimen. Note the problem on the report, and document the corrective action taken.

2. Prolonged transport. Do not process. Alert the submitter and request a repeat specimen. Note the problem on the patient's report: "Received after prolonged delay."

3. Improper or leaking container. Do not process. Call the submitter and request a repeat specimen. Document the problem and the corrective action taken.

4. Specimen unsuitable for request (e.g., anaerobic request from aerobic transport). Do not process. Contact the submitter, clarify the test request, and indicate the discrepancy. Request a proper specimen for the test requested.

5. Duplicate specimens on same day for the same request (except blood). Do not process. Place the specimen in a proper preservative at the correct storage temperature. Call the submitter and explain the duplication. Note the problem on the report.

There may be instances in which a given specimen is processed despite its quality being compromised, e.g., in a difficult or unusual case or for an infectious disease consultation. Table 4 lists specimens that provide little if any clinical information; processing of these specimens should be discouraged.

Specimen management for parasitology and virology is discussed in the relevant chapters of this Manual. In addition, excellent resources are available to assist in the development of parasitology (28, 33, 35, 38, 45) and virology (7–10, 22) specimen management manuals.

SPECIMEN MANAGEMENT

Tables 1 through 5 give the salient features of specimen management in response to requests commonly received in many clinical microbiology laboratories.

REFERENCES

1. **Baker, A. S., B. Paton, and J. Haaf.** 1989. Ocular infections: clinical and laboratory considerations. *Clin. Microbiol. Newsl.* **11:**97–101.
2. **Bannatyne, R. M., C. Clausen, and L. R. Mcarthy.** 1979. *Cumitech 10, Laboratory Diagnosis of Upper Respiratory Tract Infections.* Coordinating ed., I. B. R. Duncan. American Society for Microbiology, Washington, D.C.

3. **Bannister, E. R.** 1993. *Clostridium difficile* and toxin detection. *Clin. Microbiol. Newsl.* **15:**121–123.

4. **Bartlett, J. G., K. J. Ryan, T. F. Smith, and W. R. Wilson.** 1987. *Cumitech 7A, Laboratory Diagnosis of Lower Respiratory Tract Infections.* Coordinating ed., J. A. Washington II. American Society for Microbiology, Washington, D.C.

5. **Bartlett, R. C.** 1985. Quality control, p. 14–23. *In* E. H. Lennette, A. Balows, W. J. Hausler, Jr., and H. J. Shadomy (ed.), *Manual of Clinical Microbiology,* 4th ed. American Society for Microbiology, Washington, D.C.

6. **Carr, D. T., A. G. Karlson, and G. G. Stillwell.** 1967. A comparison of cultures of induced sputum and gastric washings in the diagnosis of tuberculosis. *Mayo Clin. Proc.* **42:**23–25.

7. **Chernesky, M. A., C. G. Ray, and T. F. Smith.** 1982. *Cumitech 15, Laboratory Diagnosis of Viral Infections.* Coordinating ed., W. L. Drew. American Society for Microbiology, Washington, D.C.

8. **Cherry, J. D., and M. J. Miller.** 1992. Use of the virology laboratory, p. 2363–2369. *In* R. D. Feigin and J. D. Cherry (ed.), *Textbook of Pediatric Infectious Diseases,* 3rd ed. The W. B. Saunders Co., Philadelphia.

9. **Chonmaitree, T., C. D. Baldwin, and H. L. Lucia.** 1989. Role of the virology laboratory in diagnosis and management of patients with central nervous system disease. *Clin. Microbiol. Rev.* **2:**1–14.

10. **Christensen, M. L.** 1989. Human viral gastroenteritis. *Clin. Microbiol. Rev.* **2:**51–89.

11. **Clarridge, J. E., M. T. Pezzlo, and K. L. Vosti.** 1987. *Cumitech 2A, Laboratory Diagnosis of Urinary Tract Infections.* Coordinating ed., A. S. Weissfeld. American Society for Microbiology, Washington, D.C.

12. **Cook, J. H., and M. Pezzlo.** 1992. Specimen receipt and accessioning, p. 1.2.1–1.2.4. *In* H. D. Isenberg (ed.), *Clinical Microbiology Procedures Handbook,* vol. 1. American Society for Microbiology, Washington, D.C.

13. **Edwards, M. S.** 1992. Infections due to human and animal bites, p. 2234–2345. *In* R. D. Feigin and J. D. Cherry (ed.), *Textbook of Pediatric Infectious Diseases,* 3rd ed. The W. B. Saunders Co., Philadelphia.

14. **Ellis, C. J.** 1991. The use and abuse of blood cultures. *Infect. Dis. Newsl.* **10:**27–30.

15. **Eschenbach, D., H. M. Pollock, and J. Schachter.** 1983. *Cumitech 17, Laboratory Diagnosis of Female Genital Tract Infections.* Coordinating ed., S. J. Rubin. American Society for Microbiology, Washington, D.C.

16. **Goldmann, D. A., and G. B. Pier.** 1993. Pathogenesis of infections related to intravascular catheterization. *Clin. Microbiol. Rev.* **6:**176–187.

17. **Goldstein, E. J. C.** 1989. Bite infections, p. 455–463. *In* S. W. Finegold and W. L. George (ed.), *Anaerobic Infections in Humans.* Academic Press, Inc., San Diego, Calif.

18. **Gray, L. D., and D. P. Fedorko.** 1992. Laboratory diagnosis of bacterial meningitis. *Clin. Microbiol. Rev.* **5:**130–145.

19. **Hagen, J. C., W. S. Wood, and T. Hashimoto.** 1977. Effect of temperature on survival of *Bacteroides fragilis* subsp. *fragilis* and *Escherichia coli* in pus. *J. Clin. Microbiol.* **6:**567–570.

20. **Haley, L. D., J. Trandel, and M. B. Coyle.** 1980. *Cumitech 11, Practical Methods for Culture and Identification of Fungi in the Clinical Microbiology Laboratory.* Coordinating ed., J. C. Sherris. American Society for Microbiology, Washington, D.C.

21. **Harris, J. C., H. L. DuPont, and R. B. Hornick.** 1972. Fecal leukocytes in diarrheal illness. *Ann. Intern. Med.* **76:**697–703.

22. **Hedberg, C. W., and M. T. Osterholm.** 1993. Outbreaks of foodborne and waterborne viral gastroenteritis. *Clin. Microbiol. Rev.* **6:**199–207.

23. **Holden, J.** 1992. Collection and transport of clinical specimens for anaerobic culture, p. 2.2.1–2.2.6. *In* H. D. Isenberg (ed.), *Clinical Microbiology Procedures Handbook,* vol. 1. American Society for Microbiology, Washington, D.C.

24. **Isenberg, H. D., F. D. Schoenknecht, and A. von Graevenitz.** 1979. *Cumitech 9, Collection and Processing of Bacteriological Specimens.* Coordinating ed., S. J. Rubin. American Society for Microbiology, Washington, D.C.

25. **Johnson, F. B.** 1990. Transport of viral specimens. *Clin. Microbiol. Rev.* **3:**120–131.

26. **Jones, D. B., T. J. Liesegang, and N. M. Robinson.** 1981. *Cumitech 13, Laboratory Diagnosis of Ocular Infections.* Coordinating ed., J. A. Washington II. American Society for Microbiology, Washington, D.C.

27. **Knoop, F. C., M. Owens, and I. C. Crocker.** 1993. *Clostridium difficile:* clinical disease and diagnosis. *Clin. Microbiol. Rev.* **6:**251–258.

28. **Krogstad, D. J., G. S. Visvesvara, K. W. Walls, and J. W. Smith.** 1991. Blood and tissue protozoa, p. 727–750. *In* A. Balows, W. J. Hausler, Jr., K. L. Herrmann, H. D. Isenberg, and H. J. Shadomy (ed.), *Manual of Clinical Microbiology,* 5th ed. American Society for Microbiology, Washington, D.C.

29. **Linares, J., A. Sitges-Serra, J. Garau, J. L. Perez, and R. Martin.** 1985. Pathogenesis of catheter sepsis: a prospective study with quantitative and semiquantitative cultures of catheter hub and segments. *J. Clin. Microbiol.* **21:**357–360.

30. **MacLowry, J. D., P. R. Murray, and L. B. Reller.** 1982. *Cumitech 1A, Blood Cultures II.* Coordinating ed., J. A. Washington II. American Society for Microbiology, Washington, D.C.

31. **Maki, D. G.** 1980. Sepsis associated with infusion therapy, p. 207–253. *In* S. Karan (ed.), *Controversies in Surgical Sepsis.* Praeger, New York.

32. **Mattia, A. R.** 1993. FDA review criteria for blood culture systems. *Clin. Microbiol. Newsl.* **15:**132–136.

33. **Melvin, D. M., and M. M. Brooke.** 1982. *Laboratory Procedures for the Diagnosis of Intestinal Parasites,* 3rd ed. Publication no. (CDC) 82-8282. U.S. Department of Health and Human Services, Atlanta.

34. **Miller, J. M.** 1985. *Handbook of Specimen Collection and Handling in Microbiology. CDC Laboratory Manual.* U.S. Department of Health and Human Services, Atlanta.

35. **Nanduri, J., and J. W. Kazura.** 1989. Clinical and laboratory aspects of filariasis. *Clin. Microbiol. Rev.* **2:**39–47.

36. **National Committee for Clinical Laboratory Standards.** 1989. *Guidelines for Laboratory Safety,* p. 11–16. CAP Environment, Safety, and Health Committee, National Committee for Clinical Laboratory Standards, Villanova, Pa.

37. **National Committee for Clinical Laboratory Standards.** 1991. *Tentative Standard M29-T2. Protection of Laboratory Workers from Infectious Disease Transmitted by Blood, Body Fluid, and Tissue,* vol. 11, no. 14, p. 31–32. National Committee for Clinical Laboratory Standards, Villanova, Pa.

38. **National Committee for Clinical Laboratory Standards.** 1992. *Tentative Standard M15-T. Slide Preparation and Staining of Blood Films for the Laboratory Diagnosis of Parasitic Diseases,* vol. 12, no. 15. National Committee for Clinical Laboratory Standards, Villanova, Pa.

39. **Ray, C. G., B. L. Wasilauskas, and R. Zabransky.** 1982. *Cumitech 14, Laboratory Diagnosis of Central Nervous System Infections.* Coordinating ed., L. R. McCarty. American Society for Microbiology, Washington, D.C.

40. **Riley, J. A., and M. P. Weinstein.** 1991. Laboratory diagnosis of bacteremia and endocarditis. *Infect. Dis. Newsl.* **10:**4–6.

41. **Ryan, M. R., and P. R. Murray.** 1993. Historical evolution of automated blood culture systems. *Clin. Microbiol. Newsl.* **15:**105–108.

42. **Sack, R. B., R. C. Tilton, and A. S. Weissfeld.** 1980. *Cumitech 12, Laboratory Diagnosis of Bacterial Diarrhea.* Coordinating ed., S. J. Rubin. American Society for Microbiology, Washington, D.C.

43. **Shea, Y. R.** 1992. Specimen collection and transport, p. 1.1.1–1.1.30. *In* H. D. Isenberg (ed.), *Clinical Microbiology Procedures Handbook,* vol. 1. American Society for Microbiology, Washington, D.C.

44. **Simor, A. E., F. J. Roberts, and J. A. Smith.** 1988. *Cumitech 23, Infections of the Skin and Subcutaneous Tissues.* Coordinating ed., J. A. Smith. American Society for Microbiology, Washington, D.C.

45. **Smith, J. W., and M. S. Bartlett.** 1991. Diagnostic parasitology: introduction and methods, p. 701–726. *In* A. Balows,

W. J. Hausler, Jr., K. L. Herrmann, H. D. Isenberg, and H. J. Shadomy (ed.), *Manual of Clinical Microbiology*, 5th ed. American Society for Microbiology, Washington, D.C.

46. **Swartz, M. N.** 1990. Cellulitis and superficial infections, p. 796–807. *In* G. L. Mandell (ed.), *Principles and Practices of Infectious Diseases*, 3rd ed. Churchill Livingstone, London.

47. **Tunkel, A. R., and W. M. Scheld.** 1993. Pathogenesis and pathophysiology of bacterial meningitis. *Clin. Microbiol. Rev.* **6:**118–136.

48. **Van Enk, R. A., and K. D. Thompson.** 1990. Microbiologic analysis of amniotic fluid. *Clin. Microbiol. Newsl.* **12:** 169–172.

Clinical Microscopy

KIMBERLE CHAPIN

4

Despite the development of rapid identification systems and molecular diagnostics, the microscopic examination of clinical material still provides the most rapid method of judging specimen quality and detecting any pathogen(s) present in clinical specimens. Staining procedures are relatively simple to perform, are cost-effective, and can be done routinely in a variety of clinical settings. In addition, the advent of numerous enhancement techniques for both staining procedures and use of equipment has further increased the sensitivity and specificity of specimen examination and the utility of the microscope in diagnosis. The complexity of the microscopic examination resides in differentiating specific organisms, evaluating the staining reactions, and determining the significance of what is detected by the examiner. These complexities become easier to resolve as the examiner acquires experience with microscopy and a clinical knowledge of disease entities.

This chapter emphasizes the basic principles of microscopy through description and illustration of the components of microscopes commonly used in the clinical microbiology laboratory, staining principles, and specific differential and enhancement techniques used for specimen examination. A discussion of use, interpretation, and significance of stained clinical material and tables describing the staining principle, clinical application, advantages, and disadvantages of each method and technique are provided as a quick reference. Finally, a series of photomicrographs illustrates points in the text. The reader is referred to the following texts for specific staining formulas and preparation procedures and for other microscopic techniques not described in detail in this chapter: *Manual of Clinical Microbiology*, 5th ed. (3), *Bailey and Scott's Diagnostic Microbiology* (4), and *Microscopy and Photomicrography—a Working Manual* (43).

HISTORICAL BACKGROUND

No single person can be credited for the discovery of the microscope. Galileo and Descartes laid the foundation with their invention and use of the telescope, but it was Hooke (1635–1703), Divini (1610–1685), Kircher (?1602–1680), and van Leeuwenhoek (1632–1723) who were among the first to bring small life-forms to the eyes of those in the 17th century (8). Hooke described his compound microscope in the 1665 treatise "Micrographia," which included illustra-tions of mold forms and of the anatomy of a flea. Leeuwenhoek was perhaps the most enthusiastic proponent of using the microscope and examining the "new world." He ground more than 500 lenses with magnifying power from ×300 to ×500 and resolving power of 1 μm, quality unsurpassed until the 19th century. In a letter submitted to the Royal Society of London in 1678, he reported his numerous observations of "very small animalcules" and included drawings of basic organism shapes and movements (8, 12, 31).

LIGHT MICROSCOPY

The modern light microscope utilizes a built-in light source and a compound lens, meaning that at least two lens systems are employed. Stained specimens are visualized by transillumination; that is, light is focused on the object, which is seen against a bright background. (Fig. 1).

The microscope has two major components: the lens systems and the mechanical parts (Fig. 2). The three lens systems include the substage condenser, the objectives, and the oculars, or eyepieces. The condenser, sitting just below the specimen stage, is the first lens of the microscope. The condenser lens does not magnify the image but concentrates the beam of light into the plane of the specimen and ultimately affects the resolution of the image. The light source, a coiled-filament tungsten lamp, starts as a horizontal beam and is projected up through the condenser via a mirror. Glare from the beam is prevented by focusing the light on the condenser just below the stage rather than directly on the specimen. This method is known as Köhler illumination and is most efficient with a properly centered light path on the specimen. This centering is accomplished by using the condenser adjustment screws on the stage. In addition, the substage diaphragm controls the diameter of the light beam projected into the object. The contrast of an image may be enhanced when the condenser aperture of the diaphragm is made smaller, because the amount of oblique light reaching the object is reduced (Fig. 3). This is most useful with wet-mount and uniformly stained preparations, but it results in decreased resolution. The condenser should not be moved for the purpose of adjusting the light intensity except when the thickness of slides varies. Instead, a knob or wheel at the base of the microscope controls the field diaphragm and alters the filtering of light from the light source. The human eye is most sensitive to

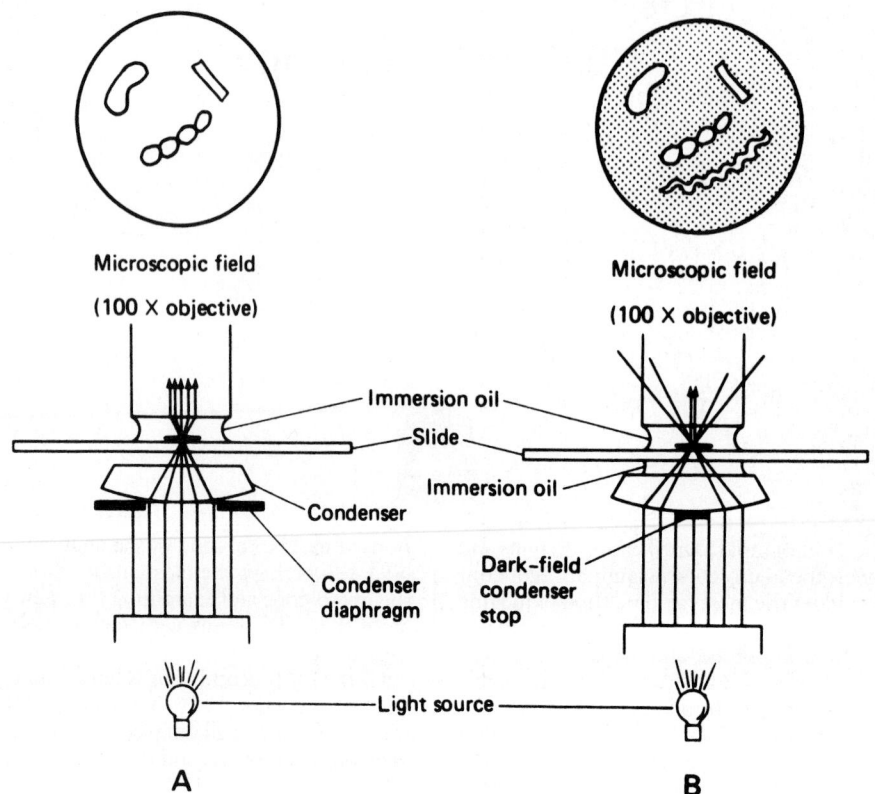

FIGURE 1 (A) Bright-field microscopy. Light with a resolution limit of 0.2 μm is focused directly on the specimen. Bacteria are visualized against a bright background. (B) Dark-field microscopy. Scattered light is viewed through the objective, and resolution is slightly enhanced to 0.1 μm, allowing spirochetes to be seen. Objects appear luminous against a black background. Illustration from reference 40, courtesy of Appleton & Lange, Norwalk, Conn.

blue wavelengths; thus, the use of a blue filter over the field diaphragm may enhance visualization and cause less fatigue when multiple slides are viewed (9, 39).

The objectives, which are the second critical component of the compound microscope lens system, may contain multiple lens elements. Two important factors independent of one another determine the performance of the objective: magnification and resolution. One purpose of the objective is to collect light dispersed from the object and form a magnified image some distance above the lens. For viewing clinical specimens, this distance is 160 to 170 mm or the length of the body tube of the microscope. Objectives used in clinical microbiology generally include a 4× scanning lens, a 10× intermediate lens, a 40× high dry lens, and a 100× immersion oil lens. All lenses used in the microscope are biconvex, and white light entering them is refracted, or split into the colors of the spectrum. Each color is refracted at a different angle with a different point of focus and projects an image surrounded by color fringes. This is known as chromatic aberration. To decrease chromatic aberration, lenses that correct for the dispersing wavelengths are used. Achromatic objectives are the simplest lens systems; they are corrected for dispersion of red and blue wavelengths. Apochromats are the most highly corrected objectives; they totally eliminate chromatic and spherical aberrations and are used in microphotography (9, 23, 39).

The second important factor for objective performance is resolution, which is the smallest distance between two structures that allows them to be viewed as distinct components. Resolution is dependent on the numerical aperture (NA) of the lens system and the light wavelength employed according to the formula $R = k \times \lambda/\mathrm{NA}$, where k is the constant 0.61, and λ is wavelength (white light = 0.55 μm). Resolution is directly proportional to the wavelength and inversely proportional to the NA. Thus, for viewing specimen slides, resolution is dependent on the NA. NA is determined by the product of two factors, the refractive index (RI) of the medium filling the space between the glass slide and the front of the objective lens (the medium being air, oil, or water) and the sine of the semiangle of the aperture of the lens (Θ). NA is calculated as NA = RI × sin Θ. Specifically, the RI is the ratio of the light path angle of incidence and angle of refraction in one medium to those in another as the light path transverses the various media. The RI is known for most transparent objects. The sin Θ is the maximum semiangle of the light path allowed to reach the objective, as represented in Fig. 3. Thus, an objective with a high NA allows for the greatest light-gathering capability and the best resolution (23, 39). The resolution limit for the compound light microscope is approximately 0.2 μm with a 100× objective and immersion oil, which allows differentiation between most bacte-

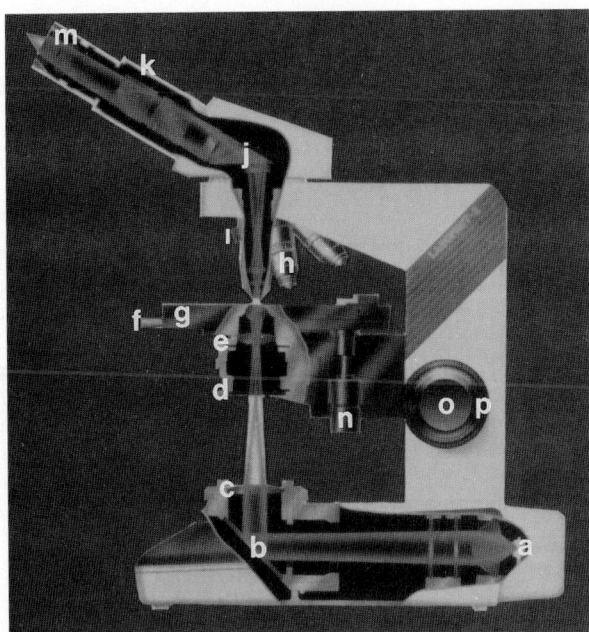

FIGURE 2 Compound light microscope components (lettered) and light path from lamp through ocular objective. Microscope photograph courtesy of Nikon, Inc. a, Light source; b, mirror; c, field diaphragm; d, substage diaphragm; e, substage condenser; f, condenser adjustment screws; g, stage; h, objective; i, revolving nosepiece; j, prism; k, body tube; m, ocular objective; n, stage position adjustment; o, fine-focus control; p, coarse-focus control.

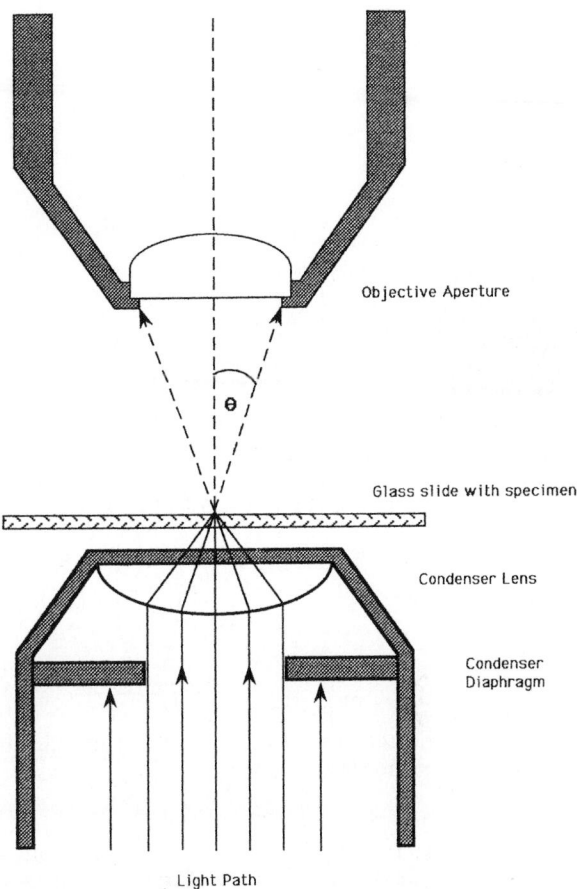

FIGURE 3 Illustration of condenser function showing focusing of light paths at the plane of the specimen. The semiangle of the objective aperture (Θ) is used to calculate the numerical aperture. The smaller the Θ value, the better the contrast.

rial species. However, treponemes, which are generally 0.1 μm wide, cannot be visualized with the light microscope.

The eyepieces, or ocular objectives, contain the last lenses of the microscope and magnify the image 10 to 15 times. The total magnification of the lens systems in the microscope is determined by multiplying the magnification of the objective at the nosepiece by the magnification of the ocular objective.

The mechanical components of the microscope include the object stage, the nosepiece, and the body tube. The specimen stage provides a stable mount for slides and contains vernier scales to allow easy relocation of a particular area of interest on a slide. The nosepiece is a revolving component that accommodates multiple objectives. The objectives should be parfocal (having the focal points in the same plane), so that when the user switches to another objective, only minor fine focusing is necessary.

DARK-FIELD MICROSCOPY

Dark-field microscopy, although not commonly used in the clinical microbiology laboratory, overcomes the limit of resolution of the light microscope (0.2 μm) and can be useful in visualizing motile spirochetes such as treponemes and *Borrelia* spp. (7). This method of microscopy excludes directly transmitted light by using a dark-field condenser and allows only oblique or scattered light to be directed onto the specimen (Fig. 1). This setup allows finer structures to be seen because the resolution improves to approximately 0.1 μm, the size of a spirochete. Bacteria appear brightly luminous against a black background.

FLUORESCENCE MICROSCOPY

Fluorescence microscopy has become commonplace in most clinical laboratories because of ease of use in smear interpretation. In addition, the use of inexpensive fluorescence lens adapters for the bright-field microscope make this technique cost-effective for all laboratories.

Fluorescence is dependent on the ability of fluorophores (naturally fluorescent substances) or fluorochromes (fluorescent dyes) to absorb the energy of nonvisible UV and short visible wavelengths, become excited, and remit the energy in the form of longer visible wavelengths. Each fluorochrome has characteristic wavelengths of absorption and emission that yield maximum fluorescence (39). High-pressure gas lamps of mercury, halogen, and xenon are capable of emitting short-wavelength light that is used with fluorochromes. A series of filters is placed between the light source and the specimen. These include a heat filter, a red stop filter that eliminates infrared waves, and a wavelength selecter or exciter filter that transmits the light of the desired wavelength but obscures other incident visible light. Most fluorescence microscopes use incident illumination, or illumination from above that passes through the objective down to the object via a dichromatic mirror beam

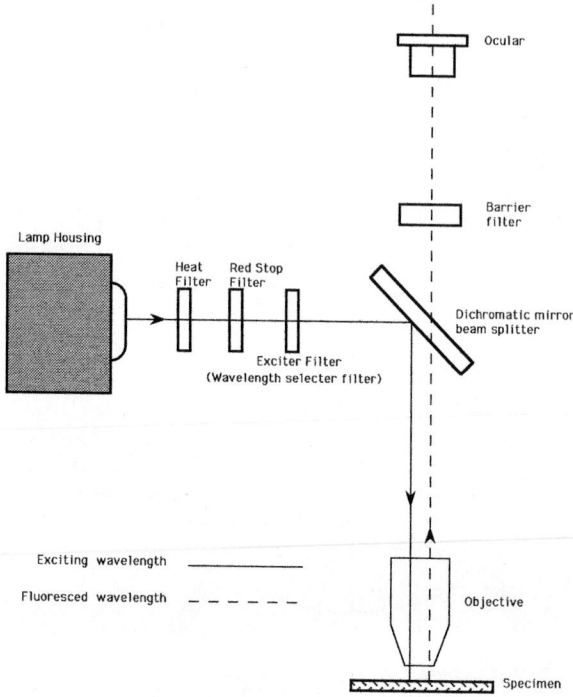

FIGURE 4 Fluorescence microscope incident illumination light path and microscope components.

splitter (Fig. 4). The dichromatic mirror transmits light of some wavelengths and reflects light of other wavelengths. The light passes through the filters, hits the mirror, and is directed through the objective onto the fluorochrome-stained smear. The longer wavelengths emitted from the fluorescing specimen pass through the objective and bounce off the mirror (light passes through because it is a different wavelength) to the ocular to form an image. Areas with bound fluorochrome fluoresce, and other areas are dark.

Looking at the equation $R = k \times \lambda/NA$, it can easily be seen why fluorescence microscopes that use shorter wavelengths and objectives of the same NA as the light microscope have better resolution.

Recently introduced into the market are two microscope adapters: the UV ParaLens Adaptor (Becton Dickinson, Franklin Lakes, N.J.) and the Makler Fluorescent Objective (Flow Inc., London, United Kingdom). These convert the bright-field microscope to a fluorescence microscope and are much less expensive than a traditional fluorescence microscope. The adapters consist of a lens with an epi-illumination system that attaches to a spare lens port on the nosepiece of any conventional microscope. All the major filter components of a traditional fluorescence microscope are contained within the epi-illumination system: the excitation filter, dichromatic mirror beam splitter, and barrier filter. The light source is provided by a high-intensity tungsten halogen lamp contained in a portable housing with fiber-optic cable. Lenses range from 10× to 100×. The adapters have successfully been used to demonstrate malarial parasites, rabies antibody, and fluorescent-antibody-stained *Pneumocystis carinii* in pulmonary specimens (32, 38). The lamp intensity is less than that of a traditional fluorescence microscope, but it did not reduce the sensitivity of detection for the above-mentioned pathogens. De-

creased lamp intensity may be a factor in using a fluorescent stain to detect acid-fast organisms. One study showed the ParaLens Adaptor to be 84% sensitive and 93% specific compared with a traditional fluorescence microscope that, when the fluorescence adapter was used, had a smear sensitivity equivalent to that of Kinyoun smear preparations (37).

ELECTRON MICROSCOPY

The electron microscope is a major capital investment and is rarely used for diagnostic purposes in microbiology. The power of the electron microscope is the significantly increased resolution resulting from use of an electric tungsten filament that emits electrons. The electron beam is aimed at the specimen, and electrons are focused by varying the electromagnetic fields, just as the light source and glass lens are adjusted in light microscopy. The image is subsequently projected onto a fluorescent screen (7).

CARE AND USE OF THE MICROSCOPE

Proper care and maintenance of the microscope will result in better interpretation of slide material. Ideally, the microscope should be covered when not in use, and the light source should be dimmed or turned off when slides are not being examined. Before examining material, check to be sure the optical components are clean and the two eyepieces match. Lower the stage when placing slides on the stage, and watch the ascent of the slide toward the objective from the side. Clean objective and ocular lenses with Kim-wipes or lens paper. However, cotton or wool swabs may be needed to reach the convex component of the lens. Remove oil on a lens with lens paper or gauze whenever possible, since these materials will not scratch the lens. If the oil cannot be easily removed, a lens cleaning solution may be needed. These solutions are typically a mixture of diluted acetic acid, isopropyl alcohol, and acetone and should be used sparingly. Alcohol or acetone alone should not be used to clean the lenses, since they dissolve mounting cements. The best preventive measures for avoiding use of these solutions are to make sure that only immersion oil is used and that care is taken not to drag other objectives through the oil. Finally, do not remove objectives, as this increases the potential for damage and of not being able to replace them in the same threading (9, 39).

DIRECT EXAMINATION OF SPECIMENS

The first step in processing virtually all clinical material is microscopic examination of the specimen. Direct examination is a rapid, cost-effective diagnostic aid. Methods for direct examination are aimed at identifying microorganisms and enumerating cells. Visible microorganisms may denote the presumptive etiological agent, guiding the laboratory in selecting appropriate isolation media and the physician in selecting empirical antibiotic therapy. The quality of the specimen and the measure of the inflammatory response can also be evaluated.

HISTOLOGIC BASIS OF STAINING

Cellular material and organisms are usually transparent and are best distinguished by the use of dyes or biological stains. Leeuwenhoek, in 1719, was the first to attempt differentiation of bacteria by using natural colored agents such as beet juice (31). With the exception of hematoxylin, natural dyes have for the most part been replaced by artificial dyes. Artificial dyes are products of chemical derivatives of coal tar, especially benzene. Members of two other important chemical groups, the chromophores and auxochromes, complete the dye compound (28).

Benzene, an aromatic organic compound, undergoes substitution reactions with radicals to form new compounds. Some of the molecular changes result in a colored product. Specific groups that react with benzene and are associated with color are called chromophores. The most important groups are C=C, C=O, C=S, C=N, N=N, N=O, and NO_2. The greater the number of chromophores in a compound, the deeper the color of the compound. Benzene plus a chromophore group is a chromogen. Although the chromogen is colored, it does not have affinity for bacteria or tissues, and washing or mechanical processes will readily remove the compound. The affinity of the dye is due to an additional group called the auxochrome. The auxochrome group gives the compound the property of electrostatic dissociation or the ability to form salt linkages with the ionizable radicals on proteins, glycoproteins, and lipoproteins on tissue or on organism cell wall components. This process can occur either directly or through the chelating action of a mordant (29). Dyes are usually sold as salts; thus, it is the auxochrome group that usually determines whether a dye is classified as cationic (basic) or anionic (acidic). Most dyes retain their cationic or anionic properties throughout the pH range of staining (pH 3 to 9) and thus reliably stain those structures that are oppositely charged. For example, DNA, which is acidic, can be stained with a basic dye. Crystal violet and safranin are typical cationic (basic) dyes, and picric acid is a typical anionic (acid) dye (9).

DIFFERENTIAL STAINING

While direct visualization of specimens in various wet mounts is useful, differentially stained specimens are the most helpful for presumptive identification of the majority of pathogens. The Gram stain and acid-fast stain are examples of differential stains. In addition, the fluorescent stains aid in identification of organisms because of the specific attachment of the fluorochromes in the dyes to organism components: auramine, calcofluor white, acridine orange, and fluorescein isothiocyanate (FITC) bound to monoclonal antibodies are examples.

In fixed differential smear preparations, four components are typically used in a progressive manner: the primary stain, a mordant, a decolorizing agent(s), and a secondary stain, or counterstain. The primary stain usually stains all cellular components and organisms in the specimen the same color, as seen in simple single-stain procedures such as that using methylene blue. The mordant acts as a fixative to anchor the dye in the specimen, making the dye insoluble. Heat, phenol, and iodine are examples of fixatives. Decolorizing agents are typically acids and alcohols such as the acetone-alcohol mixture used in the Gram stain and the sulfuric acid used in the modified acid-fast procedure (9). Removal of the primary stain with the decolorizing agent allows a secondary stain, or counterstain, to be taken up by the decolorized organisms and background. The secondary stain differentiates between types of bacteria, such as the purple and pink organisms seen in the Gram stain, or between organism and background, such as a pink acid-fast organism in a blue counterstained background.

SMEAR PREPARATION

Smears may be made from clinical material, broths, or colonies. Smears should be thin, since thick smears may peel or flake off of the slide during staining. Thick smears also make the timing of the decolorization step harder to judge. Methanol fixation is preferred for fixing clinical specimens, since heat may cause artifacts and does not fix the specimen to the slide as well (30). The smear may be stained after fixing and drying.

STAINING METHODS

The following stains and techniques are individual methods used most often in the clinical microbiology laboratory. For each staining procedure, the clinical utility, basic components (with chemical explanations where useful), expected appearance of clinical material, interpretation, variation of performance, and quality control will be described. The staining methods and significant characteristics are provided in abbreviated form in Tables 1, 2, 4, and 5.

WET MOUNTS

Clinical Utility

Wet-mount preparations (Table 1) are used to examine unstained clinical specimens. One can determine the morphology of organisms, gross structure, and biological activity including motility, reaction to certain chemicals, and serological reactivity. The specimens are examined under bright-field, phase-contrast, or dark-field microscopy depending on the microorganisms sought.

Basic Saline Wet Mount

The saline wet-mount preparation is used mainly for determining motility and gross morphology of organisms such as fungal hyphae, endospores, protozoan trophozoites, and helminth eggs and larvae.

Basic Procedure

1. Add 1 drop of 0.85% warm (37°C) aqueous NaCl to the slide.
2. Add 1 drop of specimen to the slide, and mix it with the NaCl.
3. Overlay with a coverslip.
4. Examine at ×100 to ×1,000.

Appearance of Clinical Material

The slide is examined with the light microscope for characteristic morphology and motility of organisms. Using the condenser diaphragm to decrease the circumference of the circle of light reduces the amount of transmitted light, making objects more clearly visible. Motility may not be apparent if smears are allowed to dry or cool. Wet-mount preparations can be prevented from drying by ringing the coverslips with paraffin or Vaseline before overlaying the specimen drop on the slide. Brownian movement (intrinsic molecular vibrations) can be confused with motility; thus, to avoid misinterpreting motility, the specimen should be examined for directional motility.

TABLE 1 Direct specimen examination methods

Direct examination method	Application	Principle	Time required (min)	Advantage(s)	Disadvantage(s)
Wet mount	Stool, vaginal discharge, urine sediment, aspirates	Used to detect organism motility and morphology of parasitic forms and fungal elements.	1	Rapid and specific	Limited contrast and resolution. Brownian movement may be confused with motility. Experienced microscopist required.
Modified wet mounts 10% KOH	Specimens with suspected fungi, e.g., skin scrapings, fluid aspirates	Proteinaceous host cell components are partially digested by alkali. Fungal cell wall stays intact.	5–10	Rapid detection of fungal elements	Background material may be confusing. Experienced microscopist required.
10% KOH with LPCP	Specimens with suspected fungi, e.g., skin scrapings, fluid aspirates	Adds contrast for detecting fungal forms.	5–10	Enhances detection of fungal forms.	Background material may be confusing. Experienced microscopist required.
Colloidal carbon (India ink, nigrosin)	CSF, other body fluids for *Cryptococcus neoformans*	Polysaccharide capsule excludes ink particles, giving halo appearance.	1	Rapid detection; diagnostic in CSF when present	Not as sensitive as cryptococcal antigen. Confusion with cells and artifacts. Experienced microscopist required.
Lugol's iodine	Stool	Nonspecific contrast dye to help differentiate parasitic cysts from WBCs. Cysts take up dye and appear light brown.	1	Rapid; enhances differentiation of specific forms.	Background material may be confusing. Experienced microscopist required.
Methylene blue	Stool for WBCs	WBCs in stool stain blue; their presence suggests invasive bowel disease.	1	Rapid; enhances differentiation of specific cells.	WBCs may disintegrate if stool is not examined promptly.

Modified Saline Wet Mount with 10% KOH

Wet mounts prepared in 10% KOH are used to help distinguish fungal elements in thick mucoid specimens or in specimens with keratinous material such as skin, hair, or nails. The proteinaceous components of the host cells are partially digested, leaving the polysaccharide-containing fungal cell wall intact and more apparent (14). An aliquot of specimen is simply added to a drop of 10% KOH, which can be preserved with 0.1% thimerosal (Sigma Chemical Co.). The slide remains at room temperature for 5 to 30 min after the addition of KOH, depending on specimen type, to allow digestion to occur. Digestive capabilities can be enhanced by adding 40% dimethyl sulfoxide or by gently heating the specimen.

10% KOH with LPCB

The wet mount with KOH and lactophenol cotton blue (LPCB) is used for the same purposes as the KOH preparation but incorporates the LPCB dye. LPCB enhances the visibility of fungal elements, since aniline blue stains the outer cell wall of fungi and lactic acid acts as a clearing agent (14).

Colloidal Carbon Wet Mounts (India Ink and Nigrosin)

Clinical Utility

Colloidal carbon wet mounts are used for the visualization of encapsulated microorganisms, especially *Cryptococcus neoformans*, directly in clinical specimens.

Basic Procedure

1. Use equal parts of either Pelikan India ink or nigrosin and patient spinal fluid on the slide.
2. Mix well, and add a coverslip.
3. Examine at ×100 to ×1,000.

Appearance of Clinical Material

The polysaccharide capsules of organisms will exclude the particles of ink, and the capsules will appear as clear haloes around the organisms on semiopaque backgrounds.

Interpretation

Multiple artifacts, such as erythrocytes, leukocytes (WBCs), talc particles from gloves, bubbles, and globules after a myelogram may also displace the colloidal suspension, mimicking yeast cells. These artifacts make it necessary to examine the wet mount carefully to distinguish properties consistent with the organisms: rounded shapes with buds of various sizes and double-contoured cell walls. Interpretation can also be hindered if the emulsion with the colloid suspension is too thick, blocking the transmission of light completely.

Lugol's Iodine

Lugol's iodine is a wet-mount modification incorporating iodine used in conjunction with the saline mount to examine feces or other material for intestinal protozoa and helminth ova or larvae. Iodine stains protozoan nuclei and intracytoplasmic organelles brown so that they are more easily seen. Since iodine paralyzes bacteria and protozoan trophozoites, motility cannot be interpreted (2).

Methylene Blue

Clinical Utility

Methylene blue is a simple direct stain used for a variety of purposes. The stain is used to differentiate organisms by morphological characteristics, to identify the characteristic metachromatic granules of *Corynebacterium diphtheriae*, and to detect the presence of fecal WBCs. It is also used as an adjunct to the Gram stain in detecting the presence of bacteria. The procedure is performed with a wet mount for stool specimens and with fixed smears for other sources (29).

Basic Procedure

1. Air dry and methanol fix the smear.
2. Add methylene blue stain for 30 to 60 s.
3. Examine at ×100 to ×1,000.

Appearance of Clinical Material

Methylene blue stains organisms or WBCs a deep blue in a light gray background. *C. diphtheriae* appears as blue bacilli with prominent darker blue metachromatic granules.

Interpretation

Methylene blue is not commonly used in the diagnostic laboratory but in specific circumstances can increase the sensitivity of determining whether an organism is present in a specimen. For example, the stain reveals the morphology of fusiform bacteria and spirochetes from oral infections (Vincent's angina) that may not be seen with the Gram stain. It may also establish the intracellular location of microorganisms such as *Neisseria* spp. Methylene blue is the stain of choice for identifying the metachromatic granules of diphtheria; however, one should be careful about overstaining, since this will lessen the contrast between the bacteria and the granules. This stain is also used to detect the presence of fecal WBCs, which indicate invasive diseases.

FIXED DIFFERENTIAL METHODS

See Table 2.

Gram Stain

Clinical Utility

The Gram stain is the single most useful and cost-effective test in the clinical microbiology laboratory. It is the differential stain most commonly used for direct microscopic examination of specimens and bacterial colonies because it has a broad staining spectrum. First devised by Hans Christian Joachim Gram late in the 19th century, it has re-

mained basically the same procedure and serves in dividing bacteria into two main groups: gram-positive organisms, which retain the primary crystal violet dye and appear deep blue or purple, and gram-negative organisms, which can be decolorized, thereby losing the primary stain and subsequently taking up the counterstain safranin and appearing red or pink. The staining spectrum includes almost all bacteria, many fungi, and parasites such as trichomonads, *Strongyloides* larvae, and miscellaneous protozoan cysts. The significant exceptions include those organisms, such as *Mycoplasma* spp., without a cell wall, and *Chlamydia* and *Rickettsia* spp., which are too small. *Legionella* spp. will not stain directly from clinical specimens, but once bacterial growth occurs on medium, *Legionella* bacteria can be visualized. Mycobacteria are generally not seen with the Gram stain; however, in smears illustrating heavy infections, the organism may give a beaded appearance that is somewhat similar to that of *Nocardia* spp. or may exhibit organism "ghosts" (16). The Gram stain can also be utilized to differentiate epithelial and inflammatory cells, thus providing information about the state of infection and the quality of the specimen (5, 19, 35).

Two methods of Gram staining are presented here. The first method is the conventional Gram stain used by most laboratories. The second is an altered Gram stain devised by Atkins in 1920 that uses gentian violet, a different mordant, and acetone as the decolorizing agent (1, 21).

Conventional Gram Stain Procedure

1. Fix the specimen with 95% methanol for 2 min. (Some laboratories prefer heat fixation rather than methanol fixation.)
2. For the primary stain, use crystal violet (alcoholic solution).
3. The mordant is Gram's iodine (binds the alkaline crystal violet dye to the cell wall).
4. The decolorizer is acetone and 95% ethyl alcohol (50-50 mixture).
5. The secondary stain is safranin.
6. Examine at ×100 to ×1,000.

Components of Atkin's Gram Stain

Primary Stain: Gentian Violet

Crystal violet (90%) dye	20 g
Alcohol (95%)	200 ml
Ammonium oxalate	8 g
Distilled water	800 ml

Dissolve the crystal violet in alcohol, dissolve the ammonium oxalate in distilled water, mix the two solutions together, and filter them after 24 h.

Mordant: Atkin's Iodine

Iodine (crystals)	20 g
1 N NaOH	100 ml
Distilled water	900 ml

Dissolve the NaOH with iodine before adding water, and store in a brown bottle at room temperature. The decolorizer is acetone, and the secondary stain is safranin.

There is some debate on the length of time each staining component should be left on the slide. In actual practice, the two dyes and the mordant should each be allowed to

TABLE 2 Differential fixed stained methods

Differential fixed stained methods	Application	Principle	Time required	Advantage(s)	Disadvantage(s)
Gram stain	Differential bacterial stain; stains yeast cells; stains are most often used to assess suitability of specimen for culture.	Gram-positive bacteria and yeast cells retain crystal violet and stain blue. Gram-negative organisms allow crystal violet to wash out in decolorization and subsequently take up counterstain safranin and stain pink. Quantity of cells and organisms present is used to assess quality of specimen.	3 min	Rapid; commonly performed and thus more easily interpreted. Stains bacteria differentially on basis of cell components and aids in choice of antibiotic therapy. Assessment of specimen for culture comparison of smear result to culture result.	Organisms with damaged cell walls stain unpredictably. *Nocardia* spp. and fungi may not take up crystal violet completely. Background and cellular elements stain pink, often masking gram-negative organisms.
Acid-fast stains					
Kinyoun (cold), Ziehl-Neelsen (hot), modified (for *Nocardia* spp.)	Detection of *Mycobacteria* spp., *Nocardia* spp., and some parasites	Presence of long-chain fatty acids (mycolic acids) leads cell wall to resist decolorization of basic carbol fuchsin dye when acid alcohol washed. Organisms appear pink.	15 min (depending on method)	Acid-fast characteristic specific for *Mycobacteria*, *Nocardia*, *Cryptosporidia*, and *Sarcocystis* spp. and *I. bellio*. Presence of organism generally diagnostic of infection in untreated host.	Low organism number makes slide examination tedious. Cannot differentiate different *Mycobacterium* spp. Tissue homogenates often mask presence of organism because of deeply staining background.
PAS	Detection of fungal elements in clinical specimens	Combination of acid hydrolysis and staining fungal elements stain pink-magenta.	1 h	Most fungal elements stain.	Time-consuming. Respiratory specimens must be digested. *Blastocystis* spp. stain variably.
Toluidine blue-O	Rapid examination of respiratory specimens for *P. carinii*	Background material removed by sulfation reagent and appears light blue. *P. carinii* cysts stain reddish-blue.	20 min	Rapid method for detection of *P. carinii* cysts	Differentiation of *P. carinii* from yeast cells may be difficult. Trophozoites not discernible. Not recommended for all respiratory specimens.
Wright-Giemsa	Detection of blood parasites, viral and chlamydial inclusions, toxoplasmosis, *P. carinii*, and *Rickettsia* spp.	Differential staining of basophilic and acidophilic material	10 min-1 h	Detection of multiple organisms and cellular inclusions	Not specific for inclusions (*Chlamydia* spp. the exception). Cannot determine bacterial Gram reaction.

remain on the slide for at least 15 s. More time has little effect. Most critical is the amount of time the decolorizer is used. Unfortunately, the amount of decolorizer is directly related to the thickness of the specimen on the slide. The old benchmark that the slide should continue to receive decolorizer until no more crystal violet is seen washing away is still true but difficult to attain in practice.

Gram-positive organisms are thought to retain the crystal violet dye because of the increased number of cross-linked teichoic acids and the decreased permeability of their cell walls to organic solvents, since they contain little lipid. Gram-negative organisms, because of higher lipid content associated with the cell wall, show increased permeability to decolorizer and lose the crystal violet dye (7).

Appearance of Clinical Material

1. Organisms that retain the crystal-violet iodine complex are dark blue or purple and are called gram positive.

2. Organisms that lose the primary complex and take up the secondary dye safranin appear red and are called gram negative.

3. Fungi (yeasts) appear gram positive; *Nocardia* and *Actinomyces* spp. are weakly gram positive or stain in a "beaded" fashion.

4. Inflammatory cells appear gram negative, and epithelial cells may appear gram positive and/or gram negative depending on the thickness of the smear.

5. Backgrounds of clinical specimens usually appear gram negative but may appear gram positive. Fibrin, mucus, and erythrocytes often stain gram negative.

Interpretation

A brief gross examination tells the observer where the material is located on the slide and what the possibility of underdecolorization is. Examination of many fields under low power (×10) evaluates the adequacy of the specimen and with oil immersion characterizes the bacteria (Table 3). The Gram reaction, morphology, and arrangement of the organisms, tissue, and inflammatory cells and the specimen type give the physician clues to the preliminary identification and significance of the organisms. For example, Gram-stained smears of exudate from patients with acute infection most often contain one pathogen with inflammatory

TABLE 3 Classification of Gram-stained respiratory specimens[a]

Group[c]	No. of cells/low-power field[b]	
	WBCs	Squamous epithelial cells
6 (TTA[d])	<25	<25
5	>25	<10
4	>25	10–25
3	>25	>25
2	10–25	>25
1	<10	>25

[a]See references 5 and 34.
[b]×100 magnification.
[c]Do not culture expectorated sputum falling into group 1, 2, or 3. Inform the physician that the specimen is unsuitable for culture, and request a repeat specimen.
[d]TTA, transtracheal aspirate. A specimen with no organisms and >10 epithelial cells be rejected (34).

cells. It is often possible to make an educated guess about the etiologic agent causing the infection. For example, a sputum sample with numerous gram-positive lancet-shaped diplococci and WBCs and few epithelial cells is likely to contain *Streptococcus pneumoniae* (Fig. 5). Similarly, a sample of cerebrospinal fluid (CSF) with intracellular gram-negative diplococci is likely to contain *Neisseria* species. WBCs may be scarce in early or late disease, in the leukopenic patient, or when leukotoxic organisms such as *Clostridium* spp. are present. When there is contamination of a sterile area by normal flora, a variety of organisms with or without squamous epithelial cells may be present (5, 35).

Laboratories can determine the relative clinical usefulness of a specimen by its Gram-stained appearance. Various nomograms have been generated, especially for the evaluation of sputum. Based on the ratio of oropharyngeal epithelial cells and polymorphonuclear WBCs (PMNs), the specimen can receive a rating (Table 3) (5). For example, a sputum specimen with >25 epithelial cells per ×100 magnification field, <10 PMNs per ×100 field, and multiple organisms is judged to contain primarily normal oral flora, is an inadequate specimen, and should not be cul-

FIGURE 5 *(See following pages.)* (A) Example of a poor-quality expectorated-sputum specimen showing multiple squamous epithelial cells, bacteria, and fungi. This specimen should be rejected and not cultured. (B) Blood culture Gram stain of *Streptococcus agalactiae* showing typical chaining of *Streptococcus* spp. (C) Atkin's Gram stain method. Example of a good-quality expectorated-sputum specimen showing numerous WBCs and cells of a single bacterium, *S. pneumoniae*. Probable pneumococcal pneumonia can be diagnosed from this smear preparation. Note uniform staining of pneumococci compared to that in smear in panel D. (D) Conventional Gram stain method with same specimen as in panel C. Note overdecolorization of organisms compared to that by Atkin's method. (E) Blood culture specimen smear treated with conventional Gram stain reagents showing gram-positive organisms. (F) Same blood culture specimen as in panel C, but smear was made with Gram stain and with tartrazine and fast green enhancer reagents. Polymicrobial infection with both gram-positive and gram-negative organisms is shown. Note the gray-blue of the background that allows the gram-negative organisms to be seen. (G) Fluorescent acridine stain of *S. aureus* showing bright orange-yellow fluorescence. (H) Typical appearance of pink gram-negative bacteria, *Fusobacterium nucleatum*, WBCs, and proteinaceous background. (I) Typical appearance of purple gram-positive bacteria and yeast cells with Gram stain. (J) Gram stain of sputum showing WBCs, no organisms, and "ghost" forms of *M. tuberculosis*. (K) Appearance of *M. avium-M. intracellulare* after acid-fast staining. The primary carbol fuchsin stain is retained after acid decolorization. The background stains blue as with other non-acid-fast organisms. *Nocardia* spp. look similar with a partial acid-fast stain preparation. (L) Fluorescent auramine stain of *M. tuberculosis* showing bright green fluorescence. (M) Appearance of *Nocardia* spp. in a Gram stain preparation of a brain biopsy specimen. Note the faint staining and fine-beaded appearance of the organisms. (N) Appearance of *Nocardia* spp. after acid-fast staining. The primary carbol fuchsin stain is removed in the acid decolorizing step and the organism takes up the methylene blue counterstain. (O) Acid-fast smear of stool specimen showing an obvious *Cryptosporidium* oocyst. (P) PAS stain of lung biopsy specimen showing septated hyphal elements.

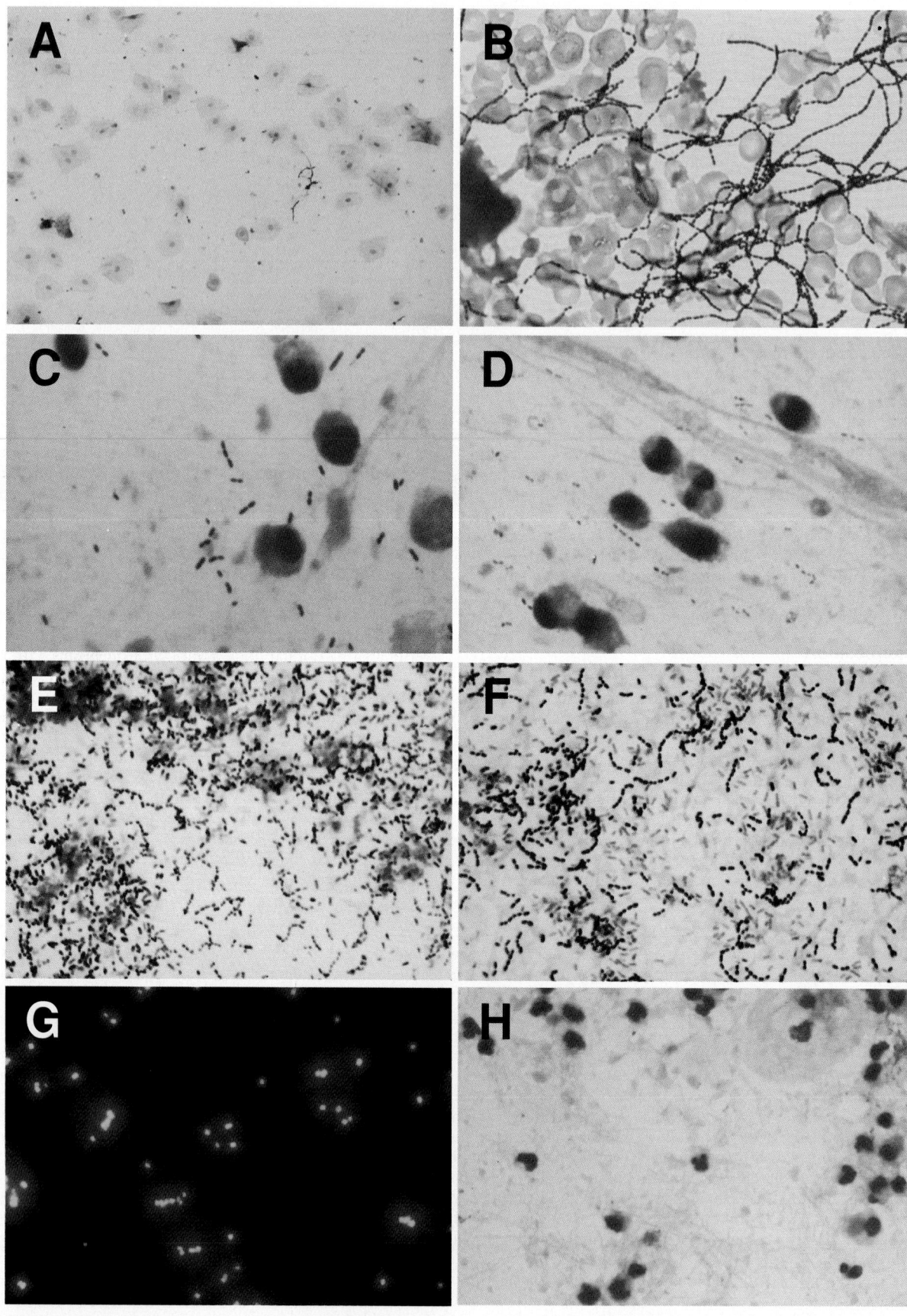

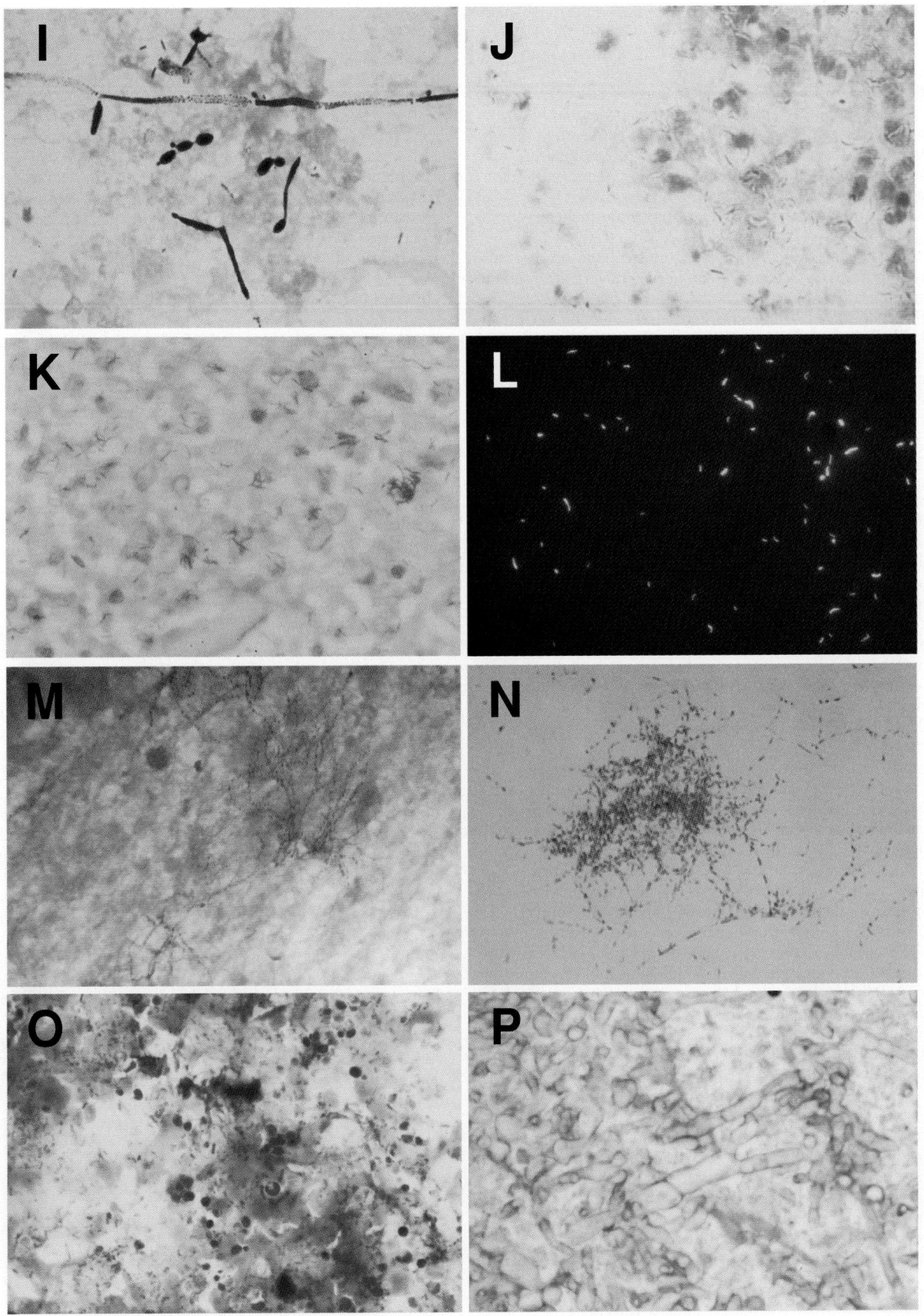

tured. A sputum specimen with >25 PMNs per ×100 field, no epithelial cells, and one predominant organism reflects infection at the site (5, 35). A sputum Gram stain with >25 PMNs and <10 epithelial cells per ×100 field but no organisms present should suggest microbes such as *Legionella* spp., *Mycoplasma* spp., or mycobacteria. Likewise, a specimen with >25 PMNs per ×100 field and a predominant organism that does not yield a positive culture may indicate significant anaerobic, fastidious, or antibiotic-damaged organisms.

Recently, criteria for evaluation of endotracheal aspirate specimens have also been proposed. In a study by Morris et al., the number of PMNs was found to have no discriminating value and was not predictive of an interpretable specimen (34). Instead, the absence of organisms and the presence of >10 squamous epithelial cells per low-power field were found to be the best criteria for rejection of endotracheal aspirate specimens. It is important to note that in that study, all specimens with PMNs but no organisms were also negative for *Legionella* spp. (22). The authors point out that the criteria for rejection are not used for determining if pneumonia is absent or present but that a specimen submitted and subsequently rejected is unlikely to provide any help in diagnosing the etiological agent causing the pneumonia. Application of the criteria resulted in rejection of up to 40% of endotracheal specimens (22).

Problems with analysis of the Gram stain generally result from errors in preparation of the slide, such as a smear that is too thick, excessive heat fixing (which can distort organisms), improper decolorization, and inexperience. Overdecolorization results in an abundance of gram-negative-appearing bacteria, while underdecolorizing results in too many gram-positive-appearing bacteria. If a chain of cocci resembling streptococci (normally gram positive) and epithelial cells appears to be gram negative, the slide is overdecolorized. Slides stained by the Atkin's Gram stain method are less sensitive to decolorization because the mordant is more effective in retaining crystal violet. This allows better visualization of gram-positive organisms, especially those very sensitive to decolorization such as *S. pneumoniae* and *Bacillus* spp. (Fig. 5). The Atkin's method does not offer a significant advantage over the conventional Gram stain in visualizing gram-negative organisms and in fact, such visualization may be more difficult with some specimens such as blood cultures. Laboratories should evaluate both Gram stain procedures and then institute one as the routine procedure. An adequate quality control measure to ensure correct staining procedure includes staining known gram-positive and gram-negative organisms such as *Staphylococcus aureus* and *Escherichia coli*, respectively. Alternatively, a swab of the inside of one's own mouth will always exhibit a variety of gram-positive and gram-negative cocci and bacilli.

A number of other factors may cause difficulty in interpretation. Specimens collected from patients receiving antibiotics, cultures transferred from antibiotic-containing media, very young or very old isolates, and bacteria containing autolytic enzyme systems may change both the Gram-staining characteristics and the typical morphology of the organism. In these cases gram-positive organisms tend to lose their cell wall integrity, lose the crystal violet more readily in the decolorization step, and appear to be gram negative. Typical gram-negative organisms may exhibit profound morphological changes; for instance, a bacillus may appear extremely elongated or globular.

Background material and artifacts can also interfere with interpretation. Precipitated gram-positive stain generally appears as irregular coccoid shapes or as asters resembling fungal hyphae. Mucus or other proteinaceous matter that stains gram negative may be misidentified as gram-negative bacilli or make visualization of slender gram-negative rods, such as *Haemophilus influenzae* or *Fusobacterium* spp., difficult (Fig. 5). Likewise, *Actinomyces* and *Nocardia* spp., which are slender, branching, and weakly gram positive, may blend into the background and be difficult to observe. However, careful examination of a clinical specimen can provide extremely valuable information that can be combined with the patient's clinical presentation to guide therapeutic and diagnostic decisions (15).

Enhancement Techniques

Some enhancement techniques are basic modifications of the Gram stain. These include the use of methylene blue, tartrazine and light green (enhanced Gram stain [Carr-Scarborough]), carbol fuchsin, and other combinations. Each of these procedures is listed in Table 4. The main purpose of these stains is to make organisms that are normally difficult to detect with the Gram stain stand out more prominently. Typically, these enhancement techniques are used for better visualizing gram-negative organisms. This is done by two methods. One method is to simply stain the organism a darker color. This is the method used with the methylene blue stain that stains all organisms blue but allows for greater contrast with the background so that the normally weakly staining gram-negative organisms are visible (13, 33). A second method is to make the background inflammatory cells and mucus a different color than the usual red-pink, which often masks gram-negative organisms. This method is used with the tartrazine-fast green stain, which makes the background gray or green and makes the organisms easily visible (24, 25, 29). In this staining method the Gram reaction of the organisms is preserved. The stain is most helpful for detection of gram-negative organisms (Fig. 5). A totally different enhancement technique is mechanical manipulation of the specimen by using a cytocentrifuge. The cytocentrifuge method employs nondisposable specimen funnels that are mounted with a slide and filter card and placed in the centrifuge. During centrifugation, the filter card absorbs the supernatant, while cells and microorganisms are centrifuged through a hole in the filter paper strip and deposited in a continuous layering fashion on a 6-mm-diameter circular area on the slide. The method is especially sensitive for detection of pathogens from sterile body fluids, especially peritoneal fluids (10, 41). The method has also been used for detection of acid-fast organisms and *P. carinii* from respiratory specimens (17). The deposition of specimen in a discrete area and the ability to lyse erythrocytes during centrifugation are particularly advantageous characteristics that allow more rapid and enhanced resolution in smear examination.

ACID-FAST STAINS

Ziehl-Neelsen Stain

The cells of certain parasites and bacteria contain long-chain (50- to 90-carbon) fatty acids (mycolic acids) that give them a coat impervious to crystal violet and other basic dyes. Heat or detergent must be used to allow penetration of the primary dye into the bacterium. Once the dye has been forced into the cell, it cannot be decolorized by the usual acid alcohol solvent. The acid-fast stain is useful

TABLE 4 Enhanced fixed stained methods

Enhanced method	Application	Principle	Time required (min)	Advantage(s)	Disadvantage(s)
Fixed stain methods Anaerobic Gram stain variation	Differential bacterial stain detects anaerobic organisms not easily seen with regular Gram stain, especially gram-negative organisms.	Uses basic carbol fuchsin as counterstain instead of safranin. Enhances gram-negative organisms.	3	Darker stain of gram-negative anaerobes such as *Fusobacterium* spp.	
Tartrazine, fast green, Gram stain variation	Differential bacterial stain enhances detection of organisms from background material.	Use of fast green and tartrazine before safranin counterstain allows significant suppression of red-pink color of background material; organisms still stain gram positive (purple) and gram negative (pink).	3	Allows excellent enhancement of small gram-negative organisms and mixed cultures to be visualized.	Slight change in color of organisms compared with regular Gram stain may be confusing.
Methylene blue	Detects metachromatic granules of C. *diphtheriae*. Enhances detection of lightly staining gram-negative rods, spirochetes, and mycoplasma colonies.	Simple stain that dyes all components blue. Organisms are dark blue against gray-blue background.	3	Rapid viewing of smears; increased contrast of organism compared with background tissue; detection of C. *diphtheriae* granules and oral spirochetes.	Does not differentiate bacterial cell wall.
Mechanical method Cytospin preparations	Detects bacteria in body fluids and *Pneumocystis* and *Mycobacteria* spp. in respiratory specimens.	Specimen is deposited in layers in discrete area during centrifugation.	5–10	Increased sensitivity for detecting organisms in body fluids. Deposits specimen in discrete 6-mm-wide area that allows more rapid examination. Lysing of erythrocytes during centrifugation yields clear specimen fields. Can be used for both small and large volumes of fluid.	Thick purulent specimens may be difficult to read. Damaged cells or bacteria may have altered morphology.

for identification of a specific group of bacteria, including *Mycobacteria* and *Nocardia* spp. (occasionally *Actinomyces* spp. with the Putt modification) and the oocysts of *Cryptosporidium* spp., *Isopora belli*, and *Sarcocystis* spp. A number of modifications of the Ziehl-Neelsen procedure are employed to differentiate these various acid-fast organisms (2).

Basic Procedure

1. Heat fix the specimen for 2 h.
2. The primary stain is basic carbol fuchsin.
3. The mordant is steam for 3 to 5 min (heat allows penetration of the primary dye through the cell wall).
4. The decolorizer is 3% H_2SO_4 in 95% ethanol.
5. The secondary stain, or counterstain, is methylene blue.
6. Examine at ×400 to ×1,000.

Appearance of Clinical Material

An acid-fast organism will stain red, and the background of cellular elements and other bacteria will be blue, the color of the counterstain (Fig. 5).

Kinyoun Modification

The only difference between the Ziehl-Neelsen and Kinyoun stains is the substitution of phenol for steam. Both stains have the same sensitivity and specificity, yet the Kinyoun (cold) stain procedure is less time-consuming and easier to perform.

Acid-Fast Modification

Another modification of the Ziehl-Neelsen stain uses a weaker decolorizing agent (0.5 to 1.0% sulfuric acid) in place of the 3% acid alcohol. This particular stain helps differentiate those organisms known as partially or weakly acid fast, particularly *Nocardia* spp. Some strains of *Actinomyces* may appear partially acid fast (i.e., speckled red) by the Putt modification. *Propionibacterium* and *Actinomycetes* spp. should be negative (2).

The acid-fast stains are important clinically and are relatively simple to use. The same careful examination techniques described for the Gram stain should be used. Definitive identification of an acid-fast organism from a clinical specimen cannot be made by staining alone, but certain clues may be helpful. Mycobacteria often appear as slender, slightly curved rods and may show darker granules that give the impression of beading. *Nocardia* spp. often branch and almost always show a speckled appearance. *Mycobacterium kansasii* may form long, often broad and banded cells (16).

Difficulty in interpretation can result from smears too thick or insufficiently decolorized, yielding an acid-fast artifact. As a quality control measure, a known acid-fast organism such as nonpathogenic *Mycobacterium tuberculosis* HRV 37 and a non-acid-fast organism such as *Streptomyces* spp. can be stained in parallel with the clinical specimen and compared.

Factors such as age, exposure to drugs, and the particular acid-fast organism itself may vary the acid-fast presentation. For example, while *M. tuberculosis* is consistently acid fast (with the Ziehl-Neelsen or Kinyoun stain [Fig. 5]), *Mycobacterium leprae*, *Nocardia* spp., and *Actinomyces* spp. are not. Therefore, use of the modified acid-fast bacillus stain utilizing a weaker decolorizer is necessary to reveal these organisms.

Small numbers of acid-fast organisms generally are clin-ically significant for most clinical specimens. However, the use of acid-fast stains for gastric aspirates in the interpretation of pulmonary disease in adults or for stool specimens from human immunodeficiency virus-positive patients in diagnosing *Mycobacterium avium-intracellulare* yields very poor specificity (false-positive smears with saprophytic organisms) as well as poor sensitivity (45). In addition, patients receiving adequate therapy may still have positive smears without positive cultures for a number of weeks. Rarely, small numbers of acid-fast organisms in thick or quality control smears may represent transferred contamination. Knowledge of specimen type and method of collection in conjunction with careful preparation and technique can aid in deciphering this problem.

MISCELLANEOUS STAINS

PAS

The periodic acid-Schiff (PAS) stain is used for observation of fungi from direct clinical specimens, especially for demonstrating yeast cells and hyphae in tissues. The procedure is a multistep method combining hydrolysis and staining. The periodic acid step hydrolyzes the cell wall aldehydes, which are then able to combine with the modified Schiff reagent coloring the cell wall carbohydrates a bright pink-magenta (14).

Basic Procedure

1. Fix the smear with formalin-ethanol for 1 min.
2. Add 5% periodic acid for 5 min.
3. The primary stain is basic fuchsin zinc (or Na) hydrosulfite (Schiff reagent).
4. The secondary stain, or counterstain, is picric acid or light green.
5. Examine at ×100 to ×400.

Appearance of Clinical Material

Fungal elements stain a bright pink-magenta or purple against an orange background if picric acid is used as the counterstain or against a green background if light green is used (14).

Interpretation

The PAS stain is one of the best general stains, since most fungal elements in clinical material will take up the stain. However, the PAS stain procedure is rather involved, requiring several different reagents and time-consuming steps and has been replaced in many laboratories by the calcofluor white stain procedure. The PAS stain cannot be used for undigested respiratory secretions, since mucin will also stain bright pink-magenta. Also, *Nocardia* spp. do not stain well, and *Blastomyces dermatitidis* appears pleomorphic. A suspension of hyphal forms of yeast cells is adequate to use with specimens when controlling the quality of the staining procedure.

Toluidine Blue-O

Clinical Utility

Toluidine blue-O is used primarily for the rapid detection of *P. carinii* from lung biopsy imprints and bronchoalveolar lavage specimens (2, 17, 36).

Basic Procedure

1. Air dry the smear.
2. Solution A is the sulfation reagent.
3. Solution B is toluidine blue-O.
4. Fix the specimen in absolute ethanol.
5. Examine at ×100 to ×1,000.

Appearance of Clinical Material

Toluidine blue-O stains the cysts of *P. carinii* reddish blue or dark purple against a light blue background. The cysts are often clumped and may be punched in, appearing crescent shaped. Trophozoites are not discernible.

Interpretation

Toluidine blue-O stain is used for the rapid examination of respiratory tract material for the presence of *P. carinii*. Although the silver stain, monoclonal antibody, and calcofluor white stains are also used, the toluidine blue-O stain is an easy and rapid stain to use and yields reliable results from appropriate specimens (44). The stain will produce the best yields when mucoid-appearing flecks are chosen from bronchial lavage specimens. Coughed sputum is not a recommended specimen. Sometimes the background may be light purple, or *P. carinii* may be confused with yeast forms. A good quality control measure is to include a positive control, preferably a positive specimen, with each specimen run (17).

Giemsa and Wright Stains

Clinical Utility

Giemsa and Wright stains are modifications of the Romanowsky stain, which is a combination of methylene blue and eosin. Giemsa and Wright stains are used primarily for the differentiation of intracellular and extracellular circulating blood parasites, particularly *Plasmodium* and *Leishmania* spp.; for demonstrating the intracellular yeast forms of *Histoplasma capsulatum* in bone marrow or peripheral blood smears; for visualizing the inclusions of viral or other infected cells; and for demonstrating *Rickettsia* spp. and *P. carinii* (2). These stains are typically used by the hematology laboratory for demonstrating the differences of nuclei and cytoplasmic features of the blood cell components. Microbiologists will often be consulted about bacteria and parasitic forms seen in smears from a variety of specimen types. The procedure and clinical utility differ slightly for Giemsa and Wright stains. The Wright stain often has the fixative in combination with the staining solution, so both processes occur at the same time, simplifying the procedure and making it convenient for daily use. The Giemsa protocol separates the fixative and stain, requiring fixing of the smear before staining. However, the Giemsa stain has superior staining characteristics and a wider spectrum of stainable entities (9).

Basic Procedure, Wright's Stain

1. Methanol fix the specimen for 2 to 3 min.
2. Allow the slide to air dry.
3. Use Wright stain for 5 min.
4. Wash the specimen gently in distilled water.
5. Examine at ×100 to ×1,000.

Basic Procedure, Giemsa Stain

1. Methanol fix the specimen for 5 min.
2. Allow the slide to air dry.
3. Use Giemsa stain for 10 to 60 min.
4. Rapidly rinse the specimen in 95% ethyl alcohol to remove excess dye.
5. Examine at ×100 to ×1,000.

Appearance of Clinical Material

Protozoan trophozoites demonstrate a red nucleus and gray-blue cytoplasm. The intracellular yeast forms of *H. capsulatum* and the inclusion bodies of infected cells typically appear blue (basophilic) and intracytoplasmic. Elementary bodies of *Chlamydia* spp. and intracystic bodies and trophozoites of *P. carinii* stain purple, and *Rickettsia* spp. appear bluish purple. The blood components stain as follows: erythrocytes, pale gray-blue; WBCs, purple nuclei and pale purple cytoplasm; eosinophilic granules, bright purple-red; and neutrophilic granules, deep pink-purple (9, 28).

Interpretation

The Giemsa stain will detect blood parasites, especially malaria, *Leishmania* spp., *Babesia* spp., and microfilaria. In bone marrow or peripheral blood smears, it will detect *H. capsulatum*, which occurs intracellularly within mononuclear cells as small round to oval yeast cells 2 to 5 μm in diameter. The stain is especially useful for visualizing inclusions of viral or other infected cells directly from clinical material, such as corneal scrapings, urine sediments, conjunctival specimens from neonates in whom *Chlamydia trachomatis* is suspected, and the bases of suspected herpetic vesicles. The stain can also be applied to infected in vitro cell cultures such as a monolayer of chlamydia-infected McCoy cells. Other parasites such as toxoplasmas in brain tissue and trophozoites of *P. carinii* from respiratory specimens may also be detected (cyst walls do not stain). A good quality control measure is the observation of a peripheral blood smear with the differentiating characteristics of the various blood cell components as described previously. Neither the tissue Wright nor the Giemsa stain reliably stains fungal or bacterial elements; each must be used in conjunction with other staining procedures if the etiology of an infection is unknown.

FLUORESCENT STAINING PROCEDURES
See Table 5.

Acridine Orange

Acridine orange is a fluorochrome that can be intercalated into nucleic acid in both the native and the denatured states (39). The stain is rapid and can differentiate between bacterial and fungal DNAs. In a number of studies the stain has been shown to be more sensitive than the Gram stain in the detection of organisms from blood culture broths, CSF, and buffy coat preparations from neonates (21, 26, 33). Acridine orange is also useful in a series of miscellaneous infectious etiologies, such as *Acanthamoeba* infections, infectious keratitis, and *Helicobacter pylori* gastritis (18, 39). The stain may be particularly helpful in thick or purulent specimens for which the Gram stain is not interpretable.

Basic Procedure

1. Methanol fix the smear.
2. Add the acridine orange solution.
3. Examine at ×100 to ×1,000.

TABLE 5 Fluorescent stain methods[a]

Fluorescence method[b]	Application	Principle	Time required	Advantage(s)	Disadvantage(s)
Staining					
Acridine orange	Detection of bacteria in blood cultures, buffy coats, and corneal scrapings; detection of fungal organisms	Fluorochrome is intercalated into nucleic acid in both native and denatured state. Bacterial and fungal DNAs fluoresce orange, and mammalian DNA fluoresces green with UV light.	3 min	Sensitive method for detection of organisms in blood, CSF, and tissue sites. Detects low organism numbers. Thick or bloody smears can be used. Can Gram stain same slide to confirm suspicious colonies.	Cellular specimens with abundance of DNA may yield difficult-to-interpret fluorescence. Interobserver variability seen with buffy coats.
Auramine-rhodamine	Detection of mycobacteria and other acid-fast organisms	Nonspecific fluorochromes that bind to mycolic acids and resist rinsing by acid alcohol (identical to acid-fast stains). Organisms fluoresce orange-yellow with UV light.	30 min	Allows rapid screening of specimens at lower magnification. May be more sensitive than acid-fast stain. Can acid-fast stain same slide to confirm suspicious colonies.	Single organism or low organism number may not be confirmable.
Calcofluor white	Detection of fungal elements in wet mounts of clinical specimens and of P. carinii in respiratory specimens	Nonspecific fluorochrome that binds to cellulose of cell walls in fungi and certain procaryotes. Fungi fluoresce blue-white or green depending on UV filter used.	KOH clearing plus 1-min stain	Can be mixed with KOH to clear specimen. Rapid screening for fungal elements at lower magnification. Rapid detection of P. carinii.	Background fluorescence may cause difficulty in interpretation with cellular or mixed bacterial and fungal specimens.
Antibody-conjugated stain					
Fluorescein conjugated	Detection of specific organisms, both directly in clinical material and in culture confirmation	Monoclonal antibodies bound to fluorochrome FITC detect antigens for specific pathogens in clinical specimens and fluoresce apple green with UV light.	1 h	Specific organism identification. Especially useful for Bordetella, Legionella, and Pneumocystis spp. and for virus identification.	Adequate clinical specimen must be submitted.
Mechanical					
Fluorescence lens attachment for bright-field microscope	Detection of organisms by fluorescence in laboratories without FM; field use.	Removable lens with fluorescence optics and portable lamp housing that attaches to nosepiece of standard light microscope.	NA	Allows fluorescence microscopy at 1/5 to 1/10 cost of FM. Background fluorescence is diminished compared to that of regular FM.	Need to use oil with highest-magnification objectives. Intensity of visualized organisms less than with FM. Sensitivity may be less than with regular FM.

[a]NA, not applicable; FM, fluorescence microscope.
[b]All stains require either a fluorescence microscope or fluorescence lens attachment.

Appearance of Clinical Material and Interpretation

Bacterial and fungal DNAs fluoresce orange under UV light, and mammalian DNA fluoresces green. Cellular specimens or heavily laden bacterial specimens may be difficult to interpret owing to excessive fluorescence, and some interobserver variability may be noted. Smears with suspicious organisms may be confirmed directly with the Gram stain. A sample of a positive blood culture broth should be fixed and stained in conjunction with the sample smear preparation as a quality control.

Auramine-Rhodamine

Clinical Utility

Auramine and rhodamine are nonspecific fluorochromes that bind to mycolic acids and are resistant to decolorization with acid alcohol (7). Staining procedures with these fluorochromes are thus equivalent to the fuchsin-based acid-fast procedures. The stain has become commonplace in laboratories that routinely perform acid-fast examinations because it allows rapid screening of specimens and because the procedure may be more sensitive than the traditional acid-fast procedures.

Basic Procedure

1. Fix the smear.
2. The primary stain is auramine-rhodamine (a single fluorochrome may be used).
3. Decolorize the specimen.
4. The secondary stain is potassium permanganate.
5. Examine at ×100 to ×400.

Appearance of Clinical Material and Interpretation

Acid-fast organisms fluoresce orange-yellow in a black background. If the secondary stain is not used, the organisms will fluoresce a yellow-green color. Smears with suspicious organisms may be confirmed directly with a Kinyoun stain. However, a single organism or a low number of organisms may be difficult to confirm. A known positive acid-fast smear should be run in conjunction with the fluorescent procedure as a quality control measure.

Calcofluor White

Clinical Utility

Calcofluor white is a nonspecific fluorochrome that binds to the β1,3-linked polysaccharides, specifically cellulose and chitin of cell walls in fungi, including yeast cells, hyphae, pseudohyphae, and spherules (7). Like the auramine-rhodamine stain, calcofluor white has become commonplace in microbiology laboratories because of the rapidity and specificity with which specimens can be observed. The fluorochrome can be mixed with KOH to clear the specimen for easier observation of fungal elements (14). The stain has also been described as a rapid detection method for *P. carinii* and *Acanthamoeba* spp. (44).

Basic Procedure

1. Use 1 drop of calcofluor white.
2. Add 1 drop of 10% KOH.
3. Add a coverslip.
4. Examine at ×100 to ×400.

Appearance of Clinical Material and Interpretation

Fungal elements and *Pneumocystis* cysts appear bright green or blue, depending on the UV filter used, against a dark background. Cellular or mixed fungal specimens that contain *P. carinii* and/or fungal elements may be difficult to interpret, especially with induced sputum specimens. Classic *P. carinii* cysts are generally 5 to 7 μm in diameter, round, and uniform in size, and they exhibit a characteristic peripheral cyst wall staining with an intense internal "double-parenthesis-like" structure. Yeast cells are differentiated from *P. carinii* by budding and intense internal staining (6).

Antibody Stain Methods

See chapter 11 of this Manual on immunoassays and Table 5 for details on antibody stain methods.

HISTOLOGIC TISSUE SPECIMEN INTERPRETATION

The clinical microbiologist is often asked to consult in smear interpretation of blood and body fluid specimens and of histologic stained sections of tissue. It should be remembered that most fungal elements, parasites, viral inclusions, and bacteria cannot be definitively identified in these preparations. The stain typically employed in the hematology laboratory for blood and body fluids is the Wright-Giemsa preparation, which uniformly stains all bacteria blue. Thus, one must not call a blue coccus a gram-positive coccus until a Gram stain can be done to help in differentiation. The same is true for tissue specimens; fungal elements that are aseptate and broad should be called zygomycetes and not a specific species such as *Mucor* spp.

In stained tissue preparations, pathogens may be significantly different in appearance owing to the staining and fixative practices used in the histology laboratory. For the best outcome in histologic diagnosis of infectious etiologies, good communication between the surgical pathologist, microbiologist, and primary physician will result in securing tissue for fixing as well as for culture, and the two methods together more often than not provide the definitive diagnosis. In typical tissue preparations, Wright-Giemsa stain does not stain bacterial or fungal elements reliably, with the important exception of *H. capsulatum*, which will stain with the Wright stain of bone marrow and peripheral blood. These preparations also stain infected cell culture monolayers and aid in visualizing viral and *Chlamydia* inclusion bodies and toxoplasmosis in tissue.

Silver stains are commonly employed for staining tissue sections. These stains are best for detection of fungal elements and *P. carinii* cyst walls in tissue (36). However, the stain also can detect bacteria and parasites. Differentiation of various yeast forms of *P. carinii* is difficult, and interpretation should be cautious and done in conjunction with other special stains.

A series of reports describe combinations of differential staining methods in tissue for the elucidation of fungal forms, specifically yeasts, and may be particularly helpful in differentiating the capsules of *Cryptococcus* and *Blastomyces* spp. (27, 46).

I express my sincere appreciation to Earl Robertson for computer graphics, David Lay for photographic assistance, Jeanne Kee for reviewing the manuscript, and Rosemary Farmer for manuscript preparation.

REFERENCES

1. **Atkins, K. N.** 1920. Report of Committee on Descriptive Chart. III. A modification of the Gram stain. *J. Bacteriol.* **5:**321–324.

2. **Balows, A., and W. Hausler.** 1988. *Diagnostic Procedures for Bacterial, Mycotic and Parasitic Infection,* 7th ed. American Public Health Association, Washington, D.C.

3. **Balows, A., W. J. Hausler, Jr., K. L. Herrmann, H. D. Isenberg, and H. J. Shadomy (ed.).** 1991. *Manual of Clinical Microbiology,* 5th ed. American Society for Microbiology, Washington, D.C.

4. **Baron, E. J., and S. M. Finegold (ed.).** 1990. *Bailey and Scott's Diagnostic Microbiology,* 8th ed. The C. V. Mosby Co., St. Louis.

5. **Bartlett, J. G., K. J. Ryan, T. F. Smith, and W. R. Wilson.** 1987. *Cumitech 7A, Laboratory Diagnosis of Lower Respiratory Tract Infections.* Coordinating ed., J. A. Washington II. American Society for Microbiology, Washington, D.C.

6. **Baselski, V. S., M. K. Robison, L. W. Pifer, and D. R. Woods.** 1990. Rapid detection of *Pneumocystis carinii* in bronchoalveolar lavage samples by using cellufluor staining. *J. Clin. Microbiol.* **28:**393–394.

7. **Berlin, O. G. W., W. L. Drew, M. A. C. Edelstein, L. S. Garcia, and G. D. Roberts.** 1990. Optical methods for laboratory diagnosis of infectious diseases, p. 64–80. *In* E. J. Baron and S. M. Finegold (ed.), *Bailey and Scott's Diagnostic Microbiology,* 8th ed. The C. V. Mosby Co., St. Louis.

8. **Boorstin, D. J.** 1983. *The Discoverers,* p. 327–332. Random House, Inc., New York.

9. **Carson, F. L.** 1990. Instrumentation, p. 43–67. *In* S. Borysewicz (ed.), *Histotechnology: a Self-Instructional Text.* American Society of Clinical Pathologists, Chicago.

10. **Chapin-Robertson, K., S. E. Dahlberg, and S. C. Edberg.** 1992. Clinical and laboratory analyses of cytospin-prepared Gram stains for recovery and diagnosis of bacteria from sterile body fluids. *J. Clin. Microbiol.* **30:**377–380.

11. **Clarridge, J. E., and J. M. Mullins.** 1987. Microscopy and staining, p. 87–103. *In* B. J. Howard (ed.), *Clinical and Pathogenic Microbiology.* The C. V. Mosby Co., St. Louis.

12. **Cohen, I. B.** 1980. The microscope, p. 111–115. *In* I. B. Cohen (ed.), *From Leonardo to Lavoisier, 1450–1800 (Album of Science).* Charles Scribner's Sons, New York.

13. **Daly, J. A., W. M. Gooch III, and J. M. Matsen.** 1985. Evaluation of the Wayson variation of a methylene blue staining procedure for the detection of microorganisms in cerebrospinal fluid. *J. Clin. Microbiol.* **21:**919–921.

14. **Emmons, C., C. Binford, K. J. Kwon-Chung, and J. Utz.** 1977. *Medical Mycology,* 3rd ed. Lea & Febiger, Philadelphia.

15. **Fine, M. J., J. J. Orloff, J. D. Rihs, R. M. Vickers, S. Kominos, W. N. Kapoor, V. C. Arena, and V. L. Yu.** 1991. Evaluation of housestaff physicians' preparation and interpretation of sputum Gram stains for community-acquired pneumonia. *J. Gen. Intern. Med.* **6:**189–198.

16. **Fisher, J. F., M. Ganapathy, B. H. Edwards, and C. L. Newman.** 1990. Utility of Gram's and Giemsa stains in the diagnosis of pulmonary tuberculosis. *Am. Rev. Respir. Dis.* **141:**511–513.

17. **Gill, V. J., N. A. Nelson, F. Stock, and G. Evans.** 1988. Optimal use of the cytocentrifuge for recovery and diagnosis of *Pneumocystis carinii* in bronchoalveolar lavage and sputum specimens. *J. Clin. Microbiol.* **26:**1641–1644.

18. **Groden, L. R., J. Rodnite, J. H. Brinser, and G. I. Genvert.** 1990. Acridine orange and Gram stains in infectious keratitis. *Cornea* **9:**122–124.

19. **Heineman, H. S., J. K. Chawla, and W. M. Lofton.** 1977. Misinformation from sputum cultures without microscopic examination. *J. Clin. Microbiol.* **6:**518–527.

20. **Hendrickson, D. A., and M. M. Krenz.** 1991. Reagents and stains. p. 1306–1314. *In* A. Balows, W. J. Hausler, Jr., K. L. Herrmann, H. D. Isenberg, and H. J. Shadomy (ed.), *Manual of Clinical Microbiology,* 5th ed. American Society for Microbiology, Washington, D.C.

21. **Henrickson, K. J., K. R. Powell, and D. H. Ryan.** 1988. Evaluation of acridine orange-stained buffy coat smears for identification of bacteremia in children. *J. Pediatr.* **112:**65–66.

22. **Isenberg, H. D.** 1993. Rejection criteria for endotracheal aspirates. *J. Clin. Microbiol.* **31:**2555. (Letter.)

23. **Junqueira, L. C., and J. Carneiro (ed.).** 1983. Methods of study, p. 1–17. *Basic Histology,* 4th ed. Lange Medical Publications, Los Altos, Calif.

24. **Kee, J. C., A. Hill, and K. C. Chapin.** 1994. Use of an enhanced Gram stain method for detection of organisms from body fluids and blood culture broths, abstr. C-242, p. 533. *Abstr. 94th Gen. Meet. Am. Soc. Microbiol. 1994.*

25. **Larson, A. M., M. J. Dougherty, D. J. Nowowiejski, D. F. Welch, G. M. Matar, B. Swaminathan, and M. B. Coyle.** 1994. Detection of *Bartonella* (*Rochalimaea*) *quintana* by routine acridine orange staining of broth blood cultures. *J. Clin. Microbiol.* **32:**1492–1496.

26. **Lauer, B. A., L. B. Reller, and S. Mirrett.** 1981. Comparison of acridine orange and Gram stains for detection of microorganisms in cerebrospinal fluid and other clinical specimens. *J. Clin. Microbiol.* **14:**201–205.

27. **Lazcano, O., V. O. Speights, Jr., J. G. Strickler, J. W. Bilbao, J. Becker, and J. Diaz.** 1993. Combined histochemical stains in the differential diagnosis of *Cryptococcus neoformans.* *Mod. Pathol.* **6:**80–84.

28. **Lillie, R. D.** 1977. The general nature of dyes and their classification, p. 19–39. *In* E. H. Stotz and V. M. Emmel (ed.), *H. J. Conn's Biological Stains,* 9th ed. The Williams & Wilkins Co., Baltimore.

29. **Lillie, R. D.** 1977. The mechanisms of staining, p. 40–59. *In* E. H. Stotz and V. M. Emmel (ed.), *H. J. Conn's Biological Stains,* 9th ed. The Williams & Wilkins Co., Baltimore.

30. **Mangels, J. I., M. E. Cox, and L. H. Lindberg.** 1984. Methanol fixation: an alternative to heat fixation of smears before staining. *Diagn. Microbiol. Infect. Dis.* **2:**129.

31. **Marti-Ibañez, F.** 1962. Baroque medicine, p. 185–195. *In* F. Marti-Ibañez (ed.), *The Epic of Medicine.* Clarkson N. Potter, Inc., New York.

32. **Mattia, A. R., M. A. Waldron, and L. S. Sierra.** 1993. Use of the UV ParaLens adapter as an alternative to conventional fluorescence microscopy for detection of *Pneumocystis carinii* in direct immunofluorescent monoclonal antibody-stained pulmonary specimens. *J. Clin. Microbiol.* **31:**720–721.

33. **Mirrett, S., B. A. Lauer, G. A. Miller, and L. B. Reller.** 1982. Comparison of acridine orange, methylene blue, and Gram stains for blood cultures. *J. Clin. Microbiol.* **15:**562–566.

34. **Morris, A. J., D. C. Tanner, and L. B. Reller.** 1993. Rejection criteria for endotracheal aspirates from adults. *J. Clin. Microbiol.* **31:**1027–1029.

35. **Murray, P. R., and J. A. Washington II.** 1975. Microscopic and bacteriologic analysis of expectorated sputum. *Mayo Clin. Proc.* **50:**339–344.

36. **Paradis, I. L., C. Ross, A. Dekker, and J. Dauber.** 1990. A comparison of modified methenamine silver and toluidine blue stains for the detection of *Pneumocystis carinii* in bronchoalveolar lavage specimens from immunosuppressed patients. *Acta Cytol.* **34:**511–516.

37. **Patterson, K. V., C. L. McDonald, B. Miller, and K. C. Chapin.** 1994. Use of the UV ParaLens® adaptor for detection of acid-fast organisms, abstr. C-122, p. 512. *Abstr. 94th Gen. Meet. Am. Soc. Microbiol. 1994.*

38. **Polsuwan, C., B. Lumlertdaecha, W. Tepsumethanon, and H. Wilde.** 1992. Using the UV ParaLens® adapter on a standard laboratory microscope for fluorescent rabies antibody detection. *Trans. R. Soc. Trop. Med. Hyg.* **86:**107–108.

39. **Rose, R. A.** 1982. Light microscopy, p. 1–19. *In* J. D. Bancroft and A. Stevens (ed.), *Theory and Practice of Histological Techniques,* 2nd ed. Churchill Livingstone, New York.

40. **Ryan, K. J., and C. G. Ray.** 1990. Laboratory diagnosis of

infectious diseases, p. 233–273. *In* J. C. Sherris (ed.), *Medical Microbiology: an Introduction to Infectious Diseases*, 2nd ed. Elsevier Science Publishing Co., Inc., New York.

41. **Shanholtzer, C. J., P. J. Schaper, and L. R. Peterson.** 1982. Concentrated Gram stain smears prepared with a cytospin centrifuge. *J. Clin. Microbiol.* **16:**1052–1056.

42. **Simor, A. E., N. B. Cooter, and D. E. Low.** 1990. Comparison of four stains and a urease test for rapid detection of *Helicobacter pylori* in gastric biopsies. *Eur. J. Clin. Microbiol. Infect. Dis.* **9:**350–352.

43. **Smith, R. F.** 1993. *Microscopy and Photomicrography*, 2nd ed. CRC Press, Inc., Boca Raton, Fla.

44. **Stratton, M., J. Hryniewicki, S. L. Aarnaes, F. Tan, L. M. De La Maza, and E. M. Peterson.** 1991. Comparison of monoclonal antibody and calcofluor white stains for the detection of *Pneumocystis carinii* from respiratory specimens. *J. Clin. Microbiol.* **29:**645–647.

45. **Strumpf, I. J., A. Y. Tsang, M. A. Schork, and J. G. Weg.** 1976. The reliability of gastric smears by auramine-rhodamine staining technique for the diagnosis of tuberculosis. *Am. Rev. Respir. Dis.* **114:**971–976.

46. **Youngberg, G. A.** 1993. Correspondence re: O. Lazcano, V. O. Speights, Jr., J. G. Strickler, J. E. Bilbao, J. Becker, and J. Diaz. Combined histochemical stains in the differential diagnosis of *Cryptococcus neoformans.* (*Mod. Pathol.* **6:**80, 1993). *Mod. Pathol.* **6:**392–393.

LABORATORY MANAGEMENT AND REGULATORY ISSUES

II

VOLUME EDITOR
FRED C. TENOVER
SECTION EDITOR
JOHN E. McGOWAN, JR.

Quality Assurance: Quality Improvement, Quality Control, and Test Validation

DAVID L. SEWELL AND RON B. SCHIFMAN

5

Quality assurance is a term used to denote a system for continuously improving reliability, efficiency, and utilization of products and services. The elements of quality assurance are structure, process, and outcome. Structure is a measure of the adequacy of the workplace and the provisions required to do the job, process is the proficiency with which the work is done, and outcome is the consequences of the work performed. Quality assurance procedures are planned and conducted by specifying measurable indicators, collecting data, and analyzing information by using thresholds, trends, or benchmarks (7). Information is used to understand and make adjustments in the system's structure and process to improve outcome (6, 9–11, 37, 38, 63).

Quality assurance in clinical microbiology spans a wide spectrum, from monitoring the performance of equipment and reagents to examining the clinical value of services and information (58). Quality assurance consists of two related but separate functional components: quality control (QC) and quality improvement (QI). QC is a process for detecting and correcting defects by establishing limits of acceptable performance. QC procedures are used to systematically monitor and correct defects after they have occurred. In contrast, the aim of QI activities is to permanently prevent deficiencies in a process before they unfold. QI is accomplished by first understanding a process through measurement and observation and then using the information to anticipate potential weaknesses and defects in order to make adjustments that bring about lasting improvement (37, 78). Continuous QI is the ongoing sequence of planning, measurement, analysis, and adjustment (10). Thus, the goals of quality assurance are to maintain reliable analytical performance (QC) and to enhance and sustain the clinical value of laboratory services (QI).

The clinical microbiology laboratory operates to provide reliable and useful patient data as an aid in the diagnosis and treatment of infectious diseases. In the current health care market and with the passage of the Clinical Laboratory Improvement Amendments of 1988 (CLIA '88), laboratories must move from primary reliance on QC procedures that detect errors after they occur to methods that prevent errors throughout the testing process, from the ordering of a test to the application of the test result. Nonproductive QC methods must be eliminated, and the laboratory's resources must be channeled to cost-effective methods that improve patient care. This chapter is intended to serve as a

reference guide to issues involving quality assurance in the clinical microbiology laboratory.

QUALITY IMPROVEMENT

In the United States, standards for quality assurance in laboratory medicine have been largely influenced by federal legislation (Clinical Laboratory Improvement Act of 1967 [CLIA '67] and CLIA '88). CLIA regulations deal primarily with internal QC, external proficiency testing, and personnel standards (21). The College of American Pathologists (CAP) administers programs that laboratories may use to satisfy federal inspection and accreditation requirements. Hospital-wide quality standards are largely driven by the Joint Commission on Accreditation of Healthcare Organizations (JCAHO) (34). The JCAHO, founded in 1951, currently evaluates and accredits more than 5,000 hospitals and more than 3,000 other health care organizations. JCAHO standards and methods for evaluation and accreditation have evolved through four major developments, starting with peer review, moving to medical audit and then systematic quality assurance, and now being guided by principles of QI. Beginning in 1992, the JCAHO incrementally revised accreditation standards to embrace the principles and goals of continuous QI. This change represents a gradual evolution from monitoring and evaluating departmental activities to a general examination of how health care systems as a whole manage the flow of patient care that is intended to reveal opportunities for improvement. As a result, the emphasis on coordinating and integrating interdepartmental activities is expanding, and more attention is being given to improving processes and preventing problems. Beginning in 1994, the JCAHO defined nine dimensions of performance for "doing the right thing well" (Table 1). These definitions serve as a useful guide for QI as described below.

The QI Process

Plan and Design

QI must be planned carefully. Planning involves a systematic approach to designing, measuring, assessing, and improving performance. Processes targeted for improvement should be selected on the basis of perceived need. Clinical microbiology activities that involve high patient risk (e.g.,

TABLE 1 JCAHO dimensions of performance

Performance dimension	Definition
Efficacy	Degree to which care accomplishes projected or desired outcome
Appropriateness	Degree to which care is relevant to patient's clinical need
Availability	Degree to which appropriate care is available to meet patient's needs
Timeliness	Degree to which care is provided at most beneficial or necessary time
Effectiveness	Degree to which care is provided correctly to achieve desired or projected outcome
Continuity	Degree to which patient care is coordinated among practitioners, among organizations, and across time
Safety	Degree to which risk of intervention and risk in environment are reduced for patients and health care providers
Efficiency	Relationship between outcomes and resources used to deliver patient care
Respect and caring	Degree to which patient is involved in care decisions and to which providers show sensitivity and respect for patient's needs and expectations

cerebrospinal fluid [CSF] cultures), are problem prone (e.g., specimen collections), or affect large numbers of patients (e.g., reporting accuracy) provide opportunities for continuous QI.

The planning process begins by defining the QI objective. This may involve expectations of patients or providers, information based on standards or practice guidelines, or comparative performance measures (benchmarking). It is important not to specify measurements (indicators) before the QI objectives are adequately defined. Early in the planning phase, integral steps in the process should be delineated and understood by observation of and discussion with those involved. Suggestions for improvement often come early in the planning stage as a result of observation and communication before any measurements are actually made. Participation by health care workers in other departments is often essential and should begin early in the development stage. Teamwork, with participation and representation by everyone who is involved in the process, should be encouraged. This has important consequences, since the most troublesome and difficult-to-understand segment of a process involves "handoffs" between departments. For complex procedures, a fish bone diagram (11) or flow chart may improve understanding of and expand insight into a process's functionality. For example, if the goal is to improve the value of sputum microbiology, it is important to involve those who collect (nurses), order and interpret (physicians), and examine (microbiologists) sputum specimens. By working together, the team describes the process and their expectations for improvement. This may involve any or all of the performance dimensions listed in Table 1.

It is also necessary that organizational leaders strongly support and encourage the pursuit of specific QI initiatives. This is important for setting priorities, allocating resources, providing training, and fostering communication and coordination between individuals and departments. Perhaps most important, support by health care leaders and managers is often needed to facilitate the modification of a process under their supervision.

Measurement and Assessment

After the QI objective is defined, it is necessary to find objective performance measures that identify opportunities for improvement and appraise how modifications are affecting the process or the outcome. Measurement should be continuous, if possible, and relevant to the QI objective. Information assessment can be devised to monitor trends,

document improvement progress, determine whether performance objectives (thresholds) are met, identify further action for improvement, or identify areas requiring more intensive intervention or readjustments. Analysis can involve statistical methods or comparisons of data to consensus standards or external databases. An example of the process is given in Fig. 1. Assessment may also entail a sentinel event that provokes action whenever it occurs (e.g., lost CSF specimen) or may involve monitoring excessive variation from an external or internal database, variation from recognized standards, or discrepancies between laboratory measurements and clinical findings (11, 58).

Improvement

Improvement usually involves modifications that are designed into an existing process. There must be sufficient resources to make the change and agreement that it will have some chance of success. The impact of any modifications must be considered. An apparently beneficial change at one step of the process should not degrade performance at other steps. Expectations of what performance improvement target can be achieved should be clarified. When action to improve a complex process is contemplated, it may be helpful to initially implement changes on a trial basis with a limited scope. If the trial is successful, wider implementation can proceed, with high expectations for improvement.

QI Procedures

Specimens

The collection of microbiology specimens is generally not under the direct control of the clinical laboratory. This circumstance makes it very important to work closely with health care workers who are responsible for obtaining specimens. At the very least, it is important that microbiologists specify procedures for obtaining the best specimens for microbiologic examinations and maintain an active training program describing these procedures. The quality of specimen collections should be monitored continuously, and collection processes should be examined regularly to identify opportunities to improve performance.

Specimen Volume

The amount of specimen submitted for microbiologic examination can have important consequences for achieving good results. For example, CSF volume may be insufficient when multiple procedures are requested. The reasons

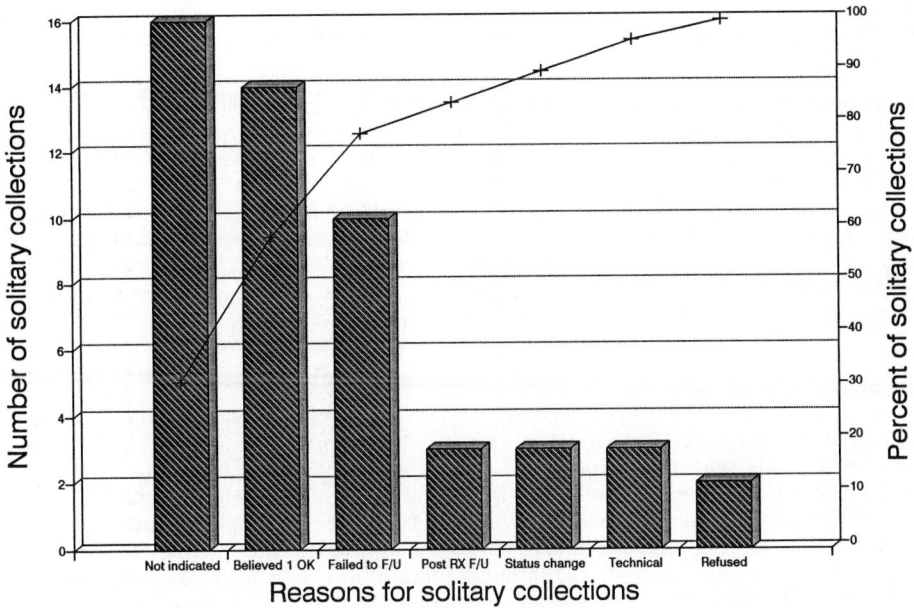

FIGURE 1 Pareto chart of solitary blood cultures for January to March 1990 (64). The Pareto chart provides a display on the horizontal axis of specific categories or groupings, beginning with the most prevalent type of event and continuing with the others in descending order. The absolute amount in each category is shown on the left axis, and a cumulative distribution (from 0 to 100%) is shown on the right axis. This type of presentation displays quantitative relationships between groups and helps set priorities for action. Making a change in the most prevalent category will have the greatest impact on overall improvement. Presenting data in this way helps anticipate the most likely sources of difficulty and sets priorities for improvement. F/U, follow-up.

for insufficient CSF volume (e.g., request for excessive testing, inappropriate guidelines) should be determined by working with those who order the tests and obtain the specimens. A change in the system (e.g., placement of easily visible directions for appropriate ordering and proper specimen volume in the lumbar puncture preparation tray or deferral of certain types of testing) may help prevent this problem permanently (2, 3, 17).

As another example, the sensitivity of blood cultures is highly dependent on blood volume. Insufficient volume for blood cultures can cause false-negative results and indicates an important, high-risk procedure that requires continuous monitoring (44). Improvements can be achieved by placing blood culture collections under the control of the institution's phlebotomy personnel, who are instructed as to the correct volume of blood to obtain.

Specimen Quantity

It is important to obtain the correct number of specimens for microbiologic examination. Excessive specimen submissions cause unnecessary work for the laboratory, while insufficient submissions may compromise the test's diagnostic reliability (40). It is important to establish guidelines for the appropriate number of specimens for each collection site and to monitor compliance. For example, solitary blood cultures for adults decrease the test's diagnostic reliability and are a mark of inappropriate test utilization (67). Blood culture collection by a phlebotomy service instructed to obtained multiple blood culture specimens can obviate this problem.

At least three morning sputum specimens for mycobacterial culture are recommended to achieve adequate test sensitivity. Insufficient specimen collections have been associated with delayed diagnosis of tuberculosis (36). Compliance with sputum collection standards should be monitored, and specimen collection procedures should be designed to ensure that an adequate number of specimens is obtained. For example, the laboratory can automatically send reminders for additional sputum collections for mycobacterial culture after the first is received, or the laboratory could wait to do this only if additional specimens are not submitted within a reasonable time.

Specimen Quality

Several techniques can be used to monitor and improve specimen quality. The diagnostic value of expectorated sputum specimens and endotracheal aspirates can be measured by cytologic examination (5, 49). Specimens with large numbers of squamous epithelial cells are presumed to contain a large volume of oropharyngeal flora that would obscure or misrepresent microbiologic findings associated with lower respiratory tract infections. Processing of specimens judged to be of poor quality can be deferred. The quality of expectorated sputum specimens depends in part on how carefully the patient is instructed about the collection process. This task generally falls upon the nursing staff or respiratory care service. It may therefore be useful to monitor the quality of sputum specimens by specific nursing unit in order to spot areas that show proportionately large numbers of sputum samples with excessive epithelial cells. This finding would indicate that more intensive educational activities, changes in collection procedures, or other corrective measures are warranted. Likewise, continued

TABLE 2 Examples of indications for restricting microbiology testing

Indication for restricting microbiology testing	Reference(s)
Routine bacterial culture and parasite examinations of enteric specimens from patients hospitalized for >3 days	19, 72, 82
CSF culture for mycobacteria of specimens with normal cell counts, glucose, and protein	3
Sputum cultures for mycobacteria from patients who have not had tuberculin skin test and chest roentgenogram examinations	73
Veneral Disease Research Laboratory and antigen testing of CSF until initial CSF findings are reviewed	2
More than one specimen received from same site by same method of collection each day (except blood cultures, CSF, and feces)	5
Bacterial culture of poor-quality specimens: from mouth, bowel contents, perirectal abscess, decubiti, pilonidal abscess, lochia, Foley catheter tip, vomitus, placenta (vaginally delivered)	5

monitoring will demonstrate the success or failure of attempts to maintain good-quality sputum collections.

Contamination rates of blood cultures and other sterile body fluids (e.g., CSF, joint fluid) are an indicator of the quality of aseptic specimen collection and laboratory processing. False-positive blood cultures can be a significant problem leading to longer hospital stays, unnecessary or prolonged therapy, and higher costs (8). It is important to monitor and minimize blood contamination rates. Procedures to decrease contamination should be designed into the processing steps (66, 74).

Another important component of specimen quality involves the length of time it takes to transport a specimen to the laboratory after it has been collected. Excessive delays may compromise results. For example, urine specimens left at room temperature for longer than 1 h may give rise to misleading false-positive results from overgrowth of organisms. Mixed urine culture results (three or more organisms) suggest a delayed or poorly collected specimen. Monitoring mixed cultures by specific nursing unit may uncover opportunities for more intensive evaluation or training to improve collection or transport procedures.

Test Utilization

The most important goal of QI procedures for test utilization should be to optimize the contribution that laboratory services make toward patient care while avoiding harmful effects. This is done by striving to ensure that testing is purposeful and efficient. Tests that are costly or have limited value may be restricted (Table 2). In other cases, such as solitary blood cultures or solitary sputum cultures for mycobacteria, as discussed above, underutilization may be a problem.

Methods to ensure that appropriate tests are ordered when indicated (as opposed to preventing inappropriate tests from being ordered when they are not indicated) and that test results are properly used are QI objectives that directly benefit the patient, especially if corrective action can be linked to the monitoring and evaluation processes. It is important to establish guidelines for appropriate microbiologic testing and to monitor compliance. It is necessary to establish these guidelines in conjunction with the medical staff and to establish a consensus definition of inappropriate testing. The guidelines may include clinical indications for testing, practice guidelines, and appropriate testing procedures for specific types of specimens. High-quality testing for an inappropriate indication will produce information that is either clinically useless or potentially misleading.

Appropriate Testing

The laboratory can specify criteria for appropriate testing and use this information to monitor test utilization practices. For example, pyuria is an important finding for requesting and interpreting urine cultures. Failure to perform a urinalysis for patients for whom a urine culture is ordered may indicate poor test selection and may compromise test interpretation. As another example, requests for inpatient stool cultures and ova and parasite examinations may not be indicated because of low diagnostic yield (19, 72, 82). Submitting stool specimens from inpatients with diarrhea for bacterial cultures and ova and parasite examinations but not testing for *Clostridium difficile* toxin may indicate poor test selection. Procedures to examine stool specimens can be designed to monitor and address these potential deficiencies in test utilization.

Some specimens are inherently of poor quality. Microbiologic examination of these specimens (e.g., from decubitus ulcer, periodontal lesion, perirectal abscess, or Foley catheter tip) produces no useful information. Consensus criteria and appropriate disposition of specimens collected from inappropriate sites should be defined in conjunction with the medical staff. Many laboratories defer processing of poor-quality specimens but hold the specimen until the physician is notified. The prevalence of inappropriate specimens should be monitored.

Utilization of Appropriate Results

The microbiology laboratory is often the earliest source of critical diagnostic information that can directly benefit the patient. Special effort should be taken to notify physicians of results that are judged to be important for immediate patient care or that are vital for clinical management. Positive blood cultures, CSF Gram smear examinations, and acid-fast bacilli in sputum smears or cultures are examples of results that many laboratories judge to be critical. However, in some cases, the clinical importance of information may not be adequately recognized because information about the patient's therapy and clinical condition is unavailable. Although all the necessary information is available to the "health care system," the patient does not immediately benefit because data are not integrated in a timely and effective fashion. This is a deficiency that can produce significant delays in appropriate therapeutic adjustments.

For example, optimal utilization of antimicrobial susceptibility test results requires more than simply making data available by written or computerized report (12, 71). Two other components are important. (i) There should be a process for looking at possible discrepancies between labo-

TABLE 3 Example of classification scheme for evaluating errors made in final report

Type of error	Example
Category A	Primary provider has responded to result by ordering next test, repeating test, or changing treatment or diagnosis
Category B	Serious error but unlikely to affect patient care; similar to category A error but primary care provider has not yet been notified or has not yet acted on reported erroneous result
Category C	Minor clerical error in reporting and "cosmetic" corrections to report

ratory results and patient therapy by reviewing current pharmaceutical and clinical data. Patients may be receiving antibiotic regimens to which an isolated pathogen is resistant, they may not be receiving treatment for an unrecognized clinically significant positive culture result, or they may be receiving treatment that is effective but more toxic or costly than necessary. (ii) There should also be a process for rapidly notifying the provider when these therapeutic discrepancies are uncovered. Without these additional components, patients with serious infections are at risk for delays or failures in treatment and, given that antimicrobial susceptibility testing data are predictive of therapeutic responses, unfavorable outcomes (23, 62, 75, 77). Computerized systems that link pharmacy information with data from the microbiology laboratory are now available from manufacturers of automated antimicrobial susceptibility testing instruments. Reports generated by these systems can be utilized to identify patients who are receiving inappropriate therapy. Application of these sources of information with effective intervention procedures may prove to be one of the most important quality improvement strategies in clinical microbiology.

Reporting Results

Two of the most important factors contributing to the quality of information in microbiology results are accuracy and timeliness. Clerical reporting errors may have adverse clinical consequences. Reporting timeliness and effectiveness may be gauged in part by how many telephone inquiries are made in proportion to the number of results reported (59). Reasons for the telephone inquiry (e.g., uncharted, lost, or delayed report?) may provide information about how to improve reporting efficiency.

It is also important to actively search for reporting mistakes and investigate their cause. This knowledge may identify reporting procedures that could be improved so that errors will not recur. Continued monitoring for errors will provide information to determine whether corrective actions are successfully reducing the frequency and clinical severity of errors (Table 3) and whether the microbiology section is able to maintain a low incidence of reporting errors (32).

When an error is detected after the final report goes out, a revised report should describe corrections but must not be modified in a way that conceals the initial erroneous result. An ongoing system for actively searching for errors in reported results should be in place. This searching can be done by retrospective comparison of results on worksheets, instrument logs, etc., with results reported on the chart. The clinical significance of errors discovered after the report is available for patient care should be classified and periodically reviewed to determine whether modifications in procedures or policies are needed.

Turnaround Time

One of the first steps in developing a turnaround time procedure is the establishment of measurable goals by con-

sensus of those involved in the process. These goals should be based on both the time constraints associated with performing the test and the realistic and medically relevant expectations of the health care providers. The consensus ensures that the goals are realistic and that everyone understands how the system works. Reaching a consensus presents an opportunity for suggestions to improve the process (61, 63). For example, a survey of over 400 clinical laboratories showed that the median turnaround time for processing, examining, and reporting CSF Gram smears was 45 min (31). The target turnaround time that laboratories established for this procedure was set at <60 min for 46%, 60 min for 47%, and >60 min for 8% of such smears. The survey demonstrated that these goals were met on average 62% of the time and that 15.3% of laboratories surveyed met the goal 100% of the time (31).

The effect that turnaround time of laboratory testing has on patient care is dependent on many factors, a large number of which are external to laboratory processing events. While easy to do, it is insufficient to track only internal laboratory turnaround times. The various components of the entire process must be examined. Whenever possible, the components examined should include both the time when specimens are collected and the time when results are utilized. For example, reporting "pleomorphic gram-negative rods with numerous inflammatory cells" in a CSF specimen would evoke another procedure to verify that the patient receives timely and appropriate therapy after a reasonable time.

Cumulative Susceptibility Reports

The laboratory should track cumulative hospital susceptibility patterns and make information on the patterns available to the medical staff. Laboratory information systems should have the capability to stratify this information by location, culture site, and time interval as needed for more specific evaluations. Such stratification aids in the empiric selection of antibiotics, formulary development by the pharmacy, and analysis of trends in endemic resistance by infection control practitioners.

Infection Control

The clinical laboratory provides valuable technical and informational support to the institution's infection control program (42). Furthermore, the JCAHO standards require the provision of adequate laboratory support for an infection control program. Review of microbiology reports is the most widely used method for case finding, and review of patient medical records is the next most widely used method (1). Properly organized laboratory information can be used for identifying clusters of infections that might be caused by a breakdown in technique or by cross infections (18, 39, 41, 65). Furthermore, total and specific nosocomial infection rates can be accurately estimated by simply calculating the ratio of positive cultures (from patients who

TABLE 4 User perceptions as measure of quality

Personnel are courteous.
Repeats usually confirm the first test.
Sample mix-ups are infrequent.
Routine results are available for rounds.
Statim results are available in time.
Results go to the right place.
Health care workers know whom and where to call for help.
Laboratory staff are readily available and eager to help.
Health care workers rarely have difficulty knowing what specimen to collect.
Health care workers know what to do to resolve problems with the laboratory.
Needs are understood, met, and exceeded.
Complaints and problems are swiftly addressed, resolved, and rarely repeated.

have been hospitalized for several days) to total hospital admissions or discharges (16, 24).

User Perceptions

A perception that the laboratory is doing a good job is evidence that QI goals are being met (57). Measurement of this perception can provide an important perspective on the quality of laboratory service (Table 4).

Q-Probes

Q-Probes is a voluntary subscription program of the CAP (30). It is an interinstitutional program comprising periodic studies dealing with quality assessment, QI, and benchmarking in pathology and laboratory medicine (Table 5). The data generated by the program are used to compare the facility's performance with that of other institutions, a process referred to as benchmarking. This information can in turn be utilized to identify opportunities for improvement.

QUALITY CONTROL

The passage of CLIA '67 (20) codified the QC measures in use today and focused attention on monitoring test performance as the means for improving laboratory test results. CLIA '67 regulated only those laboratories that provided services reimbursed under Medicare or that accepted specimens for testing from out of state. During this period, laboratories diligently recorded data related to all aspects of test performance, including staff education and training,

TABLE 5 Quality assurance issues examined by CAP interlaboratory Q-Probes program with application to microbiology in years 1989 to 1994

Blood culture utilization	Turnaround time of CSF Gram smear
Nosocomial infections	
Cumulative susceptibility results	Error reporting
	Sputum quality
Antibiotic usage	Blood culture contamination
Stool microbiology	Effects of laboratory computer downtime
Laboratory diagnosis of tuberculosis	Viral hepatitis test utilization
Quality of reference laboratories	

facilities and resources, procedure manuals, equipment, media and reagents, and proficiency testing programs, and statistically analyzed these data. The objective of the laboratory's QC program was to ensure reliable analytical test performance, but little attention was focused on the clinical value of laboratory services.

CLIA '88 (25–28) calls for a new approach and a new emphasis in the federal regulation of clinical laboratories. See chapter 6 of this Manual for a more complete discussion of CLIA '88. Although the general QC standards related to laboratory testing are similar to the standards in CLIA '67, CLIA '88 now mandates that, as discussed earlier in this chapter, laboratories develop comprehensive QA activities that monitor, evaluate, and improve the overall testing process (6).

CLIA '88 significantly expands the scope of federal regulatory authority to all laboratories that perform tests on human specimens for the purposes of diagnosis, prevention, or treatment of disease or for assessment of the health of a person. CLIA '67 regulated laboratories on the basis of their location, size, and type. CLIA '88 regulates laboratories on the basis of the complexity of the tests that they perform: waived tests (e.g., fecal occult blood), moderate-complexity tests (e.g., Gram smears of urethral or cervical exudates), or high-complexity tests (e.g., bacterial culture and antimicrobial susceptibility tests). The majority of tests performed in the clinical microbiology laboratory are classified as high-complexity tests (29).

Objective of a QC Program

The clinical microbiology laboratory is responsible for providing accurate and relevant information that is necessary for management of the patient. The accuracy and clinical value of the laboratory's analysis of the clinical specimen and microbial isolate are dependent on a QC program that assesses the quality of the specimen; documents the validity of the test method; monitors the performance of test procedures, reagents, media, instruments, and personnel; and reviews test results for errors and clinical relevance. An effective QC program is dependent on a continuous process of assessment and improvement.

Overview of CLIA '88 Standards

In addition to CLIA '88 standards, laboratories must comply with all regulations regarding the health and safety of patients and employees and the handling, storage, and disposal of hazardous waste (see also chapter 6 of this Manual). All federal regulations and standards are considered minimal and may be superseded by stricter standards imposed by states or other certifying agencies. A brief summary of some of the significant changes from CLIA '67 standards follows.

Personnel

Standards for personnel are now linked to test complexity. Most laboratories performing moderate-complexity testing are likely to be physicians' offices and primary-care clinics. These laboratories must have employees who perform in four areas of responsibility: laboratory director, technical consultant, clinical consultant, and testing personnel. High-complexity testing (includes the majority of microbiological tests) is likely to occur in hospital-based and large reference laboratories. In addition to the personnel required for moderate-complexity testing, high-complexity testing laboratories must have a technical and general supervisor. Most likely, the personnel currently employed in larger laboratories will meet CLIA '88 standards. CLIA '88 lowers the educational requirements for some personnel who work in laboratories.

Patient Test Management

Laboratories are required to implement systems that guarantee the integrity and identity of patient specimens throughout the process of collection, testing, and reporting of results. This standard mandates specimen-labeling instructions for clients, specific information on test requisitions, procedures for handling test results and records, and requirements for specimen referral. This standard will probably have an impact on smaller laboratories, since the majority of larger laboratories have these systems currently in place.

Validation of Test Method

Laboratories must establish and verify the performance of all new test methods implemented after September 1, 1992 (27). If the test method has been approved for in vitro diagnostic use by the Federal Drug Administration (FDA), the laboratory has only to demonstrate that it can obtain results comparable to those reported by the manufacturer, that the reference range is appropriate for the patient population served by the laboratory, and that the laboratory follows the manufacturer's test instructions. Otherwise, the laboratory must follow all applicable requirements for validation of the test procedure.

Proficiency Testing

Proficiency testing has changed from an educational experience to a regulatory requirement. Laboratories must participate in an approved proficiency-testing program for each specialty area in which testing is performed. Sanctions will be applied against those laboratories that fail to pass 80% of the proficiency test samples.

QA

QA activities must document active review and problem solving in the areas of patient test management, QC, proficiency testing, personnel, patient and user complaints, and relationships of patient information to patient test results. Twice yearly, test results must be compared among multiple instruments, sites, or methodologies that are used to report the same test. The validity of test methods not covered by a proficiency-testing program must be documented by the laboratory. The results of QA activities must be reviewed with the laboratory staff.

Elements of a QC Program

Personnel

The essential factor in a laboratory is the employment of highly trained individuals who work in an environment that promotes teamwork, commitment to patient care, and high standards for test performance. Employees need to be informed of and involved in decisions by their employer that affect their work and careers. CLIA '88 defines the personnel requirements for laboratories based on the complexity test model (27). In addition to redefining the titles of laboratory personnel, CLIA '88 requires that testing personnel have at least a high school diploma and appropriate training. The general supervisor must be available on site in laboratories performing high-complexity tests. All laboratory staff members who perform tests must document that they can complete the procedure competently. Their

training must include all preanalytic, analytic, and postanalytic phases of testing and must be documented twice during the first year of employment and once a year thereafter.

The employee's personnel record should contain this information in addition to information on past training and certification, the tasks and procedures the employee is authorized to perform, continuing education activities, and the annual performance appraisal. Although the quality of employee performance is difficult to measure quantitatively in a performance appraisal, the establishment of specific, clearly defined standards and the opportunity to acquire new knowledge and skills motivate most employees. Regardless of the federal requirements for personnel, excellent laboratories will continue to hire highly trained individuals.

Procedure Manual

The laboratory's procedure manual must contain all test methods performed by the laboratory. Both CLIA '88 and the CAP Commission on Laboratory Accreditation require that the laboratory's procedure manual be written in a specific format. CLIA '88 requires a 16-item format (27), while the CAP mandates that the manual be substantially in compliance with the National Committee for Clinical Laboratory Standards (NCCLS) publication GP2-A (15, 50). Compared to NCCLS publication GP2-A, the CLIA '88 format requires two additional content areas: (i) a description of the course of action to be taken in the event that a test system becomes inoperable and (ii) written criteria for the proper collection, preservation, and transportation of specimens accepted from outside clients. A significant difference between CLIA '88 and CAP requirements involves the acceptability of the manufacturer's product inserts and procedure manuals in place of a laboratory manual. The manufacturer's product insert or procedure manual is acceptable under CLIA '88 if it includes all of the required CLIA '88 items. Package inserts are not acceptable to the CAP, but a manufacturer's instrument manual that complies with NCCLS publication GP2-A is acceptable.

The original procedure manual and each change to it must be dated, approved, and signed by the current laboratory director and be readily available to the staff, preferably at the workstation. Orderly updating and review of manuals take place annually and should involve the personnel who actually use the manual on a daily basis. All replaced procedures must be retained by the laboratory for at least 2 years.

The compilation of a complete procedure manual is a difficult task but is essential to guaranteeing that all laboratory personnel perform the test in an identical fashion (4, 45). This ensures the accuracy and reproducibility of the test result. In addition to a step-by-step description of the test procedure, the manual should contain the unique laboratory policies relating to the extent of specimen workup and isolate identification.

Patient Test Management

Proper specimen handling, integrity, identification, and reporting of results must be documented throughout the testing process through written procedures that detail patient preparation, specimen collection, labeling, preservation, and transportation. Tests are performed at the written or electronic request of an authorized health care worker. Since oral test requests must be followed by a written authorization within 30 days, the laboratory should develop a policy that defines how oral requests or changes to orders are documented.

Records and Reports

The absolute requirement that detailed records be maintained for all aspects of the testing process began with CLIA '67 and continues unchanged with CLIA '88. These records document what has occurred, depict trends, serve as a data source for resolving problems, and establish the credibility of the laboratory. The standards for record keeping and the retention time for records are found in CLIA '88 (4) and the manuals of other inspection or accreditation agencies (13, 69). Examples of QC record formats are contained in reference 33a.

The test requisition should contain the following minimum information: patient's name or identifier, date the specimen was collected and received by the laboratory, name and address of the requester, test requested, and any other pertinent data (27). These records are retained by the laboratory for 2 years in a readily accessible location.

A test record or work card is assigned to each specimen to ensure reliable identification of the patient specimen as it is processed and tested by the laboratory. The procedures performed, personnel performing the procedures, notes made by personnel, communication between health care personnel, and results obtained are recorded. If applicable, instrument printouts should be affixed to the work card. In the event that the specimen does not meet the laboratory's criteria for acceptability, the reason for rejection and disposition of the specimen is documented. A work card is a legal document that can be used to reconstruct the test process and assess the accuracy of the final report. These records should also be retained for 2 years.

A test report conveys the test result to the authorized person in a timely, accurate, reliable manner that ensures patient confidentiality. The report also contains the name and address of the laboratory performing the test, reference ranges, and, if applicable, reasons for specimen rejection. If the test result indicates a life-threatening situation for the patient, the laboratory must immediately alert the person caring for the patient and document the communication. If an erroneous result was reported, contact the person utilizing the result, explain that the laboratory result was in error, and issue a clearly identified corrected report. Do not remove the original erroneous report from the patient's chart. Additionally, the laboratory must make available (if requested) the test method used to generate the result, the performance specifications of the test (if applicable), and any known test interferences. These records must be retained for 2 years.

Media, Reagents, and Commercial Systems

The laboratory must monitor and evaluate the accuracy and performance of each test. The written test procedure should define the conditions for specimen and reagent storage, reagent preparation (e.g., water quality), and test conditions (e.g., temperature, humidity). In general, outdated reagents should not be used. However, the CAP accreditation program allows laboratories to use certain costly outdated reagents but only if the laboratory has a written policy identifying the reagents, the circumstances when extended usage applies, the special control procedures to be implemented, and the person authorizing the usage (14). All reagents, solutions, culture media, and other supplies are labeled to indicate identity; strength, titer, or concentration (when significant); storage requirements; prepara-

tion and expiration dates; and other pertinent information for usage.

Beginning September 1, 1994, laboratories may follow the manufacturer's instructions for QC if the product has been cleared by the FDA as meeting the requirements of CLIA '88 (27). For each method developed in-house, modified, or not cleared by the FDA, the laboratory must evaluate instrument and reagent stability; operator variance; and number, type, and frequency of testing of control material in order to establish criteria for acceptability of test results. The test performance is monitored with QC material that is tested exactly as patient specimens are tested. For qualitative tests, negative and positive controls are used, and for quantitative tests, two controls of different strengths are included with each run of patient specimens. If applicable, statistical parameters (e.g., mean and standard deviation) are determined for each lot of control material. If control material is not available, the validity of the test method must be established by an alternative method.

Media

Commercially prepared media listed in NCCLS publication M22-A (52) are exempt from QC performed by the user (except chocolate agar, *Campylobacter* media, and selective media for the isolation of pathogenic *Neisseria* spp.) provided that the manufacturer meets the QC standards outlined in NCCLS publication M22-A. Additionally, the laboratory must ensure that the physical characteristics of the media have not been compromised during shipment (e.g., freezing, dehydration).

Each batch of user-prepared media and each shipment of commercially prepared media not listed in NCCLS publication M22-A should be evaluated for sterility, ability to support growth, and, as appropriate, selectivity, inhibition, and/or biochemical response. QC procedures, organism maintenance, and tables listing appropriate QC organisms can be found in a number of excellent references (46–48, 52, 70, 81).

Reagents

Each batch or shipment of reagents, disks, stains, antisera, and microbial identification systems is checked for positive and negative reactivity and for graded reactivity if applicable when it is prepared or opened. The reliability of the reagent must be documented before the reagent is used for patient testing. The QC strains, frequency of testing, and expected results for a variety of microbiological tests can be found in other references (46–48, 70, 81). Ideally, American Type Culture Collection strains are used for QC procedures, but other organisms that produce comparable results may be used.

The following CLIA '88 requirements for the QC of frequently used tests in bacteriology, mycobacteriology, and mycology are nearly identical to the recommendations in CLIA '67 and represent minimum standards. Regrettably, CLIA '88 did not critically review and eliminate the QC procedures that do not contribute to improved patient care but are costly to perform. In addition, CLIA '88 recommends only a positive control organism, although most references suggest both a positive and a negative control for fungal and mycobacteriology tests (46–48, 70, 81).

Reagents used for routine bacteriology tests are checked each day or week of use with a positive and a negative control organism. Catalase, coagulase, β-lactamase, and oxidase tests are monitored daily, and bacitracin, optochin, x and v strips or disks, and ONPG (*o*-nitrophenyl-β-D-galactopyranoside) reagents are checked each week of use.

The reagents and tests used for the identification of mycobacteria are checked each day of use with one *Mycobacterium* species that produces a positive reaction. However, the iron uptake test is monitored with both a positive and a negative control organism.

Reagents and tests used for the identification of fungal isolates are checked each week of use with a positive control organism. The nitrate reagent used to determine nitrate assimilation is monitored with a peptone control.

For antimicrobial susceptibility testing of bacteria, each batch of medium and each lot of disks are monitored before or concurrently with use, using the NCCLS-recommended QC strains (53). A fungal isolate susceptible to each antifungal drug is used each time the fungal susceptibility test is performed. The recommendations for the QC of mycobacterial susceptibility testing are similar except that QC is performed each week of use. Detailed procedures for the QC of these two susceptibility test methods are available in reference 33a and other references (51, 53, 55, 56).

The recommended frequency of testing antisera used for the identification of bacteria varies with the accreditation agency (4). CLIA '88 requires that antisera be checked when opened and monthly thereafter with a positive and a negative control organism.

Products for the direct detection of antigen in clinical specimens and DNA probes are checked each day of use with material that produces a positive and a negative reaction. If the test method uses an extraction step, the system must be checked with an appropriate microorganism.

Stains

Microbiologic stains are checked with positive and negative control organisms either weekly (Gram, Kinyoun, Ziehl-Neelsen, mycobacterial fluorochrome), with each use (all fluorescent stains), or with each day of use (all other stains). A detailed list of QC organisms and frequency of testing is available in other references (46–48, 70, 81).

Commercial System

QC procedures for manual and automated systems are based on the recommendations of the manufacturers. QC testing is performed on each shipment and each lot. Components of kits with different lot numbers cannot be interchanged unless specified by the manufacturer.

Facility Requirements

Adequate space, lighting, ventilation, and utilities must be available for the performance of tests. Instruments should be protected from electrical fluctuations or interruptions. Specifications, preparation, and QC of laboratory water should be defined for each test, if applicable (35, 54). The determination of adequate environmental conditions is left to individual laboratories and ultimately the person who inspects the laboratory. Safety precautions must be established, posted, and observed.

Equipment

Periodic maintenance and function checks are performed and documented for all equipment (27). If the equipment is cleared by the FDA, the laboratory may follow the frequency of maintenance and function checks specified by the manufacturer. For equipment not cleared by the FDA, developed in-house, or modified, the laboratory must establish the frequency of maintenance and function checks that

guarantees accurate test results. Maintenance logs for each instrument should be retained for the life of the equipment. Reference 33a discusses the maintenance and QC of common microbiological instruments and equipment.

Remedial Actions

Documented remedial actions must be taken for QC material that fails to meet the acceptable criteria, for equipment and test methods that do not meet performance specifications, for turnaround times over the accepted threshold, for patient results outside the reportable range, for inappropriate reference ranges, and for errors detected in reported patient results. In other words, laboratories must respond to any event that is outside what is expected and must record what action they have taken as outlined under QI.

Proficiency Testing

All laboratories must participate in an external proficiency program that is approved by the U.S. Department of Health and Human Services and reflects the specialty (e.g., mycobacteriology) and level of expertise of the laboratory (27). This program consists of at least five samples per testing event and at least three events at approximately equal intervals per year. Proficiency test samples are processed and analyzed by routine test methods and personnel, as affirmed by the person performing the test and the director of the laboratory. The laboratory's result is compared with the response that agrees with the results from 90% of 10 reference laboratories or 90% or more of participating laboratories. Laboratories must either submit correct responses for 80% of the proficiency test samples or face sanctions. The laboratory should have a plan for responding to a failure that documents the source of the problem and the corrective action taken.

VALIDATION OF METHOD PERFORMANCE

Technology assessment is used to make decisions about which tests are appropriate for a particular laboratory and to define the benefits and limitations of a new test or technology (76). This decision is often based on subjective impressions regarding the utility of the method rather than on an objective evaluation of the strengths and weaknesses of a test. As part of the process of method assessment and validation, CLIA '88 requires that laboratories verify or establish test performance specifications prior to reporting patient results (27). If a test is approved by the FDA, the laboratory need demonstrate only that it can obtain results comparable to those of the manufacturers. See references 43 and 80 for further details.

New Test Assessment

The most important consideration in the decision to implement a new test is the clinical need for or value of the test. In other words, how useful is the test for patient care? Factors that need to be considered but may be difficult to substantiate include the needs of the test requester in terms of the type of test (e.g., screening or diagnostic) and the turnaround time (e.g., statim or routine), the influence of a test on patient management, and the influence of a test on patient outcome (76, 80). The type of test and speed of testing required may be determined by simply soliciting the views of the clinicians who would use the test. The influence of a test on patient management and outcome, unfortunately, are the most difficult factors to measure quantitatively.

The next step in the assessment of a new test is to evaluate the performance of the method and compare it to those of other available techniques. The procedure for the evaluation of a test method should include, when applicable, the precision, linearity, accuracy, sensitivity, specificity, predictive value, and reference ranges for the targeted patient population. For nonculture, quantitative tests, the precision or reproducibility and the technical accuracy of the test can be assessed by the use of calibrators and control samples to determine the mean and standard deviation of replicate determinations, the linearity can be determined by using serial dilutions of external control material or elevated-titer patient samples, and interference checks can be performed by adding interfering substances to the sample or by reviewing the relevant literature.

For culture methods and techniques that yield a dichotomous positive-negative result, laboratories have generally determined the specificity and sensitivity of the method (22, 33). Test sensitivity is defined as the frequency of a positive test in patients with the disease (percentage of true positives among all positive patients). Test specificity is defined as the frequency of a negative test in patients without the disease (percentage of true negatives among all negative patients). To determine sensitivity and specificity, the true patient status must be known. The sensitivity and specificity of a test are independent of the prevalence of the disease in a population, and these two parameters are therefore not as useful to the clinician as the predictive value of a test (22).

The predictive value of a test is used to rule in or rule out a disease. The predictive value of a positive test (PVP) is the probability that a positive result indicates that the patient has the disease (percentage of true positives among all positive tests). The predictive value of a negative test (PVN) is the probability that a negative test indicates that the patient does not have the disease (percentage of true negatives among all negative tests). The PVP or PVN is dependent on the sensitivity (PVN), specificity (PVP), and prevalence of disease (PVN and PVP). The higher the prevalence of disease and specificity, the higher the PVP, since there will be fewer false-positive tests. The lower the prevalence of disease and the higher the sensitivity, the higher the PVN, since there will be fewer false-negative tests. Last, test accuracy or efficiency is the frequency with which patients are correctly identified by a test result and is dependent on both PVP and PVN (22).

These statistical parameters are helpful in the evaluation of a screening or diagnostic test in a particular disease state. In general, tests with a high PVP are useful when a false diagnosis results in therapy that is potentially harmful or has economic, social, or psychological consequences. Conversely, tests with a high PVN are useful as screening tests when you want to detect all positive patients.

The third step in the assessment of a new test is evaluating the cost of providing the test. This is most typically done by determining the direct (e.g., reagents, labor, instruments) and indirect (e.g., billing, overhead, heat, etc.) costs and comparing costs to anticipated revenues. A significant factor in cost analysis will be the number of required QC samples per run, whether the test is performed individually or in batches, and the number of repetitions required to confirm a positive result. Cost per test is useful to know when the laboratory sells an individual test to generate revenue. However, this approach is less useful with prospective payment, managed care, and health mainte-maye plans, where expense per patient admission is the ︎nsideration (76, 80). The cost of a new test must include an assessment of how the test will be used and whether it is a replacement test or an additional expense.

Last, the test assessment must address the availability of technically skilled personnel to perform the test, necessary instrumentation, and physical facilities. In addition, any special requirements for specimen collection, transportation, and result reporting may be a potential problem.

Comparability of Test Methods

As the number of new diagnostic tests increases, the laboratory is faced with choosing among multiple tests that can be used to diagnose the same disease. Typically, new tests are compared by testing samples in parallel, testing samples with known potencies, or comparing test results with clinical diagnoses. However, other factors need to be considered when multiple tests are being evaluated, especially when the evaluation is based on published results. Schwartz et al. (68) identified five methodological problems associated with test evaluations: (i) inadequate reference standard, (ii) spectrum bias, (iii) failure to use different cutoff points with immunoassays, (iv) referral bias, and (v) inadequate sample size.

In order to assess the accuracy of a new test, a reference standard is required. In microbiology, the reference standard is rarely both highly specific and sensitive and therefore will misclassify some results, which may distort the assessment of the new test (76). In fact, the new test may actually be more accurate than the reference standard. Ilstrup (33) recommends that when the true disease status of the patient is unknown, sensitivity and specificity should not be used; instead, disagreements between the new test and the reference test should be resolved by other methods.

Spectrum bias refers to the failure to consider the heterogeneity of diseased and nondiseased populations in assessing test performance (60). Patient-to-patient variation in disease severity becomes important when two diagnostic tests are compared and the test sensitivities vary with the stage of the disease (60, 76). Spectrum bias also occurs if the study population does not include patients with a clinical presentation that is similar to the disease state. Ideally, identical patient samples should be used for the comparison of two tests.

Defining the test reactivity by cutoff points is a problem, especially with immunoassays, because the results are ordinal or continuously scaled (e.g., titers, 1+, 2+, etc.). The sensitivity and specificity of the test will depend on which test values are considered positive or negative. To compare tests, plot the sensitivity on the y axis and the specificity on the x axis as the definition of a positive value is varied. This plot is a receiver operating curve analysis (33). In addition to statistical parameters, the best test will depend on the objectives of the test (diagnostic or screen), the prevalence of the disease, the cost of false-negative or false-positive results, and the availability of a confirmatory test.

The retesting of a sample that is initially nonreactive or equivocal or is initially positive or equivocal is termed referral bias (68). In the first instance (nonreactive or equivocal), selective retesting will underestimate the sensitivity, and in the second instance (positive or equivocal), it will underestimate the specificity of the first test relative to the second test.

The last problem identified by Schwartz et al. (68) involved the use of inadequate sample size for test evaluation. When the sensitivity, specificity, and predictive values of tests are being compared, the confidence intervals (CIs) should be calculated for the statistic (80). CIs give an estimate of the precision (imprecision) of the statistical parameter for the test method and vary with the sample size and level of confidence (e.g., 90, 95, or 99%). The larger the CI, the less precise the estimate of sensitivity, specificity, and predictive values of the test will be (80). Washington (79) presents an excellent discussion of the calculation and use of CI.

REFERENCES

1. **Abrutyn, E., and G. H. Talbot.** 1987. Surveillance strategies: a primer. *Infect. Control* **8:**459–463.
2. **Albright, R. E., R. H. Christenson, R. L. Habig, T. P. Mears, and K. A. Schneider.** 1988. Cerebrospinal fluid (CSF) TRAP. A method to improve CSF laboratory efficiency. *Am. J. Clin. Pathol.* **90:**707–710.
3. **Albright, R. E., Jr., C. B. Graham III, R. H. Christenson, W. A. Schell, M. C. Bledsoe, J. L. Emlet, T. P. Mears, L. B. Reller, and K. A. Schneider.** 1991. Issues in cerebrospinal fluid management: acid-fast bacillus smear and culture. *Am. J. Clin. Pathol.* **95:**418–423.
4. **August, M. J., J. A. Hindler, T. W. Huber, and D. L. Sewell.** 1990. *Cumitech 3A, Quality Control and Quality Assurance Practices in Clinical Microbiology.* Coordinating ed., A. S. Weissfeld. American Society for Microbiology, Washington, D.C.
5. **Bartlett, R. C.** 1981. How far to go—how fast to go, p. 12–44. In V. Lorian (ed.), *Significance of Medical Microbiology in the Care of Patients,* 2nd ed. The Williams & Wilkins Co., Baltimore.
6. **Bartlett, R. C., M. Mazens-Sullivan, J. Z. Tetreault, S. Lobel, and J. Nivard.** 1994. Evolving approaches to management of quality in clinical microbiology. *Clin. Microbiol. Rev.* **7:**55–88.
7. **Batalden, P. B., and E. D. Buchanan.** 1989. Industrial models of quality improvement, p. 133–159. In N. Goldfield and D. B. Nash (ed.), *Providing Quality Care.* American College of Physicians, Philadelphia.
8. **Bates, D. W., L. Goldman, and T. H. Lee.** 1991. Contaminant blood cultures and resource utilization. The true consequences of false-positive results. *JAMA* **265:**365–369.
9. **Berwick, D. M.** 1988. Measuring health care quality. *Pediatr. Rev.* **10:**11–16.
10. **Berwick, D. M.** 1989. Continuous improvement as an ideal in health care. *N. Engl. J. Med.* **320:**53–56.
11. **Berwick, D. M., A. B. Godfrey, and J. Roessner.** 1991. *Curing Health Care. New Strategies for Quality Improvement. A Report on the National Demonstration Project on Quality Improvement in Health Care.* Jossey-Bass, San Francisco.
12. **Campo, L., and J. M. Mylotte.** 1988. The use of microbiology reports by physicians in prescribing antimicrobial agents. *Am. J. Med. Sci.* **296:**392–398.
13. **College of American Pathologists.** 1988. *Professional Relations Manual,* 9th ed. College of American Pathologists, Northfield, Ill.
14. **Commission on Laboratory Accreditation.** 1990. *Inspection Checklist. Diagnostic Immunology and Syphilis Serology.* College of American Pathologists, Northfield, Ill.
15. **Commission on Laboratory Accreditation.** 1992. *Inspection Checklist. Microbiology.* College of American Pathologists, Northfield, Ill.
16. **Costel, E. E., S. Mitchell, and A. B. Kaiser.** 1985. Abbreviated surveillance of nosocomial urinary tract infections: a new approach. *Infect. Control* **6:**11–13.
17. **Crowson, T. W., E. C. Rich, B. F. Woolfrey, and D. P. Connelly.** 1984. Over-utilization of cultures of CSF for mycobacteria. *JAMA* **251:**70–72.
18. **Evans, R. S., J. P. Burke, D. C. Classen, R. M. Gardner, R. L. Menlove, K. M. Goodrich, L. E. Stevens, and S. L. Pestotnik.** 1992. Computerized identification of patients at high risk for hospital-acquired infection. *Am. J. Infect. Control* **20:**4–10.
19. **Fan, K., A. J. Morris, and L. B. Reller.** 1993. Application of rejection criteria for stool cultures for bacterial enteric pathogens. *J. Clin. Microbiol.* **31:**2233–2235.
20. **Federal Register.** 1968. Clinical Laboratories Improvement Act of 1967. *Fed. Regist.* **33:**15297–15303.
21. **Federal Register.** 1990. Medicare, Medicaid and CLIA programs: regulations implementing the Clinical Laboratory Improvement Amendments of 1988 (CLIA-88); proposed rule. *Fed. Regist.* **55:**20896–20959.
22. **Galen, R. S., and S. R. Gambino.** 1975. *Beyond Normality: the Predictive Value and Efficiency of Medical Diagnosis.* John Wiley & Sons, Inc., New York.
23. **Granato, P. A.** 1993. The impact of same-day tests versus traditional overnight testing. *Diagn. Microbiol. Infect. Dis.* **16:**237–243.
24. **Gross, P. A., A. Beaugard, and C. VanAntwerpen.** 1980. Surveillance for nosocomial infections: can the sources of data be reduced? *Infect. Control* **1:**233–236.
25. **Health Care Financing Administration.** 1988. Medicare, Medicaid, and CLIA programs; revision of the clinical laboratory regulations for the Medicare, Medicaid, and Clinical Laboratories Improvement Act of 1967 programs. *Fed. Regist.* **53:**29590–29632.
26. **Health Care Financing Administration.** 1990. Medicare, Medicaid, and CLIA programs; revision of laboratory regulations; final rule with request for comments. *Fed. Regist.* **55:**19538–19610.
27. **Health Care Financing Administration.** 1992. Clinical Laboratory Improvement Amendments of 1988; final rule. *Fed. Regist.* **57:**7137–7186.
28. **Health Care Financing Administration.** 1993. Medicare, Medicaid, and CLIA programs; CLIA fee collection; correction and final rule. *Fed. Regist.* **58:**5212–5237.
29. **Health Care Financing Administration.** 1993. Compiled list of clinical laboratory test systems, assays, and examinations categorized by complexity. *Fed. Regist.* **58:**39860–39878.
30. **Howanitz, P. J.** 1990. Quality assurance measurements in departments of pathology and laboratory medicine. *Arch. Pathol. Lab. Med.* **114:**1131–1135.
31. **Howanitz, P. J., and S. Steindel.** 1991. Intralaboratory performance and laboratorians' expectations for stat turnaround times. *Arch. Pathol. Lab. Med.* **115:**977–983.
32. **Howanitz, P. J., K. Walker, and P. Bachner.** 1992. Quantification of errors in laboratory reports: a quality improvement study of the College of American Pathologists' Q-Probes program. *Arch. Pathol. Lab. Med.* **116:**694–700.
33. **Ilstrup, D. M.** 1990. Statistical methods in microbiology. *Clin. Microbiol. Rev.* **3:**219–226.
33a.**Isenberg, H. D. (ed.).** 1992. *Clinical Microbiology Procedures Handbook,* 2 vol. American Society for Microbiology, Washington, D.C.
34. **Joint Commission on Accreditation of Healthcare Organizations.** 1992. Quality assessment and improvement, p. 137–143. In *Accreditation Manual for Hospitals.* Joint Committee on Accreditation of Healthcare Organizations, Chicago.
35. **Kempf, J. L., and D. L. Sewell.** 1992. Preparation and quality control of laboratory water, p. 13.4.1–13.4.11. In H. D. Isenberg (ed.), *Clinical Microbiology Procedures Handbook,* vol. 2. American Society for Microbiology, Washington, D.C.
36. **Kramer, F., T. Modilevsky, A. R. Valiany, J. M. Leedom, and P. F. Barnes.** 1990. Delayed diagnosis of tuberculosis in patients with human immunodeficiency virus infection. *Am. J. Med.* **89:**451–456.
37. **Kritchevsky, S. B., and B. P. Simmons.** 1991. Continuous quality improvement. Concepts and applications for patient care. *JAMA* **266:**1817–1823.
38. **Laffel, G., and D. Blumenthal.** 1989. The case for using industrial quality management science in health care organizations. *JAMA* **262:**2869–2873.
39. **Laxson, L. B., M. J. Blaser, and S. M. Parkhurst.** 1984. Surveillance for the detection of nosocomial infections and the potential for nosocomial outbreaks. *Am. J. Infect. Control* **12:**318–324.
40. **Makadon, H. J., D. Bor, and G. Friedland.** 1987. Febrile inpatients: house officer's use of bl[...] Med. **2:**293–297.
41. **McGluckin, M. B., and E. Ab[...]**

method for early detection of nosocomial outbreaks. *Am. J. Infect. Control* **7:**18–21.

42. **McGowan, Jr., J. E.** 1991. New laboratory techniques for hospital infection control. *Am. J. Med.* **3**(Suppl. B):**245S–251S**.

43. **McPherson, B. S., and C. A. Needham.** 1987. Method evaluation and test selection, p. 27–33. *In* B. B. Wentworth (ed.), *Diagnostic Procedures for Bacterial Infections*, 7th ed. American Public Health Association, Washington, D.C.

44. **Mermel, L. A., and D. G. Maki.** 1993. Detection of bacteremia in adults: consequences of culturing an inadequate volume of blood. *Ann. Intern. Med.* **119:**270–272.

45. **Miller, J. M.** 1985. Procedure manual, p. 1–11. *In* J. M. Miller and B. B. Wentworth (ed.), *Methods for Quality Control in Diagnostic Microbiology*. American Public Health Association, Washington, D.C.

46. **Miller, J. M.** 1987. *Quality Control in Microbiology*. Centers for Disease Control, Atlanta.

47. **Miller, J. M.** 1991. Quality control of media, reagents, and stains, p. 1203–1225. *In* A. Balows, W. J. Hausler, Jr., K. L. Herrmann, H. D. Isenberg, and H. J. Shadomy (ed.), *Manual of Clinical Microbiology*, 5th ed. American Society for Microbiology, Washington, D.C.

48. **Miller, J. M., and B. B. Wentworth (ed.).** 1985. *Methods for Quality Control in Diagnostic Microbiology*. American Public Health Association, Washington, D.C.

49. **Morris, A. J., D. C. Tanner, and L. B. Reller.** 1993. Rejection criteria for endotracheal aspirates from adults. *J. Clin. Microbiol.* **31:**1027–1029.

50. **National Committee for Clinical Laboratory Standards.** 1984. *Clinical Laboratory Procedure Manuals. Approved Guideline GP2-A*. National Committee for Clinical Laboratory Standards, Villanova, Pa.

51. **National Committee for Clinical Laboratory Standards.** 1990. *Antimycobacterial Susceptibility Testing. Proposed Standard M24-P*. National Committee for Clinical Laboratory Standards, Villanova, Pa.

52. **National Committee for Clinical Laboratory Standards.** 1990. *Quality Assurance for Commercially Prepared Microbiological Culture Media. Approved Standard M22-A*. National Committee for Clinical Laboratory Standards, Villanova, Pa.

53. **National Committee for Clinical Laboratory Standards.** 1990. *Performance Standards for Antimicrobial Disk Susceptibility Tests*, 4th ed. *Approved Standard M2-A4*. National Committee for Clinical Laboratory Standards, Villanova, Pa.

54. **National Committee for Clinical Laboratory Standards.** 1991. *Preparation and Testing of Reagent Water in the Clinical Laboratory*, 2nd ed. *Approved Guideline C3-A2*. National Committee for Clinical Laboratory Standards, Villanova, Pa.

55. **National Committee for Clinical Laboratory Standards.** 1992. *Performance Standards for Antimicrobial Susceptibility Testing. Fourth Informational Supplement M100-S4*. National Committee for Clinical Laboratory Standards, Villanova, Pa.

56. **National Committee for Clinical Laboratory Standards.** 1992. *Reference Method for Broth Dilution Antifungal Susceptibility Testing of Yeasts. Proposed Standard M27-P*. National Committee for Clinical Laboratory Standards, Villanova, Pa.

57. **Pedler, S. J., and A. J. Bint.** 1991. Survey of users' attitudes to their local microbiology laboratory. *J. Clin. Pathol.* **44:**6–9.

58. **Pfaller, M. A.** 1992. The microbiology laboratory, p. 493–507. *In* R. P. Wenzel (ed.), *Assessing Quality Health Care*. The Williams & Wilkins Co., Baltimore.

59. **Phillips, G., B. W. Senior, and H. McEwan.** 1992. Use of telephone inquiries to a microbiology laboratory as a proxy measure of reporting efficiency. *J. Clin. Pathol.* **45:**250–253.

60. **Rasohoff, D. F., and A. R. Feinstein.** 1978. Problems of spectrum bias in evaluating the efficacy of diagnostic tests. *N. Engl. J. Med.* **299:**926–931.

61. **Rogers, S., M. J. Bywater, and D. S. Reeves.** 1991. Audit of turn-around times in a microbiology laboratory. *J. Clin. Pathol.* **44:**257–258.

62. **Schentag, J. J., C. H. Ballow, A. L. Fritz, J. A. Paladino,** J. D. Williams, T. J. Cumbo, R. V. Ali, V. A. Galletta, M. B. Gutfeld, and M. H. Adelman. 1993. Changes in antimicrobial agent usage resulting from interactions among clinical pharmacy, the infectious disease division, and the microbiology laboratory. *Diagn. Microbiol. Infect. Dis.* **16:**255–264.

63. **Schifman, R. B.** 1990. Quality assurance goals in clinical pathology. *Arch. Pathol. Lab. Med.* **114:**1140–1144.

64. **Schifman, R. B.** 1992. Quality assessment and improvement (quality assurance), p. 13.1.1–13.1.29. *In* H. D. Isenberg (ed.), *Clinical Microbiology Procedures Handbook*, vol. 2. American Society for Microbiology, Washington, D.C.

65. **Schifman, R. B., and R. A. Palmer.** 1985. Surveillance of nosocomial infections by computer analysis of positive culture rates. *J. Clin. Microbiol.* **21:**493–495.

66. **Schifman, R. B., and A. Pindur.** 1993. The effect of skin disinfection materials on reducing blood culture contamination. *Am. J. Clin. Pathol.* **99:**536–538.

67. **Schifman, R. B., C. Strand, E. Braun, A. Louis-Charles, R. P. Spark, and M. Fried.** 1991. Solitary blood cultures as a quality assurance indicator. *Quality Assurance Utilization Rev.* **4:**132–137.

68. **Schwartz, J. S., P. E. Dans, and B. P. Kinosian.** 1988. Human immunodeficiency virus test evaluation, performance, and use: proposal to make good tests better. *JAMA* **259:**2574–2579.

69. **Sewell, D. L.** 1992. Laboratory records, p. 13.3.1–13.3.19. *In* H. D. Isenberg (ed.), *Clinical Microbiology Procedures Handbook*, vol. 2. American Society for Microbiology, Washington, D.C.

70. **Sewell, D. L.** 1992. Quality control, p. 13.2.1–13.2.35. *In* H. D. Isenberg (ed.), *Clinical Microbiology Procedures Handbook*, vol. 2. American Society for Microbiology, Washington, D.C.

71. **Sharp, S. E.** 1993. Effective reporting of susceptibility test results. *Diagn. Microbiol. Infect. Dis.* **16:**251–254.

72. **Siegel, D. L., P. H. Edelstein, and I. Nachamikin.** 1990. Inappropriate testing for diarrheal diseases in the hospital. *JAMA* **263:**979–982.

73. **Soo Hoo, G. W., D. L. Palmer, and R. L. Sopher.** 1984. Reducing tuberculosis detection costs. *Chest* **86:**860–862.

74. **Strand, C. L., R. R. Wajsbort, and K. Sturmann.** 1993. Effect of iodophor vs iodine tincture skin preparation on blood culture contamination rate. *JAMA* **269:**1004–1006.

75. **Trenholme, G. M., R. L. Kaplan, P. H. Karakusis, T. Stine, J. Fuhrer, W. Landau, and S. Levine.** 1989. Clinical impact of rapid identification and susceptibility testing of bacterial blood culture isolates. *J. Clin. Microbiol.* **27:**1342–1345.

76. **Valenstein, P.** 1993. Technology assessment for the diagnostic laboratory. Workshop no. 9107. American Society for Clinical Pathologists Workshop Meeting, Orlando, Fla.

77. **VonSeggern, R. L.** 1987. Culture and antibiotic monitoring service in a community hospital. *Am. J. Hosp. Pharm.* **44:**1358–1362.

78. **Walton, M.** 1986. *The Deming Management Method*. Dodd, Mead & Co., New York.

79. **Washington, J. A., II.** 1990. Confidence intervals: an important component of data presentations. *Clin. Microbiol. Newsl.* **12:**109–110.

80. **Washington, J. A., II, and G. V. Doern.** 1991. Assessment of new technology, p. 44–48. *In* A. Balows, W. J. Hausler, Jr., K. L. Herrmann, H. D. Isenberg, and H. J. Shadomy (ed.), *Manual of Clinical Microbiology*, 5th ed. American Society for Microbiology, Washington, D.C.

81. **Weissfeld, A. S., and R. C. Bartlett.** 1987. Quality control, p. 35–65. *In* B. J. Howard, J. Klass II, S. J. Rubin, A. S. Weissfeld, and R. C. Tilton (ed.), *Clinical and Pathogenic Microbiology*. The C.V. Mosby Co., St. Louis.

82. **Yannelli, B., I. Gurevich, P. E. Schoch, and B. A. Cunha.** 1988. Yield of stool cultures, ova and parasite tests, and *Clostridium difficile* determinations in nosocomial diarrheas. *Am. J. Infect. Control* **16:**246–249.

Addressing Regulatory Issues in the Clinical Microbiology Laboratory

JOHN E. McGOWAN, JR., AND JAMES D. MacLOWRY

6

The health care field in the United States is now "one of the most highly regulated industries in the country" (25). Laboratory personnel find themselves reviewed, surveyed, examined, and regulated to a degree never before experienced. This quantity of review has often been accompanied by problems of review quality in the form of conflicting, vague, and wasteful standards and regulations (26). Few data exist to show that the quality of results has been enhanced by these regulations.

This chapter will consider some of the reasons for the current degree of regulation, describe some of the main regulatory groups with which clinical microbiologists must deal, briefly review selected regulations of current importance, and suggest ways to make the process more efficient.

WHY HAS REGULATION INCREASED?

The introduction in 1983 of prospective reimbursement by diagnosis-related groups for Medicare patients triggered tremendous changes in health care financing. This change at the federal level created other prospective systems for reimbursement of hospitals. These have led to dramatic changes in the patterns of patient care (7, 26). Particularly prominent under such systems are a decline in patient admissions to acute-care hospitals and a drop in average length of hospital stay for those who are admitted. Outpatient diagnosis and treatment now are provided for many illnesses that before prospective reimbursement would have been managed exclusively in the hospital.

Following the implementation of these policies, there was a perception among many health care professionals and the public that the policies had led to a decline in the quality of patient care (7, 8). To the politicians who instituted the fiscal changes, this perception was undesirable. Their attempt to counter these perceptions resulted in an increasing number of laws and regulations mandating that quality of care be ensured. At the same time, anxiety in health care workers about the risk of human immunodeficiency virus infection and chemical and other environmental exposures escalated to a level of near hysteria (11, 20).

A third underlying factor was the concern of the U.S. public about the validity of certain laboratory diagnostic tests. Biased, emotional, misleading media crusades in the mid-1980s about the ways in which certain testing, especially in cytology, had been conducted created a crisis of public confidence in laboratory testing. This prompted Congress and many state health departments to look closely at the operations of clinical laboratories. This review culminated in the Clinical Laboratory Improvement Amendments of 1988 (CLIA '88), which mandated additional procedures, reviews, and proficiency testing for all laboratories that dealt with human specimens, whether or not the laboratories were associated with health care institutions (23). Ironically, while these regulations increased record keeping and mandated complex proficiency testing, they also weakened the qualification requirements for the individuals performing tests.

The consequence of all these developments has been the creation of "legions of agencies, bureaus, and administrations to review many aspects of health care" (26). Between 1983 and 1990, the increase in administrative expenditures was nearly double that for professional care or service departments (31). The total cost for health care administration in 1990 was 24.8% of each hospital's spending for health care (34). As a result, most U.S. hospitals, and clinical laboratories in particular, now find themselves reviewed, surveyed, and examined to a degree never before experienced. Various governing bodies at the federal, state, and local levels as well as several professional societies have issued guidelines, consensus statements, or directives for proper health care. Only some of these national, state, and local regulatory authorities and accrediting bodies have coordinated their rules or their protocols for inspection, and several do not accept the results of inspection by other groups (2). This quantity of review has been accompanied by problems of review quality. Few good ways to measure quality of patient care are documented at present, and the techniques to be mandated so that quality care will be ensured are equally unclear. The result is that various guidelines and regulations often are vague as well as conflicting (2). A by-product has been the need for extensive paperwork describing procedures, documenting performance, and providing responses to inspections. Some of the reporting is redundant, requiring responses to the same question for various agencies at different times.

An administration proposal for extensive reform of health care was being introduced in Congress as this chapter was written (12). A hallmark of this Health Security Act of 1993 was an increasing degree of governmental review and oversight (18). New "regional purchasing alli-

TABLE 1 Some governmental agencies involved in setting regulations or guidelines for microbiology laboratories in the United States[a]

Federal

HCFA. Regulates federal licensure (laboratories that ship specimens across state lines). Most recent update in CLIA '88 specifies extensive new procedures and standards for proficiency testing, certification, quality control. Also regulates qualification for funding from Medicare (reimbursement of health care costs for eligible participants).

CDC. Issues guidelines in several microbiology areas (but these guidelines often are adopted by agencies that can regulate and/or punish).

NIOSH. Branch of CDC that sets standards for safety and health of workers.

Environmental Protection Agency. Sets standards for safety registers and regulates chemical sterilants and disinfectants used on devices or environmental surfaces. Regulates medical waste.

OSHA. Sets regulations for employee safety (e.g., bloodborne pathogens, chemical safety) and inspects for compliance.

Food and Drug Administration. Sets standards for use of drugs, antiseptics and disinfectants, some reagents, and medical devices and equipment.

Post Office. Sets standards for shipment of specimens, organisms, etc.

Health Resource Service Administration. Sets standards for testing of organ donors and recipients in the Organ Procurement and Transplantation Network, which it administers.

Agency for Toxic Substances and Disease Registry. Sets guidelines for handling of medical waste.

Agency for Health Care Policy and Research. Sets guidelines for quality, appropriateness, and effectiveness of health services and access to such services.

Nuclear Regulatory Commission. Regulates use of radioisotopes in laboratories.

Department of Transportation. Regulates shipment of etiologic agents.

Department of Agriculture. Regulates shipment of certain infective agents, particularly those of veterinary concern.

State and/or local (county, city, etc.)

Laboratory Licensure Division. Requirements vary from state to state. Regulations concern fire, safety, etc.

Laboratory Licensure Division. Qualification for funding from Medicaid (state reimbursement of health care costs for eligible participants).

Health Department. Regulates procedures for reporting of infectious agents.

[a] This table is adapted in part from references 25 and 26.

ances" are proposed as the negotiating agents for obtaining patient health care services. These alliances also would be responsible for monitoring and publishing on a regular basis easily understood information about cost, performance of "providers" (i.e., health care workers), access to plans, and satisfaction of those enrolled. Quality information and standards for improvement of health care institutions would be determined nationally by a National Health Board in Washington, D.C., rather than by each state or each health care institution. Individual states would have to submit to the National Health Board a plan for collection of data about quality indicators and a scheme for providing these data to both individuals and the health alliances. The new legislation also includes provisions for licensure, certification, and inspection of providers, clinics, hospitals, long-term-care facilities, laboratories, and allied health personnel.

It is not likely that the act proposed by the administration will be the final one adopted, and much time and many amendments can be expected before the bill becomes law (12). Nevertheless, laboratorians should take note of the emphasis on certification, inspection, and publication of quality standards and performance. Proficiency testing results seem likely to be part of the material published, so great attention will have to be paid to the newer, more complex demands for proficiency testing embodied in CLIA '88 (see below). A review of the legislative process is exceptionally helpful in gaining an understanding of how this massive social legislation will be approached by Congress and what the problems inherent in its passage will be (17).

SPECIFIC REGULATORY GROUPS

To indicate the current level of regulation experienced by the clinical microbiology laboratory, Table 1 lists some of the governmental agencies that influence laboratory practice. Many of these agencies issue regulations and have the power to enforce compliance (25). Other nongovernmental organizations or professional societies (Table 2) provide position papers, guidelines, or standards of conduct (24, 27, 30).

IMPACT OF SELECTED REGULATORY EFFORTS ON THE MICROBIOLOGY LABORATORY

HCFA and CLIA '88

The Health Care Financing Administration (HCFA) is part of the U.S. Department of Health and Human Services. Since it was established in 1977, HCFA has been responsible, among other mandates, for the conduct and financing of the Medicare and Medicaid programs. HCFA also monitors federal quality assurance standards in laboratories and other areas of hospitals, nursing homes, and other health care facilities. Part of this activity involves oversight of the health care peer review organizations in each state.

As part of its charge for quality assurance monitoring, HCFA in 1979 was given the authority for implementing and administering the CLIA '67, which placed numerous requirements on diagnostic laboratories involved with Medicare and Medicaid reimbursement as well as those

TABLE 2 Some nongovernmental agencies involved in setting regulations or guidelines for microbiology laboratories in the United States[a]

JCAHO. Conducts "voluntary" inspection and accreditation of hospitals.

CAP. Conducts voluntary laboratory inspections.

NCCLS. Sets standards for selected laboratory procedures, e.g., susceptibility testing of antimicrobial agents (27).

American College of Physicians. Produces position papers on use of diagnostic tests.

Association for Practitioners in Infection Control. Provides guidelines for aspects of hospital infection control, e.g., use of disinfectants.

American Society for Microbiology. Provides consensus guidelines for practices in clinical laboratories, e.g., susceptibility testing.

Infectious Diseases Society of America. Provides guidelines for practices in various areas, e.g., use of antimicrobial agents (24).

[a] This table is adapted in part from references 25 and 26.

involved with interstate transport and testing of specimens (33). Prior to 1979, what is now the Centers for Disease Control and Prevention (CDC) was the federal agency responsible for enforcement of CLIA '67, an act that targeted many areas of laboratory practice and contained particularly stringent qualifications for testing personnel. New public concern about the quality of laboratory testing in the mid-1980s led Congress to take "the opportunity and responsibility to intervene more directly in the practice of medicine" and pass CLIA '88 (23). The original provisions of the draft rules for this act, published in May 1990, generated a tremendous scientific outcry that resulted in more than 60,000 comments on the draft. This response led to important revisions of the regulations that were implemented in February 1992 when published as 19 pages of rules and 727 pages of comments (13). The rule was revised again in August, September, and October 1992 and again in January and July 1993. Further revisions are expected, possibly in mid- to late 1994.

CLIA '88 was an attempt by Congress to apply comprehensive certification standards to all clinical laboratories (33). It defines the scope of laboratories included under the regulations as all those laboratories providing test data for the diagnosis, treatment, or prevention of disease in humans. This represents a sizable extension, as it includes in the regulated group the physician's office laboratory (POL). The total number of POLs in the United States is unclear; HCFA estimates the number to be about 130,000 (which would represent 85% of all laboratories). While the number of POLs is huge, they perform <10% of the more than 8 billion laboratory tests performed annually in the United States (19), so the work volume in each is clearly quite small. The remaining 95% of laboratory testing is conducted by the approximately 12,000 Medicare-certified laboratories and the 1,300 laboratories certified for interstate testing (33).

CLIA '88 addresses a number of specific areas that are relevant to microbiology laboratory operation. Seven of these will be highlighted below.

Licensure

The law has linked registration and licensure of laboratories to Medicare and Medicaid reimbursement. Filling out the proper forms and paying the required fees now permits the laboratory (not the laboratory director) to receive a "registration certificate" good for 2 years or until inspection, if inspection is required (see below). A laboratory registering with HCFA pays a registration fee of $100 to $350, depending on laboratory test volume. If applicable, the laboratory also pays a fee for inspection that ranges from $300 to $2,870, again depending on test volume. Laboratories covered under the wide definition of included laboratories must register with HCFA even if they are doing tests free. They must have a separate license for "separate addresses" but can get separate licenses for different test sections at the same site if they choose. Thus, a microbiology laboratory that is a section of a larger laboratory may be licensed under CLIA '88 separately or as part of a parent laboratory. The decision about which to obtain is guided by the tradeoff involved: with a separate license, the microbiology operation will have more paperwork and separate inspections but may be able to continue operating if other separately licensed areas of the operation fail proficiency testing and are forced to suspend operations until they pass repeat test challenges. Licensed microbiology laboratories must notify HCFA within 6 months if they change, add, or delete tests or change methods. They must notify HCFA within 30 days if they change address, name, owner, or director.

Laboratory Classification

Licensure and further requirements are tied to the categorization of the laboratory in terms of the type and complexity of tests performed and the methods used to conduct testing. The formulas used to classify each type of test and its complexity are fairly arcane and will not be considered further in this presentation (13). Approximately 12,000 tests have been listed by HCFA in detail by category (14, 16).

Four categories of testing currently are included. A small number of tests are classified as "waived" (as inspections and proficiency testing are waived for laboratories that perform only these tests). Included are a few procedures thought to be simple to perform and having minimal impact on a patient's health if performed incorrectly. No microbiological tests are included in this group. A second category is "physician-performed microscopy." This group includes wet mounts (vaginal, cervical, skin), potassium hydroxide (KOH) preparations, pinworm examinations, fern tests, postcoital vaginal or cervical mucus tests, and urine sediment tests. A laboratory performing only these tests done only by a physician, only by microscopy, only on patients covered by the physician's practice group, and only as part of a patient visit is eligible for licensure in this category. Laboratories in this classification are not subject to HCFA inspections but are still subject to CLIA '88 requirements for quality assurance and quality control (see below). Obviously, this category will also be irrelevant to microbiology laboratories.

A third category is for laboratories performing tests of moderate complexity (MC). This category includes about 75% of all listed laboratory tests, but most microbiology tests are excluded and placed in the remaining category of high complexity (HC) tests. Any laboratory that performs one or more tests defined in a list published by HCFA as HC must be licensed as an HC laboratory, and this will include most microbiology laboratories. Both MC and HC laboratories must comply with the same requirements for quality control, quality assurance, and proficiency testing that are detailed below. Both are subject to HCFA inspection, but personnel requirements are more rigid for the HC laboratory than for the laboratory licensed as MC.

At present, a number of medical groups are trying to create a new test group ("physician-waived tests") that could possibly include such tests as a Gram stain (23). The likely success of these efforts is not clear at present, but microbiologists can anticipate constant challenges to the classification of most (if not all) microbiological tests as MC or HC, and they will have to keep up with the proposed and accepted changes as they occur.

Documentation and Record Keeping

Regardless of the classification of the laboratory, CLIA '88 requires that a laboratory have written records of test requests, written procedures for specimen collection (including labeling techniques), and a current written list of all tests performed and methods used with dates of changes. A written authorization must be given from the laboratory director for specific personnel to do or supervise specific tests; a general authorization from the director must be given to those who organize quality control, quality assurance, and proficiency testing activities as applicable. Microbiology laboratories are required to keep requisitions and reports for at least 2 years; longer retention periods are required for reports and slides from surgical pathology, cytology, and histopathology.

Quality Control and Quality Assurance

The areas of quality control and quality assurance are areas of major focus for HCFA inspections (15). Most microbiology laboratories previously regulated by CLIA '67 will find few new quality control requirements that they do not already address. For example, this section requires that for each test, a written procedure defining 16 aspects listed in regulations be provided. These aspects include accuracy, precision, analytic sensitivity, reporting range, and reference ranges. Procedures and schedules for instrument maintenance and instrument recalibration (with new lots of reagents, etc., but at least every 6 months) are also required, as are procedures for running controls with each set of test samples. Documentation of achievement of the performance specifications defined as well as records of steps taken whenever the results defined as minimally acceptable were not achieved or when controls were not correct is required as well.

Quality assurance requires such activities as documenting that testing meets clinical needs. Provision is needed for listing and definitions of performance indicators (e.g., observed and expected turnaround times for results) that will reflect clinical utility in the specific setting of a given laboratory. Results of surveillance of these indicators are demanded, as is a log of problems affecting clinical care and how these were resolved. Again, there will be few rules in CLIA '88 for the average microbiology laboratory that were not previously demanded by other regulations or accreditors.

Personnel Requirements

Under CLIA '88, personnel requirements are complex, and they clearly will continue to change under pressure from several political directions (23). At present, job descriptions are needed for the laboratory director and testing personnel in all laboratory categories from waived to HC. Clinical consultants, technical consultants, and general supervisors are required in MC and HC laboratories.

Regardless of the requirements for each position, it is essential to keep careful records that detail the training and experience that justify the employment of personnel in each of the defined positions. The job description and specific authorization from the laboratory director for personnel at a given level to perform a specified test or tests should be part of the laboratory policy manual.

Proficiency Testing

CLIA '88 regulations define about 200 MC or HC test-instrument combinations for which a laboratory employing the test must demonstrate its competence with unknown challenge specimens. The microbiology laboratory will receive at least five challenges three times per year for each CLIA '88-listed analyte. The laboratory must attain an overall score of at least 80% per analyte and specialty for each proficiency testing event. If a microbiology laboratory fails two successive proficiency testing events or two of three consecutive challenges, this is considered unsatisfactory performance. For unsatisfactory performance, the laboratory must undertake appropriate training and technical assistance to correct the problems associated with the proficiency testing failure. In microbiology, two types of samples must be included. Some can have only a principal organism, and 50% of samples must be mixtures of the principal organism and appropriate normal flora. A problem at present with proficiency testing is that not all analytes have proficiency test modules available for them (23).

A more basic problem related to proficiency testing is that the initial concept originated as an educational activity to help laboratories improve their quality and compare their performance with that of their peers. The CLIA '88 legislation has converted proficiency testing to a punitive function that is the most significant parameter for maintaining certification. However, no data support its use for this purpose. The few studies designed to evaluate what guarantees quality in the laboratory have demonstrated that the most critical elements are the training and experience of the technologists, and this effect is particularly obvious in microbiology (21, 22).

Laboratories participating in proficiency testing cannot send challenge specimens out to another laboratory. They must test the proficiency challenge specimens by the same method currently used for testing of patient specimens. Results of such proficiency testing must be reported to HCFA and applicable governmental units (state, local, etc.) in an electronic format. The agencies will check on proficiency test performance and will also cross-reference to see that a laboratory that is billing Medicare or Medicaid is certified (23). If proficiency testing reveals inadequacies of proficiency testing, relevant laboratory activities must be stopped under threat of large penalties. Of concern are the strict definitions of acceptable performance in such testing, but revision of these definitions is currently being discussed by the CLIA committee.

Several groups have applied to HCFA to be designated approved vendors for proficiency testing. These include the College of American Pathologists (CAP), the American Society of Internal Medicine, the American Academy of Family Practitioners, the American Association of Bioanalysts, and several private enterprise firms. In addition, the State of Washington has been given "deemed status" for its laboratory licensing and inspection program, and a number of other states have applied for this HCFA approval as well. Before signing up with a testing program, microbiologists should carefully consider the terms and conditions of a given proficiency testing program, the likely competence of the program sponsor to provide accurate and adequate challenge specimens, and the ability of the vendor to provide interpretations of performance quickly and correctly. This need is emphasized by the difficult position in which the laboratory is placed if testing results are deemed inadequate.

Inspections

MC and HC laboratories are required to undergo inspection before relicensure will be granted. The inspection can be conducted by HCFA or by an organization or state that has been approved by HCFA as a surrogate inspector. The inspection will involve an on-site visit for interviews, observation of testing, review of documents, and review of proficiency testing results. Any test listed on the application for laboratory registration as being performed is fair game for the inspection process. If deficiencies are found, the inspected laboratory has 10 days to submit a plan of corrective action. The manual of procedures for inspection is now available and provides a good guide for the process and likely questions (15).

A prime reason for passage of CLIA '88 was concern about the quality of test results in cytology laboratories and in POLs. It is ironic that cytology laboratories are not a current priority of the CLIA oversight groups and that recent changes proposed by the Clinton administration relax the requirements for POLs (1, 32). Thus, the primary group being regulated more stringently under the new law is the laboratories that already were responsive to the demands of CLIA '67. The need to impose more bureaucratic demands on this group is questionable, but it currently is a fact of laboratory life. Thus, microbiology laboratories must add to their current administrative burdens the work required to deal with the complex and shifting requirements of this HCFA initiative as another cost of doing business.

OSHA Final Standard for Bloodborne Pathogens

The Occupational Safety and Health Administration (OSHA) is a branch of the U.S. Department of Labor. Federal OSHA activities are conducted directly by the agency or are delegated to approved plans that operate in many states (25). Regulations mandated by OSHA often result from standards and guidelines from the National Institute for Occupational Safety and Health (NIOSH). A center within CDC, NIOSH conducts research on occupational safety and health problems and conducts workplace investigations to gather data or testimony from employers and employees. NIOSH standards are developed on the basis of the agency's congressional mandate to develop methods that can provide zero risk to workers. These standards usually are adopted by OSHA as regulations. Although OSHA is charged with considering cost-effectiveness or clinical relevance, whereas NIOSH is not, OSHA mandates typically vary little from NIOSH guidelines.

The major impact of OSHA on microbiology laboratory practice to date comes from the "Final Rule on Occupational Exposure to Bloodborne Pathogens," released in December 1991 and made final in March 1992 (29). This regulation mandates several actions applicable to many areas of health care institutions, including their laboratories. Among these are the following specifications. (i) Employers must provide hepatitis B vaccination and personal protective equipment for laboratory personnel and other hospital workers who can "reasonably" be expected to have contact with blood or other potentially infectious material.

The latter category includes most body fluids and unfixed tissues or organs but not stool, urine, sputum (unless bloody), saliva (unless in dental procedures), or intact skin. (ii) Laboratories and other areas must produce an extensive written plan specifying medical evaluation and follow-up for health care workers who are exposed to the potentially infected material listed above. The plan must be reviewed and updated annually and must be available to employees as well as to OSHA inspectors when requested. (iii) Health care institutions must implement numerous engineering controls (such as containers for phlebotomy needles and other sharps) in the workplace. (iv) The employer must annually provide and require attendance at programs for education of workers about preventive techniques and required practices. Extensive directions for record keeping are specified as well.

The act of Congress creating OSHA gave this agency the authority to enforce any existing standard of practice to protect workers. OSHA conducts unannounced inspections of health care facilities both on a routine basis and when a complaint is received from a health care worker or workers. Finding that the health care facility is not complying with OSHA mandates can result in fines of up to $70,000 for each violation (25).

In the laboratory, key features to emphasize in developing the plan required by OSHA for worker protection include (i) identifying laboratory areas and procedures that reasonably can be expected to involve exposure to infectious specimens; (ii) writing guidelines for cleaning bench tops, instruments, and areas where spills have occurred; (iii) implementing safety education and training for workers on an annual basis; (iv) providing specified containers for sharps (puncture resistant, labeled or color coded, leakproof on sides and bottom, and designed so that one cannot insert a hand into the container); (v) specifying that potentially infectious materials must be in leakproof containers during collection, storage, handling, processing, transport, or shipping and must be properly labeled as biohazards; and (vi) describing the proper handling of wastes that are "regulated" in the OSHA document. The last three aspects are prime goals of the regulation, and determining whether these requirements have been implemented is a frequent target of OSHA inspectors when they visit a laboratory. Many of these points are considered in detail in chapter 7.

OSHA Enforcement Policy for Occupational Exposure to Tuberculosis

In October 1993, OSHA issued an enforcement policy dealing with occupational exposure to tuberculosis (6). This document specifies administrative actions, engineering controls, and personal protective equipment for workers in health care facilities, correctional institutions, homeless shelters, long-term-care facilities, and drug treatment centers who are potentially exposed to specimens, cultures, or patients with *Mycobacterium tuberculosis*. Early identification of potentially infected patients and worker training are aspects that are emphasized.

The CDC officially has no regulatory responsibility, but its activities include developing guidelines and recommendations on prevention and control of infections, including sterilization and disinfection; conducting research and providing consultation on methods for disease diagnosis and antimicrobial susceptibility testing; defining risk factors for transmission in health care settings; conducting surveillance of health care worker exposure incidents; and developing strategies for prophylaxis. CDC guidelines often are used as a scientific basis for OSHA regulations. The new OSHA enforcement policy on tuberculosis was issued shortly after the CDC released a draft revision of its guidelines for preventing transmission of tuberculosis in health care facilities (4). The draft CDC policy includes a 60-day period for public comment, so the OSHA document does not use this CDC update as the basis for its new enforcement actions. Instead, OSHA is enforcing compliance with prior 1990 CDC guidelines for tuberculosis control (3). This is being done under the "general duty" clause of the Occupational Safety and Health Act of 1970, which gives OSHA the authority to enforce compliance with existing industry standards.

Laboratory aspects of this document focus on use of cumbersome and expensive masks called protective respirators that contain high-efficiency particulate airflow (HEPA) filters. This requirement goes well beyond what was noted by the CDC in their 1990 guidelines (3). There seem to be few data to support this requirement, as CDC has reported that hospital outbreaks were terminated and transmission ceased when the 1990 guidelines, including use of the much less expensive dust-mist protective respirator, were implemented. In spite of the lack of scientific documentation, microbiology laboratories that are subject to OSHA routine inspection must review and implement the required actions soon, as the fines and other penalties from OSHA for noncompliance are punitive and prohibitively expensive.

Design, proper functioning, and monitoring of airflow systems and ventilation hoods in laboratories are also areas subject to OSHA inspection that have been emphasized in the new enforcement policy. In general, prevention guidelines follow those of the CDC-National Institutes of Health Task Force as recently revised (5).

AGENCIES THAT ACCREDIT AND LICENSE CLINICAL LABORATORIES

Accreditation and licensing of microbiology laboratories are carried out by a variety of governmental and nongovernmental groups. These include the federal government for participation in the Medicare and Medicaid programs and for licensure of laboratories that accept specimens from more than one state, state and city health departments, the CAP, and the Joint Commission on Accreditation of Healthcare Organizations (JCAHO). Unfortunately, the different bodies have guidelines that conflict, especially in the area of personnel standards and quality control (2). Several of these groups will be considered here.

As noted above, HCFA sets standards for performance for laboratories rendering services to Medicare patients. Inspections to document compliance with these standards and enforcement of mandates are usually accomplished by the Medicare agencies hired by contract to act as intermediaries in each state. Medicare and Medicaid rules apply to both hospital and independent laboratories, and the rules are separate for each (33). HCFA also controls licensure of laboratories that are involved in interstate commerce (i.e., accept specimens from more than one state).

Two groups, the JCAHO and the CAP, have been given special permission ("deemed status") by the Medicare program to have their inspections accepted as substitutes for some or part of federal and/or state inspections. Medicare still conducts a limited number of inspections of CAP-accredited and JCAHO-approved laboratories to verify that they comply with federal requirements.

The JCAHO began in 1915 as a standard-setting program of the American College of Surgeons. Now its governing body includes the American Hospital Association, the American College of Physicians, the American Medical Association, and the American Dental Association as well (25). The group was known as the Joint Commission on Accreditation of Hospitals for many years; the recent name change is to emphasize the variety of health care facilities now operative and the need for inspection and accreditation of many of these units. The JCAHO accredits a variety of health care facilities, and this certification serves completely or partly as a substitute for state licensure in most states. JCAHO certification is also deemed to meet HCFA's Medicare and Medicaid conditions of participation. The JCAHO also has applied for deemed status under CLIA '88 (see above). In 1993, JCAHO updated and consolidated its standards for accrediting clinical laboratories in all settings of health care.

The CAP has sponsored a voluntary inspection and certification process for laboratories since 1961 (33). This program has been supported strongly by the microbiological and laboratory community since its inception. CAP certification is accepted by the JCAHO as an alternative to most elements of its inspection and accreditation process, with the exception of the safety and quality assurance elements, which are still part of a JCAHO survey even when the laboratory is CAP accredited. CAP certification also has been given deemed status to substitute for annual Medicare inspection of hospital laboratories as long as the CAP program remains equivalent to or more stringent than the conditions in CLIA '67 (33). The CAP also has received deemed status under CLIA '88. Participation in CAP proficiency testing is mandatory for CAP accreditation, and results of such proficiency testing are reviewed as an integral part of a CAP laboratory inspection.

Microbiology laboratories may wish to develop a strategy for accreditation inspections that minimizes the number of inspectors that are encountered in a given period. CAP inspections currently are accepted as substitutes in part or totally for many state inspections of hospital laboratories for Medicare participation, and they exempt the institution from most of the elements of the JCAHO inspection.

Current proposals for health care reform include sweeping requirements for quality assurance and consumer satisfaction (9, 18). Responsibility for licensure, proficiency testing, and quality assurance enforcement is to be vested in state health regulators under the overall direction of a federal group overseeing national quality standards. The relevance of all the accrediting groups discussed above to the new health care plans now being proposed by the federal government is not clear, and some may not survive the transition to health care systems. This emphasizes the need for the microbiologist to remain conversant with the changes in health care and their regulatory fallout.

CURRENT PROBLEMS WITH THE REGULATORY PROCESS

Current regulatory attempts have two aspects that have a negative impact on the microbiologist.

"Guideline Creep"

The impact of pronouncements by governmental and other groups is not always what the titles of these papers imply. For example, the CDC issues "guidelines," and ostensibly these "are not legally enforced or regulated" (10). However, the agencies that can provide enforcement or accreditation of hospitals often use the CDC guidelines as their basis for inspection or accreditation. For example, OSHA jumped the gun and adopted some of the proposals in the CDC draft guideline revision for tuberculosis control in hospitals (6) even before the comment period on the CDC proposal had expired (4). Similarly, the JCAHO portrays its periodic accreditation reviews as a "voluntary" process, conducted at the request of (and paid for by!) the organization that has invited JCAHO to inspect. However, as discussed above, certification by the JCAHO can in part qualify a hospital for payment by federal health care programs, and failing the JCAHO certification process can lead to lack of payment for a sizable proportion of the patients cared for by many hospitals. A third example is the National Committee for Clinical Laboratory Standards (NCCLS), which sets guidelines for laboratory practice through a voluntary organization with representatives from "professions, government, and industry that proposes uniform standards" in areas such as safety of employees from bloodborne pathogens (27) or from instrument biohazards (28). In truth, however, NCCLS "approved standards" quickly become the methods followed by most manufacturers of instruments and supplies for susceptibility testing and the inspection protocols of accreditors. Once this has occurred, the doctrines assume the status of national standards of care and procedure. This "guideline creep" is a problem for the microbiologist who is faced with conflicting regulations and guidelines that supposedly are the source for regulations (26).

Medicolegal Implications of Regulation

In recent years, laboratory regulation has had more influence than before on malpractice prosecutions and defenses (26). Both the frequency of lawsuits claiming medical malpractice and the amount of money awarded in some of these suits have increased dramatically in the United States recently. One required element of medical malpractice in the United States is demonstration of negligence, which is a failure to maintain the standard of care owed to the patient. The customary standard of care is defined as the "degree of care and skill ordinarily used under similar circumstances by members of the profession." For a given situation, the standard of care is defined by objective criteria or the testimony of a qualified expert. Under this system, guidelines and standards as well as regulations have clear legal implications for health care organizations throughout the country, as they may become the basis for claims of proper or improper procedure in malpractice suits.

A further consequence of the litigious nature evoked by our social and health care systems is that liability consequences must be examined for all hospital activities. One major tactic for plaintiffs is to determine whether the hospital has a policy or procedure that is not being followed. When deviations occur, the courtroom discussion may focus on determining whether the procedure not followed was a regulation or customary standard. This potential for liability means that a hospital or laboratory must think twice before endorsing guidelines or adopting policies.

In truth, then, guidelines often begin as well-meaning attempts to provide some assistance to laboratory and other workers, but they soon may acquire the regulatory or legal clout that requires obedience. This must be kept in mind as new "guidance" is provided to microbiology laboratories by various groups.

CONCLUSION

The interaction of changes in health care in the past decade and new proposals for sweeping changes in the near future means that increased regulation of microbiology laboratories is here to stay. What is at issue is finding the most efficient ways to address this regulatory environment (26). We must campaign for close cooperation between regulatory groups, clear distinction between sociopolitical and scientific justification for regulation, and cost-benefit analysis of the regulations and their impact. Microbiology laboratory personnel are important to this effort because the diagnostic and management skills required for dealing with these regulatory issues are the same abilities used to make clinical microbiology cost-effective and efficient. It is vital for the regulated (in this case, microbiologists) to become and remain involved in all discussions of guidelines, policies, regulations, and laws that may have an impact on the microbiology laboratory if they wish the end result to be reasonable and clinically relevant.

REFERENCES

1. **Albertson, D.** 1993. Clintons float CLIA changes as part of reform strategy. *MLO* **25:**21–23.
2. **August, M. J., J. A. Hindler, T. W. Huber, and D. L. Sewell.** 1990. *Cumitech 3A, Quality Control and Quality Assurance Practices in Clinical Microbiology.* Coordinating ed., A. S. Weissfeld. American Society for Microbiology, Washington, D.C.
3. **Centers for Disease Control.** 1990. Guidelines for preventing the transmission of tuberculosis in health-care settings, with special focus on HIV-related issues. *Morbid. Mortal. Weekly Rep.* **39**(RR-17):1–39.
4. **Centers for Disease Control and Prevention.** 1993. Draft guidelines for preventing the transmission of tuberculosis in health-care facilities, second edition. Notice of comment period. *Fed. Regist.* **58:**52810–52854.
5. **Centers for Disease Control and Prevention and National Institutes of Health.** 1993. *Biosafety in Microbiological and Biomedical Laboratories,* 3rd ed. HHS publication no. CDC 93-8395. U.S. Government Printing Office, Washington, D.C.
6. **Clark, R. A.** 1993. *Enforcement Policy and Procedures for Occupational Exposure to Tuberculosis.* Occupational Safety and Health Administration, Washington, D.C.
7. **Conn, R. B.** 1989. The golden age of American medicine. *Am. J. Clin. Pathol.* **91:**725–730.
8. **Davies, N. E., and L. H. Felder.** 1990. Applying brakes to the runaway American health care system. A proposed agenda. *JAMA* **263:**73–76.
9. **Dechene, J. C.** 1993. Managed care networks: what are your options? *CAP Today* **17**(9):21–27.
10. **Garner, J. S.** 1990. The success and limitations of Centers for Disease Control (CDC) guidelines for nosocomial infection control, abstr. 31. *Proc. Joint Meet. Hosp. Infect. Soc. Br. Soc. Antimicrob. Chemother. Assoc. Acad. Microbiol.*
11. **Goldsmith, M. F.** 1990. Even "in perspective," HIV specter haunts health care workers most. *JAMA* **263:**2413–2420.
12. **Greenberg, D. S.** 1993. 1342 pages of health-care reform. *Lancet* **342:**1164.
13. **Health Care Financing Administration.** 1992. Medicare, Medicaid, and CLIA programs. Regulations implementing the Clinical Laboratory Improvement Amendments of 1988 (CLIA). *Fed. Regist.* **57:**7002–7186, 7245–7288.

14. **Health Care Financing Administration.** 1992. Specific list for categorization of laboratory test systems, assays and examinations by complexity. *Fed. Regist.* **57:**39211–39233, 40258–40296.

15. **Health Care Financing Administration.** 1992. *Surveyor's Manual for Laboratory and Laboratory Surveys.* Publication no. PB-92-146174. National Technical Information Service, Springfield, Va.

16. **Health Care Financing Administration.** 1993. Compiled list of clinical laboratory test systems, assays and examinations categorized by complexity. *Fed. Regist.* **58:**39860–39973.

17. **Iglehart, J. K.** 1993. Health care reform—the labyrinth of Congress. *N. Engl. J. Med.* **329:**1593–1596.

18. **Iglehart, J. K.** 1993. Health policy report—managed competition. *N. Engl. J. Med.* **328:**1208–1212.

19. **Koenig, A. S., III.** 1993. No need for office labs to close. *CAP Today* **17**(3):40–41.

20. **Lerman, S. E., and H. M. Kipen.** 1990. Material safety data sheets—caveat emptor. *Arch. Intern. Med.* **150:**981–984.

21. **Lunz, M. E., B. M. Castleberry, and K. James.** 1992. Laboratory staff qualifications and accuracy of proficiency test results. *Arch. Pathol. Lab. Med.* **116:**820–824.

22. **Lunz, M. E., B. M. Castleberry, K. James, and J. Stahl.** 1987. The role of certified medical technologists in providing quality laboratory results. *Lab. Med.* **18:**547–550.

23. **MacLowry, J. D.** 1993. Uncle Sam is in all labs—meet CLIA '88. *Infect. Dis. Clin. Pract.* **2:**227–230.

24. **Marr, J. J., H. L. Moffitt, and C. M. Kunin.** 1988. Guidelines for improving the use of antimicrobial agents in hospitals: a statement by the Infectious Diseases Society of America. *J. Infect. Dis.* **157:**869–876.

25. **McDonald, L. L., and G. Pugliese.** 1993. Regulatory, accreditation, and professional agencies influencing infection control programs, p. 58–69. *In* R. P. Wenzel (ed.), *Prevention and Control of Nosocomial Infections,* 2nd ed. The Williams & Wilkins Co., Baltimore.

26. **McGowan, J. E., Jr.** 1991. Regulation of clinical microbiology and infection control—is the gain worth the pain? *J. Hosp. Infect.* **18**(Suppl. A):110–116.

27. **National Committee for Clinical Laboratory Standards.** 1991. *Tentative Guideline: Protection of Laboratory Workers from Infectious Disease Transmitted by Blood, Body Fluids, and Tissue,* 2nd ed. Document M29-T2. National Committee for Clinical Laboratory Standards, Villanova, Pa.

28. **National Committee for Clinical Laboratory Standards.** 1991. *Tentative Guideline: Protection of Laboratory Workers from Instrument Biohazards.* Document 117-P. National Committee for Clinical Laboratory Standards, Villanova, Pa.

29. **Occupational Safety and Health Administration.** 1991. Occupational exposure to bloodborne pathogens: proposed rule and notice of hearing. *Fed. Regist.* **56:**64003–64182.

30. **Sahm, D. F., M. A. Neumann, C. Thornsberry, and J. E. McGowan, Jr.** 1988. *Cumitech 25, Current Concepts and Approaches to Antimicrobial Agent Susceptibility Testing.* Coordinating ed., J. E. McGowan, Jr. American Society for Microbiology, Washington, D.C.

31. **Shulkin, D. J., A. L. Hillman, and W. M. Cooper.** 1993. Reasons for increasing administrative costs in hospitals. *Ann. Intern. Med.* **119:**74–78.

32. **Voelker, R.** 1993. Cytology proficiency testing has stumped experts. *JAMA* **270:**2779–2780.

33. **Weisfeld, A.** 1990. Regulators and inspectors of concern to clinical microbiologists. *Clin. Microbiol. Newsl.* **12:**125–126.

34. **Woolhandler, S., D. U. Himmelstein, and J. P. Lewontin.** 1993. Administrative costs in U.S. hospitals. *N. Engl. J. Med.* **329:**400–403.

Laboratory Safety and Infectious Waste Management

BARBARA A. STRAIN AND DIETER H. M. GRÖSCHEL

7

The safe operation of a clinical laboratory is one of the management responsibilities of the laboratory director but requires compliance with the safety program by each laboratory worker. In one of the first textbooks on bacteriological investigation and diagnosis, published in 1894, Heim advised laboratorians to burn, boil, disinfect, and clean used materials immediately after the completion of work and to keep the work environment clean (27). Despite such warnings, laboratory-associated infections have occurred, and some have resulted in death. Well-known examples are the physicians who lost their lives while studying rickettsial diseases: H. T. Ricketts (1910), S. J. M. von Prowazek (1915), and E. Weil (1922).

In 1949, Sulkin and Pike began to collect information on laboratory-associated infections, and in 1979, Pike published a comprehensive review of the incidence, fatalities, causes, and prevention of laboratory-associated infections (46). These studies and experiences in research laboratories led to the development of laboratory safety programs (22). Despite the increased safety in microbiology laboratories (30, 59), laboratory-associated infections still occurred (29). Concerns about laboratory-associated hepatitis B virus (HBV) infections enforced the safety concepts and extended them to blood banks and other specialties of laboratory medicine in the 1970s. A major revival of interest in laboratory safety was caused by the AIDS epidemic of the 1980s and led to a series of recommendations by the then Centers for Disease Control (CDC) (5, 7–9). The Occupational Safety and Health Administration (OSHA) issued a standard on occupational exposure to bloodborne pathogens (41) that required the employer to establish "a written infection-control plan designed to minimize or eliminate employee exposure" (the Exposure Control Plan), to observe universal precautions, and to institute engineering and work practice controls. Some of these requirements had been included in previous recommendations for laboratory safety (2) and in the standards of the College of American Pathologists' laboratory accreditation program (14) and of the Joint Commission for the Accreditation of Healthcare Organizations (34). Another aspect of federally mandated occupational safety and health plans is the Hazards Communication Act of 1984, which requires all employers to inform their employees how to deal with the potential dangers arising from exposure to chemicals and other hazardous materials (40).

Several publications can be consulted for details on establishment of a laboratory safety program (12, 15, 21, 37, 38, 47, 48).

RISK ASSESSMENT

Our knowledge about the risk of acquiring laboratory-associated infections is rather limited, since there is no centralized state or federal reporting system. We base our risk assessment on data collected by Pike 15 years ago (46) and on more recent data on HBV and human immunodeficiency virus (3, 6, 20, 25). The OSHA standard (41) covers only bloodborne pathogens; other bacteria, fungi, viruses, and parasites that present risk of infection to laboratory workers are discussed in a manual published by CDC and the National Institutes of Health (NIH) (12).

Epidemiological Considerations

Transmission of disease in clinical laboratories is not well understood, and a cause-effect analysis is often not possible. The methods of transmission and the routes of infection may be quite different from those known for an organism, and the infectious dose may be much higher than in a community-acquired infection (4). Among laboratory personnel will be chronically ill persons with permanently or temporarily impaired immune defenses or allergies to chemicals or microbial components.

In the clinical laboratory, a broad spectrum of infectious agents can be introduced with patient specimens. Different levels of safety consciousness during handling of the same patient sputum specimen in a mycobacteriology or a general bacteriology laboratory can be surprising. Microbiologists are quite familiar with shigella infections occurring in laboratory technologists after they have handled a positive stool culture or with the skin test conversions of persons working with tuberculosis specimens or cultures despite supposedly excellent precautions.

Direct contact with infectious material is possible at all stages of microbiological work. Hand decontamination and the use of personal protective devices, such as gloves and gowns, interrupt this mode of transmission. Indirect contact usually occurs because surfaces of containers, work areas, or equipment are visibly or invisibly contaminated. Splashes or aerosols of potentially infectious materials are controlled by safety devices such as biological safety cabinets, closed centrifuge tubes, splash shields, and facial masks. Despite the prohibition of eating, drinking, smoking, and application of cosmetics in the laboratory and despite the separation of food storage refrigerators from laboratory refrigerators, transmission by ingestion, probably by direct or indirect contact with the organisms, occurs. Twenty-three cases of typhoid fever were contracted from teaching or proficiency-testing specimens (29, 37). The number of cases of shigella infection acquired from patient or teaching cultures is unknown.

Much of our knowledge about transmission by inoculation comes from the CDC studies of HBV infections in health care workers (6, 23). Penetration of intact skin by microorganisms is rare and usually involves organisms that can penetrate the skin in natural infection. Abraded, scratched, burned, or dermatitic skin; nasal and oral mucosa; and the conjunctiva, especially of contact lens wearers, are well-known portals of entry for a number of bacteria, fungi, and viruses. Today we are especially concerned about bloodborne viruses such as the hepatitis B and C viruses and human immunodeficiency virus, which may be contained in human blood, serum, plasma, or other body fluids. The reader is referred to the relevant literature for assessments of the risk of laboratory infection with these viruses (3, 5–12, 25, 28, 38). Other bloodborne infections (e.g., cytomegaloviral and arboviral infections, viral hemorrhagic fever, syphilis, leptospirosis, borreliosis, brucellosis, malaria, and babesiosis) that may be hazardous to laboratory workers are discussed in references 12, 37, and 41.

Universal Precautions

In 1987, CDC introduced the concept of "universal blood and body fluid precautions" or "universal precautions," to be applied consistently in the care of all patients and in the handling of blood and body fluid specimens (8). This approach, practiced for a number of years only for patients under blood and body fluid precautions (23), had been advocated for more than a decade as an effective way to reduce the risk of HBV transmission in clinical laboratories and blood banks. The arrival of AIDS was a strong motivating factor in revising existing safety programs and stimulating federal action such as the OSHA rules (9, 41). The National Committee for Clinical Laboratory Standards also has prepared a guideline for the protection of laboratory workers from infectious disease transmitted by blood, body fluids, and tissue (38). This document assists clinical laboratory scientists as well as anatomical pathologists in developing safety programs in accordance with universal precautions. It emphasizes the protection techniques and laboratory procedures that result in a safer work environment.

Blood Collection

All blood specimens should be considered potentially infectious, and care must be taken not to spill blood in patient rooms, collection areas, and laboratories. Disposable paper covers in the work area help absorb spills and are discarded as contaminated material in infectious-waste containers. Personal protective equipment such as gloves and garments should be worn during venipuncture and the handling of blood. Needles and lancets must be dealt with carefully, especially with agitated patients, to avoid accidental self-injury (32, 36). Great care must be taken when transferring blood from syringes to tubes or blood culture bottles, because the chance of an accidental needle stick is very high. This precaution applies also to subculturing samples from incubated blood culture bottles. In order to avoid popping of stoppers and spraying of contents, blood and body fluids should not be forced from syringes into tubes or bottles. Needles should never be recapped by hand; needle recapping, removal, and protection devices are available commercially. Used needles and syringes must be discarded in needle-proof containers for safe waste disposal. Many commercial designs of such containers are available, but one must ensure that they keep their integrity if they are to be decontaminated in a steam autoclave.

Bleeding of the venipuncture site should be controlled by exerting pressure with a sterile gauze pad; then the skin can be decontaminated with alcohol or iodophor, and an adhesive bandage can be applied. The venipuncturist should inspect the blood collection tubes before removing gloves and remove any outside contamination with an alcohol planchet.

For specimen handling, transport, storage, and shipment, see chapter 3 of this Manual.

Laboratory Work Practices

Each laboratory worker must be trained in safe working habits. The workplace must be kept clean and uncluttered, and each employee must be responsible for disinfecting and cleaning the work area, for proper disposal of infectious material, for the use of personal protective devices, and for frequent hand washing. Eating, drinking, smoking, application of cosmetics, nail biting, and handling of contact lenses are not permitted in the work area. Pipetting is performed only with pipetting devices; centrifugation of specimens or cultures should be performed only in plastic or uncracked glass tubes with tightly fitting closures or in safety shields. Tissue grinding should be performed with approved, closed devices, preferably in a biological safety cabinet. Care must be taken when removing caps or stoppers from tubes or bottles; a splash shield prevents exposure to splashes and aerosols.

Safety begins with the good laboratory practices of each laboratory worker. Initial orientation and training must be reinforced by continuing education and monitoring of compliance by the first-line supervisor. Unauthorized individuals, such as sales representatives and family members, should not be allowed in the work area of a microbiology laboratory. Authorized visitors, such as students, physicians, and service or repair personnel, must be acquainted with basic laboratory safety rules.

SAFETY MANAGEMENT

The responsibilities of management are to anticipate problems and to develop safety procedures and training programs based on present or potential hazards that may endanger personnel and on the behavioral factors leading to unsafe acts (21, 37). Management must be committed to the goals of health and safety of the laboratory. Improved protection and promotion of employee health and safety are proportional to the participation of laboratory management in the planning and decision making related to laboratory safety (48, 55). Employees must share the responsibility for their own safety and that of their coworkers once safety guidelines have been established.

Safety Programs

Safety awareness should become habit and a way of life in the laboratory. Laboratorians must take the time to perform a procedure, no matter how urgent, in a safe manner, for it is as easy to work safely as it is to work dangerously.

Education is the keystone to effective safety programs. Educational activities should be provided for all persons who may become exposed to potential laboratory hazards. In addition to laboratory workers, these personnel include clerical support, maintenance, supply room, housekeeping, and other employees who may come into contact with laboratory operations.

All safety information should be in a written form that is easily understood and contained in a laboratory manual that is readily available at all times. The manual should include general information (fire safety, emergency telephone numbers, proper apparel, personal hygiene, and first aid) as well as detailed policies and procedures based on the type of laboratory functions performed; it should be written according to established standards (9, 14, 38). The manual should also include other pertinent guidelines such as the following.

1. The chemical hygiene plan as part of the OSHA's right-to-know legislation (40)
2. The exposure control plan, to include universal precautions (41)
3. Spill cleanup procedures
4. Location and permitted quantities of flammable compounds for daily use
5. Labeling and handling of flammable, caustic, toxic, carcinogenic, and radioactive materials
6. Storage and handling of compressed gas cylinders
7. Electrical safety
8. Control of splashes and aerosolization
9. Use and maintenance of chemical and biological safety cabinets
10. Use of steam autoclaves for both sterilization and decontamination
11. Use of disinfectants and antiseptics

After policies and procedures have been written, they must be put into practice. Review of the safety manual and of the location and use of safety equipment should be done as part of the orientation program for new employees. If radioactive materials are used or stored in the laboratory, a radiation safety program should be included in the orientation. Further education should be an ongoing, regular process to ensure that laboratory workers are always prepared and up to date. First-line supervisors must insist that safety procedures be consistently followed and should set good examples. Performance of work in a safe manner should be a part of each employee's job description, and failure to do so should result in a formal reprimand.

The laboratory should appoint a safety officer to organize and update the safety policies and procedure manual. The officer should also chair a laboratory-wide safety committee. The committee should meet regularly to review accidents, perform inspections, and discuss laboratory safety issues. The laboratory safety officer should also be a member of the institutional safety committee. The safety officer should ensure that there are regular checks of eyewashes, safety showers, alarms, and fire extinguishers and should organize and conduct continuing safety education.

Chemical Hygiene Plan

The hazard communication standard final rule published in 1987 (40) was developed to ensure the safety of employees who come or may come into contact with hazardous chemicals in the workplace (15). Employers are responsible for disseminating information regarding the hazards of chemicals by using a comprehensive communication program called the Right to Know.

For those employees who have contact with hazardous chemicals, a comprehensive chemical hygiene plan that includes guidelines on proper container labeling, manufacturer's material safety data sheets, and a training-retraining program must be written.

The laboratory safety officer or designated personnel should be responsible for developing and implementing a training program that covers all the necessary aspects of the hazardous communication standard. Employees are trained during laboratory orientation and yearly thereafter. Written documentation of training is to be available in the laboratory.

Although the effectiveness of safety programs is difficult to measure, one should compare the number of hours involved in training with the incident rate of laboratory accidents. Decreased rates may be influenced by other factors, but such a comparison would provide a rough approximation of the effectiveness of training (25, 48).

Training could consist of a simple self-study notebook containing examples of container labels and material safety data sheets and how to interpret their information. The training notebook could also contain methods for detecting the presence of hazardous chemicals (e.g., by smell), where to find and how to use protective personal equipment, and a list of hazardous chemicals. The notebook could be combined with self-guided computer software or videotapes, both of which are commercially available. Training should also involve a tour of the laboratory, which could be part of the general safety orientation, to point out where chemicals are stored and used. The U.S. Department of Labor or the state occupational safety and health office can provide suggestions for setting up a training program.

Exposure Control Plan and Universal Precautions

OSHA standard 29 CFR part 1910 (41) requires that there be a plan to control exposures to employees through work practice controls, personal protective equipment, and proper disinfection and cleaning practices.

Universal precautions must be followed as part of the exposure plan to prevent the transmission of infectious disease by blood, body fluids, and tissue. Laboratories must also prevent transmission by microbiological cultures and infectious waste. All personnel must receive training in universal precautions before they can be assigned to work in an area with potential exposure. Annual retraining is required by OSHA, and documentation of all training must be retained for 3 years. Additionally, all personnel must be evaluated annually for compliance with universal precautions while performing job duties that have been designated in the exposure control plan as having a potential for exposure (24).

Infection Control and Employee Health Programs

The hospital microbiology laboratory must establish an effective working relationship with the hospital epidemiologist or infection control officer. A system for notification of the presence of important nosocomial isolates and reportable communicable diseases should be developed. The infection control officer should coordinate the monitoring of autoclaves used for sterilization of surgical instruments, contaminated waste, and other supplies and should use the microbiology laboratory for consultation when necessary. Instruction of hospital and laboratory staff should include

TABLE 1 Summary of recommended biosafety levels for infectious agents[a]

BSL	Practices	Safety equipment (primary barriers)	Facilities (secondary barriers)
1	Standard microbiological practices	None required	Open bench-top sink required
2	BSL-1 practice plus Limited access Biohazard warning signs Sharps precautions Biosafety manual defining any needed waste decontamination or medical surveillance policies	Class I or II BSCs or other physical containment devices used for all manipulations of agents that cause splashes or aerosols of infectious materials; PPE: laboratory coats, gloves, and face protection as needed	BSL-1 plus autoclave available
3	BSL-2 practice plus Controlled access Decontamination of all waste Decontamination of laboratory clothing before laundering Baseline serum test	Class I or II BSCs or other physical containment devices used for all manipulations of agents; PPE: protective laboratory clothing, gloves, and respiratory protection as needed	BSL-2 plus Physical separation from access corridors Self-closing, double-door access Exhausted air not recirculated Negative airflow into laboratory
4	BSL-3 practices plus Clothing change before entering Shower on exit All material decontaminated on exit from facility	All procedures conducted in class III BSCs or class I or II BSCs *in combination with* full-body, air-supplied, positive-pressure personnel suit	BSL-3 plus Separate building or isolated zone Dedicated supply-exhaust, vacuum, and decontamination systems Other requirements outlined in text

[a]From reference 12. BSC, biosafety cabinet; PPE, personal protective equipment.

the meaning of patient isolation and universal precautions and how they apply to the use of appropriate barrier protection during collection of patient samples.

All laboratory accidents should be reported to the supervisor. The employee involved should then be referred to the institution's employee health department. If the institution does not have such a department, then it must institute a mechanism for documentation of the laboratory accident and arrange for immediate and follow-up medical care by a medical practitioner, clinic, or hospital.

Forms for reporting accidents should be available. These forms should contain, in addition to the usual employee demographic data, written documentation of the nature of the accident, a statement as to whether the accident could have been prevented, and a description of measures that will be taken to prevent a recurrence. The reports should be reviewed by the safety committee at its regular meetings.

Among the basic functions of an employee health department are reducing the number of missed workdays and minimizing the number of on-the-job injuries and illnesses. In a hospital, the employee health department in cooperation with the infection control department helps curb the spread of disease between patients and staff. The department also provides instructional programs, physical examinations, tuberculosis skin tests, vaccinations, and other services for the continued good health of employees.

Immunizations
Under the OSHA plan, employers must provide HBV vaccination to all employees who have potential occupational exposure to bloodborne pathogens (41). Most hospitals already have immunization programs in place. Laboratories not affiliated with a hospital or other medical facility must arrange with a licensed physician for an immunization program; this arrangement should include postexposure management. According to OSHA estimates, HBV vaccination of all health care workers would result in 2 fewer deaths and 18 to 28 fewer cases of hepatitis and its complications per 1,000 workers exposed over a lifetime.

CONTAINMENT DESIGN, EQUIPMENT, AND TECHNIQUES
Environmental safety and safety consciousness of employees are supported by appropriate design of clinical laboratories, provision of safe equipment, and adoption of appropriate containment techniques.

Design
The design of a clinical microbiology laboratory must be guided by the scope of its operations. Applying the CDC classification of etiologic agents and the concept of containment according to four levels of safety, CDC and NIH developed guidelines for microbiological laboratories that considered the potential hazard of microbial agents and laboratory function (55).

Table 1 summarizes the recommended facilities, safety equipment, and practices (12).

The laboratory should be located away from patient care, visitor, and administrative areas and should have limited access. Ventilation and climate control systems should prevent airflow from laboratories into public corridors and nonlaboratory areas. Laboratories should be designed so that they can be easily cleaned, and the furniture should be of solid construction, with sufficient spacing to allow access for cleaning. Workbenches and equipment tables should be covered with water-impervious, heat- and chemical-resistant materials. Each laboratory requires a sink for hand washing and an eyewash station. If laboratory windows open, they should be covered with insect screens. Most laboratory operations involving routine diagnostic patient care can be performed in a basic facility under CDC-NIH biosafety level 2, using a biological safety cabinet and other practical containment for procedures that may increase the risk of exposure resulting from, for example, aerosol production (12, 38). Specialty services such as the handling and identification of mycobacteria and certain biphasic fungi require CDC-NIH biosafety level 3. Routine diagnostic procedures, however, may be performed in a biosafety level 2 facility with strict adherence to practices and the use of safety equipment as recommended for biosafety level 3. In level 3 laboratories, the interior surfaces of the rooms should be water resistant, the windows should be sealed, the access should be restricted, and there should be a negative airflow into the laboratory and no recirculation of exhausted air (12).

In addition to complying with microbiological safety recommendations, laboratories must adhere to the local fire and safety code by providing fire-fighting equipment and special storage for flammable materials and compressed gases. Other safety requirements address the use and storage of chemicals and radioactive materials. The biohazard symbol must be displayed on laboratory doors, and any equipment that may contain biohazardous material, e.g., blood culture incubators and refrigerators, must be labeled with the biohazard symbol and must give information about the type of hazard and the names and telephone numbers of persons to be contacted in cases of emergency. An evacuation plan must be posted in each room.

Containment Equipment

Containment equipment should allow the user to provide safe preparation and transfer of potentially hazardous materials without breaking the barrier. It should minimize exposure and permit the safe decontamination of effluent and the work area. The most important primary containment equipment in the clinical microbiology laboratory is the biological safety cabinet (12, 13). Two types are commonly found in the microbiology laboratory: class I, an open-fronted, negative-pressure, ventilated cabinet with HEPA-filtered air exhaust, and class II, a vertical-laminar-flow cabinet with HEPA-filtered recirculated airflow in the work area and HEPA-filtered air exhaust. With an inward face velocity of 75 ft (ca. 0.38 m/s), these cabinets are useful for partial containment of biosafety level 2 and 3 agents when used in conjunction with good laboratory practices. Class I cabinets protect the worker from contamination; class II cabinets protect the worker and the work area. Class II cabinets are most commonly used in microbiology laboratories, especially for virus and tissue culture work. All cabinets must be tested and certified after installation and after each move as well as at least annually thereafter. Laboratory workers must be trained in the proper use of the cabinets and be informed that containment, especially of aerosols, is not absolute, since interruption of the inward airflow by various means might lead to escape of particles from the work area. For a more detailed discussion of biological safety cabinets, the reader is referred to references 12 and 13.

Centrifugation has long been considered a particularly hazardous laboratory procedure. Of special concern is the centrifugation of digested sputum from tuberculosis patients. *Mycobacterium tuberculosis* has a 50% human infective dose of 10 organisms by the respiratory route (47). The use of high-speed centrifugation of mycobacterial broth cultures for DNA probe identification has brought new safety concerns (16a). Aerosol generation can be prevented to a very high degree by the use of special plug seal caps for disposable centrifuge tubes that provide a leakproof seal for the centrifugation of infectious material. Sealed centrifuge buckets and shields are recommended for work with mycobacteria and are available from all manufacturers as an additional safety feature. To reduce breakage of centrifuge tubes, only plastic tubes should be used. Each laboratory must have a procedure addressing centrifuge accidents and tube breakage.

Standard blenders and homogenizers are known as aerosol producers, and their use in a biological safety cabinet will not prevent potential worker exposure due to penetration of the front air curtain. Today, safe stainless steel containers, even for small volumes, are available for blenders (Eberbach Corp., Ann Arbor, Mich.; Waring Products Division, Hartford, Conn.). The stomacher (Seward Laboratory, London, England) holds the material to be blended in a plastic bag that can be discarded after the operation. Chatigny (13) described other laboratory-made enclosures for hand-operated devices that could be adapted to motorized grinders and sonic disruptors. If hand-held tissue grinders are used, they should be made of disposable plastic. These procedures should be performed in a biological safety cabinet.

For certain operations, especially in anaerobe and parasitology laboratories, explosion-proof centrifuges and refrigerators are required. For work with toxic chemicals and solvents, a chemical hood must be available. Flammable and corrosive chemicals must be contained in safety containers and should be kept in a safety cabinet. Special attention must be paid to the ether use policy.

Stock cultures of microorganisms must be kept in containers that resist breakage, and the containers should be stored in easily reached cabinets inaccessible to unauthorized persons.

For laboratory workers, a most important protection is the elimination of mouth pipetting and the introduction of pipetting devices. From small rubber bulbs to calibrated micropipettes with disposable, sterilizable tips to very sophisticated automated pipettes, users can select whatever is needed for the work at hand. When using automated pipettes, one should consider their potential contamination. Bacteriological inoculation loops need no longer be a source of aerosols during heat sterilization, since electrical incinerator loop sterilizers have virtually replaced the Bunsen burner. Disposable plastic loops and needles eliminate the need for heat sterilization devices.

Containment Techniques

The best design and optimal equipment cannot guarantee laboratory safety without the personal commitment of each worker and supervisor. Employers are forced by law to provide employees with personal protection devices (41). These include barrier devices such as protective gowns and coats, gloves, goggles, and face shields.

Personnel must wear a protective garment with long sleeves while working in the microbiology laboratory. In areas with possible blood or body fluid exposure, these garments should be made of fluid-resistant materials. The institution is responsible for providing clean garments to all staff. All contaminated garments should be decontaminated and sent to the laundry, as specified by established infection control protocols.

To prevent skin contamination with blood, body fluids, and other infectious materials, disposable gloves must be worn. Tightly fitted around the wrists, the gloves plus laboratory gowns protect the entire arm during work in biological safety cabinets. Disposable gloves are no absolute barrier for microorganisms, so they should be removed immediately after use, and the hands should be washed. Employees should be instructed that gloves protect only against contamination by contact and should remember that contaminated gloves can be a source of environmental transfer of infectious agents to themselves and their co-workers.

All laboratory procedures are performed in a way that minimizes the creation of aerosols (12). Procedures that have even the slightest potential of splashing, spurting, or forming an aerosol should be performed only while the employee is wearing a mask and goggles or a face shield or is protected by a bench-top safety shield. The work performed may be the unstoppering of vacuum collection tubes, the transfer of inocula from blood culture bottles with high internal gas pressures, the injection of materials through rubber stoppers, or the obtaining of an arterial blood sample from a patient. Needle sticks or other injuries from sharp instruments can be avoided by proper technique and provision of appropriate safety devices. All such injuries must be reported to the supervisor and to the person responsible for the employee health program for appropriate care and follow-up (32, 36).

Skin defects on hands, arms, neck, and face should be covered with a water-impermeable occlusive dressing (38). Under certain circumstances, the laboratory director or the employee health department may temporarily assign the employee to a nonrisk area.

Hand-washing sinks with nonirritating soap should be located in each laboratory. For accidental skin contamination with infectious agents, antiseptic soap should be available. In areas without hand-washing facilities, a waterless antiseptic preparation or alcohol planchets are useful for rapid skin decontamination after an accident or a blood or body fluid contact. Thorough hand washing should follow.

Each work area should be disinfected after completion of work and after accidental contamination. The recommendations for disinfection presented in chapter 8 of this Manual should be followed. Under no circumstances should antiseptics be used for disinfection. A procedure for the rapid treatment of spills and the appropriate materials should be available on all work stations.

Safety in the laboratory is based on strict adherence to standard microbiological practices and techniques, the provision and proper use of primary barrier equipment, and a facility that provides the laboratory with an environment adequate for its function. For the individual laboratory worker, safety is ensured not only by the employer but also by a personal commitment to protect himself or herself and all coworkers.

INFECTIOUS WASTE MANAGEMENT

General Information

Clinical microbiologists have been cognizant for many decades of the dangers of infectious waste generated in their laboratories and have instituted proper waste disposal procedures. The Hospital Infections Program of the CDC and the Joint Commission for the Accreditation of Hospitals recommended the designation of certain hospital solid waste as infectious and suggested appropriate disposal methods (22, 52). Concerns about transmission of the human immunodeficiency virus (7) led to the introduction of universal precautions (8, 9) and more definitive recommendations by voluntary and official organizations. The College of American Pathologists addresses waste disposal in its inspection check list and refers to federal, state, and local regulations (14). The Society for Hospital Epidemiologists offered in a 1992 position paper on medical waste an excellent review of the waste problem (50).

EPA and State Regulations

In 1986, the Environmental Protection Agency (EPA) published a guide for infectious waste management (17) that considered infectious waste not as a hazardous waste but as solid waste. The agency suggested that all generators of such waste develop proper management for its disposal. In subsequent years, most states developed or revised regulations on infectious waste, because the public became greatly concerned about the possibility of AIDS spreading from hospitals into communities through hospital waste. The EPA Office of Solid Waste held public hearings (18) and published standards for the tracking and management of medical waste (19) that were based on the congressional Medical Waste Tracking Act of 1988 (56). The results of these efforts were published by EPA in several interim reports, and their potential impact on the hospital and laboratory community are discussed in the Society for Hospital Epidemiology of America position paper (50).

It is beyond the scope of this chapter to discuss the various state regulations; they vary widely in scope and complexity. The microbiologist is advised to consult the local or state health or waste management departments. An example of a state regulation is the one issued by the Commonwealth of Virginia in 1988 (16).

Definition of Infectious Waste

U.S. hospitals produce about 15 lb (ca. 6.8 kg) of solid waste per bed per day (50), and about 10% is considered infectious waste according to CDC definitions (22). Whereas disposal of general solid waste costs $0.01 to $0.25/lb (1 lb = ca. 0.45 kg), usually for transportation to a landfill, commercial off-site treatment of infectious waste such as steam sterilization or incineration costs $0.30 to $1.00/lb or more. For fiscal reasons it is important that the generator of waste define infectious and noninfectious waste clearly and conservatively and design a program that will allow the personnel of hospitals and laboratories to separate the waste into categories at the site of generation.

In contrast to radioactive or chemical waste, infectious waste cannot be objectively identified (51). We could define it as waste that is capable of causing an infectious disease. In order to define a substance as causing infectious disease, one has to consider the basic epidemiological factors required for the transmission of disease, including dose, presence and virulence of a pathogen, susceptibility of the host, and portal of entry, according to Rutala and Mayhall the factor most commonly missing (50). Except for documented laboratory infections from microbiological cultures and injuries from contaminated sharp items such as needles, scalpels, and broken glass or plastic with subsequent infection, there is no epidemiological evidence that disposed hospital waste causes disease in the community (1, 49, 50), and there is no microbiological evidence that hospital waste is more infective than residential waste. Jager et al. (31) showed that hospital waste is no more contaminated than household waste.

Considering the potential public health impact of waste on coworkers and the community, one should classify as potential infectious waste only microbiological cultures and stocks and "sharps" (used sharp instruments, such as needles, lancets, and scalpels, and contaminated broken glass or plastic). Complying with the concept of universal precautions (9) for the protection of health care personnel, one can include blood, blood products, body fluids, and materials heavily contaminated with such fluids as infectious waste. Pathological waste from operating rooms, surgical pathology, and autopsies is often treated like infectious waste for esthetic reasons but not because it is infectious per se. In order to include certain noninfectious wastes in regulations, waste containing infectious material may also be called biological, biomedical, medical, or hospital waste by regulators and lawyers. This causes confusion and increases a laboratory's or hospital's amount of "infectious waste" without scientific justification (54). The lower the percentage of designated infectious waste, the cheaper the disposal cost for the institution will be. Obviously, local, state, and federal regulations must be respected.

Segregation of Infectious from Noninfectious Waste

Segregation of infectious from noninfectious waste at the site of generation is the most cost-saving part of the management program. Appropriate receptacles must be provided, and laboratory and hospital employees must be instructed. Receptacles for solid infectious waste should be used for that purpose only and not for the deposit of waste paper or coffee cups.

Liquid Infectious Waste

Liquid infectious waste such as blood, body fluids, suctioned fluids, etc., can be disposed of by the generator in a designated sink that leads directly into the sanitary sewer system. Pathological waste may be ground and flushed into the sanitary sewer with permission of the local sewer authority (7, 22). Laboratory liquid waste such as supernatants of body fluids, tissue suspensions or liquefied sputum after centrifugation, or fluids from feeding and washing of infected cell cultures is best collected in special leakproof and unbreakable receptacles. Decanting should be performed in a biological safety cabinet because of the generation of droplets and aerosols. Most laboratorians keep in the receptacle a concentration of disinfectant high enough to circumvent inactivation by proteins and dilution by liquids. Liquid laboratory waste, especially from mycobacteriology, mycology, and virology laboratories, is usually steam sterilized in the laboratory. If it is decontaminated at a site away from the laboratory, it must be contained in a closed, leakproof, unbreakable container (12) that may be placed in a contaminated-materials container for transport.

Packaging, Storage, Transportation and Treatment of Sharps

Convention and regulations use the color red on bags or containers to designate infectious waste receptacles. Plastic bags should be leakproof and capable of passing the American Society for Testing and Materials 125-lb drop weight test (16). Usually, two plastic bags are used (and required) for sturdiness and as an infection control measure. The bags must be closed tightly with tape before transport.

Venipuncture personnel and medical technologists using needles for the transfer of specimens or for subculturing blood culture media must be provided with appropriate devices for needle disposal that do not require recapping or removal by hand.

Many institutions collect solid waste, including sharps, in double-walled corrugated fiberboard boxes or equivalent rigid containers lined with a plastic bag. At one university hospital, the introduction of a new contaminated-materials container for disposal of needles and syringes as well as other contaminated materials resulted in a drop of needle stick injuries from 11.1/1,000 employees in the first year to 8.1/1,000 employees at the end of the second year (44). If sharp items, especially needles, are disposed of in such boxes, one should test the box wall for penetration by needles and scalpels. Heavy materials must be supported in appropriate containers. Needles and lancets may be deposited directly in puncture-proof rigid boxes at the site of use (laboratory bench, patient room, venipuncture tray or cart). If these containers are steam sterilized on site, one should ensure the continued integrity of the container after sterilization to avoid needle sticks occurring during removal for disposal. These containers should not be placed in compactors.

The advantages of a lined rigid container over a plastic bag are realized by the personnel responsible for transporting the waste. Frequently, broken glass or needles penetrate the plastic bags and cause injuries; they may also lead to leakage of waste. Considerations about size should include the expected weight of the waste and the dimensions of steam autoclaves and incinerators used for treatment of infectious waste.

Contaminated-materials containers shall display the international biohazard sign and a reference to the content as required by regulation or contract (infectious, medical, biomedical, biological, or hospital waste). The generator of the waste (laboratory, nursing unit, clinic) should indicate the actual content of the container (sharps, microbiological cultures, blood tubes, dialysis materials). If the containers are transported outside of the generator's premises, the name and telephone number of the responsible person or office should be printed on the box.

Areas for the storage of infectious waste must be clean and impervious to liquids. Floor drains shall discharge into the sewer system, and ventilation should consider potential exposure of humans and animals to the effluent. There should be an ongoing cleaning program as well as vermin and insect control. Access to the storage space must be limited to authorized personnel, and the doors must carry the biohazard sign with information about the content and the name and telephone number of the responsible person or office. If infectious waste is stored for longer periods, such as more than 72 h in Virginia (16) and up to 7 days in some states, or if the geographic location is associated with high ambient temperatures that may allow bacteria and fungi to grow freely, the waste should be stored in a refrigerated space at 2 to 7°C (35 to 45°F). Local regulations or ordinances may limit the storage time under refrigeration.

Transport of infectious waste in health care institutions usually occurs in two phases: collection within the institution and transport to the treatment facility. Waste containers are collected from nursing units, clinics, and laboratories by housekeeping personnel and brought either to the on-site treatment area or to the collection point for pickup for off-site treatment. Mechanical devices should not be used for the transfer of infectious waste containers (17). Carts used for the transport of waste should be designed for this purpose only and be cleaned regularly with a detergent-germicide.

For off-site transport, closed, leakproof dumpsters or trucks that display the biohazard symbol are recommended. Scattering, spilling of, or leakage from the transported material must be prevented. EPA does not consider the truck a rigid containment system; therefore, all waste must be contained in rigid or semirigid leakproof containers before being loaded on the truck (17). Present regulations and policies of the U.S. Department of Transportation and the U.S. Department of Energy about displaying the biohazard sign are confusing to institutional and commercial waste haulers (see reference 42, p. 10–11) and need to be standardized.

Under the OSHA regulations (41), employers must ensure the protection of all employees against occupational exposure to bloodborne diseases, especially HBV and human immunodeficiency virus, and must train employees in recognizing risks and using protective measures. Health care workers and their unions or organizations have been concerned about the exposure in high-risk tasks such as transporting infectious waste. Of all the needle stick or other sharps injuries reported to the employee health services of two university hospitals, the incidence was highest for the employees of environmental services (32, 36).

Another problem of concern to waste transporters is spills of infectious material from broken containers. Policies and procedures for cleanup must be available and must be part of training and continuing education.

Disposal and Treatment of Infectious Waste by Institution or Contractor

Most localities no longer permit the disposal of untreated infectious waste in landfills. Once infectious waste has been effectively treated, it may be handled and disposed of as noninfectious, general solid waste unless it is subject to other regulations as a radioactive or chemical hazard.

Incineration and steam sterilization are the most common treatment methods for infectious waste, but chemical and thermal disinfection is often practiced in clinical laboratories. (See chapters 8 and 19 of this Manual for recommended procedures of disinfection and sterilization in laboratories and hospitals.)

Incineration has the advantage of greatly reducing the volume of treated materials and thus was favored by many health care institutions for on-site treatment of both infectious and noninfectious waste. However, the increased concerns of the public and environmental control agencies about high emission rates of pollutants due to incineration of plastics and metals have made on-site incineration less desirable or even impossible. Regional cooperative or commercial incineration plants were developed to provide state-of-the-art treatment of infectious waste. Statewide or regional prohibitions to installing new incinerators or operating existing ones for treating infectious waste have caused considerable problems in the disposal of infectious waste and have increased the cost of waste handling (42, 43).

Steam sterilization is mainly practiced in clinical laboratories and occasionally in central on-site plants of hospitals. The efficacy of the process depends on the type, volume, and density of materials; their packaging; and the presence of water. These factors influence steam penetration, heat conduction, and, consequently, the degree of microbial killing as well as the time required to achieve sterility. Biological indicators are used to control the efficacy of the sterilizer. For the steam sterilization of clean instruments, dressings, and supplies, standard operating procedures were promulgated by hospital and professional organizations (45). In contrast, steam sterilization of waste is not standardized, although one often finds that the same operating standards are applied to waste as to operating room packs. Experimental studies by Lauer et al. (35) and Rutala et al. (53) clearly demonstrated the need for different operating parameters of steam sterilizers for the treatment of laboratory waste. Some of these parameters were discussed by Vesley and Lauer (see reference 58, p. 190–192). It is quite possible that the total processing time for waste may be 60 to 90 min, thereby greatly restricting the use of the autoclave for other purposes. The question of whether it is necessary to achieve overkill, that is, sterilization of *Bacillus stearothermophilus* spores or even sterilization of the waste itself when rendering infectious waste noninfectious, was raised by Rutala et al. (53) and Vesley (57). The setting of an arbitrary standard such as monitoring the efficacy of waste treatment with temperature-sensitive chemical indicators (16) or spore strips certainly is a less reliable control of successful treatment than ongoing operation and process control of the steam autoclave by a trained staffperson and supervisor. The proper performance of a steam sterilizer is more reliably controlled by an ongoing quality control (inspection-maintenance) program and occasional testing with thermocouples.

Autoclaved waste must be distinguished from non-treated infectious waste if it is transported within the facility or transferred to a landfill. If material in red bags or boxes is transported, the containers should be placed into a regular waste bag or, preferably, in a neutral transport box. The generator is responsible for marking the container as containing waste rendered noninfectious by steam treatment. Some local and state regulations have strict requirements for record keeping, identification, and tracking of steam-sterilized waste.

Alternative infectious waste treatment and disposal technologies are described in chapter 8 of this Manual and in reference 33.

After treatment, infectious waste must be noninfectious and can be disposed of like other solid waste. The need for labeling autoclaved infectious waste was mentioned earlier. Properly labeled autoclaved waste will be recognized as being noninfectious by waste haulers and landfill operators. Sometimes the waste is compacted together with regular trash and sent to the landfill. Since compacting may reexpose red plastic bags or containers, it is important that the generator reassure the waste hauler and the landfill operator, preferably in writing, that the waste was rendered noninfectious. To prevent sharps injuries of the landfill personnel, disinfected sharps should not be compacted.

Disposal of ashes in landfills has raised concerns from landfill operators and environmentalists. Heavy metals such

TABLE 2 Treatment and disposal of infectious waste in a university hospital

Source or type of infectious waste	Treatment or disposal method[a]			
	Steam	Incineration	Chemical	Sewer
Microbiological waste	+	+	+	−
Human blood, blood products, body fluids	+	+	−	+
Dialysis waste, solid	−	+	−	−
Waste from strict isolation[b]	+	+	+	+
Pathological waste	−	+	−	+
Used sharps	+	+	−	−
Cleanup of infectious-waste spill	+	+	−	−
Animal carcasses and waste	−	+	−	−
Animal carcasses and waste infected with human pathogen	+	+	−	−

[a]Approved disposal containers: sharps boxes, contaminated-material containers (23), and double plastic bags (to be replaced with larger boxes). Steam, sterilization with steam autoclave by generator or centrally and disposal by contractor; incineration, incineration in university incineration plant; chemical, disinfection with EPA-approved chemical disinfectant (also physical disinfection by boiling) and incineration or disposal by contractor; sewer, disposal of liquid waste into sanitary sewer system.
[b]Includes waste from patients isolated because of infection with methicillin-resistant *Staphylococcus aureus*.

as lead and cadmium are found not only in incinerator emissions but also in ash from hospital incinerators. The Commonwealth of Virginia (16) requires that ash be analyzed frequently for the presence of toxic substances and the total organic carbon content. If ash is found to be hazardous, it must be disposed of as hazardous waste, and the incineration unit will be closed.

Infectious Waste Management Program

Each institution needs a written plan for the proper management of infectious waste with special emphasis on microbiology waste and sharps. The Joint Commission on Accreditation of Healthcare Organizations assigns this responsibility to the infection control committee (34), the College of American Pathologists assigns it to the laboratory director (14), and the Clinical Laboratory Improvement Amendments of 1988 require protection from biohazardous materials (see reference 26, p. 7163). Table 2 shows the waste treatment and disposal plan of a university hospital as it was developed by the hospital epidemiology and microbiology services, the safety committee, and the university office of environmental health and safety in compliance with proposed state regulations. A contingency plan must be developed in case the routine plan fails.

All personnel generating, collecting, transporting, and storing infectious waste must be trained in infectious waste management, protective behavior and equipment, and procedures and policies regarding spills of liquid or solid infectious waste. Personnel specifically charged with waste handling must participate in employee health and continuing education programs. All personnel experiencing direct exposure to infectious waste must report this to their supervisor and to the employee health coordinator for appropriate immediate and follow-up care as required by OSHA (41).

Although insufficient research data are available to support all recommendations in this section, basic safety and infection control measures as well as knowledge of microbiology and epidemiology allow design of an infectious waste management system that should protect laboratory and hospital workers, waste haulers, employees of treatment facilities and landfills, and the public from infectious disease generated by laboratory waste and that comply with local and state regulations and accreditation standards (14, 26, 29, 34, 39).

REFERENCES

1. **Agency for Toxic Substances and Disease Registry.** 1990. The public health implications of medical waste: a report to Congress. Public Health Service, U.S. Department of Health and Human Services, Atlanta.
2. **Barkley, W. E.** 1981. Containment and disinfection, p. 487–503. *In* P. Gerhardt, R. G. E. Murray, R. N. Costilow, E. W. Nester, W. A. Wood, N. R. Krieg, and G. B. Phillips (ed.), *Manual of Methods for General Bacteriology*. American Society for Microbiology, Washington, D.C.
3. **Bell, D. M.** 1991. Human immunodeficiency virus transmission in health care settings: risk and risk reduction. *Am. J. Med.* 91(Suppl. 3B):294S–300S.
4. **Benenson, A. S. (ed.).** 1990. *Control of Communicable Diseases in Man*, 15th ed. American Public Health Association, Washington, D.C.
5. **Centers for Disease Control.** 1982. Acquired immune deficiency syndrome (AIDS): precautions for clinical and laboratory staffs. *Morbid. Mortal. Weekly Rep.* 31:577–580.
6. **Centers for Disease Control.** 1985. Recommendations for protection against viral hepatitis. *Morbid. Mortal. Weekly Rep.* 34:313–324, 329–335.
7. **Centers for Disease Control.** 1985. Recommendations for preventing transmission of infection with human T-lymphotropic virus type III/lymphadenopathy-associated virus in the work place. *Morbid. Mortal. Weekly Rep.* 34:681–686, 691–695.
8. **Centers for Disease Control.** 1987. Recommendations for prevention of HIV transmission in health-care settings. *Morbid. Mortal. Weekly Rep.* 36:3S–18S.
9. **Centers for Disease Control.** 1988. Update: universal precautions for prevention of transmission of human immunodeficiency virus, hepatitis B virus, and other bloodborne pathogens in health-care settings. *Morbid. Mortal. Weekly Rep.* 37:377–382, 387–388.
10. **Centers for Disease Control.** 1989. Guidelines for prevention of transmission of human immunodeficiency virus and hepatitis B virus to health-care and public-safety workers. *Morbid. Mortal. Weekly Rep.* 38(Suppl. 6):1–37.
11. **Centers for Disease Control and Prevention.** 1993. Update: acquired immunodeficiency syndrome—United States, 1992. *Morbid. Mortal. Weekly Rep.* 42:547–551, 557.
12. **Centers for Disease Control and Prevention/National Insti-**

tutes of Health. 1993. *Biosafety in Microbiological and Biomedical Laboratories*, 3rd ed. HHS publication no. (CDC)93-8395. Public Health Service, U.S. Department of Health and Human Services, Washington, D.C.

13. **Chatigny, M.** 1986. Primary barriers, p. 144–163. *In* B. M. Miller, D. H. M. Gröschel, J. H. Richardson, D. Vesley, J. R. Songer, R. D. Housewright, and W. E. Barkley (ed.), *Laboratory Safety: Principles and Practices.* American Society for Microbiology, Washington, D.C.

14. **College of American Pathologists.** 1993. *Commission on Laboratory Accreditation: 1993 Inspection Checklist*, Section I. Laboratory general, p. 19–29. College of American Pathologists, Northfield, Ill.

15. **Committee on Hazardous Substances in the Laboratory, National Research Council.** 1980. *Prudent Practices for Handling Hazardous Chemicals in Laboratories.* National Academy Press, Washington, D.C.

16. **Department of Waste Management.** 1988. *Infectious Waste Management Regulations, October 1, 1989.* Commonwealth of Virginia, Richmond, Va.

16a.**Desmond, E. P., et al.** 1993. Letter to the editor. *Clin. Microbiol. Newsl.* **15:**95.

17. **Environmental Protection Agency.** 1986. *EPA Guide for Infectious Waste Management.* EPA publication no. 1530-SW-86-014. Office of Solid Waste, Washington, D.C.

18. **Environmental Protection Agency.** 1988. 40 CFR part 261. Hazardous waste management system; identification and listing of hazardous waste; infectious waste management. *Fed. Regist.* **53:**20140–20143.

19. **Environmental Protection Agency.** 1989. 40 CFR parts 22 and 259. Standards for the tracking and management of medical waste; interim final rule and request for comments. *Fed. Regist.* **54:**12326–12395.

20. **Fahey, B. J., S. E. Beekman, J. M. Schmitt, J. M. Fedio, and D. K. Henderson.** 1993. Managing occupational exposure to HIV-1 in the healthcare workplace. *Infect. Control Hosp. Epidemiol.* **14:**405–412.

21. **Fuscaldo, A. A., B. J. Erlick, and B. Hindman.** 1980. *Laboratory Safety: Theory and Practice.* Academic Press, Inc., New York.

22. **Garner, J. S., and M. S. Favero.** 1986. CDC guideline for handwashing and hospital environmental control, 1985. *Infect. Control* **7:**231–243.

23. **Garner, J. S., and B. P. Simmons.** 1983. CDC guideline for isolation precautions in hospitals. *Infect. Control* **4:**245–325.

24. **Gauthier, D. K., J. G. Turner, L. G. Langley, C. J. Neil, and P. L. Rush.** 1991. Monitoring universal precautions: a new assessment tool. *Infect. Control Hosp. Epidemiol.* **12:**597–601.

25. **Gerberding, J. L.** 1991. Does knowledge of human immunodeficiency virus infection decrease the frequency of occupational exposure to blood? *Am. J. Med.* **91**(Suppl. 3b):308S–311S.

26. **Health Care Financing Administration, Public Health Service, Department of Health and Human Services.** 1992. Clinical laboratory improvement amendments of 1988; final rule. *Fed. Regist.* **57:**7001–7288.

27. **Heim, L.** 1894. *Lehrbuch der bakteriologischen Untersuchung und Diagnostik*, p. 488. F. Enke, Stuttgart, Germany.

28. **Hodinka, R. L., P. H. Gilligan, and M. L. Smiley.** 1988. Survival of human immunodeficiency virus in blood cultures. *Arch. Pathol. Lab. Med.* **112:**1251–1254.

29. **Holmes, M. B., D. L. Johnson, N. J. Fiumara, and W. M. McCormack.** 1980. Acquisition of typhoid fever from proficiency-testing specimens. *N. Engl. J. Med.* **303:**519–521.

30. **Huffaker, R. H.** 1974. Biological safety in the clinical laboratory, p. 871–878. *In* E. H. Lennette, E. H. Spaulding, and J. P. Truant (ed.), *Manual of Clinical Microbiology*, 2nd ed. American Society for Microbiology, Washington, D.C.

31. **Jager, E., L. Xander, and H. Rüden.** 1989. Hospital wastes. 1. Communication: microbiological investigations of hospital wastes from various wards of a big and of a smaller hospital in comparison to household refuse. *Zentralbl. Hyg.* **188:**343–364.

32. **Jagger, J., E. H. Hunt, J. Brand-Elnaggar, and R. D. Pear-**

son. 1988. Rates of needle-stick injury caused by various devices in a university hospital. *N. Engl. J. Med.* **319:**284–288.

33. **Jetté, L. P., and S. Lapierre.** 1992. Evaluation of a mechanical/chemical infectious waste disposal system. *Infect. Control Hosp. Epidemiol.* **13:**387–393.

34. **Joint Commission on Accreditation of Healthcare Organizations.** 1992. *Accreditation Manual for Hospitals, 1993.* Joint Commission on Accreditation of Healthcare Organizations, Chicago.

35. **Lauer, J. L., D. R. Battles, and D. Vesley.** 1982. Decontaminating infectious laboratory waste by autoclaving. *Appl. Environ. Microbiol.* **44:**690–694.

36. **McCormick, R. D., M. G. Meisch, F. G. Ircink, and D. G. Maki.** 1991. Epidemiology of hospital sharps injuries: a 14-year prospective study in the pre-AIDS and AIDS eras. *Am. J. Med.* **91**(Suppl. 3B):301S–307S.

37. **Miller, B. M., D. H. M. Gröschel, J. H. Richardson, D. Vesley, J. R. Songer, R. D. Housewright, and W. E. Barkley (ed.).** 1986. *Laboratory Safety: Principles and Practice.* American Society for Microbiology, Washington, D.C.

38. **National Committee for Clinical Laboratory Standards.** 1991. *Protection of Laboratory Workers from Infectious Disease Transmitted by Blood and Tissue*, 2nd ed. *Tentative Guideline M29-T2.* National Committee for Clinical Laboratory Standards, Villanova, Pa.

39. **National Committee for Clinical Laboratory Standards.** 1993. *Clinical Laboratory Waste Management. Approved Guideline GP5-A.* National Committee for Clinical Laboratory Standards, Villanova, Pa.

40. **Occupational Safety and Health Administration, U.S. Department of Labor.** 1987. 29 CFR part 1910.1200. Hazard communications, final rule. *Fed. Regist.* **52:**31852–31885.

41. **Occupational Safety and Health Administration, U.S. Department of Labor.** 1991. 29 CFR part 1910. Occupational exposure to bloodborne pathogens; final rule. *Fed. Regist.* **56:**64175–64182.

42. **Office of Technology Assessment.** 1988. *Issues in Medical Waste Management. Background Paper.* U.S. Congress, Washington, D.C.

43. **Office of Technology Assessment.** 1990. *Finding the Rx for Managing Medical Wastes.* U.S. Congress, Washington, D.C.

44. **Osterman, C. A.** 1975. Relation of new disposal unit to risk of needle puncture injuries. *Hosp. Top.* **March/April 1975.**

45. **Perkins, J. J.** 1969. *Principles and Methods of Sterilization in Health Sciences.* Charles C Thomas, Publisher, Springfield, Ill.

46. **Pike, R. M.** 1979. Laboratory-associated infections: incidence, fatalities, causes and prevention. *Annu. Rev. Microbiol.* **33:**41–66.

47. **Richardson, J. H., and W. E. Barkley.** 1985. Biological safety in the clinical laboratory, p. 138–142. *In* E. H. Lennette, A. Balows, W. J. Hausler, Jr., and H. J. Shadomy (ed.), *Manual of Clinical Microbiology*, 4th ed. American Society for Microbiology, Washington, D.C.

48. **Richardson, J. H., E. Schoenfeld, J. J. Tulis, and W. M. Wagner (ed.).** 1986. *Proceedings of the 1985 Institute on Critical Issues in Health Laboratory Practice: Safety Management in the Public Health Laboratory.* The DuPont Co., Wilmington, Del.

49. **Rutala, W. A.** 1984. Infectious waste. *Infect. Control* **5:**149–150.

50. **Rutala, W. A., and C. G. Mayhall.** 1992. Medical waste. SHEA position paper. *Infect. Control Hosp. Epidemiol.* **13:**38–48.

51. **Rutala, W. A., R. L. Odette, and G. P. Samsa.** 1989. Management of infectious waste in US hospitals. *JAMA* **262:**1635–1640.

52. **Rutala, W. A., and F. A. Sarubbi.** 1983. Management of infectious waste from hospitals. *Infect. Control* **4:**198–204.

53. **Rutala, W. A., M. M. Stiegel, and F. A. Sarubbi.** 1982. Decontamination of laboratory microbiological waste by steam sterilization. *Appl. Environ. Microbiol.* **43:**1311–1316.

54. **Rutala, W. A., and D. J. Weber.** 1991. Infectious waste—

mismatch between science and policy. *N. Engl. J. Med.* **325:** 578–582.

55. **Songer, J. R.** 1986. Management and codification of risks in the laboratory, p. 120–122. *In* B. M. Miller, D. H. M. Gröschel, J. H. Richardson, D. Vesley, J. R. Songer, R. D. Housewright, and W. E. Barkley (ed.), *Laboratory Safety: Principles and Practices.* American Society for Microbiology, Washington, D.C.

56. **U.S. Congress.** 1988. 42 USC 6901. Medical Waste Tracking Act of 1988. Public Law 100-582. U.S. Congress, Washington, D.C.

57. **Vesley, D.** Personal communication.

58. **Vesley, D., and J. L. Lauer.** 1986. Decontamination, sterilization, disinfection, and antisepsis in the microbiology laboratory, p. 182–198. *In* B. M. Miller, D. H. M. Gröschel, J. H. Richardson, D. Vesley, J. R. Songer, R. D. Housewright, and W. E. Barkley (ed.), *Laboratory Safety: Principles and Practices.* American Society for Microbiology, Washington, D.C.

59. **Wedum, A. G., W. E. Barkley, and A. Hellman.** 1972. Handling of infectious agents. *J. Am. Vet. Assoc.* **161:**1557–1567.

Sterilization, Decontamination, and Disinfection Procedures for the Microbiology Laboratory

FREDERIC JOHN MARSIK AND GERALD A. DENYS

8

The microbiology laboratory over the years has presented an ever-increasing safety challenge to microbiologists. This challenge comes not only from the many types of microorganisms that the microbiologist encounters but also from the new technologies (recombinant DNA, PCR) and equipment used in the microbiology laboratory. Every aspect of the work, from streaking a plate to cleaning up a spill to disposing of contaminated waste, has the potential to be hazardous. This chapter provides information on decontamination and waste disposal, which are key components of an overall laboratory safety program. A comprehensive discussion of safety is given in chapter 7 of this Manual.

The information in this chapter is not meant to be all-encompassing. Instead, it is meant to provide a foundation to which microbiologists can add their knowledge about disinfection, sterilization, antisepsis, and waste disposal as it relates to the work environment of the microbiology laboratory. Information on these subjects as it relates to the hospital setting can be found in chapter 19 of this Manual. Physical methods of achieving sterilization will also be reviewed. Details on how to use a particular brand of disinfectant or antiseptic will not be found in this chapter. Specific instructions for how to use a product may be obtained from the manufacturer. Reputable manufacturers have technical service representatives to contact when specific questions arise about use of their product. Any instructions given on the use of a product should be thorough and clear. Once obtained, instructions should be followed precisely. This includes not only using the material at the proper concentration for the proper amount of time but, in the case of aerosolized products, recognizing as important what seems to be a nonconsequential action such as holding the nozzle a specified distance from the surface to be sprayed.

All disinfectants, sanitizers, and sterilants must be registered with the Environmental Protection Agency (EPA) before they can be marketed. Antiseptics must be registered with the Food and Drug Administration (FDA). At least 1,200 disinfectants made by about 330 manufacturers are sold on the U.S. market, and approximately 64 gaseous and liquid sterilants are produced by more than two dozen companies in the United States. It is important to understand that the EPA does not validate any manufacturer's claims for their product. The FDA registers only those products that are sold for the purpose of sterilizing critical devices such as surgical instruments that penetrate the blood barrier and semicritical devices such as vaginal specula that contact (but do not penetrate) mucous membranes. The FDA, however, like the EPA, does not directly evaluate these products for efficacy. That a product is found acceptable for sterilizing a critical medical device does not mean that the product can be used effectively in the laboratory setting. It is prudent to purchase products from well-known manufacturers. Information on products can often be found in the literature. It may also be appropriate in some situations to test products under the conditions in which they will be used. However, such testing must be carried out with well-designed protocols and under strict test conditions in order to avoid obtaining erroneous results.

As with all chemicals, chemical products in the laboratory must be used with caution. All manufacturers must supply with their products material safety data sheets that inform the user of any potential hazards of the material. Such practices as autoclaving or leaving solutions in open containers that allow vapors to disseminate should be avoided.

The definitions of some common terms used in discussions of decontamination follow. These definitions should be agreed upon when performance claims for products are discussed with manufacturer representatives.

Antiseptic: a substance that inhibits or destroys microorganisms. No sporicidal action is implied. The term is used specifically for substances applied topically to living tissue.

Biocide: a substance that kills all living microorganisms, including spores, both pathogenic and nonpathogenic (for example, bactericides, virucides).

Biostat: an agent that prevents the growth of microorganisms but does not necessarily kill them. The removal or neutralization of a biostat may allow for the regrowth of microorganisms (for example, bacteriostatic, fungistatic).

Decontamination: the removal of microorganisms with no quantitative implication. The term is relative, and the end can be achieved by sterilization or disinfection. When a highly infectious or pathogenic organism is suspected of being present, decontamination is best achieved by a process that renders the material sterile.

Disinfection: a process that reduces or completely eliminates all pathogenic microorganisms except spores. An object that has been properly disinfected could theoretically transmit disease-producing organisms, but the possibility is greatly reduced.

Germicide: a substance that destroys microorganisms, especially pathogenic microorganisms. Technically, a germicide does not destroy spores. However, the term may be used commonly to refer to substances with sporicidal activity. It applies to agents used both on living tissue and on inanimate objects.

Sanitation: the process by which microbial contamination is brought to a "safe" level. This process refers primarily to the process of "cleaning" inanimate objects.

Sterilization: the use of physical and/or chemical procedures to completely eliminate or destroy all forms of microbial life. This term, while absolute, is relative to our ability to detect microorganisms. A solution that has been filtered appropriately to remove all bacteria and fungi and is then termed "sterile" may indeed contain other forms of microorganisms, such as viruses, and thus is technically not sterile.

PHYSICAL STERILANTS

Heat

Moist heat produced by steam autoclaving is the major sterilizing agent used in the clinical microbiology laboratory. Autoclaves are used for sterilization of both culture media and heat-stable supplies as well as for treatment of infectious waste. A steam sterilizer, or autoclave, is an insulated pressure chamber that uses saturated steam to obtain elevated temperatures. Air is removed from the chamber by either gravity displacement or prevacuum. A gravity displacement autoclave is most commonly used. Lighter steam is fed into the chamber to displace heavier air. A brief exposure to steam under pressure can destroy bacterial spores (31).

For routine sterilization of culture media and supplies, the exposure time and conditions are 15 min at 121°C (250°F) and 15 lb/in². For infectious waste, the exposure time is usually extended to 30 to 60 min (6). In addition to the correct time and temperature, direct steam contact is critical for sterilization (57). When infectious materials are treated, procedures to maximize steam penetration into the waste load should be implemented. Processing laboratory waste at 132°C (270°F) accelerates sterilization (35). Other factors that can influence treatment effectiveness are density, physical state and size, and organic content of the waste.

Materials that should not be autoclaved include antineoplastic agents, toxic chemicals, and radioisotopes, which may not be destroyed, and volatile chemicals, which could be vaporized and disseminated by heat. A double polypropylene bag with charcoal vent filter and absorbent has been developed for autoclaving solid infectious, low-level radioactive waste (65).

Dry-heat sterilization is used for materials that cannot be sterilized by steam because of the possibility of moisture damage or because the material is impermeable to steam (31). Dry heat is less efficient than moist heat and requires a longer exposure time and higher temperatures. Dry-heat sterilization is usually performed in a hot-air oven. The mechanism of destruction by dry heat is an oxidative process. Examples of materials for which dry heat is used include oils, petrolatum, powders, sharp instruments, and glassware. Dry heat or thermal inactivation-sterilization has

also been utilized as an alternative infectious waste treatment option.

Filtration

Filtration is a process used to remove microorganisms and microscopic particles from solutions, air, and other gases. Numerous materials have been used as filter media for a variety of applications (36). However, commercially available cellulose acetate and cellulose nitrate membrane filters are the types most commonly used in the laboratory. A 1.2-μm-pore-size membrane filter is used as a prefilter to prevent filter clogging. A 0.45-μm-pore-size membrane filter is used to remove bacteria, yeast cells, and molds from biological and pharmaceutical fluids. For more critical applications, such as removal of pseudomonad-like organisms, a 0.22-μm (or smaller)-pore-size filter is employed.

The most common applications of filtration sterilization in the laboratory include diagnostic drugs, growth media, tissue culture medium, serum, and serum fractions. Quality control testing for sterility of the final filtered product should be performed to test the integrity of the filtration system used. Another common application of filtration is the sterilization of air and gases. High-efficiency particulate air (HEPA) filters are used in biological safety cabinets and laminar-flow rooms. The control of microorganisms in the air is important in the food- and beverage-processing industries, pharmaceutical industry, hospitals, and laboratories.

UV Radiation

UV radiation is a form of electromagnetic wave radiation that acts on cellular nucleic acids (61). Germicidal lamps and tubes have been employed to generate UV radiation by passage of electric discharge through low-pressure mercury vapor enclosed in special glass tubes. Microorganisms are highly susceptible to UV light of 254-nm wavelength. UV has been most widely used to control airborne and surface microorganisms (51). Other applications include cold sterilization of certain chemicals and plastics for pharmaceutical use, sterilization of serum for cell cultures, and disinfection of water (61). Use of UV lamps in laboratories performing PCR tests can decrease the amount of amplified DNA deposited on surfaces (58).

The effectiveness of UV irradiation depends on intensity of the light, distance from the exposed item to the UV source, relative humidity, and microbial species. The primary disadvantage of UV irradiation as a sterilant is its inability to penetrate materials. For inactivation, the microorganism must have direct exposure to the UV light.

Ionizing Radiation

Ionizing radiation in the electromagnetic spectrum has had lethal effects on microorganisms (62). These types of radiation include microwaves, gamma rays, X ray, and electrons. The lethal effect of ionizing radiation results from direct action on a target molecule, resulting in energy being transferred within the target molecule, and from indirect action by diffusion of radicals. The primary cellular target is the DNA. These effects are produced at low temperature and are often referred to as cold sterilization.

The use of microwave energy has been investigated for a variety of applications. Whether its bactericidal action is due to nonthermal effects of microwave exposure (19) or to thermal effects of microwave heating (71) remains unclear. One commercial microwave sterilization system (CEM MikroClave Sterilization System, Matthews, N.C.) has

TABLE 1 Properties of chemical disinfectants[a]

Disinfectant	EPA registered		Use dilution	Inactivation by organic material	Bactericidal	Sporicidal	Tuberculocidal	Fungicidal	Virucidal		Major characteristics						Application		
	Disinfectant	Sterilant							Lipophilic	Hydrophilic	Flammable	Explosion potential	Corrosive	Skin irritant	Respiratory irritant	Eye irritant	Work surface	Dirty glassware	Equipment decontamination
Liquids																			
Alcohols (ethyl and isopropyl)	No	No	60–85%	+	+	−	+	+	+	+[b]	+	−	−	+	−	+	+	−	−
Chlorine	Yes	No	100–1,000 ppm chlorine	+	+	±	±[c]	+	+	+	−	−	+	+	+	+	+	+	+
Glutaraldehyde	Yes	Yes	2–5%	−	+	+	+	+	+	+	−	−	−	+	+	+	−	+	+
Iodophors	Yes	No	30–1,000 ppm iodine	−	+	−	±	±	+	+	−	−	+	+	−	+	+	+	+
Phenolic compounds	Yes	No	0.5–5.0% aqueous	−	+	−	+	+	+	±	−	−	+	+	+	+	+	+	+
Quaternary ammonium compounds	Yes	No	0.5–1.5% aqueous	+	+	−	−	±	+	−	−	−	−	−	−	+	+	+	+
Gases																			
ETO	Yes	Yes	12%	+	+	+	+	+	+	+	+[d]	+[d]	+	−	−	+	+	+	+
Formaldehyde	Yes	Yes	2–5%	+	+	+	+	+	+	+	+	+	+	−	−	+	+	+	+
Hydrogen peroxide	Yes	Yes	30%	+	+	+	+	+			+	+	+	−	−	+	+	+	+

[a]For more specific information on shelf life, cleansing action, health safety, and corrosiveness, consult manufacturer's specifications prior to use. Use all disinfectants per manufacturer's recommendation.
[b]Isopropyl alcohol is not as active as ethyl alcohol.
[c]As much as 10,000 ppm available chlorine has been reported to be necessary to kill mycobacteria (9).
[d]Neither flammable nor explosive when mixed with fluorocarbon, N_2, or CO_2.

been developed for the flash sterilization of microbiological media (38). Validation studies have demonstrated this process to be sporicidal (39).

Gamma irradiation is used to sterilize disposable medical supplies such as plastic hypodermic syringes, sutures, culture media, and containers. Gamma rays are generated from radioisotopes such as cobalt-60 and have a high degree of penetrability. However, this technology is not applicable to the laboratory.

Electron beam sterilization has been used to sterilize medical products. It requires an accelerator that generates focused beams of high-energy electron particles. This technology has been applied to the treatment of infectious waste, but because electrons do not penetrate as well as gamma rays, this technology has limited use and is not applicable to the laboratory.

Ultrasound

Ultrasonic energy at a low-kilohertz frequency range inactivates microorganisms in aqueous suspensions (59). The physical effect of ultrasound treatment appears to be transient cavitation. Ultrasonic cleaners and sonicators are often used for cleaning instruments but are not considered sterilizers. However, combining ultrasonic with chemical treatment can have a lethal effect on microorganisms (3).

CHEMICAL DISINFECTANTS

See Table 1 for a summary of the information on chemical disinfectants that follows.

Alcohol

Ethyl or isopropyl alcohol diluted to a concentration of 60 to 85% in water can be used very effectively for disinfection. Because methyl alcohol has lower bactericidal activity than either ethyl or isopropyl alcohol, its use is not recommended (69). The alcohols are bactericidal, fungicidal, and tuberculocidal, but they are not sporicidal (34). Ethyl alcohol has a broader spectrum of virucidal activity than isopropyl alcohol, being more effective against both lipophilic viruses (adenovirus, herpesvirus, influenza virus) and hydrophilic viruses (poliovirus 1, echovirus 6, coxsackie B1 virus) (32). Both ethyl and isopropyl alcohols have activity against human immunodeficiency virus (34). Alcohols work primarily by denaturing protein.

The presence of high concentrations of organic matter can diminish the activity of alcohols. In the laboratory, alcohols can be used to wipe down clean inanimate surfaces. Because of their high volatility and flammability, they must be used in well-ventilated areas away from open flames and sparks.

Formaldehyde

A solution of 37% formaldehyde, referred to as formalin, can be used as a sterilant, while concentrations of 3 to 8% can be used as disinfectants. However, because of formaldehyde's irritating fumes and carcinogenicity (16), its use as a general disinfectant in the laboratory is not recommended.

Glutaraldehyde

A saturated dialdehyde, glutaraldehyde is a relatively powerful agent that is considered a high-level disinfectant and a sterilizing agent when used appropriately. Alkaline (pH 7.4 to 8.5) 2% solutions have bactericidal, sporicidal, fungicidal, and virucidal activity against both lipophilic and hydrophilic viruses (12, 24, 32). Claims for the activity of glutaraldehyde products against mycobacteria must be evaluated carefully. Though glutaraldehyde is active against mycobacteria, adequate exposure time must be allowed (60). Glutaraldehyde is biocidal as a result of its alkylation of sulfhydryl, hydroxyl, carboxyl, and amino groups of essential chemical constituents of microorganisms.

Glutaraldehyde cannot be used as a general surface disinfectant because of its irritating vapors and required long contact time. It is an ideal material to use on contaminated equipment that can be submerged for some time. Containers of this material should be kept closed to avoid diffusion of its irritating vapors. The exposure time needed to disinfect an item will depend on the organism(s) suspected of contaminating the equipment. The manufacturer's instructions should be carefully followed.

Phenolics

Phenol itself is not used as a disinfectant because of its toxicity, carcinogenicity, and corrosiveness. Phenolic derivatives in which a functional group (e.g., chloro, bromo, alkyl, benzyl, phenyl, amyl) replaces one of the hydrogen atoms on the aromatic ring are commonly used as disinfectants. This replacement reduces the corrosive, toxic, and carcinogenic potentials of the phenol. However, skin irritation and absorption through the skin still occur, so proper precautions should be taken when these materials are used. The three phenolic derivates most commonly used as disinfectants are *ortho*-phenyl phenol, *ortho*-benzyl-*para*-chlorophenol, and *para*-tertiary amylphenol. The addition of detergents to the base formulation results in products that clean and disinfect in one step.

Phenolic compounds kill by inactivation of enzyme systems, precipitation of proteins, and disruption of the cell wall and membrane. These compounds are not considered sporicidal, but when used at their proper concentration for the correct amount of time, they are considered bactericidal, tuberculocidal, virucidal, and fungicidal. Human immunodeficiency virus is inactivated by a 0.5% phenolic solution (37). A concentration of 2% is required to inactivate fungi (68). The concentration at which to use a phenolic compound depends on the conditions under which it will be used. Generally, a concentration of 2 to 5% is used, with the lower concentration requiring a longer exposure time. Because there is a wide variety of phenolic disinfectants with various concentrations of active ingredient available on the market, it is prudent to very carefully evaluate label claims.

Quaternary Ammonium Compounds

The general chemical formula for quaternary ammonium compounds, familiarly known as "quats," is shown in Fig. 1.

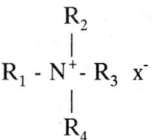

$$R_1 - N^+ - R_3 \quad x^-$$

with R_2 above and R_4 below the nitrogen.

FIGURE 1 General chemical formula for quaternary ammonium compounds (quats). R, alkyl or heterocyclic radicals that may be alike or different; x^-, usually a chloride or bromide to form the salt.

The R group is alkyl or heterocyclic radicals that may be alike or different. It is the type or types of R groups linked to the N^+ that give a quat its antimicrobial activity. The x^- is usually a chloride or bromide to form the salt. Quats are cationic surface-active detergents. Because they are strongly surface active, they have good activity against lipophilic viruses. Quats are bacteriostatic, sporostatic, fungistatic, and algistatic at fairly low concentrations. They can be bactericidal at medium to high concentrations but are not sporicidal, tuberculocidal or virucidal against hydrophilic viruses even at high concentrations (40). *Pseudomonas* spp. growing in an ammonium acetate-containing quat (1) and disease outbreaks associated with gram-negative bacterial contaminants in quats (17) have been reported. Quats kill organisms by disrupting the cell membrane, inactivating enzymes, and denaturing proteins (28). Disruption of the cell membrane appears to be the primary cause of cell death (21).

Quats are odorless, nonstaining, noncorrosive, inexpensive, and relatively nontoxic, thus attractive to use. However, the claims made by manufacturers for the bactericidal activities of their quat products have been difficult to reproduce (55). While this nonreproducibility may be related in some part to test methodology, it does not completely explain the variable results. Manufacturers' claims must be carefully scrutinized, and it must be realized that in-house evaluations may not be definitive.

The antimicrobial activities of quats are reduced in the presence of organic material, anionic detergents (soaps), and such material as gauze and cotton pads. Limiting the use of quats to ordinary environmental decontamination of noncritical surfaces may be appropriate.

Halogens

The halogen group consists of bromine, chlorine, fluorine, and iodine. Only chlorine and iodine, however, are commonly used for disinfection in the laboratory setting. The means by which chlorine and iodine kill microorganisms have not been clearly elucidated. Because it is a strong oxidizing agent, chlorine is thought to kill by a process of oxidation (20). Iodine is thought to kill by reacting with N-H and S-H groups of amino acids as well as with the phenolic group of the amino acid tyrosine and the carbon-carbon double bonds of unsaturated fatty acids (25).

As pure substances, both chlorine and iodine are unstable, corrosive, and toxic. In an attempt to ameliorate these undesirable characteristics, these halogens have been combined with other chemicals. For example, chlorine has been combined with *p*-toluene-sulfonamide, with this combination being known as chloramine T, and iodine has been combined with carriers such as polyvinylpyrrolidone, with these products being referred to as iodophors.

Chlorine

A universal disinfectant, chlorine is most commonly available as a solution of sodium hypochlorite (NaOCl) at a concentration of 1 to 5%. The solution commonly referred to as "household bleach" is often purchased with the assumption that it will contain 5.25% NaOCl. However, it is important that the concentration be specifically stated when the solution is purchased and that it be checked prior to use, since solutions containing less than 5.25% NaOCl are also sold as household bleach. Free chlorine at concentrations of <5 ppm kills vegetative bacteria. The amount of chlorine needed to kill bacterial spores is 10 to 1,000 times greater than that needed to kill vegetative cells (20). According to the recent literature, the amount of chlorine needed to kill mycobacteria ranges from 1,000 ppm (56) to 10,000 ppm (9). Viruses have been inactivated by 200 ppm available chlorine in 10 min (32). Fungal organisms have been killed in 1 h at concentrations of 100 ppm, while fungal spores have been killed at concentrations of 500 ppm (20).

A 1:10 dilution of a 5.25% stock solution of NaOCl, which equates to approximately 5,000 ppm available chlorine, is recommended by the Centers for Disease Control and Prevention for cleaning up blood spills (13, 23, 33). Because chlorine can be inactivated by blood, serum, and other proteinaceous substances, it is recommended that the surface be cleaned prior to using bleach. Under current Occupational Safety and Health Administration regulations, a disinfectant with a tuberculocidal claim may also be necessary when mycobacteria may be present (47). It is best to make up fresh bleach solutions for cleaning up spills. Fresh solutions of NaOCl made up with pH 8.0 tap water and kept in closed opaque containers are stable for up to a month. However, when these containers are opened and closed repeatedly over a period of a month, the concentration of active chlorine may decrease by as much as 50% (27). Thus, a solution intended to have a concentration of 5,000 ppm after 30 days should be diluted only 1:5 when made up from 5.25% stock solution (54).

Iodine

Commercial preparations of iodine coupled with carriers are available as general disinfectants. The manufacturer claims for these products often state that they are bactericidal, fungicidal, virucidal, and tuberculocidal. However, they may not inactivate nonlipid viruses such as poliovirus (32) and rotavirus (67). Iodophors are considered sporicidal (53), but the contact time must be prolonged (8, 41, 68). Critical to remember when using iodophors is that they must be used at the dilution stated by the manufacturer. An iodophor disinfectant meant to be diluted 1:200 but actually diluted only 1:10 may not provide enough free iodine to inactivate microorganisms. Iodophors meant to be used on the skin should not be used as disinfectants, since they may not have the amount of available iodine required when thus diluted. The use of iodine-containing compounds in the laboratory is primarily limited to those situations when items can be submerged in the iodine solution for prolonged periods. Use for surface decontamination is possible, but the specific situation needs to be evaluated carefully.

Peroxygen Compounds

Included among the peroxygen compounds are the disinfectants hydrogen peroxide and peracetic acid. Both compounds are bactericidal, virucidal, sporicidal, and fungicidal (10). The activities of these compounds against mycobacteria are not well documented. Hydrogen peroxide kills primarily by the production of hydroxyl radicals, which bring about oxidation of critical systems of the microorganism. The mode of action of peracetic acid has not been clearly elucidated, but the compound is thought to function much like hydrogen peroxide. Both compounds are widely used in industrial settings for disinfection of a wide variety of materials and equipment. Neither, however, is commonly used as a general disinfectant in the laboratory.

Heavy Metals

The use of heavy metals such as mercury in the form of mercuric chloride is not recommended because of their potential toxicity and because they are environmental pollutants.

Sanitizers

Products sold as or considered to be sanitizers or disinfectants diluted to the point at which they are considered to be sanitizers should not be used in the laboratory. These types of products produce only a reduction in the number of bacterial contaminants to "safe" levels as judged by public health requirements.

SELECTION OF DISINFECTANTS

There is no one ideal disinfectant. The chemical type and the particular brand of disinfectant need to be chosen very carefully for each application. It is not unusual for several types of chemical disinfectants to be used in the same laboratory. The following items need to be considered when a disinfectant is being chosen.

1. Type and number of microbes present
2. Type and amount of organic matter present
3. Contact time
4. Type of surface to be disinfected (porous or nonporous) and its potential for corrosion or deterioration
5. Type of water available for diluting the material: hard water may reduce the rate of kill of certain disinfectants
6. EPA registration: Only products with EPA registration dates of 1986 or later should be used, particularly if tuberculocidal claims are made
7. Ability of the manufacturer to provide clear and accurate data on the efficacy of their disinfectant. For example, that a product kills sporeforming bacteria does not necessarily mean the product is sporicidal.
8. Labels on container (should be easy to read) and the instructions for use (should be easy to understand)
9. Safety and environmental acceptability of product
10. Personnel acceptance
11. Local, state, and federal regulations
12. Cost: More expensive is not necessarily better, but do not save money if doing so jeopardizes personnel safety.
13. Shelf life: Long shelf lives of certain undiluted disinfectants may be dramatically shortened once the solutions are diluted.
14. Residual activity

The consumers of disinfectants have two choices in evaluating the efficacy of the agent against environmental bioburden. These are to accept the manufacturer's claims or to perform the validation or challenge studies for oneself. The first alternative is the easiest and for most laboratories will be the only possibility because of the time and expertise required to perform validation testing. The choice of a disinfectant manufactured by a reputable manufacturer will result in the proper choice most of the time. If validation studies are performed, these need to be done according to a well-designed protocol. One of the in vitro tests commonly used is the dilution test published by the Association of Official Analytical Chemists (7). In situ evaluation, which involves measuring viable contamination before and after the disinfectant is applied to the selected surface, can also be performed. There is, however, no standardized test for in situ evaluations. See references 14 and 54 for a more detailed discussion of what must be considered when disinfectants are evaluated.

CHEMICAL STERILANTS

Peracetic Acid

Peracetic acid has been widely used for high-level disinfection in the food and beverage industries and for cold sterilization of surgical instruments. It has also been applied as an aerosol to sterilize equipment used in raising germfree laboratory animals (10). A commercial peracetic acid sterilizer (STERIS Corp., Mentor, Ohio) is available for diagnostic and surgical instruments that can be immersed in liquid. Items to be treated are exposed to 0.2% peracetic acid, rinsed with sterile filtered water, and flushed with sterile air. The entire treatment cycle takes less than 30 min.

Peracetic acid is an oxidizing agent that attacks the sensitive sulfhydryl and sulfur bonds in proteins of the cell wall, enzymes, and other metabolites. Peracetic acid is sporicidal alone and in combination with hydrogen peroxide (4). It is effective in the presence of organic material and produces nontoxic end products (acetic acid and O_2); however, it is limited to materials that can be immersed.

Glutaraldehyde

The use of glutaraldehyde solutions as disinfectants in the laboratory was discussed above. Application of 2% glutaraldehyde as a liquid chemical sterilant has been used primarily for medical and surgical material that cannot be sterilized by heat or irradiation. Glutaraldehyde has also been used in the preparation of vaccines (50). Adequate safety precautions should be observed to minimize occupational exposure.

GAS-VAPOR STERILANTS

Gas-vapor sterilization is primarily used to sterilize medical equipment and various medical and industrial products (48). The application of gas sterilization in the laboratory has been limited to sterilization of laboratory plastics and decontamination of biological safety cabinets. Gas-vapor sterilizers may be used for the treatment of contaminated equipment. Because of the toxicity associated with ethylene oxide (ETO) and formaldehyde, alternative sterilization technologies, including vapor-phase hydrogen peroxide, plasma gas, chlorine dioxide gas, and ozone, have been developed (29).

ETO Gas

ETO is the primary gas used in hospitals to sterilize heat-sensitive supplies. In the ETO sterilization process, air is evacuated from a pressure vessel, and a sterilant mixture (12% ETO–88% Freon) is added at a specified temperature and humidity. After a specific exposure time, the gas is removed. ETO is highly effective against bacteria, molds, yeast cells, and viruses. Its bactericidal and sporicidal activity is due to alkylation of sulfhydryl, amino, carboxyl, phenolic, and hydroxyl groups. The major factors that can affect ETO sterilization are gas concentration, temperature, relative humidity, exposure time, and carrier material. Because ETO is considered a mutagen and a carcinogen, restrictions on ETO emissions have been imposed. Alternative diluent gases such as nitrogen and CO_2 are being used in vacuum cycles to replace the chlorofluorocarbon Freon.

Formaldehyde Vapor

Formaldehyde has been used in the laboratory to sterilize HEPA filters in biological safety cabinets. Sterilization with low-temperature steam-formaldehyde has also been available in Europe (45). The basic treatment cycle involves vaporizing a 2 to 5% formaldehyde solution in the presence of steam at a temperature of 60 to 80°C. The bactericidal activity of formaldehyde appears to be due to the reaction with cell protein and DNA or RNA. Formaldehyde is known to be carcinogenic (16), and residues have been detected in cellulosic materials, polyester fibers, natural rubber, and other polymeric materials.

Vapor-Phase Hydrogen Peroxide

Liquid hydrogen peroxide is a powerful oxidizing agent. The vapor phase of hydrogen peroxide is highly sporicidal. A vapor-phase hydrogen peroxide sterilization system (AMSCO Scientific, Erie, Pa.) is available for surface sterilization of instruments. An aqueous solution of 30% hydrogen peroxide is introduced into a heated vaporizer. It operates at 55 to 60°C and has a 90-min cycle. This system operates with a low concentration of sterilant, produces nontoxic end products, and provides a rapid cycle time. However, it is limited to surface sterilization. It is also corrosive to some metals, natural rubber, and nylon. Decontamination of biological safety cabinets using hydrogen peroxide vapor is also obtainable without affecting performance characteristics (VHP 1000; AMSCO Scientific).

Plasma Gas

Plasma gas sterilization has recently been applied to the treatment of medical devices (Surgikos, Inc., Arlington, Tex.). In this process, hydrogen peroxide is injected into the sterilization chamber and vaporized. The vapor diffuses throughout the chamber, and radio frequency energy is applied to the chamber to create gas plasma. The free radicals formed in the plasma gas are involved in microbial inactivation. The process temperature does not exceed 40°C. Plasma gas created outside the system can also be directed into the sterilization chamber (AbTox, Inc., Mundelein, Ill.). Because of the deep vacuum levels required for treatment, distortion or destruction of devices and surface modifications of materials may occur.

Chlorine Dioxide Gas

Chlorine dioxide is an oxidizing agent that attacks proteins but does not alter nucleic acids like ETO does. The use of chlorine dioxide gas as a sterilant of medical supplies has been investigated. Sterilization can be achieved at low concentrations and ambient temperature, and the gaseous effluent can be easily neutralized. Its reactivity is selective, however, and the toxicity effects of high-level, short-term exposure are unknown.

Ozone

Like chlorine dioxide, ozone has oxidative properties. It has potential application for sterilizing medical instruments. In the operation of the ozone gas sterilizer (Life Support, Inc., Erie, Pa.), oxygen is converted to ozone, which is then humidified and dispersed into the sterilization chamber. During a 1- to 3-h treatment cycle, ozone gas flows through the chamber. Residual ozone is then converted back to oxygen. While ozone sterilization produces nontoxic end products at low concentrations and temperature, the penetrability of ozone is marginal, the gas is corrosive to many plastics and most metals, and it degrades rubber.

SELECTION OF STERILANTS

There is no ideal sterilant. Various established and new sterilization methods have specific applications based on material compatibility and penetration requirements. Because some chemical sterilants are also used as high-level disinfectants, explicit use requirements must be complied with (26). The following items should be considered when selecting a sterilant:

1. Activity: bactericidal, sporicidal, tuberculocidal, fungicidal, and virucidal
2. Rapidity of action: sterilization achieved within a short time
3. Penetration: strong penetration of common packing material and interior space of instrumentation
4. Compatibility: produces little or no change in appearance or function of materials through repeated sterilization cycles
5. Nontoxicity: presents no health risk and poses no potential environmental hazard
6. Organic material resistance: efficacy not lost in the presence of organic material
7. Adaptability: suitable for large and small applications
8. Monitoring capability: treatment cycle monitored easily and accurately
9. Cost-effectiveness: reasonable cost for equipment, installation, and operation

ANTISEPTICS

Antiseptics are found in the laboratory primarily in products used for hand washing. The necessity for hand washing is based on the fact that the hands become contaminated with disease-producing microorganisms during laboratory procedures and then serve as the means by which laboratorians may autoinoculate themselves (70). Whether the product used for hand washing in the laboratory needs to contain an antibacterial agent has never been shown by any well-controlled studies. Thus, one can only extrapolate from studies done in patient care settings. Such studies give insufficient evidence to suggest that products containing an antibacterial agent are any better than ordinary soap for routine hand washing (64, 69). However, in those situations in which personnel perform critical care procedures on patients, the use of products containing antibacterial agents reduces the number of nosocomial infections (52, 63). Thus, one can be led to the comfortable, although perhaps incorrect, conclusion that products containing an antibacterial agent would be better in the laboratory setting. More critical, however, may actually be washing the hands in the proper manner. The proper technique involves a rigorous rubbing together of all surfaces of well-lathered hands for 10 to 15 s followed by rinsing under a stream of water (22). Particular attention should be paid to areas under the fingernails and around the cuticles if it is suspected that the hands have become extraordinarily contaminated.

Some of the same chemicals found in disinfectants are also found in hand-washing preparations. However, a hand-washing product should not be used as a disinfectant, and a disinfectant should not be used for hand washing. The more common chemicals found in hand-washing products are alcohols, chlorhexidine gluconate, iodophors, chloroxylenol, and triclosan.

Alcohols are very fast acting and effective against a broad spectrum of microorganisms, including mycobacteria, fungi, and viruses. They are safe and inexpensive. The two alcohols most commonly found in products are isopropyl and ethyl alcohols in a concentration range of 70 to 90%. Alcohols have a tendency to dry the skin, but products with emollients help minimize this drying.

Chlorhexidine gluconate is a broad-spectrum antiseptic with good activity against gram-positive and gram-negative bacteria but poorer activity against viruses, mycobacteria, and fungi. A residue is left by this material, and it may continue antibacterial activity for several hours. This antiseptic is generally used in a concentration range of 2 to 4%. Chlorhexidine gluconate activity is affected by hard water, some lotions, and soaps.

Iodophors, while having good activity against both gram-positive and gram-negative bacteria, fungi, mycobacteria, and viruses, have the disadvantage of staining and causing skin irritation. They are generally available in concentrations of 0.7% available iodine.

Chloroxylenol has good antiseptic activity but may not be as active as chlorhexidine and iodophors. It is used in concentrations ranging from 1.5 to 3.5%.

Triclosan is commonly used in deodorant soaps. It is active against both gram-positive and gram-negative organisms with the exception of *Pseudomonas* species. Its activity against other microorganisms is not as good. It is generally used in a concentration range of 0.3 to 1.0%.

Other chemicals that may be found in hand-washing products are hexachlorophene and benzalkonium chloride. Both of these have fallen into some disfavor because of potential neurotoxicity and an ability to become contaminated with bacteria, respectively.

Further information on these chemicals as used for hand washing can be found in references 5 and 33.

The type(s) of hand-washing agent to be used in the laboratory needs to be well thought out. Three key factors to consider are (i) the types of microorganisms the worker will be exposed to, (ii) the acceptance of the agent by personnel, and (iii) the cost. The facilities for hand washing should be conveniently located and convenient to use. No matter how well hand washing is done, personnel should continually be reminded to keep their hands, gloved and ungloved, away from their mouth, nose, eyes, and hair. In addition, they should be reminded that their hands, gloved and ungloved, should be washed after protective clothing is removed before leaving the laboratory and before eating.

ALTERNATIVE INFECTIOUS WASTE TREATMENT AND DISPOSAL TECHNOLOGIES

Many hospitals are reexamining their options for the treatment and disposal of infectious medical and laboratory waste because of stricter federal, state, and local regulations. It has been reported that the average hospital generates 8,400 lb (ca. 3,810 kg) of medical waste each month (2). The traditional methods of waste treatment and disposal are incineration and steam sterilization.

Incineration

The most common method of treating and disposing of both infectious medical waste and noninfectious waste is incineration. Of the current on-site incinerators available, the controlled-air type is the most widely used. The controlled-air incinerator involves a dual chamber in which two sequential combustion processes take place and a stack for venting combustion gas. Waste is fed into a primary chamber, where the combustion temperature (1,600 to 1,800°F [ca. 870 to 980°C]) is regulated by restricted airflow to volatilize and oxidize the fixed carbon in the waste. Next, in the secondary chamber, excess air is introduced in addition to the volatile gas-air mixture to achieve a higher temperature (1,800+°F [ca. 980+°C]) and complete the combustion process. In air pollution-controlled systems, scrubbers are installed after the secondary chamber to remove particulates and acid gases.

Although incineration appears to be the most suitable choice to destroy both infectious and hazardous materials and to reduce waste, it has its drawbacks. Extreme variations in waste contents such as bulk fluids or an improperly operated incinerator may result in the release of viable microorganisms via stack emissions, ash residue, or wastewater. Toxic air emissions and heavy metals in the ash pose potential health problems. Community awareness and permitting and licensing requirements, especially in populous areas in the United States, have shut down or limited the use of incineration.

Steam Sterilization

Nearly one-third of hospitals in this country treat their microbiological waste by steam sterilization. A steam sterilizer or autoclave is an insulated pressure chamber in which saturated steam is used to elevate temperature. Two means of removing air from the chamber are gravity displacement and a prevacuum cycle. The gravity displacement autoclave is the type most commonly used for treatment of infectious waste. Steam is fed into the chamber to displace heavier air. Operational conditions in the laboratory are usually a temperature of 121°C (250°F), a pressure of 15 lb/in^2, and a cycle time of 30 min. Commercial autoclaves operate at a more elevated temperature and pressure (132°C [270°F], 60 to 75 lb/in^2) and at an operational cycle time of 1 h. Types of waste that should not be autoclaved include antineoplastic agents, toxic chemicals, and radioisotopes, which may not be destroyed, and volatile chemicals, which could be vaporized and disseminated by heat. Clinical laboratories should also treat infectious waste separately from microbiological media and supplies.

The effectiveness of steam sterilization depends on time, temperature, and direct steam contact with the infectious agents. Other factors that can influence treatment efficiency include density, physical state and size, and organic content of the waste. Although biological indicators are used to control the efficacy of autoclaves, sterilization may not be achieved because waste composition is so variable. Physical destruction may also be required by a secondary mechanical process before acceptance at a landfill. See chapter 7 of this Manual for more information on infectious waste management.

Recently, several alternative treatment options have appeared on the market. Many of these methods have evolved by combining known technologies. The efficacies of these methods are dependent on contact time, number of microorganisms in the waste, organic content, volume, and physical state of the waste. Because there is no national standard, approval of each technology is based on individual state and local requirements. The minimum requirement for decontamination of infectious waste has been expressed as a 6-$\log_{10}$ reduction in vegetative bacteria, fungi, mycobacteria, and lipid-containing viruses. For bacterial endospores, a 4-$\log_{10}$ reduction must be demonstrated.

No single technology is applicable for every type of waste stream generated and for every institution. The microbiologist should be aware of the currently available treatment options and their advantages and disadvantages (18). A comparison of various waste treatment technologies and factors for selection are presented in Table 2. A brief description of each technology is given below to help in understanding the treatment process. Many new advances will no doubt occur before this information is in print because of the rapid emergence of new technology and continued improvements to existing equipment. More information on infectious waste management and treatment options can be obtained from a number of resources (11, 15, 44, 49).

Mechanical-Chemical

One of the first alternative waste treatment systems was manufactured by Medical SafeTEC, Inc. (Indianapolis, Ind.). Model Z1200 employs a high-speed hammer mill to pulverize waste material. Bags or containers of infectious waste are manually loaded onto an enclosed conveyer, transported to a feed hopper, and sprayed with a sodium hypochlorite solution. Waste and disinfectant first pass through a preshredder into a high-speed hammer mill. Unrecognizable solid materials are captured in a rotary separator, transported up a screw auger conveyer, and deposited into a waste collection cart. Liquids are either discharged directly into the sanitary sewer or recirculated through the system. The entire process is maintained under negative pressure, and air is vented through a HEPA filter system.

The Condor model HR600 medical waste treatment system (Winfield Environmental, Escondido, Calif.) is a controlled system that shreds and granulates medical waste. Before waste collection, bags are electronically scanned for detectable metal instruments or radioactive isotopes. Waste collection bags are fed into a cutting chamber, where the waste is reduced, and then treated with aqueous chlorine dioxide. A dewatering process is used to recycle most of the oxidizing solution. Fresh disinfectant is automatically metered into the treatment solution, keeping it active. The solid waste is removed through a continuous-action screw auger into a collection cart. Liquid effluent is discharged into the sanitary sewer. Air released is vented through HEPA filters and carbon filter systems.

The Mediclean IWP-1000 (Mediclean Technology, Inc., West Warwick, R.I.) medical waste processor is similar to the Condor system. However, during the granulation process, waste is continually circulated and mixed with fresh chlorine dioxide.

The STERIS-ECOMED system (STERIS Corp., and ECOMED, Inc., Indianapolis, Ind.) is designed for point-of-generation destruction. Infectious waste and STERIS 20

TABLE 2 Comparison of regulated medical waste treatment technologies[a]

Factor	Traditional method		Alternative mechanical-chemical methods of treatment[b]		
	Incineration	Steam or autoclave	Mechanical-sodium hypochlorite[c]	Mechanical-chlorine dioxide[d]	Mechanical-peracetic acid[e]
Operations					
Applicability[m]	All, mixed	Most	Most	Most[n]	Most
Equipment operation	Complex	Easy	Moderate	Easy	Easy
Operator requirement[p]	Highly skilled	Trained	Trained	Trained	Trained
Waste segregation[p]	None	Needed	Needed	Needed	Needed
Load standardization	Needed[q]	Needed[q]	Needed	Needed[q]	Needed[q]
Capacity[t]	1,200 lb/h	200 lb/h	1,000 lb/h	600 lb/h	20 lb/h
Effect of treatment on waste	Burned	Unchanged	Shredded or ground	Shredded	Pulverized
Vol reduction	85–90%	30%	Up to 85–95%	Up to 85–95%	Up to 85%
Occupational hazards	Moderate	Low	Moderate	Low	Low
QA-QC biological indicator	Bacillus subtilis	Bacillus stearothermophilus	Bacillus subtilis	Bacillus subtilis	Bacillus subtilis
QA-QC validation	Temp	Temp	Chemical	Temp, chemical	Colorimetric
Testing	Complex, expensive	Easy, inexpensive	Moderate, inexpensive	Easy, inexpensive	Moderate, inexpensive
Potential side benefits	Energy recovery	None	Yes[u]	Yes[u]	Yes[u,v]
Location, on-site, off-site	Both	Both	Both	Both, mobile	On-site
Regulatory requirements[y]					
Risk of release to air and prevention mechanisms	High, scrubber	Low, vent	Low, HEPA filter	Low, HEPA and carbon filters	Low, HEPA filter
Risk of release to water and prevention	Low, none	Low, vent	Low, drain	Low, drain	Low, drain
Disposal of residues					
Liquids	None[z]	Sanitary sewer[aa]	Sanitary sewer[aa,bb]	Sanitary sewer[aa,bb]	Sanitary sewer[aa]
Solids	Hazardous ash[cc]	Sanitary landfill[dd]	Sanitary landfill	Sanitary landfill	Sanitary landfill
Costs					
Capital	High	Low	Moderate	Moderate	Low
Labor	High	Low	Moderate	Low	Low
Operating	High	Low	Moderate	Low	Moderate
Maintenance	High	Low	Moderate	None[ee]	Low
Downtime	High	Low	Low to moderate	Low to moderate	Low

(Continued on next page)

TABLE 2 (Continued)

	Alternative mechanical-physical methods of treatment[b]						
	Mechanical-steam autoclave[f]	Mechanical-dry heat[g]	Mechanical-microwave[h]	Mechanical-radio waves[i]	Mechanical-electron beam[j]	Compaction-steam autoclave[k]	Compression-infrared[l]
	Most	Most	Most[n,o]	Most	Most	Most[n]	Most[n]
	Easy	Moderate	Easy	Complex	Easy	Easy	Easy
	Trained	Trained	Trained	Trained	Trained	Trained	Trained
	Needed	Needed	Needed	Needed	Needed	Needed	Needed
	Needed[q]	Needed[q]	Needed[q,r]	Needed[q,r]	Needed[q,s]	Needed[q]	Needed[q,r]
	300 lb/h	2,000 lb/h	550 lb/h	Not available	800 lb/h	400 lb/h	50 lb/20 min
	Shredded	Shredded or ground	Shredded	Shredded	Shredded	Compacted	Compressed
	80%	Up to 85%	Up to 85%	Up to 85%	70%	50%	80–90%
	Low	Low	Low	Low	Low	Low	Low
	Bacillus stearothermophilus	Bacillus subtilis	Bacillus subtilis	Bacillus subtilis	Bacillus subtilis	Bacillus stearothermophilus	Bacillus subtilis
	Temp, pressure	Temp	Temp	Temp, energy	Dosimeters	Temp, pressure	Temp, wt, time
	Easy, inexpensive	Easy, inexpensive	Easy, inexpensive	Easy, inexpensive	Easy, inexpensive	Easy, inexpensive	Easy, inexpensive
	Yes[u,w]	Yes[u,w]	Yes[u]	Yes[u,w,x]	Yes[u,w]	Yes[u]	Yes[u]
	On-site, mobile	Both	Both	Off-site	Both	Both	Both, mobile
	Low, HEPA and carbon filters	Low, HEPA filter	Low, HEPA filter	Low, HEPA filter	Low, via catalytic converter	Low, vent	Low, via HEPA and carbon filters
	Low, evaporation	Low, drain	Low, evaporation	None	None	Low, drain	None
	None[z]	Sanitary sewer[aa]	None[z]	None[z]	None[z]	Sanitary sewer[aa]	None
	Sanitary landfill	Sanitary landfill	Sanitary landfill	Sanitary landfill	Sanitary landfill	Sanitary landfill	Sanitary landfill
	Moderate	Moderate	High	High	Moderate	Low	Low
	Low	Not available	Low	Not available	Moderate	Low	Low
	Low	Low	Moderate	Not available	Low	Low	Low to moderate
	Moderate	Moderate	Moderate	Not available	Low	Low	Low
	Low to moderate	Low to moderate	Moderate to high	Low	Low	Low	Low

[a]Adapted from reference 49 with permission.
[b]Information provided by manufacturers.
[c]Model Z1200; Medical SafeTEC, Inc., Indianapolis, Ind.
[d]Winfield Condor model HR600; Winfield Environmental, Escondido, Calif.
[e]STERIS/ECOMED, Mentor, Ohio, and Indianapolis, Ind.
[f]G.T.H. Roland 2DA-M3; S.A.S. Systems, Inc., Houston, Tex.
[g]BioMed Waste Systems, Inc., Boston, Mass.
[h]ABB Sanitec, model HGA-250-S; ABB Environmental Services, Inc., Roseland, N.J.
[i]Stericycle, Inc., Deerfield, Ill.
[j]Nutek Corp., Palo Alto, Calif.
[k]Model MARK VII; San-I-Pak Pacific, Inc., Tracy, Calif.
[l]DISPOZ-ALL 2000; Medifor-X Corp., Redding, Conn.
[m]To infectious waste types.
[n]No more than 10% bulk fluid by weight.

[o]Metallic content must be less than 1% by weight.
[p]To eliminate pathological, chemotherapeutic, and nontreatable wastes.
[q]To eliminate bulk fluids such as dialysis fluid.
[r]To separate sharps from waste.
[s]Tote carrier.
[t]1 lb = 453.59237 g.
[u]No harmful by-products; lower volume to commercial waste streams.
[v]Point-of-generation testing.
[w]Plastic recovery for recycling.
[x]Produce fossil fuel alternative.
[y]Permits and other regulatory requirements (e.g., air emissions, wastewater discharge, ash disposal, electron beam license) are dependent on individual state and local policies.
[z]Moisture retained in solids or held within the unit.

[aa]Low-volume or intermittent drain to sanitary sewer.
[bb]Potential formation of carcinogenic compounds in chlorine-based systems.
[cc]Resource Conservation and Recovery Act-permitted landfill.
[dd]Potential problem with recognizable red-bag waste.
[ee]Included under service agreement.

sterilant concentrate (peracetic acid) are placed into a grinding chamber that simultaneously pulverizes and decontaminates. After a 10-min process cycle, the treated waste is mechanically dewatered and immediately deposited into the normal trash.

Mechanical-Thermal

The Roland biomedical waste treatment system S.A.S.-1 (S.A.S. Systems, Inc., Houston, Tex.) shreds and sterilizes infectious waste in an enclosed unit that operates as an autoclave. Infectious waste in containers is lifted by a hydraulic lift and dropped into the autoclave shredder chamber. After the waste feed door is closed, a vacuum is drawn to displace air from the treatment chamber. Steam is then injected to reach an operating pressure of 31 lb/in^2 and a temperature of 135°C (275°F). Waste is shredded while recirculating in saturated steam. A final vacuum phase dehydrates solid material prior to final disposal. The exhausted air passes through a HEPA filter and an activated-carbon filter to remove particulate matter, odor, and volatile organic compounds. The S.A.S.-1 system operates continuously by a programmed process control computer.

In the BioMed continuous thermal disinfection unit (BioMed Waste Systems, Inc., Boston, Mass.), clinical waste is shredded and then disinfected by dry heat. Infectious waste is mechanically fed into a shredder and then dropped into a disinfection unit. The disinfection unit consists of hollow screws that transport waste. This unit heats the waste to at least 120°C for 45 min. Heat is introduced by hot oil or steam that circulates within the sealed screws. The waste is then conveyed to a surge bin, where it is held at temperatures up to 120°C (248°F) before final disposal. Condensed steam is collected in a holding tank before disposal into the sanitary sewer, and processed air is discharged through a HEPA and charcoal filter system.

ABB Environmental Services, Inc. (Roseland, N.J.), manufactures the ABB Sanitec microwave disinfection system (model HGA-250-S), which shreds waste and then heats it by microwaving. Infectious waste placed in a waste container is automatically fed into a hopper, shredded, and sprayed with steam. The waste is then transported up a screw conveyer, where it is heated at 95°C (203°F) for 30 min by multiple microwaves. After heating, the solid waste is conveyed to a temperature-holding section before final waste disposal. Processed air is discharged through HEPA filters. For proper treatment, waste material should contain less than 10% liquid by weight and less than 1% metal content.

The Stericycle electrothermal deactivation system (Stericycle, Inc., Deerfield, Ill.) shreds waste and thermally inactivates it by low-frequency radio waves. The shredded waste is exposed to an oscillating, high-intensity electrical field that generates temperatures high enough to inactivate microorganisms but not destroy or change the material. After treatment, plastics are recycled. Treatment cycle time and temperature vary in accordance with state codes for microbial inactivation. The system is maintained under negative pressure, and air is released through a HEPA filter system.

Compaction-Thermal

A combination high-vacuum steam sterilizer and compactor (model MARK VII) has been developed by San-I-Pak Pacific, Inc. (Tracy, Calif.). Infectious waste is loaded into a self-operating autoclave-compactor. After a 55-min treatment cycle at 135 to 138°C (275 to 281°F), a chamber liner moves treated waste out of the sterilizer chamber. The waste then drops into a compaction chamber that compacts the waste into a roll-off container for final disposal. This system is also designed for compaction and disposal of general refuse.

Compression-Thermal

The DISPOZ-ALL 2000 (Medifor-X Corp., Redding, Conn.) is a dry-heat sterilization system that utilizes infrared and convection heat. Infectious waste is manually batched into the computerized system and heated up to 204°C (400°F) for 20 min. After the treatment cycle, waste is automatically transferred to a compressing chamber for physical impaction while still hot. Processed air is released through a HEPA and charcoal filter system.

Other Technologies

Other technologies currently under development include electron beam radiation (Nutek Corp., Palo Alto, Calif.) and plasma energy (Plasma Energy Applied Technology, Huntsville, Ala.). Plasma, an electrically charged gas, can destroy medical waste at extremely high temperatures. It does not have the emissions problems of incineration and creates recyclable by-products (30).

QUALITY ASSURANCE-QUALITY CONTROL PROGRAM

A quality assurance-quality control program is essential to ensure that established procedures for disinfection and sterilization are followed. Important components in this program can be incorporated into laboratory safety management and infectious waste management procedures. For example, laboratories must routinely disinfect work surfaces and comply with Centers for Disease Control and Prevention recommendations for universal precautions. Infectious waste must be treated by physical or chemical means before being transported to a landfill. Routine monitoring of chemical sterilant and disinfectant effectiveness is difficult, because these substances cannot be evaluated reliably by the same criteria applied to sterilization procedures such as autoclaving.

Laboratories must also document work practices, procedures, and policies that ensure employee protection in a chemical hygiene plan (47). See the *Clinical Microbiology Procedures Handbook* for more information on biohazards and safety compliance (66).

REGULATORY REQUIREMENTS

The federal, state, and local regulations under which laboratories operate are many. These regulations are often not easy to understand, often contradict one another, and require some ingenuity to implement. The publication that has most recently had an impact on the laboratory is the Occupational Safety and Health Administration's document on bloodborne pathogens (46). This document should be carefully reviewed, and a copy should be in each laboratory's library. The document places the responsibility on the employer to make sure that employees have the proper equipment, are taught proper techniques, and can actually implement what they are taught. This includes having the proper disinfectants available for use as well as being taught proper hand-washing technique. See chapters 6 and 7 of this Manual and publications from the National

Committee for Clinical Laboratory Standards (41–44) for further information on methods of meeting regulatory requirements.

REFERENCES

1. **Adair, F. W., S. G. Geftic, and J. Gelzer.** 1969. Resistance of *Pseudomonas* to quaternary ammonium compounds. Growth in benzalkonium chloride solution. *Appl. Microbiol.* **18:**299–302.

2. **Agency for Toxic Substances and Disease Registry.** 1990. The public health implications of medical waste: a report to Congress. Publication no. PB91-100271. U.S. Department of Health and Human Services, Atlanta.

3. **Ahmed, F. I. K., and C. Russell.** 1975. Synergism between ultrasonic waves and hydrogen peroxide in the killing of microorganisms. *J. Appl. Bacteriol.* **39:**31–40.

4. **Alasri, A., M. Valverde, C. Roques, G. Michel, C. Cabassud, and P. Aptel.** 1993. Sporicidal properties of peracetic acid and hydrogen peroxide, alone and in combination, in comparison with chlorine and formaldehyde for ultrafiltration membrane disinfection. *Can. J. Microbiol.* **39:**52–60.

5. **Ansari, S. A., S. A. Sattar, S. Springthorpe, G. Wells, and W. Tostowaryk.** 1989. In vivo protocol for testing efficacy of hand-washing agents against viruses and bacteria: experiments with rotavirus and *Escherichia coli. Appl. Environ. Microbiol.* **55:**3113–3118.

6. **Association for the Advancement of Medical Instrumentation.** 1988. *Steam Sterilization and Sterility Assurance.* Association for the Advancement of Medical Instrumentation, Arlington, Va.

7. **Association of Official Analytical Chemists.** 1990. *Official Methods of Analysis of the Association of Analytical Chemists,* 15th ed., p. 133–143. Association of Official Analytical Chemists, Arlington, Va.

8. **Berkelman, R. L., B. W. Holland, and R. L. Anderson.** 1982. Increased bactericidal activity of dilute preparations of povidone-iodine solutions. *J. Clin. Microbiol.* **15:**635–639.

9. **Best, M., S. A. Sattar, V. S. Springthorpe, and M. E. Kennedy.** 1990. Efficacies of selected disinfectants against *Mycobacterium tuberculosis. J. Clin. Microbiol.* **28:**2234–2239.

10. **Block, S. S.** 1991. Peroxygen compounds, p. 167–181. *In* S. S. Block (ed.), *Disinfection, Sterilization and Preservation.* Lea & Febiger, Philadelphia.

11. **Block, S. S.** 1991. Infectious medical waste: treatment and sanitary disposal, p. 730–748. *In* S. S. Block (ed.), *Disinfection, Sterilization and Preservation.* Lea & Febiger, Philadelphia.

12. **Borick, P. M., F. H. Donershine, and V. L. Chandler.** 1964. Alkalinized glutaraldehyde, a new antimicrobial agent. *J. Pharm. Sci.* **53:**1273–1275.

13. **Centers for Disease Control.** 1982. Acquired immune deficiency syndrome (AIDS): precautions for clinical and laboratory staffs. *Morbid. Mortal. Weekly Rep.* **31:**577–580.

14. **Cremieux, A., and J. Fleurette.** 1991. Methods of testing disinfectants, p. 1009–1027. *In* S. S. Block (ed.), *Disinfection, Sterilization and Preservation.* Lea & Febiger, Philadelphia.

15. **Denys, G. A.** 1992. Infectious waste management. *Encycl. Microbiol.* **2:**493–504.

16. **Department of Health, Education and Welfare.** 1981. *Formaldehyde: Evidence of Carcinogenicity.* NIOSH current intelligence bulletin 34. DHEW (NIOSH) publication no. 81-111.

17. **Dixon, R. E., R. A. Kaslow, D. C. Mackel, C. C. Fulkerson, and G. F. Mallison.** 1976. Aqueous quaternary ammonium antiseptics and disinfectants. Use and misuse. *JAMA* **236:**2415–2417.

18. **Doucet, L. G.** 1992. Which on site waste/treatment option will work best for you? *Health Facilities Management* **December:**14–19.

19. **Dreyfuss, M. S., and J. R. Chipley.** 1980. Comparison of effects of sublethal microwave radiation and conventional heating on the metabolic activity of *Staphylococcus aureus. Appl. Environ. Microbiol.* **39:**13–16.

20. **Dychdala, G. R.** 1991. Chlorine and chlorine compounds, p. 131–151. *In* S. S. Block (ed.), *Disinfection, Sterilization and Preservation.* Lea & Febiger, Philadelphia.

21. **Franklin, T. J., and G. A. Snow.** 1984. *Biochemistry of Antimicrobial Action.* Chapman & Hall, Ltd., London.

22. **Garner, J. S., and M. S. Favero.** 1985. CDC guideline for handwashing and hospital environmental control. *Infect. Control* **7:**231–243.

23. **Garner, J. S., and B. P. Simmons.** 1983. Guideline for isolation precautions in hospitals. *Infect. Control* **4:**245–325.

24. **Gorman, S. P., and E. M. Scott.** 1977. A quantitative evaluation of the antifungal properties of glutaraldehyde. *J. Appl. Bacteriol.* **47:**463–468.

25. **Gottardi, W.** 1991. Iodine and iodine compounds, p. 152–166. *In* S. S. Block (ed.), *Disinfection, Sterilization and Preservation.* Lea & Febiger, Philadelphia.

26. **Gurevich, I.** 1993. Efficacy of chemical sterilants/disinfectants: is there a light at the end of the tunnel? *Infect. Control Hosp. Epidemiol.* **14:**276–278.

27. **Hoffman, P. N., J. E. Death, and D. Coates.** 1981. The stability of sodium hypochlorite solutions, p. 77–83. *In* C. H. Collins, M. C. Allwood, S. F. Blood, and A. Fox (ed.), *Disinfectants: Their Use and Evaluation of Effectiveness.* Academic Press, London.

28. **Hugo, W. B.** 1965. Some aspects on the action of cationic surface active agents on microbial cells with special reference to their mode of action on enzymes, p. 69–82. *In* S.C.I. monograph no. 19: Surface-Active Agents in Microbiology. London Society of the Chemical Industry, London.

29. **Janssen, D. W., and P. M. Schneider.** 1992. Overview of ethylene oxide alternative methodologies in the clinical setting. *J. Healthcare Material Management* **September:**31–39.

30. **Jones, D. R.** 1993. Plasma energy technology promises large reduction in medical waste volume, weight. *Med. Waste Management* **1**(5):7–8.

31. **Joslyn, L. J.** 1991. Sterilization by heat, p. 495–526. *In* S. S. Block (ed.), *Disinfection, Sterilization and Preservation.* Lea & Febiger, Philadelphia.

32. **Klein, M., and A. DeForest.** 1963. The inactivation of viruses by germicides. *Chem. Specialists Manufacturing Assoc. Proc.* **49:**116–118.

33. **Larson, E.** 1988. Guideline for use of topical antimicrobial agents. *Am. J. Infect. Control* **16:**253–266.

34. **Larson, E. L., and H. E. Morton.** 1991. Alcohols, p. 191–203. *In* S. S. Block (ed.), *Disinfection, Sterilization and Preservation.* Lea & Febiger, Philadelphia.

35. **Lauer, J. L., D. R. Battles, and D. Vesley.** 1982. Decontaminating infectious waste by autoclaving. *Appl. Environ. Microbiol.* **44:**690–694.

36. **Levy, R. V., and T. J. Leahy.** 1991. Sterilization filtration, p. 527–552. *In* S. S. Block (ed.), *Disinfection, Sterilization and Preservation.* Lea & Febiger, Philadelphia.

37. **Martin, L. S., J. S. McDougal, and S. L. Loskoski.** 1985. Disinfection and inactivation of the human T lymphotropic virus type III/lymphadenopathy-associated virus. *J. Infect. Dis.* **152:**400–403.

38. **Meehan, Z., and J. D. Ferguson.** 1991. Flash sterilization of microbiological media in a microwave sterilization system, abstr. I-102, p. 207. *Abstr. 91st Gen. Meet. Am. Soc. Microbiol. 1991.*

39. **Meehan, Z., and R. Ramaley.** 1991. Evaluation of biological indicators in a microwave sterilization system, abstr. I-103, p. 207. *Abstr. 91st Gen. Meet. Am. Soc. Microbiol. 1991.*

40. **Merianos, J. J.** 1991. Quaternary ammonium compounds, p. 225–255. *In* S. S. Block (ed.), *Disinfection, Sterilization and Preservation.* Lea & Febiger, Philadelphia.

41. **National Committee for Clinical Laboratory Standards.** 1989. *Guidelines for Laboratory Safety. Proposed Guideline GP17-P.* National Committee for Clinical Laboratory Standards, Villanova, Pa.

42. **National Committee for Clinical Laboratory Standards.** 1991. *Protection of Laboratory Workers from Infectious Disease Transmitted by Blood, Body Fluids, and Tissue,* 2nd ed. *Tentative*

Guidelines M29-T2. National Committee for Clinical Laboratory Standards, Villanova, Pa.

43. **National Committee for Clinical Laboratory Standards.** 1991. *Protection of Laboratory Workers from Instrument Biohazards. Proposed Guidelines II 7-P*. National Committee for Clinical Laboratory Standards, Villanova, Pa.

44. **National Committee for Clinical Laboratory Standards.** 1993. *Clinical Laboratory Waste Management. Approved Guideline GP5-A*. National Committee for Clinical Laboratory Standards, Villanova, Pa.

45. **Nystrom, B.** 1991. New technology for sterilization and disinfection. *Am. J. Med.* **91:**264S–266S.

46. **Occupational Safety and Health Administration.** 1991. Occupational exposure to bloodborne pathogens; final rule. *Fed. Regist.* **56:**64003–64182.

47. **Occupational Safety and Health Administration, Department of Labor.** 1990. *Occupational Exposure to Hazardous Chemicals in the Laboratories*. 29CFR 1910.1450. U.S. Government Printing Office, Washington, D.C.

48. **Parisi, A. N., and W. E. Young.** 1991. Sterilization with ethylene oxide and other gases, p. 580–595. *In* S. S. Block (ed.), *Disinfection, Sterilization and Preservation*. Lea & Febiger, Philadelphia.

49. **Reinhardt, P. A., and J. G. Gordon.** 1991. *Infectious and Medical Waste Management*. Lewis Publishers, Boca Raton, Fla.

50. **Relyveld, E. H., and S. Ben-Efraim.** 1983. Preparation of vaccines by the action of glutaraldehyde on toxins, bacteria, viruses, allergens and cells. *Methods Enzymol.* **93:**24–60.

51. **Riley, R. L., and E. A. Nardell.** 1989. Clearing the air. The theory and application of ultraviolet air disinfection. *Am. Rev. Respir. Dis.* **139:**1286–1294.

52. **Rotter, M. L.** 1984. Hygienic hand disinfection. *Infect. Control* **5:**18–22.

53. **Russell, A. D.** 1990. Bacterial spores and chemical sporicidal agents. *Clin. Microbiol. Rev.* **3:**99–119.

54. **Rutala, W.** 1990. APIC guidelines for selection and use of disinfectants. *Infect. Control* **18:**99–117.

55. **Rutala, W. A., and E. C. Cole.** 1987. Ineffectiveness of hospital disinfectants against bacteria: a collaborative study. *Infect. Control* **8:**501–506.

56. **Rutala, W. A., E. C. Cole, N. S. Wannamaker, and D. J. Weber.** 1991. Inactivation of *Mycobacterium tuberculosis* and *Mycobacteria bovis* by 14 hospital disinfectants. *Am. J. Med.* **91**(Suppl. 3B):267S–271S.

57. **Rutula, W. A., M. M. Stiegel, and F. A. Sarubbi, Jr.** 1982. Decontamination of laboratory microbiological waste by steam sterilization. *Appl. Environ. Microbiol.* **43:**1311–1316.

58. **Sarkar, G., and S. Sommer.** 1991. Parameters affecting susceptibility of PCR contamination to UV inactivation. *BioTechniques* **10:**590–594.

59. **Scherba, G., R. M. Weigel, and W. D. O'Brien, Jr.** 1991. Quantitative assessment of the germicidal efficacy of ultrasonic energy. *Appl. Environ. Microbiol.* **57:**2079–2084.

60. **Scott, E. M., and S. P. Gorman.** 1991. Glutaraldehyde, p. 377–384. *In* S. S. Block (ed.), *Disinfection, Sterilization and Preservation*. Lea & Febiger, Philadelphia.

61. **Shechmeister, I. L.** 1991. Sterilization by ultraviolet irradiation, p. 553–565. *In* S. S. Block (ed.), *Disinfection, Sterilization and Preservation*. Lea & Febiger, Philadelphia.

62. **Silverman, G. J.** 1991. Sterilization and preservation by ionizing irradiation, p. 566–579. *In* S. S. Block (ed.), *Disinfection, Sterilization and Preservation*. Lea & Febiger, Philadelphia.

63. **Stanley, G., M. Sheetz, M. Pfaller, and R. Wenzel.** 1992. Comparative efficacy of alternative handwashing agents in reducing nosocomial infections in intensive care units. *N. Engl. J. Med.* **327:**88–93.

64. **Steer, A. C., and G. G. Mallison.** 1975. Handwashing practices for the prevention of nosocomial infections. *Ann. Intern. Med.* **83:**683–690.

65. **Stinson, M. C., M. S. Galanek, A. M. Ducatman, F. X. Masse, and D. R. Kuritzkes.** 1990. Model for inactivation and disposal of infectious human immunodeficiency virus and radioactive waste in a BL3 facility. *Appl. Environ. Microbiol.* **56:**264–268.

66. **Sulkin, I. R., and R. Gershon (ed.).** 1992. Biohazards and safety, p. 14.1.1–14.4.3. *In* H. D. Isenberg (ed.), *Clinical Microbiology Procedures Handbook*, vol. 2. American Society for Microbiology, Washington, D.C.

67. **Suttar, S. A., R. A. Raphael, H. Lochnan, and V. S. Springthorpe.** 1983. Rotavirus inactivation by chemical disinfectants and antiseptics used in hospitals. *Can. J. Microbiol.* **29:**1464–1469.

68. **Terleckyi, B., and D. D. Axler.** 1987. Quantitative neutralization assay of fungicidal activity of disinfectants. *Antimicrob. Agents Chemother.* **31:**794–798.

69. **Tilley, F. W., and J. M. Schaffer.** 1926. Relation between the chemical constitution and germicidal activity of the monohydric alcohols and phenols. *J. Bacteriol.* **12:**303–309.

70. **U.S. Public Health Service.** 1974. *NIH Biohazards Safety Guide*. GPO stock no. 1740-00383. U.S. Government Printing Office, Washington, D.C.

71. **Vela, G. R., and J. F. Wu.** 1979. Mechanism of lethal action of 2,450-MHz radiation on microorganisms. *Appl. Environ. Microbiol.* **37:**550–553.

DIAGNOSTIC TECHNOLOGIES IN CLINICAL MICROBIOLOGY

VOLUME EDITOR
FRED C. TENOVER

SECTION EDITOR
GARY V. DOERN

Introduction

GARY V. DOERN

The past 2 decades have witnessed an explosion of new technology in the clinical microbiology laboratory. As late as the early 1970s, definitive laboratory diagnoses of infectious diseases were largely accomplished only through the use of cumbersome, costly, time-consuming, often subjective techniques, most of which necessitated a highly trained and experienced technologist work force. Microorganisms were visualized directly in stained clinical material via microscopy. Alternatively or in addition, organisms were cultured by using 100-year-old methods on solid or liquid media. Organisms recovered in culture were identified by laborious, conventional, growth-based biochemical methods, and when necessary, antimicrobial susceptibility tests were nearly always performed by a manual overnight disk diffusion procedure. In selected cases but usually as an adjunct to microscopy and culture, relatively crude techniques such as agglutination, complement fixation, or neutralization were used to examine patient sera for the presence of antibody reactive either with whole cells or with defined microbial antigens. Certain levels of antibody or increasing titers with time were taken as indicating the cause of infection. In general, the practice of clinical microbiology was labor intensive, results of tests were not available for several days, the discipline was largely predicated on servicing a hospitalized patient population, and little process control existed.

The technologic changes that have taken place during the past 20 years in the clinical microbiology laboratory exceed by several orders of magnitude everything that has happened during the first 100 years of this discipline's existence. The physical structure of laboratories, staffing patterns, work flow, and turnaround time have all been profoundly influenced by technologic advances. These changes will continue, leading inevitably in the future to more decentralized testing, for example, point-of-care procedures and perhaps even in-home testing.

Five specific areas have witnessed the greatest change. Bacterial identification is now achieved largely through use of miniaturized substrate utilization systems, many of which are instrumented. Antimicrobial susceptibility testing is most commonly accomplished with a commercial broth microdilution procedure or by some form of instrumentation. Bacteremia and fungemia are now often detected by a commercial instrument-assisted blood culture system. Three such systems currently exist, two predicated on con-

tinuous monitoring of evolved CO_2 in blood culture media and the third based on repeated measurements of pressure in the headspaces of blood culture bottles. These automated systems followed lysis-centrifugation and slide blood cultures as recent innovations in the areas of blood culture microbiology. A fourth area of major technologic change in clinical microbiology is the advent of new immunodiagnostic methods for both antigen and antibody detection. First radioimmunoassay and immunofluorescence and then enzyme immunoassay, latex agglutination, and immunoblotting have emerged as important diagnostic tools. The field of immunodiagnosis has benefitted immensely from the development of monoclonal antibody technology. The advantages of monoclonal antibodies are numerous and include enhanced specificity, consistency of reagents, and inexpensive continuous sources of materials. Finally and most recently, nucleic acid probe technology has assumed an increasingly important role in the clinical microbiology laboratory both for detecting and for characterizing microorganisms. With respect to direct detection of organisms in clinical material, probe techniques initially appeared to be of limited utility owing to poor sensitivity. However, recent advances in both signal and target amplification have made possible an exciting new era in clinical microbiology, the age of molecular diagnostics.

The intent of this section of the 6th edition of the *Manual of Clinical Microbiology* is to present a comprehensive description of five diagnostic technologies: biochemical systems for the identification of bacteria and yeasts, immunoassays for use in detecting both antigens and antibodies, gas-liquid chromatography and high-pressure liquid chromatography for identifying microorganisms as well as detecting them directly in clinical specimens, molecular diagnostic methods, and cell culture systems. An effort has been made to present the fundamental principles and functional attributes of each of these technologies. The reader may anticipate obtaining an understanding of how different techniques work, what their benefits and shortcomings are, and what issues must be considered before these technologies are implemented in the laboratory. In addition, numerous examples of practical applications are provided. Many of these techniques are mentioned in subsequent chapters of this Manual but without further discussion. It is hoped that by mastering the basic descriptive information included in this section, the reader will have a clear perception of how and why these technologies are becoming more commonly utilized in today's diagnostic laboratory.

Most of the recent technologic advances in clinical microbiology share two key features. Results of tests can be defined using objective end points, often as a product of instrumentation, and test results can be obtained far more rapidly than was possible with traditional techniques. As a direct consequence of the first observation, staffing requirements in the clinical microbiology laboratory have changed dramatically. Fewer people are required, and as the subjective, interpretive nature of the work has diminished, the educational and experience requirements of workers have decreased accordingly. Furthermore, faster turnaround time has created an opportunity for enhanced clinical impact. However, this has not been achieved without added cost. Almost without exception, newer technologies cost more than their traditional, less timely counterparts. It is ironic that these newer, more costly technologies should be emerging at precisely the same time the health care industry in the United States has been confronted with the challenge of cost control to an extent that never existed in the past. Increasingly, the laboratorian is faced with the dilemma, What price technology?

When considering implementation of a new technology, numerous issues need to be addressed. Arguably the most important is how well the new procedure accomplishes its stated objective. Addressing this issue involves comparing new procedures with existing procedures. Two specific questions arise. How are the performance characteristics of a new procedure best determined, and who performs the

evaluation? Answering the first question increasingly necessitates reconsideration of existing notions as to what constitutes minimum "clinical" performance. The second question is even more perplexing. Simple perusal of representative journals in the field makes it apparent that comprehensive, well-controlled clinical laboratory evaluations of new technologies that used to be so common are not being performed today with the same frequency. There are several possible explanations for this. The sheer number of new technologies and applications precludes careful evaluations of each technique, primarily because of fiscal cutbacks in clinical laboratories that make such investigations too expensive to perform. Industry support for independent studies has also diminished. The studies themselves are frequently tedious to perform, and individuals in positions of responsibility in clinical laboratories have less time to devote to such studies. The end result is that the laboratorian confronted with a new technology is often totally reliant on manufacturer's performance claims, which in turn are usually predicated on manufacturer-organized investigations. This reality has given the Food and Drug Administration process of test system approval new importance, especially when this process is viewed in the light of another reality: most new ideas in clinical microbiology technology today emanate from industry, not the end users.

It is more important now than ever before that laboratorians have a clear understanding of the new and exciting technologies that confront them. The five chapters in this section are intended to provide that understanding.

Substrate Utilization Systems for the Identification of Bacteria and Yeasts

J. MICHAEL MILLER AND CAROLINE M. O'HARA

10

Gaining and maintaining expertise in identifying clinically important microbes has been the goal of clinical laboratorians for decades. From the early years of diagnostic methods in microbiology to the 1960s, when advances in microbial identification began to emerge, skill in interpretive judgment and the use of tubed and plated media were the basis of identification. Organisms were identified by what we now refer to as conventional procedures, which include reactions in tubed media, observation of physical characteristics such as colony morphology and odor, and results of Gram stain, agglutination tests, and susceptibility profiles. These conventional procedures eventually defined the genera and species of bacteria and yeasts and became the reference method by which we confirm isolates.

The next step in the evolution of identification methods simply miniaturized commonly used biochemical reactions into a more convenient format (4). Later, a systems approach became the industry standard, and this approach is still the basis of most currently used substrate profile systems. In a system-dependent methodology, a set of substrates is carefully selected to allow a positive- and negative-reaction pattern to emerge, creating a metabolic profile to be compared with an established database profile. In many systems, different sets of substrates are necessary to identify rapidly growing members of the family *Enterobacteriaceae*, slower-growing gram-negative non-*Enterobacteriaceae*, gram-positive cocci, gram-negative cocci, and anaerobes. Yeasts require yet another profile set.

Biochemical profiles are determined by the reactions of individual organisms with each of the substrates in the system. The accuracy of the reactions depends on the user following the manufacturers' directions regarding inoculum preparation, inoculum density, incubation times, and test interpretation. Most systems rely on pH changes resulting from utilization of substrate, enzymatic reactions that allow release of a chromogenic or fluorogenic compound, tetrazolium-based indicators of metabolic activity in the presence of a variety of carbon sources, detection of volatile or nonvolatile acids, or recognition of visible growth, as follows.

pH-based reactions (many 15- to 24-h systems). As a general rule, carbohydrate utilization by microorganisms results in an acid pH in the test medium, while protein utilization or the release of nitrogen-containing products results in an alkaline pH. A change in pH from alkalinity or neutrality to acid usually indicates a positive reaction, although negative responses are also key factors in identification schemes.

Enzyme profiles (many 4-h systems). Enzyme profile tests are usually based on preformed enzymes and require minimal microbial growth. When a colorless chromogen or fluorogen complex is hydrolyzed by an appropriate enzyme, the chromogen or fluorogen is released, providing a method of detecting the reaction either visually or automatically.

Carbon source utilization. Rather than detect metabolic by-products, this method simply measures metabolic activity. Tetrazolium-labeled carbon sources are colorless until electrons activated by metabolic activity are transferred to the dye, creating a purple color detectable visually or with instruments.

Volatile or nonvolatile acid detection. The end products of metabolism in broth cultures of organisms are analyzed by gas-liquid chromatography, and the data are analyzed by computer and compared to database libraries to arrive at a result.

Visual detection of growth (yeast identification systems). Assimilation assays often depend on the ability of bacteria or yeasts to grow in the presence of a substrate. Visual detection of growth is a positive result.

Additional tests for microbial identification that use other means of detecting a positive response for a given substrate may also be included.

Although no formal definition of "rapid" exists for describing the time required for results to be generated, most laboratorians expect rapid systems to provide usable results within 2 to 4 h of incubation. Clearly, the generation times of microbes (usually 30 min or longer) will not allow growth-dependent methods to generate detectable biochemical responses within this time. To overcome the problem of generation times, manufacturers of rapid systems utilize novel substrates on which preformed enzymes, produced by the organisms to be tested, may react to elicit responses detectable within 2 to 4 h.

CRITERIA FOR SELECTING A SYSTEM

The laboratorian must consider several issues when selecting a system to be used in the laboratory, especially when the equipment will likely provide both organism identification and antimicrobial susceptibility test results. Because the cost of some automated instruments may exceed $100,000, purchase of these identification systems represents a significant capital expenditure and a long-term commitment to that technology. Supervisors and managers in the laboratory should make such major decisions carefully and with expert consultation. Before the first technical representative is seen in the laboratory, the following important questions must be answered.

1. Why do I need a new system? Can I justify it as a benefit to the laboratory or the hospital?
2. Will management support the need for a new system? Is funding available?
3. Do I truly need a "rapid" system? Will earlier results actually get to physicians or patient records and have an impact on patient care?
4. What will I base my final decision on? What questions do I need to have answered about the system?
5. Should I buy the system outright or negotiate for a reagent rental contract?

Once these questions are answered, the next step is to begin the search for the right instrument or system to meet the needs of the laboratory and the medical staff. As a general rule, it is best not to be the first to purchase a new system without having seen in the peer-reviewed literature the results of evaluations performed by reputable clinical laboratories. If microbiology journals are unavailable, ask the representative to supply you with articles about the ability of the system to correctly identify the range of isolates usually seen in your laboratory. During conversations and demonstrations, make sure the following questions are answered.

1. Do I like the overall quality of the system?
2. How accurate is the identification? How accurate are susceptibility test results? (Get documentation to prove the answers to these questions.)
3. What is the turnaround time for a complete test?
4. How much technologist time is required for test setup? for test completion? for quality control?
5. Is the system expandable?
6. How much training time is required or suggested by the manufacturer?
7. What service contracts are recommended? What is their cost?
8. How much bench space is required? Are special electrical outlets or communication lines required?
9. What is the cost per test? What is the cost for quality control testing?
10. What is the shelf life of the test kits? What are their storage requirements, and do I have space?
11. Is the instrument protected from brownout?
12. How are software updates handled? Are they free?
13. Can I trade in my current system?
14. Is the system usable by all shifts?
15. Are epidemiology programs available with the software? Can the pharmacy be linked to the system?
16. Are the printed reports usable?
17. Can the system be interfaced with our current laboratory computer system? one- or two-way interface?
18. Who else in this geographic area uses this system?
19. Do I trust this system for my needs?

Visit other laboratories similar to yours that are now using the system, and talk with the users. Ask if they like the system, if they would buy it again, how much downtime they have experienced, and if the service from the manufacturer has been acceptable. Discuss the issues of serviceability and the mechanical reliability of the test system.

Select a system that has been fully evaluated and whose accuracy exceeds 90% in its overall ability to identify common and uncommon bacteria normally seen in your hospital or laboratory. Commonly isolated organisms should be identified with at least 95% of the accuracy of conventional methods. Do not place unwarranted expectations on the system being considered. Some systems may be unable to accurately identify the more fastidious isolates even though they are listed in the manufacturer's data base.

The accuracy of susceptibility testing for combination panels is as important as the accuracy of identification, perhaps more so. Because of the complexities of drug-microbe interactions and the novel resistance mechanisms that appear to be emerging, consultation may be necessary in order to be sure of the accuracy of susceptibility testing by a system. Section X of this book discusses the issues involved in susceptibility testing.

EVALUATING AN INSTRUMENT OR SYSTEM

Anytime an identification system is added to the laboratory, it must be documented that the system performs as described by the manufacturer. The first evidence of acceptable performance should be found in published reports by other laboratories that have evaluated the system in a sound, scientific manner (6). Read these reports carefully, looking for data that support the conclusions of the paper. Simply reading and accepting the abstract from these published studies may be misleading, especially if the study protocol was poorly conceived (5). Microbiologists who evaluate instruments against a "gold standard" or compare one system's performance to that of another must be precise in experimental design, or the resulting data interpretation could be misleading (3, 5).

The next evidence of acceptable performance by a new identification instrument should be in-laboratory verification of performance by the purchasing laboratory. The regulations implementing the Clinical Laboratory Improvement Amendments of 1988 (CLIA '88) specify that for systems placed into service after 1 September 1992,

> Prior to reporting patient test results, the laboratory must verify or establish, for each method, the performance specifications for the following performance characteristics: accuracy; precision; analytical sensitivity and specificity, if applicable; . . . . (3a)

Because identification systems provide qualitative information, only accuracy should be addressed. However, MIC tests need to have accuracy and precision measured. The CLIA regulations do not specify how accuracy is to be verified. Each laboratory is responsible for devising its own verification protocol, and verification must be done whether the laboratory is introducing a new system for the first time or is replacing an old system with a new one.

Clearly, smaller laboratories will have fewer resources than larger laboratories have with which to verify an iden-

TABLE 1 Summary of identification systems currently available[a]

System	Manufacturer	Organisms identified	Storage temp (°C)	No. of tests	Incubation	Automated
ANI	bioMérieux Vitek	Anaerobes	2–8	28	4 h; aerobic	Yes
API 20A	bioMérieux Vitek	Anaerobes	2–8	21	24 h; anaerobic	No
API 20C	bioMérieux Vitek	Yeasts	2–8	20	72 h	No
API 20E	bioMérieux Vitek	Enterobacteriaceae and nonfermenting gram-negative bacteria	2–8	21	24–48 h	No
API 20 Strep	bioMérieux Vitek	Streptococci and enterococci	2–8	20	4–24 h	No
API An-IDENT	bioMérieux Vitek	Anaerobes	2–8	21	4 h; aerobic	No
API Coryne	bioMérieux Vitek	Corynebacteria	2–8	20	24 h	No
API NFT (Rapid NFT)	bioMérieux Vitek	Gram-negative non-Enterobacteriaceae	2–8	20	24–48 h	No
API Rapid 20E	bioMérieux Vitek	Enterobacteriaceae	2–8	21	4 h	No
API Staph-IDENT	bioMérieux Vitek	Staphylococci and micrococci	2–8	10	5 h	No
API Staph (STAPH-Trac)	bioMérieux Vitek	Staphylococci and micrococci	2–8	20	24 h	No
Bacterial Identification Panel	Alamar	Enterobacteriaceae, some gram-negative nonfermenters	RT	26	18–20 h	Reader only
Crystal E/NF	BDMS	Enterobacteriaceae, some gram-negative nonfermenters	2–8	30	18–20 h	No
Crystal Rapid Stool/Enteric	BDMS	Gram-negative stool pathogens	2–8	30	18–20 h	No
Enterotube II	BDMS	Enterobacteriaceae	2–8	15	18–24 h	No
EPS (Enteric Pathogen Screen)	bioMérieux Vitek	Edwardsiella, Salmonella, Shigella, and Yersinia	2–8	10	4–8 h	Yes
ES MicroPlate	Biolog	Escherichia coli K-12, Salmonella typhimurium LT2	2–8	95	4–24 h	Reader only
Fox Dual GNI	Micro-Media Systems	Enteric and nonenteric gram-negative bacteria	−20–−40	33	18–24 h	No
GN Microplate	Biolog	Aerobic gram-negative bacteria	2–8	95	4–24 h	Reader only
GNI	bioMérieux Vitek	Enterobacteriaceae and other nonfermenting bacteria	2–8	29	4–13 h	Yes
GP Microplate	Biolog	Most gram-positive cocci and bacilli	2–8	95	4–24 h	Reader only
GPI	bioMérieux Vitek	Gram-positive cocci and bacilli	2–8–20	29	4–15 h	Yes
ID Tri-Panel	Difco/Pasco	Gram-negative and gram-positive bacteria	2–8–20	30	16–20 h; 40–44 h	Reader only
Micro-ID	Organon Teknika	Enterobacteriaceae	2–8	15	4 h	No
Minitek	BDMS	Anaerobes, Enterobacteriaceae, gram-positive organisms, Neisseria spp., nonfermenters, and yeasts	2–25	4–21, depending on need	4 h for Enterobacteriaceae and Neisseria spp. to 72 h for yeasts	No

(Continued on next page)

TABLE 1 Summary of identification systems currently available[a] *(Continued)*

System	Manufacturer	Organisms identified	Storage temp (°C)	No. of tests	Incubation	Automated
NEG ID Type 2	Baxter Diagnostics	*Enterobacteriaceae* and other fermenting and nonfermenting bacteria	2–30	34	15–42 h	Yes
Neisseria Enzyme Test	Carr-Scarborough	*Neisseria* and *Moraxella* spp.	2–8	3	30 min	No
NHI	bioMérieux Vitek	*Neisseria* and *Haemophilus* spp.	2–8	15	4 h	Yes
Oxi/Ferm	BDMS	Gram-negative, oxidase-positive fermenters and nonfermenters	2–8	9	24–48 h	No
Pos ID	Baxter Diagnostics	Gram-positive cocci and *Listeria* spp.	2–30	27	18–48 h	Yes
quadFERM +	bioMérieux Vitek	*Neisseria* and *Branhamella* spp.	2–8	4	2 h	No
RapID ANA II	IDS	Anaerobes	2–8	18	4–6 h; aerobic	No
Rapid Anaerobe	Baxter Diagnostics	Anaerobes	RT	24	4 h; aerobic	Yes
RAPIDEC STAPH	bioMérieux Vitek	Staphylococci	2–8	4	2 h	No
Rapid NEG ID 2	Baxter Diagnostics	*Enterobacteriaceae* and other fermenting and nonfermenting bacteria	2–8	36	2 h	Yes
RapID NF Plus	IDS	Nonfermenting gram-negative bacteria	2–8	17	4 h	No
RapID NH	IDS	Members of family *Neisseriaceae*, *Haemophilus* spp., and other gram-negative bacteria	2–8	13	4 h; 1 h for gonococci	No
RapID onE	IDS	*Enterobacteriaceae* and other oxidase-negative bacteria	2–8	19	4 h	No
Rapid POS ID	Baxter Diagnostics	Gram-positive cocci and *Listeria* spp.	2–8	34	2 h	Yes
RapID SS/u	IDS	Common urinary tract pathogens	2–8	11	2 h	No
RapID STR	IDS	Streptococci	2–8	14	4 h	No
Rapid Yeast ID	Baxter Diagnostics	Yeasts	RT	27	4 h	Yes
r/b Enteric Differential System	Remel	*Enterobacteriaceae*	2–8	15	18–24 h	No
Sensititre AP 80	Radiometer America	*Enterobacteriaceae* and nonfermenting gram-negative bacteria	RT	32	5–18 h	Reader only
Sensititre Microbact	Radiometer America	*Enterobacteriaceae* and selected gram-negative bacteria	2–8	12 or 24, depending on need	24–48 h	No
UID-3/UID-1	bioMérieux Vitek	Urinary tract pathogens directly from urine	2–8	9	1–13 h	Yes
Uni-N/F-Tek	Remel	Gram-negative fermenting and nonfermenting bacteria	2–8	18	24–48 h	No
UniScept 20E	bioMérieux Vitek	*Enterobacteriaceae* and nonfermenting gram-negative bacteria	2–8	20	24–48 h	Yes
UniScept 20GP	bioMérieux Vitek	Staphylococci, group D enterococci, and nonenterococci	2–8	20	18–24 h	Yes
Uni-Yeast Tek	Remel	Yeasts	2–8	13	24 h–6 days	No
YBC	bioMérieux Vitek	Yeasts	2–8	26	24–48 h	Yes

[a]Abbreviations: BDMS, Becton Dickinson Microbiology Systems; IDS, Innovative Diagnostic Systems; RT, room temperature.

TABLE 2 Comparison of features of automated identification systems[a]

Feature	Vitek Jr.	Vitek	autoSCAN-4	WalkAway-40	WalkAway-96	Sensititre	UniScept	Biolog	MIS	AutoSceptor
Capacity of system	30	120	Unlimited	40	96	Unlimited	Unlimited	Unlimited	60	15
No. of ID (no. of substrates tested)[b]										
GN	85 (29)	85 (29)	116 (35)	116 (35,[c] 36[d])	116 (35,[c] 36[d])	140 (32)	105 (20)	259 (95)	300 (NA[e])	78 (24)
GP	50 (29)	50 (29)	48 (29)	48 (29,[c] 34[d])	48 (29,[c] 34[d])	No	18 (20)	126 (95)	150 (NA)	No
ANA	No	No	58 (24)	58 (24)	58 (24)	No	83 (20)	83 (20)	600 (NA)	No
FAS	No	No	21 (18)	21 (18)	21 (18)	No	No	49 (95)	40 (NA)	No
ENV	No	No	No	No	No	No	No	360 (95)	200 (NA)	No
Yeasts	36 (26)	36 (26)	40 (27)	40 (27)	40 (27)	No	No	No	194 (NA)	No
MYCO	No	No	No	No	No	No	No	No	28 (NA)	No
Inoculation	Automated	Automated	Manual	Manual	Manual	Automated	Automated	Manual	Automated	Automated
Incubation	On-line	On-line	Off-line	On-line	On-line	On-line	Off-line	Off-line	On-line	Off-line
GN	4–18 h	4–18 h	24 or 48 h	2 or 15–42 h	2 or 15–42 h	5 or 18 h	18–24 h	4 or 24 h	30 min	18–24 h
GP	4–15 h	4–15 h	24 or 48 h	2 or 15–42 h	2 or 15–42 h		18–24 h	4 or 24 h	30 min	
ANA			4 h	4 h	4 h		4 h		30 min	
ENV								4 or 24 h	30 min	
Yeast	24 or 48 h	24 or 48 h	4 h	4 h	4 h				30 min	
MYCO									30 min	
Manual reagent addition	No	No	Yes	No	No	No	Yes	No	No	Yes
Additional tests required before incubation	Yes	Yes	Yes	Yes	Yes	No	Yes	No	No	Yes
Storage temp	4°C	4°C	RT[c,f] 4°C[g]	RT[c,f] 4°C[g]	RT[c,f] 4°C[g]	RT	4°C	4°C	RT	RT
Susceptibility testing	Yes	Yes	Yes	Yes	Yes	Yes	Yes	No	No	Yes
Urine screen or identification	Yes	Yes	No	No	No	No	No	No	No	No
DMS[h]	Yes	Yes	Yes	Yes	Yes	Yes	Yes	Yes	Yes	Yes
Computer interface	Yes	Yes	Yes	Yes	Yes	Yes	Yes	No	Yes	Yes
List price	$32,500	$70,500	$45,900	$80,000	$120,000	$42,500	$57,000	$17,950	$45,000	$39,261

[a]Modified from Stager and Davis (6) with permission.
[b]No. of ID, number of groups, genera, or species identified; GN, aerobic gram-negative bacilli; GP, aerobic gram-positive bacilli; ANA, anaerobic bacilli; FAS, fastidious bacteria; ENV, environmentally isolated bacteria; MYCO, mycobacteria.
[c]Conventional identification panel.
[d]Fluorogenic identification panel.
[e]NA, not applicable.
[f]RT, room temperature.
[g]All rapid identification panels.
[h]DMS, Data Management System.

tification system. Laboratory size, however, must have no bearing on the identification accuracy of laboratory methods or of the patient care provided by a laboratory. It is unnecessary that every laboratory reverify what has already been done by the manufacturer and by other laboratories that have published data on the accuracy of a system. A true establishment or verification of accuracy requires exhaustive testing of hundreds of strains. The role of verification by the purchasing laboratory should be to ensure that personnel using the system can make it perform at the levels of accuracy already documented by the manufacturer and published in the literature.

The laboratorian should expect a level of 95% agreement with the existing system or reference method and should accept, in the final analysis, no less than 90% agreement. This takes into account the fact that the new system may be more accurate than the old one. Try to keep the total cost of verification within a range of $250 to $1,000 depending on the size of the laboratory. In fact, one may stipulate to the manufacturer that the final purchase of a system depends on successful verification of the system's accuracy and ask that the manufacturer assist in the process by providing stock strains for that purpose. Verification protocols for identification systems will vary but may be structured around one of the following suggestions.

1. Test the quality control organisms PLUS achieve >90.0% agreement for 1 week of consecutive parallel testing (a minimum of 50 strains) with the existing method. Discrepancies must be arbitrated by a reference laboratory.

2. Test the quality control organisms PLUS two or three known reference strains (stock cultures) of each commonly isolated organism in up to 50 tests (small laboratories) or 100 tests (large laboratories).

3. Test the quality control organisms PLUS make sure that 20 to 50 organism identifications (12 to 15 different species) agree in concurrent testing with the current method or with the results of reference laboratory testing of split samples.

The Food and Drug Administration (FDA) must now approve a manufacturer-submitted quality control protocol. This clearance is unrelated to the system's premarket or 510(K) approval for commercial distribution. Systems not cleared by FDA or protocols even slightly modified from the manufacturer's recommendations must meet additional rigorous standards that establish the equivalency of the altered or uncleared method.

System Construction

Microbial identification systems are considered either manual or automated. Manual methods offer the advantages of using the analytic skills of the technologists in reading and interpreting the tests, whereas automated systems offer a hands-off approach allowing more technologist time for other duties. For all systems, the backbone of accuracy is the strength and utility of the database. Databases are constructed by using known, clinically relevant strains, and they include the type strains of most taxa in the database. In some cases, before an organism is added to the database, it is evaluated by cluster analysis to confirm its relationship to other strains in the same taxa.

The number of species included in a database may vary from around 200 species to as many as 1,200 species if clinical, environmental, and research parameters are a part of the system. For most commercial systems, database main-

tenance is a continuous process, and software upgrades incorporating major taxonomic changes are provided by the manufacturer at intervals of up to every 4 years. Some systems may allow the user to make minor changes at their local workstation.

System identifications are supported by algorithm-based decision making either through a computer or through a preprinted profile index. Bayes's theorem, or a modification of it, is often the basis of algorithm construction from data matrices.

Bayes's theorem is one of the statistical methods used by manufacturers to arrive at a certain taxon based on the reaction profile produced by the unknown clinical isolate. In this regard, Bayes's theorem allows us to consider two important issues in order to arrive at an accurate conclusion: (i) $P(t_i|R)$ is the probability that an organism exhibiting test pattern R belongs to taxon t, and (ii) $P(R|t_i)$ is the probability that members of taxon t will exhibit test pattern R. Prior to testing or observation, we make the assumptions that an unknown isolate has an equal chance of being any taxon and that each test used to identify the isolate is independent of all other tests. In this case, Bayes's theorem can be written as

$$P(t_i|R) = \frac{P(R|t_i)}{\Sigma_i P(R|t_i)}$$

By observing reference identification charts, we know the expected pattern of the population of taxon t_i (e.g., *Escherichia coli*). R is the test pattern composed of R_1, R_2, R_n, where R_1 is the result for test 1, R_2 is the result for test 2, etc., for a given taxon. We can then incorporate the percentages (likelihoods that t_i will exhibit R_1, etc.) into Bayes's theorem to arrive at an accurate taxon.

Clinical microbiologists must not, however, become dependent on these likelihoods and percentages when interpretive judgment suggests an alternative taxonomic conclusion. Bacteria isolated from clinical specimens often tend to stretch the rules of nomenclature, and they may not react as expected in a commercial system even though a legitimate result is produced. The result from the most reliable system can be wrong. In these cases, a backup method of identification must be used.

D'Amato et al. have described how the systems utilize database profiles and probability matrices to arrive at the identification of an unknown taxon (1, 2).

Commercial manufacturers of identification systems rely heavily on input from their clients and customers. Laboratorians are encouraged to communicate with the product manufacturer about problems and observations. Manufacturers depend on customer satisfaction, and most are willing to assist in problem solving or in projects that could add strength to their system. These companies, like their users, are clearly interested in the highest quality of cost-effective patient care. Tables 1 and 2 provide a summary of available identification systems and compare salient features offered by the automated and nonautomated methods.

REFERENCES

1. **D'Amato, R. F., B. Holmes, and E. J. Bottone.** 1981. The systems approach to diagnostic microbiology. *Crit. Rev. Microbiol.* **9:**1–44.
2. **D'Amato, R. F., E. J. Bottone, and D. Amsterdam.** 1991. Substrate profile systems for the identification of bacteria and

yeasts by rapid and automated approaches, p. 128–136. *In* A. Balows, W. J. Hausler, Jr., K. L. Herrmann, H. D. Isenberg, and H. J. Shadomy (ed.), *Manual of Clinical Microbiology*, 5th ed. American Society for Microbiology, Washington, D.C.

3. **Edberg, S. C., and L. S. Konowe.** 1982. A systematic means to conduct a microbiology evaluation, p. 268–299. *In* V. Lorian (ed.), *Significance of Medical Microbiology in the Care of Patients*, 2nd ed. The Williams & Wilkins Co., Baltimore.

3a. **Federal Register.** 1992. Clinical Laboratory Improvement Amendments of 1988; final rule. *Fed. Regist.* **57:**7164.

4. **Hartman, P. A.** 1968. *Miniaturized Microbiological Methods.* Academic Press, Inc., New York.

5. **Miller, J. M.** 1991. Guest commentary: evaluating biochemical identification systems. *J. Clin. Microbiol.* **29:**1559–1561.

6. **Stager, C. E., and J. R. Davis.** 1992. Automated systems for identification of microorganisms. *Clin. Microbiol. Rev.* **5:**302–327.

Immunoassays for the Diagnosis of Infectious Diseases

JOHN E. HERRMANN

11

In the past, the traditional approaches for the investigation of infectious diseases included culture techniques, serologic tests, and biochemical assays. Serodiagnosis was initially accomplished by using precipitin, agglutination, and complement fixation tests, but these techniques were labor intensive and relatively insensitive and nonspecific. Immunoassay methods came into wider usage first with the development of immunofluorescence and radioimmunoassay (RIA) techniques and then with enzyme immunoassays (EIAs). Immunofluorescence is still used in a variety of applications, but EIA techniques have largely replaced RIA, particularly for routine diagnostic applications.

Diagnosis of infectious diseases by immunoassays involves two general approaches: (i) testing for specific microbial antigens or (ii) testing for microbial-antigen-specific antibodies. Tests for antigen may be designed for direct detection of the antigen in a clinical specimen or for identification of a given agent after it has been cultivated.

Tests for antibodies may be designed to detect any antibody response regardless of immunoglobulin class, or they may be specific for immunoglobulin G (IgG) or IgM antibodies. IgM antibodies appear earlier during infection than IgG antibodies, usually peaking 7 to 10 days after infection, and persist for only a few weeks. IgG antibodies appear later, generally peaking 4 to 6 weeks after infection, and persist longer. Because the IgM antibody response is transient, the presence of this antibody indicates recent infection, and determination of an antigen-specific IgM response (and thus diagnosis) can be made with a single serum specimen. However, IgM responses do not occur solely during primary infections and may persist for long periods in some individuals. Persistence of IgM may also depend on the infectious agent. Thus, the suitability of IgM antibody responses as indications of recent infection needs to be evaluated for each infectious agent. For immunodiagnostic tests measuring total antibody, paired sera are required, one taken during the acute phase of the disease (preferably within 7 days of onset of symptoms) and one taken during convalescence (3 to 4 weeks later). A fourfold or greater increase in antibody titer is considered a positive test.

Serum IgA antibodies vary greatly as to when they first appear and how long they persist and are not routinely used for immunodiagnosis. Serum IgE may be elevated in parasitic infections but is not specifically associated with other infectious agents. The significance and diagnostic value of IgD antibodies in infections have not been determined.

The value of a particular diagnostic test can be measured in terms of sensitivity, specificity, and predictive value. Sensitivity of a test refers to the percentage of individuals with a specific disease who yield a positive test result. A negative result in a person who has the disease is referred to as a false negative. Specificity of an immunoassay refers to the percentage of persons who do not have a disease who yield a negative test result. A positive test in a person negative for disease is referred to as a false positive.

The term sensitivity applied to immunodiagnostic tests may also refer to the minimum amount of a particular antigen or antibody that can be detected by a given test. The term specificity, particularly in regard to antibodies used in an immunoassay, may also refer to the ability of the assay to distinguish the target antigen from other antigens. The context in which each term is discussed usually indicates which meaning is intended.

The predictive value of an immunodiagnostic test indicates the probability that the test will correctly indicate the presence or absence of the disease state. The predictive value of a positive test is the proportion of those who have the disease (true positives) among those who gave positive test results. The predictive value of a negative test is the proportion of those who do not have the disease (true negatives) among those who gave negative test results. The methods of calculating sensitivity, specificity, and predictive values are shown in Fig. 1.

The determination of a given individual as a true positive or a true negative for evaluating immunoassay performance may be based on clinical criteria alone or, more often, on the results of one or more other reference diagnostic tests. The standard by which other tests are measured is sometimes referred to as the "gold standard."

Sensitivity, specificity, and positive and negative predictive values are given as percentages. However, unless a particular test is both 100% sensitive and 100% specific, the suitability of the test for diagnosis is subjective. No single immunoassay has 100% sensitivity and specificity. Therefore, subjective considerations such as the benefits of rapid testing, ease of use, and cost have to be weighed against decreased diagnostic performance and the possibility of a lower quality of patient care.

$$\% \text{ Sensitivity} = \frac{\text{true positives}}{\text{true positives} + \text{false negatives}} \times 100$$

$$\% \text{ Specificity} = \frac{\text{true negatives}}{\text{false positives} + \text{true negatives}} \times 100$$

$$\% \text{ Positive predictive value} = \frac{\text{true positives}}{\text{true positives} + \text{false positives}} \times 100$$

$$\% \text{ Negative predictive value} = \frac{\text{true negatives}}{\text{true negatives} + \text{false negatives}} \times 100$$

FIGURE 1 Evaluating immunodiagnostic tests

IMMUNOASSAY FORMATS

EIA and RIA

Solid-phase EIAs evolved as a result of the findings by Nakane and Pierce (60) that antibodies could be labeled with enzymes for use in histochemical staining procedures and by Catt and Tregear (12), who described solid-phase RIA. The substitution of enzymes for radioactive labels in the solid-phase RIA resulted in a solid-phase EIA for IgG detection (21). The authors of that study, Engvall and Perlmann, coined the term enzyme-linked immunosorbent assay (ELISA) for solid-phase EIAs.

Early solid-phase EIAs were not as sensitive as the corresponding RIAs, but improvements in enzyme-labeling techniques have made the two types of assays comparable for detecting a number of antigens and antibodies. In many systems when RIA and EIA have been directly compared, there is little difference in the sensitivities or specificities of the two assays.

The advantages of EIAs versus RIAs are mainly convenience in use, in that the labeled immunoreagents used in EIAs are stable for long periods and the precautions and disposal procedures required for radioisotopes are unnecessary. In addition, the use of chromogenic substrates for the enzyme labels permits visual interpretation of test results in some cases. The only real disadvantages of EIAs are the loss of antibody reactivity that may result from conjugation to enzymes and the limits of substrate detection. For example, use of enzymes that have molecular weights higher than that of IgG molecules such as β-D-galactosidase (540,000) can cause steric hindrance of antibody activity (35). With regard to the limits of substrate detection, enzyme detection has been improved by the use of fluorogenic, luminescent, or radioactive substrates (78). This will be discussed below.

Solid-Phase Assays

The solid-phase or heterogeneous EIA requires immobilization of antigens or antibodies on a solid surface as a means of separating antigen-antibody complexes. Solid-phase surfaces used to immobilize antigens or antibodies have for most applications been polystyrene beads, tubes, and the wells of microtiter plates, including those of polyvinyl chloride. Proteins are usually coupled to these surfaces by passive adsorption. Adsorption of antigens or antibodies to nitrocellulose membranes was adapted to detection of viruses by EIA (7). Covalent linkage of antibodies or antigens to a variety of surfaces, including porous glass (50), nylon (30), cellulose (22), and agarose beads (69), has also been described.

The majority of solid-phase EIAs of diagnostic utility use plastic microtiter plates or beads, with antigen or antibody passively adsorbed to the solid phase. When a given antigen does not readily attach, then antibody to the antigen may be applied for antigen capture. For some antigens, nonspecific adsorbents such as poly-L-lysine have been used to enhance antigen adsorption (34) or the C1q component of complement has been used to capture antigen-antibody complexes (78). A number of different formats of ELISAs for antigens and antibodies are possible. These are illustrated in Fig. 2A and B, 3A and B, and 4. Most assays are of the noncompetitive type, although a number of competitive assays have been described. Examples of competitive and noncompetitive assays for antibody detection are shown in Fig. 3A and B. The disadvantage of many competitive assays when applied to antigen detection is that the labeled antigen used is usually more difficult to prepare than labeled antibody. With the advent of monoclonal antibodies (discussed below), competitive assays for specific antibodies are becoming more common. The same formats would apply to RIAs as well. For RIAs, radioactive labels are used instead of enzymes, and specific binding is determined with a gamma- or beta-ray counter, depending on the isotope used.

Homogeneous Assays

To avoid the need for separating antigen-antibody complexes, homogeneous EIAs were developed. In homogeneous immunoassays, the antigen-antibody reactions take place and are measured in solution without any separation of bound and unbound components. The assays are based on reaction of antigen with an antibody-enzyme complex. This reaction results in steric hindrance of the enzyme, which causes a decrease in product after reaction with enzyme substrate. The major advantage of homogeneous assays is that they do not require the separation and washing steps required in heterogeneous assays. The major disadvantage of this type of EIA is that it is difficult to apply to detection of high-molecular-weight antigens with sufficient sensitivity. Thus, the homogeneous assay has been used mainly for detection of hormones, drugs, and other low-molecular-weight substances. Means of improving the sensitivities of homogeneous assays have been devised but not widely used. Thus, most of the applications discussed here will be limited to EIAs of the solid-phase type.

Factors in Sensitivity and Specificity

Several factors determine how efficient any type of immunoassay is in detecting antigens and antibodies. Some fac-

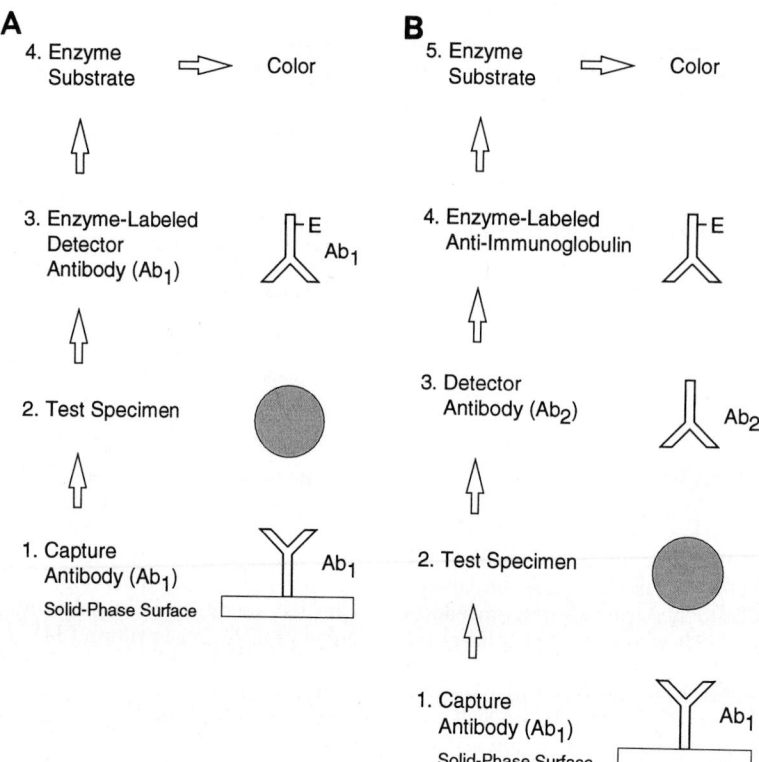

FIGURE 2 ELISA for antigen detection. (A) Direct ELISA. Steps: 1, antigen-specific antibody (Ab$_1$) is attached to a solid-phase surface; 2, test specimen is added; 3, enzyme-labeled antibody (Ab$_1$) is added; 4, a chromogenic enzyme substrate is added. The color developed is proportional to the amount of antigen present in the test specimen. (B) Indirect ELISA. Steps 1 and 2 are the same as for the direct test. Other steps: 3, specific antibody (Ab$_2$), prepared in a species different from that used for generation of Ab$_1$, is added; 4, enzyme-labeled antiglobulin specific for the species used to prepare Ab$_2$ is added; 5, chromogenic enzyme substrate is added, and results are determined as for the direct ELISA.

tors are inherent and cannot be controlled, e.g., the amount of antigen that is usually present in a positive clinical specimen. Other factors, such as test design, can be controlled. Some of the more important variables that can be controlled, primarily in EIA formats, are discussed below.

Most EIA formats require covalent coupling of enzymes to an antibody or antigen. A number of enzymes and coupling techniques have been tried. The best results have been obtained with horseradish peroxidase coupled by use of periodate (59) and alkaline phosphatase coupled by use of glutaraldehyde (4). Most of the assays found to be clinically useful in diagnosing infectious diseases use chromogenic substrates, although fluorogenic and radioactive substrates have been described (78), as have luminescent ones (3, 67).

A development in EIA methodology that has found wide application in diagnostic microbiology is the use of avidin and biotin (27), the test being based on the high-affinity complex formed between biotin and avidin. Avidin is a glycoprotein obtained from egg white, and biotin is a component of the vitamin B$_2$ complex that is inactivated by avidin. Streptavidin, a 60-kDa protein obtained from *Streptomyces avidinii*, resembles avidin obtained from egg white in its binding properties, although it has a lower affinity constant. It is often used as a substitute for egg white avidin. In immunoassays, specific antibody labeled with

biotin, which reacts with immobilized antigen, is most commonly used; the indicator system is usually enzyme-labeled avidin. Another approach is to use biotin-labeled antibody, unlabeled avidin, and biotin-labeled enzyme to indicate the antigen-antibody reaction. Soluble-avidin–biotinylated-enzyme complexes, similar in principle to the peroxidase-antiperoxidase complexes used in histochemical procedures, have also been used. An extensive variety of enzyme- and fluorescence-labeled avidin reagents are now commercially available for use in immunoassays. The enzymes commonly available are horseradish peroxidase, alkaline phosphatase, glucose oxidase, and β-galactosidase; fluorescent labels include fluorescein, rhodamine, and Texas red, which after excitation emit light of different wavelengths, thus permitting dual labels to be discernible by either instrumentation or fluorescence microscopy. Similar enzymes can also be obtained biotinylated for use in EIA. An example of this approach is the use of biotinylated β-lactamase in combination with avidin, which is effective in detecting rotavirus antigen (80).

Fluorogenic substrates measured by fluorimetry can detect smaller amounts of enzyme than chromogenic substrates and thus provide a potential means of improving assay sensitivity, and chemiluminescent substrates have also resulted in more sensitive EIAs (2). Radioactive enzyme substrates were first described for detecting cholera toxin,

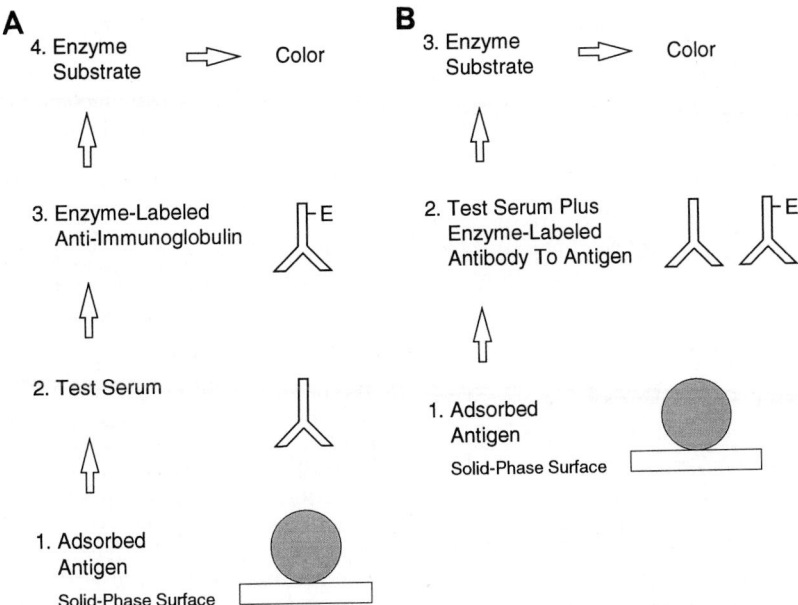

FIGURE 3 ELISA for antibody detection. (A) Noncompetitive ELISA. Steps: 1, specific antigen is attached to a solid-phase surface by passive adsorption or with antigen-specific antibody if necessary; 2, test serum containing specific antibody is added; 3, enzyme-labeled antiglobulin specific for the test serum species is added; 4, chromogenic enzyme substrate is added. The color developed is proportional to the amount of antibody present. (B) Competitive ELISA. Steps: 1, antigen is attached as for the noncompetitive assay; 2, the test serum and an enzyme-labeled antibody specific for the attached antigen are added together; 3, chromogenic enzyme substrate is added. The color developed is inversely proportional to the amount of antibody present.

with alkaline phosphatase as the enzyme and [³H]AMP as the substrate (28).

In addition to antibody affinity and the sensitivity of the indicator system used, the sensitivity of many antigen detection systems, including latex agglutination assays, depends on the amount of antibody that can be effectively immobilized on a solid phase. Methods for immobilizing antibody on plastic surfaces are usually based on simple adsorption, although covalent-linking methods have been utilized. The amount of immunoglobulin that can be immobilized on various plastics has been reported to be from 0.9 to 5.7 ng/mm² (31). With covalent coupling, the quantity of immunoglobulin immobilized may be increased from 100 to 590 ng/mm². Increasing the amount of antibody bound to a solid-phase surface should result in increased sensitivity of the immunoassay. However, increasing the concentration of antibody for coating surfaces beyond 10 μg/ml does not increase immunoassay sensitivity, apparently because of desorption of antibody from the plastic surface and steric hindrance of the antibody that is adsorbed.

The use of antibodies that are highly specific and have high affinities is the most critical aspect of many immunoassay techniques for detecting antigen. An example of how the reagents used determine the effectiveness of an EIA is the detection of *Clostridium difficile* toxin, for which the sensitivity was increased from 58.6 to 95% by changing the immunoreagents used (48). The diluents used for antigen preparation can also alter the sensitivity. Disrupting microbial agents with detergents or other chemicals may increase the sensitivity of some assays but decrease that of others. A number of diluents not usually used in EIAs were tested by

Conroy and Esen (17) for adsorbing zein, a protein present in corn, to polystyrene. These diluents included detergents, acids, alcohols, and urea. Use of alcohols or urea significantly increased the sensitivity of the EIA. Whether this approach might be applicable to some microbial antigens remains to be determined.

An increase in the sensitivity of immunoassays can sometimes be achieved by use of indirect instead of direct assays. In the direct assay, the antibody when optimally labeled usually contains two to three enzyme molecules per antibody molecule. In the indirect test, which utilizes enzyme-labeled anti-immunoglobulin, approximately five anti-immunoglobulin molecules can combine with one antibody molecule, which results in a greater number of enzyme molecules in the overall reaction.

Antibody Detection by EIA

EIAs for antibodies to microbial agents have been utilized for most of the common infectious diseases, because the sensitivity required is well within the range of an EIA. The choice of assay depends on the sensitivity required, the availability of reagents, and whether an immunoglobulin class-specific test is desired. For the detection of IgM, either the noncompetitive EIA or class capture method can be used. The principles of the class capture method are given in Fig. 4. In the noncompetitive EIA, enzyme-labeled antiglobulin (Fig. 3A) specific for IgM is used. The advantage of IgM capture methods is that there is less of a problem with sera containing rheumatoid factor. Rheumatoid factor of the IgM class can produce false-positive reactions by binding to IgG complexed with antigen, which can then bind to the enzyme-labeled anti-IgM antibody used. The

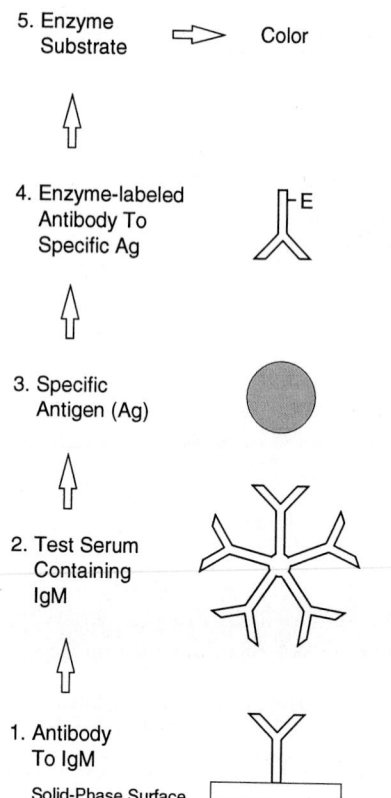

FIGURE 4 Class-specific (IgM) capture ELISA. Steps: 1, antibody specific for IgM is attached to a solid-phase surface; 2, the test serum containing IgM is added; 3, the specific antigen is added; 4, enzyme-labeled antibody specific for the antigen is added; 5, chromogenic enzyme substrate is added. The color developed is proportional to the amount of antigen-specific IgM present in the test serum.

use of IgM-specific antibody immobilized on a solid phase to capture IgM (Fig. 4) helps avoid this problem, because the enzyme-labeled indicator used is either antigen or IgG antibody.

Enzyme-labeled antigen has been used in tests for IgM in the serodiagnosis of a number of viral infections, e.g., cytomegalovirus, Epstein-Barr virus (EBV), flaviviruses, and hepatitis A virus, but the tests require purified antigen for labeling. Thus, they are somewhat limited to those agents for which production and purification of antigen are relatively simple and offer improved diagnostic capability compared with that obtained with enzyme-labeled antibodies.

Serologic EIAs have now been devised for virtually all bacterial and mycotic agents of any significance in human (and animal) health. However, developing tests for antibody that have consistent diagnostic accuracy can be difficult, and various degrees of sensitivity and specificity have been reported. In a review of 34 ELISAs or RIAs for 24 bacterial species, a sensitivity greater than 95% was reported for many assays, but the sensitivity was as low as 50% in some cases (38). Thus, each test must be carefully examined to determine how useful it is for a specific infection. In addition to problems with low sensitivity, the major drawback with many assays is the lack of standardization. Without the availability of standard serum samples for evaluation of assays, new assays require testing by a number of investigators before their validities can be assessed. In many instances, however, serologic diagnosis is the only means available to many laboratories for diagnosing some diseases when culture of the causative agent is either not possible or ineffective.

The application of antibody EIA for diagnosis of viral and rickettsial infections is also extensive. The assay is most frequently applied in screening for immune status, such as rubella testing; for cytomegalovirus antibody; for antibodies to hepatitis B antigens; and for human immunodeficiency virus (HIV) antibodies. Serologic diagnosis of infectious mononucleosis is also the diagnostic method of choice, and commercially available tests for hepatitis A virus IgM have also proven useful for diagnosis of hepatitis A virus infection. Because many viruses are difficult to isolate or have not yet been cultivated, serologic tests, such as that for Norwalk virus infection, are often the most useful for diagnosis. Recent advances in producing recombinant Norwalk virus antigens have made serologic diagnosis of this virus a practical approach (40). The use of an IgM capture EIA for determining recent viral infection is becoming more common and may provide aid in diagnosis where antigen detection methods are not available.

Antigen Detection by EIA

Methods for detecting antigen by EIA can be competitive or noncompetitive, as with the antibody assays diagrammed in Fig. 3A and B. The lower limits of sensitivity for detecting antigen by EIA in most studies are approximately 100 pg to 1 ng, although lower levels of sensitivity have been described. This level of sensitivity is sufficient to detect virtually all bacterial, viral, or parasitic agents following propagation in the laboratory but is not always sufficient to detect microbial antigens directly in clinical specimens. This is especially true in asymptomatic infections, e.g., genital herpes simplex virus or chlamydial infections, because the total antigen load available for detection may be far lower than that found in symptomatic infections. The formats most often employed for detection of antigen by ELISA are shown in Fig. 2 and 3.

The tests developed for bacterial infections are primarily for diseases whose causative agents are difficult to culture or for which rapid diagnosis will permit prompt treatment, such as for group A streptococcal infection and *Chlamydia trachomatis* genital infection. In general, there are far fewer antigen detection EIAs than antibody tests, for the reasons cited above. The efficiencies of the assays reported are variable, but most assays are not as sensitive as the corresponding culture technique.

The interest in EIA antigen detection for rapid diagnosis of viral infections remains high, because EIAs are simple to perform and require less time, expense, and expertise than isolation of the agents in cell culture. Further, some viruses either cannot be cultivated or are difficult to cultivate. Such viruses include hepatitis A and B viruses, rotavirus, Norwalk virus, and Norwalk-like viruses. Tests for antigens of hepatitis viruses have been commercially available for some time, have been extensively evaluated, and need not be elaborated on here. Tests for HIV antigens have also been extensively developed.

In addition to tests for HIV and the hepatitis viruses, including hepatitis C (46), the antigen tests most frequently developed have been for respiratory viruses, herpesviruses, and gastroenteritis viruses. Rapid diagnosis by EIA has been developed for a number of respiratory viruses.

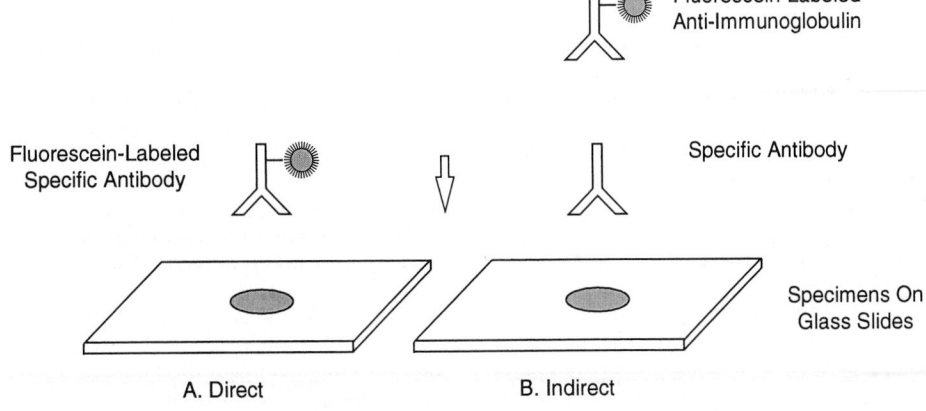

FIGURE 5 Immunofluorescence procedures. (A) In the direct test, fluorescein-labeled antibody specific for a given antigen is incubated with a specimen fixed on a glass microscope slide. If the antigen is present in the specimen, it will be bright yellow-green when viewed under a fluorescence microscope. (B) In the indirect test, unlabeled specific antibody and then fluorescein-labeled anti-immunoglobulin specific for the primary antibody are incubated with the specimen. The slides are examined with a fluorescence microscope as in the direct test.

Several have been described for diagnosis of respiratory syncytial virus. Specificity does not appear to be a problem with any EIA reported, but the sensitivity is generally less than that found by culture. Commercially produced assays for respiratory syncytial virus have been available for some time, as have immunoassays for detection of influenza A virus in respiratory secretions (41, 75).

The interest in sexually transmitted herpes simplex virus and the availability of treatment have led to the development of a number of EIAs for rapid diagnosis of herpes infections. Most of the polyclonal-antibody-based tests lack sufficient sensitivity to be used as a substitute for culture, but monoclonal antibody EIAs are more promising, giving sensitivities in the 90% range, with one report showing sensitivity greater than that of culture (16). EIAs for other herpesviruses, including varicella-zoster virus and cytomegalovirus, are also available.

Immunoassays have been developed for a number of viral gastroenteritis agents, including rotavirus, enteric adenoviruses, Norwalk virus, astrovirus, and calicivirus. Most of the polyclonal antibody assays have been replaced by monoclonal antibody assays, as discussed below.

FA Techniques

Fluorescent-antibody (FA) techniques have been widely used in the past for detection of microbial antigens and antibodies and still have many applications. The technique can be used in either direct or indirect formats, similar to the formats used for EIA and RIA. Diagrams of direct and indirect FA techniques are shown in Fig. 5A and B.

Antigens are usually detected by a direct test. Antibodies are generally detected by indirect tests, because globulin is being assayed for. To test for antibody by the FA technique, microbial antigens are dried on glass slides and treated with a fixative, usually acetone, methanol, or formalin. Dilutions of test serum are incubated with the antigen, the slides are rinsed, and fluorescein-labeled antiglobulin is added. After the slides are rinsed and air dried, a few

drops of buffered glycerol are placed over the specimens, and a coverslip is added. The test slides are examined under a fluorescence microscope and compared with controls (positive and negative sera) for bright areas of yellow-green fluorescence.

To test for antigen by the FA technique, clinical samples, e.g., sputum, nasal washings, cells from lesions, or exudates, are fixed on glass slides. Pretreatment of the samples may be required. The fixed specimens are tested for the presence of specific antigens by incubation with fluorescein-labeled antibody in a direct test (Fig. 5A) or with specific antibody first and then fluorescein-labeled antiglobulin in an indirect test (Fig. 5B). More than one antigen or microbial species can be detected in a given specimen if the specific antibodies are labeled with fluorescent compounds that appear as different colors under fluorescent light, e.g., fluorescein and rhodamine.

Detection of Antibody by FA Technique

The FA technique is commonly used for detection of antibodies to *Legionella pneumophila* and for serodiagnosis of several parasitic diseases, including malaria, leishmaniasis, African trypanosomiasis, pneumocystosis, toxoplasmosis, and schistosomiasis. Antibody levels resulting from these infections are often high, and because of the chronic nature of many parasitic diseases, detection of antibodies of any class in a single serum specimen can be presumptive evidence of an active infection.

For assay of antibodies by the FA technique, microbial antigens are fixed on glass slides, and a patient's serum is tested by the indirect method (Fig. 5B). Tests for antibodies to a variety of microbial antigens are also commercially available in the FIAX system, as discussed below.

Detection of Antigen by FA Technique

Direct detection of antigen in clinical specimens is particularly effective for diagnosis of respiratory virus infections and permits diagnosis within 2 to 3 h after the specimen is received. In comparison to isolation in cell culture, the FA technique is approximately 96% sensitive and 98% specific for diagnosis of respiratory syncytial virus, influenza virus type A, and parainfluenza viruses types 1 through 3 (24, 63). Commercially available kits that use monoclonal antibodies labeled with fluorescein can also be used for identifying *C. trachomatis* in smears from cervical swabs. FA testing of sputum and swab specimens from the nasopharynx for detection of *L. pneumophila* and *Bordetella pertussis* is also common.

The limitations on the use of the FA technique for viral diagnosis are that the assay requires cells containing viral antigen and is not useful for detection of virus in samples such as serum or feces. In combination with cell culture, the FA technique can be used to shorten the time required for virus identification. The assay has been utilized for the identification of many viruses and also of *C. trachomatis* in cell culture.

Solid-Phase Fluorescence Immunoassays (FIAX)

The solid-phase fluorescence immunoassay is similar in principle to other FA techniques except that fluorescence is measured by a fluorometer rather than by a fluorescence microscope. The most common use of this assay in clinical laboratories is with the commercial system FIAX (BioWhittaker, Inc., Walkersville, Md.). In FIAX assays, microbial antigens are immobilized on nitrocellulose disks attached to plastic strips rather than on glass microscope slides. The FIAX sampling strips are immersed in the serum samples to be tested, and specific antibodies, if present, bind to the antigen on the strip. The strips are transferred to a solution of fluorescein-labeled anti-human IgG antibody, which binds to the IgG antibodies complexed to the antigen. The amount of bound, labeled antibody is determined in a fluorometer, which quantitates the fluorescence signal emitted. FIAX tests for detecting antibodies to cytomegalovirus and EBV and to measles, mumps, rubella, herpes simplex, and varicella viruses; *Toxoplasma gondii*; *Mycoplasma pneumoniae* (IgG and IgM); *Borrelia burgdorferi* (IgG and IgM); and *Helicobacter pylori* are available.

Agglutination Assays

Because there is no need for a separation step, agglutination tests based on specific antibody immobilized on polystyrene (latex) particles offer a useful alternative to EIAs or RIAs. In addition, antibody-coated latex particles are stable for long periods, and the tests require only 15 to 30 min for completion. Agglutination assays are also simple to perform. Polystyrene beads of uniform size (generally in the 1-μm range) coated with the specific antibody or antigen desired are mixed with test specimen, and the suspension is examined for visible evidence of clumping. Agglutination assays can be performed in a tube or, more commonly, a slide format. Agglutination reactions can also be measured photometrically. In addition to the direct latex agglutination test, latex agglutination inhibition tests have also been devised. Latex agglutination tests have found the widest application in detecting soluble antigens in normally sterile body fluids such as urine, cerebrospinal fluid, and serum from patients with infection due to *Haemophilus influenzae*, *Streptococcus pneumoniae*, group A and B streptococci, *Neisseria meningitidis*, and *Cryptococcus neoformans*.

The major problem to date in the use of these tests for direct detection of viruses is the lack of sensitivity compared with other immunoassay techniques such as EIA. EIA techniques amplify antigen-antibody reactions because of substrate turnover. In theory, one enzyme molecule can be detected; hence, one antibody molecule and one molecule of analyte could also be detected. Agglutination reactions are not amplified but are based on antigen-antibody reactions forming visible complexes. This complex formation requires a greater number of antigen and antibody molecules than would be required for EIA, and thus, the sensitivity is lower. The sensitivity of an agglutination test may be satisfactory for diagnosis of some bacterial infections, but the test may not be as useful for detecting viruses, which have a much lower antigenic mass per infectious unit. For example, a latex agglutination test for diagnosis of rotavirus infection is available for rapid screening of fecal samples, but it is less sensitive than EIA (57). The judicious application of monoclonal antibodies and better techniques for immobilizing antibody on latex particles may give agglutination techniques the sensitivity required for direct detection of some viruses in clinical specimens. Techniques that combine agglutination formats with EIA have been commercially developed for some microbial antigens. However, latex agglutination techniques have not been sufficiently developed for general use in rapid viral diagnosis.

Immunoblotting of Proteins

The most common form of immunoblotting utilizes immobilized proteins on a solid support to detect protein-specific antibodies. This technique is often referred to as Western blotting, a term derived from the names for DNA blotting (Southern blotting, after its inventor) and RNA blotting (subsequently named Northern blotting, to differentiate it from DNA blotting). Proteins separated by gel electrophoresis are transferred to an immobilizing membrane, usually made from nitrocellulose, polyvinylidene difluoride, or nylon. Polyvinylidene chloride and nylon membranes are particularly useful when chemiluminescent detection systems are used. Serum specimens are applied, and antibody binds to proteins on the support matrix. Bound antibody is then detected by use of fluorescence-, enzyme-, or radionuclide-labeled anti-immunoglobulin specific for the test serum species. Western blotting has been widely applied in basic research and has been useful clinically as a confirmatory serologic test in a variety of infectious diseases.

The general components of immunoblotting as used for immunodiagnosis with human sera include (i) electrophoretic separation of protein antigen on gels (usually sodium dodecyl sulfate-polyacrylamide), (ii) electrophoretic transfer of the protein bands to a nitrocellulose or other support membrane, (iii) blocking of free-protein-binding sites on the membrane, (iv) addition of the test serum, and (v) detection of the specifically bound serum antibodies. The procedure is diagrammed in Fig. 6. The results, seen as bands formed, are compared with bands obtained with known positive and negative sera for the specific microbial antigens used in the test. The interpretation of results is thus subjective and may depend on the technical expertise of the operator if the bands are not clearly defined.

Complement Fixation Test

Complement consists of a group of heat-labile serum proteins that interact in sequence, manifesting a variety of biological activities. Among these is the ability of complement to lyse erythrocytes (RBCs) in the presence of RBC-specific antibody. This reaction is the basis for the complement fixation test. In a complement fixation test for serologic diagnosis of infectious disease, serum antibodies form complexes with specific microbial antigens, binding

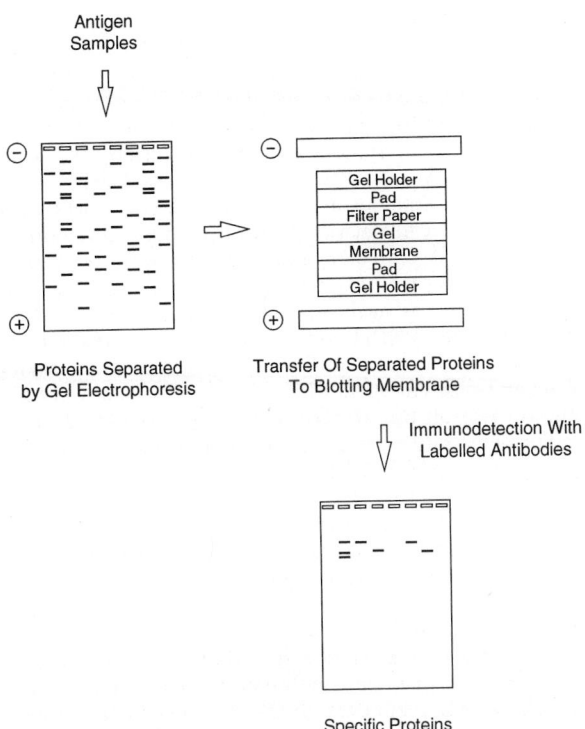

Antigen
Samples

Proteins Separated
by Gel Electrophoresis

Gel Holder
Pad
Filter Paper
Gel
Membrane
Pad
Gel Holder

Transfer Of Separated Proteins
To Blotting Membrane

Immunodetection With
Labelled Antibodies

Specific Proteins
Identified

FIGURE 6 Immunoblotting (Western blotting). Proteins separated by gel electrophoresis are electrophoretically transferred to a nitrocellulose or other blotting membrane and then detected immunologically with antibodies that are either radiolabeled or enzyme labeled. With radiolabeled antibodies, the bands specifically reacting are detected by exposure to X-ray film. With enzyme-labeled antibodies, the bands can be detected by staining with chromogenic enzyme substrates, which precipitate color directly onto the specific proteins, or with chemiluminescent enzyme substrates. Chemiluminescent bands are detected by exposure to X-ray film.

guinea pig complement. Reaction mixtures also contain sheep RBCs and rabbit antibody specifically reactive with the sheep RBCs (i.e., anti-sheep RBCs or hemolysin) in specified amounts. Sheep RBCs and hemolysin activate complement, leading to lysis of the RBCs. This is the indicator system. If the test serum contains antibody to the microbial antigen used, complement is activated (fixed) and depleted, thus preventing lysis of the RBCs in the indicator system. In other words, an inverse relationship exists between the amount of complement-fixing antibody detected and the amount of RBC lysis in the indicator system.

Because the reagents used must be carefully standardized, because the procedure is extremely labor intensive and relatively insensitive, and because such problems as anticomplementary activity can occur (substances in serum bind complement nonspecifically), the complement fixation test has been largely replaced by other immunodiagnostic techniques.

Specialized Virologic Assays

A few immunologic assays used in clinical laboratories are primarily for viral identification. These include immune electron microscopy (IEM), virus neutralization tests, and hemagglutination inhibition (HI) assays. Until other, more convenient immunoassays became available, IEM was a major diagnostic method, especially for enteric viruses. To perform IEM, immune serum was mixed with the virus-containing sample and applied to an electron microscope grid. If positive, virus particles were seen as aggregates coated with antibodies. The coating antibodies were also visible by electron microscopy.

The HI test is still used to detect antibodies to viruses that contain hemagglutinin and is particularly useful for detecting antibodies to influenza virus. In the HI test, virus-specific antibodies, if present, react with viral hemagglutinin, thus preventing agglutination of RBCs added to the assay mixture.

Identification of viruses and detection of virus-specific serum antibodies by neutralization have been the most important standard techniques for use with virtually all viruses that can be cultivated. In the neutralization test, sera or antibodies are mixed with virus prior to or at the time of adsorption to susceptible cells. If antibody specific for the virus used is present, adsorption or another stage of viral replication is blocked, and a decrease in infectivity is seen.

MONOCLONAL ANTIBODY IMMUNOASSAYS

Although polyclonal antibodies have been successfully used as diagnostic reagents in infectious disease for decades, tests using these reagents have often been difficult to standardize. This difficulty has led to variability in inter- and intralaboratory test results. There is generally less variation in commercially produced test kits utilizing polyclonal antisera, but the inherent difficulties with antisera produced in animals remain. One such difficulty is the antibodies to infectious agents that develop in animals as a result of natural exposure. For example, most animals used for antiserum production usually have antibodies to rotaviruses, which, because of the group-reactive antigen present in all rotaviruses, can react with human strains as well. Thus, if an antiserum against a viral or bacterial enteric pathogen unrelated to rotaviruses is prepared for use in an antigen-detecting immunoassay of stool specimens, there could be reactivity with any rotaviruses present. This reactivity could occur even if two species of animals were used to prepare antisera, such as may be done for a solid-phase EIA or an RIA. Another inherent difficulty is the large number of antigenic determinants in most immunogens used in infectious disease research and the resulting broad specificity of the antiserum produced, which can lead to undesired cross-reactions with other antigens. Variation in the abilities of animals of the same species to react to a given antigen is also a problem inherent in the production of antisera, and it is sometimes difficult to reproduce an antiserum with the characteristics originally obtained. This lack of reproducibility often limits the availability of a given immunoreagent.

Since the description of hybridoma technology for production of monoclonal antibodies by Kohler and Milstein in 1975 (45), the use of monoclonal antibodies has become central to both diagnosis and basic studies of infectious diseases. Because of their defined specificity and the potential for greater sensitivity when they are used in immunoassays, due in part to higher signal-to-noise ratios, mono-

TABLE 1 Advantages and limitations of monoclonal antibodies for diagnosis of infectious diseases[a]

Advantages	Limitations

1. Permanent supply	1. Intrinsic cross-reactions
	  a. Same determinant on different carriers
	  b. Similar determinants: related chemical structures
	  c. Unrelated structures: coincidental expression of determinant shape

2. Chemical reproducibility	
3. Defined specificity	
4. Overcomes problems of limited antigen supply for antiserum production.	
5. Potential for greater sensitivity due to lower background values that can often be obtained, with resulting higher signal-to-noise ratios	2. May be difficult to obtain high-affinity antibody.
	3. Cost may be greater than that of antisera against same antigen.
	4. Narrow range of reactivity may result in low sensitivity.

[a]This table is based in part on material presented in references 32, 54, and 79.

clonal antibodies appear to offer solutions to the problems associated with polyclonal antibodies.

Characteristics of Monoclonal Antibodies to Microbial Antigens

Affinity and Sensitivity

Use of monoclonal antibodies has been highly successful in some applications but limited in others, owing to the nature of the antibodies. The primary advantages of monoclonal antibodies and also some of the limiting factors in their use are listed in Table 1. For detecting microbial antigens in clinical specimens either directly or after cultivation, the utility of a monoclonal antibody prepared against a single epitope depends on the specific immunoassay format employed, the nature of the microbial antigen being detected, and certain characteristics of the monoclonal antibody itself. If a particular species does not have a group antigen or contains several variants of a group antigen, a monoclonal antibody may not be sufficiently broadly reactive. For direct detection of antigens in clinical specimens, high-affinity antibody is needed to give the sensitivity that is usually necessary for a successful assay. Production of monoclonal antibodies with high affinity constants is generally more difficult than preparation of high-affinity polyclonal antibodies. Using pools of two or more monoclonal antibodies may circumvent this difficulty, in that mixtures of monoclonal antibodies directed against different determinants on a given antigen should give greater sensitivity than one monoclonal antibody used alone.

The nature of a particular antigen in terms of percentage of the total antigenic mass, number of repeating epitopes it has, whether it is a group antigen, and how accessible the epitope is to antibody is also critical to sensitivity in detection by monoclonal antibodies. For example, a major antigen of rotaviruses, VP6, is contained in all human and animal rotavirus strains and can be detected with poly-

clonal antisera in an RIA or EIA. In an EIA format, a single monoclonal antibody to this VP6 group antigen is more sensitive than polyclonal sera in detecting human rotavirus antigen and is also far more specific (33). In contrast, there is sufficient variation in the large number of strains (isolates) of *Neisseria gonorrhoeae* that detection, and even identification after cultivation, has been difficult when monoclonal antibodies are used. In one study utilizing a commercial coagglutination test, only 86% of the clinical isolates could be correctly identified by the test (55). Others have found that this test, which uses a pool of monoclonal antibodies directed against gonococcal protein I antigens, is more effective (49). However, the overall results demonstrate that monoclonal antibodies, even in pools consisting of several different such antibodies, can be overly specific and thus result in lowered sensitivity for identification and/or detection of some microbial antigens.

Specificity

The major characteristic of monoclonal antibodies is their unique specificity. However, as monoclonal antibodies have been used more widely, it has become apparent that there may be intrinsic cross-reactions such as those listed in Table 1. These types of cross-reactions indicate that multispecificity may be found with both monoclonal antibodies and polyclonal antibodies, although it is far more limited with the former. An example of multispecificity found in infectious agents has been shown for coxsackievirus group B4: a monoclonal antibody to this virus reacted with heart tissue (65). Another example is that observed with monoclonal antibodies to *C. trachomatis* (72). Some of the monoclonal antibodies prepared against *Chlamydia*-specific lipopolysaccharide (LPS) also reacted with antigens of heat-treated *Acinetobacter* spp. and with the LPS from *Salmonella minnesota* R595, the same type of cross-reactions that have been seen with polyclonal sera to *Chlamydia* LPS (9, 11, 61). Also, monoclonal antibody in an *Aspergillus* test reacted with several other fungal antigens (43).

In the preparation of hyperimmune sera, it has been noted that as higher-affinity antibody is produced in a given antiserum, the specificity of that serum decreases owing to a polyclonal response. Studies of monoclonal antibody produced by spleen cells in vitro demonstrated that homogeneously reactive antibody from different spleen cell foci can give different affinities (44). This finding suggests that monoclonal antibodies with different affinity constants retain their specificities. However, in a study that examined the relation of avidity to specificity, an inverse relationship between the avidity and specificity of monoclonal antichlamydial antibodies was found (53). The authors of that study suggested that nonspecific monoclonal antibodies become less specific with increasing avidity because avid nonfitting bonds are not capable of differentiating between different determinants. Whether this loss of specificity will prove to be universal or will be restricted to certain antigens remains to be determined.

Another possibility that could result in false-positive results with monoclonal antibodies is binding to Fc receptors present in a given specimen, as has been found with polyclonal antibodies. Such binding is more likely to occur in monoclonal antibodies of murine origin that are not of class IgG1, as this isotype does not bind well to Fc receptors. The problem of this type of nonspecific binding in the detection of microbial antigens has not proven to be of major importance in polyclonal systems and thus is not very likely to cause difficulties with monoclonal antibodies used for diagnosis.

Compared with the potential of polyclonal sera, the potential for undesired cross-reactivity with monoclonal antibodies is minor and has not yet been shown to be a problem in immunodiagnosis. Rather, those cross-reactions that have been found and are likely to be found will be useful in studying basic mechanisms and in better defining the relatedness of similar antigens.

Production of Monoclonal Antibodies

The production and characterization of monoclonal antibodies to microbial antigens are, in general, similar to those of other antigens, and the methodology is described in numerous books and reviews (62). There are specialized techniques that may be particularly applicable to microbial species in some instances. These include use of in vitro immunization techniques and use of EBV-transformed B lymphocytes from infected patients for use alone or for fusion to either murine or human cell lines.

In general, monoclonal antibodies to microbial antigens have been produced by standard in vivo techniques, but when obtaining sufficient microbial antigen is difficult, the in vitro technique may be advantageous. An example of when this technique has been found useful is the production of antibodies to Snow Mountain virus. The virus is among the 27-nm gastroenteritis viruses that cannot be cultivated, and the only available antigen is that purified from stools of infected patients. By using in vitro immunization techniques (73), it was possible to produce monoclonal antibodies to this virus with 1 to 10 ng of antigen purified from stools. The method is particularly useful for obtaining hybridomas producing IgM monoclonal antibodies, but it can also be used to obtain those producing class IgG. Immunization of animals in vivo prior to in vitro immunization increases the likelihood of obtaining IgG-producing hybridomas. The methods for production of monoclonal antibodies by this technique have been given in detail elsewhere (8).

Other techniques that are applicable to situations in which there is limited antigen availability are the EBV transformation technique and the EBV-hybridoma technique. These techniques are generally used for producing human monoclonal antibodies with the potential for use in the treatment or prophylaxis of disease. The techniques also offer the possibility for production of monoclonal antibodies when antigen supply is limited. There are two general methods of producing human monoclonal antibodies: transformation of human B cells with EBV and fusion of these or nontransformed human B cells with either mouse or human myeloma and human lymphoblastoid cell lines. The EBV transformation technique for producing specific antibody, first described by Steinitz et al. (68), has been used to produce monoclonal antibodies to herpes simplex virus glycoprotein D (66) and influenza virus nucleoprotein (58). The major disadvantage of using the EBV transformation technique alone, without subsequent fusion, is that the cell lines may lose their abilities to produce specific antibody after long-term culture. Another disadvantage to this technique is that small quantities of antibody (1 μg/ml) are usually produced (23). Because the efficiency of direct human B-cell–human or mouse cell fusion is low, combined EBV transformation-cell fusion techniques were developed. Monoclonal antibodies to *Mycobacterium leprae* have been produced by fusion of EBV-transformed donor lymphocytes to human plasmocytomas (64) or to human ×

mouse (human B lymphocytes-mouse myeloma line) (23). Details of the methods for each of these techniques are available elsewhere (23, 64). How useful these techniques will prove to be in the production of monoclonal antibodies to human pathogens that cannot be produced by either standard fusion or in vitro techniques remains to be determined.

Much of the experience gained with polyclonal antibody immunoassays, in terms of labeling with enzymes or isotopes as discussed in the sections above, is applicable to monoclonal antibody immunoassays as well. There are differences among various monoclonal antibodies that need to be evaluated when developing an immunoassay, and some modifications from conditions used for polyclonal antibodies may be necessary. For example, in a study concerned with the binding of monoclonal antibodies to influenza virus proteins on a solid phase, it was found that binding capacity was greatly influenced by the pH of the antigen coating buffer (42).

It may be difficult to produce monoclonal antibodies to some antigens that have a sufficiently high affinities to be useful in immunoassays that require high sensitivity. One approach to circumventing this problem is to use mixtures of two or more monoclonal antibodies to increase the sensitivity of the assay (20). The increase in sensitivity may be more than additive, apparently primarily because of the formation of a circular complex consisting of one of each type of monoclonal antibody and two antigen molecules (21). Test specificity may also be influenced in the mixed monoclonal antibody technique.

Applications in Diagnosis of Infectious Diseases

The application of monoclonal antibodies to diagnosis of infectious diseases is rapidly increasing and includes both direct detection of microbial antigens in clinical specimens and identification of the agent or a specific component of the agent after cultivation. Monoclonal antibodies to many bacterial and viral antigens have been produced, and the number continues to grow. Examples of some monoclonal antibodies that have been used for direct detection in clinical samples are provided below.

Examples of monoclonal-antibody-based assays for direct detection of bacterial antigens in an EIA format include assays for *C. trachomatis* genus-specific antigen in cervical swab specimens (13, 56); *H. influenzae* serotype b in cerebrospinal fluid, serum, and urine (5, 26); *L. pneumophila* soluble antigen in urine, serum, and autopsy specimens (6, 10); *N. meningitidis* group A antigens in cerebrospinal fluids; group B streptococcal antigens in body fluids (70); and *Yersinia pestis* antigen in sera of patients with plague (77).

Fluorescent-monoclonal-antibody tests are useful for detecting *C. trachomatis* elementary bodies in cervical swabs with both genus-specific (1) and species-specific (71) antibodies, *L. pneumophila* in lung tissue (10), and *Treponema pallidum* in lesion exudates of early syphilis (39).

For detection of viral antigens in clinical specimens, there has been a rapid development of monoclonal-antibody-based assays, and many such assays are now commercially available. Commercially produced EIAs or FA tests for a variety of hepatitis viral antigens, rotavirus, respiratory syncytial virus, herpes simplex virus, adenovirus group antigen, enteric adenoviruses (serotypes 40 and 41), and HIV antigens are available in kit form.

Monoclonal-antibody-based immunoassays, in either the EIA or the immunofluorescence format, have also been devised for detection of herpes simplex virus (25); varicella-zoster virus in lesions (15); astrovirus, rotavirus, and enteric adenovirus antigens in stools (33, 36, 37); influenza virus, parainfluenza viruses (74, 76), and respiratory syncytial virus in respiratory secretions (14, 47); cytomegalovirus antigens in lung biopsy samples and bronchoalveolar lavage fluids (29, 52); and hepatitis A virus in stools (18). Innumerable monoclonal antibodies against various components of most common viruses have also been prepared for characterization, basic research, or epidemiologic studies. For the most part, these have not been applicable to routine diagnostic procedures, but many are useful for epidemiologic studies. An example of this approach is P serotyping of rotaviruses with monoclonal antibodies for the specific VP4 types (19).

CONCLUSIONS

Progress in the application of immunoassays for the diagnosis of infectious diseases has been so rapid that methods of detecting a wide variety of microbial antigens and antibodies are now available. For antigen detection, many of the assays originally developed with polyclonal antibodies have been converted and now use monoclonal antibodies. Because immunoassays are generally simple to perform and give timely results of sufficient sensitivity for routine clinical diagnosis, they will undoubtedly continue to be widely used, even as newer methods with potentially greater sensitivity, such as the PCR-based methods, become more commonly available. It is also likely that immunoassay techniques will be combined with newer technologies such as those used in molecular diagnosis to create even more useful diagnostic capabilities in the future (51).

REFERENCES

1. **Alexander, I., I. D. Paul, and E. O. Caul.** 1985. Evaluation of a genus reactive monoclonal antibody in rapid identification of *Chlamydia trachomatis* by direct immunofluorescence. *Genitourin. Med.* **61:**252–254.
2. **Avrameas, S.** 1992. Amplification systems in immunoenzymatic techniques. *J. Immunol. Methods* **150:**23–32.
3. **Avrameas, S., and J. L. Guesdon.** 1982. Immunoenzyme techniques: principles and recent developments, p. 33–54. *In* L. de la Maza and E. M. Peterson (ed.), *Medical Virology.* Elsevier Science Publishing, New York.
4. **Avrameas, S., T. Ternynck, and J. L. Guesdon.** 1978. Coupling of enzymes to antibodies and antigens. *Scand. J. Immunol.* **8:**7–23.
5. **Belmaaza, A., H. Hamel, S. Mousseau, S. Montplaisir, and B. R. Brodeur.** 1986. Rapid diagnosis of severe *Haemophilus influenzae* serotype b infections by monoclonal antibody enzyme immunoassay for outer membrane proteins. *J. Clin. Microbiol.* **24:**440–443.
6. **Bibb, W. F., P. M. Arnow, L. Thacker, and R. M. McKinney.** 1984. Detection of soluble *Legionella pneumophila* antigens in serum and urine specimens by enzyme-linked immunosorbent assay with monoclonal and polyclonal antibodies. *J. Clin. Microbiol.* **20:**478–482.
7. **Bode, L., L. Beutin, and H. Kohler.** 1984. Nitrocellulose-enzyme-linked immunosorbent assay (NC-ELISA)—a sensitive technique for the rapid visual detection of both viral antigens and antibodies. *J. Virol. Methods* **8:**111–121.
8. **Boss, B. D.** 1986. An improved in vitro immunization procedure for the production of monoclonal antibodies. *Methods Enzymol.* **121:**27–32.
9. **Brade, H., and H. Brunner.** 1979. Serological cross-reactions between *Acinetobacter calcoaceticus* and chlamydiae. *J. Clin. Microbiol.* **10:**819–822.
10. **Brown, S. L., W. F. Bibb, and R. M. McKinney.** 1985. Use of monoclonal antibodies in an epidemiological marker system: a retrospective study of lung specimens from the 1976 outbreak of Legionnaires disease in Philadelphia by indirect fluorescent-antibody and enzyme-linked immunosorbent assay methods. *J. Clin. Microbiol.* **21:**15–19.
11. **Caldwell, H. D., and P. J. Hitchcock.** 1984. Monoclonal antibody against a genus-specific antigen of *Chlamydia* species: location of the epitope on chlamydial lipopolysaccharide. *Infect. Immun.* **44:**306–314.
12. **Catt, K. J., and G. W. Tregear.** 1967. Solid phase radioimmunoassay in antibody-coated tubes. *Science* **158:**1570–1572.
13. **Caul, E. O., and I. D. Paul.** 1985. Monoclonal antibody based ELISA for detecting *Chlamydia trachomatis. Lancet* **i:**279.
14. **Chonmaitree, T., B. J. Bessette-Henderson, R. E. Helper, and H. L. Lucia.** 1987. Comparison of three rapid diagnostic techniques for detection of respiratory syncytial virus from nasal wash specimens. *J. Clin. Microbiol.* **25:**746–747.
15. **Cleveland, P. H., and D. D. Richman.** 1987. Enzyme immunofiltration staining assay for immediate diagnosis of herpes simplex virus and varicella-zoster virus directly from clinical specimens. *J. Clin. Microbiol.* **25:**416–420.
16. **Cone, R. W., P. D. Swenson, A. C. Hobson, M. Remington, and L. Corey.** 1993. Herpes simplex virus detection from genital lesions: a comparative study using antigen detection (Herpchek) and culture. *J. Clin. Microbiol.* **31:**1774–1776.
17. **Conroy, J. M., and A. Esen.** 1984. An enzyme-linked immunosorbent assay for zein and other proteins using unconventional solvents for antigen adsorption. *Anal. Biochem.* **137:**182–187.
18. **Coulepis, A. G., M. F. Veale, A. MacGregor, M. Kornitschuk, and I. D. Gust.** 1985. Detection of hepatitis A virus and antibody by solid-phase radioimmunoassay and enzyme-linked immunosorbent assay with monoclonal antibodies. *J. Clin. Microbiol.* **22:**119–124.
19. **Coulson, B. S.** 1993. Typing of human rotavirus VP4 by an enzyme immunoassay using monoclonal antibodies. *J. Clin. Microbiol.* **31:**1–8.
20. **Ehrlich, P. H., and W. R. Moyle.** 1983. Cooperative immunoassays: ultrasensitive assays with mixed monoclonal antibodies. *Science* **221:**279–281.
21. **Engvall, E., and P. Perlmann.** 1971. Enzyme-linked immunosorbent assay (ELISA). Quantitative assay of immunoglobulin G. *Immunochemistry* **8:**871–874.
22. **Ferrua, B., R. Maiolini, and R. Masseyeff.** 1979. Coupling of gamma globulin to microcrystalline cellulose by periodate oxidation. *J. Immunol. Methods* **25:**49–53.
23. **Foung, S. K. H., E. G. Engleman, and F. C. Grumet.** 1986. Generation of human monoclonal antibodies by fusion of EBV-activated B cells to a human-mouse hybridoma. *Methods Enzymol.* **121:**168–173.
24. **Gardner, P. S.** 1983. Identification of viruses. *Histochem. J.* **15:**613–620.
25. **Goldstein, L. C., L. Corey, J. K. McDougall, E. Tolentino, and R. Nowinski.** 1983. Monoclonal antibodies to herpes simplex viruses: use in antigenic typing and rapid diagnosis. *J. Infect. Dis.* **147:**829–837.
26. **Groeneveld, K., L. Van Alphen, N. J. Geelen-van den Broeck, P. P. Eijk, H. C. Zanen, and R. J. Van Ketel.** 1987. Detection of *Haemophilus influenzae* with monoclonal antibody. *Lancet* **i:**441–442.
27. **Guesdon, J. L., T. Ternynck, and S. Avrameas.** 1979. The use of avidin-biotin interaction in immunoenzymatic techniques. *J. Histochem. Cytochem.* **27:**1131–1139.
28. **Harris, C. C., R. H. Yolken, H. Krokan, and I. C. Hsu.** 1979. Ultrasensitive enzymatic radioimmunoassay: application to detection of cholera toxin and rotavirus. *Proc. Natl. Acad. Sci. USA* **76:**5336.
29. **Heckman, R. C., D. Myerson, J. D. Meyers, H. M. Shulman, G. E. Sale, L. C. Goldstein, M. Rastetter, N. Flournoy, and E. D. Thomas.** 1985. Rapid diagnosis of cytomegalovirus pneumonia by tissue immunofluorescence with a murine monoclonal antibody. *J. Infect. Dis.* **151:**325–329.

30. **Hendry, R. M., and J. E. Herrmann.** 1983. Detection and identification of influenza virus antigen by nylon-coupled enzyme linked immunoassay. *J. Virol. Methods* **6:**9–17.

31. **Herrmann, J. E.** 1981. Quantitation of antibody immobilized on plastics. *Methods Enzymol.* **73:**239–244.

32. **Herrmann, J. E.** 1989. Progress in the use of monoclonal antibodies for diagnostic microbiology, p. 679–695. *In* B. Swaminathan and G. Prakash (ed.), *Nucleic Acid and Monoclonal Antibody Probes: Applications in Diagnostic Microbiology.* Marcel Dekker, Inc., New York.

33. **Herrmann, J. E., N. R. Blacklow, D. M. Perron, G. Cukor, P. J. Krause, J. S. Hyams, H. J. Barrett, and P. L. Ogra.** 1985. Enzyme immunoassay with monoclonal antibodies for the detection of rotavirus in stool specimens. *J. Infect. Dis.* **152:**831–832.

34. **Herrmann, J. E., R. M. Hendry, and M. F. Collins.** 1979. Factors involved in enzyme-linked immunoassay and evaluation of the method for identification of enteroviruses. *J. Clin. Microbiol.* **10:**210–217.

35. **Herrmann, J. E., and S. A. Morse.** 1974. Conjugation of enzymes to antipoliovirus globulin: effect of enzyme molecular weight on virus neutralization capacity. *Immunochemistry* **11:**79–82.

36. **Herrmann, J. E., N. A. Nowak, D. M. Perron-Henry, R. W. Hudson, W. D. Cubitt, and N. R. Blacklow.** 1990. Diagnosis of astrovirus gastroenteritis by antigen detection with monoclonal antibodies. *J. Infect. Dis.* **161:**180–184.

37. **Herrmann, J. E., D. M. Perron-Henry, and N. R. Blacklow.** 1987. Antigen detection with monoclonal antibodies for the diagnosis of adenovirus gastroenteritis. *J. Infect. Dis.* **155:**1167–1171.

38. **Hill, H. R., and J. M. Matsen.** 1983. Enzyme-linked immunosorbent assay and radioimmunoassay in the serologic diagnosis of infectious diseases. *J. Infect. Dis.* **147:**258–263.

39. **Hook, E. W., III, R. E. Roddy, S. A. Lukehart, J. Hom, K. K. Holmes, and M. R. Tam.** 1985. Detection of *Treponema pallidum* in lesion exudate with a pathogen-specific monoclonal antibody. *J. Clin. Microbiol.* **22:**241–244.

40. **Jiang, X., M. Wang, K. Wang, and M. Estes.** 1993. Sequence and genomic organization of Norwalk virus. *Virology* **195:**51–61.

41. **Johnston, S. L. G., and H. Bloy.** 1993. Evaluation of a rapid enzyme immunoassay for detection of influenza A virus. *J. Clin. Microbiol.* **31:**142–143.

42. **Kammer, K.** 1983. Monoclonal antibodies to influenza A virus FM1 (H1N1) proteins require individual conditions for optimal reactivity in binding assays. *Immunology* **48:**799–808.

43. **Kappe, R., and A. Schulze-Berge.** 1993. New cause for false-positive results with the Pastorex Aspergillus latex agglutination test. *J. Clin. Microbiol.* **31:**2489–2490.

44. **Klinman, N. R.** 1969. Antibody with homogeneous antigen binding produced by splenic foci in organ culture. *Immunochemistry* **6:**757–759.

45. **Kohler, G., and C. Milstein.** 1979. Continuous cultures of fused cells secreting antibody of predefined specificity. *Nature* (London) **256:**495–497.

46. **Kuo, G., Q. L. Choo, H. J. Alter, G. L. Gitnick, A. G. Redeker, R. H. Purcell, T. Miyamura, J. L. Dienstag, M. J. Alter, and C. E. Stevens.** 1989. An assay for circulating antibodies to a major etiologic virus of human non-A, non-B hepatitis. *Science* **244:**362–364.

47. **Lauer, B. A., H. A. Masters, C. G. Wren, and M. J. Levin.** 1985. Rapid detection of respiratory syncytial virus in nasopharyngeal secretions by enzyme-linked immunosorbent assay. *J. Clin. Microbiol.* **22:**782–785.

48. **Laughon, B. E., R. P. Viscidi, S. L. Gdovin, R. H. Yolken, and J. G. Bartlett.** 1984. Enzyme immunoassays for detection of *Clostridium difficile* toxins A and B in fecal specimens. *J. Infect. Dis.* **149:**781–788.

49. **Lawton, W. D., and G. J. Battaglioli.** 1983. GonoGen coagglutination test for confirmation of *Neisseria gonorrhoeae*. *J. Clin. Microbiol.* **18:**1264–1265.

50. **Lynn, M.** 1975. Inorganic support intermediates: covalent coupling of enzymes an inorganic supports, p. 1–48. *In* H. H. Weetall (ed.), *Immobilized Enzymes, Antigens, Antibodies, and Peptides.* Marcel Dekker, Inc., New York.

51. **Mallet, F., C. Hebrard, D. Brand, E. Chapuis, P. Cros, P. Allibert, J. M. Besnier, F. Barin, and B. Mandrand.** 1993. Enzyme-linked oligosorbent assay for detection of polymerase chain reaction-amplified human immunodeficiency virus type 1. *J. Clin. Microbiol.* **31:**1444–1449.

52. **Martin, W. J., II, and T. F. Smith.** 1986. Rapid detection of cytomegalovirus in bronchoalveolar lavage specimens by a monoclonal antibody method. *J. Clin. Microbiol.* **23:**1006–1008.

53. **Matikainen, M. T., and O. P. Lehtonen.** 1984. Relation between avidity and specificity of monoclonal anti-chlamydial antibodies in culture supernatants and ascitic fluids determined by enzyme immunoassay. *J. Immunol. Methods* **72:**341–347.

54. **Milstein, C.** 1986. Overview: monoclonal antibodies, p. 107.1–107.12. *In* D. M. Weir (ed.), *Handbook of Experimental Immunology*, 4th ed., vol. 4. Blackwell Scientific Publications, Oxford.

55. **Minshew, B. H., J. L. Beardsley, and J. S. Knapp.** 1985. Evaluation of GonoGen coagglutination test for serodiagnosis of *Neisseria gonorrhoeae*: identification of problem isolates by auxotyping, serotyping, and with a fluorescent antibody reagent. *Diagn. Microbiol. Infect. Dis.* **3:**41–46.

56. **Mohanty, K. C., J. J. O'Neill, and H. H. Hambling.** 1986. Comparison of enzyme immunoassays and cell culture for detecting *Chlamydia trachomatis*. *Genitourin. Med.* **62:**175–176.

57. **Morinet, F., F. Ferchal, R. Colimon, and Y. Perol.** 1984. Comparison of six methods for detecting human rotavirus in stools. *Eur. J. Clin. Microbiol.* **3:**136.

58. **Muggeridge, M. I., D. M. Mitchell, E. D. Zanders, and P. C. L. Beverley.** 1983. Production of human monoclonal antibody to X31 influenza virus nucleoprotein. *J. Gen. Virol.* **64:**697–700.

59. **Nakane, P. K., and A. Kawaoi.** 1974. Peroxidase-labelled antibody. A new method of conjugation. *J. Histochem. Cytochem.* **22:**1084.

60. **Nakane, P. K., and G. B. Pierce, Jr.** 1966. Enzyme-labeled antibodies: preparation and application for the localization of antigens. *J. Histochem. Cytochem.* **14:**929–931.

61. **Nurminen, M., E. Wahlstrom, M. Kleemola, M. Leinonen, P. Saikku, and P. Makela.** 1984. Immunologically related ketodeoxyoctonate-containing structures in *Chlamydia trachomatis*, Re mutants of *Salmonella* species, and *Acinetobacter calcoaceticus* var. anitratus. *Infect. Immun.* **44:**609–613.

62. **Peter, J. H., and H. Baumgarten.** 1992. *Monoclonal Antibodies.* Springer-Verlag, Berlin.

63. **Ray, C. G., and L. L. Minnich.** 1987. Efficiency of immunofluorescence for rapid detection of common respiratory viruses. *J. Clin. Microbiol.* **25:**355–357.

64. **Roder, J. C., S. P. C. Cole, and D. Kozbor.** 1986. The EBV-hybridoma technique. *Methods Enzymol.* **121:**140–167.

65. **Saegusa, J., B. S. Prabkakar, K. Essani, P. R. McClintock, Y. Fukuda, V. J. Ferrans, and A. L. Notkins.** 1986. Monoclonal antibody to coxsackievirus B4 reacts with myocardium. *J. Infect. Dis.* **153:**372–373.

66. **Seigneurin, J. M., C. Desgranges, D. Seigneurin, J. Paire, J. C. Renversez, B. Jacquemont, and C. Micouin.** 1983. Herpes simplex virus glycoprotein D: human monoclonal antibody produced by bone marrow cell line. *Science* **221:**173–175.

67. **Seitz, R.** 1984. Immunoassay labels based on chemiluminescence and bioluminescence. *Clin. Biochem.* **17:**120–124.

68. **Steinitz, M., G. Klein, S. Koskimies, and O. Makela.** 1977. EB virus induced B lymphocyte cell lines producing specific antibody. *Nature* (London) **269:**420–422.

69. **Streefkerk, J. G., and A. M. Deelder.** 1975. Serodiagnostic application of immunohistoperoxidase reactions on antigen-coupled agarose beads. *J. Immunol. Methods* **7:**225–236.

70. **Sugaswara, R. J., C. M. Prato, and J. E. Sippel.** 1984.

Enzyme-linked immunosorbent assay with a monoclonal antibody for detecting group A meningococcal antigens in cerebrospinal fluid. *J. Clin. Microbiol.* **19:**230–234.

71. **Tam, M. R., W. E. Stamm, H. H. Handsfield, R. Stephens, C. C. Kuo, K. K. Holmes, K. Ditzenberger, M. Krieger, and R. C. Nowinski.** 1984. Culture-independent diagnosis of *Chlamydia trachomatis* using monoclonal antibodies. *N. Engl. J. Med.* **310:**1146–1150.

72. **Thornley, M. J., S. E. Zamze, N. D. Byrne, M. Lusher, and R. T. Evans.** 1985. Properties of monoclonal antibodies to the genus-specific antigen of chlamydia and their use for antigen detection by reverse passive haemagglutination. *J. Gen. Microbiol.* **131:**7–15.

73. **Treanor, J. J., H. P. Madore, and R. Dolin.** 1988. Development of a monoclonal antibody to the Snow mountain agent of gastroenteritis. *Proc. Natl. Acad. Sci. USA* **85:**3616–3617.

74. **Walls, H. H., K. H. Johansson, M. W. Harmon, P. Halonen, and A. Kendal.** 1986. Time-resolved fluoroimmunoassay with monoclonal antibodies for rapid diagnosis of influenza infections. *J. Clin. Microbiol.* **24:**907–912.

75. **Waner, J. L., S. J. Todd, J. Shalaby, P. Murphy, and L. V.** Wall. 1991. Comparison of Directigen FLU-A with viral isolation and direct immunofluorescence for the rapid detection and identification of influenza A virus. *J. Clin. Microbiol.* **29:**470–482.

76. **Waner, J. L., N. J. Whitehurst, T. Downs, and D. G. Graves.** 1985. Production of monoclonal antibodies against parainfluenza 3 virus and their use in diagnosis by immunofluorescence. *J. Clin. Microbiol.* **22:**535–538.

77. **Williams, J. E., M. K. Gentry, C. A. Broden, F. Laister, and R. H. Yolken.** 1984. Use of an enzyme-linked immunosorbent assay to measure antigenemia during acute plague. *Bull. W.H.O.* **62:**463–466.

78. **Yolken, R. H.** 1982. Enzyme immunoassays for the detection of infectious antigens in body fluids: current limitations and future prospects. *Rev. Infect. Dis.* **4:**35–68.

79. **Yolken, R.** 1983. Use of monoclonal antibodies for viral diagnosis. *Curr. Top. Microbiol. Immunol.* **104:**177–196.

80. **Yolken, R. H., and S. B. Wee.** 1984. Enzyme immunoassays in which biotinylated β-lactamase is used for the detection of microbial antigens. *J. Clin. Microbiol.* **19:**356–360.

Gas-Liquid and High-Performance Liquid Chromatographic Methods for the Identification of Microorganisms

ANDREW B. ONDERDONK AND MYRON SASSER

12

Phenotypic characterization of microorganisms is a significant part of clinical microbiology. The use of methods such as chromatography for identification purposes has become increasingly important. When used appropriately, such techniques are extremely useful for identifying organisms of clinical interest.

BASIC PRINCIPLES OF CHROMATOGRAPHY

Chromatography (10, 20, 60) is a separation method based on the different solubilities of compounds (analytes) between two immiscible phases. Because each analyte has a specific solubility (partition coefficient) in relation to the two immiscible phases, extraction of analytes in mixtures from one phase into the other phase occurs with different efficiencies. All chromatography methods involve a stationary phase and a mobile, or moving, phase. Separations are based on the physical partitioning of components between the stationary and mobile phases during analysis. The mobile phase may be a gas or liquid, and the stationary phase may be a liquid or solid.

Chromatographic methods are characterized according to the mobile phase utilized. Gas chromatography (GC) utilizes a gas as the mobile phase, and liquid chromatography (LC) uses a liquid as the mobile phase. Further classification of the methods for separation also include the stationary phase. For example, a method using gas as a mobile phase and a liquid as stationary phase is called gas-liquid chromatography (GLC), while a method using a liquid for both the mobile and the stationary phases is called liquid-liquid chromatography. So-called thin-layer chromatography (TLC) uses a liquid mobile phase and a solid stationary phase. Size exclusion or gel permeation chromatography is a form of LC that uses the stationary phase as a molecular sieve, so separations are based on molecular size and weight. Ion-exchange chromatography uses a solid stationary phase, and separation is due to the differential elution of ionically bound sample components with liquid buffers. High-performance liquid chromatography (HPLC) is a form of liquid-solid chromatography that forces the liquid phase over the solid phase under pressure.

SEPARATION AND DETECTION OF SPECIFIC BIOLOGIC SIGNALS

Depending on the class of compound to be separated and/or detected, one chromatographic method may be more appropriate than others (27). For example, separation of proteins from each other in a way that does not damage the separated molecules generally involves some form of liquid-solid chromatography. Complex carbohydrates and polysaccharides are normally separated by using liquid-solid chromatography, as are DNA and RNA fragments, although mono- and disaccharides may be separated by GLC methods. One of the basic components of cell membranes is long-chain fatty acids (LCFA). These are most often detected and identified by GLC techniques.

The types of detectors used to detect separated molecules vary depending on the biologic signal. Detectors range from spectrophotometric determinations of light absorbance as a function of analyte concentration (proteins and polysaccharides) to pH detectors for acids and bases. Staining of separated components, such as proteins, DNA, RNA, and carbohydrates, is often performed as a means of visually detecting components separated by TLC.

GLC

See references 10, 14, 20, 21, 32, 45, 60, and 61.

Instrumentation

Instruments designed to be used for GLC separations all have several common features, as shown in Fig. 1. Because the mobile phase (also known as a carrier) is a gas, there needs to be some way by which a sample can be entered into the carrier gas stream before the gas passes over the stationary phase (also known as the active phase). The injection port allows a sample to be injected into the moving carrier gas stream. After a sample is injected into the carrier gas stream, it is carried onto a glass or metal column containing the stationary liquid phase, which is often coated onto inert particles to increase the effective surface area. The actual separation of the various components in the sample depends on partitioning of the components between the gaseous and liquid phases as the sample travels through the column. Compounds with little solubility in the stationary phase pass through the column quickly,

GAS/LIQUID CHROMATOGRAPHY

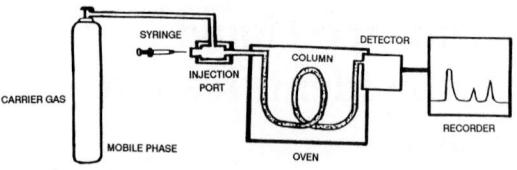

HIGH PERFORMANCE LIQUID CHROMATOGRAPHY

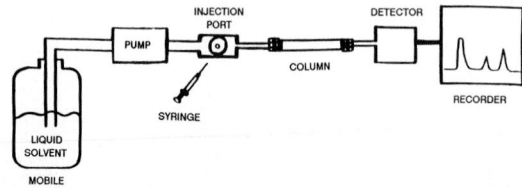

FIGURE 1 Chromatography instrumentation.

while compounds with greater solubility in the stationary phase pass through the column more slowly, at a rate consistent with their partition coefficient. The important point is that different compounds have different partition coefficients and travel at different speeds through the column; thus, similar analytes in a mixture are separated from each other.

After the separated components leave the column containing the active phase, they pass through a detector in which each compound is discerned as a discrete electrical signal. The signal from the detector is then visually recorded using a strip chart recorder. All GLC instruments are designed so that a sample in a liquid solvent can be converted to the gaseous state as the sample is injected into the carrier gas stream. This is accomplished by heating the injection port and vaporization chamber (Fig. 1) into which the sample is injected. The column and detector are also heated to prevent condensation of sample components after injection. Optimization of GLC assays takes advantage of the differences in solubilities of compounds in the active phase, depending on temperature, the rate of carrier gas flow, and the type of detector employed. GLC instruments can be adapted to different separations by simply changing the column containing the stationary phase and altering the temperatures used for analysis. This flexibility has made GLC a very useful analytic tool in the clinical laboratory.

Mobile Phases

The mobile phase, or carrier gas, used for GLC is generally an extremely pure, inert gas, which will not react with the compounds being separated. The most commonly employed carrier gases include hydrogen, helium, nitrogen, and argon.

Stationary Phases and Columns

GLC columns are referred to as packed or capillary columns depending on how the stationary phase is held in the column. Packed columns are constructed of stainless steel, aluminum, copper, or glass. These columns are usually between 1.6 and 9.5 mm in diameter and 1 to 4 m long. Packed columns contain a liquid active phase that is coated onto a solid support, such as diatomaceous-earth particles (average diameter, 100 to 300 μm), to increase surface area. Using a solid support to increase the surface area on which

the separation of analytes can occur is an important part of chromatographic technology; however, the length of such a column is limited by the resistance of the packed column to the flow of the mobile phase.

Capillary column chromatography is an alternative to packed-column chromatography that is not limited by increasing resistance to the flow of the mobile phase as the column length increases. This form of GLC takes advantage of columns with very small diameters (0.25 to 0.75 mm); hence the designation capillary column. Such columns have the liquid, active phase coated onto the inside surface of the column instead of some other solid support material. The advantage of this column is that it can be any length (20 to 40 m is common), thus increasing the efficiency with which mixtures of analytes can be separated.

Detectors

The variety of possible detectors for use with GLC systems ranges from relatively insensitive thermal conductivity detectors to extremely sensitive electron capture detectors. The two most commonly employed detectors for GLCs in clinical microbiology laboratories are thermal conductivity and flame ionization detectors. Thermal conductivity detectors sense a change in the cooling characteristics of the mobile phase passing over a heated metal filament as a change in resistance to the flow of an electric current. If an analyte is present in the mobile phase, the cooling characteristics change, and the result can be recorded as a momentary change in filament resistance. Thermal conductivity detectors can be used to detect anything that changes the cooling characteristics of the mobile phase, and they are therefore quite versatile. Flame ionization detectors use a hydrogen-oxygen flame to ionize (burn) organic compounds. As the ionized particles pass between two charged plates, a change in current between the plates can be measured and recorded. Flame ionization detectors are about 1,000 times more sensitive than thermal conductivity detectors, but they are restricted to use with organic compounds.

LC AND HPLC METHODS

HPLC is a modification of the LC methods that have been in use for many years (2, 7, 15, 22, 50, 62, 63). HPLC, like all LC methods, is not limited by the boiling point of a compound or the stability of compounds to thermal degradation, which sometimes restrict the use of GC. The analytical process involved in HPLC methods includes a mobile phase (usually called a solvent in HPLC systems) traveling over a solid phase housed within a metal or glass column. Unlike GLC instruments, HPLC instruments utilize a liquid mobile phase that is forced through the column containing the active phase by high-efficiency precision pumps. No ovens house the column, and on-line injection of sample is through a diaphragm into the stream of liquid mobile phase. Detection of separated analytes includes the use of a variety of different detectors.

The terms "normal phase" and "reverse phase" refer to the polarity of the mobile and stationary phases used for a particular separation. Normal-phase chromatography refers to a nonpolar mobile phase and a polar stationary phase. Conversely, reverse-phase separations use a polar mobile phase and a nonpolar stationary phase. Gradient elution techniques, in which the polarity of the mobile phase is continuously changing, are often employed to increase the separation efficiency for closely related compounds. One

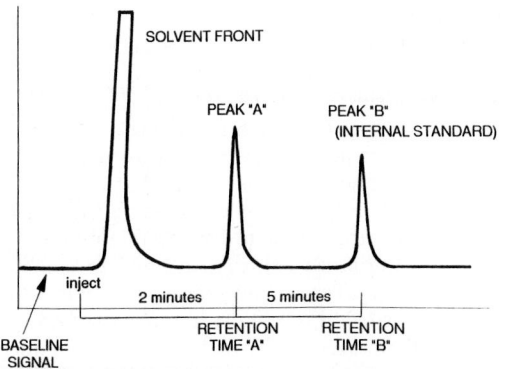

FIGURE 2 Elements of a chromatogram. Relative retention time (RR$_T$) is calculated as RR$_T$ = RT A/RT B.

TABLE 1 Volatile fatty acid products for various genera of obligate anaerobes[a]

Genus	Gram reaction	Major VFA products[b]
Bacteroides	−	A, S
Prevotella	−	A, S
Porphyromonas	−	A, S
Fusobacterium	−	a, B, s
Clostridium	+	A, B
Actinomyces	+	A, S
Propionibacterium	+	A, P
Eubacterium	+	a, b
Lactobacillus	+	a, L
Bifidobacterium	+	A, L

[a]Adapted from *Anaerobe Laboratory Manual* (28) and *Wadsworth Anaerobic Bacteriology Manual* (52).

[b]A and a, major and minor acetates; B and b, major and minor butyrates; L, major lactate; P, major propionate; S and s, major and minor succinates.

advantage of HPLC is that separated components can be collected with a fraction collector after elution from the column.

The most commonly used HPLC detectors are the UV absorption detector, the refractometer, and the fluorescence detector. The typical UV detector operates at a fixed wavelength of either 254 or 280 nm. Refractive-index detectors differentiate the sample components from the mobile solvent phase on the basis of differences in refractive index. Such detectors are sensitive to analytes at concentrations of 10^{-7} g/ml. UV detectors, in contrast, have a sensitivity of approximately 10^{-10} g/ml. Fluorescence detectors utilize a variety of light sources ranging from tungsten-halide lamps to xenon arc lamps to produce light at wavelengths that will be absorbed by analyte and reemitted at a longer wavelength. Fluorescence detectors are more sensitive than UV detectors but have a limited linear range and can be used only for analytes that fluoresce.

TLC AND OTHER TECHNIQUES

TLC methods (46) provide a rapid, cost-effective analytic tool for separating the components of sample mixtures. Commercially available TLC plates coated with appropriate stationary phases are readily available. Generally, TLC is used in conjunction with other techniques such as HPLC for qualitative determinations of sample components.

Many methods combine various instrumental techniques for more definitive analytical determinations (4, 7, 12, 25, 56, 59). These include GC-mass spectrometry, often used for drug analyses. More exotic combinations such as GC–Fourier-transformed infrared instruments are commercially available, but they are not in routine use in clinical laboratories.

CHARACTERISTICS OF CHROMATOGRAMS (GLC)

The end product for most chromatographic methods is a chromatogram, which represents the output from the detector during the period of analysis (27). If used properly, chromatograms can qualitatively identify analytes and also determine how much of a given compound is present. For methods such as GLC, the basic components of the chromatogram are the same (Fig. 2) regardless of the detector used. Components are identified on the basis of how long it takes the analyte to travel through the column and pass

through the detector. The time required for a given analyte to travel from the point of injection to the detector is called the retention time. For a particular GC and column, the time required to travel this distance will always be the same for a specific analyte as long as the rate of flow of the mobile phase and the temperature of the column are maintained. This consistency allows a chromatogram for a known mixture of compounds, such as volatile fatty acids, to be superimposed onto a chromatogram for an unknown sample and the peaks identified. An alternative to this direct approach is to include a known compound as an "internal standard" in each sample analyzed. A ratio of the retention time for the internal standard to that of any other known compound will remain the same over a wide range of temperatures and flow rates. This means that minor changes in the analytic process will not require recalibration of the chromatograph. Moreover, the internal standard can be used for quantitation as well as for judging extraction efficiency. This is the method most often employed when digital integration is used to evaluate the signal from the detector.

GLC detectors have a linear range that allows quantitation on the basis of peak area. This means that injection of twice as much analyte into the chromatograph will result in a peak that is twice as large. Peak area can be measured by hand and calculated, or more commonly, the signal can be integrated as an "electronic" area by using a measure of current deflection from baseline over a period of time (microvolt seconds).

Chromatograms are also valuable measures of instrument performance, since leaks in the system, electronic interference, and column efficiency can be assessed easily. Although digital integration does not require that a visual record of the signal be maintained, most chromatographers continue to use the chromatogram as a way to monitor system performance and separation efficiency.

MICROBIAL IDENTIFICATION

GLC Analysis of Metabolic Products as an Aid in Identifying Obligate Anaerobes

Chromatography has been used for almost 30 years as a method for the detection of short-chain organic acids (volatile fatty acids) and alcohols produced by obligate anaerobes. Obligate anaerobes are incapable of utilizing oxygen as a final electron acceptor in the electron transport chain. Consequently, a variety of alternative metabolic pathways for maximizing the amount of energy obtained during fermentation is present in obligate anaerobes. In many instances, such pathways are genus and even species specific, given that a modest amount of additional information is available. An example of the use of this information is provided in Table 1, which summarizes the short-chain-acid end products produced from glucose fermentation for several genera of obligate anaerobes and gives the organisms' Gram stain characteristics (28). Additional information on the use of short-chain fatty acids for identification of obligate anaerobes is given elsewhere (52).

The GLC system used for analyzing short-chain organic acids and alcohols produced by obligate anaerobes is usually a packed-column system (isothermal temperature, 135 to 140°C) with a thermal conductivity detector and helium as a carrier gas (mobile phase). Samples are generally obtained from peptone-yeast-glucose broth following growth. The volatile fatty acids are extracted from an acidified culture into diethyl ether, and this extract is used for analysis. In some cases, the presence of the dicarboxylic acids lactate and succinate is important. These acids are not volatile under the analytic conditions used for the other short-chain acids. In order to render lactate and succinate volatile, the culture medium is treated with reagents that will cause the formation of methyl esters of the dicarboxylic acids. These methyl esters can then be extracted into chloroform and analyzed on the same column as that used for the volatile fatty acids (28).

The individual acids present in a particular culture are identified by comparing their chromatograms to the chromatogram for a known standard containing each volatile fatty acid (C_1 to C_8) according to retention time. It is important that the column temperature and carrier gas flow rate do not change during the analytic run. Some laboratories prefer to use GLC systems equipped with flame ionization detectors and digital integration capabilities. Such systems are capable of identifying individual peaks on the basis of retention time or relative retention time compared to that of an internal standard (see above).

Composition of Microorganisms

Bacteria and yeasts are composed of four major types of biomolecules; lipids, proteins, carbohydrates, and nucleic acids. From an analytic perspective, for monomers there are 20 amino acids, five bases among the nucleic acids, approximately 10 sugars, and over 300 fatty acids. Unlike other monomeric types, the fatty acids found in microorganisms are so diverse that they provide a wide range of phenotypic characteristics for the classification of bacteria and yeasts. For any given microorganism, the fatty acids present tend to be expressed as a stable part of the structure of membranes and other lipid components and as such are always present in a predictable fashion.

While short-chain fatty acids are used as adjuncts to biochemical identification, fatty acids 9 to 20 carbons long are easily analyzed by GLC and can be used alone or with biochemical information for the identification of bacteria, yeasts, and fungi. Very large fatty acids, such as the mycolic acids, may be analyzed more effectively using HPLC or TLC. Mycolic acids are useful for the identification of mycobacteria, *Nocardia* spp., and *Rhodococcus* spp. (8, 9, 11, 35).

LCFA Analysis

A considerable mass of literature has accumulated on the utility of whole-cell fatty acid analysis for the identification of microorganisms (1, 6, 13, 34). A particularly useful reference is a paper by Welch (57). The lipid components of bacteria, overwhelmingly structural in nature, occur in the membranes and cell walls of bacteria. Yeasts and fungi often have storage lipids that can obscure these structural lipids; however, this problem can be minimized by using actively growing broth cultures. In order to compare the LCFAs of an unknown strain to those of a library of known strains, it is important to standardize the growth conditions and the analytic process. In recent years, a commercially available instrumentation and software system (Microbial Identification Systems, Newark, Del.) has made the use of LCFAs as a routine method for identification of bacteria and yeasts practical (39).

The essential elements of this system are a GC equipped with a flame ionization detector and an automatic injection system for samples, a capillary column for the separation of LCFAs, and a computer with appropriate software to perform signal analysis. Growth conditions and media have been standardized for use with a specific library of organisms depending on whether obligate anaerobes, aerobic organisms, clinical isolates, or yeasts are to be identified. Isolates are grown on a designated agar medium (or broth for obligate anaerobes) at a specified temperature, and approximately 60 mg of cells is harvested for analysis at a predetermined physiologic growth stage. The cells are then placed into a tube, and the lipid components are saponified with sodium hydroxide to release the cellular fatty acids. In order to make the LCFAs volatile under the conditions used for analysis, the LCFAs are methylated by an acid-methanol method (50% methanol in 3.25 M HCl). The LCFA-methyl esters (FAME) are then extracted into an organic solvent (hexane-methyl *tert*-butyl ether, 1:1), residual contaminants are removed with a NaOH wash (0.2 M), and the sample is placed into a small gastight vial for automated analysis. The most important part of the instrumentation is the automatic injector, which can reproducibly inject 2 μl of sample onto a column within 0.01 s. This degree of precision is necessary for the signal analysis software to correctly identify closely related FAME.

Analysis of the derivatized FAME is carried out by using a hydrogen carrier gas passing through a fused-silica capillary column containing a cross-linked phenylmethyl silicone liquid phase. The column oven temperature is programmed to increase from 170 to 270°C at a rate of 5°C/min. Detection of the separated FAME is accomplished with a flame ionization detector. Because the injection of samples onto the column is accomplished with a precision automatic injector, peak detection and identification with this system are quite reliable. The key to this system is the software that has been developed for the analysis of FAME of bacterial origin. By using known standards for each run, including calibration against internal standards and a library with thousands of entries, the chromatographic characteristics for a given isolate can be quickly matched with those of other identical (or similar) species. Variance from the established mean percent compositions for a particular species can be used to establish a similarity index. This index can be used to determine how closely an unknown isolate's FAMEs match those for a library entry.

In addition to being an automated system for microbial identification, the software provided can be used to generate a library of isolates specific to a particular laboratory or hospital. Isolates such as resistant strains of *Staphylococcus aureus* may be compared by a cluster analysis technique to determine whether isolated strains are identical or only closely related. The value of such methods for epidemiologic purposes greatly expands the utility of this method.

HPLC Analysis of Mycolic Acids

Mycolic acids are 30 to 80 carbons long and are therefore more amenable to analysis by HPLC, in which the ability to volatilize the analyte is not an issue. Just as the various FAME and percent compositions are important for identification of bacteria and yeasts as described above, the mycolic acids can be used to characterize bacteria that contain this group of compounds. A typical protocol for analysis includes saponification of the mycolic acids, derivatization into phenyl esters for UV visualization, extraction into an organic solvent, and washing of the final product to remove contaminants. Approximately 60 mg of bacterial cells is placed into an extraction tube for saponification (0.9 M NaOH in 50% ethanol at 85°C for 16 h), acidification with HCl (0.6 N), and extraction into chloroform. The chloroform is then removed under nitrogen and phenyl esters formed by using *p*-bromophenylacyl as the derivatizing agent. Analysis by HPLC is performed using a C_{18} reverse-phase column and a linear mobile-phase gradient (acetonitrile-chloroform, 90:10 to 40:60). Peaks are detected with a UV detector, and retention times are compared to those for known organisms. Profiles of unknown strains may then be compared to those for *Mycobacterium*, *Rhodococcus*, and *Nocardia* spp. (8, 9).

GLC Analysis of Carbohydrates and Other Cellular Components

Sugars that are structural in microorganisms have been less frequently used for bacterial identification. The major reasons for this are the difficulty in preparing samples for analysis and the inability to reproducibly analyze results. Complex carbohydrates must first be derivatized before chromatographic analysis. Many of these procedures may also lead to degradation of the compounds being analyzed, further complicating the process. Despite these drawbacks, sugars have been used for the phenotypic characterization of certain organisms (24, 48).

Similarly, the isoprenoid quinone compounds, ubiquitous as part of the electron transport chain of all organisms, have been used for characterization purposes. Such compounds are most often assayed by TLC or HPLC. Detection is usually accomplished by dye visualization (TLC) or by variable-wavelength UV light. As with carbohydrates, the methods for cleanly separating these compounds from each other have restricted their usefulness as commonly assayed phenotypic characteristics (26, 40, 43).

Multicomponent Analysis

Pyrolysis GC is based on the thermal decomposition of whole bacterial cells through extremely rapid heating. This process generates a variety of products that can then be separated and detected using GC techniques. Identification of peaks is via a mass spectrometer. As the signal source, the actual process generally uses a small wire to which sample has been applied. The sample is heated rapidly by passing a current through the wire, and the resultant products of pyrolysis are swept onto a separation column by the mobile phase. Although the same isolate usually yields very consistent results, sample preparation may influence the outcome of analysis. Moreover, the heating process may not always cause exactly the same biomolecular cleavage from strain to strain. Despite these problems, recent work with this technique indicates that it is a viable method for identifying bacteria (19, 24, 55).

Strategies for Identification of Bacteria Using Chromatographic Techniques

Laboratory application of chromatography for identification of bacteria and yeasts incorporates a broad array of potential methodologies. Techniques ranging from the use of TLC separation of quinones and comparison of them to those of known organisms to complex FAME pattern recognition following GLC separation are readily available (34, 37). A general introduction to the subject is provided by Novotny (43). Other useful references include articles by Moss (40) and a discussion of chemotaxonomy by Jones and Krieg in *Bergey's Manual* (31).

The application of chromatographic techniques to microbial identification involves several important considerations. Issues such as standardization of growth, extraction and derivatization methods, and compounds to be analyzed are all significant factors for a successful program. Additionally, the type of instrumentation used, method of data analysis, and algorithms used for actual identification purposes require some thought on the part of the laboratorian. Fortunately, many of the methods employed today have been standardized for use by any laboratory. Materials and methods sections from scientific publications can often be implemented directly without modification. Other issues such as peak naming and quantitation of components present are best pursued through the use of commercially available derivatizing reagents or standard mixtures such as those supplied by Supelco (Bellefonte, Pa.) and Sigma (St. Louis, Mo.). Alternatively, direct comparison with known organisms prepared for analysis in the same manner as the unknown strain may be used.

The reference literature is a useful starting point for the HPLC analysis of quinones, mycolic acids, and sugars. Similarly, references for GLC methods with volatile metabolites, short-chain fatty acids, and sugars may simplify the assay development time for these procedures (3, 5, 16–18, 23, 29, 30, 33, 36, 38, 41, 42, 44, 47, 49, 51, 53, 54, 57, 58). Data analysis may be as simple as peak identification or may include more sophisticated methods including pattern recognition and cluster analysis. The commercial vendors for chromatography supplies are often an important source of information regarding new techniques and optimum performance for specific assays and as a referral to other laboratories already using a particular procedure.

SUMMARY

Chromatographic analysis of various microbial components offers a significant alternative to more traditional identification procedures. GLC, HPLC, TLC, and other GC-based methods can supply supplemental phenotypic information that aids in identification or may serve as stand-alone identification. Many of these methods are highly automated through the use of automatic sample injection and computer analysis of detected components. The advantage of such systems is the objective manner in which identification is accomplished, the low cost per sample, and the ability to identify both asaccharolytic organisms and those with enzymatic profiles that are not distinctive. The disadvantages of these systems are the large capital expense required for initial equipment purchase, nonacceptance of alternative identification strategies, and the size of the library of known organisms that is required. However, the declining costs of computer hardware coupled to the increasing size of commercially available identification libraries have greatly facilitated the acceptance of such methods. The increasing awareness of microbiologists that there is a variety of useful biologic signals that aid in rapid and reliable identification of microorganisms indicates that chromatographic methods are being accepted as an integral part of the microbiology laboratory.

REFERENCES

1. **Athalye, M., W. C. Noble, and D. E. Minnikin.** 1985. Analysis of cellular fatty-acids by gas chromatography as a tool in the identification of medically important coryneform bacteria. *J. Appl. Bacteriol.* **58:**507–512.
2. **Belton, K. B.** 1988. Advances in LC. *Anal. Chem.* **60:**1045A–1047A.
3. **Bernard, K. A., M. Bellefeuille, and E. P. Ewan.** 1991. Cellular fatty acid composition as an adjunct to the identification of asporogenous, aerobic gram-positive rods. *J. Clin. Microbiol.* **29:**83–89.
4. **Borman, S.** 1987. Hyphenated MS at the Pittsburgh Conference. *Anal. Chem.* **59:**769A–774A.
5. **Brondz, I., and I. Olsen.** 1991. Multivariate analyses of cellular fatty acids in *Bacteroides*, *Prevotella*, *Porphyromonas*, *Wolinella*, and *Campylobacter* spp. *J. Clin. Microbiol.* **29:**183–189.
6. **Brondz, I., I. Olsen, and M. Sjostrom.** 1989. Gas chromatographic assessment of alcoholized fatty acids from yeasts: a new taxonomic method. *J. Clin. Microbiol.* **27:**2815–2819.
7. **Brown, P. R.** 1990. High performance liquid chromatography, past developments, present status, and future trends. *Anal. Chem.* **62:**995A–1007A.
8. **Butler, W. R., and J. O. Kilburn.** 1988. Identification of major slowly growing pathogenic mycobacteria and *Mycobacterium gordonae* by high-performance liquid chromatography of their mycolic acids. *J. Clin. Microbiol.* **26:**50–53.
9. **Butler, W. R., J. O. Kilburn, and G. P. Kubica.** 1987. High-performance liquid chromatography analysis of mycolic acids as an aid in laboratory identification of *Rhodococcus* and *Nocardia* species. *J. Clin. Microbiol.* **25:**2126–2131.
10. **Christian, G. D.** 1971. *Analytical Chemistry*, p. 104–140. Xerox College Publishing, Ginn and Co., Waltham, Mass.
11. **Damato, J. J., C. Kniseley, and M. T. Collins.** 1987. Characterization of *Mycobacterium paratuberculosis* by gas-liquid and thin-layer chromatography and rapid demonstration of mycobactin dependence using radiometric methods. *J. Clin. Microbiol.* **25:**2380–2383.
12. **Davies, I. A., M. W. Raynor, J. P. Kithinji, K. D. Bartle, P. T. Williams, and G. E. Andrews.** 1988. Interfacing LC/GC, SFC/GC, and SFE/GC. *Anal. Chem.* **60:**683A–702A.
13. **Eerola, E., and O.-P. Lehtonen.** 1988. Optimal data processing procedure for automatic bacterial identification by gas-liquid chromatography of cellular fatty acids. *J. Clin. Microbiol.* **26:**1745–1753.
14. **Ewing, G. W.** 1975. *Instrumental Methods of Chemical Analysis*, 4th ed., p. 364–387. McGraw-Hill Book Co., New York.
15. **Ewing, G. W.** 1975. *Instrumental Methods of Chemical Analysis*, 4th ed., p. 388–411. McGraw-Hill Book Co., New York.
16. **Ferguson, D. A., Jr., C. Li, N. R. Patel, W. R. Mayberry, D. S. Chi, and E. Thomas.** 1993. Isolation of *Helicobacter pylori* from saliva. *J. Clin. Microbiol.* **31:**2802–2804.
17. **Floyd, M. M., V. A. Silcox, W. D. Jones, Jr., W. R. Butler, and J. O. Kilburn.** 1992. Separation of *Mycobacterium bovis* BCG from *Mycobacterium tuberculosis* and *Mycobacterium bovis* by using high-performance liquid chromatography of mycolic acids. *J. Clin. Microbiol.* **30:**1327–1330.
18. **Fox, A., G. E. Black, K. Fox, and S. Rostovtseva.** 1993. Determination of carbohydrate profiles of *Bacillus anthracis* and *Bacillus cereus* including identification of o-methyl methylpentoses by using gas chromatography-mass spectrometry. *J. Clin. Microbiol.* **31:**887–894.
19. **French, G. L., I. Phillips, and S. Chinn.** 1981. Reproducible pyrolysis gas chromatography of microorganisms with solid stationary phases and isothermal oven temperatures. *J. Gen. Microbiol.* **125:**347–356.
20. **Fritz, J. S., and G. H. Schenk.** 1974. *Quantitative Analytical Chemistry*, 3rd ed., p. 367–382. Allyn and Bacon, Inc., Boston.
21. **Fritz, J. S., and G. H. Schenk.** 1974. *Quantitative Analytical Chemistry*, 3rd ed., p. 383–402. Allyn and Bacon, Inc., Boston.
22. **Fritz, J. S., and G. H. Schenk.** 1974. *Quantitative Analytical Chemistry*, 3rd ed., p. 403–420. Allyn and Bacon, Inc., Boston.
23. **Ghanem, F. M., A. C. Ridpath, W. E. C. Moore, and L. V. H. Moore.** 1991. Identification of *Clostridium botulinum*, *Clostridium argentinense*, and related organisms by cellular fatty acid analysis. *J. Clin. Microbiol.* **29:**1114–1124.
24. **Gilbart, J., A. Fox, and S. L. Morgan.** 1987. Carbohydrate profiling of bacteria by gas chromatography-mass spectrometry; chemical derivatization and analytical pyrolysis. *Eur. J. Clin. Microbiol.* **6:**715–723.
25. **Gurka, D. F., L. D. Betowski, T. A. Hinners, E. M. Heithmar, R. Titus, and J. M. Henshaw.** 1988. Environmental applications of hyphenated quadrupole techniques. *Anal. Chem.* **60:**454A–467A.
26. **Harpold, D. J., and B. L. Wasilauskas.** 1987. Rapid identification of obligately anaerobic gram-positive cocci using high-performance liquid chromatography. *J. Clin. Microbiol.* **25:**996–1001.
27. **Heftmann, E. (ed.).** 1975. *Chromatography: a Laboratory Handbook of Chromatographic and Electrophoretic Methods*, 3rd ed. Van Nostrand Reinhold Co., New York.
28. **Holdeman, L. V., E. P. Cato, and W. E. C. Moore.** 1977. *Anaerobe Laboratory Manual*, 4th ed. Virginia Polytechnic Institute and State University, Blacksburg.
29. **Hollis, D. G., R. E. Weaver, C. W. Moss, M. I. Daneshvar, and P. L. Wallace.** 1992. Chemical and cultural characterization of CDC group WO-1, a weakly oxidative gram-negative group of organisms isolated from clinical sources. *J. Clin. Microbiol.* **30:**291–295.
30. **Holst, E., J. Rollof, L. Larsson, and J. P. Nielson.** 1992. Characterization and distribution of *Pasteurella* species recovered from infected humans. *J. Clin. Microbiol.* **30:**2984–2987.
31. **Jones, D., and N. R. Krieg.** 1984. Bacterial classification V: serology and chemotaxonomy, p. 15–18. *In* N. R. Krieg and J. G. Holt (ed.), *Bergey's Manual of Systematic Bacteriology*, vol. 1. The Williams & Wilkins Co., Baltimore.
32. **Kenner, C. T., and R. E. O'Brien.** 1971. *Analytical Separations and Determinations*, p. 272–294. The Macmillan Co., New York.
33. **Kotilainen, P., P. Huovinen, and E. Eerola.** 1991. Application of gas-liquid chromatographic analysis of cellular fatty acids for species identification and typing of coagulase-negative staphylococci. *J. Clin. Microbiol.* **29:**315–322.
34. **Lambert, M. A., and C. W. Moss.** 1989. Cellular fatty acid compositions and isoprenoid quinone contents of 23 *Legionella* species. *J. Clin. Microbiol.* **27:**465–473.
35. **Levy-Frebault, V., K.-S. Goh, and H. L. David.** 1986. Mycolic acid analysis for clinical identification of *Mycobacterium avium* and related mycobacteria. *J. Clin. Microbiol.* **24:**835–839.
36. **Luquin, M., V. Ausina, F. L. Calahorra, F. Belda, M. G. Barcelo, C. Celma, and G. Prats.** 1991. Evaluation of practical chromatographic procedures for identification of clinical isolates of mycobacteria. *J. Clin. Microbiol.* **29:**120–130.
37. **Maliwan, N., R. W. Reid, S. R. Pliska, T. J. Bird, and J. R. Zvetina.** 1988. Identifying *Mycobacterium tuberculosis* cultures by gas-liquid chromatography and a computer-aided pattern recognition model. *J. Clin. Microbiol.* **26:**182–187.
38. **Merz, W. G., U. Khazan, M. A. Jabra-Rizk, L. Wu, G. J. Osterhout, and P. F. Lehmann.** 1992. Strain delineation and

epidemiology of *Candida* (*Clavispora*) *lusitaniae*. *J. Clin. Microbiol.* **30:**449–454.

39. **Miller, L. T.** 1982. Single derivatization method for routine analysis of bacterial whole-cell fatty acid methyl esters, including hydroxy acids. *J. Clin. Microbiol.* **16:**584–586.

40. **Moss, C. W.** 1985. Uses of gas-liquid chromatography and high-pressure liquid chromatography in clinical microbiology, p. 1029–1036. *In* E. H. Lennette, A. Balows, W. J. Hausler, Jr., and H. J. Shadomy (ed.), *Manual of Clinical Microbiology*, 4th ed. American Society for Microbiology, Washington, D.C.

41. **Moss, C. W., and M. I. Daneschvar.** 1992. Identification of some uncommon monounsaturated fatty acids of bacteria. *J. Clin. Microbiol.* **30:**2511–2512.

42. **Mukwaya, G. M., and D. F. Welch.** 1989. Subgrouping of *Pseudomonas cepacia* by cellular fatty acid composition. *J. Clin. Microbiol.* **27:**2640–2646.

43. **Novotny, M. V.** 1989. Recent developments in analytical chromatography. *Science* **246:**51–77.

44. **Paulsrud, J. R., S. F. Queener, M. S. Bartlett, and J. W. Smith.** 1993. Total cellular fatty acid composition of cultured *Pneumocystis carinii*. *J. Clin. Microbiol.* **31:**1899–1902.

45. **Peters, D. G., J. M. Hayes, and G. M. Hieftje.** 1974. *Chemical Separations and Measurements*, p. 560–580. The W. B. Saunders Co., Philadelphia.

46. **Poole, C. F., and S. K. Poole.** 1989. Modern thin-layer chromatography. *Anal. Chem.* **61:**1257A–1269A.

47. **Portaels, F., D. J. Dawson, L. Larsson, and L. Rigouts.** 1993. Biochemical properties and fatty acid composition of *Mycobacterium haemophilum*: study of 16 isolates from Australian patients. *J. Clin. Microbiol.* **31:**26–30.

48. **Pritchard, D. G., J. E. Coligan, S. E. Speed, and B. M. Gray.** 1981. Carbohydrate fingerprints of streptococcal cells. *J. Clin. Microbiol.* **13:**89–92.

49. **Regnery, R. L., B. E. Anderson, J. E. Clarridge III, M. C. Rodriguez-Barradas, D. C. Jones, and J. H. Carr.** 1992. Characterization of a novel *Rochalimaea* species, *R. henselae* sp. nov., isolated from blood of a febrile, human immunodeficiency virus-positive patient. *J. Clin. Microbiol.* **30:**265–274.

50. **Snyder, L. R., and J. J. Kirkland.** 1979. *Introduction to Modern Liquid Chromatography*, 2nd ed. John Wiley & Sons, Inc., New York.

51. **Stoakes, L., T. Kelly, B. Schieven, D. Harley, M. Ramos, R. Lannigan, D. Groves, and Z. Hussain.** 1991. Gas-liquid chromatographic analysis of cellular fatty acids for identification of gram-negative anaerobic bacilli. *J. Clin. Microbiol.* **29:**2636–2638.

52. **Summanen, P., E. J. Baron, D. M. Citron, C. A. Strong, H. M. Wexler, and S. M. Finegold.** 1993. *Wadsworth Anaerobic Bacteriology Manual*, 5th ed., p. 65–88. Star Publishing Co., Belmont, Calif.

53. **Thibert, L., and S. Lapierre.** 1993. Routine application of high-performance liquid chromatography for identification of mycobacteria. *J. Clin. Microbiol.* **31:**1759–1763.

54. **Tuner, K., E. J. Baron, P. Summanen, and S. M. Finegold.** 1992. Cellular fatty acids in *Fusobacterium* species as a tool for identification. *J. Clin. Microbiol.* **30:**3225–3229.

55. **Wallace, P. L., D. G. Hollis, R. E. Weaver, and C. W. Moss.** 1988. Cellular fatty acid composition of *Kingella* species, *Cardiobacterium hominis*, and *Eikenella corrodens*. *J. Clin. Microbiol.* **26:**1592–1594.

56. **Warner, M.** 1987. LC/MS and SFC/MS: will they replace GC/MS? *Anal. Chem.* **59:**855A–858A.

57. **Welch, D. F.** 1991. Applications of cellular fatty acid analysis. *Clin. Microbiol. Rev.* **4:**422–438.

58. **Welch, D. F., D. A. Pickett, L. N. Slater, A. G. Steigerwalt, and D. J. Brenner.** 1992. *Rochalimaea henselae* sp. nov., a cause of septicemia, bacillary angiomatosis, and parenchymal bacillary peliosis. *J. Clin. Microbiol.* **30:**275–280.

59. **Wilkins, C. L.** 1987. Linked gas chromatography infrared mass spectrometry. *Anal. Chem.* **59:**571A–581A.

60. **Willard, H. H., L. L. Merritt, Jr., J. A. Dean, and F. A. Settle, Jr.** 1981. *Instrumental Methods of Chemical Analysis*, 6th ed., p. 430–453. Wadsworth Publishing Co., Belmont, Calif.

61. **Willard, H. H., L. L. Merritt, Jr., J. A. Dean, and F. A. Settle, Jr.** 1981. *Instrumental Methods of Chemical Analysis*, 6th ed., p. 454–494. Wadsworth Publishing Co., Belmont, Calif.

62. **Willard, H. H., L. L. Merritt, Jr., J. A. Dean, and F. A. Settle, Jr.** 1981. *Instrumental Methods of Chemical Analysis*, 6th ed., p. 495–528. Wadsworth Publishing Co., Belmont, Calif.

63. **Willard, H. H., L. L. Merritt, Jr., J. A. Dean, and F. A. Settle, Jr.** 1981. *Instrumental Methods of Chemical Analysis*, 6th ed., p. 529–564. Wadsworth Publishing Co., Belmont, Calif.

Molecular Detection and Identification of Microorganisms

RAYMOND P. PODZORSKI AND DAVID H. PERSING

13

The intent of this chapter is twofold: (i) to provide an overview of the nucleic acid probes and nucleic acid amplification techniques that are available for use in the microbiology laboratory and (ii) to illustrate how the techniques are being applied or can be applied in the laboratory setting. Enough detail concerning these techniques and their applications will be provided so that the reader can understand how they work and begin to evaluate their strengths and weaknesses. Information is also provided on how to prepare specimens for molecular analysis, and general guidelines for the setup and management of a molecular microbiology laboratory are given. This chapter is not intended to provide specific step-by-step instructions for molecular diagnostic procedures. The reader is referred to other sources for more detailed descriptions of these procedures and their applications (86, 126).

MOLECULAR DIAGNOSTIC TESTS: BENEFITS AND LIMITATIONS

At present, DNA probes and nucleic acid amplification techniques are most useful for the characterization of microorganisms for which culture and serologic methods are difficult, extremely expensive, or unavailable. Conventional (nonamplified) DNA probe-based assays are particularly well suited for in situ hybridization in tissue in which the location and distribution of the organisms must be ascertained; for culture confirmation of slow-growing microorganisms such as *Mycobacterium tuberculosis*, *Neisseria gonorrhoeae*, and pathogenic dimorphic fungi; and for identification of M. *tuberculosis* or *Mycobacterium avium-Mycobacterium intracellulare* in broth cultures once sufficient growth has been detected radiometrically. Numerous other DNA probes are also commercially available for culture confirmation of many species of bacteria and fungi (Table 1). In addition to their use in culture confirmation, DNA probes have proved useful for identification of enterotoxin-producing strains of *Escherichia coli* (enterotoxigenic *E. coli*) because these organisms are biochemically identical to strains of *E. coli* that do not produce enterotoxins. Despite the fact that conventional DNA probe techniques are less sensitive than nucleic acid amplification techniques, probes have been successfully used for identification of microorganisms directly in clinical specimens in certain situations. DNA probes are used to detect organisms as diverse as group A streptococci in cases of strep throat and *Chlamydia trachomatis* or human papillomavirus in cervical infections.

The high level of sensitivity afforded by DNA amplification techniques makes these techniques the method of choice for direct detection of microbial DNA in clinical specimens. Even though at this writing only one commercial kit has been cleared by the U.S. Food and Drug Administration for identification of a microorganism from a patient specimen (Roche Amplicor for *C. trachomatis*), there are many reports in the literature of microorganisms being identified directly in patient specimens by nucleic acid amplification techniques. It is certain that many more commercial nucleic acid amplification-based kits will soon follow. The power of these techniques also makes them very attractive for detection of specific genes in cultured bacteria, because a specially labeled reporter probe is frequently not required. Because of the large amount of product generated during the reaction, simple ethidium bromide staining of amplified DNA is frequently all that is needed for detection.

Expensive molecular diagnostic procedures should not be used to replace procedures currently in place that have been proven to be cost-effective, rapid, sensitive, and reliable. For example, use of the shell vial and monoclonal antibodies to cytomegalovirus (CMV) to diagnose CMV viremia in immunosuppressed bone marrow transplant recipients is an excellent test in terms of sensitivity and specificity and in most cases results in a culture report 16 h from the time of specimen inoculation. It would be counterproductive to abandon the shell vial in favor of a more costly nucleic acid amplification procedure for the detection of CMV viremia without documentation of substantial benefits.

In making the decision to use a molecular diagnostic procedure, the laboratory must take into consideration all the potential benefits that the new technology may offer. The decision should not be based solely on a price comparison between the conventional assay and a molecular diagnostic assay. Instead, the decision should take into account the impact new diagnostic technology will have on clinical practice and patient management and the cost savings associated with it. For example, the rapid and reliable identification of an isolate of methicillin-resistant *Staphylococcus aureus* or the direct detection of multidrug-resistant M. *tuberculosis* in a patient specimen would lead to quicker patient isolation, a better informed choice of patient therapy, and a decrease in the time the patient is infectious. The costs saved by containing the spread of these organisms and shortening the period of patient morbidity are part of the cost-benefit equation that should be considered when the use of molecular diagnostic assays is evaluated.

It is also important to remember that despite all their sensitivity and speed, nucleic acid amplification procedures will not replace conventional culture and serologic procedures in all situations, because the results of nucleic acid amplification procedures and the results of culture or serology mean different things. Nucleic acid amplification procedures determine whether DNA or RNA from a particular organism is present in the specimen. They reveal nothing about the viability of the organism (because they can detect DNA from dead organisms) or whether the organism is involved in an infectious process. Culture, on the other hand, clearly demonstrates the viability of the organism, while a rise in titer of antibody to a specific organism agent strongly suggests involvement in infection.

MOLECULAR PROBES FOR DETECTION, IDENTIFICATION, AND CHARACTERIZATION OF MICROORGANISMS

Commercial DNA probes brought molecular biology into the clinical microbiology laboratory and have been in use for several years. The procedures for use of DNA probes are

TABLE 1 Organisms for which nucleic acid probes[a] are commercially available

Direct detection	Culture identification
Bacteria	Bacteria
Chlamydia trachomatis	Campylobacter species
Neisseria gonorrhoeae	Enterococcus species
Group A Streptococcus	Group B Streptococcus
Legionella pneumophila	Haemophilus influenzae
Gardnerella vaginalis[b]	Neisseria gonorrhoeae
Protozoa	Streptococcus pneumoniae
Trichomonas vaginalis[b]	Staphylococcus aureus
Fungi	Listeria monocytogenes
Candida species[b]	Group A Streptococcus
Virus	Mycobacterium avium
Human papillomavirus[c]	Mycobacterium intracellulare
	Mycobacterium avium complex
	Mycobacterium gordonae
	Mycobacterium tuberculosis complex
	Mycobacterium kansasii
	Fungi
	Blastomyces dermatitidis
	Coccidioides immitis
	Cryptococcus neoformans
	Histoplasma capsulatum
	Viruses
	Human papillomavirus[c]

[a]Probes have been cleared by the U.S. Food and Drug Administration. All probes are manufactured by Gen-Probe, Inc., San Diego, Calif., unless otherwise noted.
[b]Manufactured by MicroProbe Corp., Bothell, Wash. All three organisms are detected together in a single assay.
[c]Manufactured by Digene Diagnostics, Inc., Silver Spring, Md. Only specific types were detected.

TABLE 2 Comparison of detection sensitivities

Method	Approx copy no. detectable
Ethidium bromide staining	10^8
Radiolabeled oligonucleotide probes	10^6
Radiolabeled full-length probes	10^4
Enzyme-coupled probes	10^4
Chemiluminescent probes	10^4
Compound or branched probes	10^4
Nucleic acid amplification	≤ 10

now well standardized and frequently simpler than those for target or probe amplification assays. In addition, the advent of synthetic short (20 to 50 bp in length) oligonucleotide DNA probes has shortened the time required for probe hybridization assays. Large quantities of short sequences can drive hybridization reactions to completion more quickly than similar quantities of probes of 500 to 1,000 bases in length. However, the sensitivity that can be attained with conventional probes is still lower than that of nucleic acid amplification procedures, and this remains the main drawback to more widespread use of probe technology directly on patient specimens (Table 2). The niche for DNA probe technology in the clinical microbiology laboratory appears to be for in situ hybridization in tissue specimens and culture confirmation of slow-growing microorganisms. While DNA probes may be more expensive on a per test basis than conventional biochemical identification procedures, their rapid turnaround time can translate into an overall savings by providing a more timely diagnosis. In addition, using standardized methodology with DNA probes can expand the types of pathogens that a microbiology laboratory can routinely identify. The three DNA probe hybridization formats most commonly used today (solid-phase hybridization, solution-phase hybridization, and in situ hybridization) are discussed here. In addition, we also discuss probe amplification procedures and frequently used probe reporter molecules.

Solid-Phase Hybridization

The slot or dot blot is the DNA probe hybridization procedure most commonly employed in research and in some clinical laboratories (21, 46). In this procedure, intact cells are lysed, and the DNA is denatured, brought into direct contact with a nylon membrane by vacuum aspiration in a dot or slot configuration, and then fixed onto the membrane. Nucleic acids bound to solid supports such as nylon membranes are still available to participate in hybridization reactions. The membrane is immersed in a hybridization solution containing the DNA reporter probe and is allowed to hybridize. The unbound reporter probes are then washed away, and the bound probe is detected. Dot or slot blots are especially useful for the processing of multiple specimens at one time, since numerous specimens can be screened on a single filter. A variation on the dot or slot blot utilizes sandwich or capture hybridization (119, 136). This procedure uses two probes that bind to different sites on the target nucleic acid. One probe is attached to the membrane and serves to capture the target nucleic acid in the sample. The second probe is labeled and serves as the detector. This system was developed to circumvent the direct binding of target to the membrane and in some cases is more sensitive than standard membrane hybridization. Currently, slot blot assays for screening and typing several strains of human papillomaviruses are commercially available (Table 1).

The Southern blot is a solid-phase hybridization assay that allows size determination of DNA fragments bound by the reporter probe (20, 46, 153). In the Northern blot, a variation on the Southern blot, the reporter probe is used to detect RNA instead of DNA (22, 137). These procedures generally require purified sample DNA or RNA. In the Southern blot, sample DNA is isolated and digested with a restriction endonuclease, and the fragments are separated by size by agarose gel electrophoresis. In the Northern blot, total sample RNA is isolated and separated by size by denaturing agarose gel electrophoresis. In both types of blots, the nucleic acids are transferred to a nitrocellulose or nylon membrane for hybridization with a specific probe. Because of their complexity, the requirement for large amounts of sample nucleic acids, and the time required for completion of the assay, these procedures are not routinely used in microbiology laboratories.

Solution-Phase Hybridization

Many commercially available DNA probe kits are based on solution-phase hybridization, in which both the target nucleic acid and the probe are free to interact in an aqueous reaction mixture (106, 168). The free target-probe interaction in the aqueous environment speeds the rate at which hybridization occurs. The key to successful solution-phase hybridization is a single-stranded DNA probe that does not hybridize with itself. Hybridization in solution can be detected by S1 nuclease digestion of single-stranded DNA and recovery of the remaining double-stranded reporter-probe hybrids by quantitative trichloroacetic acid precipitation (17) or directly by specific binding of the double-stranded hybrids to hydroxyapatite columns, which bind double-stranded DNA selectively.

The leading method in routine use for detecting probe hybridization in solution is the hybridization protection assay (HPA) (4, 113). In HPA, an acridinium ester moiety attached to a DNA probe is hybridized to target DNA and treated with alkali. Upon addition of peroxides, the acridinium ester emits detectable light. However, in its free form, the acridinium ester-labeled DNA probe is not protected from hydrolysis and is converted to a form that does not emit light. The hybridization protection assay can be performed in a few hours and does not require removal of excess unbound single-stranded DNA or isolation of probe-bound double-stranded DNA complexes.

In Situ Hybridization

In situ hybridization assays involve the same general principles as solid- and solution-phase hybridizations. However, in situ hybridization occurs within the morphologic context of infected tissue (73, 155). When considered in conjunction with tissue morphology and host response, the information provided by these assays provides confirmation that the target organism is involved in an infectious process and gives additional information regarding distribution and abundance. In clinical settings, formalin-fixed paraffin-embedded tissue sections frequently form the starting material for in situ hybridization assays. In situ hybridization requires that target nucleic acid be made accessible for hybridization while the cellular morphology is preserved for subsequent interpretation and analysis. Therefore, the tissue processing must be delicate enough to preserve the cellular structure while denaturing the nucleic acids in the cell and making

them accessible to the probe. The sensitivity of in situ hybridization is limited by the accessibility of target nucleic acids within the cell. Because of the accessibility limitations associated with mammalian cells, relatively small probes (300 bases or less) should be used to favor tissue penetration (159).

Signal Amplification Schemes

Several ingenious signal amplification schemes have been developed in recent years to increase the sensitivities of probe-based assays (54, 171). Unlike nucleic acid target amplification or probe amplification procedures, signal amplification procedures are designed to increase the signal generated by the probe hybridized to a specific sequence of target DNA or RNA. In order to provide signal amplification, various techniques for attaching a large number of reporter molecules to the probe or the target sequence have been devised. The fact that signal amplification procedures do not involve nucleic acid target or probe amplification is a theoretical advantage, because it makes these amplification procedures less prone to the contamination problems that are of concern in enzyme-catalyzed amplification procedures. However, a relative lack of sensitivity compared to that of nucleic acid amplification procedures is still a drawback of these procedures. Signal amplification procedures can detect a minimum of 10^3 to 10^5 nucleic acid targets (162), which is well below the sensitivity of highly optimized nucleic acid target or probe amplification procedures. The chief limitation of signal amplification is the background noise due to nonspecific binding of reporter probes. Problems with high background levels have led to procedural modifications such as incorporating target capture probes to reduce the background noise levels associated with signal amplification procedures (83).

Several different types of signal amplification procedures have been developed. These systems may be as simple as using multiple labeled probes for different target sites or as elaborate as using combinations of primary and secondary probes coupled with multiple enzymes to amplify the probe signal (54, 171). The simplest signal amplification technique utilizes multiple reporter molecules attached to a single target probe. In this case, a single probe is able to produce a much greater signal than a probe with only one reporter molecule. A variation on this procedure takes a slightly different approach: instead of using multiple reporter molecules on a single probe, multiple labeled probes are used. The labeled probes are directed to different regions of the target sequence, allowing multiple hybridizations to occur in order to produce a greater signal than an individual probe could produce. More complicated signal amplification procedures have been developed to produce even greater signals. One such system includes an unlabeled target-binding probe that contains not only a region complementary to the target sequence but also a region capable of hybridizing with multiple smaller target-independent reporter probes. This system forms a compound probe network that produces a significant improvement in signal, especially when multiple reporter molecules are attached to the reporter probes.

Arguably the most powerful signal amplification system is the branched DNA (bDNA) probe system (developed by Chiron Corp.), which utilizes multiple probes together with multiple reporters (145, 162). This system utilizes a specific target capture step for isolation of the nucleic acid sequence of interest followed by hybridization with an unlabeled

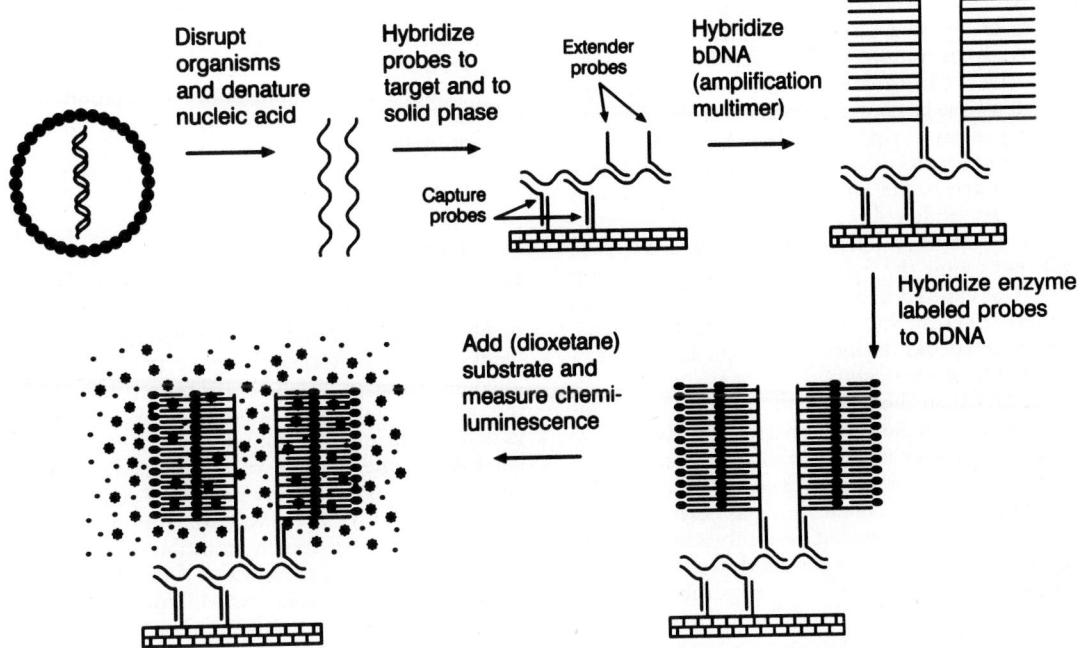

FIGURE 1 bDNA-based signal amplification. Target nucleic acid is released by disruption and capture to a solid surface via multiple contiguous capture probes. Contiguous extender probes hybridize with adjacent target sequences and contain additional sequences homologous to the branched amplification multimer. Enzyme-labeled oligonucleotides bind to the bDNA by homologous base pairing, and the enzyme-probe complex is measured by detection of chemiluminescence. All hybridization reactions occur simultaneously.

target-binding probe (Fig. 1). The target-binding probe used here is similar to the target-binding probe used in the compound probe system in that it contains two hybridization sequences. One region is complementary to the target sequence, and the other is capable of hybridizing with the bDNA amplification multimer. The amplification multimer is the key to the amplifying power of this technique. The amplification multimer is chemically synthesized with a comblike backbone made from a branching nucleoside analog that is incorporated at regular intervals along the oligonucleotide. The multimer binds many target-independent reporter probes, providing a significant boost in signal. Depending on the type of reporter probe used in this system, up to 3,000 reporter molecules can be incorporated into each target sequence. The bDNA probe system has high specificity because both capture and target probes must bind before the signal amplification can take place. Nonspecific hybridization rarely allows hybridization of both capture and target probes to the same nontarget sequence.

An advantage of bDNA probe systems is the ease with which several different capture and target-binding probes can be incorporated into a single test system that is capable of detecting nucleic acid targets with significant sequence heterogeneity, as is the case for hepatitis C virus and human immunodeficiency virus (HIV). If some specific capture or target probes fail to hybridize because of sequence variation, signal production is not lost completely because of the presence of several remaining probe complexes. bDNA probe systems are being developed for hepatitis B and C viruses (44, 172), HIV type 1 (HIV-1) (118), and CMV (151), viruses for which specific antiviral therapies are available. Because bDNA probe systems provide quantita-

tive detection over a range of several orders of magnitude, such assays may be useful for monitoring therapeutic responses during antiviral therapy. The determination of responses to antiviral therapy may be useful in assessing the need for further therapy or for adjusting doses of antiviral agents.

Reporter Molecules for Detection of Nucleic Acid Hybridization

In the past, radioactive isotopic labels (^{32}P, ^{35}S, and ^{125}I) have been the most commonly used reporter molecules. These reporters can be incorporated into the nucleic acid probe by using techniques such as end labeling or random priming (55, 56, 144). Autoradiography or beta counters are then used to detect hybridization reactions. For routine use in the clinical laboratory, radioactive isotopic detection systems have several disadvantages, including limited shelf life based on the half-life of the isotope, radiation hazard and the associated regulations that must be met by users, and disposal problems. Many nonradioactive detection systems have been developed recently, and the poor specific activities of the first-generation systems have given way to much better performance. Nonradioactive systems have the advantages of longer shelf lives, no associated special regulations, and no complicated handling or disposal problems.

The types of nonradioactive detection systems currently available utilize labels such as biotin, digoxigenin, and horseradish peroxidase (HRP). These molecules can be incorporated directly into nucleic acid probes by enzymatic or nonenzymatic methods (41, 114). These reporter groups can be detected (i) chromogenically, such as when avidin is conjugated to an enzyme (usually peroxidase or alkaline phosphatase) or by an antidigoxigenin alkaline phosphatase conjugate, or (ii) by HRP-catalyzed light production. Many of the procedures for labeling DNA-RNA probes are now available in the form of commercially prepared kits complete with chromogenic or chemiluminescent substrates. Chemiluminescence is the latest nonradioactive reporter technology to be exploited. Chemiluminescent reporters are chemical groups that release light when exposed to certain substrates after the hybridization reaction is complete (132). The emitted light can be detected by several means, i.e., autoradiography, a specially formatted scintillation counter, or a dedicated instrument called a luminometer. Chemiluminescence assays have sensitivity ranges equal to or greater than those using ^{32}P-labeled probes plus the advantages of nonradioactive reporter systems.

For probe detection by in situ hybridization, radioactive label coupled with autoradiographic detection using a silver emulsion overlaid on a specimen slide has been the most popular (14). However, assays that utilize the chromogenic procedures described above for hybridization detection are now commercially available (161). Biotin is the nonradioactive label most commonly used with in situ hybridization, with other tags such as digoxigenin used increasingly. After the conjugated enzyme is incubated with the substrate, a colored precipitate forms over the hybridization site. However, radioactive probes remain the "gold standard" for sensitivity, detecting as few as 10 copies of mRNA per cell for in situ hybridization (73).

NUCLEIC ACID AMPLIFICATION TECHNIQUES

The primary objective of nucleic acid amplification techniques is to improve the sensitivity of assays based on nucleic acids and to eventually simplify these assays by automation and the incorporation of nonradioactive detection formats. The technology behind nucleic acid amplification methods is diverse and in a constant state of flux. Because of this complexity, it is useful to assign such methods to one of three general categories: (i) target amplification systems such as PCR, self-sustaining sequence replication (3SR), or strand displacement amplification (SDA); (ii) probe amplification systems, which include Qβ replicase or the ligase chain reaction (LCR); and (iii) signal amplification (described above), in which the signal generated from each probe molecule is increased by using compound probes or branched-probe technology.

The process of in vitro nucleic acid amplification is somewhat analogous to the process of biologic amplification, or growth of an organism in culture, with enzymatic duplication and amplification of specific nucleic acid sequences. Like different types of culture media (i.e., primary isolation medium or selective medium), nucleic acid amplification-based assays can be designed to have broad or narrow specificity. Some amplification protocols are selective. They are designed to specifically amplify a nucleic acid target from a particular organism or group of organisms in the presence of other flora. Other amplification protocols are purposely designed to be nonselective. These so-called "universal" or "broad-range" systems are capable of amplifying nucleic acid sequences from a wide variety of species of organisms. In vitro nucleic acid amplification procedures have one significant advantage over conventional culture

techniques; namely, the former techniques are not limited by the ability of an organism to grow on artificial medium. The techniques described below involving target amplification and probe amplification are certain to have an impact on the practice of microbiology in the clinical laboratory in the next few years.

PCR

Since its moonlit conception in the mountains of northern California in 1983, PCR has become a technological milestone in the field of molecular biology (110). PCR is an ingenious procedure based on the ability of DNA polymerase to copy a strand of DNA by elongation of complementary strands initiated from a pair of closely spaced oligonucleotide primers (112). Theoretically, each cycle of the reaction doubles the amount of target DNA, resulting in millionfold levels of amplification. The impact that PCR has had on the practice of molecular biology rivals and perhaps exceeds that of any previously described procedure.

The fact that PCR is a very straightforward and versatile procedure has undoubtedly contributed to its widespread use (3, 50, 123, 170). As with all the nucleic acid target amplification procedures, some information about the target sequence must be available in order to design the two single-stranded DNA oligonucleotide primers that will be used in PCR to amplify target DNA (Fig. 2). These PCR primers, each about 20 to 25 nucleotides long, can be synthesized by using an automated DNA synthesizer ordered from a commercial supplier or off the shelf from some sources. PCR has high sequence specificity, in part because the two unique primers must hybridize in relatively close proximity to each other on the target DNA sequence under stringent temperature and reaction conditions before exponential amplification can occur.

A commonly used set of cycling temperatures includes denaturation of double-stranded DNA at 94°C for 15 s to 1 min, hybridization of oligonucleotide primers at 52°C for 15 s to 2 min, and extension (polymerase-mediated complementary-strand synthesis) from the primers at 72°C for 15 s to 3 min (142). These three steps are repeated for 25 to 35 cycles. In the early cycles of the reaction, the initially synthesized strands of new DNA vary in length depending on the processivity of the DNA polymerase. However, the primers begin to use the newly synthesized strands of DNA as templates, and within a few cycles, the predominant product of PCR becomes a double-stranded DNA sequence, the length of which is the sum of the lengths of the two primers plus the intervening target DNA. PCR is an exponential amplification reaction such that after n cycles there is $(1 + x)^n$ times as much target as was present initially, where x is the mean efficiency of the reaction for each cycle. Theoretically, as few as 20 cycles would yield approximately 1 million times the amount of target DNA initially present. In practice, however, the theoretical maxima are never reached, and more cycles are necessary to achieve such a level of amplification.

Despite all its power and potential applications, PCR along with all other enzyme-catalyzed procedures is a victim of its own success. Because of the incredible nucleic acid amplification power and concomitant sensitivity of these techniques, false-positive reactions caused by contaminating nucleic acids may occur. Contamination of new samples with previously amplified DNA is a significant problem that may hinder the introduction of in-house developed PCR tests into the clinical microbiology laboratory. The problem of false positives and contamination control for nucleic acid

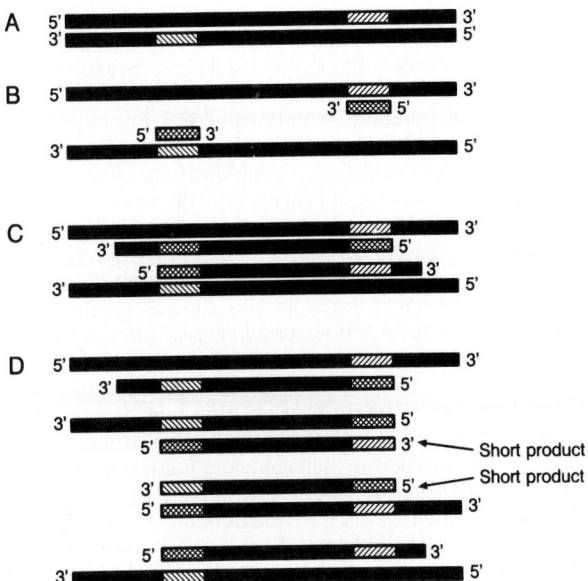

FIGURE 2 PCR. (A) In the first cycle, a double-stranded DNA target sequence is used as a template, with the primer-binding sites indicated by hatched lines. (B) These two strands are separated by heat denaturation, and the two synthetic oligonucleotide primers (cross-hatched lines) anneal to their respective recognition sequences in the 5′ → 3′ orientation. Note that the 3′ ends of the primers are facing each other. (C) A thermostable DNA polymerase initiates synthesis at the 3′ ends of the primers. Extension of the primer via DNA synthesis results in new primer-binding sites. The net result after one round of synthesis is two "ragged" copies of the original target DNA molecule. (D) In the second cycle, each of the four DNA strands shown in panel C anneals to primers (present in excess) to initiate a new round of DNA synthesis. Of the eight single-stranded products, two are of a length defined by the distance between and including the primer-annealing sites; this "short product" accumulates exponentially in subsequent cycles.

amplification procedures will be discussed at length elsewhere in this chapter.

Other problems that can occur in PCR are primer artifact formation and nonspecific hybridization of primers (142). Primer artifact formation occurs when the DNA polymerase synthesizes a template-directed product under low stringency that gains, within a few nucleotides, the primer binding site for the same or another primer. Primer artifact formation decreases the efficiency of PCR by consuming primers and competing with the true target for other necessary reaction components. Conditions that promote primer artifact formation are low-stringency annealing conditions, small target amounts, too much enzyme in early cycles, high primer concentrations, and excessive thermal cycling. Primers designed so that the last two nucleotide bases on the 3′ ends of the primers are complementary should generally be avoided in order to reduce the chances of non-template-dependent primer artifact formation. Nonspecific hybridization of one primer is usually not a major problem. Should this occur, the nontarget sequence amplifies at most 30-fold in 30 cycles and not exponentially, because the product of the first primer extension most likely would not contain a sequence complementary to that of the other primer. However, in some cases, low-stringency binding of the other primer to the nonspecific extension product of the first primer may result in amplification of nontarget sequences. Problems with nonspecific hybridization of primers are usually resolved by altering the PCR conditions or developing more specific primers.

Development of PCR-Based Assays

When PCR-based diagnostic assays are being developed, several interdependent parameters must be taken into consideration. The concentration and source of thermostable polymerase, concentration of magnesium or manganese, concentration and purity of target DNA and primers, denaturing temperature, annealing time and temperature, extension time and temperature, and number of cycles all affect the efficiency and specificity of PCR. The most critical temperatures are the denaturation and reannealing temperatures. Inadequate heating for denaturation leads to incomplete melting and hinders DNA amplification. The annealing conditions largely determine the specificity of PCR. Use of nonstringent conditions results in mispriming and amplification of nontarget sequences. Too high a stringency reduces primer annealing and results in inefficient DNA amplification. The divalent cation concentration in the reaction also has a substantial effect on the efficiency of thermostable DNA polymerases. Thermostable DNA polymerases are magnesium-dependent enzymes. Each combination of target DNA and oligonucleotide primers frequently requires a unique concentration of magnesium to maximize the efficiency of the polymerase and the specificity of the reaction.

In addition to fine-tuning the basic components and parameters associated with PCR to maximize efficiency, numerous reaction additives and procedural modifications have been reported to increase the efficiency of PCR. The addition of low concentrations of dimethyl sulfoxide has been reported to modulate the specificity of PCR, allowing previously unamplifiable targets to be amplified or allowing larger DNA targets to be amplified more efficiently (82). The use of up to 20% glycerol in PCR has also been reported to improve amplification and allow amplification of very long segments (up to 2,500 bases) of target DNA (38). However, it is important to note that dimethyl sulfoxide and glycerol are not helpful in all PCR systems; they may, in fact, be inhibitors in some cases (64).

A modification of a typical PCR that has been found to enhance the specificity of the reaction is called "hot start" (37). The standard procedure for setting up PCR involves adding all components to a microcentrifuge tube at ambient temperature or on ice, placing the tube in a thermal cycler, and raising the temperature to 94°C to denature the DNA and start the reaction. Because the thermostable polymerases have some activity at well below their optimum temperatures, any nonspecific primer-template complexes that form during reaction setup may be extended prior to denaturation. These products are amplifiable, compete for reagents in the reaction, and can result in a significant amount of nontarget amplification. If an essential component (such as the thermostable polymerase) of the PCR is omitted during reaction setup, nonspecific primer-template complexes cannot be extended. The reaction mixture can then be heated to 94°C, and the missing component can be added. At elevated temperatures, any nonspecific primer-template complexes are denatured. The reaction mixture is then cooled to an annealing temperature at which more specific priming can take place.

Modifications of PCR

Since its inception, numerous modifications of the standard PCR procedure have been developed, effectively expanding the diagnostic capabilities of PCR. A number of these modifications may eventually be used in the clinical laboratory.

Nested PCR

Nested PCR uses two sets of amplification primers, with one set targeted within the amplification product of the other set, and two rounds of amplification in order to amplify specific target DNA. This procedure is designed to increase the sensitivity of PCR by directly reamplifying the product from a primary PCR with a second PCR. Two common nested PCR procedures are in use today, a two-tube procedure and a single-tube procedure (52, 57, 74).

Two-tube procedure. In the two-tube procedure (74), primer set 1 is added to a PCR reaction mixture and run for the typical 25 to 35 cycles. The tube is then opened, and the product is transferred to a second tube containing primer set 2, which is complementary to the product from the first reaction but binds 3' internal to primer set 1. The second reaction is then run for the typical 25 to 35 cycles, and the product is analyzed by standard methods.

Single-tube procedure. The advantage of the single-tube procedure is that the tube does not have to be opened, so the chance of aerosol release of amplified DNA into the laboratory environment is reduced. However, single-tube nested PCR is much more complex than the two-tube system and may be too complicated for some laboratories. Two protocols have been developed for the single-tube nested PCR. One is based on physical separation of the two sets of reaction components in a single tube, and the other is based on differences in the annealing temperatures of the two sets of primers in the reaction. In the former, the primary reaction mixture is set up as usual with primer set 1 and all other necessary components. Internal-primer set 2 and the other necessary components required for a second reaction, less target DNA, are then inserted into a thick layer of mineral oil above the mixture containing primer set 1. After the first round of 25 to 35 cycles of PCR, the reaction tubes are spun to dilute the contents of the first reaction and to mix in the second primer set and the other components from the oil overlay. A second round of 25 to 35 cycles of PCR is then performed, and the amplification products are analyzed by standard methods.

In nested PCR based on annealing temperature differential, the reaction mixture is set up to contain both outer-primer set 1 and inner-primer set 2 (52, 57). These primer sets are designed so that primer set 1 hybridizes optimally at a substantially lower temperature than primer set 2. The PCR is run initially at the lower optimum temperature of primer set 1, which results in a mixture of products formed from both primer set 1 and primer set 2. A switch to the higher optimum temperature for primer set 2 favors amplification of the internal (nested) product from primer set 2. A variation of this method uses a higher annealing temperature for the first round of amplification (52).

Nested PCR is a very powerful amplification procedure with several advantages. Because of the high level of target amplification with nested PCR, a single target molecule can frequently be detected by direct DNA staining, thus elim-

inating the additional steps involved in postamplification hybridization with a labeled probe. Using the two-tube nested PCR procedure has the added benefit that transfer to the second set of tubes may dilute out any inhibitors of the reaction in the original sample, resulting in a successful reaction that under standard conditions may have progressed poorly. In addition, reamplification of the target DNA with a second set of primers enhances the sensitivity of the reaction by verifying the specificity of the first-round product.

However, nested PCR has some important disadvantages that may limit its use in the clinical microbiology laboratory. The most serious problem with nested PCR is specifically associated with the technically simpler and more commonly used two-tube system. Because the two-tube system involves opening a reaction tube and transferring the product to another reaction vessel, there is tremendous potential for aerosol contamination of the laboratory environment. Further complicating matters is the fact that in nested PCR, the product from the first-round PCR must be available for a second round of PCR and cannot be inactivated. Because of the high risk of laboratory contamination associated with nested PCR, multiple additional controls must be included in each experiment to ensure detection of contamination problems should they occur. For these reasons, the risk of contamination may well outweigh the benefits of nested-PCR-based assays for many laboratories, especially since well-optimized conventional reactions can usually approach the sensitivity of nested protocols.

Multiplex PCR

In multiplex PCR, two or more unique target DNA sequences in a specimen are amplified simultaneously (27). Multiplex PCR will probably be used most frequently in diagnostic assays that use one set of primers to amplify an internal control to verify the integrity of the PCR, while the second set of primers is targeted to the DNA sequence of interest. For example, in PCR analysis for the presence of the *mecA* gene from an *S. aureus* isolate, one recent multiplex application targeted a sequence of the gene coding for 16S rRNA that is ubiquitous among *S. aureus* isolates, while the other set of primers was specific for the *mecA* gene (63). A positive result with the control primers ensured that conditions for a successful PCR (i.e., successful extraction of target DNA) were present and confirmed the integrity and availability of nucleic acid for detection of the *mecA* gene. Multiplex PCR may also be used in the clinical microbiology laboratory to simultaneously probe a specimen for several different organisms in a single PCR tube (8). A PCR reaction containing a patient specimen might include a panel of primers targeted toward the signature sequences of 16S rDNA from several organisms. The primers are designed so that each amplification product is a unique size, allowing the detection and identification of a specific organism in the specimen even though primers to several different organisms are present in the reaction mixture.

A useful variation of multiplex PCR is competitive PCR (66), which allows the investigator to accurately quantitate the amount of target DNA or RNA in the starting material. The main problem in obtaining quantitative PCR data is the variation that occurs in the reaction rate. Numerous factors influence the rate of the PCR, and even when these factors are precisely controlled, tube-to-tube variation can still occur, preventing accurate quantitation of the target.

Competitive PCR compensates for the problem of variation in the PCR rate by coamplifying a competitor target nucleic acid of known concentration with the same primers used for the target nucleic acid of interest. The coamplified competitor target is designed so that it can be distinguished from the target nucleic acid of interest following PCR. Any variation in the rates of PCR in different tubes affects the yield of competitive target nucleic acid and the target nucleic acid of interest equally; relative ratios of the two nucleic acid targets should therefore be preserved with amplification. Because the concentration of the competitor target nucleic acid is known, the initial concentration of the target nucleic acid of interest can be determined by a formula adjusting for differences in the amplification efficiencies of competitor and target. Competitive PCR has been successfully used to quantitate HIV DNA and RNA in clinical specimens (129) and to accurately estimate the numbers of *Borrelia burgdorferi* spirochetes present in tissue from infected mice (177). In addition, competitor target DNA can be used as an internal control in PCR to verify the integrity of the results for the target nucleic acid of interest.

For multiplex PCR to be successful, several factors must be taken into consideration when the primers are designed. First, the primer sets should have similar annealing temperatures. A 10°C difference in annealing temperatures between the primer sets can lead to widely different amounts of amplified products or to no detectable amplification of one product or the other. The primers used should also be designed so that the products are sufficiently different in size that each product can be readily identified but not so different that amplification of one target is favored over amplification of the other. Multiplex PCR with targets that differ widely in size often favors amplification of the shorter target over the longer one, resulting in different amounts of amplified products. If wide variations in annealing temperatures or target lengths cannot be avoided, procedural modifications can be made to balance the reaction and achieve approximately equal amounts of products. Giving one set of primers an efficiency advantage by adjusting primer concentrations balances a multiplex reaction (8).

PCR Amplification of RNA

The number of applications of PCR in diagnostic microbiology would be greatly expanded if RNA targets could also be detected. For example, some viral genomic targets such as hepatitis C virus, influenza virus, and picornaviruses are present only as RNA and do not undergo reverse transcription into a DNA intermediate during their life cycles. The ability to detect RNA by using a sensitive procedure like PCR would also be useful for the identification of other organisms such as bacteria and fungi. These organisms contain thousands of copies of specific rRNA molecules that may further enhance the sensitivity of PCR. In addition, specific detection of RNA may prove useful for monitoring the effectiveness of antimicrobial therapy (because of the more rapid turnover rate of RNA) or for studying pathogenic conditions that may be associated with the expression of specific genes, such as the link between genital neoplasia and expression of the E6 and E7 viral oncogenes of human papillomaviruses.

Until recently, RNA targets could be detected by PCR only after the RNA had first been extracted and converted to cDNA by a retroviral reverse transcriptase. The newly formed cDNA was then employed as template for PCR. However, conventional reverse transcriptase reactions are fastidious and use enzymes that cannot tolerate temperatures above 42°C, which is a nonstringent hybridization temperature for most primers. This fact, coupled with the potential for single-stranded RNA templates to form stable secondary structures, leads to widely varying efficiencies of conversion of RNA to cDNA. Thus, the relative inefficiency of reverse transcription represents a major obstacle to the sensitive and specific detection of RNA targets.

A thermostable reverse transcriptase would alleviate many of the problems associated with cDNA synthesis; the elevated reaction temperature increases the stringency of primer hybridization, resulting in increased specificity, and disrupts secondary structure in the target RNA that may interfere with conventional reverse transcriptase reactions. The discovery that *E. coli* polymerase I exhibits limited reverse transcriptase activity in the presence of manganese ions led to a search for a thermostable DNA polymerase with reverse transcriptase activity. A recombinant DNA polymerase derived from the thermophilic eubacterium *Thermus thermophilus* (Tth pol) has recently become commercially available (108a). This thermostable DNA polymerase possesses efficient reverse transcriptase activity in the presence of manganese.

The Tth pol reverse transcriptase-based RNA PCR can be done under a single set of reaction conditions (178). Tth pol, two oligonucleotide primers (one used for cDNA synthesis and both used for PCR), manganese in the form of $MnCl_2$, and all the other reagents necessary for reverse transcription and PCR are mixed in a single tube. The thermal cycler is preheated to 70°C, and the reverse transcription components are incubated for 15 min. Following the reverse transcriptase reaction, the reaction mixture is heated to 95°C for 1 min to denature the RNA-DNA complexes. PCR is then begun with 2 cycles at 95°C for 15 s and 60°C for 20 s followed by 38 cycles at 90°C for 15 s and 60°C for 20 s. The sensitivity of this procedure as analyzed by gel electrophoresis has been reported to be ≤100 copies of RNA targets. The sensitivity can be improved to 10 copies of RNA targets by using a labeled hybridization probe. The ability to do the reverse transcription at 70°C is a definite improvement over older procedures and will enhance the specificity and sensitivity for detection of RNA targets.

Thermostable DNA Polymerases

In the recent past, the choice of a thermostable polymerase for a PCR-based assay was not difficult because there was only one thermostable DNA polymerase (*Taq*). Now, several different sources of thermostable DNA polymerases are commercially available (Table 3). In addition to new sources of thermostable polymerases, many genetically modified counterparts of these polymerases have appeared, and they may be superior to the original polymerase for certain applications. However, the choice of enzymes for use in PCR-based clinical tests may be limited because of legal restrictions.

The selection of a thermostable polymerase for a PCR-based assay depends on the requirements of the assay. If high denaturing temperatures and/or high numbers of amplification cycles are required for a particular assay, the use of one of the hardier thermostable DNA polymerases should be considered. If the objective of the PCR is to clone and sequence the final product, one of the DNA polymerases with 3′-to-5′ proofreading exonuclease activity can be used and will provide greater fidelity. If cycle sequencing is the basis of the assay, a thermostable polymerase without inherent 3′-to-5′ exonuclease activity would be the enzyme

TABLE 3 Common thermostable DNA polymerases[a]

Enzyme	Source	Habitat	Supplier	Stability at 95°C	Optimum activity (°C)	Cation requirement	3'-5' Exonuclease
Taq	Thermus aquaticus	Hot springs	Perkin-Elmer Corp.	$t_{1/2}$ = 40 min	50–75	$MgCl_2$	No
AmpliTaq	Recombinant Taq		Perkin-Elmer Corp.	$t_{1/2}$ = 40 min	50–75	$MgCl_2$	No
AmpliTaq Stoffel	AmpliTaq minus 289 N-terminal aa		Perkin-Elmer Corp.	$t_{1/2}$ = 80 min	75–80	$MgCl_2$	No
rTth	Thermus thermophilus	Hot springs	Perkin-Elmer Corp.	$t_{1/2}$ = 20 min	55–75	$MnCl_2$ (RT), $MgCl_2$ (PCR)	No
Ultima	Thermotoga maritima	Deep-sea thermal vents	Perkin-Elmer Corp.				Yes
Vent	Thermococcus litoralis	Deep-sea thermal vents	New England BioLabs, Inc.	>95% activity after 1 h	80	$MgSO_4$	Yes
Vent (Exo⁻)	Recombinant Vent		New England BioLabs, Inc.			$MgSO_4$	No
Deep Vent	Pyrococcus sp. strain GB-D	Deep-sea thermal vents	New England BioLabs, Inc.	>95% activity after 1 h	72–80	$MgSO_4$	Yes
Pfu	Pyrococcus furiosus	Deep-sea thermal vents	Stratagene, Inc.	>95% activity after 1 h	75	$MgCl_2$	Yes

[a]Abbreviations: $t_{1/2}$, half-life; aa, amino acids; RT, reverse transcriptase.

of choice. If the target is RNA instead of DNA, then thought should be given to evaluating a polymerase with both reverse transcriptase and DNA polymerase activities in the same tube rather then attempting separate reactions. Caution must be exercised when a thermostable polymerase with 3'-to-5' proofreading exonuclease activity is used. With the latter enzymes, it is important not to add quantities of polymerase in excess of those recommended by the manufacturer, because excess polymerase can result in nucleic acid degradation. To further decrease the chances of nucleic acid degradation, it is recommended that the enzyme be added to the reaction mixture after the primers have been annealed to the target DNA and that long incubation periods at elevated temperatures be avoided. Once a polymerase has been selected and the assay has been tailored to that particular enzyme, the polymerase probably cannot be changed without reevaluating the reaction conditions.

3SR

The first non-PCR nucleic acid amplification technique was a transcription-based amplification system (TAS) developed in 1989 by Kwoh et al. (92). In TAS, amplification of RNA is based on a repeated two-cycle reaction of cDNA synthesis followed by transcription of the cDNA template into multiple copies of RNA. A major drawback to TAS was that heat was needed to denature the RNA-DNA intermediate formed during each cycle of the reaction. Heating inactivated the enzymes used in TAS, requiring their replenishment at the end of each cycle. During the development of TAS, it was discovered that the addition of E. coli RNase H to the reaction resulted in degradation of the RNA template within the double-stranded RNA-DNA intermediates, allowing the cycling reaction to proceed under isothermal conditions (72). This nucleic acid amplification reaction has now come to be known as 3SR or nucleic acid sequence-based amplification (34).

3SR utilizes the collective activities of three enzymes

(avian myeloblastosis virus reverse transcriptase, RNase H, and bacteriophage T7 DNA-dependent RNA polymerase) to isothermally amplify an RNA target (Fig. 3). The initial steps in the reaction involve the formation of cDNAs from the target RNA by using oligonucleotide primers containing a T7 polymerase-binding site. The RNase H in the reaction degrades the initial strands of target RNA in the RNA-DNA hybrids after they have served as templates for the first primer. The second primer binds to the newly formed cDNA and is extended, resulting in the formation of double-stranded cDNAs with one or both strands capable of serving as transcription templates for T7 RNA polymerase. It is reported that up to a 10-million-fold amplification occurs within a 1- to 2-h incubation, with an initial rate of amplification of 10-fold every 2.5 min for the first 10 min.

Although 3SR involves several enzymes and a complex series of reactions, individual steps in the reaction are invisible to the user. Advantages of 3SR include very rapid reaction kinetics and the lack of a requirement for special equipment other than a temperature-controlled environment of 37 to 42°C. 3SR may be useful for the selective amplification of single-stranded RNA targets. PCR requires isolation of purified RNA and/or pretreatment of the samples with DNase prior to the reverse transcription and amplification steps. With 3SR, because reaction conditions are maintained at temperatures that do not extensively denature DNA, the DNA is largely inaccessible to the 3SR machinery, and pretreatment steps are less necessary. While 3SR is well suited for amplification of single-stranded RNA, it is a more complicated procedure when it comes to the amplification of double-stranded DNA. For DNA amplification, heat denaturation steps are required in the initial stages of the reaction in order to generate the necessary amplification intermediates. Despite this potential shortcoming, 3SR has been used successfully to detect C. trachomatis (75), azidothymidine resistance in HIV (67, 71), and human papillomavirus (19).

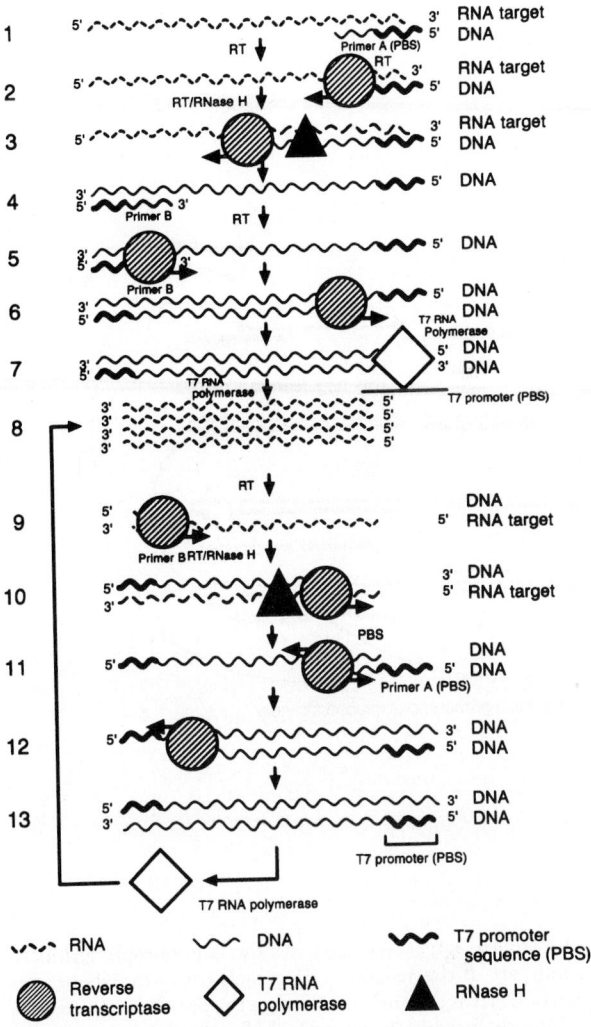

FIGURE 3 3SR. The initial steps in the reaction involve formation of cDNAs from the target RNA by using oligonucleotide primers with a T7 binding site. The RNase H in the reaction degrades the initial strands of target RNA in the RNA-DNA hybrids after they have served as templates for the initial primer. The second T7-containing primer then primes the initial single-stranded cDNAs, resulting in the formation of double-stranded cDNAs with one strand capable of serving as the transcription template for T7 RNA polymerase. This results in the synthesis of numerous copies of antisense RNA. These antisense RNAs serve as templates for one of the T7-containing oligonucleotide primers. After a round of priming, DNA polymerization by reverse transcriptase (RT), RNase H degradation of the RNA-DNA hybrid, priming, and extension of the newly formed cDNA strand by the other T7-containing oligonucleotide primer, numerous copies of double-stranded cDNA are formed again. At this stage of the reaction, both strands of the cDNA are capable of serving as transcription templates for T7 polymerase. These cDNAs can yield either sense or antisense RNAs, which then reenter the cycle of priming, DNA polymerization by reverse transcriptase, RNase H degradation of the RNA-DNA hybrid, priming, and extension of the newly formed cDNA strand by the other T7-containing oligonucleotide primer with subsequent formation of additional double-stranded cDNAs that can serve as templates for either sense or antisense RNA synthesis. PBS, promoter-based sequence.

SDA

SDA is the newest addition to the nucleic acid amplification technologies (164, 165). Developed and patented by Becton Dickinson, Inc., in 1991, SDA is an isothermal DNA amplification procedure that uses specific primers, a DNA polymerase, and a restriction endonuclease to achieve exponential amplification of target. Published reports claim ~10^7-fold amplification of target following a 2-h reaction (164). The key technology behind SDA is the generation of site-specific nicks by the restriction endonuclease *Hinc*II. The site-specific nicks are used by an exonuclease-deficient DNA polymerase (Klenow fragment) to displace the nicked strand of DNA, generating a single-stranded DNA target for synthesis primers. At the same time, the exonuclease-deficient Klenow fragment uses the remaining template strand to generate a new strand of DNA. Normally, restriction enzyme cleavage produces double-stranded DNA products, which would not be a proper template for SDA. However, if α-thio-substituted nucleotides (dATPαS) are used to synthesize a double-stranded, hemiphosphorothioated DNA recognition site for restriction enzyme cleavage, the restriction enzyme is capable of cleaving only the unmodified strand, while the hemiphosphorothioated strand remains unbroken. dATPαS is the key base in SDA because it is responsible for transforming the original target sequence into the hemiphosphorothioate form, which is resistant to restriction endonuclease cleavage. This results in a nick being formed in the unmodified primers used to initiate SDA, with the hemiphosphorothioated strand serving as a template for DNA synthesis by the exonuclease-deficient Klenow fragment at the site of the nick (Fig. 4). The latest version of SDA can be thought of as a two-part reaction (164). The initial rounds of the reaction transform the original target sequence into the hemiphosphorothioate form with nickable *Hinc*II sites at each end. The second part of the reaction involves exponential amplification of the transformed target sequence.

Although the SDA reaction is complex, individual reaction steps occur simultaneously and do not require attention once the reaction has been initiated. With the exception of the initial 95°C denaturation step, SDA is isothermal and requires no specialized laboratory equipment other than a temperature-controlled environment of 37°C. In addition, SDA can be applied to either single- or double-stranded DNA. With SDA, a limit to the size of the target DNA that can be efficiently amplified has been observed, and the level of amplification decreases ~10-fold for each 50-nucleotide increase in target length over a range of 50 to 200 nucleotides. The developers suggest that this may be related to the lower processivity of the DNA polymerase for double-stranded templates (where strand displacement is required) than for single-stranded templates (where displacement is not required). While this may not result in a serious functional limitation of SDA, it may complicate postamplification decontamination procedures. As with all enzyme-catalyzed nucleic acid amplification procedures, accidental contamination of reagents or samples with target DNA or previously amplified target is a serious problem. Compared to PCR, SDA may be more difficult to decontaminate because the shorter (50- to 100-bp) target sequences generated are more difficult to inactivate with the procedures described in this chapter.

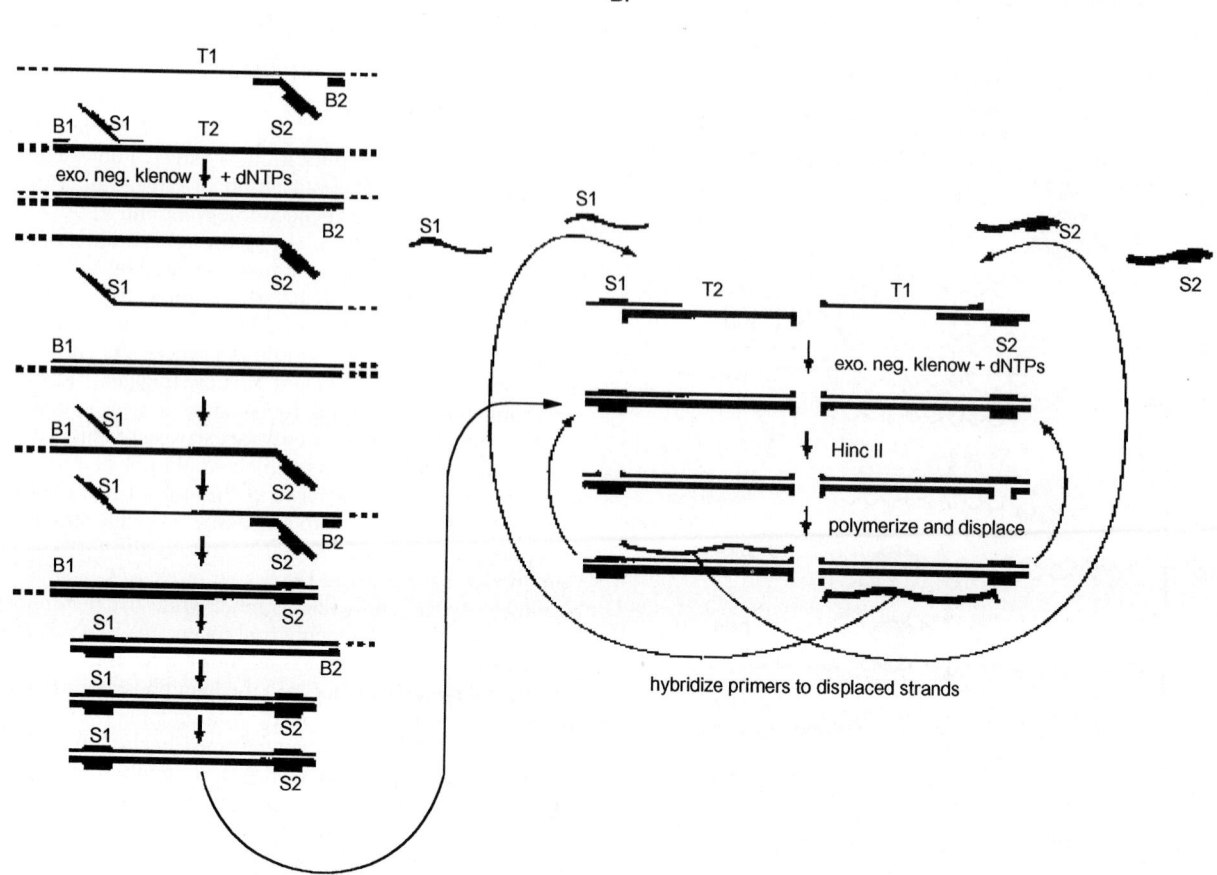

FIGURE 4 SDA. The initial rounds of the reaction (A) transform the original target sequence into the hemiphosphorothioate form with nickable *Hinc*II sites at each end that enter into the second part of the reaction (B), which involves exponential amplification of the transformed target sequence. In reaction part A, sample DNA is denatured at 95°C in the presence of an excess of four specific primers that define the target sequence. Two primers, S1 and S2, contain unmodified *Hinc*II recognition sites at their 5′ ends and specific target-binding sequences at their 3′ ends. S1 and S2 bind opposite strands of DNA flanking the target region. The other two primers, B1 and B2, are target-binding primers only, without a restriction endonuclease recognition sequence, and bind opposite strands of DNA immediately upstream of primers S1 and S2. The incorporation of primers B1 and B2 concomitantly generates a product with defined ends during the reaction and eliminates the need for restriction enzyme cleavage of the sample DNA prior to SDA. Following the addition of the primers, the reaction mixture is allowed to cool to 37°C, and exonuclease-deficient Klenow fragment (exo. neg. klenow) and *Hinc*II are added in excess together with dGTP, dCTP, TTP, and dATPαS (dNTPs). The DNA polymerase activity of the exonuclease-deficient Klenow fragment now extends all four primers simultaneously (A). Primers S1 and S2 form complementary strands of modified DNA that contain unmodified *Hinc*II sites at their 5′ ends. B1 and B2 prime the same strands and displace the newly synthesized strands primed with S1 and S2, producing new strands of DNA with defined ends that start immediately upstream of the S1 and S2 binding sites. Now S1 and B1 bind to the displaced strand initially primed with S2, while S2 and B2 bind to the displaced strand initially primed with S1 (A). Extension and displacement reactions on these templates produce two defined fragments with a hemiphosphorothioate *Hinc*II site at each end. Copies of the original target DNA that contain hemiphosphorothioate *Hinc*II ends have now been generated (A, bottom). These copies now enter the second, and prominent, part of the SDA reaction (B). Following priming and extension from S1 and S2, a double-stranded fragment of a specific size that contains a *Hinc*II site on each end that is susceptible to nicking (remember that S1 and S2 contain unmodified *Hinc*II recognition sites at their 5′ ends) is generated. Repeated cycles of nicking, DNA polymerization and strand displacement, and priming of the displaced single strands with S1 or S2 result in exponential amplification of target DNA. (Reprinted with permission from Becton Dickinson and Co., Research Triangle Park, N.C.)

Recently, SDA was used to amplify *M. tuberculosis* DNA from clinical isolates and alkali-treated sputum specimens (40). SDA readily detected the DNA target sequence in 40 different clinical isolates of *M. tuberculosis* and also exhibited excellent sensitivity for target DNA from the sputum specimens. There will undoubtedly be more reports in the future on microbiologic applications of SDA.

Qβ Replicase

The Qβ replicase system is a probe amplification procedure that was initially described in 1988 (101). This system is based on the incorporation of a single-stranded oligonucleotide probe into an RNA molecule that can be exponentially amplified after target hybridization by the enzyme Qβ replicase (51). The kinetics of the Qβ replicase reaction are very rapid. A 200-nucleotide RNA target takes ~20 s per round of replication at 37°C. Theoretically, a 200-nucleotide RNA target will yield 10^{12} progeny strands in only 13 min (89). This amount of product could easily be assayed by any one of a number of nucleic acid detection techniques.

The Qβ replicase assay itself is technically straightforward (101, 134) (Fig. 5). Target DNA is first heated to 85°C to denature it into single strands. If the target is single-stranded RNA, this step is not required. As the reaction mixture is cooling, the midivariant-1 (MDV-1) (109) reporter probe (a naturally occurring template for Qβ replicase into which the probe oligonucleotide has been inserted) and specific capture probes are added. The capture probes are used to allow removal of unbound reporter probes later in the procedure. The reaction mixture is incubated for 30 min at 37°C to allow hybridization to take place. The conditions in which hybridization occur may be adjusted to optimize the specificity of the reaction; e.g., the hybridization may be carried out under conditions that can discriminate between targets differing by only a single base. In addition, to further enhance the specificity of the assay, the sequences of the reporter and capture probes are optimized so that both must hybridize to generate a signal. After hybridization, the unbound MDV-1 reporter probes are washed from the system, while the bound MDV-1 probes are captured and remain behind. The reaction mixture is then warmed to 37°C, and Qβ replicase is added to initiate the MDV-1 amplification reaction. After 15 min, the amplification reaction is stopped by the addition of EDTA. The amplified product can be detected by any one of a number of techniques. Prichard and Stefano used a fluorescent dye that binds nucleic acids to measure the accumulation of RNA (134).

There are several potential advantages associated with the Qβ replicase amplification procedure (101). The speed with which the entire procedure can be done is remarkable: probe hybridization, amplification, and detection of amplified product can be completed in 2 to 3 h. The procedure involves isothermal reaction conditions that require no specialized laboratory equipment except a temperature-controlled environment of 37°C. The high levels of probe amplification make the procedure extremely sensitive and convenient, because such a large amount of amplified product is present that it can easily be detected by nonradioactive means. Because of the simplicity and tremendous amplifying power of Qβ replicase amplification, most likely only a few refinements would be required to adapt the procedure to automated instrumentation. In addition, the procedure lends itself to a simultaneous multiple-target detection system (101). A mixture of different MDV-1 reporter probes might be used in a single assay; after hybridization and amplification, they could be sorted out on the basis of their hybridization with a specific set of complementary DNA capture probes, permitting the simultaneous identification of several different targets.

The main disadvantage of the Qβ system is that unbound reporter probes or nonspecifically bound reporter probes serve as templates for amplification, resulting in false-positive results. For this reason, careful attention must be paid to hybridization conditions in order to achieve maximum sensitivity. In addition, an extremely efficient procedure to remove unbound or nonspecifically bound probe must be used. One procedure that has been used successfully to remove unbound MDV-1 probes in the Qβ assay system is reversible target capture (83, 134) or dual

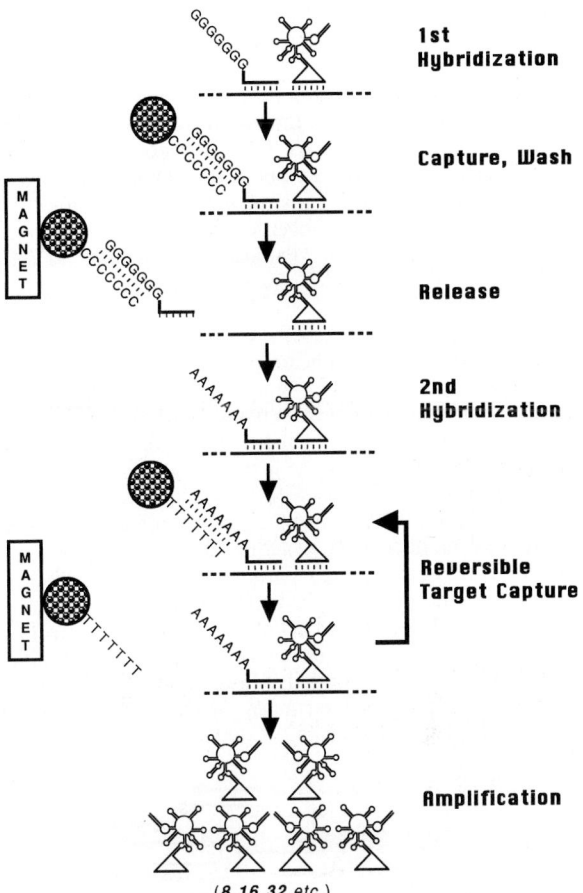

FIGURE 5 Qβ replicase dual capture assay. Both a poly(dG) target-specific capture probe and a target-specific substrate for Qβ replicase, MDV-1, are allowed to hybridize to target DNA or RNA. After hybridization, these complexes are captured onto oligo(dC)-derivatized paramagnetic beads via the poly(dG) tails on the capture probes. The beads are washed extensively to remove nonhybridized MDV-1 probes. The captured MDV-1–target hybrids are released from the capture probe-oligo(dC)-derivatized paramagnetic bead complex and hybridized to a poly(dA) target-specific probe. These complexes are captured onto oligo(dT)-derivatized paramagnetic beads via the poly(dA) tails on the capture probes. The beads are again extensively washed to removed nonhybridized MDV-1 probes. The poly(dA) capture probe–MDV-1–target hybrids are released from the oligo(dT)-derivatized paramagnetic beads, bound to fresh oligo(dT)-derivatized paramagnetic beads, and washed again to remove nonhybridized MDV-1 probes (reverse target capture). After three or more total cycles of washing, release, and recapture, the target–MDV-1 probe complexes are eluted into amplification buffer containing Qβ replicase. Qβ replicase is then added to the washed target-probe complex, and the reaction mixture is incubated at 37°C, resulting in amplification of the probe. (Reprinted with permission from GENE-TRAK Systems, Framingham, Mass.)

capture (Fig. 5). In a Qβ replicase assay developed by Prichard et al., four unique capture probes were used in the hybridization reaction (134). When the assay is fully optimized, the average efficiency of capture (or recapture) is about 65 to 75% per cycle. Because of the problem of background hybridization, the current limit of detection in commercial versions of this assay is about 600 target molecules.

If the problem of background noise can be solved, the simplicity and tremendous amplifying power of the Qβ replicase procedure will make it an attractive technology for the molecular diagnostic laboratory. The Qβ replicase system is currently being evaluated for its ability to identify many different types of pathogenic microorganisms, including *M. tuberculosis* (149), *Plasmodium falciparum*, HIV-1, CMV, *C. trachomatis*, and *N. gonorrhoeae* (91).

LCR

The LCR, or ligase amplification reaction, as it is sometimes referred to, is a probe amplification technique that was first described in 1989 by Wu and Wallace (175). LCR amplification is based on sequential rounds of template-dependent ligation of two juxtaposed oligonucleotide probes. Linear amplification of the specific probes is achieved when a single pair of oligonucleotides is ligated. Exponential amplification is achieved when two pairs of oligonucleotides, one complementary to the upper strand of target DNA and one complementary to the lower strand, are used. The procedure was modified in 1991 by Barany (6) and independently by Wallace, who cloned a thermostable DNA ligase and incorporated it into LCR. The thermostable DNA ligase resolved many of the early problems

associated with LCR, such as the need to add fresh ligase after each cycle and the relatively nonstringent annealing conditions associated with running the reaction at 30°C with bacteriophage T4 ligase.

In the first step of LCR, two sets of oligonucleotide pairs are allowed to anneal to their target DNA at 65°C (Fig. 6). When complementary base pairing exists at the ligation junction, the ligase joins the pair, forming a longer product that is bound to the template. A mismatch at the pair junction, however, prevents ligation between the two oligonucleotides. This feature of the reaction determines the specificity of the detection reaction. The reaction mixture is then heated to 94°C to denature the ligated product from the target and cooled to 65°C to allow annealing and ligation, and the cycle is then repeated. Newly formed ligation products are used as templates for ligation of still more substrates. Once ligated products form, subsequent cycles increase the amounts of products at exponential rates. The principal advantage of LCR is its ability to detect single-base-pair mismatches between target DNAs.

The early rounds of LCR are critical to the outcome of the reaction for two reasons (175). First, in the initial rounds, the original DNA sample is the primary target for ligation. Therefore, the intrinsic specificity of the reaction depends entirely on the specificity of the DNA ligase in these rounds. In later rounds, the predominant templates, which are not subject to single-base-pair discrimination like

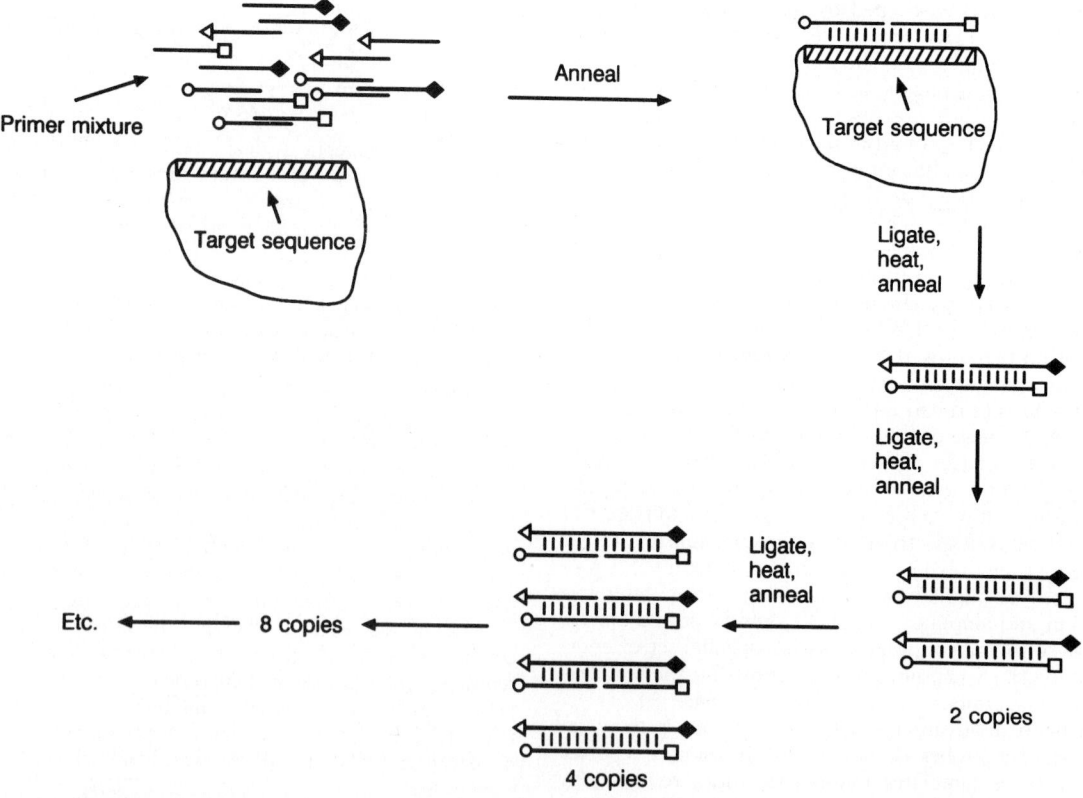

FIGURE 6 LCR. Oligonucleotide probes are annealed to template molecules in a head-to-tail fashion, with the 3′ end of one probe abutting the 5′ end of the second. DNA ligase joins the adjacent 3′ and 5′ ends to form a duplicate of one strand of the target. A second primer set, complementary to the first, then uses this duplicated strand (as well as the original target) as a template for ligation. Repeating the process results in a logarithmic accumulation of ligation products, which can be detected via the functional groups attached to the oligonucleotides.

the original template DNA, are the ligation products. Second, any undesirable template-independent ligation early on in the reaction would contribute significantly to the final signal, because these products could serve equally well as templates for amplification and are not subject to single-base-pair discrimination. It is imperative to inhibit the formation of these blunt-end ligation products, especially in the early rounds of the reaction. A modification of LCR that circumvents the problem of blunt-end ligation products involves the use of both a thermostable DNA polymerase and a ligase. Two pairs of oligonucleotide primers are annealed to opposite strands of target DNA but are spaced so that a polymerase is required to fill the gap between the primers; the ligase then connects the filled gap. This modified procedure is called gapped LCR (10).

Since 1991, several reports on the use of LCR-based probe amplification for the detection of M. tuberculosis (87), B. burgdorferi (81), and N. gonorrhoeae (9) have appeared. Abbott Laboratories, Inc., has adapted LCR to be compatible with its line of automated IMX analyzers already in place in many laboratories throughout the world (167).

PRODUCT INACTIVATION METHODS

One of the most significant problems that must be overcome before nucleic acid amplification techniques can be routinely used in the clinical microbiology laboratory is false positivity due to contaminating nucleic acids. It is ironic that one of the greatest assets of amplification procedures, their extreme sensitivity, also represents their Achilles' heel. DNA contamination of clinical specimens can come from three sources: (i) other clinical specimens when one specimen contains large numbers of target molecules of intact organisms, (ii) previously cloned DNA in the reagents used in amplification procedures, and (iii) accumulated amplification products in the laboratory resulting from repeated amplification of the same target sequence. This third type of contamination is the most serious and the most likely to occur in clinical laboratories. Initial efforts to control contamination in the laboratory should include physical separation of pre- and postamplification areas and unidirectional work flow as detailed elsewhere in this chapter. In addition, specific steps should be taken to inactivate the amplified nucleic acids so that if a carryover event does occur, the nucleic acids cannot be reamplified.

Because PCR is currently the most widely utilized amplification procedure, it has enjoyed the greatest development in the area of contamination control. However, all nucleic acid amplification procedures are subject to contamination problems. The applicability of the inactivation procedures detailed here to other amplification methods is unknown. Theoretically, the short amplification products associated with LCR or SDA may be more difficult to inactivate than the longer products of the other methods. RNA product contamination may prove simpler to control because of the inherent lability of RNA compared to DNA. Inactivation procedures will probably be necessary for all of the enzyme-catalyzed nucleic acid amplification methods before these methods can become clinically useful.

In the diagnostic molecular biology laboratory, two general approaches, one preamplification and one postamplification, can be taken to prevent false positives due to carryover of amplified DNA. Examples of pre- and postamplification techniques are discussed below.

Preamplification Inactivation

UV Light Irradiation

The first method used to inactivate contaminating DNA was UV light irradiation. In this procedure, all reagents except target DNA and Taq polymerase are combined and then briefly exposed to UV light. The rationale behind this procedure is the assumption that UV light will cause thymidine dimer formation and other covalent modifications to contaminating DNA, rendering the DNA unamplifiable. UV light decontamination is attractive because it is inexpensive and easy to perform and does not require modification of existing protocols. However, several reports question the effectiveness and reliability of UV light for inactivation of contaminating DNA (31, 146, 147). The effectiveness of this procedure is highly dependent on the length of the contaminating DNA and the G+C content of the target. Amplified products or targets of less than 300 bp are often incompletely inactivated by UV light. Even large molecules of DNA (500 or more bp) may be incompletely inactivated by UV light, presumably because of unique sequence composition (31). In addition, UV irradiation may have deleterious effects on thermostable DNA polymerases and the oligonucleotide primers, thus reducing amplification efficiency (117). Free deoxyribonucleoside triphosphates (dNTPs) present in amplification mixtures have also been reported to shield contaminating DNA and protect it from UV damage (61).

UNG

Another preamplification decontamination procedure involves modification of PCR-amplified DNA during amplification so that it can be selectively destroyed prior to amplification of unmodified target DNA. Uracil-N-glycosylase (UNG) is one of several enzymes of the excision repair system used by many species of bacteria to remove inappropriate nucleotide bases from newly synthesized DNA. The function of UNG is to recognize and cleave uracil residues that arise in DNA as a result of spontaneous deamination of cytosine residues. In a UNG-based decontamination system, dUTP replaces TTP in the PCR reaction mix and results in amplified DNA with deoxyuracil residues in place of thymidine residues (Fig. 7) (103). Thus, amplified DNA using dUTP is chemically distinct from target DNA. UNG is then added to the reaction mix prior to PCR, and the mixture is incubated briefly at room temperature and then at higher temperatures. If uracil-containing DNA (made in previous amplifications) is present in the preamplification mixture, the UNG cleaves the modified DNA at the uracil residues and destroys template activity. Following this treatment, PCR and detection of any amplified products are carried out as usual.

Although UNG-based inactivation systems have been reported to work very efficiently, there are some drawbacks. Similar to that of UV light inactivation, the efficacy of the UNG decontamination system most likely depends on the length and G+C content of the amplification product. One recent study recommends that UNG not be used on amplification products of ≤100 bp, and for maximum UNG efficiency, amplification products should be approximately 150 bp or more in length (53). Thus, it follows that the UNG system should be optimized for maximum efficiency for each application. It has also been found that partial dUTP substitution in amplified DNA can interfere with the hybridization of oligonucleotide reporter probes, decreasing the sensitivity of the assay (26). The magnitude of the

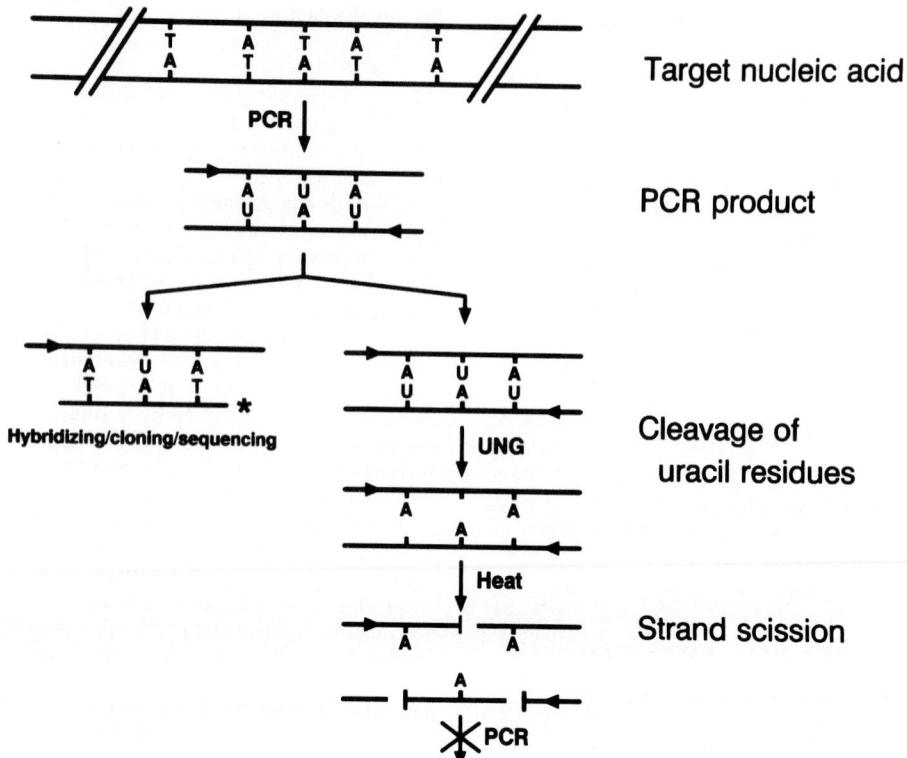

FIGURE 7 Inactivation of previously amplified DNA by selective action of UNG, which is added to reaction mixtures prior to amplification. Selective cleavage of uracil residues from any previously amplified DNA present as a contaminant results in elimination of the template activity of contaminants.

decreased hybridization signal appears to be proportional to both the dUTP concentration in the reaction mixture and the number of thymidylate residues in the probe binding site. In addition, it has been reported that some PCRs do not tolerate the substitution of dUTP for TTP (152). This fact must be taken into consideration when a DNA polymerase other than AmpliTaq is used in a PCR-based system that uses a UNG-based inactivation procedure.

Postamplification Inactivation

Photochemical Cross-Linkers

Photochemical inactivation is a simple and powerful method for postamplification inactivation of PCR products (Fig. 8). Photochemical cross-linkers derived from psoralens and isopsoralens are currently being evaluated for use as postamplification inactivation agents (32). These furocoumarin compounds represent a class of planar tricyclic reagents that are known to intercalate between base pairs of nucleic acids (Fig. 9) (30). Isopsoralen derivatives have the greatest potential use in the clinical laboratory because they do not cross-link the strands of a double-stranded DNA molecule like psoralen, so probe-based detection systems can be used following isopsoralen inactivation. Psoralen derivatives may be most useful when incorporated in a preamplification decontamination scheme. Like all inactivation systems, inactivation protocols based on psoralen and isopsoralen derivatives have certain limitations. Studies of isopsoralen inactivation have shown that for maximum efficacy, the conditions must be customized in

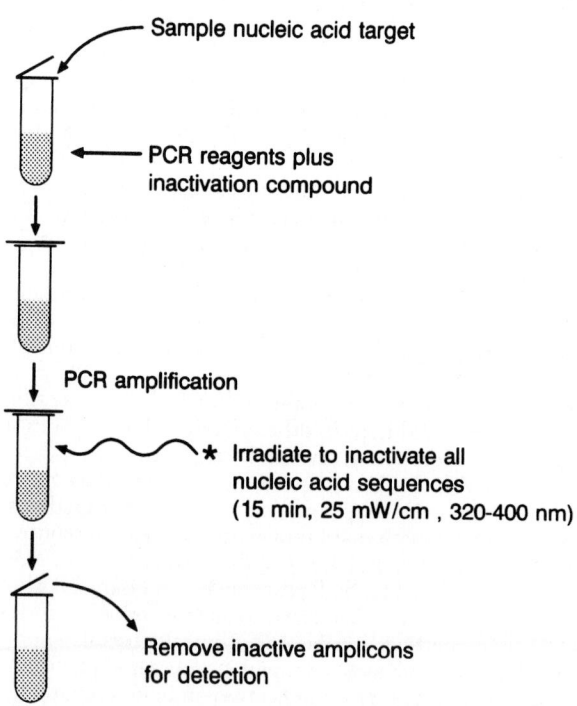

FIGURE 8 Steps in the photochemical post-PCR inactivation procedure.

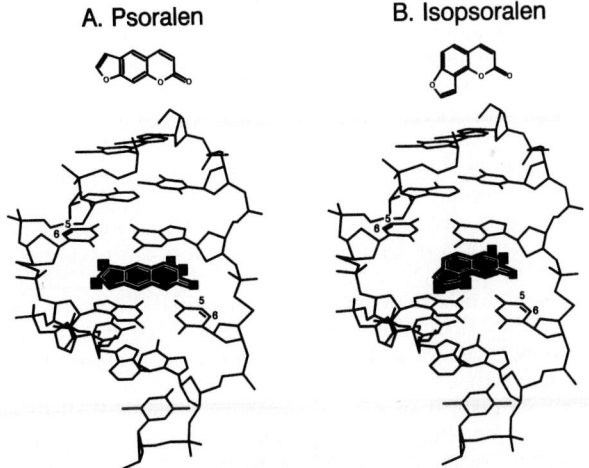

A. Psoralen **B. Isopsoralen**

FIGURE 9 Structure and reactive sites of psoralens and isopsoralens in double-stranded nucleic acids. (A) Psoralen molecule intercalated into double-stranded DNA. The intercalation site shown is created by adjacent pyrimidines on opposite strands of DNA. The geometric orientation of the psoralen permits a covalent cross-link between the two DNA strands. (B) Isopsoralen molecule intercalated into double-stranded DNA. The geometric orientation of the isopsoralen prevents covalent cross-linking between the two DNA strands. Only one strand is bound with this class of compounds.

terms of isopsoralen concentration and time of UV light exposure for each amplification product (124). Long amplification products are easier to inactivate than short products, and AT-rich products are more easily decontaminated than GC-rich products. As with UNG, a recent study suggests that amplification products should not be ≤100 for

reliable inactivation (shorter for AT-rich templates) (53, 141).

Selective Primer Cleavage

A new method of chemically modifying amplified DNA so that it can be inactivated postamplification has recently been reported (163). In this procedure, the 3′ ends of the PCR primers each contain one nucleic acid residue with ribose as the base sugar instead of deoxyribose (Fig. 10). These modifications result in the formation of an RNA-DNA duplex at one base between the primer and target DNA, but the RNA-DNA duplex does not interfere with PCR. Following PCR, the amplified DNA has incorporated a ribose linkage near the 5′ end of each strand. RNase or NaOH added postamplification cleaves the primer regions from the amplified DNA. This truncated amplification product does not serve as an efficient template for amplification with the original primers, because primer extension reactions initiated on the cleaved strand do not result in the synthesis of a binding site for the opposing primer. The result is a linear increase in the primer extension product, which does not contribute significantly to the signal. The remaining amplified DNA strands can be detected by standard methods. A drawback to this procedure as it is currently performed is that the reaction tube must be opened following amplification to add the RNase or NaOH, risking aerosolization of the amplification product and contamination of the laboratory environment. The development of reaction tubes that allow unidirectional reagent addition will make this inactivation procedure much more attractive.

Hydroxylamine Treatment

A recently described technique for postamplification inactivation of amplified DNA is hydroxylamine hydrochloride treatment (5). Hydroxylamine hydrochloride preferentially reacts with the oxygen atoms in cytosine residues, creating covalent adducts that prevent base pairing with guanine

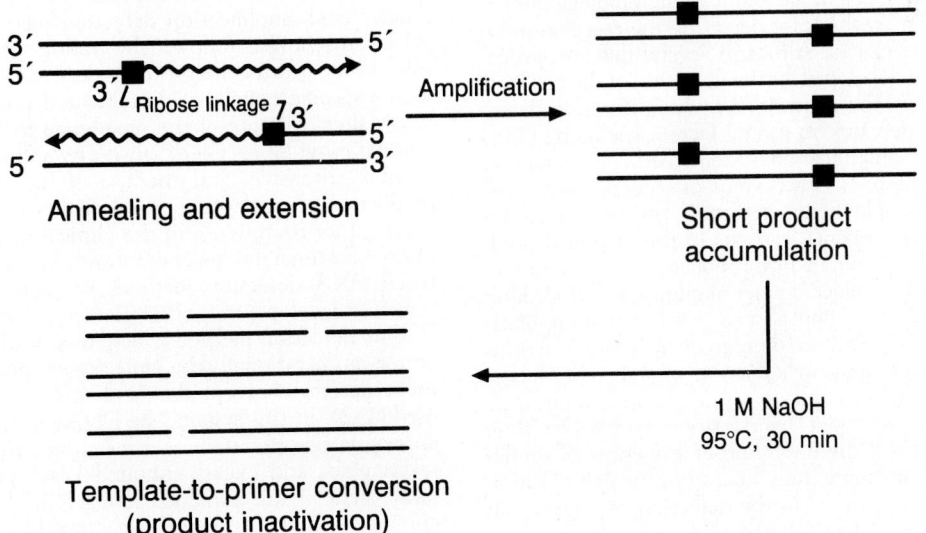

FIGURE 10 Inactivation of previously amplified DNA by selective primer cleavage. Primers containing ribose residues at their 3′ ends were used to amplify the target nucleic acid. After amplification, the ribose linkage is cleaved by alkaline (or enzymatic) hydrolysis. This results in cleavage products that can be amplified only linearly, which will not result in false positivity due to geometric amplification.

residues (18). This reaction presumably modifies primer-binding sites and the intervening target sequence so that reamplification cannot occur. Treatment of PCR products with ≥250 mM hydroxylamine hydrochloride for 30 min at room temperature effectively prevents reamplification (5). The selective modification of cytosine residues may make this method useful for inactivation of short, GC-rich amplification products, which are the most difficult products to inactivate by other methods. The main drawback of this procedure is that once the reaction product has been inactivated, hybridization with a sequence-specific probe to confirm the identity of the amplified DNA can no longer be used. In addition, as with any procedure in which the reaction tube must be opened following nucleic acid amplification, there is a risk of contaminating the laboratory environment with aerosolized templates. Like primer hydrolysis, this procedure would benefit from the development of reaction tubes that allow unidirectional reagent addition and prevent aerosolization of the contents.

Conclusions

No nucleic acid amplification inactivation procedure should be used as a substitute for good laboratory techniques and quality control. Strict separation of pre- and postamplification areas should be maintained, and specific work flow guidelines should be followed. It is important for the laboratorian to realize that amplification product inactivation procedures must be done prospectively and that no inactivation protocol will help a laboratory that has already been contaminated. Maintaining a laboratory environment free of amplification product contamination requires constant vigilance and strict adherence to protocol. It is easier and less problematic to avoid contamination at the outset than to decontaminate the laboratory once contamination has occurred.

POSTAMPLIFICATION DETECTION

Before routine nucleic acid-based amplification assays can become commonplace in the clinical microbiology laboratory, convenient methods for detecting the reaction products must be developed. Although several different procedures have been developed for the detection of the products from these assays, the choice of a postamplification detection procedure depends on several factors, including (116) (i) the type of amplification system used, i.e., target or probe amplification; (ii) the extent of sequence heterogeneity in the sample being detected; (iii) the need for detection of rare sequence variants in the amplified product; and (iv) the requirement for sequence confirmation of the amplification product. Target amplification procedures probably provide the simplest products for postamplification detection. Not only are there frequently large amounts of product, but also the sequence of product is homogeneous, reducing the complexity of detection and allowing the use of less stringent hybridization conditions than would be required if there were other sequences in similar concentrations in the mixture. Less stringent hybridization conditions may be critical to the detection of targets that contain multiple sequence variants.

Simultaneous Amplification and Ethidium Bromide Detection

An assay system in which product detection occurs concurrently with target amplification reduces the number of steps involved in detection and the potential for contamination of the laboratory environment. In one procedure, ethidium bromide is added to the amplification reaction to detect the presence of increasing concentrations of double-stranded DNA (78). Because the fluorescence of ethidium bromide increases in the presence of double-stranded DNA, an increase in fluorescence indicates positive amplification. This procedure would have the greatest applicability for specimens such as serum or plasma, which do not contain large amounts of DNA to start with. It is not necessary to open the reaction tubes to expose them to the light, because polypropylene tubes are sufficiently UV transparent that they do not interfere with the visual detection of fluorescence. An additional advantage of this type of procedure is that the progress of the amplification reaction can be continuously monitored simply by shining UV light on the reaction tube. Homogeneous assays (assays in which different reactions take place without the separation of reaction components) such as this reduce the chances of contaminating the laboratory environment, because once the amplification reaction is set up and the tube is closed, product detection does not require opening the tube. Another advantage of this type of procedure may be its potential for being easily adapted to automation. The main drawback of this type of simultaneous amplification and detection system is that there is no hybridization of a sequence-specific probe to confirm the specificity of amplified DNA. Therefore, the specificity of this procedure depends solely on whether or not DNA is amplified.

Amplification Product Detection by Agarose Gel Electrophoresis

The simplest version of product detection by agarose gel electrophoresis relies on detection of nucleic acid amplification products following agarose gel electrophoresis and ethidium bromide staining of the DNA. Although this technique is relatively easy to perform, a probe-based DNA detection system has the advantage of providing sequence specificity and increased sensitivity. Most probe-based post-nucleic acid amplification detection methods use agarose gel electrophoresis followed by transfer of the DNA to a solid phase (nitrocellulose or nylon membrane), probing with a specific radiolabeled probe, and visualization of the DNA after exposure of the membrane to X-ray film. This method provides excellent sensitivity and specificity, but it is labor intensive, and the use of radioisotopes in the production of nucleotide probes makes this approach impractical for routine use in the clinical microbiology laboratory. Fortunately, several innovative non-radioisotope-based DNA detection methods are currently in various stages of development and implementation (121, 169).

One detection method widely used in the research laboratory is operationally the same as solid-phase radioisotope probe detection except that a chemiluminescent probe is used (132). In this system, the DNA probe is linked to an enzymatic reporter such as HRP. Once hybridization has taken place and excess unbound DNA probe is washed away, peracid salt is added along with luminol and an enhancer. Light production is catalyzed by HRP bound to the DNA, and amplified DNA is detected following exposure of X-ray film for 1 h or less. While this method provides excellent sensitivity and specificity and does not require the use of radioisotopes, it is not easily automated and is probably still too labor intensive for routine use in most clinical microbiology laboratories.

Amplification Product Capture

The amplification product capture procedure is based on a solid-phase hybridization technique that incorporates the use of amplification primers biotinylated at their 5′ ends (78). Two different capture substrates are commonly used in this system. In one system, the oligonucleotide capture probes are immobilized onto nylon filters by exposure to UV light (143). The UV light activates thymine bases in DNA, which then covalently couple to the primary amines present in nylon. The capture probes are specially designed with long poly(dT) tails on their 3′ ends to facilitate their binding to the nylon. In the other commonly used system, the capture probes are attached to the wells of a 96-well microtiter plate (29, 102). To facilitate binding to the plates, the oligonucleotides are phosphorylated on their 5′ ends with T4 polynucleotide kinase before being incubated in the microtiter wells. Amplified DNA with its biotinylated 5′ ends is allowed to hybridize to the capture probes. After the unbound DNA is washed away, avidin-HRP is added, and it binds to the biotin incorporated into amplified DNA. A chromogenic substrate is then added to detect the amplified DNA colorimetrically. A prerequisite of this system is that all specific oligonucleotide binding must occur under the same hybridization conditions. This requirement can probably be met either by adjusting the length, position, and strand specificity of the oligonucleotide or by varying the amount of oligonucleotide applied to the membrane. The addition of EDTA can also minimize the differences among immobilized oligonucleotides caused by various base compositions.

DNA detection systems of this type are attractive for use in the clinical microbiology laboratory because no radioactive isotopes are involved and because when they are used with microtiter plates, they can take advantage of familiar laboratory instrumentation. In addition, these capture systems allow simultaneous detections with multiple probes, which avoids the need to use all of the amplification reaction for multiple probings (100).

Specific Product Generation during PCR Utilizing 5′→3′ Exonuclease Activity of *Taq*

As previously mentioned, it is often desirable to include a separate probe hybridization step to provide an extra level of sequence specificity and increased sensitivity over ethidium bromide staining of agarose gel-electrophoresed PCR products. The $5′ \rightarrow 3′$ exonuclease-based detection assay developed by Holland et al. incorporates this added level of product discrimination directly into PCR and allows amplification and generation of a specific product in a single tube (80). This method employs the $5′ \rightarrow 3′$ exonuclease activity of *Taq* to generate a specific detectable product concomitantly with amplification. A special oligonucleotide reporter probe designed to hybridize within the target sequence is added to the PCR. This probe is not involved in the amplification reaction and is labeled with a reporter molecule on its 5′ end and phosphorylated on its 3′ end to block extension by *Taq*. Annealing of this reporter probe to one of the PCR product strands during the course of amplification generates a target-specific substrate suitable for exonuclease cleavage. During DNA extension from a PCR primer, the $5′ \rightarrow 3′$ exonuclease activity of *Taq* cleaves 5′-terminal nucleotides of the bound reporter probe in what is essentially a nick translation reaction and releases labeled mono- and oligonucleotides that can be differentiated from undegraded probe by thin-layer chroma-

tography or polyacrylamide gel electrophoresis. Various probe labels, such as ^{32}P, fluorescent dyes, chromophores, and chemiluminescent labels, or ligands such as biotin can be used in this assay. Even in a high background of genomic DNA, labeled probe degradation occurs only in a target-specific fashion. One requirement for this detection method is that the reporter probe must bind to target before extension from a PCR primer blocks the reporter probe-binding site. If the annealing temperature of the reporter probe is close to or below the annealing temperature of the upstream amplification primer, there is a greater chance that the amplification primer will be extended before the reporter probe has bound to target. This would result in inefficient label release and could lead to a false-negative result. This problem may be overcome by varying the relative concentrations of amplification primers and reporter probe, manipulating the sequence and length of the reporter probe, or using more stable base analogs. It may be possible to directly incorporate this detection system into established clinical PCR protocols. One need only design a probe that will effectively hybridize within the target sequence and subsequently detect its specific cleavage.

HPA

Solution-phase HPA detection systems are based on the use of a chemiluminescent DNA probe and hold considerable promise (113, 132). In HPA, an acridinium ester-labeled DNA probe that is hybridized to amplified target DNA is protected from alkaline hydrolysis and, upon addition of peroxides, emits detectable light. However, in its unhybridized form, the attached acridinium ester is not protected and is hydrolyzed to a form that does not emit light. The HPA does not require the binding of amplified DNA to a solid support by DNA capture or other means, can be performed in a few hours, does not need to have the excess unbound DNA probe removed, and has a sensitivity of 10^{-16} to 10^{-19} mol (10^4 to 10^7 copies) of target DNA. However, this detection method is not widely available, because acridinium ester-labeled probes must be custom synthesized.

Conclusion

Ultimately, the detection format adopted for general use in the clinical microbiology laboratory will be one that utilizes a nonradioactive DNA probe incorporated into a system that can be easily automated. Several different types of nonradioactive reporter systems are currently available. Digoxigenin-labeled DNA probes used with an alkaline phosphatase-conjugated antidigoxigenin antibody make up a common chromogenic detection system that can be read out visually or automated with an enzyme-linked immunosorbent assay reader (114a). A more direct procedure uses fluorescence-labeled DNA probes together with a fluorogenic detection system (28). The latest entry into the nonradioactive reporter arena is the chemiluminescent DNA probe system described above.

SELECTED APPLICATIONS OF MOLECULAR METHODS IN CLINICAL MICROBIOLOGY

In numerous situations, the application of molecular methods could lead to a modification or replacement of conventional procedures in the clinical microbiology laboratory. These situations include the detection and identification of slow-growing microorganisms such as M. *tuberculosis* and other *Mycobacterium* species (13, 16, 39, 47, 76, 87, 165) and of emerging pathogens such as *B. burgdorferi* (the etiologic agent of Lyme disease [7, 68, 79, 98, 104, 105, 115, 125, 127, 128]) and *Ehrlichia* species (1, 2), which might be more rapid with molecular methods. In addition, molecular methods are beginning to be applied in the area of antimicrobial susceptibility testing. Although conventional methods of antimicrobial susceptibility testing are well standardized and generally reproducible, there are instances when these procedures fall short of predicting the true clinical response of an organism to a particular class of antibiotics or when the testing procedure requires extended periods of time. Prime examples of the shortcomings of conventional antimicrobial susceptibility testing are seen in cases of infections with methicillin-resistant *S. aureus* and in the determination of rifampin susceptibility in *M. tuberculosis*. The use of molecular methods for the genotypic determination of methicillin resistance in *S. aureus* (63, 133, 148, 160) and rifampin or isoniazid resistance in *M. tuberculosis* (33, 84, 157, 158) is currently being evaluated by several groups. A detailed account of all the applications of molecular methods in the clinical microbiology laboratory is beyond the scope of this chapter. However, two particular areas in which molecular methods would be of benefit are discussed in some detail here in order to better illustrate the application of this technology.

Detection and Identification of Uncultured Microorganisms: WAB

For infectious processes for which there are currently no known means of in vitro cultivation of the etiologic agents, nucleic acid amplification techniques are rapid and sensitive and can be used to identify these organisms directly from clinical specimens. One organism that falls into this category is the Whipple's disease-associated bacillus (WAB). Whipple's disease is a systemic disease that usually involves the gastrointestinal tract and may affect the mesentery, heart, and central nervous system (58). At present, the diagnosis of Whipple's disease is based on the clinical presentation of the patient coupled with microscopic observation of large macrophages containing diastase-resistant inclusions in periodic acid-Schiff-stained infected tissue. These procedures are time-consuming and lack sensitivity, because direct microscopic visualization of organisms is required.

Recently, PCR was used to identify the WAB (138). Broad-range primers were used to amplify a highly conserved region of the gene encoding 16S rRNA and the intervening sequence was determined directly from infected tissue. These procedures showed that the WAB is a relative of *Streptomyces* and *Actinomyces* spp. (138). This approach is possible because the 16S rRNA contains nucleic acid sequences that are highly conserved among all members of the bacterial kingdom in addition to regions that are unique to particular genera and species of bacteria, thus facilitating the phylogenetic analysis of these organisms on the basis of their 16S rRNA sequences (174). Using this approach, Relman et al. were able to identify the same organism in 5 unrelated patients with Whipple's disease but not in 10 patients without the disease (138). Use of PCR to detect the WAB may greatly improve our ability to diagnose the disease compared to diagnosis by the current procedure and may provide a means for the definitive identification of these organisms.

Universal 16S rRNA Sequencing as a General Scheme for Microbial Identification

Nucleotide sequence analysis of small-subunit (16S) rRNA or DNA has recently become extremely useful for the determination of phylogenetic relationships among members of the bacterial kingdom (174). Comparative nucleotide sequence analysis of 16S rRNAs has shown that regions of these molecules are highly conserved among all members of the bacterial kingdom, while other regions are unique to particular genera. As the already extensive library of 16S rRNA sequence data continues to accumulate, the ability to identify and classify nonviral organisms by direct sequence determination becomes more practicable.

In the past, the determination of a bacterial 16S sequence required genetic cloning of the 16S rDNA, preparation of the clone for sequencing, and DNA sequence analysis of the clone. This procedure is very lengthy, labor intensive, and costly and would never be feasible for the clinical microbiology laboratory. However, with the advent of PCR, it is now possible to amplify and sequence 16S rDNA directly from a single bacterial colony without prior DNA extraction or purification (Fig. 11) (60, 139a). Advances in nucleic acid-sequencing technology coupled with broad-range primers to highly conserved regions of 16S rDNA will very likely be used in clinical microbiology laboratories in the future for definitive identification of certain classes of difficult-to-identify and/or difficult-to-culture microorganisms. Prior to the advent of in vitro culture methods for *Mycobacterium genavense*, 16S rRNA sequencing was successfully used to characterize the organism from several AIDS patients by using isolates from patient tissue (13). In addition, the discovery of a single thermostable enzyme that can function as a reverse transcriptase and a DNA polymerase will allow amplification and sequencing directly from 16S rRNA targets that are present in 1,000 or more copies per bacterial cell, facilitating reverse transcription, DNA amplification, and nucleic acid sequence-based identification of microorganisms directly from clinical specimens (108a). On the heels of the human genome initiative, rapid advances being made in the automation of the nucleic acid sequencing process will, within the next decade, make the underlying complexity of this process invisible to the user and bring nucleic acid-sequencing technology into the clinical laboratory.

PREPARATION OF SPECIMENS FOR MOLECULAR ANALYSIS

When choosing a specimen preparation procedure, several general criteria should be met: the procedure must release the nucleic acids from the target organism, prevent degradation of the free nucleic acids, remove any substances inhibitory to nucleic acid amplification or hybridization, concentrate the nucleic acids into a small volume, and place the nucleic acids in amplification or hybridization buffer (69). In addition to these general requirements, a number of additional factors must be considered when a sample preparation procedure is being selected (69).

1. Use of the specimen. Is the specimen going to be used in a nucleic acid hybridization procedure or in a nucleic acid amplification procedure? Because of the limited sensitivity of hybridization procedures compared to that of nucleic acid amplification procedures, specimen preparation for use in hybridization must be capable of processing and concentrating larger amounts of specimens.

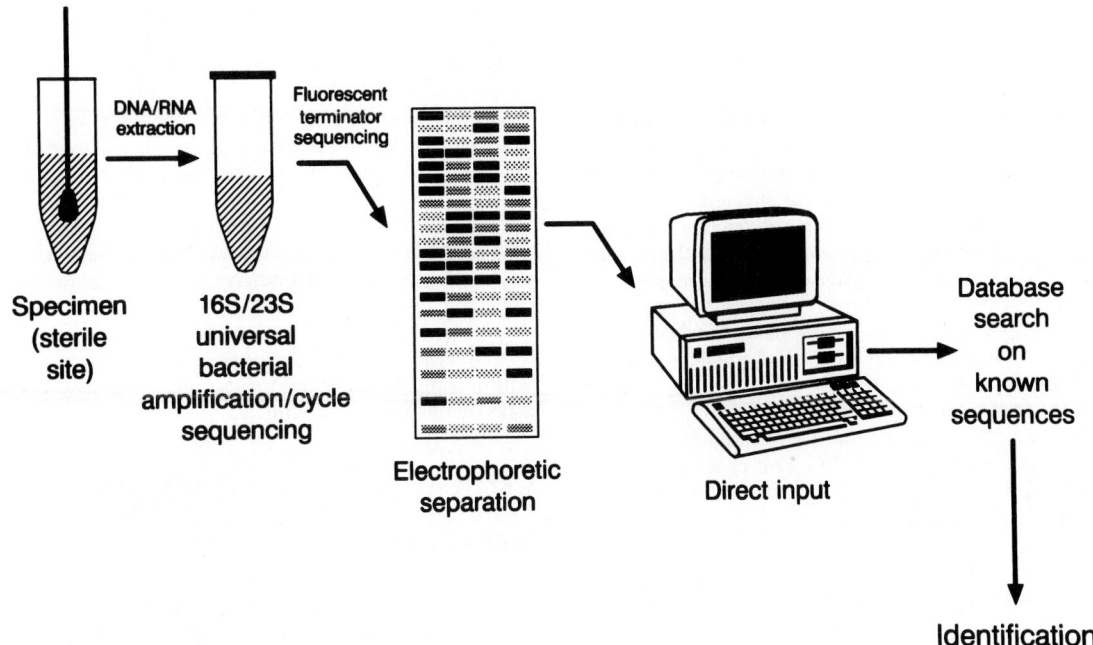

FIGURE 11 Nucleic acid sequence-based microbial identification. Broad-range (universal) primers are used in a target amplification protocol for amplification of 16S or 23S rDNA sequences. An automated sequencer determines the composition of the intervening nucleotides and directly enters sequence data into a computerized workstation. Database analysis creates a hierarchical cluster of related sequences and then generates a list of related or identical organisms.

2. Number of suspected organisms. The number of suspected organisms in a specimen dictates the sensitivity requirements for the assay and the volume of specimen required. If nucleic acid amplification is the basis of the diagnostic test, a single copy of the target DNA can theoretically be detected in an optimized procedure (11, 95). However, in nucleic acid amplification, the final reaction volume is restricted to between 20 and 100 μl, so factors such as the size of the specimen and the ease of concentration of organisms in the specimen are variables that must be considered when the optimal specimen preparation procedure is being selected.

3. Ease of lysis of the target organism. The relative ease of releasing nucleic acids from the target organism plays a large part in choosing the appropriate specimen preparation procedure. Because of their resilient cell walls, fungi and mycobacteria are considerably more difficult to release nucleic acids from than viruses or bacteria. At the other end of the spectrum are organisms that lack a cell wall, like *Mycoplasma* species, which may be prone to spontaneous lysis in specimens. In addition, gram-positive bacteria may require the use of lysing procedures more rigorous than those routinely used for the lysis of gram-negative bacteria.

4. Inhibitors of DNA polymerases. If nucleic acid amplification is the basis of the diagnostic procedure, the presence of inhibitors of DNA polymerases in the specimen must be considered. The specimen preparation procedure selected must contain steps designed to remove or neutralize potential inhibitors.

5. Characteristics of the target nucleic acid. Is the target nucleic acid DNA or RNA, double stranded or single stranded, or must the isolated nucleic acid contain only RNA without contaminating DNA? This issue is impor-

tant, because RNA is much more susceptible to degradation than DNA is (24, 36). In addition, sample preparation procedures that require boiling, alkaline lysis, or drying frequently produce partially single-stranded DNA. This single-stranded DNA can serve more readily as nonspecific templates for primers under low-stringency conditions such as during PCR setup or during isothermal amplification procedures. False priming can result in lower sensitivity and specificity for these assays (11, 59, 111).

6. Amount of extraneous nucleic acid in the specimen. How much "noise" can the assay system tolerate without compromising sensitivity and specificity?

It is clear that multiple factors must be taken into account if one is to select the most appropriate and efficient specimen preparation protocol. Fortunately, numerous procedures for preparation of specimens from several common clinical sources have been evaluated.

CSF Specimens

Small volumes (<10 μl) of cerebrospinal fluid (CSF) have been used successfully in PCR-based assays without DNA extraction (90). However, CSF is known to contain uncharacterized inhibitors of DNA polymerases (69). Therefore, nucleic acid isolation procedures that incorporate standard detergent or high-temperature cell lysis steps followed by various nucleic acid isolation and concentration steps have been evaluated. While both crude lysates and isolated nucleic acids have been used successfully in specific situations (100, 135, 140, 150), crude cell lysates are more prone to amplification problems than are isolated nucleic acids.

Blood Specimens

When whole blood or various fractions of whole blood are the specimens of choice for nucleic acid amplification procedures, certain inhibitors of DNA polymerases associated with these specimens must be dealt with. Heme and metabolic products of heme are known inhibitors of DNA polymerases (77). It has also been reported that DNA polymerases from different sources are affected by different concentrations of heme (120). The anticoagulant heparin is also known to inhibit *Taq* polymerase and the reverse transcriptase derived from murine leukemia virus (88).

Amplifiable nucleic acids have been isolated from leukocytes by using detergents to lyse the cells in whole blood and then allowing the free DNA to bind to silica matrices (176). In addition, amplifiable HIV RNA from leukocytes has been isolated by a simple procedure involving lysis of the erythrocytes in hypotonic saline followed by centrifugation to collect the leukocytes (25). This procedure is very rapid, and several specimens can be processed in only a few hours. If more rigorous isolation procedures are required, leukocytes can be isolated from whole blood by using well-established Ficoll-Hypaque isopycnic centrifugation procedures. DNA can then be isolated following lysis of the cells with detergent or following lysis with detergent and digestion with proteinase K (139, 156).

Serum or Plasma

Serum or plasma is the specimen of choice over whole blood or purified leukocytes when the target organism is found in the cell-free fraction. Here again, several standard nucleic acid isolation procedures have been used successfully to obtain amplifiable DNA following disruption of the viral capsid by detergents, proteinases, or both (97, 166). Both phenol-chloroform extraction and DNA binding to silica matrices have been successfully used (12, 107, 176). Nucleic acids have successfully been extracted from RNA viruses by using guanidine isothiocyanate lysis followed by phenol-chloroform extraction and isopropanol precipitation as well as by sodium dodecyl sulfate pronase lysis followed by phenol-chloroform extraction and ethanol precipitation (85, 116). In addition, commercially available RNA isolation kits have been used (173).

Sputum Specimens

The microorganism of most interest for molecular testing in sputum specimens is M. *tuberculosis*. Several lysis and nucleic acid extraction procedures have been evaluated for specimen preparation of M. *tuberculosis* DNA from sputum (15, 35, 48, 84, 130, 158). Because acidic polysaccharides present in sputum are known inhibitors of DNA polymerases (42, 62, 96), the basis of a successful sputum preparation procedure involves separation of the DNA from the inhibitors. Phenol-chloroform extraction of the DNA followed by ethanol precipitation has been used successfully on sputum specimens (15, 35). Use of silica matrices that bind DNA and allow the contaminants to be washed away and repetitive washing of the specimen pellet with detergents have also yielded amplifiable DNA from sputum specimens (49, 154).

Urine Specimens

Numerous lysis and nucleic acid extraction procedures have been reported for use with urine specimens (23, 65, 99). Most of these procedures are similar to those outlined for CSF. Like most specimen sources, urine is known to contain uncharacterized inhibitors of *Taq* polymerase (69).

In addition to being isolated from the common clinical specimens described above, nucleic acids have been isolated from numerous other types of specimens for molecular analysis (43, 69). Many of the protocols outlined above can be modified for use with specimens other than those for which they were originally described. Even fixed and paraffin-embedded tissue can be analyzed (70). Following the recommendations outlined above should help in selecting or developing the optimum specimen preparation procedure. As more commercially prepared kits for molecular detection and identification of microorganisms are made available, it is likely that specimen preparation procedures will become standardized for use on all pathogens found in a particular type of specimen.

CRITICAL DESIGN FEATURES AND QUALITY CONTROL IN THE DIAGNOSTIC MOLECULAR BIOLOGY LABORATORY

In the diagnostic molecular biology laboratory, in which nucleic acid amplification forms the technologic base for most assays, accuracy of testing is intimately associated with the physical layout of the facilities. Because the nucleic acid amplification techniques in use today are so powerful, the large amounts of resulting amplified products can lead to significant cross-contamination problems and can result in the generation of false-positive results. In order to address the potential problem of specimen contamination, specific laboratory designs have been developed.

General Laboratory Design Features

The design of a large, high-volume diagnostic molecular biology laboratory in which nucleic acid amplification is regularly used is based on the three major steps that are involved in testing procedures: reagent preparation, specimen preparation and reaction setup, and amplification and product analysis. Physical separation of the three areas involved in the testing procedure (i.e., a separate laboratory for each step) provides the highest level of protection against product contamination (93, 108). To further reduce the possibility of cross-contamination between the laboratories, the concept of unidirectional work flow has been integrated with physical separation of the major work areas (108). Unidirectional work flow means that each phase of the testing process proceeds from areas free of amplification products to areas in which amplification products are analyzed. This directionality lessens the possibility of amplified products being brought in from the product analysis area and contaminating reagent setup and specimen preparation areas.

While each individual laboratory has certain features unique to that area, certain features can be generally applied to all areas (108).

1. Dedicated equipment and supplies. All equipment and supplies, including not only materials directly involved in the testing procedure but also such things as laboratory coats worn by personnel working in the area and janitorial supplies for cleaning the area (dust mops, wet mops, buckets), should be dedicated to one laboratory and not used for other purposes. An easy way to keep track of equipment and supplies for a specific work area is to color code them. This makes it easier to keep materials in their assigned areas and makes it obvious when a breach of protocol has occurred.

2. Disposable supplies. Whenever possible, disposable supplies should be used, because they reduce the possibility of contamination from the products of previous assays. In addition, disposable gloves should be worn while working in all areas, and they should be changed frequently.

3. Bench-top hoods with UV lights. In the reagent preparation and specimen preparation and reaction setup laboratories, bench-top hoods are recommended for containment of specimen nucleic acids. The use of built-in UV lights to help decontaminate the hood following its use is also recommended.

4. Each laboratory in the facility should have a controlled airflow. Ideally, the facility is laid out so that reagent preparation, specimen preparation, and reaction setup and amplification laboratories are connected to the hallway via an anteroom. These laboratories and the hallway are maintained at a positive or neutral pressure, while the anteroom is maintained at a negative pressure. This airflow pattern prevents aerosols within these laboratories from escaping into the hallway and prevents aerosols in the hallway from entering the laboratories. In an alternative design, used at the Mayo Clinic, these laboratories and the anteroom all operate at a positive pressure.

The airflow for the product analysis laboratory is different. This area should be at a negative pressure, and the air from this laboratory should be exhausted directly from the building without recirculation and away from building air intake vents. Negative pressure in the product analysis area prevents aerosols from escaping into the hallway.

5. Ceiling-mounted UV light fixtures. UV lights suspended from the ceiling and controlled by timers afford another level of contamination control for the laboratory. A short 20- to 30-min exposure in the laboratory during off-hours can significantly decrease the amount of amplifiable nucleic acids found on laboratory surfaces without causing deterioration of laboratory surfaces and furnishings (108, 146, 147).

Recommendations for Individual Laboratories

See references 94 and 108 as well.

Reagent Preparation Laboratory

In the reagent preparation laboratory, all the components necessary for the nucleic acid amplification and detection procedures are prepared. Master mixes for each test procedure should contain as many of the individual reagent components as possible while meeting performance and stability requirements. Such a master mix minimizes the number of pipetting and mixing steps required for the assay, thereby decreasing the potential for introduction of contaminants and mistakes due to omission of a necessary reagent. Whenever possible, reagents should be dispensed in single-use aliquots to reduce the number of times a particular reagent needs to be opened.

Specimen Preparation and Reaction Setup Laboratory

As with any area of the clinical laboratory, universal precautions should be followed by all personnel when handling patient specimens. Strict precautions must be taken in the specimen preparation laboratory to prevent cross-contamination between specimens. Specimens should be handled with disposable plastic pipettes whenever possible. Small volumes should be pipetted with pipette tips containing aerosol-resistant barriers that are directly bonded to the walls of the tips or with positive-displacement pipettes with disposable pistons and sleeves.

It is essential that amplification reactions be set up in an amplification product-free area. To prevent cross-contamination, all pipetting in this laboratory must be done with pipette tips containing aerosol-resistant barriers that are directly bonded to the walls of the tips or with positive-displacement pipettes with disposable pistons and sleeves.

Amplification and Product Analysis Laboratory

The product analysis laboratory is the only area in which the amplification reaction vessel may be opened following the amplification procedure. To decrease the chances of product contamination of the laboratory environment, Mc-Creedy and Callaway suggest that amplification products be disposed of and not stored once a result has been determined (108). If questions concerning a test result arise, original specimens that have been saved per the policy of the institution can be reamplified.

It must be emphasized that the facility design described above is for a large, high-volume laboratory doing molecular testing that is predominantly based on nucleic acid amplification procedures. If space is limited, the procedures can be separated by setting up individual work areas within two rooms: one room for reagent preparation, specimen preparation, and reaction setup and the other room for amplification and product analysis. However, each area should have dedicated equipment and supplies and should maintain the general and individual features of each laboratory as outlined above (except for airflow). The increasing availability of commercially produced kit-based nucleic acid amplification assays will reduce the need for many of the dedicated laboratories associated with custom-designed in-house tests. In these cases, it will be the responsibility of the manufacturer to set up the kits in an amplification-free environment and to provide documentation of quality control.

Quality Control Procedures in the Diagnostic Molecular Biology Laboratory

The goal of quality control in the diagnostic molecular biology laboratory is the same as in any other area of the clinical microbiology laboratory, namely, to ensure the accuracy of test results. In addition to the routine aspects of quality control that are monitored in every section of the clinical microbiology laboratory, certain features must be stringently monitored in the diagnostic molecular biology laboratory (45).

1. Pipettors should be routinely calibrated at least once every 6 months. This is especially important because of the small volumes routinely used in nucleic acid amplification reactions.

2. Temperatures of incubators, heating blocks, and water baths should be recorded daily or at least on the day they are used. If certain incubations are performed at ambient temperature, the daily temperature of the room must be recorded.

3. The temperatures and ramping times of the thermal cycler should be regularly monitored.

4. Whenever possible, laboratory supplies should be delivered directly to the specific areas in which they will be used. This decreases the chances of possible contamination of the supplies prior to their delivery to an amplification product-free area.

Careful attention to several factors is important for maintaining the integrity of nucleic acid amplification-based procedures (45, 93, 94).

1. Use of positive and negative controls for each run of specimens. A good positive-control specimen should contain between 10 and 50 copies of well-characterized target DNA or RNA so that it reflects the level of sensitivity expected of the assay. A positive result in the negative control invalidates the run and requires that all reagents be checked for amplification, target contamination, or both.

2. Use of "spiked controls" or multiplex amplification of an internal control when suspicions arise that a specimen contains components inhibitory to the amplification reaction.

3. Use of carryover controls, which should be incorporated with each specimen whenever possible. Amplification inactivation by UNG or photochemical cross-linkers such as isopsoralen greatly reduces the chances of cross-contamination of specimens (32, 103).

4. Limited use of nested PCR. Because reamplification of an initial PCR product is the basis of nested PCR, this reaction causes serious contamination problems that cannot be dealt with easily. It is better to optimize a conventional PCR so that nesting is not required (123). If nesting must be used, the numbers of controls included with each run must be adequate to detect contamination problems.

5. Routine analysis of all single-container reagents that are repeatedly used to ensure the absence of contaminating amplification products or target nucleic acids.

6. Testing of all reagents used in specimen preparation procedures to ensure that they are functioning properly; e.g., test lysing reagents for liberating DNA or RNA from cells to ensure cell lysis without destruction of the nucleic acids.

While strict adherence to good quality control practices is important for maintaining the integrity of test results from the diagnostic molecular biology laboratory, such practices are only the beginning. Observance of what may be termed "good laboratory practice" is equally important for preventing cross-contamination of specimens and for moving work through the laboratory in a timely fashion. Good laboratory practice includes such things as maintaining a neat and organized work area and decontaminating the countertops with a solution of chlorine bleach followed by a 70% alcohol rinse, to name a few examples. Constant awareness of and monitoring for the problems that can plague nucleic acid amplification assays are the only ways to prevent small problems from growing into bigger problems.

FUTURE OF MOLECULAR TECHNIQUES IN CLINICAL MICROBIOLOGY

Nucleic acid probes and amplification technology represent some of the greatest advances in the clinical microbiology laboratory since the introduction of artificial media. Ultimately, the goal of molecular techniques in the microbiology laboratory is the direct determination of the identities and susceptibility patterns of microorganisms in clinical specimens, thus obviating the need for in vitro cultivation. While this goal may seem a bit far-fetched at present, the basic technology is at hand or, in some cases, imminent. As mentioned in this chapter, the direct identification and determination of rifampin susceptibility of M. tuberculosis in clinical specimens is being evaluated by several labora-

tories with very promising results (84, 157, 158). Because of its very slow growth and highly infectious nature, M. tuberculosis is an organism for which molecular techniques are likely to be used despite their relatively high cost. The application of molecular methods for the identification and characterization of other organisms will depend on practical technological feasibility and, ultimately, the cost-benefit ratio.

As the technology for nucleic acid amplification currently stands, application of the procedure is limited to large referral laboratories and large academic or private medical centers. Several factors will have to be addressed before these procedures become commonplace in all clinical microbiology laboratories. Probably the most significant problem that must be overcome before nucleic acid amplification techniques can be used routinely is false positivity due to contaminating nucleic acids (122, 131, 169). It is essential that the issue of contamination control be addressed and the integrity of the amplification assay results be ensured in all laboratories. "An ounce of prevention is worth a pound of cure" is especially applicable here; once clinicians have lost faith in a laboratory procedure, that faith can be very difficult to regain.

The ability to partially or fully automate these techniques is another key factor in determining how large a role these procedures will play in the average laboratory. Currently, nucleic acid amplification-based tests are very labor intensive and expensive to perform; this limits their use to large institutions and probably limits the demand for them. However, these limitations will gradually disappear as procedures are simplified and made more widely available and as clinicians realize that certain diagnoses cannot be made without this technology. Such was the experience with immunoassays. The first immunoassays were laborious and expensive and were used exclusively in the research laboratory, but as advances simplified and automated the procedures, immunoassays have become commonplace in the clinical laboratory.

Another challenge for molecular technology in the clinical microbiology laboratory is in the area of education (122). Clinical laboratory training programs at all levels will need to provide in-depth instruction in molecular diagnostic procedures. Clinical microbiologists must be willing to acquire new skills, or they may relinquish control of diagnostic molecular microbiology to more technology-oriented laboratories with little vested interest in infectious disease testing. Some recently introduced DNA probe-based tests for infectious diseases are currently being performed in clinical chemistry laboratories, partly because of the perception by manufacturers and laboratory administrators that clinical microbiologists are unfamiliar with tests based on nucleic acid chemistry. Only through education and the ability to meet new challenges will clinical microbiologists be able to oversee the introduction of molecular techniques for testing of microorganisms and to smooth the transition from conventional methods to future technologies.

REFERENCES

1. **Anderson, B. E., J. E. Dawson, D. C. Jones, and K. H. Wilson.** 1991. *Ehrlichia chaffeensis,* a new species associated with human ehrlichiosis. *J. Clin. Microbiol.* **29:**2838–2842.
2. **Anderson, B. E., J. W. Sumner, J. E. Dawson, T. Tzianabos, C. R. Greene, J. G. Olson, D. B. Fishbein, M. Olsen-Rasmussen, B. P. Holloway, E. H. George, and A. F. Azad.** 1992. Detection of the etiologic agent of human ehrlichiosis

by polymerase chain reaction. *J. Clin. Microbiol.* **30:**775–780.

3. **Arnheim, N., T. White, and W. E. Rainey.** 1990. Application of PCR: organismal and population biology. *BioScience* **40:**174–182.

4. **Arnold, L. J., Jr., P. W. Hammond, W. A. Wiese, and N. C. Nelson.** 1989. Assay formats involving acridinium-ester-labeled DNA probes. *Clin. Chem.* **35:**1588–1594.

5. **Aslanzadeh, J.** 1993. Application of hydroxylamine hydrochloride for post-PCR sterilization. *Mol. Cell. Probes* **7:**145–150.

6. **Barany, F.** 1991. Genetic disease detection and DNA amplification using cloned thermostable ligase. *Proc. Natl. Acad. Sci. USA* **88:**189–193.

7. **Barthold, S. W., D. H. Persing, A. L. Armstrong, and R. A. Peeples.** 1991. Kinetics of *Borrelia burgdorferi* dissemination and evolution of disease after intradermal inoculation of mice. *Am. J. Pathol.* **139:**263–273.

8. **Bej, A. K., M. H. Mahbubani, R. Miller, J. L. Dicesare, L. Haff, and R. M. Atlas.** 1990. Multiplex PCR amplification and immobilized capture probes for detection of bacterial pathogens and indicators in water. *Mol. Cell. Probes* **4:**353–365.

9. **Birkenmeyer, L., W. Bringer, and A. Armstrong.** 1992. Specific detection of *Neisseria gonorrhoeae* using the ligase chain reaction, abstr. C-207, p. 455. *Abstr. 92nd Gen. Meet. Am. Soc. Microbiol. 1992.* American Society for Microbiology, Washington, D.C.

10. **Birkenmeyer, L., and I. K. Mushahwar.** 1991. DNA probe amplification methods. *J. Virol. Methods* **35:**117–126.

11. **Bloch, W.** 1991. A biochemical perspective of the polymerase chain reaction. *Biochemistry* **30:**2735–2747.

12. **Boom, R., C. J. A. Sol, R. Heijtink, P. M. E. Wertheim-van Dillen, and J. van der Noordaa.** 1991. Rapid purification of hepatitis B virus DNA from serum. *J. Clin. Microbiol.* **29:**1804–1811.

13. **Bottger, E. C., A. Teske, P. Kirschner, S. Best, H. R. Chang, V. Beer, and B. Hirschel.** 1992. Disseminated infections with "*Mycobacterium genavense*" in patients with AIDS. *Lancet* **340:**76–80.

14. **Brahic, M., and A. T. Haase.** 1978. Detection of viral sequences of low reiteration frequency by in situ hybridization. *Proc. Natl. Acad. Sci. USA* **75:**6125–6129.

15. **Brisson-Noel, A., C. Aznar, C. Chureau, S. Nguyen, C. Pierre, M. Bartoli, R. Bonete, G. Pealoux, B. Gicquel, and G. Garrigue.** 1991. Diagnosis of tuberculosis by DNA amplification in clinical practice evaluation. *Lancet* **338:**364–366.

16. **Brisson-Noel, A., B. Gicquel, D. Lecossier, V. Levy-Frebault, X. Nassif, and A. J. Hance.** 1989. Rapid diagnosis of tuberculosis by amplification of mycobacterial DNA in clinical samples. *Lancet* **ii:**1069–1071.

17. **Britten, R. J., and E. H. Davidson.** 1985. Hybridization strategy, p. 3–14. *In* B. D. Hames and S. J. Higgins (ed.), *Nucleic Acid Hybridization: a Practical Approach.* IRL Press, Oxford.

18. **Brown, D. M., and P. Schell.** 1961. The reaction of hydroxylamine with cytosine and related compounds. *J. Mol. Biol.* **3:**709–710.

19. **Brown, J. T., and A. T. Wortman.** 1993. Rapid amplification of human papillomavirus type 16 and 18 E6 and E7 mRNA by 3SR, p. 414–419. *In* D. H. Persing, T. F. Smith, F. C. Tenover, and T. J. White (ed.), *Diagnostic Molecular Biology: Principles and Applications.* American Society for Microbiology, Washington, D.C.

20. **Brown, T.** 1993. Southern blotting, p. 2.9.1–2.9.15. *In* R. E. Kingston, D. D. More, J. G. Seidman, J. A. Smith, and K. Struhl (ed.), *Current Protocols in Molecular Biology*, vol. 1. Current Protocols, New York.

21. **Brown, T.** 1993. Dot and slot blotting of DNA, p. 2.9.15–2.9.20. *In* R. E. Kingston, D. D. More, J. G. Seidman, J. A. Smith, and K. Struh (ed.), *Current Protocols in Molecular Biology*, vol. 1. Current Protocols, New York.

22. **Brown, T.** 1993. Analysis of RNA by Northern and slot blot hybridization, p. 4.9.1–4.9.14. *In* R. E. Kingston, D. D. More, J. G. Seidman, J. A. Smith, and K. Struh (ed.), *Current Protocols in Molecular Biology*, vol. 1. Current Protocols, New York.

23. **Buffone, G. J., G. J. Demmler, C. M. Schimbor, and J. Greer.** 1991. Improved amplification of cytomegalovirus DNA from urine after purification of DNA with glass beads. *Clin. Chem.* **37:**1945–1949.

24. **Busch, M. P., J. C. Wilber, P. Johnson, L. Tobler, and C. S. Evans.** 1992. Impact of specimen handling and storage on detection of hepatitis C virus RNA. *Transfusion* **32:**420–425.

25. **Butcher, A., and J. Spadoro.** 1992. Using PCR for detection of HIV-1 infection. *Clin. Immunol. Newsl.* **12:**73–76.

26. **Carmody, M. W., and C. P. H. Vary.** 1993. Inhibition of DNA hybridization following partial dUTP substitution. *BioTechniques* **15:**692–699.

27. **Chamberlin, J. S., R. A. Gibbs, J. E. Rainier, P. N. Nguyen, and C. T. Caskey.** 1988. Deletion screening of Duchenne muscular dystrophy locus via multiplex DNA amplification. *Nucleic Acids Res.* **16:**11141–11156.

28. **Chehab, F. F., and Y. W. Kan.** 1989. Detection of specific DNA sequences by fluorescence amplification: a color complementation assay. *Proc. Natl. Acad. Sci. USA* **86:**9178–9182.

29. **Chevrier, D., S. R. Rasmussen, and J.-L. Guesdon.** 1993. PCR product by quantification by non-radioactive hybridization procedures using an oligonucleotide covalently bound to microwells. *Mol. Cell. Probes* **7:**187–197.

30. **Cimino, G. D., H. B. Gamper, S. T. Isaacs, and J. E. Hearst.** 1985. Psoralens as photoactive probes of nucleic acid structure and function: organic chemistry, photochemistry, and biochemistry. *Annu. Rev. Biochem.* **54:**151–193.

31. **Cimino, G. D., K. Metchette, S. T. Isaacs, and Y. S. Zhu.** 1990. More false-positive problems. *Nature* (London) **345:**773–774.

32. **Cimino, G. D., K. Metchette, J. W. Tessman, J. E. Hearst, and S. T. Isaacs.** 1991. Post-PCR sterilization: a method to control carryover contamination for the polymerase chain reaction. *Nucleic Acids Res.* **19:**99–107.

33. **Cockerill, F. R., III, J. R. Uhl, Z. Temesgen, Y. Zhang, S. T. Cole, L. Stockman, G. D. Roberts, D. L. Williams, and B. C. Kline.** Rapid identification of a point mutation of the *Mycobacterium tuberculosis* catalase-peroxidase (*katG*) gene associated with isoniazid resistance. *J. Infect. Dis.*, in press.

34. **Compton, J.** 1991. Nucleic acid sequence-based amplification. *Nature* (London) **350:**91–92.

35. **Cousins, D. V., S. D. Wilton, B. R. Francis, and B. L. Gow.** 1992. Use of polymerase chain reaction for rapid diagnosis of tuberculosis. *J. Clin. Microbiol.* **30:**255–258.

36. **Cuypers, H. T. M., D. Bresters, I. N. Winkel, H. W. Reesink, A. J. Weiner, M. Houghton, C. L. van der Poel, and P. N. Lelie.** 1992. Storage conditions of blood samples and primer selection affect the yield of cDNA polymerase chain reaction products of hepatitis C virus. *J. Clin. Microbiol.* **30:**3220–3224.

37. **D'Aquila, R. T., L. J. Bechtel, J. A. Videler, J. J. Eron, P. Gorczyca, and J. C. Kaplan.** 1991. Maximizing sensitivity and specificity of PCR by preamplification heating. *Nucleic Acids Res.* **19:**3749.

38. **Dermer, S. J., and E. M. Johnson.** 1988. Rapid DNA analysis of α_1-antitrypsin deficiency: application of an improved method for amplifying mutated gene sequences. *Lab. Invest.* **59:**403–408.

39. **De Wit, D., L. Steyn, S. Shoemaker, and M. Sogin.** 1990. Direct detection of *Mycobacterium tuberculosis* in clinical specimens by DNA amplification. *J. Clin. Microbiol.* **28:**2437–2441.

40. **Dey, M., J. Down, A. Howard, D. Howard, B. Keating, M. Little, J. Nadeau, V. Neece, C. Nycz, P. Small, and J. Wang.** 1993. Strand displacement amplification (SDA) of

M. tuberculosis DNA from clinical isolates, abstr. U-41, p. 176. *Abstr. 93rd Gen. Meet. Am. Soc. Microbiol. 1993.* American Society for Microbiology, Washington, D.C.

41. **Diamandis, E. P., and T. K. Christopoulos.** 1991. The biotin-(Strept) avidin system: principles and applications in biotechnology. *Clin. Chem.* **37:**625–636.

42. **Do, N., and R. P. Adams.** 1991. A simple technique for removing plant polysaccharide contaminants from DNA. *BioTechniques* **10:**162–166.

43. **Donofrio, J. C., J. D. Coonrod, J. N. Davidson, and R. F. Betts.** 1992. Detection of influenza A and B in respiratory secretions with the polymerase chain reaction. *PCR Methods Applic.* **1:**263–268.

44. **Douglas, D. D., J. Rakela, H. F. Taswell, D. Rabe, J. Detmer, W. Hunt, M. S. Urdea, and D. H. Persing.** 1992. *Correlation of Quantitative HCV RNA, anti-HCV IgM, and Serum ALT with Histopathological Presentation.* American Association for the Study of Liver Diseases, Chicago.

45. **Dragon, E. A., J. P. Spadoro, and R. Madej.** 1993. Quality control of polymerase chain reaction, p. 160–168. *In* D. H. Persing, T. F. Smith, F. C. Tenover, and T. J. White (ed.), *Diagnostic Molecular Biology: Principles and Applications.* American Society for Microbiology, Washington, D.C.

46. **Dyson, N. J.** 1991. Immobilization of nucleic acids and hybridization analysis, p. 111–156. *In* T. A. Brown (ed.), *Essential Molecular Biology: a Practical Approach,* vol. 2. ILR Press, Oxford.

47. **Eisenach, K. D., M. D. Cave, J. H. Bates, and T. J. Crawford.** 1990. Polymerase chain reaction amplification of a repetitive DNA sequence specific for *Mycobacterium tuberculosis. J. Infect. Dis.* **161:**977–981.

48. **Eisenach, K. D., M. D. Cave, and J. T. Crawford.** 1993. PCR detection of *Mycobacterium tuberculosis,* p. 191–196. *In* D. H. Persing, T. F. Smith, F. C. Tenover, and T. J. White (ed.), *Diagnostic Molecular Biology: Principles and Applications.* American Society for Microbiology, Washington, D.C.

49. **Eisenach, K. D., M. D. Sifford, M. D. Cave, J. H. Bates, and J. T. Crawford.** 1991. Detection of *Mycobacterium tuberculosis* in sputum samples using polymerase chain reaction. *Am. Rev. Respir. Dis.* **144:**1160–1163.

50. **Eisenstein, B. I.** 1990. The polymerase chain reaction: a new method of using molecular genetics for medical diagnosis. *N. Engl. J. Med.* **322:**178–183.

51. **Eoyang, L., and J. August.** 1971. Q beta RNA polymerase from phage Q beta infected *E. coli,* p. 829–839. *In* G. L. Cantoni and D. R. Davis (ed.), *Procedures in Nucleic Acid Research,* vol. 2. Harper & Row, Publishers, Inc., New York.

52. **Erlich, H. A., D. Gelfand, and J. J. Sninsky.** 1991. Recent advances in the polymerase chain reaction. *Science* **252:**1643–1651.

53. **Espy, M. J., T. F. Smith, and D. H. Persing.** 1993. Dependence of polymerase chain reaction product inactivation protocols on amplicon length and sequence composition. *J. Clin. Microbiol.* **31:**2361–2365.

54. **Fahrlander, P. D., and A. Klausner.** 1988. Amplifying DNA probe signals: a "Christmas tree" approach. *Bio/Technology* **6:**1165–1168.

55. **Feinberg, A. P., and B. Vogelstein.** 1983. A technique for radiolabeling DNA restriction endonuclease fragments to high specific activity. *Anal. Biochem.* **132:**6–13.

56. **Feinberg, A. P., and B. Vogelstein.** 1984. A technique for radiolabeling DNA restriction endonuclease fragments to high specific activity (addendum). *Anal. Biochem.* **137:**266–267.

57. **Feray, C., D. Samuel, V. Thiers, M. Gigou, F. Pichon, A. Bismuth, M. Reynes, P. Maisonneuve, H. Bismuth, and C. Brechot.** 1992. Reinfection of liver graft by hepatitis C after liver transplantation. *J. Clin. Invest.* **89:**1361–1365.

58. **Fleming, J. L., R. H. Wiesner, and R. G. Shorter.** 1989. Whipple's disease: clinical, biochemical, and histopathologic features and assessment of treatment in 29 patients. *Mayo Clin. Proc.* **63:**539–551.

59. **Floona, F., S. Weiss, F. Ferre, and K. Mullis.** 1990. Direct detection of HIV sequences in blood: high-gain polymerase chain reaction. *Abstr. 6th Int. Conf. AIDS* **2:**318.

60. **Frothingham, R., R. L. Allen, and K. H. Wilson.** 1991. Rapid 16S ribosomal DNA sequencing from a single colony without DNA extraction or purification. *BioTechniques* **11:**40–44.

61. **Frothingham, R., R. B. Blitchington, D. H. Lee, R. C. Greene, and K. H. Wilson.** 1992. UV absorption complicates PCR decontamination. *BioTechniques* **13:**208–210.

62. **Furukawa, K., and V. P. Bhavanandan.** 1983. Influences of anionic polysaccharides on DNA synthesis in isolated nuclei and by DNA polymerase α: correlation of observed effects with properties of the polysaccharides. *Biochim. Biophys. Acta* **740:**466–475.

63. **Geha, D. J., J. R. Uhl, C. A. Gustaferro, and D. H. Persing.** 1994. Multiplex PCR for identification of methicillin-resistant staphylococci in the clinical laboratory. *J. Clin. Microbiol.* **32:**1768–1772.

64. **Gelfand, D. H.** 1989. *Taq* DNA polymerase, p. 17–22. *In* H. A. Erlich (ed.), *PCR Technology: Principles and Applications for DNA Amplification.* Stockton Press, New York.

65. **Gerritsen, M. J., T. Olyhoek, M. A. Smits, and B. A. Bokhout.** 1991. Sample preparation method for polymerase chain reaction-based semiquantitative detection of *Leptospira interrogans* serovar Hardjo subtype hardjobovis in bovine urine. *J. Clin. Microbiol.* **29:**2805–2808.

66. **Gilliland, G., S. Perrin, and H. Franklin Bunn.** 1990. Competitive PCR for quantitation of mRNA, p. 60–69. *In* M. A. Innis, D. H. Gelfand, J. J. Sninsky, and T. J. White (ed.), *PCR Protocols: a Guide to Methods and Applications.* Academic Press, Inc., San Diego, Calif.

67. **Gingeras, T. R., P. Prodanovich, T. Latimer, J. C. Guatelli, D. D. Richman, and K. J. Barringer.** 1991. Use of self-sustained sequence replication amplification reaction to analyze and detect mutations in zidovudine-resistant human immunodeficiency virus. *J. Infect. Dis.* **164:**1066–1074.

68. **Goodman, J. L., P. Jurkovich, J. M. Kramber, and R. C. Johnson.** 1991. Molecular detection of persistent *Borrelia burgdorferi* in the urine of patients with active Lyme disease. *Infect. Immun.* **59:**269–278.

69. **Greenfield, L., and T. J. White.** 1993. Sample preparation methods, p. 122–137. *In* D. H. Persing, T. F. Smith, F. C. Tenover, and T. J. White (ed.), *Diagnostic Molecular Biology: Principles and Applications.* American Society for Microbiology, Washington, D.C.

70. **Greer, C. E., S. L. Peterson, N. B. Kiviat, and M. M. Manos.** 1991. PCR amplification from paraffin-embedded tissues: effects of fixative and fixation time. *Am. J. Clin. Pathol.* **95:**117–124.

71. **Guatelli, J. C., T. R. Gingeras, and D. D. Richman.** 1989. Nucleic acid amplification in vitro: detection of sequences with low copy numbers and application to diagnosis of human immunodeficiency virus type 1 infection. *Clin. Microbiol. Rev.* **2:**217–226.

72. **Guatelli, J. C., K. M. Whitfield, D. Y. Kwoh, K. J. Barringer, D. D. Richman, and T. R. Gingeras.** 1990. Isothermal, *in vitro* amplification of nucleic acids by multienzyme reaction modeled after retroviral replication. *Proc. Natl. Acad. Sci. USA* **87:**1874–1878.

73. **Hankin, R. C.** 1992. In situ hybridization: principles and applications. *Lab. Med.* **23:**764–770.

74. **Haqqi, T. M., G. Sarkar, C. S. David, and S. S. Sommer.** 1988. Specific amplification of a refractory segment of genomic DNA. *Nucleic Acids Res.* **16:**11844.

75. **Haydock, P. V., and S. A. Kochik.** 1993. 3SR detection of *Chlamydia trachomatis,* p. 242–246. *In* D. H. Persing, T. F. Smith, F. C. Tenover, and T. J. White (ed.), *Diagnostic Molecular Biology: Principles and Applications.* American Society for Microbiology, Washington, D.C.

76. **Hermans, P. W. M., A. R. J. Schuitema, D. van Soolingen, C. P. H. J. Verstynen, E. M. Bik, J. E. R. Thole, A. H. J. Kolk, and J. D. A. van Embden.** 1990. Specific detection of

Mycobacterium tuberculosis complex strains by polymerase chain reaction. *J. Clin. Microbiol.* **28:**1204–1213.

77. **Higuchi, R.** 1989. Simple and rapid preparation of samples for PCR, p. 31–38. *In* H. A. Erlich (ed.), *PCR Technology: Principles and Applications for DNA Amplification.* Stockton Press, New York.

78. **Higuchi, R., G. Dollinger, P. S. Walsh, and R. Griffith.** 1992. Simultaneous amplification and detection of specific DNA sequences. *Bio/Technology* **10:**413–417.

79. **Hofmeister, E. K., R. B. Markham, J. E. Childs, and R. R. Arthur.** 1992. Comparison of polymerase chain reaction and culture of *Borrelia burgdorferi* in naturally infected *Peromyscus leucopus* and experimentally infected C.B-17 *scid/scid* mice. *J. Clin. Microbiol.* **30:**2625–2631.

80. **Holland, P. M., R. D. Abramson, R. Watson, and D. H. Gelfand.** 1991. Detection of specific polymerase chain reaction product by utilizing the 5′→3′ exonuclease activity of *Thermus aquaticus* DNA polymerase. *Proc. Natl. Acad. Sci. USA* **88:**7276–7280.

81. **Hu, H., K. Elmore, I. Facey, and D. Jenderzak.** 1991. Detection of *Borrelia burgdorferi* by ligase chain reaction, abstr. D-8, p. 79. *Abstr. 91st Gen. Meet. Am. Soc. Microbiol. 1991.* American Society for Microbiology, Washington, D.C.

82. **Hung, T., K. Mak, and K. Fong.** 1990. A specificity enhancer for polymerase chain reaction. *Nucleic Acids Res.* **18:**4953.

83. **Hunsaker, W. R., H. Badri, M. Lombardo, and M. L. Collins.** 1989. Nucleic acid hybridization assays employing dA-tailed capture probes. II. Advanced multiple capture methods. *Anal. Biochem.* **181:**360–370.

84. **Hunt, J. M., G. D. Roberts, L. Stockman, T. A. Felmlee, and D. H. Persing.** 1994. Detection of a genetic locus encoding resistance to rifampin in mycobacterial cultures and in clinical specimens. *Diagn. Microbiol. Infect. Dis.* **18:**219–227.

85. **Inderlied, C. B., and K. A. Nash.** 1993. PCR detection of rubella virus, p. 394–400. *In* D. H. Persing, T. F. Smith, F. C. Tenover, and T. J. White (ed.), *Diagnostic Molecular Biology: Principles and Applications.* American Society for Microbiology, Washington, D.C.

86. **Innis, M. A., D. H. Gelfand, J. J. Sninsky, and T. J. White (ed.).** 1990. *PCR Protocols: a Guide to Methods and Applications.* Academic Press, Inc., San Diego, Calif.

87. **Iovannisci, D. M., and E. S. Winn-Deen.** 1991. Ligation amplification and fluorescence detection of *Mycobacterium tuberculosis* DNA, abstr. D-43, p. 85. *Abstr. 91st Gen. Meet. Am. Soc. Microbiol. 1991.* American Society for Microbiology, Washington, D.C.

88. **Izraeli, S., C. Pfleiderer, and T. Lion.** 1991. Detection of gene expression by PCR amplification of RNA derived from frozen heparinized whole blood. *Nucleic Acids Res.* **19:**6051.

89. **Kacian, D., D. Mills, F. Kramer, and S. Spiegelman.** 1972. A replicating RNA molecule suitable for detailed analysis of extracellular evolution and recombination. *Proc. Natl. Acad. Sci. USA* **69:**3038–3042.

90. **Kaneko, K., O. Onodera, M. Miyatake, and S. Tsuji.** 1990. Rapid diagnosis of tuberculosis meningitis by polymerase chain reaction. *Neurology* **40:**1617–1618.

91. **Klinger, J. D., and C. G. Pritchard.** 1990. Amplified probe-based assays—possibilities and challenges in clinical microbiology. *Clin. Microbiol. News* **12:**189–207.

92. **Kwoh, D. Y., G. R. Davis, K. M. Whitfield, H. L. Chappelle, L. J. DiMichele, and T. R. Gingeras.** 1989. Transcription-based amplification system and detection of amplified human immunodeficiency virus type 1 with a bead-based sandwich hybridization format. *Proc. Natl. Acad. Sci. USA* **86:**1173–1177.

93. **Kwok, S.** 1990. Procedures to minimize PCR-product carryover, p. 142–145. *In* M. A. Innis, D. H. Gelfand, J. J. Sninsky, and T. J. White (ed.), *PCR Protocols: a Guide to Methods and Applications.* Academic Press, Inc., San Diego, Calif.

94. **Kwok, S., and R. Higuchi.** 1989. Avoiding false positives with PCR. *Nature* (London) **339:**237–238.

95. **Kwok, S., D. H. Mack, K. B. Mullis, B. Poiesz, G. Ehrlich, D. Blair, A. Friedman-Kien, and J. J. Sninsky.** 1987. Identification of human immunodeficiency virus sequences by using in vitro enzymatic amplification and oligomer cleavage detection. *J. Virol.* **61:**1690–1694.

96. **Lamblin, G., H. Rahmoune, J.-M. Wierszeski, M. Lhermitte, G. Strecker, and P. Roussel.** 1991. Structure of two sulphated oligosaccharides from respiratory mucins of a patient suffering from cystic fibrosis: a fast-atom-bombardment m.s. and ¹H-n.m.r. spectroscopic study. *Biochem. J.* **275:**199–206.

97. **Lanciotti, R. S., C. H. Calisher, D. J. Gubler, G.-J. Chang, and A. V. Vornadam.** 1992. Rapid detection and typing of dengue viruses from clinical samples by using reverse transcription-polymerase chain reaction. *J. Clin. Microbiol.* **30:**545–551.

98. **Lebech, A., P. Hindersson, J. Vuust, and K. Hansen.** 1991. Comparison of in vitro culture and polymerase chain reaction for detection of *Borrelia burgdorferi* in tissue from experimentally infected animals. *J. Clin. Microbiol.* **29:**731–737.

99. **Lebech, A.-M., and K. Hansen.** 1992. Detection of *Borrelia burgdorferi* DNA in urine samples and cerebrospinal fluid samples from patients with early and late Lyme neuroborreliosis by polymerase chain reaction. *J. Clin. Microbiol.* **30:**1646–1653.

100. **Leong, D. U., and K. S. Greisen.** 1993. PCR detection of bacteria found in cerebrospinal fluid, p. 300–306. *In* D. H. Persing, T. F. Smith, F. C. Tenover, and T. J. White (ed.), *Diagnostic Molecular Biology: Principles and Applications.* American Society for Microbiology, Washington, D.C.

101. **Lizardi, P., C. Guerra, H. Lomeli, I. Tussie-Luna, and F. Kramer.** 1988. Exponential amplification of recombinant-RNA hybridization probes. *Biotechnology* **6:**1197–1202.

102. **Loeffelholz, M. J., C. A. Lewinski, S. R. Silver, A. P. Purohit, S. A. Herman, D. A. Buonagurio, and E. A. Dragon.** 1992. Detection of *Chlamydia trachomatis* in endocervical specimens by polymerase chain reaction. *J. Clin. Microbiol.* **30:**2847–2851.

103. **Longo, M. C., M. S. Berninger, and J. L. Hartley.** 1990. Use of uracil DNA glycosylase to control carry-over contamination in polymerase chain reactions. *Gene* **93:**125–128.

104. **Malawista, S. E., R. T. Schoen, T. L. Moore, D. E. Dodge, T. J. White, and D. H. Persing.** 1992. Failure of multitarget detection of *Borrelia burgdorferi*-associated DNA sequences in synovial fluids of patients with juvenile rheumatoid arthritis: a cautionary note. *Arthritis Rheum.* **35:**246–247.

105. **Malloy, D. C., R. K. Nauman, and H. Paxton.** 1990. Detection of *Borrelia burgdorferi* using the polymerase chain reaction. *J. Clin. Microbiol.* **28:**1089–1093.

106. **Matthews, J. A., and J. Kricka.** 1988. Analytical strategies for the use of DNA probes. *Anal. Biochem.* **169:**1–25.

107. **McCaustland, K. A., S. Bi, M. A. Purdy, and D. W. Bradley.** 1991. Application of two RNA extraction methods prior to amplification of hepatitis E virus nucleic acid by the polymerase chain reaction. *J. Virol. Methods* **35:**311–342.

108. **McCreedy, B. J., and T. H. Callaway.** 1993. Laboratory design and work flow, p. 149–159. *In* D. H. Persing, T. F. Smith, F. C. Tenover, and T. J. White (ed.), *Diagnostic Molecular Biology: Principles and Applications.* American Society for Microbiology, Washington, D.C.

108a.**Meyers, T. W., and D. H. Gelfand.** 1991. Reverse transcription and DNA amplification by a *Thermus themophilus* DNA polymerase. *Biochemistry* **30:**7661–7666.

109. **Miele, E. A., D. R. Mills, and F. R. Kramer.** 1983. Autocatalytic replication of a recombinant RNA. *J. Mol. Biol.* **171:**281–295.

110. **Mullis, K. B.** 1990. The unusual origin of the polymerase chain reaction. *Sci. Am.* **262:**56–65.

111. **Mullis, K. B.** 1991. The polymerase chain reaction in an anemic mode: how to avoid oligodeoxyribonuclear fusion. *PCR Methods Applic.* **1:**1–4.

112. **Mullis, K. B., and F. A. Faloona.** 1987. Specific synthesis of DNA *in vitro* via a polymerase-catalyzed reaction. *Methods Enzymol.* **155:**335–350.

113. **Nelson, N. C., and D. L. Kacian.** 1990. Chemiluminescent DNA probes: a comparison of the acridinium ester and dioxetane detection systems and their use in clinical diagnostic assays. *Clin. Chim. Acta* **194:**73–90.

114. **Nevinney-Stickel, C., H. Hinzpeter, A. Andreas, and E. D. Albert.** 1991. Non-radioactive oligotyping for HLA-DR1-DRw10 using polymerase chain reaction, digoxigenin-labeled oligonucleotides and chemiluminescence detection. *Eur. J. Immunogenet.* **18:**323–332.

114a. **Nickerson, D. A., R. Kaiser, S. Lappin, J. Stewart, L. Hood, and U. Lundegren.** 1990. Automated DNA diagnostics using an ELISA-based oligonucleotide ligation assay. *Proc. Natl. Acad. Sci. USA* **87:**8923–8927.

115. **Nocton, J. J., F. Dressler, B. J. Rutledge, P. N. Rys, D. H. Persing, and A. C. Steere.** 1994. Detection by polymerase chain reaction of *Borrelia burgdorferi* DNA in synovial fluid from patients with Lyme arthritis. *N. Engl. J. Med.* **330:**229–234.

116. **Osamu, Y., M. Tagawa, and M. Omata.** 1993. PCR detection of hepatitis B virus, p. 322–331. *In* D. H. Persing, T. F. Smith, F. C. Tenover, and T. J. White (ed.), *Diagnostic Molecular Biology: Principles and Applications.* American Society for Microbiology, Washington, D.C.

117. **Ou, C. V., J. J. Moore, and G. Schochetman.** 1991. UV irradiation to reduce false positivity in the polymerase chain reaction. *BioTechniques* **10:**442–446.

118. **Pachl, C. A., D. G. Kern, P. J. Sheridan, M. M. Stempien, J. A. Todd, Y. S. Zhu, Y. Gong, G. D. Cimino, J. C. Wilber, M. S. Urdea, and P. D. Neuwald.** 1992. Quantitative detection of HIV RNA in plasma using a signal amplification probe assay, abstr. 1247. *Program Abstr. 32nd Intersci. Conf. Antimicrob. Agents Chemother.* American Society for Microbiology, Washington, D.C.

119. **Palva, A., and M. Ranki.** 1985. Microbial diagnosis by nucleic acid sandwich hybridization. *Clin. Lab. Med.* **5:**475–490.

120. **Panaccio, M., and A. Lew.** 1991. PCR based diagnosis in the presence of 8% (v/v) blood. *Nucleic Acids Res.* **19:**1151.

121. **Persing, D. H.** 1991. Polymerase chain reaction: trenches to benches. *J. Clin. Microbiol.* **29:**1281–1285.

122. **Persing, D. H.** 1993. Diagnostic molecular microbiology: current challenges and future direction. *Diagn. Microbiol. Infect. Dis.* **16:**159–163.

123. **Persing, D. H.** 1993. In vitro nucleic acid amplification techniques, p. 51–87. *In* D. H. Persing, T. F. Smith, F. C. Tenover, and T. J. White (ed.), *Diagnostic Molecular Biology: Principles and Applications.* American Society for Microbiology, Washington, D.C.

124. **Persing, D. H., and G. D. Cimino.** 1993. Amplification product inactivation methods, p. 105–121. *In* D. H. Persing, T. F. Smith, F. C. Tenover, and T. J. White (ed.), *Diagnostic Molecular Biology: Principles and Applications.* American Society for Microbiology, Washington, D.C.

125. **Persing, D. H., B. J. Rutledge, P. N. Rys, D. S. Podzorski, P. D. Mitchell, K. D. Reed, B. Liu, E. Fikrig, and S. E. Malawista.** 1994. Target imbalance: disparity of *Borrelia burgdorferi* genetic material in synovial fluid from Lyme arthritis patients. *J. Infect. Dis.* **69:**664–668.

126. **Persing, D. H., T. F. Smith, F. C. Tenover, and T. J. White (ed.).** 1993. *Diagnostic Molecular Biology: Principles and Applications.* American Society for Microbiology, Washington, D.C.

127. **Persing, D. H., S. R. Telford III, P. N. Rys, D. E. Dodge, T. J. White, S. E. Malawista, and A. Spielman.** 1990. Detection of *Borrelia burgdorferi* DNA in museum specimens of *Ixodes dammini* ticks. *Science* **249:**1420–1423.

128. **Persing, D. H., S. R. Telford III, A. Spielman, and S. W. Barthold.** 1990. Detection of *Borrelia burgdorferi* infection in *Ixodes dammini* ticks with the polymerase chain reaction. *J. Clin. Microbiol.* **28:**566–572.

129. **Piatak, M., Jr., K.-C. Luk, B. Williams, and J. D. Lifson.** 1993. Quantitative competitive polymerase chain reaction for accurate quantitation of HIV DNA and RNA species. *BioTechniques* **14:**70–81.

130. **Pierre, C., D. Lecossier, Y. Bousougant, D. Bocart, V. Joly, P. Yeni, and A. J. Hance.** 1991. Use of a reamplification protocol improves sensitivity of detection of *Mycobacterium tuberculosis* in clinical samples by amplification of DNA. *J. Clin. Microbiol.* **29:**712–717.

131. **Podzorski, R. P., and D. H. Persing.** 1993. PCR: the next decade. *Clin. Microbiol. Newsl.* **15:**137–143.

132. **Pollard-Knight, D., C. A. Read, M. J. Downes, L. A. Howard, M. R. Leadbetter, S. A. Pheby, E. McNaughton, A. Syms, and M. A. W. Brady.** 1990. Nonradioactive nucleic acid detection by enhanced chemiluminescence using probes directly labeled with horseradish peroxidase. *Anal. Biochem.* **37:**84–89.

133. **Predari, S. C., M. Ligozzi, and R. Fontana.** 1991. Genotypic identification of methicillin-resistant coagulase-negative staphylococci by polymerase chain reaction. *Antimicrob. Agents Chemother.* **35:**2568–2573.

134. **Prichard, C. G., and J. E. Stefano.** 1990. Amplified detection of viral nucleic acid at subattomole levels using Q beta replicase. *Ann. Biol. Chem.* (Paris) **48:**492–497.

135. **Radolf, J. D.** 1993. PCR detection of *Treponema pallidum*, p. 224–229. *In* D. H. Persing, T. F. Smith, F. C. Tenover, and T. J. White (ed.), *Diagnostic Molecular Biology: Principles and Applications.* American Society for Microbiology, Washington, D.C.

136. **Ranki, M., A. Palva, M. Virtanen, M. Laaksonen, and H. Soderlund.** 1983. Sandwich hybridization as a convenient method for detection of nucleic acids in crude samples. *Gene* **21:**77–85.

137. **Reddy, K. J., and M. Gilman.** 1993. Preparation of bacterial RNA, p. 4.4.1–4.4.7. *In* R. E. Kingston, D. D. More, J. G. Seidman, J. A. Smith, and K. Struh (ed.), *Current Protocols in Molecular Biology*, vol. 1. Current Protocols, New York.

138. **Relman, D. A., T. M. Schmidt, R. P. MacDermott, and S. Falkow.** 1992. Identification of the uncultured bacillus of Whipple's disease. *N. Engl. J. Med.* **327:**293–301.

139. **Rodriguez, M., N. Prayoonwiwat, and L. R. Pease.** 1993. PCR detection of human T-cell lymphotrophic virus type 1, p. 316–321. *In* D. H. Persing, T. F. Smith, F. C. Tenover, and T. J. White (ed.), *Diagnostic Molecular Biology: Principles and Applications.* American Society for Microbiology, Washington, D.C.

139a. **Ruano, G., and K. K. Kidd.** 1991. Coupled amplification and sequencing of genomic DNA. *Proc. Natl. Acad. Sci. USA* **88:**2815–2819.

140. **Rys, P. N.** 1993. PCR detection of *Borrelia burgdorferi*, p. 203–210. *In* D. H. Persing, T. F. Smith, F. C. Tenover, and T. J. White (ed.), *Diagnostic Molecular Biology: Principles and Applications.* American Society for Microbiology, Washington, D.C.

141. **Rys, P. N., and D. H. Persing.** 1993. Preventing false positives: quantitative evaluation of three protocols for inactivation of polymerase chain reaction amplification products. *J. Clin. Microbiol.* **31:**2356–2360.

142. **Saiki, R. K.** 1989. The design and optimization of the PCR, p. 7–22. *In* H. A. Erlich (ed.), *PCR Technology: Principles and Applications for DNA Amplification.* Stockton Press, New York.

143. **Saiki, R. K., P. S. Walsh, C. H. Levenson, and H. A. Erlich.** 1989. Genetic analysis of amplified DNA with immobilized sequence-specific oligonucleotide probes. *Proc. Natl. Acad. Sci. USA* **86:**6230–6234.

144. **Sambrook, J., E. F. Fritsch, and T. Maniatis.** 1989. *Molecular Cloning: a Laboratory Manual*, 2nd ed., vol. II, p. 10.51–10.70. Cold Spring Harbor Laboratory, Cold Spring Harbor, N.Y.

145. **Sanchez-Pescador, R., M. S. Stempien, and M. S. Urdea.** 1988. Rapid chemiluminescent nucleic acid assays for detection of TEM-1 β-lactamase-mediated penicillin resistance in

Neisseria gonorrhoeae and other bacteria. *J. Clin. Microbiol.* **26:**1934–1938.

146. **Sarkar, G., and S. S. Sommer.** 1990. More light on PCR contamination. *Nature* (London) **347:**340–341.

147. **Sarkar, G., and S. S. Sommer.** 1991. Parameters affecting susceptibility of PCR contamination to UV inactivation. *BioTechniques* **10:**591–593.

148. **Serhat, U., J. Hoskins, J. E. Flokowitsch, C. Y. E. Wu, D. A. Preston, and P. L. Skatrud.** 1992. Detection of methicillin-resistant staphylococci by using the polymerase chain reaction. *J. Clin. Microbiol.* **30:**1685–1691.

149. **Shan, J., J. Liu, B. Stone, W. King, S. Popoff, J. Smith, J. Murphy, R. Nietupski, and G. Radcliffe.** 1993. A novel method for detection of *Mycobacterium tuberculosis* directly from sputum, abstr. U-43, p. 176. *Abstr. 93nd Gen. Meet. Am. Soc. Microbiol. 1993.* American Society for Microbiology, Washington, D.C.

150. **Shankar, P., N. Manjunath, K. K. Mohan, K. Prasad, M. Behari, Shriniwas, and G. K. Ahuja.** 1991. Rapid diagnosis of tuberculosis meningitis by polymerase chain reaction. *Lancet* **337:**5–7.

151. **Shen, L. P., J. A. Kolberg, R. R. Spaete, R. Miner, and W. L. Drew.** 1992. A quantitative method for detection of human cytomegalovirus DNA using a branched DNA enhanced label amplification assay, abstr. S-54, p. 408. *Abstr. 92nd Gen. Meet. Am. Soc. Microbiol. 1992.* American Society for Microbiology, Washington, D.C.

152. **Slupphaug, G., I. Alseth, I. Eftedal, G. Volden, and H. E. Krokan.** 1993. Low incorporation of dUMP by some thermostable DNA polymerases may limit their use in PCR amplifications. *Anal. Biochem.* **211:**164–169.

153. **Southern, E. M.** 1975. Detection of specific sequences among DNA fragments separated by gel electrophoresis. *J. Mol. Biol.* **98:**503–517.

154. **Sritharan, V., and R. H. Barker, Jr.** 1991. A simple method for diagnosis of M. *tuberculosis* infection in clinical samples using PCR. *Mol. Cell. Probes* **5:**385–395.

155. **Strickler, J. D., and C. D. Copenhaver.** 1990. In situ hybridization in hematology. *Am. J. Clin. Pathol.* **93**(Suppl.):544–548.

156. **Telenti, A.** 1993. PCR detection and typing of Epstein-Barr virus, p. 344–349. *In* D. H. Persing, T. F. Smith, F. C. Tenover, and T. J. White (ed.), *Diagnostic Molecular Biology: Principles and Applications.* American Society for Microbiology, Washington, D.C.

157. **Telenti, A., P. Imboden, F. Marchesi, D. Lowrie, S. Cole, M. J. Colston, L. Matter, K. Schopfer, and T. Bodmer.** 1993. Detection of rifampicin-resistance mutations in *Mycobacterium tuberculosis. Lancet* **341:**647–650.

158. **Telenti, A., P. Imboden, F. Marchesi, T. Schmidheini, and T. Bodmer.** 1993. Direct, automated detection of rifampin-resistant *Mycobacterium tuberculosis* by polymerase chain reaction and single-stranded conformation polymorphism analysis. *Antimicrob. Agents Chemother.* **37:**2054–2058.

159. **Tenover, F. C., and E. R. Unger.** 1993. Nucleic acid probes for detection and identification of infectious agents, p. 3–25. *In* D. H. Persing, T. F. Smith, F. C. Tenover, and T. J. White (ed.), *Diagnostic Molecular Biology: Principles and Applications.* American Society for Microbiology, Washington, D.C.

160. **Ubukata, K., S. Nakagami, A. Nitta, A. Yamane, S. Kawakami, M. Sugiura, and M. Konno.** 1992. Rapid detection of *mecA* gene in methicillin-resistant staphylococci by enzymatic detection of polymerase chain reaction products. *J. Clin. Microbiol.* **30:**1728–1733.

161. **Unger, E. R., L. R. Budgeon, D. Myerson, and D. J. Brigati.** 1986. Viral diagnosis by in situ hybridization: description of a rapid colorimetric method. *Am. J. Surg. Pathol.* **10:**1–8.

162. **Urdea, M. S., T. Fultz, T. J. Anderson, M. Running, J. A. Hamren, S. Ahle, and C. A. Chang.** 1991. Branched amplification multimers for the sensitive, direct detection of human hepatitis viruses. *Nucleic Acids Symp. Ser.* **24:**197–200.

163. **Walder, R. Y., J. R. Hayes, and J. A. Walder.** 1993. Use of PCR primers containing a 3′-terminal ribose residue to prevent cross-contamination of amplified sequences. *Nucleic Acids Res.* **21:**4339–4343.

164. **Walker, G. T., M. L. Fraiser, J. L. Schram, M. C. Little, J. G. Nadeau, and D. P. Malinowski.** 1992. Strand displacement amplification—an isothermal, in vitro DNA amplification technique. *Nucleic Acids Res.* **20:**1691–1696.

165. **Walker, G. T., M. C. Little, J. G. Nadeau, and D. D. Shank.** 1992. Isothermal in vitro amplification of DNA by a restriction enzyme/DNA polymerase system. *Proc. Natl. Acad. Sci. USA* **89:**392–396.

166. **Wang, J.-T., T.-H. Wang, J.-C. Sheu, S.-M. Lin, J.-T. Lin, and D.-S. Chen.** 1992. Effects of anticoagulants and storage of blood samples on efficacy of the polymerase chain reaction assay for hepatitis C virus. *J. Clin. Microbiol.* **30:**750–753.

167. **Weiss, R.** 1991. A scramble for patent rights. *Science* **254:**1293.

168. **Wetmur, J. G.** 1991. DNA probes: applications of the principles of nucleic acid hybridization. *Crit. Rev. Biochem. Mol. Biol.* **26:**227–259.

169. **White, T. J.** 1993. Amplification product detection methods, p. 138–148. *In* D. H. Persing, T. F. Smith, F. C. Tenover, and T. J. White (ed.), *Diagnostic Molecular Biology: Principles and Applications.* American Society for Microbiology, Washington, D.C.

170. **White, T. J., R. Madej, and D. H. Persing.** 1992. The polymerase chain reaction: clinical applications. *Adv. Clin. Chem.* **29:**161–196.

171. **Wiedbrauk, D. L.** 1992. Molecular methods for virus detection. *Lab. Med.* **23:**737–742.

172. **Wilber, J. C., P. J. Johnson, P. J. Daley, C. S. Chan, R. Sanchez-Pescador, P. D. Neuwald, and M. S. Urdea.** 1992. Quantitation of hepatitis C viral RNA in serum or plasma, abstr. S-44, p. 406. *Abstr. 92nd Gen Meet. Am. Soc. Microbiol. 1992.* American Society for Microbiology, Washington, D.C.

173. **Wilber, J. C., P. J. Johnson, and M. S. Urdea.** 1993. Reverse transcriptase-PCR for hepatitis C virus RNA, p. 327–331. *In* D. H. Persing, T. F. Smith, F. C. Tenover, and T. J. White (ed.), *Diagnostic Molecular Biology: Principles and Applications.* American Society for Microbiology, Washington, D.C.

174. **Woese, C. R.** 1987. Bacterial evolution. *Microbiol. Rev.* **51:**221–271.

175. **Wu, D. Y., and R. B. Wallace.** 1989. The ligation amplification reaction (LAR)—amplification of specific DNA sequences using sequential rounds of template dependent ligation. *Genomics* **4:**560–569.

176. **Yamada, O., T. Matsumoto, M. Nakashima, S. Hagare, T. Kamahora, H. Ueyama, Y. Kishi, H. Uemura, and T. Kurimaura.** 1990. A new method for extraction of DNA or RNA for polymerase chain reaction. *J. Virol. Methods* **27:**203–210.

177. **Yang, L., J. H. Weis, E. Eichwald, C. P. Kolbert, D. H. Persing, and J. J. Weis.** 1994. Heritable susceptibility to *Borrelia burgdorferi*-induced arthritis is dominant and is associated with persistence of high numbers of spirochetes in tissues. *Infect. Immun.* **62:**492–500.

178. **Young, K. K. Y., R. M. Resnick, and T. W. Meyers.** 1993. Detection of hepatitis C virus RNA by a combined reverse transcription-polymerase chain reaction assay. *J. Clin. Microbiol.* **31:**882–886.

Cell Culture Systems

LORRAINE V. JONES BRANDO

14

In 1907, when Harrison (15) published the results of his elegant experiments on the culture of explants from frog embryos, he could not have realized what a powerful and versatile tool of science had just been born. Today, large-scale cell culture allows industry to produce large lots of vaccines, monoclonal antibodies, growth factors, etc., with little variation from lot to lot. Small-scale cell culture allows clinical laboratories to detect viral, rickettsial, and chlamydial agents as well as bacterial toxins and thus assist clinicians in the diagnoses of infectious diseases. As large and diverse as this technology seems now, maturation to its current status in clinical laboratories proceeded at a fairly torpid pace owing to such menaces as microbial contamination, lack of consistent defined growth media, and lack of homogeneous cell populations. The introduction of antibiotics in culture media (3) encouraged the widespread use of cell culture as a laboratory technique, but the report in 1949 by Enders et al. (7) on growing poliovirus in cell culture may be considered the seminal experiment for the use of cell culture in the diagnostic microbiology laboratory. This chapter presents a synopsis of animal cell culture systems and their uses in the diagnostic setting.

ANIMAL CELL CULTURE SYSTEMS

Equipment and Space Requirements

The scope of work to be handled in the cell culture area of the clinical laboratory must be considered when that area is designed. Issues to consider include (i) the projected number of specimens to be handled, (ii) the number of people to be working in the area at a given time, (iii) whether culture media will be prepared on site or purchased ready to use, and (iv) whether glassware will be washed on site or disposables will be used. The answers to these and other questions dictate the size and type of equipment required.

In considering the issue of culture medium preparation, many laboratories find it more cost-effective to purchase prepared media because of the equipment, time, and effort required for on-site preparation and quality testing. Laboratories that do prepare media in house require a reliable source of high-quality water, vessels for storage of media, incubator space for quality control testing, and refrigerated space for quarantine and storage of freshly prepared media.

In general, a typical cell culture facility in a diagnostic laboratory requires several pieces of large equipment (9). Ideally, the laminar-flow hood used for medium preparation and clean, i.e., not infected, cell culture is separate from a second laminar-flow hood used for sample preparation and cell inoculation. All laminar-flow hoods should be located in clean, quiet, low-traffic areas of the laboratory away from supply air diffusers and doors. Likewise, incubators for both clean and inoculated cells should be located in low-disturbance areas, especially if large-door models with shelves near the floor are to be used. Such placement decreases the chances of dust and spores and thus contamination getting into the cultures (9). Each person should have exclusive use of an inverted microscope for daily observation of cells. Depending on the volume of samples, one or two fluorescence microscopes are required; they should be located in a dark room or a reasonable facsimile thereof. Low-speed (up to $7,300 \times g$) and superspeed (up to $50,000 \times g$) centrifuges with carriers for 96-well trays and shell vials are best located in an adjoining room because of the noise and aerosols they generate during operation. At least one refrigerator (4 to 10°C), two freezers (−20 and −70°C), and a liquid nitrogen freezer are required. The choice of the liquid nitrogen freezer will be predicated on (i) the desired capacity (number of ampoules), (ii) the frequency of access (which will have an impact on static holding time), and (iii) the convenience of access (9). Many small pieces of equipment, e.g., pipette aids, hemocytometers, coarse and fine balances, etc., may be required for individual laboratories. Storage space for consumables such as culture vessels, pipettes, glassware, etc., should be located for easy access. Finally, a wash area and sink for personnel hand washing and an eyewash station are mandatory (2, 9).

Medium Requirements

Harrison (15) grew his short-term tissue culture in its natural medium of embryo extract or lymph. From then until the 1950s, cell culture practitioners used "natural" media (embryo extracts, protein hydrolysates, lymph) for propagating cells. These undefined and unrefined media were by nature inconsistent in content and quality and were havens for microbial contaminants, all of which rendered them quite unreliable (3). The need to propagate and subculture cells in vitro in a consistent manner led to attempts to provide defined media capable of sustaining continuous cell growth. To sustain such growth, a medium must have the essential amino acids, vitamins, inorganic salts, glucose, organic supplements, hormones, and growth factors (9, 14). Eagle (6) succeeded in developing two such media, basal medium and minimum essential medium, the latter of which is today the most widely used medium (9, 14). There are now many varieties of culture media, use of which depends on the specific cells to be grown. The more common formulations employ either Earle's balanced salts for use in 5% CO_2 systems or Hanks balanced salts for closed-container systems (9). Most animal cells grow well at pH 7.4, but normal metabolites of cell growth typically drive the pH down, thus requiring that culture media be buffered. Because of its low toxicity and cost, bicarbonate is frequently used, despite its poor buffering at pH 7.4. The zwitterionic buffer HEPES (N-2-hydroxyethylpiperazine-N'-2-ethanesulfonic acid; pK_a = 7.55) has gained much popularity for its stronger buffering capacity in the 7.2 to 7.6 range and its relatively low toxicity at concentrations of 10 to 20 mM (9, 12). Concentrations of vitamins and glucose may vary according to the requirements of the cells. To increase shelf life, L-glutamine (2 mM) is often left out of the medium until just before use. Antibiotics of choice (e.g., penicillin [100 U/ml] and streptomycin [100 μg/ml]) and serum are also added before use. It is important to note here the results of Gray and Brenwald (11), who reported on the need for new antibiotics to control bacterial overgrowth of cell cultures used for diagnostic virology. Increasing numbers of samples for isolation are from immunocompromised patients who have received long-term antibiotic therapy. Consequently, bacteria in the patients' samples are often resistant to penicillin and streptomycin. Gray and Brenwald (11) found that ceftazidime and vancomycin provide adequate protection against overgrowth.

In complete medium, serum provides the "undefined" nutrients such as protein, polypeptides, hormones (insulin, hydrocortisone), and other growth factors (platelet-derived growth factor, fibroblast growth factor, epidermal growth factor, etc.) (9). Laboratories typically use heat-inactivated (56°C, 30 min) fetal bovine, horse, or human serum at 2 to 20% (vol/vol) as a supplement. Heat inactivation removes complement activity and decreases the cytotoxic action of immunoglobulins (9).

For laboratories that have a reliable source of high-quality water (meets USP specifications for water for injection) and wish to make cell culture media in house, detailed formulations are given in the literature (3, 9, 14, 27) or in the catalogs of companies that produce and sell culture media (e.g., Gibco BRL, Grand Island, N.Y.; JRH Biosciences, Lenexa, Kans.; Biowhittaker, Walkersville, Md.). It is wise for a laboratory to establish a relationship with at least two qualified vendors of cell culture media. If one of the vendors temporarily experiences quality problems or backordering, the other vendor can act as the backup.

As stated above, animal serum provides many nutrients in the cell culture medium. However, using animal serum, especially fetal bovine, as a supplement has several disadvantages, not the least of which is high cost. Additionally, sera contain minor components that may have adverse effects on cell growth. There is also natural variation from batch to batch (9). Consequently, each new lot considered for purchase must be qualified in house for adequate cell growth support as well as for virus and chlamydia growth support. It is helpful to keep the vendor-supplied certificate of analysis of each lot so that any specific attribute of qualified lots can be documented. Eventually, test lots may be preaccepted or rejected according to the certificate of

analysis. Once a serum lot has been accepted, it is prudent to purchase or reserve as many bottles as possible in an effort to cut down on testing and variability in complete media.

In recent years, scientists have developed several serum-free media that support the growth of some cells and viruses. Unfortunately, changing to such media in the clinical laboratory would require extensive testing (14), which may prove cost and time prohibitive. Several of the companies listed above that manufacture cell culture media now market serum-free media and/or serum alternatives.

All lots of media and supplements must be sterility tested before being used in the laboratory. The most convenient method is to make the medium complete with supplements and then inoculate duplicate tubes of bacterial (Trypticase soy broth) and fungal (Sabouraud) broth. Incubate one tube at 37°C and one tube at 25°C (room temperature) for 10 days. Check tubes daily for evidence of microbial growth. For cell, virus, and chlamydia growth support testing, replace the medium on the test cells (uninfected and infected) with the test lot of medium. Maintain control tubes with approved culture medium, and observe test tubes daily for any signs of toxicity or inhibition of growth.

CELLS

Categories of Adherent Cells

Cells generally may be grouped according to whether they grow as an adherent monolayer or in suspension. A few adherent lines may also be grown in suspension (e.g., HeLa S3). Cells may be further categorized by how far removed phenotypically they are from the tissue of origin and how long they last in culture. Table 1 lists the different categories of cells and the characteristics of each category (3, 9).

Large research or reference laboratories may start primary cell cultures from fresh tissue, which may be available from nearby hospitals or the regional facility of the National Cancer Institute Cooperative Human Tissue Network. This practice is not recommended for a clinical laboratory because of the amount of time and labor (not to mention cost) involved in establishing the cell cultures and the possibility that the tissues harbor unknown viruses and thus are useless for diagnostic purposes (2, 9).

Primary cells are established by simple outgrowth from a small section of tissue or by enzymatic digestion or manual mincing of tissue to separate individual cells. To boost initial growth, it is helpful to use medium with more serum (10 to 20%) in addition to antibiotics (3, 9, 27). Some primary cells (e.g., primary rhesus monkey kidney or African green monkey kidney) are exquisitely sensitive to a variety of viruses. The caveat is that, as stated earlier, fresh tissue and its resultant cells may harbor an unknown latent virus(es) whose replication or multiplication will be exacerbated by the process of making cell culture. In particular, monkey kidney cells may harbor a vacuolating virus that causes large vacuoles to form in the cell cytoplasm, thus rendering the cells useless for diagnoses of infectious agents (9, 18). When purchasing primary monkey kidney cells for use in the diagnostic laboratory, it is prudent to use cell growth medium that contains antisera to simian vacuolating viruses 5 and 40. These antisera will ordinarily prevent an exacerbation of expression of these viruses but will not interfere with the isolation and diagnosis of infectious agents (9).

TABLE 1 Characteristics of cell cultures

Characteristic	Finite (primary/cell line)	Continuous cell line
Ploidy	Diploid, euploid	Heteroploid, aneuploid
Transformation	Normal	Transformed
Tumorigenicity	Nontumorigenic	Tumorigenic
Anchor dependence	Dependent	Independent
Contact inhibition	Inhibited	Not inhibited
Mode of growth	Monolayer	Monolayer or suspension
Serum requirement	High	Low
Tissue markers	Present	Often lost
Virus susceptibility	Often retained	May be lost
Growth rate	Slow (24- to 96-h doubling time)	Rapid (12- to 24-h doubling time)

Once primary cells have been subcultured, they become a cell line and as such may last in culture for as long as 10 to 15 passages. However, susceptibility to infectious agents may change drastically over the course of several passages until the cells become resistant (18, 27). Primary cells are typically of mixed cell types. As the monolayer is subcultured, the faster-growing fibroblasts overgrow the epithelial cells until the culture consists of virtually only fibroblasts. If desired, such overgrowth can be controlled by using selective media or cultureware (9).

A more homogeneous and consistent cell culture system is found with an established cell line. Cell lines are usually of one cell type (e.g., epithelial), have a finite life span (20 to 50 population doublings), and are more consistent batch to batch than primary cells. A cell strain can be established from a cell line by cloning a particular cell type found in the culture (9).

By some mechanism, some cell lines go through what is termed in vitro transformation. Such cells have infinite growth capacity and thus are called continuous cell lines. The characteristics of a continuous line are often different from those of the parent finite cell line (Table 1). Continuous cell lines are often aneuploid, and some may lose contact inhibition and/or anchorage dependence (3, 9).

Suspension Cells

Some cells, especially transformed cells and cells of lymphoid origin, are not dependent on anchorage for growth and thus may grow in suspension. Suspension cells are somewhat easier to maintain because subculturing is just a matter of dilution with fresh growth medium rather than trypsinization. Suspension cells are grown in media specially formulated for suspension culture (e.g., RPMI 1640, Dulbecco's modified Eagle's medium high glucose, etc.) along with the usual antibiotics and serum supplements (9). Primary suspension cultures of fresh peripheral blood leukocytes (PBLs) require the cytokine interleukin-2 (IL-2) for maintenance of growth. Recombinant IL-2 or purified human IL-2 is available commercially (10). Human serum, which may be substituted for fetal bovine serum, can contribute an adequate supply of IL-2. Hospital clinical laboratories typically do not carry suspension cells, but some reference laboratories may carry them for isolation and characterization of human immunodeficiency virus.

Fresh human PBLs are isolated from whole blood by spinning the blood in a VACUTAINER CPT tube (Becton Dickinson Vacutainer Systems, Franklin Lakes, N.J.) and then removing the buffy coat layer that forms just below the plasma. The cells are washed twice in phosphate-buffered saline (PBS) and then resuspended to 1×10^6 to 2×10^6 cells per ml. To stimulate proliferation of the T lymphocytes in the cell mixture, a mitogen, in particular phytohemagglutinin-P, is added to the growth medium (3 to 5 $\mu g/ml$) for 3 days. After 3 days, the cells are gently (200 $\times$ g) pelleted out of the growth medium and then resuspended at the desired concentration in complete medium minus phytohemagglutinin-P (10). This procedure is appropriate for normal (uninfected) PBLs as well as suspected infected PBLs.

Growth and Maintenance

Many laboratories purchase cells already seeded into culture tubes or vials that are ready for inoculation with a patient's sample. The next few sections will address procedures for those laboratories that maintain cell lines and seed their own tubes, vials, etc. A variety of containers, both glass and plastic, are available for cell culture. As the cells grow and expand to fill the space allotted, they need a periodic change of culture medium ("feeding") and subculturing. Change of medium is important, because as the cells metabolize the components of the medium, the supply of these components is exhausted and they are thus not available for progeny cells (9).

The pH of the culture medium is important. Most cells stop proliferating if the pH falls to 6.5 and start dying when the pH drops from 6.5 to 6.0 (9). The pH of culture medium is easily monitored by the color of the medium, because phenol red is included in the solution. A drop in pH, indicated by a color change from red to orange or yellow, is a signal that the medium may need to be changed.

Cell concentration is also a factor in how often to change the medium and/or subculture the cells. The denser the cells, the quicker the constituents of the medium are consumed. Most transformed cells and continuous cell lines deteriorate quickly at high cell densities unless they are subcultured or the medium is changed (9). To "hold" a cell line at a high cell density, replace the growth medium with a "maintenance" or "holding" medium containing a reduced amount of serum, usually 1 to 2%.

When maintaining cells in culture, it is critical to observe the cells microscopically and be alert for deterioration. When cells become too dense, are experiencing a deficit in growth factors, become contaminated, or are experiencing senescence, some signs are easily visible: granularity around the nucleus, cytoplasmic vacuolation, or rounding up of the cells with detachment from the substratum. These signs may disappear if the medium is changed or if the cells are subcultured (9). If they do not disappear, the possibility of microbial contamination must be addressed (see below).

Cell monolayers that have reached confluency can be subcultured and expanded or seeded into tubes, vials, plates, etc. The general scheme for subculturing begins with the removal of the spent culture medium followed by a rinse with a balanced salt solution deficient in Ca^{2+} and Mg^{2+}. Adherence of cells is maintained by involvement of divalent cations and basic proteins, which form a layer between the substratum and the cells (3, 9). Rinsing will accomplish both the removal of these divalent cations and the removal of residual serum, which would inhibit the enzymatic action of trypsin. Trypsin (or another proteolytic enzyme) is added for 15 to 30 s and then removed. The cells are allowed to incubate in the residue of the dissociating agent until they round up and detach from the growth surface. Detachment can be aided by tapping the flask against the heel of the hand. Growth medium is then added, and the cells are drawn up and down (tromboned) in a pipette to disperse them into a single cell suspension. The cells are counted in a hemocytometer (27) or electronic particle counter and then dispensed at the appropriate concentration into the desired containers. A more detailed description of the procedure is given in references 3, 9, and 27.

Cryopreservation

Most mammalian cells can be stored, and their viability can be maintained for years when they are stored at liquid-nitrogen temperatures ($-196°C$, liquid; $-120°C$, vapor). A cryoprotective agent, dimethyl sulfoxide or glycerol, must be added to the growth medium of the cells before the cells are frozen. These agents protect the cells from freeze-induced damage by binding water, thereby avoiding the formation of ice crystals and the buildup of intracellular electrolytes (3, 16, 27). Suspension cells or cells that have been trypsinized are pelleted ($200 \times g$, 10 min) and then resuspended in freezing medium (growth medium plus 5 to 10% cryoprotectant) at room temperature. If lymphoid cells or lymphocytes are being frozen, higher recovery of viability is possible if the freezing medium is 90% serum–10% cryoprotectant. This suspension is dispensed as 1- to 5-ml aliquots into glass ampoules or plastic cryovials, and then the freezing process is begun. It is important for optimal cell viability and recovery to use a slow freezing process (approximately $-1°C/min$). This will aid in the formation of ice crystals extracellularly rather than intracellularly (16, 27). To achieve this slow freezing without using or purchasing expensive equipment, vials may be placed into a container such as "Mr. Frosty" (Nalge, Rochester, N.Y.). The vials are placed in a rack sitting in a foam-rubber-lined jar containing isopropanol. This jar is placed at $-70°C$ and left for at least 6 h, after which the ampoules or vials are transferred to boxes or canes in a liquid-nitrogen freezer.

To recover cells from the frozen state, vials must be retrieved from the freezer quickly and immediately put into $37°C$ water. To avoid a possible explosion of the thawing vial, this process should be carried out in a covered container, preferably inside a biosafety cabinet (2, 27).

Quality Control

Results of tests and experiments performed on contaminated cells are suspect if not entirely negated. Therefore, it is extremely important to maintain cells free from microbial and cellular contamination. Everything, excluding the patient sample, coming in contact with the cells should be sterilized by some method (gamma irradiation, filtration, steam sterilization, etc.) in order to reduce chances of microbial contamination. Antibiotics are added to cell growth medium to reduce the incidence of contamination by microbes. Even so, bacteria, mycoplasma, and fungi may be introduced via the operator, atmosphere, work surface, or solutions, and once growing, they can literally shut down a laboratory. Additionally, it is important to handle cell cultures so as to avoid cross-contamination of different cell lines (9, 16, 17, 27). A program of quality control should include the following: (i) daily gross and microscopic observation of cultures to detect overt contamination; (ii) sterility testing of all reagents before first use on cells; (iii) testing done on a regular basis for mycoplasma contamination; and (iv) a high standard of sterile technique maintained at all times (9).

For detection of bacteria and fungi, broth culture inoculation is most efficient. Mycoplasma contamination can be a more insidious problem and more difficult to detect. There are at least nine methods for detecting mycoplasma (16, 17). The American Type Culture Collection (Rockville, Md.) testing service uses both the direct culture method and the indirect Hoechst 33258 stain for DNA (4). The stain method is a quick (30-min) means of testing cell stocks on a regular basis for any mycoplasma strain. The direct culture method is more labor intensive but is the most accurate and sensitive method for detecting cultivable mycoplasma strains (9, 16, 17). Because of the need to grow a positive control, laboratories planning to test for mycoplasma by the culture method must consider the possible negative consequences of growing mycoplasma in a cell culture facility. Using a site remote from the cell culture area may be preferable (17).

To perform the Hoechst 33258 DNA stain test, indicator cells or test cells are grown on coverslips. After 24 h of incubation, test cell suspensions are added to the indicator cells. After six more days of incubation, the cells are fixed, stained for 30 min with diluted stain, rinsed, mounted, and observed for fluorescence. Positive slides show brightly stained cell nuclei with fluorescent particulate or fibrillar material over the cytoplasm (4, 16).

To avoid cross-contamination with other cell lines, handle only one cell line at a time, and decontaminate the work area thoroughly before handling a second cell line. Use separate reagents for each cell line (9). Details for testing cells for cellular contamination are given in references 16 and 17.

Finally, as they age (i.e., go through population doublings), some cell lines become resistant to virus infection (18). Therefore, new cells brought into the laboratory should be propagated to produce a low-passage-number "master" or seed stock that is tested for microbial and cellular contamination and then frozen. From this master stock, larger batches of "working" stock can be produced and frozen. Doing this provides a large supply of early-passage cells that are theoretically identical from cryovial to cryovial (17). All of these working stock cells should be identically susceptible or resistant to infectious agents.

In summary, quality control is very important in the cell culture laboratory but can be very time-consuming. Maintaining the seed stock concept and being ever vigilant in maintaining a high standard of sterile technique will save many labor hours and give the added benefits of consistent and predictable cell culture and resultant high levels of confidence in diagnostic test results.

SAFETY PRECAUTIONS

As in any kind of laboratory, the technologist in a cell culture laboratory should be cognizant of potential risks and hazards and should follow strict safety guidelines. The need for protection is defined by the source of the material and the nature of the operation. Medium preparation and serial passage of uninfected, nonhuman, nonprimate cell lines may be safely performed on the open bench or in a horizontal-flow biosafety cabinet. Procedures such as establishing primary cultures, subculturing human or primate cells, or working with infected cells or tissues or with patient samples require the use of a vertical-flow biosafety cabinet (9). Most chemicals used in a cell culture facility are used in cell culture media, which are, by their nature, considered nontoxic. However, dimethyl sulfoxide, a cryopreservative for freezing cells, is potentially dangerous because it is quickly absorbed through the skin and can carry along with it other, perhaps hazardous agents (9, 16). Cells in culture may themselves be considered potentially biohazardous because of the possibility of infecting workers with unknown viruses. Primary nonhuman primate cells are especially notorious for harboring B virus (herpesvirus) or Marburg virus, both of which are lethal for humans. Continuous cell lines should also be considered potentially hazardous because of the possibility that they harbor latent viruses (2, 18). In summary, cell cultures used in virological or bacteriological studies, as they are in the clinical laboratory, present the same degree of hazard as the virus(es) or bacterium under study; technologists should always practice universal precautions. For a more detailed discussion of laboratory safety, consult chapter 7 of this Manual.

USE OF ANIMAL CELL CULTURE IN DIAGNOSTIC MICROBIOLOGY

Virus Isolation and Detection

Since 1949, when it was reported that poliovirus multiplied in animal cell culture (7), scientists have been trying to cultivate or detect in vitro viruses that cause human disease (18). Virology sections in the clinical laboratory have become increasingly more common with the advent of effective antiviral chemotherapy (e.g., acyclovir, ganciclovir) as well as the need to document serious and/or new viral infections for transmission and epidemiology studies and for patient-contact education (e.g., hepatitis, human immunodeficiency virus). Viruses can initially be differentiated by (i) the pattern of cytopathic effect (CPE), (ii) the specific cells in which the CPE is induced, and (iii) the rapidity of the appearance of CPE. For example, two enteroviruses, poliovirus type 1 and echovirus type 11, cause CPE in rhesus monkey kidney cells. Poliovirus-induced CPE appears within 1 day after inoculation, whereas it takes 4 days for echovirus to induce the same amount of cellular destruction. Additionally, poliovirus type 1 but not echovirus type 11 will induce CPE in HEp-2 cells. Another example is the growth and production of CPE exclusively by herpes simplex virus on mink lung and rabbit kidney cells (18, 22). Typically then, specimens are inoculated into as many different cell cultures as possible to provide a susceptible host for each virus that may be present (18, 22). Table 2 lists the more common cells or cell lines used for virus isolation and characterization. The size and capacity of a laboratory and its particular needs (i.e., the predominant patient population[s]) will dictate which and how many

different cells to carry. In general, a primary monkey kidney cell culture, a human diploid cell strain (WI-38, MRC-5), and a human continuous cell line (HEp-2 or HeLa) are a good combination (18, 22, 27). Cell culture isolation remains the "gold standard" method for detection of viruses even though there is sometimes such a lengthy incubation period that results are simply for verification of diagnosis and treatment. Isolation is a very sensitive method, because theoretically, a positive result can be obtained with a single infectious virion. Unfortunately, not all viruses are capable of multiplying in vitro, or they multiply to very low titers. For these viruses, there are means of enhancing infections (e.g., rolling, centrifugation, chemical treatment) that also serve to expedite the course of infection in cell culture (19).

For detection of viruses by CPE, cell culture tubes containing fresh (3- to 8-day old) (8, 18) monolayers of indicator host cells are inoculated with 0.2 to 0.5 ml of specimen. The specimen is allowed to adsorb to the cells for 15 to 60 min at 37°C, and then fresh medium is added. The tubes are returned to the incubator and observed microscopically daily for evidence of CPE (18, 22, 27). The nature of the CPE varies by virus. Often it requires a very experienced technologist to distinguish CPE from simple cell overgrowth or senescence. Some viruses, like herpes simplex virus, cause cellular destruction with ballooning of cells (18), while respiratory syncytial virus causes syncytium (fusion) formation (22). Cytomegalovirus causes the formation of enlarged refractile cells containing intracellular inclusions (1).

Some in vitro virus infections may take several weeks to cause CPE or may not cause any overt CPE whatsoever. These viruses may cause an alteration in the infected cell surfaces so that erythrocytes adhere to the cells. This phenomenon is called hemadsorption and is usually associated with common respiratory viruses, including influenza, measles, parainfluenza, and mumps (22). The hemadsorption test, which can be done as early as 2 days postinfection, is performed by replacing the cell culture medium with a suspension (0.08 to 0.5%) of erythrocytes (usually guinea pig) and then refrigerating the culture for 20 to 30 min. When observed microscopically, a positive hemadsorption test appears as rosettes of erythrocytes adhering to infected cells (18, 22).

Many viruses associated with clinical disease exhibit surface proteins that are capable of agglutinating erythrocytes of specific animal species at defined temperatures. Accordingly, the species of erythrocytes and the incubation temperature used to assay for this agglutination will be determined by the virus suspected (18). Briefly, infected cell culture medium is mixed with a suspension of erythrocytes in a round-bottom tube or well and allowed to incubate for 0.5 to 2 h at the determined temperature. If at the end of the incubation period a button has formed in the bottom of the tube or well, no agglutination has taken place. However, if a shield or latticework of erythrocytes has formed, hemagglutination has occurred (22), implying a virus infection.

Immunostaining

For definitive identification of viruses that have been isolated, indirect or direct immunostaining is currently the method of choice. These assays may be performed as early as 16 h postinfection when herpes simplex virus or cytomegalovirus is suspected (8, 28), making these tests relatively rapid. The reporter molecules for these assays may be either fluorescent or enzymatic, the latter enabling tech-

TABLE 2 Cells used for virus or *Chlamydia* isolation

Type of cell[a]	Species or tissue of origin	Virus(es) or chlamydiae isolated[b]
Primary cells		
African green monkey	Kidney	HSV, mumps, RSV, rubella, VZV
CBMC, PBMC	Human	HIV-1, HIV-2, HTLV-I, HTLV-II, HHV-6
Chick embryo fibroblasts	Chicken	NDV, human poxviruses
Embryonic kidney, lung	Human	Adenoviruses, BK, mumps
Rabbit	Kidney	HSV
Rhesus or cynomolgus monkey	Kidney	ECHO, polioviruses, coxsackievirus groups A and B, mumps, reoviruses, influenza, measles, parainfluenza, RSV
Finite cell lines		
Foreskin fibroblasts	Human	CMV, HSV
Kidney fibroblasts	Human, fetal	Coronaviruses, HSV, rhinoviruses
Lung fibroblasts	Human, embryo	Coronaviruses, CMV, rhinoviruses, VZV
WI-38, MRC-5	Human fetal lung	Adenoviruses, CMV, polioviruses, coxsackievirus group B, enteroviruses (types 68–71), RSV, rhinoviruses
Continuous cell lines		
293	Human kidney	Adenoviruses (types 5, 40, and 41)
A549	Human lung	Adenoviruses (types 1–39)
BGMK	Buffalo green monkey kidney	Polioviruses, coxsackievirus groups A and B, reoviruses, *C. trachomatis*
HeLa	Human cervix	*C. trachomatis*, polioviruses, poxviruses, reoviruses, RSV, rhinoviruses, *C. pneumoniae*, coxsackievirus groups A and B
HEp-2	Human larynx	Adenoviruses, RSV, *C. pneumoniae*
McCoy	Mouse	*C. trachomatis*, *C. psittaci*
MDCK	Canine kidney	Influenza, parainfluenza
Mink lung	Mink	HSV
RD	Human rhabdomyosarcoma	Coronaviruses, coxsackievirus group A, polioviruses, enteroviruses (types 68–71)
RK$_{13}$	Rabbit kidney	Rubella, poxviruses
Vero, CV-1	African green monkey kidney	HSV, measles, poxviruses, BK, rubella, RSV, parainfluenza

[a]CBMC, cord blood mononuclear cells; PBMC, peripheral blood mononuclear cells.

[b]BK, human polyomavirus BK; CMV, cytomegalovirus; ECHO, enteric cytopathic human orphan viruses; HHV-6, human herpesvirus type 6; HIV-1 and HIV-2, human immunodeficiency virus types 1 and 2; HSV, herpes simplex viruses; HTLV, human T lymphotropic virus; NDV, Newcastle disease virus; RSV, respiratory syncytial virus; VZV, varicella-zoster virus.

nologists to read the assay with a light microscope. An immunostaining assay is performed on cells scraped from tube cultures or shell vials (23) (see below), on cells grown on coverslips in shell vials (8, 28), or on cells grown in multiwell trays (5, 24). The cells are fixed with acetone or alcohol, rinsed, and then incubated (30 to 60 min) with specific antibody (indirect) or conjugated specific antibody (direct). The direct assay is read after this incubation. In the indirect assay, conjugated second antibody is added and allowed to react (30 min). The slides are then rinsed and observed under the light (after addition of substrate) or fluorescence microscope. Positive slides exhibit areas of dark staining (enzymatic assay) or areas of bright green fluorescence (immunofluorescence assay) (22, 28).

Nucleic Acid Probes

The most specific method for identification of infectious agents employs nucleic acid probes. In situ hybridization of infected cells today may provide a technique complementary to immunostaining but not a replacement for it (28). Unfortunately, at this stage of development of the method, the cost of reagents and the number of labor hours involved render it less suitable than other methods for the laboratory striving for lower costs and earlier results (22).

Shell Vials

In recent years, shell vial isolation has been adapted for several viruses (8, 23, 28) as well as for chlamydia (26). This method represents a relatively rapid means of identifying agents whose identifications may ordinarily take as long as 3 weeks in traditional tube culture. Vials (with or without coverslips) are typically seeded with 5×10^4 to 1×10^5 cells in 1 ml of growth medium. When the cells are nearly confluent, the growth medium is aspirated, and then 0.2 ml of specimen is added. The vials are centrifuged (700 $\times$ g) for 30 to 60 min, after which 1 ml of growth medium is added. The vials are incubated for agent-specific times and temperatures. After appropriate incubation times, cells are rinsed once or twice with PBS. Cells can be scraped into a small volume of PBS and then transferred to a microscope slide for immunostaining. Alternatively, cells that have been grown on a coverslip can be stained directly in the shell vial. The stained coverslip is removed and placed on a slide for examination under the fluorescence microscope

(8, 23, 26, 28). Caution is advised when opening shell vial caps, as there have been reports of cross-contamination of specimens due to aerosols generated during such procedures (19).

Chlamydia Isolation

It is important to note here that *Chlamydia psittaci* is a highly infectious human pathogen whose isolation should not be attempted by laboratories lacking suitable containment facilities (biosafety level 2 or greater) and appropriately trained personnel. As stated above, chlamydia can be isolated by the shell vial technique (26). However, laboratories that handle large numbers of specimens for chlamydia isolation find it more convenient to use multiwell (48- or 96-well) plates. Cells (Table 2 and below) are grown in these plates until they are confluent. The cells (McCoy cells only) are then pretreated with DEAE-dextran (30 μg/ml) for 10 to 30 min. The pretreatment solution is then removed, and the patient sample is added. The plate is then centrifuged (1,200 $\times$ g) for 1 h, after which the specimen is allowed to adsorb at 35°C for 30 min. The sample is removed, and medium containing cycloheximide (1 μg/ml) is added. The plates are incubated for 2 to 3 days, fixed, and processed for immunofluorescence (5, 24). It is possible to isolate both *Chlamydia trachomatis* and *Chlamydia pneumoniae* on HeLa cells or McCoy cells, although recovery rates will be low and suboptimal. Even so, laboratories with limited resources may choose to use only one cell line for both. Optimal recovery of *C. trachomatis* can be achieved only by using the mouse cell line McCoy. Recovery of *C. pneumoniae* is optimal on the human cell line HEp-2 (24). *C. psittaci* is isolated on McCoy cells (26).

Bacterial Toxins

In certain cases of diarrheal disease of bacterial origin, it may be beneficial to identify not only the organism by classical bacteriological techniques but also the toxin (if any) being produced. This may be accomplished by cell culture assays. The cell culture test for the verocytotoxins (VT) of enterohemorrhagic *Escherichia coli* is commonly and easily performed in the cell culture laboratory. The VT-producing *E. coli* (VTEC) strains (20, 21), the most recently famous member being O157:H7, are now recognized as significant etiological agents of diarrhea. The VT, which traverse cell membranes and kill cells (29), are very closely related to the Shiga toxin of *Shigella dysenteriae* type 1 and therefore are also known as Shiga-like toxins. VTEC strains are strongly associated with hemorrhagic colitis and hemolytic uremic syndrome, in which the VT has direct pathogenic significance (21). VT from bacterial broth, bacterial colonies, or fecal sample filtrates is detected on Vero cell monolayers in microtiter plates. Serial dilutions of a sample (50 μl each) are added to the Vero cells. Monolayers are examined daily for 3 days for evidence of CPE (dead, detached floating cells) (20). Confirmation of specificity of CPE is accomplished by neutralization with high-titer anti-VT1 or VT2. A positive test indicates a recent VTEC infection, while a negative test implies that there has been no such infection (21).

Enterotoxigenic *E. coli* (ETEC) strains are the major cause of bacterial diarrhea in developing countries and of traveler's diarrhea. ETEC can produce a diarrhea with symptoms resembling those of cholera (21). Two plasmid-encoded enterotoxins, one heat stable and one heat labile, are produced by ETEC. The heat-labile enterotoxin as well as the functionally, structurally, and immunologically sim-

ilar enterotoxin of *Vibrio cholerae* can be detected by a relatively rapid assay on Chinese hamster ovary (CHO) cells. Sterile filtrates of cultures are added to CHO cells. After 24 h, cells are observed for elongation to three (or more) times normal width and loss of their knoblike projections (13), which indicates the presence of a toxin. Alternatively, these enterotoxins can be detected by their effect on the normally flat mouse adrenal cell line Y1. Culture filtrates are allowed to incubate in the growth medium of Y1 cells for 24 h, after which the cells are observed for the characteristic toxin-induced rounding (25). For further information about VTEC and ETEC, consult chapter 33 in this Manual.

In summary, animal cell culture has grown from its humble and difficult beginning to become a vital part of the diagnostic laboratory. Even though it is sometimes cumbersome and is being replaced by enzyme immunoassay as the assay of choice for some infectious agents, it remains the gold standard against which other tests are evaluated and to which they are compared. Cell culture is likely to be a large part of the clinical diagnostic laboratory for many years to come.

I thank Michael Forman, The Johns Hopkins Hospital; Charlotte Gaydos, The Johns Hopkins University; Judith Lovchik, University of Maryland Hospital; and Joseph Waner, The Children's Hospital of Oklahoma, for helpful discussions and guidance.

REFERENCES

1. **Arvin, A. M.** 1992. Human cytomegalovirus, p. 333–350. *In* E. H. Lennette (ed.), *Laboratory Diagnosis of Viral Infections.* Marcel Dekker, Inc., New York.
2. **Barkley, W. E.** 1979. Safety considerations in the cell culture laboratory. *Methods Enzymol.* **58:**36–43.
3. **Butler, M.** 1991. The characteristics and growth of cultured cells, p. 1–25. *In* M. Butler (ed.), *Mammalian Cell Biotechnology: a Practical Approach.* IRL Press, Oxford.
4. **Chen, T. R.** 1977. *In situ* detection of mycoplasma contamination in cell cultures by fluorescent Hoechst 33258 stain. *Exp. Cell Res.* **104:**255–262.
5. **Cles, L. D., and W. E. Stamm.** 1990. Use of HL cells for improved isolation and passage of *Chlamydia pneumoniae. J. Clin. Microbiol.* **30:**1968–1971.
6. **Eagle, H.** 1955. Nutrition needs of mammalian cells in tissue culture. *Science* **122:**501–504.
7. **Enders, J. F., T. H. Weller, and F. C. Robbins.** 1949. Cultivation of the Lansing strain of poliomyelitis virus in cultures of various human embryonic tissues. *Science* **109:**85–87.
8. **Fedorko, D. P., D. M. Ilstrup, and T. F. Smith.** 1989. Effect of age of shell vial monolayers on detection of cytomegalovirus from urine specimens. *J. Clin. Microbiol.* **27:**2107–2109.
9. **Freshney, R. I.** 1987. *Culture of Animal Cells: a Manual of Basic Technique,* 2nd ed. Alan R. Liss, Inc., New York.
10. **Gartner, S., and M. Popovic.** 1990. Virus isolation and production, p. 53–70. *In* A. Aldovini and B. D. Walker (ed.), *Techniques in HIV Research.* Stockton Press, New York.
11. **Gray, J. J., and N. P. Brenwald.** 1991. The use of antibiotics to control bacterial overgrowth of cell cultures used for diagnostic virology. *J. Virol. Methods* **32:**163–170.
12. **Gueffroy, D. E. (ed.).** 1975. *Buffers: a Guide for the Preparation and Use of Buffers in Biological Systems.* Calbiochem Corp., San Diego, Calif.
13. **Guerrant, R. L., L. L. Brunton, T. C. Schnaitman, L. I. Rebhun, and A. G. Gilman.** 1974. Cyclic adenosine monophosphate and alteration of Chinese hamster ovary cell morphology: a rapid, sensitive *in vitro* assay for the enterotoxins of *Vibrio cholerae* and *Escherichia coli. Infect. Immun.* **10:**320–327.

14. **Ham, R. G., and W. L. McKeehan.** 1979. Media and growth requirements. *Methods Enzymol.* **58:**44–93.
15. **Harrison, R. G.** 1907. Observations on the living developing nerve fiber. *Proc. Soc. Exp. Biol. Med.* **4:**140–143.
16. **Hay, R. J.** 1986. Preservation and characterisation, p. 71–112. In R. I. Freshney (ed.), *Animal Cell Culture, a Practical Approach.* IRL Press, Washington, D.C.
17. **Hay, R. J.** 1988. The seed stock concept and quality control for cell lines. *Anal. Biochem.* **171:**225–237.
18. **Hsiung, G. D., and C. K. Y. Fong.** 1982. *Diagnostic Virology.* Yale University Press, New Haven, Conn.
19. **Hughes, J. H.** 1993. Physical and chemical methods for enhancing rapid detection of viruses and other agents. *Clin. Microbiol. Rev.* **6:**150–175.
20. **Karmali, M. A.** 1987. Laboratory diagnosis of verotoxin-producing *Escherichia coli* infections. *Clin. Microbiol. Newsl.* **9:**65–70.
21. **Karmali, M. A.** 1989. Infections by verocytotoxin-producing *Escherichia coli.* *Clin. Microbiol. Rev.* **2:**15–38.
22. **Leland, D. S.** 1992. Concepts of clinical diagnostic virology, p. 3–43. In E. H. Lennette (ed.), *Laboratory Diagnosis of Viral Infections.* Marcel Dekker, Inc., New York.
23. **Olsen, M. A., K. M. Shuck, A. R. Sambol, S. M. Flor, J. O'Brien, and B. J. Cabrera.** 1993. Isolation of seven respiratory viruses in shell vials: a practical and highly sensitive method. *J. Clin. Microbiol.* **27:**2107–2109.
24. **Roblin, P. M., W. Dumornay, and M. R. Hammerschlag.** 1992. Use of HEp-2 cells for improved isolation and passage of *Chlamydia pneumoniae. J. Clin. Microbiol.* **30:**1968–1971.
25. **Sack, D. A., and R. B. Sack.** 1975. Test for enterotoxigenic *Escherichia coli* using Y1 adrenal cells in miniculture. *Infect. Immun.* **11:**334–336.
26. **Schacter, J., and C. R. Dawson.** 1979. Psittacosis-lymphogranuloma venereum agents/TRIC agents, p. 1021–1059. In E. H. Lennette and N. J. Schmidt (ed.), *Diagnostic Procedures for Viral, Rickettsial, and Chlamydial Infections,* 5th ed. American Public Health Association, Washington, D.C.
27. **Schmidt, N. J.** 1989. Cell culture procedures for diagnostic virology, p. 51–100. In N. J. Schmidt and R. W. Emmons (ed.), *Diagnostic Procedures for Viral, Rickettsial, and Chlamydial Infections,* 6th ed. American Public Health Association, Washington, D.C.
28. **Smith, T. F., A. D. Wold, and M. J. Espy.** 1992. Diagnostic virology—then and now. *Adv. Exp. Med. Biol.* **312:**191–199.
29. **Stephen, J., and R. A. Peitrowski.** 1986. *Bacterial Toxins,* 2nd ed., p. 25–39. American Society for Microbiology, Washington, D.C.

THE CLINICAL MICROBIOLOGY LABORATORY IN INFECTION CONTROL AND PREVENTION

IV

VOLUME EDITOR
MICHAEL A. PFALLER

SECTION EDITOR
RICHARD P. WENZEL

Dynamics of Hospital-Acquired Infection

LOREEN A. HERWALDT AND RICHARD P. WENZEL

15

IMPACT OF NOSOCOMIAL INFECTIONS

Between 1.75 million and 3 million (5 to 10%) of the 35 million patients admitted annually to acute-care hospitals in the United States acquire an infection that was neither present nor incubating on their admission (81, 83, 171). Such hospital-acquired (nosocomial) infections add significantly to the morbidity, mortality, and economic burden caused by the underlying diseases alone (48, 53, 63, 64, 76, 84, 119, 125, 158, 169, 177, 192). For example, approximately 100,000 to 400,000 nosocomial bloodstream infections occur each year in the United States. Of the affected patients, 40,000 to 160,000 die (crude mortality), and 25,000 to 100,000 of these die as a direct result of the infection (attributable mortality). Similarly, approximately 20,000 patients die each year as a direct result of nosocomial pneumonias (110). Some data suggest that even catheter-related urinary tract infections (UTIs) may increase a patient's risk of dying approximately threefold above that of similar patients who do not develop such infections (149). In addition to increasing mortality, nosocomial infections increase the cost of health care by $4.5 billion to $15 billion dollars annually (120).

PREVENTABLE INFECTIONS

Despite their frequency and severity, many nosocomial infections cannot be prevented. Data from the Centers for Disease Control and Prevention (CDC) Study of the Efficacy of Nosocomial Infection Control (SENIC) suggest that nosocomial infection rates could be reduced by one-third if infection control programs included at least the following: an organized, hospital-wide surveillance system; one full-time infection control nurse for every 250 hospital beds; a trained hospital epidemiologist; and a method for reporting wound infection rates to surgeons (82). During the 20 years since the CDC performed the SENIC (92, 185), patient populations have changed considerably and the practice of infection control has matured. Hence, we do not know whether a good infection control program in the 1990s can still reduce nosocomial infection rates by one-third.

Approximately 10% of nosocomial infections occur in epidemics or in clusters (85, 191), all of which, in theory, health care workers could prevent by practicing good in-fection control (186). However, the majority of nosocomial infections (90%) are endemic, and only 20 to 33% of these endemic infections may be preventable.

SITES

The incidence of nosocomial infections varies substantially by body site (Table 1). The proportion of infections at each site is consistent in all hospitals surveyed by the National Nosocomial Infections Surveillance (NNIS) system regardless of the hospital's size and medical school affiliation (57), but it varies considerably by hospital service (Fig. 1). For example, the most common infections in general surgery services are surgical site infections (SSI) (25%), but the most common infections in medical services and in nurseries are urinary tract infections (42%) and bloodstream infections (36%), respectively (57).

Bloodstream Infections

Between 1980 and 1989, the overall rate of nosocomial bloodstream infections reported by NNIS hospitals increased by 279% in small nonteaching hospitals, 196% in large nonteaching hospitals, 124% in small teaching hospitals, and 70% in large teaching hospitals (5). Also increasing substantially during that period were rates of bloodstream infections caused by coagulase-negative staphylococci (from 161 to 754%), *Staphylococcus aureus* (from 122 to 283%), enterococci (from 120 to 197%), and *Candida* species (from 75 to 487%). By 1989, the overall nosocomial bloodstream infection rates reached 1.3, 2.5, 3.8, and 6.5/1,000 discharges in small nonteaching hospitals, large nonteaching hospitals, small teaching hospitals, and large teaching hospitals, respectively (5).

Intrinsic patient factors that increase the risk of nosocomial bloodstream infections include age (≤1 or ≥60 years), malnutrition, immunosuppressive chemotherapy, loss of skin integrity, and severe underlying illness (57, 148). Treatment factors, such as indwelling devices, intensive care unit stay, and longer hospital stay, also increase the risk of nosocomial bloodstream infection (148).

The crude mortality associated with bloodstream infections is 34 to 40% for those caused by gram-negative bacteria or enterococci, 30% for those caused by coagulase-negative staphylococci, and approximately 57% for those caused by *Candida* spp. (105, 119, 187, 192). The attribut-

TABLE 1 Distribution of nosocomial infections by site

Infection	Proportion (%)		
	1975–1976 (n = 85)	1984 (n = 89)	1990–1992 (n = 57)
UTI	53	39	33
Pneumonia	13	18	15
SSI	28	17	15
Bloodstream	6	8	13
Other	NS[a]	20	24

[a]NS, not specified.

able mortality associated with bloodstream infections is approximately 25% for those caused by gram-negative bacteria or enterococci, 14% for those caused by coagulase-negative staphylococci, and approximately 38% for those caused by *Candida* spp. (105, 119, 187, 192).

Pneumonia

In the 1980s, pneumonia became the second most common nosocomial infection in the United States (40), accounting for 16% of all nosocomial infections (rate, 6 to 10/1,000 patients). More than 250,000 patients in acute-care hospitals in the United States are affected each year. Nosocomial pneumonia rates are higher in patients who have had surgery, who are hospitalized in intensive care units, and who are intubated (41).

Nosocomial pneumonias are very difficult to study because common diagnostic methods are often inadequate and because pneumonia can resemble other entities such as adult respiratory distress syndrome or congestive heart failure (7, 127). Hence, not all investigators agree on the most common etiologic agents. However, many investigators report that gram-negative rods cause over 60% of nosocomial pneumonias (11) and that *S. aureus*, particularly common in burn and surgical intensive care units, causes 15 to 20%

of cases (40, 57). *Streptococcus pneumoniae* (88) and *Haemophilus influenzae* (57), commonly isolated from patients with community-acquired pneumonias and very early (hospital days 2 to 5) nosocomial pneumonias, cause only 3 and 5% of nosocomial pneumonias, respectively. *Branhamella catarrhalis* causes nosocomial pneumonia in young children, the elderly, and patients with underlying lung disease (35, 143).

Patient factors that independently increase the risk of nosocomial pneumonia include advanced age, chronic lung disease, decreased consciousness, and large-volume aspiration (40). Treatment factors that independently increase the risk of nosocomial pneumonia include mechanical ventilation, chest surgery, and intracranial pressure monitoring (24, 39). Patients who receive antacids (16%) or intravenous ranitidine (21%) for stress ulcer prophylaxis have higher rates of late-onset nosocomial pneumonia than patients who receive sucralfate (5%; $P = 0.022$) (151). Despite those observations, the role of gastric organisms in ventilator-associated pneumonia remains controversial (11).

Nosocomial pneumonias prolong hospital stays by 8 to 9 days (64, 110) and increase duration of mechanical ventilation or intensive care unit stay threefold (38, 96). The mortality rate caused by nosocomial pneumonia has been difficult to establish. Crude mortality rates range from 20 to 50%, and approximately one-third of the deaths are attributable directly to the pneumonia. Thus, if the crude mortality is 30%, the attributable mortality is 10% (24, 39, 59, 74, 110, 172, 191). However, other investigators think that nosocomial pneumonias do not significantly increase mortality rates. By univariate analysis, Kollef found that patients with ventilator-associated pneumonia had a statistically higher mortality rate than did patients who required mechanical ventilation but did not develop pneumonia (37.2 versus 8.5%; $P < 0.001$) (102). However, in the multivariate analysis, an organ failure score of 3 or greater, a premorbid lifestyle score of 2 or greater, and supine head

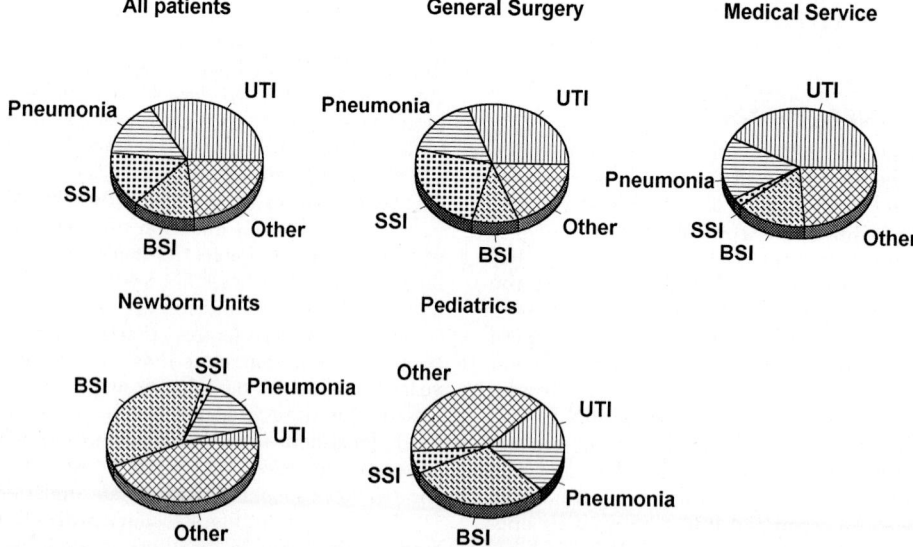

FIGURE 1 Distribution of nosocomial infections by site and major service. Data are from the hospital-wide surveillance component of the NNIS system and were obtained from 1990 through 1992. The figure is adapted from data presented in reference 57. BSI, bloodstream infections.

positioning during the first 24 h of mechanical ventilation but not nosocomial pneumonia were independently associated with mortality in an intensive care unit (102).

Nosocomial pneumonias also increase the cost of medical care by over $5,000 per infection (57). In addition, Wenzel has estimated that the annual direct cost of diagnosing and treating nosocomial pneumonia exceeds $2 billion (188).

SSIs

Each year in the United States, at least 920,000 (4%) of the 23 million patients who undergo surgery develop surgical site infections (SSIs). The proportion of nosocomial infections accounted for by SSIs has decreased from 28% in 1975 and 1976 (83) to 15% between 1990 and 1992 (57). Reports in the literature may substantially underestimate the rate of SSI, because approximately 50% of these infections develop after patients have been discharged from the hospital (20, 104). Furthermore, many surgical procedures are performed in the outpatient setting, where surveillance for infections is rarely performed (189).

Gram-positive organisms cause 56% of SSIs (19% S. aureus, 14% coagulase-negative staphylococci, 12% enterococci, and 11% other), gram-negative organisms cause 35% of SSIs, and Candida species cause 4% (57).

Investigators have identified numerous factors that predispose patients to SSI. Intrinsic patient factors that increase the risk of SSI include advanced age, obesity, and remote infection (57, 123, 139). The role of other factors, such as malnutrition and diabetes, is more controversial. Other factors that affect the risk of SSI include length of the preoperative hospital stay, preoperative shave with a razor (>12 h before the procedure), duration of the procedure, and surgical wound classification (i.e., clean, clean-contaminated, contaminated, or dirty) (123, 139). In addition, inappropriate timing of prophylactic-antibiotic administration increases the risk of SSI (32). Classen et al. documented the lowest infection rate (0.59%) in patients who received prophylactic antibiotics (half-lives, 30 to 90 min) within 2 h of the incision (32). The rate of SSI increased significantly if prophylactic antibiotics were given more than 2 h preoperatively (3.8%; odds ratio [OR] = 4.3; 95% confidence interval, 1.8 to 10.4) or at any time postoperatively (3.3%; OR = 5.8; 95% confidence interval, 2.4 to 13.8) (32).

Several groups have used multivariate analysis to develop models that predict which patients are at highest risk of developing an SSI (42, 70, 80, 139). Using data from NNIS, Culver et al. developed a risk index that included the American Society of Anesthesiologists score, the surgical wound classification, and a third variable representing the duration of surgery (42). As the risk index increased from 0 to 3, the SSI rates increased from 1.5 to 13.0% (42). However, the results of a study by Garibaldi et al. raise questions about the positive predictive value of the NNIS risk index (70). In Garibaldi's study, four factors, three of which were similar to the NNIS risk index, independently predicted the risk of SSI: the surgical wound classification, the American Society of Anesthesiologists score, the duration of surgery, and the results of intraoperative cultures. Garibaldi et al. noted that the positive predictive value of the model was only 32% and that the organisms isolated from intraoperative cultures correlated with those isolated from infections in only 41% of cases (70).

Although nearly 18% of patients who develop SSIs have a disability that persists for more than 6 months (107), few investigators have tried to determine the morbidity, mortality, and costs associated with SSI. Currently, there are no reliable data on the mortality attributable to SSI. In the era before diagnosis-related groups, Green and Wenzel evaluated the effect of SSI on the length of stay and cost of hospitalization associated with six common operations (appendectomy, cholecystectomy, colon resection, cesarian section, total abdominal hysterectomy, and coronary artery bypass grafting) (76). They determined that an SSI increased the length of stay by 62 to 113% and the cost of hospitalization by 20 to 98% (76). More recent data from the CDC suggest that each SSI adds over $3,000 to the cost of hospitalization, and Wenzel estimates that the direct cost of excess hospital stay is well over $1.5 billion (189).

UTIs

UTIs are the most common nosocomial infection, affecting 400,000 to 1 million patients in acute-care hospitals in the United States each year (68, 170). However, in 1990 to 1992, UTIs accounted for a substantially smaller proportion of nosocomial infections (33%) than they did in 1975 and 1976 (53%). Gram-negative organisms cause 59% of all nosocomial UTIs, gram-positive organisms cause 26%, and Candida species and other fungi cause 13% (57).

Intrinsic patient factors that increase the risk of UTIs include advanced age, female gender, and severe underlying disease (57, 170). Indwelling urinary catheters increase a patient's risk of developing bacteriuria and UTI; the incidence of catheter-associated bacteriuria increases by 5% each day the catheter is in place. A patient whose periurethral area is colonized with pathogenic enteric bacteria or who has diarrhea is at increased risk of developing catheter-associated bacteriuria (43, 69). While 70% of patients who develop catheter-associated bacteriuria eliminate the bacteria while the catheter is still in place, 30% develop symptomatic UTI.

Investigators have estimated that each nosocomial UTI increases the cost of hospitalization by $150 to $680 (57, 68), which is considerably less than the costs associated with other nosocomial infections. However, given the frequency of UTIs, these infections cost the medical system between $150 million and $1.8 billion dollars each year.

MICROBIOLOGY

The organisms causing nosocomial infections have changed periodically as a function of medical practice, including the use of antibiotics (Fig. 2). Recently, the interval between changes has decreased, and the number of different organisms causing nosocomial infections has increased. In the 1940s and 1950s, after penicillin and the sulfonamides were introduced, S. aureus replaced Streptococcus pyogenes and Streptococcus pneumoniae, the major nosocomial pathogens of the preantibiotic era, as the most common etiologic agent. Clinicians began using narrow-spectrum cephalosporins and aminoglycosides in the 1970s, whereupon aerobic gram-negative rods replaced S. aureus as the most common nosocomial pathogens. During the late 1970s and early 1980s, broad-spectrum cephalosporins, implantable vascular catheters, and immunosuppressive therapies such as bone marrow transplantation were introduced into clinical practice. Under these conditions, gram-positive organisms once again became important nosocomial pathogens. However, this time, allegedly nonvirulent organisms, like coagulase-negative staphylococci and the enterococci, became prominent nosocomial pathogens. Infections caused by the

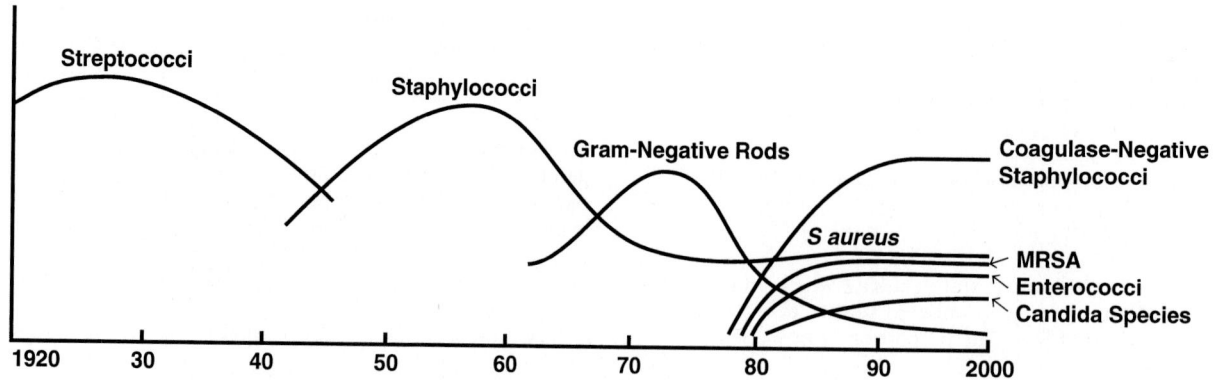

FIGURE 2 Schematic representation of the periods during which new pathogens emerged as prominent causes of nosocomial infections.

less pathogenic organisms were first identified in large university-affiliated hospitals, but they have subsequently been found in community hospitals as well (155, 164).

Since antimicrobial agents were first introduced, bacteria have found numerous ways to circumvent the lethal affects of these drugs (157). In, 1941 when physicians began treating staphylococcal infections with penicillin, all isolates were susceptible, but by the end of the decade, 60% of hospital-acquired *S. aureus* were resistant (130, 138, 195). Similarly, methicillin was introduced in 1959, and methicillin-resistant *S. aureus* (MRSA) appeared almost immediately thereafter (29, 130). During the 1980s, the trend toward resistant pathogens became more pronounced (162). MRSA has persisted, and a myriad of new resistant organisms have appeared as nosocomial pathogens.

MRSA

In the 1960s, MRSA began causing numerous nosocomial outbreaks in Europe, and by the 1970s, MRSA was causing similar problems in hospitals in the United States (28). Between 1975 and 1991, the percentage of *S. aureus* isolates in the United States that were resistant to methicillin increased dramatically, from 2.4 to 29% (57). Similarly, the number of Veterans Affairs Medical Centers (VAMCs) reporting MRSA isolates increased from 3 (2.2%) of 137 in 1975 to 111 (81%) of 137 in 1984 (150, 181). By 1987 to 1989, 248 (97%) of 256 hospitals surveyed by Boyce reported that they had seen some patients with MRSA (13).

Once in a hospital, MRSA is very difficult to eradicate (12, 34, 175). Several investigators have noted that after MRSA was introduced into a hospital, the overall rate of nosocomial *S. aureus* infections increased (18, 173). Furthermore, Muder et al. observed that patients in a long-term care facility who were colonized with MRSA were more likely to become infected than patients who carried methicillin-susceptible *S. aureus* or those who did not carry *S. aureus* (132).

Although the overall incidence of nosocomial MRSA has increased, MRSA is not uniformly distributed throughout all hospitals. In the United States, the proportion of *S. aureus* that is resistant to methicillin increases from 15% in hospitals with fewer than 200 beds to 38% in hospitals with 500 or more beds (57). Moreover, the incidence varies tremendously by location (12, 118, 179). For example, in several European countries, fewer than 2% of *S. aureus* isolates are resistant to methicillin (Denmark, 0.1%; Sweden, 0.3%; The Netherlands, 1.5%; Switzerland, 1.8%),

whereas in other countries, more than 20% of *S. aureus* isolates are resistant to methicillin (Austria, 21.6%; Belgium, 25.1%; Spain, 30.3%; France, 33.6%) (179).

The epidemiology of MRSA colonization and infection has changed as the incidence of nosocomial MRSA has increased. Previously, most patients acquired MRSA in hospitals. Now, many patients become colonized in the community or in a nursing home (28) and serve as unrecognized reservoirs after they are admitted to the hospital (112).

Risk factors for acquiring MRSA include prolonged hospital stay, treatment with broad-spectrum antibiotics, lengthy antibiotic therapy, treatment in an intensive care or burn unit, proximity to another patient with MRSA, and surgical wounds (130, 175).

Antibiotic-Resistant Enterococci

During the 1980s, the enterococci moved up in the ranks of nosocomial pathogens (129). By 1990 to 1992, these organisms caused 10% of all nosocomial infections, including 16% of UTIs, 12% of SSIs, and 9% of bloodstream infections (57).

The enterococci express intrinsic resistance to cephalosporins, semisynthetic penicillinase-resistant penicillins, and low levels of clindamycin and aminoglycosides. Compared to the streptococci, the enterococci are also relatively resistant to penicillin, ampicillin, and the ureidopenicillins (133). In addition to their intrinsic resistance, enterococci have acquired resistance to chloramphenicol, erythromycin, tetracycline, fluoroquinolones, high levels of clindamycin and aminoglycosides, and, most recently, vancomycin (133). Because they resist many antimicrobial agents, enterococci survive and even thrive in the hospital environment (72).

As early as 1973 to 1976, Calderwood et al. noted that 54 and 49% of enterococcal isolates at Massachusetts General Hospital expressed in vitro resistance to streptomycin and kanamycin, respectively (21). In 1979, investigators in Europe reported the first enterococcal isolates that expressed high-level gentamicin resistance (37, 90). These organisms now colonize and infect patients in hospitals throughout the United States (126). For example, in the mid-1980s, 16% of clinical enterococcal isolates at the University of Michigan Hospital and 55% at the Ann Arbor, Mich., VAMC expressed high-level gentamicin resistance (199).

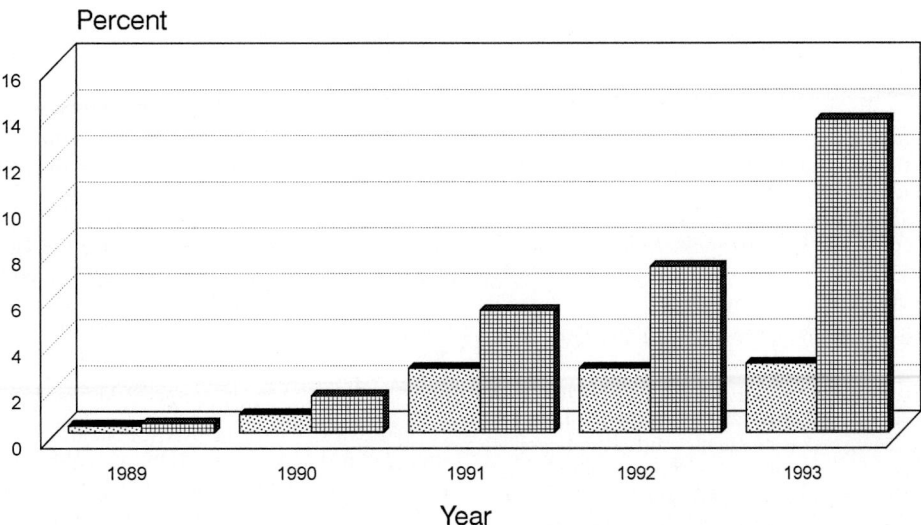

FIGURE 3 Percentages of enterococci reported as resistant to vancomycin and isolated from patients with nosocomial infections. Strains were isolated from patients in intensive care units (▦) and in non-intensive care units ( ). The figure is based on a similar figure from reference 27.

Since 1983, when Murray identified the first strain of *Enterococcus faecalis* that produced β-lactamase (134), clinical laboratories in at least six hospitals in five states have identified such strains (136). Furthermore, in 1992, Coudron et al. reported the first *Enterococcus faecium* isolate that produced β-lactamase (36). Subsequently, β-lactamase-producing, high-level-gentamicin-resistant *E. faecalis* has spread rapidly in several hospitals (154, 184), including the Richmond, Va., VAMC, where 11% of enterococci express this resistance pattern (184). Ampicillin-resistant enterococci that do not produce β-lactamase have also caused endemic and epidemic infections (15, 30, 75, 166). Chirurgi et al. noted that 9% of all enterococcal isolates in their VAMC were resistant to ampicillin (30).

In 1969, Toala et al. identified enterococcal strains with vancomycin MICs of ≥8 μg/ml (176), and in 1988, Leclercq et al. reported the first clinically important isolate of vancomycin-resistant *E. faecium* (109). Since then, those organisms have spread throughout the world (97, 168). Although called vancomycin-resistant enterococci, most of these strains in reality are resistant to most available antimicrobial agents. In the United States, vancomycin-resistant enterococci have been identified most often in intensive care units (Fig. 3), transplant units, and hematology and oncology units of hospitals on the East Coast. Data from NNIS suggest that the percentage of vancomycin-resistant isolates varies with teaching affiliation (0.6% at nonteaching hospitals and 3.0% at teaching hospitals) and the size of the hospital (0% at hospitals with fewer than 200 beds, 1.8% at hospitals with 200 to 500 beds, and 3.6% at hospitals with more than 500 beds) (27). Similarly, Jones and Sader noted that 4.4% of 1,936 enterococcal isolates from 96 sites in 46 states were resistant to vancomycin (98).

Several reports illustrate how quickly vancomycin-resistant enterococci can spread. From September 1989 until October 1991, the number of vancomycin-resistant strains isolated from patients in New York City increased exponentially from 1 to 361 cases, and the number of hospitals that had identified these organisms increased from 1 to 38

(65). Boyle et al. noted a similar dramatic increase in the number of vancomycin-resistant enterococci isolated in their hospital (19). Between 1990 and 1992, the proportion of *E. faecium* among enterococcal isolates progressively increased from 10.9 to 34.0%, and the proportion of vancomycin-susceptible strains decreased precipitously over the same period from 85.7 to 25.8%. Data from a molecular epidemiologic study by Chow et al. suggest that these organisms may be transmitted not only within a hospital or among hospitals in one city but also among hospitals in different states (31).

In general, patients acquire resistant enterococci in the hospital. Chirurgi observed that on average, ampicillin-resistant enterococci were not obtained from rectal cultures until the patient had been hospitalized for 11.3 days (30). Risk factors for colonization and infection with resistant enterococci vary with the particular antibiotic resistance and with the study (14, 15, 19, 86, 99, 113, 159). However, in more than one study, investigators have shown that prolonged hospitalization (4, 15) and prior treatment with antibiotics (4, 184) independently predict acquisition of a resistant enterococcus. In several studies, exposure to specific antibiotics such as imipenem (15), clindamycin (166), and vancomycin (99) independently predicted infection with resistant enterococci. In addition, advanced age (166), female sex (15), severe underlying illness as measured by APACHE II (184), and prior nosocomial infection (99) have been independently associated with acquisition of resistant enterococci in one hospital each.

Once in a hospital, resistant enterococci have often become endemic (19), perhaps because they can persist in the patients' gastrointestinal tracts (113), in the environment, and on the hands of health care workers (180). In numerous hospitals, investigators have isolated resistant enterococci from cultures of the environment (14, 15, 99, 113, 154, 184, 199), and some investigators have identified health care workers who were colonized with resistant strains (154, 159).

Wade et al. demonstrated that enterococci can survive

for at least 30 min on the hands of health care workers (180). Furthermore, enterococci survived after health care workers had washed their hands with bland soap but not after they had washed with clorhexidine or 60% isopropyl alcohol (180). Therefore, chlorhexidine or 60% isopropyl alcohol should replace nonmedicated soap in units where patients at high risk of acquiring resistant enterococci are treated. Moreover, housekeeping personnel should carefully clean and terminally disinfect rooms in which patients with resistant enterococci have been treated.

Resistant Gram-Negative Organisms

Many gram-negative nosocomial pathogens have developed broad-spectrum resistance to the newer β-lactam antibiotics (138). Gram-negative species such as *Enterobacter cloacae, Enterobacter aerogenes, Citrobacter freundii, Serratia* spp., *Providencia* spp., *Morganella morganii,* and *Pseudomonas aeruginosa* produce type I cephalosporinases, which are chromosomally encoded and inducible (161). These enzymes inactivate all cephalosporins and penicillins; hence, the organisms that produce them are not susceptible to the newer-generation cephalosporins. Organisms that produce type I cephalosporinases may be very difficult to detect. In particular, susceptibility testing methods that use a relatively low inoculum of organisms (e.g., microdilution methods and rapid, automated, or semiautomated methods) may not detect those enzymes before they are induced.

Some gram-negative bacteria produce extended-spectrum β-lactamases, which are plasmid-encoded enzymes that confer resistance to oxyimino-β-lactam antibiotics (e.g., cefotaxime, ceftazidime, and aztreonam). Ironically, those antibiotics were designed to resist being degraded by plasmid-encoded β-lactamases (94). Many of these β-lactamases are derivatives of TEM enzymes, but at least 16 extended-spectrum β-lactamases that are not related to that family of enzymes have been identified (94). *Klebsiella pneumoniae* and *Escherichia coli* are the organisms most likely to produce extended-spectrum β-lactamases, but *M. morganii, C. freundii, Klebsiella oxytoca,* and *P. aeruginosa* may also produce these enzymes (94). Extended-spectrum β-lactamases are very worrisome, because they inactivate most broad-spectrum β-lactam antibiotics and because organisms that produce them may also carry other resistance determinants. Although strains that produce extended-spectrum β-lactamases are becoming more frequent (94), this type of resistance may be very difficult to detect by most methods for testing antimicrobial susceptibility (100, 161).

Mycobacterium tuberculosis and Resistant *M. tuberculosis*

Most physicians once thought that tuberculosis was a disease either of the past or of the Third World. However, as the AIDS epidemic has progressed and the United States economy has declined, the incidence of tuberculosis has increased substantially. The number of tuberculosis cases reported to the CDC decreased every year but one between 1953 and the mid-1980s (10). Since 1985, however, the number of reported cases has increased 18% (10). Moreover, the number of cases in children under 5 years of age who were born in the United States increased 34% between 1987 and 1990, suggesting that transmission in the community has increased (10).

As the incidence of tuberculosis has increased, susceptible and multidrug-resistant M. *tuberculosis* (MDR-TB) strains have caused outbreaks in communities, prisons, and hospitals (8, 26, 44, 47, 55, 60, 61, 79, 91, 146). As with the vancomycin-resistant enterococci, MDR-TB has been most prevalent on the East Coast. Nosocomial outbreaks of MDR-TB have primarily affected patients infected with human immunodeficiency virus (HIV) (8, 26, 55, 60, 61, 146); however, MDR-TB has also infected individuals with healthy immune systems (8, 26, 146). In general, the period between exposure to a patient with MDR-TB and the onset of active tuberculosis has been very short (8, 61), ranging from 22 to 182 days in one outbreak (61). The mortality rate for patients infected with both HIV and MDR-TB has been very high (72 to 89%) (8, 146). The average length of survival after the diagnosis of active tuberculosis is very short. For example, in one outbreak, the median survival was 2.1 months for patients infected with MDR-TB and 14.6 months for controls infected with susceptible M. *tuberculosis* (60). Risk factors for acquiring MDR-TB have included a diagnosis of AIDS, attending a clinic for HIV patients, previous hospital admission, and previous admission on a ward for HIV patients (146).

Infected patients are more likely to transmit MDR-TB if they received aerosolized pentamidine or if their sputum smears were positive for acid-fast bacilli (8). Such patients have transmitted MDR-TB to numerous health care workers (8, 146). The risk of transmitting M. *tuberculosis* to staff appears to be greater with MDR-TB than with susceptible strains (8).

Other Resistant Organisms

Antimicrobial-resistant organisms will become more prominent during the 1990s. Organisms like *Streptococcus pneumoniae*, which in the past rarely caused outbreaks and were always susceptible to antibiotics, may become important nosocomial pathogens and may become increasingly resistant (22, 45, 73, 93, 138). Perhaps 5 to 10% of pneumococcal isolates in the United States express intermediate susceptibilities to penicillin, and up to 50% of isolates in some European countries express penicillin resistance. Thus, there is reason to be concerned about this pathogen. In addition to vancomycin-resistant enterococci, multiresistant strains of *Staphylococcus haemolyticus* have been identified by numerous investigators (67), and Schwalbe et al. isolated a strain that was resistant to vancomycin (165). Unusual organisms, including *Pediococcus acidilactici* (122), *Lactobacillus* spp., and *Leuconostoc* spp., which are intrinsically resistant to vancomycin (97, 160), or *Agrobacterium radiobacter* (56) and *Xanthomonas maltophilia* (138), which are inherently resistant to many antimicrobial agents, may cause nosocomial infections.

Impact of Resistant Organisms

The impact of resistant bacteria is very difficult to quantify. However, antibiotic resistance severely limits the number of agents that physicians can use to treat infections (128). To circumvent antibiotic resistance, clinicians may be forced to use antimicrobial agents that are more expensive, more toxic, or less efficacious than the standard drugs. Furthermore, infections caused by resistant organisms may be difficult or impossible to treat; hence, morbidity and mortality rates, lengths of hospital stays, and costs may be increased (33, 71, 72, 98, 135).

Candida Species

Once thought to be nonpathogenic, *Candida* species caused 5.6% of nosocomial bloodstream infections in 1984 (89) and 7% between 1990 and 1992 (57). Far from being nonpathogenic, these organisms cause serious infections in immunocompromised or debilitated patients. Crude death rates from candidemia range from 25 to 57% (192), and the attributable mortality rate is 38% (192). In survivors, candidemia also prolongs the length of hospital stay by 30 days above that required to treat the patients' underlying conditions (192). Underlying conditions that predispose patients to candidemia include burns, trauma, malnutrition, neutropenia, diabetes mellitus, renal failure, hematologic malignancies, and organ transplantation (193). After controlling for underlying diseases, Wey et al. identified additional independent risk factors for candidemia: numerous antibiotics received before candidemia (OR, 1.73 per antibiotic class), *Candida* species obtained from cultures of sites other than blood (OR, 10.37), prior hemodialysis (OR, 18.13), and prior use of a Hickman catheter (OR, 7.23) (193).

Viral Pathogens

The frequency of nosocomial viral infections is very difficult to quantify. These infections appear to be most common on pediatric services, where rotavirus, respiratory syncytial virus, parainfluenza virus, and adenovirus are among the most common pathogens (183). Nosocomial outbreaks of other viral diseases, such as influenza (140), measles (3), chickenpox, (78, 108), hepatitis A (50), and hepatitis B (2, 66, 117, 131, 167), may involve children and adults.

RESERVOIRS AND MODES OF TRANSMISSION

The term reservoir is used to designate the origin or niche of important pathogens. In modern hospitals, people (health care personnel, patients, and visitors) are the primary reservoirs of common hospital pathogens. Most nosocomial infections originate from patients' endogenous floras. Body sites that frequently host pathogens include the nares (*S. aureus*, including MRSA), the skin (coagulase-negative staphylococci), the gastrointestinal tract (enterococci, members of the family *Enterobacteriaceae*, and *Candida* species), and the genitourinary tract (enterococci, *Enterobacteriaceae*, *Candida* species, and *Torulopsis* spp.). Colonized or infected patients can contaminate the environment with pathogens, including the resistant enterococci (14, 99, 113, 154, 184, 199), MRSA (17), and *Clostridium difficile* (124). The contaminated environment can then serve as a secondary reservoir. In addition, the environment serves as the primary reservoir for other organisms. Environmental reservoirs include water (*Legionella* spp. [49, 87, 194] and *Pseudomonas* species) and food (*Enterobacteriaceae* and *Pseudomonas* species [23, 103, 197]).

In hospitals, health care workers, whose hands are transiently contaminated, frequently transmit pathogens (142, 190). Occasionally, health care workers who are colonized with pathogens such as *S. aureus* (16, 46, 145), *Streptococcus pyogenes* (25, 77, 95, 121, 144, 163, 182), and *Salmonella enteritidis* (101) have transmitted these organisms to patients. Contaminated vehicles such as food (178), breast milk (52), enteral nutrition solutions (111), drugs (2, 66, 115, 137), electronic thermometers (113), and intravenous fluids (116) can transmit nosocomial pathogens to patients. The hospital's air or ventilation system can transmit organisms such as *M. tuberculosis*, varicella-zoster virus (78, 108), and *Aspergillus* species (152, 153), and aerosols of contaminated water can transmit legionellae (6, 9, 51, 54, 62, 87).

Although currently not a common reservoir for nosocomial pathogens, the environment will become increasingly important as more immunosuppressed patients are treated in hospitals and as resistant organisms cause more nosocomial infections. Hence, personnel from hospital administration, infection control, engineering, housekeeping, and dietary services must work together to maintain an environment that is safe for patients, visitors, and hospital employees.

ROLE OF THE CLINICAL MICROBIOLOGY LABORATORY

The clinical microbiology laboratory has always played a key role in controlling nosocomial infections, and this role will expand in the 1990s as resistant organisms cause a larger proportion of infections (147). The first task of the laboratory is to grow and detect microbial pathogens. Although basic, that task has become more difficult as the prevalence of infections caused by M. *tuberculosis*, fungi, and unusual microbes has increased. Second, the laboratory must accurately identify causative organisms to species level. Species identification is particularly important with organisms such as the coagulase-negative staphylococci, of which some species (i.e., *Staphylococcus epidermidis* and S. *haemolyticus*) are more likely than others to cause nosocomial infections; the enterococci, of which some species (i.e., E. *faecium*) are more likely than others to resist numerous antimicrobial agents (75); and the mycobacteria, of which particular species (i.e., M. *tuberculosis*) are readily transmitted in the hospital (196).

Third, the laboratory must accurately determine the antimicrobial susceptibilities of nosocomial pathogens. In particular, the laboratory must be able to identify S. *aureus* strains that are methicillin resistant (29); enterococcal strains that express high-level resistance to aminoglycosides, β-lactams, and vancomycin; gram-negative bacteria that produce extended-spectrum β-lactamases and type I cephalosporinases (161); and M. *tuberculosis* strains that are resistant to standard therapeutic agents. At present, many laboratories cannot identify such strains, and routine testing methods may be inadequate (100, 161). For example, the proportion of laboratories that correctly identified five resistant enterococcal strains varied from 96% for high-level-vancomycin-resistant E. *faecium* to 29, 17, and 3.9% for moderate-level-vancomycin-resistant E. *faecium*, low-level-vancomycin-resistant E. *faecalis*, and β-lactamase-producing E. *faecalis*, respectively (174).

Once the laboratory has identified a resistant organism, it should immediately contact the clinician (who may need to alter therapy) and the infection control practitioner (who may need to institute isolation precautions). Furthermore, the laboratory should tabulate data on antimicrobial susceptibility and evaluate the trends in antibiotic susceptibility for common hospital pathogens.

Fourth, the laboratory and the infection control team should collaborate on epidemic investigations (156). Once infection control personnel have identified a putative cluster or outbreak, the laboratory should save all associated isolates. During outbreaks, the laboratory may also need to culture possible reservoirs: the nares of health care workers for S. *aureus*, the stools of patients and health care workers for resistant enterococci (72), and the hospital's water supply for legionellae (194). To determine whether one particular strain is causing the outbreak, the laboratory should be able to perform standard typing tests (e.g., antibiotic susceptibility tests and biotyping) and should be able either to perform molecular typing tests or to contract with experts to perform them.

Finally, because the spectrum of nosocomial pathogens is expanding and changing rapidly, the clinical microbiology laboratory must constantly review and update its methods for detecting and identifying microbial pathogens and their antimicrobial susceptibilities (147).

SUMMARY

Nosocomial infections affect at least 1.75 million to 3.5 million patients in the United States each year, add greatly to the cost of health care, and disable or kill tens of thousands of patients. The primary pathogens causing nosocomial infections have continually evolved over time, and new pathogens will be selected as the practice of medicine changes. Over the past decade, numerous resistant bacterial pathogens have appeared and spread throughout hospitals. In fact, organisms such as the vancomycin-resistant enterococci and MDR-TB have circumvented 5 decades of progress in antibiotic therapy and may herald what many have called a postantibiotic era. In the 1980s and the early 1990s, we learned the hard way that many nosocomial pathogens can become resistant to our most effective antimicrobial agents. As we approach the year 2000, we need to reemphasize classic infection control measures such as hand washing (1, 106), scrupulous care of invasive devices (114), and antibiotic control (58, 141). Moreover, we must develop new ways to diagnose, prevent, and treat these devastating infections.

REFERENCES

1. **Albert, R. K., and F. Condie.** 1981. Hand-washing patterns in medical intensive-care units. *N. Engl. J. Med.* **304:**1465–1466.
2. **Alter, M. J., J. Ahtone, and J. E. Maynard.** 1983. Hepatitis B virus transmission associated with a multiple-dose vial in a hemodialysis unit. *Ann. Intern. Med.* **99:**330–333.
3. **Ammari, L. K., L. M. Bell, and R. L. Hodinka.** 1993. Secondary measles vaccine failure in healthcare workers exposed to infected patients. *Infect. Control Hosp. Epidemiol.* **14:**81–86.
4. **Axelrod, P., and G. H. Talbot.** 1989. Risk factors for acquisition of gentamicin-resistant enterococci. A multivariate analysis. *Arch. Intern. Med.* **149:**1397–1401.
5. **Banerjee, S. N., T. G. Emori, D. H. Culver, R. P. Gaynes, W. R. Jarvis, T. Horan, J. R. Edwards, J. Tolson, T. Henderson, W. J. Martone, and the National Nosocomial Infections Surveillance System.** 1991. Secular trends in nosocomial primary bloodstream infections in the United States, 1980–1989. *Am. J. Med.* **91**(Suppl. 3B):86S–89S.
6. **Bartlett, C. L. R.** 1984. Potable water as reservoir and means of transmission, p. 210–215. *In* C. Thornsberry, A. Balows, J. C. Feeley, and W. Jakubowski (ed.), *Legionella: Proceedings of the Second International Symposium.* American Society for Microbiology, Washington, D.C.
7. **Baselski, V.** 1993. Microbiologic diagnosis of ventilator-associated pneumonia, p. 331–357. *In* R. C. Moellering, Jr., and J. A. Washington (ed.), *Infectious Disease Clinics of North America: Laboratory Diagnosis of Infectious Diseases,* vol. 7. The W. B. Saunders Co., Philadelphia.
8. **Beck-Sague, C., S. W. Dooley, M. D. Hutton, J. Otten, A. Breeden, J. T. Crawford, A. E. Pitchenik, C. Woodley, G. Cauthen, and W. R. Jarvis.** 1992. Hospital outbreak of multidrug-resistant *Mycobacterium tuberculosis* infections: factors in transmission to staff and HIV-infected patients. *JAMA* **268:**1280–1286.
9. **Best, M., J. Stout, R. R. Muder, V. L. Yu, A. Goetz, and F. Taylor.** 1983. *Legionellaceae* in the hospital water supply: epidemiological link with disease and evaluation of a method for control of nosocomial legionnaires' disease and Pittsburgh pneumonia. *Lancet* **ii:**307–310.
10. **Bloom, B. R., and C. J. L. Murray.** 1992. Tuberculosis: commentary on a reemergent killer. *Science* **257:**1055–1064.
11. **Bonten, M. J. M., C. A. Gaillard, F. H. van Tiel, H. G. W. Smeets, S. van der Geest, and E. E. Stobberingh.** 1994. The stomach is not a source for colonization of the upper respiratory tract and pneumonia in ICU patients. *Chest* **105:**878–884.
12. **Boyce, J. M.** 1989. Methicillin-resistant *Staphylococcus aureus.* Detection, epidemiology, and control measures. *Infect. Dis. Clin. N. Am.* **3:**901–913.
13. **Boyce, J. M.** 1991. Should we vigorously try to contain and control methicillin-resistant *Staphylococcus aureus? Infect. Control Hosp. Epidemiol.* **12:**46–54.
14. **Boyce, J. M., S. M. Opal, J. W. Chow, M. J. Zervos, G. Potter-Bynoe, C. B. Sherman, R. L. C. Romulo, S. Fortna, and A. A. Medeiros.** 1994. Outbreak of multidrug-resistant *Enterococcus faecium* with transferable *vanB* class vancomycin resistance. *J. Clin. Microbiol.* **32:**1148–1153.
15. **Boyce, J. M., S. M. Opal, G. Potter-Bynoe, R. G. LaForge, M. J. Zervos, G. Furtado, G. Victor, and A. A. Medeiros.** 1992. Emergence and nosocomial transmission of ampicillin-resistant enterococci. *Antimicrob. Agents Chemother.* **36:**1032–1039.
16. **Boyce, J. M., S. M. Opal, G. Potter-Bynoe, and A. A. Medeiros.** 1993. Spread of methicillin-resistant *Staphylococcus aureus* in a hospital after exposure to a health care worker with chronic sinusitis. *Clin. Infect. Dis.* **17:**496–504.
17. **Boyce, J. M., G. Potter-Bynoe, C. Chenevert, and N. L. Hughes.** 1994. Contamination of environmental surfaces with methicillin-resistant *Staphylococcus aureus* (MRSA): infection control implications. *Infect. Control Hosp. Epidemiol. Annu. Meet. Program* **15:**P22.
18. **Boyce, J. M., R. L. White, and E. Y. Spruill.** 1983. Impact of methicillin-resistant *Staphylococcus aureus* on the incidence of nosocomial staphylococcal infections. *J. Infect. Dis.* **148:**763.
19. **Boyle, J. F., S. A. Soumakis, A. Rendo, J. A. Herrington, D. G. Gianarkis, B. E. Thurberg, and B. G. Painter.** 1993. Epidemiologic analysis and genotypic characterization of a nosocomial outbreak of vancomycin-resistant enterococci. *J. Clin. Microbiol.* **31:**1280–1285.
20. **Brown, R. B., S. Bradley, E. Opitz, D. Cipriani, R. Pieczarka, and M. Sands.** 1987. Surgical wound infections documented after hospital discharge. *Am. J. Infect. Control* **15:**54–58.
21. **Calderwood, S. A., C. Wennersten, R. C. Moellering, Jr., L. J. Kunz, and D. J. Krogstad.** 1977. Resistance to six aminoglycosidic aminocyclitol antibiotics among enterococci; prevalence, evolution, and relationship to synergism with penicillin. *Antimicrob. Agents Chemother.* **12:**401–405.
22. **Cartmill, T. D. I., and H. Panigrahi.** 1992. Hospital outbreak of multiresistant *Streptococcus pneumoniae. J. Hosp. Infect.* **20:**130–132.
23. **Casewell, M., and I. Phillips.** 1978. Food as a source of *Klebsiella* species for colonisation and infection of intensive care patients. *J. Clin. Pathol.* **31:**845–849.
24. **Celis, R., A. Torres, M. Almela, R. Rodriguez-Roisin, and A. Augusti-Vidal.** 1988. Nosocomial pneumonia: a multivariate analysis of risk and prognosis. *Chest* **93:**318–324.
25. **Center for Disease Control.** 1976. Hospital outbreak of streptococcal wound infection—Utah. *Morbid. Mortal. Weekly Rep.* **25:**141.
26. **Centers for Disease Control.** 1990. Nosocomial transmission of multidrug-resistant tuberculosis to health-care workers and HIV-infected patients in an urban hospital—Florida. *Morbid. Mortal. Weekly Rep.* **39:**718–722.
27. **Centers for Disease Control.** 1993. Nosocomial enterococci resistant to vancomycin—United States, 1989–1993. *Morbid. Mortal. Weekly Rep.* **42:**597–599.

28. **Chambers, H. F.** 1988. Methicillin-resistant staphylococci. *Clin. Microbiol. Rev.* **1**:173–186.

29. **Chambers, H. F.** 1993. Detection of methicillin-resistant staphylococci, p. 425–433. *In* R. C. Moellering, Jr., and J. A. Washington (ed.), *Infectious Disease Clinics of North America: Laboratory Diagnosis of Infectious Diseases.* The W. B. Saunders Co., Philadelphia.

30. **Chirurgi, V. A., S. E. Oster, A. A. Goldberg, and R. E. McCabe.** 1992. Nosocomial acquisition of β-lactamase-negative, ampicillin-resistant enterococcus. *Arch. Intern. Med.* **152**:1457–1461.

31. **Chow, J. W., A. Kuritza, D. M. Shlaes, M. Green, D. F. Sahm, and M. J. Zervos.** 1993. Clonal spread of vancomycin-resistant *Enterococcus faecium* between patients in three hospitals in two states. *J. Clin. Microbiol.* **31**:1609–1611.

32. **Classen, D. C., R. S. Evans, S. L. Pestotnik, S. D. Horn, R. L. Menlove, and J. P. Burke.** 1992. The timing of prophylactic administration of antibiotics and the risk of surgical-wound infection. *N. Engl. J. Med.* **326**:281–286.

33. **Cohen, M. L.** 1992. Epidemiology of drug resistance: implications for a post-antimicrobial era. *Science* **257**:1050–1055.

34. **Cohen, S. H., M. M. Morita, and M. Bradford.** 1991. A seven-year experience with methicillin-resistant *Staphylococcus aureus. Am. J. Med.* **91**(Suppl. 3B):233S–237S.

35. **Cook, P. P., D. W. Hecht, and D. R. Snydman.** 1989. Nosocomial *Branhamella catarrhalis* in a paediatric intensive care unit: risk factors for disease. *J. Hosp. Infect.* **13**:299–307.

36. **Coudron, P. E., S. M. Markowitz, and E. S. Wong.** 1992. Isolation of β-lactamase-producing, aminoglycoside-resistant strain of *Enterococcus faecium. Antimicrob. Agents Chemother.* **36**:1125–1126.

37. **Courvalin, P., C. Carlier, and E. Collatz.** 1980. Plasmid-mediated resistance to aminocyclitol antibiotics in group D streptococci. *J. Bacteriol.* **143**:541–551.

38. **Craig, C. P., and S. Connelly.** 1984. Effect of intensive care unit nosocomial pneumonia on duration of stay and mortality. *Am. J. Infect. Control* **12**:233–238.

39. **Craven, D. E., L. M. Kunches, V. Kilinsky, D. A. Lichtenberg, B. J. Make, and W. R. McCabe.** 1986. Risk factors for pneumonia and fatality in patients receiving continuous mechanical ventilation. *Am. Rev. Respir. Dis.* **133**:792–796.

40. **Craven, D. E., K. A. Steger, and T. W. Barber.** 1991. Preventing nosocomial pneumonia: state of the art and perspectives for the 1990s. *Am. J. Med.* **91**(Suppl. 3B):44S–53S.

41. **Craven, D. E., K. A. Steger, and R. A. Duncan.** 1993. Prevention and control of nosocomial pneumonia, p. 580–599. *In* R. P. Wenzel (ed.), *Prevention and Control of Nosocomial Infections,* 2nd ed. The Williams & Wilkins Co., Baltimore.

42. **Culver, D. H., T. C. Horan, R. P. Gaynes, W. J. Martone, W. R. Jarvis, T. G. Emori, S. N. Banerjee, J. R. Edwards, J. S. Tolson, T. S. Henderson, and J. M. Hughes.** 1991. Surgical wound infection rates by wound class, operative procedure, and patient risk index. *Am. J. Med.* **91**(Suppl. 3B):152S–157S.

43. **Daifuku, R., and W. E. Stamm.** 1984. Association of rectal and urethral colonization with urinary tract infection in patients with indwelling catheters. *JAMA* **252**:2028–2030.

44. **Daley, C. L., P. M. Small, G. F. Schecter, G. K. Schoolnik, R. A. McAdam, W. R. Jacobs, Jr., and P. C. Hopewell.** 1992. An outbreak of tuberculosis with accelerated progression among persons infected with the human immunodeficiency virus. *N. Engl. J. Med.* **326**:231–235.

45. **Dawson, S., A. Pallett, A. Davidson, and A. Tuck.** 1992. Outbreak of multiresistant pneumococci. *J. Hosp. Infect.* **22**:328–329.

46. **Devenish, E. A., and A. A. Miles.** 1939. Control of *Staphylococcus aureus* in an operating room. *Lancet* **i**:1088–1094.

47. **DiPerri, G., M. Cruciani, M. C. Danzi, R. Luzzati, G. DeChecchi, M. Malena, S. Pizzighell, R. Mazzi, M. Solbiati, E. Concia, and D. Bassetti.** 1989. Hospital infection; nosocomial epidemic of active tuberculosis among HIV-infected patients. *Lancet* **ii**:1502–1504.

48. **Dixon, R. E.** 1987. Costs of nosocomial infections and benefits of infection control programs, p. 19–25. *In* R. P. Wenzel (ed.), *Prevention and Control of Nosocomial Infections.* The Williams & Wilkins Co., Baltimore.

49. **Doebbeling, B. N., M. A. Ishak, B. H. Wade, M. A. Pasquale, R. E. Gerszten, D. H. M. Groschel, R. J. Kadner, and R. P. Wenzel.** 1989. Nosocomial *Legionella micdadei* pneumonia: 10 years experience and a case-control study. *J. Hosp. Infect.* **13**:289–298.

50. **Doebbeling, B. N., N. Li, and R. P. Wenzel.** 1993. An outbreak of hepatitis A among health care workers: risk factors for transmission. *Am. J. Public Health* **83**:1679–1684.

51. **Doebbeling, B. N., and R. P. Wenzel.** 1987. The epidemiology of *Legionella pneumophila* infections. *Semin. Respir. Infect.* **2**:206–221.

52. **Donowitz, L. G., F. J. Marsik, K. A. Fisher, and R. P. Wenzel.** 1981. Contaminated breast milk: a source of *Klebsiella* bacteremia in a newborn intensive care unit. *Rev. Infect. Dis.* **3**:716–720.

53. **Donowitz, L. G., and R. P. Wenzel.** 1980. Endometritis following cesarean section: a controlled study of the increased duration of hospital stay and direct cost of hospitalization. *Am. J. Obstet. Gynecol.* **137**:467–469.

54. **Edelstein, P. H., R. E. Whittaker, R. L. Kreiling, and C. L. Howell.** 1982. Efficacy of ozone in eradication of *Legionella pneumophila* from hospital plumbing fixtures. *Appl. Environ. Microbiol.* **44**:1330–1334.

55. **Edlin, B. R., J. I. Tokars, M. H. Grieco, J. T. Crawford, J. Williams, E. M. Sordillo, K. R. Ong, J. O. Kilburn, S. W. Dooley, K. G. Castro, W. R. Jarvis, and S. D. Holmberg.** 1992. An outbreak of multidrug-resistant tuberculosis among patients with the acquired immunodeficiency syndrome. *N. Engl. J. Med.* **326**:1514–1521.

56. **Edmond, M. B., S. A. Riddler, C. M. Baxter, B. M. Wicklund, and A. W. Pasculle.** 1993. *Agrobacterium radiobacter*: a recently recognized opportunistic pathogen. *Clin. Infect. Dis.* **16**:388–391.

57. **Emori, T. G., and R. P. Gaynes.** 1993. An overview of nosocomial infections, including the role of the microbiology laboratory. *Clin. Microbiol. Rev.* **6**:428–442.

58. **Ena, J., R. W. Dick, R. N. Jones, and R. P. Wenzel.** 1993. The epidemiology of intravenous vancomycin usage in a university hospital: a 10-year study. *JAMA* **269**:598–602.

59. **Fagon, J.-Y., J. Chastre, A. J. Hance, P. Montravers, A. Novara, and C. Gibert.** 1993. Nosocomial pneumonia in ventilated patients: a cohort study evaluating attributable mortality and hospital stay. *Am. J. Med.* **94**:281–288.

60. **Fischl, M. A., G. L. Daikos, R. B. Uttamchandani, R. B. Poblete, J. N. Moreno, R. R. Reyes, A. M. Boota, L. M. Thompson, T. J. Cleary, S. A. Oldham, M. J. Saldana, and S. Lai.** 1992. Clinical presentation and outcome of patients with HIV infection and tuberculosis caused by multiple-drug resistant bacilli. *Ann. Intern. Med.* **117**:184–190.

61. **Fischl, M. A., R. B. Uttamchandani, G. L. Daikos, R. B. Poblete, J. N. Moreno, R. R. Reyes, A. M. Boota, L. M. Thompson, T. J. Cleary, and S. Lai.** 1992. An outbreak of tuberculosis caused by multiple-drug resistant tubercle bacilli among patients with HIV infections. *Ann. Intern. Med.* **117**:177–183.

62. **Fisher-Hoch, S. P., J. O. Tobin, A. M. Nelson, M. G. Smith, J. M. Talbot, C. L. R. Bartlett, M. G. Pritchard, R. A. Swann, and J. A. Thomas.** 1981. Investigation and control of an outbreak of Legionnaires' disease in a district general hospital. *Lancet* **ii**:932–936.

63. **Freeman, J., and J. E. McGowan, Jr.** 1978. Risk factors for nosocomial infection. *J. Infect. Dis.* **138**:811–819.

64. **Freeman, J., B. A. Rosner, and J. E. McGowan, Jr.** 1979. Adverse effects of nosocomial infection. *J. Infect. Dis.* **140**:732–740.

65. **Frieden, T. R., S. S. Munsiff, D. E. Low, B. M. Willey, G. Williams, Y. Faur, W. Eisner, S. Warren, and B. Kre-**

iswirth. 1993. Emergence of vancomycin-resistant enterococci in New York City. *Lancet* **342:**76–79.

66. **Froggatt, J. W., D. M. Dwyer, and M. A. Stephens.** 1991. Hospital outbreak of hepatitis B in patients undergoing electroconvulsive therapy, abstr. 347, p. 157. *Program Abstr. 31st Intersci. Conf. Antimicrob. Agents Chemother.*

67. **Froggat, J. W., J. L. Johnston, D. W. Galletto, and G. L. Archer.** 1989. Antimicrobial resistance in nosocomial isolation of *Staphylococcus haemolyticus. Antimicrob. Agents Chemother.* **33:**460–466.

68. **Garibaldi, R. A.** 1993. Hospital-acquired urinary tract infections, p. 600–613. *In* R. P. Wenzel (ed.), *Prevention and Control of Nosocomial Infections,* 2nd ed. The Williams & Wilkins Co., Baltimore.

69. **Garibaldi, R. A., J. P. Burke, M. L. Dickman, and C. B. Smith.** 1974. Factors predisposing to bacteriuria during indwelling urethral catheterization. *N. Engl. J. Med.* **291:**215–219.

70. **Garibaldi, R. A., D. Cushing, and T. Lerer.** 1991. Risk factors for postoperative infection. *Am. J. Med.* **91**(Suppl. 3B)**:**158S–163S.

71. **Gibbons, A.** 1992. Exploring new strategies to fight drug-resistant microbes. *Science* **257:**1036–1038.

72. **Goldman, D. A.** 1992. Vancomycin-resistant *Enterococcus faecium:* headline news. *Infect. Control Hosp. Epidemiol.* **13:**695–699.

73. **Gould, F. K., J. G. Magee, and H. R. Ingham.** 1987. A hospital outbreak of antibiotic-resistant *Streptococcus pneumoniae. J. Infect.* **15:**77–79.

74. **Graybill, J. R., L. W. Marshall, P. Charache, C. K. Wallace, and V. B. Melvin.** 1973. Nosocomial pneumonia: a continuing major problem. *Am. Rev. Respir. Dis.* **108:**1130–1140.

75. **Grayson, M. L., G. M. Eliopoulos, C. B. Wennersten, K. L. Ruoff, P. C. De Girolami, M.-J. Ferraro, and R. C. Moellering, Jr.** 1991. Increasing resistance to β-lactam antibiotics among clinical isolates of *Enterococcus faecium:* a 22-year review at one institution. *Antimicrob. Agents Chemother.* **35:**2180–2184.

76. **Green, J. W., and R. P. Wenzel.** 1977. Postoperative wound infection: a controlled study of the increased duration of hospital stay and direct cost of hospitalization. *Ann. Surg.* **185:**264–268.

77. **Gryska, P. F., and A. E. O'Dea.** 1970. Postoperative streptococcal wound infection. The anatomy of an epidemic. *JAMA* **213:**1189–1191.

78. **Gustafson, T. L., G. B. Lavely, E. T. Brawner, R. H. Hutcheson, P. F. Wright, and W. Schaffner.** 1982. An outbreak of airborne nosocomial varicella. *Pediatrics* **70:**550–556.

79. **Haley, C. E., R. C. McDonald, L. Rossi, W. D. Jones, R. W. Haley, and J. P. Luby.** 1989. Tuberculosis epidemic among hospital personnel. *Infect. Control Hosp. Epidemiol.* **10:**204–210.

80. **Haley, R. W., D. H. Culver, W. M. Morgan, J. W. White, T. G. Emori, and T. M. Hooton.** 1985. Identifying patients at high risk of surgical wound infection. *Am. J. Epidemiol.* **121:**206–215.

81. **Haley, R. W., D. H. Culver, J. W. White, W. M. Morgan, and T. G. Emori.** 1985. The nationwide nosocomial infection rate: a new need for vital statistics. *Am. J. Epidemiol.* **121:**159–167.

82. **Haley, R. W., D. H. Culver, J. W. White, W. M. Morgan, T. G. Emori, V. P. Munn, and T. M. Hooton.** 1985. The efficacy of infection surveillance and control programs in preventing nosocomial infections in United States hospitals. *Am. J. Epidemiol.* **121:**182–205.

83. **Haley, R. W., T. M. Hooten, D. H. Culver, R. C. Stanley, T. G. Emori, C. D. Hardison, D. Quade, R. H. Shachtman, D. R. Schaberg, B. V. Shah, and G. D. Schatz.** 1981. Nosocomial infections in U.S. hospitals, 1975–1976: estimated frequency by selected characteristics of patients. *Am. J. Med.* **70:**947–959.

84. **Haley, R. W., D. R. Schaberg, S. D. Von Allmen, and J. E. McGowan, Jr.** 1980. Estimating the extra charges and prolongation of hospitalization due to nosocomial infections: a comparison of methods. *J. Infect. Dis.* **141:**248–257.

85. **Haley, R. W., J. H. Tenney, J. O. Lindsey II, J. S. Garner, and J. V. Bennett.** 1985. How frequent are outbreaks of nosocomial infection in community hospitals? *Infect. Control.* **6:**233–236.

86. **Handwerger, S., B. Raucher, D. Altarac, J. Monka, S. Marchione, K. V. Singh, B. E. Murray, J. Wolff, and B. Walters.** 1993. Nosocomial outbreak due to *Enterococcus faecium* highly resistant to vancomycin, penicillin, and gentamicin. *Clin. Infect. Dis.* **16:**750–755.

87. **Helms, C. M., R. M. Massanari, R. P. Wenzel, M. A. Pfaller, N. P. Moyer, N. Hall, and the Legionella Monitoring Committee.** 1988. Legionnaires' disease associated with a hospital water system: a five year progress report on continuous hyperchlorination. *JAMA* **259:**2423–2427.

88. **Horan, T., D. Culver, W. Jarvis, G. Emori, S. Banerjee, W. Martone, and C. Thornberry.** 1988. Pathogens causing nosocomial infections. Preliminary data from the National Nosocomial Infection Surveillance System. *Antimicrob. Newsl.* **5:**65–67.

89. **Horan, T. C., J. W. White, W. R. Jarvis, T. G. Emori, D. H. Culver, V. P. Munn, C. Thornsberry, D. R. Olson, and J. M. Hughes.** 1986. Nosocomial infection surveillance 1984. Centers for Disease Control Nosocomial Infection Surveillance. *Morbid. Mortal. Weekly Rep.* **35**(1SS)**:**17SS–29SS.

90. **Horodniceanu, T., L. Bougueleret, N. El-Solh, G. Bieth, and F. Delbos.** 1979. High-level, plasmid-borne resistance to gentamicin in *Streptococcus faecalis* subsp. *zymogenes. Antimicrob. Agents Chemother.* **16:**686–689.

91. **Hutton, M. D., W. W. Stead, G. M. Cauthen, A. B. Bloch, and W. M. Ewing.** 1990. Nosocomial transmission of tuberculosis associated with a draining abscess. *J. Infect. Dis.* **161:**286–295.

92. **Iglehart, J. K.** 1982. The new era of prospective payment for hospitals. *N. Engl. J. Med.* **307:**1288–1292.

93. **Jacobs, M. R., H. J. Koornhof, R. M. Robins-Browne, C. M. Stevenson, Z. A. Vermaak, I. Freiman, G. B. Miller, M. A. Witcomb, M. Isaacson, J. I. Ward, and R. Austrian.** 1978. Emergence of multiply resistant pneumococci. *N. Engl. J. Med.* **299:**735–740.

94. **Jacoby, G. A., and A. A. Medeiros.** 1994. More extended-spectrum β-lactamases. *Antimicrob. Agents Chemother.* **35:**1697–1704.

95. **Jewett, J. F., D. E. Reid, L. E. Safon, and C. L. Easterday.** 1968. Childbed fever—a continuing entity. *JAMA* **206:**344–350.

96. **Jimenez, P., A. Torres, R. Rodriguez-Roisin, J. P. de la Bellacasa, J. Aznar, J. Gatell, and A. Agusti-Vidal.** 1989. Incidence and etiology of pneumonia acquired during mechanical ventilation. *Crit. Care Med.* **17:**882–885.

97. **Johnson, A. P., A. H. C. Uttley, N. Woodford, and R. C. George.** 1990. Resistance to vancomycin and teicoplanin: an emerging clinical problem. *Clin. Microbiol. Rev.* **3:**280–291.

98. **Jones, R. N., and H. S. Sader.** 1993. The clinical impact of enterococcal resistance. *Challenges Infect. Dis.* **1:**1–6.

99. **Karanfil, L. V., M. Murphy, A. Josephson, R. Gaynes, L. Mandel, B. C. Hill, and J. M. Swenson.** 1992. A cluster of vancomycin-resistant *Enterococcus faecium* in an intensive care unit. *Infect. Control Hosp. Epidemiol.* **13:**195–200.

100. **Katsanis, G. P., J. Spargo, M. J. Ferraro, L. Sutton, and G. A. Jacoby.** 1994. Detection of *Klebsiella pneumoniae* and *Escherichia coli* strains producing extended-spectrum β-lactamases. *J. Clin. Microbiol.* **32:**691–696.

101. **Khuri-Bulos, N. A., M. A. Khalaf, A. Shehabi, and K. Shami.** 1994. Foodhandler-associated salmonella outbreak in a university hospital despite routine surveillance cultures of kitchen employees. *Infect. Control Hosp. Epidemiol.* **15:**311–314.

102. **Kollef, M. H.** 1993. Ventilator-associated pneumonia: a multivariate analysis. *JAMA* **270:**1965–1970.

103. **Kominos, S. D., C. E. Copeland, B. Grosiak, and B. Postic.** 1972. Mode of transmission of *Pseudomonas aeruginosa* in a burn unit. *Appl. Microbiol.* **23:**309–312.

104. **Krukowski, Z. H., and N. A. Matheson.** 1988. Ten-year computerized audit of infection after abdominal surgery. *Br. J. Surg.* **75:**857–861.

105. **Landry, S. L., D. L. Kaiser, and R. P. Wenzel.** 1989. Hospital stay and mortality attributed to nosocomial enterococcal bacteremia: a controlled study. *Am. J. Infect. Control* **17:**323–329.

106. **Larson, E.** 1987. Skin cleansing, p. 250–256. *In* R. P. Wenzel (ed.), *Prevention and Control of Nosocomial Infections.* The Williams & Wilkins Co., Baltimore.

107. **Leape, L. L., T. A. Brennan, N. Laird, A. G. Lawthers, A. R. Localio, B. A. Barnes, L. Hebert, J. P. Newhouse, P. C. Weiler, and H. Hiatt.** 1991. The nature of adverse events in hospitalized patients—results of the Harvard medical practice study II. *N. Engl. J. Med.* **324:**377–384.

108. **Leclair, J. M., J. A. Zaia, M. J. Levin, R. G. Congdon, and D. A. Goldman.** 1980. Airborne transmission of chickenpox in a hospital. *N. Engl. J. Med.* **302:**450–453.

109. **Leclercq, R., E. Derlot, J. Duval, and P. Courvalin.** 1988. Plasmid-mediated resistance to vancomycin and teicoplanin in *Enterococcus faecium. N. Engl. J. Med.* **319:**157–161.

110. **Leu, H.-S., D. L. Kaiser, M. Mori, R. F. Woolson, and R. P. Wenzel.** 1989. Hospital-acquired pneumonia: attributable mortality and morbidity. *Am. J. Epidemiol.* **129:**1258–1267.

111. **Levy, J., Y. Van Laethem, G. Verhaegen, C. Perpete, J.-P. Butzler, and R. P. Wenzel.** 1989. Enteral nutrition solutions as a cause of nosocomial sepsis. *J. Parenter. Enteral Nutr.* **13:**228–234.

112. **Linnemann, C. C., P. Moore, J. L. Staneck, and M. A. Pfaller.** 1991. Reemergence of epidemic methicillin-resistant *Staphylococcus aureus* in a general hospital associated with changing staphylococcal strains. *Am. J. Med.* **91**(Suppl. 3B):238S–244S.

113. **Livornese, L. L., S. Dias, C. Samel, B. Romanowski, S. Taylor, P. May, P. Pitsakis, G. Woods, D. Kaye, M. E. Levison, and C. C. Johnson.** 1992. Hospital-acquired infection with vancomycin-resistant *Enterococcus faecium* transmitted by electronic thermometers. *Ann. Intern. Med.* **117:**112–116.

114. **Maki, D.** 1989. Risk factors for nosocomial infection in intensive care—'devices vs nature' and goals for the next decade. *Arch. Intern. Med.* **149:**30–35.

115. **Maki, D. G., B. S. Klein, R. D. McCormick, C. M. Alvarado, M. A. Zilz, S. M. Stolz, C. A. Hassemer, J. Gould, and A. R. Liegel.** 1991. Nosocomial *Pseudomonas pickettii* bacteremias traced to narcotic tampering. *JAMA* **265:**981–986.

116. **Maki, D. G., F. S. Rhame, D. C. Mackel, and J. V. Bennett.** 1976. Nationwide epidemic of septicemia caused by contaminated intravenous products. *Am. J. Med.* **60:**471–485.

117. **Maldonado, Y. A., K. Roesch, S. C. Deresinski, and R. R. Roberto.** 1989. Nosocomial hepatitis B outbreak associated with a seronegative health care worker. *Am. J. Infect. Control* **17:**99.

118. **Marples, R. R.** 1988. Methicillin-resistant *Staphylococcus aureus. Curr. Opin. Infect. Dis.* **1:**722–726.

119. **Martin, M. A., M. A. Pfaller, and R. P. Wenzel.** 1989. Coagulase-negative staphylococcal bacteremia: mortality and hospital stay. *Ann. Intern. Med.* **110:**9–16.

120. **Martone, W. J., W. R. Jarvis, D. H. Culver, and R. W. Haley.** 1992. Incidence and nature of endemic and epidemic nosocomial infections, p. 577–596. *In* J. V. Bennett and P. S. Brachman (ed.), *Hospital Infections,* 3rd ed. Little, Brown & Co., Boston.

121. **Mastro, T. D., T. A. Farley, J. A. Elliott, R. R. Facklam, J. R. Perks, J. L. Hadler, R. C. Good, and J. S. Spika.** 1990. An outbreak of surgical-wound infections due to group A streptococcus carried on the scalp. *N. Engl. J. Med.* **323:**968–972.

122. **Mastro, T. D., J. S. Spika, P. Lozno, J. Appel, and R. R. Facklam.** 1990. Vancomycin-resistant *Pediococcus acidilactici*: nine cases of bacteremia. *J. Infect. Dis.* **161:**956–960.

123. **Mayhall, C. G.** 1993. Surgical infections including burns, p. 614–664. *In* R. P. Wenzel (ed.), *Prevention and Control of Nosocomial Infections,* 2nd ed. The Williams & Wilkins Co., Baltimore.

124. **McFarland, L. V., M. E. Mulligan, R. Y. Kwok, and W. E. Stamm.** 1989. Nosocomial acquisition of *Clostridium difficile* infections. *N. Engl. J. Med.* **320:**204–210.

125. **McGowan, J. E., Jr.** 1981. Cost and benefit in control of nosocomial infection: methods for analysis. *Rev. Infect. Dis.* **3:**790–797.

126. **Mederski-Samoraj, B. D., and B. E. Murray.** 1983. High-level resistance to gentamicin in clinical isolates of enterococci. *J. Infect. Dis.* **147:**751–757.

127. **Meduri, G. U.** 1993. Diagnosis of ventilator-associated pneumonia, p. 295–329. *In* R. C. Moellering, Jr., and J. A. Washington (ed.), *Infectious Disease Clinics of North America: Laboratory Diagnosis of Infectious Diseases.* The W. B. Saunders Co., Philadelphia.

128. **Moellering, R. C., Jr.** 1991. The enterococcus: a classic example of the impact of antimicrobial resistance on therapeutic options. *J. Antimicrob. Chemother.* **28:**1–12.

129. **Moellering, R. C., Jr.** 1992. Emergence of enterococcus as a significant pathogen. *Clin. Infect. Dis.* **14:**1173–1178.

130. **Morita, M. M.** 1993. Methicillin-resistant *Staphylococcus aureus.* Past, present, and future. *Nurs. Clin. N. Am.* **28:**625–637.

131. **Moss, A. L. H.** 1981. Hospital outbreak of hepatitis B. *N. Z. Med. J.* **94:**65–66.

132. **Muder, R. R., C. Brennen, M. M. Wagener, R. M. Vickers, J. D. Rihs, G. A. Hancock, Y. C. Yee, J. M. Miller, and V. L. Yu.** 1991. Methicillin-resistant staphylococcal colonization and infection in a long-term care facility. *Ann. Intern. Med.* **114:**107–112.

133. **Murray, B. E.** 1990. The life and times of the enterococcus. *Clin. Microbiol. Rev.* **3:**46–65.

134. **Murray, B. E.** 1992. β-Lactamase-producing enterococci. *Antimicrob. Agents Chemother.* **36:**2355–2359.

135. **Murray, B. E.** 1994. Can antibiotic resistance be controlled? *N. Engl. J. Med.* **330:**1229–1230.

136. **Murray, B. E., K. V. Singh, S. M. Markowitz, H. A. Lopardo, J. E. Patterson, M. J. Zervos, E. Rubeglio, G. M. Eliopoulos, L. B. Rice, F. W. Goldstein, S. G. Jenkins, G. M. Caputo, R. Nasnas, L. S. Moore, E. S. Wong, and G. Weinstock.** 1991. Evidence for clonal spread of a single strain of β-lactamase-producing *Enterococcus* (*Streptococcus*) *faecalis* to six hospitals in five states. *J. Infect. Dis.* **163:**780–785.

137. **Nakashima, A. K., M. A. McCarthy, W. J. Martone, and R. L. Anderson.** 1987. Epidemic septic arthritis caused by *Serratia marcescens* and associated with a benzalkonium chloride antiseptic. *J. Clin. Microbiol.* **25:**1014–1018.

138. **Neu, H. C.** 1992. The crisis in antibiotic resistance. *Science* **257:**1064–1073.

139. **Nichols, R. L.** 1991. Surgical wound infection. *Am. J. Med.* **91**(Suppl. 3B):54S–64S.

140. **Pachucki, C. T., S. A. Walsh Pappas, G. F. Fuller, S. L. Krause, J. R. Lentino, and D. M. Schaaff.** 1989. Influenza A among hospital personnel and patients. Implications for recognition, prevention, and control. *Arch. Intern. Med.* **149:**77–80.

141. **Pallares, R., M. Pujol, C. Pena, J. Ariza, R. Martin, and F. Gudiol.** 1993. Cephalosporins as risk factor for nosocomial *Enterococcus faecalis* bacteremia: a matched case-control study. *Arch. Intern. Med.* **153:**1581–1586.

142. **Patterson, J. E., and M. J. Zervos.** 1990. High-level gentamicin resistance in *Enterococcus*: microbiology, genetic basis, and epidemiology. *Rev. Infect. Dis.* **12:**644–652.

143. Patterson, T. F., J. E. Patterson, B. L. Masecar, G. E. Barden, W. J. Hierholzer, Jr., and M. J. Zervos. 1988. A nosocomial outbreak of *Branhamella catarrhalis* confirmed by restriction endonuclease analysis. *J. Infect. Dis.* **157**:996–1001.

144. Paul, S. M., C. Genese, and K. Spitalny. 1990. Postoperative group A β-hemolytic streptococcus outbreak with the pathogen traced to a member of a healthcare worker's household. *Infect. Control Hosp. Epidemiol.* **11**:643–646.

145. Payne, R. W. 1967. Severe outbreak of surgical sepsis due to *Staphylococcus aureus* of unusual type and origin. *Br. Med. J.* **4**:17–20.

146. Pearson, M. L., J. A. Jereb, T. R. Frieden, J. T. Crawford, B. J. Davis, S. W. Dooley, and W. R. Jarvis. 1992. Nosocomial transmission of multidrug resistant *Mycobacterium tuberculosis*: a risk to patients and health care workers. *Ann. Intern. Med.* **117**:191–196.

147. Pfaller, M. A. 1993. Microbiology: the role of the clinical laboratory in hospital epidemiology and infection control, p. 385–405. *In* R. P. Wenzel (ed.), *Prevention and Control of Nosocomial Infections*, 2nd ed. The Williams & Wilkins Co., Baltimore.

148. Pittet, D. 1993. Nosocomial bloodstream infections, p. 512–555. *In* R. P. Wenzel (ed.), *Prevention and Control of Nosocomial Infections*, 2nd ed. The Williams & Wilkins Co., Baltimore.

149. Platt, R., B. F. Polk, B. Murdock, and B. Rosner. 1982. Mortality associated with nosocomial urinary tract infection. *N. Engl. J. Med.* **307**:637–642.

150. Preheim, L. C., D. Rimland, and M. J. Bittner. 1987. Methicillin-resistant *Staphylococcus aureus* in Veterans Administration Medical Centers. *Infect. Control* **8**:191–194.

151. Prod'hom, G., P. Leuenberger, J. Koerfer, A. Blum, R. Chiolero, M.-D. Schaller, C. Perret, O. Spinnler, J. Blondel, H. Siegrist, L. Saghafi, D. Blanc, and P. Francioli. 1994. Nosocomial pneumonia in mechanically ventilated patients receiving antacid, ranitidine, or sucralfate as prophylaxis for stress ulcer: a randomized controlled trial. *Ann. Intern. Med.* **120**:653–662.

152. Rhame, F. S. 1989. Nosocomial aspergillosis: how much protection for which patients? *Infect. Control Hosp. Epidemiol.* **10**:296–298.

153. Rhame, F. S. 1992. The inanimate environment, p. 299–333. *In* J. V. Bennett and P. S. Brachman (ed.), *Hospital Infections*. Little, Brown & Co., Boston.

154. Rhinehart, E., N. E. Smith, C. E. Wennersten, E. Gorss, J. Freeman, G. M. Eliopoulos, R. C. Moellering, Jr., and D. A. Goldman. 1990. Rapid dissemination of β-lactamase-producing, aminoglycoside-resistant *Enterococcus faecalis* among patients and staff on an infant-toddler surgical ward. *N. Engl. J. Med.* **323**:1814–1818.

155. Righter, J. 1987. Septicemia due to coagulase-negative staphylococcus in a community hospital. *Can. Med. Assoc. J.* **137**:121–125.

156. Ristuccia, P. A., and B. A. Cunha. 1987. Microbiologic aspects of infection control, p. 205–232. *In* R. P. Wenzel (ed.), *Prevention and Control of Nosocomial Infections*. The Williams & Wilkins Co., Baltimore.

157. Rockefeller University Workshop. 1994. Multiple-antibiotic-resistant pathogenic bacteria. A report of the Rockefeller University workshop. *N. Engl. J. Med.* **330**:1247–1251.

158. Rose, R., K. J. Hunting, T. R. Townsend, and R. P. Wenzel. 1977. Morbidity/mortality and economics of hospital-acquired bloodstream infections: a controlled study. *South. Med. J.* **70**:1267–1269.

159. Rubin, L. G., V. Tucci, E. Cercenado, G. Eliopoulos, and H. D. Isenberg. 1992. Vancomycin-resistant *Enterococcus faecium* in hospitalized children. *Infect. Control Hosp. Epidemiol.* **13**:700–705.

160. Ruoff, K. L., D. R. Kuritzkes, J. S. Wolfson, and M. J. Ferraro. 1988. Vancomycin-resistant gram-positive bacteria isolated from human sources. *J. Clin. Microbiol.* **26**:2064–2068.

161. Sanders, C. C., K. S. Thompson, and P. A. Bradford. 1993. Problems with detection of β-lactam resistance among nonfastidious gram-negative bacilli, p. 411–424. *In* R. C. Moellering, Jr., and J. A. Washington (ed.), *Infectious Disease Clinics of North America: Laboratory Diagnosis of Infectious Diseases*. The W. B. Saunders Co., Philadelphia.

162. Schaberg, D. R., D. H. Culver, and R. P. Gaynes. 1991. Major trends in the microbial etiology of nosocomial infection. *Am. J. Med.* **91**(Suppl. 3B):72S–75S.

163. Schaffner, W., L. B. Lefkowitz, Jr., J. S. Goodman, and M. G. Koenig. 1969. Hospital outbreak of infections with group A streptococci traced to an asymptomatic anal carrier. *N. Engl. J. Med.* **280**:1224–1227.

164. Scheckler, W. E., W. Scheibel, and D. Kresge. 1991. Temporal trends in septicemia in a community hospital. *Am. J. Med.* **91**(Suppl. 3B):90S–94S.

165. Schwalbe, R. S., J. T. Stapleton, and R. H. Gilligan. 1987. Emergence of vancomycin resistance in coagulase-negative staphylococci. *N. Engl. J. Med.* **316**:927–931.

166. Sexton, D. J., L. J. Harrell, J. J. Thorpe, D. L. Hunt, and L. B. Reller. 1993. A case-control study of nosocomial ampicillin-resistant enterococcal infection and colonization at a university hospital. *Infect. Control Hosp. Epidemiol.* **14**:629–635.

167. Shanson, D. C. 1980. Hepatitis B outbreak in operating-theatre and intensive care staff. *Lancet* **ii**:596.

168. Shlaes, D. M. 1992. Vancomycin-resistant bacteria. *Infect. Control Hosp. Epidemiol.* **13**:193–194.

169. Spengler, R. F., and W. B. Greenough III. 1978. Hospital costs and mortality attributed to nosocomial bacteremias. *JAMA* **240**:2455–2458.

170. Stamm, W. E. 1991. Catheter-associated urinary tract infections: epidemiology, pathogenesis, and prevention. *Am. J. Med.* **91**(Suppl. 3B):65S–71S.

171. Stamm, W. E., R. A. Weinstein, and R. E. Dixon. 1981. Comparison of endemic and epidemic nosocomial infections. *Am. J. Med.* **70**:393–397.

172. Stevens, R. M., D. Teres, J. J. Skillman, and D. S. Feingold. 1974. Pneumonia in an intensive care unit: a thirty-month experience. *Arch. Intern. Med.* **134**:106–111.

173. Tam, A. Y.-C., and C.-Y. Yeung. 1988. The changing pattern of severe neonatal staphylococcal infection: a 10-year study. *Aust. Paediatr.* **24**:275–278.

174. Tenover, F. C., J. Tokars, J. Swenson, S. Paul, K. Spitalny, and W. Jarvis. 1993. Ability of clinical laboratories to detect antimicrobial agent-resistant enterococci. *J. Clin. Microbiol.* **31**:1695–1699.

175. Thompson, R. L., I. Cabezudo, and R. P. Wenzel. 1982. Epidemiology of nosocomial infections caused by methicillin-resistant *Staphylococcus aureus*. *Ann. Intern. Med.* **97**:309–317.

176. Toala, P., A. McDonald, C. Wilcox, and M. Finland. 1969. Susceptibility of group D streptococcus (enterococcus) to 21 antibiotics in vitro, with special reference to species differences. *Am. J. Med. Sci.* **258**:416–430.

177. Townsend, T. R., and R. P. Wenzel. 1981. Nosocomial bloodstream infections in a newborn intensive care unit: a case-matched control study of morbidity, mortality and risk. *Am. J. Epidemiol.* **114**:73–80.

178. Villarino, M. E., D. J. Vugia, N. H. Bean, W. R. Jarvis, and J. M. Hughes. 1992. Foodborne disease prevention in health care facilities, p. 345–358. *In* J. V. Bennett and P. S. Brachman (ed.), *Hospital Infections*. Little, Brown & Co., Boston.

179. Voss, A., D. Milatovic, C. Wallrauch-Schwarz, V. T. Rosdahl, and I. Braveny. 1994. Methicillin-resistant *Staphylococcus aureus* in Europe. *Eur. J. Clin. Microbiol. Infect. Dis.* **13**:50–55.

180. Wade, J. J., N. Desai, and M. W. Casewell. 1991. Hygienic hand disinfection for the removal of epidemic vancomycin-resistant *Enterococcus faecium* and gentamicin-resistant *Enterobacter cloacae*. *J. Hosp. Infect.* **18**:211–218.

181. Wakefield, D. S., M. Pfaller, R. M. Massanari, and G. T.

Hammons. 1987. Variation in methicillin-resistant *Staphylococcus aureus* occurrence by geographic location and hospital characteristics. *Infect. Control* **8:**151–157.

182. **Walter, C. W.** 1966. The infector on the surgical team. *Clin. Neurosurg.* **14:**361–379.

183. **Welliver, R. C., and S. McLaughlin.** 1984. Unique epidemiology of nosocomial infection in a children's hospital. *Am. J. Dis. Child.* **138:**131–135.

184. **Wells, V. D., E. S. Wong, B. E. Murray, P. E. Coudron, D. S. Williams, and S. M. Markowitz.** 1992. Infections due to beta-lactamase-producing, high-level gentamicin-resistant *Enterococcus faecalis*. *Ann. Intern. Med.* **116:**285–292.

185. **Wenzel, R. P.** 1985. Nosocomial infections, diagnosis-related groups, and study on the efficacy of nosocomial infection control. Economic implications for hospitals under the prospective payment system. *Am. J. Med.* **78**(Suppl. 6B):3–7.

186. **Wenzel, R. P.** 1987. Epidemics—identification and management, p. 94–108. *In* R. P. Wenzel (ed.), *Prevention and Control of Nosocomial Infections*. The Williams & Wilkins Co., Baltimore.

187. **Wenzel, R. P.** 1988. The mortality of hospital-acquired bloodstream infections: need for a new vital statistic? *Int. J. Epidemiol.* **17:**225–227.

188. **Wenzel, R. P.** 1989. Hospital-acquired pneumonia: overview of the current state of the art for prevention and control. *Eur. J. Clin. Microbiol. Infect. Dis.* **8:**56–60.

189. **Wenzel, R. P.** 1992. Preoperative antibiotic prophylaxis. *N. Engl. J. Med.* **326:**337–339.

190. **Wenzel, R. P., M. D. Nettleman, R. N. Jones, and M. A. Pfaller.** 1991. Methicillin-resistant *Staphylococcus aureus*: implications for the 1990s and effective control measures. *Am. J. Med.* **91**(Suppl. 3B):221S–227S.

191. **Wenzel, R. P., R. L. Thompson, S. M. Landry, B. S. Russell, P. J. Miller, S. Ponce de Leon, and G. B. Miller, Jr.** 1983. Hospital-acquired infections in intensive care unit patients: an overview with emphasis on epidemics. *Infect. Control* **4:**371–375.

192. **Wey, S. B., M. Mori, M. A. Pfaller, R. F. Woolson, and R. P. Wenzel.** 1988. Hospital-acquired candidemia; attributable mortality and excess length of stay. *Arch. Intern. Med.* **148:**2642–2647.

193. **Wey, S. B., M. Mori, M. A. Pfaller, R. F. Woolson, and R. P. Wenzel.** 1989. Risk factors for hospital-acquired candidemia: a matched case-control study. *Arch. Intern. Med.* **149:**2349–2353.

194. **Winn, W. C., Jr.** 1993. *Legionella* and the clinical microbiologist, p. 377–392. *In* R. C. Moellering, Jr., and J. A. Washington (ed.), *Infectious Disease Clinics of North America: Laboratory Diagnosis of Infectious Diseases*. The W. B. Saunders Co., Philadelphia.

195. **Wise, R. I., E. A. Ossman, and D. R. Littlefield.** 1989. Personal reflections on nosocomial staphylococcal infections and the development of hospital surveillance. *Rev. Infect. Dis.* **11:**1005–1019.

196. **Witebsky, F. G., and P. S. Conville.** 1993. The laboratory diagnosis of mycobacterial diseases, p. 359–376. *In* R. C. Moellering, Jr., and J. A. Washington (ed.), *Infectious Disease Clinics of North America: Laboratory Diagnosis of Infectious Diseases*. The W. B. Saunders Co., Philadelphia.

197. **Wright, C., S. D. Kominos, and R. B. Yee.** 1976. Enterobacteriaceae and *Pseudomonas aeruginosa* recovered from vegetable salads. *Appl. Environ. Microbiol.* **31:**453–454.

198. **Zervos, M. J., S. Dembinski, T. Mikesel, and D. R. Schaberg.** 1986. High-level resistance to gentamicin in *Streptococcus faecalis*: risk factors and evidence for exogenous acquisition of infection. *J. Infect. Dis.* **153:**1075–1083.

199. **Zervos, M. J., C. A. Kauffman, P. M. Therasse, A. G. Bergman, T. S. Mikesell, and D. R. Schaberg.** 1987. Nosocomial infection by gentamicin-resistant *Streptococcus faecalis*. *Ann. Intern. Med.* **106:**687–691.

Infection Control Epidemiology and Clinical Microbiology

JOHN E. McGOWAN, JR., AND BEVERLY METCHOCK

16

Nosocomial infections (those acquired in the hospital or other health care facility) remain an important problem for hospitals today. Dealing effectively with such cross-infection in health care institutions (HCIs) requires identification of cases and their etiology, comparison of current attack rates of infection with usual baselines, characterization of epidemiologic features of the infections, development and implementation of control measures, and continuing follow-up surveillance to determine the success of the control measures. The microbiology laboratory has important responsibilities related to each step in this process. This chapter describes some of the basic epidemiologic concepts underlying the control of both epidemic and endemic nosocomial infections and the relationship of these to the microbiology laboratory.

INFECTION CONTROL EPIDEMIOLOGY AND ITS METHODS

Epidemiology is the study of occurrence, distribution, and determinants of disease (or health) in a specific human population (5). Hospital epidemiology studies these three features in patients at an HCI. A specific aspect of hospital epidemiology concerns itself with nosocomial infections and their determinants. By describing and defining occurrences of nosocomial infection and identifying reservoirs and sources, routes of transmission, likely victims, and associated factors, the infection control epidemiologist can develop procedures for control and prevention of these infections (2).

Nosocomial infections are those that develop within an HCI or are produced by organisms acquired during a stay in one of these facilities. Such infections may involve health care workers and visitors as well as patients. Patterns of occurrence are discussed in greater detail in chapter 15 of this Manual.

EPIDEMIC VERSUS ENDEMIC OCCURRENCE

Infection control epidemiology focuses both on epidemic occurrence of nosocomial infection and on baseline, or endemic, levels of disease in an attempt to minimize occurrence of each. Epidemics or outbreaks of infection are a major focus of concern for infection control programs. The need for HCI infection control programs arose from the staphylococcal epidemics of 1950 to 1970, and outbreaks still justify infection control efforts today.

Infections that occur as part of an epidemic often are amenable to control measures, so all clusters of infection should be considered within the scope of infection control activities (4). What defines an HCI outbreak (epidemic)? Many answers have been proposed, but one with general applicability, proposed by Dixon, defines an HCI outbreak as the occurrence of a greater-than-expected number of cases of a specific infection acquired in the institution (4). Usually, the cases occur in a brief interval and involve a specific patient population. By this definition, the baseline rate of occurrence of infections determines the presence or absence of an outbreak.

Classically, sharp and unexpected increases in number of cases makes recognition easy (13). However, this is not always the case. For example, cases of infection due to enterococci with high-level resistance to vancomycin have been extremely rare in most HCIs until recently (3). Even one case of infection due to a vancomycin-resistant enterococcus well might be defined as an outbreak, especially in view of the difficulty involved in treating infections due to these organisms (7). On the other hand, several cases of nosocomial infection due to *Escherichia coli* might not be considered an outbreak if the isolates all have the susceptibility pattern and biochemical reactions characteristic of most strains found in the HCI and if cases due to this organism are encountered frequently at the institution. In addition, increasing infection occurrence may represent changes in the number of patients being cared for by a HCI or changes in the patient population that reflect changes in average host susceptibility. Thus, not every cluster of cases represents an outbreak.

A study by the Centers for Disease Control reviewed occurrences of nosocomial outbreaks in seven community hospitals throughout the United States (6). Eight outbreaks involving 82 patients were confirmed in 1 year, and these accounted for 2% of all nosocomial infections and 0.09% of all hospital discharges. Extrapolated to the United States as a whole, this rate of occurrence suggests about one outbreak per year for the average community hospital with 150 beds or more. Half of the recognized outbreaks resolved spontaneously; two required that the hospital institute control measures, and the other two were resolved only after an outside investigation. The authors of that study also detailed some evidence that the organisms associated with outbreaks in larger hospitals may be "uncharacteristic" of epidemic organisms in small community hospitals (6).

Equally important, and occupying the attention of hospital epidemiologists for the great majority of their time, is attempting to minimize baseline, usual, or endemic rates of nosocomial infection. Endemic infections constitute the majority of preventable nosocomial infections (4). Thus, it is always useful to attempt to reduce these infection rates to zero. Some nosocomial infections are unavoidable using the techniques now available, so total prevention of such infections is an unattainable goal. Nevertheless, many can be prevented, and the focus of a hospital infection control program should be on identification of preventable infections, determining why they occur, and minimizing their occurrence (4). Study of endemic infections by epidemiologic techniques is the way to identify control and preventive measures that can have a measurable impact.

STEPS IN EPIDEMIOLOGIC EVALUATION

The fundamental approach to dealing with an HCI problem of nosocomial infection, whether epidemic or endemic, involves several steps and phases (4); the laboratory has a role in virtually all of these activities (Table 1). Seven steps will be considered (12, 16).

Problem Recognition

The initial step of realizing that a problem exists and defining its features is perhaps the most important action to be taken. Infection control personnel may become aware of the problem through their contact with and surveillance of the clinical services. On occasion, however, the laboratory, a follow-up clinic, or even another HCI in the area provides the first report. When a question arises about whether an endemic cross-infection problem is present or when an outbreak is clearly occurring, the investigators must begin by establishing the definition of a case. Even a rough definition will allow initial control measures to be taken, ensure that all involved agree on the nature of the problem, and help the laboratory begin to examine microbiologic aspects. This will prevent problems like an investigation of presumed staphylococcal toxic shock syndrome that turns out to be streptococcal scarlet fever, or vice versa. It will also allow the personnel to focus on potentially preventable infections rather than waste time and effort characterizing issues that cannot be changed (4).

Certain situations or types of infection are so dramatic or uncommon that even a single case may be recognized as a problem requiring immediate attention. For example, group A streptococcal infection in surgical or obstetric patients is rare enough today that occurrence of one such case would almost surely trigger investigation by the infection control team.

Case Finding and Surveillance

Once a definition has been made, attempts to identify all possible cases begin. This effort at complete case finding is crucial, because the more cases there are available for analysis, the better the chance of determining the process involved. Case finding has three major aspects.

Reliability

The first activity is ascertaining the reliability of both clinical and laboratory information. Can the clinical staff identify a case of the entity under investigation? Does the laboratory have experience in recognizing the microbiologic entity being considered? This is the stage at which "pseudo-outbreaks" or "pseudoproblems" (see below) need to be weeded out.

TABLE 1 Steps in investigating an HCI outbreak and the laboratory's role in each step[a]

Investigative step	Laboratory participation

1. Recognize the problem.
  a. Case definition
  b. Verification of clinical entity
	1. Laboratory should be surveillance and early-warning system.
	  a. Microbiologic confirmation

2. Complete case finding.
  a. Reliability of reporting
  b. Completeness of reporting
  c. Obtain additional data
	2. Do accurate microbiologic characterization.
	  a. Search database for other cases; review laboratory methods.
	  b. Complete demographics.
	  c. Process new cultures as needed.

3. Is this unusual?
  a. Calculate attack rate.
  b. Compare with baseline rates.
  c. Epidemic = higher-than-usual rate
	3. Provide archival data on occurrence.

4. Characterize the outbreak.
  a. Patient demography
  b. Location
  c. Time
	4. Is this one strain or many?
	  a. Type isolates (see chapter 17).

5. Form hypotheses about causes.
  a. Mode of spread
  b. Reservoirs
  c. Vectors
	5. Do environmental and/or personnel cultures as needed.

6. Initiate control measures.	6. Initiate laboratory aspects of control measures.
7. Do follow-up surveillance to ensure that control measures are effective.	7. Laboratory should be surveillance and early-warning system.

[a]Adapted in part from references 12 and 16.

Universality

Next, the completeness of reporting of cases must be considered (1, 2). Most nosocomial infections manifest themselves while the afflicted patient is still in the HCI. Some appear only after the patient has been discharged. For example, a number of studies have shown that up to half of all surgical wound infections become manifest after the patient has left the hospital (19). Can the reporting of cases be improved by special attention to surveillance measures? On occasion, systematic and rigorously applied surveillance may be needed to recognize subtle but important changes in pattern of infection, especially where polymicrobial etiology is present (4). This aspect is discussed in greater detail in chapter 15 of this Manual.

Information Collected

When intense search for all cases is completed, the completeness of the information obtained about each case must be reviewed. For example, infections with onset during HCI stay sometimes have been acquired in the community and have been incubating since HCI admission. Thus, not all infections with onset during HCI stay are nosocomial. This is a major reason why laboratory data alone cannot be used to define nosocomial infection and why surveillance programs for these infections must draw on other sources of data as well. Likewise, laboratory records may lack basic demographic information. For example, at Grady Memorial Hospital, our laboratory requisition records the ward on which the inpatient is found but not the clinical service of the patient. Thus, our laboratory would be unlikely to recognize a service-specific outbreak or endemic problem.

Defining Occurrence

When as many cases as possible have been identified, the question of whether this episode meets the definition of an epidemic can be considered. This requires two actions.

Rate of Occurrence

First, an attack rate must be calculated. Usual measures of infection occurrence in nosocomial infection studies are incidence or prevalence rates (2). An incidence rate is the ratio of the number of new cases of an entity (in this case, nosocomial infection) occurring in a specified population at risk of infection in a defined time period (the numerator) to the number of persons in the population (the denominator). By contrast, prevalence rate is the total number of cases of nosocomial infection in the defined population at risk at one point in time (point prevalence) or in a given interval (period prevalence). In each of these definitions, the number of episodes of interest (here, nosocomial infection) is the numerator of a fraction in which the denominator is the number of persons at risk for occurrence of the numerator event. This fraction can be adjusted for a specific characteristic of interest. For example, in a study of the occurrence of infections associated with intravascular catheters, the denominator might be the number of days such a catheter was in place or the number of patients in whom a catheter was present, depending on the focus of the study.

Comparison with Baseline

After these calculations, the attack rate of current cases is compared with the usual, baseline occurrence of the entity at the institution. The presence of an attack rate that is higher than usual then confirms that an outbreak is occurring. Note that situations may exist in which no clear increase in cases is noted from surveillance data: rates of

TABLE 2 Types of epidemiologic studies[a]

A. Descriptive epidemiology
B. Analytic epidemiology
  1. Observational studies
    a. Case-control methods
    b. Cohort methods (both prospective and retrospective)
  2. Interventional studies (clinical trials)

[a]Adapted in part from reference 8.

infection may have been excessive for extended periods, so that baseline infection rates themselves constitute a problem worthy of attention (4).

Characterizing the Episode

Description of Features

Framing the problem in terms of its location, time, persons, procedures, instruments, or other features is the next step in the systematic approach (4). The time course of an infection often is shown in the form of an epidemic curve, which is a graph of the number of cases of disease in relation to the time of onset of the nosocomial infection. This allows "eyeball" evaluation of periods before the episode in question (baseline periods) with the time the problem appeared. Such observation may be all that is needed to determine whether an outbreak exists (see next section). In addition, the characteristics of the curve itself often suggest the means by which the organism has been spread (4). Among the time trends to consider are secular (long-term) patterns, seasonal changes (important for some organisms), and acute changes from baseline (clustering) (2, 13). Location within the HCI also should be evaluated, and this analysis by place also can suggest the mode of spread of the organism. For example, HCI-acquired cases of tuberculosis in patients widely spread through the institution may suggest an airborne mode of transmission through the heating-cooling-air conditioning system of the HCI. Finally, evaluation of the demographic and clinical characteristics of the infected patients ("evaluation of person") may allow discovery of common features of the infected patients that those without infection do not share. Age, gender, type of underlying disease, prior therapy, and other such factors may be relevant or irrelevant in different situations. Any host factor that can influence the development of disease must be considered; those that are found to increase the chance of infection are defined as risk factors (2).

In the presence of enough cases, description of epidemiologic patterns may allow the problem to be understood. However, determining the important factors often requires the use of some of the analytic techniques of epidemiology that are used in other kinds of biomedical research as well. Analytic studies can be observational or interventional (Table 2). These methods more often than not involve use of appropriate control groups.

Observational Studies

Observational studies review and follow the natural course of events and are the type most frequently employed in investigations of clusters of hospital infections (8). Observational studies commonly are of two different types: case-control and cohort (Table 2). In a case-control study, a group of patients who have a nosocomial infection of interest are compared with a control (comparison) group of patients who do not have the nosocomial infection. The proportions of patients with the attribute or exposure of interest in the two groups are compared in order to identify differences that might suggest why infection occurred. Case-control studies are relatively inexpensive, quick, and reproducible compared to other options (2). This technique is used most frequently in acute investigations, such as suspected hospital outbreaks. Disadvantages of case-control studies include difficulty in choosing an appropriate control group and ruling out confounding variables. The cohort study takes a different approach. Subjects are classified as to presence or absence of some study subject of interest, and then both groups (those with and those without the feature) are monitored during their HCI stay to determine which develop nosocomial infection. The aim here also is to identify study subjects that vary between those developing and those free of nosocomial infection. Cohort studies can be retrospective (the outcome of interest has already occurred when the study is begun) or prospective (the outcome has not yet occurred when the study starts, so a period of follow-up will be involved). An advantage of cohort studies is that they permit direct estimation of the infection risk associated with a particular exposure, which cannot be done with case-control studies. On the other hand, cohort studies usually are more difficult and more expensive, and they take longer to complete than case-control methods (2). Thus, each of these types of studies has benefits and drawbacks in terms of practicality in the HCI setting, cost, and type of results that can be obtained.

Interventional Analytic Studies

Clinical trials, or interventional studies, also can be conducted in infection control epidemiology (Table 2). Here the investigators assign the exposure of given study participants to the subject of interest. Thus, one group may receive therapy through a new type of intravascular catheter while others receive it through a current standard catheter, with the choice of catheter for each patient left to the protocol of the investigator. Following patients' clinical courses to identify the occurrence of infection in each group then allows an inference to be made about the value of the intervention (here, of the catheter) as a preventive measure. Since observational studies provide no control over the interactions of different factors, it is often difficult to sort out the effect of interest from other background influences (confounders). Interventional studies offer the opportunity to make this discrimination. Again, however, this type of study method has unique advantages and disadvantages, which are discussed elsewhere (8, 18).

Developing and Testing Hypotheses

Forming postulates about the reasons for the outbreak, the reservoir of the organisms, and the mode of spread of the organism(s) to the patient is the goal of the next step in the process. Establishing these characteristics may require decisions about factors causing the episode. In developing hypotheses, the medical literature is often a good place to start: other institutions may have had similar problems with a given organism and identified the reservoir and mode of spread.

Reservoir, Source, and Mode of Spread

The reservoir of an infecting agent is the place where the organism maintains its presence in the hospital setting. The source is the place from which the infectious agent directly contacts its victim. Reservoir and source may be identical, or they may differ. For example, in an outbreak of *Pseudomonas aeruginosa* infections in dialysis outpatients at our hospital, the source of transmission to patients was the small pour bottles of iodophor antiseptic in the dialysis unit, but the reservoir of the organisms was contaminated stocks of antiseptic that were received from the manufacturer already colonized with the organism (17).

Mode of spread is the means by which the organism moves from the source to the infection victim. There are several important modes of spread; only some of these are relevant to nosocomial transmission (Table 3) (2). The most common mode of spread of HCI infectious agents is from person to person on the hands of personnel. This exemplifies indirect contact through the participation of an intermediary, in this case a health care worker. Other indirect modes of spread include instruments (e.g., an endoscope used on patient 1, inadequately decontaminated, and then used on patient 2). Direct contact between the person or environmental object with the infecting organism and the one who acquires the organism is also involved in hospital infection, and droplet transmission directly to a victim is possible as well. A special type of contact spread is a so-called common vehicle spread, where more than one person acquires the same organism(s) from a single source. Airborne transmission of infections such as tuberculosis and measles is of concern in the modern hospital, whereas vector-borne spread is rare for nosocomial infections. The distinction between direct droplet spread and airborne transmission is one of distance: droplet spread occurs by travel of droplets within a short distance, whereas airborne particles or droplet nuclei may be transmitted far beyond the physical presence of the source patient. In the HCI setting, organisms may have more than one mode of spread.

TABLE 3 Modes of spread, with examples from nosocomial infection[a]

Type of spread	Example
Indirect contact	Patient to patient on hands of health care workers; patient to patient on inadequately disinfected endoscope
Direct contact	Contaminated iodophor; contaminated food
Droplet contact	Measles
Airborne	Tuberculosis
Endogenous	Cytomegalovirus reactivation in transplant patient; mother colonized with group B streptococcus becomes infected at time of delivery
Vector-borne	Rare in hospital setting in developed countries

[a]Adapted in part from references 2 and 5.

For example, *Salmonella* infection may be foodborne, may contaminate a gastrointestinal endoscope, or may be transmitted by fecal contamination of hands of health care workers or patients (2). This versatility of the organism often complicates investigation of HCI infection episodes.

The need to distinguish the reservoirs of infection and the transmission characteristics of an episode underlies one of the major differences between the epidemiologist and the clinician in the use of microbiologic data. Colonization is the presence of a microorganism that is not associated with clinical illness. This contrasts with infection, in which the organism is producing illness. The clinician is concerned with patient welfare and so usually disregards positive cultures or other microbiologic data indicating colonization rather than infection. To the epidemiologist, however, colonization and infection can be equally helpful in tracking the movement of an organism from one HCI area to another (16).

Chance, Bias, and Confounding

Testing epidemiologic hypotheses involves deciding whether there is an association between a particular risk factor and the occurrence of nosocomial infection. A causal association is one in which change in the frequency or quantity of an exposure leads to a corresponding change in the occurrence of hospital infection (8). Associations indicate only that there is a statistical link (either positive or negative) between the presence of exposure and the occurrence of infection. This apparent connection, however, may exist only because of chance association, bias, or confounding (8). Chance association can exist when a study population provides results unlike that of an entire population. Random variation from usual behavior can produce results that appear to provide a link when there is none and can also suggest a lack of association when one truly exists. Statistical methods to estimate the probability of either of these two incorrect interpretations of epidemiologic data have been developed (18). A second factor that can interfere with recognition of associations is bias. This is a systematic error in the way that study subjects were chosen, that information was obtained from or about the subjects, or that study data were reported. Errors often occur in data collection or transcription. Epidemiologic bias is present when the errors apply differently to those with infection and those without or to those with the exposure of interest and those without. Even when systematic errors in collection or reporting of data apply equally to each study group, the result is to underestimate the degree of association between exposure and infection occurrence. Efforts to avoid bias need to be considered at the time of study design as well as at the time of analysis. The third risk in determining associations is confounding. This is present when an apparent association or lack of association is really due to the impact of a third factor that is associated with both the exposure and the infection occurrence but is independent of each. Each of these factors and methods to minimize their impact are discussed in detail elsewhere (5, 8).

Establishing Causation

Even when chance, bias, and confounding can be ruled out, the presence of a valid association does not demonstrate a cause-and-effect relationship between exposure and occurrence of infection. It is, however, one of the features of an episode that allows a judgment by the epidemiologist as to whether a causal relationship exists. Causation remains a matter of belief or judgment based on available evidence.

Unfortunately, in a field as complicated as infection control epidemiology, it is often difficult to define a clear causal relationship from observational studies. For example, nosocomial pneumonia may be associated with recent chest surgery, and this link may persist after chance, bias, and confounding are examined and ruled out. Here, however, the association merely indicates that something related to the surgery increases the risk of pneumonia, not that surgery itself causes the infection (4). In these situations, other evidence, such as the results of experimental studies (clinical trials), often is needed to establish a causal link. Thus, observational studies can be of great value in identifying the exposures, control methods, or equipment for which clinical trials that are likely to show causation may be designed.

Control Activities

Defining the reservoir, mode of spread, and associated risk factors usually points to certain control measures that can be taken to stop the progress of the epidemic or other problem under study. For example, if the data suggest that the air conditioning system of the operating suite is contaminated with the problem organism, the actions needed to minimize this problem focus on improving air cleanliness rather than on hand washing. Actions to control the outbreak now can be based on the conclusions of the analysis. At this point, new steps should be implemented, and any control efforts made earlier in the investigation should be revised as needed.

Infection control measures for the HCI usually focus on altering or removing one of the three elements requisite to infection: a pathogenic microorganism, a susceptible victim, and a mode of transmission of the organism to its victim (4). This explains the need for the development and testing of hypotheses in the steps above. The epidemiologic investigation attempts not only to characterize the nature of any and all of these determinants but also to decide which factor or factors it will be most practical to alter.

Follow-Up Surveillance

New data must be collected to make sure that the control measures achieved the desired effect. In this step, the laboratory retains a prominent role as a potential early-warning system for recurrent cases if control measures are ineffective.

ROLE OF THE LABORATORY IN EPIDEMIOLOGIC EVALUATIONS

The laboratory has an important role in each aspect of epidemiologic investigation listed in Table 1 (15).

Organism Identification

Perhaps the most important role for the laboratory is accurate identification and susceptibility testing of suspected outbreak organisms. The spectrum of organisms causing episodes or outbreaks of nosocomial infection has changed dramatically in the recent past. Still, most organisms causing outbreaks or special endemic problems of nosocomial infection can be identified in the hospital laboratory. Identification of the etiology of an infection problem can depend on the extent to which the laboratory characterizes the organism(s) in question both routinely and in special outbreak circumstances. Fortunately, technologic developments in the laboratory have increased our ability to keep pace with the changes in etiology and resistance to anti-microbial agents of HCI pathogens (15).

Some HCIs do not have the resources to identify certain organisms to species level in routine practice. For epidemiologic investigations, these institutions may have to begin to identify relevant isolates to species level or to retrieve relevant strains for identification as circumstances demand. At a minimum, the laboratory should be capable of identifying gram-positive cocci and gram-negative aerobic bacilli to the species level when special or recurring cross-infection problems make this necessary (16).

Storage of Data and Isolates

Archival information must be maintained by the laboratory to establish the baseline of occurrence of infections due to the organisms in question. Ironically, archival information required for epidemiologic investigation may be harder to retrieve now than before the advent of electronic data processing systems, as information on biochemical reactions, etc., traditionally saved on the laboratory worksheets may not be part of the record maintained on the modern laboratory information system (12).

Retrospective testing of strains is possible only when isolates of possible epidemiologic importance have been saved systematically by the laboratory. The resources to do this vary markedly from HCI to HCI, so it well might be that prior isolates are not always available when further testing is needed. Decisions about the period for which isolates should be saved routinely and in special investigative situations should be discussed with infection control personnel on a regular basis (15). The source of the isolate is an important factor in making this decision; for example, isolates from blood may be held longer than those from urine specimens. Likewise, the organism will influence this choice; for example, isolates of *Mycobacterium tuberculosis* would be more desirable for storage than most routine isolates of *E. coli*.

Timely Reporting

A crucial responsibility of the laboratory is to serve as an early-warning system for infection control problems (14). Whenever it is apparent that a problem exists, all relevant organisms should be reported to infection control personnel as soon as their isolation is recognized. At this point, arrangements should be made to save all potentially relevant organisms at least for the duration of the investigation and perhaps longer if storage facilities allow.

Supplementary Cultures

Problem investigation may require identification not only of patients but also of personnel who may be colonized with the study strain. This usually involves processing supplementary cultures of patients, personnel, or the environment. Special techniques may be required to accomplish such projects. For example, reliable detection of *Salmonella* carriage may require enhancement of growth by use of selective media. The laboratory must make careful assessment of the sites to be cultured and of the culture media and techniques to be employed (16). Likewise, different approaches may be required for processing environmental cultures to help determine or confirm the mode of spread. While environmental cultures are not recommended on a routine basis in the HCI setting, specific culture surveys to determine mode of spread in known cases of patient illness are a major contribution in many epidemiologic investigations (4). These environmental or personnel cultures

should be considered one of the costs of the HCI's infection control program and should not be billed to the individual patients affected by the problem strain. Specific culture techniques for a wide variety of inanimate items and substances have been described elsewhere (16).

Microbiologic Similarity

Determining the features of the epidemiologic problem or testing certain hypotheses about reservoir or mode of spread often requires the laboratory to define whether the outbreak strains are related or unrelated to each other. Such a determination involves conducting microbiologic studies to establish similarity or differences among isolates. To achieve this, a number of typing systems have been designed to determine organism similarity or difference; this aspect is discussed in detail in chapter 17 of this Manual.

POTENTIAL PROBLEMS RELATED TO LABORATORY ACTIVITIES IN EPIDEMIOLOGIC INVESTIGATIONS

The importance of the laboratory role in epidemiology is demonstrated by the success of investigations in which laboratory work is appropriate. It is equally well emphasized when laboratory problems lead to difficulties in characterizing a problem or potential problem in an HCI. Some of the commonest problems in laboratory support for epidemiologic activities are considered below (12).

Misdiagnosis of Outbreak ("Pseudo-Outbreak")

On occasion, the existence of an epidemic is postulated when none is actually present (10). These episodes occur when one has a clustering of apparent infections in which infection is not present or when the pattern of infections gives a false appearance of clustering.

Spurious outbreaks of nosocomial infection have been traced to a variety of sources (Table 4). For example, at our hospital, a pseudo-outbreak of *Serratia marcescens* bacteremia eventually was attributed to transfer of blood from unsterile specimen containers to blood culture collection vials (9). Another source of special importance to the laboratory is inaccurate or inconsistent microbiologic procedures. For example, 23 cases of *Legionella* infection in a single hospital apparently resulted from use of a new probe technique for diagnosis of *Legionella* infection. This test had poor specificity, and many false-positive tests were recorded (11).

Pseudo-outbreaks can take a considerable amount of time to identify. On occasion, someone in the laboratory finds a problem in a technique or procedure that brings the pseudo-outbreak to light. Usually, however, it is the clinician who recognizes the problem by noting a great disparity between the patient's clinical status and the laboratory results.

Inadequate Quality Control for Identification, Susceptibility Testing, and Typing Procedures

Organisms associated with infection control problems rarely are all isolated at the same time. Thus, laboratory characterization procedures often will be performed at different times. The need for control of batch-to-batch variation in media, phages, etc., is especially crucial with many procedures for identification, susceptibility testing, and typing. This fact explains why many of the procedures are not performed in the usual clinical laboratory in a timely fash-

TABLE 4 Sources of pseudo-outbreaks of nosocomial infection[a]

A. Related to clinical entity
1. Incorrect diagnosis of clinical entity
2. Positive cultures represent colonization rather than infection
3. Failure to distinguish community-acquired infection from nosocomial onset
B. Related to laboratory
1. Contamination during specimen collection
2. Contamination during specimen transport
3. Contamination during processing
a. Media
b. Equipment
4. Use of inadequate method or technique
C. Related to case finding
1. Increased surveillance efficiency
2. Improved laboratory techniques for identification
D. Related to chance clustering

[a]Adapted in part from references 12 and 15.

ion. In some cases, one may need to go back and identify or type isolates recovered at different periods on the same media, etc., to remove confounding variables of the testing process.

Needless Culturing, Organism Identification, or Susceptibility Testing

The overuse of supplementary culturing wastes valuable laboratory resources. It is important for the laboratory to avoid going overboard on the use of environmental cultures, personnel cultures, and organism identification and susceptibility testing. For example, use of 20 different antibiotics in situations in which one defines organisms as different if their pattern of susceptibility to these drugs varies by more than one drug almost guarantees that organisms that actually are part of the outbreak will be identified as different and thus considered unrelated.

Overuse or Overinterpretation of Typing Techniques

Typing procedures represent a great advance in the ability to characterize an outbreak. Nevertheless, certain problems are associated with their use. Interpretation is a potential pitfall; on occasion, typing is interpreted to show that organisms are identical. In reality, however, these procedures rarely can prove that organisms are identical. Instead, they usually show only that organisms are not different by the typing method(s) used. This considerable distinction focuses attention on the possibility that similar reactions can be obtained by chance alone.

One must be selective in the use of typing systems, even when an outbreak occurs. In some outbreaks, control measures can resolve the problem before or even without obtaining microbiologic data or setting up an epidemiologic investigation (15). In other situations, use of simple marker systems such as biotyping and antibiotic resistance patterns sufficiently characterizes the situation that no further testing is needed. Use of more complicated and/or less readily available typing systems should be reserved for situations in which control measures fail or in which organism or infection occurrence is being studied for academic reasons. Deal-

ing with these aspects of typing is described in detail in chapter 17 of this Manual.

CONCLUSION

Epidemiologic methods permit the efficient evaluation of many different aspects of nosocomial infection control as well as of other aspects of care in the health care setting. The steps of an epidemiologic assessment are aimed at prevention and control efforts and allow assessment of the most efficient as well as the most effective ways to achieve prevention and control. The laboratory plays a major role in epidemiologic attempts to deal with endemic nosocomial infection and faces exceptional demands for service at the beginning of and throughout a period of epidemiologic study. These activities are essential in the fight against outbreaks of HCI infection. An understanding of the usual methods and approaches of the epidemiologist can help the clinical microbiologist provide more effective support for hospital infection control activities. The investigations must be carried out rapidly and effectively yet must be as efficient and cost-effective as possible. For all of these attributes to be blended, the microbiologist must prepare in advance for the various diagnostic tasks, whether performed on site or by referral to other laboratories. After the investigation begins, it is too late to initiate a plan for the laboratory's response; contingency plans must be made for dealing with the types of cross-infection problems that have occurred most often in the past in the institution so that exceptional requests are dealt with smoothly and accurately. Fortunately, the past decade has brought improvements in epidemiologic methods and in laboratory procedures to aid infection control efforts.

REFERENCES

1. **Bartlett, C. L.** 1987. Efficacy of different surveillance systems in detecting hospital-acquired infections. *Chemioterapia* **6:**152–155.
2. **Brachman, P. S.** 1992. Epidemiology of nosocomial infections, p. 3–20. *In* J. V. Bennett and P. S. Brachman (ed.), *Hospital Infections*, 3rd ed. Little, Brown & Co., Boston.
3. **Centers for Disease Control and Prevention.** 1993. Nosocomial enterococci resistant to vancomycin—United States, 1989–1993. *Morbid. Mortal. Weekly Rep.* **42:**597–599.
4. **Dixon, R. E.** 1992. Investigation of endemic and epidemic nosocomial infections, p. 109–133. *In* J. V. Bennett and P. S. Brachman (ed.), *Hospital Infections*, 3rd ed. Little, Brown & Co., Boston.
5. **Greenberg, R. S., S. R. Daniels, W. D. Flanders, J. W. Eley, and J. R. Boring.** 1993. *Medical Epidemiology.* Appleton & Lange, Norwalk, Conn.
6. **Haley, R. W., J. H. Tenney, J. O. Lindsey, J. S. Garner, and J. V. Bennett.** 1985. How frequent are outbreaks of nosocomial infection in community hospitals? *Infect. Control* **6:**233–236.
7. **Handwerger, S., B. Raucher, D. Altarac, J. Monka, S. Marchione, K. V. Singh, B. E. Murray, J. Wolff, and B. Walters.** 1993. Nosocomial outbreak due to *Enterococcus faecium* highly resistant to vancomycin, penicillin, and gentamicin. *Clin. Infect. Dis.* **16:**750–755.
8. **Hennekens, C. H., and J. E. Buring.** 1987. *Epidemiology in Medicine.* Little, Brown & Co., Boston.
9. **Hoffman, P. C., P. M. Arnow, D. A. Goldmann, P. L. Parrott, W. E. Stamm, and J. E. McGowan, Jr.** 1976. False-positive blood cultures—association with nonsterile blood collection tubes. *JAMA* **236:**2073–2075.
10. **Kusek, J. W.** 1981. Nosocomial pseudoepidemics and pseudo-infections: an increasing problem. *Am. J. Infect. Control* **9:**70–75.

11. **Laussucq, S., D. Schuster, W. J. Alexander, W. L. Thacker, H. W. Wilkinson, and J. S. Spika.** 1988. False-positive DNA probe test for *Legionella* species associated with a cluster of respiratory illness. *J. Clin. Microbiol.* **26:**1442–1444.

12. **McGowan, J. E., Jr.** 1990. Laboratory approach to an outbreak of nosocomial infection: systems and techniques for investigation, p. 21–31. *In* K. R. Cundy, B. Kleger, E. Hinks, and L. A. Miller (ed.), *Infection Control: Dilemmas and Practical Solutions.* Plenum Press, New York.

13. **McGowan, J. E., Jr.** 1991. Abrupt changes in antibiotic resistance. *J. Hosp. Infect.* **18**(Suppl. A)**:**202–210.

14. **McGowan, J. E., Jr.** 1991. Communication with hospital staff, p. 151–158. *In* A. Balows, W. J. Hausler, Jr., K. L. Herrmann, H. D. Isenberg, and H. J. Shadomy (ed.), *Manual of Clinical Microbiology,* 5th ed. American Society for Microbiology, Washington, D.C.

15. **McGowan, J. E., Jr.** 1991. New laboratory techniques for hospital infection control. *Am. J. Med.* **91**(Suppl. 3B)**:**245S–251S.

16. **McGowan, J. E., Jr., and R. A. Weinstein.** 1992. The role of the laboratory in control of nosocomial infection, p. 187–220. *In* J. V. Bennett and P. S. Brachman (ed.), *Hospital Infections,* 3rd ed. Little, Brown & Co., Boston.

17. **Parrott, P. L., P. M. Terry, E. N. Whitworth, L. W. Frawley, R. S. Coble, I. K. Wachsmuth, and J. E. McGowan, Jr.** 1982. *Pseudomonas aeruginosa* peritonitis associated with a contaminated poloxamer-iodine solution. *Lancet* **ii:**683–685.

18. **Rothman, K. J.** 1986. *Modern Epidemiology.* Little, Brown & Co., Boston.

19. **Weigelt, J., R. Haley, and G. Siebert.** 1987. Necessity and efficiency of wound infection surveillance after discharge. *Am. J. Infect. Control* **16:**75.

Laboratory Procedures for the Epidemiologic Analysis of Microorganisms

ROBERT D. ARBEIT

17

INTRODUCTION

A considerable portion of this Manual is devoted to the process of identifying the genera and species of microorganisms (bacteria, mycobacteria, and fungi) isolated from clinical specimens. The focus of this chapter is the process of analyzing multiple isolates within a given species to determine whether they represent a single strain or multiple strains. Such analyses are useful in several clinical settings. Infection control practitioners frequently solicit laboratory support to complement intensive epidemiologic investigations, and they increasingly request microbiologic studies to help determine whether to initiate such investigations or how to focus them. Clinicians may request that isolates from an individual patient be examined to help determine whether a set of isolates represents a single infecting strain or multiple contaminants or whether a series of isolates obtained over time represents relapse of an infection due to a single strain or separate episodes of disease due to different strains. In research settings, the detailed analysis of multiple isolates can contribute to developing new insights into both the epidemiology and the pathogenesis of infection.

Strong basic clinical microbiology practices are prerequisite to further typing studies. The technologists at the bench are often the first to detect a cluster of isolates with a distinctive colonial morphology or biotype or to notice the increased prevalence of a particular species. Their observations should be encouraged and followed up on by laboratory supervisors and directors. The identification of isolates to species level and antimicrobial susceptibility testing should be completed promptly and accurately. Contemporary computer systems can greatly facilitate access to these laboratory reports and can often be programmed to alert hospital staff to the appearance of a new species or antibiotic susceptibility pattern and to provide basic routine epidemiologic analyses. Frequent, open communication among the laboratory personnel, infection control practitioners, and infectious disease clinicians is obviously critical. With the increasing sophistication and complexity of microbial typing studies, an active collaboration is required in order to interpret study results appropriately and apply them effectively in the clinical realm.

Theoretical Considerations

The central hypothesis motivating typing studies is that the isolates in a series obtained from an epidemiologic cluster or during the course of an infection in a single patient are clonally related, that is, are directly descended from a common precursor. Generations of microbiologists have observed that epidemiologically unrelated isolates of the same species often differ in several characteristics. Typing systems are based on the premise that clonally related isolates share characteristics by which they can be differentiated from unrelated isolates. In this chapter, the term isolate will be used to refer to a single colony, which is presumed to arise from a single organism. A strain is a set of isolates that, as analyzed by a typing system, are indistinguishable from each other and can be differentiated from other isolates; such isolates are thereby inferred to be clonally related to each other. (A single isolate with a distinctive characteristic[s] may also represent a strain.) The utility of a particular characteristic for typing is related to its stability within a strain and its diversity within the species. Note that the definition of a strain is operationally dependent on the resolution of the typing system, and as discussed below, clonality is more usefully considered a relative concept than an absolute condition (41).

Recent studies of microbial populations have demonstrated substantial genetic diversity within microbial species and indicate that isolates are distributed among many genetically diverging lineages. This evolutionary divergence reflects the accumulation of random, nonlethal mutations, including single-base-pair substitutions, the deletion of individual genes, or even the acquisition of DNA from other microbial species. With the development of highly sensitive molecular techniques, subtle alterations can be precisely detected, and the mechanisms underlying complex variations can be resolved.

Nevertheless, the analysis of clinically relevant isolates remains a substantial challenge. Because of the powerful selective pressure on critical microbial genetic elements, such as virulence factors and antimicrobial resistance determinants, the human pathogens within a species may represent only a subset of the genotypes within the species as a whole, which can include clinical, commensal, animal, and environmental strains. For example, the genetic diversity among invasive isolates of *Escherichia coli* cultured from patients with pyelonephritis or bacteremia is relatively limited compared to that among isolates colonizing the intestinal tract (129). Pathogenic isolates are characterized by virulence factors that enhance colonization, persistence, and invasion at the infecting site; such virulence determinants are absent from most commensal (i.e., fecal) isolates (13, 14, 80). Isolates of methicillin-resistant *Staphylococcus aureus* (MRSA) from diverse geographic locations appear to be derived from one or a very few precursor strains and consequently are genetically restricted compared to the diversity among methicillin-susceptible *S. aureus* isolates (66, 97). Thus, the most clinically relevant isolates—those with increased virulence or resistance—are often the most difficult to differentiate unambiguously.

Criteria for Evaluating Typing Systems

There is no "gold standard" by which to judge a typing method or to consistently define "true-positive" or "true-negative" results. Consequently, the terminology usually used to describe the operating characteristics of laboratory tests (e.g., sensitivity, specificity) is not strictly applicable, and the results of a typing system must be considered relative to the available epidemiologic data or to the results of other systems. Several criteria are useful in evaluating typing systems (Table 1). Typeability refers to the ability to obtain an unambiguous positive result for each isolate analyzed; nontypeable isolates are those that give either a null or an uninterpretable result. Reproducibility refers to the ability of a technique to yield the same result when the same strain is tested repeatedly. Reproducibility is influenced by both technical and biologic factors; the former may be reflected in run-to-run variations among replicate aliquots of a single isolate. In addition, biologic variation in the bacterial characteristic being examined may produce differences among independent isolates representing the same strain. Discriminatory power refers to the ability to differentiate among unrelated strains. Ideally, each unrelated isolate is detected as unique. In practice, a technique is statistically useful when the most commonly detected type represents less than 5% of the population (57, 58). For techniques that assess multiple individual characteristics (e.g., bacteriophage typing and electrophoretic methods), discriminatory power typically diminishes with decreasing reproducibility, since unrelated strains must differ in a greater number of component characteristics in order to be reliably differentiated (57).

The classic techniques for differentiating isolates (biotyping, antibiotic susceptibility patterns, serotyping, and bacteriophage typing) were based on the presence or absence of metabolic or biologic activities as expressed by whole organisms. Interpretation of such results is relatively straightforward. More recently, electrophoretic analysis of various molecular subcomponents (e.g., proteins, metabolic enzymes, plasmid DNA, chromosomal DNA, DNA produced by the PCR) has been shown to be a powerful tool for differentiating strains. Electrophoretic results are typically patterns of "bands," with each band representing a discrete bacterial product or DNA fragment. Such patterns may be extremely complex, subject to considerable technical variation, and thus difficult to analyze using logical, objective,

readily applied criteria. Consequently, ease of interpretation has become an important criterion for evaluating typing systems. Ease of performance is also critical. To be widely useful, a typing method should be applicable to a broad range of microorganisms as well as inexpensive and technically accessible, i.e., not requiring special expertise or expensive equipment or reagents. Results should be available rapidly enough to be relevant to patient management or infection control. At this time, no single typing system is optimal by all of these criteria, and no one approach is preferred for all clinical settings or infecting species.

Typing systems are often proposed or applied without adequate knowledge of their effectiveness for a broad sample of isolates. It is clearly insufficient to evaluate a typing system by comparing the results for a single well-defined outbreak with those for a handful of epidemiologically unrelated "control" strains. Almost any system has sufficient reproducibility and discriminatory power to indicate that the outbreak isolates are more similar to each other than to a few random unrelated isolates. In the absence of a gold standard, two typing systems can be formally compared only if both have been applied to the same set of isolates. Thus, rigorous evaluation of a typing method involves analyzing adequate numbers of epidemic and sporadic isolates and comparing the results directly with those of previously well-studied approaches. In determining discriminatory power, it is particularly informative to examine epidemiologically unrelated isolates that have proven to be indistinguishable by other techniques.

Classification of Typing Methods

A convenient way of classifying typing systems is to divide them into phenotypic techniques, i.e., those that detect characteristics expressed by the microorganisms, and genotypic techniques, i.e., those that involve direct DNA-based analyses of chromosomal or extrachromosomal genetic elements.

PHENOTYPIC TECHNIQUES

Typing methods that assess phenotypic differences (Table 2) are inherently limited by the capacity of microorganisms to alter the expression of the underlying genes (154). Such changes may occur unpredictably or in response to various environmental stimuli (89). In addition, point mutations representing a single nucleotide in the entire chromosome can result in the abnormal regulation or function of the gene responsible for a particular phenotype. Thus, isolates that represent the same strain and are genetically indistinguishable (or almost so) can vary in the phenotype detected. Some phenotypic approaches, such as serotyping and bacteriophage typing, require specific reagents for detecting individual types. Since the available materials may not be appropriate for all strains, a relatively large fraction of strains may give a null phenotype and consequently be nontypeable.

Biotyping

Biotyping makes use of the pattern of metabolic activities expressed by an isolate and may include specific biochemical reactions, colonial morphology, and environmental tolerances (e.g., the ability to grow on certain media or at extremes of pH or temperature). Such characteristics have classically been used for taxonomy. In most clinical microbiology laboratories, biotyping is now routinely and reliably performed by using automated systems designed for species

TABLE 1 Criteria for evaluating typing systems

Typeability
Reproducibility
Discriminatory power
Ease of interpretation
Ease of performance

TABLE 2 Characteristics of phenotypic typing systems[a]

Typing system	Proportion of strains typeable	Reproducibility	Discriminatory power	Ease of interpretation	Ease of performance
Biotyping	All	Poor	Poor	Excellent	Excellent
Antimicrobial susceptibility	All	Fair	Poor	Excellent	Excellent
Serotyping	Most	Good	Fair	Good	Fair
Bacteriophage typing	Most	Fair	Fair	Fair	Poor
Immunoblotting	All	Good	Good	Fair	Good
MLEE	All	Excellent	Good	Excellent	Good

[a]These judgments represent the views of the author; many systems remain incompletely evaluated, and characteristics may vary as applied to different species. See the text for details.

identification. The detection of multiple isolates of an unusual species can effectively indicate an outbreak (77), and occasionally, epidemic strains of common bacterial species exhibit distinctive biotypes (109). In general, however, biotyping has only limited ability to differentiate among strains within a species, and consequently, the technique has relatively poor discriminatory power.

Although variations in gene expression are the most common reason for isolates that represent the same strain to differ in one or more biochemical reactions, random mutations may also confound the interpretation of these data. One instructive example involves two colonies of *Klebsiella* spp. that were isolated from the same urine sample and that differed only in their abilities to produce indole from tryptophan (78). Since that metabolic reaction is the basis for differentiating *Klebsiella oxytoca* (indole positive) from *Klebsiella pneumoniae* (indole negative), the two isolates were considered different species. However, on detailed examination, the isolates were otherwise phenotypically and genotypically identical, suggesting that they were simply variants derived from a single clone. Although this is an extreme example, it illustrates both the limited utility of biotyping in epidemiologic studies and the hazard of making any identification system critically dependent upon a single characteristic.

Antimicrobial Susceptibility Testing

Clinical microbiology laboratories routinely test most bacterial isolates for susceptibility to a panel of antimicrobial agents. Both manual and automated methods are widely available, rigorously quality controlled, typically easily performed, and relatively inexpensive. These data are readily available to clinicians and infection control practitioners. The identification of a new or unusual pattern of antibiotic resistance among isolates cultured from multiple patients is often the first indication of an outbreak.

For many epidemiologic studies, however, antibiotic susceptibility testing has relatively limited utility not only because phenotypes vary but also because antibiotic resistance is under extraordinary selective pressure in contemporary hospitals. There are multiple genetic mechanisms by which a given strain can become abruptly resistant to a particular antibiotic. These include spontaneous point mutations (e.g., quinolone resistance in *S. aureus* [49]) and the acquisition of specific resistance genes via plasmids and transposons from other strains or even other species (e.g., the transfer of aminoglycoside-modifying enzymes among members of the family Enterobacteriaceae [86]). Since a single plasmid or compound transposon can carry multiple resistance determinants, resistance to multiple antimicrobial agents may be acquired simultaneously. On the other

hand, in the absence of specific selective pressure, such elements may be lost (41, 86, 90). As a consequence of these various genetic mechanisms, different strains may develop similar resistance patterns and, conversely, the susceptibility patterns of sequential clinical isolates representing the same strain may differ for one or more antibiotics (90, 142).

Serotyping

Serologic typing (serotyping) is based on the long-standing observation that microorganisms of the same species can differ in the antigenic determinants expressed on the cell surface. Many different surface structures, including lipopolysaccharides, capsular polysaccharides, membrane proteins, and extracellular organelles (e.g., flagella and fimbriae), exhibit such antigenic variation. Serotyping has been applied to numerous species, including both grampositive (e.g., *Streptococcus pneumoniae*) and gram-negative (e.g., *E. coli*, *Haemophilus influenzae*, and *Neisseria meningitidis*) organisms, and has been one of the classic tools for epidemiologic studies. Some distinct virulence factors and related clinical syndromes are linked to particular serotypes within a species. For example, infection with *E. coli* O157:H7 isolates is associated with the hemolytic-uremic syndrome (156).

For some organisms, such as pneumococci, legionellae, salmonellae, and shigellae, serotyping remains a primary means of evaluating isolates. The available rapid, reliable serologic assays for detecting antigen-antibody reactions use a wide variety of methods, including direct agglutination, latex coagglutination, enzyme labeling, and fluorescence labeling. The application of monoclonal antibody technology has enhanced the reproducibility of such assays. Nevertheless, as a general tool for detailed epidemiologic analyses, serotyping is subject to several critical limitations. If high-quality commercial reagents are not available, the preparation of specific typing antibodies is a difficult process and is generally restricted to reference and research laboratories. Serotyping often exhibits poor discriminatory power, even for bacterial species with large numbers of antigenic variants, because many strains may represent only a few serotypes or may be nontypeable. For example, for *Mycobacterium avium*, ~75% of clinical isolates represent only 4 of 11 defined serotypes, and 10 to 15% of strains are nontypeable because they fail to react with any available reagents or they autoagglutinate (146). Similarly, ~85% of clinical isolates of *S. aureus* express only two of eight capsular polysaccharide serotypes (7, 64).

Bacteriophage and Bacteriocin Typing

Among species for which numerous lytic bacteriophages (i.e., viruses capable of infecting and lysing bacterial cells) have been identified, strains can be characterized by their patterns of resistance or susceptibility to a standard set of phages. Until relatively recently, phage typing was the authoritative method for differentiating strains of *S. aureus* and *Salmonella* species (23, 55). Because of the need to maintain stocks of biologically active phages and control strains, phage typing is available only at reference laboratories. Even for experienced workers, the procedure is technically very demanding and suffers from considerable experimental as well as biologic variability. Since many strains are nontypeable with the standard phages, additional, experimental phages often need to be considered.

In bacteriocin typing, an isolate is assessed for susceptibility to a set of bactericidal peptides produced by selected strains (109). Although the method has been useful for certain pathogens, such as *Pseudomonas aeruginosa* (111), bacteriocin typing has limitations similar to those described for phage typing.

Electrophoretic Protein Typing and Immunoblotting

Variations in the types and structures of the proteins expressed by microorganisms can be detected by several different methods. Electrophoretic protein typing is performed by isolating materials (including proteins, glycoprotein conjugates, and lipopolysaccharides) from whole-cell or cell surface preparations, separating the materials by sodium dodecyl sulfate-polyacrylamide gel electrophoresis, and staining the proteins in the gel to determine the resulting pattern. If the proteins are radiolabeled before isolation, the pattern can be detected by autoradiography (139). In a method commonly referred to as immunoblotting, the electrophoresed bacterial products are transferred ("blotted") onto a nitrocellulose membrane and then exposed to antisera raised against specific type strains or to pooled human sera, which contain broadly reactive antibodies (92). The bound antibodies can then be detected using commercially available enzyme-labeled anti-immunoglobulins.

These techniques have been effective in epidemiologic investigations of *S. aureus* and *Clostridium difficile* (92, 93). However, because the patterns detected are very complex, comparisons among multiple strains can be difficult, and the significance of small differences is uncertain (44). The effects of technical factors such as extraction methods and bacterial growth conditions have not been fully defined. Currently, these typing methods are not widely employed.

MLEE

In multilocus enzyme electrophoresis (MLEE), isolates are analyzed for differences in the electrophoretic mobilities of a set of metabolic enzymes (128). Cell extracts containing the soluble metabolic enzymes are electrophoresed in nondenaturing starch gels. For each enzyme analyzed, a replicate gel (or slice of a gel) is stained with a specific colorimetric substrate such that the position of the enzyme is detected by the appearance of a visible reaction product (e.g., Fig. 1 in reference 128). Variations in the electrophoretic mobility of an enzyme, referred to as electromorphs, typically reflect amino acid substitutions that alter the charge of the protein and thereby identify allelic variations in the chromosomal genes encoding the enzyme. Combinations of electromorphs are designated electrophoretic types (ETs), and each distinct ET is considered to represent a multilocus genotype. Although individual enzymes may be absent (null) in particular isolates, the evaluation of multiple metabolic enzymes ensures that all isolates are typeable.

MLEE has been used most effectively to examine the population genetics of bacterial species (98). Because each individual enzyme represents a discrete, independent characteristic, the data are suitable for rigorous mathematical analysis (128). The genetic distance between any two isolates can be quantitatively estimated as the proportion of enzyme loci demonstrating different electromorphs. From such pairwise differences, the genetic diversity among large collections of isolates can be assessed, and by applying appropriate computational techniques, the genetic structure of the population can be represented graphically as a dendogram. By this approach, comprehensive population analyses of several species of pathogenic bacteria have been done (130).

Although MLEE is a very powerful tool for such population studies, it is only moderately discriminatory for the epidemiologic analysis of clinical isolates. For example, among *E. coli* isolates cultured from patients with pyelonephritis, many epidemiologically unrelated isolates, particularly those with virulence factors, represent a limited number of closely related ETs (13, 129). Moreover, MLEE requires techniques and equipment that are available in relatively few laboratories. Consequently, this method has had relatively limited application to epidemiologic studies.

GENOTYPIC TECHNIQUES

The problems associated with many of the phenotypic techniques described above have stimulated interest in DNA-based typing methods (Table 3). Initially considered the province of research settings, these technologies have matured and disseminated and are increasingly performed in clinical laboratories. However, there remain substantial complexities in interpreting the results of these methods and effectively applying them to epidemiologic studies. Additional detailed background regarding the technical aspects of several of these methods is presented in chapter 13 of this Manual.

Plasmid Analyses

Typing systems based on plasmid analyses suffer from significant limitations inherent in the fact that plasmids are mobile extrachromosomal elements and are not part of the chromosomal genotype that defines the host strain. Plasmids can be spontaneously lost from or readily acquired by a host strain; consequently, epidemiologically related isolates can exhibit different plasmid profiles (90). Since many plasmids carry antibiotic resistance determinants contained within mobile genetic elements (transposons) that can be readily acquired or deleted, the DNA composition of a plasmid can change rapidly (11, 75, 144). Further, the strong selective pressure for nosocomial organisms to express antibiotic resistance may cause such plasmids to spread rapidly among strains and even among different species and to persist for prolonged periods within an institution (11, 61, 159). Plasmids may also be involved in the spread of virulence factors and may be selected for on that basis. For example, among enterotoxigenic and enteroinvasive *E. coli* isolates, the pathogenic toxins, adhesins, and invasins are often plasmid associated (86). Many clinical isolates lack plasmids and are therefore nontypeable

TABLE 3 Characteristics of genotypic typing systems[a]

Typing system	Proportion of strains typeable	Reproducibility	Discriminatory power	Ease of interpretation	Ease of performance
Plasmid restriction digest	Most	Good	Good	Good	Excellent
REA	All	Good	Good	Poor	Excellent
Ribotyping	All	Excellent	Fair	Good	Good
PFGE	All	Excellent	Excellent	Excellent	Good
PCR restriction digest	All	Excellent	Good	Excellent	Good
Arbitrarily primed PCR	All	Good	Good	Fair	Good
Nucleotide sequence analysis	All	Excellent	Excellent	Excellent	Fair

[a] These judgments represent the views of the author; many systems remain incompletely evaluated, and characteristics may vary as applied to different species. See the text for details.

(109); others carry only one or two plasmids, a situation that provides relatively poor discriminatory power. Despite these limitations, plasmid analyses have proven useful in numerous investigations (reviewed in references 61, 86, 124, and 154) and may be quite effective in evaluating isolates obtained in a restricted time and place, for example, during acute outbreaks within a single hospital.

Plasmid profile analysis was among the earliest DNA-based techniques applied to epidemiologic studies (86). The number and sizes of the plasmids carried by an isolate can be determined by preparing a plasmid extract and subjecting it to routine agarose gel electrophoresis. However, the reproducibility of such profiles is confounded by the fact that a plasmid can exist in different molecular forms (i.e., supercoiled [closed circle], nicked [open circle], and linear), each of which migrates differently during agarose gel electrophoresis. Thus, in different preparations, each individual plasmid can be represented by a variable number of bands. Both the reproducibility and the discriminatory power of plasmid analyses can be substantially improved by digesting the plasmids with restriction enzymes (discussed below) and then electrophoretically analyzing the number and sizes of the resulting restriction fragments (86). This procedure, often referred to as restriction enzyme analysis (REA) of plasmids, is now the method of choice for plasmid studies (16, 28, 144); it is technically simple, requires only modest specialized equipment, and can be performed relatively quickly.

Restriction Endonuclease Analysis of Chromosomal DNA

A restriction endonuclease enzymatically cuts ("digests") DNA at a specific ("restricted") nucleotide recognition sequence. The number and sizes of the restriction fragments generated by digesting a given piece of DNA are influenced by both the recognition sequence of the enzyme and the composition of the DNA. Thus, an enzyme whose recognition sequence is composed of only guanine (G) and cytosine (C) will cut DNA with a low G+C content less frequently and consequently will generate fewer and larger restriction fragments than an enzyme recognizing sequences of only adenine and thymine. An enzyme that recognizes a sequence of 6 bp (a "6-bp cutter") typically has more recognition sites than an 8-bp cutter, will digest DNA more frequently, and consequently will generate more and smaller restriction fragments.

In conventional REA, bacterial DNA is digested with endonucleases that have relatively frequent restriction sites, thereby generating hundreds of fragments ranging from ~0.5 to 50 kb in length. Such fragments can be separated by size by using constant-field agarose gel electrophoresis, and the pattern can be detected by staining the gel with ethidium bromide and examining it under UV light (Fig. 1). Different strains of the same bacterial species have different REA profiles because of variations in their DNA sequences that alter the number and distribution of restriction sites. All isolates are typeable by REA, and the approach has been applied to many species; a notable example is the study of C. difficile (34, 63).

The major limitation of REA is the difficulty of comparing the complex profiles, which consist of hundreds of bands that may be unresolved and overlapping (18, 35, 104). In addition, the patterns may be confounded by restriction fragments derived from plasmids, whose DNA can readily contaminate genomic-DNA preparations. Thus, isolates that differ only in their plasmid contents may be considered different strains. The analysis of independent plasmid preparations coupled with computer-assisted subtraction techniques can compensate for this effect (18); however, such laborious procedures are rarely implemented. In general, REA has largely been supplanted by other genotypic typing methods.

Southern Blot Analysis of RFLPs

REA patterns are complex because the ethidium bromide stain detects all of the hundreds of fragments generated by the restriction enzymes used. In contrast, Southern blot analyses detect only the particular restriction fragments associated with specific chromosomal loci. Southern blots, named after the investigator who first described the technique (132), are prepared by digesting bacterial DNA, separating the restriction fragments by agarose gel electrophoresis, and then transferring ("blotting") the fragments onto a nitrocellulose or nylon membrane. The fragment(s) containing specific sequences (loci) is then detected by using a labeled piece of homologous DNA as a probe. Under the appropriate conditions, the probe binds ("hybridizes") by complementary base pair matching only to those fragments containing identical or nearly identical nucleotide sequences. Variations in the number and sizes of the fragments detected are referred to as restriction fragment length polymorphisms (RFLPs) and reflect variations in both the number of loci that are homologous to the probe and the location of restriction sites within or flanking those loci (Fig. 2). All strains carrying loci homologous to the probe are typeable, and the results are, in general, highly reproducible. The discriminatory power of an analysis is directly related to the frequency with which the fragments detected vary in number and/or in size. Thus, the choice of probes is a critical consideration.

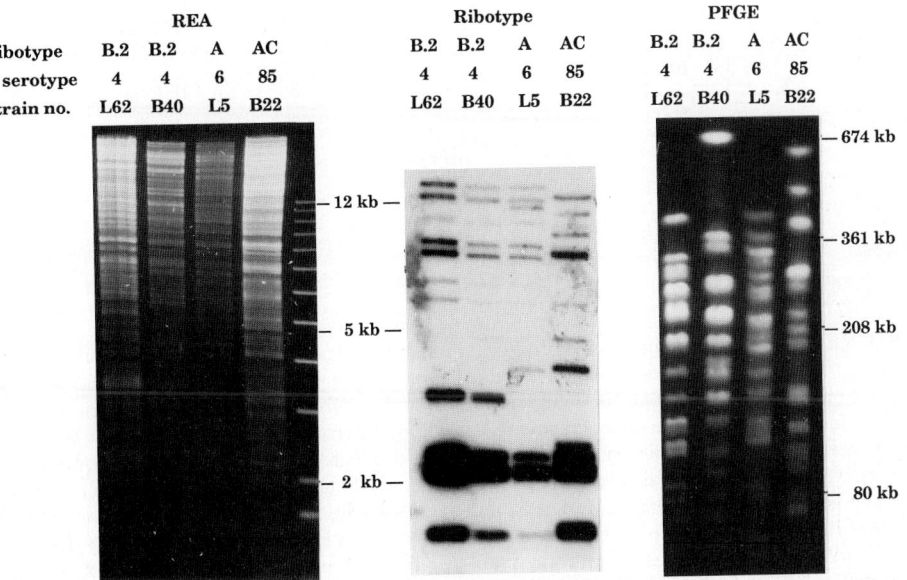

FIGURE 1 Bloodstream isolates of *E. coli* cultured from four epidemiologically unrelated patients from Long Beach, Calif. (isolates L62 and L5), and Boston, Mass. (isolates B40 and B22), and compared by three genotypic typing systems: REA, ribotyping, and PFGE. Isolates L62 and B40, from two geographically diverse sites, were indistinguishable by REA and by ribotype analysis of an *Eco*RI restriction digest; however, these isolates were distinctly different by PFGE analysis of an *Xba*I digest. Isolate L5 had some similarities to both L62 and B40 by REA and ribotyping but different by PFGE; isolate B22 was different by all three techniques. The isolates were also analyzed phenotypically. In biotyping with 30 biochemical tests (Vitek, Hazelwood, Mo.), only 4 tests differed among the isolates, with three overall patterns detected. Two isolates (L5 and L62) had the same biotype; the biotypes of strains B40 and B22 differed from that biotype by two and three tests, respectively. Of 18 antibiotics evaluated, isolates B22 and L62 were resistant only to tetracycline, whereas isolates B40 and L5 were sensitive to all antibiotics tested. The O serotypes are listed above each strain (serotypes determined by Richard Wilson, Pennsylvania State University, State College). This figure previously appeared in reference 81.

Ribotyping refers to a Southern blot analysis in which strains are characterized for the RFLPs associated with the ribosomal operon(s) (138). Operons are clusters of genes that share related functions and are often coordinately regulated; the ribosomal operons comprise nucleotide sequences coding for 16S rRNA, 25S rRNA, and one or more tRNAs. Ribosomal sequences are highly conserved, and probes prepared from isolated *E. coli* rRNA (138) or a cloned ribosomal operon (*rrn*) (13) hybridize to the chromosomal ribosomal operons of a wide range of bacterial species. All bacteria carry these operons and are therefore typeable. In general, ribotypes are stable and reproducible, with isolates from an outbreak typically having the same ribotype (115, 142).

For organisms with multiple (five to seven) ribosomal operons, such as *E. coli* and *Klebsiella*, *Haemophilus*, and *Staphylococcus* spp., ribotype patterns commonly have 10 to 15 bands and moderate discriminatory power (Fig. 1). Nevertheless, epidemiologically unrelated isolates not infrequently demonstrate the same pattern, limiting the utility of the method (13, 113, 142). For example, ~20% of 188 bloodstream isolates of *E. coli* from each of three geographically dispersed sites (Massachusetts, California, and Kenya) represented the same ribotype (84), although another genotypic technique (i.e., pulsed-field gel electrophoresis [PFGE]; discussed below) generally resolved each isolate as a distinct strain (Fig. 1). Isolates of this ribotype

typically expressed adhesin and hemolysin virulence factors; as discussed above, virulent genotypes are frequently overrepresented among collections of bacterial pathogens. For bacterial species with only a single ribosomal operon, such as mycobacteria, ribotyping typically detects only one or two bands and has limited utility for epidemiologic studies (10).

The discriminatory power of ribotyping is comparable to or slightly less than that of MLEE. In studies of isolates of *E. coli* (13), *S. aureus* (142), and *H. influenzae* type b (6), both techniques assigned isolates to the same or closely related genotypes. Such consistent, congruent results from two methods that are technically very different strongly support the hypothesis that the variations in both methods reflect the diverging genetic lineages that are characteristic of the bacterial populations. Some restriction fragments within a ribotype profile may be highly conserved and can provide a "signature" pattern that is characteristic of a particular species (51).

Southern blot analyses in which insertion sequences (IS) and transposons are used as probes have proven to be reproducible and highly discriminatory methods for strain identification (40, 66, 70, 134). These mobile genetic elements are typically present as multiple copies positioned at different chromosomal loci (75). The number and locations of IS elements, and, consequently, the RFLP(s) detected, can vary appreciably among different bacterial strains. For

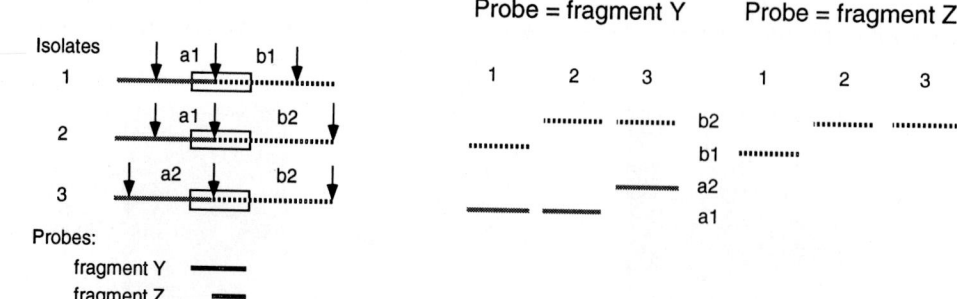

FIGURE 2 Correlation between variations in restriction sites and variations in the RFLPs detected by Southern blot analysis. As depicted, restriction enzyme A has an invariant internal restriction site within the gene (or locus) of interest and variable ("polymorphic") restriction sites to the sides ("flanks") of the gene. Fragment Y is wholly internal to the gene of interest and spans the conserved internal restriction site. A Southern blot prepared from a digest with enzyme A and probed with fragment Y will demonstrate the RFLPs shown. Fragment a is constant between isolates 1 and 2 and polymorphic between isolates 2 and 3; conversely, fragment b is polymorphic between isolates 1 and 2 and constant between 2 and 3. This situation is analogous to the RFLPs described for the *mec* locus of MRSA as described in detail by Kreiswirth et al. (66). If the same blot is probed with fragment Z, which is wholly to the right of the conserved internal restriction site, then only fragment b is detected. This situation is analogous to the methodology used to detect IS1610 elements in *M. tuberculosis* as described by van Embden et al. (150). The locus shown in this diagram would represent a single IS element; since isolates of *M. tuberculosis* typically have multiple copies of the IS element, the appearance of the blots is as depicted in Fig. 3.

example, an outbreak strain of multidrug-resistant *Mycobacterium tuberculosis* was readily differentiated from other sporadic multidrug-resistant *M. tuberculosis* strains by Southern blot analysis of IS6110, an IS present in multiple copies within the *M. tuberculosis* genome (40). This approach is currently the method of choice for epidemiologic studies of *M. tuberculosis* (Fig. 3), although its discriminatory power is relatively poor for strains with only a few copies of the IS element. To facilitate comparing the results of different studies, multiple independent laboratories have agreed to use a specific restriction enzyme to digest the chromosomal DNA and a specific subfragment of the IS to probe the Southern blots (150). By standardizing these technical variables, employing internal molecular weight markers or reference strains, and applying computerized scanners and pattern analysis, it may be possible to develop large databases of strains.

Southern blot analyses of genes encoding metabolic functions or virulence factors have been used to type isolates. There is typically only a single chromosomal copy of such genes; consequently, achieving useful strain differentiation often requires examining digests prepared with several different enzymes or probing for multiple different genetic loci (66, 104, 121, 133, 142). Such studies may be facilitated by the application of multiplexing techniques in which several different probes can be detected and distinguished simultaneously, e.g., by labeling each probe with a different fluorochrome.

The ease of interpreting Southern blots can be directly affected by the choice of enzymes for preparing the chromosomal DNA digest and the choice of probes. For example, the IS6110 blots of *M. tuberculosis* isolates are typically easier to interpret than the ribotype blots of *E. coli*. In an analysis of IS elements, essentially every band detected for a given isolate is of equal intensity; in contrast, in ribotype blots, there is considerable variability among the bands detected for each isolate (compare Fig. 1 with Fig. 3). The

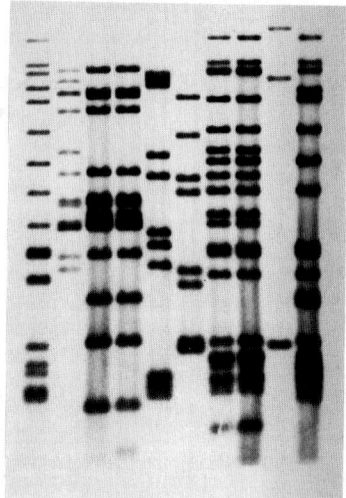

FIGURE 3 Southern blot of RFLPs associated with IS6110 among isolates of *M. tuberculosis*. The blot was prepared as described by van Embden et al. (150) from whole chromosomal DNA digested with PvuII and probed with a fragment wholly internal to IS6110 and to the right of a conserved interval PvuII site. Thus, each fragment represents a separate copy of the IS element. Because each fragment carries the same DNA sequences homologous to the probe, each fragment within an isolate is detected with similar intensity. Variation in intensity among different isolates reflects different quantities of DNA. Isolates with indistinguishable RFLP patterns are considered the same type; isolates with different patterns are considered different types. The Southern blot was generously provided by Barry Kreiswirth, Tuberculosis Center, Public Health Research Institute, New York, N.Y.

restriction enzyme and the probe used in the IS analysis were selected such that each band detected includes all of the sequences in the probe and thus hybridizes efficiently and consistently with the probe (150). In the ribotype blots shown, the restriction enzyme used (*Eco*RI) cuts at several variable sites within the ribosomal operons, generating some restriction fragments that have only short sequences homologous to the probe and consequently are detected with only a weak signal relative to the signals of fragments that have more extensive homologous sequences. Part of such variability can be eliminated by choosing a restriction enzyme (e.g., *Bam*HI) that typically cuts only outside the ribosomal operons, so that the entire operon is present on a single fragment. With that modification, all of the fragments detected generate a strong consistent signal; however, the absolute number of fragments detected decreases, and there is a concomitant decrease in the discriminatory power of the method (5). Thus, defining the optimal strategy for a Southern blot analysis can require considerable thought and effort.

Although Southern blot analyses have generally been performed by research laboratories, the procedure has been simplified by technical refinements and commercially available materials so that only a moderate level of expertise and specialized equipment is required. The preparation of probes has classically involved isolating cloned DNA fragments and labeling them with radioisotopes; however, the process has been substantially simplified by the use of the PCR (discussed below) to generate probes and the development of reliable nonradioactive detection systems (53).

PFGE of Chromosomal DNA

As noted above, a major limitation of REA with enzymes that have relatively frequent recognition sites is the difficulty of analyzing the resulting patterns composed of large numbers of overlapping, poorly resolved restriction fragments. If the bacterial genome is digested with enzymes that have relatively few restriction sites, then considerably fewer but much larger restriction fragments are generated. Until relatively recently, the characterization of large DNA fragments was limited by two major factors. First, DNA fragments of ≥25 kb are separated poorly or not at all by conventional agarose gel electrophoresis; second, DNA prepared in solution is spontaneously sheared into random fragments typically of ≤100 kb. PFGE, developed by Schwartz and Cantor in 1984 (127), is a variation of agarose gel electrophoresis in which the orientation of the electric field across the gel is changed periodically ("pulsed") rather than being kept constant as in the conventional agarose gel electrophoresis used for the REA and Southern blot studies described above. For technical reasons (21), this critical modification enables even megabase-size DNA fragments to be effectively separated by size. Suitable unsheared DNA is obtained by embedding intact organisms in agarose plugs ("inserts") and then enzymatically lysing the cell wall and digesting the cellular proteins. The isolated genomes are then digested in situ with restriction enzymes that have few recognition sites (5, 83). PFGE analysis of such digests provides a chromosomal restriction profile typically composed of 5 to 20 distinct, well-resolved fragments ranging from approximately 10 to 800 kb (Fig. 1). All bacterial isolates are theoretically typeable by PFGE, and the results are highly reproducible. The relative simplicity of the restriction profiles greatly facilitates the analysis and comparison of multiple isolates.

PFGE has been applied to a wide range of organisms (83). For several common bacterial pathogens, such as *E. coli*, enterococci, staphylococci, *P. aeruginosa*, and *M. avium* (5, 10, 94, 113, 114, 142), PFGE analysis has proven to be highly discriminatory and comparable or superior to other available techniques (Fig. 1). However, as mentioned above, some pathogens, such as MRSA, *H. influenzae* type b, and *E. coli* O157:H7, represent genetically restricted subsets of strains within a species (66, 99, 156), and consequently, epidemiologically unrelated isolates may have very similar genotypes and be indistinguishable by PFGE as well as by other typing methods (24, 142).

PFGE has two notable limitations. First, because of the need for all buffers and enzymes to be diffused into the agarose insert, the preparation of suitable DNA involves several extended incubations and takes from 2 to 4 days (83). This effort is partially offset by the fact that DNA in agarose is stable for years at 4°C and can easily be released into solution for use in other protocols. Second, PFGE requires relatively expensive, specialized equipment (21).

Typing Systems Applying the PCR

The PCR is being increasingly applied to the detection of infectious agents (108). The essential feature of PCR is the ability to rapidly and exponentially replicate ("amplify") a particular DNA sequence (the "template"). The basic procedure involves several distinct components. (i) The template should represent a relatively small fragment of DNA, typically 0.5 to 2.0 kb, because larger target sequences are difficult to amplify efficiently. Only minute quantities of the template need be present; theoretically, even a single copy is detectable. (ii) Two small oligonucleotides ("primers"), corresponding to sequences at opposite ends of the template, define the sites at which replication is initiated. The primers should be long enough to define those sites uniquely; statistical calculations indicate that 18 to 20 bp is typically sufficient. A cycle of replication involves denaturing the double-stranded DNA template, binding the primers to each strand of the template, and then synthesizing ("polymerizing") the complementary strand. (iii) A rapid, self-contained "chain reaction" is achieved by using thermostable DNA polymerases and programmable thermocyclers. An entire procedure, consisting of 20 to 30 cycles, can be conducted in a small, closed container (e.g., a microcentrifuge tube) and within a few hours will generate sufficient product ("amplicon") to be visualized and sized directly in an agarose or polyacrylamide gel.

PCR can be readily performed with commercially available reagents and thermocyclers. Since contamination with even a single copy of target DNA can produce a false-positive result, many experienced workers recommend having two physically separate work areas, one for preparing the samples and the other for performing the reactions. For diagnostic purposes, the basic procedure described can be used directly to detect the presence of the target sequences within a sample (see chapter 13 of this Manual). Several variations have been developed to provide additional information suitable for strain typing.

Restriction Digestion of PCR Products

In the most direct modification, the PCR product is digested with a restriction endonuclease, and the resulting restriction fragments are analyzed for polymorphisms by electrophoresis. This approach has been successfully used to type strains of both fungi and bacteria, including *Cryptococcus neoformans*, *S. aureus*, *Helicobacter pylori*, mycobacteria, and *Rochalimaea* spp. (43, 47, 85, 141, 152). The restriction digests are highly reproducible; however, the discriminatory power of the approach varies substantially for different species, loci, and restriction enzymes. In some instances, the digest patterns are species specific but do not identify individual strains (141), whereas in other analyses, the approach appears to be moderately discriminatory (47); strain differentiation can be increased by evaluating digests prepared with several different restriction enzymes (43, 85).

Since a highly homogeneous PCR product is required to obtain distinct restriction fragments, some investigators have employed nested PCR (47). This procedure involves a second set of primers representing sequences located inside ("nested" within) the target sequences of the initial primers. The second set of primers is added to the reaction mixture after approximately half the cycles, at which time numerous copies of the fragment specified by the initial primers have been produced. Consequently, the subfragment defined by the nested primers is amplified very rapidly. In contrast, any extraneous fragments, i.e., those generated by unanticipated binding of the first set of primers to sites other than the intended target, will not have sequences homologous to the nested primers and will not be replicated further. Thus, the final amplicon is highly enriched for the nested subfragment; this protocol obviously requires more reagents and technical effort than the routine procedure.

Rep-PCR

In PCR based on repetitive chromosomal sequences (rep-PCR), the primers used are based on short extragenic repetitive sequences, which have been identified for many members of the family *Enterobacteriaceae* and for some gram-positive bacteria as well as for fungi (148, 151, 160). Such sequences are typically present at many sites around the bacterial chromosome; when two sequences are located near enough to each other (e.g., within a few kilobases), then the DNA fragment between those sites (referred to as an "interrepeat" fragment) is effectively amplified. Since the number and locations of the repetitive sequences are quite variable, the number and sizes of the interrepeat fragments generated can similarly vary from strain to strain. The technique appears to have excellent reproducibility and moderate discriminatory power. For example, in an analysis of 29 epidemiologically unrelated isolates of *Citrobacter diversus*, 16 rep-PCR profiles were detected among isolates representing 14 MLEE types (160). When applied to higher organisms (e.g., fungi and amebae), rep-PCR profiles may also provide a rapid method for identifying particular species (102, 148).

Arbitrary Primed PCR

The variation of PCR called arbitrary primed PCR, also referred to as the random amplified polymorphic DNA assay, is based on the observation that short primers (typically 10 bp) whose sequences are not directed to any known genetic locus will nevertheless hybridize with sufficient affinity at random chromosomal sites to permit initiation of polymerization. If two such sites are located within a few kilobases of each other on opposite DNA strands and in the proper orientation, then amplification of the intervening fragment will occur (155, 158). The number and locations of these random sites vary among different strains, and thus the number and sizes of the fragments detected by electrophoresis of the amplicon also vary. The approach is attractive because it is conceptually simple and theoretically suitable for use with any organism; reports describing its application to numerous species of bacteria and fungi

have appeared in rapid succession (1, 15, 19, 31, 65, 71, 100, 116, 123, 149).

In practice, however, there are problems in achieving reproducible discriminatory results with arbitrary primed PCR. Compared with PCR that uses conventional site-specific primers, the reaction conditions (e.g., the hybridization temperature) used in arbitrary primed PCR are typically less stringent in order to facilitate initiation of the polymerization reaction at sites having one or more sequence mismatches. However, polymerization is initiated with various efficiencies at such sites, and the final quantities of DNA produced may vary widely among the different fragments amplified from a given isolate. Consequently, when the products of the reaction are visualized by ethidium bromide staining, the individual bands produced for that isolate may vary widely in intensity. Such variation is inherent in arbitrary primed PCR, is almost always present, and introduces two specific problems. First, it can be quite difficult to compare and interpret patterns whose bands demonstrate such differences in intensity (123). Second, because some of the products may represent relatively inefficient reactions, the actual fragments obtained from a single isolate may vary in different amplification reactions. Some reports indicate the need to isolate purified DNA and to quantitate DNA concentrations in order to obtain reproducible results (147). Although the PCR itself is rapid, such preparatory efforts substantially increase the time and effort required to perform the entire procedure.

In most studies, the discriminatory power of arbitrary primed PCR has not been explicitly compared with the powers of other techniques. An analysis of 26 well-characterized isolates of MRSA indicated that arbitrary primed PCR distinguished fewer distinct strains than PFGE (123). Some studies have increased the discriminatory power of the technique by performing multiple amplifications with different primers (135). However, such efforts make the approach more cumbersome to perform and the results considerably more difficult to interpret.

Nucleotide Sequence Analysis

Cloning and sequencing a particular locus in even a single isolate once represented a substantial technical effort. By using (i) the PCR to amplify a known DNA segment and (ii) automated techniques to sequence the PCR product, it is now feasible to compare multiple isolates by sequencing each one at the same locus. This approach has already been applied in analyses of natural variation within bacterial populations and clearly provides highly reliable and objective data suitable for subsequent quantitative analyses (39). The transmission of human immunodeficiency virus type 1 (HIV-1) from a dentist to his patients was confirmed by examining sequences of individual viral isolates (106).

PCR-based sequence analysis of bacterial rDNA has been used to detect and identify bacterial pathogens that cannot be recovered from infected tissues by culture techniques. Because certain ribosomal sequences are highly conserved across the bacterial kingdom, primers that will amplify ribosomal sequences from essentially any bacteria can be defined. By sequencing the amplified product and analyzing the nucleotide sequences at the relatively variable areas within the ribosomal operon, the family, genus, and species of the infecting organism can be identified. This approach has successfully detected bacteria corresponding to an unusual gram-positive actinomycete in the tissues of patients with Whipple's disease (117) and a new *Mycobac-*

terium species causing disseminated disease in patients with AIDS (25).

Particularly attractive advantages of nucleotide sequences are that the data are precise and that extensive databases can be shared with relative ease, thus facilitating comparative analyses. For example, sequence analysis of the *rpoB* chromosomal loci of multiple M. *tuberculosis* isolates has identified the distribution of different point mutations that result in rifampin resistance (140). Although nucleotide sequencing has been called "the future of molecular epidemiology" (96), several critical issues are currently unresolved. First, loci with sufficient sequence variability to permit epidemiologically useful strain differentiation must be identified for each bacterial species. In particular, the loci must be present in all isolates and have sufficient sequence variability within a span that is practical to sequence repeatedly over multiple isolates. Further, it is not clear whether sequencing at a single locus will provide reliable, unambiguous strain identification (22, 39). Finally, automated sequencers are prohibitively expensive for most settings, although the cost of such equipment will likely decline.

APPLICATION OF MICROBIAL TYPING SYSTEMS

Most requests for microbial typing reflect one apparently simple question: "Are these isolates the same or different?" As noted above, the implicit assumptions are that epidemiologically related isolates represent the clonal expansion of a single precursor, that such clonal isolates will be of the same type, and that unrelated isolates will have different types. Responding constructively to this ingenuous question requires an appreciation of the complexities inherent both in microbial biology and in obtaining and interpreting typing data.

Interpretation of Typing Results

Strictly defined, a clone is "one or more genetically identical organisms descended from a single common ancestor" (3). No typing method confirms that the entire genomes of two organisms are identical. Thus, multiple isolates representing a single type are most appropriately designated "indistinguishable," a description that emphasizes that the analysis is critically dependent on the particular typing system used. For highly discriminatory methods that consistently distinguish among epidemiologically unrelated isolates, the observation that two isolates represent the same type strongly suggests that they are both genetically and epidemiologically related (5).

The interpretation of typing results is more difficult when isolates are "similar," that is, when they differ in one or only a few of the numerous characteristics analyzed. Our experience suggests that in any typing system, a variation(s) that can be attributed to a single genetic event (e.g., a difference in a single metabolic enzyme [78] or the shift of a single band on an electrophoretic gel) is not a reliable basis for concluding that two isolates represent different strains (83). We have detected such variations, albeit infrequently, in both Southern blot and PFGE analyses of multiple isolates from the same or sequential blood cultures of individual patients (9, 10). Although such isolates are not strict clones, they are so closely related both epidemiologically and genetically that they are presumed to represent the same strain. These observations reinforce the fact

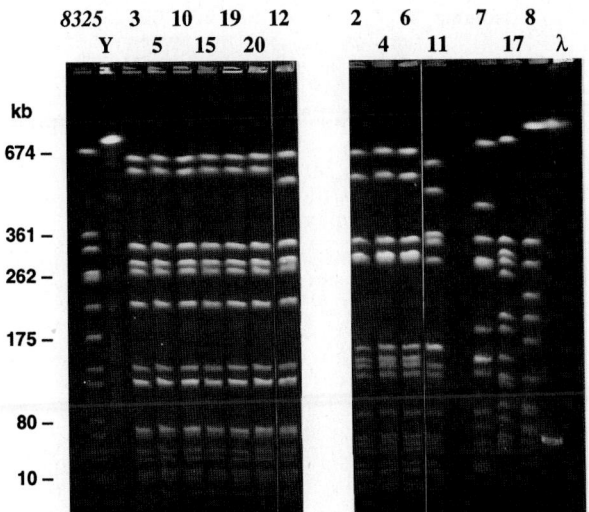

FIGURE 4 Analysis of S. *aureus* isolates by PFGE of SmaI restriction digests. Isolates are taken from set B of Tenover et al. (142). Isolates 3, 5, 10, 15, 19, 20, and 12 represent outbreak I. The PFGE profiles of the first six isolates are indistinguishable and were designated PFGE type A; the seventh isolate (isolate 12) differed by the shift of a single band, was considered a subtype of the outbreak strain, and was designated type A.1. Isolates 2, 4, 6, and 11 represent outbreak II. The PFGE profiles of the first three isolates are indistinguishable and were designated type B; the fourth isolate (isolate 11) differed by at least four bands, was considered to represent a different strain, and was designated type C. The similarities between types B and C suggest that they are genetically related; ClaI ribotypes of these isolates were indistinguishable (142). Isolates 7, 17, and 8 are epidemiologically unrelated, have distinctly different PFGE profiles, and were considered distinctly different strains. S. *aureus* ATCC 8325, which has been physically mapped, is included as an internal control and molecular weight standard. Y indicates *Saccharomyces cerevisiae* chromosomes, and λ indicates a ladder of lambda concatemers. All isolates shown were run in a single gel (9).

that clonality is a relative concept (41). In studies examining isolates collected over an extended period and/or from multiple patients, epidemiologically related isolates of the same strain are likely to demonstrate minor typing differences due to phenotypic variation or actual genotypic alterations, for example, changes in plasmid or phage content or sequence mutations altering restriction sites (142).

The inherent limitations in reproducibility due to such biologic variability as well as to technical factors also limit the discriminatory power of typing methods (57). That is, if isolates known to represent the same strain can differ to a certain degree (e.g., a single band shift), then random isolates demonstrating such differences cannot reliably be designated different strains. In many typing systems, "similar" isolates are designated types and subtypes. Isolates are assigned different types if they differ in some specified manner (e.g., two or more different antibiotic susceptibilities or bacteriophage reactions; three or more band shifts on a Southern blot analysis). Isolates that differ but not sufficiently to be designated distinct types are typically designated a subtype of the most similar type (Fig. 4). Although this process may seem reasonable, nevertheless, it is logi-

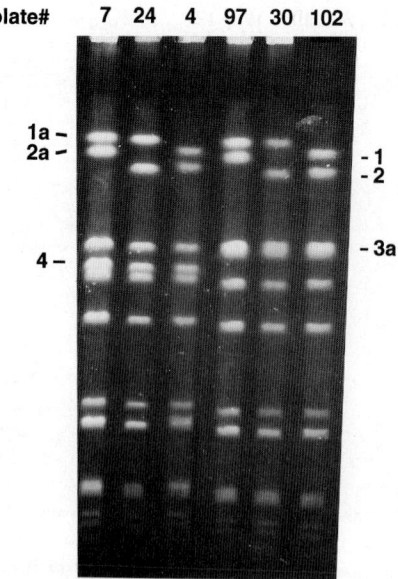

FIGURE 5 How many strains? Analysis of MRSA isolates by PFGE of *SmaI* restriction digests. All isolates were obtained from a single hospital over the course of 1 year; each pattern shown was represented by 1 to 13 isolates among more than 75 isolates examined. The set of isolates explicitly illustrates the difficulty of unambiguously designating types and subtypes, as most pairs of isolates differ by only one or two bands, but two pairs (pair 4 and 97 and pair 7 and 102) differ by three bands. In addition, Southern blot analyses (not shown) indicate that bands designated 1, 2, and 4 shift to bands 1a, 2a, and 3a, respectively, owing to the insertion of three different lysogenic bacteriophages. Thus, all six isolates have indistinguishable *SmaI* restriction fragments with respect to chromosomal DNA but differ by the presence or absence of three different lysogenic bacteriophages. All isolates shown were run in a single gel (8).

cally confounded and potentially problematic. For example, an isolate might represent a subtype to either of two different types; conversely, two subtypes of a particular type might differ sufficiently from each other to define distinct types (Fig. 5). The typing laboratory will be of greatest service to the epidemiologist or clinician if such complexities are not obscured in the process of reporting the typing data but can be shared so that the results of typing are applied collaboratively.

One approach to reconciling the biologic and technical variability inherent in molecular typing with the need for clinically and epidemiologically useful reports is to associate different classes of genetic relatedness with different probabilities of clinical and epidemiologic relatedness (142a). Isolates whose typing profiles are indistinguishable are reported as multiple isolates representing a single strain. Typically, the most common (or modal) typing profile detected among a set of isolates is considered to represent an outbreak strain. Isolates whose typing profiles differ from this modal pattern by a single genetic event are considered "closely related" genetically and "probably part of the outbreak" for clinical or epidemiologic purposes. Isolates whose profiles differ from the modal pattern by two genetic events are considered "possibly related" genetically and "possibly part of the outbreak." Isolates whose profiles differ by three or more genetic events are considered genetically "differ-

ent" and not part of the outbreak. The final decision about whether to include closely and possibly related isolates in the epidemiologic case definition of the outbreak represents the integration of molecular and clinical analyses.

In some reports, complex typing data have been analyzed by using the numerical and graphical methods (e.g., Dice coefficients and dendograms) developed in the study of evolution and population genetics. These powerful computational techniques require data sets that meet certain critical assumptions; in particular, each isolate must be described with respect to multiple distinct characteristics that vary independently. Methods generating appropriate data sets include MLEE, in which each enzyme represents a characteristic and each electromorph represents an independent variation, and nucleotide sequencing, in which each nucleotide position is a characteristic and each substitution is a variation (38, 128).

RFLP data sets typically do not satisfy these requirements. Each restriction fragment does not represent a separate, distinct characteristic; rather, each fragment is defined by two restriction sites, each comprising 4, 6, or 8 nucleotides. The simplest Southern blot analysis involves a single, well-defined chromosomal locus for which one restriction site is highly conserved and essentially invariant. In this situation, the relationships among distinct polymorphisms can be resolved, as described by Kreiswirth et al. for a set of six RFLP classes detected in association with the *mec* locus of MRSA (66). However, the quantitative analysis of more complex Southern blots (e.g., ribotypes or IS elements present at multiple loci) is highly problematic, because variations in the RFLP profile cannot be attributed to specific individual restriction sites. For PFGE profiles, which represent the entire circular chromosome, the creation or loss of a single restriction site affects two fragments. Moreover, fragments of the same molecular size cannot be assumed to represent the same chromosomal region; this can be determined only by preparing PFGE-based Southern blots and physically mapping specific loci (8, 24, 105). Conversely, restriction fragments of different sizes may represent the same chromosomal DNA and may differ because of the insertion or deletion of extrachromosomal DNA (e.g., bacteriophages) (Fig. 5) (8, 32, 72). Thus, in most instances, the reliability of numerical analyses and associated graphical representations based on RFLP data is questionable; the issue requires additional investigation.

Detection of Outbreaks

Typing methods have been widely applied in epidemiologic studies of transmissible bacterial infections. However, since true-positive and true-negative results typically cannot be established, the performance of a typing method is surprisingly difficult to express quantitatively. Nevertheless, the basic principles used in considering the performance of other laboratory tests remain applicable (60), particularly the concept of "prior probability," in this context, the likelihood that isolates were epidemiologically related independent of the results of typing.

Epidemiologic investigations are typically triggered by an increase in the prevalence of infection due to a particular bacterial species, by isolation of the same species from a cluster of patients, or by noting that isolates have a distinctive biotype or antibiotic susceptibility pattern. Thus, basic infection control surveillance and routine laboratory evaluation of isolates are practical epidemiologic screening tools. If such putative outbreaks are corroborated by detailed epidemiologic investigation, then molecular typing studies can effectively serve to verify that the isolates represent an outbreak due to a single strain. In this sequence, the typing method is being used to confirm a strong clinical and epidemiologic hypothesis. Since there is a high prior probability that an outbreak has been defined, even a moderately reproducible and discriminatory method is likely to provide useful confirmatory data. This process was well demonstrated in a recent investigation of a nosocomial outbreak of S. aureus (28); the source of the outbreak, a respiratory therapist with chronic sinusitis, was identified by case-control analysis and clearly confirmed by examining plasmid restriction digests. The molecular analysis proved particularly useful as the outbreak extended in time and the antibiotic susceptibility pattern of the strain changed. The value of performing typing studies for confirmation is dramatically illustrated when the epidemiologic hypothesis is incorrect (16, 88) or ambiguous (20, 101); in such instances, analysis of the isolates helps redirect infection control efforts.

However, to the extent that the typing results become a primary factor in defining the outbreak, then the limitations and vagaries of microbial typing may become critical and must be appreciated. This circumstance is illustrated in a recent study by Tenover et al. in which epidemic and nonepidemic isolates of S. aureus were reanalyzed by multiple different typing methods (142). One outbreak in the data set was represented by four isolates of methicillin-susceptible S. aureus related to a contaminated anesthetic; the isolates were "originally classified by bacteriophage typing as being part of the outbreak." Three isolates from the cluster were identical by 11 additional typing methods. However, the fourth isolate was considered "different" by plasmid restriction digest, PCR-based RFLPs, and PFGE (Fig. 4). The PFGE analysis indicated that the fourth isolate was genetically related to the other three, consistent with all having the same phage type, but the patterns differed at multiple fragments.

The key concepts in these examples are that epidemiologic conclusions should initially be predicated on epidemiologic data and that molecular typing is most effectively used to support those conclusions or to challenge us to reexamine them. To the extent that isolates are typed before a detailed epidemiologic study is performed, the prior probability of an outbreak is relatively low (i.e., the hypothesis motivating the typing is relatively weak), and consequently, misleading results are more likely.

Recent reports indicate that when properly applied, highly discriminatory typing methods can expand traditional concepts of nosocomial epidemiology. Classically, acute outbreaks due to a single strain are distinguished from sporadic, unrelated infections (e.g., see references 16 and 28). In contrast, several studies of nosocomial isolates representing either acute clusters or ongoing endemic persistence have documented the concurrent presence of multiple distinct strains of the same species in a hospital (42, 46, 73, 79, 101, 103, 126, 137). These experiences emphasize the value of careful strain typing in facilitating the analysis of complex epidemiologic processes.

Distinguishing Relapse from Reinfection

In addition to assisting in epidemiologic investigations, bacterial typing can contribute to the diagnosis and management of infection in the individual patient. The analysis of multiple isolates cultured from sequential episodes of infection in an individual patient can distinguish relapsing infection due to a single strain from reinfection due to a new strain and thus may help define the pathophysiology of the infection (62, 74, 79, 81, 82). Identifying reinfection may indicate that a patient is predisposed to the particular infection owing to a host defense defect (80). Culturing the same strain from a patient during separate episodes of infection suggests relapse, possibly arising from a residual focus of infection or from a site of persistent colonization (82).

Clonality of Acute Infection

Examination of multiple different isolates of the same species cultured from a patient can be useful in evaluating the presence of infection. For example, in a patient with a prosthetic valve or cerebrospinal fluid shunt, an analysis of multiple bloodstream isolates of Staphylococcus epidermidis indicating that the isolates represent a single strain suggests a true bacteremia, whereas the detection of multiple different strains is more compatible with independent skin contaminants (12). Molecular typing of S. epidermidis isolates from the skin and eyes of patients with acute postoperative endophthalmitis has implicated the patients' external tissues as the most common source of the infecting organisms (133). These interpretations are based on the assumption that acute infections due to a single species represent invasion and proliferation of a single strain (120), a hypothesis that has been formally confirmed for some other acute infections, such as bacteremic pyelonephritis (5).

In contrast, recent molecular analyses of disseminated M. avium infection among HIV-infected patients indicated that 15 to 25% of these patients have invasive disease with two distinctly different strains concurrently (10, 131). Polyclonal infection and colonization have also been demonstrated in subsets of patients infected or colonized with other pathogens, including S. aureus in hospitalized patients (79), P. aeruginosa in patients with cystic fibrosis (27), and Helicobacter pylori in patients with gastritis (43). These results suggest that different infections may have different mechanisms of pathogenesis and progression.

IMPLEMENTING A MOLECULAR EPIDEMIOLOGY LABORATORY

How a typical clinical laboratory can best meet the need for microbial typing remains an open issue. The following comments represent one approach to the process of choosing and applying a typing method and are primarily directed at those laboratories that are considering adding or expanding molecular typing capabilities.

Choosing an Appropriate Method

Several factors constrain the choice of an appropriate typing method. (i) Clinical laboratories must be prepared to analyze isolates of virtually any species. Consequently, the method should be applicable to as broad a range of organisms as possible. (ii) The method should have excellent reproducibility and discriminatory power. The results should also be reasonably easy to interpret without extensive prior experience or expertise. (iii) The feasibility and utility of an approach should have been demonstrated in multiple independent laboratories. Choosing a widely used method also facilitates obtaining consultation and assistance in the event of difficulties. (iv) Cost considerations include not only the initial capital equipment investment but also the recurrent expense of reagents and labor.

Among the phenotypic techniques (Table 2), immunoblotting and MLEE have good reproducibility and discriminatory power, but neither technique has been widely adopted. This may reflect the difficulty of interpreting the complex immunoblots and, for MLEE, the use of a relatively unfamiliar technology (nondenaturing starch gel electrophoresis) that is generally perceived as labor intensive. Among the genotypic techniques (Table 3), routine REA of chromosomal DNA is technically simple, but the complexity of the resulting patterns has dampened enthusiasm for this approach. Although Southern blot analyses are highly reproducible and easier than other methods to interpret, the need to prepare, probe, and develop the blots represents a substantial technical effort for a clinical laboratory. Only ribotyping, which provides relatively limited discriminatory power, is directly applicable to a broad range of species. Southern blot analyses of other chromosomal loci, particularly IS elements, can be highly discriminatory and for some species have proven to be the method of choice (150); however, each species analyzed requires different probes, markedly compounding the complexity of the approach. The potential of nucleotide sequence analysis is apparent, but the current cost of automated equipment (~$80,000) sharply limits its dissemination.

Plasmid restriction digests are technically reproducible, relatively simple to perform and interpret, and rapid. The only specialized equipment required is a conventional horizontal agarose gel electrophoresis box and power supply (~$1,000) and a UV transilluminator to visualize the restriction fragments after ethidium bromide staining ($1,500 to $4,000). The method has been widely and successfully applied (reviewed in references 61, 86, 124, and 154). This approach may be appropriate and useful for many clinical laboratories, provided the limitations inherent in plasmid typing are kept in mind. In particular, nontypeable isolates, i.e., those without plasmids, are inevitably encountered, and among isolates representing a single strain, plasmid carriage and structure are subject to biologic variability. This approach is most useful for evaluating sets of isolates that are circumscribed in time and place; it is much more problematic for analyzing isolates collected over extended periods. Labor and material costs are estimated at ~$15 to $20 per isolate.

The utility and reproducibility of PFGE have been confirmed in numerous laboratories; all the necessary reagents are commercially available, and the technical parameters are well described. Genomic restriction digests of many different microbial species can be analyzed by PFGE with only minor alterations in the procedure. For example, different species may require additional enzymes for cell lysis, alternative restriction endonucleases for DNA digestion, or varied pulsing conditions to resolve restriction fragments spanning different size ranges (21, 83). Specialized equipment, including an appropriately configured gel box, a programmable controller, a power supply, and a unit for cooling and recirculating the electrophoresis buffer, is required; a UV transilluminator is also necessary. Systems of coordinated components are available from commercial suppliers for $7,000 to $20,000 (21). With the development of commercially available aids to facilitate the procedure, the technical sophistication required to implement this approach is within the reach of most clinical laboratories. Labor and material costs are estimated at ~$30 per isolate.

The power and flexibility of PCR have made it a revolutionary tool in contemporary molecular biology. However, the potential speed and ease of the method must be tempered by the need to prepare samples with great care to avoid cross-contamination. Although different primers may be required for each species, custom-prepared oligonucleotides are now commercially available on request. Analysis of restriction digests of amplified fragments is the most reproducible and interpretable PCR-based approach; however, at this time, specific primers and protocols providing highly discriminatory typing have been defined and validated for only a few species. As discussed in detail above, the reproducibility and the ease of interpreting the results of arbitrary primed PCR are problematic. PCR can be readily performed with commercially available thermocyclers ($3,000 to $7,000) and reagents; a routine gel electrophoresis system and a UV transilluminator are also required. Labor and material costs are estimated at ~$15 to $20 per isolate.

Molecular Typing of Specific Organisms

In practice, the majority of hospital-based epidemiologic analyses involve a relatively limited set of organisms. Virtually every method described has been applied to these species, and a consideration of all available reports is outside the scope of this review. This section will emphasize the use of the three methods proposed above as most suitable for a clinical laboratory: plasmid typing, PFGE, and PCR-based approaches.

S. aureus

Because of their virulence, their frequent association with nosocomial outbreaks, and the emergence of multidrug-resistant strains, isolates of S. aureus have been the subject of numerous epidemiologic studies and have been examined using essentially all of the available typing techniques (91). Plasmid restriction digests have frequently proven useful (e.g., see references 16, 28, and 110). PFGE has been reported to be the most discriminatory method available, particularly for MRSA (16, 59, 114, 125, 136). Analysis of restriction digests of the PCR products of the coagulase gene appears to be moderately discriminatory (47, 142).

A recent study by Tenover et al. examined 59 isolates of S. aureus, including methicillin-susceptible strains and MRSA, by 14 different methods (142). The study set contained 29 isolates representing four distinct, previously established outbreaks plus additional epidemiologically related and unrelated isolates. This is the largest analysis available of a substantial set of isolates (of any species) evaluated by multiple techniques, and it represents a paradigm for both the insights and the complexities associated with such an investigation. The overall effectiveness of several techniques, including immunoblotting, Southern blot analyses, restriction analysis of PCR products, and PFGE, was remarkably comparable. The last two methods detected essentially all outbreak isolates but failed to distinguish an appreciable number of control strains. This result may reflect a subtle bias in that the control isolates were often selected to have the same bacteriophage type as the outbreak strain; however, epidemiologically unrelated isolates representing the same phage type are more likely to be genetically related than truly random isolates (23). In contrast, plasmid analysis, which is independent of the host strain genome, was highly effective at differentiating the outbreak strains from the controls but was less consistent at detecting the outbreak isolates. Taken together, these data suggest that in exceptionally complex or ambiguous situations, a combination of techniques that assess qualitatively different characteristics may provide the most reliable complement to epidemiologic studies.

Coagulase-Negative Staphylococci

Plasmid analyses have been very effective for studying *S. epidermidis* and other coagulase-negative streptococci (12, 29), although plasmid variability can be appreciable (11, 90). Recent studies indicate that PFGE is also effective (45, 56, 68). Although phenotypic studies have generally demonstrated extensive variation among *S. epidermidis* isolates, recent genotypic analyses suggest that, as for other pathogens, clinically significant isolates may represent a genetically restricted subset of the species. Specifically, the same ribotypes have been detected among geographically distant isolates (68), and a study of isolates cultured from the bloodstreams of infants treated in a neonatal intensive care unit demonstrated endemic persistence of the same PFGE types (56). These observations suggest that certain lineages or genotypes of *S. epidermidis* may have greater pathogenicity, perhaps related to such virulence factors as surface adhesins (145).

Enterococci

PFGE has emerged as the method of choice for analyzing isolates of the genus *Enterococcus* (37, 94, 95, 107).

E. coli and Other *Enterobacteriaceae*

PFGE is a highly effective method for differentiating among isolates of *E. coli* and is appreciably more discriminatory than MLEE or ribotyping. This was demonstrated initially in an analysis of isolates cultured from patients with pyelonephritis (5) and, more recently, in a study involving 160 bloodstream isolates cultured from patients in Massachusetts, California, and Nairobi, Kenya (80, 84). PFGE is also more discriminatory than plasmid analysis of *E. coli* isolates (48). Nevertheless, some epidemiologically unrelated isolates representing the same genetic lineage as defined by complete serotyping will have indistinguishable PFGE profiles (24, 105, 162). Although the data are limited, PFGE also appears to be highly effective for typing other *Enterobacteriaceae*, such as *Klebsiella* spp. (78) and *Enterobacter* spp. (54). Analyses of PCR products generated from repetitive sequence-based primers was comparable to MLEE for discriminating among isolates of *Citrobacter* spp. (160).

Serotyping remains the primary basis for characterizing *Salmonella* isolates and appears reasonably well correlated with distinct genetic lineages (17). Reliably differentiating among isolates of the same serotype remains difficult, perhaps reflecting restricted genetic diversity. Plasmid analysis and ribotyping appear to have a discriminatory power comparable to that of phage typing (2, 119), which is typically performed only by reference laboratories. Recently, Southern blot analysis of IS200, a species-specific multicopy IS found in the *Salmonella* chromosome, was proven to be effective for differentiating strains of the same serotype and phage type (134).

Pseudomonas spp. and Other Aerobic Gram-Negative Rods

PFGE is highly effective for analyzing isolates of *P. aeruginosa*, *Pseudomonas cepacia*, and *Acinetobacter* spp. (4, 26, 50, 52, 113, 157).

C. difficile

At this time, the application of PFGE to clinical isolates of *C. difficile* appears to be hindered by an unusual technical problem that has not been encountered with other microbial species. For 10 to 40% of nosocomial isolates of *C. difficile*, the standard protocol for in situ lysis in agarose yields genomic DNA that is partially degraded and consequently is not suitable for PFGE analysis (69). For those isolates that can be analyzed by PFGE, the discriminatory power of the technique is comparable to that of routine REA; both methods are superior to ribotyping. Plasmid analysis has proven useful in some studies, but up to 50% of *C. difficile* isolates may lack plasmids and thus be nontypeable (76, 93).

Immunoblotting and routine REA of chromosomal DNA have been effective at deciphering epidemiologic relationships among *C. difficile* isolates (33, 36, 63, 93, 122). However, interpreting the complex patterns detected by these methods requires experience, and comparing large numbers of strains is often difficult (69). Preliminary observations suggest that an implementation of arbitrary primed PCR may be effective at typing *C. difficile* (143).

Mycobacteria

With the resurgence of tuberculosis associated with the HIV pandemic and the development of outbreaks due to multidrug-resistant strains, the typing of *M. tuberculosis* has become an urgent issue. As noted above, the current method of choice is Southern blot analysis of the RFLPs associated with IS1610 (150). A major limitation of this system is that the preparation of suitable DNA requires handling high concentrations of the pathogen and consequently requires more stringent containment facilities than does work with routine bacterial isolates (118). Alternative approaches using PCR-based methods may bypass this problem (112). PFGE is also effective for typing isolates of the *M. tuberculosis* complex (67, 161).

For nontuberculous mycobacterial species, analogous IS elements have not been identified. PFGE has been successfully used to analyze isolates of the *M. avium* complex, *Mycobacterium gordonae*, and *Mycobacterium fortuitum* (10, 30, 87). For *M. avium* complex isolates in particular, PFGE provides excellent discriminatory power (10, 131, 153).

Technical Aspects of Microbial Typing

Regardless of the method used, certain issues are routinely encountered in operating a typing laboratory. A clear, concise system for uniquely identifying every isolate is required, together with a log for recording the date, source (hospital, laboratory, patient, specimen, subculture, etc.), and other pertinent information. The identification numbers routinely assigned to clinical specimens are not sufficient for a typing laboratory, since the isolate identification system needs to provide for processing multiple isolates from a single specimen or multiple different subcultures derived from the same colony. A mix-up among isolates is a potential and relatively common cause of disagreement between typing results and clinical epidemiology; such problems may also become apparent as logical inconsistencies between the results of two independent typing methods (69). Therefore, the isolate register should unambiguously identify repeated, independent samples representing the same initial organism. Finally, isolates may be provided by other institutions, either for primary analysis or for comparison.

With notable exceptions (150), there has been little formal concurrence about methodology, interpretation, or quality control in microbial typing. Pending such guidelines, one approach to monitoring the processing of isolates is to analyze isolates of a particular species in batches with species-specific reference strains as concurrent internal controls. For typing methods involving electrophoresis, molecular weight standards are generally available. Since there is no gold standard typing method, it is occasionally very useful to have the capability of analyzing isolates by two distinct methods so that the results of one system can be compared with those of the other.

Final Considerations

The application of molecular techniques to microbial typing has provided a powerful set of new tools that can facilitate both epidemiologic investigations and patient management. Typing data are most appropriately evaluated in the context of hypotheses and questions thoughtfully developed by the clinician or epidemiologist and thus used to augment rather than replace those analyses. Ideally, typing is performed independently by the laboratory to avoid any bias, but the results are considered collaboratively to ensure that both the potential insights and the unavoidable ambiguities are clearly appreciated. The manner in which the typing results are interpreted and applied is as important as the particular method employed.

REFERENCES

1. Akopyanz, N., N. O. Bukanov, T. U. Westblom, S. Kresovich, and D. E. Berg. 1992. DNA diversity among clinical isolates of *Helicobacter pylori* detected by PCR-based RAPD fingerprinting. *Nucleic Acids Res.* **20**:5137–5142.
2. Altwegg, M., F. W. Hickman-Brenner, and J. J. Farmer III. 1989. Ribosomal RNA gene restriction patterns provide increased sensitivity for typing *Salmonella typhi* strains. *J. Infect. Dis.* **160**:145–149.
3. American Heritage Dictionary of the English Language, 3rd ed. 1992. Houghton Mifflin Co., Boston.
4. Anderson, D. J., J. S. Kuhns, M. L. Vasil, D. N. Gerding, and E. N. Janoff. 1991. DNA fingerprinting by pulsed-field gel electrophoresis and ribotyping to distinguish *Pseudomonas cepacia* isolates from a nosocomial outbreak. *J. Clin. Microbiol.* **29**:648–649.
5. Arbeit, R. D., M. Arthur, R. D. Dunn, C. Kim, R. K. Selander, and R. Goldstein. 1990. Resolution of recent evolutionary divergence among *Escherichia coli* from related lineages: the application of pulsed field electrophoresis to molecular epidemiology. *J. Infect. Dis.* **161**:230–235.
6. Arbeit, R. D., R. Dunn, J. N. Maslow, R. Goldstein, and J. M. Musser. 1990. Resolution of evolutionary divergence and epidemiologic relatedness among *H. influenzae* type b (HIB) by pulse field gel electrophoresis (PFGE), abstr. 1144. *Program Abstr. 30th Intersci. Conf. Antimicrob. Agents Chemother.*
7. Arbeit, R. D., W. W. Karakawa, W. F. Vann, and J. B. Robbins. 1984. Predominance of two newly described capsular polysaccharide types among clinical isolates of *Staphylococcus aureus*. *Diagn. Microbiol. Infect. Dis.* **2**:85–91.
8. Arbeit, R. D., B. Kreiswirth, J. N. Maslow, T. Morris, and K. Murray-Leisure. Unpublished data.
9. Arbeit, R. D., and J. N. Maslow. Unpublished data.
10. Arbeit, R. D., A. Slutsky, T. W. Barber, J. N. Maslow, S. Niemczyk, J. O. Falkinham III, G. T. O'Connor, and C. F. von Reyn. 1993. Genetic diversity among strains of Mycobacterium avium causing monoclonal and polyclonal bacteremia in patients with AIDS. *J. Infect. Dis.* **167**:1384–1390.
11. Archer, G. L., D. R. Dietrick, and J. L. Johnston. 1985. Molecular epidemiology of transmissible gentamicin resistance among coagulase-negative *Staphylococci* in a cardiac surgery unit. *J. Infect. Dis.* **151**:243–251.
12. Archer, G. L., A. W. Karchmer, N. Vishniavsky, and J. L. Johnston. 1984. Plasmid-pattern analysis for the differentiation of infecting from non-infecting *Staphylococcus epidermidis*. *J. Infect. Dis.* **149**:913–920.
13. Arthur, M., R. D. Arbeit, C. Kim, P. Beltran, H. Crowe, S. Steinbach, C. Campanelli, R. A. Wilson, R. K. Selander, and R. Goldstein. 1990. Restriction fragment length polymorphisms among uropathogenic *Escherichia coli pap*-related sequences compared with *rrn* operons. *Infect. Immun.* **58**:471–479.
14. Arthur, M., C. E. Johnson, R. H. Rubin, R. D. Arbeit, C. Campanelli, C. Kim, S. Steinbach, M. Agarwal, R. Wilkinson, and R. Goldstein. 1989. Molecular epidemiology of adhesin and hemolysin virulence factors among uropathogenic *Escherichia coli*. *Infect. Immun.* **57**:303–313.
15. Aufauvre-Brown, A., J. Cohen, and D. W. Holden. 1992. Use of randomly amplified polymorphic DNA markers to distinguish isolates of *Aspergillus fumigatus*. *J. Clin. Microbiol.* **30**:2991–2993.
16. Back, N. A., C. C. Linnemann, Jr., M. A. Pfaller, J. L. Staneck, and V. Morthland. 1993. Recurrent epidemics caused by a single strain of erythromycin-resistant *Staphylococcus aureus*: the importance of molecular epidemiology. *JAMA* **270**:1329–1333.
17. Beltran, P., J. M. Musser, R. Helmuth, J. J. Farmer III, W. M. Frerichs, I. K. Wachsmuth, K. Ferris, A. C. McWhorter, J. C. Wells, A. Cravioto, and R. K. Selander. 1988. Toward a population genetic analysis of *Salmonella*: genetic diversity and relationships among strains of serotypes S. *choleraesuis*, S. *derby*, S. *dublin*, S. *enteritidis*, S. *heidelberg*, S. *infantis*, S. *newport*, S. *typhimurium*. *Proc. Natl. Acad. Sci. USA* **85**:7753–7757.
18. Bialkowska-Hobrzanska, H., D. Jaskot, and O. Hammerberg. 1990. Evaluation of restriction endonuclease fingerprinting of chromosomal DNA and plasmid profile analysis for characterization of multiresistant coagulase-negative staphylococci in bacteremic neonates. *J. Clin. Microbiol.* **28**:269–275.
19. Bingen, E., C. Boissinot, P. Desjardins, H. Cave, N. Brahimi, N. Lambert-Zechovsky, E. Denamur, P. Blot, and J. Elion. 1993. Arbitrarily primed polymerase chain reaction provides rapid differentiation of *Proteus mirabilis* isolates from a pediatric hospital. *J. Clin. Microbiol.* **31**:1055–1059.
20. Bingen, E., E. Denamur, N. Lambert-Zechovsky, N. Brahimi, M. El Lakany, and J. Elion. 1992. Rapid genotyping shows the absence of cross-contamination in *Enterobacter cloacae* nosocomial infections. *J. Hosp. Infect.* **21**:95–101.
21. Birren, B., and E. Lai. 1993. *Pulsed Field Gel Electrophoresis: a Practical Guide*. Academic Press, Inc., San Diego.
22. Bisercic, M., J. Y. Feutrier, and P. R. Reeves. 1991. Nucleotide sequences of the *gnd* genes from nine natural isolates of *Escherichia coli*: evidence of intragenic recombination as a contributing factor in the evolution of the polymorphic *gnd* locus. *J. Bacteriol.* **173**:3894–3900.
23. Blair, J. E., and R. E. O. Williams. 1961. Phage typing of staphylococci. *Bull. W.H.O.* **24**:771–784.
24. Böhm, H., and H. Karch. 1992. DNA fingerprinting of *Escherichia coli* O157:H7 strains by pulsed-field gel electrophoresis. *J. Clin. Microbiol.* **30**:2169–2172.
25. Böttger, E. C., A. Teske, P. Kirschner, S. Bost, H. R. Chang, V. Beer, and B. Hirschel. 1992. Disseminated "Mycobacterium genavense" infection in patients with AIDS. *Lancet* **340**:76–80.
26. Boukadida, J., M. de Montalembert, J.-L. Gaillard, J. Gobin, F. Grimont, D. Girault, M. Véron, and P. Berche. 1991. Outbreak of gut colonization by *Pseudomonas aeruginosa* in immunocompromised children undergoing total digestive decontamination: analysis by pulsed-field electrophoresis. *J. Clin. Microbiol.* **29**:2068–2071.
27. Boukadida, J., M. de Montalembert, G. Lenoir, P. Scheinmann, M. Véron, and P. Berche. 1993. Molecular epidemiology of chronic pulmonary colonisation by *Pseudomonas*

aeruginosa in cystic fibrosis. *J. Med. Microbiol.* **38**:29–33.

28. **Boyce, J. M., S. M. Opal, G. Potter-Bynoe, and A. A. Medeiros.** 1993. Spread of methicillin-resistant *Staphylococcus aureus* in a hospital after exposure to a health care worker with chronic sinusitis. *Clin. Infect. Dis.* **17**:496–504.

29. **Boyce, J. M., G. Potter-Bynoe, S. M. Opal, L. Dziobek, and A. A. Medeiros.** 1989. A common-source outbreak of *Staphylococcus epidermidis* infections among patients undergoing cardiac surgery. *J. Infect. Dis.* **161**:493–499.

30. **Burns, D. N., R. J. Wallace, Jr., M. E. Schultz, Y. Zhang, S. Q. Zubairi, Y. Pang, C. L. Gibert, B. A. Brown, E. S. Noel, and F. M. Gordin.** 1991. Nosocomial outbreak of respiratory tract colonization with *Mycobacterium fortuitum*: demonstration of the usefulness of pulsed-field gel electrophoresis in an epidemiologic investigation. *Am. Rev. Respir. Dis.* **144**:1153–1159.

31. **Caetano-Anollés, G., B. J. Bassam, and P. M. Gresshoff.** 1992. Primer-template interactions during DNA amplification fingerprinting with single arbitrary oligonucleotides. *Mol. Gen. Genet.* **235**:157–165.

32. **Carles-Nurit, M. J., B. Christophle, S. Broche, A. Gouby, N. Bouziges, and M. Ramuz.** 1992. DNA polymorphisms in methicillin-susceptible and methicillin-resistant strains of *Staphylococcus aureus. J. Clin. Microbiol.* **30**:2092–2096.

33. **Clabots, C. R., S. Johnson, K. M. Bettin, P. A. Mathie, M. E. Mulligan, D. R. Schaberg, L. R. Peterson, and D. N. Gerding.** 1993. Development of a rapid and efficient restriction endonuclease analysis typing system for *Clostridium difficile* and correlation with other typing systems. *J. Clin. Microbiol.* **31**:1870–1875.

34. **Clabots, C. R., S. Johnson, M. M. Olson, L. R. Peterson, and D. N. Gerding.** 1992. Acquisition of *Clostridium difficile* by hospitalized patients: evidence for colonized new admissions as a source of infection. *J. Infect. Dis.* **166**:561–567.

35. **Cleary, P. P., E. L. Kaplan, C. Livdahl, and S. Skjold.** 1988. DNA fingerprints of *Streptococcus pyogenes* are M type specific. *J. Infect. Dis.* **158**:1317–1323.

36. **Devlin, H. R., W. Au, L. Foux, and W. C. Bradbury.** 1987. Restriction endonuclease analysis of nosocomial isolates of *Clostridium difficile. J. Clin. Microbiol.* **25**:2168–2172.

37. **Donabedian, S. M., J. W. Chow, J. M. Boyce, R. E. McCabe, S. M. Markowitz, P. E. Coudron, A. Kuritza, C. L. Pierson, and M. J. Zervos.** 1992. Molecular typing of ampicillin-resistant, non-β-lactamase-producing *Enterococcus faecium* isolates from diverse geographic areas. *J. Clin. Microbiol.* **30**:2757–2761.

38. **Doolittle, R. F.** 1990. Molecular evolution: computer analysis of protein and nucleic acid sequences. *Methods Enzymol.* **183**.

39. **DuBose, R. F., D. E. Dykhuizen, and D. L. Hartl.** 1988. Genetic exchange among natural isolates of bacteria: recombination within the *phoA* gene of *Escherichia coli. Proc. Natl. Acad. Sci. USA* **85**:7036–7040.

40. **Edlin, B. R., J. I. Tokars, M. H. Grieco, J. T. Crawford, J. Williams, E. M. Sordillo, K. R. Ong, J. O. Kilburn, S. W. Dooley, K. G. Castro, W. R. Jarvis, and S. D. Holmberg.** 1992. An outbreak of multidrug-resistant tuberculosis among hospitalized patients with the acquired immunodeficiency syndrome. *N. Engl. J. Med.* **326**:1514–1521.

41. **Eisenstein, B. I.** 1990. New molecular techniques for microbial epidemiology and the diagnosis of infectious diseases. *J. Infect. Dis.* **161**:595–602.

42. **el-Adhami, W., L. Roberts, A. Vickery, B. Inglis, A. Gibbs, and P. R. Stewart.** 1991. Epidemiological analysis of a methicillin-resistant *Staphylococcus aureus* outbreak using restriction fragment length polymorphisms of genomic DNA. *J. Gen. Microbiol.* **137**:2713–2720.

43. **Fujimoto, S., B. Marshall, and M. J. Blaser.** 1994. PCR-based restriction fragment length polymorphism typing of *Helicobacter pylori. J. Clin. Microbiol.* **32**:331–334.

44. **Gaston, M. A., P. S. Duff, J. Naidoo, K. Ellis, J. I. S. Roberts, J. F. Richardson, R. R. Marples, and E. M. Cook.** 1988. Evaluation of electrophoretic methods for typing methicillin-resistant *Staphylococcus aureus. J. Med. Microbiol.* **26**:189–197.

45. **Goering, R. V., and M. A. Winters.** 1992. Rapid method for epidemiological evaluation of gram-positive cocci by field inversion gel electrophoresis. *J. Clin. Microbiol.* **30**:577–580.

46. **Goetz, M. B., M. E. Mulligan, R. Kwok, H. O'Brien, C. Caballes, and J. P. Garcia.** 1992. Management and epidemiologic analyses of an outbreak due to methicillin-resistant *Staphylococcus aureus. Am. J. Med.* **92**:607–614.

47. **Goh, S.-H., S. K. Byrne, J. L. Zhang, and A. W. Chow.** 1992. Molecular typing of *Staphylococcus aureus* on the basis of coagulase gene polymorphisms. *J. Clin. Microbiol.* **30**:1642–1645.

48. **Gordillo, M. E., G. R. Reeve, J. Pappas, J. J. Mathewson, H. L. DuPont, and B. E. Murray.** 1992. Molecular characterization of strains of enteroinvasive *Escherichia coli* O143, including isolates from a large outbreak in Houston, Texas. *J. Clin. Microbiol.* **30**:889–893.

49. **Goswitz, J. J., K. E. Willard, C. E. Fashing, and L. R. Peterson.** 1992. Detection of *gyrA* gene mutations associated with ciprofloxacin resistance in methicillin-resistant *Staphylococcus aureus*: analysis by polymerase chain reaction and automated direct DNA sequencing. *Antimicrob. Agents Chemother.* **36**:1166–1169.

50. **Gouby, A., M.-J. Carles-Nurit, N. Bouziges, G. Bourg, R. Mesnard, and P. J. M. Bouvet.** 1992. Use of pulsed-field gel electrophoresis for investigation of hospital outbreaks of *Acinetobacter baumannii. J. Clin. Microbiol.* **30**:1588–1591.

51. **Grimont, F., and P. A. D. Grimont.** 1986. Ribosomal ribonucleic acid gene restriction patterns as potential taxonomic tools. *Ann. Inst. Pasteur Microbiol.* **137B**:165–175.

52. **Grothues, D., U. Koopman, H. von der Hardt, and B. Tümmler.** 1988. Genome fingerprinting of *Pseudomonas aeruginosa* indicates colonization of cystic fibrosis siblings with closely related isolates. *J. Clin. Microbiol.* **26**:1973–1977.

53. **Gustaferro, C. A.** 1993. Chemiluminescent ribotyping, p. 584–589. *In* D. H. Persing, T. F. Smith, F. C. Tenover, and T. J. White (ed.), *Diagnostic Molecular Microbiology: Principles and Applications.* American Society for Microbiology, Washington, D.C.

54. **Haertl, R., and G. Bandlow.** 1993. Epidemiological fingerprinting of *Enterobacter cloacae* by small-fragment restriction endonuclease analysis and pulsed-field gel electrophoresis of genomic restriction fragments. *J. Clin. Microbiol.* **31**:128–133.

55. **Hickman-Brenner, F. W., A. D. Stubbs, and J. J. Farmer III.** 1991. Phage typing of *Salmonella enteritidis* in the United States. *J. Clin. Microbiol.* **29**:2817–2823.

56. **Huebner, J., G. B. Pier, J. N. Maslow, E. Muller, H. Shiro, M. Parent, A. Kropec, R. D. Arbeit, and D. A. Goldmann.** 1994. Endemic nosocomial transmission of *Staphylococcus epidermidis* bacteremia strains in a neonatal ICU over a 10-year period. *J. Infect. Dis.* **169**:526–531.

57. **Hunter, P. R.** 1990. Reproducibility and indices of discriminatory power of microbial typing methods. *J. Clin. Microbiol.* **28**:1903–1905.

58. **Hunter, P. R., and M. A. Gaston.** 1988. Numerical index of the discriminatory ability of typing systems: an application of Simpson's index of diversity. *J. Clin. Microbiol.* **26**:2465–2466.

59. **Ichiyama, S., M. Ohta, K. Shimokata, N. Kato, and J. Takeuchi.** 1991. Genomic DNA fingerprinting by pulsed-field gel electrophoresis as an epidemiological marker for study of nosocomial infections caused by methicillin-resistant *Staphylococcus aureus. J. Clin. Microbiol.* **29**:2690–2695.

60. **Ingelfinger, J. A., F. Mosteller, L. A. Thibodeau, and J. H. Ware.** 1983. *Biostatistics in Clinical Medicine.* Macmillan Publishing Co., Inc., New York.

61. **John, J. F., Jr., and J. A. Twitty.** 1986. Plasmids as epidemiologic markers in nosocomial gram-negative bacilli: experience at a university and review of the literature. *Rev. Infect. Dis.* **8**:693–704.

62. **Johnson, C. E., J. N. Maslow, D. C. Fattlar, K. S. Adams, and R. D. Arbeit.** 1993. The role of bacterial adhesins in the outcome of childhood urinary tract infections. *Am. J. Dis. Child.* **147**:1090–1093.

63. Johnson, S., C. R. Clabots, F. V. Linn, M. M. Olson, L. R. Peterson, and D. N. Gerding. 1990. Nosocomial *Clostridium difficile* colonisation and disease. *Lancet* **336:**97–100.

64. Karakawa, W. W., J. M. Fournier, W. F. Vann, R. Arbeit, R. S. Schneerson, and J. B. Robbins. 1985. Method for the serological typing of the capsular polysaccharides of *Staphylococcus aureus*. *J. Clin. Microbiol.* **22:**445–447.

65. Kersulyte, D., J. P. Woods, E. J. Keath, W. E. Goldman, and D. E. Berg. 1992. Diversity among clinical isolates of *Histoplasma capsulatum* detected by polymerase chain reaction with arbitrary primers. *J. Bacteriol.* **174:**7075–7079.

66. Kreiswirth, B., J. Kornblum, R. D. Arbeit, W. Eisner, J. N. Maslow, A. McGeer, D. E. Low, and R. P. Novick. 1993. Evidence for a clonal origin of methicillin resistance in *Staphylococcus aureus*. *Science* **259:**227–230.

67. Kristjánsson, M., P. Green, H. L. Manning, A. M. Slutsky, S. M. Brecher, C. F. von Reyn, R. D. Arbeit, and J. N. Maslow. 1993. Molecular confirmation of Bacillus Calmette-Guérin (BCG) as the cause of pulmonary infection following urinary tract installation. *Clin. Infect. Dis.* **17:**228–230.

68. Kristjánsson, M., L. Herwaldt, R. D. Arbeit, C. Cavanaugh, and J. N. Maslow. 1993. Comparison of pulsed field gel electrophoresis (PFGE) and ribotyping (RT) for differentiating isolates of coagulase-negative staphylococci (CNS), abstr. 1635. *Program Abstr. 33rd Intersci. Conf. Antimicrob. Agents Chemother.*

69. Kristjánsson, M., M. H. Samore, D. N. Gerding, P. C. DeGirolami, K. M. Bettin, A. W. Karchmer, and R. D. Arbeit. 1994. Comparison of restriction endonuclease analysis, ribotyping, and pulsed field gel electrophoresis for molecular differentiation of *Clostridium difficile* strains. *J. Clin. Microbiol.* **32:**1963–1969.

70. Lawrence, J. G., D. E. Dykhuizen, R. F. DuBose, and D. L. Hartl. 1989. Phylogenetic analysis using insertion sequence fingerprinting in *Escherichia coli*. *Mol. Biol. Evol.* **6:**1–14.

71. Lehmann, P. F., D. Lin, and B. A. Lasker. 1992. Genotypic identification and characterization of species and strains within the genus *Candida* by using random amplified polymorphic DNA. *J. Clin. Microbiol.* **30:**3249–3254.

72. Lina, B., M. Bes, F. Vandenesch, T. Greenland, J. Etienne, and J. Fleurette. 1993. Role of bacteriophages in genomic variability of related coagulase-negative staphylococci. *FEMS Microbiol. Lett.* **109:**273–278.

73. Linnemann, C. C., Jr., P. Moore, J. L. Staneck, and M. A. Pfaller. 1991. Reemergence of epidemic methicillin-resistant *Staphylococcus aureus* in a general hospital associated with changing staphylococcal strains. *Am. J. Med.* **91**(Suppl. 3B):238S–244S.

74. LiPuma, J., T. Stull, S. Dasen, K. Pidcock, D. Kaye, and O. Korzeniowski. 1989. DNA polymorphisms among *Escherichia coli* isolated from bacteruric women. *J. Infect. Dis.* **159:**526–532.

75. Lupski, J. R. 1987. Molecular mechanisms for transposition of drug-resistance genes and other movable genetic elements. *Rev. Infect. Dis.* **9:**357–368.

76. Mahony, D. E., J. Clow, L. Atkinson, N. Vakharia, and W. F. Schlech. 1991. Development and application of a multiple typing system for *Clostridium difficile*. *Appl. Environ. Microbiol.* **57:**1873–1879.

77. Maki, D. G., F. S. Rhame, D. C. Mackel, and J. V. Bennett. 1976. Nationwide epidemic of septicemia caused by contaminated intravenous products. I. Epidemiologic and clinical features. *Am. J. Med.* **60:**471–485.

78. Maslow, J. N., S. Brecher, K. S. Adams, A. Durbin, S. Loring, and R. D. Arbeit. 1993. Relationship between indole production and the differentiation of *Klebsiella* species: indole-positive and negative isolates of *Klebsiella* determined to be clonal. *J. Clin. Microbiol.* **31:**2000–2003.

79. Maslow, J. N., S. Brecher, A. Durbin, J. Gunn, M. Barlow, and R. D. Arbeit. 1993. Diversity among strains of methicillin-resistant *Staphylococcus aureus* (MRSA) from patients with long-term colonization, abstr. 143. *Annu. Meet. Infect. Dis. Soc. Am.*

80. Maslow, J. N., M. E. Mulligan, K. S. Adams, J. C. Justis, and R. D. Arbeit. 1993. Bacterial adhesins and host factors in the development and outcome of *Escherichia coli* bacteremia. *Clin. Infect. Dis.* **17:**89–97.

81. Maslow, J. N., M. E. Mulligan, and R. D. Arbeit. 1993. Molecular epidemiology: the application of contemporary techniques to typing bacteria. *Clin. Infect. Dis.* **17:**153–164.

82. Maslow, J. N., M. E. Mulligan, and R. D. Arbeit. 1994. Recurrent *Escherichia coli* bacteremia. *J. Clin. Microbiol.* **32:**710–714.

83. Maslow, J. N., A. M. Slutsky, and R. D. Arbeit. 1993. The application of pulsed field gel electrophoresis to molecular epidemiology, p. 563–572. *In* D. H. Persing, T. F. Smith, F. C. Tenover, and T. J. White (ed.), *Diagnostic Molecular Microbiology: Principles and Applications*. American Society of Microbiology, Washington, D.C.

84. Maslow, J. N., T. Whittam, R. A. Wilson, M. E. Mulligan, C. Gilks, K. S. Adams, and R. D. Arbeit. Unpublished data.

85. Matar, G. M., B. Swaminathan, S. B. Hunter, L. N. Slater, and D. F. Welch. 1993. Polymerase chain reaction-based restriction fragment length polymorphism analysis of a fragment of the ribosomal operon from *Rochalimaea* species for subtyping. *J. Clin. Microbiol.* **31:**1730–1734.

86. Mayer, L. W. 1988. Use of plasmid profiles in epidemiologic surveillance of disease outbreaks and in tracing the transmission of antibiotic resistance. *Clin. Microbiol. Rev.* **1:**228–243.

87. Mazurek, G. H., S. Hartman, Y. Shang, B. A. Brown, J. S. R. Hector, D. Murphy, and R. J. Wallace, Jr. 1993. Large DNA restriction fragment polymorphism in the *Mycobacterium avium-M. intracellulare* complex: a potential epidemiologic tool. *J. Clin. Microbiol.* **31:**390–394.

88. McGeer, A., D. E. Low, J. Penner, J. Ng, C. Goldman, and A. E. Simor. 1990. Use of molecular typing to study the epidemiology of *Serratia marcescens*. *J. Clin. Microbiol.* **28:**55–58.

89. Mekalanos, J. J. 1992. Environmental signals controlling expression of virulence determinants in bacteria. *J. Bacteriol.* **174:**1–7.

90. Mickelsen, P. A., J. J. Plorde, K. P. Gordon, C. Hargiss, J. McClure, F. D. Schoenknecht, F. Condie, F. C. Tenover, and L. S. Tompkins. 1985. Instability of antibiotic resistance in a strain of *Staphylococcus epidermidis* isolated from an outbreak of prosthetic valve endocarditis. *J. Infect. Dis.* **152:**50–58.

91. Mulligan, M. E., and R. D. Arbeit. 1991. Epidemiologic and clinical utility of typing systems for differentiating among strains of methicillin-resistant *Staphylococcus aureus*. *Infect. Control Hosp. Epidemiol.* **12:**20–28.

92. Mulligan, M. E., R. Y. Y. Kwok, D. M. Citron, J. F. John, Jr., and P. B. Smith. 1988. Immunoblots, antimicrobial resistance, and bacteriophage typing of oxacillin-resistant *Staphylococcus aureus*. *J. Clin. Microbiol.* **26:**2395–2401.

93. Mulligan, M. E., L. R. Peterson, R. Y. Y. Kwok, C. R. Clabots, and D. N. Gerding. 1988. Immunoblots and plasmid fingerprints compared with serotyping and polyacrylamide gel electrophoresis for typing *Clostridium difficile*. *J. Clin. Microbiol.* **26:**41–46.

94. Murray, B. E., K. V. Singh, J. D. Heath, B. R. Sharma, and G. M. Weinstock. 1990. Comparison of genomic DNAs of different enterococcal isolates using restriction endonucleases with infrequent recognition sites. *J. Clin. Microbiol.* **28:**2059–2063.

95. Murray, B. E., K. V. Singh, S. M. Markowitz, H. A. Lopardo, J. Evans Patterson, M. J. Zervos, E. Rubeglio, G. M. Eliopoulos, L. B. Rice, F. W. Goldstein, S. G. Jenkins, G. M. Caputo, R. Nasnas, L. S. Moore, E. S. Wong, and G. Weinstock. 1991. Evidence for clonal spread of a single strain of β-lactamase-producing *Enterococcus (Streptococcus) faecalis* to six hospitals in five states. *J. Infect. Dis.* **163:**780–785.

96. Musser, J. 1993. Multilocus enzyme electrophoresis "real time" DNA sequence analysis of nosocomial outbreaks. Paper presented at 33rd Interscience Conference on Antimicrobial Agents and Chemotherapy, Oct. 17–20, 1993, New Orleans.

97. Musser, J. M., and V. Kapur. 1992. Clonal analysis of

methicillin-resistant *Staphylococcus aureus* strains from intercontinental sources: association of the *mec* gene with divergent phylogenetic lineages implies dissemination by horizontal transfer and recombination. *J. Clin. Microbiol.* **30:**2058–2063.

98. **Musser, J. M., J. S. Kroll, D. M. Granoff, E. R. Moxon, B. R. Brodeur, J. Campos, H. Dabernat, and W. Fredericksen.** 1990. Global genetic structure and molecular epidemiology of encapsulated *Haemophilus influenzae. Rev. Infect. Dis.* **12:**75–111.

99. **Musser, J. M., J. S. Kroll, E. R. Moxon, and R. K. Selander.** 1988. Evolutionary genetics of the encapsulated strains of *Haemophilus influenzae. Proc. Natl. Acad. Sci. USA* **85:**7758–7762.

100. **Myers, L. E., S. V. P. S. Silva, J. D. Procunier, and P. B. Little.** 1993. Genomic fingerprinting of "*Haemophilus somnus*" isolates by using a random-amplified polymorphic DNA assay. *J. Clin. Microbiol.* **31:**512–517.

101. **Nicolle, L. E., H. Bialkowska-Hobrzanska, L. Romance, V. S. Harry, and S. Parker.** 1992. Clonal diversity of methicillin-resistant *Staphylococcus aureus* in an acute-care institution. *Infect. Control Hosp. Epidemiol.* **13:**33–37.

102. **Niesters, H. G. M., W. H. F. Goessens, J. F. M. G. Meis, and W. G. V. Quint.** 1993. Rapid, polymerase chain reaction-based identification assays for *Candida* species. *J. Clin. Microbiol.* **31:**904–910.

103. **Noel, G. J., B. N. Kreiswirth, P. J. Edelson, M. Nesin, S. Projan, W. Eisner, D. J. Bauer, H. de Lencastre, A. M. sa Figueiredo, and A. Tomasz.** 1992. Multiple methicillin-resistant *Staphylococcus aureus* strains as a cause for a single outbreak of severe disease in hospitalized neonates. *Pediatr. Infect. Dis. J.* **11:**184–188.

104. **Ogle, J., J. Janda, D. Woods, and M. Vasil.** 1987. Characterization and use of a DNA probe as an epidemiological marker for *Pseudomonas aeruginosa. J. Infect. Dis.* **155:**119–126.

105. **Ott, M., L. Bender, G. Blum, M. Schmittroth, M. Achtman, H. Tschäpe, and J. Hacker.** 1991. Virulence patterns and long-range genetic mapping of extraintestinal *Escherichia coli* K1, K5, and K100 isolates: use of pulsed-field gel electrophoresis. *Infect. Immun.* **59:**2664–2672.

106. **Ou, C.-Y., C. A. Ciesielski, G. Myers, C. I. Bandea, C.-C. Luo, B. T. M. Korber, J. I. Mullins, G. Schochetman, R. L. Berkelman, A. N. Economou, J. J. Witte, L. J. Furman, G. A. Satten, K. A. MacInnes, J. W. Curran, H. W. Jaffe, Laboratory Investigation Group, and Epidemiologic Investigation Group.** 1992. Molecular epidemiology of HIV transmission in a dental practice. *Science* **256:**1165–1171.

107. **Patterson, J. E., A. Wanger, K. K. Zscheck, M. J. Zervos, and B. E. Murray.** 1990. Molecular epidemiology of β-lactamase-producing enterococci. *Antimicrob. Agents Chemother.* **34:**302–305.

108. **Persing, D. H., T. F. Smith, F. C. Tenover, and T. J. White (ed.).** 1993. *Diagnostic Molecular Microbiology: Principles and Applications.* American Society for Microbiology, Washington, D.C.

109. **Pfaller, M. A.** 1991. Typing methods for epidemiologic investigation, p. 171–182. *In* A. Balows, W. J. Hausler, Jr., K. L. Herrmann, H. D. Isenberg, and H. J. Shadomy (ed.), *Manual of Clinical Microbiology,* 5th ed. American Society for Microbiology, Washington, D.C.

110. **Pfaller, M. A., D. S. Wakefield, R. Hollis, M. Fredrickson, E. Evans, and R. M. Massanari.** 1991. The clinical microbiology laboratory as an aid in infection control: the application of molecular techniques in epidemiologic studies of methicillin-resistant *Staphylococcus aureus. Diagn. Microbiol. Infect. Dis.* **14:**209–217.

111. **Pitt, T. L.** 1988. Epidemiological typing of *Pseudomonas aeruginosa. Eur. J. Clin. Microbiol. Infect. Dis.* **7:**238–247.

112. **Plikaytis, B., J. Marden, J. Woodley, W. Butler, J. Crawford, and T. Shinnick.** 1993. Multiplex PCR specific for multidrug-resistant *Mycobacterium tuberculosis* strain W, abstr. T19. *Am. Soc. Microbiol. Conf. Mol. Diagn. Therapeut.*

113. **Poh, C. L., C. C. Yeo, and L. Tay.** 1992. Genome finger-

printing by pulsed-field gel electrophoresis and ribotyping to differentiate *Pseudomonas aeruginosa* serotype O11 strains. *Eur. J. Clin. Microbiol. Infect. Dis.* **11:**817–822.

114. **Prevost, G., B. Jaulhac, and Y. Piemont.** 1992. DNA fingerprinting by pulsed-field gel electrophoresis is more effective than ribotyping in distinguishing among methicillin-resistant *Staphylococcus aureus* isolates. *J. Clin. Microbiol.* **30:**967–973.

115. **Rabkin, C., W. Jarvis, R. Anderson, J. Govan, J. Klinger, J. LiPuma, W. Martone, H. Monteil, C. Richard, S. Shigeta, A. Sossa, T. Stull, J. Swenson, and D. Woods.** 1989. *Pseudomonas cepacia* typing systems: collaborative study to assess their potential in epidemiologic investigations. *Rev. Infect. Dis.* **11:**600–607.

116. **Ralph, D., M. McClelland, J. Welsh, G. Baranton, and P. Perolat.** 1993. *Leptospira* species categorized by arbitrarily primed polymerase chain reaction (PCR) and by mapped restriction polymorphisms in PCR-amplified rRNA genes. *J. Bacteriol.* **175:**973–981.

117. **Relman, D. A., T. M. Schmidt, R. P. MacDermott, and S. Falkow.** 1992. Identification of the uncultured bacillus of Whipple's disease. *N. Engl. J. Med.* **327:**293–301.

118. **Richardson, J. H., and W. E. Barkley.** 1988. *Biosafety in Microbiological and Biomedical Laboratories* (CDC-NIH). U.S. Department of Health and Human Services publication no. (NIH) 88-8395. U.S. Government Printing Office, Washington, D.C.

119. **Rodrigue, D. C., D. N. Cameron, N. D. Puhr, F. W. Brenner, M. E. St. Louis, I. K. Wachsmuth, and R. V. Tauxe.** 1992. Comparison of plasmid profiles, phage types, and antimicrobial resistance patterns of *Salmonella enteritidis* isolates in the United States. *J. Clin. Microbiol.* **30:**854–857.

120. **Rubin, L. G.** 1987. Bacterial colonization and infection resulting from multiplication of a single organism. *Rev. Infect. Dis.* **9:**488–493.

121. **Samadpour, M., S. Moseley, and S. Lory.** 1988. Biotinylated DNA probes for exotoxin A and pilin genes in the differentiation of *Pseudomonas aeruginosa* strains. *J. Clin. Microbiol.* **26:**2319–2323.

122. **Samore, M. H., K. M. Bettin, P. C. DeGirolami, C. R. Clabots, D. N. Gerding, and A. W. Karchmer.** 1994. Wide diversity of *Clostridium difficile* types at a tertiary referral hospital. *J. Infect. Dis.* **170:**615–621.

123. **Saulnier, P., C. Bourneix, G. Prévost, and A. Andremont.** 1993. Random amplified polymorphic DNA assay is less discriminant than pulsed-field gel electrophoresis for typing strains of methicillin-resistant *Staphylococcus aureus. J. Clin. Microbiol.* **31:**982–985.

124. **Schaberg, D. R., and M. Zervos.** 1986. Plasmid analysis in the study of the epidemiology of nosocomial gram-positive cocci. *Rev. Infect. Dis.* **8:**705–712.

125. **Schlichting, C., C. Branger, J.-M. Fournier, W. Witte, A. Boutonnier, C. Wolz, P. Goullet, and G. Döring.** 1993. Typing of *Staphylococcus aureus* by pulsed-field gel electrophoresis, zymotyping, capsular typing, and phage typing: resolution of clonal relationships. *J. Clin. Microbiol.* **31:**227–232.

126. **Schoonmaker, D., T. Heimberger, and G. Birkhead.** 1992. Comparison of ribotyping and restriction enzyme analysis using pulsed-field gel electrophoresis for distinguishing *Legionella pneumophila* isolates obtained during a nosocomial outbreak. *J. Clin. Microbiol.* **30:**1491–1498.

127. **Schwartz, D. C., and C. R. Cantor.** 1984. Separation of yeast chromosome-sized DNAs by pulsed field gradient gel electrophoresis. *Cell* **37:**67–75.

128. **Selander, R. K., D. A. Caugant, H. Ochman, J. M. Muser, M. N. Gilmour, and T. S. Whittam.** 1986. Methods of multilocus enzyme electrophoresis for bacterial population genetics and systematics. *Appl. Environ. Microbiol.* **51:**873–884.

129. **Selander, R. K., D. A. Caugant, and T. S. Whittam.** 1987. Genetic structure and variation in natural populations of *Escherichia coli,* p. 1625–1648. *In* F. C. Neidhardt, K. L. Ingraham, B. Magasanik, K. B. Low, M. Schaechter, and H. E. Umbarger (ed.), *Escherichia coli and Salmonella typhi-*

murium: Cellular and Molecular Biology, vol. 2. American Society for Microbiology, Washington, D.C.

130. **Selander, R. K., J. M. Musser, D. A. Caugant, M. N. Gilmour, and T. S. Whittam.** 1987. Population genetics of pathogenic bacteria. *Microb. Pathog.* **3:**1–7.

131. **Slutsky, A. M., R. D. Arbeit, T. W. Barber, J. Rich, C. F. von Reyn, W. Pieciak, M. A. Barlow, and J. N. Maslow.** 1994. Polyclonal infection due to *Mycobacterium avium* complex in patients with AIDS detected by pulsed field gel electrophoresis of sequential clinical isolates. *J. Clin. Microbiol.* **32:**1773–1778.

132. **Southern, E. M.** 1975. Detection of specific sequences among DNA fragments separated by gel electrophoresis. *J. Mol. Biol.* **98:**503–517.

133. **Speaker, M. G., F. A. Milch, M. K. Shah, W. Eisner, and B. N. Kreiswirth.** 1991. Role of external bacterial flora in the pathogenesis of acute postoperative endophthalmitis. *Ophthalmology* **98:**639–650.

134. **Stanley, J., N. Baquar, and E. J. Threlfall.** 1993. Genotypes and phylogenetic relationships of *Salmonella typhimurium* are defined by molecular fingerprinting of IS200 and 16S rrn loci. *J. Gen. Microbiol.* **139:**1133–1140.

135. **Struelens, M. J., R. Bax, A. Deplano, W. G. V. Quint, and A. van Belkum.** 1993. Concordant clonal delineation of methicillin-resistant *Staphylococcus aureus* by macrorestriction analysis and polymerase chain reaction genome fingerprinting. *J. Clin. Microbiol.* **31:**1964–1970.

136. **Struelens, M. J., A. Deplano, C. Godard, N. Maes, and E. Serruys.** 1992. Epidemiologic typing and delineation of genetic relatedness of methicillin-resistant *Staphylococcus aureus* by macrorestriction analysis of genomic DNA by using pulsed-field gel electrophoresis. *J. Clin. Microbiol.* **30:**2599–2605.

137. **Struelens, M. J., F. Rost, A. Deplano, A. Maas, V. Schwam, E. Serruys, and M. Cremer.** 1993. *Pseudomonas aeruginosa* and *Enterobacteriaceae* bacteremia after biliary endoscopy: an outbreak investigation using DNA macrorestriction analysis. *Am. J. Med.* **95:**489–498.

138. **Stull, T. L., J. J. LiPuma, and T. D. Edlind.** 1988. A broad-spectrum probe for molecular epidemiology of bacteria: ribosomal RNA. *J. Infect. Dis.* **157:**280–286.

139. **Tabaqchali, S., R. Silman, and D. Holland.** 1987. Automation in clinical microbiology: a new approach to identifying micro-organisms by automated pattern matching of proteins labelled with ^{35}S-methionine. *J. Clin. Pathol.* **40:**1070–1087.

140. **Telenti, A., P. Imboden, F. Marchesi, D. Lowrie, S. Cole, M. J. Colston, L. Matter, K. Schopfer, and T. Bodmer.** 1993. Detection of rifampicin-resistance mutations in *Mycobacterium tuberculosis. Lancet* **341:**647–650.

141. **Telenti, A., F. Marchesi, M. Balz, F. Bally, E. C. Böttger, and T. Bodmer.** 1993. Rapid identification of mycobacteria to the species level by polymerase chain reaction and restriction enzyme analysis. *J. Clin. Microbiol.* **31:**175–178.

142. **Tenover, F. C., R. Arbeit, G. Archer, J. Biddle, S. Byrne, R. Goering, G. Hancock, G. A. Hébert, B. Hill, R. Hollis, W. R. Jarvis, B. Kreiswirth, W. Eisner, J. Maslow, L. K. McDougal, J. M. Miller, M. Mulligan, and M. A. Pfaller.** 1994. Comparison of traditional and molecular methods of typing isolates of *Staphylococcus aureus. J. Clin. Microbiol.* **32:**407–415.

142a.**Tenover, F. C., R. D. Arbeit, R. V. Goering, P. A. Mickelson, B. A. Murray, D. H. Persing, and B. Swaminathan.** Unpublished data.

143. **Tenover, F. C., and G. Killgore.** Unpublished data.

144. **Thal, L. A., J. W. Chow, J. Evans Patterson, M. B. Perri, S. Donabedian, D. B. Clewell and M. J. Zervos.** 1993. Molecular characterization of highly gentamicin-resistant *Enterococcus faecalis* isolates lacking high-level streptomycin resistance. *Antimicrob. Agents Chemother.* **37:**134–137.

145. **Tojo, M., J. Yamashita, D. A. Goldmann, and G. B. Pier.** 1988. Isolation and characterization of a capsular polysaccharide adhesin from *Staphylococcus epidermidis. J. Infect. Dis.* **157:**713–722.

146. **Tsang, A. Y., J. C. Denner, P. J. Brennan, and J. K. McClatchy.** 1992. Clinical and epidemiological importance of typing of *Mycobacterium avium* complex isolates. *J. Clin. Microbiol.* **30:**479–484.

147. **van Belkum, A., R. Bax, P. Peerbooms, W. H. F. Goessens, N. van Leeuwen, and W. G. V. Quint.** 1993. Comparison of phage typing and DNA fingerprinting by polymerase chain reaction for discrimination of methicillin-resistant *Staphylococcus aureus* strains. *J. Clin. Microbiol.* **31:**798–803.

148. **van Belkum, A., J. De Jonckheere, and W. G. V. Quint.** 1992. Genotyping *Naegleria* spp. and *Naegleria fowleri* isolates by interrepeat polymerase chain reaction. *J. Clin. Microbiol.* **30:**2595–2598.

149. **van Belkum, A., M. Struelens, and W. Quint.** 1993. Typing of *Legionella pneumophila* strains by polymerase chain reaction-mediated DNA fingerprinting. *J. Clin. Microbiol.* **31:**2198–2200.

150. **van Embden, J. D. A., M. D. Cave, J. T. Crawford, J. W. Dale, K. D. Eisenach, B. Gicquel, P. Hermans, C. Martin, R. McAdam, T. M. Shinnick, and P. M. Small.** 1993. Strain identification of *Mycobacterium tuberculosis* by DNA fingerprinting: recommendations for a standardized methodology. *J. Clin. Microbiol.* **31:**406–409.

151. **Versalovic, J., T. Koeuth, and J. R. Lupski.** 1991. Distribution of repetitive DNA sequences in eubacteria and application to fingerprinting of bacterial genomes. *Nucleic Acids Res.* **19:**6823–6831.

152. **Vilgalys, R., and M. Hester.** 1990. Rapid genetic identification and mapping of enzymatically amplified ribosomal DNA from several *Cryptococcus* species. *J. Bacteriol.* **172:**4238–4246.

153. **von Reyn, C. F., J. N. Maslow, T. W. Barber, J. O. Falkinham III, and R. D. Arbeit.** 1994. Persistent colonization of potable water as a source of *Mycobacterium avium* infection in AIDS. *Lancet* **343:**1137–1141.

154. **Wachsmuth, K.** 1985. Genotypic approaches to the diagnosis of bacterial infections: plasmid analyses and gene probes. *Infect. Control* **6:**100–109.

155. **Welsh, J., and M. McClelland.** 1990. Fingerprinting genomes using PCR with arbitrary primers. *Nucleic Acids Res.* **18:**7213–7218.

156. **Whittam, T. S., I. K. Wachsmuth, and R. A. Wilson.** 1988. Genetic evidence of clonal descent of *Escherichia coli* O157:H7 associated with hemorrhagic colitis and hemolytic uremic syndrome. *J. Infect. Dis.* **157:**1124–1133.

157. **Widmer, A. F., R. P. Wenzel, A. Trilla, M. J. Bale, R. N. Jones, and B. N. Doebbeling.** 1993. Outbreak of *Pseudomonas aeruginosa* infections in a surgical intensive care unit: probable transmission via hands of a health care worker. *Clin. Infect. Dis.* **16:**372–376.

158. **Williams, J. G. K., A. R. Kubelik, K. J. Livak, J. A. Rafalski, and S. V. Tingey.** 1990. DNA polymorphisms amplified by arbitrary primers are useful as genetic markers. *Nucleic Acids Res.* **18:**6531–6535.

159. **Woodford, N., D. Morrison, A. P. Johnson, V. Briant, R. C. George, and B. Cookson.** 1993. Application of DNA probes for rRNA and *vanA* genes to investigation of a nosocomial cluster of vancomycin-resistant enterococci. *J. Clin. Microbiol.* **31:**653–658.

160. **Woods, C. R., Jr., J. Versalovic, T. Koeuth, and J. R. Lupski.** 1992. Analysis of relationships among isolates of *Citrobacter diversus* by using DNA fingerprints generated by repetitive sequence-based primers in the polymerase chain reaction. *J. Clin. Microbiol.* **30:**2921–2929.

161. **Zhang, Y., G. H. Mazurek, M. D. Cave, K. D. Eisenach, Y. Pang, D. T. Murphy, and R. J. Wallace, Jr.** 1992. DNA polymorphisms in strains of *Mycobacterium tuberculosis* analyzed by pulsed-field gel electrophoresis: a tool for epidemiology. *J. Clin. Microbiol.* **30:**1551–1556.

162. **Zingler, G., M. Ott, G. Blum, U. Falkenhagen, G. Naumann, W. Sokolowska-Köhler, and J. Hacker.** 1992. Clonal analysis of *Escherichia coli* serotype O6 strains from urinary tract infections. *Microb. Pathog.* **12:**299–310.

Procedures To Use During Outbreaks of Food-Borne Disease

FRANK L. BRYAN

18

Investigation of an outbreak of food-borne illness is a team effort, often involving epidemiologists, laboratory personnel, and sanitarians. Each has an essential role to perform. The laboratory aids medical practitioners by testing clinical specimens to either verify diagnosis or confirm the etiological agent responsible for the illness. The laboratory group is often the first to determine that patients may be ill from a food-borne disease.

Epidemiologists interview ill persons, seek persons at risk but not ill (controls), and calculate and compare incidence rates of both groups. They make time, place, and person associations and calculate the probability that a food was the responsible vehicle. From these data, they form hypotheses that must be confirmed. This confirmation is done by the laboratory (i) finding the same type of pathogen in specimens from most of the ill persons but not from persons in the control group (or from only a few of them) and (ii) isolating the same type of pathogen from the responsible vehicle.

Sanitarians also play a key role in proving that a food is responsible for illness by making observations and measurements that relate to contamination, survival, and growth of the etiological agent. The source and mode of contamination, the manner by which the pathogen survived processing, and the degree of bacterial growth that allowed development of populations or toxins in concentrations high enough to cause the illness are sought. With such data, sanitarians explain events that led to the outbreak and initiate control and preventive measures. The laboratory provides confirmation that these events occurred by testing samples that are submitted and by performing other tests (e.g., challenge tests). Furthermore, laboratory personnel must often advise investigators on taking suitable specimens from patients, controls, and food handlers and on collecting, preserving, and transporting samples of foods that are apt to yield pathogens.

Depending on the probable disease under investigation, specimens and samples can include (i) serum, stools, vomitus, and urine from patients and controls; (ii) blood, spleen, and liver tissue and, from fatal cases, intestinal contents; (iii) stool or rectal swabs, blood, nasal, or throat swabs, and exudate or pus from lesions of persons who handled the implicated or suspect food; (iv) food from implicated lots of processed foods; (v) leftovers from an epidemiologically implicated food or meal; (vi) swabs of equipment used to process epidemiologically implicated foods; (vii) portions, scraps, or swabs of raw foods that may have introduced pathogens into the kitchen or plant environments or that show that the pathogen was present in the environment; (viii) water, air, or other environmental sites that may harbor pathogens; and (ix) when applicable, rectal or cloacal swabs of animals, swabs or portions of animal droppings, samples of feed, or swabs of environmental contacts of animals to determine reservoirs or sources of contamination.

This chapter is limited to describing procedures for collecting and processing food and environmental samples and to interpreting results of laboratory examinations of the samples. Information for collecting specimens is presented in chapter 3 of this Manual and in material by Balows et al. (1) and Lennette et al. (13). Procedures for making complete investigations of food- and waterborne illnesses are described by Bryan et al. (5, 6).

SELECTION AND COLLECTION OF SAMPLES

Sample units of foods that are suspected of being the vehicle of the etiological agent under investigation and perhaps samples of other potentially hazardous foods served at the time that the ill persons ate should be collected. Suspect foods are those that are implicated by an attack rate table, odds ratio, probability test, or other epidemiological data under investigation or that have a history of being vehicles in similar outbreaks of the food-borne disease under investigation. Potentially hazardous foods are those that support bacterial growth because of their properties (pH, nutrients, water activity [a_w]).

Collect sample units in such a way that they are not contaminated by the person collecting the sample or by utensils or containers used. Efforts should be made to minimize change in microbiological status during transport and storage. Collection of a sample several times larger than the intended analytical sample will allow for testing for several microorganisms, accidents, and retesting.

Collect sample units of suspect foods at patients' domiciles, at establishments where the suspect food or meal was ingested or from which it was purchased, or from the place of production or processing, as applicable to the situation being investigated. Ordinarily, collect sample units of any leftover food or beverage that was ingested within the last

FOOD SAMPLE COLLECTION REPORT

Form E

	Sample Number	Number of Sample Units Taken	Number of Units in Lot

Place Collected	Address		Phone

Person-In-Charge	Sample		Date/Hour Collected

Product Identification: Name and Nature	Brand		Code/Lot Number
Name of Manufacturer, Buyer, Seller, Importer (as appropriate)	Address		Container Size, Production Date or Weight
Type of Identification of Transportation Vehicle	Origin of Shipment		Date of Shipment / Arrival Date
Bill of Lading or Contract Number		Destination	

Reason for Collecting Sample:
☐ Food from alleged outbreak ☐ Ingredient of outbreak food ☐ Port of entry ☐ Routine ☐ Environmental ☐ Similar food prepared in similar manner to that involved in outbreak ☐ Special survey ☐ Other (Specify)

Method of Collecting and Shipping Sample: Method of Sterilizing: Container	Collection Utensil	Method of sampling: ☐ Random throughout lot ☐ Random throughout accessible units ☐ Other
Location Food Stored When Sampled	Temperature: Food / Storage Unit	Time Between Serving and Sampling
Shipped: ☐ Refrigerated. ☐ Frozen. ☐ Ambient	Identification Marks	Cost of Sample

Investigator/Sampler	Title	Agency

Signature of Sampler		Signature of Representative of Party Concerned

Test Requested on Basis of Epidemiologic Data	Presence/Absence	Count/Concentration	Definitive Type
☐ Staphylococci			
☐ Staphyloenterotoxin			
☐ C. perfringens			
☐ B. cereus			
☐ Salmonella			
☐ Shigella			
☐ E. coli			
☐ V. parahaemolyticus			
☐ Aerobic Colony Count			
☐ Coliform			
☐ Fecal Coliform			
☐ Enterococci			
☐ Enterobacteriacea			
☐ Yeast/Molds			
☐ Other (specify)			

Condition of Food When Received	pH	a_w	Temperature: When received

Comments and interpretations

Laboratory Analyst	Laboratory	Date/Hour: Received	Started	Completed

FIGURE 1 Sample collection form.

72 h or of ingredients used in the preparation of these foods at patients' domiciles. Obtain sample units of suspect foods or potentially hazardous foods from the suspect meal at food service establishments as well as sample units of food from an allegedly contaminated lot or at processing plants. Pay particular attention to the collection of samples before and after critical processes to determine the likelihood of contamination, survival, or growth, as applicable to the process.

Collect representative samples of foods chosen randomly from unopened, original packages during surveys or quality control evaluations (12). Sampling during outbreak investigations, however, is done quite differently. On the basis of professional judgment, one should collect sample units of food from equipment at the point of operation (such as after possible contamination, survival, or growth) that is most likely to yield food-borne pathogens. Usually, sample units should be taken from the geometric center of a food, the area where multiplication would most likely have occurred and consequently the site most likely to yield positive results and the highest counts. An exception is fermented sausage, which is sampled a short distance below the surface. Collect either top or side layers of refrigerated foods (for which cooling is more rapid than in the center of the mass) to get an indication of what the microbiological populations were soon after serving and before refrigeration.

Foods that could have been served previously should be sought in storage facilities. If no foods are left over from the suspect meal or lot, collect sample units of food prepared in a manner similar to that of the suspect food, raw foods, ingredients used in the suspect food, or foods from the same suspect source. Collect separate portions of each compo-

TABLE 1 Methods of collecting, preserving, packing, and shipping samples

Sample	Method of collecting and preserving	Method of packing and shipping
Solid food	Cut or separate portions of food with sterile knife or other implement if necessary. Using a sterile implement (e.g., spoon, scoop), aseptically collect at least 200 g of sample from geometric center or other locations as deemed necessary, transfer it to a sterile plastic bag or wide-mouth glass jar, and refrigerate or store in ice.	Label. Put refrigerant around sample container. Do not freeze or use dry ice. Take sample to laboratory, or ship it by most rapid means.
Liquid food or beverage	Stir or shake unless specific site is to be sampled. Take sample in one of the following ways.	
	1. Pour or ladle with sterile implement at least 200 ml into sterile container. Refrigerate or store in ice.	As above.
	2. Put pipette or long sterile tube into liquid, and cover top opening with finger. Transfer at least 200 ml into sterile jar or bag. Refrigerate or store in ice.	As above.
	3. Immerse Moore swab into vat of liquid food, or insert swab into pipeline and allow liquid to flow through. Keep swab in place for several hours if practicable. Transfer swab to jar containing enrichment broth for specific pathogen sought.	Label. Take sample to laboratory as soon as practicable; refrigeration may not be needed.
	4. If liquid is not viscous, pass 1 to 2 liters through membrane filter. Transfer filter pad aseptically into jar of enrichment broth for specific pathogen sought.	As above.
Frozen food	Use one of the following procedures.	
	1. Ship entire package of frozen food to laboratory without thawing or opening.	Label. Keep frozen. Use dry ice if necessary. Take or ship in insulated container.
	2. Break at least 200-g portion apart with hands encased in sterile plastic glove or inverted plastic sampling bag.	As above.
	3. Chop or saw off (with sterile or disinfected cutting surface) at least 200 g of frozen meat or seafood.	As above.
	4. Drill with large-diameter sterile auger from top of opened container diagonally through center to bottom at opposite side. Repeat from other side and other locations until at least 200 g is collected.	As above.
	5. Chip frozen material with hammer and sterile chisel so that it falls on sterile plastic sheet or foil, and collect chips with sterile implement. Transfer at least 200 g of chips into sterile container.	As above.
Canned food	Collect can or jar. Put it into plastic bag, or wrap it in foil. Open it only in laboratory under controlled conditions, where surfaces will be disinfected or other precautions will be taken before can is opened.	Label bag. Refrigeration is not necessary.
Dry or dehydrated food	Collect sample from as close to center of container as possible. Two procedures may be used. 1. Remove surface layer down to near center with sterile spoon or scoop. Withdraw at least 200 g of sample with different spoon, scoop, spatula, tongue depressor, or similar instrument.	Label. Keep in tightly sealed, moisture-resistant container. Take or ship to laboratory. Do not refrigerate.

(Continued on next page)

TABLE 1 Methods of collecting, preserving, packing, and shipping samples *(Continued)*

Sample	Method of collecting and preserving	Method of packing and shipping
	2. Insert sterile hollow tube from top of one side of container diagonally through center to bottom of opposite side. Hold top of tube, and transfer sample to sterile container. Repeat from opposite side and other locations until at least 200 g of sample is collected. Transfer material to sterile container.	
Raw meat or poultry	Sample in one of the following ways. 1. With sterile implement or sterile plastic glove, put chicken carcass, poultry part, or cut of meat into large sterile plastic bag. Add 100–300 ml of enrichment broth, shake, and massage. For turkeys or large cuts of meat (e.g., roasts), use 500 ml of enrichment broth. Remove carcass, part, or cut of meat, and close bag. 2. With hand encased in sterile plastic glove, wipe sterile sponge over large area of carcass or cut of meat. Put sponge into jar of enrichment broth. 3. Moisten swab with buffered distilled water or 0.1% peptone water. Swab large portion of carcass or cut of meat. Put tip of swab stick into tube of enrichment broth. 4. With hand encased in sterile plastic glove, wipe carcass with sterile gauze squares. Put gauze into jar of enrichment broth. 5. Aseptically cut portions of meat or skin from different areas of carcass or cut of meat, or remove portion of carcass. Put at least 200 g into sterile plastic bag or glass jar. Refrigerate or store in ice. 6. Aseptically cut 200 g of neck skin of poultry, and put into sterile container.	Same as with solid or liquid food, or if in enrichment broth, take to laboratory as soon as practicable.
Scrap material, air filters, sweepings, dust, litter, etc.	Cut filter or scrap material with sterile knife if necessary. Pick up or scoop, as applicable, material with sterile tongue depressor, spatula, spoon, or tongs, and put in sterile plastic bag or wide-mouth jar. Collect at least 200 g. Refrigerate or store food scraps in ice.	Handle like solid or dry food as applicable. If food material is included, put refrigerant around sample, and take or ship to laboratory by most rapid means available.
Water	Different types of samples may be required. Collect 1–5 liters. 1. Take historical samples, including water in bottles in refrigerators, ice cubes, and water in tanks. 2. Take line water samples after turning on tap and allowing it to run for 10 s. Hold sterile bottle under tap, and fill to 1 in. (ca. 2.5 cm) below lip. 3. Take water source samples after water has run for 5 min. Hold sterile bottle under tap, and fill to 1 in. (ca. 2.5 cm) below lip. 4. Pour through membrane filter. 5. Use Moore swabs to sample water in streams or pipelines; keep them in place for up to 48 h, and then transfer to jars of enrichment broth.	Tape collection container closed; label. Pack with absorbent material. Box, and take or ship to laboratory. Refrigeration is usually not necessary.

(Continued on next page)

TABLE 1 *(Continued)*

Sample	Method of collecting and preserving	Method of packing and shipping
Environmental or equipment surface	Use one of the following methods.	Label. Put refrigerant around sample container. Take sample to laboratory, or ship it by most rapid means.
	1. With hand encased in sterile plastic glove, wipe sterile sponge over large area of surface or measured area (if counts are to be made). Put sponge into either jar of enrichment broth or measured amount of diluent (if counts are to be made).	
	2. Moisten swab with buffered distilled water or 0.1% peptone water. Swab either large area of surface or measured area (if counts are to be made). Put swab into tube of diluent.	As above.
	3. Agar contact (Rodac) devices and similar plates are frequently used for verification of cleaning, but they have minimal use during outbreak investigations.	Tape closed, and follow above procedure.
Air	Use one of the following techniques depending on sampling objectives.	
	1. Draw air through slit sampler for measured time. Microorganisms impact on revolving agar surface.	Remove plate. Tape collection container closed. Label. Refrigerate. Ship to laboratory by most rapid means.
	2. Draw air through sieve sampler (which has cascade arrangement with successive smaller holes that captures larger particles first) for a measured time. Microorganisms impact on agar plates.	As above.
	3. Draw air through centrifugal sampler, which impinges particles onto agar surface, for measured time.	As above.
	4. Draw air through electrostatic precipitator for measured time. Microorganisms are given a charge, and they collect on rotating disc.	As above.
	5. Expose either agar plate or jar of enrichment broth for measured time. Microorganisms fall on surfaces.	As above.
	6. Draw air through membrane or fiber filter for measured time.	Put filter either in enrichment broth or on agar surface. Label. Either incubate or refrigerate, and ship to laboratory.
	7. Draw air through liquid impinger for measured time. Microorganisms are impinged in the liquid (e.g., buffered distilled water or enrichment broth).	Refrigerate. Tape collection container closed, label, and take sample to laboratory.

nent of a food (such as meringue and filling for custard pies) that was prepared by different persons or was prepared or stored in different ways before final assembly.

If the mode of spread or source of etiological agents is sought and if definitive typing of isolates is contemplated, rub sponges or swabs over food contact surfaces of equipment to collect environmental samples. Wear sterile gloves when holding the sponges. Rub sterile sponges or swabs randomly over large areas of raw foods (e.g., foods of animal origin) that are likely sources for the pathogen under investigation and over food contact surfaces of utensils and equipment used to process, store, or prepare suspect foods.

Collect sample units aseptically with sterile implements or sterile gloves, and put them into sterile containers as described in Table 1. Typical equipment for collecting,

holding, and preserving samples is listed in Table 2. Ordinarily, wrap and sterilize sampling implements in the laboratory before going to the site where the food was allegedly mishandled or mistreated. Field disinfections of sampling equipment can be accomplished as described in Table 3 when necessary. See references 5, 6, 10 to 12, 14, 15, and 19 for additional information on sample collection.

HANDLING SAMPLES AFTER COLLECTION

Mark all samples by an identifying code as soon as they have been collected. A description of the material sampled and any features that might be useful in interpreting results (such as the temperature of the food or the environment from which it was taken) should be recorded in a notebook or on a sample collection form such as that illustrated in Fig. 1 (6).

Frequently, sample units are held for a while before they are taken or sent to the laboratory. Keep frozen foods frozen either in freezers in the establishment in which they are collected or in containers in which they are in contact with dry ice until shipped. Store perishable foods at temperatures between 0 and 4.4°C either in refrigerators at the establishment in which they are collected or in an insulated container in which they are surrounded by ice or other coolant. Cool sample units of hot foods or beverages in sealed containers rapidly by holding them under cold running water or putting them in an ice bath. Then hold them at 0 to 4.4°C.

As soon as sampling has been completed, pack sample units appropriately to avoid spillage or breakage and to maintain desired temperatures. Pack frozen foods in dry ice. Keep sample units of perishable or chilled food cold with ice in plastic bags or other coolant. If dry ice is used for nonfrozen foods (which is not recommended), insulate the sample units to prevent them from freezing. When it is not possible to deliver the sample personally, send it to the laboratory by the most rapid and economical means feasible, which is often overnight mail. Advise the laboratory of the forthcoming samples, method of shipment, shipment numbers, and expected time of arrival. (See Table 1 for procedures for preserving and transporting specific types of samples.)

RECEIVING AND PREPARING SAMPLE UNITS FOR ANALYSIS

Upon arrival of the sample in the laboratory, note the physical appearance of each sample unit and its container. Also, if a temperature control sample is submitted, measure and record its temperature. Otherwise, measure and record the temperatures of perishable foods upon their arrival. This can be done by wrapping the plastic bag around a thermometer. Avoid contaminating the food during temperature measurements with thermometers or thermocouples if they are inserted directly into foods; use disinfected probes. When appropriate, measure pH and water activity (a_w) of the food. This is usually done after the analytical sample has been withdrawn.

Analyze samples as soon as practicable (but if at all possible, within 30 h) after they arrive at the laboratory. It is sometimes necessary, however, to store them for a short time (e.g., overnight) before work can begin. Those that arrive frozen should be either sampled while still frozen or thawed in a refrigerator at 0 to 4.4°C. Store low-moisture and canned foods at room temperature.

Aseptically remove sample units from their containers. Prepare and examine a Gram or other appropriate stain of a drop of liquid food or beverage or of a well-mixed (1:10) homogenate of each analytical sample of solid food. Methods for preparing homogenates and examination for specific pathogens are described by Vanderzant and Splittstoesser (18).

DETERMINATION OF APPROPRIATE TESTS

Determine the types of tests and the order in which they are to be done by considering (i) clinical signs and symptoms and incubation periods (if known) of the affected persons, (ii) results of Gram or other appropriate stain of liquid foods

or the homogenates of solid foods, and (iii) types of foods to be tested or (iv) epidemiological data. Six categories (according to signs and symptoms) of food-borne diseases are listed in Table 4, as follows: (i) upper gastrointestinal tract signs and symptoms (nausea, vomiting) occur first or predominate, (ii) lower gastrointestinal tract (abdominal cramps, diarrhea) signs and symptoms occur first and predominate, (iii) sore throat and respiratory tract signs and symptoms, (iv) neurological signs and symptoms (visual disturbances, vertigo, tingling, paralysis), (v) allergic signs and symptoms (facial flushing, itching), and (vi) signs and symptoms associated with general infection (fever, chills, malaise, prostration, aches, swollen lymph nodes).

Usually, the syndrome can readily be classified in one of these categories. If the patient's incubation period (duration between ingestion of implicated food or meal and onset of first recognized sign or symptom) is known, further classification can be done. This classification may take the form of <1, 1 to 6, 7 to 12, 13 to 72, and >72 h of incubation. Initially, one should analyze portions of the sample units for the etiological agents or toxins most likely associated with the syndrome. If the results are negative, analyze another portion for other plausible agents.

The Gram reaction and cellular morphology of predominant organisms in a sample of food provide information indicating the kind of microorganisms to seek by culture techniques. (See Table 5 for a guide to the interpretation of staining results.) An analytical sample, however, will usually have to contain a rather high number of microorganisms before they can be seen in a microscopic field. Furthermore, many spoilage and nonpathogenic bacteria are similar in appearance to the listed pathogens. Hence, this technique will not always indicate the causative agent, but it may aid in forming hypotheses about microorganisms for which testing ought to be done.

Bacteriological studies and epidemiological investigations have established the usual sources, reservoirs, and vehicles of many food-borne pathogens. Tests to run when examining foods that are suspected to be or have been epidemiologically implicated as vehicles of food-borne illness are listed in Table 6. See references 1, 2, 4, 7 to 9, 13, and 16 for additional information about food-borne illnesses.

INTERPRETATION OF RESULTS

Food products have diverse chemical components and properties. These products are processed, stored, and prepared in a variety of ways that affect the quantity and type of microorganisms that may have been introduced into, survived in, and grew in them. Microorganisms in foods are in a dynamic state, and the numbers and predominant types vary with time, with the nature of the substrate, and with environmental conditions to which the food and later the sample unit have been subjected. These same factors influence the methods used to isolate pathogens in the foods as well as those used to interpret results of the laboratory tests.

Pathogens are often easier to isolate from clinical specimens than from foods. For example, pathogenic members of the *Enterobacteriaceae* in the human gut are in an environment that provides greater enrichment than that provided by many foods. These same pathogens sometimes occur in small numbers in foods because of injury caused by processing, die-off during storage, or overgrowth by spoilage organisms. Therefore, routine clinical laboratory methods may be inadequate to detect

TABLE 2 Equipment useful for investigations[a]

Item	Examples
Sterile sample containers	Plastic bags (disposable or Whirl-Pak type), wide-mouth jars (0.3- to 1.5-liter capacity) with screw caps, water sample bottles (bottles for chlorinated water should contain enough sodium thiosulfate to provide a concn of 100 mg of this compound/ml of sample), foil or heavy wrapping paper (wrapped), large plastic bags, metal cans
Sterile and wrapped sampling implements	Spoons, scoops, tongue depressor blades, butcher knife, forceps, tongs, spatula, drill bits, metal tubes (1.23–2.5 cm in diameter, 30–60 cm long), pipettes, scissors, swabs, Moore swabs (compact pads of gauze made from strips 120 by 15 cm tied in the center with long piece of stout twine or wire; for sewer drain, stream, or pipeline samples)
Specimen-collecting equipment	Cartons (with lids) for stool specimens, bottles containing preservative and transport medium, stool specimen protective canister and cartons, sterile swabs, rectal swab outfits, sterile gauze pads (10 by 10 cm), tubes of transport medium
Temperature-measuring devices	Thermocouples (assortment, including needle points of various lengths, button or surface-registering type, welded ends) with recording potentiometer, data logger, or digital indicator; bayonet-type thermometer (5–9 in. [ca. 12.5–22.5 cm] long) with digital readout or dial (0–220°F). Optionally, thermosisters and indicator thermometers with platinum probes can be used. Reflecting thermometers can be used to measure temperatures of surfaces.
Supporting equipment	Fine-point felt-tip marking pen, pencils, note pad, roll of adhesive or masking tape, labels, waterproof cardboard tags with eyelets and wire ties, box-opening knife, cotton, flashlight, electric drill, matches, 0.1% peptone water or buffered distilled water (e.g., 5 ml in screw-cap tubes), test tube rack to fit tubes used, insulated chest, pH meter, a_w hygrometer, spare batteries for all equipment using them, investigational forms
Sterilizing agents	95% ethyl alcohol, propane torch
Coolants	Refrigerant in plastic bags, liquid in cans, rubber or heavy plastic bags that can be filled with water and frozen, heavy-duty plastic bags for ice, "canned ice"
Clothing and coverings	White laboratory-type coat, paper hats, disposable plastic gloves, disposable plastic boots

[a]At least 15 sterile plastic bags or wide-mouth jars, 15 sterile spoons, six specimen collection containers or devices, temperature-measuring devices, one each of the supporting equipment, and sterilizing agents should be preassembled in a kit that is kept by the agency responsible for investigating food-borne illness. Periodic resterilization or replacement of sterile supplies, media, and transport media is required to maintain kit in a ready-to-use condition.

pathogens in foods. (See reference 18 for appropriate procedures for foods.)

An agent thought to be responsible for an illness can be confirmed if a pathogen that causes a syndrome similar to that under investigation is isolated from appropriate specimens from patients. An agent also can be confirmed if toxins are identified in specimens from patients or if there is evidence of a rise in antibody titer in serum specimens

TABLE 3 Field disinfection of equipment

Medium	Procedure
Steam	Expose at 100°C for 1 h in enclosed chamber if proper equipment is available.
Flame	Expose contact surfaces to flame from burning alcohol, propane torch, Bunsen burner, alcohol lamp, or alcohol-soaked cotton. Wash, rinse, dry, immerse in 95% alcohol, remove, and flame until alcohol burns away. Repeat three times. (Spores may survive if procedure is inadequately performed.)
Hot water	Wash and then immerse in boiling water. (Spores may survive if not removed by washing.)
Disinfectants	Wash, rinse, and immerse for at least 1 min in solution containing disinfectant equivalent to not less than 50 ppm of hypochlorite or 12.5 ppm of available iodine (water temp at least 75°F [25°C]). Rinse in sterile water. (Spores may survive if contaminants are not removed by washing.)

TABLE 4 Guide for laboratory tests indicated by certain signs and/or symptoms and incubation periods

Category of illness	Incubation period	Predominant signs and symptoms	Specimens to analyze		Substance(s) to test for
			From patients[a]	From food workers	
Upper gastrointestinal tract (nausea, vomiting); signs and symptoms occur first or predominate	<1 h	Nausea, vomiting, unusual taste, burning of mouth	Vomitus, urine, blood, stool		Antimony, lead, arsenic, zinc, cadmium, copper, fluoride, tin
		Nausea, vomiting, retching, diarrhea, abdominal pain	Vomitus		Gastroenteritis-causing type mushrooms
	1–2 h	Nausea, vomiting, cyanosis, headache, dizziness, dyspnea, trembling, weakness, loss of consciousness	Blood		Nitrites
	1–6 h; mean, 2–4 h	Nausea, vomiting, retching, diarrhea, abdominal pain, prostration	Vomitus, stool	Nasal swab, swab of lesion on skin if *Staphylococcus aureus* is suspected	*Staphylococcus aureus* and its enterotoxins; *Bacillus cereus* and its emetic toxin
	6–24 h	Nausea, vomiting, diarrhea, thirst, dilation of pupils, collapse, coma	Urine, blood, (SGOT, SGPT enzyme tests), vomitus		Diarrhetic shellfish poison; *Amanita, Phallin,* and *Gyromitrin* toxin group of mushrooms
Lower gastrointestinal tract (abdominal cramps, diarrhea); signs and symptoms occur first or predominate	8–22 h; mean, 10–12 h	Abdominal cramps, diarrhea	Stool	Stool, rectal swab	*Clostridium perfringens, Bacillus cereus* or its diarrheagenic toxin, *Enterococcus faecalis* (?), *Enterococcus faecium* (?)
	12–72 h; mean, 10–12 h	Abdominal cramps, diarrhea, vomiting, fever, chills, malaise	Stool	Stool, rectal swab	*Salmonella, Arizona, Campylobacter jejuni, Shigella,* pathogenic *Escherichia coli,* other *Enterobacteriaceae, Vibrio cholerae, Vibrio parahaemolyticus, Yersinia enterocolitica, Aeromonas hydrophila, Plesiomonas shigelloides, Pseudomonas aeruginosa*
		As above but diarrhea is bloody	Stool	Stool	*Escherichia coli* O157:H7, *Shigella*

	Symptoms	Specimen	Specimen	Etiologic agent
1.5–3 days	Diarrhea, fever, vomiting, abdominal pain; possibly respiratory symptoms	Stool, blood	Stool, blood	Enteric viruses (e.g., poliovirus, coxsackieviruses, ECHO[b]) parvovirus-like (Norwalk-like) agents, rotaviruses, astroviruses, caliciviruses, coronaviruses
1–12 days, usually 7	Profuse watery diarrhea, abdominal pain, malaise, fever, anorexia, nausea, vomiting	Stool, intestinal or tissue biopsy sample	Stool	Cryptosporidium spp.
1–6 wk	Mucoid diarrhea (fatty stools), abdominal pain, weight loss	Stool, intestinal or tissue biopsy sample	Stool	Giardia lamblia
1 to several wk; mean, 3–4 wk	Abdominal pain, diarrhea, constipation, headache, drowsiness, ulcers, variable—often asymptomatic	Stool	Stool	Entamoeba histolytica
3–6 mo	Nervousness, insomnia, hunger pains, anorexia, weight loss, abdominal pain, sometimes gastroenteritis	Stool	Stool	Taenia saginata, Taenia solium
Sore throat and respiratory				
<1 h	Burning mouth and throat, vomiting	Vomitus		Sodium hydroxide, other strong alkalis or acids
12–72 h	Sore throat, fever, nausea, vomiting, rhinorrhea, sometimes rash	Throat swab, blood	Throat swab, swab of lesions on skin	Streptococcus pyogenes
2–5 days	Inflamed throat and nose, spreading grayish exudate, fever, chills, sore throat, malaise, difficulty swallowing, edema of cervical lymph node	Throat swab, blood	Throat swab, swab of lesions	Corynebacterium diphtheriae

(Continued on next page)

TABLE 4 Guide for laboratory tests indicated by certain signs and/or symptoms and incubation periods *(Continued)*

Category of illness	Incubation period	Predominant signs and symptoms	Specimens to analyze		Substance(s) to test for
			From patients[a]	From food workers	
Neurological (visual disturbances, vertigo, tingling, paralysis)	<1h	Tingling and numbness, giddiness, staggering, drowsiness, tightness of throat, incoherent speech, respiratory paralysis			Paralytic shellfish toxin
		Gastroenteritis, nervousness, blurred vision, chest pain, cyanosis, twitching, convulsions	Blood, urine, fat biopsy sample		Organic phosphate
		Excessive salivation, perspiration, and tearing; gastroenteritis, irregular pulse, pupils constricted, asthmatic breathing	Urine		Muscaria-type mushrooms
		Lightheadedness, drowsiness, followed by state of excitement, confusion, delirium, visual disturbances	Urine		Ibotonic acid and muscimol groups of mushrooms
		Tingling and numbness, dizziness, pallor, gastroenteritis, hemorrhage, desquamation of skin, eyes fixed, loss of reflexes, twitching, paralysis			Tetraodontoxin
	1–6h	Tingling and numbness, gastroenteritis, dizziness, dry mouth, muscular aches, dilated pupils, blurred vision, paralysis			Ciguatera toxin
		Nausea, vomiting, tingling, dizziness, weakness, anorexia, weight loss, confusion	Blood, urine, stool, gastric washings		Chlorinated hydrocarbons
	12–72h	Vertigo; double or blurred vision; loss of reflex to light; difficulty swallowing, speaking, breathing; dry mouth; weakness; respiratory paralysis	Blood, stool		*Clostridium botulinum* and its neurotoxins
	>72h	Numbness, weakness of legs, spastic paralysis, impairment of vision, blindness, coma	Urine, blood, stool, hair		Organic mercury
		Gastroenteritis, leg pain, ungainly high-stepping gait, foot and wrist drop	Biopsy of gastrocnemius muscle		Triorthocresyl phosphate

Category	Incubation period	Signs and symptoms	Specimen	Specimen	Agent
Allergic type (facial flushing, itching)	<1 h	Headache, dizziness, nausea, vomiting, peppery taste, burning of throat, facial swelling and flushing, stomach pain, itching of skin		Vomitus	Histamine
		Numbness around mouth, tingling sensation, flushing, dizziness, headache, nausea			Monosodium glutamate
		Flushing, sensation of warmth, itching, abdominal pain, puffing of face and knees		Blood	Nicotinic acid
Generalized infection (fever, chills, malaise, prostration, aches, swollen lymph nodes)	4–28 days; mean, 9 days	Gastroenteritis, fever, edema about eyes, perspiration, muscular pain, chills, prostration, labored breathing		Muscle biopsy sample	Trichinella spiralis
	7–28 days; mean, 14 days	Malaise, headache, fever, cough, nausea, vomiting, constipation, abdominal pain, chills, rose spots, bloody stools	Stool	Stool, urine, blood	Salmonella typhi
	10–13 days	Fever, headache, myalgia, rash		Lymph node biopsy sample, blood	Toxoplasma gondii
	10–50 days; mean, 25–30 days	Fever, malaise, lassitude, anorexia, nausea, abdominal pain, jaundice	Blood	Blood, feces (within 1 wk of onset of dark urine)	Serum antibody or hepatitis A virus
	Various periods (depends on specific illness)	Fever, chills, headache or joint ache, prostration, malaise, swollen lymph nodes, and other specific signs and symptoms of disease in question	Blood, stool, if warranted by disease in question	Blood, stool, urine, sputum, lymph node, gastric washings (one or more, depending on organism)	Bacillus anthracis, Brucella melitensis, Brucella abortus, Brucella suis, Coxiella burnetii, Francisella tularensis, Listeria monocytogenes, Mycobacterium tuberculosis, other Mycobacterium spp., Pasteurella multocida, Streptobacillus moniliformis, other pathogens as deemed appropriate based on syndrome (see references 3 and 17)

[a]SGOT, serum glutamic oxalacetic transaminase; SGPT, serum glutamic pyruvic transaminase.
[b]ECHO, enteric cytopathic human orphan viruses.

TABLE 5 Microscopic observations for determining morphology of predominant organisms[a] in suspect foods

Gram reaction	Cell shape	Cell grouping	Flagellum arrangement[b]	Spore	Organism to test for[c]
Positive	Large rods	Single and in pairs but frequently form tangled chains	Peritrichous	Ellipsoidal; central or paracentral spores	Bacillus cereus
Positive	Large rods	Single, in pairs or in short chains	Peritrichous	Ovoid; central, subterminal, terminal oval spores swelling cell	Clostridium botulinum
Positive	Short, almost coccal rods; sometimes filaments	Single, in pairs or in short chains	Atrichous	Ovoid; subterminal large, oval spores with flattened ends, nonswelling cells	Clostridium perfringens
Positive	Small rods with rounded ends; sometimes filaments	Single and in pairs	Peritrichous (maximum of 4), sometimes atrichous	None	Listeria monocytogenes
Positive	Cocci	Grapelike clusters but sometimes in pairs or short chains	Atrichous	None	Staphylococcus aureus
Positive	Cocci	Long chains	Atrichous	None	Streptococcus pyogenes
Positive	Cocci	Short chains	Atrichous	None	Enterococcus faecalis, Enterococcus faecium
Negative	Rods (short, plump)	Single and in pairs	Peritrichous	None	Salmonella serovars, Escherichia coli; Yersinia enterocolitica
Negative	Rods (short, plump)	Single and in pairs	Atrichous	None	Salmonella pullorum, Salmonella gallinarum, Shigella serovars, Escherichia coli (few types atrichous)
Negative	Slightly curved rods	Single and in pairs	Monotrichous (polar)	None	Vibrio parahaemolyticus
Negative	Short curved rods (comma shaped) often headed in same direction, pleomorphic	Single and in spiral chains	Monotrichous (polar)	None	Vibrio cholerae and related vibrios
Negative	Comma and S-shaped, pleomorphic	Single or in short or long spiral chains	Usually monotrichous but some bipolar or atrichous	None	Campylobacter jejuni

[a] As disclosed by Gram or other stain of food sample or food sample homogenate (1:10 dilution); large numbers of cells (frequently 10^5 or more) must be present before they can be readily detected by microscopic examination.
[b] Features are not always seen in Gram stain.
[c] See Bryan (3, 4) and Stiles (17) for other organisms to test for.

TABLE 6 Tests to examine foods alleged, suspected, or epidemiologically implicated as vehicles of food-borne illness

Food	Consider testing for
Beans, pinto and similar types	*Clostridium perfringens*, *Bacillus cereus*
Canned foods (primarily home canned)	*Clostridium botulinum* and its neurotoxins, pH
Cereals, foods containing cornstarch	*Bacillus cereus*, mycotoxins
Cheese	*Staphylococcus aureus* and its enterotoxins, *Brucella* spp., *Salmonella* serovars, pathogenic *Escherichia coli*, pH, a_w
Cheese, soft varieties	As above and *Listeria monocytogenes*
Chick peas, garbanzo beans	*Clostridium perfringens*, *Bacillus cereus*
Confectionery products	*Salmonella* serovars, *Staphylococcus aureus* and its enterotoxins
Corned beef	*Staphylococcus aureus* and its enterotoxins, *Salmonella* serovars
Cream-filled baked goods, custards	*Staphylococcus aureus* and its enterotoxins, *Salmonella* serovars, *Bacillus cereus*, pH, a_w
Crustacea	*Vibrio parahaemolyticus*, *Vibrio cholerae* O1
Beef and products containing beef	*Salmonella* serovars, *Clostridium perfringens*, *Staphylococcus aureus* and its enterotoxins, *Escherichia coli* O157:H7, *Campylobacter jejuni*, *Toxoplasma gondii*, *Taenia saginata*
Eggs and egg products	*Salmonella* serovars, *Salmonella enteritidis* of particular concern, *Streptococcus pyogenes*
Fermented meats	*Staphylococcus aureus* and its enterotoxins
Fish	*Vibrio parahaemolyticus*, *Vibrio cholerae*, *Aeromonas hydrophila*, *Plesiomonas shigelloides*, histamine, *Proteus* and *Morganella* spp., fish toxins, *Diphyllobothrium* spp., *Anisakis* spp. and other fish parasites that infect humans
Fruits, raw	*Shigella* serovars, pathogenic *Escherichia coli*
Ham	*Staphylococcus aureus* and its enterotoxins
Hamburger	*Escherichia coli* O157:H7, *Salmonella* serovars
Mayonnaise	pH, *Escherichia coli* O157:H7
Melons	*Salmonella* serovars, *Shigella* serovars, pathogenic *Escherichia coli*
Mexican-style foods	*Clostridium perfringens*, *Bacillus cereus*, *Shigella* serovars, *Salmonella* serovars, *Staphylococcus aureus* and its enterotoxins
Milk, raw (pasteurized contaminated by raw)	*Salmonella* serovars, *Campylobacter jejuni*, *Listeria monocytogenes*, *Staphylococcus aureus* and its enterotoxins, *Streptococcus pyogenes*, *Yersinia enterocolitica*
Milk, dry	*Salmonella* serovars, *Staphylococcus aureus* and its enterotoxins
Molluscan shellfish	*Vibrio parahaemolyticus*, *Vibrio cholerae* O1 and non-O1 varieties, *Vibrio vulnificus*, *Aeromonas hydrophila*, *Plesiomonas shigelloides*, shellfish toxin (saxitoxin), epidemiological and serological implications of hepatitis A or Norwalk-like viruses
Oriental-style foods	*Vibrio parahaemolyticus*, *Bacillus cereus*, monosodium glutamate
Pasta and foods containing pasta	*Bacillus cereus*, *Staphylococcus aureus* and its enterotoxins, pathogenic *Escherichia coli*
Pork	*Salmonella* serovars, *Clostridium perfringens*, *Staphylococcus aureus* and its enterotoxins, *Campylobacter aureus*, *Yersinia enterocolitica*, *Toxoplasma gondii*, *Taenia solum*
Potato	*Bacillus cereus*, *Clostridium botulinum* and its neurotoxins
Poultry, poultry products, and foods containing poultry	*Salmonella* serovars, *Campylobacter jejuni*, *Clostridium perfringens*, *Staphylococcus aureus* and its enterotoxins, *Yersinia enterocolitica*
Pulses	*Clostridium perfringens*, *Bacillus cereus*
Rice	*Bacillus cereus*
Salads containing protein ingredients (e.g., potato, egg, meat, poultry, or fish)	*Staphylococcus aureus* and its enterotoxins, *Salmonella* serovars, *Shigella* serovars, *Streptococcus pyogenes*, pathogenic *Escherichia coli*, pH, epidemiological and serological evidence of hepatitis A or Norwalk-like viruses
Smoked meat, poultry, fish products	*Salmonella* serovars, *Staphylococcus aureus* and its enterotoxins, *Clostridium botulinum* and its neurotoxins, a_w
Soft drinks, fruit juices, and concentrates (previously in metallic containers or vending machine)	Chemicals such as copper, zinc, cadmium, lead, antimony, and tin; pH
Soups, stews, chowders, gumbos	*Bacillus cereus*, *Clostridium perfringens*
Sous vide, vacuum-packed modified-atmosphere packaged foods	*Clostridium botulinum* and its neurotoxins, *Listeria monocytogenes*
Vegetables, raw, including salads	*Shigella* serovars, pathogenic *Escherichia coli*

TABLE 7 Criteria[a] for confirming an outbreak of food-borne disease when cases have typical syndrome characteristics of the disease

Disease	Criteria for confirmation of case or outbreak[b]
Bacillus cereus gastroenteritis	Isolation of same serotype of *Bacillus cereus* from stool specimen from most ill persons but not from controls Isolation of $\geq 10^5$ *Bacillus cereus*/g of epidemiologically implicated food Detection of enterotoxin by ligated rabbit ileal loop, vascular permeability reaction, immunogel-diffusion, aggregate-hemagglutination, or reverse passive latex agglutination Detection of emetic enterotoxin by monkey feeding and HEp-2 cell test
Brucellosis	Isolation of *Brucella* spp. from blood of ill persons Fourfold or greater increase in agglutination titer between blood specimens taken during acute illness and 3 to 6 wk after onset of illness
Botulism	Isolation of *Clostridium botulinum* from stool of ill person who ate epidemiologically implicated food Detection of botulinal toxin in sera, feces, or food by mouse test Typical clinical syndrome in persons known to have eaten same epidemiologically implicated foods as person with laboratory-proven cases (Suspicion is cast when patients have typical clinical syndrome and history of eating home-canned or home-fermented fish, roe, or sea mammal meat.)
Campylobacteriosis	Isolation of *C. jejuni* from stool or blood of most ill persons or from epidemiologically implicated food Same strains in patients and food according to bacterial restriction endonuclease DNA analysis Fourfold or greater rise of agglutination titer between blood specimens taken during acute illness and 2 to 4 wk after onset of illness
Cholera	Isolation of *Vibrio cholerae* from vomitus or stool of ill person who ate epidemiologically implicated foods Isolation of *Vibrio cholerae* from epidemiologically implicated food Rise of vibriocidal, antitoxin, or bacterial agglutinating antibody titer during acute or early convalescent phases of illness and fall of titer during convalescent phase in nonimmunized persons Demonstration of culture or filtrate to be enterotoxigenic by gut loop, infant mouse cell culture, or other biological technique (Suspicion is cast when patients have typical clinical syndrome and history of eating raw seafoods.)
Clostridium perfringens enteritis	Fecal spore count of $>10^6$/g in most ill persons examined within few days of illness onset Isolation of same serotype of *Clostridium perfringens* from specimens from most ill persons but not from controls Isolation of same serotype of *Clostridium perfringens* from ill persons and epidemiologically implicated food Isolation of $\geq 10^5$ *Clostridium perfringens*/g of epidemiologically implicated food Demonstration of toxin in feces by reverse passive hemagglutination and fluorescent-antibody techniques

Escherichia coli diarrhea	Isolation of same serotype of *Escherichia coli* (known to be enterotoxigenic, enteroinvasive, enterohemolytic, or enteropathogenic) from stool specimens from most ill persons but not from controls
	Isolation of same serotype of *Escherichia coli* (known to be enterotoxigenic, enteroinvasive, enterohemolytic, or enteropathogenic) from ill persons and from epidemiologically implicated food
	Demonstration of toxin to be:
	   Enterotoxigenic by ileal loop (LT), infant mouse (ST), Y1 adrenal cell culture (LT), ELISA, or T and ST genes
	   Invasive by production Sereny test, HeLa cell tissue culture, or genes for invasive proteins
	   Enterohemorrhagic by infant rabbit model, Vero or endothelial cell, tissue culture, ELISA, or other immunoassay; by enterotoxin; or by adhesion genes
	   Enteroadherent by HEp-2 cell tissue assay or aggregative-adherence genes
	   Enteropathogenic by HEp-2 cell tissue assay or focal adherence genes

(Continued on next page)

TABLE 7 *(Continued)*

Disease	Criteria for confirmation of case or outbreak[b]
Listeriosis	Isolation of *Listeria monocytogenes* from autopsy or fetal material of fatal case Isolation of same phage type (from same serogroup) of culture isolated from patients and food (Fetuses, newborns, and immunosuppressed persons are highly susceptible.) (Virulence of strains is tested by Anton test in rabbits, inoculation of mice, and inoculation of embryonated eggs.)
Salmonellosis	Isolation of *Salmonella* spp. from stool or rectal swab (urine or blood if septicemic symptoms occur) of ill persons Isolation of *Salmonella* from epidemiologically implicated food Isolation of same serovar of *Salmonella* from ill persons and from epidemiologically implicated food
Shigellosis	Isolation of *Shigella* serovars from stool or rectal swab of ill persons Isolation of *Shigella* servars from epidemiologically implicated food Isolation of same *Shigella* serovar from ill persons and from epidemiologically implicated food (and from stool of food worker)
Staphylococcal enterotoxicosis (intoxication) or food poisoning	Detection of enterotoxin in epidemiologically implicated food by serological assay Isolation of same phage type of specimen from ill person and from epidemiologically implicated food (and skin, nose, or lesion of food worker) Isolation of $\geq 10^5$ *S. aureus*/g of epidemiologically implicated food
Streptococcal sore throat or scarlet fever or diarrhea	Isolation of same M and T types of group A or G streptococci from throats of ill persons and epidemiologically implicated food
Vibrio cholerae non-O1 gastroenteritis	Isolation of *Vibrio cholerae* of same serotype from stools of ill persons Isolation of *Vibrio cholerae* from epidemiologically implicated foods
Vibrio parahaemolyticus gastroenteritis	Isolation of Kanagawa-positive *Vibrio parahaemolyticus* of same serotype from stool of most ill persons Isolation of $\geq 10^5$ *Vibrio parahaemolyticus* from epidemiologically implicated food
Vibrio vulnificus septicemia	Isolation of *Vibrio vulnificus* from blood of ill person (Patient has underlying chronic illness [e.g., liver or blood related].)
Yersiniosis	Isolation of *Yersinia enterocolitica* or *Yersinia pseudotuberculosis* from stool or blood of most ill persons Isolation of *Yersinia enterocolitica* or *Yersinia pseudotuberculosis* from epidemiologically implicated food Fourfold or greater rise of agglutination titer between blood specimens taken during acute illness and 2 to 4 wk after onset of illness
Other bacterial diseases	Variable, depending on clinical and laboratory appraisal of individual circumstances
Hepatitis A	Detection of IgM anti-hepatitis A virus from persons who ate epidemiologically implicated foods Observation of typical virus by immune electron microscopy (impractical in most laboratories) Titer rise (Fourfold rise is difficult to demonstrate unless patient is seen early in course of illness.) (Liver function tests helpful.) (Suspicion is cast when adult patients have history of recent ingestion of raw shellfish.)

(Continued on next page)

TABLE 7 Criteria[a] for confirming an outbreak of food-borne disease when cases have typical syndrome characteristics of the disease   (Continued)

Disease	Criteria for confirmation of case or outbreak[b]
Norwalk and Parvo-like viral diseases	Serological evidence of virus Fourfold or greater rise of antibody titer of sera from acute phase to convalescent phase Observation of typical small, round, structured virus by immune electron microscopy followed by confirmation with testing with reference antisera (impractical in most laboratories) (Suspicion is cast when patients have syndrome, incubation period, and duration of illness compatible with typically described illness.)
Other viral diseases	Variable, depending on clinical and laboratory appraisal of individual circumstances Observation of virus particle (that reacts with patient's convalescent-phase but not acute-phase sera) by immune electron microscopy helpful when particle reacts with reference sera
Cryptosporidiosis	Isolation of *Cryptosporidium* spp. from most ill persons plus either food association or isolation from implicated food Large no. of oocysts in stools of ill persons Developmental stages in jejunal, ileal, or colonic biopsy Serum antibody
Giardiasis	Demonstration of *Giardia lamblia* in stool, duodenal contents, or small bowel biopsy sample Detection of antigen in stool of patients
Trichinellosis	Recovery of trichinella cysts from deltoid or gastrocnemius muscle biopsy sample Demonstration of larva in epidemiologically implicated food Serological evidence of infection (Suspicion is cast when patients have typical syndrome and history of eating pork, bear, or Arctic mammals.) (Marked eosinophilia is typical.)
Diarrhetic shellfish poisoning	Detection of increased no. of *Dinophysis* spp. in water from which epidemiologically implicated mollusks were gathered Detection of toxin in epidemiologically implicated shellfish by mouse or rat test
Paralytic shellfish poisoning	Detection of large no. of toxigenic species of dinoflagellates in water from which epidemiologically implicated mollusks were harvested Detection of saxitoxin in epidemiologically implicated mollusk by mouse bioassay (Suspicion is cast when patients have history of eating raw shellfish or when area from which harvesting done has history of red tides.)
Ciguatera	Demonstration of ciguatoxin in epidemiologically implicated fish Typical clinical syndrome in persons who have eaten fish (e.g., grouper, snapper, jack) previously associated with illness in area
Puffer fish poisoning	Demonstration of tetraodontoxin in puffer fish (Suspicion cast when patients have typical syndrome and history of eating puffer fish.)
Scombroid (histamine-like) poisoning	Detection of histamine levels of ≥ 50 mg/100 g of fish muscle (Suspicion is cast when patients have typical syndrome and history of eating *Scombroidei* fish, e.g., tuna, mackerel, bluefish, mahi-mahi.)

(Continued on next page)

TABLE 7 *(Continued)*

Disease	Criteria for confirmation of case or outbreak[b]
Amanitotoxin, phallin, or gyromitrin group mushroom poisoning	Demonstration of amanitotoxin, phallin or gyromitrin in urine or in epidemiologically implicated mushrooms (Suspicion is cast when patients have history of eating gathered mushrooms.)
Gastroenteritis-causing group mushroom poisoning	Demonstration of toxic chemical in epidemiologically implicated mushrooms (Suspicion is cast when patients have history of eating gathered mushrooms.)
Ibotenic acid and muscimol groups mushroom poisoning	Demonstration of ibotenic acid or muscimol in epidemiologically implicated mushrooms (Suspicion is cast when patients have history of eating gathered mushrooms.)
Mushroom-alcohol intolerance	Demonstration of toxic chemicals in urine or in epidemiologically implicated mushrooms (Suspicion is cast when patients have history of eating gathered mushrooms and drinking alcohol within 24 h.)
Muscarine group mushroom poisoning	Demonstration of muscarine in urine or in epidemiologically implicated mushrooms (Suspicion is cast when patients have history of eating gathered mushrooms.)
Plant poisoning	Demonstration of toxic chemical in epidemiologically implicated fruit, seed, or plant (Suspicion is cast when there is history of eating known toxic species of plant.)
MSG poisoning	(Suspicion is cast when large quantities of MSG have been added to epidemiologically implicated foods.)
Heavy-metal poisoning	Demonstration of high concentration of metallic ion in epidemiologically implicated food or beverage (Suspicion is cast when patients have history of storing high-acid food or beverage in metallic container or pipelines or obtaining them from vending machines.)
Other chemical poisoning	Demonstration of high concentration of chemical substance in epidemiologically implicated food or beverage (Suspicion is cast when patients have a history of use or when suspect chemical is found in food or in environment where foods are stored or prepared.)

[a]Revised from table in previous edition on the basis of references given, Center for Disease Control guidelines, and unpublished data (11a).

[b]One or more of these criteria are demonstrated for confirmation. An outbreak is defined as two or more patients who experience the same illness after ingestion of epidemiologically implicated food. Exceptions are for single cases in situations where a chemical or toxin is recovered from the epidemiologically implicated food or where the parasite can be transmitted only by foods. LT, heat-labile toxin; ST, heat-stable toxin; ELISA, enzyme-linked immunosorbent assay; IgM, immunoglobulin M; MSG, monosodium glutamate.

from patients several days or weeks following episodes of illnesses.

To confirm actual involvement of a food, the same type of pathogen (serovar, phage type, DNA type, or other definitive type) or the same toxin as was found in specimens from patients must be found in the epidemiologically implicated food. Even when clinical specimens are not available, a food is confirmed as a vehicle if toxic substances (e.g., zinc, staphylococcal enterotoxin, verotoxin, or botulinal toxin) are detected in it. Testing isolates for their ability to be enteroinvasive, enterotoxigenic, enterohemorrhagic, enteroadherent, or enteropathogenic shows their pathogenicity (1, 9). A vehicle can be epidemiologically suspect if food-specific attack rates are high in a group of persons who have eaten a specific food and low in a group of persons who have not eaten the food, if the odds ratio is greater than 1, or if the probability of the event occurring by chance alone is less than 0.05. Further association can be made if a significant number of specific pathogens (such as

10^5 or more *Staphylococcus aureus*, *Clostridium perfringens*, or *Bacillus cereus* organisms per g of food) that cause a syndrome similar to that of the patients are isolated from the food or if enteric pathogens (such as *Salmonella* serovars or *Shigella* serovars) are recovered from the food. Criteria for classifying outbreaks of specific food-borne diseases as "confirmed" are given in Table 7.

The source or mode of spread of the causative agent can often be ascertained if the agent is isolated from raw foods, food ingredients, equipment, food workers, or live animals or their environment. Definitive typing of isolates, however, is always required to make such associations. An account of the preparation of a food that is suspected as being a vehicle must contain reference to appropriate opportunities for pathogens or toxins to contaminate the food and, where applicable, to opportunities for survival and growth of pathogens. Otherwise, the account is incomplete or in error. When necessary, field studies with appropriate laboratory backup or challenge testing should be done to provide evidence of contamination, survival of microorganisms or toxins, or growth of pathogenic bacteria.

If pathogens are not found, review results of the initial staining, and make additional attempts to isolate pathogens (or microbes not yet identified as pathogens) by using enriched media and different conditions of temperature, time, and atmosphere.

REFERENCES

1. **Balows, A., W. J. Hausler, Jr., and E. H. Lennette (ed.).** 1988. *Laboratory Diagnosis of Infectious Diseases, Principles and Practices*, vol. 1. *Bacterial, Mycotic and Parasitic Diseases.* Springer-Verlag, New York.

2. **Benenson, A. A. (ed.).** 1990. *Control of Communicable Diseases in Man*, 15th ed. American Public Health Association, Washington, D.C.

3. **Bryan, F. L.** 1979. Infections and intoxications due to other bacteria, p. 211–297. *In* H. Riemann and F. L. Bryan (ed.), *Foodborne Infections and Intoxications*, 2nd ed. Academic Press, Inc., New York.

4. **Bryan, F. L.** 1982. *Diseases Transmitted by Foods. A Classification and Summary*, 2nd ed. Centers for Disease Control, Atlanta.

5. **Bryan, F. L., H. W. Anderson, K. J. Baker, G. F. Craun, W. Duel, K. H. Lewis, T. W. McKinley, R. A. Robinson, R. C. Swanson, and E. C. D. Todd.** 1979. *Procedures to Investigate Waterborne Illness.* International Association of Milk, Food, and Environmental Sanitarians, Ames, Iowa.

6. **Bryan, F. L., H. W. Anderson, O. D. Cook, J. Guzewich, K. H. Lewis, R. C. Swanson, and E. C. D. Todd.** 1988. *Procedures to Investigate Foodborne Illness*, 4th ed., revised, 1989. International Association of Milk, Food, and Environmental Sanitarians, Ames, Iowa.

7. **Buckle, K. A., J. A. Davey, M. J. Eyles, A. D. Hocking, K. G. Newton, and E. J. Stuttard.** 1989. *Foodborne Microorganisms of Public Health Significance*, 4th ed., AIFST (New South Wales Branch) Food Microbiology Group, Commonwealth Scientific and Industrial Research Organisation, North Ryde, New South Wales, Australia.

8. **Doyle, M. P. (ed.).** 1989. *Foodborne Bacterial Pathogens.* Marcel Dekker, Inc., New York.

9. **Gorbach, S. L., J. G. Barlett, and N. R. Blacklow (ed.).** 1992. *Infectious Diseases.* The W. B. Saunders Co., Philadelphia.

10. **Grace, V., G. A. Houghty, S. E. Barnard, and J. Lindamood.** 1985. Sampling dairy and related products, p. 89–117. *In* G. H. Richardson (ed.), *Standard Methods for the Examination of Dairy Products*, 15th ed. American Public Health Association, Washington, D.C.

11. **Greenberg, A. E., L. S. Cleseri, and A. D. Eaton (ed.).** 1992. *Standard Methods for the Examination of Water and Wastewater*, 18th ed. American Public Health Association, Washington, D.C.

11a. **Griffin, P. M., N. H. Bean, B. L. Herwaldt, and R. I. Glass (Centers for Disease Control and Prevention).** Unpublished data.

12. **International Commission on Microbiological Specifications for Foods.** 1986. *Microorganisms in Foods. 2. Sampling for Microbiological Analysis: Principles and Specific Applications*, 2nd ed. University of Toronto Press, Toronto, Canada.

13. **Lennette, E. H., P. Halonen, and F. A. Murphy (ed.).** 1988. *Laboratory Diagnosis of Infectious Diseases, Principles and Practices*, vol. 2. *Viral, Rickettsial and Chlamydial Diseases.* Springer-Verlag, New York.

14. **Messer, J. W., T. F. Midura, and J. T. Peeler.** 1992. Sampling plans, sample collection, shipment, and preparation for analysis, p. 24–49. *In* C. Vanderzant and D. F. Splittstoesser (ed.), *Compendium of Methods for the Microbiological Examination of Foods*, 3rd ed. American Public Health Association, Washington, D.C.

15. **Richardson, G. H. (ed.).** 1985. *Standard Methods for the Examination of Dairy Products*, 15th ed. American Public Health Association, Washington, D.C.

16. **Riemann, H., and F. L. Bryan (ed.).** *Food-borne Infections and Intoxications*, 2nd ed. Academic Press, Inc., New York.

17. **Styles, M. E.** 1989. Less recognized or presumptive foodborne pathogenic bacteria, p. 673–733. *In* M. P. Doyle (ed.), *Foodborne Bacterial Pathogens.* Marcel Dekker, Inc., New York.

18. **Vanderzant, C., and D. Splittstoesser (ed.).** 1993. *Compendium of Methods for the Microbiological Examination of Foods*, 3rd ed. American Public Health Association, Washington, D.C.

19. **Veek, M. E., I. Weitzman, R. C. Swanson, and J. P. Lucus.** 1992. Investigation of foodborne illness outbreaks, p. 747–761. *In* C. Vanderzant and D. F. Splittstoesser (ed.), *Compendium of Methods for the Microbiological Examination of Foods*, 3rd ed. American Public Health Association, Washington, D.C.

Antisepsis, Disinfection, and Sterilization in Hospitals and Related Institutions

WILLIAM A. RUTALA

19

The correct use of antiseptics, disinfectants, and sterilization processes to inactivate or remove microorganisms is an essential component of an effective infection control program. The need for appropriate antisepsis, disinfection, and sterilization is emphasized by dozens of articles documenting infection after improper decontamination of skin or patient care items. The choice of the best liquid chemical germicide (i.e., antiseptic, disinfectant) or sterilization process depends on a variety of factors, and no single germicide or process is adequate in all circumstances. The purpose of this chapter is to facilitate the selection and use of liquid chemical germicides and sterilization processes in the health care setting by defining terms, reviewing the characteristics of liquid chemical germicides and sterilization processes, and providing recommendations on their use.

DEFINITION OF TERMS

Sterilization is the complete elimination or destruction of all forms of microbial life and is accomplished in the hospital by either physical or chemical processes. Steam under pressure, dry heat, ethylene oxide (ETO) gas, and liquid chemicals are the principal sterilizing agents used in the hospital. The term "sterilization" is intended to convey an absolute meaning, not a relative one. Unfortunately, some health professionals as well as the technical and commercial literature refer to "disinfection" as "sterilization" and items as being "partially sterile." When chemicals are used for the purpose of destroying all forms of microbiological life, including fungal and bacterial spores, they may be called chemical sterilants. These same germicides used for shorter exposure periods may also be part of the disinfection process.

Disinfection is a process that eliminates from inanimate objects many or all pathogenic microorganisms with the exception of the bacterial endospore. This elimination is usually accomplished by the use of liquid chemicals or wet pasteurization in health care settings. The efficacy of disinfection is affected by a number of factors, each of which may nullify or limit the efficacy of the process. Some of the factors that have been shown to affect disinfection efficacy are prior cleaning of the object, organic load present, type and level of microbial contamination, concentration of and exposure time to the germicide, nature of the object (e.g.,

crevices, hinges, lumens), and temperature and pH of the disinfection process.

By definition, then, disinfection differs from sterilization by its lack of sporicidal property, but this is obviously an oversimplification. A few disinfectants will kill spores if prolonged exposure times (6 to 10 h) are used; these disinfectants are called chemical sterilants. At similar concentrations but with shorter exposure periods (<30 min), these same disinfectants may kill all microorganisms with the exception of high numbers of bacterial spores, and they are then called high-level disinfectants. Other disinfectants (low level) may kill most vegetative bacteria, some fungi, and some viruses in a practical period of time (≤10 min), whereas others (intermediate level) may kill tubercle bacilli, vegetative bacteria, most viruses, and most fungi but not necessarily bacterial spores. It is apparent that the germicides differ markedly among themselves, primarily in their antimicrobial spectra and rapidity of action. Tables 1 and 2 will be discussed later and consulted in this context.

Cleaning is the removal of all foreign material (e.g., soil, organic material) from objects. It is normally accomplished by using water with detergents or enzymatic products. Cleaning must precede disinfection and sterilization procedures. Studies have shown that manual and mechanical cleaning of endoscopes achieves approximately a 5-log reduction of contaminating organisms. Thus, cleaning alone is very effective in reducing the number of microorganisms present on contaminated equipment. Decontamination is a procedure that removes pathogenic microorganisms from objects so that they are safe to handle.

Antisepsis is the process of eliminating or inhibiting the growth of microorganisms on skin or living tissue. This is accomplished by the use of antiseptic agents that are chemical germicides that prevent or arrest the growth or action of microorganisms either by inhibiting their activity or by destroying them.

Other terms that deserve discussion because they are found in the literature are words with a "-cidal" or "-cide" suffix, such as germicide. A germicide is an agent that destroys microorganisms, particularly pathogenic organisms ("germs"). It is like the word disinfectant except that germicides are applied to both living tissue and inanimate objects, whereas disinfectants are applied only to inanimate objects. Other words with the suffix "-cide" (e.g., virucide, fungicide, bactericide, sporicide, tuberculocide) refer to

TABLE 1 Methods of sterilization and disinfection[a]

Object(s)	Sterilization, critical items[b]		Disinfection procedure[c]		
	Procedure	Exposure time (h)	High level[d,e]	Intermediate level[f]	Low level[g]
Smooth, hard surface[h]	A	MR	C	I	I
	B	MR	D	K	J
	C	MR	E	L	K
	D	6	F		L
	E	6	G[i]		M
	F	MR	H		
Rubber tubing and catheters[e]	A	MR	C		
	B	MR	D		
	C	MR	E		
	D	6	F		
	E	6	G[i]		
	F	MR			
Polyethylene tubing and catheters[e,j]	A	MR	C		
	B	MR	D		
	C	MR	E		
	D	6	F		
	E	6	G[i]		
	F	MR			
Lensed instruments	B	MR	C		
	C	MR	D		
	D	6	E		
	E	6	F		
	F	MR			
Thermometers (oral and rectal)[k]				I[k]	
Hinged instruments	A	MR	C		
	B	MR	D		
	C	MR	E		
	D	6	F		
	E	6			
	F	MR			

[a]Modified from reference 97. Abbreviations: A, heat sterilization, including steam or hot air (see manufacturer's recommendations); B, ETO gas (see manufacturer's recommendations); C, glutaraldehyde-based formulations (2%; exercise caution with all glutaraldehyde formulations when further in-use dilution is anticipated); D, demand-release chlorine dioxide (will corrode aluminum, copper, brass, series 400 stainless steel, and chrome with prolonged exposure); E, stabilized hydrogen peroxide (6%; will corrode copper, zinc, and brass); F, peracetic acid (concentration variable, but ≤1% is sporicidal); G, wet pasteurization at 75°C for 30 min after detergent cleaning; H, sodium hypochlorite (1,000 ppm of available chlorine; will corrode metal instruments); I, ethyl or isopropyl alcohol (70 to 90%); J, sodium hypochlorite (100 ppm of available chlorine); K, phenolic germicidal detergent solution (follow product label for use dilution); L, iodophor germicidal detergent solution (follow product label for use dilution); M, quaternary ammonium germicidal detergent solution (follow product label for use dilution); MR, manufacturer's recommendations.

[b]Will enter tissue or vascular system, or blood will flow through them.

[c]The longer the exposure to a disinfectant, the more likely it is that all microorganisms will be eliminated. A 10-min exposure is not adequate to disinfect many objects, especially those that are difficult to clean because they have narrow channels or other areas that can harbor organic material and bacteria. A 20-min exposure is the minimum needed to reliably kill M. tuberculosis and nontuberculous mycobacteria with glutaraldehyde.

[d]Semicritical items; i.e., will come in contact with mucous membrane or nonintact skin. Exposure time, ≥20 min.

[e]Tubing must be completely filled for disinfection; care must be taken to avoid entrapment of air bubbles during immersion.

[f]Some semicritical items and noncritical items. Exposure time, ≤10 min.

[g]Noncritical items; i.e., will come in contact with intact skin. Exposure time, ≤10 min.

[h]See text for discussion of hydrotherapy.

[i]Pasteurization (washer-disinfector) of respiratory therapy and anesthesia equipment is a recognized alternative to high-level disinfection. Some data challenge the efficacy of some pasteurization units.

[j]Thermostability should be investigated when appropriate.

[k]Do not mix rectal and oral thermometers at any stage of handling or processing.

compounds that destroy the microorganism identified by the prefix. For example, a bactericide kills bacteria. For more extensive discussions of the topics of antisepsis, disinfection, and sterilization, the reader may consult several references (9, 33–35, 49, 52, 88, 89, 97, 101).

ANTISEPSIS

Most nosocomial infections are believed to be transmitted by the hands of health care workers. A specific pathogen can be transmitted from patient to patient or from the hospital environment to a patient. Thus, many nosocomial

TABLE 2 Some common disinfectants and their use dilutions, properties, and cost[a]

Germicide	Use dilution	Level of disinfection	Inactivation[b]						Important characteristics									Approx cost ($US/gal[c])	
			Bacteria	Lipophilic viruses	Hydrophilic viruses	M. tuberculosis	Mycotic agents	Bacterial spores	Shelf life >1 wk	Corrosive or deleterious effects	Residue	Inactivated by organic matter	Skin irritant	Eye irritant	Respiratory irritant	Toxic	Easily obtainable	Purchase	At use dilution
Isopropyl alcohol	60–95%	Int	+	+	−	+	+	−	+	±	−	+	±	+	−	+	+	3.70 (70%)	3.70 (70%)
Hydrogen peroxide	3–25%	CS/high	+	+	+	+	+	+	+	−	−	±	+	+	−	+	+	24.50 (6%)	24.50 (6%)
Formaldehyde	3–8%	High/int	+	+	+	+	+	±	+	−	+	−	+	+	+	+	+	38.42 (37% wt)	3.84 (3.7% wt)
Quaternary ammonium compounds	0.4–1.6% aqueous	Low	+	+	−	−	±	−	+	−	−	+	+	+	−	+	+	10.77	0.04 (0.4%)
Phenolics	0.4–5% aqueous	Int/low	+	+	±	+	±	−	+	−	+	±	+	+	−	+	+	9.70–15.70	0.06 (0.4%)–0.08 (0.8%)
Chlorine	100–1,000 ppm of free chlorine	High/low	+	+	+	+	+	±	+	+	+	+	+	+	+	+	+	1.00 (5.25%)	0.10 (0.5%)
Iodophors	30–50 ppm of free iodine	Int	+	+	+	±	±	−	+	±	+	+	±	+	−	+	+	10.10 (10%)	0.05 (0.05%)
Glutaraldehyde	2%	CS/high	+	+	+	+	+	+	+	−	+	−	+	+	+	+	+	6.50–14.00	6.50–14.00

[a]Modified from reference 107. Abbreviations and symbols: Int, intermediate; CS, chemosterilizer; +, yes; −, no; ±, variable results.
[b]With a contact time of ≤30 min, inactivates all indicated microorganisms except bacterial spores, which require 6 to 10 h of contact time.
[c]1 gal = 3.785 liters.

infections can be prevented by using hand washing to interrupt the transmission of infectious agents on the hands. Despite the simplicity and effectiveness of hand washing, several studies have documented poor compliance with recommendations concerning hand washing. Improvements in the frequency of hand washing have resulted in decreased infection rates in intensive care units. Other data suggest that nosocomial infection rates are reduced by the use of antiseptic hand-washing products in an intensive care unit setting compared with rates when plain soaps or no hand washing is used (28).

Epidemiology

The microbial flora of the skin consists of two types of microorganisms: transient and resident. Transient microflora represent recent contaminants that survive poorly on the skin and are considered noncolonizing microflora. Organisms such as *Escherichia coli* and other gram-negative bacteria can be isolated from the skin but are not repeatedly cultured from the skin of most persons (35, 49, 97). These organisms can be largely removed by hand washing with

plain (bland) soaps unless there is heavy hand contamination ($\geq 3 \log_{10}$) (30, 31).

Resident microorganisms are considered permanent residents of the skin of most persons and may not be removed by hand washing with plain soaps. Organisms considered to be resident flora include the coagulase-negative staphylococci; members of the genera *Corynebacterium*, *Propionibacterium*, and *Acinetobacter*; and certain members of the *Klebsiella-Enterobacter* group. These resident microflora usually can be killed or inhibited from growth by use of an antimicrobial hand-washing product. These antimicrobial soaps are considered drugs, because they are intended to kill or inhibit microorganisms on the skin. They are regulated by the Food and Drug Administration (FDA) under the federal Food, Drug and Cosmetic Act. Antimicrobial soaps or antiseptics should be considered for use before personnel care for newborns, between care of patients in critical care units, and before personnel care for immunocompromised patients (35, 49, 52, 97).

Studies have also demonstrated important quantitative differences in the skin flora of hospitalized patients and

TABLE 3 Characteristics of topical antimicrobial ingredients[a]

Ingredient(s)	Antiseptic effect on:					Rapidity of action	Residual activity	Concn (%)	Affected by organic matter?	Safety
	GPB	GNB	M. tuberculosis	Fungi	Viruses					
Alcohols	Excellent	Excellent	Excellent	Good	Good	Most rapid	None	70–92	No data	Drying, volatile
Chlorhexidine	Excellent	Excellent	Good	Fair	Good	Int	Excellent	4, 2, or 0.5 in alcohol	Minimally	Ototoxicity, keratitis
Iodophors	Excellent	Good	Good	Good	Good	Int	Minimal	10, 7.5, 2, 0.5	Yes	Skin irritation
PCMX	Good	Fair	Fair	Fair	Fair	Int	Good	0.5–3.75	Minimally	Need data
Triclosan (Irgasan DP-300)	Good	Good	Fair	Poor	Unknown	Int	Excellent	0.3–1.0	Minimally	Need data

[a]Modified from reference 49. Abbreviations: GPB, gram-positive bacteria; GNB, gram-negative bacteria; Int, intermediate.

nonhospitalized healthy adults as well as a higher rate of antimicrobial resistance in hospitalized patients. For example, patients had lower numbers of organisms at the skin sites sampled but significantly higher carrier rates of *Proteus*, *Pseudomonas*, and *Candida* spp. Coagulase-negative staphylococci of patients were significantly more resistant than those of controls to 8 of 10 antimicrobial agents tested. Methicillin resistance occurred in only 2.9% of the isolates from healthy controls but in 44.3% of isolates from patients (56). Other studies suggest differences in the composition and antimicrobial resistance of bacteria recovered from the hands of health care personnel compared to bacteria in a control population (40).

Antiseptic Agents

No single antiseptic agent is ideal for use in all situations. The selection of an antiseptic for use in a particular setting must take into consideration several factors, including active ingredient and concentration, user acceptability, rapidity of action, persistence, cost, inactivation by organic matter, safety and toxicity. At least five active ingredients that have antimicrobial properties are currently available in a variety of hand-washing products (66). These include alcohols, chlorhexidine gluconate (CHG), iodophors, parachlorometaxylenol (PCMX), and triclosan (also called Irgasan). Hexachlorophene, which was widely used until the 1970s, is no longer recommended for routine hand washing because of its potential neurological toxicity and lack of activity against most microorganisms except gram-positive bacteria. To assist in choosing products that are both safe and effective, each of the aforementioned products will be reviewed as to their mode of action, spectrum of activity, safety, toxicity, rapidity of action, persistence, inactivation by organic matter and available preparations (Table 3). For a more comprehensive discussion of the selection and use of antiseptics, the reader should examine the references for this section (49, 50, 52).

Alcohol

The most likely explanation for the antimicrobial action of alcohol is denaturation of proteins. Alcohols have excellent bactericidal and mycobactericidal activity. They are also active against fungi and viruses; however, they have no sporicidal activity. Alcohol is among the safest known antiseptics.

In appropriate concentrations (60 to 90% ethyl or isopropyl alcohol), alcohols rapidly reduce the microbial counts on the skin. A vigorous 1-min rubbing with alcohol in quantities sufficient to wet the hands completely is the most effective method of hand antisepsis. Although alcohols are not in common use in the United States as surgical hand scrubs, they are very effective for this use and as a hand-washing agent (53). In the past several years, a variety of alcohol-based hand rinses and foams have been introduced for use by health care workers whenever hand-washing sinks with running water and towels are unavailable (54).

The major disadvantage of alcohol for skin antisepsis is its drying effect, which is caused by the alcohol removing lipid from the skin. Newer products include emollients along with the alcohol to minimize skin drying. These products are acceptable to users and have excellent antibacterial activity (53, 54). Other disadvantages of the alcohols are their lack of residual effect; their inactivation by organic matter; and their volatile and flammable nature, which requires that they be stored in a cool, well-ventilated area. Alcohol for skin preparation must be allowed to thoroughly evaporate, particularly around ignitable devices such as those used in electrosurgery or laser surgery (49).

CHG

CHG is a cationic bisbiguanide that inactivates microorganisms by causing destruction of cell membranes and precipitation of cell contents. CHG has a broad spectrum of activity that includes fungi, viruses, bacteria, and mycobacteria.

Data from animal and human studies indicate that CHG is safe. However, ototoxicity can result if CHG is instilled directly into the middle ear, and corneal damage can result from instillation of CHG into the eye. It has a relatively low skin irritancy potential.

Although CHG acts more slowly than the alcohols, several clinical studies report good reductions in flora after a 15-s hand wash. One of the most unique attributes of CHG is its persistence, which is desirable when there is a high risk of infection and where a sustained and maximal reduction in microbial flora reduces infection risk (e.g., surgery). Studies have shown that after a few days of daily use of products that contain CHG, bacterial yields from hands are as low as after the use of alcohol-based products.

The activity of CHG is not significantly affected by blood or other organic material. However, its activity is pH dependent (5.5 to 7.0) and is reduced or neutralized in the presence of anionic- and nonionic-based moisturizers and surfactants, inorganic anions, and organic anions such as neutral soaps. For this reason, the activity of CHG is particularly formula dependent and may be influenced by individual skin differences (e.g., pH, moisture level, and secretions). Thus, it is prudent to consider all hand care products (e.g., hand-washing agents, lotions) when selecting products for the health care work environment (7). The most common formulation of CHG is a 4% concentration in a detergent base. Newer 2% formulations and foams appear to be slightly lower in antimicrobial activity than the 4% concentration. CHG is also available as an alcohol-based hand rinse (0.5% CHG) in some countries (49). Gloves containing CHG have been shown to inactivate viruses, but their role in reducing a worker's risk of exposure to infectious fluid-borne pathogens is unknown (71).

Iodophors

Iodophors are chemical complexes that consist of iodine and a carrier such as polyvinylpyrrolidone (povidone) or ethoxylated nonionic detergents (poloxamers). Since these products release free iodine from a reservoir of available iodine gradually and at a low concentration, they retain the germicidal efficacy of iodine but unlike iodine are generally nonstaining, noncorrosive to metal surfaces, and relatively free of toxicity. Iodophors are recognized as being relatively harsh on the skin. The lethal effects of iodine are the results of their ability to penetrate the cell wall of microorganisms and disrupt protein and nucleic acid structure and synthesis. Reports on the antimicrobial activity of iodophors demonstrate that they are bactericidal, mycobactericidal, and virucidal but may require longer exposure times to kill certain fungi and bacterial spores. Studies that have compared microbial reductions in skin flora have found iodophors to be comparable to or slightly less effective than CHG.

Iodophors have little if any residual effect, and the antimicrobial activity is neutralized rapidly in the presence of organic material such as blood or sputum. The iodophor most commonly used is povidone-iodine, which contains 7.5% iodophor with about 0.75% titratable iodine.

PCMX

PCMX is a chlorine-substituted xylenol that derives its antimicrobial activity by microbial cell wall destruction and enzyme inactivation. It is much less effective than the three aforementioned antiseptics. It has good activity against gram-positive bacteria but less activity against gram-negative bacteria, mycobacteria, fungi, and viruses.

PCMX is mild on the skin and produces a sustained residual activity for a few hours. Like CHG, PCMX is neutralized by nonionic surfactants, and its efficacy is highly formula dependent. PCMX is currently available in a number of hand-washing products, usually at concentrations of 0.5 to 3.75% (4, 17).

Triclosan

Triclosan (5-chloro-2-[2,4-dichlorophenoxy]phenol, or Irgasan DP-300) is a diphenyl ether. Triclosan has good activity against gram-positive bacteria and most gram-negative bacteria, but it appears to be a poor fungicide. Little information is available regarding its activity against mycobacteria and viruses. Triclosan has excellent persistence on the skin, and its activity is only minimally affected by organic matter (49). In one study, 0.3% triclosan was less effective than 2% CHG in reducing skin flora (51). Triclosan is commonly used as an antimicrobial ingredient of soaps for deodorant purposes.

CONTROL MEASURES

Hand Washing

For general patient care, a plain, nonantimicrobial soap in any convenient form (bar, leaflets, liquid, powder) is acceptable. If bar soap is used, the hospital should provide small bars that can be changed frequently and soap racks that promote drainage. If liquid soap is used, either a disposable or reusable liquid-soap container can be used; however, reusable containers should be cleaned when empty and refilled with fresh soap (35).

The efficacy of hand washing is influenced by several factors. Hands should be thoroughly covered with hand-washing product in adequate amounts (probably 3 to 5 ml per antiseptic hand wash and 1 ml per nonantiseptic hand wash [55]) and vigorously rubbed together for 10 to 15 s, generating friction on all surfaces of the hands and fingers. Cleaning under the fingernails is a crucial part of hand washing, since the majority of organisms on the hand live in the subungual region. For example, the subungual spaces have an average log_{10} CFU of 5.39, compared to 2.55 (fingernail) to 3.53 (palm) for other hand sites (68). Gram-negative rods are recovered in higher numbers from the fingertips of nursing personnel with artificial nails than from nurses with natural nails, but the impact of artificial nails on nosocomial infection is unknown (80). Hands should be thoroughly rinsed under a stream of water to remove residual soap and then dried. Either electric air driers or paper towels should be used for drying hands. Hand-drying equipment should be placed near the sink. Hand-washing sinks should be conveniently located throughout the hospital to facilitate frequent and appropriate hand washing.

Surgical Hand Scrub

A surgical hand scrub for approximately 5 min is performed to remove transient flora and reduce resident flora for the duration of a surgical procedure. Since glove perforations occur during surgical procedures, operating room staff attempt to reduce the microbial flora on the hands to the lowest level possible during surgery to minimize the risk of microbial inoculation into the wound. Agents that persist on the skin can help maintain lower bacterial counts under the gloves. Thus, agents with good antimicrobial activity and persistence, such as CHG or an iodophor, are acceptable products. An alcohol preparation yields excellent reduction in flora. This reduction is accomplished by first washing the hands and arms and cleaning the fingernails thoroughly, and then drying. Next, an alcohol solution containing emollient is applied, and the alcohol is rubbed until dry. Approximately 3 to 5 ml of alcohol per application should be used, with continuing applications for approximately 5 min, using a total of 9 to 25 ml of alcohol (49).

Preoperative Patient Skin Preparation

Antiseptics are also used for preoperative skin preparation to rapidly decrease the number of skin flora at the operative site. After the patient's skin has been physically cleaned, an antimicrobial agent that rapidly reduces the level of skin microbial flora is applied. Recommended agents include the alcohols, iodophors, or CHG. Since CHG can cause ototoxicity and eye injury, it should not be instilled into the ears or near the periorbital area (49).

Patient Skin Preparation at Catheter Site

Catheter-related bacteremias or fungemias occur in 3 to 7% of central venous catheters and 1% of arterial catheters and are most commonly caused by skin microorganisms. A trial that investigated whether antiseptics used to disinfect the insertion site reduced catheter-related infection found that catheter-related infection was higher in the 10% povidone-iodine and 70% alcohol groups than in the 2% CHG group. The results suggest that use of CHG for cutaneous disinfection helps protect against infection of the central venous and arterial catheters (61).

Gloving

Use of gloves as barriers against direct contact with blood or other potentially infectious materials has increased significantly since their use was recommended by the Centers for Disease Control and Prevention (CDC) and required by the Occupational Safety and Health Administration (OSHA). Along with this increased use of gloves in health care, there has been an increase in allergic and irritant reactions to glove ingredients. Remedies include wearing vinyl gloves or hypoallergenic gloves, which are required by OSHA for employees with latex sensitivity (73).

While gloves do provide barrier protection against blood and other potentially infectious materials, they are certainly not impermeable. Bacteria and viruses can leak through gloves, and new gloves leak about 5% of the time. This leakage rate rises to more than 50% as gloves are stressed during use. While both latex and vinyl gloves provide some barrier protection, latex gloves generally have lower leakage rates when tested under in-use conditions (24, 48, 84). Thus, the use of gloves does not replace hand washing, and hand washing after removing gloves is required by OSHA.

Gloved hands should not be washed in order to reuse gloves. Studies have shown that washing gloves may decrease the integrity of the gloves and that microorganisms adhere to gloves and are not easily removed from gloves after hand washing and drying (27). OSHA prohibits the washing or decontamination of disposable gloves for reuse. Inappropriate glove use has also been recognized as a problem. Failure to change gloves between patients and between contacts with known contaminated sites was identified as the likely source of an *Acinetobacter* outbreak (76).

Hand Lotions

Hand lotions are often recommended to minimize drying resulting from frequent hand washing. Outbreaks of bacterial infection have been caused by contaminated hand lotions. Concerns have also been expressed about the potential for oil-based lotion formulations (e.g., petroleum jelly) to weaken latex gloves and cause increased permeability. Additionally, lotions such as anionic moisturizing agents may interfere with the residual antibacterial activity of some antiseptics such as CHG (7). Thus, the interaction between lotions and antimicrobial products on antimicrobial activity as well as the effect of lotions on the permeability of gloves must be considered at the time of product selection. Hand lotions designed to protect against latex sensitivity resulting from glove use are now being marketed, but their clinical effectiveness has not been documented.

A RATIONAL APPROACH TO DISINFECTION AND STERILIZATION

About 25 years ago, a rational approach to disinfection and sterilization of patient care items or equipment was devised by Earle H. Spaulding (101). This classification scheme is so clear and logical that it has been retained, refined, and successfully used by infection control practitioners and others who plan methods for disinfection or sterilization (18, 33–35, 87–89, 97). Spaulding believed that the nature of disinfection could be understood more readily if instruments and items for patient care were divided into three categories based on the degree of risk of infection involved in the use of the items. The three categories he described were critical, semicritical, and noncritical. This terminology is employed by the 1985 CDC guideline for handwashing and hospital environmental control (35), the CDC's guidelines for the prevention of transmission of human immunodeficiency virus and hepatitis B virus to health-care and public-safety workers (18), and the Association for Practitioners in Infection Control guideline on selection and use of disinfectants (87).

Critical Items

Critical items are so called because of the high risk of infection if such an item is contaminated with any microorganism, including bacterial spores. Thus, it is critical that objects that enter sterile tissue or the vascular system be sterile. This category includes surgical instruments, cardiac and urinary catheters, and implants. Most of the items in this category should be purchased sterile or be sterilized by autoclaving if possible. If heat labile, the object may be treated with ETO or, rarely, by chemical sterilants if other methods are unsuitable. Table 1 shows several germicides categorized as chemical sterilants. These include 2% glutaraldehyde-based formulations, 6% stabilized hydrogen peroxide, peracetic acid, and demand-release chlorine dioxide. Chemical sterilization can be relied upon to produce sterility only if cleaning precedes treatment and if proper guidelines as to organic load, contact time, temperature, and pH are met.

As of June 1993, the FDA has primary responsibility for the premarket review of safety and efficacy requirements for liquid chemical germicides that are sterilants intended for use on critical and semicritical devices (70).

Semicritical Items

Objects that come in contact with mucous membranes or skin that is not intact should be free of all microorganisms with the exception of high numbers of bacterial spores and are called semicritical objects. Intact mucous membranes are generally resistant to infection by common bacterial spores but susceptible to other organisms such as tubercle bacilli and viruses. Respiratory therapy and anesthesia equipment, endoscopes, and diaphragm fitting rings are included in this category. Semicritical items minimally require high-level disinfection by wet pasteurization or chemical disinfectants. Glutaraldehyde, stabilized hydrogen peroxide, peracetic acid, and chlorine and chlorine compounds are dependable high-level disinfectants provided the factors influencing germicidal procedures are considered (Table 1). When selecting a disinfectant for use with certain patient care items, the chemical compatibility after extended use with the items must also be considered. For example, while chlorine and chlorine-releasing compounds are considered high-level disinfectants, they are generally not used for disinfecting semicritical items because of their corrosive effects.

Laparoscopes and arthroscopes are considered critical items because they enter sterile tissue, and ideally, they should be sterilized between patients. However, they commonly undergo only high-level disinfection between patients in the United States. Although limited data are available, there is no evidence to demonstrate that high-level disinfection of these scopes poses an infection risk to patients (90).

It is recommended that semicritical items be rinsed with sterile water to prevent contamination with organisms that may be present in tap water, such as nontuberculous mycobacteria and *Legionella* spp. (35, 60, 69, 87, 90, 98). In circumstances in which a sterile water rinse is not feasible, a tap water rinse should be followed by an alcohol rinse and forced-air drying (36, 90, 98). Institution of forced-air drying significantly reduces bacterial contamination of stored endoscopes, presumably by removing the wet environment favorable for bacterial growth (36). After being rinsed, items should be dried and stored (e.g., packaged) in a manner that does not recontaminate them.

Some semicritical items (i.e., hydrotherapy tanks used for patients whose skin is not intact, thermometers) may be effectively disinfected with high-level (i.e., chlorine) or intermediate-level (i.e., phenolic, iodophor, alcohol) disinfectants.

Noncritical Items

Noncritical items come in contact with intact skin but not mucous membranes. Intact skin acts as an effective barrier to most microorganisms, and sterility is not critical. Examples of noncritical items are bedpans, blood pressure cuffs, crutches, bed rails, linens, some food utensils, bedside tables, patient furniture, and floors. In contrast to critical and some semicritical items, most noncritical reusable items may be cleaned where they are used and do not need to be transported to a central processing area. There is virtually no risk of transmitting infectious agents to patients via noncritical items (105); however, these items could potentially contribute to secondary transmission by contaminating hands of health care workers or by coming into contact with medical equipment that will subsequently come in contact with patients (34). The low-level disinfectants listed for noncritical items in Table 1 may be used. The Environmental Protection Agency (EPA) has primary responsibility for premarket review of general-purpose disinfectants used on noncritical items (70).

Changes Since 1981

As a guide to the appropriate selection and use of disinfectants, a table was prepared by the CDC in 1981 and is presented here in modified form (Table 1). This table contains several significant changes from the original guideline. First, formaldehyde-alcohol has been deleted as a chemical sterilant and high-level disinfectant because it no longer has a role in disinfection strategies. It is corrosive, irritating, toxic, and not commonly used. Second, two new chemical sterilants, demand-release chlorine dioxide and peracetic acid, have been added. Third, 3% phenolic and iodophors have been deleted as high-level disinfectants because of their unproven efficacy against bacterial spores, M. *tuberculosis*, and/or some fungi. Fourth, isopropyl alcohol and ethyl alcohol have been excluded as high-level disinfectants because of their inability to inactivate bacterial spores and because of the inability of isopropyl alcohol to inactivate hydrophilic viruses. Fifth, a 1:16 dilution of 2.0% glutaraldehyde–7.05% phenol–1.20% sodium phenate (which contains 0.125% glutaraldehyde, 0.440% phenol, and 0.075% sodium phenate when diluted) has been deleted as a high-level disinfectant because of 11 scientific publications that demonstrate a lack of bactericidal activity in the presence of organic matter; a lack of fungicidal, tuberculocidal, and sporicidal activities; and reduced virucidal activity. This product was removed from the market in December 1991. Sixth, the exposure time required to achieve high-level disinfection has been changed from 10 to 30 min to 20 min or more (87–89).

PROBLEMS WITH DISINFECTION OF HOSPITAL EQUIPMENT

Concerns with Spaulding Scheme

One problem associated with the aforementioned Spaulding scheme is its oversimplification. For example, it does not consider problems in processing complicated medical equipment that is often heat labile or problems of inactivating certain microorganisms. Thus, in some situations it is still difficult to choose a method of disinfection after the categories of risk to patients have been considered. This is especially true for a few medical devices (i.e., arthroscopes, laparoscopes, biopsy forceps) in the critical category, because there is controversy about whether we should sterilize or high-level disinfect these patient care items (90). Sterilization would not be a problem if these items could be steam sterilized, but most of them are heat labile, and sterilization is achieved by using ETO, a procedure that may be too time-consuming for routine use between patients. Although the value of sterilization of these items at first seems obvious, evidence that sterilization of them improves patient care by reducing the risk of infection is lacking (42). Presumably, these reasons account for why the majority of procedures done in hospitals with arthroscopes, laparoscopes, and biopsy forceps are performed with equipment that has been processed by high-level disinfection rather than sterilization (90).

Several other problems are associated with the disinfection of patient care items. The optimal contact time and disinfection scheme are not known for all equipment. For this reason, disinfectant strategies for several semicritical items (e.g., endoscopes, applanation tonometers, cryosurgical instruments, diaphragm fitting rings) are highly variable and are further discussed below. While additional studies are needed to determine whether simplified disinfecting procedures are efficacious in a clinical setting, it is prudent to follow the CDC and Association for Practitioners in Infection Control guidelines until studies define effective alternative processes (16, 35, 58, 87).

Endoscopes

High-level disinfection can be expected to destroy all microorganisms except high numbers of bacterial spores. An immersion time of ≥20 min in 2% glutaraldehyde (or other high-level disinfectant) is required to adequately disinfect semicritical items such as endoscopes between patient procedures, particularly in view of the disputed tuberculocidal efficacy of glutaraldehyde-based disinfectants. Flexible endoscopic instruments are particularly difficult to disinfect and easy to damage because of their intricate design and delicate materials. It must be highlighted that meticulous cleaning must precede any sterilization or disinfection procedures, or outbreaks of infection may occur.

Examining the data on nosocomial infections related only to endoscopes, one finds that 281 infections were transmitted by gastrointestinal endoscopy and 96 were transmitted by bronchoscopy. The clinical spectrum of these infections ranged from asymptomatic colonization to death. *Salmonella* species and *Pseudomonas aeruginosa* were repeatedly identified as causative agents of infections transmitted by gastrointestinal endoscopy, and *Mycobacterium tuberculosis*, atypical mycobacteria, and *P. aeruginosa* were the most common causes of infections transmitted by bronchoscopy. Major reasons for transmission were inadequate cleaning, improper selection of a disinfecting agent, or failure to follow recommended cleaning and disinfection procedures (100). One multistate investigation found that 24% of the bacterial cultures from the internal channels of 71 gastrointestinal endoscopes grew 100,000 colonies or more of bacteria after completion of all disinfection-sterilization procedures and prior to use in the next patient (44). Automatic endoscope disinfection machines have also been linked to outbreaks of infections or colonization. Outbreaks involving endoscopic accessories such as suction valves and biopsy forceps support a recommendation that if such an item cannot be cleaned of all foreign matter, it should be sterilized, preferably with steam.

Clearly, there is a need for further development and redesign of automatic endoscope-washing machines and endoscopes so that they do not represent a potential source of infectious agents (12). A newly redesigned endoscope has been introduced. It includes a reusable endoscope without channels and a sterile sheath set comprising a single disposable unit: sheath; air, water, and suction channels; distal window; and cover for the endoscope control body. All contaminated surfaces including the channels are then discarded, thereby eliminating any concern for cross-transmission of infectious agents from previous patients. Further clinical trials and microbiological evaluations are needed to document the comparability, cost-effectiveness, safety, and

reduced infection risk of this system (85). Recommendations for the cleaning and disinfection of endoscopic equipment have been published and should be followed (64, 98, 106).

In general, endoscope disinfection involves five steps: cleaning, disinfecting, rinsing, drying, and storing. (i) To clean the endoscope, mechanically clean external surfaces, ports, and internal channels with water and a detergent or enzymatic cleaner. (ii) To disinfect, immerse the endoscope in high-level disinfectant, perfuse the disinfectant into the suction-biopsy channel and air-water channel, and expose the equipment for at least 20 min. (iii) To rinse the endoscope and channels, use sterile water, and if this is not feasible, tap water followed with alcohol. (iv) To dry the insertion tube and inner channels, use forced air after disinfection and prior to storage. (v) To store the endoscope, use a method that prevents recontamination (e.g., hang it vertically).

Tonometers, Diaphragm Fitting Rings, Cryosurgical Instruments

Disinfection strategies for several other semicritical items (e.g., applanation tonometers, cryosurgical instruments, and diaphragm fitting rings) are highly variable. For example, one study revealed that no uniform technique was in use for disinfection of applanation tonometers, with disinfectant contact times varying from <15 s to 20 min (90). Concern for transmission of viruses (e.g., herpes simplex virus [HSV], adenovirus type 8, human immunodeficiency virus [HIV]) by tonometer tips has prompted CDC disinfection recommendations (16). These recommendations are that the instrument be wiped clean and disinfected for 5 to 10 min with 3% hydrogen peroxide, 5,000 ppm chlorine, 70% ethyl alcohol, or 70% isopropyl alcohol. After disinfection, the device should be thoroughly rinsed in tap water and dried before use. Although these disinfectants and exposure times should kill microorganisms of relevance in ophthalmology, there are no studies that provide direct support for this belief (23). The American Academy of Ophthalmology also has developed specific guidelines for preventing infection in ophthalmology practice, but it considers only certain infectious agents (e.g., HIV, herpes, adenovirus) (1). Because a short and simple cleaning procedure is desirable in the clinical setting, the tonometer tip is sometimes swabbed with a 70% isopropyl alcohol wipe (23). Preliminary reports suggest that wiping the tonometer tip with an alcohol swab and then allowing the alcohol to evaporate may be an effective means of eliminating HSV, HIV type 1 and adenovirus (23). Since these studies involved only a few replicates and were conducted in a controlled laboratory setting, further studies are needed before this technique can be recommended. In addition, two studies have found that disinfection of pneumotonometer tips with a 70% isopropyl alcohol wipe between uses contributed to an outbreak of epidemic keratoconjunctivitis caused by adenovirus type 8 (41, 47).

No studies have evaluated disinfection techniques for other items that contact mucous membranes such as diaphragm fitting rings, cryosurgical probes, or vaginal probes used in sonographic scanning. Lettau et al. of the CDC supported a diaphragm fitting ring manufacturer's recommendation, which involved a soap-and-water wash followed by a 15-min immersion in 70% alcohol (58). This disinfection method should be adequate to inactivate HIV type 1, HBV, and HSV, even though alcohols are not classified as high-level disinfectants because their activity against picornaviruses is somewhat limited. There are no data on the inactivation of human papillomavirus by alcohol or other disinfectants, because in vitro replication of complete virions has not been achieved. Thus, while isopropyl alcohol for 15 min should kill microorganisms of relevance in gynecology, no clinical studies provide direct support for this procedure. Cryosurgical probes should be high-level disinfected, and our policy for vaginal probes used in sonographic scanning requires that a new condom be used to cover the probe for each patient, and since condoms may fail, high-level disinfection of the probe is also performed (90).

Dental Instruments

Scientific articles and increased publicity about the potential for transmitting infectious agents in dentistry have focused attention on dental instruments as possible agents of disease transmission. The American Dental Association states that surgical and other instruments that normally penetrate soft tissue or bone (e.g., forceps, scalpels, bone chisels, scalers, and surgical burs) are classified as critical and must be sterilized after each use or discarded. Instruments that are not intended to penetrate oral soft tissues or bone (e.g., amalgam condensers, air and water syringes) but may come in contact with oral tissues are classified as semicritical and should also be sterilized after each use (2). This is consistent with the recommendations from the CDC and the FDA (19, 25). Hand pieces that cannot be heat sterilized should be retrofitted to attain heat tolerance. Hand pieces that cannot be retrofitted and thus not heat sterilized should not be used (25). Chemical disinfection is not recommended for critical or semicritical dental instruments. Methods of sterilization that may be used for critical and semicritical dental instruments and materials that are heat stable include steam under pressure (autoclave), chemical vapor, and dry heat. ETO should also be an effective means of sterilization if the instrument to be sterilized is clean and dry. Consideration must be given to the effect a sterilization process may have on instruments and materials.

Uncovered operatory surfaces (e.g., countertops, chair switches, light handles) should be disinfected between patients. This can be accomplished by utilizing a disinfectant that is registered with the EPA as a "hospital disinfectant." There are several categories of such products (e.g., chlorine, phenolics) (2, 72). If waterproof surface covers are used to prevent contamination of surfaces and are carefully removed and replaced between patients, the protected surfaces need not be disinfected after each use but must be disinfected at the end of the day.

Disinfection of HBV-, HIV-, or Tuberculosis-Contaminated Devices

Should we sterilize or high-level disinfect semicritical medical devices contaminated with blood from patients infected with HIV or HBV or with respiratory secretions from a patient with pulmonary tuberculosis? The CDC recommendation for high-level disinfection is appropriate, because experiments have demonstrated the effectiveness of high-level disinfectants in inactivating these and other pathogens that may contaminate semicritical devices (11, 17, 81, 83, 88, 89, 93, 94). Nonetheless, some hospitals modify their disinfection procedures when the endoscopes are used with a patient known or suspected to be infected with HIV, HBV, or M. tuberculosis (90). This is inconsis-

tent with the concept of universal precautions, which presumes that all patients are potentially infected with blood-borne pathogens (17). Several studies have highlighted the inability to distinguish HIV- or HBV-infected patients from noninfected patients on clinical grounds (38). It is also likely that in many patients, mycobacterial infection will not be immediately clinically apparent. It should be noted that in most cases, hospitals have gas sterilized the endoscopic instruments because they believed that this practice reduced the risk of infection (90). ETO is not routinely used for endoscope sterilization because of the lengthy processing time. Endoscopes and other semicritical devices should be managed the same way whether or not the patient is infected with M. tuberculosis, HIV, or HBV.

Inactivation of Creutzfeldt-Jakob Agent

The only infectious agent that requires unique decontamination recommendations is the virus of Creutzfeldt-Jakob disease (13). Creutzfeldt-Jakob disease has been transmitted iatrogenically via implanted brain electrodes that were disinfected with ethanol and formaldehyde after use on a patient with a known case of Creutzfeldt-Jakob disease. It has been observed in recipients of contaminated human growth hormone and gonadotropin and of corneal and dura mater grafts. The need for such recommendations is due to an extremely resistant subpopulation of Creutzfeldt-Jakob disease-like viruses (3) and the protection afforded this tissue-associated virus. Steam sterilization for 1 h at a temperature of 132°C has been recommended as the preferred method for the treatment of contaminated material. Immersion in 1 N sodium hydroxide (which is caustic) for 1 h at room temperature is an alternative procedure for critical and semicritical items when autoclaving is not possible. Because noncritical patient care items or surfaces (e.g., autopsy tables, floors) have not been involved in disease transmission, these surfaces may be disinfected with either bleach (undiluted or up to 1:10 dilution) or 1 N sodium hydroxide at room temperature for 15 min or less (87–89). A formalin-formic acid procedure is required for inactivating virus infectivity in tissue samples from patients with Creutzfeldt-Jakob disease (14).

DISINFECTION

A great number of disinfectants are used in the health care setting, including alcohol, chlorine and chlorine compounds, formaldehyde, glutaraldehyde, hydrogen peroxide, iodophors, peracetic acid, phenolics, quaternary ammonium compounds, and combinations of these disinfectants. Each formulation of active and inert ingredients is considered a unique product and must be EPA or FDA registered. These disinfectants are not interchangeable, and an overview of the performance characteristics of each is intended to provide the user with sufficient information to select an appropriate disinfectant for any item and use it in the most efficient way. Excessive costs may result from incorrect concentrations and inappropriate disinfectants. Finally, occupational skin diseases among cleaning personnel have been associated with the use of several disinfectants such as formaldehyde, glutaraldehyde, chlorine, and others, and precautions (e.g., gloves, proper ventilation) should be used to minimize exposure.

Chemical Disinfectants

Alcohol

In the sphere of hospital disinfection, the word "alcohol" refers to two water-soluble chemical compounds whose germicidal characteristics are generally underrated; the two are ethyl alcohol and isopropyl alcohol. These alcohols are rapidly bactericidal rather than bacteriostatic against vegetative forms of bacteria; they are also tuberculocidal, fungicidal, and virucidal, but they do not destroy bacterial spores. Their killing activity drops sharply when they are diluted below 50% concentration, and the optimum bactericidal concentration is in the range of 60 to 90% by volume. The most feasible explanation for their antimicrobial action is denaturation of proteins.

Alcohols are not recommended for sterilizing medical and surgical materials, principally because of their lack of sporicidal action and their inability to penetrate protein-rich materials. Fatal postoperative wound infections with *Clostridium* spp. have occurred when alcohols were used to sterilize surgical instruments contaminated with bacterial spores. Thus, ethyl alcohol and isopropyl alcohol are not high-level disinfectants because of their inability to inactivate bacterial spores and because of isopropyl alcohol's inability to kill hydrophilic viruses (e.g., echovirus, coxsackie virus). Alcohols have been used effectively to disinfect oral and rectal thermometers. Alcohol towelettes have been used for years to disinfect small surfaces such as rubber stoppers of multiple-dose medication vials or vaccine bottles. Furthermore, alcohol is occasionally used to disinfect external surfaces of equipment (e.g., stethoscopes, ventilators, manual ventilation bags), cardiopulmonary resuscitation manikins, or medication preparation areas. Two studies demonstrated the effectiveness of 70% isopropyl alcohol in disinfecting reusable transducer heads in a controlled environment (79, 103). In contrast, Beck-Sague and Jarvis described three outbreaks when alcohol was used to disinfect transducer heads in an intensive care setting (6).

The documented shortcomings of alcohol use on equipment are that alcohols damage the shellac mountings of lensed instruments, tend to swell and harden rubber and certain plastic tubing after prolonged and repeated use, bleach rubber and plastic tiles, and damage tonometer tips (cause the glue to deteriorate) after the equivalent of one working year of routine use. Lingel and Coffey also found that tonometer biprisms soaked in alcohol for 4 days developed rough front surfaces that could potentially cause corneal damage. This roughening appeared to be caused by a weakening of the cementing substances used to fabricate the biprisms (59). Corneal opacification has been reported when tonometer tips were swabbed with alcohol immediately before intraocular pressure measurements were taken (99). Alcohols are flammable and consequently must be stored in a cool, well-ventilated area. They also evaporate rapidly, and this makes extended exposure time difficult to achieve unless the items are immersed.

Chlorine and Chlorine Compounds

Hypochlorites are the most widely used of the chlorine disinfectants and are available in a liquid (e.g., sodium hypochlorite) or solid (e.g., calcium hypochlorite, sodium dichloroisocyanurate) form. They have a broad spectrum of antimicrobial activity (i.e., bactericidal, fungicidal, tuberculocidal, sporicidal, and virucidal) and are inexpensive and fast acting. Their use in hospitals is limited by their corrosiveness, inactivation by organic matter, and relative instability. The microbicidal activity of chlorine is largely attributed to undissociated hypochlorous acid (HOCl). The dissociation of hypochlorous acid to the less microbicidal form (hypochlorite ion OCl⁻) is dependent on pH. The disinfecting efficacy of chlorine decreases with an increase in pH that parallels the conversion of undissociated hypochlorous acid to hypochlorite ion (29). A potential hazard is production of the carcinogen *bis*-chloromethyl when hypochlorite solutions come into contact with formaldehyde and production of the animal carcinogen trihalomethane when hot water is hyperchlorinated. A mixture of sodium hypochlorite and acid will also effect a rapid evolution of toxic chlorine gas.

Alternative compounds that release chlorine and are used in the hospital setting include demand-release chlorine dioxide and chloramine T. The advantage of these compounds over the hypochlorites is that they retain chlorine longer and so exert a more prolonged bactericidal effect. Sodium dichloroisocyanurate tablets are stable, and the microbicidal activities of solutions prepared from sodium dichloroisocyanurate tablets may be greater than those of sodium hypochlorite solutions containing the same total available chlorine for two reasons. First, with sodium dichloroisocyanurate, only 50% of the total available chlorine present is free (HOCl and OCl⁻), while the remainder is combined (mono- or dichloroisocyanurate); as free available chlorine is used up, the latter is released to restore the equilibrium. Second, solutions of sodium dichloroisocyanurate are acidic, while sodium hypochlorite solutions are alkaline, and the more microbicidal type of chlorine (HOCl) is believed to predominate (21).

The exact mechanism by which free chlorine destroys microorganisms has not been elucidated. The postulated mechanism of chlorine disinfection is inhibition of some key enzymatic reactions within the cell, protein denaturation, and inactivation of nucleic acids (29).

Inorganic chlorine solution is used for disinfecting tonometer heads and for spot disinfecting countertops and floors. A 1:10 to 1:100 dilution of 5.25% sodium hypochlorite (i.e., household bleach) (18) or an EPA-registered hospital disinfectant (87) has been recommended for decontaminating blood spills. Either of these will minimize the risk of employee exposure to blood. Since hypochlorites and other germicides are substantially inactivated in the presence of blood, the surface should be cleaned before an EPA-registered disinfectant or a 1:10 to 1:100 solution of household bleach is applied. At least 500 ppm of available chlorine for 10 min is recommended for decontamination of cardiopulmonary resuscitation training manikins (4a). Full-strength bleach is recommended for the disinfection of needles and syringes. The difference in the recommended concentrations of bleach reflects the difficulty of cleaning the interior of needles and syringes and the use of needles and syringes for parenteral injection (20). Clinicians should not alter their use of chlorine on environmental surfaces based on testing methodologies that do not simulate actual disinfection practices.

Dilute solutions of hypochlorite have been suggested for disinfection in the rooms of patients with *Clostridium difficile*-associated diarrhea or colitis as a means of preventing spread of the organism. However, studies suggest that asymptomatic patients constitute an important reservoir within the hospital and that person-to-person transmission, including transient carriage of *C. difficile* on the hands of hospital personnel, is the principal means of transmission between patients. Therefore, hand washing, barrier precautions, and meticulous environmental cleaning may be equally effective in preventing the spread of this organism.

Chlorine has long been favored as the preferred disinfectant in water treatment. Hyperchlorination of a *Legionella*-contaminated hospital water system resulted in a dramatic decrease (30 to 1.5%) in the isolation of *Legionella pneumophila* from water outlets and a cessation of nosocomial Legionnaires' disease in the affected unit (39). Chloramine T and hypochlorites have been evaluated for disinfecting hydrotherapy equipment. Hypochlorite solutions in tap water at pH ≥8.0 are stable for 1 month when stored at room temperature (23°C) in closed, opaque plastic containers. The levels of free available chlorine in solutions in opened and closed polyethylene containers are maximally reduced to 40 to 50% of the original concentration in 1 month. Thus, if a user wished to have a solution containing 500 ppm of available chlorine at day 30, a solution containing 1,000 ppm of chlorine should be prepared at time 0. There is no decomposition of sodium hypochlorite solution after 30 days when the solution is stored in a closed brown bottle (92).

Formaldehyde

Formaldehyde in both liquid and gaseous states is used as a disinfectant and sterilant. The liquid form will be considered briefly in this section. Formaldehyde is sold and used principally as a water-based solution called formalin, which is 37% formaldehyde by weight. The aqueous solution is a bactericide, tuberculocide, fungicide, virucide, and sporicide. OSHA has indicated that formaldehyde should be handled in the workplace as a potential carcinogen and has set an employee exposure standard for formaldehyde that limits an 8-h time-weighted average (TWA) exposure to a concentration of 0.75 ppm. For this reason and others, employees should have limited direct contact with formaldehyde, and these considerations limit its role in sterilization and disinfection processes.

Formaldehyde inactivates microorganisms by alkylating the amino and sulfhydryl groups of proteins and ring nitrogen atoms of purine bases (34).

Although formaldehyde-alcohol is a chemical sterilant and formaldehyde is a high-level disinfectant, its hospital uses are limited by its irritating fumes and the pungent odor that is apparent at very low levels (<1 ppm). For these reasons and others, such as carcinogenicity, this germicide is excluded from Table 1. When it is employed, there is generally limited direct employee exposure; however, significant exposures to formaldehyde have been documented for employees of renal transplant units and students in a gross anatomy laboratory. Formaldehyde is used in the health care setting to prepare viral vaccines (e.g., poliovirus, influenza), as an embalming agent, to preserve anatomical specimens, and in the past and especially when mixed with ethanol, to sterilize surgical instruments. A survey conducted in 1989 found that formaldehyde was the disinfectant used by 47% of the hemodialysis centers in the United States for reprocessing hemodialyzers, a 50% decrease from 1983. Other disinfectants that are available for dialysis systems are chlorine-based disinfectants, glutaraldehyde-based disinfectants, peracetic acid, and peracetic acid and hydrogen peroxide. Some dialysis systems use hot-water disinfection for the control of microbial contamination. If formaldehyde is used at room temperature, CDC recommends a concentration of 4% with a minimum exposure time of 24 h to disinfect disposable hemodialyzers that are reused on the same patient (32). Aqueous formaldehyde

solutions (1 to 2%) have also been used to disinfect the internal fluid pathways. To minimize the potential health hazard to dialysis patients, the dialysis equipment must be thoroughly rinsed and tested for residual formaldehyde before use.

Glutaraldehyde

Glutaraldehyde is a saturated dialdehyde that has justifiably gained wide acceptance as a high-level disinfectant and chemical sterilant. Aqueous solutions of glutaraldehyde are acidic and generally not sporicidal. Only when the solution is "activated" to pH 7.5 to 8.5 (made alkaline) by use of alkalinating agents does the solution become sporicidal. Once activated, these solutions have a shelf life of 14 days because of the polymerization of the glutaraldehyde molecules at alkaline pH levels. This polymerization blocks the active sites (aldehyde groups) of the glutaraldehyde molecules that are responsible for its biocidal activity. The in vitro inactivation of microorganisms by glutaraldehydes has been extensively investigated and reviewed (95).

Novel glutaraldehyde formulations (e.g., glutaraldehyde-phenol-sodium phenate, potentiated acid glutaraldehyde, stabilized alkaline glutaraldehyde, glutaraldehyde-phenylphenol-amylphenol) produced in the past several years have overcome the problem of rapid loss of activity (e.g., use life of 28 to 30 days) while generally maintaining excellent microbicidal activity. However, it should be recognized that antimicrobial activity is dependent not only on age but also on use conditions such as dilution and organic stress. Manufacturers' literature for these preparations suggests that the neutral or alkaline glutaraldehydes possess microbicidal and anticorrosion properties superior to those of the acid glutaraldehydes, and a few published reports substantiate these claims. The use of glutaraldehyde-based solutions in hospitals is widespread because of the advantages of these solutions, which include excellent biocidal properties; activity in the presence of organic matter (20% bovine serum); noncorrosive action on endoscopic equipment, thermometers, and rubber or plastic equipment; and noncoagulation of proteinaceous material.

The biocidal activity of glutaraldehyde is a consequence of its alkylation of sulfhydryl, hydroxyl, carboxy, and amino groups of microorganisms, which alters RNA, DNA, and protein synthesis. For an extensive review of the mechanism of action of glutaraldehydes, see Scott and Gorman (95).

Dilution of glutaraldehyde during use commonly occurs. A recent study found a glutaraldehyde concentration of 1.0 to 1.1% in manual and automatic baths used for endoscopes at the end of the 14-day reuse period (67). Others have shown that the glutaraldehyde level falls below 1% to as low as 0.27% on day 4 of reuse. This drop emphasizes the need to ensure that semicritical equipment is disinfected with an acceptable concentration of glutaraldehyde. Data suggest that 1.0% glutaraldehyde is the minimum effective concentration for use as a high-level disinfectant. Glutaraldehyde test strips are available for determining whether an effective concentration of active ingredients (e.g., glutaraldehyde) is present despite repeated use and dilution. The frequency of testing should be based on how frequently the solutions are used (e.g., if used daily, then test daily), but the strips should not be used to extend the use life beyond the expiration date. Glutaraldehyde test kits have been evaluated preliminarily for accuracy and range. The indicator should be considered unacceptable or unsafe when a dilution to 1% glutaraldehyde or lower has occurred.

Glutaraldehyde is used most commonly as a high-level disinfectant for medical equipment such as endoscopes, spirometry tubing, dialyzers, transducers, anesthesia and respiratory therapy equipment, and hemodialysis proportioning and dialysate delivery systems. In regard to disinfection of spirometers, a recent study showed that mouthpieces and spirometry tubing become contaminated with microorganisms and should be at least high-level disinfected (e.g., with glutaraldehyde) between patients. However, since there was no bacterial contamination of the surfaces inside the spirometers and cross-transmission was unlikely, the investigators concluded that it was unnecessary to clean the interior surfaces of the spirometers. Glutaraldehyde is noncorrosive to metal and does not damage lensed instruments, rubber, or plastics. Glutaraldehyde should not be used for cleaning noncritical surfaces; it is too toxic and expensive.

Proctitis believed to be due to glutaraldehyde exposure from residual endoscope solution contaminating the air-water channel has been reported and is preventable by careful endoscope rinsing. Similarly, keratopathy was caused by ophthalmic instruments that were inadequately rinsed after being soaked in 2% glutaraldehyde.

Health care workers can become exposed to elevated levels of glutaraldehyde vapor when equipment is processed in poorly ventilated rooms, when spills occur, or when there are open immersion baths. In these situations, the level of glutaraldehyde in the air could reach its ceiling limit of 0.2 ppm. Engineering and work practice controls that may be used to combat these problems include improved ventilation, tight-fitting lids on immersion baths, personal protection (e.g., gloves, goggles) to minimize skin or mucous membrane contact, and vapor containment systems that utilize activated-charcoal filters to absorb the glutaraldehyde molecules. Some workers have been fitted with half-face respirators with organic vapor filters or offered a type C supplied-air respirator with a full face piece operated in a positive-pressure mode. Even though enforcement of the ceiling limit was suspended on March 23, 1993, by the Court of Appeals, it is prudent to limit employee exposure to 0.2 ppm, since at this level, glutaraldehyde is irritating to the eyes, throat, and nose. Epistaxis, allergic contact dermatitis, asthma, and rhinitis have also been reported in health care workers exposed to glutaraldehyde. Some automatic endoscope disinfection machines reduce employee exposure to glutaraldehyde. Dosimeters are available for measuring glutaraldehyde levels in the workplace to ensure a safe work environment.

Hydrogen Peroxide

The literature contains limited accounts of the properties, germicidal effectiveness, and potential uses for stabilized hydrogen peroxide in the hospital setting. Reports ascribing good germicidal activity to hydrogen peroxide have been published and attest to its bactericidal, virucidal, sporicidal, and fungicidal properties (104). Hydrogen peroxide works by the production of destructive hydroxyl free radicals that can attack membrane lipids, DNA, and other essential cell components.

Commercially available 3% hydrogen peroxide is a stable and effective disinfectant when used on inanimate surfaces. It has been used in concentrations from 3 to 6% for the disinfection of soft contact lenses (3% for 2 to 3 h) (104), tonometer biprisms, and ventilators. Corneal damage from a hydrogen peroxide-soaked tonometer tip that was not properly rinsed has been reported. Hydrogen peroxide has also been instilled into urinary drainage bags in an attempt to eliminate the bag as a source of bladder bacteriuria and environmental contamination. While the instillation of hydrogen peroxide into the bag reduced microbial contamination of the bag, this procedure did not reduce the incidence of catheter-associated bacteriuria.

Iodophors

Iodine solutions or tinctures have long been used by health professionals primarily as antiseptics on skin or tissue. Iodophors, on the other hand, have enjoyed use as both antiseptics and disinfectants. As previously mentioned, an iodophor is a combination of iodine and a solubilizing agent or carrier; the resulting complex provides a sustained-release reservoir of iodine and releases small amounts of free iodine in aqueous solution. The best-known and most widely used iodophor is povidone-iodine, a compound of polyvinylpyrrolidone with iodine. This product and other iodophors retain the germicidal efficacy of iodine but unlike iodine are generally nonstaining and relatively free of toxicity and irritancy (37).

Several reports documenting intrinsic microbial contamination of antiseptic formulations of povidone-iodine and poloxamer-iodine caused a reappraisal of the concepts concerning the chemistry and use of iodophors. It seems that free iodine (I_2) contributes to the bactericidal activity of iodophors and that dilutions of iodophors demonstrate more rapid bactericidal action than a full-strength povidone-iodine solution. The reason that dilution increases bactericidal activity is unclear, but it has been suggested that dilution of povidone-iodine results in weakening of the iodine linkage to the carrier polymer, with an accompanying increase of free iodine in solution. Therefore, iodophors must be diluted according to the manufacturers' directions to achieve antimicrobial activity.

Iodine is able to penetrate the cell walls of microorganisms quickly, and it is thought that the lethal effects result from a disruption of protein and nucleic acid structures and synthesis.

Published reports on the in vitro antimicrobial efficacy of iodophors demonstrate that iodophors are bactericidal, mycobactericidal, and virucidal but may require prolonged contact times to kill certain fungi and bacterial spores (88, 89, 101).

Besides their use as an antiseptic, iodophors have been used for the disinfection of blood culture bottles and medical equipment such as hydrotherapy tanks, thermometers, and endoscopes. Antiseptic iodophors are not suitable for use as hard-surface disinfectants because of concentration differences. That is, iodophors formulated as antiseptics contain significantly less free iodine than those formulated as disinfectants (34).

Peracetic Acid

Peracetic acid or peroxyacetic acid is characterized by a very rapid action against all microorganisms including bacterial spores in low concentrations (0.001 to 0.2%). A special advantage of peracetic acid is that it has no harmful decomposition products (i.e., acetic acid, water, oxygen, hydrogen peroxide) and leaves no residue. It remains effective in the presence of organic matter and is sporicidal even at low temperatures. Peracetic acid can corrode copper, brass, bronze, plain steel, and galvanized iron, but these effects can be reduced by additives and pH modifications. It is considered unstable, particularly when diluted; for example, a 1% solution loses half its strength through hydrolysis in 6 days, whereas 40% peracetic acid loses only 1 to 2% of its activity per month (10).

Little is known about the mechanism of action of peracetic acid, but it is thought to function as other oxidizing agents do; that is, it denatures proteins, disrupts cell wall permeability, and oxidizes sulfhydryl and sulfur bonds in proteins, enzymes, and other metabolites (10).

The combination of peracetic acid and hydrogen peroxide has been used for disinfecting hemodialyzers. The percentage of centers using a peracetic acid-hydrogen peroxide-based disinfectant for reprocessing dialyzers increased from 5% in 1983 to 46% in 1989. Preliminary data reported by the FDA showed increased mortality in dialysis patients when dialyzers were disinfected with peracetic acid-hydrogen peroxide. Although the findings are preliminary, the agency has asked the manufacturer of this product to notify dialysis facilities about the study results and provide technical assistance on the proper use of the disinfectant. An automated machine using peracetic acid to chemically sterilize medical, surgical, and dental instruments (e.g., endoscopes, arthroscopes) is used in the United States. The manufacturer's data demonstrate that this system inactivates *Bacillus subtilis* and *Clostridium sporogenes* when the solution is heated to 50°C and the exposure time is 12 min or less; however, the efficacy of this system has not been independently verified (88). Since the design of an automatic endoscope washer, the material used, and the flow of the disinfectant-chemical sterilant and rinse water through the endoscope are critical to the effectiveness of the system in eliminating contaminating microorganisms, each machine type should be independently verified.

Phenolics

Phenol has occupied a prominent place in the field of hospital disinfection since its initial use by Lister as a germicide in his pioneering work on antiseptic surgery. In the past 30 years, however, work has been concentrated on the numerous phenol derivatives, or phenolics, and their antimicrobial properties. Phenol derivatives originate when a functional group (e.g., alkyl, phenyl, benzyl, halogen) replaces one of the hydrogen atoms on the aromatic ring. Two of the phenol derivatives that are commonly found as constituents of hospital disinfectants are *ortho*-phenylphenol and *ortho*-benzyl-*para*-chlorophenol. The antimicrobial properties of these compounds and many other phenol derivatives are much improved over those of the parent chemical. Phenolics are assimilated by porous materials, and the residual disinfectant may cause tissue irritation. In 1970, Kahn reported depigmentation of skin caused by phenolic germicidal detergents containing *para*-tertiary butylphenol and *para*-tertiary amylphenol (45).

In higher concentrations, phenol acts as a gross protoplasmic poison, penetrating and disrupting the cell wall and precipitating the cell proteins. Low concentrations of phenol and higher-molecular-weight phenol derivatives cause bacterial death by inactivating essential enzyme systems and leaking essential metabolites from the cell wall (82).

Published reports on the antimicrobial efficacies of commonly used phenolic detergents show that phenolics are bactericidal, fungicidal, virucidal, and tuberculocidal (45, 82, 88, 89, 94, 101).

Manufacturers' data using standardized AOAC methods demonstrate that commercial phenolic detergents are not sporicidal but are bactericidal, fungicidal, virucidal, and tuberculocidal at their recommended use dilutions. Generally, these efficacy claims against microorganisms have not been verified by independent laboratories or the EPA. Attempts to substantiate the bactericidal label claims of phenolic detergents using the AOAC use dilution method have failed (91). However, these same studies have shown extreme variability of test results among laboratories testing identical products.

This class of compounds is used for decontamination of the hospital environment, including laboratory surfaces, and of noncritical medical and surgical items. Phenolics are not recommended for semicritical items because of the lack of published efficacy data for many of the available formulations and because residual disinfectant on porous materials may cause tissue irritation even when the material is thoroughly rinsed.

The use of phenolics in nurseries has been justifiably questioned because of the occurrence of hyperbilirubinemia in infants placed in nurseries that use phenolic detergents. In addition, Doan and coworkers demonstrated microbilirubin level increases in phenolic-exposed infants compared to non-phenolic-exposed infants when the phenolic was prepared according to the manufacturers' recommended dilution (26). If phenolics are used to clean nursery floors, they must be diluted according to the recommendation on the product label. Phenolics should not be used to disinfect infant bassinets and incubators.

Quaternary Ammonium Compounds

The quaternary ammonium compounds are widely used as disinfectants but should not be used as antiseptics. The elimination of such solutions as antiseptics on skin and tissue (e.g., benzalkonium chloride) was recommended by the CDC (97) because of several outbreaks of infections associated with in-use contamination. There have also been a few reports of nosocomial infections associated with contaminated quaternary ammonium compounds used to disinfect patient care supplies or equipment such as cystoscopes or cardiac catheters. The quaternaries are good cleaning agents, but materials such as cotton and gauze pads may make them less microbicidal because these materials absorb the active ingredients. As with several other disinfectants (e.g., phenolics, iodophors), gram-negative bacteria have been found to survive or grow in them.

Chemically, the quaternaries are organically substituted ammonium compounds in which the nitrogen atom has a valence of 5, four of the substituent radicals (R1 through R4) are alkyl or heterocyclic radicals of a given size or chain length, and the fifth (X-) is a halide, sulfate, or similar radical. Each compound exhibits its own antimicrobial characteristics: hence the search for one compound with outstanding antimicrobial properties. Some of the chemical names of quaternary ammonium compounds that are used in hospitals are alkyl dimethyl benzyl ammonium chloride, alkyl didecyl dimethyl ammonium chloride, and dialkyl dimethyl ammonium chloride. The newer quaternary ammonium compounds, referred to as twin-chain or dialkyl quaternaries (e.g., didecyl dimethyl ammonium chloride), purportedly remain active in hard water and are tolerant of anionic residues.

The bactericidal action of the quaternaries has been attributed to the inactivation of energy-producing enzymes, denaturation of essential cell proteins, and disruption of the cell membrane. Evidence offered in support of these and other possibilities is provided by Sykes (102) and Petrocci (78).

Results from manufacturers' data sheets and from published scientific literature indicate that the quaternaries sold as hospital disinfectants are generally fungicidal, bactericidal, and virucidal against lipophilic viruses; they are not sporicidal and generally are not tuberculocidal or virucidal against hydrophilic viruses (78, 88, 89, 94, 101). Best et al. demonstrated the poor mycobactericidal activities of quaternary ammonium compounds (8). Attempts to reproduce the manufacturers' bactericidal and tuberculocidal claims using the AOAC tests with a limited number of quaternary ammonium compounds have failed (91). Studies have shown, however, an extreme variability of test results among laboratories testing identical products (91).

The quaternaries are commonly used for ordinary environmental sanitation of noncritical surfaces such as floors, furniture, and walls.

STERILIZATION

Sterilization removes or destroys all microorganisms on the surface of an article or in a fluid. However, this is an oversimplification, since the concept of what constitutes "sterile" is measured as a probability of sterility for each item to be sterilized. This probability is commonly referred to as the sterility assurance level of the product and is defined as the $\log_{10}$ number of the probability of a survivor on a single item. For example, if the probability of a spore surviving is 1 in 1 million, the sterility assurance level would be 6 (33, 74). Items that enter tissue or the vascular system and equipment (e.g., excorporeal circulator, hemodialysis coil) through which blood or sterile fluids circulate must be sterilized. Several detailed reviews (4, 5, 15, 43, 62, 77, 86) deal with principles of ETO and steam sterilization and have been used as references for this section.

Steam Sterilization

Of all the methods available for sterilization, moist heat in the form of saturated steam under pressure is the most widely used and the most dependable. Steam sterilization is nontoxic, inexpensive, and sporicidal and rapidly heats and penetrates fabrics. For these reasons, steam sterilization should be used whenever possible on all items that are not heat and moisture sensitive (e.g., steam-sterilizable respiratory therapy and anesthesia equipment) even when it is not essential to prevent disease transmission. Moist heat destroys microorganisms by the irreversible coagulation and denaturation of enzymes and structural proteins. In support of this opinion, it has been found that the presence of water significantly affects the coagulation temperature of proteins and the temperature at which microorganisms are destroyed.

The basic principle of steam sterilization, as accomplished in an autoclave, is to expose each item to direct steam contact at the required temperature and pressure for the specified time. Thus, there are four parameters of steam sterilization: steam, pressure, temperature, and time. The ideal steam for sterilization is 100% dry saturated steam with no saturated water in the form of a fine mist. Pressure serves as a means of obtaining the high temperatures necessary to kill microorganisms quickly. Specific temperature must be obtained to ensure microbicidal activity. The two common steam sterilization temperatures are 121°C (250°F) and 132°C (270°F). These temperatures must be maintained for a minimum time to kill microorganisms. *Bacillus stearothermophilus* spores are used to monitor the efficacy of steam sterilization. Recognized exposure periods for sterilization of wrapped hospital supplies are 30 min at 121°C in a gravity displacement sterilizer and 4 min at 132°C in a prevacuum sterilizer. At constant temperatures, sterilization times vary depending on the sizes and types of items as well as the sterilizer type.

The two basic types of steam sterilizers are the gravity displacement autoclave and the high-speed prevacuum sterilizer. In the former, steam is admitted at the top of the sterilizing chamber, and because the steam is lighter than air, it forces air out the bottom of the chamber through the drain vent. The gravity displacement autoclaves are primarily used to process laboratory media, water, pharmaceutical products, regulated medical waste, and nonporous articles whose surfaces have direct steam contact. For gravity displacement sterilizers, the penetration time is prolonged by incomplete air elimination. This point is illustrated with the decontamination of microbiological waste (10 lb [ca. 4.5 kg]) which requires at least 45 min at 121°C because the entrapped air remaining in a load of waste greatly retards steam permeation and heating efficiency. High-speed prevacuum sterilizers are similar to gravity displacement sterilizers except that they are fitted with a vacuum pump to ensure air removal from the sterilizing chamber and the load before the steam is admitted. The advantage of this difference is that there is nearly instantaneous steam penetration even into porous loads. The Bowie-Dick test using 100% cotton surgical towels (huckaback) is used daily in the first cycle of all vacuum-type autoclaves to evaluate the efficacy of air removal. Air that is not removed from the chamber will interfere with steam contact. Smaller disposable test packs have been devised to replace the stack of folded towels for testing the efficacy of the vacuum system in a prevacuum sterilizer.

ETO Gas Sterilization

ETO is used almost exclusively in the United States to sterilize medical products that cannot be steam sterilized. ETO is a colorless gas that is flammable and explosive; however, mixtures of ETO (10 to 12%) and carbon dioxide or the fluoridated hydrocarbons reduce the risk. Because of the implications of the effect of the halocarbons on the ozone layer, restrictions on the use of ETO are emerging (75). The effectiveness of ETO sterilization is influenced by four essential elements: gas concentration, temperature, humidity, and exposure time. The operational ranges for these four parameters are 450 to 1,200 mg/liter, 29 to 65°C, 45 to 85%, and 2 to 5 h, respectively. Within certain limitations, an increase in gas concentration and temperature may shorten the time necessary for achieving sterilization. ETO inactivates all microorganisms, although bacterial spores (especially those of *B. subtilis*) are more resistant than other microorganisms, and for this reason *B. subtilis* is the recommended biological indicator. The microbicidal activity of ETO is considered to be the result of alkylation of protein, DNA, and RNA. Alkylation, or the replacement of a hydrogen atom with an alkyl group, within cells prevents normal cellular metabolism and replication.

The main disadvantages associated with ETO are the lengthy cycle time, the cost, and the potential hazards to patients and staff; the advantage is that it can sterilize heat- or moisture-sensitive medical equipment without deleterious results. The basic ETO sterilization cycle consists of five stages (i.e., preconditioning and humidification, gas introduction, exposure, evacuation, and air washes) and takes approximately 2.5 h excluding aeration time. Mechanical aeration for 8 to 12 h at 50 to 60°C allows desorption of the toxic ETO residuum contained in exposed absorbent materials. Ambient room aeration will also achieve desorption of the toxic ETO but requires 7 days at 20°C.

In recent years, the toxicity of ETO to employees has raised considerable concerns. In June 1984, OSHA reduced the permissible exposure limit for ETO to a TWA of 1 ppm. Determination of employee exposure is made from breathing-zone air samples that are representative of the 8-h TWA for an employee in each work area. Monitoring of ETO exposure levels may be performed by passive sampling or by direct readout instruments. For <0.5 ppm, the employer can discontinue the monitoring for affected employees. If the level is >1 ppm, the employer must establish a regulated area and must then make an effort to comply through work practice alterations and by engineering controls (e.g., improved ventilation, process isolation, and/or effective equipment repair). When these modifications are not sufficient to reduce employee exposure to <1 ppm, the employer needs to supplement them by use of respiratory protection. Employees who are exposed to ETO above the action level of 0.5 ppm for at least 30 days/year are covered by a medical surveillance program (22).

Other Sterilization Methods

Dry-Heat Sterilizers

Dry-heat sterilization should be used only for materials (e.g., powders, petroleum products, sharp instruments) that might be damaged by moist heat or for materials that are impenetrable to moist heat. The advantages of dry heat are its penetrating power and lack of corrosiveness for metal and sharp instruments, but the slow heat penetration and microbial killing make this a time-consuming method. The most common time-temperature relationships for sterilization with hot-air sterilizers are 170°C (340°F) for 60 min, 160°C (320°F) for 120 min, and 150°C (300°F) for 150 min. B. subtilis spores should be used to monitor the sterilization process for dry heat, since they are more resistant to dry heat than B. stearothermophilus spores. The primary lethal process is considered to be oxidation of cell constituents.

Ionizing Radiation

Sterilization by ionizing radiation, primarily by cobalt-60 gamma rays and electron accelerators, is a low-temperature sterilization method that has been used for a host of medical products (e.g., tissue for transplantation, drugs, pharmaceuticals). Because of high sterilization costs, this method is an unfavorable alternative to ETO sterilization in hospitals but is suitable for large-scale sterilization. Several good reviews that deal with the sources, effects, and application of ionizing radiation (86, 96) may be referred to for more detail.

Liquid Chemicals

Several EPA-registered liquid chemicals are capable of producing sterile medical and surgical materials after exposure periods of 6 to 10 h (Table 1). Sterilization with liquid chemicals is recommended only for those materials that cannot be sterilized by heat or ETO.

Microwave

Recent reports have shown microwaves to be an effective microbicide. The microwaves produced by a "home-type" microwave oven (2.45 GHz) have been shown to completely inactivate bacterial cultures, viruses, and B. stearothermophilus spores within 60 s to 5 min depending on the challenge organism (57). Its use in hospitals requires further evaluation.

Other Gaseous Sterilization Methods

Reprocessing of heat-labile medical equipment is a major problem in hospitals. ETO has been the sterilant of choice for sterilizing heat-labile medical equipment. In spite of its excellent properties, it is toxic, mutagenic, and possibly carcinogenic. For this reason, alternative gaseous sterilization methods are being investigated for ETO replacement. These gaseous sterilants include chlorine dioxide, hydrogen peroxide, formaldehyde, peracetic acid, and ozone. A gas plasma sterilization system that utilizes a highly ionized gas (hydrogen peroxide), which is created under vacuum with either radio frequency or microwave energy, has recently received FDA approval for marketing. At present, it appears unlikely that any of these other gaseous sterilization methods will completely supplant ETO, primarily because the sterilants may not possess the product penetration ability of ETO (75).

Sterilizing Practices

Hospitals should perform most cleaning, disinfecting, and sterilizing of patient care supplies in a central processing department to more easily control quality. Some hospitals are able to promote the same level of efficiency and safety in the preparation of supplies for other areas such as the operating room, anesthesia, and respiratory therapy.

Cleaning

As repeatedly mentioned, items must be cleaned with water and detergents or enzymatic cleaners before they are processed. Precleaning in patient care areas may be needed on items that are heavily soiled with feces, sputum, blood, etc. Items sent to central processing without gross soil removed may be difficult to clean because of dried secretions and excretions.

Several types of mechanical cleaning machines (e.g., utensil washer-sanitizer, ultrasonic cleaner, washer-sterilizer, dishwasher) may facilitate cleaning and decontamination of most items. Delicate and intricate objects as well as heat- or moisture-sensitive articles are carefully cleaned by hand. All used items sent to the central processing area should be considered contaminated and need to be handled with gloves (forceps or tongs are sometimes needed to avoid exposure to sharps) and decontaminated by one of the aforementioned methods to render them safer to handle.

Packaging, Loading, and Storage

Once items are cleaned, dried, and inspected, objects requiring sterilization must be wrapped or packaged. The packaging material must allow penetration of the sterilant and maintain the sterility of the processed item after sterilization. Commonly used packaging material include muslin (140 to 288 thread count per in. [1 in. = 2.54 cm]), Kraft paper, nonwoven wraps, and paper or plastic peel-down packages.

All items to be sterilized should be arranged so that all surfaces will be directly exposed to the sterilizing agent. Packs should not exceed the maximum dimensions, weight, and density of 12 by 12 by 20 in., 12 lb (ca. 5.4 kg), and 7.2 lb/ft^3 [ca. 3.3 kg/(2.83 × 10^{-2} m^3)], respectively.

Safe storage times for sterile packs vary with the porosity of the wrapper and the storage conditions (e.g., open versus closed cabinets). Heat-sealed, plastic peel-down pouches and wrapped packs sealed in 3-mil (0.003-in.) polyethylene overwrap have been reported to be sterile for as long as 9 months after sterilization. The 3-mil polyethylene is applied after sterilization to extend the shelf lives of infrequently used items (63). Supplies wrapped in double-thickness muslin, comprising four layers, or the equivalent remain sterile for at least 30 days. Any item that has been sterilized should not be used after the expiration date or if the sterilized package has been dropped, wet, torn, or punctured.

Though many hospitals continue to date every sterilized product and use the time-related shelf life practice, some hospitals have switched to an event-related shelf life practice. This latter practice recognizes that the product should remain sterile until some event causes it to become contaminated (e.g., tear in packaging, packaging becomes wet or is dropped on a contaminated surface such as the floor) (65). Some data support this policy (46).

Monitoring

The sterilization procedure should be monitored routinely by a combination of mechanical, chemical, and biological parameters. These process parameters evaluate the sterilizing conditions and indirectly the microbiological status of the processed items. The mechanical techniques for steam sterilization include a daily assessment of cycle time and temperature by examining the temperature record chart as well as an assessment of pressure via the pressure gauge. Unfortunately, two other essential elements for ETO sterilization (i.e., gas concentration and humidity) cannot be monitored.

Chemical indicators are affixed on the outside of each pack to show that the package has been processed through a sterilization cycle, but these indicators do not prove that sterilization has been achieved. Preferably, a chemical indicator should also be placed on the inside of each pack to verify steam penetration. Chemical indicators are usually paper printed with either heat- or chemical-sensitive inks that change color when one or more germicidal-related parameters are present.

Biological indicators are the only process indicators that directly measure sterilization. $B. subtilis$ spores (10^6) are used to monitor ETO and dry heat, and $B. stearothermophilus$ spores (10^5) are used to monitor steam sterilization. $B. stearothermophilus$ is incubated at 55°C, and $B. subtilis$ is incubated at 35 to 37°C. Steam and ETO sterilizers should be monitored at least weekly with the appropriate commercial preparation of spores, but each load should be monitored if it contains implantable objects. When feasible, do not use implantable items until results of spore tests are known to be negative.

"Flash" steam sterilization is defined as sterilization of an unwrapped object at 132°C for 3 min at 27 to 28 lb (ca. 12 to 13 kg) of pressure in a gravity displacement autoclave. Implantables should ideally not be flash sterilized, and flash sterilization should be restricted to emergency situations. It should not be used to compensate for inadequate inventories of instruments. Flash sterilization has a smaller margin of safety, because the cycle parameters are the minimum parameters for sterilization, and deviation from the exposure time, temperature, or pressure can produce nonsterile items. In addition, the sterilized item will not be protected by packaging after sterilization, allowing for exogenous contamination.

CONCLUSION

When properly used, liquid chemical germicides (i.e., antiseptics, disinfectants) and sterilization processes can minimize the risk of nosocomial infections. However, current guidelines must be strictly adhered to.

I gratefully acknowledge Eva Clontz for her invaluable assistance in preparing this chapter, David J. Weber for reviewing the chapter, and Elaine L. Larson for her contributions to the antisepsis section of the chapter. This chapter has been extensively modified from reference 88 with permission from The Williams & Wilkins Co.

REFERENCES

1. **American Academy of Ophthalmology.** 1992. *Updated Recommendations for Ophthalmic Practice in Relation to the Human Immunodeficiency Virus.* American Academy of Ophthalmology, San Francisco.
2. **American Dental Association.** 1992. Infection control recommendations for the dental office and the dental laboratory. *J. Am. Dent. Assoc.* **123:**1–8.
3. **American Neurological Association.** 1986. Precautions in handling tissues, fluids, and other contaminated materials from patients with documented or suspected Creutzfeldt-Jakob disease. *Ann. Neurol.* **19:**75–77.
4. **American Society for Hospital Central Service Personnel.** 1982. *Ethylene Oxide Use in Hospitals.* American Hospital Association, Chicago.
4a. **Anonymous.** 1984. Recommendations for decontaminating manikins used in cardiopulmonary resuscitation training 1983 update. *Infect. Control* **5:**399–401.
5. **Association of Operation Room Nurses.** 1991. Proposed recommended practices: sterilization. *AORN J.* **54:**82–96.
6. **Beck-Sague, C. M., and W. R. Jarvis.** 1989. Epidemic bloodstream infections associated with pressure transducers; a persistent problem. *Infect. Control Hosp. Epidemiol.* **10:**54–59.
7. **Benson, L., D. LeBlanc, L. Bush, and J. White.** 1990. The effects of surfactant systems and moisturizing products on the residual activity of a chlorhexidine gluconate handwash using a pigskin substrate. *Infect. Control Hosp. Epidemiol.* **11:**67–70.
8. **Best, M., S. A. Sattar, V. S. Springthorpe, and M. E. Kennedy.** 1990. Efficacies of selected disinfectants against *Mycobacterium tuberculosis. J. Clin. Microbiol.* **28:**2234–2239.
9. **Block, S. S.** 1991. Definition of terms, p. 18–25. *In* S. S. Block (ed.), *Disinfection, Sterilization and Preservation,* 4th ed. Lea & Febiger, Philadelphia.
10. **Block, S. S.** 1991. Peroxygen compounds, p. 167–181. *In* S. S. Block (ed.), *Disinfection, Sterilization and Preservation,* 4th ed. Lea & Febiger, Philadelphia.
11. **Bond, W. W., M. S. Favero, N. J. Petersen, and J. W. Ebert.** 1983. Inactivation of hepatitis B virus by intermediate- to high-level disinfectant chemicals. *J. Clin. Microbiol.* **18:**535–538.
12. **Bond, W. W., B. J. Ott, K. A. Franke, and J. E. McCracken.** 1991. Effective use of liquid chemical germicides on medical devices: instrument design problems, p. 1097–1106. *In* S. S. Block (ed.), *Disinfection, Sterilization and Preservation,* 4th ed. Lea & Febiger, Philadelphia.
13. **Brown, P., C. J. Gibbs, H. L. Amyx, D. T. Kingsbury, R. G. Rohwer, M. P. Sulima, and D. C. Gajdusek.** 1982. Chemical disinfection of Creutzfeldt-Jakob disease virus. *N. Engl. J. Med.* **306:**1279–1281.
14. **Brown, P., A. Wolff, and C. Gajdusek.** 1990. A simple and effective method for inactivating virus infectivity in formalin-fixed tissue samples from patients with Creutzfeldt-Jakob disease. *Neurology* **40:**887–890.
15. **Caputo, R. A., and T. E. Odlaug.** 1983. Sterilization with ethylene oxide and other gases, p. 47–64. *In* S. S. Block (ed.), *Disinfection, Sterilization and Preservation,* 3rd ed. Lea & Febiger, Philadelphia.
16. **Centers for Disease Control.** 1985. Recommendations for

preventing possible transmission of human T-lymphotropic virus type III/lymphadenopathy-associated virus from tears. *Morbid. Mortal. Weekly Rep.* **34:**533–534.

17. **Centers for Disease Control.** 1987. Recommendations for prevention of HIV transmission in health-care settings. *Morbid. Mortal. Weekly Rep.* **36**(Suppl.):S3–S18.

18. **Centers for Disease Control.** 1989. Guidelines for prevention of transmission of human immunodeficiency virus and hepatitis B virus to health-care and public-safety workers. *Morbid. Mortal. Weekly Rep.* **38**(Suppl. 6):1–37.

19. **Centers for Disease Control.** 1993. Recommended infection control practices for dentistry, 1993. *Morbid. Mortal. Weekly Rep.* **41:**1–12.

20. **Centers for Disease Control.** 1993. Use of bleach for disinfection of drug injection equipment. *Morbid. Mortal. Weekly Rep.* **42:**418–419.

21. **Coates, D.** 1988. Comparison of sodium hypochlorite and sodium dichloroisocyanurate disinfectants: neutralization by serum. *J. Hosp. Infect.* **11:**60–67.

22. **Code of Federal Regulations.** 1988. Ethylene oxide. 29 CFR Part 1910.1047, July 1, p. 971–992. Department of Labor, Occupational Safety and Health Administration, Washington, D.C.

23. **Craven, E. R., S. L. Butler, J. P. McCulley, and J. P. Luby.** 1987. Applanation tonometer tip sterilization for adenovirus type 8. *Ophthalmology* **94:**1538–1540.

24. **DeGroot-Kosolcharoen, J., and J. M. Jones.** 1989. Permeability of latex and vinyl gloves to water and blood. *Am. J. Infect. Control* **17:**196–201.

25. **Department of Health and Human Services.** 1992. *Dental Handpiece Sterilization.* Food and Drug Administration, Washington, D.C.

26. **Doan, H. M., L. Keith, and A. T. Shennan.** 1979. Phenol and neonatal jaundice. *Pediatrics* **64:**324–325.

27. **Doebbeling, B. N., M. A. Pfaller, A. K. Houston, and R. P. Wenzel.** 1988. Removal of nosocomial pathogens from the contaminated glove: implications for glove reuse and handwashing. *Ann. Intern. Med.* **109:**394–398.

28. **Doebbeling, B. N., G. L. Stanley, C. T. Sheetz, M. A. Pfaller, A. K. Houston, L. Annis, N. Li, and R. P. Wenzel.** 1992. Comparative efficacy of alternative hand-washing agents in reducing nosocomial infections in intensive care units. *N. Engl. J. Med.* **327:**88–93.

29. **Dychdala, G. R.** 1991. Chlorine and chlorine compounds, p. 131–151. *In* S. S. Block (ed.), *Disinfection, Sterilization and Preservation,* 4th ed. Lea & Febiger, Philadelphia.

30. **Eckert, D. G., N. J. Ehrenkranz, and B. C. Alfonso.** 1989. Indications for alcohol or bland soap in removal of aerobic gram-negative skin bacteria: assessment by a novel method. *Infect. Control Hosp. Epidemiol.* **10:**306–310.

31. **Ehrenkranz, N. J., and B. C. Alfonso.** 1991. Failure of bland soap handwash to prevent hand transfer of patient bacteria to urethral catheters. *Infect. Control Hosp. Epidemiol.* **12:**654–662.

32. **Favero, M. S., M. J. Alter, and L. A. Bland.** 1992. Dialysis-associated infections and their control, p. 375–403. *In* J. V. Bennett and P. S. Brachman (ed.), *Hospital Infections,* 3rd ed. Little, Brown & Co., Boston.

33. **Favero, M. S., and W. W. Bond.** 1991. Sterilization, disinfection, and antisepsis in the hospital, p. 183–200. *In* A. Balows, W. A. Hausler, Jr., K. L. Herrmann, H. D. Isenberg, and H. J. Shadomy (ed.), *Manual of Clinical Microbiology,* 5th ed. American Society for Microbiology, Washington, D.C.

34. **Favero, M. S., and W. W. Bond.** 1991. Chemical disinfection of medical and surgical materials, p. 617–641. *In* S. S. Block (ed.), *Disinfection, Sterilization and Preservation,* 4th ed. Lea & Febiger, Philadelphia.

35. **Garner, J. S., and M. S. Favero.** 1986. Guideline for handwashing and hospital environmental control, 1985. *Am. J. Infect. Control* **14:**110–126.

36. **Gerding, D. N., L. R. Peterson, and J. A. Vennes.** 1982. Cleaning and disinfection of fiberoptic endoscopes: evalua-

tion of glutaraldehyde exposure time and forced-air drying. *Gastroenterology* **83:**613–618.

37. **Gottardi, W.** 1991. Iodine and iodine compounds, p. 152–166. *In* S. S. Block (ed.), *Disinfection, Sterilization and Preservation,* 4th ed. Lea & Febiger, Philadelphia.

38. **Handsfield, H. H., M. J. Cummings, and P. D. Swenson.** 1987. Prevalence of antibody to human immunodeficiency virus and hepatitis B surface antigen in blood samples submitted to a hospital laboratory: implications for handling specimens. *JAMA* **258:**3395–3397.

39. **Helms, C. M., R. M. Massanari, R. Zeitler, S. Streed, M. J. R. Gilchrist, N. Hall, W. J. Hausler, Jr., J. Sywassink, W. Johnson, L. Wintermeyer, and W. J. Hierholzer, Jr.** 1983. Legionnaires' disease associated with a hospital water system: a cluster of 24 nosocomial cases. *Ann. Intern. Med.* **99:**172–178.

40. **Horn, W. A., E. L. Larson, K. J. McGinley, and J. J. Leyden.** 1988. Microbial flora on the hands of health care personnel: differences in composition and antibacterial resistance. *Infect. Control Hosp. Epidemiol.* **9:**189–193.

41. **Jernigan, J. A., B. S. Lowry, F. G. Hayden, S. A. Kyger, B. P. Conway, D. H. M. Groschel, and B. M. Farr.** 1993. Adenovirus type 8 epidemic keratoconjunctivitis in an eye clinic: risk factors and control. *J. Infect. Dis.* **167:**1307–1313.

42. **Johnston, L. L., D. A. Shneider, M. D. Austin, F. G. Goodman, J. M. Bullock, and J. A. DeBruin.** 1982. Two percent glutaraldehyde: a disinfectant in arthroscopy and arthroscopic surgery. *J. Bone Joint Surg.* **64A:**237–239.

43. **Joslyn, L.** 1991. Sterilization by heat, p. 495–526. *In* S. S. Block (ed.), *Disinfection, Sterilization and Preservation,* 4th ed. Lea & Febiger, Philadelphia.

44. **Kaczmarek, R. G., R. M. Moore, J. McCrohan, D. A. Goldmann, C. Reynolds, C. Caquelin, and E. Israel.** 1992. Multi-state investigation of the actual disinfection/sterilization of endoscopes in health care facilities. *Am. J. Med.* **92:**257–261.

45. **Kahn, G.** 1970. Depigmentation caused by phenolic detergent germicides. *Arch. Dermatol.* **102:**177–187.

46. **Klapes, N. A., V. W. Greene, A. C. Langholz, and C. Hunstiger.** 1987. Effect of long-term storage on sterile status of devices in surgical packs. *Infect. Control* **8:**289–293.

47. **Koo, D., B. Bouvier, M. Wesley, P. Courtright, and A. Reingold.** 1989. Epidemic keratoconjunctivitis in a university medical center ophthalmology clinic; need for re-evaluation of the design and disinfection of instruments. *Infect. Control Hosp. Epidemiol.* **10:**547–552.

48. **Korniewicz, D. M., B. E. Laughon, A. Butz, and E. Larson.** 1989. Integrity of vinyl and latex procedure gloves. *Nurs. Res.* **38:**144–146.

49. **Larson, E.** 1988. APIC guidelines for use of topical antimicrobial agents. *Am. J. Infect. Control* **16:**253–266.

50. **Larson, E.** 1989. Handwashing: it's essential—even when you use gloves. *Am. J. Nurs.* **July:**934–941.

51. **Larson, E., K. Mayur, and B. A. Laughon.** 1988. Influence of two handwashing frequencies on reduction in colonizing flora with three handwashing products used by health care personnel. *Am. J. Infect. Control* **17:**83–88.

52. **Larson, E. L.** APIC guideline for handwashing in healthcare settings. *Am. J. Infect. Control,* in press.

53. **Larson, E. L., A. M. Butz, D. L. Gullette, and B. A. Laughon.** 1990. Alcohol for surgical scrubbing? *Infect. Control Hosp. Epidemiol.* **11:**139–143.

54. **Larson, E. L., P. I. Eke, and B. E. Laughon.** 1986. Efficacy of alcohol-based hand rinses under frequent-use conditions. *Antimicrob. Agents Chemother.* **30:**542–544.

55. **Larson, E. L., P. I. Eke, M. P. Wilder, and B. E. Laughon.** 1987. Quantity of soap as a variable in handwashing. *Infect. Control* **8:**371–375.

56. **Larson, E. L., K. J. McGinley, A. R. Foglia, G. H. Talbot, and J. J. Leyden.** 1986. Composition and antimicrobic resistance of skin flora in hospitalized and healthy adults. *J. Clin. Microbiol.* **23:**604–608.

57. **Latimer, J. M., and J. M. Matsen.** 1977. Microwave oven

irradiation as a method for bacterial decontamination in a clinical microbiology laboratory. *J. Clin. Microbiol.* **6:**340–342.

58. **Lettau, L. A., W. W. Bond, and J. S. McDougal.** 1985. Hepatitis and diaphragm fitting. *JAMA* **254:**752.

59. **Lingel, N. J., and B. Coffey.** 1992. Effects of disinfecting solutions recommended by the Centers for Disease Control on Goldmann tonometer biprisms. *J. Am. Optom. Assoc.* **63:**43–48.

60. **Lowry, P. W., W. R. Jarvis, A. D. Oberle, L. A. Bland, R. Silberman, J. A. Bocchini, Jr., H. D. Dean, J. M. Swenson, and R. J. Wallace, Jr.** 1988. *Mycobacterium chelonae* causing otitis media in an ear-nose-and-throat practice. *N. Engl. J. Med.* **319:**978–982.

61. **Maki, D. G., M. Ringer, and C. J. Alvarado.** 1991. Prospective randomised trial of povidone-iodine, alcohol, and chlorhexidine for prevention of infection associated with central venous and arterial catheters. *Lancet* **338:**339–343.

62. **Mallison, G. F.** 1980. Decontamination, disinfection and sterilization. *Nurs. Clin. N. Am.* **15:**757–767.

63. **Mallison, G. F., and P. G. Standard.** 1974. Safe storage times for sterile packs. *Hospitals* **48:**77–78, 80.

64. **Martin, M. A., and M. Reichelderfer.** 1993. Draft APIC guideline for the infection prevention and control in flexible endoscopy. *Am. J. Infect. Control* **21:**42A–66A.

65. **Mayworm, D.** 1984. Sterile shelf life and expiration dating. *J. Hosp. Supply Processing Distribution* **November–December:** 32–35.

66. **Mbithi, J. N., V. S. Springthorpe, and S. A. Sattar.** 1993. Comparative in vivo efficiencies of hand-washing agents against hepatitis A virus (HM-175) and poliovirus type 1 (Sabin). *Appl. Environ. Microbiol.* **59:**3463–3469.

67. **Mbithi, J. N., V. S. Springthorpe, S. A. Sattar, and M. Pacquette.** 1993. Bactericidal, virucidal, and mycobactericidal activities of reused alkaline glutaraldehyde in an endoscopy unit. *J. Clin. Microbiol.* **31:**2988–2995.

68. **McGinley, K. J., E. L. Larson, and J. J. Leyden.** 1988. Composition and density of microflora in the subungual space of the hand. *J. Clin. Microbiol.* **26:**950–953.

69. **Meenhorst, P. L., A. L. Reingold, D. G. Groothuis, G. W. Gorman, H. W. Wilkinson, R. M. McKinney, J. C. Feeley, D. J. Brenner, and R. van Furth.** 1985. Water-related nosocomial pneumonia caused by *Legionella pneumophila* serogroups 1 and 10. *J. Infect. Dis.* **152:**356–364.

70. **Memorandum of Understanding Between The Food and Drug Administration, Public Health Service, and The Environmental Protection Agency.** June 4, 1993.

71. **Modak, S., L. Sampath, H. S. S. Miller, and I. Millman.** 1992. Rapid inactivation of infectious pathogens by chlorhexidine-coated gloves. *Infect. Control Hosp. Epidemiol.* **13:** 463–471.

72. **Molinari, J. A., M. J. Gleason, J. A. Cottone, and E. D. Barrett.** 1987. Comparison of dental surface disinfectants. *Gen. Dent.* **May–June:**171–175.

73. **Occupational Safety and Health Administration.** 1991. Occupational exposure to bloodborne pathogens; final rule. *Fed. Regist.* **56:**64003–64182.

74. **Oxborrow, G. S., and R. Berube.** 1991. Sterility testing—validation of sterilization processes, and sporicide testing, p. 1047–1057. *In* S. S. Block (ed.), *Disinfection, Sterilization and Preservation*, 4th ed. Lea & Febiger, Philadelphia.

75. **Parisi, A. N., and W. E. Young.** 1991. Sterilization with ethylene oxide and other gases, p. 580–595. *In* S. S. Block (ed.), *Disinfection, Sterilization and Preservation*, 4th ed. Lea & Febiger, Philadelphia.

76. **Patterson, J. E., J. J. Vecchio, E. L. Pantelick, P. Farrel, D. Mazon, M. J. Zervos, and W. J. Hierholzer.** 1991. Association of contaminated gloves with transmission of *Acinetobacter calcoaceticus* var. *anitratus* in an intensive care unit. *Am. J. Med.* **91:**479–483.

77. **Perkins, J. J.** 1969. *Principles and Methods of Sterilization in Health Sciences*, 2nd ed. Charles C Thomas, Publisher, Springfield, Ill.

78. **Petrocci, A. N.** 1983. Surface active agents: quaternary ammonium compounds, p. 309–329. *In* S. S. Block (ed.), *Disinfection, Sterilization and Preservation*, 3rd ed. Lea & Febiger, Philadelphia.

79. **Platt, R., J. L. Lehr, S. Marino, A. Munoz, B. Nash, and D. B. Raemer.** 1988. Safe and cost-effective cleaning of pressure monitoring transducers. *Infect. Control Hosp. Epidemiol.* **9:**409–416.

80. **Pottinger, J., S. Burns, and C. Manske.** 1989. Bacterial carriage by artificial versus natural nails. *Am. J. Infect. Control* **17:**340–344.

81. **Prince, D. L., H. N. Prince, O. Thraenhart, E. Muchmore, E. Bonder, and J. Pugh.** 1993. Methodological approaches to disinfection of human hepatitis B virus. *J. Clin. Microbiol.* **31:**3296–3304.

82. **Prindle, R. F.** 1983. Phenolic compounds, p. 197–224. *In* S. S. Block (ed.), *Disinfection, Sterilization and Preservation*, 3rd ed. Lea & Febiger, Philadelphia.

83. **Resnick, L., K. Veren, Z. Salahuddin, S. Tondreau, and P. D. Markham.** 1986. Stability and inactivation of HTLV-III/LAV under clinical and laboratory environments. *JAMA* **255:**1887–1891.

84. **Richards, J. M., R. J. Sydiskis, W. M. Davidson, S. D. Josell, and D. S. Lavine.** 1993. Permeability of latex gloves after contact with dental materials. *Am. J. Orthod. Dentofac. Orthop.* **104:**224–229.

85. **Rothstein, R. I.** 1992. Disposable sterile sheathed flexible sigmoidoscopy. *Gastrointestinal Endosc.* **38:**277.

86. **Russell, A. D., W. B. Hugo, and G. A. J. Ayliffe.** 1982. *Principles and Practice of Disinfection, Preservation and Sterilisation.* Blackwell Scientific Publications, Oxford.

87. **Rutala, W. A.** 1990. APIC guideline for selection and use of disinfectants. *Am. J. Infect. Control* **58:**99–117.

88. **Rutala, W. A.** 1993. Disinfection, sterilization and waste disposal, p. 460–495. *In* R. P. Wenzel (ed.) *Prevention and Control of Nosocomial Infections*, 2nd ed. The Williams & Wilkins Co., Baltimore.

89. **Rutala, W. A.** Selection and use of disinfectants in health care. *In* C. G. Mayhall (ed.), *Infection Control and Hospital Epidemiology.* The Williams & Wilkins Co., Baltimore, in press.

90. **Rutala, W. A., E. P. Clontz, D. J. Weber, and K. K. Hoffmann.** 1991. Disinfection practices for endoscopes and other semicritical items. *Infect. Control Hosp. Epidemiol.* **12:** 282–288.

91. **Rutala, W. A., and E. C. Cole.** 1987. Ineffectiveness of hospital disinfectants against bacteria: a collaborative study. *Infect. Control* **8:**501–506.

92. **Rutala, W. A., E. C. Cole, and C. A. Thomann.** 1987. Stability and bactericidal activity of chlorine solutions, abstr. 1150, p. 297. *Abstr. 27th Intersci. Conf. Antimicrob. Agents Chemother.*

93. **Rutala, W. A., E. C. Cole, N. S. Wannamaker, and D. J. Weber.** 1991. Inactivation of *Mycobacterium tuberculosis* and *Mycobacterium bovis* by 14 hospital disinfectants. *Am. J. Med.* **91**(Suppl. 3B):267S–271S.

94. **Sattar, S. A., and V. S. Springthorpe.** 1991. Survival and disinfectant inactivation of the human immunodeficiency virus: a critical review. *Rev. Infect. Dis.* **13:**430–447.

95. **Scott, E. M., and S. P. Gorman.** 1991. Glutaraldehyde, p. 596–614. *In* S. S. Block (ed.), *Disinfection, Sterilization and Preservation*, 4th ed. Lea & Febiger, Philadelphia.

96. **Silverman, G. J.** 1991. Sterilization and preservation by ionizing irradiation, p. 566–579. *In* S. S. Block (ed.), *Disinfection, Sterilization and Preservation*, 4th ed. Lea & Febiger, Philadelphia.

97. **Simmons, B. P.** 1983. Guideline for hospital environmental control. *Am. J. Infect. Control.* **11:**97–115.

98. **Society of Gastroenterology Nurses and Associates.** 1990. *Recommended Guidelines for Infection Control in Endoscopy Settings.* Society of Gastroenterology Nurses and Associates, Rochester, N.Y.

99. **Soukiasian, S. H., G. K. Asdourian, J. S. Weiss, and H. A.**

Kachadoorian. 1988. A complication from alcohol-swabbed tonometer tips. *Am. J. Ophthalmol.* **105:**424–425.

100. **Spach, D. H., F. E. Silverstein, and W. E. Stamm.** 1993. Transmission of infection by gastrointestinal endoscopy and bronchoscopy. *Ann. Intern. Med.* **118:**117–128.

101. **Spaulding, E. H.** 1968. Chemical disinfection of medical and surgical materials, p. 517–531. *In* C. A. Lawrence and S. S. Block (ed.), *Disinfection, Sterilization and Preservation.* Lea & Febiger, Philadelphia.

102. **Sykes, G.** 1965. *Disinfection and Sterilization*, 2nd ed., p. 362–376. E & FN Spon Ltd., London.

103. **Talbot, G. H., M. Skros, and M. Provencher.** 1985. 70% alcohol disinfection of transducer heads: experimental trials. *Infect. Control* **6:**237–239.

104. **Turner, F. J.** 1983. Hydrogen peroxide and other oxidant disinfectants, p. 240–250. *In* S. S. Block (ed.), *Disinfection, Sterilization and Preservation*, 3rd ed. Lea & Febiger, Philadelphia.

105. **Weber, D. J., and W. A. Rutala.** 1993. Environmental issues and nosocomial infections, p. 420–449. *In* R. P. Wenzel (ed.), *Prevention and Control of Nosocomial Infections*, 2nd ed. The Williams & Wilkins Co., Baltimore.

106. **Working Party of the British Society of Gastroenterology.** 1988. Cleaning and disinfection of equipment for gastrointestinal flexible endoscopy; interim recommendations of a Working Party of the British Society of Gastroenterology. *Gut* **29:**1134–1151.

107. **World Health Organization.** 1983. *Laboratory Biosafety Manual.*

BACTERIOLOGY

VOLUME EDITORS
ELLEN JO BARON AND PATRICK R. MURRAY

SECTION EDITORS
MARY JANE FERRARO, PETER H. GILLIGAN,
MICHAEL A. SAUBOLLE, AND
ALICE S. WEISSFELD

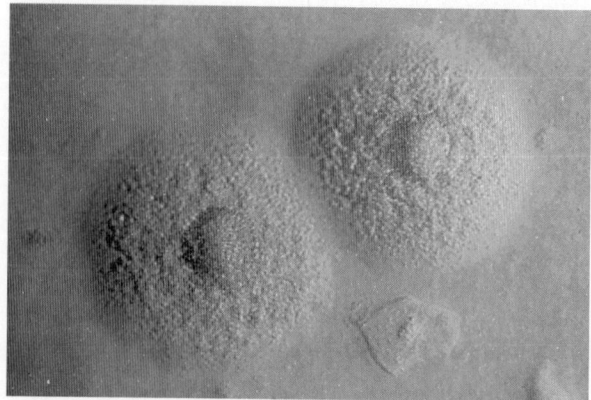

Mycoplasma hominis colonies (from a genital discharge specimen) taken by Normarski contrast bright-field microscopy
(*courtesy of Ellen Jo Baron*).

Classification and Identification of Bacteria

ELLEN JO BARON, ALICE S. WEISSFELD, PAMELA A. FUSELIER,
AND DON J. BRENNER

20

CONCEPTUAL FRAMEWORK FOR THE BASIS OF TAXONOMIC STUDIES

Definitions

The enterprise of diagnostic (and clinical) microbiology involves detecting and identifying etiologic agents of disease when they are recovered from clinical specimens or from environmental sources. The science of taxonomy includes identification, classification, and naming (nomenclature) of these microorganisms. Unfortunately, the importance of taxonomy has not been appreciated by many clinical microbiologists. Since taxonomy encompasses both classification and identification, much of the work that clinical microbiologists do falls within the definition of taxonomy. Bacterial identification and all typing schemes, whether they are based on biochemical, serologic, or other factors, are actually taxonomic classifications at or below the species level. Important applications of taxonomy include molecular or immunologic subtyping for epidemiologic purposes and the direct detection of microbial nucleic acid in specimens, with the specificity of detection dependent on the accuracy of the original classification of the species or strain. Reproducible systems for identifying microorganisms are critical, so that a particular microorganism will be identified correctly and consistently, regardless of the source or the laboratory. Sydney M. Finegold has written, "The primary purpose of nomenclature of microorganisms is to permit us to know as exactly as possible what another clinician, microbiologist, epidemiologist, or author is referring to when describing an organism responsible for infection of an individual or outbreak" (11). It is incumbent on clinical microbiologists to be aware of the close relationship between their work and taxonomy and to keep informed about taxonomic advances and the resulting changes that impact on their responsibilities in the laboratory.

The classification aspects of taxonomy serve to place microorganisms into orderly arrangements of groups so that a new isolate can more easily be characterized by comparison with known organisms. The choice of criteria for placement into groups is somewhat arbitrary, although there is agreement that the best classifications are those that reflect phylogenetic relationships (18, 33). This arbitrariness is reflected in the genetic definition of a species as strains of bacteria that exhibit 70% DNA relatedness with

5% or less divergence within related sequences, which has been recommended by a committee of experts on the classification and identification of bacteria (33). From the clinical standpoint, identification uses the attributes developed for a classification scheme to characterize and identify individual bacterial species or strains, to distinguish from all others the organisms being sought, to verify the authenticity or special properties of a microbial clone, or to recognize the etiologic agent of a disease.

Rationale for Changes in Nomenclature

On the basis of new methods that reveal more specific information about previously classified microorganisms, the classifications of such microorganisms and thus their taxonomic positions may be modified. As microorganisms are split from larger groupings into smaller, more cohesive and more delimited groups, their genus and species names must be changed to reflect their new separate status (nomenclature). Thus, nomenclature follows taxonomic classification. For example, using taxonomic methods, Shah and Collins classified certain weakly saccharolytic strains of pigmenting anaerobic gram-negative bacilli as a new genus, subsequently named *Porphyromonas*, separate from the *Bacteroides* species (26). This classification allowed microbiologists to distinguish the unique environmental niche in the gingival crevices that some of these organisms occupy and to recognize an easily performed test (susceptibility to vancomycin) that clinical laboratories could use for genus differentiation among pigmented anaerobic gram-negative rods.

Rules of Nomenclature

The choice of a microbial name is largely arbitrary, but the name must comply with a set of rules contained in the International Code of Nomenclature of Bacteria (29). A bacterial name is valid only if the manuscript proposing the change has been published in the *International Journal of Systematic Bacteriology* (IJSB) or if the proposed name appears on a list published in IJSB. The correct name of a species or a higher taxonomic designation (from lowest to highest, these designations include subspecies, species, genus, tribe, family, order, class, and kingdom) is based on validity and priority of publication and legitimacy of the name with regard to the rules of nomenclature (29). Latin or Greek versions of descriptive terms, geographic source of the first isolate, or names of distinguished scientists are usually chosen; the usage must be grammatically correct. On 1 January 1980, the first approved list of bacterial names was published in IJSB (27). Only those names considered by the editors to have adequate descriptions and available type or reference strains were included. Even if a "legitimate" species name was inadvertently omitted from that list, it had to be republished in a valid fashion to regain approval. The publication of this list was a tremendous boon to microbiologists, as many incompletely described species and synonyms were eliminated in this process. The number of names of bacteria decreased dramatically, from approximately 29,000 to just over 1,700 at the start of the approved lists.

Disagreements and Controversies Regarding Taxonomic Decisions

Taxonomic data can be interpreted differently by different investigators, which results in disagreements about grouping or splitting genera. In some cases, different methods reveal different categorizations of the same microorganisms (12). Thus, taxonomy, as this discussion implies, is not static. When two or more names exist at the genus level or above because of different interpretation of taxonomic data and all have been published validly, then all names can be used. The final choice is based on scientific usage, not rules. This has happened with *Moraxella* (*Branhamella*) *catarrhalis*, *Actinobacillus* (*Haemophilus*) *actinomycetemcomitans*, and *Legionella* (*Fluorobacter*) *dumoffii*, for example. Some of these names are still controversial. The authors of chapter 26 in this Manual prefer the genus designation *Branhamella* for the organism previously known as "*Neisseria catarrhalis*"; therefore, the organism is called *Branhamella* here, even though the species could also be correctly classified within the genus *Moraxella* by taxonomic methods. Because "*Neisseria catarrhalis*" was not included in the list of approved names, it is written with quotation marks, indicating a nonvalid name.

When two or more names exist for the same organism (species), the final name choice is governed by the rules of nomenclature: the oldest legitimate name has precedence. Thus, when *Providencia rustigianii* and *Providencia fredericiana* were both validly published but the organisms were subsequently shown to be identical species, the older name, *P. rustigianii*, had precedence. The other name (synonym) was validly published but not legitimate. In some cases, new taxonomic studies or more stringent interpretation led to publication of a new genus name for members of a larger genus grouping, such as *Helicobacter pylori* for *Campylobacter pylori*. Immediate acceptance of the new name occurred. This was not the case for *Tatlockia micdadei*; the original name, *Legionella micdadei*, prevailed through usage.

Controversial new names can be challenged by publishing in IJSB a request for an opinion from the Judicial Commission of the International Union of Microbiological Societies. An important illustration concerns the validly published observation by Bercovier et al. of the genomic homology of *Yersinia pestis* and *Yersinia pseudotuberculosis* (3). They recommended that *Y. pestis* become *Y. pseudotuberculosis* subsp. *pestis*. Williams proposed rejection of the new name, arguing that the subspecies designation would cause "confusion with dire consequences for public health" and requested a ruling from the Judicial Commission (34). The Commission agreed with Williams and retained the name *Y. pestis* because of "practical concerns for human welfare."

Importance of Recognizing and Incorporating Taxonomic Advances

Taxonomic advances have allowed us to detect differences among groups of microorganisms previously thought to be similar and to exploit those differences to our advantage. For example, taxonomic methods allowed workers from the Centers for Disease Control and Prevention to recognize that a fulminant, febrile, often fatal disease of children in Brazil was caused by strains of *Haemophilus influenzae* biogroup aegyptius (also still considered an acceptable species, *Haemophilus aegyptius*, by some authorities) that were all derived from a single clone (4). Once the relatedness of these strains was established, a simple whole-cell agglutination in specific antiserum was developed to identify those strains possessing the virulence factors for systemic disease. The common use of sorbitol-containing MacConkey agar for initial recognition of *Escherichia coli* O157:H7, the agent of severe hemorrhagic uremic syndrome, is also based on more esoteric taxonomic methods developed in reference laboratories (10). Microbiologists must make use of available references to stay abreast of new species and nomenclatural changes. Important taxonomic reference books for microbiologists include *Bergey's Manual for Systematic Bacteriology* (The Williams & Wilkins Co.), *Bergey's Manual of Determinative Bacteriology* (The Williams & Wilkins Co.), and *The Prokaryotes* (Springer-Verlag). Microbiologists can read IJSB, *Systematic and Applied Microbiology*, *Current Microbiology*, *Journal of Clinical Microbiology*, *Journal of General Microbiology*, and others to find articles pertaining to taxonomic updates. IJSB, however, remains the final arbiter.

TAXONOMIC METHODS

Taxonomic methods used to classify newly discovered or unusual microbes are not those used for routine identification in clinical microbiology laboratories. Some taxonomic tools, such as antibiograms, phage typing, and the molecular methods of plasmid analysis and ribotyping, have been adopted or adapted for characterizing microbial strains for epidemiologic purposes (Table 1). Explanations of these methods are given in chapters 13, 16, and 17 of this Manual, and some specific procedures are presented in *Diagnostic Molecular Microbiology: Principles and Applications* (20) and *Clinical Microbiology Procedures Handbook* (31). Other methods, such as DNA-DNA hybridization and rRNA sequencing, are used primarily for taxonomic studies; they are also described in chapter 13. Although methods discussed in this chapter relate to bacteria, these methods are equally valid for eukaryotic organisms and are being used increasingly for taxonomic and epidemiologic studies with fungi and parasitic organisms.

Numerical Taxonomy

Results generated by most taxonomic methods can be expressed quantitatively and then analyzed by using sophisticated computer programs that establish relative similarities among the strains studied. This discipline, called numerical taxonomy, phenetics, taxonometrics, or numerical systematics, allows a great number of traits to be examined for each strain studied. Computerization facilitates analysis of the numerous results (often more than 100 biochemical, morphologic, and other traits are entered into the equation) and generates a diagram to visually interpret the resulting relationship among strains. In its simplest form, numerical taxonomy is used to determine the relatedness

TABLE 1 Methods used for taxonomic classification of bacteria

Method	Ease of performance[a]	Applications and comments
Use of phenotypic characteristics		
Characterization of colony morphology	1	Useful for organisms that grow on solid media; used for routine identification; initial step toward more detailed methods
Gram stain reaction	1	Single most important preliminary test for correct identification of bacteria
Biochemical reactions (biotyping)	1	Useful for organisms that are metabolically active
Chromogenic enzymatic reactions	1–2	Useful for organisms that grow slowly or poorly or that are inert in traditional biochemical tests
Carbon source metabolic patterns	2	Can evaluate large no. of parameters; requires large inoculum and growth of organism
Antibiogram	1	Used for epidemiologic purposes; more helpful if patterns are quite different; can be used to separate similar species, e.g., use of O/129 for differentiating *Vibrio* and *Aeromonas* spp.
Resistogram	2	Pattern of resistance to heavy metals and toxins; used rarely in numerical taxonomy
Colicin or pyocin typing	2–3	May help differentiate among strains within a few species as preliminary assay; requires expensive and specialized reagents and extensive quality control; should be performed in reference laboratory
Phage typing	2–3	Requires extensive phage library; useful for characterizing strains within a few species; requires expensive and specialized reagents and extensive quality control; should be performed in reference laboratory
Use of serologic characteristics		
Serologic analysis	1–4, depending on sophistication of method	Can be used for species and subspecies identification (e.g., *Shigella* spp., *Streptococcus agalactiae*); requires expensive and specialized reagents (specific antisera) and extensive quality control; should be performed in reference laboratory
Immunoblotting	3–4	Useful for strain characterization or epidemiologic studies
Analytical		
Cell wall analysis (peptidoglycans, mycolic acids, etc.)	2–3	Can differentiate between otherwise similar genera (e.g., *Nocardia* and *Streptomyces*) and among species (e.g., *Mycobacterium*); commercial systems have simplified process.
Membrane or whole-cell protein analysis (PAGE)	2–3	Useful for differentiating among otherwise similar strains within a species; also used to differentiate among species and genera
Membrane or whole-cell lipid (fatty acid methyl ester) analysis	2–3	Requires expensive but easy-to-operate instrument and extensive database. Some similar species (e.g., *Bacteroides fragilis* group) may not be differentiated.
MLEE or enzyme typing	3	Useful for epidemiologic studies and for grouping strains at species level
Respiratory enzyme (quinone) analysis	3–4	Useful primarily for differentiating among genera and, less commonly, species
Dehydrogenase enzyme pattern analysis	3–4	Useful for differentiating among certain species; not necessary for routine use
Primary molecular taxonomic		
G+C ratio	3	Useful for excluding strain from known species, but different species or genera may possess identical G+C ratios
DNA hybridization	3	Current standard for designating species. It is the only method for designating genomic species. However, phenotypic data should accompany DNA hybridization data for formal naming of genomic species.
rRNA sequence analysis and RNA-DNA hybridization	4	One of the most powerful taxonomic tools for defining broad categories of bacteria; particularly useful for evolutionary studies. rRNA sequence analysis is also used to detect and classify noncultivatable organisms.
Molecular probe analysis	2–4	Usually specific for species; gaining acceptance for direct detection of organisms in clinical specimens and environment; also used to detect genes encoding antibiotic resistance

(Continued on next page)

TABLE 1 Methods used for taxonomic classification of bacteria (*Continued*)

Method	Ease of performance[a]	Applications and comments
Molecular subtyping		
Plasmid analysis	2–3	Useful for epidemiologic studies but only for strains containing plasmids; particularly useful for tracking antimicrobial resistance plasmids during nosocomial outbreaks
PFGE	3	Patterns of whole chromosomal DNA fragments may allow strain separation when other methods fail.
Restriction fragment length polymorphism	2–3	Useful for epidemiologic purposes and strain differentiation
Ribotyping	2–4	Very useful for taxonomic studies such as differentiating species within genus (e.g., *Legionella*); may discriminate among individual strains, although PFGE preferable for most species; good but tedious method for differentiating pathogenic clones
Random amplification of polymorphic DNA	3	Used to determine whether two isolates of same species are epidemiologically related; may be useful for individual organism identification

[a]4, Could be performed in only a few very sophisticated laboratories specializing in taxonomy; 3, could be performed in specialized research laboratories; 2, could be performed in larger, more sophisticated clinical laboratories; 1, could be performed in most routine clinical laboratories.

between bacterial strains (28). The similarity coefficient is determined by dividing the number of matching test results by the number of tests performed. Usually, only positive results are used in the calculation (Jaccard coefficient) to eliminate the large numbers of negative test results. Thus, two closely related strains have a large number of positive test results in common (>95%). When a strain is compared with itself, it will be 100% related (all test results will be identical). After many tests have been performed, the data can be presented visually as a similarity triangle (Fig. 1). The strains that are related at coefficients higher than a designated level (for Fig. 1, the level is 80%) can be considered to fall within the same relational group or species.

Similarity triangles are often generated with phenotypic data. Any types of results for which a quantitative answer can be generated, however, can be used to define relatedness. A dendrogram is another graphic representation of the similarity among a number of strains; it is often used with enzyme typing or ribotyping and other gel-derived data or with numerical taxonomic data. For a dendrogram, the similarity coefficients are represented along the horizontal axis, and strains are listed along the vertical axis (Fig. 2). Strains with similar characteristics (and thus high similarity coefficients) are listed together. Lines are drawn backward to the point that represents the percent similarity between each organism pair and group. Other graphical representations of relationships among strains, such as two-dimensional plots (Fig. 3), are also used to illustrate taxonomic data and to determine similarities.

Nucleic Acid Hybridization

The sequence of nucleic acids along the chromosome of a bacterial species is unique. If the DNA is separated into two complementary single strands by heat treatment, the degree to which the single-stranded DNA will reanneal with other single-stranded nucleic acids in the reaction mixture depends on the relatedness of the strands (17). Related single strands reassociate best at a temperature 25 to 30°C below that needed to denature the DNA. Under unfavorable

```
Strains
  1 |  100
  2 |  83  100
  3 |  45  39  100
  4 |  38  51  49  100
  5 |  32  45  67  92  100
    |_____
      1   2   3   4   5
              Strains
```

FIGURE 1 Similarity triangle depicting percent relatedness between strains. Strains tested against themselves show 100% relatedness. Strains 1 and 2 are >80% related to each other, strains 4 and 5 are >90% related to each other, and strain 3 is related to itself but not to any of the others even at >70%.

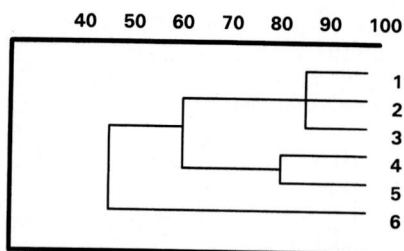

FIGURE 2 Dendrogram illustrating relatedness of strains 1, 2, and 3 at 85% and strains 4 and 5 at 80%. These two groups of strains are related to each other at around 60%, and the entire group of strains is related to strain 6 at 45%. If strains ≥80% related are considered to belong to the same species, then strains 1, 2, and 3 are one species; strains 4 and 5 are another species; and strain 6 is a third species only distantly related to the other two groups of isolates.

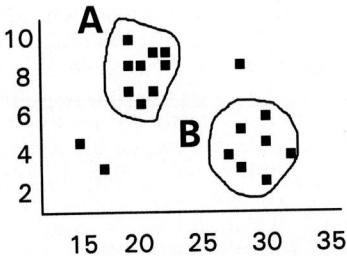

FIGURE 3 Two-dimensional plot comparing two groups (A and B) of related organisms. Strain positions are plotted on the graph on the basis of characteristics with quantitative values (e.g., number of carbon atoms in cell wall fatty acids or centimeters from the origin that an enzyme protein moves in an electrophoresis gel) that are represented on the *x* and *y* axes. Any two traits that can be quantified can be used to generate a plot. The Euclidean distance is that between the centers of each relational group (circled) or between two strains. Strains with high similarities cluster together and can be arbitrarily assigned to groups, as was done here for groups A and B.

conditions of higher temperature (10 to 15°C below the temperature needed to dissociate the strands; called stringent conditions), only very closely matched single strands reanneal. The extent of reannealing of DNA (hybridization) from different species is used to determine relatedness between any two bacterial strains and to determine whether they belong to the same species (13). A radioactive nucleotide or enzymatic label is incorporated into the known strain, which is then denatured into single-stranded DNA. The unlabeled test strain is also denatured, and the strands are combined and allowed to reassociate (hybridize). A control mixture containing only labeled and unlabeled single strands of the known species is used for comparison. All nonhybridized single strands are removed from the mixture, leaving only double-stranded hybrids. The amount of label is measured to determine the amount of hybridization and thus the relatedness (Fig. 4).

Hybridization between two single strands of nucleic acid can occur even when 15% of the nucleotide sequences are not complementary. For every 1% of unpaired nucleotides in a double-stranded nucleic acid hybrid (sequence divergence), there is approximately a 1% decrease in the thermal stability of the hybrid; i.e., the less tightly bound the two complementary strands are, the lower the temperature necessary to cause them to dissociate (16). In practice, strains belonging to the same species are usually 70 to 100% related, with less than 5% loss of thermal stability due to sequence divergence (33). However, some organisms that by DNA hybridization should be classified within a single species remain divided into separate species because of the assumed importance of the different names. *E. coli* and *Shigella* species are >70% related and are a single genomic species, but the medical significance associated with their current names has thus far maintained them as several species in two genera. *Neisseria meningitidis* and *Neisseria gonorrhoeae* fall into the same category, as do the previously mentioned *Y. pestis* and *Y. pseudotuberculosis*. On the other hand, hybridization studies may reveal the necessity for grouping strains previously thought to be separate species or genera. For example, new DNA hybridization data coupled with 16S rRNA sequence data and growth characteristics have shown that the genera *Rochalimaea* and *Bartonella* do not belong in the order *Rickettsiales*; instead, they belong in a single family (instead of two families) and even a single genus. Thus, the agent of trench fever (*Rochalimaea quintana*), an agent of cat scratch disease and bacillary angiomatosis (*Rochalimaea henselae*), and three other *Rochalimaea* species have been transferred to the genus *Bartonella*, the name that has historical precedence (5). It is anticipated that this new genus name will gain acceptance for these former *Rochalimaea* species.

rRNA Sequence Analysis

Phylogeny is the study of the evolution and development of species. Molecules, by their ordered arrangement of components, are documents of evolutionary history and can be used as tools for phylogenetic studies (36). Because the ribosome is critical to cellular function and interacts with a large number of other molecules, including mRNA and tRNAs, the sequences of the core rRNA molecules are highly constrained and have been remarkably conserved throughout evolution. The greater the time elapsed since

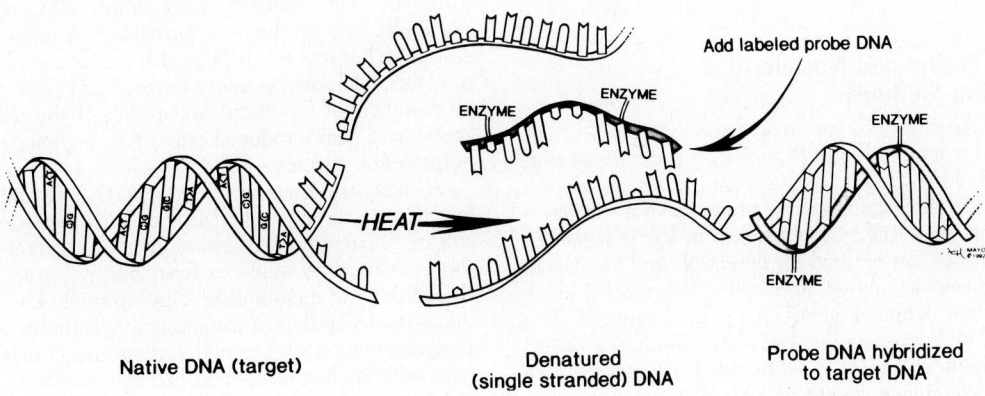

FIGURE 4 Denaturation of duplex DNA into single strands by heating, and combination of one of the single-stranded DNA segments with the labeled probe (hybridization). (From reference 20 with permission.)

divergence of two organisms from a single ancestor, the greater the number of mutational differences that will have accumulated in their molecular sequences. By examining sequence differences between organisms, it is possible to reconstruct their phylogenetic relationships. The small subunit of rRNA (16S rRNA in bacteria) has been used extensively for phylogenetic studies. The rRNA molecules for all bacteria and eukaryotic organisms, including humans, are similar enough that sequences can be easily aligned and compared and reliable phylogenies can be generated. Many bacteria have been divided into groups called rRNA superfamilies on the basis of rRNA sequence similarities, which correspond to evolutionary diversions over time (35). This allows new and reexamined strains to be categorized more easily into already familiar groups of related microbes. 16S rRNA is sequenced by using a modification of the Sanger dideoxy chain termination technique in which known primers complementary to conserved regions of the 16S rRNA are elongated by using a reverse transcriptase enzyme, described in reference 14 and in chapter 13 of this Manual. The short chains of DNA thus produced can be amplified by PCR (22) or cloning and then sequenced in an automated process.

More than 2,000 rRNA sequences are available from the Ribosomal Database Project (19), GenBank, and European Molecular Biology Laboratory (EMBL) databases. A number of methods and computer programs are available for reconstructing phylogenetic relationships from sequence data (30). rRNA sequencing is the method of choice for determining higher taxonomic relationships (above genus level); it is also extremely useful at the genus level but probably should not be the sole method used to designate genera and species (18). Because the 16S rRNA molecule contains highly variable regions, it is usually possible to find regions of 20 to 30 bases that are completely unique to any species of bacteria. DNA probes based on the sequences of these unique target regions are extremely useful in identifying bacterial species. Differentiation at the genus level is the starting point of bacterial identification. Thus, phenotypic characteristics must be correlated with phylogenetic methods for appropriate genus designations. Although 16S rRNA sequence similarity and DNA hybridization studies usually give similar results at the species level, reports in which rRNA sequence data do not agree with DNA relatedness data are accruing (3a, 12). Whether newer methods will eventually supplant DNA relatedness as the "gold standard" for designating species is still uncertain (12, 18, 33).

Molecular Probes and Nucleic Acid Amplification Methods

As long as classification is correct, every species can be differentiated from every other species on the basis of portions of its genome. Strains of the same species can also be differentiated if they differ in the presence of a virulence factor or other gene (i.e., cholera toxin in Vibrio cholerae). Molecular probes can be used for detection and identification of organisms at, above, and below the species level. Sequences that define a genus (e.g., all legionellae), a species or related complex (e.g., Mycobacterium tuberculosis complex, which includes M. tuberculosis, Mycobacterium bovis, Mycobacterium africanum, and Mycobacterium microti), or a specific clone (e.g., a specific vancomycin-resistant strain of Enterococcus faecalis) can be identified. When choosing a probe to differentiate or identify species on the basis of a "unique" gene, one must remember that any organism that contains that gene will react. Thus, strains of Vibrio mimicus that contain the cholera toxin gene will react with a probe for V. cholerae based on that gene.

The sensitivity of the method depends on the size and homology of the probe chosen and the nature of the original specimen; identification of organisms in pure cultures or from isolated colonies is usually easier than detection of a particular organism in a direct specimen. Sensitivity is enhanced by seeking a nucleic acid sequence with more copies than are found in the genome; targets such as rRNA, a repetitive sequence of DNA, or plasmid DNA present in several copies within the bacterial cell, for example, can be used. A number of commercial products are available, and some of them, such as the GenProbe rRNA system for identification of mycobacteria (8; see also chapter 31 of this Manual), have gained widespread acceptance. For samples with small numbers of target organisms and numerous confusing other substances, isolation of nucleic acid before probe testing enhances detection. Commercial systems for this purpose are also available (6).

With the exception of tests designed to detect a few well-defined pathogens, including N. gonorrhoeae and Chlamydia trachomatis in genital specimens; Gardnerella vaginalis, Candida albicans, and Trichomonas vaginalis in vaginal discharge; and Porphyromonas gingivalis and other agents of periodontal disease in tooth plaque samples (15), DNA probes have not been used widely for direct detection of microbes in clinical specimens because of low sensitivity. Nucleic acid amplification methods are being explored to address that problem, as detailed by Persing and colleagues (20). The first commercial PCR system for clinical laboratory use, for detection of Chlamydia trachomatis in genital specimens and urine, uses sequences from a cryptic plasma as the probe target (2).

Chromatographic Methods for Characterization of Bacteria

A discussion of taxonomic methods must include discussion of chromatographic separation of cellular components (e.g., fatty acids, mycolic acids, respiratory quinones) or metabolic end products (volatile and nonvolatile fatty acids). Either gas-liquid chromatography (GLC) or high-pressure liquid chromatographic (HPLC) methods are used routinely (see chapter 12 of this Manual). Anaerobic bacteria have been identified definitively in routine clinical laboratories for many years by using simple GLC to detect the metabolic end products of fermentation or proteolytic digestion (chapters 47 through 49). Evidence of unique end products, such as the major butyric acid peak produced by all members of Fusobacterium species or the major phenylacetic acid peak produced only by Porphyromonas gingivalis, helps define species.

A recently introduced system (whole-cell fatty acid methyl ester analysis [FAME]) requires digestion of a pure culture of bacterial cells, saponification, and methylation of the cellular fatty acids to form methyl esters, which are volatilized and measured by GLC that employs a very long silica-fused capillary chromatograph column. Patterns are analyzed by a computerized system and compared to patterns of known species stored in the database. A commercial form of this system (Microbial ID, Inc., Newark, Del.) has been used successfully for species or genus identification of mycobacteria (chapter 31), anaerobes, and aerobes (chapter 12). FAME patterns can be useful for taxonomic studies, often to place organisms into groups or to rule out possible identifications.

HPLC is one of the newer tools used by some laboratories for identification of mycobacteria to the species level. The pattern of mycolic acids in mycobacterial cell walls is unique for each species (chapters 12 and 31). Although the system is very reliable and results are available within hours of obtaining the isolate, the equipment is expensive, which limits its availability. HPLC is also used for characterization of the presence and relative amounts of respiratory quinone lipids in certain bacterial strains. As with mycolic acid patterns, respiratory quinone patterns are unique for separate species in many cases (7). Complex methods and expensive equipment preclude use of this method in laboratories other than those conducting research.

TOOLS USUALLY USED FOR EPIDEMIOLOGIC STUDIES

Numerous methods have been developed for molecular subtyping; these include polyacrylamide gel electrophoresis (PAGE), restriction fragment length polymorphism, ribotyping, pulsed-field gel electrophoresis (PFGE), random amplification of polymorphic DNA, karyotyping (for fungi), and others (Table 1; chapter 13) (20). Such methods, although useful for taxonomic studies, are currently used primarily for epidemiologic studies in clinical microbiology. Most of the methods are time-consuming and often expensive, and no single method works best for all organisms. Often, a simple method (such as PAGE) is used as a screening test, and more complex methods are used as reference techniques. One method not explained in other chapters that is used for both taxonomic and epidemiologic studies is multilocus enzyme electrophoresis (MLEE).

Bacteria can be differentiated by characterizing their enzymatic capabilities. A simplistic application of this principle is practiced routinely in the clinical laboratory as one type of phenotypic biochemical test (e.g., oxidase test, catalase test, chromogenic enzyme kits) for enzymes whose activities can be easily detected by using simple substrates with visible endpoints. A more sophisticated method of enzyme analysis called MLEE entails two steps: (i) separation of protein extracts of whole bacteria cells on gels or filter paper by electrophoresis and (ii) detection of the location of each enzyme on the two-dimensional gel or sheet by specific staining for the colored endpoint of enzyme activity. In this way, both the presence and the electrophoretic mobilities of the enzymes present in a bacterial clone can be compared (25). Small differences in enzyme mobility, as reflected by differences in the locations of enzyme end products, indicate strain differences. The more enzymes evaluated, the more sensitive the method becomes. A recent example used 12 enzymes to characterize strains of *Moraxella* (32). Although the test is too complex for routine use in clinical laboratories, results are useful for epidemiologic characterization of outbreak strains. MLEE can also be used to group strains at the species level.

DETECTION OF UNCULTURABLE BACTERIA

Several methods, including modifications of some of those described in this chapter, have been used to detect unculturable microbes in pathogenic processes. *Treponema pallidum*, the agent of syphilis, cannot be isolated from infected patients, but the organism can be seen on dark-field prep-

arations, and the disease can be diagnosed serologically (chapter 52). Another unculturable agent, *Spirillum minus*, the agent of one form of rat bite fever, has been visualized in blood, lymph nodes, lesion exudate, and cutaneous rash biopsy samples from infected patients by dark-field examination or by Giemsa or Wright stain. The organism is a motile, gram-negative spiral bacterium 3 to 5 μm long and 0.2 to 0.5 μm wide. It has two to six angular windings and pointed ends, with one flagellum at each pole (21). Rat bite fever caused by *Spirillum minus* is known as Sodoku and has been reported primarily from Japan. The disease is similar clinically and epidemiologically to that caused by *Streptobacillus moniliformis* (chapter 39). Approximately half of rat bite fever patients also exhibit a maculopapular rash that initially appears to be erythematous to purplish and gradually coalesces. The fatality rate is 5 to 10%.

An additional pathogenic bacterium detected exclusively by visual or serologic means is *Calymmatobacterium granulomatis*, the agent of donovanosis or granuloma inguinale. The disease is more common in tropical and developing nations than in temperate climates. Donovanosis is thought to be sexually transmitted; patients develop an indurated nodule at the site of exposure that erodes to form a beefy, aggressive, granulomatous ulcer. The lesion progresses slowly and may coalesce or form new lesions by autoinoculation. Over time, extensive genital area destruction may result. Because the organism cannot be isolated in routine culture, lesion material should be stained by Giemsa or Wright stain and examined microscopically for the presence of the organism ("Donovan bodies," clusters of blue to black bipolar-staining coccobacilli found within vacuoles in phagocytic cells). The pathognomonic forms are rarely seen in formalin-fixed, paraffin-block specimens (21).

Researchers have recently employed molecular biology tools such as PCR and probing to identify bacteria seen in tissue but difficult or impossible to isolate in culture. This approach was successful for initial characterization of *Bartonella* (*Rochalimaea*) *henselae* (23) and has been used to identify *Borrelia burgdorferi* in joint fluid and blood when cultures were negative. Universal ribosomal primers that hybridize and amplify all bacterial rRNA can be used to detect bacteria in any sterile body site. Once detected, the organism can be identified by sequencing the amplified rRNA and comparing the sequence with known rRNA sequences. Relman used such a probe to isolate the genome of the agent of Whipple's disease from infected patient tissue material (see chapter 13). Whipple's disease, a multisystemic disease with arthralgias, pleuritis, fever, weight loss, lymphadenopathy, and later cardiac and gastrointestinal tract involvement, was thought to be caused by a bacterium that was visible in stains of infected tissues but had not been grown in culture after many years of attempts. After the bacterial genome had been amplified and sequenced, the agent was found to be sufficiently unlike any known genus to warrant a new genus and species designation, *Tropheryma whippelii* (24). *T. whippelii* appears to be related most closely to the actinomycetes. Unlike *Bartonella henselae*, this agent continues to elude isolation in culture. In another modification of PCR for clinical microbiology, species- or genus-specific primers can be used to detect a specific suspected organism in patient material or mixed cultures. This method is useful both for detection of noncultivatable organisms and for more rapid identification of slow-growing organisms, such as mycobacteria.

The use of such methods has recently illuminated for microbiologists the number of uncultured or unculturable microbes present in many sites, including the periodontal crypts, fish and animal intestinal tracts, and numerous niches within the environment. The number of species thought to be present in such sites has risen dramatically with the realization of the vast amount of bacterial DNA in these materials that does not correlate with the numbers of strains recovered in cultures. One can only imagine the many new agents that will be identified with the use of the new molecular tools.

BACTERIAL IDENTIFICATION FOR THE CLINICAL LABORATORY

Principles for Identification of Bacteria

Clinical microbiology laboratories usually restrict their activities to identifying isolates on the basis of the most current taxonomic criteria available, using methods that are cost-effective and practical in the clinical setting. Edwards and Ewing (9) developed principles for characterization, classification, and identification of gram-negative bacteria based on phenotypic traits, those traits most easily determined in a clinical laboratory. These principles are paraphrased here.

1. Classification and identification of an organism should be based on its overall morphologic and biochemical pattern, not a single characteristic, no matter how important that characteristic is.
2. A large and diverse sample of strains must be tested to determine the biochemical characteristics of a given species. Results of each test should be expressed as percentages of positive or negative reactions.
3. Atypical strains, when adequately studied, are often found to be typical strains of a new species or subgroup within the species.

Flowcharts for Initial Characterization of Isolates from Clinical Specimens

Microbiologists must be up to date with the current taxonomic designations for clinically relevant isolates so that they can make the appropriate distinctions and inform clinicians when taxonomic changes are relevant to patient care. The extent to which individual isolates are identified by a given laboratory depends on the skills, interest, and time available among personnel; the goals of the laboratory and its ability to stock sufficient reagents, media, stock organisms, and other supplies necessary; the requirements of the patient population served; and the amount of money available. In all cases, a clinically significant isolate for which the physician requests definitive identification can be forwarded to a reference laboratory if the isolating laboratory does not offer the requested tests. Cost constraints on laboratories in recent years have discouraged many nonreference facilities from taking full advantage of the new technologies, which are still more costly than conventional methods and frequently supplement rather than replace older identification methods. Only when new methods can be shown to have a significant effect on patient morbidity, public health, or ultimate cost of care, e.g., by reducing length of stay, will they be more widely employed.

For the most part, clinical microbiologists continue to depend on morphologic and biochemical criteria for the identification of microorganisms isolated from patient specimens. A number of commercial systems, both automated and in kit form (chapter 10) (31), are available for defin-itive identification of the most commonly isolated species, provided they are represented in the database. A few caveats should be kept in mind. The use of computer-generated identification does not preclude the need for interpretive input from the microbiologist; i.e., the identification must make sense. For example, *Bacillus* species often resemble gram-negative bacilli and may be misidentified by an automated system as members of the family *Enterobacteriaceae* unless the microbiologist pays attention to colony morphology and the presence of spores. Anaerobes are often misidentified on the basis of kit results that use computer-generated probabilistic database tables because of incorrect Gram stain results, the first parameter in the decision matrix.

All commercial systems make identification errors. System failures include failure due to human error (wrong isolate, misnumbered module), no identification because of system inadequacies (insufficient data in database, inability to interpret atypical reactions), or incorrect identifications (the worst type of error). In some cases, such errors are obvious (a very unlikely identification is generated), but in other situations, a problem may continue for some time until an unusual event prompts its detection.

Finally, it is important to realize that results for a given test may not be equally valid among all identification systems. Thus, for the flowcharts presented here, methods may vary and may yield results that fail to match those listed. For example, the sensitivity of Christensen's urea agar for detection of the presence of urease is much greater than that of the API system (bioMérieux Vitek) urease broth microwell. In these flowcharts, we have attempted to list reactions based on conventional biochemical results and rapid tests (oxidase, catalase, etc.). However, results in individual laboratories may vary depending on the medium used. For example, different manufacturers' lots of MacConkey agar may result in growth or no growth of certain gram-negative bacilli, a key reaction in the flowcharts given here.

To help initially distinguish those isolates that are more difficult to characterize and to help the beginning microbiologist, who needs guidance in placing a new isolate into a category that will indicate which additional studies to choose, this chapter includes comprehensive flowcharts for preliminary identification of bacteria found in clinical specimens (Fig. 5 to 10 and Table 2). Based on Gram stain characteristics and a selected number of additional, easily performed enzymatic or biochemical tests, these flowcharts direct the user to the appropriate chapter(s) in this Manual for further identification procedures. They have been created as dichotomous keys; variable reactions have been taken into consideration, so the same genus may appear in several places. The charts are based on a 95% confidence level; i.e., for key reactions, 95% of the organisms yield a positive or a negative reaction. The flowcharts on the following pages are divided into the main groups of (i) aerobic and facultatively anaerobic gram-positive rods; (ii) aerobic and facultatively anaerobic cocci; (iii) aerobic and facultatively anaerobic non-glucose-utilizing gram-negative rods and coccobacilli; (iv) aerobic and facultatively anaerobic glucose-oxidizing gram-negative rods and coccobacilli; (v) aerobic and facultatively anaerobic glucose-fermenting gram-negative rods and coccobacilli; (vi) anaerobic bacteria; and (vii) organisms that require special collection, processing, culture, or staining techniques.

Choosing the appropriate chapter for further searching is not difficult with gram-positive bacteria or cocci, but the

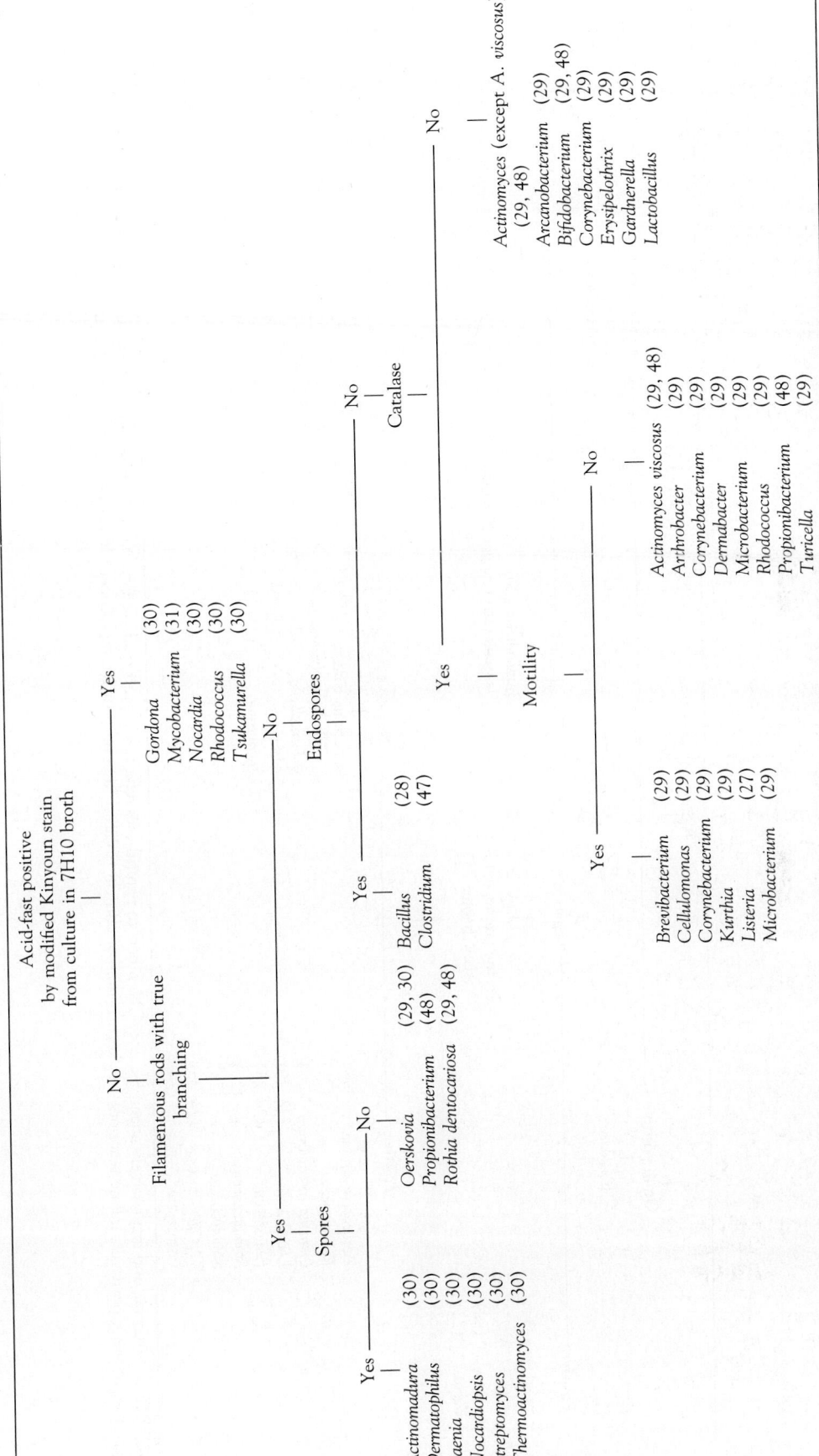

FIGURE 5 Aerobic and facultatively anaerobic gram-positive rods. Numbers in parentheses refer to chapters in this Manual.

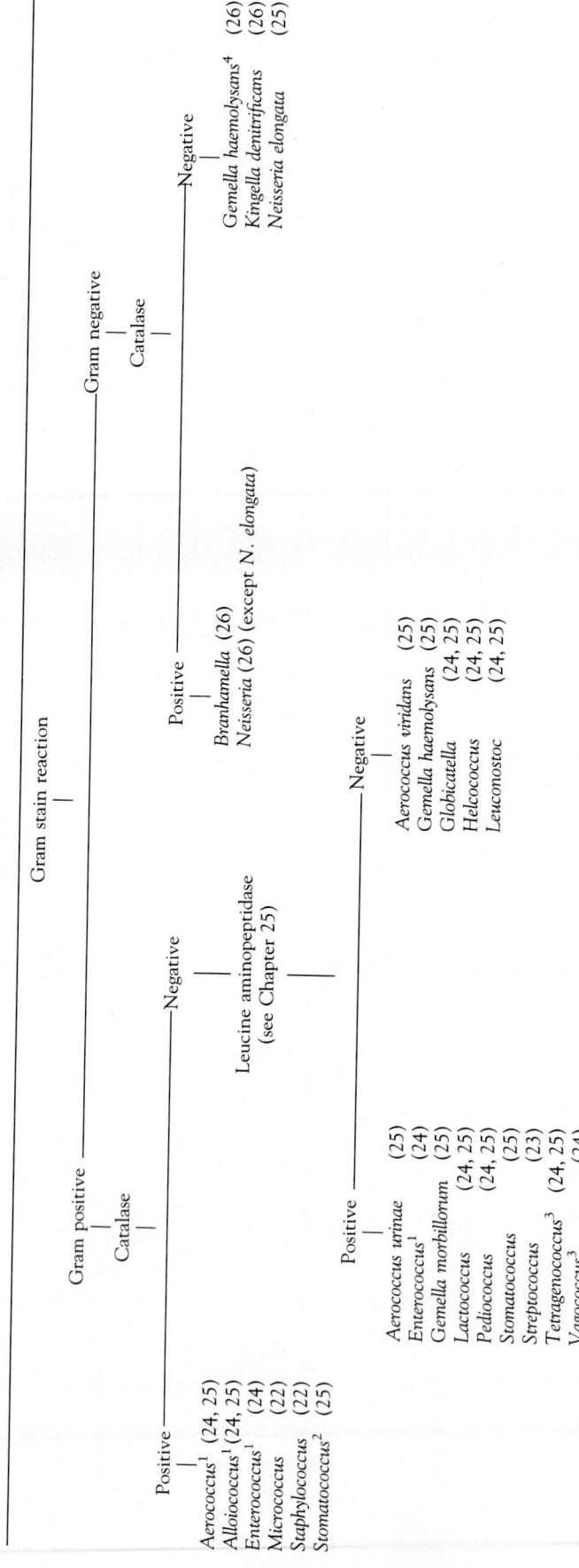

FIGURE 6 Aerobic and facultatively anaerobic cocci. Numbers in parentheses refer to chapters in this Manual. [1]May show a weak catalase (pseudocatalase) reaction; [2]may show a positive or weak positive catalase reaction; [3]no known human clinical isolates; [4]a gram-positive diplococcus that easily decolorizes to gram negative or gram variable.

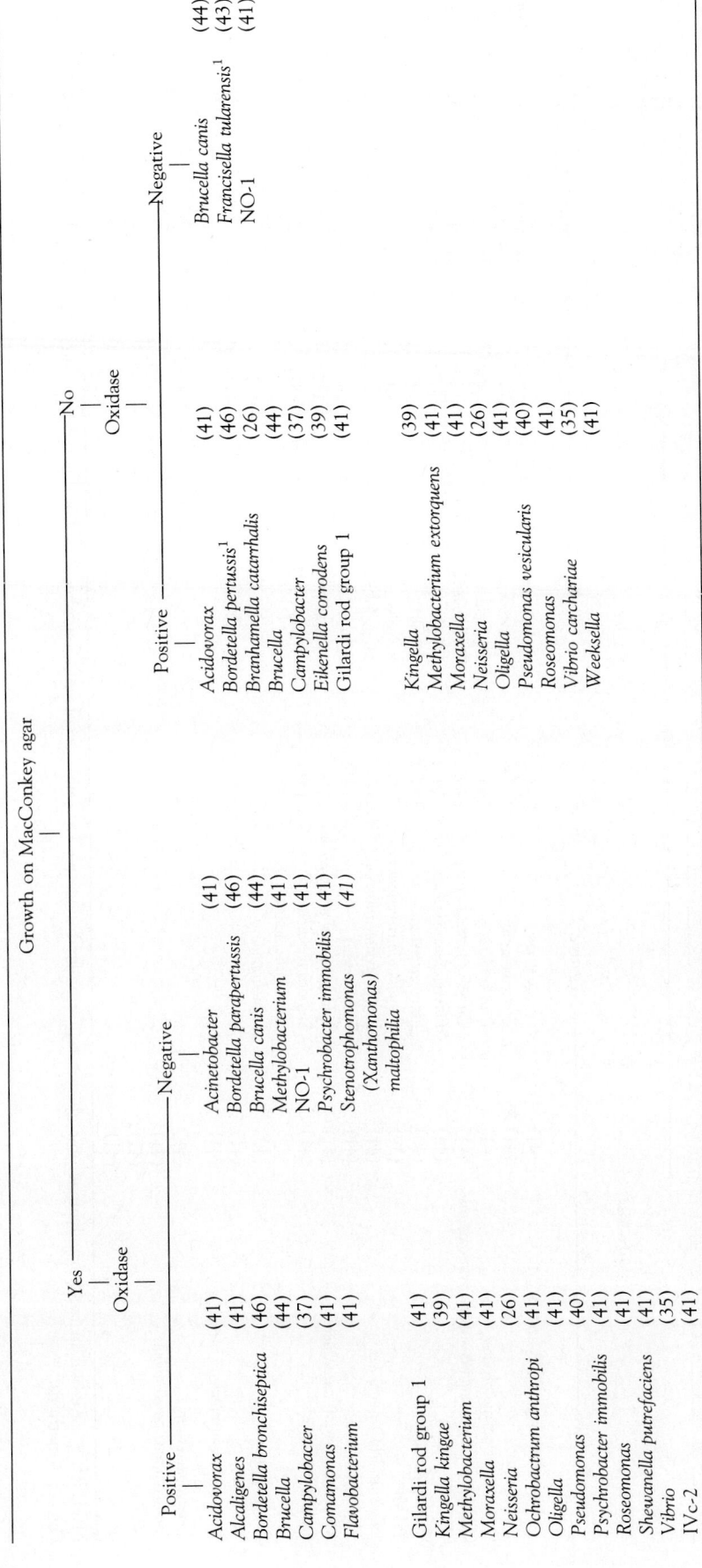

FIGURE 7 Aerobic or facultatively anaerobic non-glucose-utilizing gram-negative rods and coccobacilli. Numbers in parentheses are chapters in this Manual. Glucose metabolic activity is screened for by using triple sugar iron agar, oxidative-fermentative dextrose and oxidative-fermentative base, or Andrade's dextrose and Andrade's base. [1]Unlikely to grow on routine laboratory media.

Growth on MacConkey agar

Yes — Oxidase

Positive
- Achromobacter group B and E (41)
- Acidovorax (41)
- Agrobacterium radiobacter (41)
- Alcaligenes xylosoxidans subsp. xylosoxidans (41)
- Brucella (44)
- Burkholderia (40)
- EF-4b (39)
- EO-2 and EO-3 (41)
- Flavobacterium (41)
- Methylobacterium extorquens (41)
- Ochrobactrum anthropi (41)
- Pseudomonas (40)
- Roseomonas (41)
- Shewanella putrefaciens biovar 1 (41)
- Sphingobacterium (41)
- Sphingomonas paucimobilis (41)
- WO-1 (41)
- Ile (41)

Negative
- Acinetobacter (41)
- Brucella canis (44)
- Burkholderia (40)
- Chryseomonas luteola (41)
- Flavimonas oryzihabitans (41)
- Flavobacterium (41)
- Pseudomonas vesicularis (40)
- Psychrobacter immobilis (41)
- Roseomonas (41)
- Sphingomonas paucimobilis (41)
- Stenotrophomonas (Xanthomonas) maltophilia (41)
- Ile

No — Oxidase

Positive
- Achromobacter group F (41)
- Acidovorax (41)
- Brucella (44)
- Burkholderia mallei (40)
- EF-4b (39)
- EO-2 (41)
- Flavobacterium (41)
- Methylobacterium extorquens (41)
- Pseudomonas vesicularis (40)
- Sphingomonas paucimobilis (41)
- Sphingobacterium spiritovorum (41)
- WO-1 (41)
- Ile, IIh, and IIi (41)

Negative
- Brucella (44)
- Burkholderia mallei (40)
- Pseudomonas vesicularis (40)
- Sphingomonas paucimobilis (41)
- Ile (41)

FIGURE 8 Aerobic or facultatively anaerobic glucose-oxidizing gram-negative rods and coccobacilli. Numbers in parentheses refer to chapters in this Manual. See the legend to Fig. 7 for methods.

Growth on MacConkey agar

Yes — Oxidase

Positive

Actinobacillus[1]	(39)
Aeromonas	(36)
Chromobacterium violaceum	(39)
EF-4a[2]	(39)
Haemophilus aphrophilus	(45)
HB-5	(39)
Kingella[1,2]	(39)
Leptotrichia buccalis	(39, 49)
Pasteurella	(39)
Plesiomonas shigelloides	(36)
Vibrio[2]	(35)

Negative

Chromobacterium violaceum[2]	(39)
Enterobacteriaceae	(32–34)
Haemophilus aphrophilus	(45)
HB-5	(39)
Leptotrichia buccalis	(39, 49)
Pasteurella	(39)
Tatumella ptyseos[2]	(34)
Vibrio[2]	(35)
Yersinia[2]	(33)

No — Oxidase

Positive

Actinobacillus[1]	(39)
Aeromonas	(36)
Capnocytophaga[1,2]	(39)
Cardiobacterium hominis[1,2]	(39)
EF-4a[2]	(39)
Haemophilus aphrophilus	(45)
HB-5	(39)
Kingella[1,2]	(39)
Pasteurella	(39)
Suttonella indologenes	(39)
Vibrio[2]	(35)

Negative

Actinobacillus actinomycetemcomitans[1]	(39)
Capnocytophaga[1,2]	(39)
DF-3[1]	(29)
Gardnerella vaginalis[1,2]	(45)
Haemophilus aphrophilus	(39)
HB-5	(39)
Pasteurella	(39)
Vibrio metschnikovii[2]	(35)

FIGURE 9 Aerobic or facultatively anaerobic glucose-fermenting gram-negative rods and coccobacilli. Numbers in parentheses refer to chapters in this Manual. See the legend to Fig. 7 for methods. [1]May require the addition of 1 or 2 drops of sterile rabbit serum per 3 ml of fermentative broth medium; [2]triple sugar iron reaction does not always indicate fermentative activity.

Growth on chocolate agar in 3–5% CO_2 after 48 h of incubation

Scant growth

Actinomyces (48)
Bifidobacterium (29, 48)
Clostridium[1] (47)
Lactobacillus (48)
Leptotrichia buccalis (49)
Propionibacterium (48)

No growth → Gram stain reaction

Positive

Cocci:
Peptococcus niger (48)
Peptostreptococcus (48)
Streptococcus (48)

Rods:
Actinomyces (48)
Bifidobacterium (48)
Clostridium (47)
Eubacterium (48)
Lactobacillus (48)
Mobiluncus (48)
Propionibacterium (48)
Streptococcus (48)

Negative

Cocci:
Acidominococcus (49)
Megasphaera (49)
Veillonella (49)

Rods:
Anaerobiospirillum (49)
Anaerorhabdis (49)
Anaerovibrio (49)
Bacteroides (49)
Butyrivibrio (49)
Campylobacter (49)
Centipeda (49)
Clostridium[2] (47)
Desulfomonas (49)
Desulfovibrio (49)
Dicholebacter (49)
Fibrobacter (49)
Fusobacterium (49)
Leptotrichia (49)
Megamonas (49)
Mitsuokella (49)
Mobiluncus (48)
Porphyromonas (49)
Prevotella (49)
Rikenella (49)
Ruminobacter (49)
Sebaldella (49)
Selenomonas (49)
Succinimonas (49)
Succinivibrio (49)
Tissierella (49)

FIGURE 10 Anaerobic bacteria. Numbers in parentheses refer to chapters in this Manual. [1]*C. carnis, C. histolyticum, C. tertium,* and occasionally *C. perfringens;* [2]*C. ramosum, C. clostridioforme, C. symbiosum, C. villosum,* and *C. innocuum* almost always stain gram negative. Older cultures of other clostridia may decolorize easily and appear gram negative.

TABLE 2 Bacteria that require special collection, processing, culture, or staining techniques[a]

Acholeplasma (53)	Legionella (42)
Afipia (58)	Leptonema (50)
Arcobacter (37)	Leptospira (50)
Bartonella (Rochalimaea) (54, 58)	Mycoplasma (53)
Borrelia (51)	Rickettsia (54, 56)
Calymmatobacterium (20)	Serpulina (52)
Campylobacter (37)	Spirillum (20)
Chlamydia (55)	Streptobacillus (39)
Coxiella (54, 56)	Treponema (52)
Ehrlichia (54, 57)	Tropheryma (20)
Francisella tularensis (43)	Turneria (50)
Helicobacter (38)	Ureaplasma (53)

[a]Numbers in parentheses refer to chapters in this Manual.

diversity of gram-negative bacilli encountered in a clinical laboratory can be daunting (Fig. 7 to 9). A relatively small number of species of gram-negative aerobic rods are among the most common isolates in a clinical microbiology laboratory. To simplify initial classification, Pickett has divided them into four groups based generally on methods used for laboratory examination: (i) nonfastidious, fermentative, facultatively anaerobic, oxidase-negative rods, usually members of the Enterobacteriaceae family (enterics; chapters 32 through 34); (ii) nonfastidious, obligately aerobic, non-fermentative rods (chapters 40 and 41); (iii) fastidious hemophilic rods, primarily Haemophilus spp. (chapter 45); and (iv) unusual bacteria, including Vibrio, Aeromonas, and Pasteurella spp. and many others (chapters 35 through 39, and 42 through 46). Typically, they are encountered by clinical laboratories in proportions of 68 to 78% enterics, 12 to 16% nonfermentative rods, 8 to 15% haemophilic rods, and less than 1% unusual bacteria. Many of the unusual bacteria are discussed in chapter 39.

It is obvious that the name given to a microbe isolated from a clinical specimen conveys a great deal of information. Certain species are associated with antimicrobial resistance (e.g., Klebsiella species are resistant to ampicillin), certain species are associated with specific syndromes (Streptococcus bovis in the bloodstream almost always has a gastrointestinal origin and often signals gastrointestinal malignancy), and newly characterized genera may have unique clinical associations (Alloiococcus species seem to be involved in middle ear infections in children [chapter 25]). Therefore, it is important that microbiologists identify significant isolates to the extent necessary to allow the clinician to make correct patient care decisions. Clinicians must be informed of new taxonomic changes in a timely and considerate fashion. A consensus group that was convened to discuss the issue of adopting new taxonomy (1) proposed several actions. (i) Microbiologists should wait approximately 6 months after a new name has been validly published before changing report formats. (ii) The laboratory may send out periodic notices of new nomenclature, but the most effective method is to report both old and new names of renamed isolates for 1 to 2 years. (iii) If a newly characterized species is being reported to a physician for the first time, a telephone call or a comment on the report with some pertinent references will help the physician place the new name in context. (iv) Finally, microbiologists should make an effort to participate in discussions of individual patients and general infectious disease topics so that mutual

understanding of the roles of specific species of microbes in health and disease can be advanced.

REFERENCES

1. **Baron, E. J., and S. D. Allen.** 1993. Should clinical laboratories adopt new taxonomic changes? If so, when? *Clin. Infect. Dis.* **16**(Suppl 4):S449–S450.
2. **Bass, C. A., D. L. Jungkind, N. S. Silverman, and J. M. Bondi.** 1993. Clinical evaluation of a new polymerase chain reaction assay for detection of *Chlamydia trachomatis* in endocervical specimens. *J. Clin. Microbiol.* **31:**2648–2653.
3. **Bercovier, H., J. Ursing, D. J. Brenner, A. G. Steigerwalt, G. R. Fanning, G. P. Carter, and H. H. Mollaret.** 1980. Intra- and interspecies relatedness of *Yersinia pestis* by DNA hybridization and its relationship to *Yersinia pseudotuberculosis*. *Curr. Microbiol.* **4:**225–229.
3a. **Brenner, D. J.** Unpublished observations.
4. **Brenner, D. J., L. W. Mayer, G. M. Carlone, L. H. Harrison, W. F. Bibb, M. C. de Cunto Brandileone, F. O. Sottnek, K. Irino, M. W. Reeves, J. M. Swenson, K. A. Birkness, R. S. Weyant, S. F. Berkley, T. C. Woods, A. G. Steigerwalt, P. A. D. Grimont, R. M. McKinney, D. W. Fleming, L. L. Gheesling, R. C. Cooksey, R. J. Arko, C. V. Broome, and Brazilian Purpuric Fever Study Group.** 1988. Biochemical, genetic, and epidemiologic characterization of *Haemophilus influenzae* biogroup aegyptius (*Haemophilus aegyptius*) strains associated with Brazilian purpuric fever. *J. Clin. Microbiol.* **26:**1524–1534.
5. **Brenner, D. J., S. P. O'Connor, H. H. Winkler, and A. G. Steigerwalt.** 1993. Proposals to unify the genera *Bartonella* and *Rochalimaea*, with descriptions of *Bartonella quintana* comb. nov., *Bartonella vinsonii* comb. nov., *Bartonella henselae* comb. nov., and *Bartonella elizabethae* comb. nov., and to remove the family *Bartonellaceae* from the order *Rickettsiales*. *J. Clin. Microbiol.* **43:**777–786.
6. **Coll, P., K. Phillips, and F. C. Tenover.** 1989. Evaluation of a rapid method of extracting DNA from stool samples for use in hybridization assays. *J. Clin. Microbiol.* **27:**2245–2248.
7. **Collins, M. D.** 1985. Isoprenoid quinone analyses in bacterial classification and identification, p. 267–289. *In* M. Goodfellow and E. E. Minnikin (ed.), *Chemical Methods in Bacterial Systematics*. Academic Press, Inc., New York.
8. **Evans, K. E., A. S. Nakasone, P. A. Sutherland, L. M. de la Maza, and E. M. Peterson.** 1992. Identification of *Mycobacterium tuberculosis* and *Mycobacterium avium-M. intracellulare* directly from primary BACTEC cultures by using acridinium ester-labeled DNA probes. *J. Clin. Microbiol.* **30:**2427–2431.
9. **Ewing, W. H.** 1986. *Edwards and Ewing's Identification of Enterobacteriaceae*. Elsevier, New York.
10. **Farmer, J. J., and B. R. Davis.** 1985. H7 antiserum-sorbitol fermentation medium: a single tube screening medium for detecting *Escherichia coli* O157:H7 associated with hemorrhagic colitis. *J. Clin. Microbiol.* **22:**620–625.
11. **Finegold, S.** 1993. Introduction to summary of current nomenclature, taxonomy, and classification of various microbial agents. *Clin. Infect. Dis.* **16:**597.
12. **Fox, G. E., J. D. Wisotzkey, and P. Jurtshuk, Jr.** 1992. How close is close: 16S rRNA sequence identity may not be sufficient to guarantee species identity. *Int. J. Syst. Bacteriol.* **42:**166–170.
13. **Johnson, J. L.** 1981. Genetic characterization, p. 450–474. *In* P. Gerhardt, R. G. E. Murray, R. N. Costilow, E. W. Nester, W. A. Wood, N. R. Krieg, and G. B. Phillips (ed.), *Manual of Methods for General Bacteriology*. American Society for Microbiology, Washington, D.C.
14. **Lane, D. J., B. Pace, G. J. Olsen, D. A. Stahl, M. L. Sogin, and N. R. Pace.** 1985. Rapid determination of 16S ribosomal RNA sequences for phylogenetic analyses. *Proc. Natl. Acad. Sci. USA* **82:**6955–6959.
15. **Loesche, W. J., D. E. Lopatin, J. Stoll, N. van Poperin, and P. P. Hujoel.** 1992. Comparison of various detection methods

for periodontopathic bacteria: can culture be considered the primary reference standard? *J. Clin. Microbiol.* **30:**418–426.

16. **Marmur, J., and P. Doty.** 1961. Thermal renaturation of DNA. *J. Mol. Biol.* **3:**585–594.

17. **McCarthy, B. J., and E. T. Bolton.** 1963. An approach to the measurement of genetic relatedness among organisms. *Proc. Natl. Acad. Sci. USA* **50:**156–164.

18. **Murray, R. G. E., D. J. Brenner, R. R. Colwell, P. DeVos, M. Goodfellow, P. A. D. Grimont, N. Pfennig, E. Stackebrandt, and G. A. Zavarzin.** 1990. Report of the ad hoc committee on approaches to taxonomy within the Proteobacteria. *Int. J. Syst. Bacteriol.* **40:**213–215.

19. **Olsen, G. J., R. Overbeek, N. Larsen, T. L. Marsh, M. J. McCaughey, M. A. Maciukenas, W.-M. Kuan, T. J. Macke, Y. Xing, and C. R. Woese.** 1992. The ribosomal database project. *Nucleic Acids Res.* **20:**2199–2200.

20. **Persing, D. H., T. F. Smith, F. C. Tenover, and T. J. White (ed.).** 1993. *Diagnostic Molecular Microbiology: Principles and Applications.* American Society for Microbiology, Washington, D.C.

21. **Piot, P.** 1991. *Gardnerella, Streptobacillus, Spirillum,* and *Calymmatobacterium,* p. 483–487. *In* A. Balows, W. J. Hausler, Jr., K. L. Herrmann, H. D. Isenberg, and H. J. Shadomy (ed.), *Manual of Clinical Microbiology,* 5th ed. American Society for Microbiology, Washington, D.C.

22. **Relman, D. A.** 1993. The identification of uncultured microbial pathogens. *J. Infect. Dis.* **168:**1–8.

23. **Relman, D. A., J. S. Loutit, T. M. Schmidt, S. Falkow, and L. S. Tompkins.** 1990. The agent of bacillary angiomatosis—an approach to the identification of uncultured pathogens. *N. Engl. J. Med.* **3223:**1573–1580.

24. **Relman, D. A., T. M. Schmidt, R. P. MacDermott, and S. Falkow.** 1992. Identification of the uncultured bacillus of Whipple's disease. *N. Engl. J. Med.* **327:**293–301.

25. **Selander, R. K., D. A. Caugant, H. Ochman, J. M. Musser, M. N. Gilmourn, and T. S. Whittam.** 1986. Methods of multilocus enzyme electrophoresis for bacterial population genetics and systematics. *Appl. Environ. Microbiol.* **51:**873–884.

26. **Shah, H. N., and M. D. Collins.** 1988. Proposal for reclassification of *Bacteroides asaccharolyticus, Bacteroides gingivalis,* and *Bacteroides endodontalis* in a new genus, *Porphyromonas. Int. J. Syst. Bacteriol.* **38:**128–131.

27. **Skerman, V. B. D., V. McGowan, and P. H. A. Sneath (ed.).** 1980. Approved list of bacterial names. *Int. J. Syst. Bacteriol.* **30:**225–420.

28. **Sneath, P. H. A.** 1984. Bacterial classification. II. Numerical taxonomy, p. 5–7. *In* N. R. Krieg and J. G. Holt (ed.), *Bergey's Manual of Systematic Bacteriology,* vol. 1. The Williams & Wilkins Co., Baltimore.

29. **Sneath, P. H. A. (ed.).** 1992. *International Code of Nomenclature of Bacteria: Bacteriological Code, 1990 Revision.* American Society for Microbiology, Washington, D.C.

30. **Swofford, D. L., and G. J. Olsen.** 1990. Phylogeny reconstruction, p. 411–501. *In* D. M. Hillis and C. Moritz (ed.), *Molecular Systematics.* Sinauer Associates, Inc., Sunderland, Mass.

31. **Tenover, F. C. (ed.).** 1992. Section 10: molecular biology, p. 10.0.1–10.6.3. *In* H. D. Isenberg (ed.), *Clinical Microbiology Procedures Handbook,* vol. 2. American Society for Microbiology, Washington, D.C.

32. **Tonjum, T., D. A. Caugant, and K. Bovre.** 1992. Differentiation of *Moraxella nonliquefaciens, Moraxella lacunata,* and *Moraxella bovis* by using multilocus enzyme electrophoresis and hybridization with pilin-specific DNA probes. *J. Clin. Microbiol.* **30:**3099–3107.

33. **Wayne, L. G., D. J. Brenner, R. R. Colwell, P. A. D. Grimont, O. Kandler, M. I. Krichevsky, L. H. Moore, W. E. C. Moore, R. G. E. Murray, E. Stackebrandt, M. P. Starr, and H. G. Trüper.** 1987. Report of the Ad Hoc Committee on Reconciliation of Approaches to Bacterial Systematics. *Int. J. Syst. Bacteriol.* **37:**463–464.

34. **Williams, J. E.** 1984. Proposal to reject the new combination *Yersinia pseudotuberculosis* subsp. *pestis* for violation of the first principle of the International Code of Nomenclature of Bacteria. Request for an opinion. *Int. J. Syst. Bacteriol.* **34:**268–269.

35. **Woese, C. R.** 1987. Bacterial evolution. *Microbiol. Rev.* **51:**221–271.

36. **Zuckerkandl, E., and L. Pauling.** 1965. Molecules as documents of evolutionary history. *J. Theor. Biol.* **8:**357–366.

Processing Specimens for Bacteria

BETTY A. FORBES AND PAUL A. GRANATO

21

INTRODUCTION

General Patient Considerations

In the laboratory, the diagnosis of an infectious disease process in humans begins with proper processing and examination of the clinical specimen. Although a multitude of factors must be considered in the proper evaluation of such samples, special consideration must be given to the general patient population served by the laboratory, the incidence of certain diseases in a particular geographic area of the country, and any pertinent special patient history that may be useful in facilitating the recovery and/or detection of an etiologic agent or agents from clinical material. Each of these factors is important for making decisions about which procedures, media, and/or visual examinations will be used routinely for the timely and cost-effective laboratory diagnosis of an infectious disease. For example, if a laboratory provides services for primarily an outpatient and/or community hospital patient population, traditional approaches for the cultivation or microscopic detection of commonly occurring bacterial pathogens should be used routinely. Alternatively, if a microbiology laboratory provides services for immunocompromised patients from a tertiary care facility or oncology center and/or provides services for patients from a burn, transplant, or cystic fibrosis unit, the routine use of more elaborate processing and culture procedures for certain specimens is warranted. In addition, microbiologists must also consider staffing and budgetary resources in selecting procedures and deciding on the extent of services provided.

General Safety Considerations

The safety of laboratory personnel is of paramount importance for all clinical laboratories. Because of the inherent risks of sample aerosolization and spills associated with the handling and processing of clinical material, all patient specimens must be processed following the universal precautions guidelines. In most clinical laboratories, a special area is designated for the processing and culture of clinical samples. Ideally, these procedures should be performed in a class II biological safety cabinet or under a laminar-flow hood, with the worker wearing protective clothing, gloves, and, in some instances, a mask and protective eyewear. Previously, these safety precautions were used only for specimens identified with hazard labels. Today, every patient sample is presumed to contain a transmissible agent; therefore, special care and every safety resource available should be used when handling all specimens in order to minimize the risk of laboratory-acquired infection (see chapter 7 of this Manual).

Criteria for Specimen Acceptability

Immediately upon arrival of the specimen in the laboratory, the time and date of receipt should be noted on the requisition; many laboratories employ a time stamp device for this purpose. Once the time and date of receipt are recorded, the specimen is processed for bacterial culture by entering appropriate data into either a laboratory computer system or an equivalent paper system, examining the specimen and requisition to ensure that all criteria for acceptance are fulfilled, examining the specimen macroscopically and/or microscopically (if appropriate for the specimen type), and finally, inoculating the specimen onto primary plating media.

Laboratories must establish criteria for unacceptable specimens for culture. The requisition and specimen label should be checked for all necessary information: patient name, age, and sex; hospital number and location or address; ordering physician's name; specimen source; date and time of collection; and procedure(s) requested. If information is incomplete, the laboratory should ask a responsible person to provide the essential information or possibly to verify specific information so that further processing can be done. For any discrepancies, a written record documenting the individuals involved and how the situation was handled should be established. General criteria for specimen rejection are listed in Table 1. If a specimen is rejected a responsible individual must be notified immediately, the specimen should be saved for at least 24 h at 4°C and, again, the name(s) of the person(s) involved and the action taken should be documented.

Specimens should be processed promptly upon arrival in the laboratory. Even though this is not always possible, certain specimens, such as cerebrospinal fluids, joint fluids, genital and eye specimens, and any specimen accompanying a request for *Neisseria gonorrhoeae*, *Neisseria meningitidis*, *Haemophilus influenzae*, *Haemophilus ducreyi*, or anaerobes, should be processed immediately, regardless of other work load. Specimens of lower priority can be processed as time

TABLE 1 Criteria for specimen rejection

General	Anaerobic culture
1. Improper transport temperature 2. Improper transport medium 3. Misidentification of specimen 4. Insufficient specimen quantity 5. Leaky specimen 6. Improper transport time 7. Specimen received in fixative 8. Dried-out specimen on swabs	1. Fecal specimens include colostomy contents (except for *Clostridium difficile* or possibly other clostridia associated with gastrointestinal disease) 2. Midstream and catheterized urine 3. Throat, nose, or oropharyngeal (including gingival swab) specimens; expectorated sputa; or those obtained by naso- or orotracheal suction 4. Gastric contents 5. Vaginal and cervical swabs 6. Superficial material from skin, wounds, decubitus ulcers, eschars, and sinus tracts

permits as long as appropriate action is taken to maintain the integrity of the specimen.

Overview of Specimen Processing for Bacterial Culture

Successful laboratory diagnosis of bacterial infections depends on a number of key factors, including specimen selection such that the specimen is representative of the disease process, collection and transport, subsequent inoculation onto primary plating media, and incubation conditions for the culture. In addition, many specimens require some form of initial treatment prior to inoculation onto primary plating media. Such procedures may include dilution, concentration, tissue homogenization, or treatments such as acid washing of respiratory specimens for legionellae or decontamination of specimens for mycobacterial culture. Swab specimens received by the laboratory can be vortexed in 0.5 to 1.0 ml of saline or tryptic soy broth for about 20 s prior to inoculation. Several studies (13, 30, 82) have shown that this procedure enhances the recovery of aerobic and anaerobic bacteria from clinical specimens. This elution procedure is an alternative to direct specimen inoculation.

Direct microscopic examination of stained preparations of clinical material often provides an etiologic diagnosis so that appropriate therapy can be initiated before culture results become available. One of the cornerstones of the clinical microbiology laboratory is the Gram stain; the principle, procedure, interpretation, and reporting of the Gram stain are well described by Kruczak-Filipov and Shively (41). Although heat fixing smears to microscopic slides is common practice, methanol fixation has been reported to be superior for Gram-stained smears (47). This latter method is particularly useful for specimens containing large numbers of erythrocytes. Similarly, an acid-fast stain is the cornerstone of the mycobacteriology laboratory for the detection of mycobacteria in clinical material (18). In addition to these important stains, a large number of fluorescent-antibody stains, enzyme immunoassays, and DNA hybridization or amplification assays can be used for the direct detection of bacteria in clinical material; these assays will be addressed in the chapters that deal with specific groups of bacteria and/or in chapters 11 through 13 of this Manual. Additional information regarding microscopic procedures and stains can be found in chapter 4.

A combination of enriched, nonselective, selective, and differential media is used for the isolation of both aerobic and anaerobic bacteria from clinical specimens. A detailed summary of primary culture media by Isenberg et al. (32)

that includes each medium's type, inhibitors or indicators, purpose, and incubation atmosphere serves as an excellent reference. The various anaerobic primary plating media and their respective purposes are listed in Table 2 (50). Since anaerobic bacteria vary in their tolerance of oxygen, prolonged oxygen exposure should be avoided. However, it is not always feasible to incubate plates in an anaerobic system immediately following inoculation. Use of an anaerobic holding box (Carr-Scarborough Microbiologicals Inc., Stone Mountain, Ga.) is a practical alternative to immediate incubation if an anaerobic chamber is not available.

Finally, incubation conditions (length of incubation, temperature, and atmosphere) play a significant role in the successful recovery of bacteria from clinical specimens. Incubation conditions for any primary culture depend on the organism(s) one is attempting to recover. For example, optimal recovery of *Campylobacter jejuni* from stool culture depends not only on the use of appropriate selective media and incubation of media at 42°C but also on an atmosphere of 5% O_2 and 10% CO_2; this requirement is in contrast to the needs of other stool pathogens such as *Salmonella* and *Shigella* spp., which require incubation with ambient atmospheric conditions. Some organisms, such as *Yersinia enterocolitica* and *Listeria monocytogenes*, may require cold enrichment. To recover the large number of organisms capable of causing disease from clinical material, a variety of atmospheric conditions must be employed: ambient, CO_2 (3 and 5 to 7%) enriched, and microaerophilic, as required for *Campylobacter* spp. Although most routine cultures require incubation at 35 to 37°C, as previously mentioned, some organisms have different optimum temperatures for isolation: for example, *Y. enterocolitica* is more easily recovered with incubation at ambient temperature. Appropriate conditions can be achieved by a variety of methods, which are listed in Table 3. Finally, the length of incubation is again determined by the organisms one is attempting to culture.

General Approach

The enormous array of organisms capable of causing infection challenges the clinical microbiologist to create a rational approach to processing all specimen types. In addition, the type and size of the laboratory, the complexity of the patient population, the availability of trained personnel, and the economic resources available are among the primary factors dictating the activities of the clinical microbiology laboratory. Therefore, the primary purpose of this chapter is to recommend laboratory protocols for processing specimens for bacteriology with the understanding that these recommendations are neither all-inclusive nor

TABLE 2 Isolation media for recovery of anaerobes from clinical specimens[a]

Medium	Purpose
Anaerobic blood agar	Supports growth of strict anaerobic and facultatively anaerobic bacteria
Bacteroides bile esculin agar	Supports selective growth of *Bacteroides fragilis* group organisms; occasionally other bacteria grow, including *Fusobacterium mortiferum*, *Klebsiella pneumoniae*, *Enterococcus* species, and yeasts
Kanamycin-vancomycin laked blood agar	Supports selective growth of *Bacteroides* species and enhances pigment production in *Bacteroides melaninogenica* group. Yeasts and kanamycin-resistant gram-negative bacilli may grow on this agar.
Phenylethyl alcohol-sheep blood agar	Supports growth of most gram-positive and gram-negative anaerobes; inhibits facultative anaerobic gram-negative bacilli
Colistin-nalidixic acid blood agar	Supports growth of anaerobic and facultatively anaerobic gram-positive bacteria; inhibits most gram-negative bacteria
Thioglycolate medium (BBL-135C) supplemented with hemin and vitamin K_1	Enrichment broth used as backup to plated media
Chopped meat broth	Enrichment broth used as backup to plated media

[a]Reprinted from the *Manual of Clinical Microbiology* (50) with permission of the publisher.

TABLE 3 Options for obtaining appropriate atmospheric conditions for primary plate incubation

Atmosphere	Method
Ambient	Air incubator
3% CO_2	Candle jar
5–7% CO_2	CO_2 incubator; jar-type systems with appropriate generator envelopes; bag-type systems with appropriate generator envelopes
Specifically for *Campylobacter* spp.	Jar-type systems with appropriate generator envelopes; bag-type systems with appropriate generator envelopes; Campy Gas Tank (Adams Scientific, West Warwick, R.I., or local supplier can often provide appropriate mixture)
Anaerobic	Jar-type systems with appropriate generator envelopes; bag-type systems with appropriate generator envelopes; anaerobic chamber

where appropriate, the possible advantages and/or limitations of one method over another.

BLOOD

Although a variety of blood culture systems exists, each with its own protocols, some issues are common to all, including volume of blood, medium additives, and special protocols. Since the magnitude of bacteremia can be very low in adults, the rate of isolation of bacteria from blood is directly related to the volume of blood cultured (29, 34, 77). Collection of 20 to 30 ml of blood per culture for adults is regarded as optimal (73). In a recent study, Mermel and Maki reported a 3% increase in yield of blood cultures in adults for every milliliter of blood cultured (48); their results showed that few microbiology laboratories ensure adequate blood volumes. Upon receipt of the specimen, a laboratory should routinely document the blood volume submitted for culture and should, at the very least, begin an educational program as to the importance of culturing appropriate blood volumes for adults.

There are two major groups of blood culture medium additives: those that prevent coagulation and also inhibit the bactericidal action of blood and those that are presumed to remove exogenous components from the blood that may compromise the recovery of organisms. However, the advantages and disadvantages of each agent must be carefully considered prior to use. For example, penicillinase inactivates penicillin in the blood of the patient, but its addition carries a significant risk for contamination of the blood culture. Blood culture additives and their respective issues are reviewed by Tilton (69) and by Washington and Ilstrup (73).

The selection of a blood culture system dictates the laboratory's processing protocol. Blood culture systems include those that are manual, semiautomated, or completely automated. The choice of one system over another should be based on the number of blood cultures performed, the

applicable to all laboratory settings. The decision as to which systems to use routinely and which systems to use on special request should be made on the basis of the patient population served by the laboratory as well as other factors. The goal of the clinical microbiologist is to provide reliable and clinically relevant laboratory results in the most cost-effective and efficient manner.

Instead of providing detailed descriptions of routine processing techniques, including use of stains, reagents, or other methods, we only highlight routine protocols for processing of specimens for bacterial culture; several excellent books addressing the routine aspects of specimen processing for bacteriology are available and should serve as key references for detailed procedures and/or information on media. Examples are the *Manual of Clinical Microbiology*, 5th ed. (32), and the *Clinical Microbiology Procedures Handbook* (11). In addition, our recommendations for routine primary plating media and Gram stain according to specimen type are summarized in Tables 4 and 5. The primary focus of this chapter is on unusual or special circumstances regarding the processing of specimens for bacteriology and,

TABLE 4 Recommendations for primary plating media and Gram stains[a]

Specimen type	Stain		Medium													
	GRM	SP	EB	TV	BA	CHC	MAC, EMB	CNA, PEA	XLD, HE, SS	MAC-SOR	CPY	CIN	TBS	ANA	MI, TM, NYC	BLD-CX
Abscess	×		×		×	×								×		
Biopsy or tissue sample																
Bone marrow, brain	×	×[b]	×		×	×								×		
Liver, lymph node	×	×[b]	×		×	×	×							×		
Uterine	×	×[b]	×		×	×	×							×		
Unable to divide			×													
Body fluid																
Ascites, pleural, pericardial	×	×[c]	×		×	×	×							×		
Tympanocentesis	×		×		×	×										
CSF, synovial	×	×[c]	×		×	×										×
Effluent from CAPD patient	×															×
Amniotic	×		×		×	×	×							×		
Culdocentesis	×		×		×	×								×	×	
Prostate, testes, epididymis	×		×		×	×	×	×							×	
Ear					×	×	×	×								
Eye			×		×	×										
Foreign body																
IUD					×									×	×	
Catheter tip					×											
Genital																
Urethra					×	×									×	
Cervix, vaginal, Bartholin duct	×				×	×									×	
Ulcers	×														×	
Screen for N. gonorrhoeae	×[d]					×									×	
Screen for group B streptococci				×[e]				×								
Intestinal																
Stool and related							×		×	×[f]	×					
Bile, gallbladder					×		×		×							
Screen for Yersinia spp.												×				
Screen for Vibrio spp.													×			
Urine																
Clean catch or Foley catheter					×		×									
Catheterized					×		×									
Suprapubic					×		×									
Skin																
Deep wound	×	×			×	×	×							×		
Superficial, cellulitis					×	×	×									
Traumatized areas (burns, bites, etc.)					×	×	×									
Quantitative blood culture						×										

[a]Abbreviations: GRM, Gram stain; SP, specimen preparation (i.e., dilution, concentration, etc.); EB, enrichment broth; TV, enrichment broth for group B streptococci; BA, blood agar; CHC, chocolate agar; MAC, MacConkey agar; EMB, eosin-methylene blue agar; CNA, colistin-nalixidic acid agar; PEA, phenylethyl alcohol agar; XLD, xylose-lysine-deoxycholate agar; HE, Hektoen enteric agar; SS, salmonella-shigella agar; MAC-SOR, MacConkey agar with 1% sorbitol; CPY, selective agar for Campylobacter spp.; CIN, cefsulodin-Irgasan-novobiocin agar; TBS, thiosulfate-citrate-bile-salts-sucrose agar; ANA, anaerobic agars; ML, Martin-Lewis agar; TM, modified Thayer-Martin agar; NYC, New York City agar; BLD-CX, blood culture bottles; CSF, cerebrospinal fluid; CAPD, chronic ambulatory peritoneal dialysis; IUD, intrauterine device.

[b]Homogenize.

[c]Concentrate.

[d]Gram stain recommended for specimens obtained from male patients.

[e]Enhances detection.

[f]Depending on the geographic area, set up either routinely for all stools or only for bloody stools.

TABLE 5 Recommendations for primary plating media and stains for respiratory tract specimens[a]

Specimen type	Stain				Medium												Comments
	SP	GRM	DFA	AFB	BA	CHC	MAC, EMB	PC	ANA	CYE	SCYE	MNYC, MBR	LJ, MID, TBB	CHL	LOE, TIN	BG, RL	
Throat																	
Streptococcus pyogenes					X												
Corynebacterium diphtheriae					X										X		
Neisseria gonorrhoeae						X											Selective for GC as well
Mycoplasma pneumoniae												X					Serology
Nasopharynx																	
Routine					X	X											
Corynebacterium diphtheriae					X										X		
Bordetella pertussis					X											X	
Mycoplasma pneumoniae												X					Serology
Chlamydia spp.														X			Serology
Sputum																	
Routine	X[b]	X			X	X											
Legionella spp.	X[c]		X							X	X						Serology
Mycobacteria	X[d]			X									X				
Nocardia spp.[e,f]				X	X								X				
Mycoplasma pneumoniae												X					Serology
Chlamydia spp.														X			
Cystic fibrosis patient	X[b]	X			X	X	X	X									Screen for S. aureus as well
Bronchoalveolar lavage, washing, aspirate																	
Routine	X[g]	X			X	X	X										
Legionella spp.	X[c]		X							X	X						Serology
Mycobacteria	X[d]			X									X				
Nocardia spp.[e,f]				X	X								X				
Chlamydia spp.														X			

(Continued on next page)

TABLE 5 Recommendations for primary plating media and stains for respiratory tract specimens[a] (Continued)

Specimen type	Stain				Medium												Comments
	SP	GRM	DFA	AFB	BA	CHC	MAC EMB	PC	ANA	CYE	SCYE	MNYC MBR	LJ, MID, TBB	CHL	LOE, TIN	BG, RL	
Tracheal aspirate		×			×	×	×										
Transtracheal aspirate		×			×	×	×		×								
Lung biopsy	×[h]	×			×	×	×		×								

[a] Abbreviations: SP, specimen preparation; GRM, Gram stain; DFA, direct fluorescent-monoclonal-antibody stain for *Legionella pneumophila*; AFB, stain for acid-fast bacilli; BA, blood agar; CHC, chocolate agar; MAC, MacConkey agar; EMB, eosin-methylene blue agar; PC, *P. cepacia* selective agar; ANA, anaerobic agar; CYE, charcoal-yeast extract agar; SCYE, selective charcoal-yeast extract agar; MNYC, modified New York City agar; MBR, *M. pneumoniae* selective broth; LJ, Lowenstein-Jensen agar; MID, Middlebrook agar; TBB, liquid broth medium for the isolation of mycobacteria; CHL, *Chlamydia* culture; LOE, Loeffler serum-glucose medium; TIN, Tinsdale (cysteine-tellurite) agar; BG, Bordet-Gengou agar; RL, Regan-Lowe agar (55); GC, gonococci.
[b] Add mucolytic agent.
[c] Acid wash.
[d] Decontaminate, and concentrate by centrifugation.
[e] Set up fungus culture as well.
[f] Modified acid-fast stain.
[g] Dilutions for semiquantitative cultures.
[h] Mince for *M. pneumoniae*.

patient population served, and the laboratory's resources, including personnel, space, and finances.

If the patient population served by the laboratory includes patients immunosuppressed because of transplantation protocols, chemotherapeutic regimens, or human immunodeficiency virus infection, then laboratories should use additional blood culture systems or devices as adjuncts to their routine blood culture system. For example, use of the Isolator lysis-centrifugation system (Wampole Laboratories, Cranbury, N.J.) enhances the recovery of fungi and *Staphylococcus aureus* compared with that of broth-based systems (79). Moreover, since the lysed and centrifuged sediment is used for inoculation, different medium types can be selected to recover such organisms as mycobacteria (37) and *Bartonella* (*Rochalimaea*) spp. (65). However, owing to reports of decreased recovery of *Streptococcus pneumoniae* and *H. influenzae* with these systems compared to recovery with broth-based systems (36, 59, 74), the Isolator system should not be used as the sole blood-culturing system. A number of other options such as antibiotic-inactivating resins and other additives, various bottle sizes and media, and large volumes of specimen are provided with commercial blood culture systems. Each has its own advantages and limitations, so combinations of options are necessary in order to detect all clinically significant bacteremias.

Special Considerations

Assessment of Line Sepsis

Central venous catheters (CVCs) are valuable multipurpose devices used in patients in intensive care. These catheters serve to administer fluids, blood products, and antibiotics; monitor the patient's hemodynamics; and provide access for blood sampling. One of the major complications associated with the placement of CVCs is catheter-related sepsis; over approximately a 2-year period, 82% of bacteremias were associated with CVCs (54). A critically ill patient with a CVC who becomes febrile presents clinicians with a management dilemma, since no characteristic clinical features of catheter-related sepsis distinguish this infection from sepsis arising from other body sites. There is no consensus as to the mechanism of pathogenesis of CVC infections. Two theories have been proposed: (i) the organism gains access to the bloodstream along the outside of the catheter, originating at the skin insertion site (15, 44); and (ii) the hub is considered the source of infection, with organisms gaining access to the catheter and migrating toward the bloodstream along the inner lumen of the catheter (6, 42, 64).

Although many laboratories routinely perform semiquantitative cultures of CVC tips, this common practice can confirm the clinical suspicion of catheter-related sepsis and provide retrospective epidemiologic data only after removal of the catheter. Although some data are conflicting, a method that compares the number of bacterial colonies from blood simultaneously drawn through the catheter and obtained from a peripheral site has continued to show promise in a number of studies (16, 49, 78). On the basis of these data and until another method(s) becomes available, laboratories should provide quantitative cultures of blood drawn through the catheter for the assessment of catheter-related sepsis; one method is described in Table 6.

Unusual Organisms

Some significant bacterial pathogens require conditions that would not be met with standard blood culture systems.

TABLE 6 Collection and processing of blood specimens for assessment of catheter-related sepsis[a]

Collection	Draw peripheral blood sample into appropriately labeled 1.5-ml pediatric Isolator. Draw blood sample from catheter into appropriately labeled 1.5-ml pediatric Isolator. If catheter has multiple lumens, draw blood sample through each port into appropriately labeled 1.5-ml pediatric Isolator.
Processing	1. For each 1.5-ml pediatric Isolator received, label corresponding chocolate plate with appropriate patient information and site from which specimen was drawn.
	2. Inoculate each plate with 0.5 ml of blood.
	3. Incubate for 72 h at 37°C. Examine daily for growth.
	4. Count colonies on any plate exhibiting growth, and record no.
Interpretation	1. Negative: no growth on peripheral blood culture and on cultures of samples drawn through catheter
	2. Positive: If colony count of catheter culture(s) is five times or more that of peripheral blood cultures, results indicate probable catheter infection.
	3. If colony count of catheter culture(s) is less than five times that of peripheral blood culture, results do not indicate catheter infection with sepsis originating from undetermined site.

[a]See references 16 and 49 as well.

These organisms and references for their required processing conditions are listed in Table 7; conditions may vary in regard to collection, nutritional needs, and/or required length of incubation.

Unusual Circumstances

Occasionally, the laboratory receives for culture blood products from a possible hemolytic transfusion reaction or blood samples obtained at necroscopy. These products or blood specimens should be processed as routine blood cultures. However, the results of a recent study (80) show that postmortem blood cultures rarely, if ever, provide information that is not already known, and such cultures are of little diagnostic utility, particularly when antemortem blood cultures have been collected.

BODY FLUIDS OTHER THAN URINE AND BLOOD

Upon receipt by the laboratory, all body fluids other than blood and urine that are from normally sterile sites (ascites, cerebrospinal, pleural, pericardial, synovial, amniotic, and culdocentesis fluids, etc.) should be processed for direct microscopic examination and subsequent culture. As a general rule, if one bacterium is seen per oil immersion field, at least 10^5 bacteria per ml of specimen are present. Often, the numbers of bacteria in normally sterile body fluids are few. Therefore, it is recommended that the laboratory concentrate organisms in body fluids for microscopic examination and culture. This task is complicated by the fact that some bacteria require a significant g force in order to be sedimented by centrifugation. The sensitivity of the Gram stain may be improved by filtration, high-speed centrifugation, and cytocentrifugation. The *Manual of Clinical Microbiology*, 5th ed., recommends centrifugation of body fluids at 1,500 × g for 15 min followed by smear preparation and culture inoculation of the resulting sediment (32). A preferred practice for concentration is cytocentrifugation. According to the findings of Chapin-Robertson et al. (9), which clearly showed the superior sensitivity of cytocentrifugation compared even with that of high-speed centrifugation, clinical microbiology laboratories should adopt the routine cytocentrifugation of body fluids.

The practice of using blood culture bottles for the collection and/or culture of body fluids has not been recommended in the past. However, sufficient evidence has accumulated to recommend the use of blood culture bottles or an equivalent broth system in conjunction with direct plating, particularly for peritoneal fluids (4, 60, 61), synovial fluids (57, 63, 71), and cerebrospinal fluids (57). In general, if sufficient specimen is available, cultures should be inoculated with the same volumes specified by the manufacturer for blood specimens. If the volume is insufficient to follow the manufacturer's recommendations, as little as 0.1 ml could be inoculated.

Continuous ambulatory peritoneal dialysis (CAPD), an accepted treatment modality for end-stage renal disease, is complicated by a high incidence of peritonitis. Effluents from CAPD patients with suspected peritonitis present a particular challenge for the microbiology laboratory because of the large volume of dialysate in the peritoneal cavity, which results in a low concentration of bacteria in samples submitted for culture. A number of studies indicate that either the use of the adult Isolator tube (20, 81) or direct inoculation of dialysis effluent into blood culture bottles (20, 46, 62) is a sensitive and specific method of culture. If blood culture bottles are employed, cultures should be inoculated with the same volumes recommended by the manufacturer for blood specimens.

FOREIGN BODIES

Foley Catheters

Foley catheters are considered unacceptable for culture (26).

TABLE 7 Bacteria with special requirements for optimal recovery from blood specimens

Organism(s)	Reference chapter in this Manual
Anaerobiospirillum succiniciproducens	49
Borrelia spp.	51
Brucella spp.	44
Cardiobacterium hominis	39
Leptospira spp.	50
Mycobacterium spp.	31
Bartonella (*Rochalimaea*) spp.	58
Streptobacillus moniliformis	39

Intravenous Catheter Tips

Culture of intravenous catheters must employ some form of quantitation. Many laboratories use the semiquantitative method described by Maki et al. (45), in which the intravenous portion of the catheter is rolled across the surface of a blood agar plate. However, this method cultures only the external surface of the catheter and appears to have a low positive predictive value for catheter-related bacteremia (14). Over the last decade, several quantitative culture methods as well as other methods including staining of catheter tips have been developed, and to date, the answer to the question of which method is the most appropriate remains controversial.

Miscellaneous

Other foreign bodies, such as surgical screws, plates, rods, etc., should be inoculated into a broth such as tryptic soy.

GASTROINTESTINAL TRACT

The visual examination of stool sometimes provides a clue to the possible etiology of infection. For instance, the presence of obvious blood and mucus in stool may suggest infection by hemorrhagic *Escherichia coli* or some other invasive enteropathogen. The presence of rice water stool might indicate infection due to *Vibrio cholerae*.

Although this visual examination of stool is sometimes helpful, the microscopic examination of stool may be more useful. An applicator stick is used to mix a fleck of mucus or stool with an equal amount of Loeffler methylene blue on a glass slide. A coverslip is placed over this mixture, which is examined under the high dry objective for the presence of fecal leukocytes. The presence of numerous leukocytes is often associated with infection caused by invasive pathogens such as *Salmonella*, *Shigella*, or *Campylobacter* spp., while the absence of such cells might suggest disease caused by a noninvasive toxigenic organism (27). However, a recent study (31) has shown that the absence of fecal leukocytes, at least in children, is not a reliable predictor in ruling out the presence of invasive pathogens. Since false-positive and false-negative results occur, the fecal-leukocyte examination is most useful when the smear is strongly positive. Routine Gram stain examination of stool is not often useful unless the smear contains large numbers of gram-positive cocci in clusters, suggesting staphylococcal enterocolitis, or of curved gram-negative rods, indicating possible *Campylobacter* enteritis.

The choice of media for primary culture of stool samples is largely dependent on the bacterial enteropathogens sought routinely. In general, all stool samples should be cultured for *Campylobacter*, *Salmonella*, and *Shigella* spp. Media for detection of other pathogens may be inoculated routinely depending on the laboratory's geographic location and patient population.

The battery of primary culture media used for the recovery of *Campylobacter*, *Salmonella*, and *Shigella* spp. includes a CAMPY-blood agar plate, MacConkey or eosin-methylene blue agar, and two moderately selective enteric differential media, such as Hektoen enteric agar and xylose-lysine-deoxycholate agar. In certain circumstances, a portion of stool may be inoculated into an enteric enrichment broth such as gram-negative broth, which is used to enhance the recovery of the low numbers of *Salmonella* and *Shigella* spp. found in some fecal samples. Special enrichment broths are available for *Campylobacter* spp., but their routine use is usually not indicated.

The CAMPY-blood agar plate is incubated at 42°C for up to 72 h in a reduced-oxygen atmosphere. All other culture media are incubated in air at 35°C for 48 h and examined daily.

Yersinia spp.

Cefsulodin-Irgasan-novobiocin (CIN) agar, also known as *Yersinia* selective agar, is recommended for the recovery of *Y. enterocolitia* from stool. If CIN agar is not available, a MacConkey agar plate may be substituted. Once inoculated, the CIN or MacConkey agar is incubated at 32°C for 24 h or at 22 to 25°C (room temperature) for 48 h. Although strains of *Y. enterocolitica* grow faster at 37 than at 25°C, the lower temperature is recommended for isolation. Cold enrichment procedures are available for the improved recovery for *Yersinia* spp., but these methods are not recommended for routine use because they require prolonged incubation of up to 3 weeks.

Vibrio spp.

Thiosulfate-citrate-bile salts-sucrose (TCBS) agar is the selective-differential medium of choice for the recovery of pathogenic vibrios from stool. If a vibrio gastroenteritis is suspected and TCBS medium is not available, a MacConkey or sorbitol-MacConkey agar plate may be used for primary isolation (35). Culture plates should be incubated at 35°C. The routine use of TCBS agar as a regular part of a stool culture battery may be indicated for laboratories located near coastal areas or in areas where *Vibrio* infections are endemic, particularly during the summer, when disease incidence is highest.

Aeromonas, *Plesiomonas*, and *Edwardsiella* spp.

Aeromonas, *Plesiomonas*, and *Edwardsiella* spp. and other organisms can be recovered from stool and have been reported as infrequent causes of diarrhea. These organisms grow well on MacConkey agar incubated at 35°C, and no other specialized culture medium is needed for their isolation.

Hemorrhagic *E. coli*

Hemorrhagic *E. coli* is a relatively newly recognized enteropathogen that frequently produces a bloody diarrhea. The organism has been associated with several major outbreaks at fast-food restaurants (8, 25). Stools should be cultured routinely for hemorrhagic *E. coli* as well as for the other commonly occurring enteropathogens (*Campylobacter*, *Salmonella*, and *Shigella* spp.). Specimens should be inoculated onto sorbitol-MacConkey agar, which is used as a selective-differential screening medium for this organism. Culture plates should be incubated at 35°C.

Clostridium difficile

Clostridium difficile is a major cause of antibiotic-associated diarrhea and pseudomembranous colitis (43) as well as the leading cause of nosocomial diarrhea (38). For epidemiologic studies, stool samples may usually be submitted to a laboratory for culture of the organism and/or detection of one or both of the toxins that are responsible for the disease. Although not recommended for routine diagnosis, stool samples for *Clostridium difficile* culture should be inoculated to cycloserine-cefoxitin-egg yolk-fructose agar and incubated anaerobically for 48 h before observation. A variety of different assays are available for the detection of

Clostridium difficile toxin(s) in stool. The reader is referred to chapter 47 of this Manual for a complete discussion of the diagnosis of *Clostridium difficile*-related diarrhea.

Gastric Biopsies for *Helicobacter pylori*

Helicobacter pylori can be detected in gastric biopsy samples, brushings, and aspirates; however, gastric biopsy samples are considered the specimen of choice. The organism can be detected in stained smears of histologic sections, by culture, or through the production of urease in special medium.

Helicobacter pylori can be observed readily in histologic sections of biopsy material by using silver, hematoxylin and eosin, or Giemsa stain. A modification of the Giemsa stain technique, which is easier to perform, has been recommended for use in the routine microscopic examination of gastric biopsy samples (24).

The urease test is a rapid and inexpensive method for detecting *Helicobacter pylori* in tissue. The organism is a prolific producer of urease, which can be detected by macerating the biopsy specimen and placing it into a tube of urea broth or onto a Chistensen urea agar slant. Alternatively, a CLO (*Helicobacter pylori* was formerly called *Campylobacter*-like organism, or CLO) test is commercially available and is performed by placing the biopsy sample into a small well containing urea agar. With any of these urease tests, a positive result often develops within 15 to 30 min. See chapter 38 of this Manual for more information.

Helicobacter pylori may be cultured by mincing or homogenizing the biopsy specimen in 0.9% saline and inoculating the sample onto Marshall medium containing vancomycin, nalidixic acid, and amphotericin B (22). The culture plate is incubated microaerophilically at 35°C for up to 7 days. Alternatively, other more readily available media may be used for the cultivation of this organism, as described in chapter 38 of this Manual.

GENITAL TRACT

The bacteria most commonly associated with genital tract infections in men and women are *N. gonorrhoeae* and *Chlamydia trachomatis*. In addition to detecting these organisms, the laboratory must be able to detect other microorganisms that are unique to either men or women. For example, because of the ability of group B streptococci to cause neonatal meningitis and septicemia, pregnant women are commonly identified as to whether they are carriers of these organisms.

For the most part, direct microscopic examination of most specimen types from the genital tract provides useful information. Gram-stained smears are particularly useful for the presumptive diagnosis of gonorrhea in men and the detection of fusiform-spirochetal disease and bacterial vaginosis (BV) in women. Any surgically obtained specimen, including aspirates of abscesses and culdocentesis fluid, should be Gram stained.

Vulvovaginitis is a common clinical syndrome that is caused primarily by *Trichomonas vaginalis* and *Candida albicans*. In addition, many women present with discharge and complaints of a "fishy" vaginal odor; this condition is referred to as BV.

Since other vaginal infections closely resemble BV, an accurate differential diagnosis depends on laboratory examination of the genital specimen. First, the pH of the vaginal discharge is elevated above the normal 4.5 in about 90% of women with BV (10). Second, if 10% KOH is added to the vaginal discharge, a distinctive fishy odor will be released in

TABLE 8 Scoring system for Gram-stained vaginal smears[a]

	Code		
Lactobacillus morphotype	*Gardnerella* and *Bacteroides* morphotypes	Curved gram-variable rod	Score[b]
4+	0	0	0
3+	1+	1+ or 2+	1
2+	2+	3+ or 4+	2
1+	3+		3
0	4+		4

[a]Smears are coded 0 to 4+ on the basis of the average number of organisms with that morphotype per oil immersion field: 0, none present; 1+, <1 present; 2+, 1 to 4 present; 3+, 5 to 30 present; 4+, ≥30 present. Determine the morphotype quantity code, and then determine the score from the corresponding box in the Score column. Note that a lower score is given to curved gram-variable rods. Table is adapted from the *Journal of Clinical Microbiology* (52) with permission of the publisher.

[b]Code each morphotype separately, and add all three morphotype quantity codes together for a total score. Total score = lactobacilli + *Gardnerella* and *Bacteroides* spp. + curved rods. Interpret total score as follows: 0 to 3, normal; 4 to 6, intermediate; ≥7, BV.

about 70% of patients with BV (72); however, this test can be positive in some patients with trichomoniasis. Third, the presence on wet mount of squamous epithelial cells covered with small coccobacilli ("clue cells") is also consistent with a clinical diagnosis of BV. However, probably the best means by which to differentiate BV from other vaginal infections is direct Gram stain of vaginal fluid. Spiegel (66) described use of the Gram stain for the diagnosis of BV. In a more recent study, Nugent et al. (52) proposed a standardized method for interpretation of Gram stains for BV (Table 8). This method emphasizes the presence or absence of the morphotypes that are most reliably recognized. The bacterial morphotypes quantitated were large, gram-positive bacilli of the *Lactobacillus* morphotype; smaller gram-variable bacilli called the *Gardnerella* morphotype; and curved rods. Organisms are quantitated, and then a score is assigned to the smear and subsequently interpreted (Table 8). In general, most authors agree that the Gram stain is more sensitive than either of the two previous tests and is more specific than culture for *Gardnerella vaginalis* and the detection of clue cells; thus, it is recommended that direct Gram stains of vaginal fluids be performed as the standard test for diagnosis for BV.

Finally, preexisting lesions due to other diseases may become secondarily infected with a mixed anaerobic flora of fusobacteria and spirochetes (56). Since these infections can progress rapidly, Gram stains of vaginal specimens should be examined for the presence of these organisms. In general, examination of the smear will reveal large numbers of fusiform bacteria; if these morphotypes are observed, examine the smear carefully for the presence of faintly staining, loosely coiled spirochetes.

The primary plating media recommended for the culture of different genital specimens are listed in Table 4. For the most part, any specimen obtained at the time of pelvic surgery and specimens obtained by laparoscopy, culdocentesis, or needle aspiration of an abscess should be cultured not only for aerobic and facultative anaerobes (including *N. gonorrhoeae*, *Chlamydia trachomatis*, and *Mycoplasma* spp.) but also for strict anaerobes. Vaginal swabs are not acceptable specimens for anaerobic culture.

Group B Streptococci

Commonly, pregnant women are screened for the presence of group B streptococci between 26 and 28 weeks of gestation. Specimens submitted for culture of group B streptococci should be inoculated onto a colistin-nalidixic acid plate and into Todd-Hewitt broth containing gentamicin and nalidixic acid. The broth is subcultured onto a colistin-nalidixic acid or blood agar medium either when it becomes turbid or after 24 h of incubation.

N. gonorrhoeae

Two important factors regarding the culture of *N. gonorrhoeae* from clinical specimens must be kept in mind. First, many physician's offices and laboratories employ various types of selective media containing vancomycin and other antimicrobial agents for primary isolation of this organism. The prevalences of vancomycin-susceptible strains of *N. gonorrhoeae* vary greatly from one geographic region to another and also fluctuate over time (5). Therefore, it is recommended that choice of media for primary isolation be based on locally generated data. Chocolate agar medium may be used in geographic areas where there is an unacceptably high incidence of vancomycin-susceptible strains of gonococci.

The columnar epithelium of the endocervical canal is the primary site of infection in nonhysterectomized women and the sole site infected in 46% of women with gonorrhea (19); the anal canal is the sole site of infection in 2 to 20% of these patients. In contrast, the urethra is the primary site of infection in hysterectomized women. For greatest sensitivity, it is recommended that specimens from all sites (urethra, cervix, and anal canal) be cultured routinely.

Other Organisms

A number of other bacterial agents associated with genital infections require different collection and transport conditions, special culture conditions, or special stains for their successful recovery or detection in clinical specimens. For the isolation or detection of *S. aureus* in specimens obtained from patients with toxic shock syndrome or of *H. ducreyi, Treponema pallidum,* urogenital mycoplasmas, and *Chlamydia trachomatis* from genital specimens, please consult chapters 22, 45, 52, 53, and 55, respectively.

OCULAR SPECIMENS

Scrapings and/or Exudates

Conjunctival and corneal scrapings and conjunctival exudates submitted to the laboratory for bacteriologic examination should have a smear prepared for Gram stain and another smear prepared for Giemsa stain, if the volume of sample permits. The presence of significant numbers of organisms extracellularly or within neutrophils on the Gram stain suggests bacterial infection. Bacteria composing the indigenous microbial flora (e.g., staphylococci and diphtheroids) of the eye may occasionally be detected, but these typically appear as one to two organisms per oil immersion field in random areas. The Giemsa-stained smear is used to examine for the presence of intracytoplasmic inclusions suggestive of *Chlamydia trachomatis* infection. These inclusion bodies are most commonly found in areas containing numerous neutrophils and sheets of epithelial cells. Because the sensitivity of the Giemsa stain in detecting *Chlamydia trachomatis* is 90% in infants but only 40% in adults, the monoclonal direct fluorescent antibody (DFA)

assay (Syva, Palo Alto, Calif.) may be considered an alternative stain for specimens collected from adult patients.

Owing to the limited nature of the clinical sample, 5% sheep blood and chocolate agar plates should be routinely inoculated along with an enrichment broth if the volume of sample permits. Corneal and conjunctival scrapings are best inoculated onto media by the physician. Under such circumstances, the physician should be provided with a chocolate agar plate and several slides for smear preparation. All culture media should be incubated at 35°C in 5 to 10% CO_2.

Intraocular Fluids

The procedures for staining and examining smears of intraocular fluids are the same as those just outlined for scrapings and exudates. Intraocular fluids contain pigment granules that may be misinterpreted by the inexperienced microscopist as gram-positive cocci. The pigment granules can usually be distinguished by their larger size, oval shape, and faint brown granules. Smears of vitreous or aqueous fluid are reliable in detecting almost 70% of cases of bacterial endophthalmitis.

Eye fluids for bacterial culture are routinely inoculated onto blood and chocolate agars and an enrichment broth. Media should be incubated at 35°C in 5 to 10% CO_2 for 48 to 72 h.

The eye and its associated structures form a complex organ that is predisposed to infection by a variety of microorganisms. The reader is referred to *Cumitech 13* (33) for an excellent, comprehensive review of this subject.

LOWER RESPIRATORY TRACT

Initial Processing

The laboratory diagnosis of a bacterial lower respiratory tract infection is problematic because of the difficulty associated with obtaining a quality lower respiratory tract sample that is not contaminated to some degree with the normal flora found in upper-airway secretions. The most common lower respiratory tract sample submitted to the laboratory for bacteriologic evaluation is expectorated sputum because of the ease and convenience of collecting it. Unfortunately, if a proper specimen collection technique is not rigidly enforced, sputa can be heavily contaminated with oropharyngeal secretions that result in needless work for the laboratory and generation of laboratory reports that may be potentially misleading to the clinician.

To minimize the problem, grading systems for evaluating the quality of expectorated sputum that are based on smear examination of the sample have been developed. One such system was developed by Bartlett et al. (3). In this system, a Gram stain smear is prepared from the purulent portion of sputum. Using the low-power 10× objective, 20 to 30 separate microscopic fields are examined by counting the numbers of epithelial cells and neutrophils in each field. As shown in Table 9, a positive or negative numerical grade is assigned to each microscopic field examined. Positive numbers are assigned for the presence of neutrophils, which indicate active infection, while negative numbers are assigned for the presence of epithelial cells, which indicate oropharyngeal contamination. The average grade is then calculated by totaling the individual grades and dividing by the number of microscopic fields examined. A final score of 0 or less indicates either lack of inflammation or the presence of significant oropharyngeal contamination, suggest-

TABLE 9 Bartlett's grading system for assessing sputum quality[a]

Type of cell and no./10× low-power field	Grade
Neutrophils	
<10	0
10–25	+1
>25	+2
With mucus present	+1
Epithelial	
10–25	−1
>25	−2

[a]Reprinted from *Cumitech 7* (3) with permission of the publisher.

ing that the specimen is unacceptable for culture. In such a case, the physician should be telephoned immediately and a repeat sample should be requested.

Murray and Washington (51) proposed a similar grading system for the smear evaluation of sputum. As shown in Table 10, these authors reported that group 1 through 4 samples represent poor-quality samples owing to the large number of epithelial cells present and, as such, are not suitable for culture. They considered only group 5 specimens of sufficient quality to be worthy of culture. However, in a subsequent clinical study performed by Van Scoy (70), group 4 samples were also regarded as acceptable for culture.

Both the Bartlett et al. and the Murray and Washington grading systems are reliable for assessing the quality of expectorated sputum samples. Considerable time, labor, and cost can be saved by eliminating the need for performing culture and subsequent workup of poor-quality specimens. As such, the routine use of either of these grading systems is highly recommended. Once a sputum specimen has been judged acceptable for culture, the Gram stain smear should be examined under oil immersion to establish a possible etiology of infection based on the predominating bacterial morphotype present. These observations should be recorded and included in the final laboratory report transmitted to the physician.

Because a good-quality sputum sample contains mucin and is highly viscous by nature, a microorganism may not be evenly distributed throughout the sample. Accordingly, the recovery of a potential lower respiratory pathogen may be adversely affected by the small portion of specimen sampled for culture. Previously, sputum samples would be transferred to a sterile petri dish and teased out with a sterile forceps to locate a purulent area of specimen for culture. Aside from being esthetically unpleasing, this practice has been discon-

TABLE 10 Murray and Washington's grading system for assessing sputum quality[a]

Group	No. of cells/low-power field	
	Epithelial	Leukocytes
1	25	10
2	25	10–25
3	25	25
4	10–25	25
5	<10	25

[a]Reprinted from reference 51 with permission of the publisher.

tinued in most laboratories because of the associated health risks to the worker.

Instead, the use of a mucolytic agent, dithiothreitol (Sputolysin; Calbiochem), is recommended for the homogenization of sputum. This material is nontoxic to microorganisms and provides an even distribution of microorganisms throughout the specimen by reducing the disulfide bonds in mucin to create a specimen of waterlike consistency.

The homogenization procedure is performed by adding an equal volume of Sputolysin to the sample. *Note:* Because the addition of this reagent dilutes the sample, all smears for microscopic examination should be prepared from the original specimen before homogenization. Vortex the sample for 20 to 30 s, and allow it to sit at room temperature for 15 min. Vortex the sample again to achieve final homogenization, and the specimen is now ready for immediate culture.

A variety of different lower respiratory tract specimens may be submitted to the microbiology laboratory for bacteriologic analysis depending on the method used for specimen collection. Such samples may include expectorated sputum; induced sputum; bronchial washings, brushings, and/or biopsy samples; tracheal aspirates; transtracheal aspirates; bronchoalveolar lavage fluid; open-lung biopsy samples; needle aspirates, etc. *Note:* Solid tissue samples must be homogenized in a tissue grinder before culture (see Tissue or Surgical Biopsy Samples below for details of this tissue homogenization procedure).

The media selected for routine culture of lower respiratory tract specimens are listed in Table 5. In addition, surgically collected tissue samples, needle aspirates, and transtracheal aspirates should be cultured for anaerobes by using anaerobic agar media and an anaerobic broth (i.e., chopped meat-glucose or thioglycolate).

Special Considerations

Legionella spp.

Legionella species are not an uncommon cause of community-acquired and nosocomial pneumonias. The organisms can be detected microscopically and/or by culture from a variety of different lower respiratory tract specimens such as lung tissue, transtracheal aspirates, lavage fluids, and sputum. Sputum should not be assessed microscopically, because legionellosis does not induce a strong inflammatory response. The DFA test is the most reliable method available for the microscopic detection of *Legionella* spp. in these samples. When properly performed and carefully controlled, this rapid test is nearly 100% specific but lacks sensitivity (about 75%) owing to the relatively large number of organisms required for microscopic visualization (10^4 to 10^5 organisms per ml). Touch preparations are made from tissue samples, and thin smears are prepared from liquid specimens on microscope slides. Laboratorians are reminded that a negative smear result does not rule out the presence of disease due to species other than *Legionella pneumophila*.

Legionella species have been recovered from a variety of pulmonary and extrapulmonary specimens. Because these organisms grow slowly, their detection in culture may be obscured by faster-growing organisms in contaminated specimens, even if selective *Legionella* isolation media are used. Recovery of *Legionella* spp. from heavily contaminated specimens (i.e., sputum) can be improved by using an acid wash decontamination procedure (7). The acid wash decontamination procedure is readily performed by diluting the contaminated specimen 10-fold in KCl-HCl buffer (pH 2.2) and incubating it for no more than 4 min at room temperature. Decontamination exposure beyond 4 min must be avoided, because legionellae are also slowly inactivated by this treatment. Inoculate the sample to each of the following media: nonselective buffered charcoal-yeast extract (BCYE) agar and semiselective *Legionella* culture medium (BCYE with polymyxin B, anisomycin, and cefamandole or BCYE with polymyxin B, anisomycin, and vancomycin) as well as a 5% sheep blood agar plate. Specimens that are not likely to be heavily contaminated do not require decontamination but should be inoculated to the same culture media. Culture plates should be incubated at 35°C in a humid atmosphere (to prevent drying of the medium, which is inhibitory to growth) with or without 5% CO_2 and examined daily for a minimum of 5 to 7 days for the appearance of suspicious growth. For additional information on the legionellae, the reader is referred to chapter 42 of this Manual.

Mycobacteria

The microscopic detection and cultural recovery of mycobacteria from respiratory samples may involve the use of any combination of a host of different technologies. A detailed discussion of each system is beyond the scope of this section. Refer to chapter 31 of this Manual for a detailed description of appropriate procedures.

Mycoplasma and *Chlamydia* spp.

Smear examination of respiratory secretions is not useful for establishing the diagnosis of pulmonary infection caused by *Mycoplasma pneumoniae*, *Chlamydia psittaci*, *Chlamydia trachomatis* in infants, or *Chlamydia pneumoniae*. The laboratory diagnosis of any of these infectious processes is achieved by cultural recovery of the organism or detection of a significant rise in specific antibody titers between acute- and convalescent-phase sera. Traditionally, *M. pneumoniae* can be cultured from respiratory samples by using diphasic medium (see chapter 53 for specific details). Alternatively, New York City (NYC) medium used in conjunction with a semiselective NYC medium enrichment transport broth containing phenol red indicator and penicillin has been used successfully for the rapid (2- to 4-day) cultural recovery of *M. pneumoniae* from throat specimens (23).

Chlamydia psittaci and Chlamydia pneumoniae may also cause lower respiratory tract infections. *Chlamydia psittaci* is an uncommon cause of pneumonia that is associated with exposure to birds. At this time, serology is the most reliable method for establishing the laboratory diagnosis of infection caused by either of these *Chlamydia* spp. Neonatal *Chlamydia trachomatis* infections of the lower respiratory tract are readily diagnosed by using the cultural and noncultural methods described in chapter 55 of this Manual.

Burkholdia (*Pseudomonas*) *cepacia*

B. (*P.*) *cepacia* is often the hallmark of end-stage disease when recovered in respiratory samples from patients with cystic fibrosis (67, 68). However, the detection of *B. cepacia* in respiratory secretions is made difficult because the organism is frequently overgrown with mucoid *Pseudomonas aeruginosa*, which heavily colonizes these patients (21). Laboratories that provide services for patients with this disease should consider the routine use of *P. cepacia* agar (21) and/or OFPBL medium (75) for the culture of respiratory specimens to facilitate the detection of this important organism (also see chapter 40 of this Manual).

Quantitative Culture

A quantitative culture procedure for the reliable diagnosis of nosocomial pneumonia, particularly ventilator-associated pneumonia, has been developed. The procedure is based on the collection of lower respiratory tract samples with a fiberoptic bronchoscope using a protected specimen brush or obtaining a protected bronchoalveolar lavage specimen (39). Upon receipt in the laboratory, the protected specimen brush is placed in a tube containing 1 ml of sterile saline and vortexed vigorously. Separate 0.1-ml volumes are inoculated onto 5% sheep blood and chocolate agar plates. Each colony detected is equivalent to 10 CFU. For protected bronchoalveolar specimens, 0.01 ml of sample is inoculated to each medium, with each resultant colony being equivalent to 100 CFU/ml. The diagnostic threshold for pneumonia with a protected specimen brush specimen is $\geq 10^4$ CFU/ml.

UPPER RESPIRATORY TRACT

Gram stain smear examination of material collected from the upper respiratory tract is a useful screening procedure for establishing the presumptive diagnosis of infection but should be limited to a few select specimen types. These include sinus aspirates, tympanocentesis fluids, and any other upper-airway sample that is not contaminated with oropharyngeal secretions during the collection process. Smears of the oropharynx are not recommended for the evaluation of bacterial infection unless Vincent's angina (fusospirochetosis) or diphtheria is suspected. A Gram stain smear that shows spirochetes and fusiform bacilli is diagnostic of Vincent's angina, while a Loeffler methylene blue stain smear that shows diphtheroids with metachromatic granules suggests diphtheria.

Antigen detection tests are also available for the rapid, noncultural diagnosis of certain infections, particularly pharyngitis caused by *Streptococcus pyogenes*. Detailed descriptions of the many tests available for this noncultural screen are given elsewhere (chapters 11, 13, and 23 of this Manual). In general, most of these tests are reliable in detecting moderate to large numbers of group A streptococci in the sample but are limited in their abilities to detect low numbers of organism. Negative noncultural screening tests for *Streptococcus pyogenes* should therefore include a duplicate pharyngeal swab submitted for culture to identify patients with low-grade infection.

The upper-airway sample most commonly submitted to the laboratory for bacterial analysis is a posterior pharyngeal specimen, and the bacterial pathogen most commonly examined for is *Streptococcus pyogenes*. Accordingly, all pharyngeal specimens should be routinely inoculated onto a selective agar or a 5% sheep blood agar plate. Several cuts in the primary area of inoculation should be made with the edge of the bacteriologic loop to facilitate the detection of streptolysin O. The culture plate should be incubated at 35°C in 5% CO_2. Alternatively, the plate may be incubated anaerobically, which enhances the hemolytic activity of streptolysin O. Dykstra et al. (17) have shown that anaerobic incubation improves the detection of *Streptococcus pyogenes*, in that 98% of 149 strains of *Streptococcus pyogenes* grown anaerobically were detected, while only 63% of the same strains incubated under aerobic conditions were detected.

A variety of selective blood agar media are available for increasing the detection of group A streptococci. These media contain antibiotics, alone or in combination, that suppress the growth of pharyngeal flora, thereby facilitating the detection of small numbers of beta-hemolytic streptococci. The recovery of small numbers of group A streptococci can also be improved by preincubating the specimen swab in an enriched selective broth before inoculating it onto the appropriate culture medium.

The remainder of upper-airway specimens (i.e., nares, nasopharynx, sinus, tympanocentesis, epiglottis, etc.) should be inoculated routinely onto blood agar as well as onto a chocolate agar plate to facilitate the recovery of fastidious bacteria, such as *Haemophilus*, *Branhamella*, and *Neisseria* spp., etc., that may be pathogens when recovered from these samples. All culture plates should be incubated at 35°C in 5% CO_2. With the possible exception of sinus aspirates and tympanocentesis fluids, routine anaerobic culture is not indicated for upper-airway specimens.

Special Considerations

Diphtheria

Throat and/or nasopharyngeal swabs should be inoculated onto Loeffler's and Tinsdale (tellurite) media as well as onto a blood agar plate to rule out the presence of beta-hemolytic streptococci. Plates should be incubated at 35°C in 5% CO_2 and examined daily for 72 h for suspicious growth.

Pertussis

The laboratory diagnosis of pertussis is achieved by direct microscopic examination using a DFA test and/or by culture (see chapter 46 of this Manual). In the DFA test, the specimen of choice is a slide prepared at the bedside or a nasopharyngeal swab submitted to the laboratory in Casamino Acids transport medium. A smear can be prepared from the Casamino Acids transport medium by using a two-well slide. Because the organism is extremely susceptible to drying, heat, and cold, bedside inoculation of the specimen and immediate transport of the culture plate, protected from temperature extremes, to the laboratory is recommended.

N. gonorrhoeae

The recovery of N. *gonorrhoeae* is best achieved by inoculation of a pharyngeal specimen onto NYC or Martin-Lewis medium. Incubate the plates at 35°C in 5% CO_2 for 3 days, and examine them daily for the presence of gonococcal growth.

N. meningitidis

Nasal carriers of N. *meningitidis* may be screened by inoculating a nasopharyngeal swab specimen onto chocolate and/or NYC or Martin-Lewis media. Culture plates should be incubated at 35°C in 5% CO_2 for 3 days and examined daily for the presence of meningococcal growth.

Methicillin-Resistant S. aureus

Culture screens for detecting carriers of methicillin-resistant S. *aureus* may include swab specimens collected from the anterior nares, axilla, inguinal area, etc. Swab specimens may be inoculated onto mannitol-salt agar and/or an oxacillin-containing (Ox-Screen) medium incubated at 35°C for 72 h and examined daily for the appearance of suspicious growth.

Arcanobacterium haemolyticum

A. (formerly *Corynebacterium*) *haemolyticum* has been reported to cause pharyngitis and peritonsillar abscess (12). The organism can be recovered by inoculating a throat swab onto streptococcal selective agar or a 5% sheep blood agar plate and incubating the medium at 35°C for 48 to 72 h. Clinical significance and culture recovery are discussed completely in chapter 29 of this Manual.

SKIN

Specimens from skin include superficial specimens, open- and closed-wound specimens, and aspirates from cellulitis. Refer to Table 5 for initial processing and the recommended primary plating media.

TISSUE OR SURGICAL BIOPSY

The initial processing of tissue or biopsy samples begins with the preparation of touch preparation smears for Gram stain smear examination. If the tissue sample is large, the specimen should be examined grossly, with the assistance of a pathologist, if necessary, for areas of inflammation or necrosis. A small portion of the area of cultural interest is dissected by aseptic technique.

Tissue or biopsy touch preparations are prepared by blotting the specimen onto a sterile microscopic slide. Care must be used to prevent extraneous contamination of the specimen and to rotate the tissue during the blotting process so that more than one surface area of the sample can be examined microscopically. Allow the smear to air dry, heat or methanol fix it, and then Gram stain it per the usual procedure.

Once the smear has been prepared, the tissue or biopsy sample must be homogenized before culture. Tissue specimens should first be minced with sterile scissors. Transfer the minced tissue or biopsy to a sterile tissue grinder, mechanical device, or homogenizer containing 1 to 2 ml of sterile nutrient broth, and homogenize the sample. Alternatively, a sterile mortar and pestle with a sterile abrasive and several milliliters of sterile nutrient broth may be used. The resultant mixture is used for culture as well as for the preparation of additional smears for Gram stain examination. Note that specimens for fungal culture should be minced only; grinding may destroy hyphal elements.

Homogenized tissue or biopsy samples should routinely be inoculated for aerobic and anaerobic culture onto the media recommended in Table 4. If requested, cultures for mycobacteria that use Lowenstein-Jensen agar and other appropriate media should be inoculated with the homogenized sample. Excess suspension or unused tissue should be stored at 4°C for several weeks in case cultures must be repeated or additional cultures are indicated on the basis of the specimen's histopathology.

Special Considerations

Quantitative Tissue Cultures

Quantitative tissue cultures may be requested by a physician for evaluating the clinical significance of a bacterial isolate(s) recovered from tissue. The presence of $\geq 10^5$ organisms per g of tissue is alleged to represent bacterial infection as opposed to colonization of the sample (28, 58).

The following protocol may be used for performing quantitative bacterial cultures.

1. Weigh the sterile, empty specimen container.
2. Weigh the tissue in the container.
3. Mince the tissue with sterile scissors, and then grind it with a sterile pestle in a sterile mortar or in a tissue homogenizer containing 1 ml of nutrient broth per g of tissue.
4. Dilute the suspension as follows:

0.9 ml of nutrient broth + 0.1 ml of suspension (10^{-1})

0.9 ml of nutrient broth + 0.1 ml of 10^{-1} suspension (10^{-2})

0.9 ml of nutrient broth + 0.1 ml of 10^{-2} suspension (10^{-3})

5. Inoculate 0.1 ml of each diluted suspension onto a blood agar plate. Spread the inoculum evenly over the surface of the agar with a sterile bent ("hockey stick") glass rod.
6. Incubate the blood agar plates at 35°C for 24 h.
7. Quantitate growth as follows:

$$\text{organisms/gram of tissue} = \frac{N \times D \times F \times 10}{W}$$

where N is the number of colonies on the agar plate selected for enumeration, D is the reciprocal of the dilution inoculated onto agar, V is the volume of broth used to suspend the tissue, W is the weight of the tissue in grams, F is the dilution factor $(V + W)/W$, and 10 is the constant factor, since 0.1 ml of suspension is used as inoculum.

A rapid screening procedure for quantitating tissue on a Gram-stained smear is accomplished as follows.

1. Transfer 0.01 ml of undiluted suspension from step 3 above onto a clean glass microscope slide, and spread it over an area not exceeding 15 mm in diameter.
2. Dry in an oven (75°C) for 15 min or air dry and fix in methanol.
3. Gram stain.
4. Examine all fields under oil immersion (1,000×). Finding one or more bacteria in any field (not per oil immersion field) denotes the presence of at least 10^5 bacteria per g of tissue.

Decubitus Ulcer

The clinical evaluation of bacterial results of cultures of decubitus ulcers is made difficult by the large number of microorganisms that colonize these body surface sites without causing overt infection. To minimize this major problem, biopsy samples from decubiti should be dipped into 70% ethanol and flamed to kill colonizing surface bacteria (40). The specimen should then be processed according to the protocol described in the previous section for routine culture.

Bone Bank Specimens

Bone bank specimens may be submitted to the laboratory for evaluation of the sterile integrity of a new lot of bone bank material before its use in bone grafting. Aseptically transfer the bone chip(s) to a tube of sterile nutrient broth or an unused aerobic blood culture bottle, and incubate it at 35°C for 5 days, examining daily for the appearance of growth.

Necropsy Tissue

Postmortem tissue may be submitted for bacteriologic examination when infection is thought to be the cause of death. Tissue samples must be collected by sterile technique. Because devitalized tissue serves as an excellent growth medium for microbial flora, specimens should be collected as close to the time of death as possible to prevent overgrowth of tissue with resident microflora. Tissue samples collected from cadavers maintained at room temperature beyond 4 h or at 4°C beyond 24 h are of limited bacteriologic value. Necropsy tissues that are acceptable for bacteriologic evaluation should be processed and cultured according to the procedure outlined above in the section on routine culture of tissue.

URINE

Since most urinary pathogens grow well in urine, significant distortion in the bacterial count may occur in urine samples left at room temperature for several hours (76). For this reason, urine samples received by the laboratory beyond 2 h after collection without evidence of refrigeration (effective for up to 24 h) or without use of a urine transport tube containing a preservative to retard bacterial growth may be unsuitable for culture.

Gram stain examination of an unspun urine sample is a rapid and reliable method for detecting significant bacteriuria and pyuria. The procedure is easily performed by mixing the urine and placing 1 drop on a glass slide. Allow the drop to dry without spreading, methanol fix it, and Gram stain it. The presence of at least one bacterial cell and one leukocyte per oil immersion field correlates with significant bacteriuria ($\geq 10^5$ CFU/ml) and pyuria. In addition, the bacterial morphotype observed in positive smears may provide a useful guide in the initial selection of antimicrobial therapy.

Although recommended, routine Gram stain smear examination of urine samples may not be feasible for most laboratories because of staff limitations. In these situations, the procedure may be reserved for those patients in whom a rapid diagnosis of significant bacteriuria is indicated (i.e., a patient who has developed sepsis secondary to a possible urinary tract infection).

Even though the Gram stain is an inexpensive method for detecting bacteriuria at a level of $\geq 10^5$ CFU/ml, the test lacks sensitivity at lower levels. Therefore, the Gram stain should not be used to rule out the presence of urinary infection when specimens are collected by cystoscopy or other invasive means, where the recovery of lower numbers of bacteria may be significant.

A wide variety of chemical, automated, and culture screening tests for detecting significant bacteriuria are available. Readers should consult an excellent comprehensive review of these systems by Pezzlo (53). Although most screening methods are reliable in detecting $\geq 10^5$ CFU/ml, like the Gram stain, they lose sensitivity in identifying infected patients with low-level bacteriuria.

The two general methods for quantitating bacteria in urine are the pour plate procedure and the surface streak procedure. The pour plate procedure is the standard with which all other methods are compared. This technique, however, is rarely used in the routine clinical laboratory because it is cumbersome to perform and does not allow for the convenient workup of organisms. The surface streak procedure is easily performed and usually involves the use of a calibrated loop. These loops deliver approximately 0.01 or 0.001 ml of urine to the surface of the plate, making each resultant colony representative of 100 or 1,000 CFU/ml, respectively, in the original specimen.

The surface streak procedure is performed by dipping the calibrated loop into a well-mixed urine sample until the loop is just completely submerged below the meniscus of the liquid. It is essential that the loop enter and exit the urine sample in a vertical fashion, held perpendicular to the bottom of the specimen container to ensure proper sample loading in the loop. Any deviation from this position can make a difference of 100% in the amount of urine delivered to the surface of the agar plate (1). After the loop is removed from the urine, the sample is inoculated down the middle of the plate in a single streak from which additional spreading of the inoculum is performed. This procedure is repeated for each culture medium used.

Routine urine specimens should be plated onto a 5% sheep blood agar plate and either MacConkey or eosin-methylene blue agar. Additional types of culture plates may be added to this routine specimen if a particular situation warrants. For instance, a chocolate agar plate may be added to recover H. influenzae and Haemophilus parainfluenzae, which have been reported as rare causes of urinary tract infection, particularly in men (2). Likewise, an anaerobic agar plate may also be inoculated with urine collected by suprapubic bladder aspiration to detect the presence of an anaerobe as a rare but possible cause of urinary disease.

Patients who have persistent pyuria but sterile urine are possible candidates for uncommon causes of urinary tract infection, such as anaerobes or Mycobacterium tuberculosis. The specimen of choice for the laboratory diagnosis of an anaerobic urinary infection is a suprapubic bladder aspirate. This is also the preferred method for collecting urine from infants and from patients in whom the interpretation of results of voided specimens is difficult. Such samples should be cultured for aerobic and anaerobic bacteria as described in the previous section. All other types of urine samples (i.e., clean catch, midstream, catheterized, etc.) are unacceptable for anaerobic culture because they are invariably contaminated by the anaerobic flora of the urethra during specimen collection, which makes subsequent interpretation of the results questionable. In cases of suspected renal tuberculosis, acid-fast culture should be performed on the complete, first-voided, clean-catch morning urine sample. Twenty-four-hour pooled urine specimens are not recommended for acid-fast culture because they are more likely to be contaminated and to contain fewer numbers of viable tubercle bacilli than the first-voided sample.

Urine specimens may be used for the detection of Chlamydia trachomatis and for cultivation of N. gonorrhoeae in patients with or without urethritis, even though the organisms do not cause urinary tract infection. Such a sample is particularly useful in the asymptomatic patient who has a "dry" urethra, which makes difficult the collection of a quality urethral swab sample. The first 5 to 10 ml of clean-catch urine voided is the specimen of choice. Upon arrival in the laboratory, the sample should be centrifuged immediately, and the sediment should be used to inoculate a selective medium for the cultivation of N. gonorrhoeae or a Chlamydia transport medium for antigen detection. Specimens for gonococcal culture must be delivered to the laboratory promptly and processed immediately, because the organism does not survive for long in urine.

Individuals who have a history of occupational exposure to animals or who have been exposed to stagnant water may present with symptoms suggestive of leptospirosis. Urine is one of the specimens of choice in the cultivation of this organism on Fletcher's semisolid medium. The reader is referred to chapter 50 of this Manual for the specific details of performing this cultural recovery.

REFERENCES

1. **Albers, A. C., and R. D. Fletcher.** 1983. Accuracy of the calibrated loop technique. *J. Clin. Microbiol.* **18:**40–42.
2. **Albritton, W. L., G. W. Hammon, and A. R. Ronald.** 1978. Bacteria *Haemophilus influenzae* genitourinary tract infection in adults. *Arch. Intern. Med.* **138:**1819–1821.
3. **Bartlett, J. G., N. S. Brewer, and J. J. Ryan.** 1978. *Cumitech 7, Laboratory Diagnosis of Lower Respiratory Tract Infections.* Coordinating ed., J. A. Washington II. American Society for Microbiology, Washington, D.C.
4. **Bobadilla, M., J. Sifuentes, and G. Garcia-Tsao.** 1989. Improved method for the bacteriological diagnosis of spontaneous bacterial peritonitis. *J. Clin. Microbiol.* **27:**2145–2147.
5. **Bonin, P., T. T. Tanino, and H. H. Handsfield.** 1984. Isolation of *Neisseria gonorrhoeae* on selective and nonselective media in a sexually transmitted disease clinic. *J. Clin. Microbiol.* **19:**218–220.
6. **Brun-Buisson, C., F. Abrouk, P. Legrand, Y. Huet, L. Larabi, and M. Rapin.** 1987. Diagnosis of central venous catheter-related sepsis. Critical level of quantitative tip cultures. *Arch. Intern. Med.* **147:**873–877.
7. **Buesching, W. J., R. A. Brust, and L. W. Ayers.** 1983. Enhanced primary isolation of *Legionella pneumophila* from clinical specimens by low-pH treatment. *J. Clin. Microbiol.* **17:**1153–1155.
8. **Centers for Disease Control.** 1993. Update: multistate outbreak of *Escherichia coli* O157:H7 infections from hamburgers—western United States, 1992–1993. *Morbid. Mortal. Weekly Rep.* **42:**258–263.
9. **Chapin-Robertson, K., S. E. Dahlberg, and S. C. Edberg.** 1992. Clinical and laboratory analysis of cytospin-prepared gram stains for recovery and diagnosis of bacteria from sterile body fluids. *J. Clin. Microbiol.* **30:**377–380.
10. **Chen, K. C. S., P. S. Forsyth, T. M. Buchanan, and K. K. Holmes.** 1979. Amine content of vaginal fluid from untreated and treated patients with nonspecific vaginitis. *J. Clin. Invest.* **63:**825–835.
11. **Cintron, F.** 1992. Initial processing, inoculation, and incubation of aerobic bacteriology specimens, p. 1.4.1–1.4.19. *In* H. D. Isenberg (ed.), *Clinical Microbiology Procedures Handbook*, vol. 1. American Society for Microbiology, Washington, D.C.
12. **Clarridge, J. E.** 1989. The recognition and significance of *Arcanobacterium haemolyticum. Clin. Microbiol. Newsl.* **11:**41–45.
13. **Collee, J. G., B. Watt, R. Brown, and S. Johnstone.** 1974. The recovery of anaerobic bacteria from swabs. *J. Hyg. Camb.* **72:**339–347.
14. **Collignon, P. G., N. Soni, I. Y. Pearson, W. P. Woods, R. Munro, and T. C. Sorrell.** 1986. Is semiquantitative culture of central vein catheter tips useful in the diagnosis of catheter-associated bacteremia? *J. Clin. Microbiol.* **24:**532–535.
15. **Cooper, G. L., and C. C. Hopkins.** 1985. Rapid diagnosis of intravascular catheter-associated infection by direct Gram staining of catheter segments. *N. Engl. J. Med.* **312:**1142–1147.
16. **Douard, M. C., G. Arlet, G. Leverger, R. Paulien, C. Maintrop, E. Clementi, B. Eurin, and G. Scharson.** 1991. Quantitative blood cultures for diagnosis and management of catheter-related sepsis in pediatric hematology and oncology patients. *Intensive Care Med.* **17:**30–35.
17. **Dykstra, M. A., J. C. McLaughlin, and R. C. Bartlett.** 1979. Comparison of media and techniques for detection of group A streptococci in throat swab specimens. *J. Clin. Microbiol.* **9:**236–238.
18. **Ebersole, L. L.** 1992. Acid-fast stain procedures, p. 3.5.1–3.5.11. *In* H. D. Isenberg (ed.), *Clinical Microbiology Procedures Handbook*, vol. 1. American Society for Microbiology, Washington, D.C.

19. **Ehret, J. M., and J. S. Knapp.** 1989. Gonorrhea. *Clin. Lab. Med.* **9:**445–480.

20. **Forbes, B. A., P. A. Frymoyer, R. T. Kopecky, J. M. Wojtaszek, and D. J. Pettit.** 1988. Evaluation of the lysis-centrifugation system for culturing dialysates from continuous ambulatory peritoneal dialysis patients with peritonitis. *Am. J. Kidney Dis.* **XI:**176–179.

21. **Gilligan, P. H., P. A. Gage, L. M. Bradshaw, D. V. Schidlow, and B. T. DeCicco.** 1985. Isolation medium for the recovery of *Pseudomonas cepacia* from respiratory secretions of patients with cystic fibrosis. *J. Clin. Microbiol.* **22:**5–8.

22. **Goodwin, C. S., E. Blincow, J. R. Warren, T. E. Watters, C. R. Sanderson, and L. Easton.** 1985. Evaluation of cultural techniques for isolating *Campylobacter pyloridis* from endoscopic biopsies or gastric mucosa. *J. Clin. Pathol.* **38:**1127–1131.

23. **Granato, P. A., L. Poe, and L. B. Weiner.** 1983. Use of New York City medium for enhanced recovery of *Mycoplasma pneumoniae* from clinical specimens. *J. Clin. Microbiol.* **17:**1077–1080.

24. **Gray, S. F., J. I. Wyatt, and B. J. Rathbone.** 1986. Letter. *J. Clin. Pathol.* **39:**1279.

25. **Griffin, P. M., and R. V. Tauxe.** 1991. The epidemiology of infections caused by *Escherichia coli* O157:H7, other enterohemorrhagic *E. coli*, and the associated hemolytic uremic syndrome. *Epidemiol. Rev.* **13:**60–98.

26. **Gross, P. A., L. M. Harkavey, G. E. Barden, and M. Kerstein.** 1974. Positive Foley catheter tip cultures—fact or fancy? *JAMA* **228:**72–73.

27. **Harris, J. C., H. L. DuPont, and R. B. Hornick.** 1972. Fecal leukocytes in diarrheal illness. *Ann. Intern. Med.* **76:**697–703.

28. **Heggers, J. P., M. C. Robson, and J. D. Ristroph.** 1969. A rapid method of performing quantitative wound cultures. *Milit. Med.* **134:**666–670.

29. **Henry, N. K., C. A. McLimans, A. J. Wright, R. L. Thompson, W. R. Wilson, and J. A. Washington II.** 1983. Microbiological and clinical evaluation of the ISOLATOR lysis-centrifugation blood culture tube. *J. Clin. Microbiol.* **17:**864–869.

30. **Hoffmann, S., A. M. Jensen, and T. Justesen.** 1983. The recovery of anaerobic bacteria from swabs in three transport systems. *Acta Pathol. Microbiol. Immunol. Scand. Sect. B* **91:**23–26.

31. **Huicho, L., D. Sanchez, M. Contreras, M. Paredes, H. Murga, L. Chinchay, and G. Guevara.** 1993. Occult blood and fecal leukocytes as screening tests in childhood infectious diarrhea: an old problem revisited. *Pediatr. Infect. Dis. J.* **12:**474–477.

32. **Isenberg, H. D., E. J. Baron, R. F. D'Amato, R. C. Johnson, P. R. Murray, F. G. Rodgers, and A. von Graevenitz.** 1991. Recommendations for the isolation of bacteria from clinical specimens, p. 216–221. *In* A. Balows, W. G. Hausler, Jr., K. L. Herrmann, H. D. Isenberg, and H. J. Shadomy (ed.), *Manual of Clinical Microbiology*, 5th ed. American Society for Microbiology, Washington, D.C.

33. **Jones, D. B., T. J. Liesegang, and N. M. Robinson.** 1981. *Cumitech 13, Laboratory Diagnosis of Ocular Infections*. Coordinating ed., J. A. Washington II. American Society for Microbiology, Washington, D.C.

34. **Kellogg, J. A., J. P. Mangella, and J. H. McConville.** 1984. Clinical laboratory comparison of the 10-mL Isolator blood culture system with BACTEC radiometric blood culture media. *J. Clin. Microbiol.* **20:**618–623.

35. **Kelly, M. T., F. W. Hickman-Brenner, and J. J. Farmer III.** 1991. *Vibrio*, p. 384–395. *In* A. Balows, W. J. Hausler, Jr., K. L. Herrmann, H. D. Isenberg, and H. J. Shadomy (ed.), *Manual of Clinical Microbiology*, 5th ed. American Society for Microbiology, Washington, D.C.

36. **Kiehn, T. E.** 1988. Isolators versus broth for blood cultures. *Clin. Microbiol. Newsl.* **10:**955–958.

37. **Kiehn, T. E., F. F. Edwards, P. Brannon, A. Y. Tsang, M. Maio, J. W. M. Gold, E. Whimbey, B. Wong, J. K. McClatch, and D. Armstrong.** 1985. Infections caused by M.

38. **Kim, K.-H., R. Fekety, D. H. Batts, D. Brown, M. Cudmore, J. Silva, Jr., and D. Waters.** 1981. Isolation of *Clostridium difficile* from the environment and contacts of patients with antibiotic-associated colitis. *J. Infect. Dis.* **143:**42–44.

39. **Kirkpatrick, M. B., and J. B. Bass, Jr.** 1989. Quantitative bacterial cultures of bronchoalveolar lavage fluids and protected brush catheter specimens from normal subjects. *Am. Rev. Respir. Dis.* **139:**546–548.

40. **Krizek, T. J., and M. C. Robson.** 1975. Evaluation of quantitative bacteriology in wound management. *Am. J. Surg.* **130:**579–584.

41. **Kruczak-Filipov, P., and R. G. Shively.** 1992. Gram stain procedure, p. 1.5.1–1.5.18. *In* H. D. Isenberg (ed.), *Clinical Microbiology Procedures Handbook*, vol. 1. American Society for Microbiology, Washington, D.C.

42. **Linares, J., A. Sitges-Serra, J. Garau, J. L. Perez, and R. Martin.** 1985. Pathogenesis of catheter sepsis: a prospective study with quantitative and semiquantitative cultures of catheter hub and segments. *J. Clin. Microbiol.* **21:**357–360.

43. **Lyerly, D. M., H. C. Krivan, and T. D. Wilkins.** 1988. *Clostridium difficile*: its disease and toxin. *Clin. Microbiol. Rev.* **1:**1–18.

44. **Maki, D. G.** 1983. Infections associated with intravascular lines, p. 309–363. *In* J. S. Remington and M. N. Swartz (ed.), *Current Clinical Topics in Infectious Diseases*, vol. 3. McGraw-Hill Book Co., New York.

45. **Maki, D. G., C. E. Weiss, and H. W. Sarafin.** 1977. A semiquantitative culture method for identifying intravenous catheter-related infection. *N. Engl. J. Med.* **296:**1305–1309.

46. **Males, B. M., J. J. Walshe, L. Garringer, D. Koscinski, and D. Amsterdam.** 1986. Addi-Chek filtration, BACTEC, and 10-ml culture methods for the recovery of microorganisms from dialysis effluent during episodes of peritonitis. *J. Clin. Microbiol.* **23:**350–353.

47. **Mangels, J. I., M. E. Cox, and Lois H. Lindberg.** 1984. Methanol fixation: an alternative to heat fixation of smears before staining. *Diagn. Microbiol. Infect. Dis.* **2:**129–137.

48. **Mermel, L. A., and D. G. Maki.** 1993. Detection of bacteremia in adults: consequences of culturing an inadequate volume of blood. *Ann. Intern. Med.* **119:**270–272.

49. **Mosca, R., S. Curtas, B. Forbes, and M. M. Meguid.** 1987. The benefits of Isolator cultures in the management of suspected catheter sepsis. *Surgery* **102:**718–722.

50. **Murray, P. M., and D. M. Citron.** 1991. General processing of specimens for anaerobic bacteria, p. 488–504. *In* A. Balows, W. G. Hausler, Jr., K. L. Herrmann, H. D. Isenberg, and H. J. Shadomy (ed.), *Manual of Clinical Microbiology*, 5th ed. American Society for Microbiology, Washington, D.C.

51. **Murray, P. R., and J. A. Washington II.** 1975. Microscopic and bacteriologic analysis of expectorated sputum. *Mayo Clin. Proc.* **50:**339–344.

52. **Nugent, R. P., M. A. Krohn, and S. L. Hillier.** 1991. Reliability of diagnosing bacterial vaginosis is improved by a standardized method of Gram stain interpretation. *J. Clin. Microbiol.* **29:**297–301.

53. **Pezzlo, M.** 1988. Detection of urinary tract infections by rapid methods. *Clin. Microbiol. Rev.* **1:**268–280.

54. **Putterman, C.** 1990. Central venous catheter related sepsis: a clinical review. *Resuscitation* **20:**1–16.

55. **Regan, J., and F. Lowe.** 1977. Enrichment medium for the isolation of *Bordetella*. *J. Clin. Microbiol.* **6:**303–309.

56. **Rein, M. F.** 1990. Vulvovaginitis and cervicitis, p. 953–965. *In* G. L. Mandell, R. G. Douglas, Jr., and J. E. Bennett (ed.), *Principles and Practice of Infectious Diseases*, 3rd ed. Churchill Livingstone, New York.

avium complex in immunocompromised patients: diagnosis by blood culture and fecal examination, antimicrobial sensitivity tests, and morphological and seroagglutination characteristics. *J. Clin. Microbiol.* **21:**168–173.

57. **Reinhold, C. E., D. J. Nickolai, T. E. Piccinini, B. A. Byford, M. K. York, and G. F. Brooks.** 1988. Evaluation of broth media for routine culture of cerebrospinal and joint fluid specimens. *Am. J. Clin. Pathol.* **89:**671–674.

58. **Robson, M. C., and J. P. Heggers.** 1969. Bacterial quantitation of open wounds. *Milit. Med.* **134:**19–24.

59. **Rubin, L. G., and J. Baranowski.** 1985. Comparison of the duPont Isolator 1.5 microbial tube and Trypticase soy broth for the recovery of *Haemophilus influenzae* type b in experimental bacteremia. *J. Clin. Microbiol.* **22:**815–818.

60. **Runyon, B. A., M. R. Antillon, E. A. Akriviadis, and J. G. McHutchison.** 1990. Bedside inoculation of blood culture bottles with ascitic fluid is superior to delayed inoculation in the detection of spontaneous bacterial peritonitis. *J. Clin. Microbiol.* **28:**2811–2812.

61. **Runyon, B. A., E. T. Umland, and T. Merlin.** 1987. Inoculation of blood culture bottles with ascitic fluid: improved detection of spontaneous bacterial peritonitis. *Arch. Intern. Med.* **147:**73–75.

62. **Ryan, S., and S. Fessia.** 1987. Improved method for the recovery of peritonitis-causing microorganisms from peritoneal dialysate. *J. Clin. Microbiol.* **25:**383–384.

63. **Salminen, I., R. von Essen, K. Koota, and A. Nissinen.** 1984. A pitfall in purulent arthritis brought out in *Kingella kingae* infection of the knee. *Ann. Rheum. Dis.* **43:**656–657.

64. **Salzman, M. B., H. D. Isenberg, J. F. Shapiro, P. J. Lipsitz, and L. G. Rubin.** 1993. A prospective study of the catheter hub as the portal of entry for microorganisms causing catheter-related sepsis in neonates. *J. Infect. Dis.* **167:**487–490.

65. **Slater, L. N., D. F. Welch, D. Hensel, and D. W. Coody.** 1990. A newly recognized fastidious gram-negative pathogen as a cause of fever and bacteremia. *N. Engl. J. Med.* **323:** 1587–1593.

66. **Spiegel, C. A.** 1983. Diagnosis of bacterial vaginosis by direct Gram stain of vaginal fluid. *J. Clin. Microbiol.* **18:**170–177.

67. **Tablan, O. C., W. J. Martone, C. F. Doershuk, R. C. Stern, M. J. Thomassen, J. D. Klinger, J. W. White, L. A. Carson, and W. R. Jarvis.** 1987. Colonization of the respiratory tract with *Pseudomonas cepacia* in cystic fibrosis. Risk factors and outcomes. *Chest* **91:**527–532.

68. **Thomassen, M. J., C. A. Demko, J. D. Klinger, and R. C. Stern.** 1985. *Pseudomonas cepacia* colonization among patients with cystic fibrosis: a new opportunist. *Am. Rev. Respir. Dis.* **131:**791–796.

69. **Tilton, R. C.** 1982. The laboratory approach to the detection of bacteremia. *Annu. Rev. Microbiol.* **36:**467–493.

70. **Van Scoy, R. E.** 1977. Bacterial sputum cultures: a clinician's viewpoint. *Mayo Clin. Proc.* **52:**39–41.

71. **Von Essen, R., and A. Holtta.** 1986. Improved method of isolating bacteria from joint fluids by the use of blood culture bottles. *Ann. Rheum. Dis.* **45:**454–457.

72. **Vontver, L. A., and D. A. Eschenbach.** 1981. The role of *Gardnerella vaginalis* in non-specific vaginitis. *Clin. Obstet. Gynecol.* **24:**439–460.

73. **Washington, J. A., and D. M. Ilstrup.** 1986. Blood cultures: issues and controversies. *Rev. Infect. Dis.* **8:**792–802.

74. **Welch, D., R. K. Scribner, and D. Hensel.** 1985. Evaluation of a lysis direct plating method for pediatric blood cultures. *J. Clin. Microbiol.* **21:**955–958.

75. **Welch, D. F., M. J. Muszynski, C. H. Pai, M. J. Marcon, M. M. Hribar, P. H. Gilligan, J. M. Matsen, P. A. Ahlin, B. C. Hilman, and S. A. Chartrand.** 1987. Selective and differential medium for recovery of *Pseudomonas cepacia* from the respiratory tracts of patients with cystic fibrosis. *J. Clin. Microbiol.* **25:**1730–1734.

76. **Weldon, D. B., and M. Slack.** 1977. Multiplication of contaminant bacteria in urine and interpretation of delayed culture. *J. Clin. Pathol.* **30:**615–619.

77. **Werner, A. S., C. G. Cobbs, D. Kaye, and E. W. Hook.** 1967. Studies on the bacteremia of bacterial endocarditis. *JAMA* **202:**199–203.

78. **Whimbey, E., B. Wong, T. E. Kiehn, and D. Armstrong.** 1984. Clinical correlations of serial quantitative blood cultures determined by lysis-centrifugation in patients with persistent septicemia. *J. Clin. Microbiol.* **19:**766–771.

79. **Wilson, M. L., T. E. Davis, S. Mirrett, J. Reynolds, D. Fuller, S. D. Allen, K. K. Flint, F. Koontz, and L. B. Reller.** 1993. Controlled comparison of the BACTEC high-blood-volume fungal medium, BACTEC Plus 26 aerobic blood culture bottle, and 10-milliliter Isolator blood culture system for detection of fungemia and bacteremia. *J. Clin. Microbiol.* **31:**865–871.

80. **Wilson, S. J., M. L. Wilson, and L. B. Reller.** 1993. Diagnostic utility of postmortem blood cultures. *Arch. Pathol. Lab. Med.* **117:**986–988.

81. **Woods, G. L., and J. A. Washington.** 1987. Comparison of methods for processing dialysate in suspected continuous ambulatory peritoneal dialysis-associated peritonitis. *Diagn. Microbiol. Infect. Dis.* **7:**155–157.

82. **Yrios, J. W., E. Balish, A. Helstad, C. Field, and S. Inhorn.** Survival of anaerobic and aerobic bacteria on cotton swabs in three transport systems. *J. Clin. Microbiol.* **1:**196–200.

Staphylococcus and *Micrococcus*

WESLEY E. KLOOS AND TAMMY L. BANNERMAN

22

TAXONOMY

Members of the genera *Staphylococcus* and *Micrococcus* are catalase-positive, gram-positive cocci and are placed with *Stomatococcus* and *Planococcus* in the family Micrococcaceae (137). The family is not a phylogenetically coherent group (148). The results of DNA base composition (144), DNA-rRNA hybridization (74), and comparative oligonucleotide cataloging of 16S rRNA (98) studies have indicated that the genera *Staphylococcus* and *Micrococcus* are not closely related. The genus *Staphylococcus* is most closely related to the genera *Enterococcus*, *Bacillus*, *Listeria*, *Planococcus*, and *Brochothrix* and belongs to the broad *Bacillus-Lactobacillus-Streptococcus* cluster (99, 147). The genus *Micrococcus*, on the other hand, is most closely related to the genus *Arthrobacter* of the coryneform group (148).

DESCRIPTION OF THE GENERA

Members of the genus *Staphylococcus* are gram-positive cocci (0.5 to 1.5 μm in diameter) that occur singly and in pairs, tetrads, short chains (three or four cells), and irregular grapelike clusters. Ogston (117) introduced the name "staphylococcus" (from *staphylé*, a bunch of grapes) for the group of micrococci causing inflammation and suppuration. Rosenbach (135) used the term in a taxonomic sense and provided the first description of the genus *Staphylococcus*. Staphylococci are nonmotile, non-spore forming, and usually catalase positive, and they are usually unencapsulated or have limited capsule formation. Most species are facultative anaerobes. Except for *Staphylococcus saccharolyticus* and *S. aureus* subsp. *anaerobius*, their growth is more rapid and abundant under aerobic conditions. These exceptional organisms are also catalase negative. Some uncommon strains of staphylococci may require the presence of CO_2 or other metabolites (hemin, menadione, etc.) or a hypertonic medium for growth.

Members of the genus *Micrococcus* are gram-positive cocci (0.5 to 2.0 μm in diameter) that occur mostly in pairs, tetrads, and irregular clusters. Cohn (33) introduced the genus *Micrococcus* to represent small spherical bacteria such as staphylococci, micrococci, and streptococci as well as some other groups. Evans et al. (42) proposed separating staphylococci from micrococci on the basis of their relation to oxygen. The facultative cocci were placed in the genus

Staphylococcus, and the obligate aerobes were placed in the genus *Micrococcus*. Although we now know that there are some exceptions to this proposal, it aided in placing staphylococcal and micrococcal systematics on a fruitful course. By the mid-1960s, a clear distinction could be made between staphylococci and micrococci on the basis of their DNA base compositions (144). Members of the genus *Staphylococcus* have a DNA G+C content of 30 to 39 mol%, whereas members of the genus *Micrococcus* have a G+C content within the range of 63 to 73 mol%. The cell walls of staphylococci contain peptidoglycan and teichoic acid (136). The diamino acid present in the peptidoglycan is L-lysine, and the interpeptide bridge of the peptidoglycan consists of oligoglycine peptides (susceptible to the action of lysostaphin). Depending on the species and the relative amount of glycine present in the growth medium, some glycine residues may be substituted with L-serine or L-alanine. Cell wall teichoic acids of staphylococci may be poly(polyolphosphate), poly(glycerolphosphate-glycosylphosphate), or poly(glycosylphosphate), depending on the species. Glycerol or ribitol or both occur as typical components of poly(polyolphosphate) teichoic acids. Substituents of these teichoic acids may include *N*-acetylgalactosamine, glucose, and *N*-acetylglucosamine. Members of the genus *Micrococcus* have no glycine in the interpeptide bridge of the peptidoglycan and do not have teichoic acids. The respiratory chains of staphylococci and micrococci differ in cytochrome and menaquinone composition. Most staphylococci contain only *a*- and *b*-type cytochromes, whereas micrococci also have *c*- and *d*-type cytochromes. The exceptional species *S. caseolyticus*, *S. lentus*, *S. sciuri*, and *S. vitulus* contain *a*-, *b*-, and *c*-type cytochromes. Staphylococci contain unsaturated polyisoprenoid side chains in their menaquinones, whereas micrococci have hydrogenated menaquinones. In the clinical laboratory, staphylococci can be easily distinguished from micrococci on the basis of the former's resistance to bacitracin and susceptibility to furazolidone (10). The main characteristics used for differentiating staphylococci from micrococci and other gram-positive cocci encountered in the clinical laboratory are listed in Table 1.

The genus *Staphylococcus* is currently composed of 32 species (15, 28, 59, 66, 86, 89, 154, 167), as depicted in Table 2.

The genus *Micrococcus* is currently composed of nine

TABLE 1 Differentiation of members of the genus *Staphylococcus* from other gram-positive cocci[a]

Columns grouped under "Growth on:" are 5% NaCl agar, 6.5% NaCl agar, 12% NaCl agar, and P agar in 18 h[b]. Columns grouped under "Resistance to:" are Lysostaphin, Erythromycin, Bacitracin, and Furazolidone.

Genus and exceptional species	Mol% G+C of DNA	Strict aerobe	Facultative anaerobe or microaerophile	Strict anaerobe	Tetrad cell arrangement	Strong adherence on agar	Motility	5% NaCl agar	6.5% NaCl agar	12% NaCl agar	P agar in 18 h[b]	Catalase[c]	Benzidine test[d]	Modified oxidase test[e]	Anaerobic acid from glucose[f]	Aerobic acid from glycerol	Growth on Schleifer-Krämer agar[g]	Lysostaphin (200 µg/ml)	Erythromycin (0.04 µg/ml)	Bacitracin (0.04 U)[h]	Furazolidone (100 µg)[i]
Staphylococcus	30–39	–	d	–	d	–[j]	–	+	+	d	+	+	+	–	d	+	+	–	+	ND	–
S. aureus subsp. anaerobius		–	±	±	–	–	–	+	+	d	–	–	±	–	+	+	+	–	+	ND	–
S. saccharolyticus		–	±	±	+	–	–	+	+	±	–	±	–	±	+	+	+	–	+	+	–
S. hominis		–	±[k]	±	–	–	–	+	+	±	–	+	+	–	+	+	ND	–	+	+	–
S. auricularis		–	+	–	–	+	–	+	+	±	+	+	+	–	–	+	ND	–	+	+	–
S. saprophyticus, S. cohnii, S. xylosus		d	d	–	–	–	–	+	+	±	+	+	+	–	–	+	+	–	+	+	–
S. kloosii, S. equorum, S. arlettae		±	±	–	–	–	–	+	+	±	d	+	+	–	–	+	+	–	+	+	–
S. intermedius		–	+	–	–	–	–	+	+	+	+	+	+	–	+	+	±	–	+	+	–
S. caseolyticus, S. sciuri, S. lentus, S. vitulus		±	±	–	d	–	–	+	+	d	d	+	+	+	–	+	+	–	+	+	–
Enterococcus	34–42	–	+	–	–	–	d	+	+	(±)	±	–	–	–	+	d	(±)	+	+	+	–
Streptococcus	34–46	–	+	d	–	–	–	d	d	–	–	–	–	–	+	d		+	ND	d	–
Aerococcus	35–40	–	+	–	+	–	–	+	+	+	–	+	–	–	(+)	ND	ND	+	ND	–	–
Planococcus	39–52	+	–	–	d	–	+	+	+	+	–	+	+	ND	–	+	–	+	ND	ND	–
Stomatococcus	56–60	–	+	–	d	+	–					±			–	+	d	+	ND	ND	d
Micrococcus	66–75	+	–	–	+	–	–	+	+	d	+	+	+	+	–	–	–	+[l]	–[m]	–	+
M. kristinae		±	±	–	+	–	–	+	+	±	–	+	+	+	(+)	+	(±)	+	+	–	+
M. agilis		+	–	–	+	–	d	+	±	–	–	+	+	+	–	–	–	d	–	–	+

[a] The genera *Staphylococcus*, *Planococcus*, *Stomatococcus*, and *Micrococcus* are included in the family *Micrococcaceae* (Prévot). The genera *Enterococcus*, *Streptococcus*, and *Aerococcus* are also included in this table, as species from these genera may be found together with staphylococci in clinical specimens and may share certain properties with them. Exceptional species of these last three genera have not been singled out for comparisons. Symbols: +, 90% or more species or strains positive; ±, 90% or more species or strains weakly positive; –, 90% or more species or strains negative; d, 11 to 89% of species or strains positive; ND, not determined. Parentheses indicate a delayed reaction.

[b] Growth on P agar is under aerobic conditions and at 35 to 37°C. Positive growth is indicated for detectable colony formation of at least 1-mm diameter; ± indicates detectable colony formation of between 0.5- and 1-mm diameter. Growth on sheep or bovine blood agar is slightly greater but less discriminative between staphylococci and other genera.

[c] Sometimes a weak catalase or pseudocatalase reaction can be observed in certain strains of species designated catalase negative (136). In some species, catalase activity may be activated by hemin supplementation.

[d] Benzidine test detects the presence of cytochromes. Some strains of benzidine test-negative species can synthesize cytochromes on aerobic media supplemented with hemin.

[e] See reference 137.

[f] Standard oxidation-fermentation test (42, 137).

[g] Growth (KRAN-agar supplemented with sodium azide, potassium thiocyanate, lithium chloride, and glycine) (139) is under aerobic conditions and at 35 to 37°C for 24 to 48 h. Positive growth is indicated for a number of CFU on selective medium comparable to that on plate count agar and a colony 0.5 mm in diameter; ± indicates a significant reduction in CFU on selective medium compared to that on plate count agar, and parentheses indicate a colony of pinpoint size to 0.5-mm diameter.

[h] Disk is used. Positive indicates resistance and no zone of inhibition. *Micrococcus*, *Stomatococcus*, and *Aerococcus* spp. are susceptible and have inhibition zones 10 to 25 mm in diameter (10).

[i] Disk is used. Positive indicates resistance and from no zone of inhibition to 9 mm. Susceptible species have inhibition zones 15 to 35 mm in diameter (10).

[j] Some strains of *S. epidermidis* adhere tenaciously to the surface of agar, and this property is correlated with heavy slime production.

[k] *S. hominis* does not demonstrate growth in the anaerobic portion of a thioglycolate medium (85) within 24 h and may produce only very poor growth in this portion following 3 to 5 days of incubation. However, it will grow and ferment glucose anaerobically (standard oxidation-fermentation test). Failure to grow anaerobically in thioglycolate may be due in part to inhibition by certain of the ingredients.

[l] Some strains of *M. luteus*, *M. roseus*, *M. agilis*, and *M. sedentarius* demonstrate susceptibility to lysostaphin, presumably because of contaminating levels of endo-β-N-acetylglucosaminidase activity in some commercial sources.

[m] A few *Micrococcus* strains demonstrate a high level (MIC, ≥ 50 µg/ml) of erythromycin resistance.

species (91): *Micrococcus luteus, M. varians, M. roseus, M. lylae, M. kristinae, M. sedentarius, M. nishinomiyaensis, M. agilis,* and *M. halobius.*

NATURAL HABITATS

Staphylococci are widespread in nature, though they are mainly found living on the skin, skin glands, and mucous membranes of mammals and birds. They are sometimes found in the mouth, blood, mammary glands, and intestinal, genitourinary, and upper respiratory tracts of these hosts.

Staphylococci found on humans and other primates include *S. aureus, S. epidermidis, S. capitis, S. caprae, S. saccharolyticus, S. warneri, S. pasteuri, S. haemolyticus, S. hominis, S. lugdunensis, S. auricularis, S. saprophyticus, S. cohnii, S. xylosus,* and *S. simulans* (13, 28, 76, 77, 84, 138). Some of these species are transients on domestic animals. *S. aureus* is found on various other mammals and may be represented on some host species by different ecovars (108). *S. schleiferi, S. intermedius,* and *S. felis* are commonly found living on carnivora (66, 86). However, *S. schleiferi* may produce serious infections in humans (45, 69), and *S. intermedius* may produce infections in humans as a result of dog bites (152). *S. hyicus, S. chromogenes, S. sciuri, S. lentus, S. vitulus,* and *S. caseolyticus* are common residents of ungulates and their dairy and meat products (11, 39, 79, 86, 129).

Some *Staphylococcus* species demonstrate habitat or niche preferences on their particular hosts (77). For example, *S. capitis* is found as large populations on the adult human head, especially the scalp and forehead, where sebaceous glands are numerous and well developed. *S. auricularis* has a strong preference for the external auditory meatus. *S. hominis* and *S. haemolyticus* generally produce larger populations in areas of the skin where apocrine glands are numerous, such as the axillae and pubic areas. *S. aureus* prefers the anterior nares as a habitat, especially in the adult human.

Certain *Staphylococcus* species are found frequently as etiologic agents of a variety of human and animal infections. In this chapter, we will be concerned primarily with the identification of *S. aureus, S. epidermidis, S. haemolyticus, S. lugdunensis,* and *S. saprophyticus,* species most commonly associated with human infections, and *S. intermedius* and *S. hyicus,* species of special veterinary interest. *S. schleiferi* has been considered a significant pathogen in some European countries but has seldom been isolated from patients with infections in the United States.

Micrococci are widespread in nature and, like staphylococci, are commonly found on the skin of humans and other mammals (82, 87, 91). *M. luteus* and *M. varians* are the predominant species found on humans. *M. varians* is the predominant species living on other mammals. Micrococci have been occasionally isolated from fish, shellfish, algal blooms, natural waters, sand, and plant products (91). *M. kristinae* has been isolated from spoiled beer producing a fruity odor, and *M. varians* has been used together with pediococci as a starter culture in the processing of fermented meats (91).

CLINICAL SIGNIFICANCE

Staphylococcus Species

The coagulase-positive species *S. aureus* is well documented as a human opportunistic pathogen. As a nosocomial

pathogen, *S. aureus* has been a major cause of morbidity and mortality. *S. aureus* infections are often acute and pyogenic and, if untreated, may spread to surrounding tissue or, via bacteremia, to metastatic sites (involving other organs). Some of the infections caused by *S. aureus* involve the skin and include furuncles or boils, cellulitis, impetigo, scalded skin syndrome, and postoperative wound infections of various sites. Other major infections produced by *S. aureus* include bacteremia, pneumonia, osteomyelitis, acute endocarditis, myocarditis, pericarditis, cervicitis, cerebritis, meningitis, and abscesses of the muscle, urogenital tract, central nervous system, and various intra-abdominal organs. Food poisoning is often attributed to staphylococcal enterotoxin.

Methicillin-resistant *S. aureus* (MRSA) emerged in the 1980s as a major clinical and epidemiologic problem in hospitals (24). The widespread nature of this problem continues into the early 1990s (119). It has been suggested that hospitals of all sizes are facing the MRSA problem. Two recent reviews have discussed several means of management or control of the spread of MRSA (25, 113).

A community-acquired disease of potentially serious consequence, toxic shock syndrome (TSS), has also been attributed to infection or colonization with *S. aureus* (140), with a single clone causing the majority of cases (114). TSS is prevalent in young, menstruating females who use certain types of highly absorbent tampons (159). TSS-associated nongenital *S. aureus* has been described also in men and nonmenstruating women. TSS is associated with strains that produce and secrete the exotoxin toxic shock syndrome toxin 1 (TSST-1) (140). TSST-1 is a member of a superantigen family that has the ability to stimulate T cells (29). Aside from being a T-cell stimulator, TSST-1 is also a potent inducer of tumor necrosis factor (121) and the cytokine interleukin-1 (122). These two agents are thought to play an important role in TSS pathogenesis (109). Methods for recognizing TSST-1 production include radioimmunoassay, enzyme-linked immunosorbent assay (Toxin Technology, Inc., Madison, Wis.), and reverse passive latex agglutination (Oxoid USA, Columbia, Md.) kits.

The coagulase-negative *Staphylococcus* (CoNS) species as a group constitute a major component of the normal microflora of the human. Over the last two decades, there has been an increase in the documentation of infections due to CoNS, especially with the species *S. epidermidis.* The infection rate has been correlated with the increase in the use of prosthetic and indwelling devices and the growing number of immunocompromised patients in hospitals. Accurate identification of CoNS is needed, so that the clinical disease produced by this group of bacteria can be precisely delineated and the etiologic agent can be determined. A recent review of CoNS summarized results that are helpful in identifying the etiologic agent (79). These results included (i) the isolation of a strain in pure culture from the infected site or body fluid (most contaminated clinical specimens produce mixed cultures of different strains and/or species; however, some infections may be the consequence of more than one strain or species) and (ii) the repeated isolation of the same strain or combination of strains over the course of the infection. *S. epidermidis* has been isolated from 74 to 92% of patients with hospital-acquired CoNS bacteremia (106). Cardiac infections caused by *S. epidermidis* have occurred after procedures such as cardiac valve and cardiovascular surgery and cardiotomy (7, 9, 90, 97). *S. epidermidis* has also been implicated as the etiologic agent in infections of cerebrospinal fluid shunts (50), prosthetic joints (26), and orthopedic devices (31). *S. epidermidis* is a

TABLE 2 Differentiation of *Staphylococcus* species

Species	Colony size (large)[b]	Colony pigment[c]	Anaerobic growth[d]	Aerobic growth[e]	Staphylocoagulase	Clumping factor[f]	Heat-stable nuclease	Hemolysins[g]	Catalase[h]	Oxidase[i]	Alkaline phosphatase	Arginine arylamidase	Pyrrolidonyl arylamidase[j]	Ornithine decarboxylase	Urease[j]	β-Glucosidase[j]	β-Glucuronidase[j]	β-Galactosidase[j]	Arginine utilization[j]	Acetoin production	Nitrate reduction	Esculin hydrolysis	Novobiocin resistance[k]	Polymyxin B resistance[l]	D-Trehalose	D-Mannitol	D-Mannose	D-Turanose	D-Xylose	D-Cellobiose	L-Arabinose	Maltose	α-Lactose	Sucrose	N-Acetylglucosamine	Raffinose
S. aureus subsp. *aureus*	+	+	+	+	+	+	+	+	+	−	+	ND	ND	−	d	+	−	−	+	+	+	−	−	+	+	ND	+	+	−	−	−	+	+	+	+	−
S. aureus subsp. *anaerobius*	−	−	(+)	(±)	+	+	+	+	−	−	+w	ND	ND	ND	ND	−	−	−	ND	−	−	−	−	ND	+	−	(+)	ND	−	−	−	+	+	+	−	−
S. epidermidis	−	−	+	+	−	−	+	(d)	+	−	+	−	−	ND	+	(d)	−	−	d	d	+	−	−	+	+	+	+	(d)	−	−	−	+	d	+	+	−
S. capitis subsp. *capitis*	−	−	(+)	+	−	−	−	(d)	+	−	+	−	(d)	(d)	−	−	−	−	d	d	d	−	−	ND	−	+	+	−	−	−	−	+	d	(+)	−	−
S. capitis subsp. *ureolyticus*	d	(d)	(+)	(±)	−	−	+	(d)	+	−	(+)	−	d	−	+	(d)	d	−	d	+	+	−	−	−	−	+	+	d	−	−	−	(d)	(d)	(+)	d	−
S. caprae	−	−	+	+	−	−	+	(d)	+	−	d	−	ND	ND	+	+	+	ND	+	+	+	−	−	ND	+	d	+	ND	−	−	−	+	+	+	ND	−
S. saccharolyticus	−	d	(+)	−	−	−	−	−	−	−	−	−	ND	ND	+	ND	+	ND	d	ND	d	−	−	ND	(+)	d	(+)	ND	−	−	−	(+)	d	−	ND	−
S. warneri	−	d	+	+	−	−	−	(d)	+	−	−	−	+	ND	+	+	d	(+)	d	d	d	−	−	d	+	d	+	(d)	−	−	−	(d)	d	+	−	−
S. pasteuri[n]	−	d	+	+	−	−	−	(d)	+	−	−	−	d	ND	+	+	d	ND	d	d	d	−	−	d	+	d	+	(d)	−	−	−	(d)	d	+	d	−
S. haemolyticus	+	d	+	+	−	−	+	(+)	+	−	−	−	+	−	−	+	−	−	+	+	+	−	−	−	+	d	+	(d)	−	−	−	+	d	+	+	−
S. hominis	−	d	+	+	−	−	−	(d)	+	−	−	−	d	−	+	d	−	−	−	+	d	−	−	d	d	d	+	(d)	−	−	−	d	d	+	−	−
S. lugdunensis	+	−	+	+	−	(+)	−	(+)	+	−	+	+	+	−	d	+	+	(+)	+	+	+	−	−	d	+	d	+	+	−	−	−	+	+	+	+	+
S. schleiferi subsp. *schleiferi*	−	−	+	+	−	+	+	(+)	+	−	+	−	+	−	−	+	−	(+)	+	+	+	−	−	ND	−	−	+	−	−	−	−	−	−	−	(+)	−
S. schleiferi subsp. *coagulans*	d	−	(+)	+	+	+	+	(+)	+	−	+	−	+	−	+	+	−	ND	+	+	+	−	−	ND	+	−	+	−	−	−	−	−	−	(±)	ND	−
S. muscae	−	d	−	+	−	−	−	−	+	−	+	−	d	ND	+	d	−	ND	−	+	+	−	−	ND	−	+	+	(d)	−	−	−	−	−	+	ND	−
S. auricularis	−	−	(d)	+	−	−	−	−	+	−	−	−	−	ND	−	d	−	−	−	−	−	−	−	ND	+	−	+	+	−	−	−	(+)	−	(±)	d	−
S. saprophyticus	−	d	+	+	−	−	−	(d)	+	−	−	−	−	ND	+	+	−	+	−	d	−	d	+	−	+	d	(d)	d	+	(d)	−	(d)	+	+	d	−
S. cohnii subsp. *cohnii*	d	d	(d)	+	−	−	−	(d)	+	−	d	−	−	ND	−	−	−	+	−	d	d	d	+	ND	+	d	−	d	+	+	+	(d)	d	d	d	−
S. cohnii subsp. *urealyticum*	+	d	+	+	−	−	−	−	+	−	d	−	−	ND	+	d	−	d	−	d	+	d	+	ND	+	d	+	d	+	+	+	(+)	d	+	+	ND
S. xylosus	+	+	(+)	(+)	−	−	−	(d)	+	−	(+)	−	d	ND	+	d	−	d	−	d	+	d	+	ND	+	+	+	−	+	+	+	(±)	+	+	+	ND
S. kloosii	+	d	(d)	+	−	−	−	(d)	+	−	(+)	−	−	ND	+	ND	−	−	−	−	+	−	+	ND	+	d	+	ND	−	−	−	d	d	+	ND	+
S. equorum	+	d	(+)	+	−	−	−	−	+	−	(d)	−	d	ND	+	ND	−	−	−	−	+	−	+	ND	+	d	+	p	−	−	−	+	d	+	ND	+
S. arlettae	+	+	+	+	−	−	−	(d)	+	−	+	−	d	−	+	d	−	(d)	−	−	+	−	+	ND	+	+	+	−	(d)	−	−	d	+	+	+	−
S. gallinarum	+	d	(+)	+	−	−	−	d	+	−	+	−	d	−	+	ND	−	−	d	−	+	ND	+	ND	+	+	+	d	−	−	−	(±)	+	+	+	−
S. simulans	+	−	+	+	−	−	d	(d)	+	−	+	+	−	−	+	+	−	−	+	+	+	−	−	−	+	+	+	−	(d)	−	−	−	d	+	+	−
S. carnosus	+	−	+	+	−	−	d	−	+	−	+	d	−	−	−	+	−	−	+	−	+	−	−	−	+	+	+	d	−	−	−	d	d	+	ND	ND
S. piscifermentans	+	−	+	+	−	−	−	(d)	+	−	+	−	−	−	−	−	−	−	+	+	+	ND	−	ND	−	d	+	−	−	−	−	d	d	+	+	ND
S. felis	+	−	−	(+)	−	−	+	d	+	−	+	−	−	−	+	ND	−	(d)	+	−	+	ND	−	+	−	(d)	−	ND	−	ND	−	−	d	−	ND	−
S. intermedius	+	−	+	+	+	−	+	+	+	−	+	−	d	−	+	ND	ND	ND	d	−	+	ND	−	ND	+	+	+	p	−	−	−	(±)	+	+	ND	−
S. delphini	+	−	+	+	+	−	+	d	+	−	+	−	p	ND	+	d	−	ND	d	−	+	−	−	−	+	+	+	ND	−	ND	−	−	+	+	ND	−
S. hyicus	+	−	+	+	d	d	ND	+	+	−	+	−	d	ND	+	−	+	ND	+	−	+	−	−	−	d	d	+	p	−	−	−	−	d	d	ND	−
S. chromogenes	+	+	+	+	−	−	+	+	+	−	+	−	−	ND	+	ND	−	(d)	+	−	+	ND	−	ND	+	+	+	ND	−	−	−	(±)	+	+	ND	−
S. caseolyticus	−	d	(±)	+	−	−	−	−	+	+	+	ND	−	ND	+	d	+	ND	−	−	+	+	−	−	+	+	+	(±)	(d)	+	d	(±)	(d)	d	d	+
S. sciuri	+	d	(±)	+	−	−	ND	p	+	+	+	ND	p	ND	+	d	−	ND	−	−	+	ND	+	−	p	+	(d)	(±)	(d)	+	d	−	(d)	+	d	−
S. lentus	−	d	(±)	(+)	−	−	−	−	+	+	+	ND	−	ND	+	+	−	ND	−	−	+	d	+	ND	p	+	(+)	(±)	(±)	+	d	(d)	(d)	+	d	+
S. vitulus	−	+	+	(+)	−	−	−	d	+	+	−	ND	−	ND	d	p	d	−	−	−	+	d	−	ND	(d)	+	−	−	p	d	d	−	d	+	−	−

(Footnotes on next page)

(Footnotes to Table 2)

^aSymbols (unless otherwise indicated): +, 90% or more strains positive; ±, 90% or more strains weakly positive; −, 90% or more strains negative; d, 11 to 89% of strains positive; ND, not determined. Parentheses indicate a delayed reaction.

^bPositive is defined as a colony diameter of ≥6 mm after incubation on P agar at 34 to 35°C for 3 days and at room temperature (ca. 25°C) for an additional 2 days (84).

^cPositive is defined as the visual detection of carotenoid pigments (e.g., yellow, yellow-orange, or orange) during colony development at normal incubation or room temperatures. Pigments may be enhanced by the addition of milk, fat, glycerol monoacetate, or soaps to P agar.

^dIn a semisolid thioglycolate medium (85). Symbols: +, moderate or heavy growth down the tube within 18 to 24 h; ±, heavier growth in the upper portion of the tube and weaker growth in the lower, anaerobic portion of the tube; −, no visible growth within 48 h, but very weak diffuse growth or a few scattered, small colonies may be observed in the lower portion of tube by 72 to 96 h. Parentheses indicate delayed growth appearing within 24 to 72 h, sometimes noted as large discrete colonies in the lower portion of the tube.

^eOn P agar or bovine, sheep, or human blood agar at 34 to 37°C. *S. equorum* grows slowly at 35 to 37°C; its optimum growth temperature is 30°C. Anaerobic species *S. saccharolyticus* and *S. aureus* subsp. *anaerobius* grow very slowly in the presence of air. *S. aureus* subsp. *anaerobius* requires the addition of blood, serum, or egg yolk for growth on primary isolation medium. *S. auricularis*, *S. lentus*, and *S. vitulus* produce just detectable colonies on P agar in 24 to 36 h, and these colonies remain very small (1 to 2 mm in diameter).

^fDetected in rabbit or human plasma (slide coagulase test). Human plasma is preferred for the detection of clumping factor with *S. lugdunensis* and *S. schleiferi*. Latex agglutination is somewhat less reliable for detection of clumping factor or fibrinogen affinity factor for *S. lugdunensis*.

^gHemolysis on bovine blood agar (81). Symbols: +, wide zone of hemolysis within 24 to 36 h; (+), delayed moderate to wide zone of hemolysis within 48 to 72 h; (d), no or delayed hemolysis; −, no or only very narrow zone (≤1 mm) of hemolysis within 72 h. Some of the strains designated negative may produce a slight greening or browning of blood agar.

^hCatalase and cytochrome synthesis cannot be induced in *S. aureus* subsp. *anaerobius* by the addition of H₂O₂ or hemin to the culture medium. Catalase can be induced in *S. saccharolyticus* by hemin supplementation. In this species, cytochromes *a* and *b* are present in small quantities.

ⁱDetermined by the modified oxidase test to detect the presence of cytochrome *c*.

^jDetermined primarily by commercial rapid-identification tests.

^kPositive is defined as an MIC of ≥1.6 μg/ml or a growth inhibition zone diameter of ≤16 mm with a 5-μg novobiocin disk.

^lPositive is defined as a growth inhibition zone diameter of <10 mm with a 300-U polymyxin B disk.

^mAlkaline phosphatase activity is negative for approximately 6 to 15% of strains of *S. epidermidis*, depending on the population sampled. A low but significant number of clinical isolates have been phosphatase negative.

ⁿrRNA gene restriction site polymorphism using pBA2 as a probe can distinguish this species from other staphylococcal species, including *S. warneri* (28).

common agent in peritonitis during continuous ambulatory peritoneal dialysis (93). It has been reported in a few cases of osteomyelitis (118). Nosocomial methicillin-resistant *S. epidermidis* strains have become a serious clinical problem in the 1980s, especially in patients who have prosthetic heart valves or who have undergone other forms of cardiac surgery (7, 71).

S. saprophyticus is an important opportunistic pathogen in human urinary tract infections (UTIs), especially in young, sexually active females (105, 164). It has been proposed as an agent of nongonococcal urethritis in males or a cause of other sexually transmitted diseases (64) and prostatitis (20, 115). Isolation of the species from wound infections and septicemia has been noted (55).

Several other CoNS species have been implicated in human infections. *S. haemolyticus*, the second most frequently encountered species of this group, has been associated with various clinical infections. These include native valve endocarditis, septicemia, peritonitis, UTIs, and wound, bone, and joint infections (75, 79, 96, 106, 126, 163). *S. hominis*, *S. warneri*, *S. capitis*, *S. simulans*, *S. cohnii*, *S. xylosus*, and *S. saccharolyticus* have been associated with a variety of infections (9, 12, 23, 68, 70, 79, 103, 106, 126, 169, 171). Two relatively new species of CoNS, *S. lugdunensis* and *S. schleiferi*, appear to be significant opportunistic pathogens. *S. lugdunensis* has been implicated in native and prosthetic valve endocarditis, septicemia, brain abscesses, chronic osteoarthritis, and infections of soft tissues, bone, peritoneal fluid, and catheters (41, 45, 62, 143). *S. schleiferi* has been associated with brain empyema, wound infections, bacteremia and rachis osteitis, and infections of catheters (45, 69). In many cases, patients with infections caused by these CoNS had predisposing or underlying diseases affecting the immune system and had also experienced surgery or intravascular manipulations.

The coagulase-positive species *S. intermedius* and the coagulase-variable species *S. hyicus* are of particular importance in veterinary infections. *S. aureus* and these species are serious opportunistic pathogens of animals. *S. intermedius* has been associated with a variety of canine infections, including otitis externa, pyoderma, abscesses, reproductive tract infections, mastitis, and wound infections (40, 129, 134). *S. hyicus* has been implicated in infectious exudative epidermitis and septic polyarthritis in pigs and mastitis in cows (40, 128, 129).

CoNS are a major cause of foreign-body infections. A review of the processes involved in foreign-body infections caused by CoNS (79) suggests that the first step involves the nonspecific or specific adhesion of bacteria to biomaterials, and this step is followed by the production of slime (30, 44, 67, 124, 160). Clinical isolates of CoNS that display a polysaccharide-adhesion are generally more adherent to catheters in vitro and elaborate more biofilm (111, 112).

Micrococcus Species

Species of micrococci are generally considered harmless saprophytes that colonize the skin, mucosa, and oropharynx. However, *Micrococcus* spp. can be opportunistic pathogens in immunocompromised patients. *M. luteus* has been implicated as the causative agent in cases of intracranial abscesses (142), pneumonia (146), septic arthritis (170), and meningitis (47). *M. sedentarius* was the etiologic agent in a case of prosthetic valve endocarditis (104) and is often associated with pitted keratolysis (116). Other infections associated with *Micrococcus* spp. include bacteremia, continuous ambulatory peritoneal dialysis peritonitis, and the infection of a cerebrospinal fluid shunt (100).

COLLECTION, TRANSPORT, AND STORAGE OF SPECIMENS

The general principles of collection, transport, and storage of specimens (see chapter 3 of this Manual) are applicable to staphylococci and micrococci. No special methods or precautions are usually required for these organisms because they are easily obtained from clinical material of most infection sites and are relatively resistant to drying and moderate temperature changes. Some strains of staphylococci may require anaerobic conditions or CO_2 supplementation for satisfactory growth, but these strains survive transport and limited storage in air.

DIRECT EXAMINATION

The direct microscopic examination of normally sterile fluids such as cerebrospinal fluid, joint aspirates, or pulmonary secretions collected by transtracheal aspiration can be of great value. Direct examination of certain nonsterile fluids may also be very useful if the microscopist carefully evaluates the specimen by noting the presence of inflammatory cells versus epithelial cells. Even if large numbers of gram-positive cocci are present, only a presumptive report of "gram-positive cocci resembling staphylococci (or micrococci)" should be made. This report must be confirmed by culture and appropriate identification techniques. It must also be emphasized that microscopy by itself cannot adequately differentiate various species of staphylococci or micrococci from one another or from planococci, some streptococci, aerococci, or various anaerobic cocci.

ISOLATION PROCEDURES

Considering the widespread distribution of staphylococci and micrococci over the body surface, careful and thoughtful procedures should be used to isolate organisms from the focus or foci of infection without collecting surrounding normal flora (79).

Several isolation procedures are currently available for the diagnosis of central venus catheter infections, including quantitative blood cultures taken from blood drawn from a peripheral catheter (46) and semiquantitative counts taken from a distal segment of the catheter (8, 102).

Native valve endocarditis and prosthetic valve endocarditis are often associated with staphylococcal bacteremia. The most convincing laboratory findings include the rapid isolation of staphylococci from more than one blood culture, a high intensity of bacteremia, and the presence of the same strain(s) in sequential isolations (4, 72). Generally, in patients with suspected bacterial endocarditis, three blood cultures are sufficient to isolate the etiologic agent. It is essential that blood for culture be collected aseptically and from more than one venipuncture site in the event that a contaminant is accidently introduced at one of the sites.

For infections of hip prostheses, aspiration of the joint space and washing of the prosthesis with broth commonly yield the infecting bacteria. Ultrasonic oscillation may be used to shake off from prosthetic surfaces adherent organisms imbedded in a biofilm matrix (19, 162).

A freshly voided, midstream, clean-catch sample is usually satisfactory for making a determination of UTIs. Su-

prapubic aspiration may be indicated for patients who have a low bacterial count in clean-catch specimens, for neonates, and for young infants. Sometimes UTI due to *S. saprophyticus* is accompanied by a bacteremia with the same organism (55).

The basic procedures for culture and isolation described in chapter 21 of this Manual should be followed. Regardless of the source of the specimen, blood agar (preferably sheep blood agar) and a liquid medium such as thioglycolate broth should be inoculated. On blood agar, abundant growth of most staphylococcal species occurs within 18 to 24 h, and that of micrococci occurs within 36 to 48 h. Only individual colonies should be picked for preliminary identification testing at this time (e.g., from patients with acute infections). Since most species cannot be distinguished from each other on the basis of colony morphology with a 24-h incubation period, colonies should be allowed to grow for at least an additional 2 to 3 days before the primary isolation plate is confirmed for species or strain composition (84). This growth period is particularly important if it is necessary to sample more than one colony to obtain sufficient inocula and for determining the predominant organism or a pure culture. Failure to hold plates for 72 h can result in (i) selection of more than one species or strain if two or more colonies are sampled to produce an inoculum, (ii) selection of an organism(s) not producing the infection if the specimen contains two or more different species or strains, and (iii) incorrect labeling of a mixed culture as a pure culture. Colonies should be Gram stained, subcultured, and tested for genus, species, and, when applicable, strain properties. Most staphylococci of major medical interest produce growth in the upper as well as the lower anaerobic portions of the thioglycolate broth or semisolid agar (85).

Specimens from heavily contaminated sources such as feces should also be streaked onto a selective medium: Schleifer-Krämer agar (139), mannitol-salt agar, Columbia colistin-nalidixic acid (CNA) agar, lipase-salt-mannitol agar (Remel, Lenexa, Kans.), or phenylethyl alcohol agar (see chapter 21 of this Manual). These media inhibit the growth of gram-negative organisms but allow staphylococci and certain other gram-positive cocci to grow. On selective media, incubation should be extended to at least 48 to 72 h for discernible colony development.

IDENTIFICATION

Staphylococcus Species

Staphylococcus species can be identified on the basis of a variety of conventional phenotypic characteristics (79, 85, 86) (Table 2). The most clinically significant species can be identified on the basis of several key characteristics (Table 3). Species can be identified also on the basis of molecular phenotypic properties such as cellular fatty acids (92), multilocus enzymes (173), and whole-cell polypeptides (32) and of genotypic properties such as chromosome restriction fragments (22) and ribotypes (38, 65, 157) (also see chapter 20 of this Manual). Most molecular methods are currently confined to the reference or research laboratory.

Some laboratories may choose to restrict complete species identification of the CoNS to isolates from normally sterile sites such as blood or joint or cerebrospinal fluid and to routinely (i) distinguish *S. saprophyticus* from other CoNS isolated from urine; (ii) identify *S. epidermidis*, *S. lugdunensis*, and *S. schleiferi* isolated from colonized shunts, catheters, or prosthetic devices; and (iii) identify *S. epider-*

TABLE 3 Key tests for identification of most clinically significant *Staphylococcus* species

Species	Colony pigment[b]	Staphylocoagulase	Clumping factor[b]	Heat-stable nuclease	Alkaline phosphatase	Pyrrolidonyl arylamidase[b]	Ornithine decarboxylase	Urease[b]	β-Galactosidase[b]	Acetoin production	Novobiocin resistance[b]	Polymyxin B resistance[b]	D-Trehalose	D-Mannitol	D-Mannose	D-Turanose	D-Xylose	D-Cellobiose	Maltose	Sucrose
S. aureus	+	+	+	+	+	−	−	d	−	+	−	+	+	+	+	+	−	−	+	+
S. epidermidis	−	−	−	−	+	−	(d)	+	−	+	−	+	−	−	(+)	(d)	−	−	+	+
S. haemolyticus	d	−	−	−	−	+	−	−	−	+	−	−	+	d	−	(d)	−	−	+	+
S. lugdunensis	d	−	(+)	−	−	+	+	d	−	+	−	d	+	−	+	(d)	−	−	+	+
S. schleiferi subsp. schleiferi	−	−	+	+	+	+	−	−	(+)	+	−	−	d	−	+	−	−	−	−	−
S. saprophyticus	d	−	−	−	−	−	−	+	+	+	+	−	+	d	−	+	−	−	+	+
S. intermedius	−	+	d	+	+	−	−	+	+	−	−	−	+	(d)	+	d	−	−	(±)	+
S. hyicus	−	d	−	+	+	−	−	d	−	−	−	+	+	−	+	−	−	−	−	+

[a]Symbols: +, 90% or more strains positive; ±, 90% or more strains weakly positive; −, 90% or more strains negative; d, 11 to 89% of strains positive. Parentheses indicate a delayed reaction.

[b]Descriptions are the same as in Table 2.

midis, S. lugdunensis, S. haemolyticus, and S. warneri isolated from soft tissue infections or endocarditis.

Colonial Appearance

On nonselective blood agar, nutrient agar, tryptic soy agar, brain heart infusion agar, or P agar (84, 138), isolated colonies of most staphylococci are 1 to 3 mm in diameter within 24 h and 3 to 8 mm in diameter by 3 days of incubation in air at 34 to 37°C, depending on the species. The exceptional species S. aureus subsp. anaerobius, S. saccharolyticus, S. auricularis, S. equorum, S. caseolyticus, S. vitulus, and S. lentus grow more slowly than other staphylococci and usually require 24 to 36 h for colony development to be detectable. Colony morphology can be a useful supplementary characteristic in the identification of species. Isolated colonies that have developed for several days at 34 to 37°C and then for 2 days at room temperature should be described.

Colonies of S. aureus are usually large (6 to 8 mm in diameter), smooth, entire, slightly raised, and translucent. On P agar, they become nearly transparent by 3 to 5 days of incubation. The colonies of most strains are pigmented, ranging from cream yellow to orange. Some unusual strains of S. aureus produce dwarf colonies. Rare strains with relatively large capsules produce colonies that are smaller and more convex than those of unencapsulated strains and have a glistening, wet appearance. S. epidermidis colonies are relatively small and 2.5 to 6 mm in diameter. Pigment is not usually detected. Some of the slime-producing strains are extremely sticky and adhere to the agar surface. Colonies of S. haemolyticus are usually larger than those of S. epidermidis and S. hominis and are 5 to 9 mm in diameter. They are smooth, butyrous, and opaque like those of the related species S. hominis and may be unpigmented or cream to yellow-orange in color. Colonies of S. lugdunensis are usually 4 to 7 mm in diameter, smooth, and glossy and may be unpigmented or cream to yellow-orange in color. The edge

is entire and rather flat, while the center is slightly domed or acuminated. They are sometimes confused with colonies of S. warneri. S. schleiferi colonies are usually 3 to 5 mm in diameter and unpigmented. They are smooth, glossy, and slightly convex with entire edges. Colonies of S. saprophyticus are large (5 to 8 mm in diameter), entire, very glossy, opaque, smooth, butyrous, and more convex than colonies of the aforementioned species. Approximately one-half of the strains are pigmented, ranging from cream to yellow-orange. Colonies of S. intermedius and S. hyicus are relatively large and usually 5 to 8 mm in diameter. They are slightly convex, entire, smooth, glossy, and usually unpigmented. Colonies of S. intermedius are translucent. Those of S. hyicus are more opaque, becoming translucent with prolonged incubation.

Coagulase Production

The ability to clot plasma continues to be the most widely used and generally accepted criterion for the identification of pathogenic staphylococci associated with acute infections, i.e., S. aureus in humans and animals and S. intermedius and S. hyicus in animals. Two different coagulase tests can be performed: a tube test for free coagulase and a slide test for bound coagulase, or clumping factor (81). While the tube test is definitive, the slide test may be used as a rapid screening technique to identify S. aureus. A positive slide test may also aid in the identification of the new species S. lugdunensis and S. schleiferi (Table 3). A variety of plasmas may be used for either test; however, dehydrated rabbit plasma containing citrate or EDTA is commercially available and is most satisfactory except that human plasma is somewhat more satisfactory for the identification of S. lugdunensis and S. schleiferi. Human plasma should not be used unless it has been carefully tested for clotting capability and lack of inhibitors.

The tube coagulase test is best performed by mixing 0.1 ml of an overnight culture in brain heart infusion broth with 0.5 ml of reconstituted plasma, incubating the mixture at 37°C in a water bath or heat block for 4 h, and observing the tube for clot formation by slowly tilting the tube 90° from the vertical. Alternatively, a large, well-isolated colony on nonhibitory agar can be transferred into 0.5 ml of reconstituted plasma and incubated as describe above. Any degree of clotting constitutes a positive test. However, a flocculent or fibrous precipitate is not a true clot and should be recorded as a negative result. Incubation of the test overnight has also been recommended for *S. aureus*, since a small number of strains may require longer than 4 h for clot formation. For veterinary clinical laboratories, it is important to note that some strains of *S. intermedius* and most coagulase-producing strains of *S. hyicus* require more than 4 h for a positive coagulase test. Clot formation by these species may require 12 to 24 h of incubation. If incubation exceeds 4 h, the following points must be considered: (i) staphylokinase produced by some strains may lyse the clot after prolonged incubation, yielding false-negative results; (ii) if the plasma used is not sterile (and some is not), either false-positive or false-negative results may occur; and (iii) an inoculum from an agar-grown colony may not be pure, and a contaminant may produce false results after prolonged incubation. In this regard, plasma containing EDTA is superior to citrated plasma because citrate-utilizing organisms (e.g., some streptococci) may form clots by consuming the citrate. For those uncommon *S. aureus* strains requiring a longer clotting period, other characteristics (Table 3) should also be tested to confirm identity. Additional characteristics are required to identify rare coagulase-negative mutants and some encapsulated strains.

The slide coagulase test is performed by making a heavy uniform suspension of growth in distilled water, stirring the mixture to a homogeneous composition so as not to confuse clumping with autoagglutination, adding 1 drop of plasma, and observing for clumping within 10 s. The slide test is very rapid and more economical of plasma than the tube test. However, 10 to 15% of *S. aureus* strains may yield a negative result, which requires that the isolates be reexamined by the tube test. Slide tests must be read quickly, because false-positive results may appear with reaction times longer than 10 s. In addition, colonies for testing must not be picked from media containing high concentrations of salt (e.g., mannitol-salt agar), because autoagglutination and false-positive results may occur. Some uncommon strains of *S. intermedius* may give a positive slide test result. Alternative methods for the slide test include commercial hemagglutination slide tests for clumping factor (Staphyloslide test, Becton Dickinson Microbiology Systems, Cockeysville, Md.) and commercial latex agglutination tests that detect both clumping factor and protein A (Sero-STAT Staph latex slide test, Adams Scientific, West Warwick, R.I.; Staph Latex Kit, Remel Laboratories; Bacto Staph Latex Test, Difco Laboratories, Detroit, Mich.; Staph-a-Lex, Trinity Laboratories Inc., Raleigh, N.C.; Staphaurex, Murex Diagnostics, Norcross, Ga.; FAStaph, Carr-Scarborough Microbiologicals, Inc., Stone Mountain, Ga.; and Accu-Staph, Curtin Matheson Scientific, Houston, Tex.). Latex agglutination tests often have a higher specificity and sensitivity than the conventional slide test for the identification of *S. aureus*, though they are presently less reliable for the identification of *S. lugdunensis*. Some members of the *S. saprophyticus* and *S. sciuri* species groups may produce confusing positive results. A new latex agglutination test (Pastorex Staph-Plus; Sanofi Diagnostics Pasteur, Marnes-la-Coquette, France) that detects clumping

factor, protein A, and serotype 5 and 8 capsular polysaccharides of *S. aureus* is now available. It is reliable for the identification of both methicillin-susceptible *S. aureus* and MRSA strains (48). When the organism being tested is suspected of being *S. aureus*, negative slide tests should be confirmed by the tube coagulase test.

Heat-Stable Nuclease

A heat-stable staphylococcal nuclease (thermonuclease [TNase]) that has endo- and exonucleolytic properties and can cleave DNA or RNA is produced by most strains of *S. aureus*, *S. schleiferi*, *S. intermedius*, and *S. hyicus*. Some strains of *S. epidermidis*, *S. simulans*, and *S. carnosus* demonstrate a weak TNase activity. TNase can be detected by using a metachromatic-agar diffusion procedure and DNA-toluidine blue agar (94). A seroinhibition test has been developed to distinguish *S. aureus* TNase from those of other species (95). A commercial TNase test with toluidine blue agar is available (Remel) and can be interpreted in 4 h.

Phosphatase Activity

Phosphatase activity can be determined by using a modification of the technique of Pennock and Huddy (123), in which a 0.005 M solution of phenolphthalein diphosphate (sodium salt in 0.01 M citric acid-sodium citrate buffer, pH 5.8) is used as the substrate. Color is developed by the addition of 4-aminoantipyrine and potassium ferricyanide. Phosphatase activity is indicated by the development of a deep red color. A newer, alternative method for determining phosphatase activity based on the hydrolysis of *p*-nitrophenylphosphate into P_i and *p*-nitrophenol by alkaline phosphatase has been incorporated into several of the commercial biochemical test systems for staphylococcal species identification. Phosphatase activity is indicated by the release of yellow *p*-nitrophenol from the colorless substrate. Key Scientific Co. (Round Rock, Tex.) manufactures an alkaline phosphatase tablet that may detect activity in staphylococci, though it has not yet been widely accepted because of a small database.

Strains of *S. aureus*, *S. schleiferi*, *S. intermedius*, and *S. hyicus* and most strains of *S. epidermidis* are alkaline phosphatase positive. Phosphatase-negative strains of *S. epidermidis* can be distinguished from the related species *S. hominis* on the basis of their strong anaerobic growth in thioglycolate within 18 to 24 h or their resistance to polymyxin B (300-U disk).

Pyrrolidonyl Arylamidase Activity

Pyrrolidonyl arylamidase (pyrrolidonase) activity can be determined by the hydrolysis of pyroglutamyl-β-naphthylamide (L-pyrrolidonyl-β-naphthylamide [PYR]) into L-pyrrolidone and β-naphthylamine, the latter of which combines with a PYR reagent (*p*-dimethylaminocinnamaldehyde) to produce a red color. A commercial kit containing PYR broth and PYR reagent (Carr-Scarborough) recommended for the identification of group A streptococci and enterococci is also useful for distinguishing certain staphylococcal species (61). A slight modification in the standard procedure is required. A loopful of a 24-h agar slant culture or several well-isolated colonies are dispersed in the PYR broth (containing 0.01% PYR) to the turbidity of a McFarland no. 2 standard. The suspension is incubated at 35°C for 2 h. After incubation, 2 drops of PYR reagent are added to each tube without mixing. The development of a dark purple-red color within 2 min indicates a positive activity. Yellow, orange, or pink is considered a negative result. Alternatively, the basic features of the test have been incorporated into several of the commercial biochemical test panels for the identification of staphylococcal species. *S. haemolyticus*, *S. lugdunensis*, *S. schleiferi*, and *S. intermedius* are usually pyrrolidonase positive.

Ornithine Decarboxylase Activity

A positive ornithine decarboxylase activity can identify the species *S. lugdunensis* with considerable accuracy. Ornithine decarboxylase activity can be determined by a slight modification of the test described by Moeller (110). Decarboxylase basal medium (Becton Dickinson Microbiology Systems; Difco; GIBCO Laboratories, Grand Island, N.Y.) is prepared according to the instructions of the manufacturer, 1% (wt/vol) L-ornithine dihydrochloride is added, and the final medium is adjusted to pH 6 with 1 N sodium hydroxide before sterilization. The medium is dispensed in 3- to 4-ml amounts into small (13 by 100 mm) screw-cap tubes and autoclaved at 121°C for 10 min. A loopful of an overnight agar slant culture or several well-isolated colonies are dispersed in the test broth, and the agar in each tube is overlaid with 4 to 5 mm of sterile mineral oil. Inoculated tubes should be incubated at 35 to 37°C for up to 24 h. They can be read initially as early as 8 h for the positive identification of most strains of *S. lugdunensis*; at this time, *S. epidermidis* will produce negative results. A positive reaction is indicated by alkalinization of the medium, with a change in the initial grayish color or slight yellowing (caused by the initial fermentation of glucose) to violet (caused by decarboxylation of L-ornithine). A yellow color at 24 h indicates a negative result. The basic features of this test have been incorporated into the ID 32 STAPH (bioMérieux Vitek, Inc., Hazelwood, Mo.) identification system.

Urease Activity

A conventional urease test broth (Urea R broth; Difco) with a reduced buffer capacity can be used for detecting urease activity within 4 h in staphylococcal species. The test detects the release of ammonia from urea, resulting in an increase in pH that is shown by the phenol red indicator changing from yellow or orange to red or cerise. At present, comprehensive studies using this medium for the identification of staphylococcal species have not been reported. However, a miniaturization of this urease test has been incorporated into several of the commercial biochemical test systems for species identification of staphylococci and is represented by a large database. *S. epidermidis*, *S. intermedius*, and most strains of *S. saprophyticus* are usually urease positive.

β-Galactosidase Activity

Detection of high levels of β-galactosidase activity for the differentiation of certain staphylococcal species can be accomplished by commercial biochemical test systems that use 2-naphthol-β-D-galactopyranoside as a substrate. Fast blue BB salt in 2-methoxyethanol is added to the test well after an appropriate incubation period to detect free β-naphthol released by β-galactosidase. Positive activity is indicated by a plum purple color. By this assay, *S. intermedius* and most strains of *S. saprophyticus* are β-galactosidase positive; *S. schleiferi* is delayed or weakly positive.

Acetoin Production

Acetoin production from glucose or pyruvate is a useful alternative characteristic to distinguish *S. aureus* (positive) from another coagulase-positive species, *S. intermedius*

(negative), and coagulase-positive strains of *S. hyicus* (negative). The rapid paper disk method of Davis and Hoyling (36) is recommended for this test. The accuracy of the disk test is comparable to that of conventional Voges-Proskauer tests requiring longer incubation. Alternatively, acetoin production can be determined by a miniaturized Voges-Proskauer test incorporated into several of the commercial biochemical test systems for staphylococcal species identification.

Novobiocin Resistance

A simple disk diffusion test for estimating novobiocin susceptibility and distinguishing *S. saprophyticus* from other clinically important species can be performed by use of a 5-μg novobiocin disk on P agar (85), Mueller-Hinton agar (1), or tryptic soy sheep blood agar (54). With an inoculum suspension equivalent in turbidity to a 0.5 McFarland opacity standard and incubation at 35 to 37°C for overnight to 24 h, novobiocin resistance is indicated by an inhibition zone diameter of ≤16 mm with any of these media. Rapid disk elution procedures using either manual or automated-instrument interpretation have also been reported to reliably predict novobiocin resistance after only 4 to 5 h of incubation (2, 60). Novobiocin resistance is intrinsic to *S. saprophyticus* and several other species (Table 2), but it is uncommon in the other clinically important species.

Polymyxin B Resistance

A simple disk diffusion test for estimating polymyxin B susceptibility in order to distinguish several of the clinically important species can be done by using a 300-U polymyxin B disk (61). The test can be performed on any of the media mentioned above for estimation of novobiocin resistance. However, the largest database has been obtained with the use of tryptic soy sheep blood agar. Test conditions should be similar to those described above for novobiocin resistance. The 5-μg novobiocin disk and the 300-U polymyxin B disk can be tested on the same inoculated plate. Polymyxin B resistance is indicated by an inhibition zone diameter of <10 mm. *S. aureus*, *S. epidermidis*, *S. hyicus*, and *S. chromogenes* are usually resistant. Some strains of *S. lugdunensis* are also resistant.

Acid Production from Carbohydrates

Acid production from carbohydrates can be easily detected by using the agar plate method of Kloos and Schleifer (85). Carbohydrate reactions are also incorporated into several of the commercial biochemical test systems for staphylococcal species identification. These systems use a more acid-sensitive indicator than the bromcresol purple (pH ≤5.2) in the agar plate method; e.g., API STAPH-IDENT uses cresol red, which turns yellow at pH 6.8. For this and other reasons, results with conventional carbohydrate tests (Tables 2 and 3) may be slightly different from those obtained with rapid commercial biochemical test systems.

S. epidermidis can be distinguished from other novobiocin-susceptible species by its production of acid from maltose and sucrose and absence of acid production from trehalose and mannitol. Some uncommon strains of this species may produce acid from trehalose. These isolates can be distinguished from other species on the basis of phosphatase activity, anaerobic growth in thioglycolate, polymyxin B resistance, colony morphology, and absence of ornithine decarboxylase and pyrrolidonase activities. *S. lugdunensis* can be identified by its production of acid from trehalose, mannose, maltose, and sucrose and absence of acid production from mannitol. *S. schleiferi* produces acid from mannose and sometimes from trehalose but does not produce acid from mannitol, maltose, or sucrose. *S. saprophyticus* can be distinguished from other novobiocin-resistant species by its production of acid from sucrose and turanose and absence of acid production from mannose, xylose, cellobiose, arabinose, and raffinose.

Identification of Species by Using Commercial Biochemical or Nucleic Acid Test Systems

Several manufacturers of commercial kit identification systems (see chapter 10 of this Manual) or automated instruments have released products that can identify a number of the *Staphylococcus* species with an accuracy of 70 to >90% with relative speed and simplicity (14, 35, 80, 88, 126). Since their introduction, systems have been improved and expanded to include more species. Their reliability will continue to increase as the result of a growing database and development of more discriminating tests. *S. aureus*, *S. epidermidis*, *S. capitis*, *S. haemolyticus*, *S. saprophyticus*, *S. simulans*, and *S. intermedius* can be identified reliably by most of the commercial systems now available. For some systems, reliability depends on additional testing as suggested by the manufacturer. Additional testing might include determining coagulase, clumping factor, or ornithine decarboxylase activity; anaerobic growth in thioglycolate; or novobiocin resistance. If one or more of these key tests are not included in the particular manufacturer's product, identification could be uncertain with respect to some species. For example, in the API STAPH-IDENT (bioMérieux Vitek) system, phosphatase-negative *S. epidermidis* strains would be confused with *S. hominis* having an API profile of 2000 or 2040 (88). By determining the overnight anaerobic growth of such strains in a thioglycolate semisolid medium, one can distinguish between these species (85). Identification systems now available include the following: API STAPH-IDENT, STAPH Trac System, and ID 32 STAPH (bioMérieux Vitek); Vitek and Vitek Jr. fully automated microbiology systems that utilize a Gram Positive Identification (GPI) Card (bioMérieux Vitek); MicroScan Pos ID panel (read manually or on MicroScan instrumentation), MicroScan Rapid Pos ID panel (read by the AutoScan-W/A), and, in addition to ID panels, the Pos Combo Type 6 and Rapid Pos Combo Type 1 panels that combine species identification and antimicrobial susceptibility tests (Baxter Diagnostics Inc., MicroScan Division, West Sacramento, Calif.); Minitek Gram-Positive Set (Becton Dickinson Microbiology Systems); Sceptor *Staphylococcus* MIC/ID Panel and Sceptor Gram Positive Breakpoint/ID Panel (Becton Dickinson Instrument Systems, Towson, Md.); GP MicroPlate test panel (read manually, using Biolog computer software for interpretation, or automatically with the Biolog MicroStation) (Biolog, Haywood, Calif.); and The Microbial Identification System (MIS) that automates microbial identification by combining cellular fatty acid analysis with computerized high-resolution gas chromatography (MIDI, Newark, Del.).

Rapid detection of the species *S. aureus* can be made using the AccuProbe culture identification test for *S. aureus* (Gen-Probe, Inc., San Diego, Calif.). This test, a DNA probe assay directed against rRNA, is very accurate (100% specificity) but relatively expensive. Tube coagulase-negative and slide test-negative strains of *S. aureus* should be identified correctly by AccuProbe. *S. aureus* may be identified also by using the new immunoenzymatic assay developed by Guzmàn and coworkers (57) that is based on a monoclonal antibody, C1-10/11, prepared against the *S. aureus* endo-β-N-acetylglucosaminidase (SaG). These investigators are developing a C1-10/11-based commercial kit (56). The 100% specificity of the immunoenzymatic assay depends on the unique chemical-physical properties of the SaG. Remel manufactures a variety of individual test media and reagents that aid in identifying the genus *Staphylococcus* and certain of its species. The products include Coagulase Plasma, Coagulase Mannitol broth, Baird Parker Agar, DNase Test Agar w/Methyl Green, Staph Latex Kit, Thermonuclease Agar w/Toluidine Blue, Mannitol Salt Agar, and Vogel-Johnson Agar to aid in the presumptive identification of *S. aureus*. The uncommon coagulase-negative subspecies *S. schleiferi* subsp. *schleiferi* and the veterinary species *S. intermedius* can also produce positive reactions with several of these test products. The Lysostaphin Test Kit differentiates staphylococci from micrococci, and the Microdase Disk, a rapid oxidase test, differentiates micrococci from most staphylococci.

On occasion, it may be necessary to send isolates to a reference laboratory to confirm an identification. The Centers for Disease Control and Prevention (CDC) performs this service and has recently developed a numerical code system for the reference identification of *Staphylococcus* species that is based on a select panel of 18 conventional biochemical reactions.

Micrococcus Species

Pigment production and colony morphology may be used as simple tests in the presumptive identification of *Micrococcus* species (87). Several other phenotypic characteristics may be used to distinguish the various species (87, 91). *M. lylae* can be distinguished from *M. luteus* by its cream-white or unpigmented colonies, lack of growth on organic nitrogen agar, lysozyme resistance, and cell wall peptidoglycan. The yellow-pigmented species, *M. luteus* and *M. varians*, differ from each other in production of acid from glucose, nitrate reduction, lysozyme susceptibility, growth on inorganic nitrogen agar, color reaction on Simmons citrate agar, and oxidase reaction. *M. luteus* demonstrates oxidase activity, can grow on inorganic nitrogen sources, and is susceptible to lysozyme. *M. roseus* differs from other species in having pink colonies, nitrate reduction, and an inability to hydrolyze gelatin. *M. agilis* is the only species having flagella. It is psychrophilic and exhibits β-galactosidase activity. *M. kristinae* produces acid aerobically from glycerol and aerobically and anaerobically from glucose. It also produces acetoin and hydrolyzes esculin. *M. kristinae* produces unique wrinkled growth on purple agar (Difco) containing 1% maltose. The orange-pigmented *M. nishinomiyaensis* may be distinguished from *M. kristinae* by the former's very pale orange colonies and lack of growth on 7.5% NaCl agar, acetoin production, and esculin hydrolysis. *M. sedentarius* differs from other *Micrococcus* species by being resistant to penicillin and methicillin, producing often water-soluble expigment, growing very slowly, and reacting positively to the arginine dihydrolase test. *M. halobius* can be easy separated from other species, as it requires at least 5% NaCl for growth.

Strain Identification

Members of a bacterial strain constitute a population of cells descended from a common ancestor that existed at a relatively recent point in time. In the most recent examples, a strain represents a clonal population, with each of its members being genetically identical (isogenic) and demonstrating identical phenotypic characteristics (78). It is also reasonable to consider a strain as representing a clonal population in which some of its members differ from one another on the basis of only one or a few mutations or the loss or acquisition of an extrachromosomal element (e.g., phage, plasmid, transposon, etc.) Strain identification is very important in determining the etiologic agent in a staphylococcal infection. The clinical laboratory is now faced with the challenge of distinguishing epidemiologically significant isolates from unrelated isolates of the same species. This differentiation is important in examining isolates from individual patients as well as isolates obtained in outbreak situations. General considerations for epidemiologic typing of bacteria are discussed in chapters 13, 16, and 17 of this Manual.

The list of characteristics used for staphylococcal strain identification has grown considerably over the past 20 years. Most tests require special media, techniques, and/or instrumentation and would be better performed in a reference laboratory. Nevertheless, some approaches to strain identification can be considered by the small clinical laboratory. For example, for most *Staphylococcus* species and subspecies, the 4- to 15-digit profiles obtained with commercially available rapid species identification systems (described above) can be of some use in identifying strains, especially if a profile is relatively uncommon. Members of the same strain usually have the same profile. Colony morphology is a characteristic that can identify individual strains in many of the staphylococcal species. In general, strain colony recognition has little impact on patient care associated with the treatment of acute illness caused by *S. aureus*. However, strain colony recognition is useful when a coagulase-negative *Staphylococcus* or *Micrococcus* species is suspected in situations involving chronic infections and treatment failure. Colonies should be allowed to develop on the primary isolation medium for 3 to 4 days at 35 to 37°C and then for 2 days at room temperature before an initial screening. Colonies of the same strain generally exhibit similar size, consistency, edge, profile, luster, and color on nonselective media commonly used for the culture of staphylococci or micrococci. Certain strains may exhibit variant morphotypes, and in these situations, chromosomal analyses should help clarify the genetic relationship of each morphotype. At least one colony of each morphotype on the primary isolation plate should be selected for subsequent analyses. Antibiograms can be used to assist in staphylococcal strain identification. They are commonly determined in the laboratory, and highly standardized procedures have been established. A unique susceptibility pattern can serve as a valuable marker. The more common patterns will provide some support for identification if testing is confined to a small area or community. On occasion, a strain may demonstrate variation in pattern due to the acquisition or loss of antibiotic resistance genes or their activity, making identification more difficult.

Plasmid composition and the restriction endonuclease analysis of specific plasmids can serve as valuable molecular typing systems for strain identification (5, 83, 120). In most staphylococcal species, there is a relationship between antibiotic resistance pattern and the presence of certain plasmids carrying resistance genes. In this regard, plasmid composition may not be entirely independent of the antibiogram. Restriction endonuclease fragment analysis of plasmids can be useful in distinguishing plasmids of identical sizes. Such plasmids are considered to be different if their fragment patterns are different. However, some common plasmids are highly conserved (e.g., small tetracycline [*tetA tetB*] resistance plasmids or small erythromycin [*ermC*] resistance plasmids) and often have identical fragment patterns irrespective of the strain or species carrying them. Some strains exhibit clonal variation in their plasmid profiles. This variation is most often represented by the addition or deletion of an entire plasmid or a restriction fragment within a plasmid, though occasionally, different recombinant plasmids can be observed.

Several molecular typing methods examining complex characteristics such as cellular fatty acids (92, 168), multilocus enzymes (114, 173), pyrolysis mass spectrometry (49), and whole-cell polypeptides (101, 158) are currently being examined for strain identification. Molecular typing techniques aimed at examining the chromosomes of staphylococci have shown considerable promise in identifying strains or specific lineages. The methods include field inversion gel electrophoresis (FIGE) (51, 52) and pulsed-field gel electrophoresis (PFGE) of SmaI-digested genomic DNA (13, 133, 150) and ribotyping (21, 37, 38, 58, 132, 157). Although we would expect restriction fragment patterns to be quite stable and similar among members of the same strain, some clonal variation has been observed with certain strains with respect to the sizes and numbers of fragments present. Such variation might be explained by the acquisition or loss of prophages, transposition events, and/or recombination events with resident extrachromosomal DNA.

ANTIBIOTIC SUSCEPTIBILITIES

Nosocomial infections caused by methicillin-resistant staphylococci pose a serious problem for health care institutions. The detection of resistance in these isolates has been hampered by the variability in standard techniques used in determining methicillin resistance. The resistant strains are often heteroresistant to β-lactam antibiotics in that two subpopulations (one susceptible and the other resistant) coexist within a culture (27). Each cell in the population may carry the genetic information for resistance, but only a small fraction (10^{-8} to 10^{-4}) can actually express the resistant phenotype under in vitro testing conditions. The resistant subpopulation usually grows much more slowly than the susceptible subpopulation and therefore may be missed when in vitro testing is performed. The successful detection of heteroresistant strains depends largely on promoting the growth of the resistant subpopulations, which is favored by neutral pH, cooler temperatures (30 to 35°C), the presence of NaCl (2 to 4%), and possible prolonged incubation (up to 48 h) (34, 107). Good detection of MRSA and methicillin-resistant *S. epidermidis* can be made by using methods recommended by the National Committee for Clinical Laboratory Standards and described in chapters 113, 115, 116, and 117 of this Manual.

Several molecular techniques for the detection of methicillin resistance, including the use of a *mecA* gene probe (6, 151) or the PCR (131, 161), have been investigated. The PCR technique is as accurate as DNA hybridization, and it provides data in less than 5 h. The *mecA* gene probe is more sensitive than broth microdilution and more specific than agar dilution (6). However, care should be taken when DNA probes are used in antibiotic susceptibility testing, since the probes only indicate gene sequences and the potential of the bacterium to demonstrate antibiotic resistance (155). They do not necessarily discriminate between functional and nonfunctional genes. Two commercially available DNA probes, AccuProbe and FlashTrack

(Gen-Probe), which are species specific and bacterial generic, respectively, have been investigated for their abilities to detect oxacillin resistance in *S. aureus* on the basis of growth in the presence of oxacillin (172). They were found to be rapid (detection in 5 h) and reliable.

With the increase in methicillin resistance in *Staphylococcus* species, other antibiotics have been used in the treatment of serious infections caused by this group of bacteria. The glycopeptide vancomycin has been regarded as the drug of choice for the treatment of infections due to methicillin-resistant staphylococci. However, some patients fail to respond to treatment with vancomycin, and the side effects associated with vancomycin have limited its use in some cases. Some strains of *S. haemolyticus* (141, 163) are vancomycin resistant. The continued widespread use of this antibiotic may foster resistance in *Staphylococcus* species and other bacteria.

Alternative antibiotics investigated for the treatment of infections caused by methicillin-resistant staphylococci include (i) a glycopeptide teicoplanin, (ii) quinolones, (iii) rifampin in combination with other antibiotics, and (iv) trimethroprim-sulfamethoxazole (TMP-SMX). Teicoplanin may have fewer side effects than vancomycin, but it must be given in higher doses than was originally thought necessary (127). Decreased susceptibility to teicoplanin has been found through in vitro testing of isolates of *S. epidermidis*, *S. haemolyticus*, and *S. aureus* (16, 53, 73).

The quinolone ciprofloxacin has been used successfully alone or in combination with rifampin in treatment for MRSA (130). However, another report has shown the combination to be ineffective (125). Ciprofloxacin-resistant strains of staphylococci have been reported (17, 18). Studies have indicated that resistance to one quinolone may predispose the isolate to become resistant to other quinolones (156).

TMP-SMX has been used effectively in the treatment of staphylococcal endocarditis and meningitis (153); however, in vitro resistance is common. TMP-SMX in combination with rifampin and mupirocin alone have been used effectively to eradicate nasal carriage of resistant staphylococci (166). Staphylococci are highly susceptible to rifampin; however, since resistance develops rapidly, the drug should be used in combination therapy. Its combination with vancomycin has been clinically effective (43), although this combination may appear to be indifferent or antagonistic in vitro. Its combination with novobiocin has been effective in eradicating an MRSA carrier state (3, 165) and is more effective than its combination with TMP-SMX (165).

Micrococci appear to be susceptible to most antibiotics. Successful treatments have used vancomycin, penicillin, gentamicin, clindamycin, or a combination of these antibiotics (100).

EVALUATION, INTERPRETATION, AND REPORTING OF RESULTS

Considering the widespread reputation of the species, it is prudent to consider *S. aureus* the etiologic agent when it is isolated from a clinical specimen. Suspected isolates should be confirmed on the basis of coagulase testing and also preferably on the basis of biochemical profile by using a commercial identification system. At a significant but reasonable initial cost, the AccuProbe culture identification test for *S. aureus* (Gen-Probe) may be used to accurately identify the uncommon tube coagulase-negative and slide

test-negative strains of *S. aureus* and could be used for the routine identification of the species. Strains of *S. aureus* should be monitored in the advent of outbreaks and for the surveillance of nosocomial populations. Phage typing of *S. aureus* to monitor strains is for the most part being replaced by molecular typing methods in reference laboratories.

It is a common practice to consider *S. epidermidis* the etiologic agent when it is isolated from colonized shunts, catheters, or prosthetic devices and *S. saprophyticus* the etiologic agent when it is isolated from patients with UTI, especially if the bacteria are present in large numbers or as the predominant organism. Although these assessments are not always accurate, they are based on the known pathogenic potential of these species. Traditionally, in UTIs, colony counts of ≥100,000 CFU/ml in two or more cultures of midstream urine indicate a significant bacteriuria or UTI (145). However, since staphylococci grow relatively slowly in urine, it has been suggested that lower colony counts of 100 to 10,000 CFU/ml should be considered an appropriate range for significant bacteriuria in the presence of pyuria (63, 149). Repeated isolation of a predominant strain or a strain in pure culture is quite convincing in the attempt to determine the etiologic agent. For many of the other staphylococcal species and micrococcal species, it is imperative that individual strains be identified and monitored, preferably over the course of the infection, before their etiology can be evaluated. In the small clinic or small community hospital laboratory, CoNS species are seldom identified, though isolates of interest are sometimes sent to private or other reference laboratories for identification to the species, subspecies, or strain level. When it is deemed necessary to identify the etiologic agent, e.g., as a result of treatment failure, during an outbreak, etc., it is important that one or more aged (≥72-h) colonies of a particular morphotype be isolated from the primary isolation plate for each culture to be identified. The practice of pooling two or more young (24- to 48-h) colonies in the preparation of an inoculum or culture carries with it the risk of producing a mixed culture, resulting in erroneous identification and an erroneous accompanying antibiogram. Selecting only one young colony from a primary isolation plate carries with it the risk of missing the actual etiologic agent. For detecting individual species, subspecies, and strains and for assessing the bacterial population structure, primary isolation plates should be maintained for at least 72 h and preferably for 1 or 2 days longer before being discarded.

At many large community and teaching hospitals, state health departments, and the CDC, both conventional and molecular methods are being performed for the complete identification of staphylococci. CoNS species and subspecies are usually identified on the basis of their phenotypic characteristics by commercial rapid identification systems and some supplemental conventional methods (discussed above). Any isolation of the species *S. aureus* should be considered suspect, and the isolate should be processed for a confirmed identification together with an antibiogram. In general, a combination of two or more conventional typing techniques is employed for strain identification. Ribotyping, PFGE, and FIGE appear to be the most objective and discriminatory molecular techniques currently available. Initial guidelines for the interpretation of PFGE banding patterns from investigations in outbreaks of *S. aureus* are being established at the CDC. It is expected that in the near future, the systems described above for the complete identification of staphylococci and micrococci will become commercially available.

REFERENCES

1. **Almeida, R. J., and J. H. Jorgensen.** 1982. Use of Mueller-Hinton agar to determine novobiocin susceptibility of coagulase-negative staphylococci. *J. Clin. Microbiol.* **16:**1155–1156.

2. **Almeida, R. J., and J. H. Jorgensen.** 1983. Rapid determination of novobiocin resistance of coagulase-negative staphylococci with the MS-2 system. *J. Clin. Microbiol.* **17:**558–560.

3. **Arathoon, E. G., J. R. Hamilton, C. E. Hench, and D. A. Stevens.** 1990. Efficacy of short courses of oral novobiocin-rifampin in eradicating carrier state of methicillin-resistant *Staphylococcus aureus* and in vitro killing studies of clinical isolates. *Antimicrob. Agents Chemother.* **34:**1655–1659.

4. **Archer, G. L.** 1985. Coagulase-negative staphylococci in blood cultures: a clinician's dilemma. *Infect. Control* **6:**477–478.

5. **Archer, G. L., D. R. Dietrick, and J. L. Johnson.** 1985. Molecular epidemiology of transmissible gentamicin resistance among coagulase-negative staphylococci in a cardiac surgery unit. *J. Infect. Dis.* **151:**243–251.

6. **Archer, G. L., and E. Pennell.** 1990. Detection of methicillin resistance in staphylococci by using a DNA probe. *Antimicrob. Agents Chemother.* **34:**1720–1724.

7. **Archer, G. L., and M. J. Tenenbaum.** 1980. Antibiotic-resistant *Staphylococcus epidermidis* in patients undergoing cardiac surgery. *Antimicrob. Agents Chemother.* **17:**269–272.

8. **Aufwerber, E., S. Ringertz, and U. Ransjö.** 1991. Routine semiquantitative cultures and central venous catheter-related bacteremia. *APMIS* **99:**627–630.

9. **Baddour, L. M., T. N. Phillips, and A. L. Bisno.** 1986. Coagulase-negative staphylococcal endocarditis: occurrence in patients with mitral valve prolapse. *Arch. Intern. Med.* **146:**119–121.

10. **Baker, J. S.** 1984. Comparison of various methods for differentiation of staphylococci and micrococci. *J. Clin. Microbiol.* **19:**875–879.

11. **Ballard, D. N., W. Bowen, V. Thayer, and W. E. Kloos.** Unpublished observations.

12. **Bandres, J. C., and R. O. Darouiche.** 1992. *Staphylococcus capitis* endocarditis: a new cause of an old disease. *Clin. Infect. Dis.* **14:**366–367.

13. **Bannerman, T. L., L. W. Ayers, and W. E. Kloos.** Unpublished observations.

14. **Bannerman, T. L., K. T. Kleeman, and W. E. Kloos.** 1993. Evaluation of the Vitek System's gram-positive identification card for species identification of coagulase-negative staphylococci. *J. Clin. Microbiol.* **31:**1322–1325.

15. **Bannerman, T. L., and W. E. Kloos.** 1991. *Staphylococcus capitis* subsp. *urealyticus* subsp. nov. from human skin. *Int. J. Syst. Bacteriol.* **41:**144–147.

16. **Bannerman, T. L., D. L. Wadiak, and W. E. Kloos.** 1991. Susceptibility of *Staphylococcus* species and subspecies to teicoplanin. *Antimicrob. Agents Chemother.* **35:**1919–1922.

17. **Bannerman, T. L., D. L. Wadiak, and W. E. Kloos.** 1991. Susceptibility of *Staphylococcus* species and subspecies to fleroxacin. *Antimicrob. Agents Chemother.* **35:**2135–2139.

18. **Barry, A. L., M. A. Pfaller, and P. C. Fuchs.** 1992. Spontaneously occurring staphylococcal mutants resistant to clinically achievable concentrations of ciprofloxacin and temafloxacin. *Eur. J. Clin. Microbiol. Infect. Dis.* **11:**243–246.

19. **Bergamini, T. M., D. F. Bandyk, and D. Govostis.** 1989. Identification of *Staphylococcus epidermidis* vascular graft infections: a comparison of culture techniques. *J. Vasc. Surg.* **9:**665–670.

20. **Bergman, B., H. Wedren, and S. E. Holm.** 1989. *Staphylococcus saprophyticus* in males with symptoms of chronic prostatitis. *Urology* **34:**241–245.

21. **Bialkowska-Hobrzanska, H., V. Harry, D. Jaskot, and O. Hammerberg.** 1990. Typing of coagulase-negative staphylococci by Southern hybridization of chromosomal DNA fingerprints with a ribosomal RNA probe. *Eur. J. Clin. Microbiol. Infect. Dis.* **9:**588–594.

22. **Bialkowska-Hobrzanska, H., D. Jaskot, and O. Hammerberg.** 1990. Evaluation of restriction endonuclease fingerprinting of chromosomal DNA and plasmid profile analysis for characterization of multiresistant coagulase-negative staphylococci in bacteremic neonates. *J. Clin. Microbiol.* **28:**269–275.

23. **Bowman, R. A., and M. Buck.** 1984. *Staphylococcus hominis* septicaemia in patients with cancer. *Med. J. Aust.* **140:**26–27.

24. **Boyce, J. M.** 1990. Increasing prevalence of methicillin-resistant *Staphylococcus aureus* in the United States. *Infect. Control Hosp. Epidemiol.* **11:**639–642.

25. **Boyce, J. M.** 1991. Should we vigorously try to contain and control methicillin-resistant *Staphylococcus aureus*? *Infect. Control Hosp. Epidemiol.* **12:**46–54.

26. **Brause, B. D.** 1986. Infections associated with prosthetic joints. *Clin. Rheum. Dis.* **12:**523–535.

27. **Chambers, H. F.** 1988. Methicillin-resistant staphylococci. *Clin. Microbiol. Rev.* **1:**173–186.

28. **Chesneau, O., A. Morvan, F. Grimont, H. Labischinski, and N. El Solh.** 1993. *Staphylococcus pasteuri* sp. nov., isolated from human, animal, and food specimens. *Int. J. Syst. Bacteriol.* **43:**237–244.

29. **Choi, Y., B. Kotzin, L. Herron, J. Callahan, P. Marrack, and J. Kappler.** 1989. Interaction of *Staphylococcus aureus* toxin "superantigens" with human T cells. *Proc. Natl. Acad. Sci. USA* **86:**8941–8945.

30. **Christensen, G. D., W. A. Simpson, J. J. Younger, L. M. Baddour, F. F. Barrett, D. M. Melton, and E. H. Beachey.** 1986. Adherence of coagulase-negative staphylococci to plastic tissue culture plates: a quantitative model for the adherence of staphylococci to medical devices. *J. Clin. Microbiol.* **22:**996–1006.

31. **Clarke, A. M.** 1979. Prophylactic antibiotics for total hip arthroplasty—the significance of *Staphylococcus epidermidis*. *J. Antimicrob. Chemother.* **5:**493–502.

32. **Clink, J., and T. H. Pennington.** 1987. Staphylococcal whole-cell polypeptide analysis: evaluation as a taxonomic and typing tool. *J. Med. Microbiol.* **23:**41–44.

33. **Cohn, F.** 1872. Untersuchungen uber Bacterien. *Beitr. Biol. Pflanz. Bd. 1* **2:**127–224.

34. **Coudron, P. E., D. L. Jones, H. P. Dalton, and G. L. Archer.** 1986. Evaluation of laboratory tests for detection of methicillin-resistant *Staphylococcus aureus* and *Staphylococcus epidermidis*. *J. Clin. Microbiol.* **24:**764–769.

35. **Crouch, S. F., T. A. Pearson, and D. M. Parham.** 1987. Comparison of modified Minitek system with Staph-Ident system for species identification of coagulase-negative staphylococci. *J. Clin. Microbiol.* **25:**1626–1628.

36. **Davis, G. H. G., and B. Hoyling.** 1973. Use of a rapid acetoin test in the identification of staphylococci and micrococci. *Int. J. Syst. Bacteriol.* **23:**281–282.

37. **DeBuyser, M.-L., A. Morvan, S. Aubert, F. Dilasser, and N. El Solh.** 1992. Evaluation of ribosomal RNA gene probe for the identification of species and subspecies within the genus *Staphylococcus*. *J. Gen. Microbiol.* **138:**889–899.

38. **DeBuyser, M.-L., A. Morvan, F. Grimont, and N. El Solh.** 1989. Characterization of *Staphylococcus* species by ribosomal RNA gene restriction patterns. *J. Gen. Microbiol.* **135:**989–999.

39. **Devriese, L. A.** 1986. Coagulase-negative staphylococci in animals, p. 51–57. *In* P.-A. Mårdh and K. H. Schleifer (ed.), *Coagulase-Negative Staphylococci*. Almqvist and Wiksell International, Stockholm.

40. **Devriese, L. A., and V. Hájek.** 1980. A review. Identification of pathogenic staphylococci isolated from animals and foods derived from animals. *J. Appl. Bacteriol.* **49:**1–11.

41. **Etienne, J., B. Pangon, C. Leport, M. Wolff, B. Clair, C. Perronne, Y. Brun, and A. Bure.** 1989. *Staphylococcus lugdunensis* endocarditis. *Lancet* **i:**390.

42. **Evans, J. B., W. L. Bradford, Jr., and C. F. Niven.** 1955. Comments concerning the taxonomy of the genera *Micrococcus* and *Staphylococcus*. *Int. Bull. Bacteriol. Nomencl. Taxon.* **5:**61–66.

43. **Faville, R. J., Jr., D. E. Zaske, E. L. Kaplan, K. Crossley,**

L. D. Sabath, and P. G. Quie. 1978. *Staphylococcus aureus* endocarditis: combined therapy with vancomycin and rifampin. JAMA **240**:1963–1965.

44. **Fleer, A., and J. Verhoef.** 1989. An evaluation of the role of surface hydrophobicity and extracellular slime in the pathogenesis of foreign-body-related infections due to coagulase-negative staphylococci. *J. Invest. Surg.* **2**:391–396.

45. **Fleurette, J., M. Bes, Y. Brun, J. Freney, F. Forey, M. Coulet, M. E. Reverdy, and J. Etienne.** 1989. Clinical isolates of *Staphylococcus lugdunensis* and *S. schleiferi*: bacteriological characteristics and susceptibility to antimicrobial agents. *Res. Microbiol.* **140**:107–118.

46. **Flynn, P. M., J. L. Shenep, D. C. Stokes, and F. F. Barrett.** 1987. *In situ* management of confirmed central venous catheter-related bacteremia. *Pediatr. Infect. Dis. J.* **6**:729–734.

47. **Fosse, T., Y. Peloux, C. Granthil, B. Toga, J. Bertrando, and M. Sethian.** 1985. Meningitis due to *Micrococcus luteus*. *Infection* **13**:280–281.

48. **Fournier, J.-M., A. Bouvet, D. Mathieu, F. Nato, A. Boutonnier, R. Gerbal, P. Brunengo, C. Saulnier, N. Sagot, B. Slizewicz, and J.-C. Mazie.** 1993. New latex reagent using monoclonal antibodies to capsular polysaccharide for reliable identification of both oxacillin-susceptible and oxacillin-resistant *Staphylococcus aureus*. *J. Clin. Microbiol.* **31**:1342–1344.

49. **Freeman, R., M. Goodfellow, A. C. Ward, S. J. Hudson, F. K. Gould, and N. F. Lightfoot.** 1991. Epidemiological typing of coagulase-negative staphylococci by pyrolysis mass spectrometry. *J. Med. Microbiol.* **34**:245–248.

50. **George, R., L. Leibrock, and M. Epstein.** 1979. Long term analysis of cerebrospinal fluid shunt infections: a 25-year experience. *J. Neurosurg.* **51**:804–811.

51. **Goering, R. V., and T. D. Duensing.** 1990. Rapid field inversion gel electrophoresis in combination with an rRNA gene probe in the epidemiological evaluation of staphylococci. *J. Clin. Microbiol.* **28**:426–429.

52. **Goering, R. V., and M. A. Winters.** 1992. Rapid method for epidemiological evaluation of gram-positive cocci by field inversion gel electrophoresis. *J. Clin. Microbiol.* **30**:577–580.

53. **Goldstein, F. W., A. Coutrot, A. Sieffer, and J. F. Acar.** 1990. Percentages and distributions of teicoplanin- and vancomycin-resistant strains among coagulase-negative staphylococci. *Antimicrob. Agents Chemother.* **34**:899–900.

54. **Goldstein, J., R. Schulman, E. Kelly, G. McKinley, and J. Fung.** 1983. Effect of different media on determination of novobiocin resistance for differentiation of coagulase-negative staphylococci. *J. Clin. Microbiol.* **18**:592–595.

55. **Golledge, C. L.** 1988. *Staphylococcus saprophyticus* bacteremia. *J. Infect. Dis.* **157**:215.

56. **Guardati, M. C., C. A. Guzmàn, G. Piatti, and C. Pruzzo.** 1993. Rapid methods for identification of *Staphylococcus aureus* when both human and animal staphylococci are tested: comparison with a new immunoenzymatic assay. *J. Clin. Microbiol.* **31**:1606–1608.

57. **Guzmàn, C. A., M. C. Guardati, D. Fenoglio, G. Coratza, C. Pruzzo, and G. Satta.** 1992. A novel immunoenzymatic assay for the identification of *Staphylococcus aureus* strains negative for coagulase and protein A. *J. Clin. Microbiol.* **30**:1194–1197.

58. **Hadorn, K., W. Lenz, F. H. Kayser, I. Shalit, and C. Krasemann.** 1990. Use of a ribosomal RNA gene probe for epidemiological study of methicillin and ciprofloxacin resistant *Staphylococcus aureus*. *Eur. J. Clin. Microbiol. Infect. Dis.* **9**:649–653.

59. **Hájek, V., W. Ludwig, K. H. Schleifer, N. Springer, W. Zitzelsberger, R. M. Kroppenstedt, and M. Kocur.** 1992. *Staphylococcus muscae*, a new species isolated from flies. *Int. J. Syst. Bacteriol.* **42**:97–101.

60. **Harrington, B. J., and J. M. Gaydos.** 1984. Five-hour novobiocin test for differentiation of coagulase-negative staphylococci. *J. Clin. Microbiol.* **19**:279–280.

61. **Hébert, G. A., C. G. Crowder, G. A. Hancock, W. R. Jarvis, and C. Thornsberry.** 1988. Characteristics of coagulase-negative staphylococci that help differentiate these species and other members of the family *Micrococcaceae*. *J. Clin. Microbiol.* **26**:1939–1949.

62. **Herchline, T. E., and L. W. Ayers.** 1991. Occurrence of *Staphylococcus lugdunensis* in consecutive clinical cultures and relationship of isolation to infection. *J. Clin. Microbiol.* **29**:419–421.

63. **Hovelius, B.** 1986. Epidemiological and clinical aspects of urinary tract infections caused by *Staphylococcus saprophyticus*, p. 195–202. *In* P.-A. Mårdh and K. H. Schleifer (ed.), *Coagulase-Negative Staphylococci*. Almqvist and Wiksell International, Stockholm.

64. **Hovelius, B., I. Thelin, and P. A. Mårdh.** 1979. *Staphylococcus saprophyticus* in the aetiology of nongonococcal urethritis. *Br. J. Vener. Dis.* **55**:369–374.

65. **Hubner, R., J. Webster, J. Bruce, E. Cole, J. Neubauer, J. Wyer, R. Betts, J. Banks, T. Bannerman, D. Ballard, and W. Kloos.** 1993. Typing of bacteria through the analysis of the ribosomal RNA genes, abstr. R-10, p. 295. *Abstr. Gen. Meet. Am. Soc. Microbiol. 1993*.

66. **Igimi, S., E. Takahashi, and T. Mitsuoka.** 1990. *Staphylococcus schleiferi* subsp. *coagulans* subsp. nov., isolated from the external auditory meatus of dogs with external ear otitis. *Int. J. Syst. Bacteriol.* **40**:409–411.

67. **Jansen, B., F. Schumacher-Perdreau, G. Peters, and G. Pulverer.** 1989. New aspects in the pathogenesis and prevention of polymer-associated foreign-body infections caused by coagulase-negative staphylococci. *J. Invest. Surg.* **2**:361–380.

68. **Jansen, B., F. Schumacher-Perdreau, G. Peters, G. Reinhold, and J. Schönemann.** 1992. Native valve endocarditis caused by *Staphylococcus simulans*. *Eur. J. Clin. Microbiol. Dis.* **11**:268–269.

69. **Jean-Pierre, H., H. Darbas, A. Jean-Roussenq, and G. Boyer.** 1989. Pathogenicity in two cases of *Staphylococcus schleiferi*, a recently described species. *J. Clin. Microbiol.* **27**:2110–2111.

70. **Kamath, U., C. Singer, and H. D. Isenberg.** 1992. Clinical significance of *Staphylococcus warneri* bacteremia. *J. Clin. Microbiol.* **30**:261–264.

71. **Karchmer, A. W., G. L. Archer, and W. E. Dismukes.** 1983. *Staphylococcus epidermidis* causing prosthetic valve endocarditis: microbiologic and clinical observations as guides to therapy. *Ann. Intern. Med.* **98**:447–455.

72. **Karchmer, A. W., and G. M. Caputo.** 1986. Endocarditis due to coagulase-negative staphylococci, p. 179–187. *In* P.-A. Mårdh and K. H. Schleifer (ed.), *Coagulase-Negative Staphylococci*. Almquist and Wiskell International, Stockholm.

73. **Kenny, M. T., G. D. Mayer, J. K. Dulworth, M. A. Brackman, and K. Farrar.** 1992. Evaluation of the teicoplanin broth microdilution and disk diffusion susceptibility tests and recommended interpretive criteria. *Diagn. Microbiol. Infect. Dis.* **15**:609–612.

74. **Kilpper, R., U. Buhl, and K. H. Schleifer.** 1980. Nucleic acid homology studies between *Peptococcus saccharolyticus* and various anaerobic and facultive anaerobic Gram-positive cocci. *FEMS Microbiol. Lett.* **8**:205–210.

75. **Kleeman, K. T., T. L. Bannerman, and W. E. Kloos.** 1993. Species distribution of coagulase-negative staphylococcal isolates at a community hospital and implications for selection of staphylococcal identification procedures. *J. Clin. Microbiol.* **31**:1318–1321.

76. **Kloos, W. E.** 1980. Natural populations of the genus *Staphylococcus*. *Annu. Rev. Microbiol.* **34**:559–592.

77. **Kloos, W. E.** 1986. Ecology of human skin, p. 37–50. *In* P.-A. Mårdh and K. H. Schleifer (ed.), *Coagulase-Negative Staphylococci*. Almqvist and Wiksell International, Stockholm.

78. **Kloos, W. E.** 1990. Systematics and the natural history of staphylococci. 1. *J. Appl. Bacteriol. Symp. Suppl.* **69**:25S–37S.

79. **Kloos, W. E., and T. L. Bannerman.** 1994. Update on clinical significance of coagulase-negative staphylococci. *Clin. Microbiol. Rev.* **7**:117–140.

80. **Kloos, W. E., and C. G. George.** 1991. Identification of *Staphylococcus* species and subspecies with the MicroScan Pos

ID and Rapid Pos ID panel systems. *J. Clin. Microbiol.* **29:** 738–744.

81. **Kloos, W. E., and J. H. Jorgensen.** 1985. Staphylococci, p. 143–153. *In* E. H. Lennette, A. Balows, W. J. Hausler, Jr., and H. J. Shadomy (ed.), *Manual of Clinical Microbiology*, 4th ed. American Society for Microbiology, Washington, D.C.

82. **Kloos, W. E., and M. S. Musselwhite.** 1975. Distribution and persistence of *Staphylococcus* and *Micrococcus* species and other aerobic bacteria on human skin. *Appl. Microbiol.* **30:** 381–395.

83. **Kloos, W. E., B. S. Orban, and D. D. Walker.** 1981. Plasmid composition of *Staphylococcus* species. *Can. J. Microbiol.* **27:** 271–278.

84. **Kloos, W. E., and K. H. Schleifer.** 1975. Isolation and characterization of staphylococci from human skin. II. Descriptions of four new species: *Staphylococcus warneri, Staphylococcus capitis, Staphylococcus hominis,* and *Staphylococcus simulans. Int. J. Syst. Bacteriol.* **25:**62–79.

85. **Kloos, W. E., and K. H. Schleifer.** 1975. Simplified scheme for routine identification of human *Staphylococcus* species. *J. Clin. Microbiol.* **1:**82–88.

86. **Kloos, W. E., K. H. Schleifer, and F. Götz.** 1991. The genus *Staphylococcus*, p. 1369–1420. *In* A. Balows, H. G. Truper, M. Dworkin, W. Harder, and K. H. Schleifer (ed.), *The Prokaryotes*, 2nd ed. Springer-Verlag, New York.

87. **Kloos, W. E., T. G. Tornabene, and K. H. Schleifer.** 1974. Isolation and characterization of micrococci from human skin, including two new species: *Micrococcus lylae* and *Micrococcus kristinae. Int. J. Syst. Bacteriol.* **24:**79–101.

88. **Kloos, W. E., and J. F. Wolfshohl.** 1982. Identification of *Staphylococcus* species with the API STAPH-IDENT system. *J. Clin. Microbiol.* **16:**509–516.

89. **Kloos, W. E., and J. F. Wolfshohl.** 1991. *Staphylococcus cohnii* subspecies: *Staphylococcus cohnii* subsp. nov. and *Staphylococcus cohnii* subsp. *urealyticum* subsp. nov. *Int. J. Syst. Bacteriol.* **41:**284–289.

90. **Kluge, R. M., F. M. Calia, J. S. McLaughlin, and R. B. Hornick.** 1974. Sources of contamination in open heart surgery. JAMA **230:**1415–1418.

91. **Kocur, M., W. E. Kloos, and K. H. Schleifer.** 1991. The genus *Micrococcus*, p. 1300–1311. *In* A. Balows, H. G. Truper, M. Dworkin, W. Harder, and K. H. Schleifer (ed.), *The Prokaryotes*, 2nd ed. Springer-Verlag, New York.

92. **Kotilainen, P., P. Huovinen, and E. Eerola.** 1991. Application of gas-liquid chromatographic analysis of cellular fatty acids for species identification and typing of coagulase-negative staphylococci. *J. Clin. Microbiol.* **29:**315–322.

93. **Kraus, E. S., and D. A. Spector.** 1983. Characteristics and sequelae of peritonitis in diabetics and nondiabetics receiving chronic intermittent peritoneal dialysis. *Medicine* **62:**52–57.

94. **Lachica, R. V. F., P. D. Hoeprich, and C. Genigeorgis.** 1972. Metachromatic agar-diffusion microslide technique for detecting staphylococcal nuclease in foods. *Appl. Microbiol.* **23:**168–169.

95. **Lachica, R. V. F., S. S. Jang, and P. D. Hoeprich.** 1979. Thermonuclease seroinhibition test for distinguishing *Staphylococcus aureus* and other coagulase-positive staphylococci. *J. Clin. Microbiol.* **9:**141–143.

96. **Low, D. E., B. K. Schmidt, H. M. Kirpalani, R. Moodie, B. Kreiswirth, A. Matlow, and E. L. Ford-Jones.** 1992. An endemic strain of *Staphylococcus haemolyticus* colonizing and causing bacteremia in neonatal intensive care unit patients. *Pediatrics* **89:**696–700.

97. **Lowy, F. D., and S. M. Hammer.** 1983. *Staphylococcus epidermidis* infections. *Ann. Intern. Med.* **99:**834–839.

98. **Ludwig, W., K. H. Schleifer, G. E. Fox, E. Seewaldt, and E. Stackebrandt.** 1981. A phylogenetic analysis of staphylococci, *Peptococcus saccharolyticus* and *Micrococcus mucilaginosus. J. Gen. Microbiol.* **125:**357–366.

99. **Ludwig, W., E. Seewaldt, R. Kilpper-Bälz, K. H. Schleifer, L. Magrum, C. R. Woese, G. F. Fox, and E. Stackebrandt.** 1985. The phylogenetic position of *Streptococcus* and *Enterococcus. J. Gen. Microbiol.* **131:**543–551.

100. **Magee, J. T., I. A. Burnett, J. M. Hindmarch, and R. C. Spencer.** 1990. *Micrococcus* and *Stomatococcus* spp. from human infections. *J. Infect.* **16:**67–73.

101. **Maggs, A. F., and T. H. Pennington.** 1989. Temporal study of staphylococcal species on the skin of human subjects in isolation and clonal analysis of *Staphylococcus capitis* by sodium dodecyl sulfate-polyacrylamide gel electrophoresis. *J. Clin. Microbiol.* **27:**2627–2632.

102. **Maki, D., C. E. Weise, and H. W. Safarin.** 1977. A semiquantitative culture method for identifying intravenous-catheter-related infection. *N. Engl. J. Med.* **296:**1305–1309.

103. **Males, B. M., W. R. Bartholomew, and D. Amsterdam.** 1985. *Staphylococcus simulans* septicemia in a patient with chronic osteomyelitis and pyarthrosis. *J. Clin. Microbiol.* **21:**255–257.

104. **Manzell, J. P., J. A. Kellogg, and E. Q. Rogers.** 1989. *Micrococcus sedentarius* as a cause of prosthetic valve endocarditis, abstr. C-208, p. 428. *Abstr. Annu. Meet. Am. Soc. Microbiol. 1989.*

105. **Marrie, T. J., C. Kwan, M. A. Noble, A. West, and L. Duffield.** 1982. *Staphylococcus saprophyticus* as a cause of urinary tract infections. *J. Clin. Microbiol.* **16:**427–431.

106. **Martin, M. A., M. A. Pfaller, and R. P. Wenzel.** 1989. Coagulase-negative staphylococcal bacteremia. *Ann. Intern. Med.* **110:**9–16.

107. **McDougal, L. K., and C. Thornsberry.** 1984. New recommendations for disk diffusion antimicrobial susceptibility tests for methicillin-resistant (heteroresistant) staphylococci. *J. Clin. Microbiol.* **19:**482–488.

108. **Meyer, W.** 1967. A proposal for subdividing the species *Staphylococcus aureus. Int. J. Syst. Bacteriol.* **17:**387–389.

109. **Miethke, T., K. Duschek, C. Wahl, K. Heeg, and H. Wagner.** 1993. Pathogenesis of the toxic shock syndrome: T cell mediated lethal shock caused by the superantigen TSST-1 *Eur. J. Immunol.* **23:**1494–1500.

110. **Moeller, V.** 1955. Simplified tests for some amino acid decarboxylases and for the arginine dihydrolase system. *Acta Pathol. Microbiol. Scand.* **36:**158–172.

111. **Muller, E., J. Hübner, N. Gutierriz, S. Takeda, D. A. Goldmann, and G. B. Pier.** 1993. Isolation and characterization of transposon mutants of *Staphylococcus epidermidis* deficient in capsular polysaccharide/adhesin and slime. *Infect. Immun.* **61:**551–558.

112. **Muller, E., S. Takeda, H. Shiro, D. Goldmann, and G. B. Pier.** 1993. Occurrence of capsular polysaccharide/adhesin among clinical isolates of coagulase-negative staphylococci. *J. Infect. Dis.* **168:**1211–1218.

113. **Mulligan, M. E., K. A. Murray-Leisure, B. S. Ribner, H. C. Standiford, J. F. John, J. A. Korvick, C. A. Kauffman, and V. L. Yu.** 1993. Methicillin-resistant *Staphylococcus aureus*: a consensus review of the microbiology, pathogenesis, and epidemiology with implication for prevention and management. *Am. J. Med.* **94:**313–328.

114. **Musser, J. M., P. M. Schlievert, A. W. Chow, P. Ewan, B. N. Kreiswirth, V. T. Rosdahl, A. S. Naidu, W. White, and R. K. Selander.** 1990. A single clone of *Staphylococcus aureus* causes the majority of cases of toxic shock syndrome. *Proc. Natl. Acad. Sci. USA* **87:**225–229.

115. **Nickel, J. C., and J. W. Costerton.** 1992. Coagulase-negative *Staphylococcus* in chronic prostatitis. *J. Urol.* **147:**398–401.

116. **Nordstrom, K. M., K. J. McGinley, J. M. Zechman, and J. J. Leyden.** 1987. Similarities between *Dermatophilus congolensis* and *Micrococcus sedentarius*: identity of the etiologic agent of pitted keratolysis, abstr. R-21, p. 244. *Abstr. Annu. Meet. Am. Soc. Microbiol. 1987.*

117. **Ogston, A.** 1883. *Micrococcus* poisoning. *J. Anat. Physiol.* **17:**317–324.

118. **Paley, D., C. F. Moseley, P. Armstrong, and C. G. Prober.** 1986. Primary osteomyelitis caused by coagulase-negative staphylococci. *J. Pediatr. Orthop.* **6:**622–626.

119. **Panlilio, A. L., D. H. Culver, R. P. Gaynes, S. Banerjee, T. S. Henderson, J. S. Tolson, and W. J. Martone.** 1992.

Methicillin-resistant *Staphylococcus aureus* in U.S. hospitals, 1975–1991. *Infect. Control Hosp. Epidemiol.* **13**:582–586.

120. **Parisi, J. T., and D. W. Hecht.** 1980. Plasmid profiles in epidemiologic studies of infections of *Staphylococcus epidermidis*. *J. Infect. Dis.* **141**:637–643.

121. **Parsonnet, J., and Z. A. Gillis.** 1988. Production of tumor necrosis factor by human monocytes in response to toxic shock syndrome toxin-1. *J. Infect. Dis.* **158**:1026–1033.

122. **Parsonnet, J., R. K. Hickman, D. D. Eardley, and G. B. Pier.** 1985. Induction of human interleukin-1 by toxic shock syndrome toxin-1. *J. Infect. Dis.* **151**:514–522.

123. **Pennock, C. A., and R. B. Huddy.** 1967. Phosphatase reaction of coagulase-negative staphylococci and micrococci. *J. Pathol. Bacteriol.* **93**:685–688.

124. **Peters, G., F. Schumacher-Perdreau, B. Jansen, M. Bey, and G. Pulverer.** 1987. Biology of *S. epidermidis* extracellular slime, p. 15–32. *In* G. Pulverer, P. G. Quie, and G. Peters (ed.), *Pathogenicity and Clinical Significance of Coagulase-Negative Staphylococci.* Gustav Fischer Verlag, Stuttgart, Germany.

125. **Peterson, L. R., J. N. Quick, B. Jensen, S. Homann, S. Johnson, J. Tenquist, C. Shanholtzer, R. Petzel, L. Sinn, and D. N. Gerding.** 1990. Emergence of ciprofloxacin resistance in nosocomial methicillin-resistant *Staphylococcus aureus* isolates. *Arch. Intern. Med.* **150**:2151–2155.

126. **Pfaller, M. A., and L. A. Herwaldt.** 1988. Laboratory, clinical, and epidemiological aspects of coagulase-negative staphylococci. *Clin. Microbiol. Rev.* **1**:281–299.

127. **Phillips, G., and C. L. Golledge.** 1992. Vancomycin and teicoplanin: something old, something new. *Med. J. Aust.* **156**:53–57.

128. **Phillips, W. E., Jr., R. E. King, and W. E. Kloos.** 1980. Isolation of *Staphylococcus hyicus* subsp. *hyicus* from a pig with septic polyarthritis. *Am. J. Vet. Res.* **41**:274–276.

129. **Phillips, W. E., Jr., and W. E. Kloos.** 1981. Identification of coagulase-positive *Staphylococcus intermedius* and *Staphylococcus hyicus* subsp. *hyicus* isolates from veterinary clinical specimens. *J. Clin. Microbiol.* **14**:671–673.

130. **Piercy, E. A., D. Barbaro, J. P. Luby, and P. A. Machowiak.** 1989. Ciprofloxacin for methicillin-resistant *Staphylococcus aureus* infections. *Antimicrob. Agents Chemother.* **33**:128–130.

131. **Predari, S. C., M. Ligozzi, and R. Fontana.** 1991. Genotypic identification of methicillin-resistant coagulase-negative staphylococci by polymerase chain reaction. *Antimicrob. Agents Chemother.* **35**:2568–2573.

132. **Preheim, L., D. Pitcher, R. Owen, and B. Cookson.** 1991. Typing of methicillin resistant and susceptible *Staphylococcus aureus* strains by ribosomal RNA gene restriction patterns using a biotinylated probe. *Eur. J. Clin. Microbiol. Infect. Dis.* **10**:428–436.

133. **Prevost, G., B. Jaulhac, and Y. Piemont.** 1992. DNA fingerprinting by pulsed-field gel electrophoresis is more effective than ribotyping in distinguishing among methicillin-resistant *Staphylococcus aureus* isolates. *J. Clin. Microbiol.* **30**:967–973.

134. **Raus, J., and D. N. Love.** 1983. Characterization of coagulase-positive *Staphylococcus intermedius* and *Staphylococcus aureus* isolated from veterinary clinical specimens. *J. Clin. Microbiol.* **18**:789–792.

135. **Rosenbach, F. J.** 1884. *Mikro-organismen bei den Wund-Infections-Krankheiten des Menschen.* J. F. Bergmann, Wiesbaden, Germany.

136. **Schleifer, K. H.** 1986. Gram-positive cocci, p. 999–1002. *In* J. G. Holt, P. H. A. Sneath, N. S. Mair, and M. S. Sharpe (ed.), *Bergey's Manual of Systematic Bacteriology*, vol. 2. The Williams & Wilkins Co., Baltimore.

137. **Schleifer, K. H.** 1986. Taxonomy of coagulase-negative staphylococci, p. 11–26. *In* P.-A. Mårdh and K. H. Schleifer (ed.), *Coagulase-Negative Staphylococci.* Almqvist and Wiksell International, Stockholm.

138. **Schleifer, K. H., and W. E. Kloos.** 1975. Isolation and characterization of staphylococci from human skin. I.

139. **Schleifer, K. H., and E. Krämer.** 1980. Selective medium for isolating staphylococci. *Zentralbl. Bakteriol. Hyg. Abt. 1 Orig. Reihe C* **1**:270–280.

140. **Schlievert, P. M., K. N. Shands, B. B. Dan, G. P. Schmid, and R. D. Nishimura.** 1981. Identification and characterization of an exotoxin from *Staphylococcus aureus* associated with toxic shock syndrome. *J. Infect. Dis.* **143**:509–516.

141. **Schwalbe, R. S., J. T. Stapleton, and P. H. Gilligan.** 1987. Emergence of vancomycin resistance in coagulase-negative staphylococci. *N. Engl. J. Med.* **316**:927–931.

142. **Selladurai, B. M., S. Sivakumaran, A. Subramanian, and A. R. Mohamad.** 1993. Intracranial suppuration caused by *Micrococcus luteus*. *Br. J. Neurosurg.* **7**:205–208.

143. **Shuttleworth, R., and W. D. Colby.** 1992. *Staphylococcus lugdunensis* endocarditis. *J. Clin. Microbiol.* **30**:1948–1952.

144. **Silvestri, L. G., and L. R. Hill.** 1965. Agreement between deoxyribonucleic acid base composition and taxonomic classification of gram-positive cocci. *J. Bacteriol.* **90**:136–140.

145. **Sobel, J. D., and D. Kaye.** 1985. Urinary tract infections, p. 426–452. *In* G. L. Mandell, R. G. Douglas, Jr., and J. E. Bennett (ed.), *Principles and Practice of Infectious Disease*, 2nd ed. John Wiley & Sons, Inc., New York.

146. **Souhami, L., R. Feld, P. G. Tuffnell, and T. Fellner.** 1979. *Micrococcus luteus* pneumonia: a case report and review of the literature. *Med. Pediatr. Oncol.* **7**:309–314.

147. **Stackebrandt, E., and M. Teuber.** 1988. Molecular taxonomy and phylogenetic position of lactic acid bacteria. *Biochimie* **70**:317–324.

148. **Stackebrandt, E., and C. R. Woese.** 1979. A phylogenetic dissection of the family *Micrococcaceae*. *Curr. Microbiol.* **2**:317–322.

149. **Stamm, W. E.** 1988. Protocol for diagnosis of urinary tract infection: reconsidering the criterion for significant bacteriuria. *Urology* **32**(Suppl.):6–10.

150. **Stewart, P. R., W. El-Adhami, B. Inglis, and J. C. Franklin.** 1993. Analysis of an outbreak of variably methicillin-resistant *Staphylococcus aureus* with chromosomal RFLPs and *mec* region probes. *J. Med. Microbiol.* **38**:270–277.

151. **Suzuki, E., K. Hiramatsu, and T. Yokota.** 1992. Survey of methicillin-resistant clinical strains of coagulase-negative staphylococci for *mecA* gene distribution. *Antimicrob. Agents Chemother.* **36**:429–434.

152. **Talan, D., D. Staatz, A. Staatz, E. Goldstein, and G. Overturf.** 1988. *Staphylococcus intermedius* in canine gingiva and canine-inflicted human wound infections. A new zoonotic pathogen, abstr. C-357, p. 391. *Abstr. Annu. Meet. Am. Soc. Microbiol. 1988.*

153. **Tamer, M. A., and J. D. Bray.** 1986. Trimethroprim-sulfamethoxazole treatment of multiantibiotic-resistant staphylococci endocarditis and meningitis. *Clin. Pediatr.* **21**:125–126.

154. **Tanasupawat, S., Y. Hashimoto, T. Ezaki, M. Kozaki, and K. Komagata.** 1992. *Staphylococcus piscifermentans* sp. nov., from fermented fish in Thailand. *Int. J. Syst. Bacteriol.* **42**:577–581.

155. **Tenover, F. C.** 1988. Diagnostic deoxyribonucleic acid probes for infectious diseases. *Clin. Microbiol. Rev.* **1**:82–101.

156. **Thomson, K. S., C. C. Sanders, and M. E. Hayden.** 1991. In vitro studies with five quinolones: evidence for changes in relative potency as quinolone resistance arises. *Antimicrob. Agents Chemother.* **35**:2329–2334.

157. **Thomson-Carter, F. M., P. E. Carter, and T. H. Pennington.** 1989. Differentiation of staphylococcal species and strains by ribosomal RNA gene restriction patterns. *J. Gen. Microbiol.* **135**:2093–2097.

158. **Thomson-Carter, F. M., and T. H. Pennington.** 1989. Characterization of coagulase-negative staphylococci by sodium dodecyl sulfate-polyacrylamide gel electrophoresis and immunoblot analyses. *J. Clin. Microbiol.* **27**:2199–2203.

159. **Tierno, P. M., Jr., and B. A. Hanna.** 1989. Ecology of toxic

shock syndrome: amplification of toxic shock syndrome toxin 1 by materials of medical interest. *Rev. Infect. Dis.* **11**(Suppl 1):S182–S186.

160. **Tojo, M., M. Yamashita, D. A. Goldmann, and G. B. Pier.** 1988. Isolation and characterization of a capsular polysaccharide adhesin from *Staphylococcus epidermidis. J. Infect. Dis.* **157**:713–730.

161. **Tokue, Y., S. Shoji, K. Satoh, A. Wantanabe, and M. Motomiya.** 1992. Comparison of a polymerase chain reaction assay and a conventional microbiologic method for detection of methicillin-resistant *Staphylococcus aureus. Antimicrob. Agents Chemother.* **36**:6–9.

162. **Tollefson, D. F., D. F. Bandyk, H. W. Kaebnick, G. R. Seabrook, and J. B. Towne.** 1987. Surface biofilm disruption-enhanced recovery of microorganisms from vascular prostheses. *Arch. Surg.* **122**:38–43.

163. **Veach, L. A., M. A. Pfaller, M. Barrett, F. P. Koontz, and R. P. Wenzel.** 1990. Vancomycin resistance in *Staphylococcus haemolyticus* causing colonization and blood stream infection. *J. Clin. Microbiol.* **28**:2064–2068.

164. **Wallmark, G. I., I. Anemark, and B. Telander.** 1978. *Staphylococcus saprophyticus:* a frequent cause of urinary tract infections among female outpatients. *J. Infect. Dis.* **138**:791–797.

165. **Walsh, T. J., H. C. Standiford, A. C. Reboli, J. F. John, M. E. Mulligan, B. S. Ribner, J. Z. Montgomerie, M. B. Goetz, C. G. Mayhall, D. Rimland, D. A. Stevens, S. L. Hansen, G. C. Gerard, and R. J. Ragual.** 1993. Randomized double-blind trial of rifampin with either novobiocin or trimethoprim-sulfamethoxazole against methicillin-resistant *Staphylococcus aureus* colonization: prevention of antimicrobial resistance and effect of host factors on outcome. *Antimicrob. Agents Chemother.* **37**:1334–1342.

166. **Ward, T. T., R. E. Winn, A. F. Hartstein, and D. L. Sewell.** 1981. Observations relating to an interhospital outbreak of methicillin-resistant *Staphylococcus aureus:* role of antimicrobial therapy in infection control. *Infect. Control* **2**:453–459.

167. **Webster, J. A., T. L. Bannerman, R. J. Hubner, D. N. Ballard, E. M. Cole, J. L. Bruce, F. Fiedler, K. Schubert, and W. E. Kloos.** 1994. Identification of the *Staphylococcus sciuri* species group with EcoRI fragments containing rRNA sequences and description of *Staphylococcus vitulus* sp. nov. *Int. J. Syst. Bacteriol.* **44**:454–460.

168. **Welch, D.** 1991. Applications of cellular fatty acid analysis. *Clin. Microbiol. Rev.* **4**:422–438.

169. **Westblom, T. U., G. J. Gorse, T. W. Milligan, and A. H. Schindzielorz.** 1990. Anaerobic endocarditis caused by *Staphylococcus saccharolyticus. J. Clin. Microbiol.* **28**:2818–2819.

170. **Wharton, M., J. R. Rice, R. McCallum, and H. A. Gallis.** 1986. Septic arthritis due to *Micrococcus luteus. J. Rheumatol.* **13**:659–660.

171. **Wood, C. A., D. L. Sewell, and L. J. Strausbaugh.** 1989. Vertebral osteomyelitis and native valve endocarditis caused by *Staphylococcus warneri. Diagn. Microbiol. Infect. Dis.* **12**:261–263.

172. **Youmans, G. R., T. E. Davis, and D. D. Fuller.** 1993. Use of chemiluminescent DNA probes in the rapid detection of oxacillin resistance in clinically isolated strains of *Staphylococcus aureus. Diagn. Microbiol. Infect. Dis.* **16**:99–104.

173. **Zimmerman, R. J., and W. E. Kloos.** 1976. Comparative zone electrophoresis of esterases of *Staphylococcus* species isolated from mammalian skin. *Can. J. Microbiol.* **22**:771–779.

Streptococcus

KATHRYN L. RUOFF

23

TAXONOMY

Important changes in the classification of the streptococci have resulted from relatively recent molecular taxonomic studies of the genus. The enterococci (previously considered group D streptococci) and the lactococci (formerly classified as group N streptococci) now reside in their own genera, *Enterococcus* and *Lactococcus*, respectively (41). Although traditional phenotypic criteria (hemolytic reactions, Lancefield serology) for classification of the streptococci are still useful in certain circumstances, the older classification schemes must be tempered by new taxonomic knowledge. In the case of beta-hemolytic streptococci, we now know that unrelated species may produce identical Lancefield antigens (30) and that strains genetically related at the species level may evidence heterogeneous Lancefield antigens (18).

In spite of these exceptions to the traditional rules of streptococcal taxonomy, hemolytic reactions and Lancefield serology can still be used to divide the streptococci into broad categories as a first step in identification of clinical isolates. Beta-hemolytic isolates with Lancefield group A, C, or G antigen can be subdivided into two groups: large-colony (>0.5 mm in diameter) and small-colony (<0.5 mm in diameter) formers. Large-colony-forming group A (*Streptococcus pyogenes*) C and G strains are "pyogenic" streptococci replete with a variety of effective virulence mechanisms. The small-colony-forming beta-hemolytic strains with the same Lancefield antigens are members of a group of species ("S. milleri" group) belonging to the "viridans" streptococcal division. Although the small-colony-forming strains may participate in infection, they are also found as commensals whose pathogenic abilities appear to be much more subtle than those of the pyogenic streptococci. *S. agalactiae* is still identified reliably by its production of Lancefield group B antigen or other phenotypic traits.

Among non-beta-hemolytic streptococcal strains, alpha-reacting isolates can be separated into the species *S. pneumoniae* (optochin and bile susceptible) and the viridans division, composed of a number of species groups. The nutritionally variant streptococci, previously thought to be nutritional mutants of viridans strains, have recently been shown to constitute a separate group of streptococcal species (6). Streptococci with Lancefield group D antigen include the nonhemolytic species S. bovis. Organisms previously thought to be anaerobic streptococci have been shown to be unrelated to members of the genus *Streptococcus* (32).

DESCRIPTION OF THE GENUS

Bergey's Manual of Systematic Bacteriology (24) describes streptococci as gram-positive, catalase-negative, facultatively anaerobic bacteria that are spherical or ovoid and less than 2 μm in diameter. The reader is referred to Procedures for Initial Differentiation of Genera with Negative or Weak Catalase Reactions in chapter 25 of this Manual for a description of the catalase test. Although streptococci grow in the presence of oxygen, they are unable to synthesize heme compounds and are therefore incapable of respiratory metabolism. Some strains (S. pneumoniae, certain viridans species) require elevated levels (5%) of CO_2 for growth; the growth of many streptococcal isolates is stimulated in a CO_2-enriched atmosphere. Streptococci are nutritionally fastidious with variable nutritional requirements, and growth on complex media is enhanced by the addition of blood or serum. Glucose and other carbohydrates are metabolized fermentatively, with lactic acid as the major metabolic end product. Gas is not produced by glucose metabolism. Isolates of streptococci produce the enzyme leucine aminopeptidase (also called leucine arylamidase, or LAP); production of pyrrolidonyl arylamidase (PYR) varies depending on strain. All streptococci examined to date are susceptible to vancomycin.

NATURAL HABITATS

Streptococci are usually found as parasites of humans and other animals. While some streptococci function as virulent pathogens, other strains live harmoniously with their hosts as normally avirulent commensals. Streptococci colonize the skin and mucous membranes and can be isolated as part of the normal flora of the alimentary, respiratory, and genital tracts.

CLINICAL SIGNIFICANCE

Group A Streptococci (*S. pyogenes*)

Beta-hemolytic, large-colony-forming streptococci with Lancefield group A antigen are included in the species *S. pyogenes* and represent one of the most impressive human pathogens. The numerous virulence factors of *S. pyogenes* (M protein, lipoteichoic acid, enzymes, toxins) allow it to produce a wide array of serious infections, including pharyngitis, respiratory infection, skin (impetigo, erysipelas) and soft tissue infections, endocarditis, meningitis, puerperal sepsis, and arthritis. Infection with toxin-producing strains can result in scarlet fever or more serious toxic shock-like syndromes.

Group A streptococcal pharyngitis is characterized by pharyngeal pain, swelling, and erythema accompanied by fever and anterior cervical adenopathy. Suppurative sequelae of streptococcal pharyngitis may result from the spread of infection to contiguous tissue or from bacteremic dissemination. Nonsuppurative sequelae include rheumatic fever and acute glomerulonephritis. While either of these conditions may follow pharyngitis, only glomerulonephritis is linked with group A streptococcal infections of the skin. Recent observations suggest that the incidence of rheumatic fever and extremely severe streptococcal infections is currently on the increase in the United States (5).

Group B Streptococci

Beta-hemolytic streptococci with Lancefield group B antigen (*S. agalactiae*) are an important cause of serious neonatal infection characterized by sepsis and meningitis. Heavy colonization of the maternal genital tract is associated with colonization of infants and risk of neonatal disease. Early-onset infection occurs within the first few days after delivery and often is associated with pneumonia, while late-onset disease usually appears after 1 week of age. Group B streptococci are also associated with postpartum infections. Conditions that predispose nonpregnant adults to group B streptococcal infection include diabetes mellitus, cancer, and human immunodeficiency virus infection. Adult group B infections include bacteremia, endocarditis, skin and soft tissue infection, and osteomyelitis (44).

Other Beta-Hemolytic Streptococci (Non-Group A or B)

Group C and G streptococci that form large colonies are pyogenic streptococci similar to *S. pyogenes* and cause a wide range of serious infections such as bacteremia, endocarditis, meningitis, septic arthritis, and infections of the respiratory tract and skin. Although the clinical symptoms of pharyngeal infection caused by these streptococci are similar to those of group A pharyngitis, group C and G streptococci lack a strong association with nonsuppurative sequelae. Two reports, however, describe poststreptococcal glomerulonephritis associated with outbreaks of group C pharyngitis (3, 15).

Small-colony-forming beta-hemolytic streptococcal strains may express Lancefield group A, C, F, or G antigen or may be nongroupable. These streptococci are usually identified as members of the "*S. milleri*" group of species. Although these organisms can be isolated from pyogenic infections (notably abscesses), they appear to reside in the pharynx as commensals and are probably not agents of pharyngitis (8).

S. pneumoniae

The pneumococcus is an important agent of community-acquired pneumonia that may be accompanied by bacteremia. Oropharyngeal carriage of pneumococci is common and contributes to the difficulty of interpreting the significance of pneumococci in cultures of expectorated sputum. Other pneumococcal infections include otitis media, sinusitis, meningitis, and endocarditis. Vaccines designed to protect against infection by pneumococci with predominant capsular polysaccharide types are now available.

Viridans Streptococci

Strains of viridans streptococci are normal inhabitants of the mammalian oral cavity, gastrointestinal tract, and female genital tract. Although these streptococci often appear as contaminants in blood cultures, their presence may be associated with subacute bacterial endocarditis, especially in patients with damaged or prosthetic valves. The viridans streptococci are also isolated from infections in other organ systems and seem to be assuming an increasing role in infection in neutropenic patients (16).

S. bovis

Bacteremia caused by *S. bovis* isolates, particularly biotype I, is associated with malignancies of the gastrointestinal tract (40). These organisms are also agents of endocarditis and have been isolated from patients with meningitis (37).

Nutritionally Variant Streptococci

The nutritionally variant streptococci are normal residents of the oral cavity but have been identified as agents of endocarditis involving both native and prosthetic valves. These organisms have also been isolated from patients with ophthalmic infections (35).

COLLECTION, TRANSPORT, AND STORAGE OF SPECIMENS

Specimens suspected of harboring streptococci should be collected according to protocols outlined in chapter 3 of this Manual. A transport system need not be used if transport time is under 2 h. Although some streptococci (i.e., *S. pyogenes*) can survive desiccation and refrigeration, other strains (e.g., pneumococci) are fairly fragile, and thus, every effort should be made to process specimens as soon as possible.

ISOLATION PROCEDURES

General Procedures

Many commonly used nonselective laboratory media support the growth of streptococci; complex media enriched with blood are desirable, because the hemolytic reaction of the streptococcal isolate may be determined early in the identification process. Hemolytic patterns may vary with the source of animal blood or type of basal medium used in blood agars. Media selective for gram-positive bacteria (e.g., phenylethyl alcohol or Columbia with colistin and nalidixic acid agars) will also support growth of streptococci. Nutritionally variant streptococci usually grow on chocolate agar and brucella agar with 5% horse blood and in thioglycolate broth but not on Trypticase soy agar with 5% sheep blood. Nutritionally variant streptococci can be cultured on nonsupportive media that have been appropriately supplemented (see Tests for Satelliting Behavior below).

Cultures for isolation of streptococci should be incubated at 35 to 37°C. Ambient atmospheres are suitable for many streptococci, but since pneumococci and some viridans strains require elevated CO_2 concentrations, incubation in an atmosphere containing 5% CO_2 or in a candle jar will enhance recovery of streptococcal isolates. Streptococci grow well in anaerobic atmospheres, but their facultative nature makes anaerobic incubation unnecessary for isolation.

Special Procedures for Throat Cultures

Complex media containing 5% sheep blood are usually recommended for the culture of throat specimens because nicotinamide adenine dinucleotidase (NADase) activity in sheep blood reduces the NAD content to levels insufficient for growth of *Haemophilus haemolyticus*, a commensal that forms colonies that might be confused with those of beta-hemolytic streptococci. Throat swabs should be rolled firmly over one-sixth of the plate to deposit the specimen. A loop is employed to carefully streak the inoculum over the surface of the plate. The loop is then used (without sterilization) to stab the agar several times in an area of the plate that has not been streaked in an effort to deposit beta-hemolytic streptococci beneath the agar surface. Subsurface growth will display the most reliable hemolytic reactions owing to the activity of both oxygen-stable and oxygen-labile streptolysins. Other procedures involving overnight incubation of swabs in broth for enrichment of streptococci and a pour-streak plate method for culturing throat specimens have also been described (17).

In a review of methods designed to improve recovery of beta-hemolytic streptococci from throat specimens, Kellogg (29) concluded that 90 to 95% of *S. pyogenes* isolates from symptomatic patients would be detected by any of the following protocols: sheep blood agar incubated anaerobically for 48 h; sheep blood agar incubated aerobically without CO_2 supplementation for 48 h (a cover glass can be placed over the primary inoculation area in order to reduce oxygen tension and enhance hemolysis); sheep blood agar containing trimethoprim-sulfamethoxazole incubated aerobically in 5 to 10% CO_2 for 48 h; sheep blood agar containing trimethoprim-sulfamethoxazole incubated anaerobically for 48 h. Cultures should be examined after 18 to 24 h of incubation and reincubated if negative, with a final examination at 48 h. Incubation of throat culture plates in a CO_2-enriched atmosphere seems to encourage the recovery of non-group A beta-hemolytic streptococci. For further details, the reader is referred to Kellogg's review (29) and a study evaluating two commercially available selective media (43).

Special Procedures for Group B Streptococci

Detection of group B streptococci in the genital tracts of expectant mothers can identify infants at risk for infection and guide intrapartum administration of antibiotics. Selective enrichment broths, consisting of Todd-Hewitt broth supplemented with gentamicin and nalidixic acid (36) or with colistin and nalidixic acid (28), have been described for enhanced recovery of group B streptococci from genital specimens.

IDENTIFICATION

Direct Detection

Direct detection of streptococci via Gram stains is most useful when applied to specimens that are normally devoid of indigenous streptococcal flora. The Quellung test is a more specific method for microscopic detection of pneumococci in specimens and relies on visual enhancement of the pneumococcal capsule after reaction of the streptococci with anticapsular antisera. Details for performance of the Quellung test can be found in the previous edition of this Manual. Other specific microscopic methods for detecting group A and B streptococci have employed immunofluorescence staining (17).

Direct detection of streptococcal antigens has been used to identify group A streptococci in throat specimens. The sensitivities of these techniques depend not only on the methodology used but also on the numbers of streptococci present in the sample. In general, streptococci are transferred from the swab containing the specimen to a chemical (e.g., nitrous acid) or enzymatic (e.g., pronase) extraction solution. After a short incubation period, antigen in the suspension is detected via agglutination methods (19), enzyme immunoassay (14, 27), or other methods (20, 23). While antigen detection methods are rapid and specific, false-negative results may occur with specimens containing low numbers of streptococci (19).

Antigen detection techniques have also been used for rapid identification of group B streptococci in urogenital specimens. Neither latex agglutination assays (22) nor enzyme immunoassay methods (21, 47) have proven sensitive enough to detect low levels of colonization. These methods do, however, seem to be effective in identifying heavily colonized women. Older protocols, involving incubation of specimens in a selective medium for 5 to 20 h before antigen testing via agglutination, reveal even light colonization (31).

Normally sterile body fluids may also be tested directly for streptococcal antigens. Commercially available products that detect the Lancefield antigen of group B streptococci and the capsular polysaccharide antigen of pneumococci may be used to determine the presence of these organisms in cerebrospinal fluid and blood cultures.

Recently, a nucleic acid probe for direct detection of group A streptococci in throat swabs has become commercially available (25). Additional evaluation of this or similar products is necessary before their true utility and cost-effectiveness can be established. Use of a nucleic acid probe-based product for direct detection of *S. pneumoniae* and group B streptococci from positive blood cultures has also been described (11).

Hemolytic Reactions

Beta-hemolysis appears as complete clearing (lysis of erythrocytes) of the medium. This reaction may be obscured by the inhibition of streptolysin O by oxygen or by the production of peroxide by streptococci growing in air or in the presence of increased CO_2; thus, anaerobic incubation or observing hemolysis in the area of stabs in the agar is optimal for accurate determination of beta-hemolytic reactions. In the alpha reaction, blood cells are not completely lysed, but growth is surrounded by a greenish discoloration of the agar due to streptococcal action on hemoglobin. Nonhemolytic or gamma-hemolytic streptococci have no effect on blood agar.

TABLE 1 Differentiating characteristics of beta-hemolytic streptococci

Lancefield group	Colony size	Species	PYR	VP	CAMP[a]	Trehalose	Sorbitol
A	Large	S. pyogenes	+	−	−		
A	Small	"S. milleri" group	−	+	−		
B		S. agalactiae	−	−	+		
C	Large	S. equi[b]	−	−	−	−	−
		S. equisimilis[b]	−	−	−	+	−
		S. zooepidemicus[b]	−	−	−	−	+
C	Small	"S. milleri"	−	+	−		
F	Small	"S. milleri" group	−	+	−		
G	Large[b]		−	−	−		
G	Small	"S. milleri" group	−	+	−		
Nongroupable	Small	"S. milleri" group	−	+	−		

[a]Test for synergistic hemolysis.
[b]Traditional classification of large-colony group C streptococci is presented here. See text for a discussion of more recent taxonomic data.

Description of Colonies

Streptococcal colonies vary in color from gray to whitish and usually glisten, although dry colonies are also observed. The beta-hemolytic pyogenic streptococci of groups A, C, and G form relatively large (>0.5 mm in diameter after 24 h of incubation) colonies compared with the pinpoint colonies of the small-colony-forming beta-hemolytic strains of the "S. milleri" group. Cultures of the small-colony-forming beta-hemolytic streptococci often produce a distinct odor that has been described as buttery or caramel-like. Colonies of group B streptococci tend to be larger and have less pronounced zones of beta-hemolysis than other beta-hemolytic strains have; some group B strains are non-hemolytic.

Alpha-hemolytic colonies with depressions in their centers are characteristic of pneumococci, while viridans streptococcal colonies have a domed appearance. Pneumococci may also produce various amounts of capsular polysaccha-ride, contributing to a mucoid colonial appearance. Some viridans streptococcal strains, along with S. bovis, form nonhemolytic, grayish colonies.

Identification of Beta-Hemolytic Isolates

Serological Tests

Numerous products using rapid antigen extraction methods and agglutination techniques for antigen detection are commercially available for the Lancefield grouping of beta-hemolytic isolates. Older antigen extraction methods and a protocol for precipitin testing are detailed in the previous edition of this Manual (17). The presence of the group B antigen seems to correlate closely with a strain's identity as S. agalactiae, and group F isolates are all small-colony-forming "S. milleri" group strains. Lancefield antigens of groups A, C, and G, however, are not specific to a single type of streptococcus. Organisms with these antigens can be differentiated with biochemical tests summarized in Table 1. When nongroupable beta-hemolytic isolates that fail to react with Lancefield group A, B, C, F, or G antisera are encountered, physiological characterization may aid in identification.

Physiological Tests

PYR Test

The PYR test determines the activity of pyrrolidonyl arylamidase, an enzyme produced by S. pyogenes but not by other beta-hemolytic streptococci (Table 1). Rapid meth-ods for performance of the PYR test are commercially available. Since other organisms that may be isolated along with S. pyogenes may also give a positive PYR test, only pure cultures or isolated colonies of beta-hemolytic streptococci should be tested. Beta-hemolytic strains of enterococci might be confused with S. pyogenes, since both organisms are PYR positive. Differences in colonial size and morphol-ogy should allow for differentiation of these bacteria (see chapter 24 of this Manual).

VP Test

A positive Voges-Proskauer (VP) test for acetoin pro-duction will differentiate small-colony-forming beta-hemo-lytic "S. milleri" group strains with Lancefield group A, C, or G antigens from large-colony-forming pyogenic strains with the same Lancefield antigens (Table 1). Facklam and Washington (17) describe a version of this test for strepto-cocci in which overnight growth from an entire agar plate culture is used to inoculate 2 ml of VP broth. After 6 h of incubation at 35°C, Coblentz reagents A and B are added, and the tube is shaken and then incubated at room tem-perature for 30 min. A positive test is indicated by devel-opment of a cherry red (or even slightly pink) color.

Differentiation of Large-Colony Group C Isolates with Carbohydrate Fermentation Tests

Traditionally, large-colony-forming group C isolates have been differentiated into species based on their abilities to ferment various carbohydrates (Table 1). Studies of genetic relatedness of group C and G streptococci (18) have, however, shown that human isolates of the group C species S. equisimilis are genetically similar enough to group G large-colony human isolates to be classified in the same species. Strains traditionally classified as S. zooepidemicus were considered, on the basis of genetic data, to form a subspecies of S. equi. If traditional characterization of large-colony group C streptococci is desired, the carbohydrate fermentation broth used for viridans streptococci (see be-low) can be utilized.

CAMP Test

The majority of group B streptococci produce a diffusible extracellular protein (CAMP factor) that acts synergisti-cally with staphylococcal beta-lysin to cause lysis of eryth-rocytes. Single straight streaks of the streptococcus to be

TABLE 2 Differentiation of non-beta-hemolytic streptococci[a]

Type or species of streptococcus	Optochin susceptibility	Bile solubility	Bile esculin	Satelliting behavior
S. pneumoniae[b]	+	+	−	−
Viridans	−	−	−	−
S. bovis	−	−	+	−
Nutritionally variant[c]				+

[a]Strains of group B streptococci may be non-beta-hemolytic. Serological methods or the CAMP test should be used to rule out possible group B strains.
[b]Strains may be PYR positive.
[c]PYR positive.

tested and a beta-lysin-producing *S. aureus* strain are made perpendicular to each other and about 3 to 4 mm apart on the surface of a sheep blood agar plate. After overnight incubation in ambient atmosphere at 35°C, a positive test will appear as an arrowhead-shaped zone of complete hemolysis in the area into which both staphylococcal beta-lysin and CAMP factor have diffused (17). An alternative method employing beta-lysin-containing disks has also been described (46).

Nucleic Acid Probe Tests

Nucleic acid probes for identification of cultured isolates of group A and B streptococci are currently available (10).

Identification of Non-Beta-Hemolytic Isolates

Serological Tests

Serological testing can be useful for revealing non-beta-hemolytic strains of group B streptococci. Testing for Lancefield group D antigen may aid in identification of *S. bovis*, but this antigen is not easily demonstrated in some strains. Moreover, the group D antigen is nonspecific; it is produced by certain streptococci and by members of the genera *Enterococcus* and *Pediococcus*. Physiological testing is more reliable for identifying non-beta-hemolytic isolates.

Physiological Tests

Table 2 summarizes the physiological characteristics of non-beta-hemolytic (alpha- and nonhemolytic) streptococci. The optochin and bile solubility tests are used to differentiate pneumococci from viridans streptococci; further characterization of viridans streptococci to the species or species group level can be accomplished with physiological testing and identification kits. Nutritionally variant strains are characterized by slow growth, tiny colonies, and pleomorphic Gram stains. Growth will not occur on Trypticase soy agar with sheep blood unless the medium is supplemented with pyridoxal or cross-streaked with *S. aureus*. *S. adjacens* and *S. defectivus*, the two species of nutritionally variant streptococci so far described (6), are PYR positive.

The nonhemolytic streptococci may belong to the species *S. bovis* or to various viridans species. The ability of *S. bovis* to hydrolyze the glycoside esculin in the presence of 40% bile distinguishes it from viridans strains. Complete characterization of *S. bovis* to the biotype level can be accomplished with commercially available products (40). The CAMP test may be of use for testing isolates suspected of being non-beta-hemolytic group B streptococci.

Optochin Test

Commercially available optochin disks are applied to a quarter of a blood agar plate that has been streaked with a few colonies of the organism to be tested. After overnight incubation at 35°C in either a candle jar or a CO_2 incubator, inhibition zones are measured. Zones of ≥14 mm with a 6-mm disk or ≥16 mm with a 10-mm disk indicate inhibition and identify the isolate as *S. pneumoniae*. Isolates displaying smaller zones of inhibition should be subjected to an additional test for bile solubility to confirm their identities (17).

Bile Solubility Test

Aliquots containing 0.5 ml of a 0.5 to 1.0 McFarland standard saline suspension of the organisms to be tested are added to each of two small test tubes (13 by 100 mm). An equal amount (0.5 ml) of 2% sodium deoxycholate (bile) is added to one tube, while 0.5 ml of saline is added to the second tube to serve as a control. A positive bile solubility test (indicative of pneumococci) appears as clearing in the presence of deoxycholate but not in the control tube after incubation at 35°C for up to 2 h (17).

In a plate method for testing bile solubility, a drop of a solution of 10% sodium deoxycholate is placed directly on a colony of the strain to be tested. The plate can be kept at room temperature or placed into an aerobic incubator at 35°C for approximately 15 min until the reagent dries, but it must be kept level to prevent the reagent from running across the plate and washing away the colony. Pneumococcal colonies will disappear or be flattened, while bile-resistant streptococcal colonies will be unaffected (4).

Bile Esculin Test

Bile esculin medium (available from commercial sources) in either plates or slants should be inoculated with one to three colonies of the organism to be tested and incubated at 35°C in an ambient atmosphere for up to 48 h. A definite blackening of plated medium or blackening of at least one-half of an agar slant is considered a positive test, indicating *S. bovis* (17). Occasional viridans strains will be positive in this test or will display weakly positive reactions that are difficult to interpret (40).

Tests for Satelliting Behavior

A strain to be tested for satelliting behavior is streaked for confluent growth on a medium that fails to support growth or supports only weak growth. A single cross-streak of *S. aureus* (ATCC 25923 is suitable) is applied to the inoculated area. After incubation at 35°C in an atmosphere containing an elevated level of CO_2, nutritionally variant

TABLE 3 Characteristics of commonly isolated viridans streptococci[a]

Species	VP	Arginine	Esculin	Mannitol	Sorbitol	Urease
S. mutans group	+	−[b]	+	+	+	−
S. salivarius group[c]	+[d]	−	+	−	−	±
S. bovis[c]	+	−	+	±[e]	−	−
"S. milleri" group	+	+	±	±	−	−
S. sanguis group	−	+[f]	+[g]	−	−	−
S. mitis	−	−	−	−	−	−

[a]Information in this table is based on a review by Coykendall (9). See text for test methods.

[b]S. rattus is arginine positive.

[c]The urease test differentiates S. salivarius (about 70% positive) from mannitol-negative strains of S. bovis. Commercially available identification kits also differentiate these species accurately (40).

[d]S. vestibularis is VP variable. S. vestibularis is also alpha-hemolytic, unlike nonhemolytic S. salivarius.

[e]S. bovis biotype I is positive; biotype II is negative.

[f]S. crista may be arginine negative.

[g]S. crista is esculin negative; S. parasanguis may be negative.

streptococci will grow only in the vicinity of the staphylococcal growth. Alternatively, medium can be supplemented with pyridoxal in the form of an aqueous stock solution of filter-sterilized 0.01% pyridoxal. This solution, which can be stored frozen, should be added to medium to achieve a final concentration of 0.001% (38).

Nucleic Acid Probe Tests

A nucleic acid probe test for identification of pneumococcal isolates is now commercially available (12).

Identification of Viridans Streptococci and S. bovis: Physiological Tests and Identification Kits

In view of the changing taxonomy of the viridans streptococci and the difficulty of accurate identification when phenotypic methods alone are employed, the simplified approach to identification illustrated in Table 3 and advocated by Facklam and Washington (17) seems reasonable for the clinical laboratory. Commercially available kits provide species identifications of viridans streptococci, but the nomenclature systems used are not always uniform among these products and may not be current with respect to the latest taxonomic changes (26). Commercially available identification systems do, however, provide biotypes that may be useful with patients who present with recurring positive cultures (e.g., bacteremias in an endocarditis patient): knowing the biotypes of isolates from consecutive cultures helps assess the efficacy of treatment.

The S. mutans group contains at least seven distinct species, including the nonhemolytic cariogenic S. mutans. This organism may form rod-shaped cells under certain conditions and produces a glucose-containing polysaccharide. Other members of the S. mutans group that are iso-

lated from humans are S. sobrinus, S. rattus, and S. cricetus. S. salivarius and S. vestibularis are the two members of the S. salivarius species group that are commonly isolated from humans. Nonhemolytic S. salivarius has some phenotypic resemblance to S. bovis but may be urease positive, a unique trait among streptococcal strains. The S. sanguis group includes S. sanguis, S. gordonii, S. parasanguis, and S. crista; the group has been known in previous taxonomic schemes as S. sanguis I. Streptococci belonging to the S. mitis group have been known in the past as S. mitior, S. sanguis II, and S. oralis. The "S. milleri" group encompasses alpha-, beta-, and nonhemolytic strains with various (A, C, F, G) or no demonstrable Lancefield antigen. These organisms have been known as a single species ("S. milleri," S. anginosus) or as multiple species (S. anginosus-constellatus, S. MG-intermedius) in past classification schemes. Currently, they are divided into three separate species (S. anginosus, S. intermedius, and S. constellatus) on the basis of genetic relatedness studies. Whiley and coworkers (45) have identified phenotypic traits for distinguishing the three species (Table 4). General information on viridans streptococci can be found in references 9 and 39.

VP Test

See Identification of Beta-Hemolytic Isolates: Physiological Tests above.

Arginine Degradation Test

Moeller's decarboxylase broth containing arginine (commercially available) should be inoculated with the test organism, overlaid with mineral oil, and incubated at 35°C for up to 7 days. Degradation of arginine results in an increase in pH, indicated by development of a purple color.

TABLE 4 Current classification of "S. milleri" species group (43)

Enzyme	S. anginosus	S. constellatus	S. intermedius
β-D-Fucosidase	−	−	+
β-N-Acetylglucosaminidase	−	−	+
β-N-Acetylgalactosaminidase	−	−	+
Sialidase	−	−	+
β-Galactosidase	−	−	+
β-Glucosidase	+	−	V[a]
Hyaluronidase	−	+	+

[a]V, variable (16 to 84% of strains positive).

Negative results are indicated by a yellow color, which is due to acid accumulation from metabolism of glucose only.

Esculin Hydrolysis

Esculin agar slants (available from commercial sources) are inoculated and incubated for up to 1 week. A positive reaction appears as blackening of the medium; no change in color indicates a negative esculin hydrolysis test.

Carbohydrate Fermentation Tests

Carbohydrate fermentation broth containing the desired carbohydrate is inoculated and incubated at 35°C for up to 7 days. A positive reaction appears as a yellow color, which is due to acid production; the medium remains purple in the absence of carbohydrate utilization. The medium described below has been recommended by Facklam and Washington (17) and may be obtained commercially (Remel, Lenexa, Kans.).

Carbohydrate fermentation broth (34)

Heart infusion broth ..... 22.5 g in 900 ml of distilled
water
Carbohydrate .... 10 g in 100 ml of distilled
water
Bromocresol purple .... 1 ml of a solution consisting
of 1.6 g of bromocresol
purple in 100 ml of 95%
ethanol

Combine the above solutions, dispense the mixture into screw-cap tubes (3 ml per tube [13 by 100 mm]), and autoclave for 10 min at 121°C.

Urea Hydrolysis Test

Christensen urea agar (available commercially) is inoculated and incubated at 35°C for up to 7 days. Development of a pink color indicates a positive reaction.

SEROLOGICAL TESTS

Serological tests are available to detect immune responses in patient sera to both extracellular products (streptolysin O, hyaluronidase, DNase B, NADase, and streptokinase) and cellular components (M protein, group A antigen) of *S. pyogenes*. These tests are useful in demonstrating antecedent streptococcal infection in patients who lack documentation of recent infection but who present with nonsuppurative sequelae (rheumatic fever, glomerulonephritis). Ayoub and Harden (2) provide further discussion and methods for performing these tests.

ANTIBIOTIC SUSCEPTIBILITIES

In spite of occasional reports of resistance in certain isolates, penicillin remains the drug of choice for treatment of infections caused by streptococci, while cephalosporins and erythromycin serve as alternative choices for treatment (33). Although erythromycin resistance among group A streptococci is not common in North America, resistance has been observed in a number of locales worldwide (17). Susceptibility to tetracycline is variable among streptococcal strains.

Penicillin treatment failures have been ascribed to tolerance, a situation in which bacterial growth is inhibited by penicillin but the drug's bactericidal activity is greatly re-duced. Tolerance has been described in in vitro beta-hemolytic and viridans streptococci, but testing for this trait is prone to technical problems. Another possible explanation for treatment failure, especially for pharyngitis, is the presence of β-lactamase-producing bacteria at the site of infection (17). Concern about the efficacy of penicillin alone for effective treatment of life-threatening infections has led to the use of synergistic combinations of drugs (e.g., penicillin and gentamicin) for treatment of severe group B disease (17), endocarditis caused by nutritionally variant streptococci (42), and endocarditis due to viridans streptococci that are relatively resistant to penicillin (13). A recent report described a group B streptococcal isolate resistant to high levels of gentamicin and to synergistic drug combinations (7).

Penicillin resistance in pneumococci has been reported in areas all over the world. A recent publication (1) noted that 5 to 10% of pneumococcal isolates in the continental United States display low or high penicillin resistance, but the incidence in Alaska is higher (25.8%). Strains displaying resistance to penicillin only as well as resistance to penicillin and a variety of other antibiotics have been encountered. More recently, pneumococci for which MICs of expanded-spectrum cephalosporins (e.g., ceftriaxone or cefotaxime) were elevated have also been reported.

EVALUATION, INTERPRETATION, AND REPORTING OF RESULTS

Beta-hemolytic streptococci and pneumococci are virulent pathogens, and consequently, the detection and reporting of these organisms from all types of specimens have been emphasized in the clinical laboratory. Timely evaluation and reporting of throat specimens positive for *S. pyogenes* will allow prompt antibiotic treatment, reducing the risk of nonsuppurative sequelae. Controversy concerning the significance of small-colony-forming group A, C, and G beta-hemolytic streptococci in throat cultures still exists (8).

Attempts at identification of viridans streptococci to the species or species group level should be reserved for isolates from serious infections. It should be remembered that "*S. milleri*" group organisms recovered from abscess or wound specimens, even when present with other organisms, are likely to be present as pathogens instead of contaminants. As mentioned previously, *S. bovis* biotyping may be useful because of a correlation between biotype I and endocarditis and gastrointestinal malignancy. Nutritionally variant streptococci should be suspected in "culture-negative" endocarditis, and supplementation of media with pyridoxal or a staphylococcal cross-streak should be employed to enhance their isolation.

REFERENCES

1. **Applebaum, P. C.** 1992. Antimicrobial resistance in *Streptococcus pneumoniae*: an overview. *Clin. Infect. Dis.* **15:**77–83.
2. **Ayoub, E. M., and E. Harden.** 1992. Immune response to streptococcal antigens: diagnostic methods, p. 427–434. *In* N. R. Rose, E. Conway de Macario, J. L. Fahey, H. Friedman, and G. M. Penn (ed.), *Manual of Clinical Laboratory Immunology*, 4th ed. American Society for Microbiology, Washington, D.C.
3. **Barnham, M., T. J. Thornton, and K. Lange.** 1983. Nephritis caused by *Streptococcus zooepidemicus* (Lancefield group C). *Lancet* **i:**945–948.
4. **Baron, E. J., and S. M. Finegold.** 1990. Streptococci and related genera, p. 333–352. *In* E. J. Baron and S. M. Finegold

(ed.), *Bailey and Scott's Diagnostic Microbiology*, 8th ed. The C. V. Mosby Co., St. Louis.

5. **Bisno, A. L.** 1991. Group A streptococcal infections and acute rheumatic fever. *New Engl. J. Med.* **325:**783–793.

6. **Bouvet, A., F. Grimont, and P. A. D. Grimont.** 1989. *Streptococcus defectivus* sp. nov. and *Streptococcus adjacens* sp. nov., nutritionally variant streptococci from human clinical specimens. *Int. J. Syst. Bacteriol.* **39:**290–294.

7. **Buu-Hoi, A., C. LeBouguenec, and T. Horaud.** 1990. High-level chromosomal gentamicin resistance in *Streptococcus agalactiae* (group B). *Antimicrob. Agents Chemother.* **34:**985–988.

8. **Cimolai, N., B. J. Morrison, L. MacCulloch, D. F. Smith, and J. Hlady.** 1991. Beta-haemolytic non-group A streptococci and pharyngitis: a case-control study. *Eur. J. Pediatr.* **150:**776–779.

9. **Coykendall, A. L.** 1989. Classification and identification of the viridans streptococci. *Clin. Microbiol. Rev.* **2:**315–328.

10. **Daly, J. A., N. L. Clifton, K. C. Seskin, and W. M. Gooch III.** 1991. Use of rapid, nonradioactive DNA probes in culture confirmation tests to detect *Streptococcus agalactiae*, *Haemophilus influenzae*, and *Enterococcus* spp. from pediatric patients with significant infections. *J. Clin. Microbiol.* **29:**80–82.

11. **Davis, T. E., and D. D. Fuller.** 1991. Direct identification of bacterial isolates in blood cultures by using a DNA probe. *J. Clin. Microbiol.* **29:**2192–2196.

12. **Denys, G. A., and R. B. Carey.** 1992. Identification of *Streptococcus pneumoniae* with a DNA probe. *J. Clin. Microbiol.* **30:**2725–2727.

13. **Dinubile, M. J.** 1990. Treatment of endocarditis caused by relatively resistant nonenterococcal streptococci: is penicillin enough? *Rev. Infect. Dis.* **12:**112–117.

14. **Drulak, M., W. Bartholomew, L. LaScolea, D. Amsterdam, N. Gunnersen, J. Yong, C. Fijalkowski, and S. Winston.** 1991. Evaluation of the modified Visuwell Strep-A enzyme immunoassay for detection of group-A *Streptococcus* from throat swabs. *Diagn. Microbiol. Infect. Dis.* **14:**281–285.

15. **Duca, E., G. Teodorovici, C. Radu, A. Vita, P. Talasman-Niculescu, E. Bernescu, C. Feldi, and V. Rosca.** 1969. A new nephritogenic streptococcus. *J. Hyg. Camb.* **67:**691–698.

16. **Elting, L. S., G. P. Bodey, and B. H. Keefe.** 1992. Septicemia and shock syndrome due to viridans streptococci: a case-control study of predisposing factors. *Clin. Infect. Dis.* **14:**1201–1207.

17. **Facklam, R. R., and J. A. Washington II.** 1991. *Streptococcus* and related catalase-negative gram-positive cocci, p. 238–257. *In* A. Balows, W. J. Hausler, Jr., K. L. Herrmann, H. D. Isenberg, and H. J. Shadomy (ed.), *Manual of Clinical Microbiology*, 5th ed. American Society for Microbiology, Washington, D.C.

18. **Farrow, J. A. E., and M. D. Collins.** 1984. Taxonomic studies on streptococci of serological groups C, G and L and possibly related taxa. *Syst. Appl. Microbiol.* **5:**483–493.

19. **Gerber, M. A.** 1986. Diagnosis of group A beta-hemolytic streptococcal pharyngitis, use of antigen detection tests. *Diagn. Microbiol. Infect. Dis.* **4:**5S–15S.

20. **Gerber, M. A., M. F. Randolph, and K. K. DeMeo.** 1990. Liposome immunoassay for rapid identification of group A streptococci directly from throat swabs. *J. Clin. Microbiol.* **28:**1463–1464.

21. **Granato, P. A., and M. T. Petosa.** 1991. Evaluation of a rapid screening test for detecting group B streptococci in pregnant women. *J. Clin. Microbiol.* **29:**1536–1538.

22. **Green, M., B. Dashefsky, E. R. Wald, S. Laifer, J. Harger, and R. Guthrie.** 1993. Comparison of two antigen assays for rapid intrapartum detection of vaginal group B streptococcal colonization. *J. Clin. Microbiol.* **31:**78–82.

23. **Harbeck, R. J., J. Teague, G. R. Crossen, D. M. Maul, and P. L. Childers.** 1993. Novel, rapid optical immunoassay technique for detection of group A streptococci from pharyngeal specimens: comparison with standard culture methods. *J. Clin. Microbiol.* **31:**839–844.

24. **Hardie, J. M.** 1986. Genus *Streptococcus* Rosenbach 1884,22[AL], p. 1043–1071. *In* P. H. A. Sneath, N. S. Mair,

M. E. Sharpe, and J. G. Holt (ed.), *Bergey's Manual of Systematic Bacteriology*, vol. 2. The Williams & Wilkins Co., Baltimore.

25. **Heiter, B. J., and P. P. Bourbeau.** 1993. Comparison of the Gen-Probe group A streptococcus direct test with culture and a rapid streptococcal antigen detection assay for diagnosis of streptococcal pharyngitis. *J. Clin. Microbiol.* **31:**2070–2073.

26. **Hinnebusch, C. J., D. M. Nikolai, and D. A. Bruckner.** 1991. Comparison of API Rapid STREP, Baxter Microscan Rapid Pos ID Panel, BBL Minitek Differential Identification System, IDS RapID STR System, and Vitek GPI to conventional biochemical tests for identification of viridans streptococci. *Am. J. Clin. Pathol.* **96:**459–463.

27. **Hoffman, S.** 1990. Detection of group A streptococcal antigen from throat swabs with five diagnostic kits in general practice. *Diagn. Microbiol. Infect. Dis.* **13:**209–215.

28. **Jones, D. E., E. M. Friedl, K. S. Kanarek, J. K. Williams, and D. V. Lim.** 1983. Rapid identification of pregnant women heavily colonized with group B streptococci. *J. Clin. Microbiol.* **18:**558–560.

29. **Kellogg, J.** 1990. Suitability of throat culture procedures for detection of group A streptococci and as reference standards for evaluation of streptococcal antigen detection kits. *J. Clin. Microbiol.* **28:**165–169.

30. **Lawrence, J., D. M. Yajko, and W. K. Hadley.** 1985. Incidence and characterization of beta-hemolytic *Streptococcus milleri* and differentiation from *S. pyogenes* (group A), *S. equisimilis* (group C), and large-colony group G streptococci. *J. Clin. Microbiol.* **22:**772–777.

31. **Lim, D. V., W. J. Morales, and A. F. Walsh.** 1987. Lim group B strep broth and coagglutination for rapid identification of group B streptococci in preterm pregnant women. *J. Clin. Microbiol.* **25:**452–453.

32. **Ludwig, W., M. Weizenegger, R. Kilpper-Balz, and K. H. Schleiffer.** 1988. Phylogenetic relationships of anaerobic streptococci. *Int. J. Syst. Bacteriol.* **38:**15–18.

33. **Moellering, R. C., Jr.** 1990. Principles of anti-infective therapy, p. 206–318. *In* G. L. Mandell, R. G. Douglas, Jr., and J. E. Bennett (ed.), *Principles and Practice of Infectious Diseases*, 3rd ed. Churchill Livingstone, New York.

34. **Nash, P., and M. M. Krenz.** 1991. Culture media, p. 1226–1288. *In* A. Balows, W. J. Hausler, Jr., K. L. Herrmann, H. D. Isenberg, and H. J. Shadomy (ed.), *Manual of Clinical Microbiology*, 5th ed. American Society for Microbiology, Washington, D.C.

35. **Ormerod, L. D., K. L. Ruoff, D. M. Meisler, P. J. Wasson, J. C. Kinter, S. P. Dunn, J. H. Lass, and I. van de Rijn.** 1991. Infectious crystalline keratopathy. Role of nutritionally variant streptococci and other bacterial factors. *Ophthalmology* **98:**159–169.

36. **Persson, K. M.-S., and A. Forsgren.** 1987. Evaluation of culture methods for isolation of group B streptococci. *Diagn. Microbiol. Infect. Dis.* **6:**175–177.

37. **Purdy, R. A., B. Cassidy, and T. J. Marrie.** 1990. *Streptococcus bovis* meningitis: report of 2 cases. *Neurology* **40:**1782–1784.

38. **Ruoff, K. L.** 1991. Nutritionally variant streptococci. *Clin. Microbiol. Rev.* **4:**184–190.

39. **Ruoff, K. L.** 1993. Dealing with viridans streptococci in the clinical laboratory: the continuing challenge. *Clin. Microbiol. Newsl.* **15:**73–76.

40. **Ruoff, K. L., S. I. Miller, C. V. Garner, M. J. Ferraro, and S. B. Calderwood.** 1989. Bacteremia with *Streptococcus bovis* and *Streptococcus salivarius*: clinical correlates of more accurate identification of isolates. *J. Clin. Microbiol.* **27:**305–308.

41. **Schleifer, K. H., and R. Kilpper-Balz.** 1987. Molecular and chemotaxonomic approaches to the classification of streptococci, enterococci and lactococci: a review. *Syst. Appl. Microbiol.* **10:**1–19.

42. **Stein, D. S., and K. E. Nelson.** 1987. Endocarditis due to nutritionally deficient streptococci: therapeutic dilemma. *Rev. Infect. Dis.* **9:**908–916.

43. **Welch, D. F., D. Hensel, D. Pickett, and S. Johnson.** 1991.

Comparative evaluation of selective and nonselective culture techniques for isolation of group A beta-hemolytic streptococci. *Am. J. Clin. Pathol.* **95**:587–590.

44. **Wessels, M. R., and D. L. Kasper.** 1993. The changing spectrum of group B streptococcal disease. *N. Engl. J. Med.* **328**:1843–1844.

45. **Whiley, R. A., H. Fraser, J. M. Hardie, and D. Beighton.** 1990. Phenotypic differentiation of *Streptococcus intermedius*, *Streptococcus constellatus*, and *Streptococcus anginosus* strains within the "*Streptococcus milleri* group." *J. Clin. Microbiol.* **28**:1497–1501.

46. **Wilkinson, H. W.** 1977. CAMP-disk test for presumptive identification of group B streptococci. *J. Clin. Microbiol.* **6**:42–45.

47. **Wust, J., G. Hebisch, and K. Peters.** 1993. Evaluation of two enzyme immunoassays for rapid detection of group B streptococci in pregnant women. *Eur. J. Clin. Microbiol. Infect. Dis.* **12**:124–127.

Enterococcus

RICHARD R. FACKLAM AND DANIEL F. SAHM

24

TAXONOMY

In 1970, Kalina (37) suggested that a genus for the enterococcal streptococci *Streptococcus faecalis* and *Streptococcus faecium* and the subspecies of these two taxons based on cellular arrangement and phenotypic characteristics be named *Enterococcus*. No action on this proposal was ever taken, and the use of the genus name *Streptococcus* continued. Genetic evidence that *S. faecalis* and *S. faecium* were sufficiently different from the other members of the genus to merit a separate genus was provided by Schleifer and Kilpper-Balz in 1984 (62). It has been 10 years since this proposal, and it is generally accepted that the genus *Enterococcus* is valid. Since 1984, 17 other species have been proposed for inclusion in the genus *Enterococcus* (Table 1). Genetic evidence for these proposals has been provided by DNA-DNA and DNA-rRNA hybridizations as well as by 16S rRNA sequencing.

Recent information, including unpublished data from our laboratory, indicates that two of the proposed species should not be included in the *Enterococcus* genus. *Enterococcus solitarius* was shown by 16S rRNA to be more closely related to *Tetragenococcus halophilus* than to any species of the genus *Enterococcus* (67). DNA homology studies indicate that *E. solitarius* was equally related to the reference strains of *Enterococcus* and *Tetragenococcus*. Therefore, the true taxonomic status of this species remains in question. The rRNA sequence of *E. seriolicida* has been shown to be identical to the rRNA sequence of *Lactococcus garvieae* (18). Using whole-cell protein analysis (21), we have found that the protein profiles of *E. seriolicida* and *L. garvieae* are identical. In addition to the above evidence that *E. seriolicida* and *E. solitarius* strains do not belong in the genus *Enterococcus*, these two species also fail to react positively with the AccuProbe *Enterococcus* probe manufactured by Gen-Probe. This probe is based on a DNA oligomer having a structure complementary to a segment of enterococcal rRNA (15, 22). All strains of known *Enterococcus* species, except the type strains of *E. cecorum*, *E. columbae*, and *E. saccharolyticus*, react positively with this probe. We suggest that *E. seroiolicida* and *E. solitarius* should not be included in the genus *Enterococcus* at this time.

DESCRIPTION OF GENUS

The enterococci are gram-positive cocci that occur singly, in pairs, and in short chains. Cells are sometimes coccobacillary when Gram stains are prepared from agar plate growth. Cells are more oval and in chains when Gram stains are prepared from thioglycolate broth. The enterococci are facultatively anaerobic, and optimum growth occurs at 35°C. Most strains grow at 10 and 45°C. All strains grow in broth containing 6.5% NaCl and hydrolyze esculin in the presence of 40% bile salts (bile-esculin medium). Motility is observed with some species. Most enterococci hydrolyze pyrrolidonyl-β-naphthylamide (PYR); the exceptions are *E. cecorum*, *E. columbae*, and *E. saccharolyticus*. All strains produce leucine aminopeptidase (LAP). Enterococci do not contain cytochrome enzymes, but on occasion, the catalase test appears positive. A pseudocatalase is sometimes produced, and a weak effervescence is observed in the catalase test. Nearly all strains are homofermentive, gas is not produced, and lactic acid is the end product of glucose fermentation. Most strains produce a cell wall-associated glycerol teichoic acid antigen that is identified as the streptococcal group D antigen. Detection of the group D antigen is sometimes difficult and depends upon the extraction procedure and the quality of the antiserum used (25, 40, 64). The G+C content of the DNA ranges from 37 to 45 mol% (62).

Clinical isolates of catalase-negative gram-positive cocci can be identified as enterococci by the tests listed in Table 2. The details of how to perform the tests listed in Table 2 are given in the fifth edition of this Manual (25). The only change from the instructions given in the fifth edition is the availability of individual disks (Carr-Scarborough, Stone Mountain, Ga.; Murex Corp., Norcross, Ga.; Remel, Lenexa, Kans.) for performing the LAP test, which is performed in the same fashion as the PYR test.

The majority of other genera of catalase-negative gram-positive cocci will be discussed in chapters 23 and 25 of this Manual. Note that because strains of *Lactococcus*, *Leuconostoc*, and *Pediococcus* have been isolated from human infections, the presumptive identification of enterococci by bile-esculin reaction and growth in 6.5% NaCl broth only can be erroneous. *Tetragenococcus* (13) and *Vagococcus* (8) species, which also have phenotypic characteristics similar to those of the enterococci, have been split from the genera

308

TABLE 1 Proposals of species to be included in the genus *Enterococcus*

Species	Date	Reference
E. faecalis	1984	62
E. faecium	1984	62
E. avium	1984	11
E. casseliflavus	1984	11
E. durans	1984	11
E. gallinarum	1984	11
E. malodoratus	1984	11
E. hirae	1985	26
E. mundtii	1986	10
E. raffinosus	1989	9
E. solitarius[a]	1989	9
E. pseudoavium	1989	9
E. cecorum	1989	66
E. columbae	1990	16
E. saccharolyticus	1990	59
E. dispar	1991	12
E. sulfureus	1991	45
E. seriolicida[a]	1991	41
E. flavescens	1992	58

[a]Probably does not belong to genus *Enterococcus*.

Pediococcus and *Lactococcus*, respectively, and have not been isolated from human sources.

The most accurate presumptive identification of a catalase-negative gram-positive coccus as an enterococcus can be accomplished by demonstrating that the unknown strain is vancomycin sensitive and PYR and LAP positive and that it grows in 6.5% NaCl at 45°C. Demonstrating by extraction and serological reaction with group-specific antiserum that an unknown strain has group D antigen may be helpful in identifying some strains, but these reactions should be interpreted cautiously. Most strains of pediococci

and about half the strains of leuconostocs isolated from human infections also have group D antigens (24). Confirmation that a strain is an enterococcus requires complete identification to the species level. Positive reactions with the AccuProbe *Enterococcus* test (Gen-Probe, Inc., San Diego, Calif.) can also be used to confirm an unknown strain as an enterococcus (unpublished data).

NATURAL HABITATS

The nature of these bacteria, which allows them to grow and survive in harsh environments, allows them to persist almost everywhere. Enterococci can be found in soil, food, water, animals, birds, and insects (17, 20, 39, 50). In humans, as in other animals, enterococci inhabit the gastrointestinal tract and the genitourinary tract. *E. faecalis* is one of the most common bacteria isolated from the gastrointestinal tracts of humans. Because of changes in taxonomy, it is difficult to determine the distribution of other enterococcal species in humans. It is likely that *E. faecium* is also commonly found in the gastrointestinal tracts of humans (46).

CLINICAL SIGNIFICANCE

The ubiquitous nature of enterococci requires caution in establishing the clinical significance of a particular isolate so that unnecessary work and potentially misleading laboratory reports can be avoided whenever possible. This is especially important regarding in vitro susceptibility testing decisions (see section on antimicrobial susceptibilities below).

Murray (51) has thoroughly reviewed and summarized the variety of infections in which enterococci are involved. Urinary tract infections are the most common. Enterococci are implicated in approximately 10% of all such infections (27) and in 16% of nosocomial urinary tract infections (61). Enterococcal bacteriuria usually occurs in patients

TABLE 2 Phenotypic characteristics of facultatively anaerobic, catalase-negative, gram-positive coccus genera[a]

Genus	Cell arrangement	VAN	GAS	PYR	LAP	BE	NaCl	Growth at: 10°C	Growth at: 45°C	MOT	HEM
Enterococcus	ch	S[b]	−	+	+	+	+	+	+	V	α, β, n
Streptococcus	ch	S	−	−[c]	+	−[d]	−[e]	−	V	−	α, β, n
Globicatella	ch	S	−	+	−	−	+	−	−	−	α
Lactococcus	ch	S	−	+	+	+	V	+	−[f]	−	α, n
Vagococcus	ch	S	−	+	+	+	+	+	−	+	α, n
Leuconostoc	ch	R	+	−	−	V	V	+	V	−	α, n
Pediococcus	cl, T	R	−	−	+	+	V	−	+	−	α
Tetragenococcus	cl, T	S	−	−	+	+	+	−	+	−	α
Aerococcus	cl, T	S	−	+	−	V	+	−	+	−	α
Gemella	cl, T	S	−	+	V	−	−	−	−	−	α, n
Helcococcus	cl, T	S	−	+	−	+	−	−	−	−	n

[a]Abbreviations and symbols: ch, chains; cl, clumps; T, tetrads; VAN, susceptibility to vancomycin (30-μg disk); GAS, gas produced form glucose in Mann, Rogosa, Sharpe Lactobacillus broth (MRS); PYR, production of pyrrolidonyl arylamidase; LAP, production of leucine aminopeptidase; BE, reaction on bile-esculin medium; NaCl, growth in broth containing 6.5% NaCl; MOT, motile; HEM, hemolysis on blood agar containing 5% sheep blood; α, alpha-hemolysis; β, beta-hemolysis; n, no hemolysis; S, susceptible; R, resistant; −, ≥5% negative reaction; +, ≥95% positive reaction; V, variable reaction.
[b]Some strains are vancomycin resistant but still show a small inhibition around the disk; other strains grow right up to the disk and are vancomycin resistant under the defined criteria.
[c]Group A streptococci and nutritionally variant streptococci are PYR positive; all others are negative.
[d]Of viridans streptococci, 5 to 10% are bile-esculin positive.
[e]Some beta-hemolytic streptococci grow in 6.5% NaCl broth.
[f]Some strains of lactococci grow very slowly at 45°C.

with underlying structural abnormalities and/or in those who have undergone urologic manipulations (49). Intra-abdominal or pelvic wound infections are the next most commonly encountered infections. However, these wound cultures are frequently polymicrobial, and the role of enterococci in this setting remains controversial (56). Enterococci are being recovered from wound infections at an increasing rate, which likely results from increased antibiotic usage and emerging resistance among these organisms (49, 51, 61).

Bacteremia is the third most common type of infection, and enterococci are the third leading cause of nosocomial bacteremia (31, 33, 44, 49, 61). These infections most often occur in elderly patients who have serious underlying medical problems, and they also occur in other immunocompromised patients who have had prolonged hospitalization and have received antimicrobial therapy (1, 31). Although endocarditis is a serious enterococcal infection, it is less common than bacteremia. Enterococci are estimated to cause between 5 and 20% of bacterial endocarditis cases (47, 65). E. faecalis is the most commonly encountered species in this setting, but various other species also have been implicated as causes of endocarditis (23).

Enterococcal infections of the respiratory tract or the central nervous system may occur, but they are rare (51). The significance of isolates from these sites should be carefully evaluated before any clinical decisions are made.

Although the spectrum of infections caused by enterococci has remained relatively unchanged since the review by Murray in 1990 (51), the prevalence of these organisms as nosocomial pathogens is clearly increasing. Enterococci are second only to Escherichia coli as agents of nosocomial urinary tract infections and are third (behind Staphylococcus aureus and coagulase-negative staphylococci) as a cause of nosocomial bacteremia (61). This trend is likely to continue as the overall population ages and more people become at risk of infection owing to a generalized deterioration of health (47).

COLLECTION, TRANSPORT, AND STORAGE OF SPECIMENS

The standard methods of collecting blood, urine, wound culture specimens, and other specimens should be adequate (see chapters 3 and 21 of this Manual).

The transport of enterococci is simple and can be performed with almost any transport medium or on swabs that are kept dry. Like most clinical samples, the transported material should be cultured as soon as possible, preferably within 1 h.

Enterococci can be stored indefinitely when lyophilized. Cultures frozen at −70°C can be stored for several years. Most strains of enterococci survive several months at 4°C on ordinary agar slants.

ISOLATION PROCEDURES

Trypticase soy–5% sheep blood agar, brain heart infusion–5% sheep blood agar, or any blood agar base containing 5% animal blood supports the growth of enterococci. Some strains of E. faecalis are beta-hemolytic on agar bases containing rabbit or horse blood but non-beta-hemolytic on the same base media containing sheep blood. If the clinical sample is likely to contain gram-negative bacteria, then bile-esculin azide, Pfizer selective enterococcus, or some other commercially prepared medium containing azide is an excellent primary isolation medium. The azide inhibits the gram-negative bacteria, and the enterococci appear as black colonies because of hydrolysis of esculin. Other media such as Columbia colistin-nalidixic acid agar (CNA) or phenylethyl alcohol agar (PEA) have been used successfully to isolate enterococci. The advantage CNA has over PEA is that the hemolytic reaction can be read from CNA but not from PEA.

All enterococci grow at 35 to 37°C and do not require an atmosphere containing increased levels of carbon dioxide. However, some strains grow better in atmospheres containing increased levels of carbon dioxide.

IDENTIFICATION OF ENTEROCOCCUS SPECIES

Once it is established that the unknown catalase-negative gram-positive coccus is an enterococcus, then the tests listed in Table 3 can be used to identify the species. It is best to separate the species into four groups based on acid formation in mannitol, sorbitol, and sorbose broths and hydrolysis of arginine. Group I consists of E. avium, E. malodoratus, E. raffinosus, and E. pseudoavium. These four species form acid in all three of the aforementioned carbohydrate broths but do not hydrolyze arginine. Group II consists of E. faecalis, E. faecium, E. casseliflavus, E. mundtii, E. flavescens, and E. gallinarum. These six species form acid in mannitol broth and hydrolyze arginine but fail to form acid in sorbose broth and give variable reactions in sorbitol broth. Group III consists of E. durans, E. hirae, E. dispar, and E. faecalis variant. These four species hydrolyze arginine but do not form acid in mannitol, sorbitol, or sorbose broths.

The species of each group are then identified by specific reactions. The species in group I are identified by the reactions in arabinose and raffinose broth. E. avium forms acid in arabinose but not raffinose broth. E. malodoratus forms acid in raffinose but not arabinose broth. E. raffinosus forms acid in both, while E. pseudoavium forms acid in neither arabinose nor raffinose broth.

The species in group II are identified by specific reactions. E. faecalis is the only member of group II to tolerate tellurite and utilize pyruvate. E. faecium and E. gallinarum have similar characteristics but can be differentiated from each other by the motility test. E. gallinarum is motile, but E. faecium is not. E. casseliflavus, E. mundtii, and E. flavescens are pigmented (yellow) and similar to each other in other characteristics as well; however, E. casseliflavus and E. flavescens are motile, but E. mundtii is not. That E. flavescens fails to form acid in ribose broth but E. casseliflavus does form acid differentiates these two species.

The members of group III are easily identified by the reactions in the pyruvate, raffinose, and sucrose tests. E. durans is negative in these three tests. E. hirae is positive in one or both raffinose and sucrose tests and negative in the pyruvate test. E. dispar is positive in all three tests. E. faecalis variant strain is positive in the test for pyruvate utilization but not in that for acid formation in raffinose or sucrose broth. E. sulfureus does not form acid in mannitol, sorbitol, or sorbose broth and does not deaminate arginine. These characteristics are unique and are not shared by other enterococcal species.

The majority of clinical isolates of enterococci (80 to 90%) will be E. faecalis. E. faecium is found in 5 to 10% of

TABLE 3 Identification of *Enterococcus* species[a]

Species	MAN	SBL	SOR	ARG	ARA	RAF	TEL	MOT	PIG	SUC	PYU	RIB
Group I												
E. avium	+	+	+	−	+	−	−	−	−	+	+	+
E. malodoratus	+	+	+	−	−	+	−	−	−	+	+	+
E. raffinosus	+	+	+	−	+	+	−	−	−	+	+	+
E. pseudoavium	+	+	+	−	−	−	−	−	−	+	+	+
Group II												
E. faecalis	+	+	−	+	−	−	+	−	−	+·	+	+
E. faecium	+	−·	−	+	+	−·	−	−	−	+·	−	+
E. casseliflavus	+	−·	−	+	+	+	−·	+	+	+	−·	+
E. mundtii	+	−·	−	+	+	+	−	−	+	+	−	+
E. flavescens	+	−·	−	+	+	+	−	+	+	+	−	−
E. gallinarum	+	−	−	+	+	+	−	+	−	+	−	+
Group III												
E. durans	−·	−	−	+	−	−	−	−	−	−	−	?
E. hirae	−	−	−	+	−	+	−	−	−	+	−	?
E. dispar	−	−	−	+	−	+	−	−	−	+	+	?
E. faecalis (var)	−	−	−	+	−	−	+	−	−	−	+	?
Group IV												
E. sulfureus	−	−	−	−	−	+	−	−	+	+	−	+

[a]Abbreviations and symbols: MAN, mannitol; SBL, sorbitol; SOR, sorbose; ARG, arginine; ARA, arabinose; RAF, raffinose; TEL, 0.04% tellurite; MOT, motility; PIG, pigmented; SUC, sucrose; PYU, pyruvate, RIB, ribose; +, >90% positive; −, <10% positive; −· or +·, occasional exception (<3% of strains show aberrant reactions); ?, note tested, so results are unknown.

enterococcal infections (4, 5, 30, 32, 60). The other enterococcal species are identified less frequently. However, clusters of infections with *E. raffinosus* (6) and *E. casseliflavus* (55) have been reported. Therefore, the distribution of species varies with each clinical setting. *E. avium*, *E. gallinarum*, *E. mundtii*, *E. flavescens*, *E. durans*, *E. hirae*, and *E. faecalis* variant strains have been isolated from human sources, though rarely (23). *E. malodoratus*, *E. pseudoavium*, and *E. sulfureus* have not been isolated from human sources.

All the tests listed in Table 3 except acid formation in ribose broth have been described previously (23–25). Acid formation in ribose is determined in the same fashion as that in the other carbohydrate broths listed in Table 3.

Phenotypic characteristics of typical enterococci, the three atypical *Enterococcus* species, and *Streptococcus bovis*-type organisms are outlined in Table 4. The three atypical species differ from the typical enterococci in that they are

PYR negative. In addition, some strains grow poorly if at all in broth containing 6.5% NaCl. This latter characteristic may be medium dependent in that different results were observed in two reference laboratories (personal observation). In some respects, *E. cecorum* and *E. columbae* are more similar to *S. bovis* strains than to the typical *Enterococcus* species (Table 4). The three atypical *Enterococcus* species can be differentiated by reactions in mannitol, sorbitol, sorbose, and ribose broths (Table 4). None of the three species has been isolated from human sources; all original isolates have been obtained from bovine or avian sources (16, 17, 59, 67).

TYPING METHODS

Because enterococci are a leading cause of nosocomial infections and frequently exhibit multiple antibiotic resistance, there has been an increasing need to type and

TABLE 4 Phenotypic characteristics of three atypical *Enterococcus* species and *S. bovis* group[a]

Group or species	BE	NaCl	PYR	LAP	GAS	VAN	Growth at: 10°C	Growth at: 45°C	ARG	MAN	SBL	SOR	RIB
Typical enterococci	+	+	+	+	−	S	+	+	NA	NA	NA	NA	NA
E. saccharolyticus	+	+	−	+	−	S	+	+	−	+	+	+	+
E. cecorum	+	+[b]	−	+	−	S	+[b]	+	−	−	+	−	+
E. columbae	+	+[b]	−	+	−	S	+	+	−	−	−	−	+
S. bovis group	+	−	−	+	−	S	−	+	−	V	−	−	−

[a]For abbreviations and symbols, see Table 2, footnote a, and Table 3, footnote a. NA, not applicable in this situation.
[b]Grows very slowly; may require 7 to 10 days of incubation.

subtype isolates as a means of assisting infection control and epidemiological studies both within and among various medical institutions. Classic typing methods have included bacteriocin typing, phage typing, biochemical reaction profiles, antimicrobial resistance patterns, and serological characterization. Although these approaches have occasionally yielded useful information, they generally are time-consuming, expensive, somewhat subjective, and esoteric, and they frequently fail to adequately discriminate among strains (14, 42, 43, 53).

The advent of various molecular techniques has substantially enhanced our ability to "type" enterococci. Plasmid profiling may be helpful in some instances (2, 14), but plasmid yield is often inconsistent, and accurate interpretations of the electrophoretic patterns are frequently difficult if not impossible. Similar interpretation problems are encountered with the use of conventional electrophoresis of endonuclease-digested genomic DNA (34, 35, 38). Currently, the most useful methods include contour-clamped homogeneous electric field electrophoresis (CHEF; 53), field inversion gel electrophoresis (3, 28), and ribotyping (68) (also see chapter 13 in this Manual). Some investigators have used more than one method, such as CHEF and plasmid analysis, for their studies (2, 19). However, among these techniques, CHEF appears to be widely applied and widely useful for studying enterococcal species exhibiting a variety of antimicrobial resistance patterns (7, 19, 48, 52, 54, 57, 63), and in at least one direct comparison with ribotyping, CHEF showed definite advantages regarding strain discrimination (29).

ANTIMICROBIAL SUSCEPTIBILITIES AND EVALUATION, INTERPRETATION, AND REPORTING OF RESULTS

The unpredictable nature of enterococcal susceptibility to antimicrobial agents dictates that in vitro testing be done; when and against which antimicrobial agents it is done depend on the site of infection and the significance of a particular isolate. Additionally, because acquired resistance to penicillins, vancomycin, and high levels of aminoglycosides has been described for a variety of enterococcal species, the decision to perform susceptibility testing should not be influenced by the species of the isolate. The methods that should be used for testing enterococci are discussed in detail in chapters 113, 115, and 117 of this Manual.

Enterococci are intrinsically resistant to the inhibitory and bactericidal activities of most commonly used agents. Therefore, recommended therapy for serious infections (i.e., endocarditis, meningitis, or possibly other serious infections in immunocompromised patients) includes a cell wall-active agent, such as a penicillin or vancomycin, combined with an aminoglycoside (usually gentamicin) or sometimes streptomycin (51). These combinations overcome the intrinsic resistance exhibited by enterococci and effect synergistic killing. However, enterococci are adept at acquiring resistance to any of the components of combination therapy and in doing so eliminate that agent's contribution to synergistic killing. In vitro testing of isolates from patients with serious infections should focus on detection of this acquired resistance.

Susceptibility testing for resistance to synergy should be done with any enterococcal isolate implicated in infections for which combination therapy is indicated, e.g., isolates from blood or cerebrospinal fluid. Enterococci are also fre-

quently encountered in polymicrobial infections associated with the gastrointestinal tract or superficial wounds of hospitalized patients. Their pathogenic significance in such settings is uncertain, but susceptibility testing is warranted when predominant or heavy growth is observed (49).

Testing of enterococcal isolates from lower urinary tract infections is optional, as these infections usually respond to therapy with ampicillin. Ciprofloxacin, nitrofurantoin, or tetracycline can also be used, but the prevalence of resistance to these agents is higher (51). In cases of treatment failure, testing is always warranted. Because bactericidal activity is not usually needed for efficacious treatment of uncomplicated urinary tract infections, testing of urine isolates for resistance to components of combination therapy is rarely indicated and does not provide any additional useful information for the management of patients with such infections.

Once the need to test a particular isolate has been established, selection of the appropriate antimicrobial agents for testing must be considered. For enterococcal isolates from blood and cerebrospinal fluid, the components of combination therapy should be tested. Testing should be done for resistance to ampicillin or penicillin (includes β-lactamase testing), vancomycin, and high levels of gentamicin and streptomycin. Details regarding recommended testing methods and interpretation of results for these agents are provided in chapters 113, 115, and 117 of this Manual. Occasionally, there may be interest in an enterococcal isolate's susceptibility to beta-lactam antibiotics other than ampicillin (i.e., mezlocillin, piperacillin, azlocillin, amoxicillin-clavulanate, ampicillin-sulbactam, imipenem). Although these antibiotics do not offer any significant advantages over ampicillin when they are used against enterococci, they may be of interest for use against polymicrobial infections involving enterococci and gram-negative bacilli. Ampicillin along with a test for β-lactamase production can be used to detect resistance to these other beta-lactam antibiotics. Therefore, selection of these other drugs for testing is rarely needed. A positive test for β-lactamase production indicates resistance to ampicillin and the acylureidopenicillins (i.e., azlocillin, mezlocillin, piperacillin). Resistance to ampicillin by disk diffusion or dilution methods likely would indicate penicillin-binding-protein-mediated resistance to these same agents, beta-lactam–β-lactamase inhibitor combinations, and imipenem (36).

Testing agents to which enterococci are intrinsically resistant is contraindicated. The results of such testing are superfluous and can be dangerously misleading. The drugs that should not be tested include aztreonam, any cephalosporin, clindamycin, methicillin (or oxacillin), trimethoprim-sulfamethoxazole, and aminoglycosides (standard concentrations). For those infrequent instances when testing a urinary tract isolate is appropriate, ampicillin, ciprofloxacin, nitrofurantoin, norfloxacin, or tetracycline could be selected.

Reporting susceptibility testing results obtained with enterococci isolated from patients with deep-seated infections is relatively more complicated than reporting results obtained with other organism-antimicrobial agent combinations because testing of enterococci is essentially a screen for resistance to combination therapy. Therefore, simply reporting susceptibility or resistance results obtained with ampicillin, vancomycin, gentamicin, and streptomycin could be seriously misleading. In such a report, a "susceptible" result indicates that the antimicrobial agent may be an appropriate choice for single-agent therapy. To avoid such misinterpretation, the susceptibility report should be accompanied by an explanation (via computer, hard copy, or direct communication) that combination therapy is usually recommended and that resistance to particular agents indicates that those antimicrobial drugs would not be effective choices for combination therapy (36).

Because testing enterococci from patients with urinary tract infections involves antimicrobial agents different from those used for combination therapy and because combination therapy is not commonly used for such infections, reporting of results for isolates from this site is usually straightforward and does not require accompanying information.

REFERENCES

1. **Awada, A., P. van der Auwera, F. Meunier, D. Daneau, and J. Klastersky.** 1992. Streptococcal and enterococcal bacteremia in patients with cancer. *Clin. Infect. Dis.* **15**:33–48.
2. **Boyce, J. M., S. M. Opal, G. Potter-Bynoe, R. G. LaForge, M. J. Zervos, G. Furtado, G. Victor, and A. A. Medeiros.** 1992. Emergence and nosocomial transmission of ampicillin-resistant enterococci. *Antimicrob. Agents Chemother.* **36**:1032–1039.
3. **Boyle, J. F., S. A. Soumakis, A. Rendo, J. A. Herrington, D. G. Gianarkis, B. E. Thurberg, and B. G. Painter.** 1993. Epidemiologic analysis and genotypic characterization of nosocomial outbreak of vancomycin-resistant enterococci. *J. Clin. Microbiol.* **31**:1280–1285.
4. **Bryce, E. A., S. J. V. Zemcov, and A. M. Clarke.** 1991. Species identification and antibiotic resistance patterns of the enterococci. *Eur. J. Clin. Microbiol. Infect. Dis.* **10**:745–747.
5. **Buschelman, B. J., B. J. Bale, and R. N. Jones.** 1993. Species identification and determination of high-level aminoglycoside resistance among enterococci. Comparison study of sterile body fluid isolates. *Diagn. Microbiol. Infect. Dis.* **16**:119–122.
6. **Chirurgi, V. A., S. E. Oster, A. A. Goldberg, M. J. Zervos, and R. E. McCabe.** 1991. Ampicillin-resistant *Enterococcus raffinosus* in an acute-care hospital: case-control study and antimicrobial susceptibilities. *J. Clin. Microbiol.* **29**:2663–2665.
7. **Chow, J. W., A. Kuritza, D. M. Shlaes, M. Green, D. F. Sahm, and M. J. Zervos.** 1993. Clonal spread of vancomycin-resistant *Enterococcus faecium* between patients in three hospitals in two states. *J. Clin. Microbiol.* **31**:1609–1611.
8. **Collins, M. D., C. Ash, J. A. E. Farrow, S. Wallbanks, and A. M. Williams.** 1989. 16S ribosomal ribonucleic acid sequence analyses of lactococci and related taxa. Description of *Vagococcus fluvialis* gen. nov., sp. nov. *J. Appl. Bacteriol.* **67**:453–460.
9. **Collins, M. D., R. R. Facklam, J. A. E. Farrow, and R. Williamson.** 1989. *Enterococcus raffinosus* sp. nov., *Enterococcus solitarius* sp. nov. and *Enterococcus pseudoavium* sp. nov. *FEMS Microbiol. Lett.* **57**:283–288.
10. **Collins, M. D., J. A. E. Farrow, and D. Jones.** 1986. *Enterococcus mundtii* sp. nov. *Int. J. Syst. Bacteriol.* **36**:8–12.
11. **Collins, M. D., D. Jones, J. A. E. Farrow, R. Kilpper-Balz, and K. H. Schleifer.** 1984. *Enterococcus avium* nom. rev., comb. nov.; *E. casseliflavus* nom. rev., comb. nov.; *E. durans* nom. rev., comb. nov.; *E. gallinarum* comb. nov.; and *E. malodoratus* sp. nov. *Int. J. Syst. Bacteriol.* **34**:220–223.
12. **Collins, M. D., U. M. Rodrigues, N. E. Pigott, and R. R. Facklam.** 1991. *Enterococcus dispar* sp. nov. a new *Enterococcus* species from human sources. *Lett. Appl. Microbiol.* **12**:95–98.
13. **Collins, M. D., A. M. Williams, and S. Wallbanks.** 1990. The phylogeny of *Aerococcus* and *Pediococcus* as determined by 16S rRNA sequence analysis: description of *Tetragenococcus* gen. nov. *FEMS Microbiol. Lett.* **70**:255–262.
14. **Coudron, P. E., C. G. Mayhall, R. R. Facklam, A. C. Spadora, V. A. Lamb, M. R. Lybrand, and H. P. Dalton.** 1984. *Streptococcus faecium* outbreak in a neonatal intensive care unit. *J. Clin. Microbiol.* **20**:1044–1048.
15. **Daly, J. A., N. L. Clifton, K. C. Seskin, and W. M. Gooch III.** 1991. Use of rapid, nonradioactive DNA probes in culture confirmation tests to detect *Streptococcus agalactiae*, *Haemophilus influenzae*, and *Enterococcus* spp. from pediatric patients with significant infections. *J. Clin. Microbiol.* **29**:80–82.
16. **Devriese, L. A., K. Ceyssens, U. M. Rodrigues, and M. D. Collins.** 1990. *Enterococcus columbae*, a species from pigeon intestines. *FEMS Microbiol. Lett.* **71**:247–252.
17. **Devriese, L. A., L. Laurier, P. De Herdt, and F. Haesebrouck.** 1992. Enterococcal and streptococcal species isolated from faeces of calves, young cattle and dairy cows. *J. Appl. Bacteriol.* **72**:29–31.
18. **Domenech, A., J. Prieta, J. F. Fernandez-Garayzabal, M. D. Collins, D. Jones, and L. Dominquez.** 1963. Phenotypic and phylogenetic evidence for a close relationship between *Lactococcus garvieae* and *Enterococcus seriolicida*. *Microbiologia* **9**:63–68.
19. **Donabedian, S. M., J. W. Chow, J. M. Boyce, R. E. McCabe, S. M. Markowitz, P. E. Couldron, A. Kuritza, C. L. Pierson, M. J. Zervos.** 1992. Molecular typing of ampicillin-resistant, non-β-lactamase-producing *Enterococcus faecium* isolates from diverse geographic areas. *J. Clin. Microbiol.* **30**:2757–2761.
20. **Dutka, B. J., and K. K. Kwan.** 1978. Comparison of eight media-procedures for recovering faecal streptococci from water under winter conditions. *J. Appl. Bacteriol.* **45**:333–340.
21. **Elliott, J. A., M. D. Collins, N. E. Pigott, and R. R. Facklam.** 1991. Differentiation of *Lactococcus lactis* and *Lactococcus garvieae* from humans by comparison of whole-cell protein patterns. *J. Clin. Microbiol.* **29**:2731–2734.
22. **Enns, R. K.** 1988. DNA probes: an overview and comparison with current methods. *Lab. Med.* **19**:295–300.
23. **Facklam, R. R., and M. D. Collins.** 1989. Identification of *Enterococcus* species isolated from human infections by a conventional test scheme. *J. Clin. Microbiol.* **27**:731–734.
24. **Facklam, R. R., D. Hollis, and M. D. Collins.** 1989. Identification of gram-positive coccal and coccobacillary vancomycin-resistant bacteria. *J. Clin. Microbiol.* **27**:724–730.
25. **Facklam, R. R., and J. A. Washington II.** 1991. *Streptococcus* and related catalase-negative gram-positive cocci, p. 238–257. *In* A. Balows, W. J. Hausler, K. L. Herrmann, H. D. Isenberg, and H. J. Shadomy (ed.), *Manual of Clinical Microbiology*, 5th ed. American Society for Microbiology, Washington, D.C.
26. **Farrow, J. A. E., and M. D. Collins.** 1985. *Enterococcus hirae*, a new species that includes amino acid assay strain NCDO 1258 and strains causing growth depression in young chickens. *Int. J. Syst. Bacteriol.* **35**:73–75.
27. **Felmingham, D., A. P. R. Wilson, A. I. Quintana, and R. N. Gruneberg.** 1992. *Enterococcus* species in urinary tract infection. *Clin. Infect. Dis.* **15**:295–301.
28. **Goering, R. V., and M. A. Winters.** 1992. Rapid method for epidemiological evaluation of gram-positive cocci by field inversion gel electrophoresis. *J. Clin. Microbiol.* **30**:577–580.
29. **Gordillo, M. E., K. V. Singh, and B. E. Murray.** 1993. Comparison of ribotyping and pulsed-field gel electrophoresis for subspecies differentiation of strains of *Enterococcus faecalis*. *J. Clin. Microbiol.* **31**:1570–1574.
30. **Gordon, S., J. M. Swensen, B. C. Hill, N. E. Pigott, R. R. Facklam, R. C. Cooksey, C. Thornsberry, Enterococcal Study Group, W. R. Jarvis, and F. C. Tenover.** 1992. Antimicrobial susceptibility patterns of common and unusual species of enterococci causing infections in the United States. *J. Clin. Microbiol.* **30**:2373–2378.
31. **Graninger, W., and R. Ragette.** 1992. Nosocomial bacteremia due to *Enterococcus faecalis* without endocarditis. *Clin. Infect. Dis.* **15**:49–57.
32. **Guiney, M., and G. Urwin.** 1993. Frequency and antimicrobial susceptibility of clinical isolates of enterococci. *Eur. J. Clin. Microbiol. Infect. Dis.* **12**:362–366.
33. **Gullberg, R. M., S. R. Homann, and J. P. Phair.** 1989. Enterococcal bacteremia: analysis of 75 episodes. *Rev. Infect. Dis.* **11**:74–85.
34. **Hall, L. M. C., B. Duke, M. Guiney, and R. Williams.** 1992.

Typing of *Enterococcus* species by DNA restriction fragment analysis. *J. Clin. Microbiol.* **30:**915–919.

35. **Hall, L. M. C., B. Duke, G. Urwin, and M. Guiney.** 1992. Epidemiology of *Enterococcus faecalis* urinary tract infection in a teaching hospital in London, United Kingdom. *J. Clin. Microbiol.* **30:**1953–1957.

36. **Hindler, J. A., and D. F. Sahm.** 1992. Controversies and confusion regarding antimicrobial susceptibility testing of enterococci. *Antimicrob. Newsl.* **8:**65–74.

37. **Kalina, A. P.** 1970. The taxonomy and nomenclature of enterococci. *Int. J. Syst. Bacteriol.* **20:**185–189.

38. **Kaufhold, A., and P. Ferrieri.** 1993. Molecular investigation of clinical *Enterococcus faecium* isolates highly resistant to gentamicin. *Zentralbl. Bakteriol. Abt. 1* **278:**83–101.

39. **Kibbey, H. J., C. Hagedorn, and E. L. McCoy.** 1978. Use of fecal streptococci as indicators of pollution in soil. *Appl. Environ. Microbiol.* **35:**711–717.

40. **Knudtson, L. M., and P. A. Hartman.** 1993. Comparison of four latex agglutination kits to rapidly identify Lancefield group D enterococci and fecal streptococci. *J. Rapid Methods Autom. Microbiol.* **1:**301–304.

41. **Kusuda, R., K. Kawai, F. Salati, C. R. Banner, and J. L. Fryer.** 1991. *Enterococcus seriolicida* sp. nov., a fish pathogen. *Int. J. Syst. Bacteriol.* **41:**406–409.

42. **Luginbuhl, L. M., H. A. Rotbart, R. R. Facklam, M. H. Roe, and J. A. Elliott.** 1987. Neonatal enterococcal sepsis: case-control study and description of an outbreak. *Pediatr. Infect. Dis. J.* **6:**1022–1030.

43. **Maekawa, S., M. Yoshioka, and Y. Kumamoto.** 1992. Proposal of a new scheme for the serological typing of *Enterococcus faecalis* strains. *Microbiol. Immunol.* **36:**671–681.

44. **Maki, D. G., and W. A. Agger.** 1988. Enterococcal bacteremia: clinical features, the risk of endocarditis, and management. *Medicine* (Baltimore) **67:**248–269.

45. **Martinez-Murcia, A. J., and M. D. Collins.** 1991. *Enterococcus sulfureus*, a new yellow-pigmented *Enterococcus* species. *FEMS Microbiol. Lett.* **80:**69–74.

46. **Mead, G. C.** 1978. Streptococci in the intestinal flora of man and other non-ruminant animals, p. 245–261. *In* F. A. Skinner and L. B. Quesnel (ed.), *Streptococci*. Academic Press, Inc. (London) Ltd., London.

47. **Megran, D. W.** 1992. Enterococcal endocarditis. *Clin. Infect. Dis.* **15:**63–71.

48. **Miranda, A. G., K. V. Singh, and B. E. Murray.** 1991. DNA fingerprinting of *Enterococcus faecium* by pulsed-field gel electrophoresis may be a useful epidemiologic tool. *J. Clin. Microbiol.* **29:**2752–2757.

49. **Moellering, R. C., Jr.** 1992. Emergence of *Enterococcus* as a significant pathogen. *Clin. Infect. Dis.* **14:**1173–1178.

50. **Mundt, J. O., W. F. Graham, and I. E. McCarty.** 1967. Spherical lactic acid-producing bacteria of southern-grown raw and processed vegetables. *Appl. Environ. Microbiol.* **15:**1303–1308.

51. **Murray, B. E.** 1990. The life and times of the *Enterococcus*. *Clin. Microbiol. Rev.* **3:**46–65.

52. **Murray, B. E., H. A. Lopardo, E. A. Rubeglio, M. Frosolono, and K. V. Singh.** 1992. Intrahospital spread of a single gentamicin-resistant, β-lactamase-producing strain of *Enterococcus faecalis* in Argentina. *Antimicrob. Agents Chemother.* **36:**230–232.

53. **Murray, B. E., K. V. Singh, J. D. Heath, B. R. Skharma, and G. M. Weinstock.** 1990. Comparison of genomic DNAs of different enterococcal isolates using restriction endonucleases

with infrequent recognition sites. *J. Clin. Microbiol.* **28:**2059–2063.

54. **Murray, B. E., K. V. Singh, S. M. Markowitz, H. A. Lopardo, J. E. Patterson, M. J. Zervos, E. Rubeglio, G. M. Eliopoulos, L. B. Rice, F. W. Goldstein, S. G. Jenkins, G. M. Caputo, R. Nasnas, L. S. Moore, E. S. Wong, and G. Weinstock.** 1991. Evidence for clonal spread of a single strain of β-lactamase-producing *Enterococcus* (*Streptococcus*) *faecalis* in six hospitals in five states. *J. Infect. Dis.* **163:**780–785.

55. **Nauschuetz, W. F., S. B. Trevino, L. S. Harrison, R. N. Longfield, L. Fletcher, and W. G. Wortham.** 1993. *Enterococcus casseliflavus* as an agent of nosocomial bloodstream infections. *Med. Microbiol. Lett.* **2:**102–108.

56. **Nichols, R. L., and A. C. Muzik.** 1992. Enterococcal infections in surgical patients: the mystery continues. *Clin. Infect. Dis.* **15:**72–76.

57. **Patterson, J. E., K. V. Singh, and B. E. Murray.** 1991. Epidemiology of an endemic strain of β-lactamase-producing *Enterococcus faecalis*. *J. Clin. Microbiol.* **29:**2513–2516.

58. **Pompei, R., F. Berlutti, M. C. Thaller, A. Ingianni, G. Cortis, and B. Dainelli.** 1992. *Enterococcus flavescens* sp. nov., a new species of enterococci of clinical origin. *Int. J. Syst. Bacteriol.* **42:**365–369.

59. **Rodrigues, U., and M. D. Collins.** 1990. Phylogenetic analysis of *Streptococcus saccharolyticus* based on 16S rRNA sequencing. *FEMS Microbiol. Lett.* **71:**231–234.

60. **Ruoff, K. L., L. De La Maza, M. J. Murtagh, J. D. Spargo, and M. J. Ferraro.** 1990. Species identities of enterococci isolated from clinical specimens. *J. Clin. Microbiol.* **28:**435–437.

61. **Schaberg, D. R., D. H. Culver, and R. P. Gaynes.** 1991. Major trends in the microbial etiology of nosocomial infection. *Am. J. Med.* **91:**79S–82S.

62. **Schleifer, K. H., and R. Kilpper-Balz.** 1984. Transfer of *Streptococcus faecalis* and *Streptococcus faecium* to the genus *Enterococcus* nom. rev. as *Enterococcus faecalis* comb. nov. and *Enterococcus faecium* comb. nov. *Int. J. Syst. Bacteriol.* **34:**31–34.

63. **Thal, L. A., J. W. Chow, J. E. Patterson, M. B. Perri, S. Donabedian, D. B. Clewell, and M. J. Zervos.** 1993. Molecular characterization of highly gentamicin-resistant *Enterococcus faecalis* isolates lacking high-level streptomycin resistance. *Antimicrob. Agents Chemother.* **37:**134–137.

64. **Truant, A. L., and V. Satishchandran.** 1993. Comparison of Streptex versus PathoDx for group D typing of vancomycin-resistant *Enterococcus*. *Diagn. Microbiol. Infect. Dis.* **16:**89–91.

65. **Watanakunakorn, C., and T. Burkert.** 1993. Infective endocarditis at a large community teaching hospital, 1980–1990. *Medicine* **72:**90–102.

66. **Williams, A. M., J. A. E. Farrow, and M. D. Collins.** 1989. Reverse transcriptase sequencing of 16S ribosomal RNA from *Streptococcus cecorum*. *Lett. Appl. Microbiol.* **8:**185–189.

67. **Williams, A. M., U. M. Rodrigues, and M. D. Collins.** 1991. Intrageneric relationships of *Enterococci* as determined by reverse transcriptase sequencing of small-subunit rRNA. *Res. Microbiol.* **142:**67–74.

68. **Woodford, N., D. Morrison, A. P. Johnson, V. Briant, R. C. George, and B. Cookson.** 1993. Application of DNA probes for rRNA and *vanA* genes to investigation of a nosocomial cluster of vancomycin-resistant enterococci. *J. Clin. Microbiol.* **31:**653–658.

Leuconostoc, Pediococcus, Stomatococcus, and Miscellaneous Gram-Positive Cocci That Grow Aerobically

KATHRYN L. RUOFF

25

TAXONOMY

The bacteria included in this chapter are taxonomically diverse, but they share the characteristic of being infrequent clinical isolates that may function as opportunistic pathogens. Most of these organisms resemble other, better-known clinical isolates (i.e., streptococci, enterococci, staphylococci) and consequently may be mistaken for members of those genera. These bacteria may have been misidentified or overlooked in clinical cultures in the past or may represent emerging pathogens in compromised patient populations. Table 1 lists the organisms included here along with some of their basic characteristics. Consult the references listed in Table 1 for detailed taxonomic information.

Most organisms listed in Table 1 exhibit fairly low G+C contents (30 to 45 mol%) and are not currently affiliated with taxa above the genus level. *Stomatococcus* spp. (55 to 60 mol% G+C) and *Micrococcus* spp. (64 to 75 mol% G+C), however, are both included along with staphylococci in the family *Micrococcaceae*, but Stackebrandt (48) has noted that both of these genera probably belong, with other nonstaphylococcal organisms, in a yet-to-be-described family. Stomatococci have been classified in the past as either micrococci or staphylococci.

Organisms we now know as *Gemella morbillorum* were noted in 1917 by Tunicliff (50), who was searching for the etiologic agent of measles. The organism she isolated from the blood cultures of numerous measles patients was originally named *Diplococcus rubeolae*. This bacterium was also known as *Diplococcus morbillorum*, *Peptostreptococcus morbillorum*, and *Streptococcus morbillorum* until a proposal to include it in the genus *Gemella* was made in 1988 (34). *Gemella haemolysans* was originally classified as a *Neisseria* sp. because of its gram-variable or even gram-negative nature and its cellular morphology (diplococci with flattened adjacent sides).

The genus *Lactococcus* is composed of organisms formerly classified as Lancefield group N streptococci (46). The sole species of the genus *Alloiococcus* was originally named *Alloiococcus otitis* (1), but von Graevenitz recommended that it be renamed *Alloiococcus otitidis* in keeping with the rules of the Bacteriological Code (51). Five of the 10 genera listed in Table 1 have been described since 1982 (*Stomatococcus*, *Lactococcus*, *Globicatella*, *Helcococcus*, *Alloiococcus*). The study of gram-positive cocci by molecular taxonomic methods has encouraged the delineation of these new groups of organisms and the refinement of a genetically based taxonomy for the gram-positive cocci.

DESCRIPTION OF THE GROUP

The organisms included in this chapter form gram-positive coccoid cells, but *G. haemolysans* may appear gram variable or gram negative because of the ease with which it is decolorized. Cell shape and arrangement can aid in dividing these organisms into two broad groups, as illustrated in Table 1; those with a "streptococcal-like" Gram stain (coccobacilli in pairs and chains) and those with a "staphylococcal-like" Gram stain (i.e., more spherical cocci in pairs, tetrads, or clusters). Dividing these diverse bacteria into two groups based on cellular shape and arrangement serves only as an aid in identification; no relatedness of organisms is implied by this grouping.

Most species in these genera are facultative anaerobes, but *Micrococcus* and *Alloiococcus* spp. are obligate aerobes. *Aerococcus* spp. are classified as microaerophiles that grow poorly if at all under anaerobic conditions. The obligately aerobic genera are strongly catalase positive, while some strains of *Aerococcus* may exhibit weakly positive catalase reactions due to nonheme catalase activity. *Stomatococcus* spp., capable of heme synthesis, may evidence negative, weakly positive, or strongly positive catalase reactions. None of the genera are beta-hemolytic on routinely employed blood agars, but *G. haemolysans* is described as producing beta-hemolysis on agars supplemented with rabbit or horse blood (41).

NATURAL HABITATS

Some of the genera discussed here are well-described members of normal human oral or upper respiratory flora (*Stomatococcus* and *Gemella* spp.). Foods and vegetation are normal habitats for lactococci, pediococci, and leuconostocs; these organisms may also be normal flora of the alimentary tract, but thorough data supporting this contention are lacking. Aerococci and micrococci are environmental isolates that can also be found on human skin. Although globicatellas, helcococci, and alloiococci have been isolated from human sources, their natural habitats are not well characterized.

TABLE 1 Some characteristics of infrequently isolated gram-positive cocci that grow aerobically[a]

Genus	Catalase	Relationship to oxygen	Appearance of Gram stain	Reference
Leuconostoc	−	Facultative	cb, pr, ch	28
Lactococcus	−	Facultative	cb, pr, ch	46
Globicatella	−	Facultative	cb, pr, ch	13
Pediococcus	−	Facultative	c, pr, tet, cl	29
Aerococcus	− or w	Microaerophilic	c, pr, tet, cl	20
Gemella[b]	−	Aerobic or facultative[c]	c, pr, ch, cl[d]	41
Helcococcus	−	Facultative	c, pr, ch, cl	14
Stomatococcus	−, +, or w	Facultative	c, pr, cl	45
Micrococcus	+	Aerobic	c, cl, tet	45
Alloiococcus	+[e]	Aerobic	cb, pr, tet	1

[a]Abbreviations: w, weak; cb, coccobacilli; pr, pairs; ch, chains; c, cocci; tet, tetrads; cl, clusters. Morphology is most characteristic for organisms grown in broth.
[b]*G. haemolysans* is easily decolorized and may appear gram variable or gram negative.
[c]*G. haemolysans* prefers an aerobic growth atmosphere, while *G. morbillorum* flourishes under anaerobic conditions.
[d]*G. haemolysans* cells usually occur as diplococci with adjacent sides flattened, while *G. morbillorum* cells are found in pairs, sometimes with cells of unequal sizes in a given pair, and chains.
[e]*Alloiococcus* strains may be negative in the catalase test, especially when grown on media devoid of blood.

CLINICAL SIGNIFICANCE

The organisms included in this chapter appear either as contaminants in clinical cultures or as infrequently isolated opportunistic pathogens. These bacteria appear to be of low virulence and are normally pathogenic only in compromised hosts. Infection often occurs in previously damaged tissues (e.g., heart valves) or may be nosocomial and associated with prolonged hospitalization, antibiotic treatment, invasive procedures, and the presence of foreign bodies. Specimens likely to yield significant isolates of these bacteria are blood, cerebrospinal fluid, urine, and wound material.

Leuconostoc and Pediococcus spp.

The vancomycin-resistant organisms *Leuconostoc* and *Pediococcus* spp. have been recognized in clinical specimens since the mid-1980s. Handwerger and colleagues (31) noted host defense impairment, invasive procedures breaching the integument, gastrointestinal symptoms, and prior antibiotic treatment as common features among adult patients infected with *Leuconostoc* spp. They also observed a predisposition to *Leuconostoc* bacteremia among neonates, suggesting that infants may become colonized during delivery by leuconostocs inhabiting the maternal genital tract. In addition to being isolated from patients with bacteremia, leuconostocs have also been isolated from cerebrospinal fluid, peritoneal dialysate fluid, and wounds.

The clinical significance of *Pediococcus* spp. is less well documented. Mastro and coworkers (36) could find no clearly defined syndrome associated with isolation of *Pediococcus acidilactici* from blood cultures in nine cases they reviewed. They concluded that while this organism may be an opportunist in severely compromised patients, further data were needed to clarify its clinical significance. More recent reports have described *P. acidilactici* as an agent of septicemia and hepatic abscess (30, 47).

Stomatococcus spp.

A 1978 report by Rubin and colleagues (43) on endocarditis caused by "*Micrococcus mucilaginosus*" incertae sedis represents the first well-documented account of disease caused by the organism currently called *Stomatococcus mu-cilaginosus*. Numerous reports of infection by this organism began appearing in the mid- to late 1980s. McWhinney and coworkers (37) and Ascher et al. (3) have reviewed recently published cases of *Stomatococcus* infection (endocarditis, bacteremia, intravascular catheter infection, meningitis, peritonitis) and noted serious underlying disease, neutropenia, presence of a foreign body, cardiac valve disease, and intravenous drug use as predisposing factors. Destruction of oral mucous membranes as a result of chemo- or radiotherapy was hypothesized to play a role in dissemination of stomatococci from their normal habitat in the oral cavity.

Aerococcus spp.

Although aerococci appear as contaminants in clinical cultures, occasional reports have noted a clinically significant role for these organisms in cases of endocarditis and bacteremia (16, 32, 40). A new *Aerococcus* species, *Aerococcus urinae* (2), was recently implicated as a pathogen in patients predisposed to urinary tract infection (12).

Gemella spp.

G. haemolysans has been isolated as a pathogen in cases of endocarditis (11) and meningitis (27) and from a patient with total knee arthroplasty (17). The other currently described member of this genus, *G. morbillorum*, has been isolated (when still classified as a streptococcus; see Taxonomy above) from blood, respiratory, genitourinary, wound, and abscess cultures (21) and from infection in an arteriovenous shunt (4).

Alloiococcus sp.

Aloiococci have been isolated from the middle ear fluid of children with chronic otitis media. The observation of intracellular alloiococci and significant levels of inflammatory cells in middle ear fluids bearing this organism led Faden and Dryja (24) to suggest that *Alloiococcus* sp. plays a pathogenic role in persistent otitis media.

Micrococcus spp.

While micrococci appear in clinical cultures as contaminants, they may occasionally participate in infections such as endocarditis (33).

TABLE 2 Differentiating features of gram-positive cocci with negative or weakly positive catalase reactions[a]

Genus	PYR	VAN	LAP	Gas[b]	Esculin hydrolysis	Growth in 6.5% NaCl
Streptococcus	V[c]	S	+	−	V	V[d]
Enterococcus	+	S[e]	+	−	+	+
Lactococcus	V	S	+	−	V[f]	V
Leuconostoc	−	R	−	+	V	V
Pediococcus	−	R	+	−	V[f]	V
Gemella	+	S	V	−	−	−
Aerococcus[g]	+	S	−	−	V	+
Helcococcus	+	S	−	−	+	V
Globicatella	+	S	−	−	+	+
Stomatococcus	+[f]	S	+	−	+	−

[a]Abbreviations and symbols: +, positive; −, negative; V, variable; S, susceptible; R, resistant; VAN, vancomycin.
[b]Gas production from glucose in sealed MRS broth (Lactobacilli MRS [de Man, Rogosa, and Sharpe] broth).
[c]*S. pyogenes*, nutritionally variant streptococci, and some strains of pneumococci are PYR positive.
[d]Viridans and group D streptococci are negative; group B streptococci may be positive.
[e]Vancomycin-resistant enterococcal strains have been described, but currently, most isolates are susceptible.
[f]Most strains are positive.
[g]*A. urinae*, a newly described urinary tract pathogen, is PYR negative and LAP positive.

Lactococcus spp.

Difficulties in distinguishing lactococci from either streptococci or enterococci have probably led to the misidentification of clinical *Lactococcus* isolates in the past and may have contributed to the paucity of reports concerning the clinical role of these bacteria. Elliott and coworkers (18) studied the phenotypic characteristics of a number of lactococcal strains isolated from blood, patients with urinary tract infections, and an eye wound culture. The authors noted that three of the blood culture isolates were from patients diagnosed with prosthetic valve endocarditis.

Globicatella sp.

Globicatella sanguis, the sole species in the genus, has been isolated from patients with bacteremia, urinary tract infection, and meningitis (13).

Helcococcus sp.

Helcococcus kunzii, the only member of this newly described genus, has been isolated from wound cultures, notably foot ulcer cultures, characteristically as part of a mixture of bacteria (14). Consequently, the clinical significance of this organism is difficult to determine, since it may be present merely as a colonizer of the wound site.

COLLECTION, TRANSPORT, AND STORAGE OF SPECIMENS

The organisms described in this chapter have all been isolated from routine cultures of clinical specimens, and special requirements for collection and processing of specimens have not been described. Since these bacteria are facultatively anaerobic, microaerophilic, or obligately aerobic, aerobic collection, transport, and storage methods as described in chapter 3 of this Manual should allow for their isolation.

ISOLATION PROCEDURES

Generally, there are no special requirements for isolation of the group of bacteria discussed here; general recommendations for the culture of blood, body fluids, and other specimens should be followed. These organisms are likely to be isolated on rich, nonselective media (e.g., blood or chocolate agar, thioglycolate broth), since they are nutritionally fastidious. If selective isolation of members of the vancomycin-resistant genera *Leuconostoc* and *Pediococcus* is desired, Thayer-Martin medium may be used to inhibit normal flora or other contaminating microorganisms (44).

IDENTIFICATION

The suggestions for identification presented in this section reflect currently available information on the infrequently isolated gram-positive cocci that grow aerobically. Additional taxonomic studies and observations on clinical isolates may alter future protocols and introduce additional organisms into the identification schemes presented here. Although Gram stain morphology (best visualized from broth cultures) is prone to subjective interpretation, it has been employed as a major decision point in the identification protocols, with two general categories: Gram stain morphology resembling that of streptococci, meaning coccobacilli in pairs and chains; and staphylococcal morphology, consisting of coccoid (usually more spherical) cells arranged in pairs, clusters, or tetrads. The flow diagrams in

TABLE 3 Differentiating characteristics of catalase-positive or weakly positive gram-positive cocci[a]

Genus	Catalase	Obligately aerobic	Oxidase	Salt tolerance
Staphylococcus	+	−	−	+[b]
Stomatococcus	−, +, or w	−	−	−[b]
Micrococcus	+	+	+	+[b]
Alloiococcus	+[c]	+	−	+[d]

[a]Abbreviation and symbols: +, positive; −, negative; w, weak.
[b]Growth in the presence of 5% NaCl is tested on nutrient agar supplemented with 5% NaCl.
[c]*Alloiococcus* strains may be negative in the catalase test, especially when grown on media devoid of blood.
[d]Growth in 6.5% NaCl is tested with salt tolerance medium for gram-positive cocci. See Growth in 6.5% NaCl in the text.

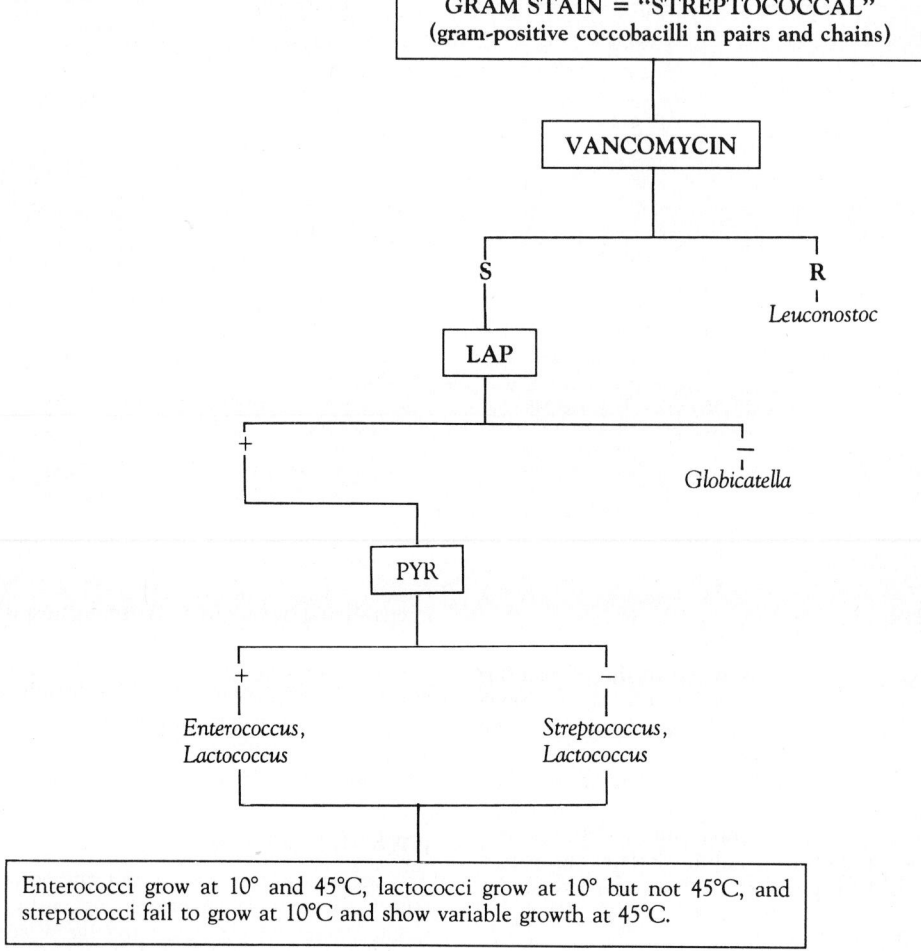

FIGURE 1 Identification of non-beta-hemolytic catalase-negative or weakly positive gram-positive cocci that grow aerobically. Gram stains of G. *morbillorum* may appear streptococcal but are distinctive in that individual cells in a given pair are often of unequal sizes. See text for further distinguishing features of this organism. S, susceptible; R, resistant.

this section should not be used for definitive identification; in most cases, additional procedures (Tables 2 and 3) are recommended before identification to the genus level is made. Identifications of unfamiliar organisms from important specimens should be confirmed by a reference laboratory.

Procedures for Initial Differentiation of Genera with Negative or Weak Catalase Reactions

Procedures for initial testing of catalase-negative or weakly positive isolates with "streptococcal" (Fig. 1) or "staphylococcal" (Fig. 2) Gram stains are represented in the flow diagrams. Note that *Gemella* strains may display either type of cellular morphology, depending on the species (see Additional Procedures for Characterization of Genera with Negative or Weak Catalase Reactions below). Descriptions of tests for these organisms follow.

Catalase Test

If a positive catalase reaction is observed with growth from blood-containing media, growth from a medium devoid of blood (e.g., brain heart infusion agar) should be used to

repeat the catalase test. A loopful of growth is transferred to a microscope slide or empty petri dish and observed for the evolution of bubbles after addition of a drop of 3% H_2O_2. It may be necessary to use a hand lens to detect weakly positive reactions.

Vancomycin Susceptibility Test

In the method of Facklam and Washington (23), a heavy inoculum (5 to 10 colonies) is spread with a loop over half of a plate containing Trypticase soy agar with 5% sheep blood. A 30-μg vancomycin disk is placed in the center of the inoculated area, and the plate is incubated in a CO_2-enriched atmosphere at 35°C overnight. Any zone of inhibition indicates susceptibility, while resistant strains exhibit no inhibition zone.

LAP Test

The LAP test determines the presence of the enzyme leucine aminopeptidase (LAP), also called leucine arylamidase. An assay for this enzyme is contained in the API Rapid Strep System (bioMérieux Vitek, Hazelwood, Mo.), and a rapid disk test for LAP is also available commercially

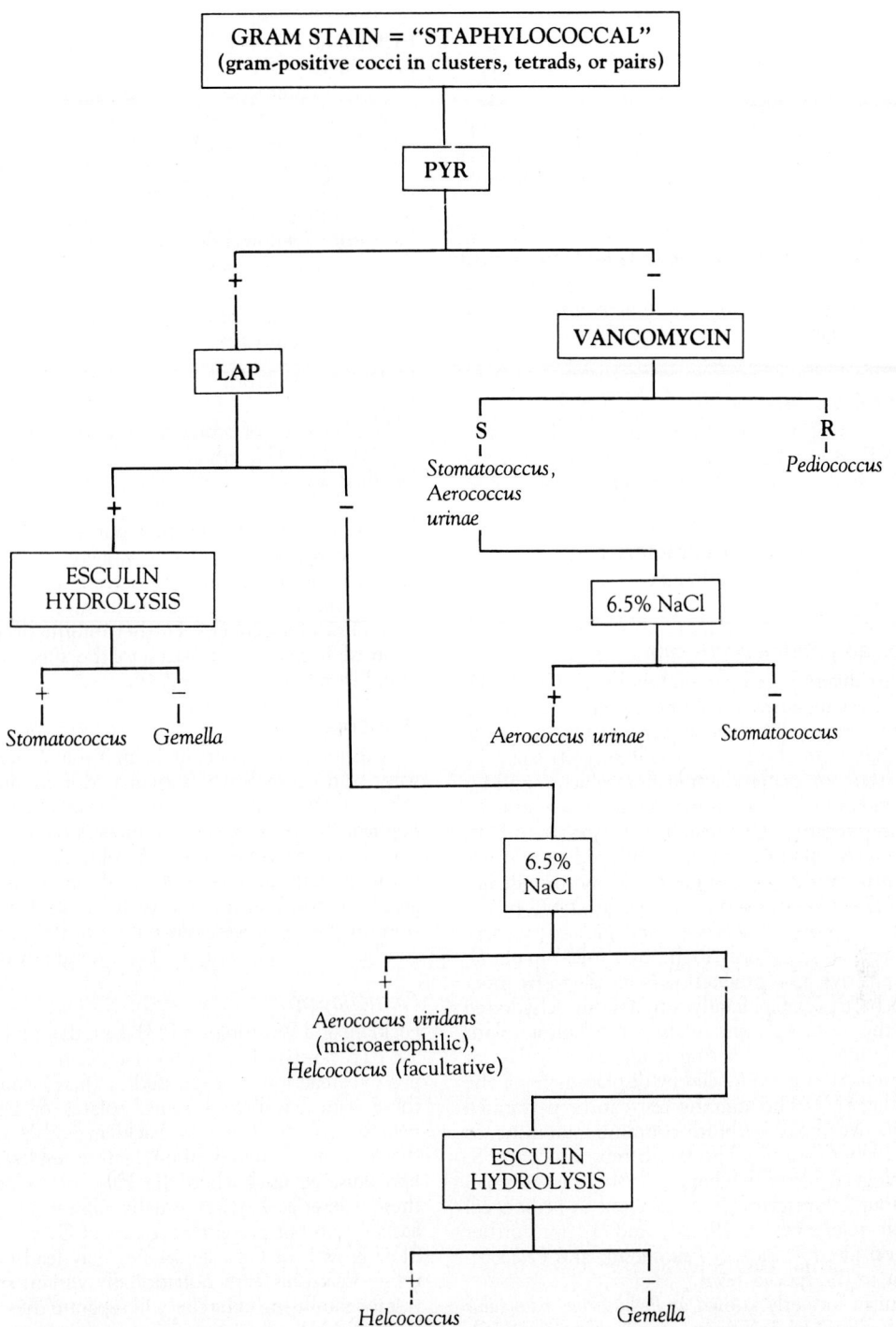

FIGURE 2 Identification of non-beta-hemolytic catalase-negative or weakly positive gram-positive cocci. S, susceptible; R, resistant.

(Carr-Scarborough Microbiologicals, Inc., Stone Mountain, Ga.). The manufacturer's instructions should be followed when the test is performed.

Growth in 6.5% NaCl

Salt-supplemented heart infusion broth with or without an acid-base indicator may be used for the NaCl growth test. According to the method of Facklam and Washington (23), salt-supplemented broth containing bromocresol purple is inoculated with two or three colonies of the organism to be tested and incubated at 35°C for up to 72 h. Turbidity with or without a color change from purple to yellow (due to production of acid) indicates growth. The medium recommended by Facklam and Washington is as follows.

Salt tolerance medium for gram-positive cocci (39)

Heart infusion broth	25 g
Sodium chloride	60 g
Indicator (1.6 g of bromocresol purple in 100 ml of 95% ethanol)	1 ml
Glucose	1 g
Distilled water..	1,000 ml

PYR Test

See Identification of Beta-Hemolytic Isolates: Physiological Tests in chapter 23 of this Manual for a description of the pyrrolidonyl arylamidase (PYR) test. Rapid disk tests are commercially available.

Esculin Hydrolysis Test

See Identification of Viridans Streptococci and S. bovis: Physiological Tests and Identification Kits in chapter 23 of this Manual for a description of the esculin hydrolysis test.

Additional Procedures for Characterization of Genera with Negative or Weak Catalase Reactions

Leuconostoc and Pediococcus spp.

In addition to differing cellular morphologies (Table 1), members of the vancomycin-resistant genera Leuconostoc and Pediococcus, along with vancomycin-resistant strains of lactobacilli that form short coccobacillary cells, may be separated by tests for gas production from glucose and for arginine degradation. Leuconostocs produce gas and are always arginine negative. Lactobacilli are variable in both tests, but a positive arginine test in a gas-producing strain rules out identity of the organism as a leuconostoc. Pediococci are gas negative and show variable reactions in the arginine test, although P. acidilactici and Pediococcus pentosaceus, the species commonly found in clinical material, are arginine positive. Gas production is measured by inoculating MRS broth (commercially available in dehydrated form) with the test organism, sealing the culture with melted petrolatum, and incubating it for up to 7 days at 35°C. Gas production is evidenced by displacement of the petrolatum plug (23). The arginine test can be performed with Moeller's decarboxylase broth containing arginine, as described in Identification of Viridans Streptococci and S. bovis: Physiological Tests in chapter 23 of this Manual. Lancefield group D antigen can be detected in pediococci (23). Consult references 5, 19, 23, and 42 for further information on identification of Leuconostoc and Pediococcus organisms to the species level.

The organism formerly known as Pediococcus halophilus was recently reclassified as Tetragenococcus halophilus (15). Although this organism shares the trait of LAP production with pediococci, it is PYR positive and vancomycin susceptible, suggesting its unrelatedness to typical pediococci. Although there are no current reports of clinical isolates of T. halophilus (22), this species should be suspected if vancomycin-susceptible, Pediococcus-like strains are encountered.

Stomatococcus spp.

See Procedures for Differentiation of Genera with Positive or Weakly Positive Catalase Reactions below.

Lactococcus spp.

Facklam and Washington (23) recommended growth temperature tests for distinguishing lactococci from streptococci and enterococci. Consult Fig. 1 for growth temperature characteristics of each of the genera. Broths are inoculated with a single colony or drop of broth culture of the test strain and incubated at 35°C for up to 7 days. A water bath is recommended for incubation of cultures at 45°C. Turbidity with or without a change in the broth's indicator to yellow indicates a positive test. The formula for the recommended medium is as follows.

Streptococcal and other gram-positive coccal growth medium (39)

Heart infusion broth	25 g
Glucose	1 g
Indicator (1.6 g of bromocresol purple in 100 ml of 95% ethanol)	1 ml
Distilled water...	1,000 ml

If it is important to rule out enterococci, suspicious isolates can be tested with a commercially available nucleic acid probe for the genus Enterococcus. Lactococcus lactis and Lactococcus garvieae are the species most commonly isolated from clinical specimens. Further information on differentiation of Lactococcus isolates to the species level may be found in references 18 and 46.

Aerococcus spp.

In addition to the traits listed in Table 2, aerococci display weak or no growth when incubated in an anaerobic atmosphere (20). This trait can be tested for by incubating duplicate blood agar plate cultures of the organism in question in an anaerobic and an aerobic atmosphere and comparing growth after 24 to 48 h. A. urinae is PYR negative and LAP positive, in contrast to A. viridans (2). Information on the heterogeneous nature of the genus Aerococcus can be found in a study by Bosley and coworkers (8).

Gemella spp.

Facklam and Washington (23) state that all Gemella species are PYR positive but that G. morbillorum displays a weakly positive reaction. Earlier studies (6, 7) generally support these data but record some isolates of both species as negative in the PYR test. Facklam and Washington note that in order to avoid false-negative results, a large inoculum must be used when the PYR test is performed with these bacteria. LAP is usually absent in isolates of G. haemolysans but present in strains of G. morbillorum (6, 7). Slow growth of Gemella species may lead to confusion of these organisms with nutritionally variant streptococci. A test for satelliting behavior will separate these two groups of bacteria (23).

Cells of G. haemolysans are easily decolorized and resemble those of neisseriae, since they occur in pairs with the adjacent sides flattened. G. haemolysans prefers an aerobic growth atmosphere. G. morbillorum cells are gram positive and arranged in pairs and short chains; individual cells in a given pair may be of unequal sizes. G. morbillorum is described as favoring anaerobic growth conditions.

Globicatella sp.

G. sanguis resembles viridans streptococci with respect to colonial and cellular morphology but differs from them by displaying a positive PYR reaction, a negative LAP reaction, and salt tolerance. Although these physiological traits are identical to those of *A. viridans*, *Globicatella*'s streptococcal cellular morphology distinguishes it from aerococci (13).

Helcococcus sp.

In addition to the characteristics shown in Table 2, most isolates of this slowly growing, tiny-colony-forming coccus produce an API Rapid Strep (bioMérieux Vitek) profile of 4100413, corresponding to an identification of "doubtful" *A. viridans*. Colonial morphology, good growth under anaerobic conditions, and stimulation of growth by addition of serum or Tween 80 to the medium differentiate *H. kunzii* from aerococci (14).

Procedures for Differentiation of Genera with Positive or Weakly Positive Catalase Reactions

Table 3 lists some distinguishing characteristics of micrococci, stomatococci, and alloiococci and compares them with those of staphylococci. *Stomatococcus* spp. are distinguished from species in the other genera by their inability to grow in the presence of 5% NaCl. Nutrient agar that has been supplemented with 5% NaCl is suitable for determining salt tolerance in stomatococci, micrococci, and staphylococci. Studies of *Alloiococcus* sp. have employed 6.5% salt broth, as described in Growth in 6.5% NaCl above. The slowly growing *Alloiococcus* sp. is differentiated from stomatococci and staphylococci by virtue of its obligately aerobic nature. A negative oxidase test distinguishes *Alloiococcus* sp. from the other obligate aerobe in this group, *Micrococcus* spp. *Alloiococcus* strains may produce a negative reaction in the catalase test, especially when grown on media devoid of blood.

Micrococcus spp. are the only oxidase-positive organisms among the group. The modified oxidase test described by Faller and Schleifer (25) should be used when these organisms are tested. It is performed by smearing a colony from sheep blood agar onto a piece of filter paper. Modified oxidase reagent, consisting of 6% tetramethylphenylenediamine hydrochloride (available from Eastman Kodak, Rochester, N.Y.) in dimethyl sulfoxide (Sigma Chemical Co., St. Louis, Mo.), is dropped onto the bacterial growth. Development of a dark blue color within 2 min indicates a positive oxidase reaction. The test is also available in the form of a ready-to-use disk impregnated with the oxidase reagent (Remel, Lenexa, Kans.). The reader is referred to the references for further details concerning the characterization of *Stomatococcus* spp. (3, 37, 38) and *Alloiococcus* sp. (1, 24).

ANTIBIOTIC SUSCEPTIBILITIES

Members of the vancomycin-resistant genera *Leuconostoc* and *Pediococcus* are also resistant to teicoplanin. Although they are usually susceptible to imipenem, minocycline, chloramphenicol, and gentamicin, their penicillin MICs correspond to the moderately susceptible category (49). Resistance to penicillin has been documented among some *Stomatococcus* strains, and susceptibility patterns to other commonly used antimicrobial agents seem to vary with the isolate (3, 37, 38). Stomatococci are uniformly susceptible to vancomycin. The observation that *Stomatococcus* spp. exhibit poor or no growth on Mueller-Hinton agar and on

Mueller-Hinton agar with sheep blood may make susceptibility testing problematic (37, 38).

Aerococcus viridans and *G. haemolysans* appear to be susceptible to penicillin and vancomycin and display a low level of resistance to aminoglycosides (9, 10). Buu-Hoi and colleagues (9) noted that while *A. viridans* seems to be naturally susceptible to macrolides, tetracyclines, and chloramphenicol, resistance to these agents has been observed. *A. urinae* has been described as susceptible to penicillin and vancomycin but resistant to sulfonamides and netilmicin (12). Buu-Hoi and coworkers (10) demonstrated a synergistic effect of penicillin and gentamicin in *G. haemolysans*.

Limited information on the antimicrobial susceptibility of *Lactococcus* spp. suggests that these organisms are moderately susceptible to penicillin and susceptible to vancomycin (26). One report documented treatment of lactococcal endocarditis with penicillin and gentamicin (35). A review of micrococcal endocarditis in cardiac surgery patients noted that MICs of penicillin ranged from 3.12 to 40.0 μg/ml and those of vancomycin spanned 1.56 to 10.0 μg/ml. All of the *Micrococcus* isolates tested were susceptible to cephalothin (33). Information on the antimicrobial susceptibilities of globicatellas, helcococci, and alloiococci is currently unavailable.

EVALUATION, INTERPRETATION, AND REPORTING OF RESULTS

Since the gram-positive cocci discussed in this chapter may appear in clinical cultures as contaminants or part of normal flora, efforts to identify them should be made only when isolates are considered clinically significant (i.e., isolated repeatedly, in pure culture, or from normally sterile sites). It should be remembered that these bacteria are opportunists; isolation from an immunocompetent patient may not have the same significance as isolation from a compromised host. Communication with clinicians should guide the microbiology laboratory in evaluating the significance of these infrequently isolated organisms.

Vancomycin susceptibility testing should be performed routinely on significant isolates. Documenting resistance to this antibiotic will not only guide therapy but also aid in identification of the isolate. The method mentioned in this chapter, which uses a nonstandardized inoculum, seems to be fairly reliable for determining susceptibility to this drug for identification purposes. Since general standardized susceptibility testing methods do not exist for these infrequently isolated gram-positive cocci, caution should be observed in interpretation of in vitro susceptibility test results. A reference laboratory should be consulted for identification or confirmation of the identity of unfamiliar organisms.

REFERENCES

1. **Aguirre, M., and M. D. Collins.** 1992. Phylogenetic analysis of *Alloiococcus otitis* gen. nov., sp. nov., an organism from human middle ear fluid. *Int. J. Syst. Bacteriol.* **42:**79–83.
2. **Aguirre, M., and M. D. Collins.** 1992. Phylogenetic analysis of some *Aerococcus*-like organisms from urinary tract infections: description of *Aerococcus urinae* sp. nov. *J. Gen. Microbiol.* **138:**401–405.
3. **Ascher, D. P., C. Zbick, C. White, and G. W. Fischer.** 1991. Infections due to *Stomatococcus mucilaginosus*: 10 cases and review. *Rev. Infect. Dis.* **13:**1048–1052.
4. **Bannatyne, R. M., and I. W. Fong.** 1992. *Gemella morbillorum*

infection in an arteriovenous shunt. *Clin. Microbiol. Newsl.* **14:**7–8.

5. **Barreau, C., and G. Wagener.** 1990. Characterization of *Leuconostoc lactis* strains from human sources. *J. Clin. Microbiol.* **28:**1728–1733.

6. **Berger, U.** 1985. Prevalence of *Gemella haemolysans* on the pharyngeal mucosa of man. *Med. Microbiol. Immunol.* **174:** 267–274.

7. **Berger, U., and A. Pervanidis.** 1986. Differentiation of *Gemella haemolysans* (Thjotta and Boe 1938) Berger 1960, from *Streptococcus morbillorum* (Prevot 1933) Holdeman and Moore 1974. *Zentralbl. Bakteriol. Hyg. Abt. A* **261:**311–321.

8. **Bosley, G. S., P. L. Wallace, C. W. Moss, A. G. Steigerwalt, D. J. Brenner, J. M. Swenson, G. A. Hebert, and R. R. Facklam.** 1990. Phenotypic characterization, cellular fatty acid composition, and DNA relatedness of aerococci and comparison to related genera. *J. Clin. Microbiol.* **28:**416–421.

9. **Buu-Hoi, A., C. LeBouguenec, and T. Horaud.** 1989. Genetic basis of antibiotic resistance in *Aerococcus viridans*. *Antimicrob. Agents Chemother.* **33:**529–534.

10. **Buu-Hoi, A., A. Sapoetra, C. Branger, and J. F. Acar.** 1982. Antimicrobial susceptibility of *Gemella haemolysans* isolated from patients with subacute endocarditis. *Eur. J. Clin. Microbiol.* **1:**102–106.

11. **Chatelain, R., J. Croize, P. Rouge, C. Massot, H. Dabernat, J. C. Auvergnat, A. Buu-Hoi, J. P. Stahl, and F. Bimet.** 1982. Isolement de *Gemella haemolysans* dans trois cas d'endocardites bacteriennes. *Med. Malad. Infect.* **12:**25–30.

12. **Christensen, J. J., H. Vibits, J. Ursing, and B. Korner.** 1991. *Aerococcus*-like organism, a newly recognized potential urinary tract pathogen. *J. Clin. Microbiol.* **29:**1049–1053.

13. **Collins, M. D., M. Aguirre, R. R. Facklam, J. Shallcross, and A. M. Williams.** 1992. *Globicatella sanguis* gen. nov., sp. nov., a new Gram-positive catalase negative bacterium from human sources. *J. Appl. Bacteriol.* **73:**433–437.

14. **Collins, M. D., R. R. Facklam, U. M. Rodrigues, and K. L. Ruoff.** 1993. Phylogenetic analysis of some *Aerococcus*-like organisms from clinical sources: description of *Helcococcus kunzii* gen. nov., sp. nov. *Int. J. Syst. Bacteriol.* **43:**425–429.

15. **Collins, M. D., A. M. Williams, and S. Wallbanks.** 1990. The phylogeny of *Aerococcus* and *Pediococcus* as determined by 16s rRNA sequence analysis: description of *Tetragenococcus* gen. nov. *FEMS Microbiol. Lett.* **70:**255–262.

16. **Colman, G.** 1967. *Aerococcus*-like organisms isolated from human infections. *J. Clin. Pathol.* **20:**294–297.

17. **Eggelmeijer, F., P. Petit, and B. A. C. Dijkmans.** 1992. Total knee arthroplasty infection due to *Gemella haemolysans*. *Br. J. Rheumatol.* **31:**67–69.

18. **Elliott, J. A., M. D. Collins, N. E. Pigott, and R. R. Facklam.** 1991. Differentiation of *Lactococcus lactis* and *Lactococcus garvieae* from humans by comparison of whole-cell protein patterns. *J. Clin. Microbiol.* **29:**2731–2734.

19. **Elliott, J. A., and R. R. Facklam.** 1993. Identification of *Leuconostoc* spp. by analysis of soluble whole-cell protein patterns. *J. Clin. Microbiol.* **31:**1030–1033.

20. **Evans, J. B.** 1986. Genus *Aerococcus* Williams, Hirch and Cowan 1953, 475[AL], p. 1080. *In* P. H. A. Sneath, N. S. Mair, M. E. Sharpe, and J. G. Holt (ed.), *Bergey's Manual of Systematic Bacteriology*, vol. 2. The Williams & Wilkins Co., Baltimore.

21. **Facklam, R. R.** 1977. Physiological differentiation of viridans streptococci. *J. Clin. Microbiol.* **5:**184–201.

22. **Facklam, R. R.** 1993. Personal communication.

23. **Facklam, R. R., and J. A. Washington II.** 1991. *Streptococcus* and related catalase-negative gram-positive cocci, p. 238–257. *In* A. Balows, W. J. Hausler, Jr., K. L. Herrmann, H. D. Isenberg, and H. J. Shadomy (ed.), *Manual of Clinical Microbiology*, 5th ed. American Society for Microbiology, Washington, D.C.

24. **Faden, H., and D. Dryja.** 1989. Recovery of a unique bacterial organism in human middle ear fluid and its possible role in chronic otitis media. *J. Clin. Microbiol.* **27:**2488–2491.

25. **Faller, A., and K.-H. Schleifer.** 1981. Modified oxidase and benzidine tests for separation of staphylococci from micrococci. *J. Clin. Microbiol.* **13:**1031–1035.

26. **Furutan, N. P., R. F. Breiman, M. A. Fischer, and R. R. Facklam.** 1991. *Lactococcus garvieae* infection in humans: a cause of prosthetic valve endocarditis, abstr. C-297, p. 391. *Abstr. 91st Gen. Meet. Am. Soc. Microbiol. 1991.*

27. **Garcia-Marcos, J. A., M. Meseguer, and F. Baquero.** 1992. Meningitis due to *Gemella haemolysans*. *Clin. Microbiol. Newsl.* **14:**142–143.

28. **Garvie, E. I.** 1986. Genus *Leuconostoc* van Tieghem 1878, 198[AL] emend mut. char. Hucker and Pederson 1930, 66[AL], p. 1071–1075. *In* P. H. A. Sneath, N. S. Mair, M. E. Sharpe, and J. G. Holt (ed.), *Bergey's Manual of Systematic Bacteriology*, vol. 2. The Williams & Wilkins Co., Baltimore.

29. **Garvie, E. I.** 1986. Genus *Pediococcus* Claussen 1903, 68[AL], p. 1075–1079. *In* P. H. A. Sneath, N. S. Mair, M. E. Sharpe, and J. G. Holt (ed.), *Bergey's Manual of Systematic Bacteriology*, vol. 2. The Williams & Wilkins Co., Baltimore.

30. **Golledge, C. L., N. Stingemore, M. Aravena, and K. Joske.** 1990. Septicemia caused by vancomycin-resistant *Pediococcus acidilactici*. *J. Clin. Microbiol.* **28:**1678–1679.

31. **Handwerger, S., H. Horowitz, K. Coburn, A. Kolokathis, and G. P. Wormser.** 1990. Infection due to *Leuconostoc* species: six cases and review. *Rev. Infect. Dis.* **12:**602–610.

32. **Kern, W., and E. Vanek.** 1987. *Aerococcus* bacteremia associated with granulocytopenia. *Eur. J. Clin. Microbiol.* **6:**670–673.

33. **Keys, T. F., and W. L. Hewitt.** 1973. Endocarditis due to micrococci and *Staphylococcus epidermidis*. *Arch. Intern. Med.* **132:**216–220.

34. **Kilpper-Balz, R., and K. H. Schleifer.** 1988. Transfer of *Streptococcus morbillorum* to the genus *Gemella* as *Gemella morbillorum* comb. nov. *Int. J. Syst. Bacteriol.* **38:**442–443.

35. **Mannion, P. T., and M. M. Rothburn.** 1990. Diagnosis of bacterial endocarditis caused by *Streptococcus lactis*. *J. Infect.* **21:**317–318.

36. **Mastro, T. D., J. S. Spika, P. Lozano, J. Appel, and R. R. Facklam.** 1990. Vancomycin-resistant *Pediococcus acidilactici*: nine cases of bacteremia. *J. Infect. Dis.* **161:**956–960.

37. **McWhinney, P. H. M., C. C. Kibbler, S. H. Gillespie, S. Patel, D. Morrison, A. V. Hoffbrand, and H. G. Prentice.** 1992. *Stomatococcus mucilaginosus*: an emerging pathogen in neutropenic patients. *Clin. Infect. Dis.* **14:**641–646.

38. **Mitchell, P. S., B. J. Huston, R. N. Jones, L. Holcomb, and F. P. Koontz.** 1990. *Stomatococcus mucilaginosus* bacteremias; typical case presentations, simplified diagnostic criteria, and a literature review. *Diagn. Microbiol. Infect. Dis.* **13:**521–525.

39. **Nash, P., and M. M. Krenz.** 1991. Culture media, p. 1226–1288. *In* A. Balows, W. J. Hausler, Jr., K. L. Herrmann, H. D. Isenberg, and H. J. Shadomy (ed.), *Manual of Clinical Microbiology*, 5th ed. American Society for Microbiology, Washington, D.C.

40. **Parker, M. T., and L. C. Ball.** 1976. Streptococci and aerococci associated with systemic infection in man. *J. Med. Microbiol.* **9:**275–302.

41. **Reyn, A.** 1986. Genus *Gemella* Berger 1960, 253[AL], p. 1081–1082. *In* P. H. A. Sneath, N. S. Mair, M. E. Sharpe, and J. G. Holt (ed.), *Bergey's Manual of Systematic Bacteriology*, vol. 2. The Williams & Wilkins Co., Baltimore.

42. **Riebel, W. J., and J. A. Washington.** 1990. Clinical and microbiologic characteristics of pediococci. *J. Clin. Microbiol.* **28:**1348–1355.

43. **Rubin, S. J., R. W. Lyons, and A. J. Murcia.** 1978. Endocarditis associated with cardiac catheterization due to a gram-positive coccus designated *Micrococcus mucilaginosus incertae sedis*. *J. Clin. Microbiol.* **7:**546–549.

44. **Ruoff, K. L., D. R. Kuritzkes, J. S. Wolfson, and M. J. Ferraro.** 1988. Vancomycin-resistant gram-positive bacteria isolated from human sources. *J. Clin. Microbiol.* **26:**2064–2068.

45. **Schleifer, K. H.** 1986. Family I. *Micrococcaceae* Prevot 1961, 31[AL], p. 1003–1035. *In* P. H. A. Sneath, N. S. Mair, M. E.

Sharpe, and J. G. Holt (ed.), *Bergey's Manual of Systematic Bacteriology*, vol. 2. The Williams & Wilkins Co., Baltimore.

46. **Schleifer, K. H., J. Kraus, C. Dvorak, R. Kilpper-Balz, M. D. Collins, and W. Fischer.** 1985. Transfer of *Streptococcus lactis* and related streptococci to the genus *Lactococcus* gen. nov. *Syst. Appl. Microbiol.* **6:**183–195.

47. **Sire, J. M., P. Y. Donnio, R. Mensard, P. Pouedras, and J. L. Avril.** 1992. Septicemia and hepatic abscess caused by *Pediococcus acidilactici*. *Eur. J. Clin. Microbiol. Infect. Dis.* **11:**623–625.

48. **Stackebrandt, E.** 1992. The genus *Stomatococcus*, p. 1320–1322. *In* A. Balows, H. G. Truper, M. Dworkin, W. Harder, and K. H. Schleifer (ed.), *The Prokaryotes*, 2nd ed., vol. 2. Springer-Verlag, New York.

49. **Swenson, J. M., R. R. Facklam, and C. Thornsberry.** 1990. Antimicrobial susceptibility of vancomycin-resistant *Leuconostoc, Pediococcus*, and *Lactobacillus* species. *Antimicrob. Agents Chemother.* **34:**543–549.

50. **Tunicliff, R.** 1917. The cultivation of a micrococcus from blood in pre-eruptive and eruptive stages of measles. *JAMA* **68:**1028–1030.

51. **von Graevenitz, A.** 1993. Revised nomenclature of *Alloiococcus otitis*. *J. Clin. Microbiol.* **31:**472.

Neisseria and Branhamella

JOAN S. KNAPP AND ROSELYN J. RICE

26

TAXONOMY

Neisseria species and *Branhamella catarrhalis* are classified with the genera *Moraxella*, *Kingella*, and *Acinetobacter* in the family *Neisseriaceae* (10). The genus *Neisseria* contains species isolated from humans and animals. The taxonomic positions of some *Neisseria* species (*Neisseria caviae*, *N. ovis*, and *N. cuniculi*) and *B. catarrhalis* are currently being debated. Proposals have been made to assign *B. catarrhalis* to the genus *Moraxella* (*Moraxella catarrhalis*) in the family *Moraxellaceae* or to its own genus, *Branhamella*, in the family *Branhamaceae* (13, 69); the name *B. catarrhalis* will be used in this chapter. The genera *Eikenella*, *Simonsiella*, and *Alysiella*; CDC groups EF-4 and M-5 (*N. weaverii* sp. nov.); and *N. iguanae* sp. nov. have also been assigned to the family *Neisseriaceae* (6, 10, 38). Descriptions of species in this chapter will be limited to those of human origin. Information relating to species of animal origin will be limited to a table of differential characteristics that should be consulted when a gram-negative diplococcus is not readily identifiable as a human *Neisseria* species, e.g., an isolate from a wound inflicted by an animal bite.

DESCRIPTION OF THE GENUS

Neisseria species (except *N. elongata*) and *B. catarrhalis* are gram-negative diplococci with adjacent sides flattened to give the characteristic kidney or coffee bean appearance observed in stained smears (77). Cells divide in two planes at right angles to each other, resulting in tetrads. Cells range in size from 0.6 to 1.5 μm, depending on the species, source of the isolate, and age of the culture. Occasionally, giant cells may be observed in some cultures. *N. elongata* and *Kingella denitrificans* are gram-negative coccobacilli that may form diplobacilli; the diplobacilli of *K. denitrificans* may resemble diplococci and thus may be confused with *N. gonorrhoeae*. Cells are nonmotile and do not produce endospores. Cells of some species are encapsulated, and some species produce a yellow-green carotenoid pigment. Cells of some species, most notably *N. gonorrhoeae*, may autolyze in culture after approximately 24 h of incubation.

Most *Neisseria*, *Branhamella*, and *Kingella* strains have complex growth requirements; some strains may be exquisitely sensitive to fatty acids, necessitating the incorporation of soluble starch in growth media. All species are aerobic and have an optimal growth temperature of 35 to 37°C; *B. catarrhalis* strains are tolerant of lower temperatures than the other species and grow well at 28°C. Growth of many *Neisseria* species is enhanced by humidity and CO_2, and strains of some species, notably *N. gonorrhoeae*, require CO_2. All species are oxidase positive and, with the exception of *N. elongata* and *K. denitrificans*, catalase positive. Species have limited metabolic activities and are differentiated by their abilities to produce catalase, produce acid from a few carbohydrates, produce polysaccharide from sucrose, produce DNase, reduce nitrate and nitrite, oxidize fatty acids such as tributyrin, and produce certain enzymes.

Neisseria species are oxidative; i.e., they produce acid from carbohydrates by oxidation, not fermentation. Consequently, the term fermentation should not be used to describe acid production by these species. Because these species are oxidative and produce less acid from carbohydrates than do fermentative organisms and because they also produce ammonia from peptones, which may neutralize any acid produced from carbohydrates, acid production must be determined in medium with a low protein/carbohydrate ratio and a sensitive indicator such as phenol red (47). The terms carbohydrate utilization and carbohydrate degradation should not be used to describe the results obtained in acid production tests. Some species, e.g., *N. cinerea*, may use glucose but rapidly overoxidize acid to carbon dioxide, with the result that acid does not accumulate in the reaction tube although the carbohydrate has been used (47). The term acid production is the most accurate term to describe the reaction observed in laboratory diagnostic tests.

An extensive discussion of the genetic relatedness among *Neisseria* and related species is beyond the scope of this chapter. However, elegant studies of the genetic relatedness between species have been done (23, 68). The relatedness between species can be summarized as follows: (i) *N. gonorrhoeae* and *N. meningitidis* (and *N. kochii*) are very closely related; (ii) *Neisseria lactamica*, *Neisseria polysaccharea*, *N. cinerea*, and *Neisseria flavescens* are related to *N. gonorrhoeae* and to each other more distantly; and (iii) strains of the "saccharolytic" *Neisseria* species (*N. subflava* bv. subflava, *N. subflava* bv. flava, *N. subflava* bv. perflava, *N. sicca*, and *N. mucosa*) are distantly related to the first two groups. As anticipated, species such as *B. catarrhalis* and

K. denitrificans are still more distantly related, as are the species of animal origin.

NATURAL HABITAT

Most *Neisseria* species are nonpathogenic and are normal inhabitants only of the oro- and nasopharyngeal mucous membranes of humans (49). On rare occasions, nonpathogenic species may be isolated from other mucous membranes, but these sites should be considered opportunistic sites of colonization rather than normal habitats. Among *Neisseria* species, only strains of *N. meningitidis* and *N. gonorrhoeae* are considered pathogens. Strains of *N. gonorrhoeae* are always considered pathogenic and may infect exposed anogenital and oropharyngeal mucous membranes. Strains of *N. meningitidis* may colonize the oro- and nasopharynx as nonpathogens in a "carrier" state and may colonize exposed anogenital mucosal membranes, particularly of homosexual men (42); some strains may cause epidemic and acute meningitis.

The normal patterns of oro- and nasopharyngeal colonization by *Neisseria* species were studied extensively in the early 1900s, but many species were misidentified owing to the use of inappropriate differential tests and media (47). Recent studies have demonstrated that both adults and children may be colonized simultaneously by several *Neisseria* species or by multiple strains of the *N. subflava* bv. perflava-*N. sicca* group (46a, 49). Patterns of colonization appear to be stable, persisting for several months in adults (49). Similarly, colonization by *N. meningitidis* may persist for many months. *N. lactamica* is isolated more frequently from children and young adolescents than from adults; the isolation of *N. polysaccharea* may vary geographically (3, 7, 8, 33, 65). *B. catarrhalis* is rarely isolated from the oropharynx of a healthy adult but may be carried more frequently in children and older adults (49, 76). The frequency and persistence of colonization by *K. denitrificans* appear to vary geographically (43, 49, 50).

CLINICAL SIGNIFICANCE

N. gonorrhoeae

Despite estimates that only one-half of cases are reported, gonorrhea remains the most frequently reported communicable disease in the United States (16). The number of reported cases of gonorrhea increased steadily from 1964 to 1977, fluctuated through the early 1980s, increased until 1987, and since 1987 has decreased annually (16). The incidence of gonorrhea is highest in high-density urban areas among persons under 24 years of age who have multiple sex partners and engage in unprotected sexual intercourse. The recent decline in the prevalence of gonorrhea, particularly among gay and bisexual men, may be attributed in part to changes in sexual behavior in response to the AIDS epidemic and to recommendations, in the United States, for the routine use of highly effective antimicrobial agents to treat gonorrhea (17). Factors that may contribute to the persistence and spread of gonorrhea are the occurrence of asymptomatic infections (particularly in women), which may be transmitted unknowingly between sexual partners, and the high-frequency spread of infections by persons with multiple sexual partners, including those who exchange sex for money or drugs.

N. gonorrhoeae is highly susceptible to adverse environmental conditions, including extreme temperatures and drying; the organism does not survive long outside its host. Gonorrhea is transmitted by direct, close, usually sexual contact between individuals. Infections are transmitted more efficiently from an infected man to a woman (in 50 to 60% of instances of one sexual exposure) than from an infected woman to a man (in 35% of instances of one sexual exposure). Transmission of gonorrhea to neonates usually occurs during birth. Transmission of gonorrhea to older infants is often associated with allegations of sexual abuse by an adult or older adolescent. Nonsexual human (skin-skin, skin-mucous membrane, or autoinoculation) or fomite transmission (excluding laboratory accidents) of *N. gonorrhoeae* has not been documented.

Uncomplicated Gonorrhea

The majority of gonococcal infections are uncomplicated lower genital tract infections caused by direct infection of the columnar epithelium of mucosal membranes. In men, acute urethritis with symptoms including a scant clear to a copious purulent discharge and urinary burning and frequency usually occurs within 7 days after infection. Asymptomatic infections are estimated to occur in 1 to 5% of infected men (20); the prevalence of asymptomatic infections may be higher in contacts of women with asymptomatic infections. If untreated, men may develop epididymitis, prostatitis, and urethral stricture. In women, the primary site of infection is usually the endocervix. Symptomatic uncomplicated infections are characterized by vaginal discharge, dysuria, and an erythematous, friable cervical os. Women with asymptomatic infections often manifest low-grade or no endocervical inflammation or discharge. In prepubertal girls, gonococcal infections may present as vulvovaginitis. Local complications may include urethral vestibular abscesses or salpingitis and pelvic inflammatory disease (PID). Asymptomatic infections may go undetected more frequently in women than in men.

Oropharyngeal and Anorectal Infections

Persons practicing receptive oral or anal intercourse may acquire oral or anorectal gonococcal infections. However, women who do not engage in anal intercourse may acquire anorectal infections by contamination from cervical secretions. Symptomatic oropharyngeal infections may manifest as a mild pharyngitis. Symptoms of anorectal infections may include a copious purulent discharge, burning or stinging pain, tenesmus, and blood in stools. Anorectal infections may also be asymptomatic.

Conjunctivitis

Ocular infections, most frequently diagnosed in newborns exposed to infected secretions during birth (ophthalmia neonatorum) but also occasionally seen in adults, must be treated promptly to prevent blindness. Conjunctival infections involve tearing and edema associated with a purulent exudate. If the infection is not treated, scarring or perforation of the cornea may occur. In industrial countries, neonatal blindness due to gonococcal infections has been largely eliminated by the routine application of prophylactic agents such as silver nitrate drops or erythromycin ointment to the eyes of newborns.

PID

If left untreated, endocervical infections may ascend locally to cause PID manifested as endometritis, salpingitis, pelvic peritonitis, and tubo-ovarian abscesses. It is estimated that 8 to 10% of women with endocervical gonococcal infections may develop PID, the most serious complication of which is permanent damage to the reproductive system. Although other organisms, including *Chlamydia trachomatis* and anaerobes, are associated with the development of PID, gonococci have been isolated in 10 to 70% of cases and are believed to play a major role in initial episodes of this disease. Anaerobic vaginal and gut flora have been associated more frequently with subsequent episodes of PID. The inflammatory response to infections may result in scarring and blockage of the fallopian tubes and, subsequently, infertility and ectopic pregnancies. Fitz-Hugh-Curtis syndrome (perihepatitis) may be associated as frequently with gonococcal PID as with chlamydial PID. A diagnosis of PID must be considered for women with lower abdominal pain and adnexal tenderness. Although laparoscopy is the best method for diagnosing PID if the inflammation has reached the fallopian tubes and the pelvic peritoneum, this procedure is rarely performed, and endocervical specimens for *N. gonorrhoeae* and *C. trachomatis* must be taken. PID should be treated empirically with recommended antimicrobial regimens with broad-spectrum activity against not only *N. gonorrhoeae* but also *C. trachomatis* and anaerobes (17).

DGI

Approximately 1 to 3% of persons with gonorrhea may develop disseminated gonococcal infection (DGI), which has often been associated with untreated asymptomatic infections, although persons deficient in complement component C7, C8, or C9 may experience recurrent DGI. Gonococci may be isolated from blood in fewer than 50% of patients. Patients may develop "dermatitis-arthritis" syndrome, with fever, chills, skin lesions, and diffuse arthralgias in the hands, feet, ankles, and elbows. Skin lesions, which occur most frequently on the extensor surfaces of the hands and feet, may be macular, pustular, hemorrhagic lesions with central necrosis. A small number of patients develop septic arthritis, and gonococci may be isolated in approximately 50% of these cases. A few patients may also develop endocarditis or meningitis. A diagnosis of DGI may be based on the identification of gonococci from synovial fluid, blood, or cerebrospinal fluid (CSF). However, DGI must be differentiated from meningococcemia.

Because gonococci are rarely observed in clinical specimens and the organism cannot be isolated from many patients, a presumptive diagnosis of DGI may be based on the isolation of gonococci from a mucosal site in the patient or the patient's sexual partner, the observation of skin lesions on the extremities, or the resolution of symptoms when the patient is treated with an appropriate antimicrobial agent. Cases of DGI are associated with highly penicillin-susceptible organisms. However, owing to recent case reports of antibiotic-resistant strains causing DGI, persons with suspected or confirmed DGI should be treated with appropriate extended-spectrum antimicrobial regimens.

N. meningitidis

Meningococci are isolated most frequently from the oro- or nasopharynges of asymptomatic carriers. Meningococcal colonization may persist for several weeks to several months and may occur in as many as 5 to 15% of individuals or more in confined populations such as military recruits. Because meningococci do not survive well outside their human host and have no alternative host, transmission of meningococcal strains usually occurs by direct contact with contaminated respiratory secretions or airborne droplets. Sexual transmission of meningococcal strains may cause lower genital tract infections in women (32) and homosexual men, resulting in anogenital carriage in addition to oro- and nasopharyngeal carriage (32, 39). Because meningococci may be isolated from anogenital sites usually infected by gonococci, it is necessary to identify neisserial isolates from these sites with confirmatory tests.

Meningococci may occasionally disseminate from the nasopharynx to cause meningococcemia and/or meningitis. The incidence of meningococcemia is highest in school-age children, adolescents, and young adults. In some individuals, the disease progresses rapidly, resulting in the fulminant death of a previously healthy person within a few hours of the onset of symptoms. Close contacts of persons with meningococcemia are at highest risk for acquiring the disease and must receive immediate prophylaxis.

Meningococcemia is usually characterized by profound vascular effects, including a petechial or purpuric skin rash that occurs in about 75% of patients. In fulminating infections (Waterhouse-Friderichsen syndrome), widespread coagulation and fulminant sepsis occur, resulting in shock and, usually, death.

Meningococcal strains have also caused arthritis—a relatively frequent complication of meningococcemia—and, rarely, purulent conjunctivitis, sinusitis, endocarditis, and primary pneumonia. Persons with inherited complement deficiencies are at greater risk for acquiring systemic meningococcal infections and may experience repeated episodes (39). On rare occasions, endocervical meningococcal infections have been associated with PID.

B. catarrhalis

Although *B. catarrhalis* has been recognized as a cause of human infections since the early 1900s, the variety of infections caused by this species has been clearly documented only in the past 10 to 15 years (26). *B. catarrhalis* causes infections ranging from acute, localized infections such as otitis media, sinusitis, and bronchopneumonia to life-threatening, systemic diseases including endocarditis and meningitis. *B. catarrhalis* causes 10 to 15% of cases of otitis media (31) and a similar proportion of cases of sinusitis. In addition, this organism causes a large proportion of cases of lower respiratory tract infections in elderly patients with chronic obstructive pulmonary diseases and is exceeded only by *Haemophilus influenzae* and *Streptococcus pneumoniae* as a causative agent of acute purulent exacerbations of chronic bronchitis (26). *B. catarrhalis* is also associated with frank pneumonia. Infections of the maxillary sinuses, middle ears, or bronchi may occur through contiguous spread of organisms from their normal habitat.

The status of *B. catarrhalis* as a member of the normal oropharyngeal flora deserves reevaluation. Descriptions in the early 1900s of *N. catarrhalis* in the normal oropharyngeal flora were probably incorrect; for many years, strains of *N. cinerea* were misidentified as *N. catarrhalis* (47). *B. catarrhalis* is not isolated frequently from the oropharynges of healthy persons (49, 76). This species has, however, been isolated from some individuals with signs of respiratory infections but only for the duration of those symptoms (46a). The species may be normal flora in the nasopharynx or upper respiratory tract rather than in the oropharynx. From a practical perspective, *B. catarrhalis* must be distinguished from *Neisseria* species in the clinical laboratory.

Other *Neisseria* Species

Most *Neisseria* species (see Table 3) are considered commensal inhabitants of the oro- and nasopharynges of

TABLE 1 Body sites and culture media for the isolation of N. gonorrhoeae, N. meningitidis, and B. catarrhalis

Species	Syndrome	Gender[a]	Site(s)	Media
N. gonorrhoeae	Uncomplicated	Female	Endocervix, (Bartholin's gland), rectum,[b] (urethra), pharynx[b]	Selective
		Male		
		Heterosexual	Urethra	Selective
		Homosexual	Urethra, rectum,[b] pharynx[b]	Selective
	PID		Endocervix, endometrium,[c] fallopian tubes[c]	Selective, nonselective
	DGI		Endocervix (female), urethra (male), skin lesions	Selective, nonselective
			Joint fluid	Nonselective, selective
			Blood	Blood culture medium
	Ophthalmia		Conjunctiva	Nonselective
N. meningitidis	Meningitis		CSF, skin lesions	Nonselective
			Blood	Blood culture medium
			Nasopharynx	Selective, nonselective
B. catarrhalis	Otitis media, sinusitis		Sputum (transtracheal aspirates)	Nonselective

[a]Only if different sites are cultured for males and females.
[b]If there is a history of oral-genital or anal-genital exposure.
[c]If a laparoscopic examination is performed.

healthy persons. N. flavescens has been isolated only once, in association with an outbreak of meningitis. Most species have been implicated as etiologic agents in one or more of the following infections: meningitis, bacteremia, endocarditis, empyema, pericarditis, and pneumonia (37). N. lactamica has been isolated from the urogenital tract (44). When isolated from normally sterile sites, these species must be identified and distinguished from N. gonorrhoeae and N. meningitidis. N. cinerea, which has been isolated on gonococcal selective media from endocervical and rectal infections, from conjunctivitis in neonates, and from lymphadenitis, has been misidentified repeatedly as N. gonorrhoeae (19, 22, 27, 53). N. lactamica and N. polysaccharea, both of which are isolated on selective media, must be differentiated from N. meningitidis. N. kochii, considered a subspecies of N. gonorrhoeae, has been isolated in Egypt from patients with conjunctivitis and urethritis (47). Strains have the colonial morphologic characteristics of N. meningitidis and the biochemical characteristics of N. gonorrhoeae. This species has not been described in the United States, but, depending on the diagnostic tests used for identification, isolates would be described as N. gonorrhoeae or a nontypeable, maltose-negative N. meningitidis. Occasionally, Neisseria species of animal origin (see Table 4) may be isolated from bite wounds inflicted by cats or dogs (36, 38). Thus, it is important to keep animal species in mind when the characteristics of a gram-negative, oxidase-positive diplococcus are not those of a described human species.

COLLECTION, TRANSPORT, AND STORAGE OF SPECIMENS

N. gonorrhoeae

Specimens for the isolation of N. gonorrhoeae may be obtained from the genital tract, urine, anal area, oropharynx, conjunctiva, Bartholin's gland, fallopian tubes, endometrium, blood, joint fluid, skin lesions, or gastric contents of neonates as described in Cumitech 4A (29), Cumitech 17A (5), and chapter 3 of this Manual (Table 1). A few special precautions and procedures mentioned here may facilitate

recovery. Although endocervical specimens are appropriate with sexually active girls and women, prepubescent females may be infected in the vaginal mucosa. A cotton swab may be rubbed against the posterior vaginal wall for 10 to 15 s to absorb secretions, or if the hymen is intact, the specimen may be collected from the vaginal orifice. Urethral specimens should be collected at least 1 h after the patient has urinated. Clean-catch, midstream urine specimens (10 to 15 ml) should be centrifuged, and the sediment should be inoculated onto a selective medium for the isolation of N. gonorrhoeae. Anorectal swabs that are contaminated with feces should be discarded, and a second specimen should be obtained.

Blood should be inoculated immediately into blood culture media. If blood is transported in tubes containing sodium polyanetholsulfonate (SPS), which may be toxic to N. gonorrhoeae, it should be inoculated into culture medium within 1 h after collection. Skin lesion specimens should be kept moist. Gonococci are more likely to be isolated from a punch biopsy of skin than from an aspirate; nonselective and selective media should be inoculated.

N. meningitidis

Specimens for the isolation of meningococci should be immediately transported to the laboratory under ambient conditions. These organisms are highly susceptible to temperature extremes and desiccation. Specimens may include CSF, blood, petechial aspirates, biopsy samples, joint fluid, conjunctival swabs, sputum or transtracheal aspirates, and nasopharyngeal swabs. Because there is some evidence that SPS is toxic to N. meningitidis, tubes containing this component should not be used for blood specimens when meningococcemia is suspected. Nasopharyngeal specimens produce a higher yield than oropharyngeal specimens.

B. catarrhalis

Tympanocentesis fluid and sinus aspirates are the ideal specimens from which to isolate the causative agents of otitis media and sinusitis, respectively. These specimens are rarely collected because they involve costly, invasive procedures that cause much discomfort to the patient. Thus, these infections are usually treated empirically. According to a comparison of Gram stain and culture results for sputum and transtracheal aspirates, sputum is a satisfactory specimen for the isolation of *B. catarrhalis* from patients with lower respiratory tract infections (2).

Media, Specimen Inoculation, and Incubation Conditions

Among the *Neisseria* species, *N. gonorrhoeae* and *N. meningitidis* are recognized as the etiologic agents of most clinically distinct and serious infections in humans. *N. gonorrhoeae* typically causes urogenital tract disease, and *N. meningitidis* causes more serious systemic, and sometimes fatal, septicemia and meningitis. Procedures for the transport of specimens and for the isolation, identification, and storage of organisms apply not only to *N. gonorrhoeae* and *N. meningitidis* but also to *B. catarrhalis* and the commensal *Neisseria* and related species that may occasionally be associated with these infections.

Selective Media

Selective media for *N. gonorrhoeae* and *N. meningitidis*, such as modified Thayer-Martin (MTM) (56) and Martin-Lewis (ML) (58) media, contain four antimicrobial agents: vancomycin (MTM, 3 μg/ml; ML, 4 μg/ml) to inhibit gram-positive bacteria; colistin (7.5 μg/ml) to inhibit gram-negative bacteria, including the commensal *Neisseria* species; trimethoprim lactate (5 μg/ml) to inhibit swarming *Proteus* species; and an antifungal agent (MTM, 13.5 μg of nystatin per ml; ML, 20 μg of anisomycin per ml). New York City (NYC) medium is a clear medium in which hemolyzed horse erythrocytes, horse plasma, and yeast dialysate are substituted for the hemoglobin and supplements contained in the chocolate agar-based media (30). NYC medium contains the antimicrobial agents vancomycin (2 μg/ml), colistin (5.5 μg/ml), amphotericin B (1.2 μg/ml), and trimethoprim lactate (3 μg/ml). When prepared properly and used while fresh, these media permit the selective isolation of *N. gonorrhoeae* and *N. meningitidis* while inhibiting the growth of most commensal *Neisseria* species. It should be noted, however, that strains of the commensal species *N. lactamica* and *K. denitrificans* are routinely isolated on selective media; occasionally, some strains of *N. subflava* bv. perflava, *N. cinerea*, and *B. catarrhalis* are also isolated on selective media. NYC medium also permits the growth of large-colony mycoplasmas and *Ureaplasma urealyticum* (T-cell mycoplasmas). Specimens for the isolation of *N. meningitidis* from sites with normal flora may be inoculated on nonselective blood or chocolate medium in addition to selective media.

Some gonococcal strains, especially AHU strains (so named because they require arginine, hypoxanthine, and uracil for growth on chemically defined media) are susceptible to the concentrations of vancomycin used in selective media. Although AHU isolates accounted for more than 50% of isolates in some geographic areas in the United States in the mid-1970s, they are now rare (71). Still other gonococci may be susceptible to trimethoprim. It is therefore advisable for laboratorians to consider periodic assessment of the adequacy of isolation of potentially more fastidious strains of *N. gonorrhoeae* through a quality assessment program that compares isolation rates from urethral and endocervical specimens from men and women, respectively, on selective and nonselective isolation media.

Nonselective Media

Ideally, specimens from sterile sites such as joint fluids, skin lesions, and conjunctiva should be inoculated onto both nonselective and selective media. A disadvantage of transporting specimens on nonselective medium is that any contaminating organism may overgrow a pathogenic species. Contaminating organisms may be introduced into the specimen during the collection process, during handling of the specimen and its container, or from the condensation that develops during the incubation process. Specimens received on nonselective medium are incubated as described above for specimens received on selective media.

Specimens for the isolation of *B. catarrhalis* and nasopharyngeal specimens for the isolation of *N. meningitidis* may be inoculated onto 5% sheep blood agar in addition to or in place of chocolate agar. Since gonococcal strains may not grow on sheep blood agar, specimens should be inoculated onto chocolate agar if there is a possibility that gonococci will be isolated from the specimen.

Blood Culture Media

Most blood culture media (tryptic soy, Columbia, and brain heart infusion) support the growth of gonococci and meningococci, provided the inoculated medium is vented and incubated in a CO_2-enriched atmosphere. However, avoid blood culture media that contain SPS, which may be toxic to some strains of *N. gonorrhoeae* and *N. meningitidis*. Media that contain gelatin (Carr Scarborough Microbiological, Decatur, Ga.) neutralize the effect of SPS (28). Lysed blood also neutralizes the toxicity of SPS; thus, cell lysis in the Isolator system (Wampole, Cranbury, N.Y.) overcomes the toxicity of SPS (73).

Incubation Conditions

Gonococci are the most fastidious of the *Neisseria* species, require complex growth media, and are highly susceptible to toxic substances such as fatty acids. In addition, although some gonococci require CO_2 for growth, the growth of all species is enhanced by a CO_2-enriched atmosphere. Thus, cultures for the isolation of *N. gonorrhoeae* or *N. meningitidis* species are always incubated at 35 to 37°C in a CO_2-enriched, humid atmosphere.

CO_2-Enriched Atmosphere

CO_2 incubators are desirable if large numbers of specimens must be processed. Alternative methods for providing a CO_2-enriched atmosphere (candle extinction jars or commercial CO_2-generating systems) may be used when specimens must be transported to a laboratory or when only small numbers of specimens are processed. A candle extinction jar produces a CO_2 concentration of 3 to 5%. It is important (i) to use a nonscented candle (the vapor from scented candles may be toxic) and (ii) to relight the candle if the jar is opened to add more plates.

Humidity

Both gonococci and meningococci require increased humidity for good growth. The humidity in incubation chambers may be increased in several ways: placing a pan of water on the bottom shelf of a CO_2 incubator; placing moistened but not dripping paper towels on the bottom the candle extinction jar (these should be replaced frequently); or placing a few drops of sterile saline or culture broth onto the surface of the culture medium, allowing it to soak in before inoculation, and sealing the plates in a zipper-locked bag with a CO_2-generating tablet.

Specimen Transport

The best method for preserving viable organisms is to inoculate specimens directly onto a nutritive medium and incubate them at 35 to 37°C in a CO_2-enriched atmosphere immediately after collection. Specimens from sites with normal flora should be inoculated onto a selective medium such as modified MTM, ML, or NYC medium. Nasopharyngeal specimens for *N. meningitidis* may be inoculated onto chocolate or blood agar medium. Joint fluid, conjunctival specimens, or specimens from sterile sites may also be inoculated onto a nonselective medium such as chocolate medium supplemented with IsoVitaleX (or an equivalent supplement). Specimens for the isolation of meningococci may be inoculated onto a blood agar medium.

If specimens must be transported to a local laboratory and it is not possible to immediately incubate the inoculated media before transport, it is more important to place the inoculated plates in a CO_2-enriched atmosphere than to incubate them at 35 to 37°C. Inoculated media may be held at room temperature in a CO_2-enriched atmosphere in either candle extinction jars or commercial CO_2-generating systems such as JEMBEC plates (57) for up to 5 h without appreciable loss of viability. If inoculated specimens must be transported between locations in very hot or very cold weather, it is recommended that containers be transported in an insulated styrofoam container. If specimens must be transported to a distant town, the inoculated media should be incubated for 18 to 24 h before being transported, and the specimen should arrive within 48 h.

Before the development of nutritive transport systems, specimens for the isolation of gonococci were shipped to laboratories in buffered, nonnutritive transport media. Although gonococci survive in these media for short periods, their viability decreases rapidly after 6 to 12 h, so that they may not be recovered after 24 h. In addition, the specimen is diluted, thus reducing the probability of isolating organisms. With the development of commercial zipper-locked, CO_2-generating systems, semisolid transport media such as Stuart's or Amies medium are no longer recommended for the transport of gonococcal specimens.

Storage

Cultures

An isolate will usually survive no longer than 48 h in culture, although some isolates may survive for 72 to 96 h. Isolates should be subcultured every 18 to 24 h to maintain maximum viability. An 18- to 24-h culture should be used to inoculate any diagnostic test that requires viable organisms. Similarly, isolates that are stored by freezing or lyophilization should also be subcultured at least once after the initial recovery culture before being used in a diagnostic test.

Frozen Storage

Long-term storage of gonococcal isolates is achieved with a −70°C freezer, liquid nitrogen (−196°C), or freeze-drying. To store frozen isolates, dense suspensions of 18- to 24-h pure cultures are prepared in tryptic soy broth containing 20% glycerin. As many as 99% of cells in a suspension may be destroyed during the freezing and thawing of the preparations. The loss of viability of gonococci in frozen storage is due to physical destruction (shearing) of cells by

crystals of the suspending medium that form during the freezing and thawing processes. For this reason, frozen preparations of gonococcal isolates may not be repeatedly thawed and refrozen. The loss of viability during freezing may be minimized by "flash" freezing the specimen in an acetone or alcohol bath containing dry ice. Some isolates may survive storage at −20°C for up to 2 weeks. The use of freezing media containing glycerin results in a "softer" product from which a sample may be taken with a sterile bacterial loop without thawing the preparation.

Lyophilization

Some laboratories have access to lyophilization (freeze-drying) facilities. To prepare lyophilates, 18- to 24-h cultures of isolates are suspended in special lyophilization media and distributed in small aliquots (usually 0.25 to 0.5 ml) in lyophilization ampoules. As with frozen storage, ca. 99% of the organisms are killed during the freezing process. Skim milk is not recommended as a suspending medium for the lyophilization of *Neisseria* species, because fatty acids in the milk may be toxic for some organisms and it is impossible to determine the density of the suspension. The suspensions are frozen at −70°C or in an ethanol-dry ice bath and dried in a vacuum for 18 to 24 h until the moisture has evaporated. The dried preparation should be powdery; if the preparation is syrupy, the vial should be discarded. One ampoule should also be opened immediately to ascertain that the preparation is viable and pure and to verify the identity of the organism and important characteristics, e.g., antimicrobial susceptibilities. Ampoules are best stored at 4 to 10 or −20°C; storage at room temperature is not recommended. Oxygen may diffuse slowly into the ampoule through the thin seal, particularly with thin-walled ampoules. Thus, one ampoule should be opened every 1 to 2 years to confirm that the preparation is viable. If the preparation is not viable after 48 h of incubation, new ampoules must be prepared.

ISOLATION PROCEDURES

Inoculated Agar Media

Incubate specimens inoculated on selective or nonselective media for up to 72 h, examining plates at 24-h intervals. If no growth is observed after 72 h, the culture should be reported as "no growth." Colonies of *N. gonorrhoeae* vary in diameter from 0.5 to 1 mm owing to the formation of different colony types (11). When examining cultures for *N. gonorrhoeae*, remember that AHU strains may produce colonies atypically small (ca. 0.25 mm in diameter) compared with those of most gonococcal strains (0.5 to 1 mm in diameter). In contrast, after incubation for 18 to 24 h, colonies of *N. meningitidis* are usually larger (1 to 2 mm in diameter) and flatter than those of *N. gonorrhoeae*. Colonies of encapsulated serogroup A and C strains may be mucoid. Among the *Neisseria* and related species, colonies of *N. lactamica*, *N. cinerea*, *N. polysaccharea*, *N. kochii*, and *K. denitrificans* are similar in size, appearance, and consistency to those of *N. gonorrhoeae* and *N. meningitidis*. Colonies of *B. catarrhalis* and the saccharolytic species *N. subflava*, *N. sicca*, and *N. mucosa* are usually 1 to 3 mm in diameter and opaque, and they vary in color from grayish pink (*B. catarrhalis*) to yellow (*N. subflava*). Colonies vary in consistency. The pinkish brown colonies of *B. catarrhalis* have a friable, "hockey puck" consistency and may be moved intact over the surface of the medium with a bacteriologic

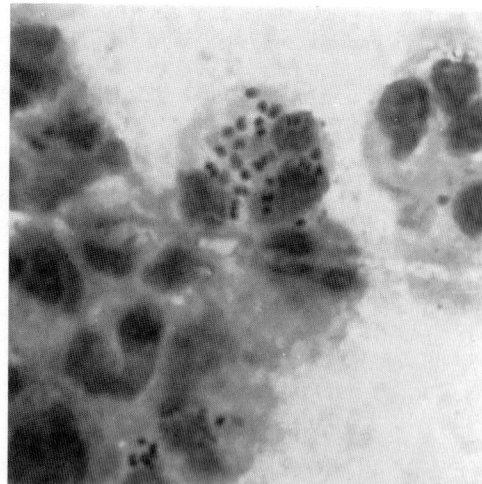

FIGURE 1 Gram stain of male urethral exudate. Some PMNs contain many diplococci; others contain none. Magnification, ×1,500.

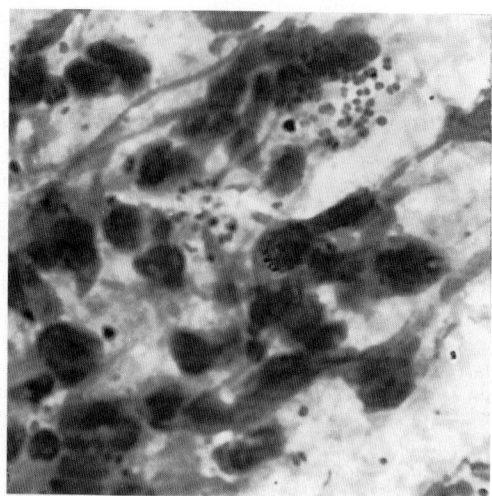

FIGURE 2 Gram stain of mucopurulent rectal exudate showing many diplococci inside PMNs. Magnification, ×1,500.

loop; colonies disintegrate in chunks when broken with a loop. Colonies of *N. subflava* bv. perflava and *N. mucosa* are convex, glistening, and butyrous; colonies of *N. subflava* bv. subflava and *N. subflava* bv. flava are low convex to flat with a matte surface and a slightly friable consistency. Colonies of *N. sicca* adhere to the agar surface and become wrinkled with prolonged incubation.

Blood Agar

Although many gonococcal strains do not grow on sheep blood agar, the more robust strains may grow quite well on these media. The colonial morphologic characteristics of *Neisseria* species and *B. catarrhalis* on blood agar are similar to those on chocolate agar except for pigmentation. Like the colonies of *N. lactamica*, *N. cinerea*, *N. polysaccharea*, and *N. kochii*, colonies of *N. gonorrhoeae* and *N. meningitidis* on blood agar are grayish to white and more opaque than those on chocolate media. Similarly, the pigmentation and opacity of colonies of the other species are more pronounced on blood agar than on chocolate agar.

DIRECT EXAMINATION

Gram Stain

A Gram stain may be performed immediately after the specimen is collected. If smears are too thick, the decolorization process may not adequately decolorize diplococci in clumps, or the smear may be overdecolorized, with the result that some gram-positive organisms may appear to be gram negative. If a Gram stain cannot be performed on site, a smear should be prepared and transported to a laboratory for staining and interpretation.

Gram-stained smears of urethral discharges from men with symptomatic gonococcal urethritis usually contain two or more intracellular gram-negative diplococci in occasional polymorphonuclear leukocytes (PMNs); many PMNs contain no diplococci (Fig. 1). Other bacteria are rarely seen in these smears. Gram-negative intracellular diplococci may not be observed in smears from men with early symptomatic infections, although many extracellular diplococci may be observed in stringy, mucoid material.

Gram-stained smears from endocervical and anorectal specimens must be interpreted carefully. If endocervical specimens have been collected properly and are not diluted with cervical mucus or contaminated with vaginal secretions, a reliable observation of gram-negative diplococci may be made. Similarly, mucopurulent material from rectal specimens collected through an anoscope may provide an acceptable specimen for a Gram stain (Fig. 2). However, because both vaginal and anorectal mucosae are colonized with gram-negative coccobacilli and bipolarly staining enteric bacteria, there is a danger of overinterpreting these smears, and results should be considered presumptive until confirmed by culture. Report the Gram stain characteristics, the cell morphology, and the number of all organisms present in the smear together with the types of cells (PMNs, squamous epithelial). Note whether intracellular gram-negative diplococci are observed in PMNs, and quantify the number either by using the terms few, moderate, and many or by scoring from 1+ to 4+.

Antigen Detection and Nucleic Acid Probe Tests for *N. gonorrhoeae*

Direct nonculture tests are available for detecting *N. gonorrhoeae* in patient specimens. These tests are useful for screening specimens from patients in geographically isolated locations where culture facilities are not available and specimens for culture confirmation cannot be transported to a laboratory within 48 h of collection. An enzyme-linked immunosorbent assay (Gonozyme; Abbott Laboratories, North Chicago, Ill.) detects gonococcal antigen. This test has been reported to be as sensitive and specific as the Gram stain for detecting gonococci in male urethral specimens and first-void urine specimens (67) but is less sensitive for detecting gonococci in endocervical specimens (72). Because the antigen detection test may cross-react with commensal *Neisseria* and related species (47), this test can be used only to make a presumptive diagnosis of a gonococcal infection.

A direct nonculture nucleic acid probe test (PACE

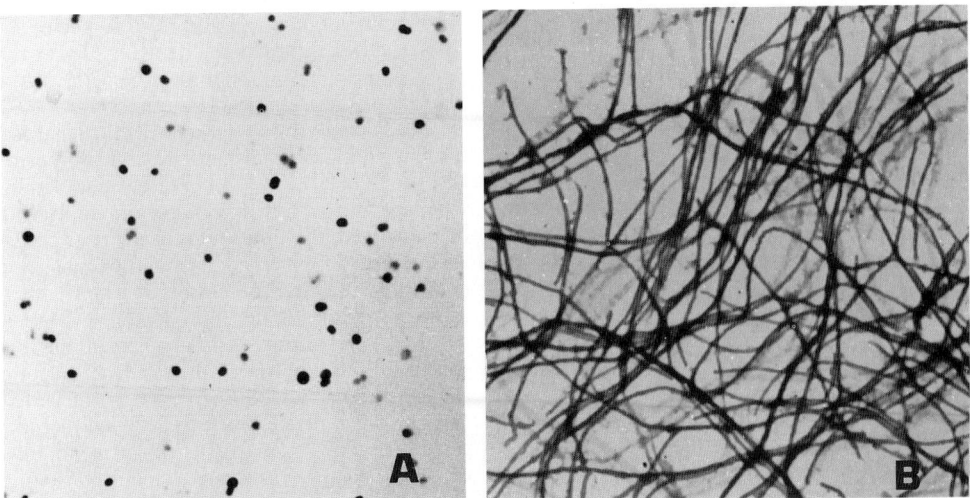

FIGURE 3 Cocci (A) and bacilli (B) exposed to subinhibitory concentrations of penicillin. Some cocci are swollen but still coccoid; bacilli form long strings. Magnification, ×1,000.

[probe assay chemiluminescence enhanced], GenProbe, San Diego, Calif.) has been used to identify *N. gonorrhoeae* in specimens and has been evaluated with patient populations at high risk for gonorrhea (35, 54, 62). This test detects gonococci directly in patient specimens in a 2-h test in which a nonisotopic DNA probe hybridizes with gonococcal rRNA. This test was highly sensitive and specific for detecting *N. gonorrhoeae* in urogenital and endocervical specimens; it was less sensitive for detecting gonococci in pharyngeal and rectal specimens from patients at high risk for gonorrhea (35, 54, 62). There have been no documented reports of cross-reactivity with nongonococcal species, and the test appears to be highly specific for *N. gonorrhoeae*. There have, however, been anecdotal, unpublished reports from diverse geographic locations of false-negative (probe-negative, culture-positive) and false-positive (probe-positive, culture-negative) results with this test. In view of these problems, laboratories using the PACE are encouraged to periodically conduct an evaluation comparing probe results with culture results. Because the PACE does not allow for isolation and confirmed identification of an etiologic agent, this test also must be used only for screening purposes, and the results may be used only to make a presumptive diagnosis of a gonococcal infection.

Latex Agglutination and Coagglutination Tests for *N. meningitidis*

Currently, there are no immunologic tests that permit the presumptive laboratory identification of gonorrhea by the detection of gonococcal antigens in patient sera. Commercial latex agglutination tests (Becton Dickinson Microbiology Systems; Wellcome Diagnostics, Research Triangle Park, N.C.) and coagglutination tests (Pharmacia, Rahway, N.J.) for detecting meningococcal antigens in body fluids are commercially available. These kits contain a polyvalent antibody reagent for serogroups A, C, Y, and W135 and a separate reagent for serogroup B that also detects the cross-reacting *Escherichia coli* K1 antigen. Antigen detection tests should always be run in conjunction with Gram stain and culture. Positive results provide a rapid presumptive diagnosis that permits early administration of appropriate ther-

apy. Negative results do not, however, exclude a diagnosis of meningococcemia.

PCR Test for *N. meningitidis*

An experimental PCR for detection of *N. meningitidis* has been developed but is not commercially available (61). Preliminary data suggest that this procedure may be useful in some clinical settings for screening of CSF for meningococcal DNA. However, the limitation that the test must be performed in an ultraclean environment to avoid contamination with nonclinical meningococcal DNA may restrict the performance of the test to reference laboratories.

IDENTIFICATION

Presumptive Identification

Identification of *Neisseria* and related species may be made at two levels, presumptive and confirmed. A presumptive identification of a neisserial infection may be made with a Gram stain and oxidase test and simply indicates that a gram-negative, oxidase-positive diplococcus has been isolated from a specimen. Although antimicrobial therapy may be initiated on the basis of a presumptive test result, additional tests must be performed to identify (confirm) an isolate.

A thin smear should be prepared by emulsifying one suspect colony in a drop of water. Care must be taken to distinguish gram-negative coccobacilli from gram-negative diplococci. Cells of some *K. denitrificans* strains may appear to be diplococci but can be identified as bacilli by using a cell elongation test (Fig. 3) (12). The suspect organism is inoculated onto a chocolate agar plate, and a 10-IU penicillin disk is placed on the inoculated area. After incubation for 18 to 24 h, growth is harvested from the edge of the zone of inhibition and stained with crystal violet or safranin. In subinhibitory concentrations of penicillin, bacilli form long bacillary forms because of their inability to divide. In contrast, true diplococci retain their characteristic cell morphology.

An oxidase test should also be performed on suspect colonies (remember, however, that oxidase reagents are

TABLE 2 Differential characteristics for genera of the family *Neisseriaceae*

Genus	Characteristic			
	Cell morphology	Oxidase	Catalase	Acid from glucose
Neisseria	Diplococcus	+	+	V[a]
Branhamella	Diplococcus	+	+	−
Kingella	Coccobacillus[b]	+	−	+
Moraxella	Rod	+	+	−

[a]Acid from glucose is species dependent.
[b]Cells may appear as diplobacilli.

toxic for gonococci; thus, if oxidase reagent is placed on colonies that must be isolated, they must be subcultured immediately). It is preferable, however, not to perform the oxidase test on the plate (particularly when few colonies are present) but to perform the test with Kovac's modification by harvesting growth with a sterile platinum (not nichrome) loop and placing it onto a filter paper freshly impregnated with oxidase reagent. Alternatively, dispense a drop of oxidase reagent onto the growth from a suspect colony harvested on a sterile swab; oxidase-positive cells will immediately turn purple. In some instances, only one suspect colony is present on a plate. Because, occasionally, a subculture from this colony may not grow, restreak and reincubate the primary culture plate after preparing a Gram stain, subculturing the colony, and performing an oxidase test by one of the alternative methods; dispersed cells will usually grow luxuriantly and permit isolation of the organism. A presumptive identification of a *Neisseria* species or *B. catarrhalis* may be made when colonies of gram-negative, oxidase-positive diplococci are detected in the culture.

Confirmatory Identification

Members of three genera (*Neisseria*, *Branhamella*, and *Kingella*) must be considered when *Neisseria* species of clinical importance are being identified; characteristics that differentiate these genera and the genus *Moraxella* are given in Table 2. Because *N. meningitidis*, commensal *Neisseria* species, and *B. catarrhalis* may occasionally be isolated from patients with infections traditionally associated with *N. gonorrhoeae*, and vice versa, tests must be performed to differentiate among these species. Characteristics that differentiate among these species and other *Neisseria* spp. are given in Table 3. A variety of commercial tests are available for the identification of *Neisseria* and related species. Tests range from those that measure one parameter of the organism, e.g., acid production or enzymes, to those that measure several parameters, e.g., acid production, enzymes, and nitrate reduction. Most tests are designed to differentiate among species isolated on selective media; thus, when organisms are isolated on nonselective media, it is important to determine whether they can grow on selective media, i.e., if they are colistin resistant. Some tests require inoculation of growth from pure subcultures of suspect colonies onto a nonselective medium (chocolate or blood agar), whereas other tests require inoculation of growth from primary culture plates. In addition, because problems have been identified with most rapid tests, it is important to conduct internal evaluations of any test that will be routinely used to identify organisms. A panel of named reference strains of *Neisseria* and related species should be used to assess the performance characteristics of the test and to select additional tests that must be performed to accurately identify isolates.

Considering the social and medicolegal implications of misidentifying a gram-negative oxidase-positive diplococcus, the clinical laboratorian must be skilled in knowing on what specimens tests may be performed, how the tests must be performed, the limitations of the tests chosen to identify an isolate, and which additional tests must be performed to confirm the identity of an isolate. The number of tests that must be performed depends on the level of identification required. For example, if a specimen has been collected specifically to confirm a clinical diagnosis of gonorrhea or meningococcal infection, tests may be performed specifically to differentiate between *N. gonorrhoeae* or *N. meningitidis*. In contrast, if the etiology of an infection is less certain, it may be necessary to identify the organism to the species level. In addition, if specimens have been collected as a result of allegations of sexual abuse or assault, it is also necessary to identify an isolate to species level; in this instance, the identification must be made with two tests that measure different parameters, e.g., acid production, or other biochemical, serologic, or nucleic acid probe tests.

Colistin susceptibility is particularly useful for differentiating between the colistin-susceptible commensal *Neisseria* species and colistin-resistant species, particularly when a specimen has been inoculated only onto a nonselective medium. Colistin resistance may be determined either by inoculating the isolate onto MTM or ML medium or by placing a 10-μg colistin disk onto inoculated chocolate agar. Strains that grow luxuriantly on the selective media or show no inhibition zone in the disk test may be interpreted as colistin resistant. Strains of *N. gonorrhoeae*, *N. meningitidis*, and *N. lactamica*; some strains of *N. polysaccharea*; and *K. denitrificans* are colistin resistant. However, remember also that some strains of *N. cinerea*, *N. subflava* bv. perflava, and *B. catarrhalis* may also be isolated on selective media and may grow when subcultured on these media. The performance of a colistin susceptibility test will permit use of confirmatory tests that are designed to differentiate among organisms isolated on selective media.

Biochemical Tests

Acid Production from Carbohydrates

Neisseria and related species may be differentiated by their patterns of acid production from glucose, maltose, lactose, sucrose, and fructose; acid production from fructose is not routinely tested (Table 3). The advantages of acid production tests are that they provide a reaction profile that permits immediate differentiation among several groups of *Neisseria* species and that their use is not limited to testing strains isolated on selective media. Although acid production from carbohydrates by *Neisseria* species has traditionally been determined in cystine Trypticase agar medium containing 1% carbohydrate, this medium is no longer recommended, because it is designed to detect acid production by fermentative organisms and is relatively insensitive for the detection of acid produced by the oxidative *Neisseria* species (47). Rapid methods, e.g., QuadFERM+ (bioMérieux Vitek, Hazelwood, Mo.), that permit detection of acid from neisseriae within several hours have been developed. Patterns of acid production by strains permit immediate confirmed identification of *N. lactamica* (the only species that produces acid from lactose) and elimination of

TABLE 3 Characteristics of *Neisseria* and related species of human origin[a]

Species	Colony morphology on chocolate medium	Growth on:			Acid from:					Reduction of NO₃	Polysaccharide from SUC	DNase
		MTM, ML, and NYC media	Chocolate or blood agar (22°C)	Nutrient agar (35°C)	GLU	MAL	LAC	SUC	FRU			
N. gonorrhoeae	Beige to gray-brown, translucent, smooth, 0.5–1 mm in diam	+	0	0	+	0	0	0	0	0	0	0
N. meningitidis	Beige to gray-brown, translucent, smooth, 1–3 mm in diam, encapsulated, mucoid	+	0	V	+	+	0	0	0	0	0	0
N. lactamica	Beige or gray-brown to yellowish, translucent, smooth, 1–2 mm in diam	+	V	+	+	+	+	0	0	0	0	0
N. cinerea[b]	Beige or gray-brown to yellowish, translucent, smooth, 1–2 mm in diam	V	0	+	0	0	0	0	0	0	0	0
N. polysaccharea	Beige or gray-brown to yellowish, translucent, smooth, 1–2 mm in diam	V	0	+	+	+	0	0	0	0	+	0
N. subflava[c]	Greenish yellow, opaque, 1–3 mm in diam, smooth to rough, sometimes adherent	V	+	+	+	+	0	V	V	0	V	0
N. sicca	White, opaque, 1–3 mm in diam, dry, adherent, wrinkled with age	0	+	+	+	+	0	+	+	0	+	0
N. mucosa	Greenish yellow, smooth, 1–3 mm in diam	0	+	+	+	+	0	+	+	+	+	0
N. flavescens	Yellow, opaque, 1–2 mm in diam, smooth	0	+	+	0	0	0	0	0	0	+	0
N. elongata[d]	Gray-brown, translucent, smooth, 1–2 mm in diam, glistening, dry, claylike consistency	0	+	+	0	0	0	0	0	0	0	0
B. catarrhalis	Pinkish brown, opaque, 1–3 mm in diam, dry, "hockey puck" consistency	V	+	+	0	0	0	0	0	+	0	+
K. denitrificans	Beige to gray-brown, translucent, smooth, 1–2 mm in diam	+	NT	+	+	0	0	0	0	+	0	0

[a]Symbols and abbreviations: +, strains typically positive but genetic mutants may be negative; 0, most strains negative; V, strain dependent; NT, not tested; GLU, glucose; MAL, maltose; LAC, lactose; SUC, sucrose; FRU, fructose.

[b]Some strains grow on selective media even though they are colistin susceptible.

[c]Includes biovars subflava, flava, and perflava. *N. subflava* bv. perflava strains produce acid from sucrose and fructose and produce polysaccharide from sucrose; *N. subflava* bv. flava and *N. subflava* bv. subflava do not produce polysaccharide from sucrose.

[d]Rod-shaped organism. Catalase test is weakly positive or negative compared with those of other *Neisseria* species. Results in table are for *N. elongata* subsp. *elongata*. Strains of *N. elongata* subsp. *glycolytica* may produce a weak acid reaction from D-glucose; are catalase positive, do not reduce nitrate, but do reduce nitrite. Strains of *N. elongata* subsp. *nitroreducans*, formerly CDC group M-6, may produce a weak acid reaction from D-glucose, are catalase negative, and reduce nitrate and nitrite.

other species, namely, *N. subflava* bv. perflava, *N. sicca*, and *N. mucosa* (which all produce acid from sucrose), from further consideration unless species identification is required. Without additional tests, strains of commensal *Neisseria* spp. may be misidentified as *N. meningitidis*, *N. gonorrhoeae*, or *B. catarrhalis*. *N. polysaccharea*, *N. subflava* bv. flava, and *N. subflava* bv. subflava, which produce acid from glucose and maltose, may be misidentified as *N. meningitidis*. *K. denitrificans*, which also produces acid from glucose, may be misidentified as *N. gonorrhoeae*. *N. cinerea* may be misidentified as *B. catarrhalis* in some tests if additional differential tests are not performed. In addition to confusion between species with identical acid production patterns, false-positive and false-negative acid reactions from glucose have resulted in misidentification of some isolates. Specifically, some gonococcal strains, particularly AHU isolates, have been described as glucose negative because they failed to produce detectable acid in some tests. Conversely, some strains of *N. cinerea*, which is considered glucose negative, may produce weak acid reactions from glucose in some tests (24, 47).

Chromogenic Enzyme Substrate Tests

Enzyme substrate tests (Gonochek II, bioMérieux Vitek; Identicult-Neisseria, IDN, Adams Scientific, West Warwick, R.I.) permit rapid differentiation among species isolated on gonococcal selective media; differentiation is limited, however, to *N. gonorrhoeae*, *N. meningitidis*, *N. lactamica*, and *B. catarrhalis*. *N. lactamica* produces β-galactosidase, *N. meningitidis* produces γ-glutamylaminopeptidase, and *N. gonorrhoeae* produces hydroxyprolylaminopeptidase; *B. catarrhalis* produce none of these enzymes (21). The major problem associated with enzyme substrate tests is that strains of *N. cinerea*, *N. polysaccharea*, *N. subflava* bv. perflava, and *K. denitrificans* are hydroxyprolylaminopeptidase positive and will be misidentified as *N. gonorrhoeae* if tested only with an enzyme substrate test. In spite of these problems, enzyme substrate tests permit differentiation among *N. meningitidis* (γ-glutamylaminopeptidase positive), *N. polysaccharea* and *N. subflava* biovars (hydroxyprolylaminopeptidase positive, γ-glutamylaminopeptidase negative), and maltose-negative variants of *N. meningitidis* (γ-glutamylaminopeptidase positive), which may be misidentified in acid production tests (70).

Multitest Identification Systems

Some commercial tests include not only acid production tests but also other biochemical tests, including tests for enzyme substrates, DNase, nitrate reduction, and β-lactamase production, and are often intended for the identification of *Haemophilus* species in addition to *Neisseria* and related species. These systems identify *N. gonorrhoeae*, *N. meningitidis*, and *B. catarrhalis*, although additional tests may be required to identify some isolates (40, 41, 66). Strictly speaking, the quadFERM+ test is a multitest system in which a DNase test (which differentiates between *N. cinerea* [DNase negative] and *B. catarrhalis* [DNase positive]) is included with acid production tests and a β-lactamase test. No problems have been reported with the acid production component of this test; *N. cinerea* isolates do not produce detectable acid, and AHU gonococcal isolates produce clearly positive acid reactions from glucose. Some problems have been encountered with the β-lactamase test, which is an acidometric test in this product.

Serologic Tests for Identification of *N. gonorrhoeae*

Fluorescent-Antibody Tests

A polyvalent test (Difco Laboratories, Detroit, Mich.) and a monoclonal-fluorescent-antibody test (*Neisseria gonorrhoeae* Culture Confirmation Test; Syva Co., Palo Alto, Calif.) are available for the culture confirmation of *N. gonorrhoeae*. Polyvalent reagents have cross-reacted with nongonococcal isolates or failed to react with some gonococcal strains (47) and are not used widely in the United States. Although this test is sensitive and specific (47), some gonococcal strains react weakly or not at all with this reagent (47), and nonspecific Fc binding with nongonococcal isolates may occur if mixed cultures are used. The advantage of this test is that it can be performed with growth from a single colony on a primary isolation plate and does not require a pure culture.

Coagglutination Tests for *N. gonorrhoeae*

Three coagglutination test are available: the Phadebact GC OMNI (Karo Bio Diagnostics AB, Huddinge, Sweden), the GonoGen I (New Horizons Diagnostics, Columbia, Md.), and the Meritec GC (Meridian Diagnostics, Cincinnati, Ohio). Another test, the GC Monoclonal Antibody test (Karo Bio Diagnostics AB), permits subgrouping of gonococcal isolates into WI (protein IA) and WII/WIII (protein IB) groups. Tests are performed in a similar manner: test and control reagents are reacted with a drop of a boiled suspension of suspect organisms. A positive reaction is recorded if agglutination occurs with the test reagent but not with the control reagent. Because coagglutination tests can be performed with growth from primary culture plates, an identification of *N. gonorrhoeae* may be obtained some 24 h earlier than is possible with tests that require a pure culture for inoculation. However, some gonococcal strains have not reacted in these tests, and cross-reactions with strains of *N. meningitidis*, *N. lactamica*, *N. cinerea*, and *K. denitrificans* have been noted in tests performed in accordance with the manufacturer's directions (47). Thus, another test should be used to confirm an identification of *N. gonorrhoeae* when (i) isolates suspected of being gonococci fail to react or give equivocal reactions with a reagent, (ii) positive reactions are obtained with isolates from extragenital sites in patients at low risk for gonorrhea, and (iii) positive reactions are obtained with isolates from genital or extragenital sites from children, whether or not sexual abuse is suspected.

Gonogen II

Gonogen II is a serologic test in which monoclonal antibodies have been conjugated to colloidal gold. Colonies of suspect organisms are suspended in a lysing solution, a drop of which is mixed with the monoclonal antibody-gold conjugate. After being mixed for 2 to 10 min, the mixture is passed through a microfilter that retains the antigen-antibody complex. When *N. gonorrhoeae* is present, a red color is seen on the filter; with nongonococcal isolates, the filter remains white or becomes pale pink.

Nucleic Acid Probe Culture Confirmation Tests

A DNA probe test (AccuProbe, Gen-Probe) has been developed as a culture confirmation test for *N. gonorrhoeae* (55). The test format is similar to that of the nonculture PACE for the direct detection of *N. gonorrhoeae* in speci-

TABLE 4 Characteristics of *Neisseria* species of animal origin (6, 38, 49)[a]

Species	Cell morphology	Acid from:				Reduction of:		Polysaccharide from SUC	Tributryin hydrolysis
		GLU	MAL	LAC	SUC	NO_3	NO_2		
N. canis	Diplococcus	0	0	0	0	+	0	0	0
N. weaverii	Rod	0	0	0	0	0	+	NT	NT
N. caviae	Diplococcus	0	0	0	0	+	+	NT	+
N. ovis	Diplococcus	0	0	0	0	+	0	NT	+
N. cuniculi	Diplococcus	0	0	0	0	0	0	NT	+
N. iguanae	Diplococcus	V	0	NT	V	+	V	+	NT

[a]Symbols and abbreviations: +, strains typically positive but genetic mutants may be negative; 0, most strains negative; V, strain dependent; NT, not tested; GLU, glucose; MAL, maltose; LAC, lactose; SUC, sucrose.

mens. This test is highly sensitive and specific for the identification of *N. gonorrhoeae* compared with rapid acid production and coagglutination tests (81).

Additional Tests

Nitrate and Nitrite Reduction Tests

The nitrate reduction test is useful for differentiating between strains of *N. gonorrhoeae* (nitrate negative) and those of *K. denitrificans* and *B. catarrhalis* (nitrate positive). The test is performed in a standard nitrate broth that has been heavily inoculated to give a dense suspension of organisms. Sulfanilic acid and α-naphthylamine reagents are added after incubation at 35 to 37°C for 48 h. The development of a red color indicates the presence of nitrite. If the medium remains colorless, zinc dust is added to catalyze the reduction of residual nitrate. The subsequent development of a red color indicates that nitrate has not been reduced; if the medium remains colorless, nitrate has been reduced beyond nitrite to nitrogenous gases. The nitrite reduction test is performed in a similar base medium containing nitrite (0.1, 0.01, and 0.001%); residual nitrite is detected with sulfanilic acid and α-naphthylamine reagents. Although the nitrite reduction test is not useful for differentiating between human neisserial species, both nitrate and nitrite reduction tests are important for differentiating between neisserial species of animal origin (Table 4).

Superoxol Test

The superoxol test is analogous to the catalase test but performed with a 30% hydrogen peroxide solution and has been recommended for differentiating between *N. gonorrhoeae* and other *Neisseria* and related species (4). The reagent is dropped onto growth on a chocolate (not blood) agar plate, and the degree of bubbling is observed. Gonococci typically give an immediate, explosive (++++) reaction. *N. meningitidis* and *B. catarrhalis* strains may also give a ++++ reaction in this test (80). Although a reevaluation of this test indicates that most commensal *Neisseria* species give weaker but nevertheless positive reactions, the superoxol test may be very useful for differentiating between *N. gonorrhoeae* strains (superoxol positive) and *N. cinerea* and *K. denitrificans* strains that do not react with this reagent (46a).

Polysaccharide from Sucrose

The polysaccharide-from-sucrose test is not commercially available but is particularly useful for differentiating between *N. polysaccharea* (polysaccharide positive) and *N. meningitidis* (polysaccharide negative). Strains of *N. flave-* scens are polysaccharide positive, as are strains of *N. subflava* bv. perflava, *N. sicca*, and *N. mucosa*; strains of *N. subflava* bv. flava and *N. subflava* bv. subflava are polysaccharide negative. The test is performed on a starch-free base medium that supports growth of the organism and that has 1% starch added. The traditional medium with 5% sucrose is inhibitory for many strains (49). Inoculated media are incubated at 35 to 37°C for 24 to 48 h. A drop of Gram's iodine is added to isolated colonies; if polysaccharide has been produced, the growth will immediately turn dark brown to purple. The reaction will fade with time but can be redeveloped by adding more iodine solution. The inoculated medium should be tested only during the 48 h of incubation, since polysaccharide-producing organisms will metabolize the polysaccharide on prolonged incubation, possibly resulting in a false-negative reaction.

Tributyrin Hydrolysis

B. catarrhalis strains may hydrolyze tributyrin, whereas other human *Neisseria* species do not metabolize this substrate. Several tests have been developed to detect this activity and provide a rapid method for confirming this species (63).

SEROLOGIC TESTS

N. gonorrhoeae

Typing systems used to differentiate between strains of *N. gonorrhoeae* include auxotyping, nutritional typing, and serotyping with monoclonal antibodies in coagglutination tests (52, 71). A dual auxotype-serovar classification system provides greater discrimination among strains than is possible with either typing system alone (71).

Auxotyping involves determining what individual metabolites strains require for growth on a chemically defined medium. If a strain fails to grow on a medium from which arginine has been omitted, the strain is recorded as arginine requiring. Thus, strains may be identified as having single or multiple growth requirements. Serotyping is performed with a panel of monoclonal antibody coagglutination reagents; monoclonal antibodies are directed against epitopes on the protein I molecule in the gonococcal outer membrane (71). Strains are divided into two major serogroups, IA and IB, on the basis of the protein I species expressed by the strain; further subdivision into serovars is made according to patterns of reaction with a panel of six protein IA- and protein IB-specific antibody reagents. Typing systems are not used specifically for the identification of isolates as *N. gonorrhoeae*. Rather, they are used to compare isolates for molecular epidemiologic investigations, which may range from studies of the dynamics of gonococcal strains in communities and the geographic spread of antimicrobial-resistant strains to comparisons between strains for medicolegal investigations (51, 71).

N. meningitidis

Meningococci are subgrouped into serologic groups on the basis of the presence of capsular polysaccharide or outer membrane protein antigens. Although 13 serogroups are currently recognized (A, B, C, D, 29E, H, I, K, L, W135, X, Y, and Z), strains belonging to groups A, B, C, Y, and W135 are most frequently implicated in systemic disease (82). Classically, group A and C strains cause epidemic meningococcal disease; group B strains are usually associated with sporadic infections during interepidemic periods, although they have been implicated in localized outbreaks. During epidemics, carriers tend to be colonized with the prevailing serogroup. Other serogroups associated with conditions of crowding (military recruits) and social disruption (refugees) are isolated sporadically from both carriers and patients with disease. In sub-Saharan Africa and certain Scandinavian countries, outbreaks of meningococcal group A and C disease are common. In North America, groups B and C are commonly associated with sporadic cases of septicemia and/or meningitis and occasionally with regionally contained outbreaks. The utility of other methods for molecular epidemiologic typing, including antigenic analyses of outer membrane proteins, multilocus enzyme electrophoresis, ribotyping, PCR, and pulsed-field electrophoresis, is being investigated (1, 46, 74, 79).

ANTIBIOTIC SUSCEPTIBILITIES

N. gonorrhoeae

Antimicrobial resistance is now widespread among strains of N. gonorrhoeae and occurs both as chromosomally mediated resistance to a variety of agents and as plasmid-mediated resistance to penicillins (penicillinase [β-lactamase]-producing N. gonorrhoeae) and to tetracycline (tetracycline-resistant N. gonorrhoeae) (45, 64). Owing to the increased frequency of antimicrobial agent-resistant strains of N. gonorrhoeae (34; Fig. 4), the Centers for Disease Control and Prevention (CDC) has recommended that selected extended-spectrum cephalosporins and newer fluoroquinolones be used as primary therapy against uncomplicated gonococcal infections (17); penicillins and tetracyclines are no longer recommended for the primary treatment of uncomplicated gonococcal infections. Although tetracycline is no longer recommended for the treatment of gonorrhea, dual therapy including tetracycline is recommended to empirically treat concurrent chlamydial infections. Owing to the use of cephalosporins and fluoroquinolones to treat gonorrhea, the CDC no longer recommends test-of-cure cultures, because treatment failures have not been documented with these agents (14, 15, 17). Recently, however, gonococcal isolates with decreased susceptibilities to fluoroquinolones (MICs of ciprofloxacin and ofloxacin, 2.0 μg/ml) have been identified (18); failure of a strain with similar MICs to respond to 500 mg of ciprofloxacin has been reported (75).

Susceptibilities of gonococcal isolates are determined by a disk diffusion method recommended by the National Committee for Clinical Laboratory Standards (18, 60). β-Lactamase production can be determined by a highly sensitive nitrocefin test (48). Ideally, susceptibilities to penicillin, tetracycline, spectinomycin, a cephalosporin, and a

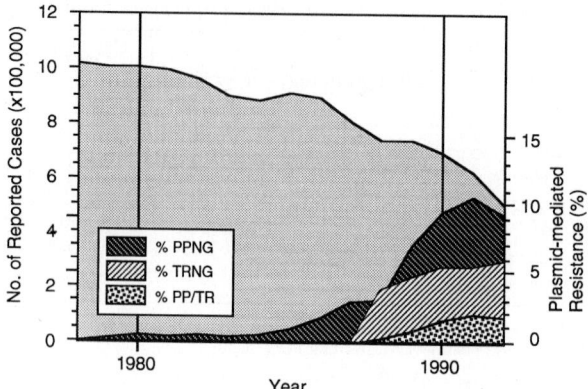

FIGURE 4 Percentage of cases of gonorrhea and frequency of gonorrhea caused by strains of N. gonorrhoeae with plasmid-mediated resistance to penicillin (PPNG), tetracycline (TRNG), and both penicillin and tetracycline (PP/TR) in the United States, 1988 to 1992, compared with the total number of reported cases. The number of cases of PPNG are based on a combination of passive surveillance for PPNG (1978 to 1987) and data provided by active surveillance in the Gonococcal Isolate Surveillance Project (GISP). TRNG and PP/TR data are provided by GISP.

fluoroquinolone should be determined. At a minimum, susceptibilities to the agents routinely used to treat gonorrhea should be determined; in clinical settings where cases of PID are treated, susceptibility to antigonococcal agents such as cefoxitin should also be tested. Because most isolates have remained highly susceptible to extended-spectrum cephalosporins, resistance to both penicillin and tetracycline may serve as an indicator of decreased susceptibility to these agents (78). Determination of spectinomycin susceptibilities is recommended if this agent is used as an alternative therapy.

Because resistance to cephalosporins and fluoroquinolones has emerged in other organisms and may be emerging in gonococci, it is important to determine the susceptibilities of clinical isolates from patients whose symptoms persist after treatment. Isolates that exhibit decreased susceptibility to the agent (zone inhibition size less than the susceptible criterion) should be submitted to a reference laboratory for confirmation. If possible, laboratories should implement a routine surveillance program to monitor the susceptibilities of gonococcal isolates in order to detect changes that may indicate the emergence of clinical resistance to therapeutic agents that may compromise treatment.

N. meningitidis

Although penicillin G has remained effective for the treatment of meningococcal infections including meningitis, meningococci exhibiting chromosomally mediated and plasmid-mediated decreased susceptibility or resistance to penicillin and tetracycline have been reported (25, 59). Chloramphenicol is still recommended as an alternative for treating meningococcal infections in patients allergic to penicillin. Because of the development of resistance over the years, sulfonamides are no longer indicated for treatment of meningococcal infections.

The narrow-spectrum cephalosporins are not recommended for therapy of meningococcal disease. In vitro susceptibility studies of newer broad-spectrum cephalosporins, e.g., cefotaxime, ceftriaxone, cefoperazone, and ceftizoxime, have demonstrated excellent activity of these agents against meningococcal isolates.

Chemoprophylaxis is recommended to reduce nasopharyngeal meningococcal carriage, particularly in contacts of patients with meningococcal infection and in potential contacts in families, military installations, schools, and closed units to prevent further transmission and disease. Rifampin, minocycline, and, more recently, some of the newer fluoroquinolones, e.g., ciprofloxacin, reduce the meningococcal carrier state.

B. catarrhalis

In the early antibiotic era, B. catarrhalis was highly susceptible to penicillin, but recently, an increasing number of resistant strains has been reported. Recent in vitro susceptibility testing of B. catarrhalis has shown that the majority of clinical isolates remain susceptible to cephamycins, cephalosporins, β-lactamase-stable penicillins (e.g., amoxicillin-clavulanate), tetracyclines, and trimethoprim-sulfamethoxazole. Limited experience with quinolones suggests that these agents are active against β-lactamase-positive and -negative strains.

Other *Neisseria* Species

Comprehensive in vitro susceptibility data for commensal *Neisseria* species are not available. Occasional reports of clinical disease such as endocarditis, meningitis, empyema, pneumonia, and ocular infections caused by other *Neisseria* species should prompt clinicians to administer antimicrobial therapy guided by empirical in vitro susceptibility testing. Generally, commensal *Neisseria* species have been reported to be susceptible in vitro to penicillin, ampicillin, and tetracyclines, although some strains may possess the TetM determinant (50). Strains of N. cinerea, however, appear to exhibit uniformly decreased susceptibility or resistance to erythromycin (MICs, 4 to 16 μg/ml) when tested by procedures for N. gonorrhoeae (46a). Hence, there is some concern that ocular infections with N. cinerea in newborns may not respond to topical prophylaxis with erythromycin ointment (9). Thus, in vitro susceptibility testing is recommended for any gram-negative, oxidase-positive diplococcus isolated from ocular and other clinically relevant specimens.

EVALUATION, INTERPRETATION, AND REPORTING OF RESULTS

The level of testing and the format for reporting laboratory results should be directed by the sociodemographic characteristics, e.g., age and gender, of the patient clientele served and by a working knowledge of the incidence and prevalence of clinically significant disease caused by *Neisseria* species, e.g., gonorrhea or meningococcal meningitis, in that population. When specimens are collected from patients at high risk for gonorrhea and there are no sociologic or medicolegal implications of a diagnosis of gonorrhea, a presumptive identification of N. gonorrhoeae may be sufficient if it is intended to facilitate early and effective treatment of infections. However, when specimens are collected to confirm a clinical diagnosis in patients, e.g., women and children at low risk for gonorrhea, special concerns apply to the laboratory processing and retention of specimens be-

cause of the medicolegal consequences that may ensue if an organism is identified as N. gonorrhoeae. Laboratorians and clinicians must know the patient population that they serve and decide to what levels of specificity identifications of *Neisseria* and related species must be made.

Special protocols should be developed for processing and reporting specimens from alleged victims of sexual abuse and assault. In these instances, organisms must be confirmed by at least two confirmatory tests that detect different characteristics of the organism. It must be emphasized that accurate communication between the clinician and the laboratory is essential to ensure that specimens of potential medicolegal importance are processed according to these criteria. In too many instances, because a specimen has not been clearly marked as being from a child, an organism has been identified with only one confirmatory test and discarded according to a routine protocol.

In general, however, with the exception of high-risk patients, for whom presumptive identifications may suffice, similar principles apply to the identification of all gram-negative, oxidase-positive diplococci. Although many tests for the identification of *Neisseria* and related species are marketed as confirmatory tests, most do not provide sufficient information to accurately identify an isolate to the species level without the performance of additional tests. As noted earlier in the chapter, several species may give identical reactions in some tests, e.g., maltose-negative N. meningitidis and K. denitrificans may give reactions identical to those of N. gonorrhoeae in some acid production tests. In addition, problems in the identification of N. gonorrhoeae and related species have been identified with most tests, e.g., false-negative acid production reactions from glucose with N. gonorrhoeae or false-positive reactions from glucose with N. cinerea.

Culture confirmation tests are preferred for the identification of *Neisseria* and related species because they require the isolation of an organism that can be examined by multiple tests if the results of the primary tests are equivocal. In general, multitest systems provide the most information about several biochemical characteristics of an isolate. These tests are most useful if identification to the species level is required, and they may help characterize isolates in the rare instances when hybrid organisms are isolated (47). Rapid identification tests, including serologic and nucleic acid probe tests, that provide a yes-no answer, i.e., an isolate either is or is not N. gonorrhoeae, may be adequate if it is necessary only to eliminate N. gonorrhoeae as the causative agent; these tests are of limited usefulness when identification to the species level is required.

Reporting

Because of the serious social and medicolegal consequences of misdiagnosing gonorrhea or misidentifying strains of N. gonorrhoeae, the CDC has recommended criteria for reporting diagnoses of gonorrhea (15). Three levels of diagnosis are defined: these levels are suggestive, which is defined on the basis of clinical findings; presumptive; and definitive, with the last two levels including the results of laboratory diagnostic tests.

A suggestive diagnosis is defined by the presence of (i) a mucopurulent endocervical or urethral exudate on physical examination and (ii) sexual exposure to a person infected with N. gonorrhoeae. A presumptive diagnosis of gonorrhea is made on the basis of two of the following three criteria: (i) typical gram-negative intracellular diplococci on microscopic examination of a smear of urethral exudates (men) or endocervical secretions (women); (ii) growth of *N. gonorrhoeae* from the urethra (men) or endocervix (women) on culture medium and demonstration of typical colonial morphology, positive oxidase reaction, and typical gram-negative morphology; and/or (iii) detection of *N. gonorrhoeae* by nonculture laboratory tests. A definitive diagnosis requires (i) isolation of *N. gonorrhoeae* from sites of exposure (e.g., urethra, endocervix, throat, rectum) by culture (usually on a selective medium) and demonstration of typical colonial morphology, positive oxidase reaction, and typical gram-negative morphology and (ii) confirmation of isolates by biochemical, enzymatic, serologic, or nucleic acid testing, e.g., carbohydrate utilization, rapid enzyme substrate tests, serologic methods such as coagglutination or fluorescent-antibody tests, or DNA probe technique.

For reporting purposes, the laboratory should perform species-level identification and confirmation in order to report a definitive result for the clinician unless otherwise requested. If an organism suspected to be *N. gonorrhoeae* is tested with rapid tests but not with additional confirmatory tests that compensate for problems associated with the primary test and is reported as "presumptive *N. gonorrhoeae*," it is important that a clinician receiving this report understand that additional tests may be required to confirm this identification. Ideally, to avoid confusion, an organism should be reported only as "gram-negative, oxidase-positive diplococcus isolated" unless it has been identified to the species level with sufficient tests to ensure the accuracy of the identification.

REFERENCES

1. **Achtman, M., B. A. Crowe, A. Olyhoek, W. Strittmatter, and G. Morelli.** 1988. Recent results on epidemic meningococcal meningitis. *J. Med. Microbiol.* **26:**172–177.
2. **Ainsworth, S. M., S. B. Nagy, L. A. Morgan, G. R. Miller, and J. L. Perry.** 1990. Interpretation of Gram-stained sputa containing *Moraxella* (*Branhamella*) *catarrhalis*. *J. Clin. Microbiol.* **28:**2559–2560.
3. **Anand, C. M., F. Ashton, H. Shaw, and R. Gordon.** 1991. Variability in growth of *Neisseria polysaccharea* on colistin-containing medium selective for *Neisseria* spp. *J. Clin. Microbiol.* **29:**2434–2437.
4. **Arko, R. J., and T. Odugbemi.** 1984. Superoxol and amylase inhibition tests for distinguishing gonococcal and nongonococcal cultures growing on selective media. *J. Clin. Microbiol.* **20:**1–4.
5. **Baron, E. J., G. H. Cassell, L. B. Duffy, D. A. Eschenbach, J. R. Greenwood, S. M. Harvey, N. E. Madinger, E. M. Peterson, and K. B. Waites.** 1993. *Cumitech 17A: Laboratory Diagnosis of Female Genital Tract Infections.* Coordinating ed., E. J. Baron. American Society for Microbiology, Washington, D.C.
6. **Barrett, S. J., L. K. Schlater, R. J. Montall, and P. H. A. Sneath.** 1994. A new species of *Neisseria* from iguanid lizards, *Neisseria iguanae* sp. nov. *Lett. Appl. Microbiol.* **18:**200–202.
7. **Blakebrough, I. S., V. M. Greenwood, H. C. Whittle, A. K. Bradley, and H. M. Gilles.** 1982. The epidemiology of infections due to *Neisseria meningitidis* and *Neisseria lactamica* in a Northern Nigerian community. *J. Infect. Dis.* **146:**626–637.
8. **Boquete, M. T., C. Marcos, and J. A. Saez-Niéto.** 1986. Characterization of *Neisseria polysacchareae* sp. nov. (Riou, 1983) in previously unidentified noncapsular strains of *Neisseria meningitidis*. *J. Clin. Microbiol.* **24:**973–975.
9. **Bourbeau, P., V. Holla, and S. Piemontese.** 1990. Ophthalmia neonatorum caused by *Neisseria cinerea*. *J. Clin. Microbiol.* **28:**1640–1641.
10. **Bøvre, K.** 1984. Family VIII. *Neisseriaceae* Prévot 1933, 119[AL], p. 288–309. *In* N. R. Krieg and J. G. Holt (ed.), *Bergey's Manual of Systematic Bacteriology*, vol. 1. The Williams & Wilkins Co., Baltimore.
11. **Brown, W. J., and S. J. Kraus.** 1974. Gonococcal colony types. *JAMA* **228:**862–863.
12. **Catlin, B. W.** 1975. Cellular elongation under the influence of antibacterial agents: way to differentiate coccobacilli from cocci. *J. Clin. Microbiol.* **1:**102–105.
13. **Catlin, B. W.** 1991. *Branhamaceae* fam. nov., a proposed family to accommodate the genera *Branhamella* and *Moraxella*. *Int. J. Syst. Bacteriol.* **41:**320–323.
14. **Centers for Disease Control.** 1987. Antibiotic-resistant strains of *Neisseria gonorrhoeae*: policy guidelines for detection, management, and control. *Morbid. Mortal. Weekly Rep.* **36**(Suppl. 5S):1S–18S.
15. **Centers for Disease Control.** 1991. *Sexually Transmitted Diseases Clinical Practice Guidelines.* U.S. Public Health Service, Atlanta.
16. **Centers for Disease Control.** 1992. *Sexually Transmitted Diseases Surveillance 1991.* U.S. Department of Health and Human Services, Atlanta.
17. **Centers for Disease Control.** 1993. Sexually transmitted diseases treatment guidelines. *Morbid. Mortal. Weekly Rep.* **42**(RR-14):1–101.
18. **Centers for Disease Control and Prevention.** 1994. Decreased susceptibility of *Neisseria gonorrhoeae* to fluoroquinolones—Ohio and Hawaii, 1992–1994. *Morbid. Mortal. Weekly Rep.* **43:**325–327.
19. **Clausen, C. R., J. S. Knapp, and P. A. Totten.** 1984. Lymphadenitis due to *Neisseria cinerea*. *Lancet* **i:**908.
20. **Dalabetta, G., and E. W. Hook III.** 1987. Gonococcal infections. *Infect. Dis. Clin. N. Am.* **1:**25–54.
21. **D'Amato, R. F., L. A. Eriquez, K. M. Thomfohrde, and E. Singerman.** 1978. Rapid identification of *Neisseria gonorrhoeae* and *Neisseria meningitidis* by using enzyme profiles. *J. Clin. Microbiol.* **7:**77–81.
22. **Denison, M. R., S. Perlman, and R. D. Anderson.** 1988. Misidentification of *Neisseria* species in a neonate with conjunctivitis. *Pediatrics* **81:**877–878.
23. **Dewhirst, F. E., B. H. Paster, and P. L. Bright.** 1989. *Chromobacterium, Eikenella, Kingella, Neisseria, Simonsiella,* and *Vitreoscilla* species comprise a major branch of the beta group *Proteobacteria* by 16S ribosomal ribonucleic acid sequence comparison: transfer of *Eikenella* and *Simonsiella* to the family *Neisseriaceae* (emend.). *Int. J. Syst. Bacteriol.* **39:**258–266.
24. **Dillon, J. R., M. Carballo, and M. Pauze.** 1988. Evaluation of eight methods for identification of pathogenic *Neisseria* species: Neisseria Kwik, RIM-N, Gonobio-Test, Minitek, Gonochek II, GonoGen, Phadebact Monoclonal GC OMNI test, and Syva MicroTrak test. *J. Clin. Microbiol.* **26:**493–497.
25. **Dillon, J. R., M. Pauzé, and K.-H. Yeung.** 1983. Spread of penicillinase-producing and transfer plasmids from the gonococcus to *Neisseria meningitidis*. *Lancet* **i:**779–781.
26. **Doern, G. V.** 1986. *Branhamella catarrhalis*—an emerging human pathogen. *Diagn. Microbiol. Infect. Dis.* **4:**191–201.
27. **Dossett, J. H., P. C. Appelbaum, J. S. Knapp, and P. A. Totten.** 1985. Proctitis associated with *Neisseria cinerea* misidentified as *Neisseria gonorrhoeae* in a child. *J. Clin. Microbiol.* **21:**575–577.
28. **Eng, J., and E. Holten.** 1977. Gelatin neutralization of the inhibitory effect of sodium polyanethol sulfonate on *Neisseria meningitidis* in blood culture. *J. Clin. Microbiol.* **6:**1–3.
29. **Evangelista, A. T., and H. R. Beilstein.** 1993. *Cumitech 4A: Laboratory Diagnosis of Gonorrhea.* Coordinating ed., C. Abramson. American Society for Microbiology, Washington D.C.
30. **Faur, Y. C., M. H. Weisburd, M. E. Wilson, and P. S. May.** 1973. A new medium for the isolation of pathogenic *Neisseria* (NYC medium). I. Formulation and comparisons with standard media. *Health Lab. Sci.* **10:**44–52.
31. **Giebink, G. S.** 1989. The microbiology of otitis media. *Pediatr. Infect. Dis. J.* **8:**518–520.
32. **Given, K. F., B. W. Thomas, and A. G. Johnston.** 1977. Isolation of *Neisseria meningitidis* from the urethra, cervix, and

anal canal: further observations. *Br. J. Vener. Dis.* **53**:109–112.

33. Gold, R., I. Goldschneider, M. L. Lepow, T. F. Draper, and M. Randolph. 1978. Carriage of *Neisseria meningitidis* and *Neisseria lactamica* in infants and children. *J. Infect. Dis.* **137**:112–121.

34. Gorwitz, R. J., A. K. Nakashima, J. S. Moran, and J. S. Knapp. 1993. Sentinel surveillance for antimicrobial resistance in *Neisseria gonorrhoeae*—United States, 1988–1991. *Morbid. Mortal. Weekly Rep.* **42**(Suppl. 3):29–39.

35. Granato, P. A., and M. R. Franz. 1990. Use of the Gen-Probe PACE system for the detection of *Neisseria gonorrhoeae* in urogenital samples. *Diagn. Microbiol. Infect. Dis.* **13**:217–221.

36. Guibourdenche, M., T. Lambert, and J.-Y. Riou. 1989. Isolation of *Neisseria canis* in mixed culture for a patient after a cat bite. *J. Clin. Microbiol.* **27**:1673–1674.

37. Herbert, D. A., and J. Ruskin. 1981. Are the "nonpathogenic" neisseriae pathogenic? *Am. J. Clin. Pathol.* **75**:739–742.

38. Holmes, B., M. Costas, S. L. W. On, P. VanDamme, E. Falsen, and K. Kersters. 1993. *Neisseria weaverii* sp. nov. (formerly CDC group M-5) from dog bite wounds of humans. *Int. J. Syst. Bacteriol.* **43**:687–693.

39. Janda, W. M., M. Bohnhoff, J. A. Morello, and S. A. Lerner. 1980. Prevalence and site-pathogen studies of *Neisseria meningitidis* and *N. gonorrhoeae* in homosexual men. *JAMA* **244**:2060–2064.

40. Janda, W. M., J. J. Bradna, and P. Ruther. 1989. Identification of *Neisseria* spp., *Haemophilus* spp., and other fastidious gram-negative bacteria with the MicroScan *Haemophilus*-*Neisseria* identification panel. *J. Clin. Microbiol.* **27**:869–873.

41. Janda, W. M., P. J. Malloy, and P. C. Schreckenberger. 1987. Clinical evaluation of the Vitek *Neisseria*-*Haemophilus* identification card. *J. Clin. Microbiol.* **25**:37–41.

42. Janda, W. M., J. A. Morello, S. A. Lerner, and M. Bohnhoff. 1983. Characteristics of pathogenic *Neisseria* spp. isolated from homosexual men. *J. Clin. Microbiol.* **17**:85–91.

43. Janda, W. M., J. G. Ulanday, M. Bohnhoff, and L. J. LeBeau. 1985. Evaluation of the RIM-N, Gonochek II, and Phadebact systems for the identification of pathogenic *Neisseria* spp. and *Branhamella catarrhalis*. *J. Clin. Microbiol.* **21**:734–737.

44. Jephcott, A. E., and R. S. Morton. 1982. Isolation of *Neisseria lactamica* from a genital site. *Lancet* **ii**:739–740.

45. Johnson, S. R., and S. A. Morse. 1988. Antibiotic resistance in *Neisseria gonorrhoeae*: genetics and mechanisms of resistance. *Sex. Transm. Dis.* **15**:217–224.

46. Kertesz, D. A., S. K. Byrne, and A. W. Chow. 1993. Characterization of *Neisseria meningitidis* by polymerase chain reaction and restriction endonuclease digestion of the *porA* gene. *J. Clin. Microbiol.* **31**:2594–2598.

46a. Knapp, J. S. Unpublished observations.

47. Knapp, J. S. 1988. Historical perspectives and identification of *Neisseria* and related species. *Clin. Microbiol. Rev.* **1**:415–431.

48. Knapp, J. S. 1988. Laboratory methods for the detection and phenotypic characterization of *Neisseria gonorrhoeae* strains resistant to antimicrobial agents. *Sex. Transm. Dis.* **15**:225–233.

49. Knapp, J. S., and E. W. Hook III. 1988. Prevalence and persistence of *Neisseria cinerea* and other *Neisseria* spp. in adults. *J. Clin. Microbiol.* **26**:896–900.

50. Knapp, J. S., S. R. Johnson, J. M. Zenilman, M. C. Roberts, and S. A. Morse. 1989. High-level tetracycline resistance resulting from TetM in strains of *Neisseria* spp., *Kingella denitrificans*, and *Eikenella corrodens*. *Antimicrob. Agents Chemother.* **32**:765–757.

51. Knapp, J. S., E. G. Sandström, and K. K. Holmes. 1985. Overview of the epidemiologic and clinical applications of the auxotype/serovar classification of *Neisseria gonorrhoeae*, p. 6–12. In G. K. Schoolnik, G. F. Brooks, S. Falkow, C. E. Frasch, J. S. Knapp, J. A. McCutchan, and S. A. Morse (ed.), *The Pathogenic Neisseriae*. American Society for Microbiology, Washington, D.C.

52. Knapp, J. S., M. R. Tam, R. C. Nowinski, K. K. Holmes, and E. G. Sandström. 1984. Serological classification of *Neisseria gonorrhoeae* using monoclonal antibodies directed against gonococcal outer membrane protein I. *J. Infect. Dis.* **150**:44–48.

53. Knapp, J. S., P. A. Totten, M. H. Mulks, and B. H. Minshew. 1984. Characterization of *Neisseria cinerea*, a nonpathogenic species isolated on Martin-Lewis medium selective for pathogenic *Neisseria* species. *J. Clin. Microbiol.* **19**:63–67.

54. Lewis, J. S., O. Fakile, E. Foss, G. Legarza, A. Leskys, K. Lowe, and D. Powning. 1993. Direct DNA probe assay for *Neisseria gonorrhoeae* in pharyngeal and rectal specimens. *J. Clin. Microbiol.* **31**:2783–2785.

55. Lewis, J. S., D. Kranig-Brown, and D. A. Trainor. 1990. DNA probe confirmatory test for *Neisseria gonorrhoeae*. *J. Clin. Microbiol.* **28**:2349–2350.

56. Martin, J. E., J. H. Armstrong, and P. B. Smith. 1974. New system for cultivation of *Neisseria gonorrhoeae*. *Appl. Microbiol.* **27**:802–805.

57. Martin, J. E., and R. L. Jackson. 1975. A biological environment chamber for the culture of *Neisseria gonorrhoeae*. *Appl. Microbiol.* **27**:802–805.

58. Martin, J. E., and J. S. Lewis. 1977. Anisomycin: improved antimycotic activity in modified Thayer-Martin medium. *Public Health Lab.* **35**:53–62.

59. Mendelman, P. M., J. Campos, D. O. Chaffin, D. A. Serfass, A. L. Smith, and J. A. Saez-Nieto. 1988. Relative penicillin G resistance in *Neisseria meningitidis* and reduced affinity of penicillin-binding protein 3. *Antimicrob. Agents Chemother.* **32**:706–709.

60. National Committee for Clinical Laboratory Standards. 1993. *Approved Standard: Performance Standards for Antimicrobial Disk Susceptibility Tests*, 5th ed. Document M2-A5. National Committee for Clinical Laboratory Standards, Villanova, Pa.

61. Ni, H., A. I. Knight, K. Cartwright, W. H. Palmer, and J. McFadden. 1992. Polymerase chain reaction for diagnosis of meningococcal meningitis. *Lancet* **340**:1432–1434.

62. Panke, E. S., L. I. Yang, P. A. Leist, P. Magevny, R. J. Fry, and R. F. Lee. 1991. Comparison of Gen-Probe DNA probe test and culture for the detection of *Neisseria gonorrhoeae* in endocervical specimens. *J. Clin. Microbiol.* **29**:883–888.

63. Pérez, J. L., A. Pulido, F. Pantozzi, and R. Martin. 1990. Butyrate esterase (tributyrin) spot test, a simple method for immediate identification of *Moraxella* (*Branhamella*) *catarrhalis*. *J. Clin. Microbiol.* **28**:2347–2348.

64. Rice, R. J., and J. S. Knapp. 1993. Comparative in vitro activities of penicillin G, amoxicillin-clavulanic acid, selected cephalosporins and quinolone antimicrobial agents against representative resistance phenotypes of *Neisseria gonorrhoeae*. *Antimicrob. Agents Chemother.* **38**:155–158.

65. Riou, J.-Y., and M. Guibourdenche. 1987. *Neisseria polysaccharea* sp. nov. *Int. J. Syst. Bacteriol.* **37**:163–165.

66. Robinson, M. J., and T. R. Oberdorfer. 1983. Identification of pathogenic *Neisseria* species with RapID NH system. *J. Clin. Microbiol.* **17**:400–404.

67. Roongpisuthipong, A., J. S. Lewis, S. J. Kraus, and S. A. Morse. 1988. Gonococcal urethritis diagnosed from enzyme immunoassay of urine sediment. *Sex. Transm. Dis.* **15**:192–195.

68. Rossau, R. G., Vandenbussche, S. Thielemans, P. Segers, H. Grosch, E. Göthe, W. Mannheim, and J. E. DeLey. 1989. Ribosomal ribonucleic acid cistron similarities and deoxyribonucleic acid homologies of *Neisseria*, *Kingella*, *Eikenella*, *Simonsiella*, *Alysiella*, and Centers for Disease Control groups EF-4 and M-5 in the emended family *Neisseriaceae*. *Int. J. Syst. Bacteriol.* **39**:185–198.

69. Rossau, R. G., A. van Landschoot, M. Gillis, and J. de Ley. 1991. Taxonomy of *Moraxellaceae* fam. nov., a new bacterial family to accommodate the genera *Moraxella*, *Acinetobacter*, and *Psychrobacter* and related organisms. *Int. J. Syst. Bacteriol.* **41**:310–319.

70. Saez-Nieto, J. A., A. Fenoll, J. Bazquez, and J. Casal. 1982.

Prevalence of maltose-negative *Neisseria meningitidis* variants during an epidemic period in Spain. *J. Clin. Microbiol.* **15:**78–81.

71. **Sarafian, S. K., and J. S. Knapp.** 1989. Molecular epidemiology of gonorrhea. *Clin. Microbiol. Rev.* **2:**S49–S55.

72. **Schachter, J., W. M. McCormick, R. F. Smith, R. M. Parks, R. Bailey, and A. C. Ohlin.** 1984. Enzyme immunoassay for diagnosis of gonorrhea. *J. Clin. Microbiol.* **19:**57–59.

73. **Scribner, R. K., and D. F. Welch.** 1984. Neutralization of the inhibitory effect of sodium polyanetholsulfonate on *Neisseria meningitidis* in blood cultures processed with the Du Pont Isolator system. *J. Clin. Microbiol.* **20:**40–42.

74. **Strathdee, C. A., S. D. Tyler, A. Ryan, W. M. Johnson, and F. E. Ashton.** 1993. Genomic fingerprinting of *Neisseria meningitidis* associated with group C meningococcal disease in Canada. *J. Clin. Microbiol.* **31:**2506–2508.

75. **Tapsall, J. W., T. R. Schultz, R. Lovett, and R. Monro.** 1992. Failure of 500 mg ciprofloxacin therapy in male urethral gonorrhoea. *Med. J. Aust.* **156:**143.

76. **Vaneechoutte, M., G. Verschraegen, G. Claeys, B. Weise, and A. M. van den Abeele.** 1990. Respiratory tract carrier rates of *Moraxella* (*Branhamella*) *catarrhalis* in adults and children and interpretation of the isolation of *M. catarrhalis* from sputum. *J. Clin. Microbiol.* **28:**2647–2680.

77. **Vedros, N. A.** 1984. Genus I. *Neisseria* Trevisan 1885, 105[AL], p. 290–296. *In* N. R. Krieg and J. G. Holt (ed.), *Bergey's Manual of Systematic Bacteriology*, vol. 1. The Williams & Wilkins Co., Baltimore.

78. **Whittington, W. L., and J. S. Knapp.** 1988. Trends in antimicrobial resistance in *Neisseria gonorrhoeae* in the United States. *Sex. Transm. Dis.* **15:**202–210.

79. **Woods, T. C., L. O. Helsel, B. Swaminathan, W. F. Bibb, R. W. Pinner, B. G. Gellin, S. F. Collin, S. H. Waterman, M. W. Reeves, D. J. Brenner, and C. V. Broome.** 1992. Characterization of *Neisseria meningitidis* serogroup C by multilocus enzyme electrophoresis and ribosomal DNA restriction profiles (ribotyping). *J. Clin. Microbiol.* **30:**132–137.

80. **Young, H., A. B. Harris, and J. W. Tapsall.** 1984. Differentiation of gonococcal and nongonococcal neisseriae by the superoxol test. *Br. J. Vener. Dis.* **60:**87–89.

81. **Young, H., and A. Moyes.** 1993. Comparative evaluation of AccuProbe culture identification test for *Neisseria gonorrhoeae* and other rapid methods. *J. Clin. Microbiol.* **31:**1996–1999.

82. **Zollinger, W. D., B. L. Brandt, and E. C. Tramont.** 1986. Immune response to *Neisseria meningitidis*, p. 346–352. *In* N. R. Rose, H. Friedman, and J. L. Fahey (ed.), *Manual of Clinical Laboratory Immunology*, 3rd ed. American Society for Microbiology, Washington, D.C.

Listeria

BALA SWAMINATHAN, JOCELYN ROCOURT, AND JACQUES BILLE

27

TAXONOMY

The genus *Listeria* was first included in the broad group of coryneform bacteria in the seventh edition of *Bergey's Manual of Determinative Bacteriology* in 1957 (41). In 1986, the use of 16S rRNA partial sequencing unambiguously located the genus *Listeria* within the *Clostridium* subbranch, together with the genera *Staphylococcus*, *Streptococcus*, *Lactobacillus*, and *Brochothrix* (25). This phylogenetic position of genus *Listeria* is consistent with the low guanine and cytosine (G+C) content of its DNA (<50%), the presence of lipoteichoic acids, and the lack of mycolic acids (49).

Listeria monocytogenes remained the only recognized species in this genus during the 2 decades after its discovery until the description of *L. denitrificans* (now *Jonesia denitrificans* [43]) in 1948, *L. grayi* in 1966, and *L. murrayi* in 1971. DNA-DNA homology and 16S rRNA sequencing results obtained during the last decade demonstrate that this genus comprises two lines of descent: one is composed of *L. monocytogenes* and genomically closely related species (*L. innocua*, *L. ivanovii* subsp. *ivanovii*, *L. ivanovii* subsp. *londoniensis*, *L. welshimeri*, and *L. seeligeri*), and the other contains *L. grayi* (8, 12, 41). This structure of the genus *Listeria* is supported by classification based on multilocus enzyme analysis results (9). Only *L. monocytogenes* causes infection in humans, while *L. ivanovii* is occasionally responsible for abortion in animals. The other species are nonpathogenic.

DESCRIPTION OF THE GENUS

Members of the genus *Listeria* are asporogenous, nonbranching, regular, short, gram-positive rods that occur singly or in short chains. Filaments 6 to 20 μm long may occur in older or rough cultures. The organisms are motile at 28°C by means of one to five peritrichous flagella. Colonies are small (1 to 2 mm after 1 or 2 days of incubation at 37°C), smooth, and blue-gray when examined with obliquely transmitted light. The optimum growth temperature is between 30 and 37°C, but growth occurs at 4°C within a few days. *Listeria* spp. are facultatively anaerobic. Catalase is produced, and the oxidase test is negative. Acid is produced from D-glucose and other sugars. The Voges-Proskauer and methyl red test reactions are positive. Esculin is hydrolyzed in a few hours. Urea and gelatin are not hydrolyzed. Neither

indole nor H_2S is produced. The cell wall contains a directly cross-linked peptidoglycan based on *meso*-diaminopimelic acid, and the major menaquinone (MK-7) contains seven isoprene units. The G+C content of the DNA is 36 to 38% (49).

NATURAL HABITATS

Listeria species are widely distributed in the environment. They have been isolated from soil, decaying vegetable matter, silage, sewage, water, animal feed, fresh and frozen poultry, fresh and processed meats, raw milk, cheese, slaughterhouse waste, and asymptomatic human and animal carriers (49). *L. monocytogenes* has been isolated from 42 species of mammals, 22 species of birds, fish, crustaceans, and insects. Nevertheless, the primary habitats of *L. monocytogenes* are considered to be the soil and decaying vegetable matter, in which it survives and grows saprophytically. Interestingly, cellobiose, a plant-derived disaccharide, represses the production by *L. monocytogenes* of two important virulence factors, namely, listeriolysin and phosphatidylinositol-specific phospholipase C (37). This fact suggests that the organism has successfully evolved to produce virulence factors only when introduced into an animal or human host. Because of its widespread occurrence, *L. monocytogenes* has many opportunities to enter food production and processing environments.

CLINICAL SIGNIFICANCE

In nonpregnant human adults, *L. monocytogenes* primarily causes meningitis, encephalitis, or septicemia (32, 47). Elderly patients or persons with predisposing conditions that lower cell-mediated immunity, such as transplants, lymphomas, and AIDS, are especially susceptible. On rare occasions, patients have no recognizable predisposing conditions. The tropism of *L. monocytogenes* for the central nervous system leads to severe disease, often with high mortality (20 to 50%) or with neurologic sequelae among survivors.

In pregnant women, *L. monocytogenes* often causes an influenzalike bacteremic illness that, if untreated, may lead to amnionitis and infection of the fetus, resulting in abortion, stillbirth, or premature birth. Early diagnosis can be made by detecting *L. monocytogenes* in maternal blood cultures; at birth, the diagnosis is made by detecting the organism in cerebrospinal fluid (CSF), blood, amniotic fluid, respiratory secretions, placental or cutaneous swabs, gastric aspirate, or meconium of the neonate. Direct microscopic visualization of gram-positive rods in these specimens could be invaluable in early diagnosis of the disease.

Focal infections occur rarely after an episode of bacteremia. However, primary cutaneous listeriosis with or without bacteremia has been reported among veterinarians and abattoir workers, who acquire the illness through contact with infected animal tissues. Endocarditis, arthritis, osteomyelitis, intra-abdominal abscesses, endophthalmitis, and pleuropulmonary infections have been described infrequently.

The incubation period and infective dose have not been firmly established, although they may be inversely related. Reported incubation times vary from a few days to 2 to 3 months. Gastrointestinal symptoms such as diarrhea have been observed in some individuals with listeriosis but are not commonly associated with ingestion of contaminated food. A transient carrier state exists in 2 to 20% of animals and humans.

Listeriosis can occur as sporadic cases or as epidemics; in both, contaminated foods are the primary vehicles of transmission. A few limited, non-food-related, nosocomial outbreaks, mainly in nurseries, have been described. A recent cluster of listeriosis cases occurring among newborn infants in a maternity ward was traced to contaminated mineral oil used to clean the infants (46). The number of sporadic cases of listeriosis in countries that report the illness is typically in the range of 0.5 to 0.8 cases per 100,000 persons; during food-borne-disease outbreaks, the incidence may rise to 5 cases per 100,000 persons (53). Foods implicated as vehicles of infection include coleslaw (cabbage), soft cheeses, poultry, turkey frankfurters, mushrooms, milk, and pork tongue in jelly. Large numbers of organisms ($>10^3$ CFU/g) were detected in foods quantitatively assayed for the organism.

The pathogenesis of *L. monocytogenes* in human infections is unclear. However, after contaminated food has been ingested, the development of an invasive infection in some individuals depends on several factors: host susceptibility, gastric acidity, inoculum size, and virulence factors of the organism. After penetrating the epithelial barrier of the intestinal tract, *L. monocytogenes* can grow within hepatic and splenic macrophages, destroying them with listeriolysin, a hemolytic protein that binds to lipids of the host cell membrane. Immunity to listeriosis relies mainly on T-cell-mediated activation of macrophages by lymphokines; the role of humoral defenses is not fully understood.

COLLECTION, TRANSPORTATION, AND STORAGE OF SPECIMENS

Laboratory Safety

The infectious dose for listeriosis has not been determined, and it may depend, in part, on the susceptibility of the host. Groups at highest risk of acquiring infection are pregnant women, neonates, immunocompromised patients, and the elderly; however, up to 30% of adults with listeriosis may be immunocompetent. Therefore, laboratorians working with *L. monocytogenes* should be made aware of this potential and advised to be particularly cautious when working with this organism (21).

Specimens

Clinical

L. monocytogenes is readily isolated from clinical specimens obtained from normally sterile sites (blood, CSF, amniotic fluid, placenta, or fetal tissue). These specimens should be transported as soon as possible or stored at 4°C for up to 48 h. Stool specimens are more productive than rectal swabs when epidemiologic studies of carriage rates are undertaken. One gram of stool (a piece the size of a grape) can be inoculated into 100 ml of a selective enrichment broth such as U.S. Department of Agriculture (USDA) primary selective enrichment broth and then shipped at room temperature by overnight mail. If this is not possible, stools should be shipped frozen on dry ice by overnight mail. Other nonsterile site specimens may be stored at 4°C for 24 to 48 h. To avoid overgrowth of *L. monocytogenes* by contaminating microflora during longer periods of storage, freezing of specimens at −20°C is recommended.

Foods

Food samples must be collected aseptically in sterile containers. Whenever possible, foods packaged in original containers must be collected. Attempts should be made to collect at least 100 g of sample. Samples may be placed in sterile Whirl-Pak bags and shipped on ice by overnight mail. Ice cream and other frozen products are best transported in the frozen state in the original container and must be thawed immediately before analysis. Although *L. monocytogenes* is relatively resistant to freezing, repeated freezing and thawing may adversely affect the viability of the bacteria.

Isolates

Cultures of *Listeria* spp. may be shipped to a distant laboratory on a non-glucose-containing agar slant (such as heart infusion agar or tryptic soy agar) packaged to conform with the requirements for interstate shipment of etiologic agents (Code of Federal Regulations, 42CFR Part 72).

ISOLATION PROCEDURES

Culture

Clinical specimens from normally sterile sites can be directly plated on tryptic soy agar containing 5% sheep, horse, or rabbit blood. Samples for blood culture can be inoculated into conventional blood culture broth.

Clinical specimens obtained from nonsterile sites and foods and specimens obtained from the environment should be selectively enriched for *Listeria* spp. before being plated. For approximately 40 years, the isolation of *L. monocytogenes* from nonsterile specimens relied on the cultural procedure of cold enrichment in a nonselective broth at 4°C for as long as 2 or more months followed by selective plating. The investigations of listeriosis outbreaks in the 1980s established foods as the major vehicles for the transmission of *L. monocytogenes* and provided the impetus for developing more rapid selective enrichment methods for isolating *L. monocytogenes* from foods.

Although several selective enrichment methods for *L. monocytogenes* have been developed, none is 100% effective. The USDA method and the Netherlands Government Food Inspection Service (NGFIS) method are used together at the Centers for Disease Control and Prevention to isolate *L. monocytogenes* from non-sterile-site clinical specimens and foods (17). Individually, the two methods are approximately 75% sensitive, but when used in conjunction with each other, they are 90% sensitive.

The USDA method involves enrichment of the specimen in University of Vermont (UVM) primary selective enrichment broth (1 part sample plus 9 parts of broth) at 30°C. After 24 h, 0.1 ml of the enrichment culture is plated on lithium chloride-phenylethanol-moxalactam (LPM) agar and Oxford or modified Oxford agar (1). Another 0.1 ml of the enrichment culture is added to 10 ml of UVM secondary selective enrichment broth, which is incubated for an additional 24 h. The secondary enrichment culture is plated as described above. The plates are incubated at 35°C and examined after 24 and 48 h. All of the media named above are described in the compendium by Atlas and Parks (1) and in the fifth edition of this Manual (31). Also, they may be purchased commercially.

The NGFIS method involves enrichment of the specimen in liquid polymyxin-acriflavine-lithium chloride-ceftazidime-esculin-mannitol (PALCAM)-egg yolk broth (51) at 30°C and plating of the enrichment culture on commercially available PALCAM agar at 24 and 48 h. PALCAM agar is incubated at 30°C for 48 h under microaerobic conditions (5% oxygen, 7.5% carbon dioxide, 7.5% hydrogen, and 80% nitrogen).

LPM agar was developed as a highly selective but nondifferential medium for the isolation of *Listeria* species. Colonies on LPM agar are examined under a stereozoom microscope (magnification, ×15 to ×25), with oblique lighting directed to the microscope stage by a concave mirror positioned at a 45° angle to the incident light (Henry illumination). *Listeria* colonies appear blue, while colonies of other bacteria appear yellowish or orange.

Oxford and PALCAM agars contain selective differential chemicals that eliminate the need for examination under oblique lighting. On Oxford and modified Oxford agars, *Listeria* colonies appear black, are 1 to 3 mm in diameter, and are surrounded by a black halo after 24 to 48 h of incubation at 37°C; the color formation is due to the hydrolysis of esculin by *Listeria* spp. and the formation of black iron-phenol compounds in the medium. On PALCAM agar, *Listeria* colonies appear gray-green, are approximately 2 mm in diameter, and have black sunken centers; esculin, ferric iron, D-mannitol, and phenol red contribute to this color formation.

Suspect colonies (3 to 10 colonies per plate) are picked to Trypticase soy agar with 5% sheep blood and incubated for 18 h at 35°C for further workup.

Rapid Detection

For more rapid determination of the presence or absence of *Listeria* species, rapid tests based on monoclonal antibodies, nucleic acid hybridization assays, and PCR can be performed. A sandwich enzyme immunoassay kit using a panel of monoclonal antibodies has been developed for use with *Listeria* spp. The kit, called Listeria-Tek, is available from Organon Teknika Corp., Durham, N.C., and is intended for the qualitative detection of *Listeria* species in meats, dairy products, and environmental samples. A nonisotopic DNA probe test for the detection of *Listeria* spp. in foods is commercially available as the Gene-Trak *Listeria* assay from Gene-Trak Systems, Framingham, Mass. The Gene-Trak assay uses a synthetic DNA probe that is complementary to specific sequences in the 16S rRNA of *Listeria* spp. Both tests are genus specific only, and both require selective enrichment of food samples before the test is applied. These

kits are not approved for the analysis of clinical specimens for diagnosis and/or treatment.

L. monocytogenes DNA in CSF and tissue (fresh or in paraffin blocks) can be specifically detected by PCR-based tests (20, 46). A frequently used target for PCR amplification is the listeriolysin gene, although other virulence-associated genes are potentially useful. The PCR assay is highly sensitive and specific and could be particularly useful when prior administration of antimicrobial agents compromises culture.

IDENTIFICATION

Genus Identification

A simplified identification is based on the following tests: Gram stain, observation of tumbling motility in wet mount, positive catalase reaction, acid production from D-glucose, esculin hydrolysis, and positive Voges-Proskauer and methyl red reactions.

Species Identification

The scheme for identification of *Listeria* species is shown in Table 1.

Identification of *Listeria* isolates to species level is crucial, because all species can contaminate foods, but only *L. monocytogenes* is of public health concern. Identification is based on a limited number of biochemical markers, among which hemolysis is essential to differentiating between *L. monocytogenes* and the most frequently isolated nonpathogenic *Listeria* species, *L. innocua*.

Hemolysis

Only three species, *L. monocytogenes*, *L. seeligeri*, and *L. ivanovii*, are hemolytic. Recent studies indicated hemolysin to be the major virulence factor of *L. monocytogenes*; however, hemolysis alone cannot be used as an indicator of the presence of a virulent species, because *L. seeligeri* is hemolytic but nonpathogenic. *L. monocytogenes* and *L. seeligeri* produce narrow zones of hemolysis that frequently do not extend much beyond the edge of the colonies, whereas *L. ivanovii* exhibits a wide zone of hemolysis (Fig. 1).

The CAMP test uses *Staphylococcus aureus* and *Rhodococcus equi* streaked in one direction on a sheep blood agar plate and test cultures of *Listeria* spp. streaked at right angles to (but not touching) the *S. aureus* and *R. equi* lines. According to *Bergey's Manual of Systematic Bacteriology* (49), hemolysis of *L. monocytogenes* and *L. seeligeri* is enhanced in the vicinity of the *S. aureus* streak, and *L. ivanovii* hemolysis is enhanced in the vicinity of *R. equi* (typical picture of a shovel [Fig. 2]). However, because many investigators have reported observing a synergistic hemolysis reaction between *L. monocytogenes* and *R. equi*, this CAMP reaction must be interpreted with caution. A Beta Lysin disk (Remel, Lenexa, Kans.) could be used to observe hemolysis enhancement of *L. monocytogenes* with beta-lysin from *S. aureus* (Fig. 3).

Acid Production from Carbohydrates

L. monocytogenes is always D-xylose negative and α-methyl-D-mannoside positive. Rare atypical strains may be L-rhamnose negative.

Miniaturized Biochemical Tests

The API-*Listeria* test (bioMérieux Vitek, Inc., Hazelwood, Mo.) was specifically designed for this genus and includes

TABLE 1 Biochemical differentiation of species in the genus *Listeria*[a]

Characteristic	*L. grayi*	*L. innocua*	*L. ivanovii*	*L. ivanovii* subsp. *londoniensis*	*L. monocytogenes*	*L. seeligeri*	*L. welshimeri*
Beta-hemolysis	−	−	++[b]	++	+	+	−
CAMP test reaction							
S. aureus	−	−	−	−	+	+	−
R. equi	−	−	+	+	−	−	−
Acid production from:							
Mannitol	+	−	−	−	−	−	−
α-Methyl-D-mannoside	+	+	−	−	+	−	+
L-Rhamnose	V	V	−	−	+	−	V
Soluble starch	+	−	−	−	−	ND	ND
D-Xylose	−	−	+	+	−	+	+
Ribose	V	−	+	−	−	−	−
N-Acetyl-β-D-mannosamine			−	+			
Hippurate hydrolysis	−	+	+	+	+	ND	ND
Reduction of nitrate	−	−	−	−	−	ND	ND
Pathogenicity for mice	−	−	+	?	+	−	−
Serotype	S	4ab, US, 6a, 6b	5	5	1/2a, 1/2b, 1/2c, 3a, 3b, 3c, 4a, 4ab, 4b, 4c, 4d, 4e, 7	1/2a, 1/2b, 1/2c, US, 4b, 4d, 6b	1/2b, 4c, 6a, 6b, US

[a]See references 8 and 49. Symbols and abbreviations: +, ≥90% of strains are positive; −, ≥90% of strains are negative; ND, not determined; V, variable; US, undesignated serotype; S, specific.

[b]Usually a wide zone or multiple zones.

10 biochemical differentiation tests in a microtube format. It includes a patented "DIM" test, quite different from hemolysis, which distinguishes between *L. monocytogenes* and *L. innocua* (6).

DNA Probe Assay for Colony Confirmation

A 30-min chemiluminescence DNA probe assay is available (Gen-Probe, San Diego, Calif.) for the rapid confir-mation of *L. monocytogenes* from colonies on primary iso-lation plates. This assay was highly specific for *L. monocytogenes* in two independent evaluations (33, 36).

Differentiation of the Genus *Listeria* from Other Genera

Because they share some characteristics, *Listeria* spp. and some other gram-positive bacteria may be confused. *Strep-*

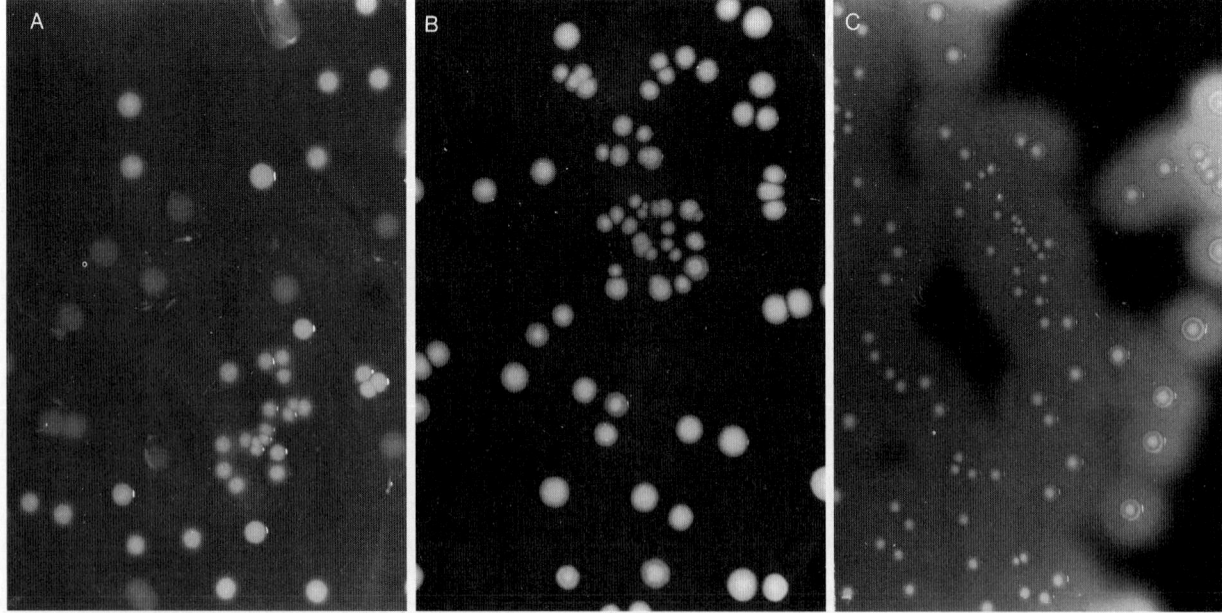

FIGURE 1 Macroscopic view of colonies on 5% human blood agar plates after 24 h of incubation. (A) *L. monocytogenes*: discrete zone of beta-hemolysis under the removed colonies. (B) *L. innocua*: no hemolysis. (C) *L. ivanovii*: wide zone of beta-hemolysis around the colonies.

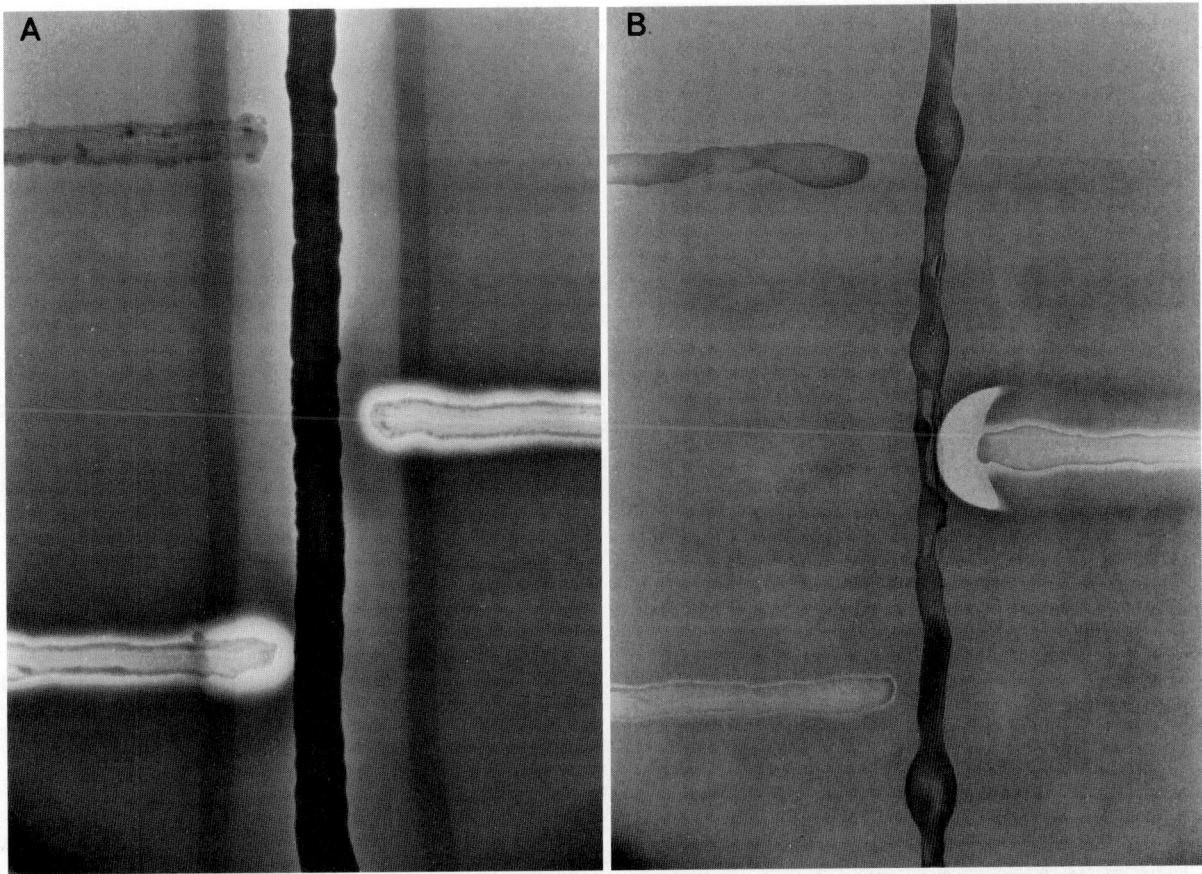

FIGURE 2 CAMP test done with *S. aureus* (A) and *R. equi* (B) after 24 h of incubation. Upper left, *L. innocua*; lower left, *L. monocytogenes*; right center, *L. ivanovii*.

tococcus spp. may be differentiated from *Listeria* spp. on the basis of Gram stain morphology, motility, and catalase activity. *Erysipelothrix* spp. differ from *Listeria* spp. in motility, catalase reaction, and ability to grow at 4°C (*Erysipelothrix* spp. do not grow at that temperature). Strictly aerobic acid production from glucose, positive Voges-Proskauer reaction, and inability to hydrolyze urea separate *Listeria* spp. (which exhibit these characteristics) from motile *Corynebacterium* spp. (which do not). Among background microflora of foods, *Lactobacillus* spp. are usually nonmotile and catalase negative, *Brochothrix* spp. are unable to grow at 37°C, and *Kurthia* spp. are strictly aerobic, oxidase negative, and esculin negative.

Determination of Pathogenicity

Methods using laboratory animals for evaluation of the virulence potential of *Listeria* isolates are available but are not used routinely, because most, if not all, natural isolates of *L. monocytogenes* are virulent. Such tests include intraperitoneal inoculation of mice, inoculation of the chorioallantoic membrane of embryonated eggs, and inoculation of the conjunctivas of rabbits (Anton test). A sensitive immunocompromised-mouse model in which relatively low numbers of virulent listeriae cause the deaths of these immunosuppressed animals within 3 days has been developed (50).

Cell culture cytotoxicity assays using the human intestinal epithelial line Caco-2 have been developed to determine the virulence potential of *Listeria* isolates in vitro.

While the results generally agree with those of animal tests, cytotoxicity assays do not provide a quantitative measure of virulence as the animal tests do (50% lethal dose). Also, some outbreak-associated *L. monocytogenes* isolates show very little cytotoxicity in the Caco-2 cell assays (39).

Serotyping and Other Typing Techniques

Serotyping

Strains of *Listeria* species are divided into serotypes on the basis of cellular (O) and flagellar (H) antigens (48). Thirteen serotypes (1/2a, 1/2b, 1/2c, 3a, 3b, 3c, 4a, 4ab, 4b, 4c, 4d, 4e, and 7) of *L. monocytogenes* are known; 4bX, a variant of serotype 4b, was implicated in a listeriosis outbreak traced to contaminated paté in England (29). Serotyping antigens are shared among *L. monocytogenes*, *L. innocua*, *L. seeligeri*, and *L. welshimeri*. Most human disease is caused by serotypes 1/2a, 1/2b, and 4b; therefore, serotyping alone is not sufficiently discriminating for subtyping purposes. Nevertheless, serotyping serves a useful purpose as a first-level discriminator. Also, unlike other subtyping methods, serotype designations are universal. Determination of the serotype also facilitates selection of appropriate controls for molecular subtyping methods.

Phage Typing

Because of the low discriminating ability of serotyping, phage typing was the only means of distinguishing strains of

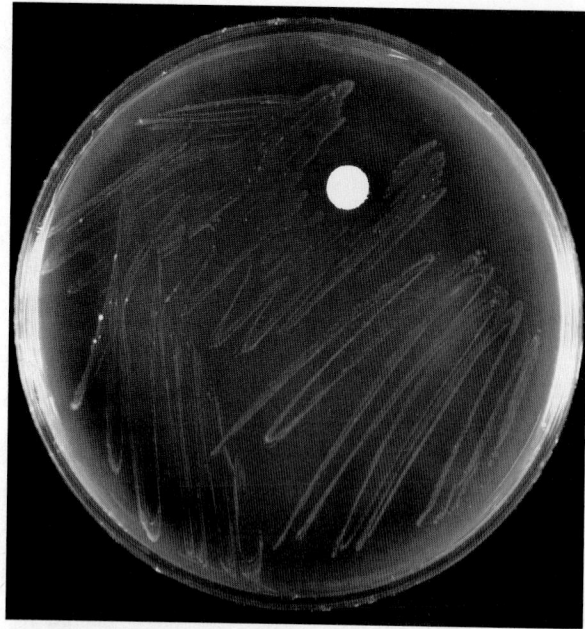

FIGURE 3 Enhancement of beta-hemolysis by *L. monocytogenes* demonstrated by using a Beta Lysin disk.

the same serotype before the introduction of molecular methods. *Listeria* phages were isolated from lysogenic strains of all *Listeria* species and from sewage (24, 42). A multicenter study between four laboratories was organized in 1981 to select an international set of phages among those isolated in France and Germany and to define a standardized method (42). Phage typing has been successfully used in clinical and epidemiologic evaluations involving recurrent infections in humans, nosocomial infections, and common source outbreaks (5, 13, 14, 22, 28, 45). The isolation of strains with the same phage type both from patients in epidemics and from the foods implicated by case-control studies confirmed the food-borne transmission of listeriosis. Despite its usefulness, phage typing is hampered by the nontypeability of some strains; the percentage of nontypeable strains may vary according to the origin of the strain (food and environmental isolates are frequently nontypeable). A recently published method of reverse phage typing may represent an interesting alternative in this regard (23, 24).

MEE

Typically, 10 to 25 enzyme loci are examined by multilocus enzyme electrophoresis (MEE) for each strain. Each unique mobility variant of an enzyme is given a unique numeric allele designation. Each strain is defined by a string of numerical values for the alleles examined (electrophoretic type). MEE has been applied with great success to epidemiologic investigations of outbreaks and sporadic disease (4, 7, 38, 40, 47). Strains implicated in three food-borne outbreaks were determined by MEE to be identical or highly related; this finding has caused some investigators to hypothesize that outbreaks are caused by a few clones that probably possess unique virulence factors. Nevertheless, MEE alone may not be adequate to subtype serotype 4b, because large numbers of epidemiologically unrelated strains cluster in a few electrophoretic types.

MEE has also been very useful in taxonomic studies and in differentiating between different *Listeria* species (9).

DNA FINGERPRINTING

DNA Microrestriction Patterns

Characterization of chromosomal DNA by restriction endonuclease analysis or ribosomal DNA gene restriction patterns (ribotyping) has been used to differentiate *L. monocytogenes* strains in different serotypes and within a given serotype, in particular serotype 4b, which is most frequently involved in outbreaks. Microrestriction patterns generated by high-frequency cutting restriction enzymes (e.g., *Eco*RI) have proved useful in epidemiologic investigations (2, 35), although the complexity of patterns makes it difficult to compare the patterns of several strains. Ribotyping simplifies the microrestriction patterns by rendering visible only the DNA fragments containing part or all of the ribosomal genes, but its discriminating ability, particularly for serotype 4b, may not be adequate (2, 15, 18, 19, 34).

DNA Macrorestriction Patterns

Pulsed-field gel electrophoresis (PFGE) distinguished 48 different patterns among 77 strains (10, 11). Analysis of strains responsible for major outbreaks that occurred during the last 2 decades shows that most of these strains belong to a small number of well-defined and very closely related clones (11). To date, this typing appears to be the most discriminatory for *L. monocytogenes*. Nevertheless, like other molecular methodologies, PFGE is not suitable at present for application to routine typing; serotyping and phage typing remain the most appropriate methods for screening large numbers of strains.

RAPD

Random amplified polymorphic DNA (RAPD) typing involves the use of a short (usually 10- to 15-bp), arbitrarily chosen primer to amplify nearly homologous sequences of the genomic DNA under low-stringency conditions. RAPD differentiates strains of various *Listeria* species, various serotypes within *L. monocytogenes*, and various subtypes within a serotype (27). Only limited data on RAPD are available, but the evidence indicates the potential of the method for subtyping *L. monocytogenes* (26).

Present Status of Subtyping

The vast majority of *L. monocytogenes* strains causing sporadic infections or outbreaks belong to the three serotypes 1/2a, 1/2b, and 4b. Strains of serotype 1/2a are highly heterogeneous and thus are easily differentiated by any of the molecular methods and by phage typing when the strains are phage typeable. In contrast, strains of serotype 4b are more closely related and probably necessitate the combined use of several methods to be optimally differentiated. An international multicenter study to evaluate the different subtyping methods and to standardize the most promising methods is currently under way.

SEROLOGIC TESTS

Serologic responses to whole-cell antigens are not diagnostic because of antigenic cross-reactivity between *L. monocytogenes* and other gram-positive bacteria such as staphylococci, enterococci, and *Bacillus* species. Further, patients with culture-confirmed listeriosis have had undetectable antibody levels (47). Determination of antibody levels to listeriolysin may be a better indicator of infection (3). The antibody response to listeriolysin is primarily immunoglobulin G, and no anti-listeriolysin immunoglobulin M has been detected in CSF or sera of patients with listeriosis.

ANTIMICROBIAL AGENT SUSCEPTIBILITIES

The pattern of antimicrobial agent susceptibility and resistance of *L. monocytogenes* has been relatively stable for many years. In vitro, the organism is susceptible to penicillin, ampicillin, gentamicin, erythromycin, tetracycline, rifampin, and chloramphenicol (44, 52). However, many of these antimicrobial agents are only bacteriostatic. Penicillin or ampicillin with or without an aminoglycoside is usually recommended for the treatment of listeriosis. Studies in vitro and in animal models have shown that an aminoglycoside enhances the antimicrobial (bactericidal) activity of penicillin against *L. monocytogenes* (30). Trimethoprim-sulfamethoxazole and aminoglycosides are among the few anti-infective agents that are bactericidal to *L. monocytogenes*; only trimethoprim-sulfamethoxazole has been used occasionally with success. Recently, resistance plasmids conferring resistance to chloramphenicol, macrolides, and tetracyclines have been found in several clinical isolates of *L. monocytogenes* and have raised concern for the future (16). Cephalosporins, which are ineffective, should never be administered when listeriosis is suspected.

EVALUATION, INTERPRETATION, AND REPORTING OF RESULTS

Colonies that show subdued beta-hemolysis on blood agar plated with blood, CSF, or other normally sterile-site specimens should be subjected to motility tests, Gram staining, and sugar utilization tests to confirm identification. The CAMP test is not necessary on a routine basis, and the Beta Lysin test may be substituted for it. Use of the API-*Listeria* test may eliminate the need for enhanced hemolysis testing altogether.

If *L. monocytogenes* is present in low numbers in CSF, the direct examination of Gram-stained clinical specimens may be of little or no value. Also, Gram-stained *Listeria* cells closely resemble other gram-positive bacteria such as streptococci and corynebacteria. If antimicrobial therapy was initiated before a CSF specimen was obtained, culture results may be negative. In these instances, Gram staining may be useful, but additional confirmation by methods such as PCR (for laboratories that have the capability) may be needed.

Serologic responses to whole-cell antigens are unreliable, although in culture-negative cases, determination of serologic response to listeriolysin may be of diagnostic value retrospectively.

REFERENCES

1. **Atlas, R. M., and L. C. Parks (ed.).** 1993. *Handbook of Microbiological Media.* CRC Press, Inc., Boca Raton, Fla.
2. **Baloga, A. O., and S. K. Harlander.** 1991. Comparison of methods for discrimination between strains of *Listeria monocytogenes* from epidemiological surveys. *Appl. Environ. Microbiol.* **57:**2324–2331.
3. **Berche, P., K. A. Reich, M. Bonnichon, J.-L. Beretti, C. Geoffroy, J. Raveneau, P. Cossart, J.-L. Gaillard, P. Geslin, H. Kreis, and M. Veron.** 1990. Detection of anti-listeriolysin O for serodiagnosis of human listeriosis. *Lancet* **ii:**624–627.
4. **Bibb, W. F., B. G. Gellin, R. Weaver, B. Schwartz, B. D. Plikaytis, M. W. Reeves, R. W. Pinner, and C. V. Broome.** 1990. Analysis of clinical and food-borne isolates of *Listeria monocytogenes* in the United States by multilocus enzyme electrophoresis and application of the method to epidemiologic investigations. *Appl. Environ. Microbiol.* **56:**2133–2141.
5. **Bille, J.** 1989. Anatomy of a listeriosis outbreak, p. 29–36. *In* Foodborne Listeriosis. Proceedings of a Symposium. B. Behr's Gmbh & Co., Hamburg, Germany.
6. **Bille, J., B. Catimel, E. Bannerman, C. Jacquet, M. N. Yersin, I. Caniaux, D. Monget, and J. Rocourt.** 1992. API-*Listeria*, a new and promising one-day system to identify *Listeria* isolates. *Appl. Environ. Microbiol.* **58:**1857–1860.
7. **Boerlin, P., and J.-C. Piffaretti.** 1991. Typing of human, animal, food, and environmental isolates of *Listeria monocytogenes* by multilocus enzyme electrophoresis. *Appl. Environ. Microbiol.* **57:**1624–1629.
8. **Boerlin, P., J. Rocourt, F. Grimont, P. A. D. Grimont, C. Jacquet, and J.-C. Piffaretti.** 1992. *Listeria ivanovii* subsp. *londoniensis* subsp. nov. *Int. J. Syst. Bacteriol.* **42:**69–73.
9. **Boerlin, P., J. Rocourt, and J.-C. Piffaretti.** 1991. Taxonomy of the genus *Listeria* by using multilocus enzyme electrophoresis. *Int. J. Syst. Bacteriol.* **41:**59–64.
10. **Brosch, R., C. Buchrieser, and J. Rocourt.** 1991. Subtyping of *Listeria monocytogenes* serovar 4b by use of low frequency cleavage restriction endonucleases and pulsed-field gel electrophoresis. *Res. Microbiol.* **142:**667–675.
11. **Buchrieser, C., R. Brosch, B. Catimel, and J. Rocourt.** 1993. Pulsed-field gel electrophoresis applied for comparing *Listeria monocytogenes* strains involved in outbreaks. *Can. J. Microbiol.* **39:**395–401.
12. **Collins, M. D., S. Wallbanks, D. J. Lane, J. Shah, R. Nietupski, J. Smida, M. Dorsch, and E. Stackebrandt.** 1991. Phylogenetic analysis of the genus *Listeria* based on reverse transcriptase sequencing of 16S rRNA. *Int. J. Syst. Bacteriol.* **41:**240–246.
13. **Fleming, D. W., S. L. Cochi, K. L. MacDonald, J. Brondum, P. S. Hayes, B. D. Plikaytis, M. B. Holmes, A. Audurier, C. V. Broome, and A. L. Reingold.** 1985. Pasteurized milk as a vehicle of infection in an outbreak of listeriosis. *N. Engl. J. Med.* **312:**404–407.
14. **Goulet, V., A. Lepoutre, J. Rocourt, A. L. Courtieu, P. Dehaumont, and P. Veit.** 1993. Epidémie de listériose en France—bilan final et résultats de l'enquête épidémiologique. *Bull. Epidemiol. Hebdom.* **4:**13–14.
15. **Graves, L. M., B. Swaminathan, M. W. Reeves, and J. Wenger.** 1991. Ribosomal DNA fingerprinting of *Listeria monocytogenes* using a digoxigenin-labeled DNA probe. *Eur. J. Epidemiol.* **7:**77–82.
16. **Hadorn, K., H. Hachler, A. Schaffner, and F. H. Kayser.** 1993. Genetic characterization of plasmid-encoded multiple antibiotic resistance in a strain of *Listeria monocytogenes* causing endocarditis. *Eur. J. Clin. Microbiol. Infect. Dis.* **12:**928–937.
17. **Hayes, P. S., L. M. Graves, B. Swaminathan, G. W. Ajello, G. B. Malcolm, R. E. Weaver, R. Ransom, K. Deaver, B. D. Plikaytis, A. Schuchat, J. D. Wenger, R. W. Pinner, C. V. Broome, and The *Listeria* Study Group.** 1992. Comparison of three selective enrichment methods for the isolation of *Listeria monocytogenes* from naturally contaminated foods. *J. Food Prot.* **55:**952–959.
18. **Jacquet, C., S. Aubert, N. El Sohl, and J. Rocourt.** 1992. Use of rRNA gene restriction patterns for the identification of *Listeria* species. *Syst. Appl. Microbiol.* **15:**42–46.
19. **Jacquet, C., J. Bille, and J. Rocourt.** 1992. Typing of *Listeria monocytogenes* by restriction fragment length polymorphism of the ribosomal ribonucleic acid gene region. *Zentralbl. Bakteriol. Hyg. A* **276:**356–365.
20. **Jaton, K., R. Sahll, and J. Bille.** 1992. Development of polymerase chain reaction assays for detection of *Listeria monocytogenes* in clinical cerebrospinal fluid samples. *J. Clin. Microbiol.* **30:**1931–1936.
21. **Jones, G. L. (ed.).** 1989. *Isolation and Identification of Listeria*

monocytogenes. Public Health Service, Centers for Disease Control, U.S. Department of Health and Human Services, Atlanta.

22. **Linnan, M. J., L. Mascola, X. D. Lou, V. Goulet, S. May, C. Salminen, D. W. Hird, M. L. Yonekura, P. Hayes, R. Weaver, A. Audurier, B. D. Plikaytis, S. L. Fannin, A. Kleks, and C. V. Broome.** 1988. Epidemic listeriosis associated with Mexican-style cheese. *N. Engl. J. Med.* **319:** 823–828.

23. **Loessner, M. J.** 1991. Improved procedure for bacteriophage typing of *Listeria* strains and evaluation of new phages. *Appl. Environ. Microbiol.* **57:**882–884.

24. **Loessner, M. J., and M. Busse.** 1990. Bacteriophage typing of *Listeria* species. *Appl. Environ. Microbiol.* **56:**1912–1918.

25. **Ludwig, W., K.-H. Schleifer, and E. Stackebrandt.** 1984. 16S rRNA analysis of *Listeria monocytogenes* and *Brochothrix thermosphacta*. *FEMS Microbiol. Lett.* **25:**199–204.

26. **MacGowan, A. P., K. O'Donaghue, S. Nicholls, J. McLauchlin, P. M. Bennett, and D. S. Reeves.** 1993. Typing of *Listeria* spp. by random amplified polymorphic DNA (RAPD) analysis. *J. Med. Microbiol.* **38:**322–327.

27. **Mazurier, S. I., and K. Wernars.** 1992. Typing of *Listeria* strains by random amplification of polymorphic DNA. *Res. Microbiol.* **143:**499–505.

28. **McLauchlin, J., A. Audurier, and A. G. Taylor.** 1986. Aspects of the epidemiology of human *Listeria monocytogenes* infections in Britain 1967–1984; the use of serotyping and phage typing. *J. Med. Microbiol.* **22:**367–377.

29. **McLauchlin, J., S. M. Hall, S. K. Velani, and R. J. Gilbert.** 1991. Human listeriosis and paté: a possible association. *Br. Med. J.* **303:**773–775.

30. **Moellering, R. C., G. Medoff, I. Leech, C. Wennersten, and L. J. Kunz.** 1972. Antibiotic synergism against *Listeria monocytogenes*. *Antimicrob. Agents. Chemother.* **1:**30–34.

31. **Nash, P., and M. M. Krenz.** 1991. Culture media, p. 1226–1288. *In* A. Balows, W. J. Hausler, Jr., K. L. Herrmann, H. D. Isenberg, and H. J. Shadomy (ed.), *Manual of Clinical Microbiology*, 5th ed. American Society for Microbiology, Washington, D.C.

32. **Nieman, R. E., and B. Lorber.** 1980. Listeriosis in adults, a changing pattern: report of eight cases and a review of the literature, 1968–1978. *Rev. Infect. Dis.* **2:**207–227.

33. **Ninet, B., E. Bannerman, and J. Bille.** 1992. Assessment of the Accuprobe *Listeria monocytogenes* culture identification reagent kit for rapid colony confirmation and its application in various enrichment broths. *Appl. Environ. Microbiol.* **58:**4055–4059.

34. **Nocera, D., M. Altwegg, G. Martinetti Lucchini, E. Bannerman, F. Ischer, J. Rocourt, and J. Bille.** 1993. Characterization of *Listeria* strains from a foodborne listeriosis outbreak by rDNA gene restriction patterns compared to four other typing methods. *Eur. J. Clin. Microbiol. Infect. Dis.* **12:**162–169.

35. **Nocera, D., E. Bannerman, J. Rocourt, K. Jaton-Ogay, and J. Bille.** 1990. Characterization by DNA restriction endonuclease analysis of *Listeria monocytogenes* strains related to the Swiss epidemic of listeriosis. *J. Clin. Microbiol.* **28:**2259–2263.

36. **Okwumabua, O., B. Swaminathan, P. Edmonds, J. Wenger, J. Hogan, and M. Alden.** 1992. Evaluation of a chemiluminescent DNA probe assay for the rapid confirmation of *Listeria monocytogenes*. *Res. Microbiol.* **143:**183–189.

37. **Park, S. F., and R. G. Kroll.** 1993. Expression of listeriolysin and phosphatidylinositol-specific phospholipase C is repressed by the plant derived molecule cellobiose in *Listeria monocytogenes*. *Mol. Microbiol.* **8:**653–661.

38. **Piffaretti, J.-C., H. Kressebuch, M. Aeschbacher, J. Bille, E. Bannerman, J. M. Musser, R. K. Selander, and J. Rocourt.** 1989. Genetic characterization of clones of the bacterium *Listeria monocytogenes* causing epidemic disease. *Proc. Natl. Acad. Sci. USA* **86:**3818–3822.

39. **Pine, L., S. Kathariou, F. Quinn, V. George, J. D. Wenger, and R. E. Weaver.** 1991. Cytopathogenic effects in enterocytelike Caco-2 cells differentiate virulent from avirulent *Listeria* strains. *J. Clin. Microbiol.* **29:**990–996.

40. **Pinner, R. W., A. Schuchat, B. Swaminathan, P. S. Hayes, K. A. Deaver, R. E. Weaver, B. D. Plikaytis, M. Reeves, C. V. Broome, J. D. Wenger, and The Listeria Study Group.** 1992. Role of foods in sporadic listeriosis. II. A microbiologic and epidemiologic investigation. *JAMA* **267:**2046–2050.

41. **Rocourt, J.** 1988. Taxonomy of the genus *Listeria*. *Infection* **16**(Suppl. 2)**:**89–91.

42. **Rocourt, J., A. Audurier, A. L. Courtieu, J. Durst, S. Ortel, A. Schrettenbrunner, and A. G. Taylor.** 1985. A multicenter study on the phage typing of *Listeria monocytogenes*. *Zentralbl. Bakteriol. Hyg. A* **259:**489–497.

43. **Rocourt, J., U. Wehmeyer, and E. Stackebrandt.** 1987. Transfer of *Listeria denitrificans* to a new genus, *Jonesia* gen. nov. as *Jonesia denitrificans* comb. nov. *Int. J. Syst. Bacteriol.* **37:**266–270.

44. **Schald, W. M.** 1983. Evaluation of rifampin and other antibiotics against *Listeria monocytogenes* in vitro and in vivo. *Rev. Infect. Dis.* **5**(Suppl. 3)**:**S593–S599.

45. **Schlech, W. F., III, P. M. Lavigne, R. A. Bortolussi, A. C. Allen, E. V. Haldane, A. J. Wort, A. W. Hightower, S. E. Johnson, S. H. King, E. S. Nicholls, and C. V. Broome.** 1983. Epidemic listeriosis—evidence for transmission by food. *N. Engl. J. Med.* **308:**203–206.

46. **Schuchat, A., C. Lizano, C. V. Broome, B. Swaminathan, C. Kim, and K. Winn.** 1991. Outbreak of neonatal listeriosis associated with mineral oil. *Pediatr. Infect. Dis.* **10:**183–189.

47. **Schuchat, A., B. Swaminathan, and C. V. Broome.** 1991. Epidemiology of human listeriosis. *Clin. Microbiol. Rev.* **4:**169–183.

48. **Seeliger, H. P. R., and K. Hohne.** 1979. Serotyping of *Listeria monocytogenes* and related species, p. 31–49. *In* T. Bergen and J. R. Norris (ed.), *Methods in Microbiology*, vol. 13. Academic Press, London.

49. **Seeliger, H. P. R., and D. Jones.** 1986. Genus *Listeria* Pirie, 1940, 383[AL], p. 1235–1245. *In* P. H. A. Sneath, H. S. Mair, M. E. Sharp, and J. G. Holt (ed.), *Bergey's Manual of Systematic Bacteriology*, vol. 2. The Williams & Wilkins Co., Baltimore.

50. **Stelma, G. N., A. L. Reyes, J. T. Peeler, D. W. Francis, J. M. Hunt, P. L. Spaulding, C. H. Johnson, and J. Lovett.** 1987. Pathogenicity test for *Listeria monocytogenes* using immunocompromised mice. *J. Clin. Microbiol.* **25:**2085–2089.

51. **van Netten, P., I. Perales, A. van de Moosdijk, G. D. W. Curtis, and D. A. A. Mossel.** 1989. Liquid and solid differential media for the detection and enumeration of *Listeria monocytogenes* and other *Listeria* spp. *Int. J. Food Microbiol.* **8:**299–316.

52. **Wiggins, G. L., W. L. Albritton, and J. C. Feeley.** 1978. Antibiotic susceptibility of clinical isolates of *Listeria monocytogenes*. *Antimicrob. Agents. Chemother.* **12:**854–860.

53. **World Health Organization.** 1988. *Report of a WHO Informal Working Group.* WHO/EHE/FOS/88.5. World Health Organization, Geneva.

Bacillus

PETER C. B. TURNBULL AND JOHN M. KRAMER

28

TAXONOMY

The family *Bacillaceae* comprises a highly diverse group of endospore-forming bacteria ranging from the anaerobic gram-negative members of the genus *Desulfotomaculum* to the aerobic gram-positive cocci of the genus *Sporosarcina*. The two principal genera of the family are *Bacillus* and *Clostridium* (see chapter 47 of this Manual).

The genus *Bacillus* includes 51 validly described species and many additional species of uncertain taxonomic status. Substantial heterogeneity within the genus is reflected in the DNA base composition of *Bacillus* strains, which ranges from 32 to 69 mol% G+C (5), and the results of at least three extensive numerical taxonomic studies have confirmed the need for a subdivision of the genus into several smaller genera (22).

DESCRIPTION OF THE GENUS

Bacillus species, familiarly known as aerobic spore-bearing bacilli, are aerobic or facultatively anaerobic and gram positive or gram variable, some species being clearly gram positive only in young cultures. The vegetative cells are straight, round- or square-ended rods ranging from 0.5 by 1.2 to 2.5 by 10 μm that occur singly or in chains that range from a few to many cells in length. The endospores (one per mother cell [sporangium] in the presence of oxygen) are cylindrical, oval, round, or occasionally kidney shaped; depending on the species, they are centrally, subterminally, or terminally placed and characteristically swelling or not swelling the sporangium. Under appropriate conditions (the presence of HCO_3^- in an anaerobic or CO_2 atmosphere), *Bacillus anthracis*, *B. subtilis*, *B. licheniformis*, and *B. megaterium* produce a polypeptide (poly-γ-D-glutamic acid) capsule visible when the organism is stained with polychrome methylene blue (M'Fadyean stain [24]). With the principal exceptions of *B. anthracis* and *B. mycoides*, almost all *Bacillus* species are motile, although there is great strain-to-strain variation in the extent of motility in any one species. Most are catalase positive.

NATURAL HABITATS

Bacillus species are mostly saprophytes widely distributed in nature, particularly in soil, whence they are spread in dust, in water, and on materials of plant and animal origins. The broad range of physiologic characteristics exhibited within the genus is reflected in a wide range of facultative variants of mesophilic species, facultative and obligate thermophiles, psychrophiles, acidophiles, halophiles, and so on that are capable of survival in spore form or even of growth at environmental extremes. The ecology of *Bacillus* species has been comprehensively reviewed elsewhere (19).

CLINICAL SIGNIFICANCE

The majority of *Bacillus* species apparently have little or no pathogenic potential and are rarely associated with disease in humans or lower animals. The principal exceptions to this are *B. anthracis*, the agent of anthrax, and *B. cereus*, but a number of other species, particularly those of the *B. subtilis* group, have been implicated in food poisoning and other human and animal infections. The resistance of the spores to heat, radiation, disinfectants, and desiccation also results in *Bacillus* species being troublesome contaminants in the operating room, on surgical dressings, in pharmaceutical products, and in foods.

The clinical conditions associated with *Bacillus* species are thoroughly reviewed by Drobniewski (7) and summarized in Table 1. *Bacillus* species other than *B. anthracis* have recently been taken more seriously in relation to infections of compromised hosts, in mixed infections, as secondary invaders, and, particularly in the case of *B. cereus*, as occasional primary pathogens.

B. cereus has long been associated with food poisoning. To a lesser extent, *B. subtilis*, *B. licheniformis*, and occasionally other *Bacillus* species are also being incriminated increasingly in food-borne illness (15).

B. cereus is the etiologic agent of two distinct food poisoning syndromes: (i) the diarrheal type, characterized by abdominal pain with diarrhea 8 to 16 h after ingestion of the contaminated food and associated with a diversity of foods from meats and vegetable dishes to pastas, desserts, cakes, sauces, and milk, and (ii) the emetic type, characterized by nausea and vomiting 1 to 5 h after ingestion of the offending food, predominantly oriental rice dishes, although other foods such as pasteurized cream, milk pudding, pastas, and reconstituted formulas have occasionally been implicated (15). Both syndromes arise as a direct result of the fact that *B. cereus* spores can survive normal

TABLE 1 *Bacillus* infections reported since 1960[a]

Species	Clinical condition(s)	Reports or occurrence[b]
B. alvei	Sepsis, meningitis	+
B. anthracis	Anthrax	
	Cutaneous (eschar, malignant pustule)	+ + + +
	Intestinal	+ +
	Pulmonary	+
	Meningitis (secondary to cutaneous)	+
B. brevis	Bacteremia	+
B. cereus	Infected wounds, mild to severe, necrotic or gangrenous	+ + + +
	Bovine mastitis	+ + + +
	Bacteremia or septicemia, including drug abuse	+ + +
	Bovine abortion	+ + +
	Pneumonia, pleurisy, empyema, meningitis, endocarditis, osteomyelitis, panophthalmitis or endophthalmitis	+ + +
	Other (burns, ear infection, peritonitis, urinary tract infection)	+
	Food poisoning	+ + + +
B. circulans	Wound infection, bacteremia, septicemia, abscesses	+ + +
	Meningitis	+
B. coagulans	Bacteremia or septicemia	+
B. licheniformis	Bacteremia or septicemia	+ + +
	Other (peritonitis, ophthalmitis, bovine toxemia)	+
	Food poisoning	+ + + +
B. macerans	Bacteremia or septicemia	+
B. pumilus	Food poisoning	+ +
	Rectal fistula	+
B. sphaericus	Bacteremia, endocarditis, meningitis, pseudotumor	+
	Food poisoning	+
B. subtilis	Bacteremia or septicemia, endocarditis, respiratory infections	+ +
	Food poisoning	+ + + +
B. thuringiensis	Wound, eye infection, bovine mastitis	+

[a] Adapted from *Topley and Wilson's Principles of Bacteriology, Virology and Immunity*, 8th ed., with permission of Edward Arnold (Publishers) Limited.

[b] +, One or two reports (of each clinical entity listed); + +, a few reports; + + +, several reports; + + + +, many reports or frequent and regular occurrence now accepted. Most of the relevant reports can be found in references 1, 4, 7, 10, 15, 19, 20, and 23.

cooking procedures; under conditions of improper storage after cooking, the spores germinate, and the vegetative cells multiply.

Although the toxin of *B. anthracis* is now well characterized (16), the toxigenic basis of *B. cereus* food poisoning and other infections has been only partially elucidated (23, 27). No toxins or other virulence factors have yet been identified to account for symptoms periodically associated with other *Bacillus* species.

Anthrax is primarily a disease of herbivores; before an effective veterinary vaccine became available in the late 1930s, anthrax was one of the foremost causes worldwide of mortality in cattle, sheep, goats, and horses. Humans almost invariably contract anthrax directly or indirectly from animals. The development and application of veterinary and human vaccines together with improvements in factory hygiene and sterilization procedures for imported animal products and increased use of synthetic alternatives to animal hides or hair have resulted in a marked decline in the incidence of the disease in both animals and humans over the past 4 decades. Nevertheless, the disease continues to be endemic in many countries, particularly those that lack an efficient vaccination policy. Even among developed countries, few are entirely free of the disease (14). Because anthrax spores remain viable in soil for many years and their persistence does not depend on animal reservoirs, *B. anthracis* is exceedingly difficult to eradicate from an area in which it is endemic; regions where it is not endemic must be constantly on the alert for the arrival of *B. anthracis* in imported products of animal origin.

In comparison with herbivores, humans are moderately resistant to anthrax. Human anthrax is traditionally classified as either (i) nonindustrial, resulting from close contact with infected animals or their carcasses after death from the disease, or (ii) industrial, acquired by those employed in processing wool, hair, hides, bones, or other animal products. Nonindustrial anthrax usually manifests itself as the cutaneous form of the disease or, far less frequently, as intestinal anthrax if the infected meat is eaten. Industrial anthrax is also usually cutaneous but has a higher probability than nonindustrial anthrax of taking the pulmonary form from inhalation of spore-laden dust (2). A few reports exist of laboratory-acquired infections (21), and direct human-to-human transmission is exceedingly rare. Direct animal-to-animal transmission is also rare; epidemics or epizootics are almost always of the point source type.

In cutaneous anthrax, which accounts for 95 to 99% of human cases worldwide, infection occurs through a break in the skin, and the lesions therefore generally occur on exposed regions of the body. Fewer than 20% of untreated cases of cutaneous anthrax are fatal; the rare fatalities seen today are due to obstruction of the airways by the accompanying extensive edema when the lesion is on the face or neck and to sequelae of secondary cellulitis or meningitis. The intestinal and pulmonary forms are more often fatal than the cutaneous type because they go unrecognized until beyond the time for effective therapy.

The incubation period in cutaneous anthrax is 2 to 3 days. A small pimple or papule appears initially; over the next 24 h, a ring of vesicles develops around the papule, which ulcerates, dries, and blackens into the characteristic eschar. The eschar enlarges, becoming thick and adherent to underlying tissues over the ensuing week. The lesion is surrounded by edema, which may be very extensive. Pus and pain are normally absent; their presence or the presence of marked lymphangitis and fever probably indicates secondary bacterial infection.

Intestinal anthrax is essentially cutaneous anthrax occurring on the intestinal mucosa. Symptoms of gastroenteritis sometimes but not invariably occur. In pulmonary anthrax, the inhaled spores are carried by macrophages from the lungs to the lymphatic system, where they germinate and multiply, ultimately leading to fatal septicemia.

In fatal cases of any of the forms, generalized symptoms may be mild (fatigue, malaise, and slight fever) or absent before sudden onset of acute illness, characterized by dyspnea, cyanosis, severe pyrexia, and disorientation followed by circulatory failure, shock, coma, and death, all within a few hours. Depending somewhat on the host species, bacteria in the blood rapidly build up over the last few hours; in the most susceptible species, levels of 10^7 to 10^9 bacilli per ml may be present at death. Behind these events is the toxin of *B. anthracis*, although the precise manner in which the toxin produces the signs and symptoms observed has yet to be elucidated (16).

COLLECTION, TRANSPORT, AND STORAGE OF SPECIMENS

Bacillus species are quite hardy and usually survive transport to the laboratory either in freshly collected specimens or in a standard transport medium.

Safety Aspects in Relation to Anthrax

Despite its bad name, anthrax is not highly contagious. Cutaneous anthrax is readily treated and is life threatening only in exceptional cases; infectious doses of the human pulmonary and intestinal forms (also treatable if recognized early enough) are generally very high (>10,000 spores). Precautions, therefore, need to be sensible but not extreme.

When specimens related to suspected anthrax are being collected, disposable gloves, disposable apron or overalls, and boots that can be disinfected after use should be worn. If dusty samples suspected of harboring high numbers of anthrax spores are involved, the use of headgear and dust masks should be considered.

Disposable items should be discarded into suitable containers for autoclaving followed by incineration. Nonautoclavable items should be immersed overnight in 10 to 30% formalin (4 to 12% formaldehyde solution, with the concentration depending on the perceived extent of contamination with anthrax spores). Items that cannot be immersed should be placed in bags for transfer to a formaldehyde fumigation chamber. The best disinfectant for specimen spillages is, again, formalin; where use of formalin is impractical, hypochlorite solutions made up of 50:50 (vol/vol) methanol-water (or ethanol-water) are more effective than solutions in water alone against anthrax spores. The strength of solution chosen will depend on (i) the perceived hazard, (ii) the particular surface or article involved, (iii) the likely duration of contact of the chlorine with the surface or article, and (iv) whether, after flooding or wiping down with the hypochlorite solutions, the contaminated materials can be transferred to bins for autoclaving. Hypochlorite solutions may range from 1,000 to 75,000 ppm of available chlorine (24).

Human Anthrax

In all cases, specimens from possible sources of the infection (carcasses, hides, hair, bones, etc.) should be sought.

Cutaneous Anthrax

Swabs are appropriate for collecting vesicular exudate found in early lesions of cutaneous anthrax. In the well-formed eschar, in which vesicular exudate is absent, the edge of the eschar should be lifted up with forceps, and fluid should be obtained by application of a capillary tube under this edge. Despite its frequently angry appearance, the lesion is not painful (unless secondarily infected). Adequate material for both culture and a M'Fadyean-stained smear should be submitted.

Intestinal Anthrax

If the intestinal anthrax patient is not severely ill, a fecal specimen may be collected. If the patient is severely ill, blood should be cultured, although isolation may not be possible after antimicrobial treatment. A M'Fadyean-stained blood smear may reveal the capsulated bacilli, or if treatment has been initiated, capsule "ghosts" may be visible.

Postmortem blood collected by venipuncture (a characteristic of anthrax is nonclotting blood at death) should be examined by M'Fadyean-stained smear and culture. Any hemorrhagic fluid from nose, mouth, or anus should be cultured. If these cultures are positive, no further specimens are needed. If they are negative, specimens of peritoneal fluid, spleen, and/or mesenteric lymph nodes, aspirated by techniques that avoid spillage of fluids, may be collected for smear and culture. Histology is optional and probably academic.

Pulmonary Anthrax

If the pulmonary anthrax patient is not severely ill, immediate specimen collection is probably unnecessary, and the person should be treated and simply observed. Paired sera (when the patient is first seen and >10 days later) may be useful for confirmation of diagnosis.

For the severely ill patient, blood smear and culture should be done. Again, results will depend somewhat on previous treatment. Postmortem, the approach given for intestinal anthrax should be followed.

Animal Anthrax

Anthrax should be considered as the possible cause of death in herbivorous animals that have died suddenly and unexpectedly, particularly if hemorrhage from the nose, mouth, or anus has occurred and if death has taken place at a site with a history of anthrax (the occurrence may be several decades previous).

Carcasses 1 to 2 Days Old

Because of the nonclotting nature of blood in victims of anthrax, it is usually possible to aspirate a few drops of blood from a vein to use for (i) a M'Fadyean-stained smear and (ii) a direct plate culture on blood agar.

Pigs frequently do not develop the enormous terminal bacteremia seen in herbivores, and the capsulated bacilli may not be visible in smears of blood. When cervical edema is present, smears and cultures should be made of fluid aspirated from the enlarged mandibular and suprapharyngeal lymph nodes. In porcine intestinal anthrax, which may be obvious only at necropsy, bacilli are usually visible in stained smears made from mesenteric lymph nodes.

Older Putrefying Carcasses

B. anthracis competes poorly with decay-causing organisms and may not be seen in smears after 2 to 3 days (the killing effect of putrefaction is one of the reasons for regulations requiring that anthrax carcasses be left intact). Culture is necessary for confirmation of diagnosis.

Sections of tissue or any blood-stained material should be collected. If the animal has been opened, spleen or lymph node specimens should be taken. The best specimens for confirmation of anthrax in putrefying carcasses are often samples of soil from beneath the nose and anus; these samples may be contaminated with the terminal hemorrhagic exudate.

Other Specimens

Tests for the presence of B. anthracis may be requested for a variety of specimens such as animal products (e.g., wool, hides, hair, bonemeal) from endemic regions, soil or other materials from old burial sites or from sites of tanneries or laboratories due for redevelopment, or other environmental materials associated with outbreaks (e.g., sewage sludge). At present, culture using selective agar techniques as described below is the only approach available, and accordingly, specimens should be collected and dispatched to the appropriate laboratory.

ISOLATION PROCEDURES

Normally Sterile Specimens

Bacillus species are usually easily recoverable on blood or nutrient agar after overnight incubation.

Specimens with Mixed Microflora

In specimens submitted during food poisoning investigations or for isolation of B. anthracis from old carcasses, animal products, or environmental specimens, the Bacillus species will mostly be present as spores. Heating (solid samples should first be emulsified in sterile deionized water [1:1, wt/vol]) at 62.5°C for 15 min will both heat shock the spores and effectively destroy non-spore-forming contaminants. Alternatively, add filter-sterilized 95 to 100% ethanol to the specimen to a final concentration of 1:1 (vol/vol), and hold for 30 to 60 min at room temperature (alcohol shock). Direct plate cultures are made on blood, nutrient, or, as appropriate, selective agars by spreading up to 250-μl volumes of undiluted and 10-, 100-, and 1,000-fold dilutions of treated sample. (Concentrating the spores

by centrifuging the specimens is contraindicated and simply results in thicker lawns of contaminants.)

Enrichment procedures are generally inappropriate for isolation of Bacillus species from clinical specimens. The few exceptions include searches for B. anthracis in old specimens or for B. cereus in stools ≥3 days after a food poisoning episode. In these circumstances, enrichment can be carried out by adding nutrient or tryptic soy broth with polymyxin (100,000 U/liter) to the heat-treated specimen (ethanol treatment is not appropriate on this occasion).

The best selective agar for B. anthracis is polymyxin-lysozyme-EDTA-thallous acetate (PLET) agar (24). For environmental specimens, animal products, or specimens from old carcasses, spread 0.25-ml aliquots of undiluted and 1:10, 1:100, and 1:1,000 dilutions of a heat- or alcohol-treated suspension of the specimen across PLET plates, which are best read after about 40 h of incubation at 37°C. Circular creamy white colonies, 1 to 3 mm diameter, with a ground-glass texture are subcultured on (i) blood agar plates to test for gamma phage, penicillin susceptibility, and hemolysis and (ii) directly or subsequently in blood to look for capsule production. A flow chart for isolation and identification of B. anthracis from various categories of specimens is given in reference 24.

Several selective agars for the isolation, identification, and enumeration of B. cereus have been designed (7, 12, 18, 29). There are no selective media for other Bacillus species.

IDENTIFICATION

Catalase production and aerobic endospore formation distinguish Bacillus species from clostridia. The few catalase-negative Bacillus species are unlikely to be encountered in the clinical laboratory.

Differential characteristics of selected species of Bacillus are shown in Tables 2 and 3. B. cereus, B. mycoides, B. thuringiensis, and, to a lesser extent, B. anthracis synthesize lecithinases, forming opaque zones of precipitation around colonies on egg yolk agar; strains of B. laterosporus, B. macerans, and B. polymyxa may produce lecithinase weakly, with the reaction being visible only beneath the colony when growth is scraped away.

Several Bacillus species, particularly those in morphologic group 1 (see below), elaborate one or more hemolysins. Some species, in particular B. subtilis, B. brevis, B. firmus, B. megaterium, and B. sphaericus, are strict aerobes; others are facultative anaerobes.

Parasporal bodies are characteristic of certain Bacillus species, and differentiation of B. thuringiensis from B. cereus is largely dependent on observation of the cuboid or diamond-shaped parasporal crystals of B. thuringiensis in sporulation agar or old (>2 days) nutrient agar cultures by phase-contrast microscopy or after staining with buffalo black or malachite green.

For differentiation and identification of the many diverse members of this very large genus, Gordon et al. (9) divided the genus Bacillus into three defined groups based on spore and sporangium morphologies, as follows.

Group 1. Sporangia not swollen. Spores ellipsoidal or cylindrical, central or terminal. Gram positive. Subgroup a (large-cell group): bacillary body width, ≥1 μm; protoplasmic globules demonstrable. Subgroup B (small-cell group): width, <1 μm; protoplasmic globules not demonstrable.

TABLE 2 Characteristics used for identification of selected *Bacillus* species[a]

Morphologic group[b]	Gram reaction	Lipid globules in protoplast	Catalase production	Spores: Ellipsoidal	Spores: Spherical	Spores: Central or paracentral	Spores: Terminal or subterminal	Spores: Swelling sporangium	Motility	Anaerobic growth	V-P reaction	pH <6.0 in V-P broth	Growth at 50°C	Growth at 60°C	Egg yolk reaction	Growth in 0.001% lysozyme	Growth at pH 5.7	Growth in 7% NaCl	AS glucose	AS arabinose	AS xylose	AS mannitol	Acid + gas from AS glucose	Starch hydrolysis	Citrate utilization	Propionate utilization	Nitrate reduction	Dihydroxyacetone production	Indole production	Phenylalanine deamination	Casein decomposition	Tyrosine decomposition
Group 1																																
Subgroup A																																
B. megaterium	+	+	+	+	—	+	—	—	a^c	—	—	a^c	—	—	—	—	+	+	+	a	a	+	—	+	v^d	n	b	n	n	a^e	+	a
B. cereus "group"[,d]	+	+	+	+	—	+	—	—	v^d	+	+	+	—	—	+	+	+	+	+	—	—	—	—	+	+	n	+	n	n	—	+	v^d
Subgroup B																																
B. licheniformis	+	—	+	+	—	+	—	—	+	+	+	+^c	+	—	—	—	+	+	+	+	+	+	—	+^c	+	+	+^c	n	n	n	+	—
B. subtilis	+	—	+	+	—	+	—	—	+	—	+	a^c	a^c	—	—	b	+	+	+	+	+	+	—	—^c	+	—	+^c	n	n	n	+	—
B. pumilus	+	—	+	+	—	+	—	—	+	—	+	+	a^c	—	—	a	+	+	+	+	+	+	—	+	+	—	—^c	n	n	n	+	—
B. firmus	+	—	+	+	—	v	v	—	a	—	—	—	—	—	—	—	—	—	+	b	b	b	—	+	—	—	+^c	n	n	n	+	b
B. coagulans	+	—	+	+	—	v	v	v	+	+	+^c	+	+	b^c	—	—	+	—	+	a	a	b	—	+	b	—	b	n	n	n	b	—
Group 2																																
B. polymyxa	v	—	+	+	—	v	v	+	+	+	+	a^c	—	—	—	a	+	—	+	+	+	+	+	+	—	n	+	+	—	n	+	—
B. macerans	v	—	+	+	—	—	+	+	+	+	—	+^c	+	—	—	—	+	—	+	+	+	+	+	+	b	n	+	—	—	n	—	—
B. circulans	v	—	a^c	+	—	v	v	+	a	a	—	+	b^c	—	—	b	b	b	+	+	+	+	—	+	b	n	b	—	—	n	b	—
B. stearothermophilus	v	—	+	+	—	—	+	+	+	—	—	+	+	+	—	—	—	—	+	b	b	b	—	+	—	n	a	—	—	n	a	—
B. alvei	v	—	+	+	—	v	v	+	+	+	—	+^c	—	—	—	+	+	—	+	—	—	—	—	—	—	n	—	+	+	n	+	b
B. laterosporus[e]	v	—	+	+	—	+	—	+	+	+	—	—	—^c	—	(+)	+	—	—	+	—	—	+	—	—	—	n	+	—	a	—	+	+
B. brevis	v	—	+	+	—	v	v	+	+	—	—	—	a^c	b^c	—	b	b	—	+	—	—	a	—	—	b	n	a	—	—	—	+	+
Group 3																																
B. sphaericus	v	—	+	—	+	—	+	+	+	+	—	—	—	—	—	a	b	b	—	—	—	—	—	—	b	n	—	—	—	+	a	—

[a] V-P, Voges-Proskauer; AS, ammonium salt; +, ≥85% of strains positive; a, 50 to 84% of strains positive; b, 15 to 49% of strains positive; —, <15% of strains positive; v, inconstant; n, not applicable; (+), under colony, which needs to be scraped off for reaction to be seen.

[b] From Gordon et al. (9).

[c] Different from the entry given in the version of this table in the fifth edition of this volume (28) as a result of errors that subsequently came to our attention.

[d] See Table 3.

[e] Spore in sporangium has characteristic canoe shape.

TABLE 3 Distinguishing characteristics of *B. cereus* and closely related species[a]

Species	Colony on blood agar	Motility	Hemolysis[b]	Lysis by gamma phage	Penicillin susceptibility	Crystalline parasporal inclusion	Citrate utilization	Growth in 7% NaCl	Tyrosine decomposition	Possession of plasmid pX01 and elaboration of anthrax toxin	Possession of plasmid pX02 and elaboration of capsule	Virulent in guinea pigs and mice with M'Fadyean reaction on blood smear at death
B. cereus	Slight green tinge	+	+	−	−	−	+[c]	+[c]	+[c]	−	−	−
B. anthracis												
Virulent	Gray-white to white; generally smaller than *B. cereus* colony	−	−	+	+	−	b	+[c]	−[c]	+	+	+
Avirulent	Like virulent *B. anthracis* colony	−	−	+	+	−	b	+	−	+/−[d]	−[d]	−
B. thuringiensis	Like *B. cereus* colony	+	+	−	−	+	+[c]	+[c]	+[c]	−	−	−
B. cereus subsp. *mycoides*	Spreading rhizoid colonies with marked tailing	−	(+)	−	−	−	a	a	a	−	−	−

[a]+, ≥85% of strains positive; a, 50 to 84% of strains positive; b, 15 to 49% of strains positive; −, <15% of strains positive; (+), usually weak positive.
[b]On sheep or horse blood agar at 24 h.
[c]From Gordon et al. (9).
[d]Naturally occurring pX01[+] pX02[−] isolates have been found (26). Naturally occurring pX01[−] pX02[−] isolates that fulfill all the other criteria of *B. anthracis* also occur naturally (26). pX01[−] pX02[+] strains have been created in the laboratory but have not been observed in the naturally occurring state.

Group 2. Sporangia swollen by ellipsoidal spores, central or terminal. Gram variable.

Group 3. Sporangia swollen. Spores spherical, subterminal, or terminal. Gram variable.

The predominant species within these groups are shown in Table 2. *Bacillus* species that did not appear to belong to any of these groups were placed in a fourth group termed unassigned strains. *B. anthracis*, *B. cereus*, *B. mycoides*, and *B. thuringiensis* are closely related; basic distinguishing features for their differentiation in the laboratory are given in Table 3.

It is generally easy to distinguish virulent *B. anthracis* from other members of the *B. cereus* group. An isolate with the correct colonial morphology, white or gray, nonhemolytic or only weakly hemolytic, nonmotile, gamma phage and penicillin susceptible, and able to produce the characteristic capsule is *B. anthracis*. Isolates showing all these characteristics except the ability to produce capsule may be avirulent forms of *B. anthracis* lacking capsule and/or toxin genes (26) and should be referred to a specialist center. Such isolates are frequently identified as *B. cereus* and discarded. Recent molecular studies have revealed unique genetic profiles among *B. anthracis* strains, justifying the view that *B. anthracis* is fully distinguishable from *B. cereus* and *B. thuringiensis* (11).

The capsule of virulent *B. anthracis* can be demonstrated on nutrient agar containing 0.7% sodium bicarbonate incubated overnight under CO_2 (candle jars perform well). Colonies of the capsulated *B. anthracis* appear mucoid, and the capsule can be visualized by staining smears with M'Fadyean's polychrome methylene blue or India ink (24). Alternatively, 2.5 ml of blood (defibrinated horse blood seems best; fetal calf serum seems quite good) can be inoc-

ulated with a pinhead quantity of growth from the initial suspect colony, incubated statically for 6 to 18 h at 37°C, and observed by M'Fadyean stain for capsules. If doubt persists, inoculation of a light suspension of the culture into a mouse or guinea pig (the guinea pig is more susceptible) may be necessary to confirm identity. A M'Fadyean-stained blood smear made at death (about 48 h after inoculation) will reveal large numbers of the typical large, capsulated, square-end rods mostly in chains.

The pX01 and pX02 plasmids carrying the genes that code for the toxin and the capsule, respectively, and the genes themselves can be detected by using molecular techniques. However, these tests are still confined to the specialist laboratory. The avirulent *B. anthracis* isolates lacking one or both plasmids are generally found in environmental samples (26); however, one such isolate was recently recovered from the blood of a child with pyrexia.

A commercially available system for *Bacillus* identification that is based on the API 50CH kit has yet to be subjected to detailed independent assessment and therefore cannot be recommended at present as a reliable, stand-alone approach. Other enzymatic and molecular methods for identification of *Bacillus* species are still in development.

SEROLOGIC TESTS

Species and Strain Differentiation

Attempts at developing simple serologic differentiation systems for *Bacillus* species over the past 4 decades were always plagued by problems of cross-reacting antigens and, in the case of spores, hydrophobic surface properties leading to autoagglutination. Monoclonal antibody tests based on specific epitopes in the spore cortex or cell wall of *B. anthracis* that might reliably distinguish *B. anthracis* from other closely related *Bacillus* species are not yet available. Similarly, *B. anthracis* strain differentiation is proving particularly difficult. Restriction fragment length polymorphism and PCR fingerprinting show promise (11).

Strain differentiation systems for *B. cereus* based on flagellar (H) antigens are available for investigations of food poisoning outbreaks or other *B. cereus*-associated clinical problems (15).

B. thuringiensis strains are classified on the basis of their flagellar antigens; 14 serotypes have been recognized, with in some cases further subdivisions based on H-antigen subfactors (6).

Toxin and Antitoxin Detection

The three protein components of anthrax toxin (protective antigen, lethal factor, and edema factor) and antibodies to them are utilizable in enzyme immunoassay systems (16). For routine confirmation of anthrax infection or for monitoring response to anthrax vaccines, protective antigen alone appears to be satisfactory (25). Serology has proved useful for epidemiologic investigations in humans and animals, but for human anthrax, early treatment sometimes prevents antibody development.

The enterotoxin complex responsible for the diarrheal type of *B. cereus* food poisoning has been partially purified and characterized (15, 23). It is the basis of at least two commercial systems undergoing evaluation for detection of the toxin in foods and feces (3, 10). An alternative to animal tests for the emetic toxin of *B. cereus* is based on the semiquantitative vacuolation response of cultured HEp-2 cells when they are exposed to the toxin in food extracts or culture filtrates (13).

ANTIMICROBIAL SUSCEPTIBILITIES

B. anthracis is almost invariably susceptible to penicillin; only three resistant isolates have been reported (17). It is susceptible to gentamicin, erythromycin, and chloramphenicol, and recent tests have shown infection to respond to ciprofloxacin and doxycycline (8); the organism is normally susceptible to streptomycin but resistant to cefuroxime. While *B. anthracis* shows in vitro susceptibility to tetracycline, veterinary experience suggests that this agent may not show clinical efficacy. Fuller discussions on the treatment of anthrax are given elsewhere (24).

B. cereus and *B. thuringiensis* produce a broad-spectrum β-lactamase and are thus resistant to penicillin, ampicillin, and cephalosporins. They are also resistant to trimethoprim, almost always susceptible to clindamycin, erythromycin, chloramphenicol, vancomycin, and the aminoglycosides, and usually susceptible to tetracycline and the sulfonamides (1, 7, 27).

INTERPRETING SIGNIFICANCE OF *BACILLUS* ISOLATIONS

Apart from *B. anthracis*, the majority of *Bacillus* species are common environmental contaminants. For this reason, the isolation of a *Bacillus* species from a single clinical specimen is generally not a sufficient basis for incriminating the organism as the etiologic agent responsible for the condition being investigated.

Low-level contamination of foodstuffs by *Bacillus* species

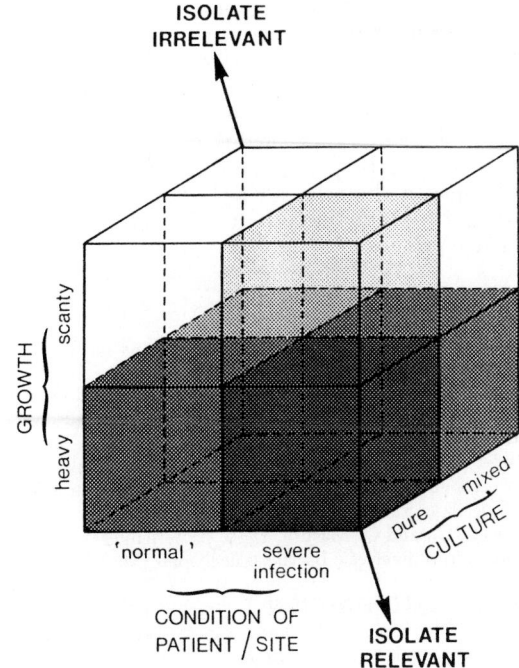

FIGURE 1 Diagrammatic representation of considerations necessary for assessing the relevance of a *Bacillus* isolate to an infection (reproduced from reference 27 with permission). In addition to the factors shown, repeated isolation of the same strain increases confidence in its significance.

is common, as is asymptomatic transient fecal carriage throughout human and animal populations. Therefore, in food poisoning investigations, qualitative isolation tests are insufficient. The ideal criteria for proving that a *Bacillus* species is the etiologic agent responsible are the isolation of significant numbers ($>10^5$ CFU/g) of the *Bacillus* species from the epidemiologically incriminated food (and in the case of suspected *B. cereus* food poisoning, detection of emetic toxin and/or enterotoxin) and the recovery of the same strain (biotype, serotype, phage type, plasmid type, etc.) in significant numbers from acute-phase specimens (feces or vomitus) from the patients.

In practice, it is rare that these ideal criteria can be met. A complete set of the appropriate food and clinical specimens is seldom available, and the epidemiologic aspects of the outbreak, such as incubation times, clinical symptoms, types of food implicated, time lapse and manner in which the food was stored between the episode and specimen collection, must all be considered along with the laboratory findings in formulating conclusions as to etiology.

Assessment of relevance in nongastrointestinal conditions is discussed elsewhere (27) and is graphically illustrated in Fig. 1. In general, greater significance should be attached to repeated isolation of the same organism, particularly from blood cultures and when found in large numbers and pure culture. Histopathology and specialist tests on the toxigenicity of the isolate may also help in the decision as to its relevance.

B. anthracis continues to be generally regarded as an obligate pathogen; its continued existence in the ecosystem appears to depend on a periodic multiplication phase within an animal host, and its environmental presence reflects contamination from an animal source at some time in the past rather than self-maintenance within the environment. In human or animal specimens, it is usually looked for only when the history of the case indicates that it is reasonable to suspect anthrax. It is rarely challenged that demonstration of capsulated *B. anthracis*, even in low numbers, from such specimens supplies confirmation of clinical suspicion; this belief is based on the knowledge that after death of the host, the bacterium is rapidly destroyed by putrefactive processes.

REFERENCES

1. **Anonymous.** 1983. *Bacillus cereus* as a systemic pathogen. *Lancet* **ii:**1469.
2. **Brachman, P. S., A. F. Kaufmann, and F. G. Dalldorf.** 1966. Industrial inhalation anthrax. *Bacteriol. Rev.* **30:**646–657.
3. **Buchanan, R. L., and F. J. Schultz.** 1992. Evaluation of the Oxoid BCET-RPLA kit for the detection of *Bacillus cereus* diarrheal enterotoxin as compared to cell culture cytotonicity. *J. Food Prot.* **55:**440–443.
4. **Chastel, C., and O. Masure.** 1986. Bactériémies, septicémies et infections diverses è *Bacillus licheniformis*. *Med. Mal. Infect.* **4:**226.
5. **Claus, D., and D. Fritze.** 1989. Taxonomy of *Bacillus*, p. 5–26. *In* C. R. Harwood (ed.), *Bacillus*. Biotechnology handbooks no. 2. Plenum Press, New York.
6. **de Barjac, H.** 1981. Insect pathogens in the genus *Bacillus*, p. 241–250. *In* R. C. W. Berkeley and M. Goodfellow (ed.), *The Aerobic Endospore-Forming Bacteria: Classification and Identification*. Academic Press Ltd., London.
7. **Drobniewski, F. A.** 1993. *Bacillus cereus* and related species. *Clin. Microbiol. Rev.* **6:**324–338.
8. **Friedlander, A. M., S. L. Welkos, M. L. M. Pitt, J. W. Ezzell, P. W. Worsham, B. E. Ivins, J. R. Lowe, G. B. Howe, P. Mikesell, and W. B. Lawrence.** 1993. Postexposure prophylaxis against experimental inhalation anthrax. *J. Infect. Dis.* **167:**1239–1242.
9. **Gordon, R. E., W. C. Haynes, and C. H.-N. Pang.** 1973. *The Genus Bacillus*. U.S. Department of Agriculture agricultural handbook no. 427. U.S. Department of Agriculture, Washington, D.C.
10. **Granum, P. E., S. Brynestad, and J. M. Kramer.** 1993. Analysis of enterotoxin production by *Bacillus cereus* from dairy products, food poisoning incidents and non-gastrointestinal infections. *Int. J. Food Microbiol.* **17:**269–279.
11. **Henderson, I., C. J. Duggleby, and P. C. B. Turnbull.** 1994. Differentiation of *Bacillus anthracis* from other *Bacillus cereus* group bacteria with the PCR. *Int. J. Syst. Bacteriol.* **44:**99–105.
12. **Holbrook, R., and J. M. Anderson.** 1980. An improved selective and diagnostic medium for the isolation and enumeration of *Bacillus cereus* in foods. *Can. J. Microbiol.* **263:**753–759.
13. **Hughes, S., B. Bartholomew, J. C. Hardy, and J. M. Kramer.** 1988. Potential application of a HEp-2 cell assay in the investigation of *Bacillus cereus* emetic-syndrome food poisoning. *FEMS Microbiol. Lett.* **52:**7–12.
14. **Hugh-Jones, M.** 1989. Global trends in the incidence of anthrax in livestock. Proceedings of the International Workshop on Anthrax. *Salisbury Med. Bull.* **68**(Suppl.):2–4.
15. **Kramer, J. M., and R. J. Gilbert.** 1989. *Bacillus cereus* and other *Bacillus* species, p. 21–70. *In* M. P. Doyle (ed.), *Foodborne Bacterial Pathogens*. Marcel Dekker, Inc., New York.
16. **Leppla, S. H.** 1991. The anthrax toxin complex, p. 277–302. *In* J. E. Alouf and J. H. Freer (ed.), *Sourcebook of Bacterial Protein Toxins*. Academic Press, Inc., New York.
17. **Lightfoot, N. F., R. J. D. Scott, and P. C. B. Turnbull.** 1989. Antimicrobial susceptibility of *Bacillus anthracis*. Proceedings of the International Workshop on Anthrax. *Salisbury Med. Bull.* **68**(Suppl.):95–98.
18. **Mossel, D. A. A., M. J. Koopman, and E. Jongerius.** 1967. Enumeration of *Bacillus cereus* in foods. *Appl. Microbiol.* **15:**650–653.
19. **Norris, J. R., R. C. W. Berkeley, N. A. Logan, and A. G. O'Donnell.** 1981. The genera *Bacillus* and *Sporolactobacillus*, p. 1711–1742. *In* M. P. Starr, H. Stolp, H. G. Truper, A. Balows, and H. G. Schlegel (ed.), *The Prokaryotes*. *A Handbook on Habitats, Isolation, and Identification of Bacteria*, vol. 2. Springer-Verlag, New York.
20. **Parry, J. M., P. C. B. Turnbull, and J. R. Gibson.** 1983. *A Colour Atlas of Bacillus Species*. Wolfe medical atlas no. 19. Wolfe Medical Publications, London.
21. **Pike, R. M., S. E. Sulkin, and M. L. Schulze.** 1965. Continuing importance of laboratory-acquired infections. *Am. J. Public Health* **55:**190–199.
22. **Priest, F. G.** 1989. Isolation and identification of aerobic endospore-forming bacteria, p. 27–56. *In* C. R. Harwood (ed.), *Bacillus*. Biotechnology handbooks no. 2. Plenum Press, New York.
23. **Turnbull, P. C. B.** 1986. *Bacillus cereus* toxins, p. 397–448. *In* F. Dorner and J. Drews (ed.), *International Encyclopedia of Pharmacology and Therapeutics*, section 119, *Pharmacology of Bacterial Toxins*. Pergamon Press, Oxford.
24. **Turnbull, P. C. B., R. Bohm, H. G. B. Chizyuka, T. Fujikura, M. E. Hugh-Jones, and J. Melling.** 1993. *Guidelines for the Surveillance and Control of Anthrax in Humans and Animals*. WHO/Zoon./93.170. World Health Organization, Geneva.
25. **Turnbull, P. C. B., M. Doganay, P. M. Lindeque, B. Aygen, and J. McLaughlin.** 1992. Serology and anthrax in humans, livestock and Etosha National Park wildlife. *Epidemiol. Infect.* **108:**299–313.
26. **Turnbull, P. C. B., R. A. Hutson, M. J. Ward, M. N. Jones, C. P. Quinn, N. J. Finnie, C. J. Duggleby, J. M. Kramer, and J. Melling.** 1992. *Bacillus anthracis* but not always anthrax. *J. Appl. Bacteriol.* **72:**21–28.
27. **Turnbull, P. C. B., and J. M. Kramer.** 1983. Non-gastrointestinal *Bacillus cereus* infections: an analysis of exotoxin production by strains isolated over a two-year period. *J. Clin. Pathol.* **36:**1091–1096.
28. **Turnbull, P. C. B., and J. M. Kramer.** 1991. *Bacillus*, p. 296–303. *In* A. Balows, W. J. Hausler, Jr., K. L. Herrmann, H. D. Isenberg, and H. J. Shadomy (ed.), *Manual of Clinical Microbiology*, 5th ed. American Society for Microbiology, Washington, D.C.
29. **van Netten, P., and J. M. Kramer.** 1992. Media for the detection and enumeration of *Bacillus cereus* in foods: a review. *Int. J. Food Microbiol.* **17:**85–99.

Corynebacterium and Miscellaneous Irregular Gram-Positive Rods, *Erysipelothrix*, and *Gardnerella*

JILL E. CLARRIDGE AND CAROL A. SPIEGEL

29

CORYNEBACTERIUM SPP. AND OTHER CORYNEFORM ORGANISMS

General Taxonomy

Recent DNA and 16S rRNA relatedness studies have made profound changes in our concepts of the gram-positive rods. As determined by rRNA sequencing, there are two major taxonomic branches of gram-positive organisms. The actinomycete group, in which most of the genera constituting the coryneform organisms are found, is characterized by a high guanine-plus-cytosine (G+C) base content of the DNA and an irregular cell shape. This group includes the genera *Corynebacterium*, *Actinomyces*, *Arcanobacterium*, *Brevibacterium*, *Mycobacterium*, and others (Table 1). The other group, containing the clinically important genera *Staphylococcus*, *Streptococcus*, *Clostridium*, *Bacillus*, *Listeria*, *Lactobacillus*, and *Erysipelothrix*, is characterized by a lower G+C content and a more regular morphology. It is important to realize that groups of rods (e.g., *Bacillus*, *Lactobacillus*, and *Corynebacterium* spp.) can be more closely related to some cocci (*Staphylococcus*, *Streptococcus*, and *Micrococcus* spp., respectively) than to each other. In this chapter, because *Erysipelothrix* spp. are more distantly related and *Gardnerella* spp. are only uncertainly related to the coryneform organisms, they will be discussed separately. Within the high-G+C group, genera that include clinically encountered organisms that may initially be difficult to distinguish from *Corynebacterium* spp. by growth or morphologic characteristics are *Mycobacterium* (rapidly growing), *Nocardia*, *Rhodococcus*, *Arcanobacterium*, *Actinomyces*, *Aureobacterium*, *Brevibacterium*, *Cellulomonas*, *Dermabacter*, *Microbacterium*, *Oerskovia*, *Rothia*, *Turicella*, and aerotolerant strains of *Propionibacterium* and *Bifidobacterium*. There are, in addition, clinically isolated strains that do not fit into any presently defined genera. Although a great variety of organisms has been included in the coryneform group, these organisms can be systematically separated and classified. The term coryneform in itself is misleading, because many of these organisms are not club shaped. However, for convenience, the term will be used instead of the more accurate description "irregular gram-positive rod," which is used in *Bergey's Manual* (47, 51). Major characteristics that are important in distinguishing relatedness of these gram-positive rods include their physiochemical compositions (cell wall constituents and cellular fatty acid [CFA] composition), bio-

chemical and metabolic characteristics (e.g., the production of catalase or the production of metabolic fatty acids), motility, atmospheric conditions required for growth, pigment production, and colonial and cellular morphologies. Some basic characteristics underlying the classification of gram-positive rods are shown in Table 1.

Because of their occasional morphologic similarity, the low-G+C genera, i.e., *Listeria*, *Erysipelothrix*, and some strains of *Lactobacillus*, *Bacillus*, and *Clostridium*, may need to be considered when one is confronted with a gram-positive rod, even though these genera are not closely related to the genus *Corynebacterium*. A *Kurthia* sp. is a rare clinical isolate and is taxonomically more related to *Bacillus* spp.; its cellular and colonial morphologies would make it more confusable with the *Bacillus cereus* group of organisms than with the coryneforms. *Listeria*, *Lactobacillus*, *Bacillus*, and *Clostridium* species are discussed in other chapters of this Manual.

Within the high-G+C group are more than 20 clinically isolated genera. Most isolates of *Mycobacterium*, *Nocardia*, and *Rhodococcus* spp. are easily distinguished from *Corynebacterium* spp. (even though the genera are closely related) by means of acid-fast staining or evidence of branching. These organisms are discussed in chapters 30 and 31 of this Manual. The strains of *Bifidobacterium*, *Actinomyces*, and *Propionibacterium* that are strict anaerobes can also be excluded from the *Corynebacterium* group and are discussed in chapter 48; however, one must be aware that there are aerotolerant or facultative strains in these three genera. Major distinguishing taxonomic characteristics of the genera discussed in this chapter are given in Table 1; some of these are new groups just now being described (21, 32–34, 71, 100).

The taxonomy of the genus *Corynebacterium* is undergoing tremendous changes at this time. Some culture collection strains have been found to consist of many different species (21, 71). Even the type strains are often not uniformly described (21, 22, 46a, 51). Thirteen species of *Corynebacterium*, excluding the plant corynebacteria, are listed in *Bergey's Manual of Determinative Bacteriology* (18, 47). There are also five newly published named species: *Corynebacterium urealyticum* (formerly CDC group D2), *C. afermentans* (CDC group ANF1), *C. propinquum* (CDC group ANF3), *C. accolans* (CDC group 6), and *C. amycolatum* (possibly synonymous with CDC groups F2 and I2) in

TABLE 1 Selected taxonomic characteristics determining grouping of gram-positive rods[a]

Group	Group CFA characteristics	Cell wall	Representative organism	Catalase	Motility	Glucose metabolism	Habitat	Comment
Irregular gram-positive rods (high G+C)	Branched	L-Lysine	*Rothia*	+	−	F	Humans, oral cavity	
		DAB	"*C. aquaticum*"	+	+/V	F	Environment	
		L-Ornithine	*Cellulomonas*	+	V	F	Soil, waste	
		L-Lysine	*Microbacterium*	+	V	F/NF	Dairy, sewage	
			Oerskovia, CDC group A	+	+/V	F	Soil, clinical	
			Arthrobacter	+	−	NF	Soil	
		meso-DAP	*Brevibacterium* (CDC group B)	+	−	NF	Skin, food	
			Dermabacter (CDC groups 3 and 5)	+	−	F	Skin	
	Straight, unsaturated	meso-DAP and l-Lysine	*Propionibacterium*	−/+	−	F	Humans, other animals	No mycolic acids
		meso-DAP, some with tuberculosteric and mycolic acids	*Turicella*	+	−	NF	Humans	
			Corynebacterium	+	−	F/NF	Humans, animals, environment	Some do not have mycolic acids
			Rhodococcus	+	−	NF	Soil, humans, horses	Weakly acid fast
			Nocardia	+	−	NF	Soil, humans, animal	Weakly acid fast
			Mycobacterium	+	−	NF	Environment, humans	Acid fast
		L-Lysine	*Actinomyces*, *Arcanobacterium*	−/+	−	F	Humans, other mammals	
			Bifidobacterium, CDC group E	−	−	F	Humans	
Regular gram-positive rods (low G+C)	Branched		*Kurthia*	+	+	NF	Food, environment	Resembles *Bacillus* sp.
			Bacillus	+	+	F/NF	Ubiquitous	
			Listeria	+	+	F	Food, environment, humans	
	Straight		*Lactobacillus*	−	−/+	F	Humans, animal food	
			Erysipelothrix	−	−	F	Animals (turkeys, swine) and humans	
Unassigned	Straight	Lysine	*Gardnerella*	−	−	F	Human genitourinary tract	

[a]Data are based on data in *Bergey's Manual* (18, 47, 51) except that CFA data are from references 8 and 92. Abbreviations: V, variable; F, fermentative; NF, nonfermentative; DAB, diaminobutyric acid.

addition to CDC groups such as F1, G1, G2, and I1. Some species (C. diphtheriae and C. afermentans) have been divided into subspecies or types, and some (C. xerosis, CDC groups F1 and G1, and C. jeikeium) have been found to consist of multiple DNA homology groups; in Table 2, these are listed as DNA groups for C. xerosis and as lipid-dependent (LD) groups for organisms that overlap with CDC groups F1, G1, and G2. Though the taxonomic positions of C. genitalium and C. pseudogenitalium are not clear, it is certain that at least some strains are synonymous with C. jeikeium and CDC group F1, respectively. With further work, some of these strains may be found to be separate species, and some may fit in with preexisting species. The type strain, C. diphtheriae, was first isolated and described in 1880.

Description of the Genus *Corynebacterium* and Other Coryneform Genera

The genus Corynebacterium is composed of organisms that are gram-positive tapered or slightly curved rods. Usually, the ends are rounded, and the sides are not parallel; instead, either the middle or the end is slightly wider, making the organisms cocoon or club shaped ("coryne" means club in Greek). The cells can also be barred or segmented. Cells are arranged singly, in palisades of parallel cells, or in pairs that have not fully separated and form V's; together, groups of these organisms may take on the appearance of Chinese letters. They are not acid fast and do not branch. The cell wall contains meso-diaminopimelic acid (meso-DAP) and most have short-chain (22- to 38-carbon) mycolic acids, called corynomycolic acids; a few species (e.g., C. amycolatum) do not possess corynomycolic acids. Some corynebacteria contain tuberculostearic acids. The G+C content is broad (between 51 and 71%), reflecting the genetic diversity. The CFAs are of the straight and unsaturated types. Corynebacteria in general are catalase positive and nonmotile and do not hydrolyze esculin or gelatin. Most of the clinically isolated strains are negative for utilization of lactose, xylose, and mannitol.

Species without these characteristics that have previously been included in the genus Corynebacterium or the coryneform group but have subsequently been reclassified to other genera include Arcanobacterium haemolyticum, Actinomyces pyogenes, Actinomyces neuii (CDC group 1), Oerskovia spp. (CDC groups A1 and A2), "C. aquaticum" (has not yet been renamed), Dermabacter hominis (CDC groups 3 and 5), and Brevibacterium spp. (CDC groups B1 and B3). CDC groups 2, 4, and 7 are known not to be Corynebacterium spp. because of their fatty acid and cell wall characteristics.

Many of the organisms classified as irregular gram-positive rods (there are over 36 genera) may have been called related coryneforms or even classified as Corynebacterium spp. in the past. They are very diverse not only morphologically but also metabolically and structurally. Some, like Oerskovia, Brevibacterium, and Arcanobacterium spp., are distinguished by being pleomorphic in the sense that they have different forms (e.g., long and/or branching filaments and coccoid or short bacilli) at different stages of their growth cycles or under different conditions of growth, but at a particular time, the individual cells are often quite uniform. Other genera, like Turicella and Bifidobacterium, are pleomorphic in the sense that individual bacterial cells from a given colony are different from one another.

These genera and species can be distinguished by their cell wall and fatty acid constituents as well as by the characteristics listed in Tables 1 and 3 (47, 51). Figure 1 presents easily observable characteristics that help differentiate organisms in these genera from Corynebacterium spp.

Natural Habitats

C. diphtheriae is most often isolated from the nasopharynx or skin lesions of patients with diphtheria. C. diphtheriae can survive for up to 6 months in dust and fomites in the environment surrounding patients who have nasopharyngeal and cutaneous diphtheria (43). C. diphtheriae can also be carried on the mucous membranes and skin of healthy humans, particularly if it is a nontoxigenic strain. Other Corynebacterium spp. are found on the skin and mucous membranes and in the gastrointestinal tracts of humans, mammals, and some other animals, often occupying a specific niche, as shown in Table 4.

The habitats of other coryneform bacteria are diverse. The genera Cellulomonas, Microbacterium, and Curtobacterium as well as the newly described genera Brachybacterium and Pimelobacter are associated with animals, soil, and food, being generally not pathogenic and only rarely isolated in the clinical laboratory. Others, such as Brevibacterium, Dermabacter, and Actinomyces spp., are commonly isolated from skin and other sites but rarely as a pathogen (2). Arcanobacterium and Rothia spp. are found in the oral cavity, while Turicella spp. are associated with the ear, and Oerskovia, "C. aquaticum," and CDC group A are associated with the environment; these organisms are isolated occasionally in the clinical laboratory, but their natural and pathogenic significances have yet to be defined (Table 4).

Clinical Significance

The most common disease caused by C. diphtheriae is diphtheria, an acute communicable disease manifested by both local infection of the upper respiratory tract and the systemic effects of a toxin, which are most notable in the heart and peripheral nerves. The signs and symptoms of the disease are a pharyngeal membrane, sore throat, dysphagia, malaise, headache, and nausea. Death can result from respiratory obstruction by the membrane or myocarditis caused by the toxin. Since active immunization became available in the 1920s, the number of reported cases of diphtheria in the United States has decreased from about 60,000 cases in 1932 to only a few per year at present. However, there is an occasional outbreak in nonimmunized or partially immunized persons. C. diphtheriae is pathogenic primarily by production of a 62,000-Da exotoxin, the gene for which is carried on a bacteriophage. If a strain is not infected with the bacteriophage and thus does not have the tox gene, it is called nontoxigenic C. diphtheriae.

C. diphtheriae also causes cutaneous diphtheria. During the 1970s, there were several outbreaks of this disease in the northwestern United States, particularly among an indigent alcoholic population living in substandard conditions (43). Cutaneous diphtheria is reportedly more common in tropical and subtropical areas. The infection occurs in previously traumatized skin, often with concomitant isolation of Streptococcus pyogenes. The lesions appear necrotic and black, with occasional formation of a membrane. Although the organisms are present in high numbers, they can be confused with other species of Corynebacterium that are normally found on the skin. It is interesting that both the toxigenic and the nontoxigenic types of C. diphtheriae were isolated from successive waves of the epidemic in Seattle. There are increasing numbers of reports of both toxigenic and nontoxigenic C. diphtheriae causing endocarditis with-

TABLE 2 Characteristics of *Corynebacterium* spp.[a]

Organism	Nitrate	Urease	Pyrazinamidase	AP	Glucose	Ribose	Maltose	Sucrose	Glycogen	Requires serum	Fermentation	Comment or API Coryne code
C. diphtheriae subsp. gravis	+	−	−	−	+	+	+	−	+	−	F	
C. diphtheriae subsp. mitis	+	−	−	−	+	+	+	−	−	−	F	
C. diphtheriae subsp. intermedius	+	−	−	−	+	(+)V	+	−	−	+	F	
C. diphtheriae subsp. belfanti	−	+	−	−	+	+	+	−	+	−	F	
"C. ulcerans"	−	+	−	V/−	+	+	+	−	−	−	F	
C. pseudotuberculosis	−	+	−	−	+	+	V	−	−	−	F	
C. striatum	+	−	+	+	+	+/−	−	+	−	−	F	3100105
C. xerosis ATCC 373 T	+	−	+	+	+	+	+	+	−	−	F	3110325
"C. xerosis" (DNA group C)	−	−	+	+	+	+	+	+		−	F	2110325
"C. xerosis" (DNA group D)	+	+	+	+	+	+	+	+		+	F	5100305
"C. xerosis" (DNA group E)	+	−	+	+	+	+	−	+		+	F	3001325
"C. xerosis" (DNA group F)	−	−	+	+	+	+	+	+		−	F	2100325
C. minutissimum	−	−	−	+	+	V	+	V		−	F	
C. renale	−	+	−	−	+	−	−	V		−/+	F	
C. cystitidis	+	+	−	−	+	−	−	−		−	F	Xylose+
C. pilosum	+	V/−	V	+	V	−	−	−		−	F	
C. bovis	−	+	+	−	+	−	−	−		+/−	NF	ONPG+
C. kutscheri	+	−	−	+	+	−	+	+		−	F	Esculin+
C. matruchotii	−	+	+ (B)	ND	V	ND	V	V	ND	−	NF	
C. pseudodiphtheriticum	−	+	+	V	−	−	−	−		−	NF	
C. urealyticum	−	+	+	+	−	−	−	−		+	NF	
C. jeikeium	−	−	+	−	+	+	−	−		+	NF	Fructose−
CDC group F1 (LD5-LD6)	V	+	+	−	+	V	+	+		+	F	The two LD groups are separated by DNA homology
CDC group G1 (LD4)	+	−	−	+	+	+	−	+		+	F	Light beige
LD3 new species (some G1)	+	−	V	+	+	+	−	V		+	F	Light beige
CDC group G2 (LD2)	−	−	V	+	+	+	V	+V		+	F	
CDC group I1	+	−	+	−	+	+/−	−	−		−	NF	3100104
C. afermentans (group ANF1)	−	−	V	V	−	−	−	−		+/−	NF	Two subsp.: lipophilum and afermentans
C. propinquum (CDC group ANF3)	+	−	V	V	−	−	−	−		−	NF	
C. accolans (CDC group 6)	+	−	−	+	+	−	−	−		−	F	
"C. asperum"	−	−/+	+	+	+	−	+	−		−	F	2101/0324
C. amycolatum	−	+	−	+	+	+	+	−		−	F	2101324
CDC group F2	V	+	+	−	+	+	+	−		−	F	3/2101324
CDC group I2	+	−	+	+	+	+	+	−		−	F	3100324

[a]Data are based on CDC charts (45, 46a) and updated literature as described in the text. All isolates are catalase positive, have a straight fatty acid type, and are negative for gelatin, esculin, lactose, xylose, and mannitol except as noted. Abbreviations: AP, alkaline phosphatase; LD, lipid-dependent species; V, variable; F, fermentative; NF, nonfermentative; ND, no data.

TABLE 3 Characteristics of other coryneform groups[a]

Test organism	Motility	Catalase	CFA type	Fermentation	Nitrate	Urease	Esculin	Gelatin	Glucose	Xylose	Mannose	Maltose	Lactose	Sucrose	Hemolysis	Pigment	Comment	API Coryne code or comment
Arcanobacterium haemolyticum	–	–	S	F	–	–	–	–	+	–	–	+	+	+/V	β		Reverse CAMP+	
Actinomyces pyogenes	–	–	S	F	–	–	–	+	+	+	–	+	+	V	β			
Actinomyces viscosus	–	+	S	F	+	–/V	V	–/V	+	–/V	+	+	+	+	–	Red	Red after 1 wk on blood agar	341073/75
Actinomyces odontolyticus	–	–	S	F	+	+	V	–	+	V	–	V	V	+	–			25/410745
Actinomyces naeslundi	–	–	S	F	V	+	+	–	+	+	+	+	+	+	–		CAMP+	
Actinomyces neuii subsp. *neuii* (CDC group 1)	–	+	S	F	+	–	–	–	+	+	+	+	+	+	–/α		CAMP+	
Actinomyces neuii subsp. *anitratus* (CDC group 1-like)	–	+	S	F	–	–	–	–	+	+	+	+	+	+				
CDC group 2	–	–	S	F	–	–	–	–	+	–	–	+	–	–			Needs serum	Club shaped
Turicella otitidis	–	+	S	NF	–	–	–		–	–	–	–	–	–	–		CAMP+	
Bifidobacterium adolescentis-group	–	–	S	F	–	–	–	–	+	+	–	+	+	+			Needs serum	
Erysipelothrix spp.	–	–	S	F	+/–	–	–	+	+/v	–	–	–	+	–	α	White or yellow	H$_2$S+	
Brevibacterium spp.- CDC group B	–	+	B	NF	–	–	–	+	–	–	–	–	–	–		White or yellow	*B. linens* yellow	DNase+
Brevibacterium casei	–	+	B	NF	–	–	+	+	–	–	–	–	–	–		White or yellow	Cheese odor	DNase+
"*Brevibacterium mcbrellneri*"	–	+	B	NF	–	–	–	+	–	–	–	–	–	–		White or tan	Body odor	
Exiguobacterium (*Brevibacterium*) *acetylicum*	+	+	B	F	V	–	+	V	+	+	+	+	+	+	V	Yellow	CDC description	
Dermabacter hominis, CDC group 3	–	+	B	F	–	–	+	–	+	+	–	+	–	+	–		Lysine+, ornithine+	45/47/ 40765
Dermabacter hominis, CDC group 5	–	+	B	F	–	–	+	–	+	+	–	+	+	+	–			45703/165
Oerskovia spp.	+	+	B	F	+/V	–	+	+	+	+	–	+	+	+	V		See text	7470765
CDC group A3	+	+	B	F	+	–	+	–	+	+	+	+	+	+	V	Yellow/–		
CDC group A4	+	+	B	F	V	–	+	V	+	+	+	+	+	+	V	Yellow/–		
CDC group A5	+	+	B	F	V	–	V	V	+	–/+	+	+	V	+	V	Yellow		

(Continued on next page)

TABLE 3 Characteristics of other coryneform groups[a] *(Continued)*

Test organism	Moti-lity	Cata-lase	CFA type	Fermen-tation	Nitrate	Urease	Esculin	Gelatin	Glu-cose	Xylose	Man-nose	Mal-tose	Lac-tose	Suc-rose	Hemol-ysis	Pigment	Comment	API Coryne code or comment
"C. aquaticum"	+	+	B	NF	V	–	V	V	+	V	V	+	–	+	–	Yellow		
Rothia spp.	+	+	B	F	+	–	+	–/V	+	–	–	+	–	+				7050/2125
CDC group 4	–	+		F	V	–	V	–	+	–	–	+	–	+		Black/gray	H$_2$S+	
CDC group 7	–	+		F	–	–	+	+	V	–	–	V	+	+				
Propionibacterium avidum subsp. granulosum	–	+	B	F	–	–	V	+	+	–	–	V	–	+	β		CAMP+	
Rhodococcus equi	–	+	S	F	V	V	V	–	+	V	V	V	–	+	–	Pink	CAMP+	
Gardnerella vaginalis	–	–	S	F	–	–	–	–	+	V	V	+	V	V	β		Hemolytic on human blood	
Erysipelothrix spp.	–	–	S	F	–	–	–	–	–	–	–	–	+	–	α		H$_2$S+	

[a]Data are based on CDC charts (45, 46a), information from A. von Graevenitz (91a), and additional information from references cited in the text. Abbreviations: S, straight; B, branched; F, fermentative; NF, nonfermentative; V, variable.

out the patient having the symptoms of classic diphtheria (3, 27, 88). It is postulated that the nontoxigenic strains occur more often in people who have been previously immunized (88).

In general, although the nondiphtheria corynebacteria are of low-level pathogenicity, they are isolated more frequently in clinical laboratories than is *C. diphtheriae*. The probable clinical significance of the nondiphtheria *Corynebacterium* spp. is highly dependent on the context in which the organism is isolated. The sites and clinical significances of various species are listed in Table 4. Guidelines for isolating and identifying the irregular gram-positive rods should be established by each laboratory. A reasonable guideline is that isolates should be considered potentially clinically significant and identified more fully if they are found (i) as a predominant or copredominant organism in adequately collected purulent sputum, wound, or abscess specimens or in a specimen from a normally sterile site; (ii) in two or more blood culture bottles; or (iii) at concentrations of $>10^4$ CFU/ml as the only isolate (care should be taken to ensure that the very small colonies do not in fact represent a mixture) or at $>10^5$ CFU/ml as the predominant isolate in a urine culture.

Other laboratory factors to be considered in determining significance include (i) the presence of the organism in the original Gram stain, (ii) the absolute numbers of organisms, (iii) the presence and relative numbers of other organisms, (iv) the pathogenicity of the other organisms, (v) the presence of inflammatory cells, and (vi) the number of specimens from which the organism is isolated. Even with these observations, additional clinical information is often needed to differentiate colonization from infection.

The important species of nondiphtheria corynebacteria, the reported diseases they cause, and their sites of isolation are given in Table 4. However, with the new definitions of some species and the inadequate descriptions of organisms in some literature, the clinical significance may be revised in the future.

Although "*C. ulcerans*" is no longer a valid species and is now considered to be in the *C. diphtheriae* group, the clinical literature generally retains the name. This organism has been associated with mastitis in cattle and pharyngitis in humans. Since it can be carried as a commensal in both humans and animals, true pathogenicity may be difficult to prove. Some strains of "*C. ulcerans*" produce diphtheria toxin and can cause a diphtherialike illness (12, 67, 99). *C. pseudotuberculosis* is primarily an animal pathogen. In sheep, goats, horses, and other warm-blooded animals, it causes ulcerative lymphangitis, abscesses, and other chronic purulent infections. On rare occasions, *C. pseudotuberculosis* is isolated from humans, usually after exposure to animals, and it may cause lymphadenitis and pneumonia.

C. xerosis, *C. striatum*, *C. minutissimum*, and *C. pseudodiphtheriticum* are normal skin and oropharyngeal flora. Asymptomatic overgrowth by these organisms can occur after antimicrobial therapy (91). However, *C. pseudodiphtheriticum* has been isolated from patients with pneumonia and from the urine of patients with urinary tract symptoms (62); two recent cases of tracheitis have also been documented (19). Both *C. striatum* and *C. xerosis* have also been associated with pneumonia (5, 96), and there is a recent report of nosocomial transmission of *C. striatum* with resulting pneumonia (56). *C. minutissimum* can cause abscess formation but probably not erythrasma of the skin.

C. jeikeium, the valid name of the CDC JK group, is the most common corynebacterial pathogen isolated in the

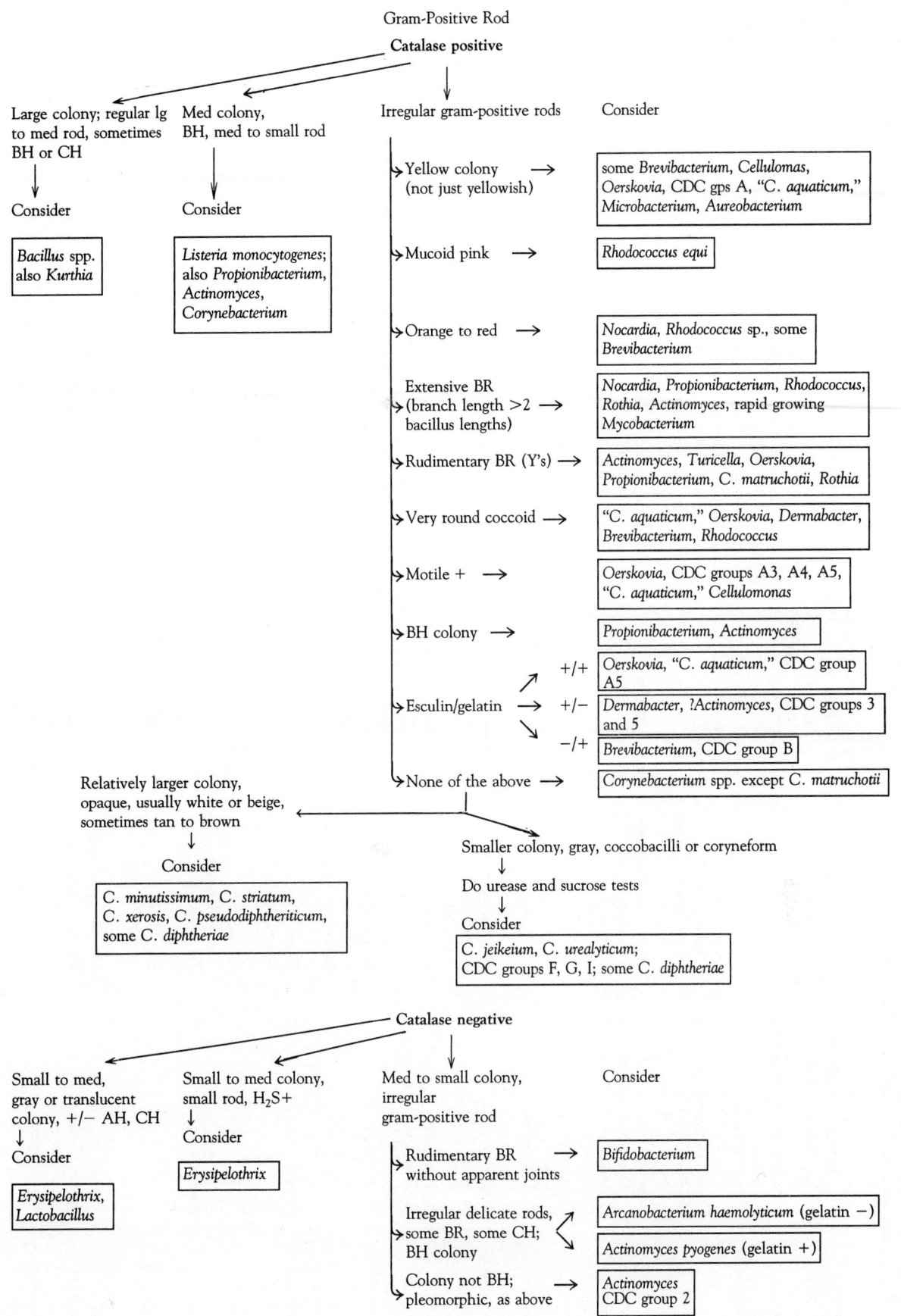

FIGURE 1 Flowchart of easily observable characteristics to be used as a guide for distinguishing genera and some species of gram-positive rods. The next step is to consult the appropriate table or chapter. Abbreviations: AH, alpha-hemolytic; BH, beta-hemolytic; BR, branching; CH, chaining; lg, large; med, medium. Colony characteristics were determined on 5% sheep blood agar–Trypticase soy agar at 24 to 48 h of incubation in 3 to 8% CO_2 with increased humidity.

TABLE 4 Clinical significance of coryneform organisms[a]

Organism	Site or specimen from which isolated			Clinical significance, disease	Reference(s)
	Not associated with human disease	Association unknown	Associated with disease		
C. diphtheriae	Rare skin and nasopharyngeal carriage		Throat, respiratory, skin, blood	Diphtheria, cutaneous diphtheria, endocarditis, septicemia, pharyngitis	3, 21, 27, 29, 43, 54, 88
C. pseudotuberculosis	Sheep, goats, horses		Purulent equine abcess	Horse abortion, horse abscesses, human necrotizing lymphadenitis	21
C. ulcerans	Horses, cattle, nasopharynx	Throat	Throat	Diphtheria, sore throat	21, 67
C. striatum	Common on skin	Miscellaneous human		Pneumonia and lung abscess in immunocompetent host	5, 21
C. xerosis	Common on skin, nasopharynx, conjunctiva	Blood, miscellaneous human	Blood, miscellaneous human	Endocarditis, septicemia, pneumonia in immunocompromised host	21
C. minutissimum	Common on skin		Blood	Septicemia, graft infection	21
C. renale group	Cattle		Cattle	? Abscesses	21
C. bovis	Cows			? Eye infections, septicemia	21
C. kutscheri	Rodents		Rodents	? Septic arthritis	54, 61
C. matruchotii	Oral cavity, primates		Eye	Eye infections	21, 46a, 97
C. mycetoides				? Tropical ulcers	18, 21
C. pseudodiphthericum	Respiratory, skin	Respiratory	Respiratory, blood	Endocarditis, tracheitis, pneumonia, lung abscess	19, 21, 62
C. urealyticum	Skin	Urine, blood, miscellaneous	Urine	Urinary tract infection, cystitis, alkaline-encrusted stones, wound infections in compromised hosts	26, 55, 80, 82
C. jeikeium	Skin, especially inguinal, axillary, and rectal areas; urine	Miscellaneous	Blood, abscess at site of i.v.	Endocarditis, septicemia, wound infections	21, 25, 26, 49
CDC group F1 (LD5 and LD6)		Urogenital	Urine, urogenital	Urinary tract infections	71, 81
CDC group F2		Miscellaneous human			
CDC group G1 (LD4)			Eye	Eye infections	46a, 54
CDC group G1 (LD3)		Blood, respiratory	Blood	Endocarditis	71
CDC group G2, G2 (LD2)		Miscellaneous human	Blood	Endocarditis	46a, 71
CDC group I1		Wound, miscellaneous	Blood	Endocarditis	30, 46a
C. afermentans ANF3	Skin	Ear, blood			46a
		Respiratory			44
C. accolans, group 6		Eye, blood, miscellaneous human			45
CDC group I2		Miscellaneous human			
Arcanobacterium spp.	Throat, pharynx	Throat, pharynx	Throat, pharynx, wounds	Pharyngitis, cellulitis, wound infections	45, 46a
					17, 21
Actinomyces pyogenes	Skin of animals		Purulent material from wounds	Soft tissue infections of cattle, swine, goats, horses, and humans	21

Organism	Natural habitat	Specimens (human source)	Specimens	Clinical significance	Reference(s)
Actinomyces neuii subsp. neuii (CDC group 1)		Eye, miscellaneous human	Eye	Endophthalmitis	33, 45, 54, 64
A. neuii subsp. anitratus					33
CDC group 2		Miscellaneous human	Not known		45, 54
Bifidobacterium adolescentis, CDC group E	Gastrointestinal tract	Miscellaneous human			41
Turicella spp.		Ear	Ear	Possible otitis media	34
Erysipelothrix spp.	Animals and animal environments		Blood, skin, ulcer	Erysipeloid (skin), septicemia, arthritis, endocarditis	9, 34, 38
Brevibacterium casei (CDC groups B1 and B3)	Skin, cheese	Blood, sterile fluids, miscellaneous human	Blood, sterile fluids	Septicemia, prosthesis and line infections in compromised patients	32, 39, 40
"Brevibacterium mcbrellneri"	Skin		Genital hair	White piedra, with Trichosporon beigelii	60
Other Brevibacterium spp. (CDC groups B1 and B3)	Skin, cheese	Blood, miscellaneous human	Blood, miscellaneous human		32, 39, 40
Dermabacter hominis (CDC groups 3 and 5)	Skin		Skin		8, 35
Oerskovia xanthineolytica	Environment, soil	Miscellaneous human			46a
Oerskovia turbata	Environment, soil	Blood, CSF, miscellaneous human	Blood, heart tissue	Endocarditis	21, 46a
CDC groups A3, A4, and A5	Natural habitat unknown			Possible endophthalmitis	4, 64
C. aquaticum	Environment, water, miscellaneous human	CSF, blood, miscellaneous human		Bacteremia, ? meningitis reported	21, 53
Rothia dentocariosa	Oral cavity	Blood	Blood	Prosthetic valve endocarditis	85
CDC group 7					45
CDC group 4	Genitourinary, miscellaneous	Genitourinary, miscellaneous			
Rhodococcus equi	Horses	Respiratory specimens (human and horse), miscellaneous human	Respiratory specimens (human and horse)	Pneumonia, especially in compromised host; soft tissue infection; osteomyelitis	21
Gardnerella vaginalis	Horses (mares)	Urogenital (male)	Urogenital (female)	Bacterial vaginosis, endometritis, neonatal scalp infections, postpartum sepsis	

[a]Abbreviations: CSF, cerebrospinal fluid; i.v., intravenous.

clinical laboratory (58, 93). It is associated with septicemia, endocarditis, skin and soft tissue infections, and, occasionally, meningitis, peritonitis, and pneumonia, particularly in the compromised and previously treated host (48, 49, 95). Although *C. jeikeium* can also be isolated from urine, it is rarely associated with disease at this site (26). Cerebrospinal fluid shunt infections, prosthetic valve endocarditis, and infections associated with prostheses have been reported (21, 37).

C. urealyticum (CDC group D2) has been isolated from the urine of patients with alkaline-encrusted cystitis and urinary tract struvite calculi, although it can be found in the urine of asymptomatic patients (26, 69, 80–82). The need to culture for this organism may depend on the patient population, as rates for significant isolations may vary from essentially zero to 8% (26, 73). *C. urealyticum* is also associated with endocarditis as well as with wound and other infections (55, 82, 94). It is usually resistant to common antimicrobial agents except vancomycin and may be confused with *C. jeikeium* if identification procedures are not done.

As lipid-dependent CDC coryneform groups (some are called LD genospecies, as defined in the taxonomy section) have been subdivided by DNA homology studies, some strong site and disease associations have appeared (71). CDC group G1 seems to contain several different DNA homology groups. One set of strains, genospecies LD3, is mainly oropharyngeal, whereas LD4, also called CDC group G1, is associated with the eye. *C. (Bacterionema) matruchotii* has also been isolated from the eye, often in association with infections (97). On the other hand, LD2 (mostly CDC group G2) is more associated with endocarditis (71). CDC group F1, composed of genospecies LD5 and LD6, is isolated from urinary tract specimens (71). Group F1 has been implicated as a urinary tract pathogen (81).

Endocarditis (often in patients with prosthetic heart valves) and infections of pacemakers or other prostheses are the most commonly reported diseases caused by nondiphtheria corynebacteria. Although all species of *Corynebacterium* can probably cause endocarditis, those specifically reported include *C. xerosis*, *C. minutissimum*, *C. pseudodiphtheriticum*, *C. jeikeium*, *C. urealyticum*, and CDC groups G2 and I1 (21, 30, 48, 55, 88).

C. mycetoides is reported to be associated with tropical ulcers. *Turicella otitidis* and lipophilic coryneform bacteria (such as *C. afermentans*) have been isolated from the ear, sometimes from patients with otitis media, although a causal relationship is unclear (34, 77).

The other coryneform organisms have varied clinical significance. *Arcanobacterium haemolyticum* is associated primarily with wound infections and pharyngitis in young adults, although cases of septicemia, endocarditis, and osteomyelitis have been reported (11, 17). The clinical significance has been difficult to assess both because the organism can be isolated without disease and because it is often isolated in association with beta-hemolytic streptococci. *Actinomyces pyogenes* is primarily an animal pathogen, but human isolates associated with wound infections and septicemia have been reported. A group of *Actinomyces pyogenes*-like organisms differing in gelatin metabolism have been isolated from human wound infections (100).

Brevibacterium spp. (CDC groups B1 and B3), some of which are found in food (especially cheese), are very common skin flora and have been associated with malodorous feet (2). A newly described strain, tentatively called "*Brevibacterium mcbrellneri*," was isolated from infected genital hairs of patients with white piedra in association with *Trichosporon* spp. (60). Strains resembling *B. epidermidis* have been isolated from patients with peritonitis who were receiving continuous ambulatory peritoneal dialysis (39). *B. casei* may be the most common clinical isolate among the brevibacteria. Most isolates are from blood and sterile fluid specimens. They are associated with septicemia in immunocompromised patients and possibly meningitis (40). *Dermabacter* spp. are also common skin bacteria (52), with isolates most often obtained from blood cultures and sometimes associated with sepsis or endocarditis but also presumed sometimes to be a contaminant (8, 35). The second most common site of isolation is wounds and abscesses, in which this organism may occasionally play a primary role (8, 35).

"*C. aquaticum*" has been reported to be associated with endophthalmitis, meningitis in a child, and septicemia as well as urinary tract infection in a child (53, 87). One group of CDC group A4-like organisms that were reported to have caused endophthalmitis were reidentified as CDC group 1 (now renamed *Actinomyces neuii* subsp. *neuii*) on the basis of esculin hydrolysis and CFA data (64). Another report of endophthalmitis in which both *Candida albicans* and CDC group A4 were isolated leaves in question the role of CDC group A4 (4). However, there are other reports of the isolation of CDC group A4 in a clinical setting (46). *Oerskovia* spp. have been isolated from blood and abscesses. *Rothia dentocariosa*, a member of the normal oral flora, can be a relatively common clinical isolate from oropharyngeal specimens. Although rarely reported as a pathogen, when it is, as with the oral streptococci, it is most often associated with endocarditis. CDC group 4 has been isolated from the genitourinary tracts of young women (45).

Collection, Transport, and Storage of Specimens

For cases of suspected diphtheria, material for culture is obtained on a swab from the inflamed areas of the membranes that are formed in the throat and nasopharynx. Swabbings of the nasopharynx and throat can also be cultured for detection of *C. diphtheriae* in suspected carriers. Material from wounds should be removed by swab or aspiration, with care being taken to avoid normal skin flora. Smears are made of the original material for subsequent Neisser or Loeffler methylene blue stain for observation of metachromatic granules and Gram stain for coryneform morphology (which is, however, indistinguishable from that of other *Corynebacterium* spp.). The specimen should be immediately transported to the laboratory or inoculated onto proper media. If this is not possible, the swab should be sent to a reference laboratory by being shipped dry in a sterile tube or in a special packet containing a desiccant such as silica gel (54).

For the nondiphtheria corynebacteria and other coryneform organisms, no special specimen handling is required. Because *Corynebacterium* organisms and some other coryneform organisms are a significant part of the normal flora of the skin and upper respiratory tract, distinguishing commensal from disease-associated organisms is a major problem. Care must be taken in collecting the specimen. A Gram stain of the original specimen should be prepared and examined for assessment of clinical significance.

Isolation Procedures

Specimens for *C. diphtheriae* should be streaked onto a blood agar plate and onto a tellurite-containing medium (e.g., cystine-tellurite agar or modified Tinsdale medium).

The tellurite-containing media have the advantage of serving as both selective and differentiation media. A Loeffler slant should also be inoculated, because C. diphtheriae grows most rapidly on this lipid-rich medium, and the Gram stains from colonies grown on Loeffler media are best for the demonstration of metachromatic granules. In addition to Tinsdale or tellurite medium, inoculation of Loeffler or Pai slants may increase the sensitivity of culture and is particularly necessary if no liquid enrichment medium was used for transport. For better recovery, specimens submitted as single swabs should be transferred to a serum- or blood-containing medium that can be subcultured after 4 days of incubation onto the media described above. When silica gel is used for transportation, it is essential that desiccated swabs be incubated overnight in a broth supplemented with plasma or blood before they are plated on the routine media recommended. All media are incubated at 35 to 37°C with or without added CO_2.

Many clinical laboratories, however, rarely have these special selective media in stock, because requests for C. diphtheriae are rare, and medium shelf life is short. Tinsdale medium has a reported shelf life of only 4 days, although storage periods of up to 4 weeks have been suggested as being appropriate (54). If the recommended media are not available, a fosfomycin disk may be placed on blood agar after inoculation as an alternative; coryneforms that grow in the surrounding zone of inhibition are selected for identification (98). Columbia colistin-nalidixic acid (CNA) agar, which selects for gram-positive organisms, may also be used but will require examination of all coryneform organisms that grow.

Two difficulties with the laboratory diagnosis of unsuspected diphtheria infection are that colonies of C. diphtheriae are not distinctive on the plates usually used for plating throat or wound cultures and that the presence of coryneform organisms isolated with other throat or skin flora is usually not reported. Thus, the diagnosis of diphtheria in a nonepidemic setting is almost always initially based on clinical observations. For example, in a recent fatal case of diphtheria in Dade County, Florida, the initial throat cultures were reported as negative. After the diagnosis was made on the clinical basis of the presence of a tonsillar membrane, cultures of autopsy material from the patient and subsequent throat swabs of contacts were plated onto appropriate media and yielded growth of C. diphtheriae (29).

Most other Corynebacterium species grow on blood agar at 35 to 37°C with or without CO_2 enrichment. Since colonies of some species are small at 24 h, for optimal isolation rates, plates should be incubated for 48 h. A significantly better rate of isolation of C. urealyticum and C. jeikeium from potentially contaminated sites can be achieved by using selective agar or CNA plates instead of sheep blood agar plates; 72 h of incubation has been recommended (26). However, we have found that the colonies on CNA are visible after a full 48 h of incubation if plates are carefully examined. It may be costly to hold urine cultures for 72 h, and some have questioned the need to specifically culture for C. urealyticum unless the urine is alkaline and struvite crystals, leukocytes, and erythrocytes are present (73).

Identification of *Corynebacterium* spp. and Coryneform Organisms

Although the recent use of whole-cell fatty acid analysis, the API Coryne strip (bioMérieux Vitek, Hazelwood, Mo.),

cell wall analysis, and classic biochemical methods allow microbiologists to correctly identify more coryneform organisms, ambiguities in the taxonomy and definition of many Corynebacterium spp. and species in related genera hamper appropriate organism identification. Even with a large number of tests, clinical laboratories would therefore probably be unable to identify many clinical isolates (16, 31, 93). Additionally, because these organisms are common clinical isolates, the time required and the expense involved in attempting to identify them can be great. Therefore, only clinically significant isolates should be identified fully.

An initial practical approach to identification is presented in Fig. 1. Note that the greatest weight is placed on Gram stain morphology from 24- to 48-h colonies on blood agar (Fig. 2), on colony color and morphology, and on the absence of some characteristics. This is an attempt to simplify initial categorization for those clinical laboratories without extensive resources. Use of these basic and morphologic criteria can also help prevent errors of identification to the genus level that may occur when standard biochemicals, the API Coryne strip, the Biolog panel, or the MIDI fatty acid analysis (see chapter 12 of this Manual) is used for identification.

All of the currently available methods useful for identifying the coryneforms suffer from a poor database. Although most yield correct and reproducible biochemical results, the name assigned to the organism is often erroneous. It is therefore important to record not just the identification name generated by the system but also the individual biochemical results or biotype number. The conventional identification protocol used at the Centers for Disease Control and Prevention (CDC) tests for nitrate reduction, urease, motility, fermentative ability, esculin hydrolysis, gelatin hydrolysis, serum requirement, and utilization (in Andrade's base or brain heart infusion) of glucose, mannitol, maltose, sucrose, lactose, and xylose by standard methods (45, 46a, 54). The organisms included in the CDC charts are Corynebacterium spp., Arcanobacterium haemolyticum, CDC coryneform groups, Actinomyces pyogenes, Oerskovia spp., and Rhodococcus spp. (45, 46a). Results are usually available in 2 to 7 days. Von Graevenitz used a scheme in which DNase, hippurate, and the CAMP reaction were found to be useful (93).

The test method for stimulation by or requirement of serum or lipid is not uniformly presented in various publications. The concept is that some strains grow better in the presence of added lipid. Such strains are called lipophilic or serum stimulated; if they require lipid, they are called lipid dependent. The test can be performed by comparing growth (turbidity) in broths with or without added Tween 80 or serum or by adding a drop of Tween 80 or serum to a Mueller-Hinton agar plate on which the isolate has been inoculated with a swab as for disk diffusion susceptibility testing. In either case, enhanced growth compared to background growth is evaluated.

The most accessible, useful commercial system available is the API Coryne system, which consists of 20 biochemical tests in cupules (31, 36, 93). Listed by the groups of three used to generate the code number, the biochemical tests include nitrate reduction, pyrazinaminidase, and pyrrolidonyl arylamidase; alkaline phosphatase, β-glucuronidase, and β-galactosidase; α-glucosidase, N-acetyl-β-glucosaminidase, and esculin; urease, gelatin hydrolysis, and blank; carbohydrate fermentation of glucose, ribose, and xylose; mannitol, maltose, and lactose; and sucrose, glyco-

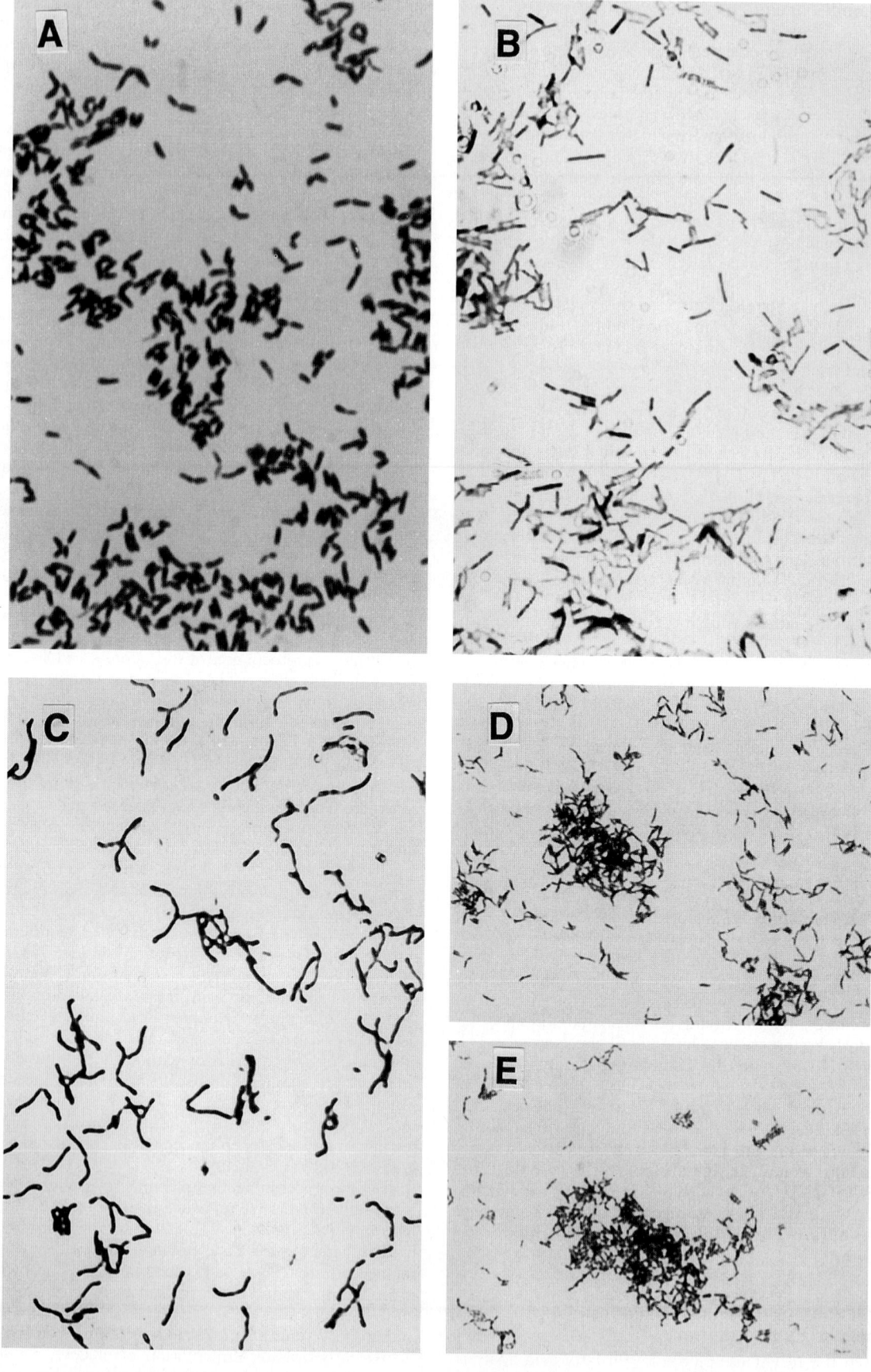

gen, and catalase. Identification numbers are generated as with the API 20E strip (see chapter 10 of this Manual). Organisms included in the database are *Corynebacterium* spp., CDC coryneform groups, *Actinomyces pyogenes*, *Arcanobacterium haemolyticum*, *Brevibacterium* spp., *Erysipelothrix rhusiopathiae*, *Gardnerella vaginalis*, *Listeria* spp., *Oerskovia* spp., and *Rhodococcus equi*. The enzymatic results are available in 24 h; the strip can then be cut, and the esculin, gelatin, urease, and sugar portions can be reincubated, with reactions read in up to 5 days. Organisms that are well identified by this system include most of the clearly defined *Corynebacterium* spp. (e.g., *C. diphtheriae*, *C. urealyticum*, *C. striatum*, *C. jeikeium*, and CDC groups G, F, and I), "*C. aquaticum*," *Erysipelothrix* spp., *Actinomyces pyogenes*, and *Arcanobacterium haemolyticum* (31). The system lumps the genospecies of *C. xerosis* and CDC groups F1, G1, and G2 together. Some *Brevibacterium* spp. are correctly identified. Organisms that could be identified by the system but are missing from or misidentified by the present database include *Rothia* spp., *Dermabacter* (which are called CDC group A by the system), *Oerskovia* spp., CDC group 1, CDC group 2, and CDC group A. Code numbers generated for some of the organisms not clearly or correctly represented in the database but published in the literature are included in the last columns of Tables 2 and 3.

Several identification methods are based on analysis of cellular structural elements. Of these, the measurement of CFAs, especially by the MIDI system (see chapter 12 of this Manual), is the most developed (7, 92). For some groups of coryneform organisms, the CFAs must be part of the defining characteristics. Although this method has the potential for a correct identification and can usually (when used in conjunction with additional information such as Gram stain and catalase reaction) place the isolate tested into the correct group, an adequate database is not yet commercially available. However, because the CFA profile defines a unique, fundamental characteristic of an organism, it is recommended that such a profile be obtained before a report on an unusual gram-positive rod is submitted for publication. General CFA data are found in Tables 1 and 3 (7, 92).

High-performance liquid chromotography (HPLC) has been used for determination of the corynomycolic acids (25). The method has detected a group of organisms that are classified as *Corynebacterium* species but lack mycolic acids (see the groups with *C. amycolatum*). Although HPLC is an excellent method of discriminating between the corynebacteria and the rhodococci, nocardiae, and mycobacteria and has been used for identification within these last three genera, the method is not sufficiently discriminatory for use with other coryneform groups.

Several systems are not very useful at this time. The standard automated Vitek system (see chapter 10) is hampered by a database that includes relatively few species of *Corynebacterium*, although *Erysipelothrix* spp., *Listeria monocytogenes*, *Arcanobacterium haemolyticum*, and *Actinomyces pyogenes* are identified accurately. The Minitek system (BBL Microbiology Systems) and the cystine tryptic agar sugar method both include too few tests to fully identify the

Corynebacterium spp. (46, 78). The Biolog system (Biolog Inc., Hayward, Calif.) utilizes 95 biochemicals in a 96-well microtiter plate format. This system has the potential to identify the coryneforms; however, many clinically isolated organisms are not yet in the database. *Arcanobacterium haemolyticum* can be correctly identified in 4 h by this system. The API 20 Strep system is useful for some organisms, but not enough different species have been evaluated (63, 89).

Although not for routine use, methods based on nucleic acid structure hybridization and 16S rRNA sequencing are used for the final definition of species relatedness (20, 35, 40, 100) (see chapter 20 of this Manual as well).

It is usually not cost-effective, clinically significant, or possible to identify all of the coryneforms to species level in a timely manner in a clinical laboratory. Even for the isolates that are significant (for example, those isolated from the blood of a patient with endocarditis), identification is often very difficult. However, certain characteristics lead one to consider the correct genera and begin identification with the right chart. For example, the clinically isolated *Corynebacterium* spp. (with the exceptions noted above) are not catalase negative, yellow, orange, or mucoid pink; do not branch; do not cleave esculin or gelatin; and are not motile, while these characteristics are found in the non-*Corynebacterium* groups (Fig. 1). Table 2 compiles some important test results for the identification of *Corynebacterium* spp., and Table 3 lists characteristics of the other genera.

C. diphtheriae, C. ulcerans, and C. pseudotuberculosis

The *C. diphtheriae* group of organisms includes the subtypes of *C. diphtheriae* (with four cultural variants or subspecies), *C. ulcerans*, and *C. pseudotuberculosis*. All form a gray-brown halo on Tinsdale medium, are pyrazinaminidase negative (*C. pseudotuberculosis* is variable), and have large amounts (23 to 29%) of a unique fatty acid, C16:1 ω7C (7). The varieties of *C. diphtheriae* differ in colony morphology; *C. diphtheriae* subsp. *intermedius* produces small colonies and requires serum, *C. diphtheriae* subsp. *mitis* produces larger colonies that are beta-hemolytic, and *C. diphtheriae* subsp. *gravis* produces larger colonies, usually without beta-hemolysis. The cultural types also have distinctive microscopic characteristics when grown on Loeffler (but not on Pai) slants. *C. diphtheriae* subsp. *mitis* meets the textbook description of *C. diphtheriae*, having long, pleomorphic, but rigid club-shaped rods. *C. diphtheriae* subsp. *intermedius* is highly pleomorphic, ranging from very long to very short rods. Cells of *C. diphtheriae* subsp. *gravis* are usually short, coccoid or pyriform rods morphologically resembling *C. pseudodiphtheriticum* cells. All strains of *C. diphtheriae* are urease negative, and all but *C. diphtheriae* subsp. *belfanti* are nitrate positive. The differentiation of types is not easily done and might have lost its relevance, since a relationship between severity of disease and biotype can no longer be recognized. *C. ulcerans* is urease positive. The biochemical

---

FIGURE 2 The very different morphologies of the coryneform organisms. Gram stains were done at 24 to 48 h on isolates from sheep blood agar (except panel E). (A) Typical *Corynebacterium* sp.: deeply staining, some V's, and palisading. (B) *Rothia* sp.: some branching, straighter. (C) *Turicella otitidis* (from reference 34): very irregular and branching. (D) *Arcanobacterium haemolyticum*: delicate, few rudimentary branches. (E) *Arcanobacterium haemolyticum* at 4 days: some are almost streptococcuslike.

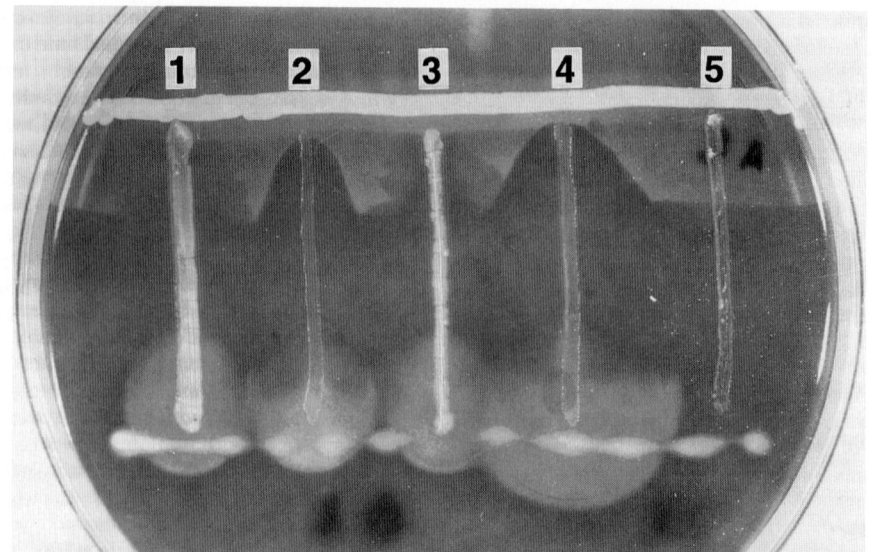

FIGURE 3 CAMP-positive and reverse-CAMP-positive tests. Rows: 1 and 3, *C. pseudotuberculosis*; 2 and 4, *Arcanobacterium haemolyticum*; 5, *Corynebacterium* sp. Top streak, *Staphylococcus aureus* ATCC 25923; bottom streak, *Rhodococcus equi*.

profiles are shown in Table 3. *C. diphtheriae* subsp. *gravis* ferments glycogen, and *C. diphtheriae* subsp. *intermedius* is stimulated by serum. *C. pseudotuberculosis* is positive in the reverse CAMP test (i.e., when streaked perpendicularly to a beta-toxin-producing *Staphylococcus aureus*), as is *Arcanobacterium haemolyticum*, but gives weaker positivity (Fig. 3). The *C. diphtheriae* group shows metachromatic granules when stained with Loeffler or Neisser stain. For optimal expression, staining must be performed from colonies on Loeffler slants. Metachromatic phosphate deposits are reddish purple in the Loeffler stain and black against the brownish cell body in the Neisser stain, resulting in an appearance of scattered matches.

    C. diphtheriae is unique (except for an occasional strain of "*C. ulcerans*" and *C. pseudotuberculosis*) in that it produces the diphtheria toxin (99). However, since toxin production is not universal among strains, all isolates should be tested for toxigenicity. At least 10 colonies should be tested, as both toxigenic and nontoxigenic varieties may be found in the same patient (54). The tests are usually performed in a reference laboratory (79). The detection of the toxins has traditionally been performed on an Elek plate on which the isolate to be tested is streaked perpendicularly to an antitoxin-containing filter paper strip. The toxin and antitoxin diffuse into the medium, forming a line of precipitation where they meet. Diphtheria toxin can also be reliably detected in tissue culture cells. Several cell lines have proved useful. The supernatant from a heavily inoculated broth culture of the test strain incubated overnight on a shaker is added to the cell line with and without added antitoxin. A cytotoxic effect may be seen under the microscope as early as 24 h: most cells are rounded and are loosely attached to the plastic surface. In contrast, the antitoxin-protected cells grow to form an even monolayer. A historical test that is rarely used at this time is the guinea pig lethality test, in which combinations of the test strain and of the test strain plus antitoxin are injected peritoneally into guinea pigs. Toxigenic strains cause death in 1 to 3 days in unprotected animals. More

recently, PCR has been used to detect the toxin gene (44, 66).

C. xerosis, C. minutissimum, C. striatum, C. jeikeium, C. urealyticum, and C. pseudodiphtheriticum

The most common *Corynebacterium* spp. isolated from clinical specimens are *C. xerosis*, *C. minutissimum*, *C. striatum*, *C. jeikeium*, *C. urealyticum*, and *C. pseudodiphtheriticum*. *C. xerosis*, *C. striatum*, and *C. minutissimum* are each a heterogeneous species. Coyle and colleagues found six different DNA homology groups among the *C. xerosis* strains in the American Type Culture Collection (ATCC), including several strains that were valid *C. striatum* species. The characteristics listed in Table 2 for *C. xerosis* are those of the type strain as described by CDC (46) and by the ATCC and Coyle et al. (20) but not by *Bergey's Manual* (18). Characteristics of other groups that may eventually be defined as new species are listed in Table 2 as DNA groups of "*C. xerosis*." Van Bousterhaut et al. describe two groups of *C. minutissimum*, a dry, waxy, more antibiotic-resistant strain and an opaque, moist, more antibiotic-susceptible strain (91).

    C. xerosis, *C. striatum*, and *C. minutissimum*, all common clinical isolates, share the characteristics of good growth on laboratory media (the colonies may initially be small, but by 48 to 72 h on sheep blood agar, they are usually more than 2 mm in diameter) and a regular medium to large coryneform cell shape. Both *C. striatum* and some isolates of *C. xerosis* can show barred forms on Gram stain. Although the colonies are often white, beige, or tan, Leonard and coworkers have described a brown-pigmented strain of *C. striatum* (56).

    Because it is the most commonly isolated corynebacterial pathogen, laboratory recognition of *C. jeikeium* is important. It is defined by its inability to reduce nitrate to nitrite and the production of acid from glucose but not from sucrose. By an initial screening for typical colony (small, gray or translucent) and cell (small, coccobacillary, not club shaped) morphologies, catalase, and sucrose utilization and urease production (in a combined test), Cartwright and colleagues claimed 100% sensitivity in identifying their clinical isolates of C. jeikeium (13). Although convenient, some caution in interpretation is needed, as CDC group G2 and C. afermentans would also be called C. jeikeium by this protocol. Riegel and coworkers have shown that biochemically identified (e.g., by API Coryne strip) C. jeikeium can be heterogeneous by DNA homology and that a negative fructose test is important in identifying valid C. jeikeium organisms; however, all C. jeikeium organisms are not fructose negative. In addition, at least one reference strain of C. genitalium is homologous to C. jeikeium (72a). Although lipid (serum) stimulates C. jeikeium growth, most isolates do not require added lipid for identification when the API Coryne system is used. C. jeikeium is noted for its resistance to beta-lactam, aminoglycoside, and some other antimicrobial agents. The resistance seems to be intrinsic, not carried on plasmids, and associated with strains most closely related to the type strain (72a).

C. urealyticum (formerly CDC group D2) resembles C. jeikeium in its resistance to most antimicrobial agents (69). Susceptible strains of C. urealyticum and C. jeikeium have been isolated, and some other Corynebacterium spp. can be resistant to antimicrobial agents; thus, antimicrobial resistance may not be an accurate identifier. C. urealyticum shares with CDC group F1 a strong urease activity.

C. matruchotii and C. mycetoides

The description of C. matruchotii (previously called Bacterionema matruchotii) in Bergey's Manual, which includes "branching" and "bacillus attached to a handle" ("whip handle") morphology as well as esculin positivity, is more in keeping with those species that have been removed from the genus Corynebacterium. The biochemical profiles from various sources vary; those listed in Table 2 are from CDC (46a). The reactions are similar (except for the maltose reaction) to those for CDC group G1, which is also isolated from the eye. Microcolonies of C. matruchotii are flat and spiderlike and are composed of filaments of various lengths. Macrocolonies can be opaque and tough, and they adhere to the agar (18).

C. mycetoides as described in Bergey's Manual are classified as corynebacteria but have unusual mycolic acids and are yellow pigmented. There are few data on isolates from clinical sources (one case report from 1942), and the organism is therefore not included in Table 2.

C. renale, C. bovis, and C. kutscheri

C. renale, C. bovis, and C. kutscheri are primarily animal isolates. All except C. bovis are urease positive. There is a question about the taxonomic position of C. bovis, since it and C. cystitidis have much higher G+C contents than the other Corynebacterium spp. C. bovis is o-nitrophenyl-β-D-galactopyranoside (ONPG) and CAMP test positive (18). Some strains in the C. renale group have been reclassified as C. pilosum and C. cystitidis on the basis of xylose and nitrate reactions.

Additional Nonfermenters and Lipophilic Organisms

C. afermentans (ANF1) and C. propinquum (ANF3) are nonfermenters. The colonies are small and translucent; the Gram stain morphology is coryneform. Within the species C. afermentans are both lipid-stimulated and lipid-indifferent strains that have been designated C. afermentans subsp.

lipophilum and C. afermentans subsp. afermentans, respectively (71, 72). The newly described C. accolans (which is synonymous with CDC group 6) resembles C. jeikeium except that it is nitrate positive.

CDC groups F1 (F2 is not closely related, as it lacks mycolic acids), G1, and G2 may require or be stimulated by the addition of lipid (see above). However, the groups are heterogeneous biochemically and by DNA hybridization. There seem to be two lipid-dependent groups within CDC group F1 and several within CDC groups G1 and G2 (71). CDC group F1 is urease positive. The characteristics of the DNA homology groups (called LD groups) are listed in Table 2. C. genitalium and C. pseudogenitalium are not valid names, as they refer to a heterogeneous group of organisms, most of which can fit into previously defined species. For example, the fructose-negative strains that have been called C. genitalium are probably C. jeikeium or CDC group ANF, and the fructose-positive organisms that have been called C. pseudogenitalium could be CDC group F1 or G1 or C. jeikeium.

Corynebacteria Lacking Mycolic Acid

C. amycolatum, "C. asperum," and CDC groups F2 and I2 lack mycolic acids, and their inclusion in the genus Corynebacterium is therefore questioned (6, 25). On the basis of the routine characterization listed in Table 2 and by the API Coryne system, the organisms would be identified as C. minutissimum or C. striatum unless a test for mycolic acids was performed (6).

Turicella otitidis

Turicella otitidis is a newly described species that resembles the corynebacteria in possessing a cell wall with meso-DAP and tuberculostearic acids, but it does not have mycolic acids. It is distinctive in its Gram stain (Fig. 2) and in being CAMP positive (34).

Arcanobacterium haemolyticum, Actinomyces pyogenes, Other Actinomyces spp., and CDC Group E (Bifidobacterium spp.)

Arcanobacterium haemolyticum, some Actinomyces spp., CDC group E, and some aerotolerant Bifidobacterium adolescentis are all catalase negative. The genus Arcanobacterium has only one species, Arcanobacterium haemolyticum. By fatty acid analysis, it is closely related to Actinomyces spp. and Corynebacterium spp. such as C. jeikeium. The varied microscopic and colonial morphologies of Arcanobacterium haemolyticum have been well described (17, 24). The colonies are beta-hemolytic with a small zone of hemolysis on sheep blood agar but a larger zone on rabbit blood. There are two colony types: smooth, mucoid, and white, and dry, friable, and grayish. Arcanobacterium haemolyticum secretes a "hemolysin" that possesses phospholipase D activity similar to that of C. pseudotuberculosis and is positive in the reverse CAMP test (in which the organism prevents hemolysis of erythrocytes instead of enhancing it) (Fig. 3). The genetic and functional similarities of the two phospholipases were recently demonstrated (23).

When Arcanobacterium haemolyticum colonies are first seen at 24 to 48 h on sheep blood agar plates, the colonies are pinpoint (0.1 mm at 24 h and 0.5 mm at 48 h) and whitish, with very small zones of beta-hemolysis. On rabbit and human blood agar plates, growth is better and zones of hemolysis are larger. The cell morphology is that of a delicate, curved rod with pointed ends and occasional rudimentary branching (Fig. 2). With age, the cells stain more unevenly, so they look like short strands of strepto-cocci (Fig. 2). This appearance, along with their catalase negativity and beta-hemolysis, allows them to be mistaken for streptococci. In addition, they will grow on streptococcal selective agars. They appear to branch more when grown in broth and may form filaments when grown under anaerobic conditions (17). *Actinomyces pyogenes* is similar to *Arcanobacterium haemolyticum* in Gram stain morphology, and *Actinomyces pyogenes* colonies resemble *Arcanobacterium haemolyticum* colonies but may be slightly larger and more hemolytic and are said to develop a brick red color by 7 days. *Actinomyces pyogenes* is usually distinguished from *Arcanobacterium haemolyticum* by being positive for gelatin and xylose. However, a group of *Actinomyces pyogenes*-like organisms that is gelatin negative has been described (100). Positivity by a rapid α-mannosidase test can differentiate *Arcanobacterium haemolyticum* and *Listeria* spp. from *Actinomyces pyogenes* and other corynebacteria (11). The API Coryne system correctly identifies *Arcanobacterium haemolyticum* and *Actinomyces pyogenes*, while the Rapid ANA does not.

Another heterogeneous catalase-negative group is composed of aerotolerant *Bifidobacterium adolescentis* and CDC group E. In serum-supplemented broth, these organisms are saccharolytic. CDC group E can be distinguished from bifidobacteria by gas chromatography of their metabolic end products.

Actinomyces neuii subsp. *neuii* and *Actinomyces neuii* subsp. *anitratus* are the new names for CDC group 1 and CDC group 1-like organisms (33, 64). They fit into the genus *Actinomyces* because of their cell wall characteristics and, like *Actinomyces viscosus*, are catalase positive. The organisms are small and slender. CDC group 2 resembles *Actinomyces* spp. and is catalase negative.

Oerskovia spp.; CDC Groups A3, A4, and A5; "C. aquaticum"; Cellulomonas

Oerskovia spp.; CDC groups A3, A4 and A5; and "C. aquaticum" constitute a particularly confusing group of organisms. All are catalase positive, have branched fatty acids, and are biochemically active. The CDC originally classified the fermentative, motile, yellow-pigmented coryneforms into five groups, A1 through A5, based on carbohydrate reactions. Members of groups A1 and A2, which were characterized by colonies with fringed edges and the ability to hydrolyze both gelatin and casein, were assigned to the genus *Oerskovia* (see chapter 30 of this Manual). Subsequently, these organisms have been shown to be closely related to members of the genus *Cellulomonas* (47). Both *Cellulomonas* and *Microbacterium* spp. can be confused with CDC group A. The two species of *Oerskovia* are separated on the basis of nitrate reduction and xanthine and hypoxanthine hydrolysis (*Oerskovia xanthineolytica* is positive for all three, and *O. turbata* is negative). The validity of CDC groups A3, A4, and A5 as separate species is not clear; although distinct in some metabolic capabilities (e.g., xylose and gelatin; Table 3), they may be taxonomically closely related. "C. aquaticum," which is not considered a *Corynebacterium* species, as it is motile and differs from the corynebacteria in cell wall characteristics, is a more common clinical isolate and can be found in the environment.

Rothia spp. and CDC Groups 4 and 7

Rothia spp. are evenly staining gram-positive organisms that can show branching. The colonies are white and raised and can be rough. *Rothia* spp. have fatty acid profiles similar to but distinct from those of the *Brevibacterium* spp., having larger amounts of 15:0, 18:0, and 18:0 fatty acids. They are fermentative and are nitrate, esculin, and glucose positive. CDC group 4 is described as a *Rothia*-like organism that has a charcoal gray colony (45). Little is known of CDC group 7. Members of this group are fermentative and distinctive in producing gelatinase and H_2S. The Gram stain morphology is that of very small coryneform organisms (45).

Brevibacterium spp. (CDC Groups B1 and B3) and Dermabacter spp.

CDC groups B1 and B3 are *Brevibacterium* spp. They are small, regular or coryneform bacilli that can show marked conversion to a coccoid form. There can be irregular or extended longer bacillary forms but rarely (if ever) branching forms. The colonies usually have a distinctive odor, which some describe as resembling cheese or body odor. CDC groups B1 and B3, which are similar organisms that could oxidize glucose (B1) or not (B3), are not valid groups according to DNA hybridization studies (40). Instead, one homology group encompasses *B. casei* and strains identified by CDC biochemical testing as both B1 and B3, while the type strains of *Brevibacterium linens* and *B. epidermidis* are more closely related to other isolates that have also been identified as groups B1 and B3. There seem to be many additional strains, some poorly defined but linked by having unusually high amounts of the fatty acids anti-iso 15:0 and anti-iso 17:0. A newly described strain, tentatively called "B. mcbrellneri," grows more slowly than do typical *Brevibacterium* strains and exhibits a gray-cream, friable, wrinled colony (60). *B. acetylicum* may be renamed "Exiguobacterium acetylicum" and differs considerably from *Brevibacterium* spp. It is motile and has very different CFAs. The major characteristics of *Brevibacterium* spp. are the production of methanethiol and a distinct cell wall with *meso*-DAP and no mycolic acids. *Brevibacterium* spp. are generally biochemically inert toward carbohydrates but proteolytically active, do not grow anaerobically, and are nonfermentative. Most stains are DNase positive, gelatin positive, and urease negative. The colonies grow well, with *B. casei* often being white, gray, and smooth and *B. linens* being yellow or orange. *B. casei* is most often associated with humans. The API Coryne system may be able to differentiate some species on the basis of ability to utilize ribose (60).

Dermabacter hominis was described in 1988 to encompass skin bacteria with unique cell wall characteristics (52). CDC fermentative groups 3 and 5 are *Dermabacter hominis* (8, 35). In addition to the characteristics given in Tables 1 and 3, *Dermabacter* and *Brevibacterium* spp. share cell wall and fatty acid characteristics, but *Dermabacter* spp. are fermentative, can grow well anaerobically, and are biochemically more active. Many *Dermabacter* isolates are resistant to aminoglycosides.

Kurthia, Rhodococcus, and Nocardia spp.

Rhodococcus and *Nocardia* spp. can be mistaken for coryneform organisms on Gram stain. Although *Rhodococcus* and *Nocardia* spp. as well as rapidly growing *Mycobacterium* spp. are closely related to *Corynebacterium* spp., they are discussed in chapters 30 and 31 of this Manual rather than here.

Kurthia spp. are found in the environment and are large, regular filamentous, gram-positive organisms resembling those of the B. cereus group except that they are nonspore-forming strict aerobes and nonfermentative. The colonies are large, nonhemolytic, and usually creamy or tan-yellow. Clinically, *Kurthia* spp. associated with endocarditis have been reported.

Antimicrobial Susceptibilities

The best preventative against diphtheria is vaccination. Diphtheria antitoxin, which is hyperimmune antiserum produced in horses, is the most important treatment for patients with diphtheria. Antimicrobial treatment does not affect the toxin already circulating; however, penicillin or erythromycin is recommended to eliminate *C. diphtheriae* from the respiratory tract in the disease or carrier state in order to stop toxin production, relieve local infection, and prevent spread of the organism. Erythromycin resistance after extensive use of erythromycin has been reported (43).

For susceptible strains of other *Corynebacterium* spp., the treatment of choice is also penicillin or erythromycin. For resistant strains, such as most *C. jeikeium* and *C. urealyticum* isolates, vancomycin is the drug of choice. Most of the noncorynebacterial coryneforms demonstrate variable susceptibilities to the beta-lactams and the aminoglycosides, being universally susceptible only to vancomycin. *Dermabacter* spp. are often resistant to aminoglycosides (35).

Since the patterns of susceptibility for all these coryneforms are not predictable, susceptibility testing should be done on clinically significant isolates. The susceptibilities of these isolates should be tested by an MIC dilution technique using Mueller-Hinton broth supplemented, if needed, with 0.01% Tween 80 and aerobic incubation or by Kirby-Bauer disk diffusion on Mueller-Hinton agar with 5% sheep blood (35, 72a, 93). The National Committee for Clinical Laboratory Standards (NCCLS) does not publish interpretive standards for penicillin for the gram-positive rods except for *Listeria* spp., but some workers have used the standards published for the staphylococci for interpreting penicillin susceptibility (35). The MICs can be reported as levels without interpretive comments. However, if interpretive results (susceptible, intermediate, and resistant) are reported, it should be noted that they are arrived at by nonstandard methods. Procedure manuals should indicate which interpretive standard (e.g., for *Staphylococcus* or *Listeria* spp.) is to be used in a particular laboratory.

ERYSIPELOTHRIX

Taxonomy and Description of the Genus

The genus *Erysipelothrix* is classified as a regular gram-positive rod related to other gram-positive organisms in the *Bacillus-Clostridium* group, which has a lower guanine-plus-cytosine content of the DNA than do the corynebacteria and others in the actinomycete group. The name is derived from the Greek words "erysipelas," meaning red skin (which is characteristic of the disease caused by the organism), and "thrix," meaning hair (which is characteristic of the organism's microscopic morphology when it forms chaining rods or filaments). *Erysipelothrix* spp. are similar to *Lactobacillus* spp. in morphology, displaying both short and long filamentous forms (up to 60 μm long), and in being catalase negative, but they are different in peptidoglycan, diamino acid, and fatty acid components. Both genera produce lactic acid as a major metabolic end product. They can most reliably be distinguished by the ability of *E. rhusiopathiae* to produce H_2S. *Erysipelothrix* strains are facultatively anaerobic, nonsporeforming, non-acid fast, and nonmotile. They do not hydrolyze esculin or urea. Until recently, the genus had but a single species, *E. rhusiopathiae*. In 1988, a second species, *E. tonsillarum*, isolated from swine and differing from *E. rhusiopathiae* only by DNA-DNA homology, was described (86). There have been several excellent reviews of *Erysipelothrix* spp. (38, 70).

Natural Habitat

E. rhusiopathiae is widely distributed in nature and can be carried by a variety of animals. It has been isolated from soil and water, probably as a result of contamination by animal products. It causes economically significant disease in swine, turkeys, ducks, and sheep but can also be carried by these and a great variety of other birds, fish, and mammals with no apparent disease. It has been isolated from the external surfaces of fish and seafood. *E. tonsillarum* has been isolated from numerous water and animal sources as well as the tonsils of healthy swine (15).

Clinical Significance

The most common disease caused by *E. rhusiopathiae* affects domestic swine and was described over 100 years ago. Swine erysipeloid can present in a variety of manifestations, including a skin infection marked by reddish purple diamond-shaped colorations, endocarditis, septicemia, and arthritis. *E. rhusiopathiae* can cause a similar spectrum of diseases in humans and can infect many other animals, including sheep, cattle, and turkeys. Human acquisition of the disease is most often subsequent to animal or animal product exposure. Thus, veterinarians, abattoir workers, fish handlers, and butchers are at particular risk. Historically, the most common form is a localized cellulitis, called erysipeloid because of the similarity to streptococcal erysipelas. The cellulitis of erysipeloid occurs about 2 to 7 days after an animal-related abrasion, often on the hand or arm. Lesions are painful and, as in swine, can spread peripherally while the central area fades.

More recently, the septicemic form has accounted for a larger proportion of the reported cases; some patients have concurrent skin rash. Endocarditis involving native valves, with or without previous damage, frequently accompanies septicemia. Arthritis is a rarely reported complication (38, 70, 75).

E. tonsillarum has not been isolated from humans and does not seem to cause disease in swine.

Collection, Transport, and Storage of Specimens

Collection of specimens is dependent on the form of the disease. Blood cultures can be collected and handled routinely for suspected cases of endocarditis or septicemia (see chapter 21 of this Manual). If there is a skin lesion, the specimen is best collected by biopsy through the full thickness of skin at the advancing edge of the infected area (9).

Isolation Procedures

E. rhusiopathiae grows on sheep blood-Trypticase soy agar and chocolate agar as well as on media selective for gram-positive organisms (CNA, phenylethanol agar). Plates should be incubated at 36°C with 5 to 10% CO_2. Although *E. rhusiopathiae* will also grow anaerobically and usually in air after subculture, there has been a report that microaerophilic or anaerobic conditions are needed for initial isolation. Commercial blood-culturing systems support the growth of *E. rhusiopathiae*. In blood culture media, the organisms can assume bizarre and irregular shapes (Fig. 4).

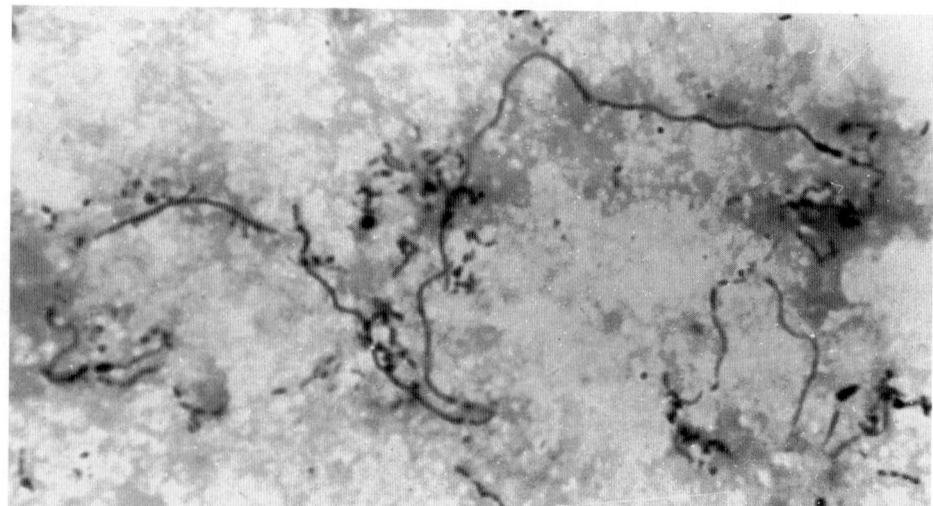

FIGURE 4 *E. rhusiopathiae* Gram stain of a blood culture. Note long and short forms.

Identification

E. rhusiopathiae colonies may be small (<0.2 mm) after 24 h of incubation; 48 to 72 h of incubation is recommended. On initial isolation, the organisms can assume two colony forms: a larger flat, rough form in which the organisms grow as long filaments (0.3 μm in diameter and up to 60 μm in length) and a smaller raised, smooth form in which the organisms are short rods (0.2 by 0.5 μm). Both are a translucent gray and can cause an alpha-hemolysis type of greening of blood-containing media. The colonies can also be smooth and contain both long and short forms; sometimes the long forms appear granular.

E. rhusiopathiae is a nonsporeforming, nonmotile, catalase-negative regular gram-positive rod; the most significant biochemical test for this species is H_2S production in a triple sugar iron agar slant. Characteristics are listed in Table 3. The organism is unusual in that it is positive for both lactose and *N*-acetyl-glucuronidase and negative for maltose. Automated systems such as Vitek and the API Coryne strip identify *E. rhusiopathiae* correctly. Of reported reactions, *E. tonsillorum* differs biochemically from *E. rhusiopathiae* only in the ability of the former to ferment sucrose. *E. rhusiopathiae* is an important pathogen to watch for when meat and slaughterhouses are inspected. Since isolation from contaminated sources can take several days, a recently developed rapid, direct, PCR-based test to detect *E. rhusiopathiae* in meat and other sources could be of economic importance (57).

Serologic Tests

Since there is apparently little serologic response to infection with *E. rhusiopathiae*, serologic tests are not useful for diagnosis. Infection does not seem to confer lasting immunity, and reinfections can occur (38). No effective vaccine for either humans or animals has been developed. Historically, *E. rhusiopathiae* has been divided into 22 serovars based on the heat-stable somatic antigens, with most isolates being serovar 1 or 2 (50). However, serotypes are not consistent with either biotypes or DNA homology groups (i.e., strains of *E. rhusiopathiae* and *E. tonsillorum* share the same serotype) and are probably not of epidemiologic use (15, 86).

Antimicrobial Susceptibilities

Most strains of *E. rhusiopathiae* are susceptible to penicillin, ampicillin, and cephalothin. In general, the beta-lactam antimicrobial agents, the quinolones (ciprofloxacin), clindamycin, and imipenem show good activity against *E. rhusiopathiae*. It is of interest that most strains are resistant to vancomycin, a characteristic shared by lactobacilli and a few other gram-positive organisms but not by the majority of gram-positive rods. Resistance to the aminoglycosides and trimethoprim-sulfamethoxazole is widespread. There is variable susceptibility to tetracycline, chloramphenicol, and erythromycin. There is no NCCLS guideline for the antimicrobial testing of *E. rhusiopathiae*. Testing by the Kirby-Bauer disk diffusion method using *Listeria* zone sizes for interpretation is a reasonable interim solution. Useful antibiotics to test would include not only the drugs for treatment (penicillin G, cephalothin, and others as appropriate) but also vancomycin for purposes of identification.

GARDNERELLA

Description of the Genus

The genus *Gardnerella* contains only one species, *Gardnerella vaginalis*. The type strain is 594 Gardner and Dukes (ATCC 14018; NCTC 10287). On the basis of DNA hybridization and phenotypic taxonomic studies, this genus was not thought to be clearly related to any other established taxon. However, more recent studies show a relationship to *Bifidobacterium* spp. (68). The *Gardnerella* cell wall resembles those of gram-positive bacteria both morphologically and in chemical composition. The organism has not been assigned to an existing family. The cells are nonmotile, pleomorphic, nonencapsulated, gram-negative to gram-variable rods that average 0.5 to 1.5 μm in length. The CFAs include laurate, myristate, stearate, oleate, and palmitate. The mole percent G+C of the DNA is 42 to 44.

G. vaginalis is facultatively anaerobic and fermentative, with acetic acid as the major end product. Catalase and oxidase are not produced. Most strains are fastidious in their growth requirements. Colonies on human blood bilayer Tween agar are 0.3 to 0.5 mm in diameter after incubation for 48 h at 35 to 37°C in an atmosphere of 5% CO_2. The optimal pH for growth ranges between 6 and 7. A distinct beta-hemolysis is produced on human and rabbit blood agars but not on sheep blood agar media. Acid is produced from a wide variety of carbohydrates.

Natural Habitat

G. vaginalis is a member of the endogenous vaginal flora of up to 69% of women of reproductive age with a normal vaginal examination and of mares (74, 90). Over 90% of male partners of women with bacterial vaginosis have urethral colonization with G. vaginalis. This organism has also been found in the rectums of heterosexual males, children, and women with and without bacterial vaginosis.

Clinical Significance

G. vaginalis is one of several species associated with bacterial vaginosis. Women with symptomatic bacterial vaginosis complain of an increased malodorous ("fishy") vaginal discharge. A clinical diagnosis is made if three or more of the following signs are present: white, homogeneous, nonadherent vaginal discharge; vaginal fluid with a pH of >4.5; clue cells (vaginal epithelial cells covered with coccobacillary rods so that the cell border is obliterated); and a fishy, aminelike odor on addition of 10% KOH (1).

The mode(s) of transmission of bacterial vaginosis is not clear. Some evidence suggests that, like urinary tract infections, it results from vaginal colonization by rectal organisms. Other studies suggest sexual transmission. In a recent study, the number of sexual partners a woman had was directly related to the occurrence of bacterial vaginosis. The reader is referred to a recent review article for more detailed information on bacterial vaginosis (83).

In pregnancy, bacterial vaginosis has been associated with preterm birth, premature rupture of membranes, and chorioamnionitis (59). G. vaginalis is a common blood isolate in postpartum and postabortal fevers. It can cause fatal and nonfatal septicemia and umbilical and fetal scalp monitor infections in neonates. It is a rare cause of urinary tract infection. Though it has been associated with balanoposthitis, its presence in the male urethra or on the glans is usually not associated with clinical manifestations.

Collection, Transport, and Storage of Specimens

Extravaginal specimens can be collected with calcium alginate or cotton-tipped swabs. One swab should be placed into a transport medium such as Amies or Stuart's. A second swab should be used to prepare a Gram stain smear. Isolation media should be inoculated as soon as possible and no later than 24 h after specimen collection. Culture for G. vaginalis is not recommended for the diagnosis of bacterial vaginosis (see below). Rather, a smear should be made as described above and stained with the Kopeloff modification of the Gram stain (84).

Because G. vaginalis is susceptible to sodium polyanethol sulfonate (SPS), an SPS-free system should be used to optimize its recovery from blood culture. Direct plating, pour plates, and supplementation of media with 1.2% gelatin have been recommended for increased isolation from blood. Lysis centrifugation has not yet been systematically evaluated.

Isolation Procedures

Direct Detection

Direct examination of vaginal secretions is more relevant for the diagnosis of bacterial vaginosis than is isolation of G. vaginalis from these specimens, because G. vaginalis is often part of the endogenous vaginal flora. The smear allows not only for a better evaluation of the vaginal flora but also for preservation of the preparation for later examination. In typical smears from patients with bacterial vaginosis, clue cells are accompanied by a mixed flora consisting of very large numbers of small gram-negative (predominantly Prevotella and Porphyromonas spp.) and gram-variable (predominantly G. vaginalis) rods and coccobacilli in the absence of larger gram-positive rods (lactobacilli). Curved gram-negative rods (Mobiluncus spp.) may be found as well. When the clinical examination is normal, Lactobacillus morphotypes are found alone or in the presence of small numbers of G. vaginalis morphotypes (84). Use of a standardized Gram stain interpretation scheme improves the reproducibility of this technique (65; see chapter 21 of this Manual as well).

Fluorescent-antibody tests for the detection and identification of G. vaginalis have been reported (42). An oligonucleotide probe for G. vaginalis has been used to diagnose bacterial vaginosis in the presence of an elevated vaginal-fluid pH (76).

Culture

Semiselective human blood bilayer Tween agar may be inoculated by rolling a swab across a sector of the plate (90). This inoculum should be streaked with a loop to allow a semiquantitative estimate of the growth of the isolate, and the plates may be incubated at 35°C in a candle jar or in a humid atmosphere containing 5% CO_2. After 48 h of incubation, the plates may be examined for colonies exhibiting diffuse beta-hemolysis. Isolates should be subcultured at least every 72 h, since older cultures are often no longer viable. For isolation of G. vaginalis from blood, SPS should be deleted from the medium, since this ingredient may inhibit the organism.

Identification

A presumptive identification of G. vaginalis may be based on the typical cell morphology on the Gram stain, beta-hemolysis with diffuse edges on human blood bilayer Tween agar, and a negative catalase test (68). When a more accurate identification is required, α-glucosidase (100% positive), β-glucosidase (0% positive), hippurate hydrolysis (90% positive), and starch hydrolysis (100% positive) tests yield maximum discriminative value (68). Inhibition tests with disks containing 50 μg of metronidazole (zone of inhibition present), 5 μg of trimethoprim (zone of inhibition present), 1 mg of sulfonamide (no zone of inhibition), and 10% bile (zone of inhibition present) are also useful in the differentiation of G. vaginalis, vaginal lactobacilli, and unclassified catalase-negative coryneform organisms. The last two groups of organisms are resistant to metronidazole, trimethoprim, and bile and are variably susceptible to sulfonamide. However, these data were developed with potato starch dextrose and H agars (68). When extragenital isolates must be fully characterized, the Haemophilus-Neisseria identification panel (American Micro Scan, Sacramento, Calif.) and another rapid system (Austin Biological Laboratories, Austin, Tex.) can be used, while API ZYM, API 20A, and Minitek have been unreliable to this date. Conventional test methods for identification of G. vaginalis have been reviewed elsewhere (14). Isolates can be subdivided into six biotypes on the basis of the production of lipase, using oleate as the substrate, beta-galactosidase, and the hydrolysis of hippurate (10).

Serologic Tests

No serologic tests for G. *vaginalis* or bacterial vaginosis are commercially available.

Antimicrobial Susceptibilities

Susceptibility testing of G. *vaginalis* is not generally recommended. Metronidazole is the drug of choice for treatment of bacterial vaginosis and extravaginal infections with bacterial-vaginosis-associated flora. The hydroxy metabolite of metronidazole is generally two to four times more active against G. *vaginalis* than is the parent compound (14). Ampicillin or amoxicillin is the drug of choice for infection with a pure culture of G. *vaginalis*, e.g., septicemia. β-Lactamase-positive G. *vaginalis* isolates have not been reported.

Helpful suggestions by Guido Funke, Dannie Hollis, and Kathryn Bernard are gratefully acknowledged.

REFERENCES

1. **Amsel, R., P. A. Totten, C. A. Spiegel, K. C. S. Chen, D. Eschenbach, and K. K. Holmes.** 1983. Nonspecific vaginitis: diagnostic criteria and microbial and epidemiologic associations. *Am. J. Med.* **874:**14–22.
2. **Anthony, R. M., W. C. Noble, and D. G. Pitcher.** 1992. Characterization of aerobic nonlipophilic coryneforms from human feet. *Clin. Exp. Dermatol.* **17:**102–105.
3. **Barakett, V., G. Bellaich, and J. C. Petit.** 1992. Fatal septicemia due to a toxigenic strain of *Corynebacterium diphtheriae* subspecies *mitis*. *Eur. J. Clin. Microbiol. Infect. Dis.* **11:**761–762. (Letter.)
4. **Barker, C., L. Leitch, N. P. Brenwald, and M. Farrington.** 1990. Mixed haematogenous endophthalmitis caused by *Candida albicans* and CDC fermentative corynebacterium group A-4. *Br. J. Ophthalmol.* **74:**247–248.
5. **Barr, J. G., and P. G. Murphy.** 1986. *Corynebacterium striatum*: an unusual organism isolated in pure culture from sputum. *J. Infect.* **13:**297–298. (Letter.)
6. **Barreau, C., F. Bimet, M. Kiredjian, N. Rouillon, and C. Bizet.** 1993. Comparative chemotaxonomic studies of mycolic acid-free coryneform bacteria of human origin. *J. Clin. Microbiol.* **31:**2085–2090.
7. **Bernard, K. A., M. Bellefeuille, and E. P. Ewan.** 1991. Cellular fatty acid composition as an adjunct to the identification of asporogenous, aerobic gram-positive rods. *J. Clin. Microbiol.* **29:**83–89.
8. **Bernard, K. A., M. Bellefeuille, D. G. Hollis, M. I. Daneshwar, and C. W. Moss.** 1994. Cellular fatty acid composition and cultural characterization of CDC fermentative coryneform groups 3 and 5. *J. Clin. Microbiol.* **32:**1217–1222.
9. **Bille, J., and M. P. Doyle.** 1991. *Listeria* and *Erysipelothrix*, p. 287–295. *In* A. Balows, W. J. Hausler, Jr., K. L. Herrmann, H. D. Isenberg, and H. J. Shadomy (ed.), *Manual of Clinical Microbiology*, 5th ed. American Society for Microbiology, Washington, D.C.
10. **Briselden, A. M., and S. L. Hillier.** 1990. Longitudinal study of the biotypes of *Gardnerella vaginalis*. *J. Clin. Microbiol.* **28:**2761–2764.
11. **Carlson, P., and S. Kontainen.** 1994. Alpha-mannosidase: a rapid test for identification of *Arcanobacterium haemolyticum*. *J. Clin. Microbiol.* **32:**854–855.
12. **Carne, H. R., and E. O. Onon.** 1982. The exotoxins of *Corynebacterium ulcerans*. *J. Hyg.* **88:**173–191.
13. **Cartwright, C. P., F. Stock, and P. M. Kruczak-Filipov.** 1993. Rapid method for presumptive identification of *Corynebacterium jeikeium*. *J. Clin. Microbiol.* **31:**3320–3322.
14. **Catlin, W.** 1992. *Gardnerella vaginalis*: characteristics, clinical considerations, and controversies. *Clin. Microbiol. Rev.* **5:**213–237.
15. **Chooromoney, K. N., D. J. Hampson, G. J. Eamens, and

M. J. Turner.** 1994. Analysis of *Erysipelothrix rhusiopathiae* and *Erysipelothrix tonsillorum* by multilocus enzyme electrophoresis. *J. Clin. Microbiol.* **32:**371–376.
16. **Clarridge, J. E.** 1986. When, why, and how far should coryneforms be identified? *Clin. Microbiol. Newsl.* **8:**32–34.
17. **Clarridge, J. E.** 1989. The recognition and significance of *Arcanobacterium haemolyticum*. *Clin. Microbiol. Newsl.* **11:**41–45.
18. **Collins, M. D., and C. S. Cummins.** 1986. Genus *Corynebacterium* Lehmann and Neumann 1896, p. 1266–1276. *In* P. H. A. Sneath, N. S. Mair, M. E. Sharpe, and J. G. Holt (ed.), *Bergey's Manual of Systematic Bacteriology*, vol. 2. The Williams & Williams Co., Baltimore.
19. **Colt, H. G., J. F. Morris, B. J. Marston, and D. L. Sewell.** 1991. Necrotizing tracheitis caused by *Corynebacterium pseudodiphtheriticum*: unique case and review. *Infect. Dis.* **13:**73–76.
20. **Coyle, M. B., R. B. Leonard, D. J. Nowowiejski, A. Malekniazi, and D. J. Finn.** 1993. Evidence of multiple taxa within commercially available reference strains of *Corynebacterium xerosis*. *J. Clin. Microbiol.* **31:**1788–1793.
21. **Coyle, M. B., and B. A. Lipsky.** 1990. Coryneform bacteria in infectious disease: clinical and laboratory aspects. *Clin. Microbiol. Rev.* **3:**227–246.
22. **Coyle, M. B., D. J. Nowowiejski, J. Q. Russell, and N. B. Groman.** 1993. Laboratory review of reference strains of *Corynebacterium diphtheriae* indicated mistyped *intermedius* strains. *J. Clin. Microbiol.* **31:**3060–3062.
23. **Cuevas, W., and G. Songer.** 1993. *Arcanobacterium haemolyticum* phospholipase D is genetically and functionally similar to *Corynebacterium pseudotuberculosis* phospholipase D. *Infect. Immun.* **61:**4310–4316.
24. **Cummings, L. A., W. K. Wu, A. M. Larson, S. Gavin, J. Fine, and M. B. Coyle.** 1993. Effects of media, atmosphere, and incubation time on colonial morphology of *Arcanobacterium haemolyticum*. *J. Clin. Microbiol.* **31:**3223–3226.
25. **De Briel, D., F. Couderc, P. Riegel, and R. Minck.** 1992. High-performance liquid chromatography of corynomycolic acids as a tool in identification of *Corynebacterium* species and related organisms. *J. Clin. Microbiol.* **30:**1407–1417.
26. **De Briel, D., J. C. Lang, G. Rougeron, P. Chabot, and A. Le Raou.** 1991. Multiresistant corynebacteria in bacteriuria: a comparative study of the role of *Corynebacterium* group D2 and *Corynebacterium jeikeium*. *J. Hosp. Infect.* **17:**38–43.
27. **Efstratiou, A., S. M. Tiley, A. Sangrador, E. Greenacre, B. D. Cookson, C. A. Sharon Chen, R. Mallon, and G. L. Gilbert.** 1993. Invasive disease caused by multiple clones of *Corynebacterium diphtheriae*. *Clin. Infect. Dis.* **17:**136.
28. **Evangelista, A. T., K. M. Coppola, and G. Furness.** 1984. Relationship between group JK corynebacteria and the biotypes of *Corynebacterium genitalium* and *Corynebacterium pseudogenitalium*. *Can. J. Microbiol.* **30:**1052–1057.
29. **Farizo, K. M., P. M. Strebel, R. T. Chen, A. Kimbler, R. J. Cleary, and S. L. Cochi.** 1993. Fatal respiratory disease due to *Corynebacterium diphtheriae*: case report and review of guidelines for management, investigation, and control. *Clin. Infect. Dis.* **16:**59–68.
30. **Farrer, W.** 1987. Four-valve endocarditis caused by *Corynebacterium* CDC group 11. *South. Med. J.* **80:**923–925.
31. **Freney, J., M. T. Duperron, C. Courtier, W. Hansen, F. Allard, J. M. Boeufgras, D. Monget, and J. Fleurette.** 1991. Evaluation of API Coryne in comparison with conventional methods for identifying coryneform bacteria. *J. Clin. Microbiol.* **29:**38–41.
32. **Funke, G., and A. Carlotti.** 1994. Differentiation of *Brevibacterium* spp. encountered in clinical specimens. *J. Clin. Microbiol.* **32:**1729–1732.
33. **Funke, G., G. Martinetti Lucchini, G. E. Pfyffer, M. Marchiani, and A. von Graevenitz.** 1993. Characteristics of CDC group 1 and group 1-like coryneform bacteria isolated from clinical specimens. *J. Clin. Microbiol.* **31:**2907–2912.

34. **Funke, G., G. E. Pfyffer, and A. von Graevenitz.** 1993. A hitherto undescribed coryneform bacterium isolated from patients with otitis media. *Microbiol. Lett.* **2:**183–190.

35. **Funke, G., S. Stubbs, G. E. Pfyffer, M. Marchiani, and M. D. Collins.** 1994. Characteristics of CDC group 3 and group 5 coryneform bacteria isolated from clinical specimens and assignment to the genus *Dermabacter*. *J. Clin. Microbiol.* **32:**1223–1228.

36. **Gavin, S. E., R. B. Leonard, A. M. Briselden, and M. B. Coyle.** 1992. Evaluation of the rapid CORYNE identification system for *Corynebacterium* species and other coryneforms. *J. Clin. Microbiol.* **30:**1692–1695.

37. **Greene, K. A., R. J. Clark, and J. M. Zabramski.** 1993. Ventricular CSF shunt infections associated with *Corynebacterium jeikeium*: report of three cases and review. *Clin. Infect. Dis.* **16:**139–141.

38. **Grieco, M. H., and C. Sheldon.** 1970. *Erysipelothrix rhusiopathiae.* *Ann. N.Y. Acad. Sci.* **174:**523–532.

39. **Gruner, E., G. E. Pfyffer, and A. von Graevenitz.** 1993. Characterization of *Brevibacterium* spp. from clinical specimens. *J. Clin. Microbiol.* **31:**1408–1412.

40. **Gruner, E., A. G. Steigerwalt, and D. G. Hollis.** Human infections caused by *Brevibacterium casei*, formerly CDC groups B-1 and B-3. *J. Clin. Microbiol.* **32:**1511–1518.

41. **Guillard, F., P. C. Appelbaum, and F. B. Sparrow.** 1980. Pyelonephritis and septicemia due to Gram-positive rods similar to *Corynebacterium* group E (aerotolerant *Bifidobacterium adolescentis*). *Ann. Intern. Med.* **92:**635–636.

42. **Hansen, W., B. Vray, K. Miller, F. Croakaert, and E. Yourassowski.** 1987. Detection of *Gardnerella vaginalis* in vaginal specimens by direct immunofluorescence. *J. Clin. Microbiol.* **25:**1934–1937.

43. **Harnish, J. P., E. Tronca, C. Nolan, M. Turck, and K. Holmes.** 1989. Diphtheria among alcoholic urban adults. *Ann. Intern. Med.* **111:**71–82.

44. **Hauser, D., M. R. Popoff, M. Kiredjian, P Boquet, and F. Bimet.** 1993. Polymerase chain reaction for diagnosis of potentially toxinogenic *Corynebacterium diphtheriae* strains: correlation with ADP-ribosylation activity assay. *J. Clin. Microbiol.* **31:**2720–2723.

45. **Hollis, D. G.** 1992. *Potential New CDC Coryneform Groups.* Paper handed out at 92nd Gen. Meet. Am. Soc. Microbiol.

46. **Hollis, D. G., F. O. Sottnek, W. J. Brown, and R. E. Weaver.** 1980. Use of the rapid fermentation test in determining carbohydrate reactions of fastidious bacteria in clinical laboratories. *J. Clin. Microbiol.* **12:**620–623.

46a. **Hollis, D. G., and R. E. Weaver.** 1981. *Gram-Positive Organisms: a Guide to Identification.* Centers for Disease Control, Atlanta.

47. **Holt, J. G., N. Krieg, P. Sneath, J. Staley, and S. Williams.** 1994. *Bergey's Manual of Determinative Bacteriology*, 9th ed. The Williams & Wilkins Co., Baltimore.

48. **Jackman, P. J. H., D. G. Pitcher, S. Pelczynska, and P. Borman.** 1987. Classification of corynebacteria associated with endocarditis (group JK) as *Corynebacterium jeikeium* sp. nov. *Syst. Appl. Microbiol.* **9:**83–90.

49. **Johnson, A., P. Hulse, and B. A. Oppenheim.** 1992. *Corynebacterium jeikeium* meningitis and transverse myelitis in a neutropenic patient. *Eur. J. Clin. Microbiol. Infect. Dis.* **11:**473–474.

50. **Jones, D.** 1986. Genus *Erysipelothrix*, p. 1245–1249. *In* P. H. A. Sneath, N. S. Mair, M. E. Sharpe, and J. G. Holt (ed.), *Bergey's Manual of Systematic Bacteriology*, vol. 2. The Williams & Wilkins Co., Baltimore.

51. **Jones, D., and M. D. Collins.** 1986. Irregular, nonsporing gram-positive rods, p. 1261–1434. *In* P. H. A. Sneath, N. S. Mair, M. E. Sharpe, and J. G. Holt (ed.), *Bergey's Manual of Systematic Bacteriology*, vol. 2. The Williams & Wilkins Co., Baltimore.

52. **Jones, D., and M. D. Collins.** 1988. Taxonomic studies on some human cutaneous coryneform bacteria: description of *Dermabacter hominis*. gen. nov., sp. nov. *FEMS Microbiol. Lett.* **51:**51–56.

53. **Kaplan, A., and F. Israel.** 1988. *Corynebacterium aquaticum* infection in a patient with chronic granulomatous disease. *Am. J. Med. Sci.* **196:**57–58.

54. **Krech, T., and D. G. Hollis.** 1991. *Corynebacterium* and related organisms, p. 277–286. *In* A. Balows, W. J. Hausler, Jr., K. Herrmann, H. D. Isenberg, and H. J. Shadomy (ed.), *Manual of Clinical Microbiology*, 5th ed. American Society for Microbiology, Washington, D.C.

55. **Langs, J. C., D. de Briel, C. Sauvage, J. G. Blickle, and H. Akel.** 1988. Endocardite a *Corynebacterium* du groupe D2 a point de depart urinaire. *Med. Mal. Infect.* **5:**293–295.

56. **Leonard, R. B., D. J. Nowowiejski, J. J. Warren, D. J. Finn, and M. B. Coyle.** 1994. Molecular evidence of person-to-person transmission of a pigmented strain of *Corynebacterium striatum* in intensive care units. *J. Clin. Microbiol.* **32:**164–169.

57. **Makino, S., Y. Okada, and T. Maruyama.** 1994. Direct and rapid detection of *Erysipelothrix rhusiopathiae* DNA in animals by PCR. *J. Clin. Microbiol.* **32:**1526–1531.

58. **Marshall, R. J., and E. Johnson.** 1990. Corynebacteria: incidence among samples submitted to a clinical laboratory for culture. *Med. Lab. Sci.* **47:**36–41.

59. **Martius, J., and D. A. Eschenbach.** 1990. The role of bacterial vaginosis as a cause of amniotic fluid infection, chorioamnionitis and prematurity—a review. *Arch. Gynecol. Obstet.* **247:**1–13.

60. **McBride, M. E., K. M. Eller, H. S. Black, J. E. Clarridge, and J. Wolf.** 1993. A new *Brevibacterium* sp. isolated from infected genital hair of patients with white piedra. *J. Med. Microbiol.* **39:**255–261.

61. **Messina, O. D., J. A. Maldonado-Cocco, A. Pescio, A. Farinati, and O. Garcia-Morteo.** 1989. *Corynebacterium kutscheri* septic arthritis. *Arthritis Rheum.* **32:**1053.

62. **Miller, R. A., A. Rompalo, and M. B. Coyle.** 1986. *Corynebacterium pseudodiphtheriticum* pneumonia in an immunologically intact host. *Diagn. Microbiol. Infect. Dis.* **4:**165–171.

63. **Morrison, J. R. A., and G. S. Tillotson.** 1988. Identification of *Actinomyces* (*Corynebacterium*) *pyogenes* with the API 20 Strep system. *J. Clin. Microbiol.* **26:**1865–1866.

64. **NaWas, T. E., D. G. Hollis, C. W. Moss, and R. E. Weaver.** 1987. Comparison of biochemical, morphologic, and chemical characteristics of Centers for Disease Control fermentative coryneform groups 1, 2, and A-4. *J. Clin. Microbiol.* **25:**1354–1358.

65. **Nugent, R. P., M. A. Krohn, and S. L. Hillier.** 1991. Reliability of diagnosing bacterial vaginosis is improved by a standardized method of Gram stain interpretation. *J. Clin. Microbiol.* **29:**297–301.

66. **Pallen, M. J.** 1991. Rapid screening for toxigenic *Corynebacterium diphtheriae* by polymerase chain reaction. *J. Clin. Pathol.* **44:**1025–1026.

67. **Pers, C.** 1987. Infection due to "*Corynebacterium ulcerans*" producing diphtheria toxin. A case report from Denmark. *Acta. Pathol. Microbiol. Immunol. Scand. Sect. B* **95:**361–362.

68. **Piot, P., E. van Dyck, P. A. Totten, and K. K. Holmes.** 1982. Identification of *Gardnerella* (*Haemophilus*) *vaginalis*. *J. Clin. Microbiol.* **15:**19–24.

69. **Pitcher, D., A. Soto, F. Soriano, and P. Valero-Guillen.** 1992. Classification of coryneform bacteria associated with human urinary tract infection (group D2) as *Corynebacterium urealyticum* sp. nov. *Int. J. Syst. Bacteriol.* **42:**178–181.

70. **Reboli, A. C., and W. E. Farrar.** 1989. *Erysipelothrix rhusiopathiae*: an occupational pathogen. *Clin. Microbiol. Rev.* **2:**354–359.

71. **Riegel, P. D., and D. De Briel.** 1993. DNA relatedness among human lipophilic diphtheroids, p. 296, R17. *Abstr. 93rd Gen. Meet. Am. Soc. Microbiol. 1993.* American Society for Microbiology, Washington, D.C.

72. **Riegel, P., D. De Briel, G. Prevost, et al.** 1993. Taxonomic study of *Corynebacterium* group ANF-1 strains: proposal of *Corynebacterium afermentans* sp. nov. containing the subspecies *C. afermentans* subsp. *afermentans* subsp. nov. and *C.*

afermentans subsp. *lipophilum* subsp. nov. *Int. J. Syst. Bacteriol.* **43:**287–292.

72a. **Riegel, P., D. deBriel, G. Prevost, F. Jehl, and H. Monteil.** 1994. Genomic diversity among *Corynebacterium jeikeium* strains and comparison with biochemical characteristics and antimicrobial susceptibilities. *J. Clin. Microbiol.* **32:**1860–1865.

73. **Ryan, M., and P. R. Murray.** 1994. Prevalence of *Corynebacterium urealyticum* in urine specimens collected at a university-affiliated medical center. *J. Clin. Microbiol.* **32:**1395–1396.

74. **Salmon, S. A., R. D. Walker, C. L. Carleton, S. Shah, and B. E. Robinson.** 1991. Characterization of *Gardnerella vaginalis* and *G. vaginalis*-like organisms from the reproductive tract of the mare. *J. Clin. Microbiol.* **29:**1157–1161.

75. **Schuster M. G., P. J. Brennan, and P. Edelstein.** 1993. Persistent bacteremia with *Erysipelothrix rhusiopathiae* in a hospitalized patient. *Clin. Infect. Dis.* **17:**783–784.

76. **Sheiness, D., K. Dix, S. Watanabe, and S. L. Hillier.** 1992. High levels of *Gardnerella vaginalis* detected with an oligonucleotide probe combined with elevated pH as a diagnostic indicator of bacterial vaginosis. *J. Clin. Microbiol.* **30:**642–648.

77. **Simonet, M., D. De Briel, I. Boucot, R. Minck, and M. Veron.** 1993. Coryneform bacteria isolated from middle ear fluid. *J. Clin. Microbiol.* **31:**1667–1668.

78. **Slifkin, M., G. Gil, and C. Engwall.** 1986. Rapid identification of group JK and other corynebacteria with the Minitek system. *J. Clin. Microbiol.* **24:**177–180.

79. **Snell, J. J. S., J. V. Demello, P. S. Gardner, W. Kwantes, and R. Brooks.** 1984. Detection of toxin production by *Corynebacterium diphtheriae*: results of a trial organised as part of the United Kingdom National External Microbiological Quality Assessment Scheme. *J. Clin. Pathol.* **37:**796–799.

80. **Soriano, F., and R. Fernandez-Roblas.** 1988. Infections caused by antibiotic-resistant *Corynebacterium* group D2. *Eur. J. Clin. Microbiol. Infect. Dis.* **7:**337–341.

81. **Soriano, F., and C. Ponte.** 1992. A case of urinary tract infection caused by *Corynebacterium urealyticum* and coryneform group F1. *Eur. J. Clin. Microbiol. Infect. Dis.* **11:**626–628.

82. **Soriano, F., C. Ponte, P. Ruiz, and J. Zapardiel.** 1993. Non-urinary tract infections caused by multiply antibiotic-resistant *Corynebacterium urealyticum*. *Clin. Infect. Dis.* **17:**890–891.

83. **Spiegel, C. A.** 1991. Bacterial vaginosis. *Clin. Microbiol. Rev.* **4:**485–502.

84. **Spiegel, C. A., R. Amsel, and K. K. Holmes.** 1983. Diagnosis of bacterial vaginosis by direct Gram stain of vaginal fluid. *J. Clin. Microbiol.* **18:**170–177.

85. **Sudduth, E. J., J. D. Rozich, and W. E. Farrar.** 1993. *Rothia dentocariosa* endocarditis complicated by perivalvular abscess. *Clin. Infect. Dis.* **17:**772–775.

86. **Takahashi, T., T. Fujisawa, Y. Tamura, S. Suzuki, M. Muramatsu, T. Sawada, Y. Benno, and T. Mitsuoka.** 1992. DNA relatedness among *Erysipelothrix rhusiopathiae* strains representing all twenty-three serovars and *Erysipelothrix tonsillarum*. *Int. J. Syst. Bacteriol.* **42:**469–473.

87. **Tendler, C., and E. J. Bottone.** 1989. *Corynebacterium aquaticum* urinary tract infection in a neonate, and concepts regarding the role of the organism as a neonatal pathogen. *J. Clin. Microbiol.* **27:**343–345.

88. **Tiley, S. M., K. R. Kociuba, L. G. Heron, and R. Muro.** 1993. Infective endocarditis due to nontoxigenic *Corynebacterium diphtheriae*: a report of seven cases and review. *Clin. Infect. Dis.* **16:**271–275.

89. **Tillotson, G., M. Arora, M. Robbins, and J. Holton.** 1988. Identification of *Corynebacterium jeikeium* and corynebacterium CDC group D2 with the API 20 Strep system. *Eur. J. Clin. Microbiol. Infect. Dis.* **7:**675–678.

90. **Totten, P. A., R. Amsel, J. Hale, P. Piot, and K. K. Holmes.** 1982. Selective differential human blood bilayer media for isolation of *Gardnerella* (*Haemophilus*) *vaginalis*. *J. Clin. Microbiol.* **15:**141–147.

91. **Van Bosterhaut, B., R. Cuvelier, E. Serruys, F. Pouthier, and G. Wauters.** 1992. Three cases of opportunistic infection caused by propionic acid producing *Corynebacterium minutissimum*. *Eur. J. Clin. Microbiol. Infect. Dis.* **11:**628–631.

91a. **von Graevenitz, A.** Personal communication.

92. **von Graevenitz, A., G. Osterhout, and J. Dick.** 1991. Grouping of some clinically relevant Gram-positive rods by automated fatty acid analysis. *APMIS* **99:**147–154.

93. **von Graevenitz, A., V. Punter, E. Gruner, G. E. Pfyffer, and G. Funke.** 1994. Identification of coryneform and other Gram-positive rods with several methods. *APMIS* **102:**381–389.

94. **Walkden, D., K. P. Klugman, S. Vally, and P. Naidoo.** 1993. Urinary tract infection with *Corynebacterium urealyticum* in South Africa. *Eur. J. Clin. Microbiol. Infect. Dis.* **12:**18–24.

95. **Waters, B. L.** 1989. Pathology of culture-proven JK *Corynebacterium* pneumonia. An autopsy case report. *Am. J. Clin. Pathol.* **91:**616–619.

96. **Watkins, D. A., A. Chahine, R. J. Creger, M. R. Jacobs, and H. M. Lazarus.** 1993. *Corynebacterium striatum*: a diphtheroid with pathogenic potential. *Clin. Infect. Dis.* **17:**21–25.

97. **Wilhelmus, K. R., N. M. Robinson, and D. B. Jones.** 1979. *Bacterionema matruchotii* ocular infections. *Am. J. Ophthalmol.* **87:**143–147.

98. **Wirsing von Konig, C. H., T. Krech, H. Finger, and M. Bergmann.** 1988. Use of fosfomycin disks for isolation of diphtheroids. *Eur. J. Clin. Microbiol. Infect. Dis.* **7:**190–193.

99. **Wong, T. P., and N. Groman.** 1984. Production of diphtheria toxin by selected isolates of *Corynebacterium ulcerans* and *Corynebacterium pseudotuberculosis*. *Infect. Immun.* **43:**1114–1116.

100. **Wust, J., G. M. Luchini, J. Luthy-Hottenstein, F. Brun, and M. Altwegg.** 1993. Isolation of gram-positive rods that resemble but are clearly distinct from *Actinomyces pyogenes* from mixed wound infections. *J. Clin. Microbiol.* **31:**1127–1135.

Nocardia, Rhodococcus, Streptomyces, Oerskovia, and Other Aerobic Actinomycetes of Medical Importance

BLAINE L. BEAMAN, MICHAEL A. SAUBOLLE, AND RICHARD J. WALLACE

30

GENERAL TAXONOMY

Historically, the actinomycetes (ray fungi) have been considered a distinct group of microorganisms defined by morphologic criteria. Thus, they were identified on the basis of an ability to grow as branching, filamentous cells that either formed spores or reproduced by fragmentation of hyphae. Because of their morphologic resemblance to filamentous fungi with no sexual stages, the actinomycetes were considered members of the Fungi Imperfecti. However, following the development of electron microscopy, chemotaxonomy, and nucleic acid methodology, it has been conclusively proven that these organisms are not eucaryotic fungi; instead, they are procaryotic, gram-positive bacteria. They consist of a large and diverse group of bacteria, and even though cellular morphology remains central to defining what constitutes an actinomycete, it is clear that many of the genera are distantly related phylogenetically (29, 36, 43). It is important to emphasize that actinomycetes are not intermediate links between the eucaryotic fungi and the procaryotic bacteria, and the claim that they are "higher bacteria" is dubious (36, 38). This chapter focuses on the medically important genera of the strictly aerobic actinomycetes, while the anaerobic species are discussed in chapter 48 of this Manual.

Even though more than 40 genera are currently described in the group of organisms collectively known as the aerobic actinomycetes, only 14 genera appear to be relevant to human and veterinary medicine. The medically important genera include the following: *Corynebacterium*, *Mycobacterium*, *Nocardia*, *Rhodococcus*, *Gordona*, *Tsukamurella*, *Streptomyces*, *Oerskovia*, *Actinomadura*, *Dermatophilus*, *Nocardiopsis*, *Thermoactinomyces*, *Saccharomonospora*, and *Saccharopolyspora*. Of these, the nocardioform actinomycetes are most prominent numerically in causing disease in humans. Nocardioform bacteria form a suprageneric group that is defined by cells growing as filaments that may exhibit branching, as irregularly shaped rods, and as cocci. All aerobic members of this group contain type IV cell walls and mycolic acids (Table 1). Aerobic nocardioform actinomycetes include the following genera: *Corynebacterium*, *Mycobacterium*, *Nocardia*, *Rhodococcus*, *Gordona*, and *Tsukamurella* (29, 36, 38, 53). Historically and traditionally, *Corynebacterium* spp. (chapter 29 of this Manual) and My-

cobacterium spp. (chapter 31) are discussed separately from other members of this group.

GENERAL DESCRIPTION OF AEROBIC ACTINOMYCETES

When rinsed by a weak mineral acid in alcohol, the bacteria that contain mycolic acids in their cell walls retain the dye carbol fuchsin longer than do cells lacking mycolic acids. The avidity of this dye for the bacterial cell increases significantly when the mycolic acids in the cell envelope become larger and more complex and when larger amounts of them occur (6). *Mycobacterium* spp. have the largest and most complex mycolic acids (C_{60} to C_{90}). As a consequence, *Mycobacterium tuberculosis* and related organisms bind carbol fuchsin so strongly that it is not removed by 3% HCl in 95% (vol/vol) ethanol. Therefore, these organisms are strongly acid-alcohol fast. In contrast, *Corynebacterium* spp. possess the smallest and simplest mycolic acids (C_{22} to C_{38}), so that most of the carbol fuchsin is removed by 1% HCl in 50% ethanol in water. Some strains of *Corynebacterium* grown under appropriate conditions do retain some dye, and they are weakly acid fast. *Nocardia*, *Rhodococcus*, *Gordona*, and *Tsukamurella* spp. possess mycolic acids that are intermediate in size and complexity between those of the mycobacteria and the corynebacteria. All of these organisms display various degrees of acid fastness, depending on the strain of organism combined with the conditions of growth as well as the methods used to remove the primary dye. These organisms are usually described as partially acid fast. The intensity of staining by all of the methods used to determine acid fastness depends on both the amount of dye bound to the mycolic acids and the relative amount of stain trapped within the cell during destaining with an acid solution. The acid fastness of nocardiae is most variable. Some strains of *Nocardia asteroides* grown under appropriate conditions are strongly acid alcohol fast in either the Ziehl-Neelsen or the auramine (fluorescent) staining procedure (7). In contrast, certain strains of *N. asteroides* grown in brain heart infusion broth are not acid fast even when the less rigorous modified Kinyoun acid-fast method with 1% H_2SO_4 in water as the decolorizing agent is used (7, 9). Growth of nocardiae and rhodococci in lipid-rich media; media supplemented with glycerol, serum, or egg; and milk-based media enhances the acid fastness of most strains (36).

TABLE 1 Selected chemotaxonomic and morphologic features of clinically significant genera of aerobic actinomycetes[a]

Feature	Nocardia	Rhodococcus	Gordona	Tsukamurella	Streptomyces	Oerskovia	Actinomadura	Dermatophilus	Nocardiopsis
Cell wall type[b]	IV	IV	IV	IV	I	VI	III	III	III
Characteristic cell wall sugar(s)	Arabinose, galactose	Arabinose, galactose	Arabinose, galactose	Arabinose, galactose	None	Galactose	Madurose or none	Madurose	None
Mycolic acids (carbon length)	44–60	34–52	48–66	64–78	None	None	None	None	None
Predominant menaquinones	MK-8 or MK-9	MK-8	MK-9	MK-9	MK-9	MK-9	MK-9	MK-8	MK-10
Mol% G+C (DNA)	64–72	63–73	63–69	67–68	69–78	70.5–75	65–77	57–59	65.7
Aerial hyphae[c]	+/-	-	-	-	+	-	+/-	+[d]/-	+
Motility	-	-	-	-	-	+/-	-	+	-

[a]Table was compiled from data in references 20, 29, 30, 38, and 41.
[b]Cell wall types: I, L-isomer of DAP; III, meso-DAP and madurose; IV, meso-DAP, arabinose, and galactose; VI, lacks DAP but has galactose.
[c]+, predominantly present; +/-, predominantly present but may at times be absent; -, predominantly absent.
[d]Present in 10% CO_2; otherwise may be absent.

The acid fastness of nocardiae and rhodococci can be distinguished from that of mycobacteria by extraction of the smear with fresh pyridine in a Coplin jar for 4 h prior to staining for acid fastness by the Kinyoun method (9). With this procedure, all *Mycobacterium* spp. except *Mycobacterium leprae* remain acid fast, while all nocardiae and rhodococci become non-acid fast.

When inoculated into environments that support growth, cells of all strains of actinomycetes elongate to form branching, filamentous forms. The rate and extent of filament elongation with lateral branching depend on the strain of actinomycete, the type of growth medium, and the temperature of incubation. Some strains of *Rhodococcus* and most species of *Mycobacterium* and *Corynebacterium* do not form branching filamentous cells during their growth (29, 36, 38). Many actinomycetes grow slowest and form the longest and most extensively branched filaments, which are often less than 0.5 mm in diameter and may exceed 100 mm in length, on tap water agar or a chemically defined minimal agar. On a more enriched medium, such as brain heart infusion agar, the organisms often grow much more rapidly, forming robust filaments (0.5 to 1.5 mm in diameter) that may fragment earlier than on minimal media. Therefore, the filaments of these actinomycetes may not be as long or as extensively branched prior to fragmentation. In some genera, the sinuous actinomycetal filaments (hyphae) grow both into the agar and along its surface (substrate hyphae). Furthermore, some of the filaments protrude away from the surface and extend into the air (aerial hyphae). Under the low power of the microscope, these aerial filaments appear darker and more refractile than the substrate cells, and they are easily recognized. In general, *Mycobacterium*, *Rhodococcus*, *Gordona*, *Tsukamurella*, *Corynebacterium*, and *Actinomyces* spp. do not exhibit aerial growth (aerial mycelia), while in contrast, various amounts of aerial filamentation can be demonstrated in most strains of *Nocardia*, *Actinomadura*, *Streptomyces*, *Dermatophilus*, *Nocardiopsis*, *Saccharomonospora*, *Saccharopolyspora*, and *Thermoactinomyces* (Tables 1 and 2) (29, 30, 38, 43, 56).

Figure 1 presents a scheme for easy and rapid tentative recognition and grouping of most aerobic actinomycetes encountered in the clinical laboratory. Once the organism has been placed within one of these groups, its identity must be confirmed by additional tests, some of which may not be readily available to the clinical laboratory. For example, to fully identify the nocardioform actinomycetes, either gas chromatographic or high-pressure liquid chromatographic (HPLC) analysis of the lipids, especially the mycolic acids, may be essential (20, 29). Furthermore, defining and separating the genera within the actinomycetes relies most often on the biochemical characterization of constituents of the cell envelope (Tables 1 and 2).

NATURAL HABITATS OF ACTINOMYCETES

Most aerobic actinomycetes represent part of the indigenous microflora found in soil, mud, and dust; on the surfaces of vegetation (e.g., thorns); within decaying vegetation; in composts; in both fresh and marine water; and in decaying fecal deposits from animals (7, 29, 37, 38, 49). On the other hand, some members of the actinomycetes, especially within the genera *Mycobacterium*, *Corynebacterium*, and *Dermatophilus*, are parasites of humans and other animal species, representing either part of the normal flora or a pathogen (6, 23, 75). Actinomycetes are distributed worldwide. Certain groups or species (e.g., *N. asteroides*) are more

TABLE 2 Selected features of clinically significant aerobic actinomycetes[a]

Feature	Nocardia	Rhodococcus	Gordona	Tsukamurella	Streptomyces	Oerskovia	Actinomadura	Dermatophilus	Nocardiopsis
Modified acid fast	+	+/−	+/−	+/−	−	−	−	−	−
Lysozyme resistance (0.05%)	+	+/−	−	+	−	−	−	−	+
Nitrate reduction	+/−	+/−	+	−	+/−	+	+	−	−
Ethylene glycol degradation	−	+	+	+	−	−	−	−	+
Urease	+	+/−	+	+	+/−	−/+	−	+	+
Anaerobic growth	−	−	−	−	−	+	−	+	−
Growth on paraffin	+	+	+	+	−	+	−	−	−

[a]Table is compiled from data in references 29, 30, 36, 38, and 41. +, predominantly positive; +/−, mostly positive with some species or strains positive; −/+, mostly negative with some species or strains positive; −, predominantly negative.

prominent in the temperate regions of the world, whereas others (e.g., *Nocardia brasiliensis*) are more frequently encountered in tropical and subtropical areas. Nevertheless, all of the actinomycetes of medical importance have been isolated from clinical and environmental sources within the United States. Most actinomycetes are mesophilic; however, thermophilic groups abound in manure and compost piles (7, 29, 36–38, 43, 49, 53).

NOCARDIA

Taxonomy

In 1888, Edmond Nocard isolated a filamentous aerobic organism from cattle with farcy. In the same year, Rabe recovered filamentous aerobic organism from a dog, and in 1890, Eppinger isolated a similar organism from a human with a fatal infection (7). In 1889, Trevisan created the genus *Nocardia* to accommodate the organism isolated by Nocard from cattle with bovine farcy. This organism was named *Nocardia farcinica*, and the human and dog isolates became *N. asteroides*. Approximately 80 years later, taxonomic studies of Nocard's original isolates revealed that two distinctly different filamentous organisms were attributed to Nocard in different culture collections. One of these belonged to the genus *Mycobacterium*, and the other was a *Nocardia* sp. Subsequent analysis of the etiology of bovine farcy revealed that a *Mycobacterium* sp., not a *Nocardia* sp., was the cause of this disease in Africa; Nocard's original isolate was therefore probably *Mycobacterium farcinogenes*. Because of the uncertainty of the taxonomic nature of *N. farcinica*, it was deleted as a valid species at that time, and *N. asteroides* was adopted as the type species for the genus (7). However, the *N. asteroides* taxon is decidedly heterogeneous, and recently, two new species, *N. farcinica* and *Nocardia nova*, were separated from it to form their own more homogeneous groups (41, 67, 71).

The nocardia are aerobic, catalase-positive, gram-positive, nonmotile, branching, filamentous bacteria (less than 1.5 mm in diameter) that fragment into irregularly shaped rods and cocci. Aerial filaments (hyphae) are always produced in variable amounts. With the exception of *Nocardia amarae*, all nocardiae are resistant to lysozyme. The cell wall contains *meso*-diaminopimelic acid (*meso*-DAP), arabinose, galactose, and a mixture of mycolic acids with carbon chain lengths varying from C_{44} to C_{64}. The cell envelope contains straight-chain saturated and unsaturated fatty acids as well as major amounts of 10-methylstearic acid (tuberculostearic acid). The phospholipids include diphosphatidylglycerol, phosphatidylethanolamine, phosphatidylinositol, and phosphatidylinositol mannosides; the menaquinones are either tetrahydrogenated with eight isoprene units [MK-8(H_4)] or dihydrogenated with nine isoprene units [MK-9(H_2)]. The G+C mole percentage of the DNA ranges from 64 to 72 (29, 38, 41, 53).

There are at least 12 currently accepted species within the genus *Nocardia*, including *N. asteroides*, *N. amarae*, *N. brasiliensis*, *N. brevicatena*, *N. carnea*, *N. farcinica*, *N. nova*, *N. otitidiscaviarum* (*N. caviae*), *N. pinensis*, *N. seriolae*, *N. transvalensis*, and *N. vaccinii* (29, 30, 38). Three of these, *N. asteroides* sensu stricto, *N. farcinica*, and *N. nova*, form the *N. asteroides* complex, which in the past was often considered a single species (29, 30, 38, 41).

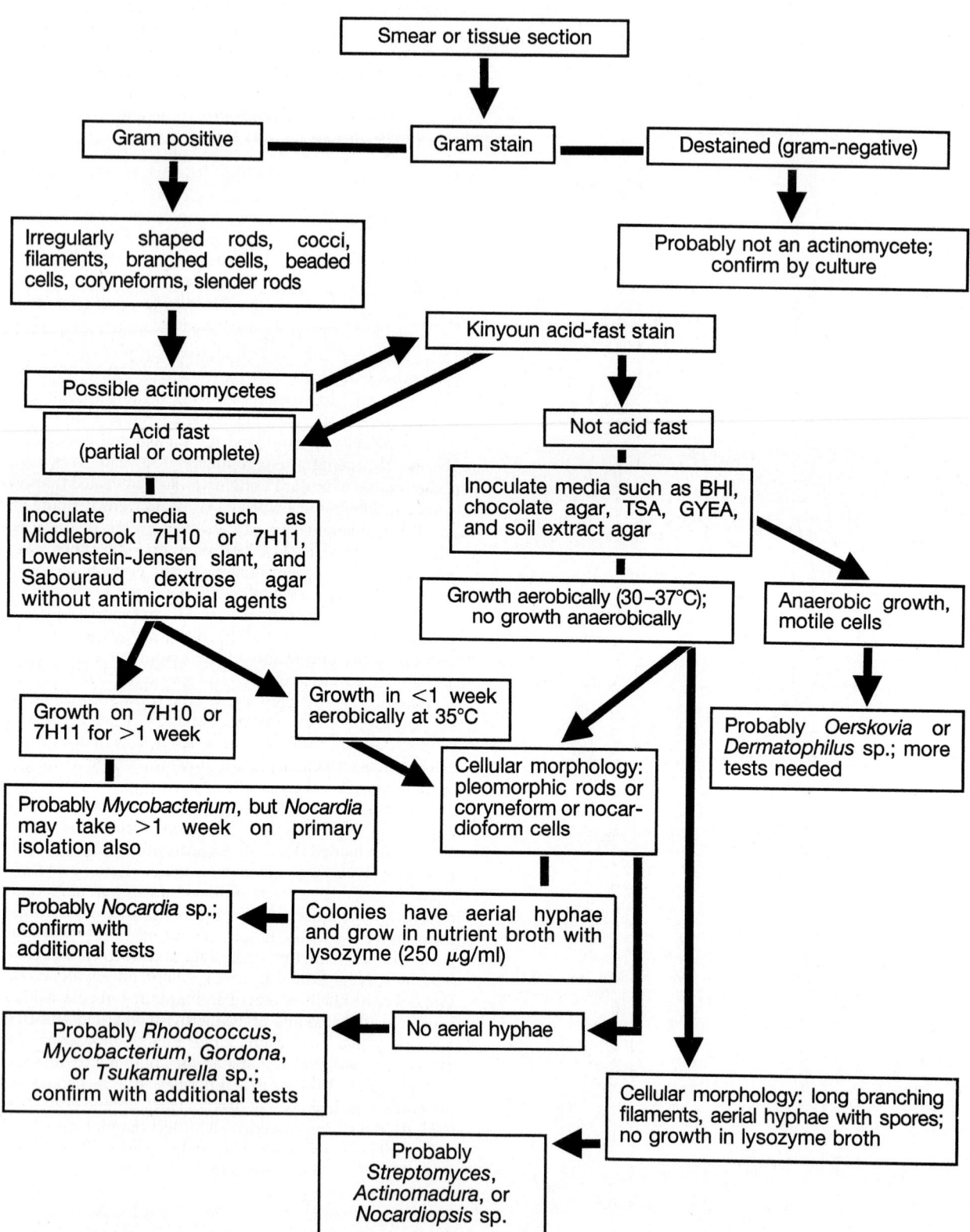

FIGURE 1 Flow diagram for the tentative differentiation and identification of medically important aerobic actinomycetes. BHI, brain heart infusion; TSA, Trypticase soy agar; GYEA, glucose-yeast extract agar.

Natural Habitat

The nocardiae are thought to be associated ubiquitously with soil and vegetation. Indeed, numerous reports describe the recovery of pathogenic species from environmental sources such as riparian soil, freshly cultivated soil, dust, plant thorns, decaying and fresh vegetable matter, fresh water, and marine environments (7, 49). There are also numerous reports on the failure to isolate nocardiae from these sources, and attempts to isolate nocardiae from defined geographic locations in the more temperate regions of the United States are not always successful. Therefore, the pathogenic species of *Nocardia* may not be as ubiquitous in the environment as previously suggested.

Clinical Significance

The first human disease caused by *N. asteroides* was described (in 1889) as "pseudotuberculosis," a systemic disease involving the brain (7). However, the first human case of "pseudotuberculosis" in the United States (in 1898) was presented as a primary pulmonary disease (7). Since these early descriptions, several thousand cases of nocardial infections of humans have been reported in the world literature. In general, all of these infections in humans can be divided into the following six categories based on the body site involved and specific clinical and pathologic features: (i) pulmonary nocardiosis; (ii) systemic nocardiosis; (iii) central nervous system nocardiosis; (iv) extrapulmonary, localized nocardiosis; (v) cutaneous, subcutaneous, and lymphocutaneous nocardioses; and (vi) nocardial mycetoma. In temperate regions of the world such as the United States, the most frequently recognized form of disease is pulmonary nocardiosis caused by *N. asteroides*. In tropical and subtropical regions, mycetoma caused by *N. brasiliensis* is most often diagnosed (7).

The incidence of nocardial infections in human and animal populations is not known. Relatively few attempts to determine the prevalence of nocardiosis have been made; however, several reports indicate that infections by these bacteria are not rare, are frequently misdiagnosed, and are underdiagnosed and that the incidence of infection is apparently increasing (7, 41).

The spectrum of disease caused by nocardiae is broad and varies from a self-limited or inapparent infection to an aggressive, destructive disease resulting in death. The clinical and pathologic features are broad and often nonspecific, mimicking other disease processes. Nevertheless, the nocardiae typically induce a pyogenic response with abscess formation. *Nocardia* spp. cause disease in every region of the body and may involve any organ system. However, there are specific target sites for most strains of nocardiae. Clinical reports on thousands of cases indicate that the regions of the body affected most often are (in descending order of prevalence) lungs, brain, skin, eyes, muscle, bone, lymph nodes, heart, kidneys, joints, bone marrow, spinal column, peritoneum, inner ear, liver, and other regions (7).

Every category of nocardial infection described above has been diagnosed in previously healthy, nonimmunocompromised humans. However, with the exceptions of mycetomas and cutaneous, subcutaneous, and lymphocutaneous infections, other forms of nocardiosis are frequently recognized in individuals who have some underlying debilitation. Therefore, nocardiae not only are primary pathogens but also are emerging as important opportunistic pathogens. The most common underlying factors that predispose individuals to nocardiosis include organ transplantation; malig-

nancies; AIDS; administration of steroids for problems such as arthritis, asthma, and emphysema; alcohol abuse; systemic lupus erythematosus; alveolar proteinosis; chronic granulomatous disease; diabetes; sarcoidosis; and a wide variety of other debilitating factors. The current increase of nocardial infections in the United States is probably due at least in part to increased numbers of immunocompromised individuals within the population as a whole (7).

A mycetoma is a chronic purulogranulomatous disease that usually begins as a painless nodule at the site of a localized injury such as a puncture wound from a thorn or splinter. With time, this lesion increases in size and usually becomes purulent and necrotic, producing sinus tracts that exude a serous fluid that may contain pus and characteristic granules. Multiple nodular lesions develop along these sinus tracts, and the entire lesion expands by direct extension through the tissues, ultimately invading muscle and adjacent bones and causing destructive osteomyelitis. Mycetomas are divided into two categories that are based on etiology. Those caused by fungi are referred to as eumycetomas, whereas those caused by actinomycetes are termed actinomycetomas. The aerobic actinomycetes causing mycetoma include members of the genera *Nocardia*, *Actinomadura*, *Streptomyces*, and *Nocardiopsis* (39, 59, 63, 64, 74). *N. brasiliensis* appears to be the most frequently recognized etiology of actinomycetoma in most regions of the world, although *Actinomadura madurae* and *Streptomyces somaliensis* are most prominent in some geographic locations. Nevertheless, all of the pathogenic species of *Nocardia* have been recovered from actinomycetomas.

Specimen Collection, Transport, and Storage

Samples should be collected aseptically from affected areas and transported to the diagnostic laboratory according to the standard protocols used for routine microbiologic studies (see chapter 21 of this Manual). Samples suspected of containing nocardiae should not be refrigerated or placed on ice prior to being plated on appropriate culture media. Some strains of *N. asteroides* and *N. brasiliensis* from clinical samples rapidly lose viability when exposed to near-freezing temperatures. Bronchial brushings are often considered more reliable for recovery of nocardiae from patients with pulmonary infections but are not always necessary, since isolation of nocardiae from sputum samples may be successful (7, 29, 36, 43, 53). Blood cultures may be taken, especially when disseminated disease is suspected; however, these cultures must be incubated for at least 2 weeks before being discarded, and they are frequently nonproductive. Sometimes invasive techniques involving needle aspirations and tissue biopsies are necessary in order to obtain a culture positive for nocardiae. This is especially true for brain infections, since the cerebrospinal fluid (CSF) is usually negative for nocardiae, even though the patient may have extensive lesions within the brain (7, 29). Nevertheless, the CSF should also be cultured for nocardiae in all cases of central nervous system infection, and these cultures should not be discarded for at least 4 weeks (7).

DIRECT EXAMINATION AND DIAGNOSTIC APPROACHES

Isolation, recognition, and identification of aerobic actinomycetes such as *Nocardia* spp. in the diagnostic laboratory can be both challenging and frustrating. The direct microscopic examination of stained preparations of exudates and

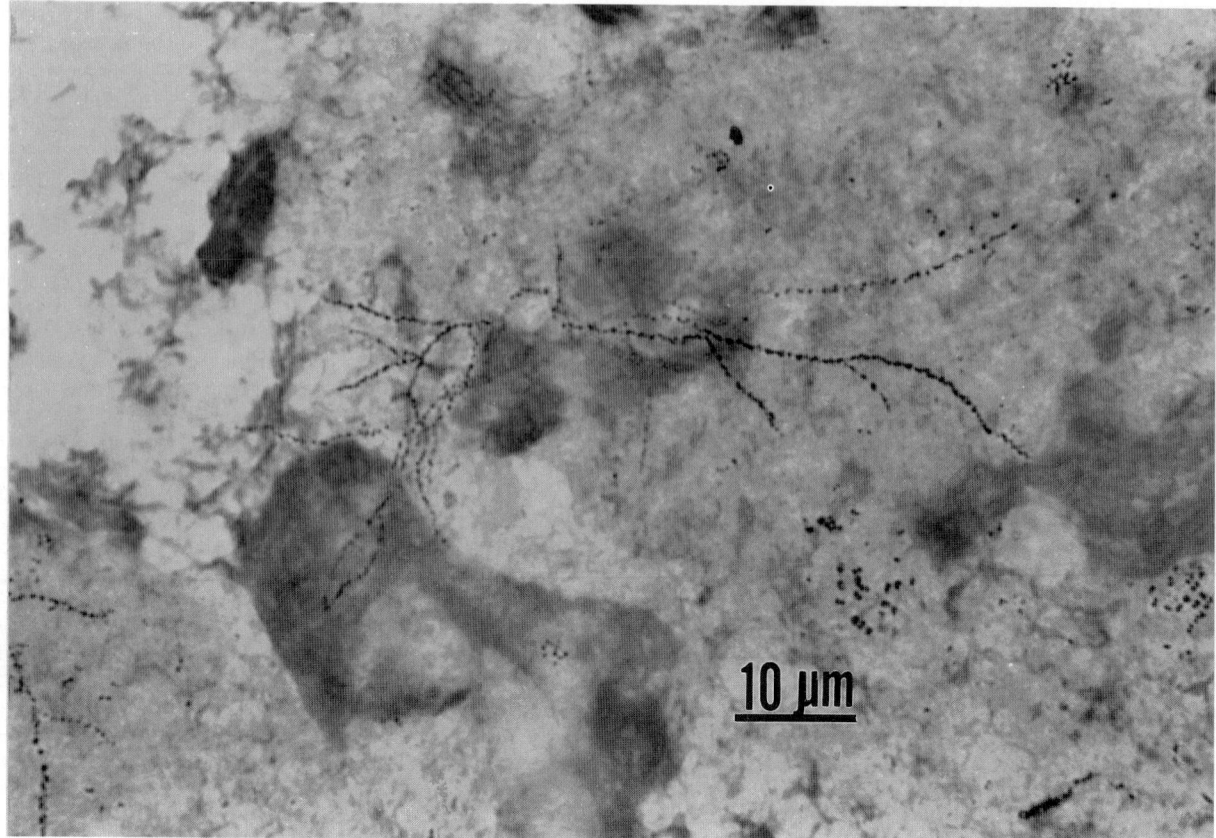

FIGURE 2 Branched, beaded filaments and coccoid cells of *N. asteroides* in the sputum of a patient with pulmonary nocardiosis (fatal case). Gram stain.

tissues may provide presumptive identification of nocardiae by the demonstration of gram-positive, beaded, branching filaments that are acid fast (Fig. 2). More often, however, this is not the case: the nocardiae may not be seen at all; they may appear as gram-positive rods, cocci, or short filaments; or they may show coryneform morphology (7, 41). These bacteria may not be acid fast by any of the methods used, or they may appear as acid-fast bacilli that resemble *Mycobacterium* spp. In smears of exudates, the bacteria can be visualized both intracellularly in phagocytic cells and extracellularly among inflammatory cells. In tissue sections, nocardiae are visualized best by the Brown and Brenn modification of the Gram stain. Nocardiae are not usually apparent in hematoxylin-and-eosin-stained sections, but they can at times be visualized by methenamine-silver or periodic acid-Schiff procedures (7, 36).

All of the pathogenic nocardiae have the ability to invade tissue as single, filamentous cells; *N. brasiliensis*, however, has a propensity for growing in tissue as a micro-colony of organisms giving rise to a granule. These granules are most commonly produced during induction of myceto-mas. Mycetomas caused by other nocardiae such as *N. asteroides*, *N. otitidiscaviarum*, and *N. transvalensis* also contain granules (36, 39, 41, 64).

Isolation Procedures

The nocardiae do not have complex growth requirements and, as a consequence, grow well on most commonly used routine bacteriologic media (41). Nevertheless, isolation

and recognition of nocardiae from clinical samples can be problematic, with frequent failure to recover nocardiae in the laboratory, even when the organisms are visible in stained preparations (7). Nocardiae usually require a minimum of 48 to 72 h before visible colonies become evident (29, 36). Therefore, if nocardiae are suspected, culture plates should be incubated for at least 2 weeks before being discarded. This lengthy incubation requires that the plates be in a humidified environment or sealed to prevent dehydration. Unfortunately, the extended period of incubation often results in overgrowth by contaminating bacteria and fungi, which may hide or inhibit the growth of nocardiae (36). Furthermore, nocardiae may manifest extremely variable colonial morphologies on different culture media, which may make recognition difficult (Fig. 3, 4A, and 4B).

There is experimental evidence that some nocardiae become physiologically and structurally altered as they grow within host tissues. Therefore, transferring these altered forms from tissue to laboratory media may be lethal for the organism or the organisms may require an extended period to adapt to the new environment (7).

To enhance recovery and inhibit bacterial and fungal overgrowth, numerous investigators have developed and studied a variety of media selective for nocardiae (3, 5, 16, 33, 44, 58, 66). One of the first of these incorporated paraffin as a sole carbon source. Nocardiae, rhodococci, and some mycobacteria can grow on paraffin, whereas most other bacteria cannot (44). More recent studies have incorporated multiple antimicrobial agents to inhibit organisms other than the nocardiae. Paraffin agar was compared to the more expensive modified Thayer-Martin medium for recovery of nocardiae from clinical samples and found to be as effective for isolation of nocardiae (5). In addition, several reports have shown that selective buffered charcoal-yeast extract medium designed for the isolation of *Legionella* spp. also enhances recovery of *Nocardia* spp. from sputum (16, 33, 41, 66).

Since nocardiae may survive the decontamination and concentration methods used for *Mycobacterium* spp., sputum or other samples that may be heavily contaminated with unwanted microorganisms can be digested with sodium hydroxide or treated with *N*-acetyl-L-cysteine and inoculated onto standard mycobacterial media such as Lowenstein-Jensen medium or Middlebrook 7H10 agar (36, 45). Since many strains of nocardiae are killed by 0.5% *N*-acetyl-L-cysteine, 2% NaOH–*N*-acetyl-L-cysteine, or benzalkonium chloride, these methods should not be relied upon entirely (45). All organisms that grow on these media after decontamination must be examined carefully with the microscope. Macroscopically, the colonies of *Nocardia* and *Mycobacterium* spp. can look the same on media designed for the cultivation of *Mycobacterium* spp. (29, 36).

Blood cultures should be incubated in the standard two-bottle culture systems, biphasic blood culture bottles, or the automated systems described in chapter 21 of this Manual (7, 36). Nocardiae grow well on blood and chocolate agar plates and sometimes form visible colonies on blood agar in 36 to 48 h. At this early stage, the colonies are small and nondescript, and they may be surrounded by a small zone of alpha-hemolysis on sheep or bovine blood agar. These colonies can be confused easily with those of other microorganisms that represent the normal flora of the upper respiratory tract. Examination of the colonies under a dissecting microscope can help one recognize the growth characteristic of nocardiae. On continued incubation, the colonies enlarge and the leading edge of the colony digs into the agar so that the colony adheres to the plate when it is manipulated with a wire loop. Within 3 to 7 days, the colonies may become surrounded by a zone of beta-hemolysis. Many clinical isolates of *N. otitidiscaviarum*, *N. brasiliensis*, and, to a lesser extent, *N. asteroides* are beta-hemolytic on sheep blood or bovine blood agar. Some isolates of *Nocardia* spp. exhibit alpha-hemolysis only, whereas most isolates of *N. asteroides* appear to be nonhemolytic on sheep blood agar (7).

Identification

From a clinical perspective, *N. asteroides*, *N. brasiliensis*, *N. farcinica*, *N. nova*, *N. otitidiscaviarum*, and *N. transvalensis* are the species of *Nocardia* that cause most diseases in humans (7, 29, 30, 38, 41). As a consequence, our discussion of identification and differentiation is limited to these species.

After the recognition of an isolate as a *Nocardia* sp., the species should be identified. This process, however, is not always simple or straightforward. Certain key characteristics are very useful as guides. Tentative identification can be accomplished by using Table 3 (7, 21, 36–38, 41, 43). The Gordon tests have proven to be very reliable for the initial identification of species (31, 32, 38); however, species within the *N. asteroides* complex (*N. farcinica*, *N. nova*, and *N. asteroides* sensu stricto) cannot be differentiated by these methods alone. Therefore, additional tests will be necessary to confirm the identification of these species (Table 3). It is

important to note that any given clinical isolate can deviate from the list of properties published for that species.

Gordon and colleagues demonstrated that the pathogenic species of nocardiae can be reliably placed within specific groupings by determining their abilities to decompose casein, xanthine, hypoxanthine, and tyrosine. Thus, tentative species assignments of pathogenic strains of nocardiae can usually, but not always, be made with only these four simple media; *N. transvalensis* and *N. otitidiscaviarum* may not be reliably differentiated with these four tests. Additional tests that can differentiate the two species and *N. brasiliensis* are presented in Table 3.

As mentioned previously, the *N. asteroides* complex is heterogeneous, and many investigators have analyzed the diversity of this group in order to identify homogeneous clusters (29, 67, 71, 72). As a consequence, the *N. asteroides* complex has been divided into the species *N. farcinica*, *N. nova*, and *N. asteroides* sensu stricto. *N. farcinica* is distinguished from *N. asteroides* sensu stricto and *N. nova* by its ability to grow on 2,3-butylene glycol (1%, vol/vol), 1,2-propylene glycol (1%, vol/vol), and rhamnose (it also produces acid when grown on this sugar) as sole carbon sources; to utilize acetamide as both a carbon and a nitrogen source; and possibly by its ability to produce opacification on Middlebrook agar (7, 29, 41, 67, 71, 72). Additionally, *N. farcinica* is resistant to the antimicrobial agents cefamandole and tobramycin, while *N. asteroides* is not (67, 71, 72). *N. nova* can be separated from *N. asteroides* and *N. farcinica* by its 2-week arylsulfatase activity, α- and β-esterase activities, lack of acid phosphatase activity, inability to use monoethanolamine as a carbon and nitrogen source, and susceptibility to cefamandole and erythromycin (38, 41, 67, 71, 72).

The mycolic acid composition of nocardiae has been reported to be characteristic for different species, and by both gas chromatography and HPLC analysis, tentative species assignments can be based on mycolic acid profiles (20, 72). It is important that mycolic acid profiles used for diagnostic purposes be interpreted cautiously. Numerous factors, including the stage of growth, type of culture media used, and temperature of incubation, may alter the mycolic acid profiles of nocardiae (10). Furthermore, induced mutations affecting the mycolic acid profile of a strain of *N. asteroides* have been reported (8). Therefore, all of these variables must be identified and rigorously controlled for comparisons among different diagnostic laboratories to be valid. Taking these considerations into account, cells of *N. asteroides* sensu stricto contain a mixture of mycolic acids varying in size from C_{42} to C_{58}, with sizes C_{50} and C_{52} being dominant; mycolic acids from *N. farcinica* vary in size from C_{44} to C_{58}, with the dominant sizes being C_{52} and C_{54}; and mycolic acids from *N. nova* vary in carbon chain length from C_{52} to C_{62}, with chain lengths of C_{56} and C_{58} being dominant (10, 29, 38, 41, 72).

It was shown that using a pool of DNA fragments from five different strains of *N. asteroides* was necessary to create a diagnostic probe that was sufficient to identify a large number of clinical isolates of *N. asteroides* (18). Unfortunately, study systems using DNA probes for diagnostic purposes need further study and evaluation, and these probes are not yet available for clinical use.

Serologic Tests

Serologic tests for nocardiae lack specificity and sensitivity, and patients exhibit low immunologic responsiveness against many nocardiae during an infectious process. At-

tempts to develop serologic tests have been complicated further by the antigenic heterogeneity of most pathogenic strains of *Nocardia* spp. (14, 15, 35, 48, 49, 54, 55, 62, 65).

Currently, diagnosis of nocardiosis relies on isolation and identification of the organism from the patient. Nevertheless, some serologic tests are useful in augmenting recognition of infection by nocardiae. For example, indirect immunofluorescence microscopy that uses dilutions of the patient's serum as the primary antibody against cells of *N. asteroides* affixed to glass slides can be an important screening process to aid in diagnosis. At a serum dilution of 1:100, most individuals (>90%) infected with nocardiae (culturally proven cases) give a positive immunofluorescence, whereas the majority of uninfected individuals are negative at this dilution of serum (35). Therefore, a negative result suggests that nocardiae are not involved in the disease process in the patient. It is important to emphasize that a positive immunofluorescence does not confirm diagnosis because of cross-reactivity with *Mycobacterium* spp. and other actinomycetes (35). Complement fixation and immunodiffusion are less frequently utilized for diagnosis of nocardial infections, but these assays may provide presumptive results similar to those of the immunofluorescence test. Unfortunately, these tests appear to be less sensitive than immunofluorescence, resulting in a greater proportion of false negatives (48, 49).

More recently, defined, purified antigens combined with either an enzyme-linked immunosorbent assay (ELISA) or a Western immunoblot analysis have been used to diagnose nocardial infections. A 55/54-kDa antigen purified from the culture filtrate of *N. asteroides* can be used for serodiagnosis of nocardial disease by using either ELISA or Western blot analysis (1, 2, 14, 15). However, even though the 55/54-kDa antigen is an important immunodiagnostic antigen for nocardiae, several clinical isolates of *N. asteroides* and *N. otitidiscaviarum* do not appear to secrete this antigen. In one study, mice infected with these strains were not seropositive for the 55/54-kDa antigen, but other characteristic antigens that could be visualized by Western blot analysis were produced (35). As a consequence, it has been recommended that in the absence of the 55/54-kDa antigen, the presence of both a 62- and a 36-kDa antigen in the Western blot profile be used for serodiagnosis of nocardial disease.

Additional key bands that may be present in Western blot profiles of sera from patients with culturally proven nocardiosis correspond to the following sizes: 90, 36, and 25 kDa (26/24 kDa) (35, 54, 55, 65). Indeed, the 26/24-kDa band is valuable for confirmation of the etiology of mycetomas caused by *N. brasiliensis* (55). Furthermore, the titer of antibody against this antigen decreases during treatment of mycetoma, and the levels of antibody return to background following complete cure of the mycetoma. These observations suggest that levels in serum of antibody against this antigen can in some instances be used for monitoring the efficacy of antinocardial therapy.

Antimicrobial Susceptibilities

The treatment of choice for nocardiosis has included the sulfonamides since the successful use of these agents in 1944 (12). Prior to 1944, pulmonary nocardiosis was usually fatal. The currently used sulfonamides include the highly soluble sulfisoxazole and sulfamethoxazole (SMX), with the latter given alone or in combination with trimethoprim (TMP), which adds to the activity of sulfonamides alone. For localized pulmonary or cutaneous nocardiosis, treatment with sulfonamides usually provides rapid resolution of disease (27, 59).

In patients with disseminated nocardiosis, therapy with sulfonamides has generally had a poor outcome (27, 59). The mortality rate for *N. asteroides* brain abscess has been approximately 50% (27). This high mortality has resulted in evaluation of a number of additional agents (28, 34, 69, 70, 73). Amikacin, ceftriaxone, cefotaxime, and imipenem have emerged as the major supplemental agents for nocardiosis. Typically, patients receive both amikacin and ceftriaxone with either a sulfonamide or the combination TMP-SMX until the patient is clinically improved and then the sulfonamide alone for 6 to 12 months.

For patients infected with *N. brasiliensis* (localized cutaneous infections or mycetoma), monotherapy with a sulfonamide or TMP-SMX has proven satisfactory (59). In patients with allergy to sulfonamides, amoxicillin-clavulanic acid is the optimal alternative oral agent, and minocycline is the second choice. For serious or disseminated disease, amikacin with either imipenem or ceftriaxone is most useful (68). Other drugs with activity against some isolates of the *N. asteroides* complex include minocycline, doxycycline, gentamicin, tobramycin, and the newer fluorinated quinolones (34, 70, 73) (Table 4).

The need for multidrug regimens for patients with serious infections due to *Nocardia* spp., the variable susceptibilities of *Nocardia* spp. to most drugs, and the frequent problem of allergy to sulfonamides or TMP-SMX (especially in the setting of AIDS) mandate that susceptibility testing be performed on clinically significant isolates of *Nocardia* (34, 69–71, 73). Although standardized methods of susceptibility testing for nocardiae have not yet been adopted by the National Committee for Clinical Laboratory Standards (NCCLS), a working group to standardize test methods for the aerobic actinomycetes has been proposed (41, 46). Problems with susceptibility testing by any method include the difficulty of obtaining a uniform suspension of organisms for inoculation, the relatively slow growth of the organism such that 2 to 3 days of incubation is required, and the limited availability of clinical data to support the laboratory interpretation of the results. Despite these problems, most clinicians agree that carefully performed susceptibility studies provide useful information that is reasonably predictive of clinical outcome.

The most commonly used methods of testing suscepti-

FIGURE 3 Variations in colony morphologies of nocardiae belonging to the *N. asteroides* complex grown on glucose-yeast extract agar at 37°C for 14 days. All of the cultures were inoculated onto the same batch of agar on the same day and incubated identically. The magnification of each colony is the same (bar = 1 cm), so sizes and appearances can be compared directly. (A) *N. asteroides* sensu stricto isolated from a nonimmunocompromised host with a pulmonary infection; (B) *N. asteroides* sensu stricto isolated from the kidney of a renal transplant patient with a fatal, disseminated infection; (C) *N. asteroides* sensu stricto ATCC 14759; (D) *N. asteroides* sensu stricto from a Hodgkin's disease patient with a fatal systemic infection; (E) *N. farcinica* from an alcoholic patient with a pulmonary infection; (F) *N. farcinica* from a patient with fatal endocarditis, taken following heart surgery to correct a defective valve; (G) *N. nova* from a dairy cow with severe mastitis.

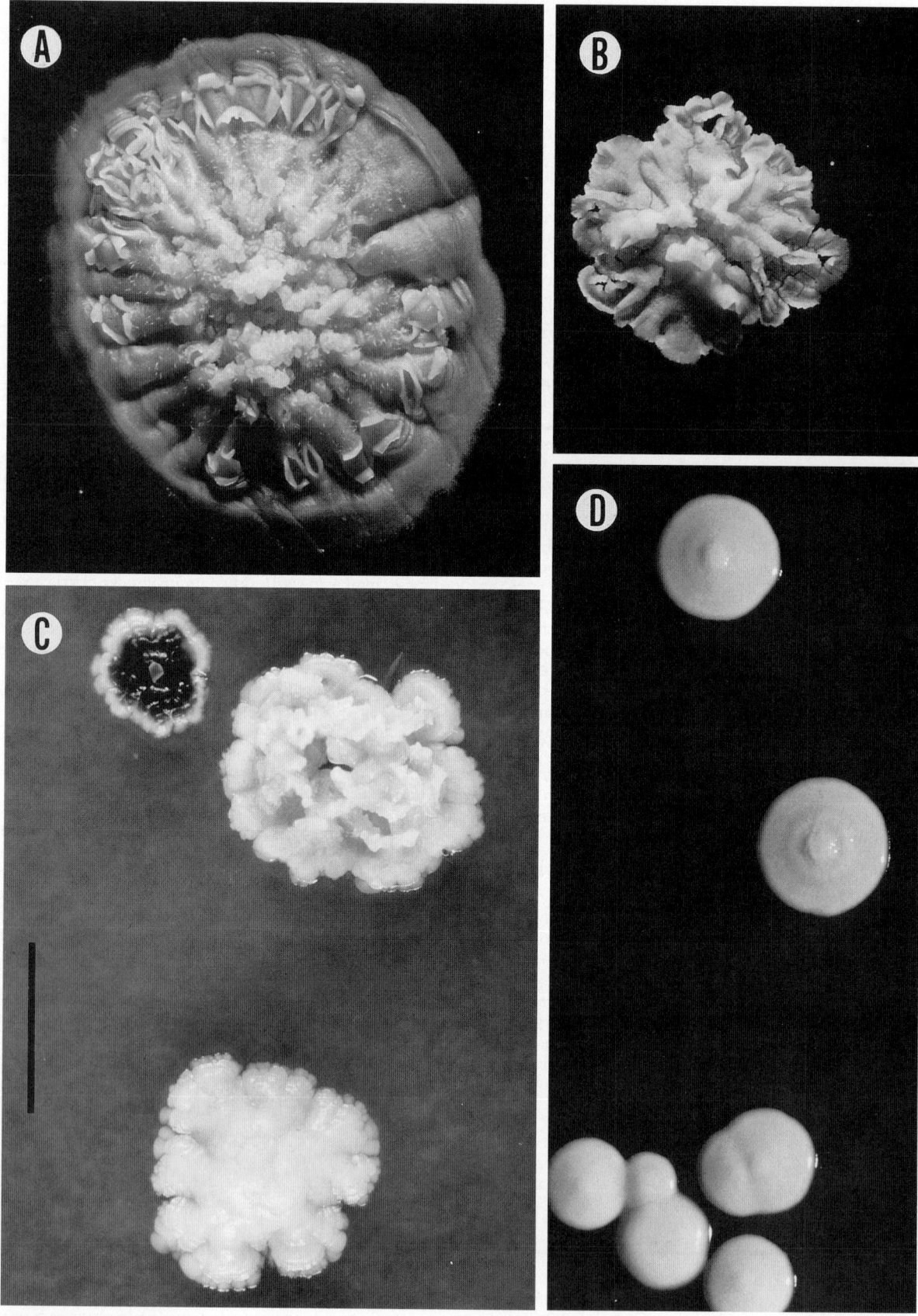

TABLE 3 Hydrolysis and selected biochemical profiles of pathogenic species of *Nocardia*[a]

| Organism | Decomposition of: | | | | | | Growth at 45°C | Arylsulfatase (14 day) | Acid from: | |
	Casein	Hypoxanthine	Tyrosine	Xanthine	Gelatin	Starch			Glucose	Rhamnose
N. asteroides	−	−	−	−	−	−	+/−	−	+	−/+
N. farcinica	−	−	−	−	−	−	+	−	+	+/−
N. nova	−	−	−	−	−	−	−	+	+	−
N. brasiliensis	+	+/−	+	−	+	−	−	−	+	−
N. otitidiscaviarum	−	+	−	+	−	−	+/−	−	+	−
N. transvalensis	−/+	+/−	−/+	+/−	−	+	−	−	+	−

[a]Table was compiled from data in references 7, 21, 36–38, and 41. +, predominantly positive; +/−, mostly positive with some species or strains negative; −/+, mostly negative with some species or strains positive; −, predominantly negative.

bility have been a disk diffusion method using Mueller-Hinton agar and a broth microdilution method using Mueller-Hinton broth (69, 70). Disk diffusion susceptibility was the first method described for susceptibility testing of *Nocardia* spp. that had general clinical applicability. Because the growth rates of *Nocardia* spp. are different from those of rapidly growing organisms such as *Staphylococcus* spp., the zone sizes have different correlations with MICs against nocardiae than with those against the more rapidly growing bacterial species. Tentative breakpoints for susceptibility and resistance have been published (69), but these are limited by the small number of isolates compared by disk diffusion and MIC determination (generally less than 100 for each drug). This technique at best provides only a crude indication of antibiotic activity.

A broth microdilution methodology has also been well described for *Nocardia* spp. (69). This utilizes standard twofold dilutions of antimicrobial agents in Mueller-Hinton broth. Drugs tested by this method generally include amikacin, amoxicillin-clavulanic acid, ampicillin, cefotaxime and/or ceftriaxone, ciprofloxacin, erythromycin, imipenem, minocycline, and a sulfonamide or TMP-SMX (69, 70). Current MIC breakpoints for susceptibility and resistance are those of the NCCLS for organisms that grow rapidly when the same antimicrobial agents are used (46). Details of this method were recently published in the *Clinical Microbiology Procedures Handbook* (17).

Recently, the Epsilometer or E Test (AB Biodisk, Solna, Sweden) was evaluated for the determination of susceptibilities of the nocardiae (13). This system uses a strip impregnated with a gradient concentration of antimicrobial agent. The analysis is performed in a manner almost identical to that used for disk diffusion studies except that in lieu of measuring diameters of zones of inhibition, the MIC of an antimicrobial agent is determined by reading the point at which the ellipsoidal zone of inhibition intersects the E Test strip. Results obtained with the E Test have excellent correlation with those obtained by either disk diffusion (>93%) or broth dilution (>96%) methods. Furthermore, there is excellent inter- and intralaboratory reproducibility (>95%) with the E Test. Unfortunately, this method presently suffers from the same problem as other in vitro susceptibility study methods by not having adequate system-

atic studies correlating results to the clinical outcomes in patients (13).

Evaluation, Interpretation, and Reporting of Results

The pathogenic species of *Nocardia* are not seen frequently in the clinical laboratory as environmental contaminants, and these organisms rarely represent part of the normal flora of the human body. Therefore, isolates of *Nocardia* spp. from a clinical sample should always be presumptively identified and reported so that the clinician can interpret the isolate's possible involvement in the disease process. Gram-stained preparations of clinical material such as sputum that reveal gram-positive, beaded filaments are presumptive evidence of nocardiae, pending culture results (Fig. 2). Even if the samples remain negative by culture, visualization of these organisms in smears should not be ignored. A modified Kinyoun acid-fast stain may differentiate the nocardiae from *Actinomyces* spp. as well as from most other actinomycetes (7, 41).

RHODOCOCCUS, GORDONA, AND TSUKAMURELLA

Taxonomy

The genus *Rhodococcus* includes a diverse group of organisms quite variable in morphology, growth patterns, biochemical characteristics, and capacity for causing disease. The cells are aerobic, catalase-positive, gram-positive, nonmotile, branching filamentous bacteria that fragment into rods and cocci. The branching filaments may be rudimentary, highly transitory, or extensive. Aerial filamentation (aerial hyphae) is not usually produced, but there are exceptions. Most strains are sensitive to lysozyme, but some are not. The cell wall contains *meso*-DAP, arabinose, galactose, and a mixture of mycolic acids with carbon chain lengths varying from C_{34} to C_{52} (C_{40} predominates). The cell envelope contains straight-chain saturated and unsaturated fatty acids as well as major amounts of tuberculostearic acid. The phospholipids include diphosphatidylglycerol, phosphatidylethanolamine, and phosphatidylinositol mannosides. The menaquinones are dihydrogenated with either eight or nine isoprene units [MK-8(H$_2$) for *Rhodo-*

FIGURE 4 Comparisons of colony morphologies of different aerobic actinomycetes grown on glucose-yeast extract agar as described in the legend to Fig. 3. Bar = 1 cm. (A) *N. otitidiscaviarum* from a patient with no identifiable underlying illness and a fatal brain infection; (B) *N. brasiliensis* from a mycetoma of the foot; (C) *A. madurae* from a mycetoma of the foot; (D) *R. equi* from an Arabian foal with pneumonia.

TABLE 4 Recommended drugs for susceptibility testing of *Nocardia* spp.

Essential	Optional
SMX or TMP-SMX	Gentamicin
Amikacin	Tobramycin
Ceftriaxone or cefotaxime	Erythromycin
Imipenem	Ciprofloxacin
Amoxicillin-clavulanic acid	Ampicillin
Minocycline	

coccus spp. or MK-9(H$_2$) for *Gordona* spp.]. The G+C mole percentage of the DNA ranges from 63 to 73 (Tables 1 and 2). Some strains of *Rhodococcus* can be difficult to differentiate from some strains of *Nocardia*. Recently, this group has undergone reorganization. Previously, 20 species (including several incertae sedis) were recognized, and these species could be divided into three stable groups on the basis of chemical constituents (Table 5) (38). One group is characterized by cell wall constituents made up of shorter C$_{34}$ to C$_{52}$ mycolic acids and eight isoprene dihydrogenated menaquinones, while a second group has the longer C$_{48}$ to C$_{66}$ mycolic acids and menaquinones with nine isoprene units (38). Members of the first group remained in the genus *Rhodococcus*, but members of the second group were transferred to the genus *Gordona*, now including *Gordona bronchialis*, *G. rubropertincta*, *G. sputi*, and *G. terrae* (Table 5) (38, 50, 61). Furthermore, on the basis of molecular and chemical methods, it was suggested that *Rhodococcus aurianticus* be replaced by a third new genus and species, *Tsukamurella paurometabola* (22, 37).

Natural Habitat

Organisms in this group are associated with a variety of environmental locations and especially with soil and farm animals. They have often been isolated from feces of herbivores and swine, with fecal contamination significantly increasing the organisms' rates of multiplication in soil. They have also been isolated from fresh and marine waters and from the guts of some arthropods (25, 38, 40).

Clinical Significance

Rhodococcus equi was first reported to cause disease in humans 40 years after its establishment as an animal pathogen in the 1920s. Numerous cases have since been reported, with the majority in immunocompromised patients, especially those with AIDS (25). Although members of the

genus *Rhodococcus* and the related genera *Gordona and Tsukamurella* are infrequently isolated from human clinical specimens, their increasing roles as opportunistic pathogens are becoming more apparent (25, 40, 41, 50).

Pathogenic primarily for livestock, *R. equi* has been the species most commonly associated with disease in humans. It is a facultative intracellular pathogen that resists phagocytosis as well as intracellular killing by macrophages (25). Pneumonia, with possible abscess and cavity formation, is the primary presentation in more than 75% of cases, and the organism is frequently associated with concomitant bacteremia in immunosuppressed patients. *R. equi* has also been the cause of bacteremia in association with central line infection, focal cervical lymph node infection, peritonitis, and endophthalmitis (26).

Other rhodococci with pathogenic potential include *Rhodococcus fascians* (a plant pathogen), *Rhodococcus luteus* and *Rhodococcus erythropolis* (agents of endophthalmitis), four species incertae sedis, and several unidentified rhodococcal species (11, 60). Included in the last group is a new proposed species reported to have contaminated erythrocyte concentrates received by three patients, with resultant septicemia; the isolates resembled but were significantly different from *R. erythropolis* and *Rhodococcus ruber* (24).

In the related genus *Gordona*, *G. bronchialis*, *G. sputi*, *G. terrae*, and a few unidentified *Gordona* species have occasionally been associated with infection of the skin, pulmonary disease, and line-associated bacteremia, the last being reported in two immunocompetent patients undergoing long-term parenteral nutrition (19, 51). *T. paurometabola* has been reported to cause meningitis, pneumonia, necrotizing fasciitis, and peritonitis in a patient undergoing ambulatory peritoneal dialysis as well as bacteremia in three patients with central lines, abscesses, and synovitis (4, 57).

Overall, 63 isolates of rhodococci and 41 isolates of other uncommon aerobic actinomycetes (16 of *T. paurometabolum*) were submitted to the Centers for Disease Control and Prevention (CDC) in Atlanta for identification over a 29-month period in the late 1980s (41, 42). Although the organisms are part of the environmental microbiota, clinically significant cases of infection caused by *Rhodococcus*, *Gordona*, and *Tsukamurella* spp. may be grossly underdiagnosed, because isolates frequently are not recognized as possible etiologic agents but rather are considered probable contaminating diphtheroids (41).

The organisms are primarily acquired via air; contaminated dust may be inhaled or may settle to contaminate medical devices such as central lines. Although the organisms have been associated with animals, a history of contact with animals or contaminated soil is not necessary and may occur with only approximately 30% of patients (25, 40). Immunocompromising conditions such as AIDS or long-term reliance on corticosteroids significantly increase patient risk for clinical infection (26, 30).

Specimen Collection, Transport and Storage

Depending on the clinical presentation, blood, lower respiratory tract secretions, catheter tips, and specimens collected from infected sites may yield rhodococci. At times, more invasive procedures, such as bronchial brushing to obtain lower respiratory tract secretions, may more reliably provide rhodococcal isolates (50). Collection and transport techniques should otherwise follow submission protocols commonly used for routine microbiologic isolation studies.

TABLE 5 Species included in the genera *Rhodococcus*, *Gordona*, and *Tsukamurella*[a]

Genus	Species
Rhodococcus	*aichiensis*, *chlorophenicolus*, *chubuensis*, *coprophilus*, *equi*, *erythropolis*, *fascians*, *globerulus*, *luteus*, *marinonascens*, *maris*, *obuensis*, *rhodnii*, *rhodochrous*, *roseus*, *ruber*
Gordona	*bronchialis*, *rubropertincta*, *sputi*, *terrae*
Tsukamurella	*paurometabolum*, *wratislaviensis*

[a]Table was compiled from data in references 38 and 41.

Direct Examination and Diagnostic Approaches

In stained preparations of exudates or tissue, the rhodococci and members of related genera are typically seen as gram-positive short rods or coccobacilli, often forming palisades, and may easily be confused with the diphtheroids. They may be located extracellularly or intracellularly (41). Unlike some of the other aerobic actinomycetes such as the streptomycetes and nocardiae, they almost never form elongated filamentous or branching forms except very early in their growth cycles (36). Although some strains are partially acid fast, acid-fast organisms are less frequently seen, and often only a small percentage of organisms seen on a smear have acid-fast characteristics.

In exudates, the rhodococci may at times form white to yellow, small, soft granules made up of loose aggregates of bacteria (36). Histopathologic evaluations of specimens can be very helpful, especially if an isolate's precise role in the infectious process is in question.

Isolation Procedures

Most species of rhodococci as well as members of the genera *Gordona* and *Tsukamurella* grow well on routine bacteriologic media, and they are frequently isolated in the routine bacteriology laboratory (41, 50). However, because of their slower growth, they may be easily overgrown by organisms of the normal microbiota that contaminate specimens. When considering the aerobic actinomycetes as a whole, selective media such as brain heart infusion agar with chloramphenicol and cycloheximide or Columbia agar with colistin and nalidixic acid are usually recommended as supplements to routine media to enhance isolation (36). Routine media should also be inoculated, and these may include blood agar, chocolate agar, brain heart infusion agar, Trypticase soy agar, or Sabouraud dextrose agar. The rhodococci will not grow on all MacConkey formulations. Digestion-decontamination procedures, such as the ones used for mycobacteria, may be detrimental to some of the aerobic actinomycete genera and should not be used (36, 41).

Although the majority of pathogenic actinomycetes grow well at 35°C, some do not grow well at 37°C, while most seem to grow at 30°C (36, 50). Since rhodococci are obligate aerobes, culture plates should be incubated in ambient air or in 5% CO_2. The plates should be examined every 2 days during the first week and every 3 to 4 days thereafter for 2 to 3 weeks. For rapid detection of minute colonies, it is useful to initially read plates with either a handheld lens or a colony-viewing microscope. Plates should be sealed to prevent drying over the longer incubation period.

Identification

On agar, rhodococci may originate as clear to white, non-hemolytic, rounded, frequently mucoid colonies, often turning salmon pink to red within 4 to 7 days (Fig. 4D). Color, however, may at times vary from red, orange, yellow, and cream to white (36, 41, 50). Growth is usually good between 15 and 40°C but is slow, often requiring several days for initial visual observation. In most instances, aerial mycelia are not produced, although *Rhodococcus coprophilus* and *G. bronchialis* have been reported to produce either feeble aerial hyphae (the former) or synnemata (the latter), which may be confused with hyphae (37). The production of aerial and vegetative mycelia is best studied on freshly prepared tap water agar (36). Additionally, the rhodococci

are usually susceptible to vancomycin, erythromycin, and gentamicin, a characteristic that preliminarily may help differentiate them from the nocardiae.

Microscopically, the organisms are gram positive, coccobacillary, and nonmotile, resembling other coryneform bacteria. A small number of cells may at times stain by the modified Kinyoun weak-acid-fast method, contributing to their being confused with fragmented nocardial cells. Acid-fast-staining properties can be enhanced by passage of an isolate through a high-lipid medium such as sterile milk, Middlebrook 7H10 agar, or Lowenstein-Jensen medium (36).

Although characteristic colonial and microscopic morphologies may be helpful in recognition, the rhodococci may be further recognized and preliminarily identified by their production of catalase and their inability to grow in lysozyme or to hydrolyze casein, hypoxanthine, tyrosine, and xanthine (31, 41). Hydrolysis of Christensen's urea and reduction of nitrate may vary by species. *R. equi*, the most common clinical isolate, is oxidase negative and otherwise quite inert, being unable to oxidize or ferment carbohydrates or alcohols; it produces lipase and phosphatase but no other enzymes such as DNase, protease, lecithinase, or elastase (37, 41, 50).

Recently, a commercial miniaturized differentiation system, the Rapid CORYNE (API-bioMérieux S.A., Marcy-l'Etoile, France), was introduced for the 24- to 48-h identification of coryneform bacteria, including *R. equi* and *Oerskovia* species. However, preliminary evaluations showed the system to have limited accuracy, with only 65% of *R. equi* and 50% of *Oerskovia* species being identified correctly without supplemental tests. Unfortunately, another problem with this system is that it is based on fermentative reactions, and all rhodococci and gordonae key out as *R. equi* (39a). *R. equi* also produces interesting equi factors, which although not unique, can be helpful characteristic markers in its identification. Equi factors produce areas of complete hemolysis on sheep blood agar when they interact with hemolysin of *Listeria monocytogenes*, beta-toxin of *Staphylococcus aureus*, or phospholipase D of *Corynebacterium pseudotuberculosis* (see chapter 29 of this Manual) (50).

At times, differentiation between species of the genera *Rhodococcus*, *Gordona*, and *Tsukamurella* can be difficult. Definitive identification for taxonomic purposes by using numerical phenetic criteria is usually impractical in the clinical laboratory, needing 107 to 129 biochemical tests (37). Less extensive batteries of biochemical evaluations, such as the 29-test panel used at the Actinomycete Laboratory at the CDC, are possible and helpful but may not identify all species in this group of organisms (37). Selected biochemical profiles to help separate the key genera and species are listed in Table 6.

Identification may require additional analysis at the chemical and molecular levels to evaluate compositions of mycolic acids, amino acids, menaquinones, and sugars as well as DNA base contents (G+C mole percentage), a requirement usually beyond the capabilities of most clinical laboratories. Presence or absence of DAP, its stereoisomeric configuration (*meso* or L form), and the cell wall chemotype are helpful in differentiating some genera of aerobic actinomycetes. Such studies usually include whole-cell hydrolysis and then end product analysis by paper, thin-layer, or liquid chromatography (36). Presence of the *meso* isomer of DAP, arabinose, and galactose defines cell wall type IV, which is characteristic of the rhodococci, nocardiae, and

TABLE 6 Hydrolysis and selected biochemical profiles of some clinically significant aerobic actinomycetes[a]

Organism	Hydrolysis of:					Nitrate to nitrite	Urease
	Casein	Hypoxanthine	Tyrosine	Xanthine	Gelatin		
Rhodococcus equi	−	−	−	−	−	+/−	+
Rhodococcus spp.	−	−	−/+	−	−	+/−	+/−
Gordona spp.	−	−	−	−	+	+/−	+/−
Tsukamurella spp. ("*R. aurianticus*")	−	+	−	+	−	+	+
Streptomyces somaliensis	+	+	−	+	+	−	+
Streptomyces anulatus (*S. griseus*)	+	+	+	+/−	−	−	−
Oerskovia turbata	+	−	−	−	−	−	+/−
Oerskovia xanthineolytica	+	+	−	+	+	+	−/+
Actinomadura spp.	+	+	+	−	+	+	−/+
Nocardiopsis spp.	+	+	+	+/−	+	+/−	+/−

[a]Table was compiled from data in references 36–38, 41, and 50.

mycobacteria and separates them from the oerskoviae (cell wall type VI) and streptomycetes (cell wall type I) (23, 30, 38).

Recently, analysis of restriction length polymorphism in rRNA genes (ribotyping) was found to be useful for further differentiation of isolates for taxonomic or epidemiologic purposes (19, 60). Gas-liquid chromatography fails to detect metabolic products from glucose in the rhodococci, but analysis of mycolic acids by HPLC was efficacious in separating some species (20, 50). Patterns were stable and could be used to separate the rhodococci from the nocardiae and other aerobic actinomycetes and to provide additional data for species identification (20). Mycolic acids of a more polar nature were identifiable in *R. equi*, *R. erythropolis*, and *Rhodococcus rhodocrous*, whereas less polar mycolates were found in *G. sputi*, *G. bronchialis*, *G. rubropertincta*, and *G. terrae*.

Serologic Tests

R. equi has at least 27 capsular serotypes associated with its polysaccharide capsule, which does not seem related to virulence (50). Although the rhodococci and related genera have antigenic variability and initiate immune responses in their hosts, serologic studies do not replace culture in the diagnosis of infections. Several immunologic assays, including a lymphocyte stimulation assay, a synergistic hemolysis inhibition assay, an agar gel immunoprecipitation test, and ELISAs, have had very limited success in diagnosing clinical disease (50). These assays have often shown, however, that subclinical infection and exposure to *R. equi* are widespread, especially in foals.

Antimicrobial Susceptibilities

Our experience with therapy of deep-seated infections due to the rhodococci is still limited. In vitro, human rhodococcal isolates are normally susceptible to erythromycin, rifampin, vancomycin, TMP-SMX, tetracycline, ampicillin-sulbactam, amoxicillin-clavulanic acid, the aminoglycosides, and imipenem (40). The broad-spectrum cephalosporins and the fluoroquinolones also have good activities, but susceptibility may be more variable and should be tested prior to long-term use. The penicillins, penicillinase-resistant penicillins, and narrow- and expanded-spectrum cephalosporins are often inactive or show only moderate activity (25, 40).

The combination of erythromycin and rifampin may be synergistic against the intracellular rhodococci and is the therapy of choice in foals, but it may be disappointing in patients with AIDS (25). Other agents with good activity as well as good penetration into macrophages include gentamicin, tobramicin, and ciprofloxacin. Vancomycin especially has been recommended as having good efficacy either alone or in combination with another agent such as imipenem (47). Duration of therapy is uncertain, although periods of many weeks up to 6 months have been suggested, because relapses may occur after shorter courses. In some instances of reported cases of relapse, review of the antimicrobial agents used has indicated decreased in vitro activity as a possible contributor to poor outcome. Unfortunately, standardized susceptibility tests are not yet available, although methods applied to the nocardiae and modifications of methods recommended by the NCCLS for rapid growing bacteria may provide insight into susceptibility patterns when necessary (23, 26, 40, 46). Surgical intervention may also be necessary in some instances, especially if implants are present.

Evaluation, Interpretation, and Reporting of Results

As clinical laboratories become more proficient at recognizing and identifying the rhodococci and related organisms, their clinical role will be better elucidated and more appreciated. These organisms should not be immediately relegated to the "contamination" category, but their significance should be carefully weighed in light of the patient's presentation and clinical status.

The rhodococci should always be at least presumptively identified and reported so that the clinician can make a careful interpretation of an isolate's possible involvement in the disease process. The isolate may be mixed with other organisms and still be a significant contributor to morbidity (50). If identification is difficult, a preliminary report with presumptive or tentative identification and susceptibilities should be sent to the clinician, and the isolate should be submitted to a reference laboratory capable of providing additional information as to its identity. The clinician should be advised of the isolate's pathogenic potential in the right clinical setting. Certainly, the clinical picture, specimen type, number of isolations, and histopathologic review of specimens may provide clues as to the role of the isolate (41).

STREPTOMYCES

Taxonomy

The genus *Streptomyces* consists of numerous poorly defined species of sporoactinomycetes, which are aerobic, catalase-positive, gram-positive, nonmotile, extensively branched filamentous bacteria that form aerial filaments (aerial hyphae) with short to long chains of nonmotile spores. They have a characteristic type I cell wall consisting of L-DAP and glycine but no characteristic sugar and no mycolic acids. The cell envelope contains saturated *iso* and *ante-iso* fatty acids but does not have tuberculostearic acid. The cells are sensitive to lysozyme. The menaquinones are hexa- or octahydrogenated with nine isoprene units [MK-9(H$_6$) or MK-9(H$_8$)]. The G+C mole percentage of the DNA ranges from 69 to 78 (36, 41, 42, 53).

Over 3,000 species of *Streptomyces* have been described, with identification often being primarily based on features of morphology and pigmentation. On the basis of phenetic studies, many species were reclustered into species groups, each being named after the first valid species in the group. Therefore, *Streptomyces griseus* was superseded (controversially) in its group by *Streptomyces anulatus*, which has precedence (41).

Natural Habitat

Being primarily soil inhabitants, the pathogenic strains of streptomycetes have diverse geographic prevalences, none being common in the United States. *S. somaliensis* prefers arid regions with sandy soil and is found most frequently in Africa, Arabia, Mexico, and portions of South America. *Streptomyces paraguayensis* is associated with soil, decaying vegetation, plants, and thorns and is found most commonly in Mexico, Central America, and South America (36). Owing to controversy over the actual identity of human isolates of *S. paraguayensis*, there is a question as to its role as a primary pathogen (36). *S. anulatus* (*S. griseus*) appears to be the most common isolate in the United States, being reported as representing 7.7% of all aerobic actinomycetes referred to the CDC over a 2.5-year period (42). Common sources included sputum, wound, blood, and brain tissue. Although cases of primary infection of the lower respiratory tract have not been confirmed, *S. anulatus* has been reported to be associated with lung disease and sepsis, brain abscess, and panniculitis as well as cervical lymphadenitis in an immunocompromised host (42). In the past, other species of the genus *Streptomyces* have commonly been considered saprophytes, being found universally as environmental contaminants (36).

Clinical Significance

The genus *Streptomyces* has not usually been recognized as being associated with human disease in the United States and has rarely been documented as a cause of disseminated disease. Although a plethora of species belongs to this genus, only three species (*S. anulatus* [*S. griseus*], *S. somaliensis*, and *S. paraguayensis* [questionable taxonomic validity]) have in the past been reported with any frequency as isolates from human infections (30, 38, 41, 43). Several recent reports, however, suggest that an increasing number of other species in the genus may play clinically significant roles (e.g., *Streptomyces albus*, *Streptomyces coelicolor*, *Streptomyces lavendulae*, *Streptomyces rimosus*, and *Streptomyces violaceoruber* [41]).

In general, streptomycetes are most commonly limited to causing actinomycotic mycetoma (39, 63). Areas of the body more prone to formation of mycetomas are those that are frequently traumatized or that come into contact with soil. The organisms probably gain access to subcutaneous tissues through abrasions in the skin. Initial formation of painless nodules is followed by fluctuance and progression to formation of draining sinuses. The sinuses begin to drain, with the exudate often containing hard, compacted, filamentous microcolonies or granules. Additional sinus tracts and nodules may form, and the involved areas eventually become deformed by fibrous tissue. The process remains characteristically pyogenic rather than granulomatous and may reach deeper connective tissue, muscle, and bone. The disease rarely spreads hematogenously, although it may occasionally involve lymph nodes via the lymphatics (39, 63). The streptomycetes rarely cause actinomycetoma in the United States, while *S. somaliensis* remains the most common species of this genus causing mycetomas worldwide.

Specimen Collection, Transport, and Storage

When granules are present, drainage from sinus tracts is adequate for isolation of actinomycetes from mycetomas. Pus or exudate may be expressed directly into a sterile tube or container or, alternatively, collected on a gauze pad laid over the drainage site for several hours. If only a small amount of drainage material can be obtained, it should be kept moist and not allowed to dry. If drainage is absent or minimal or if the surface material is contaminated or colonized, biopsy of deeper tissue is preferred, and specimens may include scrapings or aspirates from draining lesions (36, 56). Collection of other specimens such as lymph node or other tissue, lower respiratory tract secretions, or blood for culture may be indicated depending on the patient's clinical presentation.

Direct Examination and Diagnostic Approaches

Biopsy or drainage material collected from mycetomas should be examined for granules (averaging 1 mm in diameter), which may be hard. Granules of *S. somaliensis* tend to be white to yellow to brown and round to oval; those of *S. paraguayensis* are commonly black (36, 74). The granules should be washed in sterile saline several times and examined microscopically for filaments after emulsification in a drop of 10% potassium hydroxide or crushing between two slides (56, 64). Gram stain will reveal the streptomycetes as a tangled mass of gram-positive fine filaments, differentiating them from the broad, segmented hyphae of fungi. The Gram, Brown and Brenn modification of the Gram, the Gram-Weigert, and hematoxylin and eosin stains will detect *Streptomyces* spp. in tissue, while the Gomori methenamine silver stain usually fails to stain the filaments (56, 64, 74).

Isolation Procedures

The streptomycetes are not fastidious, and culture setup and incubation as well as the schedule for reading of plates should follow guidelines set forth for the rhodococci and nocardia. Some strains of streptomycetes, however, may not grow well on all of the selective media for the nocardiae, producing small colonies on such agars. It is therefore recommended that routine media be set up concurrently as outlined for the rhodococci. The addition of Sabouraud dextrose agar, both with and without chloramphenicol (0.05 mg/ml), may also be useful in the primary isolation of *Streptomyces* spp. (36, 41, 56).

Identification

Streptomycetes are slower growing than many bacteria isolated from clinical samples, and they produce glabrous or waxy heaped colonies. Colonies of *S. somaliensis* are leathery and vary in color from cream to dark brown or black; characteristic chains of spores are often present on white aerial hyphae (36). *S. paraguayensis* may produce white to gray, tough colonies with short aerial hyphae. The other nonpathogenic streptomycetes most frequently grow as chalky colonies of various colors, with white to grayish colony forms predominating. Colonial morphology is too variable to be used in differentiating the streptomycetes.

Colonies of *Streptomyces* spp. produce substrate hyphae that do not fragment easily, and they often have specialized aerial hyphae bearing curved to spiral chains of spores (36). Production, configuration, and pigmentation of spores and spore chains have been widely used in identification and classification of the streptomycetes. In one simplified scheme, three morphologic categories of chains are recognized: spore chains that are straight to flexuous (*Rectiflexibles*), those that are spiral (*Spirales*), and those that are looped (*Retinaculiaperti*). These broad types of identification schemes, however, have problems with overlap and a lack of distinct categories (38). Although substrate mycelia are non-acid fast, aerial spores may at times stain acid fast, thereby causing some confusion in separating them from the nocardiae or rhodococci (36). The production of substrate and aerial hyphae as well as the production of spores may be best evaluated on tap water agar incubated for up to 7 days.

Although many workers have applied morphologic and pigmentation studies together with numeric taxonomic data to delineate the genus *Streptomyces*, definitive identification to species level usually remains extremely difficult (38). Molecular techniques such as DNA homology evaluations or genetic probe studies may allow a better understanding and differentiation of this genus in the future.

Serologic Tests

Gel diffusion methods using antigens prepared from *S. somaliensis* have been used to distinguish actinomycetomas from eumycetomas caused by the fungi (63). It has been suggested that antigens prepared from actinomycetes and fungi known to cause mycetomas be used to screen patient sera to distinguish mycetomas caused by fungi from those caused by bacteria. However, clinical data on using the serologic approach for diagnosis are lacking.

Antimicrobial Susceptibilities

Therapeutic regimens may differ for an actinomycotic mycetoma and infection at other sites. A mycetoma may respond to therapy, although a protracted course of several months may be necessary to stop relapse. Streptomycin in combination with another agent such as dapsone or TMP-SMX has more commonly been suggested. *S. somaliensis*, however, has shown the most in vitro resistance to these agents and the most clinical failures (41). There is little experience in treating other streptomycete infections. Using a modified broth microdilution method, McNeil et al. found broad variability in the susceptibilities of strains of *S. anulatus* (*S. griseus*) in the United States (42). Amikacin had excellent activity (100% of strains susceptible); minocycline (90%), erythromycin (86%), doxycycline and imipenem (81%), and ceftriaxone (80%) were also active against *S. anulatus*. Other antimicrobial agents studied had poorer activities, ranging from 43% (ciprofloxacin) to 71% (TMP-SMX). Although not standardized, susceptibility studies can be used as a guide for developing a therapeutic protocol.

Evaluation and Interpretation of Results

The role of streptomycetes in mycetoma is well known and readily documented by culture or histopathology. Their role in other infectious processes, on the other hand, is less well appreciated and more suspect. The isolation rate of *S. anulatus* (*S. griseus*) from sputum and wounds, which is higher than those of other *Streptomyces* spp., arouses suspicion concerning its potential role as a primary pathogen (41, 42). As with rhodococci, an isolate's role should be carefully assessed along with the parameters of a patient's clinical status, specimen type, number of specimens yielding the isolate, and a histopathologic review where applicable. The isolate should be identified to genus level and submitted to a reference laboratory for further identification when necessary.

OERSKOVIA

Taxonomy

Members of the genus *Oerskovia* are facultatively anaerobic, catalase-positive (aerobic growth only), gram-positive, branching filamentous bacteria that fragment into motile pleomorphic rods (some strains are not motile). No aerial filaments are formed. The cell wall is type VI with lysine (no DAP), galactose (no arabinose), and no mycolic acids. The menaquinones are tetrahydrogenated with nine isoprene units [MK-9(H$_4$)]. The G+C mole percentage of the DNA ranges from 70 to 75. Two motile species, *Oerskovia turbata* and *Oerskovia xanthineolytica*, and two types of less well defined, nonmotile *Oerskovia*-like strains (type A and type B), have been described elsewhere (38, 41). Previously, these organisms were classified in many genera, including *Nocardia*. The presence of a type VI cell wall with lysine and galactose as major components and the absence of DAP and mycolic acids together with facultatively anaerobic growth and the ability to fragment into motile elements separates members of this genus from all other aerobic actinomycetes. Some investigators believe that this genus should be combined with the genus *Cellulomonas* (see chapter 29 of this Manual); however, the genus *Oerskovia* currently remains separated from the genus *Cellulomonas* (38, 41).

Natural Habitat

The oerskoviae are found primarily in soil, water rich in organic material, decaying plant materials, brewery sewage, and aluminum hydroxide gels (38). Because they are found primarily in soil, the oerskoviae probably cause infection by contamination of wounds or implant sites from a natural environmental reservoir (38, 52).

Clinical Significance

Previously, the genus *Oerskovia* has rarely been implicated in human disease. However, the number of cases may be underdiagnosed because of poor recognition of the etiologic agent, and the numbers of infections caused by *Oerskovia* spp. may be larger than suspected. Fewer than 40 isolates were submitted to the CDC prior to 1977, with another 5 isolates being submitted in a 2.5-year period in the late 1980s (40, 41). The most common sources of isolation have included blood, heart tissue, and CSF. Urine was also a source in an anecdotal experience. Published reports have implicated the oerskoviae as primary etiologies of bacteremia, peritonitis, endophthalmitis, endocarditis, meningitis, and pyonephrosis. Infection is frequently associated with the presence of indwelling foreign bodies (e.g., catheters, prosthetic heart valves) or penetrating injuries (52).

Specimen Collection, Transport, and Storage

Blood, CSF, or other normally sterile body fluids (depending on the patient's presenting condition) readily yield the oerskoviae if they are present. Essentially the same methods described above for the rhodococci can be used for the oerskoviae.

Direct Examination and Diagnostic Approaches

Members of the genus *Oerskovia* are seen as gram-positive rods in specimens stained by routine methods (52). Concentration of normally sterile body fluids such as CSF or peritoneal fluid by cytocentrifugation can increase the sensitivity of direct smears.

Isolation Procedures

The oerskoviae are easily isolated on most routine bacterial agar- or broth-based media, including sheep blood and chocolate agar. However, fungal media such as Sabouraud dextrose agar may fail to yield adequate growth. Colonies normally become visible after 24 to 48 h of incubation at 35°C under aerobic, 5% CO_2, or anaerobic conditions (52).

Identification

The two major species of *Oerskovia* usually form bright yellow colonies without aerial hyphae after 48 h of incubation. In younger colonies, gram-positive, non-acid-fast surface filaments with extensive branching may be seen microscopically. In older colonies, these filaments fragment into short, often motile, rodlike forms. The oerskoviae are oxidase negative and catalase positive when grown aerobically, and they are susceptible to vancomycin (38, 41, 52). Biochemically, they produce acid from glucose, lactose, sucrose, maltose, and xylose; hydrolyze gelatin and starch; and usually reduce nitrate to nitrite (38). The two most important species may be differentiated by hydrolysis studies and growth at 42°C. Thus, *O. xanthineolytica* hydrolyzes casein, xanthine, hypoxanthine, and esculin but not tyrosine, and it grows at 42°C, while in contrast, *O. turbata* fails to hydrolyze xanthine or hypoxanthine and does not grow at 42°C (38, 52). The nonmotile *Oerskovia*-like strains can be differentiated from these two species by their lack of motility and their apparent inability to hydrolyze casein (Fig. 1; 38, 41).

Serologic Tests

Serologies have not contributed to the diagnosis of infection by members of the genus *Oerskovia* and are not normally available.

Antimicrobial Susceptibilities

Experience with treatment of infections caused by members of this genus is greatly limited. Cure may be more difficult to achieve because of the common association with foreign bodies. Thus, although penicillin, ampicillin, rifampin, vancomycin, and the aminoglycosides are all effective in vitro, reports have documented failure of therapy with active agents unless the foreign body is removed (52).

Although in vitro susceptibility studies are not standardized, they may help in the choice of therapy (42).

Evaluation and Interpretation of Results

The parameters and clinical suspicion used to evaluate the role in human disease played by the rhodococci and streptomycetes may also be applied when the significance of an isolate of *Oerskovia* is evaluated.

MISCELLANEOUS AEROBIC ACTINOMYCETES: *DERMATOPHILUS, ACTINOMADURA, NOCARDIOPSIS, THERMOACTINOMYCES, SACCHAROMONOSPORA,* AND *SACCHAROPOLYSPORA (FAENIA, MICROPOLYSPORA)*

Taxonomy

Although the genus *Actinomadura* contains 26 species, only 2 (*A. madurae* and *Actinomadura pelletieri*) are clinically significant (41). The genus consists of aerobic, catalase-positive, gram-positive, branched, filamentous organisms that grow in the substrate as nonfragmented filaments. They may produce aerial filaments, but colonies without aerial filaments have leathery or cartilaginous surfaces. The cells are sensitive to lysozyme. The cell walls contain *meso*-DAP, galactose, and madurose but no mycolic acids. Tuberculostearic acid is a major fatty acid. The menaquinones are variably hydrogenated with nine isoprene units [MK-9(H_4, H_6, H_8)]. The G+C mole percentage of the DNA ranges from 65 to 77 (Tables 1, 2, and 6; 41, 75).

The genus *Dermatophilus* contains only one species, *Dermatophilus congolensis*. The cells of *D. congolensis* are aerobic or facultatively anaerobic, catalase-positive, gram-positive, branching filamentous bacteria that may form aerial filaments (aerial mycelium) when grown in an increased-CO_2 atmosphere. The filaments are long and tapering, with right-angle lateral branching. Septae develop along the filaments both horizontally and vertically, giving rise to parallel rows of coccoid cells (up to eight rows along a single filament) that become motile by a tuft of flagella. The organisms are not acid fast. The cell walls contain *meso*-DAP and madurose but not mycolic acids. The organisms have a type I phospholipid pattern. The G+C mole percentage of the DNA ranges from 57 to 59 (38, 41, 75).

Organisms in the genus *Nocardiopsis*, segregated from the genus *Actinomadura* in 1976, consist of cells that are aerobic, catalase-positive, gram-positive, non-acid-fast, nonmotile, extensively branched long filaments with well-developed aerial extensions that completely fragment into aerial spores. The cell walls are sensitive to lysozyme. The cell walls contain *meso*-DAP and no characteristic sugars. The organisms do not contain mycolic acids. They have either a type III or a type IV phospholipid pattern, and the menaquinones are either MK-10(H_4, H_6, H_8) or MK-9(H_4). The G+C mole percentage of the DNA ranges from 64 to 69 (Tables 1, 2, and 6; 38, 41). The genus contains a single clinically significant species, *Nocardiopsis dassonvillei*.

The genus *Saccharopolyspora* (also known as *Faenia* or *Micropolyspora*) consists of aerobic, catalase-positive, gram-positive, non-acid-fast, nonmotile organisms that form branched substrate filaments and aerial extensions. Both the substrate and aerial hyphae possess short unbranched lateral and terminal sporophores that give rise to as many as

20 spores in chains. The organisms are susceptible to ly-sozyme. The cell walls contain meso-DAP, arabinose, and galactose but no mycolic acids. Growth occurs on media containing 10% NaCl when incubated at 60°C. The mena-quinones are MK-9(H_4, H_6, H_8) (38).

The genus *Saccharomonospora* contains a monosporic actinomycete that has a type IV cell wall containing meso-DAP, arabinose, and galactose. The cells do not contain mycolic acids. They are moderately thermophilic, since there is a good growth at 50°C, but growth is poor above 60°C. They are sensitive to lysozyme. The G+C mole percentage of the DNA ranges from 69 to 74 (38).

Cells in the genus *Thermoactinomyces* are aerobic, cata-lase-positive, gram-positive, non-acid-fast, nonmotile, spore-forming organisms that form well-developed, branched substrate filaments that may protrude as aerial extensions. The spores, which have all of the properties of true bacterial endospores, are formed singly from sporo-phores on both substrate and aerial filaments. The organ-isms are thermophilic (growth at 55°C) and resistant to lysozyme. The cell wall has meso-DAP but no characteristic sugars. Mycolic acids are absent. The G+C mole percent-age of the DNA ranges from 52 to 55 (38).

Natural Habitats

A. madurae is found widely associated with soil, while *A. pelletieri* has been isolated only from clinical specimens. *D. congolensis* has not been recovered from the environment and has been associated only with infected material; con-taminated soil is probably its reservoir. *Saccharopolyspora rectivirgula*, *Saccharomonospora viridis*, and *Thermoactinomy-ces* spp. are thermophilic actinomycetes that grow in com-post, hay, straw, manures, and other vegetable matter that may undergo spontaneous heating up to 60°C (36, 38, 53).

Clinical Significance

D. congolensis causes an acute to chronic suppurative cuta-neous disease called dermatophilosis that affects a wide range of animals, including humans (75). Although der-matophilosis is recognized most frequently in livestock (i.e., cattle, sheep, and horses) in tropical and subtropical areas of Africa, it has been diagnosed in most regions of the world. In the United States, it has been confirmed in cattle, wild animals, and humans.

A. madurae, *A. pelletieri*, and *Nocardiopsis dassonvillei* cause actinomycetomas in humans (36, 38, 39, 53). Infec-tions caused by these organisms are recognized more fre-quently in the tropical and subtropical regions of Africa, India, and Central and South America; however, they have been diagnosed in the United States. Although these or-ganisms usually cause localized infections that result in mycetomas, they occasionally cause pulmonary or dissemi-nated disease in compromised individuals.

Spores or cellular fragments of *Saccharopolyspora rec-tivirgula*, *Saccharomonospora viridis*, and *Thermoactinomy-ces* spp. become airborne and may be inhaled following mechanical disturbance of materials containing these organisms. As a consequence, they often induce hyper-sensitivity pneumonitis (extrinsic allergic alveolitis), re-sulting in a process called "farmer's lung disease." There is little or no evidence that these thermophilic actino-mycetes cause an actual infectious process, even though they are capable of growing at 37°C. Therefore, they are not usually recovered from pulmonary secretions from individuals with this form of disease. Extrinsic allergic alveolitis is prevalent among individuals living in agri-culturally active areas in temperate regions of the world, including the United States (36, 38, 53).

Direct Examination and Diagnostic Approaches

Dermatophilosis is presumptively diagnosed by demonstra-tion of the organism in characteristic lesions through mi-croscopic examination. Using Gram, methylene blue, or Giemsa staining, *Dermatophilus* cells appear in lesions as branched filaments composed of parallel packets of coccoid cells that are due to both longitudinal and transverse sep-tation (75). The diagnosis must be confirmed by isolation and identification of the organism from the scab of the lesion.

In general, recognition of infections caused by *Actino-madura* and *Nocardiopsis* spp. relies on demonstration of the organism in mycetomatous lesions by microscopic exami-nation of characteristic granules, especially in hematoxylin-eosin-stained sections of biopsy samples from affected areas (36, 38, 41, 53). Visually, exudates from mycetomas caused by *A. madurae* contain relatively large, irregularly lobulated granules that are 1 to 5 mm in diameter and vary in color from white to yellowish or pink to red. These granules tend to be soft and easily crushed to a pastelike consistency. Sections of granules stained by hematoxylin and eosin show a poorly stained center surrounded by a basophilic mesh of filaments that may be circumscribed further by a pinkish or eosinophilic border composed of filaments and club-shaped elements. In contrast, granules produced by *A. pelletieri* are smaller than those produced by *A. madurae*. Furthermore, they are usually deep red, more regularly spherical (the granules may be lobulated or angular), and soft. Prepara-tions stained by hematoxylin and eosin tend to show frac-tured, round, sharply delineated granules with basophilic leading edges (36). *Nocardiopsis dassonvillei* forms small, moderately hard granules that are spherical and cream colored. Sections stained with hematoxylin and eosin re-veal basophilic granules that have entire, sharply defined edges with no radiating filaments. As with *D. congolensis*, confirmed diagnosis for the etiology of a mycetomatous lesion requires the isolation and identification of the spe-cific organism involved (36, 38, 53).

Diagnosis of hypersensitivity pneumonitis caused by the thermophilic actinomycetes relies on a combination of a history of exposure and appropriate clinical signs and a positive serology against specific antigens from these organisms.

Isolation Procedures

Some strains of *D. congolensis* grow well on a large variety of culture media incubated aerobically at 36°C (75). How-ever, most strains grow optimally on blood agar or brain heart infusion agar or in tryptone broth incubated under microaerophilic conditions with increased CO_2 at 37°C. *D. congolensis* does not grow on Sabouraud dextrose agar or Czapek agar, which supports the growth of many other genera of actinomycetes.

Identification

The colonial morphology of clinical isolates of *D. congo-lensis* is extremely variable. On sheep blood agar, the col-onies usually have short aerial hyphae and are surrounded by a zone of beta-hemolysis. Colonial pigmentation of *D. congolensis* varies from bright yellow to bright orange, or colonies may be pale yellow to gray-white. The organisms are gram-positive branching filaments that form septae both longitudinally and laterally, yielding cocci arranged in parallel chains. Each coccus, which fragments from the filament, generates a tuft of flagella in order to become motile. These motile cocci are called zoospores, and each zoospore develops into a new branching filament that repeats the developmental cycle (41, 75).

A. *madurae* is a strictly aerobic, branching, filamentous organism that grows optimally at 37°C. Colonies of this organism often have a leathery, membranous surface and develop a bright red pigmentation (Fig. 4C). They digest casein, gelatin, tyrosine, bromcresol purple milk, and starch but not xanthine (Tables 1, 2, and 3). Acid is produced when the organism is grown on arabinose and xylose. A. *madurae* does not utilize paraffin and will not grow in lysozyme broth. A. *pelletieri* generally grows more slowly, does not digest starch, and does not produce acid when grown on arabinose and xylose (38, 41, 53).

Nocardiopsis dassonvillei colonies have more abundant and extrinsic aerial hyphae than the colonies of *Actinomadura* spp., and bright pigmentation usually is not evident. However, the colonies may be brownish red or have a violet tint, and there may be a grayish violet diffusible pigment. Most strains digest xanthine, tyrosine, and starch. Acid is produced from cellobiose, maltose, and mannitol but not from arabinose or xylose. Urease is usually not produced, and the organisms will not grow in lysozyme media (36, 38).

Saccharomonospora spp., *Saccharopolyspora* spp., and *Thermoactinomyces* spp. cause extrinsic allergic alveolitis rather than invasive infectious disease. Therefore, these organisms are not often isolated from clinical material. Nevertheless, they can be differentiated easily from most other actinomycetes because of their abilities to grow at 55°C (36, 38).

Serologic Tests

No serologic tests for the diagnosis of infections caused by *D. congolensis*, *Actinomadura* spp., or *Nocardiopsis dassonvillei* are routinely available. Many serologic tests have been utilized experimentally for studying immune responses to these organisms, and some of these tests, including immunodiffusion tests, ELISA, and Western blot immunoassays, appear to have diagnostic value (36, 38, 41, 53, 75).

In contrast, the diagnosis of extrinsic allergic alveolitis induced by the thermophilic actinomycetes is made primarily by serologic analysis. Currently, agar gel immunodiffusion and gel immunodiffusion assays appear to be used most often, even though other methods such as counterimmunoelectrophoresis and ELISA are being investigated and may provide much earlier and more sensitive diagnoses. Reagents for these tests, including antigens and control sera, are commercially available (36, 38).

Antimicrobial Susceptibilities

Treatment of *D. congolensis* infections frequently requires topical medication combined with parenteral antimicrobial agents. These organisms are variably susceptible, depending on the strain and the specific host. As a consequence, there are no specific drugs of choice; however, high doses of various synthetic penicillins combined with streptomycin appear to be most effective. Chloramphenicol, tetracycline, erythromycin, kanamycin, and sulfonamides may also be useful (75).

Mycetomas can be refractory to chemotherapy. There are numerous reports on the ineffectiveness of chemotherapy for treatment of mycetomas caused by *Actinomadura* and *Nocardiopsis* spp., but successful therapy with combinations of surgical debridement and long-term application of streptomycin as well as of TMP-SMX or dapsone has been reported. Rifampin, sulfadoxine, and pyrimethamine may also be useful (39, 53).

Since the thermophilic actinomycetes cause a hypersensitivity pneumonitis, therapy is directed against the allergic response of the host. Thus, treatment relies on the use of glucocorticosteroids and bronchodilators (36, 38).

REFERENCES

1. **Angeles, A. M., and A. M. Sugar.** 1987. Identification of a common immunodominant protein in culture filtrates of three *Nocardia* species and use in etiologic diagnosis of mycetoma. *J. Clin. Microbiol.* **25:**2278–2280.

2. **Angeles, A. M., and A. M. Sugar.** 1987. Rapid diagnosis of nocardiosis with an enzyme immunoassay. *J. Infect. Dis.* **155:** 292–296.

3. **Ashdown, L. R.** 1990. An improved screening technique for isolation of *Nocardia* species from sputum specimens. *Pathology* **22:**157–161.

4. **Auerbach, S. B., M. M. McNeil, J. M. Brown, B. A. Lasker, and W. R. Jarvis.** 1992. Outbreak of pseudoinfection with *Tsukamurella paurometabola* traced to laboratory contamination: efficacy of joint epidemiologic and laboratory investigation. *Clin. Infect. Dis.* **14:**1015–1022.

5. **Ayyar, S., U. Tendolkar, and L. Deodhar.** 1992. A comparison of three media for isolation of *Nocardia* species from clinical specimens. *J. Postgrad. Med.* **38:**70–72.

6. **Barksdale, L., and K. S. Kim.** 1977. Mycobacterium. *Bacteriol. Rev.* **41:**217–372.

7. **Beaman, B. L., and L. Beaman.** 1994. *Nocardia*: host-parasite relationships. *Clin. Microbiol. Rev.* **7:**213–264.

8. **Beaman, B. L., A. L. Bourgeois, and S. E. Moring.** 1981. Cell wall modification resulting from *in vitro* induction of L-phase variants of *Nocardia asteroides*. *J. Bacteriol.* **148:**600–609.

9. **Beaman, B. L., and J. Burnside.** 1973. Puridine extraction of nocardial acid-fastness. *Appl. Microbiol.* **26:**426–428.

10. **Beaman, B. L., S. E. Moring, and T. Ioneda.** 1988. Effect of growth stage on mycolic acid structure in cell walls of *Nocardia asteroides* GUH-2. *J. Bacteriol.* **170:**1137–1142.

11. **Below-van, H., C. M. Wilk, K. P. Schaal, and G. O. H. Naumann.** 1991. *Rhodococcus luteus* and *Rhodococcus erythropolis* chronic endophthalmitis after lens implantation. *Am. J. Ophthalmol.* **112:**596–597.

12. **Benbow, E. P., D. T. Smith, and K. S. Grimson.** 1944. Sulfonamide therapy in actinomycosis. *Am. Rev. Tuberc.* **49:** 395–407.

13. **Biehle, J. R., S. J. Cavalieri, M. A. Saubolle, and L. J. Getsinger.** 1994. Comparative evaluation of the E Test for susceptibility testing of *Nocardia* species. *Diagn. Microbiol. Infect. Dis.* **19:**101–110.

14. **Boiron, P., and F. Provost.** 1990. Use of partially purified 54-kilodalton antigen for diagnosis of nocardiosis by Western blot (immunoblot) assay. *J. Clin. Microbiol.* **28:**328–331.

15. **Boiron, P., and D. Stynen.** 1992. Immunodiagnosis of nocardiosis. *Gene* **115:**219–222.

16. **Boquest, A. L., and F. A. Tosolini.** 1991. Isolation of *Nocardia asteroides* from respiratory specimens using *Legionella* selective media. *Aust. Microbiologist* **12:**23–24.

17. **Brown, B. A., and R. J. Wallace, Jr.** 1993. Broth microdilution MIC test for *Nocardia* spp., p. 5.12.1–5.12.9. *In* H. D. Isenberg (ed.), *Clinical Microbiology Procedures Handbook*, vol. 1. American Society for Microbiology, Washington, D.C.

18. **Brownell, G. H., and K. E. Belcher.** 1990. DNA probes for the identification of *Nocardia asteroides*. *J. Clin. Microbiol.* **28:**2082–2086.

19. **Buchman, A. L., M. M. McNeil, J. M. Brown, B. A. Lasker, and M. E. Ament.** 1992. Central venous catheter sepsis caused by unusual *Gordona* (*Rhodococcus*) species: identification with a digoxigenin-labeled rDNA probe. *Clin. Infect. Dis.* **15:**694–697.

20. **Butler, W. R., J. O. Kilburn, and G. P. Kubica.** 1987.

High-performance liquid chromatography analysis of mycolic acids as an aid in laboratory identification of *Rhodococcus* and *Nocardia* species. *J. Clin. Microbiol.* **25:**2126–2131.

21. **Collins, C. H., M. D. Yates, and A. H. C. Uttley.** 1988. Presumptive identification of nocardias in a clinical laboratory. *J. Appl. Bacteriol.* **65:**55–59.

22. **Collins, M. D., J. Smida, M. Dorsch, and E. Stackebrant.** 1988. *Tsukamurella* gen. nov. harboring *Corynebacterium paurometabolum* and *Rhodococcus aurianticus. Int. J. Syst. Bacteriol.* **38:**385–391.

23. **Coyle, M. B., and B. A. Lipsky.** 1990. Coryneform bacteria in infectious diseases: clinical and laboratory aspects. *Clin. Microbiol. Rev.* **3:**227–246.

24. **de'Clari, F., T. Menghini, G. Biaggi, and P. Magnoli.** 1992. Correspondence: septicemia due to a new species of *Rhodococcus* that contaminated closed system packed red cells—cure with imipenem. *J. Antimicrob. Chemother.* **30:**729–730.

25. **Drancourt, M., E. Bonnet, H. Gallais, Y. Peloux, and D. Raoult.** 1992. *Rhodococcus equi* infection in patients with AIDS. *J. Infect.* **24:**123–131.

26. **Ebersole, L. L., and J. L. Paturzo.** 1988. Endophthalmitis caused by *Rhodococcus equi* Prescott serotype 4. *J. Clin. Microbiol.* **26:**1221–1222.

27. **Geisler, P. J., and B. R. Andersen.** 1979. Results of therapy in systemic nocardiosis. *Am. J. Med. Sci.* **278:**188–193.

28. **Gombert, M. E., and T. M. Aulicino.** 1983. Synergism of imipenem and amikacin in combination with other antibiotics against *Nocardia asteroides. Antimicrob. Agents Chemother.* **24:**810–811.

29. **Goodfellow, M.** 1992. The family Nocardiaceae, p. 1188–1213. *In* A. Balows, H. G. Truper, M. Dworkin, W. Harder, and K. H. Schleifer (ed.), *The Prokaryotes*, 2nd ed. Springer-Verlag, New York.

30. **Goodfellow, M., and T. Cross.** 1984. Classification, p. 7–164. *In* M. Goodfellow, M. Mordarski, and S. T. Williams (ed.), *Biology of the Actinomycetes*. Academic Press, Inc. (London), Ltd., London.

31. **Gordon, R. E., and D. A. Barnett.** 1977. Resistance to rifampin and lysozyme of strains of some species of *Mycobacterium* and *Nocardia* as a taxonomic tool. *Int. J. Syst. Bacteriol.* **27:**176–178.

32. **Gordon, R. E., and J. M. Mihm.** 1959. A comparison of *Nocardia asteroides* and *Nocardia brasiliensis. J. Gen. Microbiol.* **20:**129–155.

33. **Kerr, E., H. Snell, B. L. Black, M. Storey, and W. D. Colby.** 1992. Isolation of *Nocardia asteroides* from respiratory specimens by using selective buffered charcoal-yeast extract agar. *J. Clin. Microbiol.* **30:**1320–1322.

34. **Kitzis, M. D., L. Gutman, and J. F. Acar.** 1985. *In vitro* susceptibility of *Nocardia asteroides* to 21 β-lactam antibiotics, in combination with three β-lactamase inhibitors, and its relationship to the β-lactamase content. *J. Antimicrob. Chemother.* **15:**23–30.

35. **Kjelstrom, J. A., and B. L. Beaman.** 1993. Development of a serologic panel for the recognition of nocardial infections in a murine model. *Diagn. Microbiol. Infect. Dis.* **16:**291–301.

36. **Land, G. A., M. R. McGinnis, J. L. Staneck, and A. Gatson.** 1991. Aerobic pathogenic Actinomycetales, p. 340–359. *In* A. Balows, W. J. Hausler, Jr., K. L. Herrmann, H. D. Isenberg, and H. J. Shadomy (ed.), *Manual of Clinical Microbiology*, 5th ed. American Society for Microbiology, Washington, D.C.

37. **Lasker, B. A., J. M. Brown, and M. M. McNeil.** 1992. Identification and epidemiological typing of clinical and environmental isolates of the genus *Rhodococcus* with use of a digoxigenin-labeled rDNA gene probe. *Clin. Infect. Dis.* **15:**223–233.

38. **Lechevalier, H. A., and M. A. Goodfellow.** 1994. Nocardioform actinomycetes, p. 625–652. *In* J. G. Holt, N. R. Krieg, P. H. A. Sneath, J. T. Staley, and S. T. Williams (ed.), *Bergey's Manual of Systematic Bacteriology*, 9th ed. The Williams & Wilkins Co., Baltimore.

39. **Mahgoub, E. S.** 1990. Agents of mycetoma, p. 1977–1980. *In* G. L. Mandell, G. L. Douglas, Jr., and J. E. Bennett (ed.), *Principles and Practice of Infectious Disease*, 3rd ed. Churchill Livingstone, New York.

39a.**McNeil, M.** Personal communication.

40. **McNeil, M. M., and J. M. Brown.** 1992. Distribution and antimicrobial susceptibility of *Rhodococcus equi* from clinical specimens. *Eur. J. Epidemiol.* **8:**437–443.

41. **McNeil, M. M., and J. M. Brown.** 1994. The medically important aerobic actinomycetes: epidemiology and microbiology. *Clin. Microbiol. Rev.* **7:**358–417.

42. **McNeil, M. M., J. M. Brown, W. R. Jarvis, and L. Ajello.** 1990. Comparison of species distribution and antimicrobial susceptibility of aerobic actinomycetes from clinical specimens. *Rev. Infect. Dis.* **12:**778–783.

43. **Mishra, S. K., R. E. Gordon, and D. A. Barnett.** 1980. Identification of nocardiae and streptomyces of medical importance. *J. Clin. Microbiol.* **11:**728–736.

44. **Mishra, S. K., and H. S. Randhawa.** 1969. Application of paraffin bait techniques to the isolation of *Nocardia asteroides* from clinical specimens. *Appl. Microbiol.* **18:**686–687.

45. **Murray, P. R., R. L. Heeren, and A. C. Niles.** 1987. Effect of decontamination procedures on recovery of *Nocardia* spp. *J. Clin. Microbiol.* **25:**2010–2011.

46. **National Committee for Clinical Laboratory Standards.** 1993. *Methods for Dilution Susceptibility Tests for Bacteria That Grow Aerobically*, 3rd ed. Approved standard. NCCLS document M7-A3. National Committee for Clinical Laboratory Standards, Villanova, Pa.

47. **Nordmann, P., J. Kerestedjian, and E. Ronco.** 1992. Therapy of *Rhodococcus equi* disseminated infections in nude mice. *J. Clin. Microbiol.* **36:**1244–1248.

48. **Pier, A. C., and R. E. Fichtner.** 1971. Serologic typing of *Nocardia asteroides* by immunodiffusion. *Am. Rev. Respir. Dis.* **103:**698–707.

49. **Pier, A. C., and R. E. Fichtner.** 1981. Distribution of serotypes of *Nocardia asteroides* from animal, human, and environmental sources. *J. Clin. Microbiol.* **13:**548–553.

50. **Prescott, J. F.** 1991. *Rhodococcus equi*: an animal and human pathogen. *Clin. Microbiol. Rev.* **4:**20–34.

51. **Richet, H. M., P. C. Craven, J. M. Brown, B. A. Lasker, C. D. Cox, M. M. McNeil, A. D. Tice, W. R. Jarvis, and O. C. Tablan.** 1991. A cluster of *Rhodococcus* (*Gordona*) *bronchialis* sternal-wound infections after coronary-artery bypass surgery. *N. Engl. J. Med.* **324:**104–108.

52. **Rihs, J. D., M. M. McNeil, J. M. Brown, and V. L. Yu.** 1990. *Oerskovia xanthineolytica* implicated in peritonitis associated with peritoneal dialysis: case report and review of *Oerskovia* infections in humans. *J. Clin. Microbiol.* **28:**1934–1937.

53. **Rippon, J. W.** 1988. *Medical Mycology: the Pathogenic Fungi and the Pathogenic Actinomycetes*, 3rd ed. The W. B. Saunders Co., Philadelphia.

54. **Salinas-Carmona, M. C., L. Vera, O. Welsh, and M. Rodriguez.** 1992. Antibody response to *Nocardia brasiliensis* antigens in man. *Zentralbl. Bakteriol.* **276:**390–397.

55. **Salinas-Carmona, M. C., O. Welsh, and S. M. Casillas.** 1993. Enzyme-linked immunosorbent assay for serological diagnosis of *Nocardia brasiliensis* and clinical correlation with mycetoma infections. *J. Clin. Microbiol.* **31:**2901–2906.

56. **Schaal, K. P.** 1984. Laboratory diagnosis of actinomycete disease, p. 441–456. *In* M. Goodfellow, M. Mordarski, and S. T. Williams (ed.), *The Biology of the Actinomycetes*. Academic Press, London.

57. **Shapiro, C. L., R. F. Haft, N. M. Gantz, G. V. Doern, J. C. Christenson, R. O'Brien, J. C. Overall, B. A. Brown, and R. J. Wallace, Jr.** 1992. *Tsukamurella paurometabolum*: a novel pathogen causing catheter-related bacteremia in patients with cancer. *Clin. Infect. Dis.* **14:**200–203.

58. **Shawar, R. M., D. G. Moore, and M. T. LaRocco.** 1990. Cultivation of *Nocardia* spp. on chemically defined media for selective recovery of isolates from clinical specimens. *J. Clin. Microbiol.* **28:**508–512.

59. **Smego, R. A., Jr., and H. A. Gallis.** 1984. The clinical spectrum of *Nocardia brasiliensis* infection in the United States. *Rev. Infect. Dis.* **6:**164–180.

60. **Spark, R. P., M. M. McNeil, J. M. Brown, B. A. Lasker, M. A. Montano, and M. D. Garfield.** 1993. *Rhodococcus* infection in immunocompetent host. *Arch. Pathol. Lab. Med.* **117:**515–520.

61. **Stackebrandt, E., J. Smida, and M. D. Collins.** 1988. Evidence of phylogenetic heterogeneity within the genus *Rhodococcus*: revival of the genus *Gordona* (Tsukamura). *J. Gen. Appl. Microbiol.* **34:**341–348.

62. **Sugar, A. M., G. K. Schoolnik, and D. A. Stevens.** 1985. Antibody response in human nocardiosis: identification of two immunodominant culture filtrate antigens derived from *Nocardia asteroides*. *J. Infect. Dis.* **151:**859–901.

63. **Tight, R. R., and M. S. Bartlett.** 1981. Actinomycetoma in the United States. *Rev. Infect. Dis.* **3:**1139–1150.

64. **Venugopal, T. V., and P. V. Venugopal.** 1981. Mycetoma, p. 1762–1775. *In* A. L. Braude (ed.), *Medical Microbiology and Infectious Disease*. The W. B. Saunders Co., Philadelphia.

65. **Vera-Cabrera, L., M. C. Salinas-Carmona, O. Welsh, and M. A. Rodriguez.** 1992. Isolation and purification of two immunodominant antigens from *Nocardia brasiliensis*. *J. Clin. Microbiol.* **30:**1183–1188.

66. **Vickers, R. M., J. D. Rihs, and V. L. Yu.** 1992. Clinical demonstration of isolation of *Nocardia asteroides* on buffered charcoal-yeast extract media. *J. Clin. Microbiol.* **30:**227–228.

67. **Wallace, R. J., Jr., B. A. Brown, M. Tsukamura, J. M. Brown, and G. O. Onyi.** 1991. Clinical and laboratory features of *Nocardia nova*. *J. Clin. Microbiol.* **29:**2407–2411.

68. **Wallace, R. J., Jr., D. R. Nash, W. K. Johnson, L. C. Steel,** and V. A. Steingrube. 1987. β-Lactam resistance in *Nocardia brasiliensis* is mediated by β-lactamase and reversed in the presence of clavulanic acid. *J. Infect. Dis.* **156:**959–966.

69. **Wallace, R. J., Jr., and L. C. Steele.** 1988. Susceptibility testing of *Nocardia* species for the clinical laboratory. *Diagn. Microbiol. Infect. Dis.* **9:**155–166.

70. **Wallace, R. J., Jr., L. C. Steele, G. Sumter, and J. M. Smith.** 1988. Antimicrobial susceptibility patterns of *Nocardia asteroides*. *Antimicrob. Agents Chemother.* **32:**1776–1779.

71. **Wallace, R. J., Jr., M. Tsukamura, B. A. Brown, J. Brown, V. A. Steingrube, Y. Zhang, and D. R. Nash.** 1990. Cefotaxime-resistant *Nocardia asteroides* strains are isolates of the controversial species *Nocardia farcinica*. *J. Clin. Microbiol.* **28:** 2726–2732.

72. **Yano, I., T. Imaeda, and M. Tsukamura.** 1990. Characterization of *Nocardia nova*. *Int. J. Syst. Bacteriol.* **40:**170–174.

73. **Yazawa, K., Y. Mikami, S. Ohashi, M. Miyaji, Y. Ichihara, and C. Nishimura.** 1992. *In vitro* activity of new carbapenem antibiotics: comparative studies with meropenem, L-627 and imipenem against pathogenic *Nocardia* spp. *J. Antimicrob. Chemother.* **29:**169–172.

74. **Yu, A. M., S. Zhao, and L. Y. Nie.** 1993. Mycetomas in northern Yemen: identification of causative organisms and epidemiologic considerations. *Am. J. Trop. Med. Hyg.* **48:** 812–817.

75. **Zaria, L. T.** 1993. *Dermatophilus congolensis* infection (dermatophilosis) in animals and men. An update. *Comp. Immun. Microbiol. Infect. Dis.* **16:**179–222.

Mycobacterium

FREDERICK S. NOLTE AND BEVERLY METCHOCK

31

Tuberculosis remains a major global public health problem. The World Health Organization estimates that 8 million new cases and 3 million deaths are directly attributable to the disease each year (117). This makes tuberculosis the leading cause of death due to a single infectious agent.

In the United States, the incidence of tuberculosis declined steadily until 1985, when the downward trend reversed. The number of cases of tuberculosis increased 18.4% from 1985 to 1991. The Centers for Disease Control and Prevention (CDC) estimates that during that period, 39,000 cases of tuberculosis occurred in excess of those expected if the pre-1985 downward trend had continued (31). A number of factors are thought to account for the resurgence of tuberculosis in the United States. These factors include the advent of the AIDS epidemic, immigration from areas of high endemicity, general deterioration of the health care infrastructure, transmission in high-risk environments, and increase in the number of multidrug-resistant strains of *Mycobacterium tuberculosis*.

Clinical mycobacteriology laboratories play an important role in the control of the spread of tuberculosis through the timely detection, isolation, identification, and drug susceptibility testing of *M. tuberculosis*. Substantial changes in clinical mycobacteriology have occurred since the publication of the last edition of this Manual. The availability of biphasic isolation media, the refinement of high-performance liquid chromatography (HPLC) for identification of mycobacteria, the widespread use of nonisotopic probes for culture identification, and the application of nucleic acid amplification techniques for direct detection of *M. tuberculosis* in clinical specimens are important new developments in the field. In addition, progress in our understanding of the molecular genetics of drug resistance in mycobacteria has been substantial, and sensitive growth indicator systems that could provide the basis for more rapid susceptibility testing procedures have been developed. Perhaps the biggest change, however, is the new sense of urgency regarding the reporting of results of acid-fast smears, cultures, and drug susceptibility tests to clinicians that has been prompted by the resurgence of tuberculosis and outbreaks involving multidrug-resistant strains.

There have also been major recent developments in the taxonomy and clinical conditions associated with nontuberculous mycobacteria (NTM). Recognition of M. *avium* complex (MAC) organisms as causes of disseminated disease in patients with AIDS has revived interest in these organisms and led to advances in our understanding of their biological characteristics and of many of the epidemiologic and clinical aspects of infection. M. *genavense* is a recently described species that is also capable of causing disseminated disease in human immunodeficiency virus (HIV)-infected individuals. M. *asiaticum*, M. *haemophilum*, M. *malmoense*, M. *shimoidei*, and M. *celatum* are other potentially pathogenic species for which systematics and disease associations have been recently clarified.

The methods described in this chapter enable laboratories to provide accurate results within a clinically relevant time frame. However, the level of service and the choice of methods used in an individual laboratory setting should be determined by the patient population served by the laboratory and the available resources. Considering the expense of many of the newer detection, culture, and identification methods, it seems likely that the trend toward regionalization of diagnostic mycobacteriology services will continue.

TAXONOMY

The genus *Mycobacterium* is the only genus in the family *Mycobacteriaceae*. The high G+C content of the DNA of mycobacteria (62 to 70%) is similar to that of the other mycolic acid-producing bacteria, *Nocardia* (60 to 69%), *Rhodococcus* (59 to 69%), and *Corynebacterium* (51 to 59%) spp. This similarity may support the consolidation of these genera into a single family (233).

A natural division occurs between slowly and relatively rapidly growing species of mycobacteria. Slow growers require more than 7 days to produce easily seen colonies on solid media from a dilute inoculum under ideal culture conditions. Rapid growers require less than 7 days under comparable conditions. In practice, different sets of biochemical tests are necessary to characterize slowly growing and rapidly growing species.

M. *tuberculosis* complex includes the species M. *bovis*, M. *microti*, and M. *africanum*. M. *bovis* was the name given to the bovine tubercle bacillus in 1896. The bacillus of Calmette-Guérin (BCG), which is used as a vaccine against tuberculosis in many parts of the world, conforms to the properties described for M. *bovis* except that its pathogenicity is more attenuated. M. *africanum* may represent an intermediate form between M. *tuberculosis* and M. *bovis*, and retention of M. *africanum* as a distinct species is probably not justified. M. *microti* also occupies a position along the phenotypic continuum between M. *tuberculosis* and M. *bovis*.

MAC consists of 28 serovars of two distinct species, M. *avium* and M. *intracellulare*, and has in the past also included three additional serovars of M. *scrofulaceum* (M. *avium*-M. *intracellulare*-M. *scrofulaceum* complex). Inclusion of M. *scrofulaceum* in the MAC is no longer appropriate, given recent advances in mycobacterial systematics (228). The criteria used to distinguish M. *avium* from M. *intracellulare* are now well established. Three subspecies of M. *avium* were proposed on the basis of phenotypic characteristics and genetic studies: M. *avium* subsp. *avium*, M. *avium* subsp. *paratuberculosis*, and M. *avium* subsp. *silvaticum* (205). The International Working Group on Mycobacterial Taxonomy has suggested that there is taxonomic evidence for a third species within the MAC (227).

M. *kansasii* is the name currently given to the photochromogenic "yellow bacillus" originally described in 1953. Studies of the base sequences of the 16S rRNA and of catalase serology suggest that M. *kansasii* is phylogenetically very closely related to a nonpathogenic, slowly growing, nonpigmented species, M. *gastri* (170, 226). However, the distinction between the two at the species or subspecies level should be maintained because of the clinical significance of most isolates of M. *kansasii*. The taxonomy of mycobacteria continues to evolve as molecular methods of distinguishing species become more sophisticated.

DESCRIPTION OF THE GENUS

The mycobacteria are slightly curved or straight bacilli, 0.2 to 0.6 by 1.0 to 10 μm in size, sometimes with branching. Filamentous or myceliumlike growth may occur, but it easily fragments into rods or coccoid elements. Mycobacteria have cell walls with a high lipid content that includes waxes having characteristic mycolic acids with long, branched chains.

The high lipid content of the cell wall excludes the usual aniline dyes. Mycobacteria are not readily stained by the Gram method but are considered gram positive. Special staining procedures are used to promote the uptake of dye, and once stained, mycobacteria are not easily decolorized even with acid-alcohol. This resistance to decolorization by acid-alcohol is termed acid fastness. Acid fastness may be partly or completely lost at some stage of growth by some proportion of the cells of some species. Cells of rapidly growing mycobacteria may be less than 10% acid fast.

The mycobacteria are aerobic, nonsporeforming, nonmotile bacilli. Colony morphology varies among the species. M. *tuberculosis* forms rough colonies with the bacilli compacted into curving strands (cords). In contrast, MAC usually forms smooth transparent colonies with the bacilli arranged in no definite pattern on primary cultures. M. *kansasii* is intermediate between the previous examples, forming slightly rough colonies. Diffusible pigment is rare, but colonies of some species are regularly or variably yellow, orange, or, rarely, pink. Some species require light to form pigment (photochromogens), and others form pigment in either the light or the dark (scotochromogens). Aerial filaments are rarely formed and are never visible without magnification.

Growth rates for mycobacteria are slow to very slow, with generation times varying by species and ranging from 2 to >20 h. Easily visible colonies may be produced after 2

days to 8 weeks of incubation under optimal conditions, depending on the species. Optimal temperatures for growth vary widely among species, ranging from 30 to almost 45°C. Most species adapt readily to growth on simple substrates, using ammonia or amino acids as nitrogen sources and glycerol as a carbon source in the presence of mineral salts. Some species require medium supplements such as hemin, mycobactins, or other iron transport compounds. M. *leprae* has not been cultured outside of living cells.

In 1959, Runyon established a grouping system for NTM based on pigmentation, colonial morphology, and growth rate (172). It soon became apparent that all species of mycobacteria did not fit within the four groups of the Runyon scheme. Clinical laboratories have abandoned the Runyon groups in favor of identification to species level.

The term atypical mycobacteria is of historical interest but should not be used by clinical laboratories when referring to non-M. *tuberculosis* mycobacteria. When these species were first described, they were grouped together as atypical mycobacteria because they were not typical of M. *tuberculosis*. Obviously, since these mycobacteria are not atypical but are characteristic of their own species, the term NTM is preferable when referring to these species as a group. Whenever possible, species names should be used.

CLINICAL SIGNIFICANCE AND NATURAL HABITATS

The genus *Mycobacteria* includes obligate parasites, saprophytes, and opportunistic pathogens. Most species are free living in soil and water, but the major ecological niche for others such as M. *tuberculosis* complex and M. *leprae* is diseased tissues of humans and other warm-blooded animals.

The following section contains descriptions of the diseases associated with the clinically important species of mycobacteria. Also included is a review of the diseases associated with the potentially pathogenic species. Much of the information on NTM was obtained from several recent review articles (228, 238, 240). The reader should consult these articles for more information and for access to the primary literature.

M. tuberculosis Complex

Robert Koch was the first to establish the causal relationship between the tubercle bacillus and the disease tuberculosis. The organism was named M. *tuberculosis* in 1886, presumably because it resembled a fungus in its slow growth and colony morphology.

Cases of tuberculosis are not evenly distributed throughout all segments of the U.S. population. Certain population subgroups have a higher risk of tuberculosis either because they have a higher likelihood of exposure and infection or because they are more likely to progress to active disease (29). In some cases, both factors may be important. Groups known to have higher prevalences of tuberculosis than the general population include medically underserved ethnic minorities, homeless persons, prison inmates, alcoholics, injecting drug users, the elderly, foreign-born persons from areas of high prevalence, and contacts of persons with active tuberculosis. Groups with a higher likelihood of progression from latent infection to active tuberculosis include persons with certain underlying medical conditions (compromised immune system, status postgastrectomy, body weight ≥10% below the ideal, chronic renal failure, and diabetes mellitus), persons who have been infected with tuberculosis within the past 2 years, children ≤4 years old, and persons with fibrotic lesions on chest radiograph. HIV infection is the greatest known risk factor for the progression of latent infection to active tuberculosis.

M. tuberculosis is carried in airborne particles known as droplet nuclei that are generated when patients with pulmonary tuberculosis cough. These particles are from 1 to 5 μm in size and are kept suspended in air by normal air currents. Infections occur when a susceptible person inhales the droplet nuclei containing M. tuberculosis. The infected droplet nuclei reach the terminal airways of the lung. In the alveoli, the organisms are engulfed by the alveolar macrophages and may spread throughout the body. Usually, the host's cell-mediated immune response limits further multiplication and spread of M. tuberculosis; however, some bacilli may remain viable but dormant for many years after the initial infection. Patients latently infected with M. tuberculosis usually have a positive purified protein derivative (PPD) skin test but are asymptomatic and not infectious. In general, persons with latent infections have a 10% risk for development of active tuberculosis at some time during their lives. The risk is greatest within the first 2 years after infection. Patients with HIV have a 10 to 15% risk per year for progression to infection.

Pulmonary tuberculosis in adults is a slowly progressive inflammatory process characterized by intense chronic inflammation, necrosis, and caseation. The cavities that form in the lungs may rupture into bronchi, thus allowing large numbers of organisms to spread to other areas of the lungs, to be aerosolized by coughing, and thus infect other persons. The usual clinical features of pulmonary tuberculosis include cough, weight loss, low-grade fever, dyspnea, and chest pain. Tuberculosis in AIDS patients progresses much more rapidly and often disseminates, sometimes without formation of typical granulomas. Other clinical manifestations of infection with M. tuberculosis include cervical adenitis, skin infections, pericarditis, synovitis and meningitis.

M. bovis causes tuberculosis in cattle, humans, and other primates; carnivores, including dogs and cats; swine; parrots; and some birds of prey. The disease produced in humans is virtually indistinguishable from that caused by M. tuberculosis and is treated similarly.

M. africanum is a cause of human tuberculosis in tropical Africa.

M. microti causes naturally acquired generalized tuberculosis in the vole and produces local lesions in guinea pigs, rabbits, and calves.

M. leprae

Leprosy (Hansen's disease) is a chronic granulomatuous disease caused by M. leprae. The principal manifestations of the disease include anesthetic skin lesions and peripheral neuropathy with peripheral nerve thickening. The clinical form that the disease takes in an individual depends on the degree of cell-mediated immunity expressed toward M. leprae. An intense cell-mediated immune response with elimination of the leprosy bacillus produces the tuberculoid form of the disease. Absence of cell-mediated immunity results in lepromatous leprosy. The medical complications of leprosy arise from nerve damage, immune reactions, and infiltration by the mycobacteria (21).

Recent estimates by the World Health Organization put the global prevalence of leprosy at 10 to 12 million patients, with the majority of cases occurring in Asia and Africa.

Leprosy was endemic in parts of northern Europe until the early 20th century, and small pockets of infection persist in the United States in Texas and Louisiana. The majority of infections now seen in Europe and North America are acquired abroad.

A patient with untreated lepromatous leprosy may discharge up to 8×10^8 acid-fast bacilli (AFB) in a single nose blow. Shedding from the nose is more important than shedding from skin lesions for transmission of the disease. Primary infection probably also occurs through the nose. The reservoir for M. leprae is not well established, but naturally occurring infections of the nine-banded armadillo have been documented in Texas and Louisiana.

M. leprae differs from all other mycobacteria in that it cannot be cultured in vitro. The diagnosis of leprosy is essentially clinical, based on finding one or more signs of the disease and supported by finding mycobacteria on slit skin smears or in skin biopsy samples. If the diagnosis remains in doubt, skin testing with lepromin, a preparation of leprosy-bacillus antigen, may provide further evidence of infection.

MAC

MAC organisms are ubiquitous in nature and have been isolated from water, soil, plants, house dust, and a myriad of other environmental sources. MAC organisms were recovered from 25% of water samples taken from estuaries in South Carolina, Georgia, and the Gulf Coast states (219). These organisms are of low pathogenicity and frequently colonize individuals without causing disease. This lack of disease complicates the interpretation of culture results, especially for cultures of sputum and other respiratory tract specimens.

M. avium is an important cause of disease in poultry and swine. It was first recognized as a cause of disease in chickens in the late 1800s but was not recognized as a cause of human disease until 1943. In the late 1950s, M. avium caused an outbreak of pulmonary disease at Battey State Hospital in Rome, Ga., and was referred to for some time as the Battey bacillus. MAC infections have become increasingly more common in the United States, and M. avium is the most common NTM species associated with human disease. The greatest increase in MAC infections during the past decade has been in patients with AIDS.

Prior to the advent of AIDS, pulmonary disease was the most common presentation of MAC infection. Four clinical patterns of lung disease caused by MAC have been recognized: solitary nodules; chronic bronchitis or bronchiectasis, mainly in older white women; tuberculosislike infiltrates; and diffuse infiltrates in immunocompromised patients (200). Pulmonary disease due to MAC typically occurs in white males 45 to 65 years of age with preexisting lung disease such as chronic bronchitis, emphysema, healed tuberculosis, bronchiectasis, or pneumoconiosis. The clinical presentation is similar to that of tuberculosis, with productive cough, fatigue, fever, weight loss, and night sweats. Recent reports have emphasized increasing recognition of MAC pulmonary disease in individuals, especially women, who have apparently had no predisposing disorders of the lungs or immune system (200). MAC organisms are frequently isolated from adults with cystic fibrosis, particularly in the southeastern United States, but their contribution to the disease process is not well established (114).

MAC is the leading cause of mycobacterial lymphadenitis in children less than 12 years of age. Upper anterior cervical, submandibular, submaxillary, and preauricular lymph nodes may be involved. Children more than 12 years of age are rarely infected unless they are immunocompromised or have disseminated disease.

Disseminated MAC infections in patients without AIDS usually occur in patients with underlying malignancy, inherited or therapeutic immunodeficiency (especially children or young adults with hematologic malignancies), or severe combined immunodeficiency syndrome; transplant recipients; and patients treated with corticosteroids or cytoxic drugs. The most common clinical presentation is fever of unknown origin, and dissemination may involve any organ system (92).

A role for mycobacteria in the etiology of Crohn's disease has long been a subject of speculation. Crohn's disease is an inflammatory bowel disease that occurs in humans. It is similar to Johne's disease, which occurs in cattle, sheep, and goats. M. *avium* subsp. *paratuberculosis* is the etiologic agent of Johne's disease, and mycobactin-dependent strains of M. *avium* subsp. *paratuberculosis* have been isolated from bowel tissues from a small proportion of patients with Crohn's disease. The role of mycobacteria in Crohn's disease and the relationship between Crohn's disease and Johne's disease has recently been reviewed (38).

Patients with AIDS may present with focal or disseminated MAC infections. MAC is commonly isolated from sputum or stool cultures from patients with HIV infection. It is thought that MAC is acquired from the environment and colonizes either the respiratory or the gastrointestinal tract before disseminating in HIV-infected patients (60, 85, 90). Retrospective studies have documented that 80% of HIV-infected persons with stool colonization progressed to disseminated disease (90).

Focal infections commonly involve the lungs or gastrointestinal tract and occasionally involve peripheral lymph nodes (95). In AIDS patients, the clinical presentation of focal pulmonary disease due to MAC is similar to that of the disease in other immunocompromised patients, but cavity formation is rare (<5%). There are few clinical or radiographic findings that distinguish MAC pulmonary disease from pulmonary disease due to M. *tuberculosis* or to the myriad of other opportunistic pulmonary pathogens that are of concern in this patient population. A specific etiologic diagnosis of MAC pulmonary infection requires repeated isolation of MAC from respiratory specimens and persistent signs and symptoms not attributable to another pathogen.

The gastrointestinal tract is most likely the initial site of infection in AIDS patients who develop disseminated infections. Involvement of the esophagus, duodenum, colon, sigmoid, and rectum have all been described. Histopathologic findings may include marked histiocytic and mycobacterial infiltration that when seen in the small intestine resembles Whipple's disease.

Peripheral lymphadenitis due to MAC occasionally occurs in HIV-infected individuals without disseminated disease (7). While lymphadenitis may occur in patients with disseminated MAC infections, isolated peripheral lymphadenitis is more commonly due to M. *tuberculosis* than to MAC.

Disseminated MAC infection in AIDS patients commonly causes a progressive illness characterized by intermittent fever, sweats, weakness, anorexia, and weight loss. It may be a cause of wasting syndrome experienced by many of these patients. The prevalence of this type of infection in patients with AIDS is estimated to be approximately 10% (238). The disease occurs usually a year or more after the

diagnosis of AIDS has been made, when the CD4 count is <100 cells per mm^3.

Bacteremia occurs in almost all patients, and the organism is found almost exclusively in the circulating monocytes. The magnitude of mycobacteremia in most patients ranges from 10^1 to 10^3 CFU/ml, but blood counts as high as 10^6 CFU/ml are not uncommon. In disseminated disease, almost any organ system can be involved, with levels of mycobacteria in the tissues as high as 10^{10} CFU/g of tissue. Despite the enormous tissue loads of mycobacteria, there is no consensus on the extent to which disseminated MAC infections contribute to mortality. Although disseminated infections in AIDS patients are not immediately life threatening, recent evidence indicates that these infections severely limit the quality of life and long-term survival of these patients (36, 91, 101). Antimycobacterial therapy appears to prolong survival of patients with disseminated infections, but no single agent or combination of agents is uniformly effective.

AIDS patients are infected more often with M. *avium* than with M. *intracellulare*. In one report, 98% of MAC infections in 45 AIDS patients were due to M. *avium*, whereas 40% of MAC infections in patients without AIDS were due to M. *intracellulare* (80). In addition, relatively few serovars of M. *avium* account for the majority of the infections in AIDS patients, with serovars 1, 4, and 8 predominating in the United States. Approximately 90% of mycobacterial infections in AIDS patients involve either MAC or M. *tuberculosis*, but the remaining 10% are due to almost every other known potentially pathogenic species.

M. kansasii

M. *kansasii* causes chronic pulmonary infections resembling classic tuberculosis. Most cases in the United States have been reported from California, Texas, Louisiana, Florida, Illinois, and Missouri. In years past, M. *kansasii* infections were among the most common NTM infections in the United States, but during the past 2 decades, the incidence of diseases due to this species has declined (74). Although M. *kansasii* has occasionally been cultured from water samples from various parts of the world, its natural reservoir is still largely unknown.

Chronic pulmonary disease is the most common manifestation, classically involving the upper lobe, with evidence of cavities and scarring in most cases. Pleural effusions and lymphadenopathy are rare. Pulmonary disease due to M. *kansasii* differs from that due to MAC in that fewer patients have underlying lung disease and the response to chemotherapy is much better.

Extrapulmonary infections are uncommon. They include cervical lymphadenitis in children (243), cutaneous and soft tissue infections (12, 55), and musculoskeletal system involvement (51, 130, 224). M. *kansasii* rarely disseminates except in patients with severely impaired cellular immunity and most recently in patients with AIDS (181). This organism causes the second most common opportunistic mycobacterial infection (after MAC-caused infections) associated with AIDS.

M. scrofulaceum

The species name *scrofulaceum* was derived from the word scrofula, the historical term used to describe mycobacterial infections of the cervical lymph glands. M. *scrofulaceum* is most commonly associated with cervical lymphadenitis in children (169). In one study of 16 children with mycobacterial lymphadenitis, 6 cases were caused by M. *scrofula-*

ceum, 4 were caused by M. *tuberculosis*, and 6 were caused by MAC (68).

Cervical lymphadenitis caused by M. *scrofulaceum* typically occurs in children 1 to 5 years of age. The disease is usually unilateral, involving nodes located in the submandibular or periauricular areas. The children appear healthy and are generally free of constitutional symptoms and local pain in the involved area. The nodes usually soften, rupture, and drain without any sequelae. Occasionally, the nodes spontaneously regress and heal by fibrosis and calcification. The route of infection is thought to be by way of the lymphatic vessels that drain the mouth and pharynx.

Uncommon extranodal manifestations of M. *scrofulaceum* infection include pulmonary disease, disseminated disease, and rare cases of conjunctivitis, osteomyelitis, meningitis, and granulomatous hepatitis. M. *scrofulaceum* accounted for approximately 2% of the mycobacterial infections in a series of 212 AIDS patients in the United States (73).

M. xenopi

M. *xenopi* was first isolated from an African toad in 1957 but was not recognized as a potential human pathogen until 1965. By 1979, 50 cases of M. *xenopi* infections in humans had been reported, primarily from Great Britain, France, Denmark, Australia, and the United States (237). The organism has been recovered from potable water, including water storage tanks in hospitals. Nosocomial acquisition of infection has been described (5). Birds are considered a possible natural reservoir of M. *xenopi* in Great Britain.

Pulmonary disease is the most common manifestation of infection, usually occurring in patients with underlying lung disease or other predisposing condition (lung malignancy, alcoholism, or diabetes). Pulmonary infections have been reported only in adults and occur in males more frequently than in females. The disease may be chronic, subacute, or acute. Clinically, pulmonary infections with M. *xenopi* resemble those in patients infected with M. *tuberculosis*, MAC, or M. *kansasii*. Rare extrapulmonary and disseminated infections have also been described in immunocompromised individuals, including renal transplant recipients, peritoneal dialysis patients, and patients with AIDS.

Rapid Growers

Because the three rapidly growing species M. *fortuitum*, M. *chelonae*, and M. *abscessus* share a number of characteristics and are associated with similar types of infections, they are referred to collectively as the M. *fortuitum* complex. The complex consists of two groups of species. The M. *fortuitum* group consists of M. *fortuitum*, M. *peregrinum*, and a third unnamed biovariant (125). The M. *chelonae* group consists of M. *chelonae* (formerly M. *chelonae* subsp. *chelonae*), M. *abscessus* (formerly M. *chelonae* subsp. *abscessus*), and a third biovariant known as M. *chelonae*-like organisms (125). However, these two groups of species can easily be separated on the biochemical basis of nitrate reduction and iron uptake, and since differences in susceptibility to antimicrobial agents are substantial (M. *chelonae* group is more resistant), identification to species level is important.

Most infections attributed to rapidly growing mycobacteria are associated with members of the M. *fortuitum* complex. Rare infections have also been attributed to other rapidly growing mycobacteria, including M. *smegmatis*, M. *flavescens*, M. *thermoresistibile*, M. *neoaurum*, and strains without species designations. In addition to these species,

there are at least 19 species of saprophytic rapidly growing mycobacteria (228).

M. *fortuitum* complex organisms are responsible for a number of different types of infections, including osteomyelitis, cellulitis, disseminated disease with multiple nodular soft tissue abscesses, surgical-wound infections, post-traumatic-wound infections, otitis media, corneal ulcers, and chronic pulmonary disease (221). In a large series of infections due to rapidly growing mycobacteria, M. *fortuitum* and M. *chelonae* occurred with equal frequency. The most common diseases caused by these organisms are soft tissue and skeletal infections that result from direct inoculation of contaminated materials via injections, surgery, and penetrating trauma. M. *fortuitum* has been isolated from water, soil, and dust, but the distribution of M. *chelonae* in the environment is less clear.

M. szulgai

M. *szulgai* was first described as a distinct species in 1972. Although a comparison of 16S rRNA sequences shows that M. *szulgai* is very closely related to M. *malmoense*, phenotypic distinctions between the two species can be made easily (225). The distribution of this organism appears to be worldwide, but its natural reservoirs are unknown.

It is an infrequent cause of human disease. In one review, two-thirds of the 24 cases reported prior to 1987 presented as chronic pulmonary disease that was indistinguishable from tuberculosis and occurred almost exclusively in white middle-aged males (134). The remaining case presentations included olecranon bursitis, cervical adenitis, tenosynovitis, cutaneous infections, and osteomyelitis. Two cases of M. *szulgai* infection in AIDS patients have been reported (73).

M. malmoense

The species name M. *malmoense* is derived from Malmo, the name of the city in Sweden from which the first case report originated (169). Since the original description of M. *malmoense*, disease due to this organism has been increasingly recognized in England, Scotland, Wales, Sweden, and France. M. *malmoense* has been only rarely isolated from patients in the United States. M. *malmoense* infections may be more common than suspected, because it may require 8 to 12 weeks to isolate some strains, which is longer than many laboratories in the United States hold mycobacterial cultures.

Patients with M. *malmoense* infections usually present with chronic pulmonary disease. Most of these patients are middle-aged men with previously documented pneumoconiosis. Rare extrapulmonary and disseminated infections have also been reported. The natural reservoir of this organism is unknown.

M. simiae

As the name suggests, M. *simiae* was first isolated from monkeys. It has been repeatedly isolated from tap water, but reports of human infection are relatively few. The clinical presentations included chronic pulmonary disease, osteomyelitis, and disseminated disease with renal involvement. Few cases of M. *simiae* infection in AIDS patients have been reported. M. *simiae* may be common in the environment in some geographic regions. In Tel Aviv, Israel, 399 strains were isolated from 287 patients from 1975 to 1981, making M. *simiae* the most common NTM isolated from sputum (129). Multiple positive cultures were obtained from 18 patients, all of whom were middle aged or older and most of whom had a history of tuberculosis.

M. genavense

M. *genavense* is the proposed name for a slow-growing NTM that has been isolated from the blood of AIDS patients in Switzerland, Austria, Germany, and the United States (13). Analysis of 16S rRNA sequences indicates that these organisms are members of a proposed new species most closely related to M. *simiae*. Disseminated infections with M. *genavense* are clinically indistinguishable from those with MAC but may be more amenable to antimicrobial therapy.

M. *genavense* was first isolated in BACTEC 13A medium only after extended incubation (6 to 8 weeks). It failed to grow on Lowenstein-Jensen (LJ) 7H11 agar or other media commonly used for the isolation of mycobacteria. Middlebrook 7H11 agar supplemented with mycobactin J consistently supported the growth of M. *genavense* (44). M. *genavense* may be an underrecognized cause of disseminated infections in AIDS patients because its failure to grow on solid media and its requirement for extended incubation preclude its detection in many mycobacteriology laboratories.

M. marinum

M. *marinum* causes cutaneous infections acquired as a result of trauma to skin in contact with contaminated fresh or salt water. The skin lesions may be of two types. The more typical presentation is a single papulonodular lesion confined to one extremity, usually involving the elbow, knee, foot, toe, or finger. The lesion appears 2 to 3 weeks after inoculation and with time may become verrucous or ulcerated. A second variety resembles cutaneous sporotrichosis: the primary inoculation site develops into an abscess, and secondary spread occurs centrally along the lymphatics. Recent reports emphasize deeper infections, including tenosynovitis, arthritis, bursitis, and osteomyelitis. Rare disseminated infections and visceral involvement have been reported (71, 126).

M. ulcerans

M. *ulcerans* causes an ulcerating infection that usually involves the lower extremities and is limited to the tropics, especially Africa and Australia. In Africa, the disease is known as Buruli ulcer, and in Australia, it is called Bairnsdale ulcer. The disease typically begins as a painless lump under the skin at the site of previous trauma. After a few weeks, a shallow ulcer develops at the site of the lump. The lesions may become necrotic and extend into the subcutaneous tissues. Satellite nodules may develop and ulcerate, but constitutional symptoms rarely occur unless the ulcers become superinfected.

M. haemophilum

M. *haemophilum* was first isolated in 1978 from a subcutaneous lesion of a patient with Hodgkin's disease (169). The organism is unique among mycobacteria in its requirement for hemin or ferric ammonium citrate for growth. Since the first description of the species, most of the reported cases have been female renal transplant recipients living in Australia. M. *haemophilum* has also been isolated from multiple sites in AIDS patients in the United States, from patients who have received bone marrow transplantation, and from localized lesions in pediatric patients with cervical lymphadenopathy who otherwise had no underlying immunocompromising factors (133, 171, 196).

The clinical presentation is multiple skin nodules, commonly involving the extremities, occurring in clusters or without a definite pattern. The nodules are occasionally found on the face and abdomen and in the gluteal and breast regions. Abscesses and draining fistulas may develop from the nodules. M. *haemophilum* infections may be underrecognized because of the organism's nutritional requirements and predilection for lower temperatures of incubation (i.e., 30°C).

M. shimoidei

The original description of the organism and proposal of the name M. *shimoidei* were based on seven isolates obtained from a 56-year-old male patient in Japan over an 11-year period. The patient had cavitary pulmonary disease resembling tuberculosis that ultimately proved fatal. The DNA homology studies established M. *shimoidei* as a distinct species (94). Phenotypically, M. *shimoidei* may be confused with the M. *terrae* complex, but the two species can be differentiated on the basis of catalase and β-galactosidase reactions. Only a few other cases of pulmonary disease have been reported since the first case description.

M. asiaticum

M. *asiaticum* was first isolated from monkeys in 1965 but was not recognized as a distinct species until 1971. M. *asiaticum* is phenotypically similar to M. *gordonae*, and there is only one easily determined phenotypic characteristic by which these two species can be distinguished: M. *gordonae* is scotochromogenic, whereas M. *asiaticum* is photochromogenic (228).

The first report of M. *asiaticum* human disease was published in 1983 (11). In a series of five patients from Australia, M. *asiaticum* was isolated from pulmonary specimens. Two of the patients in this series met the criteria for pulmonary mycobacteriosis. M. *asiaticum* has been only rarely isolated from patients in the United States.

M. celatum

M. *celatum* is a slowly growing nonphotochromogen species recently described by Butler et al. (26). It is similar biochemically and morphologically to a group of strains described by Brander et al. (15). M. *celatum* shares characteristics with MAC, M. *xenopi*, M. *malmoense*, and M. *shimoidei* but most closely resembles M. *xenopi*. M. *celatum* can be differentiated from M. *xenopi* by its poor growth at 45°C, production of large colonies on 7H10 agar, and production of trace amounts to 2% of the fatty acid 2-docosanol. A few strains have been misidentified as M. *tuberculosis* complex in a commercially available DNA probe test (Gen-Probe) (25).

Strains have been isolated from diverse geographic areas throughout the United States, Finland, and Somalia. Most strains have been isolated from respiratory tract specimens, but the organisms have been recovered from other sources as well, including stool, CSF, and blood. The clinical significance of these isolates was not defined. In one series, 32% of the patients from whom M. *celatum* was isolated were infected with HIV (26).

Mycobacterium Species Rarely Causing Human Disease

Several species of slowly growing mycobacteria, including M. *gordonae*, M. *gastri*, and M. *terrae* complex, and most rapidly growing mycobacteria other than M. *fortuitum* complex are so rarely associated with human disease that they are best characterized as nonpathogens. Many of the putative case reports of infections attributable to these non-pathogenic mycobacteria lack sufficient documentation of the organism's identification or its disease association. As a result, the authenticity of many of these reports is suspect.

M. gordonae is the nonpathogenic species most commonly encountered in clinical mycobacteriology laboratories. The organism is widely distributed in soil and water. In only one of the 38 published reports of M. gordonae infection was there convincing evidence that M. gordonae played a role in the patient's disease process (228). This sole clinically significant infection occurred in an AIDS patient, with M. gordonae isolated from sputum and bone marrow.

Unlike M. gordonae, M. gastri is not frequently encountered in the clinical laboratory. Wayne and Sramek found only three reports of human infection attributed to M. gastri (228). None of the reports gave sufficient details on the properties of the organisms to conclusively identify them as M. gastri.

M. terrae complex consists of the three species M. terrae, M. nonchromogenicum, and M. triviale. Clinical disease is so rarely associated with M. terrae complex that identification to species level is usually not required; however, an adequate number of tests should be performed to ensure that the potentially pathogenic species M. malmoense is not misidentified as M. terrae complex. Infections of the lung, tendon sheath, and intestine have been reported (67, 120, 136).

M. smegmatis appears to be more commonly associated with human disease than any of the other rapidly growing mycobacteria that do not belong to the M. fortuitum complex. Wallace et al. published an excellent review of the clinical conditions associated with M. smegmatis infection (220). It has most often been associated with soft tissue lesions following trauma or surgery. Rare cases of significant infections caused by M. flavescens, M. neoaurum, and M. thermoresistibile have also been reported.

COLLECTION, TRANSPORT, AND STORAGE OF SPECIMENS

Safety Procedures

M. tuberculosis is a proven hazard to anyone exposed to infectious aerosols in the laboratory (79). Nosocomial transmission of this organism from patients or specimens is of major concern to health care workers (28). Because of the low infective dose of M. tuberculosis for humans (50% infective dose, <10 bacilli) and in some laboratories a high rate of isolation of acid-fast organisms from clinical specimens (>10%), sputa and other clinical specimens from patients with suspected or known cases of tuberculosis must be considered potentially infectious and handled with appropriate precautions (35). Laboratory safety procedures aimed at minimizing the dispersal of mycobacteria into the air and protecting personnel from infection by inhalation of droplet nuclei are of the utmost importance. The control of aerosols and other forms of mycobacterial contamination is achieved by the use of appropriate biological safety cabinets (BSCs), centrifuges with safety carriers, and meticulous processing techniques. BSCs must be maintained according to the manufacturers' guidelines and certified at regular intervals.

Biosafety level 2 practices, containment equipment, and facilities are required for laboratories processing specimens

for preparation of direct acid-fast smears and performing any cultures of specimens suspected to contain mycobacteria, provided that aerosol-generating manipulations of such specimens are conducted in a class I or II BSC. Practices include wearing of gloves and protective laboratory clothing, appropriate hand washing after specimen handling, avoidance of the generation of aerosols, and if use of needles is necessary, appropriate handling and discarding of specimens into puncture-resistant containers. Liquefaction and concentration of sputum for acid-fast staining may be conducted safely on the open bench by first treating the specimens (in a class I or II BSC) with an equal volume of 5% sodium hypochlorite solution (undiluted household bleach) and waiting 15 min before centrifugation (175, 186). The sensitivity of such smears is increased by cytocentrifugation (175). The more rigorous biosafety level 3 practices, containment equipment, and facilities are required for laboratories that process specimens for mycobacterial culture and that propagate and manipulate cultures of M. tuberculosis (e.g., identification and susceptibility testing). These practices require that laboratory access be restricted, that directional airflow be used to maintain the laboratory under negative pressure, and that workers wear special laboratory clothing, masks, and gloves when appropriate. For a detailed description of safety requirements in the mycobacteriology laboratory, refer to the CDC-National Institutes of Health biosafety guide for laboratories (35).

All work involving specimens or cultures, such as making smears, inoculating media, adding reagents to biochemical tests, opening centrifuge cups, and sonicating specimens, must be performed in a BSC. All specimens suspected of containing mycobacteria (including specimens processed for other microorganisms), except for those being centrifuged for concentration purposes, must be handled within the BSC. Specimens that are to be taken out of the BSC should be covered before transport. To further decrease aerosolization, paper towels soaked with a phenolic disinfectant can be used to cover work surfaces, line discard pans, and wipe the outsides of culture tubes or containers. All work surfaces, including bench tops and the inside of the BSC, should be cleaned with an appropriate disinfectant before and after work. Effective disinfectants include Amphyl or other phenol-soap mixtures and 0.05 to 0.5% sodium hypochlorite (concentration varies according to the nature of the contaminated surface). Five percent phenol is no longer recommended as a surface disinfectant because of the documented toxicity of this compound to personnel. UV light is a useful adjunct for surface decontamination and may be used to radiate the work area when it is not in use. Centrifuges should be used with aerosol-free safety carriers to contain debris in case of tube breakage. Use of electric incinerators rather than open flames is recommended. The excess inoculum from inoculating loops, wire, or spades may be removed by dipping the tool into a container of 5% phenol or 95% ethanol in washed sand prior to insertion in an incinerator. Use syringes with permanently attached needles if needles are required. An autoclave should be available in an easily accessible area and should be used to decontaminate infectious waste before removal of the waste to disposal areas.

Personnel should be regularly monitored with the Mantoux PPD skin test (every 6 months and more often if a conversion has been documented) (35, 106) to demonstrate conversions. Those with positive skin tests should be evaluated for active tuberculosis with a chest X ray and a clinical workup. Physical examinations should be obtained when circumstances indicate that they are necessary. New converters should be referred to the Employee Health and the Infection Control Departments for epidemiologic evaluation. Laboratory personnel should work in a manner that protects themselves and others from infection. Proper training in laboratory procedures and safety must be provided for all persons who work in mycobacteriology laboratories. The laboratory should have a protocol for emergency action to be taken in case of accidents that might form aerosols. Personnel should hold their breath, make sure BSCs are on and centrifuges are off, leave the room, and keep the door closed for at least 30 min. Using appropriate respiratory protection devices, personnel can return to the accident area to cover the spill with disinfectant. PPD-negative personnel should be skin tested 3 and 6 months after the accident. Persons who are immunocompromised should be discouraged from working in the mycobacteriology laboratory (35).

Specimen Collection

Many different types of specimens may be submitted for mycobacterial culture, but the majority of the specimens submitted are from the respiratory tract. Tissue, normally sterile body fluids, urine, and gastric aspirates are other commonly submitted specimens. Blood and stool specimens may be submitted from patients with AIDS. The laboratory must provide guidance for proper collection and transport of these specimens, since the quality of specimens collected and the proper transport of those specimens to the laboratory are critical to the successful isolation of AFB.

Specimens should be collected and submitted in sterile, leakproof, disposable, appropriately labeled, laboratory-approved containers. Do not use waxed containers, as they may provide false-positive smear results (106). Initial specimens should ideally be collected prior to the initiation of antimycobacterial chemotherapy. Specimens should be collected aseptically, or the collection method should bypass areas of contamination as much as possible in order to minimize contamination with indigenous flora. Avoid contamination with tap water or other fluids that may contain either viable or nonviable environmental mycobacteria, since saprophytic mycobacteria may produce false-positive culture and/or smear results (77, 219). In general, swabs are not recommended for the isolation of mycobacteria, since they provide only a limited amount of material, and the hydrophobicity of the mycobacteria often compromises their isolation from swabs onto solid or broth media. Do not use any fixatives or preservatives, and transport specimens to the laboratory in as short a time as is practical to avoid overgrowth by contaminating bacteria and fungi. If transport to the laboratory is delayed more than 1 h, refrigerate all specimens except blood. Once specimens have been received in the laboratory, refrigerate them until processing.

Sputum

Sputum, both expectorated and induced, is the principal specimen obtained for the diagnosis of pulmonary tuberculosis. Collect an early-morning specimen, preferably 5 to 10 ml, from a deep, productive cough on at least 3 but usually not more than 5 or 6 consecutive days. Processing of additional specimens does not seem to improve recovery. For follow-up of patients on therapy, collect specimens at weekly intervals beginning 3 weeks after initiation of therapy. For expectorated sputum, patients should be instructed

to cough deeply to produce specimens distinct from saliva or nasopharyngeal discharge. The patient should also be instructed to press the rim of the container under the lower lip at the time of expectoration to minimize the chance of contaminating the outside of the container. For induced sputum, use sterile hypertonic saline, and avoid sputum contamination with nebulizer reservoir water to avoid possible false-positive culture or smear results due to saprophytic mycobacteria (219). Induction techniques should be performed by qualified personnel using appropriate respiratory protection devices in an enclosed area with appropriate airflow to minimize the risk of infectious aerosols (34). Indicate on the requisition whether the specimen is induced or expectorated to ensure proper handling, as induced sputa appear watery and much like saliva. The screening of sputum for upper respiratory tract contamination, as is done for bacteriologic workups, is not performed for AFB cultures, but specimens should preferably be lower respiratory tract secretions. Any isolate of M. *tuberculosis*, even that in saliva, is considered significant. Pooled sputum specimens are unacceptable specimens for mycobacterial culture because of increased contamination and lower test sensitivity (106). Clinicians should consider the diagnosis of pulmonary mycobacterial disease in patients with or at risk for HIV infection, even if the clinical presentation is unusual (88), and sputum samples should be collected even if results of chest X-ray examinations are negative.

Bronchoalveolar Lavage Fluids and Bronchial Washings

Invasive collection techniques may be necessary to diagnose pulmonary mycobacteriosis in some patients unable to produce sputum. Bronchial washings, bronchoalveolar lavage fluid, transbronchial biopsy specimens, and brush biopsy specimens may all be collected during bronchoscopy. Collect at least 5 ml of bronchial washing or bronchoalveolar lavage fluid in a sterile container. Avoid contaminating the bronchoscope with tap water, which may contain saprophytic mycobacteria. The material obtained from bronchial brushings can be placed in a sterile tube containing approximately 10 ml of Middlebrook 7H9 broth supplemented with bovine serum albumin (BSA; final concentration of 1 to 2%) and Tween 80 (0.5%). Bronchial washings and bronchoalveolar lavage fluids can usually be sent directly to the laboratory in the containers in which they have been collected. Frequently, bronchoscopy causes the patient to produce sputum naturally for several days after the procedure, and specimens collected a day or two after bronchoscopy enhance detection of mycobacteria. Transtracheal aspirates are not commonly submitted, but specimens obtained by invasive techniques such as fine-needle aspiration and open-lung biopsy may be submitted in difficult cases.

Gastric Lavage Fluids

Aspiration of swallowed sputum from the stomach by gastric lavage may be necessary for infants, young children, and the obtunded. On each of 3 consecutive days, collect 5 to 10 ml of fluid in a sterile container without a preservative. Fasting, early-morning specimens are recommended in order to obtain sputum swallowed during sleep. Gastric contents are initially collected with a sterile suction syringe connected to a tube inserted in the stomach. Sterile saline (20 to 30 ml) may then be introduced into the stomach and aspirated as lavage fluid. The gastric contents and lavage fluid may be pooled in a sterile container. Adjust fluid to neutral pH with 100 mg of sodium carbonate immediately following collection or within 4 h. If the specimens cannot be processed within 4 h, the laboratory should provide sterile disposable containers with 100 mg of sodium carbonate for collection. Unneutralized specimens are not acceptable, as acid is detrimental to the mycobacteria. Saprophytic mycobacteria may be recovered from gastric lavage fluid.

Blood

Cultures for the isolation of mycobacteria from blood should be reserved for immunocompromised patients, particularly those with AIDS. The majority of disseminated mycobacterial infections are due to MAC. Disseminated MAC infection occurs usually among patients with CD4 counts of <100/mm^3 (83, 85). Positive MAC blood cultures are always associated with clinical evidence of disease (83). The Isolator lysis-centrifugation system (Wampole Laboratories, Cranbury, N.J.) or the radiometric BACTEC 13A blood culture bottle (Becton-Dickinson Diagnostic Instrument Systems, Cockeysville, Md.) is recommended for mycobacterial blood cultures (1, 108, 109, 111). Isolator tubes contain saponin, which lyses cells and releases intracellular mycobacteria. Sediment from the Isolator tube can be cultured into LJ or 7H11 medium or both. The use of Isolator blood sediments to inoculate BACTEC 12B medium is contraindicated because the sediments may be inhibitory to MAC (223). The BACTEC 13A bottle contains a lysing agent and is designed specifically for the recovery of mycobacteria from blood. The 13A medium can be directly inoculated with 5 ml of blood without the potential hazards associated with the lysis-centrifugation procedure. A potential disadvantage of using BACTEC 13A medium is that this method cannot determine the number of CFU per volume of blood. However, the significance and necessity of obtaining quantitative data with each culture are unclear. The BACTEC blood culture system is discussed in more detail below. Direct inoculation of blood onto a solid medium is not recommended (1, 69, 111, 159).

For specimen collection, disinfect the site as for routine blood culture. If blood needs to be transported before inoculation of BACTEC medium, use sodium polyanethol-sulfonate (SPS) or heparin as an anticoagulant. Blood collected in EDTA or blood that is coagulated is not acceptable. MAC organisms survive for prolonged periods in Isolator tubes, so processing of Isolator tubes may be delayed for 24 h if absolutely necessary (84, 217).

Urine

Collect the first morning specimens, either by catheterization or midstream clean catch, into a sterile container on 3 consecutive days. Appropriate cleaning of genitalia should precede collection. Organisms accumulate in the bladder overnight, and the first morning void provides best results. Specimens collected at other times are dilute and thus not optimal. A minimum of 40 ml of urine is usually required for culture. For suprapubic tap, collect as much as possible in a syringe with a Luer tip cap or other sterile container. Twenty-four-hour pooled specimens, catheter bag, and small-volume (<40 ml unless a larger volume is not obtainable) specimens are unacceptable.

Stools

Stool specimens (>1 g) should be collected in sterile, wax-free, disposable clean containers or transferred from a bedpan or from plastic wrap stretched over the toilet bowl and sent directly to the laboratory. Stool specimens are recommended only for detection of MAC involvement in the gastrointestinal tracts of patients with AIDS.

The clinical utility of culture of stool for AFB remains controversial. Past recommendations have been that stool be cultured for mycobacteria only if the direct smear of unprocessed stool is positive for AFB. More recent studies showed the sensitivity of smear to be only 32 to 34% (108, 145), suggesting that fecal-smear results should not determine whether a mycobacterial culture is performed. Routine screening cultures or smears do not appear to be effective ways to identify those patients at risk for developing disseminated MAC infection (84). However, if screening cultures are positive, dissemination often follows (90, 101, 105, 161).

Body Fluids

Body fluids (cerebrospinal [CSF], pleural, peritoneal, pericardial, etc.) are aseptically collected by aspiration or surgical procedures. Collect as much as possible (10- to 15-ml minimum) in a sterile container or syringe with a Luer tip cap. Bloody specimens may be anticoagulated with SPS. Because certain body fluids (such as CSF and peritoneal dialysis effluent) may have low numbers of mycobacteria present, larger specimen volumes increase culture yields. CSF culture requires at least 2 ml. Small-volume specimens may be directly inoculated into BACTEC or Septi-Chek (Becton-Dickinson) broth.

Tissues (Lymph Node, Skin, Other Biopsy Material)

Aseptically collect at least 1 g of tissue, if possible, into a sterile container without fixative or preservative. Do not immerse in saline or other fluid or wrap in gauze. For cutaneous ulcers, collect biopsy material from the periphery of the lesion. Specimens submitted in formalin are unacceptable.

Abscess Contents, Aspirated Fluid, Skin Lesions, Wounds

Cleanse the skin with alcohol before aspirating the sample, and then aspirate as much material as possible into a syringe with a Luer tip cap. If the volume is insufficient for aspiration by needle and syringe, collect the specimen on a swab, and place the swab in transport medium (Amies or Stuart's). Negative results from swab specimens, however, are not reliable. For cutaneous lesions, aspirate material from under the margin of the lesion. The laboratory may provide 7H9 broth for transport of small volumes of aspirates. Dry swabs are not acceptable.

ISOLATION AND STAINING PROCEDURES

The laboratory diagnosis of mycobacterial diseases depends on the detection and recovery of AFB from clinical specimens. Microbiologists must be prepared to process and culture a variety of both respiratory and nonrespiratory specimens, of which some are contaminated and others are not. Appropriate culture media and conditions of incubation must be selected to facilitate optimal recovery of mycobacteria.

Processing of Specimens

Normally Sterile Specimens

Tissues or body fluids collected aseptically usually do not require the digestion and decontamination procedures used with contaminated specimens (190). Decontamination of a specimen should be attempted only if the specimen is thought to be contaminated. If the need to decontaminate a specimen is not clear, the specimen may be refrigerated until the routine bacteriologic cultures are checked the next day. If routine cultures are not available, the specimen may be initially inoculated to blood or chocolate agar plates to check for sterility before being processed for mycobacteria (169).

Normally sterile tissues may be ground in sterile 0.85% saline or 0.2% BSA and then inoculated directly to both solid and liquid media. Because body fluids commonly contain small numbers of mycobacteria, these fluids should be concentrated before inoculation of media. The fluids are centrifuged at $\geq 3,000 \times g$, and the sediment is inoculated to liquid and solid media. When the volume of fluid submitted for culture is small, the fluid may be added directly to liquid media, such as Middlebrook 7H9 (in a ratio of 1 part specimen to 5 parts broth), the BACTEC 12B bottle, or the Septi-Chek AFB system. Specimens that would normally be inoculated directly may be decontaminated at some institutions because of problems with contamination specific to those institutions. Blood and bone marrow specimens are discussed below.

Contaminated Specimens

Digestion and Decontamination Methods

Most specimens submitted for mycobacterial culture consist of a complex organic matrix contaminated with a variety of organisms that can rapidly outgrow the mycobacteria. Mucin may trap mycobacterial cells and protect contaminating bacteria from the action of decontaminating agents. Thus, mycobacteria are recovered optimally from clinical specimens through use of procedures that reduce or eliminate contaminating bacteria while releasing mycobacteria trapped in mucin and cells. Liquefaction of certain specimens, particularly sputum, is often necessary. The mycobacteria are then concentrated to enhance detection in stained smears and by culture. No one method of digestion and decontamination is ideal for all clinical specimens, all laboratories, and all circumstances. The laboratorian must be aware of the inherent limitations of the various methods.

Sodium hydroxide, the most commonly used decontaminant, also serves as a mucolytic agent but must be used cautiously, because it is only somewhat less harmful to tubercle bacilli than to the contaminating organisms. The stronger the alkali, the higher its temperature during the time it acts on the specimen, and the longer it is allowed to act, the greater will be the killing action on both contaminants and mycobacteria (119). Liquefaction is facilitated by vigorous mixing of a solution in a sealed container with a vortex-type mixer. To use the mixer properly and minimize aerosol production, the tube should be held on the vibrating base in such a way that churning, splashing, and foaming of the mixture are avoided. Homogenation should occur by centrifugal swirling, and this swirling should not be vigorous enough to allow material to rise to the cap. Wait at least 15 min after agitation before opening the tube to allow any fine aerosol droplets formed during the mixing

to settle. All such procedures should be carried out in a BSC.

Most commonly, a combination liquefaction-decontamination mixture is used. N-Acetyl-L-cysteine (NALC), dithiothreitol, and several enzymes effectively liquefy sputum. These agents have no direct inhibitory effect on bacterial cells; however, their use permits treatment with lower concentrations of sodium hydroxide, thereby indirectly improving the recovery of mycobacteria. Cetylpyridinium chloride (CPC) in specimens mailed from remote collection stations to a central processing station allows good recovery of M. *tuberculosis* without overgrowth by contaminating bacteria (186).

Specimens submitted for the culture of mycobacteria should be processed as soon as possible. Commonly used digestion-decontamination methods are the NaOH method, the Zephiran-trisodium phosphate method, and the NALC-NaOH method. The mycobacteria are somewhat protected from the effects of various chemical digestants by the lipids contained within the cell wall. However, even under the best of conditions, all currently available procedures are to some extent toxic for mycobacteria. Thus, the best yield of mycobacteria may be expected to result from the use of the mildest decontamination procedure that sufficiently controls contaminants. Strict adherence to specimen processing procedures is mandatory to ensure survival of the maximal number of mycobacteria.

The type of procedure selected by the laboratory should reflect the number and types of specimens received and the time and staff required to process them. Most laboratories process specimens in batches; current recommendations suggest that specimen batches be processed daily (203). The most widely used digestion-decontamination method is the NALC-NaOH method. A variety of specimens can be processed by the NALC-NaOH method, and the method is compatible with the BACTEC culture system. Whatever method is used, great care must be taken to prevent laboratory cross-contamination of patient specimens during processing (185). A single positive culture for M. *tuberculosis* could be the basis of a diagnosis of tuberculosis, and thus, a false-positive culture may have profound consequences for the clinical management of the patient, epidemiologic investigations, and public health control measures. Commonly used digestion-decontamination methods are described below. Further details are provided in the guide by Kent and Kubica (106), and step-by-step instructions are given in the *Clinical Microbiology Procedures Handbook* (97).

NALC-NaOH method

PRINCIPLE. NALC is a mucolytic agent that at a concentration of 0.5 to 2.0% can rapidly digest sputum. Decontamination is achieved by the addition of sodium hydroxide at a final concentration of 1%. Sodium citrate included in the digestant mixture exerts a stabilizing effect on the NALC by chelating heavy-metal ions that may be present in the specimens and may inactivate the NALC.

REAGENTS. **(i) Digestant.** For each 100 ml of digestant desired, combine 50 ml of sterile 0.1 M (2.94%) trisodium citrate with 50 ml of 4% NaOH. The NaOH and citrate mixture can be sterilized and stored for future use. To this solution add 0.5 g of powdered NALC just before use. This solution must be used within 24 h of the addition of NALC, because the mucolytic action of NALC is inactivated on exposure to air. Sputolysin (Behring Diagnostics, La Jolla, Calif.), a concentrated sterile solution of dithiothreitol (Cleland's reagent), is more stable in air than NALC and can be used as the mucolytic agent.

(ii) Phosphate buffer. Buffer is 0.067 M and pH 6.8. Mix 50 ml of solution A (0.067 M disodium phosphate and 9.47 g of anhydrous Na_2HPO_4 in 1 liter of distilled water) and 50 ml of solution B (0.067 M monopotassium phosphate and 9.07 g of KH_2PO_4 in 1 liter of distilled water). If the final buffer requires pH adjustment, add solution A to raise the pH or solution B to lower it.

(iii) BSA (optional, see below). Use a sterile 0.2% solution of BSA fraction V adjusted to pH 6.8.

PROCEDURE

1. Working in a BSC, transfer a maximum volume of 10 ml of specimen to a sterile, graduated, 50-ml plastic centrifuge tube labeled with the appropriate identification. The tube should have a leakproof and aerosol-free screw cap. Add an equal volume of the NALC-NaOH solution. Dispense the solution with a fresh, sterile pipette for each specimen, taking care to avoid contamination. If the amount of specimen received is more than 10 ml, select about 10 ml of a purulent or bloody portion of the material. The final concentration of NaOH in the tube is 1%.

2. With the cap tightened completely, invert the tube to ensure that the NALC-NaOH solution contacts all inside surfaces of the tube and cap, and then mix the contents for approximately 20 s on a Vortex mixer until they are liquefied. If liquefaction is not complete in this time, agitate the solution at intervals during the following decontamination period.

3. Allow the mixture to stand for 15 min at room temperature (20 to 25°C), with occasional gentle shaking by hand. Avoid movement that causes aeration of the specimen, since the NALC is readily inactivated by oxidation. A small pinch of crystalline NALC may be added to especially viscous specimens to effect better liquefaction. Arrange workflow so that specimens remain in contact with the decontaminating agent for only 15 min at room temperature. Overprocessing results in reduced recovery of mycobacteria. If more active decontamination is needed, increase the concentration of NaOH rather than the time the specimen is exposed to digestant.

4. Add phosphate buffer (pH 6.8) up to the 50-ml mark on the tube. Addition of this solution minimizes the continuing action of NaOH and helps lower the specific gravity of the specimen, allowing more efficient sedimentation of mycobacteria by centrifugation. Recap the tube, and swirl it by hand to mix the contents well. In order to avoid cross-contamination, do not allow the diluent container to touch the specimen tubes, or else use a separate tube of prealiquoted diluent for each specimen tube. Addition of sterile H_2O at this step is not recommended, because it does not bring down the pH as much as phosphate buffer does.

5. Centrifuge the solution for at least 15 min at ≥3,000 × g. Use aerosol-free sealed centrifuge cups. Use of a refrigerated centrifuge and higher relative centrifugal force may increase recovery of mycobacteria (166).

6. Carefully decant the supernatant fluid into a splash-proof discard container containing a suitable disinfectant. To help prevent cross-contamination of specimens, do not touch the lip of the tube to the discard container; after decanting, wipe the lip of each tube with disinfectant-soaked gauze (separate piece for each tube) to absorb drips; and recap the tubes.

7. Using a separate sterile pipette for each tube, add to the sediment 1 to 2 ml of sterile 0.2% BSA fraction V (pH 6.8) or 1 to 2 ml of phosphate buffer (pH 6.8), and resuspend the sediment with the pipette or by shaking the tube gently by hand. Some microbiologists feel that BSA has a buffering and detoxifying effect on the sediment and increases adhesion of the specimen to solid media. However, BSA is not recommended for use with the BACTEC, because BSA may delay detection times.

8. Inoculate specimens onto appropriate solid culture media (LJ, Middlebrook 7H11 and 7H10, or selective medium) and into broth media. A 1:10 dilution of the specimen (0.5 ml of the specimen in 4.5 ml of sterile water or saline) has been recommended for the reduction of any possible toxic substances in the specimen. Evidence documenting the effectiveness of this practice is lacking, however. Label the media with proper patient identification. Solid media may be inoculated by using disposable capillary pipettes to deliver 3 drops to each medium. Plates should be tilted, the drops should be streaked, or a bent sterile glass rod should be used to spread the inoculum to ensure isolated colonies and the detection of more than one colony type if present. Incubation of culture medium is discussed below.

9. Prepare a smear for acid-fast staining by using a sterile disposable pipette to place 1 drop of the undiluted sediment onto a clean, properly labeled microscopic slide, covering an area approximately 1 by 2 cm. Place the smears on an electric slide warmer at 65 to 75°C for 2 h to dry and fix. Alternatively, the smears can be air dried and fixed by passing the slide three or four times through the blue cone of a flame. Heat fixing does not always kill mycobacteria. Thus, slides should be considered potentially infectious and handled appropriately.

10. Refrigerate the remaining undiluted sediment for later use if needed (direct susceptibility testing, further treatment if specimen is contaminated, etc.).

The addition of water or buffer to the treated specimen prior to centrifugation (step 4) can be omitted without any appreciable effect on the growth or isolation rate of mycobacteria (165). After addition of NALC-NaOH, specimens are vortexed for 15 s and then centrifuged for 15 min. The supernatant is decanted, and the sediment is suspended in 1 to 2 ml of phosphate-buffered saline (pH 5.3). If this modification is used, it may be necessary to lower the concentration of NaOH from 2 to 1.5%. In addition, timing is very critical, so batch sizes should be limited.

The NALC-NaOH method can be used to process gastric lavage specimens, tissues, stools, urine, and other body fluids in addition to sputum specimens. Centrifuge neutralized gastric lavage specimens, urine, and other body fluids (10 ml or more) at ≥3,000 × g for 30 min in sterile, screw-cap, 50-ml centrifuge tubes; decant the supernatants; suspend the sediments in 2 to 5 ml of sterile distilled water; add an equal volume of NALC-NaOH; and proceed as for sputum. If a gastric lavage specimen is mucopurulent, add 50 to 100 mg of NALC powder per 50 ml of lavage fluid, and vortex before centrifugation. Tissue that is not collected aseptically can be ground, placed in a tube, homogenized by vortexing, and processed like sputum. For stool specimens, place approximately 1 g of a formed specimen or 1 to 5 ml of a liquid specimen in a total volume of 10 ml of

Middlebrook 7H9 broth, sterile water, or sterile saline; vortex vigorously for 30 s; and then allow large particles to settle to the bottom of the tube for 15 min. Remove 7 to 8 ml of supernatant, place it into a 50-ml centrifuge tube, and process like sputum. Alternatively, the diluted stool specimen can be filtered through gauze to remove particles.

Sodium hydroxide method

PRINCIPLE. Sodium hydroxide (NaOH) is both a digestant and a decontaminant. The digestant, 2 to 4% NaOH (final specimen concentration, 1 to 2%), should be used at the lowest concentration that will effectively digest and decontaminate the specimen. Because of the toxicity of NaOH to mycobacteria, strict adherence to the time limits for processing is absolutely necessary.

REAGENTS. **(i) Digestant.** Use NaOH solution (2 to 4%). Sterilize by autoclaving.
(ii) 2 N HCl. Dilute 33 ml of concentrated HCl to 200 ml with water. Sterilize by autoclaving.
(iii) Phenol red indicator. Combine 20 ml of phenol red solution (0.4% in 4% NaOH) and 85 ml of concentrated HCl with distilled water to make 1,000 ml.
(iv) Phosphate buffer (0.067 M, pH 6.8). See the NALC-NaOH procedure above for buffer preparation.

PROCEDURE. Follow the steps described above for the NALC-NaOH method, substituting sodium hydroxide (2%) for the NALC-alkali digestant.

1. Transfer a maximum volume of 10 ml of specimen to a sterile 50-ml screw-cap plastic centrifuge tube. Add an equal volume of NaOH.
2. With the cap tightened, invert the tube, and then either agitate the mixture vigorously for 15 min on a mechanical mixer, or vortex it vigorously and let it stand for 15 min. Timing must be rigidly controlled. If it is absolutely necessary to reduce excessive contamination, the NaOH concentration can be increased to 3 or 4% rather than increasing the time of exposure to the alkali.
3. Add phosphate buffer (pH 6.8) up to the 50-ml mark on the tube. Recap the tube, and swirl it by hand to mix the contents well.
4. Centrifuge the specimen at ≥3,000 × g for 15 min, decant the supernatant, and add a few drops of phenol red indicator to the sediment. Neutralize the sediment with HCl. The tube contents must be thoroughly mixed, and acid addition must be stopped when the solution becomes persistently yellow.
5. Resuspend the sediment in 1 to 2 ml of phosphate buffer or sterile 0.1% BSA fraction V.
6. Inoculate the resuspended sediment to appropriate culture media, and prepare a smear.

Zephiran-trisodium phosphate method

PRINCIPLE. Trisodium phosphate liquefies sputum readily but requires a long exposure for decontamination of the specimens when used alone. Benzalkonium chloride (Zephiran), a quaternary ammonium compound, used with trisodium phosphate as described below shortens the required exposure and selectively destroys many contaminants with little activity against tubercle bacilli. Since Zephiran is bacteriostatic to mycobacteria, the digested,

centrifuged sediment must be neutralized with buffer before being inoculated onto agar medium. The phospholipids of egg medium provide a built-in neutralizer for this compound, making neutralization unnecessary. This system can be used when the laboratory is unable to monitor the exposure time to the decontaminating agent, since timing of the digestion-decontamination process is not critical. This method is incompatible with the BACTEC system.

REAGENTS. **(i) Zephiran-trisodium phosphate digestant.** Dissolve 1 kg of trisodium phosphate ($Na_3PO_4 \cdot 12H_2O$) in 4 liters of hot distilled water. Add 7.5 ml of Zephiran concentrate (17% benzalkonium chloride; Winthrop Laboratories, New York, N.Y.), and mix. Store at room temperature.
(ii) Neutralizing buffer. The neutralizing buffer (pH 6.6) is prepared by adding 37.5 ml of 0.067 M disodium phosphate to 62.5 ml of 0.067 M monopotassium phosphate (for preparation of buffer solutions, see NALC-NaOH procedure above).

PROCEDURE

1. Transfer a maximum volume of 10 ml of specimen to a sterile, 50-ml, screw-cap plastic centrifuge tube. Add an equal volume of the Zephiran-trisodium phosphate digestant.
2. With the cap tightened, invert the tube, and then agitate the mixture vigorously for 30 min on a mechanical shaker. Permit the material to stand without shaking for an additional 30 min at room temperature.
3. Centrifuge the specimen at ≥3,000 × g for 15 min, decant the supernatant, and add 20 ml of neutralizing buffer. Vortex for 30 s to thoroughly suspend the sediment in the buffer. (The neutralizing buffer serves to inactivate traces of Zephiran in the sediment.)
4. Centrifuge the specimen again for 20 min.
5. Decant the supernatant, retaining some fluid in which to resuspend the sediment.
6. Inoculate egg-based media, and make the smear. If the sediment is to be inoculated to agar-based media, lecithin must be used as a neutralizer (118). This system cannot be used in conjunction with the BACTEC system.

Oxalic acid method

PRINCIPLE. The oxalic acid method (42) is reported to be superior to alkali methods for processing specimens consistently contaminated with *Pseudomonas* species and certain other contaminants. Specimens processed by this method may be used with the BACTEC system. The oxalic acid procedure can be used to decontaminate a previously processed sediment when cultures are contaminated with *Pseudomonas* spp. (235).

REAGENTS
5% Oxalic acid
Physiologic saline (0.85%)
4% NaOH
Phenol red indicator or pH paper

PROCEDURE

1. Add an equal volume of 5% oxalic acid to 10 ml or less of specimen in a 50-ml centrifuge tube.

2. Vortex the solution, and then allow it to stand at room temperature for 30 min with occasional shaking.

3. Add sterile physiologic saline to the 50-ml mark on the centrifuge tube. Recap the tube, and invert it several times to mix the contents.

4. Centrifuge for 15 min at ≥3,000 × g, decant the supernatant fluid, and add a few drops of phenol red indicator to the sediment. Alternatively, use pH paper to monitor the pH.

5. Neutralize with 4% NaOH.

6. Resuspend the sediment, inoculate it to media, and make the smear.

CPC-sodium chloride method

PRINCIPLE. CPC, a quaternary ammonium compound, is used to decontaminate specimens while sodium chloride effects liquefaction. Since CPC is bacteriostatic for mycobacteria inoculated onto agar-based media and since this effect is not neutralized in the digestion process, sediment from specimens treated with CPC should be inoculated only onto egg-based media. This method cannot be used in conjunction with the BACTEC system.

The CPC method is a means of digesting and decontaminating sputum specimens in transit for >24 h. After arrival in the laboratory, the digested-decontaminated specimens need only be concentrated by centrifugation, and the sediment can then be inoculated directly onto media. Mycobacteria remain viable for 8 days in the solution. Contamination rates associated with treatment of sputum specimens by the CPC method are lower than those associated the NALC-NaOH method (186).

REAGENTS. **(i) CPC digestant-decontaminant.** Dissolve 10 g of CPC and 20 g of NaCl in 1,000 ml of distilled water. The solution is self-sterilizing and remains stable if protected from light, extreme heat, and evaporation. Dissolve with gentle heat any crystals that might form in the working solution. Ingredients include sterile water and sterile saline or 0.2% sterile BSA fraction V.

PROCEDURE

1. Collect 10 ml or less of sputum in a 50-ml screw-cap centrifuge tube.

2. Inside a BSC, add an equal volume of CPC-NaCl, cap the tube securely, and shake it by hand until the specimen liquefies.

3. Package the specimen appropriately in a double mailing container as specified by current postal regulations, and send it to a processing laboratory. The specimen is digested and decontaminated in transit.

4. Upon receipt in the processing laboratory (allow at least 24 h for digestion-decontamination to be completed), dilute the digested-decontaminated specimen to the 50-ml mark with sterile distilled water, and recap the tube securely. Invert the tube several times to mix the contents.

5. Centrifuge at ≥3,000 × g for 15 min, decant the supernatant fluid, and suspend the sediment in 1 to 2 ml of sterile water, saline, or 0.2% BSA fraction V.

6. Inoculate the resuspended sediment onto egg medium, and make the smear.

Sulfuric acid

PRINCIPLE. Sulfuric acid is used to decontaminate the specimen. This method may be useful for urine and other watery body fluids that yield contaminated cultures when they are processed by one of the alkaline digestants.

REAGENTS
4% Sulfuric acid
4% Sodium hydroxide
Sterile distilled water
Phenol red indicator

PROCEDURE

1. Centrifuge the entire specimen for 30 min at 3,000 × g. This may require several tubes.

2. Decant the supernatant fluids, and pool the sediments if several tubes were used for a single specimen.

3. Add an equal volume of 4% sulfuric acid to the sediment.

4. Vortex, and let stand for 15 min at room temperature.

5. Fill the tube to the 50-ml mark with sterile water.

6. Centrifuge at ≥3,000 × g for 15 min, and decant the supernatant.

7. Add 1 drop of phenol red indicator, and neutralize the mixture with 4% NaOH until a persistent pale pink forms.

8. Inoculate media, and make the smear.

Optimizing Decontamination Procedures

Excessive contamination is defined as contamination of more than 5% of all specimens cultured. A high contamination rate (see also section on quality assurance later in this chapter) suggests either too weak a decontaminant or incomplete digestion. One or a combination of several of the following measures may be used to help decrease the decontamination rate.

1. Cautiously increase the strength of the alkali. Four percent NaOH will probably kill most tubercle bacilli.

2. Use a selective medium (one that contains antibiotics in addition to the dye malachite green) in addition to a nonselective primary culture medium to inhibit the growth of bacterial and fungal contaminants. Selective 7H11 agar (Mitchison medium), Mycobactosel agar (158) (BBL Microbiology Systems, Cockeysville, Md.), or the Gruft modification of LJ medium (which has penicillin [50 U/ml] and nalidixic acid [35 μg/ml] added to the regular formula) should be considered.

3. Use an alternative digestion-decontamination procedure for problem specimen types. Respiratory secretions from patients with cystic fibrosis, which are often overgrown with pseudomonads, can successfully be decontaminated with NALC-NaOH followed by 5% oxalic acid (235).

To determine the decontaminating capabilities of each new batch of reagents, the laboratory may wish to inoculate blood agar plates with four to six decontaminated sputum specimens in addition to inoculating mycobacterial media. The numbers of contaminants that grow after 48 h of incubation at 35°C should be minimal to none (106).

Decontamination of Urine

For urine, the entire specimen (at least 50 ml) should be centrifuged at 3,000 to 3,800 × g for 30 min. This may require several tubes. The supernatants are discarded, and the sediments are combined and suspended in 2 to 5 ml of sterile distilled water. Both the NALC-NaOH and the

TABLE 1 Acid-fast-smear evaluation and reporting[a]

Report	No. of AFB seen[b]		
	Fuchsin stain	Fluorochrome stain	
	1,000×	250×	450×
No AFB seen	0	0	0
Doubtful; repeat	1–2/300 F (3 sweeps)	1–2/30 F (1 sweep)	1–2/70 F (1.5 sweeps)
1+	1–9/100 F (1 sweep)	1–9/10 F	2–18/50 F (1 sweep)
2+	1–9/10 F	1–9/F	4–36/10 F
3+	1–9/F	10–90/F	4–36/F
4+	>9/F	>90/F	>36/F

[a]Table is adapted from Kent and Kubica (106).
[b]F, microscope fields. In all cases, one full sweep refers to scanning the full length (2 cm) of a smear 1 cm wide by 2 cm long.

sulfuric acid methods are suitable for decontamination of urine specimens.

Decontamination of Stools

The oxalic acid and NALC-NaOH decontamination methods both appear to be useful for the recovery of MAC organisms from stool specimens (242). The most useful media for recovering MAC from stool specimens are Mitchison's selective 7H11 agar and Mycobactosel LJ medium (242). These media have relatively low contamination rates and show good recovery of MAC compared with nonselective media.

Acid-Fast Stain Procedures

Mycobacteria are difficult to stain. The large amount of lipids present in their cell walls render them impermeable to the dyes used in the Gram stain, and the appearance of mycobacteria in a Gram-stained specimen may be variable. Mycobacteria may be gram invisible, they may appear as negatively stained images or "ghosts," or they may appear as beaded gram-positive rods (208). To an experienced microscopist, such an appearance may lead to the diagnosis of an unsuspected mycobacterial infection.

Mycobacteria are able to form stable complexes with certain arylmethane dyes such as fuchsin and auramine O. Although the exact nature of the acid-fast staining reaction is not completely understood, phenol in the primary stain allows penetration of the stain. The cell wall mycolic acid residues retain the primary stain even after exposure to acid-alcohol or strong mineral acids. This resistance to decolorization is required for an organism to be termed acid fast. A counterstain is employed to highlight the stained organisms for easier microscopic recognition.

The acid-fast nature of an organism can be determined by several methods. In carbol fuchsin staining procedures, such as Ziehl-Neelsen and Kinyoun staining, AFB appear red against a blue or green background, depending on the counterstain used. In fluorochrome staining procedures, such as with auramine O or auramine-rhodamine, AFB fluoresce yellow to orange (the color may vary with the filter system used). Carbol fuchsin-stained smears are examined with a 100× oil immersion objective (×1,000 magnification) and a light microscope. Fluorochrome-stained smears can be scanned at ×150 to ×200 magnification with a fluorescence microscope, with confirmation of AFB morphology at ×450 or ×1,000 magnification.

Because acid-fast artifacts may be present in a smear, it is necessary to view cell morphology carefully. AFB are approximately 1 to 10 μm long and typically are slender rods that may appear curved or bent. Individual bacilli may display heavily stained areas and areas of alternating stain, producing a beaded appearance. Some NTM species may appear pleomorphic, appearing as long filaments or coccoid forms with uniform staining properties. Rapidly growing mycobacteria may be somewhat less acid fast than other mycobacteria and may not stain with the fluorochrome stain (104). *M. kansasii* can often be recognized in stained sputum smears by its large size and cross-banding appearance. If the presence of a rapid grower is suspected and regular acid-fast stains are negative, it may be worthwhile to stain with a modified acid-fast stain and use a weaker decolorizing process. Organisms that are truly acid fast are difficult to overdecolorize. Nonmycobacterial organisms with various degrees of acid fastness include *Rhodococcus* species, *Nocardia* species, *Legionella micdadei*, and the cysts of *Cryptosporidium* species, *Isospora* species, and other microsporidia.

Each slide made from a clinical specimen should be thoroughly examined for the presence of AFB. When the Ziehl-Neelsen-stained smear is read, a minimum of 300 fields should be examined before the smear is reported as negative. This can be accomplished by performing three parallel passes along the long axis of the slide or nine passes along the short axis (106). Because the fluorochrome stain is read at a lower power than the fuchsin stain, more material can be examined in the time it would take to examine the required number of fields on a fuchsin-stained smear. At the lower magnification, the smear should be read in three parallel sweeps of the long axis, with a minimum of 30 fields of view. This requires as little time as 1.5 min. This ease of detection of AFB with the fluorochrome stain makes it the preferred staining method for clinical specimens. Positive fluorochrome smears can be restained by one of the carbol fuchsin methods to confirm the presence of AFB and to store slides for future reference (106).

Report all smears in which no AFB have been seen as negative. When acid-fast organisms are observed on a smear, report the smear as positive, and indicate the staining method. Provide information as to the quantity of AFB observed on the smear. The recommended interpretations and ways to report smear results are given in Table 1. If only one or two organisms are seen on an entire smear, this fact should be noted but not reported. A confirmation of this finding should be attempted by use of other smears from the same specimen or, if possible, smears prepared from a new specimen. Observations made with fluorochrome smears should be converted to a format that equates these observations with those made with a 100× oil immersion objective.

The direct acid-fast stain should not be used in place of culture but as an adjunct to culture. Definitive diagnosis of mycobacterial disease (except leprosy) requires growth of the microorganism. A minimum of 5×10^3 to 5×10^4 bacilli per ml of sputum is required for detection by smear, whereas culture detects as few as 10 to 100 viable organisms (106).

Although the sensitivity of the direct acid-fast smear examination for the diagnosis of mycobacterial infection is lower than that of culture methods, the acid-fast smear has an important role in early diagnosis of mycobacterial infection because of the relatively long time required for mycobacteria to be detected by culture methods. It is important to detect the presence of mycobacterial disease as rapidly as possible for implementation of appropriate patient care and public health measures. Patients with positive smears are considered more likely to spread tuberculosis. The acid-fast stain also serves to confirm the acid-fast nature of organisms recovered from culture and to monitor the effectiveness of antimycobacterial therapy. The quantitation of a positive smear is also used as an aid in the determination of appropriate dilutions of a specimen for direct susceptibility testing.

The overall sensitivity of direct smear ranges from 22 to 80% (14, 22, 132, 135, 149, 190, 195). Factors influencing sensitivity include types of specimens examined, centrifugation speed, staining technique, culture method used, and patient population being evaluated. Respiratory specimens yield the highest smear positivity rate (132, 149, 160, 175). Furthermore, if more than one respiratory tract specimen is submitted to the laboratory, up to 96% of patients with pulmonary tuberculosis may be detected by the acid-fast smear examination (132). Direct smear examination has less value in detecting the presence of infection caused by mycobacteria other than M. *tuberculosis*. The auramine-rhodamine stain is more sensitive than the carbol fuchsin stain, probably because the fluorochrome-stained smears are easier to read, and has proven to be a very reliable and specific stain (190, 198). Several reports have indicated that smear positivity is correlated with the number of colonies recovered on culture plates (132, 149, 160).

The specificity of direct smear examination is very high. Prolonged or too harsh specimen decontamination and short incubation of cultures may account for smear-positive, culture-negative results. Patients with pulmonary disease may have positive smears with negative cultures (for 2 to 10 weeks) during a course of appropriate treatment (115, 166). Cross-contamination of slides during the staining process and use of water contaminated with saprophytic mycobacteria during staining procedures are potential sources of false-positive results (54, 149). Staining jars or dishes should not be used. AFB may also be transferred in the oil used for microscopy. Specimen containers free of waxes and oils should be used, because these substances may appear on a smear as acid-fast artifacts. It is best to confirm positive smears by having them reviewed by another experienced reader.

Acid-fast stains may be used on any type of clinical materials or organisms recovered from cultures that are suspected to be mycobacteria. In laboratories that routinely culture for mycobacteria, smears from sputum and other respiratory tract secretions are usually prepared after concentration of the specimen. If rapid evaluation of a specimen is needed, a direct smear may be prepared from the purulent portion of the specimen; however, the sensitivity of this smear will be less than that of a concentrated specimen. Information regarding specific staining procedures is given in chapters 4 and 20 of this Manual.

Culture Media

Although most mycobacteria grow in simple synthetic media once they have adapted to in vitro growth, more complex media should be used for primary isolation (106). Many different media are available for the recovery of mycobacteria; they include nonselective and selective media, the latter containing one or more antimicrobial agents to prevent overgrowth by contaminating bacteria or fungi. Solid media may be egg based or agar based. All contain malachite green, a dye that suppresses the growth of contaminating bacteria. Most media used in mycobacteriology are commercially available ready to use or as powdered bases that require minimal preparation and occasional addition of supplements.

Egg-Based Media

Egg-based media contain whole eggs or egg yolks, potato flour, salts, and glycerol and are solidified by inspissation. These media have long shelf lives (several months when refrigerated, which is ideal for laboratories that process only a few specimens), resist drying for prolonged incubation periods, and support good growth of most mycobacteria. Also, materials in the inoculum and medium that are toxic to mycobacteria are neutralized. The disadvantages of these media include the difficulty in distinguishing colonies from debris and the impossibility of achieving accurate and consistent drug concentrations for susceptibility testing. The heat required for solidification of the media and components of the egg inactivate antituberculous drugs. When an egg-based medium becomes contaminated, the entire tube is involved, and the medium may completely liquefy.

Of the egg-based media, LJ medium is most commonly used in clinical laboratories. In general, LJ medium recovers M. *tuberculosis* well but is not as reliable for the recovery of other species. Petragnani medium contains about twice as much malachite green as does LJ medium and is most commonly used for recovery of mycobacteria from heavily contaminated specimens. The increased concentration of malachite green can cause partial inhibition of some strains of mycobacteria. American Trudeau Society medium has the simplest formulation of all the egg media, containing a lower concentration of malachite green than LJ, and is thus more easily overgrown by contaminants. It should be reserved for specimens that are unlikely to be contaminated, such as CSF and pleural fluid. Growth of mycobacteria on American Trudeau Society medium will be less inhibited owing to the low concentration of malachite green, resulting in earlier growth of larger colonies.

Agar-Based Media

Agar-based media are transparent and provide a ready means of detecting early growth of microscopic colonies, which are easily distinguished from inoculum debris. Colonies may be observed in 10 to 12 days, in contrast to 18 to 24 days with opaque egg-based media. Microscopic examination can be performed by simply turning the plate medium over and focusing on the agar surface through the bottom of the plate and the agar at ×14 to ×100 magnification and may provide both earlier detection of growth than unaided visual examination and presumptive identification of the species of mycobacteria present. Use of a thinly poured 7H11 agar plate (10 by 90 mm; Remel, Lenexa, Kans.) facilitates this process (234). Agar-based media can be used for susceptibility testing. Agar-based medium does not readily support growth of contaminants (106); however, the plates are expensive to prepare, and the shelf life is relatively short (1 month in the refrigerator). Care should be exercised in the preparation, incubation, and storage of the media, because excessive exposure to heat or light may result in deterioration of the media and release of formaldehyde, which is toxic to mycobacteria (141).

Middlebrook medium contains 2% glycerol, which enhances the growth of MAC. Additional supplements may be helpful for recovery of other mycobacteria and in certain situations. For example, the addition of 0.2% pyruvic acid is recommended if M. *bovis* is suspected (53), and 0.25% L-asparginine or 0.1% potassium aspartate must be added to 7H10 agar for maximal production of niacin (113). The Middlebrook 7H11 formulation is preferred over 7H10 for initial isolation from a specimen, because the addition of 0.1% enzymatic hydrolysate of casein (the only difference from 7H10) improves the recovery of isoniazid (INH)-resistant strains of M. *tuberculosis*.

Selective Media

The addition of antimicrobial agents to LJ, 7H10, or 7H11 medium may be helpful for eliminating the growth of contaminating organisms (138, 158). If selective medium is used for a particular specimen or specimen type, it should not be used alone but in conjunction with a nonselective agar- or egg-based medium. Egg-based selective media include LJ-Gruft with penicillin and nalidixic acid and LJ-Mycobactosel with cycloheximide, lincomycin, and nalidixic acid. Mitchison selective 7H11 (7H11S) medium and its modifications contain carbenicillin (especially useful for inhibiting *Pseudomonas* spp.), polymyxin B, trimethoprim lactate, and amphotericin B.

Liquid Media

Middlebrook 7H9 and Dubos Tween albumin broths are commonly used for subculturing stock strains of mycobacteria and preparing inocula for drug susceptibility tests and other in vitro tests. 7H9 broth is used as the basal medium for several identification tests. Broth media also can be used for recovering organisms from specimens with low bacterial counts, such as CSF. Tween 80 in liquid medium acts as a surfactant and allows for dispersal of clumps of mycobacterial growth, resulting in more homogeneous growth.

BACTEC AFB System

A radiometric detection system for mycobacterial growth developed by Middlebrook et al. (140) in 1977 has evolved into the BACTEC AFB system (Becton Dickinson Diagnostic Instruments, Sparks, Md.). Two types of BACTEC mycobacterial culture media are currently available. BACTEC 12B medium is a broth medium consisting of 7H9 broth base, BSA, casein hydrolysate, catalase and ^{14}C-labeled substrate (palmitic acid). Four milliliters of broth is contained in a 20-ml vial sealed with a rubber septum. For this system, specimens are processed as they would be for conventional cultures. The vial is inoculated with 0.5 ml of processed specimen. As the mycobacteria

metabolize the substrate, $^{14}CO_2$ is liberated and is detected quantitatively in the headspace gas by the BACTEC 460 instrument. The amount of $^{14}CO_2$ is translated into a growth index (GI). The rate and amount of $^{14}CO_2$ produced are directly proportional to the rate and amount of growth occurring in the medium. Every time a vial is tested, the instrument introduces fresh 5 to 10% CO_2 into the vial headspace. The radiometric technique can detect very small quantities of $^{14}CO_2$, which allows detection of mycobacterial growth at a very early stage. In many instances, detection time of positive cultures is decreased to <7 days for NTM and averages 9 to 14 days for M. *tuberculosis*, although earlier times to detection for M. *tuberculosis* can occur. A GI of >10 is considered significant; however, other bacteria may grow within the bottle. Studies have shown that the use of the BACTEC method has significantly improved the rates and times of recovery of mycobacteria from respiratory secretions and other specimens (116, 144, 168). An antimicrobial-mixture–growth-promoting supplement containing polymyxin B, amphotericin B, nalidixic acid, trimethoprim, azlocillin, and polyoxylene stearate is added to the 12B medium for suppression of contaminants and enhancement of mycobacterial growth. The semiautomated BACTEC 460 instrument can test 60 vials at a rate of one vial per 82 ± 2 s. The instrument is equipped with a hood that provides HEPA-filtered exhaust air and negative pressure in the test area. The hood also has a UV light source as a safety feature. Quality control and maintenance procedures are described in the instrument operating manual. Specimens processed by the Zephirantrisodium phosphate, benzalkonium chloride, lauryl sulfate, or CPC method cannot be used with the BACTEC system, as residual quantities of these substances in the inoculum inhibit mycobacterial growth. Highly cellular clinical specimens may give false-positive results soon after inoculation owing to metabolic activity of the cells.

BACTEC 12B medium has been reported to yield more positive cultures from clinical specimens than other media (144, 168). This is particularly significant with smear-negative specimens and specimens from treated patients with chronic disease. Improved detection of MAC in 12B medium has also been reported (87). The time to detection for positive growth is significantly shorter than that with other media. Limitations of the BACTEC system include the inability to observe colony morphology, difficulty in recognizing mixed cultures, overgrowth by contaminants, cost, radioisotope disposal, and extensive use of needles. The BACTEC TB system may be used for antimicrobial susceptibility testing (168). Generally, the initial positive vial can be used directly for identification and drug susceptibility testing. It is good practice to subculture positive 12B vials onto a fresh solid medium to obtain isolated colonies and abundant growth. Cross-contamination between vials with the BACTEC system and subsequent pseudomycobacterial infections have been reported (40, 148, 212), but careful quality control practices can minimize this problem.

BACTEC 13A medium is used for blood and bone marrow aspirate specimens. Thirty milliliters of the medium (Middlebrook 7H13 broth with the anticoagulant SPS) is contained in a 50-ml vial. Up to 5 ml of specimen can be directly inoculated into the hypotonic medium without prior concentration or lysis. The principle for the detection of growth in 13A is the same as that for detection of growth in 12B. An enrichment fluid is added to a 13A vial before or after specimen inoculation. BACTEC 13A is a sensitive medium that can detect very low concentrations of viable

AFB in blood specimens (236). The sensitivity is similar to that obtained by processing blood with the Isolator system (1). On average, positive cultures are detected in 12 to 16 days (1, 100, 109). Limitations of the system include lack of a way to quantitate the number of mycobacteria originally present in the specimen.

Biphasic Media

The Septi-Chek system (Becton Dickinson) is a mycobacterial culture system consisting of a capped bottle containing 20 ml of modified 7H9 broth under an enhanced (20%) CO_2 atmosphere and a paddle containing three types of solid media (modified LJ, Middlebrook 7H11, and chocolate agars) encased in a plastic tube. Bacterial contamination is detected on the chocolate agar. The bottles and the paddles are supplied separately. For this system, specimens are processed as they would be for conventional cultures. Cultures are inoculated by removing the bottle cap, adding the processed specimen, and then attaching the paddle to the bottle. Solid media are inoculated after 24 h of incubation in an upright position by inverting the bottles. A supplement containing glucose, glycerin, oleic acid, pyridoxal HCl, catalase, albumin, azlocillin, nalidixic acid, trimethoprim, polymyxin B, and amphotericin B is added to the culture bottle prior to inoculation. During the incubation period, the bottles are periodically tipped to reinoculate the solid media as the cultures are being read. The sensitivity of this system is similar to that of the BACTEC system according to some authors (46, 98, 182). Although the average time to detection of growth is longer than that with the BACTEC system, it is generally shorter than that with conventional media. Septi-Chek may be a viable alternative to the BACTEC system for low-volume laboratories or laboratories that wish to avoid the use of radioisotopes. The system has not been evaluated for use with blood cultures.

Heme-Containing Medium for *M. haemophilum*

M. haemophilum will grow on egg- or agar-based medium only if the medium is supplemented with hemin, hemoglobin, or ferric ammonium citrate (50). Thus, specimens from skin lesions, joints, or bone should also be inoculated onto chocolate agar; Middlebrook 7H10 agar with hemolyzed sheep erythrocytes, hemin, or an X-factor disk; or LJ medium containing 1% ferric ammonium citrate to enhance recovery of this organism (196). BACTEC media should be similarly supplemented.

Medium Selection

Medium selection for the isolation of mycobacteria and the culture-reading schedule are usually based on personal preference and/or laboratory tradition. Both should be optimized for the most rapid detection of positive cultures and identification of mycobacterial isolates. The variety of media and methods available today should be sufficient to permit laboratories to develop a system optimal for their patient population and administrative needs. Traditionally, one egg-based medium and one agar medium are used for the primary isolation of mycobacteria. Using media of two different basal compositions that complement each other makes possible the isolation of most strains of mycobacteria that cause human disease. Although clear agar medium permits earlier detection of mycobacteria, LJ medium often yields a greater number of positive cultures on prolonged incubation (119).

Currently, it is recommended that a broth-based system,

BACTEC, or Septi-Chek be used for primary mycobacterial isolation in order to favor rapid detection of positive cultures. Some laboratories, however, may face difficulty in conforming to local requirements for monitoring and disposing of radioisotopes. BACTEC 12B medium may be used as a stand-alone medium for recovery of mycobacteria from clinical specimens, but for maximum recovery and for aid in the detection of mixed mycobacterial infections, one piece of conventional solid medium should be used in addition to the 12B vial (192). The decision to use 12B medium alone or with additional conventional medium should be made after an institution has reviewed its own experience and requirements. All positive cultures, even if identified directly from the broth, need to be subcultured to solid media to detect mixed cultures and to correlate direct identification results with colony morphology. The Septi-Chek system can be used as a stand-alone system.

Detection of colonies offers several advantages over detection of growth in broth, because colonial morphology can provide clues to identification and facilitate the selection of confirmatory tests, including DNA probe tests. However, BACTEC vials can provide important microscopic clues, such as cord formation, and it is possible to use sediment from BACTEC vials for probe culture confirmation tests prior to detection of growth on solid media (see below). The reliability of the criterion of cord formation for presumptive identification of *M. tuberculosis* should be determined within each laboratory (146, 241).

Incubation

Temperature

The optimal incubation temperature for most specimens is 35 to 37°C. Exceptions to this include specimens obtained from skin and soft tissue suspected to contain M. *marinum*, M. *ulcerans*, M. *chelonae* (formerly M. *chelonae* subsp. *chelonae*), or M. *haemophilum*. In this case, since these organisms have lower optimal growth temperatures, inoculate a second set of media, and incubate them at 25 to 33°C. Temperature is critical for the BACTEC system and is different from that recommended for other systems (36 to 38°C); the system is based on metabolism of mycobacteria, which is optimal at 37 to 37.5°C for most species, and lower temperatures slow down detection time. Although it is not necessary to incubate BACTEC vials in the dark, excessive light may degrade the medium or inhibit growth. BACTEC 12B culture vials should not be shaken during incubation; however, agitation of BACTEC 13A vials may enhance the growth of MAC and lead to shorter detection times (100).

Atmosphere

An atmosphere of 5 to 10% CO_2 in air stimulates the growth of all mycobacteria grown in tubes or plates (10). Middlebrook agars require a CO_2 atmosphere to ensure growth, while it is necessary to incubate egg media under CO_2 only for the first 7 to 10 days after inoculation (the log phase of growth); subsequently, LJ cultures can be removed to ambient air incubators if space is limited. In the absence of CO_2 incubators, plates may be incubated in commercially available bags with CO_2-generating tablets. Candle extinction jars are unacceptable for use in the mycobacteriology laboratory, because the oxygen tension is reduced below that required for growth of mycobacteria. BACTEC vials and Septi-Chek bottles do not require incubation in increased CO_2. As supplied, Septi-Chek bottles contain enough CO_2 that the final concentration is 5 to 8% after inoculation. Before inoculation, all BACTEC 12B vials should be tested in the BACTEC 460 instrument to eliminate bottles with high background counts and also to establish a CO_2-enriched atmosphere within the vial. The BACTEC 13A blood culture vials should not be tested in the BACTEC 460 instrument before inoculation, as they are partially evacuated to accommodate 5 to 6 ml of blood; testing of vials immediately after inoculation is also not required, since the vials already have a CO_2 atmosphere inside.

Time

Mycobacterial cultures are generally held for 6 to 8 weeks before being discarded as negative. Specimens with positive smears that are culture negative should be held for an additional 4 weeks. Plates should be incubated medium side down until all the inoculum has been absorbed. Once the inoculum has been absorbed, plated media should be incubated inverted in CO_2-permeable polyethylene bags or sealed with CO_2-permeable shrink-seal or cellulose bands to prevent media from drying out during the long incubation period. Tubed media should be incubated in a slanted position with screw caps loose for at least a week until the sediment has been absorbed; they can then be incubated upright if space is needed. The caps on the tubes should be tightened at 2 to 3 weeks to prevent desiccation of the media. Specimens from skin lesions should be incubated for 8 to 12 weeks if *M. ulcerans* is suspected.

Reading Schedule

Mycobacteria are relatively slowly growing organisms, and cultures can thus be examined less frequently than routine bacteriologic cultures. All solid media should be examined within 3 to 5 days after inoculation to permit early detection of rapidly growing mycobacteria and to make possible prompt removal of contaminated cultures. Young cultures up to 4 weeks of age should be examined two times per week, and older cultures should be examined at weekly intervals. Use of a hand lens for opaque media and a microscope for agar media facilitates early detection of microcolonies. Conventional liquid media (7H9 or Dubos) used for primary isolation should be carefully examined weekly for particulate matter and then stained and/or subcultured prior to being discarded as negative.

The reading schedule for BACTEC vials varies according to the laboratory work load. Low-volume laboratories may read cultures three times a week for the first 2 or 3 weeks and weekly thereafter for a total of 6 weeks, while high-volume laboratories may read cultures two times a week for the first 2 weeks and weekly thereafter. Some laboratories separate smear-positive specimens from smear-negative specimens and test the former more frequently, thus diminishing the work load. In addition, separation of "probable positive" cultures from negative cultures decreases the possibility of vial cross-contamination by the instrument. With more frequent testing of all specimens, however, earlier detection of positive cultures is expected. Readings of negative cultures in 12B medium usually remain below a GI of 10; a GI of ≥10 is considered presumptively positive. At this point, the vials should be separated and tested daily. An acid-fast stain is performed when the GI is ≥50 to determine whether the culture contains mycobacteria. The morphology of mycobacteria seen in smears from 12B medium may be used to presumptively identify M. *tuberculosis* complex (241) or to decide how to progress with identification methods (146). In addition, a smear of the

broth from the vial may be Gram stained, and/or the broth may be subcultured onto a sheep blood agar plate to determine whether contamination is present. When the GI is ≥500, BACTEC antimicrobial susceptibility testing can be performed (see below). Negative BACTEC 13A vials usually read 10 to 15. When the GI reaches ≥20, the culture is considered presumptively positive. At this point, a smear is prepared and stained for AFB. If the smear is positive, the vial is incubated further or subcultured for subsequent testing. If the smear is negative, continue to incubate and test.

The Septi-Chek system may be inspected for growth four times per week or daily for the first 1 to 2 weeks, with inversion of the bottles for reinoculation of the agar medium if growth is not observed. Thereafter, both bottles and paddles are inspected weekly for growth without inversion.

IDENTIFICATION OF MYCOBACTERIA

Mycobacteria should always be identified to the species level if possible. According to traditional methods, mycobacteria are usually identified by traits such as rate of growth, colonial morphology, pigmentation, and, for differential purposes, biochemical profiles. The identification should be based on as many observations as possible, but it is prudent to select only the key biochemical tests that appear to be useful for the species suspected. Table 2 presents the characteristics of common species of mycobacteria recovered from clinical specimens. Many strains will have biochemical profiles and morphologic features compatible with the data in this table; however, variation occurs among strains, and properties may deviate from those presented here. In many instances, it is necessary to identify the organisms on the basis of a best-fit analysis. The traditional methods are well established, standardized, and relatively inexpensive but slow in providing clinically relevant information. Newer laboratory methods for mycobacterial identification, such as chromatographic analysis and nucleic acid probes, have decreased the mycobacteriologist's reliance on biochemical profiles.

Acid-Fast Staining

Microscopy is the first step in identification. An acid-fast stain is used to confirm that the isolate is a *Mycobacterium* species and to determine whether the culture is contaminated with other bacteria. The staining reactions of the mycobacteria were described in the previous section. Although certain morphologic features of mycobacteria have been associated with certain species, microscopic morphologic features should not be the basis for the final species identification.

Growth Characteristics

Growth Rate and Temperature

The growth rate is the amount of time required for mature colonies visible without magnification to form on solid media. Mycobacteria forming such colonies within 7 days are termed rapid growers, while those requiring longer periods are termed slow growers. Some rapid growers may take longer than 7 days to grow on primary isolation media, and conversely, excess inoculum may cause a slowly growing mycobacterium to appear to be a rapid grower. Thus, the growth rate should be confirmed by subculture as follows.

TABLE 2 Distinctive properties of cultivable mycobacteria encountered in clinical specimens[a]

Descriptive term	Species	Growth rate[b]	Optimal temp (°C)	Usual colony morphology[c]	Pigmentation[d]	Niacin	Growth on T2H (10 μg/ml)	Nitrate reduction
TB complex	M. tuberculosis	S	37	R	N (100)	+ (95)	+	+ (97)
	M. africanum	S	37	R	N	−	V	−
	M. bovis	S	37	Rt	N (100)	− (4)	−	− (9)
Nonchromogens	MAC	S	37	St/R	N (87)	− (0)	+	− (4)
	M. xenopi	S	42	S	S (21)	− (0)	+	− (7)
	M. haemophilum	S[f]	30	R	N	−	+	−
	M. malmoense	S	37	S	N (88)	− (0)	+	− (1)
	M. shimoidei	S	37	R	N	−	+	−
	M. genavense	S	37	St	N	−	+	−
	M. celatum	S	37	S/St	N (100)	−	+	− (0)
	M. ulcerans	S	30	R	N	−	+	−
	M. terrae complex	S	37	SR	N (93)	− (1)	+	± (67)
	M. triviale	M	37	R	N (100)	− (0)	+	+ (89)
	M. gastri	S	37	S/SR/R	N (100)	− (0)	+	− (0)
	M. nonchromogenicum	S	37	SR	N		+	
Photochromogens	M. kansasii	S	37	SR/S	P (96)	− (4)	+	+ (99)
	M. marinum	M	30	S/SR	P (100)	−/+ (21)	+	− (0)
	M. simiae	S	37	S	P (90)	± (63)	+	− (28)
	M. asiaticum	S	37	S	P (86)	− (0)	+	− (5)
Scotochromogens	M. gordonae	S	37	S	S (99)	− (0)	+	− (1)
	M. scrofulaceum	S	37	S	S (97)	− (0)	+	− (5)
	M. szulgai	S	37	S or R	S/P (93)	− (0)	+	+ (100)
	M. flavescens	M	37	S	S (100)[h]	− (0)	+	+ (92)
Rapid growers	M. fortuitum group	R	28	Sf/Rf	N (100)	−/+	+	+ (100)
	M. chelonae group	R	28	S/R	N (100)	−/+	+	− (1)
	M. smegmatis	R	28	R/S	N		+	+
	M. phlei	R	28	R	S		+	+
	M. vaccae	R	28	S	S		+	+

[a]Modified from reference 169. Plus and minus signs indicate the presence and absence, respectively, of the feature; blank spaces indicated either that the information is not currently available or that the property is unimportant. V, Variable; ±, usually present; −/+, usually absent. Percentage of CDC-tested strains positive in each test is given in parentheses, and test result is based on these percentages.

[b]S, slow; M, moderate; R, rapid.

[c]R, rough; S, smooth; SR, intermediate in roughness; t, thin or transparent; f, filamentous extensions.

[d]P, photochromogenic; S, scotochromogenic; N, nonphotochromogenic. M. szulgai is scotochromogenic at 37°C and photochromogenic at 24°C.

[e]Urease test was performed by the method of Steadham (194).

[f]Requires hemin as growth factor.

[g]Arylsulfatase reaction at 14 days is positive.

[h]Young cultures may be nonchromogenic or possess only pale pigment that may intensify with age.

[i]M. chelonae is negative; M. abscessus is positive.

1. Dilute a 7-day broth culture or a saline suspension of the organism from a freshly grown slant or plate sufficiently to obtain isolated colonies.

2. Inoculate 0.1 ml of several 10-fold dilutions of the test organism onto either egg-based or agar-based medium, and incubate at 35 to 37°C. Isolates suspected of being M. ulcerans, M. marinum, or M. haemophilum require incubation at 30 to 32°C.

3. Observe cultures at 5 to 7 days and weekly thereafter for visible colonies.

Growth in relation to temperature can usually be adequately determined by observing primary cultures or subcultures at 37 and 30°C. When more definitive identification is needed, isolates should be incubated at 24, 30, 35 to 37, and 42°C.

With the BACTEC TB system and other broth-based systems, the growth rate of mycobacteria present in a clinical specimen cannot be determined. A broth medium is used to decrease detection time and is often the first medium to demonstrate growth. The rate of growth in broth is based somewhat on the number of organisms present as well as on the growth rate of the individual organisms. The elevated GI of a positive culture in the BACTEC system gives no suggestion to the identity of the organism present in the BACTEC vial. For example, a large number of cells of M. tuberculosis may produce an elevated GI within 2 to 3 days, but the organism is still considered a slow grower and may take several weeks to grow on solid media.

Colony Morphology

Species of mycobacteria can be tentatively identified by using transmitted light to microscopically observe young

TABLE 2 *(Continued)*

Semiquantitative catalase (mm of bubbles)	68°C catalase	Tween hydrolysis	Tellurite reduction	Tolerance to 5% NaCl	Iron uptake	Arysulfatase 3 days	MacConkey agar	Urease[e]	Pyrazin-amidase, 4 days	Nucleic acid probes available
<45 (89)	− (1)	± (68)	−/+ (36)	− (0)	−	− (0)	−	± (64)	+	+
<45	−	−	−	−	−	−	−	+	−	−
<45 (69)	− (2)	− (21)		− (0)	−	− (0)	−	± (50)	−	+
<45 (98)	± (60)	− (2)	+ (81)	− (0)	−	− (1)	−/+	− (2)	+	+
>45 (85)	± (31)	− (12)	± (65)	− (0)	−	± (36)	−	− (0)	V	−
<45	−	−	−	−	−	−	−	−	+	−
<45 (99)	± (66)	+ (99)	+ (74)	− (0)	−	− (0)		− (9)	+	−
<45	−	+		−	−	−	−	−	+	−
>45	+	+			−	−		+	+	−
<45 (100)	+ (100)	− (0)	+ (100)	− (0)	− (0)	+ (100)	− (0)	− (0)	+ (100)	−
<45	+	−		−		−		V	−	−
>45 (93)	+ (92)	+ (99)	−/+ (46)	− (2)	−	− (2)	V	− (13)	V	−
>45 (100)	+ (100)	+ (100)	− (25)	+ (100)	−	± (56)	−	−/+ (33)	V	−
<45 (100)	− (11)	+ (100)	± (50)	− (0)	−	− (0)	−	−/+ (44)	−	−
>45	+	+	−	−	−	−	V	−	V	−
>45 (93)	+ (91)	+ (99)	−/+ (31)	− (0)	−	− (0)	−	−/+ (49)	−	+
<45 (98)	− (30)	+ (97)	−/+ (39)	− (0)	−	−/+ (41)[g]	−	+ (83)	+	−
>45 (93)	+ (95)	− (9)	+ (82)	− (0)	−	− (0)	−	± (69)	+	−
>45 (95)	+ (95)	+ (95)	− (20)	− (0)	−	− (0)	−	− (10)		−
>45 (90)	+ (96)	+ (100)	− (29)	− (0)	−	V (0)	−	V (31)	−/+	+
>45 (84)	+ (94)	− (2)	± (64)	− (0)	−	V (0)	−	V (31)	±	−
>45 (98)	+ (93)	−/+ (49)	± (53)	− (0)	−	V (0)	−	+ (72)	+	−
>45 (94)	+ (100)	+ (100)	−/+ (44)	± (62)	−	− (0)	−	+ (72)	+	−
>45 (93)	+ (90)	−/+ (43)	+ (92)	+ (85)	+	+ (97)	+	+ (70)	+	−
>45 (92)	± (53)	−/+ (39)	+ (89)	V[i]	−	+ (95)	+	+ (89)	+	−
>45	+	+	+	+	+	−	−	−		−
>45	+	+	+	+	+	−	−	−		−
>45	+	+	+	V	+	−	−	−		−

(5- to 14-day-old) colonies on plates inverted under the 10× objective of a stereomicroscope, according to the scheme developed by Runyon (173). This is best accomplished with a clear agar medium like Middlebrook 7H10 or 7H11 agar, so that colonies are easily visualized. Although grossly visible colonies may not yet be evident on the agar surface, they are commonly found by microscopic observation. Specific colonial morphologic features are described in Table 3. Because colonial features may provide a presumptive identification of the organisms, they can be used to suggest which biochemical tests or nucleic acid probes to use for the definitive identification. In addition, microscopic examination allows more timely detection of mixed cultures.

Pigmentation and Photoreactivity

Some mycobacteria produce carotenoid pigments without light. Others require photoactivation for pigment production. Mycobacteria are classified into three groups based on the production of pigments. Photochromogens produce nonpigmented colonies when grown in the dark and pigmented colonies only after exposure to light and reincubation. Scotochromogens produce deep yellow to orange pigmented colonies when grown in either the light or the dark, and some strains may have an increased pigment production upon continuous exposure to light. Nonphotochromogens are nonpigmented in the light and dark or have only a pale yellow, buff, or tan pigment that does not intensify after light exposure.

Pigment production should be studied in slowly growing isolates that are nonpigmented when grown in the dark and in pigmented isolates that have not been grown continuously in the dark. Pigment production should be tested for with young cultures and is usually more easily observed in cultures with isolated colonies (106). Three tubes of media are inoculated with a broth culture of the test organism diluted sufficiently to yield isolated colonies. Two tubes should be shielded from light by wrapping them with aluminum foil, and the third should be left uncovered so it will be exposed to the ambient light in the incubator. When growth is first detected in the unshielded tube, examine the foil-wrapped tubes, and if growth is detected, expose one tube with its cap loosened to light (100-W tungsten bulb placed 20 to 25 cm from the culture or fluorescent equivalent) for 3 to 5 h. The cap must be loose to allow induction of the soluble pigment, which is controlled by an oxygen-dependent, photoinducible enzyme (9). Reshield the tube, reincubate, and inspect both shielded tubes after 24, 48, and 72 h for the development of pigment. Compare colonies in the light-exposed tube with those in the shielded tube. Pigmentation patterns need to be carefully evaluated, since variations within species occur. Most important, some MAC isolates can be pigmented. All scotochromogens should be tested for photochromogenicity at 25°C to differentiate M. szulgai, which is a scotochromogen at 37°C but a photochromogen at 25°C.

Conventional Biochemical Tests

After a mycobacterial isolate has been assigned to a preliminary subgroup on the basis of growth characteristics (rate of growth, pigmentation, photoreactivity), it has traditionally been definitively identified to species or complex level by using conventional biochemical tests. Table 2 gives complete biochemical test profiles for the most commonly encountered species. The principles for these tests are described below. Detailed descriptions of methods, procedures, and controls can be found in the *Clinical Microbiology Procedures Handbook* (97). Many of the reagents and media are commercially available. Key tests performed on the isolate, i.e., those useful for identifying the suspected species, are selected on the basis of the preliminary grouping. Preference is usually given to tests that will rule in an identification by a positive result rather than rule out an identification by a negative result. Suggested groupings of key biochemical tests for identification of mycobacteria are given in Table 4. Appropriate quality control strains need to be run with each biochemical test. Alternatives to biochemical tests are also discussed below.

Arylsulfatase

Arylsulfatase hydrolyzes the bond between the sulfate group and the aromatic ring of tripotassium phenolphthalein disulfate incorporated into the growth medium to form free phenolphthalein. The presence of phenolphthalein is indicated by a red color that develops in the test medium when alkali is added. Most mycobacteria possess arylsulfatase activity, so the test conditions are varied to help identify various species. The 3-day test (229) is used mainly to identify potentially clinically significant rapid growers. With few exceptions, only M. fortuitum and M. chelonae split phenolphthalein from tripotassium phenolphthalein sulfate within 3 days. The 14-day test may be useful in the identification of slowly growing mycobacteria such as M. marinum, M. szulgai, M. xenopi, M. triviale, M. smegmatis, and M. flavescens. Two procedures for the determination of arylsulfatase production are available: the Wayne method, which utilizes an enriched agar-base medium for a 3-day test (229), and the 3- and 14-day CDC broth tests (124). The Wayne method is probably sufficient for most laboratories.

Catalase

Catalase is an intracellular soluble enzyme capable of degrading hydrogen peroxide to water and oxygen. The presence of the enzyme is detected by adding H_2O_2 to a culture of the test organisms and observing for the formation of oxygen bubbles in the reaction mixture. Virtually all mycobacteria, except M. gastri, certain INH-resistant mutants of M. tuberculosis and M. bovis, and some nonpathogenic INH-resistant strains of M. kansasii, possess catalase enzymes. Species of catalase-producing mycobacteria can be distinguished by quantitative differences in catalase activity demonstrated by intact cells in the semiquantitative catalase test and by the differences in heat stability detected by the 68°C catalase test.

Semiquantitative Catalase Test

The semiquantitative catalase test divides the mycobacteria into two groups: those producing <45 mm of bubbles (low catalase) and those producing >45 mm of bubbles (high catalase) (122). Among those that produce <45 mm of bubbles are M. tuberculosis, M. bovis, M. marinum, MAC, M. xenopi, M. malmoense, M. haemophilum, and M. gastri. M. kansasii has two recognized subgroups: one produces <45 mm of bubbles, whereas strains more commonly associated with disease produce >45 mm of bubbles. Other species of mycobacteria usually produce a column of bubbles >45 mm high.

Heat-Stable (68°C) Catalase Test

Some mycobacteria lose catalase activity when suspended in pH 7 buffer and heated to 68°C for 20 min (123). Always included in this group are M. tuberculosis, M. bovis, M. gastri, and M. haemophilum, although strains of other species may also be negative in this test. The test is valuable in conjunction with the niacin test for the identification of M. tuberculosis.

Catalase Drop Method

The catalase drop test can be used for the quick determination of significant INH resistance of M. tuberculosis (169). If a colony suspected or known to be M. tuberculosis fails to form bubbles after a drop of reagent is added, the isolate is presumptively INH resistant. However, INH resistance cannot be ruled out in catalase-producing colonies.

Growth on MacConkey Agar without Crystal Violet

Growth on MacConkey agar without crystal violet is used to distinguish rapidly growing mycobacteria (103). The potentially pathogenic members of the M. fortuitum group and the M. chelonae group usually grow on MacConkey agar without crystal violet, whereas the commonly saprophytic species are inhibited by this medium. Some strain of M. smegmatis grow on this medium, but the 3-day arylsulfatase test distinguishes these species from organisms of the M. chelonae group and the M. fortuitum group.

Iron Uptake

The iron uptake test (231) is used to detect mycobacteria capable of converting ferric ammonium citrate to an iron oxide. The iron oxide is visible as a reddish brown rust color in the colonies. The medium shows a tan discoloration. Iron uptake is useful in distinguishing M. chelonae, commonly negative, from M. fortuitum and most other rapid growers, which are positive. Slow growers and most M. flavescens are not capable of accumulating iron oxides.

Niacin Accumulation Test

Niacin (nicotinic acid) functions as a precursor in the biosynthesis of coenzymes NAD and NADP. Although all mycobacteria produce nicotinic acid, comparative studies have shown that because of a blocked metabolic pathway for conversion of free niacin to nicotinic acid mononucleotide, M. tuberculosis accumulates niacin and excretes it into the culture medium (174). Niacin-negative M. tuberculosis isolates are extremely rare. The niacin test should not be used alone to identify M. tuberculosis, however, because some strains of M. simiae, BCG, and other mycobacteria, although infrequently encountered, also accumulate niacin. Performance of the supportive nitrate reduction and 68°C catalase tests is necessary for confirming the identification of M. tuberculosis.

TABLE 3 Growth characteristics of commonly isolated mycobacteria[a]

Organism	Growth rate (days)	Pigment production		Colonial morphology on Middlebrook 7H10 agar	Features of suspension in Middlebrook 7H9 broth
		Light	Dark		
MAC	10–21	Buff to yellow	Buff to yellow	Thin, transparent, glistening or matte, smooth, entire, rounded; some colonies rough and wrinkled	Uniformly homogeneous suspension
M. bovis	25–90	Colorless to buff	Colorless to buff	Small, thin, often nonpigmented, raised, rough, later wrinkled and dry; some colonies inhibited on this medium	Heterogeneous; finely granular suspension
M. chelonae	3–7	Buff	Buff	Rounded, smooth, matte, periphery entire or scalloped, no branching filaments; some colonies rough and wrinkled	Heterogeneous; coarsely granular suspension
M. fortuitum	3–7	Buff	Buff	Circular, convex, wrinkled or matte; obvious branching filaments on periphery	Heterogeneous; coarsely granular suspension
M. gastri	10–21	Colorless to buff	Colorless to buff	Round, smooth, convex, glistening; often resembles M. avium-M. intracellulare complex	Uniformly homogeneous suspension
M. genavense[b]	5–67	Colorless	Colorless	Tiny, transparent (dysgonic) or dense and creamy or flat and dry (eugonic); resembles MAC on prolonged incubation; requires mycobactin J for growth	Heterogeneous finely granular suspension
M. gordonae	10–25	Yellow to orange	Yellow to orange	Round, smooth, convex, yellow to orange, glistening	Uniformly homogeneous suspension
M. haemophilum	14–28	Buff to gray	Buff to gray	Grayish white, smooth to rough	Usually homogeneous suspension
M. kansasii	10–21	Yellow	Buff	Raised, smooth; some rough and wrinkled; carotene crystals numerous after exposure to light	Usually heterogeneous finely granular suspension; some isolates give uniformly homogeneous suspension
M. marinum	5–14	Yellow	Buff	Round, smooth; some may be wrinkled	Uniformly homogeneous suspension
M. scrofulaceum	10–14	Yellow	Yellow	Smooth, moist, yellow, round	Uniformly yellow and homogeneous
M. simiae	7–14	Yellow	Buff	Smooth, domed, slightly pigmented	Heterogeneous coarsely granular suspension
M. smegmatis	3–7	Buff to yellow	Buff to yellow	Raised, rough to wrinkled, scalloped edges	Heterogeneous finely granular suspension
M. szulgai	14–28	Yellow to orange	Buff at 25°C; yellow at 37°C	Smooth to rough; periphery somewhat irregular	Heterogeneous finely granular suspension
M. terrae	10–21	Buff	Buff	Round, smooth, glistening, sometimes colorless	Uniformly homogeneous suspension
M. tuberculosis	12–28	Buff	Buff	Flat, rough, spreading to irregular periphery	Uniformly heterogeneous coarsely granular suspension
M. xenopi	28–42	Yellow	Yellow	Small, domed, yellow, smooth or rough; at 45°C, resembles miniature bird's nest	Uniformly homogeneous suspension

[a]Modified from reference 167.
[b]From reference 44.

TABLE 4 Suggested groupings of key biochemical tests for identification of mycobacteria

Species group	Key biochemical tests
M. tuberculosis complex	Niacin, nitrate reduction, 68°C catalase
Nonphotochromogens	Nitrate and tellurite reduction, semiquantitative and 68°C catalase, Tween 80 hydrolysis
Scotochromogens	Nitrate reduction, Tween 80 hydrolysis, urease, 5% NaCl tolerance
Photochromogens	Nitrate reduction, Tween 80 hydrolysis, semiquantitative catalase, urease
Rapid growers	Arylsulfatase, nitrate reduction, iron uptake, growth on MacConkey agar

Before the niacin test is performed, cultures should be checked for purity; be 3 to 4 weeks old, preferably on egg medium or on Middlebrook 7H10 or 7H11 medium enriched with 0.25% L-asparagine or 0.1% potassium aspartate (113); and have sufficient growth of 50 or more colonies. Do not perform this test on cultures that are mixed with other mycobacteria or bacteria, as the contaminant could metabolize the niacin and lead to a false-negative test result. The niacin test does not require the use of actively growing cultures and may be done either with chemical reagents or commercially available paper strips (245). The niacin is extracted from the medium and then detected by its reaction with a cyanogen halide in the presence of a primary amine.

Reagent-impregnated paper test strips are commercially available. The test is based on the formation of cyanogen chloride by the reaction of chloramine T and potassium thiocyanate in the presence of citric acid. Niacin then reacts with cyanogen chloride and couples with a primary aromatic amine to produce a yellow color. The paper strip method obviates the need to prepare and store the unstable chemicals used to demonstrate the presence of niacin. Directions supplied with the strips should be followed.

Nitrate Reduction

Mycobacteria differ quantitatively in their abilities to reduce nitrate. The nitrate reduction test (216) is valuable for the identification of some mycobacteria that possess similar characteristics of colony morphology, growth rate, and pigmentation. M. tuberculosis, M. kansasii, M. szulgai, some non-disease-associated strains of nonphotochromogens, and M. fortuitum are nitrate reductase positive. The test can be performed by using chemical reagents or reagent-impregnated chemical strips or by combining the nitrate reduction test with the niacin test. Reagents can be purchased from commercial sources or prepared in-house. The paper strip method yields the most consistent results with strong nitrate reducers such as M. tuberculosis (162). The combined niacin-nitrate test provides intensely positive reactions in both the niacin and the nitrate tests.

Pyrazinamidase

The enzyme pyrazinamidase hydrolyzes pyrazinamide to pyrazinoic acid. This acid is detected by the addition of ferrous ammonium sulfate to the culture medium (230). The formation of a pink ferrous-pyrazinoic acid complex indicates a positive test. This test is most useful in separating M. marinum from M. kansasii and M. bovis from M. tuberculosis. M. bovis is negative even at 7 days, whereas M. tuberculosis is positive within 4 days.

Sodium Chloride Tolerance

Few mycobacteria are able to tolerate or grow in the presence of 5% sodium chloride. M. triviale is the only slowly growing mycobacteria to do so. Pathogenic rapidly growing species, except M. mucogenicum and most isolates of M. chelonae, also grow in 5% NaCl (107, 184).

Inhibition by T2H

The thiophene-2-carboxylic acid hydrazide (T2H) test is used to distinguish niacin-positive M. bovis from M. tuberculosis and other nonchromogenic slowly growing mycobacteria. M. bovis is susceptible to low concentrations (1 to 5 μg/ml) of T2H (214), whereas M. tuberculosis is resistant.

Tellurite Reduction

Tellurite reductase reduces colorless potassium tellurite to a black metallic tellurium precipitate. MAC has the ability to reduce tellurite within 3 to 4 days (112). This distinctive property is used to separate MAC from most other nonphotochromogens. Some rapid growers can similarly reduce tellurite within this time frame.

Tween 80 Hydrolysis

Lipases produced by some mycobacterial species hydrolyze the detergent polyoxyethylene sorbitan monooleate (Tween 80) into oleic acid and polyoxyethylene sorbitol. Neutral red in the pH 7 test medium is bound by Tween 80 and has an amber color at a neutral pH. If Tween 80 is hydrolyzed, however, neutral red is no longer bound, and it reverts to its usual red color at pH 7 (232). This test is used to separate potentially pathogenic (negative) from commonly saprophytic (positive) slowly growing scotochromogens and nonphotochromogens.

Urease

The ability of an isolate to hydrolyze urea is useful in identifying both scotochromogens and nonphotochromogens. M. scrofulaceum is urease positive, whereas members of the MAC are urease negative. Thus, a negative urease test is helpful in the recognition of pigmented strains of the MAC. Three different test methods are available: the Murphy-Hawkins disk method (147), the Steadham method (194), and the Wayne method (97, 230). All are available commercially.

BACTEC NAP Test

P-Nitro-α-acetylamino-β-hydroxypropiophenone (NAP) is an intermediate compound in the synthesis of chloramphenicol that inhibits species in the M. tuberculosis complex (128, 143, 193). Other mycobacterial species are inhibited only slightly or not at all by this compound. The production of CO_2 in the BACTEC vial is monitored by the BACTEC instrument to detect the inhibition of growth. If mycobacteria are inhibited, they are unable to metabolize in the presence of NAP and thus do not produce CO_2.

Reagents

A BACTEC test vial containing a disk impregnated with 5 mg of NAP is the only reagent necessary.

Procedure

1. After a BACTEC 12B vial has become positive with a GI of ≥50 and an AFB stain of the broth has demonstrated an apparently pure culture of AFB, a portion of the well-mixed broth is removed and added to a new 4-ml 12B vial (without antimicrobial agents) according to the following scheme.

Daily GI	Amt (ml) removed and added to new 12B vial
50–100 first day	None
50–100 more than 1 day	1.0
101–200	0.8
201–400	0.6
401–600	0.4
601–800	0.3
801–999	0.2
999 more than 1 day	0.1

This diluted culture is thoroughly mixed, and 1 ml is immediately transferred to the NAP vial. The remainder of the diluted broth then serves as the growth control. Vials reading 50 to 100 for the first time are not diluted; 1.0 ml of the broth is placed directly into the NAP vial, and the original positive vial serves as the control.

2. Swab the top of the NAP vial and the control vial with an appropriate disinfectant and then an alcohol swab.

3. Test both vials on the BACTEC instrument immediately to purge with 5% CO_2.

4. Incubate at 37°C, and test daily for the next 2 to 6 days. Record the daily GI of both vials.

The daily GI of the control vial should continue to increase steadily. In the NAP vial, the decrease or increase of GI depends on the species of mycobacteria present. A decreased or unchanged GI (two consecutive ≥20% decreases in GI after inoculation or a slight increase in the first 2 days with a subsequent decrease or no increase in GI) indicates that the isolate is presumptively a member of the M. *tuberculosis* complex, whereas an increase in GI indicates probable NTM. Results should be reported only when a clear trend is achieved and should be interpreted in conjunction with the isolates' morphologic and growth characteristics. Cultures of M. *tuberculosis* contaminated with bacteria or mixed with other mycobacterial species may show resistance to NAP. The NAP test should always be confirmed by either conventional biochemical tests or DNA probe studies. M. *tuberculosis* complex should not be reported in less than 4 days. Strains of M. *kansasii*, M. *gastri*, M. *szulgai*, M. *terrae*, and M. *triviale* are partially inhibited by NAP (128). Interpretation of the NAP test in the first 2 to 4 days may be misleading with these strains because of their longer lag phases. Optimal incubation temperature is critical for the NAP test. M. *marinum* grows best at 30°C, and M. *xenopi* grows best at 42°C. NAP may inhibit the growth of these organisms at 37°C, and thus the test may give false-positive results for these organisms when they are incubated at 37°C. However, incubation at temperatures other than 37°C is not routine and usually is performed only when problems are encountered. The test results will be clear and can be reported within 4 days if the test is incubated at the optimal growth temperature for these organisms.

Mycobacteria growing on solid media may also be tested with the NAP test. Colonies are inoculated into a 12B vial and allowed to reach the appropriate GI prior to the inoculation of the NAP vial as described above.

Controls

M. *tuberculosis* serves as a positive control, and M. *avium* serves as a negative control.

Chromatographic Analysis

Analysis of lipid composition by thin-layer chromatographic, capillary gas chromatographic (CGC), and HPLC methods is recognized as a useful tool for identification of mycobacteria (17, 23, 24, 204, 206). CGC and HPLC are the chromatographic methods with the most promise for this application (see chapter 12 of this Manual for the principles of chromatographic techniques).

A CGC identification system for mycobacteria is commercially available (MIDI, Newark, Del.). The system hardware consists of an autosampler, a gas chromatograph, a flame ionization detector, an integrator, and a computer. Software includes the identification system software and a library of fatty acid profiles of well-characterized mycobacteria. Sample preparation for CGC analysis includes saponification of approximately 40 mg of bacterial cells, methylation of the fatty acids, extraction of the fatty acid methyl esters, and a sodium hydroxide wash to clean up the sample. The organic phase is transferred to a CGC vial for analysis. The fatty acid profile of an unknown mycobacterium is compared with those stored in the database by using a covariance matrix, principal component analysis, and pattern recognition software. The system will analyze about 45 samples per day with an average operator time per sample of about 6 min. Although the system is reported to be highly reliable for identification of mycobacteria, no evaluations comparing the current system configuration with other methods have been published (169).

Reverse-phase HPLC of mycolic acid esters is also a rapid, reproducible method for identification of *Mycobacterium* species (24, 204). Sample preparation for HPLC analysis is similar to that described for CGC analysis but requires the additional step of derivatizing the mycolic acids to UV-absorbing esters with *p*-bromophenacyl bromide. HPLC requires only minimal colony growth, and chromatographic profile analysis can be performed once visible growth is obtained on solid medium. The interpretation of HPLC-generated chromatographic data has been simplified by the use of a personal-computer-based library consisting of data on 45 species of *Mycobacterium* in conjunction with a commercially available pattern recognition software package, Pirouette (Infometrix, Seattle, Wash.) (70). The software processes the chromatographic data using the K-nearest neighbor algorithm. The results obtained with a set of over 1,300 strains used to evaluate the library and the pattern recognition software are given in Table 5 (70). The *Mycobacterium* library is available from the CDC free of charge. Currently, HPLC methods for the identification of mycobacteria are used in research laboratories and large reference laboratories, including the CDC and some state and county public health department laboratories.

Both CGC and HPLC are technically complex, requiring expensive equipment and software. However, the costs of consumables are relatively low. The chromatographic methods are rapid, providing definitive identification of an isolate in <2 h. Chromatographic analysis of lipids is an attractive alternative to conventional biochemical testing for large reference laboratories, and chromatography is finding increased use as the primary method of identification in those settings.

Chromatographic methods have also been used for the direct detection of mycobacterial components in clinical specimens. The most successful approach has been the detection of tuberculostearic acid (TSA) by combination of gas chromatography and mass spectroscopy with selective ion monitoring (19, 65, 127). TSA is a structural component of mycobacteria and other actinomycetes. The direct detection of TSA in CSF and sputum is a rapid and sensitive test for the diagnosis of tuberculosis; however, the cost of the equipment and its maintenance is prohibitively high at this time.

DNA Probes for Culture Confirmation

Beginning in 1987, DNA probes complementary to species-specific sequences of rRNA became commercially available for the identification of M. tuberculosis complex, MAC, M. gordonae, and M. kansasii (AccuProbe; Gen-Probe, San Diego, Calif.). These probes can be used to identify isolates that arise on solid culture media or from broth culture. The tests can be performed with the growth from primary cultures, whereas conventional biochemical testing requires inoculation of subcultures and a subsequent delay in testing until adequate growth has occurred. Results of probe assays can be obtained in about 2 h. The original ^{125}I-labeled probes were both sensitive and specific (41, 56, 61, 72, 110, 150). The current acridinium ester-labeled AccuProbes are as sensitive and specific as the radioisotopically labeled probes (76, 131, 215) and have long shelf lives and reduced biohazard, thus allowing more clinical laboratories access to the technology.

The probe test is an in-solution assay that uses the hybridization protection assay principle (4). Chemiluminescent acridinium ester is attached to the probe in such a manner that when DNA-RNA hybrids are formed, the acridinium ester is oriented within the hybrid. The rRNA is released from the test organism by the action of a lysing agent, sonication, and heat. The lysate is reacted with the probe, and if a stable DNA-RNA hybrid is formed, the acridinium label within the hybrid is protected from differential hydrolysis by a selection reagent, while unbound label is inactivated. When an alkaline hydrogen peroxide solution is added to elicit chemiluminescence, only the hybrid bound acridinium ester is available to emit light. The light produced is proportional to the amount of hybridized probe, is measured on a chemiluminometer, and is expressed in relative light units. The reader determines whether a sample is positive or negative by comparing the sample's light emission values to a standard cutoff value.

Probes can be combined with the BACTEC TB system to optimize the rapid detection and identification of mycobacteria present in clinical samples (41, 61, 110, 157), but procedural modifications are required, especially with media containing blood (41, 62). Briefly, after a GI of at least 100 is reached, a portion of the broth culture (1.0 to 1.3 ml) is removed and placed into a 1.5-ml screw-cap microcentrifuge tube, and the bacilli are concentrated by centrifugation at 9,000 to 10,000 × g for at least 5 to 7 min. The microcentrifuge used should be equipped with aerosol-con-

taining tube carriers or rotor. The supernatant is decanted, the pellet is suspended in the detection reagents, and the testing proceeds as for testing colonies from solid media. Specimens containing blood are treated with 100 μl of 10% sodium dodecyl sulfate–50 mM EDTA (pH 7.2) prior to microcentrifugation to allow for lysis of the erythrocytes and solubilization of the membranes. Pellets are washed with 1.0 to 1.5 ml of sterile water, centrifuged, and processed as described above. These protocols are not U.S. Food and Drug Administration approved and therefore should be validated with appropriate control organisms by individual users. It has been recommended that a specimen blank be included with all probe assays from BACTEC blood cultures to help in evaluating the high levels of nonspecific chemiluminescence seen with specimens containing blood (62). A positive and negative control specific for each probe should be run each time the assay is performed.

Users of the AccuProbe system should be aware of a problem with failure of the sonicating water baths that are used for lysis of the mycobacteria (222). Either transducer failure or mineral buildup in the sonicator tank can result in false-negative results, depending on the position of the sample in the water bath. If probe results do not correlate with clinical or cultural observations, the test should be repeated or confirmed by an alternative method.

Use of the probes for the identification of mycobacterial isolates can result in rapid identification of the vast majority of isolates found in clinical laboratories and eliminates the necessity of using numerous biochemical tests for the identification process. Remaining isolates require identification by biochemical reactions or chromatographic methods. Probes do not differentiate between M. tuberculosis, M. bovis, M. bovis BCG, M. africanum, and M. microti. Recent reports indicate that some isolates of M. terrae and M. celatum complex produce false-positive reactions with the M. tuberculosis complex probe (25, 179). Although probes offer a rapid and accurate means of isolate identification, poor sensitivity precludes their use for direct detection of mycobacteria in clinical specimens at this time.

Direct Detection and Identification by Nucleic Acid Amplification and Hybridization

The inherent limitations of culture-based diagnostic tests specific for M. tuberculosis and other mycobacterial species pose problems for both individual patient management and implementation of appropriate hospital infection control and public health control measures. An ideal diagnostic procedure would detect and identify mycobacteria directly from clinical specimens, thereby avoiding the relatively lengthy time required for culturing. Direct detection requires either an extremely sensitive and specific assay or a process by which a diagnostically useful component of the target organism can be amplified to a detectable level. Nucleic acid amplification (also discussed in chapter 13 of this Manual) is one such process being actively developed and evaluated, particularly for the diagnosis of tuberculosis and leprosy. Several different amplification strategies have been employed.

Many PCR assays (see chapter 13) have been evaluated for their abilities to offer a rapid diagnosis of tuberculosis or other mycobacterial disease (2, 18, 43, 52, 58, 59, 63, 82, 86, 142, 183, 189, 191). These assays differ in specimen preparation protocols, targets for amplification, primers, and methods of detection. When the assays are applied to

TABLE 5 Accuracy of HPLC profile analysis for identification of mycobacteria[a]

Species	No. of strains evaluated	No. of incorrect identifications	% Accuracy
M. tuberculosis	649	1	99
M. avium	157	13	92
M. intracellulare	56	4	93
M. scrofulaceum	40	7	83
M. gordonae	47	2	96
M. kansasii	54	2	96
M. terrae	11	1	91
M. nonchromogenicum	20	1	95
M. xenopi	35	0	100
M. fortuitum-M. peregrinum	42	3	93
M. chelonae	30	1	97
M. abscessus	54	2	96
M. marinum	26	1	96
M. chelonae-like	16	4	75
M. malmoense	10	0	100
M. simiae	11	1	91
M. szulgai	23	0	100
M. bovis BCG	11	0	100
M. haemophilum	6	0	100
M. asiaticum	7	0	100
M. shimoidei	2	1	50
M. gastri	6	0	100
M. celatum	10	1	90
T. paurometabolum	4	0	100
Non-Mycobacterium spp.	6	0	100

[a]Reprinted from reference 70 with permission.

respiratory specimens, the reported sensitivities range from 55.9 to >100% compared to culture. The procedures often employ cumbersome specimen preparation and DNA isolation methods, use radioisotopic methods for detection of amplicon, fail to assess the presence of amplification reaction inhibitors, and/or were not designed to limit false-positive results due to carryover of PCR products from previous amplification reactions. These factors present obstacles to the routine use of PCR assays in clinical laboratories. In addition, many of these assays have been studied only with selected specimens, often from patients known to have tuberculosis, and the results are often compared to results of less-than-optimal smear and culture methods. Thus, the value of PCR for the routine detection of M. tuberculosis in a clinical laboratory is unclear. Recently, two relatively large clinical trials assessed the efficiency of PCR for the routine detection of M. tuberculosis in the clinical laboratory and compared PCR results to those of fluorochrome smear and culture, including BACTEC culture as well as solid-medium culture. Clarridge et al. (39) evaluated 1,166 specimens, of which approximately half were respiratory specimens, and Forbes and Hicks (64) examined 734 respiratory specimens. The sensitivity of PCR for respiratory specimens compared to culture was 84% in both evaluations. A PCR assay for M. tuberculosis that incorporates a simple sample preparation protocol, an internal positive control, chemiluminescent detection of amplicon, and a protocol to limit false positives due to amplicon contami-

nation (154) has recently been subjected to a large clinical trial of 2,410 consecutive sputum specimens, and the results were compared with those of fluorochrome smear and culture, both conventional and BACTEC methods, to determine the clinical sensitivity of the assay for the diagnosis of pulmonary tuberculosis. While 100% specific, the assay had a disappointingly low sensitivity of 60% (139).

PCR-based assays for detection of M. tuberculosis are offered by several commercial laboratories. However, as of June 1994, none of these methods has been cleared by the Food and Drug Administration for use as an in vitro diagnostic product. The false-positive rates, false-negative rates, reproducibilities, and predictive values of these tests are not fully known, and their usefulness in patient management and public health practices has not been established (32). Laboratories vary greatly on both the sensitivities and the specificities of their assays (155). Implementation of an effective system in monitoring sensitivity and specificity is required before PCR can be used reliably in the diagnosis of tuberculosis. Laboratories that use these services or offer their own testing are cautioned to use their own quality assurance validation program and clinical evaluation to determine performance characteristics prior to using the tests and reporting patient results. Proposed guidelines for molecular diagnostic methods in clinical microbiology are available from the National Committee for Clinical Laboratory Standards. In these guidelines, the development, validation, quality assurance, and routine use of nucleic acid amplification assays are addressed in detail (153).

Commercial assays based on a variety of different amplification strategies are being developed, are undergoing clinical evaluation by several manufacturers, and will soon be available. These strategies include PCR, a transcription-based amplification system (102), ligase chain reaction (96), strand displacement amplification reaction (218), and reporter phage systems. These assay systems, from specimen procurement to interpretation of results, will need to be evaluated with rigorous clinical challenges against high-quality culture tests in the clinical laboratory setting to determine their diagnostic value. As these systems are standardized and problems with inhibition of amplification reactions by endogenous inhibitors are eliminated, the systems promise to become accepted and widely used in the clinical laboratory.

Strain-Typing Systems

Epidemiologic studies of tuberculosis can be strengthened by the application of strain typing systems (see chapters 13 and 16 of this Manual). Historically, unusual drug susceptibility patterns and phage typing (188) have been used for this purpose, but they have significant limitations. DNA fingerprinting of M. tuberculosis has proven to be a powerful epidemiologic tool (45, 57, 137). A standardized technique for restriction fragment length polymorphism (RFLP) that exploits variability in both the number and the chromosomal position of a transposable element, IS6110, has been developed (211).

The technique involves the growth of M. tuberculosis, extraction of DNA, restriction endonuclease digestion using PvuII, Southern blotting, and probing for IS6110. The sizes of the resulting IS6110-hybridizing fragments are determined by the use of molecular size markers, and the strain-specific patterns of isolates are compared. It is estimated that 10^4 to 10^5 different RFLP patterns exist for M. tuberculosis, and these patterns have proven to be stable in outbreak situations. DNA fingerprinting of M. tuberculosis has been applied to epidemiology, documenting cross-contamination in the laboratory, and resolving workman's compensation issues surrounding occupational exposure to tuberculosis. RFLP typing is offered by several regional laboratories in the United States through a program funded by the CDC.

Haas et al. (81) developed a rapid, highly sensitive, highly specific method for typing strains of M. tuberculosis that is based on the IS6110 RFLP by PCR. The key feature of this method, termed mixed-linker PCR, is the ability to amplify multiple restriction fragments containing IS6110 sequences and variable sequences adjacent to the restriction site. The main advantage of mixed-linker PCR fingerprinting over the traditional RFLP analysis is that the former requires smaller amounts of genomic DNA. Traditional RFLP typing requires a 2- to 3-week-old subculture of the isolate or heavy growth on the original slant to provide enough DNA for IS6110 analysis. Because of the capacity of PCR to amplify specific DNA fragments, it is possible to obtain a fingerprint directly from a single colony of the primary isolate without further subculture. Mixed-linker PCR can also be applied directly to fingerprinting of strongly positive smear-positive specimens (81). This may provide the opportunity for real-time epidemiology and early recognition of infection with known drug-resistant strains long before the results of susceptibility tests are available.

Serovar distinctions in the MAC are based on a seroagglutination procedure originally developed by Schaefer (180). The serologic specificities of serovar antigens are conferred by oligosaccharide residues of the C-mycoside glycopeptidolipids (16). Currently, strains are serotyped by using a combination of thin-layer chromatography (209), enzyme-linked immunosorbent assay (ELISA) analysis of species and type-specific glycolipids (244), and the conventional seroagglutination procedure. Combined use of serotyping and species-specific DNA probes has shown that serovars 1 through 6 and 8 through 11 are M. avium, while serovars 7, 12 through 17, 19, 20, and 25 are M. intracellulare (177).

IMMUNODIAGNOSTIC TESTS FOR TUBERCULOSIS

A variety of immunodiagnostic tests for tuberculosis are based on recognition of specific host responses to the infecting organism. Historically, the first immunodiagnostic test was the tuberculin skin test. The shortcomings of this test include its inability to distinguish active disease from past sensitization and an unknown predictive accuracy (187). Various in vitro tests of cell-mediated immune response to mycobacterial antigens have been described, but they are expensive and technically demanding and provide no more diagnostic information than the tuberculin skin test.

Much effort has been devoted to the development of serologic tests for tuberculosis, but no test has found widespread clinical use (8, 49). The specificities of serologic tests that use crude antigen preparations are too low for clinical application. Specificity can be increased by using purified antigens, but since not all patients respond to the same antigens, the increased specificity often results in decreased sensitivity (37, 49, 99, 213). Sensitivity and specificity remain high if ELISA results obtained with a set of purified antigens are combined. The antigens tested in serologic assays include the 38,000 antigen (48, 99) and the antigen 85 complex (176, 210).

A number of antigen capture assays based on ELISA, radioimmunoassay, or agglutination of antibody-coated latex particles have been described (47, 78). The results reported for some assays have been promising for CSF, but experience with sputum and other specimen types is limited. In six major studies, the sensitivities of immunoassays for detection of mycobacterial antigen in CSF ranged from 65.8 to 100%, and the specificities ranged from 95 to 100% (47). Another approach to detection of mycobacterial antigens is to apply an antigen detection test to a liquid culture after several days of incubation (197).

QUALITY ASSURANCE

An active quality assurance program is essential in a clinical mycobacteriology laboratory. The laboratory must have programs in place to ensure that it is generating accurate and reliable results and that the procedures used to generate those results provide relevant, timely data in the delivery of health care. Quality assurance includes but is not limited to those activities in the laboratory traditionally associated with quality control.

As the incidence of tuberculosis declined in the United States, so, too, did the number of primary mycobacteriology laboratories. Many of the functions performed by these laboratories were assumed by general bacteriology laboratories. The small number of cultures for mycobacteria and the even smaller number of positive cultures made it difficult to maintain proficiency in many of these laboratories. Because of these changes in laboratory practice, the Public Health Service introduced the levels-of-service concept in 1967. In this scheme, laboratories determine the level of service that best fits the needs of the patient population they serve, the experience of their personnel, their laboratory facilities, and the number of specimens they receive. The concept of levels of service is supported by the CDC and the American Thoracic Society (ATS) (121). The College of American Pathologists (CAP) proposed extents of service for participation in mycobacterial interlaboratory comparison surveys. ATS levels I, II, and III correlate with CAP extents 2, 3, and 4. ATS makes no provision for laboratories in which no mycobacterial procedures are performed (CAP extent 1). According to the data from the CAP mycobacterial surveys, the majority of clinical laboratories are extent 2. Only about 10% offer the full range of services described for extent 4. For performing tests that are beyond their own level of routine service, laboratories should carefully select reference laboratories.

Procedure manuals for mycobacteriology should be readily available and should conform to the guidelines provided by the National Committee for Clinical Laboratory Standards (151). Procedures should be followed by all personnel, and any changes should be approved by the laboratory director.

Personnel working in the clinical mycobacteriology laboratory must have proper training and certification in the specific functions they perform. The number of procedures performed must be high enough to ensure the continued proficiency of the personnel. Since many of the techniques used in mycobacteriology are not used elsewhere in the general bacteriology laboratory, personnel should be assigned to mycobacteriology laboratory for extended periods if rotation through the other laboratory sections is necessary.

Laboratories must participate in a recognized mycobacterial proficiency-testing survey. The CAP survey is designed for laboratories to participate at the extent or level of service that they usually offer. Proficiency-testing surveys can be used as part of a continuing education program as well as to satisfy state and federal licensure requirements. The surveys also provide the opportunity to acquire strains of mycobacteria with unusual or specific characteristics that can be used to augment the laboratory's stock culture collection.

The laboratory must maintain a collection of well-characterized mycobacteria to be used for training, continuing education, and quality control of test systems. These control organisms may be obtained from several sources, including the American Type Culture Collection, the Trudeau Mycobacterial Culture Collection based at the National Jewish Hospital and Research Center in Denver, Colo., and proficiency-testing programs. Frequently used stock cultures can be maintained on LJ slants or in 7H9 broth at 37°C or room temperature if they are subcultured monthly. Cultures on LJ slants may be held for up to 1 year if they are stored at 4°C. Rapid freezing of organisms suspended in skim milk or broth medium and storage at −20 to −70°C is the best option for long-term maintenance of stock cultures.

The National Committee for Clinical Laboratory Standards guidelines for quality control of culture media should be followed (152). Each shipment of media obtained from commercial sources should be inspected for signs of deterioration, and the manufacturer should provide documentation of the quality control evaluation of the media prior to its shipment. Laboratories that prepare their own media must document the performance characteristics of each new lot. The adequacy of mycobacterial growth can be assessed with stock cultures of M. tuberculosis, M. kansasii, M. scrofulaceum, M. intracellulare, and M. fortuitum. For media containing hemin or ferric ammonium citrate, use M. haemophilum. Escherichia coli can be used to test the inhibitory capacities of selective media. Prolonged storage (>5 weeks) or exposure to light can result in the formation of formaldehyde in Middlebrook 7H10 and 7H11 agars. Formation of formaldehyde is associated with poor growth of mycobacteria. To avoid this problem, new lots of media are put into use 1 week after inoculation of quality control organisms. If the new lot fails quality control, refrigerated aliquots of saved specimens can be reinoculated onto another lot or a different type of medium (169).

Appropriate quality control strains should be tested concurrently with patient isolates in identification systems such that positive and negative results are obtained for each component test or reaction.

The adequacy of acid-fast-staining procedures can be determined and stained smears can be microscopically evaluated by including slides containing mycobacteria and non-AFB in each run. Ideally, the positive control slides should be prepared from a concentrated sputum specimen obtained from a patient with active tuberculosis. In practice, many laboratories use suspensions of stock cultures or seeded negative sputum as positive controls for acid-fast-staining procedures. The presence of AFB in the negative controls or a >2% increase in the percentage of smear-positive but culture-negative specimens that cannot be attributed to response to antituberculosis therapy suggests that the water or reagents used in the staining procedure are contaminated with environmental mycobacteria (106). M. gordonae and M. terrae complex are most often involved. The water used

in staining may be monitored periodically for mycobacteria by filtering 1 liter of water through a 0.22-μm-pore-size filter and culturing the filter on 7H11 agar. AFB may also be carried over from one slide to another in the immersion oil used with the oil immersion lens. Oil should be removed from the lens after a positive slide has been examined.

Laboratories must monitor and document the contamination rates for cultures. Contamination rates between 3 and 5% are generally considered acceptable. Rates below 3% may indicate that the decontamination procedure is too harsh and that the procedure needs to be modified to minimize the lethal effect on mycobacteria. Contamination rates of >5% indicate too weak a decontamination procedure, which could compromise mycobacterial cultures due to overgrowth of contaminants. See the section on optimizing decontamination procedures above.

False-positive cultures may result from mislabeling, specimen carryover, or cross-contamination between culture tubes or vials. Mechanisms for minimizing false-positive cultures and rapidly recognizing their occurrence should be in place. This is particularly important in the mycobacteriology laboratory, because specimens that require digestion and decontamination are usually processed in batches. The order in which specimens are processed and the media that are inoculated should be recorded. Positive cultures that occur in a batch with previously positive cultures should alert the laboratory to the possibility of carryover or cross-contamination. The significance of an isolate may be determined by reviewing the order in which the specimens were processed, the direct smear results, the time to positivity, and the clinical history. Cross-contamination of culture vials in the BACTEC radiometric system due to inadequately sterilized sample needles has been documented (40, 148, 212). Cross-contamination in the BACTEC system is probably rare if the manufacturer's recommendations for operation and maintenance are closely followed. Unfortunately, when it occurs, it may be difficult to recognize, because it may skip several vials.

The resurgence of tuberculosis and the rapid progression of disease caused by multidrug-resistant strains in AIDS patients have focused attention on the time required to report AFB smear results and to isolate, identify, and determine the antimicrobial susceptibility of M. tuberculosis. Laboratories should document the time required to complete these studies and determine whether the results are available to health care providers in a timely fashion. The CDC has recommended that AFB smear results be available within 24 h of specimen collection and that the times required for identification and drug susceptibility testing of M. tuberculosis be no longer than 14 and 30 days, respectively, from the time of specimen collection (203). These recommendations are unrealistic for many laboratories, especially those serving patient populations in which the prevalence of tuberculosis is low and in regions where multidrug-resistant tuberculosis (MDR-TB) is rare.

ANTIMICROBIAL SUSCEPTIBILITY

The current approach to treatment of tuberculosis is based on several principles that have evolved over the past 40 years (207). First, the spontaneous mutations in M. tuberculosis that confer resistance to each antituberculosis drug preclude the use of a single drug for treatment. The treatment of tuberculosis should include at least two drugs to which the organism is susceptible to prevent the emergence of drug-resistant strains. Second, to be successful, the treatment of smear-positive tuberculosis must continue well beyond the time that AFB are no longer present in the sputum. Third, although pretreatment susceptibility studies have minimal impact on the initial treatment of tuberculosis, such testing should be done because of the potential public health benefits. The CDC recommends that initial isolates from all patients be tested for drug susceptibility to confirm the anticipated effectiveness of chemotherapy. Susceptibility testing should be repeated if the patient continues to produce culture-positive sputum after 3 months of treatment (203).

The first-line drugs in the treatment of tuberculosis are INH, rifampin, pyrazinamide, ethambutal, and streptomycin. The second-line drugs (*para*-aminosalicylic acid, ethionamide, cycloserine, capreomycin, kanamycin, amikacin, ciprofloxacin, ofloxacin, and rifabutin) should be used if resistance or toxicity occurs. Currently, two regimens, INH plus rifampin for 9 months and INH plus rifampin plus pyrazinamide for 2 months and then INH plus rifampin for 4 months, are recommended for treatment of drug-susceptible tuberculosis (3).

Drug-resistant tuberculosis has become a serious concern. In the United States, the rise in drug-resistant tuberculosis and the outbreaks of MDR-TB are manifestations of serious underlying problems with the health care infrastructure. The CDC recently conducted a nationwide survey of drug resistance among all tuberculosis cases reported during the first quarter of 1991 (30). Overall, 14.9% of cases had isolates resistant to at least one antituberculosis drug, and 3.3% had isolates resistant to both INH and rifampin, two of the most effective antituberculosis drugs. These findings show an increase in drug resistance since a previous survey conducted from 1982 to 1986. Drug resistance is a much larger problem in certain cities, most notably New York and Miami. In New York City, 33% of cases had organisms resistant to at least one drug, and 19% had organisms resistant to both INH and rifampin (66).

Clusters of MDR-TB have been reported in a number of U.S. hospitals and one prison system (57). Some organisms have been resistant to as many as seven drugs (INH, rifampin, kanamycin, ethambutol, ethionamide, streptomycin, and rifabutin). Nosocomial transmission of MDR-TB has been documented, with most transmissions occurring between AIDS patients (156). The short time between diagnosis and death (4 to 16 weeks) and the high mortality (72 to 89%) of MDR-TB underscore the need to provide physicians with timely reports of drug susceptibility studies.

The increasing problem of drug resistance has heightened interest in the study of the genetic determinants and mechanisms of drug resistance in M. *tuberculosis*. Deletion of the *katG* gene encoding both catalase and peroxidase from the chromosome of M. *tuberculosis* is associated with INH resistance (246). Mutations in another gene, *inhA*, which is involved in mycolic acid biosynthesis, also lead to INH resistance in M. *tuberculosis* (6).

Mutations in the *rpoB* gene, which encodes the beta subunit of RNA polymerase, lead to high-level resistance to rifampin in M. *tuberculosis* and M. *leprae* (89, 201). In clinical isolates of rifampin-resistant M. *tuberculosis*, a limited number of mutations were found, and all clustered within a region of 23 amino acids. Armed with this knowledge, Telenti et al. (202) developed a method for rapid detection of rifampin resistance in cultured M. *tuberculosis* and for direct detection in sputum by using amplification of a region of *rpoB* by PCR and identification of nucleotide

changes by single-strand conformation polymorphism analysis of the amplification products.

Understanding of the genetic determinants of ethambutol, fluoroquinolone, and streptomycin resistance in M. *tuberculosis* has also increased over the past several years. It may be possible in the near future to have available rapid methods for screening strains for all the medically important resistance alleles.

Strains of MAC are intrinsically resistant to antituberculosis drugs and many other antimicrobial agents owing to failure of these drugs to penetrate the lipid-rich cell wall (163, 164). There is a strong relationship between colony type and antimicrobial resistance. Transparent colony types are more resistant than opaque colony types, and colony type transition occurs at relatively high frequencies. The phenotypic heterogeneity of MAC complicates both the performance and the interpretation of susceptibility-testing studies.

Optimal regimens for treatment of either chronic pulmonary disease or disseminated infections in AIDS patients have not been defined. In addition, no therapeutic regimen has shown a sustained clinical benefit for patients with disseminated MAC. However, two randomized, placebo-controlled multicenter studies of the use of rifabutin for prevention of disseminated MAC infections showed that prophylactic rifabutin reduced the frequency of MAC bacteremia by 50% (27, 75). Patients with HIV infection and <100 CD4$^+$ lymphocytes should be given prophylaxis against MAC, and this prophylaxis should be continued for the rest of the patient's life unless multiple-drug therapy becomes necessary for disseminated disease (33). Although no optimal therapeutic regimen for disseminated MAC disease has been defined, a U.S. Public Health Service Task Force on Prophylaxis and Therapy for MAC recommended the following (33).

1. Treatment regimens should include at least two agents.

2. Every regimen should include azithromycin or clarithromycin. Many experts would use ethambutol as a second drug. One or more of the following agents can be added as second, third, and fourth agents: clofazimine, rifabutin, rifampin, ciprofloxacin, and, in some situations, amikacin (Table 6).

3. Therapy should be continued for the lifetime of the patient if clinical and microbiologic improvement is observed.

In vitro susceptibility testing of MAC with the methods and interpretive criteria used for M. *tuberculosis* has no value in guiding the selection of drugs to treat MAC disease. No studies have correlated in vitro susceptibility test results with clinical outcome, and there are no standardized methods for performing these tests. Until guidelines for performing and interpreting drug susceptibility tests for MAC are established, these tests should not be performed routinely.

Almost all strains of rapidly growing mycobacteria are resistant to the drugs used for the treatment of tuberculosis. The antimicrobial agents used often depend on the identification of the isolate and the results of drug susceptibility studies. Agar disk elution and broth dilution methods for drug susceptibility testing of rapidly growing mycobacteria have been described elsewhere (20, 20a, 20b, 199). Strains of M. *chelonae* tend to be much more resistant than either M. *fortuitum* or M. *abscessus*. Antimicrobial agents recommended for treatment of infections with rapidly growing mycobacteria and several other mycobacterioses are listed in Table 6.

EVALUATION, INTERPRETATION, AND REPORTING OF RESULTS

During the investigation of nosocomial outbreaks of MDR-TB, several problems involving clinical microbiology laboratories were identified. These included significant delays in reporting AFB smear results, identifying M. *tuberculosis*, and providing drug susceptibility results. To address these problems, and recognizing that improved technology is available for use in mycobacteriology laboratories, the CDC recommended that turnaround times from receipt of specimen to reporting of results be 24 h for the AFB smear, 14 days for detection of growth, 21 days for identification of M. *tuberculosis*, and 28 days for susceptibility testing (30). To meet these goals, laboratories must employ the most reliable and rapid methods for culture, identification, and susceptibility testing and in many cases must increase the availability of the AFB smear.

In a survey of 55 state public health laboratories, Huebner et al. (93) found that only 71% used fluorochrome rather than basic fuchsin stains for screening clinical specimens and that the results were reported, on average, 2 days after specimen receipt. Only 20% used rapid radiometric methods for culture, but 72% used rapid identification techniques (probes, NAP, or HPLC). Only 23% of laboratories performing susceptibility testing used the rapid radiometric method.

A similar survey of 1,600 clinical laboratories that participated in the CAP Mycobacteriology E Survey was conducted by Woods and Witebsky (239). Only just over half of the laboratories stained smears with the fluorochrome stain, despite its increased sensitivity and ease of reading. Only 26% of the laboratories processed contaminated specimens daily, thus potentially providing AFB smear results within 24 h. The most rapid methods available for isolation and identification of M. *tuberculosis* were used by only 30 and 35% of the respondents, respectively. Only half of the laboratories that performed susceptibility tests used the radiometric method.

These surveys indicate that considerable resources will need to be expended in both clinical and public health laboratories if the decreased turnaround-time goals suggested by the CDC are to be met. Laboratories play a pivotal role in the diagnosis and control of tuberculosis, and every effort should be made to implement sensitive and rapid methods for detection, identification, and susceptibility testing of M. *tuberculosis*. Specifically, these methods include (i) fluorochrome stain for mycobacteria in smears; (ii) radiometric, Septi-Chek, or microcolony method for culture; (iii) NAP differentiation test, DNA probes, or chromatographic analysis for identification; and (iv) direct testing of smear-positive specimen concentrates by the radiometric or agar methods for susceptibility testing (239).

The 24-h turnaround time for AFB smear results presents a challenge to most clinical laboratories. In many laboratories, the daily processing of specimens required to meet this goal would add considerable expense to the laboratory budget. Whether these expenses would be offset by savings from outside the laboratory remains to be seen. In regions where tuberculosis is uncommon, less frequent processing is acceptable. Turnaround-time goals for AFB smear results should be established for each institution after consultation with infection control practitioners and infectious disease specialists.

Nucleic acid amplification assays offer the promise of same-day detection and identification of M. *tuberculosis*. Implementation of this new technology in the clinical mycobacteriology laboratory presents several new challenges. Although the performance characteristics of many of these assays are quite good for smear-positive respiratory specimens, little information exists on the use of these tests for diagnosis of paucibacillary pulmonary or extrapulmonary disease. Unfortunately, it is in the last two categories that these tests are likely to have the greatest clinical impact. The new technology will supplement rather than replace culture. Culture will still be required to obtain organisms for susceptibility testing and to detect mycobacteria other than M. *tuberculosis*. The practicality, cost-effectiveness, and overall clinical utility of these nucleic acid amplification assays have not yet been defined.

The extreme analytical sensitivity of nucleic acid amplification methods makes false-positive results due to sample contamination and product carryover a major concern. However, the use of appropriate laboratory practices and methods for product inactivation will limit the occurrence of false positives to acceptable levels. The presence of interfering substances in specimens can produce false-negative results. In order to increase the level of confidence in negative results, the ability of each sample to support nucleic acid amplification should be assessed for specimen types known to contain interfering substances.

Nucleic acid-based assays do not depend on the presence of living organisms for detection. As a result, these tests probably should not be used to monitor patients receiving antituberculosis drugs, since these tests may be positive even after smears and cultures are negative.

The significance of the isolation of NTM may be difficult to assess, since many species are opportunistic pathogens (178, 238, 240). The following criteria are useful in establishing a role for NTM in a disease process: (i) repeated isolation of a large number of colonies of the same species from the same anatomical site over an extended period; (ii) clinical or radiographic evidence of disease; (iii) histopathologic evidence of the presence of mycobacteria in tissues; (iv) increasing numbers of mycobacteria in sequential specimens; (v) isolation from a normally sterile site; (vi) predilection of the species to cause disease at that site; (vii) presence of predisposing conditions in the patient; and (viii) absence of other identifiable causes of disease. In addition to the above criteria, accurate identification of NTM will prevent rarely encountered pathogens from being mistaken for nonpathogenic species (228).

The accurate and timely reporting of results of AFB microscopy, culture, identification, and drug susceptibility tests is essential to effective management of individual patients and to appropriate implementation of public health and infection control measures. The use of the telephone or facsimile transmission for reporting avoids the delays inherent in preparing and mailing written reports (93, 203). Physicians caring for the patients need to be notified for proper diagnosis and treatment. Infection control practitioners need to be notified in order to isolate patients with active tuberculosis and evaluate their contacts within the hospital. Public health officials must be notified to trace contacts of patients with active tuberculosis, to monitor treatment of known cases of tuberculosis, and to provide statistical data on the incidence of tuberculosis. The regulations for reporting cases of tuberculosis vary

TABLE 6 Treatment recommendations for mycobacterioses other than tuberculosis[a]

Mycobacterium	Established regimen	Additional or suggested agents
MAC (chronic pulmonary), M. scrofulaceum, M. malmoense, M. simiae	RIF, ETH, INH, STM, or AMIK	CLAR (AZI), CIP, CLOF
MAC (disseminated)[b]	CLAR (AZI), ETH	CLOF, RIFB, RIF, CIP, AMIK
M. kansasii, M. szulgai	RIF, INH, ETH	STM, CIP, CLAR
M. xenopi	RIF, INH, ETH	STM
M. marinum	ETH, RIF, DOX, or TMP-SMX	STM, CIP
M. haemophilum		RIF, CFOX, DOX, TMP-SMX
M. fortuitum	AMIK, CIP, SULF	CLOF, CLAR, CFOX, DOX, IPM
M. abscessus	AMIK	CLOF, CLAR, CLOX
M. chelonae	TOB (AMIK)	CLOF, CLAR, DOX

[a]Modified from Wolinsky (238). Abbreviations: RIF, rifampin; ETH, ethambutol; INH, isoniazid; STM, streptomycin; AMIK, amikacin; CLAR, clarithromycin; AZI, azithromycin; CIP, ciprofloxacin; CLOF, clofazimine; RIFB, rifabutin; DOX, doxycycline; TMP-SMX, trimethoprim-sulfamethoxazole; CFOX, cefoxitin; SULF, sulfamides; IPM, imipenem; TOB, tobramycin.

[b]Recommended by U.S. Public Health Service Task Force on Prophylaxis and Therapy of MAC (33).

from state to state, and laboratories need to communicate with local health departments or tuberculosis control officers about the regulations governing the reporting of AFB smears, cultures, and drug susceptibility testing.

Laboratories should verbally report the first positive smear and subsequent positive smears if numbers of organisms increase after an initial decrease. Verbal reports should also be issued for the first positive culture and for other information that may have an impact on diagnosis or treatment, including preliminary identification and resistance of M. tuberculosis to antituberculosis drugs. All telephone reports should be documented in the laboratory. Written reports of the information reported verbally should be sent to the clinician caring for the patient and the infection control office and should be added to the patients' medical record. To avoid duplication and delays in reporting, the procedures for notification need to established in cooperation with primary care providers, infection control practitioners, infectious disease specialists, respiratory medicine staff, public health officials, and other groups that may be responsible for the diagnosis and control of tuberculosis.

REFERENCES

1. Agy, M. B., C. K. Wassis, J. J. Plorde, L. C. Carlson, and M. B. Coyle. 1989. Evaluation of four mycobacterial blood culture media: BACTEC 13A, Isolator/BACTEC 12B, Isolator/Middlebrook Agar and a biphasic medium. Diagn. Microbiol. Infect. Dis. 12:303–308.
2. Altamirano, M., M. T. Kelly, A. Wong, E. T. Bessuille, W. Black, and J. A. Smith. 1992. Characterization of a DNA probe for detection of Mycobacterium tuberculosis complex in clinical samples by polymerase chain reaction. J. Clin. Microbiol. 30:2173–2176.
3. American Thoracic Society and Centers for Disease Control. 1986. Treatment of tuberculosis in adults and children. Am. Rev. Respir. Dis. 134:355–363.
4. Arnold, L. J., P. W. Hammond, W. A. Wiese, and N. C. Nelson. 1989. Assay formats involving acridinium-ester-labeled DNA probes. Clin. Chem. 35:1588–1594.
5. Ausina, V., J. Barrio, M. Luguin, M. A. Sambeat, M. Gurgui, G. Verger, and G. Prats. 1988. Mycobacterium xenopi infections in AIDS. Ann. Intern. Med. 109:927–928.
6. Banerjee, A., E. Dubnau, A. Quemard, V. Balasubramanian, K. S. Um, T. Wilson, D. Collins, G. de Lisle, and W. R. Jacobs, Jr. 1994. inhA, a gene encoding a target for isoniazid and ethionamide in Mycobacterium tuberculosis. Science 263:227–230.
7. Barbaro, D., V. Orcutt, and B. Coldiron. 1989. Mycobacterium avium-Mycobacterium intracellulare infection limited to the skin and lymph nodes in patients with AIDS. Rev. Infect. Dis. 11:625–628.
8. Bardana, E., Jr., J. McClatchy, R. Farr, and P. Minden. 1973. Universal occurrence of antibodies to tubercle bacilli in sera from nontuberculous and tuberculous individuals. Clin. Exp. Immunol. 13:65–77.
9. Barksdale, L., and K. S. Kim. 1977. Mycobacterium. Bacteriol. Rev. 21:217.
10. Beam, E. R., and G. P. Kubica. 1968. Stimulatory effects of carbon dioxide on the primary isolation of tubercle bacilli on agar containing medium. Am. J. Clin. Pathol. 50:395–397.
11. Blacklock, Z., D. Dawson, D. Kane, and D. McEvoy. 1983. Mycobacterium asiaticum as a potential pulmonary pathogen for humans. A clinical and bacteriologic review of five cases. Am. Rev. Respir. Dis. 127:241–244.
12. Bolivar, R., T. Satterwhite, and M. Floyd. 1980. Cutaneous lesions due to Mycobacterium kansasii. Arch. Dermatol. 116:207–208.
13. Bottger, E. C., A. Teske, P. Kirschner, S. Bost, H. R. Chang, V. Beer, and B. Hirschel. 1991. Disseminated "Mycobacterium genavense" infection in patients with AIDS. Lancet 340:76–80.
14. Boyd, J. C., and M. J. Marr. 1975. Decreasing reliability of acid-fast smear techniques for detection of tuberculosis. Ann. Intern. Med. 82:489–492.
15. Brander, E., E. Jantzen, R. Huttunen, A. Julkunen, and M. Katila. 1992. Characterization of a distinct group of slowly growing mycobacteria by biochemical tests and lipid analysis. J. Clin. Microbiol. 30:1972–1975.
16. Brennan, P. 1989. Structure of mycobacteria: recent devel-

opments in defining cell wall carbohydrates and proteins. *Rev. Infect. Dis.* **11:**S420–S430.

17. **Brennan, P. J., M. Heifets, and B. P. Ullom.** 1982. Thin-layer chromatography of lipid antigens as a means of identifying nontuberculous mycobacteria. *J. Clin. Microbiol.* **15:**447–455.

18. **Brisson-Noel, A., C. Azner, C. Chureau, S. Nguyen, C. Pierre, M. Bartoli, G. Pialoux, B. Cicquel, and G. Carrique.** 1991. Diagnosis of tuberculosis by DNA amplification in clinical practice evaluation. *Lancet* **338:**364–366.

19. **Brooks, J., M. Daneshvar, R. Haberberger, and I. Mikail.** 1990. Rapid diagnosis of tuberculous meningitis by frequency-pulsed electron-capture gas-liquid chromatography detection of carboxylic acids in cerebrospinal fluid. *J. Clin. Microbiol.* **28:**989–997.

20. **Brown, B. A., J. M. Swenson, and R. J. Wallace, Jr.** 1992. Agar disk elution test for rapidly growing mycobacteria, p. 5.10.1–5.10.11. *In* H. D. Isenberg (ed.), *Clinical Microbiology Procedures Handbook.* American Society for Microbiology, Washington, D.C.

20a.**Brown, B. A., J. M. Swenson, and R. J. Wallace, Jr.** 1992. Broth microdilution MIC test for rapidly growing mycobacteria, p. 5.11.1–5.11.10. *In* H. D. Isenberg (ed.), *Clinical Microbiology Procedures Handbook.* American Society for Microbiology, Washington, D.C.

20b.**Brown, B. A., and R. J. Wallace, Jr.** 1992. Broth microdilution MIC test for *Nocardia* spp., p. 5.12.1–5.12.9. *In* H. D. Isenberg (ed.), *Clinical Microbiology Procedures Handbook.* American Society for Microbiology, Washington, D.C.

21. **Bryceson, A., and R. E. Pfaltzgraft.** 1990. *Leprosy.* Churchill Livingstone, London.

22. **Burdash, N. M., J. P. Manos, D. Ross, and E. R. Bannister.** 1976. Evaluation of the acid-fast smear. *J. Clin. Microbiol.* **4:**190–191.

23. **Butler, W. R., D. G. Ahearn, and J. O. Kilburn.** 1986. High-performance liquid chromatography of mycolic acids as a tool in the identification of *Corynebacterium*, *Nocardia*, *Rhodococcus*, and *Mycobacterium* species. *J. Clin. Microbiol.* **23:**182–185.

24. **Butler, W. R., K. C. Jost, and J. O. Kilburn.** 1991. Identification of mycobacteria by high-performance liquid chromatography. *J. Clin. Microbiol.* **29:**2468–2472.

25. **Butler, W. R., S. P. O'Connor, M. A. Yakrus, and W. M. Gross.** 1994. Cross-reactivity of genetic probe for detection of *Mycobacterium tuberculosis* with newly described species *Mycobacterium celatum*. *J. Clin. Microbiol.* **32:**536–538.

26. **Butler, W. R., S. P. O'Connor, M. A. Yakrus, R. W. Smithwick, B. B. Plikaytis, C. W. Moss, M. M. Floyd, C. L. Woodley, J. O. Kilburn, F. S. Vadney, and W. M. Gross.** 1993. *Mycobacterium celatum*. *Int. J. Syst. Bacteriol.* **43:**539–548.

27. **Cameron, W., P. Sparti, N. Pietroski, D. Pearce, L. Eron, P. Sullam, M. Conant, T. Chew, S. Deresinki, B. Bernstein, J. Burnside, I. Fong, R. Hewitt, F. Smaill, S. Pomerantz, J. Schoenfelder, B. Wynne, P. Olliaro, M. Davis, P. Narang, and P. Viray.** 1992. Rifabutin prevents *Mycobacterium avium* complex bacteremia in patients with AIDS and CD4 <200, abstr. 888, p. 258. *Program Abstr. 32nd Intersci. Conf. Antimicrob. Agents Chemother.* American Society for Microbiology, Washington, D.C.

28. **Castro, K. G., and S. W. Dooley.** 1993. *Mycobacterium tuberculosis* transmission in healthcare settings: is it influenced by coinfection with human immunodeficiency virus? *Infect. Control Hosp. Epidemiol.* **14:**65.

29. **Centers for Disease Control.** 1990. Screening for tuberculosis and tuberculous infection in high-risk populations and the use of preventative therapy for tuberculous infection in the United States: recommendations of the Advisory Committee for the Elimination of Tuberculosis. *Morbid. Mortal. Weekly Rep.* **39**(RR-8):1–7.

30. **Centers for Disease Control.** 1992. National MDR-TB Task Force, national action plan to combat multidrug-resistant tuberculosis. *Morbid. Mortal. Weekly Rep.* **41:**1–48.

31. **Centers for Disease Control.** 1992. Tuberculosis morbidity—United States, 1991. *Morbid. Mortal. Weekly Rep.* **41:**240.

32. **Centers for Disease Control.** 1993. Diagnosis of tuberculosis by nucleic acid amplification methods applied to clinical specimens. *Morbid. Mortal. Weekly Rep.* **42:**686.

33. **Centers for Disease Control.** 1993. Recommendations on prophylaxis and therapy for disseminated *Mycobacterium avium* complex for adults and adolescents infected with human immunodeficiency virus. *Morbid. Mortal. Weekly Rep.* **42**(RR-9):14–20.

34. **Centers for Disease Control and Prevention.** 1993. *Draft Guidelines for Preventing the Transmission of Tuberculosis in Health Care Facilities*, 2nd ed., p. 52810–52854. U.S. Department of Health and Human Services, Washington, D.C.

35. **Centers for Disease Control-National Institutes of Health.** 1993. *Biosafety in Microbiological and Biomedical Laboratories*, 3rd ed. U.S. Department of Health and Human Services publication no. (CDC) 93-8395. U.S. Government Printing Office, Washington, D.C.

36. **Chaisson, R. E., J. Keruly, D. D. Richman, T. Creagh-Kirk, and R. D. Moore.** 1991. Incidence and natural history of *Mycobacterium avium* complex infection in advanced HIV disease treated with zidovudine. *Am. Rev. Respir. Dis.* **143:**A278.

37. **Chan, S. L., Z. Reggiardo, T. M. Daniel, D. J. Girling, and D. A. Mitchison.** 1990. Seriodiagnosis of tuberculosis using an ELISA with antigen 5 and a hemagglutination assay with glycolipid antigens. *Am. Rev. Respir. Dis.* **142:**385–390.

38. **Chiodini, R.** 1989. Crohn's disease and the mycobacterioses: a review and comparison of two disease entities. *Clin. Microbiol. Rev.* **2:**90–117.

39. **Clarridge, J., R. Shawar, T. Shinnick, and B. Plikaytis.** 1993. Large-scale use of polymerase chain reaction for detection of *Mycobacterium tuberculosis* in a routine mycobacteriology laboratory. *J. Clin. Microbiol.* **31:**2049–2056.

40. **Conville, P., and F. Witebsky.** 1989. Inter-bottle transfer of mycobacteria by the BACTEC 460. *Diagn. Microbiol. Infect. Dis.* **12:**401–405.

41. **Conville, P. S., J. F. Keiser, and F. G. Witebsky.** 1989. Comparison of three techniques for concentrating positive BACTEC 13A bottles of mycobacterial DNA probe analysis. *Diagn. Microbiol. Infect. Dis.* **12:**309–313.

42. **Corper, H. J., and N. Uyei.** 1930. Oxalic acid as a reagent for isolating tubercle bacilli and a study of the growth of acid-fast nonpathogens on different mediums with their reactions to chemical reagents. *J. Lab. Clin. Med.* **15:**348–369.

43. **Cousins, D., S. Wilton, B. Francis, and B. Gow.** 1992. Use of the polymerase chain reaction for rapid diagnosis of tuberculosis. *J. Clin. Microbiol.* **30:**255–258.

44. **Coyle, M. B., L. Carlson, C. Wallis, R. Leonard, V. Raisys, J. Kilburn, M. Samadpour, and E. Bottger.** 1992. Laboratory aspects of *Mycobacterium genavense*, a proposed species isolated from AIDS patients. *J. Clin. Microbiol.* **30:**3206–3212.

45. **Daley, C. L., P. M. Small, G. F. Schecter, G. K. Schoolnik, R. A. McAdam, R. Jacobs, Jr., and P. C. Hopewell.** 1992. An outbreak of tuberculosis with accelerated progression among persons infected with the human immunodeficiency virus: an analysis using restriction-fragment-length polymorphisms. *N. Engl. J. Med.* **326:**231–235.

46. **D'Amato, R., H. Isenberg, L. Hochstein, A. Mastellone, and P. Alperstein.** 1991. Evaluation of Roche Septi-Chek AFB system for recovery of mycobacteria. *J. Clin. Microbiol.* **29:**2906–2908.

47. **Daniel, T.** 1989. Rapid diagnosis of tuberculosis: laboratory techniques applicable in developing countries. *Rev. Infect. Dis.* **11:**S471–S478.

48. **Daniel, T., G. de Murillo, J. Sawyer, A. McLean Griffin, E. Pinto, S. Debanne, P. Espinosa, and E. Cespedes.** 1986. Field evaluation of enzyme-linked immunosorbent assay for the serodiagnosis of tuberculosis. *Am. Rev. Respir. Dis.* **134:**662–665.

49. **Daniel, T. M., and S. M. Debanne.** 1987. The serodiagnosis

of tuberculosis and other mycobacterial diseases by enzyme-linked immunosorbent assay. *Am. Rev. Respir. Dis.* **158:**678–680.

50. **Dawson, D., and F. Jennis.** 1980. Mycobacteria with a growth requirement for ferric ammonium citrate identified as *Mycobacterium haemophilum. J. Clin. Microbiol.* **11:**190–192.

51. **DeMerieux, P., E. Keystone, M. Hutcheon, and C. Laskin.** 1980. Polyarthritis due to *Mycobacterium kansasii* in a patient with rheumatoid arthritis. *Ann. Rheum. Dis.* **39:**90–94.

52. **DeWit, D., L. Steyn, S. Shoemaker, and M. Sogin.** 1990. Direct detection of *Mycobacterium tuberculosis* in clinical specimens by DNA amplification. *J. Clin. Microbiol.* **28:**2437–2441.

53. **Dixon, J. M. S., and E. H. Cuthbert.** 1967. Isolation of tubercle bacilli from uncentrifuged sputum on pyruvic acid medium. *Am. Rev. Respir. Dis.* **96:**119–122.

54. **Dizon, D., C. Mihailescu, and H. C. Bae.** 1976. Simple procedure for detection of *Mycobacterium gordonae* in water causing false-positive acid-fast smears. *J. Clin. Microbiol.* **3:**211.

55. **Dore, N., J.-P. Collins, and E. Mankiewicz.** 1979. A sporotrichoid-like *Mycobacterium kansasii* infection of the skin treated with minocycline hydrochloride. *Br. J. Dermatol.* **101:**75–79.

56. **Drake, T. A., J. A. Hindler, O. G. W. Berlin, and D. Bruckner.** 1987. Rapid identification of *Mycobacterium avium* complex in culture using DNA probes. *J. Clin. Microbiol.* **25:**1442–1445.

57. **Edlin, B. R., J. I. Tokars, M. H. Greico, J. T. Crawford, J. Williams, E. M. Sordillo, K. R. Ong, J. O. Kilburn, S. W. Dooley, K. G. Castron, W. R. Jarvis, and S. D. Holmberg.** 1992. An outbreak of multidrug-resistant tuberculosis among hospitalized patients with the acquired immunodeficiency syndrome. *N. Engl. J. Med.* **326:**1514–1521.

58. **Eisenach, K., M. Cave, J. Bates, and J. Crawford.** 1990. Polymerase chain reaction amplification of a repetitive sequence specific for *Mycobacterium tuberculosis. J. Infect. Dis.* **161:**977–981.

59. **Eisenach, K. D., M. Sifford, M. Cave, J. Bates, and J. Crawford.** 1991. Detection of *Mycobacterium tuberculosis* in sputum samples using a polymerase chain reaction. *Am. Rev. Respir. Dis.* **144:**1160–1163.

60. **Ellner, J., M. Goldberger, and D. Parenti.** 1991. *Mycobacterium avium* infection and AIDS: a therapeutic dilemma in rapid evolution. *J. Infect. Dis.* **163:**1326–1335.

61. **Ellner, P. D., T. E. Kiehn, R. Cammarata, and M. Hosmer.** 1988. Rapid detection and identification of pathogenic mycobacteria by combining radiometric and nucleic acid probe methods. *J. Clin. Microbiol.* **26:**1349–1352.

62. **Evans, K., A. Nakasone, P. Sutherland, L. de la Maza, and E. Peterson.** 1992. Identification of *Mycobacterium tuberculosis* and *Mycobacterium avium-M. intracellulare* directly from primary BACTEC cultures by using acridinium ester-labeled DNA probes. *J. Clin. Microbiol.* **30:**2427–2431.

63. **Fauville-Dufaux, M., B. Vanfleterer, L. De Wit, J. Vincke, J. Van Vooren, M. Yates, E. Serruys, and J. Content.** 1992. Rapid detection of tuberculous and non-tuberculous mycobacteria by polymerase chain reaction amplification of a 162 bp DNA fragment from antigen 85. *Eur. J. Clin. Microbiol. Infect. Dis.* **11:**797–803.

64. **Forbes, B., and K. Hicks.** 1993. Direct detection of *Mycobacterium tuberculosis* in respiratory specimens in a clinical laboratory by polymerase chain reaction. *J. Clin. Microbiol.* **31:**1688–1694.

65. **French, G., C. Chan, S. Cheung, and K. Oo.** 1987. Diagnosis of pulmonary tuberculosis by detection of tuberculostearic acid in sputum by using gas chromatography-mass spectrometry with selected ion monitoring. *J. Infect. Dis.* **156:**356.

66. **Frieden, T. R., T. Sterling, A. Pablos-Mendez, J. O. Kilburn, G. M. Cauthen, and S. W. Dooley.** 1993. The emergence of drug-resistant tuberculosis in New York City. *N. Engl. J. Med.* **328:**521–526.

67. **Fujisaa, K., H. Watanabe, K. Yamamoto, T. Nasu, Y. Kitahara, and M. Nakano.** 1989. Primary atypical mycobacteriosis of the intestine: a report of three cases. *Gut* **30:**541–545.

68. **Gill, J., E. A. Fanning, and S. Chomyc.** 1987. Childhood lymphadenitis in a harsh northern climate due to atypical mycobacteria. *Scand. J. Infect. Dis.* **19:**77–83.

69. **Gill, V., and F. Stock.** 1987. Detection of *Mycobacterium avium-Mycobacterium intracellulare* in blood cultures using concentrated and unconcentrated blood in conjunction with radiometric detection systems. *Diagn. Microbiol. Infect. Dis.* **6:**119.

70. **Glickman, S. E., J. O. Kilburn, W. R. Butler, and L. S. Ramos.** 1994. Rapid identification of mycolic acid patterns of mycobacteria by high-performance liquid chromatography using pattern recognition software and a *Mycobacterium* library. *J. Clin. Microbiol.* **32:**740–745.

71. **Gombert, M., E. Goldstein, M. Corrado, A. Shin, and K. Butt.** 1981. Disseminated *Mycobacterium marinum* infection after renal transplantation. *Ann. Intern. Med.* **94:**486–487.

72. **Gonzalez, R., and B. A. Hanna.** 1987. Evaluation of Gen-Probe DNA hybridization systems for the identification of *Mycobacterium avium-intracellulare. Diagn. Microbiol. Infect. Dis.* **8:**69–78.

73. **Good, R. C., V. A. Silcox, J. O. Kilburn, and B. D. Plikaytis.** 1985. Identification and drug susceptibility test results for *Mycobacterium* spp. *Clin. Microbiol. Newsl.* **7:**133–135.

74. **Good, R. C., and D. E. Snider.** 1982. Isolation of nontuberculous mycobacteria in the United States 1980. *J. Infect. Dis.* **146:**829–833.

75. **Gordin, F., S. Nightingale, B. Wynne, J. Kaufman Schoenfelder, D. Souder, P. Narang, R. Viray, and F. Seigal.** 1992. Rifabutin monotherapy prevents or delays *Mycobacterium avium* complex (MAC) bacteremia in patients with AIDS, abstr. 889, p. 258. *Program Abstr. 32nd Intersci. Conf. Antimicrob. Agents Chemother.* American Society for Microbiology, Washington, D.C.

76. **Goto, M., S. Oka, K. Okuzumi, S. Kimur, and K. Shimada.** 1991. Evaluation of acridinium ester-labeled DNA probes for identification of *Mycobacterium tuberculosis* and *Mycobacterium avium-Mycobacterium intracellulare* complex in culture. *J. Clin. Microbiol.* **29:**2473–2476.

77. **Graham, L., N. Warren, A. Tsang, and H. Dalton.** 1988. *Mycobacterium avium* complex pseudobacteriuria from a hospital water supply. *J. Clin. Microbiol.* **26:**1034–1036.

78. **Grange, J.** 1989. The rapid diagnosis of paucibacillary tuberculosis. *Tubercle* **70:**1–4.

79. **Grist, N., and J. Emslie.** 1985. Infections in British clinical laboratories, 1982–3. *J. Clin. Pathol.* **38:**721–725.

80. **Guthertz, L. S., B. Damsker, E. J. Bottone, E. G. Ford, T. D. Midura, and J. M. Janda.** 1989. *Mycobacterium avium* and *Mycobacterium intracellulare* infections in patients with and without AIDS. *J. Infect. Dis.* **160:**1037–1041.

81. **Haas, W., W. Butler, C. Woodley, and J. Crawford.** 1993. Mixed-linker polymerase chain reaction: a new method for rapid fingerprinting of isolates of the *Mycobacterium tuberculosis* complex. *J. Clin. Microbiol.* **31:**1293–1298.

82. **Hance, A., B. Grandchamp, V. Levy-Frebault, D. Lecossier, J. Rauzier, D. Bocart, and B. Gicquel.** 1989. Detection and identification of mycobacteria by amplification of mycobacterial DNA. *Mol. Microbiol.* **3:**843–849.

83. **Havlik, J., C. Horsburch, B. Metchock, P. William, S. Fan, and S. Thompson.** 1992. Disseminated *Mycobacterium avium* complex infection: clinical identification and epidemiologic trends. *J. Infect. Dis.* **165:**577–580.

84. **Havlik, J. A., B. Metchock, S. E. Thompson III, K. Barrett, D. Rimland, and C. R. Horsburgh, Jr.** 1993. A prospective evaluation of *Mycobacterium avium* complex colonization of the respiratory and gastrointestinal tracts of persons with human immunodeficiency virus infection. *J. Infect. Dis.* **168:**1045–1048.

85. **Hawkins, C. C., W. M. Gold, E. Whimbey, T. E. Kiehn, P.**

Brannon, R. Cammarata, A. E. Brown, and D. Armstrong. 1986. *Mycobacterium avium* complex infections in patients with the acquired immune deficiency syndrome. *Ann. Intern. Med.* **105**:184–188.

86. **Hermans, P., A. Schuitema, D. Van Soolingen, C. Verstynen, E. Bik, J. Thole, A. Kolk, and J. van Embden.** 1990. Specific detection of *Mycobacterium tuberculosis* complex strains by polymerase chain reaction. *J. Clin. Microbiol.* **28**:1204–1213.

87. **Hoffner, S.** 1988. Improved detection of M. *avium* complex with the BACTEC radiometric system. *Diagn. Microbiol. Infect. Dis.* **10**:1–6.

88. **Holzman, R. S., P. C. Hopewell, A. E. Pitchenik, L. B. Reichman, and R. L. Stoneburner.** 1987. Diagnosis and management of mycobacterial infection and disease in persons with human immune deficiency virus infection. *Ann. Intern. Med.* **106**:254–256.

89. **Honore, N., and S. Cole.** 1993. Molecular basis of rifampin resistance in *Mycobacterium leprae*. *Antimicrob. Agents Chemother.* **37**:414–418.

90. **Horsburgh, C., B. Metchock, J. McGowan, Jr., and S. Thompson.** 1992. Clinical implications of recovery of *Mycobacterium avium* complex from the stool or respiratory tract of HIV-infected individuals. *AIDS* **6**:512–514.

91. **Horsburgh, C. R., Jr., J. A. Havlik, D. A. Ellis, E. Kennedy, S. A. Fann, R. E. DuBois, and S. E. Thompson.** 1991. Survival of patients with acquired immunodeficiency syndrome and disseminated *Mycobacterium avium* complex infection with and without antimycobacterial chemotherapy. *Am. Rev. Respir. Dis.* **144**:557–559.

92. **Horsburgh, C. R., Jr., U. G. Mason III, D. C. Farhi, and M. D. Iseman.** 1985. Disseminated infection with *Mycobacterium avium-intracellulare*. A report of 13 cases and review of the literature. *Medicine* **64**:36–48.

93. **Huebner, R. E., R. C. Good, and J. I. Tokar.** 1993. Current practices in mycobacteriology: results of a survey of state public health laboratories. *J. Clin. Microbiol.* **31**:771–775.

94. **Imaeda, T., G. Broslawski, and S. Imaeda.** 1988. Genomic relatedness among mycobacterial species by nonisotopic blot hybridization. *Int. J. Syst. Bacteriol.* **38**:151–156.

95. **Inderlied, C., C. Kempler, and L. Bermudez.** 1993. The *Mycobacterium avium* complex. *Clin. Microbiol. Rev.* **6**:266–310.

96. **Iovannisci, D. M., and E. S. Winn-Deen.** 1993. Ligation amplification and fluorescence detection of *Mycobacterium tuberculosis* DNA. *Mol. Cell. Probes* **7**:35–43.

97. **Isenberg, H. D. (ed.).** 1992. *Clinical Microbiology Procedures Handbook*, vol. 1. American Society for Microbiology, Washington, D.C.

98. **Isenberg, H. D., R. F. D'Amato, L. Heifets, P. R. Murray, M. Scardamaglia, M. C. Jacobs, P. Alperstein, and A. Niles.** 1991. Collaborative feasibility study of a biphasic system (Roche Septi-Check AFB) for rapid detection and isolation of mycobacteria. *J. Clin. Microbiol.* **29**:1719–1722.

99. **Jackett, P. S., G. Bothamley, H. Batra, A. Mistry, D. Young, and J. Ivanyi.** 1988. Specificity of antibodies to immunodominant mycobacterial antigens in pulmonary tuberculosis. *J. Clin. Microbiol.* **26**:2313–2318.

100. **Jackson, K., A. Sievers, and B. Dwyer.** 1991. Effect of agitation of BACTEC 13A blood cultures on recovery of *Mycobacterium avium* complex. *J. Clin. Microbiol.* **29**:1801–1803.

101. **Jacobson, M., P. Hopewell, D. Yajko, W. Hadley, E. Lazarus, P. Mohanty, G. Modin, D. Feigal, P. Cusick, and M. Sande.** 1991. Natural history of disseminated *Mycobacterium avium* complex infection in AIDS. *J. Infect. Dis.* **164**:994–998.

102. **Jonas, V., M. J. Alden, J. I. Curry, K. Kamisango, C. Knott, R. Lankford, J. Wolfe, and D. Moore.** 1993. Detection and identification of *Mycobacterium tuberculosis* directly from sputum sediments by amplification of rRNA. *J. Clin. Microbiol.* **31**:2410–2416.

103. **Jones, W. D., and G. P. Kubica.** 1964. The use of MacCon-

key's agar for differential typing of *Mycobacterium fortuitum*. *Am. J. Med. Technol.* **30**:187–195.

104. **Joseph, S., E. Vaichulis, and V. Houk.** 1967. Lack of auramine-rhodamine fluorescence of Runyon group IV mycobacteria. *Am. Rev. Respir. Dis.* **95**:114.

105. **Kemper, C., T.-C. Meng, J. Nussbaum, J. Chiu, D. Feigal, A. Bartok, J. Leedom, J. Tilles, S. Deresinski, and J. McCutchan.** 1992. Treatment of *Mycobacterium avium* complex bacteremia in AIDS with a four-drug oral regimen: rifampicin, ethambutol, clofazimine, ciprofloxacin. *Ann. Intern. Med.* **116**:466–472.

106. **Kent, P. T., and G. P. Kubica.** 1985. *Public Health Mycobacteriology: a Guide for the Level III Laboratory*. Centers for Disease Control, U.S. Department of Health and Human Services, Atlanta.

107. **Kestle, D. G., V. D. Abbott, and G. P. Kubica.** 1967. Differential identification of mycobacteria. II. Subgroups of groups II and II (Runyon) with different clinical significance. *Am. Rev. Respir. Dis.* **95**:1041–1052.

108. **Kiehn, T. E., and R. Cammarata.** 1986. Laboratory diagnosis of mycobacterial infections in patients with acquired immunodeficiency syndrome. *J. Clin. Microbiol.* **24**:708–711.

109. **Kiehn, T. E., and R. Cammarata.** 1988. Comparative recoveries of *Mycobacterium avium-M. intracellulare* from Isolator lysis-centrifugation and BACTEC 13A blood culture systems. *J. Clin. Microbiol.* **26**:760–761.

110. **Kiehn, T. E., and F. F. Edwards.** 1987. Rapid identification using a specific DNA probe of *Mycobacterium avium* complex from patients with acquired immunodeficiency syndrome. *J. Clin. Microbiol.* **25**:1551–1552.

111. **Kiehn, T. E., F. F. Edwards, P. Brannon, A. Y. Tsang, M. Maio, J. W. M. Gold, E. Whimby, B. Wong, J. K. McClatchy, and D. Armstrong.** 1985. Infections caused by *Mycobacterium avium* complex in immunocompromised patients: diagnosis by blood culture and fecal examination, antimicrobial susceptibility tests, and morphological and seroagglutination characteristics. *J. Clin. Microbiol.* **21**:168–173.

112. **Kilburn, J. O., V. A. Silcox, and G. P. Kubica.** 1969. Differential identification of mycobacteria. V. The tellurite reduction test. *Am. Rev. Respir. Dis.* **99**:94–100.

113. **Kilburn, J. O., K. D. Stottmeier, and G. P. Kubica.** 1968. Aspartic acid as a precursor for niacin synthesis by tubercle bacilli grown on 7H10 agar medium. *Am. J. Clin. Pathol.* **50**:582–586.

114. **Kilby, J., P. Gilligan, J. Yankaskas, W. Highsmith, Jr., L. Edwards, and M. Knowled.** 1992. Nontuberculous mycobacteria in adult patients with cystic fibrosis. *Chest* **102**:70–75.

115. **Kim, T. C., R. S. Blackman, K. M. Heatwole, T. Kim, and D. F. Rochester.** 1984. Acid fact bacilli in sputum smears of patients with pulmonary tuberculosis. Prevalence and significance of negative smears pretreatment and positive smears post-treatment. *Am. Rev. Respir. Dis.* **129**:264–268.

116. **Kirihara, J. M., S. L. Hillier, and M. B. Coyle.** 1985. Improved detection times for *Mycobacterium avium* complex and *Mycobacterium tuberculosis* with the BACTEC radiometric systems. *J. Clin. Microbiol.* **22**:841–845.

117. **Kochi, A.** 1991. The global tuberculosis situation and the new control strategy of the World Health Organization. *Tubercle* **72**:1–6.

118. **Krasnow, I., and G. Kidd.** 1965. The effect of a buffered wash of sputum sediments digested with Zephiran trisodium phosphate on the recovery of acid-fast bacilli. *Am. J. Clin. Pathol.* **44**:238–240.

119. **Krasnow, I., and L. G. Wayne.** 1969. Comparison of methods for tuberculosis bacteriology. *Appl. Microbiol.* **18**:915–917.

120. **Krisher, K. K., M. C. Kallay, and F. S. Nolte.** 1988. Primary pulmonary infection caused by *Mycobacterium terrae* complex. *Diagn. Microbiol. Infect. Dis.* **11**:171–175.

121. **Kubica, G. P., W. M. Gross, J. E. Hawkins, H. M. Sommers, A. L. Vestal, and L. G. Wayne.** 1975. Laboratory

services for mycobacterial diseases. *Am. Rev. Respir. Dis.* **112**:773–787.

122. **Kubica, G. P., W. D. Jones, Jr., V. D. Abbott, R. E. Beam, J. O. Kilburn, and J. Cater, Jr.** 1966. Differential identification of mycobacteria. I. Tests on catalase activity. *Am. Rev. Respir. Dis.* **94**:400–405.

123. **Kubica, G. P., and G. L. Pool.** 1960. Studies on the catalase activity of acid-fast bacilli. I. An attempt to subgroup these organisms on the basis of their catalase activities at different temperatures and pH. *Am. Rev. Respir. Dis.* **81**:387–391.

124. **Kubica, G. P., and A. L. Rigdon.** 1961. The arylsulfatase activity of acid-fast bacilli. III. Preliminary investigation of rapidly growing acid-fast bacilli. *Am. Rev. Respir. Dis.* **83**:737–740.

125. **Kusunoki, S., and T. Ezaki.** 1992. Proposal of *Mycobacterium peregrinum* sp. nov., nom. rev., and elevation of *Mycobacterium chelonae* subsp. *abscessus* (Kubica et al.) to species status: *Mycobacterium abscessus* comb. nov. *Int. J. Syst. Bacteriol.* **42**:240–245.

126. **Lacaille, F., S. Blanche, C. Bodemer, C. Durand, Y. De Prost, and J. Gaillard.** 1990. Persistent *Mycobacterium marinum* infection in a child with probable visceral involvement. *Pediatr. Infect. Dis. J.* **9**:58–60.

127. **Larsson, L., G. Odham, G. Westerdahl, and B. Olsson.** 1987. Diagnosis of pulmonary tuberculosis by selected-ion monitoring: improved analysis of tuberculostearate in sputum using negative-ion mass spectrometry. *J. Clin. Microbiol.* **25**:893–896.

128. **Laszlo, A., and S. H. Siddiqi.** 1984. Evaluation of a rapid radiometric differentiation test for the *Mycobacterium tuberculosis* complex by selective inhibition with *p*-nitro-α-acetylamino-β-hydroxypropiophenone. *J. Clin. Microbiol.* **19**:694–698.

129. **Lavy, A., and Y. Yoshpe-Purer.** 1982. Isolation of *Mycobacterium simiae* from clinical specimens in Israel. *Tubercle* **63**:279–285.

130. **Leader, M., P. Revell, and G. Clarke.** 1984. Synovial infection with *Mycobacterium kansasii*. *Ann. Rheum. Dis.* **43**:80–82.

131. **Lebrun, L., F. Espinasse, J. Povenda, and V. Vincent-Levy-Frebault.** 1992. Evaluation of nonradioactive DNA probes for identification of mycobacteria. *J. Clin. Microbiol.* **30**:2476–2478.

132. **Lipsky, B. J., J. Gates, F. C. Tenover, and J. J. Plorde.** 1984. Factors affecting the clinical value for acid-fast bacilli. *Rev. Infect. Dis.* **6**:214–222.

133. **Males, B. M., T. E. West, and W. R. Bartholomew.** 1987. *Mycobacterium haemophilum* infection in a patient with acquired immune deficiency syndrome. *J. Clin. Microbiol.* **25**:186–190.

134. **Maloney, J. M., C. R. Gregg, D. S. Stephens, F. A. Manian, and D. Rimland.** 1987. Infections caused by *Mycobacterium szulgai* in humans. *Rev. Infect. Dis.* **9**:1120–1126.

135. **Marraro, R. V., E. M. Rodgers, and T. H. Roberts.** 1975. The acid-fast smear: fact or fiction? *Am. J. Med. Technol.* **37**:277–279.

136. **May, D. C., J. E. Kutz, R. S. Howell, M. J. Raff, and J. C. Melo.** 1983. *Mycobacterium terrae* tenosynovitis: chronic infections in a previously healthy individual. *South. Med. J.* **76**:1445–1447.

137. **Mazurek, G., M. Cave, K. Eisenach, R. Wallace, J. Bates, and J. Crawford.** 1991. Chromosomal DNA fingerprint patterns produced with IS6110 as strain-specific markers for epidemiologic study of tuberculosis. *J. Clin. Microbiol.* **29**:2030–2033.

138. **McClatchy, J. K., R. F. Waggoner, W. Kanes, M. S. Cernick, and T. L. Bolton.** 1976. Isolation of mycobacteria from clinical specimens by use of selective 7H11 medium. *Am. J. Clin. Pathol.* **65**:412–416.

139. **Metchock, B., F. S. Nolte, O. Okwumabua, L. Diem, P. S. Mitchell, C. Thurmond, J. E. McGowan, Jr., and T. Shinnick.** 1993. *Abstr. 93rd Gen. Meet. Am. Soc. Microbiol. 1993*, abstr. C-117, p. 466.

140. **Middlebrook, G., Z. Reggiardo, and W. D. Tigert.** 1977. Automatable radiometric detection of growth of *Mycobacterium tuberculosis* in selective media. *Am. Rev. Respir. Dis.* **115**:1066–1069.

141. **Miliner, R. A., K. D. Stottmeier, and G. P. Kubica.** 1969. Formaldehyde: a photothermal activated toxic substance produced in Middlebrook 7H10 medium. *Am. Rev. Respir. Dis.* **99**:603–607.

142. **Miyazaki, Y., H. Koga, S. Kohno, and M. Kaku.** 1993. Nested polymerase chain reaction for detection of *Mycobacterium tuberculosis* in clinical specimens. *J. Clin. Microbiol.* **31**:2228–2232.

143. **Morgan, M. A., K. A. Doerr, H. O. Hempel, N. L. Goodman, and G. D. Roberts.** 1985. Evaluation of the *p*-nitro-α-acetylamino-β-hydroxypropiophenone differential test for identification of *Mycobacterium tuberculosis* complex. *J. Clin. Microbiol.* **21**:634–635.

144. **Morgan, M. A., C. D. Horstmeier, D. R. DeYoung, and G. D. Roberts.** 1983. Comparison of radiometric method (BACTEC) and conventional culture media for recovery of mycobacteria from smear-negative specimens. *J. Clin. Microbiol.* **18**:384–388.

145. **Morris, A., L. B. Reller, M. Salfinger, K. Jackson, A. Sievers, and B. Dwyer.** 1993. Mycobacteria in stool specimens: the nonvalue of smears for predicting culture results. *J. Clin. Microbiol.* **31**:1385–1387.

146. **Morris, A. J., and L. B. Reller.** 1993. Reliability of cord formation in BACTEC media for presumptive identification of mycobacteria. *J. Clin. Microbiol.* **31**:2533–2534.

147. **Murphy, D. B., and J. E. Hawkins.** 1975. Use of urease test disks in the identification of mycobacteria. *J. Clin. Microbiol.* **1**:465–468.

148. **Murray, P.** 1991. Mycobacterial cross-contamination with the modified BACTEC 460 TB system. *Diagn. Microbiol. Infect. Dis.* **14**:33–35.

149. **Murray, P. R., C. Elmore, and D. J. Krogstad.** 1980. The acid-fast stain: a specific and predictive test for mycobacterial disease. *Ann. Intern. Med.* **92**:512–513.

150. **Musial, C. E., L. S. Tice, L. Stockman, and G. D. Roberts.** 1988. Identification of mycobacteria from culture by using the Gen-Probe rapid diagnostic system for *Mycobacterium avium* complex and *Mycobacterium tuberculosis* complex. *J. Clin. Microbiol.* **26**:2120–2123.

151. **National Committee for Clinical Laboratory Standards.** 1984. *Clinical Laboratory Procedure Manuals*. Approved standard GP2-A. National Committee for Clinical Laboratory Standards, Villanova, Pa.

152. **National Committee for Clinical Laboratory Standards.** 1990. *Quality Assurance for Commercially Prepared Microbiologic Culture Media*. Approved standard M-22A. National Committee for Clinical Laboratory Standards, Villanova, Pa.

153. **National Committee for Clinical Laboratory Standards.** 1994. *Specifications for Molecular Microbiology Methods for Infectious Diseases*. Proposed standard MM-P. National Committee for Clinical Laboratory Standards, Villanova, Pa.

154. **Nolte, F. S., B. Metchock, J. E. McGowan, Jr., A. Edwards, O. Okwumabua, C. Thurmond, P. S. Mitchell, B. Plikaytis, and T. Shinnick.** 1993. Direct detection of *Mycobacterium tuberculosis* in sputum by polymerase chain reaction and DNA hybridization. *J. Clin. Microbiol.* **31**:1777–1782.

155. **Noordhoek, G. T., A. H. J. Kolk, G. Bjune, D. Catty, J. W. Dale, P. E. M. Fine, P. Godfrey-Faussett, S.-N. Cho, T. Shinnick, S. B. Svenson, S. Wilson, and J. D. A. van Embden.** 1994. Sensitivity and specificity of PCR for detection of *Mycobacterium tuberculosis*: a blind comparison study among seven laboratories. *J. Clin. Microbiol.* **32**:277–284.

156. **Pearson, M. L., J. A. Jereb, T. R. Frieden, J. T. Crawford, B. J. Davis, S. W. Dooley, and W. R. Jarvis.** 1992. Nosocomial transmission of multi-drug resistant *Mycobacterium tuberculosis*. A risk to patients and health care workers. *Ann. Intern. Med.* **117**:191–196.

157. **Peterson, E. M., R. Lu, C. Floyd, A. Nakasone, G.**

Friiedly, and L. M. DeLaMaza. 1989. Direct identification of *Mycobacterium tuberculosis, Mycobacterium avium,* and *Mycobacterium intracellulare* from amplified primary cultures in BACTEC media using DNA probes. *J. Clin. Microbiol.* **27:** 1543–1547.

158. **Petran, E. L., and H. D. Vera.** 1971. Media for selective isolation of mycobacteria. *Health Lab. Sci.* **8:**225–230.

159. **Pierce, P. F., D. R. DeYoung, and G. D. Roberts.** 1983. Mycobacteremia and the new blood culture systems. *Ann. Intern. Med.* **99:**786–789.

160. **Pollock, H. M., and E. J. Wieman.** 1977. Smear results in the diagnosis of mycobacterioses using blue light fluorescence microscopy. *J. Clin. Microbiol.* **5:**329–331.

161. **Poropatich, C. O., A. M. Labriola, and C. U. Tuazon.** 1987. Acid-fast smear and culture of respiratory secretions, bone marrow and stools as predictors of disseminated *Mycobacterium avium* complex infection. *J. Clin. Microbiol.* **25:** 929–930.

162. **Quigley, H. S., and H. R. Elston.** 1970. Nitrite strips for detection of nitrate reduction by mycobacteria. *Am. J. Clin. Pathol.* **53:**663–665.

163. **Rastogi, N., C. Frehel, A. Ryter, H. Ohayon, M. Lesourd, and H. D. David.** 1981. Multiple drug resistance in *Mycobacterium avium:* is the wall architecture responsible for the exclusion of antimicrobial agents? *Antimicrob. Agents Chemother.* **20:**666–677.

164. **Rastogi, N., K. Goh, and H. David.** 1990. Enhancement of drug susceptibility of *Mycobacterium avium* by inhibitors of cell envelope synthesis. *Antimicrob. Agents Chemother.* **34:** 759–764.

165. **Ratnam, S., F. A. Stead, and M. Howes.** 1987. Simplified acetylcysteine-alkali digestion-decontamination procedure for isolation of mycobacteria from clinical specimens. *J. Clin. Microbiol.* **25:**1428–1432.

166. **Rickman, T. W., and N. P. Moyer.** 1980. Increased sensitivity of acid-fast smears. *J. Clin. Microbiol.* **11:**618–620.

167. **Roberts, G. D.** 1985. Mycobacteria and nocardia, p. 379–418. *In* J. A. Washington II (ed.), *Laboratory Procedures in Clinical Microbiology.* Springer-Verlag, New York.

168. **Roberts, G. D., N. L. Goodman, L. Heifets, H. W. Larsh, T. H. Lindner, J. K. McClatchy, M. R. McGinnis, S. H. Siddiqi, and P. Wright.** 1983. Evaluation of the BACTEC radiometric method for recovery of mycobacteria and drug susceptibility testing of *Mycobacterium tuberculosis* from acid-fast smear-positive specimens. *J. Clin. Microbiol.* **18:**689–696.

169. **Roberts, G. D., E. W. Koneman, and Y. K. Kim.** 1991. Mycobacterium, p. 304–339. *In* A. Balows, W. J. Hausler, Jr., K. L. Herrmann, H. J. Isenberg, and H. J. Shadomy (ed.), *Manual of Clinical Microbiology,* 5th ed. American Society for Microbiology, Washington, D.C.

170. **Rogall, T., J. Wolters, T. Flohr, and E. Bottger.** 1990. Towards a phylogeny and definition of species at the molecular level within the genus *Mycobacterium. Int. J. Syst. Bacteriol.* **40:**323–330.

171. **Rogers, P. L., R. E. Walker, H. C. Lane, F. G. Witebsky, J. A. Kovacs, J. E. Parrillo, and H. Masue.** 1988. Disseminated *Mycobacterium haemophilum* infection in two patients with the acquired immunodeficiency syndrome. *Am. J. Med.* **84:**640–642.

172. **Runyon, E. H.** 1959. Anonymous mycobacteria in pulmonary disease. *Med. Clin. N. Am.* **43:**273–290.

173. **Runyon, E. H.** 1970. Identification of mycobacterial pathogens using colony characteristics. *Am. J. Clin. Pathol.* **54:** 578–586.

174. **Runyon, E. H., M. J. Seline, and H. W. Harris.** 1959. Distinguishing mycobacteria by the niacin test. *Am. Rev. Tuberc. Pulm. Dis.* **79:**663–665.

175. **Saceanu, C., N. Pfeiffer, and T. McLean.** 1993. Evaluation of sputum smears concentrated by cytocentrifugation for detection of acid-fast bacilli. *J. Clin. Microbiol.* **31:**2371–2374.

176. **Sada, E. D., L. E. Ferguson, and T. M. Daniel.** 1990. An ELISA for the serodiagnosis of tuberculosis using a 30,000-Da native antigen of *M. tuberculosis. J. Infect. Dis.* **162:**928–931.

177. **Saito, H., H. Tomioka, K. Sato, H. Tasaka, and D. Dawson.** 1990. Identification of various serovar strains of *Mycobacterium avium* complex by using DNA probes specific for *Mycobacterium avium* and *Mycobacterium intracellulare. J. Clin. Microbiol.* **28:**1694–1697.

178. **Saubolle, M.** 1989. Nontuberculous mycobacteria as agents of human disease in the United States. *Clin. Microbiol. Newsl.* **11:**113–117.

179. **Saubolle, M. A., G. D. Roberts, N. L. Goodman, T. Davis, and V. Jonas.** 1993. *Abstr. 93rd Gen. Meet. Am. Soc. Microbiol.,* abstr. U-93, p. 287.

180. **Schaefer, W.** 1979. Serological identification of atypical mycobacteria, p. 323–344. *In* T. Bergan and J. R. Norris (ed.), *Methods in Microbiology.* Academic Press Ltd., London.

181. **Scherer, R., R. Sable, M. Sonnenberg, S. Cooper, P. Spencer, S. Schwimmer, F. Kocka, P. Muthuswamy, and C. Kallick.** 1986. Disseminated infection with *Mycobacterium kansasii* in the acquired immunodeficiency syndrome. *Ann. Intern. Med.* **105:**710–712.

182. **Sewell, D., A. Rashad, W. Rourke, S. Poor, J. McCarthy, and M. Pfaller.** 1993. Comparison of the Septi-Chek AFB and BACTEC systems and conventional culture for recovery of mycobacteria. *J. Clin. Microbiol.* **31:**2689–2691.

183. **Shawar, R., F. El-Zaatari, A. Nataraj, and J. Clarridge.** 1993. Detection of *Mycobacterium tuberculosis* in clinical samples by two-step polymerase chain reaction and nonisotopic hybridization methods. *J. Clin. Microbiol.* **31:**61–65.

184. **Silcox, V. A., R. A. Good, and M. M. Floyd.** 1981. Identification of clinically significant *Mycobacterium fortuitum* complex isolates. *J. Clin. Microbiol.* **14:**686–691.

185. **Small, P., N. McClenny, S. Sigh, G. Schoolnik, L. Tompkins, and P. Mickelsen.** 1993. Molecular strain typing of *Mycobacterium tuberculosis* to confirm cross-contamination in the mycobacteriology laboratory and modification of procedures to minimize occurrence of false-positive cultures. *J. Clin. Microbiol.* **31:**1677–1682.

186. **Smithwick, R. W., C. B. Stratigos, and H. L. David.** 1975. Use of cetylpyridinium chloride and sodium chloride for the decontamination of sputum specimens that are transported to the laboratory for the isolation of *Mycobacterium tuberculosis. J. Clin. Microbiol.* **1:**411–413.

187. **Snider, D.** 1982. The tuberculin skin test. *Am. Rev. Respir. Dis.* **125**(Suppl.):108–118.

188. **Snider, D., W. Jones, and R. Good.** 1984. The usefulness of phage typing *Mycobacterium tuberculosis* isolates. *Am. Rev. Respir. Dis.* **130:**1095–1099.

189. **Soini, H., M. Skurnik, K. Lippo, E. Tala, and M. Vilganen.** 1992. Detection and identification of mycobacteria by amplification of a segment of the gene coding for the 32-kilodalton protein. *J. Clin. Microbiol.* **30:**2025–2028.

190. **Sommers, H. M., J. K. McClatchy, and J. A. Monella.** 1983. *Cumitech 16, Laboratory Diagnosis of the Mycobacterioses.* Coordinating ed., H. M. Sommers and J. K. McClatchy. American Society for Microbiology, Washington, D.C.

191. **Sritharan, V., and R. Barker, Jr.** 1991. A simple method for diagnosing *M. tuberculosis* infection in clinical samples using PCR. *Mol. Cell. Probes* **5:**385–395.

192. **Stager, C., J. Libonati, S. Siddiqi, J. Davis, N. Hooper, J. Baker, and M. Carter.** 1991. Role of solid media when used in conjunction with the BACTEC system for mycobacterial isolation and identification. *J. Clin. Microbiol.* **29:**154–157.

193. **Stager, C. E., J. R. Davis, and S. H. Siddiqi.** 1988. Identification of *Mycobacterium tuberculosis* in sputum by direct inoculation of the BACTEC NAP vial. *Diagn. Microbiol. Infect. Dis.* **10:**67–73.

194. **Steadham, J. E.** 1979. Reliable urease test for identification of mycobacteria. *J. Clin. Microbiol.* **10:**134–137.

195. **Stottmeier, K. D.** 1975. Acid-fast smears and tuberculosis. *Ann. Intern. Med.* **83:**429–430.

196. **Straus, W. L., S. M. Ostroff, D. B. Jernigan, T. E. Kiehn,**

E. M. Sordillo, D. Armstrong, N. Boone, N. Schneider, J. O. Kilburn, V. A. Silcox, V. LaBombardi, and R. C. Good. 1994. Clinical and epidemiologic characteristics of *Mycobacterium haemophilum*, an emerging pathogen in immunocompromised patients. *Ann. Intern. Med.* **120**:118–125.

197. **Strauss, E., N. Wu, M. A. H. Quraishi, and S. Levine.** 1981. Clinical application of the radioimmunoassay of secretory tuberculoprotein. *Proc. Natl. Acad. Sci. USA* **78**:3214–3217.

198. **Strumpf, I. J., A. Y. Tsang, and J. W. Sayre.** 1979. Reevaluation of sputum staining for the diagnosis of pulmonary tuberculosis. *Am. Rev. Respir. Dis.* **119**:599–602.

199. **Swenson, J. M., C. Thornsberry, and V. A. Silcox.** 1982. Rapidly growing mycobacteria: testing of susceptibility to 34 antimicrobial agents by broth microdilution. *Antimicrob. Agents Chemother.* **22**:186–192.

200. **Teirstein, A., B. Damsker, P. Kirschner, D. Krellenstein, R. Robinson, and M. Chuang.** 1990. Pulmonary infection with MAI: diagnosis, clinical patterns, treatment. *Mt. Sinai J. Med.* **57**:209–215.

201. **Telenti, A., P. Imboden, F. Marchesi, D. Lowrie, S. Cole, M. Colston, L. Matter, K. Schopfer, and T. Bodmer.** 1993. Detection of rifampicin-resistance mutations in *Mycobacterium tuberculosis*. *Lancet* **341**:647–650.

202. **Telenti, A., P. Imboden, F. Marchesi, T. Schmidheini, and T. Bodmer.** 1993. Direct, automated detection of rifampin-resistant *Mycobacterium tuberculosis* by polymerase chain reaction and single-strand conformation polymorphism analysis. *Antimicrob. Agents Chemother.* **37**:2054–2058.

203. **Tenover, F., J. Crawford, R. Huebner, L. Getter, C. Horsburgh, Jr., and R. Good.** 1993. The resurgence of tuberculosis: is your laboratory ready? *J. Clin. Microbiol.* **32**:767–770.

204. **Thibert, L., and S. Lapierre.** 1993. Routine application of high-performance liquid chromatography for identification of mycobacteria. *J. Clin. Microbiol.* **31**:1759–1763.

205. **Thorel, M.-F., M. Krichevsky, and V. Levy-Frebault.** 1990. Numerical taxonomy of mycobactin-dependent mycobacteria, emended description of *Mycobacterium avium*, and description of *Mycobacterium avium* subsp. *avium* subsp. nov., *Mycobacterium avium* subsp. *paratuberculosis* subsp. nov., and *Mycobacterium avium* subsp. *silvaticum* subsp. nov. *Int. J. Syst. Bacteriol.* **40**:254–260.

206. **Tisdall, P. A., G. D. Roberts, and J. P. Anhalt.** 1979. Identification of clinical isolates of mycobacteria with gas-liquid chromatography alone. *J. Clin. Microbiol.* **10**:506–514.

207. **Toosi, Z., and J. J. Ellner.** 1992. Tuberculosis, p. 1244–1245. *In* S. L. Gorbach, J. G. Bartlett, and N. R. Blacklow (ed.), *Infectious Diseases.* The W. B. Saunders Co., Philadelphia.

208. **Trifiro, S., A.-M. Bourgault, F. Lebel, and P. Rene.** 1990. Ghost mycobacteria on Gram stain. *J. Clin. Microbiol.* **28**:146.

209. **Tsang, A., I. Drupa, M. Goldberg, J. McClatchy, and P. Brennan.** 1983. Use of serology and thin-layer chromatography for the assembly of an authenticated collection of serovars within the *Mycobacterium avium-Mycobacterium intracellulare-Mycobacterium scrofulaceum* complex. *Int. J. Syst. Bacteriol.* **33**:285–292.

210. **Turner, M., J. P. van Vooren, J. deBruyn, E. Serruysl, P. Dierckz, and J. C. Yerhault.** 1988. Humoral immune response in human tuberculosis: immunoglobulins G, A and M directed against the purified P32 protein antigen of *Mycobacterium bovis* bacillus Calmette-Guerin. *J. Clin. Microbiol.* **26**:1714–1719.

211. **van Embden, J. D. A., M. D. Cave, J. T. Crawford, J. W. Dale, K. D. Eisenach, B. Gicquel, P. Hermans, C. Martin, R. McAdam, T. M. Shinnick, and P. M. Small.** 1993. Strain identification of *Mycobacterium tuberculosis* of DNA fingerprinting: recommendations for a standardized methodology. *J. Clin. Microbiol.* **31**:406–409.

212. **Vannier, A., J. Tarrand, and P. Murray.** 1988. Mycobac-

terial cross contamination during radiometric culturing. *J. Clin. Microbiol.* **26**:1501–1505.

213. **Verbon, A., S. Kuijper, H. Jansen, P. Speelman, and A. Kolk.** 1990. Antigens in culture supernatant of *Mycobacterium tuberculosis* epitopes defined by monoclonal and human antibodies. *J. Gen. Microbiol.* **136**:955–964.

214. **Vestal, A. L., and G. P. Kubica.** 1967. Differential identification of mycobacteria. III. Use of thiacetazone, thiophen-2-carboxylic acid hydrazide and triphenyltetrazolium chloride. *Scand. J. Respir. Dis.* **48**:142–148.

215. **Viljanen, M. K., L. Olkkonen, and M. L. Katila.** 1993. Conventional identification characteristics, mycolate and fatty acid composition, and clinical significance of MAIX Accuprobe-positive isolates of *Mycobacterium avium* complex. *J. Clin. Microbiol.* **31**:1376–1378.

216. **Virtanen, S.** 1960. A study of nitrate reduction by mycobacteria. *Acta Tuberc. Scand. Suppl.* **48**:1–119.

217. **von Reyn, C. F., S. Hemmingan, S. Niemczyk, and N. J. Jacobs.** 1991. Effect of delays in processing on the survival of *Mycobacterium avium-M. intracellulare* in the Isolator blood culture system. *J. Clin. Microbiol.* **29**:1211–1214.

218. **Walker, G., M. Little, J. Nadeau, and D. Shank.** 1992. Isothermal in vitro amplification of DNA by a restriction enzyme/DNA polymerase system. *Proc. Natl. Acad. Sci. USA* **89**:392–396.

219. **Wallace, R.** 1987. Nontuberculous mycobacteria and water: a love affair with increasing clinical importance. *Infect. Dis. Clin. N. Am.* **1**:677–686.

220. **Wallace, R. J., Jr., D. R. Nash, M. Tsukamur, Z. M. Blacklock, and V. A. Silcox.** 1988. Human disease due to *Mycobacterium smegmatis*. *J. Infect. Dis.* **158**:52–59.

221. **Wallace, R. J., Jr., J. M. Swenson, V. Silcox, R. Good, J. A. Tschen, and M. S. Stone.** 1983. Spectrum of disease due to rapidly growing mycobacteria. *Rev. Infect. Dis.* **5**:657–679.

222. **Walton, D. T., and M. Valesco.** 1991. Identification of *Mycobacterium gordonae* from culture by the Gen-Probe rapid diagnostic system: evaluation of 218 isolates and potential sources of false-negative results. *J. Clin. Microbiol.* **29**:1850–1854.

223. **Wasilauskas, B., and R. Morrell, Jr.** 1994. Inhibitory effect of the Isolator blood culture system on growth of *Mycobacterium avium-M. intracellulare* in BACTEC 12B bottles. *J. Clin. Microbiol.* **32**:654–657.

224. **Watanakunakorn, C., and A. Trott.** 1973. Vertebral osteomyelitis due to *Mycobacterium kansasii*. *Am. Rev. Respir. Dis.* **107**:846–850.

225. **Wayne, L.** 1985. The "atypical" mycobacteria: recognition and disease association. *Crit. Rev. Microbiol.* **12**:185–222.

226. **Wayne, L., and G. Diaz.** 1982. Serological, taxonomic, and kinetic studies of the T and M classes of mycobacterial catalase. *Int. J. Syst. Bacteriol.* **32**:296–304.

227. **Wayne, L., R. Good, M. Krichevsky, Z. Blacklock, H. David, D. Dawson, W. Gross, J. Hawkins, V. Levy-Frebault, C. McManus, F. Portaels, S. Rusch-Gerdes, K. Schroder, V. Silcox, M. Tsukamura, L. Van den Breen, and M. Yakrus.** 1991. Fourth report of the cooperative open-ended study of slowly growing mycobacteria of the International Working Group on Mycobacterial Taxonomy. *Int. J. Syst. Bacteriol.* **41**:463–472.

228. **Wayne, L., and H. Sramek.** 1992. Agents of newly recognized or infrequently encountered mycobacterial diseases. *Clin. Microbiol. Rev.* **5**:1–25.

229. **Wayne, L. G.** 1961. Recognition of *Mycobacterium fortuitum* by means of the three-day phenolphthalein sulfatase test. *Am. J. Clin. Pathol.* **36**:185–187.

230. **Wayne, L. G.** 1974. Simple pyrazinamidase and urease tests for routine identification of mycobacteria. *Am. Rev. Respir. Dis.* **109**:147–151.

231. **Wayne, L. G., and J. R. Doubek.** 1968. Diagnostic key of mycobacteria encountered in clinical laboratories. *Appl. Microbiol.* **16**:925–931.

232. **Wayne, L. G., J. R. Doubek, and R. L. Russell.** 1964. Tests

employing Tween 80 and substrate. *Am. Rev. Respir. Dis.* **90:**588–597.

233. **Wayne, L. G., and G. P. Kubica.** 1986. Family *Mycobacteriaceae* Chester, p. 1436–1457. *In* P. H. A. Sneath, N. S. Mair, M. E. Sharpe, and J. G. Holt (ed.), *Bergey's Manual of Systematic Bacteriology*, vol. 2. The Williams & Wilkins Co., Baltimore.

234. **Welch, D., A. Guruswamy, S. Sides, C. Shaw, and M. Gilchrist.** 1993. Timely culture of mycobacteria which utilizes a microcolony method. *J. Clin. Microbiol.* **31:**2178–2184.

235. **Whittier, S., R. Hopfer, M. Knowles, and P. Gilligan.** 1993. Improved recovery of mycobacteria from respiratory secretions of patients with cystic fibrosis. *J. Clin. Microbiol.* **31:**861–864.

236. **Witebsky, F. G., J. Keiser, P. Conville, R. Bryan, C. H. Park, R. Walker, and S. H. Siddiqi.** 1988. Comparison of BACTEC 13A medium and DuPont Isolator for detection of mycobacteria. *J. Clin. Microbiol.* **256:**1501–1505.

237. **Wolinsky, E.** 1979. Nontuberculous mycobacteria and associated diseases. *Am. Rev. Respir. Dis.* **119:**107–159.

238. **Wolinsky, E.** 1992. Mycobacterial diseases other than tuberculosis. *Clin. Infect. Dis.* **15:**1–12.

239. **Woods, G., and F. Witebsky.** 1993. Current status of mycobacterial testing in clinical laboratory. Results of a questionnaire completed by participants in the College of American Pathologists mycobacteriology E survey. *Arch. Pathol. Lab. Med.* **117:**876–884.

240. **Woods, G. L., and J. A. Washington II.** 1987. Mycobacteria other than *Mycobacterium tuberculosis*: review of microbiologic and clinical aspects. *Rev. Infect. Dis.* **9:**275–294.

241. **Yagupsky, P., D. Kaminski, K. Palmer, and F. Nolte.** 1990. Cord formation of BACTEC 7H12 medium for rapid, presumptive identification of *Mycobacterium tuberculosis* complex. *J. Clin. Microbiol.* **28:**1451–1453.

242. **Yajko, D. M., P. S. Nassos, C. A. Sanders, P. C. Gonzalez, A. L. Reingold, C. R. Horsburgh, P. Hopewell, D. P. Chin, and W. K. Hadley.** 1993. Comparison of four decontamination methods for recovery of *Mycobacterium avium* complex from stools. *J. Clin. Microbiol.* **31:**302–306.

243. **Yamauchi, T., P. Ferrieri, and B. F. Anthony.** 1980. The aetiology of acute cervical adenitis in children: serological and bacteriological studies. *J. Med. Microbiol.* **13:**37–43.

244. **Yanagihara, D., V. Barr, C. Krisley, A. Tsang, J. McClatchy, and P. Brennan.** 1985. Enzyme-linked immunosorbent assay of glycolipid antigens for identification of mycobacteria. *J. Clin. Microbiol.* **21:**569–574.

245. **Young, W. D., Jr., A. Maslansky, M. S. Lefar, and D. P. Kronish.** 1970. Development of a paper strip test for detection of niacin produced by mycobacteria. *Appl. Microbiol.* **20:**939–945.

246. **Zhang, Y., B. Heym, B. Allen, D. Young, and S. Cole.** 1992. The catalase-peroxidase gene and isoniazid resistance of *Mycobacterium tuberculosis*. *Nature* (London) **358:**591–593.

Enterobacteriaceae: Introduction and Identification

J. J. FARMER III

32

In the fifth edition of this Manual, Farmer and Kelly commented that it was becoming more difficult to cover the family *Enterobacteriaceae* in a single chapter. The family includes the plague bacillus *Yersinia pestis*, the typhoid bacillus *Salmonella typhi* (48), four species that often cause diarrhea and other intestinal infections, seven species that frequently cause nosocomial infections, many other organisms that occasionally cause human or animal infections, and over 40 species that occasionally occur in human clinical specimens. In the fifth edition, all these organisms were covered in a single chapter on *Enterobacteriaceae* that was divided into six parts. In the current edition, the material is divided among three chapters. This chapter is an introduction to the family and describes the overall plan for isolation and identification; chapter 33 covers *Salmonella*, *Shigella*, *Escherichia coli*, and *Yersinia*, the four genera that include the enteric pathogens; and chapter 34 covers the remaining genera and species in the family. Because of space limitations, many topics are discussed briefly, and only a few primary literature citations are given. Several books, reviews, and chapters are recommended for more detailed information (4, 11, 12, 23, 29, 35, 64).

NOMENCLATURE AND CLASSIFICATION

The nomenclature and classification of members of the family *Enterobacteriaceae* have always been confusing (23, 29, 63, 64). Until recently, genera and species were defined by biochemical and antigenic analysis. Now, techniques such as nucleic acid hybridization and nucleic acid sequencing that measure evolutionary distance (see chapter 20 of this Manual) have better defined the relationships of all the organisms in the family (12, 29).

In this chapter, I use the nomenclature and classification used in the Enteric Bacteriology Laboratories, Centers for Disease Control and Prevention (29, 33), which may differ from other classifications (23, 64). Table 1 lists most of the genera, species, subspecies, biogroups, and unnamed Enteric Groups included in the family. Several new organisms have been added since the fifth edition of this Manual. These include *Citrobacter braakii*, *Citrobacter farmeri*, *Citrobacter sedlakii*, *Citrobacter werkmanii*, *Citrobacter youngae*, *Citrobacter* species 9, *Citrobacter* species 10, *Citrobacter* species 11 (15), *Enterobacter hormaechei* (70), *Morganella morganii* subsp. *morganii*

(60), *M. morganii* subsp. *sibonii* (60), and *Trabulsiella guamensis* (68) (Table 1).

Recently, Gavini et al. (39) proposed that a well-known species in the family, *Enterobacter agglomerans*, be moved to the genus *Pantoea*, where it would be reclassified as *Pantoea agglomerans*. Because this proposal would cause as many nomenclatural problems as it would solve and would introduce a new name that would not be familiar to most clinical microbiologists, the Enteric Bacteriology laboratories will continue to report "*Enterobacter agglomerans* group" for this heterogeneous group of organisms until a definitive classification can be proposed for the entire group. There is also a recent proposal to reclassify *Xenorhabdus luminescens* in a new genus as *Photorhabdus luminescens* (9).

Most of the newly described organisms in Table 1 are very rare in clinical specimens (29), and most clinically significant isolates belong to 20 to 25 species that have been well known for many years (23). This pattern is illustrated by the lists of organisms that most often cause bacteremia and meningitis (Table 2), nosocomial infections (Table 3), and infections of the gastrointestinal tract (Table 4). The original citations for many of the new genera (1, 10, 31, 37, 40, 41, 52–55, 58, 75, 76) and species (2, 3, 7, 8, 13–19, 21, 24–27, 30, 32, 34, 36, 38, 42–45, 47, 49–51, 56, 57, 59, 60, 66, 69, 73, 77, 79) are given in the References section.

DESCRIPTION OF THE FAMILY

Most organisms in the family *Enterobacteriaceae* share the following properties: they are gram negative and rod shaped; do not form spores; are motile by peritrichous flagella or nonmotile; grow on peptone or meat extract media without the addition of sodium chloride or other supplements, grow well on MacConkey agar, and grow both aerobically and anaerobically; ferment (rather than oxidize) D-glucose, often with gas production; are catalase positive and oxidase negative; reduce nitrate to nitrite; and have a 39 to 59% guanine-plus-cytosine (G+C) content of DNA (64).

NATURAL HABITATS

Enterobacteriaceae are widely distributed on plants and in soil, water, and the intestines of humans and animals (64).

TABLE 1 Biochemical reactions of the named species, biogroups, and Enteric Groups of the family *Enterobacteriaceae*[a]

Organism	Indole production	Methyl red	Voges-Proskauer	Citrate (Simmons')	Hydrogen sulfide (TSI)	Urea hydrolysis	Phenylalanine deaminase	Lysine decarboxylase	Arginine dihydrolase	Ornithine decarboxylase	Motility (36°C)	Gelatin hydrolysis (22°C)	Growth in KCN	Malonate utilization	D-Glucose, acid	D-Glucose, gas	Lactose fermentation	Sucrose fermentation	D-Mannitol fermentation	Dulcitol fermentation	Salicin fermentation	Adonitol fermentation	myo-Inositol fermentation	D-Sorbitol fermentation	L-Arabinose fermentation	Raffinose fermentation	L-Rhamnose fermentation	Maltose fermentation	D-Xylose fermentation	Trehalose fermentation	Cellobiose fermentation	alpha-Methyl-D-glucoside fermentation	Erythritol fermentation	Esculin hydrolysis	Melibiose fermentation	D-Arabitol fermentation	Glycerol fermentation	Mucate fermentation	Tartrate, Jordan's	Acetate utilization	Lipase (corn oil)	DNase at 25°C	Nitrate → Nitrite	Oxidase, Kovacs	ONPG test	Yellow pigment	D-Mannose fermentation
Budvicia																																															
B. aquatica[b]	0	93	0	0	0	33	0	0	0	0	27	0	0	0	100	53	87	100	60	0	0	0	0	0	80	0	100	0	93	0	0	0	0	0	100	27	0	20	27	0	0	0	100	0	93	0	0
Buttiauxella																																															
B. agrestis	0	100	0	100	0	0	0	0	0	100	100	0	80	60	100	100	100	100	100	0	100	0	0	0	100	100	100	100	100	100	100	0	0	100	100	0	60	100	60	0	0	0	100	0	100	0	100
Cedecea																																															
C. davisae[b]	0	50	50	78	0	0	0	0	67	95	95	0	86	91	100	70	19	100	100	0	99	0	0	0	100	10	0	100	100	100	100	5	0	45	0	0	0	0	44	0	91	0	100	0	90	0	100
C. lapagei[b]	0	40	80	99	0	0	0	0	80	0	95	0	100	99	100	100	60	100	100	0	100	0	0	0	100	0	0	100	100	100	100	0	0	100	0	0	0	0	75	60	100	0	100	0	99	0	100
C. neteri[b]	0	100	50	95	0	0	0	0	85	100	80	0	65	100	100	100	35	100	100	0	100	0	0	100	100	0	0	100	100	100	100	0	0	100	0	0	0	0	86	50	100	0	100	0	100	0	100
Cedecea sp. 3[b]	0	100	50	100	0	0	0	0	100	93	100	0	100	0	100	100	100	100	100	0	100	0	0	100	100	100	0	100	100	100	100	0	0	100	100	100	0	0	65	50	100	0	100	0	100	0	100
Cedecea sp. 5[b]	0	100	50	100	0	0	0	0	50	50	100	0	100	0	100	100	100	100	100	0	100	0	0	100	100	100	0	100	100	100	100	50	0	100	100	100	0	0	53	50	50	0	100	0	100	0	100
Citrobacter																																															
C. freundii[b]	33	100	0	78	78	44	0	0	67	0	89	0	89	11	100	89	78	89	100	11	0	0	0	99	100	44	100	100	89	100	44	11	0	0	100	0	100	100	100	44	0	0	100	0	89	0	100
C. diversus (koseri)[b]	99	100	0	99	0	75	0	0	80	99	95	0	0	95	100	97	50	40	100	40	15	99	0	99	99	99	99	99	99	100	99	40	0	1	99	98	99	95	96	75	0	0	100	0	97	0	100
C. amalonaticus[b]	100	100	0	95	5	85	0	0	85	95	95	0	99	1	100	98	35	9	100	0	30	0	0	99	99	5	100	99	99	100	100	2	0	5	99	0	99	96	96	86	0	0	100	0	97	0	100
C. farmeri	100	100	0	10	0	59	0	0	50	95	97	0	93	5	100	96	25	100	95	2	10	0	0	98	100	100	100	95	100	100	100	75	0	5	100	0	65	100	93	65	0	0	85	0	90	0	100
C. youngae[b]	15	100	0	75	65	80	0	0	67	5	95	0	95	5	100	75	20	20	100	85	5	0	5	100	100	10	100	100	100	100	45	0	0	0	100	5	90	100	93	53	0	0	100	0	80	0	100
C. braakii[b]	33	100	0	87	60	47	0	0	67	93	95	0	100	100	100	93	17	100	100	33	10	0	0	100	100	0	100	100	100	100	73	33	0	0	100	5	90	100	100	100	0	0	100	0	100	0	100
C. werkmanii[b]	0	100	0	100	100	100	0	0	100	0	100	0	100	100	100	100	100	0	100	0	0	0	0	100	100	0	100	100	100	100	0	0	0	0	80	0	0	100	100	83	0	0	100	0	100	0	100
C. sedlakii[b]	83	100	0	83	0	100	0	0	100	100	100	0	0	100	100	100	100	0	100	0	17	0	0	100	100	100	100	100	100	100	100	0	0	17	0	0	100	100	100	0	0	0	100	0	100	0	100
Citrobacter species 9[b]	0	100	0	0	0	0	0	0	0	0	0	0	0	100	100	100	100	100	100	100	0	0	0	100	100	0	100	100	100	100	100	0	0	0	100	0	0	0	100	0	0	0	100	0	100	0	100
Citrobacter species 10[b]	0	100	0	33	67	0	0	0	33	100	67	0	100	100	100	100	67	33	100	0	0	0	0	100	100	67	100	100	100	100	67	67	0	0	67	0	67	67	100	33	0	0	100	0	67	0	100
Citrobacter species 11[b]	100	100	0	100	67	67	0	0	67	0	100	0	100	100	100	100	67	33	100	100	33	0	0	100	100	33	100	100	100	100	100	0	0	0	33	0	100	100	100	33	0	0	100	0	100	0	100
Edwardsiella																																															
E. tarda[b]	99	100	0	1	100	0	0	100	0	100	98	0	0	0	100	50	0	0	0	0	0	0	0	0	9	0	0	100	0	0	0	0	0	0	0	0	30	0	25	0	0	0	100	0	0	0	100
E. tarda biogroup 1[b]	100	100	0	0	100	0	0	100	0	100	100	0	0	100	100	35	0	100	0	100	0	0	100	0	100	0	0	100	0	0	0	0	0	0	0	0	0	0	0	0	0	0	100	0	0	0	100
E. hoshinae[b]	50	0	0	0	0	0	0	100	0	95	100	0	100	100	100	50	0	100	100	0	50	0	0	0	0	0	0	100	0	100	0	0	0	0	0	0	30	0	0	0	0	0	100	0	0	0	100
E. ictaluri	0	100	0	0	0	0	0	100	0	65	0	0	0	0	100	0	0	0	100	0	0	0	0	0	13	0	0	100	0	0	0	0	0	0	0	0	65	0	0	0	0	0	100	0	0	0	100
Enterobacter																																															
E. aerogenes[b]	0	5	98	95	0	2	0	98	0	98	97	0	98	95	100	100	95	100	100	5	100	98	95	100	100	96	99	99	99	100	100	95	0	98	99	100	98	90	95	50	0	0	100	0	100	0	95
E. cloacae[b]	0	5	100	100	0	65	0	0	97	96	95	0	98	75	100	100	93	97	100	15	75	25	15	95	100	97	92	100	99	100	99	85	0	30	98	15	40	75	30	75	0	0	100	0	100	0	98
E. agglomerans group[b]	20	50	70	70	0	20	20	0	0	0	85	0	35	65	100	20	55	75	100	15	65	15	15	30	100	83	89	100	93	97	99	7	0	60	50	0	30	40	25	30	0	0	85	0	90	75	100
E. gergoviae[b]	0	5	100	99	0	93	20	90	0	100	90	0	0	96	100	98	100	98	100	5	99	0	0	0	100	97	99	100	99	100	100	2	0	97	97	97	100	2	97	93	0	0	99	0	97	0	100
E. sakazakii[b]	11	5	100	99	0	1	0	0	99	91	96	2	99	18	100	98	100	100	100	0	99	75	75	1	100	99	99	100	100	100	100	96	0	100	100	0	15	75	30	96	0	0	99	0	100	98	100
E. taylorae[b]	0	7	100	70	0	0	50	0	94	99	92	0	98	91	100	100	10	100	100	0	91	0	0	9	100	100	100	99	100	100	100	1	0	90	100	0	0	35	9	35	0	0	100	0	91	0	95
E. amnigenus biogroup 1[b]	0	7	100	100	0	0	0	0	9	55	100	0	100	100	100	100	70	100	100	0	100	0	0	0	100	100	100	100	100	100	100	55	0	91	100	0	0	9	9	100	0	0	100	0	100	0	100
E. amnigenus biogroup 2[b]	0	0	100	100	0	60	0	0	35	95	100	0	100	91	100	100	35	100	100	0	100	0	0	0	100	70	100	100	97	100	100	0	0	95	100	0	11	21	30	87	0	0	100	0	100	0	100
E. asburiae[b]	0	65	2	96	0	87	0	0	21	95	0	0	97	100	100	95	75	100	100	87	100	0	0	100	100	100	5	100	96	100	100	83	0	0	100	0	4	96	13	74	0	0	100	0	95	0	100
E. hormaechei[b]	0	100	100	65	0	65	0	0	78	89	52	0	100	100	100	83	9	100	100	0	44	0	0	100	100	0	100	100	100	100	100	0	0	0	100	0	100	100	100	87	0	0	100	0	100	0	100
E. intermedium	0	57	100	100	0	0	0	0	100	89	89	0	65	100	100	100	100	65	100	100	100	100	100	100	100	100	100	96	100	100	100	100	100	100	100	100	100	100	100	100	0	0	100	0	100	0	100
E. cancerogenus	0	0	100	100	0	0	0	0	100	100	100	0	100	100	100	100	0	0	100	0	100	0	0	0	100	0	100	100	100	100	0	0	0	100	100	0	0	100	0	100	0	0	100	0	100	0	100
E. dissolvens	0	0	100	100	0	0	0	0	100	100	100	0	100	100	100	100	100	100	100	0	100	0	0	100	100	100	100	100	100	100	100	100	0	100	100	0	100	100	0	0	0	0	100	0	100	0	100
E. nimipressuralis	100	100	100	100	0	100	0	100	100	100	100	0	100	100	100	100	100	100	100	0	100	0	0	100	100	100	100	100	100	100	100	100	0	100	100	0	100	100	0	0	0	0	100	0	100	0	100

(Continued on next page)

TABLE 1 Biochemical reactions of the named species, biogroups, and Enteric Groups of the family *Enterobacteriaceae*[a]   (Continued)

Organism	Indole production	Methyl red	Voges-Proskauer	Citrate (Simmons')	Hydrogen sulfide (TSI)	Urea hydrolysis	Phenylalanine deaminase	Lysine decarboxylase	Arginine dihydrolase	Ornithine decarboxylase	Motility (36°C)	Gelatin hydrolysis (22°C)	Growth in KCN	Malonate utilization	D-Glucose, acid	D-Glucose, gas	Lactose fermentation	Sucrose fermentation	D-Mannitol fermentation	Dulcitol fermentation	Salicin fermentation	Adonitol fermentation	myo-Inositol fermentation	D-Sorbitol fermentation	L-Arabinose fermentation	Raffinose fermentation	L-Rhamnose fermentation	Maltose fermentation	D-Xylose fermentation	Trehalose fermentation	Cellobiose fermentation	alpha-Methyl-D-glucoside fermentation	Erythritol fermentation	Esculin hydrolysis	Melibiose fermentation	D-Arabitol fermentation	Glycerol fermentation	Mucate fermentation	Tartrate, Jordan's	Acetate utilization	Lipase (corn oil)	DNase at 25°C	Nitrate → Nitrite	Oxidase, Kovacs	ONPG test	Yellow pigment	D-Mannose fermentation
Proteus																																															
P. mirabilis[b]	2	97	50	65	98	98	98	0	0	99	95	90	98	2	100	96	2	15	0	0	0	0	0	0	0	1	0	0	98	98	1	1	0	0	0	0	70	0	87	20	92	50	95	0	0	0	0
P. vulgaris[b]	98	95	0	15	98	95	99	0	0	0	95	95	99	0	100	85	1	97	0	0	50	0	0	0	0	1	5	0	95	30	0	60	0	50	0	0	60	0	80	25	80	80	98	0	1	0	0
P. penneri[b]	0	100	0	0	30	100	99	0	0	0	85	50	99	0	100	45	0	15	0	0	0	0	0	0	0	1	0	0	95	55	0	80	0	0	0	0	55	0	85	5	45	40	90	0	1	0	0
P. myxofaciens	0	100	0	50	0	100	100	0	0	0	100	100	100	0	100	100	0	100	0	0	50	0	0	0	0	1	0	0	100	100	0	100	0	0	0	0	100	0	100	0	100	50	100	0	0	0	0
Providencia																																															
P. rettgeri[b]	99	93	0	95	0	98	98	0	0	0	94	0	97	0	100	10	5	15	96	0	50	50	90	1	0	5	70	2	10	0	3	2	75	35	5	0	60	0	95	60	0	0	100	0	5	0	100
P. stuartii[b]	98	100	0	93	0	30	95	0	0	0	85	0	100	0	100	0	2	50	5	0	2	5	95	1	1	7	0	0	7	98	5	0	0	0	0	0	50	0	90	75	0	10	100	0	10	0	100
P. alcalifaciens[b]	99	100	0	98	0	0	99	0	0	1	96	0	100	0	100	85	0	15	100	0	1	98	0	1	1	1	0	1	1	0	0	0	0	0	0	0	15	0	50	40	0	0	100	0	0	0	100
P. rustigianii[b]	98	65	0	15	0	0	100	0	0	0	30	0	100	0	100	35	0	15	2	0	0	0	0	0	1	1	0	1	1	0	0	0	0	0	0	0	5	0	50	25	0	0	100	0	0	0	100
P. heimbachae	0	85	0	0	0	0	100	0	0	0	46	0	8	0	100	0	0	0	0	0	0	92	46	0	0	100	0	54	8	0	0	0	0	0	0	92	55	0	69	0	0	0	100	0	0	0	100
Rahnella																																															
R. aquatilis[b]	0	88	100	94	0	0	95	0	0	0	6	0	0	100	98	98	100	100	88	0	100	0	0	94	100	94	94	94	94	100	100	0	0	100	100	0	13	30	6	6	0	0	100	0	100	0	100
Salmonella																																															
DNA group 1 strains[b]																																														
Most serotypes[b]	1	100	0	95	95	1	0	98	70	97	95	0	0	0	100	96	1	1	96	90	0	0	35	95	99	2	95	97	97	99	5	2	0	5	95	95	5	90	90	90	0	2	100	0	2	0	100
S. typhi[b]	0	100	0	0	97	0	0	98	3	0	97	0	0	0	100	0	1	0	96	5	0	0	0	90	2	0	0	97	82	99	5	0	0	0	95	0	20	100	100	0	0	0	100	0	0	0	100
S. choleraesuis[b]	0	100	25	25	90	0	0	95	55	95	95	0	0	0	100	95	0	0	100	5	0	0	0	90	2	0	100	70	0	100	5	0	0	0	45	90	20	0	85	1	0	0	98	0	0	0	95
S. paratyphi A[b]	0	100	0	0	10	0	0	0	15	95	95	0	0	0	100	99	0	0	5	90	0	0	0	95	80	1	95	95	100	100	5	0	1	0	95	95	10	50	70	0	0	0	100	0	0	0	100
S. gallinarum[b]	0	100	0	0	90	0	0	90	10	1	0	0	0	0	100	0	0	0	90	90	0	0	0	95	100	10	100	5	70	50	10	1	0	0	100	0	10	50	100	0	0	0	100	0	0	0	100
S. pullorum[b]	0	90	0	0	90	0	0	100	10	95	0	0	0	0	100	90	0	0	90	0	0	0	5	90	100	10	100	5	90	90	10	0	1	0	0	90	25	0	50	0	0	0	100	0	0	0	100
DNA group 2 strains[b]	2	100	0	100	99	0	0	100	70	98	98	2	0	95	100	99	1	1	90	0	5	0	5	100	99	1	99	100	100	99	8	0	0	15	85	99	25	96	50	95	0	0	100	0	15	0	95
DNA group 3a strains[b]	1	100	0	99	99	0	0	70	70	99	99	0	0	95	100	99	15	1	100	0	5	0	0	99	99	1	99	98	100	99	1	1	0	1	95	1	10	90	50	90	0	2	100	0	100	0	100
DNA group 3b strains[b]	2	100	0	98	99	0	0	70	70	100	99	0	1	95	100	99	85	5	100	0	60	5	0	99	99	0	88	100	100	100	50	1	0	0	100	1	10	30	20	75	0	2	100	0	92	0	100
DNA group 4 strains[b]	0	100	0	94	100	2	0	100	94	100	98	0	95	100	100	94	0	5	100	0	0	5	0	100	94	0	88	100	100	100	1	0	0	0	94	5	33	88	65	70	0	2	100	0	94	0	100
DNA group 5 strains[b]	0	100	100	89	100	0	0	94	94	100	100	0	100	100	100	100	0	0	98	94	0	0	0	100	100	60	100	100	100	100	0	50	0	0	100	0	0	89	100	89	0	2	100	100	94	0	100
DNA group 6 strains[b]	0	100	100	89	100	0	0	67	67	100	100	0	0	100	100	100	22	0	67	0	0	0	0	100	100	0	100	100	94	100	0	0	0	0	89	0	33	89	100	89	0	2	100	0	44	0	100
Serratia																																															
S. marcescens[b]	1	20	98	98	0	15	0	99	0	99	97	90	95	3	100	55	2	99	99	0	95	40	75	99	0	2	0	96	7	99	5	1	1	95	0	99	95	0	75	50	98	98	98	0	95	0	99
S. marcescens biogroup 1[b]	1	100	60	30	0	0	0	55	4	65	17	30	70	0	100	10	4	100	96	0	92	30	30	92	2	0	0	70	0	100	4	0	0	96	0	92	92	0	50	4	75	82	83	0	75	0	100
S. liquefaciens group[b]	0	93	93	95	0	3	0	95	0	95	30	90	95	2	100	75	10	98	100	5	97	5	60	100	85	75	15	100	100	100	5	5	0	75	75	95	95	75	75	80	85	85	100	0	93	0	95
S. rubidaea[b]	0	20	100	95	0	2	0	55	0	0	85	90	25	94	100	30	100	99	100	0	99	99	20	1	98	99	1	99	100	100	94	1	0	94	99	20	20	70	70	80	99	85	100	0	94	0	100
S. odorifera biogroup 1[b]	60	60	50	97	0	5	0	94	0	100	94	95	60	0	100	13	70	100	100	0	95	50	100	100	100	7	95	100	100	100	100	0	7	95	100	40	40	100	100	65	35	100	100	0	100	0	100
S. odorifera biogroup 2[b]	50	60	0	100	0	0	0	0	0	100	0	60	19	0	100	40	97	100	97	0	45	55	50	65	100	94	0	94	96	100	100	0	0	81	93	50	50	65	17	55	70	20	100	0	100	0	100
S. plymuthica[b]	0	80	75	75	0	0	0	0	0	0	50	100	30	0	100	90	80	100	100	0	100	0	55	100	100	70	35	94	94	100	88	8	0	100	40	0	50	40	17	40	77	100	92	0	70	0	100
S. ficaria[b]	0	75	100	100	0	0	0	0	0	0	100	100	55	0	100	0	15	100	100	0	100	0	55	0	100	0	100	100	40	100	100	91	0	100	100	0	50	0	100	15	77	20	100	8	100	0	100
S. entomophila	0	20	100	100	0	0	0	0	0	100	100	0	70	0	100	100	0	100	100	0	100	0	30	55	100	100	76	40	40	100	6	91	0	100	40	60	88	100	58	15	20	100	100	0	44	0	100
"Serratia" fonticola[b]	0	100	9	91	0	13	0	100	97	97	91	0	0	88	100	79	97	21	91	67	100	0	30	100	100	0	76	97	85	100	6	91	0	100	89	0	33	89	100	89	0	0	100	0	100	0	100
Tatumella																																															
T. ptyseos[b]	0	0	5	2	0	0	90	0	0	0	0	0	0	0	100	0	0	98	0	0	55	0	0	0	0	11	0	0	9	93	0	0	0	0	0	0	7	0	0	0	0	0	98	0	95	0	99
Trabulsiella																																															
T. guamensis	40	100	0	100	0	0	0	50	0	100	100	0	100	100	100	100	0	0	0	0	13	0	0	0	100	0	0	0	100	100	0	0	0	40	0	0	100	0	50	88	0	0	100	0	100	0	100

(Continued on next page)

440

TABLE 1 (Continued)

Organism	Indole production	Methyl red	Voges-Proskauer	Citrate (Simmons')	Hydrogen sulfide (TSI)	Urea hydrolysis	Phenylalanine deaminase	Lysine decarboxylase	Arginine dihydrolase	Ornithine decarboxylase	Motility (36°C)	Gelatin hydrolysis (22°C)	Growth in KCN	Malonate utilization	D-Glucose, acid	D-Glucose, gas	Lactose fermentation	Sucrose fermentation	D-Mannitol fermentation	Dulcitol fermentation	Salicin fermentation	Adonitol fermentation	myo-Inositol fermentation	D-Sorbitol fermentation	L-Arabinose fermentation	Raffinose fermentation	L-Rhamnose fermentation	Maltose fermentation	D-Xylose fermentation	Trehalose fermentation	Cellobiose fermentation	alpha-Methyl-D-glucoside fermentation	Erythritol fermentation	Esculin hydrolysis	Melibiose fermentation	D-Arabitol fermentation	Glycerol fermentation	Mucate fermentation	Tartrate, Jordan's	Acetate utilization	Lipase (corn oil)	DNase at 25°C	Nitrate → Nitrite	Oxidase, Kovacs	ONPG test	Yellow pigment	D-Mannose fermentation
Escherichia-Shigella																																															
E. coli [b]	98	99	0	1	1	1	0	90	17	65	95	0	3	0	100	95	95	50	98	60	40	5	1	94	99	50	80	95	95	98	2	0	0	35	75	5	75	95	95	90	0	0	100	0	95	0	98
E. coli, inactive [b]	80	95	0	1	1	1	0	40	3	20	5	0	1	0	100	5	25	15	93	40	10	3	1	75	85	15	65	80	70	90	2	0	0	5	40	5	65	30	85	40	0	0	98	0	45	0	97
Shigella, O groups A, B, C [b]	50	100	0	0	0	0	0	0	5	1	0	0	0	0	100	0	0	0	0	2	0	0	0	30	60	50	50	30	2	80	2	0	0	0	50	0	10	0	30	2	0	0	100	0	2	0	100
S. sonnei [b]	0	100	0	0	0	0	0	0	2	98	0	0	0	0	100	0	2	1	99	0	0	0	0	2	90	3	75	90	0	96	5	0	0	0	25	0	15	10	90	96	0	0	100	0	90	0	100
E. fergusonii [b]	98	100	0	17	0	0	0	95	5	100	93	0	0	35	100	97	0	0	98	60	65	98	0	0	98	0	92	96	96	96	96	0	0	40	0	100	20	10	96	78	0	0	100	0	83	0	100
E. hermannii [b]	99	100	0	1	0	0	0	6	0	100	99	0	94	85	100	97	45	45	98	19	40	0	0	0	100	40	97	100	100	100	97	0	0	20	0	8	3	78	35	30	0	0	100	0	100	98	100
E. vulneris [b]	0	100	0	0	0	0	0	85	30	0	100	0	15	85	100	97	15	8	100	0	0	0	0	1	100	99	93	100	100	100	0	25	0	20	100	0	25	50	2	0	0	0	100	0	100	50	100
E. blattae	0	100	0	50	0	0	0	100	0	100	100	0		100	100	100	0	0	100	0	0	0	0	0	100	0	100	100	100	75	0	0	0	0	100	100	100	50	50	0	0	0	100	0	0	0	100
Ewingella																																															
E. americana [b]	0	84	95	95	0	0	0	0	0	0	60	0	5	0	100	0	70	0	100	0	80	0	0	0	0	0	0	16	13	99	10	0	0	50	0	99	24	0	35	10	0	0	97	0	85	0	99
Hafnia																																															
H. alvei [b]	0	40	85	10	0	0	0	100	6	98	85	0	95	50	100	98	5	10	99	0	13	0	0	0	95	2	97	100	98	95	15	0	0	7	0	0	95	0	70	15	0	0	100	0	90	0	100
H. alvei biogroup 1	0	85	70	0	0	0	0	100	0	45	0	0		45	100	0	0	0	55	0	55	0	0	0	0	0	0	100	70	70	0	0	0	0	0	0	0	0	30	0	0	0	100	0	30	0	100
Klebsiella																																															
K. pneumoniae [b]	10	10	98	98	0	95	0	98	0	0	0	0	98	93	100	97	98	100	99	30	100	90	95	99	100	100	99	98	99	99	100	90	2	99	99	98	97	90	95	75	0	0	99	0	99	0	99
K. oxytoca [b]	99	20	95	96	0	90	0	99	0	0	0	0	97	98	100	97	100	100	99	55	100	99	98	99	98	100	100	100	100	100	100	98	0	100	99	98	99	93	98	90	0	0	100	0	100	1	100
K. ornithinolytica [b]	100	96	70	98	0	90	0	100	0	100	0	0	100	98	100	100	100	100	100	10	100	100	95	100	100	100	100	100	100	100	100	100	0	100	100	100	100	96	100	95	0	0	100	0	100	1	100
K. planticola [b]	20	20	98	95	0	98	0	100	0	0	0	0	100	93	100	100	100	100	100	15	100	100	55	92	100	100	100	100	100	100	100	100	0	100	100	100	100	96	50	62	0	0	80	0	100	1	100
K. ozaenae [b]	0	98	0	30	0	10	0	40	6	3	0	0	88	3	100	50	30	20	100	2	97	97	65	65	90	90	55	95	95	98	92	70	0	80	97	95	65	25	50	2	0	0	80	0	80	0	100
K. rhinoscleromatis [b]	0	100	0	0	0	0	0	0	0	0	0	0	80	95	100	0	0	75	100	0	98	100	95	100	90	90	0	100	100	100	100	100	0	30	100	100	50	0	100	0	0	0	100	0	0	0	100
K. terrigena	0	60	100	40	0		0	100	0	20	0	0	100	100	100	80	100	100	100	20	100	0	80	100	100	100	100	100	100	100	100	100	0	100	100	100	100	100		20	0	0	100	0	100	0	100
Kluyvera																																															
K. ascorbata [b]	92	100	0	96	0	0	0	97	0	100	95	0	92	96	100	93	98	98	100	25	100	0	0	40	100	98	100	100	99	100	100	98	0	99	99	0	40	90	35	50	0	0	100	0	100	0	100
K. cryocrescens [b]	90	100	0	80	0	0	0	23	0	100	90	0	86	86	100	95	95	81	95	0	100	0	0	45	100	100	100	100	91	100	100	95	0	100	100	0	5	81	19	86	0	0	100	0	100	0	100
Leclercia																																															
L. adecarboxylata [b]	100	100	0	0	0	48	0	0	0	0	79	0	97	93	100	97	93	66	100	86	100	93	0	0	100	66	100	100	100	100	100	0	0	100	100	96	3	93	83	28	0	0	100	0	100	37	100
Leminorella																																															
L. grimontii [b]	0	100	0	100	100	0	0	0	0	0	0	0	0	0	100	33	0	0	0	0	0	0	0	0	0	0	0	0	83	0	0	0	0	0	0	0	17	100	100	0	0	0	100	0	0	0	0
L. richardii [b]	0	0	0	100	100	0	0	0	0	0	0	0	0	0	100	0	0	0	0	83	0	0	0	0	0	0	0	0	100	100	0	0	0	0	0	0	0	50	100	0	0	0	100	0	0	0	0
Moellerella																																															
M. wisconsensis [b]	0	100	0	80	0	0	0	0	0	100	0	0	70	60	100	60	100	100	60	0	0	100	0	0	0	0	0	30	0	0	0	0	0	0	100	75	10	0	30	10	0	0	90	0	90	0	100
Morganella																																															
M. morganii ss morganii [b]	95	95	0	0	20	95	95	1	0	95	95	0	98	1	100	90	0	0	0	0	0	0	0	0	0	0	0	0	0	0	0	0	0	0	0	0	5	0	95	0	0	0	90	0	10	0	98
M. morganii biogroup 1 [b]	100	95	0	0	15	100	100	0	0	80	80	0	90	5	100	93	0	0	0	0	0	0	0	0	0	0	0	0	0	0	0	0	0	0	0	0	100	0	100	0	0	0	90	0	20	0	100
M. morganii ss Sibonii 1 [b]	50	86	0	0	7	100	93	0	0	64	79	0	79	0	100	86	0	7	0	0	0	0	0	0	0	0	0	0	0	100	0	0	0	0	0	0	7	0	100	0	0	0	100	0	0	0	100
Obesumbacterium																																															
O. proteus biogroup 2	0	15	0	0	0	0	0	0	0	0	0	0	0	0	100	0	0	0	0	0	0	0	0	0	0	0	0	50	15	85	85	0	0	0	0	0	0	0	15	0	0	0	100	0	0	0	85
Pragia																																															
P. fontium	0	100	0	89	89	0	22	0	0	0	100	0	0	0	100	0	0	0	0	0	78	0	0	0	0	0	0	0	0	0	0	0	0	78	0	0	0	0	0	0	0	0	100	0	0	0	0

(Continued on next page)

441

TABLE 1 Biochemical reactions of the named species, biogroups, and Enteric Groups of the family *Enterobacteriaceae*[a] (Continued)

Organism	Indole production	Methyl red	Voges-Proskauer	Citrate (Simmons')	Hydrogen sulfide (TSI)	Urea hydrolysis	Phenylalanine deaminase	Lysine decarboxylase	Arginine dihydrolase	Ornithine decarboxylase	Motility (36°C)	Gelatin hydrolysis (22°C)	Growth in KCN	Malonate utilization	D-Glucose, acid	D-Glucose, gas	Lactose fermentation	Sucrose fermentation	D-Mannitol fermentation	Dulcitol fermentation	Salicin fermentation	Adonitol fermentation	myo-Inositol fermentation	D-Sorbitol fermentation	L-Arabinose fermentation	Raffinose fermentation	L-Rhamnose fermentation	Maltose fermentation	D-Xylose fermentation	Trehalose fermentation	Cellobiose fermentation	alpha-Methyl-D-glucoside fermentation	Erythritol fermentation	Esculin hydrolysis	Melibiose fermentation	D-Arabitol fermentation	Glycerol fermentation	Mucate fermentation	Tartrate, Jordan's	Acetate utilization	Lipase (corn oil)	DNase at 25°C	Nitrate → Nitrite	Oxidase, Kovacs	ONPG test	Yellow pigment	D-Mannose fermentation
Xenorhabdus																																															
X. luminescens (25°C)	50	0	0	0	0	25	0	0	0	0	0	50	0	0	75	0	0	0	0	0	0	0	0	0	0	0	0	25	0	0	0	0	0	0	0	0	0	0	50	0	0	0	0	0	0	50	100
X. luminescens DNA group 5[b]	0	0	0	50	0	60	0	0	0	0	100	80	20	0	100	0	0	0	0	0	0	0	0	0	0	0	0	0	0	0	0	0	0	0	0	0	0	0	60	20	0	0	0	0	0	60	100
X. nematophilus (25°C)	40	0	0	20	0	0	0	0	0	0	100	80	0	0	80	0	0	0	0	0	0	0	0	0	0	0	0	0	0	0	0	0	0	0	0	0	0	0	60	0	0	20	20	0	0	60	80
Yersinia																																															
Y. enterocolitica[b]	50	97	2	0	0	75	0	0	0	95	2	0	2	0	100	5	5	95	98	0	20	0	30	99	98	5	1	75	70	98	75	0	0	25	1	40	90	0	85	15	55	5	98	0	95	0	100
Y. frederiksenii[b]	100	100	2	15	0	70	0	0	0	95	5	0	0	0	100	40	40	100	100	0	92	0	20	100	100	30	99	100	100	100	100	0	0	85	100	100	85	5	55	15	55	0	100	0	100	0	100
Y. intermedia[b]	100	100	5	5	0	80	0	0	0	100	5	0	10	5	100	18	35	100	100	0	100	0	15	100	100	45	100	100	100	100	96	77	0	100	80	45	60	6	88	18	12	0	94	0	90	0	100
Y. kristensenii[b]	30	92	5	0	0	77	0	0	0	92	5	0	0	0	100	23	8	0	100	0	15	0	15	100	77	0	0	100	85	100	100	0	0	0	0	45	70	0	40	8	0	0	100	0	70	0	100
Y. rohdei[b]	0	62	0	0	0	62	0	0	0	40	0	0	0	0	100	0	0	100	100	0	0	0	0	60	60	62	0	100	38	80	25	0	0	0	50	0	38	0	100	0	0	0	88	0	50	0	100
Y. aldovae	0	80	0	0	0	60	0	0	0	80	0	0	0	0	100	0	0	0	80	0	0	0	0	100	100	0	0	100	40	100	0	0	0	0	0	0	0	0	100	0	0	0	100	0	0	0	100
Y. bercovieri[b]	0	100	0	0	0	60	0	0	0	80	0	0	0	0	100	0	20	100	100	0	20	0	0	100	100	0	0	100	60	100	100	0	0	20	0	0	20	0	100	0	0	0	100	0	80	0	100
Y. mollaretii[b]	0	100	0	0	0	20	0	0	0	80	0	0	0	0	100	0	0	100	100	0	20	0	0	100	100	0	0	100	60	100	100	0	0	0	0	0	0	0	100	0	0	0	100	0	20	0	100
Y. pestis[b]	0	100	0	0	0	5	0	0	0	0	0	0	0	0	100	0	40	0	100	0	70	0	0	50	50	15	1	60	90	100	0	0	0	50	20	0	50	0	0	0	0	0	100	0	0	0	100
Y. pseudotuberculosis[b]	0	80	0	0	0	95	0	0	5	0	0	0	0	0	100	0	0	0	100	0	25	0	0	50	5	5	70	95	90	100	0	0	0	95	70	0	50	0	50	0	0	0	85	0	70	0	100
"Yersinia" ruckeri[b]	0	97	10	0	0	0	0	5	5	100	75	30	15	0	100	5	0	0	100	0	0	0	0	0	5	0	0	95	0	95	5	0	0	0	0	0	30	0	30	30	30	0	75	0	50	0	95
Yokenella (Koserella)																																															
Y. regensburgei[b]	0	100	0	92	0	0	0	100	8	100	100	0	92	0	100	100	0	0	100	0	8	0	0	0	100	25	100	100	100	100	100	0	0	67	92	0	0	0	0	25	0	0	100	0	100	0	100
Enteric Group 58[b]	0	100	0	85	0	70	0	0	0	85	0	0	80	85	100	30	30	0	100	85	100	0	0	100	100	25	100	100	100	100	100	55	0	0	0	0	30	0	60	45	0	0	100	0	100	0	100
Enteric Group 59[b]	10	100	0	100	0	0	100	60	60	0	100	0	80	90	100	85	0	100	100	0	100	0	0	0	100	100	100	100	100	100	100	10	0	100	0	0	10	60	50	50	0	0	100	0	100	0	100
Enteric Group 60[b]	0	100	0	0	0	50	0	0	0	100	0	0	0	0	100	100	80	50	50	0	0	0	0	0	25	0	100	0	100	100	0	0	0	100	10	75	75	0	75	0	0	0	100	0	100	25	100
Enteric Group 63	0	100	0	0	0	0	100	0	50	100	65	0	0	0	100	100	0	0	100	0	100	0	100	100	100	100	0	100	100	100	100	65	0	100	100	0	0	65	50	0	0	0	100	0	100	0	100
Enteric Group 64	0	100	50	50	0	0	0	50	0	0	0	100	100	100	100	50	100	100	100	0	0	0	0	0	100	0	100	50	100	100	100	0	0	100	0	100	50	100	0	0	0	0	100	0	100	0	100
Enteric Group 68[b]	0	100	0	0	0	0	0	100	100	0	100	100	100	100	100	100	0	100	100	0	50	0	0	0	100	100	0	100	100	100	100	0	0	100	100	0	0	0	0	0	0	100	100	0	100	0	100
Enteric Group 69	0	100	100	100	0	0	0	100	100	0	100	100	100	100	100	100	100	100	100	0	100	0	0	100	100	100	0	100	100	100	100	0	0	100	100	0	0	100	0	25	0	0	100	0	100	0	100

[a] Each number gives the percentage of positive reactions after 2 days of incubation at 36°C (unless a different temperature is indicated). The vast majority of these positive reactions occur within 24 h. Reactions that become positive after 2 days are not considered. Abbreviations: TSI, triple sugar iron agar; ONPG, o-nitrophenyl-β-D-galactopyranoside.

[b] Known to occur in clinical specimens.

442

TABLE 2 Distribution of the species of *Enterobacteriaceae* and *Vibrionaceae* and other organisms in patients with bacteremia and meningitis[a]

Organism	Total no. of cases/yr[b] (no. of fatal cases)	
	Blood	Spinal fluid
Family *Enterobacteriaceae*		
Escherichia coli	5,796 (1,129)	54 (9)
Klebsiella, all species	1,260 (157)	14 (3)
K. pneumoniae	999 (129)	11 (3)
K. oxytoca	251 (25)	3
K. ozaenae	20 (4)	1
Proteus, all species	995 (123)	8 (1)
P. mirabilis	824 (92)	8 (1)
P. vulgaris	53 (10)	0
Enterobacter, all species	475 (60)	6 (1)
E. cloacae	353 (46)	5 (1)
E. aerogenes	71 (10)	2
E. agglomerans group	49 (4)	0
E. sakazakii	6 (1)	0
E. intermedium	5 (1)	0
E. gergoviae	1	0
Salmonella, all serotypes	365 (29)	4 (1)
S. typhi	103	0
S. typhimurium	60 (10)	1
S. paratyphi A	21 (1)	0
S. enteritidis	52 (10)	1
S. virchow	36 (1)	1 (1)
S. paratyphi B	12	0
S. dublin	8 (1)	0
S. panama	7 (1)	0
S. stanley	7 (1)	0
Serratia, all species	155 (16)	1
S. marcescens	93 (14)	1
S. liquefaciens group	49 (2)	0
S. odorifera	2	0
S. plymuthica	1	0
S. rubidaea	1	0
Citrobacter, all species	149 (20)	0
C. freundii group	98 (13)	0
C. diversus	35 (5)	0
Morganella morganii	124 (19)	0
Providencia, all species	30 (4)	0
P. stuartii	25 (4)	0
P. rettgeri	3	0
Yersinia enterocolitica	15 (2)	0
Yersinia pseudotuberculosis	1	0
Hafnia alvei	9 (1)	0
Shigella flexneri	2	0
Edwardsiella tarda	1	0
Family *Vibrionaceae*		
Aeromonas, all species	43 (10)	1
Plesiomonas shigelloides	1	0
Vibrio vulnificus	1	0
Other organisms[c]		
Staphylococcus aureus	3,580 (427)	50 (12)
Streptococcus pneumoniae	2,213 (338)	336 (65)
Pseudomonas aeruginosa	677 (183)	10 (2)
Haemophilus influenzae	455 (39)	510 (15)
Streptococcus spp., group B	406 (45)	71 (15)
Neisseria meningitidis	237 (38)	662 (46)

[a]Data for bacteremia (blood, bone marrow, spleen, or heart) and meningitis are from surveillance reports for England, Wales, and Ireland for 1986 and 1987 (the last 2 years these data were published). These are not reportable diseases in the United States. Data are adapted from the *Communicable Disease Report* annual tabulations for 1986 and *Communicable Disease Report* annual tabulations for 1987.

[b]Values are averages for 1986 and 1987.

[c]For comparison.

TABLE 3 Important causes of nosocomial infections in the United States[a]

Organism	No. (%) of isolates				
	Urinary tract infection	Wound	Pneumonia	Blood	Total
Enterobacteriaceae					
Escherichia coli	11,135 (26)	1,951 (10)	946 (6)	733 (6)	14,765 (16)
Enterobacter, all species	2,339 (6)	1,529 (8)	1,625 (11)	610 (5)	6,130 (7)
Klebsiella pneumoniae	2,664 (6)	618 (3)	1,042 (7)	548 (4)	4,872 (5)
Proteus mirabilis	2,312 (5)	712 (4)	503 (3)	105 (1)	3,632 (4)
Citrobacter, all species	812 (2)	321 (2)	226 (1)	82 (1)	1,441 (2)
Serratia marcescens	367 (1)	271 (1)	579 (4)	152 (1)	1,369 (2)
Other organisms[b]					
Enterococci	6,720 (16)	2,645 (13)	342 (2)	1,037 (8)	10,744 (12)
Pseudomonas aeruginosa	5,127 (12)	1,668 (8)	2,598 (17)	543 (4)	9,936 (11)
Staphylococcus aureus	823 (2)	3,439 (17)	2,401 (16)	1,984 (16)	8,647 (10)
Coagulase-negative staphycocci	1,634 (4)	2,472 (12)	293 (2)	3,384 (27)	7,783 (9)
Candida albicans	2,978 (7)	481 (2)	615 (4)	617 (5)	4,691 (5)
Streptococcus, all species	207 (0)	539 (3)	231 (1)	465 (4)	1,442 (2)
Candida, other species	853 (2)	81 (0)	109 (1)	330 (3)	1,373 (2)

[a]Based on nosocomial infection surveillance for the United States, 1986 through 1989 (74).
[b]For comparison.

Some species occupy very limited ecological niches. *Salmonella typhi* causes typhoid fever and is found only in humans (23, 48). In contrast, strains of *Klebsiella pneumoniae* are

TABLE 4 *Salmonella* and *Shigella* serotypes in the United States[a]

Serotype	No. of isolates
Salmonella typhimurium	7,894
Salmonella enteritidis	6,547
Salmonella heidelberg	2,519
Salmonella hadar	1,526
Salmonella newport	1,478
Salmonella agona	748
Salmonella thompson	689
Salmonella javiana	646
Salmonella oranienburg	595
Salmonella montevideo	558
Subtotal (10 most common serotypes)	23,200
Salmonella typhi	446
Salmonella paratyphi A	79
Salmonella paratyphi B	110
Salmonella paratyphi C	2
Salmonella choleraesuis	34
Other serotypes, or not typed	10,649
Total *Salmonella* isolates	34,520
Shigella sonnei (serogroup D)	11,094
Shigella flexneri (serogroup B)	4,015
Shigella boydii (serogroup C)	396
Shigella dysenteriae (serogroup A)	181
Shigella, not completely typed	1,573
Total *Shigella* isolates	17,259

[a]Data are from the Centers for Disease Control and Prevention's Annual *Salmonella* Surveillance Summary for 1992 and Annual *Shigella* Surveillance Summary for 1990.

distributed widely in the environment and contribute to biochemical and geochemical processes (64). However, strains of *K. pneumoniae* also cause human disease ranging from asymptomatic colonization of the intestinal, urinary, and respiratory tracts to fatal septicemia.

CLINICAL SIGNIFICANCE

Strains of *Enterobacteriaceae* are associated with abscesses, pneumonia, meningitis, septicemia, and infections of wounds, the urinary tract, and the intestine. They are a major component of the normal intestinal flora of humans but are relatively uncommon at other body sites. Several species of *Enterobacteriaceae* are very important causes of nosocomial infections (Table 3). Of clinically significant isolates, *Enterobacteriaceae* may account for 80% of gram-negative bacilli and 50% of all clinically significant isolates in clinical microbiology laboratories. They account for nearly 50% of septicemia cases (Table 2), more than 70% of urinary tract infections, and a significant percentage of intestinal infections.

Human Extraintestinal Infections

Except for the species of *Shigella*, which rarely cause infections outside the gastrointestinal tract, many species of *Enterobacteriaceae* commonly cause extraintestinal infections. However, a small number of species, including *E. coli*, *K. pneumoniae*, *Klebsiella oxytoca*, *Proteus mirabilis*, *Enterobacter aerogenes*, *Enterobacter cloacae*, and *Serratia marcescens*, account for most of these infections (Tables 2 and 3). Urinary tract infections, primarily cystitis, are the most common (74), followed by respiratory, wound, bloodstream, and central nervous system infections. Many of these infections, especially sepsis and meningitis, are life threatening, and they are often hospital acquired. Because of the severe nature of these infections, prompt isolation, identification, and susceptibility testing of *Enterobacteriaceae* isolates are essential.

Human Intestinal Infections

Several organisms in the family *Enterobacteriaceae* are also important causes of intestinal infections of humans (Table 4) and animals in the United States and worldwide. Although other species in the family have been associated with diarrhea (78) or even implicated as causes of diarrhea, only organisms in four genera (*Escherichia* [22, 28, 62, 80], *Salmonella* [20, 33, 48, 81], *Shigella* [23], and *Yersinia* [6, 61, 72]) have been clearly documented as enteric pathogens. These four genera are discussed in chapter 33 of this Manual.

SPECIMEN COLLECTION, TRANSPORT, AND PROCESSING

Extraintestinal Specimens

Enterobacteriaceae are recovered from infections at many different body sites, and normal practices (chapters 3 and 21) for collecting blood, body fluid, respiratory, wound, urine, and other specimens should be followed.

Intestinal Specimens

Stool cultures are usually submitted to the laboratory with a request to isolate and identify the cause of a possible intestinal infection, usually manifested as diarrhea. The groups of *Enterobacteriaceae* usually associated with diarrhea in the United States are essentially all strains of *Salmonella* and *Shigella* and certain pathogenic strains of *E. coli* and *Yersinia enterocolitica*. Other *Enterobacteriaceae* such as *Citrobacter*, *Hafnia*, *Morganella*, *Proteus*, *Klebsiella*, *Enterobacter*, and *Serratia* have occasionally been implicated (78). "Enterotoxin-producing strains" of these *Enterobacteriaceae* have been isolated from people with diarrhea (78), but the etiologic roles of these strains are doubtful. Some laboratories issue reports for stool cultures such as "*Klebsiella pneumoniae* in pure culture (10 of 10 colonies tested)" to reflect this drastic change in stool flora. There is no evidence that strains of these other genera are important causes of diarrhea.

Stool specimens require special attention to both collection and transportation and should be taken early in the course of illness, when the causative agent is likely to be present on primary plates in the largest numbers. At this stage, enrichment broths should be unnecessary. Except for typhoid fever, the isolation rate for enteric pathogens declines as the patient recovers. Freshly passed stool is better than rectal swabs, since there is less chance for improper collection, and mucus and blood-stained portions can be selected for culture. Stool cultures should be plated within 2 h of collection in order to recover fastidious pathogens such as *Shigella*. If rapid processing is not possible, a small portion of feces or a swab coated with feces should be placed in transport medium. Commonly used transport media include Stuart, Amies, Cary-Blair, and buffered glycerol saline. Cary-Blair is probably the best overall transport medium for diarrheal stools. Specimens held in transport media should be refrigerated until examined. In certain circumstances, such as with infants or during mass screening in outbreak investigations, rectal swabs may be useful. Mucosa within the rectal vault must be sampled for these specimens to be of optimum value. Many different procedures have been described for processing stool specimens. In cases of chronic or severe diarrhea, rare pathogens should also be considered. More information about the isolation and identification of isolates of *Salmonella*, *Shigella*, *E. coli*, and *Y. enterocolitica* is given in chapter 33.

Stool specimens should be examined visually for the presence of blood or mucus. They can also be examined microscopically for leukocytes (46). Although microscopic examination for fecal leukocytes may aid in diagnosing invasive diarrhea due to *Shigella* strains, it is not a substitute for culture and is not used routinely in most clinical laboratories (71). Definitive identification of *Enterobacteriaceae* is almost always based on biochemical reactions of pure cultures rather than on microscopic examination. Although identification by fluorescent-antibody staining is theoretically possible for all enteric pathogens, it has been of limited success because the method is difficult and there are many serologic cross-reactions among the species of *Enterobacteriaceae* (23). This technique has been limited to detection of *Salmonella* strains (primarily by the food industry) and certain serogroups of *E. coli* and to outbreak investigations.

ISOLATION

Extraintestinal Specimens

Most strains of *Enterobacteriaceae* grow readily on the plating media commonly used in clinical microbiology laboratories. Specimens from a normally sterile body site are cultured on blood or chocolate agar. Specimens such as urine, respiratory tract, or wound, which are likely to contain mixtures of organisms, are almost always cultured on selective media to enhance recovery of *Enterobacteriaceae*. MacConkey agar, generally interchangeable with eosin-methylene blue agar (EMB), is usually used because it allows a preliminary grouping of enteric and other gram-negative bacteria. The most common isolates of *Enterobacteriaceae* have a characteristic appearance on blood agar and MacConkey agar that is useful for preliminary identification (Table 5). Broth enrichment can increase the isolation rate if small numbers of *Enterobacteriaceae* are present but is not normally required.

Intestinal Specimens

Media that should be used routinely include a nonselective medium such as blood agar, a differential medium of low to moderate selectivity such as MacConkey agar, and a more selective differential medium such as XLD (xylose-lysine-deoxycholate agar) or Hektoen enteric (HE) agar. A broth enrichment such as selenite (or GN or tetrathionate) can be included, particularly if the specimen is not optimal. A highly selective medium such as brilliant green agar or bismuth sulfite agar can also be included for isolating strains of *Salmonella*. A plate of sorbitol-MacConkey agar (SMAC) can be added to enhance the isolation of Shiga-like toxin-producing strains of *Escherichia coli* O157:H7. This medium should be used if the stool is frankly bloody or if the patient has a diagnosis of hemolytic uremic syndrome and can be used for all fecal specimens if resources permit. When *Y. enterocolitica* is suspected, a selective-differential medium such as CIN (cefsulodin-irgasan-novobiocin) agar (also called *Yersinia* selective agar) can be added if resources permit. A complete stool culture procedure should also include media for isolation of *Campylobacter* and possibly *Vibrio* strains in areas where cholera and other *Vibrio* infections are common.

TABLE 5 Colonial appearances of most-common *Enterobacteriaceae* on MacConkey agar and sheep blood agar[a]

Genus or species	Appearance on and typical colony diam:	
	MacConkey agar	Sheep blood agar[b]
Salmonella and *Shigella*	Colorless, flat, 2–3 mm	Smooth, 2–3 mm
Yersinia enterocolitica	Colorless, <1 mm	Smooth, <1 mm
Escherichia coli (lactose positive)	Red, usually surrounded by precipitated bile, 2–3 mm	Smooth, 2–3 mm
Escherichia coli (lactose negative)	Colorless, 2–3 mm	Smooth, 2–3 mm
Klebsiella pneumoniae	Pink, mucoid, 3–4 mm	Mucoid, 3–4 mm
Enterobacter	Pink, not as mucoid as *Klebsiella* colonies, 2–4 mm	Smooth, 3–4 mm
Proteus vulgaris and *Proteus mirabilis*	Colorless, flat, may swarm slightly, 2–3 mm	Swarm in waves to cover plate
Other *Proteus*, *Providencia*, and *Morganella*	Colorless, flat, no swarming, 2–3 mm	Flat, 2–3 mm, no swarming

[a]Most strains appear this way, but there are exceptions.
[b]Unlike *Vibrionaceae*, most strains of *Enterobacteriaceae* are nonhemolytic (a few strains of *E. coli* produce a strong hemolysin).

IDENTIFICATION

There are many different approaches to identifying *Enterobacteriaceae* (29).

Conventional Biochemical Tests in Tubes

Tube testing was once used by clinical microbiology laboratories and is still widely used in reference and public health laboratories (23, 29). Although many laboratories prepare their own media, most are also available commercially. Growth from a single colony is inoculated into each tube, and the tests are read at 24 h and usually also at 48 h (and kept longer by reference laboratories). Unfortunately, the media and tests are not well standardized, and few laboratories use exactly the same procedures. Even with so many variables, this approach usually results in correct identification of the common species of *Enterobacteriaceae*. Table 1 gives the results for *Enterobacteriaceae* in 47 tests (for the media and methods used for this biochemical testing, see references 27 and 48 and also the fifth edition of this Manual).

Kits for Identification

A kit is defined as a series of miniaturized or standardized tests that are available commercially. This approach is similar to the conventional tube method, with the main differences being in the miniaturization, number of tests available, and method of reading results (sometimes by machine). Kits are now used by most American laboratories and are discussed in chapter 10 of this Manual.

Molecular Methods of Identification

Molecular methods have proved useful for identification to the level of family, genus, species, serotype, and strain and for differentiating pathogenic from nonpathogenic strains (chapter 20). However, few of these methods are commercially available, which has greatly limited their usefulness in clinical microbiology laboratories.

Problem Strains

Most strains of *Enterobacteriaceae* grow rapidly on plating media and on media used for biochemical identification, but occasionally, a slow-growing or fastidious strain is encountered. Some strains grow poorly on blood agar but much better on chocolate agar incubated in a candle jar. This characteristic suggests a possible nutritional requirement or a mutation involving respiration. There are slow-growing strains of *E. coli*, *K. pneumoniae*, and *Serratia marcescens*, and the typical biochemical reactions of these strains usually require extended incubation. Another type of problem organism is sometimes isolated from patients who are taking antimicrobial agents. Li et al. described such "pleiotropic" mutants of *Serratia marcescens* (67) and *Salmonella* strains. These strains react atypically in many of the standard biochemical tests and are difficult to identify. A different type of pleiotropic mutant (65) lost the ability to produce hydrogen sulfide, reduce nitrate to nitrite, and produce gas from glucose because of chlorate resistance acquired after exposure to Dakin's solution.

Laboratories occasionally isolate strains that grow rapidly but do not have biochemical reactions that fit the described species, biogroups, or Enteric Groups of *Enterobacteriaceae*. This type of culture can only be reported as "unidentified" at present. It may be an atypical strain of a described organism (Table 1), or it may belong to a new species that has not been described.

ANTIBIOTIC SUSCEPTIBILITY

Several methods are available for testing *Enterobacteriaceae* antibiotic susceptibilities, but the most popular are disk diffusion (5) and broth dilution, both described in chapter 113 of this Manual. The reader should consult a current textbook or review of infectious diseases for a description of antibiotic usage in clinical practice.

When antibiotics were first introduced, there was only slight resistance among the species of *Enterobacteriaceae*. Today, antibiotic resistance is much more common among strains isolated from humans and animals. Resistance patterns vary depending on the organism and its site of isolation. The antibiogram for individual strains is sometimes used for epidemiologic studies (chapter 16).

Intrinsic Resistance

Intrinsic resistance is a genetic property of most strains of a species and evolved long before the clinical use of antibiotics. This evolution can best be shown by studying strains isolated and stored before the antibiotic era or by studying strains from nature that presumably have not been exposed to antibiotics. For example, all strains of *Serratia marcescens* apparently have intrinsic resistance to penicillin G, colistin, and cephalothin. Table 6 lists some species of *Enterobacteriaceae* and their intrinsic resistance patterns.

TABLE 6 Intrinsic antimicrobial resistance in *Enterobacteriaceae*

Species	Drugs most strains are resistant to
Buttiauxella	Cephalothin
Cedecea spp.	Polymyxins, ampicillin, cephalothin
Citrobacter amalonaticus	Ampicillin
Citrobacter freundii	Cephalothin
Citrobacter diversus	Cephalothin, carbenicillin
Edwardsiella tarda	Colistin
Enterobacter cloacae	Cephalothin
Enterobacter aerogenes	Cephalothin
Many other *Enterobacter* spp.	Cephalothin
Escherichia hermannii	Ampicillin, carbenicillin
Ewingella americana	Cephalothin
Hafnia alvei	Cephalothin
Klebsiella pneumoniae	Ampicillin, carbenicillin
Kluyvera ascorbata	Ampicillin
Kluyvera cryocrescens	Ampicillin
Proteus mirabilis	Polymyxins, tetracycline, nitrofurantoin
Proteus vulgaris	Polymyxins, ampicillin, nitrofurantoin, tetracycline
Morganella morganii	Polymyxins, ampicillin, cephalothin
Providencia rettgeri	Polymyxins, cephalothin, nitrofurantoin, tetracycline
Other *Providencia* spp.[a]	Polymyxins, nitrofurantoin
Serratia marcescens[b]	Polymyxins, cephalothin, nitrofurantoin
Serratia fonticola	Ampicillin, carbenicillin, cephalothin
Other *Serratia* spp.	Polymyxins,[c] cephalothin

[a]Most strains of *Providencia stuartii* are also resistant to cephalothin and tetracycline.

[b]Can also be resistant to ampicillin, carbenicillin, streptomycin, and tetracycline.

[c]Resistance to polymyxins is common in *Serratia* species, but some strains have zones of 10 to 12 mm or larger.

Nonintrinsic Resistance

Much antibiotic resistance occurring today is due to the selective pressure of antibiotic usage, and many strains of *Enterobacteriaceae* that commonly cause disease have become highly antibiotic resistant through this evolutionary mechanism. Genetic studies have shown that this type of antibiotic resistance is usually plasmid mediated.

The Antibiogram as a Marker in Epidemiologic Studies

Antibiotic susceptibility testing is usually done on isolates that are clinically significant and provides an "antibiogram" useful for comparing isolates in epidemiologic studies. When the selective ecological pressure of antibiotics is changed, the resistance patterns of epidemic (or endemic) strains may also change. These changes have been documented in outbreaks that have lasted for several months or longer. Even with these limitations in stability, the antibiogram is probably the most useful and practical laboratory marker for comparing strains and recognizing infection problems.

Some of the material in this chapter was taken from or adapted from publications, including chapters on and reviews of the family Enterobacteriaceae (29, 35), by authors from the Centers for Disease Control and Prevention. I thank these authors for allowing me to use this material with a minimum of rewriting.

REFERENCES

1. **Aldová, E., O. Hausner, D. J. Brenner, Z. Kocmoud, J. Schindler, B. Potužníková, and P. Petráš.** 1988. *Pragia fontium* gen. nov., sp. nov. of the family *Enterobacteriaceae*, isolated from water. *Int. J. Syst. Bacteriol.* **38:**183–189.
2. **Aleksić, S., A. G. Steigerwalt, J. Bockemühl, G. P. Huntley-Carter, and D. J. Brenner.** 1987. *Yersinia rohdei* sp. nov. isolated from human and dog feces and surface water. *Int. J. Syst. Bacteriol.* **37:**327–332.
3. **Bagley, S. T., R. J. Seidler, and D. J. Brenner.** 1981. *Klebsiella planticola* sp. nov.: a new species of *Enterobacteriaceae* found primarily in nonclinical environments. *Curr. Microbiol.* **6:**105–109.
4. **Balows, A., H. G. Trüper, M. Dworkin, W. Harder, and K.-H. Schleifer (ed.).** 1992. *The Prokaryotes*, 2nd ed., vol. 3, p. 2673–2937. Springer-Verlag KG, Berlin.
5. **Bauer, A. W., W. M. M. Kirby, J. C. Sherris, and M. Turck.** 1966. Antibiotic susceptibility testing by a standardized single disk method. *Am. J. Clin. Pathol.* **45:**493–496.
6. **Bercovier, H., D. J. Brenner, J. Ursing, A. G. Steigerwalt, G. R. Fanning, J. M. Alonso, G. A. Carter, and H. H. Mollaret.** 1980. Characterization of *Yersinia enterocolitica sensu stricto*. *Curr. Microbiol.* **4:**201–206.
7. **Bercovier, H., A. G. Steigerwalt, A. Guiyoule, G. Huntley-Carter, and D. J. Brenner.** 1984. *Yersinia aldovae* (formerly *Yersinia enterocolitica*-like group X2): a new species of *Enterobacteriaceae* isolated from aquatic ecosystems. *Int. J. Syst. Bacteriol.* **34:**166–172.
8. **Bercovier, H., J. Ursing, D. J. Brenner, A. G. Steigerwalt, G. R. Fanning, G. P. Carter, and H. H. Mollaret.** 1980. *Yersinia kristensenii*: a new species of *Enterobacteriaceae* composed of sucrose-negative strains (formerly called *Yersinia enterocolitica* or *Yersinia enterocolitica*-like). *Curr. Microbiol.* **4:**219–224.
9. **Boemare, N. E., R. J. Akhurst, and R. G. Mourant.** 1993. DNA relatedness between *Xenorhabdus* spp. (*Enterobacteriaceae*), symbiotic bacteria of entomopathogenic nematodes, and a proposal to transfer *Xenorhabdus luminescens* to a new genus, *Photorhabdus* gen. nov. *Int. J. Syst. Bacteriol.* **43:**249–255.
10. **Bouvet, O. M. M., P. A. D. Grimont, C. Richard, E. Aldova, O. Hausner, and M. Gabrhelova.** 1985. *Budvicia acquatica* gen. nov., sp. nov.: a hydrogen sulfide-producing member of the *Enterobacteriaceae*. *Int. J. Syst. Bacteriol.* **35:**60–64.
11. **Brenner, D. J.** 1992. Additional genera of *Enterobacteriaceae*, p. 2922–2937. *In* A. Balows, H. G. Trüper, M. Dworkin, W. Harder, and K.-H. Schleifer (ed.), *The Prokaryotes*, 2nd ed. Springer-Verlag KG, Berlin.
12. **Brenner, D. J.** 1992. Introduction to the family *Enterobacteriaceae*, p. 2673–2695. *In* A. Balows, H. G. Trüper, M. Dworkin, W. Harder, and K.-H. Schleifer (ed.), *The Prokaryotes*, 2nd ed. Springer-Verlag KG, Berlin.
13. **Brenner, D. J., H. Bercovier, J. Ursing, J. M. Alonso, A. G. Steigerwalt, G. R. Fanning, G. P. Carter, and H. H. Mollaret.** 1980. *Yersinia intermedia*: a new species of *Enterobacteriaceae* composed of rhamnose-positive, melibiose-positive, raffinose-positive strains (formerly called atypical *Yersinia enterocolitica* or *Yersinia enterocolitica*-like). *Curr. Microbiol.* **4:**207–212.
14. **Brenner, D. J., B. R. Davis, A. G. Steigerwalt, C. F. Riddle, A. C. McWhorter, S. D. Allen, J. J. Farmer III, Y. Saitoh, and G. R. Fanning.** 1982. Atypical biogroups of *Escherichia coli* found in clinical specimens and description of *Escherichia hermannii* sp. nov. *J. Clin. Microbiol.* **15:**703–713.

15. Brenner, D. J., P. A. D. Grimont, A. G. Steigerwalt, G. R. Fanning, E. Ageron, and C. F. Riddle. 1993. Classification of citrobacteria by DNA hybridization: designation of *Citrobacter farmeri* sp. nov., *Citrobacter youngae* sp. nov., *Citrobacter braakii* sp. nov., *Citrobacter werkmanii* sp. nov., *Citrobacter sedlakii* sp. nov., and three unnamed *Citrobacter* genomospecies. *Int. J. Syst. Bacteriol.* **43**:645–658.

16. Brenner, D. J., A. C. McWhorter, A. Kai, A. G. Steigerwalt, and J. J. Farmer III. 1986. *Enterobacter asburiae* sp. nov., a new species found in clinical specimens, and reassignment of *Erwinia dissolvens* and *Erwinia nimipressuralis* to the genus *Enterobacter* as *Enterobacter dissolvens* comb. nov. and *Enterobacter nimipressuralis* comb. nov. *J. Clin. Microbiol.* **23**:1114–1120.

17. Brenner, D. J., A. C. McWhorter, J. K. Leete Knutson, and A. G. Steigerwalt. 1982. *Escherichia vulneris*: a new species of *Enterobacteriaceae* associated with human wounds. *J. Clin. Microbiol.* **15**:1133–1140.

18. Brenner, D. J., C. Richard, A. G. Steigerwalt, M. A. Asbury, and M. Mandel. 1980. *Enterbacter gergoviae* sp. nov.: a new species of *Enterobacteriaceae* found in clinical specimens and the environment. *Int. J. Syst. Bacteriol.* **30**:1–6.

19. Cosenza, B. J., and J. D. Podgwaite. 1966. A new species of *Proteus* isolated from larvae of the gypsy moth *Porthetria dispar* (L.). *Antonie von Leeuwenhoek J. Microbiol. Serol.* **32**:187–191.

20. Crosa, J. H., D. J. Brenner, W. H. Ewing, and S. Falkow. 1973. Molecular relationships among the *Salmonelleae*. *J. Bacteriol.* **115**:307–315.

21. Dickey, R. S., and C. H. Zumoff. 1988. Emended description of *Enterobacter cancerogenus* comb. nov. (formerly *Erwinia cancerogena*). *Int. J. Syst. Bacteriol.* **38**:371–374.

22. DuPont, H. L., S. B. Formal, R. B. Hornick, M. J. Snyder, J. P. Libonati, D. G. Sheahan, E. H. LaBrec, and J. P. Kalas. 1971. Pathogenesis of *Escherichia coli* diarrhea. *N. Engl. J. Med.* **285**:1–9.

23. Ewing, W. H. 1986. *Edwards and Ewing's Identification of Enterobacteriaceae*, 4th ed. Elsevier Science Publishing Co., New York.

24. Ewing, W. H., B. R. Davis, M. A. Fife, and E. F. Lessel. 1973. Biochemical characterization of *Serratia liquefaciens* (Grimes and Hennerty) Bascomb et al. (formerly *Enterobacter liquefaciens*) and *Serratia rubidaea* comb. nov. and designation of type and neotype strains. *Int. J. Syst. Bacteriol.* **23**:217–225.

25. Ewing, W. H., and M. A. Fife. 1972. *Enterobacter agglomerans* (Beijerinck) comb. nov. (the *herbicola-lathyri* bacteria). *Int. J. Syst. Bacteriol.* **22**:4–11.

26. Ewing, W. H., A. J. Ross, D. J. Brenner, and G. R. Fanning. 1978. *Yersinia ruckerii* sp. nov., the redmouth (RM) bacterium. *Int. J. Syst. Bacteriol.* **28**:37–44.

27. Farmer, J. J., III, M. A. Asbury, F. W. Hickman, D. J. Brenner, and The Enterobacteriaceae Study Group. 1980. *Enterobacter sakazakii*: a new species of "*Enterobacteriaceae*" isolated from clinical specimens. *Int. J. Syst. Bacteriol.* **30**:569–584.

28. Farmer, J. J., III, and B. R. Davis. 1985. H7 antiserum-sorbitol fermentation medium: a single tube screening medium for detecting *Escherichia coli* O157:H7 associated with hemorrhagic colitis. *J. Clin. Microbiol.* **22**:620–625. (Note: This paper has a misprint in the formula for MacConkey sorbitol agar. The paper says to use 22.2 g of MacConkey agar base; the correct amount is 40 g, as given in the instructions on the bottle.)

29. Farmer, J. J., III, B. R. Davis, F. W. Hickman-Brenner, A. McWhorter, G. P. Huntley-Carter, M. A. Asbury, C. Riddle, H. G. Wathen, C. Elias, G. R. Fanning, A. G. Steigerwalt, C. M. O'Hara, G. K. Morris, P. B. Smith, and D. J. Brenner. 1984. Biochemical identification of new species and biogroups of *Enterobacteriaceae* isolated from clinical specimens. *J. Clin. Microbiol.* **21**:46–76.

30. Farmer, J. J., III, G. R. Fanning, B. R. Davis, C. M. O'Hara, C. Riddle, F. W. Hickman-Brenner, M. A. Asbury, V. Lowery, and D. J. Brenner. 1984. *Escherichia fergusonii* and *Enterobacter taylorae*: two new species of *Enterobacteriaceae* isolated from clinical specimens. *J. Clin. Microbiol.* **21**:77–81.

31. Farmer, J. J., III, G. R. Fanning, G. P. Huntley-Carter, B. Holmes, F. W. Hickman, C. Richard, and D. J. Brenner. 1981. *Kluyvera*: a new (redefined) genus in the family *Enterobacteriaceae*: identification of *Kluyvera ascorbata* sp. nov. and *Kluyvera cryocrescens* sp. nov. in clinical specimens. *J. Clin. Microbiol.* **13**:919–933.

32. Farmer, J. J., III, J. H. Jorgensen, P. A. D. Grimont, R. J. Akhurst, G. O. Poinar, Jr., E. Ageron, G. V. Pierce, J. A. Smith, G. P. Carter, K. L. Wilson, and F. W. Hickman-Brenner. 1989. *Xenorhabdus luminescens* (DNA hybridization group 5) from human clinical specimens. *J. Clin. Microbiol.* **27**:1594–1600.

33. Farmer, J. J., III, A. C. McWhorter, D. J. Brenner, and G. K. Morris. 1984. The *Salmonella-Arizona* group of *Enterobacteriaceae*: nomenclature, classification and reporting. *Clin. Microbiol. Newsl.* **6**:63–66.

34. Farmer, J. J., III, N. K. Sheth, J. A. Hudzinski, H. D. Rose, and M. F. Asbury. 1982. Bacteremia due to *Cedecea neteri* sp. nov. *J. Clin. Microbiol.* **16**:775–778.

35. Farmer, J. J., III, J. G. Wells, P. M. Griffin, and I. K. Wachsmuth. 1987. *Enterobacteriaceae* infections, p. 233–296. In B. B. Wentworth (ed.), *Diagnostic Procedures for Bacterial Infections*, 7th ed. American Public Health Association, Washington, D.C.

36. Ferragut, C., D. Izard, F. Gavini, K. Kersters, J. DeLey, and H. Leclerc. 1983. *Klebsiella trevisanii*: a new species from water and soil. *Int. J. Syst. Bacteriol.* **33**:133–142.

37. Ferragut, C., D. Izard, F. Gavini, B. Lefebre, and H. Leclerc. 1981. *Buttiauxella*, a new genus of the family *Enterobacteriaceae*. *Zentralbl. Bakteriol. Parasitenkd. Infektionskr. Hyg. Abt. 1 Orig. Reihe C* **2**:33–44.

38. Gavini, F., C. Ferragut, D. Izard, P. A. Trinel, H. Leclerc, B. Lefebvre, and D. A. A. Mossel. 1979. *Serratia fonticola*, a new species from water. *Int. J. Syst. Bacteriol.* **29**:92–101.

39. Gavini, F., J. Mergaert, A. Beji, C. Mielcarek, D. Izard, K. Kersters, and J. De Ley. 1989. Transfer of *Enterobacter agglomerans* (Beijerinck 1888) Ewing and Fife 1972 to *Pantoea* gen. nov. as *Pantoea agglomerans* comb. nov. and description of *Pantoea dispersa* sp. nov. *Int. J. Syst. Bacteriol.* **39**:337–345.

40. Grimont, P. A. D., J. J. Farmer III, F. Grimont, M. A. Asbury, D. J. Brenner, and C. Deval. 1983. *Ewingella americana* gen. nov., sp. nov., a new *Enterobacteriaceae* isolated from clinical specimens. *Ann. Microbiol. (Inst. Pasteur)* **134A**:39–52.

41. Grimont, P. A. D., F. Grimont, J. J. Farmer III, and M. A. Asbury. 1981. *Cedecea davisae* gen. nov., sp. nov., and *Cedecea lapagei* sp. nov., new *Enterobacteriaceae* from clinical specimens. *Int. J. Syst. Bacteriol.* **31**:317–326.

42. Grimont, P. A. D., F. Grimont, C. Richard, B. R. Davis, A. G. Steigerwalt, and D. J. Brenner. 1978. Deoxyribonucleic acid relatedness between *Serratia plymuthica* and other *Serratia* species, with a description of *Serratia odorifera* sp. nov. (type strain: ICPB 3995). *Int. J. Syst. Bacteriol.* **28**:453–463.

43. Grimont, P. A. D., F. Grimont, C. Richard, and R. Sakazaki. 1980. *Edwardsiella hoshinae*, a new species of *Enterobacteriaceae*. *Curr. Microbiol.* **4**:347–351.

44. Grimont, P. A. D., F. Grimont, and M. P. Starr. 1979. *Serratia ficaria* sp. nov., a bacterial species associated with smyrna figs and the fig wasp *Blastophaga psenes*. *Curr. Microbiol.* **2**:277–282.

45. Grimont, P. A. D., T. A. Jackson, E. Ageron, and M. J. Noonan. 1988. *Serratia entomophila* sp. nov. associated with amber disease in the New Zealand grass grub *Costelytra zealandica*. *Int. J. Syst. Bacteriol.* **38**:1–6.

46. Harris, J. C., H. L. DuPont, and R. B. Hornick. 1971. Fecal leukocytes in diarrheal illness. *Ann. Intern. Med.* **76**:697–703.

47. Hawke, J. P., A. C. McWhorter, A. G. Steigerwalt, and D. J. Brenner. 1981. *Edwardsiella ictaluri* sp. nov., the causative agent of enteric septicemia of catfish. *Int. J. Syst. Bacteriol.* **31**:396–400.

48. Hickman, F. W., and J. J. Farmer III. 1978. *Salmonella typhi*:

identification, antibiograms, serology, and bacteriophage typing. *Am. J. Med. Technol.* **44:**1149–1159.

49. **Hickman, F. W., J. J. Farmer III, A. G. Steigerwalt, and D. J. Brenner.** 1980. Unusual groups of *Morganella* ("*Proteus*") *morganii* isolated from clinical specimens: lysine-positive and ornithine-negative biogroups. *J. Clin. Microbiol.* **12:**88–94.

50. **Hickman, F. W., A. G. Steigerwalt, J. J. Farmer III, and D. J. Brenner.** 1982. Identification of *Proteus penneri* sp. nov., formerly known as *P. vulgaris* indole negative or as *Proteus vulgaris* biogroup 1. *J. Clin. Microbiol.* **15:**1097–1102.

51. **Hickman-Brenner, F., J. J. Farmer III, A. G. Steigerwalt, and D. J. Brenner.** 1983. *Providencia rustigianii*: a new species in the family *Enterobacteriaceae* formerly known as *Providencia alcalifaciens* biogroup 3. *J. Clin. Microbiol.* **15:**1057–1060.

52. **Hickman-Brenner, F. W., G. P. Huntley-Carter, G. R. Fanning, D. J. Brenner, and J. J. Farmer III.** 1985. *Koserella trabulsii*, a new genus and species of *Enterobacteriaceae* formerly known as Enteric Group 45. *J. Clin. Microbiol.* **21:**39–42.

53. **Hickman-Brenner, F. W., G. P. Huntley-Carter, Y. Saitoh, A. G. Steigerwalt, J. J. Farmer III, and D. J. Brenner.** 1984. *Moellerella wisconsensis*, a new genus and species of *Enterobacteriaceae* found in human stool specimens. *J. Clin. Microbiol.* **19:**460–463.

54. **Hickman-Brenner, F. W., M. P. Vohra, G. P. Huntley-Carter, G. R. Fanning, V. A. Lowery III, D. J. Brenner, and J. J. Farmer III.** 1985. *Leminorella*, a new genus of *Enterobacteriaceae*: identification of *Leminorella grimontii* sp. nov. and *Leminorella richardii* sp. nov. found in clinical specimens. *J. Clin. Microbiol.* **21:**234–239.

55. **Hollis, D. G., F. W. Hickman, G. R. Fanning, J. J. Farmer III, R. E. Weaver, and D. J. Brenner.** 1981. *Tatumella ptyseos* gen. nov., sp. nov., a member of the family *Enterobacteriaceae* found in clinical specimens. *J. Clin. Microbiol.* **14:**79–88.

56. **Izard, D., C. Ferragut, F. Gavini, K. Kersters, J. DeLey, and H. Leclerc.** 1981. *Klebsiella terrigena*, a new species from soil and water. *Int. J. Syst. Bacteriol.* **31:**116–127.

57. **Izard, D., F. Gavini, and H. Leclerc.** 1980. Polynucleotide sequence relatedness and genome size among *Enterobacter intermedium* sp. nov. and the species *Enterobacter cloacae* and *Klebsiella pneumoniae*. *Zentralbl. Bakteriol. Parasitenkd. Infektionskr. Hyg. Abt. 1 Orig. Reihe C* **1:**51–60.

58. **Izard, D., F. Gavini, P. A. Trinel, and H. Leclerc.** 1979. *Rahnella aquatilis*, nouveau membre de la famille des *Enterobacteriaceae*. *Ann. Microbiol. (Inst. Pasteur)* **130A:**163–177.

59. **Izard, D., F. Gavini, P. A. Trinel, and H. Leclerc.** 1981. Deoxyribonucleic acid relatedness between *Enterobacter cloacae* and *Enterobacter amnigenus* sp. nov. *Int. J. Syst. Bacteriol.* **31:**35–42.

60. **Jensen, K. T., W. Frederiksen, F. W. Hickman-Brenner, A. G. Steigerwalt, C. F. Riddle, and D. J. Brenner.** 1992. Recognition of *Morganella* subspecies, with proposal of *Morganella morganii* subspecies *morganii* subsp. nov. and *Morganella morganii* subspecies *sibonii* subsp. nov. *Int. J. Syst. Bacteriol.* **42:**613–620.

61. **Kandolo, K., and G. Wauters.** 1985. Pyrazinamidase activity in *Yersinia enterocolitica* and related organisms. *J. Clin. Microbiol.* **21:**980–982.

62. **Karmali, M. A.** 1989. Infection by verocytotoxin-producing *Escherichia coli*. *Clin. Microbiol. Rev.* **2:**15–38.

63. **Kauffmann, F.** 1966. *The Bacteriology of Enterobacteriaceae*. The Williams and Wilkins Co., Baltimore.

64. **Krieg, N. R., and J. G. Holt (ed.).** 1984. *Bergey's Manual of Systematic Bacteriology*, vol. 1, p. 408–516. The Williams and Wilkins Co., Baltimore.

65. **Lannigan, R., and Z. Hussian.** 1993. Wound isolate of *Salmonella typhimurium* that became chlorate resistant after exposure to Dakin's solution: concomitant loss of hydrogen sulfide production, gas production, and nitrate reduction. *J. Clin. Microbiol.* **31:**2497–2498.

66. **Le Minor, L., M. Y. Popoff, B. Laurent, and D. Hermant.** 1986. Individualisation d'une septième sous-espèce de *Salmonella*: *S. choleraesuis* subsp. *indica* subsp. nov. *Ann. Microbiol. (Inst. Pasteur)* **137B:**211–217.

67. **Li, K., J. J. Farmer III, and A. Coppola.** 1974. A novel type of resistant bacteria induced by gentamicin. *Trans. N.Y. Acad. Sci.* **36:**369–396.

68. **McWhorter, A. C., R. L. Haddock, F. A. Nocon, A. G. Steigerwalt, D. J. Brenner, S. Aleksic, J. Bockemuhl, and J. J. Farmer III.** 1991. *Trabulsiella guamensis*, a new genus and species of the family *Enterobacteriaceae* that resembles *Salmonella* subgroups 4 and 5. *J. Clin. Microbiol.* **29:**1480–1485.

69. **Müller, H. E., C. M. O'Hara, G. R. Fanning, F. W. Hickman-Brenner, J. M. Swenson, and D. J. Brenner.** 1986. *Providencia heimbachae*, a new species of *Enterobacteriaceae* isolated from animals. *Int. J. Syst. Bacteriol.* **36:**252–256.

70. **O'Hara, C. M., A. G. Steigerwalt, B. C. Hill, J. J. Farmer III, G. R. Fanning, and D. J. Brenner.** 1989. *Enterobacter hormaechei*, a new species of the family *Enterobacteriaceae* formerly known as Enteric Group 75. *J. Clin. Microbiol.* **27:**2046–2049.

71. **Pickering, L. K., H. L. DuPont, J. Olarte, R. Conklin, and C. Ericsson.** 1977. Fecal leucocytes in enteric infections. *Am. J. Clin. Pathol.* **68:**562–565.

72. **Riley, G., and S. Toma.** 1989. Detection of pathogenic *Yersinia enterocolitica* by using Congo red-magnesium oxalate agar medium. *J. Clin. Microbiol.* **27:**213–214.

73. **Sakazaki, R., K. Tamura, Y. Kosako, and E. Yoshizaki.** 1989. *Klebsiella ornithinolytica* sp. nov., formerly known as ornithine-positive *Klebsiella oxytoca*. *Curr. Microbiol.* **18:**201–206.

74. **Schaberg, D. R.** 1991. Major trends in the microbial etiology of nosocomial infections. *Ann. Intern. Med.* **91**(Suppl. 3B):72S–75S.

75. **Tamura, K., R. Sakazaki, Y. Kosako, and E. Yoshizaki.** 1986. *Leclercia adecarboxylata* gen. nov., comb. nov., formerly known as *Escherichia adecarboxylata*. *Curr. Microbiol.* **13:**179–184.

76. **Thomas, G. M., and G. O. Poinar, Jr.** 1979. *Xenorhabdus* gen. nov., a genus of entomopathogenic nematophilic bacteria of the family *Enterobacteriaceae*. *Int. J. Syst. Bacteriol.* **29:**352–360.

77. **Ursing, J., D. J. Brenner, H. Bercovier, G. R. Fanning, A. G. Steigerwalt, J. Brault, and H. H. Mollaret.** 1980. *Yersinia frederiksenii*: a new species of *Enterobacteriaceae* composed of rhamnose-positive strains (formerly called atypical *Yersinia enterocolitica* or *Yersinia enterocolitica*-like). *Curr. Microbiol.* **4:**213–217.

78. **Wadstrom, T., A. Aust-Kettis, D. Habte, J. Holmgren, G. Meeuwisse, R. Mollby, and O. Soderlind.** 1976. Enterotoxin-producing bacteria and parasites in stool of Ethiopian children with diarrhoeal disease. *Arch. Dis. Child.* **51:**865–870.

79. **Wauters, G., M. Janssens, A. G. Steigerwalt, and D. J. Brenner.** 1988. *Yersinia mollaretii* sp. nov. and *Yersinia bercovieri* sp. nov., formerly called *Yersinia enterocolitica* biogroups 3A and 3B. *Int. J. Syst. Bacteriol.* **38:**424–429.

80. **Wells, J. G., B. R. Davis, I. K. Wachsmuth, L. W. Riley, R. S. Remis, R. Sokolow, and G. K. Morris.** 1983. Laboratory investigation of hemorrhagic colitis outbreaks associated with a rare *Escherichia coli* serotype. *J. Clin. Microbiol.* **18:**512–520.

81. **World Health Organization Collaborating Centre for Reference and Research on Salmonella.** 1992. *Antigenic Formulae of the Salmonella Serovars*. WHO International *Salmonella* Center, Institut Pasteur, Paris.

Escherichia, Salmonella, Shigella, and *Yersinia*

LARRY D. GRAY

33

Salmonella, Shigella, and *Yersinia* spp. and certain strains of *Escherichia coli* are enteric pathogens capable of causing severe gastroenteritis and life-threatening systemic illnesses (17).

ESCHERICHIA SPP.

E. coli is the most common bacterium isolated in clinical microbiology laboratories, the most prevalent facultative gram-negative rod in feces, the most common cause of urinary tract infection, and a common cause of both intestinal and extraintestinal infections (see chapter 32 of this Manual) (1, 2, 13, 17, 23, 27, 37).

E. coli Strains That Cause Diarrhea

E. coli is part of the healthy bowel fecal flora of both humans and lower animals; however, some strains of *E. coli* can cause severe and life-threatening diarrhea. Five distinct groups of *E. coli* cause gastrointestinal illnesses ranging from mild diarrhea to choleralike diarrhea to potentially fatal complications such as hemolytic uremic syndrome (HUS) (Table 1) (1, 2, 13, 30, 37). Enterotoxigenic *E. coli* (ETEC) produces cholera toxin-like enterotoxins that elicit profuse watery diarrhea. Enteropathogenic *E. coli* (EPEC) causes infantile diarrhea. Enteroinvasive *E. coli* (EIEC) invades intestinal epithelium and produces dysentery similar to that caused by *Shigella* spp. Enterohemorrhagic *E. coli* (EHEC) is a defined subset of Shiga-like (vero) toxin-producing *E. coli*. At least one serotype (O157:H7) of EHEC can cause hemorrhagic colitis and HUS. Enteroaggregative *E. coli* (EAggEC) is the most recently described diarrheagenic *E. coli*.

Strains of *E. coli* that cause intestinal infections resemble healthy bowel strains on common plating media and in biochemical tests (17). Unlike *Salmonella* and *Shigella* spp., pathogenic strains of *E. coli* can be lactose positive or negative. After they are subcultured, these strains can be serotyped (e.g., *E. coli* O157:H7) or tested in specific virulence assays.

ETEC

Characteristics

ETEC causes dehydrating infantile diarrhea in developing countries and traveler's diarrhea, but it is rare in the United States. ETEC can produce nausea, abdominal cramps, low fever, and a sudden-onset, profuse, watery diarrhea that is not unlike a mild case of cholera. Traveler's diarrhea caused by ETEC can be severe but is rarely fatal. ETEC produces plasmid-mediated, cell-associated heat-stable enterotoxins (ST) and heat-labile enterotoxins (LT) (13, 37).

Detection of ETEC Enterotoxins

The assays for LT and ST are relatively complicated, often involve biological systems or cell culture, and are performed primarily in reference or research laboratories. However, commercial products for the detection of ETEC LT and ST are available (Unipath [Oxoid], Ogdensburg, N.Y.). Cell culture assays for LT are described in chapter 14 of this Manual.

EPEC

Characteristics

EPEC causes enteric diseases (particularly in small children) characterized by fever, vomiting, and prominent and watery diarrhea, usually with mucus but not blood (1, 37). The first *E. coli* isolates shown to be enteric pathogens were described as such in 1955 and were named EPEC at that time. Since 1950, O55, O111, and other "enteropathogenic serotypes" have been shown to be important causes of worldwide infantile diarrhea.

Detection

EPEC can be isolated on routine enteric media. Polyvalent antisera for detecting EPEC O-antigen groups are still commercially available. In developed countries, these antisera probably do not have a place in routine laboratory procedures, but they can be used in neonatal outbreaks and cases of severe or chronic diarrhea. Five to 10 lactose-positive colonies of *E. coli* should be picked and examined for agglutination in specific antiserum. Strains that agglutinate in the serologic tests must be confirmed by titration according to the manufacturer's instructions (to eliminate cross-reactions) and should be submitted to a reference laboratory for confirmation and determination of H antigen (i.e., complete O:H typing). EPEC also has been detected by enzyme-linked immunosorbent assays (ELISA) and cell culture assays (37).

TABLE 1 Properties of *E. coli* strains that cause enteric infections

Strain	Pathogenic mechanisms	Enteric infection(s)	Common clinical presentations	Common age group	Common risk factor
ETEC	LT and ST	Diarrhea; traveler's diarrhea	Profuse watery diarrhea, cramps, nausea, dehydration	Adults, children	Foreign travel (usually Mexico)
EPEC	Adherence factor; attachment to and effacement of intestinal epithelium	Acute diarrhea	Watery diarrhea, fever, vomiting, mucus in stool	Children <2 yr old, adults	Age <2 yr
EIEC	Invasion and destruction of intestinal mucosal epithelium	Dysentery similar to *Shigella* dysentery	Dysentery; scant stool; blood, mucus, and leukocytes in stool; fever; cramps	Adults	Foreign travel (usually Mexico)
EHEC	Shiga-like toxins	Diarrhea; hemorrhagic colitis	Diarrhea (no leukocytes); abdominal cramps; blood in stool; fever, HUS, and TTP[a] may or may not be present	Children, elderly	Consumption of undercooked ground beef
EAggEC	Unknown	Chronic and acute diarrheas	Watery diarrhea, vomiting	All ages	Unknown

[a] TTP, thrombotic thrombocytopenic purpura.

EIEC

Characteristics

Like *Shigella* spp. and unlike most *E. coli* strains associated with enteritis, EIEC invades the colonic mucosa, multiplies within mucosal epithelial cells, and disrupts the epithelial cells. Also like *Shigella* spp., EIEC produces enteritis characterized by fever, abdominal cramps, watery diarrhea, or typical bacillary dysentery with leukocytes (often excreted in sheets), blood, and mucus. In most studies, EIEC appears to be relatively rare in both developing and developed countries (17).

Detection

Like *Shigella* spp., most EIEC strains are nonmotile and non- or late lactose fermenters. EIEC can be isolated on routine enteric media and can be identified by the following: Sereny test, O:H serogrouping, ELISA, or invasiveness into HEp-2 or HeLa cells. All EIEC strains are lysine decarboxylase negative, and almost all are nonmotile. The most common serotypes of EIEC are biochemically lactose positive; serotypes O152 and O124 are lactose negative. Antigenically, EIEC can cross-react with antisera to *Shigella* spp.

EHEC

Characteristics

EHEC strains are water borne and food borne. EHEC is ingested most commonly with undercooked ground beef (10, 22, 28). Beef usually is contaminated at slaughterhouses, where it contacts bovine feces. Person-to-person transmission occurs by the oral-fecal route, and appropriate infection precautions should be observed. EHEC produces two distinct lysogenic bacteriophage-encoded toxins that inhibit protein synthesis and are active against Vero and HeLa cells: Shiga-like toxin 1 (verotoxin 1) and Shiga-like toxin 2 (verotoxin 2) (22, 27). Unfortunately, simple methods and commercially available products for detecting strains that produce Shiga-like toxins do not exist; however, tests to detect strains that produce Shiga-like toxins can be performed by some reference laboratories. The Vero cell assay for Shiga-like toxins is described in chapter 14 of this Manual.

There are >50 serotypes of EHEC (17, 22). However, *E. coli* O157:H7 is the prototype EHEC. It is by far the most studied and publicized serotype, the only serotype that can be easily isolated and identified in most clinical microbiology laboratories, and the only EHEC that is a public health problem. Almost all published information regarding EHEC applies only to *E. coli* serotype O157:H7. In most geographic areas, *E. coli* O157:H7 is the most frequently isolated EHEC. Non-O157 EHEC strains are not common causes of diarrhea and HUS and have not been shown to cause bloody diarrhea or outbreaks of diarrhea in the United States or Canada (22, 28).

In many parts of North America, *E. coli* O157:H7 often is the second or third most commonly isolated enteric bacterial pathogen (usually isolated more often than *Shigella* and *Yersinia* spp.). It is usually the most common bacterial pathogen isolated from bloody stools and has been isolated from as many as 40% of all bloody stools (22). The occurrence of *E. coli* O157:H7 infections is greatest in June, July, and August in temperate climates (22).

Clinical Manifestations

After an incubation period of 3 to 5 days (range, 1 to 8 days), *E. coli* O157:H7 can cause an asymptomatic infection, mild diarrhea, or a diarrheal illness that is characterized by nonbloody (progressing to bloody) diarrhea and abdominal cramps (together known as hemorrhagic colitis), few leukocytes in stools, and lack of significant fever (10, 22). In adults, the illness usually is self-limited, lasts 5 to 8 days, and resolves without sequelae. Of patients with *E. coli* O157:H7 diarrhea, 2 to 7% develop HUS, the most important complication of EHEC infection. HUS is characterized by hemolytic anemia, thrombocytopenia, and acute renal failure (27). *E. coli* O157:H7 is the leading cause of acute renal failure in children less than 4 years old (27). Persons who develop HUS have mortality rates of 3 to 10% and severe or chronic renal, cardiac, and neurologic complication rates of 4 to 30% (10, 27).

Detection and Identification

In 1993, the Centers for Disease Control and Prevention (CDC) recommended that all laboratories culture routinely for *E. coli* O157:H7 (10, 28). Ideally, each laboratory should culture for *E. coli* O157:H7 routinely for a few months to a year to determine the prevalence of the bacterium and then decide whether routine culture for *E. coli* O157:H7 is cost-effective. This suggestion will not be practical for many laboratories. Alternatively, other surveillance plans can be implemented: only patients with bloody stools could be cultured for *E. coli* O157:H7 routinely, all stools could be cultured routinely for *E. coli* O157:H7 but only during the months with the highest incidence of the disease, etc. (28).

Stools from adults and children should be cultured within 7 and 30 days, respectively, of the onset of illness. Ideally, stool from any patient who reports having bloody diarrhea should be cultured for *E. coli* O157:H7, because stool from a patient who has had bloody diarrhea does not always appear bloody in the laboratory.

Approximately 80% of most healthy bowel fecal flora *E. coli* ferment D-sorbitol in ≤24 h (28). On the other hand, virtually all *E. coli* O157:H7 do not ferment (or ferment very slowly) D-sorbitol, a characteristic used to differentiate *E. coli* O157:H7 on MacConkey-sorbitol agar medium (SMAC). Therefore, *E. coli* colonies that do not ferment D-sorbitol (colorless colonies after 24 to 48 h at 35 to 37°C) can be screened directly from SMAC or after subculture with one of two latex agglutination tests for *E. coli* O157 antigen (Unipath [Oxoid]; Pro-Lab Diagnostics, Round Rock, Tex.) (31, 42). If SMAC is used for primary culture and selection, only three colonies need to be screened for the presence of O157 antigen.

Unlike 92 to 96% of all *E. coli*, most Shiga-like-toxin-producing isolates of *E. coli* O157:H7 do not produce β-D-glucuronidase and cannot cleave 4-methylumbelliferyl-β-D-glucuronide (MUG) to an end product that is visible under 366-nm UV light; therefore, they are MUG negative (28, 41). A MUG test is commercially available (Remel, Lenexa, Kans.).

Before any report is issued, the isolate must be proven to be *E. coli* by standard biochemical tests. Sorbitol-negative *E. coli* O157:H7 isolates do not need to be tested for toxin production, because virtually all of these isolates produce toxin.

Susceptibility to Antimicrobial Agents

There is no evidence that use of antimicrobial agents to treat *E. coli* O157:H7 disease changes the course of the disease or is beneficial in any way (21, 27).

EAggEC

The term enteroadherent *E. coli* refers to *E. coli* strains that do not produce LT or ST; are not invasive; usually are not grouped according to O:H serotypes as ETEC, EPEC, EIEC, or EHEC; and adhere to HEp-2 and HeLa cells in characteristic patterns (35, 37).

EAggEC is the best-studied enteroadherent *E. coli*. EAggEC has been associated with chronic diarrhea in many parts of the world. In children, EAggEC produces intestinal illness characterized by watery diarrhea, vomiting, dehydration, and, occasionally, abdominal pains, fever, and bloody stools.

Detection

The diagnosis of EAggEC-associated diarrhea can be accomplished only by a liquid-culture clump aggregation test, a test for adherence to cells, or a DNA probe test (4, 5, 35).

Species Other Than *E. coli*

In addition to *E. coli*, the genus *Escherichia* includes *Escherichia hermanii* (former CDC enteric group 11), *Escherichia vulneris* (former CDC enteric group 1), *Escherichia fergusonii* (former CDC enteric group 10), and *Escherichia blattae*. Although uncommon, these species occasionally are isolated from human specimens and appear to be potential pathogens (16). *Escherichia adecarboxylata/agglomerans* has been assigned to a new genus (*Leclercia adecarboxylata*).

SALMONELLA SPP.

Salmonella spp. have been isolated from humans and almost all animals throughout the world. Some serotypes of *Salmonella* are virtually species specific. For example, humans are the only known natural reservoir for serotype *Salmonella typhi* and (for all practical purposes) serotypes *Salmonella paratyphi* A, B, and C.

Salmonella spp. cause many types of infections, from mild self-limiting gastroenteritis to life-threatening typhoid fever. The most common form of *Salmonella* disease is self-limiting gastroenteritis with fever lasting less than 2 days and diarrhea lasting less than 7 days. Typhoid fever, the best-studied enteric fever, is characterized by fever, headache, diarrhea, and abdominal pain and can produce fatal respiratory, hepatic, spleen, and/or neurologic damage. Bacteremia, meningitis, respiratory disease, cardiac disease, osteomyelitis, and other local infections caused by *Salmonella* spp. have also been reported (25).

S. typhi and *S. paratyphi* A and B cause gastroenteritis, bacteremia, and enteric fever; *Salmonella choleraesuis* causes gastroenteritis and enteric fever, especially in children; and *Salmonella typhimurium* is the most frequently isolated serotype in the United States. The first four serotypes are not common in the United States. Serotypes other than the first four are most likely to cause uncomplicated gastroenteritis.

Nomenclature and Classification

Nomenclature and classification of these bacteria have changed many times and still are not stabilized (15, 17, 18). The genera *Salmonella* and (the former) *Arizona* are so closely related in evolutionary lines and degree of DNA homology that they should be considered a single genus: *Salmonella* (18). Most experts agree that there are seven distinct subgroups of *Salmonella*, each with its own phenotypic characteristics and its own historical progression of nomenclature. See the previous edition of this Manual (10) for a useful and practical tabular presentation of the characteristics of the seven subgroups of *Salmonella*. Most salmonellae (>99%) isolated in clinical microbiology laboratories belong to subgroup 1.

Serotypes

The members of each of the seven subgroups of *Salmonella* can be serotyped according to somatic O, surface Vi, phase 1 flagellar, and phase 2 flagellar antigens (17). The detailed antigenic description of each serotype is always given in a notation in which major antigenic groups are separated by colons (O antigens:Vi antigen:phase 1 antigens:phase 2 antigens). For simplicity and convenience, each of the most important serotypes usually is named for the location where it was first isolated or for something else with which it is associated (i.e., each serotype is treated and named as if it were a species). For example, *Salmonella* subgroup 1 (choleraesuis) serotype 1,4,5,12:i:1,2 can be called "*Salmonella,* subgroup 1, serotype typhimurium" or "*Salmonella,* serotype typhimurium"; however, it is usually called simply *S. typhimurium.* Experts at the CDC have stated that this simple nomenclature is widely accepted, practical, and clinically informative (18).

Most laboratories use commercially available polyvalent antisera to determine the O-antigen groups of isolates of *Salmonella* (serogroup A, B, C1, C2, D, etc.) and to provide rapid and preliminary identification (e.g., *Salmonella* serogroup B) to physicians. Preliminary serogrouping is important, because certain results of serogrouping can suggest that an isolate might be one of the five important serotypes. Specifically, the well-recognized isolates *S. paratyphi* A, *S. typhimurium, S. paratyphi* B, *S. choleraesuis,* and *S. typhi* are members of serogroups A, B, B, C, and D, respectively. Most state laboratories perform complete O:H serotyping of *Salmonella* isolates. If an isolate fails to agglutinate in polyvalent antiserum, a glass test tube containing a very heavy suspension (in saline) should be placed into boiling water for 15 to 30 min (longer times can be required) and cooled, and the suspension should be retested with the antiserum. Boiling destroys the surface Vi capsular antigen that can block agglutination.

Three commercially available latex agglutination tests can be used for the rapid detection of *Salmonella* and *Shigella* spp. after growth on solid media and in broth (Wellcolex Colour *Salmonella* and Wellcolex Colour *Shigella,* Murex, Kent, England; Bactigen *Salmonella-Shigella* Test, Wampole Laboratories, Cranbury, N.J.) (8, 19, 24, 33, 38).

Isolation

Routine selective and differential enteric agar media usually are sufficient for isolating *Salmonella* and *Shigella* spp. (12). Both a differential enteric agar medium (MacConkey or eosin-methylene blue) and a moderately selective agar medium (*Salmonella-Shigella,* xylose-lysine-deoxycholate, or Hektoen) should be used. Highly selective enteric agar media (brilliant green and bismuth sulfate) can be reserved for use only during outbreaks. Brilliant green agar is especially good for isolating *Salmonella* spp. (except *S. typhi* and *S. paratyphi*), and bismuth sulfate agar is effective for isolating *S. typhi.*

An enrichment broth (selenite or gram negative [GN]) can be added to the primary media to facilitate the recovery of small numbers of *Salmonella* and *Shigella* spp. However, enrichment broths are not cost-effective enough for routine

use except during outbreaks, when screening for carriers, and in other clinically warranted situations.

Identification

Suspicious colonies of *Salmonella* spp. can be identified by manual or commercial biochemical methods. A less expensive (but usually 1-day-longer) screening test for *Salmonella* spp. uses triple sugar iron (TSI) agar and lysine iron agar. The typical biochemical pattern on TSI is alkaline/acid, gas positive, and H_2S positive. Some strains are lactose positive and acid/acid, but these strains usually are H_2S negative. The typical pattern on lysine iron agar is alkaline/alkaline and H_2S positive; however, lysine- and H_2S-negative strains exist. *S. paratyphi* A is both lysine and H_2S negative. To reduce laboratory expenses without decreasing the quality of clinically relevant results, some laboratories no longer serogroup *Salmonella* and *Shigella* spp. but instead send isolates to reference laboratories or to local or state public health laboratories for serogrouping. Isolates that give appropriate preliminary biochemical reactions should be serogrouped with commercially available polyvalent antisera, and a preliminary report should be issued. Subsequently, the isolate must be fully identified biochemically and fully serotyped by a reference laboratory before a final report is issued.

The 4-methylumbelliferyl caprilate fluorescence disk test (Remel) for detecting C8-esterase can be used to screen stool isolates for *Salmonella* spp. False positives can be caused by other members of the family *Enterobacteriaceae* (3, 34, 36, 39).

Serodiagnosis

The Widal test for antibodies to the O antigens of *Salmonella* serotypes most likely to cause typhoid fever, usually *S. typhi* and *S. paratyphi* A and B, can be useful in helping diagnose typhoid fever when other methods have failed (40).

SHIGELLA SPP.

Shigella spp. cause classic bacillary dysentery (shigellosis), a descending intestinal illness characterized by abdominal pain, fever, large volumes of watery stools, and, 1 to 2 days later, smaller volumes of stools that often contain much blood and mucus and many leukocytes. The pathogenicity of shigellosis involves invasion and inflammation of the colonic epithelium, destruction of the superficial mucosa, sloughing of the mucosa, and production of mucosal ulcers. *Shigella* spp. rarely invade beyond the mucosa; recovery of the bacterium in blood is rare. *Shigella dysenteriae* can cause a particularly severe form of dysentery that has been reported to have fatality rates of up to 20%. Infection with *Shigella* spp. other than *S. dysenteriae* usually is self-limited and rarely fatal except in the elderly and in undernourished children.

Most cases of shigellosis are individual cases and are due to person-to-person transmission. When associated with outbreaks, the disease usually is transmitted by contaminated food and/or water.

The four named species-serogroups of *Shigella* are *S. dysenteriae* (serogroup A), *Shigella flexneri* (serogroup B), *Shigella boydii* (serogroup C), and *Shigella sonnei* (serogroup D). Serotyping *Shigella* spp. beyond the serogroup level is usually done only by reference laboratories and is extremely useful and even imperative in epidemiologic investigations.

Shigella spp. first showed resistance to sulfonamides in

the 1950s and have developed multiple resistances since that time. *Shigella* isolates became resistant to tetracycline, then to ampicillin in the 1970s and 1980s, and recently to trimethoprim-sulfamethoxazole. This development of multiple resistance is especially prominent and problematic outside the United States. However, multiply resistant *Shigella* isolates are becoming increasingly common in the United States.

Specimen, Media, and Identification

In order of decreasing productivity, rectal swab specimens, stool, and anal swab specimens are the specimens of choice (14). Streaks and collections of blood, mucus, and pus in stool specimens are extremely productive and should be cultured.

For optimal chances of isolating *Shigella* spp., a differential enteric agar medium and a moderately selective agar medium should be used (20). Xylose-lysine-deoxycholate medium is especially good for isolating *Shigella* spp.

Suspect colonies can be identified directly in many commercial identification systems. Alternatively, suspect colonies can be screened for *Shigella* spp. by subculturing the colonies onto TSI or Kligler iron agar (41). Isolates that are alkaline/acid on TSI or Kligler iron agar, H_2S negative, and gas negative can be further identified to genus and species. *Shigella* spp. other than *S. sonnei* should be sent to a state or reference laboratory for serotyping.

Occasionally, biochemically identified isolates of *Shigella* will not agglutinate in *Shigella* antisera. Suspensions of these isolates should be heated in a water bath at 100°C for 15 to 30 min and retested for agglutination. Such isolates could be EIEC.

Differentiation of *Shigella* spp. and *E. coli*

Taxonomically, *Shigella* spp. and *E. coli* are essentially the same genus and species. Their DNA relatedness is very high, they are often difficult to differentiate biochemically, and they cross-react serologically (see chapter 32 of this Manual). However, they have remained separate species for clinical reasons. The following are suggested guidelines for identifying *Shigella* spp. (17).

1. Isolates are always nonmotile and lysine negative.
2. With the exception of a few strains of a few serotypes (*S. flexneri* 6, *S. boydii* 13 and 14, and *S. dysenteriae* 3), gas is not produced during carbohydrate fermentation.
3. Isolates that ferment mucate or are alkaline on acetate or Christensen citrate agar are likely to be *E. coli*.
4. *S. boydii* and *S. dysenteriae* are very rare in the United States, and isolates identified as one of these should be retested or confirmed in some way before a final report is issued.

YERSINIA SPP.

All 11 species of *Yersinia* are well established members of the family *Enterobacteriaceae*, and all have been isolated from clinical specimens. At least three species are unquestionably human pathogens: *Yersinia pestis*, *Yersinia enterocolitica*, and *Yersinia pseudotuberculosis*. Yersinioses are zoonotic infections that usually affect rodents, small animals, and birds; humans are accidental hosts. Both *Y. pestis* and *Y. pseudotuberculosis* are rarely isolated in the United States.

Y. pestis

Y. pestis causes both urban and sylvatic plague. Rats are the natural reservoirs for *Y. pestis* in urban plague (the devas-tating "city" disease of the Middle Ages), and small animals such as ground squirrels, field and wood rats, rabbits, and domestic cats are the natural reservoirs in sylvatic plague (the "country" plague as it exists today in many countries, including the United States). Humans become infected with the bacterium when they are bitten by fleas of natural reservoirs, when they handle infected animals or tissues, and, much less commonly, when they inhale aerosolized *Y. pestis* generated by a person with pneumonic plague (12). The last reported case of human-to-human transmission of plague in the United States occurred in 1924.

Infections due to *Y. pestis* are rare in the United States; most (89%) of the cases have been reported from New Mexico, Arizona, California, and Colorado (12). From 1970 to 1991, there were 295 reported cases of indigenous plague in the United States (high year, 1983 [40 cases]).

Y. pestis is not a fastidious bacterium; it grows well on blood agar and many other enteric media. However, after 24 h on blood agar and many enteric agars, the colonies are only pinpoint, much smaller than those of other *Enterobacteriaceae*. After 48 h, the colonies are 1 to 1.5 mm in diameter and gray-white, and they can appear slightly mucoid.

Y. pestis is inactive in routine biochemical tests but has a typical pattern that closely resembles that of *Y. pseudotuberculoisis* (see chapter 32 of this Manual). Broth cultures of *Y. pestis* have a characteristic "stalactite pattern" in which clumps of cells adhere to the side of the tube and settle to the bottom if the tube is disturbed. Many commercial identification systems do not include *Y. pestis* in their data banks; therefore, *Y. pestis* must be suspected from its appearance on primary media and from clinical and epidemiologic information. Isolates suspected of being *Y. pestis* should be reported immediately by telephone to the state health department, which will request that the isolate be sent to a specialized reference laboratory. In the United States, isolates should be sent to Plague Section, Bacterial Zoonoses Branch, Centers for Disease Control and Prevention, P.O. Box 2087, Fort Collins, CO 80522; telephone, (303) 221-6450.

Y. enterocolitica

In the last several years, *Y. enterocolitica* has been isolated from many kinds of clinical and nonclinical specimens in many countries, particularly in Europe, Scandinavia, and Canada. Serotype O:3 is reported to be an increasingly significant enteric pathogen in the United States, especially in California, New York City, and some pediatric populations (6, 7, 29, 32). The natural reservoirs of the bacterium are many kinds of animals, especially pigs, rodents, livestock, and rabbits. *Y. enterocolitica* is transmitted by ingestion of contaminated food (often milk and pork) and water, probably by the fecal-oral route, and perhaps by contact with infected animals.

Clinical Manifestations

Y. enterocolitica is a significant and invasive enteric pathogen that causes several well-recognized diseases, especially in younger persons, and several uncommon postinfection syndromes (11). The most common diseases caused by the bacterium are (hemorrhagic) enterocolitis, terminal ileitis, mesenteric lymphadenitis (pseudoappendicular syndrome), septicemia, and focal infections in many extraintestinal sites.

Enterocolitis caused by *Y. enterocolitica* is characterized by diarrhea, low fever, and abdominal pain. Leukocytes and, often, blood can be present in stools. The illness usually is self-limiting, but many complications have been reported (11). Patients (commonly teenagers) with pseudoappendicular syndrome usually present with fever, right lower quadrant tenderness, and abdominal pain. The syndrome is clinically similar to and often misdiagnosed as appendicitis but is actually mesenteric lymphadenitis and terminal ileitis. *Y. enterocolitica* can be cultured from the lymph nodes and ileum. Many cases of fatal bacteremia and septic shock have been associated with *Y. enterocolitica*-contaminated blood transfusion products (9, 26).

Biogroups and Serotypes

Most isolates can be serotyped as one of >50 serotypes according to somatic O antigens. Only certain biotypes and serotypes of *Y. enterocolitica* are generally considered enteric and invasive pathogens. In the United States, serotypes O:8 and O:5,27 cause mesenteric lymphadenitis and the most invasive forms of *Y. enterocolitica* disease.

Isolation

Although strains of *Y. enterocolitica* grow faster at 37 than at 25°C, the lower temperature is recommended for primary isolation. Strains of *Y. enterocolitica* grow well on MacConkey agar incubated at 37°C, but the colonies are much smaller than those of other species of *Enterobacteriaceae*. Cefsulodin-irgasan-novobiocin (CIN) agar is specifically designed to isolate *Yersinia* spp. Cultures on CIN agar are incubated at 32°C for 24 h or at 22 to 25°C for 48 h. After 18 to 24 h on CIN agar, colonies of *Y. enterocolitica* are translucent without dark red centers; by 48 h, the colonies are dark pink with translucent borders and are occasionally surrounded by a zone of precipitated bile. Both pathogenic and nonpathogenic strains of *Y. enterocolitica* and other *Yersinia* species grow on CIN agar, but most other bacteria (except *Citrobacter* spp.) are inhibited. Strains of *Y. enterocolitica* usually are lactose negative, but lactose-positive strains exist.

If a dedicated medium for *Yersinia* spp. is not used, cultures on MacConkey agar can be examined after 24 h at 35 to 37°C for small colorless colonies that become much larger after an additional 24 h of incubation at room temperature.

Many laboratories culture for *Y. enterocolitica* only on request, because the routine use of special selective media for *Y. enterocolitica* (CIN agar) is a low-yield procedure and is not cost-effective.

Identification

On TSI agar, most isolates of *Y. enterocolitica* typically produce an acid/acid reaction with no gas or H_2S. Some strains produce an alkaline reaction because of slow sucrose fermentation or the production of alkaline products from peptones in the medium. The typical acid/acid reaction on TSI agar is the same as that of some *Enterobacteriaceae* such as *E. coli*; therefore, *Y. enterocolitica* can be missed if TSI agar is used to screen for *Salmonella* and *Shigella* spp. On Kligler's iron agar, *Y. enterocolitica* will produce an alkaline/acid reaction similar to that of *Salmonella* and *Shigella* spp. Most isolates of *Y. enterocolitica* are urea positive (often at 24 h; sometimes several days are required) and motile at 25 but not 37°C.

Serodiagnosis

In culture-negative cases of suspected *Y. enterocolitica* or *Y. pseudotuberculosis* disease, the results of serologic assays (mi-crohemagglutination, complement fixation, and enzyme immunoassay) for antibody to these bacteria might be helpful. Such assays are available in some large commercial clinical laboratories.

Y. pseudotuberculosis

Y. pseudotuberculosis is biochemically similar to *Y. pestis*, causes illnesses (usually in persons 5 to 15 years old) similar to those caused by *Y. enterocolitica*, and is isolated (usually from blood) only rarely in the United States. The disease caused by this bacterium is a zoonosis. The natural reservoirs of the bacterium are rodents, wild animals, and game birds throughout the world.

Other Species of *Yersinia*

The other eight species of *Yersinia* (*Yersinia frederiksenii*, *Yersinia intermedia*, *Yersinia kristensenii*, *Yersinia aldovae*, *Yersinia bercovieri*, *Yersinia mollaretti*, *Yersinia rohdei*, and "*Yersinia ruckeri*") are also found in both intestinal and extraintestinal specimens and are biochemically similar to each other. These species can grow at 4°C and on CIN agar, and they can multiply in refrigerated foods. The pathogenic potential of these eight species is controversial, but at this time, isolates should not be disregarded.

REFERENCES

1. **Abbott, S. L., and J. M. Janda.** 1992. Bacterial gastroenteritis. I. Incidence and etiological agents. *Clin. Microbiol. Newsl.* **14:**17–21.
2. **Abbott, S. L., and J. M. Janda.** 1992. Bacterial gastroenteritis. II. Pathogenesis and laboratory identification. *Clin. Microbiol. Newsl.* **14:**25–28.
3. **Aguirre, P. M., J. B. Cacho, L. Folgurira, M. Lopez, J. Garcia, and A. C. Valasco.** 1990. Rapid fluorescence method for screening *Salmonella* spp. from enteric differential media. *J. Clin. Microbiol.* **28:**148–149.
4. **Albert, M. J., F. Quadri, A. Hague, and N. A. Bhuiyan.** 1993. Bacterial clump formation at the surface of liquid culture as a rapid test for identification of enteroaggregative *Escherichia coli. J. Clin. Microbiol.* **31:**1397–1399.
5. **Baudry, B., S. J. Saravino, P. Vial, J. B. Kaper, and M. M. Levine.** 1990. A sensitive and specific DNA probe to identify enteroaggregative *Escherichia coli*, a recently discovered diarrhea pathogen. *J. Infect. Dis.* **161:**1249–1257.
6. **Bisset, J., L. C. Powers, S. L. Abbott, and J. M. Janda.** 1990. Epidemiologic investigations of *Yersinia enterocolitica* and related species: sources, frequency, and serogroup distribution. *J. Clin. Microbiol.* **28:**910–912.
7. **Bottone, E. J., C. R. Gallans, and M. F. Sierra.** 1987. Disease spectrum of *Yersinia enterocolitica* serogroup O:3, the predominant source of human infection in New York City. *Contrib. Microbiol. Immunol.* **9:**50–60.
8. **Bouvet, P. J. M., and S. Jeanjean.** 1992. Evaluation of two colored latex kits, the Wellcolex Colour *Salmonella* test and the Wellcolex Colour *Shigella* test, for serological grouping of *Salmonella* and *Shigella* species. *J. Clin. Microbiol.* **30:**2184–2186.
9. **Centers for Disease Control.** 1991. Update. *Yersinia enterocolitica* bacteremia and endotoxin shock associated with red blood cell transfusions—United States, 1991. *Morbid. Mortal. Weekly Rep.* **40:**176–178.
10. **Centers for Disease Control and Prevention.** 1993. Update: Multistate outbreak of *Escherichia coli* O157:H7 infections from hamburgers—western United States, 1993. *Morbid. Mortal. Weekly Rep.* **42:**257–253.
11. **Cover, T. L., and R. C. Aber.** 1989. *Yersinia enterocolitica. N. Engl. J. Med.* **321:**16–24.
12. **Crook, L. D., and B. Tempest.** 1992. Plague. A review of 27 cases. *Arch. Intern. Med.* **152:**1253–1256.

13. **Eisenstein, B. I.** 1990. Enterobacteriaceae, p. 1658–1673. In G. L. Mandell, R. G. Douglas, and J. E. Bennett (ed.), *Principles and Practice of Infectious Diseases*. Churchill Livingstone, New York.

14. **Escheverria, P., O. Sethabuts, and C. Pitarangsi.** 1991. Microbiology and diagnosis of infections with *Shigella* and enteroinvasive *Escherichia coli*. *Rev. Infect. Dis.* **13**(Suppl. 4): S220–S225.

15. **Ewing, W. H.** 1986. *Edwards and Ewing's Identification of Enterobacteriaceae*, 4th ed. Elsevier Science Publishing Co., New York.

16. **Farmer, J. J., III, B. R. Davis, F. W. Hickman-Brenner, A. McWhorter, G. P. Huntley-Carter, M. A. Asbury, C. Riddle, H. G. Wathen, C. Elias, G. R. Fanning, A. G. Steigerwalt, C. M. O'Hara, G. K. Morris, P. B. Smith, and D. J. Brenner.** 1985. Biochemical identification of new species and biogroups of *Enterobacteriaceae* isolated from clinical specimens. *J. Clin. Microbiol.* **21**:46–76.

17. **Farmer, J. J., III, and M. T. Kelly.** 1990. Enterobacteriaceae, p. 360–383. In A. Balows, W. J. Hausler, Jr., K. L. Herrmann, H. D. Isenberg, and H. J. Shadomy (ed.), *Manual of Clinical Microbiology*, 5th ed. American Society for Microbiology, Washington, D.C.

18. **Farmer, J. J., III, A. C. McWhorter, D. J. Brenner, and G. D. Morris.** 1984. The *Salmonella-Arizona* group of *Enterobacteriaceae*: nomenclature, classification and reporting. *Clin. Microbiol. Newsl.* **6**:63–66.

19. **Fedorko, D. P., S. M. Lehman, P. K. W. Yu, J. J. Germer, and J. P. Anhalt.** 1989. Increased efficiency of stool culture for the detection of *Salmonella* and *Shigella*. *Diagn. Microbiol. Infect. Dis.* **12**:463–466.

20. **Grasmick, A.** 1992. Processing and interpretation of bacterial fecal cultures, p. 1.10.1–1.10.25. In H. D. Isenberg (ed.), *Clinical Microbiology Procedures Handbook*. American Society for Microbiology, Washington, D.C.

21. **Griffin, P. M., S. M. Sotroff, R. V. Tauxe, K. D. Green, J. G. Wells, J. H. Lewis, and P. A. Blake.** 1988. Illnesses associated with *Escherichia coli* O157:H7 infections. *Ann. Intern. Med.* **109**:705–712.

22. **Griffin, P. M., and R. V. Tauxe.** 1991. The epidemiology of infections caused by *Escherichia coli* O157:H7, other enterohemorrhagic *E. coli*, and associated hemolytic uremic syndrome. *Epidemiol. Rev.* **13**:60–98.

23. **Guerrant, R. L.** 1990. Principals and syndromes of enteric infections, p. 837–851. In G. L. Mandell, R. G. Douglas, and J. E. Bennett (ed.), *Principles and Practice of Infectious Diseases*. Churchill Livingstone, New York.

24. **Hadfield, S. G., A. Lane, and M. B. McIllmurray.** 1987. A novel colored latex test for the detection and identification of more than one antigen. *J. Immunol. Methods* **97**:153–158.

25. **Hook, E. W.** 1990. *Salmonella* species (including typhoid fever), p. 1700–1715. In G. L. Mandell, R. G. Douglas, and J. E. Bennett (ed.), *Principles and Practice of Infectious Diseases*. Churchill Livingstone, New York.

26. **Jacobs, J., D. Jamaer, J. Vandevan, M. Wouters, C. Vermylen, and J. Vandepitte.** 1989. *Yersinia enterocolitica* in donor blood: a case report and review. *J. Clin. Microbiol.* **27**:1119–1121.

27. **Karmali, M. A.** 1989. Infection by verocytotoxin-producing *Escherichia coli*. *Clin. Microbiol. Rev.* **2**:15–38.

28. **Kay, B. A., P. M. Griffin, N. A. Strockbine, and J. G. Wells.** 1994. Too fast food: bloody diarrhea and death from *Escherichia coli* O157:H7. *Clin. Microbiol. Newsl.* **16**:17–19.

29. **Lee, L. A., J. Taylor, G. P. Carter, B. Quinn, J. J. Farmer III, R. V. Tauxe, and the *Yersinia enterocolitica* Collaborative Study Group.** 1990. *Yersinia enterocolitica* O:3: an emerging cause of pediatric gastroenteritis in the United States. *J. Infect. Dis.* **163**:660–663.

30. **Levine, M. M.** 1987. *Escherichia coli* that cause diarrhea: enterotoxigenic, enteropathogenic, enteroinvasive, enterohemorrhagic, and enteroadherent. *J. Infect. Dis.* **155**:377–389.

31. **March, S. B., and S. Ratnam.** 1989. Latex agglutination test for detection of *Escherichia coli* serotype O157:H7. *J. Clin. Microbiol.* **27**:1675–1677.

32. **Metchock, B., D. L. Lonsway, G. P. Carter, L. A. Lee, and J. E. McGowan, Jr.** 1991. *Yersinia enterocolitica*: a frequent seasonal stool isolate from children at an urban hospital in the southeast United States. *J. Clin. Microbiol.* **29**:2868–2869.

33. **Metzler, J., and I. Nachamkin.** 1988. Evaluation of a LA test for detection of *Salmonella* and *Shigella* spp. by using broth enrichment. *J. Clin. Microbiol.* **26**:2501–2504.

34. **Munoz, P., M. D. Diaz, E. Cercenado, J. Rodriguez-Creixems, J. Berenguer, and E. Bouza.** 1993. Rapid screening of *Salmonella* species from stool cultures. *Am. J. Clin. Pathol.* **100**:404–406.

35. **Nataro, J. P., J. B. Kaper, R. Robins-Browne, V. Prado, P. Vial, and M. M. Levine.** 1987. Patterns of adherence of diarrheagenic *Escherichia coli* to HEp-2 cells. *Pediatr. Infect. Dis. J.* **6**:829–831.

36. **Olsson, M., A. Syk, and R. Wollin.** 1992. Identification of salmonellae with 4-methylumbelliferyl caprylate fluorescence test. *J. Clin. Microbiol.* **29**:2631–2632.

37. **Raj, P.** 1993. Pathogenesis and laboratory diagnosis of *Escherichia coli*-associated enterocolitis. *Clin. Microbiol. Newsl.* **15**: 89–93.

38. **Rohner, P., S. Dharan, and R. Auckenthaler.** 1992. Evaluation of the Wellcolex Colour *Salmonella* test for detection of *Salmonella* in enrichment broths. *J. Clin. Microbiol.* **30**:3274–3276.

39. **Ruiz, J., M. A. Sempere, M. C. Varela, and J. Gomez.** 1992. Modification of the methodology of stool culture for *Salmonella* detection. *J. Clin. Microbiol.* **30**:525–526.

40. **Sack, R. B., and D. A. Sack.** 1992. Immunologic methods for the diagnosis of infections by *Enterobacteriaceae* and *Vibrionaceae*, p. 482–488. In N. R. Rose, E. C. de Macario, J. L. Fahey, H. Friedman, and G. M. Penn (ed.), *Manual of Clinical Laboratory Immunology*, 4th ed. American Society for Microbiology, Washington, D.C.

41. **Shigei, J.** 1992. Test methods used in the identification of commonly isolated aerobic gram-negative bacteria, p. 1.19.1–1.19.111. In H. D. Isenberg (ed.), *Clinical Microbiology Procedures Handbook*. American Society for Microbiology, Washington, D.C.

42. **Sowers, E. G., and N. A. Strockbine.** 1991. A comparison of two commercial agglutination tests for identification of *Escherichia coli* O157:H7, abstr. C-281, p. 389. *91st Annu. Meet. Am. Soc. Microbiol. 1991.*

Enterobacteriaceae: Opportunistic Pathogens and Other Genera

MARY J. R. GILCHRIST

34

In the family *Enterobacteriaceae*, the genera *Salmonella*, *Shigella*, *Escherichia*, and *Yersinia* hold the greatest historical significance for human disease, especially in association with enteric illness. These genera are described in chapter 33 of this Manual. The balance of the genera, the subject of this chapter, are introduced in chapter 32. The more prominent of these other genera (*Citrobacter*, *Enterobacter*, *Klebsiella*, *Morganella*, *Proteus*, *Providencia*, and *Serratia*) have long been known to be involved in human infections, e.g., of the urinary tract, and are increasingly being associated with nosocomial infections at a variety of sites. Among these genera has arisen a subset of organisms that possess novel antibiotic resistance mechanisms, including inducible or extended-spectrum β-lactamases, aminoglycoside-modifying enzymes, and the like. Antibiotic-resistant strains of *Klebsiella pneumoniae* or *Enterobacter* spp. now colonize many intensive care units, and the infection control community is seeking means to diminish their impact on morbidity and mortality of hospitalized patients. The less prominent genera are newly isolated or renamed (*Budvicia*, *Buttiauxella*, *Cedecea*, *Edwardsiella*, *Hafnia*, *Kluyvera*, *Leclercia*, *Leminorella*, *Moellerella*, *Pantoea*, *Pragia*, *Rahnella*, *Tatumella*, *Trabulsiella*, *Xenorhabdus*, and *Yokenella*). The newer genera are not necessarily "new" but merely newly recognized, a result of our increased ability to accurately identify differences in organisms (see chapter 20 of this Manual). In many cases, the clinical significance of these organisms is as yet unknown. As the taxonomically significant biochemical properties of these organisms are increasingly incorporated into the databases of the automated bacterial identification systems that are employed in routine, primary clinical laboratories (Table 1), we will come to understand their association with disease or colonization. For the most part, when these organisms are associated with human diseases, it is as opportunistic pathogens.

The biochemical properties, differential reactions, and unique characteristics of the other genera of *Enterobacteriaceae* are given in Table 1 of chapter 32. To simplify matters, some generalizations can be made about these organisms. As members of the *Enterobacteriaceae*, they are oxidase negative and catalase positive, except that some *Xenorhabdus* species are catalase negative. When motile, they have peritrichous flagella, with the exception of *Tatumella*, which elaborates lateral flagella. With the exception of *Pantoea* spp., *Klebsiella ozaenae*, *Xenorhabdus* spp.,

and *Erwinia* spp., they reduce nitrate to nitrite. In addition to several H_2S^+ *Proteus* species, *Budvicia* spp., some *Edwardsiella tarda* strains, *Leminorella* spp., *Pragia* spp., and some *Citrobacter* spp. produce hydrogen sulfide.

Some of these species can be recognized by their unique colony morphology on blood or nutrient agar. Swarming is a well-recognized characteristic of the genus *Proteus*. Swarming should not be a sole identifying characteristic, however, since other organisms swarm. New information (53a) based on virulence properties of fresh isolates from animal spleens suggests that *Salmonella* species are particularly pathogenic when they present as swarming colonies. It would be unfortunate to miss such a *Salmonella* strain in a laboratory that employs a screening test protocol, e.g., for swarming and indole, for the identification of *Proteus* spp. Growth on carbohydrate-rich media allows some species to produce capsular material, which leads to mucoid colonies. While *K. pneumoniae* is most recognized for this characteristic, other species of *Klebsiella* and *Enterobacter* are also positive. A number of species produce pigmented colonies (Table 2). A gram-negative bacillus that grows as a yellow colony may be a potential *Enterobacter sakazakii* or *Enterobacter agglomerans*; a positive oxidase test instead suggests *Flavobacterium* species, not a member of the *Enterobacteriaceae*. This distinction is particularly relevant in cases of neonatal sepsis and meningitis, in which all of these organisms have been reported. With the exception of luminescence in *Xenorhabdus* spp., a characteristic that is not readily recognized, the balance of the species of other *Enterobacteriaceae* have undistinguished colony morphology and must be recognized by multiple biochemical tests. Table 1 lists the commercial identification systems that currently include these organisms in their databases. Individuals seeking extensive tabular information on the distinguishing features of these organisms are encouraged to consult chapter 32 of this Manual, the 44 tables in *Bergey's Manual of Determinative Bacteriology* (43), and other classic sources of information (15, 24).

BUDVICIA SP.

Described in 1985 (6), the genus *Budvicia* contains a single species, *Budvicia aquatica*. Most of the strains have been isolated from water, but a few have been isolated from human feces. *B. aquatica* produces hydrogen sulfide.

TABLE 1 Inclusion of unusual *Enterobacteriaceae* in the databases of commercially available identification systems

Organism	Inclusion in database of:				
	API 20E	VITEK GNI	MicroScan Conventional Panel	Biolog	MIDI
Budvicia aquatica				×	
Buttiauxella agrestis[a]				×	
Cedecea davisae	×	×	×	×	×
Cedecea lapagei	×	×	×	×	×
Cedecea neteri	×		×	×	×
Citrobacter amalonaticus[b]	×	×	×	×	×
Citrobacter freundii group[b]	×	×	×	×	×
Citrobacter diversus (*C. koseri*)[b]	×	×	×	×	×
Edwardsiella hoshinae	×			×	×
Edwardsiella ictaluri				×	×
Edwardsiella tarda	×	×	×	×	×
Enterobacter aerogenes	×	×	×	×	×
Enterobacter agglomerans	×	×	×	×	×
Enterobacter amnigenus		×	×	×	×
Enterobacter asburiae	×	×	×	×	×
Enterobacter cancerogenus[a]					×
Enterobacter cloacae	×	×	×	×	×
Enterobacter gergoviae	×	×	×	×	×
Enterobacter hormaechei		×		×	
Enterobacter intermedium(s)[a]	×	×	×	×	×
Enterobacter sakazakii	×	×	×	×	×
Enterobacter taylorae	×	×	×	×	×
Erwinia species				8 species	10 species
Ewingella americana	×	×	×	×	
Hafnia alvei	×	×	×	×	×
Klebsiella ornithinolytica	—[c]		×	×	
Klebsiella oxytoca	×	×	×	×	×
Klebsiella ozaenae	×	×	×	—[d]	—[d]
Klebsiella planticola				×	×
Klebsiella pneumoniae	×	×	×	×	×
Klebsiella rhinoscleromatis	×	×	×	—[d]	—[d]
Klebsiella terrigena				×	×
Klebsiella trevisanii					×[e]
Kluyvera ascorbata	—[f]	—[f]	×	×	×
Kluyvera cryocrescens	—[f]	—[f]	×	×	×
Koserella (see *Yokenella*)					
Leclercia adecarboxylata	×	×	×	×	×
Leminorella grimontii			×	×	×
Leminorella richardii			×	×	×
Moellerella wisconsensis	×		×	×	
Morganella morganii	×	×	×	×	×
Pantoea agglomerans	×[g]	×[g]	×[g]	×[g]	×[g]
Pantoea dispersa				×	
Pragia fontium[a]				×	
Proteus mirabilis	×	×	×	×	×
Proteus myxofaciens[a]				×	×
Proteus penneri	×	×	×	×	×
Proteus vulgaris	×	×	×	×	×
Providencia alcalifaciens	×	×	×	×	×
Providencia heimbachae[a]				×	

(Continued on next page)

TABLE 1 (*Continued*)

Organism	Inclusion in database of:				
	API 20E	VITEK GNI	MicroScan Conventional Panel	Biolog	MIDI
Providencia rettgeri	×	×	×	×	×
Providencia rustigianii			×	×	×
Providencia stuartii	×	×	×	×	×
Rahnella aquatilis				×	×
Serratia entomophila[a]				×	
Serratia ficaria				×	
Serratia fonticola		×	×	×	×
Serratia grimesii				—[h]	×
Serratia liquefaciens	×	×	×	×	×
Serratia marcescens	×	×	×	×	×
Serratia odorifera	×	×	×	×	×
Serratia plymuthica	×	×	×	×	×
Serratia proteomaculans				×	×
Serratia rubidaea	×	×	×	×	×
Tatumella ptyseos			×	×	
Trabulsiella guamensis				×	×
Xenorhabdus luminescens				×	×
Xenorhabdus nematophilus				×	×
Yokenella regensburgii	—[i]		×	×	

[a]Isolates are from environmental or animal (nonhuman) sources.
[b]Proposed new species: *Citrobacter farmeri* sp. nov., *Citrobacter youngae* sp. nov., *Citrobacter braakii* sp. nov., *Citrobacter werkmanii* sp. nov., *Citrobacter sedlakii* sp. nov., *Citrobacter* genomospecies 9 through 11.
[c]Listed as *Klebsiella* group 47.
[d]Listed as a subspecies of *K. pneumoniae*.
[e]Synonym of *K. planticola*.
[f]Listed as "species."
[g]Listed as *E. agglomerans*.
[h]A member of the *S. liquefaciens* group.
[i]Listed as *Koserella* spp.

TABLE 2 Colony color of pigmented species of other *Enterobacteriaceae*[a]

Organism	Colony color (temp of incubation [°C])	% Pigmented if <100%
Enterobacter agglomerans	Yellow	76–89
Enterobacter sakazakii	Yellow	
Erwinia ananas	Yellow (27)	
Erwinia percicinus	Red (27)	
Erwinia stewartii	Yellow (27)	90–100
Erwinia uredovora	Yellow (27)	90–100
Escherichia hermannii	Yellow	90–100
Escherichia vulneris	Yellow	26–75
Leclercia adecarboxylata	Yellow	11–25
Pantoea agglomerans	Yellow	90–100
Pantoea dispersa	Yellow	26–75
Serratia marcescens	Red	26–75
Serratia plymuthica	Red	26–75
Serratia rubidaea	Red	90–100
Xenorhabdus beddingii	Brown (25)	90–100
Xenorhabdus bovienii	Yellow (25)	90–100
Xenorhabdus luminescens	Yellow, orange, or red (25)	90–100
Xenorhabdus nematophilus	Yellow (25)	26–75
Xenorhabdus poinarii	Brown (25)	90–100

[a]Compiled from reference 43.

BUTTIAUXELLA SP.

Described in 1981 (26), the genus *Buttiauxella* includes a single species, *Buttiauxella agrestis*. Biochemically similar to *Kluyvera* spp. and to enteric groups 63 and 64, the known strains of *Buttiauxella agrestis* have been isolated from environmental sources, especially water.

CEDECEA SPP.

Described in 1981, the genus *Cedecea* comprises three named species (*Cedecea davisae* [30], *Cedecea lapagei* [30], and *Cedecea neteri* [23]) and two unnamed species. Although strains of *Cedecea* have biochemical reactions similar to those of the genus *Serratia*, *Cedecea* spp. may be distinguished by their lack of DNase enzyme activity. Although strains of *Cedecea* are rare in clinical specimens, there are a number of reports of infections due to these organisms. About half of the human clinical isolates are from respiratory tract specimens (43). A case of bacteremia due to *C. neteri* has been reported (23).

CITROBACTER SPP.

The genus *Citrobacter* until recently comprised three species. *Citrobacter koserii* (synonyms, *Citrobacter diversus* and *Levinea malonatica* in some countries) is a cause of neonatal meningitis and sepsis (16) and has been described in nursery outbreaks associated with colonization of patients. *Citrobacter freundii* is well characterized as a cause of sepsis and infections of a number of tissues. Isolated from fecal specimens and possessing cytotoxic activity, *Citrobacter freundii* has been implicated as a cause of gastrointestinal infection. Because some strains of *Citrobacter freundii* resemble *Salmonella* spp. biochemically and agglutinate in *Salmonella* polyvalent O antiserum, they may be incorrectly identified as *Salmonella* spp. *Citrobacter amalonaticus* (synonym, *Levinea amalonatica*), named for its nonreactivity with malonate, is occasionally isolated from feces but rarely from extraintestinal sources. A new publication (7) proposed the introduction of five named (*Citrobacter farmeri*, *Citrobacter youngae*, *Citrobacter braakii*, *Citrobacter werkmanii*, *Citrobacter sedlakii*) and three as yet unnamed *Citrobacter* genomospecies. As these are not yet in the databases of commercial identification systems, it will be some time before routine laboratories begin to identify them.

EDWARDSIELLA SPP.

Described in 1965, the genus *Edwardsiella* is composed of three species and a biogroup: *Edwardsiella tarda*, *Edwardsiella hoshinae*, *Edwardsiella ictaluri*, and *Edwardsiella tarda* biogroup 1. In nature, these organisms reside in the intestines of cold-blooded animals and in water and are pathogens for eels and catfish (35, 43). Of the three species, *Edwardsiella tarda* is most prominently associated with serious human infections (61). Associated with an asymptomatic carrier state, mild diarrheal disease, and many extraintestinal infections (including soft tissue infections, meningitis, osteomyelitis, cholangitis, and sepsis), *E. tarda* was recently shown to also cause multiple hepatic abscesses (62). *Edwardsiella tarda* is easily recognized if media that demonstrate its production of hydrogen sulfide are employed; *Edwardsiella tarda* biogroup 1, however, at this time isolated only from reptiles, is H$_2$S negative. *Edwardsiella hoshinae* is rarely isolated from human clinical specimens (32). Associated with fish, *Edwardsiella ictaluri* is an uncommon isolate in the clinical setting.

ENTEROBACTER SPP.

Two species frequently isolated from clinical specimens, *Enterobacter cloacae* and *Enterobacter aerogenes*, are prominent causes of nosocomial infections. *E. agglomerans*, occasionally isolated from clinical specimens (18, 28), is best referred to as *E. agglomerans* complex, as it is actually a heterogeneous group of organisms composed of a dozen DNA hybridization groups. One of these DNA hybridization groups has been proposed for classification as *Pantoea agglomerans*. *E. agglomerans* shares with *E. sakazakii* the characteristic yellow pigmentation of its colonies and the tendency to cause life-threatening neonatal meningitis and sepsis (19, 53). A recent report on a neonatal outbreak of *E. sakazakii* sepsis and meningitis implicated dried infant formula as the source of the organism (11). Rare in human clinical specimens (8, 9, 20, 47), the other species of *Enterobacter* can be difficult to identify because they are biochemically similar to one another. The clinical relevance of the newly described species of *Enterobacter* is just beginning to be appreciated. There is a report of the isolation of *Enterobacter amnigenus* from the blood and an intravenous catheter of a heart transplant recipient, but it was associated with *Pseudomonas aeruginosa* bacteremia, so its clinical impact was not clear (4). *Enterobacter taylorae* has been reported as a cause of serious nosocomial infection (56). *Enterobacter cancerogenus*, *Enterobacter intermedius*, *Enterobacter dissolvens*, and *Enterobacter nimipressuralis* have not been isolated from human clinical specimens. In a review (58) of the clinical relevance of *Enterobacter* infections, it was suggested that avoidance of broad-spectrum cephalosporins for surgical prophylaxis and therapy in patients with *Enterobacter* infections would lower the prevalence of bacteremias due to the organism and reduce the incidence of multiresistance.

ERWINIA SPP.

Members of the genus *Erwinia* are a diverse group of organisms that primarily colonize plants and have rarely been isolated from humans (43).

EWINGELLA SP.

Described in 1983 (29), the genus *Ewingella* is composed of a single species, *Ewingella americana*. Biochemically, the genus *Ewingella* is similar to the genus *Cedecea*. In a recent report of an immunocompromised patient with septicemia due to *E. americana*, the human clinical experience with the agent was reviewed (13). The species has been isolated from a variety of sites in cases where clinical significance could not be established. It has also been reported in a common-source outbreak in a surgery and in an epidemic of pseudobacteremia due to contaminated blood collection tubes.

HAFNIA SP.

The genus *Hafnia* is composed of a single species, *Hafnia alvei*, that was formerly classified as *Enterobacter hafniae*. Some organisms formerly classified as atypical *Hafnia* spp. have been reclassified as *Yokenella* spp. (see below). The organisms are found in feces of humans and animals (particularly birds), sewage, soil, and dairy products (43).

KLEBSIELLA SPP.

K. pneumoniae is an important cause of nosocomial and community-acquired infections. With the acquisition of new antibiotic resistance mechanisms, the organism is achieving notoriety as an ominous nosocomial pathogen. Biochemically similar to *K. pneumoniae* except for its indole positivity, *Klebsiella oxytoca* also shares a tendency to produce capsules that manifest as mucoid colonies when grown on carbohydrate-rich media. However, mucoid colonies may be misleading; capsule formation may also occur in organisms such as *E. aerogenes* and *E. cloacae*. *K. ozaenae* and *Klebsiella rhinoscleromatis*, both of which are classified by some as subspecies of *K. pneumoniae*, are associated with two chronic diseases of the nasal mucosa (ozaena and rhinoscleroma) that are found primarily in the tropics. *Klebsiella planticola* (synonym, *Klebsiella trevisanii*) (3, 25), *Klebsiella terrigena* (45), and *Klebsiella* group 47, which is also known as *Klebsiella ornithinolytica* (57), are rare in human clinical specimens. A recent report (54) that specifically sought *K. terrigena* in clinical specimens found that most isolates are from the respiratory tract but did not definitively establish a role for the organism in pulmonary disease.

KLUYVERA SPP.

Described in 1981 (21), the genus *Kluyvera* includes two species, *Kluyvera ascorbata* and *Kluyvera cryocrescens*. Biochemically, these organisms are similar to *Buttiauxella* sp. These organisms are generally associated with opportunistic infections, but there are now reports of infections in otherwise healthy hosts, including pyelonephritis (14) and soft tissue infection (49).

KOSERELLA SPP.

See *Yokenella* sp. below.

LECLERCIA SP.

Described in 1986 (58) and previously known as *Escherichia adecarboxylata* and as enteric group 41, the genus *Leclercia* is composed of a single species, *Leclercia adecarboxylata*, which has been isolated from a variety of human clinical specimens (including blood and sputum [43]), food, water, and the environment.

LEMINORELLA SPP.

Described in 1985 (41), the genus *Leminorella*, formerly known as enteric group 57, comprises the species *Leminorella grimontii* and *Leminorella richardii*, both of which are characteristically positive for the production of hydrogen sulfide. Most of the human isolates have been from feces, but a few have been from urine. There is as yet little published to document the types of infections that these agents may cause.

MOELLERELLA SP.

Described in 1984 (40), the genus *Moellerella* comprises a single species, *Moellerella wisconsensis*. The organism is biochemically inactive. Isolates are primarily from stool specimens (43), but a role in diarrhea has not been established, and there is no evidence for such a role.

MORGANELLA SP.

The genus *Morganella* includes a single species, *Morganella morganii*, originally classified in the genus *Proteus* as *Proteus morganii*. A group of M. *morganii* strains designated biogroup 1 (36) were characterized as nonmotile, lysine negative, and able to ferment glycerol within 24 h. Recently, it was proposed (48) that the species be reconstituted as two subspecies, M. *morganii* subsp. *morganii* and M. *morganii* subsp. *sibonii*. M. *morganii* is a well-known cause of urinary tract and other extraintestinal infections. M. *morganii* has been mentioned as a possible enteric pathogen, but its etiologic role is doubtful. There is a recent report (10) of chorioamnionitis and associated sepsis due to M. *morganii* complicated by acute respiratory distress syndrome.

PANTOEA SPP.

Proposed in 1989, the genus *Pantoea* would be composed of two species, *P. agglomerans* (from species now classified as *Erwinia herbicola*, *Erwinia milletiae*, and one of the DNA hybridization groups of *Enterobacter agglomerans*) and *Pantoea dispersa*. The former organism is isolated from plants and water and has been found in human wounds, blood, and urine as an opportunistic pathogen (28), but the latter proposed species, *P. dispersa*, has not been found in human specimens.

PRAGIA SP.

Described in 1988 (2), the genus *Pragia* includes a single species, *Pragia fontium*. Another of the newly described members of the family *Enterobacteriaceae* that produce hydrogen sulfide, *Pragia fontium* has been isolated from water sources. Although there is no evidence of human pathogenicity (43), the organism is mentioned so that it can be distinguished from other hydrogen sulfide-producing members of the family *Enterobacteriaceae*.

PROTEUS SPP.

In 1978, the description of the genus *Proteus* was altered, and two additional genera, *Providencia* and *Morganella*, were established for species formerly classified as *Proteus* spp. The remaining two primary species, *Proteus mirabilis* and *Proteus vulgaris*, frequently cause urinary tract and other extraintestinal infections. They are easily recognized on nonselective media such as blood agar because they swarm and have a characteristic odor. They may be differentiated from one another by their differing indole and ornithine reactions (see Table 1 in chapter 32 of this Manual). *Proteus penneri* (37) is occasionally isolated from clinical specimens and has been described as a nosocomial pathogen. *Proteus myxofaciens* has not yet been described in human infections.

PROVIDENCIA SPP.

The genus *Providencia* was originally established for urease-negative organisms that were biochemically similar to species of *Proteus*. This distinction is no longer relevant, since *Providencia stuartii* is urease positive. This property of the organism means that it shares with *Proteus* spp. the tendencies to alkalinize urine and to promote the development of crystals and urinary calculi. *Providencia rettgeri* and *Providencia stuartii* are well-known causes of urinary tract and other extraintestinal infections and have caused many nosocomial outbreaks. *Providencia alcalifaciens*, a rare isolate in most clinical microbiology laboratories, has been isolated from feces. There have been only a few human isolates of *Providencia rustigianii* (38). *Providencia heimbachae* has been isolated from animal specimens (52).

RAHNELLA SP.

Described in 1979 (46), the genus *Rahnella* comprises a single species, *Rahnella aquatilis*. Most strains have been isolated from water, but there are descriptions of catheter-related bacteremia in a bone marrow transplant recipient (44) and urinary tract infection in a renal transplant patient (1).

SERRATIA SPP. (60)

Serratia marcescens is an important cause of extraintestinal infections, having caused many nosocomial outbreaks associated with blood transfusions, surgery, and the urinary tract. A number of other *Serratia* species have been described. These include *Serratia ficaria* (33), *Serratia fonticola* (27), *Serratia liquefaciens* (17), *Serratia odorifera* (31), and *Serratia plymuthica* (31). *S. marcescens* does not ferment

L-arabinose, which easily differentiates it from other species except *Serratia entomophila* (34), which does not occur in human clinical specimens. *S. odorifera* has been described recently in a case of nosocomial sepsis (51). The role of *S. plymuthica* in community-acquired bacteremia was recently reviewed elsewhere (55). *Serratia proteomaculans* subsp. *quinovora* was recently described; it was involved in a fatal case of pneumonia (5).

TATUMELLA SP.

Described in 1981 (42), the genus *Tatumella* is composed of a single species, *Tatumella ptyseos*. Although the majority of the isolates have been from the respiratory tract, where they have doubtful clinical significance, several isolates have been from blood.

TRABULSIELLA SPP.

The name *Trabulsiella* was proposed in 1991 (50) as the provisional name for enteric group 90. These organisms biochemically resemble *Salmonella* spp. Their clinical significance has yet to be established.

XENORHABDUS SPP.

Described in 1979 (59), the genus *Xenorhabdus* has a number of species, two of which (*Xenorhabdus nematophilus* and *Xenorhabdus luminescens*) are pathogenic for nematodes. Several strains isolated from human blood and wounds were identified as *X. luminescens* DNA hybridization group 5 (22). The organism has a unique set of biochemical reactions. Colonies glow in the dark, but the bioluminescence is very weak; after 15 min of dark adaptation, one can observe the faint light produced by the growth. Subsequent studies of an isolate from a wound have shown that luminescence is best when incubation is at 33°C (12).

YOKENELLA SP.

Described in 1985 (39), the genus *Yokenella* was formerly called enteric group 45. The organism in this enteric group was subsequently classified as *Koserella trabulsii*. When it was found to be indistinguishable from a previously described organism, the two agents were classified as one, *Yokenella regensburgei*, the only species in the genus (43). Biochemically similar to *Hafnia* spp., the organisms have been isolated from the respiratory tract, wounds, urine, and feces.

I acknowledge the contribution of J. J. Farmer, author of the chapter on Enterobacteriaceae in the previous edition of this Manual, who contributed greatly to the planning, writing, and editing of this chapter.

REFERENCES

1. **Alballaa, S. R., S. M. Hussain Qadri, O. al-Furayh, and K. al-Qatary.** 1992. Urinary tract infection due to *Rahnella aquatilis* in a renal transplant patient. *J. Clin. Microbiol.* **30:**2948–2950.
2. **Aldová, E., O. Hausner, D. J. Brenner, Z. Kocmoud, J. Schindler, B. Potuzníková, and P. Petráš.** 1988. *Pragia fontium* gen. nov., sp. nov. of the family *Enterobacteriaceae*, isolated from water. *Int. J. Syst. Bacteriol.* **38:**183–189.
3. **Bagley, S. T., R. J. Seidler, and D. J. Brenner.** 1981. *Klebsiella planticola* sp. nov.: a new species of *Enterobacteriaceae* found primarily in nonclinical environments. *Curr. Microbiol.* **6:**105–109.
4. **Bollet, C., A. Elkouby, P. Pietri, and P. deMicco.** 1991. Isolation of *Enterobacter amnigenus* from a heart transplant recipient. *Eur. J. Clin. Microbiol. Infect. Dis.* **10:**1071–1073.
5. **Bollet, C., P. Grimont, M. Gainnier, A. Geissler, P. DeMicco, and J. M. Sainty.** 1993. Fatal pneumonia due to *Serratia proteomaculans* subsp. *quinovora. J. Clin. Microbiol.* **31:**444–445.
6. **Bouvet, O. M. M., P. A. D. Grimont, C. Richard, E. Aldová, O. Hausner, and M. Gabrhelova.** 1985. *Budvicia aquatica* gen. nov., sp. nov.: a hydrogen sulfide-producing member of the *Enterobacteriaceae. Int. J. Syst. Bacteriol.* **35:**60–64.
7. **Brenner, D. J., P. A. D. Grimont, A. G. Steigerwalt, G. R. Fanning, E. Ageron, and C. F. Riddle.** 1993. Classification of citrobacteria by DNA hybridization: designation of *Citrobacter farmeri* sp. nov., *Citrobacter youngae* sp. nov., *Citrobacter braakii* sp. nov., *Citrobacter werkmanii* sp. nov., *Citrobacter sedlakii* sp. nov., and three unnamed *Citrobacter* genomospecies. *Int. J. Syst. Bacteriol.* **43:**645–658.
8. **Brenner, D. J., A. C. McWhorter, A. Kai, A. G. Steigerwalt, and J. J. Farmer III.** 1986. *Enterobacter asburiae* sp. nov., a new species found in clinical specimens, and reassignment of *Erwinia dissolvens* and *Erwinia nimipressuralis* to the genus *Enterobacter* as *Enterobacter dissolvens* comb. nov. and *Enterobacter nimipressuralis* comb. nov. *J. Clin. Microbiol.* **23:**1114–1120.
9. **Brenner, D. J., C. Richard, A. G. Steigerwalt, M. A. Asbury, and M. Mandel.** 1980. *Enterobacter gergoviae* sp. nov.: a new species of *Enterobacteriaceae* found in clinical specimens and the environment. *Int. J. Syst. Bacteriol.* **30:**1–6.
10. **Carmona, F., F. Fabregues, R. Alvarez, J. Vila, and V. Cararach.** 1992. A rare case of chorioamnionitis by *Morganella morganii* complicated by sepsis and ARDS. *Eur. J. Obstet. Gynecol. Reprod. Biol.* **45:**67–70.
11. **Clark, N. C., B. C. Hill, C. M. O'Hara, O. Steingrimsson, and R. C. Cooksey.** 1990. Epidemiologic typing of *Enterobacter sakazakii* in two neonatal nosocomial outbreaks. *Diagn. Microbiol. Infect. Dis.* **13:**467–472.
12. **Colepicolo, P., K. W. Cho, G. O. Poinar, and J. W. Hastings.** 1989. Growth and luminescence of the bacterium *Xenorhabdus luminescens* from a human wound. *Appl. Environ. Microbiol.* **55:**2601–2606.
13. **Devreese, K., G. Claeys, and G. Verschraegen.** 1992. Septicemia with *Ewingella americana. J. Clin. Microbiol.* **30:**2746–2747.
14. **Dollberg, S., A. Gandacu, and A. Klar.** 1990. Acute pyelonephritis due to a *Kluyvera* species in a child. *Eur. J. Clin. Microbiol. Infect. Dis.* **9:**281–283.
15. **Ewing, W. H.** 1986. *Edwards and Ewing's Identification of Enterobacteriaceae*, 4th ed. Elsevier Science Publishing Co., New York.
16. **Ewing, W. H., and B. R. Davis.** 1972. Biochemical characterization of *Citrobacter diversus* (Burkey) Werkman and Gillen and designation of the neotype strain. *Int. J. Syst. Bacteriol.* **22:**12–18.
17. **Ewing, W. H., B. R. Davis, M. A. Fife, and E. F. Lessel.** 1973. Biochemical characterization of *Serratia liquefaciens* (Grimes and Hennerty) Bascomb et al. (formerly *Enterobacter liquefaciens*) and *Serratia rubidaea* (Stapp) comb. nov. and designation of type and neotype strains. *Int. J. Syst. Bacteriol.* **23:**217–225.
18. **Ewing, W. H., and M. A. Fife.** 1972. *Enterobacter agglomerans* (Beijerinck) comb. nov. (the *herbicola-lathyri* bacteria). *Int. J. Syst. Bacteriol.* **22:**4–11.
19. **Farmer, J. J., III, M. A. Asbury, F. W. Hickman, D. J. Brenner, and The Enterobacteriaceae Study Group.** 1980. *Enterobacter sakazakii*: a new species of "*Enterobacteriaceae*" isolated from clinical specimens. *Int. J. Syst. Bacteriol.* **30:**569–584.
20. **Farmer, J. J., III, G. R. Fanning, B. R. Davis, C. M. O'Hara, C. Riddle, F. W. Hickman-Brenner, M. A. Asbury, V. A. Lowery III, and D. J. Brenner.** 1984. *Escherichia fergusonii* and *Enterobacter taylorae*: two new species of *Enterobacteriaceae* isolated from clinical specimens. *J. Clin. Microbiol.* **21:**77–81.

21. Farmer, J. J., III, G. R. Fanning, G. P. Huntley-Carter, B. Holmes, F. W. Hickman, C. Richard, and D. J. Brenner. 1981. *Kluyvera*, a new (redefined) genus in the family *Enterobacteriaceae*: identification of *Kluyvera ascorbata* sp. nov. and *Kluyvera cryocrescens* sp. nov. in clinical specimens. *J. Clin. Microbiol.* **13**:919–933.

22. Farmer, J. J., III, J. H. Jorgensen, P. A. D. Grimont, R. J. Akhurst, G. O. Poinar, Jr., E. Ageron, G. V. Pierce, J. A. Smith, G. P. Carter, K. L. Wilson, and F. W. Hickman-Brenner. 1989. *Xenorhabdus luminescens* (DNA hybridization group 5) from human clinical specimens. *J. Clin. Microbiol.* **27**:1594–1600.

23. Farmer, J. J., III, N. K. Sheth, J. A. Hudzinski, H. D. Rose, and M. F. Asbury. 1982. Bacteremia due to *Cedecea neteri* sp. nov. *J. Clin. Microbiol.* **16**:775–778.

24. Farmer, J. J., III, J. G. Wells, P. M. Griffin, and I. K. Wachsmuth. 1987. *Enterobacteriaceae* infections, p. 233–296. In B. B. Wentworth (ed.), *Diagnostic Procedures for Bacterial Infections*, 7th ed. American Public Health Association, Washington, D.C.

25. Ferragut, C., D. Izard, F. Gavini, K. Kersters, J. DeLey, and H. Leclerc. 1983. *Klebsiella trevisanii*: a new species from water and soil. *Int. J. Syst. Bacteriol.* **33**:133–142.

26. Ferragut, C., D. Izard, F. Gavini, B. Lefebre, and H. Leclerc. 1981. *Buttiauxella*, a new genus of the family *Enterobacteriaceae*. *Zentralbl. Bakteriol. Parasitenkd. Infectionskr. Hyg. Abt. 1 Orig. Reihe C* **2**:33–44.

27. Gavini, F., C. Ferragut, D. Izard, P. A. Trinel, H. Leclerc, B. Lefebvre, and D. A. A. Mossel. 1979. *Serratia fonticola*, a new species from water. *Int. J. Syst. Bacteriol.* **29**:92–101.

28. Gavini, F., J. Mergaert, A. Beji, C. Mielcarek, D. Izard, K. Kersters, and J. De Ley. 1989. Transfer of *Enterobacter agglomerans* (Beijerinck 1888) Ewing and Fife 1972 to *Pantoea* gen. nov. as *Pantoea agglomerans* comb. nov. and description of *Pantoea dispersa* sp. nov. *Int. J. Syst. Bacteriol.* **39**:337–345.

29. Grimont, P. A. D., J. J. Farmer III, F. Grimont, M. A. Asbury, D. J. Brenner, and C. Deval. 1983. *Ewingella americana* gen. nov., sp. nov., a new *Enterobacteriaceae* isolated from clinical specimens. *Ann. Microbiol. (Inst. Pasteur)* **134A**:39–52.

30. Grimont, P. A. D., F. Grimont, J. J. Farmer III, and M. A. Asbury. 1981. *Cedecea davisae* gen. nov., sp. nov., and *Cedecea lapagei* sp. nov., new *Enterobacteriaceae* from clinical specimens. *Int. J. Syst. Bacteriol.* **31**:317–326.

31. Grimont, P. A. D., F. Grimont, C. Richard, B. R. Davis, A. G. Steigerwalt, and D. J. Brenner. 1978. Deoxyribonucleic acid relatedness between *Serratia plymuthica* and other *Serratia* species, with a description of *Serratia odorifera* sp. nov. (type strain: ICPB 3995). *Int. J. Syst. Bacteriol.* **28**:453–463.

32. Grimont, P. A. D., F. Grimont, C. Richard, and R. Sakazaki. 1980. *Edwardsiella hoshinae*, a new species of *Enterobacteriaceae*. *Curr. Microbiol.* **4**:347–351.

33. Grimont, P. A. D., F. Grimont, and M. P. Starr. 1979. *Serratia ficaria* sp. nov., a bacterial species associated with smyrna figs and the fig wasp *Blastophaga psenes*. *Curr. Microbiol.* **2**:277–282.

34. Grimont, P. A. D., T. A. Jackson, E. Ageron, and M. J. Noonan. 1988. *Serratia entomophila* sp. nov. associated with amber disease in the New Zealand grass grub *Costelytra zealandica*. *Int. J. Syst. Bacteriol.* **38**:1–6.

35. Hawke, J. P., A. C. McWhorter, A. G. Steigerwalt, and D. J. Brenner. 1981. *Edwardsiella ictaluri* sp. nov., the causative agent of enteric septicemia of catfish. *Int. J. Syst. Bacteriol.* **31**:396–400.

36. Hickman, F. W., J. J. Farmer III, A. G. Steigerwalt, and D. J. Brenner. 1980. Unusual groups of *Morganella* ("*Proteus*") *morganii* isolated from clinical specimens: lysine-positive and ornithine-negative biogroups. *J. Clin. Microbiol.* **12**:88–94.

37. Hickman, F. W., A. G. Steigerwalt, J. J. Farmer III, and D. J. Brenner. 1982. Identification of *Proteus penneri* sp. nov., formerly known as *Proteus vulgaris* indole negative or as *Proteus vulgaris* biogroup 1. *J. Clin. Microbiol.* **15**:1097–1102.

38. Hickman-Brenner, F., J. J. Farmer III, A. G. Steigerwalt, and D. J. Brenner. 1983. *Providencia rustigianii*: a new species in the family *Enterobacteriaceae* formerly known as *Providencia alcalifaciens* biogroup 3. *J. Clin. Microbiol.* **15**:1057–1060.

39. Hickman-Brenner, F. W., G. P. Huntley-Carter, G. R. Fanning, D. J. Brenner, and J. J. Farmer III. 1985. *Koserella trabulsii*, a new genus and species of *Enterobacteriaceae* formerly known as enteric group 45. *J. Clin. Microbiol.* **21**:39–42.

40. Hickman-Brenner, F. W., G. P. Huntley-Carter, Y. Saitoh, A. G. Steigerwalt, J. J. Farmer III, and D. J. Brenner. 1984. *Moellerella wisconsensis*, a new genus and species of *Enterobacteriaceae* found in human stool specimens. *J. Clin. Microbiol.* **19**:460–463.

41. Hickman-Brenner, F. W., M. P. Vohra, G. P. Huntley-Carter, G. R. Fanning, V. A. Lowery III, D. J. Brenner, and J. J. Farmer III. 1985. *Leminorella*, a new genus of *Enterobacteriaceae*: identification of *Leminorella grimontii* sp. nov. and *Leminorella richardii* sp. nov. found in clinical specimens. *J. Clin. Microbiol.* **21**:234–239.

42. Hollis, D. G., F. W. Hickman, G. R. Fanning, J. J. Farmer III, R. E. Weaver, and D. J. Brenner. 1981. *Tatumella ptyseos* gen. nov., sp. nov., a member of the family *Enterobacteriaceae* found in clinical specimens. *J. Clin. Microbiol.* **14**:79–88.

43. Holt, J. G., N. R. Krieg, P. H. A. Sneath, J. T. Staley, S. T. Williams (ed.). 1994. *Bergey's Manual of Determinative Bacteriology*, 9th ed. The Williams & Wilkins Co., Baltimore.

44. Hoppe, J. E., M. Herter, S. Aleksic, T. Klingebiel, and D. Niethammer. 1993. Catheter-related *Rahnella aquatilis* bacteremia in a pediatric bone marrow transplant recipient. *J. Clin. Microbiol.* **31**:1911–1912.

45. Izard, D., C. Ferragut, F. Gavini, K. Kersters, J. DeLey, and H. Leclerc. 1981. *Klebsiella terrigena*, a new species from soil and water. *Int. J. Syst. Bacteriol.* **31**:116–127.

46. Izard, D., F. Gavini, P. A. Trinel, and H. Leclerc. 1979. *Rahnella aquatilis*, nouveau membre de la famille des *Enterobacteriaceae*. *Ann. Microbiol. (Inst. Pasteur)* **130A**:163–177.

47. Izard, D., F. Gavini, P. A. Trinel, and H. Leclerc. 1981. Deoxyribonucleic acid relatedness between *Enterobacter cloacae* and *Enterobacter amnigenus* sp. nov. *Int. J. Syst. Bacteriol.* **31**:35–42.

48. Jensen, K. T., W. Fredericksen, F. W. Hickman-Brenner, A. G. Steigerwalt, C. F. Riddle, and D. J. Brenner. 1992. Recognition of *Morganella* subspecies with proposal of M. *morganii* subsp. *morganii* and M. *morganii* subsp. *sibonii*. *Int. J. Syst. Bacteriol.* **42**:613–620.

49. Luttrell, R. E., G. A. Rannick, J. L. Soto-Hernandez, and A. Verghese. 1988. *Kluyvera* species soft tissue infection: case report and review. *J. Clin. Microbiol.* **26**:2650–2651.

50. McWhorter, A. C., R. L. Haddock, F. A. Nocon, A. G. Steigerwalt, D. J. Brenner, S. Aleksic, J. Bockemuhl, and J. J. Farmer III. 1991. *Trabulsiella guamensis*, a new genus and species of the family *Enterobacteriaceae* that resembles *Salmonella* subgroups 4 and 5. *J. Clin. Microbiol.* **29**:1480–1485.

51. Mermel, L. A., and C. A. Spiegel. 1992. Nosocomial sepsis due to *Serratia odorifera* biovar 1. *Clin. Infect. Dis.* **14**:208–210.

52. Muller, H. E., C. M. O'Hara, G. R. Fanning, F. W. Hickman-Brenner, J. M. Swenson, and D. J. Brenner. 1986. *Providencia heimbachae*, a new species of *Enterobacteriaceae* isolated from animals. *Int. J. Syst. Bacteriol.* **36**:252–256.

53. Muytjens, H. L., H. C. Zanen, H. J. Sonderkamp, L. A. Kollée, I. K. Wachsmuth, and J. J. Farmer III. 1983. Analysis of eight cases of neonatal meningitis and sepsis due to *Enterobacter sakazakii*. *J. Clin. Microbiol.* **18**:115–120.

53a. Petter, J. (Southeast Poultry Research Laboratory). Unpublished data.

54. Podschun, R., and U. Ullmann. 1992. Isolation of *Klebsiella terrigena* from clinical specimens. *Eur. J. Clin. Microbiol. Infect. Dis.* **11**:349–352.

55. Reina, J., N. Borrell, and I. Llompart. 1992. Community acquired bacteremia caused by *Serratia plymuthia*. Case report and review of the literature. *Diagn. Microbiol. Infect. Dis.* **15**:449–452.

56. **Rubinstien, E. M., P. Klevjer-Anderson, C. A. Smith, M. T. Drouin, and J. E. Patterson.** 1993. *Enterobacter taylorae*, a new opportunistic pathogen: report of four cases. *J. Clin. Microbiol.* **31:**249–254.

57. **Sakazaki, R., K. Tamura, Y. Kosako, and E. Yoshizaki.** 1989. *Klebsiella ornithinolytica* sp. nov., formerly known as ornithine-positive *Klebsiella oxytoca*. *Curr. Microbiol.* **18:**201–206.

58. **Shlaes, D. M.** 1993. The clinical relevance of *Enterobacter* infections. *Clin. Ther.* **15**(Suppl. A)**:**21–28.

59. **Thomas, G. M., and G. O. Poinar, Jr.** 1979. *Xenorhabdus* gen. nov., a genus of entomopathogenic nematophilic bacteria of the family *Enterobacteriaceae. Int. J. Syst. Bacteriol.* **29:**352–360.

60. **von Graevenitz, A., and S. J. Rubin (ed.).** 1980. *The Genus Serratia.* CRC Press, Inc., Boca Raton, Fla.

61. **Wilson, J. P., R. R. Waterer, J. D. Wofford, Jr., and S. W. Chapman.** 1989. Serious infections with *Edwardsiella tarda.* A case report and review of the literature. *Arch. Intern. Med.* **149:**108–210.

62. **Zighelboim, J., T. W. Williams, Jr., M. W. Bradshaw, and R. L. Harris.** 1992. Successful medical management of a patient with multiple hepatic abscesses due to *Edwardsiella tarda. Clin. Infect. Dis.* **14:**117–120.

Vibrio

JAMES C. McLAUGHLIN

35

TAXONOMY

The genus *Vibrio* is in the family *Vibrionaceae*, which also includes the genera *Aeromonas*, *Plesiomonas*, and *Photobacterium*. *Aeromonas* and *Plesiomonas* are discussed in chapter 36 of this Manual. *Vibrio* species isolated in clinical laboratories share features with members of the families *Enterobacteriaceae* and *Pseudomonadaceae*. Excellent reviews of clinically significant *Vibrio* species (40) and the whole *Vibrio* genus (24) are available.

DESCRIPTION OF THE GENUS

Members of the genus *Vibrio* are straight or curved gram-negative bacilli 0.5 to 0.8 μm in diameter and 1.4 to 2.6 μm in length. Usually motile by a single polar flagellum, some strains produce multiple lateral flagella when grown on solid media. *Vibrio* spp. grow in the presence or absence of oxygen. All species except *Vibrio metschnikovii* produce oxidase, all ferment glucose, and some produce gas. Sodium stimulates growth, and some species are halophilic. Sensitivity to the vibriostatic compound O/129 (2,4-diamino-6,7-diisopropylpteridine) varies. The genus contains more than 30 species, and 12 of these are human pathogens or have been isolated from human clinical specimens. Seven of the 12 species have been named since 1979. *V. cholerae* Pacini 1854 is the type species for the genus (8).

NATURAL HABITATS

Vibrio species are natural inhabitants of brackish and salt water worldwide. Human disease is associated with ingestion of contaminated water or consumption of contaminated shellfish or seafood. Wound and systemic infections have developed following contact with water. Some *Vibrio* species are pathogenic for marine animals and fish. This chapter focuses only on species isolated from human clinical specimens. *Vibrio* species and their associated human diseases are listed in Table 1.

CLINICAL SIGNIFICANCE

V. cholerae

Seven cholera pandemics have been reported since 1817. All except the current, the seventh, have swept out of the Ganges River delta in India. The present pandemic began in Sulawesi, Indonesia, in 1961. Over the next several years, it spread across Asia and the subcontinent and into Africa, where by 1991, more than 19 countries were affected. After an absence of over a century, cholera returned to Latin America in 1991. The history of cholera and other aspects of the disease are well covered in two excellent books, one edited by Barua and Greenough (7) and the other edited by Wachsmuth, Blake, and Ølsvik (93).

V. cholerae is the etiologic agent of a secretory diarrhea spread by the fecal-oral route. Two biotypes, El Tor and Classical, are associated with human disease. Infections with *V. cholerae* may be asymptomatic, mild, or severe. If not treated, patients with severe cholera may die within 5 h as a result of massive fluid and electrolyte loss. Fewer than 100,000 cases of cholera were reported to the World Health Organization in 1990. This number had been constant for years. However, in January 1991, cholera returned to Latin America, and in 1991 and 1992, 745,309 cases and 6,403 deaths were reported to the World Health Organization from Latin American countries alone (95). Because of the ability of *V. cholerae* El Tor to survive in water once introduced and because of socioeconomic conditions in Latin America, cholera is likely to remain there for years.

In the United States, no cases of cholera had occurred since 1911. Then in 1973, a man living in Port Lavaca, Tex., who had undergone a partial gastrectomy developed severe cholera. No further cases were detected in the United States until 1978. Subsequent studies have found that Gulf Coast waters are natural reservoirs for a unique strain of *V. cholerae*. Cases of cholera occur in Gulf Coast states annually and are usually associated with the ingestion of *V. cholerae*-contaminated shellfish or seafood. The Latin American epidemic strain has been isolated from U.S. Gulf Coast waters (21), but no cholera cases due to this strain as a result of consuming shellfish from U.S. waters have been reported (22).

The potential for imported cases of cholera in the United States is great. In 1992, two outbreaks of cholera occurred in individuals living in New Jersey and New York who ate crabs transported in a suitcase from Latin America (27). Three cases of cholera in Maryland were traced to the consumption of frozen coconut milk imported from Asia (46). More than 75 passengers developed cholera after

TABLE 1 Association of *Vibrio* species with different clinical syndromes[a]

Species	Clinical syndrome[b]			
	Gastroenteritis	Wound infection	Ear infection	Septicemia
V. cholerae O1	+++			
V. cholerae non-O1	+++	++	+	+
V. mimicus	++		+	
V. fluvialis	++			
V. parahaemolyticus	+++			+
V. alginolyticus	(+)	++	++	+
V. cincinnatiensis				+
V. hollisae	++			+
V. vulnificus	(+)	++		++
V. furnissii	(+)			
V. damsela		++		+
V. metschnikovii	(+)			+
V. carchariae		+		

[a]Table is adapted from Pavia et al. (68).
[b]Symbols: +++, frequently reported; ++, less common (6 to 100 reports); +, rare (1 to 5 reports); (+), occurs at this site, but etiologic role as a cause of intestinal infection has not been established.

eating contaminated seafood cocktails on an international flight from Latin America to Los Angeles (53).

Treatment

Although antibiotics reduce the number of days of diarrhea, replacement of fluids and electrolytes remains the most important aspect of cholera treatment. The most efficacious intravenous and oral replacement fluids have been developed at the International Center for Diarrhoeal Diseases Research/Dhaka in Dhaka, Bangladesh. The impact of oral rehydration fluids on preventing mortality due to cholera and other diarrheal diseases worldwide is enormous. The rapid availability of oral rehydration fluids played a major role in the low death rate in the Latin American epidemic that began in early 1991.

Non-O1 *V. cholerae*

Non-O1 *V. cholerae* have also been designated NAG, for nonagglutinable *V. cholerae*, and, erroneously, NCV, for noncholera vibrio. These organisms are biochemically indistinguishable from O1 *V. cholerae*. A variety of taxonomic approaches have demonstrated that O1 and non-O1 *V. cholerae* are the same species. Several serotyping systems have been developed for these organisms, but it is likely that molecular techniques will replace these.

In 1935, Gardner and Venkatraman published a classification scheme for *V. cholerae* (30). *V. cholerae* isolated from cholera patients was designated O1, and all other strains were designated non-O1. Non-O1 *V. cholerae* cause both gastroenteritis and systemic infections. Some strains produce cholera enterotoxin. Non-O1 *V. cholerae* have been isolated from blood, wounds, ears, sputum, cerebrospinal fluid, and urine (29, 58). Extraintestinal infections caused by non-O1 strains usually occur in immunocompromised patients (78). Cases of non-O1 *V. cholerae* occur inland (29, 61, 88) as well along the Gulf Coast and in the Great Lakes region of the United States. Non-O1 *V. cholerae* have been isolated from surface waters in western Colorado (76), Chesapeake Bay (35), the Pacific Ocean, and the Gulf of Mexico. Aquatic birds serve as carriers (15, 63).

V. cholerae O139

In March 1993, the first reports of an outbreak of epidemic cholera due to a new serogroup of non-O1 *V. cholerae* appeared (9, 71). Because the strain does not belong to any of the 138 previously described serogroups of *V. cholerae*, it has been designated *V. cholerae* O139 and given the synonym Bengal, because it was originally isolated from patients living along the Bay of Bengal. It was reported first from Madras and then from Bangladesh. By late 1993, it had already been reported from Thailand, suggesting a very rapid spread (16). The seventh pandemic strain of El Tor *V. cholerae* O1 took more than 3 years to reach Thailand from Celebes in Indonesia. In some parts of India, O139 has nearly replaced the El Tor seventh pandemic strain. A case of imported cholera due to *V. cholerae* O139 has been reported in a traveler returning to the United States from India (89). The patient recovered after four days of watery diarrhea without rehydration therapy. Infection with the O139 strain is clinically indistinguishable from the cholera produced by O1 *V. cholerae*. Patients develop severe secretory diarrhea, vomiting, and rapid onset of dehydration. An important difference is that the majority of patients with O139 diarrhea are adults, indicating that there is no preexisting immunity to this serotype; previous infection with O1 *V. cholerae* does not appear to be protective (2). Vaccines against *V. cholerae* O1 are not expected to be protective against *V. cholerae* O139 (84).

Toxin production by this new serotype organism was demonstrated in Y1 adrenal cells and the rabbit ileal loop assay (4). Toxin homology studies demonstrate relatedness of the amino-terminal sequences of *V. cholerae* O139 to El Tor *V. cholerae* O1 *ctxA1*, *ctxA1*, and *ctxB* sequences, suggesting that the new serotype is an O antigen mutant of an El Tor strain (31). The modifications of the O antigen and the genetic alterations between El Tor *V. cholerae* and *V. cholerae* O139 have been discussed elsewhere (52).

V. cholerae O139 is isolated from water in Bangladesh. Of 92 water samples collected between February and June 1993, 12% were positive for the organism, and all strains produced cholera toxin (39). In contrast, during epidemic periods, *V. cholerae* O1 is normally isolated from less than 1% of water samples.

The organism is sensitive to tetracycline, ampicillin, chloramphenicol, erythromycin, and ciprofloxacin (4) but resistant to trimethoprim-sulfamethoxazole and furazolidone (94). Specific diagnosis can be made with O139 antiserum, which is available from several state health departments; the Centers for Disease Control and Prevention (CDC), Atlanta, Ga.; and the Department of Microbiology, Kyoto University, Kyoto, Japan (94).

V. mimicus

V. mimicus appears to be a rare cause of human diarrhea. Of 21 strains of *V. mimicus* studied at CDC in 1981, 19 were from patients who developed diarrhea following the consumption of seafood (typically, raw oysters), and 2 were from patients with otitis media (80). During 1989, four isolates from patients with gastroenteritis were reported by the Gulf Coast *Vibrio* Working Group (50).

V. parahaemolyticus

V. parahaemolyticus was first reported in 1950 in Japan, where it was responsible for 20 deaths among 272 patients with acute gastroenteritis who had eaten half-dried sardine, called *shirasu* in Japanese (56). It has been isolated in waters and from patients worldwide. It has a 12-min generation time, and infection is a result of eating fish or seafood. *V. parahaemolyticus* is a common cause of infection in Japan and has been accountable for a number of common-source outbreaks (6) and sporadic cases in the United States. It was the *Vibrio* species most frequently isolated by the Gulf Coast *Vibrio* Working Group in 1989 (50). It is responsible most commonly for a short-lived acute diarrhea, which may be bloody, that lasts an average of 3 days (6).

V. fluvialis

V. fluvialis was isolated from more than 500 patients with diarrhea in Bangladesh and designated *Vibrio*-like group EF-6 (38). The organism was subsequently named *V. fluvialis* in a numerical taxonomic study (48). *V. fluvialis* has been isolated in the United States from a patient with a fatal case of diarrhea (86) and in Bangladesh from the blood of a patient with shigellosis and *V. fluvialis* bacteremia who died (3). Seven patients with diarrhea, three requiring hospitalization, were identified in the Gulf Coast states in 1989 (50).

V. furnissii

In 1983, strains of *V. fluvialis* producing gas from glucose were shown by DNA relatedness studies to be a new species and were designated *V. furnissii* (14). Although *V. furnissii* has been isolated from patients with acute gastroenteritis, its role as a cause of diarrhea has not been established. *V. furnissii* has been isolated together with *V. fluvialis* in the stool of a 1-month-old infant with diarrhea that resolved without treatment (33).

V. alginolyticus

V. alginolyticus has been isolated from patients with ear infections, conjunctivitis (49), intracranial infection (65), wounds following burns (23), wounds following an air crash (54), and simple exposure to seawater (70). In the collection at CDC, most of the clinical isolates have been from the ear (24). Infections due to *V. alginolyticus* are extraintestinal, and this organism has not been reported as a cause of diarrhea.

V. carchariae

V. carchariae, from the Greek *carcharias*, meaning shark, was isolated from the bite wound of a child following a shark attack (68).

V. damsela

First named in 1981, when it was found to be the cause of ulcers on the damselfish *Chromis punctipinnis* (51), *V. damsela* has been associated almost exclusively with wound infections. Six strains of *V. damsela* originally designated EF-5 at CDC were isolated from wounds of otherwise healthy persons (60). Five of these cases were associated with contact with salt or brackish water. A case of fulminant septicemia developed in a patient who had cut his hand while filleting bluefish (69). *V. damsela* has been described as the cause of a rapidly progressive necrotizing infection (19). The organism produces a cytolysin that is lethal in mice. The toxin is distinct from those produced by *V. vulnificus*, *V. parahaemolyticus*, and *V. cholerae* (45). *V. damsela* isolated from a fatal wound infection demonstrated two hemolytic phenotypes, one of which inhibited the hemolysis caused by *Staphylococcus aureus* (18).

V. cincinnatiensis

A *Vibrio* species isolated from the blood and cerebrospinal fluid of a 70-year-old man with meningitis was named *V. cincinnatiensis* (13). The patient had no recent history of exposure to seawater or consumption of seafood. Moxalactam therapy was successful (11).

V. hollisae

V. hollisae (original CDC designation, EF-13) has been associated with diarrhea in individuals who have eaten raw seafood (32). A case of *V. hollisae* septicemia in a man with chronic active hepatitis and presenting with gastrointestinal illness has been reported (75). The Gulf Coast *Vibrio* Working Group reported eight isolates of *V. hollisae* during 1989 from patients with gastroenteritis and one from a patient with a wound infection (50).

V. metschnikovii

V. metschnikovii was first described in 1888 and redefined in 1978 (47). Although it has been isolated from environmental sources, only one case report has been published (41). The organism was isolated from blood cultures of an 82-year-old woman who presented with acute cholecystitis. She had no recent history of travel or seafood consumption. Other human isolates are reported by Farmer and Hickman-Brenner (24).

V. vulnificus

V. vulnificus is responsible for wound infections and primary septicemia following ingestion of raw seafood or contact with seawater. The epidemiology, biochemical characteristics, and antibiotic susceptibilities of *V. vulnificus* strains were first described in 1979 on the basis of 39 strains submitted to CDC (10). Seventy-five percent of the patients with primary septicemia also had hepatic disease, 95% were over 40 years old, and 90% were male. The organism was isolated on both coasts of the United States. Liver disease is a risk factor for invasive infection with this organism (85). *V. vulnificus* has also been recovered from the stools of three patients with diarrhea who had eaten raw oysters and were taking medication that reduced gastric acidity (42), but its etiologic role in diarrhea has not been proven. Of 18 isolates of *V. vulnificus* from four Gulf Coast states in 1989, only 3 were from patients with gastroenteritis, and in 2 of these cases, *V. parahaemolyticus* was simultaneously isolated (50). Observations of large series of patients with *V. vulnificus* infections in Taiwan and Korea have been published (17, 67).

COLLECTION, TRANSPORT, AND STORAGE OF SPECIMENS

For *Vibrio* species associated with diarrheal disease, a stool specimen collected in the acute stage of illness before antibiotics are given is the specimen of choice. If any delay in culturing is anticipated, addition of the specimen to a transport medium is essential. Cary-Blair transport medium is recommended. Inoculation into alkaline-peptone water (44) is acceptable if subculture will be within 6 to 8 h. Rectal swabs are adequate if they are immediately placed into a transport medium or alkaline-peptone water to prevent drying. *Vibrio* spp. are particularly susceptible to drying. Because of the public health importance of the isolation of *Vibrio* spp., inoculation to an enrichment broth is recommended. Alkaline-peptone water and tellurite-taurocholate-peptone broth are both good enrichment broths. When attempting to culture *Vibrio* spp. from extraintestinal sites, routine collection and transport methods are adequate.

DIRECT EXAMINATION

Rapid Detection

Dark-field microscopy of a stool specimen from a cholera patient usually shows the rapid, darting, or shooting-star motility that is characteristic of *V. cholerae*. A drop of *V. cholerae* O1 antiserum added to the slide rapidly immobilizes the organisms. This test can be used for the tentative laboratory diagnosis of *V. cholerae* O1 cholera. A dark-field microscope is not essential. The mobility can be seen by lowering the condenser, increasing the light intensity, and placing one's finger partially in the path of the light.

A latex agglutination test (Denka Seiken Co., Tokyo, Japan) (79) and an antigen capture test (New Horizons Diagnostics Corp., Columbia, Md.) for the direct detection of *V. cholerae* O1 in stool specimens are available, although they have not been evaluated extensively.

Molecular Methods

A variety of molecular approaches to the laboratory diagnosis of cholera and to toxin detection have been developed. PCR has been used for the detection of the cholera toxin A subunit gene (*ctxA*) (26). A 564-bp fragment of *ctxA* was used to identify 140 isolates of toxigenic O1 *V. cholerae* from food, water, and patients during the Latin American epidemic. A nonradioactive DNA oligonucleotide probe specific for *ctxA* is capable of detecting toxigenic *V. cholerae* on nonselective medium (96). PCR amplification of the *ctx* operon in rice water stools without DNA extraction has been reported to be effective. The sensitivity of this method was reported to be three viable *V. cholerae* cells (82). PCR amplification of the cholera toxin gene was shown to be a more sensitive method than a bead enzyme-linked immunosorbent assay (ELISA) for detecting cholera toxin (72). A detailed PCR protocol for the differentiation of toxigenic from nontoxigenic *V. cholerae* O1 colonies has

TABLE 2 Appearance of *Vibrio* species on TCBS[a]

Organism	Colony Color on TCBS	Growth and plating efficiency
V. alginolyticus	Yellow	Good
V. carchariae	Yellow	Good
V. cholerae	Yellow	Good
V. cincinnatiensis	Yellow	Very poor
V. fluvialis	Yellow	Good
V. furnissii	Yellow	Good
V. metschnikovii	Yellow	May be reduced
V. damsela	Green[b]	Reduced at 36°C
V. hollisae	Green	Very poor
V. mimicus	Green	Good
V. parahaemolyticus	Green[c]	Good
V. vulnificus	Green[d]	Good

[a]Adapted from reference 44.
[b]5% yellow.
[c]1% yellow.
[d]10% yellow.

been published (64). None of these molecular methods are commercially available.

CULTURE AND ISOLATION

Definitive laboratory diagnosis of cholera requires isolation of *V. cholerae* from the stool of a patient with diarrhea. All of the *Vibrio* species pathogenic for humans except *V. hollisae* will grow on thiosulfate-citrate-bile salts-sucrose agar (TCBS), which is recommended for the isolation of *Vibrio* spp. from stool specimens. The colors of colonies on this medium vary (Table 2). TCBS is very selective for *Vibrio* spp. Media from different manufacturers show variations in their abilities to select for *Vibrio* spp. and inhibit the growth of other bacteria (55). MacConkey agar is not recommended, because not all *Vibrio* spp. grow well on this medium. Those that do grow on MacConkey agar appear as lactose-negative colonies. *Vibrio* spp. from sites other than stools can be recovered on 5% sheep blood agar, which is not selective enough for use alone for isolation of *Vibrio* species from stools. With the exception of hemolytic strains of *V. cholerae* O1 and *V. damsela*, *Vibrio* species are nonhemolytic on sheep blood agar. A modified taurocholate-tellurite-gelatin agar containing 4-methylumbelliferyl-β-D-galactoside has been proposed as an alternative to TCBS for differentiation and presumptive identification of *Vibrio* species (62).

The routine use of TCBS for plating of stool specimens is not cost-effective in the United States even in areas where *Vibrio* infections are endemic (50). Therefore, the Gulf Coast *Vibrio* Working Group recommends using TCBS when patients present with a compatible diarrheal illness and a history of eating raw seafood. Additionally, specimens from wound infections following exposure to seawater and fish or marine animals should be cultured on TCBS.

Because of the wide range of salt requirements among *Vibrio* species, isolates from human clinical material, including blood specimens, may appear as aberrant forms not recognizable as *Vibrio* spp. A stool isolate of *V. parahaemolyticus* subcultured from TCBS demonstrated aberrant biochemical and morphologic features until it was inoculated onto medium supplemented with sodium chloride (12).

TABLE 3 Biochemical and other characteristics of 12 *Vibrio* species found in human clinical specimens[a]

Test[b]	% Positive[c]											
	V. cholerae	*V. mimicus*	*V. metschnikovii*	*V. cincinnatiensis*	*V. hollisae*	*V. damsela*	*V. fluvialis*	*V. furnissii*	*V. alginolyticus*	*V. parahaemolyticus*	*V. vulnificus*	*V. carchariae*
Indole production[d] (HIB, 1% NaCl)	99	98	20	8	97	0	13	11	85	98	97	100
Methyl red (1% NaCl)	99	99	96	93	0	100	96	100	75	80	80	100
Voges-Proskauer[d] (1% NaCl, Barritt)	75	9	96	0	0	95	0	0	95	0	0	50
Citrate, Simmons	97	99	75	21	0	0	93	100	1	3	75	0
H₂S on TSI	0	0	0	0	0	0	0	0	0	0	0	0
Urea hydrolysis	0	1	0	0	0	0	0	0	0	15	1	0
Phenylalanine deaminase	0	0	0	0	0	0	0	0	1	1	35	NG
Arginine, Moeller[d] (1% NaCl)	0	0	60	0	0	95	93	100	0	0	0	0
Lysine, Moeller[d] (1% NaCl)	99	100	35	57	0	50	0	0	99	100	99	100
Ornithine, Moeller[d] (1% NaCl)	99	99	0	0	0	0	0	0	50	95	55	0
Motility (36°C)	99	98	74	86	0	25	70	89	99	99	99	0
Gelatin hydrolysis (1% NaCl, 22°C)	90	65	65	0	0	6	85	86	90	95	75	0
KCN test (% that grow)	10	2	0	0	0	5	65	89	15	20	1	0
Malonate utilization	1	0	0	0	0	0	0	11	0	0	0	0
D-Glucose, acid production[d]	100	100	100	100	100	100	100	100	100	100	100	50
D-Glucose, gas production[d]	0	0	0	0	0	10	0	100	0	0	0	0
Acid production from:												
D-Adonitol	0	0	0	0	0	0	0	0	1	0	0	0
L-Arabinose[d]	0	1	0	100	97	0	93	100	1	80	0	0
D-Arabitol[d]	0	0	0	0	0	0	65	89	0	0	0	0
Cellobiose	8	0	9	100	0	0	30	11	3	5	99	50
Dulcitol	0	0	0	0	0	0	0	0	0	3	0	0
Erythritol	0	0	0	0	0	0	0	0	0	0	0	0
D-Galactose	90	82	45	100	100	90	96	100	20	92	96	0
Glycerol	30	13	100	100	0	0	7	55	80	50	1	0
myo-Inositol	0	0	40	100	0	0	3	0	0	0	85	50
Lactose[d]	7	21	50	0	0	0	3	0	0	1	85	0
Maltose[d]	99	99	100	100	0	100	100	100	100	99	100	100
D-Mannitol[d]	99	99	96	100	0	0	97	100	100	100	45	50
D-Mannose	78	99	100	100	100	100	100	100	99	100	98	50
Melibiose	1	0	0	7	0	0	3	11	1	1	40	0
α-Methyl-D-glucoside	0	0	25	57	0	5	0	0	1	0	0	0
Raffinose	0	0	0	0	0	0	0	11	0	0	0	0
L-Rhamnose	0	0	0	0	0	0	0	45	0	1	1	0
Salicin[d]	1	0	9	100	0	0	0	0	4	1	95	0

(Continued on next page)

TABLE 3 Biochemical and other characteristics of 12 *Vibrio* species found in human clinical specimens[a] (*Continued*)

Test[b]	% Positive[c]											
	V. cholerae	*V. mimicus*	*V. metschnikovii*	*V. cincinnatiensis*	*V. hollisae*	*V. damsela*	*V. fluvialis*	*V. furnissii*	*V. alginolyticus*	*V. parahaemolyticus*	*V. vulnificus*	*V. carchariae*
D-Sorbitol	1	0	45	0	0	0	3	0	1	1	0	0
Sucrose[d]	100	0	100	100	0	5	100	100	99	1	15	50
Trehalose	99	94	100	100	0	86	100	100	100	99	100	50
D-Xylose	0	0	0	43	0	0	0	0	0	0	0	0
Mucate, acid production	1	0	0	0	0	0	0	0	0	0	0	0
Tartrate, Jordan	75	12	35	0	65	0	35	22	95	93	84	50
Esculin hydrolysis	0	0	60	0	0	0	8	0	3	1	40	0
Acetate utilization	92	78	25	14	0	0	70	65	0	1	7	0
Nitrate → nitrate[d]	99	100	0	100	100	100	100	100	100	100	100	100
Oxidase[d]	100	100	0	100	100	95	100	100	100	100	100	100
DNase, 25°C	93	55	50	79	0	75	90	100	95	92	50	100
Lipase (corn oil)[d]	92	17	100	36	0	0	40	89	85	90	92	0
ONPG test[d]	94	90	50	86	0	0	0	35	0	5	75	0
Yellow pigment at 25°C	0	0	0	0	0	0	0	0	0	0	0	0
Tyrosine clearing	13	30	5	0	3	0	65	45	70	77	75	0
Growth in nutrient broth with:												
0% NaCl[d]	100	100	0	0	0	0	0	0	0	0	0	0
1% NaCl[d]	100	100	100	100	99	100	99	99	99	100	99	100
6% NaCl[d]	53	49	78	100	83	95	96	100	100	99	65	100
8% NaCl[d]	1	0	44	62	0	0	71	78	94	80	0	0
10% NaCl[d]	0	0	4	0	0	0	4	0	69	2	0	0
12% NaCl	0	0	0	0	0	0	0	0	17	1	0	0
Swarming (marine agar, 25°C)	-	-	-	+	-	-	-	-	+	+	-	100
String test	100	100	100	80	100	80	100	100	91	64	100	100
O/129, zone of inhibition[e]	99	95	90	25	40	90	31	0	19	20	98	100
Polymyxin B, zone of inhibition	22	88	100	92	100	85	100	89	63	54	3	100

[a]From reference 44.

[b]HIB, heart infusion broth; 1% NaCl, 1% NaCl added to the standard medium to enhance growth; TSI, triple sugar iron agar.

[c]After 48 h of incubation at 36°C (unless other conditions are indicated). Most positive reactions occur during the first 24 h. NG, no growth (probably because NaCl concentration is too low); +, most strains (generally about 90 to 100%) positive; −, most strains negative (generally about 0 to 10% positive).

[d]Test is recommended as part of the routine set for *Vibrio* identification.

[e]Disk potency, 150 μg.

TABLE 4 Eight key differential tests for dividing the 12 clinically significant *Vibrio* species into six groups[a]

| | Reaction of species in[b]: | | | | | | | | | | | |
| | Group 1 | | Group 2, *V. metschnikovii* | Group 3, *V. cincinnatiensis* | Group 4, *V. hollisae* | Group 5 | | | Group 6 | | | |
Test	*V. cholerae*	*V. mimicus*				*V. damsela*	*V. fluvialis*	*V. furnissii*	*V. alginolyticus*	*V. parahaemolyticus*	*V. vulnificus*	*V. carchariae*
Growth in nutrient broth:												
With no NaCl added	+	+	–	–	–	–	–	–	–	–	–	–
With 1% NaCl added	+	+	+	+	+	+	+	+	+	+	+	+
Oxidase	+	+	–	+	+	+	+	+	+	+	+	+
Nitrate → nitrite	+	+	–	+	+	+	+	+	+	+	+	+
myo-Inositol fermentation	–	–	v	+	–	–	–	–	–	–	–	–
Arginine dihydrolase	–	–	v	–	–	+	+	+	–	–	–	–
Lysine decarboxylase	+	+	+	+	–	+	–	–	+	+	+	+
Ornithine decarboxylase	+	+	–	–	–	–	–	–	–	+	+	+

[a]From reference 44.
[b]Key test results are boxed. All data are for reactions within 2 days at 35 to 37°C unless otherwise specified. Symbols: +, most strains (generally about 90 to 100%) positive; –, most strains negative (generally about 0 to 10% positive); v, strain-to-strain variation.

DETECTION OF CHOLERA TOXINS

Secretory diarrhea, the hallmark of cholera, is caused by enterotoxins. Cholera toxin is composed of A and B subunits. The B portion, composed of five identical subunits, is responsible for binding to ganglioside receptors on the cell surface. The A subunit functions to increase the level of cyclic AMP in interstitial cells, leading to the loss of fluid and electrolytes into the intestinal lumen. Finkelstein reviews details of the structure and function of cholera enterotoxin (28). Additionally, *V. cholerae* produces a zonula occludens toxin that loosens the tight junctions between intestinal epithelial cells (25). Classically, the *V. cholerae* toxins have been detected by using a variety of in vivo models, including the ligated rabbit ileal loop. Tissue culture assays using Y1 adrenal cells or Chinese hamster kidney cells are also used.

An ELISA using GM_1 ganglioside as the capture molecule has been developed for detection of the heat-labile enterotoxins of both *V. cholerae* O1 and *Escherichia coli* (77). Additionally, a toxin bead ELISA was used successfully to rapidly detect cholera toxin from samples in a cholera outbreak (73). In comparison to a standard ELISA, a latex agglutination kit (VET-RPLA; Oxoid, U.S.A.) that detects cholera toxin has a sensitivity of 97% and a specificity of 100% (5). These systems offer the advantage of making possible a diagnosis of cholera in the field within a short time of specimen receipt.

PCR was compared with the GM_1 ELISA in the determination of toxigenicity of *V. cholerae* O1 collected in Argentina during 1991. When primers for both ctxA1 and ctxA2 were used, the correlation between the two tests was excellent (91).

IDENTIFICATION

Classic methods can be used for the identification of *Vibrio* species (Table 3). The addition of 1% NaCl to medium without salt is recommended to enhance the positivity of biochemical reactions (24). Fewer positive reactions are seen above or below the optimum NaCl concentration for growth (92). Kelly et al. (44) present a scheme (Table 4) for separating *Vibrio* species into six groups. These groupings provide a convenient starting point for further species identification.

Vibrio species are included in the databases of several commercial identification systems. The reliability of these databases, however, has not been evaluated for large numbers of strains.

General Tests for Initial Characterization of *Vibrio* Species

Vibriostatic Test

First described in 1954 (81) for the differentiation of *Pseudomonas* from *Vibrio* species, O/129 (2,4-diamino-6,7-diisopropylpteridine) has been used for separating *Vibrio* species from *Aeromonas* spp. and for intraspecies differentiation. Disks and powder are available from Oxoid, U.S.A., and the phosphate salt is available from Sigma (catalog no. D0781). *V. cholerae* O1 and non-O1 have been reported to be 99 and 95% susceptible to O/129, respectively. A 0.5 MacFarland suspension of the test organism is prepared and inoculated onto a Trypticase soy agar plate as if performing a Kirby-Bauer disk diffusion test. Disks of O/129 (10 and

TABLE 5 Differentiation of arginine dihydrolase-positive *Vibrio* species and comparison with *Aeromonas* species[a]

Test or property	Reaction[b]			
	V. damsela	*V. fluvialis*	*V. furnissii*	*Aeromonas* spp.
Growth in nutrient broth with:				
No added NaCl	−	−	−	+
1% NaCl	+	+	+	+
Voges-Proskauer	+	−	−	v
Citrate, Simmons	−	+	+	v
Fermentation of:				
D-Galacturonic acid	−	+	+	−
L-Arabinose	−	+	+	v
D-Mannitol	−	+	+	+
Sucrose	−	+	+	(+)
Gas production during fermentation	v	−	+	v

[a]From reference 44.
[b]See Table 6, footnote b, for explanation of symbols.

150 μg) are placed on the plate, which is then incubated overnight. Any zone of inhibition is considered positive.

Reports of *V. cholerae* resistant to O/129 have now been published. Thirty-two (67%) of 51 strains isolated from water, plankton, and rectal swabs collected in Bangladesh were determined to be resistant to O/129 (37). Over 80% of 110 clinical isolates of *V. cholerae* O1 collected in India in 1989 and 1990 were resistant to 150 μg of O/129 (74). One O/129-resistant non-O1 *V. cholerae* strain isolated in California from a patient with gastroenteritis has been reported (1). O/129 susceptibility as a diagnostic test should now be used with caution.

String Test

The string test was first described by Smith in 1970 (83). In this test, a suspension of a suspected *Vibrio* colony in 0.5% sodium deoxycholate in saline on a glass slide is stirred with a loop. *Vibrio* species lyse, releasing DNA, which can be pulled up into a string with a loop. *Aeromonas* spp. are negative in the string test. A 0.5% solution of the reagent can be prepared with deoxycholic acid and sodium salt (ICN Biochemicals, catalog no. 101496; Sigma, catalog no. D6750).

Salt Requirement and Tolerance Tests

The salt requirement and tolerance tests are important for differentiating *Vibrio* species. Nutrient broth without NaCl in the formulation must be used to prepare the medium. Nutrient broths without added NaCl and with 1, 2, 4, 6, 8, 10, and 12% NaCl are used to determine salt tolerance and requirement (24).

Biochemical Identification

The rapid identification of *V. cholerae* is of major epidemiologic importance. *V. cholerae* grows well on TCBS, forming yellow colonies in 18 to 24 h. It can also be isolated on MacConkey agar as colorless colonies. Yellow colonies on TCBS that are oxidase positive on nonselective medium should be promptly evaluated by performing agglutination tests with specific *V. cholerae* antiserum and confirmatory biochemical tests. A positive string test can be used to rapidly differentiate *V. cholerae* from *Aeromonas* spp., which are also oxidase positive and can appear on TCBS as yellow colonies.

V. cholerae is composed of two biotypes of epidemiologic

significance, Classical and El Tor. These biotypes can be differentiated by a series of tests, including agglutination of chicken and sheep erythrocytes, Mukerjee phage susceptibility, polymyxin B susceptibility, hemolysis on blood agar, and others (66). El Tor strains cause a higher ratio of asymptomatic cases to cholera gravis (severe disease) cases than does the Classical biotype. The seventh pandemic strain is El Tor, either Inaba or Ogawa, except in the southern part of Bangladesh, the ancestral home of cholera, where Classical strains still predominate.

The two common serotypes of *V. cholerae* O1 are designated Inaba and Ogawa. They share an A antigen. Ogawa has the antigenic formula AB, and Inaba has the formula AC. Hikojima, a rare serotype, carries all three antigens, ABC.

V. mimicus, so named for its resemblance to *V. cholerae* O1, was shown in 1981 to be a distinct species (20). It is not a halophile and can be differentiated from *V. cholerae* by a negative sucrose test, negative Voges-Proskauer test, and sensitivity to polymyxin B and colistin. Few isolates of *V. mimicus* produce cholera toxin. A thermostable hemolysin-like toxin has been detected in 16 of 17 clinical isolates in Japan (90).

V. fluvialis is in the arginine dihydrolase-positive group of *Vibrio* species (Table 5). It is a halophile, and most strains tolerate 8% but not 10% NaCl. It can be confused with *Aeromonas* species, but the salt requirement differentiates these organisms. *V. furnissii* is also arginine dihydrolase positive. *V. furnissii* and *V. fluvialis* can be differentiated on the basis of gas production, esculin hydrolysis, and carbon source utilization (44).

V. damsela is a halophile in group 5 that tolerates 6% but not 8% NaCl. It differs from other *Vibrio* spp. in being indole negative and methyl red and Voges-Proskauer positive. It is only weakly motile. On sheep blood agar, it produces a zone of beta-hemolysis.

Some strains of *V. hollisae*, the only species in group 4, fail to grow on TCBS or MacConkey agar but will grow and demonstrate weak hemolysis on blood agar.

V. parahaemolyticus (group 6) grows readily as green colonies on TCBS and as nonhemolytic colonies on sheep blood agar. It does not reliably grow on the highly selective media routinely used for the isolation of enteric pathogens. *V. parahaemolyticus* is in the arginine dihydrolase-negative, lysine decarboxylase-positive group of pathogenic *Vibrio*

TABLE 6 Differentiation of arginine dihydrolase-negative, lysine decarboxylase-positive *Vibrio* species[a]

Test or property	Reaction[b]			
	V. alginolyticus	*V. parahaemolyticus*	*V. vulnificus*	*V. carchariae*
Voges-Proskauer	+	−	−	−
Growth in nutrient broth with:				
8% NaCl	+	(+)	−	−
10% NaCl	v	−	−	−
Fermentation of:				
Sucrose	+	−	(−)	+
Salicin	−	−	+	−
Cellobiose	−	−	+	v
Lactose	−	−	(+)	−
L-Arabinose	−	(+)	−	−
Swarming (marine agar, 25°C)	+	+	−	+
Zone of inhibition around:				
Colistin	Large	Large	Small	Small
Ampicillin	Small	Small	Large	Small
Carbenicillin	Small	Small	Large	Small

[a]From reference 44.
[b]Symbols: +, most strains (generally about 90 to 100%) positive; −, most strains negative (generally about 0 to 10% positive); (+), many strains (generally about 75 to 90%) positive; v, strain-to-strain variation (generally about 25 to 75% positive); (−), many strains negative (generally about 10 to 25% positive).

species (Tables 4 and 6). Many isolates of *V. parahaemolyticus* have now been shown to be urease positive. A hemolysin can be demonstrated on Watgasuma medium, and the production of this hemolysin, the Kanagawa phenomenon, is related to pathogenicity (57). A serotyping system is based on 12 O antigens and 55 K antigens.

V. vulnificus is a halophile in the same biochemical group as *V. parahaemolyticus*. It is most readily differentiated from *V. parahaemolyticus* by its abilities to produce *o*-nitrophenyl-β-D-galactopyranoside (ONPG) and to ferment lactose. It requires 0.5% sodium chloride for growth and tolerates 6% but not 8% NaCl. The biochemical characteristics useful in differentiating *Vibrio* species in the arginine dihydrolase-negative, lysine decarboxylase-positive group are given in Table 6. *V. vulnificus* has been detected in artificially contaminated oysters by PCR (34) and on nonselective media with a labeled probe against the cytolysin gene (97).

V. alginolyticus is one of the group 6 halophilic *Vibrio* species, which are arginine dihydrolase negative and lysine decarboxylase positive (Table 6). A *Proteus*-like swarming due to the action of peritrichous flagella is usually seen on blood agar. *V. alginolyticus* grows in the presence of 8% salt, and some strains grow in 10% salt. *V. carchariae* is also in the arginine dihydrolase-negative, lysine decarboxylase-positive group of *Vibrio* species. The one clinical isolate reported required salt for growth and grew in 3, 6, and 9% NaCl (68).

V. cincinnatiensis and *V. metschnikovii* are the only *Vibrio* species isolated from human clinical specimens that ferment *myo*-inositol.

Unlike other *Vibrio* species isolated from human clinical material, *V. metschnikovii* does not produce oxidase, nor does it reduce nitrate to nitrite.

ANTIBIOTIC SUSCEPTIBILITY

Susceptibility testing of *Vibrio* species should be performed with Mueller-Hinton broth or agar without added NaCl (36). The addition of NaCl to Mueller-Hinton agar inter-

fered with the activity of gentamicin in a study of the susceptibilities of *V. parahaemolyticus* and *V. alginolyticus* strains (43). The susceptibilities of 1,025 isolates of the 12 *Vibrio* species isolated from human clinical specimens and submitted to the CDC have been reported elsewhere (24). Testing was done by the Kirby-Bauer method. Most of the isolates were susceptible to chloramphenicol, gentamicin, and nalidixic acid.

Norfloxacin, a quinolone, has good activity against 40 strains of O1 and non-O1 *V. cholerae*, 22 strains of *V. parahaemolyticus*, and 19 strains of *V. vulnificus* tested. The susceptibility of smaller numbers of isolates of other pathogenic species was also reported (59).

The first widespread outbreak of cholera due to a multiply resistant strain of *V. cholerae* was reported from Tanzania in 1977. Plasmid-encoded multiple-antibiotic resistance in El Tor strains was detected in Bangladesh in 1980 (87). Resistance to trimethoprim-sulfamethoxazole and ampicillin has been linked to resistance to the vibriostatic compound O/129 (37). Tetracycline resistance of *V. cholerae* O1 is common in areas where the organism is endemic.

Although it has been reported that the susceptibilities of *Vibrio* species can be determined by using semiautomated methods, comparative studies of results obtained by Kirby-Bauer disk diffusion and semiautomated susceptibility testing methods have not been published (88).

REFERENCES

1. Abbott, S. L., W. K. W. Cheung, B. A. Portoni, and J. M. Janda. 1992. Isolation of vibriostatic agent O/129-resistant *Vibrio cholerae* non-O1 from a patient with gastroenteritis. *J. Clin. Microbiol.* 30:1598–1599.
2. Albert, M. J., M. Ansaruzzaman, P. K. Bardhan, A. S. G. Faruque, S. M. Faruque, M. S. Islam, D. Mahalanabis, R. B. Sack, M. A. Salam, A. K. Siddique, M. D. Yunus, and K. Zaman. 1993. Large epidemic of cholera-like disease in Bangladesh caused by *Vibrio cholerae* O139 synonym Bengal. *Lancet* 342:387–390.
3. Albert, M. J., M. A. Hossain, K. Alam, I. Kabir, P. K. B. Neogi, and S. Tzipori. 1991. A fatal case associated with

shigellosis and *Vibrio fluvialis* bacteremia. *Diagn. Microbiol. Infect. Dis.* **14**:509–510.

4. **Albert, M. J., A. K. Siddique, M. S. Islam, A. S. G. Faruque, M. Ansaruzzaman, S. M. Faruque, and R. B. Sack.** 1993. Large outbreak of clinical cholera due to *Vibrio cholerae* non-O1 in Bangladesh. *Lancet* **341**:704. (Letter.)

5. **Almeida, R. J., F. W. Hickman-Brenner, E. G. Sowers, N. D. Puhr, J. J. Farmer III, and I. K. Wachsmuth.** 1990. Comparison of a latex agglutination assay and an enzyme-linked immunosorbent assay for detecting cholera toxin. *J. Clin. Microbiol.* **28**:128–130.

6. **Barker, W. H., Jr.** 1974. *Vibrio parahaemolyticus* outbreaks in the United States. *Lancet* **i**:551–554.

7. **Barua, D., and W. B. Greenough III (ed.).** 1992. *Current Topics in Infectious Disease: Cholera.* Plenum Publishing Co., New York.

8. **Baumann, P., A. L. Furniss, and J. V. Lee.** 1984. Genus I. *Vibrio* Pacini 1854, 411AL, p. 518–538. *In* N. R. Krieg and J. G. Holt (ed.), *Bergey's Manual of Systematic Bacteriology*, vol. 1. The Williams & Wilkins Co., Baltimore.

9. **Bhattacharya, M. K., S. K. Bhattacharya, S. Garg, P. K. Saha, D. Dutta, G. B. Nair, B. C. Deb, and K. P. Das.** 1993. Outbreak of *Vibrio cholerae* non-O1 in India and Bangladesh. *Lancet* **341**:1346–1347. (Letter.)

10. **Blake, P. A., M. H. Merson, R. E. Weaver, D. G. Hollis, and P. C. Heublein.** 1979. Disease caused by a marine *Vibrio*. Clinical characteristics and epidemiology. *N. Engl. J. Med.* **300**:1–5.

11. **Bode, R. B., P. R. Brayton, R. R. Colwell, F. M. Russo, and W. E. Bullock.** 1986. A new *Vibrio* species, *Vibrio cincinnatiensis*, causing meningitis: successful treatment in an adult. *Ann. Intern. Med.* **104**:55–56.

12. **Bottone, E. J., and T. Robin.** 1978. *Vibrio parahaemolyticus*: suspicion of presence based on aberrant biochemical and morphological features. *J. Clin. Microbiol.* **8**:760–763.

13. **Brayton, P. R., R. B. Bode, R. R. Colwell, M. T. MacDonell, H. L. Hall, D. J. Grimes, P. A. West, and T. N. Bryant.** 1986. *Vibrio cincinnatiensis* sp. nov., a new human pathogen. *J. Clin. Microbiol.* **23**:104–108.

14. **Brenner, D. J., F. W. Hickman-Brenner, J. V. Lee, A. G. Steigerwalt, G. R. Fanning, D. G. Hollis, J. J. Farmer III, R. E. Weaver, and R. J. Seidler.** 1983. *Vibrio furnissii* (formerly aerogenic biogroup of *Vibrio fluvialis*), a new species isolated from human feces and the environment. *J. Clin. Microbiol.* **18**:816–824.

15. **Buck, J. D.** 1990. Isolation of *Candida albicans* and halophilic *Vibrio* spp. from aquatic birds in Connecticut and Florida. *Appl. Environ. Microbiol.* **56**:826–828.

16. **Chongsa-nguan, M., W. Chaicumpa, P. Moolasart, P. Kandhasingha, T. Shimada, H. Kurazono, and Y. Takeda.** 1993. *Vibrio cholerae* O139 Bengal in Bangkok. *Lancet* **342**:430–431. (Letter.)

17. **Chuang, Y.-C., C.-Y. Yuan, C.-Y. Liu, C.-K. Lan, and A. H.-M. Huang.** 1992. *Vibrio vulnificus* infection in Taiwan: report of 28 cases and review of clinical manifestations and treatment. *Clin. Infect. Dis.* **15**:271–276.

18. **Clarridge, J. E., and S. Zighelboim-Daum.** 1985. Isolation and characterization of two hemolytic phenotypes of *Vibrio damsela* associated with a fatal wound infection. *J. Clin. Microbiol.* **21**:302–306.

19. **Coffey, J. A., Jr., R. L. Harris, M. L. Rutledge, M. W. Bradshaw, and T. W. Williams, Jr.** 1986. *Vibrio damsela*: another potentially virulent marine *Vibrio*. *J. Infect. Dis.* **153**:800–802.

20. **Davis, B. R., G. R. Fanning, J. M. Madden, A. G. Steigerwalt, H. B. Bradford, Jr., H. L. Smith, Jr., and D. J. Brenner.** 1981. Characterization of biochemically atypical *Vibrio cholerae* strains and designation of a new pathogenic species, *Vibrio mimicus. J. Clin. Microbiol.* **14**:631–639.

21. **DePaola, A., G. M. Capers, M. L. Motes, Ø. Olsvik, P. I. Fields, J. Wells, I. K. Wachsmuth, T. A. Cebula, W. H. Koch, F. Khambaty, M. H. Kothary, W. L. Payne, and B. A. Wentz.** 1992. Isolation of Latin American epidemic strain of *Vibrio cholerae* O1 from US Gulf Coast. *Lancet* **339**:624. (Letter.)

22. **Eichold, B. H., II, J. R. Williamson, C. H. Woernle, and R. M. McPhearson.** 1993. Isolation of *Vibrio cholerae* O1 from oysters—Mobile Bay, 1991–1992. *Morbid. Mortal. Weekly Rep.* **42**:91–93.

23. **English, V. L., and R. B. Lindberg.** 1977. Isolation of *Vibrio alginolyticus* from wounds and blood of a burn patient. *Am. J. Med. Technol.* **44**:989–993.

24. **Farmer, J. J., III, and F. W. Hickman-Brenner.** 1992. The genera *Vibrio* and *Photobacterium*, p. 2952–3011. *In* A. Balows, H. G. Trüper, M. Dworkin, W. Harder, and K.-H. Schleifer (ed.), *The Prokaryotes: a Handbook on the Biology of Bacteria*, vol. 3. *Ecophysiology, Isolation, Identification, Applications*. Springer-Verlag, New York.

25. **Fasano, A., B. Baudry, D. W. Pumplin, S. S. Wasserman, B. D. Tall, J. M. Ketley, and J. B. Kaper.** 1991. *Vibrio cholerae* produces a second enterotoxin, which affects intestinal tight junctions. *Proc. Natl. Acad. Sci. USA* **88**:5242–5246.

26. **Fields, P. I., T. Popovic, K. Wachsmuth, and Ø. Olsvik.** 1992. Use of polymerase chain reaction for detection of toxigenic *Vibrio cholerae* O1 strains from the Latin American cholera epidemic. *J. Clin. Microbiol.* **30**:2118–2121.

27. **Finelli, L., D. Swerdlow, K. Mertz, H. Ragazzoni, and K. Spitalny.** 1992. Outbreak of cholera associated with crab brought from an area with epidemic disease. *J. Infect. Dis.* **166**:1433–1435.

28. **Finkelstein, R. A.** 1992. Cholera enterotoxin (choleragen): a historical perspective, p. 155–187. *In* D. Barua and W. B. Greenough III (ed.), *Current Topics in Infectious Disease: Cholera*. Plenum Publishing Co., New York.

29. **Florman, A. L., A. H. Cushing, T. Byers, and S. Popejoy.** 1990. *Vibrio cholerae* bacteremia in a 22-month-old New Mexican child. *Pediatr. Infect. Dis. J.* **9**:63–65. (Letter.)

30. **Gardner, A. D., and K. V. Venkatraman.** 1935. The antigens of the cholera group of vibrios. *J. Hyg.* **35**:262–282.

31. **Hall, R. H., F. M. Khambaty, M. Kothary, and S. P. Keasler.** 1993. Non-O1 *Vibrio cholerae*. *Lancet* **342**:430. (Letter.)

32. **Hickman, F. W., J. J. Farmer III, D. G. Hollis, G. R. Fanning, A. G. Steigerwalt, R. E. Weaver, and D. J. Brenner.** 1982. Identification of *Vibrio hollisae* sp. nov. from patients with diarrhea. *J. Clin. Microbiol.* **15**:395–401.

33. **Hickman-Brenner, F. W., D. J. Brenner, A. G. Steigerwalt, M. Schreiber, S. D. Holmberg, L. M. Baldy, C. S. Lewis, N. M. Pickens, and J. J. Farmer III.** 1984. *Vibrio fluvialis* and *Vibrio furnissii* isolated from a stool sample of one patient. *J. Clin. Microbiol.* **20**:125–127.

34. **Hill, W. E., S. P. Keasler, M. W. Trucksess, P. Feng, C. A. Kaysner, and K. A. Lampel.** 1991. Polymerase chain reaction identification of *Vibrio vulnificus* in artificially contaminated oysters. *Appl. Environ. Microbiol.* **57**:707–711.

35. **Hoge, C. W., D. Watsky, R. N. Peeler, J. P. Libonati, E. Israel, and J. G. Morris, Jr.** 1989. Epidemiology and spectrum of *Vibrio* infections in a Chesapeake Bay community. *J. Infect. Dis.* **160**:985–993.

36. **Hollis, D. G., R. E. Weaver, C. N. Baker, and C. Thornsberry.** 1976. Halophilic *Vibrio* species isolated from blood cultures. *J. Clin. Microbiol.* **3**:425–431.

37. **Huq, A., M. Alam, S. Parveen, and R. R. Colwell.** 1992. Occurrence of resistance to vibriostatic compound O/129 in *Vibrio cholerae* O1 isolated from clinical and environmental samples in Bangladesh. *J. Clin. Microbiol.* **30**:219–221.

38. **Huq, M. I., A. K. M. J. Alam, D. J. Brenner, and G. K. Morris.** 1980. Isolation of *Vibrio*-like group, EF-6, from patients with diarrhea. *J. Clin. Microbiol.* **11**:621–624.

39. **Islam, M. S., M. K. Hasan, M. A. Miah, F. Qadri, M. Yunus, R. B. Sack, and M. J. Albert.** 1993. Isolation of *Vibrio cholerae* O139 Bengal from water in Bangladesh. *Lancet* **430**:342. (Letter.)

40. **Janda, J. M., C. Powers, R. G. Bryant, and S. L. Abbott.** 1988. Current perspectives on the epidemiology and pathogenesis of clinically significant *Vibrio* spp. *Clin. Microbiol. Rev.* **1**:245–267.

41. **Jean-Jacques, W., K. A. Rajashekaraiah, J. J. Farmer III, F. W. Hickman, J. G. Morris, and C. A. Kallick.** 1981. *Vibrio metschnikovii* bacteremia in a patient with cholecystitis. *J. Clin. Microbiol.* **14:**711–712.

42. **Johnston, J. M., S. F. Becker, and L. M. McFarland.** 1986. Gastroenteritis in patients with stool isolates of *Vibrio vulnificus*. *Am. J. Med.* **80:**336–338.

43. **Joseph, S. W., R. M. DeBell, and W. P. Brown.** 1978. In vitro response to chloramphenicol, tetracycline, ampicillin, gentamicin, and beta-lactamase production by halophilic vibrios from human and environmental sources. *Antimicrob. Agents Chemother.* **13:**244–248.

44. **Kelly, M. T., F. W. Hickman-Brenner, and J. J. Farmer III.** 1992. *Vibrio*, p. 384–395. *In* A. Balows, W. J. Hausler, Jr., K. L. Herrmann, H. D. Isenberg, and H. J. Shadomy (ed.), *Manual of Clinical Microbiology*, 5th ed. American Society for Microbiology, Washington, D.C.

45. **Kreger, A. S.** 1984. Cytolytic activity and virulence of *Vibrio damsela*. *Infect. Immun.* **44:**326–331.

46. **Lacey, C., R. Talbot, J. Taylor, D. Dwyer, B. Jolbitado, C. Morrison, E. Butler-Senkel, S. Strauss, D. Murphy-Baxam, J. Libonati, E. Israel, N. Ridley, M. Smith, J. Zingeser, G. Miller, Jr., and K. Ungchusak.** 1991. Cholera associated with imported frozen coconut milk—Maryland, 1991. *Morbid. Mortal. Weekly Rep.* **40:**844–845.

47. **Lee, J. V., T. J. Donovan, and A. L. Furniss.** 1978. Characterization, taxonomy, and emended description of *Vibrio metschnikovii*. *Int. J. Syst. Bacteriol.* **28:**99–111.

48. **Lee, J. V., P. Shread, A. L. Furniss, and T. N. Bryant.** 1981. Taxonomy and description of *Vibrio fluvialis* sp. nov. (synonym group F vibrios, group EF6). *J. Appl. Bacteriol.* **50:**73–94.

49. **Lessner, A. M., R. M. Webb, and B. Rabin.** 1985. *Vibrio alginolyticus* conjunctivitis. First reported case. *Arch. Ophthalmol.* **103:**229–230.

50. **Levine, W. C., P. M. Griffin, and the Gulf Coast *Vibrio* Working Group.** 1993. *Vibrio* infections on the Gulf Coast: results of first year of regional surveillance. *J. Infect. Dis.* **167:**479–483.

51. **Love, M., D. Teebken-Fisher, J. E. Hose, J. J. Farmer III, F. W. Hickman, and G. R. Fanning.** 1981. *Vibrio damsela*, a marine bacterium, causes skin ulcers on the damselfish *Chromis punctipinnis*. *Science* **214:**1139–1140.

52. **Manning, P. A., U. H. Stroeher, and R. Morona.** 1994. Molecular basis for O-antigen biosynthesis in *Vibrio cholerae* O1: Ogawa-Inaba switching, p. 77–94. *In* I. K. Wachsmuth, P. A. Blake, and Ø. Olsvik (ed.), *Vibrio cholerae and Cholera: Molecular to Global Perspectives*. American Society for Microbiology, Washington, D.C.

53. **Mascola, L., M. Tormey, D. Ewert, S. Fannin, T. Prendergast, H. Meyers, M. Ginsberg, F. Taylor, S. Abbott, S. B. Werner, G. W. Rutherford, O. Ravenholt, L. Empey, D. Kwalick, R. Salcido, and D. Brus.** 1992. Cholera associated with an international airline flight, 1992. *Morbid. Mortal. Weekly Rep.* **41:**134–135.

54. **Matsiota-Bernard, P., and C. Nauciel.** 1993. *Vibrio alginolyticus* wound infection after exposure to sea water in an air crash. *Eur. J. Clin. Microbiol. Infect. Dis.* **12:**474–475.

55. **McCormack, W. M., W. E. DeWitt, P. E. Bailey, G. K. Morris, P. Soeharjono, and E. J. Gangarosa.** 1974. Evaluation of thiosulfate-citrate-bile salts-sucrose agar, a selective medium for the isolation of *Vibrio cholerae* and other pathogenic vibrios. *J. Infect. Dis.* **129:**497–500.

56. **Miwatani, T., and Y. Takeda.** 1976. Discovery of *Vibrio parahaemolyticus*, p. 1. *In* T. Miwatani and Y. Takeda (ed.), *Vibrio parahaemolyticus: a Causative Bacterium of Food Poisoning*. Saikon Publishing Co., Ltd., Tokyo.

57. **Miyamoto, Y., T. Kato, Y. Obara, S. Akiyama, K. Takizawa, and S. Yamai.** 1969. In vitro hemolytic characteristic of *Vibrio parahaemolyticus*: its close correlation with human pathogenicity. *J. Bacteriol.* **100:**1147–1149.

58. **Morris, J. G., Jr.** 1990. Non-O group 1 *Vibrio cholerae*: a look at the epidemiology of an occasional pathogen. *Epidemiol. Rev.* **12:**179–191.

59. **Morris, J. G., Jr., J. H. Tenney, and G. L. Drusano.** 1985. In vitro susceptibility of pathogenic *Vibrio* species to norfloxacin and six other antimicrobial agents. *Antimicrob. Agents Chemother.* **28:**442–445.

60. **Morris, J. G., Jr., R. Wilson, D. G. Hollis, R. E. Weaver, H. G. Miller, C. O. Tacket, F. W. Hickman, and P. A. Blake.** 1982. Illness caused by *Vibrio damsela* and *Vibrio hollisae*. *Lancet* **i:**1294–1297.

61. **Mulder, G. D., T. M. Ries, and T. R. Beaver.** 1989. Nontoxigenic *Vibrio cholerae* wound infection after exposure to contaminated lake water. *J. Infect. Dis.* **159:**809–811. (Letter.)

62. **O'Brien, M., and R. Colwell.** 1985. Modified taurocholate-tellurite-gelatin agar for improved differentiation of *Vibrio* species. *J. Clin. Microbiol.* **22:**1011–1013.

63. **Ogg, J. E., R. A. Ryder, and H. L. Smith, Jr.** 1989. Isolation of *Vibrio cholerae* from aquatic birds in Colorado and Utah. *Appl. Environ. Microbiol.* **55:**95–99.

64. **Olsvik, Ø., T. Popovic, and P. I. Fields.** 1993. PCR detection of toxin genes in strains of *Vibrio cholerae* O1, p. 266–276. *In* D. H. Persing, T. F. Smith, F. C. Tenover, and T. J. White (ed.), *Diagnostic Molecular Microbiology: Principles and Applications*. American Society for Microbiology, Washington, D.C.

65. **Opal, S. M., and J. R. Saxon.** 1986. Intracranial infection by *Vibrio alginolyticus* following injury in salt water. *J. Clin. Microbiol.* **23:**373–374.

66. **Pal, S. C.** 1992. Laboratory diagnosis, p. 229–251. *In* D. Barua and W. B. Greenough III (ed.), *Current Topics in Infectious Disease: Cholera*. Plenum Publishing Co., New York.

67. **Park, S. D., H. S. Shon, and N. J. Joh.** 1991. *Vibrio vulnificus* septicemia in Korea: clinical and epidemiologic findings in seventy patients. *J. Am. Acad. Dermatol.* **24:**397–403.

68. **Pavia, A. T., J. A. Bryan, K. L. Maher, T. R. Hester, Jr., and J. J. Farmer III.** 1989. *Vibrio carchariae* infection after a shark bite. *Ann. Intern. Med.* **111:**85–86.

69. **Perez-Tirse, J., J. F. Levine, and M. Mecca.** 1993. *Vibrio damsela*: a cause of fulminant septicemia. *Arch. Intern. Med.* **153:**1838–1840.

70. **Pezzlo, M., P. J. Valter, and M. J. Burns.** 1979. Wound infection associated with *Vibrio alginolyticus*. *Am. J. Clin. Pathol.* **71:**476–478.

71. **Ramamurthy, T., S. Garg, R. Sharma, S. K. Bhattacharya, G. B. Nair, T. Shimada, T. Takeda, T. Karasawa, H. Kurazano, A. Pal, and Y. Takeda.** 1993. Emergence of novel strain of *Vibrio cholerae* with epidemic potential in southern and eastern India. *Lancet* **341:**703–704. (Letter.)

72. **Ramamurthy, T., A. Pal, P. K. Bag, S. K. Bhattacharya, G. B. Nair, H. Kurozano, S. Yamasaki, H. Shirai, T. Takeda, Y. Uesaka, K. Horigome, and Y. Takeda.** 1993. Detection of cholera toxin gene in stool specimens by polymerase chain reaction: comparison with bead enzyme-linked immunosorbent assay and culture method for laboratory diagnosis of cholera. *J. Clin. Microbiol.* **31:**3068–3070.

73. **Ramamurthy, T., A. Pal, G. B. Nair, S. C. Pal, T. Takeda, and Y. Takeda.** 1990. Experience with toxin bead ELISA in cholera outbreak. *Lancet* **336:**375–376.

74. **Ramamurthy, T., A. Pal, S. C. Pal, and G. B. Nair.** 1992. Taxonomical implications of the emergence of high frequency of occurrence of 2,4-diamino-6,7-diisopropylpteridine-resistant strains of *Vibrio cholerae* from clinical cases of cholera in Calcutta, India. *J. Clin. Microbiol.* **30:**742–743.

75. **Rank, E. L., I. B. Smith, and M. Langer.** 1988. Bacteremia caused by *Vibrio hollisae*. *J. Clin. Microbiol.* **26:**375–376.

76. **Rhodes, J. B., H. L. Smith, Jr., and J. E. Ogg.** 1986. Isolation of non-O1 *Vibrio cholerae* serovars from surface waters in western Colorado. *Appl. Environ. Microbiol.* **51:**1216–1219.

77. **Sack, D. A., S. Huda, P. K. B. Neogi, R. R. Daniel, and W. M. Spira.** 1980. Microtiter ganglioside enzyme-linked immunosorbent assay for *Vibrio* and *Escherichia coli* heat-labile enterotoxins and antitoxin. *J. Clin. Microbiol.* **11:**35–40.

78. **Safrin, S., J. G. Morris, Jr., M. Adams, V. Pons, R. Jacobs, and J. E. Conte, Jr.** 1988. Non-O:1 *Vibrio cholerae* bacteremia: case report and review. *Rev. Infect. Dis.* **10:**1012–1017.

79. **Shaffer, N., E. S. do Santos, P.-A. Andreason, and J. J. Farmer III.** 1989. Rapid laboratory diagnosis of cholera in the field. *Trans. R. Soc. Trop. Med. Hyg.* **83:**119–120.

80. **Shandera, W. X., J. M. Johnston, B. R. Davis, and P. A. Blake.** 1983. Disease from infection with *Vibrio mimicus,* a newly recognized *Vibrio* species. *Ann. Intern. Med.* **99:**169–171.

81. **Shewan, J. M., W. Hodgkiss, and J. Liston.** 1954. A method for the rapid differentiation of certain non-pathogenic, asporogenous bacilli. *Nature* (London) **153:**208–209.

82. **Shirai, H., M. Nishibuchi, T. Ramamurthy, S. K. Bhattacharya, S. C. Pal, and Y. Takeda.** 1991. Polymerase chain reaction for detection of the cholera enterotoxin operon of *Vibrio cholerae.* *J. Clin. Microbiol.* **29:**2517–2521.

83. **Smith, H. L., Jr.** 1970. A presumptive test for vibrios: the "string" test. *Bull. W.H.O.* **42:**817–818.

84. **Swerdlow, D. L., and A. A. Ries.** 1993. *Vibrio cholerae* non-O1—the eighth pandemic? *Lancet* **342:**382–383.

85. **Tacket, C. O., F. Brenner, and P. A. Blake.** 1984. Clinical features and an epidemiological study of *Vibrio vulnificus* infections. *J. Infect. Dis.* **149:**558–561.

86. **Tacket, C. O., F. Hickman, G. V. Pierce, and L. F. Mendoza.** 1982. Diarrhea associated with *Vibrio fluvialis* in the United States. *J. Clin. Microbiol.* **16:**991–992.

87. **Threlfall, E. J., B. Rowe, and I. Huq.** 1980. Plasmid-encoded multiple antibiotic resistance in *Vibrio cholera* El Tor from Bangladesh. *Lancet* **i:**1247–1248. (Letter.)

88. **Tison, D. L., and M. T. Kelly.** 1984. *Vibrio* species of medical importance. *Diagn. Microbiol. Infect. Dis.* **2:**263–276.

89. **Tormey, M., L. Mascola, L. Kilman, P. Nagami, E. DeBess, S. Abbott, and G. W. Rutherford.** 1993. Imported cholera associated with a newly described toxigenic *Vibrio cholerae* O139 strain—California, 1993. *Morbid. Mortal. Weekly Rep.* **42:**501–503.

90. **Uchimura, M., K. Koiwai, Y. Tsuruoka, and H. Tanaka.** 1993. High prevalence of thermostable direct hemolysin (TDH)-like toxin in *Vibrio mimicus* strains isolated from diarrhoeal patients. *Epidemiol. Infect.* **11:**49–53.

91. **Varela, P., M. Rivas, N. Binsztein, M. L. Cremona, P. Herrmann, O. Burrone, R. A. Ugalde, and A. C. C. Frasch.** 1993. Identification of toxigenic *Vibrio cholerae* from the Argentine outbreak by PCR for *ctx* A1 and *ctx* A2-B. *FEBS Lett.* **315:**74–76.

92. **Varnam, A. H., and M. G. Evans.** 1991. *Foodborne Pathogens: an Illustrated Text,* p. 159. Mosby-Yearbook, Inc., St. Louis.

93. **Wachsmuth, I. K., P. A. Blake, and Ø. Olsvik (ed.).** 1994. *Vibrio cholerae and Cholera: Molecular to Global Perspectives.* American Society for Microbiology, Washington, D.C.

94. **World Health Organization.** 1993. Epidemic diarrhoea due to *Vibrio cholerae* non-O1. *Weekly Epidemiol. Rec.* **68(20):**141–142.

95. **World Health Organization.** 1993. Cholera in 1992. *Weekly Epidemiol. Rec.* **68(21):**149–155.

96. **Wright, A. C., Y. Guo, J. A. Johnson, J. P. Nataro, and J. G. Morris, Jr.** 1992. Development and testing of a nonradioactive DNA oligonucleotide probe that is specific for *Vibrio cholerae* cholera toxin. *J. Clin. Microbiol.* **30:**2302–2306.

97. **Wright, A. C., G. A. Miceli, W. L. Landry, J. B. Christy, W. D. Watkins, and J. G. Morris, Jr.** 1993. Rapid identification of *Vibrio vulnificus* on nonselective media with an alkaline phosphatase-labeled oligonucleotide probe. *Appl. Environ. Microbiol.* **59:**541–546.

Aeromonas and *Plesiomonas*

J. MICHAEL JANDA, SHARON L. ABBOTT, AND AMY M. CARNAHAN

36

TAXONOMY

The genera *Aeromonas* and *Plesiomonas* are in the first edition of *Bergey's Manual of Systematic Bacteriology* as members of the family *Vibrionaceae* because, like the genus *Vibrio*, they consist of gram-negative straight rods that are generally motile by polar flagella and are facultative anaerobes (7). Further, all three genera are oxidase positive, utilize D-glucose as a sole or principal carbon and energy source, and have a guanine-plus-cytosine content within the range 38 to 63 mol% (37). However, a growing body of molecular genetic evidence strongly suggests that aeromonads should constitute their own family, *Aeromonadaceae*, and that plesiomonads are actually closer to the genus *Proteus* within the family *Enterobacteriaceae* (33).

The genus *Aeromonas* has gradually expanded from a single mesophilic species, *Aeromonas hydrophila*, and a single psychrophilic species, *A. salmonicida*, to include at least 14 genomospecies or DNA hybridization groups (HGs) from which 11 phenotypic species have been named (22). The genus *Plesiomonas* still includes the single species *Plesiomonas shigelloides* (8). New *Aeromonas* species isolated from clinical specimens since the fifth edition of the *Manual of Clinical Microbiology* (47) include *A. jandaei* (11), *A. trota* (10), and *A. allosaccharophila* (34) (Table 1). *A. veronii* is now considered to contain two biovars, *A. veronii* biovar veronii (esculin hydrolysis and ornithine decarboxylase positive) and *A. veronii* biovar sobria (esculin hydrolysis and ornithine decarboxylase negative). These two biovars are actually members of the same DNA hybridization group (HG 8/10), and what has previously been identified in the clinical laboratory as *A. sobria* is actually *A. veronii* biovar sobria.

DESCRIPTION OF THE GENERA

Aeromonas species are straight gram-negative rods that measure 0.3 to 1.0 by 1.0 to 3.5 μm but can occur as coccobacilli and filamentous rods. All species, except most *A. salmonicida* and *A. media* strains, are motile, are catalase and oxidase positive, and reduce nitrate to nitrite. They ferment D-glucose as well as many other carbohydrates, and most also produce gas from D-glucose. They grow over a wide temperature range, with most mesophilic strains growing from 0 to 45°C and psychrophiles, especially *A. salmo-*

nicida, growing only below 35 to 37°C (37, 47). Aeromonads are capable of producing a number of exoenzymes; several cellular fatty acids (19), such as hexadecanoic acid (16:0); and two unsaturated acids, hexadecenoic acid (16:1) and octadecenoic acid (18:1). The guanine-plus-cytosine content of the DNA for *Aeromonas* species ranges from 57 to 64 mol% (37).

Plesiomonas species (formerly known as C27 or *A. shigelloides*) are straight gram-negative rods that measure 0.8 to 1.0 by 3.0 μm and can occur singly, in pairs, or in short chains. They are usually motile by two to five lophotrichous polar flagella and are positive for catalase, oxidase, and nitrate reduction. They ferment D-glucose and a few other carbohydrates but without production of gas. They grow over a slightly higher temperature range than *Aeromonas* spp. (8 to 45°C), and their main cellular fatty acids differ from those of aeromonads, with 3-hydroxylauric acid replacing 3-hydroxypentadecanoic acid (30). The guanine-plus-cytosine content of *Plesiomonas* DNA is 51 mol% (47).

NATURAL HABITATS

Aeromonas and *Plesiomonas* species are indigenous to aquatic environments worldwide. Aeromonads have been isolated from freshwater, chlorinated water, polluted water, and occasionally marine environments, although they seem to prefer brackish or estuarine water, and their numbers are highest in the warmer months of May through November (13, 20). Plesiomonads are generally found only in tropical aquatic climates, because they do not grow from 0 to 8°C. They also appear to prefer freshwater and estuarine waters over a marine environment (8). Several different *Aeromonas* species have also been recovered in significant numbers from grocery store produce and meats and from environmental and seafood sources (36).

Aeromonas species are associated with a wide variety of diseases in warm- and cold-blooded vertebrates, including frogs, fish, reptiles, snakes, and birds (16). Studies indicate a distinct distribution pattern of certain HGs in clinical *Aeromonas* isolates from diverse geographic origins (5). The predominant species appear to be *A. hydrophila* (HG1), *A. caviae* (HG4), and *A. veronii* biovar sobria (HG8). The rarer genomospecies (HGs 2, 3, 5, 6, 7, and 11) seem to occur more frequently in environmental and veterinary specimens, but there are few studies of large groups of

TABLE 1 Clinical distribution of *Aeromonas* spp. involved in human infections

Genomospecies	DNA group	Clinical data		Frequency of isolation[a]			
		Frequency[b]	Significance[c]	Feces	Wounds	Blood	Other
A. hydrophila	1	+++	1	+++	+++	+++	+++
A. salmonicida	3	++	3	+	−	−	−
A. caviae	4	+++	1	++	+	+++	+++
A. media	5	++	3	+	−	−	−
A. veronii biovar sobria	8	+++	1	++	+	+++	+++
A. veronii biovar veronii	10	+	3	+	+	−	+
A. jandaei	9	++	2	+	+	−	+
A. schubertii	12	++	2	−	+	+	−
A. trota	13	+	3	+	−	+	−
A. allosaccharophila		+	3	+	−	−	−

[a]+++, commonly isolated from this site; +, occasionally to rarely isolated from this site; −, not isolated from this site to date.
[b]+++, commonly isolated in the laboratory; ++, occasionally isolated in the laboratory; +, rarely isolated in the laboratory.
[c]1, major pathogenic species of humans; 2, proven cause of disease in rare instances; 3, occasionally isolated from humans (feces) but significance unknown.

environmental aeromonads to substantiate this distribution.

CLINICAL SIGNIFICANCE

Aeromonas spp.

Aeromonads are chiefly isolated from the gastrointestinal tract and from extraintestinal sites (23) (Table 1). In the case of gastrointestinal infections, definitive evidence linking these microorganisms to infectious diarrhea is still lacking (not all of Koch's postulates have been fulfilled). For extraintestinal sources such as bacteremia and wound infections, their role as bona fide human pathogens has been firmly established over the past two decades; most *Aeromonas* strains (>70%) recovered from wounds appear to be clinically significant (23). Other less common infections caused by *Aeromonas* spp. include endocarditis, meningitis, pneumonia, osteomyelitis, peritonitis, conjunctivitis, thrombophlebitis, and cholecystitis. Anatomic sites where aeromonads are of questionable significance include the respiratory tract (throat, sputum, bronchial washes), the genitourinary tract, and wounds with indwelling catheters and drains.

Aeromonas-associated gastroenteritis principally affects children under the age of 5 (39). Cases of adult-associated gastroenteritis have also been described (17). The highest seasonal incidence of *Aeromonas* diarrhea in the United States occurs during the summer months, when warmer temperatures presumably lead to higher infectious doses in contaminated foods (meats, produce, dairy) and potable water sources (35). Some studies indicate that *Aeromonas* infection is the second or third leading cause of bacterial gastroenteritis during these months. In its mildest form, *Aeromonas* gastroenteritis resembles a secretory (watery) diarrhea, typical of that caused by many other enteric pathogens; under rare circumstances, the diarrhea has been known to mimic cholera. In its more severe form, it resembles shigellosis or bacillary dysentery, in which the stool specimen is bloody in appearance and numerous fecal leukocytes are present. Although not as common as enterotoxigenic *Escherichia coli*, *Aeromonas* spp. have occasionally been associated with traveler's diarrhea. A predisposition to developing *Aeromonas*-associated gastroenteritis has also been reported in adults with hypochlorhydric conditions (including gastrectomy) (17).

Wound infections caused by aeromonads usually occur when abraded mucosal surfaces come into contact with contaminated water, soil, or marine products (fish fins or hooks) during recreational or occupational activities (25). Most of the illnesses resulting from such traumas resemble a typical cellulitis with localized inflammation (redness, swelling) of the subcutaneous tissues. A more fulminant form of this disease, in which rapid destruction of soft tissue is accompanied by crepitation and gangrene, resembling clostridial myonecrosis, has been reported. Such infections are uniformly fatal in individuals who simultaneously present with positive *Aeromonas* blood cultures. In rare instances, septicemic individuals may develop ecthyma gangrenosa lesions on the extremities or trunk that are more typically observed in persons suffering from *Pseudomonas aeruginosa*- or *Vibrio*-associated bacteremias (23). One newly described infection due to *Aeromonas* spp. involves the use of medicinal leeches (43). Such leeches, used to relieve venous congestion after reconstructive surgery, often harbor in their gut aeromonads that are introduced into open wounds during the attachment process. The aeromonads then multiply in these infected wounds, causing mild to moderately severe cellulitis that may require antimicrobial therapy and in some instances debridement.

Aeromonas bacteremia is most commonly seen in individuals with hepatobiliary disease such as Laennec's cirrhosis and in those suffering from hematologic malignancies or solid tumors (23). Although still rare, *Aeromonas* septicemia in immunocompetent persons has been documented by an increasing number of reports (18). Almost all of these infections are thought to originate from ingestion of these bacteria from food sources, with subsequent invasion through the gastrointestinal epithelium. The fatality rate associated with *Aeromonas* sepsis ranges from 30 to 70% in selected surveys (23).

Plesiomonas spp.

Infections attributed to *P. shigelloides* are almost exclusively restricted to two clinical settings (8). The most common presentation is a watery diarrheal illness most often found in individuals with a history of freshwater contact, seafood consumption, exposure to amphibia or reptiles, or travel to developing countries. The peak season of *P. shigelloides*-associated diarrheal disease appears to be similar to that of *Aeromonas*-associated diarrhea (warmer months). An age predilection for developing *P. shigelloides*-associated gastroenteritis has not been reported. As for *Aeromonas* spp., definitive studies linking this species to infectious gastroenteritis are lacking, although two outbreaks of diarrheal disease caused by this organism in Japan have been well described (46). The second well-recognized syndrome associated with *P. shigelloides* is septicemia, often accompanied by meningitis. Most published cases of *Plesiomonas* meningitis have been in infants whose deliveries have been complicated by various medical conditions, including prolonged rupture of the mother's membranes (8). The fatality rate in such instances approaches 70%. Several cases of *P. shigelloides* septicemia in splenectomized adults have also been reported, all of which have had fatal outcomes. On rare occasions, *P. shigelloides* has been isolated from anatomic sources other than feces, blood, or cerebrospinal fluid. These include bile, wounds, joint fluid, and lymph nodes.

COLLECTION, TRANSPORT, AND STORAGE OF SPECIMENS

Both *Aeromonas* and *Plesiomonas* spp. are hardy organisms that multiply under a variety of physiologic conditions. Since almost all aeromonads and plesiomonads recovered from sterile or extraintestinal body sites grow well on routine selective and differential agars, their recovery from such anatomic sites and identification are usually not problems. For gastrointestinal specimens, Cary-Blair transport medium is satisfactory for *Aeromonas* and *Plesiomonas* spp.; however, stool is always preferable to rectal swabs for the recovery of either agent.

ISOLATION

For primary isolation, either a semispecific medium for *Aeromonas* spp. or a nonselective agar should be used, since most aeromonads are sucrose positive and many *A. caviae* strains are also lactose positive and thus will not be recognized easily on media such as Hektoen enteric and MacConkey agars. Blood agar (with or without ampicillin) or cefsulodin-irgasan-novobiocin agar may be useful in the recovery of aeromonads from fecal specimens (26). With blood agar, spot oxidase and indole tests can be performed on individual colonies, and hemolytic phenospecies (*A. hydrophila*, *A. veronii*) can also be observed. Because *P. shigelloides* is sucrose negative and delayed lactose positive, most commonly used selective and differential agars are satisfactory. When the anticipated recovery of aeromonads or plesiomonads is poor (subacute or chronic diarrhea), an enrichment procedure (alkaline peptone water, pH 8.5) may be required.

IDENTIFICATION

To separate these genera, presumptive *Aeromonas*, *Plesiomonas*, and *Vibrio* isolates should be screened for susceptibility to the vibriostatic agent O/129 (2,4-diamino-6,7-diisopropylpteridine) at both 10- and 150-μg concentrations, growth on nutrient broth in the absence (0%) or presence (6%) of NaCl, and growth on thiosulfate citrate bile salts sucrose agar (TCBS). Determination of the correct genus is becoming increasingly difficult because of emerging world-

wide resistance of *Vibrio cholerae* to O/129 and lot-to-lot variability in the specificity and selectivity of commercially prepared TCBS. Despite these difficulties, correct genus assignment is very important in light of the continuing epidemic of *V. cholerae* O1 in the Western Hemisphere and the recent isolation of the epidemic *V. cholerae* O139 strain in the United States (45). For strains whose genus (*Aeromonas*, nonhalophilic *Vibrio*) cannot be immediately determined, additional tests should be performed, including tests for gas production from D-glucose and for ornithine decarboxylase activity. Identification of *P. shigelloides* is generally not a problem, as its unusual biochemical properties (lysine decarboxylase, ornithine decarboxylase, and arginine dihydrolase positive and fermentation of *m*-inositol) help distinguish this group from either aeromonads or vibrios (Table 2).

With the expansion of legitimate species within the genus *Aeromonas*, identification of individual strains to species continues to be very difficult. Currently, most commercially available kit systems lack both the specificity and the sensitivity to accurately identify aeromonads past the genus level. Aeromonads may also be identified to the most common phenospecies (*A. hydrophila*, *A. veronii*, *A. caviae*) by using abbreviated schemes (9, 23). Such identifications may provide important information on both the invasive and the enterotoxigenic capabilities of a strain and on general susceptibility profiles. For definitive identification of other genomospecies, a more complete biochemical profile that may include a number of laboratory tests not available in commercial systems is required (1). Biochemical reactions useful in such identifications are listed in Table 2. Commercial systems frequently identify *A. caviae* as *Vibrio fluvialis*. Such problems can be avoided by using the screening reactions listed above (e.g., salt tolerance) and esculin hydrolysis (*V. fluvialis* is esculin negative).

Serogrouping or serotyping of *Aeromonas* spp. or *P. shigelloides* may be useful for epidemiologic purposes or in order to determine strain relatedness within a given genomospecies. Currently, there are over 90 recognized or provisional serogroups within the genus *Aeromonas* (National Institutes of Health, Tokyo, Japan; Central Public Health Laboratory, London, England); serogroup designations are not genomospecies specific (41, 44). For *P. shigelloides*, a serotype system based on the presence of somatic and flagellar antigens has been developed (3, 42). Over 100 different serotypes or serovars can be distinguished in this manner by using one of two international typing systems (National Institutes of Health, Tokyo, Japan; Institute of Hygiene and Epidemiology, Prague, Czech Republic).

Aeromonas and *Plesiomonas* spp. have also been investigated for enzymatic activities (12, 31, 38) by using either chromogenic substrates (API ZYM, API esterase, and arylamidase galleries) or carbohydrate utilization in the presence of a redox indicator (BIOLOG GN microplate). *Aeromonas* strains can be identified to genomospecies or can be epidemiologically fingerprinted by using such molecular technologies as multilocus enzyme electrophoresis (4), ribotyping (32), chromosomal restriction enzyme analysis (15, 28), and protein (outer membrane) electrophoresis (28). Plasmid analysis is of very limited value owing to the low frequency of strains (20 to 30%) carrying extrachromosomal elements (the exception is *A. salmonicida*). Recent molecular investigations have used cloned gene products (21) or PCR-amplified regions of 16S rRNA (6) to accurately identify some of the more infrequently encountered aeromonads (*A. schubertii*, *A. jandaei*, *A. trota*).

TABLE 2 Relevant biochemical properties of *Aeromonas* spp. and *P. shigelloides*[a]

Characteristic	Response							
	A. hydrophila	*A. caviae*	*A. veronii* biovar sobria	*A. veronii* biovar veronii	*A. jandaei*	*A. schubertii*	*A. trota*	*P. shigelloides*
Screening reactions								
Oxidase	+	+	+	+	+	+	+	+
O/129, 10 µg/150 µg	R/R	R/R	R/R	R/R	R/R	R/R	R/R	S/S
Growth on:								
TCBS	−	−	−	−	−	−	−	−
Nutrient agar–0% NaCl	+	+	+	+	+	+	+	+
Nutrient agar–6% NaCl	−	−	−	−	−	−	−	−
Phenotypic properties								
Indole	+	+	+	+	+	−	+	+
Glucose (gas)	+	−	+	+	+	−	+	−
Arginine dihydrolase	+	+	+	−	+	+	+	−
Lysine decarboxylase	+	−	+	+	+	+	+	+
Ornithine decarboxylase	−	−	−	+	−	−	−	+
Voges-Proskauer	+	−	+	+	+	−	−	+
Acid from:								
L-Arabinose	+	+	−	−	−	−	−	−
Lactose	−	+	−	−	−	−	−	−
Sucrose	+	+	+	+	−	−	−	−
m-Inositol	−	−	−	−	−	−	−	+
D-Mannitol	+	+	+	+	+	−	+	−
Salicin	+	+	−	+	−	−	−	V
Esculin hydrolysis	+	+	−	+	−	−	−	−
Beta-hemolysis on sheep blood	+	−	+	+	+	V	V	−

[a]R, resistant; S, susceptible; V, variable (20 to 80% positive).

SEROLOGIC TESTS

Serogroup results can be combined with species designations to trace clusters of infections or reputed outbreaks. Some serogroups of *Aeromonas* show partial or complete antigenic identity with some serogroups of non-O1 *V. cholerae*, *V. fluvialis*, and *P. shigelloides*. In some cases, when the clinical significance of the *Aeromonas* isolate is questionable, demonstration of an immune response to the organism may be beneficial. The simplest technique to perform is a bacterial (tube) agglutination; because of the low sensitivity of this assay and the large number of low agglutination titers to *Aeromonas* spp. found in asymptomatic individuals, titers of ≤1:64 are probably not of diagnostic value. Paired sera (acute and convalescent phases) demonstrating a fourfold rise or decline in immune response are always preferable. Other immunologic assays useful in assessing the clinical significance of aeromonads include detection of fecal (secretory immunoglobulin A to bacterial lipopolysaccharide (24), detection of immunoglobulin G to specific bacterial surface proteins by immunoblotting, and enzyme-linked immunosorbent assay for whole-cell envelopes (29).)

P. shigelloides can cross-react with *Shigella* grouping sera, particularly with group D (*Shigella sonnei*), in the clinical laboratory, sometimes leading to the initial misidentification of *P. shigelloides* as a shigella (2). As with *Aeromonas* spp., plesiomonads can be serotyped for epidemiologic purposes. No major studies on the immune response of individuals reputedly infected with *P. shigelloides* have been reported to date (24).

ANTIBIOTIC SUSCEPTIBILITIES

Most *Aeromonas* spp. are resistant to ampicillin and susceptible to other antimicrobial agents such as broad-spectrum and broad-spectrum-like cephalosporins, the aminoglycosides, quinolones, tetracycline, chloramphenicol, carbapenems, and trimethoprim-sulfamethoxazole (27). Susceptibility to narrow- and expanded-spectrum cephalosporins is species associated, since most *A. veronii* are susceptible to agents such as cephalothin, while *A. hydrophila* and *A. caviae* are resistant. *A. trota* is unusual in its uniform susceptibility to ampicillin (10). Most aeromonads produce inducible β-lactamases; recent investigations using commercial microdilution tests indicate that very major errors concerning susceptibility to β-lactam antibiotics (ticarcillin, cephalothin, piperacillin, etc.) can be made when results are compared with those of agar dilution (40). This suggests that rapid systems should not be relied upon when β-lactam compounds are tested against *Aeromonas* spp.

For *P. shigelloides*, the consensus results of several studies indicate overall susceptibility to all major classes of antibiotics mentioned for aeromonads, including narrow- and expanded-spectrum cephalosporins (14).

EVALUATION, INTERPRETATION, AND REPORTING OF RESULTS

Probably the most difficult decision the laboratory faces regarding *Aeromonas* spp. concerns the role of these agents in diarrheal disease. During the acute phase of a gastrointestinal illness, the laboratory should report both the phenospecies (*A. hydrophila*, *A. caviae*, *A. veronii*) and the relative quantity of *Aeromonas* organisms present on either differential or nonselective medium. For cases of chronic diarrhea, the isolation of *Aeromonas* spp., while of potential significance, is not definitive. Either pathologic evidence (from tissue biopsy and culture) or the demonstration of an immune response by one of the above-described techniques is necessary as well. None of the serologic tests previously mentioned have been standardized. However, from published data it appears that immunoblot studies aimed at detecting an immune response against the bacterial lipopolysaccharide or cell-associated proteins offer the most promise. The use of PCR or probes to detect virulence factors or specific genomospecies is not appropriate until the role of aeromonads and their microbial products in gastroenteritis is clarified.

REFERENCES

1. **Abbott, S. L., W. K. W. Cheung, S. Kroske-Bystrom, T. Malekzadeh, and J. M. Janda.** 1992. Identification of *Aeromonas* strains to the genospecies level in the clinical laboratory. *J. Clin. Microbiol.* **30:**1261–1266.
2. **Abbott, S. L., R. P. Kokka, and J. M. Janda.** 1991. Laboratory investigations on the low pathogenic potential of *Plesiomonas shigelloides*. *J. Clin. Microbiol.* **29:**148–153.
3. **Aldova, E.** 1992. Comparison of Shimada and Sakazaki's and Aldova's antigenic schemes for *Plesiomonas shigelloides*. *Syst. Appl. Microbiol.* **15:**70–75.
4. **Altwegg, M., M. W. Reeves, R. Altwegg-Bissig, and D. J. Brenner.** 1991. Multilocus enzyme analysis of the genus *Aeromonas* and its use for species identification. *Zentralbl. Bakteriol.* **275:**28–45.
5. **Altwegg, M., A. G. Steigerwalt, R. Altwegg-Bissig, J. Luthy-Hottenstein, and D. J. Brenner.** 1990. Biochemical identification of *Aeromonas* genospecies isolated from humans. *J. Clin. Microbiol.* **28:**258–264.
6. **Ash, C., A. J. Martinez-Murcia, and M. D. Collins.** 1993. Identification of *Aeromonas schubertii* and *Aeromonas jandaei* by using a polymerase chain reaction-probe test. *FEMS Microbiol. Lett.* **108:**151–156.
7. **Baumann, P., and R. H. W. Schubert.** 1984. Family II. *Vibrionaceae* Veron, 1965,5245, p. 516–517. *In* N. R. Krieg and J. G. Holt (ed.), *Bergey's Manual of Systematic Bacteriology*, vol. 1. The Williams & Wilkins Co., Baltimore.
8. **Brenden, R. A., M. A. Miller, and J. M. Janda.** 1988. Clinical disease spectrum and pathogenic factors associated with *Plesiomonas shigelloides* infections in humans. *Rev. Infect. Dis.* **10:**303–316.
9. **Carnahan, A. M., S. Behram, and S. W. Joseph.** 1991. Aerokey II: a flexible key for identifying clinical *Aeromonas* species. *J. Clin. Microbiol.* **29:**2843–2849.
10. **Carnahan, A. M., T. Chakraborty, G. R. Fanning, D. Verma, A. Ali, J. M. Janda, and S. W. Joseph.** 1991. *Aeromonas trota*, sp. nov., an ampicillin-susceptible species isolated from clinical specimens. *J. Clin. Microbiol.* **29:**1206–1210.
11. **Carnahan, A. M., G. R. Fanning, and S. W. Joseph.** 1991. *Aeromonas jandaei* (formerly genospecies DNA group 9 *A. sobria*), a new sucrose-negative species isolated from clinical specimens. *J. Clin. Microbiol.* **29:**560–564.
12. **Carnahan, A. M., S. W. Joseph, and J. M. Janda.** 1989. Species identification of *Aeromonas* strains based on carbon substrate oxidation profiles. *J. Clin. Microbiol.* **27:**2128–2129.
13. **Chowdbury, M. A. R., H. Yamanaka, S. Miyoshi, and S. Shinoda.** 1990. Ecology of mesophilic *Aeromonas* spp. in aquatic environments of a temperate region and relationship with some biotic and abiotic environmental parameters. *Zentralbl. Hyg.* **190:**344–356.
14. **Clark, R. B., P. D. Lister, L. Arneson-Rotert, and J. M. Janda.** 1990. In vitro susceptibilities of *Plesiomonas shigelloides*

15. **de la Morena, M. L., R. Van, K. Singh, M. Brian, B. E. Murray, and L. K. Pickering.** 1993. Diarrhea associated with *Aeromonas* species in children in day care centers. *J. Infect. Dis.* **168:**215–218.
16. **Farmer, J. J., III, M. J. Arduino, and F. W. Hickman-Brenner.** 1992. The genera *Aeromonas* and *Plesiomonas*, p. 3012–3028. *In* A. Balows, H. G. Truper, M. Dworkin, W. Harder, and K. H. Schleifer (ed.), *The Prokaryotes*, 2nd ed. Springer-Verlag, New York.
17. **George, W. L., M. M. Nakata, J. Thompson, and M. L. White.** 1985. *Aeromonas*-related diarrhea in adults. *Arch. Intern. Med.* **145:**2207–2211.
18. **Golik, A., Y. Leonov, F. Schlaeffer, I. Gluskin, and G. Lewinsohn.** 1990. *Aeromonas* species bacteremia in nonimmunocompromised hosts: two case reports and a review of the literature. *Isr. J. Med. Sci.* **26:**87–90.
19. **Hansen, W., J. Freney, M. Labbe, F. Renaud, E. Yourassowsky, and J. Fleurette.** 1991. Gas-liquid chromatographic analysis of cellular fatty acid methyl esters in *Aeromonas* species. *Zentralbl. Bakteriol.* **275:**1–10.
20. **Hazen, T. C., C. B. Fleirmans, R. P. Hirsch, and G. W. Esch.** 1978. Prevalence and distribution of *Aeromonas hydrophila* in the United States. *Appl. Environ. Microbiol.* **36:**731–738.
21. **Husslein, V., T. Chakraborty, A. Carnahan, and S. W. Joseph.** 1992. Molecular studies on the aerolysin gene of *Aeromonas* species and discovery of a species-specific probe for *Aeromonas trota*. *Clin. Infect. Dis.* **14:**1061–1068.
22. **Janda, J. M.** 1991. Recent advances in the study of the taxonomy, pathogenicity, and infectious syndromes associated with the genus *Aeromonas*. *Clin. Microbiol. Rev.* **4:**397–410.
23. **Janda, J. M., and P. S. Duffey.** 1988. Mesophilic aeromonads in human disease: current taxonomy, laboratory identification, and infectious disease spectrum. *Rev. Infect. Dis.* **10:**980–997.
24. **Jiang, Z. D., A. C. Nelson, J. J. Mathewson, C. D. Ericsson, and H. L. DuPont.** 1991. Intestinal secretory immune response to infection with *Aeromonas* species and *Plesiomonas shigelloides* among students from the United States in Mexico. *J. Infect. Dis.* **164:**979–982.
25. **Kelly, K. A., J. M. Koehler, and L. R. Ashdown.** 1993. Spectrum of extraintestinal disease due to *Aeromonas* species in tropical Queensland, Australia. *Clin. Infect. Dis.* **16:**574–579.
26. **Kelly, M. T., E. M. D. Stroh, and J. Jessop.** 1988. Comparison of blood agar, ampicillin blood agar, MacConkey-ampicillin-Tween agar, and modified cefsulodin-irgasan-novobiocin agar for isolation of *Aeromonas* from stool specimens. *J. Clin. Microbiol.* **30:**1598–1599.
27. **Koehler, J. M., and L. R. Ashdown.** 1993. In vitro susceptibilities of tropical strains of *Aeromonas* species from Queensland, Australia, to 22 antimicrobial agents. *Antimicrob. Agents Chemother.* **37:**905–907.
28. **Kuijper, E. J., L. van Alphen, E. Leenders, and H. C. Zanen.** 1989. Typing of *Aeromonas* strains by DNA restriction endonuclease analysis and polyacrylamide gel electrophoresis of cell envelopes. *J. Clin. Microbiol.* **27:**1280–1285.
29. **Kuijper, E. J., L. van Alphen, M. F. Peeters, and D. J. Brenner.** 1990. Human serum antibody response to the presence of *Aeromonas* spp. in the intestinal tract. *J. Clin. Microbiol.* **28:**584–590.
30. **Lambert, M. A., F. W. Hickman-Brenner, J. J. Farmer III, and C. W. Moss.** 1983. Differentiation of *Vibrionaceae* species by their cellular fatty acid composition. *Int. J. Syst. Bacteriol.* **33:**777–792.
31. **Manafi, M., and M.-L. Rotter.** 1992. Enzymatic profile of *Plesiomonas shigelloides*. *J. Microbiol. Methods* **16:**175–180.
32. **Martinetti Lucchini, G., and M. Altwegg.** 1992. rRNA gene restriction patterns as taxonomic tools for the genus *Aeromonas*. *Int. J. Syst. Bacteriol.* **42:**384–389.
33. **Martinez-Murcia, A. J., S. Benlloch, and M. D. Collins.** 1992. Phylogenetic interrelationships of members of the gen-

era *Aeromonas* and *Plesiomonas* as determined by 16S ribosomal DNA sequencing: lack of congruence with results of DNA-DNA hybridizations. *Int. J. Syst. Bacteriol.* **42:**412–421.

34. **Martinez-Murcia, A. J., C. Esteve, E. Garay, and M. D. Collins.** 1991. *Aeromonas allosaccharophila* sp. nov., a new mesophilic member of the genus *Aeromonas*. *FEMS Microbiol. Lett.* **91:**199–206.

35. **Moyer, N. D.** 1987. Clinical significance of *Aeromonas* species isolated from patients with diarrhea. *J. Clin. Microbiol.* **25:** 2044–2048.

36. **Nishikawa, Y., and T. Kishi.** 1988. Isolation and characterization of motile *Aeromonas* from human, food, and environmental specimens. *Epidemiol. Infect.* **101:**213–223.

37. **Popoff, M.** 1984. Genus III. *Aeromonas* Kluyver and Van Niel 1936,398, p. 545–548. *In* N. R. Krieg and J. G. Holt (ed.), *Bergey's Manual of Systematic Bacteriology*, vol. 1. The Williams & Wilkins Co., Baltimore.

38. **Renaud, F., J. Freney, J. M. Boeufgra, D. Monget, A. Sedaillan, and J. Bleurett.** 1988. Carbon substrate assimilation patterns of clinical and environmental strains of *Aeromonas hydrophila*, *Aeromonas sobria*, and *Aeromonas caviae* observed with a micromethod. *Zentralbl. Bakteriol. Hyg. A* **269:**323–330.

39. **San Joaquin, V. H., and D. A. Pickett.** 1988. *Aeromonas*-associated gastroenteritis in children. *Pediatr. Infect. Dis. J.* **7:**53–57.

40. **Schadow, K. H., D. K. Giger, and C. C. Sanders.** 1993.

Failure of the Vitek system to detect beta-lactam resistance in *Aeromonas* species. *Am. J. Clin. Pathol.* **100:**308–310.

41. **Shimada, T., and Y. Kosako.** 1991. Comparison of two O-serogrouping systems for mesophilic *Aeromonas* spp. *J. Clin. Microbiol.* **29:**197–199.

42. **Shimada, T., and R. Sakazaki.** 1985. New O and H antigens and additional serovars of *Plesiomonas shigelloides*. *Jpn. J. Med. Sci. Biol.* **38:**73–76.

43. **Snower, D. P., C. Ruef, A. P. Kuritza, and S. C. Edberg.** 1989. *Aeromonas hydrophila* infection associated with the use of medicinal leeches. *J. Clin. Microbiol.* **27:**1421–1422.

44. **Thomas, L. V., R. J. Gross, T. Cheasty, and B. Rowe.** 1990. Extended serogrouping scheme for motile, mesophilic *Aeromonas* species. *J. Clin. Microbiol.* **28:**980–984.

45. **Torrney, M., L. Mascola, L. Kilman, P. Nagami, E. DeBess, S. Abbott, and G. W. Rutherford.** 1993. Imported cholera associated with a newly described toxigenic *Vibrio cholerae* O139 strain—California. *Morbid. Mortal. Weekly Rep.* **42:** 501–504.

46. **Tsukamoto, T., Y. Kinoshita, T. Shimada, and R. Sakazaki.** 1978. Two epidemics of diarrhoeal disease possibly caused by *Plesiomonas shigelloides*. *J. Hyg. Camb.* **80:**275–280.

47. **von Graevenitz, A., and M. Altwegg.** 1991. *Aeromonas* and *Plesiomonas*, p. 396–401. *In* A. Balows, W. J. Hausler, Jr., K. L. Herrmann, H. D. Isenberg, and H. J. Shadomy (ed.), *Manual of Clinical Microbiology*, 5th ed. American Society for Microbiology, Washington, D.C.

Campylobacter and Arcobacter

IRVING NACHAMKIN

37

TAXONOMY

Two closely related genera, *Campylobacter* and *Arcobacter*, are now included in the family *Campylobacteraceae* (82, 84). The family *Campylobacteraceae* includes 18 species and subspecies within the genus *Campylobacter* and 4 species in the genus *Arcobacter* (Table 1). Two species that were formerly classified in the genus *Wolinella* (*Wolinella curva*, *Wolinella recta*) have been reclassified as *Campylobacter* (84). Several species formerly classified as *Campylobacter* are now classified in the genus *Arcobacter* and include *Arcobacter cryaerophilus*, *A. nitrofigilis*, and *A. butzleri* (84, 85). An additional species, *A. skirrowii*, was also recently described (85). *Campylobacter cinaedi* and *C. fennelliae* are now classified in the genus *Helicobacter* (84) and are described in chapter 38 of this Manual. *Bacteroides gracilis* and *Bacteroides ureolyticus* are thought to be closely related to the genus *Campylobacter*; however, this classification is still uncertain.

DESCRIPTION OF THE GENUS

Campylobacters are curved, S-shaped, or spiral rods that are 0.2 to 0.9 μm wide and 0.5 to 5 μm long. Campylobacters are gram-negative, non-spore-forming rods that may form spherical or coccoid bodies in old cultures or in cultures exposed to air for prolonged periods. Organisms are motile by means of a single polar unsheathed flagellum at one or both ends. Species are microaerophilic and have a respiratory type of metabolism. Some strains grow aerobically or anaerobically. An atmosphere containing increased hydrogen may be required by some species for microaerobic growth (82).

Arcobacters are gram-negative, slightly curved, curved, S-shaped, or helical non-spore-forming rods that are 0.2 to 0.9 μm wide and 1 to 3 μm long. Organisms are motile, with a single polar unsheathed flagellum. Arcobacters grow at 15, 25, and 30°C but have variable growth at 37 and 42°C. Organisms are microaerophilic and do not require hydrogen for growth. Arcobacters may grow aerobically at 30°C and anaerobically at 35 to 37°C. Most strains are nonhemolytic. *A. skirrowii* may be alpha-hemolytic. Most strains are susceptible to nalidixic acid but show a variable response to cephalothin (85).

CLINICAL SIGNIFICANCE

C. jejuni and C. coli

C. jejuni subsp. *jejuni* (referred to as *C. jejuni* throughout this chapter) and *C. coli* have been recognized since the late 1970s as agents of gastrointestinal infection. Many clinical and epidemiologic investigations have established *C. jejuni* as one of the most common causes of sporadic bacterial enteritis in the United States. Studies by the Centers for Disease Control and Prevention using limited surveillance of campylobacter infection in addition to estimates based on outbreak- and hospital-based studies in the United States suggest that the overall infection rate is 1,000/100,000 population. Over 2 million cases are estimated to occur annually, an incidence of infection similar to that in the United Kingdom and other developed nations (73). *Campylobacter* infections usually occur sporadically in the summer months and usually follow ingestion of improperly handled or improperly cooked food, primarily poultry products. The incidence of infection follows a bimodal age distribution, with the highest incidence in infants and young children and a second peak in young adults 20 to 40 years old. Outbreaks usually occur in the spring and fall months and have been associated with ingestion of contaminated milk and water (73, 92).

The incidence of campylobacter infections in developing countries such as Mexico and Thailand may be orders of magnitude higher than that in the United States (75). In developing countries, *Campylobacter* spp. are frequently isolated from individuals who may or may not have diarrheal disease. Most symptomatic infections occur in infancy and early childhood, and their frequency decreases with age. Travelers to developing countries are at risk for developing *Campylobacter* infection, with isolation rates from 0 to 39% reported in different studies (73, 75).

C. jejuni and *C. coli* are the most common *Campylobacter* species associated with diarrheal illness and are clinically indistinguishable. Most laboratories do not routinely distinguish these organisms. Approximately 5 to 10% of cases in the United States that are reported as being due to *C. jejuni* are probably due to *C. coli*, and this percentage may be higher in other parts of the world (75). A spectrum of illness is seen during *C. jejuni-C. coli* infection: patients range from asymptomatic to severely ill. Symptoms and signs usually include fever, abdominal cramping, and diar-

TABLE 1 Phenotypic properties of *Campylobacter* species[a]

Organism	Catalase	Nitrate reduction	Nitrite reduction	H₂ required	Urease	H₂S (TSI)	Hippurate hydrolysis	Indoxyl acetate hydrolysis	Growth at: 15°C	25°C	42°C	Growth in or on: 3.5% NaCl	1% glycine	MacConkey agar	Susceptibility to: Nalidixic acid	Cephalothin
C. jejuni subsp. *jejuni*	+	+	−	−	−	−	+	+	−	−	+	−	+	+	S	R
C. jejuni subsp. *doylei*	V	−	−	−	−	−	V	+	−	−	−	−	+	−	S	S
C. coli	+	+	−	−	−	−	−	+	−	−	+	−	+	+	S	R
C. fetus subsp. *fetus*	+	+	−	−	−	−	−	−	−	+	−	−	+	+	V	S
C. fetus subsp. *venerealis*	+	+	−	−	−	−	−	−	−	+	−	−	−	+	R	S
C. lari	+	+	−	−	V	−	−	−	−	−	+	−	+	+	R	R
C. upsaliensis	W	+	−	−	−	−	−	+	−	−	+	−	V	−	S	S
C. hyointestinalis	+	+	−	V	−	+	−	−	−	−	+	−	+	−	R	S
C. sputorum biovar sputorum	−	+	+	−	−	+	−	−	−	−	+	−	+	+	S	S
C. sputorum biovar bubulus	−	+	+	−	−	+	−	−	−	−	+	+	+	−	R	S
C. sputorum biovar fecalis	+	+	+	−	−	+	−	−	−	−	+	−	+	+	R	S
C. helveticus	−	+	+	+	−	+	−	+	−	−	+	V	V	−	S	S
C. mucosalis	−	+	+	+	−	+	−	−	−	−	+	−	+	+	R	S
C. concisus	−	+	+	+	−	+	−	−	−	−	+	−	+	+	R	R
C. curvus	−	+	+	+	−	+	−	+	−	−	+	−	+	−	S	R
C. rectus	−	+	+	+	−	+	−	+	−	−	W	−	+	−	S	ND
"*C. showae*"	+	+	+	+	−	+	−	+	−	−	+	−	V	−	R	ND
A. cryaerophilus group 1A	+	V	−	−	−	−	−	−	+	+	+	−	−	−	V	R
A. cryaerophilus group 1B	+	V	−	−	−	−	−	+	+	+	−	−	−	+	S	V
A. nitrofigilis	+	+	−	−	+	−	−	−	+	+	−	+	−	−	S	S
A. butzleri	W	+	−	−	−	−	−	+	+	+	V	V	+	+	S	R
A. skirrowii	+	+	−	−	−	−	−	+	+	+	V	V	V	−	S	S

[a]TSI, triple sugar iron; W, weak reaction; S, susceptible; R, resistant; V, variable reaction; ND, not determined. Adapted from references 17, 33, 58, 71, 82, and 85.

rhea (with or without blood and fecal leukocytes) that lasts several days to more than 1 week. Symptomatic infections are usually self-limited, but relapses may occur in 5 to 10% of untreated patients (6). *Campylobacter* infection may mimic acute appendicitis and result in unnecessary surgery. Extraintestinal infections do occur and include bacteremia, reactive arthritis, bursitis, urinary tract infection, meningitis, endocarditis, peritonitis, erythema nodosum, pancreatitis, abortion, and neonatal sepsis (6). Bacteremia has been reported to occur at a rate of 1.5/1,000 intestinal infections, with the highest rate in the elderly (69). Persistent diarrheal illness and bacteremia may occur in immunocompromised hosts, such as patients with human immunodeficiency virus infection (56), and may be difficult to treat. Deaths attributable to *C. jejuni* infection have been reported but occur rarely (73).

Guillain-Barré syndrome (GBS) appears to be strongly associated with *Campylobacter* infection as one of the sequelae of infection (45). *C. jejuni* has been recovered from Japanese patients with GBS, and most of these isolates belong to a single serotype, O:19, suggesting that only certain *Campylobacter* serotypes are associated with disease (36). Recent studies have also shown an association of *Campylobacter* spp. with an illness clinically similar to GBS known as acute motor axonal neuropathy that occurs primarily in northern China and may occur in other parts of the world (43). The pathogenesis of GBS and GBS-like illness is not understood but may include both host factors and/or antigens of *Campylobacter* spp. that cross-react with nerve tissue components.

Little is known about the pathogenesis of campylobacter infection. The infective dose of *Campylobacter* spp. does not appear to be high, with 1,000 organisms being capable of causing illness (5). The signs and symptoms of infection suggest an invasive mechanism of disease. In vitro and in vivo studies suggest that campylobacter first colonizes the intestinal mucous layer, mediated by motility, and then invades and/or translocates across the epithelial surface to the underlying tissue, where other putative virulence factors are then elaborated (35). *C. jejuni* has been reported to produce a cholera-like enterotoxin; however, other studies have refuted the significance of these findings (34). The identification of useful in vivo models and advances in studying *Campylobacter* virulence factors at the molecular level will be important for our future understanding of the pathogenesis of infection.

C. fetus

C. fetus subsp. *fetus* (referred to here as *C. fetus*) is primarily associated with bacteremia and extraintestinal infections in patients with underlying diseases, and infection with this organism may have a poor outcome in some patients (6). *C. fetus* is also associated with septic abortions, septic arthritis, abscesses, meningitis, endocarditis, mycotic aneurysm, thrombophlebitis, peritonitis, and salpingitis (46). Although gastroenteritis does occur with this species, the incidence is probably underestimated because the organism does not grow well at 42°C and is usually susceptible to cephalothin, an antimicrobial agent used in some common selective media for stool culture (13). *C. fetus* produces a surface protein capsule composed of a high-molecular-weight surface array protein that is essential for the virulence of the organism. This high-molecular-weight protein confers resistance to serum-mediated killing and phagocytosis and may explain why *C. fetus* is able to cause systemic infection (8). *C. fetus* subsp. *venerealis* is not associated with human infection.

C. upsaliensis

C. upsaliensis is a recently described thermotolerant species that appears to be an important cause of diarrhea and bacteremia (22, 46). *C. upsaliensis* is susceptible to many antimicrobial agents present in *C. jejuni*-selective media, and thus, most *C. upsaliensis* isolates are recovered by using a filtration method (22). *C. upsaliensis* may occasionally be recovered on some selective media.

C. lari

C. lari (formerly *C. laridis*) is a thermophilic species that was isolated first from gulls (of the genus *Larus*) and has since been isolated from other avian species, dogs, cats, and chickens. *C. lari* has been infrequently reported from humans with bacteremia and with gastrointestinal and urinary tract infections. An outbreak of waterborne infection that occurred in 1985 affected over 100 individuals (46, 74).

Other Species

Other *Campylobacter* species have been isolated from clinical specimens of patients with a variety of diseases, but their pathogenic roles have not been determined. *C. jejuni* subsp. *doylei* is a nitrate-negative subspecies of *C. jejuni* rarely recovered from patients with upper gastrointestinal tract infections and gastroenteritis (22, 72). *C. hyointestinalis* has been occasionally associated with proctitis and diarrhea in human infection (14). *C. concisus* is associated primarily with periodontal disease but has also been isolated from patients with bacteremia, foot ulcer, and upper and lower gastrointestinal tract infections (30, 83). *C. sputorum* biovar sputorum and *C. sputorum* biovar bubulus have been associated with lung, axillary, scrotal, and groin abscesses (46). *C. mucosalis* was isolated from two children with enteritis (19). *C. helveticus* (71) is a newly described species recovered from domestic cats and dogs and has not been reported from human sources. *C. rectus* is primarily isolated from patients with active periodontal infections (61) but has also been isolated from a patient with pulmonary infection (70). "*C. showae*" is a recently described species isolated from human gingival crevice (17) that is somewhat different from other *Campylobacter* species in its morphologic characteristics, appearing as straight rods with multiple flagella.

Arcobacter spp.

Arcobacter are aerotolerant, *Campylobacter*-like organisms frequently isolated from cattle and pigs suffering from abortion and enteritis (90). Two of the four *Arcobacter* species have been associated with human infection. *A. butzleri* has been isolated from patients with bacteremia, endocarditis, peritonitis, and diarrhea (33, 77, 85). *A. cryaerophilus* group 1B has been isolated from patients with bacteremia and diarrhea (33, 85).

SPECIMEN COLLECTION, TRANSPORT, AND STORAGE

Fecal Samples

Fecal specimens are the preferred samples for isolating *Campylobacter* species from patients with gastrointestinal infections; however, rectal swabs are acceptable for culture. Multiple samples may increase the isolation rate of campylobacters. A transport medium should be used for transporting fecal specimens to the laboratory if a delay of more than 2 h is anticipated and for transporting rectal swabs. Cary-Blair medium containing reduced agar (1.6 g/liter) appears to be the most suitable single transport medium for *Campylobacter* spp. as well as for other enteric pathogens (88). Specimens received in Cary-Blair medium should be stored at 4°C if processing is not performed immediately.

Blood

Campylobacter species, primarily *C. fetus*, *C. jejuni*, and *C. upsaliensis*, have been isolated from blood; however, in only a few studies have optimal conditions for isolating *Campylobacter* spp. from blood culture systems been evaluated. Both the BACTEC system (aerobic bottles) and the Septi-Chek system appear to support the growth of the common *Campylobacter* species (32, 39, 87). Other systems, such as anaerobic broth (Trypticase soy broth; Difco) or Wampole Isolator systems, may not be as sensitive (32).

DIRECT EXAMINATION

Campylobacter spp. are not easily visualized with the safranin counterstain commonly used in the Gram stain procedure; carbol fuchsin or basic fuchsin should be used as the counterstain for smears of stools or pure cultures (53, 66). Because of their characteristic microscopic morphology, *Campylobacter* spp. may be detected by direct Gram stain examination of stools obtained from patients with acute enteritis. The sensitivity of the stain ranges from 66 to 94%, and its specificity is very high (53, 66).

Fecal leukocytes may be present during *Campylobacter* infection and have been reported in from 25 to 80% of culture-proven cases (18, 20). Some authors suggest that, in order to improve the cost-effectiveness of diagnosis, a test for the presence of leukocytes in stools be used to indicate the need for culture (24). While the likelihood of *Campylobacter* spp. or other enteroinvasive pathogens may be higher in the presence of fecal leukocytes, the absence of fecal leukocytes does not rule out disease. Thus, examination of stool samples for fecal leukocytes is not recommended as a test for predicting bacterial infection or as a marker for selective culturing for *Campylobacter* spp. or other stool pathogens (18).

CULTURE AND ISOLATION

Most *Campylobacter* species require a microaerobic atmosphere containing approximately 5% O_2, 10% CO_2, and 85% N_2 for optimal recovery. Several manufacturers produce microaerobic gas generator packs that are suitable for routine use. A Tri-Gas incubator (80) or evacuation and replacement of an anaerobic incubator with the approximate gas mixture may also be used for routine cultures (48). The concentration of oxygen generated in candle jars is suboptimal for the isolation of *Campylobacter* spp. and should not be used for routine laboratory isolation procedures (41). Some species of *Campylobacter*, such as *C. sputorum*, *C. concisus*, *C. mucosalis*, *C. curvus*, *C. rectus*, and *C. hyointestinalis*, may require hydrogen for primary isolation and growth. These species may not be recovered when the conventional microaerobic conditions are used, since the amount of hydrogen generated in properly used commercial gas packs is <2%. A gas mixture of 10% CO_2, 6% H_2, and the balance N_2 used in an evacuation-replacement jar is sufficient for isolating hydrogen-requiring species (32a).

Primary plating on selective media in combination with a filtration method (described below) is the optimal method for recovering the variety of *Campylobacter* species from clinical samples. A number of selective media are recommended for isolating *C. jejuni* and *C. coli*. These include the blood-containing Skirrow medium (68) and Campylobacter–cefoperazone-vancomycin-amphotericin (CVA) medium (64) and the blood-free charcoal cefoperazone deoxycholate agar (CCDA) (28), charcoal-based selective medium (CSM) (31), and selective semisolid motility medium (23). To achieve the highest yield of *Campylobacter* spp. from stool samples, a combination of media including either CCDA or CSM as one of the media appears to be the optimal method (16, 25) and may increase the recovery of *Campylobacter* spp. by as much as 10 to 15% over the use of a single medium.

Most of these selective media have one or more antimicrobial agents, mainly cefoperazone, as the primary inhibitor of enteric bacterial flora. The antimicrobial agents such as cephalothin, colistin, and polymyxin B that are present in some selective medium formulations may be inhibitory to some strains of *C. jejuni* and *C. coli* (21, 50) and are inhibitory to *C. fetus* subsp. *fetus*. *C. jejuni* subsp. *doylei*, *C. upsaliensis*, and *A. butzleri* generally will not grow on cephalothin-containing media. When choosing selective media for primary isolation of *Campylobacter* spp. from fecal samples, laboratories should use cefoperazone-containing media and discontinue using cephalothin-containing formulations.

If *Campylobacter* infection is suspected at the time blood specimens are drawn, subculture broth medium to a nonselective blood agar medium, and incubate plates under microaerobic conditions at 37°C. This treatment will allow the isolation of thermophilic and nonthermophilic species. Similarly, blood drawn in Isolator tubes for bacterial culture should be cultured on a nonselective blood plate incubated under microaerobic conditions at 37°C. If a curved, gram-negative rod is observed upon Gram stain examination of a positive blood culture bottle, a sample should be cultured on a nonselective blood plate and incubated under microaerobic conditions at 37°C. An alternative staining method such as acridine orange may also be useful for detecting campylobacters in blood culture bottles if the Gram stain is negative.

Optimal conditions for recovery of *Arcobacter* spp. from clinical specimens have not been determined. What we now know as *Arcobacter* spp. were first isolated on semisolid media designed to isolate *Leptospira* spp. (85, 90). *Arcobacter* species are aerotolerant and have been recovered on certain selective media such as Campy-CVA (2) incubated under microaerobic conditions at 37°C and on nonselective media used in the filtration method. Vandamme et al. designed a medium specifically for the recovery of *Arcobacter* species (85), but the medium has not been evaluated in a clinical setting. Borczyk et al. (10) found that *A. cryaerophila* grew well on cefsulodin-irgasan-novobiocin (CIN) agar than on Mueller-Hinton blood agar, CSM, and Skirrow medium. However, Campy-CVA was found by Anderson et al., who used fecal samples, to be more sensitive than CIN agar (2, 32a).

Species within the genera *Campylobacter* and *Arcobacter* have different optimal temperatures for growth. The choice of incubation temperature for routine stool cultures, therefore, is critical in determining the spectrum of species that will be isolated. By convention, most laboratories use 42°C as the primary incubation temperature for campylobacters. This temperature allows growth of *C. jejuni* and *C. coli* on selective media. *C. upsaliensis* also grows well at 42°C but usually is not recovered on selective media. *C. fetus* grows poorly at this temperature and generally will not be recovered. *Arcobacter* species are not thermophilic and will generally not be recovered at 42°C.

In contrast, most *Campylobacter* and *Arcobacter* species grow well at 37°C. Several of the selective media, such as Skirrow medium and semisolid motility medium, were devised for use at 42°C and have poor selective properties at 37°C, whereas CCDA and CSM show good selective properties at 37°C (16). Ideally, stool samples should be inoculated to selective plates incubated at both 37 and 42°C. Plates should be incubated for 72 h before being reported as negative.

Because of the expense of including several types of media and filtration in the initial workup for campylobacters, one reasonable approach would be to include two selective plates (a blood-free and a blood-containing formulation) incubated at 42°C. If cultures are negative and no other organism is identified, additional samples should be plated on selective media, processed by the filtration method, and incubated at 37°C under microaerobic conditions.

Enrichment Cultures

A number of enrichment broths have been formulated to enhance the recovery of *Campylobacter* spp. from stool samples. These include Campy-thio (7), *Campylobacter* enrichment broth (42), and Preston enrichment (9). Enrichment cultures may be beneficial when low numbers of organism are expected because of delayed transport to the laboratory or because the acute stage of disease has passed.

Filtration

Various investigators have cultured stools by using a filtration method with a nonselective medium to isolate antibiotic-susceptible *Campylobacter* spp. (21, 22) and *Arcobacter* spp. (33). The method is based on the principal that campylobacters can pass through a membrane filter (0.45- to 0.65-μm pore size) with relative ease (because of their motility), while other stool flora are retained during the short processing time.

Cellulose acetate membrane filters with a 0.65-μm pore size are recommended for routine use (89). The filtration technique is performed by placing a sterile 0.65-μm-pore-size cellulose acetate filter on a plate containing antibiotic-free CCDA, CSM, or blood-containing medium. Five to 15 drops of fecal suspension are placed on the filter, and the plate is incubated at 37°C for 1 h. The filter is then removed, and the plate is incubated at 37°C under microaerobic conditions, preferably with an atmosphere containing increased hydrogen (for the hydrogen-requiring species). Stool samples containing ≥10^5 CFU of *Campylobacter* spp. per ml will be detected with this method, and the filtration method is thus not as sensitive as primary culture with selective medium (22). Filtration should be used to complement direct culture on selective plating media and not as a replacement.

IDENTIFICATION

Depending on the medium used, *Campylobacter* spp. may have different colonial appearances. In general, *Campy-*

lobacter spp. produce gray, flat, irregular, spreading colonies, particularly on freshly prepared media. As the moisture content decreases, colonies may become round, convex, and glistening, with little spreading. Hemolysis on blood agar is not observed. *Arcobacter* spp. are morphologically similar to *Campylobacter* spp. (85). For initial analysis, a Gram stain examination of the colony and an oxidase test should be performed. Oxidase-positive colonies exhibiting a characteristic Gram stain appearance (e.g., gram negative, curved to S-shaped rods) that are isolated from selective media incubated at 42°C under microaerobic conditions can be reliably reported as *Campylobacter* spp. until other biochemical tests are performed. The Gram stain appearance of *Arcobacter* spp. may differ from that of typical *Campylobacter* spp. *A. butzleri* is only slightly curved, and *A. cryaerophila* tends to be much more helical.

COMMON PHENOTYPIC TESTS

Several phenotypic tests have been described for identifying *Campylobacter* spp. The most routinely useful tests for initial identification include growth temperature studies (e.g., growth at 25, 37, and 42°C), presence of catalase, hippurate hydrolysis, indoxyl acetate hydrolysis, nitrate reduction, production of H_2S, and antibiotic sensitivity by the disk method (4, 29, 48). The H_2S reaction on triple sugar iron medium works best if the medium is freshly prepared. The nitrate reduction test should be performed using media described by Barrett et al. (4). Unfortunately, *Campylobacter* species are difficult to differentiate from *Arcobacter* species on the basis of phenotypic tests. However, an aerotolerant species (i.e., growth under aerobic conditions) that grows on MacConkey agar (under microaerobic conditions) can be presumptively identified as an *Arcobacter* species. The failure to grow on MacConkey agar, however, does not rule out the presence of *Arcobacter* species.

To obtain consistent and reproducible results, a standardized suspension and inoculum should be used for phenotypic tests. When performing growth temperature and oxygen tolerance studies, prepare a suspension of organism in heart infusion broth or tryptic soy broth to the turbidity of a McFarland no. 1 standard. Using a fiber-tipped swab dipped in the broth suspension, make a single streak across the plate (Mueller-Hinton agar with 5% sheep blood is a suitable medium), and incubate the plates at the desired temperature and/or atmospheric conditions (4, 48).

Hippurate Hydrolysis

Hydrolysis of sodium hippurate is the major test for distinguishing *C. jejuni* subsp. *jejuni* (and also *C. jejuni* subsp. *doylei*) from other *Campylobacter* species. Strains isolated on selective medium that grow at 42°C, are oxidase positive, show characteristic microscopic morphology, and give a positive result with hippurate hydrolysis should be reported as *C. jejuni* subsp. *jejuni*, and for routine clinical purposes, no other tests need to be performed. The method is described in the *Clinical Microbiology Procedures Handbook* (60). Commercially available disk methods for rapid hippurate hydrolysis testing compare favorably with the tube method (12). Use a large inoculum (i.e., full loop) when performing this test. Occasional strains of *C. jejuni* may be hippurate negative (47, 81). Gas-liquid chromatography for detecting benzoic acid (liberated by hydrolysis of sodium hippurate) is the most sensitive assay for hippurate hydrolysis (47) and can be used for more definitive determination.

Nalidixic Acid-Cephalothin Disk Tests

Inhibition by (susceptibility to) or resistance of *Campylobacter* spp. to nalidixic acid and cephalothin has routinely been used as an aid for species identification. Tests can be performed using standard disk diffusion techniques with any nonselective medium that supports the growth of the *Campylobacter* strain and 30-μg nalidixic acid and 30-μg cephalothin disks. Incubate the cultures for 1 to 2 days at 37°C for nonthermophilic campylobacters and at 42°C for thermophilic strains. Any zone of inhibition is considered to indicate susceptibility. Some variability does exist with this test, and even with *C. jejuni* and *C. coli*, some strains may not give the expected results. Occasional strains of *C. jejuni* and *C. coli* may be cephalothin susceptible. Nalidixic acid-resistant *C. jejuni* and *C. coli* strains have been noted by a number of investigators, and in Europe, increasing resistance to fluoroquinolones has been noted (15, 62). Since there is cross-resistance between the fluoroquinolones and nalidixic acid, the finding of nalidixic acid resistance does not exclude the identification of *C. jejuni* or *C. coli*. Hippurate-positive strains should be reported as *C. jejuni*, regardless of the nalidixic acid disk results. Detection of a nalidixic acid-resistant strain of *C. jejuni* should also alert the laboratory and suggest to the physician that the strain may be resistant to fluoroquinolones and that further susceptibility testing may be warranted if antimicrobial therapy with this class of drugs is used.

Indoxyl Acetate Hydrolysis

Indoxyl acetate hydrolysis (44, 58) is useful for differentiating some species of thermophilic *Campylobacter*. The disk method is rapid (5 to 30 min) and easy to perform. Disks are prepared by making a 20% solution of indoxyl acetate in acetone and adding 25 μl of this solution to a blank paper disk (0.6 cm; BBL). Dry the disks at room temperature, and store them at 4°C in a brown tube with desiccant. Disks may also be available from commercial sources. Place a large loopful of growth from a plate on the disk, and then add a drop of sterile distilled water. Hydrolysis of indoxyl acetate is indicated by the appearance of a dark blue or blue-green color. Weak positives show a pale blue color in 10 to 30 min. Lack of a color change indicates a negative reaction.

NONCULTURE METHODS

A number of commercial systems have been developed as an aid to identifying *Campylobacter* spp. at the genus level. Two immunologic assays (Meritec Campy jcl, Meridian Diagnostics, Cincinnati, Ohio; Campyslide, BBL Microbiology Systems, Cockeysville, Md.) (26, 49) can detect *C. jejuni* and *C. coli* but cannot differentiate between the two. At the time of evaluation, the Meritec Campy jcl assay could not reliably identify *C. lari* (49). A DNA probe directed against *Campylobacter* RNA sequences (Accuprobe) that is produced by Gen-Probe, Inc. (San Diego, Calif.), detects *C. jejuni* subsp. *jejuni* subsp. *C. jejuni* subsp. *doylei*, *C. coli*, and *C. lari* and was 100% sensitive for all isolates tested (59, 79). However, the probe also hybridized with 2 of 17 *C. hyointestinalis* strains (59). Thus, these methods may be useful for confirming *Campylobacter* spp. if other tests are not conclusive.

Although a specific test for *Campylobacter* spp. in fecal samples is not commercially available, several investigators have evaluated using the PCR to directly detect *Campy-*lobacter spp. in stool samples (52, 86). The major obstacle in the development of a PCR assay of stools for the clinical laboratory is the lack of a simple and practical DNA extraction procedure for stools that eliminates PCR inhibitors.

EPIDEMIOLOGIC TYPING SYSTEMS

Many typing systems have been devised to study the epidemiology of *Campylobacter* infections; they vary in complexity and ability to discriminate between strains. These methods include serotyping, biotyping, bacteriocin sensitivity, detection of preformed enzymes, auxotyping, lectin binding, phage typing, multilocus enzyme electrophoresis, and molecular-based methods such as restriction endonuclease analysis, ribotyping, and PCR (54). The most frequently used system is serotyping. Two major serotyping schemes are used worldwide; they detect heat-labile (40) and O (55) antigens. The heat-labile serotyping scheme originally described by Lior et al. (40) can detect over 100 serotypes of *C. jejuni*, *C. coli*, and *C. lari*. Uncharacterized bacterial surface antigens and, in some serotypes, flagellar antigens are the serodeterminants for this serotyping system (1). The Penner and Hennessy O serotyping scheme (55) detects 60 types of *C. jejuni* and *C. coli* (54) and is based on detection of lipopolysaccharide antigens. Serotyping is performed in only a few reference laboratories because of the time and expense needed to maintain quality serotyping antisera. Commercially available heat-labile antigen serotyping reagents are generally of very poor quality (51). A combination of serotyping and molecular methods should be used for studying campylobacter infections.

SEROLOGY

Serum immunoglobulin G (IgG), IgM, and IgA levels rise in response to infection, but serum and fecal IgA levels rise during the first few weeks of infection and then fall rapidly (6). Serum antibody assays vary in both sensitivity and specificity for detecting campylobacter infection, and test performance appears to be population dependent. Patients with campylobacter infection may give false-positive legionella antibody tests (11). Serologic testing appears to be useful for epidemiologic investigations but is not recommended for routine diagnosis.

TREATMENT AND ANTIMICROBIAL SUSCEPTIBILITY TESTING

C. jejuni and *C. coli* are susceptible to a variety of antimicrobial agents, including macrolides, fluoroquinolones, aminoglycosides, cholamphenicol, and tetracycline (6). Erythromycin has been the drug of choice for treating *C. jejuni* gastrointestinal infections, and ciprofloxacin has been used as an alternative drug (6). Early therapy of *Campylobacter* infection with erythromycin or ciprofloxacin is effective in eliminating the organism from stool and may also reduce the duration of symptoms associated with infection (20, 57, 65).

C. jejuni is generally susceptible to erythromycin, with resistance rates of less than 5% (63, 67). Rates of erythromycin resistance in *C. coli* vary considerably, with up to 80% of strains showing resistance in some studies (27, 38, 63, 76). Although ciprofloxacin has been effective in treating *Campylobacter* infections, emergence of ciprofloxacin resistance during therapy has been reported (20, 57, 93). Several in vitro studies suggest an increasing resistance to fluoroquinolones (15, 62, 91), and thus, the effectiveness of these drugs may be diminished in the future. C. *jejuni* and C. *coli* may also produce a β-lactamase that appears to be active against amoxicillin, ampicillin, and ticarcillin (37).

Parenteral therapy is used to treat systemic C. *fetus* infections. Drugs used include erythromycin, ampicillin, aminoglycosides, and chloramphenicol, depending on the type of infection (6).

Susceptibility tests for *Campylobacter* spp. are not standardized. Consequently, the literature contains some variability in the susceptibility data reported. Recommendations for the agar dilution method include using Mueller-Hinton agar supplemented with 5% horse or sheep blood and incubated for 16 to 18 h under microaerobic conditions (78). The E-test (PDM Epsilometer; AB Biodisk, Solna, Sweden), which uses Mueller-Hinton agar with 5% sheep blood, compares favorably with agar dilution methods (3, 27) but is not recommended for testing clindamycin.

REFERENCES

1. **Alm, R. A., P. Guerry, M. E. Power, H. Lior, and T. J. Trust.** 1991. Analysis of the role of flagella in the heat-labile Lior serotyping scheme of thermophilic campylobacters by mutant allele exchange. *J. Clin. Microbiol.* **29:**2438–2445.
2. **Anderson, K. F., J. A. Kiehlbauch, D. C. Anderson, H. M. McClure, and I. K. Wachsmuth.** 1993. *Arcobacter* (*Campylobacter*) *butzleri*-associated diarrheal illness in a nonhuman primate. *Infect. Immun.* **61:**2220–2223.
3. **Baker, C. N.** 1992. The E-test and *Campylobacter jejuni*. *Diagn. Microbiol. Infect. Dis.* **15:**469–472.
4. **Barrett, T. J., C. M. Patton, and G. K. Morris.** 1988. Differentiation of *Campylobacter* species using phenotypic characterization. *Lab. Med.* **19:**96–102.
5. **Black, R. E., M. M. Levine, M. L. Clements, T. P. Hughs, and M. J. Blaser.** 1988. Experimental *Campylobacter jejuni* infections in humans. *J. Infect. Dis.* **157:**472–480.
6. **Blaser, M. J.** 1990. *Campylobacter* species, p. 1649–1658. *In* G. L. Mandell, R. G. Douglas, and J. E. Bennett (ed.), *Principles and Practice of Infectious Diseases.* Churchill Livingstone, New York.
7. **Blaser, M. J., I. D. Berkowitz, F. M. LaForce, J. Craven, L. B. Reller, and W. L. Wang.** 1979. Campylobacter enteritis: clinical and epidemiologic features. *Ann. Intern. Med.* **91:**179–185.
8. **Blaser, M. J., and Z. Pei.** 1993. Pathogenesis of *Campylobacter fetus* infections: critical role of high-molecular weight S-layer proteins in virulence. *J. Infect. Dis.* **167:**372–377.
9. **Bolton, F. J., and L. Robertson.** 1982. A selective medium for isolating *Campylobacter jejuni/coli*. *J. Clin. Pathol.* **35:**462–467.
10. **Borczyk, A., S. D. Rosa, and H. Lior.** 1991. Enhanced recognition of *Campylobacter cryaerophila* in clinical and environmental specimens, abstr. C-267, p. 386. *Abstr. Gen. Meet. Am. Soc. Microbiol. 1991.*
11. **Boswell, T. C. J., and G. Kudesia.** 1992. Serological cross-reactions between *Legionella pneumophila* and *Campylobacter* in the indirect fluorescent antibody test. *Epidemiol. Infect.* **109:**291–295.
12. **Cacho, J. B., P. M. Aguirre, A. Hernanz, and A. C. Velasco.** 1989. Evaluation of a disk method for detection of hippurate hydrolysis by *Campylobacter* spp. *J. Clin. Microbiol.* **27:**359–360.
13. **Edmonds, P., C. M. Patton, T. J. Barrett, G. K. Morris, A. G. Steigerwalt, and D. J. Brenner.** 1985. Biochemical and genetic characteristics of atypical *Campylobacter fetus* subsp. *fetus* isolated from humans in the United States. *J. Clin. Microbiol.* **21:**936–940.
14. **Edmonds, P., C. M. Patton, P. M. Griffin, T. J. Barrett, G. P. Schmid, C. N. Baker, M. A. Lambert, and D. J. Brenner.** 1987. *Campylobacter hyointestinalis* associated with human gastrointestinal disease in the United States. *J. Clin. Microbiol.* **25:**685–691.
15. **Endtz, H. P., G. J. Ruijs, B. van Klingeren, W. H. Jansen, T. van der Reyden, and R. P. Mouton.** 1991. Quinolone resistance in campylobacter isolated from man and poultry following the introduction of fluoroquinolones in veterinary medicine. *J. Antimicrob. Chemother.* **27:**199–208.
16. **Endtz, H. P., G. J. Ruijs, A. H. Zwinderman, T. van der Reijden, M. Biever, and R. P. Mouton.** 1991. Comparison of six media, including a semisolid agar, for the isolation of various *Campylobacter* species from stool specimens. *J. Clin. Microbiol.* **29:**1007–1010.
17. **Etoh, Y., F. E. Dewhirst, B. J. Paster, A. Yamamoto, and N. Goto.** 1993. *Campylobacter showae* sp. nov., isolated from the human oral cavity. *Int. J. Syst. Bacteriol.* **43:**631–639.
18. **Fan, K., A. J. Morris, and L. B. Reller.** 1993. Application of rejection criteria for stool cultures for bacterial enteric pathogens. *J. Clin. Microbiol.* **31:**2233–2235.
19. **Figura, N., P. Guglielmetti, A. Zanchi, N. Partini, D. Armellini, P. F. Bayeli, M. Bugnoli, and S. Verdiani.** 1993. Two cases of *Campylobacter mucosalis* enteritis in children. *J. Clin. Microbiol.* **31:**727–728.
20. **Goodman, L. J., G. M. Trenholme, R. L. Kaplan, J. Segreti, D. Hines, R. Petrak, J. A. Nelson, K. W. Mayer, W. Landau, and G. W. Parkhurst.** 1990. Empiric antimicrobial therapy of domestically acquired acute diarrhea in urban adults. *Arch. Intern. Med.* **150:**541–546.
21. **Goossens, H., M. De Boeck, H. Coignau, L. Vlaes, C. Van den Borre, and J. Butzler.** 1986. Modified selective medium for isolation of *Campylobacter* spp. from feces: comparison with Preston medium, a blood-free medium, and a filtration system. *J. Clin. Microbiol.* **24:**840–843.
22. **Goossens, H., L. Vlaes, M. De Boeck, B. Pot, K. Kersters, J. Levy, P. de Mol, J. P. Butzler, and P. Vandamme.** 1990. Is "*Campylobacter upsaliensis*" an unrecognised cause of human diarrhoea? *Lancet* **335:**584–586.
23. **Goossens, H., L. Vlaes, I. Galand, C. Van den Borre, and J. P. Butzler.** 1989. Semisolid blood-free selective-motility medium for the isolation of campylobacters from stool specimens. *J. Clin. Microbiol.* **27:**1077–1080.
24. **Guerrant, R. L., and D. A. Bobak.** 1991. Bacterial and protozoal gastroenteritis. *N. Engl. J. Med.* **325:**327–340.
25. **Gun-Monro, J., R. P. Rennie, J. H. Thornley, H. L. Richardson, D. Hodge, and J. Lynch.** 1987. Laboratory and clinical evaluation of isolation media for *Campylobacter jejuni*. *J. Clin. Microbiol.* **25:**2274–2277.
26. **Hodinka, R. L., and P. H. Gilligan.** 1988. Evaluation of the Campyslide agglutination test for confirmatory identification of selected *Campylobacter* species. *J. Clin. Microbiol.* **26:**47–49.
27. **Huang, M. B., C. N. Baker, S. Banerjee, and F. C. Tenover.** 1992. Accuracy of the E test for determining antimicrobial susceptibilities of staphylococci, enterococci, *Campylobacter jejuni*, and gram-negative bacteria resistant to antimicrobial agents. *J. Clin. Microbiol.* **30:**3243–3248.
28. **Hutchinson, D. N., and F. J. Bolton.** 1984. Improved blood free selective medium for the isolation of *Campylobacter jejuni* from faecal specimens. *J. Clin. Pathol.* **37:**956–957.
29. **Isenberg, H. (ed.).** 1992. *Clinical Microbiology Procedures Handbook.* American Society for Microbiology, Washington, D.C.
30. **Johnson, C. C., and S. M. Finegold.** 1987. Uncommonly encountered motile, anaerobic gram-negative bacilli associated with infection. *Rev. Infect. Dis.* **9:**1150–1162.
31. **Karmali, M. A., A. E. Simor, M. Roscoe, P. C. Flemming, S. S. Smith, and J. Lane.** 1986. Evaluation of a blood-free, charcoal-based, selective medium for the isolation of *Campylobacter* organisms from feces. *J. Clin. Microbiol.* **23:**456–459.
32. **Kasten, M. J., F. Allerberger, and J. P. Anhalt.** 1991. *Campylobacter* bacteremia: clinical experience with three different blood culture systems at Mayo Clinic 1984–1990. *Infection* **19:**88–90.

32a.**Kiehlbauch, J.** Personal communication.

33. **Kiehlbauch, J. A., D. J. Brenner, M. A. Nicholson, C. N. Baker, C. M. Patton, A. G. Steigerwalt, and I. K. Wachsmuth.** 1991. *Campylobacter butzleri* sp. nov. isolated from humans and animals with diarrheal illness. *J. Clin. Microbiol.* **29:**376–385.

34. **Konkel, M. E., Y. Lobet, and W. Cieplak.** 1992. Examination of multiple isolates of *Campylobacter jejuni* for evidence of cholera toxin-like activity, p. 193–198. *In* I. Nachamkin, M. J. Blaser, and L. S. Tompkins (ed.), *Campylobacter jejuni: Current Status and Future Trends.* American Society for Microbiology, Washington, D.C.

35. **Konkel, M. E., D. J. Mead, S. F. Hayes, and W. Cieplak, Jr.** 1992. Translocation of *Campylobacter jejuni* across human polarized epithelial cell monolayer cultures. *J. Infect. Dis.* **166:**308–315.

36. **Kuroki, S., T. Saida, M. Nukina, T. Haruta, M. Yoshioka, Y. Kobayashi, and H. Nakanishi.** 1993. *Campylobacter jejuni* strains from patients with Guillain-Barre syndrome belong mostly to Penner serogroup 19 and contain β-N-acetylglucosamine residues. *Ann. Neurol.* **33:**243–247.

37. **Lachance, N., C. Gaudreau, F. Lamothe, and L. A. Lariviëre.** 1991. Role of the β-lactamase of *Campylobacter jejuni* in resistance to β-lactam agents. *Antimicrob. Agents Chemother.* **35:**813–818.

38. **Lachance, N., C. Gaudreau, F. Lamothe, and F. Turgeion.** 1993. Susceptibilities of β-lactamase-positive and -negative strains of *Campylobacter coli* to β-lactam agents. *Antimicrob. Agents Chemother.* **37:**1174–1176.

39. **Lastovica, A. J., E. Le Roux, and J. L. Penner.** 1989. "*Campylobacter upsaliensis*" isolated from blood cultures of pediatric patients. *J. Clin. Microbiol.* **27:**657–659.

40. **Lior, H., D. L. Woodward, J. A. Edgar, L. J. Laroche, and P. Gill.** 1982. Serotyping of *Campylobacter jejuni* by slide agglutination based on heat-labile antigenic factors. *J. Clin. Microbiol.* **15:**761–768.

41. **Luechtefeld, N. W., L. B. Reller, M. J. Blaser, and W. L. Wang.** 1982. Comparison of atmospheres of incubation for primary isolation of *Campylobacter fetus* subsp. *jejuni* from animal specimens: 5% oxygen versus candle jar. *J. Clin. Microbiol.* **15:**53–57.

42. **Martin, W. T., C. M. Patton, G. K. Morris, M. E. Potter, and N. D. Puhr.** 1983. Selective enrichment broth for isolation of *Campylobacter jejuni. J. Clin. Microbiol.* **17:**853–855.

43. **McKhann, G. M., D. R. Cornblath, J. W. Griffin, T. W. Ho, C. Y. Li, Z. Jiang, H. S. Wu, G. Zhaori, Y. Liu, L. P. Jou, T. C. Liu, C. Y. Gao, J. Y. Mao, M. J. Blaser, B. Mishu, and A. K. Asbury.** 1993. Acute motor axonal neuropathy: a frequent cause of acute flacid paralysis in China. *Ann. Neurol.* **33:**333–342.

44. **Mills, C. K., and R. L. Gherna.** 1987. Indoxyl acetate hydrolysis. *J. Clin. Microbiol.* **25:**1560–1561.

45. **Mishu, B., and M. J. Blaser.** 1993. Role of infection due to *Campylobacter jejuni* in the initiation of Guillain-Barre syndrome. *Clin. Infect. Dis.* **17:**104–108.

46. **Mishu, B., C. M. Patton, and R. V. Tauxe.** 1992. Clinical and epidemiological features of non-jejuni, non-coli *Campylobacter* species, p. 31–41. *In* I. Nachamkin, M. J. Blaser, and L. S. Tompkins (ed.), *Campylobacter jejuni: Current Status and Future Trends.* American Society for Microbiology, Washington, D.C.

47. **Morris, G. K., M. R. El Sherbeeny, C. M. Patton, H. Kodaka, G. L. Lombard, P. Edmonds, D. G. Hollis, and D. J. Brenner.** 1985. Comparison of four hippurate hydrolysis methods for identification of thermophilic *Campylobacter* spp. *J. Clin. Microbiol.* **22:**714–718.

48. **Morris, G. K., and C. M. Patton.** 1985. *Campylobacter,* p. 302–308. *In* E. H. Lennette, A. Balows, W. J. Hausler, Jr., and H. J. Shadomy (ed.), *Manual of Clinical Microbiology,* 4th ed. American Society for Microbiology, Washington, D.C.

49. **Nachamkin, I., and S. Barbagallo.** 1990. Culture confirmation of *Campylobacter* spp. by latex agglutination. *J. Clin. Microbiol.* **28:**817–818.

50. **Ng, L., D. E. Taylor, and M. E. Stiles.** 1988. Characterization of freshly isolated *Campylobacter coli* strains and suitability of selective media for their growth. *J. Clin. Microbiol.* **26:**518–523.

51. **Nicholson, M. A., and C. M. Patton.** 1993. Evaluation of commercial antisera for serotyping heat-labile antigens of *Campylobacter jejuni* and *Campylobacter coli. J. Clin. Microbiol.* **31:**900–903.

52. **Oyofo, B. H., S. A. Thornton, D. H. Burr, T. J. Trust, O. R. Pavlovskis, and P. Guerry.** 1992. Specific detection of *Campylobacter jejuni* and *Campylobacter coli* by using polymerase chain reaction. *J. Clin. Microbiol.* **30:**2613–2619.

53. **Park, C. H., D. L. Hixon, A. S. Polhemus, C. B. Ferguson, S. L. Hall, C. C. Risheim, and C. B. Cook.** 1983. A rapid diagnosis of campylobacter enteritis by direct smear examination. *Am. J. Clin. Pathol.* **80:**388–390.

54. **Patton, C. M., and I. K. Wachsmuth.** 1992. Typing schemes: are current methods useful?, p. 110–128. *In* I. Nachamkin, M. J. Blaser, and L. S. Tompkins (ed.), *Campylobacter jejuni: Current Status and Future Trends.* American Society for Microbiology, Washington, D.C.

55. **Penner, J. L., and J. N. Hennessy.** 1980. Passive hemagglutination technique for serotyping *Campylobacter fetus* subsp. *jejuni* on the basis of soluble heat-stable antigens. *J. Clin. Microbiol.* **12:**732–737.

56. **Perlman, D. M., N. M. Ampel, R. B. Schifman, D. L. Cohn, C. M. Patton, M. L. Aguirre, W. L. Wang, and M. J. Blaser.** 1988. Persistent *Campylobacter jejuni* infections in patients infected with human immunodeficiency virus (HIV). *Ann. Intern. Med.* **108:**540–546.

57. **Petruccelli, B. P., G. S. Murphy, J. L. Sanchez, S. Walz, R. DeFraites, J. Gelnett, R. L. Haberberger, P. Echeverria, and D. N. Taylor.** 1992. Treatment of traveler's diarrhea with ciprofloxacin and loperamide. *J. Infect. Dis.* **165:**557–560.

58. **Popovic-Uroic, T., C. M. Patton, M. A. Nicholson, and J. A. Kiehlbauch.** 1990. Evaluation of the indoxyl acetate hydrolysis test for rapid differentiation of *Campylobacter, Helicobacter,* and *Wolinella* species. *J. Clin. Microbiol.* **28:**2335–2339.

59. **Popovic-Uroic, T., C. M. Patton, I. K. Wachsmuth, and P. Roeder.** 1991. Evaluation of an oligonucleotide probe for identification of *Campylobacter* species. *Lab. Med.* **22:**533–539.

60. **Pratt-Ripplin, K., and M. Pezzlo.** 1992. Identification of commonly isolated aerobic gram-positive bacteria, p. 1.20.21–1.20.22. *In* H. Isenberg (ed.), *Clinical Microbiology Procedures Handbook,* vol. 1. American Society for Microbiology, Washington, D.C.

61. **Rams, T. E., D. Feik, and J. Slots.** 1993. *Campylobacter rectus* in human periodontitis. *Oral Microbiol. Immunol.* **8:**230–235.

62. **Rautelin, H., O. V. Renkonen, and T. U. Kosunen.** 1991. Emergence of fluoroquinolone resistance in *Campylobacter jejuni* and *Campylobacter coli* in subjects from Finland. *Antimicrob. Agents Chemother.* **35:**2065–2069.

63. **Reina, J., N. Borrell, and A. Serra.** 1992. Emergence of resistance to erythromycin and fluoroquinolones in thermotolerant *Campylobacter* strains isolated from feces 1987–1991. *Eur. J. Clin. Microbiol. Infect. Dis.* **11:**1163–1166.

64. **Reller, L. B., S. Mirrett, and L. G. Reimer.** 1983. Controlled evaluation of an improved selective medium for isolation of *Campylobacter jejuni,* abstr. C-274, p. 357. *Abstr. Annu. Meet. Am. Soc. Microbiol.* 1983.

65. **Salazar-Lindo, E., R. B. Sack, E. Chea-Woo, B. A. Kay, I. Piscoya, and R. Y. Leon-Barua.** 1986. Early treatment with erythromycin of *Campylobacter jejuni* associated dysentery in children. *J. Pediatr.* **109:**3555–3560.

66. **Sazie, E. S. M., and A. E. Titus.** 1982. Rapid diagnosis of Campylobacter enteritis. *Ann. Intern. Med.* **96:**62–63.

67. **Sjogren, E., B. Kaijser, and M. Werner.** 1992. Antimicrobial susceptibilities of *Campylobacter jejuni* and *Campylobacter coli* isolated in Sweden: a 10-year follow-up report. *Antimicrob. Agents Chemother.* **36:**2847–2849.

68. **Skirrow, M. B.** 1977. Campylobacter enteritis: a "new" disease. *Br. Med. J.* **ii:**9–11.

69. **Skirrow, M. B., D. M. Jones, E. Sutcliffe, and J. Benjamin.** 1993. Campylobacter bacteremia in England and Wales, 1981–1991. *Epidemiol. Infect.* **110:**567–573.

70. **Spiegel, C. A., and G. Telford.** 1984. Isolation of *Wolinella recta* and *Actinomyces viscosus* from an actinomycotic chest wall mass. *J. Clin. Microbiol.* **20:**1187–1189.

71. **Stanley, J., A. P. Burnens, D. Linton, S. L. W. On, M. Costas, and R. J. Owen.** 1992. *Campylobacter helveticus* sp. nov., a new thermophilic species from domestic animals: characterization, and cloning of a species-specific DNA probe. *J. Gen. Microbiol.* **138:**2293–2303.

72. **Steele, T. W., and R. J. Owen.** 1988. *Campylobacter jejuni* subspecies *doylei* (subsp. nov.), a subspecies of nitrate-negative campylobacters isolated from human clinical specimens. *Int. J. Syst. Bacteriol.* **38:**316–318.

73. **Tauxe, R. V.** 1992. Epidemiology of *Campylobacter jejuni* infections in the United States and other industrialized nations, p. 9–19. *In* I. Nachamkin, M. J. Blaser, and L. S. Tompkins (ed.), *Campylobacter jejuni: Current Status and Future Trends.* American Society for Microbiology, Washington, D.C.

74. **Tauxe, R. V., C. M. Patton, P. Edmonds, T. J. Barrett, D. J. Brenner, and P. A. Blake.** 1985. Illness associated with *Campylobacter laridis*, a newly recognized *Campylobacter* species. *J. Clin. Microbiol.* **21:**222–225.

75. **Taylor, D. N.** 1992. *Campylobacter* infections in developing countries, p. 20–30. *In* I. Nachamkin, M. J. Blaser, and L. S. Tompkins (ed.), *Campylobacter jejuni: Current Status and Future Trends.* American Society for Microbiology, Washington, D.C.

76. **Taylor, D. N., M. J. Blaser, P. Echeverria, C. Pitarangsi, L. Bodhidatta, and W. L. Wang.** 1987. Erythromycin-resistant *Campylobacter* infections in Thailand. *Antimicrob. Agents Chemother.* **31:**438–442.

77. **Taylor, D. N., J. H. Kiehlbauch, W. Tee, C. Pitarangsi, and P. Echeverria.** 1991. Isolation of group 2 aerotolerant *Campylobacter* species from Thai children with diarrhea. *J. Infect. Dis.* **163:**1062–1067.

78. **Tenover, F. C., C. N. Baker, C. L. Fennell, and C. A. Ryan.** 1992. Antimicrobial resistance in *Campylobacter* species, p. 66–73. *In* I. Nachamkin, M. J. Blaser, and L. S. Tompkins (ed.), *Campylobacter jejuni: Current Status and Future Trends.* American Society for Microbiology, Washington, D.C.

79. **Tenover, F. C., L. Carlson, S. Barbagallo, and I. Nachamkin.** 1990. DNA probe culture confirmation assay for identification of thermophilic *Campylobacter* species. *J. Clin. Microbiol.* **28:**1284–1287.

80. **Thompson, J. S., D. S. Hodge, D. E. Smith, and Y. A. Yong.** 1990. Use of tri-gas incubator for routine culture of *Campylobacter* species from fecal specimens. *J. Clin. Microbiol.* **28:**2802–2803.

81. **Totten, P. A., C. M. Patton, F. C. Tenover, T. J. Barrett,** W. E. Stamm, A. G. Steigerwalt, J. Y. Lin, K. K. Holmes, and D. J. Brenner. 1987. Prevalence and characterization of hippurate-negative *Campylobacter jejuni* in King County, Washington. *J. Clin. Microbiol.* **25:**1747–1752.

82. **Vandamme, P., and J. De Ley.** 1991. Proposal for a new family, *Campylobacteraceae*. *Int. J. Syst. Bacteriol.* **41:**451–455.

83. **Vandamme, P., E. Falsen, B. Pot, B. Hoste, K. Kersters, and J. De Ley.** 1989. Identification of EF group 22 campylobacters from gastroenteritis cases as *Campylobacter concisus*. *J. Clin. Microbiol.* **27:**1775–1781.

84. **Vandamme, P., E. Falsen, R. Rossau, B. Hoste, P. Segers, R. Tytgat, and J. De Ley.** 1991. Revision of *Campylobacter*, *Helicobacter*, and *Wolinella* taxonomy: emendation of generic descriptions and proposal of *Arcobacter* gen. nov. *Int. J. Syst. Bacteriol.* **41:**81–103.

85. **Vandamme, P., M. Vancanneyt, B. Pot, L. Mels, B. Hoste, D. Dewettinck, L. Vlaes, C. Van den Borre, R. Higgins, and J. Hommez.** 1992. Polyphasic taxonomic study of the emended genus *Arcobacter* with *Arcobacter butzleri* comb. nov. and *Arcobacter skirrowii* sp. nov., and aerotolerant bacterium isolated from veterinary specimens. *Int. J. Syst. Bacteriol.* **42:**344–356.

86. **Waegel, A., and I. Nachamkin.** 1993. Polymerase chain reaction for detecting *Campylobacter jejuni* in stool samples, abstr. C-442, p. 524. *Abstr. 93rd Gen. Meet. Am. Soc. Microbiol. 1993.*

87. **Wang, W. L., and M. J. Blaser.** 1986. Detection of pathogenic *Campylobacter* species in blood culture systems. *J. Clin. Microbiol.* **23:**709–714.

88. **Wang, W. L., L. B. Reller, B. Smallwood, N. W. Luechtefeld, and M. J. Blaser.** 1983. Evaluation of transport media for *Campylobacter jejuni* in human fecal specimens. *J. Clin. Microbiol.* **18:**803–807.

89. **Wells, J. G., N. D. Puhr, C. M. Patton, M. A. Nicholson, M. A. Lambert, and R. C. Jerris.** 1989. Comparison of selective media and filtration for the isolation of *Campylobacter* from feces, abstr. C-231, p. 432. *Abstr. Annu. Meet. Am. Soc. Microbiol. 1989.*

90. **Wesley, I. V.** *Arcobacter* infections. *In* G. W. Beran (ed.), *CRC Handbook of Zoonosis*, in press. CRC Press, Boca Raton, Fla.

91. **Wistrom, J., M. Jertborn, E. Ekwall, K. Norlin, B. Soderquist, A. Stromberg, R. Lundholm, H. Hogevik, L. Lagergren, G. Englund, and S. Norrby.** 1992. Empiric treatment of acute diarrheal disease with norfloxacin. *Ann. Intern. Med.* **117:**202–208.

92. **Wood, R. C., K. L. MacDonald, and M. T. Osterholm.** 1992. Campylobacter enteritis outbreaks associated with drinking raw milk during youth activities. A 10-year review of outbreaks in the United States. *JAMA* **268:**3228–3230.

93. **Wretlind, B., A. Stromberg, L. Ostlund, E. Sjogren, and B. Kaijser.** 1992. Rapid emergence of quinolone resistance in *Campylobacter jejuni* in patients treated with norfloxacin. *Scand. J. Infect. Dis.* **24:**685–686.

Helicobacter

ROBERT C. JERRIS

38

DESCRIPTION OF THE GENUS

The genus *Helicobacter* was established in 1989 during taxonomic restructuring of the species *Campylobacter pylori* and *Campylobacter mustelae*. Convincing data from analysis of cellular fatty acids, ultrastructure, respiratory quinones, antimicrobial susceptibility profiles, growth requirements, enzyme capabilities, and 16S rRNA sequences clearly supported the establishment of a new genus (21).

Organisms in the genus are helical, curved, or straight unbranched gram-negative rods 0.3 to 1.0 μm wide and 1.5 to 5.0 μm long. The organisms have rounded ends, and when the cells are curved, they demonstrate spiral periodicity. In older cultures, the organisms appear as spheroid or coccoid bodies. They are motile by means of a single, polar sheathed flagellum (*Helicobacter cinaedi* and *H. fennelliae*) or multiple unipolar or bipolar lateral sheathed flagella (*H. pylori*). The organisms are microaerophilic, with optimum growth at 37°C. Catalase and oxidase reactions are positive; H_2S production (in triple sugar iron) and hippurate hydrolysis are negative. The organisms are asaccharolytic. The G+C content of the DNA ranges from 35 to 44 mol% (62).

There are at least nine species within the genus *Helicobacter*, as detailed in Table 1. The feces of birds and pigs also contain helicobacter-like organisms, and it is predicted that with further evaluation, these strains may be designated new species.

Three species are significant human pathogens: *H. pylori* (previously named *C. pylori* and *C. pyloridis*), *H. fennelliae* (previously named *C. fennelliae*), and *H. cinaedi* (previously named *C. cinaedi*). The remainder of this chapter will deal with these three species.

NATURAL HABITAT

H. pylori

The major habitat of *H. pylori* is the human gastric mucosa. The enzyme urease, produced in abundance by the organism, converts urea to ammonia, which enables the organism to survive in the mucosa despite the high acidity of the lumen. Additionally, the spiral shape and active motility of *H. pylori* make it able to resist peristalsis. The microaerophilic nature of the organism is suited to the environment of the mucosal gel, and specific adhesins produced by the organism are probably responsible for long-term residency.

H. pylori has also been detected in dental plaque by culture (29, 33) and by PCR (3). It has been successfully cultured from the feces of Gambian children (55) and detected in gastritis patients' feces by PCR (34). The organism has also been cultured from the saliva of a patient with gastritis (17).

H. pylori has worldwide distribution, and although the mode(s) of acquisition and transmission is not entirely clear, it appears to be acquired by the fecal-oral or the oral-oral route.

H. cinaedi and H. fennelliae

H. cinaedi's natural habitat is the intestinal tract in rodents. In humans, *H. cinaedi* and *H. fennelliae* have been isolated most frequently from rectal cultures (56). *H. cinaedi* has also been cultured from the feces of adult women and children (61). Although no comprehensive epidemiological studies exist, it can be inferred from case reports and animal models (19) that these organisms may be resident in the gastrointestinal tract or may exist as part of the transient fecal flora.

CLINICAL SIGNIFICANCE

H. pylori

Since its discovery in 1982 and publication of the landmark paper by Marshall and Warren (36), *H. pylori* has received much attention as an agent of gastritis. In developing countries, the organism is acquired early in childhood, and up to 90% of children are infected by age 5 (55). In the United States, few infections occur during childhood, and the organism has a projected incidence of about 0.5 to 1.0% per year, with about 50% incidence at age 60 (6). The highest incidence in both developing and developed countries is in poorer, lower socioeconomic groups, for which crowding and poor sanitation are risk factors.

On acquisition, *H. pylori* resides primarily in the mucous layer of the gastric mucosa without directly invading the gastric epithelium. The host mounts a humoral immune response that can be detected serologically but appears to have little effect in eradicating the organism (6). A number of potentially pathogenic and adaptive substances, including urease, cytotoxins, mucinase, lipase, phospholipase A, adhesins, and a hemolysin, are elaborated by the organism (12, 15). The organism colonizes the stomach for many

TABLE 1 *Helicobacter* species and related organisms

Species	Hosts	Source or habitat
H. pylori	Humans	Gastric mucosa
H. mustelae	Ferrets	Gastric mucosa
H. felis	Cats, dogs	Gastric mucosa
H. nemestrinae	Macaque monkeys	Gastric mucosa
H. muridarum	Rats, mice	Intestinal mucosa
H. acinonyx	Cheetahs	Gastric mucosa
H. cinaedi	Humans, rodents	Intestinal mucosa
H. fennelliae	Humans	Intestinal mucosa
"*H. rappini*"[a]	Sheep, dogs, humans	Liver (sheep), stomach (dogs), feces (humans)
Gastrospirillum hominis[b]	Cheetahs, humans	Gastric mucosa

[a]Previously *Flexispira rappini*. A proposal to place the organism within the genus *Helicobacter* has been made.

[b]Unculturable helix-shaped bacteria found by McNulty et al. (37) in stained gastric biopsy specimens. RNA sequencing places the organism in the genus *Helicobacter*, and the name "*H. heilmanii*" has been proposed (23).

years or decades and causes a persistent low-grade gastric inflammation. In the majority of cases, this chronic, superficial gastritis is asymptomatic. Certain host or organism factors, not yet characterized, are responsible for progression of the inflammatory response. Candidate factors include a specific 120- to 128-kDa high-molecular-mass antigen that is linked to cytotoxin production (57), a specific vacuolating toxin (13), and the ability of substances elaborated by the organism to recruit and activate inflammatory cells (32).

The association between chronic gastritis, *H. pylori* infection, and peptic ulceration is strong. In idiopathic peptic ulcer disease, *H. pylori* is present at rates significantly higher than those in age-matched controls (7). In addition, *H. pylori* is associated with duodenal ulcers (4, 6). Longitudinal studies over a 10-year period have shown that patients with chronic gastritis have a 13-fold increase in the risk of developing duodenal ulcers (51). Furthermore, clinical trials have shown that specific therapy against *H. pylori* results in long-lasting remission or cure of the diseases (6). As there are no satisfactory models of peptic ulcer disease, our knowledge about the pathogenesis of the disease is incomplete.

Chronic superficial gastritis may progress or lead to chronic gastric atrophy with the risk of gastric adenocarcinoma. Thus, *H. pylori* may be a risk factor for gastric cancer. Both epidemiological and prospective studies support this hypothesis (39, 43, 44).

H. cinaedi and *H. fennelliae*

H. cinaedi and *H. fennelliae* were first isolated from homosexual men who presented with proctocolitis, proctitis, enteritis, and bacteremia (56). The organisms were also found colonizing the gastrointestinal tract without clinical symptomatology. Since then, *H. cinaedi* has been noted as a cause of bacteremia primarily in patients with AIDS (1, 10, 45). It may have a role in gastroenteritis in other patient populations (54, 61). Additionally, *H. cinaedi* has been reported from a neonate with septicemia and meningitis (41). The patient's mother cared for pet hamsters during the first two trimesters of her pregnancy. Successive bacte-

remias with *H. cinaedi* and *H. fennelliae* in a bisexual male have been described (38).

Complete epidemiological profiles for *H. cinaedi* and *H. fennelliae* remain undefined.

SPECIMEN COLLECTION AND TRANSPORT

H. pylori

For detection of *H. pylori*, obtain multiple biopsy specimens, including at least one from the gastric antrum and one from the corpus, because of the patchy distribution of the organism (26). Substances (such as cimetidine and benzocaine, but not lidocaine) and procedures that inhibit the growth of *H. pylori* should be avoided (9). Contamination of biopsy forceps with glutaraldehyde also inhibits growth.

Biopsy material should be placed directly into a transport medium, because *H. pylori* is extremely susceptible to desiccation. Physiological sterile saline is appropriate if the specimen can be transported and processed without delay. If a delay in transport is anticipated, Stuart's transport medium held at 4°C for up to 24 h is preferred (28, 63). Other transport media shown to be effective include thioglycolate broth, nutrient broth, and brucella broth (24).

H. cinaedi and *H. fennelliae*

Conventional methods for obtaining rectal swabs, diarrheic stools, and blood for culture have yielded the organisms. It appears that incubation with the use of appropriate selective media in the proper atmosphere is the most critical aspect in cultivation of *H. cinaedi* and *H. fennelliae*.

DIRECT EXAMINATION AND DETECTION

For *H. pylori*, direct smears from touch preparations of minced tissue are preferred, as organisms are difficult to see in homogenized tissue. The routine Gram stain is generally adequate for staining the organisms. Alternatively, commercially available basic fuchsin (4%) may be used as a counterstain.

H. pylori may be detected in histological sections stained by the hematoxylin and eosin method. Organisms appear adjacent to the gastric epithelium as faintly staining, eosinophilic, curved to spiral rods measuring 3.0 by 0.5 μm, often with patchy distribution, as seen in Fig. 1. On rare occasions, a special stain like Warthin-Starry or Giemsa may be required to visualize low numbers of organisms.

Indirect methods such as the commercially available campylobacter-like organism (CLO) test rely on the strong urease activity of *H. pylori* and its ability to break down urea to ammonia and bicarbonate. Small pieces of biopsied tissue are placed into a tube containing a pellet of modified Christensen's urea agar. The presence of the organism is detected by an indicator change from yellow to pink, indicating urease activity.

A second test, also based on *H. pylori*'s efficient hydrolysis of urea, is the labeled-carbon breath test. Patients are fasted overnight and given a test meal plus a solution of ^{13}C- or ^{14}C-labeled urea. The presence of *C. pylori* is detected by the liberation of labeled CO_2 (as a by-product of the degradation of urea) in the patient's breath.

Serological tests (enzyme immunoassay and latex agglutination tests) to detect antibodies to *H. pylori* are commercially available (25, 50, 53), as discussed below under Evaluation, Interpretation, and Reporting of Results.

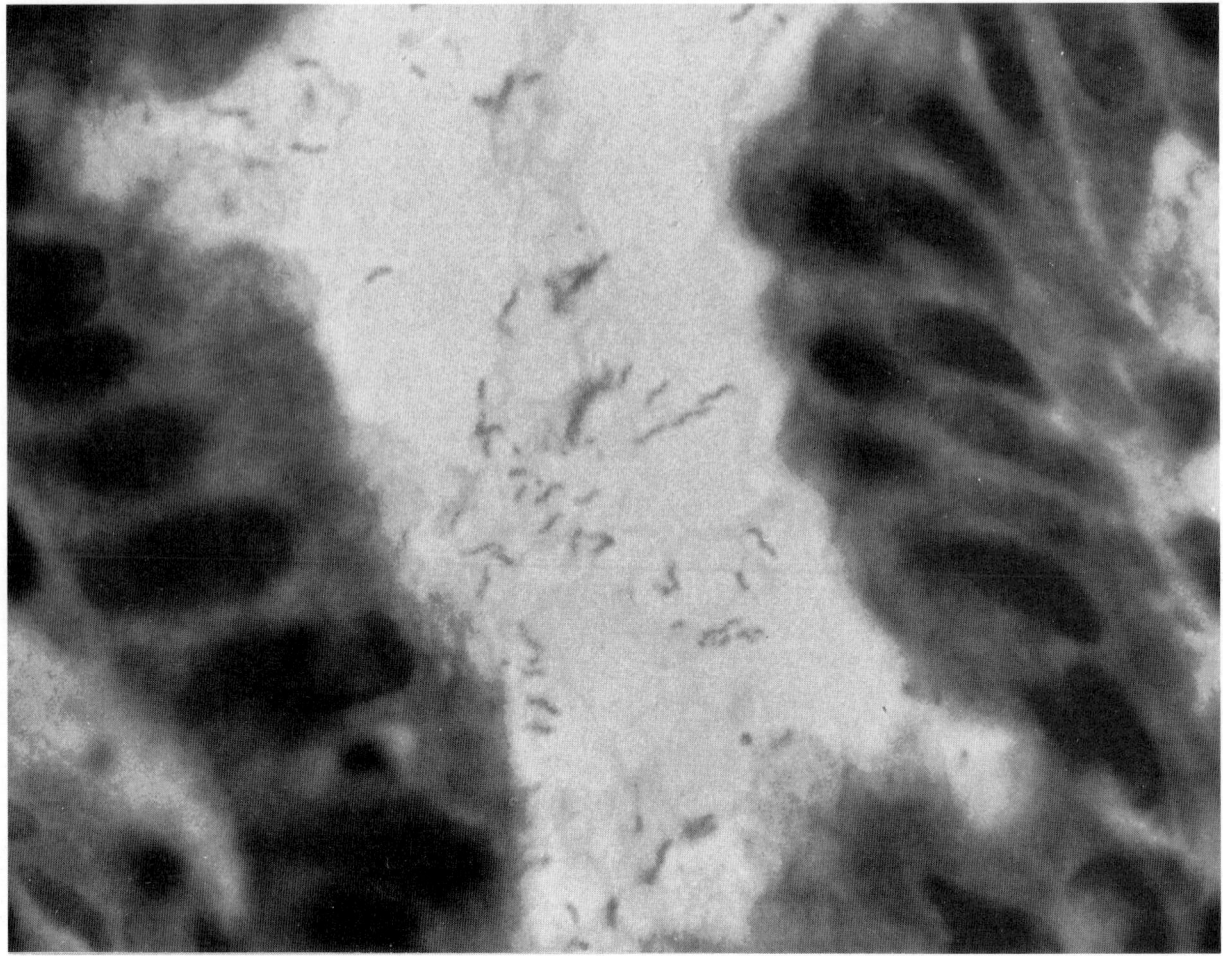

FIGURE 1 Hematoxylin and eosin stain of a gastric biopsy sample with *H. pylori* present in the gastric mucosa.

The PCR has been used successfully for identification of *H. pylori*. Details of the primers and conditions for amplification and detection are available (59).

CULTURE AND ISOLATION

H. pylori

Biopsy specimens should be minced and gently homogenized in physiological sterile saline. The homogenate should be plated immediately, and both selective and nonselective media have been recommended for use (22). Traditional chocolate agar and nutrient-rich media such as brain heart infusion or brucella agar supplemented with 5 to 7% horse or rabbit blood are adequate as nonselective media (23). An egg yolk emulsion agar with Columbia agar base has also been shown to yield excellent nonselective growth of *H. pylori* (64). Skirrow's and other routine *Campylobacter* media are too inhibitory for *H. pylori* isolation (14, 47). Dent's medium (Oxoid, through Unipath, Ogdensberg, N.Y.), a modification of Skirrow's medium made by adding amphotericin B and replacing polymyxin with cefsulodin (14), has been recommended as a selective medium. Pylori agar (bioMérieux, Marcy l'Etoile, France) has also been used for selective isolation (24). The optimal selective medium has yet to be developed.

Plates must be incubated in a microaerobic environment (5 to 7% O_2) incorporating increased CO_2 (5 to 10%) and a high relative humidity. This environment may be created by using the Campy Pak system (BBL Microbiology Systems, Cockeysville, Md.) in an anaerobic jar with a moistened towel in the bottom. A single evacuation of an anaerobic jar to 220 mm Hg (1 mm Hg = 133.322 Pa) and replacement with an anaerobic gas mixture (10% CO_2, 10% H_2, and 80% N_2) yields an acceptable atmosphere of 5% O_2, 7% CO_2, 8% H_2, and 80% N (22). The Campy Pouch (BBL), which holds one or two plates, has also been effective for subculture. Because an acceptable atmosphere is generated in the pouch, primary isolation is predicted even though no formal evaluation has been published. All biochemical and susceptibility tests for these organisms require microaerobic conditions.

The optimal temperature of incubation is 35 to 37°C. Colonies from primary isolation are generally observed by 3 to 4 days of incubation; however, a rare strain will take up to 7 days. Colonies appear as small, pinpoint, 1- to 2-mm, translucent, nonhemolytic colonies. Gram-stained smears reveal gram-negative uniform to slightly curved rods, with rare spiral shapes. This morphology differs from the predominant spiral shapes seen in histopathologic sections. Isolates may not survive even limited exposure to atmospheric bench-top environments (for some strains as little as 45 min) (24), and subculturing at 2- to 3-day intervals is required for many isolates to retain viability. In older cultures, *H. pylori* cells become coccoid, with an associated decrease in ability to subculture.

H. cinaedi and H. fennelliae

Rectal swabs and diarrheic specimens should be plated on a selective medium such as brucella agar base with 10% sheep blood, vancomycin (10 mg/liter), polymyxin (2,500 IU/ liter), trimethoprim (5 mg/liter), and amphotericin B (2 mg/liter) (16, 56). Nonselective media for subculture of blood cultures include brucella agar base supplemented with vitamin K, hemin, and 5% sheep blood; brucella agar base plus 10% sheep blood; heart infusion agar plus 5% rabbit blood; and Columbia agar base plus 5% sheep or horse blood.

Organisms are frequently not discernible in Gram-stained isolates from blood cultures unless basic fuchsin is used as a counterstain. Dark-field examination and acridine orange staining may also be used.

Plates should be incubated in a microaerobic atmosphere as described for *H. pylori*. The optimum temperature for growth is 37°C, and no growth occurs at 25 or 42°C or in ambient air, in a candle jar, or under strict anaerobic conditions. Growth is demonstrable in 4 to 7 days. Colonies are pinpoint and translucent at 72 h and develop into gray to gray-white, flat, irregular, watery colonies measuring about 1 mm in diameter. Heavy inoculation on fresh medium often yields flat, spreading growth with no distinct colonies.

On Gram stain with basic fuchsin as a counterstain, organisms appear predominantly as slender, wavy, gram-negative bacilli. Older cultures yield coccoid forms (with some strains, in as little as 48 h).

IDENTIFICATION

H. pylori

Presumptive identification of *H. pylori* can be made with typical Gram stain properties, a positive catalase test, a positive oxidase test, and a positive rapid urease reaction. Table 2 details the biochemical characteristics of this species.

Several typing schemes for *H. pylori* have been described. Among seven strains, Lior detailed five provisional serovars differentiated by heat-labile antigens (31). *H. pylori* is extremely genetically diverse, and ribopatterns occurring with *Hae*III digestion contain between five and seven fragments of up to 10 kb and appear to be an excellent means of typing. Although *Hae*III ribopatterns are highly discriminatory, up to 25% of strains have DNA with *Hae*III site-specific methylation and are not fragmented (42).

H. cinaedi and H. fennelliae

Both *H. cinaedi* and *H. fennelliae* are catalase and oxidase positive, urease negative, and susceptible to nalidixic acid (30-mg disk) when tested on brucella agar base supplemented with 10% sheep blood. *H. cinaedi* shows intermediate zones of inhibition to cephalothin (range, 12 to 16 mm; mean, 14.2 mm), whereas *H. fennelliae* is susceptible (zone size range, 28 to 30 mm; mean, 29.5 mm) (16, 27). *H. fennelliae* has an odor of hypochlorite (16). Table 2 details the biochemical characteristics of the species.

TABLE 2 Differential reactions and characteristics of species of genus *Helicobacter*[a]

Characteristic	*H. pylori*	*H. cinaedi*	*H. fennelliae*
Oxidase	+	+	+
Catalase	+	+	+
Urease	+	−	−
Hippurate hydrolysis	−	−	−
H$_2$S production on triple sugar iron agar	−	−	−
γ-Glutamyl transpeptidase	+	NA	NA
Nitrate reduction	−	+	−
Microaerobic growth at:			
25°C	−	−	−
37°C	+	+	+
42°C	−	−	−
Susceptibility to:			
Nalidixic acid (30-μg disk)	R	S	S
Cephalothin (30-μg disk)	S	I	S
Growth in the presence of:			
1% bile	−	+	+
1% glycine	−	+	+
Indoxyl acetate hydrolysis	−	−	+
Cellular fatty acid[b]	Lambert gpG	Lambert gpD	Lambert gpE

[a]Data were obtained from references 8, 16, 21, 23, 47, 48, 56, and 62. +, >90% positive; −, >90% negative; NA, not available; S, susceptible; R, resistant; I, intermediate.

[b]Lambert group G is characterized by the presence of 19:0 cyc, 3-OH-16:0, and 3-OH-18:0 and the absence of 16:1 and 3-OH-14:0 (30). Lambert group D is characterized by the presence of 12:0, 3-OH-12:0, and 3-OH-16:0 and the absence of 16:1 and 3-OH-14:0 (30). Lambert group E is characterized by the presence of a 16-carbon aldehyde and a 16-carbon dimethylacetyl and the absence of 16:1 (30).

ANTIMICROBIAL SUSCEPTIBILITY

H. pylori

Because of the inactivity of certain antimicrobial agents in the acid environment of the stomach, most laboratory assays to predict antimicrobic activity against *H. pylori* do not predict in vivo results. Routine testing of *H. pylori* susceptibility to metronidazole has been recommended, however, as in vitro results do predict clinical outcome. Agar and broth dilution methods for susceptibility testing have been used (18, 46). A substantial decrease in efficacy is noted with triple-therapy regimens when metronidazole-resistant strains are detected, with success rates dropping from 90 to 40% with resistant organisms (5). *H. pylori* is susceptible to bismuth-containing compounds at levels achievable with oral dosing and is also sensitive to the proton pump inhibitors omeprazole and lansoprazole (6).

For treatment of *H. pylori* infection, a 14-day triple-therapy regimen is generally effective (35). Bismuth (eight tablets of Pepto-Bismol or 4 tablets of De-Nol per day), metronidazole (250 mg four times a day [q.i.d.]) and either tetracycline (500 mg q.i.d.) or amoxicillin (500 mg q.i.d.) have been used with success (4, 35). Follow-up at 28 days posttreatment is suggested for eradication of the organism (35). Inappropriate therapy may result in relapse (11).

In the treatment of children, to whom bismuth is difficult to administer and for whom tetracycline is contraindi-

TABLE 3 Summary of tests for detection of *H. pylori*

Test	Endoscopy required	Cost[a]	Sensitivity (%)[b]	Specificity (%)[b]	Comments
Culture	Yes	$ $ $	77–94	100	"Gold standard"; optimal recovery, with two biopsies; requires expertise in clinical microbiology; required for sensitivity testing and epidemiological typing
Histology	Yes	$ $ $	93–99	95–99	Demonstrates host response and integrity of mucosa
Rapid urease test (e.g., CLO test)	Yes	$	86–97	86–98	Can be performed rapidly
^{13}C breath test (nonradioactive)	No	$ $	90–100	80–99	Preferred for pregnant women and children; not widely available; excellent for early posttreatment tracking
^{14}C breath test	No	$ $	90–100	92–100	Small amount of radiation exposure; excellent for posttreatment tracking
Serology	No	$	83–98	56–100	Readily available; useful seroepidemiological tool; limited value in short-term tracking of therapy

[a]$, least expensive; $$, moderately expensive; $$$, most expensive.
[b]Data are from references 50, 52, and 53.

cated, amoxicillin (50 mg/kg of body weight per day) with tinadazole (20 mg/kg/day) has been used (40). Alternative trials with different regimens and agents are ongoing in both children and adults.

H. cinaedi and *H. fennelliae*

Disk diffusion testing shows *H. cinaedi* to be susceptible to ampicillin and gentamicin (41). The agar dilution method has indicated that both *H. cinaedi* and *H. fennelliae* are susceptible to ampicillin, gentamicin, doxycycline, tetracycline, ceftriaxone, rifampin, spectinomycin, nalidixic acid, and chloramphenicol (18). A modified BACTEC procedure demonstrated inhibition of *H. fennelliae* by ampicillin, chloramphenicol, gentamicin, and ciprofloxacin (38).

Optimal therapy for infection with *H. cinaedi* and *H. fennelliae* has not been established, but ampicillin in combination with gentamicin has been used as effective therapy against *H. cinaedi* (38), as has a 2-week course of ciprofloxacin (49). Gentamicin has been effective against *C. fennelliae* (38).

EVALUATION, INTERPRETATION, AND REPORTING OF RESULTS

A standard biopsy sample in which the organism is detected by rapid urease detection methods (such as the CLO test), histology, or culture is recommended for diagnosis of *H. pylori*. Rapid urease tests are simple and cost-effective and give same-day results. Histology offers the advantage that the host inflammatory response may be visualized but has the limitation that detection of the organism may be missed when the distribution of the bacteria is irregular or patchy. Culture requires specific laboratory expertise and is necessary for performing susceptibility testing and typing of isolates.

Noninvasive breath tests for detection of *H. pylori* are alternative diagnostic procedures useful both in diagnosing disease and in determining therapeutic outcome. These methods require instrumentation for analysis but do not require a biopsy.

Serology may be useful as adjunct testing for screening purposes and epidemiological studies. At present, only limited long-term studies can firmly establish the role of serology in diagnosis and tracking of *H. pylori* disease. Data are currently being gathered with recently marketed commercial kits for serological detection. Although product inserts show sensitivities and specificities of more than 90 to 95%, the need for independent evaluation of these kits has recently been underscored by the failure of many of the products to achieve stated performance levels (50, 53). Serology obviates the need for biopsy and is relatively inexpensive. Serological assays may be limited in their value for monitoring early posttreatment tracking, as antibody levels persist for several months postinfection, falling to approximately 50% of pretreatment levels at about 6 months (2). Minor decreases in antibody levels are noted within a month of treatment, and levels continue to fall over time with successful therapy (58). If the organism is not eliminated, titers remain the same or become elevated (60). See Table 3 for a summary of tests designed to identify *H. pylori*.

Clinical suspicion of infection with *H. cinaedi* and *H. fennelliae* must be communicated to the laboratory, since appropriate laboratory methods are required for isolation. A specific humoral antibody response to *H. cinaedi* and *H. fennelliae* has been described and may prove useful as a diagnostic and epidemiologic tool (20), but tests are not available commercially. The spectrum of clinical disease caused by these organisms will undoubtedly continue to expand as routine procedures for detection and isolation are implemented in the clinical laboratory.

REFERENCES

1. **Aboulafia, D., G. Mathisen, and R. Mitsuyasu.** 1991. Case report: aggressive Kaposi's sarcoma and *Campylobacter* bacteremia in a female with transfusion associated AIDS. *Am. J. Med. Sci.* **301:**256–258.
2. **Andersen, L. P.** 1993. The antibody response to *Helicobacter pylori* infection, and the value of serologic tests to detect *H. pylori* and for post-treatment monitoring, p. 285–305. *In* C. S. Goodwin and B. W. Worsley (ed.), *Helicobacter pylori: Biology and Clinical Practice.* CRC Press, Inc., Boca Raton, Fla.
3. **Banatvala, N., C. R. Lopez, R. Owen, Y. Ardi, G. Davies, J. Hardia, and R. Feldman.** 1993. *Helicobacter pylori* in dental plaque. *Lancet* **341:**380.
4. **Bell, G. D.** 1993. Clinical aspects of infection with *Helicobacter pylori. Communicable Dis. Rep.* **3:**R59–R62.
5. **Bell, G. D., K. U. Powell, S. M. Burridge, A. Pallecaros, P. H. Jones, P. W. Gant, G. Harrison, and J. E. Trowell.**

1992. Experience with 'triple' anti-*Helicobactor pylori* eradication therapy: side effects and the importance of testing the pre-treatment isolate for metronidazole. *Aliment. Pharmacol. Ther.* **6:**427–435.

6. Blaser, M. J. 1992. *Helicobacter pylori:* its role in disease. *Clin. Infect. Dis.* **15:**386–393.

7. Blaser, M. J., G. I. Perez-Perez, J. Lindenbaum, D. Schneidman, G. Van Deventer, M. Marin-Sorensen, and W. M. Weinstein. 1991. Association of infection due to *Helicobacter pylori* with specific upper gastrointestinal pathology. *Rev. Infect. Dis.* **13**(Suppl. 8):S704–S708.

8. Burnens, A. P., J. Stanley, U. B. Schaad, and J. Nicolet. 1993. Novel *Campylobacter*-like organism resembling *Helicobacter fennelliae* from a boy with gastroenteritis and from dogs. *J. Clin. Microbiol.* **31:**1916–1917.

9. Carr, H., H. Estrin, V. Hupertz, and S. Czinn. 1987. Inhibitory effects of topical anesthetic agents on *Campylobacter pyloridis* culture results. *Gastroenterology* **92:**1338.

10. Cimolai, N., M. J. Gill, A. Jones, B. Flores, W. E. Stamm, W. Laurie, B. Madden, and M. S. Shahrabadi. 1987. "*Campylobacter cinaedi*" bacteremia: case report and laboratory findings. *J. Clin. Microbiol.* **25:**942–943.

11. Coughlin, J. G., D. Gilligan, H. Humphreys, D. McKenna, C. Dooley, E. Sweeney, C. Keane, and C. O'Morain. 1987. *Campylobacter pylori* and recurrence of duodenal ulcers—a twelve month follow-up study. *Lancet* **ii:**1109–1111.

12. Covacci, A., S. Censini, M. Bugnoli, R. Petracca, D. Burroni, G. Macchia, A. Massone, E. Papini, Z. Xiang, N. Figura, and R. Rappuoli. 1993. Molecular characterization of the 128 k-Da immunodominant antigen of *Helicobacter pylori* associated with cytotoxicity and duodenal ulcer. *Proc. Natl. Acad. Sci. USA* **90:**5791–5795.

13. Cover, T. L., and M. J. Blaser. 1992. Purification and characterization of the vacuolating toxin from *Helicobacter pylori*. *J. Biol. Chem.* **267:**10570–10575.

14. Dent, J. C., and C. A. M. McNulty. 1988. Evaluation of a new selective media for *Campylobacter pylori*. *Eur. J. Clin. Microbiol. Infect. Dis.* **7:**555–568.

15. Dunn, B. E. 1993. Pathogenic mechanisms of *Helicobacter pylori*. *Gastroenterol. Clin. N. Am.* **22:**43–57.

16. Fennell, C. L., P. A. Totten, T. C. Quinn, D. L. Patton, K. K. Holmes, and W. E. Stamm. 1984. Characterization of *Campylobacter*-like organisms from homosexual men. *J. Infect. Dis.* **149:**58–66.

17. Ferguson, D. A., L. Chuanfu, N. R. Patel, W. R. Mayberry, D. S. Chi, and E. Thomas. 1993. Isolation of *Helicobacter pylori* from saliva. *J. Clin. Microbiol.* **31:**2802–2804.

18. Flores, B. M., C. L. Fennell, K. K. Holmes, and W. E. Stamm. 1985. In vitro susceptibilities of *Campylobacter*-like organisms to 20 antimicrobial agents. *Antimicrob. Agents Chemother.* **28:**188–191.

19. Flores, B. M., C. L. Fennell, L. Kuller, M. A. Bronsdon, W. R. Morton, and W. E. Stamm. 1990. Experimental infection of pig-tailed macaques (*Macaca nemestrina*) with *Campylobacter cinaedi* and *Campylobacter fennelliae*. *Infect. Immun.* **58:**3947–3953.

20. Flores, B. M., C. L. Fennell, and W. E. Stamm. 1989. Characterization of *Campylobacter cinaedi* and *C. fennelliae* antigens and analysis of the human immune response. *J. Infect. Dis.* **159:**635–640.

21. Goodwin, C. S., J. A. Armstrong, T. Chilvers, M. Peters, M. D. Collins, L. Sly, W. McConnell, and W. E. S. Harper. 1989. Transfer of *Campylobacter pylori* and *Campylobacter mustelae* to *Helicobacter* gen. nov. as *Helicobacter pylori* comb. nov. and *Helicobacter mustelae* comb. nov., respectively. *Int. J. Syst. Bacteriol.* **39:**397–405.

22. Goodwin, C. S., E. D. Blincow, J. R. Warren, T. E. Waters, C. R. Sanderson, and L. Easton. 1985. Evaluation of cultural techniques for isolation of *Campylobacter pyloridis* from endoscopic biopsies of the gastric mucosa. *J. Clin. Pathol.* **38:**1127–1131.

23. Goodwin, C. S., and B. W. Worsley. 1993. The *Helicobacter* genus: the history of *H. pylori* and taxonomy of current spe-

cies, p. 1–13. *In* C. S. Goodwin and B. W. Worsley (ed.), *Helicobacter pylori: Biology and Clinical Practice.* CRC Press, Inc., Boca Raton, Fla.

24. Goodwin, C. S., and B. W. Worsley. 1993. Microbiology of *Helicobacter pylori*. *Gastroenterol. Clin. N. Am.* **22:**5–19.

25. Goosens, H., Y. Glupczynski, A. Burette, C. van den Borre, and J. P. Butzler. 1992. Evaluation of available second-generation immunoglobulin G enzyme immunoassays for detection of *Helicobacter pylori* infection. *J. Clin. Microbiol.* **30:**176–180.

26. Hazell, S. L., W. B. Hennessey, T. J. Borody, J. Carrick, M. Ralston, L. Brady, and A. Lee. 1987. *Campylobacter pyloridis* gastritis. II. Distribution of bacteria and associated inflammation in the gastrointestinal environment. *Am. J. Gastroenterol.* **82:**297–301.

27. Kemper, C. A., P. Mickelsen, A. Morton, B. Walton, and S. C. Deresinski. 1993. *Helicobacter* (*Campylobacter*) *fennelliae*-like organisms as an important but occult cause of bacteremia in a patient with AIDS. *J. Infect. Dis.* **26:**97–101.

28. Kjoller, M., A. Fischer, and T. Justesen. 1991. Transport conditions and number of biopsies necessary for culture of *Helicobacter pylori*. *Eur. J. Clin. Microbiol. Infect. Dis.* **10:**166–167.

29. Krajden, S., M. Fuksa, J. Anderson, J. Kempston, A. Boccia, C. Petrea, C. Babida, M. Karmali, and J. L. Penner. 1989. Examination of human stomach biopsies, saliva, and dental plaque for *Campylobacter pylori*. *J. Clin. Microbiol.* **27:**1397–1398.

30. Lambert, M. A., C. M. Patton, T. J. Barrett, and C. W. Moss. 1987. Differentiation of *Campylobacter* and *Campylobacter*-like organisms by cellular fatty acid composition. *J. Clin. Microbiol.* **25:**706–713.

31. Lior, H. 1991. Serological characterization of *H. pylori*—a provisional serotyping scheme. *Ital. J. Gastroenterol.* **23**(Suppl. 2):42.

32. Mai, U. E., G. I. Perez-Perez, J. B. Allen, S. M. Wahl, M. J. Blaser, and P. D. Smith. 1992. Surface proteins from *Helicobacter pylori* exhibit chemotactic activity for human leukocytes and are present in gastric mucosa. *J. Exp. Med.* **175:**517–525.

33. Majmudar, P., S. M. Shah, K. R. Dhunjibhoy, and H. G. Desai. 1990. Isolation of *Helicobacter pylori* from dental plaque in healthy volunteers. *Indian J. Gastroenterol.* **9:**271–272.

34. Mapstone, N. P., D. A. F. Lyncy, F. A. Lewis, A. T. R. Axon, D. S. Tompkins, M. F. Dixon, and P. Quirke. 1993. PCR identification of *H. pylori* in feces from gastritis patients. *Lancet* **341:**447.

35. Marshall, B. J. 1993. Treatment strategies for *Helicobacter pylori* infection. *Gastroenterol. Clin. N. Am.* **22:**183–198.

36. Marshall, B. J., and J. R. Warren. 1984. Unidentified curved bacilli in the stomach of patients with gastritis and peptic ulceration. *Lancet* **i:**1311–1315.

37. McNulty, C. A. M., J. Dent, A. Curry, J. S. Uff, G. A. Ford, M. G. L. Gear, and S. P. Wilkinson. 1989. New spiral bacterium in gastric mucosa. *J. Clin. Pathol.* **42:**585–591.

38. Ng, V., W. K. Hadley, C. L. Fennell, B. M. Flores, and W. E. Stamm. 1987. Successive bacteremias with "*Campylobacter cinaedi*" and "*Campylobacter fennelliae*" in a bisexual male. *J. Clin. Microbiol.* **25:**2008–2009.

39. Nomura, A., G. N. Stremmermann, P. Chyou, I. Kato, G. I. Perez-Perez, and M. J. Blaser. 1991. *Helicobacter pylori* infection and gastric carcinoma in a population of Japanese-Americans in Hawaii. *N. Engl. J. Med.* **325:**1132–1136.

40. Ordenda, G., D. Vaira, J. Holton, A. Colin, F. Altare, and N. Ansaldi. 1989. Amoxicillin plus tinidazole for *Campylobacter pylori* gastritis in children: assessment by IgG antibody, pepsinogen I, and gastrin levels. *Lancet* **i:**690–692.

41. Orlicek, S. L., D. F. Welch, and T. L. Kuhls. 1993. Septicemia and meningitis caused by *Helicobacter cinaedi* in a neonate. *J. Clin. Microbiol.* **31:**569–571.

42. Owens, R. J. 1993. Microbiological aspects of *Helicobacter pylori* infection. *Communicable Dis. Rep.* **3:**R51–R56.

43. Parsonnet, J. 1993. *Helicobacter pylori* and gastric cancer. *Gastroenterol. Clin. N. Am.* **22:**89–104.

44. **Parsonnet, J., G. D. Friedman, P. P. Vandersteen, Y. Chang, J. H. Vogelman, N. Orentresch, and R. K. Sibley.** 1991. *Helicobacter pylori* infection and the risk of gastric carcinoma. *N. Engl. J. Med.* **325:**1127–1131.

45. **Pasternak, J., R. Bolivar, R. L. Hopfer, V. Fainstein, K. Mills, A. Rios, G. P. Body, C. L. Fennell, P. A. Totten, and W. E. Stamm.** 1984. Bacteremia caused by *Campylobacter*-like organisms in two male homosexuals. *Ann. Intern. Med.* **101:**339–341.

46. **Pavicic, M. J. A. M. P., F. Namavar, T. Verboom, A. J. van Winkelhoff, and J. de Graaf.** 1993. In vitro susceptibility of *Helicobacter pylori* to several antimicrobial combinations. *Antimicrob. Agents Chemother.* **37:**1184–1186.

47. **Penner, J. L.** 1991. *Campylobacter, Helicobacter,* and related organisms, p. 402–409. *In* A. Balows, W. J. Hausler, Jr., K. L. Herrmann, H. D. Isenberg, and H. J. Shadomy (ed.), *Manual of Clinical Microbiology,* 5th ed. American Society for Microbiology, Washington, D.C.

48. **Popovic-Uroic, T., C. M. Patton, M. A. Nicholson, and J. A. Kiehlbauch.** 1990. Evaluation of indoxyl acetate hydrolysis test for rapid differentiation of *Campylobacter, Helicobacter,* and *Wolinella* species. *J. Clin. Microbiol.* **28:**2335–2339.

49. **Sacks, L. V., A. M. Lambriola, V. J. Gill, and F. M. Gordin.** 1991. Use of ciprofloxacin for successful eradication of bacteremia due to *Campylobacter cinaedi* in a human immunodeficiency virus-infected person. *Rev. Infect. Dis.* **13:**1066–1068.

50. **Schembri, M. A., S. K. Lin, and J. R. Lambert.** 1993. Comparison of commercial diagnostic tests for *Helicobacter pylori* antibodies. *J. Clin. Microbiol.* **31:**2621–2624.

51. **Sipponen, P., K. Varis, O. Fraki, V. M. Korri, K. Seppala, and M. Sivrala.** 1990. Cumulative 10-year risk of symptomatic duodenal and gastric ulcer in patients with or without chronic gastritis. A clinical follow-up of 454 outpatients. *Scand. J. Gastroenterol.* **25:**966–973.

52. **Smoot, D. T.** 1994. *Helicobacter pylori* diagnostic tests: benefits, sensitivity, and specificity, p. 77–79. *In National Institutes for Health Consensus Development Conference on Helicobacter pylori in Peptic Ulcer Disease.* National Institutes of Health, Bethesda, Md.

53. **Talley, N. J., L. Kost, A. Haddad, and A. R. Zinsmeister.** 1992. Comparison of commercial serological tests for detection of *Helicobacter pylori* antibodies. *J. Clin. Microbiol.* **30:**3145–3150.

54. **Tee, W., B. N. Anderson, B. C. Ross, and B. Dwyer.** 1987. Atypical campylobacter associated with gastroenteritis. *J. Clin. Microbiol.* **25:**1248–1252.

55. **Thomas, J. E., G. R. Gibson, M. K. Darboe, A. Dale, and L. T. Weaver.** 1992. Isolation of *H. pylori* from human feces. *Lancet* **34:**1194–1195.

56. **Totten, B. A., C. L. Fennell, F. C. Tenover, J. M. Wegenberg, P. L. Perine, W. E. Stamm, and K. K. Holmes.** 1985. *Campylobacter cinaedi* (sp. nov.) and *Campylobacter fennelliae* (sp. nov.): two new *Campylobacter* species associated with enteric disease in homosexual men. *J. Infect. Dis.* **151:**131–139.

57. **Tummuru, M. K., T. L. Cover, and M. J. Blaser.** 1993. Cloning and expression of a high-molecular-mass major antigen of *H. pylori*: evidence of linkage to cytotoxin production. *Infect. Immun.* **61:**1799–1809.

58. **Vaira, D., J. Holton, M. Cairns, A. Polydorou, J. F. Dowsett, and P. R. Salmon.** 1988. Antibody response to *Campylobacter pylori* after treatment for gastritis. *Br. Med. J.* **297:**397.

59. **Valentine, J. L.** 1992. PCR detection of *Helicobacter pylori,* p. 282–287. *In* D. H. Persing, T. F. Smith, F. C. Tenover, and T. J. White (ed.), *Diagnostic Molecular Microbiology: Principles and Applications.* American Society for Microbiology, Washington, D.C.

60. **Van Bohemen, C. G., M. L. Langenberg, E. A. J. Rauws, J. Oudbier, E. Weterings, and H. C. Zanen.** 1989. Rapidly decreased serum IgG to *Campylobacter pylori* following elimination of *Campylobacter*-positive gastritis. *Immunol. Lett.* **20:**59.

61. **Vandamme, P., E. Falsen, B. Pot, K. Kersters, and J. DeLey.** 1990. Identification of *Campylobacter cinaedi* isolated from blood and feces of children and adult females. *J. Clin. Microbiol.* **28:**1016–1020.

62. **Vandamme, P., E. Falsen, R. Rossau, B. Hoste, P. Segers, R. Tytgat, and J. DeLey.** 1991. Revision of *Campylobacter, Helicobacter,* and *Wolinella* taxonomy: emendation of generic descriptions and proposal of *Arcobacter* gen. nov. *Int. J. Syst. Bacteriol.* **41:**88–103.

63. **West, A. P., M. R. Millar, and D. S. Thompkins.** 1990. Survival of *Helicobacter pylori* in water and saline. *J. Clin. Pathol.* **43:**609.

64. **Westblom, T. U., E. Madan, and B. R. Midkiff.** 1991. Egg yolk emulsion agar, a new medium for cultivation of *Helicobacter pylori*. *J. Clin. Microbiol.* **29:**819–821.

Unusual Gram-Negative Bacteria, Including *Capnocytophaga, Eikenella, Pasteurella,* and *Streptobacillus*

BARRY HOLMES, M. JOHN PICKETT, AND DANNIE G. HOLLIS

39

The term unusual bacteria (UB) is used loosely to denote gram-negative rods other than enteric species, *Haemophilus* spp. and the nonfermentative bacteria (NFB). Some UB (species of *Aeromonas, Plesiomonas,* and *Vibrio*) are fermentative and nonfastidious like enteric species but differ from them in being oxidase positive. Some (*Campylobacter* and *Eikenella* species) fail to acidify glucose like some NFB (*Moraxella* species and some pseudomonads) but, unlike most NFB, are fastidious. Several UB (*Capnocytophaga* species, *Cardiobacterium hominis,* and some *Brucella* species) differ from both enteric species and NFB in presenting little or no growth on blood agar (BA) plates incubated aerobically for 48 h. All UB have two features in common: they are encountered infrequently in clinical specimens, and they may also easily escape detection. This chapter deals with some of the UB (Table 1); several of the better known genera are dealt with in other chapters.

MATERIALS AND METHODS

The media and procedures of the Centers for Disease Control and Prevention (CDC) (11, 58) are generally applicable to clinical laboratories. Most such laboratories use commercial BA plates prepared with 5% sheep blood; these plates may be slightly less sensitive to bacterial hemolysins than 5% rabbit blood agar prepared with heart infusion agar. BA plates become less sensitive with age. Any isolate that may be capnophilic should be inoculated onto duplicate BA plates for incubation at 35°C, one plate aerobically and the other in a candle jar (3 to 5% carbon dioxide and saturated humidity).

The oxidase test should be performed with the more sensitive tetramethyl, not dimethyl, reagent. With quite fastidious bacteria, growth in the peptone basal medium (enteric fermentation base) used for sugar fermentation tests may fail. In such instances, each tube should be supplemented with 2 drops of healthy rabbit serum. Alternatively, when growth fails in peptone basal medium, large inocula (0.5 to 1 mm³ of cell paste) with buffered sugars (such as for neisseriae) or aerobic low-peptone sugars (as for NFB; 39) can be used. No growth is required by these two methods. Similarly, a buffered amino acid broth, large inoculum, and ninhydrin reagent (to test for decarboxylation of lysine and ornithine) obviate the need for growth of the more fastidious rods. Motility medium can be that used for

pseudomonads, the motility nitrate agar used for NFB, or the oxidative-fermentative basal medium used for NFB. For indole tests, the tryptone medium should be incubated for 2 days, extracted with xylene, and then tested for indole with Ehrlich reagent. A more sensitive procedure, at least for flavobacteria (39), is use of a buffered tryptophan broth medium and modified Kovács reagent. Gram stains should be prepared from both solid and liquid media. Growth on the BA plate should be used for an oxidase test. Tryptone-glucose-yeast extract agar slants are used to determine growth at 25, 35, and 42°C, and one of these slants is used for a catalase test (3% hydrogen peroxide); if an organism does not grow on this medium, then heart infusion agar slants are used.

ACTINOBACILLUS SPP.

The genera *Actinobacillus* and *Pasteurella* are similar in the mole percentages of G+C in their chromosomal DNA, in their phenotypic features, and in being animal reservoirs for human disease. Organisms in both genera ferment glucose, reduce nitrate to nitrite, and are nonmotile. No single phenotypic feature definitively distinguishes the two (33), but an isolate can usually be identified by performance of several biochemical tests (Table 2). DNA-DNA hybridization studies indicated that the former *Pasteurella ureae* should be transferred to the genus *Actinobacillus* as *Actinobacillus ureae* (34). Certain species currently in the genus *Actinobacillus* belong in separate genera according to their 16S rRNA sequences or DNA-rRNA hybridizations. These species are *Actinobacillus actinomycetemcomitans, Actinobacillus seminis,* and *Actinobacillus capsulatus* (13, 16). Weaver et al. (58) suggested that the 33 strains tabulated by the CDC as *Actinobacillus suis* may represent more than one taxon and that some of the mannitol-fermenting strains may be the biochemically similar *Actinobacillus hominis.*

All isolates are nonmotile coccobacilli or small rods; they are oxidase positive (except that A. *actinomycetemcomitans* is often negative) and acidify both slants and butts of triple sugar iron agar (TSIA) (Table 1). Most strains of *Actinobacillus equuli* and A. *suis* but none of *Actinobacillus lignieresii* ferment melibiose and trehalose. All strains of A. *suis* but none of A. *equuli* and A. *lignieresii* hydrolyze esculin. A. *ureae* differs from most other species in being

TABLE 1 Unusual gram-negative rods[a]

Organism	No. of strains	Capnophilic	% of strains positive		TSIA, acid		Table(s) containing further data
			Catalase	Oxidase	Slant	Butt	
Actinobacillus actinomycetemcomitans	120	+	99	19[b]	100	100	2, 3
Actinobacillus equuli	19	−	73	100	100	100	2
Actinobacillus equuli-like bacterium	1	−	100	100	100	100	2
Actinobacillus lignieresii	30	−	89	100	100	100	2
Actinobacillus suis	33	−	85	100	100	100	2
Actinobacillus (Pasteurella) ureae	97	−	63	99	100	99	2
Capnocytophaga spp.							
Capnocytophaga gingivalis, C. ochracea, C. sputigena (DF-1)	155	+	7[c]	7	73	55	3
Capnocytophaga canimorsus, C. cynodegmi (DF-2)	27	+	100	96	17	15	3
Cardiobacterium hominis (IId)	65	+	1	100	93	84	3
Chromobacterium violaceum	37	−	97	67	8	94	4
DF-3	21	+	0	0	95	100	3
DF-3-like	6	+	0	0	100	100	3
EF-4a	97	−	100	100	3	73	4
EF-4b	34	−	100	100	0	6	4
Eikenella corrodens	506	+	8	100	0	0	3
Kingella denitrificans	60	−	10	100	2	0	4
Kingella kingae	33	−	0	100	32	12	4
Kingella oralis	11	−	9	100	ND[d]	ND	4
Pasteurella aerogenes	16	−	100	100	94	100	2
Pasteurella bettyae (HB-5)	44	+	2	54	100	98	3
Pasteurella dagmatis (*Pasteurella* "n. sp. 1"; includes *Pasteurella* taxon 16)	91	−	96	98	100	100	2
Pasteurella haemolytica (includes *P. trehalosi*)	67	−	86	95	100	100	2
Pasteurella multocida (includes *P. canis* and *P. stomatis*)	306[e]	−	98	97	99	99	2
Pasteurella pneumotropica	107	−	100	99	100	97	2
Streptobacillus moniliformis	16	+	0	0	71	57	3
Suttonella indologenes	6	−	0	100	100	100	4

[a]Exclusive of the genera *Aeromonas*, *Plesiomonas* (chapter 36 in this Manual), *Bordetella* (chapter 46), *Brucella* (chapter 44), *Campylobacter* (chapter 37), *Francisella* (chapter 43), *Gardnerella* (chapter 29), *Haemophilus* (chapter 45), and *Vibrio* (chapter 35) and members of the family *Bartonellaceae*. Data are mainly from Clark et al. (11). See also Dewhirst et al. (14), Mutters et al. (33), Schlater et al. (48), and Table 2 for more recently described species of pasteurellae. Not all strains were tested for some features.

[b]Of the strains tested, 95% were either oxidase negative or only weakly positive.

[c]Of the strains tested, 98% were either catalase negative or only weakly positive.

[d]ND, no data.

[e]*P. multocida* of the CDC charts (11) contains more than one species (33).

lactose and xylose negative. Unlike *A. hominis*, *A. suis* ferments D-arabinose, D-mannose, and cellobiose.

Actinobacilli were first associated with cattle, but they have since been recovered from other animals and humans (11, 37). Several human infections have followed animal bites. *A. ureae* has been recovered from blood, cerebrospinal fluid, and the respiratory tract (55). *A. hominis* was originally recovered from patients with chronic lung disease; it is also a causative agent of septicemia in hepatic failure (60). An *A. equuli*-like bacterium has been isolated from an infected horse bite wound (37). Other species, such as *Actinobacillus rossii* and *A. seminis* (52), *A. capsulatus*, *Actinobacillus muris*, and *Actinobacillus (Haemophilus) pleuropneumoniae*, are not described here, as they have so far not been isolated from clinical specimens.

On 1-day BA plates, colonies are translucent and 1 to 2 mm in diameter.

All isolates are susceptible to chloramphenicol and tetracycline (27).

DNA-DNA hybridizations and examination of surface antigens led to the proposal that *A. actinomycetemcomitans* be transferred to the genus *Haemophilus* as *Haemophilus actinomycetemcomitans* (41). Although this proposal was not favorably received (25a), the organism is certainly misplaced in the genus *Actinobacillus* (see above). Differential features are presented in Tables 1, 2, and 3.

A. actinomycetemcomitans was first recovered from actinomycotic lesions in cattle and humans. This bacterium is now particularly associated with endocarditis and periodontitis but has also been associated with other focal infections, including animal bites.

A. actinomycetemcomitans, like brucellae, *Haemophilus aphrophilus*, and *Francisella tularensis*, appears as small coccobacilli or short rods upon Gram stain. The organisms are

TABLE 2 Differential features of *Actinobacillus* and *Pasteurella* species occurring in clinical specimens[a]

Feature	A. actinomycetemcomitans	A. equuli[b]	A. hominis	A. lignieresii	A. suis	A. ureae	[P.] aerogenes	P. bettyae (HB-5)	P. canis	P. dagmatis[c]	P. haemolytica sensu stricto	P. multocida[d]	[P.] pneumotropica	P. stomatis	P. trehalosi	P. volantium
Acidified																
L-Arabinose	−	(−)	−	(−)	(+)	−	(+)	−	−	−	+	d	(−)	−	−	−
Cellobiose	+	−	−	+	+	−	−	−	−	−	(−)	−	−	−	d	NT
D-Galactose	+	d	+	d, L	(+)	−	+	−	+	+	+	+	+	+	−	NT
Lactose	+	+	+	+	+	−	+	d	−	−	d	−	d	−	−	d, w
Maltose	(+)	+	+	+	+	(+)	d	−	−	+	+	(−)	d	−	+	+
D-Mannitol	+	+	+	+	−	+	d	d	+	−	+	(+)	−	+	+	+
D-Mannose	+	+	−	+	+	d	d	−	+	+	(−)	+	+	+	+	NT
Melibiose	−	+	+, L	−	+	−	d	−	−	−	−[e]	(−)	d	−	−[e]	NT
Raffinose	−	+	+	d	+	−	−	−	−	+, w	d	−	−	−	−	NT
Salicin	(−)[f]	−	d	−	+	−	(−)	−	−	−	(−)	−	d	−	d	d
Sorbitol	−	−	−	(−)	−	(−)	+	−	d	−	+	d	(−)	−	+	NT
Sucrose	−	+	+	+	+	+	+	−	+	+	+	+	−	+	+	+
D-Trehalose	−	+	+	+	+	d	−	d	(−)	+	−	d	+	+	+	d
D-Xylose	d	+	+	+	+	−	+	−	+	−	+	d	d	−	−	−
Beta-hemolysis (sheep cells)	−	−	−	−	−	−	−	−	(−)	−	(+)	−	(+)	−	(+)	−
Catalase	+	d	+	d	+	d	+	−	+	+	+	+	+	+	+	+
Esculin hydrolysis	−	−	d	d	+	d	−	d	−	−	d	−	d	−	d	NT
Fermentation[g]	+	+	+	+	+	+	+	+	+	+	+	+	+	+	+, L	NT
Gas production from D-glucose[g]	(+)	−	−	−	−	−	(+)	d, w	−	+, w	−	−	d	−	−	−
Growth on MacConkey agar[h]	d	+	d	+	−	d	−	d	−	−	+	−	d	−	+	−
Indole production	−	−	−	−	−	−	(−)	+	d, w	+	−	+	+	+, w	−	−
NAD requirement	−	−	+	−	−	−	d	−	−	−	−	−	−	−	−	+
ONPG reaction[i]	−	d	d	d	(+)	−	d	−	+	−	d	(+)	+	−	−	+
Ornithine decarboxylase	−	−	−	−	−	−	d	+	+	−	−	+	+	+	−	−
Oxidase	+	(+)	+	(+)	(+)	(+)	d	+	+	+	+	d	+	+	+	d
Urease	−	+	+	+	+	+	+	−	−	+	−	−	+	−	−	+

[a] Data are from Sneath and Stevens (52); these data were summarized from 19 different references. Symbols and abbreviations: +, ≥90% of strains are positive; (+), 80 to 89% of strains are positive; d, 21 to 79% of strains are positive; (−), 11 to 20% of strains are positive; −, ≤10% of the strains are positive; L, late reaction, >24 h for acidification of carbohydrates or >48 h for fermentation in oxidative-fermentative medium; NT, not tested; w, weak reaction.

[b] A. equuli-like bacterium differs from A. equuli in producing gas from glucose (37).

[c] Strains that are similar except for failing to produce urease are probably Pasteurella species taxon 16 (5).

[d] Three subspecies: P. multocida subsp. gallicida is dulcitol and sorbitol positive, P. multocida subsp. multocida is dulcitol negative and sorbitol positive, and P. multocida subsp. septica is dulcitol and sorbitol negative.

[e] Positive according to Carter (9) and negative according to Kilian and Frederiksen (29) and Sneath and Stevens (51).

[f] Positive according to Sneath and Stevens (51) and negative according to Kilian and Frederiksen (29) and Phillips (38).

[g] In Hugh-Leifson medium containing glucose with incubation for 48 h at 37°C.

[h] Different formulations of this medium may explain discrepancies in the reported results; the results for P. volantium are for medium supplemented with NAD.

[i] Hydrolysis of ONPG.

TABLE 3 Features of capnophilic rods

Feature	% of strains positive (no. of strains examined)										
	Actino-bacillus actinomy-cetemcomitans[a] (120)	Capno-cytophaga canimorsus (DF-2)[b] (75)	Capno-cytophaga cynodegmi (DF-2-like)[b] (9)	Capno-cytophaga gingivalis (DF-1)[c] (25)	Capno-cytophaga ochracea (DF-1)[c] (27)	Capno-cytophaga sputigena (DF-1)[c] (6)	Cardio-bacterium hominis[a] (65)	DF-3[a] (21)	Eikenella corrodens[a] (506)	Pasteurella bettyae (HB-5)[a] (44)	Strepto-bacillus monili-formis[a] (16)
Acidified											
D-Galactose	100	89	67	0	83	0	0	ND[d]	ND	0	ND
D-Glucose	99	93	100	100	100	100	100	100	0	100	94
Lactose	0	100	100	8	92	40	0	95	0	0	0
Maltose	95	95	100	100	100	100	100	100	0	0	100
D-Mannitol	82	0	0	0	0	0	95	0	0	0	0
Raffinose	0	0	89	24	70	17	0	ND	ND	0	ND
Sucrose	0	0	89	100	100	100	98	95	0	0	7
D-Xylose	42	0	0	0	0	0	0	100	0	0	0
Arginine dihydrolase	0	99	100	0	0	0	0	0	0	0	ND
Catalase	99	100	100	0	0	0	1	0	8	2	0
Citrate (Simmons)	0	0	0	0	0	0	0	0	0	0	0
Esculin	0	64	100	75	96	83	2	100	0	0	ND
Gelatin, hydrolyzed	0	0	0	17[e]	14[e]	60	0	0	0	0	0
Indole	0	0	0	0	0	0	100	85	0	100	0
Lysine decarboxylase	0	0	0	9	5	0	0	9	82	0	ND
MacConkey agar, growth	5	2	0	0	0	0	0	0	0	58	0
Motility	0	0	0	0	0	0	0	0	0	0	0
Nitrate reduced	100	0	17	4	8	83	0	0	99	100	ND
Ornithine decarboxylase	0	0	0	0	0	0	0	0	98	0	0
Oxidase	19	97	100	0	0	0	100	0	100	54	ND
Urease	0	0	0	12	14	0	0	4	0	0	0

[a]Most data are from Clark et al. (11). Not all strains were tested for some features.
[b]Data are from Brenner et al. (8). Not all strains were tested for some features.
[c]Data are from Socransky et al. (53).
[d]ND, no data.
[e]Gelatin is not hydrolyzed according to Holt and Kinder (25).

relatively fastidious obligate capnophiles. In broth, the cells adhere to the wall of the tube, and except for a slight sediment, the medium remains clear. Colonies on 1-day-old plates may be less than 0.5 mm in diameter but enlarge to 2 to 3 mm, sometimes with rough surfaces and pitting, after several days of incubation.

Therapy has frequently involved ampicillin or penicillin, but resistance to these agents is common. In vitro, aminoglycosides, cefazolin, cefotaxime, ceftriaxone, and chloramphenicol show good activity (27), as does the azalide azithromycin; cefixime and ciprofloxacin in combination with metronidazole may be useful for patients allergic to penicillin (36).

CAPNOCYTOPHAGA SPP.

Clinical isolates of *Capnocytophaga* spp. fall into two groups, DF-1 and DF-2 (Table 3), which are based on habitat, pathogenicity, and cultural features. Three species within the DF-1 group (*Capnocytophaga gingivalis*, *Capnocytophaga ochracea*, and *Capnocytophaga sputigena*) and two within the DF-2 group (*Capnocytophaga canimorsus* and *Capnocytophaga cynodegmi*) have been described. The five species of *Capnocytophaga* are clearly separable by DNA-DNA hybridizations (8).

These organisms ferment carbohydrates except mannitol and xylose. Positive results are best obtained with relatively large inocula in small volumes of fluid basal medium or on aerobic low-peptone basal medium, particularly with the DF-2 group. The DF-1 group is catalase, oxidase, and usually arginine dihydrolase negative; in contrast, most strains of the DF-2 group are positive in these three tests. The three species of the DF-1 group cannot always be identified by conventional tests, and the CDC tabulates these as a single group (11). However, *Capnocytophaga ochracea* strains are o-nitrophenyl-β-D-galactopyranoside (ONPG) positive, usually hydrolyze starch, and may produce acid from galactose; *Capnocytophaga sputigena* strains are ONPG positive and may hydrolyze gelatin but not starch; and *Capnocytophaga gingivalis* strains are ONPG negative and do not hydrolyze gelatin or starch (25). These three species can also be separated by their protein profiles (28). Isolates may be confused with aerotolerant strains of *Leptotrichia buccalis* unless unknown strains are also tested for metabolic end products of fermentation and cellular fatty acid composition (2).

DF-1 is associated with colonization and disease of the oral cavity, particularly periodontitis, and also with septicemia; rarely, stroke can follow endocarditis. Many strains of DF-2 have been associated with septicemia or meningitis following dog bites; complications include acute renal failure and infective endocarditis. Splenectomy and alcoholism appear to be predisposing factors. There is some evidence that the lower respiratory tract could be the primary site of infection in rare cases.

All *Capnocytophaga* strains are capnophilic, facultatively anaerobic, gram-negative slender or filamentous rods with ends usually tapered. Most display gliding motility (twitching and translocation; 53), but this may not be evident on some agar media. Gliding can be detected by inoculating cultures onto heart infusion agar containing 5% rabbit serum (8). Growth on BA is usually favored when the basal medium contains heart infusion. Colonies of DF-1 may scarcely be visible on 1-day-old BA plates but usually enlarge to several millimeters after 2 to 4 days and display flat, spreading, fingerlike edges.

Most *Capnocytophaga* strains are resistant to aminoglycosides but susceptible to expanded-spectrum cephalosporins, clindamycin, chloramphenicol, the quinolones, and penicillin; occasional strains produce β-lactamases (45).

CARDIOBACTERIUM HOMINIS

C. hominis is one of the few capnophilic, nonmotile UB that are oxidase positive but catalase negative.

All strains ferment glucose and maltose but not lactose and xylose; all are indole positive, weakly so in some strains. Other characteristics are presented in Tables 1 and 3.

The normal habitat of *C. hominis* is the upper respiratory tract. It is an etiologic agent of endocarditis (58) and has also been the cause of abdominal abscesses.

In Gram stains prepared from BA cultures, pleomorphic rod-shaped cells with bulbous ends are seen; they frequently appear in clusters resembling rosettes. However, when the growth medium is supplemented with yeast extract, gram-negative rods with little evidence of pleomorphism are seen. On 1-day-old BA plates, the colonies are minute, but at 2 days, they are 1 mm in diameter, smooth, and butyrous. The colonies of some strains may pit the agar.

For endocarditis, cefazolin therapy may be effective in patients hypersensitive to penicillin.

CHROMOBACTERIUM VIOLACEUM

Chromobacterium violaceum is a facultatively anaerobic motile rod. Strains do not decarboxylate lysine and ornithine but may be arginine dihydrolase positive. Nonpigmented strains are frequently indole positive. Oxidase-negative strains could be mistaken for enteric species such as *Xenorhabdus* spp., and oxidase-positive strains could be mistaken for *Aeromonas* or *Vibrio* species. However, distinctions can usually be made by noting the above characteristics and those presented in Tables 1 and 4.

Chromobacterium violaceum is a normal inhabitant of soil and water and is only rarely associated with human disease (20). Cutaneous lesions are followed by an overwhelming septicemia; the mortality rate is high. There is also a similarity between *Chromobacterium violaceum* infection and the septicemic form of melioidosis (54). Most cases have occurred in tropical and subtropical climates. In the United States, human infections have been reported from Florida, Louisiana, and South Carolina.

Most of the 37 strains processed by the CDC produced a violet pigment (11), violacein, that is soluble in ethanol but insoluble in chloroform and water; this pigment is probably unique to this species. Some strains exhibit both pigmented and nonpigmented colonies on BA plates. Nonpigmented strains resemble *Aeromonas* species (but are maltose and mannitol negative). Colonies on 1-day-old BA plates are 0.5 to 1.5 mm in diameter, convex, and smooth. Some strains are hemolytic.

Chromobacterium violaceum is susceptible to aminoglycosides, chloramphenicol, and tetracycline but resistant to penicillin and cephalosporins (20, 58).

DF-3

DF-3 is a small, facultatively anaerobic, nutritionally fastidious coccobacillus. Its cellular fatty acids present a unique profile that is similar to those of *Capnocytophaga* species (57) and serves to distinguish it from aerotolerant

TABLE 4 *Chromobacterium* sp., *Kingella* spp., *Suttonella* sp., and group EF-4[a]

Feature	% of strains positive (no. of strains examined)						
	Chromobacterium violaceum (37)	EF-4a (97)	EF-4b (34)	*Kingella denitrificans* (60)	*Kingella kingae* (33)	*Kingella oralis* (11)	*Suttonella indologenes* (1)
Acidified							
D-Glucose	100	100	96	92	90	100	100
Lactose	0	0	0	0	0	0	0
Maltose	3	0	0	0	100	0	100
D-Mannitol	0	0	0	0	0	0	0
Sucrose	26	0	0	0	0	0	100
D-Xylose	0	0	0	0	0	0	0
Arginine dihydrolase	100	79	0	ND	ND	ND	0
Citrate (Simmons)	77	4	20	0	0	ND	0
Esculin, hydrolyzed	5	0	0	0	0	0	0
Gelatin, hydrolyzed	86	60	9	0	0	ND	0
Hemolysis	48	0	0	0	84	0	0
Indole	21	0	0	0	0	0	100
MacConkey agar, growth	100	50	71	0	36	ND	0
Motility	100	0	0	0	0	ND	0
Nitrate reduced	97	97	97	93	6	0	0
Gas from nitrate	ND	62	0	88	0	ND	ND
Urea (Christensen)	19	0	0	0	0	ND	0
42°C, growth	85	70	69	48	4	ND	0

[a]Adapted from Clark et al. (11); the data on *K. oralis* are from Dewhirst et al. (14). Not all strains were tested for some features. ND, no data.

strains of *L. buccalis*, with which it might otherwise be confused (2).

DF-3 is capnophilic and both catalase and oxidase negative (Table 1). It can be distinguished from other UB that share these three features by tests for acidification of lactose and xylose and reduction of nitrate (Table 3).

It has been recovered from an abscess, blood, feces, urine, wounds, and other body sites (11, 56). Examination of feces of patients who are immunocompromised or who have severe underlying disease reveals a spectrum of disease involving DF-3, ranging from asymptomatic carriage to chronic diarrhea (6).

The DF-3-like group, a group of six clinical isolates from blood and wounds, is most similar to group DF-3 but can be distinguished by the presence of beta-like-hemolysis and usually proteolytic activity, by a distinct cellular fatty acid profile, and by the absence of acid production from sucrose and xylose (12).

EF-4

EF-4, a nonmotile coccobacillus, resembles pasteurellae in several features, including being the cause of infections in humans following animal contact, particularly animal bites. EF-4b strains do not acidify the butts but grow well on the slants of Kligler iron agar and TSIA; consequently, they may be treated as NFB. However, comparisons of their whole-cell protein patterns by sodium dodecyl sulfate-polyacrylamide gel electrophoresis and genomic data indicate that EF-4a and EF-4b are separate species within the same genus (23). Studies of rRNA-DNA relatedness (46) showed that these two organisms, together with the M-5 group (recently proposed as *Neisseria weaveri*; see chapter 26 of this Manual), are members of an emended family, *Neisseriaceae*.

EF-4a differs from pasteurellae in fermenting only glu-

cose, in being arginine dihydrolase positive, and in having some strains that are weakly pigmented (yellow to tan). Nitrate is usually reduced to gas, and a few strains reduce both nitrate and nitrite without formation of gas. EF-4b also differs from pasteurellae in acidifying only glucose; gas is not produced from nitrate. Most strains of EF-4a produce arginine dihydrolase, whereas strains of EF-4b never do so; also, a greater proportion of strains of the former produce gelatinase and gas from nitrate. Other features are shown in Tables 1 and 4.

EIKENELLA SP.

The genus *Eikenella* contains only one species, *Eikenella corrodens*.

Cells are straight, unbranched slender rods 1.5 to 4 μm in length. *Eikenella* strains are nonmotile, but twitching may occur on some media. They are oxidase positive, usually catalase negative, capnophilic, facultatively anaerobic, and nonsaccharolytic. Hemin is usually required for aerobic growth. Ornithine and usually lysine are decarboxylated, but the arginine dihydrolase test is negative. Other features are shown in Tables 1 and 3. *E. corrodens* can be confused with another pitting or corroding bacterium, the obligate anaerobe *Bacteroides ureolyticus* (formerly *Bacteroides corrodens*). However, they differ in that *E. corrodens* is facultatively aerobic, lysine decarboxylase positive, and urease and gelatin negative.

E. corrodens is part of the normal flora of mucous membranes. It can also be etiologically significant, singly or in mixed infections, in human bite wounds, appendicitis, arthritis, empyema, endocarditis, meningitis, pleuropulmonary infections, pneumonia, postsurgical infection, and abscesses in soft tissue, including the brain (18). It may rarely cause chorioamnionitis.

On 1-day-old BA plates, colonies may be minute or

inapparent, and several days of incubation may be required before colonies of usable size are present. Such colonies have moist, clear centers surrounded by flat, spreading growth. Pitting of the medium may occur. A slight yellow pigmentation may be present in older cultures.

Strains of *Eikenella* are susceptible to several antimicrobial agents, including amoxicillin-clavulanic acid, the penicillins, quinolones, and tetracycline, but are resistant to clindamycin (10).

KINGELLA SPP.

Two organisms, one designated TM-1 and one designated "*Moraxella* new species 1," are similar to each other. Along with another new species, *Kingella indologenes* (now *Suttonella indologenes* [15], described separately below), they were eventually placed in the new genus *Kingella* with the names *Kingella denitrificans* and *Kingella kingae*, respectively. *K. kingae* has also been referred to as *Moraxella kingii*.

Kingella species are coccobacillary or straight rods 2 to 3 μm in length that commonly appear in pairs or short chains. They are nonmotile but may be piliated and show twitching motility. They are facultatively anaerobic, fermentative, oxidase positive, and usually catalase negative. They are nutritionally fastidious and hence, as noted above for other fastidious bacteria, may require special procedures for fermentation tests. *Kingella* strains usually mimic gonococci in microscopic morphology, failure to acidify any sugar other than glucose, and occasional failure to grow on TSIA. The organism is, however, a rod and also differs from gonococci in reducing nitrate and in being usually catalase negative (Tables 1 and 4). Salient features for the identification of *K. kingae* are hemolysis (usually weak), failure to produce catalase, and acidification of both glucose and maltose. Other features are presented in Tables 1 and 4. The more recently described *Kingella oralis* (14), from the human oral cavity, pits agar and is phenotypically similar to *K. kingae* except in failing to produce acid from maltose and not showing hemolysis (see also Tables 1 and 4).

The normal habitat of these organisms appears to be the upper respiratory tracts of humans. *K. kingae* is an opportunistic pathogen, particularly in young children, and has been recovered from blood, bone, corneal ulcer, intervertebral disks, joint fluid, and the nasopharynx. The BACTEC blood culture system enhances recovery of the organism from joint fluid and improves the bacteriologic diagnosis of pediatric septic arthritis (61). *K. denitrificans* is rarely if ever pathogenic, but it has been described as a cause of granulomatous disease in a patient with AIDS.

On 1-day-old BA plates, colonies are 0.5 mm or less in diameter and convex. The colonies of some strains have flat peripheries and pit the agar.

Most strains are susceptible to many antimicrobial agents, including penicillin (32, 58).

PASTEURELLA SPP.

As already noted, phenotypic features of *Pasteurella* species are similar to those of *Actinobacillus* species. Indeed, DNA-DNA hybridization studies (33) have shown that several taxa previously assigned to the genus *Pasteurella* (*Pasteurella aerogenes*, *Pasteurella haemolytica*, and *Pasteurella pneumotropica*) are more closely related to members of the genus *Actinobacillus*. Comparisons of 16S rRNA sequences (16) confirm that these three species are not closely related to each other or to *Pasteurella* spp. Hybridization (33) and

subsequent (52) studies detected more than 13 taxa in the genus *Pasteurella*; the differential phenotypic features of those taxa that occur in clinical specimens are shown in Table 2, while the characteristics of species occurring only in veterinary material are given in the fifth edition of this Manual (40).

Pasteurella species are coccobacilli or rods 1 to 2 μm in length. They are nonmotile, facultatively anaerobic, and saccharolytic. Most strains recovered from clinical specimens are catalase, oxidase, indole, and sucrose positive; most decarboxylate ornithine; some are encapsulated. Oxidase tests should be performed from cultures grown on BA or chocolate agar medium; negative results may be obtained with other growth media (21). The principal species are shown in Table 1.

Pasteurellae have been recovered from lesions in many parts of the human body, especially animal bite wounds. In animals, they cause fowl cholera, hemorrhagic septicemia, mastitis, septic pleuropneumonia, snuffles, and other focal infections. Both healthy and diseased wild and domestic animals are the reservoirs for most human infections.

Most strains are susceptible to chloramphenicol, penicillin, and tetracycline (58).

The biochemical profile of [*P.*] *aerogenes* is distinct from that of other pasteurellae (Table 2). This species has been recovered from aborted fetuses of swine and from animal bites, urine, and peritoneal fluid in humans. Colonies on 1-day-old BA plates are 0.5 to 1 mm in diameter, convex, smooth, translucent, and nonhemolytic.

The capnophilic group HB-5 has more recently been described as a species of *Pasteurella* and was given the name *Pasteurella bettii* (52), which was subsequently changed to *Pasteurella bettyae* (50). Although HB-5 and *P. bettyae* are the same organism, different results are given by different authors (11, 52) for their catalase, oxidase, maltose fermentation, and especially indole reactions (see Tables 1, 2, and 3). HB-5 exhibits coccobacilli and rods upon Gram stain. The organism is facultatively anaerobic, and all strains are weakly aerogenic. Of the usual battery of six sugars used by the CDC, only glucose is acidified. HB-5 has been recovered from amniotic fluid, blood (usually from newborn babies), finger lesions, leg abscesses, placenta, rectal sites, surgical incisions, and urogenital specimens (particularly those of females), including exudates of genital ulcers, especially Bartholin gland abscesses (1, 58). On 1-day-old BA plates, colonies are 0.5 to 1.0 mm in diameter, convex, and smooth. HB-5 is susceptible to many antimicrobial agents, including penicillin, but several strains produce β-lactamases (7).

Another more recently described species, *Pasteurella caballi* (48), differs from most other pasteurellae in being catalase negative. All of the 29 strains recovered from horses were aerogenic, failed to grow on MacConkey agar, acidified neither L-arabinose nor trehalose, and were both urease and indole negative. Although not included in Table 2, the species is mentioned here, as a strain has been isolated from a veterinary surgeon's infected wound (4).

Some *Pasteurella* strains assigned to *Pasteurella multocida* in the CDC charts (11) have been reassigned to *Pasteurella canis* on the basis of results of hybridization studies (33). Biotype 1 strains of *P. canis*, recovered from dogs (mouth) and dog bites in humans, are indole positive; biotype 2, from cattle, is indole negative. The principal test for differentiating *P. canis* and *P. multocida* is acidification of mannitol (Table 2).

The name *Pasteurella dagmatis* was proposed (33) for a

group of pasteurellae referred to by the CDC as *Pasteurella* sp. "n. sp. 1" (*Pasteurella* "gas") (11). Differential tests include acidification of maltose and xylose, production of indole, decarboxylation of ornithine, and hydrolysis of urea (Table 2). Similar strains that are urea negative are probably *Pasteurella* sp. taxon 16 (5). Many of the 91 strains processed at the CDC were recovered from human focal wounds after animal contact, particularly cat and dog bites. On 1-day-old BA plates, colonies are 1 to 2 mm in diameter and nonhemolytic.

Two biotypes were long recognized within [*P.*] haemolytica: biotype A was xylose positive, esculin negative, and usually mannose negative, while biotype T was mannose positive, xylose negative, and usually esculin positive. However, these biotypes exhibit only low levels of DNA-DNA hybridization to each other (3), and each has now been accorded species status: biotype A is *P. haemolytica*, and biotype T is *Pasteurella trehalosi* (52). Both organisms are indole and urea negative (Table 2). Freshly isolated strains are hemolytic but may lose this feature upon subculture. *P. haemolytica* is relatively common as an agent of septicemia and mastitis in domestic animals but rarely causes disease in humans (58, 62). *P. trehalosi* causes septicemia in older lambs and is retained in Table 2 only because of its historical association with *P. haemolytica* sensu stricto.

Tests for acidification of dulcitol and sorbitol delineate three subspecies within *P. multocida*: *P. multocida* subsp. *gallicida*, *P. multocida* subsp. *multocida*, and *P. multocida* subsp. *septica* (33). Like other pasteurellae and actinobacilli, *P. multocida* acidifies both the slant and the butt of TSIA. An occasional strain may give a weak or even negative result in the oxidase test, but tests with the tetramethyl reagent only rarely give negative results. Results from tests for indole, ornithine decarboxylase, and acidification of maltose and sucrose indicate the salient features of this species. *P. multocida* is apparently a commensal in the upper respiratory tracts of fowl, mammals, and possibly humans (58). In humans, it is associated with focal infections following animal bites, chronic pulmonary disease, and systemic disease, including meningitis. Colonies on 1-day-old BA plates are 1 to 2 mm in diameter and nonhemolytic. Typing methods for epidemiologic purposes include use of the API ZYM system (22), bacteriophages (35), and DNA fingerprinting (59) as well as serology.

Salient features for identification of [*Pasteurella*] pneumotropica come from tests for ornithine decarboxylase, hydrolysis of urea, and production of gas from glucose (Table 2). The organism was initially associated with pneumonic lesions in laboratory mice (11, 58). It appears to be less common than other species of pasteurellae as an agent of either animal or human disease. Strains isolated from humans and reported to belong to this species are more likely to have been strains of *P. dagmatis*. On 1-day-old BA plates, colonies are 0.5 to 1.5 mm in diameter, low convex, and nonhemolytic.

Important differential tests of *Pasteurella stomatis* are for acidification of maltose and mannitol, production of indole, and decarboxylation of ornithine (Table 2). Strains have been recovered from the respiratory tracts of cats and dogs (33); the organism has been reported to cause human infection following cat or dog bites (24, 42).

Most strains of *Pasteurella volantium* were recovered from fowl, but one strain was from a human tongue (33). All 10 strains required V factor (NAD) for growth.

STREPTOBACILLUS SP.

The genus *Streptobacillus* contains only one species, *Streptobacillus moniliformis*. It is a facultatively anaerobic, highly pleomorphic, asporogenous gram-negative rod. The cells are less than 1 μm wide by 1 to 5 μm long, but curved and looping unbranched filaments as long as 100 to 150 μm may be formed. In young cultures or in smears from clinical material, cells usually appear to be more uniform, with occasional homogeneous filaments. Irregular coccobacilli are found in older cultures, with granules and bands appearing in the filaments. L forms occur easily during cultivation. *S. moniliformis* is largely an inactive species with neither catalase nor oxidase produced. Acid without gas is produced from some sugars (26, 44; Table 3). Fermentation reactions may be determined in cysteine tryptic agar base (30). Arginine dihydrolase and H_2S may be produced (44).

The natural habitat of *S. moniliformis* is the nasopharynges of wild and laboratory rats. The organism is associated with two clinical conditions: rat bite fever and Haverhill fever. Most human cases of rat bite fever are acquired through rat bites, but one patient developed an acute arthropathy without a previous history of rat bite (19). Haverhill fever, on the other hand, is acquired by ingesting milk, or rarely other food or water, contaminated with *S. moniliformis* (31).

Rat bite fever is characterized mainly by a relapsing fever with sudden onset, chills, headache, vomiting, and skin rash. Diarrhea and weight loss are often found in infants and young children (43). The incubation period is usually less than 10 days. The bite wound usually heals rapidly without severe inflammation. Complications include hepatic and splenic congestion, pneumonia, endocarditis (47), myocarditis, various abscesses, chorioamnionitis, hepatitis, meningitis, and nephritis. Rarely, infection can prove fatal (49).

Blood and joint fluid (preferably in substantial quantities) are the major sources of *S. moniliformis*. These fluids should be mixed with equal volumes of 2.5% sodium citrate (44) to prevent clotting before microscopic examination and culture.

Media enriched with whole blood, serum, or ascitic fluid are required for the isolation of *S. moniliformis*. Only blood culture media without sodium polyanetholesulfonate should be used, as this compound may be inhibitory (30). Use of sedimented blood cells to inoculate isolation media has been advocated (44). Some microbiologists recommend the use of a medium that facilitates the isolation of the more fastidious L-phase variants, which can be recovered on heart infusion enriched with horse serum and yeast extract (44).

Specimens should be inoculated onto both solid and liquid media and then incubated at 35°C in a humidified 7 to 8% CO_2 atmosphere. In broth media, characteristic "puff balls" appear after 2 to 6 days. Colonies on agar are round, grayish, smooth, and glistening, with a diameter of 1 to 2 mm, but L-phase colonies with the typical fried-egg appearance can also appear. The organisms should be subcultured as soon as they are detected, since the marked decrease of pH in broth can be lethal to them.

Penicillin is the drug most active against this organism in vitro and is the drug of choice for treating the infection (17); treatment with penicillin plus chloramphenicol or with erythromycin has also been successful.

SUTTONELLA SP.

Suttonella indologenes (15) differs from the *Kingella* species in being both indole and sucrose positive (Table 4). The organism is only rarely encountered in clinical specimens. The few strains described were all associated with eye infections. It is susceptible to penicillin.

REFERENCES

1. **Baddour, L. M., M. S. Gelfand, R. E. Weaver, T. C. Woods, M. Altwegg, L. W. Mayer, R. A. Kelley, and D. J. Brenner.** 1989. CDC group HB-5 as a cause of genitourinary infections in adults. *J. Clin. Microbiol.* **27:**801–805.
2. **Bernard, K., C. Cooper, S. Tessier, and E. P. Ewan.** 1991. Use of chemotaxonomy as an aid to differentiate among *Capnocytophaga* species, CDC group DF-3, and aerotolerant strains of *Leptotrichia buccalis. J. Clin. Microbiol.* **29:**2263–2265.
3. **Bingham, D. P., R. Moore, and A. B. Richards.** 1990. Comparison of DNA:DNA homology and enzymatic activity between *Pasteurella haemolytica* and related species. *Am. J. Vet. Res.* **51:**1161–1166.
4. **Bisgaard, M., O. Heltberg, and W. Frederiksen.** 1991. Isolation of *Pasteurella caballi* from an infected wound on a veterinary surgeon. *APMIS* **99:**291–294.
5. **Bisgaard, M., and R. Mutters.** 1986. Characterization of some previously unclassified "*Pasteurella*" spp. obtained from the oral cavity of dogs and cats and description of a new species tentatively classified with the family *Pasteurellaceae* Pohl 1981 and provisionally called taxon 16. *Acta Pathol. Microbiol. Immunol. Scand. Sect. B* **94:**177–184.
6. **Blum, R. N., C. D. Berry, M. G. Phillips, D. L. Hamilos, and E. W. Koneman.** 1992. Clinical illnesses associated with isolation of dysgonic fermenter 3 from stool samples. *J. Clin. Microbiol.* **30:**396–400.
7. **Bogaerts, J., J. Verhaegen, W. M. Tello, S. Allen, L. Verbist, E. Van Dyck, and P. Piot.** 1990. Characterization, in vitro susceptibility, and clinical significance of CDC group HB-5 from Rwanda. *J. Clin. Microbiol.* **28:**2196–2199.
8. **Brenner, D. J., D. G. Hollis, G. R. Fanning, and R. E. Weaver.** 1989. *Capnocytophaga canimorsus* sp. nov. (formerly CDC group DF-2), a cause of septicemia following dog bite, and *C. cynodegmi* sp. nov., a cause of localized wound infection following dog bite. *J. Clin. Microbiol.* **27:**231–235.
9. **Carter, G. R.** 1984. Genus I. *Pasteurella* Trevisan 1887, 94[AL], nom. cons. Opin. 13, Jud. Comm. 1954, 153, p. 552–557. *In* N. R. Krieg and J. G. Holt (ed.), *Bergey's Manual of Systematic Bacteriology,* vol. 1. The Williams & Wilkins Co., Baltimore.
10. **Chen, C. K. C., and M. E. Wilson.** 1992. *Eikenella corrodens* in human oral and non-oral infections: a review. *J. Periodontol.* **63:**941–953.
11. **Clark, W. A., D. G. Hollis, R. E. Weaver, and P. Riley.** 1984. *Identification of Unusual Pathogenic Gram-Negative Aerobic and Facultatively Anaerobic Bacteria.* Centers for Disease Control, Atlanta.
12. **Daneshvar, M. I., D. G. Hollis, and C. W. Moss.** 1991. Chemical characterization of clinical isolates which are similar to CDC group DF-3 bacteria. *J. Clin. Microbiol.* **29:**2351–2353.
13. **De Ley, J., W. Mannheim, R. Mutters, K. Piechulla, R. Tytgat, P. Segers, M. Bisgaard, W. Frederiksen, K.-H. Hinz, and M. Vanhoucke.** 1990. Inter- and intrafamilial similarities of rRNA cistrons of the *Pasteurellaceae. Int. J. Syst. Bacteriol.* **40:**126–137.
14. **Dewhirst, F. E., C.-K. C. Chen, B. J. Paster, and J. J. Zambon.** 1993. Phylogeny of species in the family *Neisseriaceae* isolated from human dental plaque and description of *Kingella orale* sp. nov. *Int. J. Syst. Bacteriol.* **43:**490–499.
15. **Dewhirst, F. E., B. J. Paster, S. La Fontaine, and J. I. Rood.** 1990. Transfer of *Kingella indologenes* (Snell and Lapage 1976) to the genus *Suttonella* gen. nov. as *Suttonella indologenes* comb. nov.; transfer of *Bacteroides nodosus* (Beveridge 1941) to the genus *Dichelobacter* gen. nov. as *Dichelobacter nodosus* comb. nov.; and assignment of the genera *Cardiobacterium, Dichelobacter,* and *Suttonella* to *Cardiobacteriaceae* fam. nov. in the gamma division of *Proteobacteria* on the basis of 16S rRNA sequence comparisons. *Int. J. Syst. Bacteriol.* **40:**426–433.
16. **Dewhirst, F. E., B. J. Paster, I. Olsen, and G. J. Fraser.** 1992. Phylogeny of 54 representative strains of species in the family *Pasteurellaceae* as determined by comparison of 16S rRNA sequences. *J. Bacteriol.* **174:**2002–2013.
17. **Edwards, R., and R. G. Finch.** 1986. Characterisation and antibiotic susceptibilities of *Streptobacillus moniliformis. J. Med. Microbiol.* **21:**39–42.
18. **Flesher, S. A., and E. J. Bottone.** 1989. *Eikenella corrodens* cellulitis and arthritis of the knee. *J. Clin. Microbiol.* **27:**2606–2608.
19. **Fordham, J. N., E. McKay-Ferguson, A. Davies, and T. Blyth.** 1992. Rat bite fever without the bite. *Ann. Rheum. Dis.* **51:**411–412.
20. **Georghiou, P. R., G. M. O'Kane, S. Siu, and R. J. Kemp.** 1989. Near-fatal septicaemia with *Chromobacterium violaceum. Med. J. Aust.* **150:**720–721.
21. **Grehn, M., and F. Müller.** 1989. The oxidase reaction of *Pasteurella multocida* strains cultured on Mueller-Hinton medium. *J. Microbiol. Methods* **9:**333–336.
22. **Grehn, M., F. Müller, and R. Hugelshofer.** 1991. The API ZYM system as a tool for typing of *Pasteurella multocida* strains from humans. *J. Microbiol. Methods* **13:**201–206.
23. **Holmes, B., M. Costas, and A. C. Wood.** 1990. Numerical analysis of electrophoretic protein patterns of group EF-4 bacteria, predominantly from dog-bite wounds of humans. *J. Appl. Bacteriol.* **68:**81–91.
24. **Holst, E., J. Rollof, L. Larsson, and J. P. Nielsen.** 1992. Characterization and distribution of *Pasteurella* species recovered from infected humans. *J. Clin. Microbiol.* **30:**2984–2987.
25. **Holt, S. C., and S. A. Kinder.** 1989. Genus II. *Capnocytophaga* Leadbetter, Holt and Socransky 1982, 266[VP] (effective publication: Leadbetter, Holt and Socransky 1979, 13), p. 2050–2058. *In* J. T. Staley, M. P. Bryant, N. Pfennig, and J. G. Holt (ed.), *Bergey's Manual of Systematic Bacteriology,* vol. 3. The Williams & Wilkins Co., Baltimore.
25a. **International Committee on Systematic Bacteriology Subcommittee on *Pasteurellaceae* and Related Organisms.** 1987. Minutes of the meetings, 6 and 10 September 1986, Manchester, England. *Int. J. Syst. Bacteriol.* **37:**474.
26. **Josephson, S. L.** 1988. Rat-bite fever, p. 443–447. *In* A. Balows, W. J. Hausler, Jr., M. Ohashi, and A. Turano (ed.), *Laboratory Diagnosis of Infectious Disease,* vol. 1. Springer-Verlag, New York.
27. **Kaplan, A. H., D. J. Weber, E. Z. Oddone, and J. R. Perfect.** 1989. Infection due to *Actinobacillus actinomycetemcomitans:* 15 cases and review. *Rev. Infect. Dis.* **11:**46–63.
28. **Khwaja, K. J., P. Parish, M. J. Aldred, and W. G. Wade.** 1990. Protein profiles of *Capnocytophaga* species. *J. Appl. Bacteriol.* **68:**385–390.
29. **Kilian, M., and W. Frederiksen.** 1981. Identification tables for the *Haemophilus-Pasteurella-Actinobacillus* group, p. 281–290. *In* M. Kilian, W. Frederiksen, and E. L. Biberstein (ed.), *Haemophilus, Pasteurella and Actinobacillus.* Academic Press, Inc. (London), Ltd., London.
30. **Lambe, D. W., Jr., A. W. McPhedran, J. A. Mertz, and P. Stewart.** 1973. *Streptobacillus moniliformis* isolated from a case of Haverhill fever: biochemical characterization and inhibitory effect of sodium polyanethol sulfonate. *Am. J. Clin. Pathol.* **60:**854–860.
31. **McEvoy, M. B., N. D. Noah, and R. Pilsworth.** 1987. Outbreak of fever caused by *Streptobacillus moniliformis. Lancet* **ii:**1361–1363.
32. **Morrison, V. A., and K. F. Wagner.** 1989. Clinical manifestations of *Kingella kingae* infections: case report and review. *Rev. Infect. Dis.* **11:**776–782.
33. **Mutters, R., P. Ihm, S. Pohl, W. Frederiksen, and W. Mannheim.** 1985. Reclassification of the genus *Pasteurella* Trevisan 1887 on the basis of deoxyribonucleic acid homology, with proposals for the new species *Pasteurella dagmatis,*

Pasteurella canis, Pasteurella stomatis, Pasteurella anatis, and *Pasteurella langaa. Int. J. Syst. Bacteriol.* **35**:309–322.

34. **Mutters, R., S. Pohl, and W. Mannheim.** 1986. Transfer of *Pasteurella ureae* Jones 1962 to the genus *Actinobacillus* Brumpt 1910: *Actinobacillus ureae* comb. nov. *Int. J. Syst. Bacteriol.* **36**:343–344.

35. **Nielsen, J. P., and V. T. Rosdahl.** 1990. Development and epidemiological applications of a bacteriophage typing system for typing *Pasteurella multocida. J. Clin. Microbiol.* **28**:103–107.

36. **Pavicic, M. J. A. M. P., A. J. van Winkelhoff, and J. de Graaff.** 1992. In vitro susceptibilities of *Actinobacillus actinomycetemcomitans* to a number of antimicrobial combinations. *Antimicrob. Agents Chemother.* **36**:2634–2638.

37. **Peel, M. M., K. A. Hornidge, M. Luppino, A. M. Stacpoole, and R. E. Weaver.** 1991. *Actinobacillus* spp. and related bacteria in infected wounds of humans bitten by horses and sheep. *J. Clin. Microbiol.* **29**:2535–2538.

38. **Phillips, J. E.** 1984. Genus III. *Actinobacillus* Brumpt 1910, 849[AL], p. 570–575. *In* N. R. Krieg and J. G. Holt (ed.), *Bergey's Manual of Systematic Bacteriology*, vol. 1. The Williams & Wilkins Co., Baltimore.

39. **Pickett, M. J.** 1989. Methods for identification of flavobacteria. *J. Clin. Microbiol.* **27**:2309–2315.

40. **Pickett, M. J., D. G. Hollis, and E. J. Bottone.** 1989. Miscellaneous gram-negative bacteria, p. 410–428. *In* A. Balows, W. J. Hausler, Jr., K. L. Herrmann, H. D. Isenberg, and H. J. Shadomy (ed.), *Manual of Clinical Microbiology*, 5th ed. American Society for Microbiology, Washington, D.C.

41. **Potts, T. V., J. J. Zambon, and R. J. Genco.** 1985. Reassignment of *Actinobacillus actinomycetemcomitans* to the genus *Haemophilus* as *Haemophilus actinomycetemcomitans* comb. nov. *Int. J. Syst. Bacteriol.* **35**:337–341.

42. **Pouëdras, P., P. Y. Donnio, Y. Le Tulzo, and J. L. Avril.** 1993. *Pasteurella stomatis* infection following a dog bite. *Eur. J. Clin. Microbiol. Infect. Dis.* **12**:65.

43. **Raffin, B. J., and M. Freemark.** 1979. Streptobacillary ratbite fever: a pediatric problem. *Pediatrics* **64**:214–217.

44. **Rogosa, M.** 1985. *Streptobacillus moniliformis* and *Spirillum minus*, p. 400–406. *In* E. H. Lennette, A. Balows, W. J. Hausler, Jr., and H. J. Shadomy (ed.), *Manual of Clinical Microbiology*, 4th ed. American Society for Microbiology, Washington, D.C.

45. **Roscoe, D. L., S. J. V. Zemcov, D. Thornber, R. Wise, and A. M. Clarke.** 1992. Antimicrobial susceptibilities and β-lactamase characterization of *Capnocytophaga* species. *Antimicrob. Agents Chemother.* **36**:2197–2200.

46. **Rossau, R., G. Vandenbussche, S. Thielemans, P. Segers, H. Grosch, E. Göthe, W. Mannheim, and J. De Ley.** 1989. Ribosomal ribonucleic acid cistron similarities and deoxyribonucleic acid homologies of *Neisseria, Kingella, Eikenella, Simonsiella, Alysiella,* and Centers for Disease Control groups EF-4 and M-5 in the emended family *Neisseriaceae. Int. J. Syst. Bacteriol.* **39**:185–198.

47. **Rupp, M. E.** 1992. *Streptobacillus moniliformis* endocarditis: case report and review. *Clin. Infect. Dis.* **14**:769–772.

48. **Schlater, L. K., D. J. Brenner, A. G. Steigerwalt, C. W. Moss, M. A. Lambert, and R. A. Packer.** 1989. *Pasteurella caballi,* a new species from equine clinical specimens. *J. Clin. Microbiol.* **27**:2169–2174.

49. **Sens, M. A., E. W. Brown, L. R. Wilson, and T. P. Crocker.** 1989. Fatal *Streptobacillus moniliformis* infection in a two-month-old infant. *Am. J. Clin. Pathol.* **91**:612–616.

50. **Sneath, P. H. A.** 1992. Correction of orthography of epithets in *Pasteurella* and some problems with recommendations on latinization. *Int. J. Syst. Bacteriol.* **42**:658–659.

51. **Sneath, P. H. A., and M. Stevens.** 1985. A numerical taxonomic study of *Actinobacillus, Pasteurella* and *Yersinia. J. Gen. Microbiol.* **131**:2711–2738.

52. **Sneath, P. H. A., and M. Stevens.** 1990. *Actinobacillus rossii* sp. nov., *Actinobacillus seminis* sp. nov., nom. rev., *Pasteurella bettii* sp. nov., *Pasteurella lymphangitidis* sp. nov., *Pasteurella mairi* sp. nov., and *Pasteurella trehalosi* sp. nov. *Int. J. Syst. Bacteriol.* **40**:148–153.

53. **Socransky, S. S., S. C. Holt, E. R. Leadbetter, A. C. R. Tanner, E. Savitt, and B. F. Hammond.** 1979. *Capnocytophaga*: new genus of gram-negative gliding bacteria. III. Physiological characterization. *Arch. Microbiol.* **122**:29–33.

54. **Ti, T.-Y., W. C. Tan, A. P. Y. Chong, and E. H. Lee.** 1993. Nonfatal and fatal infections caused by *Chromobacterium violaceum. Clin. Infect. Dis.* **17**:505–507.

55. **Verhaegen, J., H. Verbraeken, A. Cabuy, J. Vandeven, and J. Vandepitte.** 1988. *Actinobacillus* (formerly *Pasteurella*) *ureae* meningitis and bacteraemia: report of a case and review of the literature. *J. Infect.* **17**:249–253.

56. **Wagner, D. K., J. J. Wright, A. F. Ansher, and V. J. Gill.** 1988. Dysgonic fermenter 3-associated gastrointestinal disease in a patient with common variable hypogammaglobulinemia. *Am. J. Med.* **84**:315–318.

57. **Wallace, P. L., D. G. Hollis, R. E. Weaver, and C. W. Moss.** 1989. Characterization of CDC group DF-3 by cellular fatty acid analysis. *J. Clin. Microbiol.* **27**:735–737.

58. **Weaver, R. E., D. G. Hollis, and E. J. Bottone.** 1985. Gram-negative fermentative bacteria and *Francisella tularensis,* p. 309–329. *In* E. H. Lennette, A. Balows, W. J. Hausler, Jr., and H. J. Shadomy (ed.), *Manual of Clinical Microbiology*, 4th ed. American Society for Microbiology, Washington, D.C.

59. **Wilson, M. A., M. J. Morgan, and G. E. Barger.** 1993. Comparison of DNA fingerprinting and serotyping for identification of avian *Pasteurella multocida* isolates. *J. Clin. Microbiol.* **31**:255–259.

60. **Wüst, J., J. Gubler, W. Mannheim, and A. von Graevenitz.** 1991. *Actinobacillus hominis* as a causative agent of septicemia in hepatic failure. *Eur. J. Clin. Microbiol. Infect. Dis.* **10**:693–694.

61. **Yagupsky, P., R. Dagan, C. W. Howard, M. Einhorn, I. Kassis, and A. Simu.** 1992. High prevalence of *Kingella kingae* in joint fluid from children with septic arthritis revealed by the BACTEC blood culture system. *J. Clin. Microbiol.* **30**:1278–1281.

62. **Yaneza, A. L., H. Jivan, P. Kumari, and M. S. Togoo.** 1991. *Pasteurella haemolytica* endocarditis. *J. Infect.* **23**:65–67.

Pseudomonas and *Burkholderia*

PETER H. GILLIGAN

40

TAXONOMY

In the fifth edition of the *Manual of Clinical Microbiology* (21), the genus *Pseudomonas* was divided into five groups based on rRNA homology (52). Subsequent study has led to either the reclassification or the general acceptance of new genus designations for three of these five groups. It has been proposed that organisms in homology group II are sufficiently different from organisms in homology group I (*Pseudomonas aeruginosa* and related organisms) to belong to the genus *Burkholderia*. Seven species (*Burkholderia cepacia, B. mallei, B. pseudomallei, B. pickettii, B. gladioli, B. caryophylli,* and *B. solanacearum*) formerly considered *Pseudomonas* spp. have been assigned to this recently described genus (83). The first five of these seven have been recovered from humans. Organisms in group III are now classified in a new family, the *Comamonadaceae*, which includes the genera *Comamonas* and *Acidovorax* (79). *Xanthomonas*, the genus designation for the group V organisms, is now widely accepted. *Xanthomonas maltophila*, however, has been assigned to the new genus *Stenotrophomonas* (see chapter 41 of this Manual). The organisms in group IV, *Pseudomonas diminuta* and *Pseudomonas vesicularis*, have significant genetic divergence from the group I organisms but remain in the genus *Pseudomonas*. Other organisms previously classified as *Pseudomonas* are now reclassified as *Shewanella* (*Pseudomonas*) *putrefaciens* and *Sphingomonas* (*Pseudomonas*) *paucimobilis* (39, 84). Although these name changes have been validly published, some have not been recognized by the International Committee on Nomenclature (Table 1). Use of these new genus names will depend on their acceptance by the clinical microbiology community. The genera *Pseudomonas* and *Burkholderia* are discussed in this chapter. The other organisms listed in Table 1 along with other non-glucose fermenters, including *Pseudomonas*-like group 2, are covered in chapter 41 of this Manual.

GENERAL DESCRIPTION

Pseudomonas and *Burkholderia* spp. are aerobic, non-spore-forming, gram-negative rods that are straight or slightly curved. They are 1 to 5 μm long and 0.5 to 1.0 μm wide. Some isolates grow under anaerobic conditions by using nitrate as a terminal electron acceptor. With the exception of *B. mallei*, these organisms are motile owing to the presence of one or more polar flagella (51). These bacteria are catalase positive, and most are oxidase positive. With the exception of certain strains of *P. vesicularis*, these organisms grow on MacConkey agar and appear as nonfermenters. Most strains degrade glucose oxidatively, and most species degrade nitrate to either nitrite or nitrogen gas. Certain species have distinctive colony morphologies or pigmentation. They are nutritionally quite versatile, with different species being able to utilize a variety of simple and complex carbohydrates, alcohols, and amino acids as carbon sources. Certain species can multiply at 4°C, but most are mesophilic, with optimal growth temperatures between 30 and 37°C (51).

NATURAL HABITATS

Pseudomonas spp. and *Burkholderia* spp. are environmental organisms found in water and soil and on plants, including fruits and vegetables. They are distributed worldwide. Both genera are well recognized as phytopathogens, and many species were first described in that context. Because of their abilities to survive in aqueous environments, these organisms have become particularly problematic in the hospital environment.

P. aeruginosa is the best studied of these organisms because of its importance as a nosocomial pathogen. *P. aeruginosa* has been found in a variety of aqueous solutions, including disinfectants, ointments, soaps, irrigation fluids, eyedrops, and dialysis fluids, and in dialysis equipment (26, 46). *P. aeruginosa* is frequently found in the aerators and traps of sinks; baby, whirlpool, and hydrotherapy baths; respiratory therapy equipment; and showerheads. *P. aeruginosa* is found on the surfaces of many types of raw fruits and vegetables; therefore, profoundly immunosuppressed individuals are not allowed these foods, because ingestion of the pseudomonads with subsequent colonization and translocation may lead to bacteremia. In addition to its nosocomial sources, *P. aeruginosa* may be found in swimming pools, hot tubs, contact lens solutions, cosmetics, illicit injectable drugs, and the innersoles of sneakers. All have been sources of infection.

P. aeruginosa is infrequently found as part of the human microflora of healthy individuals. The bowel, the most likely site, is colonized after ingestion of the organism.

TABLE 1 Recent taxonomic changes

Old designation	New designation
Pseudomonas cepacia	*Burkholderia cepacia*
Pseudomonas gladioli	*Burkholderia gladioli*
Pseudomonas mallei	*Burkholderia mallei*
Pseudomonas pickettii	*Burkholderia pickettii*
Pseudomonas pseudomallei	*Burkholderia pseudomallei*
Pseudomonas acidovorans	*Comamonas acidovorans*
Pseudomonas testosteroni	*Comamonas testosteroni*
Pseudomonas delafieldii	*Acidovorax delafieldii*
Pseudomonas facilis	*Acidovorax facilis*
Pseudomonas temperans	*Acidovorax temperans*
Pseudomonas putrefaciens	*Shewanella putrefaciens*
Pseudomonas paucimobilis	*Sphingomonas paucimobilis*
Pseudomonas mesophilica	*Methylobacterium extorquens*
Pseudomonas luteola	*Chrysemonas luteola*
Pseudomonas oryzihabitans	*Flavimonas oryzihabitans*

Colonization of the respiratory tract is common in hospitalized individuals, especially those who are intubated. The likelihood of colonization increases the longer the patient is hospitalized (58).

The distribution of other species of *Pseudomonas* and *Burkholderia* is similar to that of *P. aeruginosa* and depends on these organisms' abilities to survive in hostile environments. Contamination of aqueous solutions such as distilled water, soaps, disinfectants, and injectable medicines with these organisms has led to pseudoinfections, most commonly pseudobacteremia, as well as true bacteremia and other infections (24, 33, 40, 50, 53, 68).

Certain species have limited geographic distributions or unusual environments. *B. pseudomallei* is found primarily in tropical and subtropical areas; the disease (melioidosis) is most prevalent in northern Australia and Southeast Asia. It is particularly prevalent in the rice-growing regions of northern Thailand because of high concentrations of the organism in rice paddy surface water (15, 38).

Pseudomonas pseudoalcaligenes has a highly unusual habitat. It has been found at concentrations of $>10^8$ organisms per ml in metalworking fluid, a mixture of water and petroleum products. Metalworkers are exposed to aerosols containing 10^5 organisms per m^3 with no apparent ill effects (43).

CLINICAL SIGNIFICANCE

P. aeruginosa

P. aeruginosa is the most important human pathogen in the genera *Pseudomonas* and *Burkholderia* with respect to both the numbers and types of infections caused and their associated morbidity and mortality (58). It successfully combines adaptability to a variety of moist environments with a collection of potent virulence factors. The spectrum of disease caused by this agent ranges from superficial skin infections to fulminant sepsis.

Community-acquired *P. aeruginosa* infections in nonimmunocompromised individuals tend to be localized and frequently associated with contaminated water or solutions. Probably the most superficial infection associated with this organism is folliculitis, acquired in swimming pools, whirlpools, or hot tubs (27). In one outbreak, all party guests who had bathed in a wooden hot tub developed folliculitis

and otitis due to *P. aeruginosa*. High concentrations of *P. aeruginosa* were found on the surface of the hot tub (9).

Superficial infections of the ear canal due to *P. aeruginosa* frequently develop in those involved in aquatic sports, such as competitive swimmers. This condition is aptly named "swimmer's ear." This condition should not be confused with a much more severe ear infection called malignant otitis externa, an infection seen primarily in diabetics and the elderly in which *P. aeruginosa* can invade the underlying tissues, damaging cranial nerves and causing a temporal bone and basilar skull osteomyelitis. Meningitis may result (58). Successful treatment of this infection requires surgical debridement and antimicrobial therapy.

P. aeruginosa infection of the eye usually follows minor trauma to the cornea. These infections are frequently associated with contact lens use. Contaminated contact lens solution and the use of tap water during lens care have been implicated as sources of infection (30). *P. aeruginosa* infection of the eye can cause corneal ulcers, which may progress to loss of ocular function if not promptly treated.

P. aeruginosa is a common cause of osteomyelitis of the calcaneus in children (18). A puncture wound, usually caused by a nail penetrating a sneaker, occurs within the month preceding development of this infection. The inner pad of the sneaker is the source of *P. aeruginosa* (19).

The most severe community-acquired infection caused by *P. aeruginosa* is endocarditis in intravenous drug users. The drugs being injected are probably mixed with water contaminated by *P. aeruginosa*, leading to bacteremia, which may result in endocarditis. Replacement of the infected valve is usually necessary (58). These individuals may also develop osteomyelitis of a variety of bones.

An unusual mucoid phenotype of *P. aeruginosa* is associated with chronic lung infection in individuals with cystic fibrosis (CF). Approximately 70 to 80% of adolescent and adult CF patients are infected with this phenotype (22). The unusual lung environment of the CF patient is believed to switch on a cluster of genes that encode for the production of the large quantities of alginate, a polysaccharide polymer that give this organism its mucoid appearance (11, 72). Alginate appears to be antiphagocytic and is believed to induce a significant immune response in the lungs of these patients. This immune response is partially responsible for the lung damage caused by this chronic infection, which leads to the abbreviated life spans of these patients (44). Bacteremia, however, is rare, probably because of the high level of circulating antibodies to various *P. aeruginosa* virulence factors in these patients (35). Mucoid *P. aeruginosa* is rarely seen in other patient populations.

P. aeruginosa is a major cause of nosocomial infection. It is the leading cause of nosocomial respiratory tract infection and can be especially serious in the intubated patient in the intensive care unit (65). Colonization of the upper airways and endotracheal tubes must be distinguished from true infection in this setting (3), because patients with pneumonia have a high mortality and require aggressive antimicrobial therapy. On the other hand, unnecessary use of antimicrobial agents in patients who are colonized rather than infected may result in increasing antimicrobial resistance and antimicrobial side effects. Gram stains of secretions obtained by endotracheal suction that reveal large quantities of gram-negative rods and polymorphonuclear leukocytes support the diagnosis of nosocomial pneumonia. The absence of gram-negative rods and the presence of squamous epithelial cells indicate that the specimen is not useful for determining whether the patient has pneumonia; such specimens should be rejected (45).

P. aeruginosa also causes nosocomial urinary tract infections, wound infections, and bacteremia (58). Wound infections due to *P. aeruginosa* are particularly troublesome in burn patients. The incidence of *P. aeruginosa* wound infections has declined in burn patients, but the high rate of sepsis following wound infections in this setting is responsible for high mortality rates (29).

The pathogenicity of *P. aeruginosa* is explained by an impressive array of virulence factors. Polar pili mediate attachment to epithelial cells (60). Once attached, the bacteria produce tissue-damaging proteases, hemolysins, exotoxins, and endotoxin. The role of elastase, one of the proteases produced by *P. aeruginosa*, has been documented in the pathogenesis of keratitis, burn wounds, and chronic lung disease in CF patients (29, 30, 44). Exotoxin A, like diphtheria toxin, inhibits protein synthesis. The precise role of exotoxin A in the pathogenesis of *P. aeruginosa* infection is unknown, but the toxin may be associated with the dissemination of this organism and its systemic toxicity (58).

Other *Pseudomonas* Species

Pseudomonas species other than *P. aeruginosa* infrequently cause infection, in large part because of their lack of virulence factors. They may frequently colonize aqueous hospital environments. Because of the low virulence of these species, the infections they cause are often iatrogenic and are associated with the administration of contaminated solutions, medicines, and blood products or the presence of indwelling catheters (24, 28, 66, 77).

P. fluorescens has the ability to grow at 4°C and can be isolated from the skin of a small proportion of blood donors (62, 66). These two factors are responsible for occasional transfusion reactions. One patient transfused with blood contaminated with *P. fluorescens* died of septic shock as a result (66). Contamination of bone marrow transplants by *Pseudomonas* spp. during processing (37) may become more common as the numbers of such procedures increase.

Both *P. fluorescens* and *P. stutzeri* have been implicated in outbreaks of pseudobacteremia (33, 68). The recovery of these infrequently encountered organisms from blood cultures for more than one patient over a brief period suggests an outbreak of either pseudobacteremia or true bacteremia due to contaminated injectable solutions. These situations require prompt investigation by infection control practitioners.

P. stutzeri is an unusual cause of human infection, but it has been reported to cause bacteremia in immunosuppressed individuals (59) and pneumonia in alcoholics (8). It has also caused bacteremia in patients undergoing hemodialysis with machines containing *P. stutzeri*-contaminated dialysis fluid (24). *P. stutzeri* has also been recovered from wounds, the respiratory tracts of intubated patients, and the urinary tract, although its pathogenic role in those settings is unclear (31).

Other *Pseudomonas* species are found even less frequently in human infection. *P. alcaligenes* was the cause of catheter-related endocarditis in a bone marrow transplant recipient (42). *P. putida* was the agent of catheter-related bacteremia in cancer patients (1). *P. vesicularis* bacteremia was reported in a hemodialysis patient and an immunosuppressed individual (56, 77). It has also been recognized in cervical specimens because of its ability to produce bright orange colonies on Thayer-Martin agar (49). Other *Pseudo-*

monas species are infrequently encountered in clinical specimens, and their pathogenic roles are uncertain.

Burkholderia spp.

The genus *Burkholderia* contains two organisms well recognized as human pathogens, *B. pseudomallei* and *B. cepacia* (formerly *Pseudomonas pseudomallei* and *Pseudomonas cepacia*, respectively).

B. pseudomallei is the etiologic agent of melioidosis. The organism is acquired either by inhalation or by contact of cut or abraded skin with contaminated soil or water (15, 38). There appears to be a broad spectrum of disease, ranging from asymptomatic infection (as judged by serologic surveys) to fulminant sepsis (32, 38). Mortality in patients with fulminant sepsis approaches 90%. The organism frequently produces abscesses in organs of the reticuloendothelial system such as the lungs, liver, spleen, and lymph nodes. Abscesses can occur in skin, soft tissue, joints, and bones. Multiple rather than single lesions are common. Manifestations of this organism can mimic those of *Mycobacterium tuberculosis*. Like *M. tuberculosis*, this organism can survive within phagocytes (61), produce nodular lesions visible on chest radiograph and granulomatous lesions in a variety of tissues, and lie dormant for years and then reactivate (38). Infection of the central nervous system is unusual. When it occurs, encephalitis with peripheral neuropathy is more characteristic than a pyogenic meningitis (80). Relapse following appropriate antimicrobial therapy occurred in 25% of patients with bacteremic disease studied over a 5-year period (10). Molecular genotyping indicates that relapse rather than reinfection occurs in these patients (17).

Melioidosis is most prevalent in Southeast Asia and northern Australia but occurs in tropical and subtropical regions throughout the world. Concern exists that a significant number of Vietnam War veterans are latently infected with this organism; reactivation may become increasingly common as that population ages. This organism should be considered in the differential diagnosis of any individual with a fever of unknown origin or a tuberculosislike disease who has a travel history to a region of endemicity, even if travel preceded the illness by decades (5, 36).

B. cepacia is well recognized as a nosocomial pathogen, causing infections associated with contaminated disinfectants, equipment, and medications. These infections include bacteremia (particularly in patients with indwelling catheters), urinary tract infection, septic arthritis, peritonitis, and respiratory tract infection (50, 53). Because *B. cepacia* is of low virulence, the morbidity and mortality associated with these infections are low. *B. cepacia* pseudobacteremia due to contaminated disinfectants has been described (14, 50). When *B. cepacia* is isolated from the blood of multiple patients over a short period of time, the possibility of pseudobacteremia should be considered.

B. cepacia is emerging as an important pathogen in two patient populations with genetic disease: those with CF and those with chronic granulomatous disease (22, 48). CF patients chronically infected with *B. cepacia* have decreased survival rates compared to those who are not infected. Patients with the "cepacia syndrome" frequently have mild lung disease prior to *B. cepacia* infection but show rapid decline in lung function, frequent bacteremia (an unusual occurrence in CF patients), and death due to lung failure. Autopsy results reveal multiple lung abscesses containing

B. cepacia, a pathologic finding similar to that seen in fulminant *B. pseudomallei* disease (22).

Molecular techniques have demonstrated that person-to-person transmission of *B. cepacia* occurs (25, 69) and that certain strains of *B. cepacia* may be highly transmissible while others are much less so (74).

Because of their profound immunosuppression, CF patients who have received lung transplants are particularly vulnerable to infection with *B. cepacia*. *B. cepacia* nosocomial infections have occurred in this patient population with a mortality of 80% (70). Nonetheless, genotype analysis suggests that most CF patients infected posttransplant with *B. cepacia* harbored the same organism prior to transplantation (73).

The other *Burkholderia* (formerly *Pseudomonas*) species, *B. mallei*, *B. pickettii*, and *B. gladioli*, do not appear to play significant roles in human disease. *B. mallei* is the etiologic agent of glanders, a disease of livestock, particularly horses, mules, and donkeys. Human disease due to this organism has not been reported in the United States within the last 50 years (64).

B. gladioli has recently been isolated from the respiratory secretions of CF patients. This organism was recognized by its ability to grow on a selective medium used to isolate *B. cepacia*. This organism does not appear to have a role in the lung disease of these patients (12).

B. pickettii has been recovered from a variety of clinical specimens, including urine, wound exudates, blood, and cerebrospinal fluid (63). This organism has contaminated bone marrow used for transplantation during processing (37). An outbreak of nosocomial bacteremia occurred after the adulteration of narcotics with distilled water contaminated with *B. pickettii* (40). *B. pickettii* may be recovered from the respiratory tracts of CF patients. Because of biochemical similarities, *B. pickettii* may be confused with *B. cepacia*. Like *B. gladioli*, *B. pickettii* does not appear to exacerbate pulmonary disease in CF patients.

COLLECTION, TRANSPORT, AND STORAGE

Pseudomonas and *Burkholderia* species can survive in a variety of hostile environments and at temperatures found in clinical settings. Therefore, standard collection, transport, and storage techniques as outlined in chapter 3 of this Manual are sufficient to ensure recovery of these organisms from clinical specimens.

DIRECT EXAMINATION

Pseudomonas and *Burkholderia* species have similar Gram stain morphologies and are not easily distinguished from other non-glucose-fermenting gram-negative bacilli. No direct fluorescent-antibody or DNA probe reagents for the detection of any species of these two genera are currently available.

CULTURE AND ISOLATION

Organisms from the genera *Pseudomonas* and *Burkholderia* have been recovered from a variety of clinical specimens. These organisms grow well on standard laboratory media such as 5% sheep blood agar and chocolate agar. Such media can be used to recover the organisms from clinical specimens such as cerebrospinal fluid, joint fluid, or peritoneal dialysis fluid, in which a mixed flora is not anticipated. All members of the genera *Pseudomonas* and *Burkholderia*,

including *B. pseudomallei* (81), grow in broth blood culture systems, so special blood culture techniques such as the lysis-centrifugation system are not required.

Isolation of these organisms from specimens with mixed flora is facilitated by selective media. MacConkey agar is a useful selective medium for the isolation of most *Pseudomonas* species (except *P. vesicularis*). Other selective media such as cetrimide or eosin-methylene blue agars can be used to isolate *P. aeruginosa* from clinical and environmental sources.

Burkholderia species will grow on MacConkey agar, but specific selective media with the ability to inhibit *P. aeruginosa* are preferred for the isolation of *B. cepacia* and *B. pseudomallei*. Two media, PC (for *P. cepacia*) (23) and OFPBL (for oxidative-fermentative base-polymyxin B-bacitracin-lactose) (78) agars, are useful for the recovery of *B. cepacia* from respiratory secretions of CF patients. Both media contain antimicrobial agents that inhibit the growth of *P. aeruginosa*, which grows on these media infrequently. Each medium has distinct disadvantages. Because of its intense yellow pigment on OFPBL, *B. gladioli* may be confused with *B. cepacia*, colonies of which appear yellow on this medium because of lactose oxidation. OFPBL is less selective than PC agar and allows a wider variety of organisms to grow, but the inability of most of these isolates to oxidize lactose distinguishes them from *B. cepacia*. PC medium may be too selective, inhibiting the growth of an occasional *B. cepacia* clinical isolate (7), and it is not truly differential. Proficiency surveys indicate that both media are more effective than MacConkey agar for the isolation of *B. cepacia* (75).

Ashdown medium is effective for the isolation of *B. pseudomallei*; crystal violet and gentamicin act as selective agents. This medium is superior to MacConkey agar or MacConkey agar supplemented with colistin for the recovery of *B. pseudomallei* from clinical specimens containing mixed bacterial flora, such as throat, rectal, and sputum specimens (2, 82).

IDENTIFICATION AND INTERPRETATION OF CULTURE RESULTS

P. aeruginosa, P. fluorescens, and *P. putida*

P. aeruginosa isolates are easily recognized on primary isolation media on the basis of characteristic colonial morphology, production of diffusible pigments, and a grapelike or corn taco-like odor. The colonies are usually spreading and flat, have serrated edges, and show a metallic sheen. Other colonial variants, including smooth, coliform, gelatinous, dwarf, and mucoid, exist, with mucoid colonies being particularly prevalent in respiratory tract specimens from CF patients (22). *P. aeruginosa* produces a number of water-soluble pigments, including the yellow-green or yellow-brown fluorescent pigment pyoverdin (also produced by *P. fluorescens* and *P. putida*). When pyoverdin combines with the blue water-soluble phenazine pigment pyocyanin, the bright green color characteristic of *P. aeruginosa* is created. The organism may also produce one of two other water-soluble pigments: pyorubin (red) or pyomelanin (brown to brown-black). *P. aeruginosa* can be confidently identified on the basis of a positive oxidase test; a triple sugar iron (TSI) agar reaction of alkaline over no change; growth at 42°C; and production of bright blue to bluish green, red, or brown diffusible pigments on Mueller-Hinton or other non-dye-containing agar.

Occasional *P. aeruginosa* strains do not produce pigment or produce a yellowish green pigment that does not differentiate them from *P. fluorescens* or *P. putida*. Among the pyoverdin-producing organisms, the ability of *P. aeruginosa* to grow at 42°C differentiates it from *P. fluorescens* and *P. putida*. Nonpigmented *P. aeruginosa* occur; they are frequently highly mucoid strains recovered from the respiratory secretions of CF patients. In practice, the finding of a highly mucoid, non-glucose-fermenting, gram-negative rod in respiratory specimens from CF patients is sufficient to identify the organism as *P. aeruginosa*. However, when a nonpigmented isolate is recovered in a clinical setting other than this, key biochemical characteristics of *P. aeruginosa* besides those already mentioned include nitrate reduction, oxidation of glucose but not disaccharides, and arginine dihydrolase activity (Table 2).

P. fluorescens and *P. putida* are the other members of the fluorescent pseudomonad group (pyoverdin producers) that are recovered from clinical specimens. Unlike *P. aeruginosa*, they do not posses a distinctive colonial morphology or odor. In most clinical settings, there is little need to differentiate among these organisms, since they are of low virulence and usually not of clinical significance. Because of the well-known association between *P. fluorescens* and contaminated blood products, however, accurate identification may be important for isolates from patients who have recently received blood products and developed bacteremia. *P. fluorescens* can be differentiated from *P. putida* by the abilities of the former to grow at 4°C and degrade gelatin; *P. putida* can do neither. Negative gelatin results with standard tube assays should be interpreted with caution, since a significant number of *P. fluorescens* isolates may require incubation for 4 to 7 days for accurate detection of gelatin degradation (55).

Other fluorescent pseudomonads may occasionally be encountered in clinical or environmental specimens. Identification as "*Pseudomonas* species not *aeruginosa*" and susceptibility testing of the isolates, when appropriate, are sufficient in most circumstances. When necessary, these isolates can be referred to reference laboratories, where more extensive biochemical batteries and cell wall fatty acid analysis can be used to establish a definitive identification.

P. stutzeri Group

The *P. stutzeri* group consists of three organisms: *P. stutzeri*, CDC group Vb-3, and *P. mendocina*. Because of the biochemical and morphologic similarities between the two, CDC group Vb-3 should be considered a biovar of *P. stutzeri*. Of this group, *P. stutzeri* is most frequently encountered in clinical specimens. *P. stutzeri* isolates are easily recognized on primary isolation media by their distinctive dry, wrinkled colonial morphology. These colonies, which can pit or adhere to the agar, are buff to brown. The adherence can make removal of colonies from agar medium difficult. Frequently, the only way to remove this organism from agar is to lift an entire colony. Because of the difficulty in making suspensions of specific turbidity, commercial identification and susceptibility systems may not work well with this organism. Not all isolates of *P. stutzeri* produce wrinkled colonies, and such strains are not readily distinguishable from other pseudomonads. *P. mendocina* is rarely recovered as a human pathogen. It might be occasionally recovered from environmental cultures, since it can be found, like other pseudomonads, in water and soil (21). The colonies are smooth, nonwrinkled, and flat, producing a brownish yellow pigment; they are not readily distinguished from those of other pseudomonads.

Key biochemical characteristics of the *P. stutzeri* group are the ability to reduce nitrates to nitrogen gas, a positive oxidase test, and the ability to oxidize glucose but not lactose. Both *P. mendocina* and CDC group Vb-3 demonstrate arginine dihydrolase activity, while *P. stutzeri* does not. Otherwise, *P. stutzeri* and CDC group Vb-3 are indistinguishable by commonly used biochemical tests. Most strains of *P. stutzeri* and CDC group Vb-3 oxidize maltose and mannitol, while *P. mendocina* cannot metabolize these sugars (Table 2).

P. alcaligenes, *P. pseudoalcaligenes*, and *Pseudomonas* sp. Strain CDC Group 1

P. alcaligenes, *P. pseudoalcaligenes*, and *Pseudomonas* sp. strain CDC group 1 are rarely encountered in clinical and environmental samples; they do not have distinctive colonial morphologies, nor do they produce pigments. Compared to other pseudomonads, they are biochemically inert. Characteristics that distinguish them from other biochemically inert gram-negative rods are a positive oxidase reaction, motility due to a polar flagellum, and growth on MacConkey agar. Most strains of *P. pseudoalcaligenes* and many strains of *P. alcaligenes* reduce nitrates to nitrites. *Pseudomonas* sp. strain CDC group 1 can be distinguished from the other two species by its ability to reduce nitrates to gas. *P. alcaligenes* is distinguished from *P. pseudoalcaligenes* by its inability to oxidize sugars, while *P. pseudoalcaligenes* weakly oxidizes fructose (Table 2). Isolates of these organisms are difficult to identify, and many laboratories, especially those using commercial systems, may appropriately call these isolates "*Pseudomonas* sp. not *aeruginosa*." If the clinical situation requires a definitive identification, assistance from reference laboratories should be sought.

P. diminuta and *P. vesicularis*

P. diminuta and *P. vesicularis* are rarely encountered in clinical and environmental specimens. Unlike most pseudomonads, these organisms have growth requirements for specific vitamins, including pantothenate, biotin, and cyanocobalamin. In addition, *P. diminuta* requires cysteine for growth (21).

On primary isolation media, *P. diminuta* colonies are chalk white, while many strains of *P. vesicularis* produce an orange intracellular pigment. Most strains of *P. diminuta* grow on MacConkey agar, while only approximately 25% of *P. vesicularis* strains grow on this medium.

Key characteristics of this group of organisms are a positive oxidase test, motility due to a single polar flagellum, weak oxidation of glucose that is strain variable, and failure of >90% of strains to reduce nitrate to nitrite. These organisms can produce acid from ethanol, but this test is not available in most clinical laboratories.

P. vesicularis is much more likely to oxidize glucose and maltose than *P. diminuta*. The most reliable test for differentiating these two infrequently encountered organisms is the test for esculin hydrolysis. All strains of *P. vesicularis* are reported to hydrolyze this substrate, while *P. diminuta* does not (20, 21).

Burkholderia spp.

Burkholderia spp. can be recovered on most aerobically incubated primary isolation media used in the clinical laboratory to isolate enteric gram-negative bacilli. Some mem-

TABLE 2 Characteristics of *Pseudomonas* spp. found in clinical specimens[a]

Test	*P. aeruginosa*[b] (n = 93)	*P. fluorescens* (n = 233)	*P. putida* (n = 335)	*P. stutzeri* (n = 181)	*P. mendocina* (n = 6)	*P. pseudo-alcaligenes* (n = 80)	*P. alcaligenes* (n = 61)	*Pseudomonas* sp. strain CDC group 1[c] (n = 31)	*P. diminuta* (n = 52)	*P. vesicularis* (n = 55)
Oxidase	100	100	100	100	100	100	100	100	100	100
Growth:										
MacConkey	99	100	100	100	100	93	98	97	96	26
Cetrimide	88	96	93	3	100	64	15	13	0	0
6.5% NaCl	7	3	11	100	100	4	10	14	0	0
42°C	100	0	0	90	100	75	48	48	19	0
Nitrate reduction	74	19	0	100	100	93	61	100	4	7
Gas from nitrate	60	4	0	100	100	5	0	100	0	0
Pyoverdin	69	91	82	0	0	0	0	0	0	0
Arginine dihydrolase	99	99	99	0[d]	100	36	7	33	0	0
Lysine decarboxylase	0	0	0	0	0	0	ND	0	0	0
Phenylalanine deaminase	8	3	2	55	50	21	20	ND	16	6
Indole	0	0	0	0	0	0	0	0	0	0
Hemolysis	38	14	0	0	0	0	11	0	0	0
Hydrolysis										
Urea	66	44	43	17	50	8	21	3	0	0
DNA	9	0	0	0	0	0	0	ND	12	0
Lecithin	8	89	0	8	0	0	0	ND	0	0
Gelatin	46	100	0	0	0	3	2	4	58	38
Acetamide	37	1	3	0	0	4	0	ND	0	0
Esculin	0	0	0	0	0	0	ND	0	0	100
Acid from:										
Glucose[e]	98	100	100	100	100	19	0	0	29	57
Fructose	89	99	99	95	100	100	0	0	0	0
Galactose	81	98	99	91	100	4	0	0	0	15
Mannose	79	99	99	89	100	1	0	0	0	0
Rhamnose	22	43	28	23	0	0	0	0	0	38
Xylose	85	97	98	94	100	8	0	0	0	11
Lactose	0	11	13	0	0	0	0	0	0	0
Sucrose	0	47	10	0	0	0	0	0	0	0
Maltose	12	31	19	99	0	11	0	0	2	62
Mannitol	68	93	17	70	0	3	0	0	0	0
Lactose (10%)	14	61	42	0	0	0	0	0	0	0
Motile	96	100	100	100	100	90	98	100	100	98
No. of flagella	1	>1	>1	1	1	1	1	1	1	1

[a] Numbers in table are percentages of strains positive. ND, no data.
[b] Data are for apyocyanogenic isolates only.
[c] Data are from reference 13; all other data are from reference 20.
[d] Organisms similar to *P. stutzeri* but positive for arginine dihydrolase are Vb-3.
[e] Oxidative-fermentative basal medium with 1% carbohydrate.

bers of the genus *Burkholderia* have distinctive characteristics. *B. pseudomallei* colonies vary from smooth and mucoid to dry and wrinkled. The dry, wrinkled colonial morphotype allows this organism to be recognized in specimens with a mixed bacterial flora. This wrinkled colonial morphotype is similar to that of *P. stutzeri*, and because of the clinical significance of *B. pseudomallei*, it is important to distinguish the two organisms. In addition to its distinctive colonial morphology, *B. pseudomallei* has a pungent, earthy odor. However, colonies of this organism should not be sniffed deliberately, as is the practice of some microbiologists, because of the risk of laboratory-acquired infections. On Ashdown medium, the colonies are typically dry, wrinkled, and violet-purple. These characteristics are sufficiently distinct to allow a presumptive identification of *B. pseudomallei* (82).

B. cepacia, especially when recovered from the respiratory tracts of CF patients, may require 3 days of incubation before colonies are seen on selective media. On MacConkey agar, these colonies may be punctate and tenacious, while on blood agar or selective medium such as PC or OFPBL, the colonies are smooth and slightly raised. On MacConkey agar, colonies of *B. cepacia* frequently become dark pink to red because of oxidation of lactose after extended incubation (4 to 7 days). Most clinical isolates are nonpigmented, but on iron-containing media such as a TSI slant, many strains produce a bright yellow pigment. Like *B. pseudomallei*, *B. cepacia* produces a pungent, earthy odor.

B. gladioli produces a bright yellow pigment. Production of this pigment on OFPBL medium has resulted in *B. gladioli* being confused with *B. cepacia* (12).

B. pickettii grows slowly on primary isolation medium and, like *B. cepacia*, may require ≥72 h of incubation before colonies become visible.

B. mallei is the most easily identified member of this genus because it is nonmotile. Other characteristics helpful in identifying *B. mallei* are reduction of nitrate to nitrite, arginine dihydrolase activity, oxidation of glucose, and variable oxidation of maltose. It fails to oxidize sucrose or xylose (13).

Because of the importance of *B. pseudomallei* and *B. cepacia* as human pathogens, accurate identification of these organisms is important. Key characteristics of *B. pseudomallei* include positive oxidase reaction, production of gas from nitrate, multitrichous polar flagella, arginine dihydrolase and gelatinase activities, and oxidation of a wide variety of carbohydrates. An acid slant on TSI results from oxidation of both lactose and sucrose. *B. pseudomallei* can be distinguished from *P. stutzeri* by arginine dihydrolase and gelatinase activities, lactose oxidation, and number of flagella (Tables 2 and 3).

Key characteristics of *B. cepacia* include a positive oxidase reaction in >90% of strains, lysine decarboxylase activity by >90% of strains, multitrichous polar flagella, and the utilization of a wide variety of mono- and disaccharides, including lactose. Many strains of *B. cepacia* give a weak oxidase reaction. *B. cepacia* can be distinguished from *B. gladioli* on the basis of oxidase activity; lysine decarboxylase activity; and lactose, maltose, and sucrose oxidation reactions.

Distinguishing *B. pickettii* from *B. cepacia* can be difficult. Reduction of nitrate to gas is characteristic of many but not all *B. pickettii* isolates (Table 3). When this characteristic is seen, it can be useful in distinguishing these two species, because *B. cepacia* isolates cannot reduce nitrate to gas, and only a third of the isolates reduce nitrate to nitrite. Most

strains of *B. cepacia* have lysine decarboxylase activity. When *B. cepacia* does not decarboxylate lysine, differentiation of *B. cepacia* and *B. pickettii* bv. 3 strains (also referred to as *Pseudomonas thomasii*) becomes difficult. *B. pickettii* bv. 3 organisms are usually unable to reduce nitrates, and both this biovar and *B. cepacia* attack the same carbohydrates with the exception of sucrose.

Two methods can distinguish *B. pickettii* from *B. cepacia* when biochemical results are not helpful. *B. pickettii* has a single polar flagellum, while *B. cepacia* has a tuft of three to eight polar flagella. Cell wall fatty acid analysis is also useful in distinguishing these organisms (47).

Use of Commercial Kit Systems for Identification of *Pseudomonas* and *Burkholderia* spp.

Many laboratories use commercial kit systems rather than conventional biochemical tests to identify *Pseudomonas* and *Burkholderia* spp. The general principles of these kits are described in chapter 10 of this Manual. Commercial systems have been developed for the specific identification of non-glucose-fermenting rods in the clinical laboratory. These include the N/F system (Remel Laboratories, Lenexa, Kans.), the AutoMicrobic System Gram-Negative Identification card and Rapid NFT (bioMérieux Vitek, Hazelwood, Mo.), the RapID NF Plus (Innovative Diagnostics Systems, Inc., Atlanta, Ga.), and the autoSCAN W/A Rapid Neg Combo (RNC) 1 panel (Baxter MicroScan, West Sacramento, Calif.). It may take from 4 to 48 h to obtain an identification with these systems depending on whether the system identifies the organisms on the basis of preformed enzymes (rapid) or on the basis of substrate utilization.

Evaluations suggest an accuracy of more than 90% for each system in identifying *Pseudomonas* and *Burkholderia* spp. However, these evaluations are often flawed by testing only small numbers of isolates from certain species. Studies using over 50 isolates of *P. aeruginosa* and 10 or fewer isolates of less frequently encountered *Pseudomonas* and *Burkholderia* spp. are common (6, 34, 41, 54, 57, 76). These evaluations do little more than prove that the systems adequately identify *P. aeruginosa*. In addition, because companies frequently modify their databases, comparison of one evaluation with another may be meaningless. Despite these problems, certain statements can be made concerning the value of systems in identifying different *Pseudomonas* and *Burkholderia* species.

Most systems can adequately identify *P. aeruginosa*, although the use of an expensive commercial system is unwarranted when a few simple tests can accurately identify this organism. Neither the Walk-Away Rapid Neg Combo 1 panel nor the RapID NF Plus can distinguish between *P. fluorescens* and *P. putida* (34, 54, 76). Differentiating these organisms is clinically relevant only if a patient develops bacteremia with a blood transfusion as a potential source. Owing to lack of adequate evaluations, isolates identified as *P. stutzeri*, *P. mendocina*, *P. alcaligenes*, *P. pseudoalcaligenes*, *P. vesicularis*, *P. diminuta*, *B. pickettii*, *B. mallei*, and *B. gladioli* should be confirmed by alternative methods if clinically indicated.

The NFT is the only system that has been adequately evaluated for accuracy in identifying *B. pseudomallei*. One study using 400 isolates showed an overall accuracy of >99% (16). The identification of a clinical isolate as *B. pseudomallei* by any other kit system requires confirmation, preferably by conventional biochemical analysis. Reporting of this identification should be delayed until confirmation is obtained.

TABLE 3 Characteristics of *Burkholderia* spp. from clinical specimens[a]

Test	B. pseudomallei (n = 7)	B. mallei[b] (n = 8)	B. cepacia (n = 93)	B. gladioli (n = 3)	B. pickettii bv. 1 (n = 31)	B. pickettii bv. 2 (n = 25)	B. pickettii bv. 3 (n = 10)
Oxidase	100	25	93	0	100	100	100
Growth:							
MacConkey	100	88	95	100	77	100	100
Cetrimide	0	0	71	0	0	0	0
42°C	100	0	60	0	26	60	60
Nitrate reduction	100	100	37	33	87	100	20
Gas from nitrate	100	0	0	0	84	100	10
Arginine dihydrolase	100	100	0	0	0	0	0
Lysine decarboxylase	0	0	92	0	0	0	0
Ornithine decarboxylase	0	0	66	0	0	0	0
Phenylalanine deaminase	0	ND	2	0	3	40	0
Hemolysis	43	0	4	0	0	0	0
Hydrolysis							
Urea	43	12	45	100	100	100	100
Lecithin	86	ND	51	66	0	0	0
Gelatin	100	0	51	66	0	0	0
Acetamide	0	ND	74	100	77	40	80
Esculin	57	0	63	0	3	0	0
ONPG	0	ND	67	100	0	0	0
Acid from:			79				
Glucose[c]	100	100	100	100	100	100	100
Fructose	100	ND	100	100	100	100	100
Galactose	100	ND	100	100	100	100	100
Mannose	100	ND	100	100	100	100	100
Rhamnose	71	ND	0	0	0	0	0
Xylose	86	12	99	100	100	100	100
Lactose	100	12	99	0	100	0	100
Sucrose	86	0	83	0	0	0	0
Maltose	100	0	98	0	100	0	100
Mannitol	100	62	100	100	0	0	100
Lactose (10%)	100	ND	98	0	81	0	100
Motile	100	0	99	100	94	96	100
No. of flagella	>1	0	>1	1	1	1	1

[a]Numbers in table are percentages of strains positive. ND, no data; ONPG, o-nitrophenyl-β-D-galactopyranoside.
[b]Data are from reference 13; all other data are from reference 20.
[c]Oxidative-fermentative basal medium with 1% carbohydrates.

Most commercial systems identify *B. cepacia* with an accuracy of ≥90%. In most clinical settings, this level of accuracy is acceptable. However, for a CF patient, 100% accuracy is essential, because CF patients with *B. cepacia* often become medical and social outcasts owing to concern about spread of the organism to other individuals with CF. When *B. cepacia* is identified in a CF patient by a commercial system, the identity of the isolate should be confirmed with conventional biochemical testing and, if necessary, fatty acid analysis. In addition, isolates recovered from CF patients and identified by a commercial kit system as *B. pseudomallei*, *B. gladioli*, or *B. pickettii* should be confirmed by conventional biochemical and fatty acid analyses to ensure that *B. cepacia* isolates are not being misidentified.

ANTIMICROBIAL SUSCEPTIBILITY

Community-acquired isolates of *P. aeruginosa* are usually susceptible to the antipseudomonal penicillins (ticarcillin-clavulanic acid, piperacillin, piperacillin-tazobactam, and mezlocillin), the aminoglycosides (gentamicin, tobramycin, and amikacin), ciprofloxacin, cefoperazone, ceftazidime, and imipenem. Susceptibility is less predictable for ticarcillin, carbenicillin, other broad-spectrum cephalosporins (ceftriaxone and cefotaxime), fluoroquinolones other than ciprofloxacin, and the monobactam aztreonam. The organism is uniformly resistant to antistaphylococcal penicillins, ampicillin, amoxicillin-clavulanic acid, ampicillin-sulbactam, tetracyclines, macrolides, rifampin, chloramphenicol, trimethoprim-sulfamethoxazole, first- and second-generation cephalosporins, and oral broad-spectrum cephalosporins (cefixime and cefpodoxime).

Nosocomially acquired *P. aeruginosa* isolates tend to be more resistant than community-acquired strains to antimicrobial agents, frequently displaying resistance to multiple classes of antimicrobial agents. This multidrug resistance is of particular concern in chronically infected CF patients. Development of resistance during monotherapy with either cell wall-active agents (antipseudomonal penicillin, cefoperazone, and ceftazidime) or ciprofloxacin occurs frequently (4, 67) and is usually due to mutation.

B. pseudomallei is usually susceptible to the antipseudomonal penicillins, cefoperazone, ceftazidime, ampicillin-sulbactam, amoxicillin-clavulanic acid, chloramphenicol, and tetracyclines. Susceptibility to the fluoroquinolones is variable. Most isolates are resistant to the aminoglycosides and trimethoprim-sulfamethoxazole (71, 85). The organism is uniformly resistant to antistaphylococcal penicillins, ampicillin, rifampin, and the narrow- and expanded-spectrum cephalosporins.

B. cepacia is one of the most antimicrobial agent-resistant organisms encountered in the clinical laboratory. The organism is usually susceptible only to piperacillin, azlocillin, cefoperazone, ceftazidime, chloramphenicol, and trimethoprim-sulfamethoxazole. Susceptibility to imipenem is variable, and a small percentage of isolates may be susceptible to kanamycin. Strains recovered from CF patients who have received repeated antimicrobial courses are frequently resistant to all known antimicrobial agents.

REFERENCES

1. **Anaissie, E., V. Fainstein, P. Miller, K. Hassamali, S. Pitlik, G. P. Bodey, and K. Rolston.** 1987. *Pseudomonas putida*: newly recognized pathogen in patients with cancer. *Am. J. Med.* **82:**1191–1194.
2. **Ashdown, L. R.** 1979. An improved screening technique for isolation of *Pseudomonas pseudomallei* from clinical specimens. *Pathology* **11:**293–297.
3. **Baselski, V.** 1993. Microbiologic diagnosis of ventilator-associated pneumonia. *Infect. Dis. Clin. N. Am.* **7:**331–357.
4. **Bell, S. M., J. N. Pham, and J. Y. M. Lanzarone.** 1985. Mutation of *Pseudomonas aeruginosa* to piperacillin resistance mediated by β-lactamase production. *J. Antimicrob. Chemother.* **15:**665–670.
5. **Bouvy, J. J., J. E. Degener, C. Stijnen, M. P. W. Gallee, and B. van der Berg.** 1986. Septic melioidosis after a visit to Southeast Asia. *Eur. J. Clin. Microbiol.* **5:**655–656.
6. **Burdash, N. M., E. R. Bannister, J. P. Manos, and M. E. West.** 1980. A comparison of four commercial systems for the identification of non-fermentative Gram-negative bacilli. *Am. J. Clin. Pathol.* **73:**564–569.
7. **Burns, J. L., D. K. Clark, and G. L. Morlin.** 1993. Effect of specimen storage temperature on the ability to recover *Pseudomonas cepacia* from sputum. *Pediatr. Pulmonol. Suppl.* **9:**256–257.
8. **Carratala, J., A. Salazar, J. Mascaro, and M. Santin.** 1992. Community-acquired pneumonia due to *Pseudomonas stutzeri*. *Clin. Infect. Dis.* **14:**792.
9. **Centers for Disease Control.** 1982. Otitis due to *Pseudomonas aeruginosa* 0:10 associated with a mobile redwood tub system—North Carolina. *Morbid. Mortal. Weekly Rep.* **31:**541–542.
10. **Chaowagul, W., Y. Suputtamongkol, D. A. B. Dance, A. Rajchanuvong, J. Pattra-arechachai, and N. J. White.** 1993. Relapse in melioidosis: incidence and risk factors. *J. Infect. Dis.* **168:**1181–1185.
11. **Chitnis, C. E., and D. E. Ohman.** 1993. Genetic analysis of the alginate biosynthetic gene cluster of *Pseudomonas aeruginosa* shows evidence of an operonic structure. *Mol. Microbiol.* **8:**583–590.
12. **Christenson, J. C., D. F. Welch, G. Mukwaya, M. J. Muszynski, R. E. Weaver, and D. J. Brenner.** 1989. Recovery of *Pseudomonas gladioli* from respiratory tract specimens of patients with cystic fibrosis. *J. Clin. Microbiol.* **27:**270–273.
13. **Clark, W. A., D. G. Hollis, R. E. Weaver, and P. Riley.** 1984. *Identification of Unusual Pathogenic Gram-Negative Aerobic and Facultatively Anaerobic Bacteria*, p. 1–40, 72, 79, 244–287. U.S. Public Health Service, Atlanta.
14. **Craven, D. E., B. Moody, M. G. Connolly, N. R. Kollisch,**

K. D. Stottmeier, and W. R. McCabe. 1981. Pseudobacteremia caused by povidone-iodine solution contaminated with *Pseudomonas cepacia*. *N. Engl. J. Med.* **305:**621–623.
15. **Dance, D. A. B.** 1991. Melioidosis: the tip of the iceberg? *Clin. Microbiol. Rev.* **4:**52–60.
16. **Dance, D. A. B., V. Wuthiekanun, P. Naigowit, and N. J. White.** 1989. Identification of *Pseudomonas pseudomallei* in clinical practice: use of simple screening tests and API 20NE. *J. Clin. Pathol.* **42:**645–648.
17. **Desmarchelier, P. M., D. A. B. Dance, W. Chaowagul, Y. Suputtamongkol, N. J. White, and T. L. Pitt.** 1993. Relationships among *Pseudomonas pseudomallei* isolates from patients with recurrent melioidosis. *J. Clin. Microbiol.* **31:**1592–1596.
18. **Faden, H., and M. Grossi.** 1991. Acute osteomyelitis in children: reassessment of etiologic agents and their clinical characteristics. *Am. J. Dis. Child.* **145:**65–69.
19. **Fisher, M. C., J. F. Goldsmith, and P. H. Gilligan.** 1985. Sneakers as a source of *Pseudomonas aeruginosa* in children with osteomyelitis following puncture wounds. *J. Pediatr.* **106:**607–609.
20. **Gilardi, G. L.** 1989. *Identification of Glucose Non-fermenting Gram-Negative Rods*. Department of Laboratories, North General Hospital, New York.
21. **Gilardi, G. L.** 1991. *Pseudomonas and Related Genera*, p. 429–441. *In* A. Balows, W. J. Hausler, Jr., K. L. Herrmann, H. D. Isenberg, and H. J. Shadomy (ed.), *Manual of Clinical Microbiology*, 5th ed. American Society for Microbiology, Washington, D.C.
22. **Gilligan, P. H.** 1991. Microbiology of airway disease in patients with cystic fibrosis. *Clin. Microbiol. Rev.* **4:**35–51.
23. **Gilligan, P. H., P. A. Gage, L. M. Bradshaw, D. V. Schidlow, and B. T. DeCicco.** 1985. Isolation medium for the recovery of *Pseudomonas cepacia* from respiratory secretions of patients with cystic fibrosis. *J. Clin. Microbiol.* **22:**5–8.
24. **Goetz, A., V. L. Yu, J. E. Hanchett, and J. D. Rihs.** 1983. *Pseudomonas stutzeri* bacteremia associated with hemodialysis. *Arch. Intern. Med.* **143:**1909–1912.
25. **Govan, J. R. W., P. H. Brown, J. Maddison, C. J. Doherty, J. W. Nelson, M. Dodd, A. P. Greening, and A. K. Webb.** 1993. Evidence for transmission of *Pseudomonas cepacia* by social contact in cystic fibrosis. *Lancet* **342:**15–19.
26. **Grundmann, H., A. Kropec, D. Hartung, R. Berner, and F. Daschner.** 1993. *Pseudomonas aeruginosa* in a neonatal intensive care unit: reservoirs and ecology of the nosocomial pathogen. *J. Infect. Dis.* **168:**943–947.
27. **Gustafson, T. L., J. D. Band, R. H. Hutcheson, Jr., and W. Schaffner.** 1983. *Pseudomonas* folliculitis: an outbreak and review. *Rev. Infect. Dis.* **5:**1–8.
28. **Heard, S., S. Lawrence, B. Holmes, and M. Costas.** 1990. A pseudo-outbreak of *Pseudomonas* on a special care baby unit. *J. Hosp. Infect.* **16:**59–65.
29. **Holder, I. A.** 1993. *Pseudomonas aeruginosa* virulence-associated factors and their role in burn wound infections, p. 235–245. *In* R. B. Fick, Jr. (ed.), *Pseudomonas aeruginosa: the Opportunist*. CRC Press, Inc., Boca Raton, Fla.
30. **Holland, S. P., J. S. Pulido, T. K. Shires, and J. W. Costerton.** 1993. *Pseudomonas aeruginosa* ocular infections, p. 159–176. *In* R. B. Fick, Jr. (ed.), *Pseudomonas aeruginosa: the Opportunist*. CRC Press, Inc., Boca Raton, Fla.
31. **Holmes, B.** 1986. Identification and distribution of *Pseudomonas stutzeri* in clinical material. *J. Appl. Bacteriol.* **60:**401–411.
32. **Kanaphun, P., N. Thirawattanasuk, Y. Suputtamongkol, P. Naigowit, D. A. B. Dance, M. D. Smith, and N. J. White.** 1993. Serology and carriage of *Pseudomonas pseudomallei*: a prospective study in 1000 hospitalized children in northeast Thailand. *J. Infect. Dis.* **167:**230–233.
33. **Keys, T. F., L. J. Melton III, M. D. Maker, and D. M. Ilstrup.** 1983. A suspected hospital outbreak of pseudobacteremia due to *Pseudomonas stutzeri*. *J. Infect. Dis.* **147:**489–493.
34. **Kitch, T. T., M. R. Jacobs, and P. C. Appelbaum.** 1992.

Evaluation of the 4-hour RapID NF Plus method for identification of 345 gram-negative nonfermentative rods. *J. Clin. Microbiol.* **30:**1267–1270.

35. **Klinger, J. D., D. C. Straus, C. B. Hilton, and J. A. Bass.** 1978. Antibodies to proteases and exotoxin A of *Pseudomonas aeruginosa* in patients with cystic fibrosis: demonstration by radioimmunoassay. *J. Infect. Dis.* **138:**49–58.

36. **Koponen, M. A., D. Zlock, D. L. Palmer, and T. L. Merlin.** 1991. Melioidosis: forgotten, not gone! *Arch. Intern. Med.* **151:**605–608.

37. **Lazarus, H. M., M. Magalhaes-Silverman, R. M. Fox, R. J. Creger, and M. Jacobs.** 1991. Contamination during *in vitro* processing of bone marrow for transplantation: clinical significance. *Bone Marrow Transplant.* **7:**241–246.

38. **Leelarasamee, A., and S. Bovornkitti.** 1989. Melioidosis: review and update. *Rev. Infect. Dis.* **11:**413–425.

39. **MacDonell, M. T., and R. R. Colwell.** 1985. Phylogeny of the Vibrionaceae, and recommendation for two new genera, *Listonella* and *Shewanella*. *Syst. Appl. Microbiol.* **6:**171–182.

40. **Maki, D. G., B. S. Klein, R. D. McCormick, C. J. Alvarado, M. A. Zilz, S. M. Stolz, C. A. Hassemer, J. Gould, and A. R. Liegel.** 1991. Nosocomial *Pseudomonas pickettii* bacteremias traced to narcotic tampering: a case for selective drug screening of health care personnel. *JAMA* **265:**981–986.

41. **Martin, R., F. Siavoshi, and D. L. McDougal.** 1986. Comparison of rapid NFT system and conventional methods for identification of nonsaccharolytic gram-negative bacteria. *J. Clin. Microbiol.* **24:**1089–1092.

42. **Martino, P., A. Micozzi, M. Venditti, C. Gentile, C. Girmenia, R. Raccah, S. Santilli, N. Alessandri, and F. Mandelli.** 1990. Catheter-related right-sided endocarditis in bone marrow transplant recipients. *Rev. Infect. Dis.* **12:**250–257.

43. **Mattsby-Baltzer, I., L. Edebo, B. Järvholm, B. Lavenius, and T. Soderstrom.** 1990. Subclass distribution of IgG and IgA antibody response to *Pseudomonas pseudoalcaligenes* in humans exposed to infected metal-working fluid. *J. Allergy Clin. Immunol.* **86:**231–238.

44. **McCubbin, M., and R. B. Fick, Jr.** 1993. Pathogenesis of *Pseudomonas* lung disease in cystic fibrosis, p. 189–211. *In* R. B. Fick, Jr. (ed.), *Pseudomonas aeruginosa: the Opportunist.* CRC Press, Inc., Boca Raton, Fla.

45. **Morris, A. J., D. C. Tanner, and L. B. Reller.** 1993. Rejection criteria for endotracheal aspirates from adults. *J. Clin. Microbiol.* **31:**1027–1029.

46. **Morrison, A. J., Jr., and R. P. Wenzel.** 1984. Epidemiology of infections due to *Pseudomonas aeruginosa*. *Rev. Infect. Dis.* **6**(Suppl. 3):S627–S642.

47. **Mukwaya, G. M., and D. F. Welch.** 1989. Subgrouping of *Pseudomonas cepacia* by cellular fatty acid composition. *J. Clin. Microbiol.* **27:**2640–2646.

48. **O'Neil, K. M., J. H. Herman, J. F. Modlin, E. R. Moxon, and J. A. Winkelstein.** 1986. *Pseudomonas cepacia*: an emerging pathogen in chronic granulomatous disease. *J. Pediatr.* **108:**940–942.

49. **Otto, L. A., B. S. Deboo, E. L. Capers, and M. J. Pickett.** 1978. *Pseudomonas vesicularis* from cervical specimens. *J. Clin. Microbiol.* **7:**341–345.

50. **Pallent, L. J., W. B. Hugo, D. J. W. Grant, and A. Davies.** 1983. *Pseudomonas cepacia* as contaminant and infective agent. *J. Hosp. Infect.* **4:**9–13.

51. **Palleroni, N. J.** 1984. Genus I. *Pseudomonas* Migula 1894, 237^AL, p. 141–199. *In* N. R. Krieg and J. G. Holt (ed.), *Bergey's Manual of Systematic Bacteriology*, vol. 1. The Williams & Wilkins Co., Baltimore.

52. **Palleroni, N. J., R. Kunisawa, R. Contopoulou, and M. Doudoroff.** 1973. Nucleic acid homologies in the genus *Pseudomonas*. *Int. J. Syst. Bacteriol.* **23:**333–339.

53. **Pegues, D. A., L. A. Carson, R. L. Anderson, M. J. Norgard, T. A. Agent, W. R. Jarvis, and C. H. Woernle.** 1993. Outbreak of *Pseudomonas cepacia* bacteremia in oncology patients. *Clin. Infect. Dis.* **16:**407–411.

54. **Pfaller, M. A., D. Sahm, C. O'Hara, C. Ciaglia, M. Yu, N. Yamane, G. Scharnweber, and D. Rhoden.** 1991. Compari-

55. **Pickett, M. J., J. R. Greenwood, and S. M. Harvey.** 1991. Tests for detecting degradation of gelatin: comparison of five methods. *J. Clin. Microbiol.* **29:**2322–2325.

56. **Planes, M., A. Ramirez, F. Fernandez, J. A. Capdevila, and C. Tolosa.** 1992. *Pseudomonas vesicularis* bacteremia. *Infection* **20:**367–368.

57. **Plorde, J. J., J. A. Gates, L. G. Carlson, and F. C. Tenover.** 1986. Critical evaluation of the AutoMicrobic system gram-negative identification card for identification of glucose-non-fermenting gram-negative rods. *J. Clin. Microbiol.* **23:**251–257.

58. **Pollack, M.** 1990. *Pseudomonas aeruginosa*, p. 1673–1691. *In* G. L. Mandell, R. G. Douglas, Jr., and J. E. Bennett (ed.), *Principles and Practice of Infectious Diseases*, 3rd ed. Churchill Livingstone, New York.

59. **Potvliege, C., J. Jonckheer, C. Lenclud, and W. Hansen.** 1987. *Pseudomonas stutzeri* pneumonia and septicemia in a patient with multiple myeloma. *J. Clin. Microbiol.* **25:**458–459.

60. **Prince, A.** 1992. Adhesions and receptors of *Pseudomonas aeruginosa* associated with infection of the respiratory tract. *Microb. Pathog.* **13:**251–260.

61. **Pruksachartvuthi, S., N. Aswapokee, and K. Thankerngpol.** 1990. Survival of *Pseudomonas pseudomallei* in human phagocytes. *J. Med. Microbiol.* **31:**109–114.

62. **Puckett, A., G. Davison, C. C. Entwistle, and J. A. J. Barbara.** 1992. Post-transfusion septicaemia 1980–1989: importance of donor arm cleansing. *J. Clin. Pathol.* **45:**155–157.

63. **Riley, P. S., and R. E. Weaver.** 1975. Recognition of *Pseudomonas pickettii* in the clinical laboratory: biochemical characterization of 62 strains. *J. Clin. Microbiol.* **1:**61–64.

64. **Sanford, J. P.** 1990. *Pseudomonas* species including melioidosis and glanders, p. 1692–1700. *In* G. L. Mandell, R. G. Douglas, Jr., and J. E. Bennett (ed.), *Principles and Practice of Infectious Diseases*, 3rd ed. Churchill Livingstone, New York.

65. **Schaberg, D. R., D. H. Culver, and R. P. Gaynes.** 1991. Major trends in the microbial etiology of nosocomial infection. *Am. J. Med.* **91**(Suppl. 3B):72S–75S.

66. **Scott, J., F. E. Boulton, J. R. W. Govan, R. S. Miles, D. B. L. McClelland, and C. V. Prowse.** 1988. A fatal transfusion reaction associated with blood contaminated with *Pseudomonas fluorescens*. *Vox Sang.* **54:**201–204.

67. **Scully, B. E., M. F. Parry, H. C. Neu, and W. Mandell.** 1986. Oral ciprofloxacin therapy of infections due to *Pseudomonas aeruginosa*. *Lancet* **i:**819–822.

68. **Simor, A. E., J. Ricci, A. Lau, R. M. Bannatyne, and L. Ford-Jones.** 1985. Pseudobacteremia due to *Pseudomonas fluorescens*. *Pediatr. Infect. Dis.* **4:**508–512.

69. **Smith, D. L., L. B. Gumery, E. G. Smith, D. E. Stableforth, M. E. Kaufmann, and T. L. Pitt.** 1993. Epidemic of *Pseudomonas cepacia* in an adult cystic fibrosis unit: evidence of person-to-person transmission. *J. Clin. Microbiol.* **31:**3017–3022.

70. **Snell, G. I., A. de Hoyos, M. Krajden, T. Winton, and J. R. Maurer.** 1993. *Pseudomonas cepacia* in lung transplant recipients with cystic fibrosis. *Chest* **103:**466–471.

71. **Sookpranee, T., M. Sookpranee, M. A. Mellencamp, and L. C. Preheim.** 1991. *Pseudomonas pseudomallei*, a common pathogen in Thailand that is resistant to the bactericidal effects of many antibiotics. *Antimicrob. Agents Chemother.* **35:**484–489.

72. **Speert, D. P., S. W. Farmer, M. E. Campbell, J. M. Musser, R. K. Selander, and S. Kuo.** 1990. Conversion of *Pseudomonas aeruginosa* to the phenotype characteristic of strains from patients with cystic fibrosis. *J. Clin. Microbiol.* **28:**188–194.

73. **Steinbach, S., L. Sun, R.-Z. Jiang, P. Gilligan, P. Flume, and R. Goldstein.** 1993. Epidemiology of *P. cepacia* infection at the molecular genetic level. I. Clonal progeny of pre-transplant strains as the source of post-surgical infection. *Pediatr. Pulmonol. Suppl.* **9:**260.

74. **Sun, L., R.-Z. Jiang, S. Steinbach, P. Gilligan, J. Fortsner, P. Flume, and R. Goldstein.** 1993. Epidemiology of *P. cepacia* infection at the molecular level. II. Evidence for variable transmissibility. *Pediatr. Pulmonol. Suppl.* **9:**260–261.

75. **Tablan, O. C., L. A. Carson, L. B. Cusick, L. A. Bland, W. J. Martone, and W. R. Jarvis.** 1987. Laboratory proficiency test results on use of selective media for isolating *Pseudomonas cepacia* from simulated sputum specimens of patients with cystic fibrosis. *J. Clin. Microbiol.* **25:**485–487.

76. **Tenover, F. C., T. S. Mizuki, and L. G. Carlson.** 1990. Evaluation of autoSCAN-W/A automated microbiology system for the identification of non-glucose-fermenting gram-negative bacilli. *J. Clin. Microbiol.* **28:**1628–1634.

77. **Vanholder, R., E. Vanhaecke, and S. Ringoir.** 1992. *Pseudomonas* septicemia due to deficient disinfectant mixing during reuse. *Int. J. Artif. Organs* **15:**19–24.

78. **Welch, D. F., M. J. Muszynski, C. H. Pai, M. J. Marcon, M. M. Hribar, P. H. Gilligan, J. M. Matsen, P. A. Ahlin, B. C. Hilman, and S. A. Chartrand.** 1987. Selective and differential medium for recovery of *Pseudomonas cepacia* from the respiratory tracts of patients with cystic fibrosis. *J. Clin. Microbiol.* **25:**1730–1734.

79. **Willems, A., J. De Ley, M. Gillis, and K. Kersters.** 1991. *Comamonadaceae*, a new family encompassing the acidovorans rRNA complex, including *Variovorax paradoxus* gen. nov., comb. nov., for *Alcaligenes paradoxus* (Davis 1969). *Int. J. Syst. Bacteriol.* **41:**445–450.

80. **Woods, M. L., II, B. J. Currie, D. M. Howard, A. Tierney, A. Watson, N. M. Anstey, J. Philpott, V. Asche, and K.** Withnall. 1992. Neurological melioidosis: seven cases from the Northern Territory of Australia. *Clin. Infect. Dis.* **15:**163–169.

81. **Wuthiekanun, V., D. Dance, W. Chaowagul, Y. Suputtamongkol, Y. Wattanagoon, and N. White.** 1990. Blood culture techniques for the diagnosis of melioidosis. *Eur. J. Clin. Microbiol. Infect. Dis.* **9:**654–658.

82. **Wuthiekanun, V., D. A. B. Dance, Y. Wattanagoon, Y. Supputtamongkol, W. Chaowagul, and N. J. White.** 1990. The use of selective media for the isolation of *Pseudomonas pseudomallei* in clinical practice. *J. Med. Microbiol.* **33:**121–126.

83. **Yabuuchi, E., Y. Kosako, H. Oyaizu, I. Yano, H. Hotta, Y. Hashimoto, T. Ezaki, and M. Arakawa.** 1992. Proposal of *Burkholderia* gen. nov. and transfer of seven species of the genus *Pseudomonas* homology group II to the new genus, with the type species *Burkholderia cepacia* (Palleroni and Holmes 1981) comb. nov. *Microbiol. Immunol.* **36:**1251–1275.

84. **Yabuuchi, E., I. Yano, H. Oyaizu, Y. Hashimoto, T. Ezaki, and H. Yamamoto.** 1990. Proposals of *Sphingomonas paucimobilis* gen. nov. and comb. nov., *Sphingomonas parapaucimobilis* sp. nov., *Sphingobacterium yanoikuyae* sp. nov., *Sphingobacterium adhaesiva* sp. nov., *Sphingomonas capsulata* comb. nov., and two genospecies of the genus *Sphingomonas*. *Microbiol. Immunol.* **34:**99–119.

85. **Yamamoto, T., P. Naigowit, S. Dejsirilert, D. Chiewsilp, E. Kondo, T. Yokota, and K. Kanai.** 1990. In vitro susceptibilities of *Pseudomonas pseudomallei* to 27 antimicrobial agents. *Antimicrob. Agents Chemother.* **34:**2027–2029.

Acinetobacter, Alcaligenes, Moraxella, and Other Nonfermentative Gram-Negative Bacteria

ALEXANDER von GRAEVENITZ

41

The organisms covered in this chapter belong to the group of taxonomically diverse nonfermentative gram-negative rods. Chemotaxonomy and phylogeny, however, are deemphasized in favor of features that can be used for routine laboratory diagnosis. Characteristically, these organisms fail to acidify the butt of Kligler or triple sugar iron agar or of oxidative-fermentative media and grow significantly better under aerobic than under anaerobic conditions; many strains even fail to grow anaerobically. Some of these "nonfermenters" are covered in the chapters in this Manual on the genera *Pseudomonas, Bordetella, Brucella,* and *Neisseria,* and on unusual gram-negative bacteria. The rest are initially characterized by routine tests, i.e., microscopic morphology, oxidase, acidification of carbohydrates, indole production, and motility. With the exception of the animal pathogen *Moraxella bovis,* all these bacteria are catalase positive (the catalase-negative *Neisseria elongata* is discussed in chapter 26 of this Manual).

Methods used to grow and identify this group are those used for *Pseudomonas* spp. (see chapter 40). Initial incubation should be at 35 to 37°C, although many pink-pigmented strains grow only at ≤30°C and can be detected only on plates left at room temperature after the initial readings. Growth on certain selective primary media (e.g., MacConkey or salmonella-shigella agar) is variable; there can be significant lot-to-lot variations in the media. Nonfermenters growing on MacConkey agar generally form colorless colonies.

Identification media are best incubated at 30°C unless growth does not ensue. Tables 1 to 8 do not give a complete repertory of biochemical reactions; additional data can be found in other publications (10, 23). Traditional diagnostic systems, e.g., those based on oxidative-fermentative media, aerobic low peptone media, or buffered single substrates (26), have now been replaced in many laboratories by commercial kits. If such kits are used, the laboratory must be familiar with the extent of the database; organisms not included will, of course, not be identified correctly. Because assimilation test results often depend on the basal medium used, most of those results are not included in the tables. The minimal basal medium mentioned is the one used by Gilardi (23, 64). The indole test should be the one recommended by Pickett with a modified Kovács reagent and a buffered tryptophan medium (64, 70).

Identification of nonfermenters by automated fatty acid analysis has also been attempted (88; see chapter 12 of this Manual). In view of the difficulties inherent in this approach (65), I recommend that fatty acid profiles be used only in conjunction with traditional or commercial diagnostic systems. The main fatty acids of each species are listed in the tables.

OXIDASE-NEGATIVE GROUP

See Table 1.

Acinetobacter spp.

The genus *Acinetobacter* consists of oxidase-negative, nonmotile, nitrate-negative (3% positive) nonfermentative gram-negative rods that are 1 to 1.5 by 1.5 to 2.5 μm in size, sometimes difficult to decolorize, and frequently arranged in pairs. In the stationary growth phase and on nonselective agars, coccobacillary forms predominate, while early growth in fluid media and growth on plates containing cell wall-active antimicrobial agents yield mostly rods. Colonies are smooth, opaque, and slightly smaller than those of members of the family *Enterobacteriaceae.* Many strains grow on MacConkey agar as either colorless or slightly pinkish colonies. Seven named species and 12 genospecies (hybridization groups) have been outlined (3, 5, 84); they replace the old biovars *Acinetobacter anitratus, Acinetobacter lwoffii, Acinetobacter haemolyticus,* and *Acinetobacter alcaligenes.* They can be differentiated by means of biochemical and growth tests (Table 2). Genospecies 1, 2, 3, and 13 of Tjernberg and Ursing (84) may be difficult to separate in the clinical laboratory and have been referred to as the *Acinetobacter calcoaceticus-Acinetobacter baumannii* complex (21). Most glucose-oxidizing nonhemolytic clinical strains are *Acinetobacter baumannii,* most glucose-negative nonhemolytic ones are *Acinetobacter lwoffii,* and most hemolytic ones are *Acinetobacter haemolyticus.* An assay transforming a stable competent auxotroph to prototrophy (44) can confirm the genus diagnosis vis-à-vis group NO-1 but needs at least 24 h to do so. Selective media have been designed (39).

Acinetobacter species are widely distributed in nature and in the hospital environment, are the second most commonly isolated nonfermenters (*Pseudomonas aeruginosa* being the first), are able to survive on moist and dry surfaces (22), and may be present on healthy human skin. The species most frequently isolated is *Acinetobacter baumannii*

TABLE 1 Oxidase-negative, indole-negative, nonfermentative gram-negative rods[a]

Test[b]	Acinetobacter spp.	Group NO-1	Chryseomonas luteola	Flavimonas oryzihabitans	Xanthomonas (Stenotrophomonas) maltophilia
Motility	−	−	+	+	+
Pigmentation	−	−	y	y	br-tan
Growth:					
On MacConkey agar	V	V	+	+	+
At 42°C	V	V	V	V	V
On MBM + acetate	+	ND	+	+	−
Nitrate to nitrite	−	+	V	−	V
Esculin hydrolysis	−	−	+	−	+
Gelatin hydrolysis	V	−	V	−	+
Urease (Christensen)	V	−	V	V	−
Arginine dihydrolase	−	−	V	−	−
DNase	−	ND	−	−	+
ONPG	−	−	+	−	+
Starch hydrolysis	−	ND	V	V	−
Acid from:					
Glucose	V	−	+	+	+
Maltose	V	−	+	+	+
Sucrose	−	−	−	V	V
Mannitol	−	−	+	+	−
Xylose	V	−	+	+	V
CFA[c]	C18:1ω9c, C16:1ω7c, C16:0[d] (46)	C16:0, C16:1ω7c, C18:1ω7c (29)	C18:1ω7c, C16:1ω7c, C16:0 (52)	C18:1ω7c, C16:1ω7c, C16:0 (52)	i-C15:0, ai-C15:0, C16:1ω9c (99)

[a]Data are from references 10, 23, 29, and 52. For reactions of *Bordetella parapertussis*, see chapter 46 of this Manual. Symbols and abbreviations: +, ≥90% of strains positive; −, ≥90% of strains negative; ND, no data available; V, variable; br, brown; y, yellow.

[b]MBM, minimal basal medium; ONPG, o-nitrophenyl-β-D-galactopyranoside; CFA, main cellular fatty acids.

[c]Numbers in parentheses are references.

[d]*Acinetobacter lwoffii* and genospecies 15 (84), in contrast to other genospecies, do not have 2-OH C12:0 (46).

(with 19 biotypes identified by assimilation tests [4] and 34 serovars [85]), followed by *Acinetobacter lwoffii*, *Acinetobacter haemolyticus*, *Acinetobacter johnsonii*, genospecies 3 (with 26 serovars [85]), and genospecies 6 (46, 79, 84). Disease, most often hospital-acquired and most frequently observed during the warm season, may involve the respiratory tract, urinary tract, and wounds (including catheter sites) and may progress to septicemia (24). Risk factors are antibiotic treatment and/or surgery, instrumentation, and stay in intensive care units; clinical isolates, however, are more often colonizers than infecting agents (24). Hospital outbreaks have been investigated by various typing methods (17, 85). Cephalothin is ineffective and trimethoprim-sulfamethoxazole, imipenem, amoxicillin-clavulanic acid, and doxycycline are effective against most strains (24, 79), but susceptibility testing is required for each clinically significant strain. Multiply resistant strains occur mostly in the *Acinetobacter calcoaceticus-Acinetobacter baumannii* complex and in *Acinetobacter haemolyticus* (79, 97).

Group NO-1

Group NO-1 bacteria (29) are oxidase-negative, asaccharolytic, nonmotile, coccoid to medium-sized gram-negative rods forming small colonies on sheep blood agar (SBA) that can be transferred in toto by a needle. Other differential features are shown in Table 1. Since approximately 3% of asaccharolytic *Acinetobacter* spp. reduce nitrate, the definite distinguishing test between the two groups is the transformation assay (44) which is negative with group NO-1. Molecular genetic data are not extant at this time. The cellular fatty acid composition also differs from that of *Acinetobacter* (Table 1). Most strains have been isolated from dog or cat bites and are susceptible to antibiotics used for gram-negative infections.

Chryseomonas luteola and Flavimonas oryzihabitans

C. luteola and *F. oryzihabitans*, formerly called Ve-1 and Ve-2, respectively (17, 52), are oxidase-negative, short to medium-length, thick, sometimes paired nonfermentative gram-negative rods with polar flagella that produce either wrinkled and adherent or (more rarely) smooth colonies exhibiting an intracellular, water-insoluble yellow pigment. They are closely related to the genus *Pseudomonas* but differ from each other in number of flagella (more than one versus only one) and several biochemical characteristics (Table 1). Both species are found in nature and in the hospital environment and have been associated with catheter infections, septicemia, peritonitis associated with continuous ambulatory peritoneal dialysis (CAPD), and mixed infections at other sites (27). They are generally susceptible to most broad-spectrum cephalosporins, aminoglycosides, tetracycline, and quinolones (18, 23).

TABLE 2 Phenotypic characteristics of 17 *Acinetobacter* genospecies

Test	% Positive strains in genospecies[a]															
	1 (8)	2 (121)	3 (15)	4 (23)	5 (17)	6 (3)	7 (23)	8/9 (34)	10 (4)	11 (4)	12 (3)	13 (9)	14 (3)	15 (2)	16 (4)	17 (2)
Growth at:																
37°C	100	100	100	100	100	100	0	100	100	100	100	89	100	100	75	100
41°C	0	100	100	0	90	0	0	0	0	0	0	0	0	0	0	0
44°C	0	100	0	0	0	0	0	0	0	0	0	0	0	0	0	0
Gelatin hydrolysis[b]	0	0	0	96	0	100	0	0	0	0	0	0	0	0	0	0
Acid from glucose[b]	100	95	100	52	0	66	0	6	100	0	33	100	100	100	100	100
Hemolysis on SBA	0	0	0	100	0	100	0	0	0	0	0	0	100	100	100	100
Utilization of[b]:																
trans-Aconitate	100	99	100	52	0	0	0	0	0	0	0	11	67	0	0	50
β-Alanine	100	95	94	0	0	0	0	0	100	100	0	0	100	0	75	100
DL-4-Aminobutyrate	100	100	100	100	88	0	35	40	100	100	100	11	100	0	25	100
L-Arginine	100	98	100	96	95	100	35	0	0	0	100	100	100	50	100	100
Azelate	100	90	100	0	0	0	0	100	50	25	100	0	100	0	0	0
Citrate	100	100	100	91	82	100	100	0	100	100	0	100	100	50	100	100
Glutarate	100	100	100	0	0	0	0	0	100	100	100	0	100	0	0	100
L-Histidine	100	98	94	96	100	100	0	0	100	100	0	100	100	100	100	100
DL-Lactate	100	100	100	0	100	0	100	100	100	100	100	100	100	100	100	100
D-Malate	0	98	100	96	100	66	22	76	100	100	0	100	66	50	100	100
Malonate	100	98	87	0	0	0	13	0	0	0	100	11	100	0	50	50

[a]Data are from Bouvet and Grimont (3) for genospecies 1 to 12 and from Bouvet and Jeanjean (5) for genospecies 13 to 17. See those papers for methodology. *Acinetobacter* genospecies: 1, *A. calcoaceticus*; 2, *A. baumannii*; 3, unnamed; 4, *A. haemolyticus*; 5, *A. junii*; 6, unnamed; 7, *A. johnsonii*; 8 and 9 (phenotypically inseparable), *A. lwoffii*; 10 and 11, unnamed; 12, *A. radioresistens*; 13 to 17, unnamed. Genospecies 13 and 15 of Tjernberg and Ursing (84) do not correspond to genospecies 13 and 15 of Bouvet and Jeanjean (5), while genospecies 14 of the former authors corresponds to genospecies 13 of the latter authors (46). Numbers in parentheses after genospecies numbers indicate numbers of strains.

[b]Incubation at 30°C. This table lists only 11 utilization tests, while Gerner-Smidt et al. (21) recommend 14 tests for diagnostic purposes.

Xanthomonas (Stenotrophomonas) maltophilia

The transfer of the former species *Pseudomonas maltophilia* to the genus *Xanthomonas* was proposed on the basis of DNA-rRNA hybridizations, DNA guanine-plus-cytosine content, type of ubiquinones, cellular fatty acid composition, comparative enzymology, growth parameters, phage typing, and ecological niches (82). In contrast to other *Xanthomonas* spp., *X. maltophilia* shows lophotrichous flagellation, absence of xanthomonadins (yellow aryl-polyene pigments) and plant pathogenicity, and growth at 37°C. For this and other reasons, a new genus, *Stenotrophomonas*, with the single species *Stenotrophomonas maltophilia*, has recently been proposed (67).

X. (S.) maltophilia is a short to medium-sized straight gram-negative rod with a polar tuft of flagella. It forms large, smooth, glistening colonies with uneven margins and lavender-green to light purple pigmentation. In heart infusion agar with tyrosine, a water-soluble brown pigment, and on blood agar, a greenish discoloration appears underneath the growth. The smell of ammonia may be very pronounced. With a few exceptions (42), most strains require methionine (or cystine plus glycine) for growth. The organism does not produce oxidase; oxidizes maltose faster than glucose; hydrolyzes esculin, DNA, gelatin, and Tween 80; and decarboxylates lysine (Table 1). A selective medium containing imipenem has been described (92).

X. (S.) maltophilia is ubiquitous in nature and has also been isolated from the hospital environment. It is the third most frequently isolated nonfermentative gram-negative rod in the clinical laboratory. Strains may be colonizers (e.g., in cystic fibrosis) or infecting agents. Risk factors for acquisition of both are stay in intensive care units, mechan-ical ventilation, previous antimicrobial treatment (89), and, possibly, malignancies (63). Septicemia (often associated with intravenous catheters), pneumonia, wound infection, and, rarely, other infections (endocarditis, meningitis) have been reported (55). In hospital outbreaks, typing by serology (89), multilocus enzyme electrophoresis (78), and restriction fragment length polymorphisms (1) have been applied. *X. (S.) maltophilia* is frequently resistant to antimicrobial agents. National Committee for Clinical Laboratory Standards criteria regarding media, temperature, and zone sizes have to be applied strictly when susceptibility testing is done. Even then, disk interpretative errors with ciprofloxacin, aminoglycosides, and beta-lactams may be considerable (20, 28, 56, 57). Most strains are still susceptible to trimethoprim-sulfamethoxazole, moxalactam, doxycycline, and chloramphenicol; a minority of strains are susceptible to other antibiotics (18, 23).

OXIDASE-POSITIVE, INDOLE-NEGATIVE, ASACCHAROLYTIC COCCOID NONFERMENTERS

See Table 3.

Moraxella spp.

Members of the genus *Moraxella* are oxidase-positive, nonmotile, asaccharolytic coccobacilli that are often plump, occur predominantly in pairs and sometimes in short chains, and have a tendency to resist decolorization.

Moraxellae are parasitic on human skin and mucous membranes. The most frequently isolated species is *Moraxella nonliquefaciens*, which forms smooth, translucent to

TABLE 3 Oxidase-positive, indole-negative, coccoid asaccharolytic gram-negative rods[a]

Test	Moraxella lacunata	Moraxella nonliquefaciens	Moraxella canis	Moraxella lincolnii	Moraxella osloensis	Moraxella atlantae	Moraxella phenylpyruvica	Oligella urethralis	Oligella ureolytica
Motility	–	–	–	–	–	–	–	–	+[b]
Catalase	+	+	+	+	+	+	+	–	+
Hemolysis (SBA)	α/–	–	–	–	–	–	ly/–	ly/–	ly/–
Growth									
On MacConkey agar	–	–	+	–	V	+	V	V	V
At 42°C	–	V	ND	ND	D	V	V	+	–
On MBM + acetate	–	–	+	ND	+	V	V	V	+
Urease	D	–	–	ND	–	–	+	–	+
Phenylalanine deaminase (agar)	+[b]	–	–	–	–	–	+	+	+
Gelatin hydrolysis	+	–	+	–	–	–	–	–	–
Nitrate to nitrite	+	+	V	–	V	–	V	–	+
Nitrite reduction	–	–	–	V	–	V	–	+	+
DNase	–	–	+	–	–	–	–	–	–
CFA (43, 60, 76, 87)	C18:1ω9c, C16:0, C16:1ω7c	C18:1ω9c, C16:1ω7c	C18:1ω9c, C16:1ω7c	C18:1ω9c, C16:1ω7c, C16:0	C18:1ω9c, C18:0, C16:1ω7c	C16:0, C18:2, C18:0, C18:1ω9c	C18:2, C18:1ω9c, C16:0, C16:1ω7c	C18:1ω7c, C16:0	C18:1ω7c, C16:0, C19:0cyc

[a]Data are from references 6, 10, 23, 43, 60, 70a, 76, and 87. For reactions of *Brucella* spp., see chapter 44 of this Manual. Abbreviations: MBM, minimal basal medium (see text); CFA, cellular fatty acids; ly, diffuse lysis underneath colonies on SBA; D, divergent data in the literature; ND, no data; V, variable.
[b]May be delayed or difficult to demonstrate.

TABLE 4 Oxidase-positive, indole-negative, asaccharolytic, noncoccoid gram-negative rods[a]

Test	Alcaligenes piechaudii[b]	Alcaligenes faecalis[c]	"Alcaligenes faecalis type II"	Alcaligenes xylosoxidans subsp. denitrificans	Comamonas acidovorans[d]
Growth on:					
MacConkey agar	+	+	+	+	+
MBM + acetate	V	V	V	V	+
Nitrate to nitrite	+	−	−	+	+
Nitrite to gas	−	+	−	+	−
Urease	−	−	−	−	−
Flagella	pe	pe	pe	pe	>1 po
Utilization of mesaconate and itaconate	+	−	−	V	+
Acetamide hydrolysis	V	+	−	V	+
Phenylalanine deaminase	−	−	V	V	+
CFA[f]	ND	C16:0, C17:0cyc, C16:1ω7c, C18:1ω7c (14)	C16:0, C17:0cyc, C16:1ω7c, C18:1ω7c (14)	C16:0, C17:0cyc, C16:1ω7c (14)	C16:1ω7c, C16:0, C18:1ω7c (95)

[a]Data are from references 10, 16, 23, 51, 58, 95, and 96. For reactions of *N. elongata*, *N. weaveri*, asaccharolytic *Pseudomonas* spp., and *Bordetella* spp., see chapters 26, 40, and 46 of this Manual. Abbreviations: MBM, minimal basal medium; CFA, cellular fatty acids; ND, no data; V, variable; pe, peritrichous; po, polar.
[b]Formerly *Alcaligenes faecalis* type I.
[c]Formerly *Alcaligenes odorans* (see text).
[d]Produces acid from fructose and mannitol.
[e]The only species listed in this table that is DNase positive and resistant to polymyxin B and that fairly regularly forms a yellow insoluble pigment.
[f]See Table 1, footnote c.
[g]Less C17:0cyc and 3-OH C10:0 than in *Comamonas* spp., otherwise like *Comamonas* spp.

semiopaque colonies 0.1 to 0.5 mm in diameter after 24 h and 1 mm in diameter after 48 h of growth on blood agar plates. Occasionally, these colonies spread and pit the agar. The colonial morphologies of *Moraxella lincolnii* (87), *Moraxella osloensis*, and *Moraxella phenylpyruvica* are similar, but pitting is rare. It is, on the other hand, common in *Moraxella lacunata*, whose colonies are smaller and form dark haloes on chocolate agar. Colonies of *Moraxella atlantae* are small (usually 0.5 mm in diameter) and show pitting and spreading (6). Most *Moraxella canis* colonies resemble those of the *Enterobacteriaceae* (43). Microscopically, *M. canis* resembles *Moraxella catarrhalis* (*Branhamella catarrhalis*), which is discussed in chapter 26 of this Manual.

Biochemical reactions are listed in Table 3. Most laboratories do not determine the species of moraxellae because of the similarity in pathogenic significance of the species. A separation between *M. lacunata* and nonspreading *M. nonliquefaciens* may prove difficult, because gelatin hydrolysis (with any method) and liquefaction of Loeffler slants may take more than 1 week (70a). In some instances, fatty acid analysis may help determine the species (60; Table 3); in other cases, quantitative transformation of a high-level streptomycin resistance marker can be used (45). The differential diagnosis of *M. phenylpyruvica* and *Brucella* spp. is of great practical importance (68) and requires microscopy and tests for phenylalanine deaminase and acid formation from xylose. The former is positive for *M. phenylpyruvica* and the latter is positive for *Brucella* spp. The tributyrin test may be positive for several *Moraxella* spp. and therefore cannot be used to separate them from *M. (B.) catarrhalis* (69). Likewise, γ-glutamyl aminopeptidase occurs not only in *M. canis* but also in some strains of other moraxellae (43). *N. weaveri* is nitrite and phenylalanine positive.

At the least, the species *M. osloensis*, *M. nonliquefaciens*, and *M. lincolnii* are part of the normal flora of the human respiratory tract. Moraxellae are rare agents of infections (conjunctivitis, meningitis, septicemia, endocarditis, arthritis, and otolaryngologic infections) (25). Most strains are susceptible to penicillin and its derivatives, cephalosporins, tetracyclines, and aminoglycosides (18, 23).

Oligella spp.

The genus *Oligella*, a group of small coccoid gram-negative rods, includes the former *Moraxella urethralis* and CDC group IVe (now *Oligella ureolytica*) (76). *Oligella urethralis* is nonmotile, while most strains of *O. ureolytica* are motile by peritrichous flagella. Colonies are smaller than those of *M. osloensis* and are opaque to whitish. Among the biochemical characteristics of *O. ureolytica*, positive phenylalanine and urease reaction are conspicuous. In *O. ureolytica*, the urease reaction often turns positive within minutes after inoculation. Both organisms have been isolated chiefly from the human urinary tract and occasionally from other sources (10, 71). While *O. urethralis* is generally susceptible to most antibiotics, including penicillin, *O. ureolytica* exhibits variable susceptibility patterns (18, 23).

OXIDASE-POSITIVE, INDOLE-NEGATIVE, ASACCHAROLYTIC TO SACCHAROLYTIC ROD-SHAPED NONFERMENTERS
See Table 4.

Alcaligenes spp.

Members of the genus *Alcaligenes* are rods (0.5 by 1 to 0.5 by 2.6 μm) with peritrichous flagella. Both phylogenetically and biochemically, they are closely related to members of the genus *Bordetella* (see chapter 46 of this Manual). They occur mainly in the environment (91) and show limited

TABLE 4 (Continued)

Test	Comamonas terrigena	Comamonas testosteroni	Acidovorax delafieldii	Flavobacterium odoratum[e]	CDC group IVc-2	Gilardi rod group 1
Growth on:						
MacConkey agar	+	+	ND	+	+	+
MBM + acetate	−	+	V	−	+	−
Nitrate to nitrite	+	+	+	−	V	−
Nitrite to gas	−	−	V	V	−	−
Urease	−	−	−	+	+	−
Flagella	>1 po	>1 po	1 po	−	pe	
Utilization of mesaconate and itaconate	V	V	−	ND	ND	ND
Acetamide hydrolysis	−	−	−	−	−	−
Phenylalanine deaminase	V	V	ND	V	−	+
CFA[f]	C16:1ω7c, C16:0, C18:1ω7c (95)	C16:1ω7c, C16:0, C18:1ω7c (95)	—[g]	i-C15:0 (16)	C16:1ω7c, C16:0, C18:1ω7c, C17:0cyc (60a)	C18:1ω7c, C14:0, C16:0, C19:0cyc (58)

action on carbohydrates. Colonies are nonpigmented and similar in size to those of *Acinetobacter* spp. Medically important species are (i) the asaccharolytic species, including *Alcaligenes faecalis*, *Alcaligenes piechaudii* (51), and *Alcaligenes xylosoxidans* subsp. *denitrificans* (formerly *Alcaligenes denitrificans* subsp. *denitrificans*, *Achromobacter denitrificans*, and *Alcaligenes denitrificans*), and (ii) the saccharolytic species *Alcaligenes xylosoxidans* subsp. *xylosoxidans* (formerly *Alcaligenes denitrificans* subsp. *xylosoxidans* and *Achromobacter xylosoxidans*). *Alcaligenes faecalis* consists of two colonial types: the former "*Alcaligenes odorans*," with whitish spreading colonies that have a fruity smell and cause strong greening of the underlying blood agar, and "*Alcaligenes faecalis* type II," which lacks those characteristics but mostly forms colonies with a spreading edge. The asaccharolytic species are rarely observed as human pathogens (51, 91), but *Alcaligenes xylosoxidans* subsp. *xylosoxidans* is a relatively frequent agent of infection, particularly of septicemia in nosocomial settings (9, 91). Susceptibilities are unpredictable except for susceptibility to piperacillin and ticarcillin-clavulanic acid (2, 18, 23, 91).

Comamonas and Acidovorax spp.

The genus *Comamonas* includes the species *Comamonas* (formerly *Pseudomonas*) *acidovorans*, *Comamonas testosteroni*, and *Comamonas terrigena* (96); the genus *Acidovorax* includes the species *Acidovorax* (formerly *Pseudomonas*) *delafieldii*, *Acidovorax* (P.) *facilis*, and *Acidovorax* (P.) *temperans* (95). Both genera are mainly environmental. They are polarly flagellated. Only a few species oxidize carbohydrates. Their cells accumulate poly-β-hydroxybutyrate. *Comamonas* species are occasionally spirillar and have polar tufts of one to six flagella with a mean wavelength of 3.1 μm, while *Acidovorax* spp. are monotrichous with a smaller wavelength. *Comamonas* spp. often and *Acidovorax* spp. always reduce nitrate to nitrite (Table 4). Further differences relate to cellular fatty acids and utilization tests (95, 96).

Comamonas terrigena requires as growth factors methionine and nicotinamide. *Comamonas acidovorans*, the only

saccharolytic species, may produce a soluble, yellow to tan, or, rarely, a weakly fluorescing pigment and oxidizes fructose and mannitol but no other carbohydrates. In tryptone broth, anthranilic acid and kynurenine are produced; they yield an orange color with Kovács reagent. *Comamonas testosteroni* biochemically resembles *Comamonas terrigena* except for certain utilization tests (e.g., acetate) (96). Reports of infection with comamonads are rare (74). *Comamonas acidovorans* is generally resistant to more antimicrobial agents (particularly the aminoglycosides) than are the other two species. All species are at present susceptible to piperacillin, cefoxitin, cefotaxime, moxalactam, imipenem, and ciprofloxacin (18, 23).

Acidovorax spp. have been isolated very rarely and apparently never with clinical significance from human samples. Gilardi's description of one single clinical isolate of *Acidovorax delafieldii* (23) differs substantially from the description by Willems et al. (95) of 14 clinical isolates that did not produce acids from carbohydrates in oxidative-fermentative media and possessed few of the enzymes listed for the single strain. Differences between *Acidovorax* species refer chiefly to utilization of organic acids, amino acids, and sugars (95). Susceptibility studies are not available.

CDC Group IVc-2

The CDC group IVc-2 short to medium-sized gram-negative rod is asaccharolytic and motile by peritrichous flagella. Cells may stain irregularly. The urease test may rapidly turn positive. Reactions resemble those of *Bordetella bronchiseptica* (Table 4) except for nitrate reduction; also, colonies of IVc-2 are larger. The organism has been isolated from a variety of human sources (10) and as an agent of septicemia and peritonitis (62, 72); it is often resistant to ampicillin, cephalothin, and aminoglycosides (18, 23).

Gilardi Rod Group 1

Gilardi rod group 1 organisms, which are medium-length asaccharolytic gram-negative rods, resemble *Neisseria weaveri* (CDC group M-5) in many respects except that most of the former grow on MacConkey agar, do not reduce nitrite,

TABLE 5 Oxidase-positive, indole-negative, saccharolytic nonfermentative gram-negative rods[a]

Test	Agrobacterium radiobacter	Alcaligenes xylosoxidans subsp. xylosoxidans	EO-2	EO-3	Psychrobacter immobilis	Ochrobactrum anthropi	Achromobacter group(s)[b]		Flavobacterium multivorum
							B and E	F	
Growth on:									
MacConkey agar	+	+	V	+	+	+	+	–	D
MBM + acetate	+	+	+	ND	V	V	ND	ND	–
Pigment	–	–	y/–	y	–	–	–	–	sl y/–
DNase	–	–	–	ND	–	–	–	–	V
ONPG	+	–	V	ND	–	–	–	–	+
Arginine dihydrolase	–	V	–	–	–	V	+	–	–
Flagella	pe	pe	–	–	–	pe	B–, E+	–	–
Nitrate to nitrite	+	+	V	–	V	+	pe	pe	–
Nitrate to gas	–	V	–	–	–	+	+	+	–
Urease	+	–	V	+	V	+	B+, E–	+	+
Esculin hydrolysis	+	–	–	–	–	V	+	+	+
Gelatin hydrolysis	–	–	–	–	–	V	+	+	+
Acid from:									
Glucose	+	+	+	+	V	+	+	+	+
Maltose	+	–	V	V	–	V	+	+	+
Sucrose	+	–	–	–	–	V	+	+	+
Mannitol	+	–	V	+w	–	V	B+, E–	+	–
Xylose	+	+	+	+	+	+	+	+	+
Polymyxin B	S	S	S	ND	ND	S	ND	ND	R
CFA[d]	C18:1ω7c, C19:0cyc (77)	C17:0cyc, C16:0 (15)	C18:1ω7c (60)	C18:1ω7c (60)	C18:1ω9c (60)	C18:1ω7c, C19:0cyc, C18:0 (34)	C18:1ω7c (34)	ND	i-C15:0, i-2-OH-C15:0, C16:1Δ9 (12, 83)

[a] Data are from references 10, 23, 30, 35, 36, 39, 40, 50, and 60. For reactions of saccharolytic *Pseudomonas* spp., see chapter 40 of this Manual; for EF-4b, see chapter 39; for *Shewanella*, *Methylobacterium*, and *Roseomonas* spp., see Tables 6 and 7. For abbreviations, see Table 1, footnote *a*, and Table 4, footnote *a*. Other abbreviations: sly, slightly; S, susceptible; R, resistant (no zone around 10-μg disk); w, weak.
[b] See text for differences (35).
[c] Reference 83.
[d] Approximately 6% of strains are oxidase negative (23).
[e] See Table 1, footnote *c*.

and are phenylalanine deaminase positive (Table 4; 58). Strains of this group have been isolated from a variety of human sources and are susceptible to many antimicrobial agents (23).

OXIDASE-POSITIVE, INDOLE-NEGATIVE, SACCHAROLYTIC NONFERMENTERS
See Table 5.

Agrobacterium radiobacter

Human strains of *Agrobacterium* belong to the species *Agrobacterium radiobacter*, which, unlike *Agrobacterium tumefaciens*, does not carry a tumor-inducing plasmid causing crown gall or hairy root disease in plants. Cells are 0.6 to 1.0 by 1.5 to 3.0 μm long and occur singly and in pairs. Colonies on 48-h SBA have diameters of 2 mm. Agrobacteria are motile by peritrichous (occasionally monotrichous) flagella and are active sugar oxidizers. Most human strains belong to biovar 1, which is distinguished by its ability to oxidize lactose to 3-ketolactose (10). They have caused infections associated with plastic material (intravenous and peritoneal catheters) and, rarely, septicemia (41). Most strains are susceptible to broad-spectrum cephalosporins, carbapenems, tetracyclines, and gentamicin but not to tobramycin (23, 91).

Groups EO-2 and EO-3 and Psychrobacter immobilis

The classification of groups EO-2 and EO-3 and *Psychrobacter immobilis* is incomplete. They are nonmotile saccharolytic coccobacilli and grow, sometimes poorly, on MacConkey agar. In contrast to the two EO (eugonic oxidizer) groups, *Psychrobacter immobilis* grows best at 20°C and only occasionally at 37°C; cultures smell like phenylethyl alcohol agar (roses). Microscopically, EO-2 is characterized by frequently vacuolated O-shaped cells, and *Psychrobacter immobilis* is characterized by paired organisms. *Psychrobacter immobilis* can be confirmed by transformation studies (60). These three groups have all been isolated from clinical specimens, and *Psychrobacter immobilis* has also been isolated from food (10, 40, 60). Clinical data on EO-2 and EO-3 are not extant; there are a few reports of disease due to *Psychrobacter immobilis* (53).

Ochrobactrum anthropi and "Achromobacter" Groups B, E, and F

The organism formerly called CDC group Vd (biotypes 1 and 2) is the only species in the genus *Ochrobactrum*, *Ochrobactrum anthropi* (36). It is a medium-length rod with peritrichous flagella, but an individual cell may have a single flagellum only. Colonies on SBA resemble those of

TABLE 5 (Continued)

Test	Flavobacterium spiritivorum	Flavobacterium thalpophilum	Flavobacterium mizutaii	Flavobacterium yabuuchiae[c]	WO-1	Pseudomonas-like group 2	Sphingomonas paucimobilis[d]
Growth on:							
MacConkey agar	+	+	−	+	V	+	V
MBM + acetate	−	−	−	ND	ND	+	+
Pigment	sl y/−	sl y/−	sl y	sl y	y/−	ND	y
DNase	+	+	−	+	ND	−	−
ONPG	+	+	+	+	ND	+	+
Arginine dihydrolase	−	−	−	−	−	−	−
Flagella	−	−	−	−	≥1 po	≥1 po	1 po
Nitrate to nitrite	−	+	−	−	+	−	−
Nitrate to gas	−	−	V	−	V	−	−
Urease	+	+	V	+	V	+	−
Esculin hydrolysis	+	+	+	+	−	−	+
Gelatin hydrolysis	−	V	V	+	−	−	−
Acid from:							
Glucose	+	+	+	+	+[D]	+	+
Maltose	+	+	+	+	−	−	+
Sucrose	+	+	+	+	−	−	+
Mannitol	+	−	−	+	+[D]	+	−
Xylose	+	+	+	+	−	+	+
Polymyxin B	R	R	R	ND	ND	V	V
CFA[e]	i-C15:0, i-2-OH-C15:0, C16:1Δ9 (12, 83)	i-C15:0, i-2-OH-C15:0, C16:1Δ9 (12, 83)	i-C15:0, i-2-OH-C15:0, C16:1Δ9 (12, 83)	i-C15:0, i-2-OH-C15:0, C16:1Δ9 (12, 83)	C16:1ω7c, C16:0, C18:1ω7c (30)	C16:0, C16:1Δ9, C18:1 (13)	C18:1ω7c, C16:0, 2-OHC14:0 (98)

Enterobacteriaceae except that the former are smaller. The organism is identical to "*Achromobacter*" groups A, C, and D of Holmes et al. (35). *Ochrobactrum anthropi* has been isolated from various environmental and human sources, predominantly from patients with catheter-related bacteremias (36, 48). Multiple antibiotic resistance is the rule, with most strains susceptible only to trimethoprim-sulfamethoxazole, fluoroquinolones, aminoglycosides, and imipenem (23, 48, 91).

"*Achromobacter*" groups B and E constitute biotypes of a single species whose taxonomy has yet to be fully investigated (32). They are difficult to separate biochemically from *Ochrobactrum anthropi*. According to Holmes et al. (35), groups A, C, and D (identical to *Ochrobactrum anthropi*) are esculin positive, but other authors (10) give variable esculin reactions for *Ochrobactrum anthropi*. Group F, although difficult to separate phenotypically from *Ochrobactrum anthropi* and groups B and E, is genotypically different from them (32). *Achromobacter* group B has been isolated from patients with septicemia; susceptibility to chloramphenicol, ciprofloxacin, gentamicin, and tobramycin (but not amikacin) was observed in four strains (33).

Group WO-1

Members of group WO-1, reported recently (30), show one or two polar flagella and oxidize glucose weakly and perhaps late. Other characteristics are listed in Table 5. Human isolates have come from blood, urine, cerebrospinal fluid, and other sources (30). No genetic and susceptibility data have been published.

Pseudomonas-Like Group 2

The organisms in *Pseudomonas*-like group 2, included in the heterogeneous group "IVd" (50), have polar tufts of flagella and are phenylalanine deaminase positive. Other characteristics are listed in Table 5. Colonies tend to stick to the agar. Isolates have come from the respiratory tract, blood, and CAPD fluids (23). Clinical data are sparse. Susceptibility to most antibiotics used against gram-negative rods except polymyxin B is the rule (23).

Sphingomonas spp.

On the basis of nucleotide sequence analysis of 16S rRNA and the presence of unique sphingoglycolipid and ubiquinone types, the genus *Sphingomonas*, with several species, among them the type species *Sphingomonas paucimobilis* (formerly *Pseudomonas paucimobilis*), has been proposed (98). Only the type species is important clinically. It is

TABLE 6 Human biovars of *Shewanella putrefaciens*[a]

Test	Biovar 1	Biovar 2
Acid from:		
Sucrose	+	−
Maltose	+	−
Growth:		
In 6.5% NaCl	−	+
At 42°C	−	+

[a]For both biovars, cellular fatty acids are C17:1ω9c, C16:1ω9c, and i-C15:0 (61).

characterized by medium to long rods that form large, butyrous, occasionally mucoid colonies exhibiting a non-diffusible yellow carotenoid pigment. Motility (by a single polar flagellum) may be slow. Although *S. paucimobilis* is isolated frequently in clinical laboratories, clinically significant strains constitute a minority (73). The organism is widely distributed in the environment, including water. Most strains are susceptible to tetracycline, chloramphenicol, trimethoprim-sulfamethoxazole, the aminoglycosides, and fluoroquinolones; susceptibility to other antimicrobial agents varies (18, 23).

Shewanella putrefaciens

The organism formerly called *Pseudomonas putrefaciens* and *Alteromonas putrefaciens* has now been placed in the genus *Shewanella* (54), family *Vibrionaceae*, although this classification does not overcome the problem of heterogeneity of strains (66) (Table 6). Colonies on SBA are convex, circular, smooth, and occasionally mucoid; produce a brown to tan soluble pigment; and cause a green discoloration of the medium. Cells are long, short, or filamentous. Motility is due to a single polar flagellum. Ornithine decarboxylase, nitrate reductase, and DNase are always produced, and, with few exceptions (90), hydrogen sulfide is produced in Kligler and triple sugar iron agars (the alkaline butt differentiates such strains from *Salmonella* spp.). Urease and esculin are not hydrolyzed.

Biovar 1 (group III of Owen et al. [66]) is primarily found in foodstuffs and the environment and only occasionally in human specimens. Biovar 2 (group IV of Owen et al. [66]) is mainly found in clinical samples. Other biovars or groups have been described (23), but their incidences in clinical samples are very low. *Shewanella putrefaciens* strains have caused lower limb cellulitis, otitis media, and septicemia (49). They are generally susceptible to most antimicrobial agents effective against gram-negative rods except penicillin and cephalothin (18, 23, 90, 91).

Methylobacterium and Roseomonas spp.

Members of the genus *Methylobacterium* are able to utilize methanol as a sole source of carbon and energy, although this characteristic may be lost on subculture (Table 7). They occur mostly on vegetation but may also be found in the hospital environment. Most human strains belong to the species *Methylobacterium mesophilicum* (*Pseudomonas mesophilica*, *Pseudomonas extorquens*) (94). This species is motile by one polar or lateral flagellum, is oxidase positive, grows best at 25°C (many strains do not grow at 37°C), and forms a pink to coral nondiffusible pigment. Cells are large, vacuolated, and pleomorphic and may resist decolorization. Approximately 90% of strains grow within 4 to 5 days to 1-mm-diameter colonies on SBA and Thayer-Martin, Sabouraud, buffered charcoal-yeast extract, and Middlebrook-Cohn 7H11 agar but only weakly or not at all on MacConkey agar. Oxidation of sugars (xylose and sometimes glucose) is weak; urea and starch are hydrolyzed.

Methylobacterium spp. have caused septicemia, CAPD peritonitis, and other infections (47). They are best tested for susceptibility by agar or broth dilution at 30°C for 48 h (8) and show unpredictable patterns (23).

Members of the new genus *Roseomonas* (75) are also pink pigmented but differ in morphologic and biochemical characteristics from *Methylobacterium* spp. (Table 7). They are unvacuolated and rather plump and coccoid and form mostly pairs and short chains. They grow on blood agar and Thayer-Martin agar at 37°C. Colonies are often mucoid.

The oxidase test may be delayed positive (≤30 s). So far, three named species (*Roseomonas gilardii*, *Roseomonas cervicalis*, and *Roseomonas fauriae*) and three more unnamed genomospecies have been outlined. They differ from each other with regard to several biochemical tests and motility (by one polar flagellum). All have come from clinical sources (including patients with bacteremia) and have been susceptible to aminoglycosides, tetracycline, and imipenem (75).

OXIDASE-POSITIVE, NONMOTILE, PIGMENTED, INDOLE-POSITIVE OR -NEGATIVE NONFERMENTERS

See Tables 4, 5, and 8.

Flavobacterium spp. and CDC Groups IIe, IIh, and IIi

The taxonomy of bacteria belonging to the genus *Flavobacterium* and to CDC groups IIe, IIh, and IIi has not been settled. At present, the following groups and species of medical importance are recognized (31).

Group A. Saccharolytic, indole positive; includes *Flavobacterium meningosepticum* (with two genomic groups and four serogroups), *Flavobacterium* group IIb (with several genomic groups clustering around the type strains of *Flavobacterium indologenes* and *Flavobacterium gleum*), and *Flavobacterium breve* (with two genomic groups [86]).

Group B. Asaccharolytic, indole negative; includes *Flavobacterium odoratum*.

Group C. Saccharolytic, indole negative; includes *Flavobacterium yabuuchiae*, *Flavobacterium thalpophilum*, *Flavobacterium mizutaii*, *Flavobacterium multivorum*, and *Flavobacterium spiritivorum* (31, 83). They have been proposed for a new genus, *Sphingobacterium*, on account of their high content of sphingophospholipids and other taxonomic features, and the synonymy of *Flavobacterium yabuuchiae* and *Flavobacterium spiritivorum* has been postulated (83).

Group D. Asaccharolytic, indole positive (formerly CDC groups IIf and IIj); now in the genus *Weeksella* (see below).

CDC groups IIe, IIh, and IIi. Saccharolytic, indole positive (10).

Colonies of *Flavobacterium meningosepticum* are smooth and fairly large (1 to 2 mm in diameter after 24 h) but show only weak (if any) production of yellow pigment. Colonies of *Flavobacterium multivorum*, *Flavobacterium spiritivorum*, *Flavobacterium thalpophilum*, *Flavobacterium mizutaii*, and *Flavobacterium yabuuchiae* look similar. In contrast, *Flavobacterium* IIb colonies are deep yellow, owing to water-insoluble flexirubin production (70). *Flavobacterium breve* colonies are pale yellow, and those of *Flavobacterium odoratum* are yellowish to tan, produce a fruity odor, and show a tendency to spread. Microscopically, cells of *Flavobacterium meningosepticum*, *Flavobacterium* group IIb, and groups IIe, IIh, and IIi are thinner in their central than in their peripheral portions and include filamentous forms; IIh cells are significantly smaller than those of other species. Pigment formation in these groups is variable (Table 5). It should be emphasized that several results (e.g., DNase, indole, urea, starch hydrolysis) are dependent on the choice

TABLE 7 Biochemical reactions of *Methylobacterium* and *Roseomonas* spp.[a]

Test[b]	Methylo-bacterium spp.	Roseomonas spp.
Oxidase	+	+
Oxidation of methanol	+	−
Growth:		
On MacConkey agar	−	+
At 42°C	−	+
Urease	+	+
Starch hydrolysis	+	+
MBM acetate	+	−
UV absorption of colonies[c]	+	−
CFA[d]	C18:1ω7c (94)	C16:0, C18:1ω7c, C19:0Δ 11, 12, 2-OH C19:0Δ 11, 12 (94)

[a]Data are from references 23, 75, and 94.
[b]MBM, minimal basal medium; CFA, cellular fatty acids.
[c]Colonies appear dark when exposed to long-wave UV light.
[d]See Table 1, footnote c.

of medium, reagents, and length of incubation (70). Phenotypic separation between *Flavobacterium indologenes* and *Flavobacterium gleum* has been difficult; at least acid production from xylose and growth at 41°C are consistently positive in DNA groups clustering around the type strain *Flavobacterium gleum* (86).

The natural habitats of these bacteria are soil, plants, foodstuffs, and water sources, including those in hospitals. *Flavobacterium* group IIb is the most frequent human isolate, although clinical significance is rare (91). The latter is also true for the other species (19), except *Flavobacterium meningosepticum*, which has been an agent of neonatal meningitis, nosocomial miniepidemics (81; verifiable by ribotyping [11]), and, rarely, adult septicemia (80) and respiratory colonization and infection following aerosolized polymyxin B treatment (7). *Flavobacterium meningosepticum*, *Flavobacterium* group IIb, and *Flavobacterium odoratum* are resistant to many antimicrobial agents (aminoglycosides, beta-lactam antibiotics, tetracyclines, chloramphenicol) but are often susceptible to rifampin, clindamycin, erythromycin, ciprofloxacin, and trimethoprim-sulfamethoxazole (18, 23, 91). The other species are often susceptible to more antimicrobial agents (23). Since a consider-

TABLE 8 Oxidase-positive, indole-positive, nonmotile, nonfermentative gram-negative rods[a]

Test	Flavobacterium meningosepticum	Flavobacterium group IIb	Flavobacterium breve	CDC group IIe[b]	CDC group IIh	CDC group IIi	Weeksella virosa	Weeksella zoohelcum
Flexirubin pigment (insoluble)	V	+	+	−	−	−	−	−
Light yellow pigment	V	−	+	−	−	V	−	−
Other pigments (soluble)	V[c]	V[d]	+[c]	V[d]	+[c]	V[d]	+[c]	+[d]
Nitrate to nitrite	−	V	−	−	−	+	+	+
Catalase	+	+	+	V	+	+	+	+
Growth:								
On MacConkey agar	V	V	D	−	−	−	−	−
At 42°C	V	V	−	−	−	D	V	−
ONPG	+	V	−	−	−	+	−	−
Starch hydrolysis[e]	−	+	V	+	+	V	−	−
Urease[e]	−	−	−	−	−	−	−	+
DNase[e]	+	V	+	D	V	−	−	−
Esculin hydrolysis	+	+	−	−	+	+	−	−
Gelatin hydrolysis	+	+	+	−	D	−	+	+
Acid from:								
Glucose	+	+	V	+	+	+	−	−
Maltose	+	+	+	+	+	+	−	−
Sucrose	−	V	−	−	−	+	−	−
Mannitol	+	−	−	−	−	−	−	−
Xylose	−	V	−	−	−	+	−	−
Polymyxin B	R	R	R	S	V	R	S	R
CFA[f]	i-C15:0, i-2-OH-C15:0, i-3-OH-C17:0, i-C17:1 (59)	i-C15:0, i-2-OH-C15:0, i-3-OH-C17:0, i-C17:1 (59)	i-C15:0, C16:1Δ9, i-3-OHC17:0 (16)	i-C15:0, i-C17:1, i-3OH-C17:0, a-C15:0 (16)	i-C15:0, i-C17:1, i-3OH-C17:0, a-C15:0 (16)	ND	i-C15:0>i-C17:1 (37, 38)	i-C15:0>i-C17:1 (37, 38)

[a]Data are from references 10, 16, 37, and 38. For abbreviations, see Table 1, footnote a, and Table 5, footnote a.
[b]One of 15 strains was oxidase negative (10).
[c]Tan to brown (late).
[d]Tan to yellow.
[e]See text.
[f]See Table 1, footnote c.

able number of "very major" errors may occur when disk tests alone are used (93), MIC determinations are recommended for clinically significant strains.

Weeksella spp.

The former CDC groups IIf (*Weeksella virosa*) and IIj (*Weeksella zoohelcum*) have been placed into the genus *Weeksella*. These organisms are 0.6 to 2 to 3 μm long, with parallel sides and rounded ends. Colonies are 2 to 3 mm in diameter after 48 h. *W. virosa* colonies are butyrous to mucoid and adherent, and they show a tan to brown pigmentation; *W. zoohelcum* colonies are sticky and tan to yellow. *W. virosa* is urease negative and polymyxin B susceptible; *W. zoohelcum* is urease positive and polymyxin resistant. Both organisms hydrolyze gelatin and show characteristic enzyme patterns (37, 38). *W. virosa* occurs mainly in urine and vaginal samples, while *W. zoohelcum* is found most frequently in bite wounds and rarely in other specimens (10). Both organisms are susceptible to most antibiotics except aminoglycosides (23).

REFERENCES

1. **Bingen, E. H., E. Denamur, N. Y. Lambert-Zechovsky, A. Bourdois, P. Mariani-Kurkdjian, J.-P. Cezard, J. Navarro, and J. Elion.** 1991. DNA restriction fragment length polymorphism differentiates crossed from independent infections in nosocomial *Xanthomonas maltophilia* bacteremia. *J. Clin. Microbiol.* **29:**1348–1350.

2. **Bizet, C., F. Tekaia, and A. Philippon.** 1993. In-vitro susceptibility of *Alcaligenes faecalis* compared with those of other *Alcaligenes* spp. to antimicrobial agents including seven β-lactams. *J. Antimicrob. Chemother.* **32:**907–910.

3. **Bouvet, P. J. M., and P. A. D. Grimont.** 1986. Taxonomy of the genus *Acinetobacter* with the recognition of *Acinetobacter baumannii* sp. nov., *Acinetobacter haemolyticus* sp. nov., *Acinetobacter johnsonii* sp. nov., and *Acinetobacter junii* sp. nov. and emended descriptions of *Acinetobacter calcoaceticus* and *Acinetobacter lwoffii*. *Int. J. Syst. Bacteriol.* **36:**228–240.

4. **Bouvet, P. J. M., and P. A. D. Grimont.** 1987. Identification and biotyping of clinical isolates of *Acinetobacter*. *Ann. Inst. Pasteur/Microbiol.* **138:**569–578.

5. **Bouvet, P. J. M., and S. Jeanjean.** 1989. Delineation of new proteolytic genomic species in the genus *Acinetobacter*. *Res. Microbiol.* **140:**291–299.

6. **Bøvre, K., J. E. Fuglesang, N. Hagen, E. Jantzen, and L. O. Frøholm.** 1976. *Moraxella atlantae* sp. nov. and its distinction from *Moraxella phenylpyruvica*. *Int. J. Syst. Bacteriol.* **26:**511–521.

7. **Brown, R. B., D. Phillips, M. J. Barker, R. Pieczarka, M. Sands, and D. Teres.** 1989. Outbreak of nosocomial *Flavobacterium meningosepticum* respiratory infections associated with use of aerosolized polymyxin B. *Am. J. Infect. Control* **17:**121–125.

8. **Brown, W. J., R. L. Sautter, and A. E. Crist, Jr.** 1992. Susceptibility testing of clinical isolates of *Methylobacterium* species. *Antimicrob. Agents Chemother.* **36:**1635–1638.

9. **Cheron, M., E. Abachin, E. Guerot, M. El-Bez, and M. Simonet.** 1994. Investigation of hospital-acquired infections due to *Alcaligenes denitrificans* subsp. *xylosoxidans* by DNA restriction fragment length polymorphism. *J. Clin. Microbiol.* **32:**1023–1026.

10. **Clark, W. A., D. G. Hollis, R. E. Weaver, and P. Riley.** 1984. *Identification of Unusual Pathogenic Gram-Negative Aerobic and Facultatively Anaerobic Bacteria.* Centers for Disease Control, Atlanta.

11. **Colding, H., J. Bangsborg, N.-E. Fiehn, T. Bennekov, and B. Bruun.** 1994. Ribotyping for differentiating *Flavobacterium meningosepticum* isolates from clinical and environmental sources. *J. Clin. Microbiol.* **32:**501–505.

12. **Dees, S. B., G. M. Carlone, D. Hollis, and C. W. Moss.** 1985. Chemical and phenotypic characteristics of *Flavobacterium thalpophilum* compared with those of other *Flavobacterium* and *Sphingobacterium* species. *Int. J. Syst. Bacteriol.* **35:**16–22.

13. **Dees, S. B., D. G. Hollis, R. E. Weaver, and C. W. Moss.** 1983. Cellular fatty acid composition of *Pseudomonas marginata* and closely associated bacteria. *J. Clin. Microbiol.* **18:**1073–1078.

14. **Dees, S. B., and C. W. Moss.** 1975. Cellular fatty acids of *Alcaligenes* and *Pseudomonas* species isolated from clinical specimens. *J. Clin. Microbiol.* **1:**414–419.

15. **Dees, S. B., and C. W. Moss.** 1978. Identification of *Achromobacter* species by cellular fatty acids and by production of keto acids. *J. Clin. Microbiol.* **8:**61–66.

16. **Dees, S. B., C. W. Moss, D. G. Hollis, and R. E. Weaver.** 1986. Chemical characterization of *Flavobacterium odoratum*, *Flavobacterium breve*, and *Flavobacterium*-like groups IIe, IIh, and IIf. *J. Clin. Microbiol.* **23:**267–273.

17. **Dijkshoorn, L., J. L. Wubbels, A. J. Beunders, J. E. Degener, A. L. Boks, and M. F. Michel.** 1989. Use of protein profiles to identify *Acinetobacter calcoaceticus* in a respiratory care unit. *J. Clin. Pathol.* **42:**853–857.

18. **Fass, R. J., and J. Barnishan.** 1980. In vitro susceptibility of nonfermentative gram-negative bacilli other than *Pseudomonas aeruginosa* to 32 antimicrobial agents. *Rev. Infect. Dis.* **2:**841–853.

19. **Freney, J., W. Hansen, C. Ploton, H. Meugnier, S. Madier, N. Bornstein, and J. Fleurette.** 1987. Septicemia caused by *Sphingobacterium multivorum*. *J. Clin. Microbiol.* **25:**1126–1128.

20. **Garcia-Rodriguez, J. A., J. E. Garcia Sanchez, M. I. Garcia Garcia, E. Garcia Sanchez, and J. L. Munoz Bellido.** 1991. Antibiotic susceptibility profile of *Xanthomonas maltophilia*. In vitro activity of β-lactam/β-lactamase inhibitor combinations. *Diagn. Microbiol. Infect. Dis.* **14:**239–243.

21. **Gerner-Smidt, P., I. Tjernberg, and J. Ursing.** 1991. Reliability of phenotypic tests for identification of *Acinetobacter* species. *J. Clin. Microbiol.* **29:**277–282.

22. **Getchell-White, S. I., L. G. Donowitz, and D. H. M. Gröschel.** 1989. The inanimate environment of an intensive care unit as a potential source of nosocomial bacteria: evidence for long survival of *Acinetobacter calcoaceticus*. *Infect. Control Hosp. Epidemiol.* **10:**402–407.

23. **Gilardi, G. L.** 1990. *Identification of Glucose-Nonfermenting Gram-Negative Rods.* North General Hospital, New York.

24. **Glew, R. H., R. C. Moellering, Jr., and L. J. Kunz.** 1977. Infections with *Acinetobacter calcoaceticus* (*Herellea vaginicola*): clinical and laboratory studies. *Medicine* **56:**79–97.

25. **Graham, D. R., J. D. Band, C. Thornsberry, D. G. Hollis, and R. E. Weaver.** 1990. Infections caused by *Moraxella*, *Moraxella urethralis*, *Moraxella*-like groups M-5 and M-6, and *Kingella kingae* in the United States, 1953–1980. *Rev. Infect. Dis.* **12:**423–431.

26. **Greenwood, J. R.** 1985. Methods of isolation and identification of glucose-nonfermenting gram-negative rods, p. 1–16. *In* G. L. Gilardi (ed.), *Nonfermentative Gram-Negative Rods: Laboratory Identification and Clinical Aspects.* Marcel Dekker, Inc., New York.

27. **Hawkins, R. E., R. A. Moriarty, D. E. Lewis, and E. C. Oldfield.** 1991. Serious infections involving the CDC group Ve bacteria *Chryseomonas luteola* and *Flavimonas oryzihabitans*. *Rev. Infect. Dis.* **13:**257–260.

28. **Hohl, P., R. Frei, and P. Aubry.** 1991. In vitro susceptibility of 33 clinical case isolates of *Xanthomonas maltophilia*. Inconsistent correlation of agar dilution and of disk diffusion test results. *Diagn. Microbiol. Infect. Dis.* **14:**447–450.

29. **Hollis, D. G., C. W. Moss, M. I. Daneshvar, L. Meadows, J. Jordan, and B. Hill.** 1993. Characterization of Centers for Disease Control group NO-1, a fastidious, nonoxidative, gram-negative organism associated with dog and cat bites. *J. Clin. Microbiol.* **31:**746–748.

30. **Hollis, D. G., R. E. Weaver, C. W. Moss, M. I. Daneshvar, and P. L. Wallace.** 1992. Chemical and cultural characterization of CDC group WO-1, a weakly oxidative gram-negative

group of organisms isolated from clinical sources. *J. Clin. Microbiol.* **30:**291–295.

31. **Holmes, B.** 1992. Recent developments in *Flavobacterium* taxonomy, p. 6–15. *In* P. J. Jooste (ed.), *Advances in the Taxonomy and Significance of Flavobacterium, Cytophaga and Related Bacteria.* University Press, Bloemfontein, Republic of South Africa.

32. **Holmes, B., M. Costas, A. C. Wood, R. J. Owen, and D. D. Morgan.** 1990. Differentiation of *Achromobacter*-like strains from human blood by DNA restriction endonuclease digest and ribosomal RNA gene probe patterns. *Epidemiol. Infect.* **105:**541–551.

33. **Holmes, B., R. Lewis, and A. Trevett.** 1992. Septicemia due to *Achromobacter* group B: a report of two cases. *Med. Microbiol. Lett.* **1:**177–184.

34. **Holmes, B., C. W. Moss, and M. I. Daneshvar.** 1993. Cellular fatty acid compositions of "*Achromobacter* groups B and E." *J. Clin. Microbiol.* **31:**1007–1008.

35. **Holmes, B., C. A. Pinning, and C. A. Dawson.** 1986. A probability matrix for the identification of Gram-negative, aerobic, non-fermentative bacteria that grow on nutrient agar. *J. Gen. Microbiol.* **132:**1827–1842.

36. **Holmes, B., M. Popoff, M. Kiredjian, and K. Kersters.** 1988. *Ochrobactrum anthropi* gen. nov., sp. nov. from human clinical specimens and previously known as group Vd. *Int. J. Syst. Bacteriol.* **38:**406–416.

37. **Holmes, B., A. G. Steigerwalt, R. E. Weaver, and D. J. Brenner.** 1986. *Weeksella virosa* gen. nov., sp. nov. (formerly group IIf), found in human clinical specimens. *Syst. Appl. Microbiol.* **8:**185–190.

38. **Holmes, B., A. G. Steigerwalt, R. E. Weaver, and D. J. Brenner.** 1986. *Weeksella zoohelcum* sp. nov. (formerly group IIj), from human clinical specimens. *Syst. Appl. Microbiol.* **8:**191–196.

39. **Holton, J.** 1983. A note on the preparation and use of a selective and differential medium for the isolation of the *Acinetobacter* spp. from clinical sources. *J. Appl. Bacteriol.* **66:**24–26.

40. **Hudson, M. J., D. G. Hollis, R. E. Weaver, and C. G. Galvis.** 1987. Relationship of CDC group EO-2 and *Psychrobacter immobilis. J. Clin. Microbiol.* **25:**1907–1910.

41. **Hulse, M., S. Johnson, and P. Ferrieri.** 1993. *Agrobacterium* infections in humans: experience at one hospital and review. *Clin. Infect. Dis.* **16:**112–117.

42. **Ikemoto, S., K. Suzuki, T. Kaneko, and K. Komagata.** 1980. Characterization of strains of *Pseudomonas maltophilia* which do not require methionine. *Int. J. Syst. Bacteriol.* **30:**437–447.

43. **Jannes, G., M. Vaneechoutte, M. Lannoo, M. Gillis, M. Vancanneyt, P. Vandamme, G. Verschraegen, H. van Heuverswyn, and R. Rossau.** 1993. Polyphasic taxonomy leading to the proposal of *Moraxella canis* sp. nov. for *Moraxella catarrhalis*-like strains. *Int. J. Syst. Bacteriol.* **43:**438–449.

44. **Juni, E.** 1972. Interspecies transformation of *Acinetobacter:* genetic evidence for a ubiquitous genus. *J. Bacteriol.* **112:**917–931.

45. **Juni, E., G. A. Heym, M. J. Maurer, and M. L. Miller.** 1987. Combined genetic transformation and nutritional assay for identification of *Moraxella nonliquefaciens. J. Clin. Microbiol.* **25:**1691–1694.

46. **Kämpfer, P.** 1993. Grouping of *Acinetobacter* genomic species by cellular fatty acid composition. *Med. Microbiol. Lett.* **2:**394–400.

47. **Kaye, K. M., A. Macone, and P. H. Kazanjian.** 1992. Catheter infections caused by *Methylobacterium* in immunocompromised hosts: report of three cases and review of the literature. *Clin. Infect. Dis.* **14:**1010–1014.

48. **Kern, W. V., M. Oethinger, A. Kaufhold, E. Rozdzinski, and R. Marre.** 1993. *Ochrobactrum anthropi* bacteremia: report of four cases and short review. *Infection* **21:**306–310.

49. **Kim, J. H., R. A. Cooper, K. E. Welty-Wolf, L. J. Harrell, P. Zwadyk, and M. E. Klotman.** 1989. *Pseudomonas putrefaciens* bacteremia. *Rev. Infect. Dis.* **11:**97–104.

50. **King, A., B. Holmes, I. Phillips, and S. P. Lapage.** 1979. A

taxonomic study of clinical isolates of *Pseudomonas pickettii*, "*P. thomasii*" and "group IVd" bacteria. *J. Gen. Microbiol.* **114:**137–147.

51. **Kiredjian, M., B. Holmes, K. Kersters, I. Guilvout, and J. de Ley.** 1986. *Alcaligenes piechaudii*, a new species from human clinical specimens and the environment. *Int. J. Syst. Bacteriol.* **36:**282–287.

52. **Kodama, K., N. Kimura, and K. Komagata.** 1985. Two new species of *Pseudomonas: P. oryzihabitans* isolated from rice paddy and clinical specimens and *P. luteola* isolated from clinical specimens. *Int. J. Syst. Bacteriol.* **35:**467–474.

53. **Lloyd-Puryear, M., D. Wallace, T. Baldwin, and D. G. Hollis.** 1991. Meningitis caused by *Psychrobacter immobilis* in an infant. *J. Clin. Microbiol.* **29:**2041–2042.

54. **MacDonell, M. T., and R. R. Colwell.** 1985. Phylogeny of the Vibrionaceae, and recommendation for two new genera, *Listonella* and *Shewanella. Syst. Appl. Microbiol.* **6:**171–182.

55. **Marshall, W. F., M. R. Keating, J. P. Anhalt, and J. M. Steckelberg.** 1989. *Xanthomonas maltophilia:* an emerging nosocomial pathogen. *Mayo Clin. Proc.* **64:**1097–1104.

56. **McGeer, A., C. Poulos, S. R. Scriver, D. Hoban, P. Kibsey, R. Devlin, and D. E. Low.** 1993. Evaluation of disk diffusion susceptibility testing for *Xanthomonas maltophilia* (XM), abstr. C-210, p. 483. *Abstr. 93rd Gen. Meet. Am. Soc. Microbiol. 1993.*

57. **Metchock, B., B. Hill, K. Maher, and C. Thornsberry.** 1988. Susceptibility of *Xanthomonas* (*Pseudomonas*) *maltophilia* to antimicrobial agents: methodological problems, abstr. 1220, p. 327. *Program Abstr. 28th Intersci. Conf. Antimicrob. Agents Chemother.*

58. **Moss, C. W., M. I. Daneshvar, and D. G. Hollis.** 1993. Biochemical characteristics and fatty acid composition of Gilardi rod group 1 bacteria. *J. Clin. Microbiol.* **31:**689–691.

59. **Moss, C. W., and S. B. Dees.** 1978. Cellular fatty acids of *Flavobacterium meningosepticum* and *Flavobacterium* species group IIb. *J. Clin. Microbiol.* **8:**772–774.

60. **Moss, C. W., P. L. Wallace, D. G. Hollis, and R. E. Weaver.** 1988. Cultural and chemical characterization of CDC groups EO-2, M-5, and M-6, *Moraxella* (*Moraxella*) species, *Oligella urethralis, Acinetobacter* species, and *Psychrobacter immobilis. J. Clin. Microbiol.* **26:**484–492.

60a.**Moss, C. W.** Personal communication.

61. **Moule, A. L., and S. G. Wilkinson.** 1987. Polar lipids, fatty acids, and isoprenoid quinones of *Alteromonas putrefaciens* (*Shewanella putrefaciens*). *Syst. Appl. Microbiol.* **9:**192–198.

62. **Musso, D., M. Drancourt, J. Bardot, and R. Legré.** 1994. Human infection due to the CDC group IVc-2 bacterium: case report and review. *Clin. Infect. Dis.* **18:**482–484.

63. **Nagai, T.** 1984. Association of *Pseudomonas maltophilia* with malignant lesions. *J. Clin. Microbiol.* **20:**1003–1005.

64. **Nash, P., and M. M. Krenz.** 1991. Culture media, p. 1226–1288. *In* A. Balows, W. J. Hausler, Jr., K. L. Herrmann, H. D. Isenberg, and H. J. Shadomy (ed.), *Manual of Clinical Microbiology,* 5th ed. American Society for Microbiology, Washington, D.C.

65. **Osterhout, G. J., V. H. Shull, and J. D. Dick.** 1991. Identification of clinical isolates of gram-negative nonfermentative bacteria by an automated cellular fatty acid identification system. *J. Clin. Microbiol.* **29:**1822–1830.

66. **Owen, R. J., R. M. Legros, and S. P. Lapage.** 1978. Base composition, size and sequence similarities of genome deoxyribonucleic acids from clinical isolates of *Pseudomonas putrefaciens. J. Gen. Microbiol.* **104:**127–138.

67. **Palleroni, N. J., and J. F. Bradbury.** 1993. *Stenotrophomonas,* a new bacterial genus for *Xanthomonas maltophilia* (Hugh 1980) Swings et al. 1983. *Int. J. Syst. Bacteriol.* **43:**606–609.

68. **Peiris, V., S. Fraser, M. Fairhurst, D. Weston, and E. Kaczmarski.** 1992. Laboratory diagnosis of *Brucella* infection: some pitfalls. *Lancet* **339:**1415–1416.

69. **Pérez, J. L., A. Pulido, F. Pantozzi, and R. Martin.** 1990. Butyrate esterase (tributyrin) spot test, a simple method for immediate identification of *Moraxella* (*Branhamella*) *catarrhalis. J. Clin. Microbiol.* **28:**2347–2348.

70. **Pickett, M. J.** 1989. Methods for identification of flavobacteria. *J. Clin. Microbiol.* **27:**2309–2315.

70a.**Pickett, M. J.** Personal communication.

71. **Pugliese, A., B. Pacris, P. E. Schoch, and B. A. Cunha.** 1993. *Oligella urethralis* urosepsis. *Clin. Infect. Dis.* **17:**1069–1070.

72. **Ramos, J. M., F. Soriano, M. Bernacer, J. Esteban, and J. Zapardiel.** 1993. Infection caused by the nonfermentative gram-negative bacillus CDC group IVc-2: case report and literature review. *Eur. J. Clin. Microbiol. Infect. Dis.* **12:**456–458.

73. **Reina, J., A. Bassa, I. Llompart, D. Portela, and N. Borrell.** 1991. Infections with *Pseudomonas paucimobilis*: report of four cases and review. *Rev. Infect. Dis.* **13:**1072–1076.

74. **Reina, J., I. Llompart, and P. Alomar.** 1991. Acute suppurative otitis caused by *Comamonas acidovorans*. *Clin. Microbiol. Newsl.* **13:**38–39.

75. **Rihs, J. D., D. J. Brenner, R. E. Weaver, A. G. Steigerwalt, D. G. Hollis, and V. L. Yu.** 1993. *Roseomonas*, a new genus associated with bacteremia and other human infections. *J. Clin. Microbiol.* **31:**3275–3283.

76. **Rossau, R., K. Kersters, E. Falsen, E. Jantzen, P. Segers, A. Union, L. Nehls, and J. de Ley.** 1987. *Oligella*, a new genus including *Oligella urethralis* comb. nov. (formerly *Moraxella urethralis*) and *Oligella ureolytica* sp. nov. (formerly CDC group IVe): relationship to *Taylorella equigenitalis* and related taxa. *Int. J. Syst. Bacteriol.* **37:**198–210.

77. **Sawada, H., Y. Takikawa, and H. Ieki.** 1992. Fatty acid methyl ester profiles of the genus *Agrobacterium*. *Ann. Phytopathol. Soc. Jpn.* **58:**46–51.

78. **Schable, B., M. E. Villarino, M. S. Favero, and J. M. Miller.** 1991. Application of multilocus enzyme electrophoresis to epidemiologic investigations of *Xanthomonas maltophilia*. *Infect. Control Hosp. Epidemiol.* **12:**163–167.

79. **Seifert, H., R. Baginski, A. Schulze, and G. Pulverer.** 1993. Antimicrobial susceptibility of *Acinetobacter* species. *Antimicrob. Agents Chemother.* **37:**750–753.

80. **Sheridan, R. I., C. M. Ryan, M. S. Pasternack, J. M. Weber, and R. G. Tompkins.** 1993. Flavobacterial sepsis in massively burned pediatric patients. *Clin. Infect. Dis.* **17:**185–187.

81. **Siegman-Igra, Y., D. Schwartz, G. Soferman, and N. Konforti.** 1987. *Flavobacterium* group IIb bacteremia: report of a case and review of *Flavobacterium* infections. *Med. Microbiol. Immunol.* **176:**103–111.

82. **Swings, J., P. de Vos, M. van den Mooter, and J. de Ley.** 1983. Transfer of *Pseudomonas maltophilia* Hugh 1981 to the genus *Xanthomonas* as *Xanthomonas maltophilia* (Hugh 1981) comb. nov. *Int. J. Syst. Bacteriol.* **33:**409–413.

83. **Takeuchi, M., and A. Yokota.** 1992. Proposals of *Sphingobacterium faecium* sp. nov., *Sphingobacterium piscium* sp. nov., *Sphingobacterium heparinum* comb. nov., *Sphingobacterium thalpophilum* comb. nov. and two genospecies of the genus *Spingobacterium*, and synonymy of *Flavobacterium yabuuchiae* and *Sphingobacterium spiritivorum*. *J. Gen. Appl. Microbiol.* **38:**465–482.

84. **Tjernberg, I., and J. Ursing.** 1989. Clinical strains of *Acinetobacter* classified by DNA-DNA hybridization. *APMIS* **97:**595–605.

85. **Traub, W. H., and B. Leonhard.** 1994. Serotyping of *Acinetobacter baumannii* and genospecies 3: an update. *Med. Microbiol. Lett.* **3:**120–127.

86. **Ursing, J., and B. Bruun.** 1991. Genotypic heterogeneity of *Flavobacterium* group IIb and *Flavobacterium breve* demonstrated by DNA-DNA hybridization. *APMIS* **99:**780–786.

87. **Vandamme, P., M. Gillis, M. Vancanneyt, B. Hoste, K. Kerster, and E. Falsen.** 1993. *Moraxella lincolnii* sp. nov., isolated from the human respiratory tract, and reevaluation of the taxonomic position of *Moraxella osloensis*. *Int. J. Syst. Bacteriol.* **43:**474–481.

88. **Veys, A., W. Callewaert, E. Waelkens, and K. van den Abbeele.** 1989. Application of gas-liquid chromatography to the routine identification of nonfermenting gram-negative bacteria in clinical specimens. *J. Clin. Microbiol.* **27:**1538–1542.

89. **Villarino, M. E., L. E. Stevens, B. Schable, G. Mayers, J. M. Miller, J. P. Burke, and W. R. Jarvis.** 1992. Risk factors for epidemic *Xanthomonas maltophilia* infection/colonization in intensive care unit patients. *Infect. Control Hosp. Epidemiol.* **13:**201–206.

90. **von Graevenitz, A.** 1976. Detection of unusual strains of Gram-negative rods through the routine use of a deoxyribonuclease-indole medium. *Mt. Sinai J. Med.* **43:**727–735.

91. **von Graevenitz, A.** 1985. Ecology, clinical significance, and antimicrobial susceptibility of infrequently encountered glucose-nonfermenting gram-negative rods, p. 181–232. *In* G. L. Gilardi (ed.), *Nonfermentative Gram-Negative Rods: Laboratory Identification and Clinical Aspects*. Marcel Dekker, Inc., New York.

92. **von Graevenitz, A., and C. Bucher.** 1983. Isolation of *Pseudomonas maltophilia* from human stools with thienamycin. *Zentralbl. Bakteriol. Hyg. Abt. 1 Orig. Reihe A* **254:**403–404.

93. **von Graevenitz, A., and M. Grehn.** 1977. Susceptibility studies on *Flavobacterium* II-b. *FEMS Microbiol. Lett.* **2:**289–292.

94. **Wallace, P. L., D. G. Hollis, R. E. Weaver, and C. W. Moss.** 1990. Biochemical and chemical characterization of pink-pigmented oxidative bacteria. *J. Clin. Microbiol.* **28:**689–693.

95. **Willems, A., E. Falsen, B. Pot, E. Jantzen, B. Hoste, P. Vandamme, M. Gillis, K. Kersters, and J. de Ley.** 1990. *Acidovorax*, a new genus for *Pseudomonas facilis*, *Pseudomonas delafieldii*, E. Falsen (EF) group 13, EF group 16, and several clinical isolates, with the species *Acidovorax facilis* comb. nov., *Acidovorax delafieldii* comb. nov., and *Acidovorax temperans* sp. nov. *Int. J. Syst. Bacteriol.* **40:**384–398.

96. **Willems, A., B. Pot, E. Falsen, P. Vandamme, M. Gillis, K. Kersters, and J. de Ley.** 1991. Polyphasic taxonomic study of the emended genus *Comamonas*: relationship to *Aquaspirillum aquaticum*, E. Falsen group 10, and other clinical isolates. *Int. J. Syst. Bacteriol.* **41:**427–444.

97. **Wood, C. A., and A. C. Reboli.** 1993. Infections caused by imipenem-resistant *Acinetobacter calcoaceticus* biotype *anitratus*. *J. Infect. Dis.* **168:**1602–1603.

98. **Yabuuchi, E., I. Yano, H. Oyaizu, Y. Hashimoto, T. Ezaki, and H. Yamamoto.** 1990. Proposals of *Sphingomonas paucimobilis* gen. nov. and comb. nov., *Sphingomonas parapaucimobilis* sp. nov., *Sphingomonas yanoikuyae* sp. nov., *Sphingomonas adhaesiva* sp. nov., *Sphingomonas capsulata* comb. nov., and two genospecies of the genus *Sphingomonas*. *Microbiol. Immunol.* **34:**99–119.

99. **Yang, P., L. Vauterin, M. Vancanneyt, J. Swings, and K. Kersters.** 1993. Application of fatty acid methyl esters for the taxonomic analysis of the genus *Xanthomonas*. *Syst. Appl. Microbiol.* **16:**47–71.

Legionella

WASHINGTON C. WINN, JR.

42

The medical significance of the genus *Legionella* was first recognized after an epidemic of pneumonia among members of the Pennsylvania American Legion during the United States bicentennial celebration in Philadelphia (37, 66). In fact, members of the genus had been isolated by inoculation of guinea pigs and characterized as "rickettsia-like organisms" as long ago as 1943 (46, 65). A combination of epidemic disease, media attention, and the resultant marshalling of full scientific resources caused a rapid transformation of these bacterial curiosities into established pathogens.

TAXONOMY

In many respects, *Legionella* spp. and legionellosis are children of the technology age. Technology has facilitated many of the epidemics by producing either means for efficient transmission of bacteria or medical advances that reduce the resistance of patients to infection. Similarly, the taxonomy has been based from the outset on genetic analysis, tempered by recognition of phenotypic characteristics and practical considerations. The newly isolated pathogen from Philadelphia was unrelated to previously described pathogens. For that reason, a new family (*Legionellaceae*), a new genus (*Legionella*), and a new species (*pneumophila*) were established (11). This new family is a part of the gamma subgroup of the class *Proteobacteria* (38). Subsequently, a burgeoning number of additional species has been defined, primarily on the basis of genetic and serological analysis (Table 1). Some of the newly recognized species have been isolated from both human and environmental sources; others have been isolated only from the environment to date. A proposal to divide the family into three genera (39) has not been supported by analysis of 16S rRNA (38) or, perhaps more important, by usage in the microbiology community. With increasing sophistication of analysis, there has been recognition of genetic diversity in a single serogroup (40) and phenotypic variation among genetically homogeneous strains (43, 44). Serological cross-reactions between genetically disparate species have also been described. The implications of this dichotomy between phenotypic and genotypic characteristics are great for epidemiologic studies, but taxonomists have only begun to address the issue. Brenner and colleagues have suggested the creation within the species *Legionella pneumophila* of three subspecies based on genomic relationships and with crossing of serological groups (10).

DESCRIPTION OF GENUS

The genus *Legionella* consists of thin, faintly staining, gram-negative, non-spore-forming bacilli (Fig. 1A). They generally measure 0.3 to 0.9 μm by 1.5 to 5 μm in clinical specimens but may become filamentous in culture. *Legionella* spp. require L-cysteine for growth on initial isolation from clinical and environmental sources, and growth is enhanced by inclusion of iron salts in the growth medium. The members of the genus are aerobic, stimulated by 5% CO_2, nonsaccharolytic, and relatively inert biochemically. Most species are weakly oxidase and catalase positive, gelatinase positive, and motile by means of one or more polar or subpolar flagella. Expression of flagella is inconstant, however, and may be temperature dependent (73). The fatty acid composition of *Legionella* spp. can be used to identify individual isolates and to group related species (58). *L. micdadei* is weakly acid fast in tissues and clinical specimens (Fig. 1B), but this characteristic is lost after isolation on agar media (70). Characteristics of the genus *Legionella* are summarized in Table 2.

NATURAL HABITATS

Legionella spp. are widespread in the environment, being almost exclusively associated with surface and potable waters or moist environments. The potable sources may be easily recognized drinking facilities, such as taps and faucets, or less obvious distribution sites, such as showerheads and public fountains. *L. longbeachae* serogroup 1 has been isolated from potting soil stored at room temperature for 7 months but not from soils that had thoroughly dried out (85). Isolation of *L. micdadei* from water sources has been considerably more difficult to accomplish than recovery of *L. pneumophila*.

It is clear that environmental microbes are important for survival and growth of legionellae in nature, because inclusion of the complex microbiota found in tap water is necessary for growth of *L. pneumophila* in water suspensions in vitro (97). Cyanobacteria (84) and other non-*Legionella* bacteria (90) stimulate growth of *L. pneumophila* in vitro, but stimulatory activity in samples of potable water has not

TABLE 1 Selected characteristics of *Legionella* species[a]

Legionella sp.	Serogroup	Isolated from:		Autofluorescence	Gelatin liquefaction	Motility, flagellum	Brown pigment	Hippurate hydrolysis
		Humans	Environment					
L. pneumophila[b] subsp. *pneumophila* subsp. *fraseri* subsp. *pascullei*	14	Yes	Yes	−	+	+	+[c]	+
L. micdadei[b]	1	Yes	Yes		−	+	−	−
L. bozemanii	2	Yes	Yes	BW	+	+	+	−
L. dumoffii	1	Yes	Yes	BW	+	+	+[c]	−
L. feelei[b]	2	Yes	Yes	−	−	+	+[c]	+/−
L. gormanii	1	Yes	Yes	BW	+	+	+	−
L. hackeliae	2	Yes	No	−	+	+	+	−
L. israelensis	1	Yes	Yes	−	+[c]	+	−	−
L. jordanis	1	Yes	Yes	−	+	+	+	−
L. sainthelensi	2	Yes	Yes	−	+	+	+[c]	−
L. longbeachae	2	Yes	Yes	−	+	+	+	−
L. maceachernii	1	Yes	Yes	−	+	+	+	−
L. oakridgensis	1	Yes	Yes	−	+	−	+	−
L. wadsworthii	1	Yes	No	YG	+	+	−	−
L. birminghamensis	1	Yes	No	YG	+	+	−	−
L. cincinnatiensis	1	Yes	No	−	+	+	+	−
L. anisa[b]	1	Yes	Yes	BW[c]	+	+	+	−
L. tucsonensis	1	Yes	No	BW	+	+	−	−
L. lansingensis	1	Yes	No	−	−	+	−	−
L. cherrii	1	No	Yes	BW	+	+	+	−
L. erythra	2	No	Yes	R[c]	+	+	+	−
L. jamestownensis	1	No	Yes	−	+	+	+	−
L. parisiensis	1	No	Yes	BW	+	+	+	−
L. shakespearei	1	No	Yes	−	+	+	+	−
L. santicrucis	1	No	Yes	−	+	+	+	−
L. steigerwaltii	1	No	Yes	BW	+	+	+	−
L. adelaidensis	1	No	Yes	−	+	+	−	−
L. fairfieldensis	1	No	Yes	−	−	−	−	−
L. brunensis	1	No	Yes	−	+	+	+[c]	−
L. moravica	1	No	Yes	−	+	+	+[c]	−
L. quinlivianii	2	No	Yes	−	±	+	+	−
L. gratiana	1	No	Yes	−	+	+	−	−
L. quateirensis	1	No	Yes	−	+	+	+	−
L. nautarum	1	No	Yes	−	−	−	−	−
L. worsleiensis	1	No	Yes	−	+	+	+	−
L. londiniensis	1	No	Yes	−	+	−	+	+/−[c]
L. geestiana	1	No	Yes	−	+	+	+[c]	+[c]
L. rubrilucens	1	No	Yes	R[c]	+	+	+	−
L. spiritensis	2	No	Yes	−	+	+	+	+[c]

[a] Characteristics are derived from multiple references. +, ≥90% positive; −, ≥90% negative; ±, most strains positive; V, variable; YG, yellow-green fluorescence; BW, blue-white fluorescence; R, red fluorescence.
[b] Associated with outbreaks of Pontiac fever.
[c] Rare strains are negative or weakly positive only.

been clearly demonstrated. Some bacteria produce bacteriocins or other substances that inhibit the growth of *L. pneumophila* (36) or lyse bacterial cells (77).

In contrast, an increasing body of evidence suggests the importance of protozoa for survival and growth of legionellae in nature (79). A variety of free-living amoebae and ciliated protozoa has been isolated along with *Legionella* spp. from environmental water sites suspected as sources of *Legionella* infection (5). *Legionella* spp. multiply intracellularly within the amoebae, much as they do within human monocytes and macrophages (5, 79). In fact, successful isolation of *Legionella* spp. on agar media may on occasion be accomplished only after cocultivation of water samples with amoebae (32) or preincubation of water samples that contain amoebae before subculture onto agar (82). Wadowsky and colleagues demonstrated that the growth-promoting activity of tap water was eliminated by filtration through a 1-μm-pore-size membrane filter, a procedure that removed amoebae but not heterotrophic bacteria from the system (89).

The environmental niches in which *Legionella* spp. find themselves may also have practical implications for attempts at eradication and control. *Legionella* spp. in tap water (57) and adherent to plumbing surfaces (96) may be more resistant to chlorine and other biocides than bacteria cultured in vitro. Growth within the protected intracellular environment of free-living amoebae may be one of the mechanisms for increased resistance to chemicals in nature (53).

Although *Legionella* spp. are widespread in the environment, colonization of natural and artificial water sources depends heavily on the local environment. In one study, the frequency of water samples positive for *Legionella* spp. over a 6-month period ranged from 0 to 72%; there were no institutional factors that could be associated with positive cultures. The same strain of *Legionella* tends to persist in a given institution, and it has been suggested that each water outlet serves as a microecologic unit (63). It is likely that one or more of the environmental cofactors discussed above is operative in the microenvironments. Colonization of peripheral water outlets in homes (2) and hospitals (3) has been correlated with reduced hot-water temperature.

Animals may possess antibodies to *Legionella* spp., but there is no evidence that human infection results from contact with animals. Colonization of humans with *Legionella* spp. occurs rarely, if ever. Isolation of this genus from clinical specimens is usually associated with overt clinical disease.

CLINICAL SIGNIFICANCE

Legionella infections can be classified into four categories: (i) subclinical infection, (ii) nonpneumonic disease, (iii) pneumonia, and (iv) extrapulmonary inflammatory disease. The 1976 Philadelphia epidemic was referred to as Legionnaires' disease. A more general term is legionellosis or *Legionella* infection of a specific type, such as pneumonia. Cases in each category may occur as part of an epidemic, may occur sporadically or nosocomially, or may be community acquired. Nosocomial *Legionella* pneumonia and community-acquired pneumonia that require hospitalization have been estimated to represent 1 to 4% of all pneumonias in each category (9). Extrapolations from a recent study of sporadic *Legionella* pneumonia acquired in the community yield an estimate of 11,000 cases per year nationally. Failure to utilize available diagnostic tools may result in the mistaken impression that *Legionella* infections are not occurring in a hospital or a community (67). For *Legionella* infections in particular, national extrapolations are potentially misleading because of the critical importance of the local microenvironment. Even a demonstration that *Legionella* spp. contaminate natural or potable-water systems does not predict the subsequent occurrence of human disease.

Although the number of species within the genus *Legionella* continues to increase, the vast majority of human infections are caused by *L. pneumophila*, particularly serogroups 1 and 6, or by *L. micdadei* (76). Most of the clinical information therefore derives from experience with these two species. It is very likely, however, that any species is capable of causing serious human disease, particularly if the patient is sufficiently immunosuppressed. As judged by case reports, the epidemiology, clinical presentation, and pathology of infections caused by other *Legionella* spp. are similar to those caused by *L. pneumophila* and *L. micdadei*.

Subclinical infection can be inferred from the frequency of occurrence of antibody to *Legionella* spp. (up to 30%; commonly 5 to 10%) in the absence of recognized episodes of pneumonia. Even if a portion of the positive serological tests represent cross-reactions, the magnitude and reproduc-

ibility of the prevalence of antibodies cannot be dismissed. Although most subclinical infections occur in healthy hosts, the phenomenon has been described even for transplant patients (20, 49).

Epidemic, nonpneumonic illness is also known as Pontiac fever, after the location of the first recognized episode. The infection is characterized by a very high attack rate (>95%) among those exposed, a short incubation period (hours to several days), a self-limited illness, and the absence of pulmonary infiltrates on chest radiographs. Prominent symptoms include fever, malaise, myalgias, and cough, a nonspecific flu-like syndrome. The initial episode was caused by *L. pneumophila* serogroup 1 (42). Subsequently, outbreaks have been attributed to *L. feeleii*, *L. anisa*, and *L. micdadei*. Nosocomial outbreaks of Pontiac fever have not been described. Pontiac fever is by definition an epidemic disease. Sporadic cases of nonpneumonic illness, which are indistinguishable clinically from Pontiac fever, may represent a bridge between subclinical illness and pneumonia (41).

Pneumonia is the most frequent manifestation of *Legionella* infection. The onset is usually abrupt, with high fever, malaise, myalgias, headache, and a nonproductive cough (33, 54). When sputum is produced, it often contains inflammation, but bacteria are not usually demonstrated by Gram's stain. Extrapulmonary manifestations, particularly confusion and diarrhea, are frequently prominent, but investigators have not been able to differentiate *Legionella* infection from other bacterial pneumonias in multiple prospective studies. The nonproductive cough suggests atypical pneumonia, to be differentiated especially from *Mycoplasma pneumoniae* and *Chlamydia pneumoniae* infections in outpatients. The rapid progression of pulmonary infiltrates and the frequent extension to other lobes and to the contralateral lung suggest infection with other gram-negative bacilli, particularly if colonizing enteric bacilli contaminate sputum specimens. Failure of the pneumonia to respond to therapy with expanded-spectrum cephalosporins and aminoglycosides may suggest the correct diagnosis. Abscesses are commonly documented pathologically in fatal cases (95) and are recognized radiographically in a small minority of cases (31). Convalescence is often prolonged, and radiographic resolution may be protracted. Chronic residua with scarring have been described in rare cases (15).

Dissemination of bacteria through the bloodstream occurs frequently in severe *Legionella* pneumonia. In a prospective study of *Legionella* infection, 6 (38%) of 16 patients with culture-proved infection had positive blood cultures (78). The frequency of bacteremia in mild disease is unclear. Occasionally, a secondary focus of metastatic infection will occur after direct extension from the infected lung or after bacteremic spread. Pleural empyema, pericarditis, myocarditis, endocarditis, pancreatitis, pyelonephritis, peritonitis, cellulitis, hepatic abscess, abscesses related to the gastrointestinal tract, and infections of intravascular prosthetic devices have been described. The list can be expected to expand as clinical experience grows, but it is important to document the microbial etiology of unusual occurrences rigorously. In addition, noninfectious complications, including skin rashes, encephalitis, arthritis, acute renal failure, and myoglobinuria, have been associated with *Legionella* infection.

Although most extrapulmonary lesions develop concurrently with or shortly after an episode of pneumonia, occasional infections have been described in the absence of obvious pulmonary infection. Most dramatically, an out-

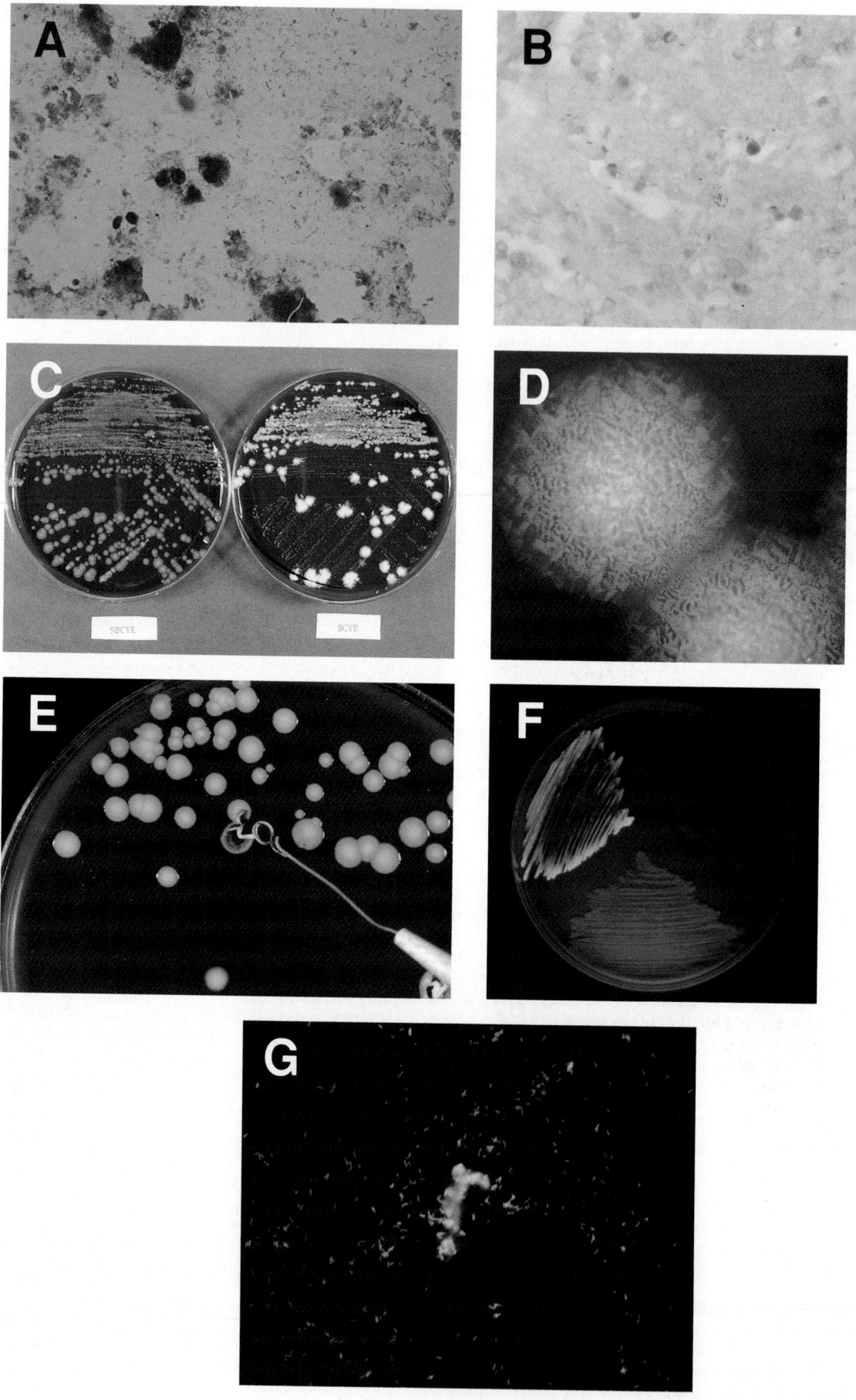

break of primary sternal-wound infections was traced to direct introduction of bacteria into the wounds during bathing (62).

Pathogenesis

Legionella infections are a result of interactions among the traditional triad of factors: bacterium, environment, and human host. The preponderance of certain species and serotypes in human infection suggests the presence of virulence factors, which have not yet been clearly defined. An antigen recognized by a single monoclonal antibody (MAb 2) is disproportionately represented on strains isolated from human infections (19) compared to environmental isolates. A 24-kDa protein, called macrophage infectivity potentiator (Mip), is required for full expression of intracellular infection (30), and mutations in the responsible gene result in diminished virulence (16). Immunization of guinea pigs with a major secretory protein results in immunoprotection, but the protein is not required for production of lethal bacterial infection (8). This major secretory protein has properties of a metalloprotease, analogous to the elastase of *Pseudomonas aeruginosa*. A major cytoplasmic membrane protein, a heat shock protein, also confers immunoprotection in a guinea pig model (7). Other potential virulence determinants have been described elsewhere (21). It is likely that virulence is a multifactorial phenomenon; to date, no single factor that can account for lethal infection has been identified (47).

Although *Legionella* spp. may cause serious disease in apparently healthy individuals, severe infection and fatal outcomes occur most commonly in patients with diminished host defenses. The most commonly documented risk factors have been cigarette smoking and chronic obstructive lung disease, chronic cardiovascular disease, renal failure, and immunosuppression caused by underlying disease or by therapy (29). Transplantation of organs and therapy with high doses of steroids have been implicated as risk factors with particular frequency. Neutropenia and infection with human immunodeficiency virus have been notably absent from the list of prominent risk factors.

The connecting link between virulent bacteria in the environment and an immunosuppressed patient is often difficult to establish. The best-documented route for transmission of infection is the generation of an infective aerosol from a *Legionella*-contaminated water source. The most direct documentation of cooling tower water as a source of infectious aerosol occurred when two maintenance workers entered an operating cooling tower to do repairs (41). Although questioned by some authorities (68), the epidemiologic evidence is convincing that water drift from evaporative condensers and cooling towers is responsible for

TABLE 2 Characteristics of genus *Legionella*

Faintly staining, thin, gram-negative, non-spore-forming bacilli
Requirement for L-cysteine on primary isolation
Mottled, cut-glass colonies on BCYEα agar under low
   magnification (dissecting microscope)
Growth stimulation by iron compounds
Utilization of amino acids for growth
Nonutilization of carbohydrates for growth
High concentrations of branched-chain fatty acids
Isoprenoid quinones (ubiquinones) with >10 isoprene units in
   side chain

epidemics of *Legionella* infection. Other open water sources that have been implicated as sources of infectious aerosols include whirlpool spas and public water fountains.

Most sporadic *Legionella* infections and some epidemics are caused by contaminated potable water that has been aerosolized by vehicles such as humidifiers, nebulizers, misters, and showerheads. In addition to inhalation of aerosols, aspiration of contaminated potable water is a probable mechanism for infection of the lower respiratory tract. Entry through the gastrointestinal tract has been suggested to explain abdominal infections, although this portal of entry has not been proved. The mechanism for transmission of bacteria in most sporadic cases is unknown, aspiration being a leading candidate. As described above, direct entry of bacteria into a fresh wound may cause nosocomial *Legionella* infection (62).

Legionella spp. are facultative intracellular pathogens and have a close relationship with monocytes and macrophages. The resident alveolar macrophage is the first phagocyte encountered by bacteria in the airspaces. This major cellular defense mechanism effectively kills many inhaled bacteria, but *Legionella* spp. multiply unimpeded in nonactivated macrophages (47). After interaction with complement receptors on the surface of macrophages, the bacteria are phagocytized into a phagosome. Fusion of phagosome and lysosome is blocked, and the bacteria multiply within the phagosome until the cell is killed, releasing the bacteria.

The inflammatory exudate in human lungs is composed of predominantly macrophages, predominantly neutrophils, or, most commonly, a mixture of polymorphonuclear neutrophils and monocytes-macrophages (95). The intense inflammatory exudate results in necrotizing infection and macroscopically visible abscesses in approximately 20% of cases. It is ironic that this often intense cellular exudate in the airspaces is so poorly mobilized into expectorated sputum.

FIGURE 1 (A) Gram's stain of *L. pneumophila* serogroup 1 from scraping of formalin-fixed lung tissue. Numerous thin gram-negative bacilli of various lengths are well demonstrated. Intracellular bacteria are evident in polymorphonuclear neutrophils. The counterstain is safranin with 0.05% basic fuchsin. Although *Legionella* spp. are not usually demonstrated in sputum, they may be seen with Gram's stain if sufficient numbers are present. Magnification, ×1,000. (B) Acid fastness of *L. micdadei* in tissue; Ziehl-Neelsen stain. The bacteria are not acid fast when isolated on agar. Magnification, ×1,250. (C) Growth of *L. pneumophila* serogroup 1 on BCYEα agar with (left) and without (right) antibiotics. Heavy growth of this isolate from sputum was obtained on both plates. Colonies on nonselective medium are small, presumably inhibited by normal flora including yeasts. (D) *L. pneumophila* serogroup 1. The faceted, cut-glass appearance of the colonies under reflected light is evident with use of a dissecting microscope. (E) *L. pneumophila* serogroup 1 on BCYEα agar. The "sticky" appearance of the colonies is evident. (F) Autofluorescence of *Legionella* spp. The blue-white fluorescence of *L. bozemanii* under long-wavelength UV light is apparent. *L. pneumophila* and *L. micdadei* do not fluoresce. (G) Direct immunofluorescence of *L. pneumophila* serogroup 1. Fluorescein-isothiocyanate conjugate polyclonal antiserum. Magnification, ×400.

TABLE 3 Diagnostic tests for *Legionella* infection

Test	Commercially available	Sensitivity (%)	Specificity (%)	Comment
Culture	Yes	70	100	Preferred method
Direct immunofluorescence	Yes	25–70	>95	Not recommended for routine use
Radioimmunoassay	Yes	70–90	>99	*L. pneumophila* serogroup 1 only
Enzyme immunoassay	No	70–90	>99	
Latex agglutination	Yes	55–90	85–99	Commercial products perform at low end
Serology (indirect immunofluorescence)	Yes	70–80	>95	Ancillary to culture

The cellular immune system is of critical importance for control of *Legionella* infections, whereas the role of antibody is unclear. The kinetics of bacterial growth in experimental models suggest that a combination of neutrophils and activated monocytes-macrophages is required for suppression of bacterial growth in vivo.

The explanation for the two very different clinical manifestations of infection, Pontiac fever and *Legionella* pneumonia, is still unclear. A strain of *L. anisa* that was associated with Pontiac fever did not multiply in human macrophages, suggesting that this biologic block might be responsible for the nonpneumonic presentation (34). In contrast, the Pontiac strain of *L. pneumophila* serogroup 1, which caused the original epidemic, produced acute pneumonia when aerosolized into guinea pigs, as did water from the evaporative condenser at the site of the epidemic (51). The mystery therefore remains unsolved.

Biohazard and Safety

The only risk for *Legionella* infection is from exposure to an infectious water source, especially when aerosols are produced. There have been no evidence of human-to-human transmission and no documented laboratory infections. Usual laboratory safety precautions are adequate with *Legionella* spp. unless a procedure is likely to produce significant aerosols.

COLLECTION, TRANSPORT, AND STORAGE OF SPECIMENS

Specimens should be obtained from the site of infection if localized disease is present. Lower respiratory tract secretions should be collected by usual methods and submitted to the laboratory as promptly as possible. There are no data to suggest the appropriate number or timing of sputum specimens. If clinical suspicion is high, several sputum samples should be submitted, because small numbers of organisms may be present. Careful induction of sputum by nebulization has the potential for increasing the yield from expectorated sputum, but there are no clinical trials on the subject. Although suspensions of NaCl reduce the recovery of *Legionella* spp., there is little alternative to procedures such as bronchoscopy or sputum induction, which use nebulized saline. Specimens should not be rejected because of delay, because *Legionella* spp. have been recovered after storage of specimens for several days at room temperature. Microscopic screening protocols, commonly used when bacterial pneumonia is suspected, are not useful for selecting samples to be cultured (48). If the presence of urinary antigen is to be tested for, a clean-catch sample should be obtained.

Blood cultures should be collected from hospitalized patients with documented pneumonia, because a positive culture is definitive evidence of infection. Clinical experience has been greatest with the BACTEC blood culture system. *L. pneumophila* has been isolated from both aerobic and anaerobic bottles but only after blind subculture onto buffered charcoal-yeast extract agar supplemented with α-ketoglutarate (BCYEα) medium (78). Lysis-centrifugation has been used to isolate *L. pneumophila* from seeded blood (18), but clinical experience has been limited. *Legionella* spp. have also been isolated in biphasic BCYE agar-broth bottles (27). The laboratory should be notified that *Legionella* spp. are suspected, because special handling is required.

Serum for antibody studies should be collected as soon as possible. A convalescent-phase specimen should be collected at least 6 weeks after the onset of infection. When infection is undifferentiated, as in Pontiac fever, documentation of the etiology must depend on serological studies.

Environmental water samples should be collected from all potential sources in clean screw-cap containers. For documentation of low levels of *Legionella* colonization, volumes of at least 1 liter should be collected. Sampling of potable-water systems should include sediment from hot-water tanks and swab samples of faucets and showerheads. If the water has recently been chlorinated, 0.5 ml of 0.1 N sodium thiosulfate should be added to each liter of water (14). Faucets and showers should be sampled with a wooden shaft swab after removal of the aerator or showerhead. The swab is then placed in 3 to 5 ml of water from the same fixture. Careful epidemiologic investigation is essential in documenting the significance of any colonized environmental sites.

DIRECT DETECTION

Investigators have attempted to develop procedures for early diagnosis by detection of antigens or nucleic acids in clinical specimens (Table 3). The initial approach, direct immunofluorescence of respiratory secretions, has a reported sensitivity of 25 to 70% (13, 28, 74). The specificity of the test is greater than 95%. False-positive results occur both because of cross-reactions with other bacteria and because of contamination with environmental *Legionella* spp. Pseudoepidemics have been caused by false-positive direct immunofluorescence tests (35). A monoclonal fluorescein-conjugated antibody to a species-specific outer membrane protein antigen produces cleaner staining and fewer cross-reactions (87). In areas of low disease prevalence, the positive predictive value of direct immunofluorescence is unacceptably low. The combination of specificity problems and low sensitivity indicates that this test should not be performed in most clinical situations.

Detection of *Legionella* antigen in urine and other body fluids has also been used successfully to diagnose *Legionella* infections expeditiously by radioimmunoassay, enzyme immunoassay, or latex agglutination (83). A commercially available radioimmunoassay for *L. pneumophila* serogroup 1 has a sensitivity of 80% or greater and a very high specificity (1). The drawbacks of this reagent are the limitation to a single serogroup, albeit the most frequent pathogen in the genus, and the need for radiolabeled reagents with a short shelf life. Cross-reactions have been demonstrated between urinary antigens of several *L. pneumophila* serogroups (55). Antigen may be excreted for months, so a positive urinary assay does not equate precisely with acute infection (56). Enzyme immunoassays and latex agglutination assays either are not commercially available or have not performed acceptably for clinical use.

A genus-specific DNA probe assay is also commercially available. After positive cutoff values were adjusted to obtain an acceptable specificity (>99%), the sensitivity of the assay was approximately 70%, which is equivalent to the sensitivity of direct immunofluorescence in the same study (74). At least one pseudoepidemic was traced to false-positive probe tests (59). This end point for the probe assay is objective, an advantage over direct immunofluorescence, and the methodology can be employed for other infectious agents. The decision to add the test will depend on local prevalence of disease. None of the direct detection techniques can substitute for culture.

ISOLATION

The classic triad of diagnostic approaches is available for diagnosis of *Legionella* infections: isolation of the bacterium in culture, direct detection of bacterial antigens or nucleic acids in clinical specimens, and documentation of a serological response to the bacterium. The overwhelmingly preferred diagnostic method is culture (100). The sensitivity of culture is comparable to that of other methods, and the specificity is 100%, a significant advantage when the prevalence of disease in the test population is low. The ubiquity of *Legionella* spp. in water systems even produced a false-positive culture result when a postmortem sample was apparently contaminated with water from the autopsy room (61). Although the results of culture are not available immediately, most isolates are recovered within 72 h, a clinically relevant time frame for assessment of therapy and clinical response.

The standard for isolation of *Legionella* spp. from clinical specimens is BCYEα, a medium developed by the pioneering efforts of the late James Feeley (71). Yeast extract is included as a rich source of nutrients; L-cysteine, iron compounds, and α-ketoglutarate are added to stimulate growth. Activated charcoal serves as a scavenger of a variety of compounds, such as superoxide radicals and peroxides, that are produced in the medium, especially after exposure to light. Other methods used to minimize the effects of toxic substances include washing the yeast extract agar and adding albumin, either as a substitute for charcoal or as a second protective measure. ACES [N-(2-acetamido)-2-aminoethanesulfonic acid] buffer incorporated in

BCYEα agar provides an optimal pH for *Legionella* growth without inhibitory effects. Although BCYEα agar is the standard, it is worth noting that *L. pneumophila* has been isolated on commercially prepared chocolate agar (22).

When specimens from nonsterile sources are cultured, selective media or treatments should be used to inhibit other bacteria (Fig. 1C). The most commonly used selective medium consists of BCYEα medium supplemented with polymyxin B, anisomycin, and either cefamandole or vancomycin. It is essential that a noninhibitory medium be inoculated also, because some strains, particularly of species other than *L. pneumophila*, are inhibited by the antibiotics, with the cephalosporin mixtures being more inhibitory than those containing vancomycin (60). Brief treatment of sputum specimens with heat or acid (13) may enhance recovery of *Legionella* spp. from clinical specimens, but the techniques have not been widely used.

When environmental samples are cultured, it is useful to add glycine to the basic BCYEα medium in addition to the vancomycin-polymyxin B-anisomycin mixture (14). With the addition of bromothymol blue and bromocresol purple for differential coloration of some *Legionella* spp., the agar becomes modified Wadowsky-Yee medium (71). Acidification of the sample for 15 min at pH 2.0 to 2.2 by addition of acid buffer (0.2 M HCl, 0.2 M KCl) followed by neutralization to pH 7.0 with 0.1 N KOH is also useful for environmental samples. Potable-water specimens should be concentrated by filtration (14). Centrifugation at $6,000 \times g$ for 10 min is comparable to filtration (12) but difficult to apply to large-volume samples. After concentration, sediment is resuspended or filters are vortexed in 10 to 20 ml of the original water. Swab specimens are expressed into their water-suspending medium before inoculation.

Preincubation of water samples at 35°C may increase the yield from lightly contaminated samples, putatively by allowing multiplication of bacteria within environmental amoebae (82). The relationship between *Legionella* spp. and free-living amoebae has been exploited for diagnosis of both human infection and environmental colonization. Isolation of *Legionella*-like bacteria from sputum specimens and of *L. micdadei* from water samples only after incubation of the samples with amoebae has been reported (32, 80). These enrichment techniques may be appropriate for use by specialized reference laboratories.

Procedures for processing water samples are available from several sources (12, 14). Commercially available media are of variable quality (52, 60), but several manufacturers have maintained production of high-quality media. If the media are made in-house, it is important to be aware of variable performance of the agar bases (24). Cost-effective alternatives for the buffering agent have been described elsewhere (26). Quality control should be performed with bacterial strains that have been passaged minimally on agar. Laboratory-adapted strains, including some available from the American Type Culture Collection, will grow well on lots of media that do not support the growth of fresh clinical isolates (52). If commercial quality control data are used, documentation of the procedures can be requested.

Inoculated plates should be streaked for isolation, incubated for a minimum of 5 days in a humid atmosphere, and observed daily for growth. CO_2 at a concentration of 2 to 5% may stimulate growth of some species, but higher concentrations of CO_2 may be inhibitory. Young colonies can be detected perhaps 12 to 24 h earlier if a dissecting microscope is used for examination.

IDENTIFICATION

Colonies on BCYEα or selective BCYEα agar should be selected for Gram stain if they are not represented on enriched medium such as sheep blood agar or chocolate agar or if they have a characteristic macroscopic experience. Viewed through a dissecting microscope (Fig. 1D), the colonies have a "mottled" surface, an iridescent red-blue-green sheen, or a faceted "cut-glass" appearance. Although the colonies themselves are round and entire, the play of light on the surface produces an appearance of irregularity or complex internal structure. As colonial growth continues, the iridescence can be appreciated with the naked eye, particularly where growth is confluent. Another clue is the sticky consistency that colonies of *L. pneumophila* often have when manipulated with a glass rod (Fig. 1E).

Legionella spp. stain very poorly with the usual safranin counterstain. The staining time of the safranin counterstain should be prolonged. Alternatively, 0.05% basic fuchsin can be added to the safranin stain to increase the intensity of the staining. If a Gram stain of the suspect colonies reveals thin, palely staining gram-negative bacilli, the colonies should be subcultured for demonstration of cysteine dependence. The most rigorous test of dependence is to subculture the isolate to two agar media that differ only in the presence of L-cysteine (71). For laboratories in which *Legionella* spp. are infrequently isolated, it may be inconvenient and costly to stock the additional cysteine-deficient media. An acceptable alternative is subculture of the isolate onto BCYEα agar and sheep blood agar in parallel. Microbiologic judgment must be used, because dichotomous growth on blood agar and BCYEα agar could result from differences in nutrient composition. Similarly, bacteria other than *Legionella* spp., including thermophilic spore-forming bacilli (88), *Francisella tularensis* (81), and *Bordetella pertussis* (72), are dependent on L-cysteine for growth or on charcoal for removal of inhibitory substances. In the case of *Bordetella* and *Francisella* spp., the problem is complicated by similar morphology in Gram stains, cross-reactions with some *Legionella* antisera (6, 72, 81), and potential clinical presentation as lower respiratory tract infection. Careful analysis of colonial morphology may suggest the correct identification, which can be confirmed by application of appropriate serological or genetic analysis.

Biochemical tests are of little use in the differentiation of *Legionella* spp. from other genera or in the differentiation of species within the genus. The genus is asaccharolytic, and other reactions are weak, variable, or temperature dependent. The accompanying table of reactions, used judiciously and with caution, may provide confirmatory information (Table 1). The hippurate hydrolysis test is useful for biochemical confirmation of the most common pathogenic species, *L. pneumophila* (45); the diagnostic utility of this test has been diminished by the variability of more recently isolated species, however. Although autofluorescence of colonies under long-wavelength UV light (Wood's lamp) differentiates certain *Legionella* spp. (Fig. 1F), the infrequency of occurrence of the fluorescent species in human disease compromises the clinical utility of the test. Browning of tyrosine-containing medium is more variable, and rare isolates of *L. pneumophila* do not produce the expected result. Catalase and oxidase reactions and bacterial motility may be weak and difficult to detect.

For practical purposes, serological typing of putative *Legionella* isolates is the simplest approach and is sufficient for characterization of most isolates in clinical laboratories. The power of serological analysis has been eroded by cross-reactions among *Legionella* species and serogroups, which are being increasingly described. A positive reaction provides a reasonably certain identification at the genus level. The positive predictive value is probably acceptably high if the isolate is identified as a commonly isolated pathogen such as *L. pneumophila* serogroup 1 or 6, *L. micdadei*, or *L. dumoffii*.

Although agglutinating antisera are commercially available, the most common procedure has been direct immunofluorescence with polyclonal or monoclonal antibodies (Fig. 1G). Commercially available polyclonal antibodies are directed at serogroup-specific antigens, probably lipopolysaccharide, and must be used for serotyping (25). Bacterial suspensions should be sparse to minimize quenching of fluorescence by the antigenic mass, which may produce a weakly positive reaction. The polyclonal reagents cross-react with some common environmental organisms and human pathogens, in particular *Pseudomonas fluorescens*, *P. aeruginosa*, *Bacteroides fragilis*, *F. tularensis*, and *B. pertussis* (see also Serological Tests below). A monoclonal fluoresceinated antiserum against a species-specific outer membrane protein antigen of *L. pneumophila* (86) produces a cleaner background and does not cross-react with the common non-*Legionella* pathogens (23, 87).

Definitive identification of isolates requires genetic analysis by techniques that are beyond the reach of most clinical laboratories. A commercially available gene probe correctly identifies a spectrum of *Legionella* spp., the exceptions being four strains of *L. bozemanii*, and does not react with non-*Legionella* bacteria (94). This procedure brings definitive identification at the genus level within the capability of many laboratories if sufficient numbers of isolates are recovered to justify the expense of the reagents. Characterization of fatty acid composition and ubiquinone content have also been used to identify *Legionella* spp. (58), but these techniques are also restricted to highly specialized laboratories. Isolates should be sent to a reference laboratory for definitive identification if (i) serological analysis indicates an infrequently isolated species or serotype, (ii) phenotypic characteristics suggest a *Legionella* sp. but available antisera do not react with the isolate, (iii) the isolate reacts with antisera to more than one serogroup or species, or (iv) phenotypic characteristics are atypical.

SEROLOGICAL TESTS

Serological tests are useful for epidemiologic studies of prevalence and for diagnosis of suspect cases in which the patient does not have pneumonia or is not able to produce adequate lower respiratory secretions. Such tests are an adjunct to microbiologic diagnosis and should not be used as a substitute for it. The disadvantages of serological tests are that the diagnosis is usually retrospective and that cross-reactions among *Legionella* spp. and with other bacteria make the result presumptive.

A variety of methodologies, including microagglutination, enzyme immunoassay, and counterimmunoelectrophoresis, has been developed to measure the serological response. The most frequently used test is indirect immunofluorescence, which is also commercially available and has been extensively evaluated. The immunologic response includes antibodies of immunoglobulin G (IgG), IgM, and, to a lesser extent, IgA classes (93). It is important to use an antiglobulin reagent that detects all classes. IgM antibody can be measured to provide an early provisional serological diagnosis with a single serum specimen. This antibody class can persist for many months in some patients, however (99).

The sensitivity of serological diagnosis is approximately 80% (28). Seroconversion occurs in most patients within 3 weeks, but patients should be monitored for 6 weeks (28), and some patients never seroconvert. The most common criterion for seroconversion is an increase in antibodies of at least fourfold to a titer of 1:128 (93). A single titer cannot be used to diagnose individual infections and is only presumptive in studies of epidemic disease.

The specificity of serological diagnosis has been estimated at 96 to 99% (92, 93). In areas where there is a moderate incidence of infection, demonstration of seroconversion will result in an acceptable predictive value for *Legionella* infection. Where disease is rare (1% of patients tested or less), the predictive value will be considerably lower, resulting in a more presumptive diagnosis. Serological cross-reactions have been demonstrated with *P. aeruginosa*, *P. fluorescens*, *Pseudomonas pseudomallei*, *Pseudomonas alcaligenes*, *Xanthomonas* spp., *Flavobacterium* spp., *Bacillus cereus* spores, *Bacteroides fragilis*, *B. pertussis*, *F. tularensis*, *Campylobacter jejuni*, and members of the family *Enterobacteriaceae*, such as *Citrobacter freundii*. Heat-killed and formalin-inactivated bacteria have been most frequently used in indirect immunofluorescence tests, with the former producing higher titers (91). The immunologic response is directed against lipopolysaccharide and protein antigens. The genus-specific 60-kDa protein antigen, for instance, has epitopes that react with 39 non-*Legionella* bacterial species (75). Some patients respond with antibodies that react with a single serogroup; others develop a more diverse response (93). To detect the spectrum of *Legionella* infections, therefore, it is important to test multiple antigens, but the precise identity of the infecting species cannot be determined with certainty. Absorption of patients' sera with cross-reacting antigens provides greater specificity but makes the assay more complicated and decreases sensitivity (75, 92).

ANTIMICROBIAL SUSCEPTIBILITIES

In vitro antimicrobial susceptibility results do not correlate well with clinical response, so routine testing is not appropriate. It has been suggested that the intracellular location of legionellae in phagocytes is responsible for the dichotomy. In addition, some antibiotics are bound by the activated charcoal in commonly used media. The standard therapy for *Legionella* infection is erythromycin, although no prospective, randomized clinical trials have ever been performed. Rifampin is often added for severe, life-threatening disease. Natural resistance to these antibiotics has not yet occurred. Trimethoprim-sulfamethoxazole, several quinolone antibiotics, and new macrolides such as clarithromycin and azithromycin have been used successfully, but clinical experience is limited.

EVALUATION, INTERPRETATION, AND REPORTING OF RESULTS

Respiratory secretions or other clinically indicated sites should be cultured. Urine should be tested for *L. pneumophila* serogroup 1 antigen if feasible. Blood cultures should be collected for *Legionella* culture for hospitalized patients.

Positive results should be called in to the caregiver and the hospital epidemiologist if appropriate. An acute-phase serum for serological analysis should be collected and stored; if a specific diagnosis is not made during the acute infection, convalescent-phase sera should be obtained 3 weeks later and again at 6 weeks if seroconversion has not occurred. An at least fourfold increase in antibody titer to a level of 1:128 or greater is presumptive evidence of acute infection.

The clinical microbiologist may also be called upon to participate in investigations of epidemic disease and in decisions about control of *Legionella* infection. A variety of molecular techniques is now available for characterizing isolated strains and matching isolates from clinical and environmental sources (4). It is no longer adequate to employ serological characterization alone.

It is potentially possible, although not easy, to prevent *Legionella* infections by controlling exposure to bacteria in the aquatic environment. There are several schools of thought on the proper approach to monitoring for the presence of *Legionella* spp. and *Legionella* infections (17, 50, 98). It is clear that attention should be concentrated on the individuals at highest risk, particularly immunosuppressed patients in hospitals. In addition, common-sense measures should be instituted. Only sterile water should be used in situations in which aerosols will be generated or when direct contact with sick patients will ensue. Attention should be paid to the proximity of air intake vents to potential generators of aerosols. Environmental and potable-water systems should be maintained according to established protocols and manufacturers' instructions.

Monitoring of clinical disease can be focused most efficiently on high-risk patients, using culture of respiratory secretions (64). Microbiologic characterization of fatal cases of pneumonia from the autopsy room represents another approach to surveillance of *Legionella* infection. Routine microbiologic surveillance of the environment in the absence of documented human infection is controversial. Once clinical infection has been documented, several methods of environmental control are available (69). Each has advantages and drawbacks, and none is permanent, so continued environmental surveillance is required.

REFERENCES

1. **Aquero-Rosenfeld, M. E., and P. H. Edelstein.** 1988. Retrospective evaluation of the Du Pont radioimmunoassay kit for detection of *Legionella pneumophila* serogroup 1 antigenuria in humans. *J. Clin. Microbiol.* **26:**1775–1778.
2. **Alary, M., and J. R. Joly.** 1991. Risk factors for contamination of domestic hot water systems by legionellae. *Appl. Environ. Microbiol.* **57:**2360–2367.
3. **Alary, M., and J. R. Joly.** 1992. Factors contributing to the contamination of hospital water distribution systems by legionellae. *J. Infect. Dis.* **165:**565–569.
4. **Barbaree, J. M.** 1993. Selecting a subtyping technique for use in investigations of legionellosis epidemics, p. 169–172. *In* J. M. Barbaree, R. F. Breiman, and A. P. Dufour (ed.), *Legionella: Current Status and Emerging Perspectives.* American Society for Microbiology, Washington, D.C.
5. **Barbaree, J. M., B. S. Fields, J. C. Feeley, G. W. Gorman, and W. T. Martin.** 1986. Isolation of protozoa from water associated with a legionellosis outbreak and demonstration of intracellular multiplication of *Legionella pneumophila. Appl. Environ. Microbiol.* **51:**422–424.
6. **Benson, R. F., W. L. Thacker, B. B. Plikaytis, and H. W. Wilkinson.** 1987. Cross-reactions in *Legionella* antisera with *Bordetella pertussis* strains. *J. Clin. Microbiol.* **25:**594–596.
7. **Blander, S. J., and M. A. Horwitz.** 1993. Major cytoplasmic

membrane protein of *Legionella pneumophila*, a genus common antigen and member of the hsp 60 family of heat shock proteins, induces protective immunity in a guinea pig model of Legionnaires' disease. *J. Clin. Invest.* **91:**717–723.

8. **Blander, S. J., L. Szeto, H. A. Shuman, and M. A. Horwitz.** 1990. An immunoprotective molecule, the major secretory protein of Legionella pneumophila, is not a virulence factor in a guinea pig model of Legionnaires' disease. *J. Clin. Invest.* **86:**817–824.

9. **Breiman, R. F.** 1993. Modes of transmission in epidemic and nonepidemic *Legionella* infection: directions for further study, p. 30–35. *In* J. M. Barbaree, R. F. Breiman, and A. P. Dufour (ed.), *Legionella: Current Status and Emerging Perspectives.* American Society for Microbiology, Washington, D.C.

10. **Brenner, D. J., A. G. Steigerwalt, P. Epple, W. F. Bibb, R. M. McKinney, R. W. Starnes, J. M. Colville, R. K. Selander, P. H. Edelstein, and C. W. Moss.** 1988. *Legionella pneumophila* serogroup Lansing 3 isolated from a patient with fatal pneumonia, and descriptions of *L. pneumophila* subsp. *pneumophila* subsp. nov., *L. pneumophila* subsp. *fraseri* subsp. nov., and *L. pneumophila* subsp. *pascullei* subsp. nov. *J. Clin. Microbiol.* **26:**1695–1703.

11. **Brenner, D. J., A. G. Steigerwalt, and J. E. McDade.** 1979. Classification of the legionnaires' disease bacterium: *Legionella pneumophila*, genus novum, species nova, of the family Legionellaceae, familia nova. *Ann. Intern. Med.* **90:**656–658.

12. **Brindle, R. J., P. J. Stannett, and R. N. Cunliffe.** 1987. *Legionella pneumophila*: comparison of isolation from water specimens by centrifugation and filtration. *Epidemiol. Infect.* **99:**241–247.

13. **Buesching, W. J., R. A. Brust, and L. W. Ayers.** 1983. Enhanced primary isolation of *Legionella pneumophila* from clinical specimens by low-pH treatment. *J. Clin. Microbiol.* **17:**1153–1155.

14. **Centers for Disease Control.** 1992. *Procedures for the Recovery of Legionella from the Environment.* U.S. Department of Health and Human Services, Atlanta.

15. **Chastre, J., G. Raghu, P. Soler, P. Brun, F. Basset, and C. Gibert.** 1987. Pulmonary fibrosis following pneumonia due to acute Legionnaires' disease. Clinical, ultrastructural, and immunofluorescent study. *Chest* **91:**57–62.

16. **Cianciotto, N. P., B. I. Eisenstein, C. H. Mody, and N. C. Engleberg.** 1990. A mutation in the mip gene results in an attenuation of *Legionella pneumophila* virulence. *J. Infect. Dis.* **162:**121–126.

17. **Dennis, P. J.** 1990. Reducing the risk of Legionnaires' disease. *Ann. Occup. Hyg.* **34:**189–193.

18. **Dorn, G. L., and W. R. Barnes.** 1979. Rapid isolation of *Legionella pneumophila* from seeded donor blood. *J. Clin. Microbiol.* **10:**114–115.

19. **Dournon, E., W. F. Bibb, P. Rajagopalan, N. Desplaces, and R. M. McKinney.** 1988. Monoclonal antibody reactivity as a virulence marker for *Legionella pneumophila* serogroup 1 strains. *J. Infect. Dis.* **157:**496–501.

20. **Dowling, J. N., A. W. Pascule, F. N. Frola, M. K. Zaphyr, and R. B. Yee.** 1984. Infections caused by *Legionella micdadei* and *Legionella pneumophila* among renal transplant recipients. *J. Infect. Dis.* **149:**703–713.

21. **Dowling, J. N., A. K. Saha, and R. H. Glew.** 1992. Virulence factors of the family Legionellaceae. *Microbiol. Rev.* **56:**32–60.

22. **Dumoff, M.** 1979. Direct in-vitro isolation of the legionnaires' disease bacterium in two fatal cases. Cultural and staining characteristics. *Ann. Intern. Med.* **90:**694–696.

23. **Edelstein, P. H., K. B. Beer, J. C. Sturge, A. J. Watson, and L. C. Goldstein.** 1985. Clinical utility of a monoclonal direct fluorescent reagent specific for *Legionella pneumophila*: comparative study with other reagents. *J. Clin. Microbiol.* **22:**419–421.

24. **Edelstein, P. H., and M. A. Edelstein.** 1991. Comparison of different agars used in the formulation of buffered charcoal yeast extract medium. *J. Clin. Microbiol.* **29:**190–191.

25. **Edelstein, P. H., and M. A. C. Edelstein.** 1989. Evaluation of the Merifluor-*Legionella* immunofluorescent reagent for identifying and detecting 21 *Legionella* species. *J. Clin. Microbiol.* **27:**2455–2458.

26. **Edelstein, P. H., and M. A. C. Edelstein.** 1993. Comparison of three buffers used in the formulation of buffered charcoal yeast extract medium. *J. Clin. Microbiol.* **31:**3329–3330.

27. **Edelstein, P. H., R. D. Meyer, and S. M. Finegold.** 1979. Isolation of *Legionella pneumophila* from blood. *Lancet* **i:**750–751.

28. **Edelstein, P. H., R. D. Meyer, and S. M. Finegold.** 1980. Laboratory diagnosis of Legionnaires' disease. *Am. Rev. Respir. Dis.* **121:**317–327.

29. **England, A. C., III, D. W. Fraser, B. D. Plikaytis, T. F. Tsai, G. Storch, and C. V. Broome.** 1981. Sporadic legionellosis in the United States: the first thousand cases. *Ann. Intern. Med.* **94:**164–170.

30. **Engleberg, N. C.** 1993. Genetic studies of *Legionella* pathogenesis, p. 63–68. *In* J. M. Barbaree, R. F. Breiman, and A. P. Dufour (ed.), *Legionella: Current Status and Emerging Perspectives.* American Society for Microbiology, Washington, D.C.

31. **Fairbank, J. T., A. C. Mamourian, P. A. Dietrich, and J. C. Girod.** 1983. The chest radiograph in Legionnaires' disease. Further observations. *Radiology* **147:**33–34.

32. **Fallon, R. J., and T. J. Rowbotham.** 1990. Microbiological investigations into an outbreak of Pontiac fever due to *Legionella micdadei* associated with use of a whirlpool. *J. Clin. Pathol.* **43:**479–483.

33. **Fang, G. D., V. L. Yu, and R. M. Vickers.** 1989. Disease due to the *Legionellaceae* (other than *Legionella pneumophila*). Historical, microbiological, clinical, and epidemiological review. *Medicine* (Baltimore) **68:**116–132.

34. **Fields, B. S., J. M. Barbaree, G. N. Sanden, and W. E. Morrill.** 1990. Virulence of a *Legionella anisa* strain associated with Pontiac fever: an evaluation using protozoan, cell culture, and guinea pig models. *Infect. Immun.* **58:**3139–3142.

35. **Finkelstein, R., W. A. Palutke, B. B. Wentworth, J. G. Geiger, and G. D. Bostic.** 1993. Colonization of the respiratory tract with *Legionella* species. *Isr. J. Med. Sci.* **29:**277–279.

36. **Flesher, A. R., D. L. Kasper, P. A. Modern, and E. O. Mason, Jr.** 1980. *Legionella pneumophila*: growth inhibition by human pharyngeal flora. *J. Infect. Dis.* **142:**313–317.

37. **Fraser, D. W., T. R. Tsai, W. Orenstein, W. E. Parkin, H. J. Beecham, R. G. Sharrar, J. Harris, G. F. Mallison, S. M. Martin, J. E. McDade, C. C. Shepard, and P. S. Brachman.** 1977. Legionnaires' disease: description of an epidemic of pneumonia. *N. Engl. J. Med.* **297:**1189–1197.

38. **Fry, N. K., S. Warwick, N. A. Saunders, and T. M. Embley.** 1991. The use of 16S ribosomal RNA analyses to investigate the phylogeny of the family Legionellaceae. *J. Gen. Microbiol.* **137:**1215–1222.

39. **Garrity, G. M., A. Brown, and R. M. Vickers.** 1980. *Tatlockia* and *Fluoribacter.* Two new genera of organisms resembling *Legionella pneumophila. Int. J. Syst. Bacteriol.* **30:**609–614.

40. **Garrity, G. M., E. M. Elder, B. Davis, R. M. Vickers, and A. Brown.** 1982. Serological and genotypic diversity among serogroup 5-reacting environmental *Legionella* isolates. *J. Clin. Microbiol.* **15:**646–653.

41. **Girod, J. C., R. C. Reichman, W. C. Winn, Jr., D. N. Klaucke, R. L. Vogt, and R. Dolin.** 1982. Pneumonic and nonpneumonic forms of legionellosis. The result of a common-source exposure to *Legionella pneumophila. Arch. Intern. Med.* **142:**545–547.

42. **Glick, T. H., M. B. Gregg, B. Berman, G. Mallison, W. W. Rhodes, Jr., and I. Kassanoff.** 1978. Pontiac fever. An epidemic of unknown etiology in a health department. I. Clinical and epidemiologic aspects. *Am. J. Epidemiol.* **107:**149–160.

43. Harrison, T. G., N. A. Saunders, A. Haththotuwa, N. Doshi, and A. G. Taylor. 1992. Further evidence that genotypically closely related strains of *Legionella pneumophila* can express different serogroup specific antigens. *J. Gen. Microbiol.* **37:**155–161.

44. Harrison, T. G., N. A. Saunders, A. Haththotuwa, G. Hallas, R. J. Birtles, and A. G. Taylor. 1990. Phenotypic variation amongst genotypically homogeneous *Legionella pneumophila* serogroup 1 isolates: implications for the investigation of outbreaks of Legionnaires' disease. *Epidemiol. Infect.* **104:**171–180.

45. Hebert, G. A. 1981. Hippurate hydrolysis by *Legionella pneumophila*. *J. Clin. Microbiol.* **13:**240–242.

46. Hebert, G. A., C. W. Moss, L. K. McDougal, F. M. Bozeman, R. M. McKinney, and D. J. Brenner. 1980. The rickettsia-like organisms TATLOCK (1943) and HEBA (1959): bacteria phenotypically similar to but genetically distinct from *Legionella pneumophila* and the WIGA bacterium. *Ann. Intern. Med.* **92:**45–52.

47. Horwitz, M. A. 1993. Toward an understanding of host and bacterial molecules mediating *Legionella pneumophila* pathogenesis, p. 55–62. *In* J. M. Barbaree, R. F. Breiman, and A. P. Dufour (ed.), *Legionella: Current Status and Emerging Perspectives.* American Society for Microbiology, Washington, D.C.

48. Ingram, J. G., and J. F. Plouffe. 1994. Danger of sputum purulence screens in culture of *Legionella* species. *J. Clin. Microbiol.* **32:**209–210.

49. Jacobs, F., C. Liesnard, J. P. Goldstein, M. J. Struelens, G. Primo, J. L. Leclerc, and J. P. Thys. 1990. Asymptomatic *Legionella pneumophila* infections in heart transplant recipients. *Transplantation* **50:**174–175.

50. Joly, J. R. 1993. Monitoring for the presence of *Legionella*: where, when, and how?, p. 211–216. *In* J. M. Barbaree, R. F. Breiman, and A. P. Dufour (ed.), *Legionella: Current Status and Emerging Perspectives.* American Society for Microbiology, Washington, D.C.

51. Kaufmann, A. F., J. E. McDade, C. M. Patton, J. V. Bennett, P. Skaliy, J. C. Feeley, D. C. Anderson, M. E. Potter, V. F. Newhouse, M. B. Gregg, and P. S. Brachman. 1981. Pontiac fever: isolation of the etiologic agent (*Legionella pneumophila*) and demonstration of its mode of transmission. *Am. J. Epidemiol.* **114:**337–347.

52. Keathley, J. D., and W. C. Winn, Jr. 1985. Comparison of media for recovery of clinical isolates of *Legionella pneumophila*. *Am. J. Clin. Pathol.* **83:**498–499.

53. King, C. H., E. B. Shotts, R. E. Wooley, and K. G. Porter. 1988. Survival of coliforms and bacterial pathogens within protozoa during chlorination. *Appl. Environ. Microbiol.* **54:**3023–3033.

54. Kirby, B. D., K. M. Snyder, R. D. Meyer, and S. M. Finegold. 1980. Legionnaires' disease: report of sixty-five nosocomially acquired cases of review of the literature. *Medicine* (Baltimore) **59:**188–205.

55. Kohler, R. B., L. J. Wheat, M. L. French, P. L. Meenhorst, W. C. Winn, Jr., and P. H. Edelstein. 1985. Cross-reactive urinary antigens among patients infected with *Legionella pneumophila* serogroups 1 and 4 and the Leiden 1 strain. *J. Infect. Dis.* **152:**1007–1012.

56. Kohler, R. B., W. C. Winn, Jr., and L. J. Wheat. 1984. Onset and duration of urinary antigen excretion in Legionnaires' disease. *J. Clin. Microbiol.* **20:**605–607.

57. Kuchta, J. M., S. J. States, J. E. McGlaughlin, J. H. Overmeyer, R. M. Wadowsky, A. M. McNamara, R. S. Wolford, and R. B. Yee. 1985. Enhanced chlorine resistance of tap water-adapted *Legionella pneumophila* as compared with agar medium-passaged strains. *Appl. Environ. Microbiol.* **50:**21–26.

58. Lambert, M. A., and C. W. Moss. 1989. Cellular fatty acid compositions and isoprenoid quinone contents of 23 *Legionella* species. *J. Clin. Microbiol.* **27:**465–473.

59. Laussucq, S., D. Schuster, W. J. Alexander, W. L. Thacker, H. W. Wilkinson, and J. S. Spika. 1988. False-

60. Lee, T. C., R. M. Vickers, V. L. Yu, and M. M. Wagener. 1993. Growth of 28 *Legionella* species on selective culture media: a comparative study. *J. Clin. Microbiol.* **31:**2764–2768.

61. Lightfoot, N. F., I. R. Richardson, J. Shrimanker, and D. J. Farrell. 1991. Post-mortem isolation of *Legionella* due to contamination. *Lancet* **337:**376.

62. Lowry, P. W., R. J. Blankenship, W. Gridley, N. J. Troup, and L. S. Tompkins. 1991. A cluster of *Legionella* sternal-wound infections due to postoperative topical exposure to contaminated tap water. *N. Engl. J. Med.* **324:**109–113.

63. Marrie, T. J., D. Haldane, G. Bezanson, and R. Peppard. 1992. Each water outlet is a unique ecological niche for *Legionella pneumophila*. *Epidemiol. Infect.* **108:**261–270.

64. Marrie, T. J., S. Macdonald, K. Clarke, and D. Haldane. 1991. Nosocomial legionnaires' disease: lessons from a four-year prospective study. *Am. J. Infect. Control* **19:**79–85.

65. McDade, J. E., D. J. Brenner, and F. M. Bozeman. 1979. Legionnaires' disease bacterium isolated in 1947. *Ann. Intern. Med.* **90:**659–661.

66. McDade, J. E., C. C. Shepard, D. W. Fraser, T. R. Tsai, M. A. Redus, and W. R. Dowdle. 1977. Legionnaires' disease: isolation of a bacterium and demonstration of its role in other respiratory disease. *N. Engl. J. Med.* **297:**1197–1203.

67. Muder, R. R., V. L. Yu, J. K. McClure, F. J. Kroboth, S. D. Kominos, and R. M. Lumish. 1983. Nosocomial Legionnaires' disease uncovered in a prospective pneumonia study. *JAMA* **249:**3184–3188.

68. Muder, R. R., V. L. Yu, and A. H. Woo. 1986. Mode of transmission of *Legionella pneumophila*. A critical review. *Arch. Intern. Med.* **146:**1607–1612.

69. Muraca, P. W., V. L. Yu, and A. Goetz. 1990. Disinfection of water distribution systems for *Legionella*: a review of application procedures and methodologies. *Infect. Control Hosp. Epidemiol.* **11:**79–88.

70. Myerowitz, R. L., A. W. Pasculle, J. N. Dowling, G. J. Pazin, M. Puerzer, R. B. Yee, C. R. Rinaldo, Jr., and T. R. Hakala. 1979. Opportunistic lung infection due to "Pittsburgh Pneumonia Agent." *N. Engl. J. Med.* **301:**953–958.

71. Nash, P., and M. M. Krenz. 1991. Culture media, p. 1226–1288. *In* A. Balows, W. J. Hausler, Jr., K. L. Herrmann, H. D. Isenberg, and H. J. Shadomy (ed.), *Manual of Clinical Microbiology*, 5th ed. American Society for Microbiology, Washington, D.C.

72. Ng, V. L., M. York, and W. K. Hadley. 1989. Unexpected isolation of *Bordetella pertussis* from patients with acquired immunodeficiency syndrome. *J. Clin. Microbiol.* **27:**337–338.

73. Ott, M., P. Messner, J. Heesemann, R. Marre, and J. Hacker. 1991. Temperature-dependent expression of flagella in *Legionella*. *J. Gen. Microbiol.* **137:**1955–1961.

74. Pasculle, A. W., G. E. Veto, S. Krystofiak, K. McKelvey, and K. Vrsalovic. 1989. Laboratory and clinical evaluation of a commercial DNA probe for the detection of *Legionella* spp. *J. Clin. Microbiol.* **27:**2350–2358.

75. Plikaytis, B. B., G. M. Carlone, C. P. Pau, and H. W. Wilkinson. 1987. Purified 60-kilodalton *Legionella* protein antigen with *Legionella*-specific and nonspecific epitopes. *J. Clin. Microbiol.* **25:**2080–2084.

76. Reingold, A. L., B. M. Thomason, B. J. Brake, L. Thacker, H. W. Wilkinson, and J. N. Kuritsky. 1984. *Legionella* pneumonia in the United States: the distribution of serogroups and species causing human illness. *J. Infect. Dis.* **149:**819.

77. Richardson, I. R. 1990. The incidence of *Bdellovibrio* spp. in man-made water systems: coexistence with legionellas. *J. Appl. Bacteriol.* **69:**134–140.

78. Rihs, J. D., V. L. Yu, J. J. Zuravleff, A. Goetz, and R. R. Muder. 1985. Isolation of *Legionella pneumophila* from blood with the BACTEC system: a prospective study yielding positive results. *J. Clin. Microbiol.* **22:**422–424.

positive DNA probe test for *Legionella* species associated with a cluster of respiratory illnesses. *J. Clin. Microbiol.* **26:**1442–1444.

79. **Rowbotham, T. J.** 1980. Preliminary report on the pathogenicity of *Legionella pneumophila* for freshwater and soil amoebae. *J. Clin. Pathol.* **33:**1179–1183.

80. **Rowbotham, T. J.** 1983. Isolation of *Legionella pneumophila* from clinical specimens via amoebae, and the interaction of those and other isolates with amoebae. *J. Clin. Pathol.* **36:**978–986.

81. **Roy, T. M., D. Fleming, and W. H. Anderson.** 1989. Tularemic pneumonia mimicking Legionnaires' disease with false-positive direct fluorescent antibody stains for *Legionella. South. Med. J.* **82:**1429–1431.

82. **Sanden, G. N., W. E. Morrill, B. S. Fields, R. F. Breiman, and J. M. Barbaree.** 1992. Incubation of water samples containing amoebae improves detection of legionellae by the culture method. *Appl. Environ. Microbiol.* **58:**2001–2004.

83. **Sathapatayavongs, B., R. B. Kohler, L. J. Wheat, A. White, and W. C. Winn, Jr.** 1983. Rapid diagnosis of Legionnaires' disease by latex agglutination. *Am. Rev. Respir. Dis.* **127:** 559–562.

84. **States, S. J., L. F. Conley, J. M. Kuchta, B. M. Oleck, M. J. Lipovich, R. S. Wolford, R. M. Wadowsky, A. M. McNamara, J. L. Sykora, G. Keleti, et al.** 1987. Survival and multiplication of *Legionella pneumophila* in municipal drinking water systems. *Appl. Environ. Microbiol.* **53:**979–986.

85. **Steele, T. W., J. Lanser, and N. Sangster.** 1990. Isolation of *Legionella longbeachae* serogroup 1 from potting mixes. *Appl. Environ. Microbiol.* **56:**49–53.

86. **Tenover, F. C., L. Carlson, L. Goldstein, J. Sturge, and J. J. Plorde.** 1985. Confirmation of *Legionella pneumophila* cultures with a fluorescein-labeled monoclonal antibody. *J. Clin. Microbiol.* **21:**983–984.

87. **Tenover, F. C., P. H. Edelstein, L. C. Goldstein, J. C. Sturge, and J. J. Plorde.** 1986. Comparison of cross-staining reactions by *Pseudomonas* spp. and fluorescein-labeled polyclonal and monoclonal antibodies directed against *Legionella pneumophila. J. Clin. Microbiol.* **23:**647–649.

88. **Thacker, L., R. M. McKinney, C. W. Moss, H. M. Sommers, M. L. Spivack, and T. F. O'Brien.** 1981. Thermophilic sporeforming bacilli that mimic fastidious growth characteristics and colonial morphology of *Legionella. J. Clin. Microbiol.* **13:**794–797.

89. **Wadowsky, R. M., L. J. Butler, M. K. Cook, S. M. Verma, M. A. Paul, B. S. Fields, G. Keleti, J. L. Sykora, and R. B. Yee.** 1988. Growth-supporting activity for *Legionella pneumophila* in tap water cultures and implication of hartmannellid amoebae as growth factors. *Appl. Environ. Microbiol.* **54:** 2677–2682.

90. **Wadowsky, R. M., and R. B. Yee.** 1985. Effect of non-Legionellaceae bacteria on the multiplication of *Legionella pneumophila* in potable water. *Appl. Environ. Microbiol.* **49:** 1206–1210.

91. **Wilkinson, H. W., and B. J. Brake.** 1982. Formalin-killed versus heat-killed *Legionella pneumophila* serogroup 1 antigen in the indirect immunofluorescence assay for legionellosis. *J. Clin. Microbiol.* **16:**979–981.

92. **Wilkinson, H. W., D. D. Cruce, and C. V. Broome.** 1981. Validation of *Legionella pneumophila* indirect immunofluorescence assay with epidemic sera. *J. Clin. Microbiol.* **13:**139–146.

93. **Wilkinson, H. W., A. L. Reingold, B. J. Brake, D. L. McGiboney, G. W. Gorman, and C. V. Broome.** 1983. Reactivity of serum from patients with suspected legionellosis against 29 antigens of Legionellaceae and *Legionella*-like organisms by indirect immunofluorescence assay. *J. Infect. Dis.* **147:**23–31.

94. **Wilkinson, H. W., J. S. Sampson, and B. B. Plikaytis.** 1986. Evaluation of a commercial gene probe for identification of *Legionella* cultures. *J. Clin. Microbiol.* **23:**217–220.

95. **Winn, W. C., Jr., and R. L. Myerowitz.** 1981. The pathology of the *Legionella* pneumonias. A review of 74 cases and the literature. *Hum. Pathol.* **12:**401–422.

96. **Wright, J. B., I. Ruseska, and J. W. Costerton.** 1991. Decreased biocide susceptibility of adherent *Legionella pneumophila. J. Appl. Bacteriol.* **71:**531–538.

97. **Yee, R. B., and R. M. Wadowsky.** 1982. Multiplication of *Legionella pneumophila* in unsterilized tap water. *Appl. Environ. Microbiol.* **43:**1330–1334.

98. **Yu, V. L., T. R. J. Beam, R. M. Lumish, R. M. Vickers, J. Fleming, C. McDermott, and J. Romano.** 1987. Routine culturing for *Legionella* in the hospital environment may be a good idea: a three-hospital prospective study. *Am. J. Med. Sci.* **294:**97–99.

99. **Zimmerman, S. E., M. L. French, S. D. Allen, E. Wilson, and R. B. Kohler.** 1982. Immunoglobulin M antibody titers in the diagnosis of Legionnaires disease. *J. Clin. Microbiol.* **16:**1007–1011.

100. **Zuravleff, J. J., V. L. Yu, J. W. Shonnard, B. K. Davis, and J. D. Rihs.** 1983. Diagnosis of Legionnaires' disease. An update of laboratory methods with new emphasis on isolation by culture. *JAMA* **250:**1981–1985.

Francisella

SCOTT J. STEWART

43

TAXONOMY

Described first in humans in 1907 and later in rodents (15), the causative agent of tularemia was called *Bacterium tularensis*. Beginning his work in 1920, Edward Francis dedicated his career to describing the clinical manifestations, diagnosis, and histopathology of tularemia (8). In 1974, the causative organism was renamed *Francisella tularensis* in recognition of his contributions. Early colloquial names, such as rabbit fever and deerfly fever in this country, hare fever in Japan, and trappers' ailment in Europe and Asia, attest to its long-recognized association with wild animals. Humans of any age, sex, or race are universally susceptible. The annual incidence of tularemia in humans in the United States has declined from several thousand in the 1930s to a few hundred in the 1990s in concert with the disappearance of market hunting and trapping. In China, where wild rabbits are still processed for consumption, the association of hunting and disease continues (18).

DESCRIPTION OF GENUS

F. tularensis is the etiologic agent of tularemia, which is, with rare exception, the only disease produced by this genus. The organism is a small, gram-negative, pleomorphic, nonmotile, non-spore-forming coccobacillus measuring 0.2 by 0.2 to 0.7 μm. It is a strict aerobe, is nonpiliated, and possesses a thin capsule composed predominantly of lipid. In nature, it is a rather hardy organism, persisting for many weeks in mud, water, and decaying animal carcasses. The organism is extremely infectious. Biosafety level 2 is recommended for clinical laboratory work with suspected material, and biosafety level 3 is required for culturing the organism in large quantities (32).

NATURAL HABITATS

Tularemia is primarily a disease of wild animals that is perpetuated in nature by ectoparasites, contaminated environment, cannibalism, and acute and chronic carriers. Human infection is incidental and is usually the result of interaction with wild animals or their environs. Tularemia is the third most commonly reported tick-borne disease in the United States.

This organism is extremely widespread (Fig. 1). Literally hundreds of species of wild and domestic animals and birds, including common house pets, have been found infected with *F. tularensis*. Dozens of biting and blood-sucking insects serve as vectors. Wild hares and ticks are the source for most of the human cases in the areas where the disease is endemic (Missouri, Arkansas, and Oklahoma), which annually report more than 50% of the cases in the United States (5). Biting flies are the most common vectors in Utah, Nevada, and California, whereas ticks are the most common vectors in the Rocky Mountain states. Sheep, beavers, and muskrats have also been responsible for human outbreaks.

Two main biovars, type A (*F. tularensis* biovar tularensis) and type B (*F. tularensis* biovar palearctica), are found in this country. Type A is found only on the North American continent. This biovar produces the more serious disease in humans, with an untreated fatality rate of about 5%. This biovar is usually associated with wild hares, sheep, and ticks. Type B is widespread in most countries of the temperate zone in the Northern Hemisphere, including North America. It produces a milder, often subclinical disease. This biovar is usually associated with water and aquatic mammals. Only a few poorly documented cases have been reported from the Southern Hemisphere.

CLINICAL SIGNIFICANCE

Tularemia often starts with the sudden onset of flu-like symptoms: chills, fever, headache, and generalized aching. The infection usually starts when the organism penetrates the skin. Following an incubation period of 2 to 10 days, an ulcer forms at the site of penetration and becomes the local focus of infection. This ulcer may persist for several months. The organism is transported via the lymphatics to the regional nodes, which enlarge and frequently become necrotic. When the organism enters the bloodstream, widespread dissemination and typical endotoxemia result.

The ulceroglandular form of the disease constitutes about 80% of the reported cases. The ulcer is erythematous, indurated, and nonhealing and has a "punched-out" appearance at 1 to 3 weeks. Ulcers on the upper extremities usually result from exposure to infected mammals and occur mostly on hunters and ranchers. Lesions on the lower extremities, head, and back are usually from the bite of a blood-sucking arthropod such as a tick, deerfly, or mosquito.

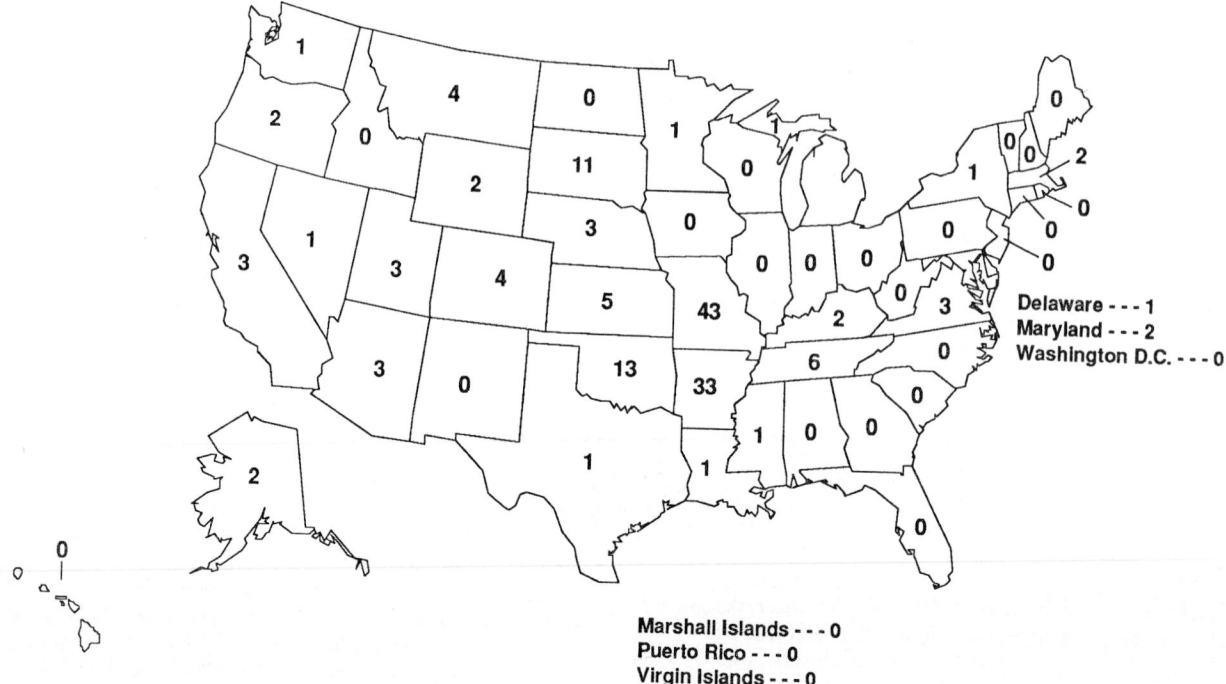

FIGURE 1 Average number of tularemia cases per year by state, 1988 to 1992 (5).

Oculoglandular tularemia, most frequently encountered by an ophthalmologist, is an unusual variation of the ulceroglandular form. The conjunctiva, usually having been infected by being rubbed with contaminated fingers, is painfully inflamed and shows numerous yellowish nodules and pinpoint ulcers. Because of the debilitating pain of this form of the infection, the patient usually seeks medical attention before regional lymphadenopathy is evident (16). Painful preauricular adenopathy is unique for tularemia and separates it from cat scratch disease, tuberculosis, sporotrichosis, and syphilis (10).

Oropharyngeal tularemia, which is manifested by a sore throat that is painful beyond its appearance and does not respond to penicillin or its analogs, is primarily a pediatric disease in this country (9, 19). The buccal or pharyngeal area is considered the portal of entry because the classic ulcer is frequently found there. The tonsils become enlarged and covered with a yellow-white pseudomembrane similar to that described for diphtheria (17).

Glandular tularemia is the term used when an ulcer is not found. On occasion, patients with unexplained regional lymphadenopathy have undergone lymph node biopsy for suspected malignancy (7). Surgical manipulations of otherwise untreated tularemic lymph nodes have initiated severe infections owing to the release of large numbers of organisms from within the node (22).

In systemic or typhoidal tularemia, which is the acute form of this infection, the organism appears to overrun the host's protective barriers. It has a mortality rate of 30 to 60% and is usually associated with a massive inoculation or a patient who is otherwise compromised. This form appears as an acute septicemia and lacks the classic external ulcer and lymphadenopathy. Toxemia, severe headache, and continuous high fever are common findings. The patient may be delirious and develop prostration and shock. If this disease is suspected, specific antibiotic therapy must be initiated immediately, as the course of disease can be so rapidly fulminating that the patient's death may precede a diagnostic workup.

Pleuropulmonary tularemia can result from either inhalation of an infectious aerosol or colonization of the pleural cavity following dissemination via the bloodstream (21). Patients with pulmonary tularemia are more likely to be older, less likely to have a history of vector involvement, more likely to have a typhoidal illness, more likely to have a positive culture, and more likely to die than those without pulmonary involvement (25).

Gastrointestinal tularemia is almost always traced to the consumption of contaminated food or water (1). This form of the infection is usually associated with additional abdominal or low-back pain and persistent diarrhea. The course of this type of infection can vary from mild, unexplained, persistent diarrhea with no other symptoms to a rapidly fulminating fatal disease. When patients with fatal cases have been autopsied, extensive ulceration throughout the bowel has been found, suggesting a massive inoculum (26).

TRANSPORT AND STORAGE OF SPECIMENS

Specimens suspected of containing *F. tularensis* should be collected and manipulated with extreme caution. Tularemia is currently listed as the third most commonly reported laboratory-associated bacterial infection. Biosafety level 2 precautions are essential (32).

DIRECT EXAMINATION

Conventional microscopy of polychromatic stained tissue smears or sections shows the organism to occur singularly and in groups, both intra- and extracellularly. Gram stain of

biopsy material is of little value, as the small, weakly staining organism cannot readily be distinguished from the background. An indirect fluorescent-antibody test can be performed with commercially available antiserum (BBL 40808;
BBL, Cockeysville, Md.; Difco 2241-47-0; Difco, Detroit,
Mich.). False-positive fluorescent-antibody stain for *Legionella* spp. has been reported in a specimen harboring *F.
tularensis* (20). Indeed, guinea pigs vaccinated with live
tularemia vaccine are protected against challenge with *Legionella* spp.

CULTURE AND ISOLATION

Bacterial isolation is difficult. In a study of over 1,000
human cases, 84% of which were laboratory confirmed by
serology, the organism was isolated in slightly more than
10% of the cases (30).

The historical medium of choice is cystine glucose blood
agar (27); however, good growth has been achieved with
modified Mueller-Hinton broth, chocolate agar supplemented with IsoVitaleX, and modified charcoal yeast extract agar (2, 6, 34). A heavy inoculum on appropriate
medium will yield a growth mass in 18 h, while individual
colonies may require 2 to 4 days of incubation to reach 2
mm in size. Overgrowth by contaminating organisms can be
reduced by incorporating 100 to 500 U of penicillin per ml
into the medium.

Direct isolation can be achieved from infected ulcer
scrapings, lymph node biopsy specimens, gastric washings,
and sputum. Blood cultures seldom reveal the organism.
Isolations from urine and feces have been extremely rare.

IDENTIFICATION

Colonies are blue-gray, round, smooth, and slightly mucoid.
On medium containing blood, a small zone of alpha hemolysis usually surrounds the colony. A slide agglutination test
using commercially available antiserum and a positive control antigen (Difco 2240-56-9) can be performed on a
culture suspension for identification. Biochemical reactions
are of no particular value and do not justify the additional
risk to the diagnostician. Fermentation of glycerol or evidence of citrulline ureadase production serve to differentiate biovar type A from type B (28). PCR has been used to
demonstrate the infection, with blood used as a source of
DNA. However, this test has not been shown to be more
sensitive than direct culturing (14).

SEROLOGY

Tularemia is most frequently confirmed by serology. Commercially available antigens (BBL 4087, Difco 2240-56-5)
can be used with the standard tube agglutination test. A
titer of less than 1:20 is not considered diagnostic, because
nonspecific cross-reactions are common at this level. A
fourfold increase in titer during illness or a single titer of
1:160 or greater is considered diagnostic. Titers into the
thousands are common late in the infection, and titers of
1:20 to 1:80 can persist for years. A microagglutination test
that is about 100-fold more sensitive than the standard tube
agglutination test has been described; it should be of great
value if the reagents become commercially available (4,
23). Enzyme-linked immunosorbent assays have proven
useful for detecting both antibodies and antigens (3, 29,
35). Although extremely rare, *Francisella philomiragia* and *F.*

tularensis biovar novicida have been associated with human
disease and can cause some serologic confusion (11).

ANTIMICROBIAL SUSCEPTIBILITIES OF *F. TULARENSIS* AND PREVENTION AND CONTROL OF DISEASE

The standardized antimicrobial susceptibility test cannot be
performed for *F. tularensis* because this organism will not
grow on the media used for this assay. For treatment, the
aminoglycoside antibiotics are the drugs of choice, with
streptomycin being the most commonly used, although
amikacin has recently been shown to be more effective
(31). These agents are indicated for the potential life-
threatening pleuropneumonia or other typhoidal infections. Imipenem-cilastatin has been used when the aminoglycosides are contraindicated because of renal complications (13). Fluoroquinolones have shown promise because of their low toxicity and their potential for oral and
topical applications (24). Chloramphenicol and tetracyclines are associated with high relapse rates. The beta-
lactams, except for imipenem, and the sulfanilamides are
considered ineffective (20). The macrolide antibiotics are
not recommended (33).

Since there is no effective control of tularemia in nature,
public awareness of this organism's ubiquitous presence and
potential for human infection should be maintained. In
areas of endemicity, one should avoid handling dead or
moribund animals and try to reduce the possibility of insect
bites by wearing protective clothing, using insect repellents,
and removing ticks promptly. The chlorination of municipal drinking water has virtually eliminated epidemics
springing from that source, but untreated water may serve as
a vector.

An investigational live attenuated vaccine is available
through the U.S. Army Medical Research and Development Command, Fort Detrick, Frederick, MD 21701. It is
recommended for persons with a high probability of exposure either to the organism or to infected animals (32).

REFERENCES

1. **Amos, H. L., and D. H. Sprunt.** 1936. Tularemia. Review of literature of cases contracted by ingestion of rabbit. JAMA **106:**1078–1080.
2. **Baker, C. N., D. G. Hollis, and C. Thornsberry.** 1985. Antimicrobial susceptibility testing of *Francisella tularensis* with a modified Mueller-Hinton broth. *J. Clin. Microbiol.* **22:**212–215.
3. **Bavanger, L., J. A. Maeland, and A. I. Naess.** 1988. Agglu-tinins and antibodies to *Francisella tularensis* outer membrane antigens in the early diagnosis of disease during an outbreak of tularemia. *J. Clin. Microbiol.* **26:**433–437.
4. **Brown, S. L., F. T. McKinney, G. C. Klein, and W. L. Jones.** 1980. Evaluation of a safranin-O-stained antigen mi-croagglutination test for *Francisella tularensis* antibodies. *J. Clin. Microbiol.* **11:**146–148.
5. **Centers for Disease Control.** 1988–1992. Summary—cases of specified notifiable diseases, United States, *Morbid. Mortal. Weekly Rep.* **36**–42.
6. **Clark, W. A., D. G. Hollis, R. E. Weaver, and P. Riley.** 1984. Identification of unusual pathogenic gram-negative aer-obic and facultatively anaerobic bacteria, p. 164–165. U.S. Department of Health and Human Services publication no. 017-023-00149. U.S. Government Printing Office, Washing-ton, D.C.
7. **Evans, M. E., D. W. Gregory, W. Schaffner, and Z. A.**

McGee. 1985. Tularemia: a 30 year experience with 88 cases. *Medicine* **64**:251–269.

8. **Francis, E.** 1925. Tularemia. *JAMA* **250**:3216–3224. [Reprint.]

9. **Fulginiti, V. A., and C. Hoyle.** 1966. Oropharyngeal tularemia. *Rocky Mount. Med. J.* **63**:41–42, 67.

10. **Halperin, S. A., T. Gast, and P. Ferrieri.** 1985. Oculoglandular syndrome caused by *Francisella tularensis. Clin. Pediatr.* **24**:520–522.

11. **Hollis, D. G., R. E. Weaver, A. G. Steigerwalt, J. D. Wenger, C. W. Moss, and D. J. Brenner.** 1989. *Francisella philomiragia* comb. nov. (formerly *Yersinia philomiragia*) and *Francisella tularensis* biogroup novicida (formerly *Francisella novicida*) associated with human disease. *J. Clin. Microbiol.* **27**:1601–1608.

12. **Konyukhov, V. F., I. S. Tartakovsky, V. V. Petrosov, Y. G. Suchkov, and S. V. Prozorovsky.** 1991. The nonspecific protection of guinea pigs immunized with a live tularemia vaccine against lung infection by *Legionella pneumophila. Zh. Mikrobiol. Epidemiol. Immunobiol.* **8**:50–51.

13. **Lee, H. C., E. Horowitz, and W. Linder.** 1991. Treatment of tularemia with imipenem/cilastatin sodium. *South. Med. J.* **84**:1277–1278.

14. **Long, G. W., J. J. Oprandy, R. B. Narayanan, A. H. Fortier, K. R. Porter, and C. A. Nacy.** 1993. Detection of *Francisella tularensis* in blood by polymerase chain reaction. *J. Clin. Microbiol.* **31**:152–154.

15. **McCoy, G. W., and C. W. Chapin.** 1912. *Bacterium tularense* the cause of a plague-like disease of rodents. *U.S. Public Health Hosp. Bull.* **53**:17–23.

16. **Nohinek, B., and J. J. Marr.** 1983. Tularemia: two manifestations in modern man. *Mo. Med.* **80**:687–689, 698.

17. **Ohara, H.** 1934. Oto-rhino-laryngologia. *Fukuoka* **7**:432–437.

18. **Pang, Z. C.** 1987. The investigation of the first outbreak of tularemia in Shandong Peninsula. *Chung Hua Liu Hsing Ping Hsueh Tsa Chih* **8(5)**:261–263.

19. **Parkhurst, J. B., and V. H. San Joaquin.** 1990. Tonsillopharyngeal tularemia: a reminder. *Am. J. Dis. Child.* **144**:1070. (Letter.)

20. **Roy, T. M., D. Flemming, and W. H. Anderson.** 1989. Tularemic pneumonia mimicking Legionnaires' disease with false-positive direct fluorescent antibody stains for *Legionella. South. Med. J.* **82**:1429–1431.

21. **Rubin, S. A.** 1978. Radiographic spectrum of pleuropulmonary tularemia. *Am. J. Roentgenol.* **131**:277–281.

22. **Saslaw, S., H. T. Eiglesbach, H. E. Wilson, J. A. Prior, and S. Carhart.** 1961. Tularemia vaccine study. *Arch. Intern. Med.* **107**:689–701.

23. **Sato, T., H. Fujita, Y. Ohara, and M. Homma.** 1990. Microagglutination test for early and specific serodiagnosis of tularemia. *J. Clin. Microbiol.* **28**:2372–2374.

24. **Scheel, O., R. Reiersen, and T. Hoel.** 1992. Treatment of tularemia with ciprofloxacin. *Eur. J. Microbiol. Infect. Dis.* **11**:447–448.

25. **Scofield, R. H., E. J. Lopez, and S. J. McNabb.** 1992. Tularemia pneumonia in Oklahoma. 1982–1987. *J. Okla. State Med. Assoc.* **85**:165–170.

26. **Simpson, W. P.** 1928. Tularemia (Francis' disease). *Ann. Intern. Med.* **1**:1007.

27. **Stewart, S. J.** 1981. Tularemia, p. 705–714. *In* A. Balows and W. J. Hausler, Jr. (ed.), *Diagnostic Procedures for Bacterial, Mycotic and Parasitic Infections,* 6th ed. American Public Health Association, Washington, D.C.

28. **Stewart, S. J.** 1988. Tularemia, p. 519–524. *In* A. Balows, W. J. Hausler, Jr., M. Ohashi, and A. Turano (ed.), *Laboratory Diagnosis of Infectious Diseases: Principles and Practice.* Springer-Verlag, New York.

29. **Tarnvik, A., S. Lofgren, L. Ohlund, and G. Sandstrom.** 1987. Detection of antigen in urine of a patient with tularemia. *Eur. J. Clin. Microbiol.* **6**:318–319.

30. **Taylor, J. P., G. R. Istre, T. C. McChesney, F. T. Satalowich, R. L. Parker, and L. M. McFarland.** 1991. Epidemiologic characteristics of human tularemia in the southwest-central states, 1981–1987. *Am. J. Epidemiol.* **133**:1032–1038.

31. **Tynkevich, N. K., N. V. Pavlovich, and I. V. Ryzhko.** 1990. Comparative study of the effectiveness of amikacin and streptomycin in experimental tularemia. *Antibiot. Khimioter.* **8**:35–37.

32. **U.S. Department of Health and Human Services.** 1988. *Biosafety in Microbiological and Biomedical Laboratories,* 2nd ed. U.S. Department of Health and Human Services publication no. 88–8395. U.S. Government Printing Office, Washington, D.C.

33. **Vasilyev, N. T., V. A. Oborin, P. G. Vasilyev, O. V. Glushkoba, I. D. Kravets, and B. A. Levchuk.** 1989. Sensitivity spectrum of *Francisella tularensis* to antibiotics and synthetic antibacterial drugs. *Antibiot. Khimioter.* **34**:662–665.

34. **Westerman, E. L., and J. McDonald.** 1986. Tularemia pneumonia mimicking Legionnaires' disease: isolation of organism on CYE agar and successful treatment with erythromycin. *South. Med. J.* **76**:1169–1170.

35. **Yurov, S. V., S. Y. Pchelintsev, S. S. Afanasyev, A. A. Vorobyev, N. N. Urakov, G. V. Cherenkova, K. K. Fedortsov, L. I. Krasnoproshina, G. S. Vlasov, and N. B. Denisova.** 1991. The use of microdot immunoenzyme analysis with visual detection for the determination of tularemia antibodies. *Zh. Mikrobiol. Epidemiol. Immunobiol.* **3**:61–64.

Brucella

NELSON P. MOYER AND LARRY A. HOLCOMB

TAXONOMY

According to DNA-DNA hybridization studies, all brucellae consist of a single genospecies, *Brucella melitensis* (31). DNA probes for molecular subtyping have shown marked homogeneity within the genus by DNA-DNA hybridization and ribotyping (28). When the low-cleavage-frequency restriction enzyme XbaI is used with pulsed-field gel electrophoresis, the type strains of the current nomen species produce biovar-specific rRNA gene restriction patterns (3). The traditional nomenclature corresponding to natural host preference and supported by cultural, biochemical, and serologic phenotypic characteristics is widely accepted in clinical microbiology and will be retained in this chapter.

DESCRIPTION OF THE GENUS

Brucellae are small, nonmotile, nonsporulating, nonencapsulated, gram-negative coccobacilli or short rods (0.5 to 0.7 by 0.6 to 1.5 μm) arranged singly, in pairs, or in short chains. Growth occurs aerobically, often enhanced by CO_2, but no growth occurs under strict anaerobic conditions. The organism grows in a minimal medium containing sodium chloride, sodium thiosulfate, ammonium sulfate, glucose, nicotinic acid, thiamine, panthotenic acid, and biotin (41). Optimal growth occurs at 37°C (10 to 40°C) at pH 6.6 to 7.4. Colonies appear on agar surfaces after 2 to 3 days of incubation and reach a 2- to 3-mm diameter after 4 or 5 days. Growth is slower on selective media. Colonies are nonhemolytic and nonpigmented and have smooth, glistening surfaces. *B. ovis* and *B. canis* occur normally in the rough form, whereas the other four species have been encountered only in the smooth form. Smooth to rough dissociation readily occurs in the laboratory. Brucellae are catalase-positive, oxidase-positive (except *B. ovis* and *B. neotomae*), urease-variable organisms that reduce nitrate to nitrite. They are citrate and methyl red negative, do not produce acetylmethylcarbinol or gelatinase, and do not release *o*-nitrophenol from *o*-nitrophenol-β-D-galactoside. Metabolism is oxidative. Production of H_2S, hydrolysis of urea, and resistance to thionin and basic fuchsin help differentiate the biovars (20). Biotyping requires oxidative metabolism and bacteriophage studies (4).

NATURAL HABITATS

Brucellosis is a zoonotic disease with a domestic-animal reservoir. It is an occupational disease of veterinarians, microbiologists, shepherds, farmers, abattoir workers, and butchers. The biovars infective for humans are *B. abortus* (cattle), *B. suis* (swine), *B. melitensis* (goats and sheep), and *B. canis* (dogs). *B. neotomae* (desert wood rat) and *B. ovis* (sheep) are not known to cause disease in humans. Transmission by milk, milk products, meat, and direct contact with infected animals is documented. Routes of infection are genital, nasopharyngeal, gastrointestinal, conjunctival, respiratory, and through abraded skin (50). The seasonal incidence in countries in which the disease is endemic is highest in spring and summer and coincides with parturition, abortion, and increased milk production of animals, commonly goats, sheep, and camels (2). Claims of person-to-person or sexual transmission are rare and improbable, since delayed onset makes it difficult to rule out a food or environmental exposure with certainty (26).

CLINICAL SIGNIFICANCE

Brucellae are intracellular parasites that cause epizootic abortions in animals and septicemic febrile illness or localized infection of bone, tissue, or organ systems in humans (36, 45). Brucellae localize and replicate within the rough endoplasmic reticulum of nonphagocytic host cells. *B. melitensis* is the most virulent of the biovars. Brucellosis in humans has a variable incubation period, an insidious or abrupt onset, and no pathognomic symptoms or signs. While the average onset (depending on the biovar) occurs in 3 to 4 weeks following exposure, the amount and route of inoculum and the nutritional and immunological status of the host play important roles in the establishment of active infection. Incubation periods of 7 to 10 months have been reported. The immune response to human brucellosis is characterized by an initial rise in titer of immunoglobulin M (IgM) antibody followed within a few weeks by a rise in titer of IgG antibody. Levels of IgM and IgG decline after treatment, and the decline indicates therapeutic response. An alteration in the decline of IgG antibody or an increase in IgG antibody is associated with incomplete recovery or relapse (6, 40). Because of inadequate therapy or intermittent exposure to brucellae, IgM may persist at moderate to low levels for months to years in asymptomatic patients with underlying infection. Occupationally exposed patients with background IgM titers develop a detectable rise in IgG level during acute episodes (25, 52). Symptoms at onset include fever, arthralgia, malaise, weight loss, anorexia, arthritis, splenomegaly, hepatomegaly, chills, and sweating. Approximately 70% of patients with acute cases in areas of endemicity are blood culture positive, and almost all of these patients have a tube agglutination titer of $\geq 1{:}160$. In 10 to 15% of patients with brucellosis, complications of an articular, osseous, visceral, or neurological nature occur. Osteomyelitis is the most frequent complication in humans (42, 50). Neurological involvement is rare, developing in 5% or fewer of cases, usually presenting as acute or chronic meningoencephalitis, which may be misdiagnosed as tuberculosis. In the United States, most of the 1,654 reported cases during the period 1978 to 1987 occurred in persons with occupational or avocational exposure to animals or laboratory cultures of *Brucella* spp. Approximately 100 cases per year are now reported in the United States (15). *B. canis* has accounted for more than 30 cases of brucellosis in humans (50). Sporadic episodes of food-associated brucellosis have been caused by *B. melitensis*-contaminated goat cheese (7, 14). Brucellosis has occurred in hunters who had butchered elk, moose, and bison in the United States and in persons who had ingested raw meat, bone marrow, or liver of reindeer, moose, and caribou in Alaska. Accidental self-inoculation and corneal contamination with live vaccine strains of *B. abortus* 19 and *B. melitensis* Rev. 1 may cause infection.

COLLECTION, TRANSPORT, AND STORAGE OF SPECIMENS

Multiple blood cultures should be obtained when brucellosis is suspected. Acute-phase serum should be obtained as soon as possible after onset of the disease, and convalescent-phase serum should be collected 14 to 21 days after onset. Additional sera may be required for establishing a diagnosis when blocking antibody is present or when IgG studies are indicated. Infected tissues, abscesses, bone marrow, and spleen and liver biopsies should be cultured (21). Rarely, cerebrospinal fluid (CSF), pleural fluid, peritoneal fluid, urine, and other specimens may be collected for isolation of brucellae. Specimens should be cultured as soon as possible after collection or refrigerated if delays are unavoidable.

DIRECT EXAMINATION

Brucellae are intracellular pathogens of the reticuloendothelial system with granulomatous foci in various organs. Examination of stained smears of tissue or exudates is rarely instructive.

CULTURE AND ISOLATION

Laboratory Safety

Infection with *Brucella* spp. is readily acquired by laboratory workers through skin contact, inhalation of aerosols, eye and mouth contact, and accidental percutaneous inoculation. *B. melitensis* has caused laboratory-acquired infections even without apparent lapses in technique while a certified biological safety cabinet (BSC) was being used (47). Procedures associated with laboratory-acquired brucellosis are

preparation of antigen suspensions, performance of slide agglutination tests, sniffing petri plates, and removing petri plates from the BSC for closer observation. *Brucella* spp. are classified as biosafety level 3 organisms, and all manipulations with live cultures and antigens must be confined to a class II BSC (16).

Blood and Bone Marrow

Conventional blood culture media require extended incubation. Potential exposure and clinical history are indications that the laboratory should extend incubation of blood culture bottles and perform blind subcultures. When blood and other body fluids are cultured, a biphasic medium with 5 to 10% CO_2 is recommended (4). If biphasic bottles are not available, commercial blood culture bottles can be vented and incubated at 35 to 37°C. However, subcultures must be made every 4 to 5 days for 30 days. Brucellae may be detected in 7 days in bone marrow aspirate cultures and in 5 days in blood cultures when specimens are inoculated into BACTEC blood culture bottles and subcultured to solid media (53). When recovery of brucellae from blood and bone marrow was compared in 50 patients with a diagnosis of clinical brucellosis, bone marrow cultures were positive in 92% and blood culture was positive in 70% (27). The average time required for growth was 4.32 days for bone marrow cultures and 6.65 days for blood cultures. Bone marrow cultures were positive for patients with acute, subacute, and chronic brucellosis, whereas blood cultures were positive only for patients with acute cases. Most laboratories report longer incubation periods for isolation of brucellae from BACTEC bottles, with positive cultures detectable after 5 to 29 days (mean, 14 days) of incubation (32). A lysis-centrifugation method has been effective for recovery of *Brucella* spp. from blood, bone marrow, and spinal fluid (22). BACTEC produces higher recovery rates, while lysis-centrifugation methods allow earlier recognition of positive cultures, with a mean recovery in 3 days versus the 14 days required for BACTEC (32, 38). *B. melitensis* has been isolated from blood in 2 to 8 days by using BacT/Alert (46).

Exudates and Tissues

Minced, blended, cubed, and stomached exudate and tissue samples produce similar results, and no single method is consistently superior. Fibrous clots, exudates, and tissue are aseptically ground in a mortar with sterile sand, and the resulting material is inoculated onto 5% sheep blood agar, brucella agar containing 5% serum, or serum glucose agar (Oxoid USA Ltd., Columbia, Md.).

Culture Media

Tryptose agar with 5% bovine serum and with or without antibiotics remains a standard plating medium for the isolation of brucellae. When significant contamination is likely, a selective medium should be inoculated in addition to the basal medium. Skirrow agar has been used under field conditions in veterinary practices to isolate *B. abortus*, *B. suis*, *B. melitensis*, *B. canis*, and *B. ovis* from contaminated semen, vaginal exudates, and milk (48). Brucellae grow readily on many of the specialized culture media already available in the clinical laboratory. Selective BCYE (polymyxin, 80 U/ml; anisomycin, 80 ng/ml; cefamandole, 4 μg/ml) is commercially available as CYE-PAC from Remel Laboratories, Lenexa, Kans. (43). Brucellae also grow on Thayer-Martin medium or chocolate agar containing VCNT (BBL), although some strains of *B. ovis* may be inhibited (11). *B. melitensis* has been isolated from CSF inoculated onto Lowenstein-Jensen egg base medium after 1 week of incubation (24). The colonies were initially disregarded as contaminants because they were not acid fast.

Incubation

Inoculated plates are incubated at 35 to 37°C in an atmosphere of 5 to 10% CO_2 for 10 days.

Examination of Cultures

After 4 to 5 days, *Brucella* colonies are 2 to 7 mm in diameter, spheroidal, moist, slightly opalescent, translucent, and bluish white in reflected light. These characteristics may vary somewhat with pH and available moisture. Transfer isolated colonies to several tryptose agar slants or slants of similar media, and incubate them under 10% CO_2 tension at 37°C for 48 h. If sufficient growth is not obtained, reincubate for another 24 h. *Brucella* spp. yield a fine, translucent growth with a slight amber tinge. Gram stain the colonies, and examine for gram-negative pleomorphic coccobacilli. Since brucellae take the counterstain poorly, apply counterstain for 1 to 3 min instead of the usual 30 s.

IDENTIFICATION

Preliminary Identification in Clinical Laboratories

Nonhemolytic colonies of oxidase-positive, gram-negative coccobacilli that do not ferment lactose or glucose and are obligate aerobes are tested for agglutination in unabsorbed anti-smooth *Brucella* serum (Difco Laboratories). Smooth strains of *Brucella* share common surface antigens with *Francisella tularensis*, *Vibrio cholerae*, *Escherichia coli* O157:H7, *Salmonella* O:30, and *Yersinia enterocolitica* serotype O:9 (13). Nonsmooth strains of brucellae cross-react with some *Pseudomonas* species. *B. ovis*, *B. canis*, and rough variants of other *Brucella* species cross-react when unabsorbed anti-rough *Brucella* sera are used, but no cross-reactions occur when these organisms are agglutinated with unabsorbed anti-smooth *Brucella* serum. Suspected *Brucella* colonies are emulsified in two separate drops of saline. A drop of anti-smooth *Brucella* serum is added to the first drop, and normal serum is added to the second. The suspensions and sera are mixed and examined for agglutination, which should be rapid and complete unless dissociation has occurred. *B. canis* and *B. ovis* will not agglutinate in anti-smooth *Brucella* serum. Control cultures of known *Brucella* spp. should be used in conjunction with the suspected culture. These morphological, biochemical, and serologic results are sufficient to place the isolate in the genus *Brucella* (35). *Brucella*-negative cultures do not rule out brucellosis. Preliminary identification is sufficient laboratory evidence for the physician to initiate therapy.

Differentiation of Biovars and Biotypes in Reference Laboratories

Six biovars and 15 biotypes constitute the genus *Brucella*. Common criteria for the identification of the biovars and biotypes are (i) production of H_2S, (ii) requirement of increased CO_2 for growth, (iii) production of urease, (iv) agglutination in monospecific sera, and (v) growth on solid media containing basic fuchsin and thionin (Table 1).

H_2S Production

Production of H_2S is detected by suspending a lead acetate paper strip directly over but not touching the inoculated surface of a tryptose agar slant. The strip is examined daily for 4 days and is replaced with a fresh strip each day. Hydrogen sulfide production blackens the strip.

Requirements for Additional CO_2

The strain under examination is inoculated on duplicate tryptose agar slants; one slant is incubated in air, and the other is incubated in an atmosphere of 5 to 10% CO_2. Since mutants that are no longer dependent on additional CO_2 occur, this test should be carried out immediately after initial isolation of the organism. Good growth in CO_2 and scanty or no growth in air indicate a requirement for CO_2.

Urease Production

A semiquantitative urease test can be performed by preparing a dense suspension of the organisms and placing a loopful of the suspension on a slant of Christensen urea agar. *B. suis* and *B. canis* turn the indicator pink in less than 1 min, whereas *B. abortus* usually takes 5 to 10 min to do so.

Slide Agglutination Test

Monospecific antisera prepared by differential absorption with *B. melitensis* and *B. abortus* can be used to determine the predominant lipopolysaccharide, M or A (13). A dense suspension of the organism to be tested is prepared in 0.5% phenolized saline and heated at 60°C for 1 h. A drop of the suspension is added to a drop of each monospecific antiserum and mixed. Agglutination should occur within 1 min with one of the sera. Control cultures of *B. abortus* biotype 1, *B. melitensis* biotype 1, and *B. ovis* or *B. canis* should be used for this test. These control cultures should agglutinate in their homologous serum within 1 min without agglutinating in the other sera.

Tube Agglutination Test

A culture suspension is prepared by harvesting a fresh slant culture with 0.5% phenolized saline. The harvest is then heated at 60°C for 1 h, diluted, and standardized to 78% light transmission at a wavelength of 650 nm in a spectrophotometer. The monospecific tube test is performed in duplicate, i.e., one set of tubes for each of the monospecific sera. The serum is diluted just beyond its known titer by the double-dilution method, starting at 1:5. An equal amount of the antigen suspension, prepared as described above, is then added to each tube and the saline control. The tubes are incubated for 24 h at 37°C and then read. Usually the strain being studied will be agglutinated to its known titer by one of the sera but not at all by the other serum. *B. melitensis* biotype 3 agglutinates in both monospecific sera at various titers (4).

Growth in Presence of Thionin and Basic Fuchsin

The concentrations of basic fuchsin and thionin (Allied Chemical and Dye Co., New York, N.Y.) used in these tests are between 20 and 40 μg of dye per ml of medium. The actual concentration of dye is that which will differentiate among control cultures of *B. melitensis*, *B. abortus*, and *B. suis* biotypes. This concentration will depend on the basal medium used as well as the bacteriostatic action and purity of the dye and must be determined by using control cultures. The medium is prepared by heating a 0.1% dye solution in a boiling-water bath for 20 min and then adding

TABLE 1 Differential characteristics of biovars and biotypes in genus *Brucella*[a]

Biovar	Biotype	CO$_2$ required	H$_2$S produced	Appearance of urease activity	Growth in presence of dyes[b]			Agglutination in sera[c]		
					Basic fuchsin, 20 µg	Thionin		A	M	R
						20 µg	40 µg			
B. melitensis	1	−	−	Variable	+	+	+	−	+	−
	2	−	−	Variable	+	+	+	+	−	−
	3	−	−	Variable	+	+	+	+	+	−
B. abortus[d]	1	(+)	+	1–2 h	+	−	−	+	−	−
	2	(+)	+	1–2 h	−	−	−	+	−	−
	3	(+)	+	1–2 h	+	+	+	+	−	−
	4	(+)	+	1–2 h	(+)	−	−	−	+	−
	5	−	−	1–2 h	+	+	−	−	+	−
	6	−	(+)	1–2 h	+	+	−	+	−	−
	9	−	+	1–2 h	+	+	−	−	+	−
B. suis	1	−	+	0–30 min	(−)	+	+	+	−	−
	2	−	−	0–30 min	−	+	−	+	−	−
	3	−	−	0–30 min	+	+	−	+	−	−
	4	−	−	0–30 min	(−)	+	+	+	+	−
	5	−	−	0–30 min	−	+	+	−	+	−
B. canis		−	−	0–30 min	−	+	+	−	−	+
B. neotomae		−	+	0–30 min	−	−	−	+	−	−
B. ovis		+	−	−	(−)	+	+	−	−	+

[a]Data courtesy of the Diagnosis Bacteriology Laboratory, National Veterinary Services Laboratory, Ames, Iowa.
[b]Biotype differentiation is obtained on Trypticase soy or tryptose agar with dye concentrations of 20 and 40 µg/ml. Other concentrations may be preferable with other growth media. Interpretation of results should be controlled with the reference strains of each biotype. Conduct tests in CO$_2$ for strains requiring CO$_2$. +, Positive; −, negative; (+), most strains positive; (−), most strains negative.
[c]A, monospecific *B. abortus* antiserum; M, monospecific *B. melitensis* antiserum; R, anti-rough serum.
[d]*B. abortus* biovars 7 and 8 were deleted by the International Committee of Systematic Bacteriology Subcommittee on Taxonomy of *Brucella*.

the required amount to the melted agar base (tryptose agar or Trypticase soy agar). The agar is mixed and poured into petri dishes. The surfaces of the plates should be dry before inoculation. From each pure culture and control strain to be tested, a suspension is made by suspending a loopful of the organism in 1.0 ml of sterile saline. The suspensions should be of equal density. A sterile cotton swab can be used to inoculate the medium. Six cultures, including reference strains, may be tested on each plate. The plates are incubated under 10% CO$_2$ at 37°C for 3 to 4 days and examined for growth (4).

Phage Typing

The phage test is particularly useful in distinguishing *B. abortus* biotypes 3 to 6 and 9 from *B. melitensis*. The routine test dilution of bacteriophage Tbilisi will completely lyse smooth cultures of *B. abortus*, but *B. suis* and *B. melitensis* cultures are not affected by this phage dilution (4). *B. suis* is partially lysed by a phage concentration of 10^4 times the routine test dilution, whereas most strains of *B. melitensis* are not lysed by this concentration.

Oxidative Metabolic Tests

Clinical isolates of *B. melitensis*, *B. abortus*, and *B. suis* can be differentiated by using GN MicroPlates (Biolog, Haywood, Calif.). Biolog GN plates are less sensitive than the Warburg respirometric method for determining oxidative potential, and biotypes cannot be differentiated (49).

Genetic Probes

Randomly cloned DNA fragments have been used as probes to type brucellae. Four probes identify the most commonly encountered biovars (*B. melitensis* 1 and 3, *B. abortus* 1 and 3, *B. suis* 2, and *B. ovis*). The four acetylaminofluorene-labeled probes are available to reference laboratories (28). This typing method could circumvent the problems associated with mailing *Brucella* cultures, since purified DNA or membranes carrying DNA restriction fragments could be mailed to reference laboratories using the probes. DNA amplification methods with sufficient sensitivity and specificity for clinical diagnosis are under development (8, 30).

SEROLOGIC TESTS

Since isolation of the organisms from infected patients is difficult, serologic tests are relied on for the routine diagnosis of brucellosis. In culture-negative patients, serologic information alone cannot provide a definitive diagnosis of brucellosis. Exposure history, epidemiological features, and clinical presentation must all be considered for reliable interpretation of serologic tests. Seropositivity resulting from long-term exposure to low levels of organisms may occur in the absence of disease in veterinarians, abattoir workers, and herdsmen. Seroprevalence studies show that 5 to 10% of the population living in areas of endemicity have agglutination titers of ≥1:160 with no evidence of disease.

The tube agglutination test is the principal test used clinically in the United States (34, 39, 52). The standard antigen is prepared from *B. abortus* 1119-3 and is available to state and federal laboratories from the National Animal Disease Laboratory, U.S. Department of Agriculture, Ames, IA 50010. Laboratories receiving an occasional request for this test should refer the specimen to a reference laboratory. The microagglutination test correlates well with the tube agglutination test and is used by public health laboratories as a quantitative screening test for large numbers of sera (12, 34). Serum titers equal to or greater than 1:80 should be retested by the tube agglutination test. The 2-mercaptoethanol test for IgG is useful in evaluating response to antimicrobial therapy and in diagnosing chronic brucellosis (52). Other tests currently used for serodiagnosis of brucellosis include the radioimmune assay, complement fixation, and enzyme-linked immunoassay (5, 52).

SUSCEPTIBILITY TO ANTIMICROBIAL AGENTS

Antibiotic Susceptibility Testing

A low MIC does not correspond to a low MBC for *Brucella* spp.; thus, in vitro susceptibility tests are secondary in importance to clinical trials in establishing the efficacy of a therapeutic regimen and in evaluating new antibiotics. Relapses in patients with brucellosis and the progression of some acute cases to chronic brucellosis have not been due to the emergence of resistant strains. According to the results of in vitro drug susceptibility tests, most strains of *B. melitensis* are resistant to penicillins, cephalosporins, trimethoprim-sulfamethoxazole (TMP-SMX), erythromycin, and tobramycin while being susceptible to chloramphenicol, streptomycin, rifampin, tetracycline, ofloxacin, kanamycin, and gentamicin (44).

Treatment

Since intracellular bacteria are relatively inaccessible to antibiotics, prolonged treatment is advised. Treatment of brucellosis must effectively control acute illness and prevent complications and relapse. A combination of tetracycline or doxycycline with streptomycin for 4 to 6 weeks is currently regarded as the best treatment for brucellosis, since the relapse rate for patients receiving this regimen is significantly lower than that with other regimens (17, 29). Streptomycin and rifampin produce relapse rates of 0 to 13% after 6 weeks of therapy (18). TMP-SMX or rifampin alone is associated with relapse. The combinations of TMP-SMX with rifampin or tetracycline and of streptomycin with rifampin are effective for complicated cases (29). Kanamycin and gentamicin can replace streptomycin if resistance develops (33, 51). TMP-SMX plus streptomycin plus rifampin is recommended for treating childhood brucellosis (2, 18). Fluoroquinolones show a low MIC in vitro but a variable clinical response. Clinical trials demonstrated no difference between 6-week regimens of doxycycline-rifampin and of ofloxacin-rifampin for treatment of acute brucellosis when patients were monitored for relapse every 3 months for 1 year after therapy (1).

EVALUATION, INTERPRETATION, AND REPORTING OF RESULTS

Few clinical laboratory findings are indicative of brucellosis. Peripheral blood films may demonstrate moderate leukopenia with relative lymphocytosis. Erythrocyte sedimentation rate is slightly increased or normal. CSF protein level is elevated, and glucose level is normal or low in patients with brucellar meningitis. A C-reactive protein value of >1 μg/dl indicates unfavorable response to therapy for acute brucellosis (37). The lymphocyte proliferation test may be the only immunological evidence of brucellosis in cases of brucellar osteomyelitis or colitis (42). Acute brucellosis is diagnosed on positive culture or serologic findings of ≥1:160 for the tube agglutination test and ≥1:320 for the Coombs test. The 2-mercaptoethanol test is useful in diagnosing chronic disease. In the absence of a previously known infection, a single tube agglutination titer of <1:160 does not exclude the possibility of *Brucella* infection, and titers of ≥1:160 lose diagnostic significance in groups repeatedly exposed to brucellae, such as abattoir employees and veterinarians. The tube agglutination test is positive for only 27% of patients with chronic brucellosis with neurologic complications. False-positive tube agglutination tests may be caused by salmonellosis, tularemia, cholera, lupus erythematosus, and myeloma. The Coombs test and enzyme-linked immunosorbent assay for IgG and IgA antibodies may provide more useful information than the tube agglutination and 2-mercaptoethanol tests for distinguishing between inactive and active brucellosis (40). Blood cultures are positive in 10 to 70% of cases depending on geographic prevalence. Misidentification of *B. melitensis* as *Moraxella phenylpyruvica* has occurred when commercial identification systems, which typically do not include *Brucella* spp. in the database, were used (39). Differentiation of brucellar vaccine strains from field strains is important in determining vaccine efficacy and in epidemiological studies. *B. abortus* 19 has a reduced requirement for erythritol, is susceptible to penicillin, and is resistant to streptomycin and dyes. *B. melitensis* Rev. 1 is resistant to streptomycin and susceptible to penicillin and dyes. *B. suis* S2 tolerates a temperature range of 15 to 44°C, while field strains prefer a range of 18 to 42°C (10). Strains of *B. melitensis* biovar 1 that are sensitive to dyes and penicillin have been isolated from humans and sheep (9, 23). As the number of strains with anomalous dye sensitivities increases, the taxonomic usefulness of this marker decreases and the validity of established biovars becomes questionable (19). Thionin blue has become difficult to find, so that its usefulness is limited.

REFERENCES

1. Akova, M., O. Uzun, H. E. Akalin, M. Hayran, S. Ünal, and D. Gür. 1993. Quinolones in treatment of human brucellosis: comparative trial of ofloxacin-rifampin versus doxycycline-rifampin. *Antimicrob. Agents Chemother.* **37:**1831–1834.
2. Al-Eissa, Y. A., A. M. Kambal, M. N. Al-Nasser, S. A. Al-Habib, I. M. Al-Fawaz, and F. A. Al-Zamil. 1990. Childhood brucellosis: a study of 102 cases. *Pediatr. Infect. Dis. J.* **9:**74–79.
3. Allardet-Servent, A., G. Bourg, M. Ramuz, M. Pages, M. Bellis, and G. Roizes. 1988. DNA polymorphism in strains of the genus *Brucella. J. Bacteriol.* **170:**4603–4607.
4. Alton, G. G., L. M. Jones, R. D. Angus, and J. M. Verger. 1988. *Techniques for the Brucellosis Laboratory.* Institut National de La Recherche Agronomique, Paris.
5. Araj, G. F., and A. F. Kaufmann. 1989. Determination by enzyme-linked immunosorbent assay of immunoglobulin G (IgG), IgM, and IgA to *Brucella melitensis* major outer membrane proteins and whole-cell heat-killed antigens in sera of patients with brucellosis. *J. Clin. Microbiol.* **27:**1909–1912.

6. Ariza, J., T. Pellicer, R. Pallarés, A. Foz, and F. Gudiol. 1992. Specific antibody profile in human brucellosis. *Clin. Infect. Dis.* **14**:131–140.

7. Arnow, P. M., M. Smaron, and V. Ormiste. 1984. Brucellosis in a group of travelers to Spain. *J. Am. Med. Assoc.* **251**:505–507.

8. Baily, G. G., J. B. Krahn, B. S. Drasar, and N. G. Stoker. 1992. Detection of *Brucella melitensis* and *Brucella abortus* by DNA amplification. *J. Trop. Med. Hyg.* **95**:271–275.

9. Banai, M., I. Mayer, and A. Cohen. 1990. Isolation, identification and characterization in Israel of *Brucella melitensis* biovar 1 atypical strains susceptible to dyes and penicillin, indicating the evolution of a new variant. *J. Clin. Microbiol.* **28**:1057–1059.

10. Bosseray, N., and M. Plommet. 1990. *Brucella suis* S2, *Brucella melitensis* Rev. 1 and *Brucella abortus* S19 living vaccines: residual virulence and immunity induced against three *Brucella* species challenge strains in mice. *Vaccine* **8**:462–468.

11. Brown, G. M., C. R. Ranger, and D. J. Kelley. 1970. Selective medium for the isolation of *Brucella ovis*. *Cornell Vet.* **61**:265–280.

12. Brown, S. L., G. C. Klein, F. T. McKinney, and W. L. Jones. 1981. Safranin O-stained antigen microagglutination test for detection of *Brucella* antibodies. *J. Clin. Microbiol.* **13**:398–400.

13. Bundle, D. R., J. W. Cherwonogrodzky, M. Caroff, and M. B. Perry. 1987. The lipopolysacchrides of *Brucella abortus* and *B. melitensis*. *Ann. Inst. Pasteur Microbiol.* **138**:92–98.

14. Centers for Disease Control. 1983. Brucellosis—Texas. *Morbid. Mortal. Weekly Rep.* **32**:548–553.

15. Centers for Disease Control. 1993. Summary of notifiable diseases in the United States, 1992. *Morbid. Mortal. Weekly Rep.* **41**:67.

16. Centers for Disease Control and National Institutes of Health. 1988. *Biosafety in Microbiological and Biomedical Laboratories*, 2nd ed. Publication no. 17-40-508-3. U.S. Government Printing Office, Washington, D.C.

17. Cisneros, J. M., P. Viciana, J. Colmenero, J. Pachon, C. Martinez, and A. Alacon. 1990. Multicenter prospective study of treatment of *Brucella melitensis* brucellosis with doxycycline for 6 weeks plus streptomycin for 2 weeks. *Antimicrob. Agents Chemother.* **34**:881–883.

18. Colmenero Castillo, J. D., S. H. Marquez, J. M. R. Iglesias, F. C. Franquelo, F. R. Diaz, and A. Alonso. 1989. Comparative trial of doxycycline plus streptomycin versus doxycycline plus rifampin for the therapy of human brucellosis. *Chemotherapy* **35**:146–152.

19. Corbel, M. J. 1991. Identification of dye-sensitive strains of *Brucella melitensis*. *J. Clin. Microbiol.* **29**:1066–1068.

20. Corbel, M. J., and W. J. Brinley-Morgan. 1984. *Brucella*, p. 377–388. *In* J. G. Holt and N. R. Krieg (ed.), *Bergey's Manual of Systematic Bacteriology*, vol. 1. The Williams & Wilkins Co., Baltimore.

21. Daugherty, M. P., J. Dolter, G. C. Evans, M. E. Griffith, B. A. Hummert, D. Lindquist, M. M. Struthers, and J. D. Wong. 1992. Processing of specimens for isolation of unusual organisms. Part 6. *Brucella* spp., p. 1.18.23–1.18.27. *In* H. D. Isenberg (ed.), *Clinical Microbiology Procedures Handbook*, vol. 1. American Society for Microbiology, Washington, D.C.

22. Etemadi, H., A. Raissadat, M. J. Pickett, Y. Zafari, and P. Vahedifar. 1984. Isolation of *Brucella* spp. from clinical specimens. *J. Clin. Microbiol.* **20**:586.

23. Ewalt, D. R., and L. B. Forbes. 1987. Atypical isolates of *Brucella abortus* from Canada and the United States characterized as dye sensitive with M antigen dominant. *J. Clin. Microbiol.* **25**:698–701.

24. Friedrich, I., S. Schonfeld, and Y. Keness. 1992. The ability of *Brucella melitensis* to grow on Loewenstein-Jensen egg medium: presentation of a case with *Brucella* meningitis. *Isr. J. Med. Sci.* **28**:806–807.

25. Gazapo, E., J. G. Lohoz, J. L. Subiza, M. Baquero, J. Gil, and E. G. de la Concha. 1989. Changes in IgM and IgG antibody concentrations in brucellosis over time: importance for diagnosis and follow-up. *J. Infect. Dis.* **159**:219–225.

26. Georghiou, P. R., and E. J. Young. 1991. Prolonged incubation in brucellosis. *Lancet* **337**:1543.

27. Gotuzzo, E., C. Carrillo, J. Guerra, and L. Llosa. 1986. An evaluation of diagnostic methods for brucellosis—the value of bone marrow culture. *J. Infect. Dis.* **153**:122–125.

28. Grimont, F., J. M. Verger, P. Cornelis, J. Limet, M. Lefèvre, M. Grayon, B. Régnault, J. Van Broeck, and P. A. D. Grimont. 1992. Molecular typing of *Brucella* with cloned DNA probes. *Res. Microbiol.* **143**:55–65.

29. Hall, W. H. 1990. Modern chemotherapy for brucellosis in humans. *Rev. Infect. Dis.* **12**:1060–1099.

30. Herman, L., and H. De Ridder. 1992. Identification of *Brucella* spp. by using the polymerase chain reaction. *Appl. Environ. Microbiol.* **58**:2099–2101.

31. International Committee on Systematic Bacteriology, Subcommittee on Taxonomy of *Brucella*. 1988. Minutes of the meeting-September 5, 1986. *Int. J. Syst. Bacteriol.* **38**:450–452.

32. Kolman, S., M. C. Maayan, G. Gotesman, L. A. Rozenszajn, B. Wolach, and R. Lang. 1991. Comparison of the Bactec and lysis concentration methods for recovery of *Brucella* species from clinical specimens. *Eur. J. Clin. Microbiol. Infect. Dis.* **10**:647–648.

33. Montejo, J. M., I. Alberola, P. Glez-Zarate, A. Alvarez, J. Alonso, A. Canovas, and C. Aguirre. 1993. Open, randomized therapeutic trial of six antimicrobial regimens in the treatment of human brucellosis. *Clin. Infect. Dis.* **16**:671–676.

34. Moyer, N. P., G. M. Evans, N. E. Pigott, J. D. Hudson, C. E. Farshy, J. C. Feeley, and W. J. Hausler, Jr. 1987. Comparison of serologic screening tests for brucellosis. *J. Clin. Microbiol.* **25**:1969–1972.

35. Moyer, N. P., and W. J. Hausler, Jr. 1992. The genus *Brucella*, p. 2384–2400. *In* A. Balows, H. G. Trüper, M. Dworkin, W. Harder, and K. H. Schleifer (ed.), *The Prokaryotes*, vol. 3, 2nd ed. Springer-Verlag, New York.

36. Moyer, N. P., and L. A. Holcomb. 1988. Brucellosis, p. 143–154. *In* A. Balows, W. J. Hausler, Jr., M. Ohashi, and A. Turano (ed.), *Laboratory Diagnosis and Infectious Diseases: Principles and Practice*, vol. 1. Springer-Verlag, New York.

37. Navarro, J. M., J. Mendoza, J. Leiva, R. Rodríguez-Contreras, and M. de la Rosa. 1990. C-reactive protein as a prognostic indicator in acute brucellosis. *Diagn. Microbiol. Infect. Dis.* **13**:269–270.

38. Navas, E., A. Guerrero, J. Cobo, and E. Loza. 1993. Faster isolation of *Brucella* spp. from blood by Isolator compared with Bactec NR. *Diagn. Microbiol. Infect. Dis.* **16**:79–81.

39. Peiris, V., S. Fraser, M. Fairhurst, D. Weston, and E. Kaczmarski. 1992. Laboratory diagnosis of *Brucella* infection: some pitfalls. *Lancet* **339**:1415.

40. Pellicer, T., J. Ariza, A. Foz, R. Pallares, and F. Gudiol. 1988. Specific antibodies detected during relapse of human brucellosis. *J. Infect. Dis.* **157**:918–924.

41. Plommet, M. 1991. Minimal requirements for growth of *Brucella suis* and other *Brucella* species. *Zentralbl. Bakteriol.* **275**:436–450.

42. Potasman, I., L. Even, M. Banai, E. Cohen, D. Angel, and M. Jaffe. 1991. Brucellosis: an unusual diagnosis for a seronegative patient with abscesses, osteomyelitis, and ulcerative colitis. *Rev. Infect. Dis.* **13**:1039–1042.

43. Raad, I., K. Rand, and D. Gaskins. 1990. Buffered charcoal-yeast extract medium for the isolation of brucellae. *J. Clin. Microbiol.* **28**:1671–1672.

44. Rubinstein, E., R. Lang, B. Shasha, B. Hagar, L. Diamanstein, G. Joseph, M. Anderson, and K. Harrison. 1991. In vitro susceptibility of *Brucella melitensis* to antibiotics. *Antimicrob. Agents Chemother.* **35**:1925–1927.

45. Smith, L. D., and T. A. Ficht. 1990. Pathogenesis of *Brucella*. *Crit. Rev. Microbiol.* **17**:209–230.

46. Solomon, H. M., and D. Jackson. 1992. Rapid diagnosis of

Brucella melitensis in blood: some operational characteristics of the BacT/Alert. *J. Clin. Microbiol.* **30:**222–224.

47. **Staszkiewicz, J., C. M. Lewis, J. Colville, M. Zervos, and J. Band.** 1991. Outbreak of *Brucella melitensis* among microbiology laboratory workers in a community hospital. *J. Clin. Microbiol.* **29:**287–290.

48. **Terzolo, H. R., F. A. Paolicchi, A. R. Moreira, and A. Homse.** 1991. Skirrow agar for simultaneous isolation of *Brucella* and *Campylobacter* species. *Vet. Rec.* **129:**531–532.

49. **Wong, J. D, J. M. Janda, and P. S. Duffey.** 1992. Preliminary studies on the use of carbon substrate utilization patterns for identification of *Brucella* species. *Diagn. Microbiol. Infect. Dis.* **15:**109–113.

50. **Young, E. J.** 1983. Human brucellosis. *Rev. Infect. Dis.* **5:**821–842.

51. **Young, E. J.** 1989. Treatment of brucellosis in humans, p. 127–141. *In* E. J. Young and M. J. Corbel (ed.), *Brucellosis: Clinical and Laboratory Aspects.* CRC Press, Inc., Boca Raton, Fla.

52. **Young, E. J.** 1991. Serologic diagnosis of human brucellosis: analysis of 214 cases by agglutination tests and review of the literature. *Rev. Infect. Dis.* **13:**359–372.

53. **Zimmerman, S. J., S. Gillikin, N. Sofat, W. R. Bartholomew, and D. Amsterdam.** 1990. Case report and seeded blood culture study of *Brucella* bacteremia. *J. Clin. Microbiol.* **28:**2139–2141.

Haemophilus

JOSEPH M. CAMPOS

45

TAXONOMY

The organism we now know as *Haemophilus influenzae* was first described in 1892 by Pfeiffer, who found it in sputa of several patients with influenza. The genus name *Haemophilus*, however, was not assigned to the organism until 1920 (99). Pittman described the six capsular serotypes of *H. influenzae* in 1931 and recognized that members of serotype b were most likely to cause invasive infections (74). Clinically encountered species of *Haemophilus* include *H. influenzae*, *H. parainfluenzae*, *H. ducreyi*, *H. aphrophilus*, *H. paraphrophilus*, *H. haemolyticus*, *H. parahaemolyticus*, and *H. segnis*, although evidence is accumulating that *H. ducreyi* may be only distantly related to the *Pasteurellaceae* family, including members of the genus *Haemophilus* (25). Although the incidence of *H. influenzae* type b infections in the United States was drastically reduced by the introduction of the first of several effective vaccines in 1985 (35), *Haemophilus* species remain important causes of a wide spectrum of human infections.

DESCRIPTION OF THE GENUS

Haemophilus species are gram-negative coccobacillary bacteria. They are facultatively anaerobic, meaning that they generate their energy for growth either oxidatively (preferred) or fermentatively. Growth is optimal for most species when they are incubated in a humid atmosphere containing 5 to 10% CO_2 maintained at a temperature between 33 and 37°C. Some *Haemophilus* cultures produce a characteristic "mouse nest" odor, which is a clue to experienced technologists of their presence. Most strains exhibit the aforementioned coccobacillary morphology under the microscope, but occasional organisms form short filaments, giving rise to the well-known phrase "pleomorphic bacilli." Most species have fastidious nutritional requirements and grow in the laboratory only when provided with complex, well-supplemented media (e.g., chocolate agar) or with an exogenous source of essential nutrients (e.g., satelliting growth around colonies of *Staphylococcus aureus*). The specific growth requirements of clinically significant species are discussed later in this chapter.

NATURAL HABITATS

Most *Haemophilus* species are normal inhabitants of the upper respiratory tracts of humans and other animals.

CLINICAL SIGNIFICANCE

A wide variety of *Haemophilus* infections may occur in humans, ranging from infections that are frequent and easily managed (e.g., conjunctivitis and otitis media) to those that are rare and potentially life threatening (e.g., meningitis, pericarditis, and Brazilian purpuric fever). The species of *Haemophilus* that most frequently cause human infections are *H. influenzae* (including biogroup aegyptius), *H. parainfluenzae*, *H. ducreyi*, and *H. aphrophilus*. The species *H. haemolyticus*, *H. parahaemolyticus*, *H. segnis*, and *H. paraphrophilus* uncommonly cause infection and will be discussed only briefly here.

H. influenzae

Spread of *H. influenzae* occurs by inhalation of respiratory droplets. Respiratory tract colonization rates by all strains of *H. influenzae* may reach as high as 50% of the population (95), but colonization by type b strains is fairly uncommon in healthy infants (0.7%) and children (3 to 5%) and is quite rare in adults (59).

Possession of a polysaccharide capsule is a major virulence factor for strains of *H. influenzae* that cause systemic infection. Such strains can be serotyped on the basis of the antigenic specificities of their capsules, with serotypes a through f originally having been characterized by Pittman in 1931 (74). The majority of encapsulated strains causing infection belong to serotype b. Collectively the encapsulated strains are referred to as typeable strains. Unencapsulated (nontypeable) strains may also cause infection, but these infections usually occur at sites contiguous with the upper respiratory tract and typically are not spread to other sites via the bloodstream. Recent molecular analysis suggests that the nontypeable strains of *H. influenzae* derive from encapsulated ancestors (87).

Until recently, *H. influenzae* type b was the leading cause of bacterial meningitis and epiglottitis; a major cause of pericarditis, pneumonia, septic arthritis, osteomyelitis, and facial cellulitis; and an occasional cause of urinary tract infection and peritonitis in children younger than 5 years of age (17). Approximately 1 in 200 children in the United States experienced invasive infection with this organism before the age of 5 years, with a peak incidence of infection at 6 to 7 months of age (15). Attendance of preschool-age children at day-care centers was a risk factor for development of primary *H. influenzae* disease (71). Sharp reductions in the incidence of invasive *H. influenzae* were observed beginning in the mid-1980s; they were probably due to a combination of vaccine administration, rifampin prophylaxis of disease contacts, and use of more efficacious therapeutic agents (58). The introduction and widespread availability of efficacious vaccines in the United States beginning in 1985 further accelerated the reduction in incidence of *H. influenzae* type b infections, so that today these infections have become quite rare. The Immunization Practices Advisory Committee of the Centers for Disease Control currently recommends administration of a licensed *H. influenzae* type b conjugate vaccine to all children starting at 2 months of age (12).

H. influenzae, predominantly nontypeable strains, causes a variety of infections in adults and children beyond the age of 9 years (47). Lower respiratory tract infections (febrile tracheobronchitis and pneumonia), frequently with concomitant bacteremia, are reported most often. In the postvaccine era, adults now represent 25% of all cases of invasive *H. influenzae* infection (30). The majority of these patients suffer underlying medical conditions that predispose them to infection. Such conditions include malignancy, chronic obstructive pulmonary disease, alcoholism, infection with human immunodeficiency virus type 1, and pregnancy. Approximately 80% of invasive isolates have been nontypeable (47). Of the typeable strains recovered from bacteremic patients, the large majority have been type b. Type f strains have constituted approximately 85% of the non-type-b typeable strains (85).

Nonencapsulated strains of *H. influenzae* tend to cause noninvasive respiratory infections in healthy children; community-acquired pneumonia and exacerbations of chronic bronchitis in adults; and occasionally, invasive infections in neonates, children, adults, and immunocompromised patients (31, 32, 64, 98). In children, these organisms are the most common cause of purulent bacterial conjunctivitis (formerly referred to as *H. aegyptius* or the Koch-Weeks bacillus) (23, 94) and, after *Streptococcus pneumoniae*, the second most frequent cause of otitis media (6).

During the late spring of 1984, an outbreak of severe childhood illness characterized by high fever, abdominal pain, vomiting, hemorrhagic skin lesions, vascular collapse, and death occurred in São Paulo State, Brazil (7). Laboratory investigation revealed that strains of *H. influenzae* biogroup aegyptius belonging to a single clone were the cause of the infection, later named Brazilian purpuric fever (8, 10). Conventional biochemical testing is unable to differentiate the Brazilian purpuric fever clone of *H. influenzae* biogroup aegyptius from *H. influenzae* biogroup III (90). Outer membrane protein profiles of the two biogroups are different, however, and can be used for differentiation (11).

H. ducreyi

H. ducreyi is the cause of chancroid, a sexually transmitted disease manifested by shallow genital ulcerations that may be accompanied by inguinal lymphadenopathy. Most cases occur in tropical climates among lower socioeconomic groups. In recent years, however, miniepidemics of disease,

including a number of outbreaks in the United States, have occurred at sites far removed from the tropics. The lesion itself begins as a tender papule that becomes pustular and then ulcerated over the course of 2 days. Lesions may merge to form larger ulcers. The ulcers in most patients are painful and thus easily distinguished from the chancres of syphilis. Tender unilateral inguinal lymphadenitis (bubo formation) is found in 50% of cases. Recent data suggest that the presence of genital ulcerations, such as those resulting from chancroid, increases the risk of acquiring human immunodeficiency virus type 1 infection during homosexual or heterosexual contact (73, 86).

H. parainfluenzae

H. parainfluenzae is ordinarily regarded as part of the commensal flora of the human upper respiratory tract. Situations arise, nevertheless, in which this organism is a serious pathogen. *H. parainfluenzae* endocarditis is such an example and can be difficult to recognize in the laboratory owing to the fastidious nature of the organism (14). The organism is also an occasional cause of secondary bacteremia and urethritis in adults (70, 88).

H. aphrophilus

H. aphrophilus is another upper respiratory tract inhabitant that can cause serious infection. Endocarditis and brain abscess are the most common presentations, with pneumonia, meningitis, and secondary bacteremia occurring less often (5). The species name aphrophilus derives from the enologic term aphros, referring to the froth formed during fermentation of grape juice. The name is an indirect reference to the bacterium's requirement for added CO_2 in its incubation atmosphere. Kilian suggested in 1976 that *H. aphrophilus* and the related organisms *H. paraphrophilus* and *Actinobacillus actinomycetemcomitans* may not meet the criteria for membership in the genus *Haemophilus* (45).

Other Species

An unnamed genospecies of *Haemophilus* genetically related to *H. influenzae* and *H. haemolyticus* has been recovered from patients with serious urogenital, neonatal, and mother-infant infections (76). Characterization of these isolates by currently used criteria identifies them as nontypeable *H. influenzae* biotype IV. However, it is quite likely that these organisms belong to a distinct, thus far unnamed species.

Other species of *Haemophilus*, e.g., *H. paraphrophilus*, *H. haemolyticus*, and *H. segnis*, are part of the human oral microbiota and very rare causes of infection.

COLLECTION, TRANSPORT, AND STORAGE OF SPECIMENS

Specimens harboring *Haemophilus* spp. must be as free as possible of contaminating flora because of the fastidious growth requirements of the organism and the ease with which colonies of *Haemophilus* can be obscured by those of other microorganisms. The importance of careful selection and adequate disinfection (where appropriate) of specimen sampling sites cannot be overemphasized.

Blood

Haemophilus bacteremia of childhood tends to be high magnitude (>100 CFU/ml), obviating the need to collect as large a quantity of blood for culture as is collected from adults (4). However, since 27% of pediatric bacteremias exhibit organism densities of <10 CFU/ml (9) and since 10 to 20% of low-magnitude occult bacteremias in children are caused by *H. influenzae* (37), a minimum of 0.5 ml and preferably several milliliters of blood should be submitted for culture. The recommendations published in *Cumitech 1A* for adults should be followed for patients beyond the age of 7 years (77).

Body Fluids

A >1-ml quantity of cerebrospinal fluid (CSF), so as to make conventional centrifugation prior to Gram stain and culture worthwhile, is strongly recommended. Similarly, partial antimicrobial treatment of patients with possible meningitis may require collection of sufficient CSF for both culture and microbial antigen detection (discussed below). Other body fluids likely to contain *Haemophilus* spp. (e.g., synovial fluid, pericardial fluid, pleural fluid) should be collected in amounts that make both Gram stain and culture possible.

Ocular Specimens

Conjunctivitis caused by *H. influenzae* usually results in copious amounts of purulent discharge. Pus should be collected on the tip of a calcium alginate flexible-shaft swab and placed in modified Stuart's transport medium for delivery to the laboratory.

Respiratory Specimens

The major difficulty in documenting *Haemophilus* spp. as causes of lower respiratory tract infection is interference from commensal upper respiratory flora. Specimen collection techniques that avoid contamination with upper respiratory secretions are recommended. Specimen collection from patients with epiglottitis should not be attempted unless emergency airway maintenance procedures can be initiated rapidly should laryngospasm occur. The utility of nasopharyngeal cultures in identifying the etiology of otitis media has been a controversial topic. Most recently, these cultures were shown to have minimal positive predictive value (71%) for nontypeable *H. influenzae* disease and very little positive predictive value for *Streptococcus pneumoniae* and *Branhamella* (*Moraxella*) *catarrhalis* infection (29). Schwartz et al. found their value highest when 2+ or more growth of a single pathogen was recovered (82).

Genital Specimens

The specimen of choice for laboratory diagnosis of chancroid is material collected from the exposed base or margin of the lesion. Needle aspirates of pus from buboes of patients manifesting inguinal lymphadenitis augment but should not replace lesional smears and cultures.

DIRECT EXAMINATION

Microscopic Examination

Visualization of *Haemophilus* spp. under the microscope is hampered by the relatively small size of these organisms. Use of biological stains such as the Gram and acridine orange (AO) stains facilitates organism detection by enhancing the contrast between bacteria and background. Methanol fixation of smears, a standard step during AO staining, enhances the appearance of *Haemophilus* spp. in Gram stains as well (see chapter 21 of this Manual). The Quellung reaction, primarily associated with identification

and serotyping of *Streptococcus pneumoniae*, can also be used for recognition of encapsulated (typeable) strains of *H. influenzae*.

Gram Stain

Microscopic examination of Gram-stained body fluid smears can make possible early detection of life-threatening *H. influenzae* infection. Although definitive identification of *H. influenzae* cannot be made, presumptive identification with a high degree of accuracy is possible for an experienced examiner. The main limitation of Gram stain, as is the case with other microscopic assays, is lack of sensitivity. Approximately 85% of cases of culture-proven *H. influenzae* meningitis yield positive Gram-stained smears of CSF (33). Organisms in specimens containing fewer than 5×10^4 CFU/ml of body fluid stand little chance of being detected. Conventional centrifugation and preparation of smears from the resuspended sediment of such specimens increase the sensitivity by 5- to 10-fold, provided there is sufficient quantity of the specimen (≥ 1.0 ml) to make centrifugation worthwhile. Cytocentrifugation of specimens, in which sediments are deposited directly onto microscope slides, reportedly increases the sensitivity of Gram stain by as much as 100-fold (83).

Gram stain has also been advocated by some as a means for presumptive diagnosis of chancroid. The cells of *H. ducreyi* characteristically exhibit the "school of fish" arrangement, although this may be more a property of broth culture suspensions than lymph node aspirates. Published studies report the sensitivity of Gram stain as less than 50% (68).

Wayson Stain

Another limitation of the Gram stain is the lack of contrast between some stained microorganisms and stained cellular debris. The tiny gram-negative bacilli of *H. influenzae* can be camouflaged very effectively by the gram-negative proteinaceous strands that frequently appear in infected body fluids. A solution to this problem is use of a staining method that differentially stains microorganisms and other components of the smear. Wayson-stained smears yield dark blue bacteria, light blue proteinaceous strands, and light blue to purple inflammatory cells (19). These smears cannot be restained with Gram reagents, however; new smears are necessary to determine the Gram reactions of organisms.

AO Stain

AO dye has a high affinity for polymerized nucleic acids, particularly double-stranded DNA. AO-stained smears are examined under a fluorescence microscope equipped to provide UV illumination. Use of this dye as a primary stain enables detection of 10-fold-fewer *Haemophilus* spp. than Gram stain can detect (49). The reason is the marked contrast between the fluorescent orange stained bacilli and the greenish black unstained background. However, the added sensitivity of the AO stain is at least partially negated when specimens contain large numbers of inflammatory cells. The fluorescence emanating from cellular nuclei can interfere greatly with detection of tiny fluorescent bacteria. AO-positive smears can be restained with Gram reagents to determine the Gram reactions of bacteria present (56).

Antigen Detection

Assay methods for detection of bacterial capsular polysaccharide in body fluids are described in chapter 11 of this

Manual. Most laboratories offering this service utilize the rapid particle agglutination assay format, which is available from several commercial sources.

Particle Agglutination

Commercially available particle agglutination immunoassays for *H. influenzae* type b antigen are of two types: those based on staphylococcal protein A-mediated coagglutination (SPAMC) of immunoglobulin G (IgG)-coated staphylococcal cells and those based on agglutination of IgG-sensitized latex particles (LPA). Both assays furnish results in less than 10 min and require no specialized equipment, although use of a mechanical rotator makes both assays more convenient to perform. More important, these particle agglutination assays are 10- to 100-fold more sensitive than counterimmunoelectrophoresis (22, 81). Head-to head comparisons of LPA and SPAMC found them to be equivalent at detecting *H. influenzae* type b antigen in CSF (16, 53), but LPA was superior to SPAMC when non-CSF body fluids from patients with culture-proven nonmeningeal *H. influenzae* type b infection were tested (53). Neither assay is sufficiently sensitive to substitute for culture.

LPA can detect *H. influenzae* type b antigen in CSF (21) and urine (38) as long as 21 days after immunization with *H. influenzae* type b vaccine. Laboratorians should alert clinicians to this possibility when unexpectedly positive test results are obtained from children of vaccine age.

Other Methods

Enzyme immunoassay, radioimmunoassay, and indirect immunofluorescence assay for *H. influenzae* type b antigen have been described, but their use has been limited primarily to research applications (50, 51, 84). Although their sensitivities may equal or even exceed that of particle agglutination, they require specialized equipment and longer assay times, and any improvement in sensitivity is likely clinically irrelevant.

Sputa from cystic fibrosis patients can be washed, smeared, and stained with an immunoperoxidase conjugate of monoclonal antibody 8BD9 directed against outer membrane protein P6 of *H. influenzae* (61). This procedure has been reported to be more sensitive than culture for detection of *H. influenzae* because of the frequency with which *Pseudomonas aeruginosa* overgrows respiratory cultures.

Immunofluorescence (80) and enzyme immunoassays (79) for detection of *H. ducreyi* antigen have been described. Evaluation of the former assay yielded a sensitivity and a specificity of 50 and 100%, respectively. The latter assay demonstrated a sensitivity and specificity of 100% during a very limited evaluation of lesion swab specimens from 30 patients.

DNA Probe

A PCR-based DNA probe assay for noncultural detection of *H. ducreyi* rRNA in genital ulcer specimens was described recently (13). The assay utilizes broad-specificity primers that amplify rRNAs from a wide variety of bacteria and specific probes labeled with [^{32}P]ATP for identification of amplified *H. ducreyi* rRNA sequences. Validation of assay performance with clinical specimens revealed a sensitivity as high as 98% but a specificity of only 51%. Most of the false-positive PCR results were felt to reflect infected patients not detected by culture.

CULTURE AND ISOLATION

Attempts to grow *Haemophilus* spp. in culture must take into account the nutritionally fastidious nature of these microorganisms. All species, with the exception of some laboratory-adapted strains of *H. aphrophilus*, require either exogenous hemin (X factor), NAD (V factor), or both. Strains of some species (e.g., *H. influenzae* biogroup aegyptius, *H. ducreyi*) have requirements for additional nutrients that when met, facilitate their detection by culture. Most species grow well on standard chocolate agar and not at all on conventional 5% sheep blood agar. The latter medium may have adequate amounts of available X factor, but intact V factor is found only intraerythrocytically. The original recipe for chocolate agar included a heating step during which sheep erythrocytes were lysed and the V-factor-destroying enzyme NADase was inactivated. Most commercial chocolate agar today is prepared by adding hemoglobin and vitamin supplements to a nutritional agar base. In a comparison of six medium bases and seven growth supplements, Rennie et al. reported that a medium comprising GC agar base plus 5% chocolatized sheep blood and 1% yeast autolysate promoted the best growth of *H. influenzae*, *H. parainfluenzae*, and other less commonly encountered species of *Haemophilus* (excluding *H. ducreyi*) (78).

Growth of *Haemophilus* species on 5% sheep blood agar can be achieved by cross-streaking the medium with a *Staphylococcus* sp. or by applying a filter paper disk or strip saturated with X+V factor to the surface of the medium. *Haemophilus* spp. form satellite colonies along the length of the staphylococcal growth or around the X+V-factor disk or strip.

Another difficulty encountered during culture of *Haemophilus* spp. is the ease with which their growth may be obscured by that of other organisms. Colonies of *Haemophilus* spp. tend to be small and translucent after overnight incubation on chocolate agar. Their presence among the mixed flora in respiratory cultures can be easily overlooked. This is a particularly vexing problem when culturing sputa from cystic fibrosis patients, most of whose cultures exhibit high counts of spreading *P. aeruginosa* colonies. *H. influenzae* type b is a major pulmonary pathogen in these patients, and its detection is necessary. Inoculation and anaerobic incubation on chocolate agar containing 300 μg of bacitracin per ml have been carried out with some success. Similar success has been reported for NAG medium (blood agar base, N-acetyl-D-glucosamine, hemin, NAD, and bacitracin) (62). Alternatively, inoculation of specimens to Levinthal agar containing bacitracin and high-titer antiserum to *H. influenzae* type b (60) or Columbia agar containing yeast extract, hemin, NAD, bacitracin, and *H. influenzae* type b antiserum (3) can be performed. On the last two media, *H. influenzae* type b colonies are recognizable by the surrounding halos of antigen-antibody precipitate.

Body fluid specimens likely to harbor *Haemophilus* spp. should be inoculated to 5% sheep blood agar, chocolate agar, and a suitable enrichment broth (e.g., modified eugonic broth, Fildes broth). Agar media should be incubated at 35 to 37°C for 48 to 72 h in an aerobic atmosphere containing 5 to 10% CO_2. As mentioned previously, *Haemophilus* colonies tend to be small and translucent and exude a mouse nest odor, owing to their production of indole from tryptophan. Encapsulated strains of *H. influenzae* form larger, glistening colonies that are somewhat more noticeable than those of nonencapsulated strains after overnight incubation. Enrichment broths should be incubated for 7 days at 35 to 37°C and carefully examined daily for evidence of growth. Because of their small dimensions, cells of *Haemophilus* spp. cannot be depended on to impart obvious turbidity to broths even when they are present at densities exceeding 10^9 CFU/ml.

Lower respiratory tract secretions collected without obvious contamination by upper respiratory flora and aspirates of pus from localized infections should be inoculated to 5% sheep blood agar and chocolate agar and incubated as described above. Inclusion of an enrichment broth is optional, owing to the probability of specimen contamination with small numbers of respiratory or cutaneous flora.

Conjunctival swabs for detection of nontypeable *H. influenzae* and members of biogroup aegyptius should be inoculated to 5% sheep blood agar and chocolate agar. If available, enriched chocolate agar (chocolate agar plus 1% IsoVitaleX) is preferred, but the majority of strains will grow on conventional chocolate agar. Cultures should be incubated at 35 to 37°C for 72 h in an aerobic atmosphere containing 5 to 10% CO_2.

Lymph node aspirates and lesional scrapings from patients with suspected chancroid should be inoculated to selective nutritionally rich media such as Hammond gonococcal medium (gonococcal agar base containing 5% fetal calf serum, 1% bovine hemoglobin, 1% CVA (cofactors-vitamins-amino acids) enrichment, and 3 μg of vancomycin per ml), Fildes enriched gonococcal medium (gonococcal agar base containing 5% Fildes reagent, 5% horse blood, and 3 μg of vancomycin per ml), or enriched Mueller-Hinton agar (Mueller-Hinton agar base containing 5% chocolatized horse blood, 1% CVA enrichment, and 3 μg of vancomycin per ml) (20, 52). Vancomycin is present in these media to suppress the growth of contaminating gram-positive cutaneous flora. Use of two different media augments the recovery of *H. ducreyi* over that from a single medium (63). Cultures should be incubated at 33 to 35°C for 72 h in a candle extinction jar.

IDENTIFICATION

X- and V-Factor Requirements

The standard workup for an isolate that exhibits microscopic and colonial properties suggestive of *Haemophilus* spp. is to determine the organism's X- and V-factor requirements.

A long-standing, still popular method for determining an isolate's X- and V-factor requirements utilizes factor-impregnated filter paper strips. A McFarland 0.5 standardized suspension of the isolate is prepared in sterile 0.85% saline, the suspension is swabbed onto the surface of a 150-mm Mueller-Hinton agar plate, and then the X, V, and X+V filter paper strips are applied to the medium surface. Care must be taken not to prepare too heavy a suspension and thus have X- or V-factor carryover from the primary growth medium and not to place the strips too close to one another and thus obtain confusing results. The plate is examined after overnight incubation to determine which of the growth factor strips promoted satelliting growth.

Inzana et al. described a broth medium alternative to the agar-based X- and V-factor strip test (36). The broth system consists of three tubes of brain heart infusion broth supplemented with X factor, V factor, or X+V factor. Tubes are inoculated from an organism suspension prepared in sterile 0.85% saline and incubated with aeration at 37°C. In most instances, tubes can be read for turbidity or clumping after 4 to 5 h of incubation, with some V-factor strains requiring overnight incubation.

Perhaps the most convenient method for ascertaining X- and V-factor growth requirements is the commercially available quadrant plate approach. Laboratories can purchase prepoured plates in which each quadrant of the 100-mm plate contains Mueller-Hinton agar supplemented with X factor, V factor, X+V factor, or X+V factor and horse erythrocytes (for determination of hemolytic capability). An organism suspension in sterile 0.85% saline is prepared and lightly inoculated to each of the four quadrants. As noted above, heavy organism suspensions must be avoided to eliminate X- and V-factor carryover from the primary growth medium. After overnight incubation, the plate is examined to determine the isolate's X- and V-factor requirements as well as its hemolytic potential.

Porphyrin Test

The porphyrin test of Kilian determines an isolate's X-factor requirement while avoiding the problem of X-factor carryover from primary culture media and X-factor contamination of test media (45). *Haemophilus* species that require X factor do not excrete porphobilinogen and porphyrins during growth because of enzymatic deficiencies in the hemin biosynthetic pathway. To perform the test, heavy suspensions of isolates are prepared in 0.5-ml quantities of 2 mM δ-aminolevulinic acid and 0.8 mM $MgSO_4$ in 0.1 M phosphate buffer (pH 6.9) and incubated for 4 h at 37°C. The tube is illuminated with UV light (~360 nm) and examined for red fluorescence. Fluorescence indicates enzymatic conversion of the δ-aminolevulinic acid substrate to porphyrins and, thus, X-factor independence. According to Kilian, coupling the results of the porphyrin test with those of one of the V-factor tests described above averts misidentification of approximately 18% of *Haemophilus* spp. (46).

Biochemical Tests

As a supplement to determining growth requirements for X and/or V factors, key biochemical tests can be performed to identify isolates to the species level (Table 1). Hemolysis of horse blood is a test useful for distinguishing *H. influenzae* from *H. haemolyticus*, and *H. parainfluenzae* from *H. parahaemolyticus*. Fermentation of 1% glucose, sucrose, lactose, and/or mannose in phenol red broth base supplemented with X and V factors is useful for confirming identification of all *Haemophilus* species. Biotyping of *H. influenzae* and *H. parainfluenzae* (34) as well as identification of *H. segnis* and the aegyptius biogroup of *H. influenzae* can be accomplished with growth-independent rapid tests for indole, urease, and ornithine decarboxylase production (Table 2).

Several commercially available kits identify and biotype *Haemophilus* species as well. Recent evaluations of API 10E and API 20E (bioMérieux, La Balme-les-Grottes, France), HNID System (Baxter Healthcare, McGaw Park, Ill.), RIM-H 1/RIM-H 2 (Ortho Diagnostic Systems, Raritan, N.J.), and API NH (bioMérieux) have been published (2, 72, 75). All were reported as acceptably accurate compared to conventional biochemical test systems.

DNA Probes

A commercially available DNA probe for identification of *H. influenzae* culture isolates is marketed under the Accuprobe product name (Gen-Probe, San Diego, Calif.). The principle of the method is hybridization of a chemilumi-

TABLE 1 Differential characteristics of *Haemophilus* species[a]

Haemophilus sp.	Factor requirement		Hemolysis of horse blood	Fermentation of:				Presence of catalase	CO_2 enhancement of growth
	X[b]	V		Glucose	Sucrose	Lactose	Mannose		
H. influenzae[c]	+	+	−	+	−	−	−	+	+
H. haemolyticus	+	+	+	+	−	−	−	+	−
H. ducreyi	+	−	−	−	−	−	−	−	−
H. parainfluenzae	−	+	−	+	+	−	+	D	D
H. parahaemolyticus	−	+	+	+	+	−	−	+	−
H. segnis	−	+	−	w	w	−	−	D	−
H. paraphrophilus	−	+	−	+	+	+	+	−	+
H. aphrophilus	−	−	−	+	+	+	+	−	+

[a] Table is adapted from reference 46a. d, Differences encountered; w, weak fermentation reaction.
[b] As determined by the porphyrin test.
[c] Includes biogroup aegyptius.

nescent acridinium ester-labeled DNA probe with complementary *H. influenzae* rRNA. A published assessment of the product found it to be extremely accurate compared to conventional test methods (18).

SEROLOGIC TESTS

Serotyping Methods

Within the *Haemophilus* genus, serotyping is relevant only for encapsulated strains of *H. influenzae*. Rapid agglutination of organisms in type-specific antisera has been used for many years but requires a quantity of growth sufficient to produce visible clumping of bacterial cells. An equally rapid method that requires fewer organisms and produces easily interpreted results is the commercially available Phadebact

TABLE 2 Differentiation of *H. influenzae* and *H. parainfluenzae* biotypes[a]

Species and biotype	Indole	Urease	Ornithine decarboxylase
H. influenzae			
I	+	+	+
II	+	+	−
III[b]	−	+	−
IV	−	+	+
V	+	−	+
VI	−	−	+
VII	+	−	−
VIII	−	−	−
H. parainfluenzae[c]			
I	−	−	+
II	−	+	+
III	−	+	−
IV	+	+	+
VI	+	−	+
VII	+	+	−
VIII	+	−	−

[a] Table is adapted from reference 46a.
[b] *H. influenzae* biotype III and *H. influenzae* biogroup aegyptius can be differentiated by analysis of outer membrane protein profiles.
[c] V-factor-dependent strains showing negative reactions in all three tests used for biotyping have been referred to as *H. parainfluenzae* biotype V (89). However, it is unclear whether such strains are *H. parainfluenzae* or are strains of *H. segnis* or *H. paraphrophilus*.

Haemophilus Test (Karo Bio Diagnostics, Huddinge, Sweden) staphylococcal coagglutination kit. The limitation of this kit is that it identifies type b strains only; non-type-b encapsulated isolates are referred to collectively as types a, c, d, e, and f. The classic slide agglutination method mentioned above must be used to identify these serotypes specifically.

Detection of Antibodies to *Haemophilus* spp.

Chancroid is the only *Haemophilus* infection for which antibody detection has been used for diagnostic and epidemiologic purposes. Enzyme immunoassays for *H. ducreyi* IgG antibodies and IgM antibodies (24, 65) have been reported.

Documentation of immunologic response to *H. influenzae* type b vaccines can also be accomplished serologically. Assay methods that have been used successfully include radioimmunoassay and enzyme immunoassay (97).

ANTIMICROBIAL SUSCEPTIBILITIES

As recently as 20 years ago, *Haemophilus* infections uniformly responded to treatment with ampicillin. Since then, ampicillin-penicillin resistance has become commonplace, and chloramphenicol resistance is not unusual in areas of the world where this drug is still administered frequently. Trimethoprim-sulfamethoxazole-, tetracycline-, and rifampin-resistant strains are also being detected with alarming frequency (28, 39).

Resistance of *H. influenzae* to ampicillin was reported first from Europe in 1972 (55) and then in the United States in 1974 (44, 93). The mechanism of resistance was later shown to be production of a plasmid-mediated β-lactamase (91). Resistance of *H. influenzae* to ampicillin by a mechanism other than production of β-lactamase was reported in 1980 (54). The mechanism appears to be structural alteration of penicillin-binding proteins, which results in elevated MICs of all β-lactam agents, including the cephalosporins (40, 57).

Resistance of *H. influenzae* to chloramphenicol was recognized in 1977 and found to be due to microbial production of an inactivating enzyme, chloramphenicol acetyltransferase (96). Shortly thereafter, *H. influenzae* strains resistant to both ampicillin and chloramphenicol by means of enzymatic inactivation were reported (43).

Antimicrobial-Agent-Inactivating Enzymes

The β-lactamase test is a rapid means of establishing whether an isolate produces the enzyme β-lactamase. A positive result indicates resistance to ampicillin-penicillin, but a negative result does not ensure susceptibility, because of the possibility of alternate modes of resistance. All three β-lactamase assay methods (acidimetric, iodometric, and chromogenic cephalosporin) perform well with Haemophilus spp., but most laboratories use convenient, commercially available derivatives of the chromogenic cephalosporin test (69).

The chloramphenicol acetyltransferase test is another rapid assay for determining antimicrobial resistance in Haemophilus spp. As with the β-lactamase assay, a positive result indicates resistance to chloramphenicol, but a negative result does not ensure susceptibility (1). The test is available commercially in the form of a reagent-impregnated disk (Remel, Lenexa, Kans.) but does not appear to be as sensitive as the original tube test (27).

Susceptibility Test Media

The National Committee for Clinical Laboratory Standards (NCCLS) recommendations for agar and broth Haemophilus susceptibility test media have changed in the last few years (66, 67). The previous recommendations of chocolatized Mueller-Hinton agar and Mueller-Hinton broth supplemented with 3 to 5% lysed horse blood have been supplanted by Haemophilus Test Medium (HTM) in its agar or broth formulation. HTM was selected because of its defined chemical composition, its lack of thymidine and thymidine analogs, its transparency, and its ready commercial availability (26). Although HTM was designed expressly for susceptibility testing of H. influenzae, further testing of both the agar and the broth formulations has found them to be acceptable for testing the less commonly encountered Haemophilus species with the same test conditions and interpretive guidelines (41).

Agar Diffusion Methods

Recent changes in the NCCLS agar disk diffusion procedure for susceptibility testing of Haemophilus spp. include careful preparation of inocula in 0.9% saline or Mueller-Hinton broth directly from a 20- to 24-h chocolate agar culture and incubation of plates in a 5 to 7% CO_2-enriched atmosphere for 16 to 18 h (67). Use of a bench-top nephelometer for standardization of inoculum density is advised in order to avoid erroneous results due to over- or under-inoculation of test media (48, 92). Further information regarding agar diffusion testing of Haemophilus spp. can be found in chapter 114 of this Manual.

The E-test (AB Biodisk, Solna, Sweden) is an agar diffusion method in which calibrated plastic strips coated with an antimicrobial concentration gradient are utilized. The point of intersection of the antimicrobial zone of growth inhibition with the calibrated strip indicates the MIC of the antimicrobial agent. The method was evaluated recently with 100 strains of H. influenzae and found to yield accurate results in a simple and reproducible fashion, especially when HTM agar was used (42).

Broth Dilution Methods

The new NCCLS broth dilution susceptibility test procedure for Haemophilus spp. calls for use of microdilution trays containing 100 μl of HTM per well. Test suspensions are prepared in 0.9% saline or Mueller-Hinton broth and ad-

justed to a turbidity equivalent to a 0.5 McFarland standard, verified by using a nephelometer. Trays are inoculated to a final cell density of 5×10^5 CFU/ml and incubated for 20 to 24 h in ambient air at 35°C (66). Additional information can be found in chapter 114 of this Manual.

EVALUATION, INTERPRETATION, AND REPORTING OF RESULTS

Laboratory diagnosis of Haemophilus infections can be approached from several testing directions. Culture has been the mainstay for diagnosis of serious infections for a long time, and that situation is likely to remain unchanged for the next several years. Rapid methods such as microscopy, antigen detection, and detection of nucleic acid polymers are important adjuncts to culture but are neither sensitive nor specific enough today to replace it. Serologic tests for Haemophilus infection are handicapped not only by their sensitivity and specificity but also by the length of time that must pass before antibodies reach detectable titers.

Culture results are extremely reliable, provided the laboratory is employing acceptable techniques during organism isolation and identification. Routine cultures should always include inoculation of a chocolate agar or other nutritionally adequate plate. Even more enriched media must be used to recover H. ducreyi from genital ulcer sites. Selective media may be necessary for recovery of Haemophilus spp. from body sites colonized with normal flora.

Isolation of Haemophilus spp. from sites not ordinarily harboring respiratory flora is almost always clinically significant. Of greatest concern in young children are the typeable (encapsulated) H. influenzae strains, especially those belonging to type b. Fortunately, routine immunization of children in many countries of the world has sharply reduced the incidence of type b infections. Both typeable and nontypeable H. influenzae strains, H. parainfluenzae, and H. aphrophilus are causes of potentially life-threatening infections in older children and adults. Detection of H. ducreyi in genital ulcers and lymph node aspirates is always meaningful.

Microscopic assays inherently lack sensitivity, requiring $\geq 10^3$ CFU of bacteria per ml to yield positive results. Except when the Quellung reaction occurs, one is unable to definitively identify viewed organisms as Haemophilus spp. Accordingly, the use of these assays should be limited to determining the presence or absence of microorganisms and to providing a guide for selection of initial antimicrobial therapy.

Antigen detection immunoassays for H. influenzae type b also have restricted utility. The present-day practice of administering ampicillin plus a broad-spectrum cephalosporin (e.g., cefotaxime, ceftriaxone) as empiric therapy for childhood bacterial meningitis has eliminated much of the urgency of identifying the disease etiology. This antimicrobial combination has fewer side effects than ampicillin plus chloramphenicol and is effective against β-lactamase-producing H. influenzae and virtually all other causes of bacterial meningitis. Most clinicians today are inclined to leave patients on this empiric therapy until culture and antimicrobial susceptibility results are available. Consequently, the clinical utility of antigen detection results for patients with bacterial meningitis is low enough that the cost-effectiveness of routine testing has been questioned.

One situation in which antigen detection has retained its clinical value is management of patients with partially treated meningitis. Gram-stained smears and cultures of CSF for such patients may very well be negative, and a positive antigen detection test may be the only indication of the etiologic agent.

Direct testing of genital specimens for *H. ducreyi* antigen or nucleic acid polymers holds much promise in that cultures for this microorganism are difficult to do and are therefore insensitive. Standardization and commercial availability of such assays are lacking, however, at this time.

Because of the unpredictability of *Haemophilus* susceptibility to antimicrobial agents, routine testing of clinically significant isolates is recommended. β-Lactamase and chloramphenicol acetyltransferase assays furnish results rapidly but have an important limitation. Positive results indicate resistance to ampicillin-penicillin or chloramphenicol, but negative results do not ensure susceptibility. Consequently, isolates yielding negative results should be tested by more definitive methods (e.g., agar diffusion, broth dilution) before they are reported as susceptible. Now that *Haemophilus* susceptibility test methods have been defined and standardized by NCCLS, these are the assays that should be performed by clinical laboratories. It is anticipated that the E-test method will be included in a future iteration of the standards.

REFERENCES

1. **Azemun, P., T. Stull, M. Roberts, and A. L. Smith.** 1981. Rapid detection of chloramphenicol resistance in *Haemophilus influenzae*. *Antimicrob. Agents Chemother.* **20:**168–170.
2. **Barbé, G., M. Babolat, J. M. Boeufgras, D. Monget, and J. Freney.** 1994. Evaluation of API NH, a new 2-hour system for identification of *Neisseria* and *Haemophilus* species and *Moraxella catarrhalis* in a routine clinical laboratory. *J. Clin. Microbiol.* **32:**187–189.
3. **Barbour, M. L., D. W. Crook, and R. T. Mayon-White.** 1993. An improved antiserum agar method for detecting carriage of *Haemophilus influenzae* type b. *Eur. J. Clin. Microbiol. Infect. Dis.* **12:**215–217.
4. **Bell, L. M., G. Alpert, J. M. Campos, and S. A. Plotkin.** 1985. Routine quantitative blood cultures in children with *Haemophilus influenzae* or *Streptococcus pneumoniae* bacteremia. *Pediatrics* **76:**901–904.
5. **Bieger, R. C., N. S. Brewer, and J. A. Washington II.** 1978. *Haemophilus aphrophilus*: a microbiologic and clinical review and report of 42 cases. *Medicine* **57:**345–355.
6. **Bluestone, C. D., and J. O. Klein.** 1983. Otitis media with effusion, atelectasis and eustachian tube dysfunction, p. 356–512. *In* C. D. Bluestone and S. E. Stool (ed.), *Pediatric Otolaryngology.* The W. B. Saunders Co., Philadelphia.
7. **Brazilian Purpuric Fever Study Group.** 1987. Brazilian purpuric fever: epidemic purpura fulminans associated with antecedent purulent conjunctivitis. *Lancet* **ii:**761–763.
8. **Brenner, D. J., L. W. Mayer, G. M. Carlone, L. H. Harrison, W. F. Bibb, M. C. de Cunto Brandileone, F. O. Sottnek, K. Irino, M. W. Reeves, J. M. Swenson, K. A. Birkness, R. S. Weyant, S. F. Berkley, T. C. Woods, A. G. Steigerwalt, P. A. D. Grimont, R. M. McKinney, D. W. Fleming, L. L. Gheesling, R. C. Cooksey, R. J. Arko, C. V. Broome, and the Brazilian Purpuric Fever Study Group.** 1988. Biochemical, genetic, and epidemiologic characterization of *Haemophilus influenzae* biogroup aegyptius (*Haemophilus aegyptius*) strains associated with Brazilian purpuric fever. *J. Clin. Microbiol.* **26:**1524–1534.
9. **Campos, J. M., and J. R. Spainhour.** 1985. Comparison of the Isolator 1.5 Microbial Tube with a conventional blood culture broth system for detection of bacteremia in children. *Diagn. Microbiol. Infect. Dis.* **3:**167–174.
10. **Carlone, G. M., L. Gorelkin, L. L. Gheesling, A. L. Erwin, S. K. Hoiseth, M. H. Mulks, S. P. O'Connor, R. S. Weyant,** J. Myrick, K. Rubin, R. S. Munford III, E. H. White, R. J. Arko, B. Swaminathan, L. M. Graves, L. W. Mayer, M. K. Robinson, S. P. Caudill, and the Brazilian Purpuric Fever Study Group. 1989. Potential virulence-associated factors in Brazilian purpuric fever. *J. Clin. Microbiol.* **27:**609–614.
11. **Carlone, G. M., F. O. Sottnek, and B. D. Plikaytis.** 1985. Comparison of outer membrane protein and biochemical profiles of *Haemophilus aegyptius* and *Haemophilus influenzae* biotype III. *J. Clin. Microbiol.* **22:**708–713.
12. **Centers for Disease Control.** 1991. *Haemophilus* b conjugate vaccines for prevention of *Haemophilus influenzae* type b disease among infants and children two months of age and older: recommendations of the Immunization Practices Advisory Committee (ACIP). *Morbid. Mortal. Weekly Rep.* **40**(RR-1): 1–7.
13. **Chui, L., W. Albritton, B. Paster, I. Maclean, and R. Marusyk.** 1993. Development of the polymerase chain reaction for diagnosis of chancroid. *J. Clin. Microbiol.* **31:**659–664.
14. **Chunn, C. J., S. R. Jones, J. A. McCutchan, E. J. Young, and D. N. Gilbert.** 1977. *Haemophilus parainfluenzae* infective endocarditis. *Medicine* **56:**99–113.
15. **Cochi, S. L., C. V. Broome, and A. W. Hightower.** 1985. Immunization of US children with *Haemophilus influenzae* type b polysaccharide vaccine: a cost effectiveness model of strategy assessment. *JAMA* **253:**521–529.
16. **Collins, J. K., and M. T. Kelly.** 1983. Comparison of Phadebact coagglutination, Bactogen latex agglutination, and counterimmunoelectrophoresis for detection of *Haemophilus influenzae* type b antigens in cerebrospinal fluid. *J. Clin. Microbiol.* **17:**1005–1008.
17. **Dajani, A. S., B. I. Asmar, and M. C. Thirumoorthi.** 1979. Systemic *Haemophilus influenzae* disease: an overview. *J. Pediatr.* **94:**355–364.
18. **Daly, J. A., N. L. Clifton, K. C. Seskin, and W. M. Gooch III.** 1991. Use of rapid, nonradioactive DNA probes in culture confirmation tests to detect *Streptococcus agalactiae*, *Haemophilus influenzae*, and *Enterococcus* spp. from pediatric patients with significant infections. *J. Clin. Microbiol.* **29:**80–82.
19. **Daly, J. A., W. M. Gooch III, and J. M. Matsen.** 1985. Evaluation of the Wayson variation of a methylene blue staining procedure for the detection of microorganisms in cerebrospinal fluid. *J. Clin. Microbiol.* **21:**919–921.
20. **Dangor, Y., S. D. Miller, H. J. Koornhof, and R. C. Ballard.** 1992. A simple medium for the primary isolation of *Haemophilus ducreyi*. *Eur. J. Clin. Microbiol. Infect. Dis.* **11:**930–934.
21. **Darville, T., R. F. Jacobs, R. A. Lucas, and B. Caldwell.** 1992. Detection of *Haemophilus influenzae* type b antigen in cerebrospinal fluid after immunization. *Pediatr. Infect. Dis. J.* **11:**243–244.
22. **Daum, R. S., G. R. Siber, J. S. Kamon, and R. R. Russell.** 1982. Evaluation of a commercial latex particle agglutination test for rapid diagnosis of *Haemophilus influenzae* type b infection. *Pediatrics* **69:**466–471.
23. **Davis, D. J., and M. Pittman.** 1950. Acute conjunctivitis caused by *Haemophilus*. *Am. J. Dis. Child.* **79:**211–219.
24. **Desjardins, M., C. E. Thompson, L. G. Filion, J. O. Ndinya-Achola, F. A. Plummer, A. R. Ronald, P. Piot, and D. W. Cameron.** 1992. Standardization of an enzyme immunoassay for human antibody to *Haemophilus ducreyi*. *J. Clin. Microbiol.* **30:**2019–2024.
25. **Dewhirst, F. E., B. J. Paster, I. Olsen, and G. J. Fraser.** 1992. Phylogeny of 54 representative strains of species in the family *Pasteurellaceae* as determined by comparison of 16S rRNA sequences. *J. Bacteriol.* **174:**2002–2013.
26. **Doern, G. V.** 1992. In vitro susceptibility testing of *Haemophilus influenzae*: review of new National Committee for Clinical Laboratory Standards recommendations. *J. Clin. Microbiol.* **30:**3035–3038.
27. **Doern, G. V., G. S. Daum, and T. A. Tubert.** 1987. In vitro chloramphenicol susceptibility testing of *Haemophilus influenzae*: disk diffusion procedures and assays for chloramphenicol acetyltransferase. *J. Clin. Microbiol.* **25:**1453–1455.
28. **Doern, G. V., J. H. Jorgensen, C. Thornsberry, T. Tubert,**

J. S. Redding, and L. A. Maher. 1988. National collaborative study of the prevalence of antimicrobial resistance among clinical isolates of *Haemophilus influenzae*. *Antimicrob. Agents Chemother.* **32:**180–185.

29. **Faden, H., J. Stanievich, L. Brodsky, J. Bernstein, and P. L. Ogra.** 1990. Changes in nasopharyngeal flora during otitis media of childhood. *Pediatr. Infect. Dis. J.* **9:**623–626.

30. **Farley, M. M., D. S. Stephens, P. S. Brachman, Jr., C. Harvey, J. D. Smith, J. D. Wenger, and CDC Meningitis Surveillance Group.** 1992. Invasive *Haemophilus influenzae* disease in adults: a prospective, population-based surveillance. *Ann. Intern. Med.* **116:**806–812.

31. **Friesen, C. A., and C. T. Cho.** 1986. Characteristic features of neonatal sepsis due to *Haemophilus influenzae*. *Rev. Infect. Dis.* **8:**777–780.

32. **Gilsdorf, J. R.** 1987. *Haemophilus influenzae* non-type b infections in children. *Am. J. Dis. Child.* **141:**1063–1065.

33. **Greenlee, J. L.** 1990. Approach to diagnosis of meningitis. Cerebrospinal fluid evaluation. *Infect. Dis. Clin. N. Am.* **4:**583–597.

34. **Harper, J. J., and M. H. Tilse.** 1991. Biotypes of *Haemophilus influenzae* that are associated with noninvasive infections. *J. Clin. Microbiol.* **29:**2539–2542.

35. **Immunization Practices Advisory Committee.** 1985. Polysaccharide vaccine for prevention of *Haemophilus influenzae* type b disease. *Morbid. Mortal. Weekly Rep.* **34:**201–205.

36. **Inzana, T. J., J. Clarridge, and R. P. Williams.** 1987. Rapid determination of X/V growth requirements of *Haemophilus* species in broth. *Diagn. Microbiol. Infect. Dis.* **6:**93–100.

37. **Jaffe, D. M.** 1994. Occult bacteremia in children. *Adv. Pediatr. Infect. Dis.* **9:**237–258.

38. **Jones, R. G., J. W. Bass, M. E. Weisse, and J. M. Vincent.** 1991. Antigenuria after immunization with *Haemophilus influenzae* oligosaccharide CRM197 conjugate (HbOC) vaccine. *Pediatr. Infect. Dis. J.* **10:**557–559.

39. **Jorgensen, J. H., G. V. Doern, L. A. Maher, A. W. Howell, and J. S. Redding.** 1990. Antimicrobial resistance among respiratory isolates of *Haemophilus influenzae, Moraxella catarrhalis,* and *Streptococcus pneumoniae* in the United States. *Antimicrob. Agents Chemother.* **34:**2075–2080.

40. **Jorgensen, J. H., G. V. Doern, C. Thornsberry, J. S. Redding, L. A. Maher, and T. Tubert.** 1988. Susceptibility of multiply resistant *Haemophilus influenzae* to newer antimicrobial agents. *Diagn. Microbiol. Infect. Dis.* **9:**27–32.

41. **Jorgensen, J. H., A. W. Howell, and L. A. Maher.** 1990. Antimicrobial susceptibility testing of less commonly isolated *Haemophilus* species using *Haemophilus* test medium. *J. Clin. Microbiol.* **28:**985–988.

42. **Jorgensen, J. H., A. W. Howell, and L. A. Maher.** 1991. Quantitative antimicrobial susceptibility testing of *Haemophilus influenzae* and *Streptococcus pneumonae* by using the E-test. *J. Clin. Microbiol.* **29:**109–114.

43. **Kenny, J. F., G. D. Isburg, and R. H. Michaels.** 1980. Meningitis due to *Haemophilus influenzae* type b resistant to both ampicillin and chloramphenicol. *Pediatrics* **66:**14–16.

44. **Khan, W., S. Ross, W. Rodriguez, G. Controni, and A. K. Saz.** 1974. *Haemophilus influenzae* type B resistant to ampicillin: a report of two cases. *JAMA* **229:**298–301.

45. **Kilian, M.** 1976. A taxonomic study of the genus *Haemophilus*, with the proposal of a new species. *J. Gen. Microbiol.* **93:**9–62.

46. **Kilian, M.** 1985. *Haemophilus*, p. 387–393. *In* E. H. Lennette, A. Balows, W. J. Hausler, Jr., and H. J. Shadomy (ed.), *Manual of Clinical Microbiology,* 4th ed. American Society for Microbiology, Washington, D.C.

46a.**Kilian, M.** 1991. *Haemophilus*, p. 463–470. *In* A. Balows, W. J. Hausler, Jr., K. L. Herrmann, H. D. Isenberg, and H. J. Shadomy (ed.), *Manual of Clinical Microbiology,* 5th ed. American Society for Microbiology, Washington, D.C.

47. **Kostman, J. R., B. L. Sherry, C. L. Fligner, S. Egaas, P. Sheeran, L. Baken, J. E. Bauwens, C. Clausen, D. M. Sherer, J. J. Plorde, T. L. Stull, and P. M. Mendelman.** 1993.

48. **Laferriere, C., M. I. Marks, and D. F. Welch.** 1983. Effect of inoculum size on *Haemophilus influenzae* type b susceptibility to new and conventional antibiotics. *Antimicrob. Agents Chemother.* **24:**287–289.

49. **Lauer, B. A., L. B. Reller, and S. Mirrett.** 1981. Comparison of acridine orange and Gram stains for detection of microorganisms in cerebrospinal fluid and other clinical specimens. *J. Clin. Microbiol.* **14:**201–205.

50. **Leinonen, M., and H. Käyhty.** 1978. Comparison of countercurrent immunoelectrophoresis, latex agglutination, and radioimmunoassay in detection of soluble capsular polysaccharide antigens of *Haemophilus influenzae* type b and *Neisseria meningitidis* of group A or C. *J. Clin. Pathol.* **31:**1172–1176.

51. **Lim, L. C. L., D. R. Pennell, and R. F. Schell.** 1990. Rapid detection of bacteria in cerebrospinal fluid by immunofluorescence staining on membrane filters. *J. Clin. Microbiol.* **28:**670–675.

52. **MacDonald, K., D. W. Cameron, G. Irungu, L. J. D'Costa, F. A. Plummer, L. A. Slaney, J. O. Ndinya-Achola, and A. R. Ronald.** 1989. Comparison of Sheffield media with standard media for the isolation of *Haemophilus ducreyi*. *Sex. Transm. Dis.* **16:**88–90.

53. **Marcon, M. J., A. C. Hamoudi, and H. J. Cannon.** 1984. Comparative laboratory evaluation of three antigen detection methods for diagnosis of *Haemophilus influenzae* type b disease. *J. Clin. Microbiol.* **19:**333–337.

54. **Markowitz, S. M.** 1980. Isolation of an ampicillin-resistant, non-beta-lactamase-producing strain of *Haemophilus influenzae*. *Antimicrob. Agents Chemother.* **17:**80–83.

55. **Mathies, A. W., Jr.** 1972. Penicillins in the treatment of bacterial meningitis. *J. R. Coll. Physicians Lond.* **6:**139–146.

56. **McCarthy, L. R., and J. E. Senne.** 1980. Evaluation of acridine orange stain for detection of microorganisms in blood cultures. *J. Clin. Microbiol.* **11:**281–285.

57. **Mendelman, P. M., D. O. Chaffin, T. L. Stull, C. E. Rubens, K. D. Mack, and A. L. Smith.** 1984. Characterization of non-β-lactamase-mediated ampicillin resistance in *Haemophilus influenzae*. *Antimicrob. Agents Chemother.* **26:**235–244.

58. **Michaels, R. H., and O. Ali.** 1993. A decline in *Haemophilus influenzae* type b meningitis. *J. Pediatr.* **122:**407–409.

59. **Michaels, R. H., C. S. Poziviak, F. E. Stonebraker, and C. W. Norder.** 1976. Factors affecting pharyngeal *Haemophilus influenzae* type b colonization rates in children. *J. Clin. Microbiol.* **4:**413–417.

60. **Michaels, R. H., F. E. Stonebraker, and J. B. Robbins.** 1975. Use of antiserum agar for detection of *Haemophilus influenzae* type b in the pharynx. *Pediatr. Res.* **9:**513–516.

61. **Möller, L. V. M., G. J. Ruijs, H. G. M. Heijerman, J. Dankert, and L. van Alphen.** 1992. *Haemophilus influenzae* is frequently detected with monoclonal antibody 8BD9 in sputum samples from patients with cystic fibrosis. *J. Clin. Microbiol.* **30:**2495–2497.

62. **Möller, L. V. M., L. van Alphen, H. Grasselier, and J. Dankert.** 1993. *N*-Acetyl-D-glucosamine medium improves recovery of *Haemophilus influenzae* from sputa of patients with cystic fibrosis. *J. Clin. Microbiol.* **31:**1952–1954.

63. **Morse, S. A.** 1989. Chancroid and *Haemophilus ducreyi*. *Clin. Microbiol. Rev.* **2:**137–157.

64. **Murphy, T. F., and M. A. Apicella.** 1987. Nontypable *Haemophilus influenzae*: a review of clinical aspects, surface antigens, and the human immune response to infection. *Rev. Infect. Dis.* **9:**1–15.

65. **Museyi, K., E. Van Dyck, T. Vervoot, D. Taylor, C. Hoge, and P. Piot.** 1988. Use of an enzyme immunoassay to detect serum IgG antibodies to *Haemophilus ducreyi*. *J. Infect. Dis.* **157:**1039–1043.

66. **National Committee for Clinical Laboratory Standards.** 1993. *Methods for Dilution Antimicrobial Susceptibility Tests for Bacteria That Grow Aerobically,* 3rd ed. Approved standard M7-A3. National Committee for Clinical Laboratory Standards, Villanova, Pa.

67. **National Committee for Clinical Laboratory Standards.** 1993. *Performance Standards for Antimicrobial Disk Susceptibility Tests,* 5th ed. Approved standard M2-A5. National Committee for Clinical Laboratory Standards, Villanova, Pa.

68. **Nsanze, H., M. V. Fast, L. J. D'Costa, P. Tukei, J. Curran, and A. R. Ronald.** 1981. Genital ulcers in Kenya: a clinical and laboratory study. *Br. J. Vener. Dis.* **59**:378–381.

69. **O'Callaghan, C. H., A. Morris, S. M. Kirby, and A. H. Shingler.** 1972. Novel method for detection of β-lactamases by using a chromogenic cephalosporin substrate. *Antimicrob. Agents Chemother.* **1**:283–288.

70. **Oill, P. A., A. W. Chow, and L. B. Guze.** 1979. Adult bacteremic *Haemophilus parainfluenzae* infections. *Arch. Intern. Med.* **139**:985–988.

71. **Osterholm, M. T.** 1990. Invasive bacterial diseases and child day care. *Semin. Pediatr. Infect. Dis.* **1**:222–233.

72. **Palladino, S., B. J. Leahy, and T. L. Newall.** 1990. Comparison of the RIM-H rapid identification kit with conventional tests for the identification of *Haemophilus* spp. *J. Clin. Microbiol.* **28**:1862–1863.

73. **Piot, P., and M. Laga.** 1989. Genital ulcers and other sexually transmitted diseases as cofactors for the sexual transmission of HIV. *Br. Med. J.* **298**:623–624.

74. **Pittman, M.** 1931. Variation and type specificity in the bacterial species *Hemophilus influenzae*. *J. Exp. Med.* **53**:471–495.

75. **Quentin, R., I. Dubarry, C. Martin, B. Cattier, and A. Goudeau.** 1992. Evaluation of four commercial methods for identification and biotyping of genital and neonatal strains of *Haemophilus* species. *Eur. J. Clin. Microbiol. Infect. Dis.* **11**:546–549.

76. **Quentin R., C. Martin, J. M. Musser, N. Pasquier-Picard, and A. Goudeau.** 1993. Genetic characterization of a cryptic genospecies of *Haemophilus* causing urogenital and neonatal infections. *J. Clin. Microbiol.* **31**:1111–1116.

77. **Reller, L. B., P. R. Murray, and J. D. MacLowry.** 1982. *Cumitech 1A, Blood Cultures II.* Coordinating ed., J. A. Washington II. American Society for Microbiology, Washington, D.C.

78. **Rennie, R., T. Gordon, Y. Yaschuk, P. Tomlin, P. Kibsey, and W. Albritton.** 1992. Laboratory and clinical evaluations of media for the primary isolation of *Haemophilus* species. *J. Clin. Microbiol.* **30**:1917–1921.

79. **Roggen, E. L., R. Pansaerts, E. Van Dyck, and P. Piot.** 1993. Antigen detection and immunological typing of *Haemophilus ducreyi* with a specific rabbit polyclonal serum. *J. Clin. Microbiol.* **31**:1820–1825.

80. **Schalla, W. O., L. L. Sanders, G. P. Schmid, M. R. Tam, and S. A. Morse.** 1986. Use of dot-immunobinding and immunofluorescence assays to investigate clinically suspected cases of chancroid. *J. Infect. Dis.* **153**:879–887.

81. **Scheifele, D. W., J. I. Ward, and G. R. Siber.** 1981. Advantage of latex agglutination over countercurrent immunoelectrophoresis in the detection of *Haemophilus influenzae* type b antigen in serum. *Pediatrics* **68**:888–891.

82. **Schwartz, R., W. J. Rodriguez, R. Mann, W. Khan, and S. Ross.** 1979. The nasopharyngeal culture in acute otitis media. *JAMA* **241**:2170–2173.

83. **Shanholtzer, C. J., P. J. Schaper, and L. R. Peterson.** 1982. Concentrated Gram-stained smears prepared with a cytospin centrifuge. *J. Clin. Microbiol.* **16**:1052–1056.

84. **Sippel, J. E., C. M. Prato, N. I. Girgis, and E. A. Edwards.** 1984. Detection of *Neisseria meningitidis* group A, *Haemophilus influenzae* type b, and *Streptococcus pneumoniae* antigens in cerebrospinal fluid specimens by antigen capture enzyme-linked immunosorbent assays. *J. Clin. Microbiol.* **20**:259–265.

85. **Slater, L. N., J. Guarnaccia, S. Makintubee, and G. R. Istre.** 1990. Bacteremic disease due to *Haemophilus influenzae* capsular type f in adults: report of five cases and review. *Rev. Infect. Dis.* **12**:628–635.

86. **Stamm, W. E., H. H. Handsfield, A. M. Rompalo, R. L. Ashley, P. L. Robertsa, and L. Corey.** 1987. The association between genital ulcer disease and acquisition of HIV infection in homosexual men. *JAMA* **260**:1429–1433.

87. **St. Geme, J. W., III, A. Takala, E. Esko, and S. Falkow.** 1994. Evidence for capsule gene sequences among pharyngeal isolates of nontypeable *Haemophilus influenzae*. *J. Infect. Dis.* **169**:337–342.

88. **Sturm, A. W.** 1986. *Haemophilus influenzae* and *Haemophilus parainfluenzae* in nongonococcal urethritis. *J. Infect. Dis.* **153**:165–167.

89. **Sturm, A. W.** 1986. Isolation of *Haemophilus influenzae* and *Haemophilus parainfluenzae* from genital-tract specimens with a selective medium. *J. Med. Microbiol.* **21**:349–352.

90. **Swaminathan, B., L. W. Mayer, W. F. Bibb, G. W. Ajello, K. Irino, K. A. Birkness, C. F. Garon, M. W. Reeves, M. Cristina de Cunto Brandileone, F. O. Sottnek, D. J. Brenner, A. G. Steigerwalt, and the Brazilian Purpuric Fever Study Group.** 1989. Microbiology of Brazilian purpuric fever and diagnostic tests. *J. Clin. Microbiol.* **27**:605–608.

91. **Sykes, R. B., M. Matthew, and C. H. O'Callaghan.** 1975. R-factor mediated beta-lactamase production by *Haemophilus influenzae*. *J. Med. Microbiol.* **8**:437–441.

92. **Syriopoulou, V. P., D. W. Scheifele, C. M. Sack, and A. L. Smith.** 1979. Effect of inoculum size on the susceptibility of *Haemophilus influenzae* b to beta-lactam antibiotics. *Antimicrob. Agents Chemother.* **16**:510–513.

93. **Tomeh, M. O., S. E. Starr, J. E. McGowan, Jr., P. M. Terry, and A. J. Nahmias.** 1974. Ampicillin-resistant *Haemophilus influenzae* type b infection. *JAMA* **229**:295–297.

94. **Trottier, S., K. Stenberg, I. A. Von Rosen, and C. Svanborg.** 1991. *Haemophilus influenzae* causing conjunctivitis in day-care children. *Pediatr. Infect. Dis. J.* **10**:578–584.

95. **Turk, D. C.** 1984. The pathogenicity of *Haemophilus influenzae*. *J. Med. Microbiol.* **18**:1–16.

96. **van Klingeren B., J. D. A. van Embden, and M. Dessens-Kroon.** 1977. Plasmid-mediated chloramphenicol resistance in *Haemophilus influenzae*. *Antimicrob. Agents Chemother.* **11**:383–387.

97. **Vella, P. P., J. M. Staub, J. Armstrong, K. T. Dolan, C. M. Rusk, S. Szymanski, W. E. Greer, S. Marburg, P. J. Kniskern, T. L. Schofield, R. L. Tolman, F. Hartner, S. Pan, R. J. Gerety, and R. W. Ellis.** 1990. Immunogenicity of a new *Haemophilus influenzae* type b conjugate vaccine (meningococcal protein conjugate) (PedvaxHIB™). *Pediatrics* **85**(Suppl.):668–675.

98. **Wallace, R. J., D. Musher, E. J. Septimus, J. E. McGowan, Jr., F. J. Quinones, K. Wiss, P. H. Vance, and P. A. Trier.** 1981. *Haemophilus influenzae* infections in adults: characterization of strains by serotypes, biotypes, and β-lactamase production. *J. Infect. Dis.* **144**:101–106.

99. **Winslow, C. E., J. Broadhurst, and R. E. Buchanan.** 1920. The families and genera of the bacteria: final report of the committee of the Society of American Bacteriologists on characterization and classification of bacterial types. *J. Bacteriol.* **5**:191–229.

Bordetella

MARIO J. MARCON

46

TAXONOMY

The genus *Bordetella* consists of four species: *Bordetella pertussis*, *B. parapertussis*, *B. bronchiseptica*, and *B. avium* (77). The first three species are closely related genetically, showing a G+C content of 66 to 70 mol%, a DNA homology of 72 to 94% (47, 48, 66), and very limited diversity by multilocus enzyme analysis (63). Although there are phenotypic differences (morphology, growth characteristics, nutritional requirements, biochemical reactions) among these species, some investigators believe they should be considered a single species or three subspecies within a single species (48). Not all investigators are in agreement with this viewpoint (45, 46). *B. avium* is clearly a distinct species, having a G+C content of 62 mol% (44). It has been proposed that, along with members of the genus *Alcaligenes*, the genus *Bordetella* should be included in a new family called the *Alcaligenaceae* (10).

DESCRIPTION OF THE GENUS

Members of the genus *Bordetella* are small, gram-negative coccobacilli or short rods ranging in size from 0.2 by 0.5 to 0.5 by 2.0 μm (66). They are obligate aerobes with an optimum growth temperature of 35 to 37°C. The bordetellae are somewhat inactive metabolically, and no acid is produced from carbohydrates. *B. pertussis* and *B. parapertussis* are nonmotile, while *B. bronchiseptica* and *B. avium* are motile by peritrichous flagella (77).

The bordetellae display heterogeneity in their growth requirements and fastidiousness to primary isolation. *B. pertussis* is the most fastidious of the group, requiring complex growth factors; fresh clinical isolates will not grow on chocolate agar or common agar bases supplemented with 5% sheep blood. Moreover, a number of substances, including fatty acids, heavy metal ions, sulfides, and peroxides, inhibit growth. Media for the primary isolation of *B. pertussis* generally include substances such as starch, charcoal, ion-exchange resins, or a high percentage of blood to inactivate the inhibitory substances. *B. parapertussis* is somewhat less fastidious than *B. pertussis*, but isolation from clinical specimens is usually accomplished only when specialized medium for *B. pertussis* is employed. *B. bronchiseptica* and *B. avium* are generally nonfastidious and can be recovered on routine laboratory agars including infusion agars with or without blood, MacConkey agar, and salmonella-shigella agar.

NATURAL HABITATS

All members of the genus *Bordetella* are respiratory pathogens of warm-blooded animals, sharing a tropism for ciliated respiratory epithelial cells. *B. pertussis* and *B. parapertussis* are uniquely human pathogens, the former being the major cause of whooping cough or pertussis and the latter being generally associated with a milder, less frequently occurring form of the disease (53). Pertussis is a highly contagious disease with attack rates of more than 90% reported in unimmunized populations (5). Person-to-person transmission occurs by the aerosol route. *B. pertussis* and *B. parapertussis* adhere to, multiply among, and remain localized in the ciliated epithelial cells of the respiratory tract. *B. bronchiseptica* and *B. avium* are commensals and opportunistic pathogens of the respiratory tracts of a variety of animals (80). For example, *B. bronchiseptica* causes infectious tracheobronchitis (kennel cough) in dogs, and *B. avium* causes rhinotracheitis (coryza) in turkeys. *B. bronchiseptica* is an opportunistic human pathogen associated with both respiratory infections (bronchitis, pneumonia) and nonrespiratory infections (wound infection, bacteremia, endocarditis, meningitis, peritonitis). Some of these infections have occurred in patients with close contact with animals. *B. bronchiseptica* has not been reported to cause pertussis. There have been no reports of the recovery of *B. avium* from humans; however, this organism closely resembles *Alcaligenes faecalis*, and it may have been misidentified in the past.

CLINICAL SIGNIFICANCE

Classical pertussis due to *B. pertussis* in an unimmunized individual usually occurs after a 7- to 10-day (range, 6 to 20 days) incubation period (75). It manifests as a disease of three stages: the catarrhal, paroxysmal, and convalescent stages. The catarrhal (flowing-discharge) stage is characterized by nonspecific symptoms that mimic a cold or other viral upper respiratory tract infection. These symptoms include rhinorrhea, sneezing, mild cough, conjunctivitis, and malaise, with no or minimal fever. During this stage, the disease is highly communicable, organisms are most readily isolated from cultures of the posterior nasopharynx, and appropriate antibiotic therapy is effective in modifying the course of disease. The catarrhal stage lasts about 1 to 2 weeks, with the cough increasing in intensity and frequency. The paroxysmal stage is marked by sudden attacks (paroxysms) of severe, repetitive coughing, often culminating with the characteristic whoop and frequently being followed by vomiting. The whooping sound is caused by the rapid inspiration of air after the clearance of mucus-blocked airways. In infants younger than 6 months, apnea is a common manifestation and the whoop may be absent. During the coughing episodes, which may occur 10 to 20 times or more during a 24-h period, the patient may become cyanotic to the point of requiring intermittent airway support. Patients typically appear normal between paroxysms. A marked lymphocytosis, both relative and absolute, usually accompanies this stage of the disease; lymphocyte counts may exceed 50,000/mm^3 and represent 70% or more of total circulating leukocytes. At this time, it becomes increasingly difficult to recover the organism from the respiratory tract, and antibiotics have little or no effect on modifying the course of disease; however, administration of antibiotics is important in eradicating the organism and preventing further spread of disease. During the paroxysmal stage, which lasts 1 to 4 weeks, a number of important complications can occur. These include secondary bacterial infections, including pneumonia and otitis media; toxic central nervous system manifestations, including convulsions and encephalopathy; hemorrhage; hernia; pneumothorax; and rectal prolapse, the latter four occurring as a result of the severe coughing. The beginning of the convalescent (recovery) stage is marked by a reduction in frequency and severity of coughing spells. Complete recovery may require weeks or months, with cough persisting during this period. Following clinical pertussis, immunity to disease is lifelong.

Although diagnosis of pertussis on clinical grounds is easily accomplished when a patient presents with a whoop and other characteristic features, diagnosis of patients without classical signs and symptoms is problematic (34). Symptoms are often absent in partially immune individuals such as (i) infants with low levels of passively transferred maternal antibody, (ii) young children who have been only partially immunized, and (iii) older children and adults with waning immunity or clinical conditions that render them immunosuppressed. This may also be the case with infection due to *B. parapertussis*, which generally causes a milder form of disease.

Differences in manifestation of disease due to *Bordetella* spp. are likely due to genetic makeup and differential expression of toxins and other virulence factors coupled with host-specific responses (16). Three virulence factors of *B. pertussis* have been most extensively studied; these are pertussis toxin (PT), filamentous hemagglutinin (FHA), and adenyl cyclase toxin (ACT) (77). PT is responsible for a wide array of biological activity, including promotion of lymphocytosis and many of the signs and symptoms of pertussis. The structural gene for PT can also be found in *B. parapertussis* and *B. bronchiseptica*, but gene expression occurs only in *B. pertussis* (77). Antibody to PT is believed to be protective against clinical pertussis. All three closely related *Bordetella* spp. produce FHA and ACT. The FHA promotes adherence of the organism to ciliated epithelial cells; antibody to FHA confers some protection by inhibition of attachment. It has been suggested that ACT promotes infection by inducing high levels of cyclic AMP in phagocytic cells, thus inhibiting their activity and reducing host defenses in the respiratory tract. The expression of these and other virulence genes of *B. pertussis* is highly regulated by a single gene referred to as *vir*, for virulence regulating (77). Mutations in the *vir* locus are responsible for phenotypic phase variations of the organism. These mutants are unable to partially or fully express the *vir*-regulated genes and may be avirulent, nontoxigenic, and nonfastidious in their growth requirements. Expression of virulence factors may be restored by cloning and transfer of the wild-type *vir* gene (57). Expression of the *vir*-regulated genes may also be modulated by growth conditions. How the loss or modulation of a specific gene or group of genes relates to alteration of virulence remains to be elucidated. It has been suggested that conversion of *B. pertussis* to *B. parapertussis* or vice versa may be a function of up-or-down regulation at the *vir* gene locus (60).

Despite the availability of an effective (80 to 90% protective) whole-cell vaccine as well as newer acellular vaccines (72), pertussis remains a disease of worldwide distribution because many developing nations do not have the resources for vaccinating their populations (81). Even in developed nations such as Great Britain and Sweden, major outbreaks have occurred, often in association with a decrease in vaccine usage because of safety concerns (14). In the United States, pertussis is endemic, with most disease occurring as isolated cases; for reasons that are unclear, sporadic epidemics occur approximately every 4 years (13). Between 2,000 and 5,000 cases have been reported annually in the United States over the past 10 years (9), but this is surely an underestimation of disease incidence. In the prevaccine era, pertussis occurred most commonly in children in the 1- to 5-year age group. Recently, however, there has been a shift of disease occurrence to children less than 1 year of age (5). This shift has probably resulted from a decrease in passively transferred maternal antibody, because adults do not receive booster vaccinations. Severity of disease is particularly great in the young infant, and a majority of fatalities occur in patients less than 1 year of age. It is still unclear how the organism is maintained in the human population. A prolonged carrier state is not known, and the organism is very infrequently recovered from healthy individuals (51). It has been suggested that asymptomatically infected adults or adults experiencing mild respiratory symptoms provide the reservoir for disease. Immunization protects against disease but is not as effective in preventing infection (54).

COLLECTION, TRANSPORT, AND STORAGE OF SPECIMENS

The literature contains many evaluations of methods for the collection, transport, and culture of *Bordetella* spp. from clinical specimens (14, 17, 18, 35, 65, 82). Although significant improvements have been made since the time of the "cough plate" technique, it should be remembered that isolation of the organism is probably at best 50% sensitive for the laboratory diagnosis of pertussis compared with comprehensive serologic testing to demonstrate seroconversion. Whichever methods are selected for a particular clinical setting, it is imperative that detailed instructions along with appropriate collection materials be readily available to medical personnel involved in the process. Readers should refer to the fifth edition of this Manual (18) or to the *Clinical Microbiology Procedures Handbook* (8, 70).

The specimen of choice for recovery of B. pertussis and B. parapertussis from the respiratory tract is secretions collected from the posterior nasopharynx (18). Specimens collected from the throat are not recommended (56). Poor sensitivity in isolation of Bordetella spp. from throat specimens reflects a greater growth competition from indigenous oropharyngeal flora than from nasopharyngeal (NP) flora as well as the predilection of the bordetellae for ciliated epithelial cells. B. pertussis has also been recovered from NP, bronchoalveolar lavage, and transbronchial specimens in human immunodeficiency virus type 1-infected or otherwise immunosuppressed patients (11, 55, 64).

NP specimens may be collected as aspirates obtained by suction or as pernasal swab specimens. The technique of NP aspiration involves connection of a small-bore catheter tube to a vacuum source via a mucus trap device such as a tracheal trap container. The other end of the catheter is then inserted through the external nares to the posterior pharynx. Suction is applied while the catheter is slowly withdrawn. Material in the catheter is then flushed into the trap with a small volume of phosphate-buffered saline or other suitable wash fluid. The main advantages of NP aspiration are that it generally yields sufficient material for other diagnostic procedures (viral cultures) and is well tolerated by patients; however, aspiration is practical in the hospital setting only (27). NP swab specimens are a practical and appropriate alternative to aspirates for recovery of Bordetella spp. The swab handle should be of flexible wire (aluminum). The fiber composition of the swab tip is critical, as some fibers contain substances that are inhibitory to or likely to trap the organisms. Both calcium alginate and Dacron have been recommended, with the former preferred over the latter (40). Generally, two pernasal swab specimens are collected by passing the swabs through the nares as far as possible into the posterior nasopharynx (until resistance is met), rotating the swabs for a few seconds, and then gently withdrawing them. The swabs may remain in the nasopharynx for up to 30 s or until a coughing spell is induced (18).

Following collection, NP specimens may be directly plated to culture media at the bedside or transferred to a transport system for delivery to the laboratory. Many prefer direct bedside or patient site plating; however, this method requires a conscientious effort by medical personnel who are collecting the specimens and may not be available or practical outside the hospital setting or at locations removed from the processing laboratory. If the specimen cannot be plated immediately, the selection of a transport system should be based on the length of time the specimen will be in transit. Some have suggested that for short transport times (a few hours), the specimen may be expressed from the swab into a solution of 1% casein hydrolysate (Casamino Acids, Difco 0230; Difco, Detroit, Mich.) and transported at room temperature (8, 19, 70). Plates can be inoculated and smears for direct fluorescent-antibody (DFA) tests (see below) can be prepared from the suspension. For those specimens to be plated on the same day of collection, some have used a swab system consisting of nonnutritive Amies transport medium with charcoal along with room temperature transport (8, 17, 18, 70). When laboratory plating will be delayed until the following day or longer, the use of half-strength charcoal-horse blood agar transport medium is recommended (8, 70). This nutritive transport medium, developed by Regan and Lowe (68), is prepared by using half the amount of charcoal agar (Oxoid CM119; Oxoid, Columbia, Md.) required for preparation of

isolation media but is similarly supplemented with 10% defibrinated horse blood and made selective with the addition of cephalexin at 40 mg/liter. The medium may be prepared as slants or deeps in screw-cap glass or plastic tubes, small jars, or flasks. The swab specimen is streaked onto the agar surface, then submerged into the agar, and left in place (swab shaft is cut off). Some data suggest that the recovery of Bordetella spp. is improved if the transport medium is incubated at 35°C at the collection site for 1 or 2 days prior to transport (35, 40, 42, 49). This initial incubation step allows the organisms to multiply, while cephalexin in the medium prevents overgrowth by normal flora bacteria. Several broth media have also been studied for enrichment of B. pertussis before plating (41, 78), but the data are insufficient to allow a recommendation for broth enrichment. Following initial incubation, Regan-Lowe transport medium is forwarded to the testing laboratory at room temperature; total time in transit should not exceed 2 or 3 days. One report documents the successful recovery of B. pertussis from Regan-Lowe transport medium held at 4°C from the time of collection; however, this latter study used only seeded NP specimens (61). In the laboratory, growth (if present) on Regan-Lowe transport medium may be tested directly for Bordetella spp., and the swab can be removed and used for inoculation of isolation medium. Jones-Kendrick transport medium (43), which is similar to Regan-Lowe transport medium but contains no blood, may also be used with only a modest decrease in efficiency of transport. Jones-Kendrick transport medium has the advantage of a longer shelf life.

DIRECT EXAMINATION

DFA Test

The DFA test has been used for a long time with various degrees of success for the rapid, direct detection of B. pertussis and B. parapertussis in NP specimens (12, 31, 65, 73). Although a number of factors have contributed to discrepancies in the reported sensitivity and specificity of DFA, the two most important factors probably are the sensitivity of the culture method used as a "gold standard" for comparison and the technical skill and experience of the technicians who perform and read the test. For example, prior to 1984, when my laboratory switched from Bordet-Gengou to charcoal-horse blood isolation medium, more cases of pertussis were detected by DFA test than by culture, and the sensitivity of DFA compared with culture was over 85%. Since 1984, we have cultured over 300 B. pertussis isolates, and the yearly DFA sensitivity has ranged from 40 to 75% (mean, 61%), with specificity approaching 95% compared with culture. Thus, DFA should always be used in conjunction with and not as a replacement for culture (20).

Slides for DFA testing can be prepared at the bedside directly from the NP swab specimens after plating or alternatively from the Casamino Acids transport medium containing the swab. It is convenient to use glass slides with dual circular rings, and at least two slides (four smears) should be prepared. Slides should be dried, heat fixed, and stained on the day of receipt or stored at −70°C and heat fixed immediately before staining. Although most recommend DFA staining for both B. pertussis and B. parapertussis, my laboratory has isolated only eight strains of the latter over the last 10 years, and we do not stain for this organism routinely. Laboratories in geographic areas reporting a larger percentage of *B. parapertussis* isolates may want to include this procedure in their protocol (53, 58).

Preparation of reagents, quality control, test performance, and subjectivity of interpretation of the DFA test present some challenges to even good laboratories. The fluorescent conjugates must be prepared from lyophilized material and titrated. Because they are highly specific, the *B. pertussis* conjugate may be used as a nonspecific or negative staining control for *B. parapertussis* and vice versa. Control slides may be prepared from stock cultures of these organisms maintained on isolation medium by weekly subculture. Alternatively, subculture plates may be sealed and held for up to 1 month at 5°C. Control slides should be prepared by initial suspension and dilution in buffered saline to contain low numbers of organisms (10 to 100) per oil immersion field. Slides should be prepared with a coverslip and fluorescent antibody mounting fluid (pH 7.2; Difco 2329-57-2) and viewed under oil immersion with a fluorescence microscope at a magnification between ×600 and ×1,000 (my laboratory uses ×630). On microscopic examination, the organisms appear as small coccobacillary or "football"-shaped rods with strong peripheral apple green fluorescence and dark centers. Both morphology and staining pattern must be consistent. It is not unusual to find cocci that stain homogeneously with weak to moderate intensity; these should be ignored. Although there are no absolute guidelines, we accept as few as five typical organisms as a positive DFA test. In a well-prepared slide, particularly one prepared directly from the swab at the bedside, a large number of leukocytes and ciliated epithelial cells along with mucous strands is visible. If very little cellular material is visible, DFA tests are reported as inconclusive owing to insufficient material. If the interpretation of the DFA test is in doubt, it is prudent to report it as negative.

It has been suggested that laboratories proficient in DFA and culture for pertussis should obtain a DFA sensitivity of 60% or better and a specificity of at least 90% over time compared with culture (18). This suggestion is in line with our experience and represents a reasonable objective for most laboratories offering diagnostic services for pertussis. With this in mind, clinicians should be instructed to use positive DFA results as strong presumptive evidence of pertussis; negative DFA results and negative cultures do not rule out pertussis.

Antigen Detection

DFA detection of *B. pertussis* requires that whole, intact bacterial cells in relatively large numbers be present in the clinical specimen. Thus, DFA tests are generally not as sensitive as good culture techniques, even late in the disease process, when culture sensitivity is known to be poor. Monoclonal antibodies for detecting soluble *B. pertussis* antigens or toxins have been developed and used in assays for detection of the enzymatic activity of *B. pertussis* ACT (79); enzyme-linked immunosorbent assays (ELISAs) for PT (15), FHA (25), or *B. pertussis*-specific lipopolysaccharide (26); and a PT neutralization assay using Chinese hamster ovary cells in vitro (30). These assays have not been thoroughly evaluated but merit further investigation.

Nucleic Acid Detection

Detection of *B. pertussis*-specific nucleic acids in NP specimens shows tremendous potential in the diagnosis of pertussis. Although detection of DNA using direct or culture-amplified gene probe technology has been applied to a number of organisms, including *Bordetella* spp. (69), most current research focuses on DNA amplification methods based on the PCR. Such methods are particularly applicable to diagnosis of diseases like pertussis because of limitations with the diagnostic methods presently available coupled with the fact that the organism is found only in individuals with active infection (with or without disease); there is no known prolonged carrier state. Furthermore, an isolate is not required for antimicrobial susceptibility testing because the organism is predictably susceptible to erythromycin. Several regions of the *B. pertussis* genome have been targets for PCR-based amplification and detection; studies that attempted to amplify highly conserved, reiterated chromosomal DNA sequences of *B. pertussis* or *B. parapertussis* showed the highest sensitivity and specificity of the PCR assay when compared with culture or serology (12, 21, 33, 76). Before PCR can be recommended as the test method of choice for diagnosis of pertussis, guidelines need to be developed through the consensus process for specimen collection and transport, preparation, amplification, detection, and quality control. In addition, PCR methods will need to be developed and kits must be manufactured by commercial companies before widespread application of this technology occurs.

CULTURE AND ISOLATION

A number of different media for the culture isolation of *B. pertussis* and *B. parapertussis* have been developed and evaluated. These include Bordet-Gengou potato infusion agar with glycerol and sheep blood (14, 18), Jones-Kendrick charcoal agar (43, 71), charcoal-horse blood agar (also referred to as Regan-Lowe isolation medium) (74), and a modified Stainer-Scholte agar medium supplemented with heptakis cyclodextrin, a growth stimulant for *B. pertussis* (1, 14). All of these have been further modified and tested with the addition of a variety of antimicrobial agents, including penicillin, methicillin, and cephalexin, and the substitution of one type of mammalian blood for another. Each of these media and variations have been used successfully by a number of investigators. Most current data support the use of charcoal agar supplemented with 10% horse blood and 40 mg of cephalexin per liter (18, 35, 39). This medium has the advantage of a longer shelf life (4 to 8 weeks) than Bordet-Gengou medium (1 to 7 days), but it is not as long as that of Jones-Kendrick or modified Stainer-Scholte medium (up to 3 months). Sheep blood may be substituted for horse blood without a significant decrease in plating efficiency (37). Because of the reported inhibition of some strains of *B. pertussis* by 40 mg of cephalexin per liter, some have recommended using lower concentrations of this antimicrobial agent, a second charcoal-horse blood plate prepared without cephalexin, or a combination of Bordet-Gengou and charcoal-horse blood agar plates (18, 49). If medium is prepared in-house, it is advisable to use a deep-pour plate (25 to 30 ml/100-mm plate) to help ensure adequate moisture. We have successfully used a single in-house-prepared charcoal-sheep blood agar plate as well as commercially obtained (Remel 50-2500 [Remel, Lenexa, Kans.]; BBL 4397883 [BBL, Cockeysville, Md.]) charcoal-horse blood agar, all containing 40 μg of cephalexin per ml. Medium without cephalexin is very susceptible to bacterial overgrowth and does not increase diagnostic yield (55, 82).

It is important that both commercial and in-house-prepared charcoal-blood agar media undergo rigorous quality control. We have detected deficiencies and rejected lots of both types of media. It is important to use a fresh clinical isolate for quality control, because repeated subculture of *B. pertussis* strains leads to loss of fastidiousness and laboratory adaptation to a variety of media. In order to have a fresh isolate available for quality control, an isolate can be frozen in small aliquots in sheep blood at −70°C. Upon retrieval from storage, the organism is passaged once on charcoal-horse blood agar, adjusted to a turbidity standard, diluted in saline, and plated in duplicate onto newly prepared and previously quality-controlled medium to yield approximately 100 CFU per plate. The plates are incubated at 35°C and read at days 3 and 5 for colony count and size. Only media that display a plating efficiency of at least 50% with similar colony size when compared with the previously controlled media are accepted.

Once plated to solid medium, *Bordetella* spp. should be cultivated at 35°C (not 37°C) in air without elevated CO_2. Some have used incubation in air with 5% CO_2 with apparent success (20, 56), but the bordetellae do not require CO_2 and one study reported relative inhibition of growth in air with 5 to 10% CO_2 compared with growth in air alone (37). It is important that plates remain moist during incubation. They can be kept moist by incubating them in a loosely sealed device such as an anaerobe jar with a wet sponge on the bottom or by wrapping them with a sealing tape that allows air exchange (Shrink Seals; Scientific Device Laboratory, Inc., Glenview, Ill.). The plates should be incubated for a minimum of 7 days before being reported as negative. It is advantageous to selectively hold plates for 10 days when patients are known to be on an antimicrobial agent at the time of collection, when specimens are DFA positive, or when one is forced to use medium whose characteristics for *Bordetella* growth are marginal.

Most isolates of *B. pertussis* are detected in 3 to 5 days, and most *B. parapertussis* isolates are detected in 2 to 4 days. The use of a stereomicroscope with angular lighting will facilitate early detection of colonies. Young colonies of both *B. pertussis* and *B. parapertussis* on charcoal-horse blood agar are small, shiny, round, and mercury-silver in color; the colonies become whitish grey with age. Colonies that develop within 24 to 48 h and are atypical for *Bordetella* spp. (large, irregular, rough, flat, pigmented) may be ignored. It is generally possible to identify colonies of *Bordetella* spp. within a background of upper respiratory tract flora, but an occasional culture will be overgrown by organisms such as *Pseudomonas* spp. and rendered uninterpretable for *Bordetella* spp.; final reports should reflect this fact. Colonies of filamentous fungi may also be a problem, but if only a few are present, the plate can often be salvaged by carefully cutting out and discarding the offending colony or colonies. Amphotericin B can be incorporated into the agar at the time of preparation if fungal contamination is a consistent problem.

IDENTIFICATION

Colonies suggestive of *Bordetella* spp. should be Gram stained, and the safranin counterstain should be applied for 1 to 2 min to enhance the intensity of staining. Gram-negative coccobacilli or short rods should be subjected to serologic identification with either fluorescein-labeled (Difco 2359-56-6 and 2378-56-3) or agglutinating (Difco 2309-50-3 and 2310-50-0) polyclonal antisera specific for *B. pertussis* and *B. parapertussis*. Only a few colonies are necessary for the fluorescein-labeled-antibody procedure, and these reagents can also be used in the DFA test on NP specimens. When this procedure is used on plate organisms,

take care to prepare a very dilute suspension of the organisms in phosphate-buffered saline (Difco FA buffer 2314-33-8) before applying to glass slides; testing of heavy suspensions may be associated with false-negative results owing to dulling of the fluorescence. The agglutination test may also be used when there is sufficient growth on the primary isolation plate. In this case, a heavy suspension of the organism should be prepared in saline by repeatedly forcing it through a pipette to break up clumps; the test can then be performed on glass or cardboard slides. Both serologic identification tests need to be controlled with known strains of *B. pertussis* and *B. parapertussis*; there should be little or no cross-reactivity between strains.

When serologic identification test results are clear cut, no additional confirmatory testing is needed. Whenever initial growth is limiting or serologic identification testing is equivocal, colonies should be subcultured to routine blood and chocolate agar plates as well as to a medium designed for bordetellae. *B. pertussis* will not grow on blood or chocolate agar, whereas isolates of *B. parapertussis* will grow on blood and sometimes on chocolate agar. Growth on chocolate agar alone is helpful in identifying the occasional strain of *Haemophilus* sp. that may be recovered on selective media for *Bordetella* spp. Repeat serologic identification testing should be performed on growth from the medium designed for *Bordetella* isolation. Additional tests that are helpful in confirming the identification of these and other *Bordetella* spp. include oxidase and urease production (Table 1).

SEROLOGIC TESTS

Because of the known lack of sensitivity of both culture and DFA tests for *B. pertussis*, a number of serologic tests for antibodies have been developed and evaluated (24, 28, 31, 52, 59, 67). Although test methods have included agglutination, complement fixation, indirect hemagglutination, toxin neutralization, and bactericidal assays, recent interest has focused on ELISAs (7, 22, 23). At least one ELISA is commercially available (Lab Products International Ltd., Raleigh, N.C.); it is formatted as two separate tests, one for immunoglobulin G (IgG) and one for IgA and IgM antibody combined. It has not been extensively evaluated in the United States and is presently not licensed for clinical use in this country. The results of these ELISA evaluations can be summarized as follows. (i) High sensitivities in terms of detection of antibody rises in paired sera can be achieved only when several ELISAs (e.g., IgG to FHA and PT along with IgA to FHA) are used in combination, and (ii) ELISAs for IgA, particularly to FHA, show promise as rapid diagnostic tests that can be performed on a single serum sample or NP aspirate specimen because IgA antibodies are produced early in infection but not following immunization with whole-cell vaccine as antigen. Furthermore, the presence of IgA in secretions is short-lived (29). Western immunoblotting techniques using crude whole-cell antigen preparations of *B. pertussis* may also differentiate between antibody responses resulting from natural infection and immunization (67).

A number of issues relating to the serologic diagnosis of pertussis make it difficult to evaluate serologic tests. These issues include the lack of a gold standard test for comparison, the problems associated with collection of appropriately timed paired sera, the difficulty in interpreting a single antibody titer in a number of clinical situations, and the relative lack of serologic test data for the diagnosis of

TABLE 1 Characteristics of *Bordetella* spp.[a]

Characteristic	B. pertussis	B. parapertussis	B. bronchiseptica	B. avium
Catalase	+	+	+	+
Oxidase	+	−	+	+
Motility	+	−	+	+
Nitrate reduction	−	−	+	−
Urease production	−	+(24 h)	+(4 h)	−
Growth on:				
Regan-Lowe agar	3–6 days	2–3 days	1–2 days	1–2 days
Blood agar	−	+	+	+
MacConkey agar	−	±	+	+
Salmonella-shigella agar	−	−	+	+
Browning of peptone agar	−	+	−	±

[a]Modified from references 18 and 77. Responses: +, activity or growth present; −, not present; ±, may or may not be present.

sporadic cases of pertussis as opposed to cases occurring during epidemics. For these and other reasons, it is presently impossible to recommend a single test method for the serologic diagnosis of pertussis. Serologic testing for multiple antibody classes to multiple antigens will likely continue to play an important role in epidemiologic surveys of disease prevalence and vaccine efficacy as well as complement culture diagnosis of pertussis in those laboratories devoting research effort to establish and validate these tests.

ANTIMICROBIAL SUSCEPTIBILITY

Erythromycin remains the drug of choice for both treatment and prophylaxis (4). Surveys performed over the past 30 years suggest that MICs of erythromycin for *B. pertussis* have not increased; no in vitro resistance has been detected (3, 4). Most studies report mean MICs in the 0.03 to 0.25 μg/ml range or MICs for 90% of strains (MIC$_{90}$s) of $\leq$0.12 μg/ml. MICs of erythromycin for *B. parapertussis* are generally higher but within the range of susceptibility (50).

Because of the gastrointestinal side effects (abdominal pain and loose stools) of erythromycin, a number of other oral antimicrobial agents against *B. pertussis* have been evaluated for in vitro activity, bacteriologic eradication from the respiratory tract, and/or clinical efficacy. Ampicillin or amoxicillin, tetracyclines, rifampin, and trimethoprim-sulfamethoxazole have moderate in vitro activities against *B. pertussis*, but their rates of bacteriologic eradication have been inferior to that of erythromycin (4). A number of fluoroquinolones have very good in vitro activities against both *B. pertussis* and *B. parapertussis*. Ciprofloxacin appears to be among the most active, with an MIC$_{90}$ of 0.06 μg/ml (2, 38). Comparison of MICs of fluoroquinolones with known achievable concentrations in respiratory secretions suggests excellent potential for bacteriologic eradication and clinical efficacy by these drugs, but clinical trials have not yet been performed. Although these drugs are presently not approved for use in children under 18 years, they could be used in adults with clinical pertussis or to limit transmission. A number of newer macrolides, including clarithromycin and azithromycin, are very active against *B. pertussis*; MIC$_{90}$s in the 0.03 to 0.06 μg/ml range are reported (32). Again, there have been no reports of clinical trials for these agents against pertussis. It should be appreciated that antimicrobial susceptibility testing of the fastidious *Bordetella* spp. is not standardized. Although most studies report the use of agar dilution methods on media containing blood, it may be difficult to compare studies. At present, laboratories need not perform routine susceptibility testing of *B. pertussis* and *B. parapertussis*, because erythromycin remains active against them. Patients are treated for 14 days and are generally considered to be noninfectious by 7 days. Although several studies reported superior bacteriologic eradication rates with use of erythromycin estolate instead of the ethylsuccinate or stearate form of the drug, two recent studies suggest that any of the available erythromycin preparations may be effective if compliance is ensured and if dosing regimens are altered to account for bioavailability (6, 36). These studies also reported clinical improvement (decreased coughing severity and frequency) in patients started on erythromycin after they had reached the early paroxysmal stage.

B. bronchiseptica and *B. avium* have antimicrobial susceptibility profiles similar to those of a number of nonfermentative gram-negative rods (62, 80). They are generally susceptible to the aminoglycosides, extended-spectrum penicillins (including mezlocillin and piperacillin) and antipseudomonal cephalosporins (ceftazidime and cefoperazone). Because MICs of these and other agents are not predictable, isolates of clinical significance should be tested in vitro by standardized disk diffusion or dilution methods.

REFERENCES

1. Aoyama, T., Y. Murase, T. Iwata, A. Imaizumi, Y. Suzuki, and Y. Sato. 1986. Comparison of blood-free medium (cyclodextrin solid medium) with Bordet-Gengou medium for clinical isolation for *Bordetella pertussis*. J. Clin. Microbiol. **23:** 1046–1048.
2. Appleman, M. E., T. L. Hadfield, J. K. Gaines, and R. E. Winn. 1987. Susceptibility of *Bordetella pertussis* to five quinoline antimicrobic drugs. Diagn. Microbiol. Infect. Dis. **8:**131–133.
3. Bannatyne, R. M., and R. Cheung. 1982. Antimicrobial susceptibility of *Bordetella pertussis* strains isolated from 1960–1981. Antimicrob. Agents Chemother. **21:**666–667.
4. Bass, J. W. 1986. Erythromycin for treatment and prevention of pertussis. Pediatr. Infect. Dis. **5:**154–157.
5. Bass, J. W., and S. R. Stephenson. 1987. The return of pertussis. Pediatr. Infect. Dis. J. **6:**141–144.
6. Bergquist, S.-O., S. Bernander, H. Dahnsjo, and B. Sundelöf. 1987. Erythromycin in the treatment of pertussis: a study of bacteriologic and clinical effects. Pediatr. Infect. Dis. J. **6:**458–461.
7. Campbell, P. B., P. L. Masters, and E. Rohwedder. 1988. Whooping cough diagnosis: a clinical evaluation of comple-

menting culture and immunofluorescence with enzyme-linked immunosorbent assay of pertussis immunoglobulin A in nasopharyngeal secretions. *J. Med. Microbiol.* 27:247–254.

8. **Daughterty, M. P., J. Dolter, G. C. Evans, M. E. Griffith, B. A. Hummert, D. Linquist, M. M. Struthers, and J. D. Wong.** 1992. Processing of specimens for isolation of unusual organisms, p. 1.18.13–1.18.17. *In* H. Isenberg (ed.), *Clinical Microbiology Procedures Handbook,* vol. 1. American Society for Microbiology, Washington, D.C.

9. **Davis, S. F., P. M. Strebel, S. L. Cochi, E. R. Zell, and S. C. Hadler.** 1992. Pertussis surveillance—United States. *Morbid. Mortal. Weekly Rep.* 41:11–19.

10. **DeLey, J., P. Segers, K. Kersters, W. Mannheim, and A. Lievens.** 1986. Intra- and intergeneric similarities of the *Bordetella* ribosomal ribonucleic acid cistrons: proposal for a new family, *Alcaligenaceae. Int. J. Syst. Bacteriol.* 36:405–414.

11. **Doebbeling, B. N., M. L. Feilmeier, and L. A. Herwaldt.** 1990. Pertussis in an adult man infected with the human immunodeficiency virus. *J. Infect. Dis.* 161:1296–1298.

12. **Ewanowich, C. A., L. W.-L. Chui, M. G. Paranchych, M. S. Peppler, R. G. Marusyk, and W. L. Albritton.** 1993. Major outbreak of pertussis in northern Alberta, Canada: analysis of discrepant direct fluorescent-antibody and culture results by using polymerase chain reaction methodology. *J. Clin. Microbiol.* 31:1715–1725.

13. **Farizo, K. M., S. L. Cochi, E. R. Zell, E. W. Brink, S. G. Wassilak, and P. Patriarca.** 1992. Epidemiological features of pertussis in the United States, 1980–1989. *Clin. Infect. Dis.* 14:708–719.

14. **Friedman, R. L.** 1988. Pertussis: the disease and new diagnostic methods. *Clin. Microbiol. Rev.* 1:365–376.

15. **Friedman, R. L., S. Paulaitis, and J. W. McMillan.** 1989. Development of a rapid diagnostic test for pertussis: direct detection of pertussis toxin in respiratory secretions. *J. Clin. Microbiol.* 27:2466–2470.

16. **Gilchrist, M. J. R.** 1990. Pertussis: pathophysiology and prevention. *Clin. Microbiol. Newsl.* 12:17–20.

17. **Gilchrist, M. J. R.** 1990. Laboratory diagnosis of pertussis. *Clin. Microbiol. Newsl.* 12:49–53.

18. **Gilchrist, M. J. R.** 1991. *Bordetella,* p. 471–477. *In* A. Balows, W. J. Hausler, Jr., K. L. Herrmann, H. D. Isenberg, and H. J. Shadomy (ed.), *Manual of Clinical Microbiology,* 5th ed. American Society for Microbiology, Washington, D.C.

19. **Gilligan, P.** 1983. Laboratory diagnosis of *Bordetella pertussis* infection. *Clin. Microbiol. Newsl.* 5:115–117.

20. **Gilligan, P. H., and M. C. Fisher.** 1984. Importance of culture in laboratory diagnosis of *Bordetella pertussis* infections. *J. Clin. Microbiol.* 20:891–893.

21. **Glare, E. M., J. C. Paton, R. R. Premier, A. J. Lawrence, and I. T. Nisbet.** 1990. Analysis of repetitive DNA sequence from *Bordetella pertussis* and its application to the diagnosis of pertussis using the polymerase chain reaction. *J. Clin. Microbiol.* 28:1982–1987.

22. **Goodman, Y. E., A. J. Wort, and F. L. Jackson.** 1981. Enzyme-linked immunosorbent assay for detection of pertussis IgG nasopharyngeal secretions as an indicator of recent infection. *J. Clin. Microbiol.* 13:286–292.

23. **Granström, G., P. Askelöf, and M. Granström.** 1988. Specific immunoglobulin A to *Bordetella pertussis* antigens in mucosal secretion for rapid diagnosis of whooping cough. *J. Clin. Microbiol.* 26:869–874.

24. **Granström, G., B. Wretlind, C. R. Salenstedt, and M. Granström.** 1988. Evaluation of serologic assays for diagnosis of whooping cough. *J. Clin. Microbiol.* 26:1818–1823.

25. **Gustafsson, B., and P. Askelof.** 1988. Monoclonal antibody-based sandwich enzyme-linked immunosorbent assay for detection of *Bordetella pertussis* filamentous hemagglutinin. *J. Clin. Microbiol.* 26:2077–2082.

26. **Gustafsson, B., U. Lindquist, and M. Andersson.** 1988. Production and characterization of monoclonal antibodies directed against *Bordetella pertussis* lipopolysaccharide. *J. Clin. Microbiol.* 26:188–193.

27. **Hallander, H. O., E. Reizenstein, B. Renemar, G. Rasmu-** son, **L. Mardin, and P. Olin.** 1993. Comparison of nasopharyngeal aspirates with swabs for culture of *Bordetella pertussis. J. Clin. Microbiol.* 31:50–52.

28. **Hallander, H. O., J. Storsaeter, and R. Möllby.** 1991. Evaluation of serology and nasopharyngeal cultures for diagnosis of pertussis in a vaccine efficacy trial. *J. Infect. Dis.* 163:1046–1054.

29. **Halperin, S. A.** 1991. Interpretation of pertussis serologic tests. *Pediatr. Infect. Dis. J.* 10:791–792.

30. **Halperin, S. A., R. Bortolussi, A. Kasina, and A. J. Wort.** 1990. Use of a Chinese hamster ovary cell cytotoxicity assay for the rapid diagnosis of pertussis. *J. Clin. Microbiol.* 28:32–38.

31. **Halperin, S. A., R. Bortolussi, and A. J. Wort.** 1989. Evaluation of culture, immunofluorescence and serology for the diagnosis of pertussis. *J. Clin. Microbiol.* 27:752–757.

32. **Hardy, D. J., D. M. Hensey, J. M. Beyer, C. Vojtko, E. J. McDonald, and P. B. Fernandes.** 1988. Comparative in vitro activities of new 14-, 15-, and 16-member macrolides. *Antimicrob. Agents Chemother.* 32:1710–1719.

33. **He, Q., J. Mertsola, H. Soini, M. Skurnik, O. Ruuskanen, and M. K. Viljanen.** 1993. Comparison of polymerase chain reaction with culture and enzyme immunoassay for diagnosis of pertussis. *J. Clin. Microbiol.* 31:642–645.

34. **Herwaldt, L. A.** 1991. Pertussis in adults: what physicians need to know. *Arch. Intern. Med.* 151:1510–1512.

35. **Hoppe, J. E.** 1988. Methods for isolation of *Bordetella pertussis* from patients with whooping cough. *Eur. J. Clin. Microbiol.* 7:616–620.

36. **Hoppe, J. E.** 1992. Comparison of erythromycin estolate and erythromycin ethylsuccinate for treatment of pertussis. *Pediatr. Infect. Dis. J.* 11:189–193.

37. **Hoppe, J. E., and M. Schlagenhaur.** 1989. Comparison of three kinds of blood and two incubation atmospheres for cultivation of *Bordetella pertussis* on charcoal agar. *J. Clin. Microbiol.* 27:2115–2117.

38. **Hoppe, J. E., and C. G. Simon.** 1990. In vitro susceptibilities of *Bordetella pertussis* and *Bordetella parapertussis* to seven fluoroquinolones. *Antimicrob. Agents Chemother.* 34:2287–2288.

39. **Hoppe, J. E., and R. Vogl.** 1986. Comparison of three media for culture of *Bordetella pertussis. Eur. J. Clin. Microbiol.* 5:361–363.

40. **Hoppe, J. E., and A. Weib.** 1987. Recovery of *Bordetella pertussis* from four kinds of swabs. *Eur. J. Clin. Microbiol.* 6:203–205.

41. **Hoppe, J. E., A. Weiss, and S. Woerz.** 1988. Failure of charcoal-horse blood broth with cephalexin to significantly increase the rate of *Bordetella* isolation from clinical specimens. *J. Clin. Microbiol.* 26:1248–1249.

42. **Hoppe, J. E., S. Woerz, and K. Botzenhart.** 1986. Comparison of specimen transport systems for *Bordetella pertussis. Eur. J. Clin. Microbiol.* 5:671–673.

43. **Jones, G. L., and P. L. Kendrick.** 1969. Study of a blood-free medium for transport and growth of *Bordetella pertussis. Health Lab. Sci.* 6:40–45.

44. **Kersters, K., K.-H. Hinz, A. Hertle, P. Segers, A. Lievens, O. Siegmann, and J. DeLey.** 1984. *Bordetella avium* sp. nov. isolated from the respiratory tracts of turkey and other birds. *Int. J. Syst. Bacteriol.* 34:56–70.

45. **Khattak, M. N., and R. C. Matthews.** 1993. Genetic relatedness of *Bordetella* species as determined by macrorestriction digests resolved by pulsed-field gel electrophoresis. *Int. J. Syst. Bacteriol.* 43:659–664.

46. **Khelef, N., B. Danve, M. J. Quentin-Millet, and N. Guiso.** 1993. *Bordetella pertussis* and *Bordetella parapertussis:* two immunologically distinct species. *Infect. Immun.* 61:486–490.

47. **Kloos, W. E., W. J. Dobrogosz, J. W. Ezzell, B. R. Kimbro, and C. R. Manclark.** 1979. DNA-DNA hybridization, plasmids, and genetic exchange in the genus *Bordetella,* p. 70–85. *In* C. R. Manclark and J. C. Hill (ed.), *International Symposium on Pertussis.* Department of Health, Education and Welfare publication no. 79-1830. U.S. Government Printing Office, Washington, D.C.

48. **Kloos, W. E., N. Mohapatra, W. J. Dobrogosz, J. W. Ezzell, and C. R. Manclark.** 1981. Deoxyribonucleotide sequence relationships among *Bordetella* species. *Int. J. Syst. Bacteriol.* **31:**173–176.

49. **Kurzynski, T. A., D. M. Boehm, J. A. Rott-Petri, R. F. Schell, and P. E. Allison.** 1988. Comparison of modified Bordet-Gengou and modified Regan-Lowe media for the isolation of *Bordetella pertussis* and *Bordetella parapertussis. J. Clin. Microbiol.* **26:**2661–2663.

50. **Kurzynski, T. A., D. M. Boehm, J. A. Rott-Petri, R. F. Schell, and P. E. Allison.** 1988. Antimicrobial susceptibilities of *Bordetella* species isolated in a multicenter pertussis surveillance project. *Antimicrob. Agents Chemother.* **32:**137–140.

51. **Lambert, H. P.** 1986. The carrier state: *Bordetella pertussis. J. Antimicrob. Chemother.* **18**(Suppl. A)**:**13–16.

52. **Lawrence, A. J., and J. C. Paton.** 1987. Efficacy of enzyme-linked immunoabsorbent assay for rapid diagnosis of *Bordetella pertussis* infection. *J. Clin. Microbiol.* **25:**2102–2104.

53. **Linneman, C. C., and E. B. Pery.** 1977. *Bordetella parapertussis:* recent experience and a review of the literature. *Am. J. Dis. Child.* **131:**560–563.

54. **Long, S. S., H. W. Lischner, A. Deforest, and J. L. Clark.** 1990. Serologic evidence of subclinical pertussis in immunized children. *J. Pediatr. Infect. Dis.* **9:**700–704.

55. **Marcon, M. J.** Unpublished data and observations.

56. **Marcon, M. J., A. C. Hamoudi, H. J. Cannon, and M. M. Hribar.** 1987. Comparison of throat and nasopharyngeal swab specimens for culture diagnosis of *Bordetella pertussis* infection. *J. Clin. Microbiol.* **25:**1109–1110.

57. **McGillivray, D. M., J. G. Coote, and R. Parton.** 1989. Cloning of the virulence regulatory (vir) locus of *Bordetella pertussis* and its expression in *B. bronchiseptica.* FEMS Microbiol. Lett. **65:**333–338.

58. **Mertsola, J.** 1985. Mixed outbreak of *Bordetella pertussis* and *parapertussis* in an elementary school. *Eur. J. Clin. Microbiol.* **4:**123–128.

59. **Mertsola, J., O. Ruuskanen, T. Kuronen, O. Meurman, and M. K. Viljanen.** 1990. Serologic diagnosis of pertussis: evaluation of pertussis toxin and other antigens in enzyme-linked immunosorbent assay. *J. Infect. Dis.* **161:**966–971.

60. **Monack, D., J. J. Munoz, M. G. Peacock, W. J. Black, and S. Falkow.** 1989. Expression of pertussis toxin correlates with pathogenesis in *Bordetella* species. *J. Infect. Dis.* **159:**205–210.

61. **Morrill, W. E., J. M. Barbaree, B. S. Fields, G. N. Sanden, and W. T. Martin.** 1988. Effects of transport temperature and medium on recovery of *Bordetella pertussis* from nasopharyngeal swabs. *J. Clin. Microbiol.* **26:**1814–1817.

62. **Mortensen, J. E., A. Burmbach, and T. R. Shryock.** 1989. Antimicrobial susceptibility of *Bordetella avium* and *Bordetella bronchiseptica* isolates. *Antimicrob. Agents Chemother.* **33:**771–772.

63. **Musser, J. M., E. L. Hewlett, M. S. Peppler, and R. K. Selander.** 1986. Genetic diversity and relationships in populations of *Bordetella* species. *J. Bacteriol.* **166:**230–237.

64. **Ng, V. L., M. York, and W. K. Hadley.** 1989. Unexpected isolation of *Bordetella pertussis* from patients with acquired immunodeficiency syndrome. *J. Clin. Microbiol.* **27:**337–338.

65. **Onorato, I. M., and S. G. F. Wassilak.** 1987. Laboratory diagnosis of pertussis: the state of the art. *Pediatr. Infect. Dis. J.* **6:**145–151.

66. **Pittman, M.** 1984. Genus *Bordetella* Moreno-Lopez 1952, 178,

p. 388–393. *In* N. R. Krieg and J. G. Holt (ed.), *Bergey's Manual of Systematic Bacteriology,* vol. 1. The Williams & Wilkins Co., Baltimore.

67. **Redd, S. C., H. S. Rumschlag, R. J. Biellik, G. N. Sanden, C. R. Reiner, and M. L. Cohen.** 1988. Immunoblot analysis of humoral immune responses following infection with *Bordetella pertussis* or immunization with diphtheria-tetanus-pertussis vaccine. *J. Clin. Microbiol.* **26:**1373–1377.

68. **Regan, J., and F. Lowe.** 1977. Enrichment medium for the isolation of *Bordetella. J. Clin. Microbiol.* **6:**303–309.

69. **Reizenstein, E., E. Morfeldt, G. Granström, M. Granström, and S. Löfdahl.** 1990. DNA hybridization for diagnosis of pertussis. *Mol. Cell. Probes* **4:**299–306.

70. **Sneed, J. O.** 1992. Laboratory diagnosis of pertussis, p. 1.14.9–1.14.13. *In* H. D. Isenberg (ed.), *Clinical Microbiology Procedures Handbook,* vol. 1. American Society for Microbiology, Washington, D.C.

71. **Stauffer, L. R., D. R. Brown, and R. D. Sandstrom.** 1983. Cephalexin-supplemented Jones-Kendrick charcoal agar for selective isolation of *Bordetella pertussis:* comparison with previously described media. *J. Clin. Microbiol.* **17:**60–62.

72. **Storsaeter, J., and P. Olin.** 1992. Relative efficacy of two acellular pertussis vaccines during three years of passive surveillance. *Vaccine* **10:**142–144.

73. **Strebel, P. M., S. L. Cochi, K. M. Farizo, B. J. Payne, S. D. Hanauer, and A. L. Baughman.** 1993. Pertussis in Missouri: evaluation of nasopharyngeal culture, direct fluorescent antibody testing, and clinical case definitions in the diagnosis of pertussis. *Clin. Infect. Dis.* **16:**276–285.

74. **Sutcliffe, E. M., and J. D. Abbott.** 1972. Selective medium for the isolation of *Bordetella pertussis* and *parapertussis. Br. Med. J.* **6:**732–733.

75. **Thomas, M. G.** 1989. Epidemiology of pertussis. *Rev. Infect. Dis.* **11:**255–262.

76. **van der Zee, A., C. Agterberg, M. Peeters, J. Schellekens, and F. R. Mooi.** 1993. Polymerase chain reaction assay for pertussis: simultaneous detection and discrimination of *Bordetella pertussis* and *Bordetella parapertussis. J. Clin. Microbiol.* **31:**2134–2140.

77. **Weiss, A. A.** 1992. The genus *Bordetella,* vol. 3, p. 2530–2543. *In* A. Balows, H. G. Truper, M. Dworkin, W. Harder, and K.-H. Schleifer (ed.), *The Prokaryotes: a Handbook on the Biology of Bacteria: Ecophysiology, Isolation, Identification, Applications,* 2nd ed. Springer-Verlag, Inc., New York.

78. **Wirsing von Köenig, C. H., A. Tacken, and H. Finger.** 1988. Use of supplemented Stanier-Scholte broth for the isolation of *Bordetella pertussis* from clinical material. *J. Clin. Microbiol.* **26:**2558–2560.

79. **Wirsing von Köenig, C. H., A. Tacken, and E. L. Hewlett.** 1989. Detection of *Bordetella pertussis* by determination of adenylate cyclase activity. *Eur. J. Clin. Microbiol. Infect. Dis.* **8:**633–636.

80. **Woolfrey, B. F., and J. A. Moody.** 1991. Human infections associated with *Bordetella bronchiseptica. Clin. Microbiol. Rev.* **4:**243–255.

81. **Wright, P. F.** 1991. Pertussis in developing countries: definition of the problem and prospects for control. *Rev. Infect. Dis.* **13:**S228–S234.

82. **Young, S. A., G. L. Anderson, and P. D. Mitchell.** 1987. Laboratory observations during an outbreak of pertussis. *Clin. Microbiol. Newsl.* **9:**176–179.

Clostridium

ANDREW B. ONDERDONK AND STEPHEN D. ALLEN

47

Spore-forming, catalase-negative, anaerobic bacilli with gram-positive cell wall components belong to the genus *Clostridium*. Detailed descriptions of 85 species are provided in the 1986 edition of *Bergey's Manual of Systematic Bacteriology* (19). There are currently more than 100 published species of *Clostridium*. Only a limited number of species are commonly encountered as agents of infection in properly collected clinical specimens from humans (Table 1). Although botulism, tetanus, and certain other clostridial diseases may arise exogenously, most *Clostridium* species associated with disease are part of the patient's own microflora, especially that present in the large bowel. Diseases caused by members of the genus *Clostridium* generally fall into one of three categories: (i) noninvasive disease in which toxin(s) is responsible for all symptoms, (ii) invasive (histotoxic) disease in which a progressive infectious process and tissue destruction occur, and (iii) purulent disease in which a closed-space mixed infection involving multiple organisms is present, usually within the peritoneal cavity. The most common toxigenic disease is antibiotic-associated colitis, caused by *Clostridium difficile*. The most common pathogen associated with a variety of invasive infectious processes in humans is *C. perfringens*.

DESCRIPTION OF THE GENUS

The vegetative cells of most species are rod shaped and straight or curved, but cells vary from short coccoid rods to long filamentous forms. Rod-shaped cells may be rounded, tapered, or blunt ended. Cells may occur singly, in pairs, or in chains of various lengths. In certain species (e.g., *C. cocleatum* and *C. spiroforme*), many rods may be joined to form tight coils or spiral configurations (19). Most species stain gram positively during the early stage of growth; however, some species, such as *C. ramosum* and *C. clostridioforme*, almost always appear gram negative after overnight culture. Several species (e.g., *C. tetani*) appear gram negative by the time spores have formed. All but a few species are motile by means of peritrichous flagella. Nonmotile species isolated from clinical specimens include *C. perfringens*, *C. ramosum*, and *C. innocuum*. The spores of clostridia are ovoid to spherical and distend the vegetative cells. Certain species, e.g., *C. perfringens*, produce spores only under special culture conditions. Although the majority of *Clostridium* species are obligate anaerobes, there is considerable species variation with respect to oxygen toxicity; some species (e.g., *C. haemolyticum* and *C. novyi* type B) are strict obligate anaerobes and will not grow when exposed to even trace amounts of oxygen. A few aerotolerant strains (*C. tertium*, *C. carnis*, *C. histolyticum*, and occasional strains of *C. perfringens*) show scant growth on plating medium incubated in a 5 to 10% CO_2 incubator in air or in a candle jar.

Since aerotolerant clostridia may grow on the surface of fresh agar medium under aerobic conditions, it is possible to confuse these species with certain facultative anaerobic *Bacillus* species. However, members of the genus *Clostridium* usually form spores only under anaerobic conditions and almost never produce catalase. Also, aerotolerant clostridia show much better growth (i.e., they form larger colonies) under anaerobic conditions than in air, whereas *Bacillus* species often form larger colonies on aerobically incubated medium than on medium incubated anaerobically.

Catalase is not produced except in rare instances. When it is produced, reactions are typically weakly positive. In addition, clostridia lack a cytochrome system and thus react negatively in the cytochrome oxidase test. They do not have a mechanism for phosphorylation via an electron transport system. Energy is obtained via ATP by substrate-level phosphorylation (19). Clostridia are usually fermentative, proteolytic, or both, but some are asaccharolytic and nonproteolytic. Many clostridia produce a range of short-chain fatty acids (e.g., acetate and butyrate) when grown in peptone-yeast extract-glucose or chopped-meat–carbohydrate medium, and many produce a variety of other fermentation products (e.g., acetone, butanol, and other alcohols). The metabolism and energy-yielding mechanisms of clostridia are extremely diverse and interesting (17).

NATURAL HABITATS

Members of the genus *Clostridium* are distributed widely in nature and are found in soil as well as in freshwater and marine sediments throughout the world. Some species are psychrophilic or thermophilic, but most are mesophilic. *Clostridium* is the only genus of spore-forming anaerobes associated with humans, either as nonpathogens at a variety of anatomic locations or at infected sites. Several *Clostridium* species reside in the lower intestinal tracts of humans and other animals as part of the normal flora (33). With

TABLE 1 *Clostridium* species most frequently encountered in clinical specimens at Indiana University Medical Center Anaerobe Laboratory, 1979 through 1988[a]

Species	Isolates	
	No.	% of total
C. perfringens	472	26
C. ramosum	257	14
C. innocuum	235	13
C. clostridioforme	200	11
C. difficile	130	7
C. butyricum	90	5
C. cadaveris	83	5
C. bifermentans	64	4
C. sporogenes	63	3
C. septicum	61	3
C. tertium	50	3
C. paraputrificum	42	2
C. glycolicum	25	1
C. subterminale	20	1
Other recognized species[b]	50	3

[a]From Allen and Baron (1a). The total of 1,842 isolates does not include 252 isolates (12% of 2,094 isolates) that did not belong to a recognized species. All of the isolates were from properly collected specimens. No isolates were from feces, intestinal materials, or other contaminated sources.

[b]Includes 1 to 10 isolates of each of the following: *C. sordellii, C. barati, C. malenominatum, C. sphenoides, C. putrificum, C. symbiosum, C. indolis, C. celatum, C. tetani, C. carnis, C. leptum,* and *C. hastiforme.*

certain notable exceptions, most species are considered harmless saprophytes.

Among the pathogenic species, both exogenous and endogenous sources exist. Strictly toxigenic diseases such as botulism, *C. perfringens* food-borne illness, gas gangrene associated with traumatic wound infections, and tetanus are well known as arising from exogenous sources. Although usually acquired from the patient's own gastrointestinal tract, *C. difficile* may be spread exogenously from person to person during outbreaks of hospital-acquired diarrhea or colitis (50, 67). Endogenous infections involving clostridia that are a part of the host's own microflora are much more common, however. As is the case for other endogenous infections involving anaerobes (endocarditis, brain abscess, aspiration pneumonia, intra-abdominal abscess, etc.), special circumstances are usually associated with the development of disease. Common predisposing factors include trauma, operative procedures, vascular stasis, obstruction, treatment with immunosuppressive agents or chemotherapeutic agents used in the treatment of malignancy, prior treatment with antimicrobial agents (as in pseudomembranous colitis), and underlying illness such as leukemia, carcinoma, or diabetes mellitus. Under the right conditions, clostridia can invade and multiply in essentially any tissue of the body.

CLINICAL SIGNIFICANCE

C. perfringens

C. perfringens is the species most commonly isolated from human clinical specimens excluding stool. It is encountered in a wide variety of clinical settings ranging from simple contamination of wounds to traumatic or nontraumatic myonecrosis, clostridial cellulitis, intra-abdominal sepsis,

gangrenous cholecystitis, postabortion infection with devastating septicemia, intravascular hemolysis, bacteremia in various clinical settings, aspiration pneumonia, necrotizing pneumonia, thoracic empyema, subdural empyema, and brain abscess (11, 29, 30, 79). The organism is ubiquitous, occurring in soil, water, sewage, and the intestinal tracts of animals and many humans.

C. perfringens, also called *C. welchii* by some investigators, is commonly considered to be one of a group of organisms referred to as the histotoxic or tissue-destroying clostridia, the most common form of tissue destruction being gas gangrene. The histotoxic clostridia most commonly involved in gas gangrene (myonecrosis) are *C. perfringens* (80%), *C. novyi*, and *C. septicum*, followed by *C. histolyticum* and *C. bifermentans*. Other species such as *C. sordellii, C. fallax, C. sporogenes,* and *C. tertium* have been encountered in patients with myonecrosis, but their pathogenic significance is not certain.

Clostridial myonecrosis (gas gangrene) is a toxin-mediated breakdown of muscle tissue associated with growth of the organism. It is a rapidly progressive, life-threatening condition with liquefactive necrosis of muscle, gas (primarily insoluble hydrogen and nitrogen) formation, and associated clinical signs of toxemia. Blood cultures are positive in about 15% of patients (38). Many excellent descriptions of the clinical and pathological features of gas gangrene have been published elsewhere (4, 11, 33, 38, 69, 86, 88, 94).

C. perfringens has also been isolated from the vaginal vaults or cervixes of approximately 1 to 9% of healthy pregnant and nonpregnant women (23). Gas gangrene of the uterus, which is now rare in the United States, has occurred most frequently as a consequence of illegal or self-induced abortions. It has also followed spontaneous abortion or vaginal delivery in which the postabortal or puerperal uterus contained blood clots and fragments of necrotic placenta or fetal tissue along with a minimum number of clostridia that could proliferate within the endometrium and subsequently invade the uterine wall. Clostridial myonecrosis of the uterus is associated with fulminant and overwhelming septicemia, marked intravascular hemolysis with a sudden drop in hemoglobin, hemolysis, profound anemia, jaundice, hemoglobinuria, and excretion of urine that is burgundy wine colored. Patients become hypotensive and not uncommonly develop cardiovascular collapse, cardiac failure, pulmonary edema, renal failure, and coma, and they may die within hours after the onset of the infection (94). A dirty, red-brown, foul-smelling vaginal discharge containing gas bubbles and a Gram-stained smear showing numerous gram-positive bacilli should suggest clostridial myonecrosis of the uterus.

The presence of gas at an infected site does not always signal clostridial myonecrosis. Various gas-producing bacteria may form gas in tissue; those that do so most frequently, particularly after laceration-type wounds involving soft tissue other than muscle, are the clostridia, especially *C. perfringens* (3, 87, 94). Crepitant cellulitis caused by clostridia, also called anaerobic cellulitis, characteristically involves subcutaneous tissues or retroperitoneal tissues (3, 4). In contrast to clostridial myonecrosis, the muscle is usually not involved to a significant extent and remains viable during crepitant cellulitis. Although both gas gangrene and crepitant cellulitis are serious, the outlook for patients who have clostridial infections confined to subcutaneous tissues is usually not so ominous as that for patients with myonecrosis, provided the correct diagnosis and treatment are initiated early in the course of illness. Pertinent findings in crepitant cellulitis include abundant gas in the tissue, often more than in myonecrosis; tissue swelling without much discoloration of the overlying skin; minimal pain; and a thin, sweet-smelling or sometimes foul-smelling exudate that may contain numerous polymorphonuclear leukocytes and bacteria (3, 11). On occasion, however, polymorphonuclear leukocytes may be absent because of the activity of the leukolytic toxins of the clostridia. The presence of boxcar-shaped, gram-variable rods and no leukocytes in a Gram stain of infected tissue should lead one to suspect clostridial infection. These infections may involve only clostridia but are frequently polymicrobial, also containing Escherichia coli, Peptostreptococcus spp., Bacteroides spp., or other bacteria (3, 69). In some patients whose initial presentation is that of focal clostridial crepitant cellulitis, there may be a fulminant clinical course with rapid spread of bacteria, gas, and cellulitis through fascial planes, and signs of toxemia, septicemia, shock, and intravascular hemolysis, progressing to renal failure and death.

Clostridium species are commonly encountered in a variety of polymicrobial infections involving the abdomen, including peritonitis, intra-abdominal abscesses, and septicemia in patients with obstructive or perforating lesions of the terminal ileum or large bowel (39). C. perfringens and C. septicum have been documented in patients with overwhelming sepsis and gangrenous necrosis of the small intestine or large bowel. Prior abdominal surgery of the bowel, penetrating wounds involving the small intestine or large bowel, and carcinoma of the colon or other malignancy have been observed in these patients. Although it is clear that clostridia play a major pathogenic role in some life-threatening necrotizing infections of the intestinal tract, the role of clostridia during intra-abdominal infections must be assessed on an individual patient basis (53).

C. perfringens produces a variety of biologically active proteins, or toxins, that play an important role in pathogenicity (reviewed by Smith [77] and more recently by Hatheway [42]). On the basis of mouse lethality assays and specific neutralization of four toxins produced in culture fluids, five toxin types of C. perfringens have been identified (types A through E) (42, 82; Table 2). In reference laboratories such as those of the Centers for Disease Control and Prevention (CDC), toxin typing by toxicity assays is employed for epidemiologic purposes (26). Most infections of humans involving C. perfringens are caused by type A strains. C. perfringens type C is the only other toxin type encountered in human illness (79). Type C is associated with a disease called enteritis necroticans (discussed below). All five lethal toxin types of clostridia produce a potent phospholipase, or alpha-toxin. Also referred to as phospholipase C, the alpha-toxin has several additional properties. It produces the opaque zone of lecithin hydrolysis products that surrounds colonies of C. perfringens growing on egg yolk agar plates. It is also a hemolysin active against erythrocyte membranes. It gives rise to an outer zone of partial hemolysis that encircles a smaller zone of complete hemolysis produced by theta-toxin (a heat-labile, oxygen-labile hemolysin easily detected around colonies of C. perfringens growing on sheep blood agar). The alpha-toxin is also active against the membranes of muscle cells, leukocytes, and platelets, and it has necrotizing activity that leads to the death of a variety of host cells and tissues. In addition to producing alpha-toxin, type B strains of C. perfringens produce beta and epsilon lethal toxins, type D strains produce epsilon lethal toxin, and type E strains

uniquely produce an iota-toxin as well as alpha-toxin (Table 2). Beta-toxin is a major lethal factor produced by type C strains of C. perfringens. This toxin is important in the pathogenesis of certain forms of necrotizing enteropathy (42).

A variety of apparently less important bioactive proteins, or minor toxins, which may or may not serve as virulence factors, are produced by C. perfringens (42, 77, 82). These include lambda-toxin (a protease), kappa-toxin (a collagenase), mu-toxin (a hyaluronidase), a neuraminidase, and a DNase (79, 94).

Although C. perfringens is commonly considered one of a group of organisms referred to as the histotoxic clostridia, it may also contribute to purely toxigenic disease and has been reported to be present in purulent infections as well.

C. perfringens Food Poisoning

C. perfringens generally ranks behind Salmonella spp. and Staphylococcus aureus as the third most common cause of food poisoning in the United States and elsewhere (6, 15). Almost all U.S. outbreaks and cases of C. perfringens food-borne gastroenteritis appear to be due to type A strains (73). In C. perfringens type A food-borne disease, the food vehicle is almost always an improperly cooked meat (e.g., beef, turkey, chicken, or pork) or a meat product, such as gravy, that has cooled slowly after cooking or may have been inadequately reheated (6). Spores surviving the initial cooking germinate, and vegetative cells proliferate during slow cooling or insufficient reheating. C. perfringens type A food-borne illness should be suspected when there is an outbreak of diarrhea with crampy abdominal pain within about 7 to 15 h after the consumption of a suspected food (73). However, the incubation period may range up to 30 h. Most patients are afebrile; nausea and vomiting occur in less than a third of patients, and the stools are frequently foamy and foul smelling. Illness results from the ingestion of food with about 10^8 viable vegetative cells, which in the alkaline environment of the small intestine undergo sporulation, producing an enterotoxin in the process (73). The enterotoxin, a 34,000-molecular-weight protein (a structural component of C. perfringens spores), induces fluid and electrolyte secretion in the ilea of mice and other animal models (96). Usually, the illness is mild, and most patients recover within 2 to 3 days after onset. The diagnosis is confirmed by the culture from epidemiologically implicated food of at least 10^5 organisms per g and by the demonstration, using a quantitative spore selection technique, of median spore counts of at least 10^6 C. perfringens spores per g of feces collected within 24 h after the onset of illness (6, 73). Latex reagents for detecting type A enterotoxin, although under development (66), are not available for clinical laboratory use. Laboratory testing for C. perfringens food-borne illness is a public health laboratory function that should be performed concomitantly with provision of epidemiologic support. Ideally, isolates of the same serotype can be cultured from epidemiologically incriminated food and ill persons but not from control subjects. Unfortunately, the experience with serotyping of C. perfringens at the CDC has not been highly successful (73). C. perfringens may produce a severe necrotizing disease of the small bowel known as enteritis necroticans, which can occur sporadically or in an epidemic form. The syndrome has been called Darmbrand in Germany and pig-bel in Papua, New Guinea (11, 68, 79, 92, 94). In this condition, seen mostly in children, there is evidence that suggests that C. perfringens type C, which either is present as part of the normal

TABLE 2 Distribution of major toxins among types of *C. perfringens*[a]

Toxin type	Occurrence and country where originally found	Alpha-toxin		Beta-toxin		Epsilon-toxin		Iota-toxin	
		Presence or absence	Characteristics	Presence or absence	Characteristics	Presence or absence	Characteristics	Presence or absence	Characteristics
A[b]	Gas gangrene of humans and animals Intestinal flora of humans and animals Putrefactive processes in soil, etc. (United States) Food poisoning (United Kingdom)	+	Lethal Lecithinase Hemolytic Necrotizing	−	Lethal Necrotizing	−	Lethal Permease	−	Lethal Dermonecrotic ADP ribosylating
B	Lamb dysentery Enterotoxemia of foals (United Kingdom) Enterotoxemia of sheep and goats (Iran)	+		+		+		−	
C	Enterotoxemia of sheep (struck) (United Kingdom) Enterotoxemia of calves and lambs (United States) Enterotoxemia of piglets (United Kingdom) Enteritis necroticans of humans (Darmbrand) (Germany) Enteritis necroticans of humans (pig-bel) (Papua New Guinea)	+		+		−		−	
D	Enterotoxemia of sheep, lambs, goats, cattle, possibly humans (Australia)	+		−		+		−	
E	Sheep and cattle, pathogenicity doubted (United Kingdom)	+		−		−		+	

[a]From Allen and Baron (1a).
[b]Also called phospholipase C.

intestinal flora or is ingested with contaminated pork or other meat during a feast, proliferates in the small intestine and produces beta-toxin. Toxin production probably leads to focal paralysis, inflammation, hemorrhage, and segmental gangrenous necrosis of the intestine, particularly the jejunum. Among the factors likely to be involved in the pathogenesis of enteritis necroticans are (i) overeating, which might distend the bowel and cause partial obstruction; (ii) poor nutrition, which leads to low levels of production of the pancreatic proteases (particularly trypsin, which ordinarily destroys beta-toxins); and (iii) a diet rich in trypsin inhibitors (such as the semicooked sweet potatoes often eaten at pig feasts) (91).

C. botulinum and Botulism

C. botulinum, the cause of a relatively rare but often fatal food poisoning, is widely distributed in nature and can be isolated from soil samples at virtually any location. The seven toxigenic types of *C. botulinum* (A through G) are based on antigenically distinct toxins produced by different strains of the organism. Type C can be subdivided into two toxin types, C_1 and C_2; C_1 is a neurotoxin, and C_2 is not a neurotoxin but causes increased vascular permeability. Types A, B, E, and F are the principal causes of botulism in humans; types C and D have been associated with botulism in birds and mammals (42). Type G organisms, now called *C. argentinense* (83), have been isolated from soil in Argentina and from autopsy material from five individuals who died suddenly, but the type G organisms have not been clearly implicated in cases of botulism (80). The potent neurotoxins of *C. botulinum* are protoplasmic proteins that are synthesized during growth of the organisms and released during lysis (25).

Botulism is a neuroparalytic disease produced by the neurotoxins of *C. botulinum*, currently classified into four categories: (i) classical food-borne botulism, an intoxication caused by the ingestion of preformed botulinal toxin in contaminated food; (ii) wound botulism, which is the rarest form of botulism (40 cases have been reported), which results from elaboration of botulinal toxin in vivo after growth of *C. botulinum* in an infected wound; (iii) infant botulism, in which botulinal toxin is elaborated in vivo in the intestinal tract of an infant who has been colonized with *C. botulinum*; and (iv) a classification of undetermined for those cases involving individuals older than 12 months in whom no food or wound source is implicated (5, 42). Although botulism is not strictly a food-borne intoxication in every instance, its association with food or, in the case of infants, the ingestion of certain spore-containing foods warrants consideration of this species as primarily a food-borne disease problem.

The mechanisms of action of the botulinal neurotoxins are not completely known, but a three-step process has been hypothesized: (i) the toxin binds irreversibly to a receptor site at the neuromuscular junction and other peripheral autonomic synaptic sites (the receptor sites have yet to be identified); (ii) a portion of the toxin molecule probably penetrates the plasma membrane, possibly by receptor-mediated endocytosis, and thus is internalized into the nerve cell; and (iii) the neurotoxin molecule acts within the nerve terminal to prevent release of acetylcholine (42, 74). The characteristic clinical hallmark of botulism is an acute flaccid paralysis, which begins with bilateral cranial nerve impairment involving muscles of the face, head, and pharynx and then descends symmetrically to involve muscles of the thorax and extremities. Death may result from respiratory failure caused by paralysis of the tongue or muscles of the pharynx that occlude the upper airway or from paralysis of the diaphragm and intercostal muscles. Patients diagnosed with food-borne or wound botulism should immediately receive trivalent (type ABE) antitoxin and promptly receive intensive respiratory care. For further information on botulism in general, see Hatheway (41, 42), Smith and Sugiyama (78), the CDC handbook (5), and Dowell (25).

Infant Botulism

Infant botulism is the most frequently encountered form of botulism; 760 cases were reported in the United States between 1976 and 1988 (6, 7). Infant botulism has been reported from North and South America, Europe, Asia, and Australia (8). In the United States in general, the geographical distribution of toxin types in infant botulism cases has paralleled the distribution of *C. botulinum* toxin types in soil sampled from across the country (76). Type A has been the most frequent botulinal toxin type in cases of infant botulism in states west of the Mississippi River, whereas type B cases have predominated east of the Mississippi (8). Interestingly, one infant from Hawaii had two different strains of *C. botulinum*; one strain was type A, and the other was type B (41). Two other cases were caused by a strain(s) of *C. botulinum* that produced toxins that required both type B antitoxin and type F antitoxin for neutralization (41). In another interesting case, type F infant botulism was caused by an organism that most closely resembled *C. barati* in its cultural and biochemical characteristics (40). Type E botulism, caused by toxigenic strains of *C. butyricum*, was confirmed in two infants from Italy (41). Infected infants have ranged from 6 days to 11.7 months old (81). The ingestion of spores by the infant during the first weeks of life is probably a prerequisite for development of the disease. Preformed toxin has not been detected in any food or liquid ingested by the babies. To date, the only clearly defined risk factors have been breastfeeding and exposure to honey, which is a potential source of spores. The CDC has recommended that honey not be fed to infants less than 1 year old (81). Another concern is that spores have also been found in a limited number of samples of corn syrup, which is often given to infants for treatment of decreased frequency of bowel movement (81). However, *C. botulinum* spores in corn syrup actually consumed by an infant with botulism have yet to be demonstrated. Whatever the sources of the spores, the ingested spores of *C. botulinum* germinate within the intestinal tract, and the vegetative cells multiply and produce the neurotoxin, which is then absorbed into the bloodstream. Decreased frequency of bowel movements, which may also be a sign of decreased intestinal motility, may be an additional risk factor for infant botulism (81). The decreased motility could lead to spread of *C. botulinum* or its toxin from the colon to the small intestine, where the toxin is probably absorbed (81). The first sign of illness is invariably constipation, although this decreased frequency of bowel movements is often overlooked. Patients who are ultimately hospitalized usually develop lethargy and mild weakness with feeding difficulties, pooled oral secretions, and an altered cry. The baby eventually becomes floppy, loses head control, and may go on to develop ophthalmoplegia, ptosis, flaccid facial expression, dysphagia, other signs of cranial nerve deficits, and generalized muscular weakness. Respiratory insufficiency necessitating respiratory therapy also may occur, as in other forms of botulism (8, 9). There is a spectrum of clinical features in infant botulism, ranging from mild illness not requiring hospitalization to sudden death, and this syndrome accounts for a small percentage of cases of sudden infant death syndrome (8, 10). The differential diagnosis of infant botulism has included sepsis, myasthenia gravis, failure to thrive, benign congenital hypotonia, and a variety of other conditions.

C. difficile and Antibiotic Colitis

C. difficile is the major cause of antibiotic-associated diarrhea and pseudomembranous colitis (62). The organism has been isolated from diverse natural habitats, including soil, hay, sand, dung from various large mammals (cows, donkeys, and horses), and from the feces of dogs, cats, rodents, and humans (32). C. difficile is carried asymptomatically as part of the gastrointestinal flora of as many as half of all healthy neonates during the first year of life, decreasing to the adult carrier rate of less than 4% in children older than 2 years (22, 90, 95). Among adults who have received antibiotic therapy, carriage rates as high as 46% have been reported (34). Hospitalized patients frequently become colonized with this organism (34, 89, 90). Antimicrobial agents of all classes and several anticancer chemotherapeutic agents have been implicated in the development of C. difficile-associated diarrhea or pseudomembranous colitis (62). The most commonly reported agents are ampicillin, clindamycin, and cephalosporins (32).

C. difficile produces at least three potential virulence factors: an enterotoxin (toxin A) that induces a positive fluid response in the rabbit ligated ileal loop model, a cytotoxin (toxin B) that induces cytopathic effects in numerous tissue culture cell lines (61, 84), and a substance that inhibits bowel motility (49). Toxin A also produces cytopathic effect in cell cultures but is 1,000 times less potent than toxin B in the cell lines that have been tested (84). Most strains tested for toxin in vitro produce either both toxins or no detectable toxin, although there are reports of one toxin being present without the other (62); it is thought that toxins A and B are both important in the pathogenesis of C. difficile-associated disease, with toxin A being more important. Toxin A seems to interfere with the cytoskeletons of intestinal epithelial cells, thereby rendering the cells nonfunctional. Complete pathogenic mechanisms for C. difficile-associated disease have yet to be defined, despite active research in the area (24). Organisms other than C. difficile, including S. aureus and C. perfringens, are rarely agents of antibiotic-associated diarrhea (16, 64).

C. difficile is now one of the most commonly detected enteric pathogens and an important cause of nosocomial infections in hospitals and nursing homes (20, 37, 50, 67). Despite the common perception that C. difficile colitis is a problem only when antibiotics are used, patients with bowel stasis, those who have had bowel surgery, and those with no known risk factors can also contract C. difficile-induced gastrointestinal disease, although this form of the disease is less common than that associated with antibiotic use. Clinical symptoms range from mild diarrhea to toxic megacolon (36). In the most severe form, this disease may include bowel perforation and death, although these events are unlikely to occur in appropriately treated patients.

C. septicum and Bacteremia

C. septicum is isolated only rarely from the feces of healthy individuals, and it is not recovered often from blood cultures of otherwise healthy individuals. Increasingly, the isolation of this organism from blood cultures has been associated with other underlying disease processes (see below). Finding of this swarming organism on blood agar plates incubated under anaerobic conditions is often a precursor to diagnosis of a serious neoplastic disease. The portal of entry for C. septicum into the bloodstream is believed to be the ileocecal region of the bowel. Whether or not C. septicum is part of the indigenous microflora of this site, at least in low concentrations, has not been established. C. septicum has, however, been found in the lumens of 10 to 68% of normal appendixes and in none of 30 nonlumen appendiceal tissue samples from gangrenous or perforated appendices (12, 33).

Of particular interest has been the association of C. septicum in the blood with malignancies, especially leukemia and lymphoma or carcinoma of the large bowel. As many as 70 to 85% of patients whose blood cultures are positive for this organism have some underlying malignancy (14, 52, 54). Another clinically important association has been observed between C. septicum bacteremia, neutropenia, and enterocolitis involving the terminal ileum or cecum (13, 51, 59, 60, 70). The neutropenia has been related to both chemotherapy for leukemia or other neoplastic disease and to cyclic neutropenia and drug-induced agranulocytosis.

Not all patients with C. septicum bacteremia have malignancy or neutropenic enterocolitis. Patients with diabetes mellitus, severe atherosclerotic cardiovascular disease, and gas gangrene may also develop C. septicum bacteremia (57). The clinical importance of recognizing C. septicum bacteremia and starting appropriate treatment without delay cannot be overemphasized. Patients with this syndrome are usually acutely and gravely ill, frequently have high temperatures, and often show metastatic spread of myonecrosis to distant anatomic sites. Mortality rates are significant; however, appropriate antibiotic therapy with β-lactam antibiotics and aggressive surgical intervention early in the course of the illness may aid in avoiding a devastating outcome (57, 58).

C. tetani and Tetanus

The strictly toxigenic disease caused by C. tetani is often associated with puncture wounds that do not appear to be serious infections, particularly in animals. The organism and its spores can be isolated from a variety of sources, including soil and the intestinal contents of numerous animal species. Toxin is elaborated at the site of apparent minor trauma and rapidly binds to neural tissue, provoking a characteristic paralysis and spasms (see below). Tetanus is now largely a disease of nonimmunized animals and humans, since an effective toxoid has been in use for many years.

The clinical syndrome known as tetanus is an extremely dramatic illness produced by the action of a potent neurotoxin, tetanospasmin, which is elaborated by C. tetani. Tetanospasmin is a heat-labile protoplasmic protein that is released after the autolysis of C. tetani. The toxin binds to gangliosides in the central nervous system and blocks inhibitory impulses to the motor neurons, thus producing prolonged muscle spasms of both flexor and extensor muscles. Like the botulinal toxins, tetanospasmin acts at myoneural junctions to inhibit the release of acetylcholine. The binding sites for tetanospasmin and botulinal toxins differ; spasticity is characteristic of tetanus, and flaccid paralysis is characteristic of botulism (25). Tetanus is still largely a disease of the unimmunized in the United States and is reported most frequently from areas of the rural South. It would be unusual for the clinical laboratory to be requested to isolate *C. tetani* from a wound, since tetanus usually presents few diagnostic problems for the clinician. For more details on tetanus, see Furste et al. (31), Smith and Williams (79), Willis (94), and the excellent review by Dowell (25).

OTHER CLOSTRIDIAL DISEASES

Clostridium spp. and Bacteremia

At most tertiary care facilities (including Brigham and Women's Hospital and Indiana University Hospital), approximately 10 to 12% of all blood cultures contain clinically significant bacteria, and about 0.5 to 2% of the total isolates are *Clostridium* species (18). The clostridial species encountered most frequently in positive blood cultures is *C. perfringens*, which accounts for 20 to ≥50% of the clostridial isolates (1). Other *Clostridium* species found less frequently in positive blood cultures include *C. septicum*, *C. ramosum*, *C. bifermentans*, *C. sordellii*, *C. difficile*, *C. tertium*, *C. cadaveris*, *C. innocuum*, *C. sporogenes*, *C. sphenoides*, *C. butyricum*, and others (1, 2, 18, 39). Underlying conditions commonly associated with clostridial bacteremia include chronic alcoholism, sepsis following intra-abdominal surgery, necrosis of the small or large bowel, genitourinary tract disorders (including septic abortions), cardiovascular disease, pulmonary diseases, underlying malignancy, diabetes, and decubitus ulcers (18, 21, 39). In many instances of bacteremia involving *C. perfringens* and certain other clostridia, isolates are of doubtful significance and may represent contaminants. Isolates deposited transiently in the perianal area may be spread to a venipuncture site, or isolates may reflect transient bacteremia of no clinical significance. One large retrospective study found that half of all clostridia isolated from blood represented contaminants.

Additional Clostridial Species of Interest

Numerous other clostridial species are occasionally associated with disease. Species such as *C. novyi*, *C. histolyticum*, *C. bifermentans*, *C. sordellii*, *C. fallax*, *C. sporogenes*, and *C. tertium* have all been associated with histotoxic clostridial disease (see *C. perfringens* above). *C. sordellii* has also been cited as producing a cytotoxin similar to that of *C. difficile* (see *C. difficile* above).

COLLECTION, TRANSPORT, AND STORAGE OF CLINICAL SPECIMENS

As with other anaerobic bacteria, the proper selection, collection, and transport of clinical specimens are extremely important for the laboratory diagnosis of clostridial infections. For recommended collection and transport procedures in general, refer to chapter 3 of this Manual. Several tissue specimens should be taken from the active site of infections when gas gangrene is suspected, because clostridia are often not distributed uniformly in pathological lesions.

Specimens for Confirmation of *C. perfringens* Food-Borne Illness

For a laboratory confirmation of *C. perfringens* food-borne illness, most clinical laboratories need to use the services of a reference laboratory (e.g., local or state public health laboratory). Food and fecal specimens to be shipped to a reference laboratory should be collected in sterile, leakproof containers and transported at 4°C. Rectal swabs should be placed in an anaerobic transport container (chapter 3) and shipped at 4°C (73).

Specimens for *C. difficile* Culture and Toxin Assay

A single, freshly passed fecal specimen (ideally 10 to 20 ml of watery stool; minimum of 5.0 ml or 5 g) is the preferred specimen for *C. difficile* culture and toxin assay (35). To lessen the chance of obtaining positive culture results from patients merely colonized with the organism, we recommend that only liquid or unformed stool specimens be processed. Results of testing solid, formed stools are not likely to contribute to the diagnosis of *C. difficile*-associated disease. An exception could be made for epidemiologic surveys in which the objective is to determine the degree of *C. difficile* carriage in a population. Swab specimens are inadequate for the toxin assay because the volume of sample obtained is too small, although swabs have been used successfully to detect carriers during epidemiologic investigations (65). Other appropriate specimens include lumen contents and surgical or autopsy samples of the large bowel. Specimens should be transported in tightly sealed, leak-proof plastic or glass containers. To yield optimal recovery, stool specimens should be cultured within 2 h of collection; although spores will survive in refrigerated stool for several days, there will probably be a large decrease in the number of viable vegetative cells of *C. difficile* in refrigerated specimens. Stools may be placed in an anaerobic environment (anaerobic transport vial or swab) if culture must be performed after storage. Adequate recovery of *C. difficile* may be expected from stools stored at 5°C for up to 2 days. Specimens for toxin assay may be stored at 5°C for up to 3 days and should be frozen at −70°C if a longer delay before the assay is performed is anticipated. Freezing at −20°C results in a dramatic loss of cytotoxin activity.

Specimens for Suspected Neutropenic Enterocolitis Involving *C. septicum*

The specimens of choice for suspected neutropenic enterocolitis involving *C. septicum* are three blood cultures collected from three different venipuncture sites, stool (at least 25 g, or 25 ml if liquid), and lumen contents or tissue from the involved ileocecal area collected at surgery or autopsy and transported in tightly sealed leakproof containers. In addition, a biopsy sample of muscle (or an aspirate of fluid from the involved area, taken by needle and syringe) should be collected if the patient is also suspected of having myonecrosis.

Specimens for *C. botulinum* Culture and Toxin Assay

The clinical diagnosis of food-borne botulism can be confirmed by the demonstration of botulinal toxin in serum, feces, gastric contents, or vomitus or by the recovery of *C. botulinum* from the feces of the patient. The organism has been isolated only rarely from asymptomatic individuals during food-borne outbreaks (5). The demonstration of botulinal toxin and *C. botulinum* in suspected foods aids in determining the food item responsible for an outbreak, but it provides only indirect evidence to support a clinical diagnosis of botulism. Ideally, 15 to 20 ml of serum (not whole blood), 25 to 50 g of stool, and the suspect food(s) should be collected. Specimens to collect from patients with suspected wound botulism include serum, feces, tissue, exudate, or swab samples from the wound. When infant botulism is suspected, collect serum (2 ml) and as much stool as possible. In most instances, the diagnosis of infant botulism has been confirmed by the detection of the toxin or *C. botulinum* or both in feces. Toxin has been detected in serum only infrequently in infants with this diagnosis.

Most hospital laboratories are not properly equipped to process specimens from patients suspected of having botulism. The CDC, Atlanta, Ga., provides epidemiologic aid and emergency laboratory services 24 h/day every day of the week. The attending physician or state epidemiologist should notify the CDC immediately when there is a suspected case of botulism so that appropriate action can be taken to establish the diagnosis, initiate treatment, and investigate the potential outbreak. The appropriate CDC telephone number (24 h/day) for enteric disease epidemiologic aid is (404) 639-3753.

DIRECT EXAMINATION

The direct microscopic examination of clinical materials can provide extremely useful information for the physician in the diagnosis and treatment of clostridial infections. Gas gangrene is an extremely urgent situation, requiring a rapid clinical diagnosis. The direct examination of a Gram-stained smear of the wound may be of special aid to the clinician in establishing the diagnosis. Characteristic findings are the absence of inflammatory cells and other cellular outlines and the presence of clostridia in smears prepared from the central areas of the lesion. It is important to distinguish clostridial myonecrosis, which requires radical surgical excision or amputation of a limb, from anaerobic cellulitis, nonclostridial anaerobic cellulitis, or spreading fascitis (11, 30). Nonclostridial anaerobic cellulitis may involve anaerobic cocci, facultatively anaerobic cocci, *Bacteroides* spp., *Fusobacterium* spp., and/or other microbial species. Cell outlines of striated muscle cells and granulocytes remain intact in the latter condition.

The Gram stain is satisfactory for the direct examination of a specimen. Special note should be taken of gram-positive rods with or without spores, because sporulation in tissue is not common with the two most frequently encountered species in wound and abscess materials, *C. perfringens* and *C. ramosum*. *C. perfringens* usually appears as large, relatively short, fat, gram-positive rods in tissue smears; the cells of *C. ramosum* are more slender, longer, and often curved (55). *C. perfringens* may or may not be encapsulated in smears from wounds; capsules usually are present in smears of endometrial specimens from postabortion *C. perfringens* infections. Special spore stains offer no advantage over Gram stains for the demonstration of spores. Examination with a phase microscope may be helpful if the spores are close to maturity. If spores are present, note their shapes (spherical or oval) and positions (terminal, subterminal, or central) in the cells. An excellent medium for the demonstration of spores is a chopped-meat agar slant. Incubate the culture anaerobically at 5 to 7°C below the optimum temperature for growth of clostridia. For most species, 30°C is satisfactory, but 37°C is better for inducing sporulation of *C. perfringens*. If spores are not visible, their presence may be deduced by subjecting a suspension of the isolate to heat (80°C) or ethanol treatment (described below).

Tests for Diagnosis of *C. difficile* Disease

There is considerable controversy about the optimal method of detecting *C. difficile*-associated disease. Currently, the three major assay methods include enzyme immunoassay for toxins A and B, latex agglutination assays for toxins A and B, and cell culture cytotoxicity assays with specific neutralization for toxin B. Culture of the organism from feces, although possible, is not considered diagnostic, because a significant portion of the adult population carries this organism. The cytotoxicity assay, coupled to neutralization of cytopathic effects with antisera directed at toxin B, is considered the "gold standard." Most laboratories that perform cell cultures for other purposes continue to use this assay for detecting *C. difficile*. The newer latex and enzyme immunoassay methods appear to be reasonably sensitive, are easy to use, but vary in their specificity and correlation with the cytotoxicity assay. For details of the various toxin assays available for toxins A and B, consult the package inserts for the latex agglutination and enzyme immunoassay kits and the *Clinical Microbiology Procedures Handbook* (48). The standard cytotoxicity assay is also commercially available, along with neutralizing antisera and positive control material (Bartels/Scientific Products).

CULTURE AND ISOLATION

A summary of useful procedures is provided below. For a more detailed treatment of the topic, the *Clinical Microbiology Procedures Handbook* (48) for anaerobic bacteriology provides detailed methodologies applicable to most clinical situations. Clostridia usually produce good growth on commercially available anaerobe blood agar and phenylethyl alcohol blood agar (PEA) after 1 to 2 days of incubation. Brucella 5% sheep blood agar, Columbia agar, or brain heart infusion agar supplemented with yeast extract, vitamin K_1, and hemin may also be used as the nonselective blood agar medium; colonial characteristics vary on these different media. A few species, such as *C. perfringens*, form colonies after overnight incubation. When clostridia are suspected in wound or abscess specimens (e.g., from gas gangrene), it is recommended that egg yolk agar (modified McClung-Toabe formula [26]) or neomycin egg yolk (NEY) agar be inoculated in addition to blood agar and PEA. To prepare NEY medium, neomycin is added to achieve a final concentration of 100 μg/ml. Neomycin is heat stable and can be added before the medium is autoclaved. Neomycin can also be added to anaerobe blood agar at the same concentration. Neomycin is added to inhibit some of the facultatively anaerobic gram-negative bacilli; thus, NEY medium is moderately selective.

After incubation, examine the blood agar and PEA cultures with a dissecting microscope, noting particularly the hemolysis pattern, colony structure, and any evidence of swarming or of motile colonies. Examine the egg yolk agar or NEY culture for evidence of lecithinase (phospholipase C) or lipase production. Lecithinase activity is indicated in either medium by the development of an insoluble, opaque, whitish precipitate within the agar. An iridescent sheen or oil-on-water appearance (pearly layer) on the surface growth indicates lipase activity. Proteolysis, the third reaction that can be seen on egg yolk agar or NEY medium, is indicated by a zone of translucent clearing in the medium around the colonies. In addition to the modified McClung-Toabe egg yolk agar formulation (26), these same reactions can be determined on the hemin-supplemented egg yolk agar formulation recommended by Summanen et al. (85) or on Lombard-Dowell egg yolk agar (Presumpto plate; 27, 55).

If swarming growth, such as that associated with swarming *Proteus* strains, has covered the surface of the agar medium, inoculate another blood agar plate and incubate it anaerobically for only 18 to 24 h. Subculture from the colonies as soon as the plates are taken from the anaerobe jar. If swarming is again observed, subculture the isolate to a PEA plate to inhibit swarming. Alternatively, anaerobic blood agar made up with 4% (or higher-concentration) agar may be useful. This mixture is known as stiff blood agar (26). When isolated colonies can be picked, subculture them to chopped-meat medium, incubate the culture overnight, and use it for the inoculation of differential media. In addition, inoculate chopped-meat–carbohydrate broth (47) for gas-liquid chromatography.

Spore Selection Techniques

Most clinical specimens from which clostridia might be isolated, other than blood (feces, material from wounds and abscesses, muscle and other soft tissue, surgical specimens, necropsy materials, etc.), yield a mixture of non-spore-forming bacteria. A spore selection technique with heat or alcohol treatment is a useful technique for isolating spore-formers while inhibiting nonsporeformers. Spores of clostridia (or of *Bacillus* spp.) resist the heat or alcohol treatment, whereas vegetative cells are killed. After treatment, spores germinate in appropriate broth medium and produce growth under appropriate conditions. Heat treatment alone should not be relied upon, because the spores of various *Clostridium* spp. vary in their degrees of resistance to heat. Treatment of specimens with at least 50% ethanol (final concentration) for 1 h aids in the selective isolation of clostridia from specimens obtained during mixed infections and circumvents the problem of different spore tolerances to heat (56). The major problem to avoid with alcohol treatment is the presence of solid specimen particles that are not adequately penetrated by the alcohol. When the specimen is not sufficiently homogenized, the vegetative cells of nonsporeformers may not be inhibited.

Alcohol Treatment

To a 1-ml sample of a fecal suspension, homogenate of a wound or exudate, etc., in a sterile screw-cap tube, add an equal volume of absolute (or 95%) ethyl alcohol (56). Mix the specimen gently at room temperature (22 to 25°C) for 1 h. An Ames Aliquot Mixer (Miles Laboratories, Inc., Elkhart, Ind.) is a convenient way to provide continuous mixing. Subculture the treated material, and inoculate chopped-meat–glucose (or chopped-meat–glucose–starch) medium, anaerobe blood agar, or egg yolk agar. Incubate the culture, and inspect it for growth. For stool specimens, it is often advantageous to alcohol treat separate 1-ml samples from a series of 1:10 dilutions. This treatment helps the alcohol penetrate solid particles.

Heat Treatment

For heat treatment (26), preheat a tube of chopped-meat–glucose–starch medium in an 80°C water bath for 5 min, and then add 1 ml of sample suspension. Heat the culture for 10 min, remove the tube, and cool it in cold water. Subculture the treated sample suspension into an unheated tube of chopped-meat–glucose–starch medium and onto an anaerobe blood agar and egg yolk agar plate. Incubate the culture anaerobically, and examine it for growth.

Laboratory Investigation of *C. perfringens* Food-Borne Illness

Methods for the enumeration of *C. perfringens* in foods and *C. perfringens* spores in feces with egg yolk-free tryptose-sulfite-cycloserine agar are described in detail by Hauschild and colleagues (43–45). Although methods for detection of *C. perfringens* enterotoxin in feces have been described previously and are of considerable interest, these assays are still considered experimental and are generally not available in clinical laboratories (63, 73, 75).

IDENTIFICATION

There are several alternatives to the phenotypic characterization of clostridial species in the laboratory. One of the traditional methods has been to use prereduced anaerobically sterilized (PRAS) media for determination of fermentation profiles and other characteristics. To identify clostridia by this method, inoculate the PRAS tubed media (available commercially; as listed in Table 3) prepared as described by Holdeman et al. (47). Although the CHO-based media of Dowell and Hawkins (26) provide an excellent system for characterizing clostridia and can be used, Table 3 is based on results obtained with PRAS media, and a few of the reactions differ. Thus, if CHO-based media are used, the CDC publications on anaerobic bacteriology should be consulted (26). Incubate the cultures for 24 to 72 h at 35 to 37°C; overnight incubation is often sufficient for many clostridia in PRAS media if growth is rapid and abundant. Examine Gram stains of the chopped-meat culture to determine the presence, positions, and shapes of the spores. If spores are not found, inoculate a tube of chopped-meat medium, heat the culture at 70°C for 10 min, and incubate the tube. Growth in this heated tube usually indicates the presence of spores, although none may be apparent microscopically. Alternatively, an alcohol spore selection technique may be helpful for those clostridia with heat-sensitive spores (56).

Determine whether carbohydrate fermentation has occurred by measuring the pH (47). Determine the metabolic products from chopped-meat–carbohydrate or peptone-yeast-glucose broth culture by gas chromatography (see chapter 12 of this Manual). Results will not be identical if peptone-yeast extract-glucose is the culture medium (26, 47). The identification of *Clostridium* species can be made without gas-liquid chromatography, but such identification usually involves more time. The information listed in Table 3 will serve to identify most of the clostridial species commonly isolated from clinical specimens. Information on additional tests and descriptions of differential characteristics of additional species can be found elsewhere (19, 26, 47, 55, 79, 85).

An alternative to the PRAS medium described above is one of the commercially available preformed enzyme or microfermentation systems described in the *Clinical Microbiology Procedures Handbook* (48). These commercially packaged systems are useful for the identification of many clostridial species of clinical interest. Similarly, long-chain free fatty acid analysis, described in chapter 12 of this Manual, is an accurate and sensitive method for the identification of isolates by using cellular fatty acid composition as a means for characterization.

Toxin tests are necessary for the identification of a few species. *C. sporogenes* cannot be differentiated with certainty from the proteolytic group I strains of *C. botulinum* unless toxin tests are used. A few strains of group III *C. botulinum* produce lecithinase as well as lipase and are difficult to distinguish from *C. novyi* type A except by toxin tests or by the use of a *C. novyi* fluorescent-antibody conjugate (79). To test for toxin, inoculate two tubes of

TABLE 3 Differential characteristics of commonly encountered clostridia[a]

Species	Spores	Egg yolk agar LEC	Egg yolk agar LIP	Growth on aerobic blood agar	Gelatin hydrolysis	Milk digestion	Indole production	Glucose	Maltose	Lactose	Sucrose	Salicin	Mannitol	Principal metabolic products on PYG or CMC	Other
C. bifermentans	OS	+	-	-	+	+	+	+	w/-	-	-	-	-	A, (p), (ib), (b), (iv), (ic)	Urease negative
C. botulinum[b]															
Group I[c]	OS	-	+	-	+	+	-	+	-/w	-	-	-	-	A, (p), ib, B, IV, (v), (ic)	
Group II[c]	OS	-/+	+	-	+	+	-/+	+	V	-	+/w	-	-	A, P, B, (v)	
Group III[c]	OS	-	+	-	+	-	-	+	+/w	+	+	+	-/+	A, B	
C. butyricum	OS	-	-	-	-	-	-	+	+	+	+	+	-	A, B	
C. cadaveris	OT	-	-	-	+	+	+	+	-	-	-	-	-/+	A, (p), B (from PY = A, p, ib, v, iv)	
C. chauvoei[d]	OS	-	-	-	+	-	-	+	+/w	+/w	+/w	-	-	A, B	Spores seldom observed; usually gram negative
C. clostridioforme	OS	-	-	-	-	-	-/+	+	+/w	+/-	+	+/-	-	A	
C. difficile	OS	-	-	-	+	-	-	+	-	-	-	-/w	+/-	A, (p), ib, B, iv, (v), ic	
C. histolyticum	OS	-	-	V	+	+	-	-	-	-	-	+	-	A, B	Aerotolerant
C. innocuum[e]	OT	-	-	-	-	-	-	+	-	-	+	+	+	A	
C. limosum	OS	+	-	-	+	+	-	+	-	-	-	-	-	A, P, B	
C. novyi A[e]	OS	+	+	-	+	-	-	+	V	+	-	-/+	-	A, P, B	
C. novyi B[e]	OS	+	+	-	+	+	+/-	+	V	+	-	-	-	A, P, B	
C. paraputrificum	OT	-	-	-	-	-	-	+	+	+	+	+	-	A, B	
C. perfringens[e]	OS	+	-	-	+	+	-	+	+	+	+	-	-	A, (p), B	Spores seldom observed; double zone of hemolysis
C. ramosum[e]	R/OT	-	-	-	-	-	-	+	+	+	+	+	+/-	A	Spores seldom observed; frequently gram negative
C. septicum	OS	-	-	-	+	+	-	+	+	+	-	V	-	A, (p), B	Spreading colony
C. sordellii	OS	+	-	-	+	-	+	+	w/+	-	-	-	-	A, (p), (ib), (iv), (ic)	Usually urease positive
C. sphenoides	RS/T	-	-	-	+	-	+	+	+	w/+	w/-	w/+	w/+	A	Usually gram negative
C. sporogenes	OS	-	+	-	+	+	-	+	-/w	-	-	-/+	-	A, (p), ib, B, iv, (v), (ic)	
C. subterminale	OS	-/+	-	-	+	+	-	-	+	+	-	-	-	A, (p), ib, B, IV, (ic)	
C. tertium	OT	-	-	+	-	-	-	+	+	+	+	+	+/w	A, B	Aerotolerant
C. tetani	RT	-	-	-	+	+/-	V	-	-	-	-	-	-	A, p, B	

[a]From Allen and Baron (1a). Based on the use of PRAS media as described by Holdeman et al. (47). Key: +, positive reaction; -, negative reaction; V, variable reaction; w, weak reaction; /, either/or; O, oval; R, round; S, subterminal; T, terminal; LEC, lecithinase production; LIP, lipase production. Fermentation products: A, acetate; B, butyrate; F, formate; IB, isobutyrate; IC, isocaproate; IV, isovalerate; P, propionate; V, valerate. (), May or may not be present. Capital letters indicate major peaks; lowercase letters indicate minor peaks. PYG, peptone-yeast extract-glucose medium; CMC, chopped-meat–carbohydrate medium.

[b]Group I contains proteolytic strains (A, B, and F), group II contains nonproteolytic strains (B, E, and F), and group III contains types C and D, and group III contains saccharolytic strains (B, E, and F).

[c]Toxin neutralization test required for identification.

[d]Pathogenic for herbivores.

[e]C. innocuum, C. perfringens, and C. ramosum are nonmotile.

chopped-meat–glucose medium. Incubate one tube at 37°C overnight, and incubate the other tube at 37°C for 3 days. Test the overnight culture first; if no toxin is found, test the 3-day culture. Centrifuge the culture, and place 1.2-ml amounts of the supernatant in several tubes. Add 0.3 ml of appropriate antiserum per tube for the various species suspected. Let the well-mixed suspensions stand for 30 min at room temperature or at 37°C, and then inject 0.5-ml portions of control supernatant (without antiserum) and antiserum mixtures intraperitoneally into each of two mice. Observe the mice for 3 days, and record deaths that occur. Only specific sera for laboratory testing should be used for toxin identification; therapeutic sera are often unsatisfactory because they may contain antibodies to toxins of species other than those listed on the label. Diagnostic clostridial antisera are available from Wellcome Reagents Ltd., Wellcome Research Laboratories, Beckenham, England BR3 3BS. The analysis of long-chain free fatty acids provides an alternative to the method described above, since the various types of *C. botulinum* can be identified on the basis of differences in their cellular fatty acid profiles.

SUSCEPTIBILITY TO ANTIMICROBIAL AGENTS

Penicillin G shows excellent activity against most but not all strains of *C. perfringens* and has traditionally been considered the antibiotic of choice for the clostridia in general (71, 72). However, resistance is increasing among *C. perfringens* to the extent that alternative antimicrobial agents may need to be considered (11). β-Lactamase has not been demonstrated in *C. perfringens*. As referred to by Hecht et al., resistance to penicillin in *C. perfringens* involves a decreased affinity of penicillin-binding protein 1 for penicillin (46). Resistance to penicillin is especially common in *C. ramosum*, *C. clostridioforme*, and *C. butyricum* (28, 71); these species produce β-lactamases that are induced by β-lactam antibiotics (46).

Although clindamycin is still highly active against most species of commonly encountered anaerobic bacteria in the United States, a number of clostridial species are frequently resistant to it. Resistance to this antibiotic occurs in some strains of *C. ramosum*, *C. difficile*, *C. tertium*, *C. subterminale*, *C. innocuum*, *C. sporogenes*, and *C. perfringens* (93).

Chloramphenicol, piperacillin, metronidazole, imipenem, and combinations of β-lactam drugs with β-lactam inhibitors (e.g., ampicillin-sulbactam) are active against nearly all of the clostridia, with only a few exceptions (11, 28, 71). The clostridia have shown variable resistance to the cephalosporins and tetracyclines, and they are usually resistant to the aminoglycosides. The quinolones, of which ciprofloxacin is the most active, have not shown remarkable activity against the clostridia (71). Many clostridia other than *C. perfringens* are resistant to cefoxitin, cefotaxime, ceftazidime, ceftizoxime, cefoperazone, and other third-generation β-lactam drugs (28, 71).

Severe *C. difficile*-associated intestinal disease is usually treated with oral vancomycin or metronidazole, although most strains of *C. difficile* are susceptible to a number of antimicrobial agents in vitro (including penicillins, tetracycline, and quinolones). For patients unable to tolerate oral antibiotics but requiring therapy, parenteral vancomycin or metronidazole is recommended (32). Antibiotic therapy often results in relapse of disease, so discontinuation of the offending agent or change to an agent less likely to

cause diarrhea should be considered the primary intervention of choice (32).

REFERENCES

1. **Allen, S. D.** Unpublished data.
1a. **Allen, S. D., and E. J. Barron.** 1991. *Clostridium*, p. 505–521. In A. Balows, W. J. Hausler, Jr., K. L. Herrmann, H. D. Isenberg, and H. J. Shadomy (ed.), *Manual of Clinical Microbiology*, 5th ed. American Society for Microbiology, Washington, D.C.
2. **Alpern, R. J., and V. R. Dowell, Jr.** 1971. Nonhistotoxic clostridial bacteremia. *Am. J. Clin. Pathol.* **55:**717–722.
3. **Altemeier, W. A., and W. R. Culbertson.** 1948. Acute nonclostridial crepitant cellulitis. *Surg. Gynecol. Obstet.* **87:**206–212.
4. **Altemeier, W. A., and W. L. Furste.** 1947. Gas gangrene. *Surg. Gynecol. Obstet.* **84:**507–523.
5. **Anonymous.** 1979. *Botulism in the United States, 1899–1977. Handbook for Epidemiologists, Clinicians and Laboratory Workers.* Center for Disease Control, Atlanta.
6. **Anonymous.** 1983. *Foodborne Disease Outbreaks Annual Summary—1981.* Centers for Disease Control, Atlanta.
7. **Anonymous.** 1989. Summary of notifiable diseases, United States—1988. *Morbid. Mortal. Weekly Rep.* **33:**165–166.
8. **Arnon, S. S.** 1989. Infant botulism, p. 601–609. In S. M. Finegold and W. L. George (ed.), *Anaerobic Infections in Humans.* Academic Press, Inc., New York.
9. **Arnon, S. S., and J. Chin.** 1979. The clinical spectrum of infant botulism. *Rev. Infect. Dis.* **1:**614–624.
10. **Arnon, S. S., K. Damus, and J. Chin.** 1981. Infant botulism: epidemiology and relation to sudden infant death syndrome. *Epidemiol. Rev.* **3:**45–66.
11. **Bartlett, J. G.** 1990. Gas gangrene (other *Clostridium*-associated diseases), p. 1850–1860. In G. L. Mandell, R. G. Douglas, Jr., and J. E. Bennett (ed.), *Principles and Practice of Infectious Diseases*, 3rd ed. Churchill Livingstone, Inc., New York.
12. **Bennion, R. S., E. J. Baron, J. E. Thompson, Jr., J. Downes, P. Summanen, D. A. Talan, and S. M. Finegold.** 1990. The bacteriology of gangrenous and perforated appendicitis—revisited. *Ann. Surg.* **221:**165–171.
13. **Bignold, L. P., and H. P. B. Harvey.** 1979. Necrotizing enterocolitis associated with invasion by *Clostridium septicum* complicating cyclic neutropaenia. *Aust. N.Z. J. Med.* **9:**426–429.
14. **Bodey, G. P., S. Rodriguez, V. Fainstein, and L. S. Elting.** 1991. Clostridial bacteremia in cancer patients. A 12-year experience. *Cancer* **67:**1928–1942.
15. **Borriello, S. P., F. E. Barclay, A. R. Welch, M. F. Stringer, G. N. Watson, R. K. T. Williams, D. V. Seal, and K. Sullens.** 1985. Epidemiology of diarrhoea caused by enterotoxigenic *Clostridium perfringens*. *J. Med. Microbiol.* **20:**363–372.
16. **Borriello, S. P., H. E. Larson, A. R. Welch, F. Barclay, M. F. Stringer, and B. A. Bartholomew.** 1984. Enterotoxigenic *Clostridium perfringens*: a possible cause of antibiotic-associated diarrhea. *Lancet* **i:**305–307.
17. **Brock, T. D., and M. T. Madigan.** 1988. *Biology of Microorganisms*, 5th ed. Prentice-Hall, Inc., Englewood Cliffs, N.J.
18. **Brook, I.** 1989. Anaerobic bacterial bacteremia: 12-year experience in two military hospitals. *J. Infect. Dis.* **160:**1071–1075.
19. **Cato, E. P., W. L. George, and S. M. Finegold.** 1986. Genus *Clostridium* Prazmowski, p. 1141–1200. In P. H. A. Sneath, N. S. Mair, M. E. Sharpe, and J. G. Holt (ed.), *Bergey's Manual of Systematic Bacteriology*, vol. 2. The Williams & Wilkins Co., Baltimore.
20. **Clabots, C. R., S. Johnson, M. M. Olson, L. R. Peterson, and D. N. Gerding.** 1992. Acquisition of *Clostridium difficile* by hospitalized patients: evidence for colonized new admission as a source of infection. *J. Infect. Dis.* **166:**561–567.
21. **Coleman, N., G. Spiers, J. Khan, V. Broadbent, D. G. Wight, and R. E. Warren.** 1993. Neutropenic enterocolitis

associated with *Clostridium tertium*. *J. Clin. Pathol.* **42:**180–183.

22. **Cooperstock, M.** 1988. *Clostridium difficile* in infants and children, p. 45–64. *In* R. D. Rolfe and S. M. Finegold (ed.), *Clostridium difficile: Its Role in Intestinal Disease.* Academic Press, Inc., New York.

23. **Decker, W. H., and W. Hall.** 1966. Treatment of abortions infected with *Clostridium welchii*. *Am. J. Obstet. Gynecol.* **95:**394–399.

24. **Donta, S. T.** 1988. Mechanism of action of *Clostridium difficile* toxins, p. 169–181. *In* R. D. Rolfe and S. M. Finegold (ed.), *Clostridium difficile: Its Role in Intestinal Disease.* Academic Press, Inc., New York.

25. **Dowell, V. R., Jr.** 1984. Botulism and tetanus: selected epidemiologic and microbiologic aspects. *Rev. Infect. Dis.* **6**(Suppl. 1):S202–S207.

26. **Dowell, V. R., Jr., and T. M. Hawkins.** 1977. *Laboratory Methods in Anaerobic Bacteriology: CDC Laboratory Manual.* Department of Health, Education, and Welfare publication no. (CDC) 78-8272. Center for Disease Control, Atlanta.

27. **Dowell, V. R., Jr., and G. L. Lombard.** 1982. Differential agar media for identification of anaerobic bacteria, p. 258–262. *In* R. C. Tilton (ed.), *Rapid Methods and Automation in Microbiology.* American Society for Microbiology, Washington, D.C.

28. **Finegold, S. M.** 1989. Therapy of anaerobic infections, p. 793–818. *In* S. M. Finegold and W. L. George (ed.), *Anaerobic Infections in Humans.* Academic Press, Inc., New York.

29. **Finegold, S. M., and W. L. George (ed.).** 1989. *Anaerobic Infections in Humans.* Academic Press, Inc., New York.

30. **Finegold, S. M., W. L. George, and M. E. Mulligan.** 1986. *Anaerobic Infections.* Year Book Medical Publishers, Inc., Chicago.

31. **Furste, W., A. Aquirre, and D. J. Knoepfler.** 1989. Tetanus, p. 611–627. *In* S. M. Finegold and W. L. George (ed.), *Anaerobic Infections in Humans.* Academic Press, Inc., New York.

32. **George, W. L.** 1989. Antimicrobial agent-associated diarrhea and colitis, p. 661–678. *In* S. M. Finegold and W. L. George (ed.), *Anaerobic Infections in Humans.* Academic Press, Inc., New York.

33. **George, W. L., and S. M. Finegold.** 1985. Clostridia in the human gastrointestinal flora, p. 1–37. *In* S. P. Borriello (ed.), *Clostridia in Gastrointestinal Disease.* CRC Press, Inc., Boca Raton, Fla.

34. **George, W. L., R. D. Rolfe, G. M. Harding, R. Klein, C. W. Putman, and S. M. Finegold.** 1982. *Clostridium difficile* and cytotoxin in feces of patients with antimicrobial agent-associated pseudomembranous colitis. *Infection* **10:**205–207.

35. **Gerding, D. N., and J. S. Brazier.** 1993. Optimal methods for identifying *Clostridium difficile* infections. *Clin. Infect. Dis.* **16**(Suppl. 4):S439–S442.

36. **Gerding, D. N., R. L. Gebhard, H. W. Sumner, and L. R. Peterson.** 1988. Pathology and diagnosis of *Clostridium difficile* disease, p. 259–286. *In* R. D. Rolfe and S. M. Finegold (ed.), *Clostridium difficile: Its Role in Intestinal Disease.* Academic Press, Inc., New York.

37. **Gilligan, P. H., L. R. McCarthy, and V. M. Genta.** 1981. Relative frequency of *Clostridium difficile* in patients with diarrheal disease. *J. Clin. Microbiol.* **14:**26–31.

38. **Gorbach, S. L.** 1979. Other *Clostridium* species (including gas gangrene), p. 1876–1885. *In* G. L. Mandell, R. G. Douglas, Jr., and J. E. Bennett (ed.), *Principles and Practice of Infectious Diseases.* John Wiley & Sons, Inc., New York.

39. **Gorbach, S. L., and H. Thadepalli.** 1975. Isolation of *Clostridium* in human infections: evaluation of 114 cases. *J. Infect. Dis.* **131:**S81–S85.

40. **Hall, J. D., L. M. McCroskey, B. J. Pincomb, and C. L. Hatheway.** 1985. Isolation of an organism resembling *Clostridium barati* which produces type F botulinal toxin from an infant with botulism. *J. Clin. Microbiol.* **21:**654–655.

41. **Hatheway, C. L.** 1988. Botulism, p. 111–133. *In* A. Balows,

W. J. Hausler, Jr., M. Ohashi, and A. Turano (ed.), *Laboratory Diagnosis of Infectious Diseases: Principles and Practice,* vol. 1. Springer-Verlag, New York.

42. **Hatheway, C. L.** 1990. Toxigenic clostridia. *Clin. Microbiol. Rev.* **3:**66–98.

43. **Hauschild, A. H. W.** 1975. Criteria and procedures for implicating *Clostridium perfringens* in food-borne outbreaks. *Can. J. Public Health* **66:**388–392.

44. **Hauschild, A. H. W., and R. Hilsheimer.** 1974. Enumeration of food-borne *Clostridium perfringens* in egg yolk-free tryptose-sulfite-cycloserine agar. *Appl. Microbiol.* **27:**521–526.

45. **Hauschild, A. H. W., R. Hilsheimer, and D. W. Griffith.** 1974. Enumeration of fecal *Clostridium perfringens* spores in egg yolk-free tryptose-sulfite-cycloserine agar. *Appl. Microbiol.* **27:**527–530.

46. **Hecht, D. W., M. H. Malany, and F. P. Tally.** 1989. Mechanisms of resistance and resistance transfer in anaerobic bacteria, p. 755–769. *In* S. M. Finegold and W. L. George (ed.), *Anaerobic Infections in Humans.* Academic Press, Inc., New York.

47. **Holdeman, L. V., E. P. Cato, and W. E. C. Moore (ed.).** 1977. *Anaerobe Laboratory Manual,* 4th ed. Virginia Polytechnic Institute and State University, Blacksburg.

48. **Isenberg, H. D. (ed.).** 1992. *Clinical Microbiology Procedures Handbook,* vol. 1, p. 2.0.1–2.12.6. American Society for Microbiology, Washington, D.C.

49. **Justus, P. G., J. L. Martin, D. A. Goldberg, N. S. Taylor, J. G. Bartlett, R. W. Alexander, and J. R. Mathias.** 1982. Myoelectric effects of *Clostridium difficile*: motility-altering factors distinct from its cytotoxin and enterotoxin in rabbits. *Gastroenterology* **83:**836–843.

50. **Kim, K.-H., R. Fekety, D. H. Batts, D. Brown, M. Cudmore, J. Silva, Jr., and D. Waters.** 1981. Isolation of *Clostridium difficile* from the environment and contacts of patients with antibiotic-associated colitis. *J. Infect. Dis.* **143:**42–44.

51. **King, A., A. Rampling, D. G. D. Wright, and R. E. Warren.** 1984. Neutropenic enterocolitis due to *Clostridium septicum* infection. *J. Clin. Pathol.* **37:**335–343.

52. **Kirchner, J. T.** 1991. *Clostridium septicum* infection. Beware of associated cancer. *Postgrad. Med.* **90:**157–160.

53. **Kliegman, R. M., and A. A. Fanaroff.** 1984. Necrotizing enterocolitis. *N. Engl. J. Med.* **310:**1093–1103.

54. **Kolbeinsson, M. E., W. D. Holder, Jr., and S. Aziz.** 1991. Recognition, management, and prevention of *Clostridium septicum* abscess in immunosuppressed patients. *Arch. Surg.* **125:**642–645.

55. **Koneman, E. W., S. D. Allen, V. R. Dowell, Jr., W. M. Janda, H. M. Sommers, and W. C. Winn, Jr.** 1988. *Color Atlas and Textbook of Diagnostic Microbiology,* 3rd ed. J. B. Lippincott Co., Philadelphia.

56. **Koransky, J. R., S. D. Allen, and V. R. Dowell, Jr.** 1978. Use of ethanol for selective isolation of sporeforming microorganisms. *Appl. Environ. Microbiol.* **35:**762–765.

57. **Koransky, J. R., M. D. Stargel, and V. R. Dowell, Jr.** 1979. *Clostridium septicum* bacteremia: its clinical significance. *Am. J. Med.* **66:**63–66.

58. **Kornbluth, A. A., J. B. Danzig, and L. H. Bernstein.** 1989. *Clostridium septicum* infection and associated malignancy. *Medicine* **68:**30–37.

59. **Kudsk, K. A.** 1992. Occult gastrointestinal malignancies producing metastatic *Clostridium septicum* infections in diabetic patients. *Surgery* **112:**765–772.

60. **Lev, R., and K. G. Sweeney.** 1993. Neutropenic enterocolitis. Two unusual cases with review of the literature. *Arch. Pathol. Lab. Med.* **117:**524–527.

61. **Lyerly, D. M., D. E. Lockwood, S. H. Richardson, and T. D. Wilkins.** 1982. Biological activities of toxins A and B of *Clostridium difficile*. *Infect. Immun.* **35:**1147–1150.

62. **Lyerly, D. M., D. E. Lockwood, S. H. Richardson, and T. D. Wilkins.** 1988. *Clostridium difficile*: its disease and toxins. *Clin. Microbiol. Rev.* **1:**1–18.

63. **Mahony, D. E., E. Gilliatt, S. V. Dawson, E. Stockdale, and S. H. S. Lee.** 1989. Vero cell assay for rapid detection of

Clostridium perfringens enterotoxin. *Appl. Environ. Microbiol.* **55**:2141–2143.

64. **McDonald, M., P. Ward, and K. Harvey.** 1982. Antibiotic-associated diarrhoea and methicillin-resistant *Staphylococcus aureus. Med. J. Aust.* **1**:462–464.

65. **McFarland, L. V., M. B. Coyle, W. H. Kremer, and W. E. Stamm.** 1987. Rectal swab cultures for *Clostridium difficile* surveillance studies. *J. Clin. Microbiol.* **25**:2241–2242.

66. **McIane, B. A., and J. T. Snyder.** 1987. Development and preliminary evaluation of a slide latex agglutination assay for detection of *Clostridium perfringens* type A enterotoxin. *J. Immunol. Methods* **100**:131–136.

67. **Mulligan, M. E., L. R. Peterson, R. Y. Y. Kwok, C. R. Clabots, and D. N. Gerding.** 1988. Immunoblots and plasmid fingerprints compared with serotyping and polyacrylamide gel electrophoresis for typing *Clostridium difficile. J. Clin. Microbiol.* **26**:41–46.

68. **Murrell, T. G. C.** 1989. Enteritis necroticans, p. 639–659. *In* S. M. Finegold and W. L. George (ed.), *Anaerobic Infections in Humans.* Academic Press, Inc., New York.

69. **Nichols, R. L., and J. W. Smith.** 1975. Gas in the wound; what does it mean? *Surg. Clin. N. Am.* **55**:1289–1296.

70. **Rifkin, G. D.** 1980. Neutropenic enterocolitis and *Clostridium septicum* infection in patients with agranulocytosis. *Arch. Intern. Med.* **140**:834–835.

71. **Rosenblatt, J. E.** 1989. Antimicrobial susceptibility of anaerobic bacteria, p. 731–753. *In* S. M. Finegold and W. L. George (ed.), *Anaerobic Infections In humans.* Academic Press, Inc., New York.

72. **Schwartzman, J. D., L. B. Reller, and W.-L. Wang.** 1977. Susceptibility of *Clostridium perfringens* isolated from human infection to twenty antibiotics. *Antimicrob. Agents Chemother.* **11**:695–697.

73. **Shandera, W. X., C. O. Tacket, and P. A. Blake.** 1983. Food poisoning due to *Clostridium perfringens* in the United States. *J. Infect. Dis.* **147**:163–170.

74. **Simpson, L. L.** 1989. Peripheral actions of the botulinum toxins, p. 153–178. *In* L. L. Simpson (ed.), *Botulinum Neurotoxin and Tetanus Toxin.* Academic Press, Inc., New York.

75. **Sisson P. R., J. M. Kramer, M. M. Brett, R. Freeman, R. J. Gilbert, and N. J. Lightfoot.** 1992. Application of pyrolysis mass spectrometry to the investigation of outbreaks of food poisoning and non-gastrointestinal infection associated with *Bacillus* species and *Clostridium perfringens. Int. J. Food Microbiol.* **17**:57–66.

76. **Smith, L. D.** 1978. The occurrence of *Clostridium botulinum* and *Clostridium tetani* in the soil of the United States. *Health Lab. Sci.* **15**:74–80.

77. **Smith, L. D.** 1979. Virulence factors of *Clostridium perfringens. Rev. Infect. Dis.* **1**:254–260.

78. **Smith, L. D., and H. Sugiyama.** 1988. *Botulism: the Organism, Its Toxins, the Disease*, 2nd ed. Charles C Thomas, Publisher, Springfield, Ill.

79. **Smith, L. D., and B. L. Williams.** 1984. *The Pathogenic Anaerobic Bacteria*, 3rd ed. Charles C Thomas, Publisher, Springfield, Ill.

80. **Sonnabend, O., W. Sonnabend, R. Heinzle, T. Sigrist, R. Dirnhofer, and U. Krech.** 1981. Isolation of *Clostridium bot-ulinum* type G and identification of type G botulinal toxin in humans: report of five sudden unexpected deaths. *J. Infect. Dis.* **143**:22–27.

81. **Spika, J. S., N. Shaffer, N. Hargrett-Bean, S. Collin, K. L. MacDonald, and P. A. Blake.** 1989. Risk factors for infant botulism in the United States. *Am. J. Dis. Child.* **143**:828–832.

82. **Sterne, M., and G. W. Warrack.** 1964. The types of *Clostridium perfringens. J. Pathol. Bacteriol.* **88**:279–283.

83. **Suen, J. C., C. L. Hatheway, A. G. Steigerwalt, and D. J. Brenner.** 1988. *Clostridium argentinense,* sp. nov.: a genetically homogeneous group composed of all strains of *Clostridium botulinum* toxin type G and some nontoxigenic strains previously identified as *Clostridium subterminale* or *Clostridium hastiforme. Int. J. Syst. Bacteriol.* **38**:375–381.

84. **Sullivan, N. M., S. Pellett, and T. D. Wilkins.** 1982. Purification and characterization of toxins A and B of *Clostridium difficile. Infect. Immun.* **35**:1032–1040.

85. **Summanen, P., E. J. Baron, D. M. Citron, C. A. Strong, H. M. Wexler, and S. M. Finegold.** 1993. *Wadsworth Anaerobic Bacteriology Manual*, 5th ed. Star Publishing Co., Belmont, Calif.

86. **Swartz, M. N.** 1990. Myositis, p. 812–818. *In* G. L. Mandell, R. G. Douglas, Jr., and J. E. Bennett (ed.), *Principles and Practice of Infectious Diseases*, 3rd ed. Churchill Livingstone, Inc., New York.

87. **Swartz, M. N.** 1990. Subcutaneous tissue infections and abscesses, p. 808–812. *In* G. L. Mandell, R. G. Douglas, Jr., and J. E. Bennett (ed.), *Principles and Practice of Infectious Diseases*, 3rd ed. Churchill Livingstone, Inc., New York.

88. **VanBeek, A., E. Zook, P. Yaw, R. Gardner, R. Smith, and J. L. Glover.** 1974. Nonclostridial gas-forming infections. *Arch. Surg.* **108**:552–557.

89. **Varki, N. M., and T. I. Aquino.** 1982. Isolation of *Clostridium difficile* from hospitalized patients without antibiotic-associated diarrhea or colitis. *J. Clin. Microbiol.* **16**:659–662.

90. **Viscidi, R., S. Willey, and J. G. Bartlett.** 1981. Isolation rates and toxigenic potential of *Clostridium difficile* isolates from various patient populations. *Gastroenterology* **81**:5–9.

91. **Walker, P. D., T. G. C. Murrell, and L. K. Nagy.** 1980. Scanning electron microscopy of the jejunum in enteritis necroticans. *J. Med. Microbiol.* **13**:445–450.

92. **Watson, D. A., J. H. Andrew, S. Banting, J. R. Markay, R. G. Stillwell, and M. Merrett.** 1991. Pig-bel but no pig: enteritis necroticans acquired in Australia. *Med. J. Aust.* **155**:47–50.

93. **Wilkins, T. B., and T. Thiel.** 1973. Resistance of some species of *Clostridium* to clindamycin. *Antimicrob. Agents Chemother.* **3**:136–137.

94. **Willis, A. T.** 1969. *Clostridia of Wound Infection.* Butterworths, London.

95. **Wilson, K. H.** 1993. The microecology of *Clostridium difficile. Clin. Infect. Dis.* **16**(Suppl. 4):S214–S218.

96. **Yamamoto, K., I. Ohishi, and G. Sakaguchi.** 1979. Fluid accumulation in mouse ligated intestine inoculated with *Clostridium perfringens* enterotoxin. *Appl. Environ. Microbiol.* **37**:181–186.

Peptostreptococcus, Propionibacterium, Eubacterium, and Other Nonsporeforming Anaerobic Gram-Positive Bacteria

SHARON L. HILLIER AND BERNARD J. MONCLA

48

The anaerobic gram-positive cocci and nonsporeforming rods constitute a genetically and phenotypically diverse group of bacteria. These organisms are, for the most part, components of the normal flora of the skin or mucosal surfaces of humans. Because of this, they are most frequently recovered from clinical specimens taken from urogenital or oral sites or from patients with systemic infections originating from these sites. These opportunistic pathogens are seldom recovered as single isolates. This group of microorganisms is most likely to be recovered from abscess, body fluid, or wound specimens in mixed cultures, usually along with anaerobic gram-negative rods or facultative organisms even when care is taken to avoid contamination during specimen collection. The recovery of these organisms from some clinical specimens is summarized in Table 1.

Isolates of these microorganisms are often undetected or unrecognized in many clinical laboratories. Many of the species discussed in this chapter have complex growth requirements or require extended anaerobic incubation for growth. Specimen handling procedures and growth conditions that are adequate for recovery of *Bacteroides fragilis* or *Clostridium* species will not be suitable for recovery of *Peptostreptococcus anaerobius* or *Mobiluncus curtisii*. Specimens from oral or pelvic abscesses will in all likelihood be negative for most anaerobes after 48 h of incubation, and extended anaerobic incubation will be required to accurately define the etiologic agents of these infections. If these organisms are not detected and reported, the clinicians ordering the cultures may be misled into believing that no anaerobic microorganisms are present and may act upon this erroneous information by discontinuing antimicrobic therapy directed against these pathogens. The most common error in handling wound, abscess, and body fluid specimens from urogenital and oral sites is the failure to isolate these fastidious and slow-growing microorganisms. Extra attention to extended (and uninterrupted) anaerobic incubation will guarantee an opportunity of detecting these organisms.

After their recovery and isolation, members of this group of microorganisms are a challenge to identify accurately. Some of the commercial rapid identification systems that accurately identify some of the fast-growing anaerobic rods from the gut are woefully inadequate for identification of the organisms we are concerned with here. However, cor-rect identification of these bacteria to the genus level usually provides the clinician with useful information about the probable susceptibility of the pathogen to various antimicrobic agents and hints as to the source of the infectious agents. For example, recovery of anaerobic *Lactobacillus* species from the blood suggests a probable genital or rectal source, whereas the recovery of a *Propionibacterium* species may suggest a skin source.

ANAEROBIC GRAM-POSITIVE COCCI

Taxonomy

The anaerobic gram-positive cocci include the genera *Peptostreptococcus* and *Peptococcus* as well as some species of *Streptococcus*. Major revisions in the taxonomy of gram-positive cocci occurred in 1983, when many species formerly in the genus *Peptococcus* were transferred to the genus *Peptostreptococcus* because of the similarities in the G+C contents of their DNAs (32). The new species of *Peptostreptococcus* of human origin that have been described since 1990 are *Peptostreptococcus hydrogenalis*, *P. vaginalis*, *P. lacrimalis*, and *P. lactolyticus* (30, 54). Two species of peptostreptococci have been observed exclusively in animals. *P. heliotrinreducens*, which was transferred from the genus *Peptococcus* in 1986, is recovered from sheep rumens, and *P. barnesae*, an obligately purine-utilizing species, is found in chicken feces (31, 74). The genus *Peptococcus* contains only one species, *Peptococcus niger* (94). *Peptococcus saccharolyticus* is now classified as *Staphylococcus saccharolyticus* (50).

The species of *Streptococcus* currently included as anaerobic streptococci include three strictly anaerobic species, *Streptococcus hansenii*, *S. phemorphus*, and *S. parvulus*, that are rarely isolated from clinical specimens. The taxonomic positions of these species remain unclear. However, analyses of the 16S RNA, DNA base composition, and peptidoglycan type suggest that *S. hansenii* shows relatedness to *Clostridium sphenoides* and *Clostridium amniovalericum*, while *S. pleomorphus* appears to be closely related to *Clostridium innocuum* (56). *S. parvulus*, which can be recovered from the gingival crevice, is more closely related to *Peptostreptococcus* species (17). Further studies will be required to establish the taxonomic position of these species. *Gemella morbillorum* (formerly *Streptococcus morbillorum*) is occasionally isolated from clinical specimens but is considered

TABLE 1 Recovery of anaerobic gram-positive rods and cocci from various specimen types

Site	Reference	No. of patients	% of patients yielding isolates from genus:				
			Peptostreptococcus	*Actinomyces*	*Eubacterium*	*Lactobacillus*[a]	*Propionibacterium*
Appendix	5	71	23	11	34	21	6
Kidney abscess	8	6	83	0	17	0	17
Bladder abscess	8	2	100	0	50	0	0
Periurethral abscess	8	7	71	0	0	0	0
Bartholin gland abscess	8	26	42	0	8	4	0
Penile abscess	8	7	86	0	0	0	0
Scrotal or testicular abscess	8	21	48	5	5	0	0
Periapical abscess	95	22	68	14	9	23	0
Orofacial infection	38	55	33	0	13	33	0
Penile wound	8	6	66	0	0	0	0
Periodontal pockets[b]		85	60	75	43	NA[c]	15
Breast abscess	11	41	51	0	2	0	27
Peritonsillar abscess	47	42	36	0	0	2	10
Blood	10	587	11	0	2	0.2	36
Nostril (normal)	9	25	16	0	0	0	75
Vagina	41	230	87	7	3	14	22

[a]Obligately anaerobic strains only.
[b]Moncla, unpublished data from University of Washington Periodontal clinic, 1984 through 1989.
[c]NA, no attempts were made to recover *Lactobacillus* spp. in these studies.

microaerophilic and is covered in chapter 23 of this Manual (51).

Anaerobic gram-positive cocci of other genera, including *Ruminococcus* and *Coprococcus*, can be isolated from the rumens of animals and the stomachs and bowels of human beings. However, their isolation from clinical specimens is exceedingly rare, and they will not be discussed further here. For information on their isolation and identification, consult volume 2 of *Bergey's Manual of Systematic Bacteriology* (82).

Description of the Group

The genera *Peptostreptococcus* and *Peptococcus* include gram-positive obligately anaerobic nonsporeforming cocci. Cells may occur in pairs, tetrads, irregular masses, or chains. These species are chemoorganotrophs that metabolize peptones and amino acids to acetic acid and often produce isobutyric, butyric, isovaleric, or isocaproic acid. Obligately anaerobic cocci belonging to the genera *Streptococcus* and *Gemella* produce lactic acid as the sole major product. Members of the genera *Coprococcus* and *Ruminococcus* differ in that they require a fermentable carbohydrate for growth.

Natural Habitat

The anaerobic gram-positive cocci are widely distributed as normal flora in humans and animals. They can be recovered routinely from the skin, upper respiratory tract, gingiva, gut, and urogenital tract. Anaerobic gram-positive cocci can be recovered from the vaginas of 90% of pregnant women, with *P. prevotii*, *P. tetradius*, *P. magnus*, and *P. asaccharolyticus* being the most common species (41). Likewise, these organisms can be recovered from 60% of periodontal pocket specimens (61a) but are relatively uncommon in the saliva.

Peptococcus niger is recovered from the vaginas of 20 to 30% of pregnant women but has been only infrequently recovered from clinical specimens (41). *Peptococcus niger* has also been recovered from sheep with foot rot (67).

Clinical Significance

The role of "anaerobic streptococci" in human infections has been recognized since the early 1900s. Schwarz and Dieckman published a report in 1926 on postpartum endometritis in 165 women (78). They recovered *P. anaerobius* from 28% of blood and endometrial cultures and other "anaerobic streptococci" from 13%. The importance of anaerobic gram-positive cocci in the etiology of postpartum endometritis has since been confirmed in many studies. One recent report listed a wide variety of species recovered from the endometria of women with postpartum fever, including *P. anaerobius* (19%), *P. asaccharolyticus* (20%), *P. magnus* (41%), *P. prevotii* (15%), and *P. tetradius* (26%). *Peptococcus niger* is a less frequent isolate and was recovered from only 4% of the endometrial specimens (42). These organisms are also frequently recovered from tubo-ovarian abscesses, the fallopian tubes or endometria of women with pelvic inflammatory disease, septic abortions, and patients with amnionitis and infection of the placental membranes (chorioamnionitis) (Table 1) (40). The isolation of *Peptostreptococcus* spp. from placental membranes is associated with preterm birth (40). In pelvic infections, these organisms are often coisolated with *Bacteroides* species (usually not *Bacteroides fragilis* group), *Prevotella* species, or facultative bacteria such as *Escherichia coli* or viridans streptococci.

Anaerobic gram-positive cocci are isolated from patients with a wide variety of head and neck infections, including periodontitis, chronic otitis media, chronic sinusitis, and brain abscess (9, 10, 12, 13). The source of anaerobic cocci in brain abscess is probably related to the presence of these bacteria in otitis and sinusitis and their subsequent spread into the central nervous system. Anaerobic gram-positive cocci present in the gingiva can spread hematogenously following dental manipulations or extractions to cause brain abscess. The organisms can also spread by aspiration from the oral cavity to cause pulmonary disease, including pneumonitis, lung abscess, empyema, or necrotizing pneumonia. As with obstetric and gynecologic infections, these types typically include other anaerobes, facultative streptococci, and *E. coli*.

Spillage of fecal contents into the peritoneum following appendicitis, diverticulitis, surgery, penetrating trauma, or cancer can lead to intra-abdominal infections involving anaerobic gram-positive cocci, *Bacteroides* and *Clostridium* species, and facultative bacteria. Intestinal perforation or malignancy may also lead to liver abscess, which involves obligate anaerobes in at least half of all cases (6). In one recent report, *Peptostreptococcus* species were isolated from one in four people with perforated appendicitis with peritonitis (5).

Bacteremia due to anaerobic gram-positive cocci most commonly follows obstetric or gynecologic infections, including postpartum endometritis and amnionitis. The most commonly occurring isolates include *P. magnus*, *P. asaccharolyticus*, and *P. anaerobius* (10). Infection of bone, joints, and grafts may also occur, with *P. magnus* being of principal importance in these sites.

Collection, Transport, and Storage of Specimens

The appropriate specimen collection and transport methods for anaerobic gram-positive cocci depend on the specimen site to be sampled. In general, aspirates are considered the optimal samples for culture of obligate anaerobes, because direct aspirates are less likely to be exposed to the effects of oxygen and desiccation. For the best recovery of anaerobic cocci, appropriate samples (amniotic fluids obtained by amniocentesis, cul-de-sac fluid obtained by culdocentesis, etc.) should be transported to the laboratory within 30 min. Because this is often impossible or the culture site does not yield an appropriate aspirate, a swab specimen is often obtained by the clinician.

When swab specimens only are obtained for culture of anaerobic cocci, it is imperative that an anaerobic transport system be used. This group of microorganisms is quickly rendered nonviable when exposed to O_2 or when held for any length of time on a dry transport swab. Two anaerobic transport media that have been evaluated for use with anaerobic cocci are the B-D Port-a-Cul anaerobic transport tube (Becton-Dickinson Microbiology Products, Cockeysville, Md.) and the Anaerobic Specimen Collector (Becton-Dickinson Vacutainer Systems, Rutherford, N.J.). Brook demonstrated that *Peptostreptococcus* species had 100% viability after 24 h at room temperature in the Port-a-Cul but less than 60% viability after 24 h in the Anaerobic Specimen Collector (7). A modified Stuart transport medium has also been reported to provide adequate protection for *P. magnus* and *P. anaerobius* for up to 48 h (88). Baron et al. compared the Accu-Cul-Shure (Technology for Medicine, Pleasantville, N.Y.), a porous plastic swab substitute, to the Port-a-Cul for recovery of bacteria from decubitus ulcers (3). The two systems were equivalent with respect to recovery of *P. asaccharolyticus*, *P. magnus*, and *P. prevotii*. The Accu-Cul-Shure has also been successfully used as a microbiologic sampling device for the endometrium, since the swab substitute is housed in a protective sleeve.

Specimens for culture of anaerobic cocci should always be transported to the laboratory as soon as possible. However, storage of specimens for up to 24 h is possible if an appropriate anaerobic transport medium is used. The optimal temperature for storage is somewhat controversial, since some authors advise against refrigeration, others recommend refrigeration, and others have found variable results depending on the type of transport medium used. In our experience, optimal transport of these organisms is most reliable at room temperature (39a).

Isolation Procedures

The anaerobic cocci grow well on most nonselective plating media suitable for anaerobic isolation. Blood, brucella, or Schaedler agar base supplemented with 5% sheep blood, vitamin K, and hemin generally yields good growth in 48 to 72 h. One report has suggested that Centers for Disease Control and Prevention (CDC) agar base gives better recovery of *P. tetradius*, *P. anaerobius*, and *P. asaccharolyticus* than brucella agar-based blood agar (80).

Acceptable broth media include peptone-yeast extract-glucose, chopped meat-glucose, and thioglycolate supplemented with hemin, vitamin K, and rabbit serum (5%). Supplementation with Tween (final concentration, 0.02%) may stimulate growth in broth media.

Identification and Identifying Characteristics

While it seems self-evident that anaerobic gram-positive cocci should stain as gram-positive cocci, this is not always true. For instance, *P. productus* and *P. anaerobius* may be elongated and resemble gram-positive coccobacilli. Any of the gram-positive cocci may lose gram positivity rapidly with age. To add to the confusion, anaerobic gram-negative cocci, including *Veillonella* and *Megasphaera* spp., may at times appear to be gram positive. Preparation of Gram stains from a variety of media, both broth and solid, and examination of subcultures at different ages often help in assessing the correct cellular morphology and Gram stain reaction. Gram-positive cocci that resemble rods while on agar medium will assume a more coccoid appearance when grown in broth. Vancomycin susceptibility (5-µg disk on brucella blood agar) is useful for establishing that a microorganism is gram positive in those instances in which gram-variable or gram-negative staining is observed.

Assignment of an anaerobic gram-positive coccus to a specific genus can also be somewhat difficult, especially if a gas chromatograph is not available. Any identification of anaerobic gram-positive cocci that is made without the aid of chromatographic analysis of growth by-products must be considered presumptive. Even though the rapid identification strips can readily identify many species of anaerobic gram-positive cocci, the accuracy of the identification is directly related to the knowledge and experience of the individual reading the test (15, 49, 64, 77). For example, the reported accuracy of the RapID-ANA II for the genus *Peptostreptococcus* ranges from 15% for *P. prevotii* to 100% for *P. micros* (18, 60). The microbiologist should take into account all of the information available when using any identification system. In a recent evaluation of the rapid ID32A kit (bioMérieux Vitek, Inc., Hazelwood, Mo.), the system correctly identified only 1 of 12 *P. prevotii* isolates (65). Those authors noted that *P. prevotii* is heterogeneous according to preformed enzyme profiles, and the taxonomy may require further clarification. By contrast, *P. asaccharolyticus* is identified well by most commercial systems. The difficulty in correctly identifying these organisms may be due in part to our lack of understanding of the true taxonomic status of some *Peptostreptococcus* species. The capabilities of several commercial systems to identify gram-positive anaerobic cocci are summarized in Table 2. Tables 3, 4, and 5 can be used to further distinguish the genera and species.

The three species of *Peptostreptococcus* most commonly seen in clinical specimens are *P. magnus*, *P. anaerobius*, and

TABLE 2 Identification of anaerobic gram-positive cocci by commercial systems

Organism	RapID-ANA II[a]		AN-Ident[b]	
	Tested	Identified to species	Tested	Identified to species
Peptostreptococcus spp.				
P. asaccharolyticus	53	49 (92)	4	4
P. magnus	40	39 (98)	5	5
P. anaerobius	31	22 (71)	5	4
P. micros	26	26 (100)	1	1
P. tetradius	4	3 (75)	9	5
P. prevotii	27	4 (15)	3	3
Peptococcus niger	0	NA	1	0 (0)

(No. (%) of isolates)

[a]Adapted from references 18 and 60. NA, not applicable.
[b]Adapted from reference 68.

P. asaccharolyticus. Colonies of P. magnus are minute to 0.5 mm, raised, dull, smooth, and nonhemolytic. Cells are 0.7 to 1.2 μm in diameter and appear in a tightly packed arrangement. P. micros colonies are minute to 1 mm, convex, and dull. Cells are smaller than those of P. magnus, being 0.3 to 0.7 μm. However, strain variability under different growth conditions makes differentiation of these two species on the basis of morphology subjective. While most colonies of P. micros are smooth, a rough variant from periodontitis patients has also been described (89). The colonies of P. anaerobius are usually somewhat larger (1 mm) and nonhemolytic, while the individual cells are smaller (0.5 to 0.6 μm in diameter) than those of P. magnus. Very young cultures of P. anaerobius may have elongated cells in chains. The colonies of P. anaerobius may have a pungently sweet odor. P. asaccharolyticus colonies are minute to 2 mm in diameter and may have a slightly yellow pigment on blood agar. The individual cells are 0.5 to 1.5

TABLE 4 Flow chart for identification of anaerobic gram-positive cocci

I. Black pigment (may be olive green to mustard), catalase positive: Peptococcus niger
II. No black pigment
  A. Sensitive to SPS[a]: Peptostreptococcus anaerobius
  B. Not sensitive to SPS
    1. Indole positive
      a. Nitrate positive, coagulase positive: Peptostreptococcus indolicus
      b. Nitrate negative, coagulase negative
        i. Acid from glucose: Peptostreptococcus hydrogenalis
        ii. No acid from glucose: Peptostreptococcus asaccharolyticus
    2. Indole negative
      a. Urease positive
        i. Acid from glucose and lactose: Peptostreptococcus lactolyticus
        ii. Acid from glucose and sucrose, no acid from lactose: Peptostreptococcus tetradius
        iii. Weak or no acid from glucose, no acid from lactose or sucrose: Peptostreptococcus prevotii
      b. Urease negative
        i. Esculin positive: Peptostreptococcus productus
        ii. Esculin negative: Peptostreptococcus magnus, Peptostreptococcus micros

[a]SPS, sodium polyanethol sulfonate (93).

μm in diameter and are arranged in pairs, tetrads, or irregular clumps.

The three newly described species of Peptostreptococcus are related to P. prevotii. These species, P. vaginalis, P. lacrimalis, and P. lactolyticus, appear as short chains or in masses. P. hydrogenalis cells are 0.7 to 1.8 μm and occur in short chains or masses.

Peptococcus niger is rarely recovered from clinical speci-

TABLE 3 Identifying characteristics of anaerobic cocci[a]

Organism	Coagulase	Indole	Nitrate	Urease	Fermentation of:					GLC analysis[b]
					Cellobiose	Glucose	Lactose	Maltose	Sucrose	
Peptococcus niger[c]	−	−	−	−	−	−	−	−	−	B, C, iv, a
Peptostreptococcus spp.										
P. anaerobius	−	−	−	−	−	+ (W)	−	+	−	A, IC, ib, b, iv
P. magnus	−	−	−	−	−	−	−	−	−	A, (l), (s)
P. micros	−	−	−	−	−	−	−	−	−	A, (l), (s)
P. indolicus	+	+	+	−	−	−	−	−	−	A, B, p, (l), (s)
P. asaccharolyticus	−	+	−	−	−	−	−	−	−	A, B, (l), (p), (s)
P. prevotii	−	−	±	+	−	+ (W)	−	+ (W)	−	B, C, iv, a
P. tetradius	−	−	−	+	−	+	−	+	+	B, L, (a), (p)
P. productus	−	−	−	−	+	+	+	+	+	A, (l), (s)
P. lactolyticus	−	−	−	+	−	+	+	+	−	B, L, a
P. vaginalis	−	−	−	−	−	+ (W)	−	+ (W)	−	B, l, a
P. lacrimalis	−	−	−	−	−	−	−	−	−	B, a, p, l
P. hydrogenalis	−	+	−	−	−	+	−	+	+	B, a, p, l

[a]Symbols: +, 90% or more positive; −, 90% or more negative; ±, usually negative but some strains positive; V, variable reaction; W, weak.
[b]GLC, gas-liquid chromatography. Analysis was of fatty acid products in peptone-yeast broth with glucose. Capital letters indicate major metabolic products, whereas lowercase letters indicate minor products. Parentheses indicate a variable reaction. A, acetic acid; P, propionic acid; IB, isobutyric acid; B, butyric acid; IV, isovaleric acid; V, valeric acid; IC, isocaproic acid; C, caproic acid; L, lactic acid; S, succinic acid.
[c]On initial cultures, usually pigmented olive green to black.

TABLE 5 Biochemical characteristics of the anaerobic streptococci and *Gemella* sp.[a]

Characteristic	Reaction[b]			
	G. morbillorum	S. hansenii	S. pleomorphus	S. parvulus
Aerotolerance	+	−	−	−
Acid from:				
Cellobiose	−	−	−	+
Fructose	−	−	+	+
Galactose	−	+	−	+
Inulin	−	−	NT	+
Lactose	−	+	−	+
Maltose	W	+	−	+
Mannose	W	−	d	+
Salicin	−	−	−	+
Sucrose	W	−	−	+
Raffinose	−	+	NT	−
Production of H₂S	−	+	W	−
Esculin hydrolysis	−	d	NT	+

[a]Adapted from reference 82.
[b]+, positive; −, negative; W, weak; d, differs among strains; NT, not tested. All species produce lactic acid from fermentation and fail to ferment mannitol, sorbitol, and starch.

mens, although it is likely that this isolate has gone unnoticed within mixed anaerobic cultures from clinical specimens. On initial isolation, the colonies are black to olive green. The pigment may not be visible to the naked eye but is evident under a dissecting microscope. The black pigment fades quickly upon exposure to oxygen. After subculture or after the culture is a week old, the colony can take

on a mustard yellow pigment. The colonies are convex and circular with a shiny smooth appearance. They are weakly catalase positive. They are indole, urease, and coagulase negative; nitrate is not reduced; and esculin and starch are not hydrolyzed. Hydrogen sulfide (in sulfide-indole-motility medium) and ammonia are produced. The only definitive means of differentiating *Peptococcus* from *Peptostreptococcus* spp. is by analyzing G+C ratios. Unfortunately, this is not practical, and the peptococci must be presumptively identified on the basis of pigment production and the catalase reaction.

Antimicrobic Susceptibilities

Most *Peptostreptococcus* isolates (96%) are susceptible to beta-lactams and most cephalosporin-type antimicrobics (Table 6), although MICs as high as 64 μg/ml have been reported for ticarcillin and cefotaxime. Clindamycin and metronidazole are slightly less active (84 and 88% of isolates, respectively, are susceptible). However, there is evidence of inducible macrolide-lincosamide resistance, with *P. asaccharolyticus* being the most resistant to erythromycin (69). Some strains that are clindamycin susceptible and erythromycin resistant have inducible clindamycin resistance.

Evaluation, Interpretation, and Reporting of Results

The significance of finding anaerobic gram-positive cocci in clinical specimens depends on the quality of the specimen and the likelihood that it was contaminated by oral, genital, or intestinal flora. These are the microorganisms most frequently recovered in mixed culture, so it may be necessary to speak to the clinician who obtained the specimen to determine the likelihood that the cocci represent contamination. For example, amniotic fluid obtained by intrauter-

TABLE 6 Antimicrobic susceptibilities[a] of *Peptostreptococcus* species[b] and *Peptococcus niger*

Antimicrobial agent	Breakpoint (μg/ml)	Peptostreptococcus species[c]				Peptococcus niger		
		No. of strains tested	MIC$_{50}$ (μg/ml)	MIC$_{90}$ (μg/ml)	MIC range (μg/ml)	No. of strains tested	MIC$_{50}$ (μg/ml)	MIC range (μg/ml)
Penicillin G	16	60	0.2	8	≤0.06–128	4	0.1	0.1–0.5
Ampicillin	4	98	2	32	≤0.06–128	4	0.1	≤0.06–2
Piperacillin	64	35	0.5	8	≤0.06–32	4	0.2	≤0.06–0.5
Mezlocillin	64	35	0.5	8	≤0.06–128	4	0.2	0.2–1
Cefuroxime	16	35	0.5	16	≤0.06–≥256	4	0.2	≤0.06–1
Cefpodoxime		63	1	32	≤0.12–264	0		
Cefixime		63	8	≥64	0.5–≥64	0		
Cefoxitin	32	123	1	8	≤0.06–64	4	0.5	0.2–4
Cefoperazone	32	60	0.5	4	≤0.06–64	4	0.2	≤0.06–8
Cefotaxime	32	35	2	8	≤0.06–16	4	1	≤0.06–2
Imipenem	8	60	1	16	≤0.06–≥256	4	1	0.5–8
Clindamycin	4	88	2	16	≤0.06–128	0		
Lincomycin	4	35	4	≥256	≤0.06–≤256	4	8	0.5–32
Ofloxacin	4	83	2	16	≤0.06–32	0		
Norfloxacin	4	39	4	32	≤0.06–128	4	1	0.5–8
Ciprofloxacin	4	139	2	4	≤0.06–32	0		
Tinidazole	16	35	2	32	≤0.06–≥256	4	1	1–16
Metronidazole	16	147	2	64	≤0.06–≥256	4	1	0.5–8
Tetracycline	4	35	2	16	≤0.06–32	4	1	0.2–8
Doxycycline		63	4	16	≤0.06–32	0		

[a]Includes data compiled from references 37, 57, 67, 90, and 91.
[b]Includes *P. anaerobius*, *P. prevotii*, *P. asaccharolyticus*, and *P. micros*.
[c]MIC$_{50}$ and MIC$_{90}$, MICs at which 50 and 90% of strains, respectively, are inhibited.

TABLE 7 Differentiation of gram-positive anaerobic rods

Characteristic	Reaction[a]							
	Actinomyces spp.	Propioni-bacterium propionicus	Bifidobacterium spp.	Lactobacillus spp.	Rothia sp.	Propioni-bacterium spp.	Eubacterium spp.	Mobiluncus spp.
Strictly anaerobic	V	−	+	V	−	V	+	+
Motility	−	−	−	−	−	−	V	+
Catalase production	±[b]	−	−	−	+	V	−	−
Metabolic products[c]	S, L, a	A, P, l, s[d]	A, L[e]	L, (a), (s)	A, L, (s), (p)	A, P, iv, s, l	A, B[f]	S, L, A
Indole production	−	V	−	−	−	−	−	−
Nitrate reduction	V	+	−	−	+	V	V	V
Mol% G+C of DNA	58–68	63–75	57–64	35–53	65–69	59–66	30–40	49–52

[a]Data were compiled from references 22, 35, 63, 72, 75, 82, and 84. Abbreviations and symbols: V, variable results occur for different species of the same genus; A, acetic acid; S, succinic acid; B, butyric acid; L, lactic acid; P, propionic acid; IV, isovaleric acid variable or, if produced, usually present only in trace amounts; lowercase letters represent products usually produced in small amounts; +, most strains positive; −, most strains negative.

[b]A. viscosus is the only member of the genus that produces catalase.

[c]Determined on peptone-yeast extract-glucose broth cultures.

[d]Under anaerobic conditions, Propionibacterium propionicus ferments glucose to CO_2, acetic acid, propionic acid, and small amounts of lactic and succinic acids. In air, glucose is converted to CO_2 and acetic acid.

[e]Acetic and lactic acids are produced in a molar ratio of 3:2.

[f]Metabolic products formed by Eubacterium species vary. See Table 8.

ine pressure catheter after rupture of fetal membranes could be contaminated by normal vaginal flora, including peptostreptococci. On the other hand, recovery of anaerobic cocci from amniotic fluid obtained by amniocentesis or at the time of cesarean section in a woman with intact fetal membranes indicates an infection that could place the mother at increased risk of postpartum endometritis. When it is unclear how a specimen was obtained, it is advisable to consult with the referring physician.

ANAEROBIC NONSPOREFORMING GRAM-POSITIVE RODS

Taxonomy

The group of anaerobic nonsporeforming gram-positive rods includes the genera Propionibacterium, Eubacterium, Bifidobacterium, Lactobacillus, Rothia, Actinomyces, and Mobiluncus. These genera are taxonomically quite diverse while sharing many phenotypic characteristics. In the past, bacterial taxonomy has relied heavily on descriptive aspects of these organisms, which has resulted in the grouping of genera into families that were probably not appropriate taxonomically.

The gram-positive anaerobic bacteria may be separated into two major subdivisions on the basis of mole percentage G+C, which appears to represent separate phylogenetic lines. The low-G+C subdivision represents a more ancient line and includes the genera Lactobacillus, Clostridium, Eubacterium, Peptostreptococcus, and Peptococcus. The high-G+C subdivision includes the genera Bifidobacterium, Actinomyces, Propionibacterium, Rothia, Bacterionema, and Mobiluncus (22, 53, 73, 96). These genera form part of the so-called Actinomycetales branch (22).

Traditionally, inclusion in a group of microorganisms was based primarily on morphologic and biochemical criteria. Although considerable progress has been made in understanding the taxonomy of these bacteria, reliable and accessible tests for identification of many of these species

are still not available. As a result, identification in the clinical laboratory is difficult. The genera fall into discrete units based on amino acid content of cell wall, major end products of glucose fermentation, and mole percentage G+C content (73, 75, 76, 82). Actinomyces cell walls contain lysine, aspartic acid, and ornithine, similar to the cell walls of Bifidobacterium species. Bacterionema cell walls have DL-diaminopimelic acid (DL-DAP), whereas Rothia cell walls contain lysine. Propionibacterium cell walls may contain meso-DAP, LL-DAP, or lysine in place of DAP. Propionibacterium propionicus (formerly Arachnia propionica) has LL-DAP in the cell walls. There is some overlap in the mole percentage G+C contents of all of these organisms (Table 7).

Several new species of Actinomyces have been described: Actinomyces georgiae (previously designated Actinomyces D08); A. gerencseriae (A. israelii serotype II); A. neuii (CDC group 1-like coryneform bacterium), with two subspecies, neuii and anitratus; and A. hyovaginalis (20, 34, 46). The genus Eubacterium contains one new species, Eubacterium yurii, which has three subspecies, yurii, margaretiae, and schtitka (58, 59). New Lactobacillus species include Lactobacillus rimae and Lactobacillus uli (66). Arachnia propionica has been reclassified as Propionibacterium propionicus (19, 23).

Mobiluncus spp. (motile anaerobic curved rods) were first recovered in 1913 from the uterine discharge of a woman with postpartum endometritis. While other workers reported the recovery of similar anaerobic rods from vaginal specimens, this group of microorganisms was not recognized and given the name Mobiluncus until 1984 (84). The genus Mobiluncus is composed of obligately anaerobic, gram-variable or gram-negative, curved, nonsporeforming rods with tapered ends. The rods occur singly or in pairs and may have a gull wing appearance. They are motile by multiple subpolar flagella. Even though single organisms stain gram negative to gram variable, they possess a multilayered gram-positive type of cell wall lacking the lipopolysaccharide, 2-keto-3-deoxyoctulosonic acid, and hydroxylated fatty acids typically found in the cell walls of gram-negative organisms (16). These organisms are susceptible to vancomycin and resistant to colicin, which is consistent with the susceptibility patterns of other gram-positive organisms. The genus *Mobiluncus* does not conform to other families of anaerobic bacteria. Because the family *Bacteroidaceae* includes other genera of curved, anaerobic, rod-shaped bacteria that produce succinic acid, the genus *Mobiluncus* was tentatively placed into the family *Bacteroidaceae* in 1984. Partial reverse transcriptase sequencing of the 16S rRNA has demonstrated that the genus *Mobiluncus* is more closely related to the genus *Actinomyces* than to the *Bacteroidaceae* (53).

Description of the Group

Members of this group are gram-positive asporogenous rods, obligately anaerobic or facultatively anaerobic, motile or nonmotile chemoorganotrophs, and saccharoclastic or non-saccharoclastic.

Natural Habitat

With the exception of *A. humiferus*, which is found exclusively in soil, the natural habitat of *Actinomyces* spp. appears to be the mucous membranes of humans and animals. *A. bovis*, *A. denticolens*, *A. howelli*, *A. hordeovulneris*, *A. hyovaginalis*, *A. slackii*, and *A. suis* have not been isolated from humans. *Rothia* sp. is a normal inhabitant of the human oral cavity. It is present in saliva and both supragingival and subgingival plaque.

Bifidobacterium species are found in human clinical material, sewage, and the intestines of humans and animals. *Eubacterium* species are found in the intestines and oral cavities (periodontia) of humans and other animals, in various plant products, and in clinical material obtained from soft tissue infections.

Propionibacterium species are isolated from skin, intestinal abscesses, and the oral cavity.

Although most isolates of *Lactobacillus* species are facultative anaerobes, approximately 20% of human isolates are obligately anaerobic. *Lactobacillus* species are found in the healthy mouth (in both saliva and plaque), intestinal tracts, and vaginas of human beings and other mammals. They are also found in a variety of food products.

The natural habitat of *Mobiluncus* spp. is the reproductive tracts and recta of humans and other primates. *Mobiluncus* species have been reported in 50 to 65% of vaginal specimens from women with bacterial vaginosis (*Gardnerella vaginalis* vaginitis, nonspecific vaginitis) (45). In our experience, *Mobiluncus* species are recovered from fewer than 10% of specimens from women with normal, *Lactobacillus*-predominant flora. *Mobiluncus* spp. have also been recovered from rectal swabs from women with bacterial vaginosis and from rectal and urethral specimens from the male sex partners of women with bacterial vaginosis (45). In our laboratory, we have also recovered *Mobiluncus* spp. and other bacteria associated with bacterial vaginosis from the vaginas of children with no history of sexual abuse. Vaginal specimens from rhesus macaque monkeys have also yielded *Mobiluncus* spp. (25).

Clinical Significance

Information on the occurrence of anaerobic gram-positive rods in clinical materials is difficult to obtain, since many data on anaerobic infections are not reliable (33). Both methodologic and taxonomic parameters contribute to our inability to obtain such data. However, data from individual laboratories specializing in anaerobes indicate that the prevalence of these anaerobes in various infections is much greater than is reflected by their isolation frequencies in many clinical laboratories. For example, Finegold determined that *Actinomyces* spp. were isolated in his laboratory from 14% of the infections involving anaerobes, while other workers reported a much lower incidence (33). Recovery of anaerobic gram-positive rods from clinical specimens is summarized in Table 1.

A. israelii, *A. naeslundii*, *A. viscosus*, *A. odontolyticus*, *A. pyogenes*, and *A. meyeri* are causative agents of disease in humans. These organisms are important constituents of the normal flora of mucous membranes and therefore must be considered opportunistic pathogens. They have on occasion been isolated from stool but are most likely transient colonizers of the gut. All except *A. pyogenes* and *A. meyeri* are found in high numbers in saliva and subgingival plaques. *A. viscosus* and *A. naeslundii* have been studied extensively and are known to play significant roles in dental caries and periodontal disease (62). *A. bovis* is the causative agent of lumpy jaw in cattle.

Actinomycosis is the most frequently encountered disease entity involving *Actinomyces* spp. The incidence of actinomycosis in the United States is difficult to determine, since this disease is not reportable. While there are a few recent reports on this subject, Slack and Gerencser concluded that "Actinomycosis occurs throughout the world and is neither a rare nor a common disease" (81). Actinomycosis is observed twice as frequently in men as in women, and the anatomical distribution is about 60% cervicofacial, 15% thoracic, 20% abdominal, and 5% other types.

Actinomycosis is characterized as a chronic granulomatous lesion that becomes suppurative and forms abscesses and draining sinuses (48, 70). A purulent actinomycotic discharge, usually containing macroscopic sulfur granules that appear as whitish, yellow, or brown granular bodies, may be present. *A. israelii* is the most common cause of actinomycosis in humans, but the other species of *Actinomyces*, as well as *Propionibacterium propionicum*, may also be etiologic agents. These infections are usually polymicrobic in nature. Frequently, *Fusobacterium* species, *Eikenella corrodens*, *Capnocytophaga* species, *Actinobacillus actinomycetemcomitans*, black-pigmented *Prevotella* species, *Porphyromonas asaccharolytica*, *Porphyromonas gingivalis*, and streptococci in various combinations are isolated along with *Actinomyces* species or *Propionibacterium propionicum*. *Actinomyces* species are almost never recovered in pure cultures. It is believed that these associated microorganisms contribute to the pathogenesis of actinomycosis.

Actinomyces species may also be involved in pelvic inflammatory disease associated with intrauterine devices (87, 97). Evans estimated that between 1.6 and 11.6% of intrauterine device users have infection with *A. israelii* (29). *Actinomyces* species are also reported to cause pyogenic liver abscess, with the majority of cases occurring in patients without preexisting oral disease or intra-abdominal infections (61). Similarly, *A. pyogenes*, an organism usually considered to be of low pathogenicity, has been recovered from infected humans in the absence of predisposing illness (26).

Among animals, the incidence of actinomycosis is greatest in cattle, although infections have also been reported in some dogs, sheep, goats, horses, cats, deer, moose, and antelope (81). Bovine actinomycosis (lumpy jaw) occurs when *A. bovis* gains entrance to the mandibular tissue following traumatic injury, after ingestion of rough plant material such as straw or silage, or by a broken tooth. As the infection develops, there is a simultaneous destruction of existing bone and stimulation of bone growth, resulting in a proliferative osteitis. The infections are eventually spread to the contiguous tissue, resulting in impairment of mastication or breathing. The yellow discharge from the sinus tracts frequently contains sulfur granules.

A. hordeovulneris is of veterinary importance because it is an agent of actinomycosis (thoracic or abdominal) of dogs (14). Infections are often associated with an irritation of the dog's skin by members of the grass genus *Hordeum* (foxtail). The awns of *Hordeum* spp. have sharp tips with barbed flaring hairs that cause the awns to become trapped in the coat of the animal. When the dog licks itself, the sharp structures of the foxtail are contaminated by the saliva and penetrate the skin to form a wound. These structures ultimately migrate into the tissues, leaving sites of inflammation and necrosis. The awns may also be inhaled or swallowed. In many cases, *A. viscosus* is isolated with *A. hordeovulneris*, suggesting that the natural habitat of the organism is the canine oral cavity. *A. suis* is believed to be involved in swine mammary gland actinomycosis. The pathogenicities of *A. denticolens*, *A. howelii*, and *A. slackii* are unknown.

Propionibacteria have been recovered from mixed infections of the skin (12). The role of *Propionibacterium acnes* in acne vulgaris is widely accepted. Recent reports of propionibacterial infections have been linked to operative procedures or foreign bodies. It is the predominant anaerobic flora in conjunctival cultures of adults and children. The organism has also proven to be pathogenic in infections of bone, joints, and the central nervous system (27). Thus, though *Propionibacterium* species have often been considered contaminants from normal flora, these organisms can act as primary pathogens and should not be immediately dismissed as contaminants.

Propionibacterium propionicus is a recognized agent of lacrimal canaliculitis and actinomycosis, and it produces abscesses in animals when injected. *Propionibacterium propionicus* was originally described as *Actinomyces propionica* because of its similarity to *Actinomyces* species (19).

Rothia is a monotypic genus whose only member is considered an opportunistic pathogen (also mentioned in chapter 29 of this Manual). It has been isolated from patients with abdominal infections, infectious endocarditis, and other infections. Its habitat appears to be human supragingival plaque (35).

Eubacterium species are not frequently encountered in clinical specimens, probably because of the difficulty in recognizing such species. *E. lentum* is the species most frequently observed. *Eubacterium* species are usually isolated in mixed culture from abscesses and wounds and very seldom from blood cultures. Several recently described species have been associated with periodontal disease: *E. nodatum*, *E. timidum*, and *E. brachy* (43, 63). A study by Hill et al. suggests that these species have been involved in numerous infections at other sites such as head and neck, thorax, bones, skin, and pelvis (39). *Eubacterium* species are found in the oral cavities of human beings and lower animals, in soils, and in animal plant products. At least 34 species are currently recognized, and at least 19 other distinct groups of unnamed species are known, but only a few are encountered in the clinical laboratory (Table 8).

Bifidobacterium species are seldom encountered in clinical materials. *Bifidobacterium dentium*, formerly *Actinomyces eriksonii* and later *B. eriksonii*, appears to be the only *Bifidobacterium* species with pathogenic potential. It has

been isolated from dental caries, feces, and vaginas of humans and from various clinical materials such as lower respiratory tract specimens. *B. longum* and *B. breve* are found only occasionally in human clinical materials.

Most *Lactobacillus* species recovered from clinical specimens are microaerophilic, but obligately anaerobic isolates can also be recovered. These organisms are found as part of the normal flora of the human mouth, intestinal tract, and vagina and may therefore be recovered from specimens contaminated by normal flora. They are rarely pathogenic. However, lactobacilli have been reported to cause endocarditis, neonatal meningitis, chorioamnionitis, amnionitis, pleuropulmonary infections, and bacteremia (4, 21, 24, 55). Identification of anaerobic lactobacilli to the species level is extremely difficult and, given the low pathogenicity of the organism, is rarely indicated.

The pathogenic potential of *Mobiluncus* spp. is not well understood. However, it is unclear whether the rare isolation of *Mobiluncus* spp. from clinical specimens is due to the inability of many laboratories to isolate and identify this fastidious microorganism or to the low pathogenicity of this organism. While *Mobiluncus* spp. are present in the vaginas of many women with bacterial vaginosis, the role of this organism in the etiology of this syndrome is not known. Curved rods resembling *Mobiluncus* spp. can easily be identified in Gram-stained vaginal smears from women with bacterial vaginosis (Fig. 1a).

The pathogenic potential of *Mobiluncus* spp. has been supported by its isolation from breast abscesses, umbilical discharge, breast wounds, and blood cultures (36, 85). *Mobiluncus* spp. can also be recovered from the endometrial aspirates of women with pelvic inflammatory disease and from the chorioamnion of the placenta from preterm deliveries. While *Mobiluncus* spp. have been recovered in pure cultures from some sterile-site specimens, they are most often recovered with *Prevotella* and *Peptostreptococcus* species.

Collection, Transport, and Storage of Specimens

This group includes microorganisms that are anaerobic, facultatively anaerobic, and aerotolerant. Therefore, samples should be treated as anaerobic and handled with the consideration outlined above for anaerobic cocci. *Propionibacterium* species, many obligately anaerobic *Lactobacillus* spp., and *Eubacterium* spp. may be lost if exposed to oxygen. The most important parameter is the elapsed time between specimen collection and incubation of the inoculated media under strict anaerobic conditions.

Detection and Isolation Procedures

Physicians must indicate clearly which specimens are to be cultured for the presence of nonsporeforming anaerobic gram-positive rods, since several additional steps are required for proper setup, incubation, and holding times.

Direct microscopic examination of clinical materials is an excellent presumptive method for detection of anaerobic gram-positive rods and should be used whenever possible. In oral pathology laboratories, specimens are sectioned, stained with Brown-Brenn stain, and examined for sulfur granules, which are considered diagnostic for actinomycosis (Fig. 2). In general, samples should be sent for microbial analysis, and such materials should be stained to guide the workup of the specimen.

A variety of materials is suitable for examination for the presence of sulfur granules: surgically removed tissues, autopsy tissues, bronchial washes, body fluids, purulent exu-

TABLE 8 Characteristics of nonsaccharoclastic species of *Eubacterium*[a]

Eubacterium species	Morphologic properties — Cellular[c]	Morphologic properties — Colonial	Metabolic products[b]	Utilization of pyruvate	Nitrate reduced	Gelatin hydrolyzed	Esculin hydrolyzed	Hydrogen produced
E. nodatum	Branched, filamentous, or club-shaped cells; nonmotile; 0.5–0.9 by 2.0–12 mm in PY	Molar tooth, heaped, or raspberry; 0.5–1.0 mm in diam	a, B, (f), (l), (s)	+	–	–	–	–
D-6 group[d]	Regular, variable length, occasional chaining, beading, or diphtheroidal	Circular, entire, convex	a, B, paa, (f), (l), (s)	–	–	–	–	+
E. brachy	Short or coccoidal, chaining, 0.4–0.8 by 0.1–3.0 mm in PY	Circular, entire, low convex, occasionally rough	ib, iv, ic (a), (f), (l), (s), (hc)	–, (+)	–	–	–	+
E. timidum	Short, regular to diphtheroidal, occasional clumps, 0.8–1.6 by 1.6–3.1 mm in PY	Circular, entire, low convex	paa (a), (f), (l), (s)	–	–	–	–	–
E. lentum	Short or coccoidal, occasional chains, 0.2–0.4 by 0.2–2.0 mm in PY	Circular, entire, low convex	(a), (f), (l), (s)	–	+	–	–	–
E. dolichum	Thin rods in long chains, slightly tapered, 0.4–0.6 by 1.6–6.0 mm in PYG	Fails to grow on blood agar plate	b (l), (a)	–	–	W	–	–
E. combesii	Singles, pairs, short chains, and palisade arrangements; motile; 0.6–0.8 by 3.0–10 mm in PYG	Circular, entire to irregular convex, semiopaque, whitish yellow, shiny, smooth	A, B, iv, l, ib, (p), (f)	+	–	+	d	+
E. yurii	Straight rods with slightly rounded edges, form unique three-dimensional brushlike aggregates	Colonies spread on blood agar, measuring about 10 mm in 48 h. Pale yellow pigment may be evident.	B, A, p			–	–	

[a]Data were compiled from references 39, 44, and 70.

[b]ic, isocaproate; iv, isovalerate; ib, isobutyrate; a, acetate; b, butyrate; l, lactate; s, succinate; f, formate; p, propionate; v, valerate; paa, phenylacetate; hc, hydrocinnamate. Capital letters represent ≥1 meq of product per 100 ml of culture; lowercase letters represent <1 meq of product per 100 ml of culture; products in parentheses are not produced uniformly. +, most strains positive; –, most strains negative; (+), usually positive but some strains give negative reaction; W, reaction weak; d, reaction variable.

[c]PY, peptone-yeast extract; PYG, peptone-yeast extract-glucose.

[d]An undescribed *Eubacterium* species that appears to be of clinical significance.

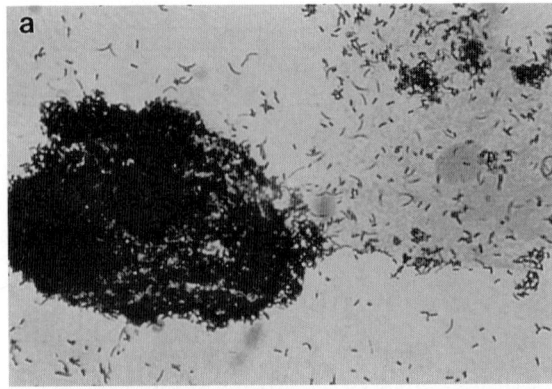

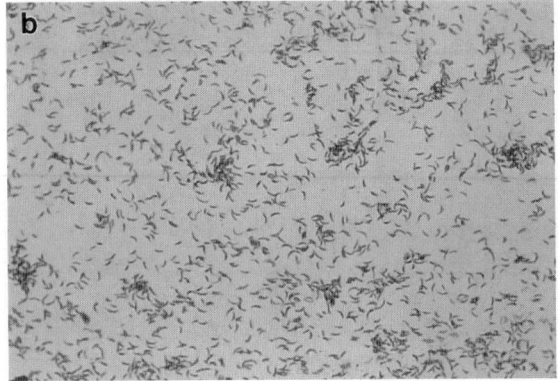

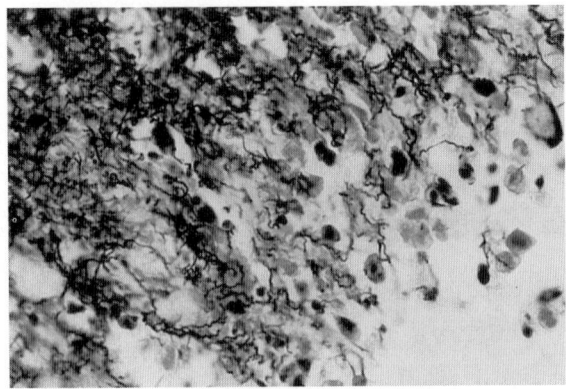

FIGURE 2 Microscopic morphology of *Actinomyces* species in a sulfur granule taken from an oral soft tissue section stained with the Brown-Brenn modification of the Gram stain. Note the characteristic appearance of the club-shaped filaments and the numerous polymorphonuclear leukocytes. Magnification, ×100. (Courtesy of Dolphine Oda, Department of Oral Biology, University of Washington, Seattle.)

FIGURE 1 (a) Gram-stained vaginal smear from a woman with bacterial vaginosis. Note the clue cell (left) and clear squamous epithelial cell (right). Smear contains numerous small gram-negative to gram-variable rods (*Gardnerella* sp.) and curved rods (*Mobiluncus* sp.). (b) Microscopic morphology of a 72-h subculture of *M. curtisii*.

dates, intrauterine devices, Papanicolaou smears, and gauze from draining wounds. Sulfur granules are irregular in shape and size (ranging from 0.1 to 5 mm), hard, and usually yellowish. The sulfur granules may be large enough to see with the unaided eye; granules from patients with oral actinomycosis sometimes resemble popcorn husks. If sulfur granules are observed, one granule should be removed with an inoculating needle, loop, or sterile forceps; placed in a drop of water on a microscope slide; and then gently crushed with a second slide. A wet mount should be examined under low power (100×). The granules have distinctly irregular edges. Upon reduction of the light intensity, the peripheries of the granules give the appearance of numerous club-shaped masses of filaments radiating from the granule, which is usually difficult to distinguish. The clubs should become very distinct at higher magnification (×1,000). The clubs should be refractile, and their appearance has been described as hyaline. Having made these initial observations, one must confirm the presence of filaments. Remove the coverslip, and dry, heat fix, and Gram stain the specimen. Gram-positive branched and unbranched filaments should be easily discernible.

A number of agar plate media are available for isolation and cultivation of nonsporeforming anaerobic gram-positive rods. Blood agars (5% sheep blood or rabbit blood) supplemented with hemin and vitamin K are most often recommended; other media include CDC anaerobe blood

agar, brain heart infusion, brucella agar, modified Schaedler agar, phenylethyl alcohol blood agar, and heart infusion agar. Media containing vancomycin should not be used. Anaerobic blood culture media, enriched thioglycolate medium, and chopped meat-glucose medium will support growth of these microorganisms.

The usual procedures for anaerobes (see chapter 21 of this Manual) should be used. We have found that fresh medium (less than 7 to 9 days old) works best for primary isolation of these anaerobes, since many of these organisms require high moisture content for optimal growth. Other authors state that plates should be wrapped in a plastic wrap to prevent desiccation and may be stored for 4 to 6 weeks. Blood agar plates can be stored inside a dark plastic-garbage-bag-lined box at 5°C. This practice prevents desiccation and protects the medium from the generation of oxidizing components by light. Plate media can be reduced overnight in either an anaerobic glove box or an anaerobic jar before use. This routine has eliminated the need to use anaerobically sterilized medium. Laboratories unable to prepare their own media may wish to consider the use of commercially prepared prereduced anaerobically sterilized blood agar such as is available from Anaerobe Systems (San Jose, Calif.). These media have an extended shelf life of up to 6 months and yield results comparable to those obtained with fresh media.

It is usually necessary to fix, section, and stain tissue specimens to demonstrate sulfur granules. If stained with hematoxylin-eosin, the eosinophilic clubs should be visible at the peripheries of the granules. Diphtheroidal cells and filaments should be apparent at the bases of the clubs. Visualization of the filaments in the granules is easier with a modified Gram stain such as MacCallum-Goodpasteur or Brown-Brenn (28). Fixed and sectioned tissues can also be stained with fluorescent-antibody reagents. Fluorescent antibodies to *Actinomyces*, *Bifidobacterium*, and *Propionibacterium* species are available from the CDC and may be used on tissue sections and clinical materials. Other organisms that may produce granules with clubs are *Nocardia*, *Streptomyces*, and *Staphylococcus* species.

The presence of *Staphylococcus* species can be ruled out

by morphology and Gram reaction. To distinguish *Actinomyces* from *Nocardia* isolates, granules should be stained for acid fastness by the method of Kinyoun or Ziehl-Neelsen, since other stains may yield unreliable results (81).

Sulfur granules, if present, should be placed in a sterile petri dish and rinsed with thioglycolate broth. They should then be transferred with a sterile Pasteur pipette to a sterile tube or preferably a Ten Broeck grinder (available from various sources) with approximately 0.5 ml of thioglycolate broth. The granules should be crushed with the tip of a sterile glass rod and immediately inoculated to one plate each of anaerobic blood agar and phenylethyl alcohol blood agar (1 drop of inoculum each). Plates should be streaked to achieve well-isolated colonies. Thioglycolate broth medium should be inoculated with 2 drops of inoculum near the bottom of the tube; the screw caps should be left loose to allow exchange of gases and exposure to anaerobic atmospheres. Media should be incubated in an anaerobic jar or anaerobic chamber at 35 to 37°C. Cultures should be examined after 48 h for growth and then reincubated for 5 to 7 days. It may be necessary to hold plates for as long as 2 to 4 weeks. Incubation of plate media in anaerobic glove boxes may lead to serious desiccation; we therefore prefer to incubate primary cultures in anaerobic jars. However, when anaerobic glove boxes are used, plates may be placed in plastic containers such as Tupperware to prevent drying. A paper towel should be placed on top of the plates in the containers to absorb the moisture that collects from condensation.

Isolation of *Actinomyces* species from specimens rich in microflora may be particularly troublesome. Selective media and methods for the isolation of *Actinomyces* species have been described elsewhere (1, 52). Cervical swabs or intrauterine devices may present a considerable problem, since the number of organisms may be quite large, and many different species are present. Traynor et al. described a method in which swabs are soaked in 5 ml of thioglycolate broth and the samples are diluted 10-fold in the same medium (10^{-1} to 10^{-4} dilutions) (87). Dilutions are plated on Columbia blood agar with and without 2.5 mg of metronidazole per liter; although this method would not inhibit many organisms such as streptococci and lactobacilli, it appears to give results comparable to visual results obtained by using fluorescent antibody in a direct microscopic examination.

The importance of *Actinomyces* species in periodontal diseases and caries has spurred workers in the dental research field to develop selective media for recovery of these species from dental plaque (52). However, the usefulness of these media with clinical materials has yet to be tested, and it would probably not be cost-effective to maintain a supply of such media in the clinical laboratory.

Identification and Distinguishing Characteristics

Colony and cell morphologies are perhaps the best means of presumptively separating anaerobic gram-positive rods into genera; however, these characteristics can be both variable and diverse. For example, *Actinomyces* species on blood agar yield numerous colony types, ranging from "smooth" to "molar tooth." The microcolony morphologies of *Actinomyces*, *Propionibacterium*, and *Rothia* species have been used by some laboratories and are very helpful, but their use does not lend itself to general clinical microbiology. The colony and cellular morphologies of these genera are quite variable, depending on culture medium used (liquid or solid) and ages and conditions of culture.

The key for the presumptive genus identification of anaerobic gram-positive rods is shown in Table 7. Many genera may be eliminated from consideration by relatively simple means. The presence of bacterial endospores indicates *Bacillus* or *Clostridium* species. Several filamentous clostridia (e.g., *Clostridium clostridioforme*, *C. perfringens*, and *C. ramosum*) fail to produce endospores on laboratory media. Growth on egg yolk agar with phospholipase C

TABLE 9 Differential characteristics of members of the genus *Actinomyces* and *Propionibacterium propionicus*[a]

Characteristic	A. viscosus	A. pyogenes	A. meyeri	A. israelii	A. naeslundii	A. odontolyticus	Propionibacterium propionicus	A. georgiae	A. gerencseriae
Catalase production	+	−	−	−	−	−		−	−
Nitrate reduction	d	−	−	d	+	+	+	−	−
Gelatin hydrolysis	−	+	−	−		−	d	d	−
H₂S production on triple sugar iron agar	+	−	−	+	+	+	d	c	
Esculin hydrolysis	d	−	−	+	+	d		+	+
Pink pigment on blood agar	−	−	−	−	−	+[c]	−	−	−
Urease	d	−	d	−	+		−	−	−
Acid from:									
meso-Inositol	d	d	−	+	+	−	d	d	d
Glycerol	d	−	d	−	d	d	d[d]	−	−
Xylose	−	d	+	+	d	d	−	−	−
Raffinose	+	−	−	+	+	−	+	−	+
Trehalose	d	d	−	−	+	−	d	+	+
Mannitol	−	−	−	d		−	+		+
Mol% G+C of DNA	59–69	56–58	64–67	57–65	63–68	62	63–65	65–69	70–71

[a]Data were adapted from references 22, 35, 73, 75, 81, and 82.
[b]Abbreviations and symbols: +, positive reactions for 90 to 100% of strains tested; −, negative reactions for 90 to 100% of strains tested; d, variable reactions.
[c]Pigmentation of *A. odontolyticus* may be variable and may take 7 to 10 days to develop. Exposure to oxygen may enhance pigmentation.
[d]Serovar 1, variable reactions; serovar 2, usually negative.

TABLE 10 Flow chart for identification of selected *Eubacterium* species[a]

I. Butyric acid produced
  A. Caproic acid produced: *E. alactolyticum*
  B. Caproic acid not produced, acid from glucose
    1. Indole produced: *E. saburreum*
    2. Indole not produced
      a. Nitrate reduced
        i. Esculin hydrolyzed: *E. multiforme*
        ii. Esculin not hydrolyzed: *E. monoforme*
      b. Nitrate not reduced
        i. Major lactic acid from glucose: *E. rectale*
        ii. No major lactic acid from glucose
          aa. Acid from glucose and esculin hydrolyzed: *E. limosum*
          bb. No acid from glucose or maltose, gelatin hydrolyzed: *E. combesii*
          cc. Weak acid from glucose and maltose, gelatin not hydrolyzed: *E. nodatum*
II. Butyric acid not produced
  A. Acid from glucose
    1. Indole produced: *E. tenue*
    2. Indole not produced
      a. Major lactic acid from glucose: *E. aerofaciens*
      b. No lactic acid from glucose: *E. contortum*
  B. No acid from glucose: *E. lentum, E. brachy, E. timidum, E. yurii*

[a]Modified from reference 63.

(lecithinase) production indicates *C. perfringens*. For identification of *C. clostridioforme* and *C. ramosum*, see chapter 47 of this Manual.

Some organisms appear more rodlike or diphtheroidal on solid medium than in broth culture. *Streptococcus mutans*, *Streptococcus intermedius*, *G. morbillorum*, and others form short coccoid forms on solid media more rapidly than in broth and may on initial inspection be mistaken for rods, for example, *Propionibacterium*, *Eubacterium*, *Bifidobacterium*, and *Lactobacillus* species. *Nocardia*, *Corynebacterium*, *Erysipelothrix*, and *Listeria* species may be eliminated from the anaerobic gram-positive nonsporeforming bacilli by testing oxygen tolerance. The observation of rods with clubs or bifurcated ends should suggest a species of *Bifidobacterium*, but this morphology may also be observed in other genera. Non-acid-fast branched filaments are good indicators for *Actinomyces* species.

The *Actinomyces* species are facultatively anaerobic except for *A. bovis*, *A. israelii*, and *A. meyeri*, which are strict anaerobes. All species grow best in primary cultures under anaerobic conditions. It may be difficult to obtain isolates in pure culture; therefore, several subcultures are recommended to ensure purity before biochemical tests are carried out.

The macroscopic appearance of *Actinomyces* colonies on blood agar and the cellular morphology are the features most frequently observed, but it should be remembered that these characteristics are extremely variable. Excellent photographs and drawings may be found in references 44, 81, and 82. Differential characteristics for species of *Actinomyces* and *Propionibacterium propionicus* are presented in Table 9.

Differentiation of several species may be difficult. *A. naeslundii* and *A. viscosus* are very similar but differ in catalase production, *A. viscosus* being positive. *A. israelii* and *Propionibacterium propionicus* are differentiated by the production of propionic acid by *Propionibacterium propionicus*. Morphologically, *A. israelii* and other *Actinomyces* species are very similar to *E. nodatum*, but the production of butyric acid by some *Eubacterium* species is a differential characteristic (39, 44). It may be quite difficult to distinguish *A. israelii* from *Propionibacterium propionicus* by the standard bench biochemical tests; however, although both organisms produce acetic and succinic acids as end products of carbohydrate metabolism, *Propionibacterium propionicus* produces propionic acid, whereas *A. israelii* produces lactic acid. Some *Actinomyces* spp. actually require CO_2 in order to produce succinate.

Eubacterium spp. (Tables 8 and 10) are easily confused with other anaerobic gram-positive rods; however, gas-liquid chromatographic analysis of metabolic end products is very useful in differentiation (39, 44, 63). Organisms may be either uniform or pleomorphic and are obligately anaerobic. They are chemoorganotrophs and may or may not be saccharoclastic. The mole percentage G+C for the species ranges from 30 to 55%, demonstrating that the genus is a heterogeneous group. This conclusion is supported by the

TABLE 11 Selected characteristics of *Bifidobacterium* species encountered in normal human flora and clinical materials[a]

Species	Fermentation of:					Occurrences in human material[b]			
	Arabinose	Cellobiose	Glycogen	Melezitose	Sucrose	Clinical	Mouth	Intestines-feces	Cervix-vagina
B. dentium	+	+	+	+[−]	+	+	+	+	+
B. bifidum	−	−	−	−	−	+		+	+
B. infantis	−	−[+]	−	−	+			+	+
B. globosum	+	V	+	−	+			+	
B. breve	−	+	+[−]	V	+	+		+	+
B. longum	+	−[+]	−	+[w]	+	+		+	
B. catenulatum	V	+	V	−	+			+	+
B. adolescentis	V	+	+	−	+			+	

[a]Taken from reference 1. Fermentation patterns of *Bifidobacterium* species overlap considerably; polyacrylamide gel electrophoresis or DNA-DNA homology data or both are required for the definitive identification of some species. All *Bifidobacterium* species produce acid from glucose. +, pH below 5.5, or positive; w, pH 5.5 to 5.9, or weak; −, pH above 5.9, or negative; V, variable; superscript, reaction of some strains of a species.
[b]The only documented pathogenic species listed is *B. dentium* (formerly *B. eriksonii*). Few other species have been isolated from clinical specimens; none are commonly encountered.

TABLE 12 Identifying characteristics of *Mobiluncus* species[a]

Characteristic	M. mulieris	M. curtisii	M. curtisii group SLH-29
Length (μm)	2.9	1.7	1.3–1.7
Gram reaction	−	V	−
Hippurate hydrolyzed	−	+	V (27%)
α-D-Galactosidase	−	+	V (7%)
Arginine dihydrolase	−	+	+
Proline aminopeptidase	+	+	+
α-D-Glucosidase	+	+	+
β-D-Glucosidase	−	−	−

[a]Adapted from references 79 and 84. V, variable; −, negative; +, positive.

fact that the metabolic end products fall into three groups: (i) butyric acid plus other short-chain fatty acids and alcohols; (ii) a combination of lactic, acetic, and formic acids with H_2; and (iii) few or no detectable acids.

The cellular and colonial morphologic characteristics of *Rothia dentocariosa* closely resemble those of *Actinomyces* spp. and to some extent those of *Nocardia* spp.; differentiation cannot be based on colonial and cellular morphologies. Biochemically, *R. dentocariosa* is fermentative, whereas *Nocardia* spp. are not; fermentation of glucose by *R. dentocariosa* does not produce succinic acid, whereas fermentation by *Actinomyces* spp. does. When grown aerobically, *R. dentocariosa* forms 1-mm colonies that are smooth or granular in appearance in 18 to 24 h. After 3 to 7 days, the colonies may be smooth or convex with highly convoluted surfaces. When grown anaerobically, the colonies are smaller and highly filamentous, resembling the "spider" colonies of *Actinomyces* spp. By contrast, *Actinomyces* spp. are preferentially anaerobic, also requiring CO_2 for maximal growth.

The key characteristic (Table 11) of *Bifidobacterium* spp. is the production of acetate and lactate at a molar ratio of 2:3, which distinguishes these organisms from other lactic acid- and acetic acid-producing gram-positive anaerobic rods.

Lactobacillus spp. can usually be identified on the basis of Gram stain, a negative catalase test, and gas chromatography showing a major lactic acid peak from glucose.

There are currently two recognized species of *Mobiluncus*: *M. curtisii* and *Mobiluncus mulieris*. They are oxidase, catalase, and indole negative and weakly saccharolytic, and they produce succinic and acetic acid during fermentation,

with or without lactic acid. Growth is not stimulated by formate and fumarate. Strains are motile by multiple subterminal flagella. H_2S is not produced. Colonies are colorless, translucent, smooth, and convex and may have a watery appearance, especially on very moist or fresh media. Colonies have a maximum diameter of 2 to 3 mm after 5 days of uninterrupted anaerobic incubation. Colonies may be less than 1 mm in diameter after 2 days of incubation.

M. curtisii can be differentiated from *M. mulieris* most easily by its length and its gram-variable rather than gram-negative staining. *M. curtisii* is shorter (mean length, 1.7 μm) than *M. mulieris* (mean length, 2.9 μm). Typical strains of *M. curtisii* hydrolyze hippurate, but up to 73% of atypical *M. curtisii* isolates (*M. curtisii* group SLH-29) identified by DNA homology lack the ability to hydrolyze hippurate (79). Gas chromatographic analysis of peptone-yeast broth supplemented with starch and 2% serum is of some usefulness in distinguishing the two species. *M. curtisii* produces a major succinate and a minor acetate peak, while *M. mulieris* generally produces major succinate and acetate peaks with or without a minor lactate peak. Both species of *Mobiluncus* are positive for proline aminopeptidase and α-D-glucosidase and negative for phosphatases and β-D-glucosidase. Typical strains of *M. curtisii* are also positive for α-D-galactosidase and arginine dihydrolase, while *M. mulieris* is negative for these two enzymes (Table 12). Several commercial systems for the identification of anaerobic gram-positive rods have been evaluated. While the number of isolates evaluated in each study is relatively small, the overall accuracy of these systems is low (Table 13).

Antimicrobic Susceptibilities

New information on antimicrobic susceptibilities of anaerobic gram-positive rods is sparse compared with the information available on other anaerobic species. Many reports fail to give results for specific species, opting instead to combine data for the group. However, it is generally accepted that the anaerobic gram-positive nonsporeforming rods are susceptible to penicillin G, carbenicillin, and chloramphenicol (92). Other antimicrobics have been reported to have various efficiencies. Clindamycin, erythromycin, and tetracycline were active against 94, 88, and 60% of isolates tested, respectively (71, 86, 90). *M. curtisii* is uniformly resistant to metronidazole, but the clinical significance of this fact is doubtful (83). Women with bacterial vaginosis who are treated with oral or intravaginal metronidazole rarely have persistent *M. curtisii* immediately after therapy (39a).

The eradication of these organisms from sites of infec-

TABLE 13 Identification of anaerobic gram-positive rods by commercial systems

Organism	% Correctly identified to genus and species (no. of isolates tested)[a]				
	API 20A	Vitek ANI	PRAS II	RapID-ANA	AN-Ident
Actinomyces spp.	14 (7)	50 (6)	NT	40 (5)	86 (7)
Propionibacterium spp.	90 (10)	89 (18)	66 (6)	NT	63 (16)
P. acnes	NT	NT	NT	96 (25)	NT
P. granulosum	NT	NT	NT	100 (2)	NT
E. lentum	100 (4)	33 (6)	100 (2)	69 (13)	100 (6)
E. limosum	NT	NT	NT	33 (3)	NT
Lactobacillus spp.	0 (1)	25 (4)	100 (1)	NT	100 (2)
Bifidobacterium spp.	0 (1)	75 (4)	NT	66 (3)	0 (1)

[a]Data were compiled from references 2, 15, 49, 64, and 77. NT, not tested.

.tion may be difficult, particularly in actinomycotic or acti-nomycosislike infections, since blood supplies to the infected site(s) may be inadequate, multiple abscesses may be present, and surrounding granulation tissue may impede antimicrobic penetration. There is considerable difficulty in obtaining antimicrobic penetration of granules with clubs, and prolonged therapy is therefore usually recommended (81).

Evaluation, Interpretation, and Reporting of Results

As with the anaerobic gram-positive cocci, interpretation of culture results may require additional clinical information as to how the specimen was obtained. Specimens yielding anaerobic gram-positive rods will yield two or three other organisms as well, so one should not automatically dismiss mixed cultures as contaminated. On the other hand, since these microorganisms are endogenous flora of the skin, oral cavity, genital tract, and gut, contamination by normal flora during specimen acquisition is often a problem. Therefore, communication with the physician submitting the specimen for culture is desirable and sometimes essential. The microbiologist handling the specimen should inquire how the specimen was obtained and determine from the physician how complete an identification of the isolates is required.

REFERENCES

1. **Allen, S. D.** 1985. Gram-positive, nonsporeforming anaerobic bacilli, p. 461–472. *In* E. H. Lennette, A. Balows, W. J. Hausler, Jr., and H. J. Shadomy (ed.), *Manual of Clinical Microbiology*, 4th ed. American Society for Microbiology, Washington, D.C.
2. **Appelbaum, P. C., C. S. Kaufmann, and J. W. Depenbusch.** 1985. Accuracy and reproducibility of a four-hour method for anaerobe identification. *J. Clin. Microbiol.* **21:**894–898.
3. **Baron, E. J., M. L. Vaisänen, M. McTeague, C. A. Strong, and S. M. Finegold.** 1993. Comparison of the Accu-Cul-Shure System and a swab placed in a B-D Port-a-Cul tube for specimen collection and transport. *Clin. Infect. Dis.* **16**(Suppl. 4):S325–S327.
4. **Bayer, A. S., A. W. Chow, D. Betts, and L. B. Guze.** 1978. Lactobacillemia—report of nine cases. Important clinical and therapeutic considerations. *Am. J. Med.* **64:**808–813.
5. **Bennion, R. S., J. E. Thompson, E. J. Baron, and S. M. Finegold.** 1990. Gangrenous and perforated appendicitis with peritonitis: treatment and bacteriology. *Clin. Ther.* **12:**31–44.
6. **Bjornson, H. S.** 1989. Biliary tract and hepatic infections, p. 333–347. *In* S. M. Finegold and W. L. George (ed.), *Anaerobic Infections in Humans*. Academic Press, Inc., San Diego, Calif.
7. **Brook, I.** 1987. Comparison of two transport systems for recovery of aerobic and anaerobic bacteria from abscesses. *J. Clin. Microbiol.* **24:**2020–2022.
8. **Brook, I.** 1988. Anaerobic bacteria in suppurative genitourinary infections. *J. Urol.* **141:**889–893.
9. **Brook, I.** 1988. Aerobic and anaerobic bacteriology of purulent nasopharyngitis in children. *J. Clin. Microbiol.* **26:**592–594.
10. **Brook, I.** 1988. Recovery of anaerobic bacteria from clinical specimens in 12 years at two military hospitals. *J. Clin. Microbiol.* **26:**1181–1188.
11. **Brook, I.** 1988. Microbiology of non-puerperal breast abscesses. *J. Infect. Dis.* **157:**377–379.
12. **Brook, I., and E. H. Frazier.** 1993. Significant recovery of nonsporulating anaerobic rods from clinical specimens. *Clin. Infect. Dis.* **16:**476–480.
13. **Brook, I., R. B. Kiehlich, and S. Grimm.** 1981. Bacteriology of acute periapical abscess in children. *J. Endodontol.* **7:**378–380.
14. **Buchanan, A. M., J. L. Scott, M. A. Gerencser, B. L. Beaman, S. Jang, and L. Biberstein.** 1984. *Actinomyces hordeovulneris* sp. nov., an agent of canine actinomycosis. *Int. J. Syst. Bacteriol.* **34:**439–443.
15. **Burlage, R. S., and P. D. Ellner.** 1985. Comparison of the PRAS II, AN-Ident, and RapID ANA systems for identification of anaerobic bacteria. *J. Clin. Microbiol.* **22:**32–35.
16. **Carlone, G. M., M. L. Thomas, R. J. Arko, G. O. Guerrant, C. W. Moss, J. M. Swenson, and S. A. Morse.** 1986. Cell wall characteristics of *Mobiluncus* species. *Int. J. Syst. Bacteriol.* **36:**288–296.
17. **Cato, E. P.** 1983. Transfer of *Peptostreptococcus parvulus* (Weinberg, Nativelle, and Prevot 1937), Smith 1957 to the genus *Streptococcus: Streptococcus parvulus* (Weinberg, Nativelle, and Prevot 1937) comb. nov., nom. rev., emend. *Int. J. Syst. Bacteriol.* **33:**82–84.
18. **Celig, D. M., and P. C. Schreckenberger.** 1991. Clinical evaluation of the RapID-ANA II panel for identification of anaerobic bacteria. *J. Clin. Microbiol.* **29:**457–462.
19. **Charfreitag, O., M. O. Collins, and E. Stackebrandt.** 1988. Reclassification of *Arachnia propionica* as *Propionibacterium propionicus* comb. nov. *Int. J. Syst. Bacteriol.* **38:**354–375.
20. **Collins, M. S., S. Stubbs, J. Hommez, and L. A. Devriese.** 1993. Molecular taxonomic studies of *Actinomyces*-like bacteria isolated from purulent lesions in pigs and description of *Actinomyces hyovaginalis* sp. nov. *Int. J. Syst. Bacteriol.* **43:**471–473.
21. **Cox, S. M., L. E. Phillips, L. J. Mercer, C. E. Stager, S. Waller, and S. Faro.** 1986. Lactobacillemia of amniotic fluid origin. *Obstet. Gynecol.* **68:**134–135.
22. **Cummins, C. S., and J. S. Johnson.** 1986. Genus I. *Propionibacterium* Orla-Jensen 1909, 337^{AL}, p. 1346–1363. *In* P. H. A. Sneath, N. S. Mair, M. E. Sharpe, and J. G. Holt (ed.), *Bergey's Manual of Systematic Bacteriology*, vol. 2. The Williams & Wilkins Co., Baltimore.
23. **Cummins, C. S., and C. W. Moss.** 1990. Fatty acid composition of *Propionibacterium propionicus* (*Arachnia propionica*). *J. Clin. Microbiol.* **40:**307–308.
24. **Davis, A. J., P. A. James, and P. M. Hawkey.** 1986. *Lactobacillus* endocarditis. *J. Infect.* **12:**169–174.
25. **Doyle, L., C. L. Young, S. S. Jang, and S. L. Hillier.** 1991. Normal vaginal aerobic and anaerobic bacterial flora of the rhesus macaque (*Macaca mulatta*). *J. Med. Primatol.* **20:**409–413.
26. **Drancourt, M., O. Oules, V. Bouche, and Y. Peloux.** 1993. Two cases of *Actinomyces pyogenes* infection in humans. *Eur. J. Clin. Microbiol. Infect. Dis.* **12:**55–57.
27. **Edmiston, C. E.** 1991. *Arachnia* and *Propionibacterium*: causal commensals or opportunistic diphtheroids. *Clin. Microbiol. Newsl.* **13:**57–59.
28. **Emmons, C. W., C. H. Binford, and J. P. Utz.** 1970. *Medical Mycology*, 2nd ed. Lea & Febiger, Philadelphia.
29. **Evans, D. T. P.** 1993. *Actinomyces israelii* in the female genital tract: a review. *Genitourin. Med.* **69:**54–59.
30. **Ezaki, T., S. L. Liu, Y. Hashimoto, and E. Yabuuchi.** 1990. *Peptostreptococcus hydrogenalis* sp. nov. from human fecal and vaginal flora. *Int. J. Syst. Bacteriol.* **40:**305–306.
31. **Ezaki, T., and E. Yabuuchi.** 1986. Transfer of *Peptococcus heliotrinreducens* corrig. to the genus *Peptostreptococcus: Peptostreptococcus heliotrinreducens* Lanigan 1983a comb. nov. *Int. J. Syst. Bacteriol.* **36:**107–108.
32. **Ezaki, T., N. Yamamoto, K. Ninomiya, S. Suzuki, and E. Yabuuchi.** 1983. Transfer of *Peptococcus indolicus, Peptococcus asaccharolyticus, Peptococcus prevotii*, and *Peptococcus magnus* to the genus *Peptostreptococcus* and proposal of *Peptostreptococcus tetradius* sp. nov. *Int. J. Syst. Bacteriol.* **33:**683–698.
33. **Finegold, S. M.** 1989. General aspects of anaerobic infection, p. 135–153. *In* S. M. Finegold and W. L. George (ed.), *Anaerobic Infections in Humans*. Academic Press, Inc., San Diego, Calif.
34. **Funke, G., S. Stubbs, A. von Graevenitz, and M. O. Collins.** 1994. Assignment of human-derived CDC group 1 coryneform bacteria and CDC group I-like bacteria to the genus *Actinomyces* as *Actinomyces neuii* subsp. *neuii* sp. nov., subsp.

nov., and *Actinomyces neuii* subsp. *anitratus* subsp. nov. *Int. J. Syst. Bacteriol.* **44:**167–171.

35. **Gerencser, M. A., and G. H. Bowden.** 1986. Genus *Rothia*. Georg and Brown 1967, 68^AL, p. 1342–1346. *In* P. H. A. Sneath, N. S. Mair, M. E. Sharpe, and J. G. Holt (ed.), *Bergey's Manual of Systematic Bacteriology*, vol. 2. The Williams & Wilkins Co., Baltimore.

36. **Glupczynski, Y., M. Labbe, F. Crockaert, P. Pepersack, P. Van Der Auwera, and E. Yourassowsky.** 1984. Isolation of *Mobiluncus* in four cases of extragenital infections in adult women. *Eur. J. Clin. Microbiol.* **3:**433–435.

37. **Goldstein, E. J. C., and D. M. Citron.** 1991. Susceptibility of anaerobic bacteria isolated from intra-abdominal infections to ofloxacin and interaction of ofloxacin with metronidazole. *Antimicrob. Agents Chemother.* **35:**2447–2449.

38. **Heimdahl, A., L. V. Konow, T. Satoh, and C. E. Nord.** 1985. Clinical appearance of orofacial infections of odontogenic origin in relation to microbiologic findings. *J. Clin. Microbiol.* **22:**299–302.

39. **Hill, G. B., O. M. Ayers, and A. P. Kohan.** 1987. Characterization and sites of infection of *Eubacterium nodatum*, *Eubacterium timidum*, *Eubacterium brachy*, and other asaccharolytic eubacteria. *J. Clin. Microbiol.* **25:**1540–1545.

39a. **Hillier, S. L.** Unpublished data.

40. **Hillier, S. L., M. A. Krohn, N. B. Kiviat, D. H. Watts, and D. A. Eschenbach.** 1991. Microbiologic causes and neonatal outcomes associated with chorioamnion infection. *Am. J. Obstet. Gynecol.* **165:**955–961.

41. **Hillier, S. L., M. A. Krohn, L. K. Rabe, S. J. Klebanoff, and D. A. Eschenbach.** 1993. Normal vaginal flora, H$_2$O$_2$-producing lactobacilli and bacterial vaginosis in pregnant women. *Clin. Infect. Dis.* **16**(Suppl. 4)**:**S273–S281.

42. **Hillier, S. L., D. H. Watts, M. F. Lee, and D. A. Eschenbach.** 1990. Etiology and treatment of post cesarean section endometritis after cephalosporin prophylaxis. *J. Reprod. Med.* **35:**322–328.

43. **Holdeman, L. V., E. P. Cato, J. A. Burmeister, and W. E. C. Moore.** 1980. Descriptions of *Eubacterium timidum* sp. nov., *Eubacterium brachy* sp. nov., and *Eubacterium nodatum* sp. nov. isolated from human periodontitis. *Int. J. Syst. Bacteriol.* **30:**163–169.

44. **Holdeman, L. V., E. P. Cato, and W. E. C. Moore (ed.).** 1977. *Anaerobe Laboratory Manual*, 4th ed. Virginia Polytechnic Institute and State University, Blacksburg.

45. **Holst, E., B. Wathne, B. Hovelius, and P.-A. Mårdh.** 1987. Bacterial vaginosis: microbiological and clinical findings. *Eur. J. Clin. Microbiol.* **6:**536–541.

46. **Johnson, J. L. I., L. V. H. Moore, B. Kaneko, and W. E. C. Moore.** 1990. *Actinomyces georgiae* sp. nov., *Actinomyces gerencseriae* sp. nov., designation of two genospecies of *Actinomyces naeslundii*, and inclusion of *A. naeslundii* serotypes II and III and *Actinomyces viscosus* serotype II in *A. naeslundii* genospecies 2. *Int. J. Syst. Bacteriol.* **40:**273–286.

47. **Jokipii, A. M., L. Jokipii, P. Sipila, and K. Jokinen.** 1988. Semiquantitative culture results and pathogenic significance of obligate anaerobes in peritonsillar abscesses. *J. Clin. Microbiol.* **26:**957–961.

48. **Juhl, G., and W. A. Brzezinski.** 1984. Disseminated actinomycosis associated with infection by *Capnocytophaga* species. *J. Infect. Dis.* **149:**654.

49. **Karachewski, N. O., E. L. Busch, and C. L. Wells.** 1985. Comparison of PRAS II, RapID ANA, and API 20A systems for identification of anaerobic bacteria. *J. Clin. Microbiol.* **21:**122–126.

50. **Kilpper-Bälz, R., and K. H. Schleifer.** 1981. Transfer of *Peptococcus saccharolyticus* (Foubert and Douglas) to the genus *Staphylococcus: Staphylococcus saccharolyticus* (Foubert and Douglas) comb. nov. *Zentralbl. Bakteriol. Parasitenkd. Infektionskr. Hyg. Abt. 1 Orig. Reihe C* **2:**324–331.

51. **Kilpper-Bälz, R., and K. H. Schleifer.** 1988. Transfer of *Streptococcus morbillorum* to the genus *Gemella* as *Gemella morbillorum* comb. nov. *Int. J. Syst. Bacteriol.* **38:**442–443.

52. **Kornman, K. S., and W. J. Loesche.** 1978. New medium for

isolation of *Actinomyces viscosus* and *Actinomyces naeslundii* from dental plaque. *J. Clin. Microbiol.* **7:**514–518.

53. **Lassnig, C., M. Dorsch, J. Wolters, E. Schaber, G. Stöffler, and E. Stackebrandt.** 1989. Phylogenetic evidence for the relationship between the genera *Mobiluncus* and *Actinomyces*. *FEMS Microbiol. Lett.* **65:**17–22.

54. **Li, N., Y. Hashimoto, S. Adnan, H. Miura, H. Yamamoto, and T. Ezaki.** 1992. Three new species of the genus *Peptostreptococcus* isolated from humans: *Peptostreptococcus vaginalis* sp. nov., *Peptostreptococcus lacrimalis* sp. nov., and *Peptostreptococcus lactolyticus* sp. nov. *Int. J. Syst. Bacteriol.* **42:**602–605.

55. **Lorenz, R. P., P. C. Appelbaum, R. M. Ward, and J. J. Botti.** 1982. Chorioamnionitis and possible neonatal infection associated with *Lactobacillus* species. *J. Clin. Microbiol.* **16:**558–561.

56. **Ludwig, W., R. Weizenegger, R. Kilpper-Bälz, and K. H. Schleifer.** 1988. Phylogenetic relationships of anaerobic streptococci. *Int. J. Syst. Bacteriol.* **38:**15–18.

57. **Madinger, N. E., J. A. McGregor, P. J. McKinney, S. T. Dembeck, C. S. Haskell, and Z. Johnson.** 1993. Comparative antibiotic susceptibilities of anaerobes associated with infection of the female reproductive tract. *Clin. Infect. Dis.* **16**(Suppl. 4)**:**S349–S352.

58. **Margaret, B. S., and G. N. Krywolap.** 1986. *Eubacterium yurii* subsp. *yurii* sp. nov. and *Eubacterium yurii* subsp. *margaretiae* subsp. nov.: test brush bacteria from subgingival dental plaque. *Int. J. Syst. Bacteriol.* **36:**145–149.

59. **Margaret, B. S., and G. N. Krywolap.** 1988. *Eubacterium yurii* subsp. *schtitka* subsp. nov.: test tube brush bacteria from subgingival dental plaque. *Int. J. Syst. Bacteriol.* **38:**207–208.

60. **Marler, L. M., J. A. Siders, L. C. Wolters, Y. Pettigrew, B. L. Skitt, and S. D. Allen.** 1991. Evaluation of the new RapID-ANA II system for the identification of clinical anaerobic isolates. *J. Clin. Microbiol.* **29:**874–878.

61. **Miyamoto, M. I., and F. C. Fang.** 1993. Pyogenic liver abscess involving *Actinomyces*: case report and review. *Clin. Infect. Dis.* **16:**303–309.

61a. **Moncla, B. J.** Unpublished data.

62. **Moore, L. V. H., W. E. C. Moore, E. P. Cato, R. M. Smibert, J. A. Burmeister, and A. M. Best.** 1987. Bacteriology of human gingivitis. *J. Dent. Res.* **66:**989–995.

63. **Moore, W. E. C., and L. V. H. Moore.** 1986. Genus *Eubacterium* Prevot 1938, 294^AL, p. 1353–1373. *In* P. H. A. Sneath, N. S. Mair, M. E. Sharpe, and J. G. Holt (ed.), *Bergey's Manual of Systematic Bacteriology*, vol. 2. The Williams & Wilkins Co., Baltimore.

64. **Murray, P. R., C. J. Weber, and A. C. Niles.** 1985. Comparative evaluation of three identification systems for anaerobes. *J. Clin. Microbiol.* **22:**52–55.

65. **Ng, J., L.-K. Ng, A. W. Chow, and J. R. Dillon.** 1994. Identification of five *Peptostreptococcus* species isolated predominantly from the female genital tract by using the rapid ID32A system. *J. Clin. Microbiol.* **32:**1302–1307.

66. **Olsen, I., J. L. Johnson, L. V. Moore, and W. E. Moore.** 1991. *Lactobacillus uli* sp. nov. and *Lactobacillus rimae* sp. nov. from the human gingival crevice and emended descriptions of *Lactobacillus minutus* and *Streptococcus parvulus. Int. J. Syst. Bacteriol.* **41:**261–266.

67. **Piriz, S., R. Cuenca, J. Valle, and S. Vadillo.** 1992. Susceptibilities of anaerobic bacteria isolated from animals with ovine foot rot to 28 antimicrobial agents. *Antimicrob. Agents Chemother.* **36:**198–201.

68. **Quentin, C., M.-A. Desailey-Chanson, and C. Bebear.** 1991. Evaluation of AN-Ident. *J. Clin. Microbiol.* **29:**231–235.

69. **Reig, M., A. Moreno, and F. Baquero.** 1992. Resistance of *Peptostreptococcus* spp. to macrolides and lincosamides: inducible and constitutive phenotypes. *Antimicrob. Agents Chemother.* **36:**662–664.

70. **Reiner, S. L., J. M. Harrelson, S. E. Miller, G. B. Hill, and H. A. Gallis.** 1987. Primary actinomycosis of an extremity: a case report and review. *Rev. Infect. Dis.* **9:**581–589.

71. **Rolfe, R. O., and S. M. Finegold.** 1982. Comparative in vitro

activity of ceftriaxone against anaerobic bacteria. *Antimicrob. Agents Chemother.* **22**:338–341.

72. **Schaal, K. P.** 1986. Genus *Actinomyces* Harz 1877, 133^AL, p. 1383–1418. *In* P. H. A. Sneath, N. S. Mair, M. E. Sharpe, and J. G. Holt (ed.), *Bergey's Manual of Systematic Bacteriology*, vol. 2. The Williams & Wilkins Co., Baltimore.

73. **Schaal, K. P., G. M. Schofield, and G. Pulverer.** 1980. Taxonomy and clinical significance of Actinomycetaceae and Propionibacteriaceae. *Infection* **8**(Suppl. 2):S122–S130.

74. **Schiefer-Ullrich, H., and J. R. Andreesen.** 1985. *Peptostreptococcus barnesae* sp. nov., a gram-positive, anaerobic, obligately partial purine utilizing coccus from chicken feces. *Arch. Microbiol.* **143**:26–31.

75. **Schofield, G. M., and K. P. Schaal.** 1980. Carbohydrate fermentation patterns of facultatively anaerobic actinomycetes using micromethods. *FEMS Microbiol. Lett.* **8**:67–69.

76. **Schofield, G. M., and K. P. Schaal.** 1981. A numerical taxonomic study of members of the Actinomycetaceae and related taxa. *J. Gen. Microbiol.* **127**:237–259.

77. **Schreckenberger, P. C., D. M. Celig, and W. M. Janda.** 1988. Clinical evaluation of the Vitek ANI card for identification of anaerobic bacteria. *J. Clin. Microbiol.* **26**:225–232.

78. **Schwarz, D., and W. J. Dieckman.** 1926. Anaerobic streptococci: their role in puerperal infection. *South. Med. J.* **19**:470–479.

79. **Schwebke, J. R., S. A. Lukehart, M. C. Roberts, and S. L. Hillier.** 1991. Identification of two new antigenic subgroups within the genus *Mobiluncus. J. Clin. Microbiol.* **29**:2204–2208.

80. **Sheppard, A., C. Cammarata, and D. H. Martin.** 1990. Comparison of different medium bases for the semiquantitative isolation of anaerobes from vaginal secretions. *J. Clin. Microbiol.* **28**:455–457.

81. **Slack, J. M., and M. A. Gerencser.** 1975. *Actinomyces, Filamentous Bacteria: Biology and Pathogenicity.* Burgess Publishing Co., Minneapolis.

82. **Sneath, P. H. A., N. S. Mair, M. E. Sharpe, and J. G. Holt (ed.).** 1986. *Bergey's Manual of Systematic Bacteriology*, vol. 2. The Williams & Wilkins Co., Baltimore.

83. **Spiegel, C. A.** 1987. Susceptibility of *Mobiluncus* species to 23 antimicrobial agents and 15 other compounds. *Antimicrob. Agents Chemother.* **31**:249–252.

84. **Spiegel, C. A., and M. Roberts.** 1984. *Mobiluncus* gen. nov., *Mobiluncus curtisii* subsp. *curtisii* sp. nov., *Mobiluncus curtisii* subsp. *holmesii* subsp. nov., and *Mobiluncus mulieris* sp. nov., curved rods from the human vagina. *Int. J. Syst. Bacteriol.* **34**:177–184.

85. **Sturm, A. W.** 1989. *Mobiluncus* species and other anaerobic bacteria in nonpuerperal breast abscess. *Eur. J. Clin. Microbiol.* **8**:789–792.

86. **Sutter, V. L., and S. M. Finegold.** 1976. Susceptibility of anaerobic bacteria to 23 antimicrobial agents. *Antimicrob. Agents Chemother.* **10**:736–752.

87. **Traynor, R. M., D. Pavatt, H. L. D. Duguid, and I. D. Duncan.** 1981. Isolation of actinomycetes from cervical specimens. *J. Clin. Pathol.* **34**:914–916.

88. **Tvede, M., and N. Høiby.** 1992. Experimental studies of survival of anaerobic bacteria at 4°C and 22°C in two different transport systems. *APMIS* **100**:1048–1052.

89. **van Dalen, P. J., T. J. M. van Steenbergen, M. M. Cowan, H. J. Busscher, and J. de Graaff.** 1993. Description of two morphotypes of *Peptostreptococcus micros. Int. J. Syst. Bacteriol.* **43**:787–793.

90. **Wexler, H. M., and S. M. Finegold.** 1988. In vitro activity of cefotetan compared with that of other antimicrobial agents against anaerobic bacteria. *Antimicrob. Agents Chemother.* **32**:601–604.

91. **Wexler, H. M., and S. M. Finegold.** 1988. In vitro activity of cefoperazone plus sulbactam compared with that of other antimicrobial agents against anaerobic bacteria. *Antimicrob. Agents Chemother.* **32**:403–406.

92. **Whiting, J. L., N. Cheng, and A. W. Chow.** 1987. Interactions of ciprofloxacin with clindamycin, metronidazole, cefoxitin, cefotaxime, and mezlocillin against gram-positive and gram-negative anaerobic bacteria. *Antimicrob. Agents Chemother.* **31**:1379–1382.

93. **Wideman, P. A., V. L. Vargo, D. Citronbaum, and S. M. Finegold.** 1976. Evaluation of the sodium polyanethol sulfonate disk test for the identification of *Peptostreptococcus anaerobius. J. Clin. Microbiol.* **4**:330–333.

94. **Wilkins, T. D., W. E. C. Moore, S. E. H. West, and L. V. Holdeman.** 1975. *Peptococcus niger* (Hall) Kluyver and van Niel 1936: emendation of description and designation of neotype strain. *Int. J. Syst. Bacteriol.* **25**:47–49.

95. **Williams, B. L., G. F. McCann, and F. D. Schoenknecht.** 1983. Bacteriology of dental abscesses of endodontic origin. *J. Clin. Microbiol.* **18**:770–774.

96. **Woese, C. R.** 1987. Bacterial evolution. *Microbiol. Rev.* **51**:221–271.

97. **Yoonessi, M., K. Crickard, I. S. Cellino, S. K. Satchidanand, and W. Fett.** 1985. Association of *Actinomyces* and intrauterine contraceptive devices. *J. Reprod. Med.* **30**:48–52.

Bacteroides, Porphyromonas, Prevotella, Fusobacterium, and Other Anaerobic Gram-Negative Bacteria

HANNELE R. JOUSIMIES-SOMER, PAULA H. SUMMANEN, AND
SYDNEY M. FINEGOLD

49

TAXONOMY

The anaerobic gram-negative bacteria are part of the normal flora of the mouth, upper respiratory tract, intestinal tract, and urogenital tract of humans and animals. Anaerobic spirochetes are covered in chapter 52, and *Campylobacter* spp. other than *Campylobacter rectus* and *Campylobacter curvus* (*Wolinella recta* and *Wolinella curva*) are discussed in chapter 37 of this Manual; the rest are discussed here. The initial differentiation of these genera is based on cellular morphology, motility, flagellar arrangement, and an analysis of metabolic end products by gas-liquid chromatography (GLC) (Table 1) (27, 53, 62). Species definition is based on biochemical characteristics, nucleic acid base composition, and homology (chapter 20 of this Manual). *Acidaminococcus*, *Megasphaera*, and *Veillonella* are the genera of anaerobic, gram-negative cocci (Table 1). In the majority of clinical specimens, only organisms of the genera *Bacteroides*, *Porphyromonas*, *Prevotella*, *Fusobacterium*, and *Veillonella* are encountered.

Bacteroides spp., *Porphyromonas* spp., *Prevotella* spp., *Fusobacterium* spp., and *Bilophila* sp. are presumptively characterized on the basis of colonial and cellular morphology, pigment production, fluorescence under long-wave UV light, susceptibility to special-potency antibiotic disks, and certain rapidly determined biochemical characteristics. A definitive species-level identification requires a battery of biochemical tests. Definitive identification is not feasible for all anaerobic isolates because of financial constraints and is not ordinarily important for clinical purposes. Generally, for clinical purposes (especially for initiating empiric therapy), it is sufficient to know the broad groupings of isolates and their usual patterns of susceptibility to antimicrobial agents.

In recent years, the taxonomy of anaerobic gram-negative bacilli has been in a state of great change, and this trend will continue. The genus *Bacteroides* has been restricted to the *Bacteroides fragilis* group only (including *Bacteroides eggerthii*) (53); the taxonomic positions of other species still included in the genus *Bacteroides* remain uncertain, but all of these species will ultimately be transferred to other genera. The genera *Prevotella* (saccharolytic) and *Porphyromonas* (pigmented asaccharolytic organisms) were previously included in the genus *Bacteroides* (52, 54). The asaccharolytic, formate- and fumarate-requiring gram-neg-

ative rods *Bacteroides ureolyticus*, *Bacteroides gracilis*, *C. curvus*, and *C. rectus* are microaerophiles instead of true anaerobic bacteria (24). *Campylobacter* spp., *W. recta*, *W. curva*, *B. gracilis*, and *B. ureolyticus* form a tight homology group, and *W. recta* and *W. curva* have been renamed *C. rectus* and *C. curvus*, respectively (65). *Wolinella succinogenes*, isolated from the bovine rumen, is currently the only species remaining in the genus *Wolinella*. The taxonomic positions of *B. gracilis* and *B. ureolyticus* remain uncertain; additional studies are required to determine whether to transfer these organisms to the genus *Campylobacter*.

The anaerobic gram-negative cocci are identified presumptively on the basis of colonial and cellular morphology, fluorescence under long-wave UV light, susceptibility to special-potency antibiotic disks, and some rapidly determined biochemical characteristics. A definite identification relies on carbohydrate fermentation test results and fatty acid profiles of metabolic end products determined by GLC.

For recent taxonomic changes, see Table 2 and reference 61.

DESCRIPTION OF THE GROUP

The anaerobic gram-negative bacteria are differentiated from the facultatively anaerobic bacteria by their inability to grow in the presence of oxygen and their susceptibility to metronidazole. Some of the species included here are indeed microaerophiles and are often resistant to metronidazole; the appropriate atmospheric requirements should be determined for all isolates.

NATURAL HABITATS AND CLINICAL SIGNIFICANCE

Gram-negative anaerobic bacilli are the anaerobes most commonly encountered in clinical infections; they are found in more than half of the specimens yielding anaerobes (6, 16, 62). The bile-resistant *B. fragilis* group (see Table 4) is the most commonly recovered anaerobe in clinical specimens and is more resistant to antimicrobial agents than are most other anaerobes. *B. fragilis* and *Bacteroides thetaiotaomicron* are of the greatest clinical significance. They are recovered from most intra-abdominal infections and may occur in infections at other sites. Members of the *B. fragilis* group are major constituents of the normal

603

TABLE 1 Differentiation of genera of gram-negative anaerobic bacteria

Characteristic	Genus
Rod-shaped cells or coccobacilli	
I. Nonmotile or peritrichous flagella	
A. Produce butyric acid (without isobutyric and isovaleric acids)	*Fusobacterium*
B. Produce major lactic acid	*Leptotrichia*
C. Produce acetic acid and hydrogen sulfide; reduce sulfate..	*Desulfomonas*
D. Not as above (A, B, or C)	*Anaerorhabdus*
	Bacteroides
	Bilophila
	Dichelobacter
	Fibrobacter
	Megamonas
	Mitsuokella
	Porphyromonas
	Prevotella
	Rikenella
	Ruminobacter
	Sebaldella
	Tissierella
II. Polar flagella	
A. Fermentative	
1. Produce butyric acid..	*Butyrivibrio*
2. Produce succinic acid	
a. Spiral cells...	*Succinivibrio*
b. Ovoid cells...	*Succinimonas*
3. Produce propionic and acetic acids...	*Anaerovibrio*
B. Nonfermentative; produce succinic acid from fumarate	*Campylobacter* (*Wolinella*)
III. Tufts of flagella on concave side of curved cells; fermentative...	*Selenomonas*
IV. Flagella in spiral arrangement along cell body; fermentative..	*Centipeda*
V. Bipolar tufts of flagella..	*Anaerobiospirillum*
Spherical or kidney bean-shaped cells	
I. Produce propionic and acetic acids..	*Veillonella*
II. Produce butyric and acetic acids..	*Acidaminococcus*
III. Produce isobutyric, butyric, isovaleric, valeric, and caproic acids...	*Megasphaera*

colonic flora, are also found in smaller numbers in the female genital tract, but are not common in the mouth or upper respiratory tract. *Bacteroides splanchnicus*, formerly included in the *B. fragilis* group, is frequently isolated from patients with intra-abdominal infections.

The pigmented anaerobic gram-negative bacilli are composed of saccharolytic and asaccharolytic species of the genera *Prevotella* and *Porphyromonas* (52, 54), respectively. Nine species of these genera are found in human clinical material (see Table 7). *Prevotella corporis*, *Prevotella denti-*

TABLE 2 Recent taxonomic changes (since the fifth edition of *Manual of Clinical Microbiology*)

New nomenclature	Previous nomenclature	Reference
Dicholebacter nodosus[a]	*Bacteroides nodosus*	11
Campylobacter curvus	*Wolinella curva*	65
Campylobacter rectus	*Wolinella recta*	65
Fusobacterium necrophorum subsp. *funduliforme*	*Fusobacterium necrophorum* biovar B	57
Fusobacterium necrophorum subsp. *necrophorum*	*Fusobacterium necrophorum* biovar A	57
Fusobacterium nucleatum subsp. *animalis*[a]	None (new subspecies)	19
Fusobacterium nucleatum subsp. *fusiforme*	None (new subspecies)	19
Fusobacterium nucleatum subsp. *nucleatum*	None (new subspecies)	12
Fusobacterium nucleatum subsp. *polymorphum*	None (new subspecies)	12
Fusobacterium nucleatum subsp. *vincentii*	None (new subspecies)	12
Fusobacterium pseudonecrophorum	*Fusobacterium necrophorum* biovar C	58
Porphyromonas canoris[a]	None (new species)	38
Porphyromonas circumdentaria[a]	None (new species)	37
Porphyromonas salivosa[a]	*Bacteroides salivosus*	37
Prevotella nigrescens	None (new species; previously part of *Prevotella intermedia*)	55

[a]Of animal origin.

cola, *Prevotella intermedia, Prevotella loescheii, Prevotella mela-ninogenica, Prevotella nigrescens* (a new species [55]), *Porphyromonas endodontalis,* and *Porphyromonas gingivalis* are found in the human oral cavity. Some are important pathogens in oral, dental, and bite infections and may produce infections of the head, neck, and lower respiratory tract. Some of the above-named pigmented organisms plus *Prevotella asaccharolytica* are also prevalent in the urogenital and intestinal tracts and are important in infections arising from these sources. Three *Porphyromonas* spp. of animal origin (*Porphyromonas canoris, Porphyromonas circumdentaria,* and *Porphyromonas salivosa*) have been named (37, 38); in addition, two new groups of isolates phenotypically similar to *Porphyromonas* species have been described (32). *Porphyromonas* spp. of animal origin have been recently encountered in humans with animal bite infections. The pigmented species still remaining in the genus *Bacteroides, Bacteroides levii* and *Bacteroides macacae,* are also of animal origin and are possibly related to *Porphyromonas* spp. (53). Strains phenotypically similar to *B. levii* have been isolated from patients with various types of human clinical infections, including pleuropulmonary, skin, and soft tissue infections (Wadsworth Laboratory) and bacterial vaginosis (25).

The bile-sensitive, nonpigmented, saccharolytic gram-negative bacilli (see Table 5) are found in the same settings as the pigmented gram-negative rods (34). *Prevotella bivia* and *Prevotella disiens* are found in female patients with genital tract infections and less frequently in patients with oral infections; these strains are often resistant to the β-lactam antibiotics, including penicillin, aminopenicillins, and cephalosporins. *Prevotella oris* and *Prevotella buccae* are found in a variety of oral (22), pleuropulmonary (43), and other infections. The less frequently encountered *Prevotella oralis* group is now represented by *P. oralis, Prevotella veroralis, Prevotella buccalis,* and *Prevotella oulorum* (54).

Prevotella zoogleoformans (indole negative) is rarely isolated from human clinical specimens, whereas *Prevotella heparinolytica* (indole positive) is found often in the oral cavity and in patients with oral-associated infections. *Mitsuokella dentalis* (a new species) has been isolated from infected root canals and periodontal pockets (18, 23); we have also encountered this organism in mandibular and gum abscesses.

The nonpigmented, asaccharolytic, or weakly fermentative species (see Table 6) *Anaerorhabdus furcosus, Bacteroides capillosus, Bacteroides coagulans, Bacteroides pneumosintes, Bacteroides putredinis, Desulfomonas pigra, Desulfovibrio* spp., and *Tissierella praeacuta* inhabit the intestinal tract and have occasionally been recovered from patients with miscellaneous infections (29, 34). *Bacteroides forsythus,* a fusiform gram-negative rod, is a putative pathogen recovered from subgingival sites in patients with periodontitis and is often isolated together with *C. rectus* and *Fusobacterium nucleatum.* Our group has encountered *Bacteroides tectum,* of feline oral flora and associated infections, in humans who have sustained cat bites. *Bilophila wadsworthia,* a bile-resistant organism, has been recovered from patients with acute appendicitis, gangrenous and perforated appendicitis, and related abscesses. It has also been isolated from various other clinical specimens including blood, joint fluid, and pleural fluid as well as from human feces and vaginal and oral secretions (5, 33). *Bilophila wadsworthia* is easily overlooked in cultures owing to its fastidious growth.

Members of the formate-fumarate-requiring *B. ureolyticus*-like group, including *B. ureolyticus, B. gracilis, C. curvus,* and *C. rectus,* have been isolated from patients with various

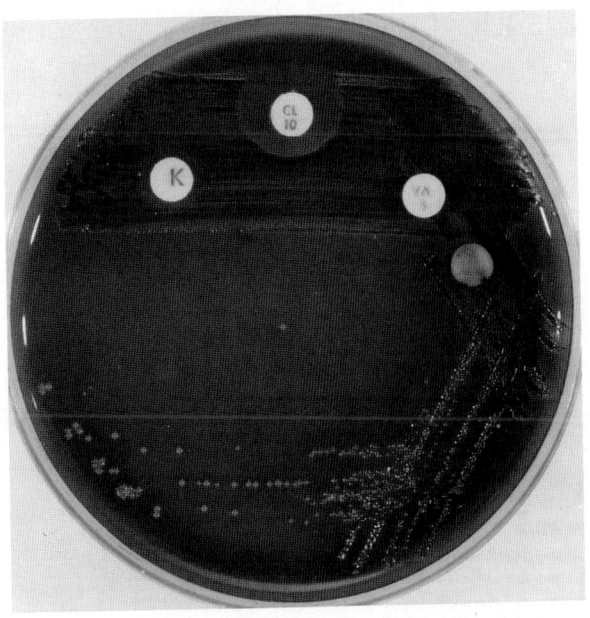

FIGURE 1 Placement of special-potency antibiotic disks. A blank disk for indole testing may be added (usually after growth has occurred).

types of infections. *B. ureolyticus* has been recovered from patients with pulmonary, head and neck, intra-abdominal, urogenital, bone, and soft tissue infections. *B. gracilis* has been recognized as an important pathogen in serious visceral or head and neck infections (30). Our recent studies suggest that the organisms currently identified as *B. gracilis* are a heterogeneous group in terms of bile resistance, composition of cellular fatty acids, atmospheric requirements, and susceptibility patterns and will probably be placed in at least two different genera (60). One group within the *B. gracilis*-related strains is more resistant to antimicrobial agents than even the *B. fragilis* group. *Campylobacter* (*Wolinella*) spp. are primarily oral isolates found in patients with periodontitis; *C. rectus* has been implicated as a putative pathogen at sites of active periodontal breakdown (13).

F. nucleatum is the *Fusobacterium* sp. most commonly encountered (see Table 8) in clinical infections (62). This organism is found in the mouth, genital, gastrointestinal, and upper respiratory tracts. It is often involved in the same types of infections as the pigmented *Prevotella* spp. and *Porphyromonas* spp. Currently, *F. nucleatum* of human origin has been divided into four different subspecies (Table 2) (12, 19, 20). *Fusobacterium necrophorum* is a very virulent anaerobe that may cause severe infection, usually in children or young adults, originating from pharyngotonsillitis. It was the most common anaerobe isolated from peritonsillar abscesses in young adults (31). In addition to peritonsillar abscess, local complications include neck space infections and jugular vein septic thrombophlebitis. There may also be multiple metastatic abscesses (most frequently in the lungs, pleural space, liver, and large joints) related to bacteremia (postanginal sepsis syndrome or Lemierre's disease) (49). *F. necrophorum* is encountered much less often in serious infections now than in the era before antimicrobial agents, but this makes it more treacherous, because many clinicians may not be familiar with the problems.

TABLE 3 Grouping of anaerobic gram-negative rods[a]

Group or organism	Kanamycin (1,000 μg)	Vancomycin (5 μg)	Colistin (10 μg)	Growth in 20% bile	Catalase	Indole	Lipase
B. fragilis group	R	R	R	+	V	V	−
Other *Bacteroides* spp.	R	R	V	$-^{+}$	$-^{+}$	V	−
Pigmented species	R	V	V	−	$-^{+}$	V	V
Porphyromonas spp.	R	S	R	−	V	+	$-^{+}$
Prevotella spp.	R^{S}	R	V	−	−	V	V
P. intermedia-P. nigrescens	R^{S}	R	S	−	−	+	$+^{-}$
P. loescheii	R	R	V	−	−	−	V
Other *Prevotella* spp.	R	R	V	−	$-^{+}$	$-^{+}$	$-^{+}$
B. ureolyticus-like group[f]	S	R	S	$-^{+}$	−	−	−
B. ureolyticus	S	R	S	−	−	−	
B. gracilis	S	R	S	$-^{+}$	−	−	
Campylobacter spp.[g]	S	R	S	−	−	−	
Bilophila sp.[h]	S	R	S	+	+	−	−
Desulfomonas pigra[i]	S	R	R	V	−	−	−
Fusobacterium spp.	S	R	S	V	−	V	V
F. nucleatum	S	R	S	−	−	+	−
F. necrophorum	S	R	S	$-^{+}$	−	+	$+^{-}$
F. varium-F. mortiferum	S	R	S	+	−	V	−

Erythromycin, commonly used in upper respiratory tract infections such as tonsillitis, is not active against fusobacteria. *F. necrophorum* has been recently separated into two subspecies: *F. necrophorum* subsp. *necrophorum* contains the lipase-positive, hemagglutinin-producing biovar A, and *F. necrophorum* subsp. *funduliforme* contains the lipase-negative, non-hemagglutinin-producing biovar B (57). Furthermore, *Fusobacterium pseudonecrophorum* was proposed to

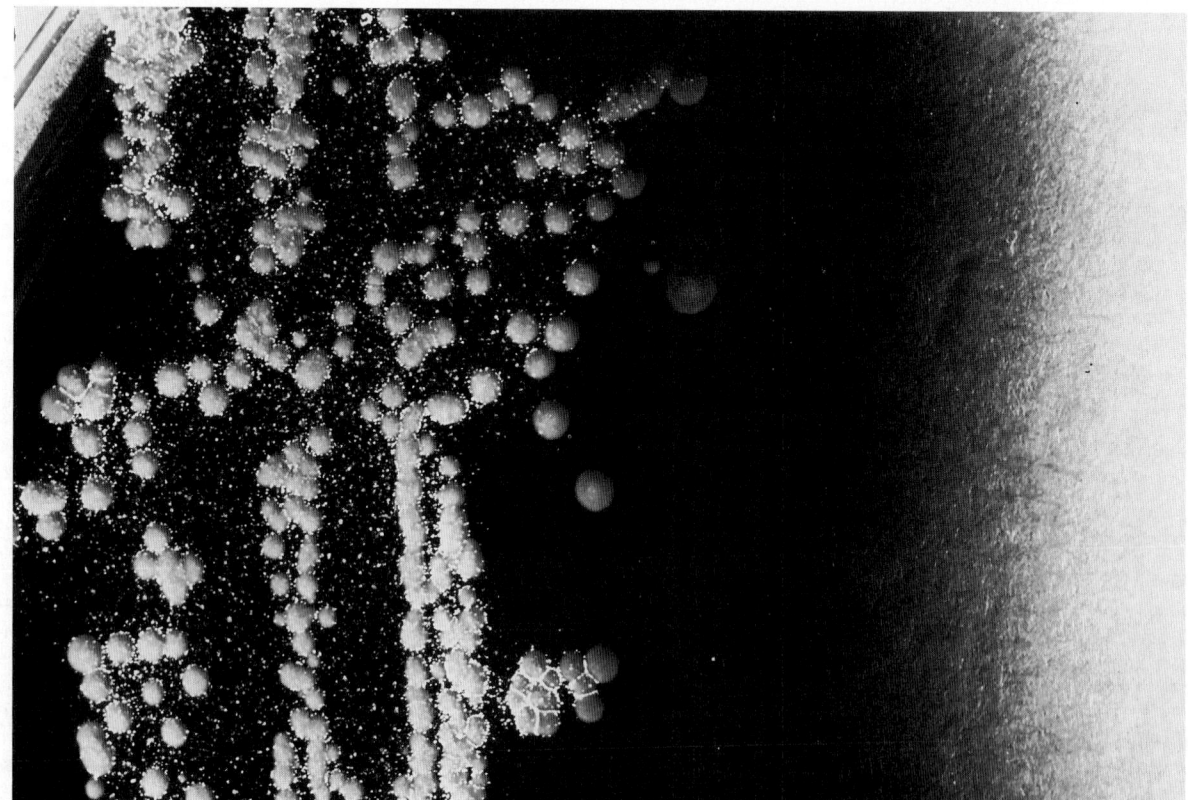

FIGURE 2 Colonies of *B. fragilis* on BBE agar. Note the blackening of the agar and colonies due to esculin hydrolysis and bile precipitation.

TABLE 3 *(Continued)*

Group or organism	Slender cells with pointed ends	Growth stimulated by formate-fumarate[b]	Nitrate reduction[c]	Urease[d]	Motility[e]	Pitting of agar	Pigment	Brick red fluorescence
B. fragilis group								−
Other Bacteroides spp.							+	+⁻
Pigmented species							+	+⁻
Porphyromonas spp.							+	+
Prevotella spp.							+	+
P. intermedia-P. nigrescens							+	+
P. loescheii							−	−
Other Prevotella spp.								
B. ureolyticus-like group[f]		+	+	V	V	V		
B. ureolyticus		+	+	+	−	V		
B. gracilis		+	+	−	−	V		
Campylobacter spp.[g]		+	+	−	+	V		
Bilophila sp.[h]	−		+	+⁻	−	−		
Desulfomonas pigra[i]	−		V	V	−	−		
Fusobacterium spp.	V							
F. nucleatum	+							
F. necrophorum	−							
F. varium-F. mortiferum	−							

[a]R, Resistant; S, susceptible; R^S, most strains resistant, some strains susceptible; V, variable; +, positive reaction for majority of strains; −, negative reaction; +⁻, most strains positive; −⁺, most strains negative, some strains positive.

[b]Compare the growth of the organism in an unsupplemented thioglycolate broth with growth in a broth supplemented with formate and fumarate. Additive: dissolve 3 g of sodium formate, 3 g of fumaric acid, and 20 pellets of sodium hydroxide in 50 ml of distilled water; adjust the pH to 7; and filter sterilize. Add 0.5 ml of additive to 10 ml of culture broth (26, 62).

[c]Use the spot nitrate disk test (see references 24a and 62).

[d]Make a heavy suspension of the organism in 0.5 ml of sterile urea broth (Difco Laboratories, Detroit, Mich.) or in sterile water, and insert a urea tablet (Rosco). Incubate the tubes aerobically for up to 24 h. A bright pink or red is positive; this color usually appears within 15 to 30 min. If the indicator becomes reduced, add Nessler reagent (24a) to determine the ammonia production (ammonia indicates a positive reaction).

[e]Check motility with a young broth culture supplemented with formate and fumarate.

[f]Formate and fumarate should be added to broth media for this group of organisms.

[g]C. (Wolinella) rectus and C. curvus.

[h]Growth stimulated by 1% pyruvate (final concentration).

[i]Desulfoviridin positive. Inoculate a tube of liquid medium supplemented with 1% pyruvate and 0.25% magnesium sulfate. Incubate until turbidity indicates good growth. Centrifuge the tube to obtain a pellet. Pipette a heavy drop of the pellet onto a slide. Add a drop of 2 N NaOH to the pellet, and immediately observe under long-wave (366-nm) UV light. Positive: red fluorescence; negative: no fluorescence.

include strains belonging to biovar C (58); however, a recent study suggested that *F. pseudonecrophorum* is a synonym for *Fusobacterium varium* (3). *Fusobacterium mortiferum* and *F. varium* are mainly encountered in patients with intra-abdominal infections. *Fusobacterium alocis*, *Fusobacterium sulci*, and *Fusobacterium periodonticum* are primarily isolated from subgingival sites in patients with gingivitis and periodontitis (64). *Fusobacterium russii* has been implicated in animal bite infections (21). *Fusobacterium ulcerans* is found in tropical ulcer (1).

Leptotrichia buccalis is a common mouth organism and may also be found in the vagina and intestinal tract. It has been involved in cases of septicemia in immunocompromised patients with oral mucosal infections (4).

Selenomonas sputigena, *Selenomonas artemidis*, *Selenomonas dianae*, *Selenomonas flueggei*, *Selenomonas infelix*, and *Selenomonas noxia* are all oral organisms, as is *Centipeda periodontii*, and are found in subgingival sites in patients with periodontitis (36, 47).

The motile *Succinivibrio dextrinosolvens* and *Butyrivibrio fibrisolvens* are normal colonic flora but may occasionally be encountered in patients with clinical infections (29). The sites of normal carriage of *Anaerobiospirillum succiniciprodu-*

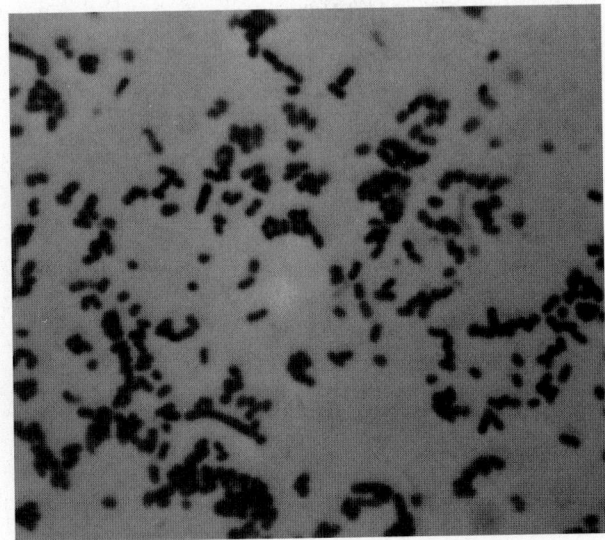

FIGURE 3 Coccobacillary cells of *Porphyromonas asaccharolytica*.

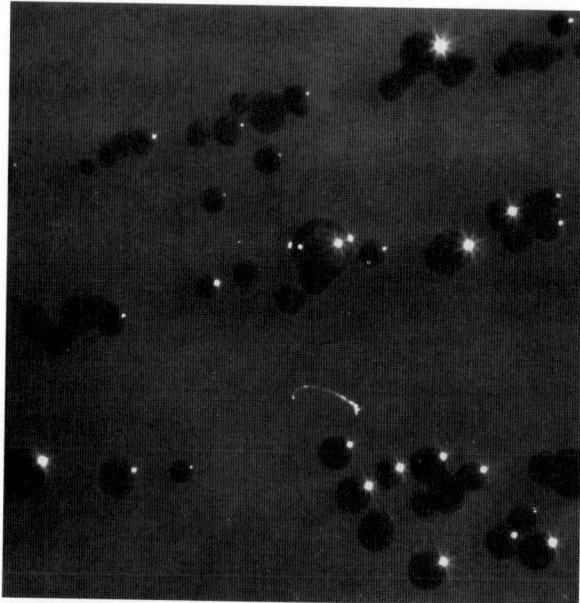

FIGURE 4 Black-pigmented colonies of *Porphyromonas asaccharolytica*.

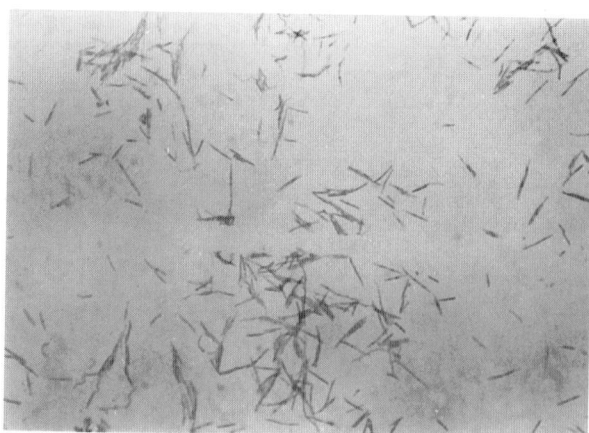

FIGURE 6 Cells of *F. nucleatum*. Note the slender shape with pointed ends.

Veillonella spp. have been encountered in patients with oral, bite wound, head, neck, and miscellaneous soft tissue infections (62).

COLLECTION, TRANSPORT, AND STORAGE OF SPECIMENS

General guidelines for collection, transport, and storage of specimens are discussed in chapter 3 of this Manual. Sites normally harboring a rich indigenous flora, such as the intestinal tract or vagina, should not be sampled and cultured for anaerobes except under special circumstances and in special ways (e.g., quantitative study of upper-small-bowel flora in the blind-loop syndrome). Lower respiratory tract specimens and endometrial samples are especially difficult to obtain without contaminating the sample with indigenous flora. Double-lumen-catheter bronchial brush-

cens in humans is unknown at present, but *Anaerobiospirillum* species are common in the fecal flora of cats and dogs (39). Strains of *Anaerobiospirillum* have been isolated from blood cultures of compromised patients and from fecal specimens of patients with diarrhea (40, 44). In the latter case, a zoonotic role for *Anaerobiospirillum* spp. has been proposed (39).

Acidaminococcus fermentans, *Megasphaera elsdenii*, and *Veillonella parvula* (see Table 9) are part of the normal human fecal flora, and *V. parvula*, as well as *Veillonella atypica* and *Veillonella dispar*, is part of the normal oral flora. Anaerobic gram-negative cocci are rare pathogens. *Veillonella* spp. are isolated more frequently from clinical specimens than are *Megasphaera* spp.; *Acidaminococcus* spp. and

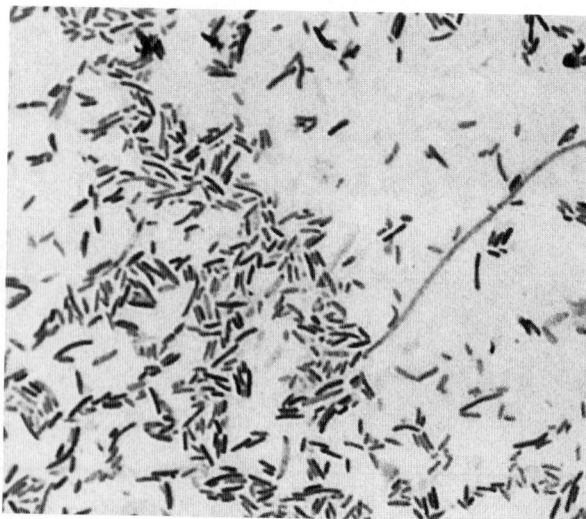

FIGURE 5 Microscopic morphology of *Bilophila wadsworthia*.

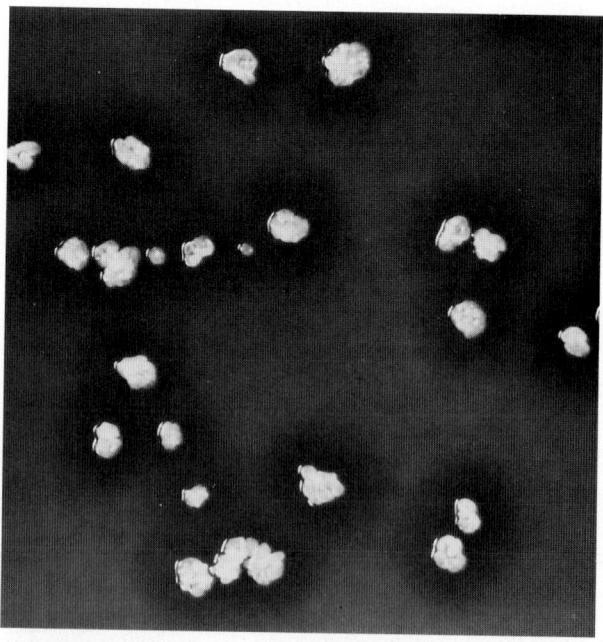

FIGURE 7 Bread crumb-shaped colonies of *F. nucleatum*.

ings and bronchoalveolar lavage fluid, both cultured quantitatively, and pleural fluid represent good specimens for the former, and an endometrial suction curette (Pipelle; Unimar, Wilton, Conn.) biopsy provides a good sample from the latter site (17). Instructions for collection of specimens from various body sites and by various methods are given in more detail elsewhere (15, 28, 62).

Ulcers should be carefully debrided, and proper samples should be collected from the base or progressive edge, where bacteria actively multiply, rather than from unremoved crust or surface pus, which are often contaminated by other bacteria not reflecting the true infecting flora.

Pus, when present, is best aspirated into a syringe through a needle and injected into an anaerobic transport vial containing an oxidation-reduction indicator. Syringes used for aspiration should not be used as transporters because of the potential danger of needle stick injuries and because oxygen diffuses through plastic syringes. Pieces of infected tissue obtained by excision or biopsy are always preferable to pus, which is, in turn, preferable to a specimen obtained by swab. Tissue samples are best transported in loosely capped containers sealed in anaerobic gas-impermeable bags. For small tissue and biopsy specimens and for subgingival and root canal samples, a semisolid anaerobic transport medium in which the specimen can be submerged may be used. Swabs are prone to drying and may carry too small a volume of sample to be cultured on several media or quantitatively. More detailed information on transport systems can be found elsewhere (15, 28, 62).

Conditions and time of transport should not affect the viability or relative proportions of bacteria present in the specimen. Fast transport is important when Gram stain and culture results are needed early for guidance of therapy. Specimens should not be refrigerated, as oxygen diffuses better at lower temperatures.

DIRECT DETECTION

The following methods of direct examination may be useful for the detection of gram-negative, non-spore-forming anaerobic bacteria: macroscopic examination, Gram stain, dark-field or phase-contrast microscopy, and GLC.

The gross appearance (purulence, necrotic tissue), fluorescence under long-wave (366-nm) UV light, and odor of the specimen can give the laboratory valuable clues to the presence of anaerobes. Fetid or putrid odor due to volatile short-chain fatty acids and amines is always associated with the presence of anaerobes in the sample. Black, necrotic tissue and/or red fluorescence of the sample may indicate the presence of pigmented gram-negative rods.

Nucleic acid probes, used in the direct demonstration of some anaerobic oral and nonoral gram-negative bacilli, are not yet standardized and produced for commercial distribution.

Of the conventional methods, Gram stain is by far the simplest and the most likely to yield significant information. It is recommended that direct smears be fixed in methanol for 30 s to preserve the host and bacterial cell morphologies (41). Standard Gram stain procedures and reagents are used, except that 0.5% basic fuchsin, which enhances the staining of gram-negative anaerobes, is substituted for safranin as the counterstain (62). In thick films from exudates and bloody fluids, recognition of organisms may be facilitated by staining with acridine orange. Dark-field and phase-contrast microscopy may be helpful in the detection of small, poorly staining organisms (B. pneumo-

sintes); for the direct observation of motility (Campylobacter spp.); for noting spores (Clostridium spp.); and for the recognition of morphotypes not cultivable on ordinary media (spirochetes).

Direct gas chromatographic analysis of specimens other than blood cultures does not add relevant information to what is obtainable from Gram-stained smears.

ISOLATION PROCEDURES: MEDIA

The use of selective media along with nonselective media will increase the yield and save time in terms of recognition and isolation of colonies. Nonselective media made from different basal media differ in their abilities to support the growth of certain groups of anaerobes; brucella base is superior to Trypticase soy (CDC base agar) and Schaedler base for isolation of gram-negative bacilli, but CDC base supports the growth of anaerobic gram-positive cocci better. Brain heart infusion base is superior to Trypticase soy in isolation efficiency for Eubacterium species but worse for pigmented gram-negative rods from subgingival and other samples (48, 56). Fastidious anaerobe agar (Lab M, Bury, England) produces luxuriant growth of fusobacteria and can be used as a base medium with or without selective agents. In academic centers performing large-scale anaerobic bacteriology, it would be ideal to use two different basal media to maximize isolation efficiency.

Isolation methods are outlined in the Clinical Microbiology Procedures Handbook (8, 50). The minimum medium setup includes (i) a nonselective, enriched, brucella-base sheep blood agar plate supplemented with vitamin K_1 and hemin (BAP); (ii) a kanamycin-vancomycin laked sheep blood agar for the selection of Bacteroides and Prevotella spp. (it should be noted, however, that the 7.5-μg/ml concentration of vancomycin in the original formulation of this medium is inhibitory for asaccharolytic Porphyromonas spp. [66], whereas a reduced concentration [2.0 μg/ml] allows the growth of most Porphyromonas spp.); (iii) a Bacteroides bile-esculin agar plate (BBE) for the selection and presumptive identification of the B. fragilis group and Bilophila sp.; and (iv) a phenylethyl alcohol-sheep blood agar plate to prevent overgrowth by aerobic gram-negative rods and swarming of some clostridia. When fusobacteria are clinically suspected as the cause of infection, special selective media may be used (62). Also, a selective medium for culturing Anaerobiospirillum spp. from fecal specimens has been described (40). The inoculated anaerobic plates are immediately placed in an anaerobic environment. After incubation for 48 h at 36°C, the plates are examined. In routine clinical microbiology, a total incubation period of at least 7 days for primary plates is recommended. With shorter incubation times, several anaerobic species may not be detected.

IDENTIFICATION

Different colony types are subcultured to a brucella BAP to which special-potency antibiotic disks (colistin, 10 μg; kanamycin, 1,000 μg; and vancomycin, 5 μg) are added (a nitrate disk may also be added) (Fig. 1), to a chocolate agar plate that is incubated in 5 to 10% CO_2, and to blood agar that is incubated in air (for aerotolerance testing). A laked rabbit blood agar plate (LRBA) for the rapid demonstration of pigment production (most reliable and effective medium) and an egg yolk agar plate (EYA) for the demonstration of lipase activity may also be inoculated at this point. The

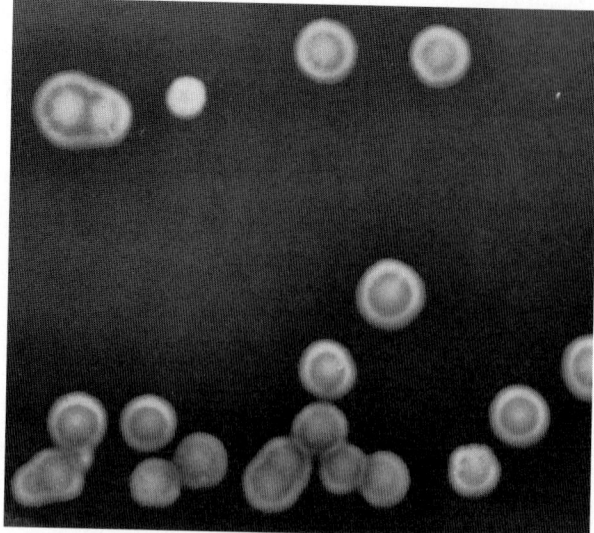

FIGURE 8 Umbonate colonies of *F. necrophorum*.

FIGURE 10 Pitting colonies of *B. ureolyticus*.

primary plates are reincubated along with the purity and test plates.

Preliminary Examination of Isolates

The characteristics that are noted and the tests that are made with BAP are detailed Gram stain and colony morphology, pigment, fluorescence (long-wave UV-light) (the last two tests are also recorded from LRBA), hemolysis, greening, pitting, spot indole reaction (*para*-dimethylaminocinnamaldehyde reagent), nitrate-nitrite reduction, special-potency antibiotic-disk susceptibility, catalase (15% H_2O_2, preferably testing colonies from EYA), and motility from broth culture or plate (hanging drop). Motile isolates are studied further as indicated in Table 1; most often, these will be *Campylobacter* spp. Another approach is the use of Lombard-Dowell Presumpto plates (Presumpto I, II, III;

Carr Scarborough, Remel Laboratories) developed at the Centers for Disease Control and Prevention; these may give additional important information for the presumptive identification of the anaerobic gram-negative bacilli (14). Good growth is necessary for the proper interpretation of test results on this medium.

The primary plates are reinspected after 4 or more days to detect slow growers, new morphotypes, or late pigmenters. In oral microbiology, two rapid in situ tests have been used. The rapid differentiation of lactose-fermenting from non-lactose-fermenting species is determined by applying 4-methylumbelliferyl-D-galactoside reagent (Sigma Chemical Co., St. Louis, Mo.; catalog no. M-1633) to the colonies and screening for fluorescent (lactose-positive) colonies under long-wave UV light (MUG test) (2). The carboxy-L-arginine-7-amino-4-methylcoumarin amide HCl (CAAM) test demonstrates trypsin-like activity of suspected colonies of *Porphyromonas gingivalis*; the colonies are treated with the CAAM reagent (Sigma; catalog no.

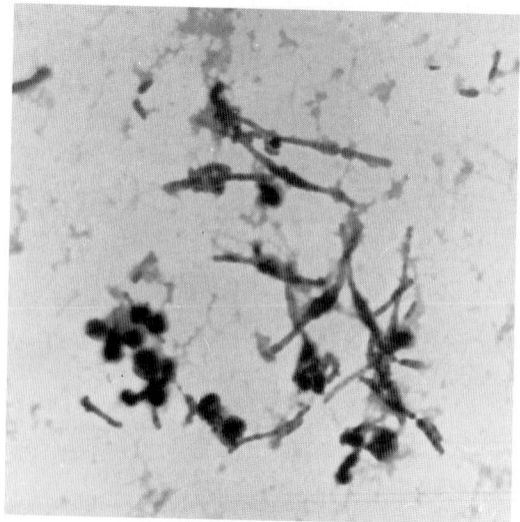

FIGURE 9 Microscopic morphology of *F. mortiferum*. There is marked pleomorphism and irregularity of staining. Note the filaments with swellings along their course.

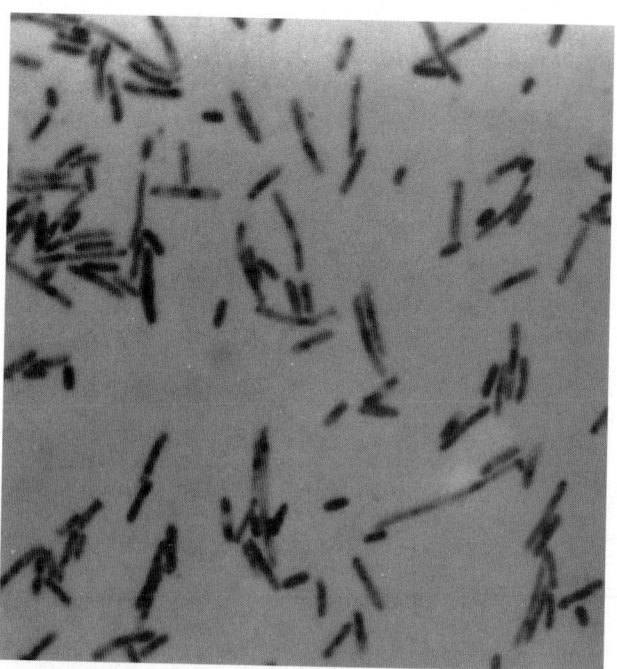

FIGURE 11 Pleomorphic, irregularly staining cells of *B. fragilis*.

TABLE 4 Characteristics of B. fragilis group[a]

Species	Growth in 20% bile	Indole	Catalase	Esculin hydrolysis	Arabinose	Cellobiose	Rhamnose	Salicin	Sucrose	Trehalose	Xylan	α-Fucosidase[b]	Fatty acids from PYG[c]
B. caccae	+	−	−+	+	+−	+−	+−	−+	+	+	−	+	A, p, S (iv)
B. distasonis	+	−	+−	+	−+	+−	V	+	+	+	−	−	A, p, S (pa, ib, iv, l)
B. eggerthii	+	+	−	+	+	+−	+−	−	−	−	+	−	A, p, S (ib, iv, l)
B. fragilis	+	−	+−	+	−	+−	−	−	+	+	−	+	A, p, S, pa (ib, iv, l)
B. merdae	+	−	−+	+	−+	V	−	+	+	−+	−	−	A, p, S (ib, iv)
B. ovatus	+	+	+−	+	+	+	+	+	+	+	+−	+	A, p, S, pa (ib, iv, l)
B. stercoris	+	+	+−	+−	−+	+	−	−+	+	−	V	V	A, p, S, f (ib, iv)
B. thetaiotaomicron	+	+	+−	+	+	+	+−	+−	+	+−	−	+	A, p, S, pa (ib, iv, l)
B. uniformis	W+	+	−+	+	+	+	−+	−	+	−w	V	+	a, p, l, S (ib, iv)
B. vulgatus	+	−	−+	+−	+	−	+	−	+	−	−+	+	A, p, S

[a]+, Positive reaction for the majority of strains; −, negative reaction; V, variable reaction; W, weak reaction; +−, most strains positive; −+, most strains negative, some strains positive; −, most strains negative, some strains weakly positive. Sugars: +, pH <5.5; W, pH 5.5 to 5.8; −, pH >5.8.
[b]Make a heavy suspension (heavier than a McFarland no. 2 standard) of the organism in 0.25 ml of sterile saline. Insert a tablet of substrate (Rosco). Incubate for 4 h (or overnight). Yellow color, positive; colorless, negative.
[c]Capital letters indicate major metabolic products from peptone-yeast-glucose (PYG), lowercase letters indicate minor products, and parentheses indicate a variable reaction for the following fatty acids: A, acetic; P, propionic; F, formic; IB, isobutyric; B, butyric; IV, isovaleric; V, valeric; L, lactic; S, succinic; PA, phenylacetic. Note that isoacids are primarily from carbohydrate-free media (e.g., peptone-yeast extract) in the case of saccharolytic organisms.

C-9396) and screened for fluorescence under long-wave UV light (59).

Presumptive Identification of Species

Most of the clinically significant gram-negative rods can be placed into broad groups with relatively few tests, and some can be presumptively identified with ease (Table 3). The special-potency antibiotic-disk pattern can be used to separate the gram-negative rods into several groups. A zone size equal to or greater than 10 mm is considered susceptible.

Most of the *Bacteroides* and *Prevotella* spp. are resistant to vancomycin and kanamycin and variable in susceptibility to colistin; *Porphyromonas* spp. are generally susceptible to vancomycin and resistant to colistin. The *B. fragilis* group can be identified presumptively by their special-potency antibiotic-disk pattern (resistant to all three disks) and by growth equal to or greater than control growth in 20% bile as determined by a tube test, by a bile disk test (62), or on a BBE agar plate (Fig. 2). Coccobacillary organisms that fluoresce red or produce black colonies are in the pigmented *Prevotella* spp.-*Porphyromonas* spp. group (Fig. 3 and 4).

Fusobacterium spp., the *B. ureolyticus*-like group, and *Bilophila* sp. are resistant to vancomycin but susceptible to both colistin and kanamycin. *B. ureolyticus*, *B. gracilis*, *Campylobacter*, and *Bilophila* colonies are usually much smaller and more translucent than those of fusobacteria. An anaerobic, gram-negative, catalase-negative rod that requires formate and fumarate for growth in broth culture or that pits agar may be presumptively identified as a *B. ureolyticus*-like organism. To document formate-fumarate requirement, inoculate two tubes of peptone yeast (or thioglycolate) broth medium, with one containing the additive and the other not containing it (control). Compare the intensity of growth (Table 3, footnote b). A strongly catalase-positive gram-negative rod (Fig. 5) that is stimulated by bile and pyruvate can be presumptively identified as a *Bilophila* sp.

Most organisms not fitting the above-described groupings are *Bacteroides* spp. or *Prevotella* spp. but occasionally are representatives of the other genera listed in Table 1.

Gram-negative cocci are sensitive to colistin and usually to kanamycin but resistant to vancomycin. Small, gram-negative cocci reducing nitrate or nitrite and growing as small, grayish white, translucent colonies that fluoresce red under UV light can be presumptively identified as *Veillonella* spp.

Rapid Identification

The anaerobic gram-negative rods that can be rapidly identified are shown in Table 3. *F. nucleatum* is a thin rod with tapering ends (Fig. 6) and is indole positive. The needle shape morphology is shared with the microaerophilic *Capnocytophaga* spp. *F. nucleatum* fluoresces chartreuse under UV light and often produces greening of the agar after exposure to air. At least three different colony morphotypes of *F. nucleatum* exist; these colonial morphologies coincide with the subspecies designations. *F. nucleatum* subsp. *nucleatum* forms small, grayish white, smooth colonies; *F. nucleatum* subsp. *fusiforme* colonies are also small (<0.5 to 1 mm), granular, and irregular (bread crumb) (Fig. 7); and *F. nucleatum* subsp. *polymorphum* colonies are large, speckled, smooth, and translucent (20). *F. necrophorum* subsp. *necrophorum* is lipase positive and usually bile sensitive. It is a pleomorphic long rod with round ends and often has bizarre forms. It produces indole, fluoresces chartreuse, pro-

TABLE 5 Characteristics of nonpigmented saccharolytic *Bacteroides* spp., *Mitsuokella* spp., and *Prevotella* spp.[a]

Subgroup and species	Growth in 20% bile	Indole	Esculin hydrolysis	Arabinose	Cellobiose	Glucose	Lactose	Salicin	Sucrose
Pentose fermenters									
B. splanchnicus	+	+	+	+	−	+	+	−	−
Mitsuokella dentalis	−	−	V	+	+	+	+	−	W
Mitsuokella multiacida	+	−	+	+	+	+	+	+	+
P. buccae	−	−	+	+	+	+	+	+	+
P. oris	−	−	+	+−	+	+	+	+	+
P. heparinolytica[e]	−	+	+	+	+	+	+	+	+
P. zoogleoformans[e]	−	−	+	V	+	+	+	V	+
Not pentose fermenters									
P. buccalis	−	−	+	−	+	+	+	−	+
P. oralis	−	−	+	−	+	+	+	+	+
P. oulorum	−	−	−	−	−	+	+	−	−
P. veroralis	−	−	+	−	+	+	+	−	+
Proteolytic									
P. bivia	−	−	−	−	−	+	+	−	−
P. disiens	−	−	−	−	−	+	−	−	−

[a] See Table 4, footnote *a*.
[b] See Table 4, footnote *b*.
[c] ONPG, *o*-nitrophenyl-β-D-galactopyranoside.
[d] See Table 4, footnote *c*.
[e] Produces viscous sediment in broth.

duces greening of the agar, and often demonstrates beta-hemolysis around the dull, umbonate colonies (Fig. 8). Lipase-negative strains require further biochemical testing. *F. mortiferum* is extremely pleomorphic, with filaments containing swollen areas with large, round bodies and exhibiting irregular staining (Fig. 9). *F. necrophorum* may have similar morphology but usually fewer round bodies. A bile-resistant fusobacterium isolated from BBE agar may be identified presumptively as *F. mortiferum* or *F. varium*; however, other fusobacteria (e.g., *F. necrophorum* and *F. ulcerans*) may grow in 20% bile. Therefore, further testing is required to confirm the presumptive identification.

B. ureolyticus, *B. gracilis*, and *Campylobacter* spp. are thin gram-negative rods with rounded ends that produce the *Fusobacterium* disk pattern. The colonies are small and translucent or transparent and may produce greening of the agar. Three colony morphotypes exist: smooth and convex, pitting (Fig. 10), and spreading. All colony types can occur in the same culture. These organisms are asaccharolytic, catalase negative, and nitrate or nitrite reducing, and they require supplementation of broth media with formate and fumarate for growth (Table 3, footnote *b*). *Campylobacter* spp. are motile and oxidase positive. *B. ureolyticus* is urease positive, and *B. gracilis* is nonmotile and urease negative

TABLE 6 Characteristics of nonpigmented weakly saccharolytic or nonsaccharolytic gram-negative bacilli[a]

Species	Growth in 20% bile	Glucose	Catalase	Indole	Nitrate	Motility	Formate-fumarate required[b]
Anaerorhabdus furcosus	+−	W	−	−	−	−	−
Bilophila wadsworthia	+	−	+	−	+	−	−
Bacteroides capillosus	−+	W−	−	−	−	−	−
Bacteroides coagulans	+	−	−	+	−	−	−
Bacteroides forsythus	−	−	−	−	−	−	+
Bacteroides gracilis	−+	−	−	−	+	−	+
Bacteroides pneumosintes	−	−	−	−	−	−	−
Bacteroides putredinis	+−	−	+−	+	−	−	−
Bacteroides tectum[f]	+	W	−	−	−	−	−
Bacteroides ureolyticus	−	−	−+	−	+	−	+
Campylobacter spp.	−+	−	−	−	+	+	+
Desulfomonas pigra	+−	−	−	−	V	−	−
Tissierella praeacuta	+	−	−	−	V	+	−

[a] See Table 4, footnote *a*.
[b] See Table 3, footnote *b*.
[c] See Table 3, footnote *i*.
[d] See Table 3, footnote *d*.
[e] See Table 4, footnote *c*.
[f] Animal origin; also isolated from cat bite infections in humans.

TABLE 5 (Continued)

Subgroup and species	Xylose	Xylan	α-Fucosidase[b]	ONPG[c] (β-galactosidase)[b]	N-Acetyl-β-glucosaminidase[b]	β-Xylosidase[b]	Fatty acids from PYG[d]
Pentose fermenters			+	+	+	-	A, p, S, ib, iv, pa (l)
B. splanchnicus	-					V	A, S
Mitsuokella dentalis	V	-		+		+	A, L, S
Mitsuokella multiacida	+			+	-	+	A, S (p, ib, b, iv, l)
P. buccae	+		-	+		+	A, S (p, ib, iv)
P. oris	+		+	+	+	+	A, p, S (iv)
P. heparinolytica[e]	+		+	+	+	+	A, P, S (ib, iv)
P. zoogleoformans[e]	V		+	+			
Not pentose fermenters			+	+	+	-	a, iv, S
P. buccalis	-	-	+	+	+	-	A, f, S (l)
P. oralis	-	-	+	+	+	-	A, S
P. oulorum	-	-	-	+	+	-	a, S
P. veroralis		+	+	+	+	-	
Proteolytic			+	+	+	-	A, iv, S (f, ib)
P. bivia	-						A, S (f, p, ib, iv)
P. disiens	-	-		-	-	-	

(Table 3, footnotes d, e, and f). Bilophila sp., which phenotypically resembles B. ureolyticus, is distinguished from the former three species by its resistance to bile and strong catalase reaction.

An indole- and lipase-positive coccobacillus that forms black-pigmented colonies or fluoresces red may be identified as belonging to the P. intermedia-P. nigrescens group. However, we have also encountered a few Porphyromonas asaccharolytica strains that are lipase positive. A rapid enzyme test for α-glucosidase (Rosco, Taastrup, Denmark) is helpful; P. intermedia-P. nigrescens is positive, and Porphyromonas asaccharolytica is negative. Any lipase-negative strains must be identified further by other biochemical tests. A lipase-positive, indole-negative, pigmented, gram-negative rod may be identified as P. loescheii or as Virginia Polytechnic Institute "Prevotella sp. strain DIC-20" genospecies (67).

Small gram-negative cocci that fluoresce red under UV light and reduce nitrate or nitrite are probably Veillonella spp. Megasphaera cells are large (>1.5 μm). Megasphaera and Acidaminococcus colonies do not fluoresce.

Definitive Identification

The definite identification of most species requires certain additional biochemical tests, metabolic end product analysis, and/or cell wall fatty acid profiling by GLC. Even in good research or reference laboratories, a percentage of strains will not be definitively identifiable. Curved rods that are not Campylobacter spp. should be checked for motility and identified according to Table 1. Organisms with small,

TABLE 6 (Continued)

Species	Desulfoviridin[c]	Urease[d]	Esculin hydrolysis	Gelatin hydrolysis	Fatty acids from PYG[e]
Anaerorhabdus furcosus	-	-	+	-[w]	a, l (s)
Bilophila wadsworthia	W[-]	+[-]	-	-	A (s)
Bacteroides capillosus	-	-	+	-[w]	a, s (p, l)
Bacteroides coagulans	-	-	-	+	a (p, l, s)
Bacteroides forsythus	-	-	+	+	A, S, pa
Bacteroides gracilis	-	-	-	-[w]	a, S
Bacteroides pneumosintes	-	-	-	+	a (l, s)
Bacteroides putredinis	-	-	-	+	A, P, ib, b, IV, S (l)
Bacteroides tectum[f]	-	-	+	+	A, p, iv, S, pa
Bacteroides ureolyticus	-	+	-	-	A, S
Campylobacter spp.	-	-	-	-	a, S
Desulfomonas pigra	+	V	-	-	A
Tissierella praeacuta	-	-	-	+	A, p, ib, B, IV, s (l)

TABLE 7 Characteristics of pigmented *Prevotella* spp., *Porphyromonas* spp., and related *Bacteroides* species[a]

Subgroup and species	Indole	Lipase	Catalase	Fermentation of:				Esculin hydrolysis	α-Fucosidase[b]
				Glucose	Cellobiose	Lactose	Sucrose		
Asaccharolytic									
Porphyromonas asaccharolytica	+	−		−	−	−	−	−	+
Porphyromonas canoris[d]	+	−	+	−	−	−	−	−	−
Porphyromonas circumdentaria[d]	+	−	+	−	−	−	−	−	−
Porphyromonas endodontalis	+	−	−	−	−	−	−	−	
Porphyromonas gingivalis[e]	+	−	−	−	−	−	−	−	
Porphyromonas salivosa[d]	+	+	+	W^-	−	W^-	−	−	
Bacteroides levii[f]	−	−	−	W	−	W	−	−	
B. macacae[d]	+	−	+	W	−	W	−	−	
Saccharolytic									
Prevotella corporis	−	−	−	+	−	−	−	−	
Prevotella denticola	−	−	−	+	$-^+$	+	+	$+^-$	+
Prevotella intermedia	+	$+^-$	−	+	−	−	$+^-$	−	+
Prevotella loescheii	−	V	−	+	+	+	+	$+^-$	+
Prevotella melaninogenica	−	−	−	+	$-^+$	+	+	$-^+$	+
Prevotella nigrescens[g]	+	$+^-$	−	+	−	−	+	−	+

[a] See Table 4, footnote *a*. W^-, most strains weakly positive, some strains negative.
[b] Reaction by API ZYM system or Rosco diagnostic tablets (see Table 4, footnote *b*).
[c] See Table 4, footnote *c*.
[d] Of animal origin; may be isolated from patients with animal bite infections.
[e] *Porphyromonas gingivalis* does not show fluorescence; *B. levii*-like organisms may show weak or no fluorescence.
[f] Negative by the API ZYM System; positive by the Rosco *o*-nitrophenyl-β-D-galactopyranoside test.
[g] *P. intermedia* shares the same characteristics. Differentiation is based on enzyme electrophoresis (55).

translucent, spreading colonies that are not *B. ureolyticus*-like should also be checked for motility. Very large fusiform rods that are isolated from the mouth or urogenital tract and have one pointed end, one blunt end, and convoluted colonies are suggestive of *Leptotrichia* spp. The characteristic GLC pattern shows major amounts of lactic acid. Tables 5 to 10 are based on reactions in prereduced, anaerobically sterilized (PRAS) liquid media (26, 46, 62). Recently, a shortened and simplified scheme (arabinose, trehalose, rhamnose, and xylan) for the identification of the *B. fragilis* group by utilizing commercial PRAS biochemicals and 24-h postincubation addition of bromothymol blue has been described (10). **Do not interpret the results from other systems with the tables given here.** Gas chromatographic analysis may be performed on broth that shows good growth of the organism (see chapter 12 of this Manual). Each lot of uninoculated broth must be assayed in parallel with samples to determine the background amounts of acetic and succinic acids, and if chopped-meat broth is used, an uninoculated broth is assayed for lactic acid. Fermentation end products vary depending on the substrate available to the organism. This may lead to misinterpretation of the GLC pattern and misidentification of the organism. For instance, saccharolytic organisms produce greater amounts of isoacids in the absence of a fermentable carbohydrate, and a fermentable carbohydrate is required for the detection of lactic acid.

Colonies of the *B. fragilis* group on brucella BAP are 2 to 3 mm in diameter, circular, entire, convex, and gray to white. The cells may be uniform, bipolarly stained, or pleomorphic (some with vacuoles) (Fig. 11); this difference is medium and age dependent. Ovoid cells suggest *Bacteroides ovatus*. Good growth in or stimulation by 20% bile (2% oxgall) is characteristic of the *B. fragilis* group; the exception to this rule is the poor growth of *Bacteroides*

uniformis in bile. Some organisms not belonging to the *B. fragilis* group (non-*B. fragilis* group) are bile resistant. These include *B. splanchnicus*, *B. tectum*, *Mitsuokella multiacida*, *Bilophila* sp., and some fusobacteria. Not all of these, however, grow on BBE agar. Morphologic characteristics, special-potency identification disks, and some biochemical reactions differentiate these species. Most of the *B. fragilis* group organisms blacken the BBE agar (esculin hydrolyzed), although *B. vulgatus* may not hydrolyze esculin. Table 4 is a key for the differentiation of members of the bile-resistant *B. fragilis* group.

Members of the non-*B. fragilis* group, nonpigmented anaerobic gram-negative bacilli, form three major subgroups: (i) saccharolytic, (ii) saccharolytic and proteolytic, and (iii) asaccharolytic. Tables 5 and 6 list the more commonly encountered or clinically important species in this group. The rapid identification of *B. ureolyticus*-like group and *Bilophila* sp. was also discussed in Rapid Identification above and Table 3.

The saccharolytic organisms fall into two categories, i.e., pentose fermenters and nonfermenters (arabinose and xylose are usually tested). Pentose fermentation may be difficult to demonstrate owing to suboptimal growth in the test medium; therefore, a screening test for the presence of the preformed enzyme β-xylosidase is recommended (for procedure, see Table 4, footnote *b*). *P. oris* and *P. buccae* are pentose fermenters (Table 5). They are phenotypically very similar but can be differentiated by the α-fucosidase- and N-acetyl-β-glucosaminidase tests (Table 4, footnote *b*) (22); furthermore, *P. buccae* is usually sensitive to the special-potency colistin disk, whereas *P. oris* is not. *P. zoogleoformans* and *P. heparinolytica* may also ferment pentoses. Both produce viscous material in broth cultures. The positive indole reaction of *P. heparinolytica* differentiates these two species. The indole production is sometimes very

TABLE 7 *(Continued)*

Subgroup and species	α-Galacto-sidase[b]	β-Galacto-sidase[b]	N-Acetyl-β-glucosaminidase[b]	Trypsin[b]	Chymotrypsin[b]	Fatty acids from PYG[c]
Asaccharolytic						
Porphyromonas asaccharolytica	−	−	−	−	−	A, P, ib, B, IV, s
Porphyromonas canoris[d]	−	+	+	−	+	A, P, ib, b, IV, s
Porphyromonas circumdentaria[d]	−	−	−	−	−	A, P, ib, b, IV, s, pa
Porphyromonas endodontalis	−	−	−	−	−	A, P, ib, B, IV, s
Porphyromonas gingivalis[e]	−	−[f]	+	+	−	A, P, ib, B, IV, s, pa
Porphyromonas salivosa[d]	+	−[f]	+	+	+	A, P, ib, B, IV, s, pa
Bacteroides levii[f]	−	+	+	−	+	A, P, ib, B, IV, s
B. macacae[d]	+	−[f]	+	+	+	A, P, ib, B, IV, s, pa
Saccharolytic						
Prevotella corporis	−	−	−	−	+[−]	A, ib, iv, S (b)
Prevotella denticola	+	+	+	−	−	A, S, (f, ib, iv, l)
Prevotella intermedia	−	−	−	−	−	A, iv, S (f, p, ib)
Prevotella loescheii	+	+	+	−	−	a, S (f, l)
Prevotella melaninogenica	+	+	+	−	−	A, S, (f, ib, iv, l)
Prevotella nigrescens[g]	−	−	−	−	−	A, iv, S (f, p, ib)

difficult to demonstrate and should be tested from a pure culture on EYA and/or from an old culture, as it tends to be a weak reaction. Other pentose fermenters include bile-resistant *Mitsuokella multiacida* and *B. splanchnicus* and bile-susceptible *Mitsuokella dentalis*. *Mitsuokella dentalis* forms characteristic, "water drop," viscous colonies on blood agar. The far better growth of *Mitsuokella* on CDC blood agar than on brucella blood agar is an additional feature that facilitates the identification of the species.

Salicin, cellobiose, and xylan are the key sugars in the differentiation of *P. oralis*, *P. buccalis*, *P. veroralis*, and *P. oulorum* (Table 5). Furthermore, *P. oulorum* produces catalase and is lipase positive (46). Certain strains of the saccharolytic pigmented *Prevotella* spp. require more than 21 days to develop pigment, and these strains (especially *P. loescheii*) closely resemble *P. veroralis*. Darker, more opaque colonies and salicin and cellobiose reactions may aid in differentiating the strains. According to a recent report (67), *P. loescheii*, "*Prevotella* sp. strain DIC-20," and some strains of *P. melaninogenica*, *P. denticola*, and *P. veroralis* cannot be differentiated by conventional biochemical tests; cellular fatty acid analysis was the only reliable method for separating these species.

P. bivia and *P. disiens* are both saccharolytic and strongly proteolytic (Table 5). Gelatin and milk are usually digested within 2 or 3 days (milk may take longer). Differentiation is based on lactose fermentation; *P. bivia* is lactose positive, and *P. disiens* is lactose negative. Under long-wave UV light, *P. bivia* and *P. disiens* colonies may fluoresce light orange to pink (coral), and *P. disiens* also may produce a brown pigment on LRBA that makes differentiation from the phenotypically similar *P. corporis* difficult.

Asaccharolytic, nonpigmented (Table 6) *Anaerorhabdus* spp., *Bacteroides* spp., and *Tissierella* spp. are infrequently isolated from clinical specimens. *B. capillosus* coagulates milk and may grow better with Tween 80-supplemented media. *B. pneumosintes* is a very tiny rod best seen in dark field (almost resembles *Veillonella* spp. in Gram stain) and forms minute colonies that may require magnification to be seen. *B. forsythus* is a fusiform bacillus that exhibits trypsin-

like activity. It requires N-acetylmuramic acid and therefore is often seen as satellite colonies around other organisms, especially fusobacteria. *Desulfomonas* sp. is a sulfate-reducing bacterium that can be confirmed with the desulfoviridin test (Table 3, footnote *i*) and differentiated from *Desulfovibrio* by negative motility.

The pigmented *Prevotella* spp. and *Porphyromonas* spp. (Table 7) vary greatly in degree and rapidity of pigment production, depending primarily on the type of blood and the composition of the base medium used in the agar. A period of 2 to 21 days may be required even on LRBA to detect pigmentation, which ranges from buff to tan to black. The identity of strains not showing pigmentation within 21 days must be established by other biochemical tests to avoid confusion with the *P. oralis* group organisms. The pigmented *Prevotella* and *Porphyromonas* spp. fluoresce pink, orange, or brick red under UV light. Fluorescence is best demonstrated in young cultures; in older cultures, especially on laked blood agar, the fluorescence is more or less masked depending on the intensity of pigment production.

The unusual special-potency antibiotic-disk pattern (susceptibility to vancomycin) in addition to asaccharolytic properties separates the *Porphyromonas* spp. from the other pigmenters. *Porphyromonas asaccharolytica*, *Porphyromonas endodontalis*, and *Porphyromonas gingivalis* are all asaccharolytic and phenotypically very similar. The key differential tests include phenylacetic acid production, trypsin-like activity, and N-acetyl-β-glucosaminidase and α-fucosidase activities (Table 7). The *Porphyromonas* spp. of animal origin are differentiated from human strains by a positive catalase reaction.

B. macacae and *B. levii* phenotypically closely resemble *Porphyromonas* spp. Because the *B. levii*-like organisms seem to be isolated more frequently from human clinical specimens than was previously appreciated, the characteristics of these two are included in Table 7.

Table 8 characterizes the more commonly isolated fusobacteria. The conversion of threonine and lactate to propionic acid is important in the differentiation of these species. Bizarre

TABLE 8 Characteristics of *Fusobacterium* species[a]

Fusobacterium sp.	Distinctive cellular morphology	Indole	Growth in 20% bile	Lipase	Gas in glucose agar[b]	Fermentation of:			Esculin hydrolysis	Lactate converted to propionate	Threonine converted to propionate	Fatty acids from PYG[c]
						Glucose	Fructose	Mannose				
F. gonidiaformans	Gonidial forms	+	−	−	4^2	−	−	−	−	−	+	A, p, B (f, l, s)
F. mortiferum[d]	Bizarre; round bodies	−	+	−	4	$+^w$	$+^w$	$+^w$	+	−	+	a, p, B (f, v, l, s)
F. naviforme	Boat shape	+	−	−	$-^2$	w^-	−	−	−	−	−	a, B, L (f, p, s)
F. necrophorum	Large, pleomorphic	+	$-^+$	$+^{-e}$	4^2	$-^w$	$-^w$	−	−	+	+	a, p, B (l, s)
F. nucleatum	Slender with pointed ends	+	−	−	$-^2$	$-^w$	$-^w$	−	−	−	+	a, p, B (F, L, s)
F. pseudonecrophorum		$+^-$	$-^+$	−	4^2	$-^w$	$-^w$	−	−	+	+	a, p, B (l, s)
F. russii[f]		−	−	−	2^-	−	−	−	+	−	−	a, B, L (f)
F. varium		$+^-$	+	−	4	w^+	w^+	$+^w$	−	−	+	a, p, B, L (s)
F. ulcerans	Pointed ends, some with central round swellings	−	+	−	2	+	−	$+^-$	−	−	+	a, p, B, l (s)

[a]See Table 4, footnote a. $+^w$, Most strains positive, some strains weakly positive; $-^w$, most strains negative, some strains weakly positive; w^-, most strains weakly positive, some strains negative; w^+, most strains weakly positive, some strains positive.

[b]Gas in PYG agar deep: −, no gas detected; 2, splits agar horizontally; 4, agar is displaced to top of tube.

[c]See Table 4, footnote c.

[d]o-Nitrophenyl-β-D-galactopyranoside positive. See Table 4, footnote b.

[e]Lipase-positive strains of *F. necrophorum* subsp. *necrophorum* and lipase-negative strains of *F. necrophorum* subsp. *funduliforme*.

[f]*F. alocis* and *F. sulci* share the same characteristics; *F. russii* colonies are larger.

TABLE 9 Characteristics of gram-negative cocci[a]

Species	Nitrate reduction	Catalase	Glucose	Fatty acids from PYG[b]
Veillonella spp.	+	V	−	A, p
Acidaminococcus fermentans	−	−	−	A, B
Megasphaera elsdenii	−	−	+	a, ib, b, iv, v, C

[a]+, Positive reaction for most strains; −, negative reaction for most strains; V, variable reaction.
[b]See Table 4, footnote c.

pleomorphic rods with very large round bodies suggest *F. mortiferum* (Fig. 9). This organism may grow on BBE agar and turn the agar black. A new species, *F. ulcerans*, isolated from tropical ulcer, closely resembles *F. mortiferum* but is esculin negative. *F. periodonticum*, an oral isolate, is indole positive and bile sensitive and converts threonine but not lactate to propionate. Contrary to earlier reports, we have failed to demonstrate glucose, fructose, or galactose fermentation of the type strain. *F. alocis*, *F. sulci*, and *F. russii* are indole negative and do not convert lactate or threonine to propionate; *F. russii* forms larger colonies (9).

Veillonella spp. are nonfermentative and produce acetic and propionic acids (Table 9). *Acidaminococcus fermentans* produces acetic and butyric acids. *Megasphaera elsdenii* ferments glucose and produces multiple fatty acids, including caproic acid.

Other Approaches to Identification

Several microsystems for the identification of anaerobes are currently available. The API 20A (Analytab Products, Inc.) and Minitek (BBL) are microtube biochemical systems that produce results in 24 to 48 h and have comput-erized databases. API 20A and Minitek are best suited for identification of saccharolytic, fast-growing organisms such as the *B. fragilis* group and many clostridia. Most asaccharolytic organisms cannot be identified by these systems, and some fastidious organisms (e.g., some *Prevotella* spp.) fail to grow in them. Even for many saccharolytic organisms, supplemental tests, including GLC, are often required for definitive identification. The color reactions in both of these systems are not always clear-cut, as shades of brown or no color (API 20A) and yellow-orange (Minitek) can make interpretation of the test results difficult.

The rapid (4-h) identification systems based on detection of preformed enzymes include RapID ANA II (Innovative Diagnostic Systems, Inc.), Rapid ID 32A (bioMérieux, Marcy l'Etoile, France), ANI Card (Vitek Systems), AN-Ident and API ZYM (bioMérieux and Analytab), and MicroScan (American MicroScan), as mentioned in chapter 10 of this Manual. Overall performance of these systems has varied from moderate to good; 60 to 90% of the isolates are identified to the species level (7, 35, 51, 63). The performance is, of course, affected by the source and nature of the isolates; the accuracy of these systems can

TABLE 10 Activities of various drugs against anaerobic gram-negative bacteria according to the Wadsworth Agar Dilution Procedure[a]

Drug	NCCLS breakpoint (μg/ml)	% Susceptible at breakpoint						
		B. fragilis	B. fragilis group[b]	B. gracilis[c]	Porphyromonas and Prevotella spp.	Fusobacterium spp.[d]	Bilophila sp.[e]	Veillonella spp.
Cefoperazone	32	50–69	50–69	50–69	>95	85–95	<50	
Cefotaxime	32	50–69	<50	50–69	>95	85–95	70–84	
Cefotetan	32	85–95	50–69	50–69	85–95	85–95	70–84	
Cefoxitin	32	85–95	70–84	50–69	>95	85–95	>95	
Ceftizoxime	64, 32[f]	70–84	70–84	50–69	>95	>95	<50	
Chloramphenicol	16	>95	>95	>95	>95	>95	>95	>95
Clindamycin	4	85–95	70–84	50–69	>95	>95	85–95	>95
Imipenem	8	>95	>95	>95	>95	>95	>95	>95
Metronidazole	16	>95	>95	85–95	>95	>95	>95	>95
Penicillin G[g]	4	<50	<50	50–69	70–84[h]	>95	<50	70–90
Ticarcillin	64				>95		>95	>95
Ticarcillin-clavulanate[i]	64/2	>95	>95	50–69	>95	85–95	>95	>95

[a]National Committee for Clinical Laboratory Standards (NCCLS)-approved method. Data are from Wadsworth Anaerobic Bacteriology Laboratory.
[b]Includes B. fragilis.
[c]Use of formate-fumarate additive is presently recommended (see Table 3, footnote b). This antibiogram is questionable, as formate-fumarate was not included in testing.
[d]Use of whole blood in test medium enhances growth.
[e]Addition of 1% pyruvate to test medium is recommended.
[f]Here, 32 μg/ml is the breakpoint for the broth microdilution test.
[g]Strains producing β-lactamase should be considered resistant. A more appropriate breakpoint for β-lactamase producers needs to be set.
[h]The breakpoint separating β-lactamase-positive from β-lactamase-negative strains is 0.5 μg/ml (31).
[i]Other beta-lactam–β-lactamase inhibitor combinations have equivalent activities.

be further increased by certain simple supplemental tests and by GLC. The API ZYM system, which allows the determination of 19 preformed enzymes in 4 h, does not have a database, but it is a useful supplement for identification of clinically encountered anaerobic bacteria, especially the *Porphyromonas* spp. (Table 7).

Individually available enzyme tests (Rosco, Taastrup, Denmark) are much cheaper than commercial kits; they can be applied in a number of situations and allow flexibility in tailoring the set to best suit special needs. The use of 4-methylumbelliferone derivatives of many substrates permits rapid spot tests based on fluorescence (42, 45).

A recently developed method for identification of anaerobes is based on analysis of whole-cell fatty acids by capillary column GLC (see chapter 12 of this Manual). An extensive database for anaerobes has been compiled, largely by the VPI Anaerobe Laboratory, and it is updated frequently (MIDI, Inc., Newark, Del.). An option for creating a cumulative database is also available.

SEROLOGICAL TESTS

Serological procedures are not practical for the identification of anaerobic bacteria from colonies. Furthermore, there are no standardized tests available for antibody or antigen detection in clinical specimens that would be useful for this group of organisms.

ANTIBIOTIC SUSCEPTIBILITIES

Susceptibility testing of anaerobes is discussed in chapter 115 of this Manual. Susceptibility patterns obtained in the Wadsworth Anaerobic Bacteriology Laboratory for the most commonly encountered gram-negative rods of clinical significance and for *Veillonella* spp. are noted in Table 10. Resistance is increasingly common among anaerobic gram-negative bacilli. This is particularly true for members of the *B. fragilis* group, which are not uncommonly resistant to expanded- and broad-spectrum cephalosporins (including β-lactamase-resistant drugs such as cefoxitin) and clindamycin. Strains of the *B. fragilis* group with resistance to imipenem and metronidazole are also encountered. Whether the reported resistance of *B. gracilis* to several antimicrobial agents is real or is due to methodological factors remains to be established (Table 10, footnote c). Metronidazole resistance is confined to the microaerophilic strains within this group.

β-Lactamase production (nitrocefin test positivity) is common among the *B. fragilis* group and in *Bilophila wadsworthia*; almost all strains are β-lactamase positive. Approximately one-third of *Prevotella* spp. are also β-lactamase producers. Occasional strains of *B. gracilis, B. coagulans, B. splanchnicus, Desulfomonas pigra, F. mortiferum, F. nucleatum, F. varium, Megamonas hypermegas, Mitsuokella multiacida,* and *Porphyromonas* spp. may produce β-lactamase.

Mechanisms of resistance other than production of β-lactamase (including high-level metalloenzyme) include changes in penicillin-binding proteins and outer membrane porin channels. Plasmids conferring resistance are also encountered.

EVALUATION, INTERPRETATION, AND REPORTING OF RESULTS

Interpretation of results of a mixed culture containing multiple isolates is difficult. Rough quantitation of cultures

together with Gram stain results (provided the specimen was properly taken and transported) is helpful because bacteria present in pure culture or in large numbers most probably are of major importance, as are organisms recovered again on repeat culture and organisms isolated from normally sterile sites. The nature of the bacteria found can also give clues to their importance in the infectious process. Certain taxa are much more important clinically in terms of frequency of occurrence, severity of infection produced, and antimicrobial resistance. Organisms of the *B. fragilis* group, especially *B. fragilis* and *B. thetaiotaomicron*, are more virulent and/or more resistant to antimicrobial drugs than are many other anaerobes. *F. necrophorum* is an exceptionally virulent organism, producing serious infections such as postanginal sepsis syndrome. *F. nucleatum* may be found as the sole infecting agent in pleuropulmonary infections (43). *Bilophila wadsworthia* is present in healthy bowel flora in low numbers, yet it is the third most common anaerobe recovered from gangrenous or perforated appendices and is a common constituent of the microbiota in other intra-abdominal infections. Furthermore, the majority of the strains are β-lactamase producers. *B. gracilis* is often associated with severe infection and may have a high degree of resistance to certain antimicrobial agents. *Porphyromonas gingivalis* is the putative key pathogen in aggressive forms of adult periodontitis and, together with *Porphyromonas endodontalis*, is often involved in root canal infections and complications of these such as odontogenic sinusitis.

Microbiologists must be willing to consult with the clinician on interpretation of the relevance of the findings. The laboratory should provide the clinician with a gradual introduction to new taxonomy by reporting both new and old names in parallel for at least 1 to 2 years. Reports must be readily available for phone or direct contact. Statements of specimen quality and possible limitations of methods used serve as most important feedback to the clinician. Dialogue between the clinician and the microbiologist should be frequent.

REFERENCES

1. **Adriaans, B., and H. Shah.** 1988. *Fusobacterium ulcerans* sp. nov. from tropical ulcers. *Int. J. Syst. Bacteriol.* **38:**447–448.
2. **Alcoforado, G. A., T. L. McKay, and J. Slots.** 1987. Rapid method for detection of lactose fermenting oral microorganisms. *Oral Microbiol. Immunol.* **2:**35–38.
3. **Bailey, G. D., and D. N. Love.** 1993. *Fusobacterium pseudonecrophorum* is a synonym for *Fusobacterium varium. Int. J. Syst. Bacteriol.* **43:**819–821.
4. **Baquero, F., J. Fernández, F. Dionda, A. Erice, J. Pérez de Oteiza, J. A. Reguera, and M. Reig.** 1990. Capnophilic and anaerobic bacteremia in neutropenic patients: an oral source. *Rev. Infect. Dis.* **2**(Suppl. 12):S157–S160.
5. **Baron, E. J., M. Curren, G. Henderson, H. Jousimies-Somer, K. Lee, K. Lechowitz, C. A. Strong, P. Summanen, K. Túner, and S. M. Finegold.** 1992. *Bilophila wadsworthia* isolates from clinical specimens. *J. Clin. Microbiol.* **30:**1882–1884.
6. **Brook, I.** 1988. Recovery of anaerobic bacteria from clinical specimens in 12 years at two military hospitals. *J. Clin. Microbiol.* **26:**1181–1188.
7. **Burlage, R. S., and P. D. Ellner.** 1985. Comparison of the PRAS II, AN-Ident, and RapID ANA systems for identification of anaerobic bacteria. *J. Clin. Microbiol.* **22:**32–35.
8. **Byrd, L.** 1992. Examination of primary culture plates for anaerobic bacteria, p. 2.4.1–2.4.6. *In* H. D. Isenberg (ed.), *Clinical Microbiology Procedures Handbook.* American Society for Microbiology, Washington, D.C.
9. **Cato, E. P., L. V. H. Moore, and W. E. C. Moore.** 1985.

Fusobacterium alocis sp. nov. and *Fusobacterium sulci* sp. nov. from the human gingival sulcus. *Int. J. Syst. Bacteriol.* **35:**475–477.

10. **Citron, D. M., E. J. Baron, S. M. Finegold, and E. J. C. Goldstein.** 1990. Short prereduced anaerobically sterilized (PRAS) biochemical scheme for identification of clinical isolates of bile-resistant *Bacteroides* species. *J. Clin. Microbiol.* **28:**2220–2223.

11. **Dewhirst, F. E., B. J. Paster, S. La Fontaine, and J. I. Rood.** 1990. Transfer of *Kingella indologenes* (Snell and Lapage 1976) to the genus *Suttonella* gen. nov.; transfer of *Bacteroides nodosus* (Beveridge 1941) to the genus *Dichelobacter* gen. nov. as *Dichelobacter nodosus* comb. nov., and assignment of the genera *Cardiobacterium, Dichelobacter,* and *Suttonella* to *Cardiobacteriaceae* fam. nov. in the gamma division of the *Proteobacteria* on the basis of 16S rRNA sequence comparisons. *Int. J. Syst. Bacteriol.* **40:**426–433.

12. **Dzink, J. L., M. T. Sheenan, and S. S. Socransky.** 1990. Proposal of three subspecies of *Fusobacterium nucleatum* Knorr 1922: *Fusobacterium nucleatum* subsp. *nucleatum* subsp. nov.; *Fusobacterium nucleatum* subsp. *polymorphum* subsp. nov., nov. rev., comb. nov.; *Fusobacterium nucleatum* subsp. *vincentii* subsp. nov., nov. rev., comb. nov. *Int. J. Syst. Bacteriol.* **40:**74–78.

13. **Dzink, J. L., S. S. Socransky, and A. D. Haffajee.** 1988. The predominant cultivable microbiota of active and inactive lesions of destructive periodontal diseases. *J. Clin. Periodontol.* **15:**316–323.

14. **Engelkirk, P. G., J. Duben-Engelkirk, and V. R. Dowell, Jr.** 1992. *Principals and Practice of Clinical Anaerobic Bacteriology.* Star Publishing Co., Belmont, Calif.

15. **Finegold, S. M., E. J. Baron, and H. M. Wexler.** 1991. *A Clinical Guide to Anaerobic Infections.* Star Publishing Co., Belmont, Calif.

16. **Finegold, S. M., and W. L. George (ed.).** 1989. *Anaerobic Infections in Humans.* Academic Press, Inc., San Diego, Calif.

17. **Finegold, S. M., H. R. Jousimies-Somer, and H. M. Wexler.** 1993. Current perspectives on anaerobic infections: diagnostic approaches, p. 257–275. *In* J. A. Washington II (ed.), *Laboratory Diagnosis of Infectious Diseases,* vol. 7. *Infectious Disease Clinics of North America.* W. B. Saunders Co., Philadelphia.

18. **Flynn, M. J., G. Li, and J. Slots.** 1994. *Mitsuokella dentalis* in human periodontitis, abstr. D-257, p. 141. *Abstr. 94th Gen. Meet. Am. Soc. Microbiol. 1994.*

19. **Gharbia, S. E., and H. N. Shah.** 1992. *Fusobacterium nucleatum* subsp. *fusiforme* subsp. nov. and *Fusobacterium nucleatum* subsp. *animalis* subsp. nov. as additional subspecies within *Fusobacterium nucleatum. Int. J. Syst. Bacteriol.* **42:**296–298.

20. **Gharbia, S. E., H. N. Shah, P. A. Lawson, and M. Haapasalo.** 1990. The distribution and frequency of *Fusobacterium nucleatum* subspecies in the human oral cavity. *Oral Microbiol. Immunol.* **5:**324–327.

21. **Goldstein, E. J. C., D. M. Citron, and S. M. Finegold.** 1984. Role of anaerobic bacteria in bite wound infections. *Rev. Infect. Dis.* **6**(Suppl. 1):S177–S183.

22. **Haapasalo, M.** 1986. *Bacteroides buccae* and related taxa in necrotic root canal infections. *J. Clin. Microbiol.* **24:**940–944.

23. **Haapasalo, M., H. Ranta, H. Shah, K. Ranta, K. Lounatmaa, and R. M. Kroppenstedt.** 1986. *Mitsuokella dentalis* sp. nov. from dental root canals. *Int. J. Syst. Bacteriol.* **36:**566–568.

24. **Han, Y.-H., R. M. Smibert, and N. R. Krieg.** 1991. *Wolinella recta, Wolinella curva, Bacteroides ureolyticus,* and *Bacteroides gracilis* are microaerophiles, not anaerobes. *Infect. Immun.* **41:**218–220.

24a. **Hendrickson, D. A., and M. M. Krenz.** 1991. Reagents and stains, p. 1297. *In* A. Balows, W. J. Hausler, Jr., K. L. Herrmann, H. D. Isenberg, and H. J. Shadomy (ed.), *Manual of Clinical Microbiology,* 5th ed. American Society for Microbiology, Washington, D.C.

25. **Hillier, S. H., M. A. Krohn, L. K. Rabe, S. J. Klebanoff, and D. A. Eschenbach.** 1993. The normal vaginal flora, H_2O_2-producing lactobacilli, and bacterial vaginosis in pregnant women. *Clin. Infect. Dis.* **16**(Suppl. 4):273–281.

26. **Holdeman, L. V., E. P. Cato, and W. E. C. Moore (ed.).** 1977. *Anaerobic Laboratory Manual,* 4th ed. Virginia Polytechnic Institute and State University, Blacksburg, Va.

27. **Holdeman, L. V., R. W. Kelley, and W. E. C. Moore.** 1984. Anaerobic gram-negative straight, curved and helical rods. Family 1. *Bacteroidaceae* Pribram 1933, 10^AL, p. 602–662. *In* N. R. Krieg and J. G. Holt (ed.), *Bergey's Manual of Systematic Bacteriology,* vol. 1. The Williams & Wilkins Co., Baltimore.

28. **Holden, J.** 1992. Collection and transport of clinical specimens for anaerobic culture, p. 2.2.1–2.2.7. *In* H. D. Isenberg (ed.), *Clinical Microbiology Procedures Handbook.* American Society for Microbiology, Washington, D.C.

29. **Johnson, C. C., and S. M. Finegold.** 1987. Uncommonly encountered, motile, anaerobic gram-negative bacilli associated with infection. *Rev. Infect. Dis.* **9:**1150–1162.

30. **Johnson, C. C., J. F. Reinhardt, M. A. C. Edelstein, M. E. Mulligan, W. L. George, and S. M. Finegold.** 1985. *Bacteroides gracilis,* an important anaerobic bacterial pathogen. *J. Clin. Microbiol.* **22:**799–802.

31. **Jousimies-Somer, H., S. Savolainen, A. Mäkitie, and J. Ylikoski.** 1993. Bacteriologic findings in peritonsillar abscesses in young adults. *Clin. Infect. Dis.* **16**(Suppl. 4):292–298.

32. **Karjalainen, J., A. Kanervo, M.-L. Väisänen, B. Forsblom, E. Sarkiala, and H. Jousimies-Somer.** 1993. *Porphyromonas*-like gram-negative rods in naturally occurring periodontitis in dogs. *FEMS Immunol. Med. Microbiol.* **6:**207–212.

33. **Kasten, M. J., J. E. Rosenblatt, and D. R. Gustafson.** 1992. *Bilophila wadsworthia* bacteremia in two patients with hepatic abscess. *J. Clin. Microbiol.* **30:**2502–2503.

34. **Kirby, B. D., W. L. George, V. L. Sutter, D. M. Citron, and S. M. Finegold.** 1980. Gram-negative anaerobic bacilli: their role in infection and patterns of susceptibility to antimicrobial agents. I. Little-known *Bacteroides* species. *Rev. Infect. Dis.* **2:**914–951.

35. **Kitch, T. T., and P. C. Appelbaum.** 1989. Accuracy and reproducibility of the 4-hour ATB 32A method for anaerobe identification. *J. Clin. Microbiol.* **27:**2509–2513.

36. **Lai, C.-H., B. M. Males, P. A. Dougherty, P. Berthold, and M. A. Listgarten.** 1983. *Centipeda periodontii* gen. nov., sp. nov. from human periodontal lesions. *Int. J. Syst. Bacteriol.* **33:**628–635.

37. **Love, D. N., G. D. Bailey, S. Collings, and D. A. Briscoe.** 1992. Description of *Porphyromonas circumdentaria* sp. nov. and reassignment of *Bacteroides salivosus* (Love, Johnson, Jones, and Calverley 1987) as *Porphyromonas* (Shah and Collins 1988) *salivosa* comb. nov. *Int. J. Syst. Bacteriol.* **42:**434–438.

38. **Love, D. N., J. Karjalainen, A. Kanervo, B. Forsblom, E. Sarkiala, G. D. Bailey, D. I. Wigney, and H. Jousimies-Somer.** 1994. *Porphyromonas canoris* sp. nov., an asaccharolytic, black-pigmented species from the gingival sulcus of dogs. *Int. J. Syst. Bacteriol.* **44:**204–208.

39. **Malnick, H., A. Jones, and J. C. Vickers.** 1989. *Anaerobiospirillum:* cause of a "new" zoonosis? Lancet **i:**1145–1146.

40. **Malnick, H., K. Williams, J. Phil-Ebosie, and A. S. Levy.** 1990. Description of a medium for isolating *Anaerobiospirillum* spp., a possible cause for zoonotic disease, from diarrheal feces and blood of humans and use of the medium in a survey of human, canine, and feline feces. *J. Clin. Microbiol.* **28:**1380–1384.

41. **Mangels, J. I., M. Cox, and L. H. Lindberg.** 1984. Methanol fixation: an alternative to heat fixation of smears before staining. *Diagn. Microbiol. Infect. Dis.* **2:**129–137.

42. **Mangels, J. I., I. Edvalson, and M. Cox.** 1993. Rapid presumptive identification of *Bacteroides fragilis* group organisms with use of 4-methylumbelliferone-derivative substrates. *Clin. Infect. Dis.* **16**(Suppl. 4):S319–S321.

43. **Marina, M., C. A. Strong, R. Civen, E. Molitoris, and S. M. Finegold.** 1993. Bacteriology of anaerobic pleuropulmonary infections: preliminary report. *Clin. Infect. Dis.* **16**(Suppl. 4):256–262.

44. **McNeil, M. M., W. J. Martone, and V. R. Dowell, Jr.** 1987. Bacteremia with *Anaerobiospirillum succiniciproducens*. *Rev. Infect. Dis.* **9:**737–742.

45. **Moncla, B. J., P. Braham, L. K. Rabe, and S. L. Hillier.** 1991. Rapid presumptive identification of black-pigmented gram-negative anaerobic bacteria by using 4-methylumbelliferone derivatives. *J. Clin. Microbiol.* **29:**1955–1958.

46. **Moore, L. V. H., E. P. Cato, and W. E. C. Moore.** 1991. *Anaerobe Laboratory Manual Update: a Supplement to the VPI Anaerobe Laboratory Manual*, 4th ed. Virginia Polytechnic Institute and State University, Blacksburg, Va.

47. **Moore, L. V. H., J. L. Johnson, and W. E. C. Moore.** 1987. *Selenomonas noxia* sp. nov., *Selenomonas flueggei* sp. nov., *Selenomonas infelix* sp. nov., *Selenomonas dianae* sp. nov., and *Selenomonas artemidis* sp. nov., from the human gingival crevice. *Int. J. Syst. Bacteriol.* **36:**271–280.

48. **Moore, W. E. C.** 1987. Microbiology of periodontal disease. *J. Periodontal Res.* **22:**335–341.

49. **Moreno, S., J. G. Altozano, B. Pinilla, J. C. Lopez, B. de Quiros, A. Ortega, and E. Bouza.** 1989. Lemierre's disease: postanginal bacteremia and pulmonary involvement caused by *Fusobacterium necrophorum*. *Rev. Infect. Dis.* **11:**319–324.

50. **Reischelderfer, C., and J. I. Mangels.** 1992. Culture media for anaerobes, p. 2.3.1–2.3.8. *In* H. D. Isenberg (ed.), *Clinical Microbiology Procedures Handbook*. American Society for Microbiology, Washington, D.C.

51. **Schreckenberger, P. C., D. M. Celig, and W. M. Janda.** 1988. Clinical evaluation of the Vitek ANI card for identification of anaerobic bacteria. *J. Clin. Microbiol.* **26:**225–230.

52. **Shah, H. N., and M. D. Collins.** 1988. Proposal for reclassification of *Bacteroides asaccharolyticus*, *Bacteroides gingivalis*, and *Bacteroides endodontalis* in a new genus, *Porphyromonas*. *Int. J. Syst. Bacteriol.* **38:**128–131.

53. **Shah, H. N., and M. D. Collins.** 1989. Proposal to restrict the genus *Bacteroides* (Castellani and Chalmers) to *Bacteroides fragilis* and closely related species. *Int. J. Syst. Bacteriol.* **39:**85–87.

54. **Shah, H. N., and M. D. Collins.** 1990. *Prevotella*, a new genus to include *Bacteroides melaninogenicus* and related species formerly classified in the genus *Bacteroides*. *Int. J. Syst. Bacteriol.* **40:**205–208.

55. **Shah, H. N., and S. E. Gharbia.** 1992. Biochemical and chemical studies on strains designated *Prevotella intermedia* and proposal of a new pigmented species, *Prevotella nigrescens* sp. nov. *Int. J. Syst. Bacteriol.* **42:**542–546.

56. **Sheppard, A., C. Cammarata, and D. H. Martin.** 1990. Comparison of different bases for the semiquantitative isolation of anaerobes from vaginal secretions. *J. Clin. Microbiol.* **28:**455–457.

57. **Shinjo, T., T. Fujisawa, and T. Mitsuoka.** 1991. Proposal of two subspecies of *Fusobacterium necrophorum* (Flugge) Moore and Holdeman: *Fusobacterium necrophorum* subsp. *necrophorum* subsp. nov., nom. rev. (ex Flugge 1886), *Fusobacterium necrophorum* subsp. *funduliforme* subsp. nov., nom. rev. (ex Halle 1898). *Int. J. Syst. Bacteriol.* **41:**395–397.

58. **Shinjo, T., K. Hiraiwa, and S. Miyazato.** 1990. Recognition of biovar C of *Fusobacterium necrophorum* (Flugge) Moore and Holdeman as *Fusobacterium pseudonecrophorum* sp. nov., nom. rev. (ex Prevot 1940). *Int. J. Syst. Bacteriol.* **40:**71–73.

59. **Slots, J.** 1987. Detection of colonies of *Bacteroides gingivalis* by a rapid fluorescence assay for trypsin-like activity. *Oral Microbiol. Immunol.* **2:**139–141.

60. **Strong, C. A., M. McTeague, M.-L. Väisänen, J. Herman, and S. M. Finegold.** 1993. Characterization of an unusual microaerophilic organism closely related to *Bacteroides gracilis*. *Clin. Infect. Dis.* **16**(Suppl. 4):188–189.

61. **Summanen, P.** 1993. Microbiology terminology update: clinically significant anaerobic gram-positive and gram-negative bacteria (excluding spirochetes). *Clin. Infect. Dis.* **16:**606–609.

62. **Summanen, P., E. J. Baron, D. M. Citron, C. Strong, H. M. Wexler, and S. M. Finegold.** 1993. *Wadsworth Anaerobic Bacteriology Manual*, 5th ed. Star Publishing Co., Belmont, Calif.

63. **Summanen, P., and H. Jousimies-Somer.** 1988. Comparative evaluation of RapID ANA and API 20A for identification of anaerobic bacteria. *Eur. J. Clin. Microbiol. Infect. Dis.* **7:**771–775.

64. **Tanner, A., C.-H. Lai, and M. Maiden.** 1992. Characteristics of oral gram-negative species, p. 299–341. *In* J. Slotts and M. A. Taubman (ed.), *Contemporary Oral Microbiology and Immunology*. Mosby-Year Book, Inc., St. Louis.

65. **Vandamme, P., E. Falsen, R. Rossau, B. Hoste, P. Segers, R. Tytgat, and J. De Ley.** 1991. Revision of *Campylobacter*, *Helicobacter*, and *Wolinella* taxonomy: emendation of generic descriptions and proposal of *Arcobacter* gen. nov. *Int. J. Syst. Bacteriol.* **41:**88–103.

66. **van Winkelhoff, A. J., and J. de Graaff.** 1983. Vancomycin as a selective agent for isolation of *Bacteroides* species. *J. Clin. Microbiol.* **18:**1282–1284.

67. **Wu, C.-C., J. L. Johnson, W. E. C. Moore, and L. V. H. Moore.** 1992. Emended descriptions of *Prevotella denticola*, *Prevotella loescheii*, *Prevotella veroralis*, and *Prevotella melaninogenica*. *Int. J. Syst. Bacteriol.* **42:**536–541.

Leptospiraceae

ARNOLD F. KAUFMANN AND ROBBIN S. WEYANT

50

TAXONOMY

The family *Leptospiraceae* and the family *Spirochaetaceae* make up the order *Spirochaetales*. The taxonomy of the *Leptospiraceae* is currently in a state of transition. In the classic taxonomy of leptospires, the basic taxon is the serovar, and pathogenicity is the key criterion for differentiating between species (13). Pathogenic serovars, which cause the disease leptospirosis, are included in the species *Leptospira interrogans*, and free-living nonpathogenic serovars are included in the species *L. biflexa*. More than 210 serovars of *L. interrogans* and 63 serovars of *L. biflexa* have been officially described (15). The serovars of each species are organized into serogroups based on shared major antigens. Serogroup designation has no official taxonomic status and is primarily intended for laboratory use.

By convention, *Leptospira* serovar designations are italicized or underlined. Genus and species of the serovar should be designated at first use; thereafter, the serovar name alone is used. Serogroup designations are not italicized or underlined, and the first letter of the serogroup name is capitalized. An illustration of correct nomenclature is *Leptospira interrogans* serovar *icterohaemorrhagiae* strain RGA, with *icterohaemorrhagiae* being a member of the Icterohaemorrhagiae serogroup.

Molecular taxonomic studies of the *Leptospiraceae* have shown a high degree of genetic heterogeneity within the classic species *L. interrogans* and *L. biflexa* (33). This heterogeneity has led to the development of a proposed phylogenetic system of leptospire taxonomy. In the phylogenetic taxonomic system, pathogenicity is not a criterion for species differentiation, and the genus *Leptospira* has been expanded to include 11 species (*L. interrogans*, *L. noguchii*, *L. weilii*, *L. santarosai*, *L. borgpetersenii*, *L. kirschneri*, *L. inadai*, *L. biflexa*, *L. meyeri*, *L. wolbachii*, and *L. parva*) (11, 25, 33). In addition, a newer genus, *Leptonema* (containing the single species *Leptonema illini*), has been described (10), and another genus, *Turneria* (containing strains currently identified as *L. parva*), has been suggested but not officially described (17).

DESCRIPTION OF FAMILY

Members of the family *Leptospiraceae* are flexible, helical rods 0.1 μm in diameter and 6 to 12 μm in length. The helical conformation is right-handed, with more than 18 coils per cell. Usually, one or both ends of a cell are hooked. Individual cells stain faintly gram negative but are best viewed by dark-field or phase-contrast microscopy. *Leptospiraceae* are motile, with two subterminal periplasmic flagella, and are able to pass through 0.2-μm-pore-size filters. These organisms are aerobic and utilize long-chain fatty alcohols as carbon and energy sources. Amino acids and carbohydrates are not utilized. Purines but not pyrimidines are utilized. Organisms may be free living or may live in association with human or animal hosts.

NATURAL HABITAT

Leptospirosis is a zoonotic disease, having its reservoir in wild, domestic, and peridomestic animals (7). Once an animal is infected, it may become an asymptomatic carrier of leptospires within the renal tubules. Some *Leptospira* serovars are host adapted for preferential but not exclusive survival in specific animal species. For example, *canicola* is most often associated with dogs, *pomona* is associated with cattle and swine, *ballum* is associated with mice, and *icterohaemorrhagiae* is associated with rats. Humans are dead-end hosts in the chain of transmission, with possible incidents of person-to-person transmission being rarely reported.

Pathogenic leptospires can survive for many days to months in soil and fresh water with a pH in the neutral range. In salt water, survival time is only a few hours.

CLINICAL SIGNIFICANCE

Infection usually results from direct or indirect exposure to the urine of leptospiruric animals (7). Indirect exposure through contaminated water and soil accounts for most sporadic cases, common-source outbreaks in swimmers, and cases in occupational groups such as rice farmers, sugarcane workers, sewer workers, and military personnel. Direct exposure occurs in pet owners, veterinarians, and persons working with livestock. Other modes of infection, such as animal bites, handling infected animal tissues, and ingestion of contaminated food and water, have less importance.

Leptospires can enter the body through breaks in skin, through the mucosa of the mouth and upper respiratory tract, and through the conjunctiva. *Leptospira* organisms can also directly penetrate skin that has been immersed in contaminated water for prolonged periods (7, 12). Thus, swimmers and persons who work with their bare feet or hands in contaminated water have a higher risk of infection.

The incubation period for leptospirosis ranges from 2 to 20 days, with a mean of 10 days (7, 14). Illness usually begins abruptly with fever, chills, rigors, intense headache, myalgia (most marked in the legs and back), nausea, and vomiting. Conjunctival injection, pharyngitis, hepatomegaly, and skin rash may also occur in the first several days of illness.

Leptospirosis is typically a biphasic illness (7, 14). The initial, or septicemic, phase lasts for 4 to 7 days. Leptospires disseminate widely throughout the body during this stage, and blood cultures are most productive for establishing the diagnosis. Following a 1- to 3-day quiescent period that is thought to reflect the initial onset of an immune response, the second, or immune, phase begins. In anicteric leptospirosis, the less severe form of the disease, the immune phase is generally of lesser duration and severity than the septicemic phase, and its most important features are meningitis (up to 40% of patients) and leptospiruria (>95% of patients). Urine cultures and serology are recommended for laboratory diagnosis during and after this phase of the illness.

About 10% of patients develop icteric leptospirosis. The hallmarks of icteric leptospirosis (jaundice and azotemia) begin late in the septicemic phase and tend to obscure the biphasic nature of the disease. Icteric leptospirosis can be clinically severe, with a mortality rate of about 10%.

COLLECTION, TRANSPORT, AND STORAGE OF SPECIMENS

Blood, cerebrospinal fluid (CSF), and urine are the specimens of choice for the recovery of leptospires from patients with leptospirosis. The most appropriate sources to culture during the first 10 days of illness are blood and CSF. These specimens should be collected prior to antibiotic treatment and while the patient is febrile. If culture media are not immediately available, blood should be collected for transport into tubes containing heparin or sodium oxalate. Tubes containing citrate solutions should be avoided, as citrate may be inhibitory (32). Blood and CSF specimens should be stored and transported at 5 to 20°C and inoculated into culture medium within 1 week of collection.

After the first week of illness, the optimal source for isolation of leptospires is urine. Extreme care should be taken to collect an uncontaminated specimen, since most urine contaminants will overgrow leptospires in culture. Cultures should be inoculated as soon as possible, especially if the urine is acidic. Leptospires survive only a few hours in acidic urine. If culture media are not immediately available, the urine can be diluted 1:10 in 1% bovine serum albumin and stored for a few days at 5 to 20°C (5).

In fatal cases of leptospirosis, viable leptospires can be isolated from multiple tissues, particularly the liver, kidney, and brain (5). Samples should be collected and processed within 4 h of death, since leptospires will not survive in autolytic tissues.

For epidemiologic investigations, freshwater ponds represent potential sources of pathogenic leptospires. Water should be collected in sterile containers and transported to the laboratory at room temperature within 72 h of collection (12).

DIRECT DETECTION OF LEPTOSPIRES

A presumptive identification of leptospires can be made by direct microscopic methods. Blood, CSF, and tissues may be screened microscopically for the presence of leptospires by either dark-field or direct fluorescent-antibody methods. Blood is prepared for analysis by mixing 5 ml of freshly drawn whole blood with either 0.5 ml of 1% sodium oxalate or 1.0 ml of 1% heparin. The mixture is then centrifuged at $500 \times g$ for 15 min, and the supernatant containing the organisms is transferred to another tube and centrifuged at $1,500 \times g$ for 30 min. The second supernatant is discarded, and a wet mount is prepared from the sediment. For urine and CSF, the sample is centrifuged at $1,500 \times g$ for 30 min, the supernatant is discarded, and a wet mount is prepared from the sediment.

Tissues may be prepared for analysis by disruption in a sterile Ten Broek tissue grinder or in a mortar and pestle and may then be suspended in sterile phosphate-buffered saline (PBS) (pH 7.2 to 7.8) (1 part tissue to 9 parts PBS). The resulting suspension is centrifuged at $500 \times g$ for 15 min. The sediment is discarded, and the supernatant containing the leptospires is poured into another tube and centrifuged at $1,500 \times g$ for 30 min. As previously described, the second supernatant is discarded, and a wet mount is prepared from the sediment.

Dark-field microscopic examination of specimens processed as described above can be useful in establishing a rapid diagnosis but should not be relied on as the sole diagnostic criterion. In specimens with small numbers of leptospires, elucidation of organisms may be difficult, even after centrifugation of the specimen. Experience and skill are required to differentiate leptospires from artifacts such as fibrils or cellular extrusions.

Numerous silver staining techniques have been described for visualizing leptospires in tissues (27). As with other microscopic techniques, the sensitivity of the technique is poor in samples with small numbers of leptospires.

Fluorescence microscopy is slightly more sensitive than dark-field microscopy or silver staining for detecting leptospires in specimens. The National Veterinary Service Laboratory, Animal and Plant Health Inspection Service (APHIS), U.S. Department of Agriculture, Ames, Iowa, produces multivalent fluorescein-conjugated antibodies for the veterinary profession. These conjugates are available through local APHIS representatives.

For fluorescence microscopy, urine or CSF sediment is placed on a slide, air dried, and fixed in absolute alcohol for 10 min. The fixed specimen is reacted with the leptospiral conjugate for 2 h at 37°C. After being washed in PBS or saline, a coverslip is mounted with 10% PBS in glycerol (pH 7.5). A fluorescence microscope with a filter combination that is usually recommended for bacterial identification, such as a BG 12 excitor and an OG1 ocular filter, is recommended for examination of the slide (27).

The fastidious nature of leptospires makes them good candidates for the development of rapid and sensitive nucleic acid hybridization-based assays. Numerous methods of detecting leptospires in specimens by direct hybridization with genomic DNA probes have been described elsewhere (20, 22, 29). In general, these methods are no more sensitive than fluorescence microscopy, dark-field microscopy, or silver staining, but they may be more specific. In vitro nucleic acid amplification approaches to detection and preliminary identification have also been described and, although still under development, look promising (8, 21, 30).

They are more sensitive than the microscopic methods described above and, when performed with the appropriate controls, may be as specific. Further studies with larger numbers of serovars will be required before these methods achieve universal acceptance.

ISOLATION PROCEDURES

Using aseptic techniques, inoculate the specimens within 24 h of collection into tubes containing 5 ml of a semisolid medium, such as Fletcher's, Ellinghausen's, or polysorbate 80 (7, 27). For blood, the inoculum volume should be kept to a minimum because of the presence of inhibitory substances. Only 1 or 2 drops (approximately 50 μl/drop) should be added to 5 ml of medium. Inoculate three to five tubes per specimen, as the bacteremia tends to be low titer. For CSF, an inoculum of 0.5 ml of undiluted specimen per 5 ml of medium is recommended.

To minimize problems with bacterial overgrowth, the following urine culture technique is recommended (27). Draw about 1 ml of urine from the specimen container into a sterile 2-ml syringe with an attached 2.5-cm, 20-gauge needle. Inoculate two tubes, each containing 5 ml of semisolid medium, with 1 drop of the undiluted urine per tube. Discharge all but 0.1 ml of the undiluted urine from the syringe, and draw 0.9 ml of sterile buffered saline into the syringe. Expel a few drops of the diluted urine before inoculating 1 drop of the 1:10 dilution into each of two tubes of semisolid medium. Repeat the dilution procedure two more times, ending up with a set of four paired tubes of inoculated medium. Add a single 30-μg antimicrobial susceptibility test disk of neomycin to one tube of each dilution pair.

To culture organs, use aseptic technique to remove a small piece of the organ, and place this piece in the barrel of a sterile 5-ml syringe. Insert the plunger, and crush the specimen through the tip of the syringe into a tube of semisolid culture medium. The inoculum should be roughly the size of a pea (6 mm in diameter). Two tubes of medium should be inoculated per specimen, and a single neomycin disk should be added to one of the tubes.

To culture pond water, laboratory animal inoculation is used as a first step to overcome problems posed by overgrowth by saprophytic leptospires and other bacteria. Inject 3 ml of the water sample intraperitoneally into a weanling guinea pig. Obtain a heart blood specimen 4 h after inoculation for culture and a baseline serum sample. Kill the animal 4 weeks later, and at that time, obtain blood, kidney, and brain for serologic tests and culture.

Hold all leptospiral culture specimens at room temperature (5 to 20°C). Examine a drop of the culture medium by dark-field microscopy once a week for 5 weeks and then twice a month for 4 months before calling a culture negative. Although leptospires are usually detectable in positive cultures within 1 to 2 weeks, holding apparently negative cultures for up to 4 months ensures that the patient's diagnosis is accurate for prognostic purposes.

When leptospires are cultivated in semisolid media, leptospiral growth initially appears as a diffuse zone near the top of the tube. As the culture matures, the leptospires condense into a well-differentiated ring at the level of the tube corresponding to the optimum oxygen tension for the organism (Dinger's ring). In general, the saprophytic serovars form rings closer to the top of the tube than the pathogenic serovars do. Leptospiral growth can be con-

firmed by examining a sample of the growth ring by dark-field microscopy.

Cultures in semisolid media should remain viable for at least 8 weeks at room temperature. For permanent storage, cultures in semisolid medium can be frozen in liquid nitrogen without added cryoprotectants.

IDENTIFICATION

Once grown in culture, leptospires can be identified to serogroup by the microscopic agglutination test (MAT) with reference rabbit sera made against the type serovars of all recognized serogroups (27). The MAT is performed on isolates that have been passaged at least three times in a liquid medium, such as polysorbate 80 liquid. Such passage effectively dilutes out agar granules that might produce agglutination artifacts and reduces nonspecific agglutination. The test is conducted in a tissue culture plate with 96 flat-bottom wells, with results being read on the stage of a dark-field microscope by using a long-working-distance 10× objective. Reference sera are diluted with PBS in twofold steps from 1:25 through 1:3,200. Then, 50 μl of each serum dilution is mixed with 50 μl of an antigen suspension standardized at 100 to 200 organisms per high-power field (an antigen density equivalent to a McFarland 0.5 or a Roessler 20 turbidity unit). The plates are gently shaken to mix the contents, covered, and incubated at room temperature for 2 to 4 h before being read at 450× on a dark-field microscope. The titer is defined as the greatest antiserum dilution at which at least 50% of the leptospires are agglutinated (a 2+ agglutination).

Final identification of an isolate to the serovar level requires the use of the agglutinin adsorption technique (27). This technique involves raising isolate-specific hyperimmune rabbit antiserum and requires the maintenance of many reference serovars and antisera. Final identification to the species level by using the phylogenetic taxonomic approach requires nucleic hybridization studies. The complexity of these techniques limits their use to reference laboratories (26).

Neither serogrouping nor serotyping is appropriately done in a clinical laboratory. Once a culture is determined to be positive, it should be shipped to a reference laboratory for further identification.

Less complex alternatives for identification are currently under development. These methods include analysis of whole-chromosome restriction endonuclease patterns and cellular fatty acid profiles (2, 4, 16, 18, 24, 25, 34). Some of these methods are limited by the need to grow large numbers of leptospire cells for analysis. As with molecular detection methods, further evaluation with a broad range of serovars and a better understanding of the molecular taxonomy of leptospires will determine the utility of these methods in the diagnostic laboratory.

Nonpathogenic serovars reportedly differ from pathogenic serovars in their ability to grow at 13°C and in the presence of 225 μg of 8-azaguanine per ml. However, we have found these characteristics as well as the ability to grow in the presence of either 2,6-diamino-purine or copper sulfate and the production of lipase to be of no value in predicting pathogenic potential (33).

SEROLOGIC TESTS

The MAT with live antigen is the standard reference laboratory procedure for detection of leptospiral antibodies

(6, 27). Both highly sensitive and specific, the MAT is the standard against which other serodiagnostic test procedures must be measured. This test, however, is time-consuming, difficult to standardize, and hazardous because of its use of live antigens. Some laboratories use formalin-killed antigens, because they are less hazardous and are stable for about 2 weeks. Although killed-antigen titers are somewhat lower and cross-reactivity is greater, results generally compare favorably with those obtained with live antigens (9).

The great antigenic diversity of leptospiral serovars requires the use of a battery of antigens. We currently use 23 antigens (*andaman, australis, autumnalis, djasiman, ballum, bataviae, canicola, celledoni, butembo, cynopteri, grippotyphosa, borincana, georgia, wolffi, copenhageni, mankarso, javanica, panama, pomona, alexi, pyrogenes, shermani,* and *tarassovi*). The antigens are live 4- to 7-day-old cultures in liquid culture medium. The antigen concentration is adjusted to 100 to 200 cells per high-power field (450×). The MAT is performed in 96-well flat-bottom tissue culture plates as described above. One drop (50 μl) of PBS is placed in each well exclusive of the first row. Two drops (100 μl) of a 1:25 serum dilution are added to the first well. A series of 1:2 dilutions for use through row 11 is prepared by use of 50-μl microdiluters. Row 12 is used as a saline control for each antigen used. One drop (50 μl) of antigen is added to each well, with a separate column for each individual antigen. The plates are gently shaken to mix the contents, covered, and incubated at room temperature for 2 h before being read at 450× on the dark-field microscope. The end point is the highest dilution at which a 2+ or greater agglutination reaction is observed.

A variety of other serologic tests have been developed, including an indirect hemagglutination test using a genus-specific antigen derived from *andaman*, a slide agglutination test using pooled formalin-fixed antigens, several enzyme-linked immunosorbent assay formats, and a macroscopic agglutination test that uses microencapsulated antigens. None of these alternative methods have the sensitivity or specificity necessary for reliable clinical diagnostic tests.

ANTIBIOTIC SUSCEPTIBILITIES

Standardized procedures have not been developed for antimicrobial susceptibility testing of leptospires. However, clinical experience, experimental animal studies, and in vitro susceptibility testing have generally correlated (1, 14, 19, 23, 28, 31). Results of controlled clinical trials support the therapeutic efficacies of doxycycline and penicillin (19, 31). In a hamster model, ampicillin, bacampicillin, cyclacillin, piperacillin, mezlocillin, doxycycline, chlortetracycline, cefotaxime, and moxalactam were effective in preventing death of infected animals (1). In vitro testing has demonstrated strain variability in susceptibility to penicillins and tetracyclines (3, 23). Further methodologic development and evaluation will be necessary before in vitro susceptibility testing can be recommended for choosing therapeutic protocols.

Evaluation, Interpretation, and Reporting of Results

Evaluating serologic findings supplemented with clinical and epidemiologic information is usually the first step in the diagnosis of leptospirosis. If the appropriate antigens are used, clinically significant MAT titers to *Leptospira* spp. should help rule out other febrile illnesses with similar clinical pictures. Using this technique, a serologically confirmed case of leptospirosis is defined by a MAT that shows

either seroconversion from a titer of less than 1:50 to at least 1:200 or a fourfold rise in titer between acute- and convalescent-phase serum specimens studied at the same laboratory. Typically, more than one antigen will be reactive, and laboratory reports of antibody titers to six or more antigens are not unusual. A MAT titer of at least 1:200 for a single specimen obtained after the onset of symptoms or a titer of at least 1:100 for consecutive specimens is suggestive but not diagnostic of leptospirosis. A titer of at least 1:800 in the presence of compatible symptoms is strong evidence of recent or current leptospirosis (7, 14). Delayed seroconversions are common. We estimate that up to 10% of patients will not seroconvert within 30 days of clinical onset. We have observed cross-reactive antibody, sometimes with significant seroconversion, associated with syphilis, relapsing fever, Lyme disease, and legionellosis.

Negative results in direct examinations of specimens for leptospires in patients with elevated MAT titers do not necessarily rule out acute disease, since MAT serology is more sensitive than microscopic or direct DNA probe hybridization tests. The isolation of leptospires in culture confirms the diagnosis and differentiates between current infection and past exposure, which may not be clearly differentiated by serology. About 50% of blood cultures will be positive if samples are collected within 7 days of clinical onset and prior to initiation of antimicrobial therapy. Urine cultures are also highly successful, with positive cultures often being obtained as long as 30 days after apparent clinical recovery.

Limited availability of laboratories with the appropriate capabilities for *Leptospira* diagnostics is a problem in all areas of the world. However, modern communication networks and the ability to rapidly ship specimens from even the remotest areas make the services of these laboratories much more widely available than they were even a few years ago.

REFERENCES

1. **Alexander, A. D., and P. L. Rule.** 1986. Penicillins, cephalosporins, and tetracyclines in treatment of hamsters with fatal leptospirosis. *Antimicrob. Agents Chemother.* **30:**835–839.
2. **Bao, L., B. Dai, W. K. Terpstra, and G. J. Van Eys.** 1990. DNA hybridization of *Leptospira* serogroup Icterohaemorrhagiae with *Leptospira* recombinant probes. *Hua-Hsi I Ko Ta Hsueh Hsueh Pao* **21:**357–361.
3. **Broughton, E. S., and L. E. Flack.** 1986. The susceptibility of a strain of *Leptospira interrogans* serogroup Icterohaemorrhagiae to amoxicillin, erythromycin, lincomycin, tetracycline, oxytetracycline, and minocycline. *Zentralbl. Bakteriol. Hyg. A* **261:**425–431.
4. **Cacciapouti, B., L. Ciceroni, and D. A. Barbini.** 1991. Fatty acid profiles, a chemotaxonomic key for the classification of strains of the family *Leptospiraceae.* *Int. J. Syst. Bacteriol.* **41:**295–300.
5. **Cole, J. R.** 1990. Spirochetes, p. 41–60. *In* G. R. Carter and J. R. Cole (ed.), *Diagnostic Procedures in Veterinary Bacteriology and Mycology,* 5th ed. Academic Press, Inc., New York.
6. **Cole, J. R., C. R. Sulzer, and A. R. Pursell.** 1973. Improved microtechnique for the leptospiral microscopic agglutination test. *Appl. Microbiol.* **25:**976–980.
7. **Faine, S. (ed.).** 1982. *Guidelines for the Control of Leptospirosis.* W.H.O. offset publication no. 67. World Health Organization, Geneva.
8. **Garretsen, M. J., T. Olyhoek, M. A. Smits, and B. A. Bokhout.** 1991. Sample preparation method for polymerase chain reaction-based semiquantitative detection of *Leptospira interrogans* serovar *hardjo* subtype *hardjobovis* in bovine urine. *J. Clin. Microbiol.* **29:**2805–2808.

9. **Gochenour, W. S., C. A. Gleiser, and M. K. Ward.** 1958. Laboratory diagnosis of leptospirosis. *Ann. N.Y. Acad. Sci.* **70:**421–426.

10. **Hovind-Hougen, K.** 1979. *Leptospiraceae*, a new family to include *Leptospira* Noguchi 1917 and *Leptonema* gen. nov. *Int. J. Syst. Bacteriol.* **29:**245–251.

11. **Hovind-Hougen, K., W. A. Ellis, and A. Birch-Andersen.** 1981. *Leptospira parva* sp. nov.: some morphological and biological characters. *Zentralbl. Bakteriol. Parasitenkd. Infektionskr. Hyg. Abt. 1 Orig. Reihe A* **250:**343–354.

12. **Jackson, L. A., A. F. Kaufmann, W. G. Adams, M. B. Phelps, C. Andreasen, C. W. Langkop, B. J. Francis, and J. D. Wenger.** 1993. Outbreak of leptospirosis associated with swimming. *Pediatr. Infect. Dis. J.* **12:**48–54.

13. **Johnson, R. C., and S. Faine.** 1984. Family II *Leptospiraceae* Hovind-Hougen 1979, 245[AL], p. 62–67. *In* N. R. Kreig and J. G. Holt (ed.), *Bergey's Manual of Systematic Bacteriology*, vol. 1. The Williams & Wilkins Co., Baltimore.

14. **Kelley, P. W.** 1992. Leptospirosis, p. 1295–1301. *In* S. L. Gorbach, J. G. Bartlett, and N. R. Blacklow (ed.), *Infectious Diseases*. The W. B. Saunders Co., Philadelphia.

15. **Kmety, E., and H. Dikken.** 1988. Revised list of *Leptospira* serovars accepted by the Subcommittee on the Taxonomy of *Leptospira* at the Manchester meeting, 6/7 September 1986. University Press, Groningen, The Netherlands.

16. **Lefebvre, R. B., J. W. Foley, and A. B. Theirmann.** 1985. Rapid and simplified protocol for isolation and characterization of leptospiral chromosomal DNA for taxonomy and diagnosis. *J. Clin. Microbiol.* **22:**606–608.

17. **Marshall, R.** 1992. International Committee on Systemic Bacteriology, Subcommittee on the Taxonomy of *Leptospira*, minutes of the meetings, 13 and 15 September 1990, Osaka, Japan. *Int. J. Syst. Bacteriol.* **42:**330–334.

18. **Marshall, R. B., B. E. Wilton, and A. J. Robinson.** 1981. Identification of leptospiral serovars by restriction endonuclease analysis. *J. Med. Microbiol.* **14:**163–166.

19. **McClain, J.B.L., W. R. Ballou, S. M. Harrison, and D. L. Steinweg.** 1984. Doxycycline therapy for leptospirosis. *Ann. Intern. Med.* **100:**696–698.

20. **McCormick, B. M., B. D. Millar, R. P. Monckton, R. T. Jones, R. J. Chappel, and B. Adler.** 1989. Detection of leptospires in pig kidney using DNA hybridization. *Res. Vet. Sci.* **47:**134–135.

21. **Merien, F., P. Amouriax, P. Perolat, G. Baranton, and I. St. Girons.** 1992. Polymerase chain reaction detection of *Leptospira* spp. in clinical samples. *J. Clin. Microbiol.* **30:**2219–2224.

22. **Miller, B. C., R. J. Chappel, and B. Adler.** 1987. Detection of leptospires in biological fluids using DNA hybridization. *Vet. Microbiol.* **15:**71–78.

23. **Oie, S., K. Hironaga, A. Koshiro, H. Konishi, and Z. Yoshii.** 1983. In vitro susceptibilities of five *Leptospira* strains to 16 antimicrobial agents. *Antimicrob. Agents Chemother.* **24:**905–908.

24. **Pacciarini, M. L., M. L. Savio, S. Tagliabue, and C. Rossi.** 1992. Repetitive sequences cloned from *Leptospira interrogans* serovar *hardjo* genotype hardjoprajitno and their application to serovar identification. *J. Clin. Microbiol.* **30:**1243–1249.

25. **Ramadass, P., B. D. W. Jarvis, R. J. Corner, D. Penny, and R. B. Marshall.** 1992. Genetic characterization of pathogenic *Leptospira* species by DNA hybridization. *Int. J. Syst. Bacteriol.* **42:**215–219.

26. **Stallman, N. D.** 1987. Minutes of the meeting, International Committee on Systematic Bacteriology, Subcommittee on the Taxonomy of *Leptospira*. *Int. J. Syst. Bacteriol.* **37:**472–473.

27. **Sulzer, C. R., and W. L. Jones.** 1978. *Leptospirosis. Methods in Laboratory Diagnosis*, revised ed. Publication no. (CDC) 74-8275. U.S. Department of Health, Education, and Welfare, Washington, D.C.

28. **Takafuji, E. T., J. W. Kirkpatrick, R. N. Miller, J. J. Karwacki, P. W. Kelley, M. W. Gray, K. M. McNeill, H. L. Timboe, R. E. Kane, and J. L. Sanchez.** 1984. An efficacy trial of doxycycline chemoprophylaxis against leptospirosis. *N. Engl. J. Med.* **310:**497–500.

29. **Terpestra, W. J., G. J. Schoone, G. S. Lighart, and J. Ter Schegget.** 1987. Detection of *Leptospira interrogans* in clinical specimens by in situ hybridization using biotin-labeled DNA probes. *J. Gen. Microbiol.* **133:**911–914.

30. **Van Eyes, G. J. J. M., C. Gravekamp, M. J. Gerritsen, W. Quint, M. T. E. Cornelissen, J. Ter Schegget, and W. J. Terpstra.** 1989. Detection of leptospires in urine by polymerase chain reaction. *J. Clin. Microbiol.* **27:**2258–2262.

31. **Watt, G., L. P. Padre, M. L. Tuazon, C. Calubaquib, E. Santiago, C. P. Ranoa, and L. W. Laughlin.** 1988. Placebo-controlled trial of intravenous penicillin for severe and late leptospirosis. *Lancet* **i:**433–435.

32. **Wolff, J. W.** 1954. *The Laboratory Diagnosis of Leptospirosis*. Charles C Thomas, Publisher, Springfield, Ill.

33. **Yasuda, P. H., A. G. Steigerwalt, K. R. Sulzer, A. F. Kaufmann, F. Rogers, and D. J. Brenner.** 1987. Deoxyribonucleic acid relatedness between serogroups and serovars in the family *Leptospiraceae* with proposals for seven new *Leptospira* species. *Int. J. Syst. Bacteriol.* **37:**407–415.

34. **Zuerner, R. L., and C. A. Bolin.** 1990. Nucleic acid probe characterizes *Leptospira interrogans* serovars by restriction fragment length polymorphisms. *Vet. Microbiol.* **24:**355–366.

Borrelia

TOM G. SCHWAN, WILLY BURGDORFER, AND PATRICIA A. ROSA

51

TAXONOMY

Bacteria of the genus *Borrelia* belong to one of two separate lineages of spirochetes (order *Spirochaetales*) that are taxonomically distinguished at the family level (90). The type species is *Borrelia anserina* (Sakharoff 1891). Species of this genus form a tight phylogenetic group within the family *Spirochaetaceae*. Other genera in this family include *Treponema*, *Spirochaeta*, *Serpulina*, and *Cristispira*. The second family of spirochetes is *Leptospiraceae*, which encompasses the genera *Leptospira* and *Leptonema*. Spirochetes are one of the few major bacterial groups for which classical morphologic criteria and 16S rRNA sequence analyses agree in predicting the phylogenetic relationships among members of the group and their position as a distinct entity within the eubacterial kingdom (90). They are neither gram-positive nor gram-negative organisms.

DESCRIPTION OF THE GENUS

Borreliae are highly motile helical organisms that are 5 to 25 μm long and 0.2 to 0.5 μm wide (11). An outer membrane encloses the periplasmic flagella and the protoplasmic cylinder (41). The flagella (7 to 20 per terminus) are found beneath the outer membrane, attached subterminally to opposite ends of the protoplasmic cylinder, and they mediate both the motility and the shape of borreliae (37).

The protoplasmic cylinder consists of the cell wall and the cytoplasmic membrane enclosing the protoplasmic contents of the cell. Microtubules have not been detected in the cytoplasm of borreliae, a trait that, along with unsheathed flagella, serves to morphologically distinguish them from the treponemes (11).

It is currently accepted that borreliae are microaerophilic (6), but evidence indicates that they can grow anaerobically (68). Borreliae require long-chain fatty acids for growth and produce lactic acid through glucose fermentation.

Borrelial DNA has a G+C mole content of approximately 30% (38). Borreliae differ from other spirochetes, as well as from almost all other bacteria, by the unique structure of their genome; they contain a linear chromosome (22, 32) and numerous linear and circular plasmids (7, 81).

Unlike pathogenic spirochetes of other genera, all bor-

reliae are arthropod borne. Individual species of *Borrelia* differ in the arthropod vectors by which they are transmitted, their reservoir hosts, the nonhost species that can be infected, and the diseases they cause. Borreliae are transmitted by a large variety of ticks and in one instance by the human body louse. They can be pathogenic for humans, domestic animals, rodents, and birds.

Recent studies have divided the Lyme disease spirochetes, *B. burgdorferi* sensu lato, into three separate species, *B. burgdorferi* sensu stricto, *B. garinii*, and *B. afzelii* (5). There appears to be no vector specificity that distinguishes them, nor is their separation universally accepted. It is unclear whether they cause separate or identical diseases; however, recent preliminary results demonstrate that in Europe, different clinical manifestations may be associated with the three currently recognized species (89).

NATURAL HABITATS

Except for *B. recurrentis* and *B. duttonii*, all species of *Borrelia* are maintained in nature by cycling through wild animals and the ticks that feed upon them. *B. recurrentis* and *B. duttonii* do not have a wild-animal reservoir (31). These two species of spirochetes infect only humans and the human body louse (*B. recurrentis*) or *Ornithodoros moubata* ticks (*B. duttonii*) that feed on human blood. The ecological components that maintain *Borrelia* spp. in nature are quite diverse and widespread throughout the world (Table 1).

The majority of currently recognized species of *Borrelia* are transmitted by "soft" ticks of the family Argasidae (31). This group includes (i) the borreliae that cause tick-borne relapsing fever, which are transmitted by soft ticks of the genus *Ornithodoros*; (ii) *Borrelia coriaceae*, the putative agent of epidemic bovine abortion of cattle, which may be transmitted by *Ornithodoros coriaceus*; and (iii) *B. anserina*, the agent of fowl spirochetosis, which is transmitted by several species of soft ticks of another genus, *Argas* (27). *B. recurrentis* also causes relapsing fever in humans but is unique among all the borreliae by not being transmitted by a tick. The sole vector of this spirochete is the human body louse, *Pediculus humanus humanus*. Although *B. recurrentis* infects mice and monkeys in the laboratory, the strict host preference of the human body louse for only humans is likely a

TABLE 1 Characteristics and distribution of arthropod-borne borreliae

Borrelia spp.	Arthropod vector	Animal reservoir	Distribution	Disease
B. recurrentis (syn., *B. obermeyeri*, *B. novyi*)	*P. humanus humanus*	Humans	Worldwide	Louse-borne, epidemic relapsing fever
B. duttonii	*Ornithodoros moubata*	Humans	Central, eastern, and southern Africa	East African tick-borne, endemic relapsing fever
B. hispanica	*Ornithodoros erraticus* (large variety)	Rodents	Spain, Portugal, Morocco, Algeria, Tunisia	Hispano-African tick-borne relapsing fever
B. crocidurae, *B. merionesi*, *B. microti*, *B. dipodilli*	*Ornithodoros erraticus* (small variety)	Rodents	Morocco, Libya, Egypt, Iran, Turkey, Senegal, Kenya	North African tick-borne relapsing fever
B. persica	*Ornithodoros tholozani* (syn., *Ornithodoros papillipes*, *Ornithodoros crossi?*)	Rodents	From west China and Kashmir to Iraq and Egypt, former USSR, India	Asiatic-African tick-borne relapsing fever
B. caucasica	*Ornithodoros verrucosus*	Rodents	Caucasus to Iraq	Caucasian tick-borne relapsing fever
B. latyschewii	*Ornithodoros tartakowskyi*	Rodents	Iran, central Asia	Caucasian tick-borne relapsing fever
B. hermsii	*Ornithodoros hermsi*	Rodents, chipmunks, tree squirrels	Western United States	American tick-borne relapsing fever
B. turicatae	*Ornithodoros turicata*	Rodents	Southwestern United States	American tick-borne relapsing fever
B. parkeri	*Ornithodoros parkeri*	Rodents	Western United States	American tick-borne relapsing fever
B. mazzottii	*Ornithodoros talaje* (*Ornithodoros dugesi?*)	Rodents	Southern United States, Mexico, Central and South America	American tick-borne relapsing fever
B. venezuelensis	*Ornithodoros rudis* (syn., *Ornithodoros venezuelensis*)	Rodents	Central and South America	American tick-borne relapsing fever
B. burgdorferi, *B. garinii*, *B. afzelii*	*Ixodes scapularis*	Rodents	Eastern and midwestern United States	Lyme disease
	Ixodes pacificus	Rodents	Western United States	Lyme disease
	Ixodes ricinus	Rodents	Europe	Lyme disease
	Ixodes persulcatus	Rodents	Asian countries	Lyme disease
B. japonica	*Ixodes ovatus*	Rodents	Japan	Lyme disease?
	?	?	Australia	Lyme disease
B. theileri	*Rhipicephalus* spp., probably other ixodid ticks	Cattle, horses, sheep	South Africa, Australia, North America, Europe	Tick spirochetosis
B. coriaceae	*Ornithodoros coriaceus*	Deer (?) Cattle (?)	Western United States	Epizootic bovine abortion (?)
B. anserina	*Argas* spp. (mites?)	Fowl	Worldwide	Avian borreliosis
B. brasiliensis	*Ornithodoros brasiliensis*	?	South America (Brazil)	?[a]
B. graingeri	*Ornithodoros graingeri*	?	East Africa (Kenya)	One laboratory case[a]
B. tillae	*Ornithodoros zumpti*	Rodents	South Africa	?[a]
B. queenslandica	*Ornithodoros gurneyi*	Rodents	Australia	?[a]
B. armenica	*Ornithodoros alactagalis*	Rodents	Armenia	?[a]

[a]Spirochete-tick associations of unknown or little health significance.

primary factor restricting this spirochete's distribution in nature.

The remaining known species of *Borrelia* are transmitted by various species of "hard" ticks of the family Ixodidae. *B. burgdorferi* sensu stricto, *B. garinii*, and *B. afzelii*, causative agents of Lyme disease and related disorders, are transmit-

ted by numerous species of ticks in the genus *Ixodes* (4, 50) and possibly by ticks in other genera as well (73).

Diverse vertebrate hosts are infected with borreliae in nature. With only a few exceptions, however, various species of rodents are the natural vertebrate hosts, and these rodent hosts, along with ticks, maintain most of the species of *Borrelia* in foci throughout many parts of the world (31). Except for *B. duttonii*, for which humans remain the only natural vertebrate hosts, tick-borne relapsing fever spirochetes all naturally infect small rodents. In western North America, *B. hermsii*, *B. turicatae*, and *B. parkeri* all naturally infect small rodents, primarily in the squirrel family. These three species appear to be maintained and transmitted only by their specific tick vectors, *Ornithodoros hermsi*, *Ornithodoros turicata*, and *Ornithodoros parkeri*, respectively. The unique ecological requirements of each of these species of tick therefore also influences the distribution of borreliae, the mammalian hosts infected, and the probability of humans entering the maintenance cycle and becoming infected.

The diversity of vertebrate hosts that are naturally infected with Lyme disease spirochetes (*B. burgdorferi* sensu lato) is far greater than the types of hosts infected by any one species of relapsing fever spirochete (3, 4, 50). This diversity is probably influenced not so much by what species of mammals the spirochetes are able to colonize but rather by the very different life cycles and behaviors displayed by the *Ornithodoros* and *Ixodes* ticks. *Ornithodoros* ticks spend their entire lives in protected microhabitats closely associated with the nest or resting places of their host and source of blood. These ticks feed rapidly, usually in 10 to 45 min and usually at night, and for the rest of the time remain off their hosts. All stages of these ticks, including larvae, multiple nymphs, and adults, most likely feed on the same species of mammal and are unlikely to be dispersed widely by the movements of their hosts.

The *Ixodes* ticks that transmit Lyme disease spirochetes have life cycles and behaviors very different from those of the relapsing fever ticks. The specificity displayed by many of the relapsing fever borreliae for a single species of tick is not true for the Lyme disease spirochetes. *Ixodes* ticks are called "three-host" ticks, referring to the fact that each of the three stages in the life cycle (larva, single nymph, and adult) feeds on a different host (84). For the *Ixodes* species associated with Lyme disease, the larvae and nymphs feed primarily on small rodents. Many species of mammals as well as birds may be hosts for these ticks, and adult ticks feed primarily on a variety of mammal species, including deer, raccoons, domestic and wild carnivores, and larger domestic mammals such as horses. These *Ixodes* ticks all feed for prolonged periods, thus contributing significantly to their own dispersal in association with their host's movement. They thereby encounter a greater diversity of hosts to subsequently feed upon. As a result, many species of small to large mammals play a role both in maintaining Lyme disease spirochetes in nature and in serving as hosts for the ticks that transmit these bacteria.

All vectors become infected by ingesting spirochetes while feeding on the blood of an infected mammal or bird. Some species of ticks may also pass spirochetes to the next generation via eggs, a phenomenon referred to as vertical or transovarial transmission. This type of transmission has the potential to maintain spirochetal infections in ticks without the necessity for the ticks to feed on infectious vertebrate hosts. Although transovarial transmission of *B. burgdorferi* has been reported (48, 66), it occurs rarely and has no importance in maintaining the spirochetes in nature. Another unusual form of direct horizontal transmission of some relapsing fever spirochetes is venereal transmission of the spirochetes in the spermatophores of the male tick, but the importance of such transmission for maintaining these spirochetes in nature is not known.

The transmission of spirochetes from the arthropod vector to susceptible vertebrate hosts varies somewhat among each of the three types of vectors and is influenced by the type of infection established in the arthropod. *B. recurrentis*, the etiologic agent of louse-borne relapsing fever, is unique both in its vector, the human body louse, and in how the spirochete is transmitted by this insect (21). In the louse, the spirochetes establish an infection restricted to the hemolymph and are absent from the salivary glands; thus, they are not transmitted by bite via saliva. Humans become infected only when body lice carrying the spirochetes are crushed, rupturing the louse's outer skeleton and releasing the spirochetes onto the skin, where they will enter the body through either broken or intact skin.

The soft ticks that transmit the relapsing fever spirochetes do so primarily by bite via infected saliva. After ingestion by a soft tick, the borreliae penetrate the gut wall and disseminate to several organs, including the ovaries, central ganglion, salivary glands, and sometimes also the coxal glands. Although *Ornithodoros* and *Argas* ticks feed rapidly, transmission results either from the direct inoculation of infected saliva or from contamination of the bite wound with infected coxal fluid, which is excreted from pores near the base of each front leg of the tick (hard ticks do not have coxal glands).

In *Ixodes* ticks infected with *B. burgdorferi*, the spirochetes are usually restricted to the midgut during periods when the ticks are off their hosts (18). However, during tick engorgement, spirochetes multiply and penetrate the midgut to invade the hemolymph and spread to the salivary glands, from which they may be transmitted via saliva that is secreted during the later stages of tick feeding (67, 91). Several studies have demonstrated that transmission of *B. burgdorferi* seldom occurs during the first 24 to 48 h of tick feeding. However, as ticks feed for longer periods, the frequency of *Ixodes* ticks infecting mammalian hosts increases dramatically (67). The regurgitation of spirochetes in gut fluids directly back out through the mouth is also considered a potential mode of transmission.

CLINICAL SIGNIFICANCE

Relapsing Fever in Humans

Relapsing fever in humans is a febrile, septicemic disease with sudden onset after an incubation period of 2 to 15 days. Fever persists for 3 to 7 days and is followed by an afebrile interval of several days to several weeks. Thereafter, as many as 13 relapses may occur, especially in untreated patients, as a result of antigenic variations in the causative borreliae. Detailed clinical descriptions of relapsing fever in humans have been presented elsewhere (17, 85). During the acute, febrile episodes, patients may also experience shaking chills, severe headache, myalgias, arthralgias, nausea, and vomiting. A skin rash that is quite variable in form and may resemble the erythema migrans lesion associated with Lyme disease has also been reported for 28% of relapsing fever patients. Long-term or chronic clinical manifestations may involve the central nervous, respiratory, and cardiovascular systems. Such chronic cases of relapsing fever may resemble and be confused with chronic Lyme disease (51). Pregnant women who contract relapsing fever may abort their fetus or give birth to an infant infected via transplacental transmission.

According to medical history, more than 50 million persons contracted louse-borne relapsing fever during the first half of this century, with epidemics occurring throughout Europe, Africa, Asia, and South America. Since 1967, however, louse-borne relapsing fever has been reported primarily from African countries, although more recent outbreaks have occurred also in South America (World Health Organization Weekly Epidemiological Record, 1967–1978).

Because of the sporadic occurrence of tick-borne relapsing fevers, extremely little is known about their incidence worldwide. More than 840 cases, including at least 6 deaths, have been recorded from Jordan, Rwanda, and Iran since 1959 (25, 26, 40).

In the United States during the period 1964 to 1993, 512 cases of tick-borne relapsing fever were recognized. Most occurred within the distributional area of *O. hermsi*, and a few occurred also within that of *O. turicata*. Outbreaks usually are sporadic and rarely involve more than two persons. Outbreaks involving a larger number of cases have been recorded occasionally, with reports from the Pacific Northwest (88) as well as the Southwest, including the Grand Canyon (16). Tick-borne relapsing fever is underreported in the United States because it is seldom recognized. Most patients are unaware of a tick bite, and unless a history of wilderness exposure, camping, or spending nights in old, rodent-infested cabins is given, the disease is rarely suspected during the initial period of fever (33).

Lyme Disease in Humans

Lyme disease is an endemic inflammatory disorder that usually begins in summer with the distinctive skin lesion erythema migrans accompanied by headache, stiff neck, myalgias, arthralgias, malaise, fatigue, or the swelling of the lymph nodes. Weeks to months later, some patients may develop meningoencephalitis, myocarditis, or migrating musculoskeletal pain. Still later, patients may develop intermittent attacks of oligoarticular arthritis or chronic arthritis in the large joints, particularly the knees. First described after an outbreak among children in Lyme, Conn., in 1975 (87), the disease appears to be more severe than erythema chronicum migrans, a tick-borne associated syndrome observed as early as 1908 in Europe (2). The etiologic agent, *B. burgdorferi*, remained obscure until 1981, when it was discovered in *Ixodes scapularis* (= *Ixodes dammini*) from New York (19) and later in the European tick vector *Ixodes ricinus*. Other clinical syndromes in Europe that appear to be related to the same agent include lymphocytoma (lymphadenosis benigna cutis), acrodermatitis chronica atrophicans, tick-borne meningoradiculitis (Garin-Bujadoux-Bannwarth's syndrome), and myositis (2, 69, 89).

In the United States from 1975 through 1979, about 500 cases of Lyme disease, the majority from the Northeast, were reported. Since 1982, the year the discovery of the causative agent of Lyme disease and related disorders was reported, through 1992, approximately 50,000 cases have come to the attention of the Centers for Disease Control and Prevention (CDC). Increased incidence is thought to be due to improved awareness and recognition of the disease as well as to actual geographic spread. The number of states with Lyme disease has also increased from 14 in 1979 to 48 by 1992.

Borrelioses in Animals

Borrelioses in rodents and lagomorphs may be similar to the diseases observed in humans. There may be one or more relapses accompanied by spirochetemias of various degrees.

Not all animals are equally susceptible to the various *Borrelia* species, and even differences in susceptibility to various isolates of a single species of spirochete may be seen. Young animals are generally more susceptible and may die as the result of infection.

Avian borreliosis, caused by *B. anserina*, affects geese, ducks, turkeys, and chickens throughout the world and in certain countries is of great economic importance because of the severe losses to the poultry industry it causes. Clinically, the disease begins with a high fever after an incubation period of about 4 days. The birds become cyanotic and have yellowish green diarrhea. During the early stages of the febrile reaction, spirochetes can be readily detected in the blood. Surviving birds recover after about 2 weeks and have long-lasting immunity.

Tick spirochetosis caused by *B. theileri* in cattle, horses, and sheep in South Africa and recently also in Australia is a benign disease characterized by one to two attacks of fever, inappetence, weight loss, weakness, and anemia.

The Lyme disease spirochete, *B. burgdorferi*, does not adversely affect wild animals. However, in dogs, cows, and horses, it has been reported to cause arthritic manifestations with lameness, stiffness, and swollen joints (20, 56, 57). White-footed mice (*Peromyscus leucopus*) once infected probably remain so for their entire lives, although there are fluctuations in the percentage of ticks that become infected while feeding upon them (18, 53). New Zealand White rabbits, hamsters, gerbils, and certain genetic lines of mice and rats are also susceptible to *B. burgdorferi* and appear to be useful for studies related to the pathogenicity of this agent.

A borrelia spirochete has also been isolated from *O. coriaceus*, the soft tick implicated as a vector of epizootic bovine abortion, which is a major disease of rangeland cattle in the western United States (49). Genetic and phenotypic characteristics revealed that this spirochete is a new species, named *B. coriaceae* (42). As yet, there is only circumstantial evidence that this organism is causally related to epizootic abortion, and as of 1993, no additional isolates of this spirochete have been obtained.

COLLECTION, TRANSPORT, AND STORAGE OF SPECIMENS

The louse-borne relapsing fever spirochete *B. recurrentis* can be collected either from infected human body lice or from human blood of patients with acute disease. Similarly, the African tick-borne relapsing fever spirochete *B. duttonii* can be collected either from infected *O. moubata* ticks or from infected human blood. All of the other tick-borne relapsing fever spirochetes can be collected from the appropriate species of infected *Ornithodoros* tick, the wild mammalian reservoir, or the blood of patients acutely ill with relapsing fever. Given the difficulty of collecting *Ornithodoros* ticks and the short time that wild mammals have spirochetes circulating in their peripheral blood, human patients acutely ill with relapsing fever are often the best sources for acquiring new isolates of relapsing fever spirochetes.

The Lyme disease spirochetes *B. burgdorferi* sensu stricto, *B. garinii*, and *B. afzelii* can all be acquired from infected *Ixodes* ticks, various infected tissues of mammalian reservoirs, or human patients. Because this group of *Borrelia* spp. does not frequently produce detectable spirochetemias in mammals, blood does not yield spirochetes as often as other

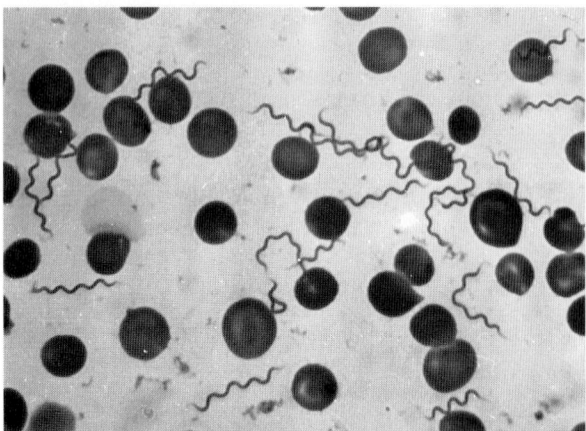

FIGURE 1 *B. hermsii* in a thin smear of rodent blood. Giemsa stain. Magnification, ×1,100.

tissues such as the skin, urinary bladder, heart, and spleen (44, 76, 83). Although these spirochetes have been isolated from human blood, skin biopsy samples taken just outside the periphery of an erythema migrans lesion have been the best sources. Lyme disease spirochetes have also been isolated from human cerebrospinal fluid and only rarely from other tissues and fluids.

The maintenance of infectious spirochetes requires the use of laboratory animals such as rats or mice or live colonies of infected ticks. Relapsing fever spirochetes can be kept alive for several months in clotted whole blood collected from the infected mammal and held at 4°C. Repeated reinoculation of mice and subsequent storage of frozen blood can maintain spirochetes in the laboratory for long periods. Vector ticks infected with their associated spirochetes (by feeding on infective mice) can be held for either one generation (*Ixodes* ticks) or for months to several years (*Ornithodoros* ticks), thus maintaining infections of these bacteria that can then be passed into mice by inoculation or tick feeding. Lyme disease spirochetes can also be inoculated into laboratory mice, hamsters, or white-footed mice, which will likely remain infected for their entire lives and can be the source of *B. burgdorferi* through culture of their infected skin or urinary bladders or by allowing ticks to feed upon them (76).

DETECTION AND ISOLATION PROCEDURES

Detection of Spirochetes in Blood by Staining

Diagnosis of most clinical cases of borrelioses except Lyme disease is based primarily on the detection of spirochetes in the peripheral blood of febrile persons and animals. During the febrile period of relapsing fever, high numbers of spirochetes may circulate in the blood and can be detected by light or dark-field microscopy of wet preparations made from a drop of whole blood mixed with a small drop of sodium citrate and covered with a coverslip. Leishman, Giemsa, May-Grunwald, Wright, and other combinations of Romanowsky stains are used to stain thin- and thick-drop films for examination by conventional light microscopy (Fig. 1). Spirochetes in the blood of mildly infected persons are more readily detected by a microhematocrit concentration technique previously described elsewhere than by microscopy (34).

Lyme disease spirochetes usually do not produce a microscopically detectable spirochetemia in humans or wild animals.

Detection of Spirochetes in Tissues by Staining

B. burgdorferi can be detected in tissue sections by light microscopy with various staining methods. This species and other members of the genus have an affinity for acid dyes, and they stain with nearly all aniline dyes. The Warthin-Starry silver stain has been used often (14), as have immunologic stains that use fluoresceinated polyclonal antiserum. Staining indirectly with a monoclonal antibody followed by a secondary fluoresceinated antibody has the potential to both detect and identify spirochetes in clinical samples (63). Numerous monoclonal antibodies that bind to a diversity of epitopes presented by several species of *Borrelia* have now been described (12, 15, 77).

Detection of *B. burgdorferi* Antigens

Considerable interest has been directed toward developing alternatives to the detection of intact Lyme disease spirochetes because of their paucity in clinical samples. Approaches have included detection of antigens in urine, blood, and other tissues by a dot blot immune assay and an antigen capture assay using colloidal gold-immune electron microscopy (28, 39). Although the initial results were encouraging and one urine-based antigen test was marketed commercially for a brief time, no such tests are currently available.

Detection of Spirochetes by Using DNA Probes and PCR

A recent strategy for detection of *Borrelia* spp. has been to demonstrate the presence of DNA specific to certain species of spirochete. Hybridization probes were developed first to specifically identify the purified DNAs of *B. burgdorferi* and *B. hermsii* as well as whole spirochetes (78, 80). Although the probes were highly specific, their sensitivities were unsatisfactory, requiring a minimum of 2,800 to 10,000 spirochetes for detection. Subsequently, assays specific for *B. burgdorferi* were developed by using the PCR. These assays are capable of detecting only one to five spirochetes (71), of classifying spirochetes from North America and Eurasia into two typing groups (70), and of identifying the three currently recognized species, *B. burgdorferi*, *B. garinii*, and *B. afzelii* (60). Other workers have further developed PCR assays, amplifying other regions of the spirochete's genome to identify *B. burgdorferi* DNA (59, 62) and to detect the spirochete in ticks (64) or in the urine of human patients with Lyme disease (35).

Detection of Spirochetes by Culture

Borreliae can be cultivated readily in their arthropod vectors or in a large variety of vertebrate hosts, particularly small rodents. Cultivation in embryonated chicken eggs has also been successful.

In vitro cultivation of borreliae in liquid medium is also possible (46). The medium currently used by most workers to isolate and maintain many species of *Borrelia* in the laboratory is Barbour-Stoenner-Kelly II (BSK II) medium (6, 10). A commercial preparation of the complete medium, BSK-H, is available from Sigma Chemical Co., St. Louis, Mo. The spirochetes are cultured most successfully when the medium has a neutral pH and is incubated in a microaerophilic environment at 30 to 37°C. The culture medium is monitored for spirochetes by dark-field micros-copy for 4 to 6 weeks; cultures may be positive in less than a week or may require many months if the medium lacks gelatin and rabbit serum and is incubated at temperatures lower than 30°C (54). Once isolated in the medium, the spirochetes may be identified by reactivity with species-specific monoclonal antibodies, although hybridization with specific DNA probes or amplification of specific DNA by PCR can be done. Cultures of borreliae can be stored for several years at −70°C by adding a cryopreservative such as glycerol (10 to 20% final concentration) to the medium.

Various modifications of BSK medium have also been used successfully for the cultivation of borreliae from ticks as well as from human and other animal tissues. Borreliae are resistant to rifampin and phosphomycin, and these antibiotics can be added to the BSK-II medium (50 and 100 μg/ml, respectively) to reduce contamination of cultures by other bacteria. Amphotericin B (10 μg/ml) is also useful at reducing fungal contamination, especially when one is attempting to isolate borreliae from triturates of entire ticks. Increased concentrations of agarose have been used to successfully grow *B. burgdorferi* on a solidified medium, allowing production of colonies from single organisms (47). Continuous serial passage for even short periods, however, may effect many biological changes in the spirochetes, thereby altering their phenotypic and genotypic character-istics (74, 75).

Animal Inoculation

Visualization or isolation in liquid medium of spirochetes from human blood may at times be very difficult if possible at all; in these circumstances, inoculation of susceptible laboratory animals for recovery of spirochetes may be given consideration. Such methods are not routinely available, and specialized reference laboratories, such as the CDC, should be consulted.

IDENTIFICATION

Historically, the taxonomic identification of borreliae, par-ticularly the relapsing fever spirochetes, has depended heavily on their geographical distribution and the natural arthropod vectors (24). This assumed specificity, however, does not always hold true. Thus, for most cases of tick-borne relapsing fevers, taxonomic identification of the eti-ologic agent is presumptive and is based on a history of exposure to a particular vector.

Morphologic criteria are not sufficient to distinguish among the species of *Borrelia*. Comparisons of total DNA by hybridization studies discriminate between the relapsing fever and Lyme disease spirochetes but are of limited utility in distinguishing among the various species of relapsing fever spirochetes (38, 43). The phenomenon of antigenic phase variations makes a precise serologic identification of relapsing fever spirochetes difficult if not impossible. The development of methods that permit the identification of different relapsing fever species is hindered by the paucity of such isolates from clinical or natural sources.

Several molecular procedures to identify *B. burgdorferi* and to differentiate it from other pathogenic spirochetes are available (38, 78, 80). *B. burgdorferi* can be directly iden-tified in tick midguts by fluorescent staining with species-specific monoclonal antibodies that recognize spirochetal outer surface proteins. It is also possible to identify *B. burgdorferi* in ticks by using the PCR and specific primers. Direct identification of *B. burgdorferi* in clinical material by using PCR has proven unreliable (except for synovial fluid) owing to the paucity of organisms in such samples. Methods to distinguish between *B. burgdorferi* and other borreliae include DNA probes and PCR assays based on sequences encoding flagellin (65), 16S rRNA (64), outer surface pro-teins (64), or other targets (35) as well as on the reactivity of bacterial isolates on immunoblots with species-specific monoclonal antibodies recognizing the major outer surface proteins (11) or a conserved 39-kDa protein (82).

SEROLOGIC TESTS

Serologic confirmation of borrelioses is accomplished by a diagnostic change in the titer of antibodies specific to the agent when a patient's sequential sera are compared. In the absence of acute-phase serum, a single convalescent-phase serum sample with an antibody titer above a specified level may be acceptable for confirmation of the diagnosis. How-ever, many species of relapsing fever spirochetes have not yet been successfully isolated and maintained in artificial culture and are therefore not available for use as antigens in serologic tests. Spontaneous change of surface antigens in these spirochetes also makes serodiagnosis difficult and ex-plains the inefficiency and failure of many serologic proce-dures, including agglutination, adhesion, opsonic activity, immobilization, and borreliolysin, in the detection of anti-bodies. Nevertheless, promising results have been obtained with the indirect immunofluorescence assay (IFA) with cultured spirochetes as antigens. This test may also be used for the characterization of antigenic types of relapsing fever borreliae and for the detection of mixed populations and serotypes of these spirochetes.

Lyme disease spirochetes are difficult to detect in human patients, and serologic tests are therefore the most practical and readily available methods for confirming Lyme disease in the diagnostic laboratory. The history of Lyme serology dates only to late 1981, when the newly isolated spirochete was used in an IFA study screening convalescent-phase sera of Lyme patients from the northeast United States (19). Since then, numerous types of serologic tests for Lyme disease have been described in the scientific literature, and new test kits have inundated an increasingly competitive commercial market. Numerous reviews and editorials that describe some of these tests, their potential problems, and the lack of standardization have been published (8, 9, 29, 30).

The IFA uses Lyme disease spirochetes as a substrate fixed onto a glass microscope slide. While this test is easy to do, there are problems in the test's specificity, reproducibil-ity, and usefulness when large numbers of samples need to be tested as well as in its subjective interpretation and lack of quantification. Owing to the presence of nonspecific antibodies in some normal sera, most workers require that a positive sample be reactive when diluted 1:256 or more, although not everyone adheres to this threshold for a pos-itive result. Unfortunately, a titer of 256 or higher may occur if the patient has been exposed to other species of spirochetes or to even more distantly related bacteria that share antigens with the Lyme disease spirochete (8). There-fore, a patient who has not been exposed to *B. burgdorferi* but who has had tick-borne relapsing fever, syphilis, lepto-spirosis, or periodontal disease may have antibodies that react positively in a Lyme test (58). Visual determination of the titer end point is also somewhat subjective.

The enzyme-linked immunosorbent assay (ELISA), or enzyme immunoassay, is probably the most widely used serologic test for confirming Lyme disease in laboratories where large numbers of samples are routinely tested (8, 13, 55). The advantages of this type of assay over the IFA are the ability to quickly test many samples and the use of a spectrophotometric determination that is quantifiable and subject to statistical analysis. The assay in its simplest form uses microtiter plates in which wells are coated with either intact spirochetes or a suspension of spirochetal antigens produced by disrupting the bacteria by sonication. While this method is fast, efficient for testing many samples, and quantifiable with less subjective error, it has the same weaknesses of specificity that the IFA has (58), and the ELISA has become even less standardized than the IFA. As mentioned above, sera from patients exposed to some other pathogenic bacteria may react falsely. Also, there is considerable variation in how the ELISA is done and how the results are reported.

Because of the poor specificity of both the IFA and the ELISA, the Western immunoblot technique, considered by some investigators to be the "gold standard," could be more beneficial than these assays, especially in the diagnosis of patients with atypical clinical manifestations who live in areas where the disease is currently considered nonendemic. Done properly, a negative serum can be identified with a high degree of confidence. Multiple banding resulting from antibodies binding to one or more of the antigens raises the concern of what is really positive. Considerable variations in the reagents and methods used subject this procedure to the same types of interlaboratory variation that exist for the ELISA.

However, the potential lack of specificity for many of these tests as described above has led investigators to search for components of the bacterium that can be used as diagnostic antigens to improve the specificities and sensitivities of serologic tests. Specific antigens have been obtained either by lysing the spirochetes and then selectively purifying the target antigen (23, 36, 45) or by using recombinant DNA techniques to clone genes, thereby expressing a specific spirochetal antigen in a foreign bacterial cell like *Escherichia coli* (52, 79). An antigen that looks quite promising is the 39-kDa protein (P39), which is conserved among all Lyme disease spirochetes, is specific to this group of borreliae, and generates a strong antibody response in infected humans and experimental animals (79, 82). This recombinant antigen has been incorporated into several new commercially available serologic test kits.

ANTIBIOTIC SUSCEPTIBILITIES

There is general agreement that antibiotics are preferable to the arsenical compounds that were originally used to treat borrelial infections (31). However, choice of antibiotic, dosage, and duration and route of treatment are controversial.

Although borreliae are susceptible to many antibiotics, tetracyclines are the drugs of choice in treating relapsing fever (31). They reduce the relapse rate and rid the central nervous system of spirochetes. However, the rapid destruction of organisms may provoke a severe Jarisch-Herxheimer reaction. Good results have been obtained with 0.5 g of tetracycline given orally every 6 h for 4 to 5 days or with a single oral dose of 100 mg of doxycycline. To avoid the Jarisch-Herxheimer reaction, a combined penicillin and tetracycline therapy is recommended. It consists of 400,000 U of procaine penicillin administered intramuscularly fol-

lowed the next day by 500 mg of tetracycline given orally every 6 h for 7 days (72).

Although all stages of Lyme disease may respond to antibiotic therapy, treatment regimens depend on the nature and severity of clinical manifestations (1, 86). Antibiotics given early in disease shorten duration of the rash and prevent later illness, although some patients with severe early disease develop manifestations later despite recommended courses of antibiotics.

Men, nonpregnant women, and children with early disease, mild neurological symptoms, and/or minor cardiac involvement have been treated orally with doxycycline (100 mg twice daily) or tetracycline HCl (250 to 500 mg thrice daily) given for 10 to 30 days. Amoxicillin (250 to 500 mg thrice daily, or 20 to 40 mg/kg of body weight per day for children) given for 10 to 30 days is also effective and is preferred for children under 8 years of age and for pregnant and lactating women. For patients who cannot take tetracycline and are allergic to penicillin, erythromycin at a dosage of 250 mg four times daily (or 30 mg/kg/day for children) is recommended.

Patients with late neurological (focal central nervous system involvement) complications, severe Lyme arthritis, or severe cardiac involvement may require intravenous application of penicillin G or ceftriaxone.

Very few therapeutic trials for Lyme disease have been conducted in a prospective randomized fashion (61). MICs and MBCs for various antibiotics are inconsistent in published studies, and it is not always possible to correlate the outcome of therapy for human infections with in vitro testing. It is also difficult to assess whether "treatment failure" of some cases of late-stage Lyme disease actually reflects "true failure" or results instead from an initial misdiagnosis of Lyme disease.

Avian borreliosis has been treated successfully with 100,000 U of procaine penicillin given intramuscularly or with oxytetracycline at 2 mg/kg of body weight (31). Borrelial infections in cattle are generally so mild as not to require treatment. However, spirochetes are effectively treated with tetracycline (2 mg/kg of body weight) given intravenously.

Borreliae are resistant to rifampin, phosphomycin, sulfonamides, and 5-fluorouracil (11). The molecular basis for these resistances is unknown; however, as discussed above, they have been useful when isolating borreliae from samples that are contaminated with other bacteria.

EVALUATION, INTERPRETATION, AND REPORTING OF RESULTS

Tick-borne relapsing fever in the United States is not a reportable disease except in California and Texas. Therefore, the disease is certainly underrecognized, both in the United States and in other regions of the world where it is endemic. A presumptive diagnosis of human cases is based primarily on recurrent febrile episodes following exposure to habitats where infected *Ornithodoros* ticks (or infected human body lice) are known to occur. Cases are confirmed by demonstrating spirochetes in a blood smear made during a febrile episode. Retrospective serologic confirmation of relapsing fever is not generally done. Antibiotic treatment early in the course of the disease may also prevent the development of detectable antibodies. For these reasons, there is little interest in acquiring, evaluating, or reporting serologic test results for relapsing fever. Additionally, no specific serologic test that will identify antibodies to only one species of relapsing fever spirochete has been developed. For patients who are considered to possibly have chronic relapsing fever, testing the patient's serum with a panel of different *Borrelia* spp. may reveal greater reactivity to the species having caused the infection, but such results are also difficult to interpret given the antigenic heterogeneity within a single *Borrelia* species.

Lyme disease is now reportable in all of the United States except Oregon (and this will likely change soon). Case recognition and reporting rely heavily on positive serologic test results (in lieu of detecting or isolating spirochetes) and clinical manifestations involving organ systems that are considered consistent for this disease. Therefore, unlike procedures for relapsing fever, serologic testing, evaluation, and reporting for Lyme disease have placed heavy demands on diagnostic laboratories. Because of the lack of standardization of the commonly used serologic tests (IFA, ELISA) and the significantly high frequency of false-positive results due to cross-reactive antibodies to other bacteria, the interpretation of a positive Lyme disease titer is very difficult without the patient's history. Thus, for many clinicians in areas where the disease is endemic, the diagnosis of Lyme disease is often based solely on clinical presentation with a possible exposure to or known history of *Ixodes* tick bite. A greater problem arises in regions where Lyme disease spirochetes and a proven vector are not yet known to occur. In such areas, caution must be taken not to place too much significance on a borderline or moderately positive titer to *B. burgdorferi* in the sera of patients who present with clinical manifestations similar to one of the many conditions that have been described for Lyme disease. Better serologic tests that use single or multiple recombinant antigens and have much greater specificities than the tests that use entire spirochetes as antigens are becoming available. During the next several years, as these new tests become established and accepted, one can hope that greater reliability will occur and greater confidence can be placed in serologic tests that either confirm or rule out a presumptive diagnosis of Lyme disease.

REFERENCES

1. **Abramowicz, M.** 1989. Treatment of Lyme disease. *Med. Lett.* **31:**57–59.
2. **Afzelius, A.** 1921. Erythema chronicum migrans. *Acta Derm. Venereol.* **2:**120–125.
3. **Anderson, J. F.** 1988. Mammalian and avian reservoirs of *Borrelia burgdorferi. Ann. N.Y. Acad. Sci.* **539:**180–191.
4. **Anderson, J. F.** 1989. Epizootiology of *Borrelia* in Ixodes tick vectors and reservoir hosts. *Rev. Infect. Dis.* **11:**1451–1459.
5. **Baranton, G., D. Postic, I. Saint-Girons, P. Boerlin, J.-C. Piffaretti, M. Assous, and P. A. D. Grimont.** 1992. Delineation of *Borrelia burgdorferi* sensu stricto, *Borrelia garinii* sp. nov., and group VS461 associated with Lyme borreliosis. *Int. J. Syst. Bacteriol.* **42:**378–383.
6. **Barbour, A. G.** 1984. Isolation and cultivation of Lyme disease spirochetes. *Yale J. Biol. Med.* **57:**521–525.
7. **Barbour, A. G.** 1988. Plasmid analysis of *Borrelia burgdorferi*, the Lyme disease agent. *J. Clin. Microbiol.* **26:**475–478.
8. **Barbour, A. G.** 1988. Laboratory aspects of Lyme borreliosis. *Clin. Microbiol. Rev.* **1:**399–414.
9. **Barbour, A. G.** 1989. The diagnosis of Lyme disease: rewards and perils. *Ann. Intern. Med.* **110:**501–502.
10. **Barbour, A. G., W. Burgdorfer, S. F. Hayes, O. Peter, and A. Aeschlimann.** 1983. Isolation of a cultivable spirochete from *Ixodes ricinus* ticks of Switzerland. *Curr. Microbiol.* **8:**123–126.
11. **Barbour, A. G., and S. F. Hayes.** 1986. Biology of *Borrelia* species. *Microbiol. Rev.* **50:**381–400.
12. **Barbour, A. G., S. F. Hayes, R. A. Heiland, M. E. Schrumpf, and S. L. Tessier.** 1986. A *Borrelia*-specific monoclonal antibody binds to a flagellar epitope. *Infect. Immun.* **52:**549–554.
13. **Berardi, V. P., K. E. Weeks, and A. C. Steere.** 1988. Serodiagnosis of early Lyme disease: analysis of IgM and IgG antibody responses by using an antibody-capture enzyme immunoassay. *J. Infect. Dis.* **158:**754–760.
14. **Berger, B. W.** 1989. Dermatologic manifestations of Lyme disease. *Rev. Infect. Dis.* **11:**S1475–S1481.
15. **Blanchard-Channell, M., and J. L. Stott.** 1991. Characterization of *Borrelia coriaceae* antigens with monoclonal antibodies. *Infect. Immun.* **59:**2790–2798.
16. **Boyer, K. M., R. S. Munford, G. O. Maupin, C. P. Pattison, M. D. Fox, A. M. Barnes, W. L. Jones, and J. E. Maynard.** 1977. Tick-borne relapsing fever: an interstate outbreak originating at Grand Canyon National Park. *Am. J. Epidemiol.* **105:**469–479.
17. **Bryceson, A. D. E., E. H. O. Parry, P. L. Perine, D. A. Warrell, D. Vukotich, and C. S. Leithead.** 1970. Louse-borne relapsing fever. A clinical and laboratory study of 62 cases in Ethiopia and a reconsideration of the literature. *J. Med.* **39:**129–170.
18. **Burgdorfer, W.** 1989. Vector/host relationships of the Lyme disease spirochete, *Borrelia burgdorferi. Rheum. Dis. Clin. N. Am.* **15:**775–787.
19. **Burgdorfer, W., A. G. Barbour, S. F. Hayes, J. L. Benach, E. Grunwaldt, and J. P. Davis.** 1982. Lyme disease—a tick-borne spirochetosis. *Science* **216:**1317–1319.
20. **Burgess, E. C.** 1988. *Borrelia burgdorferi* infection in Wisconsin horses and cows. *Ann. N.Y. Acad. Sci.* **539:**235–243.
21. **Buxton, P. A.** 1946. *The Louse: an Account of the Lice Which Infest Man, Their Medical Importance and Control*, p. 1–164. The Williams & Wilkins Co., Baltimore.
22. **Casjens, S., and W. M. Huang.** 1993. Linear chromosomal physical and genetic map of *Borrelia burgdorferi*, the Lyme disease agent. *Mol. Microbiol.* **8:**967–980.
23. **Coleman, J. L., and J. L. Benach.** 1987. Isolation of antigenic components from the Lyme disease spirochete: their role in early diagnosis. *J. Infect. Dis.* **155:**756–765.
24. **Davis, G. E.** 1956. The identification of spirochetes from human cases of relapsing fever by xenodiagnosis with comments on local specificity of tick vectors. *Exp. Parasitol.* **5:**271–275.
25. **deClercq, A. G., A. Z. Meheus, E. de Pierpont, and C. Nyirashema.** 1975. Single-dose doxycycline treatment of tick-borne relapsing fever. *East Afr. Med. J.* **8:**428–429.
26. **deZulueta, J., S. Nasrallah, J. S. Karam, A. R. Anani, G. K. Sweatman, and D. A. Muir.** 1971. Finding of tick-borne relapsing fever in Jordan by the malaria eradication service. *Ann. Trop. Med. Parasitol.* **65:**491–495.
27. **Diab, F. M., and Z. R. Soliman.** 1977. An experimental study of *Borrelia anserina* in four species of *Argas* ticks. I. Spirochete localization and densities. *Zentralbl. Parasitenkd.* **53:**201–212.
28. **Dorward, D. W., T. G. Schwan, and C. F. Garon.** 1991. Immune capture and detection of *Borrelia burgdorferi* antigens in urine, blood, or tissues from infected ticks, mice, dogs, and humans. *J. Clin. Microbiol.* **29:**1162–1170.
29. **Duffy, J., L. E. Mertz, G. H. Wobig, and J. A. Katzmann.** 1988. Diagnosing Lyme disease: the contribution of serologic testing. *Mayo Clin. Proc.* **63:**1116–1121.
30. **Eichenfield, A. H., and B. H. Athreya.** 1989. Lyme disease: of ticks and titers. *J. Pediatr.* **114:**328–332.
31. **Felsenfeld, O.** 1971. Borrelia. *Strains, Vectors, Human and Animal Borreliosis*, p. 180. Warren H. Green, Inc., St. Louis.
32. **Ferdows, M. S., and A. G. Barbour.** 1989. Megabase-sized linear DNA in the bacterium *Borrelia burgdorferi*, the Lyme disease agent. *Proc. Natl. Acad. Sci. USA* **86:**5969–5973.
33. **Fihn, S., and E. B. Larson.** 1980. Tick-borne relapsing fever in the Pacific Northwest: an underdiagnosed illness? *West. J. Med.* **133:**203–209.

34. **Goldschmid, J. M., and K. Mahomed.** 1972. The use of the microhematocrit technic for the recovery of *Borrelia duttonii* from the blood. *Am. J. Clin. Pathol.* **58**:165–169.

35. **Goodman, J. L., P. Jurkovich, J. M. Kramber, and R. C. Johnson.** 1991. Molecular detection of persistent *Borrelia burgdorferi* in the urine of patients with active Lyme disease. *Infect. Immun.* **59**:269–278.

36. **Hansen, K., K. Pii, and A.-M. Lebech.** 1991. Improved immunoglobulin M serodiagnosis in Lyme borreliosis by using a μ-capture enzyme-linked immunosorbent assay with biotinylated *Borrelia burgdorferi* flagella. *J. Clin. Microbiol.* **29**:166–173.

37. **Holt, S. C.** 1978. Anatomy and chemistry of spirochetes. *Microbiol. Rev.* **42**:114–160.

38. **Hyde, F. W., and R. C. Johnson.** 1984. Genetic relationship of Lyme disease spirochetes to *Borrelia, Treponema,* and *Leptospira* spp. *J. Clin. Microbiol.* **20**:151–154.

39. **Hyde, F. W., R. C. Johnson, T. J. White, and C. E. Shelburne.** 1989. Detection of antigens in urine of mice and humans infected with *Borrelia burgdorferi,* etiologic agent of Lyme disease. *J. Clin. Microbiol.* **27**:58–61.

40. **Janbakhsh, B., and A. Ardelan.** 1977. The nature of sporadic cases of relapsing fever in Kazeroun area, southern Iran. *Bull. Soc. Pathol. Exot.* **70**:587–589.

41. **Johnson, R. C.** 1977. The spirochetes. *Annu. Rev. Microbiol.* **31**:89–106.

42. **Johnson, R. C., W. Burgdorfer, R. S. Lane, A. G. Barbour, S. F. Hayes, and F. W. Hyde.** 1987. *Borrelia coriaceae* sp. nov.: putative agent of epizootic bovine abortion. *Int. J. Syst. Bacteriol.* **328**:454–456.

43. **Johnson, R. C., F. W. Hyde, and C. M. Rumpel.** 1984. Taxonomy of the Lyme disease spirochete. *Yale J. Biol. Med.* **57**:529–537.

44. **Johnson, R. C., N. Marek, and C. Kodner.** 1984. Infection of Syrian hamsters with Lyme disease spirochetes. *J. Clin. Microbiol.* **20**:1099–1101.

45. **Karlsson, M., G. Stiernstedt, M. Granstrom, E. Asbrink, and B. Wretlind.** 1990. Comparison of flagellum and sonicate antigens for serological diagnosis of Lyme borreliosis. *Eur. J. Clin. Microbiol. Infect. Dis.* **9**:169–177.

46. **Kelly, R.** 1971. Cultivation of *Borrelia hermsi. Science* **173**:443–444.

47. **Kurtti, T. J., U. G. Munderloh, R. C. Johnson, and G. G. Ahlstrand.** 1987. Colony formation and morphology in *Borrelia burgdorferi. J. Clin. Microbiol.* **25**:2054–2058.

48. **Lane, R. S., and W. Burgdorfer.** 1987. Transovarial and transstadial passage of *Borrelia burgdorferi* in the western black-legged tick, *Ixodes pacificus. Am. J. Trop. Med. Hyg.* **37**:188–192.

49. **Lane, R. S., W. Burgdorfer, S. F. Hayes, and A. G. Barbour.** 1985. Isolation of a spirochete from the soft tick, *Ornithodoros coriaceus:* a possible agent of epizootic bovine abortion. *Science* **230**:85–87.

50. **Lane, R. S., J. Piesman, and W. Burgdorfer.** 1991. Lyme borreliosis: relation of its causative agent to its vectors and hosts in North America and Europe. *Annu. Rev. Entomol.* **36**:587–609.

51. **Lange, W. R., T. G. Schwan, and J. D. Frame.** 1991. Can protracted relapsing fever resemble Lyme disease? *Med. Hypotheses* **35**:77–79.

52. **LeFebvre, R. B., G.-C. Perng, and R. C. Johnson.** 1990. The 83-kilodalton antigen of *Borrelia burgdorferi* which stimulates immunoglobulin M (IgM) and IgG responses in infected hosts is expressed by a chromosomal gene. *J. Clin. Microbiol.* **28**:1673–1675.

53. **Levine, J. F., M. L. Wilson, and A. Spielman.** 1985. Mice as reservoirs of the Lyme disease spirochete. *Am. J. Trop. Med. Hyg.* **34**:355–360.

54. **MacDonald, A. B., B. W. Berger, and T. G. Schwan.** 1990. Clinical implications of delayed growth of *Borrelia burgdorferi. Acta Trop.* **48**:89–94.

55. **Magnarelli, L. A., and J. F. Anderson.** 1988. Enzyme-linked immunosorbent assays for the detection of class-specific immunoglobulins to *Borrelia burgdorferi. Am. J. Epidemiol.* **127**:818–825.

56. **Magnarelli, L. A., J. F. Anderson, A. B. Schreier, and C. M. Ficke.** 1987. Clinical and serologic studies of canine borreliosis. *J. Am. Vet. Med. Assoc.* **191**:1089–1094.

57. **Magnarelli, L. A., J. F. Anderson, E. Shaw, J. E. Post, and P. C. Palka.** 1988. Borreliosis in equids in northeastern United States. *Am. J. Vet. Res.* **49**:359–362.

58. **Magnarelli, L. A., J. N. Miller, J. F. Anderson, and G. R. Riviere.** 1990. Cross-reactivity of nonspecific treponemal antibody in serologic tests for Lyme disease. *J. Clin. Microbiol.* **28**:1276–1279.

59. **Malloy, D. C., R. K. Nauman, and H. Paxton.** 1990. Detection of *Borrelia burgdorferi* using the polymerase chain reaction. *J. Clin. Microbiol.* **28**:1089–1093.

60. **Marconi, R. T., and C. F. Garon.** 1992. Development of polymerase chain reaction primer sets for diagnosis of Lyme disease and for species-specific identification of Lyme disease isolates by 16S rRNA signature nucleotide analysis. *J. Clin. Microbiol.* **30**:2830–2834.

61. **Nawakowski, J., and G. P. Wormser.** 1993. Treatment of early Lyme disease: infection associated with erythema migrans, p. 149–162. In P. K. Coyle (ed.), *Lyme Disease.* Mosby Year Book, St. Louis.

62. **Nielsen, S. L., K. K. Y. Young, and A. G. Barbour.** 1990. Detection of *Borrelia burgdorferi* DNA by the polymerase chain reaction. *Mol. Cell. Probes* **4**:73–79.

63. **Park, H. K., B. E. Jones, and A. G. Barbour.** 1986. Erythema chronicum migrans of Lyme disease: diagnosis by monoclonal antibodies. *J. Am. Acad. Dermatol.* **15**:406–410.

64. **Persing, D. H., S. R. Telford III, A. Spielman, and S. W. Barthold.** 1990. Detection of *Borrelia burgdorferi* infection in *Ixodes dammini* ticks with the polymerase chain reaction. *J. Clin. Microbiol.* **28**:566–572.

65. **Picken, R. N.** 1992. Polymerase chain reaction primers and probes derived from flagellin gene sequences for specific detection of Lyme disease and North American relapsing fever. *J. Clin. Microbiol.* **30**:99–114.

66. **Piesman, J., J. G. Donahue, T. N. Mather, and A. Spielman.** 1986. Transovarially acquired Lyme disease spirochetes (*Borrelia burgdorferi*) in field-collected larval *Ixodes dammini* (Acari: Ixodidae). *J. Med. Entomol.* **23**:219.

67. **Piesman, J., T. N. Mather, R. J. Sinsky, and A. Spielman.** 1987. Duration of tick attachment and *Borrelia burgdorferi* transmission. *J. Clin. Microbiol.* **25**:557–558.

68. **Preac-Mursic, V., B. Wilske, and S. Reinhardt.** 1991. Culture of *Borrelia burgdorferi* on six solid media. *Eur. J. Clin. Microbiol. Infect. Dis.* **10**:1076–1079.

69. **Reimers, C. D., D. E. Pongratz, U. Neubert, A. Pilz, G. Hubner, M. Naegele, B. Wilske, P. H. Duray, and J. de Koning.** 1989. Myositis caused by *Borrelia burgdorferi:* report of four cases. *J. Neurol. Sci.* **91**:215–226.

70. **Rosa, P. A., D. Hogan, and T. G. Schwan.** 1991. Polymerase chain reaction analyses identify two distinct classes of *Borrelia burgdorferi. J. Clin. Microbiol.* **29**:524–532.

71. **Rosa, P. A., and T. G. Schwan.** 1989. A specific and sensitive assay for the Lyme disease spirochete, *Borrelia burgdorferi,* using the polymerase chain reaction. *J. Infect. Dis.* **160**:1018–1029.

72. **Salih, S. Y., and D. Mustafa.** 1977. Louse-borne relapsing fever. II. Combined penicillin and tetracycline therapy in 160 Sudanese patients. *Trans. R. Soc. Trop. Med. Soc.* **71**:49–51.

73. **Schulze, T. L., G. S. Bowlen, E. M. Bosler, M. F. Laket, W. E. Parkin, R. Altman, B. G. Ormiston, and J. K. Shisler.** 1984. *Amblyomma americanum:* a potential vector of Lyme disease in New Jersey. *Science* **224**:601–603.

74. **Schwan, T. G., and W. Burgdorfer.** 1987. Antigenic changes of *Borrelia burgdorferi* as a result of in vitro cultivation. *J. Infect. Dis.* **156**:852–853.

75. **Schwan, T. G., W. Burgdorfer, and C. F. Garon.** 1988. Changes in infectivity and plasmid profile of the Lyme disease spirochete, *Borrelia burgdorferi,* as a result of in vitro cultivation. *Infect. Immun.* **56**:1831–1836.

76. **Schwan, T. G., W. Burgdorfer, M. E. Schrumpf, and R. H.**

Karstens. 1988. The urinary bladder, a consistent source of *Borrelia burgdorferi* in experimentally infected white-footed mice (*Peromyscus leucopus*). *J. Clin. Microbiol.* **26:**893–895.

77. **Schwan, T. G., K. L. Gage, R. H. Karstens, M. E. Schrumpf, S. F. Hayes, and A. G. Barbour.** 1992. Identification of the tick-borne relapsing fever spirochete, *Borrelia hermsii*, with a species-specific monoclonal antibody. *J. Clin. Microbiol.* **30:**790–795.

78. **Schwan, T. G., W. J. Simpson, M. E. Schrumpf, and R. H. Karstens.** 1989. Identification of *Borrelia burgdorferi* and *B. hermsii* using DNA hybridization probes. *J. Clin. Microbiol.* **27:**1734–1738.

79. **Simpson, W. J., W. Burgdorfer, M. E. Schrumpf, R. H. Karstens, and T. G. Schwan.** 1991. Antibody to a 39-kilodalton *Borrelia burgdorferi* antigen (P39) as a marker for infection in experimentally and naturally inoculated animals. *J. Clin. Microbiol.* **29:**236–243.

80. **Simpson, W. J., C. F. Garon, and T. G. Schwan.** 1990. *Borrelia burgdorferi* contains repeated DNA sequences that are species specific and plasmid associated. *Infect. Immun.* **58:**847–853.

81. **Simpson, W. J., C. F. Garon, and T. G. Schwan.** 1990. Analysis of supercoiled circular plasmids in infectious and non-infectious *Borrelia burgdorferi*. *Microb. Pathog.* **8:**109–118.

82. **Simpson, W. J., M. E. Schrumpf, and T. G. Schwan.** 1990. Reactivity of human Lyme borreliosis sera with a 39-kilodalton antigen specific to *Borrelia burgdorferi*. *J. Clin. Microbiol.* **28:**1329–1337.

83. **Sinsky, R. J., and J. Piesman.** 1989. Ear punch biopsy method for detection and isolation of *Borrelia burgdorferi* from rodents. *J. Clin. Microbiol.* **27:**1723–1727.

84. **Sonenshine, D. E.** 1991. *Biology of Ticks.* Oxford University Press, New York.

85. **Southern, P. M., and J. P. Sanford.** 1969. Relapsing fever: a clinical and microbiological review. *Medicine* **48:**129–149.

86. **Steere, A. C.** 1989. Lyme disease. *N. Engl. J. Med.* **321:**586–596.

87. **Steere, A. C., S. E. Malawista, J. A. Hardin, S. Ruddy, P. W. Askenase, and W. A. Andiman.** 1977. Erythema chronicum migrans and Lyme arthritis: the enlarging clinical spectrum. *Ann. Intern. Med.* **86:**685–698.

88. **Thompson, R. S., W. Burgdorfer, R. Russell, and B. J. Francis.** 1969. Outbreak of tick-borne relapsing fever in Spokane County, Washington. *J. Am. Med. Assoc.* **210:**1045–1050.

89. **vanDam, A. P., H. Kuiper, K. Vos, A. Widjojokusumo, B. M. deJongh, L. Spanjaard, A. C. P. Ramselaar, M. D. Kramer, and J. Dankert.** 1993. Different genospecies of *Borrelia burgdorferi* are associated with distinct clinical manifestations of Lyme borreliosis. *Clin. Infect. Dis.* **17:**708–717.

90. **Woese, C. R.** 1987. Bacterial evolution. *Microbiol. Rev.* **51:**221–271.

91. **Zung, J. L., S. Lewengrub, M. A. Rudzinska, A. Spielman, S. R. Telford, and J. Piesman.** 1989. Fine structural evidence for the penetration of the Lyme disease spirochete *Borrelia burgdorferi* through the gut and salivary tissues of *Ixodes dammini*. *Can. J. Zool.* **67:**1737–1748.

Treponema and Other Host-Associated Spirochetes

STEVEN J. NORRIS AND SANDRA A. LARSEN

52

GENERAL TAXONOMY

The genus *Treponema* (order *Spirochaetales*, family *Spirochaetaceae*) includes four human pathogens (Table 1) and at least six human nonpathogens. In volume 1 of *Bergey's Manual of Systematic Bacteriology* (84), the nomenclature was restructured so that the species *Treponema pallidum* includes three of the human pathogens: *T. pallidum* subsp. *pallidum* (venereal syphilis), *T. pallidum* subsp. *endemicum* (endemic syphilis), and *T. pallidum* subsp. *pertenue* (yaws). *T. carateum* (pinta) remains as a separate species owing to the lack of genetic information. The pathogenic treponemes in this group are very closely related, to the extent that they are distinguished primarily by their patterns of pathogenesis in humans and experimentally infected animals. They are morphologically indistinguishable and, where examined, have ≥95% DNA homology by hybridization (22), a high degree of sequence identity of known genes (58–60), nearly identical protein profiles (see reference 65 for a review), and shared reactivity with monoclonal antibodies (57). According to structure, host dependence, and protein content, *T. paraluiscuniculi* appears to be closely related to the human pathogens; however, it causes venereal spirochetosis of rabbits and is not known to cause human disease.

In contrast to the close relationship among the human pathogens, *T. pallidum* subsp. *pallidum* has ≤5% DNA homology with other, nonpathogenic spirochetes such as *T. phagedenis*, *T. refringens*, and *Serpulina* (formerly *Treponema*) *hyodysenteriae* (22, 87, 88). However, rRNA sequence similarities (66) indicate an evolutionary relationship between *T. pallidum*, *T. phagedenis*, *T. denticola*, and other treponemes. Because of the extensive differences between treponemes pathogenic to humans and other host-associated spirochetes, these two groups will be discussed separately below.

SPIROCHETES ASSOCIATED WITH HUMAN TREPONEMATOSES

Description of the Genus

T. pallidum is a spirochete ~0.18 μm in diameter and ranging in length from 6 to 20 μm (Fig. 1). Suspensions of *T. pallidum* are best visualized by dark-field microscopy, although the bacterium can also be seen by phase-contrast microscopy. Unstained organisms are not visible by standard bright-field microscopy because of the small cell diameter.

Morphologically, *T. pallidum* has regular helices (6 to 14 per cell) with a wavelength of 1.1 μm and an amplitude of ~0.3 μm. The ends are pointed and lack the hook shape characteristic of some commensal human spirochetes. Fresh preparations of the organism exhibit rapid rotation about the axis (the characteristic corkscrew motility) due to the action of flagella inserted in both ends and extending down the cell body within the periplasmic space (Fig. 1). Flexing and reversal of rotation can also occur, but translational motion is not observed unless *T. pallidum* is in a viscous medium. Numerous treponemes are often present in exudates from primary or secondary syphilitic lesions, and detection of organisms with the characteristic morphology and motility is still considered the only definitive means of diagnosis (see below).

Habitat and Distribution

The *T. pallidum* subspecies and *T. carateum* are obligate parasites of humans and are not known to have any animal or environmental reservoirs. Venereal syphilis (*T. pallidum* subsp. *pallidum*) has a worldwide distribution, with incidence varying widely according to geographic location and socioeconomic groups. Endemic syphilis is restricted to desert and temperate regions of North Africa and the Middle East. Yaws occurs most commonly in tropical or desert regions of Africa, South America, and Indonesia. Pinta is found primarily in tropical areas of Central and South America.

Clinical Significance

Syphilis is still a common sexually transmitted disease in many areas of the world despite the availability of effective therapy. In the United States, 112,581 cases were reported in 1992, including 33,973 cases of primary and secondary syphilis and 3,850 cases of congenital syphilis (11).

Transmission of the treponematoses occurs through direct contact with active lesions. In venereal syphilis, the primary and secondary lesions are infectious and contain large numbers of organisms until the healing stage; for epidemiologic purposes, early latent cases (duration, ≤1 year) are also considered potentially infectious. *T. pallidum* and related organisms apparently penetrate mucous mem-

636

TABLE 1 Characteristics of the human treponematoses[a]

Organism	Disease	Distribution	Predominant age of onset	Transmission	Congenital infection
T. pallidum subsp. *pallidum*	Venereal syphilis	Worldwide	Adolescents, adults	Sexual contact	Yes
T. pallidum subsp. *pertenue*	Yaws (frambesia, pian)	Tropical areas, Africa, South America, Caribbean, Indonesia	Children	Skin contact	No
T. pallidum subsp. *endemicum*	Endemic syphilis (bejel, dichuchwa)	Arid areas, Africa, Middle East	Children to adults	Mucous membrane	Rarely
T. carateum	Pinta (carate, cute)	Semiarid, warm areas; Central and South America	Children, adolescents	Skin contact	No

[a]See references 20, 21, 68, 69, and 95.

branes and abrasions in the epidermis, establish a local infection in the dermis, and disseminate rapidly from the primary site.

Venereal syphilis exhibits a wide variety of clinical manifestations (Table 2). Early syphilis consists of the primary, secondary, and early latent stages. The primary stage is

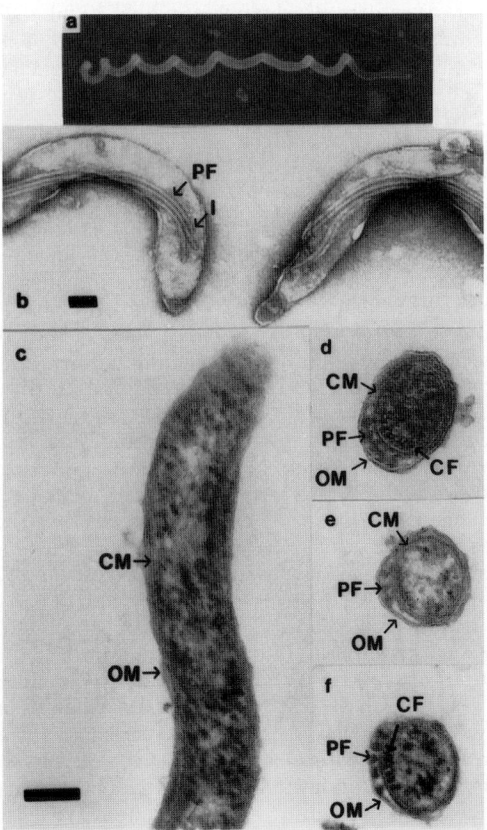

FIGURE 1 Morphology of *T. pallidum*. (a) Scanning electron micrograph showing spiral shape. (b) Negatively stained view of the tips of two organisms. Note the insertion points (I) of periplasmic flagella (PF) near the ends. (c through f) Electron micrographs of ultrathin sections, showing the outer membrane (OM), cytoplasmic membrane (CM), periplasmic flagella (PF), and locations of cytoplasmic filaments (CF). Bars = 0.1 μm. (Reprinted with permission from reference 65.)

characterized by the chancre, a firm, indurated lesion with an ulcerated or crusted surface ranging in size from a few millimeters to 1 to 2 cm. Although disseminated infection occurs in virtually all infected individuals, multiple secondary lesions arise in ~25% of patients. Lesions in these early stages of syphilis contain large numbers of organisms and should routinely be examined by dark-field microscopy.

About two-thirds of the syphilis cases reported in the United States are diagnosed at the latent (asymptomatic) stage, in which diagnosis is based solely on serologic reactivity and history. Longitudinal observations of untreated syphilis patients at the beginning of the 20th century indicated that approximately one-third remain latently infected for life, one-third undergo "biological cure" (i.e., lose serologic reactivity), and the remaining third develop tertiary manifestations (29). Tertiary syphilis, consisting of neurologic, cardiovascular, and gummatous forms, can develop decades after the initial infection (95). Cardiovascular and gummatous syphilis are now rarely encountered, apparently because of surveillance measures and antimicrobial therapy administered for other infections. Recent studies have shown that viable *T. pallidum* is present in the cerebrospinal fluid (CSF) of primary and secondary stage patients, indicating that central nervous system infection occurs very early in the course of disease (47). Neurologic manifestations, including syphilitic meningitis, may occur as early as 3 months postinfection; thus, neurosyphilis should not be considered solely a late manifestation of the disease (33). Patients, particularly those infected with human immunodeficiency virus (HIV), may have neurologic or ocular relapses following treatment with benzathine penicillin owing to the antibiotic's poor penetration of the blood-brain barrier and failure to eradicate central nervous system infection (3, 50, 55).

Congenital syphilis results from the transmission of *T. pallidum* across the placenta during the later stages of pregnancy. Early congenital syphilis tends to be manifested when the mothers are afflicted with early syphilis during the course of pregnancy, resulting in stillbirth or fulminant infection of the newborn (Table 2). Late manifestations of congenital syphilis are the outcome of chronic untreated infection and can result in multiple stigmata (Table 2) that are often not obvious until the second decade of life.

Yaws, endemic syphilis, and pinta were common infections in areas of endemicity prior to the World Health Organization eradication program beginning in 1948; in the early 1950s, it was estimated that 200 million people

TABLE 2 Common manifestations of treponematoses[a]

Venereal syphilis (T. pallidum subsp. pallidum)
  Primary (local): 10–90 days postinfection (avg, 21 days)
    Chancre (single or multiple, skin or mucous membranes)
    Regional lymphadenopathy
  Secondary (disseminated): 6 wk–6 mo postinfection
    Multiple secondary lesions (skin or mucous membranes)
    Generalized lymphadenopathy, fever, malaise
    Condylomata lata
    Alopecia
    Asymptomatic or symptomatic CNS involvement
      (meningitis)
  Latent (early, ≤1-yr duration; late, >1-yr duration)
    Reactive serologic tests for syphilis
    Asymptomatic
  Tertiary (late, chronic): mo to yr postinfection
    Gummatous (monocytic infiltrates, tissue destruction, any
      organ)
    Cardiovascular (aortic aneurysm)
    Neurosyphilis (paresis, tabes dorsalis, meningovascular
      syphilis)
  Congenital
    Early (onset, <2 yr): fulminant, disseminated infection;
      mucocutaneous lesions; osteochondritis; anemia;
      hepatosplenomegaly; CNS involvement
    Late (persistence, >2 yr): interstitial keratitis, bone and
      tooth deformities, eighth-nerve deafness, neurosyphilis,
      other tertiary manifestations
Yaws (T. pallidum subsp. pertenue)
  Early: onset, 9–90 days postinfection (avg, 21 days)
    Primary lesion (mother yaw): papular, nontender, often
      pruritic, crusted or ulcerated
    Disseminated lesions (daughter yaws): often resemble
      primary lesion or appearance may vary
    Malaise, fever, lymphadenopathy
    Osteitis, periostitis, other bone and joint manifestations
  Latent: positive serologic tests, no other signs of infection
  Late: 10% of patients
    Destructive lesions of bone and cartilage (e.g., ulcerative
      rhinopharyngitis)
    Hyperkeratotic skin lesions
Endemic syphilis (T. pallidum subsp. endemicum)
  Early
    Primary: lesion not usually detected
    Secondary: multiple oropharyngeal, cutaneous lesions
    Generalized lymphadenopathy
    Periostitis
  Latent: positive serologic tests, no other signs of infection
  Late: destructive skin, bone, and cartilage lesions
Pinta (T. carateum): restricted to skin
  Early
    Initial lesion: hyperkeratotic, pigmented papule or plaque
    Disseminated skin lesions (pintids)
    Regional lymphadenopathy
  Late: pigmentary changes in skin (hyper- and
    hypopigmentation)

[a]CNS, central nervous system.

were exposed to yaws during their lifetimes (20, 21, 68, 69). Areas where the diseases are endemic are now more restricted, but decreased surveillance has led to a recrudescence of incidence in many areas (52). Transmission of the endemic treponematoses occurs through direct contact with early lesions or with contaminated fingers or drinking or eating utensils. Yaws and endemic syphilis commonly are transmitted among children (2 to 10 years of age), whereas pinta usually has a later onset (ages 15 through 30 years). Many of the manifestations of yaws and endemic syphilis are similar to those of venereal syphilis (Table 2); the lesions of pinta appear to be restricted to the skin. Congenital transmission of the endemic treponematoses is rare.

Diagnostic Parameters and Interpretation of Laboratory Tests

The following is a brief discussion of the parameters involved in the diagnosis of syphilis and other treponematoses. Procedures for the direct detection of T. pallidum and the serologic diagnosis of syphilis are described in detail in A Manual of Tests for Syphilis (42).

Diagnostic Criteria for Venereal Syphilis

For the laboratory diagnosis of syphilis, each stage has a particular testing requirement (Table 3). Because T. pallidum cannot be readily cultured (17, 23, 63), other laboratory methods of identifying infection have been developed. The current tests for syphilis fall into three categories: direct microscopic examination, used when lesions are present; nontreponemal tests, used for screening; and treponemal tests, which are confirmatory. In the United States, the routine testing scheme is direct microscopic examination of lesion exudates followed by a nontreponemal test that if reactive is confirmed with a treponemal test (Fig. 2). Criteria for the diagnosis of syphilis are divided into three categories: definitive, presumptive, and suggestive (10).

During the primary stage, serous fluids from the lesion contain numerous treponemes detectable either by dark-field microscopy or by the direct fluorescent-antibody (DFA) test for T. pallidum (DFA-TP) (42). Humoral antibodies, as detected by the standard nontreponemal and treponemal serologic tests for syphilis, usually do not appear until 1 to 4 weeks after the chancre has formed. By the secondary stage of syphilis, the organism has invaded every organ of the body and virtually all body fluids. At this stage, with few if any exceptions, all serologic tests for syphilis are reactive, and treponemes may be found in lesions by direct microscopic examination. Nontreponemal and treponemal serologic tests are consistently reactive in the early latent stage, but the patient is asymptomatic. During the late latent stage (duration of >1 year), the proportion of patients exhibiting reactive nontreponemal tests and the reactive titers for the nontreponemal tests decrease.

Symptoms of late or tertiary syphilis may occur 10 to 20 years after the initial infection. Approximately 71% of patients with late-stage syphilis have reactive nontreponemal tests; however, treponemal tests are almost always reactive and may be the only basis for diagnosis (Table 3). Diagnosis of cardiovascular syphilis is based on the presence of symptoms that indicate aortic insufficiency or aneurysm, reactive treponemal test results, and no known history of treatment for syphilis. Neurologic forms of syphilis (e.g., syphilitic meningitis) may develop during the secondary stage. Neurosyphilis as a complication of late syphilis may occur as early as 2 years after initial infection. Neurosyphilis may take many forms, yet symptoms consistent with neurosyphilis may not always be present (33, 80). CSF examinations such as the Venereal Disease Research Laboratory (VDRL) CSF slide test and total protein and leukocyte counts should be performed on the spinal fluids of patients

TABLE 3   Criteria for diagnosis of syphilis

Early syphilis
  Primary
    Definitive: direct microscopic identification of *T. pallidum* in lesion material, lymph node aspirate, or biopsy section
    Presumptive (requires 1 and either 2 or 3)
      1. Typical lesion
      2. Reactive nontreponemal test and no history of syphilis
      3. For persons with history of syphilis, fourfold increase in most recent quantitative nontreponemal test titer compared with results of past tests
    Suggestive (requires 1 and 2)
      1. Lesion resembling chancre
      2. Sexual contact within preceding 90 days with person who has primary, secondary, or early latent syphilis
  Secondary
    Definitive: direct microscopic identification of *T. pallidum* in lesion material, lymph node aspirate, or biopsy section
    Presumptive (requires 1 and either 2 or 3)
      1. Skin or mucous membrane lesions typical of secondary syphilis
        a. Macular, papular, follicular, papulosquamous, or pustular
        b. Condylomata lata (anogenital region or mouth)
        c. Mucous patches (oropharynx or cervix)
      2. Reactive nontreponemal test titer of ≥1:8 and no previous history of syphilis
      3. For persons with history of syphilis, fourfold increase in most recent nontreponemal test titer compared with previous test results
    Suggestive (requires 1 and 2 and is made only when serologic test results are not available)
      1. Presence of clinical manifestations as described above
      2. Sexual exposure within past 6 mo to person with early syphilis
  Early latent
    Definitive: definitive diagnosis does not exist because lesions are not present in latent stage
    Presumptive (requires 1, 2, and 3 or 4)
      1. Absence of signs and symptoms
      2. Reactive nontreponemal and treponemal test results
      3. Nonreactive nontreponemal test within preceding yr
      4. Fourfold increase in nontreponemal test titer compared with previous test results for persons with history of syphilis or of symptoms compatible with early syphilis
    Suggestive (requires 1 and 2)
      1. Reactive nontreponemal test result
      2. History of sexual exposure within preceding yr
Late syphilis
  Benign and cardiovascular
    Definitive: observation of treponemes in tissue sections by direct microscopic examination with DFAT-TP
    Presumptive
      1. Reactive treponemal test
      2. No known history of treatment for syphilis
      3. Characteristic symptoms of benign or cardiovascular syphilis
  Neurosyphilis
    Definitive (requires 1 and either 2 or 3)
      1. Reactive serum treponemal test
      2. Reactive VDRL CSF test on spinal fluid sample
      3. Identification of *T. pallidum* in CSF or tissue by microscopic examination or animal inoculation
    Presumptive (requires 1 and either 2 or 3)
      1. Reactive serum treponemal test
      2. Clinical signs of neurosyphilis
      3. Elevated CSF protein (>40 mg/dl) or leukocyte count (>5 mononuclear cells/ml) in absence of other known causes
Neonatal congenital syphilis
    Definitive: demonstration of *T. pallidum* by direct microscopic examination of umbilical cord, placenta, nasal discharge, or skin lesion material
    Presumptive (requires 1, 2, and 3)
      1. Determination that infant was born to mother who had untreated or inadequately treated syphilis at delivery regardless of findings in infant
      2. Infant with reactive treponemal test result
      3. One of following additional criteria:
        a. Clinical sign or symptoms of congenital syphilis on physical examination
        b. Abnormal CSF finding without other cause
        c. Reactive VDRL CSF test result
        d. Reactive IgM antibody test specific for syphilis

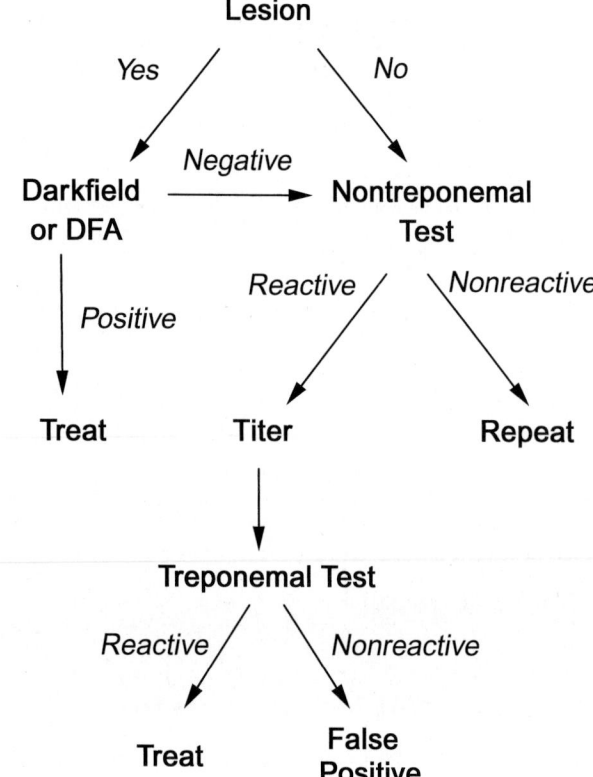

FIGURE 2 Routine screening scheme for early syphilis in the United States.

with late latent syphilis and of those with clinical symptoms and signs consistent with neurosyphilis (33, 80). Diagnosis of neurosyphilis is based on a combination of these criteria (Table 3).

Diagnostic Criteria for Congenital Syphilis

A primary stage does not occur in congenital syphilis, because the organisms directly infect the fetal circulation (98). Treponemes or the effects thereof are detectable in almost every tissue of the infant. The standard serologic tests for syphilis, based on the measurement of immunoglobulin G (IgG), reflect antibodies passively transferred from the mother to the infant rather than IgM antibodies produced during gestation. Currently, the diagnosis of neonatal congenital syphilis depends on a combination of results from physical, radiographic, serologic, and direct microscopic examinations (Table 3). Clinical signs of congenital syphilis include hepatosplenomegaly, cutaneous lesions, osteochondritis, and snuffles (40, 95). Although some clinical manifestations may be present at birth, they are more often seen at 3 weeks to 6 months. At birth, up to 50% of the infants with congenital syphilis are asymptomatic; other stigmata that may develop later include tooth and bone malformations, deafness, blindness, and learning disabilities.

Diagnostic Criteria for Syphilis among HIV-Infected Persons

In the 1980s, studies showed that 70% of serum specimens from homosexual men with AIDS reacted in treponemal

tests for syphilis (76). Studies in a heterosexual population indicated that approximately 60% of HIV-infected females had reactive syphilis serologic test results (8). Thus, there is an extremely high prevalence of syphilis within some groups of HIV-infected individuals, in part because of shared risk factors for the acquisition of these two infections. Most HIV-infected persons who also are infected with *T. pallidum* appear to respond normally in the serologic tests for syphilis (e.g., exhibit true-positive reactions) and have typical clinical signs. However, several exceptions have been published or reported to the Centers for Disease Control and Prevention (CDC) (9, 43, 50, 55, 77). Reported problems in the diagnosis of syphilis in HIV-infected patients include (i) unusual clinical signs and symptoms, (ii) lack of serologic response in a clinically confirmed case of active syphilis, (iii) failure of nontreponemal test titers to decline after treatment with standard regimens, (iv) unusually high titers in nontreponemal tests, (v) rapid progression to late stages of syphilis and neurologic involvement even after treatment of primary or secondary syphilis, and (vi) disappearance of treponemal test reactivity over time. Whether these problems in the diagnosis of syphilis occur more frequently in HIV-seropositive persons than in HIV-seronegative persons with syphilis is unclear (77, 78).

Recommendations for the diagnosis and treatment of syphilis in HIV-infected persons have been made by the CDC (9). The recommendations include increased use of direct microscopic examinations of lesion or biopsy material when clinical findings suggest syphilis but serologic test results are nonreactive; careful serologic and physical examination follow-up to ensure adequacy of treatment; CSF examination of patients in whom syphilis is latent and in whom the duration, if known, is >1 year; treatment of neurosyphilis among HIV-infected patients with at least 10 days of aqueous crystalline penicillin G or aqueous procaine penicillin G plus probenecid; and reporting of unusual manifestations of syphilis to state epidemiologists.

Diagnostic Criteria for Other Treponematoses

To date, no laboratory method has been devised to distinguish the other pathogenic treponematoses from each other or from syphilis. The standard serologic tests for syphilis are uniformly reactive with yaws, pinta, and nonvenereal endemic syphilis (bejel) (2). Western blotting (immunoblotting) assays do not differentiate the antibodies formed in response to syphilis from those formed in response to yaws or pinta (27, 58). Most molecular approaches, such as DNA sequencing, DNA probes, and PCR techniques, have also failed to individualize the pathogenic treponemes (58). Therefore, the clinical appearance of the lesions formed by *T. pallidum* subsp. *pertenue*, *T. pallidum* subsp. *pallidum*, *T. pallidum* subsp. *endemicum*, and *T. carateum*; the anatomical location of the lesion; the mode of transmission; the age of the individual; and the geographic location of the infected individual are the only criteria that can be used to diagnose these infections as separate entities (2).

Because yaws, pinta, and nonvenereal syphilis are often childhood diseases, the nontreponemal test titers of adults from geographic regions in which endemic treponematoses were virtually eliminated by mass campaigns in the 1950s and 1960s are expected to be ≤1:8 (1). Therefore, any titer of >1:8 for adults from these regions is indicative of venereal syphilis. Likewise, titers of ≥1:8 in children indicates a possible resurgence of yaws in the populations of these areas (69).

Detection and Identification of *T. pallidum* in Tissues and Tissue Exudates

General Principles

In addition to being difficult to culture, treponemes do not readily stain with common laboratory stains but can be stained with silver stains (91). Therefore, when lesions are present, the most specific and easiest means of diagnosing syphilis is by direct detection of the organism. Currently, dark-field microscopy and DFA tests are commonly used to detect *T. pallidum*. A positive result on microscopic examination is definitive for syphilis if infection with other pathogenic treponemes can be excluded. When dark-field microscopy is used, the presence of morphologically similar parasitic spirochetes within and near the genitalia requires that *T. pallidum* be viewed in its "living state" in order to observe its characteristic morphology and motility.

The newest technique for direct detection of *T. pallidum* is the PCR (6, 30, 60, 61, 82, 96). PCR as a tool for detecting *T. pallidum* is used by only a few laboratories, although commercial test manufacturers are developing tests based on PCR for use in genital ulcer diseases.

Specimen Collection

When collecting, preparing, and examining specimens, observe universal safety precautions (42). Briefly, a specimen for dark-field microscopy should consist of serous fluid free of erythrocytes, other organisms, and tissue debris. The lesion should be cleansed only if it is encrusted or obviously contaminated, and only tap water or physiologic saline (without antibacterial additives) should be used. It may be necessary to gently abrade the lesion surface to yield a clear serous fluid, which is then collected on a glass slide and covered with a coverslip. Ideal specimens for direct microscopic examination can be obtained if chancres, secondary lesions, condyloma latum, or mucous patches are present. Because even the experienced observer may find it difficult or impossible to differentiate *T. pallidum* from other parasitic spirochetes of the mouth, dark-field microscopy should not be used for the examination of samples from oral lesions. It is essential that exudates be examined as soon as possible (ideally within 20 min) to ensure retention of motility. *T. pallidum* is very sensitive to exposure to oxygen, heat, nonphysiologic pH, and desiccation.

Lesion samples for direct fluorescence assay examination are collected in the manner described for dark-field microscopy except that the samples are air dried. Motility of the organism is not required in the DFA-TP. Because the conjugates used are specific for pathogenic strains of *Treponema*, the DFA-TP is applicable to samples collected from both oral and rectal lesions. The use of the DFA-TP test has been extended to include the staining of tissue sections (DFAT-TP) (37, 42). Any tissue can be used, but tissues for paraffin-embedded sections are most frequently collected from the brain, gastrointestinal tract, placenta, umbilical cord, or skin. Often, DFAT-TP is used to diagnose late-stage or congenital syphilis or to distinguish skin lesions of secondary or late syphilis from those of Lyme disease.

Sample source for PCR is controversial (62). For example, one study found whole blood and lesion material, both exudate and tissue, to be satisfactory samples for PCR, whereas the appropriateness of serum and CSF as sample sources for PCR is still under consideration (62, 96). Contamination of the samples collected for PCR with extraneous DNA can be a major problem and may account for some of the variable results noted by some researchers.

Any specimen source can be used for rabbit infectivity testing as long as the material is less than 1 h old or was flash frozen immediately after collection and maintained in liquid nitrogen or at temperatures of −78°C or below; 10% glycerol or 5% (vol/vol) dimethyl sulfoxide should be added to liquid specimens immediately prior to freezing.

Dark-Field Microscopy

Because of their narrow width, treponemes cannot be observed with the ordinary light microscope. Microscopes equipped with a double-reflecting or single-reflecting dark-field condenser, a 40× to 45× objective and a 90× to 100× objective with a funnel stop are needed to perform the dark-field examination (42). Illumination for dark-field microscopy is obtained when light rays strike the object in the field at an oblique angle so that no direct light rays but only rays reflected from the object enter the microscope. The object itself then appears to be illuminated against a dark background. The slide is first scanned with the 40× to 45× objective. Once an organism is located, a small drop of immersion oil is placed on the coverslip, the objective is switched to the 90× to 100× objective, and the organism is brought into focus. *T. pallidum* is distinguished from other spiral organisms by the tightness and regularity of the spirals and by the characteristic corkscrew movement. Positive findings on dark-field examination are reported as "organisms found that have the characteristic morphology and motility of *T. pallidum*" and allow a specific and immediate diagnosis of syphilis. Also, primary syphilis can be diagnosed by dark-field microscopy before the appearance of reactive serologic tests. However, a negative dark-field finding does not exclude the diagnosis of syphilis, since the organisms may be too few to be demonstrated if the lesion is in the healing stage or if the infection has been altered by systemic or topical treatment. Adequate training and experience are necessary to make an accurate diagnosis by dark-field microscopy. The untrained observer may be deceived by artifacts such as cotton fibers and Brownian motion. When the direct microscopic results are negative, other sexually transmitted diseases causing lesions, such as herpes and chancroid, should be considered.

DFA-TP Test

The easier-to-perform and more specific DFA-TP method has been used to detect the presence of *T. pallidum* subspecies in tissues, body fluids, secretions, and lesion exudates (42). A UV fluorescence microscope equipped with a dark-field condenser or epifluorescence illumination and the standard set of filters is required. The test detects pathogenic treponemes and differentiates them from nonpathogenic treponemes by an antigen-antibody reaction. Initially, samples are air dried, and immediately before staining, smears are fixed with either acetone for 10 min or 100% methanol for 10 s, or slides are gently heat fixed. Smears are stained with fluorescein isothiocyanate (FITC)-labeled anti-*T. pallidum* globulins obtained from the sera of humans or rabbits with syphilis; these globulins have been absorbed with Reiter treponemes. More recently, a mouse monoclonal antibody to *T. pallidum* has been used in the DFA-TP test (37). When fluorescing treponemes displaying typical morphology are seen by microscopic examination, results are reported as "treponemes, immunologically specific for *T. pallidum*, were observed by direct immunofluorescence." Negative results are reported as "treponemes were not observed by direct immunofluorescence." As with all tests for syphilis, proper controls of the technique and reagents used in these methods are mandatory if the results are to be meaningful.

DFAT-TP Test

A combination of the DFAT-TP test and histologic stains may be used to examine biopsy and autopsy material for the presence of pathogenic *Treponema* spp. (42). To perform the DFAT-TP test, tissues from punch biopsy or surgical excision are initially fixed for 24 h in 10% neutral-buffered formalin and then embedded in paraffin blocks. Sections (2 μm) are deparaffinized and then pretreated with either NH_4OH or trypsin. Next, slides are stained with a monoclonal conjugate as for the DFA-TP. A *T. pallidum*-infected rabbit testicular tissue section is used as the control. Results are reported as for the DFA-TP test.

PCR

The methods used for PCR have not been standardized, but one group of investigators (30, 82) has used alkaline lysis extraction for removal of inhibitory components from the sample. The primers and probe for detection of *T. pallidum* were prepared from the 47-kDa gene. After a 40-cycle series of denaturing, annealing, and extension, the PCR products were visualized by dot blot or electrophoresis and Southern blot transfer, hybridization with a ^{32}P-labeled probe, and autoradiography.

Isolation Procedures

The *T. pallidum* subspecies and *T. carateum* cannot be cultivated in vitro by using standard bacteriologic techniques. The oldest method for detecting infection with *T. pallidum* is animal infectivity testing. The pathogenic treponemes (except for *T. carateum*) can be propagated by inoculation of appropriate laboratory animals (94). This technique is capable of detecting one or two infectious treponemes under optimal conditions (49) and is used as the "gold standard" for measuring the sensitivity of other methods, such as the PCR (30, 82). Rabbit inoculation was used recently to demonstrate the presence of viable *T. pallidum* in the CSF of early-syphilis patients (47). This method is not practical for routine diagnostic use because of the expense and technical difficulty.

Limited multiplication of *T. pallidum* subsp. *pallidum* has been obtained over a 10- to 20-day period in a complex tissue culture system, but continuous in vitro culture has not been achieved (17, 23, 63). The growth rate both in vivo and in vitro is very slow, with an estimated doubling time in vivo of 30 to 33 h. Recent evidence indicates that *T. pallidum* subsp. *pallidum* is a microaerophile rather than an anaerobe (16, 17). It requires molecular oxygen for energy production through an oxidative phosphorylation mechanism but is extremely sensitive to oxygen radicals and will become nonviable within minutes to hours of exposure to atmospheric levels of oxygen. Therefore, survival is best in an atmosphere containing 1 to 5% O_2 and in media containing reducing agents. Glucose appears to provide the principal energy source, and serum is required for survival and growth.

Serologic Tests

General Principles

Because lesions are not present in latent and late-stage syphilis and because *T. pallidum* subspecies cannot be readily isolated and grown in vitro, serologic tests for syphilis are performed (43). As mentioned earlier, the serologic

TABLE 4 Standard status tests for syphilis

Type of test	Name (abbreviation) of test
Nontreponemal	Venereal Disease Research Laboratory (VDRL) slide
	Unheated-serum reagin (USR)
	Rapid plasma reagin (RPR) 18-mm circle card
	Toluidine red unheated-serum test (TRUST)
Treponemal	Fluorescent-treponemal-antibody absorption (FTA-ABS)
	FTA-ABS double staining (DS)
	Microhemagglutination assay for antibodies to *T. pallidum* (MHA-TP)

tests used fall into one of two categories, nontreponemal or treponemal. Four nontreponemal tests and three treponemal tests are currently considered standard tests (Table 4).

Nontreponemal tests used for screening have the advantage of being widely available, inexpensive, convenient to perform on large numbers of specimens, and useful for determining the efficacy of treatment. Limitations of the nontreponemal serologic tests include their lack of sensitivity in early dark-field-positive primary cases and in late syphilis and the possibilities of a prozone reaction or of false-positive results. Treponemal tests employ *T. pallidum* subsp. *pallidum* as the antigen and detect specific treponemal antibodies that are usually related only to treponemal infections. Treponemal tests are primarily used to verify reactivity in the nontreponemal tests. The treponemal tests also may be used to confirm a clinical impression of syphilis in which the nontreponemal test is nonreactive but there is evidence of syphilis, such as might occur in late syphilis. Unfortunately, treponemal tests are technically more difficult and costly to perform than nontreponemal tests and cannot be used to monitor treatment. For 85% of persons successfully treated, treponemal test results remain reactive for years if not a lifetime.

Specimen Collection

When collecting, preparing, and examining specimens (42) for the serologic tests, observe universal safety precautions. Serum is the specimen of choice for both nontreponemal and treponemal tests. However, the rapid plasma reagin card test and the toluidine red unheated-serum test may also be performed with plasma samples. The technician must check the product insert to be sure that the plasma sample has not exceeded the recommended storage time and that the blood was collected in the specified anticoagulant. Plasma cannot be used in the VDRL test, since the sample must be heated before testing, and plasma cannot be used in the treponemal tests for syphilis. The VDRL test is the only test that can be used for testing CSF. The CSF is not heated before the test is performed.

When screening for congenital syphilis, the CDC recommends testing the mother's serum rather than cord blood. Recent studies compared the reactivities of the mother's serum, cord blood, and infant's serum and found that the maternal sample is the best indicator of infection, neonatal serum is second best, and cord blood is the least reactive (13, 70). Infant's serum is the specimen of choice for IgM-specific tests.

TABLE 5 Sensitivities and specificities of serologic tests for syphilis[a]

Test[b]	Sensitivity (%)				Specificity (%), nonsyphilis
	Primary syphilis	Secondary syphilis	Latent syphilis	Late syphilis	
Nontreponemal					
VDRL	78 (74–87)	100	95 (88–100)	71 (37–94)	98 (96–99)
RPR card	86 (77–100)	100	98 (95–100)	73	98 (93–99)
USR	80 (72–88)	100	95 (88–100)		99
RST[b]	82 (77–86)	100	95 (88–100)		97
TRUST	85 (77–86)	100	98 (95–100)		99 (98–99)
Treponemal					
FTA-ABS	84 (70–100)	100	100	96	97 (94–100)
MHA-TP	76 (69–90)	100	97 (97–100)	94	99 (98–100)
FTA-ABS DS	80 (69–90)	100	100		98 (97–100)

[a]Numbers in parentheses are ranges in CDC studies.
[b]RST, reagin screen test. See Table 4 for definitions of other abbreviations.

Nontreponemal Tests

All available nontreponemal tests are based on an antigen composed of an alcoholic solution containing measured amounts of cardiolipin, cholesterol, and sufficient purified lecithin to produce standard reactivity (43, 80). The VDRL and unheated-serum reagin tests are flocculation tests requiring microscopic examination, whereas the rapid plasma reagin and toluidine red unheated-serum tests produce agglutination reactions visible without magnification. The nontreponemal (reagin) tests measure IgM and IgG antibodies to lipoidal material released from damaged host cells as well as those to lipids and lipoproteins produced by the treponemes (51). The antilipoidal antibodies are produced not only as a consequence of syphilis and the other treponemal diseases but also in response to nontreponemal diseases of an acute and chronic nature in which tissue damage occurs. Nontreponemal tests may be used either as qualitative or quantitative tests. As qualitative tests, the nontreponemal tests are used in screening for syphilis. In the quantitative nontreponemal tests, serial twofold dilutions are made, and the serum is diluted until an endpoint is reached. Quantitative reactions are reported in terms of the highest (last) dilution (dil) in which the specimen is fully reactive. Quantitative tests are more informative than qualitative tests alone. Quantitative tests establish a baseline of reactivity from which change can be measured; this baseline allows evaluation of recent infection by demonstrating a fourfold rise in titer and helps detect reinfection or relapse among persons with a persistently reactive (serofast) test for syphilis. All of the nontreponemal tests have approximately the same sensitivities and specificities (Table 5) but may differ in reactivity levels as a result of the variation in antigen preparation. The different levels of reactivity are reflected in the different endpoint titers obtained when the same serum is tested in the four tests. Because success or failure of treatment is based on just a fourfold decrease in titer, the serum sample used as the baseline should be drawn the day treatment is begun. Also, because of the variation in reactivity levels among the test, the same test used in the initial testing should be used to monitor treatment.

The interpretation of the results depends on the population being tested. When the nontreponemal tests are used as a screening test in a low-risk population, all reactive results should be confirmed with a treponemal test. In some low-risk populations, every reactive result may be a false-positive result. Nontreponemal test results must also be interpreted according to the stage of syphilis suspected. Approximately 30% of those with early primary syphilis have nonreactive nontreponemal test results on the initial visit. The nontreponemal test usually becomes reactive 1 to 4 weeks after the chancre appears. In secondary syphilis, nearly all patients have nontreponemal test endpoint titers of ≥1:8. For patients with atypical lesions and/or nontreponemal test titers of <1:8, the nontreponemal tests should be repeated and a confirmatory treponemal test should be performed. Patients with late latent syphilis, i.e., with reactive nontreponemal and treponemal tests, no clinical and historical findings, and an unknown history of treatment or nonreactive serologic results, should be evaluated for potential asymptomatic neurosyphilis. Even without treatment, the nontreponemal tests may be only weakly reactive or even nonreactive in the late stages of syphilis.

When a patient is treated for neurosyphilis, the CSF cell count usually returns to normal levels first; this is followed by the CSF protein concentration and, finally, the nontreponemal serologic test results (43, 80). Like the results of quantitative serum serologic tests, the results of serial VDRL slide quantitative tests on CSF can be used to monitor response to treatment. However, the response of the CSF quantitative VDRL to treatment is less predictable; it may take many years for the CSF nontreponemal test to become nonreactive, although the titer should drop progressively.

Congenital syphilis can be controlled by treating pregnant women. False-positive nontreponemal test results for samples from pregnant women have been reported, and reactive results should be confirmed with a treponemal test. Nontreponemal titers remaining after treatment of syphilis tend to increase nonspecifically during pregnancy. This increase in titer may be confused with reinfection or relapse. This reaction may be considered nonspecific if effective therapy can be documented and if development of new lesions, a fourfold increase in titer, and a history of recent sexual exposure to a person with infectious syphilis can be ruled out. Previously, the difference between the mother's nontreponemal test titer and the infant's titer at delivery was thought to be a means of distinguishing infected from uninfected infants. If the infant's titer was higher than that of the mother's, then the infant had congenital syphilis. However, the converse is not true. A lower titer in the infant's serum than in the mother's does not rule out congenital syphilis. Examination of serum sample pairs from mothers and infants in cases of congenital syphilis indicated that in only 22% of the cases did the infant have a titer higher than that of the mother (90).

When nontreponemal tests are used for monitoring therapy, quantitative tests should be performed on patient's serum samples drawn at 3-month intervals for at least 1 year. Following adequate therapy for primary and secondary syphilis, there should be at least a fourfold decline in titer by the third or fourth month and an eightfold decline in titer by the sixth to eighth month (5). For most patients treated in early syphilis, the titer declines until little or no reaction is detected after the first year (24, 26). Patients treated in the latent or late stages may show a more gradual decline in titer (25). Low titers persist in approximately 50% of these patients after 2 years of observation. This persistent seropositivity does not signify treatment failure or reinfection, and these patients are likely to remain serofast even if they are retreated.

Prozone reactions occur in the nontreponemal tests in 1 to 2% of patients with secondary syphilis. Sera exhibiting a weakly reactive, atypical, or "rough" negative reaction in undiluted serum should be diluted as in the quantitative assay, in which case the reactivity will increase and then decrease as the endpoint titer is approached. Dilution of the antibody to 1:16 is usually adequate to obtain the proper optimal concentration and a readily detectable reaction.

False-positive reactions can occur either acutely (<6 months' duration), as an outcome of viral infections, immunizations, or pregnancy, or chronically (>6 months' duration), as a result of multisystem autoimmune diseases such as systemic lupus erythematosus or other conditions including narcotic addiction, aging, malignancy, or leprosy. False-positive reactions are generally low in titer (<1:8), but high-titer false-positive nontreponemal tests can be obtained in a high proportion (up to 10%) of intravenous drug abusers.

Treponemal Tests

Three tests are currently considered standard treponemal tests; all use *T. pallidum* as the antigen and are based on the detection of antibodies directed against cellular components (42). The treponemal tests are not intended for routine use or as screening procedures. Their greatest value is in distinguishing positive nontreponemal results from false-positive results and in establishing the diagnosis of late latent or late syphilis. Problems arise when these tests are used as screening procedures, because about 1% of the general population will have false-positive results. However, a reactive treponemal test result on a sample that is also reactive in a nontreponemal test is highly specific.

If used appropriately as confirmatory tests, the treponemal tests have few limitations; their greatest limitation is cost. Treponemal tests vary in their reactivities in early primary syphilis (Table 5); the varied sensitivities of the treponemal tests in primary syphilis are related to the time of serum collection after lesion development. With primary syphilis, the microhemagglutination assay for *T. pallidum* antibodies (MHA-TP) is less sensitive than the fluorescent-treponemal-antibody absorption (FTA-ABS) test and probably is less sensitive than the nontreponemal tests.

In the secondary stage of syphilis, the sensitivity of the MHA-TP test is equal to that of the nontreponemal and FTA-ABS tests. In secondary and latent stages, all of the treponemal tests are 100% reactive. Even though the sen-

sitivities of the treponemal tests decline somewhat in the late stage of syphilis (Table 5) because approximately 30% of patients with late syphilis are nonreactive in the nontreponemal tests, treponemal test results should be obtained if syphilis in these stages is suspected and the nontreponemal tests are nonreactive. The laboratory should be informed that late syphilis is suspected; otherwise, according to laboratory policy, a treponemal test may not be performed in the absence of a reactive nontreponemal test.

As with the nontreponemal tests, all standard treponemal tests are reactive for congenital syphilis. Passively transferred treponemal antibodies should be catabolized and undetectable among noninfected infants between the ages of 12 and 18 months.

Although false-positive results in the treponemal tests are often transient and their cause is unknown, a definite association has been made between false-positive FTA-ABS test results and the diagnosis of systemic, discoid, and drug-induced varieties of lupus erythematosus (43, 80). Patients with systemic lupus erythematosus can give false-positive FTA-ABS tests that exhibit an "atypical beading" fluorescence pattern. Patients whose serum specimens exhibit this type of reaction should be screened for autoimmune diseases; testing should include a test for serum anti-DNA antibodies. Unexplained reactive serologic results also may occur, particularly in elderly patients. Some false-positive reactions may be due to the failure of the sorbent used in the tests to remove all cross-reacting groups, genera, or families of treponemal antibodies, e.g., antibodies to Lyme disease (35, 48). In these instances, absorption with Reiter treponeme or the hemagglutination test may be the only means of differentiating between syphilis and a false-positive reaction.

Of the two treponemal test methods currently in use, the MHA-TP gives fewer false-positive test results. In general, the occurrence of false-positive hemagglutination tests is rare in "healthy" persons (<1%). Inconclusive hemagglutination tests have been reported for patients with infectious mononucleosis, especially in the presence of a high heterophile antibody. Presumably, false-positive hemagglutination test results also occur in samples from drug addicts, patients with collagen disease, patients with leprosy, and patients with other miscellaneous conditions. In some cases, it is difficult to assess whether syphilis actually coexists with these other infections. If both the FTA-ABS and the MHA-TP tests are reactive, the sample is most likely (95%) from a person who has or has had syphilis (72). However, the final decision depends on clinical judgment.

FTA-ABS Tests

The FTA-ABS test is an indirect fluorescent-antibody technique (42, 43). In this procedure, the antigen used is *T. pallidum* subsp. *pallidum* (Nichols strain). The patient's serum is first diluted 1:5 in sorbent (an extract from cultures of the nonpathogenic Reiter treponeme) to remove group treponemal antibodies that are produced in some persons in response to nonpathogenic treponemes. Next, the serum is layered on a microscope slide to which *T. pallidum* has been fixed. If the patient's serum contains antibody, the antibody coats the treponeme. FITC-labeled anti-human immunoglobulin is added, and it combines with the patient's antibodies adhering to *T. pallidum*, resulting in FITC-stained spirochetes that are visible when examined by fluorescence microscopy. A recent modification of the standard FTA-ABS test is the FTA-ABS double-staining test. The FTA-ABS double-staining technique employs a tetramethylrhodamine isothiocyanate-labeled anti-human IgG globulin and a counterstain with FITC-labeled anti-*T. pallidum* conjugate. The counterstain was developed for use with microscopes with incident illumination to eliminate the need to locate the treponemes by dark-field microscopy when the patient's serum does not contain antibodies to *T. pallidum*. Counterstaining the organism ensures that the nonreactive result is due to the absence of antibodies and not to the absence of treponemes on the slide with incident illumination.

Results of both FTA-ABS tests are reported as reactive, reactive minimal, nonreactive, or atypical fluorescence observed. Specimens initially read as 1+ (reactive minimal) should always be retested. If the results are again read as 1+, they should be reported as such with the statement that "in the absence of historical or clinical evidence of treponemal infection, this test result should be considered equivocal." A second specimen should be obtained 1 to 2 weeks after the initial specimen and submitted to the laboratory for serologic testing. Because beaded fluorescence (atypical staining) has been observed in sera from patients with active systemic lupus erythematosus and from patients with other autoimmune diseases, this observation should be reported as well.

The sources for errors are numerous with the FTA-ABS tests because they are multicomponent tests and each component must be matched with the other. Conjugates must be properly titrated, and controls for reactive, reactive minimal, and nonreactive samples and for nonspecific staining and sorbent must be included. Slides must be evaluated to ensure that the antigen is adhering. In addition, the microscope must be in proper operating condition with the appropriate filters in place.

MHA-TP Test

Passive hemagglutination of erythrocytes sensitized with antigen is an extremely simple method for the detection of antibody (42). The antigen used in the procedure is formalinized, tanned sheep erythrocytes sensitized with ultrasonicated material from *T. pallidum* (Nichols strain). The patient's serum is first mixed with absorbing diluent made from nonpathogenic Reiter treponemes and other absorbents and stabilizers. The serum is then placed in a microtiter plate, and sensitized sheep erythrocytes are added. Serum containing antibodies reacts with these cells to form a smooth mat of agglutinated cells in the microtiter plate. Unsensitized cells are used as a control for nonspecific reactivity.

Results are reported as reactive, nonreactive, or inconclusive. Reactive results are reported over a range of agglutination patterns, from a smooth mat of agglutinated cells surrounded by a smaller red circle of unagglutinated cells with hemagglutination outside the circle to a smooth mat of agglutinated cells covering the entire bottom of the well. With highly reactive serum samples, the edges of the mat are sometimes folded. Nonreactive results are reported when a definite compact button with or without a very small "hole" in its center forms in the center of well. A button of unagglutinated cells having a small hole in the center is initially read as ±, and the test should be repeated. If the same pattern is again observed, then the report should be "nonreactive." If agglutination is observed with the unsensitized cells, then the test is repeated as follows. Twofold dilutions of the absorbed serum are prepared in absorbing diluent in two rows of the microtiter plate wells. Sensitized cells are added to each well in one row, and

unsensitized cells are added to each well in the other row. A reactive result is reported, without reference to titer, if (i) the hemagglutination with sensitized cells is at least two doubling dilutions (four times greater than) that with unsensitized cells and (ii) the first dilution showing no hemagglutination with unsensitized cells has a 3+ or 4+ reaction with sensitized cells. Otherwise, the test results are reported as inconclusive.

Because the test is based on agglutination, quantitation of treponemal antibody is possible but has not proven to be worthwhile. Most studies demonstrate no practical relationship between the titer and either the progression of the disease or the clinical stage of syphilis diagnosed, and unlike the quantitative nontreponemal tests, the quantitative MHA-TP test does not seem useful in posttreatment evaluation (43, 80). The sources of error with the MHA-TP test are usually associated with the use of dusty or improper plates, pipetting, and vibrations in the laboratory.

Newer Tests for Syphilis

When medical management decisions are to be made, the laboratory should use only procedures that have been thoroughly evaluated. In the United States, the tests for syphilis are classified as standard, provisional, or investigational. Tests that are standard or provisional can be used in medical management if the results obtained are comparable to those previously reported (standard status) or if test results of the new tests are comparable to results obtained with the standard test currently being used when the two tests are performed in parallel (provisional status). Tests with investigational status should not be used as the sole criterion for diagnosis of syphilis in any stage.

Several tests using the enzyme-linked immunosorbent assay (ELISA) format have been developed as tests for syphilis (36, 42, 45, 56, 67, 97). Two of these tests are currently considered provisional tests: the VISUWELL reagin test, which is a nontreponemal test, and the Captia M test, which is an IgM capture ELISA that detects anti-*T. pallidum* antibodies and is meant for use in diagnosing congenital syphilis in newborns (36, 45, 67). In the VISUWELL reagin test, a VDRL antigen coats the well of the microtiter plate. Anticardiolipin antibodies in the patient's serum adhere to the antigen in the wells, and this antibody reaction is detected by an anti-human IgG antibody labeled with urease as the enzyme. For untreated syphilis, the test has a sensitivity of 97% and a specificity of 97% (67). Like the other nontreponemal tests, the reactivity of the VISUWELL reagin test declines with treatment of the patient.

Because infected infants can produce IgM in utero after 3 months of gestation and because the fetus can be infected with *T. pallidum* at any time during gestation, an IgM ELISA has been developed. The test is based on the use of anti-human IgM antibody to capture IgM in the patient's serum followed by use of a purified *T. pallidum* antigen to detect those IgM antibodies in the patient's serum that are directed toward *T. pallidum* (36). Next, a horseradish peroxidase-conjugated monoclonal antibody to *T. pallidum* is used to detect the antigen-patient antibody reaction through the enzyme's reaction with a substrate. One study found that the IgM capture ELISA was more sensitive than the FTA-ABS 19S IgM test in detecting probable cases of congenital syphilis (90); however, another study found the test to be equal in sensitivity to the IgM Western blot for neonatal congenital syphilis but less sensitive than the Western blot for delayed-onset congenital syphilis (4).

Of the commercial ELISAs designed to replace the

TABLE 6 Recommended treatment for syphilis[a]

Stage of syphilis	Treatment[b]
Primary, secondary, or early latent (<1-yr duration)	Benzathine penicillin G, 2.4 U i.m. in one dose; for patients allergic to penicillin (nonpregnant), 100 mg of doxycycline orally b.i.d. for 14 days or 500 mg of tetracycline orally q.i.d. for 14 days[c]
Late latent (>1-yr duration) or of unknown duration	Benzathine penicillin G, 7.2×10^6 U total, 2.4×10^6 U/wk for 3 consecutive wk; for patients allergic to penicillin (nonpregnant), 100 mg of doxycycline orally b.i.d. for 28 days or 500 mg of tetracycline orally q.i.d. for 28 days[c]
Tertiary, excluding neurosyphilis	As for late latent disease, with appropriate management of complications
Neurosyphilis	Aqueous crystalline penicillin G, $12–24 \times 10^6$ U/day i.v. ($2–4 \times 10^6$ U every 4 h) for 10–14 days, or procaine penicillin, 2.4×10^6 U/day i.m. plus 500 mg of probenecid orally q.i.d. for 10–14 days
During pregnancy	Penicillin regimen appropriate for woman's stage of syphilis; tetracycline, doxycycline, and erythromycin are contraindicated[d]
Congenital	Aqueous crystalline penicillin G, 100,000 to 150,000 U/kg/day i.v. (50,000 U/kg i.v. every 8 to 12 h) for 14 days

[a]Adapted from references 12, 32, and 99; refer to these resources for further information.
[b]i.m., intramuscularly; b.i.d., twice a day; q.i.d., four times a day; i.v., intravenously.
[c]Patients treated for syphilis with regimens that do not include penicillin may be at increased risk for treatment failure. Assurance of compliance with therapy and follow-up are recommended for such patients.
[d]Women treated in the second half of pregnancy are at risk for premature labor and/or fetal stress due to a Jarisch-Herxheimer reaction. Tetracycline and doxycycline are contraindicated because of possible fetal bone and tooth damage, and erythromycin poses a high risk of treatment failure.

FTA-ABS and MHA-TP tests as confirmatory tests for syphilis, initial evaluations have found all to have sensitivities and specificities similar to those of the other treponemal tests, but more extensive evaluation is necessary (45, 56, 97).

Another test format that has been used frequently in the research laboratory is the Western blot for T. pallidum. The test using IgG conjugate appears to be at least as sensitive and specific as the FTA-ABS tests, and efforts have been made to standardize the procedure (7, 28). To date, many investigators agree that detecting antibodies to the immunodeterminants with molecular masses of 15.5, 17, 44.5, and 47 kDa appears to be diagnostic for acquired syphilis (65). The Western blot for T. pallidum has its greatest value as a diagnostic test for congenital syphilis, when an IgM-specific conjugate is used (4, 19, 46, 81, 82). The IgM Western blot for congenital syphilis appears to have a greater specificity and sensitivity than the FTA-ABS 19S IgM test. Unfortunately, the Western blot test is not yet commercially available.

Treatment and Antimicrobial Susceptibilities

The syphilis treatment guidelines recommended by the CDC in 1993 are provided in Table 6 (12). Penicillin and its derivatives are the preferred antimicrobial agents for the treatment of syphilis and the other treponematoses. Early syphilis in immunologically normal patients can be treated effectively with a single dose of benzathine penicillin G. Neurosyphilis should be treated with either aqueous crystalline penicillin G or procaine penicillin, because benzathine penicillin does not achieve treponemacidal concentrations in the central nervous system. In one study, viable T. pallidum was present in the CSF of 12 of 40 untreated primary and secondary syphilis patients (47). In addition, central nervous system infection persisted in some HIV-positive patients following benzathine penicillin therapy (47). In immunologically normal patients, the remaining organisms are apparently cleared by the host, because the incidence of relapse is low in these patients. However, HIV-infected individuals may be susceptible to neurologic

or ocular complications of syphilis following benzathine penicillin treatment (3). Therefore, some authors have recommended that therapy appropriate for neurosyphilis (aqueous penicillin G or procaine penicillin with probenicid) be used for all patients coinfected with T. pallidum and HIV and that HIV-positive patients be examined carefully for clinical signs of neurologic involvement (e.g., CSF examination) prior to and after therapy (3, 32). Late or late latent syphilis patients with evidence of neurosyphilis or other forms of active syphilis should have a CSF examination; if neurologic involvement is not indicated, prolonged treatment with benzathine penicillin is recommended (Table 6).

Effective therapy of early syphilis may result in a Jarisch-Herxheimer reaction, consisting of fever and local intensification of lesions. This reaction, thought to result from the rapid release of treponemal antigens, occurs within the first 12 h of the initiation of therapy and usually resolves within 24 h.

Doxycycline and tetracycline can be used as alternative therapies for the treatment of penicillin-allergic individuals. Erythromycin is no longer recommended because of the occurrence of treatment failures and the possible existence of erythromycin-resistant strains (89), and the macrolides are contraindicated in pregnant women. In these cases, it is suggested that penicillin allergy be documented by thorough questioning regarding the patient's history and by skin testing. If necessary, the patient may be desensitized to permit use of penicillin therapy (12). A Jarisch-Herxheimer reaction occurring in pregnant women may cause premature labor, fetal distress, or (rarely) stillbirth. However, the high probability of fetal damage resulting from syphilitic infection in the absence of adequate therapy outweighs this risk.

Antimicrobial susceptibility testing is not straightforward, owing to the lack of a method for continuous culture of T. pallidum. A number of approaches have been used for determining the susceptibilities of representative strains (e.g., the Nichols strain) to antimicrobial agents, including in vitro loss of motility or infectivity, treatment of experi-

TABLE 7 Other human-host-associated spirochetes

Type of spirochete	Habitat	Species or group	Reference(s)
Oral	Dental plaque in gingival crevices	*T. denticola*	53, 54, 84
		T. socranskii	86
		T. vincentii	53, 84
		T. pectinovorum	85
		T. skoliodontum	53, 84
		PROS	73–75, 83
Skin associated	Sebaceous secretions in genital region	*T. phagedenis*	53, 84
		T. refringens	
		T. minutum	
Intestinal	Colon, rectum, feces	*Serpulina*-related organisms	15, 18, 31, 38, 39, 41, 44, 79, 92, 93
		B. aalborgii	34

mental animal infections, human trials, and examination of the susceptibility of nonpathogenic, cultivable treponemes (71). Recently, inhibition of protein synthesis and of in vitro multiplication in a tissue culture system have been shown to be effective means of determining susceptibility (64, 64a, 89). MICs (in milligrams per liter) determined by the in vitro culture method were 0.0005 for penicillin G, 0.2 for tetracycline, 0.0007 for ceftriaxone, 0.04 for cefetamet, 0.004 for ceftcram, and 0.007 for ceftazidime. Like other spirochetes, *T. pallidum* was relatively insensitive to seven quinolones tested (MIC range, 1 to 10 mg/liter) and to rifampin (MIC, 44 mg/liter). MBCs as determined by loss of infectivity during in vitro culture were one- to fivefold higher than MICs (64).

T. pallidum strains appear to be uniformly susceptible to penicillin and other β-lactams. There is no evidence for the development of resistance to penicillin, as has occurred with *Neisseria gonorrhoeae* and other pathogens. Penicillin G provides greater inhibitory activity than do cephalosporins and other β-lactams tested. Ceftriaxone, amoxicillin, and ampicillin are active against *T. pallidum* in vitro and also appear to be effective in the treatment of early syphilis. Amoxicillin and ampicillin offer no theoretical advantage over benzathine penicillin G. Ceftriaxone has a long serum half-life and reaches high levels in the central nervous system. However, the cost is high, multiple doses are required, and there is a 3 to 7% risk of adverse reactions in penicillin-allergic individuals; in addition, the data on efficacy in patients are limited. Therefore, ceftriaxone is not currently recommended as an alternative therapy for early syphilis.

OTHER HUMAN-HOST-ASSOCIATED SPIROCHETES

In addition to *T. pallidum* and related pathogens, a number of anaerobic treponemes and other spirochetes are parasites of humans. These include separate groups of spirochetes found in the oral cavity (particularly the gingival crevices), in sebaceous secretions in the genital region, and in the colon and rectum (Table 7). In general, these spirochetes are considered commensal organisms. However, there is evidence that spirochetes are involved in gingivitis and periodontal disease and that overgrowth of intestinal spirochetes may correlate with the occurrence of diarrhea or other bowel disorders.

Oral Spirochetes

A variety of spirochetes inhabit supragingival and subgingival plaque in humans; only a small proportion of these have been cultured in vitro and characterized (14). Those that have been cultivated and identified to the species level are within the genus *Treponema* (Table 7). They are spiral-shaped organisms ranging from 0.15 to 0.30 μm in diameter and from 5 to 16 μm in length (variable within each species). Although the species differ from one another slightly in terms of cell diameter and helical configuration, it is generally not possible to identify them on morphologic grounds; biochemical parameters such as growth requirements, carbohydrate fermentation, and enzymatic activities are used for identifying species (84).

The oral spirochetes listed in Table 7 are difficult to isolate from healthy gingiva, but they increase both in prevalence and in numbers of organisms present in patients with gingivitis or periodontal disease. Treponemes are detected in subgingival plaque of 88 to 97% of patients with periodontal disease (54). *T. socranskii* is the most common isolate, followed by *T. denticola* and *T. pectinovorum* (54, 85, 86).

The procedures for isolation and characterization of oral spirochetes have been described previously (53, 54, 84–86). Briefly, treponemes may be cultured from samples of supragingival or subgingival material removed with a sterile dental scaler having a detachable tip or paper points. Samples should be examined by dark-field microscopy for the presence and morphologic characteristics of spirochetes. The plaque material is placed into tubes with sterile anaerobic dilution broth and small glass beads and is dispersed by vortexing for a few seconds. Tenfold dilutions are made and inoculated onto agar plates containing oral treponeme isolation medium, a complex peptone-yeast extract medium containing clarified rumen fluid and serum as sources of short- and long-chain fatty acids, respectively (53, 85). Selection of treponemes can be achieved by addition of rifampin (2 μg/ml) and polymyxin B (800 U/ml) to the medium or by use of a membrane filter technique. In the latter, a sterile 45-mm-wide nitrocellulose filter with an average pore diameter of 0.15 to 0.3 μm is placed on the surface of the agar isolation medium. A sterile O ring is sealed onto the filter with 3% molten agarose, and a few drops of the diluted sample are placed in the resulting well. Treponemes are capable of translocating through the filter into the underlying medium, whereas other bacteria are trapped on the surface. Growth is indicated by a white, hazy appearance of the medium when the filter is removed after 1 to 2 weeks of incubation at 37°C. Subculture and single-colony isolation can be achieved by removing a plug from the area of growth, placing it in broth or semisolid medium, and streaking plates with 2- to 3-day cultures. At all stages, care should be taken to minimize exposure to air and to maintain strict anaerobic conditions during culture through the use of prereduced medium and an anaerobic chamber or anaerobic jars gassed with carbon dioxide. Anaerobic jars with platinum catalyst may be evacuated and refilled three times with a nitrogen-hydrogen-carbon dioxide gas mixture; alternatively, a GasPak envelope (BBL, Baltimore, Md.) may be used.

Pathogen-related oral spirochetes (PROS) were found in plaque samples from 11 of 17 patients with necrotizing ulcerative gingivitis and from 10 of 19 patients with periodontitis but were absent from 24 healthy subjects (73–75, 83). PROS have not been cultured in vitro and are identified by their reactivity with monoclonal antibodies specific for *T. pallidum* antigens, including the 37-kDa flagellar sheath protein (FlaA) and a major 47-kDa lipoprotein. Gingivitis and periodontitis patients also expressed IgG antibodies reactive with 47-, 37-, 14-, and 12-kDa *T. pallidum* subsp. *pallidum* antigens, whereas a smaller proportion of healthy subjects had detectable antibodies against only the 47- and 37-kDa antigens (73). PROS were found in periodontal tissues of ulcerative gingivitis patients (73, 75) and were capable of migrating through a mouse abdomen wall barrier, an invasive property shared with *T. pallidum* but not with the cultivable oral treponemes (74). Oral spirochetes reactive with the anti-*T. pallidum* monoclonal antibodies were also detected in gorillas and dogs (83). Further studies to establish the taxonomic relationship between PROS and other spirochetes and to determine whether PROS represent frank pathogens are in progress.

Nonpathogenic Spirochetes of the Genital Region

T. phagedenis, *T. refringens*, and *T. minutum* are treponemal species that inhabit smegma (sebaceous secretions and desquamated epithelial cells), which is found beneath the prepuce and in other epithelial folds of the genital region. *T. phagedenis* and *T. refringens* are 0.20 to 0.25 μm in diameter, whereas *T. minutum* tends to be smaller (0.15 to 0.20 μm). Although *T. phagedenis* and *T. refringens* have been shown by DNA-DNA hybridization to have ≤5% homology with *T. pallidum* (22), comparison of 16S rRNA sequences indicates that they are related to the pathogenic treponemes. These harmless healthy flora could potentially be misidentified as *T. pallidum* in dark-field preparations from skin sites. Careful cleansing of the site as described above under Diagnostic Parameters and Interpretation of Laboratory Tests obviates this possibility. The skin-associated treponemes may be collected with sterile moistened swabs and placed in anaerobic transport medium or directly into selective medium. They are readily cultured in peptone-yeast extract-glucose medium or in thioglycolate medium with 10% heat-inactivated rabbit serum under anaerobic conditions at 37°C (53).

Intestinal Spirochetes

The presence of spirochetes in the colon, rectum, and feces of humans has been recognized for over a hundred years (79). Since the description of "human intestinal spirochetosis" in 1967 (31), there has been a resurgence of interest in the possible involvement of intestinal spirochetes in

diarrhea and other gastrointestinal diseases as well as in the apparently high prevalence of these organisms in homosexual males and HIV-infected individuals (79). Despite this history, the taxonomy, prevalence, distribution, and possible disease associations of these organisms are still unclear.

Morphologically, most human intestinal spirochetes are relatively short (3 to 6 μm on average) and are 0.2 to 0.4 μm in diameter with one to two helices per cell. Longer cells may also be present. They have pointed ends and four to six periplasmic flagella attached subterminally in a single row at each end of the cell.

The human intestinal spirochetes have not as yet been classified and appear to be a heterogeneous group rather than a single species. Although it has been suggested (15) that the human spirochetes are related to *S. hyodysenteriae* (the causative agent of swine dysentery) and *Serpulina innocens* (an apathogenic swine organism), recent multilocus enzyme electrophoresis, DNA-DNA hybridization, and morphologic analyses (44) have indicated that human isolates are more closely related to a different group of pig intestinal isolates represented by strains P43/6/78 and M1. Similar organisms are also present in mice, rats, and dogs (41). The human intestinal spirochete *Brachyspira aalborgi* (34) appears to be distinct from the majority of human and animal strains (41). Restriction fragment polymorphism indicates sequence heterogeneity both among the human isolates and in comparison with isolates from other animals. Further analysis of the taxonomy and pathobiology of these organisms will be necessary to clarify their phylogenetic relationships and possible role in human disease.

When observed in association with rectal or colonic tissue, these organisms are typically attached by the tip to the apical surface of the columnar epithelial cells. Although the spirochetes can form a dense layer visible by light microscopy (using silver staining or other staining techniques), signs of tissue damage or inflammation are usually absent.

A clear association between intestinal spirochetes and disease has not been established (79). Spirochetes can be present in healthy individuals as well as in those with diarrhea or other gastrointestinal symptoms. In the early 1900s, several studies examined spirochetes in human stool specimens by light microscopy. In those studies with over 100 patients, the percentage of subjects with spirochetes ranged from 3.3 to 61%. Recent light and electron microscopic examinations and culture of rectal biopsy samples from heterosexual patients have indicated the presence of spirochetes in 1.9 to 6.9% of subjects, and helical organisms were also found to be associated with healthy and inflamed appendixes. Higher proportions of homosexual males have been reported to have intestinal spirochetes. At present, there is no clear evidence for an association between intestinal spirochetosis and HIV infection.

For isolation, fresh stool specimens or rectal swabs should be examined by dark-field microscopy for the presence of spirochetes. Rectal biopsy samples can also serve as sources of material for culture or for histologic examination for spirochetes (79). Positive samples can be cultured under anaerobic conditions at 37°C by streaking on Trypticase soy agar medium with 5 to 10% horse or calf blood and 400 μg of spectinomycin per ml (±5 μg of polymyxin B per ml) to inhibit the growth of other bacteria. After 5 to 14 days of incubation, growth will appear as a thin film or as discrete, pinpoint colonies. Weak beta-hemolysis is typically observed. No serologic procedures for detection of intestinal spirochetosis are available, and no data on antimicrobial susceptibility or treatment have been reported. Clearly, more information is needed to establish the potential medical importance of these organisms.

REFERENCES

1. **Antal, G. M., and G. Causse.** 1985. The control of endemic treponematoses. *Rev. Infect. Dis.* **7**(Suppl. 2):S220–S226.
2. **Benenson, A. S. (ed.).** 1990. *Control of Communicable Diseases in Man*, 15th ed., p. 323–324, 425–426, 483–486. American Public Health Association, Washington, D.C.
3. **Berry, C. D., T. M. Hooten, A. C. Collier, and S. A. Lukehart.** 1987. Neurologic relapse after benzathine penicillin therapy for secondary syphilis in a patient with HIV infection. *N. Engl. J. Med.* **316**:1587–1589.
4. **Bromberg, K., S. Rawstron, and G. Tannis.** 1993. Diagnosis of congenital syphilis by combining *Treponema pallidum*-specific IgM detection with immunofluorescent antigen detection for *T. pallidum*. *J. Infect. Dis.* **168**:238–242.
5. **Brown, S. T., A. Zaidi, S. A. Larsen, and G. H. Reynolds.** 1985. Serological response to syphilis treatment: a new analysis of old data. *JAMA* **253**:1296–1299.
6. **Burstain, J. M., E. Grimprel, S. A. Lukehart, M. V. Norgard, and J. D. Radolf.** 1991. Sensitive detection of *Treponema pallidum* by using the polymerase chain reaction. *J. Clin. Microbiol.* **29**:62–69.
7. **Byrne, R. E., S. Laske, M. Bell, D. Larson, J. Phillips, and J. Todd.** 1992. Evaluation of a *Treponema pallidum* Western immunoblot assay as a confirmatory test for syphilis. *J. Clin. Microbiol.* **30**:115–122.
8. **Castro, K. G., S. Lieb, H. W. Jaffe, J. P. Narkunas, C. H. Calisher, T. J. Bush, J. J. Witte, and the Belle Glade Field Study Group.** 1988. Transmission of HIV in Belle Glade, Florida: lessons for other communities in the United States. *Science* **239**:193–197.
9. **Centers for Disease Control.** 1988. Recommendations for diagnosing and treating syphilis in HIV-infected patients. *Morbid. Mortal. Weekly Rep.* **37**:600–602, 607–608.
10. **Centers for Disease Control.** 1991. *Sexually Transmitted Diseases Clinical Practices Guidelines*. Centers for Disease Control, Atlanta.
11. **Centers for Disease Control.** 1992. Summary of notifiable diseases, United States 1992. *Morbid. Mortal. Weekly Rep.* **41**(55):9.
12. **Centers for Disease Control.** 1993. Sexually transmitted diseases treatment guidelines. *Morbid. Mortal. Weekly Rep.* **42**(RR-14):27–46.
13. **Chhabra, R. S., L. P. Brion, M. Castro, L. Freundlich, and J. H. Glaser.** 1993. Comparison of maternal sera, cord blood and neonatal sera for detection of presumptive congenital syphilis: relationship with maternal treatment. *Pediatrics* **91**:88–91.
14. **Choi, B. K., B. J. Paster, F. E. Dewhirst, and U. B. Göbel.** 1994. Diversity of cultivable and uncultivable oral spirochetes from a patient with severe destructive periodontitis. *Infect. Immun.* **62**:1889–1895.
15. **Coene, M., A. M. Agliano, A. T. Paques, P. Cattani, G. Dettori, A. Sanna, and C. Cocito.** 1989. Comparative analysis of the genomes of intestinal spirochetes of human and animal origin. *Infect. Immun.* **57**:138–145.
16. **Cox, C. D.** 1983. Metabolic activities, p. 57–70. In R. F. Schell and D. M. Musher (ed.), *Pathogenesis and Immunology of Treponemal Infection*. Marcel Dekker, Inc., New York.
17. **Cox, D. L., B. Riley, P. Chang, S. Sayahtaheri, S. Tassell, and J. Hevelone.** 1990. Effects of molecular oxygen, oxidation-reduction potential, and antioxidants upon in vitro replication of *Treponema pallidum* subsp. *pallidum*. *Appl. Environ. Microbiol.* **56**:3063–3072.
18. **Dettori, G., R. Burioni, R. Grillo, and P. Cattani.** 1992. Molecular cloning and characterization of DNA from human intestinal spirochetes. *Eur. J. Epidemiol.* **8**:198–205.
19. **Dobson, S. R. M., L. H. Taber, and R. E. Baughn.** 1988. Recognition of *Treponema pallidum* antigens by IgM and IgG antibodies in congenitally infected newborns and their mothers. *J. Infect. Dis.* **157**:903–910.
20. **Engelkens, H. J. H., J. Judanarso, A. P. Oranje, V. D. Vuzevski, P. L. A. Niemel, J. J. van der Sluis, and E. Stolz.** 1991. Endemic treponematoses. I. Yaws. *Int. J. Dermatol.* **30**:77–83.
21. **Engelkens, H. J. H., P. L. A. Niemel, J. J. van der Sluis, A. Meheus, and E. Stolz.** 1991. Endemic treponematoses. II. Pinta and endemic syphilis. *Int. J. Dermatol.* **30**:231–238.
22. **Fieldsteel, A. H.** 1983. Genetics, p. 39–54. In R. F. Schell and D. M. Musher (ed.), *Pathogenesis and Immunology of Treponemal Infection*. Marcel Dekker, Inc., New York.
23. **Fieldsteel, A. H., D. L. Cox, and R. A. Moeckli.** 1981. Cultivation of virulent *Treponema pallidum* in tissue culture. *Infect. Immun.* **32**:908–915.
24. **Fiumara, N. J.** 1978. Treatment of early latent syphilis of less than one year's duration. *Sex. Transm. Dis.* **5**:85–88.
25. **Fiumara, N. J.** 1979. Serologic responses to treatment of 128 patients with late latent syphilis. *Sex. Transm. Dis.* **6**:243–246.
26. **Fiumara, N. J.** 1980. Treatment of primary and secondary syphilis; serological response. *JAMA* **243**:2500–2502.
27. **Fohn, M. J., F. S. Wignall, S. A. Baker-Zander, and S. A. Lukehart.** 1988. Specificity of antibodies from patients with pinta for antigens of *Treponema pallidum* subspecies *pallidum*. *J. Infect. Dis.* **157**:32–37.
28. **George, R. W., V. Pope, and S. A. Larsen.** 1991. Use of the Western blot for the diagnosis of syphilis. *Clin. Immunol. Newsl.* **8**:124–128.
29. **Gjestland, T.** 1955. The Oslo study of untreated syphilis: an epidemiologic investigation of the natural course of syphilis infection based upon a re-study of the Boeck-Bruusgaard material. *Acta Dermato. Vener.* **35**(Suppl. 34):1–368.
30. **Grimprel, E., P. J. Sanchez, G. D. Wendel, J. M. Burstain, G. H. McCracken, J. D. Radolf, and M. V. Norgard.** 1991. Use of polymerase chain reaction and rabbit infectivity testing to detect *Treponema pallidum* in amniotic fluid. *J. Clin. Microbiol.* **29**:1711–1718.
31. **Harland, W. A., and F. D. Lee.** 1967. Intestinal spirochaetosis. *Br. Med. J.* **3**:718–719.
32. **Hook, E. W., III.** 1989. Treatment of syphilis: current recommendations, alternatives, and continuing problems. *Rev. Infect. Dis.* **11**(Suppl. 6):S1511–S1517.
33. **Hook, E. W., III, and C. M. Marra.** 1992. Acquired syphilis in adults. *N. Engl. J. Med.* **326**:1060–1069.
34. **Hovind-Hougen, K., A. Birch-Andersen, R. Henrik-Nielsen, M. Orholm, J. O. Pedersen, P. S. Teglbjærg, and E. H. Thaysen.** 1982. Intestinal spirochetosis: morphological characterization and cultivation of the spirochete *Brachyspira aalborgi* gen. nov., sp. nov. *J. Clin. Microbiol.* **16**:1127–1136.
35. **Hunter, E. F., H. Russell, C. E. Farshy, J. S. Sampson, and S. A. Larsen.** 1986. Evaluation of sera from patients with Lyme disease in the fluorescent treponemal antibody-absorption tests for syphilis. *Sex. Transm. Dis.* **13**:232–236.
36. **Ijsselmuiden, O. E., J. J. van der Sluis, A. Mulder, E. Stolz, K. P. Bolton, and R. V. W. van Eijk.** 1989. An IgM capture enzyme-linked immunosorbent assay to detect IgM antibodies to treponemes in patients with syphilis. *Genitourin. Med.* **65**:79–83.
37. **Ito, F., R. W. George, E. F. Hunter, S. A. Larsen, and V. Pope.** 1992. Specific immunofluorescent staining of pathogenic treponemes with a monoclonal antibody. *J. Clin. Microbiol.* **30**:831–838.
38. **Jones, M. J., J. N. Miller, and W. L. George.** 1986. Microbiological and biochemical characterization of spirochetes isolated from the feces of homosexual males. *J. Clin. Microbiol.* **24**:1071–1074.
39. **Käsbohrer, A., H. R. Gelderblom, K. Arasteh, W. Heise, G. Grosse, M. L'age, A. Schönberg, M. A. Koch, and G. Pauli.** 1990. Intestinale Spirochätose bei HIV-Infektion: Vorkommen, Isolierung und Morphologie der Spirochäten. *Dtsch. Med. Wochenschr.* **115**:1499–1509.
40. **Kaufman, R. E., O. G. Jones, J. H. Blount, and P. J.**

Wiesner. 1977. Questionnaire survey of reported early congenital syphilis, problems in diagnosis prevention and treatment. *Sex. Transm. Dis.* **4:**135–139.

41. **Koopman, M. B. H., A. Käsbohrer, G. Beckmann, B. A. M. van der Zeijst, and J. G. Kusters.** 1993. Genetic similarity of intestinal spirochetes from humans and various animal species. *J. Clin. Microbiol.* **31:**711–716.

42. **Larsen, S. A., E. F. Hunter, and S. J. Kraus (ed.).** 1990. *A Manual of Tests for Syphilis*, 8th ed. American Public Health Association, Washington, D.C.

43. **Larsen, S. A., B. M. Steiner, and A. H. Rudolph.** Laboratory diagnosis and interpretation of tests from syphilis. *Clin. Microbiol. Rev.*, in press.

44. **Lee, J. I., A. J. McLaren, A. J. Lymbery, and D. J. Hampson.** 1993. Human intestinal spirochetes are distinct from *Serpulina hyodysenteriae. J. Clin. Microbiol.* **31:**16–21.

45. **Lefèvre, J. C., M. A. Bertrand, and R. Bauriaud.** 1990. Evaluation of the Captia enzyme immunoassays for detection of immunoglobulins G and M to *Treponema pallidum* in syphilis. *J. Clin. Microbiol.* **28:**1704–1707.

46. **Lewis, L. L., L. H. Taber, and R. E. Baughn.** 1990. Evaluation of immunoglobulin M Western blot analysis in the diagnosis of congenital syphilis. *J. Clin. Microbiol.* **28:**296–302.

47. **Lukehart, S. A., E. W. Hook III, S. A. Baker-Zander, A. C. Collier, C. W. Critchlow, and H. H. Handsfield.** 1988. Invasion of the central nervous system by *Treponema pallidum*: implications for diagnosis and treatment. *Ann. Intern. Med.* **109:**855–862.

48. **Magnarelli, L. A., J. N. Miller, J. F. Anderson, and G. R. Riviere.** 1990. Cross-reactivity of nonspecific treponemal antibody in serologic tests for Lyme disease. *J. Clin. Microbiol.* **28:**1276–1279.

49. **Magnuson, H. J., H. Eagle, and R. Fleischman.** 1948. The minimal infectious inoculum of *Spirochaeta pallida* (Nichols strain) and a consideration of its rate of multiplication in vivo. *Am. J. Syph.* **32:**1–18.

50. **Marra, C. M.** 1992. Syphilis and human immunodeficiency virus infection. *Semin. Neurol.* **12:**43–50.

51. **Matthews, H. M., T. K. Yang, and H. M. Jenkin.** 1979. Unique lipid composition of *Treponema pallidum* (Nichols virulent strain). *Infect. Immun.* **24:**713–719.

52. **Meheus, A., and G. M. Antal.** 1992. The endemic treponematoses: not yet eradicated. *World Health Stat. Q.* **45:**228–237.

53. **Miller, J. N., R. M. Smibert, and S. J. Norris.** 1992. The genus *Treponema*, p. 3537–3559. *In* A. Balows, H. G. Trüber, M. Dworkin, W. Harder, and K.-H. Schleifer (ed.), *The Prokaryotes. A Handbook on the Biology of Bacteria, Ecophysiology, Isolation, Identifications, and Applications*, 2nd ed. Springer Verlag, New York.

54. **Moore, L. V. H., W. E. C. Moore, E. P. Cato, R. M. Smibert, J. A. Burmeister, A. M. Best, and R. R. Ranney.** 1987. Bacteriology of human gingivitis. *J. Dent. Res.* **66:**989–995.

55. **Musher, D. M., R. J. Hamill, and R. E. Baughn.** 1990. Effect of human immunodeficiency virus (HIV) infection on the course of syphilis and on the response to treatment. *Ann. Intern. Med.* **113:**872–881.

56. **Nayar, R., and J. M. Campos.** 1993. Evaluation of the DCL Syphilis-G enzyme immunoassay test kit for the serologic diagnosis of syphilis. *Am. J. Clin. Pathol.* **99:**282–285.

57. **Noordhoek, G. T., A. Cockayne, L. M. Schouls, R. H. Meloen, E. Stolz, and J. D. A. van Embden.** 1990. A new attempt to distinguish serologically the subspecies of *Treponema pallidum* causing syphilis and yaws. *J. Clin. Microbiol.* **28:**1600–1607. (Erratum, **28:**2853.)

58. **Noordhoek, G. T., H. J. H. Engelkens, J. Judanarso, J. van der Stek, G. N. M. Aelbers, J. J. van der Sluis, J. D. A. van Embden, and E. Stolz.** 1991. Yaws in West Sumatra, Indonesia: clinical manifestations, serological findings, and characterisation of new treponema isolates by DNA probes. *Eur. J. Clin. Microbiol. Infect. Dis.* **10:**12–19.

59. **Noordhoek, G. T., P. W. M. Hermans, A. N. Paul, L. M. Schouls, J. J. van der Sluis, and J. D. A. van Embden.** 1989. *Treponema pallidum* subspecies *pallidum* (Nichols) and *Treponema pallidum* subspecies *pertenue* (CDC 2575) differ in at least one nucleotide: comparison of two homologous antigens. *Microb. Pathog.* **6:**29–42.

60. **Noordhoek, G. T., B. Wieles, J. J. van der Sluis, and J. D. A. van Embden.** 1990. Polymerase chain reaction and synthetic DNA probes—a means of distinguishing the causative agents of syphilis and yaws? *Infect. Immun.* **58:**2011–2013.

61. **Noordhoek, G. T., E. C. Wolters, M. E. J. De Jonge, and J. D. A. van Embden.** 1991. Detection by polymerase chain reaction of *Treponema pallidum* DNA in cerebrospinal fluid from neurosyphilis patients before and after antibiotic treatment. *J. Clin. Microbiol.* **29:**1976–1984.

62. **Norgard, M. V.** 1993. Clinical and diagnostic issues of acquired and congenital syphilis encompassed in the current syphilis epidemic. *Curr. Opin. Infect. Dis.* **6:**9–16.

63. **Norris, S. J., and D. G. Edmondson.** 1987. Factors affecting the multiplication and subculture of *Treponema pallidum* subsp. *pallidum* in a tissue culture system. *Infect. Immun.* **53:**534–539.

64. **Norris, S. J., and D. G. Edmondson.** 1988. In vitro culture system to determine the MICs and MBCs of antimicrobial agents against *Treponema pallidum* subsp. *pallidum* (Nichols strain). *Antimicrob. Agents Chemother.* **32:**68–74.

64a. **Norris, S. J., and N. J. Farley.** Personal observation.

65. **Norris, S. J., and the *Treponema pallidum* Polypeptide Research Group.** 1993. Polypeptides of *Treponema pallidum*: progress toward understanding their structural, functional and immunologic roles. *Microbiol. Rev.* **57:**750–779.

66. **Paster, B. J., F. E. Dewhirst, and W. G. Weisburg.** 1991. Phylogenetic analysis of the spirochetes. *J. Bacteriol.* **173:**6101–6109.

67. **Pedersen, N. S., O. Orum, and S. Mouritsen.** 1987. Enzyme-linked immunosorbent assay for detection of antibodies to Venereal Disease Research Laboratory (VDRL) antigen in syphilis. *J. Clin. Microbiol.* **25:**1711–1716.

68. **Perine, P. L., D. R. Hopkins, P. L. A. Niemel, R. K. St. John, G. Causse, and G. M. Antal.** 1984. *Handbook of Endemic Treponematoses: Yaws, Endemic Syphilis, and Pinta.* World Health Organization, Geneva.

69. **Perine, P. L., J. W. Nelson, J. O. Lewis, S. Liska, E. F. Hunter, S. A. Larsen, V. K. Agadzi, F. Kofi, J. A. K. Ofori, M. R. Tam, and M. A. Lovett.** 1985. New technologies for use in the surveillance and control of yaws. *Rev. Infect. Dis.* **7**(Suppl. 2):S295–S299.

70. **Rawstron, S. A., and K. Bromberg.** 1991. Comparison of maternal and newborn serologic tests for syphilis. *Am. J. Dis. Child.* **145:**1383–1388.

71. **Rein, M. F.** 1976. Biopharmacology of syphilotherapy. *J. Am. Vener. Dis. Assoc.* **3:**109–127.

72. **Rein, M. F., G. W. Banks, L. C. Logan, S. A. Larsen, J. C. Feeley, D. S. Kellogg, and P. J. Wiesner.** 1980. Failure of the *Treponema pallidum* immobilization test to provide additional diagnostic information about contemporary problem sera. *Sex. Transm. Dis.* **7:**101–105.

73. **Riviere, G. R., M. A. Wagoner, S. A. Baker-Zander, K. S. Weisz, D. F. Adams, L. Simonson, and S. A. Lukehart.** 1991. Identification of spirochetes related to *Treponema pallidum* in necrotizing ulcerative gingivitis and chronic periodontitis. *N. Engl. J. Med.* **325:**539–543.

74. **Riviere, G. R., K. S. Weisz, D. F. Adams, and D. D. Thomas.** 1991. Pathogen-related oral spirochetes from dental plaque are invasive. *Infect. Immun.* **59:**3377–3380.

75. **Riviere, G. R., K. S. Weisz, L. G. Simonson, and S. A. Lukehart.** 1991. Pathogen-related spirochetes identified within gingival tissue from patients with acute necrotizing ulcerative gingivitis. *Infect. Immun.* **59:**2653–2657.

76. **Rogers, M. F., D. M. Morens, J. A. Stewart, R. M. Kaminski, T. J. Spira, P. M. Feorino, S. A. Larsen, D. P. Francis, M. Wilson, L. Kaufman, and the Task Force on Acquired Immune Deficiency Syndrome.** 1983. National case control

study of Kaposi's sarcoma and *Pneumocystis carinii* pneumonia in homosexual men. II. Laboratory results. *Ann. Intern. Med.* **99**:151–158.

77. **Rolfs, R. T., E. Hindershot, K. Hackett, M. Augenbraun, G. Bolan, S. Johnson, K. Wagner, W. Brady, M. Joesoef, S. Larsen, and the Syphilis & HIV Study Group.** 1993. Treatment of early syphilis in HIV-infected and HIV-uninfected patients—preliminary results of the syphilis and HIV study. Paper presented at Int. Soc. Sex. Transm. Dis. Res. Meet.

78. **Romanowski, B., R. Sutherland, F. H. Fick, D. Mooney, and E. J. Love.** 1991. Serologic response to treatment of infectious syphilis. *Ann. Intern. Med.* **114**:1005–1009.

79. **Ruane, P. J., M. M. Nakata, J. F. Reinhardt, and W. Lance George.** 1989. Spirochete-like organisms in the human gastrointestinal tract. *Rev. Infect. Dis.* **11**:184–196.

80. **Rudolph, A. H., and S. A. Larsen.** 1993. Laboratory diagnosis of syphilis, unit 16-22A, p. 1–16. *In* D. J. Demis (ed.), *Clinical Dermatology*. J. B. Lippincott Co., Philadelphia.

81. **Sanchez, P. J., G. H. McCracken, G. D. Wendel, K. Olsen, N. Threlkeld, and M. V. Norgard.** 1989. Molecular analysis of the fetal IgM response to *Treponema pallidum* antigen: implications for improved serodiagnosis of congenital syphilis. *J. Infect. Dis.* **159**:508–517.

82. **Sanchez, P. J., G. D. Wendel, E. Grimprel, M. Goldberg, M. Hall, O. Arencibia-Mireles, J. D. Radolf, and M. V. Norgard.** 1993. Evaluation of molecular methodologies and rabbit infectivity testing for the diagnosis of congenital syphilis and neonatal central nervous system invasion by *Treponema pallidum*. *J. Infect. Dis.* **167**:148–157.

83. **Simonson, L., L. Braswell, W. Falkler, A. Laws, D. Lloyd, S. Lukehart, and G. Riviere.** 1993. Human oral spirochete antigens in certain animal populations, abstr. D-133, p. 118. *Abstr. 93rd Gen. Meet. Am. Soc. Microbiol. 1993.*

84. **Smibert, R. M.** 1984. Genus III. *Treponema* Schaudinn 1905, 1728[AL], p. 49–57. *In* N. R. Krieg and J. G. Holt (ed.), *Bergey's Manual of Systematic Bacteriology*, vol. 1. The Williams & Wilkins Co., Baltimore.

85. **Smibert, R. M., and J. A. Burmeister.** 1983. *Treponema pectinovorum* sp. nov. isolated from humans with periodontitis. *Int. J. Syst. Bacteriol.* **33**:852–856.

86. **Smibert, R. M., J. L. Johnson, and R. R. Ranney.** 1984. *Treponema socranskii* sp. nov., *Treponema socranskii* subsp. *socranskii* subsp. nov., *Treponema socranskii* subsp. *buccale* subsp. nov., and *Treponema socranskii* subsp. *paredis* subsp. nov.

isolated from the human periodontia. *Int. J. Syst. Bacteriol.* **34**:457–462.

87. **Stanton, T. B.** 1992. Proposal to change the genus designation *Serpula* to *Serpulina* gen. nov. containing the species *Serpulina hyodysenteriae* comb. nov. and *Serpulina innocens* comb. nov. *Int. J. Syst. Bacteriol.* **42**:189–190.

88. **Stanton, T. B., N. S. Jensen, T. A. Casey, L. A. Tordoff, F. E. Dewhirst, and B. J. Paster.** 1991. Reclassification of *Treponema hyodysentariae* and *Treponema innocens* in a new genus, *Serpula*, gen. nov., as *Serpula hyodysenteriae* comb. nov. and *Serpula innocens* comb. nov. *Int. J. Syst. Bacteriol.* **41**:50–58.

89. **Stapleton, J. T., L. V. Stamm, and P. J. Bassford, Jr.** 1985. Potential for development of antibiotic resistance in pathogenic treponemes. *Rev. Infect. Dis.* **7**:S314–S317.

90. **Stoll, B. J., F. K. Lee, S. A. Larsen, E. Hale, D. Schwartz, R. J. Rice, R. Ashby, R. Holmes, and A. J. Nahmias.** 1993. Improved serodiagnosis of congenital syphilis with combined assay approach. *J. Infect. Dis.* **167**:1093–1099.

91. **Swisher, B. L.** 1987. Modified Steiner procedure for microwave staining of spirochetes and nonfilamentous bacteria. *J. Histochnol.* **10**:241–243.

92. **Tompkins, D. S., S. J. Foulkes, P. G. R. Godwin, and A. P. West.** 1986. Isolation and characterisation of intestinal spirochetes. *J. Clin. Pathol.* **39**:535–541.

93. **Tompkins, D. S., M. A. Waugh, and E. M. Cooke.** 1981. Isolation of intestinal spirochaetes from homosexuals. *J. Clin. Pathol.* **34**:1385–1387.

94. **Turner, T. B., and D. H. Hollander.** 1957. *Biology of the Treponematoses*. World Health Organization, Geneva.

95. **U.S. Public Health Service.** 1968. *Syphilis: a Synopsis*. Public Health Service publication no. 1660. U.S. Government Printing Office, Washington, D.C.

96. **Wicher, K., G. T. Noordhoek, F. Abbruscato, and V. Wicher.** 1992. Detection of *Treponema pallidum* in early syphilis by DNA amplification. *J. Clin. Microbiol.* **30**:497–500.

97. **Young, H., A. Moyes, A. McMillan, and J. Patterson.** 1992. Enzyme immunoassay for antitreponemal IgG: screening or confirmatory test? *J. Clin. Pathol.* **45**:37–41.

98. **Zenker, P. N., and S. M. Berman.** 1990. Congenital syphilis: reporting and reality. *Am. J. Public Health* **80**:271–272.

99. **Zenker, P. N., and R. T. Rolfs.** 1990. Treatment of syphilis, 1989. *Rev. Infect. Dis.* **12**(Suppl. 6):S590–S608.

Mycoplasma and *Ureaplasma*

DAVID TAYLOR-ROBINSON

53

TAXONOMY AND NOMENCLATURE

Organisms in the class *Mollicutes* ("soft skin") have been derived from ancestral anaerobic bacteria (clostridia) by gene deletion. The class contains four orders, five families, and eight genera (Table 1). The mycoplasmas detected in humans belong to 12 species (and possibly 1 other) in the genus *Mycoplasma*, 1 species (*Ureaplasma urealyticum*) in the genus *Ureaplasma*, and 2 species in the genus *Acholeplasma* (Table 2). The term mollicutes is sometimes used trivially to describe any of the organisms in the class, as is the term mycoplasmas. The latter term is used in this chapter where convenient but is best used for organisms within the genus *Mycoplasma* only. Likewise, the terms ureaplasmas, spiroplasmas, and acholeplasmas refer in a trivial fashion to organisms within the genera *Ureaplasma*, *Spiroplasma*, and *Acholeplasma*, respectively.

DESCRIPTION OF MOLLICUTES

Mollicutes are the smallest free-living microorganisms. They are pleomorphic, varying in size from 0.2 to 0.3 μm, thus being about the size of poxviruses. They are contained only by a cell membrane, with no evidence of a cell wall; hence, they are not susceptible to β-lactams and do not take up Gram stain. Individual organisms, which assume various shapes from round to filamentous depending on the mycoplasmal species and the constituents of the broth medium, can be seen by dark-field and phase-contrast microscopy. These techniques have been used to demonstrate the motility of some mycoplasmal species. On agar media, mollicutes multiply to form colonies that vary in diameter from 15 to $\geq$300 μm; the largest colonies can be seen by the naked eye and often have a typical "fried-egg" appearance (Fig. 1) due to the contrast between central growth in the depth of the agar and peripheral shallow growth on the surface. In contrast, the smallest colonies, which are produced by the ureaplasmas, require low-power magnification to be seen.

Because of their small genomes, mollicutes are fastidious in their growth requirements. They need preformed nucleic acid precursors (from yeast extract) and, except for the mesoplasmas and acholeplasmas, enriched media containing cholesterol (from serum supplement). Growth rates in broth media vary, with generation times from 1 h for some ureaplasmas to 6 h for *Mycoplasma pneumoniae*. The yield of organisms from broth is small (1 to 20 mg of protein per liter), and cultures show at most a faint haze rather than frank turbidity. Colonies on agar are detectable in 2 (M. *hominis*) to 20 or more (M. *pneumoniae*) days. Most mollicutes, including those of human origin, can be categorized by their abilities to ferment glucose, utilize arginine, or hydrolyze urea. Thus, of the human *Mycoplasma* species, M. *pneumoniae* and M. *genitalium*, for example, are glycolytic, slow-growing, aerobic species that show serologic cross-reactions; M. *fermentans* and M. *penetrans* utilize both arginine and glucose and grow best under anaerobic conditions; and the remaining species utilize arginine but not glucose. Ureaplasmas are distinguished by their unique ability to hydrolyze urea. *Acholeplasma laidlawii* is glycolytic and grows in media without serum.

NATURAL HABITATS

Mollicutes have been isolated from practically all mammalian and avian species in which they have been sought and from a variety of plants and insects (spiroplasmas) (47, 74, 80). In animals, they colonize mucous membranes and are therefore found in the ororespiratory and genitourinary tracts, often behaving as commensals. Mollicutes also infect the conjunctivae, the external ear, and the mammary glands of some animals (for example, goats), but there are no reports of human mammary gland involvement. Mollicutes tend to be host specific, but A. *laidlawii* is ubiquitous, being found in various animal species and also in soil and on plants (74). From their initial site of colonization, mollicutes may disseminate to various other body sites, particularly in immunodeficient patients (19, 39, 66).

In humans who are immunocompetent, some mycoplasmas are isolated almost entirely from a particular site, for example, M. *pneumoniae* from the ororespiratory tract. On the other hand, other species have a preference for one site but are found also in another (Table 2). Thus, M. *hominis* is found most frequently in the genitourinary tract and less often in the oropharynx, presumably because of transfer through orogenital contact.

TABLE 1 Classification and some distinguishing features of mycoplasmas (class *Mollicutes*)

Classification of class *Mollicutes*	Distinguishing features			
	Sterol required	Genome size (kbp)	Mol% G+C of DNA	Other
Order I: *Mycoplasmatales*				
Family I: *Mycoplasmataceae*				
Genus I: *Mycoplasma* (~100 species)	Yes	580–1,380	23–41	
Genus II: *Ureaplasma* (5 species; *U. urealyticum* has at least 14 serotypes)	Yes	730–1,160	27–30	Urea metabolized
Order II: *Entomoplasmatales*				
Family I: *Entomoplasmataceae*				
Genus I: *Entomoplasma* (5 species)	Yes	790–1,140	27–29	
Genus II: *Mesoplasma* (4 species)	No	870–1,100	27–30	Requires 0.04% Tween 80
Family II: *Spiroplasmataceae*				
Genus I: *Spiroplasma* (≥11 species, including *Spiroplasma citri*)	Yes	940–2,200	25–31	Helical structure
Order III: *Acholeplasmatales*				
Family I: *Acholeplasmataceae*				
Genus I: *Acholeplasma* (9 species)	No	1,500–1,690	27–36	
Order IV: *Anaeroplasmatales*				
Family I: *Anaeroplasmataceae*				Obligate anaerobes
Genus I: *Anaeroplasma* (4 species)	Yes	~1,600	29–33	
Genus II: *Asteroleplasma* (1 species)	No	~1,600	40	

CLINICAL SIGNIFICANCE

Ororespiratory Tract

Eleven mycoplasmas have been found in the oral cavity (Table 2) (72). M. salivarium is a common inhabitant of the gingival crevice, especially in persons with periodontal dis-ease, and is not observed in edentulous persons. Several others, including M. orale, M. buccale, M. faucium, and M. lipophilum, also appear to be constituents of the normal oral flora. As a result of using the PCR, M. fermentans has been detected recently in the throat (30) and has been associated strongly with adult respiratory distress syndrome with or

TABLE 2 Primary sites of colonization, metabolism, and pathogenicity of mollicutes of human origin

Species	Primary site of colonization		Metabolism of:		Pathogenicity
	Oropharynx	Genitourinary tract	Glucose	Arginine	
M. salivarium	+	−	−	+	−
M. orale	+	−	−	+	−
M. buccale	+	−	−	+	−
M. faucium	+	−	−	+	−
M. lipophilum	+	−	−	+	−
M. pneumoniae	+	−	+	−	+
M. hominis	+	+	−	+	+
M. genitalium	?+	+	+	−	+
M. fermentans	+	+	+	−	?
M. primatum	−	+	−	+	−
M. spermatophilum	−	+	−	+	−
M. pirum[a]	?	?	+	+	−
M. penetrans	−	+	+	+	?
U. urealyticum[b]	+	+	−	−	+
A. laidlawii	+	−	+	−	−
A. oculi	?	−	+	−	−

[a]Human origin debatable.
[b]Metabolizes urea.

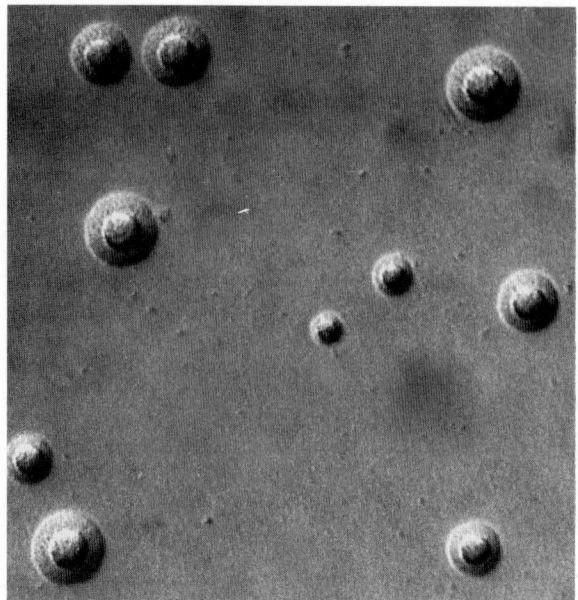

FIGURE 1 Colonies of M. hominis (human origin) up to 110 μm in diameter, showing the typical fried-egg appearance. Oblique illumination. Magnification, ×132.

without systemic disease (37). Historically, M. pneumoniae is known as a major cause of primary atypical pneumonia (8) and accounts for 15 to 20% of cases of X-ray-proven pneumonia (14). This organism may persist in the respiratory tract for several months after infection and sometimes for years in hypogammaglobulinemic patients (69). Subclinical infections are common, and the disease, called walking pneumonia, is ordinarily mild. Unlike some other respiratory infections, M. pneumoniae infections tend not to be seasonal. In large populations, disease is endemic year-round with periodic increases in incidence, but in smaller populations, outbreaks appear as epidemics (8, 14). Infection spreads slowly in families, transmission apparently requiring close contact.

Genitourinary Tract

Of the seven mycoplasmas found in the genitourinary tract (Table 2), M. hominis and U. urealyticum have been the subjects of most studies, although M. genitalium is now assuming importance. Difficulty in accepting M. hominis and U. urealyticum as causes of disease has arisen either because samples cannot be obtained easily from the affected site (for example, the fallopian tube) or because the organisms are recovered from asymptomatic individuals. Nevertheless, there is evidence that these species play etiologic roles in some genital tract diseases of both men and women. However, the organisms reach the upper tract and cause disease only in a small subpopulation of women who are colonized in the lower tract. Furthermore, ureaplasmas and M. hominis are recognized increasingly as causes of extragenital disease in immunocompromised patients (19, 39, 66) and in newborn infants, particularly those who are born preterm (76).

The results of human and animal inoculation studies and observations of immunocompromised patients are supportive of ureaplasmas being a cause of nonchlamydial, nongonococcal urethritis (NGU) in men, and antibiotic

and serologic studies provide further evidence (59, 65). There is no evidence to indicate that M. hominis is a cause of male urethritis (59). On the other hand, M. genitalium has been detected by PCR significantly more often in urethral specimens from men with acute NGU (about 20%) than in those without urethritis (about 6%) (25, 28). Antibody responses have been detected in some men with acute disease, and this mycoplasma has produced urethritis in subhuman primates. Also, evidence for persistence, obtained by DNA probe (24) and PCR, has been seen in men with chronic disease. In women, there is no evidence that M. hominis is a cause of the urethral syndrome, but U. urealyticum might be involved (56).

M. hominis and ureaplasmas have not been detected by culture of prostatic biopsy samples from patients with chronic abacterial prostatitis (11), nor has M. genitalium been found by means of the PCR technique. In contrast, ureaplasmas have been recovered from an epididymal aspirate of a patient suffering from nonchlamydial, nongonococcal acute epididymo-orchitis accompanied by a specific antibody response (27) and may be an infrequent cause of this disease.

U. urealyticum produces urease and induces crystallization of struvite and calcium phosphates in urine in vitro (21) and calculi in animal models (70). In addition, ureaplasmas have been found in the urine and calculi of patients with the infection-type stone more frequently than in those of patients with the metabolic-type stone, suggesting a possible causal association (21). M. hominis has been isolated from the upper urinary tract only in patients with symptoms of acute pyelonephritis, often with an antibody response (71), and causes possibly about 5% of cases of such disease. Obstruction or instrumentation of the urinary tract may be predisposing factors. U. urealyticum has not been implicated in the same way.

Female Reproductive Tract and Neonates

Mollicutes do not cause vaginitis, although M. hominis might do so rarely before puberty. Mollicutes are among various microorganisms that proliferate in patients with bacterial vaginosis (BV) and may contribute to the condition (68). Thus, M. hominis organisms occur in large numbers ($\geq 10^5$) in many cases, in contrast to the smaller numbers found in only about 10% of healthy women. U. urealyticum behaves similarly except that the number of organisms is usually smaller.

BV, and, hence, the microorganisms associated with it, including mollicutes, may in some women lead to pelvic inflammatory disease. M. hominis has been isolated from the endometrium and fallopian tubes of about 10% of women with salpingitis diagnosed by laparoscopy, and specific antibody responses have been recorded (68). However, the significance of this mycoplasma is difficult to judge in an individual case when several microorganisms are present. U. urealyticum has been isolated directly but rarely from affected fallopian tubes, but not alone. This, together with the negative results of serologic tests and of inoculating subhuman primates and fallopian tube organ cultures (68), does not support a causal role. In the case of M. genitalium, however, serologic data (41) and the results of subhuman primate inoculation (68) suggest that this organism may sometimes have such a role.

It is unknown whether mollicutes cause infertility in women. That U. urealyticum might cause involuntary infertility, a notion first raised about 25 years ago, remains speculative (60), as does the possibility that ureaplasmas could affect sperm (20, 60).

Ureaplasmas have been isolated from spontaneously aborted fetuses, sometimes from internal organs, and from stillborn and premature infants more often than from induced abortions or normal full-term infants (67). The results of some serologic and therapeutic (20) studies have also supported a role for these organisms in fetal morbidity. However, whether the fetus dies because of invasion by these organisms or whether it does so for some other reason and is then invaded remains unknown, as does the truth about whether ureaplasmas act alone or together with other organisms involved in BV. BV is also a possible confounding factor that needs to be taken into account in the weak association seen between ureaplasmas in the chorioamnion and low birth weight. Ureaplasmas at this site have been associated with inflammation (13), but proof that they alone cause chorioamnionitis and that their presence might lead to impaired fetal nutrition and reduced weight has not been forthcoming.

The notion that M. *hominis* organisms cause fever in some women after abortion or after a normal delivery is based on their isolation, presumably in pure culture, from the blood of about 10% of such women but not from afebrile women who have abortions or from healthy pregnant women (20, 67). In addition, antibody responses have been detected in about half of febrile aborting women but in few of those who remain afebrile (20, 67). Similar observations have been made for the isolation of U. *urealyticum* (12).

In the newborn, M. *hominis* and U. *urealyticum* acquired in utero or during birth have occasionally been reported to cause respiratory disease (44, 75). Thus, chronic lung disease and even death have occurred in very low birth weight infants (<1,000 g) (6) infected with ureaplasmas. However, the extent to which these problems are a reflection of concomitant BV in the mother is not clear. M. *hominis* and U. *urealyticum* invade the cerebrospinal fluid occasionally, particularly that of premature infants within the first few days of life (76). Either mild subclinical meningitis without sequelae or neurologic damage with permanent handicaps may ensue.

Other Body Sites and Immunocompromised Hosts

Extrapulmonary and extragenital mycoplasmal infections occur probably more often than is recognized because the organisms are not sought routinely. Most infections with M. *hominis* have been discovered fortuitously, because it grows, although not optimally, on blood agar and in routine blood cultures. Apart from postabortal and postpartum fever, septicemia due to M. *hominis* has been demonstrated (39) after renal transplantation, trauma, and genitourinary manipulations, and this mycoplasma has also been found in wound infections, brain abscesses, and osteomyelitis lesions.

A small proportion of individuals with hypogammaglobulinemia develop suppurative arthritis, and mycoplasmas and ureaplasmas, particularly those in the genitourinary tract, are responsible for many of these infections (18, 39, 66). In some patients, the arthritis responds to antibiotic therapy, whereas in others, disease and organisms persist for many months despite antibiotic and anti-inflammatory treatment and gamma globulin replacement. In some of the cases involving ureaplasmas, the arthritis has been associated with subcutaneous abscesses, persistent urethritis, and chronic urethrocystitis or cystitis (65, 66).

Septicemia by M. *hominis* may occur after organ transplantation and in patients on immunosuppressive therapy. Sternal wound infections caused by M. *hominis* in heart-lung transplant patients appear particularly common. Furthermore, polyarthritis with recovery of both M. *hominis* and ureaplasmas has been seen in a kidney allograft patient on an immunosuppressive regimen (39). The significance of M. *fermentans* (30), M. *penetrans* (77, 78), and other mollicutes in human immunodeficiency virus-positive patients with or without AIDS is currently a matter of debate.

TYPE, COLLECTION, TRANSPORT, AND STORAGE OF SPECIMENS

The considerations that apply to specimens from the genitourinary tract also apply to specimens from the ororespiratory tract.

Type and Collection

Well-taken throat swabs are at least as suitable as sputum samples for detecting M. *pneumoniae* and other respiratory mycoplasmas because of the adherence of the organisms to cells. The same applies to the recovery of mycoplasmas and ureaplasmas from the male genitourinary tract, where urethral swabbing is the usual approach, although a urine specimen may be preferred because of its noninvasiveness. Quantitative information can be provided by examination of urine (61), the numbers of organisms detected being usually 10-fold less than those detected by swabbing. Furthermore, examination of specimens taken at different stages during urination or after prostatic massage is a means of determining the location of mycoplasmas within the genitourinary tract, something that cannot be achieved by swabbing.

Swab and other specimens should be taken without contact with antiseptics, analgesics, or lubricants (sometimes used in obstetrical and gynecologic practice), since some of these substances inhibit the growth of mycoplasmas. Swabs from any site should be expressed immediately in mycoplasmal or other transport medium and not be allowed to dry. In other words, taking the medium to the patient is a sound policy. The swab should not be broken off into the medium because of possible inhibitors in the swab stick but should instead be agitated in the medium, expressed against the side of the container, and then discarded. Other samples should also be inoculated into medium soon after collection. This is important, particularly if the quantity of specimen is small (for example, prostatic fluid), in order to prevent the specimen from drying. Likewise, blood samples should be processed soon after taking, since the recovery of genital mycoplasmas declines appreciably over 1 h; 2 ml of heparinized blood in 18 ml of blood culture medium or mycoplasmal growth medium is an appropriate volume.

Transport and Storage

The following media may be used for transport and storage: (i) growth medium (see below) with or without the substrates metabolized by mollicutes; (ii) nutrient broth enriched with, for example, 10 to 20% horse serum; and (iii) sucrose-phosphate transport medium (designated 2SP) containing 10% heat-inactivated (56°C for 30 min) fetal calf serum without antibiotics. The last is primarily for the transport of chlamydiae but is useful when chlamydiae and mollicutes are being sought because both may be recovered from the medium (55), obviating the need for two urethral swabs. Transport medium is also available commercially.

If transport medium is not available or if the specimen comprises a large volume of urine, the specimen should be taken to the laboratory as rapidly as possible; if transportation takes several hours, the specimen should be kept at 4°C. Once mycoplasmal medium has been inoculated, it should be transported to the laboratory as soon as possible, preferably within 24 h, and the medium containing the organisms should be kept at 4°C in the meantime. If transportation cannot be undertaken within a few days, the medium containing the specimen should be frozen to −70°C or placed in liquid nitrogen and transported in the frozen state. Concern over rapidity of transport is greater if specimens are to be cultured than if they are to be examined by noncultural methods.

Specimens that are received frozen may be stored, if necessary, for long periods at −70°C (17) or in liquid nitrogen. When the specimens are to be examined, they should be thawed rapidly in a 37°C water bath. Specimens that have never been frozen are preferable, because there is always some loss of viability through freezing. If immediate examination is not possible, the specimens should be kept at 4°C for no longer than 72 h. If it is foreseeable that they cannot be tested within this time, they should be stored at −70°C or in liquid nitrogen but not at −20°C.

DIRECT DETECTION

The type of specimen (swab, urine, etc.) to be tested by noncultural procedures may be the same as for cultural methods, but transportation does not require maintenance of viability. Direct detection procedures have a particular place in detecting mollicutes that are otherwise almost impossible to detect (for example, M. genitalium) or for which culture is slow (for example, M. pneumoniae). The disadvantages are the difficulty of quantification and the inability to undertake further characterization or antibiotic susceptibility testing.

Antigen Detection Techniques

Antigen detection techniques have been applied mostly in a search for M. pneumoniae. Thus, antigen in respiratory exudates has been detected by direct immunofluorescence, counterimmunoelectrophoresis, immunoblotting with monoclonal antibodies, and several enzyme immunoassays (EIAs) (61). The methods have the virtue of speed, but sensitivity seems to be a problem. Thus, in one EIA (33), 10^4 CFU of M. pneumoniae was needed per ml of sample for a positive result. As a consequence, the methods have not found general acceptance.

DNA Probes and PCR

DNA probes have been insufficiently sensitive to foster widespread use. This was seen with probes for M. pneumoniae (22), U. urealyticum, and M. hominis. With the last two, only 63 and 56%, respectively, of culture-positive male urethral specimens were probe positive (51). Despite 30% of urethral specimens from homosexual men proving positive for M. genitalium with a nick-translated whole-genome DNA probe (24), which detected as little as 50 to 100 pg of DNA per blot, the general problem of insensitivity has seen a move toward the use of the PCR technique, in which there is massive amplification of DNA.

DNA primers specific for M. pneumoniae (3), M. genitalium (29, 43), M. fermentans (79), and U. urealyticum (4)

have been developed and used for DNA amplification by the PCR. The method has proved to be rapid, sensitive, specific, and clinically useful. This is so for M. pneumoniae (10, 38), a sensible strategy being to continue cultural isolation attempts only for those specimens that prove to be PCR positive. In the cases of M. genitalium (10, 25, 28) and M. fermentans (30, 37), the much greater sensitivity of the PCR compared to that of culture has shown for the first time that M. genitalium exists primarily in the genitourinary tract and M. fermentans exists primarily in the throat.

ISOLATION PROCEDURES

Growth Media and Inoculation

A medium used widely for the isolation of mollicutes comprises beef heart infusion broth, available commercially as PPLO broth, supplemented with 10% (vol/vol) fresh yeast extract (25% [wt/vol]) and 20% (vol/vol) horse serum (15). Other sera, such as fetal calf serum, may improve the medium. Furthermore, media of other formulations have been used satisfactorily for the isolation of both mycoplasmas and ureaplasmas (52). Thus, SP4 medium improves the isolation not only of the more fastidious mollicutes, such as M. pneumoniae (73), but also of the more easily isolatable ones, such as M. hominis. While medium formulations are important, the quality of the components may be even more so. The development of a successful medium is through trial and error, and components need to be pretested for their abilities to support growth. Quality control with a fastidious isolate is important.

Penicillin (1,000 IU/ml) or a synthetic penicillin having a wider antibacterial spectrum is usually added to all media. Thallium acetate and methylene blue are not recommended, because they are partially inhibitory for some mycoplasmas.

Inoculation of specimens into liquid medium, which is diluted serially, followed by subculture to liquid or agar media provides the most sensitive method for the isolation of most mollicutes and certainly of U. urealyticum (61). Color changes in liquid medium due to the metabolic activities of mollicutes may indicate growth when colonies fail to develop on agar. The clinical material (for example, expressed swab, urine, or urine deposit) is diluted 10-fold (for example, 0.2 ml of specimen or swab agitated in 1.8 ml of medium). The medium is supplemented with phenol red (0.002%) and glucose (0.1%) contained in a screw-cap vial of 2.5-ml capacity. Further serial 10-fold dilutions can be made as deemed necessary (see below). Likewise, the specimen is diluted in medium containing arginine (0.1%) and, again, in medium containing urea (0.1%). The caps of the vials are tightened, and the vials are incubated at 37°C. M. pneumoniae and some of the other mollicutes (Table 2) metabolize glucose to lactic acid by the glycolytic pathway, reducing the pH of the medium, set initially at 7.5 to 7.8, to less than 7.0 and thus changing the color from pink to yellow. M. hominis and several other mycoplasmas metabolize arginine to ammonia via ornithine by a three-enzyme system and raise the pH of the medium, set initially at 7.0, so that the color changes from yellow to pink. Ureaplasmas multiply best at a pH of 6.0 or less, and since they possess a urease that breaks down urea to ammonia, a similar change in the color of the medium occurs (61).

Dilution of specimens, as described above, is valuable for various reasons that have been detailed elsewhere (61). Among these reasons is the dilution of inhibitors and contaminating bacteria. Diluting the specimen up to 10^{-3} is useful in this regard. Further serial dilution, that is, titration, is helpful in estimating the number of organisms in the original specimen; the highest dilution at which a color change is seen may be regarded as containing one color-changing unit, assuming that there is even distribution and little aggregation of organisms. Thus, a change up to the 10^{-4} dilution suggests that the original specimen contained 10^4 color-changing units (or 10^4 organisms).

Incubation Period and Subculture

The speed of a color change and therefore the length of incubation are influenced at least partially by the number of organisms in the original specimen. M. *pneumoniae* may take several weeks to produce a change, and M. *genitalium* may take months, if it produces a change at all. On the other hand, M. *hominis* and some of the other arginine-metabolizing mycoplasmas produce a change much more rapidly, usually well within a week, and the ureaplasmas do so usually within 24 to 48 h or less and rarely thereafter.

Subculturing is performed by taking aliquots of medium (0.1 or 0.2 ml) from the cultures that are just changing color or from those that have changed most recently and introducing the aliquots into fresh broth medium and/or onto agar medium. In glucose-containing medium at pH 7.5 to 7.8, minced tissue, cellular deposits, and particularly blood produce a yellow color that may mask any specific change. This yellow color demands routine subculture, but otherwise it is usually not profitable to undertake a "blind" passage if color changes have not occurred initially.

Development of Colonies

On agar, colonies of M. *pneumoniae*, usually without a fried-egg appearance, develop best in an atmosphere of air–5% CO_2, although the fried-egg appearance may develop on subculture. Colonies of the genital mycoplasmas develop best in an atmosphere of 95% N_2 plus 5% CO_2. A strict anaerobic atmosphere with medium that has been prereduced is, however, not conducive to the development of such colonies. Many assume a diameter of about 200 to 300 μm and have the classic fried-egg appearance. Colonies of M. *genitalium* are often much smaller, and many do not have the typical appearance. U. *urealyticum* organisms were originally termed T strains or T mycoplasmas (T for tiny) (52) because of the very small colonies they produce (15 to 30 μm in diameter). Usually, these colonies do not have a fried-egg morphology because they lack the peripheral surface growth. As an aid to detecting these colonies, an aliquot of urea-containing broth medium that has just changed color should be subcultured to agar medium containing urea, 0.05 M HEPES (*N*-2-hydroxyethylpiperazine-*N'*-2-ethanesulfonic acid) buffer, and a sensitive indicator of ammonia, that is, manganous sulfate (52) or calcium chloride (61). On such medium, ureaplasmas form dark brown colonies that are slightly larger than usual and more easily recognizable. It should be emphasized, however, that apart from the colonial features of the ureaplasmas, the colonial features of any of the other mollicutes are alone insufficient to allow specific identification. Furthermore, the colonies of some bacteria on mycoplasmal agar mimic those of mollicutes.

On conventional blood agar, M. *hominis* organisms but not ureaplasmas sometimes produce nonhemolytic pinpoint colonies. Also, M. *hominis* multiplies in most routine blood culture media without changing the appearances of the media, and the occurrence of minute colonies on blood agar

following blind subculture may be the first hint of bloodstream invasion for specimens tested in a routine laboratory. The mycoplasma-inhibiting effect of sodium polyanethol sulfonate, included in blood culture media as an anticoagulant, can be overcome by the addition of gelatin (1% [wt/vol]) (45).

Commercial Culture Kits

Kits designed to isolate and identify M. *hominis* and U. *urealyticum* are available commercially. Successful use of these kits has been reported (48), and they may be of particular value where the need to detect these microorganisms arises infrequently. Specialist laboratories are likely to have developed their own media of superior performance.

IDENTIFICATION

Mollicutes are distinguishable from conventional bacteria by the former's lack of reaction with Gram stain and lack of a thick cell wall observable by electron microscopy. These features apart, colony morphology may indicate the existence of a mollicute, and nonspecific tests followed by more specific tests will lead to precise identification.

Various biochemical reactions are common to too many different mycoplasmal species to be helpful in making a diagnosis. However, certain biologic features determined routinely during the course of isolation provide some indication of species identification. Thus, a specimen from the respiratory tract that rather slowly acidifies medium containing glucose (thus turning it yellow) and produces colonies that hemadsorb is most likely to contain M. *pneumoniae*. Such a color change produced more rapidly and colonies that are nonhemadsorbing suggest M. *fermentans*, although negative hemadsorption should be interpreted with caution. An alkaline color change without turbidity in medium containing urea is almost certainly due to U. *urealyticum*, and a specimen from the genital tract that produces a similar change in medium containing arginine is most likely to contain M. *hominis*.

The preceding observations indicate the specific antisera that are required to provide definitive identification when used in one or more of several available tests (Table 3). These need to be specific although not necessarily very sensitive, since high-titer antisera are usually employed. Inhibition of colony development around a filter paper disk containing specific antiserum (agar growth inhibition) (7) is a widely used method; more than one antiserum against a particular species may be required to encompass various strains of that species (34, 35). The difficulty sometimes encountered in identification because of the close serologic relationship between species is illustrated by that between M. *genitalium* and M. *pneumoniae* (36). Certainty in identifying either of these species may require several techniques, more than one antiserum (34), and perhaps the use of monoclonal antibodies allied to Western blotting (immunoblotting) (42). Agar growth inhibition has also been used to identify the serovars of U. *urealyticum* (53), although metabolism inhibition (MI) and complement-dependent mycoplasmacidal tests have been employed more often (57). However, epi-immunofluorescence or immunoperoxidase techniques (57) enable colonies on agar to be identified directly, so that mixtures of different ureaplasmal serovars or, indeed, mycoplasmal species can be detected.

TABLE 3 Relative sensitivities and specificities of serologic tests for mollicutes[a]

Serologic technique	Sensitivity	Specificity
Radioimmunoprecipitation	++++	+++?
Complement-dependent bactericidal	+++±	+++?
EIA	+++	+++
Metabolism inhibition	+++	+++
Inhibition of growth in liquid medium	+++	+++
Indirect hemagglutination	+++	++
Micro-IF	+++	+++
Inhibition of cytopathic effect in tissue culture	++	+++
IF (colonies on agar)	++	+++
Single radial hemolysis	++	+++
CF	++	+
Cumulative hemagglutination inhibition	++	+?
Agglutination	++	+
Latex agglutination	++	+?
Hemadsorption inhibition	+	+++
Colony inhibition on agar (disk method)	+	+++
Immunodiffusion and immunoelectrophoresis	+	++

[a]+, ++, +++, and ++++ indicate poor, moderate, good, and excellent, respectively.

SEROLOGIC TESTS

The various tests that have been used to study mycoplasmas are listed in Table 3 in decreasing order of sensitivity. Sensitive tests are not required for identification of isolates, as mentioned previously, but they are needed for the quantitative measurement of antibody in seroepidemiology and in the serologic diagnosis of infection in vivo. Those used most often in diagnosis are discussed below.

Respiratory Infections

Since culture of M. pneumoniae is slow and not always successful, diagnosis of M. pneumoniae infections has usually relied on serology. Cold agglutinins, detected by agglutination of type O Rh-negative erythrocytes at 4°C, develop in about half of the patients. These agglutinins are induced occasionally by a number of other conditions, but a fourfold or greater rise in titer or a titer of 64 to 128 or more in a single serum suggests a recent M. pneumoniae infection, assuming that the serologically related M. genitalium is not able to stimulate cold agglutinins or does so rarely. Of the various more specific methods, complement fixation (CF) has been used most often, the chloroform-methanol-extractable lipid antigen being advocated (31). A fourfold or greater rise in titer, mainly to immunoglobulin M (IgM), with a peak at about 3 to 4 weeks, occurs in about 80% of cases and has been regarded as indicating a recent infection. Also, a single antibody titer of 64 to 128 or more in a suggestive clinical setting should be regarded as sufficient reason to institute therapy. However, in the older patient, a negative CF test does not exclude M. pneumoniae disease (26). A fourfold or greater M. pneumoniae antibody response or a titer of 64 to 128 or more in an indirect hemagglutination test, for which kits are available commercially, is diagnostically suggestive in an appropriate clinical setting. A modification of the test (38), an IgM antibody capture assay (9), and a commercially available particle

agglutination test (1) have all been tried. The sensitivities of these tests may be a little greater than that noted with the CF test, but the specificities of all tests, including the CF test, are a problem (26) that is highlighted by the antigenic cross-reactivity between M. pneumoniae and M. genitalium (36). More specific, perhaps, is the microimmunofluorescence (micro-IF) test in which IgM antibody is sought; the presence of this antibody provides some confidence in the accuracy of the diagnosis of a current or recent infection, but sensitivity is less than that seen with the CF test (54). There seems to be no sense in advocating a change from one test to another in search of greater specificity. A sensible and realistic approach to serodiagnosis seems to be continued use of a familiar test, for example, CF, with Western blotting used for a check on specificity in dubious cases (31). In the future, it is possible that a sensitive enzyme-linked immunosorbent assay (ELISA) (31) in which purified P1 adhesin protein provides greater specificity (26) will find a place.

Genital Infections

No single serologic test has gained favor for identifying genital infections. The relative lack of both sensitivity and specificity makes the CF test unsuitable. The indirect hemagglutination technique is too complex, difficult to reproduce, and not entirely specific for widespread use. MI tests in which inhibition of mycoplasmal growth by antibody results in lack of metabolic activity, which is detectable in various ways (58), have been used to detect, for example, antibody responses to ureaplasmas in various clinical settings. Persons tested include male subjects after experimental intraurethral inoculation of ureaplasmas, male patients with NGU (62) and epididymitis (27), infants, and mothers with spontaneous pregnancy loss (46).

EIAs or ELISAs have been used to study antibody responses to genital mollicutes. For example, in the case of U. urealyticum, a cell lysate antigen and commercially available alkaline phosphatase conjugates were used to detect ureaplasmal responses in patients with NGU (5), and a modified EIA has been used to measure antibodies to M. hominis in women with acute salpingitis (40). None of the assays, however, has found routine clinical application.

More recently, an indirect micro-IF test was used to measure antibody to M. genitalium (16), with responses being detected in men with NGU (61) and in women with salpingitis (41). The test is rapid, reproducible, and quite sensitive and specific, there being less cross-reactivity with M. pneumoniae than is seen with some other methods.

ANTIBIOTIC SUSCEPTIBILITIES

Methods Used

The antibiotic susceptibilities of mollicutes may be determined in vitro by two methods: the agar dilution method (50), in which the end point is the prevention of colony development on agar, and an MI method (50, 63), in which the antibiotic in broth medium inhibits growth of the organisms and hence their metabolism. The MI method is a simple modification of the method used for measuring antibody, with the antibody replaced by antibiotic. The MI method is preferable for many investigators, particularly when ureaplasmas are being tested, since color changes are easier to demonstrate than colony development. The end point (initial or minimal inhibitory concentration [MIC] in the MI method is the highest dilution of antibiotic that

TABLE 4 Susceptibilities of M. *pneumoniae*, M. *hominis*, M. *fermentans*, and U. *urealyticum* to various antibiotics[a]

Antibiotic(s)	M. pneumoniae	M. hominis	M. fermentans	U. urealyticum
Tetracyclines	+	+	+	+
Erythromycin	+	−	−	+
Clarithromycin	+	−	?	+
Azithromycin	+	−	?	±
Streptomycin	+	−	−	±
Spectinomycin	+	±	±	±
Gentamicin	±	±	−	±
Chloramphenicol	±	±	±	±
Clindamycin	±	+	+	−
Lincomycin	±	+	+	−
Ciprofloxacin	±	+	+	±
Difloxacin	±	+	?	±
Nalidixic acid	−	−	−	−
Cephalosporins	−	−	−	−
Penicillins	−	−	−	−
Rifampin	−	−	−	−

[a]+, susceptible (MICs, <1 μg/ml); ±, partially susceptible (MICs, 1 to 10 μg/ml); −, resistant (MICs, >10 μg/ml).

inhibits the color change when the change in the control without antibiotic has just developed (50, 63). Apart, perhaps, from the quinolones, antibiotics active against mollicutes tend not to be bactericidal, at least in concentrations that can be achieved in vivo. Lack of killing may be seen by a creeping increase in the MIC on continued incubation of the MI test; in effect, a final inhibitory concentration could be recorded after the initial reading (63), and differences in results from one laboratory to another probably reflect readings taken at various points within the range. Particular problems may be experienced in testing the susceptibility of ureaplasmas to erythromycin (32). Bactericidal activity is sought by removing the mixture of organisms and antibiotic, at whatever concentration it is considered inhibitory, and (i) determining whether there is growth in medium when the antibiotic is diluted beyond its inhibitory concentration (63) or (ii) passing the mixture through a 0.2-μm-pore-size filter, washing the filter, and then testing it by culture for viable organisms (63).

Susceptibility Profiles

The susceptibilities of M. *pneumoniae*, M. *hominis*, M. *fermentans*, and U. *urealyticum* to a range of antibiotics are shown in Table 4. The concise representation hides the fact that the susceptibilities depicted are drawn from numerous studies in which a wide range of MICs is seen for any particular antibiotic (50). Some investigators may find that the MIC of a particular antibiotic does not fall accurately within the category presented (Table 4, footnote a), but overall, the representation of the antibiotic susceptibility profiles is likely to be correct. It is noteworthy that antibiotics other than tetracyclines and erythromycin, particularly some of the newer macrolides, are active against M. *pneumoniae* (2). In contrast, M. *hominis* is not susceptible to erythromycin and some of the other macrolides but is susceptible to clindamycin and lincomycin, whereas the reverse is true for U. *urealyticum*.

The antibiotic susceptibility profiles of other mollicutes of human origin are not available in such detail. However, M. *genitalium* is similar to M. *pneumoniae* in susceptibility, being susceptible to the tetracyclines and highly susceptible

to a range of macrolides and streptogramins (49); M. *fermentans*, like M. *hominis*, is resistant to erythromycin (23).

Antibiotic Resistance

Mollicutes are innately resistant to some antibiotics (50). For example, resistance of all mollicutes to penicillins and cephalosporins is due to the absence of target antibiotic-binding proteins through lack of a cell wall. Innate resistance to erythromycin is seen in M. *hominis* and in some but not all other mollicutes. Resistance may develop through chromosomal mutations (notably to streptomycin) or through acquisition of antibiotic resistance transposons, an example of which is the *tet*M gene (50). *tet*M codes for a protein that binds to the ribosomes, protecting them from the actions of the tetracyclines. Tetracycline resistance mediated in this way has been noted particularly with M. *hominis* and U. *urealyticum*. At least 10% of ureaplasma strains isolated from patients attending sexually transmitted disease clinics are resistant (64); a few of these strains are also resistant to erythromycin (64).

Resistance of mollicutes to multiple antibiotics has been seen to develop in hypogammaglobulinemic patients (64), and eradication of these mollicutes, even of those that appear susceptible in in vitro tests, may be extremely difficult. This difficulty highlights the facts that mollicutes are inhibited but not killed by antibiotics in concentrations achievable in vivo and that a functioning immune system plays an integral part in elimination.

EVALUATION, INTERPRETATION, AND REPORTING OF RESULTS

The more important mollicutes are considered here, and reference to the preceding text is recommended.

M. pneumoniae

Detection of M. *pneumoniae* by culture is slow, but isolation of the organisms is significant clinically. Detection by PCR is fast and more sensitive and is gaining acceptance. A fourfold rise in antibody titer or a titer of 64 to 128 in a single serum in the CF test is suggestive, but confirmation by Western blotting is required because of cross-reactivity with *M. genitalium*. The development of specific ELISAs should overcome this problem.

M. hominis

Detection of *M. hominis* by culture is relatively easy and rapid (<1 week). If the test is not practiced routinely, commercial kits may be helpful. Isolation from joints, blood, and the upper genital tract is associated significantly with disease; significance is attributed only to ≥10^5 organisms in the female lower genital tract (a correlation with BV). The most useful serologic tests are MI, micro-IF, and ELISA.

U. urealyticum

Detection of *U. urealyticum* by culture is relatively easy and very rapid (1 to 2 days). Comments about commercial tests and the significance of detection are the same as for *M. hominis*. Detection of <10^4 organisms in the male urethra is unlikely to be significant. The most useful serologic tests are MI and ELISA.

M. genitalium and *M. fermentans*

The PCR is required to detect *M. genitalium* and is helpful in detecting *M. fermentans*. Detection of *M. genitalium*, without quantitation, is likely to be significant. Serologically, micro-IF is useful for detecting antibody responses to *M. genitalium*, because cross-reactivity with *M. pneumoniae* is less when tested by this method than when tested by other procedures.

REFERENCES

1. **Barker, C. E., M. Sillis, and T. G. Wreghitt.** 1990. Evaluation of Serodia Myco II particle agglutination test for detecting *Mycoplasma pneumoniae* antibody: comparison with μ-capture ELISA and indirect immunofluorescence. *J. Clin. Pathol.* **43:**163–165.
2. **Bebear, C., M. Dupon, H. Renaudin, and B. de Barbeyrac.** 1993. Potential improvements in therapeutic options for mycoplasmal respiratory infections. *Clin. Infect. Dis.* **17**(Suppl. 1):202–207.
3. **Bernet, C., M. Garret, B. de Barbeyrac, C. Bebear, and J. Bonnet.** 1989. Detection of *Mycoplasma pneumoniae* by using the polymerase chain reaction. *J. Clin. Microbiol.* **27:**2492–2496.
4. **Blanchard, A., J. Hentschel, L. Duffy, K. Baldus, and G. H. Cassell.** 1993. Detection of *Ureaplasma urealyticum* by polymerase chain reaction in the urogenital tract of adults, in amniotic fluid, and in the respiratory tract of newborns. *Clin. Infect. Dis.* **17**(Suppl. 1):148–153.
5. **Brown, M. B., G. H. Cassell, D. Taylor-Robinson, and M. C. Shepard.** 1983. Measurement of antibody to *Ureaplasma urealyticum* by an enzyme-linked immunosorbent assay and detection of antibody responses in patients with nongonococcal urethritis. *J. Clin. Microbiol.* **17:**288–295.
6. **Cassell, G. H., K. B. Waites, D. T. Crouse, P. T. Rudd, K. C. Canupp, S. Stagno, and G. R. Cutter.** 1988. Association of *Ureaplasma urealyticum* infection of the lower respiratory tract with chronic lung disease and death in very-low-birth-weight infants. *Lancet* **ii:**240–244.
7. **Clyde, W. A.** 1964. Mycoplasma species identification based upon growth inhibition by specific antisera. *J. Immunol.* **92:**958–965.
8. **Clyde, W. A.** 1993. Clinical overview of typical *Mycoplasma pneumoniae* infections. *Clin. Infect. Dis.* **17**(Suppl. 1):32–36.
9. **Coombs, R. R. A., G. Easter, P. Matejtschuk, and T. G. Wreghitt.** 1988. Red-cell IgM-antibody capture assay for the detection of *Mycoplasma pneumoniae*-specific IgM. *Epidemiol. Infect.* **100:**101–109.
10. **de Barbeyrac, B., C. Bernet-Poggi, F. Febrer, H. Renaudin, M. Dupon, and C. Bebear.** 1993. Detection of *Mycoplasma pneumoniae* and *Mycoplasma genitalium* in clinical samples by polymerase chain reaction. *Clin. Infect. Dis.* **17**(Suppl. 1):83–89.
11. **Doble, A., B. J. Thomas, P. M. Furr, M. M. Walker, J. R. W. Harris, R. O. Witherow, and D. Taylor-Robinson.** 1989. A search for infectious agents in chronic abacterial prostatitis utilising ultrasound guided biopsy. *Br. J. Urol.* **64:**297–301.
12. **Eschenbach, D. A.** 1986. *Ureaplasma urealyticum* as a cause of post-partum fever. *Pediatr. Infect. Dis.* **5**(Suppl):258–261.
13. **Eschenbach, D. A.** 1993. *Ureaplasma urealyticum* and premature birth. *Clin. Infect. Dis.* **17**(Suppl. 1):100–106.
14. **Foy, H. M.** 1993. Infections caused by *Mycoplasma pneumoniae* and possible carrier state in different populations of patients. *Clin. Infect. Dis.* **17**(Suppl. 1):37–46.
15. **Freundt, E. A.** 1983. Culture media for classic mycoplasmas, p. 127–135. *In* S. Razin and J. G. Tully (ed.), *Methods in Mycoplasmology*, vol. 1. Academic Press, Inc., New York.
16. **Furr, P. M., and D. Taylor-Robinson.** 1984. Microimmunofluorescence technique for detection of antibody to *Mycoplasma genitalium*. *J. Clin. Pathol.* **37:**1072–1074.
17. **Furr, P. M., and D. Taylor-Robinson.** 1990. Long-term viability of stored mycoplasmas and ureaplasmas. *J. Med. Microbiol.* **31:**203–206.
18. **Furr, P. M., D. Taylor-Robinson, and A. D. B. Webster.** 1994. Mycoplasmas and ureaplasmas in patients with hypogammaglobulinaemia and their role in arthritis: microbiological observations over 20 years. *Ann. Rheum. Dis.* **53:**183–187.
19. **Gelfand, E. W.** 1993. Unique susceptibility of patients with antibody deficiency to mycoplasma infection. *Clin. Infect. Dis.* **17**(Suppl. 1):250–253.
20. **Glatt, A. E., W. M. McCormack, and D. Taylor-Robinson.** 1989. Genital mycoplasmas, p. 279–293. *In* K. K. Holmes, P.-A. Mardh, P. F. Sparling, P. J. Wiesner, W. Cates, S. M. Lemon, and W. E. Stamm (ed.), *Sexually Transmitted Diseases*, 2nd ed. McGraw Hill Book Co., New York.
21. **Grenabo, L., H. Hedelin, and S. Pettersson.** 1988. Urinary infection stones caused by *Ureaplasma urealyticum*: a review. *Scand. J. Infect. Dis.* **53**(Suppl.):46–49.
22. **Harris, R., B. P. Marmion, G. Varkanis, T. Kok, B. Lunn, and J. Martin.** 1988. Laboratory diagnosis of *M. pneumoniae* infection. II. Comparison of methods for direct detection of specific antigens or nucleic acid sequences in respiratory exudates. *Epidemiol. Infect.* **101:**685–694.
23. **Hayes, M. M., D. J. Wear, and S.-C. Lo.** 1991. In vitro antimicrobial susceptibility testing for the newly identified AIDS-associated *Mycoplasma: Mycoplasma fermentans* (incognitus strain). *Arch. Pathol. Lab. Med.* **115:**464–466.
24. **Hooton, T. M., M. C. Roberts, P. L. Roberts, K. K. Holmes, W. E. Stamm, and G. R. Kenny.** 1988. Prevalence of *Mycoplasma genitalium* determined by DNA probe in men with urethritis. *Lancet* **i:**266–268.
25. **Horner, P. J., C. B. Gilroy, B. J. Thomas, R. O. M. Naidoo, and D. Taylor-Robinson.** 1993. Association of *Mycoplasma genitalium* with acute non-gonococcal urethritis. *Lancet* **342:**582–585.
26. **Jacobs, E.** 1993. Serological diagnosis of *Mycoplasma pneumoniae* infections: a critical review of current procedures. *Clin. Infect. Dis.* **17**(Suppl. 1):79–82.
27. **Jalil, N., A. Doble, C. Gilchrist, and D. Taylor-Robinson.** 1988. Infection of the epididymis by *Ureaplasma urealyticum*. *Genitourin. Med.* **64:**367–368.
28. **Jensen, J. S., R. Ørsum, B. Dohn, S. Uldum, A.-M. Worm, and K. Lind.** 1993. *Mycoplasma genitalium*: a cause of male urethritis? *Genitourin. Med.* **69:**265–269.
29. **Jensen, J. S., S. A. Uldum, J. Sondergard-Andersen, J. Vuust, and K. Lind.** 1991. Polymerase chain reaction for detection of *Mycoplasma genitalium* in clinical samples. *J. Clin. Microbiol.* **29:**46–50.
30. **Katseni, V. L., C. B. Gilroy, B. K. Ryait, K. Ariyoshi, P. D.**

Bieniasz, J. N. Weber, and D. Taylor-Robinson. 1993. *Mycoplasma fermentans* in individuals seropositive and seronegative for HIV-1. *Lancet* **341:**271–273.

31. Kenny, G. E. 1992. Serodiagnosis, p. 505–512. *In* J. Maniloff, R. N. McElhaney, L. R. Finch, and J. B. Baseman (ed.), *Mycoplasmas: Molecular Biology and Pathogenesis.* American Society for Microbiology, Washington, D.C.

32. Kenny, G. E., and F. D. Cartwright. 1993. Effect of pH, inoculum size, and incubation time on the susceptibility of *Ureaplasma urealyticum* to erythromycin *in vitro. Clin. Infect. Dis.* **17**(Suppl. 1):215–218.

33. Kok, T.-W., G. Varkanis, B. P. Marmion, J. Martin, and A. Esterman. 1988. Laboratory diagnosis of *Mycoplasma pneumoniae* infection. I. Direct detection of antigen in respiratory exudates by enzyme immunoassay. *Epidemiol. Infect.* **101:**669–684.

34. Leach, R. H., A. Hales, P. M. Furr, D. L. Michelmore, and D. Taylor-Robinson. 1987. Problems in the identification of *Mycoplasma pirum* isolated from human lymphoblastoid cell cultures. *FEMS Microbiol. Lett.* **44:**293–297.

35. Lin, J.-S. L., S. Alpert, and K. M. Radnay. 1975. Combined type-specific antisera in the identification of *Mycoplasma hominis. J. Infect. Dis.* **131:**727–730.

36. Lind, K. 1982. Serological cross-reaction between *Mycoplasma genitalium* and M. *pneumoniae. Lancet* **ii:**1158–1159.

37. Lo, S.-C., D. J. Wear, S. L. Green, P. G. Jones, and J. F. Legier. 1993. Adult respiratory distress syndrome with or without systemic disease associated with infections due to *Mycoplasma fermentans. Clin. Infect. Dis.* **17**(Suppl. 1):259–263.

38. Marmion, B. P., J. Williamson, D. A. Worswick, T.-W. Kok, and R. J. Harris. 1993. Experience with newer techniques for the laboratory detection of *Mycoplasma pneumoniae* infection: Adelaide, 1978–1992. *Clin. Infect. Dis.* **17**(Suppl. 1):90–99.

39. Meyer, R. D., and W. Clough. 1993. Extragenital *Mycoplasma hominis* infections in adults: emphasis on immunosuppression. *Clin. Infect. Dis.* **17**(Suppl. 1):243–249.

40. Miettinen, A., J. Paavonen, E. Jansson, and P. Leinikki. 1983. Enzyme immunoassay for serum antibody to *Mycoplasma hominis* in women with acute pelvic inflammatory disease. *Sex. Transm. Dis.* **10**(Suppl.):289–293.

41. Møller, B. R., D. Taylor-Robinson, and P. M. Furr. 1984. Serologic evidence implicating *Mycoplasma genitalium* in pelvic inflammatory disease. *Lancet* **i:**1102–1103.

42. Morrison-Plummer, J., D. H. Jones, K. Daly, J. G. Tully, D. Taylor-Robinson, and J. B. Baseman. 1987. Molecular characterization of *Mycoplasma genitalium* species-specific and cross-reactive determinants: identification of an immunodominant protein of M. *genitalium. Isr. J. Med. Sci.* **23:**453–457.

43. Palmer, H. M., C. B. Gilroy, P. M. Furr, and D. Taylor-Robinson. 1991. Development and evaluation of the polymerase chain reaction to detect *Mycoplasma genitalium. FEMS Microbiol. Lett.* **61:**199–203.

44. Payne, N. R., S. S. Steinberg, P. Ackerman, B. A. Chrenka, S. M. Sane, K. T. Anderson, and J. J. Fangman. 1993. New prospective studies of the association of *Ureaplasma urealyticum* colonization and chronic lung disease. *Clin. Infect. Dis.* **17**(Suppl. 1):117–121.

45. Pratt, B. 1990. Automated blood culture systems: detection of *Mycoplasma hominis* in SPS-containing media, p. 778–781. *In* G. Staneck, G. H. Cassell, J. G. Tully, and R. F. Whitcomb (ed.), *Recent Advances in Mycoplasmology.* Gustav Fischer Verlag, Stuttgart.

46. Quinn, P. A., A. B. Shewchuk, J. Shuber, K. I. Lie, E. Ryan, M. Sheu, and M. L. Chipman. 1983. Serologic evidence of *Ureaplasma urealyticum* infection in women with spontaneous pregnancy loss. *Am. J. Obstet. Gynecol.* **145:**245–250.

47. Razin, S., and M. F. Barile (ed.). 1985. *The Mycoplasmas,* vol. 4. Academic Press, Inc., New York.

48. Renaudin, H., and C. Bebear. 1990. Evaluation des systemes *Mycoplasma* PLUS et SIR *Mycoplasma* pour la detection quantitative et l'etude de la sensibilite aux antibiotiques des mycoplasmes genitaux. *Pathol. Biol.* **38:**431–435.

49. Renaudin, H., J. G. Tully, and C. Bebear. 1992. In vitro susceptibilities of *Mycoplasma genitalium* to antibiotics. *Antimicrob. Agents Chemother.* **36:**870–872.

50. Roberts, M. C. 1992. Antibiotic resistance, p. 513–523. *In* J. Maniloff, R. N. McElhaney, L. R. Finch, and J. B. Baseman (ed.), *Mycoplasmas: Molecular Biology and Pathogenesis.* American Society for Microbiology, Washington, D.C.

51. Roberts, M. C., M. Hooton, W. Stamm, K. K. Holmes, and G. E. Kenny. 1987. DNA probes for the detection of mycoplasmas in genital specimens. *Isr. J. Med. Sci.* **23:**618–620.

52. Shepard, M. C. 1983. Culture media for ureaplasmas, p. 137–146. *In* S. Razin and J. G. Tully (ed.), *Methods in Mycoplasmology,* vol. 1. Academic Press, Inc., New York.

53. Shepard, M. C., and C. D. Lunceford. 1978. Serological typing of *Ureaplasma urealyticum* isolates from urethritis patients by an agar growth inhibition method. *J. Clin. Microbiol.* **8:**566–574.

54. Sillis, M. 1990. The limitation of IgM assays in the serological diagnosis of *Mycoplasma pneumoniae* infection. *J. Med. Microbiol.* **33:**253–258.

55. Smith, T. F., L. A. Weed, G. R. Pettersen, and J. W. Segura. 1977. Recovery of chlamydia and genital mycoplasma transported in sucrose phosphate buffer and urease color test medium. *Health Lab. Sci.* **14:**30–34.

56. Stamm, W. E., K. Running, J. Hale, and K. K. Holmes. 1983. Etiologic role of *Mycoplasma hominis* and *Ureaplasma urealyticum* in women with the acute urethral syndrome. *Sex. Transm. Dis.* **10**(Suppl.):318–322.

57. Taylor-Robinson, D. 1983. Serological identification of ureaplasmas from humans, p. 57–63. *In* J. G. Tully and S. Razin (ed.), *Methods in Mycoplasmology,* vol. 2. Academic Press, Inc., New York.

58. Taylor-Robinson, D. 1983. Metabolism inhibition test, p. 411–417. *In* J. G. Tully and S. Razin (ed.), *Methods in Mycoplasmology,* vol. 1. Academic Press, Inc., New York.

59. Taylor-Robinson, D. 1985. Mycoplasmal and mixed infections of the human male urogenital tract and their possible complications, p. 27–63. *In* S. Razin and M. F. Barile (ed.), *The Mycoplasmas,* vol. 4. Academic Press, Inc., New York.

60. Taylor-Robinson, D. 1986. Evaluation of the role of *Ureaplasma urealyticum* in infertility. *Pediatr. Infect. Dis.* **5**(Suppl.):262–265.

61. Taylor-Robinson, D. 1989. Genital mycoplasma infections. *Clin. Lab. Med.* **9:**501–523.

62. Taylor-Robinson, D., and G. W. Csonka. 1981. Laboratory and clinical aspects of mycoplasmal infections of the human genitourinary tract, p. 151–186. *In* J. R. W. Harris (ed.), *Recent Advances in Sexually Transmitted Diseases,* no. 2. Churchill Livingstone, Edinburgh.

63. Taylor-Robinson, D., and P. M. Furr. 1982. The static effect of rosaramicin on *Ureaplasma urealyticum* and the development of antibiotic resistance. *J. Antimicrob. Chemother.* **10:**185–191.

64. Taylor-Robinson, D., and P. M. Furr. 1986. Clinical antibiotic resistance of *Ureaplasma urealyticum. Pediatr. Infect. Dis.* **5**(Suppl.):335–337.

65. Taylor-Robinson, D., P. M. Furr, and A. D. B. Webster. 1985. *Ureaplasma urealyticum* causing persistent urethritis in a patient with hypogammaglobulinaemia. *Genitourin. Med.* **61:**404–408.

66. Taylor-Robinson, D., P. M. Furr, and A. D. B. Webster. 1986. *Ureaplasma urealyticum* in the immunocompromised host. *Pediatr. Infect. Dis.* **5**(Suppl.):236–238.

67. Taylor-Robinson, D., and W. M. McCormack. 1979. Mycoplasmas in human genitourinary infections, p. 307–366. *In* J. G. Tully and R. F. Whitcomb (ed.), *The Mycoplasmas,* vol. 2. Academic Press, Inc., New York.

68. Taylor-Robinson, D., and P. E. Munday. 1988. Mycoplasmal infection of the female genital tract and its complications, p. 228–247. *In* M. J. Hare (ed.), *Genital Tract Infection in Women.* Churchill Livingstone, Edinburgh.

69. Taylor-Robinson, D., A. D. B. Webster, P. M. Furr, and G. L. Asherson. 1980. Prolonged persistence of *Mycoplasma pneumoniae* in a patient with hypogammaglobulinaemia. *J. Infect.* **2:**171–175.

70. Texier-Maugein, J., M. Clerc, A. Vekris, and C. Bebear. 1987. *Ureaplasma urealyticum*-induced bladder stones in rats and their prevention by flurofamide and doxycycline. *Isr. J. Med. Sci.* **23:**565–567.

71. Thomsen, A. C. 1978. Mycoplasmas in human pyelonephritis: demonstration of antibodies in serum and urine. *J. Clin. Microbiol.* **8:**197–202.

72. Tully, J. G. 1993. Current status of the mollicute flora of humans. *Clin. Infect. Dis.* **17**(Suppl. 1)**:**2–9.

73. Tully, J. G., D. L. Rose, R. F. Whitcomb, and R. P. Wenzel. 1979. Enhanced isolation of *Mycoplasma pneumoniae* from throat washings with a newly modified culture medium. *J. Infect. Dis.* **139:**478–482.

74. Tully, J. G., and R. F. Whitcomb (ed.). 1979. *The Mycoplasmas*, vol. 2. Academic Press, Inc., New York.

75. Unsworth, P. F., D. Taylor-Robinson, E. E. Shoo, and P. M. Furr. 1985. Neonatal mycoplasmaemia: *Mycoplasma hominis* as a significant cause of disease? *J. Infect.* **10:**163–168.

76. Waites, K. B., P. T. Rudd, D. T. Crouse, K. C. Canupp, K. G. Nelson, C. Ramsey, and G. H. Cassell. 1988. Chronic *Ureaplasma urealyticum* and *Mycoplasma hominis* infections of central nervous system in preterm infants. *Lancet* **i:**17–21.

77. Wang, R. Y.-H., J. W.-K. Shih, T. Grandinetti, P. F. Pierce, M. M. Hayes, D. J. Wear, H. J. Alter, and S.-C. Lo. 1992. High frequency of antibodies to *Mycoplasma penetrans* in HIV-infected patients. *Lancet* **340:**1312–1316.

78. Wang, R. Y.-H., J. W.-K. Shih, S. H. Weiss, T. Grandinetti, P. F. Pierce, M. Lange, H. J. Alter, D. J. Wear, C. L. Davies, R. K. Mayur, and S.-C. Lo. 1993. *Mycoplasma penetrans* infection in male homosexuals with AIDS: high seroprevalence and association with Kaposi's sarcoma. *Clin. Infect. Dis.* **17:**724–729.

79. Wang, R. Y.-H., W. S. Wu, M. S. Dawson, J. W.-H. Shih, and S.-C. Lo. 1992. Selective detection of *Mycoplasma fermentans* by polymerase chain reaction and by using a nucleotide sequence within the insertion sequence-like element. *J. Clin. Microbiol.* **30:**245–248.

80. Whitcomb, R. F., and J. G. Tully (ed.). 1989. *The Mycoplasmas*, vol. 5. Academic Press, Inc., New York.

CHLAMYDIA, RICKETTSIA, AND RELATED BACTERIA

VI

VOLUME EDITOR
PATRICK R. MURRAY

SECTION EDITORS
MARY JANE FERRARO AND
PETER H. GILLIGAN

Classification and Identification of *Chlamydia, Rickettsia,* and Related Bacteria

DAVID H. WALKER AND GREGORY A. DASCH

54

Genera included in the orders *Chlamydiales* and *Rickettsiales* in section 9 of *Bergey's Manual of Systematic Bacteriology* (19) were grouped together largely because of their smallness and usual intimate association with eukaryotic cells and because of the similarity of techniques required for their identification and study. Table 1 summarizes some of the properties of the major species and groups of clinical importance. The rapid application of molecular methods for their classification and identification, the continuous addition of new species, and the alterations in generic assignments of species addressed in the fifth edition of this Manual (36) have continued apace. Only recently have the phylogenetic relationships among most of the principal agents of human and veterinary disease in these orders been fully established by determination of sequences of DNA coding for 16S rRNA (rDNA) (8, 17, 22, 23). General molecular methods such as DNA sequencing and restriction fragment length polymorphism (RFLP) analysis by PCR-amplified genus-specific genes are also being rapidly applied to the identification and classification of both new species and serotypes (4–6, 8, 12, 14, 16, 18, 26). However, investigations of biochemical phenotypic traits among these species, with the notable exception of the *Bartonella* (*Rochalimaea*) species, which can be grown on axenic media (9, 37), have been quite limited because of the difficulties involved in studying obligately intracellular bacteria. Moreover, preparation and availability of specific monoclonal or polyclonal antibody reagents have lagged greatly behind the addition of new species and serotypes (with the notable exception of *Chlamydia* and some *Rickettsia* species), largely because relatively few laboratories have the specialized containment facilities or resources required for investigating many of these agents.

Four species of *Chlamydia* have been defined. By DNA-DNA hybridization, these species generally exhibit less than 30% homology and can be distinguished by RFLP analysis of their rRNA gene loci with cloned rDNA probes (14). The 16S rDNA sequences of *Chlamydia pneumoniae, C. trachomatis,* and *C. psittaci* have been determined and exhibit greater than 94% homology (17). *Chlamydia* spp. clearly do not belong to the *Archaea* domain and are only remotely related to other bacterial groups. By one algorithm, they were grouped with *Planctomyces staleyi,* a free-living aquatic bacterium that also lacks cell wall peptidoglycan (23, 35), but an alternative analysis with seven

Planctomyces species suggests no close relationships to this genus (22) but rather a distant association with clusters including flavobacteria and relatives, green sulfur bacteria, and group ε purple bacteria (proteobacteria), which includes *Campylobacter, Helicobacter,* and *Wolinella* spp. PCR-RFLP and DNA sequence analysis of variable domains of the major outer membrane protein gene *ompA* have provided rapid and precise means for distinguishing serovars of each of the species (16). Monoclonal antibodies are also available for microimmunofluorescence typing of many serovars of *C. trachomatis, C. psittaci,* and *C. pecorum,* while *C. pneumoniae* exhibits little variability (20). Serotyping with polyclonal sera prepared in a precise manner has been the standard method for classifying rickettsiae and chlamydiae (25). Although capable of distinguishing novel species, this method is cumbersome and of limited utility in resolving the current problems in rickettsial classification.

Sequence analysis of the 16S rDNA of the *Rickettsiales* has confirmed the phylogenetic diversity long suspected in the group (8, 34). In particular, *Coxiella burnetii,* an obligate intracellular parasite of the acidified phagolysosome, and *Wolbachia persica,* a membrane-enclosed transovarially passed tick symbiont, are both γ-group *Proteobacteria* and glycolytic, while the other *Rickettsiales* are all α-group *Proteobacteria* and nonglycolytic. Although the genus *Coxiella* contains only a single species, six types have been defined on the basis of genomic RFLP, plasmid content, and lipopolysaccharide type (21). Whether these genetic types correlate with various clinicopathologic features of Q fever, either acute or chronic (especially endocarditis), the latter of which probably occurs in fewer than 1% of those infected with *C. burnetii,* has been questioned by recent studies (27, 30).

Among the α-proteobacteria group bacteria originally included in the order *Rickettsiales,* 16S rDNA sequence analysis has consistently dovetailed with distinct differences among the species in relationships to host cells and antigenic composition. Both *Bartonella bacilliformis* and four species of *Rochalimaea* can be cultivated axenically on various blood agar media (8, 9, 37). Their host cell associations are most readily defined as epicellular. They are closely related to α-2 group species like *Agrobacterium* spp., members of the family *Rhizobiaceae,* and *Brucella* spp. The five species exhibit DNA-DNA hybridization homologies of 40 to 70% and differ in cellular fatty acid composition. The

TABLE 1 Comparison of epidemiologic, pathogenetic, and microbial characteristics of chlamydiae, rickettsiae, and related organisms[a]

Organism	Natural host	Mode of transmission	Proteobacterial subdivision	Target cell(s)	Subcellular location
Chlamydia trachomatis	Humans	Human to human	None	Conjunctival and genital epithelium, macrophages	Phagosome
Chlamydia psittaci	Birds, reptiles, mammals	Aerosol	None	Respiratory epithelium	Phagosome
Chlamydia pneumoniae	Humans	Aerosol	None	Respiratory epithelium	Phagosome
Spotted fever group rickettsiae[b]	Ticks, mites	Tick or mite bite	α	Endothelium	Cytosol
Typhus group rickettsiae[c]	Fleas, lice, mammals	Flea or louse feces	α	Endothelium	Cytosol
Rickettsia tsutsugamushi	Chiggers	Chigger bite	α	Poorly documented	Cytosol
Ehrlichia sennetsu group[d]	Unknown	Unknown, infected salmon fluke	α	Mononuclear phagocytes	Phagosome
Ehrlichia chaffeensis group[e]	Dogs, ticks	Tick bite	α	Mononuclear phagocytes	Phagosome
Ehrlichia phagocytophila group[f]	Herbivores, ticks	Tick bite	α	Polymorphonuclear phagocytes	Phagosome
Bartonella (*Rochalimaea*) spp.	Sand flies, humans, cats, lice	Sand fly bite, cat scratch, louse feces	α subgroup 2	Erythrocytes extracellular	Cytosol and epicellular
Afipia spp.	Unknown, ?soil	Wound contamination	α subgroup 2	Poorly defined	Poorly defined
Coxiella spp.	Mammals, ticks	Aerosol	γ	Mononuclear phagocytes	Phagolysosome

(Continued)

close phenotypic and genetic relationship of *Bartonella* and *Rochalimaea* species has resulted in their recent consolidation in the genus *Bartonella*; concurrently, the genus *Bartonella* was removed from the order *Rickettsiales*, which now contains only obligate intracellular bacteria (8). Both *Bartonella henselae* and *B. bacilliformis* cause chronic epicellular infections associated with vascular proliferative lesions (8, 37). *B. henselae* has also been strongly implicated in cases of cat scratch disease (11). By 16S rDNA sequence analysis, *Afipia* species (7), including *Afipia felis*, an agent initially thought to be the etiologic agent of cat scratch disease, also belong to the α-2 Proteobacteria but have only 90% 16S rDNA similarity to *Bartonella* spp. (22). *Afipia* spp. have very strong similarity to the symbiotic nitrogen-fixing bacterium *Bradyrhizobium japonicum* and the budding bacterium *Blastobacter denitrificans*, whose genera may require redefinition (38).

The remaining groups of *Rickettsiales* for which 16S rDNA sequences are available include six distinct genetic groups. The large genus *Rickettsia* (Table 1) can be split into the highly related typhus and spotted fever groups and the genetically diverse species *Rickettsia tsutsugamushi*, for which a separate genus, *Orientia*, has been proposed (31). These bacteria all replicate in the cytoplasm of their host cells. Three related but distinct genetic groups of species in the nonacidified phagosome-associated genus *Ehrlichia* have been discovered by 16S rDNA sequence analysis (1, 2). Since *Cowdria ruminantium* and *Neorickettsia helminthoeca* share cross-reacting antigens with some ehrlichiae (28, 29), it was not surprising that their 16S rDNA sequences (8, 33) are similar to those of the parasites of phagocytes, the *Ehrlichia chaffeensis-Ehrlichia ewingii-Ehrlichia canis* group and the *Ehrlichia risticii-Ehrlichia sennetsu* group, respectively.

Somewhat more surprisingly, the important hemoparasite *Anaplasma marginale* was closely related to the neutrophil-platelet parasites in the *Ehrlichia phagocytophila-Ehrlichia platys-Ehrlichia equi* group. All three genetic groups contain pathogens of both animals and humans. Finally, the widespread agent of invertebrate cytoplasmic incompatibility, *Wolbachia pipientis*, commonly found in mosquitoes and *Drosophila simulans*, also belongs to this branch of α-proteobacteria, but its pathogenicity for humans remains uncertain (24).

Because rickettsiae and ehrlichiae may be difficult to cultivate from clinical specimens and because for some organisms no tissue culture method for propagating them has been developed (10), PCR-RFLP and sequence analysis have gained preeminence both for the detection and the identification of specific agents. 16S rDNA sequences have been used primarily for PCR amplification of the ehrlichiae (1, 2) and *Bartonella* spp. (8), while several genes are used for the rickettsiae. The DNA sequences of the 22-kDa, 47-kDa, GroEL (60-kDa), and, particularly, the major variable 50- to 63-kDa protein genes are used for typing *R. tsutsugamushi* (15). Many new species as well as long-standing species of typhus and spotted fever group rickettsiae have been characterized recently by PCR-RFLP, pulsed-field gel electrophoresis, and antigenic analysis by polyacrylamide gel electrophoresis and Western blotting (immunoblotting) (4–6, 12, 13, 32). The conserved 17-kDa antigen gene and citrate synthase genes and the heat-modifiable variable surface protein antigen genes have been widely used for typing new isolates (3–6, 12, 18, 26).

The remaining questions are the practical implications of the taxonomic identification of these bacteria for the clinical microbiologist. These molecular tools have pro-

TABLE 1 (Continued)

Developmental cycle	Axenic cultivation	G+C content (%)	Lipopoly-saccharide	Peptido-glycan	Size of protein antigen(s) (kDa)	ATP source	Plasmid
+	−	42–45	+	−	40, 60, 68	Obligate parasitism	+
+	−	39–43	+	−	40, 60, 68	Obligate parasitism	+
+	−	40	+	−	40, 60, 68, 98	Obligate parasitism	−
−	−	32–33	+	+	190, 135	Presumed independent synthesis and parasitism	−
−	−	29–30	+	+	135	Independent synthesis and parasitism	−
−	−	28.5–30	−	−	70, 50–63, 47	Synthesis	−
Undefined	−	Unknown	−	+	91, 70, 52, 44, 20–28	Synthesis	−
Undefined	−	Unknown	−	+	110, 64, 52, 42, 33, 20–28	Presumed synthesis	−
Undefined	−	Unknown	Unknown	Unknown	Unknown	Presumed synthesis	−
−	+	39–41	+	+	100, 75, 60, 35, 17	Synthesis	+
−	+	62.5	+	+	Unknown	Synthesis	Unknown
Proposed	−	43	+	+	28	Synthesis	+

[a] +, Present; −, absent.
[b] The spotted fever group of the genus *Rickettsia* includes the human pathogens *R. rickettsii*, *R. conorii*, *R. sibirica*, *R. akari*, *R. japonica*, *R. australis*, and *R. africae*; the named species of undetermined pathogenicity *R. parkeri*, *R. montana*, *R. rhipicephali*, *R. helvetica*, and *R. massiliae*; and at least eight unnamed distinct serotypes of tick origin from North America, Europe, and Asia.
[c] The typhus group of rickettsiae includes the human pathogens *R. prowazekii*, *R. typhi*, ELB (a cat flea-associated rickettsia), and *R. canada* of uncertain pathogenicity.
[d] The *E. sennetsu* group also includes the equine pathogen *E. risticii* and the canine pathogen *N. helminthoeca*.
[e] The *E. chaffeensis* group also includes the canine pathogens *E. canis* and *E. ewingii* as well as the bovine pathogen *C. ruminantium*.
[f] The *E. phagocytophila* group also includes the closely related human granulocytic ehrlichiae and the animal pathogens *E. equi*, the canine agent *E. platys*, and *A. marginale*.

vided new means of detecting these fastidious bacteria in ill patients and in distinguishing them from the causes of other bacterial and viral infections with similar presentations. The number of "new" agents that have been discovered in the last few years is remarkable. The identification of new foci and agents of disease is greatly dependent on application of these taxonomic tools. In particular, the choice of appropriate public health disease control measures is entirely dependent on the identification of the agents involved. It behooves the clinical microbiologist concerned for the patient and the community to establish laboratory methods for the diagnosis of rickettsial diseases. With ever-increasing domestic and international travel, broadly applicable methods for establishing the diagnosis of infections caused by diverse agents are essential.

This investigation was supported in part by the Naval Medical Research and Development Command Research Task 61102A 3M161102BS13.AC.1293 and by grants AI21242 and AI31431 from the National Institute of Allergy and Infectious Diseases.

REFERENCES

1. Anderson, B. E., J. E. Dawson, D. C. Jones, and K. H. Wilson. 1991. *Ehrlichia chaffeensis*, a new species associated with human ehrlichiosis. *J. Clin. Microbiol.* **29:**2838–2842.
2. Anderson, B. E., C. E. Greene, D. C. Jones, and J. E. Dawson. 1992. *Ehrlichia ewingii* sp. nov., the etiologic agent of canine granulocytic ehrlichiosis. *Int. J. Syst. Bacteriol.* **42:** 299–302.
3. Anderson, B. E., and T. Tzianabos. 1989. Comparative sequence analysis of a genus-common rickettsial antigen gene. *J. Bacteriol.* **171:**5199–5201.
4. Azad, A. F., J. B. Sacci, Jr., W. M. Nelson, G. A. Dasch, E. T. Schmidtmann, and M. Carl. 1992. Genetic characterization and transovarial transmission of a typhus-like rickettsia found in cat fleas. *Proc. Natl. Acad. Sci. USA* **89:**43–46.
5. Baird, R. W., M. Lloyd, J. Stenos, B. C. Ross, R. S. Stewart, and B. Dwyer. 1992. Characterization and comparison of Australian human spotted fever group rickettsiae. *J. Clin. Microbiol.* **30:**2896–2902.
6. Beati, L., J.-P. Finidori, B. Gilot, and D. Raoult. 1992. Comparison of serologic typing, sodium dodecyl sulfate-polyacrylamide gel electrophoresis protein analysis, and genetic restriction fragment length polymorphism analysis for identification of rickettsiae: characterization of two new rickettsial strains. *J. Clin. Microbiol.* **30:**1922–1930.
7. Brenner, D. J., D. G. Hollis, C. W. Moss, C. K. English, G. S. Hall, J. Vincent, J. Radosevic, K. A. Birkness, W. F. Bibb, F. D. Quinn, B. Swaminathan, R. E. Weaver, M. W. Reeves, S. P. O'Connor, P. S. Hayes, and F. C. Tenover. 1991. Proposal of *Afipia* gen. nov., with *Afipia felis* sp. nov. (formerly the cat scratch disease bacillus), *Afipia clevelandensis* sp. nov. (formerly the Cleveland Clinic Foundation strain),

Afipia broomeae sp. nov., and three unnamed genospecies. *J. Clin. Microbiol.* **29:**2450–2460.

8. **Brenner, D. J., S. P. O'Connor, H. H. Winkler, and A. G. Steigerwalt.** 1993. Proposals to unify the genera *Bartonella* and *Rochalimaea*, with descriptions of *Bartonella quintana* comb. nov., *Bartonella vinsonii* comb. nov., *Bartonella henselae* comb. nov., and *Bartonella elizabethae* comb. nov., and to remove the family *Bartonellaceae* from the order *Rickettsiales. Int. J. Syst. Bacteriol.* **43:**777–786.

9. **Daly, J. S., M. G. Worthington, D. J. Brenner, C. W. Moss, D. G. Hollis, R. S. Weyant, A. G. Steigerwalt, R. E. Weaver, M. I. Daneshvar, and S. P. O'Connor.** 1993. *Rochalimaea elizabethae* sp. nov. isolated from a patient with endocarditis. *J. Clin. Microbiol.* **31:**872–881.

10. **Dasch, G. A., and E. Weiss.** 1992. The genera *Rickettsia, Rochalimaea, Ehrlichia, Cowdria,* and *Neorickettsiae,* p. 2407–2470. *In* A. Balows, H. G. Truper, M. Dworkin, W. Harder, and K.-H. Schleifer (ed.), *The Prokaryotes. A Handbook on the Biology of Bacteria: Ecophysiology, Isolation, Identification, Applications,* 2nd ed. Springer-Verlag, New York.

11. **Dolan, M. J., M. T. Wong, R. L. Regnery, J. H. Jorgensen, M. Garcia, J. Peters, and D. Drehner.** 1993. Syndrome of *Rochalimaea henselae* adenitis suggesting cat scratch disease. *Ann. Intern. Med.* **118:**331–336.

12. **Eremeeva, M. E., N. M. Balayeva, V. F. Ignatovich, and D. Raoult.** 1993. Proteinic and genomic identification of spotted fever group rickettsiae isolated in the former USSR. *J. Clin. Microbiol.* **31:**2625–2633.

13. **Eremeeva, M. E., V. Roux, and D. Raoult.** 1993. Determination of genome size and restriction pattern polymorphism of *Rickettsia prowazekii* and *Rickettsia typhi* by pulsed field gel electrophoresis. *FEMS Microbiol. Lett.* **112:**105–112.

14. **Fukushi, H., and K. Hirai.** 1993. *Chlamydia pecorum*—the fourth species of genus *Chlamydia. Microbiol. Immunol.* **37:**515–522.

15. **Furuya, Y., Y. Yoshida, T. Katayama, S. Yamamoto, and A. Kawamura, Jr.** 1993. Serotype-specific amplification of *Rickettsia tsutsugamushi* DNA by nested polymerase chain reaction. *J. Clin. Microbiol.* **31:**1637–1640.

16. **Gaydos, C. A., L. Bobo, L. Welsh, E. W. Hook III, R. Viscidi, and T. C. Quinn.** 1992. Gene typing of *Chlamydia trachomatis* by polymerase chain reaction and restriction endonuclease digestion. *Sex. Transm. Dis.* **19:**303–308.

17. **Gaydos, C. A., L. Palmer, T. C. Quinn, S. Falkow, and J. J. Eiden.** 1993. Phylogenetic relationship of *Chlamydia pneumoniae* to *Chlamydia psittaci* and *Chlamydia trachomatis* as determined by analysis of 16S ribosomal DNA sequences. *Int. J. Syst. Bacteriol.* **43:**610–612.

18. **Gilmore, R. D., Jr.** 1993. Comparison of the *rompA* gene repeat regions of *Rickettsiae* reveals species-specific arrangements of individual repeating units. *Gene* **125:**97–102.

19. **Krieg, N. R., and J. G. Holt (ed.).** 1984. *Bergey's Manual of Systematic Bacteriology,* vol. 1, p. 687–739. The Williams & Wilkins Co., Baltimore.

20. **Kuroda-Kitagawa, Y., C. Suzuki-Muramatsu, T. Yamaguchi, H. Fukushi, and K. Hirai.** 1993. Antigenic analysis of *Chlamydia pecorum* and mammalian *Chlamydia psittaci* by use of monoclonal antibodies to the major outer membrane protein and a 56- to 64-kd protein. *Am. J. Vet. Res.* **54:**709–712.

21. **Mallavia, L. P., J. E. Samuel, and M. E. Frazier.** 1991. The genetics of *Coxiella burnetii*: etiologic agent of Q fever and

chronic endocarditis, p. 259–284. *In* J. C. Williams and H. A. Thompson (ed.), *Q Fever: the Biology of Coxiella burnetii.* CRC Press, Inc., Boca Raton, Fla.

22. **Neefs, J.-M., Y. Van de Peer, P. De Rijk, S. Chapelle, and R. De Wachter.** 1993. Compilation of small ribosomal subunit RNA structures. *Nucleic Acids Res.* **21:**3025–3049.

23. **Olsen, G. J., and C. R. Woese.** 1993. Ribosomal RNA: a key to phylogeny. *FASEB J.* **7:**113–123.

24. **O'Neill, S. L., R. Giordano, A. M. E. Colbert, T. L. Karr, and H. M. Robertson.** 1992. 16S rRNA phylogenetic analysis of the bacterial endosymbionts associated with cytoplasmic incompatibility in insects. *Proc. Natl. Acad. Sci. USA* **89:**2699–2702.

25. **Philip, R. N., E. A. Casper, W. Burgdorfer, R. K. Gerloff, L. E. Hughes, and E. J. Bell.** 1978. Serologic typing of rickettsiae of the spotted fever group by microimmunofluorescence. *J. Immunol.* **121:**1961–1968.

26. **Regnery, R. L., C. L. Spruill, and B. D. Plikaytis.** 1991. Genotypic identification of rickettsiae and estimation of intraspecies sequence divergence for portions of two rickettsial genes. *J. Bacteriol.* **173:**1576–1589.

27. **Reimer, L. G.** 1993. Q fever. *Clin. Microbiol. Rev.* **6:**193–198.

28. **Rikihisa, Y.** 1991. The tribe *Ehrlichieae* and ehrlichial diseases. *Clin. Microbiol. Rev.* **4:**286–308.

29. **Rikihisa, Y.** 1991. Cross-reacting antigens between *Neorickettsia helminthoeca* and *Ehrlichia* species, shown by immunofluorescence and western immunoblotting. *J. Clin. Microbiol.* **29:**2024–2029.

30. **Stein, A., and D. Raoult.** 1993. Lack of pathotype specific gene in human *Coxiella burnetii* isolates. *Microb. Pathog.* **15:**177–185.

31. **Tamura, A., H. Urakami, and N. Ohashi.** 1991. A comparative view of *Rickettsia tsutsugamushi* and the other groups of rickettsiae. *Eur. J. Epidemiol.* **7:**259–269.

32. **Uchida, T., T. Uchiyama, K. Kumano, and D. H. Walker.** 1992. *Rickettsia japonica* sp. nov., the etiological agent of spotted fever group rickettsiosis in Japan. *Int. J. Syst. Bacteriol.* **42:**303–305.

33. **van Vliet, A. H. M., F. Jongejan, and B. A. M. van der Zeijst.** 1992. Phylogenetic position of *Cowdria ruminantium* (*Rickettsiales*) determined by analysis of amplified 16S ribosomal DNA sequences. *Int. J. Syst. Bacteriol.* **42:**494–498.

34. **Weisburg, W. G., M. E. Dobson, J. E. Samuel, G. A. Dasch, L. P. Mallavia, O. Baca, L. Mandelco, J. E. Sechrest, E. Weiss, and C. R. Woese.** 1989. Phylogenetic diversity of the rickettsiae. *J. Bacteriol.* **171:**4202–4206.

35. **Weisburg, W. G., T. P. Hatch, and C. R. Woese.** 1986. Eubacterial origin of chlamydiae. *J. Bacteriol.* **167:**570–574.

36. **Weiss, E.** 1991. Taxonomy, p. 1033–1035. *In* A. Balows, W. J. Hausler, Jr., K. L. Herrmann, H. D. Isenberg, and H. J. Shadomy (ed.), *Manual of Clinical Microbiology,* 5th ed. American Society for Microbiology, Washington, D.C.

37. **Welch, D. F., D. M. Hensel, D. A. Pickett, V. H. San Joaquin, A. Robinson, and L. N. Slater.** 1993. Bacteremia due to *Rochalimaea henselae* in a child: practical identification of isolates in the clinical laboratory. *J. Clin. Microbiol.* **31:**2381–2386.

38. **Willems, A., and M. D. Collins.** 1992. Evidence for a close genealogical relationship between *Afipia* (the causal organism of cat scratch disease), *Bradyrhizobium japonicum* and *Blastobacter denitrificans. FEMS Microbiol. Lett.* **96:**241–246.

Chlamydia

JULIUS SCHACHTER AND WALTER E. STAMM

55

INTRODUCTION

The chlamydiae are among the more common pathogens throughout the animal kingdom (19). They are nonmotile, gram-negative, obligate intracellular bacteria. Their unique developmental cycle differentiates them from all other microorganisms (23). They replicate within the cytoplasm of host cells, forming characteristic intracellular inclusions that can be seen by light microscopy. They differ from viruses by possessing both RNA and DNA and have cell walls quite similar in structure to those of gram-negative bacteria. They are susceptible to many broad-spectrum antibiotics, possess a number of enzymes, and have a restricted metabolic capacity. None of these metabolic reactions results in the production of energy. Thus, they have been considered energy parasites that use the ATP produced by the host cell for their own requirements.

TAXONOMY

Chlamydiae are presently placed in their own order, the *Chlamydiales*, family *Chlamydiaceae*, with one genus, *Chlamydia* (23). There are four species: *Chlamydia trachomatis*, *C. psittaci*, *C. pecorum*, and *C. pneumoniae*. *C. trachomatis* includes the organisms causing trachoma, inclusion conjunctivitis, lymphogranuloma venereum (LGV), and genital tract diseases. The three biovars within the species are trachoma, LGV, and murine. *C. trachomatis* strains are sensitive to the action of sulfonamides and produce a glycogenlike material within the inclusion vacuole that stains with iodine. *C. psittaci* strains infect many avian species and mammals, producing such diseases as psittacosis, ornithosis, feline pneumonitis, and bovine abortion (34). They are resistant to the action of sulfonamides and produce inclusions that do not stain with iodine. *C. pneumoniae* has less than 10% DNA relatedness to the other species and has pear-shaped rather than round elementary bodies (EBs) (15, 16). It appears to be exclusively a human pathogen. *C. pneumoniae* has been identified as the cause of a variety of respiratory tract diseases and is distributed worldwide. A fourth species, *C. pecorum*, has also been described, but its role as a pathogen is not clear, and specialized reagents are required for its identification.

DESCRIPTION OF GENUS

Growth Cycle

C. trachomatis attaches to a heparan sulfate-like molecule on the surface of susceptible host cells (40). This molecule apparently functions as a bridge between a specific receptor on the surface of the epithelial cell and a receptor on the EB. Chlamydiae are ingested by susceptible host cells by a mechanism that is not yet completely defined but is likely to be similar to receptor-mediated endocytosis. The uptake process is directly influenced by the chlamydiae, and ingestion of chlamydiae is specifically enhanced (6). After attachment, the EB enters the cell in an endosome, within which the entire growth cycle is completed. The chlamydiae prevent phagolysosomal fusion. Once the EB (diameter, 0.25 to 0.35 μm) has entered the cell, it reorganizes into a reticulate body that is larger than the EB (0.5 to 1 μm) and richer in RNA. After approximately 8 h, the reticulate body begins dividing by binary fission. Approximately 18 to 24 h after infection, the reticulate bodies become EBs by a poorly understood reorganization or condensation process. The EBs are then released to initiate another cycle of infection. The EBs are specifically adapted for extracellular survival and are the infectious form of chlamydiae. The metabolically active and replicating form, the reticulate body, does not survive well outside the host cell and seems adapted for an intracellular milieu.

Antigenic Relationships

The chlamydiae possess group (or genus)-specific, species-specific, and type-specific antigens. Although the organisms are antigenically complex, only a few antigens play a role in diagnosis and pathogenesis. The group complement fixation (CF) antigen, shared by all members of the genus, is the lipopolysaccharide (LPS), with a ketodeoxyoctanoic acid as the reactive moiety. It may be analogous to the LPS of certain gram-negative bacteria (26). One-way cross-reactions have been reported between chlamydiae and some bacteria, but these do not appear to influence serodiagnosis. The major outer membrane protein (MOMP) contains both species- and subspecies-specific antigens (7). The MOMP is responsible for most of the reactivity seen in the microimmunofluorescence (micro-IF) test, which was used to make the original description of the 15 serovars of *C. trachomatis* (37, 38). Recent studies using monoclonal antibodies for serotyping have confirmed these originally described serovars and have also identified new serovariants (18, 35). Studies employing DNA sequencing of the MOMP gene have localized the serotyping epitopes to the variable portions of the MOMP gene (33). A 60-kDa cysteine-rich structural protein has a highly immunogenic species-specific epitope (24). A 60-kDa heat shock protein that shares sequence homology with analogous human genes may play an important role in the generation of immunopathology (22).

Serovars of *C. psittaci* can be demonstrated by neutralization tests and micro-IF (1, 13). Only one serovar of *C. pneumoniae* has been demonstrated.

CLINICAL SIGNIFICANCE

C. trachomatis is almost exclusively a human pathogen (28) and is the most common sexually transmitted bacterial agent. Within this species, serotypes A, B, Ba, and C have been associated with endemic trachoma, the most common preventable form of blindness, while serotypes L1, L2, and L3 are associated with LGV. When sexual transmission of *C. trachomatis* strains other than LGV is studied, serotypes D through K are the major identifiable causes of nongonococcal urethritis in men and also cause epididymitis and Reiter's syndrome. Proctitis may occur in either sex. In women, cervicitis, urethritis, endometritis, and salpingitis result from chlamydial infection with these types. Salpingitis may produce tubal scarring, infertility, and ectopic pregnancy. The agent in the cervix may be transmitted to the neonate passing through the infected birth canal, and an eye disease (inclusion conjunctivitis of the newborn) and a characteristic chlamydial pneumonia of infants may develop (4). Vaginal, pharyngeal, and enteric infections in neonates are also recognized.

Human psittacosis is a zoonosis usually contracted from exposure to an infected avian species. *C. psittaci* is ubiquitous among avian species, and infection in birds usually involves the intestinal tract. The organism is shed in the feces, contaminates the environment, and is spread by aerosol. *C. psittaci* is also common in domestic mammals. In some parts of the world, these infections have important economic consequences, as *C. psittaci* is a cause of a number of systemic and debilitating diseases in domestic mammals and, most important, can cause abortions (34). Human chlamydial infections resulting from exposure to infected domestic mammals occur but seem to be relatively uncommon.

C. pneumoniae, like *C. trachomatis*, appears to be a human infection without an animal reservoir (14). Seroepidemiological studies indicate that infections are very common worldwide. Age-specific prevalence and incidence rates suggest that infections are commonly acquired in later childhood, adolescence, and early adulthood, resulting in seroprevalences of 40 to 50% in 30- to 40-year-old people. Manifestations of infection include pharyngitis, bronchitis, and mild pneumonia. Within households, schools, and workplace environments and among military personnel in close living quarters, transmission occurs via respiratory secretions. In seroepidemiological studies, these infections have been linked with coronary artery disease, and their role in atherosclerosis is currently under intense scrutiny.

COLLECTION, TRANSPORT, AND STORAGE

C. trachomatis is not a particularly labile organism. Maximal infectivity is achieved by keeping specimens cold and minimizing the time between collecting the specimen and processing it in the laboratory.

Collect swabs, scrapings, and small tissue samples in a special transport medium (23a). Because *Chlamydia* spp. are bacteria, the selection of antibiotics that can be used to prevent other bacterial contamination is restricted. Do not use broad spectrum antibiotics such as tetracyclines, macrolides, or penicillin. Aminoglycosides and fungicides are the mainstays. Refrigerate chlamydial specimens if they can be processed within 24 h after collection; otherwise, freeze them at −60°C.

C. psittaci strains are generally more stable. Some may persist in a contaminated environment for months without losing viability. The stability of *C. pneumoniae* has not been well studied, but it loses <50% of its infectivity in 24 h at 4°C (17).

When specimens are collected for enzyme immunoassay (EIA) or direct fluorescent-antibody (DFA) procedures, follow the descriptions and procedure instructions given in the commercial product's package inserts. When manufacturers supply swabs or specific transport media, use those materials, as use of other materials may impair the sensitivity and/or specificity of the test.

For cytological studies, air dry impression smears of tissues or scrapings of involved epithelial cell sites, and fix them appropriately (cold acetone or methanol for immunofluorescence, methanol for Giemsa stain, and heat for Macchiavello or Gimenez stain).

For most *C. trachomatis* infections of humans, it is imperative that samples be collected by vigorous swabbing or scraping of the involved epithelial cell sites. Purulent discharges are inadequate and should be cleaned from the site before sampling is done. Appropriate sites include the conjunctiva for trachoma-inclusion conjunctivitis and the anterior urethra (several centimeters into the urethra) or the cervix (within the endocervical canal) for genital infection. Because the trachoma biovar appears to infect only columnar and squamocolumnar cells, cervical specimens must be collected at the transitional zone or within the os. The organism can also infect the urethra of the female; organism recovery rates may be improved if a second sample is collected from the urethra and sent to the laboratory for testing in the same tube with the cervical sample.

Many different swab types can be used, but toxicity related to materials in swabs is not uncommon. It is useful to test swab types for toxicity in cell cultures or interference in nonculture assays when proprietary swabs are not provided by the manufacturer. As a general rule, avoid swabs with wooden sticks. Cotton, Dacron, and alginate swabs may all be used successfully, although toxicity has been noted with specific lots of each. The cytobrush has also been used for collection of endocervical specimens. It appears to collect more cells than swabs and has been associated in some investigators' experience with higher recovery rates of chlamydiae or higher rates of antigen detection by DFA (21). In other laboratories, there has been no advantage. Most likely, if clinicians are well trained to collect adequate specimens with other swabs, the cytobrush confers no particular advantage. However, with less well trained personnel, the cytobrush may result in improved specimen quality.

For women with salpingitis, the samples may be collected by needle aspiration of the involved fallopian tube. Endometrial specimens may yield the agent. Rectal mucosa, the nasopharynx, and the throat may also be sampled. For infants with pneumonia, swabs may be collected from the posterior nasopharynx or the throat, although nasopharyngeal or tracheobronchial aspirates collected by intubation appear to be a superior source of agent. For LGV, the likely specimens will be bubo pus, rectal or urethral swabbings, or biopsy samples.

For some nonculture tests, urine specimens are appropriate for diagnostic evaluation. For men, an effort should be made to obtain first-voided catch urine (the first 10 to 30 ml of urine). Specimens should be obtained at least 2 h after last micturition. Subsequent processing of the specimen will depend on the manufacturer's instructions. For some tests, 1 to 4 ml of urine is taken from the whole sample and processed further. With other tests, the entire urine specimen is processed. In either instance, it is typical to centrifuge the specimen, resuspend the sediment into the proprietary diluent, and follow the manufacturer's testing procedures for detection of chlamydial antigens or nucleotide sequences. Urine specimens from women have not been useful for antigen detection (9).

Throat swabs are collected for *C. pneumoniae*. For *C. psittaci* infection in humans (classic psittacosis), the specimens include sputum and blood specimens. *C. psittaci* has been recovered from a variety of involved anatomic sites sampled by biopsy or at necropsy. For all culture methods, the sites sampled are likely to be contaminated, and the specimen should be collected into a medium that contains appropriate antibiotics to inhibit the growth of bacteria.

ISOLATION PROCEDURES

Biosafety Considerations

C. trachomatis is not considered a particularly dangerous pathogen to handle in the laboratory, but a number of laboratory infections, usually manifested as follicular conjunctivitis, have occurred. However, the LGV biovar is a more invasive organism, and severe cases of pneumonia have occurred when research workers were exposed to aerosols created by laboratory procedures such as sonication (5). *C. psittaci* must be considered a potentially dangerous organism to handle in the laboratory. For many years it was a major cause of laboratory-acquired infections. These infections usually resulted from exposure to aerosols, but the stability of the organism in the environment is also a potential problem. The organism should not be handled in laboratories without appropriate containment facilities. Laboratory infections with *C. pneumoniae* have also occurred.

Specimen Processing

General guidelines for processing specimens are listed below. Fresh samples are preferred, but frozen material (−60°C) is acceptable.

Ocular and Genital Tract Specimens

For ocular and genital tract sites, the laboratory usually receives swabs in antibiotic-containing transport medium. Refrigerate the material until it is inoculated into cell culture if inoculation can be accomplished in less than 24 h. Otherwise, freeze it at −60°C.

Bubo Pus

To prepare bubo pus, grind the viscous material, and then suspend it in nutrient broth or cell culture medium to at least 20% by weight. Even when the pus is not viscous, dilution is advisable. If the bubo is not fluctuant, sterile saline may be injected and then aspirated for isolation attempts. Test for bacterial contaminants, and inoculate the material into cell cultures.

Blood

If there is a blood clot, grind it, and add sufficient beef heart broth or cell culture medium to make a 10% suspension. Inoculate the suspension directly into cell culture. It is usually advisable to inoculate with several further dilutions as well, because the concentrated material may be toxic to the cells.

Sputum or Throat Washings

To prepare an emulsion of sputum or throat washings, suspend sputum in 2 to 10 times (depending on its consistency) its volume of sterile antibiotic-containing broth (pH 7.2 to 7.4) or cell culture medium; emulsify thoroughly by shaking with glass beads in a sterile, tightly stoppered container. Inoculate into the cell culture system after 1 to 2 h of treatment with antibiotics at room temperature. It may be advisable to centrifuge extracts for 20 to 30 min at $100 \times g$ to remove coarse material.

Fecal Samples

Suspend cloacal or rectal swabs, droppings from caged birds, or fecal pellets in antibiotic broth or cell culture medium. Shake the suspension thoroughly. After centrifugation at $300 \times g$ for 10 min, remove the supernatant fluid. It may be further diluted (1:2 and 1:20) with antibiotic solution and held for 1 h at room temperature before inoculation into cell culture.

Tissues

Thaw frozen tissue in a refrigerator at about 4°C for 18 to 24 h. Weigh the specimen, mince it with sterile scissors, and grind it to a paste with mortar and pestle or homogenizer. After the tissue has been ground thoroughly, add the volume of antibiotic-containing diluent required to make a 10 to 20% emulsion, and mix the suspension thoroughly. For cell culture, use antibiotic-containing collection medium, and inoculate 10^{-1} and 10^{-2} dilutions.

Isolation in Cell Culture

All known chlamydiae grow in the yolk sac of the embryonated hen egg. However, with centrifugation of the inoculum, it also appears that all chlamydiae (with some variability) will grow in cell culture; psittacosis and LGV agents are capable of serial growth in cell culture without centrifugation. A number of different cell lines have been used to support the growth of chlamydiae. It does not appear that any single cell line is markedly superior to others, since successful isolation has been performed with monkey kidney, HeLa, L, and McCoy cells among others. McCoy and HeLa cells are most commonly used. There is less experience with *C. pneumoniae*. This organism grows poorly in McCoy and HeLa cells, especially in primary isolation from clinical specimens. HL cells and Hep-2 cells may be more sensitive for the recovery of *C. pneumoniae* (12).

The recommended procedure for primary isolation of chlamydiae is cell culture. The most common technique involves inoculation of clinical specimens into cyclohex-imide-treated McCoy cells (27). The basic principle involves centrifugation of the inoculum onto the cell mono-layer, incubation of monolayers for 48 to 72 h, and staining. Use of fluorescein-conjugated monoclonal antibodies represents the most sensitive method for detecting *C. trachomatis* inclusions in cell culture (32) and also allows earlier detection of inclusions. Alternatively, iodine staining can be used. It is less expensive but also less sensitive. For *C. psittaci* and *C. pneumoniae*, the inclusions can be demonstrated with genus-specific monoclonal antibodies or by the Giemsa stain.

If the coverslip-and-vial method is used, plate McCoy cells onto 13-mm coverslips contained in 15-mm-diameter (1-dram [1 fluidram = 3.697 ml]) disposable glass vials. Cell concentration (approximately 1×10^5 to 2×10^5) is selected to give a light, confluent monolayer after 24 to 48 h of incubation at 37°C. For optimal results, use the cells within 24 to 72 h after they reach confluency.

Shake the clinical specimens with glass beads before inoculation. This procedure is safer and more convenient than sonication. Standard inoculation procedure involves removing medium from the cell monolayer and replacing it with the inoculum in a volume of 0.1 to 1 ml. The specimen is then centrifuged onto the cell monolayer at approximately 3,000 × g at room temperature for 1 h. Hold the vials at 35°C for 2 h before washing the cells or changing the medium to medium containing 1 to 2 µg of cyclohex-imide per ml (this must be titrated for each batch used). Then incubate the cells at 35°C for 48 to 72 h, and examine one coverslip for inclusions by use of immunofluorescence, iodine, or Giemsa staining. The use of immunofluorescence can speed up the process, since inclusions can clearly be seen (although they are smaller) at 24 h postinfection. Giemsa stain is more sensitive than iodine stain, but the microscopic evaluation is more difficult. Slide reading can be facilitated by examining the Giemsa-stained coverslip by dark-field microscopy.

If passage of positive material or blind passage of negative material is desired, pass the material at 72 to 96 h postinoculation. Disrupt the cell monolayer by shaking it with glass beads on a Vortex mixer; treat the material by low-speed centrifugation to remove cell debris, and inoculate the supernatant as described above. For symptomatic patients, at least 90% of specimens positive for *C. trachomatis* are identified in the first passage. In screening asymptomatic patients, who often have less agent present at the infected site, more of the positive specimens require passage for detection.

With trachoma, inclusion conjunctivitis, and genital tract infections, the technique is as described above. For LGV, dilute the aspirated bubo pus (10^{-1} and 10^{-2}) before inoculation. Always make second passages, because detritus from the inoculum may make it difficult to read the slides. For many *C. psittaci* isolation attempts, it may be convenient to lengthen the incubation period to 5 to 10 days before examining the coverslips for inclusions. These organisms do not require mechanical assistance for cell-to-cell infection (30).

Many laboratories processing large numbers of specimens use flat-bottom 48- or 96-well microtiter plates rather than vials (39). In this method, cells are plated onto coverslips or, more commonly, directly onto the plates. Processing and incubation are as described above, but microscopy is modified to use either long working objectives or inverted microscopes. This procedure offers considerable savings of reagents and time and is particularly useful when mostly symptomatic patients are being screened. The microtiter technique may be less sensitive than the vial technique for screening largely asymptomatic groups of patients who have less organism present.

Ongoing quality control is important in maintaining a sensitive and specific culture system. To test the validity of specimen collection, periodic evaluation of slides by DFA permits evaluation of the cellularity of specimens being obtained. To test the adequacy of specimen transport, periodic transport of specimens with known dilutions of chlamydiae can be used to assess loss of viability or inclusion-forming units during transport. Daily inoculation of positive controls (known dilutions of chlamydia) are used to determine the overall sensitivity of the cell culture system. Laboratories using the microtiter method should evaluate possible episodes of cross-contamination by reinoculation of any positive specimen occurring in wells adjacent to a high-titer positive specimen.

Yolk Sac Isolation

Although it is not often used for evaluation of clinical specimens, the yolk sac is still employed for preparing antigens for the micro-IF test. Hold the specimen for 1 h at room temperature before using a 3.2-cm 22-gauge needle to inoculate 0.25 ml into the yolk sac. Before inoculation, incubate the fertile eggs at 38.5 to 39°C in a moist atmosphere. The eggs to be used must be obtained from a flock fed an antibiotic-free diet. The eggs should be free from mycoplasma. When they are 7 days old, candle the embryonated hen eggs for viability, mark the locations of air sacs, paint the shell with tincture of iodine, and gently punch a hole in the shell. Inoculate the specimen at a slight angle away from the embryo; label three or four eggs with a pencil or marking pen. After inoculation, swab the shell with iodine again, and seal the hole (with glue or tape). Then incubate the eggs in a moist environment at 35°C, and candle them daily for 13 days. Discard eggs that die in the first 3 days after inoculation.

Harvest the yolk sacs of eggs that die thereafter. Harvesting entails painting the shell with iodine, cracking and removing the shell over the air sac, teasing away the shell and chorioallantoic membranes, and removing the yolk sac with forceps. Strip away excess yolk material. It is important that all instruments be sterile and that fresh instruments be used for each specimen. Make and stain impression smears (Gimenez or the modified Macchiavello method). Perform sterility tests on the yolk sac in thioglycolate broth. If the embryos are still viable at 13 days postinoculation, chill the eggs for several hours, harvest the yolk sacs, grind them in nutrient broth, and centrifuge them lightly. Pass the supernatant in another group of four 7-day-old embryonated hen eggs (1 ml of 50% yolk sac per egg). After two blind passages, terminate attempts and consider the outcome negative.

The generally acceptable criteria for positive isolation are the presence of EBs in the impression smears, serially transmissible egg mortality, the presence of group antigen in the yolk sac, and the absence of contaminating bacteria.

IDENTIFICATION

Since most laboratories now use cell culture isolation systems, the basic procedure for identification of chlamydiae involves demonstration of typical intracytoplasmic inclusions by appropriate immunofluorescence, iodine, or Giemsa staining procedures. Fluorescent-antibody staining provides both a morphological and an immunological means of identification and thus may be preferable for less experienced laboratories.

C. *trachomatis* strains can be serotyped using type-specific and subspecies-specific batteries of monoclonal antibodies. These antibodies have been used in several assay formats, but the most readily adaptable, sensitive, and specific method appears to be the recently described microwell typing system, in which inclusions in microtiter wells are stained using pools of type-specific and subspecies-specific monoclonal antibodies (35). Although serotyping of chlamydial strains may be of use in epidemiological studies, it is of little clinical use unless medicolegal issues are involved.

In addition to typing of C. *trachomatis* with commercially available monoclonal antibodies, other methods of typing have been developed. These include the use of restriction endonuclease patterns of PCR-amplified DNA and direct sequencing of the variable domains in the chlamydial MOMP. These variable regions include the peptides responsible for species, serogroup, and serovar specificity. Variability in the amino acid sequence of the variable-region peptides may predict antigenic variance (33). As these tests have been applied, more subtypes have been identified, and it is likely that this process will continue, particularly if, as is expected, these variants originate from immune selection.

DIRECT CYTOLOGICAL EXAMINATION

C. *trachomatis* infections of the conjunctivae, urethra, or cervix can be diagnosed by demonstrating typical intracytoplasmic inclusions on cytological exam. The Giemsa stain was most often used in the past, but more-sensitive immunofluorescence procedures have now largely replaced it. Cytology to detect inclusions is particularly useful in diagnosing acute inclusion conjunctivitis of the newborn, in whom the sensitivity of this method exceeds 90%. The ability to detect intracellular diplococci in infants with gonococcal ophthalmia neonatorum is another benefit of this approach, and obviously, direct microscopy is much faster than isolation procedures. Cytology is relatively insensitive in diagnosing adult conjunctival and genital tract infections; other tests should be used in these situations.

Fluorescent-Antibody Technique

In the 1980s, fluorescein-conjugated monoclonal antibodies became available for cytological staining. A test using these antibodies is based on detecting EBs in smears, in contrast to previous efforts to detect inclusions (36). The early commercial DFA reagents were plagued with problems of cross-reactions with *Staphylococcus aureus* and other bacteria. These reagents have now been dramatically improved, however, and current experience indicates that the DFA test has approximately 80 to 90% sensitivity and 98 to 99% specificity compared with culture when both tests are performed under ideal circumstances.

The test requires a trained microscopist who can distinguish between fluorescing chlamydial particles and nonspecific fluorescence. Several DFA configurations are commercially available, as are a variety of monoclonal antibodies directed against MOMP or against LPS. Monoclonal antibodies to the LPS will stain all *Chlamydia* spp., but the quality of the fluorescence is somewhat mitigated by uneven distribution of LPS on the chlamydial particle. The anti-MOMP monoclonal antibodies are prepared against C.

trachomatis; they are species specific and therefore will not stain C. *psittaci* or C. *pneumoniae*. Further, the quality of fluorescence is better, because MOMP is evenly distributed on the chlamydial particle. Thus, if C. *trachomatis* is being sought, the anti-MOMP monoclonal antibodies are preferred. This procedure offers the possibility of rapid diagnosis, since the technique takes only 30 min to perform.

Giemsa Staining Technique

For Giemsa staining, air dry the smear, fix it with absolute methanol for at least 5 min, and dry it again. Then cover it with the diluted Giemsa stain (freshly prepared the same day) for 1 h. Rinse the slide rapidly in 95% ethyl alcohol to remove excess dye and enhance differentiation, and then dry it and examine it microscopically. Longer staining periods (1 to 5 h) may be preferable with heavy tissue culture monolayers. EBs stain reddish purple. The initial bodies are more basophilic, staining bluish, as do most bacteria.

NEWER NONCULTURE DIAGNOSTIC TESTS

EIA

A number of commercially available products can detect chlamydial antigens in clinical specimens by using EIA procedures (9, 31). Most of these products detect chlamydial LPS, which is more soluble than MOMP. The tests utilize either monoclonal or polyclonal antibodies to LPS and thus theoretically could detect all chlamydiae. However, they have not been well evaluated in the diagnosis of infections with C. *psittaci* or C. *pneumoniae*. For C. *trachomatis*, these tests appear to be slightly less sensitive and slightly less specific than the DFA test. Most EIAs take several hours to perform, are suitable for batch processing, and allow a laboratory to test many specimens.

The performance profiles of the commercially available EIA tests vary considerably. All the tests are less sensitive than culture, with some tests being less sensitive than others. Without confirmation, the tests have a specificity on the order of 97%. Thus, they are not amenable to screening low-prevalence populations because of the low predictive value of a positive result in such groups. To address this problem, confirmatory tests have been developed by many manufacturers. In these assays, all positive results are repeated in the presence of a monoclonal antibody directed against the *Chlamydia*-specific epitope on the LPS. This results in blocking of the specific reactions but not of the false-positive results. The appropriate application of confirmatory tests increases the specificity to well above 99.5% (20), thus reducing the number of false-positive tests in low prevalence populations. However, note that the blocking confirmatory tests have not been validated with specimens from many of the sites that are commonly tested for presence of chlamydial antigen. Another approach to confirmation is to test a specimen with a second test based on a different principle, for example, a DFA based on MOMP detection to confirm an LPS-based EIA.

Nucleic Acid Probes

Several nucleic acid probes are now commercially available, and many others have been developed and evaluated by research laboratories. One commercially available probe test (GenProbe) utilizes DNA-RNA hybridization in an effort to increase sensitivity by detecting chlamydial RNA. While this test has not been as extensively evaluated as some of the popular antigen detection methods, available data suggest that it is about as sensitive as the best antigen detection methods and is relatively specific (10).

The second class of nucleic acid probes involves use of amplification methods such as ligase chain reaction (LCR) and PCR, which are described elsewhere in this Manual. With *Chlamydia* spp., the probes are aimed at nucleotide sequences on the cryptic plasmid to take advantage of the fact that there are 7 to 10 copies of plasmid DNA contained in each *C. trachomatis* EB. At present, these tests are available only for *C. trachomatis.* Theoretically, given the multiplicity of target sites for the amplification procedures being used, these techniques should be able to detect less than one EB in purified suspensions of chlamydial particles. However, actual sensitivity with clinical specimens may be considerably less because of sampling problems or potential inhibition of the enzymes or reactions involved.

These amplification tests have received sufficient evaluation to allow them to be approved for marketing, and in general, their sensitivities are better than those seen with other nonculture procedures. Either PCR or LCR for detection of chlamydial genes in urine specimens appears to be the most sensitive diagnostic test available for chlamydial urethritis in men, providing an improved sensitivity compared with either urethral swab cultures or EIAs used on urine (2, 8). With cervical specimens, there has been some variability in results obtained in evaluations, with some suggestions that inhibitors in the specimen reduce the sensitivity of the PCR test (3). Current data suggest that PCR is more sensitive to inhibition than LCR is. The LCR results on urine specimens and on cervical and male urethral swabs have sensitivities in excess of 85% and specificities greater than 99.5%. If these preliminary results can be validated and if costs are competitive, it is likely that the amplified DNA probe tests will become the tests of choice for diagnosis of chlamydial infection in routine clinical laboratories. When organisms are needed for further study, isolation in cell culture will continue to be used.

Advice on Use of Nonculture Tests

Chlamydial infections are common. Thus, their diagnosis represents a large market, and many manufacturers provide chlamydial tests. Laboratorians are therefore faced with a bewildering and ever-increasing array of commercially available diagnostic tests. At this writing, a few fairly simple recommendations can be made.

Because cell culture is a relatively demanding technique and many laboratories want to establish routine chlamydial testing, nonculture methods of detecting the organism have become popular. The relative performances of the many EIA and DFA procedures available are difficult to assess, as only two (the Syva MicroTrak DFA procedure and the Abbott Chlamydiazyme EIA procedure) have had long-term independent assessment of performance, and there have been few head-to-head comparisons in a single patient population of the many tests available versus a single culture system. In general, these tests achieve a sensitivity of 60 to 90% and are approximately 97 to 98% specific in detecting chlamydial infection. The more recently introduced EIAs and the Syva DFA appeared to be superior in one study in which the tests were evaluated simultaneously (25), but more comparative evaluations are needed.

There are advantages and disadvantages to either test format. The DFA has the advantage of allowing the microscopist to confirm the adequacy of the specimen. Being more labor intensive, the DFA is less well suited to processing large numbers of specimens. Anti-MOMP mono-

clonal antibodies provide superior staining of *C. trachomatis* EBs but will not detect the EBs of other species. Anti-LPS monoclonal antibodies will stain all chlamydiae but tend to be less bright and also stain the particles unevenly (11). The EIA procedure is more amenable to batch processing and becomes relatively expensive when small numbers of specimens are being tested. Most EIAs detect the LPS and thus should react with all chlamydial species. They have been widely studied only for *C. trachomatis* and therefore cannot be recommended for use in *C. psittaci* or *C. pneumoniae* infection.

Both of these procedures have some advantages over culture. They are more rapid; results can be available in 30 min to 24 h. Because viability is not an issue, it is not necessary to maintain a cold chain. Specimens can be processed when cell culture is not available or when there are problems in transporting specimens to a laboratory. There is one major caveat: nonculture tests should not be used when evidence of sexual abuse is being sought. False-positive results occur with these tests, and the tests have not been extensively evaluated on some of the specimens that are tested in cases of child abuse. It is imperative that culture be used to prove the existence of chlamydial infection for legal purposes.

It is still clear that under ideal circumstances, culture remains the diagnostic method of choice. With adequate control over maintenance, storage, and transport of specimens and with well-trained specimen collectors, cell culture is more sensitive than any of the currently available antigen detection tests or direct nucleic acid probe tests. Unfortunately, these ideal circumstances do not exist in many settings. When nonculture tests are compared with less-than-ideal culture systems, their specificities may suffer because apparent false positives are really culture misses. When there are transportation problems that may affect chlamydia viability, the nonculture tests are preferred. Judging from current data, the amplification-based tests (PCR and LCR) will likely offer the most sensitive and most specific alternative in the near future.

SEROLOGICAL TESTS

The most widely used serological test for diagnosing chlamydial infections is the CF test. This test is useful in diagnosing psittacosis, in which paired sera often show fourfold or greater increases in titer. The same seems to be true for many *C. pneumoniae* infections. Approximately 50% of these infections are CF positive, although it may take 24 weeks to detect seroconversion (14). CF may also be useful in diagnosing LGV, in which single-point titers greater than 1:64 are highly supportive of this clinical diagnosis. With LGV, it is difficult to demonstrate rising titers, since the nature of the disease is such that the patient is seen by the physician after the acute stage. Any titer above 1:16 is considered significant evidence of exposure to chlamydiae. The genus-specific and relatively insensitive CF test is not particularly useful in diagnosing trachoma, inclusion conjunctivitis, or the related genital tract infections, and it plays no role in diagnosing neonatal chlamydial infections.

The micro-IF method is a much more sensitive procedure for measuring antichlamydial antibodies. It may be used in diagnosing psittacosis, in which paired sera will show rising immunoglobulin G (IgG) titers. With LGV, it is again difficult to demonstrate rising titers, but single-point titers in active cases usually have relatively high levels of IgM (>1:32) and IgG (>1:2,000) antibody. Trachoma, inclusion conjunctivitis, and genital tract infections may be diagnosed by the micro-IF technique if appropriately timed paired acute- and convalescent-phase sera can be obtained. However, it is often difficult to demonstrate rising antibody titers, particularly in sexually active people. Many of these individuals will be seen for chronic or repeat infections. The background rate of seroreactors in venereal disease clinics is >60%, making it particularly difficult to demonstrate seroconversion. In general, first attacks of chlamydial urethritis have been regularly associated with seroconversion. Individuals with systemic infection (epididymitis or salpingitis) usually have much higher antibody levels than do those with superficial infections, and women tend to have higher antibody levels than men.

Serology is particularly useful in diagnosing chlamydial pneumonia in neonates. In this case, high levels of IgM antibody are regularly found in association with disease (29). IgG antibodies are less useful, because the infants are being seen at a time when they have considerable levels of maternal IgG (all of these infections are acquired from the infected mother, who is almost always seropositive). It takes between 6 and 9 months for maternal antichlamydial antibodies to disappear. Infants older than 9 months may be tested for determination of prevalence of chlamydial infection without fear of confounding effects of maternal antibody. Infants with inclusion conjunctivitis or respiratory tract carriage of chlamydiae without pneumonia usually have very low levels of IgM antibodies. Thus, a single IgM titer of >1:32 may support the diagnosis of chlamydial pneumonia.

C. pneumoniae TWAR infections are usually diagnosed by micro-IF. The diagnostic criterion has been (i) a fourfold rise in titer, (ii) an IgM titer of ≥1:16, or (iii) an IgG titer of ≥1:512 (14).

The micro-IF technique uses many serotypes of chlamydiae, and the procedure as simplified by Wang et al. (38) is recommended. Since serology is particularly useful in diagnosing neonatal infection and since the IgM antibody responses tend to be very serovar specific, the use of single broadly reacting antigens will miss at least 15 to 25% of the infections that other procedures can prove to be caused by chlamydiae or that would be positive by a multiple-antigen micro-IF. The single-antigen tests may involve either yolk sac suspensions of agent or identification of fluorescent inclusions in tissue monolayers. Serotypes of the DEL serogroup are commonly chosen for this purpose.

Research workers should be warned that seroreactions that are specific for type A are, at least in the United States, liable to be spurious. These antibodies are usually transient and do not result in the persistent high levels of IgG antibodies that usually follow chlamydial infections.

EIA techniques that measure antichlamydial antibodies have been described. While most of these procedures have been successful in measuring IgG antibody, they have not been as reliable in measuring IgM antibody. Such tests are commercially available, but there is little published experience with them. It is likely that the EIAs are less sensitive than the micro-IF, miss some C-complex reactors, and cannot be readily applied to IgM antibody. The procedure may be of some use in selected instances and for serosurveys in laboratories where micro-IF techniques are not available.

Single serological tests are of very little value in the diagnosis of uncomplicated lower genital tract infections, which represent the majority of tests. Although manufacturers may claim that their tests are specific for *C. tracho-*

matis and that titers of a specific level are diagnostic of current infection, neither of those claims is likely to be accurate. If the tests involve partially solubilized EBs for EIA or inclusion detection by immunofluorescence, there is undoubtedly LPS present. The test will detect antibodies not only to *C. trachomatis* but also to *C. psittaci* or *C. pneumoniae* because the LPS is a genus-specific antigen. Because of the high prevalence of antibody in high-risk populations, single positive results are seldom diagnostic for current genital infection. In most research evaluations, positive predictive values have been on the order of 30 to 50%.

CF

The CF test may be performed in either the tube system or the microtiter system. Reagents should be standardized in the tube system regardless of which test system will be used. The microtiter systems are most useful in screening large numbers of sera, but it is preferable to retest all positive sera in the tube system. Occasionally, sera giving titers in the range of 1:4 to 1:8 in the microdilution system are positive at 1:16 (taken as the significant level) in the tube system. Because guinea pigs may have naturally occurring chlamydial infection, the complement must be tested for antichlamydial antibodies.

All reagents are available commercially except for high-titer group antigen, which can be prepared as follows. Inoculate yolk sacs of 7-day-old embryonated eggs with chlamydiae (e.g., psittacosis isolate 6BC) at a dose estimated to result in death of about 50% of the inoculated eggs in 5 to 7 days. Candle the eggs daily, and discard those that die early. When the 50% death end point is approached, refrigerate the remaining eggs (recently dead or live) for 3 to 24 h. Then harvest the yolk sacs. If examination of random samples shows large numbers of particles, pool the yolk sacs. This preparation may be stored at −20°C until further processing. Grind the yolk sacs in a mortar with sterile sand. Add beef heart broth (pH 7.0) to make a 20% suspension, and culture the material to determine whether it is free of bacterial contamination. Place the suspension in a flask containing sterile glass beads, and store it at 4°C for 3 to 6 weeks with daily shaking. Then centrifuge it at ca. 500 × g to remove coarse particles, transfer it to a heavy sterile flask, and steam it at 100°C or immerse it in boiling water for 30 min. After it has cooled, add liquefied phenol to 0.5%. The antigen should then be refrigerated for at least 1 week before being used. It is stable for at least 1 year if not contaminated and should have an antigen titer of 1:256 or greater. A similar preparation from uninfected yolk sacs must be included as one of the controls.

Micro-IF

The micro-IF test is usually performed with chlamydial organisms grown in yolk sac. Tissue culture-grown agent can be used, but it may be necessary to concentrate the EBs and add some normal yolk sac to improve contrast for microscopy. The individual yolk sacs are selected for EB richness and pretitrated to give an even distribution of particles. A 1 to 3% yolk sac suspension (phosphate-buffered saline, pH 7.0) is generally satisfactory. The antigens may be stored as frozen samples; after being thawed, they should be well mixed on a Vortex mixer before use. Micro-IF antigens for research purposes are available through the Washington Research Foundation. Place antigen dots on a slide in a specific pattern, with separate pen markings for each antigen. Each cluster of dots includes all of the antigenic types to be tested. Air dry the antigen dots, and fix them on slides with acetone (15 min at room temperature). Slides may be stored frozen. They may sweat when thawed for use, but they can be conveniently dried (as can the original antigen dots) with the cool airflow of a hair dryer. Place serial dilutions of serum (or tears or exudate) placed on the different antigen clusters on the slides. Separate the clusters of dots sufficiently to avoid running of serum from cluster to cluster. After the serum dilutions have been added, incubate the slides for 0.5 to 1 h in a moist chamber at 37°C. Then place them in a buffered saline wash for 5 min and then in a second 5-min wash. Dry the slides, and stain them with fluorescein-conjugated antihuman globulin. Pretitrate conjugates in a known positive system to determine appropriate working dilutions. This reagent may be prepared against any class of globulin being considered (IgA or secretory piece for secretions, IgG, or IgM). Counterstains such as bovine serum albumin conjugated with rhodamine may be included. Then wash the slides twice more, dry them, and examine them by standard fluorescence microscopy. Use of a monocular tube is recommended to allow greater precision in determining fluorescence of individual EB particles. Read the end points as the dilution giving bright fluorescence clearly associated with the well-distributed EBs throughout the antigen dot. Identification of the type-specific response is based on dilution differences reflected in the end points for different prototype antigens.

For each run of either CF or micro-IF, known positive and negative sera should always be included. These sera should always duplicate their titers as previously observed within the experimental (+1 dilution) error of the system.

ANTIMICROBIAL SUSCEPTIBILITY TESTING

Since strain-to-strain variation in antimicrobial susceptibility profiles and newly acquired drug resistance are both very infrequent among chlamydiae, susceptibility testing generally has little clinical utility. Among the drugs most active in vitro against *C. trachomatis* and *C. psittaci* are tetracycline, doxycycline, erythromycin, azithromycin, rifampin, ofloxacin, and clindamycin. Thus, the tetracyclines and macrolides have generally been the mainstays of therapy for infections due to chlamydiae.

Antimicrobial susceptibility testing for chlamydiae has not been standardized, and methodological differences doubtlessly contribute to different MICs reported from different laboratories. In general, organisms for testing are grown for at least two passages in antibiotic-free media before being harvested for use in testing. An adjusted inoculum of ~100 inclusion-forming units per microtiter well is then used to infect antibiotic-free monolayers. After centrifugation of the inoculum onto the monolayer, serial dilutions of the test antibiotic can be added either immediately or at various time intervals over the next 24 h. After 48 h, fluorescein-conjugated monoclonal antibodies are used to identify the highest antibiotic dilution that prevents inclusion formation (MIC). Generally, monolayers are also disrupted and passaged to define the highest dilution that prevents viable chlamydiae from being detected in passage (MBC).

REFERENCES

1. Banks, J., B. Eddie, M. Sung, N. Sugg, J. Schachter, and K. F. Meyer. 1970. Plaque reduction technique for demonstrating neutralizing antibodies for *Chlamydia*. *Infect. Immun.* **2**:443–447.
2. Bauwens, J. E., A. M. Clark, M. J. Loeffelholz, S. A. Herman, and W. E. Stamm. 1993. Diagnosis of *Chlamydia trachomatis* urethritis in men by polymerase chain reaction assay of first-catch urine. *J. Clin. Microbiol.* **31**:3013–3016.
3. Bauwens, J. E., A. M. Clark, and W. E. Stamm. 1993. Diagnosis of *Chlamydia trachomatis* endocervical infections by a commercial polymerase chain reaction assay. *J. Clin. Microbiol.* **31**:3023–3027.
4. Beem, M. O., and E. M. Saxon. 1977. Respiratory-tract colonization and a distinctive pneumonia syndrome in infants infected with *Chlamydia trachomatis*. *N. Engl. J. Med.* **296**:306–310.
5. Bernstein, D. I., T. Hubbard, W. Wenman, B. L. Johnson, Jr., K. K. Holmes, H. Liebhaber, J. Schachter, R. Barnes, and M. A. Lovett. 1984. Mediastinal and supraclavicular lymphadenitis and pneumonitis due to *Chlamydia trachomatis* serovars L1 and L2. *N. Engl. J. Med.* **311**:1543–1546.
6. Byrne, G. I., and J. W. Moulder. 1978. Parasite-specified phagocytosis of *Chlamydia psittaci* and *Chlamydia trachomatis* by L and HeLa cells. *Infect. Immun.* **19**:598–606.
7. Caldwell, H. D., and J. Schachter. 1982. Antigenic analysis of the major outer membrane protein of Chlamydia spp. *Infect. Immun.* **35**:1024–1031.
8. Chernesky, M., S. Castriciano, J. Sellors, I. Stewart, I. Cunningham, S. Landis, W. Seidelman, L. Grant, C. Devlin, and J. Mahony. 1990. Detection of *Chlamydia trachomatis* antigens in urine as an alternative to swabs and cultures. *J. Infect. Dis.* **161**:124–126.
9. Chernesky, M. A., J. B. Mahony, S. Castriciano, M. Mores, I. O. Stewart, S. F. Landis, W. Seidelman, E. J. Sargeant, and C. Leman. 1986. Detection of *Chlamydia trachomatis* antigens by enzyme immunoassay and immunofluorescence in genital specimens from symptomatic and asymptomatic men and women. *J. Infect. Dis.* **154**:141–148.
10. Clarke, L. M., M. F. Sierra, B. J. Daidone, N. Lopez, J. M. Covino, and W. M. McCormack. 1993. Comparison of the Syva MicroTrak enzyme immunoassay and Gen-Probe PACE 2 with cell culture for diagnosis of cervical *Chlamydia trachomatis* infection in a high-prevalence female population. *J. Clin. Microbiol.* **31**:968–971.
11. Cles, L. D., K. Bruch, and W. E. Stamm. 1988. Staining characteristics of six commercially available monoclonal immunofluorescence reagents for direct diagnosis of *Chlamydia trachomatis* infections. *J. Clin. Microbiol.* **26**:1735–1737.
12. Cles, L. D., and W. E. Stamm. 1990. Use of HL cells for improved isolation and passage of *Chlamydia pneumoniae*. *J. Clin. Microbiol.* **28**:938–940.
13. Eb, F., and J. Orfila. 1981. Serotyping of *Chlamydia psittaci* by microimmunofluorescence test. *Ann. Microbiol.* (Paris) **A132**:18.
14. Grayston, J. T. 1989. *Chlamydia pneumoniae*, strain TWAR. *Chest* **95**:664–669.
15. Grayston, J. T. 1992. Infections caused by *Chlamydia pneumoniae* strain TWAR. *Clin. Infect. Dis.* **15**:757–763.
16. Grayston, J. T., C. C. Kuo, L. A. Campbell, and S. P. Wang. 1989. *Chlamydia pneumoniae* sp. nov. for *Chlamydia* sp. strain TWAR. *Int. J. Syst. Bacteriol.* **39**:88–90.
17. Kuo, C.-C., and J. T. Grayston. 1988. Factors affecting viability and growth in HeLa 229 cells of *Chlamydia* sp. strain TWAR. *J. Clin. Microbiol.* **26**:812–815.
18. Lampe, M. F., R. J. Suchland, and W. E. Stamm. 1993. Nucleotide sequence of the variable domains within the major outer membrane protein gene from serovariants of *Chlamydia trachomatis*. *Infect. Immun.* **61**:213–219.
19. Meyer, K. F. 1967. The host spectrum of psittacosis-lymphogranuloma venereum (PL) agents. *Am. J. Ophthalmol.* **63**:1225–1246.

20. **Moncada, J., J. Schachter, G. Bolan, J. Engelman, L. Howard, I. Mushahwar, G. Ridgway, G. Mumtaz, W. Stamm, and A. Clark.** 1990. Confirmatory assay increases specificity of the Chlamydiazyme test for *Chlamydia trachomatis* infection of the cervix. *J. Clin. Microbiol.* **28:**1770–1773.

21. **Moncada, J., J. Schachter, M. Shipp, G. Bolan, and J. Wilber.** 1989. Cytobrush in collection of cervical specimens for detection of *Chlamydia trachomatis. J. Clin. Microbiol.* **27:**1863–1866.

22. **Morrison, R. P., K. Lying, and H. D. Caldwell.** 1989. Chlamydial disease pathogenesis ocular hypersensitivity elicited by a genus-specific 57 Kd protein. *J. Exp. Med.* **169:**663–675.

23. **Moulder, J. W., T. P. Hatch, C. C. Kuo, J. Schachter, and J. Storz.** 1984. Order II. Chlamydiales Storz and Page 1971, 334, p. 729–739. *In* N. R. Krieg and J. G. Holt (ed.), *Bergey's Manual of Systematic Bacteriology,* vol. 1. The Williams & Wilkins Co., Baltimore.

23a.**Nash, P., and M. M. Krenz.** 1991. Culture media, p. 1226–1288. *In* A. Balows, W. J. Hausler, Jr., K. L. Herrmann, H. D. Isenberg, and H. J. Shadomy (ed.), *Manual of Clinical Microbiology,* 5th ed. American Society for Microbiology, Washington, D. C.

24. **Newhall, W. J., B. Batteiger, and R. B. Jones.** 1982. Analysis of the human serological response to proteins of *Chlamydia trachomatis. Infect. Immun.* **38:**1181–1189.

25. **Newhall, W. J., S. DeLisle, D. Fine, R. E. Johnson, A. Hadgu, B. Matsuda, D. Osmond, J. Campbell, and W. E. Stamm.** 1994. Head-to-head evaluation of five different nonculture chlamydia tests relative to a quality-assured culture standard, abstr. 211. *Sex. Transm. Dis.* **21:**S165–S166.

26. **Nurminen, M., M. Leinonen, P. Saikku, and P. H. Makela.** 1983. The genus-specific antigen of Chlamydia: resemblance to the lipopolysaccharide of enteric bacteria. *Science* **220:**1279–1281.

27. **Ripa, K. T., and P.-A. Mardh.** 1977. Cultivation of *Chlamydia trachomatis* in cycloheximide-treated McCoy cells. *J. Clin. Microbiol.* **6:**328–331.

28. **Schachter, J.** 1978. Chlamydial infections. *N. Engl. J. Med.* **298:**428–435, 490–495, 540–549.

29. **Schachter, J., M. Grossman, and P. H. Azimi.** 1982. Serology of *Chlamydia trachomatis* in infants. *J. Infect. Dis.* **146:**530–535.

30. **Schachter, J., N. Sugg, and M. Sung.** 1978. Psittacosis: the reservoir persists. *J. Infect. Dis.* **137:**44–49.

31. **Stamm, W. E., L. A. Koutsky, J. K. Benedetti, J. L. Jourden, R. C. Brunham, and K. K. Holmes.** 1984. *Chlamydia trachomatis:* urethral infection in men. *Ann. Intern. Med.* **100:**47.

32. **Stamm, W. E., M. Tam, M. Koester, and L. Cles.** 1983. Detection of *Chlamydia trachomatis* inclusions in McCoy cell cultures with fluorescein-conjugated monoclonal antibodies. *J. Clin. Microbiol.* **17:**666–668.

33. **Stephens, R. S., E. A. Wagar, and G. K. Schoolnik.** 1988. High-resolution mapping of serovar-specific and common antigenic determinants of the major outer membrane protein of *Chlamydia trachomatis. J. Exp. Med.* **167:**817.

34. **Storz, J.** 1971. *Chlamydia and Chlamydia-Induced Diseases.* Charles C Thomas, Publisher, Springfield, Ill.

35. **Suchland, R. J., and W. E. Stamm.** 1991. Simplified microtiter well cell culture method for rapid immunotyping of *Chlamydia trachomatis. J. Clin. Microbiol.* **29:**1333–1338.

36. **Tam, M. R., W. E. Stamm, H. H. Handsfield, R. Stephens, C.-C. Kuo, K. K. Holmes, K. Ditzenberger, M. Crieger, and R. C. Nowinski.** 1984. Culture-independent diagnosis of *Chlamydia trachomatis* using monoclonal antibodies. *N. Engl. J. Med.* **310:**1146–1150.

37. **Wang, S.-P., and J. T. Grayston.** 1970. Immunologic relationship between genital TRIC, lymphogranuloma venereum and related organisms in a new microtiter indirect immunofluorescence test. *Am. J. Ophthalmol.* **70:**367–374.

38. **Wang, S.-P., J. T. Grayston, E. R. Alexander, and K. K. Holmes.** 1975. Simplified microimmunofluorescence test with trachoma lymphogranuloma venereum (*Chlamydia trachomatis*) antigens for use as a screening test for antibody. *J. Clin. Microbiol.* **1:**250–255.

39. **Yoder, B. L., W. E. Stamm, M. C. Koester, and E. R. Alexander.** 1981. Microtest procedure for isolation of *Chlamydia trachomatis. J. Clin. Microbiol.* **13:**1036–1039.

40. **Zhang, J. P., and R. S. Stephens.** 1992. Mechanism of *C. trachomatis* attachment to eukaryotic host cells. *Cell* **89:**881–889.

Rickettsia and Coxiella

JAMES G. OLSON AND JOSEPH E. McDADE

56

TAXONOMY

The order *Rickettsiales* includes a very diverse group of microorganisms; morphology, association with various organs of arthropods (primarily lice, fleas, ticks, or mites), obligate intracellular parasitism, and serologic relatedness are among the phenotypic criteria that have been used heretofore to form families and genera and to separate species within this order. Although many rickettsial species have adapted to existence within arthropods, they are also frequently capable of infecting vertebrates, including humans, usually as accidental hosts (74).

Recent characterizations of member species at the molecular level are causing reevaluations of phylogenetic relationships within the order *Rickettsiales*. Application of molecular taxonomic techniques has proven problematic, however, because with the introduction of these new technologies, rickettsiologists have not reached a consensus on the definition of a species. Serologic criteria continue to be utilized by some investigators (69); others use genomic sequencing (2, 46, 73). Restriction fragment length polymorphism analysis of the rickettsial genome (62), occasionally in combination with antigenic identification, has also been proposed as describing a species (78). Genomic analyses have been confounded by the utilization of different genes as standards and the unusually close interrelatedness of previously described rickettsial species. Genomic differences also have not been interpreted uniformly. For example, as much as 29% difference in nucleotide sequence of an antigen-encoding gene was considered to distinguish serotypes of *Rickettsia tsutsugamushi* (50), but considerably less diversity in antigen genes, as revealed by restriction fragment length polymorphism analysis, has been interpreted as designating distinct species of spotted fever group rickettsiae (39, 78). Standardization of these various approaches is needed.

Fortunately, phylogenetic and taxonomic relationships of human pathogens within the genera *Rickettsia* and *Coxiella* have been largely unaffected by these recent events, thus simplifying a discussion of clinical, epidemiologic, and diagnostic aspects of those microorganisms. Exceptions will be noted in the following discussion. *Ehrlichia* spp. (chapter 57 of this Manual) and *Bartonella* spp. (chapter 58) will be described separately.

DESCRIPTION OF GENERA AND NATURAL HABITATS

Rickettsia spp.

Members of the genus *Rickettsia* are morphologically and biochemically similar to other gram-negative bacteria. They are short, rod-shaped or coccobacillary organisms, usually 0.8 to 2.0 μm long and 0.3 to 0.5 μm in diameter. Most species grow luxuriantly in tissue culture or in the yolk sacs of embryonated eggs. Guinea pigs or mice are the most commonly used experimental hosts, although meadow voles (*Microtus pennsylvanicus*) have proven useful for isolation of spotted fever group rickettsiae (9).

Species in the genus *Rickettsia* have been subdivided into three groups of antigenically related microorganisms: spotted fever, typhus, and scrub typhus (Table 1). The spotted fever and typhus groups are composed of multiple species, whereas the scrub typhus group currently consists of different serovars of only one species, *R. tsutsugamushi*. Vector associations also help distinguish the various groups. For example, most members of the spotted fever group are found in ticks, scrub typhus rickettsiae are found in mites, and typhus rickettsiae have fleas or lice as vectors (Table 1). Limited DNA-to-DNA hybridization studies have confirmed these phenotypic groupings (47). With the exception of *R. japonica*, a recently described member of the spotted fever group that has been well documented as causing human illness in Japan (69), no new established pathogens have been described in this genus. Additional rickettsial isolates have been obtained in recent years, mostly from ticks, but their precise taxonomic positions and public health significance remain uncertain.

All members of the genus *Rickettsia* are unstable outside the host cell and are inactivated by temperatures of $\geq$56°C or by standard disinfectants.

Coxiella spp.

The genus *Coxiella* is composed of a single species, *Coxiella burnetii*. It is a short, rod-shaped microorganism, 0.2 to 0.4 μm in diameter and 0.4 to 1.0 μm in length. Although *C. burnetii* is found in numerous species of ticks, it is maintained in nature primarily by aerosol transmission. *C. burnetii* typically infects cattle, sheep, and goats; multiplies to large numbers in the placentas of pregnant animals; and is shed during parturition.

TABLE 1 Features of the pathogenic rickettsiae

Biogroup	Species	Disease in humans	Distribution	Means of transmission to humans
Spotted fever	*R. rickettsii*	RMSF	Western Hemisphere	Tick bite
	R. conorii	MSF (also called Boutonneuse fever)	Primarily Mediterranean countries, Africa, India, southwest Asia	Tick bite
	R. sibirica	Siberian tick typhus	Siberia, Mongolia, northern China	Tick bite
	R. australis	Australian tick typhus	Australia	Tick bite
	R. akari	Rickettsialpox	United States, Russia	Mite bite
	R. japonica	Oriental spotted fever	Japan	Presumably tick bite
Typhus	*R. prowazekii*	Epidemic typhus	Primarily highland areas of South America and Africa	Infected louse feces
		Recrudescent typhus (Brill-Zinsser disease)	Worldwide: follows distribution of persons with primary infections	Reactivation of latent infection
		Sporadic typhus	United States	Contact with flying squirrels (*Glaucomys volans*)
	R. typhi	Murine typhus	Worldwide	Infected flea feces
Scrub typhus	*R. tsutsugamushi*	Scrub typhus	Asia, northern Australia, Pacific Islands	Chigger bite
Q fever	*C. burnetii*	Q fever	Worldwide	Infectious aerosols

C. burnetii displays an antigenic phase variation that is unique among the rickettsiae. It exists in two antigenic phases (I and II), which are analogous to the smooth and rough forms observed with some species of bacteria. *C. burnetii* exists in antigenic phase I in nature, but it changes to phase II after continuous passage in tissue cultures or embryonated eggs. Available data indicate that the phase I antigen is a polysaccharide component of the *Coxiella* lipopolysaccharide and that the transition from phase I to phase II occurs when one or more carbohydrate components are deleted from the lipopolysaccharide moiety (25). Plasmids have been identified in *C. burnetii*, but their precise functions remain cryptic. Limited data suggest a relationship between plasmid type and clinical illness, i.e., acute versus chronic disease (3), but the influence of host factors on clinical manifestations is difficult to evaluate and precludes firm conclusions about the role of plasmids in pathogenesis.

In contrast to species in the genus *Rickettsia*, *C. burnetii* resists inactivation by physical and chemical treatment (63). It has reportedly survived for a year or more attached to wool or other fomites, and it is incompletely inactivated when held at 63°C for 30 min. Furthermore, *C. burnetii* is only partially inactivated when exposed to 1% formalin or 1% phenol for 24 h. However, it can be inactivated by 0.05% hypochlorite, 5% H_2O_2, or a 1:100 dilution of Lysol (O-phenylphenyl, 2.8%; O-benzyl-p-chlorophenyl, 2.7%; alcohol, 1.8%; xylenes, 1.5%; isopropyl alcohol, 0.9%; tetrasodium ethyl-diamine tetraacetate, 0.76%; soap, 16.5%).

CLINICAL SIGNIFICANCE

Table 1 summarizes the various rickettsioses, their respective etiologic agents, and their salient epidemiologic features. From a clinical perspective, rickettsial diseases have many features in common. The incubation period for most rickettsioses ranges from 3 to 14 days. Most patients develop nonspecific symptoms and signs. Onset of disease is sudden in about half of the cases. Fever and headache are the most commonly reported symptoms, but chills, myal-

gias, arthralgias, malaise, and anorexia also are noted. Fever increases during the first week of illness, often reaching 104°F (40°C) or higher.

Rash is a hallmark of rickettsial infection, but it usually follows systemic symptoms. Its absence should not rule out a possible rickettsial etiology, especially during the first week of illness. Conjunctivitis and pharyngitis are common. Photophobia is also observed, but evidence of more serious central nervous system impairment (confusion, stupor, delirium, seizures, and coma) is found in only about 25% of patients with Rocky Mountain spotted fever (RMSF) (38) or typhus (48) and is virtually never seen in the other rickettsial diseases.

Mild pulmonary involvement, manifested by cough and infiltrates on the chest roentgenogram, is common in epidemic typhus and is found in about half of Q fever and scrub typhus patients and about one-third of patients with RMSF. Hepatomegaly or splenomegaly is found in only about 20% of patients with rickettsioses other than Q fever, in which hepatomegaly and hepatitis may dominate the clinical picture in as many as half of the patients. In scrub typhus, regional lymphadenopathy is observed in about 20% of the patients early in the infection. Later, generalized lymphadenopathy that may be mistaken for mononucleosis is seen in about 80% of patients.

The symptoms of murine typhus (45) and flying-squirrel-associated typhus fever (18, 44) are similar to but milder than those of epidemic typhus.

Chronic Q fever and recrudescent typhus (Brill-Zinsser disease) are illnesses that can appear years after initial infection with *C. burnetii* and *R. prowazekii*, respectively. Chronic Q fever is an uncommon illness that usually affects patients with preexisting valvular heart disease; it develops as a culture-negative endocarditis accompanied by liver function abnormalities and granulomatous hepatitis. Chronic Q fever infections are frequently fatal (20, 67). The incidence of Brill-Zinsser disease is difficult to obtain. The disease occurs sporadically in persons who have lived in areas where typhus is endemic and have contracted primary cases of louse-borne typhus months to years before the onset of the recrudescent infection. The clinical manifestations of Brill-Zinsser disease are generally milder than those of primary louse-borne typhus (66).

The courses and outcomes of rickettsial diseases vary. The primary determinants are the specific infectious agent and the promptness of effective antibiotic treatment. Fatalities are common with RMSF, epidemic typhus, and scrub typhus but rare with murine typhus; no fatalities have been reported for rickettsialpox.

COLLECTION, TRANSPORT, AND STORAGE OF SPECIMENS

There are three general approaches to the laboratory diagnosis of rickettsial diseases: direct detection of rickettsiae in patient tissues, isolation of rickettsiae from tissues, and serologic tests for rickettsial antibodies. Although no technique reliably provides a diagnosis early enough to affect the outcome of the disease, several new rapid diagnostic techniques have shown promise in research laboratories and may soon allow early diagnosis in a clinical setting. Specifically, PCR technology, detection of antigen in peripheral blood cells, and immunoglobulin M (IgM) antibody capture immunoassays have shown some utility in providing laboratory confirmation early in the clinical course of disease. Serodiagnosis is the preferred diagnostic approach, however, and testing of serum specimens collected during the acute and convalescent phases of illness is still recommended. The first specimen should be obtained during the acute phase of illness, and the second specimen should be obtained 1 to 3 weeks later, depending on the test employed.

Because of the hazards of working with living rickettsiae, isolation attempts are usually limited to situations in which the outcome was fatal and postmortem tissues are the only specimens available for testing. Even then, direct fluorescent-antibody tests of formalin-fixed, paraffin-embedded tissues are faster and safer than rickettsial isolation for diagnosis.

Most pathogenic rickettsiae are hazardous microorganisms (biosafety level 3; see chapter 7 of this Manual) that have been responsible for numerous laboratory infections (54, 58); specimens obtained from patients suspected of having rickettsial diseases should be handled with appropriate care. Rickettsiae are usually present in infected blood at a relatively low concentration of approximately 10 to 100 viable organisms per ml (38) and lose viability after several hours at room temperature (Q fever rickettsiae excepted). However, viable rickettsiae may still be present in blood that has been kept at room temperature for several days. Processing freshly drawn blood to obtain serum does not pose a serious threat provided that aerosolization is minimized and surgical gloves are worn. Special care should be exercised when specimens that might contain Q fever rickettsiae are handled. *C. burnetii* is highly infectious, resists desiccation and chemical inactivation, and is frequently shed in the urine and feces of infected animals (6). Although sera from suspected Q fever patients present no unusual hazards when handled as described above, infected tissues from Q fever patients should be processed under strict biosafety level 3 conditions and then only by highly qualified personnel. The obvious caveats apply to those who attempt to isolate *C. burnetii* in laboratory animals.

Blood can be used for rickettsial isolation attempts or direct detection of rickettsiae provided that it is collected during the acute, febrile period, before antibiotic therapy. Either clotted or heparinized blood is satisfactory for rickettsial isolation. If isolation attempts cannot be started immediately, blood should be stored at or below −70°C. Blood that must be forwarded to a reference laboratory for rickettsial isolation should be shipped on liquid nitrogen or dry ice in unbreakable, leakproof shipping containers. EDTA-anticoagulated blood is preferred for direct detection. Anticoagulated blood for direct detection requires no refrigeration during shipment and may be stored at 4°C until tested. Postmortem tissues (lung, spleen, and lymph nodes) are also suitable for rickettsial isolation attempts. Thumbnail-size pieces of fresh tissues should be collected and stored frozen until shipped. Tissues obtained postmortem contain far greater numbers of organisms than does infected blood, however, and require additional care in handling, including the wearing of respirators fitted with filters of appropriate particle-size masks and back-fastening gowns.

DIRECT EXAMINATION

Immunofluorescence

Rickettsiae have been detected by immunofluorescence in tissue biopsy samples obtained from the site of tick attachment (42) and from cutaneous lesions (60, 70). Technical improvements in the biopsy assay have resulted in a highly specific assay for confirming Mediterranean spotted fever (MSF) (15). Evaluations of this technique (23, 60, 71, 76) indicate that it will detect *R. rickettsii* or *R. conorii* in about 50% of patients with RMSF or MSF, respectively. All of the factors that contribute to the lack of sensitivity are not known, although most false-negative results were for patients who had received specific antibiotic therapy before the biopsy was performed (60, 71).

Rickettsiae are also present in circulating endothelial cells. Because of their relatively low numbers, ready detection by direct fluorescence microscopy was not feasible until Drancourt and his colleagues (17) utilized monoclonal antibody-coated beads to concentrate circulating endothelial cells from patients with MSF and then detected rickettsiae in the cells by direct immunofluorescence. The procedure had a sensitivity of 66% in initial evaluation. It takes less than 3 h to perform and shows promise as a rapid diagnostic assay for MSF and other rickettsial infections.

Direct fluorescent-antibody testing of tissues collected postmortem is a useful approach for the retrospective diagnosis of rickettsial diseases. This technique was first applied to the rickettsiae by Walker and Cain (70), who successfully detected rickettsiae in kidney tissues of 7 of 10 patients who had died of suspected RMSF. Our laboratory has had its best success with lung and spleen, both highly vascularized tissues, although rickettsiae can also be seen in heart and liver.

PCR

Studies by Tzianabos et al. (68) showed that rickettsial DNA can be detected in infected blood early in rickettsial infections. Using nucleotide primers corresponding to a portion of a defined rickettsial gene, they successfully detected *R. rickettsii* DNA in seven of nine patients with confirmed cases of RMSF by the PCR technique (see chapter 13 of this Manual). Additional testing with clinical specimens indicated that the technique is highly specific but lacks sensitivity; PCR detects rickettsiae best in pa-

TABLE 2 Highlights of various serologic techniques for diagnosis of rickettsial infections

Technique	Minimum positive titer	Time after onset to usual antibody detection	Comments	References
IFA	16–64, depending on investigator	2–3 wk	Relatively sensitive and requires little antigen; can distinguish immunoglobulin isotypes	5, 19, 35, 37, 49, 52, 57
CF	8 or 16, depending on investigator	3–4 wk	Less sensitive than IFA or ELISA but very specific	19, 37, 49, 51, 57
ELISA	Optical density, 0.25 > control	1 wk in some instances	IgM capture assay promising for early diagnosis	12, 13, 16, 21, 26, 43
Latex agglutination	64	1–2 wk	Lacks sensitivity for late-convalescent-stage sera	14, 30, 32, 33, 37
IHA	40?	1–2 wk	Sensitivity that of IFA and > that of CF	1, 37, 57, 65
Immunoperoxidase	20	7 days	Not evaluated for all rickettsiae; useful in field situations	59, 77
Microagglutination	≥8	1–2 wk	Requires considerable antigen; less sensitive than IFA	49, 52, 57
Radioimmunoassay	10?	7–10 days	IgM capture assay may be useful for early diagnosis	16, 40
Weil-Felix	40–320, depending on investigator	2–3 wk	Lacks sensitivity and specificity	5, 8, 34, 37, 52, 57

tients who are seriously ill with RMSF and is of potential importance in the early diagnosis of RMSF and other rickettsial infections.

CULTURE AND ISOLATION

Isolation of rickettsiae is not recommended for routine diagnosis of rickettsial diseases. Only research laboratories that have biosafety level 3 containment and qualified personnel with extensive experience in cultivating rickettsiae should consider isolating rickettsiae from clinical specimens. Postmortem tissues are usually contaminated with bacteria; therefore, suspensions of these tissues are inoculated into susceptible animals for attempted isolation of rickettsiae. In most cases, the guinea pig is the animal of choice for rickettsial isolation. These animals are susceptible to infection by *R. typhi*, *R. rickettsii*, and *C. burnetii* (51), and infection can be monitored by measuring their body temperature. Moreover, male guinea pigs often present with scrotal swelling when infected with *R. typhi* or *R. rickettsii*. Meadow voles (*M. pennsylvanicus*) are very susceptible to *R. rickettsii* infection, but they are not readily available. Mice are the species of choice for isolation of *R. akari* and *R. tsutsugamushi*; however, some inbred mice are resistant to infection by scrub typhus rickettsiae, and outbred mice should be used for isolation of these agents.

Blood specimens are collected from animals that develop overt signs of illness following inoculation. Ill animals are humanely killed, target tissues are harvested aseptically, and blood and tissues are repassaged in animals, embryonated eggs, or tissue cultures. Gimenez stain or direct immunofluorescence may be used to confirm the isolation of rickettsiae. Further passage in tissue culture or embryonated eggs is necessary to increase the yield of organisms.

Surviving animals are bled 28 days postinoculation, and the sera are tested for rickettsial antibodies. Animals that were seronegative before inoculation and are positive after 28 days are considered to have been infected with rickettsiae. Seroconversion in the absence of illness likely indicates the presence of a strain of reduced virulence for the animal of choice.

SEROLOGIC TESTING

Various serologic procedures have been used for diagnosing rickettsial diseases (Table 2). Details of the different techniques can be found in the appropriate references and in chapter 11 of this Manual. Relatively few techniques are used regularly by most laboratories. Some tests lack sensitivity (complement fixation [CF], microagglutination, and Weil-Felix) or specificity compared with newer tests (5, 34, 37, 49, 52, 57), and reagents are not commercially available for many procedures. Antigens for CF tests for Q fever are available commercially, and the Centers for Disease Control and Prevention (CDC) distributes indirect immunofluorescence assay (IFA) reagents for the diagnosis of spotted fever, typhus, and Q fever infections to qualified public health laboratories. Otherwise, reagents can be acquired only through research laboratories.

Recommending a preferred serologic technique is not entirely straightforward. Enzyme-linked immunosorbent assay (ELISA) techniques, particularly IgM capture assays, are among the most sensitive procedures available for rickettsial diagnosis, but they require large quantities of purified antigens that are unavailable commercially. ELISAs also are quite amenable to the large-scale screening of serum specimens, but this advantage is usually lost for rickettsial diseases, because with the exception of epidemic typhus, rickettsioses usually occur sporadically at a relatively low frequency. The IFA technique remains the most popular technique for serodiagnosis of rickettsial diseases, primarily because of the availability of reagents and the ease and economy with which the assay can be incorporated into existing antibody-screening systems.

The indirect hemagglutination (IHA) test has received only limited evaluations, but it apparently has a sensitivity equal to or better than that of IFA. An erythrocyte-sensitizing substance (ESS), which can be obtained from typhus and spotted fever group rickettsiae by alkali extraction, is adsorbed onto sheep or human group O erythrocytes, and the coated cells are then used as antigens for simple agglutination tests (10, 11, 55). Unfortunately, rickettsial ESS is not available commercially. The respective ESSs exhibit group-specific antigenic reactivity. The convenience of the IHA technique makes it ideal for a clinical setting, although the relatively short shelf life (6 months) of sensitized erythrocytes (65) is a disadvantage. Nonetheless, this assay is potentially valuable as a bedside technique, particularly in field situations.

In recent years, the latex agglutination test has been used with some success in a number of public health laboratories (14, 29–33). Latex spheres are coated with ESS, and the sensitized particles are used as antigens in an agglutination test. The latex agglutination test is simple and convenient, but because it is primarily an IgM assay, it occasionally lacks sensitivity for late-convalescent-phase sera. Additionally, because some epitopes may be masked or destroyed when the ESS is adsorbed onto latex particles, some positive sera may go undetected. This assay is recommended for use in the hospital setting instead of the Weil-Felix test, which is still used in similar circumstances.

The Weil-Felix test became popular in the 1920s after it was observed that certain *Proteus* strains would agglutinate early-convalescent-phase sera from patients with suspected rickettsial disease (72). The test fell into disfavor because it lacks both sensitivity and specificity (5, 34), but it has managed to survive because of its convenience in a clinical setting. Despite its simplicity, it does not reliably provide either early or specific diagnosis of rickettsioses.

The IFA procedure for the rickettsiae is similar to conventional IFA techniques, with inactivated yolk sac or tissue culture suspensions of rickettsiae being used as antigens (49). Antigen is applied to 8- or 10-well Teflon-coated glass slides with capillary tubes. Each well should contain the requisite combination of rickettsial antigens. For example, in the United States, spotted fever and typhus group antigens should be included in all tests, with Q fever antigen added as indicated. The antigen spots are then air dried and fixed in cold acetone for 15 min. A series of twofold dilutions of each serum is then prepared, starting at a 1:16 dilution. Known positive and negative control sera should be included in each test. Sera are usually diluted in a 1% suspension of normal yolk sac prepared in phosphate-buffered saline (pH 7.2) to remove any naturally occurring antibodies to egg proteins that might react with the substrate. The remainder of the test is performed by standard methods as described in chapter 11. IFA tests are reasonably sensitive, although the subjectivity of endpoint determinations is an obvious disadvantage.

The indirect immunoperoxidase technique has been extensively utilized for routine diagnosis of scrub typhus and is useful in field situations and where fluorescent microscopes are not available.

ANTIBIOTIC SUSCEPTIBILITY

In vitro sensitivities of rickettsiae are not routinely determined because isolation is attempted in only a few research laboratories and because intracellular rickettsiae require cell culture for antibiotic susceptibility testing. Tetracyclines, chloramphenicol, and fluoroquinilones have shown activity against rickettsiae in vitro (61). Clinical studies demonstrate that tetracyclines and chloramphenicol are the antibiotics of choice; however, ciprofloxacin therapy may be an alternative (4). Combination therapy is recommended for chronic Q fever (41). In the preantibiotic era, the reported case-fatality ratio of RMSF ranged from 23 to 70% (28, 75). With the advent of antibiotics, case-fatality ratios quickly fell below 10% and have remained at approximately 3 to 7% in the United States (22).

EVALUATION, INTERPRETATION, AND REPORTING OF RESULTS

With the exception of Q fever, rickettsial diseases usually present as febrile exanthemous illnesses with protean clinical manifestations. In a clinical setting, rickettsioses must be distinguished from several viral and bacterial illnesses, including meningococcemia, measles, enteroviral exanthems, leptospirosis, typhoid fever, rubella, dengue fever, and ehrlichiosis. The interpretation of serologic tests is frequently confounded by the lack of specificity of available rickettsial reagents. Whole rickettsiae are used as antigens for many antibody assays, and since rickettsiae are antigenically related to some species of bacteria (e.g., *Proteus* species), sera from nonrickettsial patients can react at low titers with rickettsial antigens. This has necessitated the designation of minimum positive titers (Table 2) to confirm recent rickettsial infections. The minimum titers for each procedure should not be considered absolute, however, because in many instances they were assigned somewhat arbitrarily. Borderline titers obtained with single specimens collected at nonoptimum times, i.e., very early or very late in convalescence, are the most difficult to interpret. The traditional way to avoid such problems is to collect acute- and convalescent-phase serum specimens 1 to 3 weeks apart, so that rising or falling titers can be demonstrated.

CDC has established criteria for positivity for the various serologic tests. Fourfold rises in titer detected by any technique (Weil-Felix technique excepted) are considered evidence of rickettsial infections. With single serum specimens, CF titers of ≥16 in a clinically compatible case are also considered positive. With the IFA test, single titers of ≥64 are considered positive for typhus and spotted fever group infections, whereas single IFA titers of 256 are considered minimal for confirmation of Q fever. Although these titers are somewhat higher than those recommended by others (Table 2), they usually present no problem in identification of recent infections if appropriately timed sera are available for testing.

The rickettsial species responsible for an infection is difficult to identify by conventional serologic tests. There is such extensive antigenic relatedness among members of each rickettsial biogroup that a convalescent-phase serum from a patient infected with one species in the biogroup cross-reacts extensively with all other species in the same biogroup. For example, both murine and louse-borne typhus are endemic in certain areas of the world, and routine IFA tests cannot distinguish between these two infections. Antibody adsorption (24) or toxin neutralization (27) tests can distinguish patients who have epidemic or murine typhus infections, but these tests can be performed only by specialty laboratories. In addition, convalescent-phase sera from patients with RMSF or typhus occasionally cross-react with typhus or spotted fever group rickettsiae, respectively (53). However, such cross-reactions are infrequent, and the heterologous titers are routinely much lower than in homologous reactions. CDC has developed an ELISA utilizing lipopolysaccharide antigens that eliminates cross-reaction between typhus group and spotted fever group infections (36).

Determining the specific etiology of spotted fever group infections can be more or less problematic depending on the country of origin. In the Western Hemisphere, RMSF is by far the most prevalent spotted fever group infection. However, in the United States, RMSF must occasionally be differentiated from rickettsialpox; differences in the respective clinical and epidemiologic features are valuable for determining the likely etiology. Furthermore, in contrast to sera from patients with RMSF, sera from patients with rickettsialpox are uniformly negative in Weil-Felix tests. Recent data from Europe and Asia suggest that the distributions of *R. conorii* and *R. sibirica* may overlap more than previously known and could confound attempts to identify specific etiologic agents by conventional serologic testing. However, because all rickettsiae are susceptible to the tetracyclines, it is not necessary to identify the specific etiologic agent to ensure that individual patients receive proper treatment.

Antigenic phase variation must be considered for the interpretation of serologic results for Q fever. In acute, self-limited Q fever infections, antibodies to the phase II antigen appear first and dominate the humoral immune response. With chronic Q fever infections, however, phase I titers eventually equal or exceed phase II titers. Peacock et al. (56) concluded that the ratio of phase II to phase I antibodies was a useful indicator for distinguishing acute from chronic Q fever infections. The ratios were >1 for primary Q fever, ≥1 for patients with granulomatous hepatitis, and ≤1 for endocarditis patients. This generalization is useful, but it may not apply in every situation.

Our laboratory has had little experience with scrub typhus serology. The most important consideration is an awareness of the antigenic diversity of *R. tsutsugamushi* strains in a given area. Unless an appropriate combination of strains of *R. tsutsugamushi* is included in the battery of test antigens, the titers of some serum specimens could appear falsely low, and some infections could even go undetected (7, 64).

Finally, although laboratory testing is central to the diagnosis of rickettsial diseases, clinical findings can contribute to diagnosis, particularly in the face of equivocal serologic results. Although there are no unique pathognomonic features for the rickettsioses, eschars (at the site of arthropod attachment) are useful indicators of RMSF and scrub typhus in areas of endemicity; the vesicular rash of rickettsialpox is unique among the rickettsioses; and the triad of fever, headache, and rash is a useful indicator of RMSF.

REFERENCES

1. **Anacker, R. L., R. N. Philip, L. A. Thomas, and E. A. Casper.** 1979. Indirect hemagglutination test for detection of antibody for *Rickettsia rickettsii* in sera from humans and common laboratory animals. *J. Clin. Microbiol.* **10:**677–684.

2. **Anderson, B. E., J. E. Dawson, D. C. Jones, and K. H. Wilson.** 1991. *Ehrlichia chaffeensis,* a new species associated with human ehrlichiosis. *J. Clin. Microbiol.* **29:**2838–2842.

3. **Baca, O. G.** 1991. Pathogenesis of rickettsial infections; emphasis on Q fever. *Eur. J. Epidemiol.* **7:**222–228.

4. **Beltran, R. R., and J. I. H. Herrero.** 1992. Evaluation of ciprofloxacin and doxycycline in the treatment of Mediterranean spotted fever. *Eur. J. Clin. Microbiol. Infect. Dis.* **11:**427–431.

5. **Berman, S. J., and W. D. Kundin.** 1973. Scrub typhus in South Vietnam. *Ann. Intern. Med.* **79:**26–30.

6. **Bernard, K. W., G. L. Parham, W. G. Winkler, and C. G. Helmick.** 1982. Q fever control measures: recommendations for research facilities using sheep. *Infect. Control* **3:**461–465.

7. **Bourgeois, A. L., J. G. Olson, R. C. Y. Fang, and P. F. D. Van Peenan.** 1977. Epidemiological and serological study of scrub typhus among Chinese military in the Pescadores Islands of Taiwan. *Trans. R. Soc. Trop. Med. Hyg.* **71:**338–342.

8. **Brown, G. W., A. Shirai, C. Rogers, and M. G. Groves.** 1983. Diagnostic criteria for scrub typhus: probability values for immunofluorescent antibody and *Proteus* OXK agglutinin titers. *Am. J. Trop. Med. Hyg.* **32:**1101–1107.

9. **Burgdorfer, W., D. J. Sexton, R. K. Gerloff, R. L. Anacker, R. N. Philip, and L. A. Thomas.** 1975. *Rhipicephalus sanguineus:* vector of a new spotted fever group rickettsia in the United States. *Infect. Immun.* **12:**205–210.

10. **Chang, R. S., E. S. Murray, and J. C. Snyder.** 1954. Erythrocyte-sensitizing substances from rickettsiae of the Rocky Mountain spotted fever group. *J. Immunol.* **73:**8–15.

11. **Chang, R. S., J. C. Snyder, and E. S. Murray.** 1953. A serologically active erythrocyte-sensitizing substance from typhus rickettsiae. II. Serological properties. *J. Immunol.* **70:**215–221.

12. **Crum, J. W., S. Hanchalay, and C. Eamsila.** 1980. New paper enzyme-linked immunosorbent technique compared with microimmunofluorescence for detection of human serum antibodies for *Rickettsia tsutsugamushi. J. Clin. Microbiol.* **11:**584–588.

13. **Dasch, G. A., S. Halle, and A. L. Bourgeois.** 1979. Sensitive microplate enzyme-linked immunosorbent assay for detection of antibodies against the scrub typhus rickettsiae. *J. Clin. Microbiol.* **9:**38–48.

14. **de la Fuente, L., P. Anda, I. Rodriguez, K. E. Hechemy, D. Raoult, and J. Casal.** 1989. Evaluation of latex agglutination test for *Rickettsia conorii* antibodies in seropositive patients. *J. Med. Microbiol.* **28:**69–72.

15. **DeMicco, C., D. Raoult, and M. Toga.** 1986. Diagnosis of Mediterranean spotted fever by using an immunofluorescence technique. *J. Infect. Dis.* **153:**136–138.

16. **Doller, G., P. C. Doller, and H. J. Gerth.** 1984. Early diagnosis of Q fever: detection of immunoglobulin M by radioimmunoassay and enzyme immunoassay. *Eur. J. Clin. Microbiol.* **3:**550–553.

17. **Drancourt, M., F. George, P. Brouqui, J. Sampol, and D. Raoult.** 1992. Diagnosis of Mediterranean spotted fever by indirect immunofluorescence of *Rickettsia conorii* in circulating endothelial cells isolated with monoclonal antibody-coated immunomagnetic beads. *J. Infect. Dis.* **166:**660–663.

18. **Duma, R. J., D. E. Sonenshine, F. M. Bozeman, J. M. Veazy, Jr., B. L. Elisberg, D. P. Chadwick, N. I. Stocks, T. M. McGill, G. B. Miller, and J. N. MacCormack.** 1981. Epidemic typhus in the United States associated with flying squirrels. *J. Am. Med. Assoc.* **25:**2318–2323.

19. **Dupuis, G., O. Peter, M. Peacock, W. Burgdorfer, and E. Haller.** 1985. Immunoglobulin responses in acute Q fever. *J. Clin. Microbiol.* **22:**484–487.

20. **Ellis, M. E., C. C. Smith, and M. A. J. Moffat.** 1982. Chronic or fatal Q fever infection: a review of 16 patients seen in north-east Scotland (1967–80). *Q. J. Med. New Ser.* **205:**54–66.

21. **Field, P. R., J. G. Hunt, and A. M. Murphy.** 1983. Detection and persistence of specific IgM antibodies to *Coxiella burnetii* by enzyme-linked immunosorbent assay: a comparison with

immunofluorescence and complement fixation tests. *J. Infect. Dis.* **148:**477–487.

22. **Fishbein, D. B., J. E. Kaplan, K. W. Bernard, and W. G. Winkler.** 1984. Surveillance of Rocky Mountain spotted fever in the United States, 1981–1983. *J. Infect. Dis.* **150:**609–611.

23. **Fleisher, G., E. T. Lennette, and P. Honig.** 1979. Diagnosis of Rocky Mountain spotted fever by immunofluorescent identification of *Rickettsia rickettsii* in skin biopsy tissue. *J. Pediatr.* **95:**63–65.

24. **Goldwasser, R. A., and C. C. Shepard.** 1959. Fluorescent antibody methods in the differentiation of murine and epidemic typhus fever; specificity changes resulting from previous immunization. *J. Immunol.* **82:**373–380.

25. **Hackstadt, T., M. G. Peacock, P. J. Hitchcock, and R. L. Cole.** 1985. Lipopolysaccharide variation in *Coxiella burnetii:* intrastrain heterogeneity in structure and antigenicity. *Infect. Immun.* **48:**359–365.

26. **Halle, S., G. A. Dasch, and E. Weiss.** 1977. Sensitive enzyme-linked immunosorbent assay for detection of antibodies against typhus rickettsiae, *Rickettsia prowazekii* and *Rickettsia typhi. J. Clin. Microbiol.* **6:**101–110.

27. **Hamilton, H. L.** 1945. Specificity of toxic factors associated with epidemic and murine strains of typhus rickettsiae. *Am. J. Trop. Med. Hyg.* **25:**391–395.

28. **Harrell, G. T.** 1949. Rocky Mountain spotted fever. *Medicine* **28:**333–370.

29. **Hechemy, K. E., R. L. Anacker, N. L. Carlo, J. A. Fox, and H. A. Gaafar.** 1983. Absorption of *Rickettsia rickettsii* antibodies by *Rickettsia rickettsii* antigens in four diagnostic tests. *J. Clin. Microbiol.* **17:**445–449.

30. **Hechemy, K. E., R. L. Anacker, R. N. Philip, K. T. Kleeman, J. N. MacCormack, S. J. Sasovoski, and E. E. Michaelson.** 1980. Detection of Rocky Mountain spotted fever antibodies by a latex agglutination test. *J. Clin. Microbiol.* **12:**144–150.

31. **Hechemy, K. E., E. E. Michaelson, R. L. Anacker, M. Zdeb, S. J. Sasowski, K. T. Kleeman, J. M. Joseph, J. Patel, J. Kudlac, L. B. Elliott, J. Rawlings, C. E. Crump, J. D. Folds, H. Dowda, Jr., J. H. Barrick, J. R. Hindman, G. E. Killgore, D. Young, and R. H. Altieri.** 1983. Evaluation of latex-*Rickettsia rickettsii* test for Rocky Mountain spotted fever in 11 laboratories. *J. Clin. Microbiol.* **18:**938–946.

32. **Hechemy, K. E., D. Raoult, C. Eismann, Y. Han, and J. A. Fox.** 1986. Detection of antibodies to *Rickettsia conorii* with a latex agglutination test in patients with Mediterranean spotted fever. *J. Infect. Dis.* **153:**132–135.

33. **Hechemy, K. E., and B. B. Rubin.** 1983. Latex *Rickettsia rickettsii* test reactivity in seropositive patients. *J. Clin. Microbiol.* **17:**489–492.

34. **Hechemy, K. E., R. W. Stevens, S. Sasowski, E. E. Michaelson, E. A. Casper, and R. N. Philip.** 1979. Discrepancies in Weil-Felix and microimmunofluorescence test results for Rocky Mountain spotted fever. *J. Clin. Microbiol.* **9:**292–293.

35. **Hunt, J. G., P. R. Field, and A. M. Murphy.** 1983. Immunoglobulin responses to *Coxiella burnetii* (Q fever): single-serum diagnosis of acute infection, using an immunofluorescence technique. *Infect. Immun.* **39:**977–981.

36. **Jones, D., B. Anderson, J. Olson, and C. Greene.** 1993. Enzyme-linked immunosorbent assay for detection of human immunoglobulin G to lipopolysaccharide of spotted fever group rickettsiae. *J. Clin. Microbiol.* **31:**138–141.

37. **Kaplan, J. E., and L. B. Schonberger.** 1986. The sensitivity of various serologic tests in the diagnosis of Rocky Mountain spotted fever. *Am. J. Trop. Med. Hyg.* **35:**840–844.

38. **Kaplowitz, L. G., J. V. Lange, J. J. Fischer, and D. H. Walker.** 1983. Correlation of rickettsial titers, circulating endotoxin, and clinical features in Rocky Mountain spotted fever. *Arch. Intern. Med.* **143:**1149–1151.

39. **Kelly, P., L. Matthewman, L. Beati, D. Raoult, P. Mason, M. Dreary, and R. Makombe.** 1992. African tick-bite fever: a new spotted fever group rickettsiosis under an old name. *Lancet* **340:**982–983.

40. **Lackman, D. B., G. Gilda, and R. N. Philip.** 1964. Appli-

cation of the radioisotope precipitation test to the study of Q fever in man. *Health Lab. Sci.* **1:**21–28.

41. **Levy, P. Y., M. Drancourt, J. Etienne, J. C. Auvergnat, J. Beytout, J. M. Sainty, F. Goldstein, and D. Raoult.** 1991. Comparison of different antibiotic regimes for therapy of 32 cases of Q fever endocarditis. *Antimicrob. Agents Chemother.* **35:**533–537.

42. **Mansueto, S., G. Tringali, R. Di Leo, M. Maniscalco, M. Montenegro, and D. Walker.** 1984. Demonstration of spotted fever group rickettsiae in the tache noire of a healthy person in Sicily. *Am. J. Trop. Med. Hyg.* **33:**479–482.

43. **Mansueto, S., G. Vitale, C. Mocciari, R. Librizzi, I. Friscia, V. Usticano, G. Gambino, and G. Reina.** 1989. Laboratory diagnosis of boutonneuse fever by enzyme-linked immunosorbent assay. *Trans. R. Soc. Trop. Med. Hyg.* **83:**855–857.

44. **McDade, J. E., C. C. Shepard, M. A. Redus, V. F. Newhouse, and J. D. Smith.** 1980. Evidence of *Rickettsia prowazekii* infections in the United States. *Am. J. Trop. Med. Hyg.* **29:**277–284.

45. **Miller, E. S., and P. B. Beeson.** 1946. Murine typhus fever. *Medicine* **25:**1–15.

46. **Minnick, M. F., and G. L. Stiegler.** 1993. Nucleotide sequence and comparison of the 5s ribosomal RNA genes of *Rochalimaea henselae, R. quintana,* and *Brucella abortus. Nucleic Acids Res.* **21:**2518.

47. **Myers, W. F., and C. L. Wisseman, Jr.** 1980. Genetic relatedness among the typhus group of rickettsiae. *Int. J. Syst. Bacteriol.* **30:**143–150.

48. **National Research Council, Division of Medical Sciences, Committee on Pathology.** 1953. Pathology of epidemic typhus. Report of fatal cases studied by the United States of America Typhus Commission in Cairo, Egypt, during 1943–1945. *Arch. Pathol.* **56:**397–435, 512–553.

49. **Newhouse, V. F., C. C. Shepard, M. D. Redus, T. Tzianabos, and J. E. McDade.** 1979. A comparison of the complement fixation, indirect fluorescent antibody and microagglutination test for the serological diagnosis of rickettsial diseases. *Am. J. Trop. Med. Hyg.* **28:**387–395.

50. **Ohashi, N., H. Nashimoto, H. Ikeda, and A. Tamura.** 1992. Diversity of immunodominant 56-kDa type-specific antigen (TSA) of *Rickettsia tsutsugamushi. J. Biol. Chem.* **267:**12728–12735.

51. **Ormsbee, R., M. Peacock, R. Gerloff, G. Tallent, and D. Wike.** 1978. Limits of rickettsial infectivity. *Infect. Immun.* **19:**239–245.

52. **Ormsbee, R., M. Peacock, R. Philip, E. Casper, J. Plorde, T. Gabre-Kidan, and L. Wright.** 1977. Serologic diagnosis of epidemic typhus fever. *Am. J. Epidemiol.* **105:**261–271.

53. **Ormsbee, R., M. Peacock, R. Philip, E. Casper, J. Plorde, T. Gabre-Kidan, and L. Wright.** 1978. Antigenic relationships between the typhus and spotted fever groups of rickettsiae. *Am. J. Epidemiol.* **108:**53–59.

54. **Oster, C. N., D. S. Burke, R. H. Kenyon, M. S. Ascher, P. Harber, and C. E. Pedersen, Jr.** 1977. Laboratory-acquired Rocky Mountain spotted fever. *N. Engl. J. Med.* **297:**859–862.

55. **Osterman, J. V., and C. S. Eisemann.** 1978. Rickettsial indirect hemagglutination test: isolation of erythrocyte sensitizing substance. *J. Clin. Microbiol.* **8:**189–196.

56. **Peacock, M. G., R. N. Philip, J. C. Williams, and R. S. Faulkner.** 1983. Serological evaluation of Q fever in humans: enhanced phase I titers of immunoglobulins G and A are diagnostic for Q-fever endocarditis. *Infect. Immun.* **41:**1089–1098.

57. **Philip, R. N., E. A. Casper, J. N. MacCormack, D. L. Sexton, L. A. Thomas, R. L. Anacker, W. Burgdorfer, and S. Vick.** 1977. A comparison of serologic methods for diagnosis of Rocky Mountain spotted fever. *J. Epidemiol.* **105:**56–67.

58. **Pike, R. M.** 1976. Laboratory-associated infections: summary and analysis of 3921 cases. *Health Lab. Sci.* **13:**105–114.

59. **Raoult, D., C. DeMicco, H. Chaudet, and J. Tamalet.** 1985. Serological diagnosis of Mediterranean spotted fever by the

immunoperoxidase reaction. *Eur. J. Clin. Microbiol.* **4:**441–442.

60. **Raoult, D., C. DeMicco, H. Gallais, and M. Toga.** 1984. Laboratory diagnosis of Mediterranean spotted fever by immunofluorescent demonstration of *Rickettsia conorii* in cutaneous lesions. *J. Infect. Dis.* **150:**145–148.

61. **Raoult, D., M. Yeaman, and O. Baca.** 1989. Susceptibility of *Rickettsia* and *Coxiella* to quinolones. *Rev. Infect. Dis.* **11:**S986.

62. **Regnery, R. L., C. L. Spruill, and B. D. Plikaytis.** 1991. Genotypic identification of rickettsiae and estimation of intraspecies sequence divergence for portions of two rickettsial genes. *J. Bacteriol.* **173:**1576–1589.

63. **Scott, G. H., and J. C. Williams.** 1990. Susceptibility of *Coxiella burnetii* to chemical disinfectants. *Ann. N.Y. Acad. Sci.* **590:**291–296.

64. **Shiral, A., J. C. Coolbaugh, E. Gan, T. C. Chan, D. L. Huxsoll, and M. C. Groves.** 1982. Serological analysis of scrub typhus isolates from the Pescadores and Philippine Islands. *Jpn. J. Med. Sci. Biol.* **35:**255–259.

65. **Shiral, A., J. W. Dietal, and J. V. Osterman.** 1975. Indirect hemagglutination test for human antibody to typhus and spotted fever group rickettsiae. *J. Clin. Microbiol.* **2:**430–437.

66. **Snyder, J. C.** 1965. Typhus fever rickettsiae, p. 1059–1094. *In* F. L. Horsfall and I. Tamm (ed.), *Viral and Rickettsial Infections of Man.* J. B. Lippincott Co., Philadelphia.

67. **Turck, W. P. G., G. Howitt, L. A. Turnberg, H. Fox, M. Longson, M. B. Matthews, and R. Das Gupta.** 1976. Chronic Q fever. *Q. J. Med. New Ser.* **45:**193–217.

68. **Tzianabos, T., B. E. Anderson, and J. E. McDade.** 1989. Detection of *Rickettsia rickettsii* DNA in clinical specimens by using polymerase chain reaction technology. *J. Clin. Microbiol.* **27:**2866–2868.

69. **Uchida, T., T. Uchiyama, K. Kumano, and D. H. Walker.** 1992. *Rickettsia japonica* sp. nov., the etiological agent of spotted fever rickettsiosis in Japan. *Int. J. Syst. Bacteriol.* **42:**303–305.

70. **Walker, D. H., and B. G. Cain.** 1978. A method for specific diagnosis of Rocky Mountain spotted fever on fixed, paraffin-embedded tissue by immunofluorescence. *J. Infect. Dis.* **137:**206–209.

71. **Walker, D. H., B. G. Cain, and P. M. Olmstead.** 1978. Laboratory diagnosis of Rocky Mountain spotted fever by immunofluorescent demonstration of *Rickettsia rickettsii* in cutaneous lesions. *Am. J. Clin. Pathol.* **69:**619–623.

72. **Weil, E., and A. Felix.** 1916. Zur serologischen Diagnosis des Fleckfiebers. *Wien. Klin. Wochenschr.* **29:**33–35.

73. **Weisburg, W. G., M. E. Dobson, J. E. Samuel, G. A. Dasch, L. P. Mallavia, O. Baca, L. Mandelco, J. E. Sechrest, E. Weiss, and C. R. Woese.** 1989. Phylogenetic diversity of the rickettsiae. *J. Bacteriol.* **171:**4202–4206.

74. **Weiss, E., and J. W. Moulder.** 1984. Rickettsiales, p. 687–704. *In* N. R. Krieg and J. G. Holt (ed.), *Bergey's Manual of Systematic Bacteriology*, vol. 1. The Williams & Wilkins Co., Baltimore.

75. **Wilson, L. B., and W. M. Chowning.** 1904. Studies in *Pyroplasmosis hominis* ("spotted fever" or "tick fever" of the Rocky Mountains). *J. Infect. Dis.* **1:**31–57.

76. **Woodward, T. E., C. D. Pedersen, Jr., C. N. Oster, L. R. Bagley, J. Romberger, and M. J. Snyder.** 1976. Prompt confirmation of Rocky Mountain spotted fever: identification of rickettsiae in skin tissues. *J. Infect. Dis.* **134:**297–301.

77. **Yamamoto, S., and Y. Minamishima.** 1982. Serodiagnosis of tsutsugamushi fever (scrub typhus) by the indirect immunoperoxidase technique. *J. Clin. Microbiol.* **15:**1128–1132.

78. **Yu, X., Y. Jin, M. Fan, G. Xu, O. Liu, and D. Raoult.** 1993. Genotypic and antigenic identification of two new strains of spotted fever group rickettsiae isolated from China. *J. Clin. Microbiol.* **31:**83–88.

Ehrlichia

JAMES G. OLSON AND JACQUELINE E. DAWSON

57

TAXONOMY

Ehrlichia species are members of the order *Rickettsiales*, family *Rickettsiaceae*, which includes the tribes *Rickettsieae*, *Wolbachieae*, and *Ehrlichieae* (39). The tribe *Ehrlichieae* comprises three genera: *Ehrlichia*, *Cowdria*, and *Neorickettsia*. Four species are assigned to the genus *Ehrlichia*: *Ehrlichia canis*, *E. phagocytophila*, *E. equi*, and *E. sennetsu* (39). Three new species, *E. risticii*, the etiologic agent of Potomac horse fever; *E. chaffeensis*, the etiologic agent of human ehrlichiosis in the United States; and *E. ewingii*, the etiologic agent of canine granulocytic ehrlichiosis, have been subsequently accepted by the *International Journal of Systematic Bacteriology* (2, 3, 26). In addition, there are several species incertae sedis (see chapter 54 of this Manual).

DESCRIPTION OF THE GENUS

Ehrlichiae, first described in 1935, are obligate, intracytoplasmic, gram-negative rickettsiae with a tropism for mononuclear or polymorphonuclear leukocytes (39). The organisms exist in membrane-lined vacuoles within the cytoplasm of infected host cells. Ehrlichial inclusions, called morulae, contain variable numbers of organisms. Single organisms, wrapped in vacuolar membrane, have also been observed in the cytoplasm (10). Individual organisms are extremely pleomorphic, varying from 0.2 to 0.4 μm; morulae range up to 4 μm. *Ehrlichia* spp. occur in small electron-dense and large electron-lucent forms, but a clear life cycle has not been elucidated (35). Entry of *Ehrlichia* spp. into new host cells in vitro may occur by a coupled exocytosis in one cell and endocytosis in an adjacent cell (35). The natural histories of *Ehrlichia* spp. are given in Table 1.

CLINICAL SIGNIFICANCE

Animal Diseases

Canine ehrlichiosis is caused by *E. canis* infection of domestic and wild canidae (25). The acute stage begins 10 to 15 days after infection and lasts a few days to 3 weeks. Dogs commonly develop fever, anorexia, generalized depression, mucopurulent ocular and nasal discharges, weight loss, lymphadenopathy, corneal opacities, pneumonitis, and vomiting (34). Leukopenia, thrombocytopenia, and anemia

(27) frequently occur, and the bone marrow reveals a normal to increased number of myeloid cells and an increased number of megakaryocytes. Biochemical abnormalities include hyperglobulinemia and elevated transaminase levels (34). Although most dogs appear to recover from the acute stage, they usually remain infected. A subclinical phase follows, during which the dogs appear healthy but mild hematologic abnormalities (especially thrombocytopenia) may persist (27).

Chronic canine ehrlichiosis often begins a few months after acute infection. The severity of the chronic stage varies among different breeds. German shepherds develop tropical canine pancytopenia (27), the most severe form, often first manifested by epistaxis, petechial or ecchymotic hemorrhages, fever, lymphadenopathy, edema, anorexia, and dyspnea (8).

E. phagocytophila causes tick-borne fever in sheep, goats, cattle, and deer in Europe, Asia, and South Africa (46). Most animals recover within 2 weeks after the onset of illness, although spontaneous relapse may occur following splenectomy, other infectious disease, or injury (21). This disease is characterized by fever, reduced milk production, reduced growth rate, and leukopenia (22). Stillbirths and abortion occur in infected pregnant cattle (45). After clinical recovery, *E. phagocytophila* may persist for up to 25 months in sheep (21).

E. equi causes equine granulocytic ehrlichiosis and is found in the cytoplasm of granulocytes, primarily neutrophils (23). The disease is manifested by fever, depression, partial anorexia, limb edema, petechiation, icterus, ataxia, and reluctance to move (41). Thrombocytopenia, elevated plasma icterus index, decreased packed-cell volume, and marked leukopenia (involving first lymphocytes and then granulocytes) are observed. The disease is usually mild and is fatal only with injury resulting from ataxia.

Cases of Potomac horse fever, caused by *E. risticii*, peak in midsummer, suggesting vector transmission (26). Attempts to identify specific vectors have been unsuccessful. Potomac horse fever is manifested by fever, anorexia, depression, colic, laminitis, lymphadenopathy, leukopenia, and diarrhea, which may progress to dehydration, hypovolemia, and shock (26). Transplacental transmission of *E. risticii* has been reported (14), and the organism may induce abortion or resorption of the fetus.

Canine granulocytic ehrlichiosis is caused by *E. ewingii*

TABLE 1 Natural history of *Ehrlichia* spp.

Group and organism	Disease	Natural host	Experimental host (reference)	Vector (reference)
Monocytic				
E. chaffeensis	Human ehrlichiosis	Human	None identified	*Amblyomma americanum?* (4), *Dermacentor variabilis?*
E. sennetsu	Sennetsu ehrlichiosis	Human	Mouse (32), nonhuman primates (40)	None identified
E. canis	Canine ehrlichiosis, tropical canine pancytopenia	Canid	None identified	*Rhipicephalus sanguineus* (29)
E. risticii	Equine monocytic ehrlichiosis, Potomac horse fever	Horse	Mouse (35), nonhuman primates (42), cat (12), dog (37)	None identified
Granulocytic				
E. phagocytophila	Tick-borne fever	Sheep, goat, cow, bison, deer	Mouse, guinea pig (22)	*I. ricinus*
E. equi	Equine ehrlichiosis	Horse	Nonhuman primates (28), cat (28), dog (28), sheep and goat (41)	None identified
E. ewingii	Canine granulocytic ehrlichiosis	Canid	None identified	*Amblyomma americanum?* (6)

(3) and is milder than classic (monocytic) canine ehrlichiosis (18). Chronic canine granulocytic ehrlichiosis has been associated with polyarthritis (18).

Human Diseases

Sennetsu Ehrlichiosis
E. sennetsu, first isolated from a human in 1953 (32), is the causative agent of sennetsu fever, sennetsu rickettsiosis, or sennetsu ehrlichiosis. The disease has been described in southwestern Japan since the late 19th century and is characterized by sudden onset of fever, chills, headache, malaise, insomnia, diaphoresis, sore throat, and anorexia (31). Lymphadenopathy is generalized, characterized by tenderness, and most prominent in the postauricular and posterior cervical regions. Hepatomegaly and splenomegaly occur in about one-third of cases (43). Geographic distribution of confirmed cases includes Japan (31, 32, 43) and Malaysia (36, 44). Laboratory findings include leukopenia with increased neutrophil levels during the acute phase and relative and absolute pleocytosis accompanied by the appearance of atypical cells during the late febrile and convalescent phases of the illness. The disease is self-limiting, and serious complications or fatalities have not been reported. Differential diagnoses include mononucleosis and scrub typhus.

Human Ehrlichiosis in the United States
The first case of human ehrlichiosis in the United States was described in 1986 (30). The etiologic agent was isolated from a patient (10) and shown to be a distinct species, *E. chaffeensis* (1). The disease is serologically and clinically distinct from sennetsu ehrlichiosis. Clinical signs and symptoms are similar to those of Rocky Mountain spotted fever and include fever, headache, arthralgia, myalgia, anorexia, nausea, vomiting, chills, pneumonia, and, infrequently, rash. Laboratory findings include leukopenia, thrombocytopenia, and elevated levels of liver enzymes. More than 300 laboratory-confirmed cases have been reported, with nine deaths. Cases are concentrated in western, south central,

and South Atlantic states. The true incidence of human ehrlichiosis in the United States is most certainly underestimated; there is no requirement to report the disease to state health departments or to the Centers for Disease Control and Prevention (CDC). Epidemiological investigations (17, 20, 24, 33) have shown that human ehrlichiosis occurs at or above the rate of Rocky Mountain spotted fever.

COLLECTION, TRANSPORT, AND STORAGE OF SPECIMENS
Specimens for the confirmation of ehrlichiosis include anticoagulated blood (heparinized or treated with EDTA) (10) for isolation attempts and for PCR detection (5), blood films fixed in methanol for microscopic examination (30), serum for detection of antibodies (11), and spinal fluid for direct microscopic examination and PCR detection (16) and for serologic testing for antibodies.

Recommendations for optimal transport conditions of blood and tissue for recovery of ehrlichiae are difficult, because *E. chaffeensis* has been recovered from a human only once. Specimens for isolation should be transported immediately on ice packs. Blood and tissue for detection of ehrlichiae by the PCR technique may be refrigerated or frozen. Sera and spinal fluids may be stored at 4 or −20°C until tested.

ISOLATION PROCEDURES
The isolation of *E. canis*, *E. risticii*, *E. sennetsu*, and *E. chaffeensis* is accomplished by inoculating the leukocyte fraction of peripheral blood of affected mammals into cell culture. *E. canis* (13) and *E. chaffeensis* (10) are primarily propagated in a continuous canine macrophage cell line (DH82); *E. risticii* (38) and *E. sennetsu* (9) are propagated in a continuous murine macrophage cell line (P388D1). Even if no ehrlichial organisms are observed in buffy coat

smears of suspects, ehrlichial organisms can be isolated by these procedures (35).

To date, none of the granulocytic *Ehrlichia* spp. has been cultured in a continuous cell line (35), and temporary culture is limited to *E. phagocytophila* in granulocytes from infected sheep (46). Similar to *E. phagocytophila,* however, *E. equi* and *E. ewingii* can be observed in peripheral granulocytes.

IDENTIFICATION

Direct microscopic examination of Giemsa- or Diff-Quik-stained peripheral blood buffy coat smears may reveal clusters of round, dark-purple-stained, small dots or clusters of dots (morulae) in the cytoplasm of leukocytes. This type of examination is most useful during the febrile stage of the disease for detection of *E. phagocytophila, E. equi,* and *E. ewingii,* all of which primarily infect granulocytes (35). *E. canis* and *E. chaffeensis* are more frequently observed in agranulocytes by this technique than is *E. sennetsu* or *E. risticii.*

E. canis has been demonstrated in tissues by an immunohistology procedure (7). *E. risticii* has been observed in intestinal epithelial cells and macrophages in paraffin-embedded tissue specimens by an immunoperoxidase procedure and silver stain (35). Immunohistology for *E. chaffeensis* has been performed on tissue sections, aspirated bone marrow, and peripheral blood smears (15). *E. chaffeensis* was detected primarily within histiocytes, but morulae were occasionally present within lymphocytes (15).

Methods of identification by the PCR procedure have been developed and are available by special request at the CDC. Positive results have been achieved using whole-blood specimens collected in EDTA from febrile patients prior to antibiotic treatment. DNA extracted from the blood specimens is subsequently amplified with specific primers and electrophoresed through an agarose gel (5). PCR-amplified products can then be confirmed by detection with a specific probe or by direct sequencing.

SEROLOGIC TESTS

Specific humoral immune responses are the basis for the diagnosis of ehrlichial diseases. Serologic diagnosis of infection can be determined either by the preferred indirect fluorescent-antibody (IFA) test (35) or by enzyme-linked immunosorbent assay. The positive cutoff titer has not been standardized and currently varies with the laboratory conducting tests. At CDC, human ehrlichiosis is diagnosed by a fourfold change in immunoglobulin G antibody levels. Until 1990, when *E. chaffeensis* was isolated, *E. canis* was used as the diagnostic antigen. Comparative tests, however, have shown that *E. chaffeensis* is more sensitive than *E. canis* during the early stages of the disease (8a).

Ehrlichia-infected macrophages are used as the antigen for *E. canis, E. risticii, E. sennetsu,* and *E. chaffeensis* IFA. Peripheral blood leukocyte smears for an experimentally infected animal are used for serodiagnosis of *E. equi* and *E. phagocytophila.* The diagnosis of canine granulocytic ehrlichiosis is based on serologic cross-reaction with *E. canis.*

ANTIBIOTIC SUSCEPTIBILITIES

Tetracyclines and their derivatives are the preferred antibiotics for treatment of ehrlichial infections in all species (35). Human ehrlichiosis patients treated with oxytetracy-cline, doxycycline, or chloramphenicol defervesced more rapidly and had significantly shorter hospitalizations than patients treated with other antibiotics (19).

EVALUATION, INTERPRETATION, AND REPORTING OF RESULTS

Human ehrlichiosis is a potentially fatal illness that presents with nonspecific clinical signs and symptoms and is often confused with Rocky Mountain spotted fever. Serologic test results are useful for the clinician in confirming a clinical diagnosis but frequently are not available until after the patient has been discharged.

Serologic confirmation of a case of human ehrlichiosis by IFA depends on demonstrating a fourfold change in antibody titer against *E. chaffeensis,* with a minimum titer of 1:64. Ideally, the initial serum specimen is obtained during the acute phase of illness and the second is obtained at least 2 to 3 weeks later (11). Detection of ehrlichiae in blood (5) by PCR is currently the only test that allows species-level identification of *E. chaffeensis.* However, the PCR assay is not available as a routine diagnostic test.

REFERENCES

1. **Anderson, B. E., J. E. Dawson, D. C. Jones, and K. H. Wilson.** 1991. *Ehrlichia chaffeensis,* a new species associated with human ehrlichiosis. *J. Clin. Microbiol.* **29:**2838–2842.
2. **Anderson, B. E., J. E. Dawson, D. C. Jones, and K. H. Wilson.** 1992. Validation of the publication of new names and new combinations previously effectively published outside the IJSB: list no. 41. *Int. J. Syst. Bacteriol.* **42:**327–329.
3. **Anderson, B. E., C. E. Greene, D. C. Jones, and J. E. Dawson.** 1992. *Ehrlichia ewingii* sp. nov., the etiologic agent of canine granulocytic ehrlichiosis. *Int. J. Syst. Bacteriol.* **42:**299–302.
4. **Anderson, B. E., K. G. Sims, J. G. Olson, J. E. Childs, J. F. Piesman, C. M. Happ, G. O. Maupin, and B. J. Johnson.** 1993. *Amblyomma americanum:* a potential vector of human ehrlichiosis. *Am. J. Trop. Med. Hyg.* **49:**239–244.
5. **Anderson, B. E., J. W. Sumner, J. E. Dawson, T. Tzianabos, C. R. Greene, J. G. Olson, D. B. Fishbein, M. Olsen-Rasmussen, B. P. Holloway, E. H. George, and A. F. Azad.** 1992. Detection of the etiologic agent of human ehrlichiosis by polymerase chain reaction. *J. Clin. Microbiol.* **30:**775–780.
6. **Anziani, D. S., S. A. Ewing, and R. W. Barker.** 1990. Experimental transmission of a granulocytic form of the tribe Ehrlichieae by *Dermacentor variabilis* and *Amblyomma americanum* to dogs. *Am. J. Vet. Res.* **51:**929–931.
7. **Aronson, J., J. Scimeca, D. Harris, and D. H. Walker.** 1990. Immunohistologic demonstration of *Ehrlichia canis. Ann. N.Y. Acad. Sci.* **590:**148–156.
8. **Buhles, W. C., D. L. Huxsoll, and M. Ristic.** 1974. Tropical canine pancytopenia: clinical, hematologic, and serologic response of dogs to *Ehrlichia canis* infection, tetracycline therapy, and challenge inoculation. *J. Infect. Dis.* **130:**357–367.
8a.**Centers for Disease Control and Prevention.** Unpublished data.
9. **Cole, A. I., M. Ristic, G. E. Lewis, and G. Rapmund.** 1985. Continuous propagation of *Ehrlichia sennetsu* in murine macrophage cell culture. *Am. J. Trop. Med. Hyg.* **34:**774–780.
10. **Dawson, J. E., B. E. Anderson, D. B. Fishbein, J. L. Sanchez, C. S. Goldsmith, K. H. Wilson, and C. W. Duntley.** 1991. Isolation and characterization of an *Ehrlichia* sp. from a patient diagnosed with human ehrlichiosis. *J. Clin. Microbiol.* **29:**2741–2745.
11. **Dawson, J. E., D. B. Fishbein, T. R. Eng, M. A. Redus, and N. R. Greene.** 1990. Diagnosis of human ehrlichiosis with the indirect fluorescent antibody test: kinetics and specificity. *J. Infect. Dis.* **162:**91–95.

12. **Dawson, J. E., C. J. Holland, and M. Ristic.** 1988. Susceptibility of cats to infection with *E. risticii*, causative agent of equine monocytic ehrlichiosis. *Am J. Vet. Res.* **49:**2096–2100.

13. **Dawson, J. E., Y. Rikihisa, S. Ewing, and D. B. Fishbein.** 1991. Serologic diagnosis of human ehrlichiosis using two *Ehrlichia canis* isolates. *J. Infect. Dis.* **163:**564–567.

14. **Dawson, J. E., M. Ristic, C. J. Holland, R. H. Whitlock, and J. E. Sessions.** 1987. Isolation of *Ehrlichia risticii*, the causative agent of Potomac horse fever, from the fetus of an experimentally infected mare. *Vet. Rec.* **121:**232.

15. **Dumler, J. S., J. E. Dawson, and D. H. Walker.** 1993. Human ehrlichiosis: hematopathology and immunohistologic detection of *Ehrlichia chaffeensis*. *Hum. Pathol.* **24:**391–396.

16. **Dunn, B. E., T. P. Monson, J. S. Dumler, C. C. Morris, A. B. Westbrook, J. L. Duncan, J. E. Dawson, K. G. Sims, and B. E. Anderson.** 1992. Identification of *Ehrlichia chaffeensis* morulae in cerebrospinal fluid mononuclear cells. *J. Clin. Microbiol.* **30:**2207–2210.

17. **Eng, T. R., J. R. Harkess, D. B. Fishbein, J. E. Dawson, C. N. Greene, M. A. Redus, and F. T. Satalowich.** 1990. Epidemiologic, clinical, and laboratory findings of human ehrlichiosis in the United States, 1988. *JAMA* **264:**2251–2258.

18. **Ewing, S. A., W. R. Roberson, R. G. Buckner, and C. S. Hayat.** 1971. A new strain of *Ehrlichia canis*. *J. Am. Vet. Med. Assoc.* **159:**1771–1774.

19. **Fishbein, D. B., J. E. Dawson, and L. Robinson.** 1994. Human ehrlichiosis in the United States, 1985–1990. *Ann. Intern. Med.* **120:**736–743.

20. **Fishbein, D. B., A. Kemp, J. E. Dawson, N. R. Greene, M. A. Redus, and D. H. Fields.** 1989. Human ehrlichiosis: prospective active surveillance in febrile hospitalized patients. *J. Infect. Dis.* **160:**803–809.

21. **Foggie, A.** 1951. Studies on the infectious agent of tick-borne fever in sheep. *J. Pathol. Bacteriol.* **63:**1–15.

22. **Foggie, A., and C. S. Hood.** 1961. Adaptation of the infectious agent of tick-borne fever to guinea pigs and mice. *J. Comp. Pathol.* **71:**414–427.

23. **Gribble, D. H.** 1969. Equine ehrlichiosis. *J. Am. Vet. Med. Assoc.* **155:**462–469.

24. **Harkess, J. R., S. A. Ewing, J. M. Crutcher, J. Kudlac, G. McKee, and G. R. Istre.** 1989. Human ehrlichiosis in Oklahoma. *J. Infect. Dis.* **159:**576–579.

25. **Harvey, J. W., C. F. Simpson, J. M. Gaskin, and J. H. Sameck.** 1979. Ehrlichiosis in wolves, dogs, and wolf-dog crosses. *J. Am. Vet. Med. Assoc.* **175:**901–905.

26. **Holland, C. J., E. Weiss, W. Burgdorfer, A. I. Cole, and I. Kakoma.** 1985. *Ehrlichia risticii* sp. nov.: etiological agent of equine monocytic ehrlichiosis (synonym, Potomac horse fever). *Int. J. Syst. Bacteriol.* **35:**524–526.

27. **Huxsoll, D. L., P. K. Hildebrandt, R. M. Nims, and J. S. Walker.** 1970. Tropical canine pancytopenia. *J. Am. Vet. Med. Assoc.* **157:**1627–1632.

28. **Lewis, G. E., D. L. Huxsoll, M. Ristic, and A. J. Johnson.** 1975. Experimentally induced infection of dogs, cats, and nonhuman primates with *Ehrlichia equi*, etiologic agent of equine ehrlichiosis. *Am. J. Vet. Res.* **36:**85–88.

29. **Lewis, G. E., M. Ristic, R. D. Smith, T. Lincoln, and E. H. Stephenson.** 1977. The brown dog tick *Rhipicephalus sanguineus* and the dog as experimental hosts of *Ehrlichia canis*. *Am. J. Vet. Res.* **38:**1953–1955.

30. **Maeda, K., N. Markowitz, R. C. Hawley, M. Ristic, D. Cox, and J. E. McDade.** 1987. Human infection with *Ehrlichia canis*, a leukocytic rickettsia. *N. Engl. J. Med.* **316:**853–856.

31. **Misao, T., and K. Katsuta.** 1956. Epidemiology of infectious mononucleosis. *Jpn. J. Clin. Exp. Med.* **33:**73–82.

32. **Misao, T., and Y. Kobayashi.** 1954. Studies on infectious mononucleosis. I. Isolation of etiologic agent from blood, bone marrow, and lymph node of a patient with infectious mononucleosis by using mice. *Tokyo Iji Shinshi* **71:**683–686.

33. **Peterson, L. R., L. A. Sawyer, D. B. Fishbein, P. W. Kelley, R. J. Thomas, L. A. Magnarelli, M. Redus, and J. E. Dawson.** 1989. An outbreak of ehrlichiosis in members of an Army reserve unit exposed to ticks. *J. Infect. Dis.* **159:**562–568.

34. **Reardon, M. J., and K. R. Pierce.** 1981. Acute experimental canine ehrlichiosis. I. Sequential reaction of the hemic and lymphoreticular systems. *Vet. Pathol.* **18:**48–61.

35. **Rikihisa, Y.** 1991. The tribe *Ehrlichieae* and ehrlichial diseases. *Clin. Microbiol. Rev.* **4:**286–308.

36. **Ristic, M.** 1990. Current strategies in research on ehrlichiosis, p. 136–153. *In* J. C. Williams and I. Kakoma (ed.), *Ehrlichiosis: a Vector-Borne Disease of Animals and Humans.* Kluwer Academic Publishers, Boston.

37. **Ristic, M., J. E. Dawson, C. J. Holland, and A. Jenny.** 1988. Susceptibility of dogs to infection with *Ehrlichia risticii*, causative agent of equine monocytic ehrlichiosis (Potomac horse fever). *Am. J. Vet. Res.* **49:**1497–1500.

38. **Ristic, M., C. J. Holland, J. E. Dawson, J. Sessions, and J. Palmer.** 1986. Diagnosis of equine monocytic ehrlichiosis (Potomac horse fever) by indirect immunofluorescence. *J. Am. Vet. Med. Assoc.* **189:**39–46.

39. **Ristic, M., and D. L. Huxsoll.** 1984. Tribe II. *Ehrlichieae*, p. 704–711. *In* N. R. Krieg and J. G. Holt (ed.), *Bergey's Manual of Systematic Bacteriology*, vol. 1. The Williams & Wilkins Co., Baltimore.

40. **Shishido, A., S. Honjo, M. Suganuma, S. Ohtaki, M. Hikita, T. Fujiwara, and M. Takasaka.** 1965. Studies on infectious mononucleosis induced in the monkey by experimental infection with *Rickettsia sennetsu*. I. Clinical observations and etiological investigations. *Jpn. J. Med. Sci. Biol.* **18:**73–83.

41. **Stannard, A. A., K. H. Gribble, and R. S. Smith.** 1969. Equine ehrlichiosis, a disease with similarities to tick-borne fever and bovine petechial fever. *Vet. Rec.* **84:**149–150.

42. **Stephenson, E. H.** 1990. Experimental ehrlichiosis in nonhuman primates, p. 93–99. *In* J. C. Williams and I. Kakoma (ed.), *Ehrlichiosis: a Vector-Borne Disease of Animals and Humans.* Kluwer Academic Publishers, Boston.

43. **Tachibana, N.** 1986. Sennetsu fever: the disease, diagnosis, and treatment, p. 205–208. *In* H. Winkler and M. Ristic (ed.), *Microbiology—1986.* American Society for Microbiology, Washington, D.C.

44. **Weiss, E., G. A. Dasch, J. C. Williams, and Y. H. Kang.** 1990. Biological properties of the genus *Ehrlichia*: substrate utilization and energy metabolism, p. 59–67. *In* J. C. Williams and I. Kakoma (ed.), *Ehrlichiosis: a Vector-Borne Disease of Animals and Humans.* Kluwer Academic Publishers, Boston.

45. **Wilson, J. C., A. Foggie, and M. A. Carmichael.** 1964. Tick-borne fever as a cause of abortion and stillbirths in cattle. *Vet. Rec.* **76:**1081–1084.

46. **Woldehiwet, Z.** 1983. Tick-borne fever: a review. *Vet. Res. Commun.* **6:**163–175.

Bartonella

DAVID F. WELCH AND LEONARD N. SLATER

58

TAXONOMY

A recent phylogenetic analysis of the order *Rickettsiales* and members of the α subgroup of the *Proteobacteria* has confirmed that species known heretofore as *Rochalimaea* are more closely related to the genera *Brucella* and *Agrobacterium* than to members of the family *Rickettsiaceae* (3). Most highly related to *Rochalimaea* species, however, with 16S rRNA similarity values ranging from 98.7 to 98.8%, is *Bartonella bacilliformis*. By DNA hybridization, there is more than 40% DNA relatedness between *B. bacilliformis* and members of the genus *Rochalimaea*, which argues that all of these species belong in a single genus. Therefore, proposals to combine the genera *Bartonella* and *Rochalimaea* and to adopt the genus name *Bartonella* because of its nomenclatural priority have been made. This relationship between *B. bacilliformis* and the former *Rochalimaea* species was established after the recent recognition of new clinical syndromes associated with AIDS and the identification of the novel human pathogens *B.* (formerly *Rochalimaea*) *henselae* and *B.* (formerly *Rochalimaea*) *elizabethae*. The phylogenetic tree based on 16S rRNA sequence data (see chapter 54 of this Manual) reflects the proposal to classify members of the genus *Rochalimaea* and the genus *Bartonella* together and to separate them from the family *Rickettsiaceae*.

To a limited extent, this chapter also deals with members of the genus *Afipia*, because *Afipia felis* shares some clinical similarities with *B. henselae* and is somewhat related genetically to the *Bartonella* spp. (11). The described species in addition to *A. felis* are *A. clevelandensis* and *A. broomeae*.

DESCRIPTION OF GENUS

Bartonella spp. are small (0.6 by 1.0 μm) gram-negative rods that are often slightly curved. They are oxidase negative, aerobic, and highly fastidious. They do not produce acid from carbohydrates. All members of the genus can be cultured on enriched (blood-containing) bacteriologic culture medium in the presence of 5% CO_2. The optimal temperature varies from 25 to 30°C for *B. bacilliformis* to 35 to 37°C for the other species. Motility is also a variable feature. *B. bacilliformis* possesses polar flagella, while *B. henselae* and, to lesser extent, *B. quintana* display twitching motility due to the presence of pili. Pili are also associated

with marked cytoadherence. *Afipia* spp. differ from *Bartonella* spp. in that they are urease and oxidase positive. Most *Afipia* strains produce acid from D-xylose. Their generation time is usually shorter and they are less fastidious nutritionally than *Bartonella* spp. *A. felis* reduces nitrates.

NATURAL HABITATS

B. bacilliformis was described in 1909 by Barton as the erythrocyte-adherent agent associated with a group of illnesses geographically confined to intermediate altitudes of the South American Andes. Its strictly regional occurrence is likely the result of the limited distribution of its sand fly (*Lutzomyia verrucarum*) vector. As a group, these maladies have borne the eponyms bartonellosis and Carrión's disease, the latter designation honoring the Peruvian medical student whose 1885 self-inoculation from a patient's late-stage skin lesion, the verruga peruana, caused his carefully chronicled acute infection and death.

B. quintana, previously known as *Rickettsia quintana*, was first identified as a cause of debilitating febrile illness among battlefield troops in World War I, hence the eponym trench fever. The organism is globally endemic. Epidemics of bacteremic illness are associated with conditions of poor sanitation and personal hygiene, which predispose humans to exposure to the only known vector, the body louse.

B. henselae lacks a proven arthropod vector, although ticks and fleas are both candidates, but its transmission has been linked by epidemiologic and serologic studies to cats. It has also been cultured from the blood of domestic felines and from patients with human lymphadenitis consistent with cat scratch disease (CSD) and has been identified by immunohistochemical labeling in additional cases of CSD lymphadenitis. In addition, PCR amplification of DNA extracted from different preparations of CSD skin test antigens has yielded sequences consistent with *B. henselae*. CSD's worldwide distribution suggests that *B. henselae* is globally endemic.

CLINICAL SIGNIFICANCE

After a typical incubation period of about 3 weeks, acute *B. bacilliformis* infection results in bacteremic illness (Oroya fever) that may be insidious or abrupt in onset. Fever, chills, diaphoresis, headache, and mental status changes can be severe and are often associated with a rapid and profound anemia. During this stage, there can be associated lymphadenopathy and thrombocytopenia, severe myalgias and arthralgias, and complications of tissue hypoxia such as delirium, coma, dyspnea, or angina. For survivors, convalescence is associated with a decline of fever and a disappearance of erythrocyte-adherent bacteria, but an apparent increased susceptibility to opportunistic infection occurs. Without antimicrobial therapy, there can be a high fatality rate among persons lacking endemic exposure. Usually within months after acute infection, but sometimes longer, the late-stage manifestation of verruga peruana may become evident. In addition to nodular skin lesions of a variety of shapes and hues, mucosal and internal lesions can also occur. Their histology reveals neovascular proliferation, with bacteria evident in the affected tissue. Such lesions may develop at one site while receding at another, persist for months to years, and eventually become fibrotic with involution.

Bacteremia due to *B. quintana* or *B. henselae*, even when persistent, is rarely associated with short-term mortality (22). Trench fever's natural course includes several self-limited clinical patterns. Incubation may span 3 days to over 5 weeks before the sudden onset of fever, which may be brief (4 to 5 days) or uninterrupted and prolonged (2 to 6 weeks). More commonly, however, three to five and up to eight febrile paroxysms occur, each about 5 days in duration. Numerous nonspecific symptoms and signs may occur. To date, the limited number of reports of *B. quintana* bacteremia in human immunodeficiency virus (HIV)-infected hosts does not allow generalization (22). *B. henselae* was also first recognized as an agent of fever and bacteremia, initially in immunocompromised and then in immunocompetent hosts (14, 22). Recognition of its etiologic role in bacillary angiomatosis and bacillary peliosis antedated that of *B. quintana* (8, 19). It is now also recognized to cause nonvascular inflammatory response (18), of which the most important clinically is CSD.

Symptoms associated with bacteremia due to *B. henselae* probably begin at least a week after inoculation. *B. henselae* bacteremia in HIV-infected persons is characterized by insidious development of fatigue, malaise, body aches, weight loss, progressively higher and longer recurring fevers, and sometimes headache. Hepatomegaly may also be found. In contrast, *B. henselae* bacteremia in HIV-uninfected persons more often presents with abrupt onset of fever, which may persist or become relapsing. Associated arthralgias, myalgias, and headache appear to be more common, and aseptic meningitis has been documented, but other localizing symptoms or physical findings remain unusual (22). *B. henselae* bacteremia can evolve into long-term asymptomatic persistence. *B. elizabethae* has been isolated only once, as the cause of endocarditis and bacteremia. *B. quintana* and *B. henselae* have also been reported to cause endocarditis.

Bacillary Angiomatosis and Peliosis

Bacillary angiomatosis is a neovascular proliferative disorder originally described as involving the skin (causing vascular nodules) and regional lymph nodes of HIV-infected persons and since demonstrated to involve a variety of internal organs (6, 7, 19). It occurs in both immunocompromised and immunocompetent hosts. *B. henselae* and *B. quintana* have been found in these conditions by culture (8, 19), by amplification from tissue of specific 16S rRNA gene sequences (15), and by immunohistochemical labeling of bacteria in tissue (13). Cutaneous lesions often arise in crops, but both the temporal pattern and the morphologic characteristics can vary tremendously. They can be remarkably similar to the lesions of verruga peruana, but since most cases have been identified in HIV-infected persons outside the region of endemicity for *B. bacilliformis*, the major differential diagnoses clinically become Kaposi's sarcoma and pyogenic granuloma. The histologic distinction from other neovascular tumors, including the presence in bacillary angiomatosis lesions of Warthin-Starry-staining bacilli, has been clearly defined (9).

In bacillary peliosis, which involves usually the liver and sometimes the spleen in HIV-infected persons and other immunosuppressed persons, the affected tissues contain numerous blood-filled, partially endothelial-cell-lined cystic structures often separated from surrounding parenchymal cells by fibromyxoid stroma containing a mixture of inflammatory cells, dilated capillaries, and clumps of bacilli identified by Warthin-Starry staining (12). Inflammatory reactions to *B. henselae* infection without associated angiomatosis or peliosis but involving liver, spleen, lymph nodes, heart, and bone marrow have been reported (18). These reactions are characterized by nodular collections of lymphocytes and nonepithelioid histiocytes that may become centrally necrotic, containing aggregates of neutrophils and karyorrhexic debris suggestive of microscopic abscess formation.

CSD

Typical CSD (89% of cases) begins as a cutaneous papule or pustule that usually develops within a week after an animal (most commonly a juvenile cat or dog) contact at a site (usually a scratch or bite) of inoculation (4). Regional lymphadenopathy develops in 1 to 7 weeks. About one-third of patients have fever, and about one-sixth develop suppuration of nodes. The histopathology of lymph nodes includes a mixture of nonspecific inflammatory reactions including granulomata and stellate necrosis. Bacilli may be demonstrable by Warthin-Starry staining (20) and less effectively by tissue Gram staining. Atypical CSD includes Parinaud's oculoglandular syndrome (a granulomatous conjunctivitis associated with preauricular lymphadenopathy, probably occurring after primary ocular inoculation) and various other localized or systemic presentations. Spontaneous resolution usually occurs in 2 to 4 months.

Among the *Bartonella* spp., only *B. henselae* has been associated with CSD. The relative roles in the pathogenesis of CSD played by *B. henselae* and *A. felis* remain to be clearly elucidated. Both are best detected in tissue by Warthin-Starry staining (20). The organism reportedly cultured from 10 of 19 CSD tissue specimens (5) was later designated *A. felis* and deemed the cause of CSD (2). Rabbit polyclonal antibodies to the first isolate of *A. felis* reportedly reacted with bacilli in CSD lesions in immunohistochemical studies and were later also claimed to react with bacilli demonstrated in bacillary angiomatosis. However, this antiserum also reacted with an organism causing bacteremia that was subsequently proven to be *B. henselae*. Attempts to demonstrate specific antibody responses to *A. felis* in persons with CSD have been unsuccessful (1, 16). Thus, evidence implicating *B. henselae* in the etiology of CSD is much more convincing than evidence supporting *A. felis*. While the histopathologic responses of CSD could conceivably be elicited by more than one pathogen, there is a growing consensus that *B. henselae* is probably the predominant if not the sole pathogen.

COLLECTION, TRANSPORT, AND STORAGE
OF SPECIMENS

The specimen source of most isolates of *Bartonella* and *Afipia* spp. is blood or tissue. Approaches typically used for recovery of other pathogens from these sites are generally suitable, although the fastidious nature of these organisms requires that precautions be taken to minimize the interval from collection to processing. If storage of specimens is necessary, they should be kept frozen ($-20°C$) or at 4°C. *B. henselae* is inhibited by concentrations of sodium polyanethol sulfonate (SPS) that are relevant to blood culture systems, so adding agents that neutralize SPS toxicity is another precaution that should be taken when blood is cultured. Alternatively, lytic blood culture systems that combine the protective effect of free hemoglobin against SPS toxicity with the release of closely cell-associated organisms may be used.

DIRECT EXAMINATION

Although stained blood films have been used to detect *B. bacilliformis* in patients with Oroya fever, the magnitude of bacteremia associated with the other species is usually too

low for this technique to be practical. Other means of direct examination that may prove useful include Warthin-Starry silver staining of fixed tissue and detection of organisms in tissue by using an immunocytochemical labeling technique (13). Successful staining of organisms in tissue by the Warthin-Starry method depends on careful attention to details of the procedure. Critical steps include those of washing glassware in potassium dichromate-sulfuric acid solution, making the developer from solutions heated separately at 54 to 56°C (and maintained at this temperature after they are mixed) for 30 min, and then using developer as soon as it is mixed.

ISOLATION PROCEDURES

Since the majority of isolates require more than 7 days of incubation before they can be detected, routine bacterial culture protocols usually do not allow *Bartonella* spp. to be detected. Attempts to isolate *Bartonella* spp. may be driven by the clinical picture, but throughout the spectrum of manifestations, many symptoms are not distinctive enough in their clinical appearance to guide the diagnosis and thus to have the laboratory informed that special cultures are needed. If clinically directed culturing is relied upon to detect *Bartonella* spp., it is indicated in the settings of CSD, fever of unknown origin with history of cat exposure, fever of unknown origin in the immunocompromised patient, and bacillary angiomatosis. Culture protocols designed to yield other slowly growing organisms (for example, *Histoplasma capsulatum* or *Mycobacterium avium* complex on noninhibitory media) can also result in recovery of *Bartonella* spp.

A. *felis* has been most successfully isolated (from lymph node biopsy samples or aspirates) by using HeLa cells (1). Primary cell cultures of human monocytes also support growth of *A. felis*. In both systems, elongated pleomorphic organisms become visible in Gimenez-stained preparations 72 h after infection of the cell cultures.

Blood

The two fundamentally different blood culture practices of direct plating and using broth-based systems can be adapted

for detection of *Bartonella* spp. Isolator (lysis-centrifugation; Wampole, Cranbury, N.J.)-processed blood should be plated on enriched (chocolate- or blood-containing) medium incubated at 35 to 37°C under conditions of elevated CO_2 and humidity. For optimal growth, the medium should be as freshly prepared as possible. Plates can be sealed with Parafilm or Shrink Seal after the first 24 h of incubation to preserve the moisture content of the medium. Such plates usually can be incubated for up to 30 days without noticeable deterioration. Considering the length of time required for detection of *Bartonella* spp. allows other slowly growing pathogens, including *Mycobacterium tuberculosis*, to grow on the same plates, appropriate safety precautions should be taken with positive cultures.

Alternative approaches to use of the Isolator include use of a broth-based or biphasic culture system. Improved growth has been obtained in brucella broth supplemented with hemin (250 fg/ml) and peptic digest of blood (8% Fildes reagent), but this technique has primarily been applied to propagation rather than primary isolation of isolates (17). Similarly, isolates can be propagated on buffered charcoal yeast extract medium, but this is not recommended for primary isolation. *B. henselae* has been isolated in the biphasic Septi-Chek system (Roche Diagnostics, Nutley, N.J.) after prolonged incubation in excess of 40 days. Evidence of growth, if any, in the broth phase is a pellicle or adherent film on the glass surface. Biphasic media reportedly serve to isolate *A. felis* (5), but experience to date is so limited that methods of choice, with the possible exception of the tissue culture protocol of Birkness et al. (1), cannot be stated with certainty.

Bartonella spp. grow best on solid or semisolid media. In broth, they do not produce turbidity or convert enough oxidizable substrate to CO_2 for CO_2 detection blood culture systems to detect growth. Several isolates have been detected, however, by using BACTEC and resin-containing media combined with acridine orange staining at the termination of a 7-day incubation period.

Tissue

B. henselae and *B. quintana* have been isolated from liver, spleen, lymph node, and skin after homogenization either by direct plating or by cocultivation with an endothelial cell line (8). While the cocultivation method may be more successful in recovering organisms from specimens such as tissue, it is not practical for most microbiology laboratories. Freshly prepared heart infusion agar containing 5 or 10% defibrinated rabbit or horse blood supports better growth of most strains than do other media such as chocolate or 5% sheep blood, although the last two have been used successfully. Other necessary conditions include a humid atmosphere and 35 to 37°C incubation for 3 to 4 weeks. *B. bacilliformis* and *Afipia* spp. have a lower (25 to 30°C) optimal temperature for growth. Selective culture techniques have not been developed, so recovery of isolates from certain specimens such as skin may be impossible if indigenous or contaminating flora are present.

IDENTIFICATION

Colonies of *Bartonella* spp. are of two morphologic types: (i) irregular, raised, whitish, rough ("cauliflower" or "molar tooth" or "verrucous"), and dry in appearance or (ii) smaller, circular, tan, and moist in appearance, tending to pit and adhere to the agar (Fig. 1). Both types are usually present in the same culture. The degree of colonial heter-

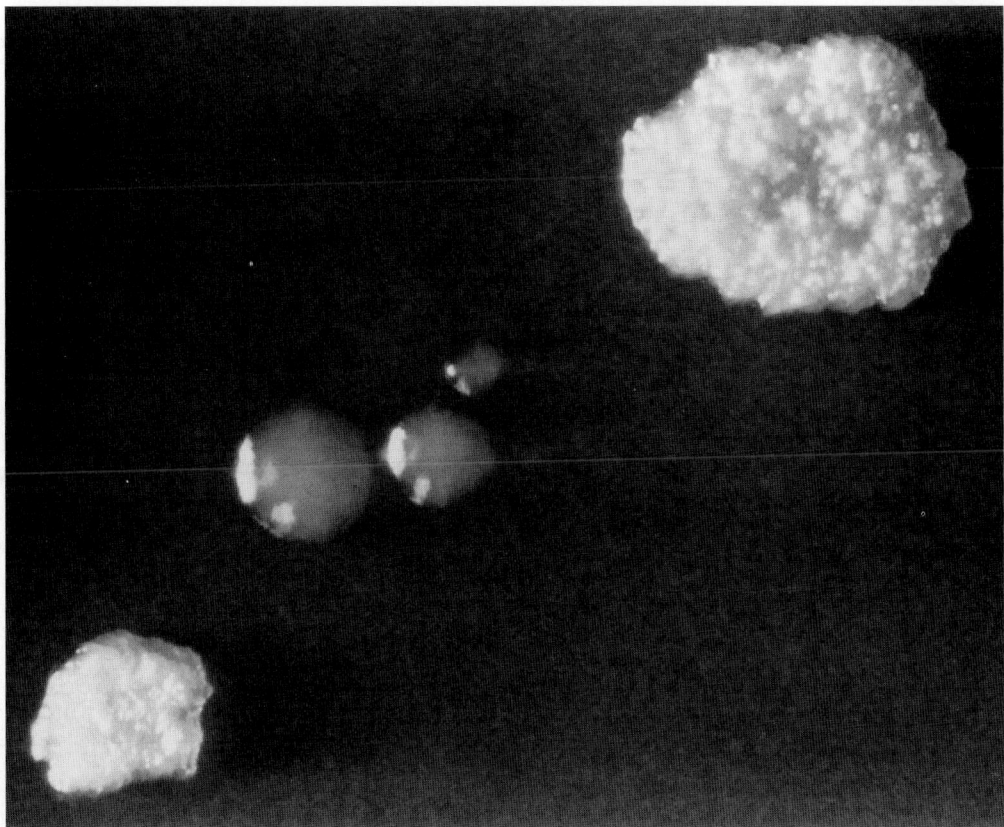

FIGURE 1 *B. henselae* after 7 days of incubation, showing heterogeneity of colonies. Magnification, ×40.

ogeneity varies by species and by strain, with *B. henselae* typically characterized by a greater proportion of rough colonies than *B. quintana*, which may even appear as uniformly smooth in primary cultures. Repeated subcultures of *B. henselae* tend to have increasing proportions of smooth colonies. The Gram stain of a colony reveals small, gram-negative, slightly curved rods. Cells, especially *B. henselae*, are very autoadherent, as can be demonstrated by attempting to scrape colonies off of a culture plate with a loop. In a wet mount, there is twitching motility of cells. These features, plus a lengthy (>7-day) incubation before their appearance and negative catalase and oxidase reactions, are sufficient for presumptive identification of *B. henselae* or *B. quintana*.

Definitive Identification

Distinguishing between *B. henselae* and *B. quintana* by simple means is not as easy, because they are phenotypically and genotypically very similar. Though not widely available, a reliable means is immunofluorescence with antisera monospecific for each of these two species. Characterization using conventional tests that are commercially available to laboratories produces results that identify the *Bartonella* spp. and distinguish them from *A. felis* (Table 1).

Motility

Isolates most likely to be encountered, i.e., *B. henselae* or *B. quintana*, do not possess flagella but do produce a twitching motion when suspended in saline and examined microscop-

ically under a coverslip. This motility is presumably interrelated with adherence, both features being mediated by fine fimbriae (pili) that are visible in negative stained preparations with the electron microscope.

Cellular Fatty Acid Composition

Determination of cellular fatty acid composition by gas-liquid chromatography is outlined in chapter 12 of this Manual. Fatty acid methyl esters are prepared from cells harvested after 7 days of incubation at 35°C in 5% CO_2 on plates containing 5% rabbit blood heart infusion agar. This approach is useful in identifying and distinguishing *Bartonella* spp. in that they have relatively simple gas-liquid chromatography profiles consisting mainly of $C_{18:1}$, $C_{18:0}$, and $C_{16:0}$ acids, with the newest member, *B. elizabethae*, containing a greater amount of $C_{17:0}$ than the other species. An unusual branched-chain fatty acid found in *Afipia* spp. is absent in *Bartonella* spp.

Identification Kits

Although none of the various commercially available identification systems contain *Bartonella* spp. in their databases, several of these systems yield a pattern of reactions that distinguishes *Bartonella* spp. from species that are in the database. Using the MicroScan rapid anaerobe panel and careful adjustment of inoculum size, it is possible to distinguish *B. henselae* and *B. quintana* on the basis of biotype codes derived from the reactions in this panel (21). If the inoculum size used in performing the Microscan tests is

TABLE 1 Differential characteristics of *Bartonella* species and *A. felis*[a]

Characteristic	B. bacilliformis	B. quintana	B. henselae	B. elizabethae	A. felis
Gram reaction	−	−	−	−	−
Catalase	+	−	−/+	−	−
Oxidase	−	−/+	−	−	+
Nitrate reduction	−	−	−	−	+
Indole	−	−	−	−	−
Urease	−	−	−	−	+
Acid from carbohydrates[b]	−	−	−	−	−
Optimal temp	25–30°C	35–37°C	35–37°C	35–37°C	25–30°C
Growth in nutrient broth	−	−	−	−	+
Hemolysis	−	−	−	−	+
Flagella	+	−[c]	−[c]	−	+
Cellular fatty acids constituting >10% of total	$C_{18:1\omega7C}$, $C_{16:0}$, $C_{16:1\omega7C}$	$C_{18:1\omega7C}$, $C_{16:0}$, $C_{18:0}$	$C_{18:1\omega7C}$, $C_{18:0}$, $C_{16:0}$	$C_{18:1\omega7C}$, $C_{17:0}$, $C_{16:0}$	$C_{18:1\omega7C}$, $C_{BR19:1}$, $C_{19:0cyc}$, $C_{18:0}$

[a]+, positive reaction; −, negative reaction; −/+, negative or weakly positive.
[b]Glucose, lactose, maltose, mannitol, and sucrose.
[c]May demonstrate twitching motility in wet mounts.

equivalent to a McFarland no. 3 standard, the biotype codes usually obtained are 10077640 (*B. henselae*), 10073640 (*B. quintana*), and 10077240 (*B. bacilliformis*). Difficulty separating *B. quintana* from *B. henselae* occurs when heavier inocula are used in these panels.

Subtyping

A PCR-based restriction fragment length polymorphism method has been applied by Matar et al. (10) to subtyping of isolates. Genomic DNA making up the spacer region between the 16S and 23S rRNA genes plus a portion of the 23S rRNA gene are amplified and then subjected to restriction endonuclease analysis. Different patterns are seen among the *Bartonella* spp., and at least seven subtypes within *B. henselae* have been determined, but no studies to date have made any epidemiologic correlations.

SEROLOGIC TESTS

Since culturing of *Bartonella* spp. is difficult and time-consuming, alternative means of identifying infectious agents are important. Human antibody responses often have been substantially cross-reactive between *B. quintana* and *B. henselae*. Nevertheless, the differential magnitude of antibody reactions to either species may allow an inference as to which may be culpable in a particular infection. This would be especially important when they may cause overlapping syndromes such as fever with bacteremia or bacillary angiomatosis. Human antibody responses to *Bartonella* species have been measured by a variety of techniques.

B. bacilliformis

Enzyme immunoassay (EIA) is the most sensitive test for detecting immunoglobulin G (IgG) antibodies among persons from the region of endemicity. IgM antibodies can be detected by immunofluorescence in persons with active Oroya fever as well as in some persons without blood smear evidence of acute bacteremic infection.

B. quintana

EIA and radioimmunoprecipitation have proven comparable, and both are superior to hemagglutination or immunofluorescence assays in sensitivity. When the same EIA is used, all cases in a small series of acute primary or relapsed

trench fever patients had measurable, although often low, levels of anti-*B. quintana* antibodies.

B. henselae

Immunofluorescence and EIA have been described. They have been used primarily to provide one of the lines of evidence implicating *B. henselae* in the pathogenesis of CSD. The tests reported to date have undergone different degrees of scrutiny and corroboration, but most appear to have high sensitivity and specificity. Persons with CSD develop IgM antibodies that can be measured early in the course of disease and IgG antibodies whose levels rise somewhat later and then decline over time.

Results of the immunofluorescence test and one EIA for *B. henselae* have demonstrated a good correlation with each other and with findings of CSD skin testing. Since skin test antigen for CSD is not commercially available or standardized and since it carries with it the potential risk of transmitting occult infection (being prepared from pus from suppurative lymph nodes), serologic testing will likely replace skin testing in aiding the clinical diagnosis of CSD. Serologic testing is also likely to play a role in corroborating the clinical suspicion of bacteremia due to *B. quintana* or *B. henselae* when an agent cannot be successfully isolated and in implicating one or the other pathogen as the cause of bacillary angiomatosis or peliosis when culture of the lesions is not attempted or is unsuccessful.

ANTIMICROBIAL SUSCEPTIBILITY

Antimicrobial susceptibility testing can be performed by incorporation of antimicrobial agents into either blood or chocolate agars or into *Haemophilus* test medium by using a broth microdilution technique. Testing of isolates is problematic for those strains displaying the most fastidious growth characteristics. Preliminary experience suggests that the E test (AB Biodisk, Solna, Sweden), which uses heart infusion rabbit blood agar, may be a useful alternative in assessing quantitative susceptibility to agents. Generally, isolates are susceptible in vitro to most antibacterial agents tested, including β-lactams, tetracyclines, macrolides, aminoglycosides, fluoroquinolones, vancomycin, rifampin, chloramphenicol, and co-trimoxazole, but resistant to nalidixic acid. In vitro resistance to penicillin and ampicillin, tetracycline, or vancomycin has been noted. *B. quintana* is similar in its in vitro susceptibility pattern except for resistance to aminoglycosides. Growing clinical experience with *B. henselae* and *B. quintana* in the setting of HIV infection suggests that therapy with β-lactams or sulfonamides may be suboptimal and supports the use of tetracyclines, erythromycin, or chloramphenicol. The antimicrobial susceptibility profiles of *Afipia* spp. display more resistance than those of *Bartonella* spp.

EVALUATION, INTERPRETATION, AND REPORTING OF RESULTS

Limited experience prevents an accurate comparison of different techniques for assessing the sensitivities of cultures. Serologic data from immunocompetent individuals have supported the diagnosis of infection when a *Bartonella* sp. has been cultured and a serologic assay was performed. Given the larger experience with CSD and other manifestations of *Bartonella* infection, for which only serologic data have been collected, the implied sensitivity of serology is 85 to 90%. The specificity of serology is probably around 95% according to screening in other populations presumed to be uninfected and for other diseases. While the value of culturing for *Bartonella* spp. is limited in certain cases by practicality issues of fresh-medium availability, length of incubation, and unknown efficiency, the use of cultures in conjunction with serology is justified in selected cases.

REFERENCES

1. **Birkness, K. A., V. G. George, E. H. White, D. S. Stephens, and F. D. Quinn.** 1992. Intracellular growth of *Afipia felis*, a putative etiologic agent of cat scratch disease. *Infect. Immun.* **60:**2280–2287.
2. **Brenner, D. J., D. G. Hollis, C. W. Moss, C. K. English, G. S. Hall, V. Judy, J. Radosevic, K. A. Birkness, W. F. Bibb, F. D. Quinn, B. Swaminathan, R. E. Weaver, M. W. Reeves, S. P. O'Connor, P. S. Hayes, F. C. Tenover, A. G. Steigerwalt, B. A. Perkins, M. I. Daneshvar, B. C. Hill, J. A. Washington, T. C. Woods, S. B. Hunter, T. L. Hadfield, G. W. Ajello, A. F. Kaufman, D. J. Wear, and J. D. Wenger.** 1991. Proposal of *Afipia* gen. nov., with *Afipia felis* sp. nov. (formerly the cat scratch disease bacillus), *Afipia clevelandensis* sp. nov. (formerly the Cleveland Clinic Foundation strain), *Afipia broomeae* sp. nov., and three unnamed genospecies. *J. Clin. Microbiol.* **29:**2450–2460.
3. **Brenner, D. J., S. P. O'Connor, H. H. Winkler, and A. G. Steigerwalt.** 1993. Proposals to unify the genera *Bartonella* and *Rochalimaea*, with descriptions of *Bartonella quintana* comb. nov., *Bartonella vinsonii* comb. nov., *Bartonella henselae* comb. nov., and *Bartonella elizabethae* comb. nov., and to remove the family *Bartonellaceae* from the order *Rickettsiales*. *Int. J. Syst. Bacteriol.* **43:**777–786.
4. **Carithers, H. A.** 1985. Cat-scratch disease. An overview based on a study of 1,200 patients. *Am. J. Dis. Child.* **139:** 1124–1133.
5. **English, C. K., D. J. Wear, A. M. Margileth, C. R. Lissner, and G. P. Walsh.** 1988. Cat-scratch disease. Isolation and culture of the bacterial agent. *J. Am. Med. Assoc.* **269:**1347–1354.
6. **Kemper, C. A., C. M. Lombard, S. C. Deresinski, and L. S. Tompkins.** 1990. Visceral bacillary epithelioid angiomatosis: possible manifestations of disseminated cat scratch disease in the immunocompromised host: a report of two cases. *Am. J. Med.* **89:**216–222.
7. **Koehler, J. E., P. E. LeBoit, B. M. Egbert, and T. G. Berger.** 1988. Cutaneous vascular lesions and disseminated cat-scratch disease in patients with the acquired immunodeficiency syndrome (AIDS) and AIDS-related complex. *Ann. Intern. Med.* **109:**449–455.
8. **Koehler, J. E., F. D. Quinn, T. G. Berger, P. E. LeBoit, and J. W. Tappero.** 1992. Isolation of *Rochalimaea* species from cutaneous and osseous lesions of bacillary angiomatosis. *N. Engl. J. Med.* **327:**1625–1632.
9. **LeBoit, P. E., T. G. Berger, B. M. Egbert, J. H. Beckstead, Y. S. B. Yen, and M. H. Stoler.** 1989. Bacillary angiomatosis. The histology and differential diagnosis of a pseudoneoplastic infection in patients with human immunodeficiency virus disease. *Am. J. Surg. Pathol.* **13:**909–920.
10. **Matar, G. M., B. Swaminathan, S. B. Hunter, L. N. Slater, and D. F. Welch.** 1993. Polymerase chain reaction-based restriction fragment length polymorphism analysis of a fragment of the ribosomal operon from *Rochalimaea* species for subtyping. *J. Clin. Microbiol.* **31:**1730–1734.
11. **O'Connor, S. P., M. Dorsch, A. G. Steigerwalt, D. J. Brenner, and E. Stackebrandt.** 1991. 16S rRNA sequences of *Bartonella bacilliformis* and cat scratch disease bacillus reveal phylogenetic relationships with the alpha-2 subgroup of the class *Proteobacteria*. *J. Clin. Microbiol.* **29:**2144–2150.
12. **Perkocha, L. A., S. M. Geaghan, T. S. B. Yen, S. L. Nishimura, S. P. Chan, R. Garcia-Kennedy, G. Honda, A. C. Stoloff, H. Z. Klein, R. L. Goldman, S. V. Meter, L. D. Ferrell, and P. E. LeBoit.** 1990. Clinical and pathological features of bacillary peliosis hepatis in association with human immunodeficiency virus infection. *N. Engl. J. Med.* **323:** 1581–1586.
13. **Reed, J., D. J. Brigati, S. D. Flynn, N. S. McNutt, K.-W. Min, D. F. Welch, and L. N. Slater.** 1992. Immunocytochemical identification of *Rochalimaea henselae* in bacillary (epithelioid) angiomatosis, parenchymal bacillary peliosis, and persistent fever with bacteremia. *Am. J. Surg. Pathol.* **16:**650–657.
14. **Regnery, R. L., B. E. Anderson, J. E. Clarridge III, M. C. Rodriquez-Barradas, D. C. Jones, and J. H. Carr.** 1992. Characterization of a novel *Rochalimaea* species, *R. henselae* sp. nov., isolated from blood of a febrile, human immunodeficiency virus-positive patient. *J. Clin. Microbiol.* **30:**265–274.
15. **Relman, D. A., J. S. Loutit, T. M. Schmidt, S. Falkow, and L. S. Tompkins.** 1990. An approach to the identification of uncultured pathogens: the agent of bacillary angiomatosis. *N. Engl. J. Med.* **323:**1573–1580.
16. **Schwartzman, W. A.** 1992. Infections due to *Rochalimaea*: the expanding clinical spectrum. *Clin. Infect. Dis.* **15:**893–902.
17. **Schwartzman, W. A., C. A. Nesbit, and E. J. Baron.** 1993. Development and evaluation of a blood-free medium for determining growth curves and optimizing growth of *Rochalimaea henselae*. *J. Clin. Microbiol.* **31:**1882–1885.
18. **Slater, L. N., J. V. Pitha, L. Herrera, M. D. Hughson, K.-W. Min, and J. A. Reed.** 1994. *Rochalimaea henselae* infection in AIDS causing inflammatory disease without angiomatosis or peliosis: demonstration by immunocytochemistry and corroboration by DNA amplification. *Arch. Pathol. Lab. Med.* **118:**33–38.
19. **Slater, L. N., D. F. Welch, and K.-W. Min.** 1992. *Rochalimaea henselae* causes bacillary angiomatosis and peliosis hepatis. *Arch. Intern. Med.* **152:**602–606.
20. **Wear, D. J., A. M. Margileth, G. W. Fischer, C. J. Schlagel, and F. M. King.** 1983. Cat scratch disease: a bacterial infection. *Science* **221:**1403–1405.
21. **Welch, D. F., D. M. Hensel, D. A. Pickett, V. H. San Joaquin, A. Robinson, and L. N. Slater.** 1993. Bacteremia in a child due to *Rochalimaea henselae*: a practical identification of isolates in the clinical laboratory. *J. Clin. Microbiol.* **31:**2381–2386.
22. **Welch, D. F., D. A. Pickett, L. N. Slater, A. G. Steigerwalt, and D. J. Brenner.** 1992. *Rochalimaea henselae* sp. nov., a cause of septicemia, bacillary angiomatosis, and parenchymal bacillary peliosis. *J. Clin. Microbiol.* **30:**275–280.

MYCOLOGY

VII

VOLUME EDITOR
MICHAEL A. PFALLER

SECTION EDITOR
ROBERT A. FROMTLING

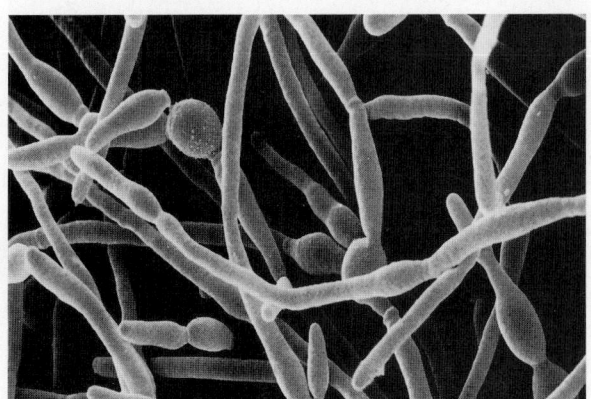

Candida albicans (courtesy of Visuals Unlimited/© David M. Phillips).

Morphology, Taxonomy, and Classification of the Fungi

DENNIS M. DIXON AND ROBERT A. FROMTLING

59

INTRODUCTION

Mycology has emerged into the microbiological mainstream. The list of opportunistic fungal pathogens is increasing at an impressive rate that is related to the expanding size of the immunocompromised patient population. This results in a need for the clinical laboratory technologist to be able to recognize an increasingly large group of potential fungal pathogens. Organisms once thought to be "contaminants" are now confirmed pathogens in compromised patients. A further need is to recognize that even though a given isolate may not be a documented fungal pathogen, its isolation from a normally sterile site and its ability to grow at 37°C require that it be considered a possible pathogen. With few exceptions, all of the fungi that infect humans share the ability to grow at 37°C. Beyond this, it is difficult to identify traits or factors associated with virulence in the pathogenic fungi.

Mycological identification can be frustrating because of the importance placed upon morphology and the need to become familiar with certain structures and terms. However, investing a small amount of time to learn a few basic structures and principles of classification can result in the ability to recognize and properly identify many medically important fungi.

MORPHOLOGY, TAXONOMY, AND CLASSIFICATION OF FUNGI

Fungi were among the first microorganisms recognized because some of the fruiting structures, such as the mushrooms, are large enough to be seen without a microscope. The word "mycology," in fact, is derived from *mykes*, the Greek word for mushroom. Fungi were initially classified with the plants, and much of the botanical influence is still seen, even though the organisms have been transferred to a separate, fifth kingdom on the basis of cell structure (Table 1). The nomenclature of the fungi is governed by the International Code of Botanical Nomenclature (adopted by the 15th International Botanical Congress, Tokyo, 1993). Morphology retains an important role in the identification of most fungi. This places demands upon the observer to become familiar with the organisms through experience. Morphology is more important for identification of moulds than of yeasts, but even with the latter, morphology on a

medium such as cornmeal agar is a key first step in the identification process.

Fungi are extremely successful organisms, as evidenced by their ubiquity in nature. They are an important component in the energy cycle, where they function as decomposers. Thus, fungi are valuable as saprophytes in nature. Of the estimated 250,000 species, fewer than 150 are known to be primary pathogens of humans. The infection of humans seems to be an accident of nature, since it represents a "dead end" for the fungus; i.e., most fungal infections are not contagious but are acquired through exposure to a point source in nature, where the organism exists as a saprophyte. This fact has practical implications in the laboratory and is why moulds, in particular, should be manipulated in a biological safety cabinet (see Laboratory Safety below).

Fungi are eukaryotic, chemoheterotrophic organisms with cell walls containing chitin and/or cellulose. They may be unicellular or multicellular; there is a tendency to be multinucleate. The body of a fungus is termed the thallus. Fungi can be grouped simply on the basis of morphology as either yeasts or moulds. A yeast can be defined morphologically as a cell that reproduces by budding, a process whereby a progenitor cell pinches off a portion of itself to produce a progeny cell. In this connotation yeasts are generally unicellular and produce circular, restricted, pasty or mucoid colonies. In contrast, moulds are multicelled, filamentous forms of fungi consisting of thread-like filaments termed hyphae that interweave to form a mycelium. The resulting colonies are "fuzzy." (For a discussion of "mould" versus "mold," see reference 5.) Many medically important fungi, such as *Blastomyces dermatitidis*, can exist in each of these two morphologies and are thus termed dimorphic. When additional forms are present, the term polymorphic (pleomorphic) is applied (Fig. 1).

Fungi reproduce by the formation of spores, which may be either asexual (involving mitosis only) or sexual (involving meiosis; preceded by fusion of the protoplasm and nuclei of two cells). One fungus can produce both sexual and asexual spores. Specialized structures (fruiting bodies, or fructifications) may be associated with either sexual or asexual spores. Information on these fruiting bodies is summarized in Table 2. Recognition of a particular fruiting structure can be a useful step in the identification process. It is important to determine the type of spore contained in

TABLE 1 Simplified taxonomic scheme illustrating major groups of Kingdom Fungi (Myceteae) in which medically important fungi are classified[a]

Taxonomic designation	Representative genera	Human disease
Division: Amastigomycota		
Subdivision: Zygomycotina		
Class: Zygomycetes		
Order: Mucorales	*Rhizopus, Mucor, Rhizomucor, Absidia, Cunninghamella, Saksenaea*	Zygomycosis; opportunistic in patients with diabetes, leukemia, severe burns, or malnutrition; rhinocerebral infections
Order: Entomophthorales	*Basidiobolus, Conidiobolus*	Zygomycosis; subcutaneous infections
Subdivision: Ascomycotina		
Class: Ascomycetes		
Subclass: Hemiascomycetidae		
Order: Endomycetales	*Saccharomyces, Pichia* (teleomorphs of some *Candida* spp.)	Numerous mycoses
Subclass: Plectomycetidae		
Order: Onygenales	*Arthroderma* (teleomorphs of *Trichophyton* and *Microsporum* spp.)	Dermatophytoses
Order: Eurotiales	*Ajellomyces* (teleomorphs of *Histoplasma* and *Blastomyces* spp.)	Systemic mycoses
	Teleomorphs of some *Aspergillus* and *Penicillium* spp.	Aspergillosis; hyalohyphomycosis
Subdivision: Basidiomycotina		
Class: Basidiomycetes		
Subclass: Holobasidiomycetidae		
Order: Agaricales	*Amanita, Agaricus*	Mushroom poisoning from consumption of poisonous species (e.g., *Amanita* spp.)
Subclass: Heterobasidiomycetidae		
Order: Filobasidiales	*Filobasidiella* (teleomorph of *Cryptococcus neoformans*)	Cryptococcosis
Subdivision[b]: Deuteromycotina		
Class: Deuteromycetes		
Subclass: Blastomycetidae (Blastomycetes[c])		
Order: Cryptococcales	Imperfect yeasts: *Candida, Cryptococcus, Trichosporon, Pityrosporum*	Numerous mycoses
Subclass: Hyphomycetidae (Hyphomycetes[c])		
Order: Moniliales		
Family: Moniliaceae	*Epidermophyton, Coccidioides, Paracoccidioides, Sporothrix, Aspergillus*	Numerous mycoses
Family: Dematiaceae	*Phialophora, Fonsecaea, Exophiala, Wangiella, Xylohypha, Bipolaris, Exserohilum, Alternaria*	Chromoblastomycosis, mycetoma, and phaeohyphomycosis
Subclass: Coelomycetidae (Coelomycetes[c])		
Order: Sphaeropsidales	*Phoma*	Phaeohyphomycosis
Division: Mastigomycota		
Subdivision: Diplomastigomycotina		
Class: Oomycetes[d]	*Pythium*	Pythiosis

[a]Modified from information in reference 1.
[b]This subdivision and all of its taxa are form taxa.
[c]Some taxonomists delete Deuteromycetes as a rank and replace this class with Blastomycetes, Hyphomycetes, and Coelomycetes rather than using these names as subclass designations.
[d]Some taxonomists place these organisms in Kingdom Protista (Protoctista).

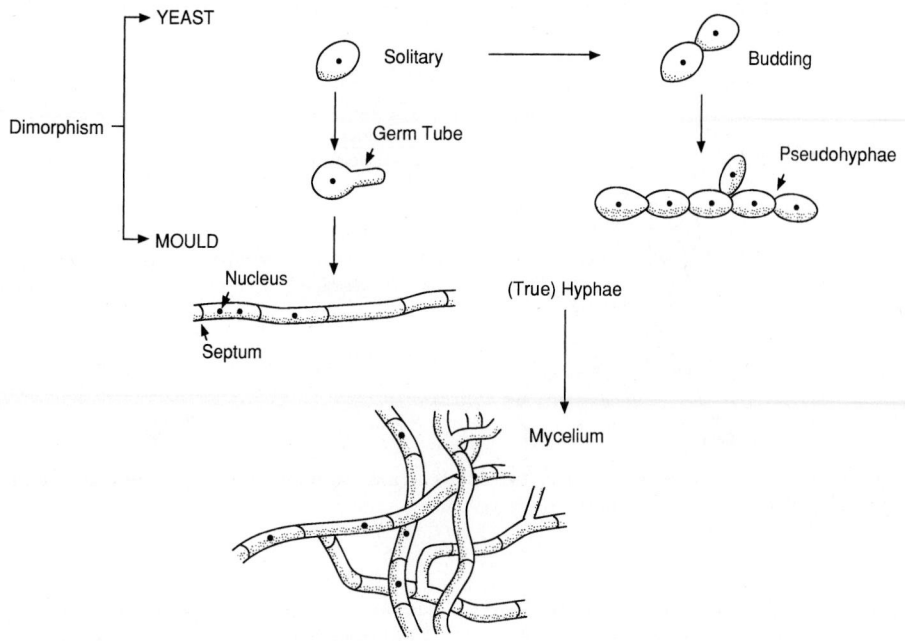

FIGURE 1 Simplified representations of the fungal thallus. Figure courtesy of J. Shockey, Merck Research Laboratories.

order to establish both the type of fruiting structure and the ultimate identification.

Asexual spores are of two general types: sporangiospores and conidia (Fig. 2). Sporangiospores are asexual spores produced within a containing structure (sporangium); they are characteristic of the lower fungi classified in the Zygomycetes, such as *Rhizopus* and *Mucor* spp. Typically, these fungi have broad, sparingly septate hyphae. In the class Oomycetes, classified in Kingdom Protista by some taxonomists, the sporangiospores (zoospores) are often motile. Conidia are asexual spores that are borne naked and are nonmotile, as evidenced in *Aspergillus* and *Penicillium* spp. and the dermatophytes. The conidial fungi are classified together as the Deuteromycetes (Fungi Imperfecti).

Sexual spores are the result of meiosis (Fig. 3). The meiotic nuclei are packaged into cells that function as spores. These spores and their mode of production serve as

the basis for taxonomic grouping. The oomycetes undergo a mating process characterized by the production of specialized structures (antheridia and oogonia; Fig. 1), resulting in oospores contained in an oogonium. The only medically important fungus in this group thus far is *Pythium*. The Zygomycetes produce zygospores, which typically are large, ornamented cells easily recognized, such as those of *Rhizopus* spp. However, zygospores are rarely seen in the clinical laboratory, since most of the medically important species are heterothallic; that is, two different strains of compatible mating types are required for the production of sexual spores.

The higher fungi are represented by the classes Ascomycetes, Basidiomycetes, and Deuteromycetes. The sexual spore of the Ascomycetes is the ascospore, characterized by its production within a containing sac (ascus). The sexual spore of the Basidiomycetes is the basidiospore, character-

TABLE 2 Fruiting structures helpful in classifying some medically important fungi

Fruiting structure	Characteristics	Taxonomic group	Representative genera
Sexual (ascomata)			
Perithecium	Spherical or flask-shaped ascoma with opening	Ascomycotina	*Chaetomium, Leptosphaeria*
Cleistothecium	Spherical ascoma; no opening	Ascomycotina	*Pseudallescheria, Emericella, Ajellomyces,*[a] *Arthroderma*[a]
Asexual (conidiomata)			
Pycnidium	Spherical or flask-shaped conidioma, usually with opening	Deuteromycotina[b]	*Hendersonula (Nattrassia), Phoma, Pyrenochaeta*
Synnema	Rope-like tangle of conidiophores, usually with apical fertile region	Deuteromycotina	*Graphium*
Sporodochium	Clump of conidiophores	Deuteromycotina	*Fusarium, Epicoccum*

[a]The term "gymnothecium" occasionally is used to refer to the characteristic, loosely woven cleistothecium produced by these genera.
[b]Specifically, the coelomycetes.

ASEXUAL FUNGAL SPORES

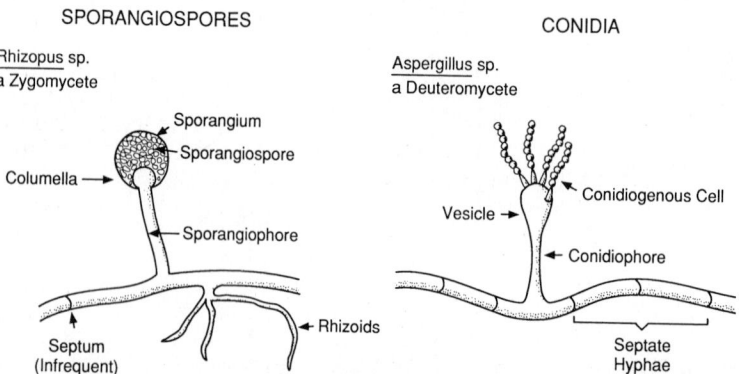

FIGURE 2 Stylized representation of two kinds of asexual fungal spores. Figure courtesy of J. Shockey, Merck Research Laboratories.

ized by extrusion from a club-shaped structure (basidium). The Deuteromycetes produce no known sexual spores. They are included with the higher fungi because they are presumed to represent the conidial stages of Ascomycetes and, more rarely, Basidiomycetes that have lost the ability to produce sexual spores. It is also possible that certain Deuteromycetes never had the ability to produce sexual spores. Since sexual spores hold taxonomic precedence, the Deuteromycetes represent a provisional taxonomic group.

Recognition of a given spore as sexual or asexual is largely dependent upon knowledge of the life cycle of the fungus (in order to determine whether the spore is the result of meiosis or mitosis) and recognition of the characteristic morphology of the spore. Of particular importance

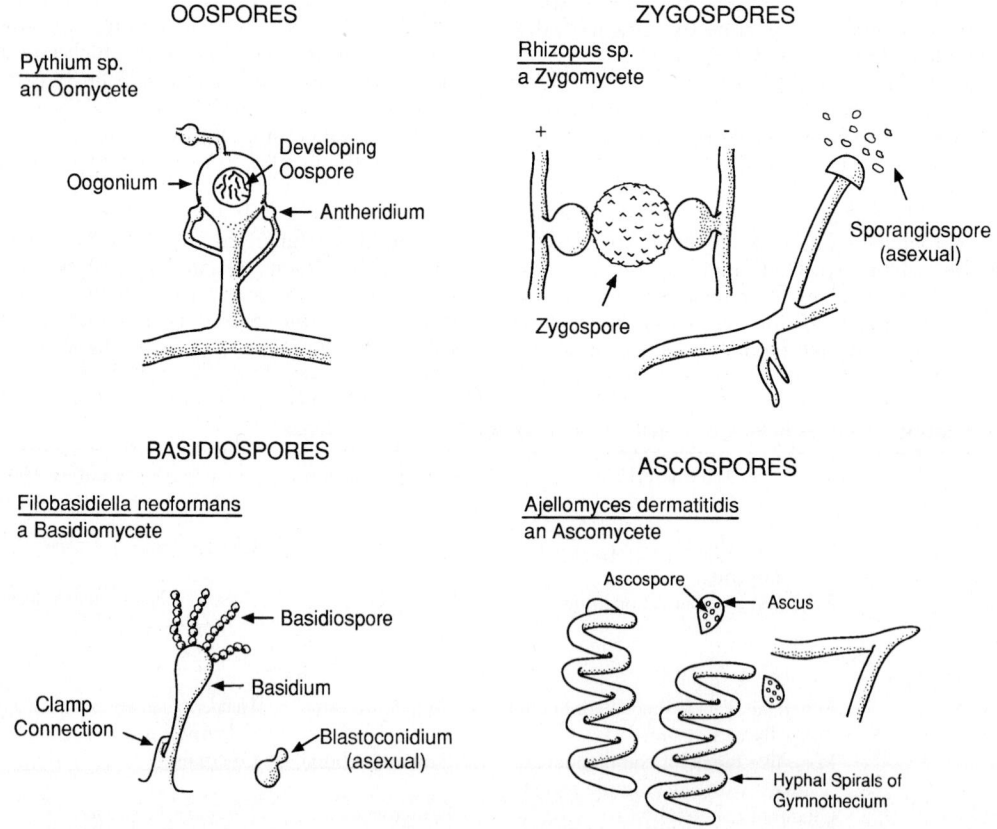

FIGURE 3 Stylized representation of four kinds of sexual fungal spores, using medically important examples. Figure courtesy of J. Shockey, Merck Research Laboratories.

with the conidial fungi is the complication imposed by finding that the sexual stage of a fungus has been named on the basis of its conidial (asexual) form. For example, *Blastomyces dermatitidis* was named in 1898 on the basis of its asexual characteristics: a fungus (*myces*) that reproduced by budding (*Blasto*) recovered from a case of dermatitis (*dermatitidis*). The fungus grows as a yeast in tissue (37°C) and as a mould in culture (25°C). It produces lateral solitary spores meeting the definition of conidia. In 1968, the fungus was found to complete a life cycle by mating with an opposite mating type to yield a diploid zygote that undergoes meiosis to yield ascospores. Since *B. dermatitidis* was the name chosen to describe the entity known only in the asexual form, a new name, *Ajellomyces dermatitidis*, was chosen for the ascomycetous form. The form isolated in clinical microbiology laboratories is *B. dermatitidis*, and this name is retained for that description. However, we now know that *B. dermatitidis* represents the conidial form of an ascomycete; the asexual form of a conidial fungus is known as the anamorph, and the sexual form is known as the teleomorph. The entire fungus is collectively termed the holomorph. The name of the teleomorph also serves as the name for the holomorph.

The Deuteromycetes can be segregated into the Blastomycetes, Coelomycetes, and Hyphomycetes. The Blastomycetes contain anamorphs that reproduce by budding, the Coelomycetes contain moulds that produce conidia within a cavity of fungal tissue conidioma (e.g., pycnidium; Table 2), the Hyphomycetes contain moulds that produce conidia without fruiting structures or with synnemata (Table 2) or that produce only sterile hyphae. The Hyphomycetes are often further described as being either dematiaceous (darkly pigmented) or hyaline, although these terms may be used to describe any fungus. Useful separations among the Deuteromycetes are made on the basis of morphology and ontogeny of the conidia and conidiogenous cells (cells that produce conidia). Conidia may develop either by a blastic process (budding) or a thallic process (fragmentation). A summary of some different types of conidiogenesis is presented in Table 3. Since different fungi can produce conidia that look similar, it is often helpful to study a variety of microscopic fields in order to evaluate the mode of development of the conidium from the conidiogenous cell.

There can be confusion in the use of the term "yeast." Some mycologists restrict this term to a specific group of Ascomycetes that produce ascospores in a free ascus, are primarily unicellular, and reproduce asexually by either budding (e.g., *Saccharomyces* spp.) or fission (e.g., *Schizosaccharomyces* spp.). In such a context, other fungi with budding morphologies but without the characteristically borne ascospores are referred to as yeast-like. Thus, although *B. dermatitidis* reproduces by budding and has a teleomorph that produces ascospores, the ascus is contained in a fruiting body (and therefore is not free). Because of this, the budding morphology would be termed "yeast-like" by those holding to a strict taxonomic usage and "yeast" by those using a morphological definition. In short, there is confusion because "yeast" is not a formal taxon.

LABORATORY PROCEDURES

Basic bacteriological techniques are applicable to the fungi. However, since fungi have longer generation times than most bacteria, cultures need to be kept longer, and provisions need to be made to prevent media from excessive drying. This can be done by pouring plates with more media

than bacteriological plates, sealing them with paraffin tape, incubating them in moist chambers, or using tubed media where possible.

Yeasts are manipulated in much the same way as bacteria. Moulds, however, require some specialized procedures. Inflexible, straight needles rather than loops are generally preferred, and colonies are transferred by cutting a small portion of growth to be used as a point inoculum. It is possible to streak moulds for isolation if there is sufficient spore production. This can be done with a standard bacteriological loop after using a straight needle to place a portion of inoculum on the plate to be streaked; this is useful in separating mixed cultures. To separate fungi mixed with bacteria, cultures can be streaked on media containing antibacterial antibiotics. If this fails, 1 drop of a broth or saline suspension of the mixed culture can be used to inoculate each of four tubes of Sabouraud broth, to which is added in series either 1, 2, 3, or 4 drops of concentrated HCl (9).

Microscopy is very important in fungal identification. Bright-field optics and a magnification factor of about 500 are sufficient to resolve the features of medically important fungi mounted in lactophenol (phenol is listed as a hazardous chemical; solutions containing phenol should be prepared, stored, and used in an approved chemical safety cabinet) (for pigmented fungi), or lactophenol cotton blue (Poirrier's blue; for unpigmented fungi). Often, however, critical evaluation of conidial ontogeny requires greater magnification (×1,000). Some structures, such as the annellations present on annellides or the collarettes present on phialides, are clearer with phase-contrast optics. Fluorescent brighteners (calcofluor white) used in conjunction with a UV light source and the proper filter combination are useful not only in detecting fungi in clinical material but also in elucidating fungal structure. One study described the use of a 330- to 380-μm excitation filter, a 420-μm barrier filter, and a 515W eyepiece side absorption filter (26).

Specimen mounts for microscopic examination of moulds include teased, mashed, and slide culture preparations. The simplest mount is the teased preparation. Sterile, straight dissecting needles are used to remove a portion of growth that is teased apart in 95% ethanol; a drop of either lactophenol or lactophenol cotton blue is added, and the mount is sealed with a coverslip. A modification of the teased preparation involves cutting a small section of growth and the uppermost surface of adherent medium from the plate and mashing this in a drop of mounting medium by using a coverslip. In more critical studies to determine conidial ontogeny, it may be necessary to set up a slide culture. This requires sterile glass slides and coverslips and a sterile moist chamber. It is convenient to autoclave a supply of glass petri dishes containing U-shaped bent glass rods. Slides can be labeled, dipped in 95% ethanol, flame sterilized, and placed on the bent rods in the petri dishes. Sporulation agar (potato glucose, cornmeal, cereal) is then cut out of plates into squares approximately one-half the size of the coverslip and placed on the slides within the petri dish. These manipulations can be performed with flame-sterilized spatulas or scalpels. The edges of the agar are then inoculated, and a flame-sterilized coverslip is placed on top of the agar. The result is a glass incubation chamber that is kept moist by the addition of water; the fungus should grow on both glass surfaces in contact with the agar. Thus, it is possible to remove the coverslip for mounting in lactophenol on a fresh slide and, after removal

TABLE 3 Groupings of conidia based on ontogeny

Conidial group	Characteristic	Representative genera
Thallic development	Hyphal tip ceases to elongate; wall growth depolarizes. Entire segment begins to differentiate, and cell wall growth occurs by thickening process.	
Arthroconidia	Thallic production whereby hyphal segments fragment into individual cells delimited by prior development of septum. All other modes of ontogeny listed below are modifications of blastic production.	*Coccidioides, Geotrichum*
Blastic development	Essentially a budding process whereby localization (polarization) of growth occurs at region of mother cell that "blows out" to give rise to conidium	
Blastoconidia	Conidium originates as blown-out portion of mother cell. Developing cell is recognizable before being delimited by septum. Daughter cell may separate completely or remain attached to form chains.	*Candida, Cryptococcus,* and most other yeasts; *Cladosporium, Xylohypha*
Phialoconidia	Blastic production from tubular, often flask-shaped conidiogenous cell (phialide) that does not increase in length or width	*Aspergillus, Fusarium, Lecythophora, Malassezia, Penicillium, Phialophora*
Annelloconidia	Blastic production from flask-shaped conidiogenous cell (annellide) that extends via percurrent growth (inner layers grow through outer layers) and often tapers	*Exophiala, Phaeoannellomyces, Scedosporium, Scopulariopsis*
Poroconidia	Blastic production through channel in cell wall of conidiogenous cell	*Alternaria, Bipolaris, Curvularia, Drechslera, Exserohilum*
Sympoduloconidia	Successive blastic conidial production characterized by continued growth of conidiogenous cell to one side of base of conidium. Typically results in zigzag or distorted ampulliform appearance of conidiogenous cell. Poroconidia of *Curvularia, Drechslera, Bipolaris,* and *Exserohilum* spp. are produced on sympodial conidiophores.	*Sporothrix, Dactylaria*
Aleurioconidia	Older but useful term describing conidia that typically have broad bases of attachment to conidiogenous cell and separate by lysis of conidiogenous hyphal walls, leaving remnant attached to conidium as annular frill. *Epidermophyton, Microsporum,* and *Trichophyton* spp. now are recognized to produce conidia by a thallic process, and their conidia are now termed holothallic conidia. *Blastomyces, Chrysosporium,* and *Histoplasma* spp. produce holoblastic conidia.	*Blastomyces, Chrysosporium, Histoplasma, Epidermophyton, Trichophyton, Microsporum*

of the agar block, to place a drop of lactophenol cotton blue and another coverslip on the original slide.

Any of the preparations described above can be kept as reference slides. Nail polish can be added to the corners of a coverslip to fix it to a slide; after excessive mounting medium evaporates or is blotted dry, the edges of the coverslip can be completely sealed to the slide to make a semipermanent mount.

The taxonomy of the medically important yeasts is based not only upon morphology but also upon extensive physiological characterization as used in clinical bacteriology. Commercial products are now available to assist in compartmentalizing the key physiological reactions necessary to identify these organisms (see chapters 60 and 61 of this Manual).

Although not as important as in the taxonomy of the yeasts, certain physiological reactions and special tests can be helpful in the identification of the moulds. Examples of selected tests for this purpose are shown in Table 4.

MOLECULAR APPROACHES TO MEDICAL MYCOLOGY

The advent of molecular biology and its technology has begun to facilitate clinical laboratory mycology. Specific examples of advances in the identification of fungi in clinical specimens should be apparent in this edition of the Manual (see chapters 62 and 63). Two examples of molecular technology applied to medically important fungi isolated in culture are the use of DNA probes for species identification of selected pathogens and the use of DNA fingerprinting for the subspecies typing of selected species.

Nonradioactive molecular probes are now commercially available for the identification of cultures of *B. dermatitidis*, *Coccidioides immitis*, *Cryptococcus neoformans*, and *Histoplasma capsulatum* (12). Use of the Accuprobe (Gene-Probe, San Diego, Calif.) provides for the culture identification of the preceding species in less than 2 h. Given that the procedure can be applied to either yeast or mould forms of very young colonies, it has reduced the time necessary to identify blood isolates of *H. capsulatum* to as little as 5 days after incubation of a lysis-centrifugation-processed blood culture specimen (12).

Molecular epidemiology is becoming commonplace in the research and reference mycology laboratories. Combinations of electrophoretic karyotyping using pulsed-field electrophoresis, restriction fragment length polymorphism analysis using restriction endonucleases, and species-specific DNA probes have resulted in the ability to conduct strain-typing studies with the medically important fungi (12, 18). Much of the work has been done with *Candida*, *Aspergillus*, and *Cryptococcus* spp., so it is now possible in specialty laboratories to track the movement of pathogens through populations and to begin to address the question of clonality of fungal drug resistance. It is predictable that such procedures will see more widespread implementation in the near future.

LABORATORY SAFETY

Biosafety has to be a top priority in any clinical laboratory. All clinical laboratory personnel should be trained in the principles of biosafety and should know how to follow established safety procedures when working with any potential human pathogen. The two most common means of acquiring fungal infections in the laboratory are inhalation of conidia that are aerosolized and accidental inoculation with sharps (e.g., needles, scalpels, broken glassware) that are associated with clinical specimens.

The term "containment" is used in describing safe methods for managing infectious agents in the laboratory environment where they are being handled and/or maintained (20). Primary containment is the protection of personnel and the immediate laboratory environment from exposure to infectious agents. This type of containment is accomplished by maintaining good microbiological technique and using appropriate safety equipment. Secondary containment is the protection of the external laboratory environment from exposure to infectious materials. A combination of facility design and operational practices is used to attain secondary containment. The purpose of containment is to reduce exposure of laboratory workers, other persons, and the outside environment to potentially hazardous agents. Appropriate containment includes a combination of laboratory practice and technique, safety equipment, and facility design.

The most important element of containment is strict adherence to standard microbiological practices and techniques. Persons working with infectious materials must be aware of potential hazards and must be trained and proficient in the practices and techniques required for handling such material safely. The laboratory supervisor is responsible for providing or arranging for appropriate training of personnel. When standard laboratory practices are judged insufficient to control the hazard associated with a particular agent or laboratory procedure, additional measures may be needed. The laboratory supervisor is responsible for selecting additional safety practices, which must be in keeping with the risk associated with the agent or procedure (see discussion of biosafety levels below).

Safety equipment includes biological safety cabinets and a variety of enclosed containers. The biological safety cabinet is the principal device used to provide containment of infectious aerosols generated by many microbiological procedures. Three types of biological safety cabinets (classes I, II, and III) are used in microbiological and clinical laboratories (20). Class I and II cabinets are open front but provide proper protection to laboratory personnel and the environment when used with appropriate techniques. Class III safety cabinets are gastight units that provide the most effective protection of the investigator and the laboratory. In addition to safety cabinets, items such as gloves, coats, gowns, shoe covers, safety glasses, boots, respirators, and face shields also serve as important safety equipment.

Biosafety levels, including laboratory practices and techniques, safety equipment, and appropriate laboratory facilities, have been defined to guide those working with infectious agents (14, 20, 23). Biosafety level 1 requires basic laboratory facilities and the use of standard microbiological practices. This level applies to the use of well-characterized microorganisms not known to cause disease in healthy adult humans or not known to colonize humans. Biosafety level 2 practices, equipment, and facilities are applicable to work with a broad spectrum of indigenous moderate-risk agents present in the community and associated with human diseases of various severities. With good microbiological techniques, these agents can be worked with safely in activities conducted on the open workbench, provided the potential for producing aerosols or splashes is low. Primary hazards to personnel working with these agents relate to accidental autoinoculation or ingestion of infectious materials. Procedures with high aerosol potential increase the risk of exposure of personnel to infectious agents and must be conducted in primary containment equipment or devices. Biosafety level 3 is used when the potential of infection from aerosols, autoinoculation, or ingestion exists and when personnel are handling microorganisms that may cause severe or lethal infections.

Clinical specimens that may contain fungi pathogenic

TABLE 4 Special tests to aid in identification of selected moulds

Test or procedure	Application	Examples	Reference(s)
Aspergillus flavus differential medium	Differentiates A. *flavus* group from most other aspergilli by production of yellow-orange colony reverse.	A. *flavus* +, A. *fumigatus* −	24
Christensen urea agar	Differentiates dermatophytes.	*Trichophyton memtagrophytes* +, *Trichophyton rubrum* −	Chapter 60, this Manual
Exoantigen test	Specifically identifies various fungi by immunodiffusion.	*Histoplasma capsulatum, Blastomyces dermatitidis, Coccidioides immitis*	8
Mycosel agar	Differentiates certain pathogens from morphologically similar saprophytes. In general, cycloheximide in Mycosel agar inhibits saprophytic fungi to greater degree than pathogens, but there are notable exceptions, such as that listed here and *Cryptococcus neoformans*, pathogenic aspergilli, and Zygomycetes.	*Dactylaria constricta* var. *gallopava* −, *D. constricta* var. *constricta* +	22
Nitrate assimilation	Differentiates *Wangiella* spp. from most other dematiaceous pathogens.	*Wangiella dermatitidis* −, *Exophiala jeanselmei* +	19, 25
Proteolysis (e.g., 12% gelatin in broth)	May differentiate some pathogenic dematiaceous fungi from saprophytes, but test is medium, fungus, test, and laboratory dependent.	*Fonsecaea* spp. −, *Cladosporium* spp. +	3
Sodium chloride tolerance	Differentiates selected dermatophytic and dematiaceous fungi.	*Exophiala werneckii*, growth in presence of ≥15% NaCl; *Exophiala jeanselmei*, no growth	7
Thermotolerance	Differentiates various fungi. Growth maxima (°C) are:		
	55	*Rhizomucor* spp.	10
	50–52	*Rhizopus rhizopodiformis*	4, 27
	45	*Aspergillus fumigatus*	21
	45	*Dactylaria constricta* var. *gallopava*	22
	42	*Xylohypha bantiana*	2
	40	*Wangiella dermatitidis*	16
	40	*Exserohilum mcginnisii*	15
	40	*Bipolaris spicifera*	15
	37	*Exophiala jeanselmei*	16
	37	*Cladosporium carrionii*	6
Trichophyton agars	Nutritionally differentiates dermatophytes.	*Trichophyton tonsurans*, greater growth on T_4 than on T_1	Chapter 60, this Manual
Wickerham assimilation and fermentations	Differentiates selected Zygomycetes.	Selected Zygomycetes	27

for humans should be handled by using biosafety level 2 practices, containment equipment, and facilities as outlined in reference 20. Standard precautions include the use of a biological safety cabinet when working with any clinical material suspected of containing fungi. This precaution protects laboratory personnel and the laboratory environment from aerosols of conidia that may form in certain clinical samples containing fungi that have a filamentous phase. Personnel must be properly trained in the use of the cabinets. An additional advantage of using biological safety cabinets is that they protect the clinical specimens from extraneous airborne contaminants.

Biosafety level 2 procedures are specifically recommended for personnel working with clinical specimens that may contain *B. dermatitidis*, *C. immitis*, *H. capsulatum*, *Cryptococcus neoformans*, *Sporothrix schenckii*, and pathogenic members of the genera *Microsporum*, *Trichophyton*, and *Epidermophyton*, which cause dermatophytoses. Biosafety level 3 should be used when propagating or manipulating cultures of *B. dermatitidis*, *C. immitis*, and *H. capsulatum* or when handling soil or other specimens that may contain infectious conidia from these three fungi. Conidia from these fungi have great potential for causing infection by inhalation. In addition to the fungi listed, the Centers for Disease Control and Prevention recommends that *Penicillium marnefei*, *Cladosporium trichoides* (*Xylohypha bantianum*), *Wangiella* (*Exophiala*) *dermatitidis*, *Fonsecaea pedrosoi*, and *Dactylaria gallopava* (*Ochroconis gallopavum*) also be handled at biosafety level 2, since these organisms have caused infection in immunocompromised people by presumed inhalation or accidental subcutaneous inoculation from environmental sources (20).

Additional information on the safe handling and processing of clinical specimens that may contain these and other fungi is outlined in chapters 60 through 68 of this Manual.

STOCK CULTURES

Stock cultures of fungi are an essential source of material for quality control organisms, for reference cultures to be used in comparative identification, for production of known metabolites, and for teaching collections. Numerous methods of preserving fungal cultures are available (17); these vary according to the organisms represented; the size of the collection; and considerations of cost, practicality, and personal preference. Two of the simplest methods involve either freezing actively growing agar slant cultures at −10°C (caution: *Epidermophyton* spp. and many Zygomycetes may not survive freezing) or preparing sterile-water suspensions of cultures. For the latter method, cultures are grown on potato glucose agar until mature (several days to 2 weeks), and mycelium, spores, or yeast cells are scraped or washed from the surface of slants with sterile distilled water and a pipette (10). The resulting fungal suspensions can be sealed in sterile vials to prevent evaporation and stored at room temperature. A time-tested but less commonly used technique is to cover an agar slant culture of a fungus with sterile mineral oil.

For long-term preservation, preparation of lyophilized cultures and storage of cultures at −70°C or in liquid nitrogen are well-established procedures, but they require expensive equipment such as lyophilizers or liquid-nitrogen freezers. However, cultures stored by these techniques may be kept for many years without loss of viability or alteration of biochemical or morphological characteristics.

CONCLUSION

The descriptions, procedures, and guidelines outlined in chapters 60 through 68 of this Manual represent current knowledge in specialized areas of medical mycology that are relevant for specialists working in clinical microbiology. In addition to the cited literature, a list of selected references appears at the end of this chapter for those interested in learning more about basic and medical mycology and related applications. The American Society for Microbiology, the Centers for Disease Control and Prevention, and several universities frequently conduct workshops and training programs in medical mycology, including specialty courses in clinical mycology. Information on these programs may be found in *ASM News* and in the newsletters of the Medical Mycological Society of the Americas, Mycological Society of America, and International Society for Human and Animal Mycology.

REFERENCES

1. **Alexopoulos, C. J., and C. W. Mims.** 1979. *Introductory Mycology,* 3rd ed. John Wiley & Sons, Inc., New York.
2. **Dixon, D. M., T. J. Walsh, W. G. Merz, and M. R. McGinnis.** 1989. Infections due to *Xylohypha bantiana* (*Cladosporium trichoides*). *Rev. Infect. Dis.* **11:**515–525.
3. **Espinel-Ingroff, A., P. R. Goldson, M. R. McGinnis, and T. M. Kerkering.** 1988. Evaluation of proteolytic activity to differentiate some dematiaceous fungi. *J. Clin. Microbiol.* **26:**301–307.
4. **Gartenberg, G., E. J. Bottone, G. T. Keusch, and I. Weitzman.** 1978. Hospital-acquired mucormycosis (*Rhizopus rhizopodiformis*) of skin and subcutaneous tissue. Epidemiology, mycology and treatment. *N. Engl. J. Med.* **299:**1115–1118.
5. **Illman, W. I.** 1970. On the use of mould versus mold for a mycelial fungus. *Mycologia* **62:**1214.
6. **Iwatsu, T.** 1984. A new species of *Cladosporium* from Japan. *Mycotaxon* **20:**521–533.
7. **Kane, J., and R. C. Summerbell.** 1987. Sodium chloride as an identification aid of *Phaeoannellomyces werneckii* and other medically important dematiaceous fungi. *J. Clin. Microbiol.* **25:**944–946.
8. **Kaufman, L., and P. G. Standard.** 1987. Specific and rapid identification of medically important fungi by exoantigen detection. *Annu. Rev. Microbiol.* **41:**209–225.
9. **McGinnis, M. R.** 1980. *Laboratory Handbook of Medical Mycology.* Academic Press, Inc., New York.
10. **McGinnis, M. R., A. A. Padhye, and L. Ajello.** 1974. Storage of stock cultures of filamentous fungi, yeasts, and some aerobic actinomycetes in sterile distilled water. *Appl. Microbiol.* **28:**218–222.
11. **McGinnis, M. R., and I. F. Salkin.** 1986. Identification of molds commonly used in proficiency tests. *Lab. Med.* **17:**138–142.
12. **Merz, W. G.** 1994. The clinical mycology laboratory: meeting the challenges of the 90s. *Infect. Dis. Clin. Practice* (Suppl. 2) **3:**560–567.
13. **Monaghan, R. L., and S. Currie.** 1985. Preservation of antibiotic production by representative bacteria and fungi. *Dev. Ind. Microbiol.* **26:**787–792.
14. **National Academy of Sciences.** 1989. *Biosafety in the Laboratory: Prudent Practices for the Handling and Disposal of Infectious Materials.* National Academy Press, Washington, D.C.
15. **Padhye, A. A., L. Ajello, M. A. Wieden, and K. K. Steinbronn.** 1986. Phaeohyphomycosis of the nasal sinuses caused by a new species of *Exserohilum*. *J. Clin. Microbiol.* **24:**245–249.
16. **Padhye, A. A., M. R. McGinnis, and L. Ajello.** 1978. Thermotolerance of *Wangiella dermatitidis*. *J. Clin. Microbiol.* **8:**424–426.
17. **Pasarell, L., and M. R. McGinnis.** 1992. Viability of fungal cultures maintained at −70°C. *J. Clin. Microbiol.* **30:**1000–1004.
18. **Pfaller, M. A.** 1992. Epidemiological typing methods for mycoses. *Clin. Infect. Dis.* **14**(Suppl. 1):S4–S10.
19. **Pincus, D. H., I. F. Salkin, N. J. Hurd, I. L. Levy, and M. E. Kemna.** 1988. Modification of potassium nitrate assimilation test for identification of clinically important yeasts. *J. Clin. Microbiol.* **26:**366–368.

20. **Richmond, J. Y., and R. W. McKinney (ed.).** 1993. *Biosafety in the Microbiological and Biomedical Laboratories*, 3rd ed. U.S. Department of Health and Human Services, publication no. (CDC) 93-8395. U.S. Government Printing Office, Washington, D.C.

21. **Rippon, J. W.** 1988. *Medical Mycology: the Pathogenic Fungi and the Pathogenic Actinomycetes*, 3rd ed. The W. B. Saunders Co., Philadelphia.

22. **Salkin, I. F., and D. M. Dixon.** 1987. *Dactylaria constricta*: description of two varieties. *Mycotaxon* **29:**377–381.

23. **Salkin, I. F., and R. Gershon.** 1992. Biohazards and safety, p. 14.1.1–14.1.6. *In* H. D. Isenberg (ed.), *Clinical Microbiology Procedures Handbook*. American Society for Microbiology, Washington, D.C.

24. **Salkin, I. F., and M. A. Gordon.** 1975. Evaluation of *Aspergillus* differential medium. *J. Clin. Microbiol.* **2:**74–75.

25. **Salkin, I. F., M. E. Kenma, and M. G. Rinaldi.** 1988. *Abstr. X Congr. Int. Soc. Hum. Anim. Mycol.*, p. 52.

26. **Salkin, I. F., M. R. McGinnis, M. J. Dykstra, and M. G. Rinaldi.** 1988. *Scedosporium inflatum*, an emerging pathogen. *J. Clin. Microbiol.* **26:**498–503.

27. **Scholer, H. J., E. Mueller, and M. A. A. Schipper.** 1983. Mucorales, p. 9–59. *In* D. H. Howard (ed.), *Fungi Pathogenic for Humans and Animals*. Marcel Dekker, Inc., New York.

SELECTED REFERENCES FOR FURTHER STUDY

1. **Ainsworth, G. C., F. K. Sparrow, and A. S. Sussmann.** 1973. *The Fungi: an Advanced Treatise*, vol. 4A and 4B. Academic Press, Inc., New York.

2. **Bullock, W., T. Kozel, S. Scherer, and D. M. Dixon.** 1993. Medical mycology in the 1900s: involvement of NIH and the wider community. *ASM News* **59:**182–185.

3. **Cole, G. T.** 1986. Models of cell differentiation in conidial fungi. *Microbiol. Rev.* **50:**95–132.

4. **Cole, G. T., and R. A. Samson.** 1979. *Pattern of Development in Conidial Fungi*. Fearon Pitman Publishers, Inc., Belmont, Calif.

5. **Dolan, C. T., J. W. Funkhoser, E. W. Koneman, N. Y. Miller, and G. D. Roberts.** 1976. *Atlases of Medical Mycology*, vol. 1–6. American Society for Clinical Pathology, Chicago. (Kodachrome slides and explanatory manual.)

6. **Haley, L. D., J. Trandel, and M. B. Coyle.** 1980. *Cumitech 11, Practical Methods for Culture and Identification of Fungi in the Clinical Microbiology Laboratory*. Coordinating ed., J. C. Sherris. American Society for Microbiology, Washington, D.C.

7. **Hawksworth, D. L.** 1974. *Mycologist's Handbook*. Commonwealth Mycological Institute, Kew, Surrey, England.

8. **Hawksworth, D. L., B. C. Sutton, and G. C. Ainsworth.** 1983. *Ainsworth & Bisby's Dictionary of the Fungi*, 7th ed. Commonwealth Mycological Institute, Kew, Surrey, United Kingdom.

9. **Kendrick, B.** 1985. *The Fifth Kingdom*. Mycologue Publications, Waterloo, Ontario, Canada.

10. **Koneman, E. W., G. D. Roberts, and S. F. Wright.** 1978. *Practical Laboratory Mycology*, 2nd ed. The Williams & Wilkins Co., Baltimore.

11. **Kwon-Chung, K. J., and J. E. Bennett.** 1992. *Medical Mycology*. Lea & Febiger, Philadelphia.

12. **Larone, D. H.** 1987. *Medically Important Fungi: a Guide to Identification*. Elsevier, New York.

13. **Phillips, G. D. (ed.).** *Review of Medical and Veterinary Mycology*. CAB International Mycological Institute, Kew, Surrey, United Kingdom.

14. **Salkin, I. F., and B. E. Robinson.** 1988. Mycology, p. 1–411. *In* B. B. Wentworth (ed.), *Diagnostic Procedures for Mycotic and Parasitic Infections*. American Public Health Association, Washington, D.C.

Detection and Recovery of Fungi from Clinical Specimens

WILLIAM G. MERZ AND GLENN D. ROBERTS

60

INTRODUCTION

The laboratory diagnosis of fungal infections requires awareness by clinicians, collection of proper specimens, and appropriate mycologic laboratory procedures. This chapter presents guidelines for collection and transport of specimens, preparation and interpretation of direct microscopic examination of specimens, and culture and incubation of mycologic cultures.

SPECIMEN COLLECTION, TRANSPORT, AND STORAGE

Specimen collection, transport, and storage are extremely important components in the provision of rapid, accurate results for the diagnosis and management of mycoses. Specimens must be collected under aseptic conditions or after appropriate hygienic preparation to optimize the significance of the mycologic results. Communication between physicians and the laboratory is essential. Specimens need to be accompanied by important information, including proper patient identifiers, time of collection, and presumptive diagnosis, which might help the laboratory.

Because of our increased awareness of safety issues and concern for health care providers at all levels, all specimens for mycologic studies should considered potentially hazardous and handled accordingly. Although blood and other body fluids or specimens contaminated with blood require special care, it is suggested that all specimens should be treated with the same care; specimens should be transported in double leakproof containers sealed at the bedside. For example, cerebrospinal fluid (CSF) should be placed in a sterile tube with a leakproof lid, and the tube should be placed in a second container, i.e., either a sealed carry container or a sealed bag. The second container prevents hazardous contact if the tube leaks in transit. Specimens that leak in transit should be judged individually. The best option is to re-collect the specimen. If the specimen is not re-collectable (e.g., collection requires an invasive procedure), exercise judgment and add a comment to the results if the specimen is processed.

Specimens for fungal cultures are best sent in a sterile container. Although there is not an extensive literature on the subject, pathogenic fungi can be recovered from specimens sent in most anaerobic bacterial transport media or other transport media. Direct examination from these transport systems should not be attempted.

Transport to the laboratory should be rapid and safe (in double containers). Fortunately, delayed processing of specimens for fungal culture is not as detrimental as it is with virologic, parasitologic, or certain bacteriologic specimens. In general, if specimen processing is delayed, storage at 4°C is appropriate. There are rare exceptions for specific types of specimens. Specimens submitted on Culturette swabs should not be stored before culturing, since *Histoplasma capsulatum*, *Blastomyces dermatitidis*, and *Cryptococcus neoformans* may be inhibited.

Hair, Skin, and Nails

Specimens for the diagnosis of tinea capitis should include both representative abnormal hairs removed with forceps and scalp scales collected by scraping. Obtain skin specimens for both dermatophytosis or primary cutaneous candidiasis by scraping the active borders of the lesions with a scalpel or glass microscope slide after wiping the affected area with an alcohol swab.

Obtain nail specimens by clipping a generous portion of the affected area of the nail, and include scrapings of the excess keratin produced beneath the nail.

Place all hair, skin, and nail specimens in an envelope or sterile culture dish for transport to the laboratory. Should storage be required, storage at room temperature is optimal. Storage at 4°C is usually acceptable; however, an occasional dermatophyte might not tolerate refrigeration. Discourage the use of swabs for collection and transport of these specimens, as direct examinations cannot be performed on such specimens, and cultures may not be optimal. If a specimen is received on a swab and a new specimen cannot be obtained, use the swab for culture only. Streak the cotton portion onto the surface of the agar, and leave it in contact with the agar surface if an agar slant is used.

External Eye Specimens

Specimens from patients with presumed mycotic keratitis should be obtained with care by an ophthalmologist, usually in an operating room. After the surface of the cornea is scraped several times, streak the Kimura scalpel on X's or C's on the bottom of appropriate fungal media in petri dishes (6). Fix another portion of the scraping on glass microscope slides for examination by the clinical and/or pathology laboratory. Keep inoculated plates at room temperature if transit time to the laboratory may be delayed.

Sterile Body Fluids (CSF; Pericardial, Peritoneal, and Synovial Fluids; Vitreous Humor)

Obtain all fluids under aseptic conditions by using a sterile syringe. The fluid can be transferred to a sterile tube for safe transport to the laboratory. Store specimens at 4°C.

Urine

Collect all urine specimens in sterile containers, and immediately send them to the laboratory. Specimens may be stored at 4°C for up to 12 h. Twenty-four-hour urine specimens and specimens collected from catheter bags are not acceptable for mycologic studies.

Vaginal Secretions

Diagnosis of vaginal candidiasis is better established by clinical characteristics and a positive direct examination of the secretions. Cultures may be misleading, since yeast cells are part of the normal vaginal flora of up to 20% of healthy women. However, cultures may be helpful in monitoring therapy, managing individuals with chronic recurring disease, or determining the etiology of infectious vaginitis when other tests have been unrewarding. Appropriate specimens are either two swabs or one swab and one smear (transported in a closed container) in order to perform both a culture and a direct examination. Specimens may be stored at 4°C if needed.

Stool Specimens

Diagnoses of many fungal infections of the gastrointestinal tract are better established by biopsy of pathologic tissue. Positive cultures may be misleading, since up to 40% of healthy individuals and up to 75% of compromised patients are colonized with yeasts. Stool cultures should be submitted in a sterile container or on two rectal swabs. Storage at 4°C is appropriate.

Respiratory Secretions

Since most respiratory specimens are collected through the upper respiratory tract, proper specimen collection procedures are extremely important in order to recover fungal pathogens in the presence of normal bacterial and fungal flora. Proper oral hygiene procedures (brushing the teeth and rinsing) should be performed prior to specimen collection. Some patients may require that sputum be induced by experienced personnel. First morning specimens are usually preferred. Twenty-four-hour collections are inappropriate; they should be rejected, and new specimens should be requested. Collect all specimens in wide-mouth containers with leakproof lids. Specimens may be stored at 4°C if necessary. A total of 5 to 10 ml is more than adequate; specimens grossly contaminated by saliva, as evidenced by numerous squamous epithelial cells, need not be rejected (9); however, specimens or parts of specimens with more macrophages or polymorphonuclear cells may help increase the recovery of pathogenic fungi. Transit time is not as critical for mycologic cultures as for bacteriologic cultures; fungi have been recovered after delays of up to 2 weeks (5).

Pneumocystis

Laboratory diagnosis of *Pneumocystis carinii* infections by direct microscopic detection requires proper specimens. Examination should be performed only on (i) induced sputum specimens that have been collected by trained personnel after the patient's teeth and mucosal surfaces have been brushed and after mouth washing or (ii) specimens collected during bronchoscopy. Avoid storage if at all possible.

Upper Respiratory Tract

Since fungi may be normal flora of the upper respiratory tract or contaminants from environmental aeroflora, cultures may be misleading. Therefore, direct examination of the specimen is extremely important. Material obtained by curetting is optimal; however, swabs are acceptable, although two are necessary for direct examination. Specimens may be stored at 4°C.

Tissues

Tissue specimens should be collected by experienced personnel using strict aseptic procedures. Specimens may be divided by the physician or the laboratory. They should be placed in a small sterile container and sent to the laboratory immediately. If transport is to be delayed, the specimen may be covered with a minimal amount of sterile saline without preservative.

Exceptions are tissue specimens so minute that they are better cultured by the physician. A corneal specimen is an example of this type of exception.

Blood

Blood specimens should be collected by experienced personnel using strict aseptic procedures. If blood culture bottles (radiometric or nonradiometric) are used, inoculate 5 or 10 ml of blood at the bedside. If the lysis-centrifugation method is used, use 10-ml tubes (1). Inoculated blood bottles or lysis tubes may be kept at room temperature but need to be taken immediately to the laboratory. Delayed detection will occur with blood bottles and recovery will be affected in lysis tubes (16) if they are not processed within 8 to 9 h of collection. Collection times are extremely important.

SPECIMEN HANDLING

The laboratory staff must take the time to verify that all pertinent information about a specimen has been provided, that the specimen has not leaked in transit, that it was delivered by an appropriate method, and that the time between collection and initial mycologic processing is acceptable. Addressing potential problems when the specimen is received is far superior to attempting to correct a problem later, when a culture might turn positive.

Safe handling of specimens for both protection of the laboratory staff and minimization of laboratory specimen contamination is extremely important. All blood or body fluid specimens and all specimens contaminated by blood or body fluids must be processed by staff wearing gloves and laboratory coats or gowns. For optimal safety, it is essential that all clinical specimens be processed in a biologic safety cabinet by trained personnel wearing laboratory coats and gloves. (Refer to chapter 59 in this section.)

Non-Culture-Dependent Methods

Rapid, accurate, non-culture-dependent methods for the diagnosis of life-threatening mycoses are needed. Methodologies with the potential for providing rapid information directly from analysis of the clinical specimen include detection of fungal structure by direct examination of the specimen, detection of circulating antigen, chemical detection of fungal constitutive macromolecules, detection of fungus-specific metabolites, and detection of fungus-specific nucleic acid sequences. However, only detection of fungal structures by direct examination and detection of antigen in clinical specimens are currently available for rapid diagnosis. The last methodologies (detection of fungus-specific metabolites and nucleic acid sequences in clinical specimens) have the potential to revolutionize the diagnosis of fungal infections. Recent studies describe a quantitative enzymatic serum assay for detection of D-arabinitol produced by most *Candida* species that is useful for both diagnosing and monitoring invasive candidiasis. Similarly, recent studies have supported the feasibility of detecting fungus- or species-specific DNA sequences in clinical specimens after in vitro amplification for the diagnosis of candidiasis, aspergillosis, and pneumocytis infections. Large prospective studies following commercialization of these assays are needed to establish the role of methodologies in the clinical mycology laboratory. All of these are important, since the sooner a serious fungal infection is diagnosed, the earlier specific antifungal therapy can be initiated, and thus the chance for a favorable outcome is increased. Accurate detection and recognition of specific elements can save days to weeks in the recovery and identification of pathogenic fungi.

Antigen Detection

Cryptococcal and histoplasma antigen tests are the only assays that are currently available and widely accepted. Experimental assays or tests for the diagnosis of aspergillosis, candidiasis, and coccidioidomycosis that are now available in Europe may in the future become available in the United States.

Detection of cryptococcal heteropolysaccharide capsular antigen is perhaps the best means of diagnosing an infectious disease. Currently, both the reference latex method and an enzyme-linked immunosorbent assay method are commercially available. Monitoring the titers of CSF specimens can be helpful in monitoring infection or diagnosing relapses. See the *Manual of Clinical Laboratory Immunology* (7) for details. The assay is highly sensitive and specific when pretreatment, careful specimen handling, and proper controls are employed. The assay can be performed on serum and CSF specimens.

Detection of the circulating polysaccharide antigen of *H. capsulatum* can be helpful in the diagnosis of life-threatening infections caused by this dimorphic fungus. The assay is a semiquantitative radioimmunoassay method that can be used with serum, urine, and CSF specimens. Monitoring titers can be helpful in monitoring therapeutic efficacy and establishing relapses. Currently, the assay is available in only one reference laboratory, that of L. Joseph Wheat, Wishard Memorial Hospital, Indianapolis, Ind.

Direct Examination

A definitive diagnosis of many mycotic infections can be made by trained personnel when they detect fungal elements specific for certain fungi pathognomonic of infection. Some examples include the detection of *H. capsulatum* in blood or bone marrow, cysts of *P. carinii*, *B. dermatitidis*, and *Coccidioides immitis* by direct examination.

Detection of specific fungal elements that are part of our normal flora or our environment may be crucial in making a decision as to whether a specific fungus recovered is a contaminant or an opportunistic pathogen. Detection of wide, wavy hyphae morphologically compatible with the

agent of zygomycosis is necessary for specific diagnosis of the acute fulminant infection in a patient with diabetes mellitus in acidosis. A culture positive for *Rhizopus* spp. may indicate a contaminant or a pathogen, and a delay in initiating aggressive antifungal therapy and surgical procedures can be crucial. Similar examples include aspergillosis or other agents of hyalohyphomycosis and phaeohyphomycosis: the detection of hyphae in a clinical specimen is strongly associated with the presence of these diseases.

In addition, the detection of specific fungal elements allows the laboratory to add specialized media or to increase the incubation time for optimal recovery of fungi. An example is the addition of mycobacterial media for better recovery of a *Nocardia* sp. when the typical thin, branching filaments are seen in a direct examination of a specimen. Flagging cultures with a positive direct examination for *B. dermatitidis*, *Coccidioides immitis*, or *H. capsulatum* has permitted recovery of these pathogens by extending the incubation period beyond the laboratory's normal incubation time.

However, the limitations and potential problems of direct examinations must be considered. First, a negative direct examination never rules out a fungal infection. The sensitivity varies with anatomical site, amount of specimen examined, number of organisms, site of pathology, type of patient, and quality of the examiner. Second, false-positive findings unfortunately do occur. Lysed lymphocytes in an India ink preparation of CSF have been mistaken for *C. neoformans*, collagen fibers have been mistaken for *Nocardia* filaments, fat droplets have been confused with budding yeast cells, etc. Equivocal findings should always be reviewed by more than one reader, or a second procedure should be performed to aid in this type of problem. A false-positive direct examination usually is more harmful than a false-negative direct examination.

Since direct examinations are less sensitive than cultures, cultures should be performed on specimens, and direct examinations should be performed only if a sufficient volume of specimen is provided. As with all policies, exceptions exist: a positive direct examination of a mouth lesion suspected to be candidiasis may be more important than a positive culture.

A number of stains or procedures can be used to detect fungal elements in direct examination of clinical specimens (Table 1). Each method has its own advantages and disadvantages. The wet preparation, KOH, calcofluor, Gram stain, Wright stain, Giemsa stain, and monoclonal fluorescent-antibody assays for detection of *P. carinii* are often performed in clinical laboratories.

The Papanicolaou stain is usually performed in a cytopathology laboratory, and the periodic acid-Schiff and methenamine silver stains are usually performed in anatomic or surgical pathology laboratories. Laboratory personnel may occasionally be called upon to aid in interpretation, so familiarity with these stains may be helpful.

As a group, fungi exhibit great diversity and polymorphism, exemplified by the various morphologies and structures seen in pathologic tissue. Recognition of these diverse structures in clinical specimens can provide very specific identification of many fungal pathogens or at least a list of possible pathogens. The characteristic fungal elements seen in clinical specimens are presented in Table 2. Most fungi appear the same in tissue and in other clinical specimens. Photomicrographs 1 through 29 illustrate these fungal elements as seen in clinical specimens.

TABLE 1 Methods available for direct microscopic detection of fungal elements in clinical specimens

Method (reference)[a]	Use	Time required	Advantages	Disadvantages
Alcian blue	Detection of C. neoformans in CSF	2 min	When positive in CSF, diagnostic of meningitis.	Not commonly used; like India ink, does not detect all cases.
Acid fast	Detection of mycobacteria and Nocardia spp.	12 min	Detects Nocardia spp. and B. dermatitidis.	Tissue specimens are difficult to interpret owing to strong background staining.
Calcofluor white (4, 8)	Detection of fungi, including P. carinii	1 min	Can be mixed with KOH; detects fungi rapidly owing to bright fluorescence.	Requires fluorescence microscope; background fluorescence prominent, but fungi exhibit more intense fluorescence. Vaginal secretions are difficult to interpret.
Giemsa	Examination of bone marrow and peripheral blood smears	15 min	Detects intracellular H. capsulatum.	Detection is usually limited to H. capsulatum.
Giemsa and rapid modifications	Examination of induced sputum and bronchoscopy specimens	13 min	Detects intracystic bodies and trophozoites of P. carinii.	Does not stain cysts but does stain organisms other than P. carinii.
Gram stain	Detection of bacteria	3 min	Commonly performed on most clinical specimens submitted for bacteriology and detects most fungi present.	Some fungi stain well; however, others (e.g., Cryptococcus spp.) stain weakly in some instances and exhibit only stippling. Some Nocardia isolates fail to stain or stain weakly. Common GS artifacts appear as yeast cells.
India ink	Detection of C. neoformans in CSF	1 min	When positive in CSF, diagnostic of meningitis.	Negative in many cases of meningitis; not reliable
Potassium hydroxide (KOH)	Clearing of specimens to make fungi more readily visible	5 min; if clearing is not complete, an additional 5–10 min is necessary.	Rapid detection of fungal elements	Experience required, since background artifacts are often confusing. Clearing of some specimens may require extended time.
Methylene blue	Detection of fungi in skin scrapings	2 min	Usually added to KOH; provides contrast for detection of fungal elements.	Background staining of cells makes reading difficult.
Methenamine silver stain	Detection of fungi in histologic sections	1 h	Best stain for detecting fungal elements	Requires specialized staining method not usually available in microbiology laboratories.
Papanicolaou stain	Examination of secretions for presence of malignant cells	30 min	Cytotechnologist can detect fungal elements.	Requires specialized staining and reader familiar with this stain.
PAS stain	Detection of fungi	20 min; 5 min additional if counterstain is employed	Stains fungal elements well; hyphae and yeasts can be readily distinguished.	Nocardia spp. do not stain well; B. dermatitidis appears pleomorphic. PAS-positive artifacts can appear as yeast cells.
Toluidine blue	Examination of induced sputum and bronchial specimens for P. carinii	25 min	Reveals cyst walls of P. carinii with purple color.	Background stains; does stain other fungi.
Fluorescent monoclonal antibody (10)	Examination of induced sputum and bronchial specimens for P. carinii	45 min	Detects cysts of P. carinii.	Specific for P. carinii
Wright stain	Examination of bone marrow and peripheral blood smears	7 min	Detects intracellular H. capsulatum.	Detection is usually limited to H. capsulatum.

[a]PAS, periodic acid-Schiff.

TABLE 2 Characteristic fungal elements seen in direct examination of clinical specimens

Morphologic fungal elements found	Organism(s)	Diam range (μm)	Characteristic features
Yeast forms	*Histoplasma capsulatum*	2–5	Small; oval to round budding cells; often found clustered within histocytes; difficult to detect when present in small numbers; often intracellular; Fig. 1 and 2
Yeast forms	*Sporothrix schenckii*	2–6	Small; oval to round to cigar shaped; single or multiple buds present; uncommonly seen in clinical specimens; Fig. 3
Yeast forms	*Cryptococcus neoformans*	2–15	Cells vary in size; usually spherical but may be football shaped; buds usually single and "pinched off"; capsule may or may not be evident; rarely, pseudohyphal forms with or without capsule may be seen; Fig. 4–7
Yeast forms	*Blastomyces dermatitidis*	8–15	Cells usually large and spherical, double refractile; buds usually single, but several may remain attached to parent cells; buds connected by broad base; Fig. 8 and 9
Yeast forms	*Paracoccidioides brasiliensis*	5–60	Cells usually large and surrounded by smaller buds around periphery ("mariner's wheel appearance"); smaller cells (2–5 μm) that resemble *H. capsulatum* may be present; buds have "pinched-off" appearance; Fig. 10 and 11
Cysts and trophozoites	*Pneumocystis carinii*	5–12 (cysts) 2–4 (trophozoites)	Cysts are round to cup shaped, possibly with four intracystic bodies; trophozoites are free smaller stages; Fig. 27 and 28
Spherules	*Coccidioides immitis*	10–200	Spherules vary in size; some contain endospores, others are empty. Adjacent spherules may resemble *B. dermatitidis*; endospores may resemble *H. capsulatum* but show no evidence of budding. Spherules may produce multiple germ tubes if direct preparation is kept in moist chamber for ≥24 h; hyphae may be found in cavitary lesions; Fig. 12 and 13
"Sporangium"	*Rhinosporidium seeberi*	6–300	Large, thick-walled sporangia containing sporangiospores; mature sporangia are larger than spherules of *C. immitis*.
Yeast forms and pseudohyphae or true hyphae	*Candida* spp.	3–4 (yeast forms) 5–10 (pseudohyphae)	Cells usually exhibit single budding; pseudohyphae, when present, are constricted at ends and remain attached like links of sausage; true hyphae, when present, have parallel walls and are septate; Fig. 14 and 15
Yeast forms and hyphae	*Malassezia furfur*	3–8 (yeast forms) 2.5–4 (hyphae)	Short, curved hyphal elements usually present along with round yeast cells that retain their spherical shapes in compacted clusters; Fig. 16
Wide nonseptate hyphae	Zygomycetes: *Mucor* spp., *Rhizopus* spp., and other genera	10–30	Hyphae are large, ribbon-like, often fractured or twisted. Occasionally, septa may be present, branching usually at right angles. Smaller hyphae overlap those of *Aspergillus* spp., particularly *Aspergillus flavus*; Fig. 17–19
Hyaline septate hyphae	Dermatophytes: skin and nails	3–15	Hyaline septate hyphae commonly seen; chains of arthroconidia may be present. Fig. 20
Hyaline septate hyphae	Hair	3–15	Arthroconidia on periphery of hair shaft that produce sheath indicate ectothrix infection. Arthroconidia formed by fragmentation of hyphae within hair shaft indicate endothrix infection.

(Continued on next page)

TABLE 2 Characteristic fungal elements seen in direct examination of clinical specimens (*Continued*)

Morphologic fungal elements found	Organism(s)	Diam range (μm)	Characteristic features
Hyaline septate hyphae	Hair	3–15	Long hyphal filaments or channels within hair shaft indicate favus hair infection.
Hyaline septate hyphae	*Aspergillus* spp.	3–12	Hyphae are septate and exhibit dichotomous, 45° angle branching; larger hyphae, often disturbed, may resemble those of zygomycetes. Fig. 21–23
Hyaline septate hyphae	*Geotrichum* spp.	4–12	Hyphae and rectangular arthroconidia are present and are sometimes rounded. Irregular forms may be present. Fig. 24
Hyaline septate hyphae	*Trichosporon* spp.	2–4 by 8	Hyphae and rectangular arthroconidia are present and are sometimes rounded. Blastoconidia may be difficult to observe.
Hyaline septate hyphae	*Pseudallescheria boydii* (cases other than mycetoma)	3–12	Hyphae are septate and are impossible to distinguish from those of other hyaline molds, e.g., *Aspergillus* spp.
Hyaline septate hyphae	*Fusarium* spp.	3–12	Hyphae are septate and are impossible to distinguish from those of other hyaline molds, e.g., *Aspergillus* spp.
Dematiaceous septate hyphae	*Bipolaris* spp., *Curvularia* spp., *Exserohilum* spp., *Exophiala* spp., *Phialophora* spp., *Wangiella dermatitidis*, *Xylohypha bantiana*	2–6	Dematiaceous polymorphous hyphae are seen; budding cells with single septa and chains of swollen rounded cells may be present. Occasionally, aggregates may be present when infection is caused by *Phialophora* and *Exophiala* spp. Fig. 29
	Phaeoannellomyces werneckii	1.5–5	Usually, large numbers of frequently branched hyphae are present along with budding cells.
Sclerotic bodies	*Cladosporium carrionii*, *Fonsecaea compacta*, *Fonsecaea pedrosoi*, *Phialophora verrucosa*, *Rhinocladiella aquaspersa*	5–20	Brown, round to pleomorphic, thick-walled cells with transverse septa. Commonly, cells contain two fission plates that form tetrad of cells. Occasionally, branched septate hyphae may be found in addition to sclerotic bodies. Fig. 25
Granules	*Acremonium falciforme*, *Acremonium kiliense*, *Acremonium recifei*	200–300	White, soft granules without cement-like matrix
Granules	*Curvularia geniculata*, *Curvularia lunata*	500–1,000	Black, hard grains with cement-like matrix at periphery
Granules	*Aspergillus nidulans*	65–160	White, soft granules without cement-like matrix
Granules	*Exophiala jeanselmei*	200–300	Black, soft granules, vacuolated, without cement-like matrix, made of dark hyphae and swollen cells
Granules	*Fusarium* spp.	200–500	White, soft granules without cement-like matrix
	Fusarium moniliforme, *Fusarium solani*	300–600	
Granules	*Leptosphaeria* spp.	400–600	Black, hard granules; cement-like matrix; periphery composed of polygonal swollen cells and center composed of hyphal network
	Leptosphaeria senegalensis, *Leptosphaeria tompkinsii*	500–1,000	
Granules	*Madurella grisea*	350–500	Black, soft granules without cement-like matrix; periphery composed of polygonal swollen cells and center composed of hyphal network
Granules	*Madurella mycetomatis*	200–900	Black to brown hard granules of two types: (i) rust brown, compact, and filled with cement-like matrix; (ii) deep brown, filled with numerous vesicles, 6–14 μm in diam, cement-like matrix in periphery, and central area of light-colored hyphae

(Continued on next page)

TABLE 2 (*Continued*)

Morphologic fungal elements found	Organism(s)	Diam range (μm)	Characteristic features
Granules	*Neotestudina rosatti*	300–600	White, soft granules with cement-like matrix at periphery
Granules	*Pseudallescheria boydii*	200–300	White, soft granules composed of hyphae and swollen cells at periphery in cement-like matrix
Granules	*Pyrenochaeta romeri*	300–600	Black, soft granules composed of polygonal swollen cells at periphery; center is network of hyphae; no cement-like matrix

CULTURE GUIDELINES

Blood

Several blood culture systems for recovery of fungi are available. The lysis-centrifugation method provides a sensitive method for detection of fungemia (18). Recent improvements in broth and biphasic methods by agitation and lysis have improved the recovery of fungi compared to that of classic broth technologies and approach the sensitivity of the lysis-centrifugation method (13). Presently, several totally automated blood systems are commercially available. Recovery of yeasts by these systems has been comparable to or better than recovery by nonradiometric and radiometric methods (11, 19, 20). A concern noted in a study of radiometric versus BacT/Alert systems was that more than 25% of fungemias were not detected metabolically but were detected only by blind terminal blood cultures. Similarly, an additional 11% of positive blood cultures would have gone undetected if incubation time had been reduced to 5 days, a possibility being considered in many laboratories. The majority of these cultures (19) were positive for *C. neoformans*. Subculturing of broth systems or use of the lysis-centrifugation method should be considered with patients at high risk for cryptococcosis or any fungal infection. Recovery of *H. capsulatum* still requires the use of lysis-centrifugation.

Other Clinical Specimens

Optimal recovery of medically important fungi from clinical specimens is related to multiple factors. The first is the specimen itself, which must be freshly collected and appropriate for the mycotic infection being considered. Table 3 lists guidelines for the selection of clinical specimens that may be submitted for mycologic cultures for the major mycoses.

The quality of the specimen to be cultured is also important. In general, there are fewer fungal cells at the site of an infection than there are bacterial cells in a bacterial infection. Therefore, enough specimen must be cultured to ensure optimal recovery. Although pretreatment of specimens may not be done routinely, Table 4 presents guidelines that may be an aid in the recovery of fungi. As a general rule, at least 1 to 2 ml of a fluid specimen or enough tissue to properly inoculate three or four media with ~0.5 ml each is sufficient. Fluids, e.g., peritoneal fluid, pericardial fluid, and even CSF in volumes of >1 to 2 ml, should be concentrated by centrifugation and culturing of the pellet. Alternatively, the specimen may be filtered through a 0.45-μm-pore-size membrane filter, and the filter can then be cultured. Most fungi will not survive the 2% NaOH treatment used for recovery of mycobacteria (12).

The value of quantitating by colony counts or semiquantitating by light, moderate, or heavy growth is equivocal. Even with urine cultures, the value of such procedures is not clear. It has been suggested that the presence of $>10^4$ colonies of yeast cells per ml is associated with pyelonephritis and that the presence of fewer colonies is associated with contamination (3). Others suggest that any colony count is abnormal (2). *Caution*: Both false positives and false negatives occur, and the physician needs to make a clinical decision on the significance of a positive urine culture for *Candida* spp. on an individual basis.

The next important factor in specimen culture is the medium used, since no one fungal medium is best for all medically important fungi. Using at least one medium that does not contain antifungal or antibacterial agents has been suggested, although generally this is not helpful. Brain heart infusion agar (BHI), inhibitory mold agar, and Sabhi are examples. To ensure maximum recovery of fastidious dimorphic fungi, one medium (BHI) containing blood (5 to 10% sheep blood) should be used. Media containing antibacterial agents should be used with any specimen that might contain normal bacterial flora. Inhibitory mold agar, Sabhi, or BHI plus gentamicin (5 to 100 μg/ml), chloramphenicol (16 μg/ml), and penicillin (20 μg/ml)-streptomycin (40 μg/ml) have commonly been used. Incorporation of norfloxacin (5 μg/ml) or ciprofloxacin (5 μg/ml) has also proven effective. The eukaryotic protein synthesis inhibitor cycloheximide (0.5 μg/ml), alone or together with antibacterial agents, should be used in one medium. In addition, specialized media for the recovery of specific fungi may be employed, e.g., media containing substrates (L-dopa, caffeic acid, etc.) for detection of *C. neoformans* by its phenol oxidase activity, yeast extract phosphate agar with addition of NH_4OH (15) for inhibition of bacteria and recovery of dimorphic fungi from specimens, media containing or overlaid with a source of long-chain fatty acids (e.g., olive oil) for the recovery of *Malassezia furfur*, DTM medium (17) specifically formulated for culture of dermatophytes, and mycobacterial medium (Lowenstein-Jensen) or Middlebrook medium for the recovery of *Nocardia* spp. Sabouraud dextrose (2%) agar containing cycloheximide and chloramphenicol is best used for recovery of dermatophytes.

Proper incubation temperature, containers, and duration are also important in optimizing recovery of fungi from clinical specimens. It is generally recommended that cultures be incubated at 30°C, since this temperature is near optimal for growth for many fungi, although culturing at

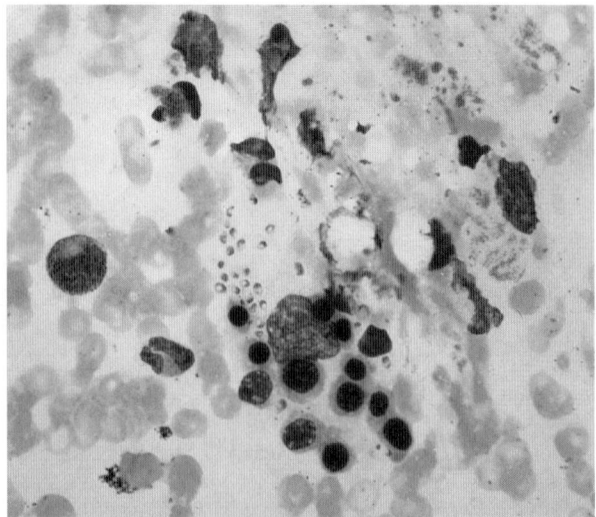

FIGURE 1 *H. capsulatum* in bone marrow. Small intracellular yeast cells are apparent in this Wright-Giemsa preparation. Magnification, ×2,360.

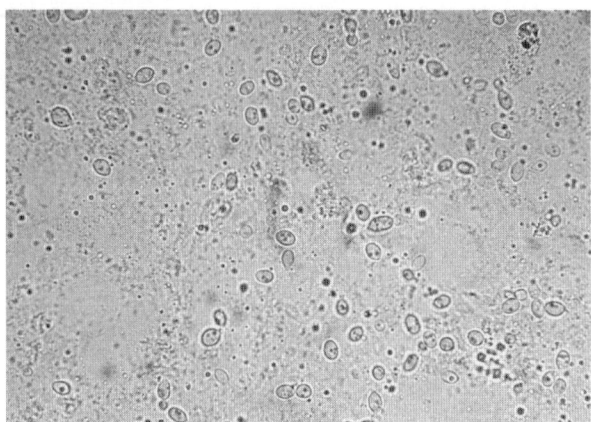

FIGURE 2 Numerous small budding yeast cells of *H. capsulatum* present in sputum as seen by bright-field microscopy. Magnification, ×1,852.

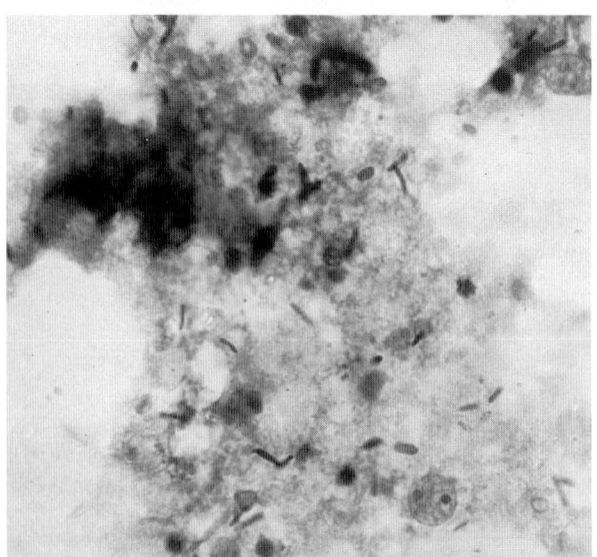

FIGURE 3 Periodic acid-Schiff stain of exudate from a cutaneous ulcer, showing numerous elongated yeast cells (cigar bodies) of *Sporothrix schenckii*. Magnification, ×1,915.

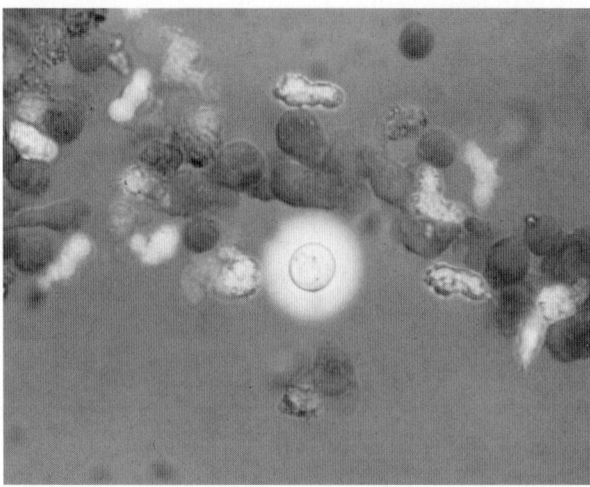

FIGURE 4 India ink preparation of CSF showing a single encapsulated, spherical yeast cell of *C. neoformans*. Magnification, ×2,385.

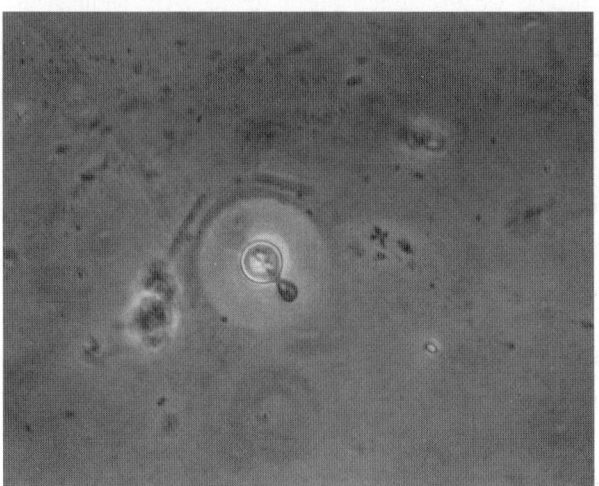

FIGURE 5 Phase-contrast photomicrograph of *C. neoformans* in sputum. Note the spherical yeast cell with a "narrow-necked" bud attached. Magnification, ×2,385.

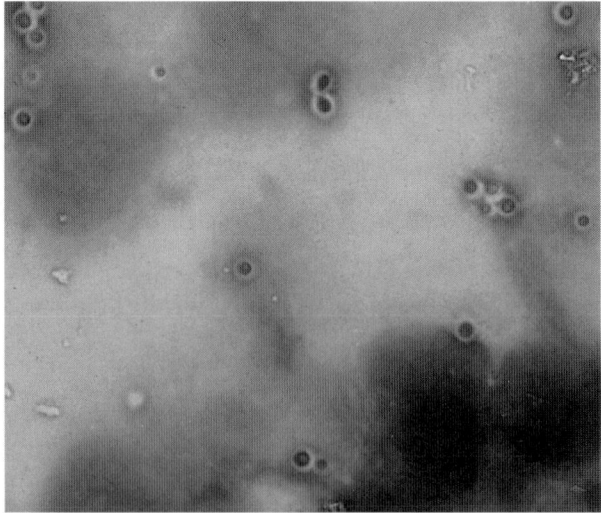

FIGURE 6 Periodic acid-Schiff stain of sputum, showing spherical yeast cells of *C. neoformans*. Note the presence of a capsule and the variation in sizes of cells. Magnification, ×1,590.

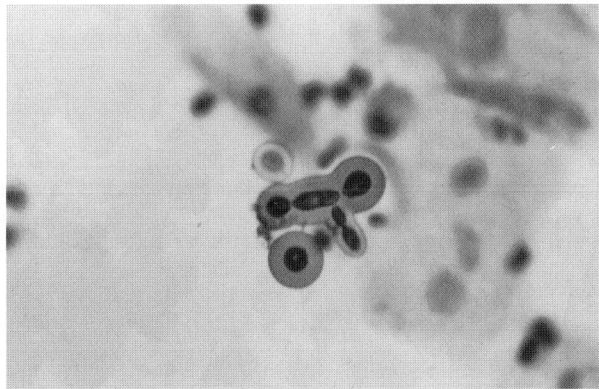

FIGURE 7 Papanicolaou stain of sputum showing the rare pseudohyphal form of *C. neoformans*. Note the capsule surrounding all cells. Magnification, ×2,210.

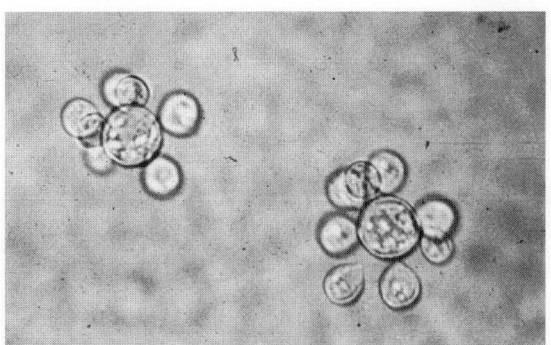

FIGURE 10 Bright-field photomicrograph of *Paracoccidioides brasiliensis*, showing multiple budding yeast cells resembling mariner's wheels. Magnification, ×1,590.

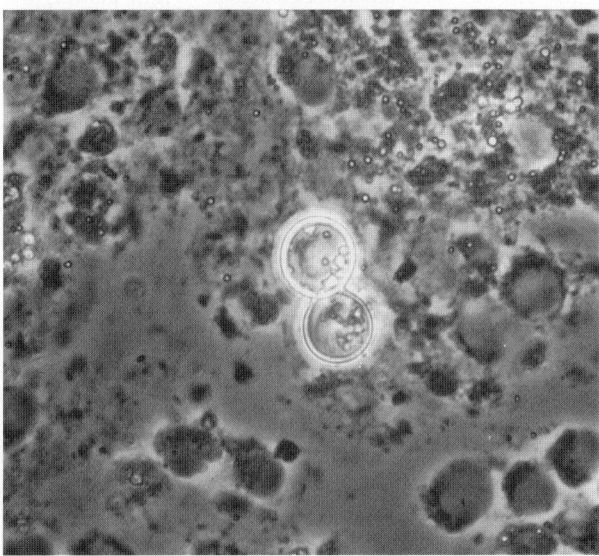

FIGURE 8 Phase-contrast photomicrograph of *B. dermatitidis* in sputum. The presence of a large, broad-based, budding yeast cell with a "double-contoured" wall is characteristic. Magnification, ×3,040.

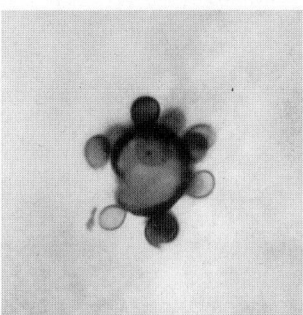

FIGURE 11 Methenamine silver stain of *P. brasiliensis* yeast form. Magnification, ×980.

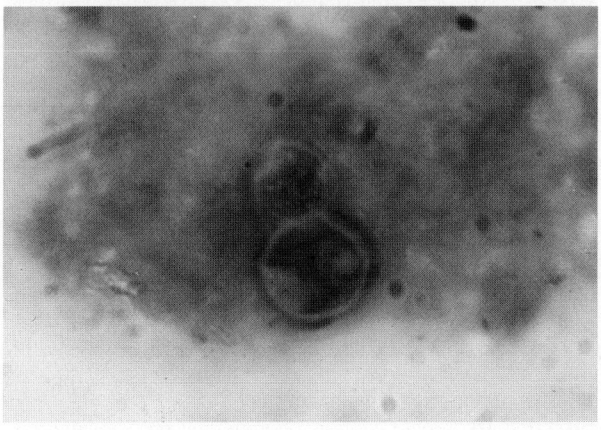

FIGURE 9 Gram stain of *B. dermatitidis* yeast form in sputum. Magnification, ×1,852.

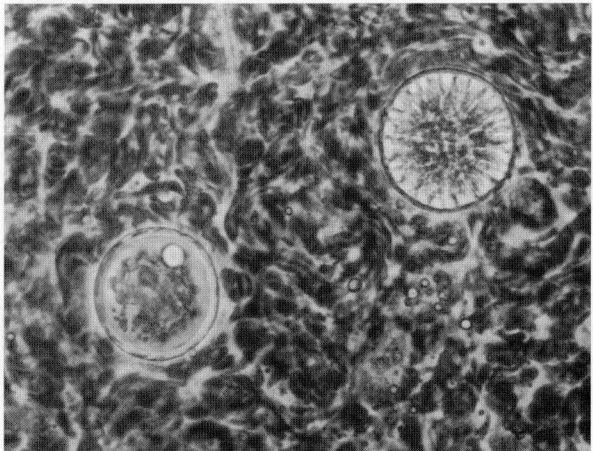

FIGURE 12 Phase-contrast photomicrograph of *Coccidioides immitis* spherules in sputum. Note the absence of endospores and the presence of cleavage furrows on one spherule. Magnification, ×2,400.

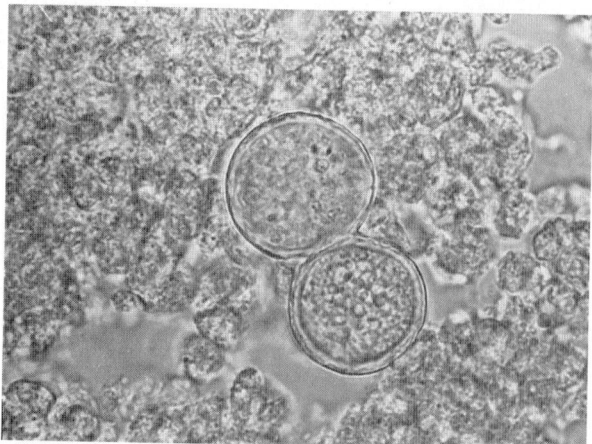

FIGURE 13 Bright-field photomicrograph of two adjacent spherules of *Coccidioides immitis* that morphologically resemble *B. dermatitidis*. Note the presence of endospores in one spherule. Magnification, ×1,228.

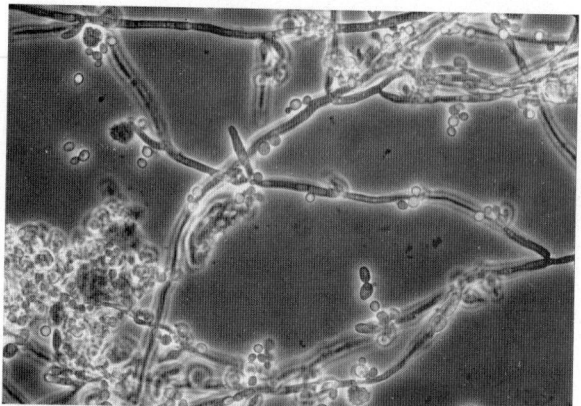

FIGURE 14 Phase-contrast photomicrograph of *Candida* sp. in peritoneal fluid. Note the presence of blastoconidia and pseudohyphae characteristic of the genus. Magnification, ×1,493.

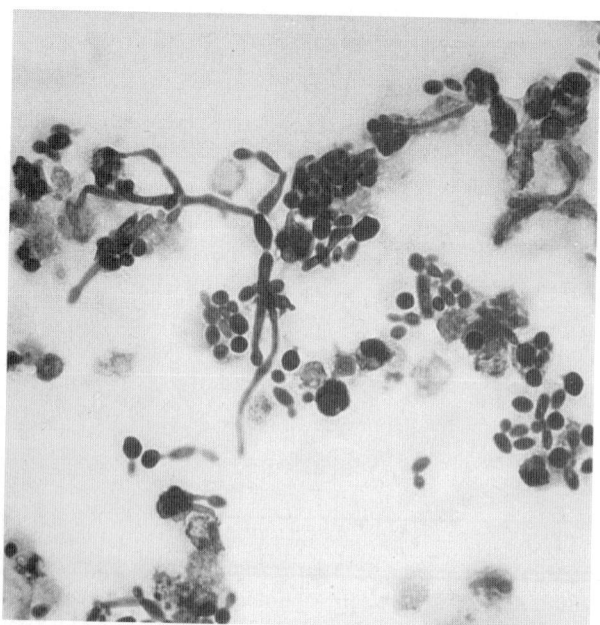

FIGURE 15 Periodic acid-Schiff stain of *Candida* sp. Magnification, ×1,390.

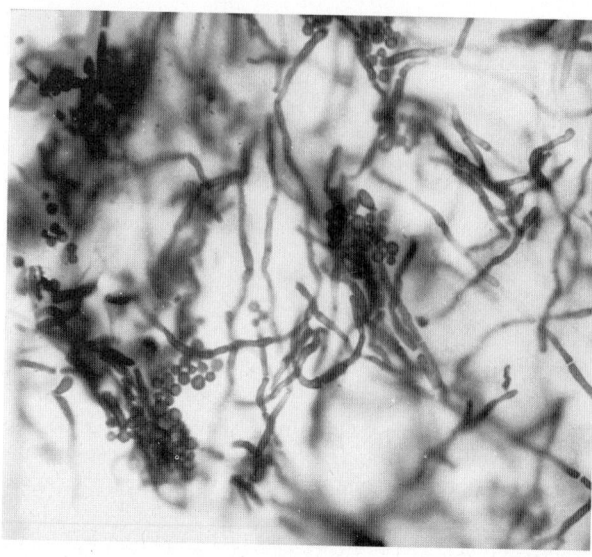

FIGURE 16 Periodic acid-Schiff stain of skin showing hyphal and yeast elements characteristic of M. *furfur*. Magnification, ×1,590.

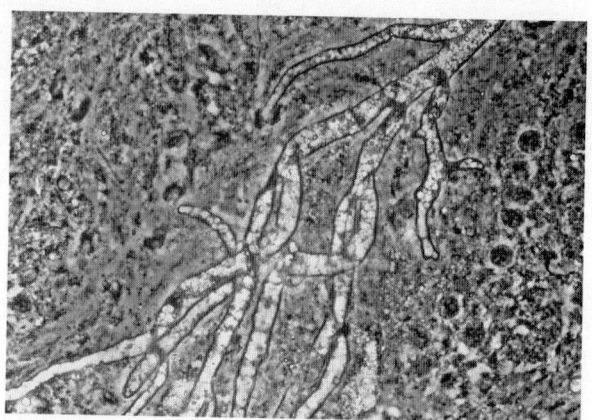

FIGURE 17 Phase-contrast photomicrograph of wound exudate, showing large, aseptate, twisted hyphae characteristic of a zygomycete. Magnification, ×1,493.

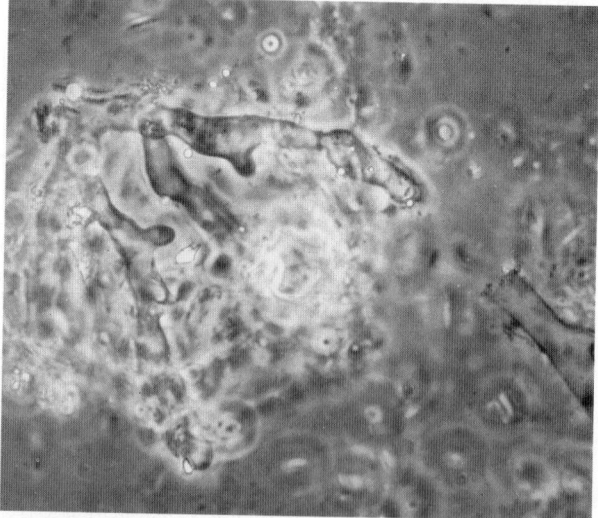

FIGURE 18 Phase-contrast photomicrograph showing fractured pieces of large aseptate hyphae characteristic of a zygomycete. Magnification, ×1,990.

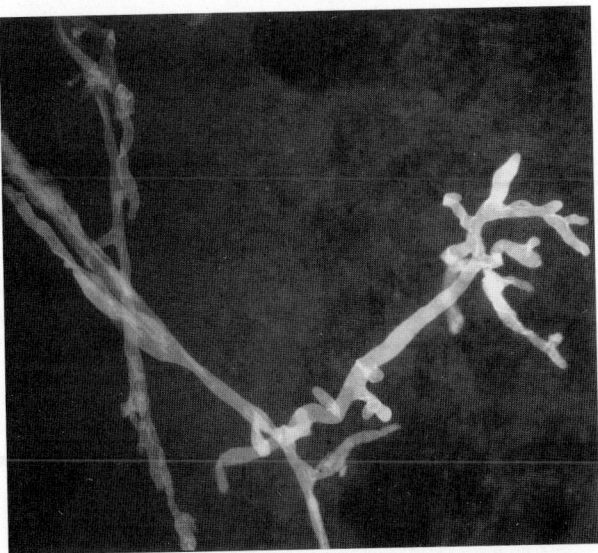

FIGURE 19   Calcofluor white-stained sputum showing characteristic hyphae of a zygomycete. Magnification, ×928.

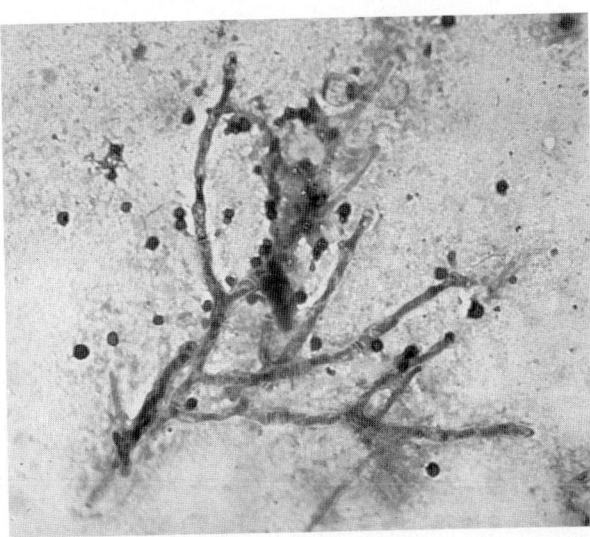

FIGURE 22   Gram stain of sputum, showing dichotomously branching, septate hyphae consistent with those of *Aspergillus* sp. Magnification, ×1,860.

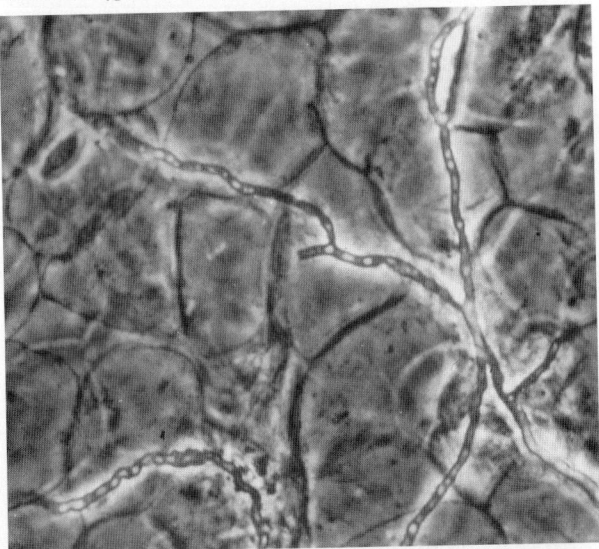

FIGURE 20   Phase-contrast photomicrograph of hyphae of a dermatophyte intertwined with squamous cells. Magnification, ×1,535.

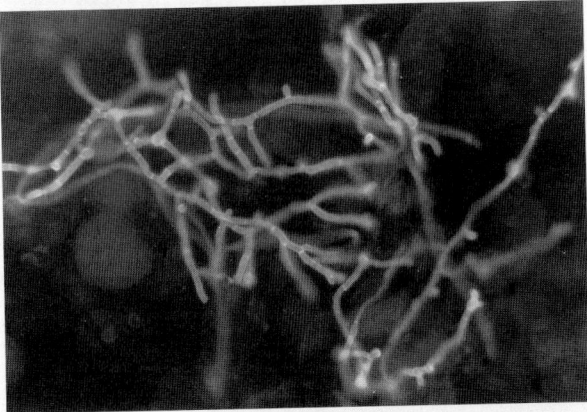

FIGURE 23   Calcofluor white stain of hyphae characteristic of *Aspergillus* sp. Magnification, ×1,200.

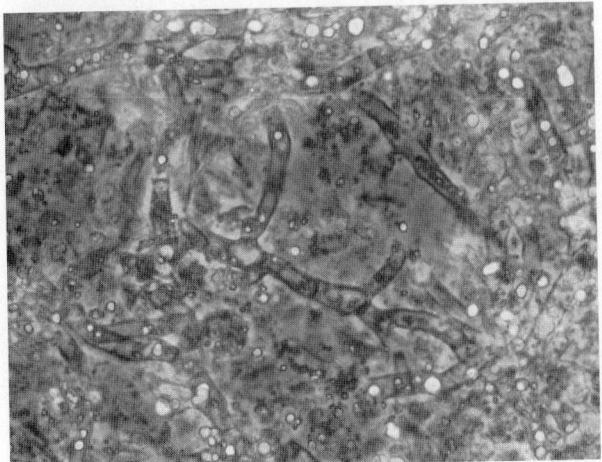

FIGURE 21   Phase-contrast photomicrograph of exudate from lung tissue, showing septate hyphae. *Aspergillus* sp. was recovered in culture. Magnification, ×3,040.

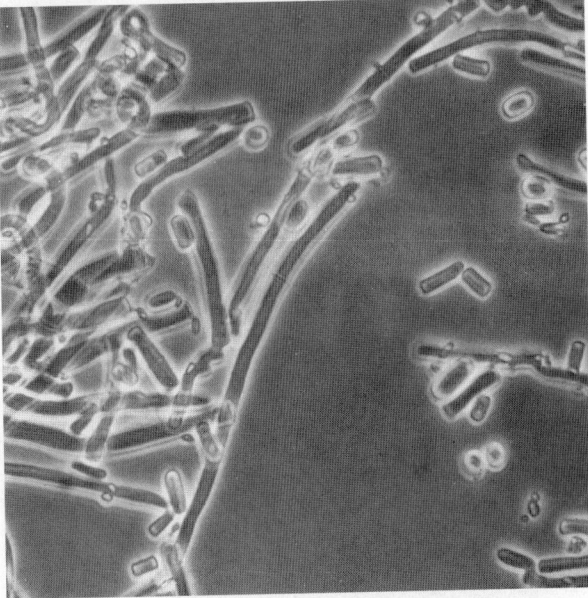

FIGURE 24   Phase-contrast photomicrograph showing hyphae and arthroconidia in sputum. *Geotrichum* sp. was recovered in culture. Magnification, ×1,597.

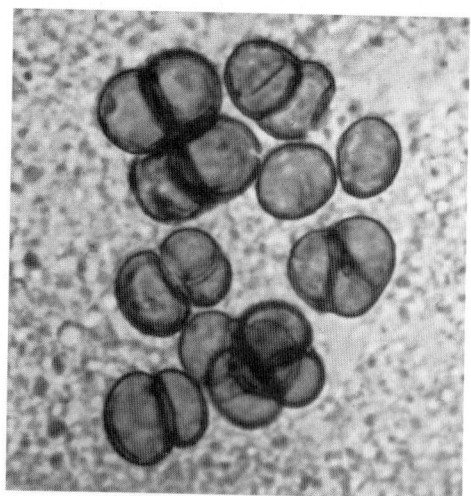

FIGURE 25 KOH preparation of sclerotic cells in tissue that are characteristic of the tissue form of the etiologic agents of chromoblastomycosis. Magnification, ×1,535.

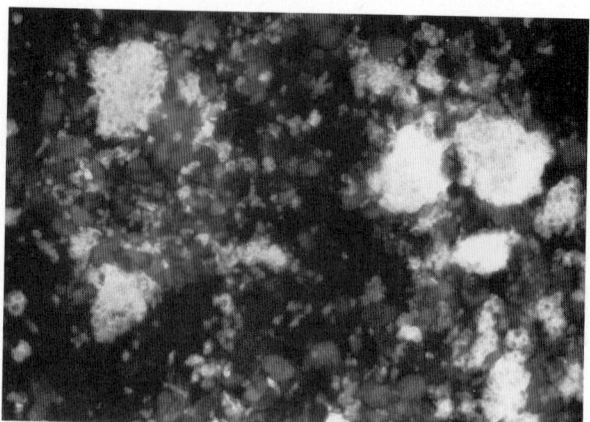

FIGURE 28 Monoclonal fluorescent-antibody stain of *P. carinii* in an induced sputum specimen. Magnification, ×1,200.

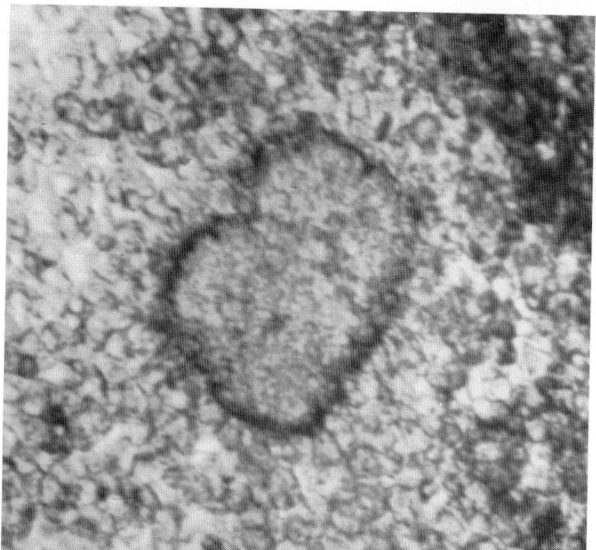

FIGURE 26 KOH preparation of a granule found in wound exudate that is characteristically seen in mycetoma. Magnification, ×500.

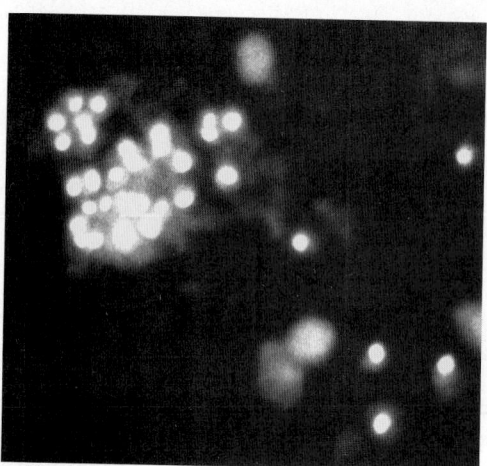

FIGURE 27 Calcofluor white stain of *P. carinii* cysts in an induced sputum specimen. Magnification, ×1,589.

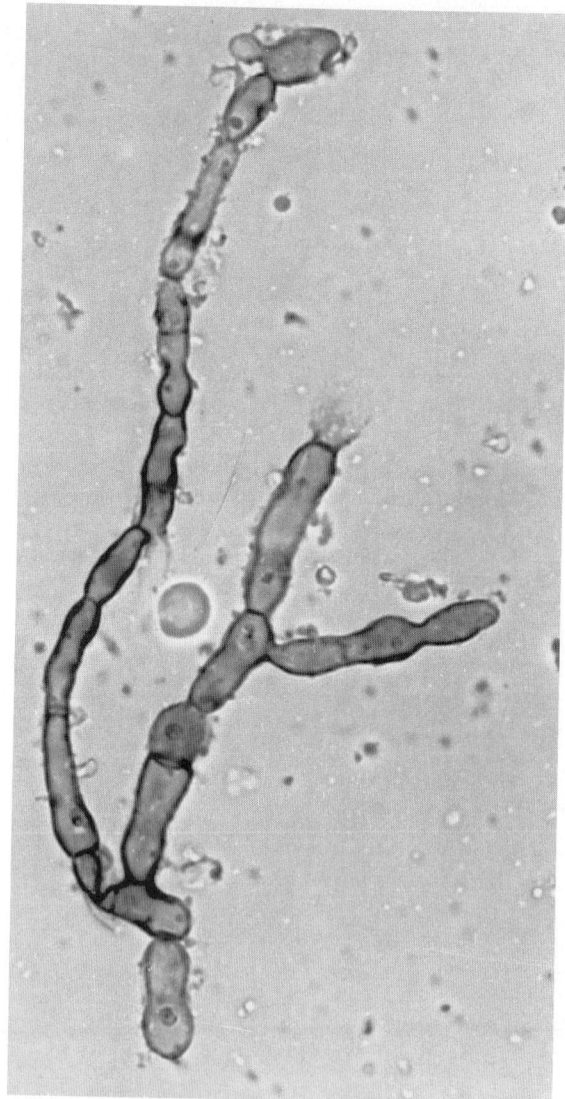

FIGURE 29 KOH stain of dematiaceous hyphae in an aspirate of a subcutaneous cyst. Magnification, ×2,000.

TABLE 3   Selection of clinical specimens for fungal culture

Specimen	Order of preference for site selection[a]													
	Blasto-mycosis	Coccidioido-mycosis	Histo-plasmosis	Paracoccidi-oidomycosis	Candi-diasis	Crypto-coccosis	Asper-gillosis	Zygo-mycosis	Pneumo-cystis	Fusa-riosis	Dermato-phytosis	Chromoblas-tomycosis	Sporo-trichosis	Myce-toma
Lower respiratory tract	1	1	1	2	X	1	1	1	1	3			3	
Blood		6	2		1	2		X		2			X	
Bone	4	X		X		X	X	X	X	X				
Bone marrow		X	3	X	X	X	X	3	X	X		X	X	
Brain	X	X	X	X	X		X	X						
CSF	X	X	X		X	5	X			5				
Eye				X	X		2	2		4			4	
Nose-nasal sinus	X	X			X		X	X						
Prostate	3	2	5	3	4	X				X			X	
Mucous membrane					X			X				2	2	
Subcutaneous tissue	X	X			X		X			X			1	1
Joints		5			2			4						
Urine	2	3	4		X	3	X			6				
Skin			X	1	X	X	X			1	1	1	X	2
Hair and nails											2	X		
Multiple systemic sites during disseminated infection	6	4	6	4	3	4	3	5	X	X			X	

[a] Predominant sites for recovery of organism are ranked in order of importance (based on most common clinical presentations). X indicates other sites from which organisms have been recovered.

TABLE 4   Pretreatment of clinical specimens prior to plating

Specimen	Pretreatment	Comments
Respiratory secretions	Lysis with mucolytic agents	Necessary for induced system specimens for *P. carinii*; optional for other mycoses; may improve yield but also may increase numbers of other organisms, i.e., yeasts and bacteria
Body fluids	Centrifugation or filtration	Necessary for optimal recovery; for all specimens >1–2 ml
Urine	Centrifugation	Necessary for optimal recovery, especially of agents of deep mycoses
Nails	Macerated in mortar or broken into very thin sections	Necessary for optimal recovery of dermatophytic and nondermatophytic agents
Tissue	Macerated in mortar or cut into small pieces	Necessary for optimal recovery of all fungi

room temperature is also acceptable. Use of culture dishes is best, since they provide large surface areas on which to spread the specimens. However, dehydration may occur. Use of 40 ml of specimen per dish may help prevent surface dehydration. Test tubes (especially larger-diameter tubes) can be used and usually do not have the dehydration problem. Large tubes are preferred, and they should be incubated horizontally in racks throughout the incubation or for at least 24 h to ensure that the inoculum is spread over a large surface area. *Caution:* All screw caps must be left loosened for proper aeration, or else growth will be either inhibited or very abnormal, leading to delayed identification.

Incubate all specimens for at least 4 weeks. Incubation for 6 weeks may be required for maximum recovery of fastidious fungi, e.g., *H. capsulatum* and *B. dermatitidis*. If a single medium (plate or tube) becomes positive for an equivocal pathogen, the remaining media should be incubated and the culture may not finalized until the end of the incubation period. Some laboratories may perform surveillance cultures for highly compromised patient populations (14). Multiple skin sites, stool or rectal swabs, urine (centrifuged or not centrifuged), throat, nares, and nasopharynx have been used. These specimens need to be inoculated on media containing antibacterial agents and then incubated for 5 to 10 days. All yeast cells and filamentous fungi should then be identified. Results may help predict what organism is causing an infection.

Regularly examine all cultures for growth. Discard any specimens overgrown by a bacterium or a fungus suspected to be a laboratory contaminant, and request a new specimen as rapidly as possible. Handle all filamentous fungi (and all fungi if possible) in a certified, laminar-flow biologic safety cabinet.

REFERENCES

1. **Bille, J., L. Stockman, G. D. Roberts, C. D. Horstmeier, and D. M. Ilstrup.** 1983. Evaluation of a lysis-centrifugation system for recovery of yeast and filamentous fungi from blood. *J. Clin. Microbiol.* **18:**469–471.
2. **Fisher, J. F., W. H. Chew, S. Shadomy, R. J. Duma, C. G. Mayhall, and W. C. House.** 1982. Urinary tract infections due to *Candida albicans*. *Rev. Infect. Dis.* **4:**1107–1118.
3. **Goldberg, P. K., P. J. Kozinn, G. J. Wise, N. Nouri, and R. B. Brooks.** 1979. Incidence and significance of candiduria. *JAMA* **241:**582–584.
4. **Hageage, G. J., and B. J. Harrington.** 1984. Use of calcofluor white in clinical mycology. *Lab. Med.* **15:**109–112.
5. **Hariri, A. R., H. O. Hempel, C. L. Kimberlin, and N. L. Goodman.** 1982. Effects of time lapse between sputum collection and culturing on isolation of clinically significant fungi. *J. Clin. Microbiol.* **15:**425–428.
6. **Jones, D. B., T. J. Liesegang, and N. M. Robinson.** 1981. *Cumitech 13, Laboratory Diagnosis of Ocular Infections.* Coor-

dinating ed., J. A. Washington II. American Society for Microbiology, Washington, D.C.
7. **Kaufman, L., and E. Reiss.** 1992. Serodiagnosis of fungal diseases, p. 506–528. *In* N. R. Rose, E. C. DeMacario, J. L. Fahey, H. Friedman, and G. M. Penn (ed.), *Manual of Clinical Laboratory Immunology,* 4th ed. American Society for Microbiology, Washington, D.C.
8. **Kim, Y. K., S. Parulekar, P. K. Yu, R. J. Pisani, T. F. Smith, and J. P. Anhalt.** 1990. Evaluation of calcofluor white stain for detection of *Pneumocystis carinii. Diagn. Microbiol. Infect. Dis.* **13:**307–310.
9. **Murray, P. R., R. E. VanScoy, and G. D. Roberts.** 1977. Should yeasts in respiratory secretions be identified? *Mayo Clin. Proc.* **52:**42–45.
10. **Ng, L., N. A. Virani, R. E. Chaisson, D. M. Yajko, H. T. Spahr, K. Cabrian, N. Rollins, P. Charache, M. Krieger, W. K. Hadley, and P. C. Hopewell.** 1990. Rapid detection of *Pneumocystis carinii* using a direct fluorescent monoclonal antibody stain. *J. Clin. Microbiol.* **28:**2228–2233.
11. **Nolte, F., J. M. Williams, R. C. Jerris, J. A. Morello, C. D. Leitch, S. Matushek, L. D. Schwabe, F. Dorigan, and F. E. Kocka.** 1993. Multicenter clinical evaluation of a continuous monitoring blood culture system using fluorescent-sensor technology (BACTEC 9240). *J. Clin. Microbiol.* **31:**552–557.
12. **Roberts, G. D., A. G. Karison, and D. R. DeYoung.** 1976. Recovery of pathogenic fungi from clinical specimens submitted for mycobacteriological culture. *J. Clin. Microbiol.* **3:**47–48.
13. **Roberts, G. D., M. A. Pfaller, E. Gueho, T. R. Rogers, C. DeVroey, and W. G. Merz.** 1992. Developments in the diagnostic mycology laboratory. *J. Med. Vet. Mycol.* **30**(Suppl. 1):241–248.
14. **Sandford, G. R., W. G. Merz, J. R. Wingard, P. Charache, and R. Saral.** 1980. The value of fungal surveillance cultures as predictors of systemic fungal infections. *J. Infect. Dis.* **142:**503–509.
15. **Smith, C. D., and N. L. Goodman.** 1975. Improved culture method for the isolation of *Histoplasma capsulatum* and *Blastomyces dermatitidis* from contaminated specimens. *Am. J. Clin. Pathol.* **68:**276–280.
16. **Stockman, L., G. D. Roberts, and D. M. Ilstrup.** 1984. Effect of storage of the Dupont lysis-centrifugation system on recovery of bacteria and fungi in a prospective clinical trial. *J. Clin. Microbiol.* **19:**283–285.
17. **Taplin, D.** 1965. The use of gentamicin in mycology. *J. Invest. Dermatol.* **45:**549–550.
18. **Telenti, A., and G. D. Roberts.** 1989. Fungal blood cultures. *Eur. J. Clin. Microbiol. Infect. Dis.* **8:**825–831.
19. **Wakefield, T., D. Wagner, N. Antik, and W. G. Merz.** 1993. Importance of >5 day incubation or terminal subculture of BacT/Alert and BACTEC 460 blood culture systems, abstr. C-72, p. 458. *Abstr. 93rd Gen. Meet. Am. Soc. Microbiol. 1993.*
20. **Wilson, M. L., M. P. Weinstein, L. G. Reiner, S. Mirrett, and L. B. Reller.** 1992. Controlled comparison of the BacT/Alert and BACTEC 660/730 nonradiometric blood culture systems. *J. Clin. Microbiol.* **30:**323–329.

Candida, Cryptococcus, and Other Yeasts of Medical Importance

NANCY G. WARREN AND KEVIN C. HAZEN

61

NATURAL HABITAT AND CLINICAL SIGNIFICANCE

Yeasts are ubiquitous in our environment, being found on fruits, vegetables, and other plant materials (exogenous). Some live as normal inhabitants in and on our bodies (endogenous). Therefore, they may be found in specimens without having any clinical significance. Yeasts are considered opportunistic pathogens and as such may be cultured from specimens of patients debilitated in some fashion, e.g., by hormone imbalance; by the administration of immunosuppressive agents such as corticosteroids, anticancer drugs, and the newer anti-AIDS drugs; or by the overuse of broad-spectrum antibiotics. Patients who undergo transplant surgery and the concomitant administration of immunosuppressive drugs often acquire yeast infections. Diseases such as AIDS and others that cause a diminution or depletion of the immunological system are also predisposing factors for yeast infections. Endocarditis caused by yeasts has been associated with the use of nonsterile equipment by drug addicts. Fungemia may occur when indwelling catheters are not changed or removed at frequent intervals; it also has been reported in infants supported by lipid-supplemented hyperalimentation (17).

Yeasts are by far the most common fungi isolated from human patients. The decision as to the significance of their presence in a specimen ultimately rests with the physician, but accurate, complete information from the laboratory is essential for reaching a reasonable conclusion. Type of specimen (e.g., closed, normally sterile sites rather than sputum or urine), number of specimens positive with the same organism from the same patient, and number of colonies formed are all critical pieces of information and should be carefully noted.

CHARACTERISTICS OF YEASTS

Yeasts are unicellular, eukaryotic, budding cells that are generally round to oval or, less often, elongate or irregular in shape. They multiply principally by the production of blastoconidia (buds). When blastoconidia are produced one after the other in a linear fashion without separation, a structure termed a pseudohypha is formed. Under certain circumstances, some yeasts may produce true septate hyphae. Such circumstances are associated with the diminu-

tion of oxygen, e.g., in host tissues or for submerged colonies in agar medium, at the bottom of broth medium in a test tube, or in the presence of 5 to 10% CO_2.

Cultures of yeasts are moist, creamy, or glabrous to membranous in texture. Several produce a capsule that may make the colony mucoid. With rare exceptions, aerial hyphae are not produced. Colonies may be hyaline, brightly colored, or darkly pigmented because of the presence of melanins. Organisms with darkly pigmented colonies, found in the family Dematiaceae, are discussed in chapter 67 of this Manual. Dimorphic fungal pathogens possessing a yeast phase in tissue are also discussed in other chapters.

Although there are many yeast genera and hundreds of yeast species, relatively few of these produce disease in humans and animals. Yeasts generally are identified (classified) by observing the macroscopic and microscopic features mentioned above. Usually, biochemical tests are also required for definitive identification to species level. In most instances, the pathogenic yeasts are found among the *Deuteromycetes,* or Fungi Imperfecti, that is, fungi that do not exhibit the sexual or teleomorphic state in culture.

Yeasts also may be classified as *Ascomycetes* or *Heterobasidiomycetes* depending on their method of sexual reproduction. A few genera may produce the sexual state on standard mycological media over time. In some instances, it may be necessary to identify the teleomorphic state of a yeast culture. This is best accomplished by submitting the culture to a reference laboratory for further study.

DIRECT EXAMINATION

The appropriate examination of a clinical specimen is essential prior to proper processing of the material. Additionally, examination often will aid the laboratorian and the physician in a preliminary identification, either ruling in or ruling out certain pathogenic yeasts. Certain methods are universal to the preliminary observation of fungi in a specimen, e.g., Gram stain, calcofluor, and 20% KOH. However, others, such as India ink preparation for the demonstration of a capsule, are used for the yeasts and yeast-like fungi only. This preparation is used with cultures and on specimens of urine, cerebrospinal fluid, etc., that have been centrifuged. It generally is not useful on primary specimens such as sputum or on other materials that do not allow even distribution of the ink. It must be remembered that if 20%

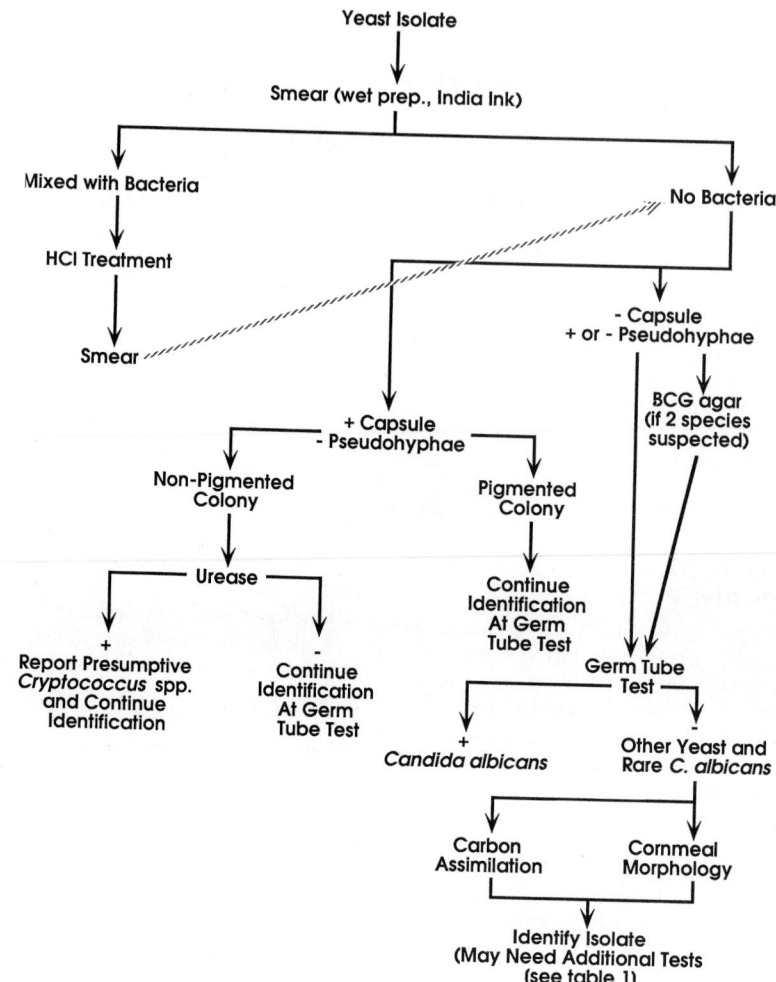

FIGURE 1 Scheme for identification of yeasts from clinical specimens.

KOH is used on a preparation, neither the India ink nor the Gram stain can subsequently be added.

During microscopic examination of a specimen for yeasts, factors that may aid in identification may be observed: (i) size and shape of the organism, (ii) mode of attachment of bud(s), (iii) presence or absence of a capsule, (iv) thickness of cell wall, (v) presence of pseudohyphae, and (vi) presence of arthroconidia. Unusual structures should also be noted.

IDENTIFICATION

A scheme for yeast identification is presented in Fig. 1. When viewed as a general guideline, it offers a basic approach to yeast identification but also is flexible enough for adaptation to individual laboratory situations. Morphologic and physiologic properties of the most commonly encountered yeasts in clinical specimens are given in Table 1. Particularly salient characteristics of individual yeasts will be reemphasized in the sections dealing with specific yeast genera. A sound, systematic approach to yeast identification cannot be stressed enough. With the numerous yeast identification systems available today, most yeast species can be identified easily; however, repeat testing sometimes

is necessary, and determining morphologic characteristics on cornmeal or similar agars should be mandatory.

Macroscopic Characteristics of Yeasts

Most yeasts (*Malassezia furfur* being an exception) grow well on common mycologic and bacteriologic media. Growth is usually detected in 48 to 72 h, and subcultures or laboratory-adapted strains may grow more rapidly. Colonies have a smooth to wrinkled, creamy appearance; some pigment may be observed initially or may intensify with age. Heavily encapsulated yeasts will have a very moist, mucoid appearance.

The ability of yeasts to grow at 37°C is a very important characteristic. Most pathogenic species grow readily at 25 and 37°C, while saprophytes usually fail to grow at the higher temperature. Additionally, some yeast species, e.g., *Candida albicans*, will grow in the presence of cycloheximide (Acti-Dione), which is found in Mycosel, Mycobiotic, and similar agars, while most other yeasts, including *Cryptococcus neoformans*, will be inhibited.

Pellicle growth on the surface of liquid media such as Sabouraud dextrose broth or malt extract broth has been used in the past to assist with yeast identification. More recent evidence suggests that this characteristic can be

TABLE 1 Cultural and biochemical characteristics of yeasts frequently isolated from clinical specimens[a]

Columns 7–18 ("Glucose" through "Dulcitol") are **Assimilation of**[b]; columns 19–24 ("Glucose" through "Trehalose") are **Fermentation of**.

Species	Growth at 37°C	Pellicle in broth	Pseudo- or true hyphae	Chlamydospores	Germ tubes	Capsule, India ink	Glucose (a)	Maltose (a)	Sucrose (a)	Lactose (a)	Galactose (a)	Melibiose (a)	Cellobiose (a)	Inositol (a)	Xylose (a)	Raffinose (a)	Trehalose (a)	Dulcitol (a)	Glucose (f)	Maltose (f)	Sucrose (f)	Lactose (f)	Galactose (f)	Trehalose (f)	Urease	KNO₃ utilization	Phenol oxidase	Ascospores
Candida albicans	+	−	+	+	+	−	+	+	+*	−	+	−	−	−	+	−	+	−	F*	F	−	−	F	F	−	−	−	−
C. catenulata	+*	−	+	−	−	−	+	+	−	−	+	−	−	−	+	−	−	−	F	F	−	−	F*	−	−	−	−	*
C. guilliermondii	+	−	+	−	−	−	+	+	+	−	+	+	+	−	+	+	+	−	F	−	F	−	F	F	−	−	−	*
C. kefyr	+	−	+	−	−	−	+	−	+	+	+	−	+*	−	+*	+	+*	−	F	−	F	F*	−	−	+*	−	−	−
C. krusei[c]	+*	+	+	−	−	−	+	−	−	−	−	−	−	−	+	−	−	−	F	−	−	−	−	−	−	−	−	*
C. lambica	+	+	+	−	−	−	+	−	−	−	−	−	+	−	+	−	−	+	−	−	−	−	−	−	+	−	−	*
C. lipolytica[c]	+	+	+	−	−	−	+	−	−	−	−	−	−	−	−	−	−	−	−	−	−	−	−	−	−	−	−	*
C. lusitaniae[d]	+	−	+	−	−	−	+	+	+	−	+	−	−	−	+	−	+	−	F	−	F	−	−	−	−	−	−	*
C. parapsilosis[e]	+	−	+	−	−	−	+	+	+	−	+	−	+	−	+	−	+	−	F	−	−	−	−	−	−	−	−	−
C. rugosa	+	−	+	−	−	−	+	−	+	−	+	−	+*	−	+*	−	+	−	−	−	−	−	−	−	−	−	−	−
C. tropicalis[d,e]	+	+	+	−[f]	−	−	+	+	+	−	+	−	+	−	+	−	+	−	F	−	−	−	−	−	−	−	−	−
C. zeylanoides	−	−	+	−	−	−	+	+	+	−	+*	−	+	−	+	−	+	−	−	−	−	−	−	−	−	−	−	−
Torulopsis glabrata	+	−	−	−	−	−	+	−	−	−	−	−	−	−	−	−	+	−	W	−	W	−	−	W	−	−	−	*
T. candida	+	−	+	−	−	−	+	+	+	−	+	−	+	−	+	−	+	−	F	−	−	−	−	F	−	−	−	−
T. pintolopesii[g]	+*	−	R	−	−	−	+	+	−	−	+	−	−	−	−	−	+*	−	F	−	−	−	−	−	−	−	−	−
Cryptococcus neoformans	+	−	−	−	−	+	+	+	+	−	+	−	+	+	+	−	+	+*	−	−	−	−	−	−	+	−	+	−
Cryptococcus albidus	−*	−	−	−	−	−	+	+	+	−	+*	−	+	+	+	+*	+	+	−	−	−	−	−	−	+	+	−	−
Cryptococcus laurentii	+*	−	−	−	−	−	+	+	+	+*	+	+*	+	+	+	+	+	+*	−	−	−	−	−	−	+	−	−	−
Cryptococcus luteolus	−	−	−	−	−	−	+	+	+	+	+	−	+	+	+	+	+*	−	−	−	−	−	−	−	+	−	−	−
Cryptococcus terreus	−*	−	−	−	−	+*	+	+*	+	+*	+*	−	+*	+	+	+*	+	+*	−	−	−	−	−	−	+	+	−	−
Cryptococcus uniguttulatus	−	−	−	−	−	+*	+	+	+	−	+	+*	+	+	+	−	+*	+	−	−	−	−	−	−	+	+	−	−
Rhodotorula glutinis	+	−	−	−	−	−	+	+	+	−	−*	+*	−	−	+	+	+	−	−	−	−	−	−	−	+	+	−	−
Rhodotorula rubra	+	−	−	−	−	−	+	+*	+	+*	+	+*	+*	−	+	+	+*	+*	−	−	−	−	−	−	+	−	−	−
S. cerevisiae	+*	+	−*	−	−	−	+	+	+	−	+	−	−	−	−	+	+	−	F	F	F	−	F	F*	−	−	−	+
H. anomala	+*	+	+	−	−	−	+	+	+	−	+	−	−	−	−	+*	+	−	F	F	F	−	F	−	−	+	−	+
Trichosporon beigelii	+*	+	+	−	−	−	+	+	+	+*	+	−	+*	+*	+*	+*	+	+	−	−	−	−	−	−	+*	−	−	−
T. pullulans	−*	+	+	−	−	−	+	+	+	+	+	−	+	+	+	+	+	−	−	−	−	−	−	−	+	+	−	−
G. candidum[h]	−	+	+	−	−	−	+	−	−	−	+	−	+	−	+	−	−	−	−	−	−	−	−	−	−	−	−	−
B. capitatus	+	+	+	−	−	−	+	+	+	−	+	−	+	−	+	−	+	−	−	−	−	−	−	−	−	−	−	−
P. wickerhamii[h]	+	−	−	−	−	−	+	−	−	−	+	−	−	−	−	−	+	−	−	−	−	−	−	−	−	−	−	−

a*, Strain variation; R, rare; F, sugar is fermented (i.e., gas is produced); W, weak fermentation. Based on data from Barnett et al. (6) and Kreger-van Rij (44).

[b] In these columns, + indicates growth greater than that of the negative control.

[c] *C. lipolytica* assimilates erythritol; *C. krusei* does not. Maximum growth temperatures are 43 to 45°C for *C. krusei* and 33 to 37°C for *C. lipolytica.*

[d] *C. lusitaniae* assimilates rhamnose; *C. tropicalis* usually does not.

[e] *C. parapsilosis* assimilates L-arabinose; *C. tropicalis* usually does not.

[f] Rare strains of *C. tropicalis* produce teardrop-shaped chlamydospores.

[g] *Torulopsis pintolopesii* is a thermophilic yeast capable of growth at 40 to 42°C.

[h] Not yeasts but may be confused with several yeast genera.

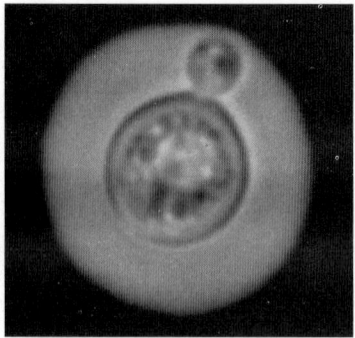

FIGURE 2 India ink preparation of *Cryptococcus neoformans*. Magnification, ×400. (Courtesy of H. J. Shadomy.)

variable; however, as an ancillary test, it may be helpful for identifying *Candida tropicalis*, *Candida krusei*, and *Trichosporon* spp.

Microscopic Characteristics of Yeasts

Upon isolation of suspected yeast cells from clinical specimens, the first examination should be of a wet preparation of a colony. A small amount of growth is emulsified in a drop of sterile distilled water and examined microscopically with a reduced light source. Observations should include size and shape of the yeast cell, method of bud attachment, and presence or absence of pseudohyphae, true hyphae, or arthroconidia. Any round or slightly oval budding yeast with rare or no pseudohyphae should be examined further for the presence of a capsule.

The same wet preparation used for the initial microscopic examination can be used for an India ink examination. Add a small drop of India ink close to the edge of the coverslip, and allow the drop to diffuse underneath the coverslip. The preparation then can be examined for the presence of encapsulated yeasts (Fig. 2). The outline of the yeast cell, surrounded by a clear area which is the mucopolysaccharide capsule, will be obvious against a dark India ink background. False-positive preparations usually are the result of contaminated ink or of artifacts that can exhibit ragged edges. Capsular size cannot be used for identification purposes, as this characteristic may be influenced by culture age, medium composition, and strain variation. The presence of a capsule does not automatically identify the yeast cell as *Cryptococcus neoformans*, since other cryptococci, *Rhodotorula* spp., and rare *Torulopsis* spp.

also produce capsules. In practice, any nonpigmented, round, encapsulated yeast cell recovered from cerebrospinal fluid should be considered *Cryptococcus neoformans* until proven otherwise.

Purity of Cultures

Before any additional physiologic tests are performed, it is essential to ensure that one has a pure culture. A Gram stain of the culture can verify purity, but bacterial contamination can often be detected during the wet preparation examination. If the culture is mixed with bacteria, the isolate should be inoculated to a blood agar plate and individual colonies should be picked or, alternatively, treated with hydrochloric acid (Fig. 1). The HCl procedure is performed by inoculating a colony into three tubes of Sabouraud dextrose broth (each containing 5 ml). Use a capillary pipette to add 4 drops of 1 N HCl to the first tube, 2 drops to the second tube, and 1 drop to the third tube. After incubation at 25°C for 24 to 48 h, subculture 0.1 ml of each broth onto fresh Sabouraud dextrose agar plates.

It is possible that more than one yeast species will be recovered from a clinical specimen, especially if the specimen is from a normally nonsterile site (95). Careful attention to colonial morphology and microscopic characteristics can offer clues to a mixed population. Subculturing individual isolates to additional media can be helpful, and the use of *Candida* bromcresol green agar or Pagano-Levin agar (Difco) (33, 95) may reveal the presence of more than one yeast species.

Morphology Studies

While examination of a wet preparation gives a primary indication of the yeast species involved, more extensive study of morphology on cornmeal agar or similar agars will afford an opportunity to correlate morphologic characteristics with results of biochemical testing. Microscopic examination should reveal thick-walled chlamydospores of *C. albicans* or other structures of the other yeasts (Table 2). Special attention also should be given to the size and shape of the pseudohyphae and the arrangement of blastoconidia along the pseudohyphae. Figure 3 illustrates several morphologic characteristics. Using known control organisms can ensure quality of the medium and allow one to compare morphologies of the known standard organisms with that of the culture in question. Experience is needed to detect some of the more subtle characteristics, and these findings should be considered preliminary, i.e., one of the first steps in complete yeast identification.

TABLE 2 Microscopic appearance of several yeasts and yeast-like fungi

Organism	Pseudohyphae	True hyphae	Blastoconidia	Arthroconidia	Annelloconidia	Chlamydospores	Ascospores
B. capitatus	×	×	×		×		
C. albicans	×	×	×			×	
Other Candida spp.	×[a]	×[a]	×				×[a]
Cryptococcus spp.			×				
Geotrichum spp.		×		×			
Hansenula spp.	×[a]		×				×
Rhodotorula spp.			×				
Saccharomyces spp.	×[a]		×				×
Torulopsis spp.			×				
Trichosporon spp.	×	×	×	×			

[a]Strain variation.

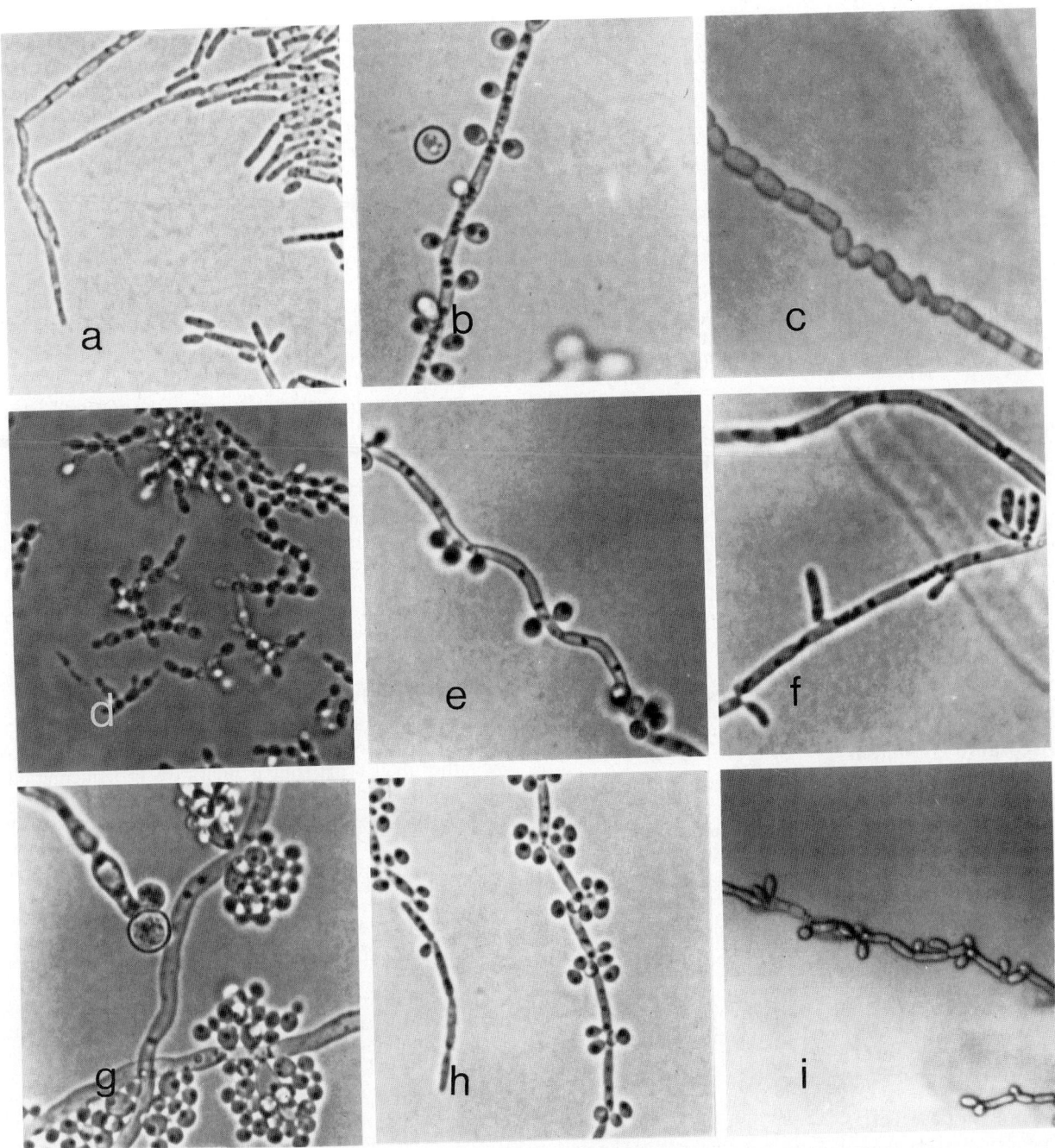

FIGURE 3 Morphologic features of some yeast and yeast-like organisms on cornmeal agar at 24 to 48 h with ambient temperature. (a) *C. krusei.* Extremely elongated, rarely branched pseudohyphae; few blastoconidia. (b) *C. tropicalis.* Blastoconidia are formed at and between septa. (c) G. *candidum* arthroconidia. (d) *C. guilliermondii.* Chains of blastoconidia forming sparse pseudohyphae in a young culture. (e) *C. lusitaniae.* Short, distinctly curved pseudohyphae with blastoconidia formed at and occasionally between septa. (f) *B. capitatus.* True hyphae and annelloconidia resembling arthroconidia. (g) *C. albicans.* Blastoconidia, chlamydospores, true hyphae, and pseudohyphae. (h) *C. parapsilosis.* Elongated, delicately curved pseudohyphae with blastoconidia at septae. (i) *T. beigelii.* Blastoconidia formed at the corners of arthroconidia. Magnification, ×370. (Courtesy of B. A. Davis.)

Germ Tube Test

One of the simplest and most valuable tests for the rapid presumptive identification of *C. albicans* is the germ tube test (Fig. 4a). The test is considered presumptive, because

not all isolates of *C. albicans* will be germ tube positive, and false positives may be obtained despite well-trained staff (18). Microscopic observation of the preparation will reveal that the short hyphal initials produced by *C. albicans* are

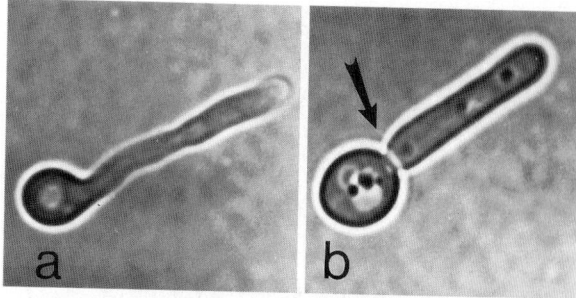

FIGURE 4 Germ tube test. (a) Germ tube formation by *C. albicans*; (b) blastoconidial germination with constriction (arrow) of *C. tropicalis* not seen with true germ tubes of *C. albicans*. Magnification, ×400. (Courtesy of B. A. Davis.)

not constricted at the junction of the blastoconidium and the germ tube. Frequently, so many *C. albicans* blastoconidia produce germ tubes that the germ tubes become entwined with each other, producing clumps of cells. *C. tropicalis* can also produce hyphal initials, but the blastoconidia are larger than those of *C. albicans*, and there is a definite constricture where the hyphal initial joins a blastoconidium (Fig. 4b). In addition to using a known culture of *C. albicans* as a positive test control, negative controls using *C. tropicalis* and *Torulopsis glabrata* should also be included. The preparation should not be incubated longer than 4 h, as other hypha-producing yeasts will begin to germinate after this time.

Ascospore Formation

Several yeasts recovered from clinical specimens may be present in the teleomorphic (sexual) state. In order to enhance production of ascospores, cultures should be inoculated onto media such as Fowell's acetate agar, incubated at room temperature for 2 to 5 days, and examined by wet preparation for the presence of ascospores within asci. Some mycologists prefer to perform special stains to detect ascospores. The Ziehl-Neelsen stain routinely employed in mycobacteriology can be used if needed; however, most ascospores can be detected easily in a drop of sterile distilled water. Ascospore production in *Saccharomyces cerevisiae* is evidenced by the presence of one to four globose spores (Fig. 5c); *Hansenula anomala* produces one to four hat-shaped spores (Fig. 5a).

Phenol Oxidase Test

The phenol oxidase screening procedure detects the ability of *Cryptococcus neoformans* to produce phenol oxidase on substrates containing caffeic acid. The most frequently used medium is birdseed agar (containing niger or thistle seeds), and a number of formulations that show various degrees of success have been described by several investigators. Apparently, test performance is related to the glucose content of the medium: the more glucose in the medium, the less likely a valid test will be obtained. When *Cryptococcus neoformans* is subcultured on the medium, colonies turn a dark brown in 2 to 5 days. Because of the wide variation in medium formulations, the test must be subjected constantly to quality control measures.

Urease

The urease test detects a yeast's ability to produce the enzyme urease. In the presence of suitable substrates, urease

splits urea and produces ammonia, which raises the pH and causes a color shift in the phenol red indicator from amber to pinkish red. The urease test aids in the identification of *Cryptococcus* spp. and *Rhodotorula* spp., which are all urease positive. Most strains of *Trichosporon* spp. are positive, while *Geotrichum* spp. and *Blastoschizomyces capitatus* are negative. Nearly all *Candida* spp. encountered in clinical specimens are urease negative, exceptions being *Candida lipolytica* and some strains of *C. krusei*.

Carbohydrate Assimilation Tests

The mainstay of yeast identification to the species level is the carbohydrate assimilation test (Table 1), which measures the ability of a yeast cell to utilize a specific carbohydrate as the sole source of carbon in the presence of oxygen. Several reliable kits and automated and semiautomated systems are on the market; they make the classic Wickerham and Burton method unnecessary for routine clinical isolates.

Nitrate Assimilation Test

The nitrate assimilation test is similar to the carbon assimilation test. It tests the ability of a yeast cell to utilize nitrate as a sole nitrogen source. While it is not necessary to perform nitrate assimilation studies on every yeast isolate, this information can be helpful with certain groups of yeasts. The test is most beneficial when trying to identify *Cryptococcus*, *Rhodotorula*, *Trichosporon*, and *Hansenula* spp.

Carbohydrate Fermentation Tests

Fermentative yeasts recovered from clinical specimens produce carbon dioxide and alcohol; therefore, production of gas rather than a pH shift is indicative of fermentation. Rarely are fermentation studies needed to identify most of the commonly isolated yeasts if the mycologist is familiar with typical morphology on cornmeal agar. The test is most helpful in differentiating the various species of *Candida*; *Cryptococcus* and *Rhodotorula* spp. are nonfermentative.

Rapid Identification of Yeasts

Short turnaround times for yeast identification, as for other activities in the clinical microbiology laboratory, allow valuable information to be quickly provided to physicians for patient management. This is particularly important given that various yeast species (such as *C. krusei*, *Candida parapsilosis*, *Candida lusitaniae*, *C. tropicalis*, and *Cryptococcus neoformans*) are inherently or potentially resistant to amphotericin B and the newer azole agents (9, 27, 88). Institution of effective antifungal therapy as early as possible can only improve patient outcome.

While the goal is rapid identification, the definition of rapid is controversial. The "gold standard" for yeast identification is assimilation and fermentation as developed by Wickerham and Burton (91). The test requires up to 28 days before identification can be provided. Rapid, therefore, could be simply defined as anything faster than the Wickerham-Burton method. With the development of identification tests that provide species information within 48 to 72 h (39) (see earlier discussion), a rapid identification test could be defined as one that yields the final result on the same day that the test is set up. Few tests that meet the criteria of definitive identification and same-day results have been developed.

A third criterion, the ability to identify a broad range of yeasts with one test, severely limits the development of rapid tests. While single tests that discriminate between

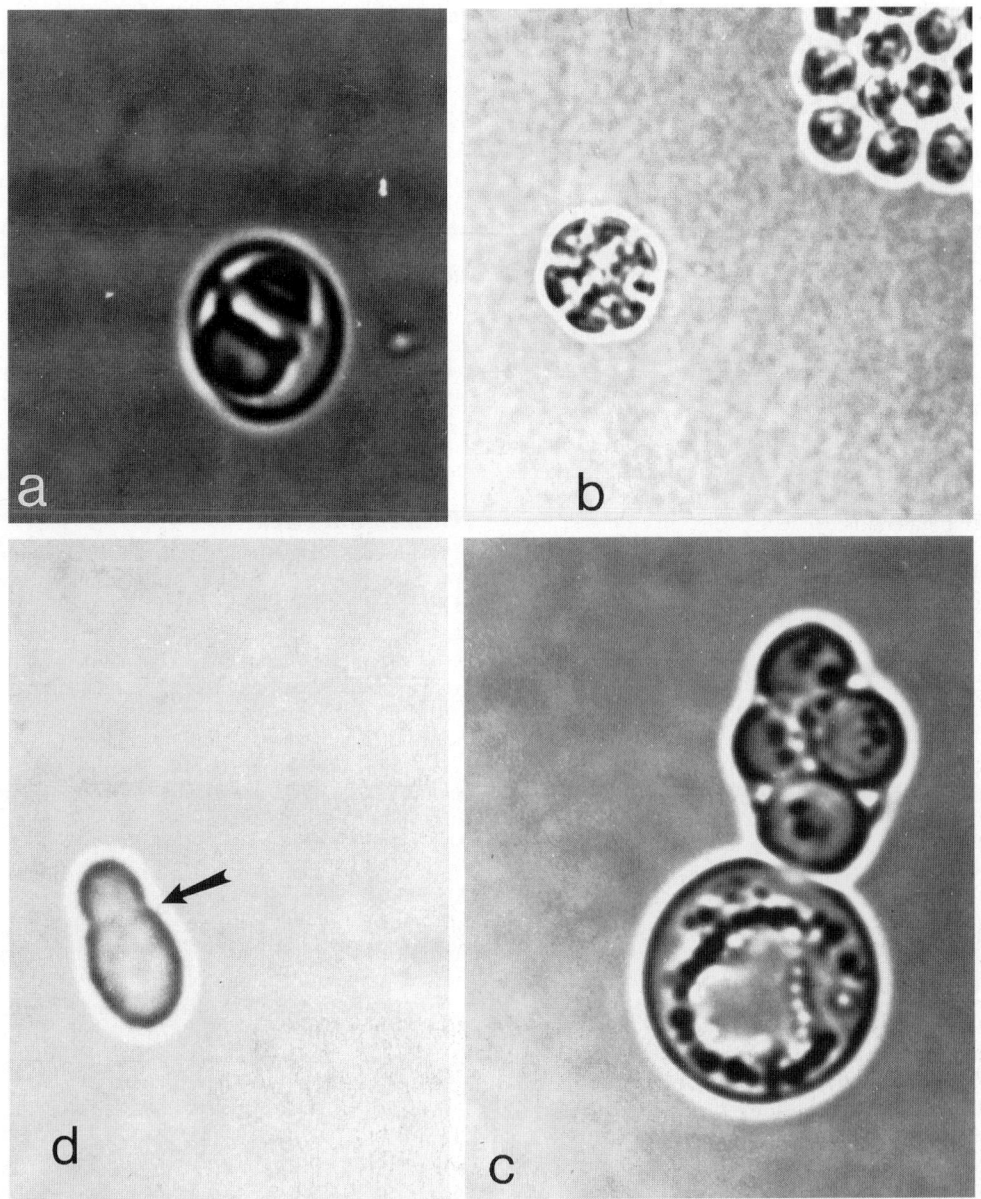

FIGURE 5   Diagnostic features of selected yeasts. (a) Ascus of *H. anomala* containing hat-shaped ascospores; (b) sporangium of *Prototheca wickerhamii* containing sporangiospores; (c) *S. cerevisiae* with vegetative cell and ascus containing four globose ascospores; (d) bottle-shaped, budding yeast demonstrating phialoconidium and collarette (arrow) of *M. furfur*. Magnification, ×1,000. (Courtesy of B. A. Davis.)

two species or confirm a presumptive identification have been developed, rapid (same-day) tests that apply to a broad range of species have only recently become available. At best, the tests that identify multiple species are limited to the more common species seen in the clinical laboratory (81). Examples of such tests are the MicroScan Yeast Identification Panel (Baxter-MicroScan, West Sacramento, Calif.), the API ZYM (Analytab Products, Plainview, N.Y.), and the API Yeast Ident (Analytab) (62, 81).

Only a limited number of yeast features have been advantageously utilized for rapid tests. These features include cell antigens (detected with latex agglutination immunoassays), DNA-RNA sequences (via direct probes or

PCR assays), lipids, and preformed enzymes (exoenzymes). Tests that utilize the first two features are described elsewhere in this Manual and will not be discussed further here.

Preformed enzymes provide one system for identification of multiple species with a single test panel. Such tests may be acutely affected by incubation temperature (36) and are effectively limited to the more common yeast species isolated in the clinical laboratory. However, detection of specific exoenzymes can be extremely useful for resolution of the confusion between two or three possible species. In particular, several rapid tests are available for this purpose, particularly for detection of *C. albicans* (for example, the *C. albicans* screen [Carr-Scarborough Microbiologicals, Decatur, Ga.] and the Albistrip [Lab M Ltd., Bury, United Kingdom]). These tests are excellent for confirmation of a germ tube-positive organism such as *C. albicans* and could be used in place of longer-term identification tests for this purpose. Using these tests for confirmation purposes can provide a financial saving to the clinical laboratory. Both tests depend on the production of two enzymes instead of a single enzyme; single-enzyme tests can lead to spurious results (80).

Lipid profiles can also be used to advantage for rapid identification of yeasts (MIDI, Newark, Del.). Procedures for these tests are somewhat complex and require use of a gas chromatograph and a computerized data station. Thus, the initial costs for this capability are high (>$30,000). However, once the equipment is available, the cost per sample is low and complete identification can be accomplished on the same day the culture is mature. The available database includes both common and rare yeasts. This technology appears very promising, but further development of the reference database is needed.

Maintenance of Yeast Cultures

Well-characterized yeast isolates should be retained for use as quality control organisms to ascertain the performance of media and to provide examples of known positive and negative biochemical and morphologic reactions. Yeasts can be maintained easily in a culture collection by heavily inoculating a tube of sterile distilled water with a 48- to 72-h-old yeast culture grown on Sabouraud dextrose agar. The yeast suspension should remain viable for at least 2 years if the cap is tightened securely on the tube and the tube is stored at room temperature. To retrieve the culture, 1 or 2 drops of the suspension are removed aseptically from the tube by using a sterile capillary pipette, and the suspension is inoculated onto Sabouraud dextrose agar.

ORGANISMS RESEMBLING YEASTS

Occasionally, organisms such as moulds, algae, etc., grow on mycologic media and produce colonies that resemble those produced by yeasts. Careful attention to morphologic characteristics will differentiate these organisms from each other and from yeasts. Following are several examples of such organisms that are recovered from clinical specimens and superficially resemble yeasts.

Genus *Geotrichum*

Members of the genus *Geotrichum* may be confused with yeasts because they produce a white to cream-colored, mealy, subsurface colony on fungal media. Occasionally, an aerial mycelium may be produced, giving a colony that superficially resembles *Coccidioides immitis*. Microscopic examination reveals septate hyaline hyphae that break into arthroconidia upon maturation (Fig. 3c). No other structures are observed. *Geotrichum candidum* may be separated from *Trichosporon beigelii* by carbohydrate assimilations and the lack of urease and blastoconidium production. *Geotrichum* spp. are occasionally isolated from patient specimens but are rarely considered pathogens except in extremely debilitated individuals.

Genus *Ustilago*

The genus *Ustilago* represents a heterobasidiomycetous fungus that is parasitic on the seeds and flowers of many cereals and grasses. It may be inhaled and may therefore be isolated from sputum specimens. The colonies are slow growing and, when young, are moist, compact, and white. With age, short hyphae are produced and the colonies become velvety or powdery. Microscopically, the cells are irregular and yeast-like or elongate and spindle shaped and may resemble arthroconidia. With the production of short hyphae, clamp connections are formed. Physiologically, most isolates are nitrate positive and have a carbohydrate assimilation pattern similar to that of *Cryptococcus albidus*.

Black Yeasts

Occasional isolates of *Sporothrix schenckii*, *Aureobasidium pullulans*, and agents of phaeohyphomycosis initially produce white to tan yeast-like colonies on primary isolation media. Identification methods for these organisms are quite different from yeast identification methods and are discussed in chapter 67 of this Manual.

Genus *Prototheca*

Prototheca spp. are ubiquitous achlorophyllous algae that produce disease in humans and animals only rarely (41). Human infection usually involves the skin and underlying tissues or olecranon bursa. *Prototheca wickerhamii* is recovered most often from human specimens, while *Prototheca zopfii* is usually associated with animal infections. Colonies are white to cream colored, yeast-like, and dull or moist to mucoid in appearance. The optimal growth temperature is 30°C, and growth is inhibited by medium containing cycloheximide. Microscopically, the cells are variable in size and shape and do not bud (Fig. 5b). Asexual reproduction is by release of sporangiospores from sporangia. Generally, sporangia of *Prototheca zopfii* (14 to 16 μm in diameter) are larger than those of *Prototheca wickerhamii* (7 to 13 μm in diameter), but this characteristic can be influenced by environmental conditions. Cells of *Prototheca wickerhamii* are round, while most strains of *Prototheca zopfii* exhibit oval to cylindrical cells. Carbon assimilation studies can be used for identification purposes. Both species assimilate glucose and galactose; additionally, *Prototheca wickerhamii* assimilates trehalose. Members of the genus *Prototheca* are nonfermentative (65).

CHARACTERISTICS OF SELECTED MEDICALLY IMPORTANT YEAST GENERA

Genus *Blastoschizomyces*

B. capitatus, a newly proposed combination for *Blastoschizomyces pseudotrichosporon* and *Trichosporon capitatum* (73), has recently been recognized as an emerging cause of invasive fungal disease in leukemic patients (59). It is widely distributed in nature and has occasionally been recovered as normal skin flora. Disseminated disease usually is associated with immunosuppressive conditions (4). Macroscopically, colonies are glabrous with radiating edges, white to cream colored, and shiny. Microscopically, isolates produce true hyphae, pseudohyphae, and annelloconidia resembling arthroconidia (Fig. 5f). On the basis of morphologic features alone, *B. capitatus* can be difficult to separate from *Trichosporon* spp., and physiologic tests are needed. *B. capitatus* is nonfermentative and can be separated from *T. beigelii* by growth on Sabouraud dextrose agar at 45°C, growth on cycloheximide-containing agar at room temperature, and failure to hydrolyze urea.

Genus *Candida*

The heterogeneous genus *Candida* belongs to the family *Cryptococcaceae* within the division Deuteromycetes (Fungi Imperfecti). The genus contains approximately 200 species. This number is not immutable. Technologic advances that affect apparent taxonomic relationships will continually result in reassignments of present species and discovery of new species. These advances may also lead to the designation of new genera.

The genus *Candida* is a form genus. As such, the taxonomic relationships of species within the genus are not well defined. Teleomorphs of several genera have been demonstrated for different species of *Candida*. The teleomorphic genera include *Clavispora, Debaryomyces, Issatchenkia, Kluyveromyces,* and *Pichia*. This large number of teleomorphic relationships demonstrates that the genus *Candida* is a mixture of unrelated species. One reason for the wide variety of species is the definition of *Candida*. The genus designation is used for any asexual yeast cell that does not have one of the following features: (i) acetic acid production; (ii) visually detectable red, pink, or orange pigments; (iii) arthroconidia; (iv) unipolar or bipolar budding on a broad base; (v) blastoconidia formed on sympodulae; (vi) buds formed on stalks; (vii) needle-shaped terminal conidia; (viii) triangular cells; (ix) enteroblastic-basipetal budding, usually with mucoid colonies and the ability to grow on inositol as a sole carbon source; and (x) ballistoconidia (44).

The genus *Candida* no longer contains species that are diazonium blue B (DBB) positive, a characteristic that is associated with yeasts having a basidiomycetous affinity (such as *Cryptococcus* spp.). DBB-positive organisms (e.g., *Candida humicola, Candida curvata, Candida diffluens,* and *Candida scottii*) have been reassigned to either genus *Cryptococcus* or genus *Rhodotorula*.

Another change in species epithets includes the controversial fusion of *Candida stellatoidea* with *C. albicans*. *Candida claussenii* and *Candida langeronii* have also been merged into *C. albicans* (92). Genetic evidence supporting the fusion of *C. stellatoidea* indicates that two types (designated I and II) of *C. stellatoidea* exist (68). Type II, which does not assimilate sucrose, appears to be a sucrose-negative mutant of *C. albicans* (50). A definitive association of type I *C. stellatoidea* with *C. albicans* has not been established.

One significant occurrence has been the suggestion that species of *Torulopsis* belong to the genus *Candida* (96). This suggestion is debatable. The original separation of the genera is based on the inability of *Torulopsis* spp. to produce pseudomycelium. However, this criterion appears to have been misapplied (96), leading to incorrect generic epithets. Furthermore, there appear to be no strong arguments for or against the merger (61). As this issue has not been definitively settled, the two genera will be treated separately here. In the future, it is likely that the merger will be accepted despite the lack of official sanction. Two recent important reference books for yeast identification (6, 44) merged species of *Torulopsis* with *Candida*. Individuals who are interested in these species should be aware of the effect of the merger on genus-species epithets. In particular, the medically important species *Torulopsis candida* has the new epithet *Candida famata*.

Candida species are ubiquitous, being found on many plants and as normal flora of the alimentary tracts of mammals and the mucocutaneous membranes of humans (61). Essentially all areas of the gastrointestinal (GI) tracts of humans can harbor *Candida* spp. The overall carriage rate in healthy individuals has been estimated to reach 80%, but this value is likely an underestimate owing to methodologic problems associated with detecting low levels of yeasts in GI specimens. The most commonly isolated species (50 to 70% of yeast isolates) from the GI tracts of humans is *C. albicans*. Other yeast species that have been isolated include *C. tropicalis, C. parapsilosis, C. krusei, Candida kefyr, Candida guilliermondii, T. glabrata,* and *Rhodotorula* spp. (61). Other species may inhabit the GI tract, but they were not identified in surveillance cultures. After *C. albicans,* the three species most commonly isolated from GI sources are *C. tropicalis, C. parapsilosis,* and *T. glabrata*.

Candida spp. can be present in clinical specimens as a result of environmental contamination, colonization, or actual disease processes. An accurate diagnosis requires proper handling of clinical material. For example, a few *C. albicans* cells will multiply rapidly in sputum left at room temperature, giving the inaccurate impression that large numbers of yeasts are present.

Candida spp. that are normal flora can invade tissue and produce life-threatening pathology in patients whose immune defenses have been altered by disease or iatrogenic intervention. *C. albicans* is the species most commonly isolated from patients with nearly all forms of candidiasis (61). Contributing to its high association with disease is its high prevalence in the healthy population, as described above. In addition, *C. albicans* appears to possess a number of virulence attributes that may promote successful parasitism. These attributes include relatively rapid germination upon seeding tissue from the bloodstream (34), protease production (51), surface integrin-like molecules for adhesion to extracellular matrix proteins (31, 84), complement protein-binding receptor (12, 29), phenotypic switching (79), and surface variation and hydrophobicity (35). These attributes have been recently reviewed elsewhere (11, 16). Only *C. tropicalis* appears to be more virulent than *C. albicans* when present in patients with leukemia or lymphoreticular malignant disease (93). Other medically important *Candida* spp. include *Candida catenulata, Candida ciferrii, C. guilliermondii, Candida haemulonii, C. kefyr, C. krusei, C. lipolytica, C. lusitaniae, Candida norvegensis, C. parapsilosis, Candida pulcherrima, Candida rugosa, Candida utilis, Candida viswanathii,* and *Candida zeylanoides* (69). This list is not exclusive, as other rare agents will certainly be added in the future. The species that are emerging opportunistic pathogens include *C. lipolytica, C. lusitaniae,* and *C. krusei*. These three species have been isolated from patients with fungemia. *C. lusitaniae,* which is generally a low-virulence organism, may be either innately or potentially resistant to amphotericin (32). *C. krusei* isolates appear to be susceptible to ketoconazole but not fluconazole (94). An extensive review of the genus and of the diseases produced by *Candida* spp. has been presented by Odds (61). Regardless of the species and of whether it comes from an immunocompromised or an immunocompetent patient, a single isolate of *Candida* from blood should be considered significant (21). Taken together with the reported rate of 10% of nosocomial bloodstream infections being caused by *Candida* species (5), the incidence of *Candida* fungemia represents a serious patient management problem.

Oral candidiasis is considered a defining illness for AIDS. It may be present in the prodromal stages of AIDS or may accompany late stages of AIDS (CD4 T-cell count of less than 400 cells per μl) (63, 83). Oral candidiasis is a marker for esophageal candidiasis in the late stages of AIDS

(83). Mucocutaneous forms of candidiasis are often related to defects in cell-mediated immunity, while systemic spread is generally associated with neutropenia (19, 25). Thus, despite the presence of oral or esophageal candidiasis, systemic spread of *Candida* spp. in AIDS patients is uncommon. When it occurs, systemic spread is associated with a drop in the neutrophil count (20).

There is a serotype preference in the oral candidiasis of AIDS and other immunocompromised patients. While the distribution of serotypes (A and B) is approximately equal in healthy individuals, serotype B is more prevalent in immunocompromised patients (10). The importance of this observation is unclear. However, serotype B isolates are physiologically different from serotype A isolates in that B serotypes exhibit greater karyotype variability and are more likely to develop resistance to 5-fluorocytosine (2, 3).

Blastoconidia of *Candida* spp. vary in shape from round to oval to elongate. Occasional initial isolates, especially from patients receiving antimicrobial agents, may be highly pleomorphic. Asexual reproduction is by multilateral budding, and a true mycelium may be present. If sexual reproduction occurs, the yeasts are classified in their teleomorphic state. Appearance of pseudohyphae and attachment of blastoconidia are important characteristics to observe when identifying *Candida* spp. Figure 3 illustrates these morphologic features. Observation of germ tubes and chlamydospores is also helpful in identifying *C. albicans*. Growth on fungal media can be detected as early as 24 h; however, colonies usually are visible in 48 to 72 h as white to cream colored or tan. They are creamy in texture and may become more membranous and convoluted with age. Occasionally, initial isolates of *C. albicans* on Sabouraud dextrose agar are wrinkled or rugose but revert to smooth colonies on subculture. In our experience, many isolates of *C. albicans* produce "colonies with feet" (i.e., colonies with short marginal extensions) on blood agar, while other yeasts do not. Most *Candida* spp. grow well aerobically at 25 to 30°C, and many will grow at 37°C or above. Several species of *Candida*, most notably *C. albicans*, are diploid (70, 90). These species apparently have lost the ability to undergo meiosis.

Carbon assimilation and occasionally fermentation studies are needed to differentiate the species (Table 1), but rapid confirmation tests for particular species are becoming available (see Rapid Identification of Yeasts above). Of the *Candida* spp. usually recovered from clinical specimens, *C. guilliermondii* is the only one to assimilate dulcitol, and *C. kefyr* (previously *Candida pseudotropicalis*) assimilates lactose. The assimilation of rhamnose can be helpful in separating *C. lusitaniae* from the biochemically similar but rhamnose-negative variants of *C. tropicalis* (76). For atypical isolates of *C. lusitaniae*, which can be confused with *C. tropicalis*, good growth on Trichophyton agar 1 with biotin but weak growth on vitamin-free medium indicates *C. tropicalis* (82). Certain rare strains of *C. tropicalis* may assimilate cellobiose weakly and exhibit an assimilation pattern similar to that of *C. parapsilosis*. The inclusion of arabinose is helpful, since *C. parapsilosis* readily assimilates this carbohydrate. Most strains of *C. tropicalis* do not. Urease is generally not produced, nor is KNO_3 utilized by the *Candida* species listed in Table 1.

Genus *Cryptococcus*

The genus *Cryptococcus* contains many species, six of which are noted in Table 1. Of these, *Cryptococcus neoformans* is considered the only human pathogen, although *Cryptococcus albidus* (45) and a few others have rarely been impli-

cated in disease in severely debilitated individuals. *Cryptococcus* spp. are round to oval yeast-like fungi ranging greatly in size (3.5 to 8 μm or more in diameter) with single budding and a narrow neck between parent and daughter cells (Fig. 2). Unusually large yeast cells (up to 60 μm) have been observed, and this size appears to be associated with higher incubation temperatures (56). Occasionally, several buds may be seen; pseudohyphae are observed rarely. The cell wall is quite fragile, and it is not unusual to find collapsed or crescent-shaped cells, especially in stained tissue sections. Cells are characterized by the presence of a mucopolysaccharide capsule varying from a wide halo to a nearly undetectable lighter zone around the cells, depending on the strain and the medium used. Use of India ink with wet preparations or the mucicarmine stain for mucopolysaccharide with fixed preparations may be helpful in making the capsule visible. Colonies are typically mucoid owing to the presence of capsular material, become dry and duller with age, and exhibit a wide range of color (cream, tan, pink, yellow). The color may darken with age. Strains possessing only a slight capsule may appear similar to colonies of *Candida* spp. All members of the genus produce urease, utilize various carbohydrates, and are nonfermentative.

Cryptococcus neoformans was the first serious fungal pathogen of humans to be identified as a heterobasidiomycetous yeast. Mating studies (47) have demonstrated two separate mating pairs that correlate with serotype. At least four serotypes of *Cryptococcus neoformans* have been identified: A, B, C, and D. Serotypes A and D produce the teleomorphic state *Filobasidiella neoformans*, and serotypes B and C produce the teleomorph originally named *Filobasidiella bacillispora* (49). The latter species was found to be identical to *Cryptococcus neoformans* var. *gattii* (85). Although there are certain differences besides serotype, e.g., utilization of creatinine and certain dicarboxylic acids, temperature tolerance, and virulence for mice, all serotypes are now considered varieties of *F. neoformans*. Standard laboratory tests do not differentiate between the serotypes, and though media for separating serotypes A and D from serotypes B and C have been recommended, these media are not available commercially (52, 74).

The different serotypes of *Cryptococcus neoformans* exhibit epidemiologic differences also (7). *Cryptococcus neoformans* var. *neoformans* (serotypes A and D) is the most common worldwide, with the highest prevalence of serotype A in Japan and in tropical and subtropical countries or areas such as Southeast Asia, Brazil, Australia, and southern California. Serotype B is much more prevalent than C, and most serotype C strains have been found in southern California (48).

Cryptococcus neoformans was first detected in the environment in the late 19th century, when Sanfelice recovered the yeast organism from peach juice. Since then, however, *Cryptococcus neoformans* var. *neoformans* has been most frequently associated with aged pigeon (and other bird) droppings and soils contaminated with these droppings. The organism is usually not found in fresh droppings but is most evident in bird excreta that have accumulated over long periods on window ledges, vacant buildings, and other roosting sites (48). It is only recently that the habitat of *Cryptococcus neoformans* var. *gattii* has been identified. Recovery of *Cryptococcus neoformans* var. *gattii* (serotype B) from the environment coincides with the flowering of the river red gum tree (*Eucalyptus camaldulensis*). Likewise, the global distribution of cryptococcal disease due to *Crypto-*

coccus neoformans var. *gattii* correlates with the distribution of this tree (22).

Initial cryptococcal infection begins by inhalation of the fungus into the lungs, usually followed by hematogenous spread to the brain and meninges. Involvement of the skin, bones, and joints is seen, and *Cryptococcus neoformans* is often cultured from the urine of patients with disseminated infection. In patients without human immunodeficiency virus infection, cryptococcosis, particularly cryptococcal meningitis, usually is seen in association with underlying conditions such as lupus erythematosis, sarcoidosis, leukemia, lymphomas, and Cushing's syndrome.

Cryptococcosis is one of the defining diseases associated with AIDS. Patients with cryptococcosis and serologic evidence of human immunodeficiency virus infections are considered to have AIDS. In nearly 45% of AIDS patients, cryptococcosis was reported as the first AIDS-defining illness. Because none of the presenting signs or symptoms of cryptococcal meningitis (such as headache, fever, and malaise) are sufficiently characteristic to distinguish it from other infections that occur in patients with AIDS, determining cryptococcal antigen titers and culturing blood and cerebrospinal fluid are useful in making a diagnosis (14).

All *Cryptococcus* species are nonfermentative aerobes. Separation of species is based upon assimilation of various carbohydrates and KNO_3 (Table 1). *Cryptococcus neoformans* may be distinguished from other *Cryptococcus* species by these biochemical studies as well as by growth at 37°C, which may be seen with *Cryptococcus albidus* and *Cryptococcus laurentii*. *Cryptococcus neoformans* may be differentiated from other yeasts and from other *Cryptococcus* species by its production of brown colonies on birdseed agar (80a), although with occasional isolates (particularly serotype C), the production of phenol oxidase may have to be induced. Differentiation from the genus *Rhodotorula*, whose members also form capsules, produce urease, and are nonfermentative, is usually accomplished by noting the utilization of inositol (Table 1) and the absence of the carotenoid pigments characteristically present in *Rhodotorula* spp. *Cryptococcus neoformans* can be distinguished from other yeasts in fixed tissue by the use of a Fontana-Masson stain that detects melanin precursors in the cell wall.

Genus *Hansenula*

Two species in the genus *Hansenula* have been reported to cause disease in humans. *H. anomala* has been associated with catheter-related infections (43), and *Hansenula polymorpha* has been recovered from mediastinal lymph nodes of a child with chronic granulomatous disease (60). Variations in colony morphology may cause confusion with *Cryptococcus* and *Candida* spp. The texture may be smooth to wrinkled, and the color may be white, cream, or tan. Microscopically, multilateral budding cells are observed. Pseudohyphae and true hyphae have been reported as characteristics of the genus; however, they have not been observed with the two species presented here. Individual species are either homothallic or heterothallic. Although *H. anomala* is heterothallic, the diploid form is the one usually recovered from clinical specimens. The asci contain one to four hat-shaped spores (Fig. 5a). All members of the genus are nitrate positive. Carbohydrate assimilation and fermentation studies are needed to identify the individual species (46).

Genus *Malassezia*

Three *Malassezia* species have been recovered from human specimens. The most common clinical species is M. *furfur* (this epithet includes the previously named species *Pityrosporum ovale* and *Pityrosporum orbiculare*). No teleomorphic state has been described; however, cell wall structure and a positive DBB staining reaction suggest that this form genus may be closely related to the *Basidiomycetes* (77). M. *furfur* is the agent of the superficial disease pityriasis versicolor (previously tinea versicolor). It has been documented to cause catheter-associated sepsis related to hyperalimentation with lipid emulsions in neonates and immunocompromised patients (17, 28, 40, 58). Pneumonia may be present in the septic picture of all age groups (58). In addition, M. *furfur* has been reported from patients with peritonitis who had received continuous peritoneal dialysis (86) and is implicated as a possible cause or associated agent in seborrheic dermatitis, dandruff, atopic dermatitis, and confluent and reticulate papillomatosis (58, 71). However, Ingham and Cunningham (40) note that the etiologic significance of M. *furfur* in these diseases is controversial. Cutaneous infections caused by *Malassezia* species are treatable with a wide number of agents. Relapse rates are high with topical agents, and prophylactic use with oral agents (itraconazole or ketoconazole) may be necessary to prevent relapse. Catheter-associated disease is best treated by removal of the catheter and, if necessary, application of a systemic antifungal agent (17, 67).

M. *furfur* is normally a commensal of human skin. *Malassezia pachydermatis*, which is typically found on animals, has only rarely been reported to cause sepsis, particularly in low-birth-weight children receiving lipid emulsions through a central venous catheter (75). Two isolates of *Malassezia sympodialis* from human sources have been reported (78). The role of this organism in human disease may be underestimated because of its similarity to M. *furfur* in routine identification tests used in the clinical laboratory (see below).

Culture of routine skin scrapings for the diagnosis of pityriasis versicolor is not usually necessary. The clinical laboratory should be prepared to isolate M. *furfur* from patients with fungemia or other serious infections. M. *furfur* is an obligate lipophile, requiring long-chain fatty acids (C_{12} to C_{24}) for growth (Table 2). M. *pachydermatis*, in contrast, is not an obligate lipophile. Initial cultures of M. *furfur* from blood specimens may result in tiny colonies on solid media or weak growth in broth cultures. Upon subculture to media without lipid supplementation, no growth will be seen.

Lipid supplementation is usually accomplished by overlaying the medium with several drops of sterile olive oil. Alternative media, such as Sabouraud dextrose agar with olive oil and Tween 80 incorporated into the agar and GYP-S agar (glucose-yeast extract-peptone plus olive oil, Tween 80, and glycerol monostearate) devised by Faergemann (24), work well and afford an opportunity for colony quantitation. Other media for supporting *Malassezia* growth have been developed (8, 55).

After incubation of the organism at 35 to 37°C under normal atmospheric conditions, growth usually appears as creamy colonies in 2 to 4 days. Microscopically, *Malassezia* species in culture appear as small, bottle-shaped, budding yeasts (Fig. 5d). Hyphal forms may also be observed in tissue. Careful observation (facilitated by differential interference contrast) will reveal that the junction between the

TABLE 3 Physiologic and antigenic characteristics of *Malassezia* spp.[a]

Malassezia species	Common habitat	Obligatory lipophilic	Lipid requirement supplied by oleic acid or Tween 80	DBB staining	Urea hydrolysis	Unipolar budding	Sympodial budding	No. of serovars	Size range (μm)
M. furfur	Human skin	+	+	+	+	+	0	3[b]	1.5–4.5 by 2.0–6.5
M. sympodialis	Human skin	+	0	+	+	+	+	ND	2.5–7.5 by 2.0–8.0
M. pachydermatis	Canine skin	0	NA	+	+	+	0	ND	2.5–5.5 by 3.0–6.5

[a]Adapted from Marcon and Powell (58). ND, not determined; NA, not applicable.
[b]All antigenically distinct from M. pachydermatis.

bud and the mother cell (which is described as a reduced phialide with a collarette) is rather broad and not as constricted as those in *Torulopsis* and *Candida* species. Routine biochemical testing for the identification of *Malassezia* spp. usually is not attempted. However, organisms that are identified with the API 20C as *C. lipolytica* (glucose, glycerol, and sorbitol positive) may actually be *M. pachydermatis* (58). In this case, morphology is helpful. Identification to the genus level is accomplished by demonstrating typical microscopic and macroscopic morphology and good growth at 37°C but no or poor growth at 25°C. Growth on Sabouraud dextrose agar with and without lipid additives allows differentiation of *M. furfur* and *M. sympodialis* from *M. pachydermatis*. It should be noted that *T. glabrata* can behave like *M. furfur*. In this case, morphology must be studied. To distinguish *M. furfur* from *M. sympodialis*, since both are obligate lipophiles and grow as phialidic yeasts in young cultures, organisms should be subcultured onto media overlaid with either oleic acid or Tween 80 (Table 3). *Malassezia* spp. are nonfermentative and urease positive.

Genus *Pichia*

The teleomorphic genus *Pichia* encompasses several species of *Candida* (for example, *Candida lambica* and *C. guilliermondii*). Generally, the anamorphic (asexual) state is the one recovered from clinical specimens, as mating studies usually are needed to produce the teleomorphic (sexual) form. The biochemical and physiologic reactions of the teleomorph should be identical to those of the anamorph as listed in Table 1.

Genus *Rhodotorula*

Rhodotorula spp. are normal inhabitants of moist skin and can be recovered from such environmental sources as shower curtains, bathtub grout, and tooth brushes (15). In rare instances, *Rhodotorula* spp. have been reported to cause septicemia (64), meningitis (66), systemic infection (72), and sepsis related to complications from indwelling central venous catheters (42).

Rhodotorula spp. and *Cryptococcus* spp. have many similar physiologic and morphologic properties. Both are round to oval multilateral budding yeasts with capsules, produce urease, and fail to ferment carbohydrates. *Rhodotorula* spp. differ from cryptococci by their inability to assimilate inositol and their obvious carotenoid pigment.

Genus *Saccharomyces*

S. cerevisiae is the most common *Saccharomyces* species recovered in the clinical laboratory. Usually thought to be nonpathogenic, it has been reported occasionally to cause thrush, vulvovaginitis (15), and fungemia (23).

Multilateral budding yeast cells are round to oval, and short rudimentary (occasionally well-developed) pseudohyphae may be formed. Ascospore production can be enhanced easily by growing the yeast on Fowell's acetate agar for 2 to 5 days at room temperature. Asci contain one to four round, smooth ascospores (Fig. 5c). Other physiologic properties are listed in Table 1. Assimilation of raffinose by *S. cerevisiae* is noteworthy. Very few yeasts encountered in the clinical laboratory utilize this carbon source, so when one does, *S. cerevisiae* should be considered.

Genus *Torulopsis*

T. glabrata is regarded as a symbiont of humans and can be isolated routinely from the oral cavity and the genitourinary, alimentary, and respiratory tracts of most individuals. As an agent of serious infection, it has been associated with endocarditis (13), meningitis (1), and multifocal disseminated disease (37). It is recovered often from urine specimens and has been estimated to account for as many as 21% of urinary yeast isolates (26).

T. glabrata consists of oval multilateral budding yeasts 2 to 4 μm in diameter that do not form pseudohyphae. *T. glabrata* typically assimilates glucose and trehalose only. Lack of pseudohyphae, growth at 37°C, and fermentation of glucose separate it from *C. zeylanoides*.

Genus *Trichosporon*

The genus *Trichosporon* has recently undergone extensive taxonomic reevaluation. Distinctive morphologic and physiologic patterns for invasive clinical, superficial clinical, and environmental isolates have been recognized among isolates of *T. beigelii* (54). Partial sequencing of the 26S rRNA in addition to studies of ultrastructure, DNA, and ubiquinone systems has led researchers to recommend a revision of the genus to include a large number of species possessing marked ecological preferences (30). For the purposes of this Manual, previously established and accepted *Trichosporon* nomenclature will be maintained while investigators continue their careful scrutiny to establish taxonomic stability. *Trichosporon* spp. belong to the family Cryptococcaceae. *T. beigelii* is the usual species seen in the clinical laboratory; *T. capitatum* has been renamed *B. capitatus* (73). *T. beigelii*, previously known as *Trichosporon cutaneum*, is the etiologic agent of white piedra, which is a superficial infection on the distal portion of the scalp, beard, and axillary and pubic hair (see chapter 65 of this Manual). The true nature of the invasive capability of *T. beigelii* was not recognized until 1970 (89). Disseminated disease due to *Trichosporon* spp. is uncommon but increasingly reported (53, 57). Infection of the immunocompromised host, especially the granulocytopenic patient, frequently produces a fatal disease that is refractory to amphotericin B therapy (38, 87).

Trichosporon spp. produce blastoconidia of various shapes, well-developed hyphae, pseudohyphae, and arthroconidia (Fig. 3J). When a *Trichosporon* isolate produces only a few blastoconidia, differentiation from *Geotrichum* spp. may be difficult. Inoculation of malt extract broth at room temperature will encourage blastoconidia production in *Trichosporon* spp., usually within 48 to 72 h. Growth usually is observed within 1 week on solid medium, and young colonies are cream colored, smooth and shiny, later becoming dry, membranous, and cerebriform. *T. beigelii* and *Trichosporon pullulans* are urease positive; *T. pullulans* is also nitrate positive. Most *Trichosporon* spp. are nonfermentative.

REFERENCES

1. Anhalt, E., J. Alvarez, and R. Bert. 1986. *Torulopsis glabrata* meningitis. South. Med. J. **79:**916.
2. Asakura, K., S.-I. Iwaguchi, M. Homma, T. Sukai, K. Higashide, and K. Tanaka. 1991. Electrophoretic karyotypes of clinically isolated yeasts of *Candida albicans* and *C. glabrata.* J. Gen. Microbiol. **137:**2531–2538.
3. Auger, P., C. Dumas, and J. Joly. 1979. A study of 666 strains of *Candida albicans:* correlation between serotype and susceptibility to 5-fluorocytosine. J. Clin. Microbiol. **139:**590–594.
4. Baird, D. R., M. Harris, R. Menon, and R. Stoddart. 1985. Systemic infection with *Trichosporon capitatum* in two patients with leukemia. Eur. J. Clin. Microbiol. **4:**62–64.
5. Banerjee, S. N., T. G. Emori, D. H. Culver, R. P. Gaynes, W. R. Jarvis, T. Horan, J. R. Edwards, J. Tolson, T. Henderson, W. J. Marton, and The National Nosocomial Infections Surveillance System. 1991. Secular trends in nosocomial primary bloodstream infections in the United States, 1980–1989. Am. J. Med. **91**(Suppl. 3B):86S–89S.
6. Barnett, J. A., R. W. Payne, and D. Yarrow. 1983. *Yeasts: Characteristics and Identification.* Cambridge University Press, Cambridge.
7. Bennett, J. E., K. J. Kwon-Chung, and D. H. Howard. 1977. Epidemiological differences among serotypes of *Cryptococcus neoformans.* Am. J. Epidemiol. **105:**582–586.
8. Bezjak, V., T. M. Al-Nakib, and R. Chandy. 1989. New oil-free media for *Malassezia furfur* (*Pityrosoporum orbiculare*), abstr. F-51, p. 466. Abstr 89th Annu. Meet. Am. Soc. Microbiol. 1989.
9. Bodey, G. P. 1992. Azole antifungal agents. Clin. Infect. Dis. **14:**S161–S169.
10. Brawner, D. L., G. L. Anderson, and K. Y. Yuen. 1992. Serotype prevalence of *Candida albicans* from blood culture isolates. J. Clin. Microbiol. **30:**149–153.
11. Calderone, R. A., and P. C. Braun. 1991. Adherence and receptor relationships of *Candida albicans.* Microbiol. Rev. **55:**1–20.
12. Calderone, R. A., L. Linehan, E. Wadsworth, and A. L. Sandberg. 1988. Identification of C3d receptors on *Candida albicans.* Infect. Immun. **56:**252–258.
13. Carmody, T. J., and K. K. Kane. 1986. *Torulopsis* (*Candida*)

14. Chuck, S. L., and M. A. Sande. 1989. Infections with *Cryptococcus neoformans* in the acquired immunodeficiency syndrome. N. Engl. J. Med. **321:**794–799.
15. Cooper, B. H., and M. Silva-Hutner. 1985. Yeasts of medical importance, p. 526–541. In E. H. Lennette, A. Balows, W. J. Hausler, Jr., and H. J. Shadomy (ed.), *Manual of Clinical Microbiology*, 4th ed. American Society for Microbiology, Washington, D.C.
16. Cutler, J. E. 1991. Putative virulence factors of *Candida albicans.* Annu. Rev. Microbiol. **45:**187–218.
17. Danker, W. M., S. A. Spector, J. Fierer, and C. E. Davis. 1987. *Malassezia* fungemia in neonates and adults: complication of hyperalimentation. Rev. Infect. Dis. **9:**743–753.
18. Dealler, S. F. 1991. *Candida albicans* colony identification in 5 minutes in a general microbiology laboratory. J. Clin. Microbiol. **29:**1081–1082.
19. Domer, J. E., and R. I. Kehrer. 1993. Introduction to *Candida:* systemic candidiasis, p. 49–116. In J. W. Murphy, H. Friedman, and M. Bendinelli (ed.), *Fungal Infections and Immune Responses.* Plenum Press, New York.
20. Drouhet, E., and B. Dupont. 1991. Candidosis in heroin addicts and AIDS: new immunologic data on chronic mucocutaneous candidosis, p. 61–72. In E. Tümbay, H. P. R. Seeliger, and O. Ang (ed.), *Candida and Candidamycosis.* Plenum Press, New York.
21. Edwards, J. E., Jr., and S. G. Filler. 1992. Current strategies for treating invasive candidiasis: emphasis on infections in nonneutropenic patients. Clin. Infect. Dis. **14**(Suppl. 1):S106–S113.
22. Ellis, D. H., and T. J. Pfeiffer. 1990. Natural habitat of *Cryptococcus neoformans* var. *gattii.* J. Clin. Microbiol. **28:**1642–1644.
23. Eschete, M. L., and B. C. West. 1980. *Saccharomyces cerevisiae* septicemia. Arch. Intern. Med. **140:**1539.
24. Faergemann, J. 1984. Quantitative culture of *Pityrosporum orbiculare.* Int. J. Dermatol. **23:**330–333.
25. Filler, S. G., and J. E. Edwards, Jr. 1993. Chronic mucocutaneous candidiasis, p. 117–133. In J. W. Murphy, H. Friedman, and M. Bendinelli (ed.), *Fungal Infections and Immune Responses.* Plenum Press, New York.
26. Frye, K. R., J. M. Donovan, and G. W. Drach. 1988. *Torulopsis glabrata* urinary infections: a review. J. Urol. **139:**1245–1249.
27. Galgiani, J. N., J. Reiser, C. Brass, A. Espinel-Ingroff, M. A. Gordon, and T. M. Kerkering. 1987. Comparison of relative susceptibilities of *Candida* species to three antifungal agents as determined by unstandarized methods. Antimicrob. Agents Chemother. **31:**1343–1347.
28. Garcia, C. R., B. L. Johnston, G. Corvi, L. J. Walker, and W. L. George. 1987. Intravenous catheter-associated *Malassezia furfur* fungemia. Am. J. Med. **83:**790–792.
29. Gilmore, B. J., E. M. Retsinas, J. S. Lorenz, and M. K. Hostetter. 1988. An iC3b receptor on *Candida albicans:* structure, function, and correlates for pathogenicity. J. Infect. Dis. **257:**38–46.
30. Guého, E., M. T. Smith, G. S. de Hoog, G. Billon-Grand, R. Christen, and W. H. Batenburg-van der Vegte. 1992. Contributions to a revision of the genus *Trichosporon.* Antonie van Leeuwenhoek **61:**289–316.
31. Gustafson, K. S., G. M. Vercellotti, C. M. Bendel, and M. K. Hostetter. 1991. Molecular mimicry in *Candida albicans.* J. Clin. Invest. **87:**1896–1902.
32. Hadfield, T. L., M. B. Smith, R. E. Winn, M. G. Rinaldi, and C. Guerra. 1987. Mycoses caused by *Candida lusitaniae.* Rev. Infect. Dis. **9:**1006–1012.
33. Haley, L. D., and C. S. Callaway. 1978. Isolation and identification of yeasts of medical importance, p. 112–113. In *Laboratory Methods in Medical Mycology.* U.S. Department of Health, Education and Welfare publication no. (CDC) 78-8361. Center for Disease Control, Atlanta.

34. Hazen, K. C., D. L. Brawner, M. H. Riesselman, J. E. Cutler, and M. A. Jutila. 1991. Differential adherence of hydrophobic and hydrophilic *Candida albicans* yeast cells to mouse tissues. *Infect. Immun.* **59:**907–912.

35. Hazen, K. C., and P. M. Glee. Cell surface hydrophobicity and medically important fungi. *Curr. Top. Med. Mycol.*, in press.

36. Hazen, K. C., and B. W. Hazen. 1987. Temperature-modulated physiological characteristics of *Candida albicans*. *Microbiol. Immunol.* **31:**497–508.

37. Hickey, W. F., L. H. Sommerville, and F. J. Schoen. 1983. Disseminated *Candida glabrata*: report of a uniquely severe infection and a literature review. *Am. J. Clin. Pathol.* **80:**724–727.

38. Hoy, J., K.-C. Hsu, K. Rolston, R. L. Hopfer, M. Luna, and G. P. Bodey. 1986. *Trichosporon beigelii* infection: a review. *Rev. Infect. Dis.* **8:**959–967.

39. Huppert, M., G. Harper, S. H. Sun, and V. Delanerolle. 1975. Rapid methods for identification of yeasts. *J. Clin. Microbiol.* **2:**21–34.

40. Ingham, E., and A. C. Cunningham. 1993. *Malassezia furfur*. *J. Med. Vet. Mycol.* **31:**265–288.

41. Kaplan, W. 1978. Protothecosis and infections caused by morphologically similar green algae, p. 218–232. *In The Black and White Yeasts: Proceedings of IV International Conference on Mycoses.* Pan American Health Organization publication no. 356. Pan American Health Organization, Washington, D.C.

42. Kiehn, T. E., E. Gorey, A. E. Brown, F. F. Edwards, and D. Armstrong. 1992. Sepsis due to *Rhodotorula* related to use of indwelling central venous catheters. *Clin. Infect. Dis.* **14:**841–846.

43. Klein, A. S., G. T. Tortora, R. Malowitz, and W. H. Greene. 1988. *Hansenula anomala*: a new fungal pathogen. *Arch. Intern. Med.* **148:**1210–1213.

44. Kreger-van Rij, N. J. W. (ed.). 1984. *The Yeasts: a Taxonomic Study*, 3rd ed. Elsevier Science Publishers B.V., Amsterdam.

45. Krumholz, R. A. 1972. Pulmonary cryptococcosis. A case due to *Cryptococcus albidus*. *Am. Rev. Respir. Dis.* **105:**421–424.

46. Kurtzman, C. P. 1984. Genus 11. *Hansenula* H. et P. Sydow, p. 165–213. *In N. J. W. Kreger-van Rij (ed.), The Yeasts: a Taxonomic Study*, 3rd ed. Elsevier Science Publishers B. V., Amsterdam.

47. Kwon-Chung, K. J. 1975. A new genus, *Filobasidiella*, the perfect state of *Cryptococcus neoformans*. *Mycologia* **67:**1197–1200.

48. Kwon-Chung, K. J., and J. E. Bennett. 1992. *Medical Mycology*, p. 397–446. Lea & Febiger, Philadelphia.

49. Kwon-Chung, K. J., J. E. Bennett, and T. S. Theodore. 1978. *Cryptococcus bacillisporus* sp. nov. serotype B-C of *Cryptococcus neoformans*. *Int. J. Syst. Bacteriol.* **28:**616–620.

50. Kwon-Chung, K. J., J. B. Hicks, and P. N. Lipke. 1990. Evidence that *Candida stellatoidea* type II is a mutant of *Candida albicans* that does not express sucrose-inhibitable α-glucosidase. *Infect. Immun.* **58:**2804–2808.

51. Kwon-Chung, K. J., D. Lehman, C. Good, and P. T. Magee. 1985. Genetic evidence for the role of extracellular proteinase in virulence of *Candida albicans*. *Infect. Immun.* **49:**571–575.

52. Kwon-Chung, K. J., I. Polacheck, and J. E. Bennett. 1982. Improved diagnostic medium for separation of *Cryptococcus neoformans* var. *neoformans* (serotypes A and D) and *Cryptococcus neoformans* var. *gattii* (serotypes B and C). *J. Clin. Microbiol.* **15:**535–537.

53. Leblond, V., O. Saint-Jean, A. Datry, G. Lecso, C. Frances, S. Bellefigh, M. Gentilini, and J. L. Binet. 1986. Systemic infections with *Trichosporon beigelii* (*cutaneum*). *Cancer* **58:**2399–2405.

54. Lee, J. W., G. A. Melcher, M. G. Rinaldi, P. A. Pizzo, and T. J. Welsh. 1990. Patterns of morphologic variation among isolates of *Trichosporon beigelii*. *J. Clin. Microbiol.* **28:**2823–2827.

55. Leeming, J. P., and F. H. Notman. 1987. Improved methods for isolation and enumeration of *Malassezia furfur* from human skin. *J. Clin. Microbiol.* **25:**2017–2019.

56. Love, G. L., G. D. Boyd, and D. L. Greer. 1985. Large *Cryptococcus neoformans* isolated from brain abscess. *J. Clin. Microbiol.* **22:**1068–1070.

57. Manzella, J. P., I. J. Berman, and M. D. Kubrika. 1982. *Trichosporon beigelii* fungemia and cutaneous dissemination. *Arch. Dermatol.* **118:**343–345.

58. Marcon, M. J., and D. A. Powell. 1992. Human infections due to *Malassezia* spp. *Clin. Microbiol. Rev.* **5:**101–119.

59. Martino, P., M. Venditti, A. Micozzi, G. Morace, L. Polonelli, M. P. Mantovani, M. C. Petti, V. L. Burgio, C. Santini, P. Serrra, and F. Mandelli. 1990. *Blastoschizomyces capitatus*: an emerging cause of invasive fungal disease in leukemia patients. *Rev. Infect. Dis.* **18:**579–582.

60. McGinnis, M. R., D. H. Walker, and J. D. Folds. 1980. *Hansenula polymorpha* infection in a child with chronic granulomatous disease. *Arch. Pathol. Lab. Med.* **104:**290–292.

61. Odds, F. C. 1988. *Candida and Candidosis*, 2nd ed. Bailliere Tindall, London.

62. Pfaller, M. A., T. Preston, M. Bale, F. P. Koontz, and B. A. Body. 1988. Comparison of the Quantum II, API Yeast Ident, and AutoMicrobic Systems for identification of clinical yeast isolates. *J. Clin. Microbiol.* **26:**2054–2058.

63. Phelan, J. A., B. R. Saltzman, G. H. Friedland, and R. S. Klein. 1987. Oral findings in patients with acquired immunodeficiency syndrome. *Oral Surg.* **64:**50–56.

64. Pien, F. D., R. L. Thompson, D. Deye, and G. D. Roberts. 1980. *Rhodotorula* septicemia: two cases and a review of the literature. *Mayo Clin. Proc.* **55:**258–260.

65. Pore, R. S. 1985. *Prototheca* taxonomy. *Mycopathologia* **90:**129–135.

66. Pore, R. S., and J. Chen. 1976. Meningitis caused by *Rhodotorula*. *Sabouraudia* **14:**331–335.

67. Redline, R. W., S. S. Redline, B. Boxerbaum, and B. B. Dahms. 1985. Systemic *Malassezia furfur* infection in patients receiving intralipid therapy. *Hum. Pathol.* **16:**815–822.

68. Rikkerink, E. H. A., B. B. Magee, and P. T. Magee. 1990. Genomic structure of *Candida stellatoidea*: extra chromosomes and gene duplication. *Infect. Immun.* **58:**949–954.

69. Rinaldi, M. G. 1993. Biology and pathogenicity of *Candida* species, p. 1–20. *In G. P. Bodey (ed.), Candidiasis: Pathogenesis, Diagnosis and Treatment*. Raven Press, New York.

70. Rine, J., W. Hansen, E. Hardeman, and R. W. Davis. 1983. Targeted selection of recombinant clones through gene dosage effects. *Proc. Natl. Acad. Sci. USA* **80:**6750–6754.

71. Ring, J., D. Abeck, and K. Neuber. 1992. Atopic eczema: role of microorganisms on the skin surface. *Allergy* **47:**265–269.

72. Rusthoven, J. J., R. Feld, and P. J. Tuffnell. 1984. Systemic infection by *Rhodotorula* spp. in the immunocompromised host. *J. Infect.* **8:**244–249.

73. Salkin, I. F., M. A. Gordon, W. M. Samsonoff, and C. L. Rieder. 1985. *Blastoschizomyces capitatus*, a new combination. *Mycotaxon* **22:**373–380.

74. Salkin, I. F., and N. J. Hurd. 1982. New medium for differentiation of *Cryptococcus neoformans* serotype pairs. *J. Clin. Microbiol.* **15:**169–171.

75. Sanguinetti, V., M. P. Tampieri, and L. Morganti. 1984. A survey of 120 isolates of *Malassezia* (*Pityrosporum*) *pachydermatis*. *Mycopathologia* **85:**93–95.

76. Schlitzer, R. L., and D. G. Ahearn. 1982. Characterization of atypical *Candida tropicalis* and other uncommon clinical yeast isolates. *J. Clin. Microbiol.* **15:**511–516.

77. Simmons, R. B., and D. G. Ahearn. 1987. Cell wall ultrastructure and DBB reaction of *Sporopachydermia quercuum*, *Bullera tsugae* and *Malassezia* spp. *Mycologia* **79:**38–43.

78. Simmons, R. B., and E. Guého. 1990. A new species of *Malassezia*. *Mycol. Res.* **94:**1146–1149.

79. Soll, D. R. 1992. High-frequency switching in *Candida albicans*. *Clin. Microbiol. Rev.* **5:**183–203.

80. Spicer, A. D., and K. C. Hazen. 1992. Rapid confirmation of

Candida albicans identification by combination of two presumptive tests. *Med. Microbiol. Lett.* **1:**284–289.

80a. **Staib, F., M. Seibold, E. Antweiler, B. Fröhlich, S. Weber, and A. Blisse.** 1987. The brown colour effect (BCE) of Cryptococcus neoformans in the diagnosis, control and epidemiology of C. neoformans infections in AIDS patients. *Zentralbl. Bakteriol. Hyg.* A **266:**167–177.

81. **St.-Germain, G., and D. Beauchesne.** 1991. Evaluation of the MicroScan rapid yeast identification panel. *J. Clin. Microbiol.* **29:**2296–2299.

82. **Summerbell, R. C.** 1992. *Candida lusitaniae* confirmed by a simple test, abstr. F-90, p. 513. *Abstr. 92nd Gen. Meet. Am. Soc. Microbiol. 1992.*

83. **Tavitian, A., J.-P. Raufman, and L. E. Rosenthal.** 1986. Oral candidiasis as a marker for esophageal candidiasis in the acquired immunodeficiency syndrome. *Ann. Intern. Med.* **104:**54–55.

84. **Tronchin, G., J. P. Bouchara, V. Annaix, R. Robert, and J. M. Senet.** 1991. Fungal cell adhesion molecules in *Candida albicans. Eur. J. Epidemiol.* **7:**23–33.

85. **Vanbreuseghem, R., and M. Takashio.** 1970. An atypical strain of *Cryptococcus neoformans* (San Felice) Vuillemin 1894. II. C. neoformans var. *gattii* var. nov. *Ann. Soc. Belge Med. Trop.* **50:**695–702.

86. **Wallace, M., H. Bagnall, D. Glen, and S. Averill.** 1979. Isolation of lipophilic yeasts in "sterile" peritonitis. *Lancet* **ii:**956.

87. **Walsh, T. J., G. P. Melcher, M. G. Rinaldi, J. Lecciones, D. A. McGough, P. Kelly, J. Lee, D. Callender, M. Rubin, and P. A. Pizzo.** 1990. *Trichosporon beigelii,* an emerging pathogen resistant to amphotericin B. *J. Clin. Microbiol.* **28:**1616–1622.

88. **Walsh, T. J., and A. Pizzo.** 1988. Treatment of systemic fungal infections: recent progress and current problems. *Eur. J. Clin. Microbiol. Infect. Dis.* **7:**460–475.

89. **Watson, K. C., and S. Kallichurum.** 1970. Brain abscess due to *Trichosporon cutaneum. J. Med. Microbiol.* **3:**191–193.

90. **Whelan, W. L., and K. J. Kwon-Chung.** 1988. Auxotrophic heterozygosities and the ploidy of *Candida parapsilosis* and *Candida krusei. J. Med. Vet. Mycol.* **26:**163–171.

91. **Wickerham, L. J., and K. A. Burton.** 1948. Carbon assimilation tests for the classification of yeasts. *J. Bacteriol.* **56:**363–371.

92. **Wickes, B. L., J. B. Hicks, W. G. Merz, and K. J. Kwon-Chung.** 1992. The molecular analysis of synonymy among medically important yeasts within the genus *Candida. J. Gen. Microbiol.* **138:**901–907.

93. **Wingard, J. R., J. D. Dick, W. G. Merz, G. R. Sanford, R. Saral, and W. H. Burns.** 1982. Differences in virulence of clinical isolates of *Candida tropicalis* and *Candida albicans* in mice. *Infect. Immun.* **37:**833–836.

94. **Wingard, J. R., W. G. Merz, M. G. Rinaldi, T. R. Johnson, J. E. Karp, and R. Saral.** 1991. Increase in *Candida krusei* infection among patients with bone marrow transplantation and neutropenia treated prophylactically with fluconazole. *N. Engl. J. Med.* **325:**1274–1277.

95. **Yamane, N., and Y. Saitoh.** 1985. Isolation and detection of multiple yeasts from a single clinical sample by use of Pagano-Levin agar medium. *J. Clin. Microbiol.* **21:**276–277.

96. **Yarrow, D., and S. A. Meyer.** 1978. Proposal for amendment of the diagnosis of the genus *Candida* Berkhout nom. cons. *Int. J. Syst. Bacteriol.* **28:**611–615.

Pneumocystis

W. KEITH HADLEY AND VALERIE L. NG

62

DESCRIPTION OF THE AGENT

Pneumocystis carinii is a unicellular eukaryotic organism with a tropism for growth on respiratory surfaces of mammals (64, 66, 78, 79, 88). Its phylogeny has not been determined with certainty. *P. carinii* was originally classified with the protozoa, but recent observations suggest that it is phylogenetically more closely related to fungi (21, 59, 64, 76, 78). Features of pneumocystis shared with fungi include (i) cyst wall ultrastructure similar to that of fungal cell wall, (ii) lamellar cristae in the mitochondria (protozoans have tubular christae), (iii) formation of intracystic bodies resembling the formation of ascospores by the ascomycetes (66), (iv) highest homology of the more conserved domains of the 16S rRNA subunit with that of ascomycetes (20, 75, 76), (v) homology of the 5S rRNA with that of primitive zygomycetes (86), (vi) homology of the protein synthesis elongation factor EF-3 with that of *Saccharomyces cerevisiae* (an ascomycete), (vii) separate proteins for thymidylate synthase and dihydrofolate reductase (protozoa produce a single bifunctional protein) (19, 21, 22), and (viii) homology of the sequence encoding *Pneumocystis* thymidylate synthase with that of ascomycetes (22).

In contrast, ergosterol, the sterol found in the membranes of most fungi, has not been detected in *Pneumocystis* cells. Lack of cell membrane ergosterol and presence of other sterols may explain the lack of clinical efficacy of commonly used antifungal agents that are dependent either on binding to or inhibiting synthesis of ergosterol (i.e., amphotericin B or imidazole and triazole antifungal agents, respectively) (21, 61).

P. carinii cannot be propagated continuously in host cell-free (axenic) culture systems. Study of the organism in vivo is limited to analysis of specimens either directly obtained from infected mammals or cocultured with feeder cells. Rat-derived *Pneumocystis* cells cocultured with mammalian-lung-derived cells is most successful and has been studied the most (1, 2, 71). Short-term host cell-free culture systems have been developed recently. These maintain metabolism and may increase the numbers of *Pneumocystis* cells over several days. Although viability can and has been measured by incorporation of radiolabel from a substrate into its product (14, 16), enumeration of *Pneumocystis* cells (polymorphic organisms that are strongly self-adherent as well as adherent to host cells) and determination of viabil-

ity remain problematic for all currently used *P. carinii* culture systems (1, 37, 40).

Animal models of *P. carinii* infection have been highly predictive of human response to infection, therapy, and prophylaxis. The corticosteroid treatment, latent-infection rat model that was studied and developed by Frenkel et al. has been used to study therapeutic or prophylactic agents before human trials are done (25). The transtracheally inoculated corticosteroid-treated rat model established a standardized inoculum and more uniform infection (3, 72). Other models include immunodeficient mice (nude mouse, *scid* mouse, T-cell-deficient mouse) and other mammals.

Molecular biology-based evidence of different varieties and possibly different species of *Pneumocystis* is emerging. Although *Pneumocystis* strains recovered from the rat, humans, and other mammals appear to be morphologically identical, differences between surface antigens, karyotypes, and nucleic acid sequences of a variety of cloned genes can be demonstrated in *Pneumocystis* strains recovered from rats and humans (30, 41, 43, 70, 74). Rats can harbor at least two *Pneumocystis* strains with marked genetic differences (18). These differences may be useful for tracking the epidemiology of *Pneumocystis* infection and must be considered when drugs and immunologic and nucleic acid diagnostic reagents are being designed.

The proposed sexual and asexual life cycles of *Pneumocystis* sp. are as depicted in Fig. 1 and discussed below (24, 64). *Pneumocystis* cells have been observed as both small (1.5- to 2-μm) and large (3.0- to 5.0-μm) trophic forms. The nucleus of each trophic form is small (0.5 to 1 μm) and can be visualized by light microscopic examination of Giemsa-stained preparations (17, 65, 78, 79). The mechanism of nuclear division is not clear; some large trophic forms have notably larger nuclei and apparent binary fission, and binucleate cells and apparent cytokinesis have been documented, but there have been only two documented observations of karyokinesis involving a mitotic spindle (63, 64). For sexual reproduction (i.e., meiosis), it has been proposed that mating of haploid gametic cells, formation of a diploid zygote, encystment, and sporogenesis occur in *Pneumocystis* cells (similar to spore formation in ascomycetous fungi). The precyst, a commonly observed large form (4.0 to 5.0 μm) with a large nucleus and beginning wall formation, has been proposed for the large parent form or zygote. In support of this hypothesis, complexes suggesting meiotic

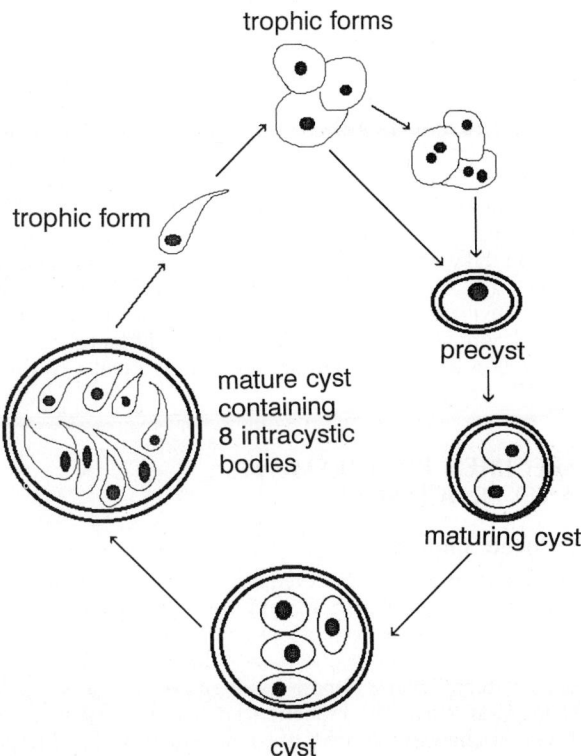

trophic forms

trophic form

mature cyst
containing
8 intracystic
bodies

precyst

maturing cyst

cyst

FIGURE 1   Suggested life cycle of *P. carinii*.

TABLE 1   Clinical conditions associated with PCP

Prematurity of infant
Malnutrition, protein and caloric (marasmus)
Corticosteroid therapy
Cytotoxic therapy
Advanced malignancy
Immunosuppressive therapy
Preparation for organ transplantation
Cushing's syndrome
Hodgkin's disease
Non-Hodgkin's lymphoma
Chronic lymphocytic leukemia
Solid tumors
Congenital immunodeficiency
HIV-1 infection and/or AIDS
Old age

carinii in 94% of children aged 2.5 years. The prevalence of antibodies detected by this immunoblot was similar in different geographic areas (56).

P. CARINII INFECTION

P. carinii is recognized as an opportunistic pathogen for patients with a wide variety of iatrogenic or naturally occurring diseases, including therapies of malignancies and various rheumatologic disorders with corticosteroids and cytotoxic drugs, immunosuppressive treatment prior to solid organ or bone marrow transplantation, congenital immunodeficiency, or Cushing's syndrome (Table 1) (32, 85, 89). *P. carinii* pneumonia is the most common opportunistic infection for individuals infected with human immunodeficiency virus type 1 (HIV-1) (10, 28, 45).

The respiratory tract is thought to be the portal of entry for *P. carinii*, since the apparent primary infection is in the lung. Airborne spread of the organism has been demonstrated in experiments with susceptible rats, and human disease is thought to occur similarly (35). Despite demonstration of middle ear infection by *P. carinii*, there is no evidence for nasopharyngeal colonization by *P. carinii* in humans.

Successful colonization and infection of a host by a parasite are usually dependent on some parasitic adherence mechanism. The trophic form of *P. carinii* adheres to the alveolar type 1 pneumocyte. By binding to mannose residues on the *P. carinii* surface glycoprotein and thus serving as a bridge between the principal surface antigen of *P. carinii* and host cells and the mannose receptor on alveolar macrophages, host fibronectin may facilitate adherence (23, 60, 87).

The infectious form of *P. carinii* is not known. The freshly released intracystic body or the small trophic form of a *Pneumocystis* cell, however, is within the size range (i.e., 1 to 3 μm) of other pulmonary pathogens successfully spread deep into the lung via aerosolization (i.e., tubercle bacillus).

P. carinii pneumonia (PCP) is thought to occur in the immunodeficient patient by reactivation of a latent infection; the healthy host cellular, and perhaps humoral, immune function is thought to be responsible for maintaining *P. carinii* in a latent state. This reasoning has been the basis for the practice of not isolating *P. carinii*-infected patients in the hospital. An alternative hypothesis for the develop-

prophase have been seen in the precyst (48). The precyst undergoes encystment with the formation of the commonly observed mature cyst (approximately 5 μm in diameter), which can contain up to eight intracystic bodies (spherical, oval, or fusiform) (65, 79, 88). The intracystic bodies are thought to form by meiosis and partitioning of the cytoplasm. After release of the intracystic bodies from the cyst (excystation), the cyst wall may remain intact or may indent and form cyclic structures (64, 79).

Ruffolo has proposed that a nomenclature in concordance with the fungal classification be used, i.e., that sporogenesis be substituted for the sexual formation of reproductive cysts, spore case be substituted for cyst, sporocyst be substituted for precyst, and spore be substituted for intracystic body (64).

The natural reservoir of *P. carinii* remains unknown. It is unlikely that infected rodents serve as a reservoir and transmit *P. carinii* as a zoonosis, since the surface antigens and nucleic acid sequences of various genes of rat-derived *P. carinii* are distinct from those of human-derived *P. carinii* (70). Serologic studies suggest that *P. carinii* is ubiquitous in the environment. It has been detected in the lungs of a variety of mammals (especially immunodeficient ones), including humans, primates, rodents, hares, ferrets, cats, dogs, and horses (34, 35). Human contact with *P. carinii* and development of anti-*Pneumocystis* antibodies occurs at an early age in most humans. Pifer et al., using an indirect fluorescent-antibody test, observed that 83% of 4-year-old children had antibody to *P. carinii* (57). Peglow et al., using immunoblotting techniques, observed antibody directed against the 40-kDa major antigen of human-derived *P.*

ment of PCP is that *P. carinii* stays in the healthy host for only brief periods and that the host is frequently exposed to sources of the organism throughout life. The conflict between the two hypotheses cannot be resolved until epidemiologic markers for *Pneumocystis* infection are readily available (84). The CD4$^+$ lymphocyte count has proven to be a useful predicter of opportunistic pneumonia in HIV-1-infected individuals. PCP is more likely to occur in adults and older children as the CD4$^+$ count falls below 200/mm^3, while in young children, who normally have higher CD4$^+$ counts, PCP occurs when the CD4$^+$ count is depressed below 450/mm^3 (11, 12, 46). The exact host immunologic defects that permit *P. carinii* proliferation are not known.

The sequential development of the commonly observed PCP pathology in humans is best described by what has been studied in the corticosteroid-mediated immunosuppressed experimental rat model system (33, 84). Attachment of *P. carinii* trophozoites to the type 1 pneumocyte is followed by diffuse alveolar injury and leakage of exudate through the basement membrane. Adherent clusters of *P. carinii* of various stages accumulate in the alveoli concomitant with a small mononuclear infiltration, edema, thickening and fibrosis of the alveolar septa, and a decrease in surfactant phospholipids. There is impaired gas exchange and altered lung compliance. Hypoxemia, increased alveolar-arterial oxygen gradient, respiratory alkalosis, reduced vital capacity, and diffusing capacity (or "alveolar-capillary block") usually occur. A temporary worsening of hypoxemia is often observed following initial appropriate treatment for PCP. *P. carinii* can occasionally be found within a noncaseating granuloma in the lung. The profuse lymphocytic plasma cell infiltrate characteristic of pneumocystosis in the marasmic, malnourished child of less than 4 months of age is seldom seen in PCP associated with immunodeficiency. *Pneumocystis* infection may spread from the lung to the pleural space and adjacent lymph nodes. Hematologic spread to organs with highly developed vascularity can occur. Reports of extrapulmonary pneumocystosis describe infections occurring in lymph nodes, spleen, liver, bone marrow, endocrine organs, gastrointestinal and genitourinary tracts, heart, eyes, and ears (31, 54, 62, 77).

DIAGNOSIS OF PCP

Diagnosis of PCP is at present dependent on the morphologic identification of the organism. The organism can be detected in a variety of respiratory specimens, i.e., sputum, induced sputum, tracheal aspirate fluid, bronchoalveolar lavage fluid, tissue obtained by transbronchial biopsy (TBBx), cellular material obtained by bronchial brush, pleural fluid, and tissue obtained by open-thorax lung biopsy. The microbiology laboratory is equipped to process these specimens and use them for diagnosis of a variety of opportunistic infections.

The diagnostic yields of the various specimens are critically dependent on the underlying disease of the patient, the expertise of the staff evaluating the patient and obtaining the specimen, and the expertise of the laboratory staff processing and examining the specimen. For AIDS patients, 80% of PCP diagnoses can be made from induced sputum specimens (42, 49, 50). In contrast, for patients with an underlying immunosuppression other than AIDS, *P. carinii* is only rarely observed in induced sputum, and diagnosis is dependent on examination of specimens obtained by bronchoscopy or biopsy (33, 36, 85, 89). The difference in diagnostic yield of induced sputum specimens

has been attributed to a presumed higher burden of *P. carinii* in AIDS patients than in other immunosuppressed patients. For infants and young children who are unable to produce sputum, tracheal aspirate fluid, open-thorax lung biopsy specimens, or respiratory specimens obtained by a special pediatric bronchoscopy service must be used for diagnosis (6, 36).

At San Francisco General Hospital, AIDS patients selected to undergo further laboratory evaluation for PCP must meet the following criteria: (i) belong to a group at risk for HIV transmission or other cause of immunodeficiency, (ii) show respiratory symptoms, and (iii) show objective evidence of lung disease (at least one of the following: abnormal chest X-ray film, reduced diffusing capacity for carbon monoxide, and abnormal lung uptake of gallium-67 citrate) (49).

SPECIMEN REQUIREMENTS AND PROCESSING

Induced Sputum

Sputum induction does not require expensive equipment or a physician's participation, and the patient experiences little discomfort and rarely any complication. Sputum induction is best done centrally by pulmonary function laboratory technicians, respiratory therapists, or a specially trained assistant. The induction facility should have adequate air changes, negative-pressure ventilation, and respirators to protect personnel from other pathogens that may be concomitantly present (e.g., *Mycobacterium tuberculosis*). Alternatively, sputum inductions can be done within a chamber with HEPA-filtered exhaust.

For induced sputum specimens, adequate preparation of the patient is of utmost importance. Vigorous brushing of the teeth, tongue, and gums with a toothbrush and 0.85% saline for 5 to 10 min before sputum induction followed by thorough rinsing is critical in order to remove as much cellular debris of oral origin as possible. Toothpaste should not be used, as it can interfere with subsequent processing and staining of the specimen. Deep inhalation of nebulized 3% sodium chloride solution by the patient will result in osmotic accumulation of fluid in and irritation of the respiratory passages, with subsequent coughing and expectoration of bronchoalveolar contents. The induced sputum specimen is usually mucoid and translucent; only rarely is it purulent. When *Pneumocystis* clumps are present, they are 0.1 to 0.2 mm in diameter and cream to light tan in appearance. In early studies of induced sputum for diagnosis of *Pneumocystis* pneumonia, the sputum was smeared directly (5, 58). This method proved less sensitive than procedures using mucolysis and concentration of the specimen (39, 49).

In the laboratory, the induced sputum specimen is mucolysed by the addition of an equal volume of freshly made 0.0065 M dithiothreitol (Stat-Pak Sputolysin, Behring Diagnostics, La Jolla, Calif.; diluted to the manufacturer's specifications) or 0.5% *N*-acetyl-L-cysteine and incubation on a rotary shaker at 35°C with intermittent vigorous vortexing until the specimen is almost completely liquefied (complete liquefication will completely disperse the *Pneumocystis* cells and make microscopic detection difficult). The mostly liquefied specimen is concentrated by centrifugation at 1,300 × g for 5 min, and the sediment is smeared on glass slides, which are then air dried and heat fixed (49). Prolonged heat fixation (i.e., 7 to 10 passes through the

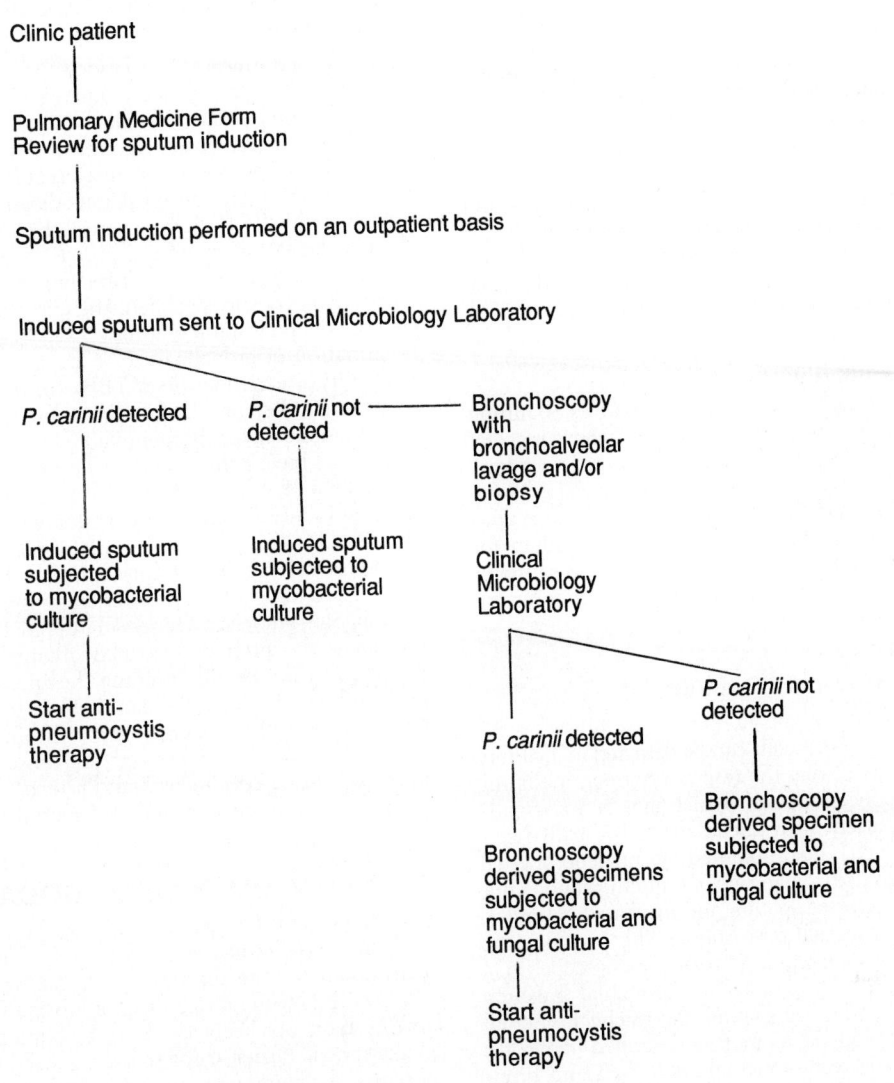

Clinic patient

Pulmonary Medicine Form
Review for sputum induction

Sputum induction performed on an outpatient basis

Induced sputum sent to Clinical Microbiology Laboratory

P. carinii detected

P. carinii not detected

Bronchoscopy with bronchoalveolar lavage and/or biopsy

Induced sputum subjected to mycobacterial culture

Induced sputum subjected to mycobacterial culture

Clinical Microbiology Laboratory

Start anti-pneumocystis therapy

P. carinii not detected

P. carinii detected

Bronchoscopy derived specimen subjected to mycobacterial and fungal culture

Bronchoscopy derived specimens subjected to mycobacterial and fungal culture

Start anti-pneumocystis therapy

FIGURE 2   Algorithm for evaluation of patient suspected of having PCP.

hottest portion of a Bunsen burner flame) is important in fixing the material to the slide, since most of the natural cellular adhesins are removed during mucolysis. Slides containing fixed material are then stained (see below) and examined microscopically for the presence of pneumocystis.

Bronchoscopy with Bronchoalveolar Lavage

Bronchoscopy is performed by a pulmonary physician and requires special facilities in case emergency resuscitation of the patient is necessary; the facility must also provide adequate ventilation to protect personnel from other respiratory pathogens. The procedure uses a fiber-optic bronchoscope to examine the respiratory tract (6, 9, 55). To obtain lavage fluid, the bronchoscope is wedged into a segmental bronchus, and 100 to 150 ml of 0.85% saline is used to flush the alveoli in the lung segment attached to the bronchus; approximately 50 to 60 ml of lavage fluid is usually recovered. This lavage fluid is liquid, does not need mucolysis,

and is easily concentrated by centrifugation (49). The centrifuged pellet is used to make smears on glass slides, and the smears are fixed in absolute methanol or acetone while the specimen is still moist. After air drying and heat fixation, slides can be stained and microscopically examined for *Pneumocystis* cells.

TBBx

A TBBx sample is obtained during bronchoscopy by using a cutting tool passed through the fiber-optic bronchoscope to punch through the bronchial wall and obtain adjacent lung tissue. The TBBx procedure can, on occasion, result in serious complications to the patient, including hemorrhage and pneumothorax.

If a TBBx sample is obtained, the biopsy specimen should be collected on Telfa or sterile gauze barely moistened with nonbacteriostatic saline and folded around the TBBx specimen. It is important to keep the specimen moist but not wet; excess saline dilutes tissue adhesins such that touch imprints will wash away during staining. The specimen should be kept moist in a small sterile container with a tightly fitted lid. If handled sterilely, the specimen can first be used to make imprint or touch preparations on clean, alcohol-flamed, and cooled slides, and the remainder can be ground or minced and subjected to various microbiologic cultures. If the biopsy tissue is large enough, touch preparations can be made from additional sterilely cut surfaces. The touch preparations should be fixed in absolute methanol or acetone while still moist; slides should be air dried and heat fixed before being stained. Alternatively, the biopsy sample can be ground in a tissue grinder, and smears can be made on microscope slides, fixed, and stained (7).

Open-Thorax Lung Biopsy

The open-thorax lung biopsy procedure involves obtaining a lung biopsy sample after the chest has been surgically opened (8). The biopsy can be visually directed to a lung lobe or segment involved by the disease process. The biopsy specimen should be handled by the operating room staff as described above for TBBx specimens. The specimen for microbiologic examination must not be placed in saline or fixative, which precludes examination of touch preparations or microbiologic culture.

Other Specimens Derived from the Respiratory Tract

On occasion, *Pneumocystis* cells can be detected in a variety of specimens obtained from the respiratory tract, including spontaneously produced sputum (which may be principally saliva, since the patient with pneumocystosis usually has a nonproductive cough), fluid obtained by tracheobronchial aspiration (perhaps the only means of sampling young children who cannot produce sputum and intubated patients), fluid obtained by bronchial wash (obtained by washing the bronchial wall without wedging the bronchoscope in a bronchus), or specimens obtained by bronchial brushing (38). Spontaneously derived sputum and tracheobronchial aspiration fluid should be treated as described above for induced sputum, whereas bronchial wash fluid and bronchial brushing can be handled as described for bronchoalveolar lavage fluid. None of the above-mentioned specimens have as high a diagnostic yield as induced sputum or specimens obtained by bronchoscopy (i.e., lavage and/or biopsy).

DIAGNOSTIC YIELDS OF INDUCED SPUTUM AND BRONCHOSCOPY-DERIVED SPECIMENS

In studies performed at San Francisco General Hospital, *P. carinii* was detected in approximately 55% of all induced sputa obtained from individuals suspected of PCP and meeting the aforementioned clinical criteria. The sensitivity of induced sputum (i.e., detection of *P. carinii* in induced specimens from patients ultimately diagnosed with PCP) is approximately 75 to 80%, whereas the sensitivity of bronchoalveolar lavage is virtually 100% (5, 9). Although earlier studies suggested that TBBx might produce the most sensitive diagnostic specimen, the experience at San Francisco General Hospital spanning 1990 to 1994 suggests that examination of bronchoalveolar lavage fluid has the same sensitivity as examination of TBBx touch preparations. All stains (e.g., Giemsa, Giemsa-like, toluidine blue, Gomori's methenamine silver, immunofluorescence, Gram-Weigert, calcofluor) have comparable sensitivities for the detection of *P. carinii* (4, 15, 51, 53, 73). By definition, the specificity of either induced sputum or bronchoalveolar lavage fluid for PCP is 100% (i.e., there are no false-positive results).

On the basis of this information, the following algorithm was devised for evaluation of patients suspected of PCP and seen at San Francisco General Hospital (Fig. 2). Induced sputum is the primary specimen obtained for evaluation because of the ease of obtaining the specimen on an outpatient basis by nonphysician staff with little attendant patient morbidity. Bronchoscopy with bronchoalveolar lavage and/or TBBx is reserved for those patients whose induced sputum specimens lack *Pneumocystis* cells or who are unable to produce an induced sputum specimen (52).

POTENTIAL DIAGNOSTIC MODALITIES

Although useful for epidemiologic studies, serologic assays to detect anti-*Pneumocystis* antibodies have not been clinically useful in the diagnosis of *Pneumocystis* pneumonia (56, 57). Likewise, antigen capture assays for *Pneumocystis* antigen have not been proven to be clinically useful. Amplification of *Pneumocystis*-specific genes by PCR has the potential for being more sensitive than the currently practiced method of staining and microscopic examination of specimens. However, as presently constituted, PCR systems are labor intensive and expensive (80, 81). The current microscopy-based method has been convincingly demonstrated to produce results concordant with clinical diagno-

---

FIGURE 3 (Opposite) *P. carinii*. (A) Giemsa-like and Diff-Quik stains of induced sputum. (A1) Single mature cyst (8 μm) containing eight fusiform intracystic bodies with purple nuclei and light blue cytoplasms. Note clear unstained wall of cyst at its periphery. Cyst is attached to alveolar macrophage. Magnification, ×1,148. (A2) Thin clump of mostly small trophic forms (2 to 3 μm) with small purple nuclei and light blue cytoplasms. One precyst (6 μm) with a large nucleus and clear (unstained) wall is visible. Magnification, ×1,148. (A3) Thick clump of precysts and cysts with various numbers of intracystic bodies and trophic forms overlying each other. Magnification, ×1,148. (A4) Thick clump with overlying *P. carinii* cells at various stages. Note squamous epithelial cells and other host cells in mucolysed concentrate. Magnification, ×276. (B) Direct fluorescent-antibody stain of induced sputum. Note the apple green fluorescein stain on wall and wall fold of cyst. Structures inside cyst are unstained (black). Trophic forms that are confluent are stained on the surface. Magnification, ×1,106. (C) Gomori's methenamine silver stain (Grocott) of bronchoalveolar lavage specimen. Cyst walls and thickenings (double comma) are stained black. Trophic forms in clump are not stained. Magnification, ×1,120. (D) Toluidine blue O stain of bronchoalveolar lavage specimen. Cyst walls and thickenings (double comma) are stained purple. Trophic forms in clump are not stained. Magnification, ×896. (E) Calcofluor stain of bronchoalveolar lavage specimen. Cyst walls with thickenings (double comma) are highly fluorescent (color varies with barrier filter). Trophic forms may be lightly stained. Magnification, ×552. (F) Papanicolaou's stain of pleural fluid. *Pneumocystis* trophic forms are phagocytized by macrophage. Nuclei are dark, and cytoplasm is green. Magnification, ×896.

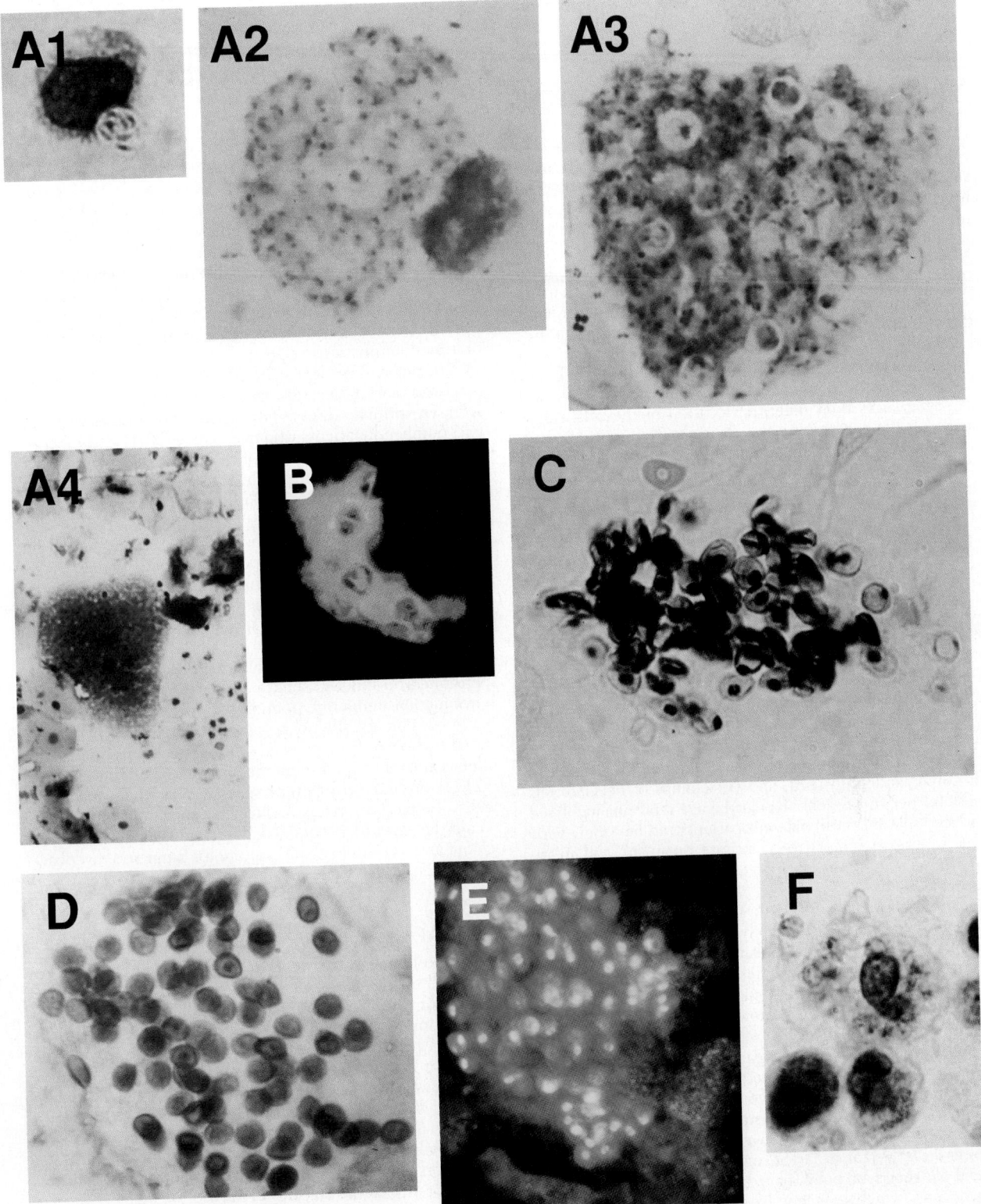

sis, patient therapy, and subsequent clinical course. A clinical role for more sensitive diagnostic methods may be to identify individuals who would be prevented from developing *P. carinii* by receiving prophylaxis; the clinical study to validate this hypothesis, however, has not been performed.

STAINING CHARACTERISTICS OF *P. CARINII*

A variety of stains have been used for the identification of *P. carinii*. The *Pneumocystis* cyst wall contains polysaccharides commonly present in fungal cell walls, chitin, and yeast glucan polymers (47). Stains commonly used to detect fungal walls (e.g., periodic acid-Schiff, Gomori's methenamine silver, toluidine blue O, and calcofluor) have been used for the identification of *Pneumocystis* cysts (4, 15, 27, 29, 44, 53, 73). In addition to cyst wall staining, two focal thickenings ("double commas") in the cyst wall that are apparently unique to *Pneumocystis* cysts also stain with these stains. The role of these focal thickenings is unknown.

Giemsa and Giemsa-like stains do not stain the cell wall but instead stain the nuclei of the cellular stages purple and the cytoplasms light blue; the location of the cell wall is made readily apparent by the clear negative stain (5, 7, 9, 13, 15, 49). Direct fluorescein-conjugated monoclonal anti-*Pneumocystis* antibodies used for immunofluorescent staining stain the walls of the cysts and trophic forms (15, 26, 51, 53).

The staining characteristics, advantages, and disadvantages of the various stains used for the detection of *P. carinii* are shown in Fig. 3 and discussed in Table 2.

Quantitation of the burden of *P. carinii* from any of the above-mentioned specimens is problematic because (i) disease may occur in discrete regions of the lung not readily accessible to sputum induction or lavage, (ii) different patients exert different efforts in producing an induced sputum specimen, (iii) the quantity of fluid aspirated back after bronchoalveolar lavage differs among patients, (iv) trophic forms and cysts in induced sputum specimens may be difficult to detect and enumerate depending on the degree of mucolysis and stain used, and (v) clumps of *Pneumocystis* cells (either in bronchoalveolar lavage fluid or mucolysed induced sputum) from different patients can be widely variable in size. Attempts to quantitate organism load in response to appropriate anti-*Pneumocystis* therapy have thus been problematic (1).

The microbiology laboratory at San Francisco General Hospital receives approximately 400 induced sputum specimens and 120 bronchoalveolar lavage fluid specimens per year for evaluation for the presence of *P. carinii* or other pulmonary pathogens. The rapid Giemsa-like stain is used for diagnosis because of the low cost, ease and rapidity of staining, and capacity to show the trophic forms and nucleated structures within cysts (precysts and intracystic bodies). This stain also permits judgment of specimen quality by demonstration of host alveolar macrophages. In addition, the distinctive Giemsa-stained morphologic appearances of *Cryptococcus neoformans*, *Histoplasma capsulatum*, *Strongyloides stercoralis* larvae, and *Toxoplasma gondii* permit rapid diagnosis of pulmonary infections caused by these pathogens, which may not be detected with other stains (49, 52). Because background host cells also stain, training and expertise in interpreting cellular elements in Giemsa-stained preparations are required to screen such preparations for *Pneumocystis* cells. Laboratories that have a lower volume of *Pneumocystis* specimens may prefer to use immunofluorescent staining or one of the other stains described in Table 2.

ORGANIZATION OF THE CLINICAL MICROBIOLOGY LABORATORY FOR DIAGNOSIS OF PCP

The clinical microbiology laboratory often receives specimens for the diagnosis of a variety of opportunistic infections by either direct stain or culture; this same specimen can also be used to diagnose PCP. By use of rapid processing and staining techniques, a laboratory diagnosis of PCP can usually be made within a few hours. Statim specimen processing is not necessary, because *Pneumocystis* trophozoites and cysts remain clearly identifiable in specimens collected at least 3 to 4 days after appropriate anti-*Pneumocystis* therapy has been instituted (69). If PCP is clinically suspected, treatment should be started without awaiting laboratory confirmation.

The algorithm adopted by an individual hospital for the diagnosis of PCP depends on the (i) age of the population seen, (ii) primary disease process that predisposes to PCP, (iii) number of patients who may have PCP, (iv) number of physicians and facilities available for performing bronchoscopy, and (v) availability of inpatient or clinic facilities.

The algorithm presented earlier (Fig. 2) is in use at San Francisco General Hospital, where the majority of the patients who receive a diagnosis of PCP have AIDS. In addition to *P. carinii*, other pulmonary pathogens may be concomitantly present. For this reason, if an induced sputum sample contains *P. carinii*, it is also subjected to mycobacterial culture, since a specimen obtained from the normally sterile lower respiratory tract of the patient under consideration will not be obtained. Fungal cultures are not performed on induced sputa because *Candida* spp. and environmental fungi are found, and disseminating fungal pathogens are usually not recovered (49). Recovery of bacterial pathogens from induced sputum specimens is virtually identical to that from spontaneously produced sputum (49, 52). If an induced sputum specimen lacks *Pneumocystis* cells, the patient undergoes bronchoscopy, and bronchoalveolar lavage fluid is obtained. The fluid is examined for *P. carinii* and also subjected to culture for fungi and mycobacteria (52).

THERAPY

The development of anti-*Pneumocystis* therapy has been hindered by the lack of a clinically relevant in vitro culture system for testing susceptibility. Many of the drugs effective for treating PCP have been developed with the corticosteroid treatment rat model (3, 25, 72). The most uniform inoculum and the most reliable model for studying anti-*Pneumocystis* drug activity remain the transtracheal inoculation of *P. carinii* into rats or mice (3, 72). Recent developments in this arena include screening for antimicrobial susceptibility by monitoring radioisotope-labeled substrate utilization by rat *P. carinii* maintained in feeder cell cultures or short-term axenic cultures (1, 2, 14, 16, 72).

Intravenous trimethoprim-sulfamethoxazole (TMP-SMX) is usually the first treatment selected for patients with acute PCP, since oral TMP-SMX can be given after the patient has responded favorably. Only a few patients, however, are able to complete the TMP-SMX regimen because of the high frequency of side effects. Parenteral

TABLE 2 Comparison of stains used to detect *P. carinii*[a]

Stain (reference)	Time to perform stain	Cyst wall	T and IB	Advantages	Disadvantages
Giemsa (7)	30–60 min	Unstained. Appears as clear (unstained) ring around IB.	Nuclei stain red-purple. Cytoplasm stains light blue.	Inexpensive; simple to perform; stains cells in all stages of *P. carinii*; stains most other pathogens (e.g., bacteria, parasites, fungi); stains host cells.	Experienced reader required to distinguish *Pneumocystis* clumps from stained host cells in background.
Giemsa-like rapid stains (e.g., Diff-Quik) (39, 49, 51, 53)	<5 min				
Fluorescein-conjugated monoclonal antibody (direct fluorescent-antibody stain) (26, 39, 51, 53)	30 min	Stains and fluoresces apple green. Cyst contents unstained (i.e., appear black). Fold in cyst wall sometimes apparent.	Stained. Appears as small polygon outlined in fluorescent apple green. Nuclei not stained.	Easy for experienced reader to screen; immunofluorescent staining is sensitive and specific.	Requires fluorescence microscope; reagent is expensive.
Methenamine silver (Gomori-Grocott) (15, 29, 53, 68)	6–24 h (conventional), 1–2 h (rapid)	Stains brown to black. Thickenings (double comma) and fold in cyst wall stain dark brown to black.	Unstained	Easy to screen for cysts.	Prolonged staining time (conventional); moderate cost (time); strong acid used; T and IB unstained; host cells unstained.
Toluidine blue O (27, 39, 53)	1–6 h	Stains violet to purple. Thickenings (double comma) and fold in cyst wall stain darker violet to purple.	Unstained	Easy to screen for cysts.	Prolonged staining time; moderate cost (time); strong acid used; T and IB unstained; host cells unstained.
Calcofluor white (4, 73)	<5 min	Stains blue-white or green depending on filter. Cyst wall and thickenings intensely fluorescent.	Unstained	Cysts fluoresce brilliantly; simple to perform; inexpensive (time).	Requires fluorescence microscope; strong alkali used; T and IB unstained; also stains fungi; needs experienced reader to distinguish morphology of *Pneumocystis* cysts from that of other fungi.
Gram-Weigert	<5 min	Unstained	IB stain purple. T faintly visible.	Commonly available in cytopathology laboratories.	Faint staining overcome by experienced observer.
Papanicolaou (68)	1–6 h	Unstained	IB stain purple. T faintly visible.	Commonly available in cytopathology laboratories.	Faint staining overcome by experienced observer.

[a] T, trophic forms; IB, intracystic bodies.

pentamidine isethionate has been used for primary therapy or as an alternate drug if TMP-SMX is too toxic. Pentamidine may also be quite toxic and not tolerated. Aerosolized, inhaled pentamidine has fewer side effects than TMP-SMX but is less effective. Other therapeutic agents or combinations (e.g., dapsone with TMP, clindamycin with primaquine, atovaquone [BW566C80] with trimetrexate) have been used for mild to moderately severe acute PCP, although they may be less effective (67, 82, 83). Both TMP-SMX and pentamidine are used for PCP prophylaxis.

MONOGRAPHS ON *P. CARINII*

1. **Hughes, W. T.** 1987. *Pneumocystis carinii Pneumonitis*, vol. 1 and 2. CRC Press, Inc., Boca Raton, Fla.
2. **Walzer, P. D. (ed.).** 1994. *Pneumocystis carinii Pneumonia*, 2nd ed., revised and expanded. Marcel Dekker, Inc., New York.

REFERENCES

1. **Armstrong, M. Y. K., and M. T. Cushion.** 1994. In vitro cultivation, p. 3–24. *In* P. D. Walzer (ed.), *Pneumocystis carinii Pneumonia*, 2nd ed. Marcel Dekker, Inc., New York.
2. **Bartlett, M. S., R. Eichholtz, and J. W. Smith.** 1985. Antimicrobial susceptibility of *Pneumocystis carinii* in culture. *Diagn. Microbiol. Infect. Dis.* **3**:381–387.
3. **Bartlett, M. S., J. A. Fishman, M. M. Durkin, S. F. Queener, and J. W. Smith.** 1990. *Pneumocystis carinii*: improved models to study efficacy of drugs for treatment or prophylaxis of pneumocystis pneumonia in the rat. *Exp. Parasitol.* **70**:100–106.
4. **Baselski, V. S., M. K. Robinson, L. W. Pifer, and D. R. Woods.** 1990. Rapid detection of *Pneumocystis carinii* in bronchoalveolar lavage samples by using calcofluor staining. *J. Clin. Microbiol.* **28**:393–394.
5. **Bigby, T. D., D. Margolskee, J. L. Curtis, P. F. Michael, D. Sheppard, W. K. Hadley, and P. C. Hopewell.** 1986. The usefulness of induced sputum in the diagnosis of *Pneumocystis carinii* pneumonia in patients with the acquired immunodeficiency syndrome. *Am. Rev. Respir. Dis.* **133**:515–518.
6. **Birriel, J. A., J. A. Adams, M. A. Saldana, K. Mavunda, S. Goldfinger, D. Vernon, B. Holzman, and R. M. McKay.** 1991. Role of flexible bronchoscopy and bronchoalveolar lavage in the diagnosis of pediatric acquired immunodeficiency syndrome-related pulmonary diseases. *Pediatrics* **87**:897–899.
7. **Blumenfeld, W., E. Wagar, and W. K. Hadley.** 1984. Use of transbronchial biopsy for diagnosis of opportunistic pulmonary infections in acquired immunodeficiency syndrome (AIDS). *Am. J. Clin. Pathol.* **81**:1–5.
8. **Bonfils-Roberts, E. A., A. Nickodem, and T. F. Nealon.** 1990. Retrospective analysis of the efficacy of open lung biopsy in acquired immunodeficiency syndrome. *Ann. Thorac. Surg.* **49**:115–117.
9. **Broaddus, C., M. D. Dake, M. S. Stulbarg, W. Blumenfeld, W. K. Hadley, J. A. Golden, and P. C. Hopewell.** 1985. Bronchoalveolar lavage and transbronchial biopsy for the diagnosis of pulmonary infections in the acquired immunodeficiency syndrome. *Ann. Intern. Med.* **102**:747–752.
10. **Centers for Disease Control.** 1986. Update: acquired immunodeficiency syndrome—United States. *Morbid. Mortal. Weekly Rep.* **35**:757–766.
11. **Centers for Disease Control.** 1991. Guidelines for prophylaxis against *Pneumocystis carinii* pneumonia for children infected with human immunodeficiency virus. *Morbid. Mortal. Weekly Rep.* **40**(RR2):1–13.
12. **Centers for Disease Control.** 1992. Recommendations for prophylaxis against *Pneumocystis carinii* pneumonia for adults and adolescents infected with human immunodeficiency virus. *Morbid. Mortal. Weekly Rep.* **41**(RR4):1–11.
13. **Chagas, C.** 1909. Nova tripanozomiaze humana. *Mem. Inst. Oswaldo Cruz* **1**:159–218.
14. **Comley, J. C. W., R. J. Mullin, L. A. Wolfe, M. H. Hanlon, and R. Ferone.** 1991. A microculture screening assay for the primary in vitro evaluation of drugs against *Pneumocystis carinii*. *Antimicrob. Agents Chemother.* **35**:1965–1974.
15. **Cregan, P., A. Yamamoto, A. Lum, T. van der Heide, M. MacDonald, and L. Pulliam.** 1990. Comparison of four methods for rapid detection of *Pneumocystis carinii* in respiratory specimens. *J. Clin. Microbiol.* **28**:2432–2436.
16. **Cushion, M. T., and D. Ebbets.** 1991. Growth and metabolism of *Pneumocystis carinii* in axenic culture. *J. Clin. Microbiol.* **28**:1385–1394.
17. **Cushion, M. T., J. J. Ruffolo, and P. D. Walzer.** 1988. Analysis of the developmental stages of *Pneumocystis carinii* in vitro. *Lab. Invest.* **58**:324–331.
18. **Cushion, M. T., J. Zhang, M. Kaselis, D. Giuntoli, S. L. Stringer, and J. R. Stringer.** 1993. Evidence for two genetic variants of *Pneumocystis carinii* coinfecting laboratory rats. *J. Clin. Microbiol.* **31**:1217–1223.
19. **Edman, J. C., U. Edman, M. Cao, B. Lundgren, J. A. Kovacs, and D. V. Santi.** 1989. Isolation and expression of the *Pneumocystis carinii* dihydrofolate reductase gene. *Proc. Natl. Acad. Sci. USA* **86**:8625–8629.
20. **Edman, J. C., J. A. Kovacs, H. Masur, D. V. Santi, H. J. Elwood, and M. L. Sogin.** 1988. Ribosomal RNA sequences show *Pneumocystis carinii* to be a member of the fungi. *Nature* (London) **334**:519–522.
21. **Edman, J. C., and M. L. Sogin.** 1994. Molecular phylogeny of *Pneumocystis carinii*, p. 91–105. *In* P. D. Walzer (ed.), *Pneumocystis carinii Pneumonia*, 2nd ed. Marcel Dekker, Inc., New York.
22. **Edman, U., J. C. Edman, B. Lundgren, and D. V. Santi.** 1989. Isolation and expression of the *Pneumocystis carinii* thymidylate synthase gene. *Proc. Natl. Acad. Sci. USA* **86**:6503–6507.
23. **Ezekowitz, R. A. B., D. J. Williams, H. Koziel, M. Y. K. Armstrong, A. Warner, F. F. Richards, and R. M. Rose.** 1991. Uptake of *Pneumocystis carinii* mediated by the macrophage mannose receptor. *Nature* (London) **351**:155–158.
24. **Filice, G., G. Carnevale, P. Lanzarini, F. Castelli, P. Olliaro, G. Carosi, and E. G. Rondanelli.** 1985. Life cycle of *Pneumocystis carinii* in the lung of immunocompromised hosts: an ultrastructural study. *Microbiologica* **8**:319–328.
25. **Frenkel, J. K., J. T. Good, and J. A. Schultz.** 1966. Latent pneumocystis infection of rats, relapse and chemotherapy. *Lab. Invest.* **15**:1559–1577.
26. **Gill, V. J., G. Evans, F. Stock, J. E. Parrillo, H. Masur, and J. A. Kovacs.** 1987. Detection of *Pneumocystis carinii* by fluorescent-antibody stain using a combination of three monoclonal antibodies. *J. Clin. Microbiol.* **25**:1837–1840.
27. **Gosey, L. L., R. M. Howard, F. G. Witebsky, F. P. Ognibene, T. C. Wu, V. J. Gill, and J. D. MacLowry.** 1985. Advantages of modified toluidine blue O stain and bronchoalveolar lavage for the diagnosis of *Pneumocystis carinii* pneumonia. *J. Clin. Microbiol.* **22**:803–807.
28. **Gottlieb, M. S., R. Schroff, H. M. Schanker, J. D. Weisman, P. T. Fan, R. A. Wolf, and A. Saxon.** 1981. *Pneumocystis carinii* pneumonia and mucosal candidiasis in previously healthy homosexual men: evidence of a new acquired cellular immunodeficiency. *N. Engl. J. Med.* **305**:1425–1431.
29. **Grocott, R. G.** 1955. A stain for fungi in tissue sections and smears using Gomori's methenamine-silver nitrate technique. *Am. J. Clin. Pathol.* **25**:975–979.
30. **Hong, S.-T., P. E. Steele, M. T. Cushion, P. D. Walzer, S. L. Stringer, and J. R. Stringer.** 1990. *Pneumocystis carinii* karyotypes. *J. Clin. Microbiol.* **28**:1785–1795.
31. **Hoover, D. R., A. J. Saah, H. Bacellar, J. Phair, R. Detels, R. Anderson, and R. A. Kaslow for the Multicenter AIDS Cohort Study.** 1993. Clinical manifestations of AIDS in the era of pneumocystis prophylaxis. *N. Engl. J. Med.* **329**:1922–1926.
32. **Hughes, W. T.** 1987. *Pneumocystis carinii Pneumonitis*, vol. 1, p. 1–7. CRC Press, Inc., Boca Raton, Fla.
33. **Hughes, W. T.** 1987. *Pneumocystis carinii Pneumonitis*, vol. 2, p. 1–15. CRC Press, Inc., Boca Raton, Fla.

34. **Hughes, W. T.** 1987. *Pneumocystis carinii Pneumonitis*, vol. 1, p. 57–69. CRC Press, Inc., Boca Raton, Fla.

35. **Hughes, W. T.** 1987. *Pneumocystis carinii Pneumonitis*, vol. 1, p. 97–104. CRC Press, Inc., Boca Raton, Fla.

36. **Hughes, W. T.** 1994. Clinical manifestations in children, p. 319–329. *In* P. D. Walzer (ed.), *Pneumocystis carinii Pneumonia*, 2nd ed. Marcel Dekker, Inc., New York.

37. **Kaneshiro, E. S., Y.-P. Wu, and M. T. Cushion.** 1991. Assays for testing *Pneumocystis carinii* viability. *J. Protozool.* **38:**85S–87S.

38. **Karpel, J. P., D. Prezant, D. Appel, and G. Bezahler.** 1986. Endotracheal lavage for the diagnosis of *Pneumocystis carinii* pneumonia in intubated patients with acquired immune deficiency syndrome. *Crit. Care Med.* **14:**741.

39. **Kovacs, J. A., V. L. Ng, H. Masur, G. Leoung, W. K. Hadley, G. Evans, H. C. Lane, F. P. Ognibene, J. Shelhamar, J. E. Parrillo, and V. J. Gill.** 1988. Diagnosis of *Pneumocystis carinii* pneumonia: improved detection in sputum with use of monoclonal antibodies. *N. Engl. J. Med.* **318:**589–593.

40. **Lapinsky, S. E., D. Glencross, N. G. Car, J. M. Kallenbach, and S. Zwi.** 1991. Quantification and assessment of viability of *Pneumocystis carinii* organisms by flow cytometry. *J. Clin. Microbiol.* **29:**911–915.

41. **Lee, C.-H., J.-J. Lu, M. S. Bartlett, M. M. Durkin, T.-H. Liu, J. Wang, B. Jiang, and J. W. Smith.** 1993. Nucleotide sequence variation in *Pneumocystis carinii* strains that infect humans. *J. Clin. Microbiol.* **31:**754–757.

42. **Levine, S. J., H. Masur, and V. J. Gill.** 1991. Effect of aerosolized pentamidine prophylaxis on the diagnosis of *Pneumocystis carinii* pneumonia by induced sputum examination in patients infected with human immunodeficiency virus. *Am. Rev. Respir. Dis.* **144:**760–764.

43. **Lundgren, B., R. Cotton, J. D. Lundgren, J. C. Edman, and J. A. Kovacs.** 1990. Identification of *Pneumocystis carinii* chromosomes and mapping of five genes. *Infect. Immun.* **58:**1705–1710.

44. **Mahan, C. T., and G. E. Sale.** 1978. Rapid methenamine silver stain for pneumocystis and fungi. *Arch. Pathol. Lab. Med.* **102:**351–352.

45. **Masur, H., M. A. Michelis, J. B. Greene, I. Onorato, R. A. V. Stouwe, R. S. Holzman, G. Wormser, L. Brettman, M. Lange, H. W. Murray, and S. Cummingham-Rundles.** 1981. An outbreak of community-acquired *Pneumocystis carinii* pneumonia: initial manifestation of cellular immune dysfunction. *N. Engl. J. Med.* **305:**1431–1438.

46. **Masur, H., F. P. Ognibene, R. Yarchoan, J. H. Shelhamer, B. F. Baird, W. Travis, A. F. Suffredini, L. Deyton, J. A. Kovacs, J. Falloon, R. Davey, M. Polis, J. Metcalf, M. Baseler, R. Wesley, V. J. Gill, A. S. Fauci, and H. C. Lane.** 1989. CD4 counts as predictors of opportunistic pneumonias in human immunodeficiency virus (HIV) infection. *Ann. Intern. Med.* **111:**223–231.

47. **Matsumoto, Y., S. Matsuda, and T. Tegoshi.** 1989. Yeast glucan in the cyst wall of *Pneumocystis carinii*. *J. Protozool.* **36:**21S–22S.

48. **Matsumoto, Y., and Y. Yoshida.** 1984. Sporogony in *Pneumocystis carinii*: synaptonemal complexes and meiotic nuclear divisions observed in precysts. *J. Protozool.* **31:**420–428.

49. **Ng, V. L., I. Gartner, L. A. Weymouth, C. D. Goodman, P. C. Hopewell, and W. K. Hadley.** 1989. The use of mucolysed induced sputum for the identification of pulmonary pathogens associated with human immunodeficiency virus infection. *Arch. Pathol. Lab. Med.* **113:**488–493.

50. **Ng, V. L., S. M. Geaghan, G. Leoung, S. Shiboski, J. Fahy, L. Schnapp, D. M. Yajko, P. C. Hopewell, and W. K. Hadley.** 1993. Lack of effect of prophylactic aerosolized pentamidine on the detection of *Pneumocystis carinii* in induced sputum or bronchoalveolar lavage specimens. *Arch. Pathol. Lab. Med.* **117:**493–496.

51. **Ng, V. L., N. A. Virani, R. E. Chaisson, D. M. Yajko, H. T. Sphar, K. Cabrian, N. Rollins, P. Charache, M. Krieger, W. K. Hadley, and P. C. Hopewell.** 1990. Rapid detection of *Pneumocystis carinii* using a direct fluorescent monoclonal antibody stain. *J. Clin. Microbiol.* **28:**2228–2233.

52. **Ng, V. L., D. M. Yajko, and W. K. Hadley.** 1993. Update on laboratory tests for the diagnosis of pulmonary disease in HIV-1-infected individuals. *Semin. Respir. Infect.* **8:**86–95.

53. **Ng, V. L., D. M. Yajko, L. W. McPhaul, I. Gartner, B. Byford, C. D. Goodman, P. S. Nassos, C. A. Sanders, E. L. Howes, G. Leoung, P. C. Hopewell, and W. K. Hadley.** 1990. Evaluation of an indirect fluorescent antibody stain for detection of *Pneumocystis carinii* in respiratory specimens. *J. Clin. Microbiol.* **28:**975–979.

54. **Northfelt, D. W., M. J. Clement, and S. Safrin.** 1990. Extrapulmonary pneumocystosis: clinical features in human immunodeficiency virus infection. *Medicine* **69:**392–398.

55. **Ognibene, F. P., J. Shelhamer, V. Gill, A. M. Macher, D. Loew, M. M. Parker, E. Gelman, A. S. Fauci, J. E. Parrillo, and H. Masur.** 1984. The diagnosis of *Pneumocystis carinii* pneumonia in patients with the acquired immunodeficiency syndrome using segmental bronchoalveolar lavage. *Am. Rev. Respir. Dis.* **129:**929–932.

56. **Peglow, S. L., A. G. Smulian, M. J. Linke, C. L. Pogue, S. Nurre, J. Crisler, J. Phair, J. W. M. Gold, D. Armstrong, and P. D. Walzer.** 1990. Serologic responses to specific *Pneumocystis carinii* antigens in health and disease. *J. Infect. Dis.* **161:**296–306.

57. **Pifer, L. L., W. T. Hughes, S. Stago, and D. Woods.** 1978. *Pneumocystis carinii* infection: evidence for high prevalence in normal and immunosuppressed children. *Pediatrics* **61:**35–41.

58. **Pitchenik, A. E., P. Ganjei, A. Torres, D. A. Evans, E. Rubin, and H. Baier.** 1986. Sputum examination for the diagnosis of *Pneumocystis carinii* pneumonia in the acquired immunodeficiency syndrome. *Am. Rev. Respir. Dis.* **133:**226–229.

59. **Pixley, F. J., A. E. Wakefield, S. Baerji, and J. M. Hopkin.** 1991. Mitochondrial gene sequences show fungal homology for *Pneumocystis carinii*. *Mol. Microbiol.* **5:**1347–1351.

60. **Pottratz, S. T., J. Paulsrud, J. S. Smith, and W. J. Martin.** 1991. *Pneumocystis carinii* attachment to cultured lung cells by pneumocystis gp 120, a fibronectin binding protein. *J. Clin. Invest.* **88:**403–407.

61. **Ragan, M. A.** 1989. Biochemical pathways and the phylogeny of the eukaryotes, p. 145–160. *In* B. Fernholm, K. Bremer, and H. Jörnvall (ed.), *The Hierarchy of Life*. Elsevier Science Publishers, Amsterdam.

62. **Raviglione, M. C.** 1990. Extrapulmonary pneumocystosis: the first 50 cases. *Rev. Infect. Dis.* **12:**1127–1138.

63. **Richardson, J. D., S. F. Queener, M. S. Bartlett, and J. W. Smith.** 1989. Binary fission of *Pneumocystis carinii* trophozoites grown in vitro. *J. Protozool.* **36:**27S–29S.

64. **Ruffolo, J. J.** 1994. *Pneumocystis carinii* cell structure, p. 25–43. *In* P. D. Walzer (ed.), *Pneumocystis carinii Pneumonia*, 2nd ed. Marcel Dekker, Inc., New York.

65. **Ruffolo, J. J., M. T. Cushion, and P. D. Walzer.** 1986. Techniques for examining *Pneumocystis carinii* in fresh specimens. *J. Clin. Microbiol.* **23:**17–21.

66. **Ruffolo, J. J., M. T. Cushion, and P. D. Walzer.** 1989. Ultrastructural observations on life cycle stages of *Pneumocystis carinii*. *J. Protozool.* **36:**53S–54S.

67. **Safrin, S.** 1994. New developments in the management of *Pneumocystis carinii* disease, p. 95–112. *In* P. Volberding and M. A. Jacobson (ed.), *AIDS Clinical Review 1993/1994*. Marcel Dekker, Inc., New York.

68. **Schumann, G. B., and J. J. Swensen.** 1991. Comparison of Papanicolaou's stain with Gomori methenamine silver (GMS) stain for cytodiagnosis of *Pneumocystis carinii* in bronchoalveolar lavage (BAL) fluid. *Am. J. Clin. Pathol.* **95:**583–586.

69. **Shelhamer, J. H., F. P. Ognibene, A. M. Macher, C. Tuazon, R. Steiss, D. Longo, J. A. Kovacs, M. M. Parker, C. Natanson, H. C. Lane, A. S. Fauci, J. E. Parrillo, and H. Masur.** 1984. Persistence of *Pneumocystis carinii* in lung tissue of acquired immunodeficiency syndrome patients treated for pneumocystis pneumonia. *Am. Rev. Respir. Dis.* **130:**1161–1165.

70. Sinclair, K., A. E. Wakefield, S. Benerji, and J. M. Hopkin. 1991. *Pneumocystis carinii* organisms derived from rat and human hosts are genetically distinct. *Mol. Biochem. Parasitol.* **45:**183–184.

71. Sloand, E. 1993. The challenge of *Pneumocystis carinii* culture. *J. Eukaryot. Microbiol.* **40:**188–195.

72. Smith, J. W., M. S. Bartlett, and S. F. Queener. 1994. Development of models and their use to discover new drugs for therapy and prophylaxis of *Pneumocystis carinii* pneumonia, p. 487–509. *In* P. D. Walzer (ed.), *Pneumocystis carinii Pneumonia*, 2nd ed. Marcel Dekker, Inc., New York.

73. Stratton, N., J. Hryniewicki, S. L. Aarnaes, G. Tan, L. de le Maza, and E. M. Peterson. 1991. Comparison of monoclonal antibody and calcofluor white stains for detection of *Pneumocystis carinii* from respiratory specimens. *J. Clin. Microbiol.* **29:**645–647.

74. Stringer, J. R., S. L. Stringer, J. Zhang, R. Baughman, A. G. Smulian, and M. L. Cushion. 1993. Molecular genetic distinction of *Pneumocystis carinii* from rats and humans. *J. Eukaryot. Microbiol.* **40:**733–741.

75. Stringer, S. L., K. Hudson, M. A. Blase, P. D. Walzer, M. T. Cushion, and J. R. Stringer. 1989. Sequence from ribosomal RNA of *Pneumocystis carinii* compared to those of four fungi suggests an ascomycetous affinity. *J. Protozool.* **36:**14S–16S.

76. Swofford, D. L., and G. J. Olsen. 1991. Phylogeny reconstruction, p. 411–501. *In* D. M. Hillis and C. Moritz (ed.), *Molecular Systematics.* Sinauer Associates, Sunderland, Mass.

77. Telzak, E. E., R. J. Cote, J. W. M. Gold, S. W. Campbell, and D. Armstrong. 1990. Extrapulmonary *Pneumocystis carinii* infections. *Rev. Infect. Dis.* **12:**380–386.

78. ul-Haque, A., S. B. Plattner, R. T. Cook, and M. N. Hart. 1987. *Pneumocystis carinii* taxonomy as viewed by electron microscopy. *Am. J. Clin. Pathol.* **87:**504–510.

79. Vavra, J., and K. Kucera. 1970. *Pneumocystis carinii* Delanoe: its ultrastructure and ultrastructural affinities. *J. Protozool.* **17:**463–483.

80. Wakefield, A. E., L. Guiver, R. F. Miller, and J. M. Hopkin. 1991. DNA amplification on induced sputum samples for diagnosis of *Pneumocystis carinii* pneumonia. *Lancet* **337:**1378–1379.

81. Wakefield, A. E., and J. M. Hopkin. 1994. A molecular approach to the diagnosis of *Pneumocystis carinii* pneumonia, p. 403–414. *In* P. D. Walzer (ed.), *Pneumocystis carinii Pneumonia*, 2nd ed. Marcel Dekker, Inc., New York.

82. Walker, R. S., and H. Masur. 1994. Current regimens of therapy and prophylaxis, p. 439–466. *In* P. D. Walzer (ed.), *Pneumocystis carinii Pneumonia*, 2nd ed. Marcel Dekker, Inc., New York.

83. Walzer, P. D. 1994. Development of new anti-*Pneumocystis carinii* drugs, p. 511–543. *In* P. D. Walzer (ed.), *Pneumocystis carinii Pneumonia*, 2nd ed. Marcel Dekker, Inc., New York.

84. Walzer, P. D. 1994. Pathogenic mechanisms, p. 251–265. *In* P. D. Walzer (ed.), *Pneumocystis carinii Pneumonia*, 2nd ed. Marcel Dekker, Inc., New York.

85. Walzer, P. D., D. P. Perl, D. J. Krogstead, P. G. Rawson, and M. G. Schultz. 1974. *Pneumocystis carinii* pneumonia in the United States: epidemiologic, diagnostic and clinical features. *Ann. Intern. Med.* **80:**83–93.

86. Watanabe, J., H. Hori, K. Tanabe, and Y. Nakamura. 1989. Phylogenetic association of *Pneumocystis carinii* with the "Rhizopoda/Myxomycota/Zygomycota group" indicated by comparison of 5S ribosomal RNA sequences. *Mol. Biochem. Parasitol.* **32:**163–167.

87. Yoneda, K., and P. D. Walzer. 1983. Attachment of *Pneumocystis carinii* to type I alveolar cells studied by freeze-fracture electron microscopy. *Infect. Immun.* **40:**812–815.

88. Yoshida, Y. 1989. Ultrastructural studies of *Pneumocystis carinii*. *J. Protozool.* **36:**53–60.

89. Young, L. S. 1984. Clinical aspects of pneumocystosis in man: epidemiology, clinical manifestations, diagnostic approaches, and sequelae, p. 139–174. *In* L. S. Young (ed.), *Pneumocystis carinii Pneumonia.* Marcel Dekker, Inc., New York.

Histoplasma, Blastomyces, Coccidioides, and Other Dimorphic Fungi Causing Systemic Mycoses

THOMAS J. WALSH, THOMAS G. MITCHELL, AND DAVISE H. LARONE

63

Fungal dimorphism may be defined as growth of a mould form in the natural environment or in the laboratory at 25 to 30°C and as growth in the yeast or spherule form in tissues or when incubated on enriched media at 37°C (55, 62, 68). The dimorphic fungi that cause endemic systemic mycoses are *Histoplasma capsulatum* var. *capsulatum, H. capsulatum* var. *duboisii, Blastomyces dermatitis, Coccidioides immitis,* and *Paracoccidioides brasiliensis.* These fungi have also been termed primary or systemic fungal pathogens. In the environment, they all produce a filamentous or mycelial form with hyaline, branching, septate hyphae. The conidia or hyphae of *B. dermatitidis, P. brasiliensis,* and both varieties of *H. capsulatum* convert to budding yeast cells in tissue or on enriched media at 37°C in the laboratory. *C. immitis* produces spherules in tissue or in vitro under the appropriate conditions. Temperature is a critical variable in inducing dimorphism in these fungi. The mycelial form at 25 to 30°C undergoes morphologic conversion to the yeast or spherule form at 37°C. *B. dermatitidis* and *P. brasiliensis* may convert to their yeast forms in response to temperature alone (42). *H. capsulatum* requires elevated temperatures and the presence of reducing conditions, such as free thiol groups from cysteine (38). The conversion of *C. immitis* from arthroconidia to spherules requires changes in temperature, CO_2 content, and nutrients (8). The mechanisms that regulate dimorphism are the subjects of active investigations (7).

Although *Sporothrix schenckii* is classically regarded as a dimorphic fungus, this fungus will be discussed elsewhere (see chapter 67 of this Manual). *Penicillium marneffei,* which also undergoes a hypha-to-yeastlike transformation from environment to host tissue, is discussed in chapter 64. The term dimorphism may be more broadly defined to encompass morphologic transformations of other fungi, such as *Candida albicans, Wangiella dermatitidis,* and *Phialophora verrucosa* (as well as other fungi causing chromoblastomycosis) (68).

This chapter provides a guide to the laboratory identification of the dimorphic fungi that cause systemic mycoses (Table 1). Discussions of taxonomy, mycology, ecology, and clinical manifestations are provided to facilitate understanding of the dimorphic fungi and their diseases as they pertain to the clinical microbiology laboratory.

TAXONOMY AND MYCOLOGY

Dimorphic fungi causing systemic mycoses are assigned to the form class *Deuteromycetes.* The names *H. capsulatum* var. *capsulatum, H. capsulatum* var. *duboisii, B. dermatitidis, C. immitis,* and *P. brasiliensis* designate the anamorphs (asexual stages) of these fungi. The teleomorphs (sexual stages) of *H. capsulatum* var. *capsulatum, H. capsulatum* var. *duboisii,* and *B. dermatitidis* are classified in the subdivision *Ascomycotina* and family *Gymnoascaceae* (34, 35, 37, 39, 40). The teleomorphs of *C. immitis* and *P. brasiliensis* are not known.

On routine fungal culture at 20 to 30°C in the laboratory, the five species usually grow as moulds and produce colonies that may be indistinguishable from each other as well as from the colonies of many saprophytic species. Colonies usually develop within days to a week, although some specimens do not become positive for many weeks. After colony formation, conidia usually appear in 1 to 2 weeks. The geographic distribution, morphologic features, and teleomorphs are summarized in Table 2.

H. capsulatum Darling 1906

The name *H. capsulatum* was suggested by the histological appearance of the yeast cells within the macrophages of a patient who had died of an infection thought to be due to a protozoan (9). Despite its name, which was based on early descriptions of the histopathologic lesions, *H. capsulatum* is neither protozoan nor encapsulated. The fungus was first cultured from the blood of an infected child (11). The sexual stage or teleomorph of *H. capsulatum* was demonstrated by Kwon-Chung et al. and named *Emmonsiella capsulata* (37). The new term, *Ajellomyces capsulatus,* was proposed by McGinnis and Katz on taxonomic grounds (41).

H. capsulatum grows at 25 to 30°C on Sabouraud glucose agar (SGA) as a white to buff colony. The mycelium is characterized by hyphae with slender conidiophores and spherical or pyriform tuberculate and nontuberculate macroconidia measuring 8 to 16 μm in diameter. Finely roughened macroconidia 2 to 5 μm in diameter may be abundant in fresh clinical isolates.

When the mould phase of *H. capsulatum* is incubated at 37°C on enriched medium, such as brain heart infusion (BHI) agar with blood and cysteine, conidia will germinate and convert to yeast cells. After subculture, blastoconidia

TABLE 1 General epidemiologic features of the primary mycoses

Feature	Histoplasmosis	Blastomycosis	Coccidioidomycosis	Paracoccidioidomycosis
Saprobic form (<35°C) with hyaline septate hyphae	Yes	Yes	Yes	Yes
Tissue form	Yeasts	Yeasts	Spherules	Yeasts
High infection rate in areas of endemicity	Yes	?	Yes	Yes
≥90% of infections initiated via respiratory tract	Yes	Yes	Yes	Yes
≥90% of infections asymptomatic	Yes	?	Yes	Yes
≥90% of infections self-limiting	Yes	?	Yes	Yes
≥90% of infections involve immunocompetent hosts	Yes	Yes	Yes	Yes
Approx % of males among patients with disease	80–90	50–90	75–90	95

that are ovoid to spherical and measure 2 to 4 μm in diameter develop. This same yeast morphology is present in tissue. Although *H. capsulatum* now should be technically termed *Histoplasma capsulatum* var. *capsulatum*, the former term will be retained throughout this chapter for simplicity.

H. capsulatum var. *duboisii* Drouhet 1957

Originally designated *Histoplasma duboisii* by Vanbreuseghem, *H. capsulatum* var. *duboisii* was later recognized by Drouhet as a variant of *H. capsulatum* and is thus designated *H. capsulatum* var. *duboisii* (56). Mating studies have confirmed that *H. capsulatum* and *H. capsulatum* var. *duboisii*

are variants of the same species (35). The in vitro morphologic features of the mycelial and yeast phases of *H. capsulatum* var. *duboisii* are virtually identical to those of *H. capsulatum* (51). In vivo, however, *H. capsulatum* var. *duboisii* develops larger, thick-walled, oval yeast cells 10 to 15 μm in diameter. This organism will hereafter be referred to as *H. capsulatum* var. *duboisii*.

B. dermatitidis Gilchrist et Stokes 1898

B. dermatitidis was originally isolated from a cutaneous lesion (16). The teleomorph of *B. dermatitidis* was identified as *Ajellomyces dermatitidis* (39, 40). At 25 to 30°C, *B.*

TABLE 2 Characteristics of dimorphic fungi causing systemic infections

Anamorph (teleomorph)	Ecology	Mycelial form	Tissue form
H. capsulatum (*A. capsulatus*)	Alkaline soil, bird and bat guano; Ohio, Missouri, and Mississippi River valleys	Hyphae, globose microconidia 3–5 μm in diam; tuberculate and nontuberculate macroconidia (8–16 μm)	Small oval yeasts; 2–4 μm in diam
H. capsulatum var. *duboisii* (*A. capsulatus*)	Central Africa	Hyphae, microconidia, tuberculate and nontuberculate macroconidia identical to those of *H. capsulatum*	Thick-walled yeasts; narrow-based budding; 10–15 μm in diam
B. dermatitidis (*A. dermatitidis*)	Riverbanks? Ohio and Mississippi River valleys	Hyphae and oval, pyriform, to globose terminal and lateral conidia (2–10 μm)	Thick-walled yeasts; wide-based singly budding; 8–15 μm in diam
C. immitis	Soil; semiarid regions of southwestern United States; Mexico, Central and South America	Hyphae and arthroconidia (3–6 μm)	Spherules (20–60 μm in diam) containing endospores 2–4 μm in diam
P. brasiliensis	Soil? Central and South America	Hyphae; rare oval to globose terminal and lateral microconidia; intercalary chlamydospores	Thick-walled multiply budding yeasts; 15–30 μm in diam

dermatitidis grows as a mould with conidia and septate hyphae. The mould form of *B. dermatitidis* produces abundant conidia from aerial hyphae and lateral conidiophores. These conidia are spherical, ovoid, or pyriform and 2 to 10 μm in diameter, resembling the macroconidia of *H. capsulatum*. However, unlike the macroconidia of *H. capsulatum*, the conidia of *B. dermatitidis* are smooth. Thick-walled chlamydospores 7 to 18 μm in diameter also may be observed in *B. dermatitidis*. Isolates may vary in rate of growth, colony appearance, and degree and type of conidiation. Colonies usually require 2 weeks or more for full development.

On enriched media at 37°C, *B. dermatitidis* grows as a yeast form, with colonies that are folded, pasty, and moist. The yeast cells of *B. dermatitidis* in vitro or in tissue are thick walled and spherical; they produce single buds with a characteristically broad base of attachment between the bud and the parent cell.

C. immitis Rixford et Gilchrist 1896

The life cycle of *C. immitis* involves development of hyphae, arthroconidia, and spherules (7). In native soil or on routine laboratory culture, *C. immitis* produces branching, septate hyphae and arthroconidia that usually, but not invariably, develop in alternate hyphal cells. As the arthroconidia mature, the hyphal cells between them disintegrate, and the arthroconidia are released as unicellular structures. The arthroconidia are characteristically barrel shaped, measuring approximately 3 by 6 μm, and often bear remnants of cell wall material from the disrupted adjacent hyphal cells.

In tissue or on special media, the arthroconidia become spherical, enlarge, and develop into spherules that may contain endospores. Spherules range in size up to 80 μm. At maturity, the spherules rupture to release the endospores, which may themselves develop into spherules.

P. brasiliensis (Splendore) Almeida 1930

P. brasiliensis at 25 to 30°C initially produces a nonspecific mycelial colony that may later bear a variety of conidia (e.g., chlamydospores, arthroconidia, and single conidia) (53). In tissue or on enriched media at 37°C, the characteristic yeast form develops. The budding yeast cells are globose to pyriform; they may become quite large (10 to 30 μm in diameter), produce numerous small buds (2 to 10 μm), and resemble a mariner's wheel or pilot wheel.

SYSTEMIC MYCOSES DUE TO DIMORPHIC FUNGI

Systemic fungal infections have become increasingly important problems with the advent of new and expanding populations of immunocompromised patients (4, 5, 15, 28, 31, 43, 66, 71, 72, 76, 78). Patients with AIDS represent the newest and most rapidly expanding group at risk for systemic mycoses (4, 43, 72). Improved transportation and mobility of contemporary society no longer restrict the presentation of histoplasmosis, blastomycosis, coccidioidomycosis, and paracoccidioidomycosis to their respective areas of endemicity. Newer advances in antifungal therapy have expanded therapeutic options and reduced the toxicity of treatments for these mycoses (69). The successful diagnosis of a systemic mycosis requires coordination of the expertise of the clinical microbiology laboratory and the physicians caring for the patients. Clear communication of a patient's medical condition and of the laboratory data is

essential for an accurate and efficient mycologic diagnosis. A basic understanding of the pathogenesis, epidemiology, and clinical manifestations of the systemic mycoses caused by the dimorphic fungi will permit a better approach to laboratory identification and clinical diagnosis.

Pathogenesis

In most cases, human systemic mycoses due to the dimorphic fungi are initiated when aerosolized conidia are inhaled. The conidia are sufficiently small to reach the lower respiratory tract, including the alveolar air spaces. Upon entry of the conidia into the lower respiratory tract, the interaction of host defenses and various fungal factors determines the outcome of infection (28). For *H. capsulatum* and *P. brasiliensis*, the conidia are initially contained by alveolar macrophages, which are modulated by T lymphocytes, resulting in localized granulomatous inflammation (10, 46, 58, 60). By comparison, a combined acute (pyogenic) and chronic (mononuclear or macrophage) inflammatory response is often observed with *C. immitis* and *B. dermatitidis* (3, 15, 19). More than 95% of cases of histoplasmosis, coccidioidomycosis, and paracoccidioidomycosis are estimated to be self-limiting and produce minimal symptoms. In most cases, the only evidence of infection is the development of an immune response, which is manifested by conversion to a positive delayed-type skin reaction and the production of specific precipitins and complement-fixing antibodies (44, 59). The small percentage of these episodes that advances to progressive pulmonary infection or clinically overt disseminated infection is often associated with predisposing risk factors, particularly underlying defects in cell-mediated immunity.

Epidemiology and Clinical Manifestations

Histoplasmosis

H. capsulatum has been isolated from soil with high nitrogen concentrations, especially those related to droppings of starlings, chickens, and bats (17). Outbreaks of pulmonary histoplasmosis have been associated with exposure to these reservoirs (18, 56, 58, 73). *H. capsulatum* is found principally along the Ohio, Mississippi, and St. Lawrence rivers but is also found in other areas of the United States and throughout the world. Histoplasmosis has been discussed in depth in several recent reviews (10, 24, 43, 72, 73).

The clinical manifestations of histoplasmosis may be classified according to site (pulmonary, extrapulmonary, or disseminated infection), duration of infection (acute, subacute, or chronic), and pattern of infection (primary versus reactivation). Primary acute pulmonary histoplasmosis may develop in a healthy, immunocompetent host who is exposed to a heavy inoculum (18, 73). Yeast cells of *H. capsulatum* may be observed on direct examination of sputum.

Histoplasmosis may reactivate years later in isolated tissues, particularly the central nervous system, adrenal glands, and mucocutaneous surfaces (73, 74). This pattern of histoplasmosis, which often occurs in elderly and immunocompromised patients, must be differentiated from the patterns of other mycoses, tuberculosis, and neoplastic disease (25, 72). Tissue from any of these sites may be submitted for culture and histopathologic studies.

Disseminated histoplasmosis may develop in immunocompromised patients with cellular immunodeficiencies (20, 43, 72, 76). However, disseminated histoplasmosis of infancy may develop in otherwise apparently healthy infants less than 2 years of age. Specimens for culture include blood, urine, bone marrow, and sputum. Patients with AIDS and disseminated histoplasmosis may have multiple cutaneous lesions. Biopsy and culture of these cutaneous lesions may reveal *H. capsulatum*.

African Histoplasmosis

African histoplasmosis is caused by *H. capsulatum* var. *duboisii* and is limited to equatorial Africa between 20°N and 10°S. African histoplasmosis is distinguished from histoplasmosis due to *H. capsulatum* var. *capsulatum* by (i) larger, thick-walled yeast cells (10 to 15 μm in diameter) in tissue biopsy samples; (ii) diminished pulmonary involvement; (iii) greater frequency of skin and bone lesions; and (iv) pronounced giant cell formation (13, 32, 56). Isolated lesions may develop in cutaneous and subcutaneous tissues or in bone (13, 32). Another presentation is multiple, disseminated lesions that may involve the skin, subcutaneous tissues, bone, lymph nodes, or abdominal organs.

Blastomycosis

Human and canine cases of blastomycosis occur principally in the Ohio and Mississippi River valleys but are not limited to these regions (14, 45). *B. dermatitidis* has been isolated from riverbank soil and beaver dams associated with outbreaks of blastomycosis (29, 30). Although the disease is commonly known as North American blastomycosis, it has been reported from other continents, though only rarely (6).

B. dermatitidis may cause self-limited or localized pulmonary lesions. Chronically progressive blastomycosis may involve one or more organs, most commonly the lungs, followed by skin, genitourinary tract, bone, or central nervous system in immunocompromised patients (3, 45, 63).

Sputum samples, bronchial lavage fluid, or lung biopsy specimens may be submitted for microscopy and culture. Sputum cytology samples collected to identify malignant cells in patients with chronic pulmonary infiltrates may reveal unsuspected yeast cells of *B. dermatitidis*. Lung biopsy may reveal a pyogranulomatous reaction with marked fibrosis. Cutaneous lesions may be ulcerative or verrucous and resemble the lesions of a variety of chronic infections or skin cancer. Biopsy demonstrates pseudoepitheliomatous hyperplasia, acanthosis, and intraepidermal and dermal abscesses containing blastoconidia of *B. dermatitidis*. Osteomyelitis develops in up to one-third of patients with blastomycosis. The genitourinary tract, especially the prostate and epididymis, is another target of blastomycosis. Urine collected for culture after prostate massage may also reveal *B. dermatitidis*. Meningitis due to blastomycosis is uncommon and difficult to diagnose by culture of lumbar cerebrospinal fluid; recovery of *B. dermatitidis* may be improved with culture of ventricular or cisternal fluid (33).

Coccidioidomycosis

C. immitis is distributed in southwestern United States, northwestern Mexico, Argentina, and other areas of Central and South America (48, 64). Clinical manifestations of coccidioidomycosis have been classified into three general groups: (i) initial pulmonary infection, which is usually self-limiting; (ii) pulmonary complications; and (iii) extrapulmonary disease (4, 15, 31). Primary infections in healthy hosts usually resolve spontaneously without antifungal therapy. However, primary pulmonary infection, particularly in immunocompromised patients, may evolve into one of several complications: pulmonary nodules, thin-walled cavities, progressive pneumonia, pyopneumothorax, and bronchopleural fistula. Certain patient populations with defective cellular immunity are apparently more susceptible to progressive pneumonia, complicated pneumonia, and dissemination.

Dissemination to extrapulmonary sites may result in cutaneous and soft tissue infection, osteomyelitis, arthritis, and meningitis. Cerebrospinal fluid, other body fluids, and biopsy specimens of tissues infected by *C. immitis* may be submitted to the clinical microbiology laboratory for microscopic examination and culture. More detailed discussion of the clinical manifestations of coccidioidomycosis can be found in several informative reviews (4, 31, 48, 65, 77).

Paracoccidioidomycosis

P. brasiliensis is restricted to Central and South America, but within this vast area, the endemicity varies considerably (5, 52). More than 95% of patients who progress to symptomatic paracoccidioidomycosis are males, possibly because of estrogen-mediated inhibition of mycelium-to-yeast form transformation (54, 65). Paracoccidioidomycosis may be considered to have three patterns of infection: acute pneumonia, chronic pneumonia, and disseminated infection (66). These infections may be further classified as primary infection or reactivation. Fever, cough, sputum production, chest pain, dyspnea, hemoptysis, malaise, and weight loss may occur with pneumonia and disseminated infection. Extrapulmonary lesions often develop on the face and oral mucosa. Other sites include lymph nodes, spleen, liver, gastrointestinal tract, and adrenal glands. The epidemiology and clinical manifestations of paracoccidioidomysosis are discussed in greater detail elsewhere (56, 66).

COLLECTION, TRANSPORT, AND STORAGE OF SPECIMENS

Methods of collection, transport, and storage of specimens are detailed in chapter 60 of this Manual. Since the respiratory tract is the most common portal of entry of the dimorphic fungi, most specimens from patients with suspected endemic mycoses are sputum, bronchoalveolar lavage fluid, transtracheal aspirates, or lung biopsy. From extrapulmonary sites that also may be infected, tissue specimens (e.g., skin, liver, and bone) as well as body fluids (e.g., blood, urine, and synovial fluid) may be submitted.

Several aspects of collection and transport of specimens are particularly relevant to the dimorphic fungi. Sputum often is contaminated by bacteria, saprophytic yeasts endogenous to the oral cavity, and airborne conidia of saprophytic moulds. Care must be taken to transport, process, and properly culture such contaminated specimens promptly to avoid overgrowth by more rapidly growing bacteria or saprophytic fungi.

Blastomycosis and coccidioidomycosis may cause suppurative cutaneous and visceral lesions. For example, the prostate may be the site of an abscess due to *B. dermatitidis*. Specimens consisting of pus or exudate should be submitted, whenever possible, in syringes. Swabs should be avoided, but when they are used to collect specimens, they should be transported directly to the laboratory, preferably in commercial transport systems to prevent drying.

Normally sterile fluids and tissues should also be transported to the laboratory promptly, especially if their collection required an invasive procedure. Extreme care should be given to such specimens, because the repetition of an invasive procedure to replace a lost or damaged specimen subjects the patient to further discomfort and risk of complications. Tissue should be divided and submitted for histopathologic and mycologic examinations. Where available and appropriate, special tissue stains, such as the Grocott-Gomori methenamine silver (GMS), periodic acid-Schiff (PAS), and Giemsa stains, should be requested. Direct fluorescent-antibody methods may identify fungal cells that are sparsely present or atypical in morphology, but such methods are not generally available.

BIOSAFETY

The conidia of the mycelial forms of dimorphic fungi are highly infectious and easily transmissible by aerosolization. The arthroconidia of *C. immitis* have a propensity for airborne transmission. The yeast forms of dimorphic fungi in tissue are also infectious if aerosolized (e.g., during specimen processing) and inhaled or if accidentally injected by direct percutaneous inoculation. Laboratory-acquired infections due to the systemic dimorphic fungi have been well documented (71); overall, laboratory-acquired fungal infections are more common than bacterial, parasitic, and chlamydial infections and second in prevalence only to viral infections (50). Coccidioidomycosis and histoplasmosis constitute 46% of these fungal infections.

All plating of specimens and study of cultures should be performed within a biosafety cabinet (see chapters 59 and 60 of this Manual). Plates with a suspicious hyaline mould should be opened only in a biosafety cabinet and handled with extreme care. Slants are preferable to plates for isolation of dimorphic fungi. Slants are safer, require less space, and are more resistant to desiccation during protracted incubation. However, plates are appropriate for initial recovery of dimorphic fungi from lysis-centrifugation blood cultures. Shrink seals are advised for plates in order to prevent accidental opening and to retard drying. If colonies of dimorphic fungi, such as *C. immitis*, are identified on plates, the plates should be sealed with tape in the biosafety cabinet, properly decontaminated, and disposed of. Slide cultures should *not* be set up for an isolate suspected of being a systemic dimorphic fungus.

LABORATORY DIAGNOSIS

General Guidelines

The dimorphic fungi that cause systemic mycoses are identified by direct microscopic examination of specimens, isolation and characterization of the fungus in culture, DNA probing of isolates, or demonstration of specific exoantigens produced in culture (Table 3).

The systemic dimorphic fungi produce hyaline, septate hyphae at 25 to 30°C in the laboratory. In the laboratory, the conidia or hyphae of most isolates of *B. dermatitidis, P. brasiliensis*, and both varieties of *H. capsulatum* convert to budding yeast cells in tissue or on enriched media at 37°C. Mycelium-to-yeast form conversion appears to be more reliable at 37°C than at 35°C. Conversion from the mycelial to the yeast form usually requires 7 to 14 days or longer. During the early phases of conversion, a mixture of mycelial and yeast forms is found. *C. immitis* produces spherules in vitro under the appropriate conditions; however, this conversion is not a routine procedure in most clinical microbiology laboratories.

On routine culture at 25 to 30°C in the laboratory, the

TABLE 3 Summary of methods for microbiologic identification of dimorphic fungi that cause systemic mycoses

Direct microscopic examination
 Wet preparations of fresh clinical specimens[a]
  Calcofluor white or KOH
  Combination of calcofluor white and KOH

Preparations of fixed specimens
 Wright or Giemsa stain (especially of bone marrow aspirates or buffy coat smears to detect *H. capsulatum* within monocytes or macrophages)
 Cytopathology (*B. dermatitidis* or *H. capsulatum* or spherules of *C. immitis*) by Papanicolaou staining methods
 Indirect fluorescent-antibody staining
 Staining of paraffin-embedded clot section of bone marrow aspirate (by H&E, PAS, GMS, Giemsa, and Wright stains), especially for *H. capsulatum*
 Histopathologic examination of paraffin-embedded tissue specimens (by H&E, PAS, GMS, Giemsa, and Wright stains)

Culture
 Nonsterile specimens
  Primary isolation on media[b] containing antibacterial agents (e.g., chloramphenicol, gentamicin, streptomycin, or penicillin) with and without cycloheximide to inhibit saprophytic fungi
  Yeast extract agar with concentrated ammonium hydroxide (inhibits growth of *Candida* spp. but not of systemic dimorphic fungi)
  Cultures incubated at 30°C under aerobic conditions
  Tubes or plates to be held for 4–8 wk or longer[c]
 Sterile specimens[d]
  Primary isolation on media[b]
  Cultures incubated at 30°C under aerobic conditions
  Subculture suspicious mycelial colonies to promote sporulation[e]
  Tubes or plates to be held for 4–8 wk or longer[c]
 Conversion of mycelial form to yeast form

Exoantigen identification
 Detection of cell-free antigens produced by mycelium from cultures

Nucleic acid probes
 Identification of cultured isolates

---

[a] Touch preparations of freshly resected tissues may also be stained with calcofluor white.
[b] Such as blood agar, BHI agar, inhibitory mould agar, SGA, enriched broth, and BHI broth.
[c] Subculture suspicious colonies. Laboratories in regions of endemicity also may consider incubating cultures directly at 37°C.
[d] Tissues should be minced or homogenized before being plated. Mincing is preferable to homogenization.
[e] Potato-glucose agar or SGA.

five species of dimorphic fungi discussed in this chapter grow as moulds and produce colonies that may be indistinguishable from each other as well as from the colonies of many saprophytic species. These colonies do not exhibit reliably consistent characteristics with regard to texture, pigmentation, or growth rate. Colonies may develop within days to a week, although some species, such as *H. capsulatum*, may not grow until after 4 to 6 weeks of incubation or longer. Detection of circulating antibodies and fungal antigens in serum or other normally sterile body fluids of infected patients is useful in establishing a clinical diagnosis of certain endemic mycoses, particularly coccidioidomycosis and histoplasmosis. Recent studies have demonstrated that *Histoplasma* polysaccharide antigen can be detected in bronchoalveolar lavage specimens from AIDS patients (75). The reader is referred to the *Manual of Clinical Laboratory Immunology* for complete coverage of serologic tests (26).

Direct Examination

Careful direct microscopic examination of specimens may provide rapid presumptive diagnosis of a systemic mycosis, which is advantageous, since these dimorphic fungi tend to grow slowly. Fresh, wet preparations of sputum, centrifuged cerebrospinal fluid or urine, pus, skin scrapings, tissue impression smears, and similar specimens should be examined directly with calcofluor white, KOH, or both. Calcofluor white binds nonspecifically to fungal cell wall polysaccharides such that the walls fluoresce brightly under UV illumination (21). Calcofluor white is a sensitive stain that offers excellent visualization of the morphology of pathogenic fungi. Its use is described further in chapter 60 of this Manual.

Additional procedures for examining sputum for fungi are provided in chapter 60. Sputum may be concentrated and stained directly with Wright or Giemsa stain to detect *H. capsulatum* within monocytes or macrophages. Patients with pulmonary mycoses due to dimorphic fungi have chronic pulmonary infiltrates often resembling lung cancer. Papanicolaou staining methods used by cytopathologists are valuable for detection of the spherules of *C. immitis* and the yeast cells of *B. dermatitidis* and *H. capsulatum* (57). Indeed, routine Papanicolaou staining of sputum from patients with suspected endemic pulmonary mycoses may be an effective method of rapid diagnosis. Indirect fluorescent-antibody staining is another rapid, sensitive, specific method of diagnosis by direct examination.

Bone marrow aspirates, buffy coat smears, and peripheral blood smears are valuable for the early detection of *H. capsulatum* in patients with disseminated histoplasmosis (20, 43). Giemsa or Wright stain reveals the yeast cells within circulating monocytes or tissue macrophages. Staining of a paraffin-embedded clot section of bone marrow aspirate by PAS, GMS, Giemsa, and Wright stains is another method that allows detection of *H. capsulatum* in granulomas (70).

Touch preparations of bone marrow biopsy samples, lymph nodes, and other tissues are an efficient means of detecting fungi in these tissues. The same group of special stains, as well as calcofluor white, may be applied to slides to which the freshly cut surface of a tissue specimen has been pressed and which have been dried and then fixed with heat or ethanol. Histopathologic examinations of specimens are usually completed within 24 h and may provide definitive diagnostic information. Immunohistologic methods employing fluorescent antibody, immunoperoxidase, and gold-silver staining are performed in a few specialized laboratories. Their use in clinical laboratories may become more routine as sensitive and specific reagents become commercially available (25).

Culture

Several considerations pertain to the systemic dimorphic fungi. As indicated previously, nonsterile specimens such as sputum or skin are often contaminated with bacteria and

TABLE 4 Species-specific exoantigens for identification of systemic dimorphic fungal pathogens

Organism	Exoantigen(s)[a]
H. capsulatum	h, m
B. dermatitidis	A
C. immitis	HS, F, HL
P. brasiliensis	1, 2, 3

[a]Exoantigens are detected by precipitin lines of identity in immunodiffusion tests of concentrated culture supernatant fluids versus reference antigens and antisera (27).

saprophytic fungi that can overgrow the slower-growing dimorphic fungi. Therefore, primary isolation media should contain antibacterial agents (e.g., chloramphenicol, gentamicin, streptomycin, or penicillin) and cycloheximide to inhibit saprophytic fungi. Media without cycloheximide should also be included, because this compound inhibits many opportunistic pathogens, such as *Cryptococcus neoformans*, *Candida* spp., *Aspergillus* spp., and zygomycetes. Alternatively, such specimens can be plated on yeast extract agar onto which a drop of concentrated ammonium hydroxide is allowed to diffuse from the edge of the plate; this compound inhibits the growth of *Candida* species but not of the systemic dimorphic fungi (61).

Normally sterile specimens may be inoculated directly onto blood agar, BHI agar, inhibitory mould agar, SGA, and enriched broth such as BHI broth. Tissues should be minced or homogenized before being plated. All cultures for systemic dimorphic fungi should be incubated at 25 to 30°C under aerobic conditions. Since tubes or plates are to be held for 4 to 8 weeks or longer, the incubator should be well humidified to prevent desiccation of the agar. Culture tubes are preferable; however, if plates are used, 25 ml of agar should be used per plate. Lids should be fastened with shrink seals or two tabs of tape to prevent accidental opening. Laboratories in regions of endemicity may also consider incubating cultures simultaneously at 37°C.

Exoantigen Identification

The mycelial forms that are suspected of being *H. capsulatum*, *B. dermatitidis*, *C. immitis*, or *P. brasiliensis* require confirmation of identification. The exoantigen technique is a simple diagnostic method that detects the presence of cell-free antigens, known as exoantigens, that are produced by the mycelial-form cultures. A specific exoantigen detected in the aqueous extract of a mycelial culture may identify any of the systemic dimorphic fungi (Table 4). This procedure is relatively rapid and is applicable to nonsporulating cultures. Exoantigens are demonstrated by immunodiffusion of specific antigens in either a concentrated aqueous extract of the colony on solid medium or the supernatant fluid of a broth culture of the isolate. Reference antisera identify specific antigens in the isolate and the control antigen.

Methods to detect exoantigens have been developed by Kaufman and Standard (27). To test a culture on solid medium, the mature slant culture is overlaid with a solution of Merthiolate (1:5,000) at room temperature. After overnight incubation, the fluid is aspirated, and a 5-ml sample is concentrated 50-fold (Minicon B-15; Amicon Corp.) or tested unconcentrated for *C. immitis*. The concentrate is placed on a microimmunodiffusion well opposite reference antigens and antisera. Note that antiserum is placed in the

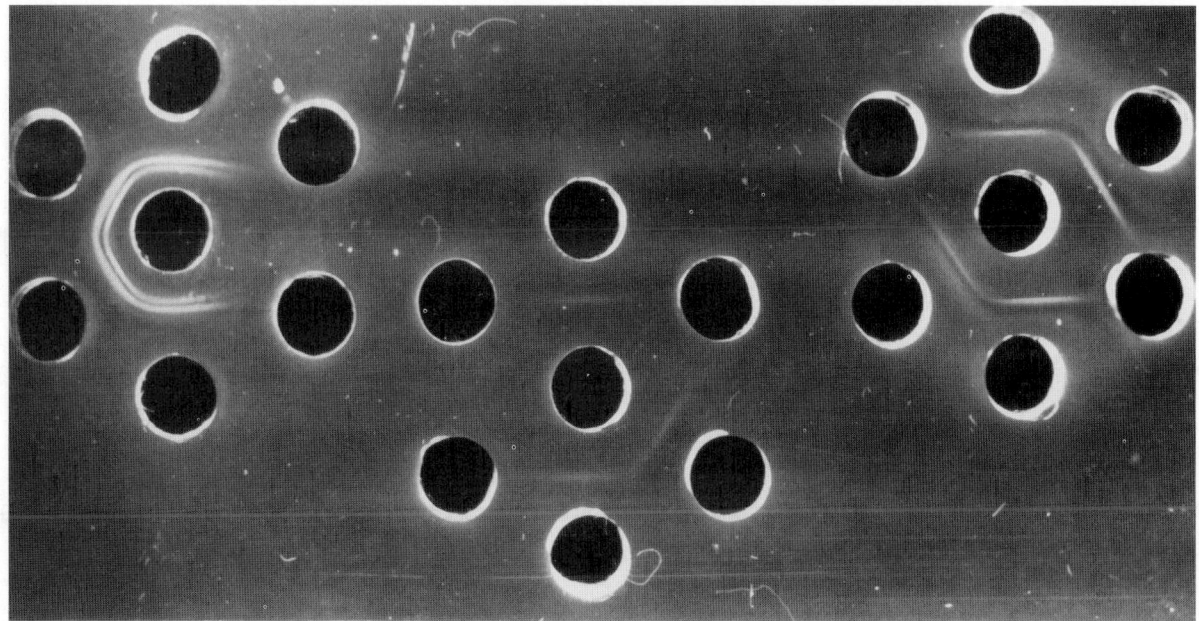

FIGURE 1   Immunodiffusion plate indicating detection of exoantigens. (Left well cluster) Center well, reference antiserum to *H. capsulatum*; top and bottom wells, reference antigen to *H. capsulatum*, indicating h (inner band) and m (outer band) precipitin lines; wells on left side, aqueous extracts of unknown isolates indicating lines of identity with bands specific for *H. capsulatum*. (Middle well cluster) Center well, reference antiserum to *B. dermatitidis*; top and bottom wells, reference antigen to *B. dermatitidis*, indicating A precipitin line; well on lower right side, aqueous extract of unknown isolate indicating line of identity with A band specific for *B. dermatitidis*. (Right well cluster) Center well, reference antiserum to *C. immitis*; top and bottom wells, reference antigen to *C. immitis*, indicating F precipitin line; wells on lower left and upper right sides, aqueous extracts of unknown isolates indicating line of identity with F band specific for *C. immitis*.

center well 1 h before the control antigens and concentrate are added to adjacent wells. The microimmunodiffusion plate is incubated for 24 h at room temperature, the template is removed, the surface of the agarose plate is washed and covered with distilled water, and the plate is read over indirect light for lines of identity (Fig. 1). Any line of identity with control antigen-antibody is significant.

Some isolates produce more readily detectable antigen when grown in liquid medium at 25°C for 3 or more days. For this procedure, one or more 30-ml BHI cultures are established in 125-ml flasks, which are incubated on a gyratory shaker at 150 rpm. After 3 days, all or a portion of the culture is removed, and 1% Merthiolate is added to give a final concentration of 1:5,000. The Merthiolate-treated sample is shaken for another day and centrifuged. Five-milliliter samples of the supernatant fluid are concentrated 10- and 25-fold as described above and tested for the presence of specific exoantigen. Note that *C. immitis* does not require concentration. If 3-day cultures are unproductive, concentrated culture supernatants from older cultures may be tested.

Nucleic Acid Probes

A recent advance in the identification of dimorphic fungi from cultures is the development of nucleic acid probes. Probes for the identification of *H. capsulatum*, *B. dermatitidis*, and *C. immitis* are commercially available in a nonisotopic kit format (AccuProbe; Gen-Probe Inc., San Diego,

Calif.). The procedure takes approximately 1 h and consists of three major steps.

1. Sample preparation. A suspension of the test organism in the mould or yeast phase is sonicated to lyse the fungal cells and release target rRNA. The suspension is then heat inactivated to ensure death of the isolate.

2. Hybridization. The organism lysate is incubated with a DNA probe labeled with acridinium ester. The probe hybridizes with the target rRNA (if present), forming a stable double-stranded hybrid. The acridinium ester is in a protected position in the double-stranded hybrid of a positive sample but remains exposed if target rRNA is not present.

3. Selection and detection. Selection reagent is added to hydrolyze acridinium ester on single, unhybridized strands of the probe. The solution tube is placed in a luminometer, into which hydrogen peroxide and sodium hydroxide are injected. In the presence of intact acridinium ester bound in the DNA probe-target rRNA hybrid, chemiluminescence occurs instantaneously and is measured by the luminometer. The amount of light generated is proportional to the amount of target nucleic acid in the test suspension. A predetermined number of relative light units must be generated for a test to be considered positive.

Studies with *H. capsulatum* have shown that isolates of any age cultured on any medium may be used successfully with the AccuProbe system (22, 23, 47). Dark-colored fungi

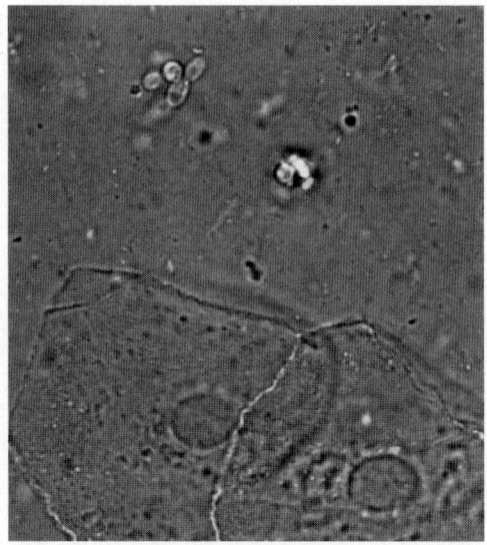

FIGURE 2 KOH wet mount of sputum showing blasto-conidia of *H. capsulatum* adjacent to epithelial cells. Original magnification, ×400.

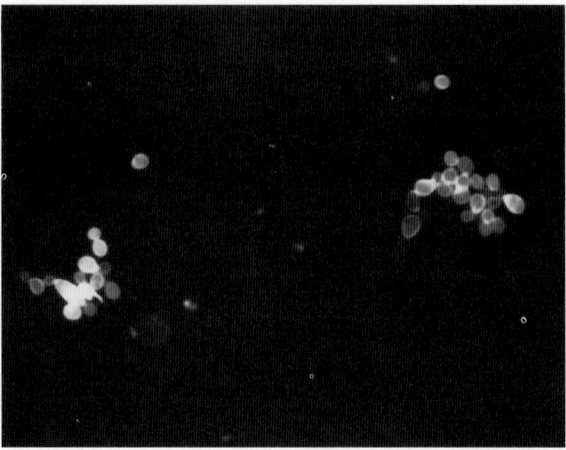

FIGURE 3 Calcofluor white wet mount of sputum showing blastoconidia of *H. capsulatum*. Original magnification, ×400.

sometimes produce false-positive results that may be due to chemiluminescence of pigments such as melanin. Since the dimorphic fungi do not produce melanin, dark colonies need not be tested, and the potential problem can be avoided.

Although the probe method is more expensive to perform than exoantigen testing, the probes offer the advantages of early testing on very young cultures, rapid processing, easy interpretation of results, and a high level of accuracy for the identification of *H. capsulatum*, *B. dermatitidis*, and *C. immitis*.

Laboratory Identification of Specific Dimorphic Fungi

H. capsulatum

Direct examination of specimens for *H. capsulatum* is best accomplished with special stains. The budding yeast cells of *H. capsulatum* (2 to 4 μm) on a calcofluor white or KOH preparation of sputum may be too small for reliable detection and may be confused with *Torulopsis glabrata*, which is similar in size and shape and often colonizes the human oropharynx (Fig. 2 and 3). The small yeast cells of *H. capsulatum* are frequently observed within the cytoplasm of macrophages. In contrast, the yeast cells of *T. glabrata* are seldom found within macrophages. Nevertheless, the only reliable methods of distinguishing between these two species are culture and indirect fluorescent-antibody staining. Giemsa and hematoxylin and eosin (H&E) stains reveal the intracellular yeast cells of *H. capsulatum* more readily, especially in sputum, blood smears, bone aspirates, and biopsy specimens. The GMS stain delineates the yeast cells but not the cellular detail of the host inflammatory cells.

Histopathologic examination of paraffin-embedded specimens by H&E and PAS stains reveals that *H. capsulatum* elicits a granulomatous inflammatory response. Different patterns of inflammation may be evident, depending on the duration and severity of infection. Large numbers of the tiny yeasts pack the cytoplasm of macrophages in acute

pulmonary or disseminated histoplasmosis (Fig. 4). The yeast cells of *H. capsulatum* must be distinguished from cells of the intracellular parasites *Leishmania donovani* and *Toxoplasma gondii*. *L. donovani* contains a kinetoplast, which is not present in the yeast cells of *H. capsulatum*. The tachyzoites of *Toxoplasma gondii* are not stained by GMS. Chronic lesions of histoplasmosis are characterized by epitheloid granulomas and fewer budding yeast cells within the macrophages. As lesions become fibrotic and calcified, the number of yeasts continues to diminish. The GMS stain is preferable for detection of small number of yeasts. Budding

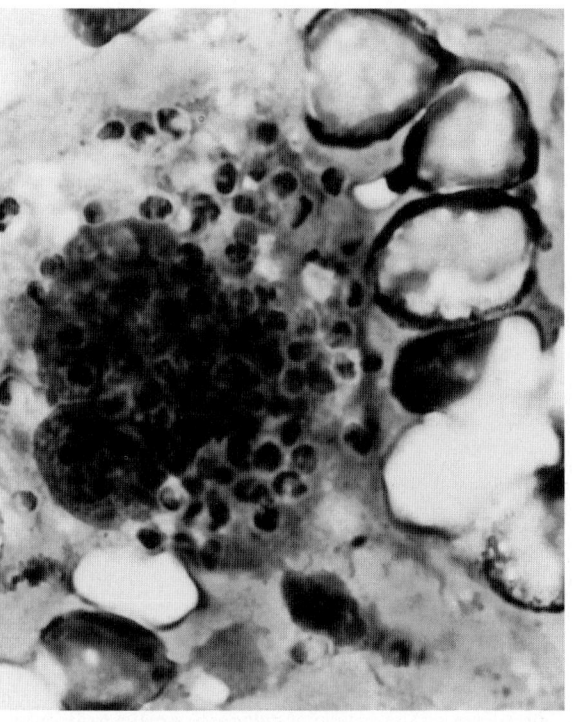

FIGURE 4 Bone marrow aspirate with macrophage containing numerous blastoconidia of *H. capsulatum*. Giemsa stain; original magnification, ×1,250.

FIGURE 5 Mycelial phase of *H. capsulatum* demonstrating tuberculate and nontuberculate macroconidia. Original magnification, ×400.

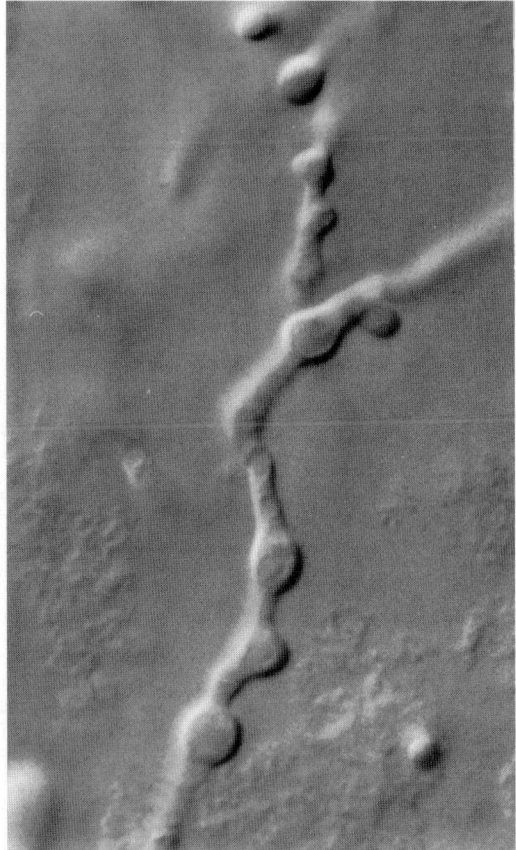

FIGURE 6 Transitional phase of *H. capsulatum* during conversion from mycelial phase to yeast phase at 35°C. Original magnification, ×400.

may not be observed in the chronic lesions of histoplasmosis.

H. capsulatum also should be distinguished histologically from the small form of *B. dermatitidis*, the endospores and young spherules of *C. immitis*, and the yeast cells of *Cryptococcus neoformans*. A yeast cell of *B. dermatitidis* has a broader base of attachment between the bud and the parent cell than does one of *H. capsulatum*. The presence of spherules of various sizes distinguishes *C. immitis* in tissue. Alcian blue or Mayer mucicarmine stain stains the polysaccharide capsule of *Cryptococcus neoformans* but does not stain *H. capsulatum*, which lacks a capsule.

Culture of specimens infected with *H. capsulatum* at 25 to 30°C reveals a fluffy, slowly growing colony with an aerial mycelium that varies in color from white to buff to brown (1). During early growth of the mycelial culture, spherical to oval to pyriform microconidia (2 to 5 μm in diameter) are present. These microconidia may be sessile on the sides of hyphae and attached to short lateral conidiophores. With continued growth, the mould develops slender conidiophores and characteristic globose and pyriform, tuberculate and nontuberculate macroconidia measuring 8 to 16 μm in diameter (Fig. 5). Since these macroconidia may resemble those of the saprophytic genus *Sepedonium*, a sus-

picious isolate must be converted to the yeast form, be shown to produce the h or m exoantigen, or give a positive reaction when tested with a specific nucleic acid probe in order to identify it as *H. capsulatum*. Furthermore, *H. capsulatum* grows on media with cycloheximide, but the monomorphic *Sepedonium* spp. are inhibited. *Chrysosporium* spp. develop conidia that resemble those of *H. capsulatum*; however, *Chrysosporium* spp. are not dimorphic, do not produce exoantigens, and do not react with *Histoplasma*-specific probes. The *Chrysosporium* state of *Renispora flavissima* may also resemble the mycelium of *H. capsulatum*.

The mycelial form of *H. capsulatum* is converted to the yeast form by incubating the mycelial culture at 37°C on an enriched medium such as BHI agar with cysteine. This conversion may be difficult and is best performed at 37°C instead of 35°C. When the mycelium is incubated at 37°C, spherical to oval budding yeast cells (2 to 5 μm in diameter) develop. Hyphal cells may form buds directly or may develop enlarged, transitional cells that subsequently begin to bud (Fig. 6). The microconidia may also convert to budding yeast cells. Complete conversion is rarely achieved, but the presence of a mixture of typical yeast cells with hyphal elements is sufficient to confirm the identification. Yeast cells of *H. capsulatum* in vitro or in vivo are small and ellipsoidal, approximately 1 to 3 by 3 to 5 μm. This conversion usually requires at least 7 to 10 days. Such yeast cells also may be isolated directly on blood agar plates or other enriched media incubated at 37°C. Buds often are formed at the smaller ends of ellipsoidal yeast cells and attached by a narrow connection.

The lysis-centrifugation technique (Isolator; Wampole Laboratories, Cranbury, N.J.) is the most effective method of recovering *H. capsulatum* and other dimorphic fungi from blood specimens (2, 49). The tube in which the blood is collected for culture contains a mixture (saponin, propylene glycol, sodium polyanetholesulfonate, and EDTA) that lyses leukocytes, prevents coagulation, and inhibits complement. After the tube is centrifuged at 3,000 × g for 30 min, the supernatant is withdrawn from the tube, and the concentrate is transferred via pipette from the bottom of the tube to culture media. A variety of media may be used, including inhibitory mould agar, SGA, BHI, or chocolate agar plates. *H. capsulatum* may be isolated as a mixture of yeast cells and hyphae directly from lysis-centrifugation blood cultures, which are routinely incubated at 35°C. The application of lysis-centrifugation and biphasic media to blood cultures, especially of blood from patients with AIDS, has increased the recovery of *H. capsulatum* from blood (49).

H. capsulatum var. *duboisii*

H. capsulatum var. *duboisii* differs reliably from *H. capsulatum* var. *capsulatum* only in its tissue form. Direct examination of purulent material or tissue biopsy specimens (usually skin, lymph nodes, or both) treated with calcofluor white or KOH reveals large, thick-walled, budding yeasts of *H. capsulatum* var. *duboisii* that measure 10 to 15 μm in diameter. The spherical to ellipsoidal yeast cells in infected tissue are found within the abundant multinucleate giant cells in a fibrogranulomatous inflammatory reaction (Fig. 7). Retraction of the cytoplasm of the phagocytes from the yeasts produces an artifactual "capsule" on H&E stain. With GMS stain, the narrow attachment between the buds and yeasts creates a figure eight or double-cell budding configuration. The yeast cells may occasionally be connected in short chains. *B. dermatitidis* may also infect skin and bone. However, *H. capsulatum* var. *duboisii* is distinguishable from *B. dermatitidis* in tissue by the presence of narrow-based budding in the former and broad-based budding in the latter. Moreover, the abundance of many intracellular thick-walled yeasts in numerous multinucleate giant cells is more typical of *H. capsulatum* var. *duboisii* than of *B. dermatitidis*.

The colonial morphology and microscopic appearance of the mould form of *H. capsulatum* var. *duboisii* grown at 25 to 30°C are the same as those of *H. capsulatum*, including the typical microconidia and macroconidia. Distinguishing between the in vitro yeast phases of *H. capsulatum* var. *duboisii* and *H. capsulatum* var. *capsulatum* may be difficult. When the mould is incubated at 37°C, the fungus converts to yeast cells that are similar in size and shape to those of *H. capsulatum* var. *capsulatum*, especially in early cultures. Some cells of *H. capsulatum* var. *duboisii* may be larger and more thick walled; however, these cells are not the same as the characteristic larger, thick-walled yeast forms present in tissues of patients with African histoplasmosis (51). Thus, histopathologic documentation of the characteristic tissue forms of *H. capsulatum* var. *duboisii* is a critical step in establishing the laboratory diagnosis of this organism.

B. dermatitidis

Direct calcofluor white or KOH mounts of sputum, exudates, and tissues can demonstrate the yeast cells of *B.*

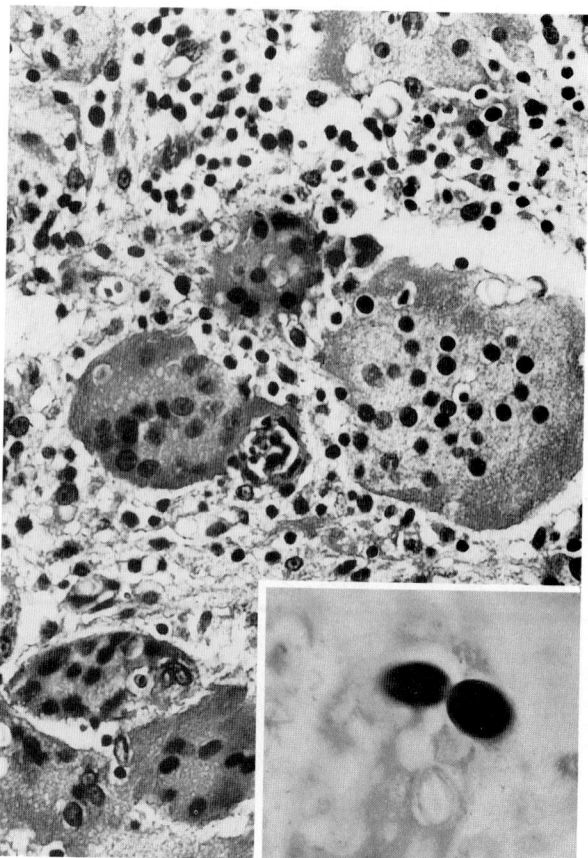

FIGURE 7 Bone biopsy specimen showing multinucleate giant cells and intracytoplasmic blastoconidia of *H. capsulatum* var. *duboisii*. Original magnification, ×250. (Insert) Bone biopsy specimen revealing budding yeast form of *H. capsulatum* var. *duboisii* (methenamine silver stain). Original magnification, ×1,250.

dermatitidis, which are large, spherical, and thick walled and measure approximately 8 to 15 μm in diameter (Fig. 8 and 9). The yeast cells bud singly and have wide bases of attachment between the buds and the parent yeast cells. This characteristic can be used to distinguish the yeast cells

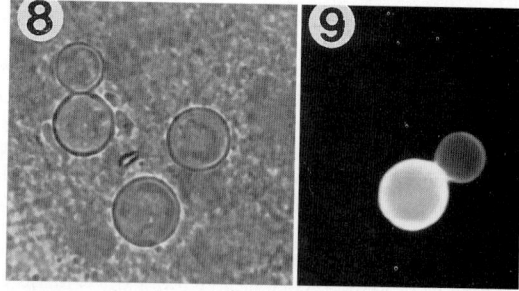

FIGURE 8 KOH wet mount of sputum showing budding yeast cell of *B. dermatitidis*. Original magnification, ×400.

FIGURE 9 Calcofluor white wet mount of sputum showing budding yeast cell of *B. dermatitidis*. Original magnification, ×400.

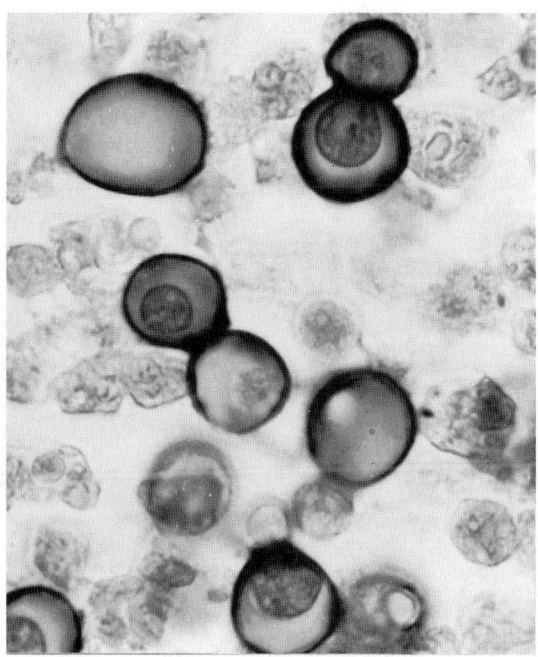

FIGURE 10 Lung biopsy specimen demonstrating blasto-conidia of *B. dermatitidis* (GMS stain). Original magnification, ×1,250.

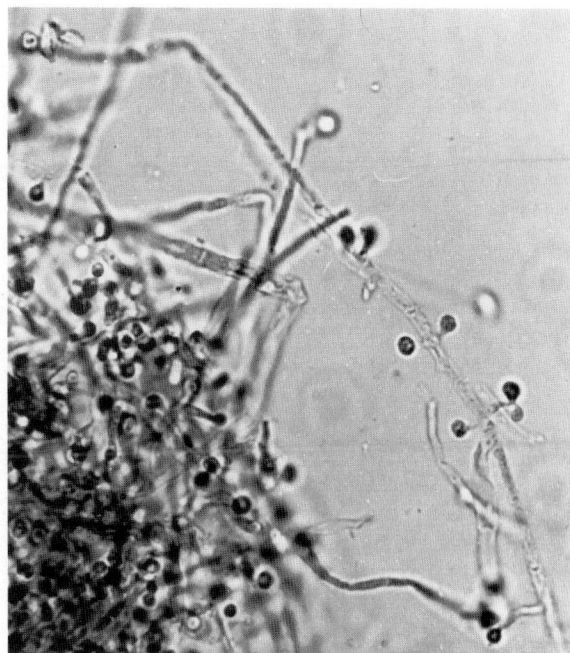

FIGURE 11 Mycelial phase of *B. dermatitidis*. Original magnification, ×400.

of *B. dermatitidis* from those of *H. capsulatum* var. *duboisii*, which have a narrow base of attachment and are usually found in multinucleate giant cells. The bud of *B. dermatitidis* often attains the same size as the parent yeast cell before becoming detached. Infected tissues stained with GMS reveal these characteristic yeast forms (Fig. 10). These yeast cells also have been recognized on cytologic specimens of sputum (often submitted to rule out primary lung cancer) treated with Papanicolaou stain. Communication with the patient's physician concerning travel history greatly facilitates the distinction between *B. dermatitidis* and *H. capsulatum* var. *duboisii*.

When specimens are cultured at 25 to 30°C on routine mycologic media, *B. dermatitidis* initially produces a fluffy white colony. Some strains develop tan, glabrous colonies without conidia, and others may produce light brown colonies with concentric rings. The mould form of *B. dermatitidis* produces conidia 2 to 10 μm in diameter that are located on long or short terminal or lateral hyphal branches (Fig. 11). These conidia are typically spherical, ovoid, or pyriform. Thick-walled chlamydospores 7 to 18 μm in diameter may also be observed in older cultures. The colony and conidia resemble those of *Chrysosporium* spp. and may not be distinguishable from an early culture of *H. capsulatum* having only hyphae, conidiophores, and microconidia. The identification is confirmed by conversion to the yeast form by growth at 37°C, a positive test with a nucleic acid probe, or detection of exoantigen A. However, confirmation of culture identification is probably not necessary if a tissue diagnosis of blastomycosis is also established.

At 37°C, the yeast form grows as a white to light brown, wrinkled colony. In vitro or in tissue, the yeast cells of *B. dermatitidis* are thick walled and spherical; they produce single buds with characteristically wide bases of attachment between the buds and the parent cells. The microscopic

morphology of *B. dermatitidis* isolates may vary. A small form produces yeast cells in tissue that resemble *H. capsulatum* (56). Although most of these small forms of *B. dermatitidis* possess the characteristic features of broadly attached buds, cells with narrow attachments may be observed. Rarely, hyphae are also seen in tissue along with yeast cells of *B. dermatitidis*.

C. immitis

Because of the risks to laboratory personnel who work with the mould form of *C. immitis*, direct examinations of sputum, exudates, and tissue are highly recommended. Mature spherules are thick walled, usually 20 to 60 μm in diameter, and easily recognized on wet mounts with KOH or calcofluor white (Fig. 12 to 15). Larger spherules may measure up to 80 μm in diameter. Endospores (2 to 4 μm) can be observed in intact or recently disrupted spherules. During maturation, spherules undergo progressive endosporulation. Immature or smaller spherules (10 to 20 μm in diameter) lacking endospores may resemble phagocytic cells, artifacts, or other fungi. Mature spherules rupture to release the endospores. Hyphae may develop in chronic cavitary and granulomatous lesions of pulmonary coccidioidomycosis or in a pleural space having a low CO_2 content (12).

The histopathology of coccidioidomycosis presents a variable inflammatory response ranging from an acute pyogenic reaction to a chronic granulomatous reaction. This variability may be due to an acute inflammatory reaction to endospores after rupture of spherules. A granulomatous response is observed in association with intact spherules. Foci of calcification and fibrosis may be present in resolving granulomatous lesions. Spherules are sparse in tissues from patients with resolving infection, but they are numerous during progressive disease. Spherules of *C. immitis* are easily identified in tissue by routine H&E, GMS, and PAS stains, particularly the last (Fig. 16). Endospores within the spher-

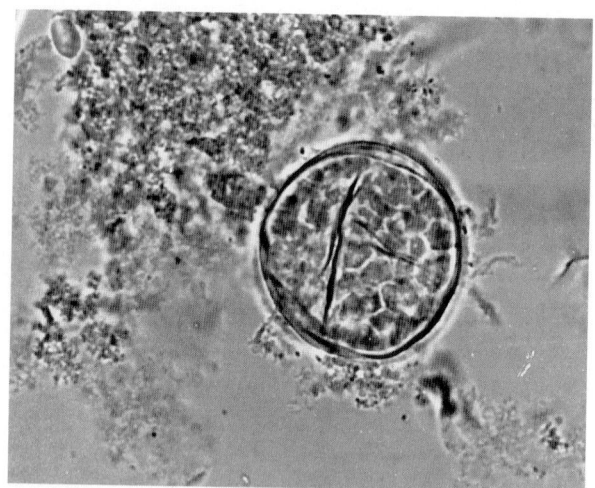

FIGURE 12 KOH wet mount of sputum showing spherule of *C. immitis*. Original magnification, ×400.

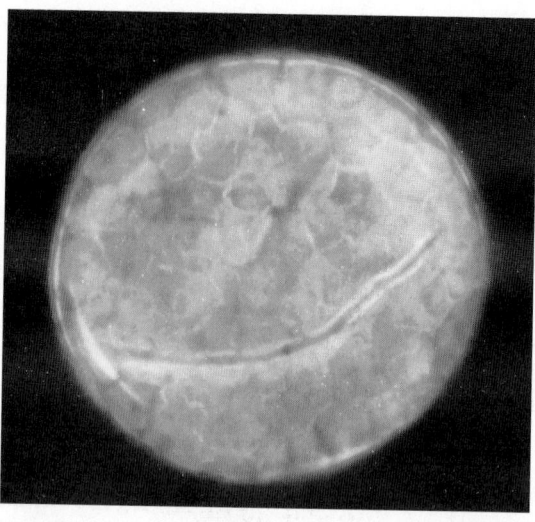

FIGURE 14 Calcofluor white wet mount of sputum showing spherule of *C. immitis*. Original magnification, ×400.

ules of infected tissue may be observed histologically by H&E and PAS stains.

Endospores and small spherules may be confused with atypical forms of *B. dermatitidis* and nonbudding yeasts of *H. capsulatum*, *P. brasiliensis*, *T. glabrata*, and *Cryptococcus neoformans*. *Prototheca wickerhamii* may resemble small spherules, and *Rhinosporidium seeberi* may simulate larger spherules of *C. immitis*. Some pollen grains found in sputum also look like spherules.

When spherules of *C. immitis* cannot be identified on wet mounts, specimens should be cultured on slants instead of plates. *C. immitis* grows readily on conventional media at 25 to 30°C, usually within 1 week, as a floccose colony composed of hyaline, septate hyphae with arthroconidia (Fig. 17). The pigmentation and texture of colonies are highly variable. Colors range from buff to yellow to tan; the texture may be floccose to powdery, depending on the degree of fragmentation of the hyphae into arthroconidia. Slide cultures should *not* be set up for isolates suspected of being *C. immitis*.

The arthroconidia of *C. immitis* develop initially in the lateral hyphal branches and are thick-walled, barrel-shaped cells 2 to 4 by 3 to 6 μm that alternate with empty,

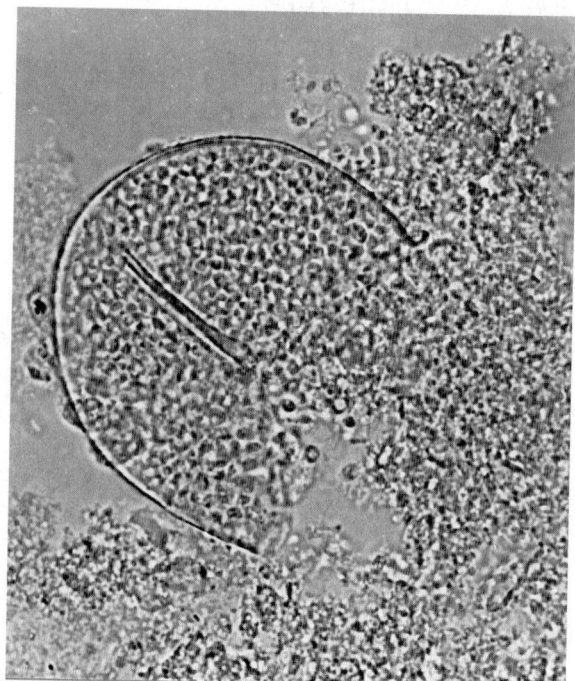

FIGURE 13 KOH wet mount of sputum showing disrupted spherule and endospores of *C. immitis*. Original magnification, ×400.

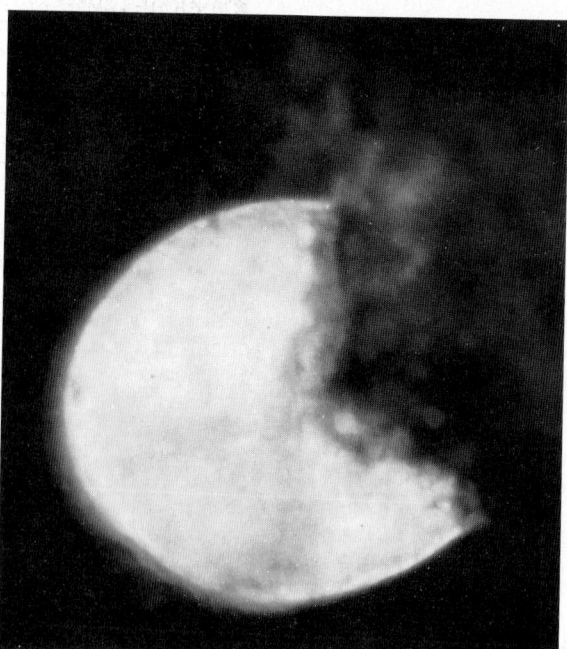

FIGURE 15 Calcofluor white wet mount of sputum showing disrupted spherule and faintly visible endospores of *C. immitis*. Original magnification, ×400.

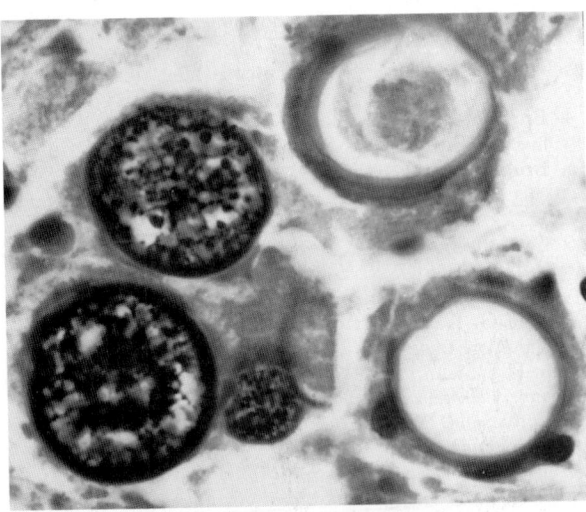

FIGURE 16 Lung biopsy specimen demonstrating empty spherules and spherules filled with endospores of *C. immitis* (PAS stain). Original magnification, ×763.

thin-walled disjunctor cells. As the walls of these disjunctor cells deteriorate, the arthroconidia become detached and dispersed. The mycelium of *C. immitis* must be distinguished from those of other genera that produce hyphae and arthroconidia, including those of genera *Malbranchea*, *Uncinocarpus*, *Arthroderma*, *Auxarthron*, *Geotrichum*, and *Oidiodendron*.

The mould form of *C. immitis* is not routinely converted to the tissue form in clinical microbiology laboratories, although a synthetic broth can be used to generate viable endospores (8). An agar medium for cultivation of spherules may also be used for rapid in vitro conversion and identification of *C. immitis* (67). Alternatively, spherules

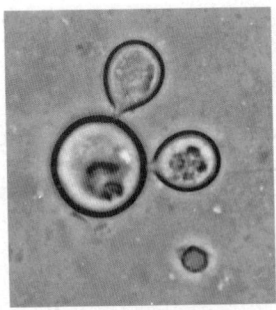

FIGURE 18 KOH wet mount of sputum showing blastoconidia of *P. brasiliensis*. Original magnification, ×400.

may be produced by intraperitoneal injection of hyphae and arthroconidia into laboratory animals in approved facilities. However, nucleic acid probes and demonstration of exoantigen HS production are the safest and fastest methods of confirmation.

P. brasiliensis

P. brasiliensis may be identified by direct examination of sputum, bronchoalveolar lavage fluid, pus from draining lymph nodes, scrapings from ulcer, or tissue biopsy samples. A wet mount of such material may reveal variations in the characteristic thick-walled pilot wheel or mariner's wheel configuration of the *P. brasiliensis* yeast form (Fig. 18 and 19). The parent yeast cell measures 15 to 30 μm in diameter, whereas the buds are 2 to 10 μm in diameter and have narrow bases of attachment. Some yeast cells may measure as much as 60 μm in diameter. The presence of multiple budding distinguishes this yeast from *Cryptococcus neoformans* and *B. dermatitidis*. Histopathologic examination using H&E, GMS, and PAS stains of tissue infected with *P. brasiliensis* reveals a pyogranulomatous process with infiltrating polymorphonuclear leukocytes, mononuclear cells, macrophages, and multinucleate giant cells. Singly budding forms that resemble *B. dermatitidis* are occasionally observed; however, the multiply budding forms seen elsewhere in the specimen distinguish *P. brasiliensis* from *B. dermatitidis*.

At 25 to 30°C, isolates of *P. brasiliensis* grow slowly and produce colonies that vary in gross morphology, ranging from glabrous brown colonies to wrinkled, floccose, beige or

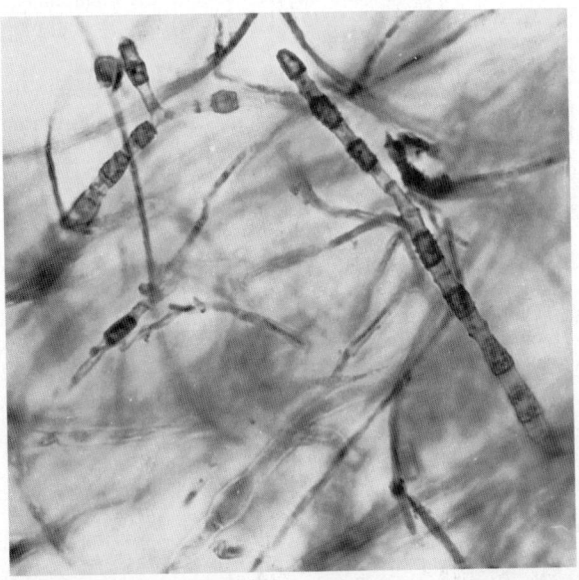

FIGURE 17 Mycelial phase of *C. immitis* demonstrating hyphae with thick-walled arthroconidia alternating with thin-walled disjunctor cells on potato glucose agar at 25°C. Original magnification, ×400.

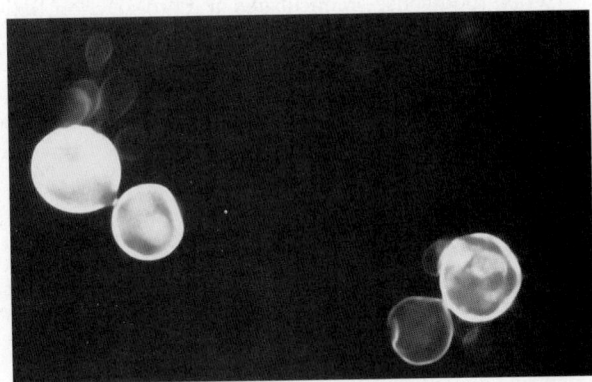

FIGURE 19 Calcofluor white wet mount of sputum showing blastoconidia of *P. brasiliensis*. Original magnification, ×400.

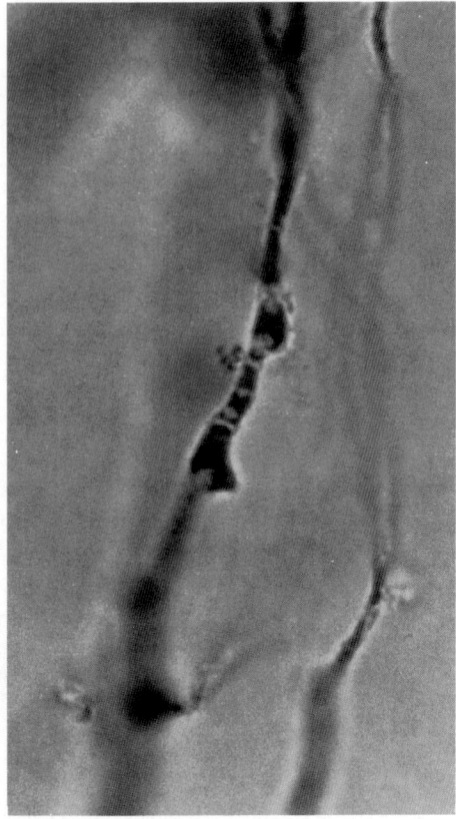

FIGURE 20 Mycelial phase of *P. brasiliensis*. Original magnification, ×400.

white colonies. Primary isolation is improved on yeast extract agar (53). The mould form of *P. brasiliensis* may require growth for several weeks before conidia develop. These conidia may be absent to infrequent. When present, the microconidia appear laterally along the hyphae and may resemble those of *B. dermatitidis*. Chlamydoconidia may predominate in the mycelium. Rectangular arthroconidia and intercalary chlamydospores may also be produced (Fig. 20). Since the microscopic features of the mycelial form of *B. brasiliensis* are not specific, conversion to the yeast form or detection of specific antigen is necessary for definitive identification.

When the hyphae are incubated at 35 to 37°C on BHI or Kelly medium, the yeast form develops slowly. The yeast colony is folded, friable, and white to gray and consists of singly and multiply budding cells. The multiply budding yeast cells demonstrate the characteristic mariner's wheel configuration, similar to that seen in tissue. The parent yeast cell measures 10 to 25 μm in diameter, and the progeny buds, which are narrowly attached, are up to 10 μm in diameter.

REFERENCES

1. **Berliner, M.** 1968. Primary subcultures of *Histoplasma capsulatum*. I. Macro- and micromorphology of the mycelial phase. *Sabouraudia* **6**:111–118.
2. **Bille, J. L., L. Stockman, G. D. Roberts, C. D. Horstmeier, and D. M. Ilstrup.** 1983. Evaluation of a lysis-centrifugation system for recovery of yeasts and filamentous fungi from blood. *J. Clin. Microbiol.* **18**:469–471.
3. **Bradsher, R. W.** 1990. Blastomycosis: fungal infections of the lung update: 1989. *Semin. Respir. Infect.* **5**:105–110.
4. **Bronnimann, D. A., and J. N. Galgiani.** 1989. Coccidioidomycosis. *Eur. J. Clin. Microbiol. Infect. Dis.* **8**:466–473.
5. **Brummer, E., E. Castaneda, and A. Restrepo.** 1993. Paracoccidioidomycosis: an update. *Clin. Microbiol. Rev.* **6**:89–117.
6. **Carman, W. F., J. A. Frean, H. H. Crewe-Brown, G. A. Culligan, and C. N. Young.** 1989. Blastomycosis in Africa. *Mycopathologia* **107**:25–32.
7. **Cole, G. T., and S. H. Sun.** 1985. Arthroconidium-spherule-endospore transformation in *Coccidioides immitis*, p. 281–333. In P. J. Szaniszlo (ed.), *Fungal Dimorphism with Emphasis on Fungi Pathogenic for Humans*. Plenum Press, New York.
8. **Converse, J. L.** 1956. Effect of physico-chemical environment on spherulation of *Coccidioides immitis* in a chemically defined medium. *J. Bacteriol.* **72**:784–792.
9. **Darling, S. T. A.** 1906. A protozoan general infection producing pseudotubercles in the lungs and focal necrosis in the liver, spleen, and lymph nodes. *JAMA* **46**:1283–1285.
10. **Davies, S. F.** 1988. Diagnosis of pulmonary fungal infections. *Semin. Respir. Infect.* **3**:162–171.
11. **de Monbreun, W. A.** 1934. The cultivation and cultural characteristics of Darling's *Histoplasma capsulatum*. *Am. J. Trop. Med. Hyg.* **14**:93–135.
12. **Dolan, M. J., C. P. Lattuda, G. P. Melcher, R. Zellmer, R. Allendoerfer, and M. G. Rinaldi.** 1992. Coccidioides immitis presenting as a mycelial pathogen with empyema and hydropneumothorax. *J. Med. Vet. Mycol.* **30**:249–255.
13. **Drouhet, E.** 1989. African histoplasmosis. *Bailliere's Clin. Trop. Med. Communicable Dis. Int. Pract. Res.* **4**:221–247.
14. **Furcolow, M. L., J. F. Busey, R. W. Menges, et al.** 1970. Prevalence and incidence studies of human and canine blastomycosis. II. Yearly incidence studies in three states, 1960–1967. *Am. J. Epidemiol.* **92**:121–131.
15. **Galgiani, J. N., and N. M. Ampel.** 1990. *Coccidioides immitis* in patients with human immunodeficiency virus infections. *Semin. Respir. Infect.* **5**:151–154.
16. **Gilchrist, T. C., and W. R. Stokes.** 1898. A case of pseudolupus vulgaris caused by a *Blastomyces*. *J. Exp. Med.* **3**:53–78.
17. **Goodman, N. L., and H. W. Larsh.** 1967. Environmental factors and growth of *Histoplasma capsulatum* in soil. *Mycopathol. Mycol. Appl.* **33**:145.
18. **Goodwin, R. A., J. E. Loyd, and R. M. Des Prez.** 1981. Histoplasmosis in normal hosts. *Medicine* (Baltimore) **60**:231.
19. **Graham, A. R., R. E. Sobonya, D. A. Bronnimann, and J. N. Galgiani.** 1988. Quantitative pathology of coccidioidomycosis in acquired immunodeficiency syndrome. *Hum. Pathol.* **19**:800–806.
20. **Graybill, J. R.** 1988. Histoplasmosis and AIDS. *J. Infect. Dis.* **158**:623–626.
21. **Hageage, G. J., and B. J. Harrington.** 1984. Use of calcofluor white in clinical mycology. *Lab. Med.* **15**:109–112.
22. **Hall, G. S., K. Pratt-Rippin, and J. A. Washington.** 1992. Evaluation of a chemiluminescent probe assay for identification of *Histoplasma capsulatum* isolates. *J. Clin. Microbiol.* **30**:3003–3004.
23. **Huffnagle, K. E., and R. M. Gander.** 1993. Evaluation of Gen-Probe's *Histoplasma capsulatum* and *Cryptococcus neoformans* AccuProbes. *J. Clin. Microbiol.* **31**:419–421.
24. **Johnson, P. C., and G. A. Sarosi.** 1989. Community-acquired fungal pneumonias. *Semin. Respir. Infect.* **4**:56–63.
25. **Kaufman, L.** 1992. Immunohistologic diagnosis of systemic mycosis: an update. *Eur. J. Epidemiol.* **8**:377–382.
26. **Kaufman, L., and E. Reiss.** 1992. Serodiagnosis of fungal diseases, p. 506–528. In N. R. Rose, E. Conway de Macario, J. L. Fahey, H. Friedman, and G. M. Penn (ed.), *Manual of Clinical Laboratory Immunology*, 4th ed. American Society for Microbiology, Washington, D.C.

27. **Kaufman, L., and P. G. Standard.** 1987. Specific and rapid identification of medically important fungi by exoantigen detection. *Annu. Rev. Microbiol.* **41:**209–225.

28. **Khardori, N.** 1989. Host-parasite interaction in fungal infections. *Eur. J. Clin. Microbiol. Infect. Dis.* **8:**331–351.

29. **Klein, B. S., J. M. Vergeront, A. F. DiSalvo, L. Kaufman, and J. P. Davis.** 1987. Two outbreaks of blastomycosis along rivers in Wisconsin. Isolation of *Blastomyces dermatitidis* from riverbank soil and evidence of its transmission along waterways. *Am. Rev. Respir. Dis.* **136:**1333–1338.

30. **Klein, B. S., J. M. Vergeront, R. J. Weeks, U. N. Kumar, G. Mathai, B. Varkey, L. Kaufman, R. W. Bradsher, J. F. Stoebig, and J. P. Davis.** 1986. Isolation of *Blastomyces dermatitidis* in soil associated with a large outbreak of blastomycosis in Wisconsin. *N. Engl. J. Med.* **314:**529–534.

31. **Knoper, S. R., and J. N. Galgiani.** 1988. Systemic fungal infections: diagnosis and treatment. I. Coccidioidomycosis. *Infect. Dis. Clin. N. Am.* **2:**861–875.

32. **Kotloff, K. L., P. A. Vial, J. W. R. Young, and A. G. Smith.** 1987. *Histoplasma duboisii* infection in a Liberian girl. *Pediatr. Infect. Dis. J.* **6:**202–205.

33. **Kravitz, G. R., S. F. Davies, M. R. Eckman, et al.** 1981. Chronic blastomycotic meningitis. *Am. J. Med.* **71:**501–505.

34. **Kwon-Chung, K. J.** 1972. Sexual stage of *Histoplasma capsulatum. Science* **175:**326–328.

35. **Kwon-Chung, K. J.** 1973. Studies on *Emmonsiella capsulata.* I. Heterothallism and development of the ascocarp. *Mycologia* **65:**109–121.

36. **Kwon-Chung, K. J.** 1975. Perfect state (*Emmonsiella capsulata*) of the fungus using large-form African histoplasmosis. *Mycologia* **67:**980.

37. **Kwon-Chung, K. J., R. J. Weeks, and H. W. Larsh.** 1974. Studies on *Emmonsiella capsulata* (*Histoplasma capsulatum*). II. Distribution of the two mating types in 13 endemic states of the United States. *Am. J. Epidemiol.* **99:**44–49.

38. **Maresca, B., and G. S. Kobayashi.** 1989. Dimorphism in *Histoplasma capsulatum*: a model for the study of cell differentiation in pathogenic fungi. *Microbiol. Rev.* **53:**186–209.

39. **McDonough, E. S., and A. L. Lewis.** 1967. *Blastomyces dermatitidis*: production of the sexual stage. *Science* **156:**528–529.

40. **McDonough, E. S., and A. L. Lewis.** 1968. The ascigerous stage of *Blastomyces dermatitidis. Mycologia* **60:**76–83.

41. **McGinnis, M. R., and B. Katz.** 1979. *Ajellomyces* and its synonym *Emmonsiella. Mycotaxon* **8:**157–164.

42. **Medoff, G., A. Painter, and G. S. Kobayashi.** 1987. Mycelial-to yeast-phase transitions of the dimorphic fungi *Blastomyces dermatitidis* and *Paracoccidioides brasiliensis. J. Bacteriol.* **169:**4055–4060.

43. **Minamoto, G., and D. Armstrong.** 1988. Fungal infections in AIDS. Histoplasmosis and coccidioidomycosis. *Infect. Dis. Clin. N. Am.* **2:**447–456.

44. **Mitchell, T. G.** 1988. Serodiagnosis of mycotic infections, p. 303–323. *In* B. B. Wentworth (ed.), *Diagnostic Procedures for Mycotic and Parasitic Infections.* American Public Health Association, Washington, D.C.

45. **Mitchell, T. G.** 1988. Blastomycosis, p. 1927–1938. *In* R. D. Feigin and J. D. Cherry (ed.), *Textbook of Pediatric Infectious Diseases.* The W. B. Saunders Co., Philadelphia.

46. **Moscardi-Bacchi, M., A. Soares, R. Mendes, S. Marques, and M. Franco.** 1989. In situ localization of T lymphocyte subsets in human paracoccidioidomycosis. *J. Med. Vet. Mycol.* **27:**149–158.

47. **Padhye, A. A., G. Smith, D. McLaughlin, P. G. Standard, and L. Kaufman.** 1992. Comparative evaluation of a chemiluminescent DNA probe and an exoantigen test for rapid identification of *Histoplasma capsulatum. J. Clin. Microbiol.* **30:**3108–3111.

48. **Pappagianis, D.** 1988. Epidemiology of coccidioidomycosis. *Curr. Top. Med. Mycol.* **2:**199–238.

49. **Paya, C. V., G. D. Roberts, and F. R. Cockerill III.** 1987. Laboratory methods for the diagnosis of disseminated histoplasmosis: clinical importance of the lysis-centrifugation blood culture technique. *Mayo Clin. Proc.* **62:**480–485.

50. **Pike, R. M.** 1979. Laboratory-associated infections: incidence, fatalities, cases, and prevention. *Annu. Rev. Microbiol.* **33:**41–66.

51. **Pine, L., E. Drouhet, and G. Reynolds.** 1964. A comparative morphological study of the yeast phases of *Histoplasma capsulatum* and *Histoplasma duboisii. Sabouraudia* **3:**211–224.

52. **Restrepo, A.** 1985. The ecology of *Paracoccidioides brasiliensis*: a puzzle still unresolved. *Sabouraudia* **23:**323–334.

53. **Restrepo, A., and I. Correa.** 1973. Comparison of two culture media for primary isolation of *Paracoccidioides brasiliensis* from sputum. *Sabouraudia* **10:**260–265.

54. **Restrepo, A., M. E. Salazar, L. E. Cano, E. P. Stover, D. Feldman, and D. A. Stevens.** 1984. Estrogens inhibit mycelium-to-yeast transformation in the fungus *Paracoccidioides brasiliensis*: implications for resistance of females to paracoccidioidomycosis. *Infect. Immun.* **46:**346–353.

55. **Rippon, J. W.** 1980. Dimorphism in pathogenic fungi. *Crit. Rev. Microbiol.* **8:**49–97.

56. **Rippon, J. W.** 1988. *Medical Mycology. The Pathogenic Fungi and the Pathogenic Actinomycetes,* 3rd ed. The W. B. Saunders Co., Philadelphia.

57. **Sanders, J. S., G. A. Sarosi, D. J. Nollett, and J. L. Thompson.** 1977. Exfoliative cytology in the rapid diagnosis of pulmonary blastomycosis. *Chest* **72:**193–196.

58. **Schwarz, J.** 1981. *Histoplasmosis.* Plenum Press, New York.

59. **Segal, G. P.** 1987. Serodiagnostic procedures in the systemic mycoses. *Semin. Respir. Med.* **9:**136–144.

60. **Singer-Vermes, L. M., E. Burger, M. F. Franco, M. M. Di-Bacchi, M. J. Mendes-Giannini, and V. L. Calich.** 1989. Evaluation of the pathogenicity and immunogenicity of seven *Paracoccidioides brasiliensis* isolates in susceptible inbred mice. *J. Med. Vet. Mycol.* **27:**71–82.

61. **Smith, C. D., and N. L. Goodman.** 1975. Improved culture method for the isolation of *Histoplasma capsulatum* and *Blastomyces dermatitidis* from contaminated specimens. *Am. J. Clin. Pathol.* **62:**276–280.

62. **Soll, D. R.** 1985. *Candida albicans,* p. 167–195. *In* P. J. Szaniszlo (ed.), *Fungal Dimorphism with Emphasis on Fungi Pathogenic for Humans.* Plenum Press, New York.

63. **Steck, W. D.** 1989. Blastomycosis. *Dermatol. Clin.* **7:**241–250.

64. **Stevens, D. A. (ed.).** 1980. *Coccidioidomycosis. A Text.* Plenum Press, New York.

65. **Stevens, D. A.** 1989. The interface of mycology and endocrinology. *J. Med. Vet. Mycol.* **27:**133–140.

66. **Sugar, A. M.** 1988. Paracoccidioidomycosis. *Infect. Dis. Clin. N. Am.* **2:**913–924.

67. **Sun, S. H., M. Huppert, and K. R. Vukovich.** 1976. Rapid in vitro conversion and identification of *Coccidioides immitis. J. Clin. Microbiol.* **3:**186–190.

68. **Szaniszlo, P. J., and J. L. Harris (ed.).** 1985. *Fungal Dimorphism—with Emphasis on Fungi Pathogenic for Humans.* Plenum Press, New York.

69. **Walsh, T. J.** 1993. Management of immunocompromised patients with evidence of an invasive mycosis. *Hematol. Clin. N. Am.* **7:**1003–1026.

70. **Walsh, T. J., R. Catchatourian, and H. Cohen.** 1983. Disseminated histoplasmosis complicating bone marrow transplantation. *Am. J. Clin. Pathol.* **79:**509–511.

71. **Walsh, T. J., and P. A. Pizzo.** 1988. Nosocomial fungal infections: a classification for hospital-acquired fungal infections and mycoses arising from endogenous flora or reactivation. *Annu. Rev. Microbiol.* **42:**517–545.

72. **Wheat, L. J.** 1988. Systemic fungal infections: diagnosis and treatment. I. Histoplasmosis. *Infect. Dis. Clin. N. Am.* **2:**841–859.

73. **Wheat, L. J.** 1989. Diagnosis and management of histoplasmosis. *Eur. J. Clin. Microbiol. Infect. Dis.* **8:**480–490.

74. **Wheat, L. J., B. E. Batteiger, and B. Sathapatayavongs.** 1990. *Histoplasma capsulatum* infections of the central nervous system. A clinical review. *Medicine* (Baltimore) **69:**244–260.

75. **Wheat, L. J., P. Connolly-Stringfield, B. Williams, K. Con-**

nolly, R. Blair, M. Bartlett, and M. Durkin. 1992. Diagnosis of histoplasmosis in patients with acquired immunodeficiency syndrome by detection of Histoplasma capsulatum polysaccharide antigen in bronchoalveolar lavage fluid. *Am. Rev. Respir. Dis.* **145:**1421–1424.

76. Wheat, L. J., T. G. Slama, J. A. Horton, et al. 1982. Risk factors for disseminated or fatal histoplasmosis. Analysis of a large urban outbreak. *Ann. Intern. Med.* **96:**159–163.

77. Williams, P. L., R. Johnson, D. Pappagianis, H. Einstein, U. Slager, F. T. Koster, J. J. Eron, and J. Morrison. 1992. Vasculitic and encephalitic complications associated with Coccidioides immitis infection of the central nervous system in humans: report of 10 cases and review. *Clin. Infect. Dis.* **14:**673–682.

78. Zeluff, B. J. 1990. Fungal pneumonia in transplant recipients. *Semin. Respir. Infect.* **5:**80–89.

Aspergillus, *Fusarium*, and Other Opportunistic Moniliaceous Fungi

MICHAEL J. KENNEDY AND LYNNE SIGLER

64

The incidence of infections caused by moniliaceous fungi that typically exist as ubiquitous saprophytic and plant parasitic moulds has risen dramatically in recent years (42). The increased use of antimicrobial agents and more aggressive and prolonged chemotherapeutic regimens for treatment of malignant diseases have contributed to this trend (7, 62, 113). Similarly, the advent of routine organ transplantation and implantation of prosthetic devices, the introduction of highly technical methods for advanced life support, and the management of newer diseases, especially AIDS, and physiologic disorders have put patients at greater risk of infection by fungi historically thought of as "laboratory contaminants" (54, 140). Moreover, as patient longevity increases with better management of AIDS or neoplasmas, there is greater risk for infection by uncommon fungal pathogens (42, 124). This chapter focuses on those fungi that grow in tissue in the form of hyaline or lightly colored septate hyphal elements and that produce a varied clinical spectrum.

CLINICAL MANIFESTATIONS AND DISEASE TERMINOLOGY

The spectrum of disease caused by hyaline moulds is diverse, is largely determined by the local and general immunologic and physiologic state of the host, and may be symptomatic or asymptomatic. In most instances, the portal of entry for fungal propagules is either through a break in the epidermis or by way of the lungs. Noted exceptions to this include introduction into the body by means of contaminated surgical instruments, intraocular lens, and prosthetic devices or other contaminated materials or solutions associated with surgery and/or routine health care (90, 136, 140).

Although these opportunistic moulds grow in most body tissues and fluids, colonization or invasion is commonly, but not solely, associated with subcutaneous soft tissue and mucous membranes. Individuals whose resistance is lowered as a result of a severe debilitating disease or immunosuppressive therapy typically suffer from invasive pulmonary or paranasal sinus infection, but in some instances, the infecting fungus may spread to surrounding tissues or disseminate to virtually any organ (90, 113, 136). Noninvasive forms of infection also have been noted in debilitated individuals as well as in individuals with apparently normal defense mechanisms. In such cases, the fungus colonizes a preexisting cavity in the lungs such as an ectatic bronchus, a tuberculous cavity, or a lung cyst (20). Other clinical syndromes are also recognized. Infections of the ear canal, nail, cutaneous and subcutaneous tissues, cornea, and other sites manifesting as cellulitis, bursitis, nephritis, endocarditis, peritonitis, fungemia, and, rarely, cerebral infections and involvement of the central nervous system have been documented (8, 121).

The use of the term aspergillosis to define infections caused by species of *Aspergillus* is well established, but the practice of coining disease names based on the genus of the fungus involved has disadvantages for infections caused by uncommon or rare fungal pathogens. The wide variety of fungi involved has made it difficult to place the organisms into accessible groups, and problems arise when fungus names are changed. To avoid the coining and proliferation of unnecessary new disease names based on the genus of the fungus involved, two major disease groups have been proposed: hyalohyphomycosis and phaeohyphomycosis (2, 3, 105). Each group encompasses a variety of clinical syndromes but is based on the presence of septate hyphal elements without or with pigmentation or melanin in the cell wall (2, 3, 105). Phaeohyphomycosis, as originally defined, is a heterologous group of infections characterized by the presence of dark (brown to black)-pigmented fungal elements in tissue (see chapter 67 of this Manual); however, there has been some confusion about the use of the term. If melanin pigmentation in tissue is the primary criterion for inclusion under the broad umbrella of this mycosis, several fungi that are dematiaceous in vitro should be excluded from this group, since they grow as hyaline elements in tissue. Species of *Scedosporium*, for instance, form pigmented conidia in vitro and hyaline hyphae in tissue that are indistinguishable from those of *Aspergillus* species. Another example of this anomaly is the strongly pigmented fungus *Nattrassia mangiferae* (*Scytalidium dimidiatum*), which usually forms hyaline or lightly pigmented hyphal elements in tissue. Although special stains (e.g., the Masson-Fontana silver stain) may help detect melanin in fungal elements in tissue, the results are not always decisive. Some fungi with variable pigmentation, such as *Sporothrix schenckii*, may stain faintly or inconsistently (91). Recently, a subcommittee of the International Society for Human and Animal Mycology (ISHAM) added to the confusion by defining phaeohyphomycosis as infection caused by a dematiaceous fungus (117). The term "hyalohyphomycosis" was proposed as a complement to phaeohyphomycosis for all subcutaneous and systemic infections in which the fungi involved have a tissue form consisting of hyaline, septate, branched, and, occasionally, toruloid mycelial elements (121). Although it was not intended to replace disease names such as aspergillosis, the term received qualified endorsement by the ISHAM subcommittee (117) only as a general term for infections caused by "*unusual* hyaline pathogens." This committee recommended that, rather than "generic" disease names, fungal diseases be given names that provide a specific description of the pathology and name the causative agent, e.g., subcutaneous cyst caused by fungus X. This approach may resolve the current difficulties in trying to categorize infections by the color of fungal elements in tissue.

ETIOLOGIC AGENTS

The opportunistic hyaline moulds that have been reported to cause infection in humans and animals are listed in Table 1. Verification of authenticity of reports can be difficult if the fungus is inadequately described and illustrated and if representative cultures are not maintained in culture collections. This chapter describes the more commonly occurring opportunists and briefly comments on some rarer agents.

Classification and Morphology

The opportunistic hyaline moulds are distributed throughout the kingdom Fungi and belong to genera of the Ascomycotina, Basidiomycotina, and Fungi Imperfecti. The Fungi Imperfecti is a special division or "form" division consisting of fungi for which no sexual stage (teleomorph) is known and fungi that are the asexual stages (anamorphs) of ascomycetes and basidiomycetes. Today, the teleomorphs of many of the opportunistic fungi are known. Although teleomorphs occasionally are isolated in culture, they may be difficult to obtain by routine culture methods. The Fungi Imperfecti includes both yeasts (unicellular growth) and moulds (filamentous growth) and is divided into three form-classes: (i) Blastomycetes, including unicellular organisms reproducing by budding or fission and sometimes producing hyphae (yeasts and yeastlike fungi; see chapter 61 of this Manual); (ii) Hyphomycetes; and (iii) Coelomycetes. Reproduction in mould fungi is characterized by the production of conidia, which are reproductive propagules produced following mitotic division of the nucleus. Conidia are formed when the nucleus migrates into the propagule from a specialized cell called a conidiogenous cell. The conidiogenous cell(s) may be borne on an erect simple or branched structure known as a conidiophore (Hyphomycetes) or within a specialized fruiting body called a conidioma (Coelomycetes).

Most of the pathogenic moulds are classified as Hyphomycetes. Identification of Hyphomycetes is based on morphology of the conidia and the mechanisms by which conidia are formed (conidiogenesis). Three basic tools are necessary for practical observation of these features. (i) An ocular micrometer (see chapter 67) is essential for recording the sizes of conidia or of sexual spores when present. Identification of the less common opportunists requires comparison with published taxonomic descriptions in which size is often a key criterion for species distinction. (ii) A dissecting microscope with magnification up to ×60 and basal illumination is useful for examining petri plate cultures or culture tubes for the presence of conidia in chains or slimy heads, specialized structures such as hülle cells, sclerotia, conidiomata, or sexual fruiting bodies forming under the aerial mycelium or embedded in the agar. (iii) Microscopic mounts that allow observation of how a fungus forms its conidia are also necessary. Slide culture preparations are excellent for many fungi (155), but conidial structures of some species of *Aspergillus* or *Penicillium* may not be typical in slide culture conditions.

Morphologic features of importance for identification of conidial fungi include (i) conidium size, shape, and pattern of septation; (ii) color of conidia and conidiophore, whether light (hyaline or moniliaceous) or dark (dematiaceous); (iii) developmental aspects of conidiogenesis, including nature of the conidiogenous cell; (iv) mechanism of conidium liberation or dehiscence; and (v) structure of the conidioma (if present). Differences in conidial shape and septation, which distinguish groupings known today as Saccardo spore groups (119), are readily observable characteristics convenient for preliminary distinction. Conidia may be single celled (amerosporae) or have one (didymosporae) or more (phragmosporae) septa. If a fungus produces both nonseptate and septate conidia, the conidia are often referred to as micro- and macroconidia. Conidia also vary in shape, being long and narrow (scolecosporae; as in *Fusarium* species), spirally coiled (helicosporae), or star shaped (staurosporae). Development of a conidium may occur by conversion of an existing cell or several cells (thallic-arthric) or may involve new wall building or blowing out of a portion of the wall (blastic). Conidiogenesis usually occurs at a particular location on a conidiogenous cell. If development occurs at a site that remains fixed and gives rise to more than one conidium, then the site is stable or determinate. If development occurs at new points on the conidiogenous cell (or axis), then the site is unstable or indeterminate. New sites of conidial development may occur as the axis lengthens (progressive) or shortens (retrogressive). The conidiogenous cell may produce only a single conidium or multiple conidia. Sympodial development involves the development of a single conidium at successive sites on a lengthening axis. Conidiogenous cells that are specialized to produce multiple conidia include the phialide and the annellide. Although it is often difficult to distinguish between these two types of cells, the annellide elongates, and sometimes narrows, during the formation of each new conidium, leaving an often imperceptible series of rings or scars on the conidiogenous cell. Scrutiny of the cell under an oil immersion objective may be necessary to make this distinction. Phialides and annellides produce conidia that accumulate either in slimy masses or in chains with the youngest at the base of the chain (basipetal). Conidia may also form in acropetal chains, with the youngest conidium at the tip of the chain. Distinction between acropetal and basipetal chains is possible by careful scrutiny of the size and wall morphology of the top and bottom conidia of the chain. The youngest conidium is recognizable by its smaller size, lighter color if the conidia are pigmented, and differences in wall ornamentation if the conidia are roughened. Some conidiogenous cells are specialized to simultaneously form multiple conidia over the surface of the swollen cell. When mature, conidia detach by fission of a double septum (schizolytic dehiscence) or by sacrifice of a supporting cell (rhexolytic dehiscence) either by fracture of a thin-walled region of the supporting cell or by lysis. Lytic dehiscence

TABLE 1 Currently recognized agents of aspergillosis and hyalohyphomycosis[a]

Acremonium spp.
    A. alabamensis
    A. falciforme
    A. kiliense
    A. potroni
    A. recifei
    A. roseo-griseum
    A. strictum

Aphanoascus fulvescens
    See also Chrysosporium spp.

Arthrographis kalrae

Aspergillus spp.
    Also listed under Eurotium spp.,
        Emericella spp., and Neosartorya spp.[b]
    A. candidus
    A. carneus
    A. clavatus
    A. deflectus
    A. fischeri group
    A. flavipes
    A. flavus
    A. fumigatus group
    A. glaucus group
    A. nidulans group
    A. niger
    A. ochraceus
    A. oryzae
    A. parasiticus
    A. restrictus
    A. sydowii
    A. terreus
    A. ustus
    A. versicolor

Beauveria spp.
    B. alba
    B. bassiana

Chrysosporium spp.
    Chrysosporium parvum (dimorphic)
        (Emmonsia parva)
    Chrysosporium sp.[c]
    See also Aphanoascus fulvescens[b]

Coprinus delicatulus

Cylindrocarpon spp.
    C. destructans[d]
    C. lichenicola
    C. tonkinense
    C. vaginae

Emericella spp.
    E. nidulans
    E. quadrilineata
    Also listed under Aspergillus
        spp.

Emmonsia parva (dimorphic)
    Chrysosporium parvum
        (dimorphic) (=Emmonsia
        parva)

Eurotium spp.
    Also listed under A. glaucus
        group
    E. amstelodami
    E. repens
    E. rubrum

Fusarium spp.
    F. anthophilum
    F. chlamydosporum
    F. dimerum
    F. moniliforme (=F.
        verticillioides sensu stricto)
    F. oxysporum
    F. proliferatum
    F. roseum
    F. sacchari
    F. solani

Gymnascella dankaliensis

Lecythophora spp.
    L. hoffmannii[d]
    L. mutabilis[f]

Microascus spp.
    M. cinereus[e]
    M. cirrosus[d]
    Also listed under
        Scopulariopsis spp.

Myceliophthora thermophila[d]

Myriodontium keratinophilum[c]

Neosartorya fischeri var. spinosa
    Also listed under A. fumigatus
        group

Onychocola canadensis

Paecilomyces spp.
    P. javanicus
    P. lilacinus
    P. marquandii
    P. variotii
    P. viridis

Penicillium spp.
    P. chrysogenum
    P. citrinum
    P. commune
    P. decumbens
    P. expansum
    P. marneffei (dimorphic)

Phialemonium spp.
    P. curvatum
    P. dimorphosporum

Scedosporium spp.
    S. apiospermum[f]
    S. prolificans[f] (formerly S.
        inflatum)

Schizophyllum commune

Scopulariopsis spp.
    S. acremonium
    S. brevicaulis
    S. brumptii[e]
    Also listed under Microascus
        spp.[b]

Scytalidium spp.
    S. hyalinum
    S. dimidiatum (synanamorph,
        Nattrassia mangiferae)[f]

Trichoderma viride

Tritirachium oryzae

Volutella cinerescens

[a]Modified from Rogers and Kennedy (140).
[b]Names for teleomorphic stages.
[c]Insufficiently documented to confirm species identification.
[d]Colonies are moniliaceous, not dark brown, gray, or black.
[e]Dematiaceous in culture. Listed here for comparison.
[f]Dematiaceous in culture. Hyphal elements in tissue are often hyaline.

involves enzymatic disintegration of the supporting cell(s) and typically occurs in the dermatophytes and related fungi.

Identification of some moulds may be difficult when different spore states are present. Subculture from the different states is necessary to confirm that the associated states belong to the same fungus and that the isolate is not contaminated. Some moulds have associated yeast forms, but maintenance of the yeast form is often dependent on the type of agar medium used, and this state may be lost upon continuous subculture. Conidiomata are formed in isolates of some moulds. The conidiomata of Hyphomycetes are sporodochia (conidiophores borne on a compact mass of hyphae or a hyphal stroma) and synnemata (conidiophores aggregated into a compound stalk). The conidia formed on the conidiomata may be the same as or different from the conidia formed on simple conidiophores. For example, the macroconidia of *Fusarium* species produced in sporodochia are generally similar, while the conidia formed on the synnemata of the *Graphium* state of *Pseudallescheria boydii* differ in size and shape from those formed on simple conidiophores. Different conidial states are often referred to as synanamorphs, and moulds having more than one independent form or spore stage are called pleomorphic, or occasionally polymorphic, fungi.

Collection, Transport, Storage, and Processing of Specimens

Clinical specimens from patients with suspected fungal infections are to be collected with prudence and then transported to the laboratory and processed as soon as possible by using the standard procedures described in chapter 60 of this Manual. Because of the diverse clinical manifestations, various sites may need to be examined for fungal elements. Biopsy material, transtracheal aspirates, and sputum samples collected in the early morning all may be useful specimens for the isolation and detection of hyaline moulds, as are infected nails. Swabs taken from mucous membranes and skin lesions are generally of little help in diagnosing infections caused by hyaline moulds and are not recommended unless duplicate specimens are taken. The reliability of determining whether an isolate is a possible pathogen is increased if the two specimens contain the same organism. Blood cultures are of little use in the isolation of *Aspergillus* species (178), but *Pseudallescheria boydii* or *Fusarium* species may be reliably detected (113, 133).

Media for isolation usually contain antibacterial agents such as chloramphenicol either alone or in combination with gentamicin, penicillin, or streptomycin. Opportunistic moulds are variably sensitive to cycloheximide, so media containing this selective agent, such as Mycobiotic agar (Difco) or Mycosel agar (Becton Dickinson), should be used cautiously. For optimum recovery, the specimen should be inoculated onto several types of media and incubated at 28 to 30°C. Suspicious isolates should be tested for the ability to grow at 37°C.

Significance of Isolation

Although the majority of saprophytic and plant parasitic moulds are not considered pathogenic and appear unlikely to be able to adapt to or take advantage of risk factors predisposing to opportunistic infection, those capable of growing at or near body temperature must be considered to have latent pathogenic capability. The variety of fungi capable of colonizing or invading human tissue has increased dramatically in recent years, as reflected by new reports of proven infection. However, the need for definitive evidence of infection should be stressed in establishing a diagnosis of infection due to a normally saprophytic mould. The laboratory procedure for confirming fungal etiology includes (i) detection in the specimen of hyphal elements that are compatible with the morphology of the isolated mould, (ii) isolation of several colonies of a fungus or repeat isolation of the same fungus from multiple specimens, (iii) accurate identification of the isolated mould, and (iv) confirmation of the mould's ability to grow at or near body temperature. Species of *Aspergillus*, *Scedosporium*, or *Fusarium* isolated from all deep tissues or body fluids must be considered colonizers or potential invasive pathogens. No fungal isolate should be discarded as a contaminant without thorough examination of the clinical specimen. Quality control measures to ensure that isolation media are

not contaminated and careful inspection of the location on a slant or plate where growth of a fungus occurs in relation to where a specimen was placed are also important factors in determining whether an isolate is involved in disease. Close communication between mycologists and physicians is also essential.

ASPERGILLUS SPECIES

Aspergillosis is now considered the second most common fungal infection requiring hospitalization in the United States (135). In patients with positive fungus cultures, *Aspergillus* species are the second most common isolate after *Candida* species (52). The pathologic responses caused by members of the genus *Aspergillus* vary in severity and clinical course and may occur as both primary and secondary infections. The clinical manifestations are largely determined by the local or general immunologic and physiologic state of the host, but various forms of the disease may integrate and overlap (134, 135). The various clinical entities of aspergillosis are described in Table 2. Risk factors for invasive aspergillosis (IA) include granulocytopenia in

TABLE 2 Classification of *Aspergillus* infections[a]

I. **Disease in the healthy host**
   A. Toxicosis or mycotoxicosis
      1. Ingestion of mycotoxins
      2. Ingestion of other metabolites
   B. Allergic manifestations
      1. Allergic asthma
      2. Allergic rhinitis
      3. Allergic sinusitis
      4. Extrinsic allergic alveolitis
      5. Hypersensitivity pneumonia
      6. Allergic bronchopulmonary aspergillosis
   C. Superficial infections
      1. Cutaneous infection
      2. Otomycosis
      3. Sinusitis
      4. Saprophytic bronchopulmonary aspergillosis
      5. Tracheobronchitis
   D. Invasive infection
      1. Single organ
      2. Multiple organs (disseminated)

II. **Infections associated with tissue damage or foreign body**
   A. Keratitis and endophthalmitis
   B. Burn wound infection
   C. Osteomyelitis
   D. Prosthetic valve endocarditis
   E. Vascular graft infection
   F. Aspergilloma (fungus ball)
   G. Empyema and pleural aspergillosis
   H. Peritonitis

III. **Infections in the compromised host**
   A. Primary cutaneous aspergillosis
   B. Sino-orbital infection
   C. Pulmonary aspergillosis
      1. Invasive tracheobronchitis
      2. Chronic necrotizing pulmonary aspergillosis
      3. Acute invasive pulmonary aspergillosis
   D. Central nervous system aspergillosis
   E. Invasive (disseminated) aspergillosis

[a]Modified from Bodey and Vartivarian (13).

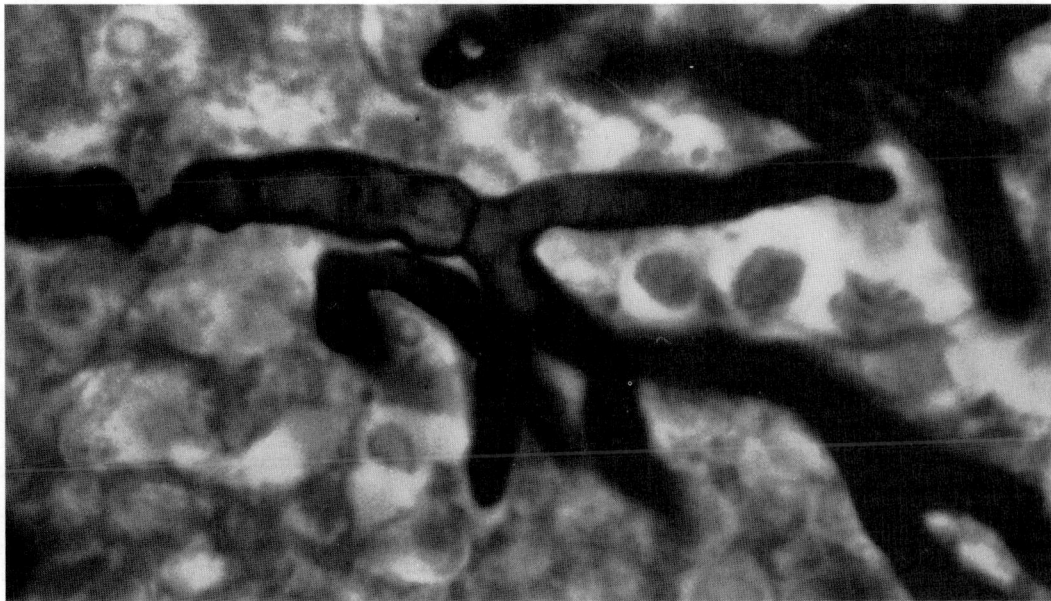

FIGURE 1 Invasive aspergillosis. Dichotomous branching and septation of hyphae. Sometimes there is a slight constriction of the hyphal wall at the point of septal formation. GMS stain. Original magnification, ×500. (Reprinted from reference 136.)

leukemic patients, neutropenia following bone marrow or organ transplantation, and high-dose corticosteroid or cytotoxic drug therapy (113). Although pulmonary aspergillosis among patients with AIDS is uncommon, several cases have been reported in patients with other risk factors, including drug-induced neutropenia, treatment with corticosteroids, and smoking of marijuana (30, 54).

Direct Examination

Assessment of the pathologic importance of any species of *Aspergillus* isolated from clinical specimens can be difficult, because aspergilli are routinely isolated from respiratory secretions, skin scrapings, and other specimens. Stimlam et al. (170), for instance, reported that when isolated, aspergilli are significant in only 10% of the cases. Thus, it is imperative that the organism be (i) demonstrated by direct microscopic examination in fresh clinical material and (ii) repeatedly isolated from the same specimen or multiple body sites to demonstrate clinical significance.

Hyphal elements and the details of hyphal morphology of aspergilli may be observed readily in routine KOH preparations with or without a fluorescent compound such as calcofluor white (113) or in tissue section with special fungal stains such as Gomori methenamine silver (GMS) stain (20, 21). *Aspergillus* hyphae stain with hematoxylin and eosin if the tissue is properly fixed, is not understained with hematoxylin, and is not necrotic. Viable hyphae are often basophilic to amphophilic, whereas hyphae in macerated or necrotic tissue tend to be eosinophilic (20).

The appearance of *Aspergillus* hyphae may vary with the type of infection. In acute invasive infection, aspergilli typically are seen as hyaline, closely septate hyphae that are 3 to 6 μm in diameter, branch dichotomously at acute (45°) angles, and have smooth parallel walls with no or slight constriction at the septa (Fig. 1) (20). In IA, hyphae proliferate extensively throughout the tissue, often in parallel or radial arrays. Aspergilli colonizing pulmonary cav-

itary lesions grow as tangled masses of hyphae and in chronic infection may exhibit atypical hyphal features such as swellings measuring up to 12 μm in diameter and/or absence of conspicuous septa. When hyphal elements typical of aspergilli are present in histologic sections, a presumptive diagnosis can be made, but it should always be confirmed by culture or immunologic techniques. Hyphae of other opportunistic hyaline moulds such as *Fusarium* species and *Pseudallescheria boydii* may resemble those of aspergilli, both in their propensity to grow in the lumens of blood vessels and in their formation of dichotomously branched hyphae. The presence of sporulating structures or conidial heads in tissue section aids in diagnosis and, if such elements are typical, may allow a preliminary assessment of the group to which the fungus belongs. Conidial heads usually occur in lung cavities or in the ear canal rather than in deep tissue. Free conidia may resemble other structures such as yeast cells, or cysts of *Pneumocystis carinii*, when stained with GMS, which masks the conidium color, but conidia of the aspergilli are often roughened. Distorted and broader hyphae may be mistaken for those of zygomycetes, but the hyphae of the latter are generally broader (up to 15 μm), have nonparallel walls that often appear collapsed and acutely twisted, and are infrequently septate (20). Problems in differential diagnosis under atypical or typical conditions when aspergilli are suspect can usually be resolved by direct immunofluorescence.

Immunodiagnosis

Immunologic tests serve as important aids in the rapid and specific diagnosis of various clinical forms of aspergillosis (147, 184). This is especially true in patients yielding negative cultures or when certain forms of aspergillosis are masked by or mistaken for other diseases (58, 77, 140, 141). The detection of circulating antibodies or *Aspergillus* antigens, for instance, has proven useful in defining the etiology of pulmonary infections in the absence of cultures or when two or more fungi coexist (77, 88, 180). Moreover, immunologic tests are extremely important for the rapid diagnosis of IA, for which early treatment is required in order to resolve such a potentially fatal form of the disease (55, 185).

A variety of serologic tests with disparate sensitivities has been developed to detect circulating antibodies. Immunodiffusion (ID) remains the most widely used technique, primarily because it is simple and easy to perform (88). Although the ID test lacks sensitivity and gives no quantitative information on antibody concentrations, a number of reports have documented its usefulness as a diagnostic tool (88, 140). Kurup and Kumar (88) noted that a micromodification of Wadsworth's agar gel diffusion against a pool of antigens correlated well with other diagnostic methods.

Other tests for detecting circulating antibodies to aspergilli that have achieved some success for diagnosis of various forms of aspergillosis include enzyme-linked immunosorbent assay (ELISA), biotin-avidin-linked immunosorbent assay (BALISA), radioimmunoassay (RIA), and indirect immunofluorescence (147). Of these, various ELISAs are the most amenable to use in the clinical laboratory. These assays can be, in some cases, highly sensitive, reliable, and versatile, although discrepancies between results from different laboratories have been reported (88). Variations in results are primarily due to the selection of *Aspergillus* antigens employed (82). For instance, both mycelial and culture filtrate antigens have been used in ELISA (75, 153), and batch-to-batch and laboratory-to-laboratory variations have been noted (75, 89). It is important to note that some antigens have limited binding specificities and bind only to certain immunoglobulin classes or isotypes within those classes (89). Also, the ELISA may not be sensitive enough to detect relatively low levels of antigen-specific immunoglobulin E or subclasses of immunoglobulin G (83, 87, 88). A variation of the conventional ELISA technique has been developed to overcome these problems (86). This method, the BALISA, takes advantage of the high-affinity binding between biotin and avidin by covalently linking protein and polysaccharide antigens of aspergilli (16). It has been reported to be highly reproducible and to have a sensitivity comparable to that of RIA (88). RIA has also been used as a reliable tool for detecting antibody (9, 122, 123); however, this assay has several obvious disadvantages associated with the use of radioisotopes that preclude its routine use in many clinical laboratories (88).

In patients with suspected allergic bronchopulmonary aspergillosis, skin test reactivity to *Aspergillus* antigen extracts has been used. Two types of reaction are observed with skin testing: type I (immediate) and type II (Arthrus). Type I reactions consist of cutaneous wheal-and-flare reactions that occur within 15 to 20 min of skin test challenge, while type II reactions usually take longer to develop, i.e., ~4 to 10 h after skin testing, and last longer. A delayed-type skin reaction also may be present in some patients, particularly those with hypersensitivity pneumonia (88).

Although detection of circulating antibodies is considered an important aid in the diagnosis of some forms of aspergillosis, demonstrable antibodies against aspergilli may be absent in some patients. Young and Bennett (196) reported, for instance, that very little or no antibody response to *Aspergillus fumigatus* was detected by ID, complement fixation, and indirect immunofluorescence in immunosuppressed patients who had IA and underlying hematologic malignancies. In a review of a number of studies, Kurup and Kumar (88) noted various degrees of success in detecting specific antibodies to aspergilli in sera of patients with IA. Similarly, antifungal therapy or the acute and fulminant courses of some *Aspergillus* infections may limit the detection of circulating antibodies to aspergilli (189). In such cases, antigen detection systems may provide information that is helpful in establishing an early, accurate diagnosis.

A variety of tests have been developed to detect soluble antigens of *Aspergillus* spp. in serum, urine, or other body fluids and even within host phagocytic cells (37, 127). Of these, RIA, inhibition ELISA and BALISA, and immunoblotting have been the most commonly used methods (61, 67, 82, 86, 176, 197). These methods appear promising but may not provide sensitivity sufficient to allow early diagnosis in some cases (88). Few are commercially available (183). Other methods and variations of these techniques are currently being evaluated in an effort to improve assay sensitivity. The development of immunologic tests employing monoclonal antibody and previously undescribed antigens for the detection of *Aspergillus* antigenemia are in progress and hold great promise for rapid antigen detection for early diagnosis and greater sensitivity (63). Methods for the detection of antigenemia have recently been summarized by Kurup and Kumar (88). Regardless of the test employed, success in detecting antigenemia is directly related to the frequency of monitoring of samples (61, 186, 194). As with the detection of circulating antibody, false-negative results in antigen assays may be ascribed, in part, to antifungal therapy, the course of infection, and the timing of sample collection.

Molecular Techniques

Rapid confirmation of the diagnosis of aspergillosis continues to be a serious problem for the clinician, particularly in IA, which is associated with a high rate of mortality in immunocompromised patients (45). The serious nature of the disease frequently leads to early initiation of therapy, but this action may impede obtaining an accurate diagnosis in patients in whom suspected aspergillosis has not been unequivocally confirmed by both culture and histopathology (140). The severity of the situation has led to a continued search for improved direct and indirect tests that are specific and reproducible in routine laboratory practice and that allow the rapid and reliable diagnosis of aspergillosis (88, 140, 185).

Molecular approaches to detecting DNA from clinical specimens offer promise where other diagnostic methods are limited (45, 166). Such methods may be highly sensitive, and results could be obtained on the same day (167). As one example, Spreadbury and coworkers (167) developed oligonucleotide primers that specifically amplify DNA from *A. fumigatus* and demonstrated both the utility and validity of PCR for the laboratory diagnosis of IA. In their study, oligonucleotide primers were used to amplify a 401-bp fragment spanning the 26S-intergenic spacer region of the *A. fumigatus* ribosomal DNA complex. These primers were highly sensitive and specific, detecting as little as 1 pg of genomic DNA from *A. fumigatus* but not from other fungal, bacterial, viral, or human DNA tested. Using PCR, these workers were able to detect *A. fumigatus* DNA in clinical samples from three of three patients with proven IA. The validity of detecting aspergillosis by PCR was confirmed in animal studies, which showed a 93% correlation between culture and PCR results. Further study is required, however, to determine the usefulness of PCR as a routine diagnostic procedure. Nonetheless, the sensitivity and specificity of molecular techniques make them attractive as alternative approaches for the early diagnosis of aspergillosis.

Another important application of molecular techniques is the identification and typing of isolates. Detailed epidemiologic studies have been hampered by the lack of simple, reproducible typing methods. Because phenotypic variability has been demonstrated for several species of aspergilli, a number of typing schemes for identification and epidemiologic study have been reported. These include alteration of gross morphology (94), susceptibility to "killer toxins" (130), isoenzyme electrophoresis with respect to esterase and phosphatase mobility (84), and immunoblot fingerprinting (18). However, none of these techniques has been developed as a formal typing system, because either the degree of discrimination is low or standardization is difficult (97).

Knowledge of the genetic characteristics of aspergilli allows for epidemiologic analysis (29) and is not dependent on gene expression as are immunoblot fingerprinting and other immunologic methods. Genetic fingerprinting of aspergilli by restriction fragment length polymorphism (RFLP) of genomic DNA has been shown to provide a measure of strain relatedness (17, 29, 31, 109). Similar results have been obtained with randomly amplified polymorphic DNA (RAPD typing) patterns (10) and Southern blots probed with ribosomal DNA (166) or nonribosomal genomic sequences (51).

More recently, the PCR has been adapted for fingerprinting aspergilli by using single primers, alone or in combination, with arbitrary (97) or repetitive (182) DNA sequences. Loudon et al. (97), for instance, used PCR to amplify polymorphic DNA patterns from 19 A. fumigatus isolates from six patients with aspergilloma. Two of five primers examined gave adequate discrimination between isolates, generating five and six types. Typeability and reproducibility compared favorably with those of immunoblot fingerprinting and XbaI-generated RFLP of the same isolates. RAPD typing was less labor-intensive than immunoblot fingerprinting, and the results suggested that aspergillomas sometimes contain isolates of more than one type. Van Belkum and coworkers (182) found that PCR amplification of repetitive DNA motifs also allowed discrimination of Aspergillus species. In addition, when a primer deduced from a prokaryotic repeat motif was used, A. fumigatus isolates originating from different patients or different anatomical locations could be typed individually. These studies imply that species- and isolate-specific probes can be developed, and they establish the usefulness of molecular techniques for detection-identification systems and epidemiologic studies.

Taxonomy, Mycology, and Identification

The genus Aspergillus is classified in the Hyphomycetes, but many species have teleomorphs (sexual stages) that are classified in the Trichocomaceae of the Ascomycotina. Early monographic treatments (131) included teleomorphic features within descriptions of Aspergillus species, a practice not consistent with the International Code of Botanical Nomenclature, and this has caused problems in the nomenclature of some well-known species. Approximately 190 taxa of Aspergillus and associated teleomorphs are known. About 20 taxa have been reliably reported from human or animal infections (90), and of these, A. fumigatus, Aspergillus flavus and Aspergillus niger are the most common pathogenic species worldwide (90, 135, 136). When isolated on

media such as Sabouraud dextrose agar, aspergilli tend to reproduce in the asexual form. Subculture to standard Czapek-Dox, Czapek-Dox with added glucose (20 to 30%) and 2% malt extract agar (131), or modifications of these media (80) stimulates development of a teleomorph and allows comparison of colonial and microscopic features with those given in published monographs and taxonomic keys (36, 80, 119, 131). Some species are osmophilic and grow poorly on media with low concentrations of sugar.

Because the large number of species within the genus makes identification to the species level difficult for even the experienced microbiologist, the aspergilli were placed into distinct, accessible groups (131). Except for the glaucus group, each name comes from a representative species within the group. Members of the glaucus group produce teleomorphs that are classified in the genus Eurotium. Many of the anamorphic names in this group are considered invalid owing to the inclusion of the teleomorph in the original description. More modern treatments of Aspergillus have formally defined the groups as sections within subgenera (36, 80). A key to the common groups of Aspergillus is provided in Fig. 2, and the diagnostic features of the more common pathogens and contaminants are described in Table 3. Although isolates of the genus Aspergillus are usually identified on the basis of conidial characteristics, the presence of sexual structures (cleistothecia and ascospores; Fig. 3) allows ready identification of members of some groups. Species of medical importance have teleomorphs in the genera Eurotium, Emericella, and Neosartorya, and they are described under these names in Table 3, since the sexual structures are important in making an accurate identification.

Growth in culture consists of septate, hyaline, branched hyphae that give rise to upright conidiophores or stipes terminating in a swollen cell (vesicle) (Fig. 2). The conidiophore varies in color, length, and wall ornamentation in different species; is usually nonseptate; and arises either directly from the vegetative hyphae or from a specialized hyphal cell called a foot cell. The vesicle is globose, subglobose, hemispherical, and pyriform (pear shaped) or clavate, and either the entire vesicle or a portion of it is covered with phialides, which form the conidia. Phialides arise simultaneously directly from the vesicle (uniseriate; Fig. 4), from an intermediate series of cells called metulae (biseriate; Fig. 5), or by a combination of both processes, as occurs in species such as A. flavus (Fig. 2). The conidia are typically ellipsoidal or globose and vary in size, color, wall markings, and thickness depending on the species. Some aspergilli produce other structures that may be helpful in the identification of an isolate. Sclerotia are firm, fruiting-body-like structures composed of swollen hyphal cells but lacking internal spores. Hülle cells are globose or variable in shape and have thick and highly refractive walls (Fig. 6). Hülle cells commonly occur in clusters, often near the center of a colony immersed in the vegetative mycelium, and their presence may be indicated by droplets of exudate on the colony obverse. They are often associated with cleistothecia such as in the Aspergillus nidulans group (Table 3). Cleistothecia are spherical structures whose outer walls are composed of interwoven hyphae and whose interiors are filled with asci and ascospores. Cleistothecium color and wall texture and ascospore shape, color, size, and wall ornamentation are important characteristics in differentiating teleomorphs of Aspergillus species.

Both macroscopic and microscopic characteristics are required for identification of Aspergillus species. Colony

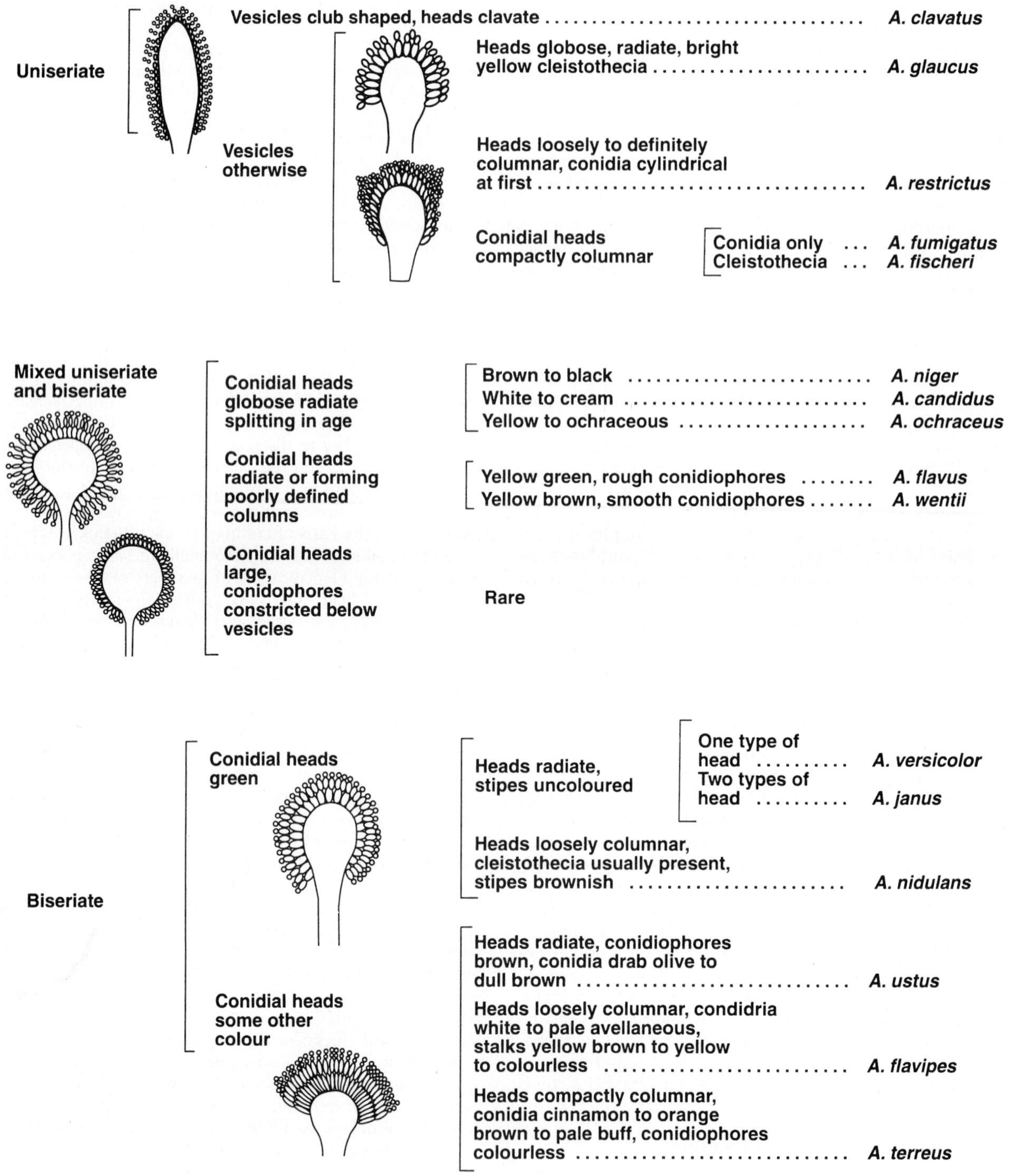

Uniseriate

Vesicles club shaped, heads clavate .... *A. clavatus*

Vesicles otherwise

Heads globose, radiate, bright yellow cleistothecia .... *A. glaucus*

Heads loosely to definitely columnar, conidia cylindrical at first .... *A. restrictus*

Conidial heads compactly columnar — Conidia only ... *A. fumigatus* / Cleistothecia ... *A. fischeri*

Mixed uniseriate and biseriate

Conidial heads globose radiate splitting in age

Brown to black .... *A. niger*
White to cream .... *A. candidus*
Yellow to ochraceous .... *A. ochraceus*

Conidial heads radiate or forming poorly defined columns

Yellow green, rough conidiophores .... *A. flavus*
Yellow brown, smooth conidiophores .... *A. wentii*

Conidial heads large, conidophores constricted below vesicles

Rare

Biseriate

Conidial heads green

Heads radiate, stipes uncoloured — One type of head .... *A. versicolor* / Two types of head .... *A. janus*

Heads loosely columnar, cleistothecia usually present, stipes brownish .... *A. nidulans*

Conidial heads some other colour

Heads radiate, conidiophores brown, conidia drab olive to dull brown .... *A. ustus*

Heads loosely columnar, condidria white to pale avellaneous, stalks yellow brown to yellow to colourless .... *A. flavipes*

Heads compactly columnar, conidia cinnamon to orange brown to pale buff, conidiophores colourless .... *A. terreus*

FIGURE 2 Key to the common "group series" of *Aspergillus*. (Reprinted with permission from reference 119.)

growth rates, obverse and reverse colors, texture, topography, and presence of exudate droplets or diffusible pigments should be recorded at 7 days on identification media (36, 80, 119). Isolates not immediately identifiable should be retained for 3 weeks or longer for observation of development of teleomorph or other structures. Examination of a colony under a dissecting microscope allows observation of conidial chains to determine whether they are borne in a single column (columnar) or whether the columns are split, with some arising at right angles to the stipe (radiate). The shapes and sizes of vesicles, metulae, phialides, conidia, ascocarps, and ascospores and the length, color, and wall ornamentation of the conidiophore are important microscopic features.

TABLE 3 Characteristics of some medically important *Aspergillus* species grown on identification media[a]

Group[b] and organism	Seriation Uni-seriate	Seriation Bi-seriate	Colony	Microscopic features	Comment	Selected references[c]
Fumigatus group, *A. fumigatus*	+		Dark blue-green to grayish turquoise; slate gray with age; reverse variable in color	Conidiophore mostly up to 300 μm long and 5–8 μm wide, smooth walled, uncolored or greenish; vesicle dome shaped, 20–30 μm in diam, phialides on upper half only; head strongly columnar; conidia subglobose to globose, occasionally ellipsoidal, smooth to echinulate, 2–3.5 μm in diam	Distinguished by blue-green colonies, thermotolerance (grows well at 45–50°C); columnar heads with single layer of phialides. Mutant forms include sterile isolates and slow-growing types with septate phialides; may be confirmed by thermotolerance and exoantigen test. Common airborne mould often associated with compost piles and found in soil of potted plants; most common pathogenic species.	24, 36, 80, 90, 119, 135, 136, 172
Neosartorya pseudofischeri	+		White to gray-green; exudate uncolored or pale brown	Conidiophore mostly 150–500 μm long and 5–8 μm wide, smooth walled, uncolored; vesicle dome shaped, 18–20 μm in diam, phialides on upper half; columnar; conidia globose or ellipsoidal, smooth or echinulate, 2–4 μm in diam; cleistothecia globose, mostly 150–350 μm; ascospores with two crests, 4.5–6 μm long, convex surface with nonanastomosing ridges	Grows at up to 50°C; conidial characteristics strongly similar to those of *A. fumigatus* but colonies usually whiter and sporulating less heavily; species distinguished by ascospore morphology; rarely reported as pathogen; some clinical isolates reported as *N. fischeri* have been reidentified as *N. pseudofischeri*.	36, 90, 119, 125, 171
Flavus group, *A. flavus*	+	+	Yellow to dark yellowish green	Conidiophore mostly 400–850 μm long and 20 μm wide, uncolored, coarsely roughened; vesicle subglobose or globose, commonly 25–45 μm in diam; loosely radiate or splitting into columns in age; conidia globose, rarely ellipsoidal, roughened, 3–6 μm in diam	Some strains produce toxins and brown to black sclerotia; growth usually enhanced at 37°C; heads vary in size, and some are uniseriate; colony color may be influenced by culture medium additives such as yeast extract; second most common human pathogen.	36, 80, 90, 119, 135, 136
Niger group, *A. niger*		+	Rapidly becoming black with broad white margin and often showing areas of yellow; reverse usually light colored or pale yellow	Conidiophore 400–3,000 μm long and 15–20 μm wide, smooth, uncolored to brownish at upper end; vesicle globose, commonly 30–75 μm in diam; radiate, then splitting into columns in age; conidia globose with thick walls, brown to black, irregularly roughened, commonly 4–5 μm in diam	Frequent cause of otomycosis; sometimes associated with colonization of cavities	36, 80, 90, 119, 135, 136
Terreus group, *A. terreus*		+	Tan to cinnamon brown, rarely orange-brown, reverse yellow or tan	Conidiophore 100–250 μm long and 4.5–6 μm wide, smooth walled, uncolored; vesicle dome shaped, 10–16 μm in diam, phialides on upper half to two-thirds; head columnar; conidia globose or subglobose, smooth, 2 μm in diam; globose or oval solitary conidia, commonly formed sessile on submerged hyphae	Cinnamon brown colonies, columnar heads, and solitary accessory conidia are highly distinctive.	36, 77, 80, 90, 119

(Continued on next page)

773

TABLE 3 Characteristics of some medically important *Aspergillus* species grown on identification media[a] *(Continued)*

Group[b] and organism	Seriation Uni- seriate	Seriation Bi- seriate	Colony	Microscopic features	Comment	Selected references[c]
Glaucus group, *Eurotium amstelodami*	+		Deep green mixed with bright yellow, reverse uncolored or pale yellow	Conidiophore mostly 200–350 μm long and 7–12 μm wide, smooth walled, uncolored or pale brown; vesicle globose, 15–30 μm in diam; conidial heads large, radiate; conidia subglobose, echinulate, 4 μm in diam; cleistothecia yellow, globose, 75–150 μm; ascospores roughened with furrow and two crests, 4.5–6 μm long and 3.5–4 μm wide	Species of glaucus group are osmophilic; cleistothecial production enhanced on high-sugar media; growth poor at 37°C; species difficult to differentiate by features of conidial state but readily identified if ascospores present; *Eurotium repens*, *E. ruber*, and *E. amstelodami* rarely reported as pathogens.	36, 80, 90, 119
Restrictus group, *A. restrictus*	+		Dull olive green to brownish green, very slow growing	Conidiophore 80–200 μm long and 4–8 μm wide, smooth or finely roughened, uncolored; vesicle hemispherical, 8–20 μm in diam, phialides on upper surface only; head strongly columnar; conidia cylindrical to ellipsoidal, roughened, 4–7 μm long and 3 to 4 μm wide	Osmophilic, growth slightly enhanced on high-sugar media; conidial heads similar to those of *A. fumigatus* but distinguished by cylindrical conidia developing in long and closely packed columns, very slow growth, and no growth at 37°C.	36, 80, 90, 119
Versicolor group *A. versicolor*		+	Variable colors ranging from pale green to gray-green, tan with patches of pink or yellow; reverse variable and often deep red	Conidiophore mostly 200–400 μm long and 5 μm wide, smooth walled, uncolored, yellowish or pale brown; vesicle ovate to elliptical, 9–16 μm in diam, fertile over most of surface; radiate to loosely columnar; conidia globose, echinulate, 2.5–3 μm in diam; hülle cells globose	Slow-growing, greenish tan but often variably colored colonies and small biseriate vesicles distinguish this species. *A. sydowii* is similar microscopically, but colonies are blue-green. Reduced structures resembling penicilli sometimes present.	36, 70, 80, 119
A. sydowii		+	Blue-green, often resembling *Penicillium* spp.	Conidiophore mostly 150–350 μm long and 3–8 μm wide, smooth and thick walled, uncolored, or pale brown; vesicle globose or elliptical, 7–13 μm in diam, fertile over most of surface; radiate; conidia globose, very rough, 2.5–4 μm in diam; reduced structures resembling penicilli common	Slow-growing blue-green colonies distinguish this species from *A. versicolor*; diminutive heads resembling penicilli common.	36, 70, 80, 90, 119
Nidulans group *Emericella nidulans*		+	Dark green in conidial forms; buff to purplish brown in primarily cleistothecial forms; reverse deep red to purple	Conidiophore mostly 70–150 μm long and 3–6 μm wide, smooth walled, becoming brown; vesicle hemispherical, 8–12 μm in diam, phialides on upper surface, columnar; conidia globose, rough, 3–4 μm in diam; cleistothecia dark red-brown, globose, 100–250 μm, enmeshed in globose hulle cells; ascospores lenticular with two longitudinal crests, about 5 μm long and 4 μm wide, reddish purple	Distinguished by reddish brown cleistothecia, abundant hülle cells, reddish purple ascospores with two crests, short conidiophores, and "stout" metulae.	36, 80, 90, 119

					References
Emericella quadrilineata	+	Olive green to grayish purple, reverse purple	Conidiogenous structures, ascocarps, and hülle cells similar to those of Emericella nidulans. Ascospores red, lenticular, 4.5–5.5 by 3.5–4 μm, with 2 major and 2 minor equatorial crests	Distinguished from E. nidulans by 4 crests on ascospores.	80, 119, 129
Flavipes group, A. flavipes	+	White with patches of yellow, becoming pale or grayish buff; reverse yellow to golden brown	Conidiophore mostly 150–400 μm long and 4–8 μm wide, uncolored to pale brown, smooth to slightly roughened; vesicle subglobose, commonly 10–20 μm in diam; radiate to loosely columnar in age; conidia globose, smooth, 2–3 μm in diam; cleistothecia and hülle cells rarely produced.	Distinguished from A. terreus by slower-growing colonies, metulae usually formed over entire vesicle, and radiate to loosely columnar heads.	80, 119
Ustus group, A. ustus	+	Brownish gray or olive gray; reverse yellow, dull reddish, or purplish	Conidiophore 75–400 μm long and 4–7 μm wide, smooth, becoming brown; vesicle globose or subglobose, 7–16 μm in diam, fertile over upper two-thirds; radiate to loosely columnar; conidia globose, rough, 3–4.5 μm in diam; irregular or elongate hülle cells present in some isolates	Distinguished by dull gray-green colonies, small vesicles, and brown conidiophores.	36, 80, 90, 119, 187
A. deflectus	+	Slow growing, becoming mouse gray with pinkish margins or patches of yellow	Conidiophore 40–125 μm long and 2.5–3.5 μm wide, smooth walled, reddish brown; vesicle hemispherical, 5–7 μm in diam, typically bent at right angle to stipe, phialides on upper surface; columnar; conidia globose, 3–3.5 μm in diam, smooth to irregularly roughened; ovate or irregular hülle cells present in some isolates	Distinguished by vesicle bent almost at right angle to conidiophore; rarely reported as pathogen in humans and dogs.	66
Candidus group, A. candidus	+	White to cream	Conidiophore mostly 200–500 μm long and 7–10 μm wide, smooth to finely roughened, uncolored; vesicle globose to subglobose, 17–35 μm in diam, fertile over entire surface; radiate; conidia globose, smooth, 3–4 μm in diam; reddish purple to black sclerotia sometimes present	Distinguished from all colored aspergilli by white, slow-growing colonies; from A. niveus by larger vesicles, metulae covering entire surface, and failure to form a teleomorph; predominantly biseriate, sometimes uniserate on smaller heads.	36, 80, 90, 119

[a] Modified from Rogers and Kennedy (140), with data from references 36 and 80.

[b] Modern concepts have replaced group names with subgenera and sections (80).

[c] Refer to reference manuals by Domsch et al. (36), Klich and Pitt (80), Kwon-Chung and Bennett (90), or Onions et al. (119) for species descriptions and illustrations.

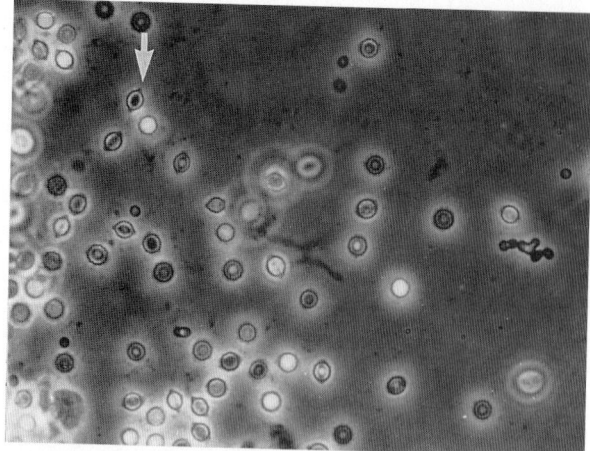

FIGURE 3 Ascospores of *Neosartorya* species. Note the two crests. Magnification, ×610.

Occasionally, nonconidiating isolates of aspergilli are isolated from specimens. A sterile isolate, especially one that is from the respiratory tract and grows at 45°C, may be suspected of being *A. fumigatus*. Exoantigen tests have proved useful for rapid identification to the genus level and in the case of *A. fumigatus* to the species level (71, 151).

SCEDOSPORIUM SPECIES

Scedosporium species are soil-, manure-, and water-inhabiting moulds with virulence and a clinical spectrum remarkably similar to those of aspergilli. Two species are known to cause human and animal infection. *Scedosporium apiospermum* is the anamorph of the ascomycete *Pseudallescheria boydii*. *Scedosporium prolificans* was originally described under the name *Scedosporium inflatum*, but morphologic and molecular evidence suggested that the latter was closely related to *Lomentospora prolificans*, a fungus originally isolated from greenhouse soil (59). No teleomorph is known. While *Pseudallescheria boydii* is the leading cause of myce-

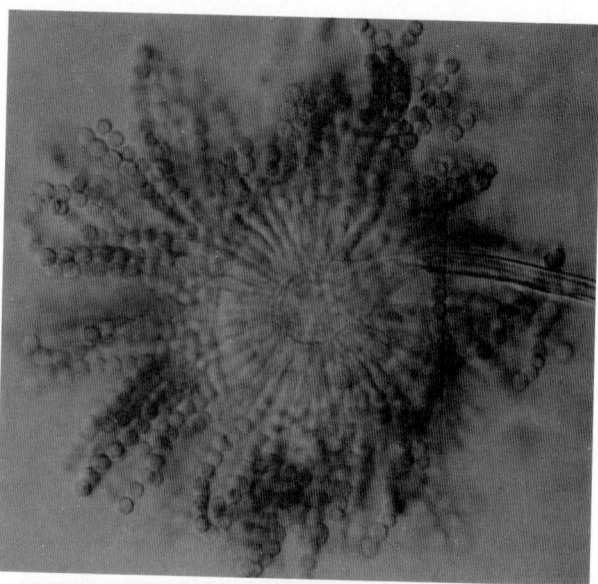

FIGURE 5 Biseriate conidial head of *Aspergillus versicolor*. Magnification, ×770.

toma in North America (discussed in chapter 68), it is recognized today as an important agent of infection from virtually any site (90, 136), including brain (38, 48) (see also chapter 67). Often causing osteomyelitis in both humans and animals (98, 144), *S. prolificans* has been isolated from a variety of other sites and from patients with disseminated infection (100, 145, 193, 195). As with aspergilli, these species are uncommonly documented as causing infections in patients with AIDS (54, 124).

The similarities between species of *Aspergillus* and *Scedosporium* in both clinical syndromes and appearance in tissue has led to the treatment of *Scedosporium* spp. in this chapter, even though species of *Scedosporium* are dematiaceous in culture. In tissue sections, the hyphae of *P. boydii* are hyaline, septate, branched, and hematoxylinophilic and can be reliably distinguished from hyphae of the aspergilli and other hyaline moulds only by direct immunofluores-

FIGURE 4 Uniseriate conidial heads of *Neosartorya fischeri* var. *spinosa*. Magnification, ×610.

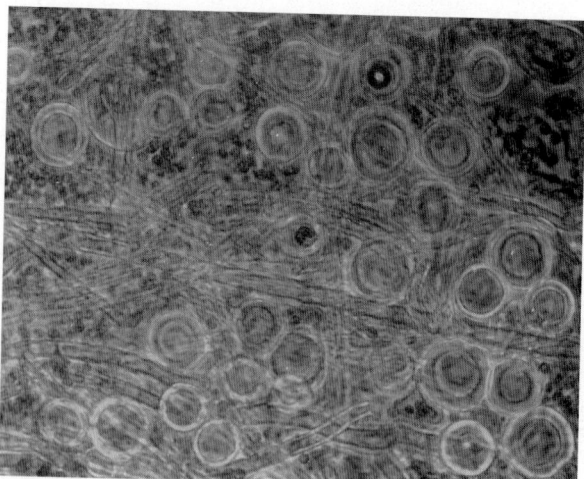

FIGURE 6 Hülle cells of *Aspergillus versicolor*. Magnification, ×460.

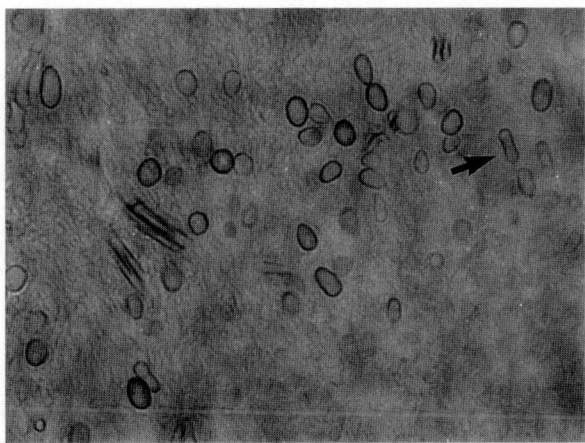

FIGURE 7 Fungus ball in sinus showing conidia and hyphae of *Scedosporium apiospermum*, GMS stain. Note that cylindrical conidia typical of the *Graphium* stage are also present (arrow). Magnification, ×770.

cence. Occasionally, conidia that are pale brown and ovoid or cylindrical are formed, generally in and around fungus balls or in the sinuses. When present, the conidia allow the ready recognition of this fungus (Fig. 7). Nevertheless, culture and serologic analysis are necessary to establish the diagnosis, especially since both aspergilli and *Pseudallescheria boydii* have been found to coexist in the same infection (136). Serologic procedures may aid in prognostic evaluation and culture identification (90).

The cultural characteristics separating *S. apiospermum* from *S. prolificans* (*S. inflatum*) are the shape and arrangement of the conidiogenous cells, the development of associated states (sexual and asexual), and a tolerance of cycloheximide. The annellides of *S. apiospermum* are usually solitary and more uniformly cylindrical in shape than those of *S. prolificans*, which are basally swollen (1.5 to 3 μm in width) and narrow abruptly to the neck. The annellides of *S. prolificans* may be solitary but typically are borne at the apices of conidiophores (Fig. 8) in clusters of two to five. Although isolates of *S. apiospermum* grow on media containing cycloheximide, most isolates of *S. prolificans* are completely inhibited by cycloheximide. Although *S. apio-*

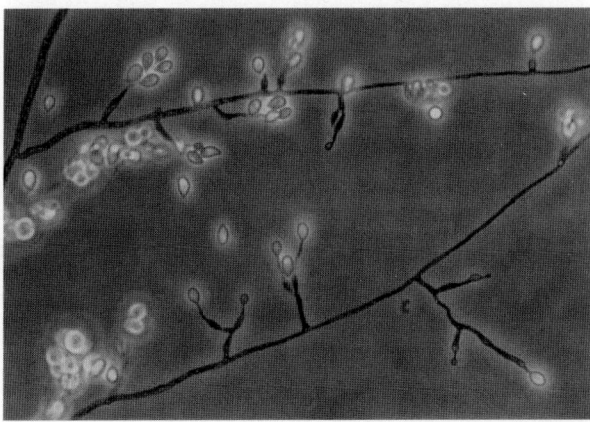

FIGURE 8 Annellides of *Scedosporium prolificans* (*S. inflatum*) showing swollen bases. Magnification, ×440.

spermum can produce three states, all states may not be produced by an individual isolate. A second type of conidial state, called the *Graphium* state, occurs in some isolates. It forms more commonly on sporulation media such as cornmeal agar, pea agar, potato-carrot agar, potato dextrose agar (PDA), or plain water agar and may be induced in some isolates by scarring the surface of the colony with a transfer needle. The conidia of the *Graphium* state are produced in the same manner as in the *Scedosporium* state, i.e., from annellidic conidiogenous cells, but the conidiogenous cells are borne on erect synnemata (a type of conidioma). Also, the conidia are cylindrical and longer than the ovoid conidia formed on the solitary annellides. *Pseudallescheria boydii* is the name given to the ascomycetous teleomorph (sexual stage) of *S. apiospermum*. It is distinguished by the formation of round, yellowish to dark brown cleistothecia containing ellipsoidal, pale brown, smooth-walled ascospores. The ascospores resemble the *Scedosporium* conidia, but they are often distinguished by the presence of a central refractile "de Bary bubble," and they lack an abscission scar. Ascocarps may be induced by culturing isolates on sporulation agar and typically are produced within the agar or beneath the mycelium near the center of the colony. So far, neither a teleomorph nor a *Graphium* stage has been found in any isolate of *S. prolificans*.

Species of *Scedosporium* have been compared with dark *Scopulariopsis* species in the differential diagnostic, but the latter may be readily distinguished by the formation of conidia in chains. Some dark *Scopulariopsis* species are associated with *Microascus* teleomorphs, the ascocarps of which differ in having short or long necks.

FUSARIUM SPECIES

Fusarium species are ubiquitous soil saprophytes and plant pathogens that can cause infection or toxicosis in humans and animals (114a, 133). Infections reported include keratitis (90), onychomycosis (70), colonization of burned or necrotic skin of wounds (191), leg ulcers (93), maxillary sinus infection (85), and infection of the deeper tissues, including mycetoma (see chapter 68), endophthalmitis, facial granuloma, osteomyelitis, and brain abscess (136). Increased incidence of pneumonia, fungemia, and disseminated infection has been seen in patients with prolonged neutropenia due to therapy for leukemia or lymphoma or following bone marrow transplantation (6, 7, 46, 90, 107, 113, 133, 136, 137). Multiple skin lesions, rhinocerebral involvement, fungemia, or all of these are common clinical findings in disseminated fusarial infection (46, 133, 188). In contrast to the low rate of recovery of *Aspergillus* species in blood cultures in cases of IA, there is a high rate of recovery (59%) from blood in cases of disseminated fusarial infection (46, 90, 133).

In direct examination, the hyphae of *Fusarium* species are haphazardly dispersed throughout the lesion and resemble the hyphae of aspergilli or *Pseudallescheria boydii* in size (3 to 6 μm in width), septation, branching pattern, and predilection for vascular invasion (20, 188). Dichotomous branching is common, branches are often constricted at their sites of origin from the parent hyphae (20, 21), and hyphae may show areas of collapse. Swellings or chlamydosporelike structures may be found, especially in vascular lesions (20), and conidia were found in the lumens of the bronchioles in an infected animal (43). Immunofluorescence and immunosorbent assays may help distinguish between infections caused by some hyaline moulds, but definitive diagnosis requires isolation and identification of the fungus.

The most common pathogenic species are *Fusarium solani*, *F. oxysporum*, and *F. moniliforme*, and there are rare reports of infection due to *F. anthophilum* (118), *F. chlamydosporum* (76), *F. dimerum* (200), *F. napiforme* (107), *F. proliferatum* (133, 173), *F. roseum* (90), and *F. sacchari* (199). Identification of species of *Fusarium* can be difficult and requires careful observation of both cultural and microscopic features (114a, 115). Although fusaria grow well on most mycological media, the medium used can have a profound influence both on colonial topography and color and on conidium development. Identification schemes usually require observation of growth rates, color of colony obverse and reverse, color of conidial masses, and microscopic features on specific media such as PDA and tap water agar supplemented with either sterilized carnation leaves or potassium chloride (115). In contrast to *Acremonium* colonies, colonies of *Fusarium* species are often fast growing, usually reaching more than 7 cm in diameter after 10 to 14 days. Colony obverse and reverse colors range from white or cream to orange, tan, brown, carmine red, reddish brown, pink, purple, or blue to blue-green (115). Aerial mycelium may be abundant, with colonies appearing cottony, or it may be sparse, with colonies initially appearing almost shiny and yeastlike. Cultural variation and loss of ability to produce macroconidia are common occurrences in cultures of *Fusarium*, especially if they are maintained on media rich in carbohydrates.

Microscopically, *Fusarium* species are characterized by the production of two types of hyaline, phialidic conidia. The macroconidia (Fig. 9A) are hyaline, two- to multi-celled (commonly with three to five septa), fusiform or sickle shaped, and often with a distinct, notched basal cell, and they typically occur in sporodochia (conidiophores are borne crowded on a compact mass of hyphae). Although macroconidia may also be produced in the aerial mycelium (Fig. 9B), the macroconidia produced in sporodochia (Fig. 9A) are the most uniform for the species. The microconidia (Fig. 9B and 10) are oval, ellipsoidal, club shaped, kidney shaped, globose, lemon shaped, pear shaped, or spindle shaped; one or two celled; and borne in slimy heads or in short chains in the aerial mycelium. Both types of conidia are born from awl-shaped phialides that usually each have a single opening (monophialides) (Fig. 9). In some species, such as *F. proliferatum*, phialides in the aerial mycelium have more than one opening (polyphialides) (Fig. 10). Other microscopic features of importance include the presence of chlamydospores and their formation in chains or clusters (Fig. 9B). Some species produce few or no microconidia, whereas other species rarely produce macroconidia. In the latter case, the distinction from *Acremonium* spp. is less clear, but colonies of *Acremonium* species are usually slower growing (<3 cm in 10 days), the hyphae are narrower (mostly <2 μm in width), and the phialides are generally more needlelike and less robust. *Cylindrocarpon* species are reported rarely as agents of infection in mycotic keratitis (90) and mycetoma (25, 201) and are recovered occasionally as contaminants from specimens. The macroconidia are distinguished by the more rounded shape of the terminal cells (36). The microscopic and colonial features of the three most common pathogenic *Fusarium* species are summarized in Table 4.

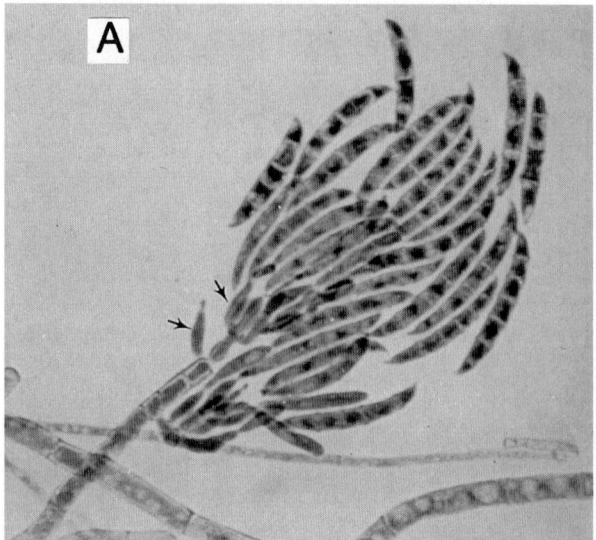

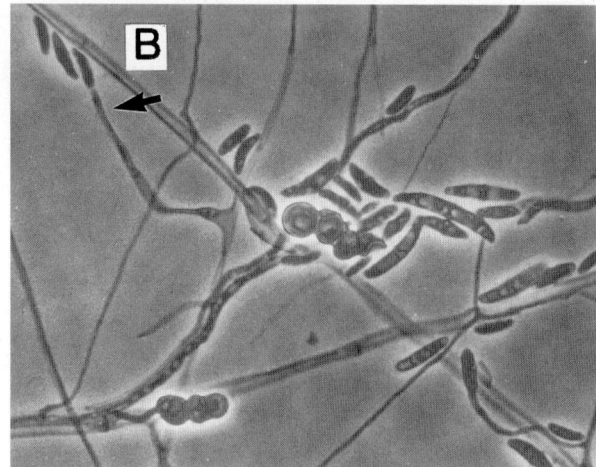

FIGURE 9 Microscopic morphology of *Fusarium* species. (A) Macroconidia produced in sporodochia from phialides (arrows). Original magnification, ×1,200. (Reprinted from reference 136.) (B) *F. solani*, showing microconidia and macroconidia formed in the aerial mycelium and presence of chlamydospores. Microconidia are formed from monophialides (arrow). Magnification, ×770.

PAECILOMYCES SPECIES

Species of *Paecilomyces* occur worldwide as soil saprophytes, insect parasites, and agents of biodeterioration. The most common pathogenic species are *Paecilomyces variotii* and *P. lilacinus*; other species rarely reported from infections include *P. marquandii*, *P. javanicus* (4), and *P. viridis* (149), originally described from infection in a chameleon (90, 140). Although *P. viridis* was implicated in a case of endophthalmitis (139), the etiologic agent was subsequently determined to be *P. variotii* (19). Most of the cases involving *P. lilacinus* have involved the eye, including keratitis, endophthalmitis, corneal ulcer, and orbital granuloma (1, 53, 90, 108, 116, 168). Predisposing factors include use of extended-wear contact lenses, implantation of inadequately sterilized lenses, and use of topical corticosteroids before onset of clinically apparent keratitis (168). *P. variotii* was the etiologic agent in several cases of endocarditis following

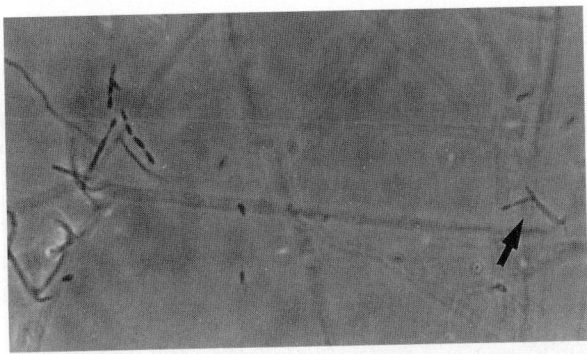

FIGURE 10 Microconidia in chains formed from mono- and polyphialides (arrow) in *Fusarium proliferatum*. Magnification, ×420.

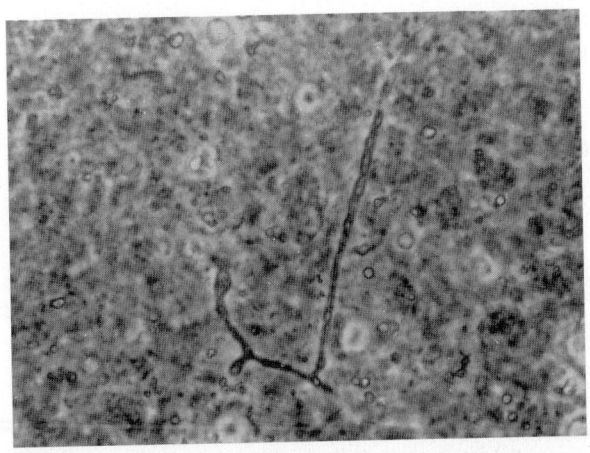

FIGURE 11 KOH preparation of necrotic tissue of draining chest lesion, showing branched hypha of *Paecilomyces variotii*. Magnification, ×770.

valve replacement (68, 90, 103, 162). Surgery and prosthetic implants appear to be risk factors predisposing to infection, since the conidia of both species appear to be relatively resistant to most sterilizing techniques (19, 36, 101, 136). Cutaneous and subcutaneous infection (19, 192), pulmonary infection (19, 26, 33), pyelonephritis (154), sinusitis (120, 138), cellulitis (60, 65), peritonitis (101), and fungemia (177) have been reported in both immunocompetent and immunosuppressed patients.

In tissues, hyphae of *Paecilomyces* species are hyaline, septate, and about 2 to 4 μm wide (Fig. 11). Hyphal and rounded, budding cellular forms have been noted in tissue sections from which *P. marquandii* or *P. viridis* was grown

(140, 146). In an animal infection, mycelia were observed to project into the lumens of necrotic tissue and to form conidiophores bearing flask-shaped phialides (20).

Samson (146) discussed 31 species in the genus *Paecilomyces* and divided them into two groups. Section *Paecilomyces* includes the type species *P. variotii* and the anamorphs of some *Byssochlamys*, *Thermoascus*, and *Talaromyces* species. Colonies of these species are usually yellowish brown or buff to brownish, and most are

TABLE 4 Morphologic features of three commonly isolated *Fusarium* spp.[a]

Characteristic	*F. solani*	*F. oxysporum*	*F. moniliforme*[b]
Growth (cm) on PDA (4 days)	3–3.5	4.5–6.5	4.5–6.5
Colonies	Grayish white with cream slimy areas, developing blue-green to bluish brown colors, never orange, reverse uncolored, rarely violet	White to peach, usually with purple or violet tints, sporodochia orange, slimy, reverse uncolored to dark blue or purple	White or pale peach to purple, sporodochia orange, reverse uncolored to dark purple
Phialides	Mono, long, slender, sometimes verticillately branched	Mono, short, stout, mostly unbranched	Mono, long, branched and unbranched
Microconidia	Cylindrical to oval, nonseptate	Oval to kidney shaped, single celled, in heads	Oval to club shaped with flattened base, mostly single celled, in chains[c] or heads
Macroconidia	Mostly 3 septate (1–5 septate), 28–42 by 4–6 μm, slightly curved, rounded at each end or basal cell notched	Mostly 3 (or up to 5) septate, moderately curved, 27–60 by 3–4.5 μm, pointed at each end	Present or rare, 3–5 (or up to 7) septate, sickle shaped to almost straight, 30–60 by 3–4 μm, basal cell foot shaped
Chlamydospores[d]	Present, formed singly and in pairs, commonly produced	Present, formed singly and in pairs, usually abundant	Absent

[a]Modified from Rogers and Kennedy (140) with data from references 36 and 115.
[b]Description of *Fusarium moniliforme* sensu stricto, also known under the earlier name *Fusarium verticillioides*.
[c]Best observed on carnation leaf agar or KCL medium (115).
[d]Best observed on carnation leaf agar or in sterile distilled water (115).

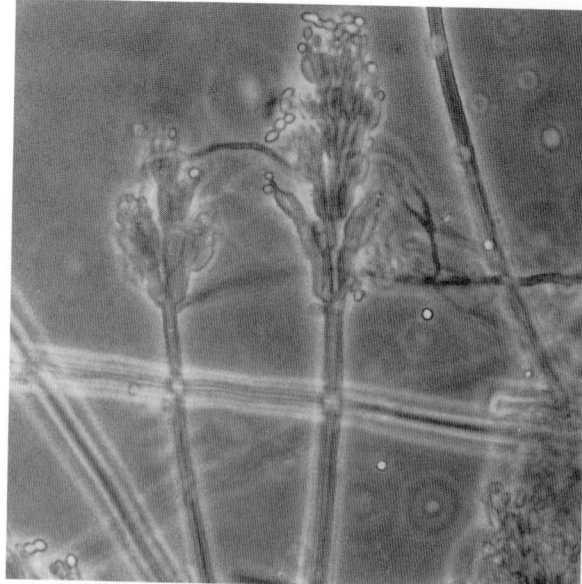

FIGURE 12 Verticillate conidiophores of *Paecilomyces lilacinus* bearing whorls of phialides. Magnification, ×750.

thermotolerant. Section *Isarioidea* is a heterogeneous assemblage of species that produce white or brightly colored colonies. Microscopically, the genus *Paecilomyces* (see Fig. 12) is similar to the genus *Penicillium* in forming verticillate conidiophores bearing whorls of branches and phialides. It is distinguished by the lack of bright green or blue-green colonies and by the basally swollen phialides that taper toward the neck and are bent away from the main axis. The conidia are single celled and produced in long, tangled, or divergent chains (36, 146).

All species of *Paecilomyces* grow well on routine media. The colonial and microscopic features of medically important *Paecilomyces* species (Fig. 12 through 14) are described in Table 5. *Byssochlamys fulva*, anamorph *P. fulvus*, is most similar to *P. variotii* but can be distinguished by its production of cleistothecia; its conidia, which are truncate at each end; and its lack of solitary conidia.

PENICILLIUM SPECIES

Members of the genus *Penicillium* are the common blue-green moulds that exist ubiquitously in nature, are among the most common of all laboratory contaminants, and can easily be isolated from respiratory specimens and body surfaces. Although *Penicillium* species grow over a wide range of temperatures from 5 to 45°C, many species are completely or strongly inhibited at 37°C and thus have low potential for causing human infection. Even repeated isolation of a *Penicillium* species from patient specimens does not necessarily indicate an etiologic role, unless the isolate is accompanied by typical fungal elements in tissue specimens or smears of lesional exudate (136). Other factors to consider when assessing the significance of repeated isolation include the isolation of any other fungal species, the patient's possible exposure to airborne conidia, and a diagnosis of bronchiectasis, since conidia may remain viable for prolonged periods without invasion or colonization (90).

The only true pathogen among members of the genus is *Penicillium marneffei*. This species is unique among the pen-

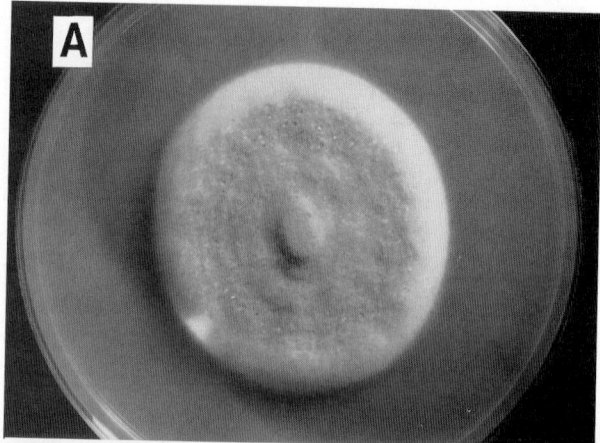

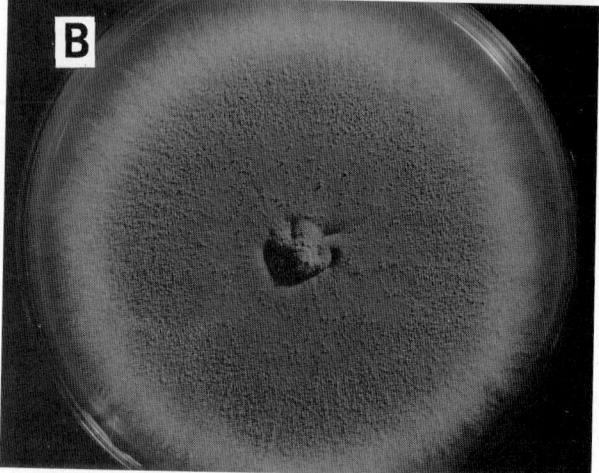

FIGURE 13 Colonies of *Paecilomyces lilacinus* after 11 days (A) and *Paecilomyces variotii* after 7 days (B) on PDA.

icillia in being dimorphic, forming in tissue a unicellular yeastlike organelle that reproduces by planate division. It forms a grayish green colony with an intense brownish red reverse and a red diffusible pigment, and it occurs predom-

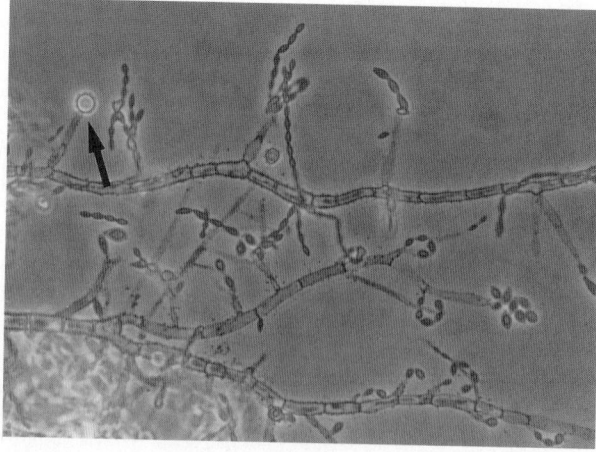

FIGURE 14 Chains of conidia of various sizes and a solitary conidium (arrow) of *Paecilomyces variotii*. Magnification, ×430.

TABLE 5 Morphologic features of selected *Paecilomyces* species[a]

Characteristic	*P. lilacinus*	*P. marquandii*	*P. variotii*
Colonies on PDA after 7 days at 25°C	3–3.5 cm in diam, initially white becoming dull red or reddish gray,[b] cottony, reverse colorless or vinaceous	2 cm in diam, violet brown[b] with white margin, cottony, often with radial folds, yellow reverse, slight yellow diffusible pigment	7–8 cm in diam, yellowish to olive brown, powdery
Temp tolerance	Strongly restricted at 37°C	No growth at 37°C	Up to 50°C (some isolates to 60°C)
Conidiophores	Rough walled, pigmented yellow or purple, with densely clustered metulae bearing whorls of 2–4 phialides (verticillate), rarely synnematous	Smooth walled, hyaline, verticillately branched, metulae with 2–4 phialides, often forming loose synnemata	Verticillately or irregularly branched, metulae bearing 2–7 phialides
Phialides	Taper abruptly to narrow neck about 1 μm wide	Taper abruptly to narrow neck about 1 μm wide	Phialides robust, tapering gradually to form long neck about 2 μm wide
Conidia	Smooth to slightly rough, ellipsoidal to fusiform, hyaline to purple in mass, 2.5–3 by 2–2.2 μm	Smooth to slightly rough, ellipsoidal to fusiform, hyaline to purple in mass, 3–3.5 by 2–2.2 μm	Variable in size and shape, mostly 3.2–5 by 2–4 μm or up to 15 μm long, ellipsoidal to cylindrical smooth, hyaline to yellow-brown
Solitary conidia	Absent	Present, thin walled, often in submerged mycelium	Terminal or intercalary, brown, thick walled

[a]Data are from references 36, 121, and 146.
[b]Color names are from A. Kornerup and J. H. Wanscher, *Methuen Handbook of Colour*, Eyre Methuen Ltd., London, 1983.

inantly in southeastern or far eastern Asia, where it was first reported to be associated with the bamboo rat. Infections caused by *Penicillium marneffei* are usually disseminated, with multiple organ involvement, manifesting as lymphadenitis, subcutaneous abscesses, bone lesions, arthritis, enlarged spleen, or lung, liver, or bowel lesions (90). Characteristic skin and mucocutaneous lesions may also be present (22). As noted by Kwon-Chung and Bennett (90), more than 30 reports of naturally occurring infection were published by 1990, and there is a rising incidence of invasive infections due to this fungus in patients with AIDS (22, 28, 54, 124, 181). The potential for laboratory-acquired infection was demonstrated by the individual who first described *Penicillium marneffei*, who accidently punctured his right index finger while inoculating laboratory rodents. The fungus could be isolated from a small nodule that developed at the puncture site (148). In direct examination, the oval, yeastlike cells of *Penicillium marneffei* may be difficult to distinguish from those of *Histoplasma capsulatum*, especially when they are found within phagocytes. However, the yeast cells of *H. capsulatum* may show evidence of budding rather than division by cross walls, and they stain with specific fluorescent-antibody conjugates of *H. capsulatum* (72). The yeastlike cells of *Penicillium marneffei* stain well with GMS and are oval and about 3 μm wide or elongate and up to 8 μm long, sometimes with a single septum (28). Colonies with sparse green aerial mycelium, reddish brown vegetative mycelium, and red diffusible pigment; compact biverticillate or irregular penicilli borne on short, smooth or finely roughened conidiophores (Fig. 15); and growth at 37°C in a yeastlike form distinguish

Penicillium marneffei from all other species of *Penicillium* (28, 90, 128). In addition, exoantigen tests have proved useful in identification (150).

Other species of *Penicillium* have been implicated in human infection, but Rippon (136) has suggested that in the vast majority of these cases, the mould isolated was a

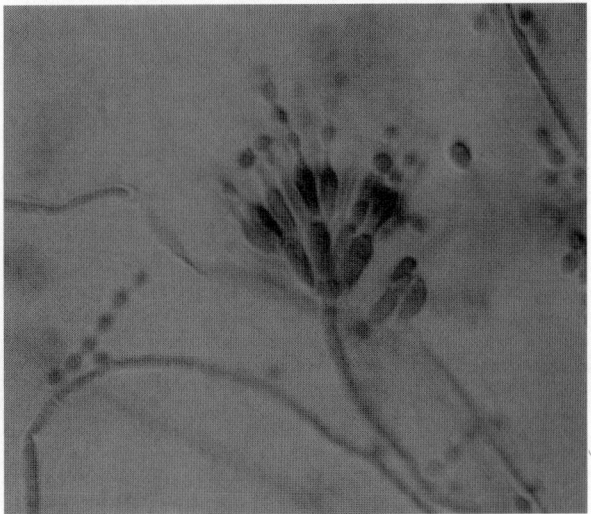

FIGURE 15 Compact biverticillate penicillus of *Penicillium marneffei*. Magnification, ×1,280.

contaminant and not associated with the pathology. Nevertheless, several reports appear to be genuine cases of infection. *Penicillium chrysogenum*, *Penicillium decumbans*, *Penicillium citrinum*, *Penicillium commune*, *Penicillium expansum*, and unidentified species have been reported to cause urinary tract infection (50), endocarditis following mitral or aortic valve replacement (27, 78), mycotic keratitis (41), pulmonary infections (49, 96), hypersensitivity pneumonitis (164), fungemia (5), and disseminated infection (64).

Identification of *Penicillium* species is a complex task requiring growth of an isolate on a variety of media at different temperatures and careful microscopic examination. The use of Sabouraud dextrose agar alone is not recommended. Pitt's (128) identification scheme employs Czapek yeast extract agar, malt extract agar, and 25% glycerol nitrate agar and incubation at 5, 25, and 37°C. Microscopically, species of *Penicillium* are distinguished by development of the classic penicillus, a simple to complex structure consisting of a slender conidiophore and zero to three or four levels of branching. The branches supporting the phialides are called metulae. Branching patterns among the penicilli include simple (not branched, or monoverticillate; subgenus *Aspergilloides*), one or two staged and symmetrical (biverticillate; subgenus *Biverticillium*) (Fig. 15), one or two staged and asymmetrical (biverticillate; subgenus *Furcatum*), and three or four staged (terverticillate or quaterverticillate; subgenus *Penicillium*) (128). The conidia of *Penicillium* species are oval or round with smooth or finely roughened surfaces, hyaline, dark green or blue green, and 2 to 5 μm in diameter, and they are formed in basipetal chains. Sclerotia and sexual fruiting bodies may form in some species of penicillia, since the genus includes anamorphs of *Talaromyces* and *Eupenicillium* (36, 128).

SCOPULARIOPSIS SPECIES

Members of the genus *Scopulariopsis* are common soil fungi and agents of deterioration, especially of cellulosic substrates. The genus *Scopulariopsis* contains both moniliaceous and dematiaceous species (112). Some of the dematiaceous species have teleomorphs in the genus *Microascus* of the Microascaceae. Other dematiaceous species can form complex conidiomata called synnemata in culture, and these species are usually classified in the closely related genus *Cephalotrichum* (also known as *Doratomyces*). Of the species reported from human infections, *Scopulariopsis brevicaulis*, *Scopulariopsis candidum*, and *Scopulariopsis acremonium* form white to buff-colored colonies, whereas *Scopulariopsis koningii*, *Scopulariopsis brumptii*, *Scopulariopsis fusca*, and the *Scopulariopsis* anamorphs of *Microascus* species form graybrown to black colonies.

Most cases of human infection are caused by *Scopulariopsis brevicaulis*, and this species is often recovered as a laboratory contaminant, especially from cutaneous specimens. Like species of *Fusarium* and *P. variotii*, it may be found growing on necrotic tissue. Infections with *Scopulariopsis brevicaulis* are usually confined to the toenails and, less commonly, the fingernails (70). Direct examination in KOH of scrapings or clippings of infected nails characteristically shows clusters of the round, thick-walled, yellowish brown conidia typical of the species or hyphal elements atypical of a dermatophyte. Invasion of deeper tissues is rare. *Scopulariopsis brevicaulis* has been isolated from a deep lesion in the soft tissues of the ankle (152) and from the toe with involvement of the bone (8, 126). Unidentified *Scopulariopsis* species have been reported to cause pneumonia in

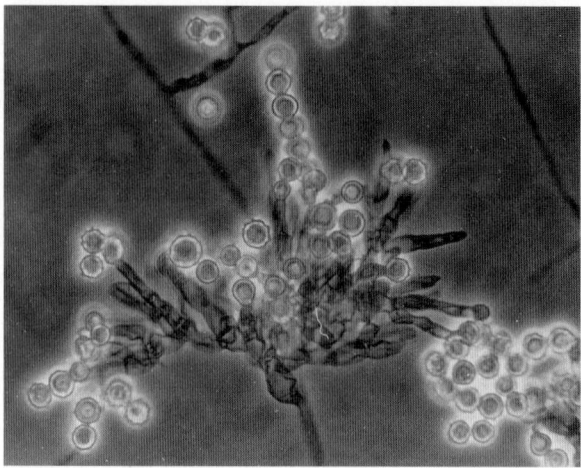

FIGURE 16 Rough-walled conidia in chains formed on annellides in *Scopulariopsis brevicaulis*. Note the branched conidiogenous apparatus. Magnification, ×650.

a patient with acute myeloblastic leukemia (190) and invasive infection in leukemic patients following chemotherapy or bone marrow transplant (114). The dark-colored *Scopulariopsis fusca*, *Scopulariopsis koningii*, and *Microascus* species *Microascus cirrosus* and *Microascus cinereus* as well as moniliaceous *Scopulariopsis acremonium* and *Scopulariopsis candidum* are recovered mostly from toenails (32, 70, 90). *Scopulariopsis brumptii* has been isolated from pulmonary nodules in an intravenous drug abuser with hypersensitivity pneumonitis (57). *Microascus cinereus* has been reported together with *Aspergillus* species in the lung (92) and in the maxillary sinus (11). Perithecia revealed by histologic examination of biopsy material from the maxillary sinus (11) suggested that the pathology was due to *Microascus cinereus*.

Although there are approximately 30 species of *Scopulariopsis*, *Scopulariopsis brevicaulis* may be identified with confidence. Isolates of this species grow well on all laboratory media, including media containing cycloheximide. Colonies of *Scopulariopsis brevicaulis* typically grow moderately rapidly and are buff or tan and coarsely powdery. In contrast, colonies of *Scopulariopsis candidum* and *Scopulariopsis acremonium* remain white or creamy white, and the conidia of *Scopulariopsis acremonium* are larger and are oval to egg shaped with a pointed tip. In common with *Paecilomyces* and *Penicillium* spp., *Scopulariopsis* species develop a complex branched conidiogenous apparatus with one or two levels of branching, but the conidiogenous cells differ in being annellides rather than phialides. Conidia are formed in basipetal chains (Fig. 16), and mature conidia are thick walled, round with a flattened base and sometimes with a small point at the tip, smooth to coarsely roughened, hyaline to tan in mass, and 5 to 8 μm wide by 5 to 7 μm long.

Dematiaceous *Scopulariopsis* species are more difficult to identify. *Scopulariopsis brumptii* has been compared with *Scedosporium* species (145), but conidia of *Scopulariopsis* species are produced in chains, whereas those of *Scedosporium* spp. are slimy. Occasional isolates grown on sporulation media may produce sexual fruiting bodies with short necks. These isolates are placed in the genus *Microascus*. At maturity, the ascospores come out of the neck of the fruiting body in long, reddish brown chains. Ascospores are

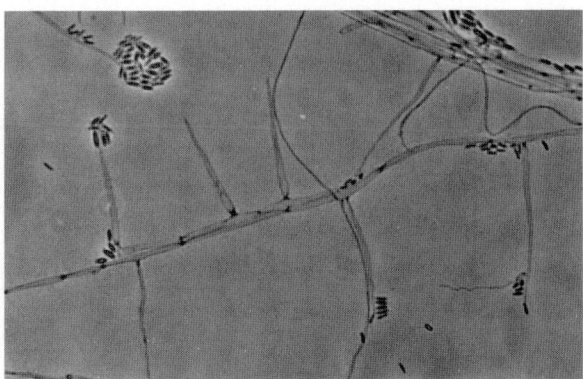

FIGURE 17 Slender, awl-shaped phialides of *Acremonium kiliense*. Magnification, ×460.

FIGURE 18 Fasciculate (showing spiky aggregations of hyphae) colony of *Acremonium kiliense* on PDA after 22 days.

single celled and are either lens shaped with one side slightly convex or almost heart shaped.

ACREMONIUM, LECYTHPHORA, AND PHIALEMONIUM SPECIES

The genus *Acremonium*, formerly called *Cephalosporium*, includes approximately 100 species associated with soil, insects, sewage, rhizospheres of plants, and other environmental substrates. It includes the anamorphs of the ascomycetous genera *Nectria*, *Emericellopsis*, and *Mycoarachis* among others (36). The genus is distinguished by formation of narrow hyphae bearing solitary, slender (usually around 2 μm wide), mostly unbranched, awl (needle)-shaped phialides (Fig. 17), with or without shallow collarettes, that form conidia in slimy masses, rarely in chains. The genera *Phialemonium* and *Lecythophora* are similar to *Acremonium* but have been separated by their formation of short, stumpy phialides without basal septa (called adelophialides) in addition to more typical awl-shaped phialides (47). Although these genera have been described with the dematiaceous medically important fungi (see chapter 67), colonies are usually white, salmon, or yellowish to greenish. Colonies of *Lecythophora mutabilis* become dark because of the development of dematiaceous chlamydospores. Hyphal elements in tissue are usually reported as hyaline (79, 106).

Most reports of infection due to *Acremonium* species are cases of mycetoma (see chapter 68), onychomycosis, mycotic keratitis, or colonization of soft contact lens (90, 136). Well-documented reports of invasive infection are rare and include reports of invasive pulmonary disease in a boy with chronic granulomatous disease; midline granuloma with eroding destruction of the hard palate, maxilla, and mandible; cerebritis in an intravenous drug abuser; endocarditis in a prosthetic valve; and osteomyelitis of the calvarium following trauma (90, 140). Recent reports include infection in arthritis (175), mycotic esophagitis (163), and fungemia in a neutropenic patient (15). Species documented as etiologic agents include *Acremonium alabamensis*, *Acremonium falciforme*, *Acremonium kiliense*, *Acremonium roseo-griseum*, and *Acremonium strictum*, but many reports are based on unidentified species (90).

Species of *Acremonium* grow well on Sabouraud dextrose agar, and some species can tolerate cycloheximide. Growth is slower than that of *Fusarium* species, i.e., usually less than 3 cm in 10 days compared with more than 7 cm for most

Fusarium species. Colonies are often white, velvety, cottony, or fasciculate (having spiky aggregations of hyphae) (Fig. 18) and flat or slightly raised in the center. Species such as *Acremonium strictum*, a common soil and occasional airborne fungus, may appear moist and pink or salmon colored and then resemble *Lecythophora hoffmannii* (47). Although the absence of adelophialides in *Acremonium strictum* should distinguish this species from *Lecythophora* species, these distinctions are not always readily observed (25, 35). *Acremonium kiliense* (Fig. 17 through 19), the most common medically important species, is distinguished by formation of solitary hyaline chlamydospores, especially on oatmeal agar (Fig. 19), and cylindrical conidia. *Acremonium recifei* also forms chlamydospores, but its conidia are sickle shaped or slightly curved. Some species of *Fusarium* fail to form macroconidia or produce them tardily, but isolates of *Fusarium* species are generally faster growing and have broader hyphae with stouter phialides.

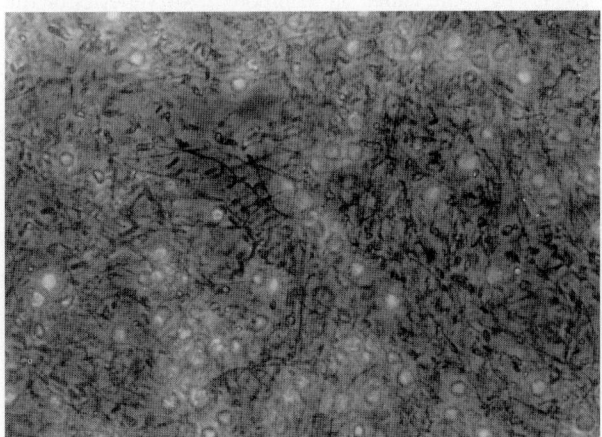

FIGURE 19 Hyaline chlamydospores of *Acremonium kiliense* formed on oatmeal agar. Magnification, ×460.

MISCELLANEOUS FUNGI

Beauveria Species

Beauveria bassiana is a soil saprophyte well known as an insect pathogen, causing muscardine in silkworms and systemic infection in a variety of insects. Although *B. bassiana* was identified as the etiologic agent in a case of lung infection in a 22-year-old female who was diagnosed originally as having tuberculosis, examination of Freour and Lahourcade's report (44) suggests that this species was not the causative agent. Their mould isolate was described as having greenish colonies, and its microscopic features also were not consistent with *B. bassiana*. *B. bassiana* and *Beauveria alba* have been reported in cases of mycotic keratitis (104, 143).

The genus *Beauveria* is characterized by basally swollen conidiogenous cells that are formed in whorls (Fig. 20A) or singly (Fig. 20B) on hyaline hyphae. The conidiogenous cell proliferates sympodially in a zigzag (rachiform or geniculate) fashion (Fig. 20A, arrow). In *B. bassiana*, the conidiogenous cells are often aggregated into sporodochia, while in *B. alba*, they are predominantly solitary. Conidia are small, one celled, hyaline, globose, and 2 to 3 μm in diameter in *B. bassiana*, and oval (1.5 μm wide by 2 μm long) in *B. alba*. Colonies of both species are fairly slow growing (usually 2 cm or less in 7 days), white, and floccose to velvety, developing tufts if sporodochia are produced.

Chrysosporium-Emmonsia-Myceliophthora-Myriodontium Species

Members of the genera *Chrysosporium*, *Emmonsia*, *Myriodontium*, and *Myceliophthora* produce solitary, usually single-celled conidia that are called aleurioconidia because of their lytic method of conidium dehiscence (70). Some species of *Chrysosporium* as well as members of the genus *Emmonsia* are closely related to the dermatophytes or to the systemic pathogens, sharing a tolerance of cycloheximide and having teleomorphs in the ascomycete order Onygenales (70, 156). In contrast, teleomorphs of *Myceliophthora* species are placed in the order Sordariales.

Emmonsia parva, also known as *Chrysosporium parvum*, is the best-known pathogen, causing adiaspiromycosis mainly in rodents and occasionally in humans. *E. parva* is dimorphic, and the infection is named after the tissue form, which is a large, thick-walled, chlamydosporelike, nonreproducing spore called an adiaspore. Two varieties of the fungus are distinguished by their maximum growth temperatures and the different sizes of their adiaspores. Most cases of human infection have been attributed to *E. parva* var. *crescens*, which has a maximum growth temperature of ~37°C and larger adiaspores (often 100 μm or more in diameter). Kwon-Chung and Bennett (90) reported on 21 cases, primarily pulmonary infections, involving this variety. In some instances, the organism could not be cultured but was recognized from histopathology alone. *E. parva* var. *parva* was isolated from biopsy specimens of bone, sputum, and bone marrow from a patient with AIDS (39). The adiaspores of this variety are smaller (usually 10 to 25 μm in diameter), and the maximum growth temperature is 40°C.

The validity of *Chrysosporium* species or *Myriodontium keratinophilum* as etiologic agents is questionable, since in the few reports of infection (46, 95, 99, 169, 179), the isolated organism has not been identified to species or described well enough to confirm the etiology. Indeed, the

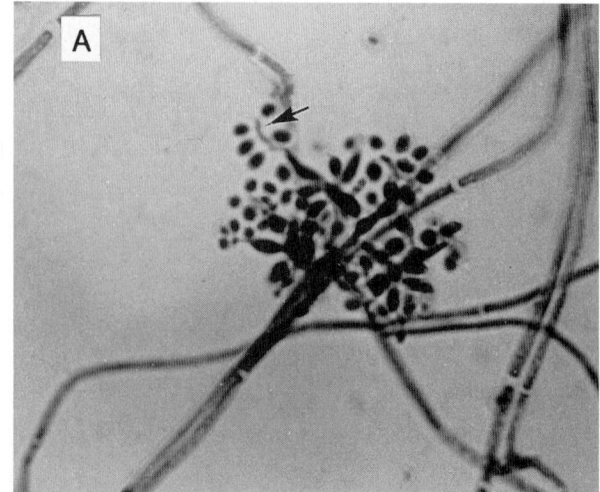

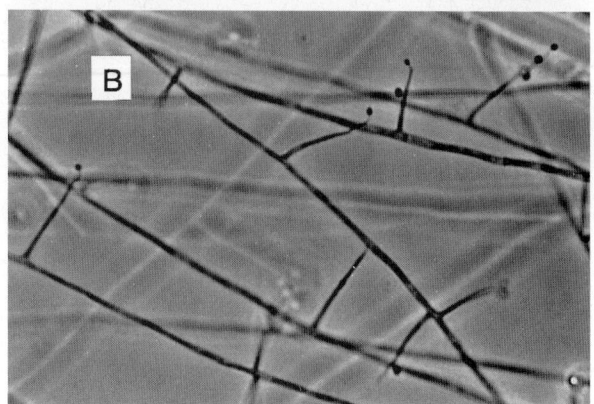

FIGURE 20 *Beauveria* species, showing basally swollen conidiogenous cells that proliferate sympodially in a zigzag (rachiform or geniculate) fashion (arrow, panel A) to form conidia singly from small denticles. (A) Conidiogenous cells of *Beauveria bassiana* are usually produced in sporodochialike clusters. Original magnification, ×1,000. (B) Conidiogenous cells of *Beauveria alba* are solitary. Magnification, ×720. (Panel A is reprinted from reference 136.)

case of osteomyelitis reported by Stillwell et al. (169) has been reported in several subsequent papers as being due to *E. parva*, even though the original authors neither mentioned this species nor reported the presence of adiaspores. Several cases of dermatophytelike infections due to *Aphanoascus fulvescens*, an ascomycete with a *Chrysosporium* anamorph, have been reported (90). Species of *Chrysosporium* and the similar *Geomyces* spp. as well as *Myriodontium keratinophilum* are occasional contaminants of cutaneous specimens, especially from feet, and respiratory specimens (70).

Members of the genus *Myceliophthora* are distinguished by the formation of solitary conidia on conidiogenous cells that are slightly swollen. The thermophilic species *Myceliophthora thermophila* was shown to cause disseminated infection in a leukemic patient (14). The conidia of most species of *Myceliophthora* are hyaline or yellow, but the conidia of *Myceliophthora thermophila* are reddish brown (70).

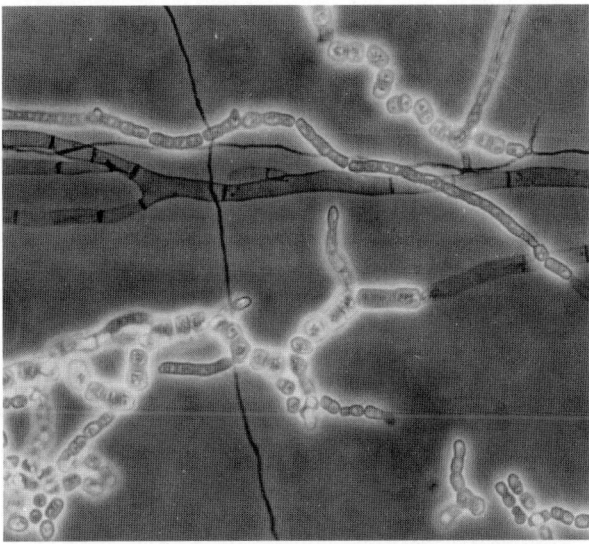

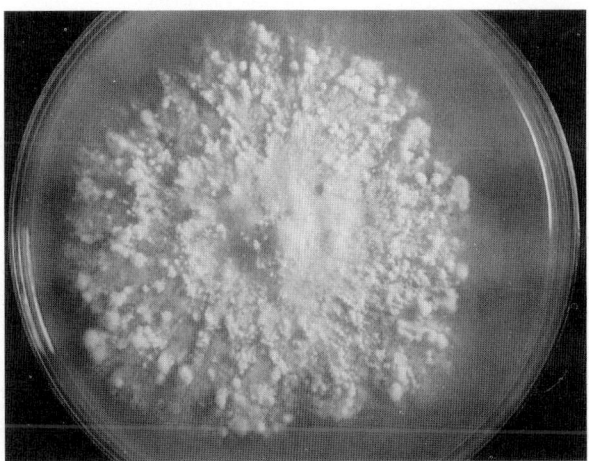

FIGURE 22 Colony of *Scytalidium hyalinum* on phytone yeast extract agar after 9 days at 37°C.

FIGURE 21 Hyphae dividing by schizolytic dehiscence to form arthroconidia in *Scytalidium hyalinum*. Magnification, ×460.

Scytalidium-Onychocola-Arthrographis Species

Scytalidium hyalinum and *Scytalidium dimidiatum* are best known as agents of onychomycosis, especially of the toenails (111), but invasive infections have been rarely reported. *Scytalidium hyalinum* caused multiple cysts in the ankle of an immunocompromised patient (198), and fungemia and subcutaneous lesions due to *Scytalidium dimidiatum* in a granulocytopenic child were recently reported (12). Another case of deep infection was attributed to *Scytalidium lignicola* (34), but the isolate involved in this case was a misidentified *Scytalidium dimidiatum*. *Scytalidium* species are characterized by the formation of schizolytic arthroconidia (158) (Fig. 21). In *Scytalidium hyalinum* and *Scytalidium dimidiatum*, the arthroconidia are formed from hyphae of various widths, ranging from 2 to 10 (or up to 15) μm wide (158, 161). The species differ in the color of the hyphae, which are hyaline to pale yellow in *Scytalidium hyalinum* and light to dark brown in *Scytalidium dimidiatum*; however, in tissue, hyphae of both species are hyaline, pale yellow, or, rarely (in the case of *S. dimidiatum*), brown. Colonies of both species are rapid growing, usually reaching the edge of the petri dish by 7 to 10 days (Fig. 22). They grow well on routine media but are susceptible to cycloheximide. On sporulation media, *Scytalidium dimidiatum* may form pycnidia of the coelomycete stage, which is called *Nattrassia mangiferae* (formerly known as *Hendersonula toruloidea*). Another arthroconidial hyphomycete rarely causing onychomycosis or infections of the glabrous skin is *Onychocola canadensis* (157, 160). Compared with *Scytalidium hyalinum*, this mould is very slow growing and tolerant of cycloheximide. *Arthrographis kalrae* is a thermotolerant arthroconidial fungus which has been recorded only rarely as a human pathogen from skin, lung, and corneal ulcer (102, 158, 159). Initial growth is often yeastlike, but, with age, colonies become tan and powdery due to the development of aerial hyphae. *A. kalrae* forms arthroconidia by fragmentation of fertile hyphae which are typically borne on slender conidiophores.

Basidiomycetes

Allergic reactions due to inhalation of fungal spores and mycetismus due to the ingestion of mushrooms have been frequently reported, but human infections caused by members of the Basidiomycotina are very rare. Several reports have described the isolation of basidiomycetes from various anatomical sites. *Coprinus cinereus* was reported from a case of prosthetic valve endocarditis (165) in which numerous septate hyphae were observed in the aortic valve, within myocardial tissue, and in the lungs. *Schizophyllum commune* has been documented as a pathogen in both immunosuppressed and immunocompetent hosts in cases of chronic lung disease (23) and allergic bronchopulmonary mycosis (69), onychomycosis (81), meningitis (56), and sinusitis (73, 142). In another case, the fungus caused ulcerative lesions that perforated the hard palate of a 4-year-old girl who presented with no underlying disease (132). Hyphae in tissue may be of two types: narrow (5 to 6 μm wide) and thin walled with rare clamp connections or broader (15 to 18 μm) and irregular with numerous clamps (56) (Fig. 23). *Schizophyllum commune* grows on routine media without cycloheximide, appearing as fast-growing, white, woolly or cottony colonies, and is thermotolerant, remaining viable at 45°C. An isolate may be suspected to be *Schizophyllum*

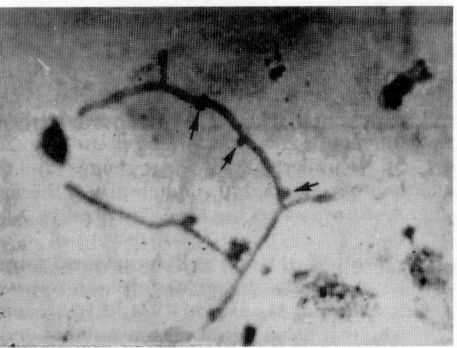

FIGURE 23 *Schizophyllum commune*. Hyphae and clamp connections (arrow) are evident in a KOH mount of scrapings from the roof of a mouth. (Reprinted from reference 140.)

commune if microscopic features of clamps and short, thin, sterigmatalike projections are seen on the vegetative hyphae. On sporulation media after 3 weeks or more of incubation, most isolates develop white to tan fan-shaped fruiting bodies, on the undersides of which are gills. Species of pasidiomycetes are occasionally isolated without evidence of tissue invasion. Emmons (40) observed mycelial strands on direct examination of sputum and reported repeated isolation of *Coprinus micaceus* from one patient. Khan et al. (74) repeatedly recovered *Sporotrichum pruinosum*, the anamorph of *Phanerochaete chrysosporium*, from respiratory secretions of three patients with chronic bronchopulmonary disorders. Like *Schizophyllum commune*, *Sporotrichum pruinosum* is thermotolerant, and it is well known as causing white rot in wood. Although meningitis was attributed to a species of *Ustilago* (110), a genus of smuts, the fungus was not isolated in culture, and the identity could not be confirmed.

REFERENCES

1. **Agrawal, P. K., B. Lal, S. Waheb, O. P. Srivastava, and S. C. Misra.** 1979. Orbital paecilomycosis due to *Paecilomyces lilacinus* (Thom) Samson. *Sabouraudia* **17:**363–370.

2. **Ajello, L.** 1982. Hyalohyphomycosis: a disease entity whose time has come. *Newsl. Med. Mycol. Soc. N.Y.* **10:**3–5.

3. **Ajello, L.** 1986. Hyalohyphomycosis and phaeohyphomycosis: two global diseases entities of public health importance. *Eur. J. Epidemol.* **2:**243–251.

4. **Allevato, P. A., J. M. Ohorodnik, E. Mezger, and J. F. Eisses.** 1984. *Paecilomyces javanicus* endocarditis of native and prosthetic aortic valve. *Am. J. Clin. Pathol.* **82:**247–252.

5. **Alvarez, S.** 1990. Systemic infection caused by *Penicillium decumbans* in a patient with acquired immunodeficiency syndrome. *J. Infect. Dis.* **162:**283.

6. **Ammari, L. K., J. M. Puck, and K. L. McGowan.** 1993. Catheter-related *Fusarium solani* fungemia and pulmonary infection in a patient with leukemia in remission. *Clin. Infect. Dis.* **16:**148–150.

7. **Anaissie, E., G. P. Bodey, H. Kantarjian, J. Ro, S. E. Vartivarian, R. Hopfer, J. Hoy, and K. Rolston.** 1989. New spectrum of fungal infections with cancer. *Rev. Infect. Dis.* **11:**369–378.

8. **Anaissie, E. J., G. P. Bodey, and M. G. Rinaldi.** 1989. Emerging fungal pathogens. *Eur. J. Clin. Microbiol. Infect. Dis.* **8:**323–331.

9. **Arbesman, C. E., K. Wicher, J. I. Wypych, R. E. Reisman, H. Dickie, and C. E. Reed.** 1974. IgE antibodies in sera of patients with allergic bronchopulmonary aspergillosis. *Clin. Allergy* **4:**349–358.

10. **Aufauvre-Brown, A., J. Cohen, and D. W. Holden.** 1992. Use of randomly amplified polymorphic DNA markers to distinguish isolates of *Aspergillus fumigatus*. *J. Clin. Microbiol.* **30:**2991–2993.

11. **Aznar, C., C. de Bievre, and C. Guiguen.** 1989. Maxillary sinusitis from *Microascus cinereus* and *Aspergillus repens*. *Mycopathologia* **105:**93–97.

12. **Benne, C. A., C. Neeleman, M. Bruin, G. S. de Hoog, and A. Fleer.** 1993. Disseminating infection with *Scytalidium dimidiatum* in a granulocytopenic child. *Eur. J. Clin. Microbiol. Infect. Dis.* **12:**118–121.

13. **Bodey, G. P., and S. Vartivarian.** 1989. Aspergillosis. *Eur. J. Clin. Infect. Dis.* **8:**413–437.

14. **Bourbeau, P., D. A. McGough, H. Fraser, N. Shah, and M. G. Rinaldi.** 1992. Fatal disseminated infection caused by *Myceliophthora thermophila*, a new agent of mycosis: case history and laboratory characteristics. *J. Clin. Microbiol.* **30:**3019–3023.

15. **Brown, N. M., E. L. Blundell, S. R. Chown, D. W. Warnock, J. A. Hill, and R. R. Slade.** 1992. *Acremonium* infection in a neutropenic patient. *J. Infect.* **25:**73–76.

16. **Brummund, W., A. Resnic, J. N. Fink, and V. P. Kurup.** 1987. *Aspergillus fumigatus*-specific antibodies in allergic broncho-pulmonary aspergillosis and aspergilloma: evidence for a polyclonal antibody response. *J. Clin. Microbiol.* **25:**5–9.

17. **Burnie, J. P., A. Coke, and R. C. Matthews.** 1992. Restriction endonuclease analysis of *Aspergillus fumigatus* DNA. *J. Clin. Pathol.* **45:**324–327.

18. **Burnie, J. P., R. C. Matthews, I. Clark, and L. J. R. Milne.** 1989. Immunoblot fingerprinting *Aspergillus fumigatus* DNA. *J. Immunol. Methods* **118:**179–186.

19. **Castro, L. G. M., A. Salebian, and M. N. Sotto.** 1990. Hyalohyphomycosis by *Paecilomyces lilacinus* in a renal transplant patient and a review of human *Paecilomyces* species infections. *J. Med. Vet. Mycol.* **28:**15–26.

20. **Chandler, F. W., W. Kaplan, and L. Ajello.** 1980. *Color Atlas and Text of the Histopathology of Mycotic Diseases.* Year Book Medical Publishers, Inc., Chicago.

21. **Chandler, F. W., and J. C. Watts.** 1987. *Pathologic Diagnosis of Fungal Infections.* American Society of Clinical Pathologists, Inc., Chicago.

22. **Chiewchanvit, S., P. Mahanupab, P. Hirunsri, and N. Vanittanakom.** 1991. Cutaneous manifestations of disseminated *Penicillium marneffei* mycosis in five HIV-infected patients. *Mycoses* **34:**245–249.

23. **Ciferri, R., A. Chavez Batista, and S. Campos.** 1956. Isolation of *Schizophyllum commune* from sputum. *Atti Int. Bot. Lab. Crittogam. Univ. Pavia* **14:**118–120.

24. **Conrad, D. J., M. Warnock, P. Blanc, M. Cowan, and J. A. Golden.** 1992. Microgranulomatous aspergillosis after shoveling wood chips: report of a fatal outcome in a patient with chronic granulomatous disease. *Am. J. Ind. Med.* **22:**411–418.

25. **de Hoog, G. S., A. Buiting, C. S. Tan, A. B. Stroebel, C. Ketterings, E. J. de Boer, B. Naafs, R. Brimicombe, M. K. E. Nohlmans-Paulssen, G. T. J. Fabius, A. H. Klokke, and L. G. Visser.** 1993. Diagnostic problems with imported cases of mycetoma in the Netherlands. *Mycoses* **36:**81–87.

26. **Dekhkan-Khodzhaeva, N. A., and S. Shamsiev.** 1982. Role of *Paecilomyces* in the etiology of protracted recurrent bronchopulmonary disease in children. *Pediatriya* **9:**12–14.

27. **DelRossi, A. J., D. Morse, P. M. Spagna, and G. M. Lemole.** 1980. Successful management of *Penicillium* endocarditis. *J. Thorac. Cardiovasc. Surg.* **80:**945–947.

28. **Deng, Z., J. L. Ribas, D. W. Gibson, and D. H. Connor.** 1988. Infections caused by *Penicillium marneffei* in China and Southeast Asia: review of eighteen published cases and report of four more Chinese cases. *Rev. Infect. Dis.* **10:**640–652.

29. **Denning, D. W., K. V. Clemons, L. H. Hanson, and D. A. Stevens.** 1990. Restriction endonuclease analysis of total cellular DNA of *Aspergillus fumigatus* isolates of geographically and epidemiologically diverse origin. *J. Infect. Dis.* **162:**1151–1158.

30. **Denning, D. W., S. E. Follansbee, M. Scolaro, S. Norris, H. Edelstein, and D. A. Stevens.** 1991. Pulmonary aspergillosis in the acquired immunodeficiency syndrome. *N. Engl. J. Med.* **324:**654–662.

31. **Denning, D. W., G. S. Shankland, and D. A. Stevens.** 1991. DNA fingerprinting of *Aspergillus fumigatus* isolates from patients with aspergilloma. *J. Med. Vet. Mycol.* **29:**339–342.

32. **De Vroey, C., A. Lasagni, E. Tosi, F. Schroder, and M. Song.** 1992. Onychomycoses due to *Microascus cirrosus* (syn. *M. desmosporus*). *Mycoses* **35:**193–196.

33. **Dharmasena, F. M., G. S. R. Davies, and D. Catovsky.** 1985. *Paecilomyces variotii* pneumonia complicating hairy cell leukemia. *Br. Med. J.* **290:**967–968.

34. **Dickinson, G. M., T. J. Cleary, T. Sanderson, and M. R. McGinnis.** 1983. First case of subcutaneous phaeohyphomycosis caused by *Scytalidium lignicola* in a human. *J. Clin. Microbiol.* **17:**155–158.

35. **Dixon, D. M., and A. Polak-Wyss.** 1990. The medically

important dematiaceous fungi and their identification. *Mycoses* **34**:1–18.

36. **Domsch, K. H., W. Gams, and T.-H. Anderson.** 1980. *Compendium of Soil Fungi.* IHW-Verlag, Eching, Germany.

37. **Dupont, B., M. Huber, S. J. Kim, and J. E. Bennett.** 1987. Galactomannan antigenemia and antigenuria in aspergillosis: studies in patients and experimentally infected rabbits. *J. Infect. Dis.* **155**:1–11.

38. **Dworzack, D. L., R. B. Clark, W. J. Borkowski, Jr., D. L. Smith, M. Dykstra, M. P. Pugsley, E. A. Horowitz, T. L. Connolly, D. L. McKinney, M. K. Hostetler, J. F. Fitzgibbons, and M. Galant.** 1989. *Pseudallescheria boydii* brain abscess: association with near-drowning and efficacy of high-dose, prolonged miconazole therapy in patients with multiple abscesses. *Medicine* **68**:218–224.

39. **Echavarria, E., E. L. Cano, and A. Restrepo.** 1993. Disseminated adiaspiromycosis in a patient with AIDS. *J. Med. Vet. Mycol.* **31**:91–97.

40. **Emmons, C. W.** 1954. Isolation of *Myxotrichum* and *Gymnoascus* from lungs of animals. *Mycologia* **46**:334–338.

41. **Eschete, M. L., J. W. King, C. West, and A. Aberle.** 1981. *Penicillium chrysogenum* endophthalmitis. *Mycopathologia* **74**:125–127.

42. **Fox, J. L.** 1993. Fungal infection rates are increasing. *ASM News* **59**:515–518.

43. **Frelier, P. F., L. Sigler, and P. E. Nelson.** 1985. Mycotic pneumonia caused by *Fusarium moniliforme* in an alligator. *J. Med. Vet. Mycol.* **23**:399–402.

44. **Freour, P., and M. Lahourcade.** 1965. Une mycose nouvelle. Etude clinique et mycologique d'une localisation pulmonaire de "Beauveria." *Bull. Soc. Med. Hop. Paris* **117**:197–206.

45. **Gabal, M. A.** 1989. Development of a chromosomal DNA probe for the laboratory diagnosis of aspergillosis. *Mycopathologia* **106**:121–129.

46. **Gamis, A. S., T. Gudnason, G. S. Giebink, and N. K. C. Ramsay.** 1991. Disseminated infection with *Fusarium* in recipients of bone marrow transplant. *Rev. Infect. Dis.* **13**:1077–1088.

47. **Gams, W., and M. R. McGinnis.** 1983. *Phialemonium*, a new anamorphic genus intermediate between *Phialemonium* and *Acremonium. Mycologia* **75**:977–987.

48. **Gari, M., J. Fruit, P. Rousseau, J. M. Garnier, C. Trichet, J. C. Baudrillart, P. Compte, P. Feucheres, and J. M. Pinon.** 1985. *Scedosporium (Monosporium) apiospermum*: multiple brain abscesses. *J. Med. Vet. Mycol.* **23**:371–376.

49. **Gelfand, M. S., F. Hammond Cole, Jr., and R. C. Baskin.** 1990. Invasive pulmonary penicilliosis: successful therapy with amphotericin B. *South. Med. J.* **83**:701–704.

50. **Gillium, J., and S. A. Vest.** 1951. *Penicillium* infection of the urinary tract. *J. Urol.* **65**:485–489.

51. **Girardin, H., J.-P. Latage, T. Srikantha, B. Morrow, and D. R. Soll.** 1993. Development of DNA probes for fingerprinting *Aspergillus fumigatus. J. Clin. Microbiol.* **31**:1547–1554.

52. **Goodwin, S. D., J. Fiedler-Kelly, T. H. Grasela, W. A. Schell, and J. R. Perfect.** 1992. A nationwide survey of clinical laboratory methodologies for fungal infections. *J. Med. Vet. Mycol.* **30**:153–160.

53. **Gordon, M. A., and S. A. Norton.** 1985. Corneal transplant infection by *Paecilomyces lilacinus. Sabouraudia* **23**:295–301.

54. **Gradon, J. D., J. G. Timpone, and S. M. Schnittman.** 1992. Emergence of unusual opportunistic pathogens in AIDS: a review. *Clin. Infect. Dis.* **15**:134–157.

55. **Graybill, J. R.** 1992. Future directions of antifungal chemotherapy. *Clin. Infect. Dis.* **14**(Suppl. 1):S170–S181.

56. **Greer, D. L.** 1977. Basidiomycetes as agents of human infections: a review. *Mycopathologia* **65**:133–139.

57. **Grieble, H. G., J. W. Rippon, N. Maliwan, and V. Daun.** 1975. *Scopulariopsis* and hypersensitivity pneumonitis in an addict. *Ann. Intern. Med.* **83**:326–329.

58. **Griffin, T. D., J. P. McFarland, and W. C. Johnson.** 1991. Hyalohyphomycosis masquerading as squamous cell carcinoma. *J. Cutan. Pathol.* **18**:116–119.

59. **Gueho, E., and G. S. de Hoog.** 1991. Taxonomy of the medical species of *Pseudallescheria* and *Scedosporium. J. Mycol. Med.* **1**:3–9.

60. **Harris, L. F., B. M. Dan, and A. W. Lefkowitz.** 1979. *Paecilomyces* cellulitis in a renal transplant patient; successful treatment with intravenous miconazole. *South. Med. J.* **72**:897–898.

61. **Haynes, K. Y., J. P. Latge, and T. R. Rogers.** 1990. Detection of *Aspergillus* antigens associated with invasive infection. *J. Clin. Microbiol.* **28**:2040–2044.

62. **Henwick, S., and S. V. Hetherington.** 1992. Aspergillosis, p. 557–572. *In* C. C. Patrick (ed.), *Infections in Immunocompromised Infants and Children.* Churchill Livingstone, Inc., New York.

63. **Hetherington, S. V., S. Henwick, D. M. Parham, and C. C. Patrick.** 1994. Monoclonal antibodies against a 97-kilodalton antigen from *Aspergillus flavus. Clin. Diagn. Lab. Immunol.* **1**:63–67.

64. **Huang, S. N., and L. S. Harris.** 1963. Acute disseminated penicilliosis. *Am. J. Clin. Pathol.* **39**:167–174.

65. **Jade, K. B., M. F. Lyons, and J. W. Gnan.** 1986. *Paecilomyces lilacinus* in an immunocompromised patient. *Arch. Dermatol.* **122**:1169–1170.

66. **Jang, S. S., T. E. Dorr, E. L. Biberstein, and A. Wong.** 1986. *Aspergillus deflectus* infection in four dogs. *J. Med. Vet. Mycol.* **24**:95–104.

67. **Johnson, T. M., V. P. Kurup, A. Resnick, R. C. Ash, J. N. Fink, and J. Kalbfleisch.** 1989. Detection of circulating *Aspergillus fumigatus* antigen in bone marrow transplant patients. *J. Clin. Lab. Med.* **114**:700–707.

68. **Kalish, S. B., and R. Goldschmidt.** 1982. Infective endocarditis caused by *Paecilomyces variotii. J. Clin. Pathol.* **78**:249–252.

69. **Kamei, K., H. Unno, K. Nagao, T. Kuriyama, K. Nishimura, and M. Miyaji.** 1994. Allergic bronchopulmonary mycosis caused by the basidiomycetous fungus *Schizophyllum commune. Clin. Infect. Dis.* **18**:305–309.

70. **Kane, J., R. C. Summerbell, L. Sigler, S. Krajden, and G. Land.** *Handbook of Dermatophytes. A Clinical Guide and Laboratory Manual of Dermatophytes and Other Filamentous Fungi from Skin, Hair and Nails,* in press. Star Publishing Co., Belmont, Calif.

71. **Kaufman, L.** 1981. Current methods for serodiagnosing systemic fungus infections and identifying their etiologic agents. *Estr. L'Ig. Mod.* **76**:426–442.

72. **Kaufman, L., and W. Kaplan.** 1961. Preparation of a fluorescent antibody specific for the yeast phase of *Histoplasma capsulatum. J. Bacteriol.* **82**:729–735.

73. **Kern, M. E., and F. A. Uecker.** 1986. Maxillary sinus infection caused by the homobasidiomycetous fungus *Schizophyllum commune. J. Clin. Microbiol.* **23**:1001–1005.

74. **Khan, Z. U., H. S. Randhawa, T. Kowshik, S. N. Gaur, and G. A. de Vries.** 1988. The pathogenic potential of *Sporotrichum pruinosum* isolated from the human respiratory tract. *J. Med. Vet. Mycol.* **26**:145–151.

75. **Khan, Z. U., M. D. Richardson, D. L. Crutcher, and D. W. Warnock.** 1984. Use of *Aspergillus fumigatus* culture filtrate fractions in ELISA for the serological diagnosis of allergic aspergillosis. *Mykosen* **27**:327–339.

76. **Kiehn, T. E., P. E. Nelson, E. M. Bernard, F. F. Edwards, B. Koziner, and D. Armstrong.** 1985. Catheter-associated fungemia caused by *Fusarium chlamydosporum* in a patient with lymphocytic lymphoma. *J. Clin. Microbiol.* **21**:501–504.

77. **Kimura, M., S. I. Udagawa, A. Shoji, H. Kume, M. Iimori, T. Satou, and S. Hashimoto.** 1990. Pulmonary aspergillosis due to *Aspergillus terreus* combined with staphylococcal pneumonia and hepatic candidiasis. *Mycopathologia* **111**:47–53.

78. **Kinare, S. G., and A. P. Chaukar.** 1978. Fungal endocarditis after cardiac valve surgery. *J. Postgrad. Med.* **24**:164–170.

79. **King, D., L. Pasarell, D. M. Dixon, M. R. McGinnis, and W. G. Merz.** 1993. A phaeohyphomycotic cyst and perito-

nitis caused by *Phialemonium* species and a reevaluation of its taxonomy. *J. Clin. Microbiol.* **31:**1804–1810.

80. **Klich, M. A., and J. I. Pitt.** 1988. *A Laboratory Guide to Common Aspergillus Species and Their Teleomorphs.* Commonwealth Scientific and Industrial Research Organization, North Ryde, Australia.

81. **Kligman, A. M.** 1950. A basidiomycete probably causing onychomycosis. *J. Invest. Dermatol.* **14:**67–70.

82. **Knight, F., and D. W. R. Mackenzie.** 1992. *Aspergillus* antigen latex test for diagnosis of invasive aspergillosis. *Lancet* **339:**188.

83. **Knutsen, A. P., P. S. Hutcheson, K. R. Mueller, and R. G. Slavin.** 1990. Serum immunoglobulins E and G anti-*Aspergillus fumigatus* antibody in patients with cystic fibrosis who have allergic bronchopulmonary aspergillosis. *J. Lab. Clin. Med.* **116:**724–727.

84. **Kulik, M. M., and A. G. Brooks.** 1970. Electrophoretic studies of soluble proteins from *Aspergillus* spp. *Mycologia* **62:**365–376.

85. **Kurien, M., V. Anandi, R. Raman, and K. N. Brahmadathan.** 1992. Maxillary sinus fusariosis in immunocompetent hosts. *Laryngol. Otol.* **106:**733–736.

86. **Kurup, V. P.** 1986. Enzyme-linked immunosorbent assay in the detection of specific antibodies against *Aspergillus* in patient sera. *Zentralbl. Bakteriol. Mikrobiol. Hyg. Reihe A* **261:**509–516.

87. **Kurup, V. P.** 1988. Production and characterization of a murine monoclonal antibody to *Aspergillus fumigatus* antigen having IgG- and IgE-binding activity. *Int. Arch. Allergy Appl. Immunol.* **86:**400–406.

88. **Kurup, V. P., and A. Kumar.** 1991. Immunodiagnosis of aspergillosis. *Clin. Microbiol. Rev.* **4:**439–456.

89. **Kurup, V. P., M. Ramasamy, P. A. Greenberger, and J. N. Fink.** 1988. Isolation and characterization of a relevant *Aspergillus fumigatus* antigen with IgG and IgE binding activity. *Int. Arch. Allergy Appl. Immunol.* **86:**176–182.

90. **Kwon-Chung, K. J., and J. E. Bennett.** 1992. *Medical Mycology.* Lea & Febiger, Philadelphia.

91. **Kwon-Chung, K. J., W. B. Hill, and J. E. Bennett.** 1981. New, special stain for histopathological diagnosis of cryptococcosis. *J. Clin. Microbiol.* **13:**383–387.

92. **Lacey, J.** 1986. *Microascus cinereus* (Emile-Weil & Gaudin) Curzi: a human pathogen? *Mycopathologia* **96:**137–142.

93. **Landau, M., A. Srebrnik, R. Wolf, E. Bashi, and S. Brenner.** 1992. Systemic ketoconazole treatment for *Fusarium* leg ulcers. *Int. J. Dermatol.* **31:**511–512.

94. **Leslie, C. E., B. Flannigan, and L. J. R. Milne.** 1988. Morphological studies on clinical isolates of *Aspergillus fumigatus. J. Med. Vet. Mycol.* **26:**335–341.

95. **Levy, F. E., J. T. Larson, E. George, and R. H. Maisel.** 1991. Invasive *Chrysosporium* infection of the nose and paranasal sinuses in an immunocompromised host. *Otolaryngol. Head Neck Surg.* **104:**384–388.

96. **Liebler, G. A., G. J. Magovern, P. Sadighi, S. B. Park, and W. J. Cushing.** 1977. *Penicillium* granuloma of the lung presenting as a solitary pulmonary nodule. *JAMA* **234:**671.

97. **Loudon, K. W., J. P. Burnie, A. P. Coke, and R. C. Matthews.** 1993. Application of polymerase chain reaction to fingerprinting *Aspergillus fumigatus* by random amplification of polymorphic DNA. *J. Clin. Microbiol.* **31:**1117–1121.

98. **Malloch, D., and I. F. Salkin.** 1984. A new species of *Scedosporium* associated with osteomyelitis in humans. *Mycotaxon* **21:**247–255.

99. **Maran, A. G. D., K. Kwong, L. J. R. Milne, and D. Lamb.** 1985. Frontal sinusitis caused by *Myriodontium keratinophilum. Br. Med. J.* **290:**207.

100. **Marin, J., M. A. Sanz, G. F. Sanz, J. Guarro, M. L. Martinez, M. Prieto, E. Gueho, and J. L. Menezo.** 1991. Disseminated *Scedosporium inflatum* infection in a patient with acute myeloblastic leukemia. *Eur. J. Clin. Microbiol. Infect. Dis.* **10:**759–761.

101. **Marzec, A., L. G. Heron, R. C. Pritchard, R. H. Butcher, H. R. Powell, A. P. S. Disney, and F. A. Tosolini.** 1993.

102. **McAleer, R., J. H. Froudist, and G. Cherian.** 1988. A dimorphic fungus *Arthrographis kalrae*, implicated in two diseases, abstr. P172. X *Congr. Int. Soc. Hum. Anim. Mycol.*

103. **McClellen, J. R., J. D. Hamilton, J. A. Alexander, W. G. Wolfe, and J. B. Reed.** 1976. *Paecilomyces variotii* endocarditis on a prosthetic aortic valve. *J. Thorac. Cardiovasc. Surg.* **71:**472–475.

104. **McDonnell, P. J., T. P. Werblin, L. Sigler, and W. R. Green.** 1985. Mycotic keratitis due to *Beauveria alba. Cornea* **3:**213–216.

105. **McGinnis, M. R., L. Ajello, and W. A. Schell.** 1985. Mycotic diseases: a proposed nomenclature. *Int. J. Dermatol.* **24:**9.

106. **McGinnis, M. R., W. Gams, and M. N. Goodwin.** 1986. *Phialemonium obovatum* infection in a burned child. *J. Med. Vet. Mycol.* **24:**51–55.

107. **Melcher, G. P., D. A. McGough, A. W. Fothergill, C. Norris, and M. G. Rinaldi.** 1993. Disseminated hyalohyphomycosis caused by a novel human pathogen, *Fusarium napiforme. J. Clin. Microbiol.* **31:**1461–1467.

108. **Miller, G. P., and G. Rebell.** 1978. Intravitreal antimycotic therapy and the cure of mycotic endophthalmitis caused by a *Paecilomyces lilacinus* contaminated pseudophakos. *Ophthalmic Surg.* **9:**54–63.

109. **Moody, S. F., and B. M. Tyler.** 1990. Use of nuclear DNA restriction fragment length polymorphisms to analyze the diversity of the *Aspergillus flavus* group: A. *flavus*, A. *parasiticus*, and A. *nomius. Appl. Environ. Microbiol.* **56:**2453–2461.

110. **Moore, M., and W. O. Russel.** 1946. Chronic leptomeningitis and ependymitis caused by *Ustilago*, probably U. *zeae*: ustilagomycosis, the second reported instance of human infection. *Am. J. Pathol.* **22:**761–773.

111. **Moore, M. K.** 1992. The infection of human skin and nail by *Scytalidium* species, p. 1–42. *In* M. Borgers, R. Hay, and M. G. Rinaldi (ed.), *Current Topics in Medical Mycology.* Springer-Verlag, New York.

112. **Morton, F. J., and G. Smith.** 1963. *The Genera Scopulariopsis Bainier, Microascus Zukal, and Doratomyces Corda.* Mycological papers no. 86. Commonwealth Mycological Institute, Kew, Surrey, England.

113. **Musial, C. A., F. R. Cokerill III, and G. D. Roberts.** 1988. Fungal infections of the immunocompromised host: clinical and laboratory aspects. *Clin. Microbiol. Rev.* **1:**349–364.

114. **Neglia, J. P., D. D. Hurd, P. Ferrieri, and D. C. Snover.** 1987. Invasive *Scopulariopsis* in the immunocompromised host. *Am. J. Med.* **83:**1163–1166.

114a. **Nelson, P. E., M. C. Dignani, and E. J. Anaissie.** 1994. Taxonomy, biology, and clinical aspects of *Fusarium* species. *Clin. Microbiol. Rev.* **7:**479–504.

115. **Nelson, P. E., T. A. Toussoun, and W. F. O. Marasas.** 1983. *Fusarium Species: an Illustrated Manual of Identification.* Pennsylvania State University Press, State College.

116. **O'Day, D. M.** 1977. Fungal endophthalmitis caused by *Paecilomyces lilacinus* after intraocular lens implantation. *Am. J. Ophthalmol.* **83:**130–131.

117. **Odds, F., T. Arai, A. F. DiSalvo, E. G. V. Evans, R. J. Hay, H. S. Randhawa, M. G. Rinaldi, and T. J. Walsh.** 1992. Nomenclature of fungal diseases: a report and recommendations from a sub-committee of the International Society for Human and Animal Mycology (ISHAM). *J. Med. Vet. Mycol.* **30:**1–10.

118. **Okuda, C., M. Ito, Y. Sato, K. Oka, and M. Hotchi.** 1987. Disseminated cutaneous *Fusarium* infection with vascular invasion in a leukemic patient. *J. Med. Vet. Mycol.* **25:**177–186.

119. **Onions, A. H. S., D. Allsopp, and H. O. W. Eggins.** 1981. *Smith's Introduction to Industrial Mycology.* Edward Arnold, London.

120. **Otcenasek, M., Z. Jirousek, Z. Nozicka, and K. Mencl.** 1984. Paecilomycosis of the maxillary sinus. *Mykosen* **27:**242–248.

Paecilomyces variotii in peritoneal dialysate. *J. Clin. Microbiol.* **31:**2392–2395.

121. **Padhye, A. A.** 1988. Hyalohyphomycosis, p. 654–662. *In* A. Balows, W. J. Hausler, Jr., M. Ohashi, and A. Turano (ed.), *Laboratory Diagnosis of Infectious Diseases. Principles and Practice*, vol. 1. *Bacterial, Mycotic and Parasitic Diseases.* Springer-Verlag, New York.

122. **Patterson, R., and M. Roberts.** 1974. IgE and IgG antibodies against *Aspergillus fumigatus* in sera of patients with bronchopulmonary allergic aspergillosis. *Int. Arch. Allergy* **46:**150–160.

123. **Patterson, R., M. Roberts, A. C. Ghory, and P. A. Greenberger.** 1980. IgA antibody against *Aspergillus fumigatus* antigen in patients with allergic bronchopulmonary aspergillosis. *Clin. Exp. Immunol.* **42:**395–398.

124. **Perfect, J. R., W. A. Schell, and M. G. Rinaldi.** 1993. Uncommon invasive fungal pathogens in the acquired immunodeficiency syndrome. *J. Med. Vet. Mycol.* **31:**175–179.

125. **Peterson, S. W.** 1992. *Neosartorya pseudofischeri* sp. nov. and its relationship to other species in *Aspergillus* section *Fumigati.* *Mycol. Res.* **96:**547–554.

126. **Phillips, P., W. S. Wood, G. Phillips, and M. G. Rinaldi.** 1989. Invasive hyalohyphomycosis caused by *Scopulariopsis brevicaulis* in a patient undergoing allogenic bone marrow transplant. *Diagn. Microbiol. Infect. Dis.* **12:**429–432.

127. **Pierard, G. E., J. Arrese-Estrada, C. Pierard-Franchimont, A. Thiry, and D. Stynen.** 1991. Immunohistochemical expression of galactomannan in the cytoplasm of phagocytic cells during invasive aspergillosis. *Am. J. Clin. Pathol.* **96:**373–376.

128. **Pitt, J. I.** 1979. *The Genus Penicillium and Its Teleomorphic States Eupenicillium and Talaromyces.* Academic Press, London.

129. **Polachek, I., A. Nagler, E. Okon, P. Drakos, J. Plaskowitz, and K. J. Kwon-Chung.** 1992. *Aspergillus quadrilineatus,* a new causative agent of fungal sinusitis. *J. Clin. Microbiol.* **30:**3290–3293.

130. **Polonelli, L., S. Conti, W. Magliani, and G. Moraci.** 1989. Biotyping of pathogenic fungi by the killer system and with monoclonal antibodies. *Mycopathologia* **107:**17–23.

131. **Raper, K. B., and D. I. Fennell.** 1965. *The Genus Aspergillus.* The Williams & Wilkins Co., Baltimore.

132. **Restrepo, A., D. L. Greer, M. Robledo, O. Osorio, and H. Mondragon.** 1973. Ulceration of the palate caused by a basidiomycete *Schizophyllum commune. Sabouraudia* **9:**201–204.

133. **Richardson, S. E., R. M. Bannatyne, R. C. Summerbell, J. Milliken, R. Gold, and S. S. Weitzman.** 1988. Disseminated fusarial infection in the immunocompromised host. *Rev. Infect. Dis.* **10:**1171–1181.

134. **Rinaldi, M. G.** 1983. Invasive aspergillosis. *Rev. Infect. Dis.* **5:**1061–1066.

135. **Rinaldi, M. G.** 1988. Aspergillosis, p. 559–572. *In* A. Balows, W. J. Hausler, Jr., M. Ohashi, and A. Turano (ed.), *Laboratory Diagnosis of Infectious Diseases. Principles and Practice*, vol. 1. *Bacterial, Mycotic and Parasitic Diseases.* Springer-Verlag, New York.

136. **Rippon, J. W.** 1988. *Medical Mycology: the Pathogenic Fungi and the Pathogenic Actinomycetes*, 3rd ed. The W. B. Saunders Co., Philadelphia.

137. **Rippon, J. W., R. A. Larson, D. M. Rosenthal, and J. Clayman.** 1988. Disseminated cutaneous and peritoneal hyalohyphomycosis caused by *Fusarium* species: three cases and review of the literature. *Mycopathologia* **101:**105–111.

138. **Rockhill, R. C., and M. D. Klein.** 1980. *Paecilomyces lilacinus* as the cause of chronic maxillary sinusitis. *J. Clin. Microbiol.* **11:**737–739.

139. **Rodrigues, M. M., and D. MacLeod.** 1975. Exogenous fungal endophthalmitis caused by *Paecilomyces. Am. J. Ophthalmol.* **79:**687–690.

140. **Rogers, A. L., and M. J. Kennedy.** 1991. Opportunistic hyaline hyphomycetes, p. 659–673. *In* A. Balows, W. J. Hausler, Jr., K. L. Herrmann, H. D. Isenberg, and H. J. Shadomy (ed.), *Manual of Clinical Microbiology*, 5th ed. American Society for Microbiology, Washington, D.C.

141. **Rogers, T. R., K. A. Haynes, and R. A. Barnes.** 1990. Value of antigen detection in predicting invasive pulmonary aspergillosis. *Lancet* **336:**1210–1213.

142. **Rosenthal, J., R. Katz, D. B. DuBois, A. Morrissey, and A. Machicao.** 1992. Chronic maxillary sinusitis associated with the mushroom *Schizophyllum commune* in a patient with AIDS. *Clin. Infect. Dis.* **14:**46–48.

143. **Sachs, S. W., J. Baum, and C. Mies.** 1985. *Beauveria bassiana* keratitis. *Br. J. Ophthalmol.* **69:**548–550.

144. **Salkin, I. F., C. R. Cooper, J. W. Bartges, M. E. Kemna, and M. G. Rinaldi.** 1992. *Scedosporium inflatum* osteomyelitis in a dog. *J. Clin. Microbiol.* **30:**2797–2800.

145. **Salkin, I. F., M. R. McGinnis, M. J. Dykstra, and M. G. Rinaldi.** 1988. *Scedosporium inflatum,* an emerging pathogen. *J. Clin. Microbiol.* **26:**498–503.

146. **Samson, R. A.** 1974. *Paecilomyces* and some allied hyphomycetes. *Stud. Mycol.* **6:**1–117.

147. **Schonheyder, H.** 1987. Pathogenetic and serological aspects of pulmonary aspergillosis. *Scand. J. Infect. Dis. Suppl.* **51:**1–62.

148. **Segretain, G.** 1959. *Penicillium marneffei* n. spp. agent d'une mycose du systeme reticuloendothelial. *Mycopathologia* **11:**327–353.

149. **Segretain, G., H. Fromentin, P. Destombes, E. R. Brygoo, and A. Dodin.** 1964. *Paecilomyces virdis* n. sp. champignon dimorphque, agent d'une mycose generalisee de *Chameleo lateralis* Gray. *C.R. Acad. Sci.* (Paris) **259:**258–261.

150. **Sekhon, A. S., J. S. K. Li, and A. K. Garg.** 1982. Penicilliosis marneffei: serological and exoantigen studies. *Mycopathologia* **77:**51–57.

151. **Sekhon, A. S., P. G. Standard, L. Kaufman, A. K. Garg, and P. Cifuentes.** 1986. Grouping of *Aspergillus* species with exoantigens. *Diagn. Immunol.* **4:**112–116.

152. **Sekhon, A. S., D. J. Williams, and J. H. Harvey.** 1974. Deep *Scopulariopsis:* a case report and sensitivity studies. *J. Clin. Pathol.* **27:**837–843.

153. **Sepulvedda, R., J. L. Longbottom, and J. Pepys.** 1979. Enzyme-linked immunosorbent assay (ELISA) for IgG and IgE antibodies to protein and polysaccharide antigens of *Aspergillus fumigatus. Clin. Allergy* **9:**359–371.

154. **Sherwood, J. A., and A. S. Dansky.** 1983. *Paecilomyces* pyelonephritis complicating nephrolithiasis and a review of *Paecilomyces* infections. *J. Urol.* **130:**526–528.

155. **Sigler, L.** 1992. Preparing and mounting slide cultures, p. 6.12.1–6.12.4. *In* H. D. Isenberg (ed.), *Clinical Microbiology Procedures Handbook*, vol. 1. American Society for Microbiology, Washington, D.C.

156. **Sigler, L.** 1993. Perspective on Onygenales and their anamorphs by a traditional taxonomist, p. 161–168. *In* D. R. Reynolds and J. W. Taylor (ed.), *The Fungal Holomorph: a Consideration of Mitotic, Meiotic and Pleomorphic Speciation.* CAB International, Wallingford, United Kingdom.

157. **Sigler, L., S. P. Abbott, and A. Woodgyer.** 1994. New records of nail and skin infection due to *Onychocola canadensis* and description of its teleomorph *Arachnomyces nodosetosus* sp. nov. *J. Med. Vet. Mycol.* **32:**275–285.

158. **Sigler, L., and J. W. Carmichael.** 1976. Taxonomy of *Malbranchea* and some other Hyphomycetes with arthroconidia. *Mycotaxon* **4:**349–488.

159. **Sigler, L., and J. W. Carmichael.** 1983. Redisposition of some fungi referred to *Oidium microspermum* and a review of *Arthrographis. Mycotaxon* **18:**495–507.

160. **Sigler, L., and L. Congly.** 1990. Toenail infection caused by *Onychocola canadensis* gen. et sp. nov. *J. Med. Vet. Mycol.* **28:**405–417.

161. **Sigler, L., and C. J. K. Wang.** 1990. *Scytalidium circinatum* sp. nov., a hyphomycete from utility poles. *Mycologia* **82:**399–404.

162. **Silver, M. D., P. G. Tuffinell, and W. G. Bigelow.** 1971. Endocarditis caused by *Paecilomyces variotii* affecting an aortic valve allograft. *J. Thorac. Cardiovasc. Surg.* **61:**278–281.

163. **Simon, G., G. Rakoczy, J. Galgoczy, T. Verebely, and J.**

Bokay. 1991. *Acremonium kiliense* in oesophagus stenosis. *Mycoses* **34:**257–260.

164. **Solley, G. O., and R. E. Hyatt.** 1980. Hypersensitivity pneumonitis induced by *Penicillium* species. *J. Allergy Clin. Immunol.* **65:**65–70.

165. **Speller, D. C. E., and A. C. MacIves.** 1971. Endocarditis caused by a *Coprinus* species. A fungus of the toadstool group. *J. Med. Microbiol.* **4:**370–374.

166. **Spreadbury, C. L., B. W. Bainbridge, and J. Cohen.** 1990. Restriction fragment length polymorphisms in isolates of *Aspergillus fumigatus* probed with part of the intergenic spacer region from the ribosomal RNA gene complex of *Aspergillus nidulans. J. Gen. Microbiol.* **136:**1991–1994.

167. **Spreadbury, C., D. Holden, A. Aufauvre-Brown, B. Bainbridge, and J. Cohen.** 1993. Detection of *Aspergillus fumigatus* by polymerase chain reaction. *J. Clin. Microbiol.* **31:**615–621.

168. **Starr, M. B.** 1987. *Paecilomyces lilacinus* keratitis: two case reports in extended wear contact lens wearers. *CLAO J.* **13:**95–101.

169. **Stillwell, W. T., B. D. Rubin, and J. L. Axelrod.** 1984. *Chrysosporium,* a new causative agent in osteomyelitis. *Orthopedics* **184:**190–192.

170. **Stimlam, C. V., D. E. Dines, R. F. Rodgers-Sullivan, G. D. Roberts, and W. C. Sheehan.** 1980. Respiratory tract *Aspergillus*—clinical significance. *Minn. Med.* **63:**25–30.

171. **Summerbell, R. C., L. de Repentigny, C. Chartrand, and G. St.-Germain.** 1992. Graft-related endocarditis caused by *Neosartorya fischeri* var. *spinosa. J. Clin. Microbiol.* **30:**1580–1582.

172. **Summerbell, R. C., S. Krajden, and J. Kane.** 1989. Potted plants in hospitals as reservoirs of pathogenic fungi. *Mycopathologia* **106:**13–22.

173. **Summerbell, R. C., S. E. Richardson, and J. Kane.** 1988. *Fusarium proliferatum* as an agent of disseminated infection in an immunocompromised patient. *J. Clin. Microbiol.* **26:**82–87.

174. **Swatek, F., C. Halde, M. J. Rinaldi, and H. J. Shadomy.** *Aspergillus* and other opportunistic saprophytic hyaline hyphomycetes, p. 584–594. *In* E. H. Lennette, A. Balows, W. J. Hausler, Jr., and H. J. Shadomy (ed.), *Manual of Clinical Microbiology,* 4th ed. American Society for Microbiology, Washington, D.C.

175. **Szombathy, S. P., M. G. Chez, and R. M. Laxer.** 1988. Acute septic arthritis due to *Acremonium. J. Rheumatol.* **15:**714–715.

176. **Talbot, G. H., M. H. Weiner, S. L. Gerson, M. Provencher, and S. Hurwitz.** 1987. Serodiagnosis of invasive aspergillosis in patients with hematologic malignancy: validation of the *Aspergillus fumigatus* antigen radioimmunoassay. *J. Infect. Dis.* **155:**12–27.

177. **Tan, T. Q., A. K. Ogden, J. Tillman, G. J. Demmler, and M. G. Rinaldi.** 1992. *Paecilomyces lilacinus* catheter-related fungemia in an immunocompromised patient. *J. Clin. Microbiol.* **30:**2479–2483.

178. **Teltenti, A., and G. D. Roberts.** 1989. Fungal blood cultures. *Eur. J. Clin. Microbiol. Infect. Dis.* **8:**825–831.

179. **Toshniwal, R., S. Goodman, S. A. Ally, V. Ray, C. Bodino, and C. A. Kallick.** 1986. Endocarditis due to *Chrysosporium* species: a disease of medical progress? *J. Infect. Dis.* **153:**638–639.

180. **Treger, T. R., D. W. Bisscher, M. S. Bartlett, and J. W. Smith.** 1985. Diagnosis of pulmonary infection caused by *Aspergillus:* usefulness of respiratory cultures. *J. Infect. Dis.* **152:**572–576.

181. **Tsang, D. N. C., P. C. K. Li, M. S. Tsui, Y. T. Lau, and E. K. Yeoh.** 1991. *Penicillium marneffei:* another pathogen to consider in patients infected with human immunodeficiency virus. *Rev. Infect. Dis.* **13:**766–777.

182. **van Belkum, A., W. G. V. Quint, B. E. de Pauw, W. J. G. Melchers, and J. F. Meis.** 1993. Typing of *Aspergillus* species and *Aspergillus fumigatus* isolates by interrepeat polymerase chain reaction. *J. Clin. Microbiol.* **31:**2502–2505.

183. **Van Cutsem, J., L. Meulemans, F. Van Gerven, and D. Stynen.** 1990. Detection of circulating galactomannan by Pastorex Aspergillus in experimental invasive aspergillosis. *Mycoses* **33:**61–69.

184. **Wallenbeck, I., S. Dreborg, O. Zetterstrom, and R. Einarsson.** 1991. *Aspergillus fumigatus* specific IgE and IgG antibodies for diagnosis of Aspergillus-related lung diseases. *Allergy* **46:**372–378.

185. **Warnock, D. W., A. B. Foot, E. M. Johnson, S. B. Mitchell, J. M. Cornish, and A. Oakhill.** 1991. *Aspergillus* antigen latex test for diagnosis of invasive aspergillosis. *Lancet* **338:**1023–1024.

186. **Weiner, M. H.** 1980. Antigenemia detected by radioimmunoassay in systemic aspergillosis. *Ann. Intern. Med.* **92:**793–796.

187. **Weiss, L. M., and W. A. Thiemke.** 1983. Disseminated *Aspergillus ustus* infection following cardiac surgery. *Am. J. Clin. Pathol.* **80:**408–411.

188. **Weitzman, I.** 1988. Saprophytic molds as agents of cutaneous and subcutaneous infection in the immunocompromised host. *Arch. Dermatol.* **122:**1161–1168.

189. **Wheat, L. J.** 1986. The role of the serologic diagnostic laboratory and the diagnosis of fungal disease, p. 43–68. *In* G. A. Sarosi and S. F. Davies (ed.), *Fungal Diseases of the Lung.* Grune and Stratton, Inc., New York.

190. **Wheat, L. J., M. Bartlett, M. Ciccarelli, and J. W. Smith.** 1984. Opportunistic *Scopulariopsis* pneumonia in an immunocompromised host. *South. Med. J.* **77:**1608–1609.

191. **Wheeler, M. S., M. R. McGinnis, W. A. Schell, and D. H. Walker.** 1981. *Fusarium* infection in burned patients. *Am. J. Clin. Pathol.* **75:**304–311.

192. **Williamson, P. R., K. J. Kwon-Chung, and J. J. Gallin.** 1992. Successful treatment of *Paecilomyces variotii* infection in a patient with chronic granulomatous disease and a review of *Paecilomyces* species infections. *Clin. Infect. Dis.* **14:**1023–1026.

193. **Wilson, C. M., E. J. O'Rourke, M. R. McGinnis, and I. F. Salkin.** 1990. *Scedosporium inflatum:* clinical spectrum of a newly recognized pathogen. *J. Infect. Dis.* **161:**102–107.

194. **Wilson, E. V., V. M. Hearn, and D. W. R. Mackenzie.** 1987. Evaluation of a test to detect circulating *Aspergillus fumigatus* antigen in a survey of immunocompromised patients with proven or suspected invasive disease. *J. Med. Vet. Mycol.* **25:**365–374.

195. **Wood, G. M., J. G. McCormack, D. B. Muir, D. H. Ellis, M. F. Ridley, R. Pritchard, and M. Harrison.** 1992. Clinical features of human infection with *Scedosporium inflatum. Clin. Infect. Dis.* **14:**1027–1033.

196. **Young, R. C., and J. E. Bennett.** 1971. Invasive aspergillosis. Absence of detectable antibody response. *Am. Rev. Respir. Dis.* **104:**710–716.

197. **Yu, B., Y. Niki, and D. Armstrong.** 1990. Use of immunoblotting to detect *Aspergillus fumigatus* antigen in sera and urine of rats with experimental invasive aspergillosis. *J. Clin. Microbiol.* **28:**1575–1579.

198. **Zaatari, G. S., G. Reed, and R. Morewessel.** 1984. Subcutaneous hyphomycosis caused by *Scytalidium hyalinum. Am. J. Clin. Pathol.* **83:**252–256.

199. **Zapater, R. C.** 1986. Opportunistic fungus infections—*Fusarium* infections (keratomycosis by *Fusarium*). *Jpn. J. Med. Mycol.* **27:**68–69.

200. **Zapater, R. C., A. Arrechea, and V. H. Guevara.** 1972. Queratomicosis por *Fusarium dimerum. Sabouraudia* **10:**274–275.

201. **Zoutman, D. E., and L. Sigler.** 1991. Mycetoma of the foot caused by *Cylindrocarpon destructans. J. Clin. Microbiol.* **29:**1855–1859.

Trichophyton, Microsporum, Epidermophyton, and Agents of Superficial Mycoses

IRENE WEITZMAN, JULIUS KANE, AND RICHARD C. SUMMERBELL

65

TAXONOMY

The etiologic agents of the dermatophytoses are classified in three anamorphic genera: *Trichophyton, Microsporum,* and *Epidermophyton.* Those dermatophytes capable of reproducing sexually, i.e., producing ascomata with asci and ascospores, are classified in the teleomorphic genus *Arthroderma* (71), family *Arthrodermataceae* of the *Onygenales* (12), phylum *Ascomycota.* The teleomorph-anamorph states of the dermatophytes and similar species are given in Table 1.

CHARACTERIZATION

Physiologically, the dermatophytes have in common the ability to digest keratin (an insoluble scleroprotein), a tolerance for cycloheximide, and the ability to produce a rise in pH into the alkaline range as a consequence of growth in medium containing glucose and peptone.

NATURAL HABITAT

Dermatophytes usually are grouped into three categories based on host preference and natural habitat (2). Anthropophilic species almost exclusively infect humans; animals are rarely infected. Geophilic species are soil-inhabiting organisms, and soil is a source of infection for both humans and lower animals. Zoophilic species are essentially pathogens of lower animals, although animal-to-human transmission is not uncommon. This grouping (Table 2) may be helpful in determining the source of infection; e.g., human infections caused by *Microsporum canis* are often the result of contact between susceptible children and stray kittens (39).

Geographically, the dermatophytes may vary in distribution (50); some, e.g., *Trichophyton rubrum,* have a global distribution, whereas others, e.g., *Trichophyton concentricum,* are geographically limited, being found only in the Pacific Islands and regions in Central and South America.

CLINICAL MANIFESTATIONS

The dermatophytoses (tinea or ringworm) are fungal infections of the keratinized tissues (hair, nails, skin, etc.) of humans and lower animals by a group of closely related keratinophilic fungi collectively referred to as dermato-

phytes, whose anamorphs (asexual forms) belong to the genera *Epidermophyton, Microsporum,* and *Trichophyton.* Cutaneous infections resembling dermatophytoses may be caused by yeasts or other unrelated filamentous fungi that are normally saprophytes or plant pathogens; these infections are often referred to as opportunistic dermatomycoses (74).

Tissue invasion is essentially cutaneous because of the inability of these fungi to penetrate deeper tissues or organs as a result of nonspecific inhibitory factors in serum (37) and inhibition of fungal keratinases (14). Infection may range from mild to severe as a consequence of the reaction of the host to the metabolic products of the fungus, the virulence of the infecting strain, the anatomic location of the infection, and local environmental factors. Occasionally, subcutaneous tissue may be invaded, e.g., in Majocchi's granuloma, kerion, mycetoma-like processes (8, 76), or, more rarely, a generalized systemic infection (6).

Anatomic Location

Traditionally, infections caused by dermatophytes have been named according to the anatomic location involved, e.g., tinea barbae (beard and moustache), tinea capitis (scalp, eyebrows, and eyelashes), tinea corporis (face and trunk), tinea cruris (groin, perineal, and perianal areas), tinea pedis (feet), and tinea unguium (nails). Different dermatophyte species may produce clinically identical lesions; conversely, a single species may infect many anatomic sites.

Tinea barbae, usually caused by zoophilic fungi, e.g., *Trichophyton verrucosum* and *Trichophyton mentagrophytes* var. *mentagrophytes,* are typically highly inflammatory and may present as acute pustular folliculitis that can progress to suppurative boggy lesions (kerion). A less severe form that appears as dry, erythematous, scaly lesions also occurs. Tinea capitis may vary in its presentation from highly erythematous, patchy, scaly areas with dull gray hair stumps to highly inflammatory with folliculitis, kerion formation, alopecia, and scarring. *Trichophyton tonsurans* and *M. canis* are the most common agents (50). Favus (tinea favosa), most commonly caused by *Trichophyton schoenleinii* and rarely caused by *Trichophyton violaceum* or *Microsporum gypseum,* is a chronic infection of the scalp and glabrous skin characterized by the formation of cup-shaped crusts resembling honeycombs (scutula). Tinea corporis, which can be

TABLE 1 Teleomorph-anamorph states of dermatophytes and similar species[a]

Anamorph	Teleomorph
Trichophyton spp.	*Arthroderma* spp.
T. ajelloi	A. uncinatum
T. flavescens	A. flavescens
T. georgiae	A. ciferrii
T. gloriae	A. gloriae
T. mentagrophytes	A. benhamiae, A. vanbreuseghemii
T. simii	A. simii
T. terrestre	A. quadrifidum, A. insingulare, A. lenticularum
T. vanbreuseghemii	A. gertlerii
Microsporum spp.	*Arthroderma* spp.
M. amazonicum	A. borellii
M. boullardii	A. corniculatum
M. cookei	A. cajetani
M. canis var. canis	A. otae
M. canis var. distortum	A. fulvum, A. incurvatum, A. gypseum
M. gypseum	A. gypseum
M. nanum	A. obtusum
M. persicolor	A. persicolor
M. racemosum	A. racemosum
M. vanbreuseghemii	A. grubyi
Microsporum sp.	A. cookiellum

[a]Data are from reference 38.

caused by any dermatophyte, is classically manifested as circular, erythematous lesions with scaly, raised, active, often vesicular borders. Chronic lesions on the trunk and extremities usually are caused by *T. rubrum*. Tinea cruris ("jock itch"), usually caused by *T. rubrum* or *Epidermophyton floccosum*, typically appears as scaly, erythematous to tawny brown, bilateral and asymmetric lesions extending down to the inner thigh and exhibiting a sharply marginated border frequently studded with small vesicles. Tinea

pedis varies in appearance; the most common manifestation is maceration, peeling, itching, and painful fissuring between the fourth and fifth toes. An acute inflammatory condition with vesicles, pustules, or a hyperkeratotic chronic type ("moccasin foot") are other manifestations. *T. mentagrophytes* var. *mentagrophytes* (granular) frequently causes the more inflammatory type of infections, whereas *T. rubrum* or *T. mentagrophytes* var. *interdigitale* (velvety) causes the more chronic type. Tinea unguium, most often caused by *T. rubrum*, usually appears as thickened, deformed, friable, discolored nails with accumulated subungual debris.

Transmission and Contagion

Anthropophilic fungi usually are transmitted by close human contact or indirectly by sharing of clothes, combs, brushes, towels, bedsheets, etc. Tinea capitis is highly contagious and may spread rapidly within a family, institution, or school. Transmission of tinea cruris is associated with shared clothing, towels, and sanitary facilities. The transmission of tinea pedis and tinea unguium is controversial (54, 74). Geophilic infections involving transmission from a soil source to humans or lower animals are rare, but outbreaks originating from infected soil with secondary human transmission have been reported (5). Infections involving zoophilic species result from animal-to-human contact (cats, dogs, cattle, laboratory animals, etc.) or from indirect transmission involving fomites.

Detection

Patients with suspected tinea capitis should be examined with a Wood's lamp (filtered UV light peak of 365 nm) in a darkened room for the presence of bright green fluorescent hairs. Such hairs, considered Wood's light positive, typically occur in the small-spored ectothrix type of hair invasion caused by *Microsporum audouinii*, *M. canis*, and *Microsporum ferrugineum*. Hairs infected with *T. schoenleinii* may show a dull green color (52).

A Wood's lamp can also be used to differentiate between infections of the skin that may be similar clinically, e.g., erythrasma (caused by a bacterium, *Corynebacterium minutissimum*) from dermatophytosis. In the former, the skin

TABLE 2 Grouping of dermatophytes on basis of host preference and natural habitat

Anthropophilic	Geophilic	Zoophilic
Epidermophyton floccosum	*Microsporum* spp.	*Microsporum* spp.
Microsporum spp.	M. fulvum	M. canis var. canis
M. audouinii	M. gypseum	M. canis var. distortum
M. ferrugineum	M. nanum	M. equinum
Trichophyton spp.	M. persicolor	M. gallinae
T. concentricum	M. praecox	*Trichophyton* spp.
T. kanei	M. racemosum	T. equinum
T. gourvilii	M. vanbreuseghemii	T. mentagrophytes var. erinacei, T. mentagrophytes var. mentagrophytes, T. mentagrophytes var. quinckeanum
T. megninii		
T. mentagrophytes var. interdigitale		
T. raubitschekii		T. simii
T. rubrum		T. verrucosum
T. schoenleinii		
T. soudanense		
T. tonsurans		
T. violaceum		
T. yaoundei		

fluoresces orange to coral red, whereas in the latter, the skin is not fluorescent.

Direct microscopic examination of skin, hair, and nails is the most rapid method of determining fungal etiology and may be accomplished readily by examining the clinical material in 10% potassium (KOH) or 10% sodium hydroxide (NaOH), 25% sodium hydroxide (NaOH) with 5% glycerin, or calcofluor white (53).

LABORATORY METHODS

Collection and Transport of Specimens

Whenever feasible, aseptic technique should be used to minimize contamination. Sufficient clinical material for direct microscopic examination and culture should be collected. The following equipment should be available for collection and transport of specimens: sterile nail clippers, scissors, forceps for epilating hairs, scalpels, sterile gauze squares, 70% alcohol for disinfection, sterile water for cleaning painful areas, surface-sterilized clean glass slides, clean pill packets or clean paper envelopes to contain and transport the clinical specimens, and appropriate culture media when feasible. Black photographic paper may be useful for collecting and better visualizing scrapings. Closed tubes are not recommended for specimens, since they retain moisture, which may result in an overgrowth of contaminants. Disposable brushes have been recommended for collection of specimens from the scalp or fur of animals (40).

Hairs from the scalp should be epilated with sterile forceps. If the specimen is Wood's light positive, epilate only fluorescent hairs. In favus, the scutulum at the mouth of the hair follicle is suitable for culture and microscopic examination. Hairs invaded by endothrix fungi may need to be dug out with the tip of a sterile scalpel blade because the hairs often break off flush at scalp level, making it difficult to grasp them with forceps. Rubbing with a sterile moistened swab has been successful with pediatric patients (22). After disinfection with alcohol or cleansing with sterile water, active borders of skin lesions should be scraped with a scalpel to collect epidermal scales. In vesicular tinea pedis, the tops of the vesicles can be removed with sterile scissors for direct examination and culture. Nails should be disinfected with alcohol gauze squares. The most desirable material for culture is the waxy subungual debris, which contains the viable fungal elements. In order to remove contaminating saprophytic fungi and bacteria, the crumbly debris directly underneath the nail near the tips should be removed with a scalpel before material is collected for culture. Similarly, if the dorsal nail plate is diseased, scrape and discard the outer surface before the nail is cultured.

Direct Microscopic Examination

Skin scrapings or hairs are placed in 1 or 2 drops of a KOH or NaOH aqueous solution on a clean glass slide. A coverslip is placed on top, and the preparation is heated gently (short of boiling) by being passed rapidly over a Bunsen burner three or four times and then allowed to sit at room temperature for a few minutes for clearing. Nails may require a stronger alkali solution (up to 25% KOH or NaOH) and a longer clearing time. Fungi may also be demonstrated by use of calcofluor white (53). All preparations should be examined under low power and confirmed under high power.

Infected skin and nails may reveal hyaline hyphal frag-

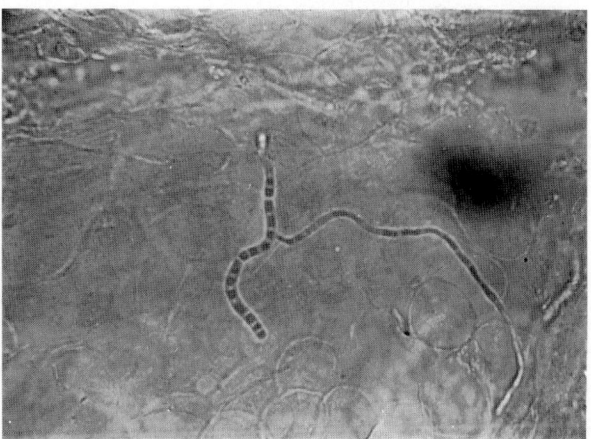

FIGURE 1 Dermatophyte hyphae in skin scraping. NaOH mount. Magnification, ×400.

ments; septate, often branched hyphae; and chains of arthroconidia (Fig. 1).

The appearance of infected hairs depends on the invading dermatophyte species. Three main types of hairs (ectothrix, endothrix, and favic) are observed by the direct microscopic examination. The terms "ectothrix" and "endothrix" refer to the location of the arthroconidia in relation to the hair shaft. Arthroconidia are formed by fragmentation of the invading hyphae. The appearance, sizes, and locations of the arthroconidia may suggest the infecting genera or species (Table 3). The appearances described below are based on descriptions given by Rippon (52).

Ectothrix Hairs

Arthroconidia appear as a mosaic sheath around the hair or as chains on the surface of the hair shaft (Fig. 2).

Arthroconidia 1 to 3 μm in diameter appearing as a mosaic sheath around the hair suggest M. audouinii, M. canis, or M. ferrugineum. Infected hairs are Wood's lamp positive.

Arthroconidia 3 to 4 μm in diameter, few in number, and scattered around the outside of the hair suggest M. gypseum and Microsporum fulvum. Hairs are Wood's lamp negative.

Arthroconidia 3 to 4 μm in diameter in chains on the hair surface suggest T. mentagrophytes complex. Hairs are Wood's lamp negative.

Large arthroconidia 8 to 12 μm in diameter in large

TABLE 3 Hair invasion by dermatophytes

Ectothrix	Endothrix	Favic
T. megninii	T. gourvilii	T. schoenleinii
T. mentagrophytes	T. soudanense	
T. verrucosum	T. tonsurans	
M. audouinii	T. violaceum	
M. canis	T. yaoundei	
M. ferrugineum		
M. fulvum		
M. gypseum		
M. praecox		

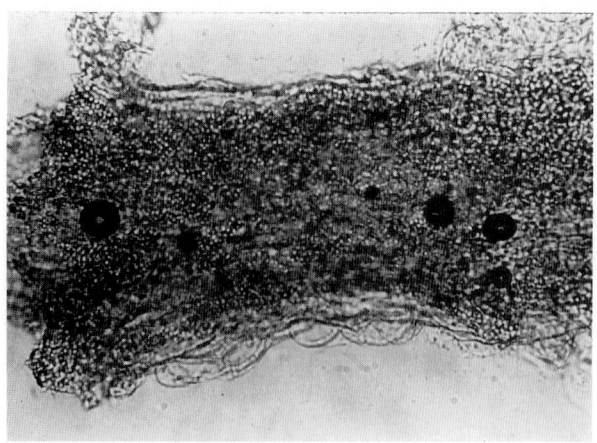

FIGURE 2 M. audouinii, ectothrix type of hair invasion. Magnification, ×400.

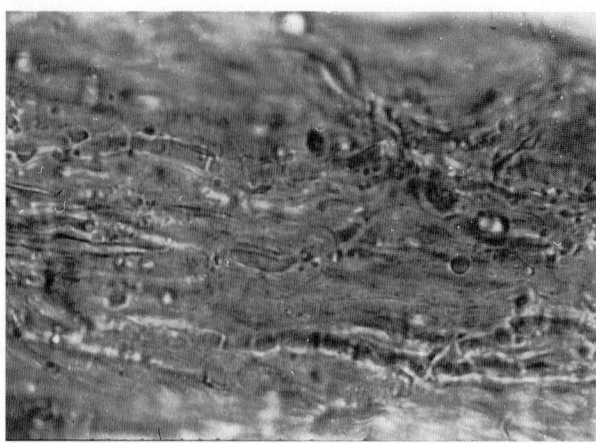

FIGURE 4 Hair infected by T. schoenleinii from a patient with favus. Magnification, ×1,000.

dense chains around the hair are characteristic of T. verrucosum. Hairs are Wood's lamp negative.

Endothrix Hairs

Endothrix hair invasion produced by T. tonsurans, T. violaceum, and Trichophyton gourvilii is observed as chains of arthroconidia 3 to 4 μm in diameter filling the insides of shortened hair stubs (Fig. 3). Hairs are Wood's lamp negative. T. rubrum, which rarely infects hair, has been described as both ectothrix and endothrix (44, 52).

Favic Hairs

Hyphal filaments, air bubbles, or tunnels and fat droplets are observed intrapilar with favic hairs (Fig. 4). These hairs are dull green under the Wood's lamp. In general, all infected hairs show hyphae within the hair shaft at some time during the course of infection, usually during the early stages.

T. concentricum does not invade hair (52), and E. floccosum invades hair only rarely (57).

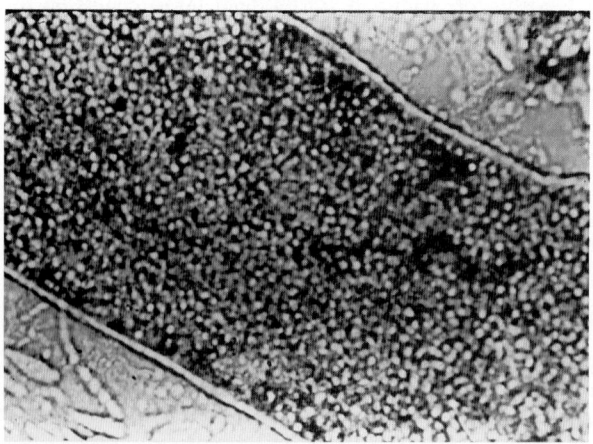

FIGURE 3 T. tonsurans, endothrix type of hair invasion. Magnification, ×1,000.

Isolation Media

In the United States, the most common media used for the isolation of dermatophytes are Sabouraud glucose agar (SDA) (pH 5.6) and Emmons' modification, with less glucose (2 instead of 4%) and a pH between 6.8 and 7.0, paired with SDA containing chloramphenicol and cycloheximide to inhibit bacterial and saprophytic fungal contamination (available commercially as Mycobiotic agar [Difco Laboratories, Detroit, Mich.; Remel, Lenexa, Kans.] or Mycosel [BBL Microbiology Systems, Cockeysville, Md.]). SDA with chloramphenicol (Remel, Difco) and inhibitory mould agar (BBL) are recommended for the isolation from contaminated specimens of opportunistic fungi that cause clinical infections resembling dermatophytosis but are sensitive to cycloheximide (53). The addition of gentamicin is recommended for specimens heavily contaminated by bacteria (65). SDA with cycloheximide, chloramphenicol, and gentamicin (CCG) is routinely used as an isolation medium in the laboratory of the junior author (R.C.S.) in addition to Casamino Acids (Difco)-erythritol-albumin agar medium plus CCG for positive skin and nail specimens. This medium discourages the growth of Candida albicans and encourages the more slowly growing dermatophytes (16). In addition, Littman oxgall agar (Difco, Remel) with added gentamicin and chloramphenicol is used routinely with nail scrapings for the isolation of yeasts and moulds sensitive to cycloheximide (63). For specimens from patients from regions where cattle are commonly raised, SDA plus CCG is replaced by bromcresol purple (BCP)-milk solids agar plus CCG and 0.5% yeast extract to enhance growth and reveal the characteristic casein hydrolysis of T. verrucosum (35).

Some laboratories use dermatophyte test medium (available commercially from Difco and Remel). This selective medium screens for the presence of dermatophytes in heavily contaminated material (from feet, nails, etc.). The growth of dermatophytes causes a rise in pH, thus changing the phenol red indicator from yellow to red. The antibiotics gentamicin and chlortetracycline inhibit bacteria, whereas cycloheximide inhibits saprophytic fungi (66). Dermatophyte test medium should be considered only for screening purposes, since fungi other than dermatophytes can grow and turn the medium red (43, 56).

Cultures are routinely incubated at 25 to 30°C and examined weekly for up to 4 weeks. Cultures of specimens suspected for *T. verrucosum* also should be incubated at 37°C, since this temperature enhances growth of this species.

IDENTIFICATION

Identification of the dermatophyte species is often based on (i) colony characteristics in pure culture on SDA and (ii) microscopic morphology. However, these criteria alone may be insufficient, since colonial appearance may vary or be similar for different species, characteristic pigmentation may fail to appear, and some isolates may not sporulate. This situation is especially pertinent for the genus *Trichophyton*. Special media may be required to stimulate pigment production; it may be necessary to use sporulation and physiologic tests in conjunction with morphology to identify the species correctly.

Colony Characteristics

The following characteristics are important in observing gross colony morphology: color of the surface and the reverse of the colony, texture of the surface (powdery, granular, woolly, cottony, velvety, or glabrous), topography (elevation, type of folding, margins, etc.), and rate of growth.

Microscopic Morphology

Microscopic morphology, especially the appearance and arrangement of the conidia (macroconidia or microconidia) and other structures, may be determined by teased mounts or slide culture preparations mounted in lactophenol cotton blue (Poirrer's blue), in lactophenol aniline blue (phenol, an ingredient of lactophenol cotton blue and lactophenol aniline blue, is listed as a hazardous chemical; therefore, solutions containing phenol should be prepared, stored, and used in an approved chemical safety cabinet), or in more permanent mounting fluids (69). Sometimes a special medium such as cornmeal or cornmeal-glucose agar, potato-glucose agar, SDA plus 3 to 5% NaCl (29, 34), or lactrimel agar (9, 27) may be required to stimulate sporulation.

Physiologic Tests

In Vitro Hair Perforation Test

The in vitro hair perforation tests devised by Ajello and Georg to distinguish between atypical isolates of *T. mentagrophytes* and *T. rubrum* (4) may also be used to distinguish *M. canis* from *Microsporum equinum* (49). Hairs exposed to *T. mentagrophytes* and *M. canis* show wedge-shaped perforations perpendicular to the hair shaft (a positive test result), whereas *T. rubrum* and *M. equinum* do not form these perforating structures. This test may also aid in identifying other species (49).

Place short strands of human hair in petri dishes, and autoclave the dishes at 121°C for 10 min; add 25 ml of sterile distilled water and 2 or 3 drops of 10% sterilized yeast extract. Inoculate these plates with several fragments of the test fungus that has been grown on SDA; incubate the plates at 25°C, and examine them at regular intervals over a period of 21 days. Hairs may be examined microscopically for perforations by removing a few segments and placing them in a drop of lactophenol cotton blue mounting fluid. Gently heating the lactophenol cotton blue mounts aids in the detection of the fungus.

Special Nutritional Requirements

Nutritional tests were originally described by Georg and Camp as an aid in the routine identification of *Trichophyton* species that seldom produce conidia or that resemble each other morphologically (18). Certain species have distinctive nutritional requirements, whereas others do not. The method employs a casein basal medium that is vitamin free (trichophyton agar 1 [T1]) to which various vitamins are added, i.e., inositol (T2), thiamine and inositol (T3), thiamine (T4), and nicotinic acid (T5), and an ammonium nitrate basal medium (T6) to which histidine is added (T7). These media are available commercially in dehydrated form from Difco and in prepared form from Remel. A small fragment (about the size of the head of a pin) from the culture to be tested is placed on the surface of the basal medium (controls) and the media containing the vitamin and amino acid additives. Care must be taken to avoid transferring agar from the fungus inoculum to the nutritional media. Cultures are incubated at room temperature (or 37°C if *T. verrucosum* is suspected) and read after 7 and 14 days. The amount of growth is graded from 0 to 4+. Charts indicating the nutritional requirements are available from the manufacturer as package inserts, from the original paper (18), and from various texts (52, 74).

Urea Hydrolysis

The ability to hydrolyze urea provides additional data to aid in the differentiation of *T. rubrum* (urease negative) from *T. mentagrophytes* (typically urease positive), *T. rubrum* from *Trichophyton raubitschekii* (urease positive) (32, 64), and *T. rubrum* from *Trichophyton megninii* (urease positive) (55, 58). Christensen urea agar and broth may both be used; however, some consider the broth to be more sensitive for this purpose (30). After the urea medium is inoculated, it is incubated at 25 to 30°C up to 7 days. The tubes should be examined every 2 to 3 days for the color change from orange or pale pink to purple-red that indicates the presence of urease, a positive test result. Negative and positive controls should always be included and read first.

Growth on BCP-Milk Solids-Glucose Medium

Type of growth (profuse versus restricted) and a change in the pH indicator (BCP) indicating alkalinity are especially useful for differentiating *T. rubrum* from *T. mentagrophytes* and *T. mentagrophytes* from *Microsporum persicolor* (34, 64). *T. rubrum* shows restricted growth and produces no alkaline reaction on BCP-milk solids-glucose medium, whereas *T. mentagrophytes* typically shows profuse growth and an alkaline reaction. Although *M. persicolor* shows profuse growth, it does not result in an alkaline reaction. Other tests for differentiating *T. mentagrophytes* from *M. persicolor* are described elsewhere (45).

Cultures to be tested are inoculated onto slants of BCP-milk solids-glucose medium and examined for pH change and growth characteristics at the end of a 7-day incubation at 25°C. A color change from pale blue to violet purple indicates an alkaline reaction.

Growth on Polished Rice Grains

Unlike most dermatophytes, *M. audouinii* grows poorly on rice grains and produces a brownish discoloration of the rice (11). *M. equinum* recently was reported to have a growth pattern resembling that of *M. audouinii* (1). This is a useful test for differentiating these species from *M. canis* and from other dermatophytes that typically grow and sporulate on rice grains.

The medium is prepared in 125-ml flasks by mixing 1 part raw unfortified rice grains and 3 parts water (11) or 8.0 g of rice grains and 25 ml of distilled water. Autoclave at 15 lb/in^2 for 15 min. Inoculate the surface of the rice, and incubate the sample for 2 weeks at 25 to 30°C.

Temperature Tolerance and Temperature Enhancement

Tests for temperature tolerance and enhancement are useful for differentiating *T. mentagrophytes* from *Trichophyton terrestre* (46), *T. mentagrophytes* from *M. persicolor* (34), *T. verrucosum* from *T. schoenleinii* (52), and *Trichophyton soudanense* from *M. ferrugineum* (73). At 37°C, *T. mentagrophytes* shows good growth, whereas *T. terrestre* does not grow, and *M. persicolor* grows poorly or not at all; growth of *T. verrucosum* and *T. soudanense* is enhanced, but that of *T. schoenleinii* and *M. ferrugineum* is not.

Inoculate two slants of SDA with an equivalent fragment of the culture. Incubate the slants at room temperature (25 to 30°C) and at 37°C. Compare the growth at both temperatures when mature colonies appear at room temperature. Appropriate controls are recommended and should be compared first.

DESCRIPTION OF ETIOLOGIC AGENTS

For practical purposes, identification of dermatophytes in the routine clinical laboratory is based on the anamorph, i.e., conidia in culture, rather than on the teleomorph (sexual form). However, Takashio's medium has been reported to differentiate between the teleomorphic species of the *M. gypseum* complex on the basis of colonial and microscopic features (15).

Two types of hyaline conidia may be produced by the dermatophytes: large multicellular, smooth or rough, thin- or thick-walled macroconidia and smaller unicellular, smooth-walled microconidia. The three genera are grouped according to the presence or absence of these two types of conidia and the appearance of the surface of the macroconidia, i.e., rough versus smooth. However, this rigid morphologic distinction has become a morphologic continuum based on overlapping characteristics (48). Identification of these species is based on the microscopic appearance and

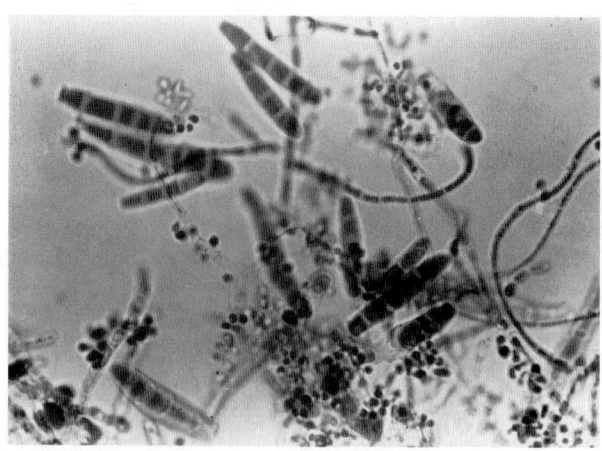

FIGURE 6 Macroconidia and microconidia of *T. mentagrophytes* on SDA–5% NaCl. Magnification, ×400.

arrangement of the conidia, colonial morphology on SDA (Fig. 5 through 20), and physiological tests (Table 4).

Trichophyton Species

Macroconidia have smooth thin to thick walls, are variable in shape (clavate, fusiform to cylindrical), vary in number of septa (1 to 12) and in size (8 to 86 by 4 to 14 μm), and are borne singly or in clusters. Microconidia, which are usually present and more numerous than macroconidia, may be globose, pyriform, or clavate and are borne singly or in grapelike clusters. Although the production of microconidia is more characteristic of this genus, three species lacking microconidia have been described: *Trichophyton (Keratinomyces) longifusum* (3, 17), *Trichophyton kanei* (62), and a variant of *T. tonsurans* (48).

Microsporum Species

Microsporum species produce macroconidia and microconidia that may be rare or numerous, depending on the species and the substrate. The distinguishing characteristic is the macroconidium, which is typically rough walled (varying from minutely roughed, echinulate to verrucose). Macroconidia also vary in shape (obovate, fusiform to cy-

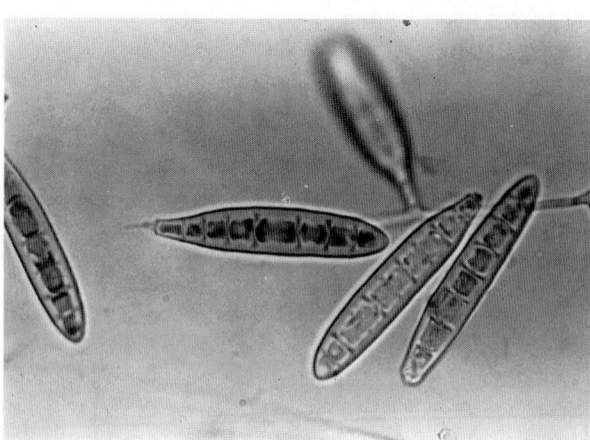

FIGURE 5 Smooth-walled macroconidia of *Trichophyton ajelloi*. Magnification, ×400.

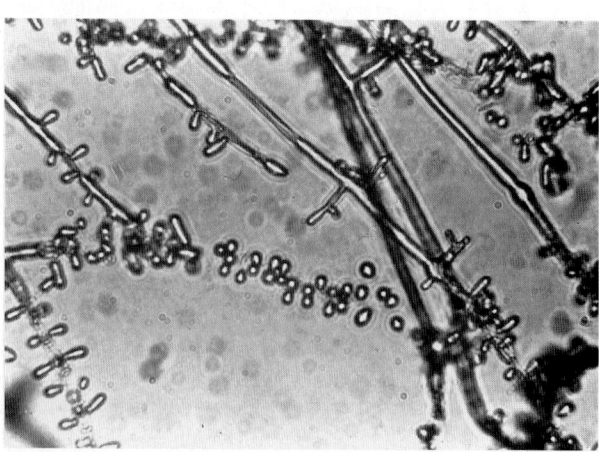

FIGURE 7 Clavate and subspherical microconidia of *T. raubitschekii*. Magnification, ×400.

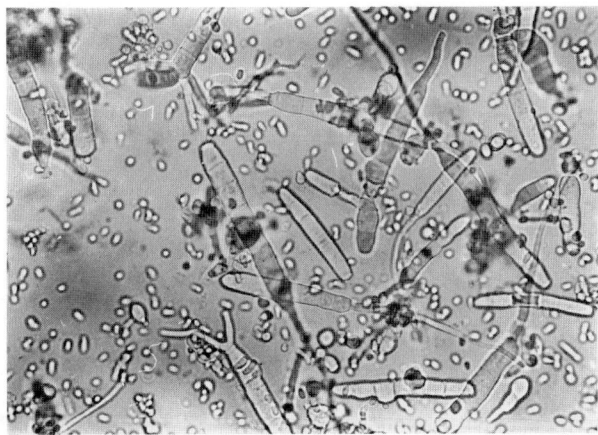

FIGURE 8 Smooth-walled macroconidia of *T. raubitschekii* from primary isolate on SDA. Magnification, ×400.

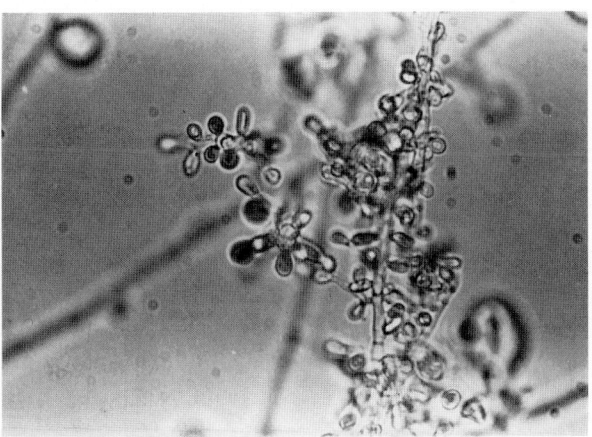

FIGURE 11 Microconidia with typical condensed cytoplasm of *T. tonsurans*. Magnification, ×400.

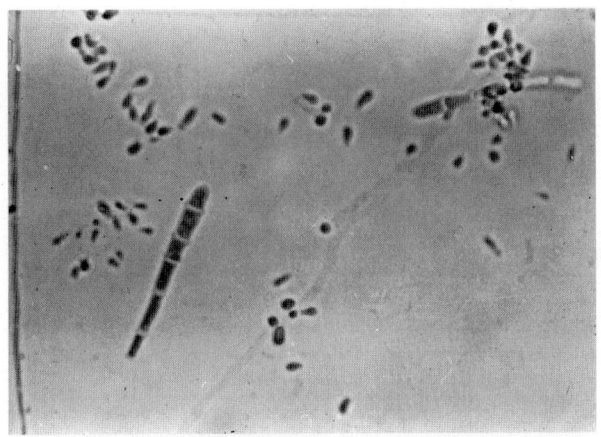

FIGURE 9 Long, narrow macroconidium and clavate to pyriform microconidia of *T. rubrum*. Magnification, ×400.

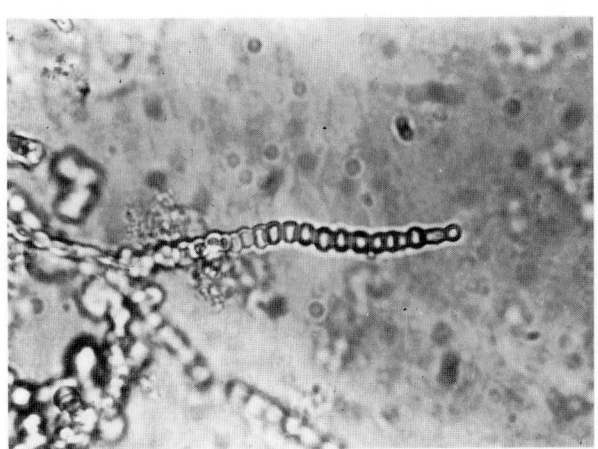

FIGURE 12 Characteristic chlamydospores produced by *T. verrucosum* or BCP-milk solids-yeast extract agar. Magnification, ×400.

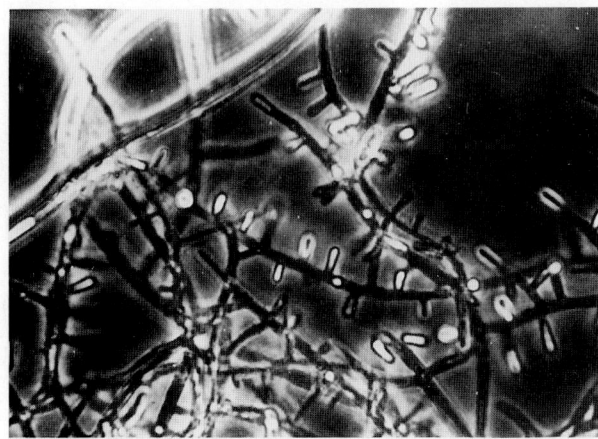

FIGURE 10 Clavate macroconidium, microconidia, and intermediate conidia of *T. terrestre*. Phase contrast; magnification, ×400.

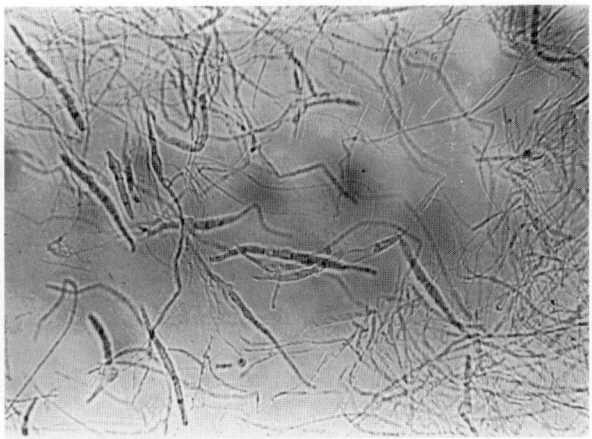

FIGURE 13 Rat-tailed macroconidia of *T. verrucosum* on BCP-milk solids-yeast extract agar. Magnification, ×400.

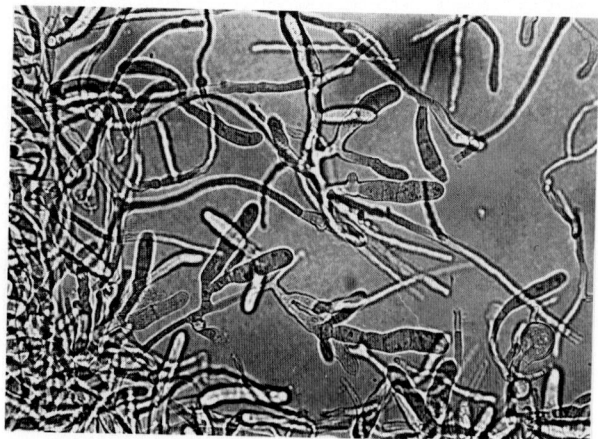

FIGURE 14 Macroconidia of *E. floccosum* on SDA. Note the absence of microconidia. Magnification, ×400.

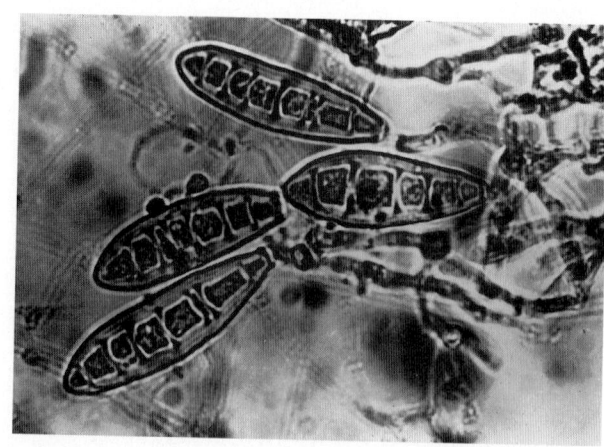

FIGURE 17 Macroconidia of *Microsporum cookei*, showing thick walls and pseudosepta. Magnification, ×400.

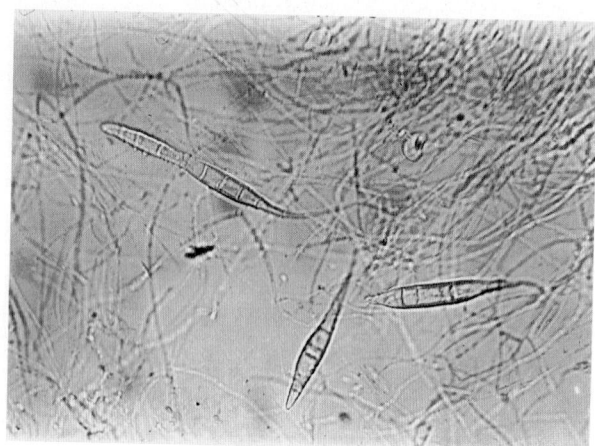

FIGURE 15 Macroconidia of *M. audouinii* on SDA–3% NaCl. Magnification, ×400.

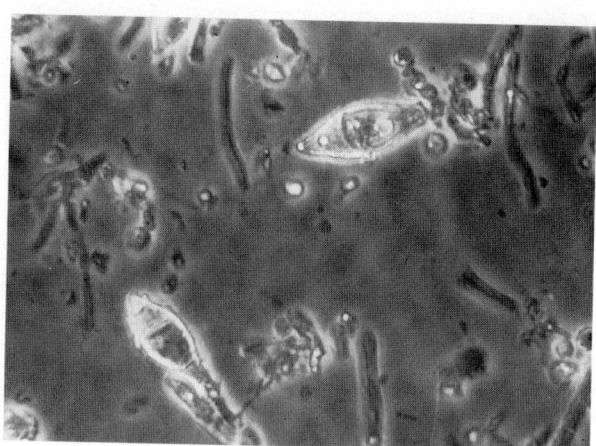

FIGURE 18 Fusiform two- to three-septate macroconidia of *M. equinum*. Phase contrast; magnification, ×400.

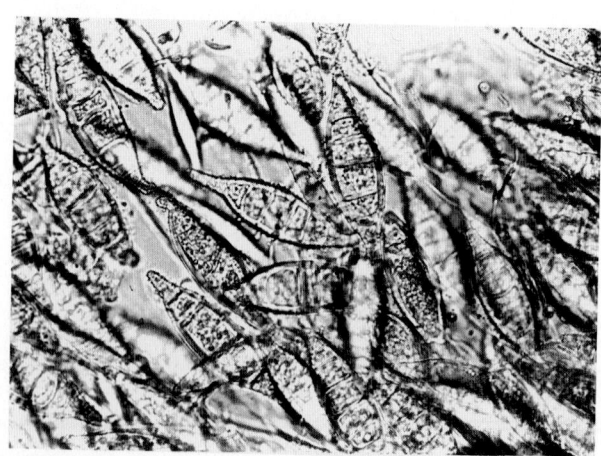

FIGURE 16 Macroconidia of *M. canis* with rough thick walls. Magnification, ×400.

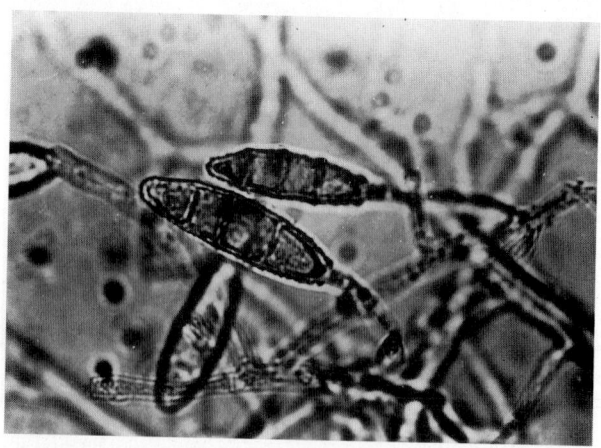

FIGURE 19 Macroconidia of *M. gypseum*. Magnification, ×400.

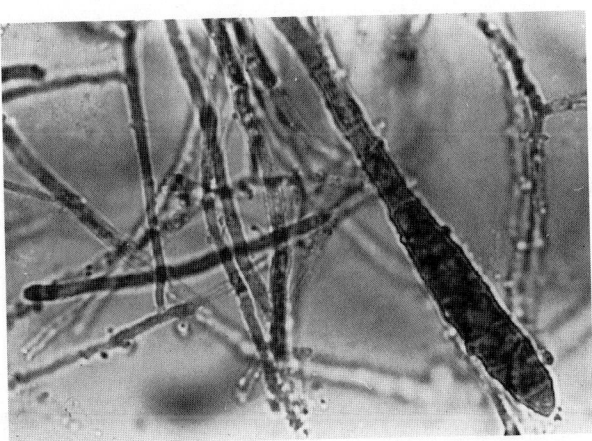

FIGURE 20 Rough-walled macroconidium of *M. persicolor* on SDA–3% NaCl. Magnification, ×1,000.

lindrofusiform), number of septa (1 to 15), size (6 to 160 by 6 to 25 µm), and width of the cell wall. Microconidia are pyriform or clavate and usually are arranged singly along the sides of the hyphae. *Microsporum* species invade skin, hair, and rarely, nails.

Epidermophyton Species

In *Epidermophyton* species, microconidia are lacking; only smooth-walled, broadly clavate macroconidia are produced. They have one to nine septa, are 20 to 60 µm long by 4 to 13 µm wide, and are borne singly or in clusters of two or

TABLE 4 Sequence of procedures for identification of dermatophytes in pure culture[a]

1. Examine colony for colors of surface and reverse, topography, texture, and rate of growth. Proceed to step 2.
2. Prepare teased mounts, and search for identifying microscopic morphology, especially presence, appearance, and arrangement of macroconidia and microconidia (consult Fig. 5–20 and Tables 5 and 6). If results are inconclusive, proceed to step 3.
3. Prepare slide cultures, and examine for characteristic morphology as indicated above if teased mounts do not provide sufficient information. Consider special media if sporulation is absent (potato-glucose agar, SDA with 3–5% NaCl, lactrimel). If results are inconclusive, proceed to step 4.
4. Perform as many of the following physiologic tests as necessary for identification.
   a. Urease.
   b. Nutritional requirements if *Trichophyton* sp. is suspected
   c. Growth on rice grains if *Microsporum* sp. is suspected
   d. In vitro hair perforation test
   e. Temperature tolerance and/or optimum temperature of growth
   f. Special media to differentiate *T. mentagrophytes* from *M. persicolor* (34), *T. rubrum* from *T. mentagrophytes* (64), and *T. soudanense* from *T. ferrugineum* (73)
   g. Mating studies (to be performed in reference laboratories)

[a]It may be necessary to incubate cultures on brain heart infusion agar or similar medium to determine absence of bacterial contamination before proceeding to step 4. Procedures are adapted from Weitzman et al. (70, 74).

three. *E. floccosum* is the only pathogen (the genus has only two species); it invades skin, nails, and, rarely, hair (57).

Characteristic features of the pathogenic dermatophytes are presented in Tables 5 and 6. These tables also include data on some similar but rarely pathogenic geophilic *Microsporum* and *Trichophyton* species and on *Trichophyton fischeri*, which must be differentiated from *T. rubrum*.

NAIL AND SKIN MYCOSES MIMICKING DERMATOPHYTOSES

Over 30 nondermatophytic mould species and a few yeast-like fungi have been reliably reported to cause onychomycosis (nail infection) (36). In North America, the prevalent agents of nondermatophytic onychomycosis are *Scopulariopsis brevicaulis*, *Aspergillus sydowii*, and, in some areas, the *Scytalidium dimidiatum* state of *Nattrassia mangiferae* (63). Various other *Aspergillus* and *Scopulariopsis* species as well as some *Fusarium* species and others may be implicated. The *Scytalidium* species causing onychomycosis, *Scytalidium dimidiatum* and *Scytalidium hyalinum*, also regularly cause infections resembling tinea pedis and tinea manuum. A recent survey in Toronto, Canada, showed that nondermatophytic filamentous fungi caused 3.3% of all onychomycoses but less than 1% of skin infections of feet and hands (63). It must be stressed that many of the nondermatophytes are inhibited by cycloheximide, so a cycloheximide-free medium such as Littman oxgall agar with chloramphenicol and gentamicin, SDA with chloramphenicol, or inhibitory mould agar must be used in their isolation.

Verification of a nondermatophytic infection may be a problem. In North America, the isolation of a known pathogenic *Scytalidium* species from nail or skin specimens positive for fungal filaments may generally be taken to indicate infection. Confirmation of the finding in a repeat sample is ideal but not essential, because these fungi are virtually never obtained as contaminants of the body surface in temperate regions. The other nondermatophytic agents of onychomycosis are more common as contaminants than as infectious agents. Confusingly, they may grow as contaminants from dermatophytic lesions, where dermatophyte filaments are seen in direct examination but no dermatophyte culture grows. The nondermatophyte may be mistakenly associated with the filaments and proposed as an infectious organism. True cases of nondermatophytic infection can be verified only by showing constancy of association in repeat samples (61): (i) elements compatible with the suspected agent must be visualized in direct (KOH or NaOH) examination; (ii) the suspected agent must be isolated in culture; (iii) the same organism must be repeatedly cultured from the suspected site and must remain culturable as long as active infection persists; and (iv) a more likely causal agent, such as a dermatophyte, should not be isolated despite repeated, adequate sampling. Thus, nondermatophytic onychomycosis caused by moulds other than *Scytalidium* spp. and nail infection caused by yeasts other than *Candida albicans* cannot normally be diagnosed on the basis of a single skin or nail sample. Exceptions to this occur rarely when fungi such as *Scopulariopsis* or *Aspergillus* spp. conidiate heavily in the nail tissue itself and reproductive structures are evident in direct examination. Mixed infections by a dermatophyte and a nondermatophyte rarely occur, and in such cases the nondermatophyte may consistently predominate. Even if the consistent presence of a nondermatophyte is verified in a lesion, sampling

TABLE 5 Important characteristics of pathogenic *Trichophyton* species[a]

Species	Colony on SDA	Microscopic morphology	Comments[b]
T. ajelloi	Cream or orange-tan; flat; powdery; reverse is blackish purple, sometimes nonpigmented; rapid growth	Microconidia rare; macroconidia numerous, fusiform to cylindrical, thick-walled, multiseptate, with 5–12 cells (Fig. 5)	Geophilic species; rarely pathogenic
T. concentricum	Beige, brown, or reddish; elevated and convoluted; glabrous to velvety; no undersurface color; slow growing	Micro- and macroconidia usually absent; chlamydospores may be present	Geographically restricted; 50% of isolates stimulated by thiamine; others are autotrophic
T. equinum	Cream colored; flat; fluffy; reverse is yellow becoming reddish brown	Elongate, clavate or subglobose, stalked microconidia; macroconidia rare, similar to those of *T. mentagrophytes*	Requires nicotinic acid; autotrophic variety has been described (59)
T. fischeri	White; velvety to cottony; reverse is brownish red to wine red; closely resembles *T. rubrum*	Pyriform and subglobose microconidia abundant along unbranched hyphae, macroconidia long and sinuous, cylindrical to clavate; thin closterospore-like hyphal projections often produced	Geophilic species not pathogenic, growth at 37°C; urease test negative; in vitro hair perforation test negative; growth restricted on BCPCDA and no pH change within 7 days; no red undersurface pigment on CEAA (28); must be differentiated from *T. rubrum* and *T. raubitschekii* (32)
T. gourvilii	Pink to red; heaped up; convoluted; glabrous; becoming velvety	Typical *Trichophyton*-type macroconidia and microconidia usually found	No special nutritional requirements
T. kanei	White; velvety to granular; reverse is brownish red to wine red	Macroconidia predominant, cylindrical to clavate, often with T-shaped bases; microconidia small, mostly cylindrical, pyriform when formed	Urease test weakly positive; in vitro hair test negative; growth restricted on BCPCDA and no pH change within 7 days
T. megninii	Pink to rose; radially folded; suede-like; reverse is wine red	Pyriform to clavate microconidia; macroconidia rare, similar to those of *T. rubrum*	Requires L-histidine; urease positive
T. mentagrophytes	White, cream, tan, yellowish or pink; flat; powdery, granular, or velvety; reverse is light tan, yellow, red, or reddish brown; sometimes produces diffusible yellow or melanoid pigment	Globose to pyriform microconidia in clusters or singly along hyphae; clavate macroconidia present in some strains (Fig. 6); coiled hyphae (spirals), nodular bodies, and antler-like hyphae may be observed	Urease positive; perforates hair in vitro; grows at 37°C; growth profuse on BCPCDA, with alkalinity within 7 days
T. raubitschekii	Buff; raised center with radial grooves; velvety to granular; reverse is blood red	Microconidia clavate, globose, or subglobose, sessile or on short stalks along unbranched hyphae; macroconidia abundant on primary isolation; thin, elongate with blunt ends, 5–9 cells (Fig. 7 and 8)	Urease positive; in vitro hair perforation test negative; brown pigmentation on CMD agar (32); growth restricted on BCPCDA and no pH change within 7 days

Species	Colony morphology	Microscopic morphology	Comments
T. rubrum	White; velvety, seldom powdery; reverse is wine red, sometimes yellow, orange, or with diffusible melanoid pigment; growth and color variants are described elsewhere (77)	Pyriform microconidia usually along unbranched hyphae; macroconidia absent to rare, thin, cylindrical to clavate in granular cultures (Fig. 9)	Urease test negative; in vitro hair perforation test negative; red undersurface pigment on CMD agar and CEAA except in yellow hyaline variants (16); restricted on BCPCDA with no pH or growth change within 7 days; must be differentiated from T. raubitschekii (32) and T. fischeri (28)
T. schoenleinii	White to tan; heaped and convoluted; glabrous or waxy, becoming velvety on subculture; reverse lacks pigment; slow growth	Micro- and macroconidia rarely seen; chlamydospores often numerous; hyphal tips often show "nailhead" morphology and branch to form antler-like structures (favic chandeliers)	Autotrophic for vitamins, which differentiates it from T. verrucosum
T. simii	White to pale buff; flat or slightly convoluted; powdery; reverse is straw to salmon colored	Numerous macroconidia; some fragment or develop swellings resembling chlamydospores; pyriform microconidia and spirals may be found	
T. soudanense	Yellow-orange (like dried apricots); flat with convolutions; suede-like texture; fringed (eyelash) periphery; reverse is yellow to orange-yellow; slow growing	Pyriform microconidia rare; macroconidia very rare; reflex branching characteristic	Growth stimulated at 37°C (73)
T. terrestre	White, pale yellow, or red; flat; granular to downy; reverse is pale yellow, yellowish brown, or red; rapid growth	Microconidia clavate to pyriform, single or clustered, short, intermediate, or fully extended to attain size of macroconidia; clavate to cylindrical, thin walled, 2–6 cells (Fig. 10)	Geophilic; rarely pathogenic; usually no growth at 37°C; must be differentiated from T. mentagrophytes
T. tonsurans	Color varies with isolate (yellow, cream, white, pink, brown, gray, etc.); convoluted; raised or flat; velvety to powdery; reverse is dark brown to mahogany red; slow growing	Clavate to elongate microconidia, some swollen into balloon forms and attached to branched conidiophores by short stalks (Fig. 11); macroconidia rare	Stimulated by thiamine
T. verrucosum	Cream to tan; flat; discoid; velvety (T. verrucosum var. discoides); white, heaped, folded, glabrous to downy (T. verrucosum var. album), yellow-ocher; convoluted; glabrous (T. verrucosum var. ochraceum); slow growing	Usually no micro- or macroconidia; chlamydospores usually numerous and in chains (Fig. 12)	All strains require thiamine; most require inositol as well; growth stimulated at 37°C, differentiating it from T. schoenleinii; growth and hydrolysis within 6 days on BCPCYA medium (35); characteristic rat-tailed macroconidia also stimulated on BCPCYA (Fig. 13)

(Continued on next page)

TABLE 5 Important characteristics of pathogenic *Trichophyton* species[a] (*Continued*)

Species	Colony on SDA	Microscopic morphology	Comments[b]
T. violaceum	Violet or lavender, rarely white; heaped and convoluted; glabrous or velvety; purple undersurface; slow growing	Micro- and macroconidia usually lacking; chlamydospores may be found	Growth and sporulation stimulated by thiamine
T. yaoundei	Buff or light yellow developing brown pigment diffusing into medium; glabrous, leathery; flat, becoming heaped and folded; reverse is light tan; slow growing	Pyriform microconidia rare; macroconidia not found; chlamydospores seen	Geographically limited to Africa or persons with African ancestry (23); growth slightly stimulated on thiamine and inositol (23)

[a]Adapted from Weitzman et al. (70, 74). Note that some recently described pathogenic dermatophytes (*Trichophyton krajdenii* [33], *Trichophyton sarkisovii* [25]) are not included in this table.
[b]CEAA, Casamino Acids-erythritol-albumin agar; CMD, cornmeal-glucose agar; BCPCDA, BCP-milk solids-dextrose agar; BCPCYA, BCP-milk solids-yeast extract agar.

should still be adequate to rule out the possibility of dermatophyte coinfection.

SUPERFICIAL MYCOSES

Superficial mycoses include diseases in which the causative fungi colonize the cornified layers of the epidermis or the suprafollicular portion of the hair. There is little tissue damage, and cellular response from the host generally is lacking. The disease is essentially cosmetic, involving changes in the pigmentation of the skin (pityriasis versicolor or tinea nigra) or formation of nodules along the hair shaft distal from the follicle) (black piedra and white piedra).

In contrast to agents of the dermatophytoses, the etiologic agents are diverse and unrelated.

Pityriasis Versicolor (Tinea Versicolor)

Pityriasis versicolor is an infection of the stratum corneum caused by the lipophilic yeast *Malassezia furfur* (synonyms: *Pityrosporum furfur*, *Pityrosporum orbiculare*, and *Pityrosporum ovale*). Lesions appear as scaly, discrete or concrescent, hypopigmented or hyperpigmented (fawn, yellow-brown, brown, or red) patches chiefly on the neck, torso, and limbs. The infection is largely cosmetic, becoming apparent when the skin fails to tan normally. The disease has a worldwide distribution; it is common in temperate zones and very prevalent in the tropics. *Malassezia furfur* is found on the normal skin and elicits disease only under conditions, local or systemic, that favor the overgrowth of the organism.

Malassezia furfur has been associated with folliculitis (7), obstructive dacryocystitis (51), systemic infections in patients receiving intralipid therapy (13), and seborrheic dermatitis, especially in patients with AIDS (19). Two excellent reviews of human infections caused by *Malassezia* spp. and characteristics of the genus have recently been published (24, 41).

Direct Examination

The fungus is observed readily when scrapings are mounted in 10% KOH plus ink (10), 25% NaOH plus 5% glycerin, or Kane's formulation (glycerol, 10 ml; Tween 80, 10 ml; phenol, 2.5 g; methylene blue, 1.0 g; distilled water, 480 ml). The presence of short, septate, occasionally branching filaments 2.5 to 4 μm in diameter and of variable lengths along with clusters of small, unicellular, oval or round budding yeast cells (Fig. 21) with a collarette between mother and daughter cells (budding is phialidic) averaging 4 μm (up to 8 μm) in size is diagnostic.

Isolation and Culture

Culture is not essential for identification unless the findings of direct microscopic examination are atypical. Also, *Malassezia furfur* is part of the normal flora of the adult skin. If culture is desired, olive oil added to the medium is essential. Scrapings may be inoculated on SDA plus cycloheximide and chloramphenicol (Mycosel; BBL), Mycobiotic agar (Difco), or Littman oxgall agar (Difco or Remel); overlaid with sterile oil (we recommend Gallo or Pompeian); and incubated at 37°C. Growth is slow; colonies are cream colored, glossy, and raised (Fig. 22), later becoming dull, dry, and tan to brownish. Only budding yeast cells appear in culture (Fig. 23); hyphae are not found in routine isolation media, although short germ tubes may be produced.

MYCOLOGY

TABLE 6 Important characteristics of pathogenic *Epidermophyton* and *Microsporum* species[a]

Species	Colony on SDA	Microscopic morphology	Comments
Epidermophyton floccosum	Yellowish green to khaki; flat to radially folded; powdery to velvety; yellow-brown on reverse; white tufts common on surface of older cultures; slow growth	Abundant, widely clavate, smooth-walled macroconidia, 20–40 by 6–8 μm, single or in clusters, with 0–4 septa (Fig. 14); chlamydospores common in older cultures	Invades skin, nails, and, rarely, hair (57); no microconidia
Microsporum spp. *M. audouinii*	Grayish white, cream to tan; flat, spreading; velvety; light salmon pink to light reddish brown on reverse; moderate growth	Usually no conidia; apiculate terminal chlamydospores are only characteristic features; pectinate hyphae may be present	Prepubertal tinea capitis and tinea corporis; poor growth and brownish discoloration of rice grains; sporulating strains have irregular fusiform and elongated macroconidia with few septa at irregular intervals (Fig. 15); microconidia rare to moderate; does not perforate hair in vitro (49)
M. canis	White to pale buff; wooly; yellow-orange to orange-brown reverse; rarely nonpigmented on reverse; rapid growth	Numerous fusiform macroconidia with thick walls and up to 15 septa; 18–125 by 5–25 μm, with asymmetric knobbed apex (Fig. 16); few microconidia	Good growth and sporulation on rice grains; perforates hair in vitro (49)
M. canis var. *distortum*	White to pale buff, usually with radial groves; colorless to yellowish on reverse; rapid growth	Macroconidia distorted, bizarre in shape, 20–60 by 7–27 μm; microconidia abundant	
M. cookei	Yellowish to reddish tan; flat; powdery, granular, or downy; dark purple-red on reverse; rapid growth	Macroconidia numerous, some resembling those of M. gypseum, but most with thicker walls (1–5 μm) (Fig. 17); microconidia abundant	Geophilic species; rarely pathogenic
M. equinum	White, pale buff to pale salmon; folding; velvety to finely powdery; buff to salmon on reverse	Macroconidia infrequent on SDA, elliptical to fusiform, 18–60 by 5–15 μm, 2–4 cells; thick walls may resemble those of M. canis (Fig. 18)	Macroconidia stimulated by niger seed medium 8 agar (31); does not perforate hair in vitro (49); growth on rice grains resembles that of M. audouinii (1)
M. ferrugineum	Yellowish to rust colored; folded; waxy; white, velvety variants found in Balkans (52); very slow grower	Usually no conidia; numerous chlamydospores, irregular hyphae, and long, straight, coarse hyphae with prominent septa ("bamboo hyphae")	Geographically restricted to parts of Africa, Asia, and Eastern Europe; light yellow colonies on Lowenstein-Jensen medium differentiates it from dark, reddish brown colonies of T. soudanense (67, 73)
M. gallinae	White tinged with pink; slightly folded, downy; raspberry red diffusing pigment on reverse; rapid to moderate growth	Macroconidia fairly abundant, blunt tipped, 6–8 by 15–50 μm, 2–10 cells; cell walls usually smooth, sometimes echinulate; pyriform microconidia	Conidia stimulated by yeast extract or thiamine in medium

(Continued on next page)

TABLE 6 Important characteristics of pathogenic *Epidermophyton* and *Microsporum* species[a] (*Continued*)

Species	Colony on SDA	Microscopic morphology	Comments
M. gypseum complex (M. gypseum, M. fulvum, and M. boullardii)	Pale buff, rosy buff to light cinnamon, white border; flat; powdery, granular to floccose; buff to reddish brown on reverse; rapid growth	Abundant macroconidia, 25–60 by 7.5–15 μm; ellipsoidal to fusiform, up to 6 septa, thin walled (Fig. 19); microconidia moderately abundant	Definitive identification of species is obtained by mating tester strains on appropriate media to induce teleomorph (47, 75) or by examining colonial and microscopic features on Takashio's medium (15)
M. nanum	Cream to buff; powdery reddish brown on reverse; moderate growth	Abundant obovate to clavate macroconidia, 10.5–30 by 6.5–13 μm, usually 2 cells; microconidia few to moderately abundant	Grows more slowly than members of M. gypseum complex; must be differentiated from *Trichothecium roseum*
M. persicolor	Yellowish buff becoming peach to pink; flat; powdery to downy; reddish brown on reverse; rapid growth	Abundant microconidia, spherical to pyriform (few clavate), stalked, borne mostly in grapelike clusters but also singly along the sides of hyphae; thin-walled macroconidia (Fig. 20) often smooth; spiral hyphae common	Resembles *T. mentagrophytes*; can be differentiated on BCP-milk solids–glucose medium (34); rough-walled macroconidia on SDA–3% NaCl (34); peach- to rose-colored colonies on cereal agar (45); absence of good growth at 37°C (34)
M. praecox	Cream to yellowish tan; folded; powdery; pale yellow to orange on reverse; moderate growth	Numerous long fusiform macroconidia, some with apical appendages, 40–90 by 7–17 μm, 2–8 septa, thin walled; microconidia absent	Does not perforate human hair in vitro (72)
M. racemosum	White, cream, or buff; flat; finely granular; dark purple-red on reverse; rapid growth	Macroconidia abundant, fusiform to ellipsoidal, 41–77 by 9 μm, 3–8 septa, moderately thick walls; numerous microconidia mostly stalked and produced in grapelike clusters	Macroconidia resemble those of M. gypseum
M. vanbreuseghemii	Pink to deep rose, light buff, or yellowish; flat; coarsely granular to downy; cream to pale yellow on reverse; rapid growth	Abundant cylindrofusiform macroconidia, 43.7–87.5 μm, thick walled, up to 12 septa; numerous pyriform to obovate microconidia borne singly along sides of hyphae	

[a]Adapted from Weitzman et al. (70, 74).

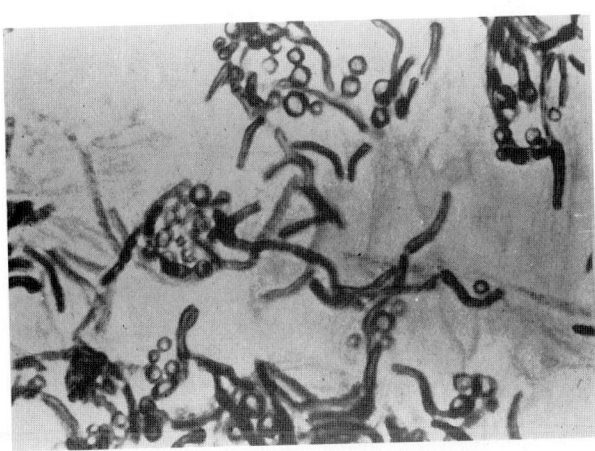

FIGURE 21 *Malassezia furfur* in skin scrapings from a lesion of pityriasis versicolor (Kane's stain). Magnification, ×1,000.

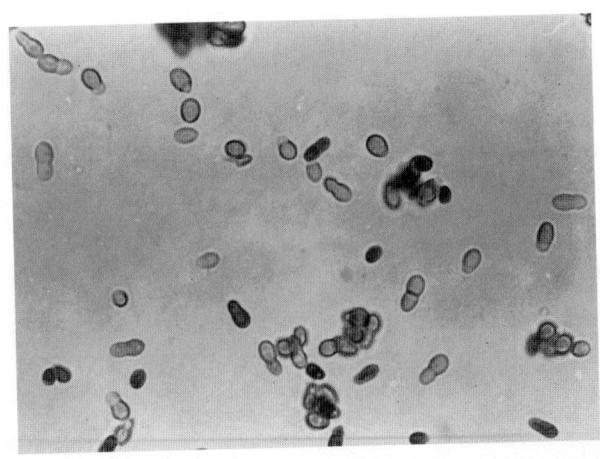

FIGURE 23 Microscopic appearance of *Malassezia furfur* yeast cells on Littman oxgall overlaid with olive oil. Magnification, ×400.

Tinea Nigra

Tinea nigra is characterized by the appearance, primarily on the palms of the hands and less commonly on the dorsa of the feet, of flat, sharply marginated, brownish black, non-scaly macules.

The disease is most common in the tropical areas of Central and South America, Africa, and Asia (51) and had been reported to be prevalent in North Carolina and in coastal areas of the southeastern United States (68). Cases diagnosed outside the areas of endemicity have resulted from travel to the American tropics or the Caribbean islands (51). The etiologic agents are *Phaeoannellomyces* (*Exophiala*) *werneckii* (42) and *Stenella araguata*, which is very rare and will not be described here.

Direct Microscopic Examination

Microscopic examination of skin scrapings in KOH or NaOH reveals numerous light brown, frequently branching, septate filaments 1.5 to 5 μm in diameter; short sinuous septate fragments (Fig. 24); and budding cells, some septate.

Isolation, Culture, and Identification

On SDA with or without antibiotics, *Phaeoannellomyces werneckii* grows slowly and usually appears within 2 to 3 weeks as moist, shiny, olive to greenish black yeast-like colonies occasionally only hyphal (42). Microscopic observation of the yeast-like colonies reveals one- to two-celled pale brown to deeply pigmented cylindrical to spindle-shaped cells with some budding. The cells are rounded at one end and tapered toward the other (Fig. 25), with annellations (rings) at the tapered end (42). With prolonged incubation, the colonies develop velvety, greenish gray to grayish black aerial mycelium over portions of the yeast-like growth. As the colonies become more mycelial, microscopic observation reveals olivaceous, thin or broad, often thick-walled hyphae septate at frequent intervals. Along these hyphae, one- or two-celled hyaline to olivaceous annelloconidia arise from intercalary annellides or from annellides integrated within the hyphae. These annelloconidia tend to accumulate in balls, eventually sliding

FIGURE 22 Culture of *Malassizia furfur* on Littman oxgall agar overlaid with olive oil.

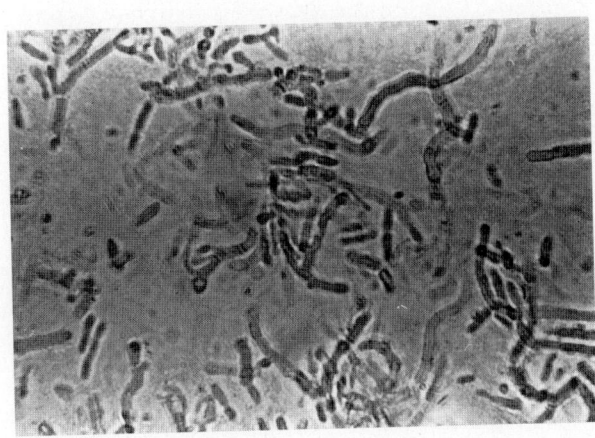

FIGURE 24 *Phaeoannellomyces werneckii* in NaOH mount of skin from lesion of tinea nigra. Magnification, ×400.

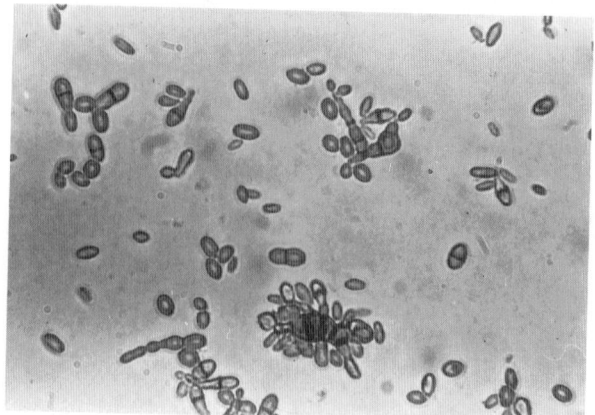

FIGURE 25 *Phaeoannellomyces werneckii* after 2 weeks on SDA. Note dematiaceous unicellular and bicellular annelloconidia. Magnification, ×400.

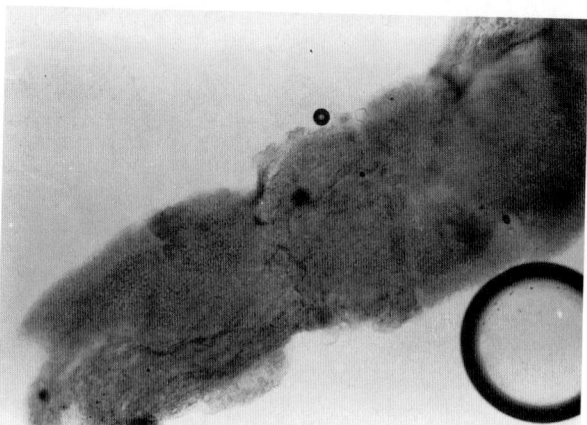

FIGURE 27 White piedra nodule on hair from the groin. Magnification, ×1,000.

down along the sides of the hyphae. Seceded conidia may produce new conidia by budding.

Black Piedra

Black piedra is a fungal infection of the scalp hair, less commonly of the beard or moustache, and rarely of axillary or pubic hairs. The disease is characterized by the presence of discrete, hard, gritty, dark brown to black nodules adhering firmly to the hair shaft (Fig. 26). It is found mostly in tropical regions in Africa, Asia, and Central and South America. Humans as well as primates are infected.

The etiologic agent in humans is *Piedraia hortae*, an ascomycete whose nodules serve as ascostromata containing locules that harbor the asci and ascospores.

Direct Microscopic Examination

Hair fragments containing the nodules are mounted in 25% NaOH or KOH, heated gently, and carefully squashed without breaking the coverslip (the nodules are very hard). The squashed preparation of a mature nodule should reveal compact masses of dark, septate hyphae around the surface of the hair and round to oval asci containing hyaline,

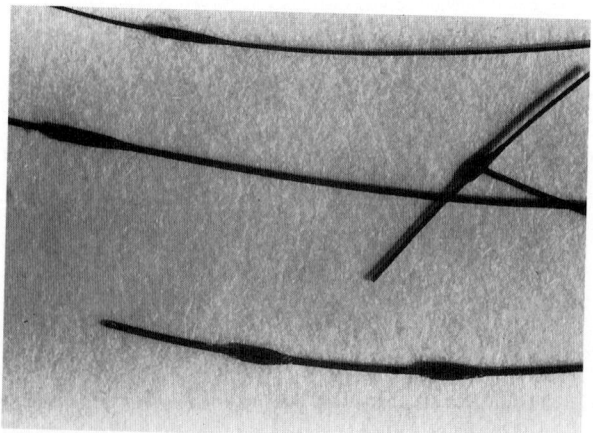

FIGURE 26 Black piedra nodules on scalp hair. NaOH mount; magnification, ×100.

curved, fusiform, aseptate ascospores that bear one or more appendages.

Isolation, Culture, and Identification

SDA with chloramphenicol and SDA with chloramphenicol and cycloheximide should be used for isolation. Some reports have indicated that cycloheximide may be inhibitory; however, others have used this antibiotic successfully.

Colonies are very slow growing, dark brown to black, and heaped in the center with a flat periphery. A dark brown to black, short, aerial mycelium eventually covers the young glabrous colony. Some colonies produce a reddish brown diffusible pigment on the agar. Microscopic examination reveals only highly septated dark hyphae and swollen intercalary cells. Conidia and ascospores are usually not found on routine mycological media.

White Piedra

White piedra is a fungal infection of the hair shaft characterized by the presence of soft white, yellowish, beige, or greenish nodules found chiefly on facial, axillary, or genital hairs (Fig. 27) and less commonly on scalp, eyebrows, and eyelashes. Nodules may be discrete or more often coalescent, forming an irregular transparent sheath.

The infection occurs sporadically in North America and Europe and more commonly in South America and the Orient (51). Although white piedra is an uncommon infection, genital white piedra is more frequent in certain populations (26, 60).

The etiologic agents of white piedra, formerly called *Trichosporon beigelii* or *Trichosporon cutaneum*, are now known to be separate species with different ecologies. The name *Trichosporon beigelii* is considered doubtful and can no longer be maintained (20). The genus *Trichosporon* was revised by Guého et al. (21) on the basis of morphology, ultrastructure, physiology, ubiquinone systems, mole percent G+C of DNA, DNA-DNA reassociation, and 26S rRNA partial sequences. Instead of *Trichosporon beigelii*, the organisms named *Trichosporon ovoides* and *Trichosporon inkin* are the most frequent isolates for white piedra, with *Trichosporon ovoides* associated mostly with white piedra of the head and *Trichosporon inkin* confined largely to pubic white piedra. *Trichosporon asteroides* and *Trichosporon cutaneum* are isolated less frequently from superficial lesions, probably as contaminants in most cases. *Trichosporon asahii*

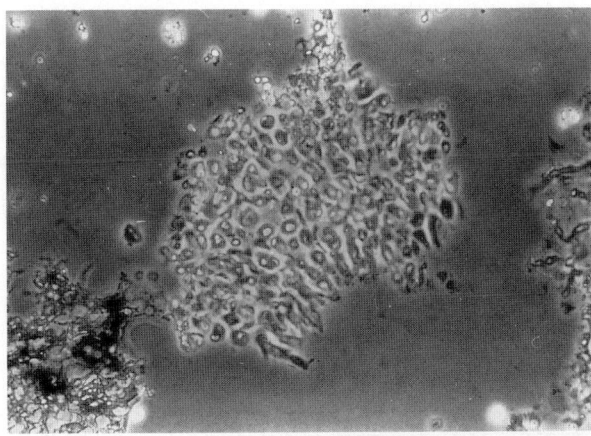

FIGURE 28 Arthroconidia from a crushed nodule of white piedra. Magnification, ×1,000.

is the major species responsible for the systemic disseminated human mycoses; less frequently, opportunistic infection is caused by *Trichosporon mucoides*, which has been isolated from superficial and systemic sources (21).

Microscopic examination of hairs containing the adherent nodules mounted in 10% KOH or 25% NaOH–5% glycerin and squashed under a coverslip will reveal intertwined hyaline septate hyphae, hyphae breaking up into oval or rectangular arthroconidia 2 to 4 μm in diameter (Fig. 28), occasional blastoconidia, and bacteria that may surround the nodule as a zoogloea.

The isolates formerly described as *Trichosporon beigelii* but now known by other species names (see above) may be readily isolated on SDA with chloramphenicol or other isolation media containing antibacterial antibiotics. The isolation medium should not contain cycloheximide, since this drug is inhibitory to most of the species. Growth is rapid, forming white to cream-colored colonies that exhibit a variety of colonial morphologies depending on the species. A description of the genus and characteristics of the species are given in chapter 61 of this Manual and in papers by Guého et al. (20, 21).

REFERENCES

1. **Aho, R.** 1987. Mycological studies on *Microsporum equinum* isolated in Finland, Sweden and Norway. *J. Med. Vet. Mycol.* **25:**255–260.
2. **Ajello, L.** 1960. Geographic distribution and prevalence of the dermatophytes. *Ann. N.Y. Acad. Sci.* **89:**30–38.
3. **Ajello, L.** 1968. A taxonomic review of the dermatophytes and related species. *Sabouraudia* **6:**147–159.
4. **Ajello, L., and L. K. Georg.** 1957. In vitro cultures for differentiating between atypical isolates of *Trichophyton mentagrophytes* and *Trichophyton rubrum*. *Mycopathol. Mycol. Appl.* **8:**3–7.
5. **Alsop, J., and A. P. Prior.** 1961. Ringworm infection in a cucumber greenhouse. *Br. Med. J.* **1:**1081–1083.
6. **Araviysky, A. N., R. A. Araviysky, and G. A. Eschkov.** 1975. Deep generalized trichophytosis. *Mycopathologia* **56:**47–65.
7. **Back, O., J. Faergemann, and R. Hornquist.** 1985. *Pityrosporum* folliculitis: a common disease of the young and middle-aged. *J. Am. Acad. Dermatol.* **12:**56–61.
8. **Barson, W. J.** 1985. Granuloma and pseudogranuloma of the skin due to *Microsporum canis*. *Arch. Dermatol.* **121:**895–897.
9. **Borelli, D.** 1962. Medios caseros para micologia. *Arch. Venez. Med. Trop. Parasitol. Med.* **4:**301–310.
10. **Cohen, M. M.** 1954. A simple procedure for staining tinea versicolor (*M. furfur*) with fountain pen ink. *J. Invest. Dermatol.* **22:**9–10.
11. **Conant, N. F.** 1936. Studies on the genus *Microsporum*. I. Cultural studies. *Arch. Dermatol.* **33:**665–683.
12. **Currah, R. S.** 1985. Taxonomy of the Onygenales: Arthrodermataceae, Gymnoascaceae, Myxotrichaceae and Onygenaceae. *Mycotaxon* **24:**1–216.
13. **Danker, W. M., S. A. Spector, J. Fierer, and C. E. Davis.** 1987. *Malassezia* fungemia in neonates and adults: complication of hyperalimentation. *Rev. Infect. Dis.* **9:**743–753.
14. **Dei Cas, E., and A. Vernes.** 1986. Parasitic adaptation of pathogenic fungi to mammalian hosts. *Crit. Rev. Microbiol.* **13:**173–218.
15. **DeMange, C., N. Contet-Andonneau, M. Kombila, M. Miegeville, M. Berthonneau, C. DeVroey, and G. Percebois.** 1992. *Microsporum gypseum* complex in man and animals. *J. Med. Vet. Mycol.* **30:**301–308.
16. **Fischer, J. B., and J. Kane.** 1974. The laboratory diagnosis of dermatophytosis complicated with *Candida albicans*. *Can. J. Microbiol.* **20:**167–182.
17. **Flórián, E., and J. Galgóczy.** 1964. *Keratinomyces longifusus* sp. nov. from Hungary. *Mycopathol. Mycol. Appl.* **24:**73–80.
18. **Georg, L. K., and L. B. Camp.** 1957. Routine nutritional tests for the identification of dermatophytes. *J. Bacteriol.* **74:**113–121.
19. **Groisser, D., E. J. Bottone, and M. Lebwohl.** 1989. Association of *Pityrosporum orbiculare* (*Malassezia furfur*) with seborrheic dermatitis in patients with acquired immunodeficiency syndrome (AIDS). *J. Am. Acad. Dermatol.* **20:**770–773.
20. **Guého, E., G. S. de Hoog, and M. T. Smith.** 1992. Neotypification of the genus *Trichosporon*. *Antonie van Leeuwenhoek* **61:**285–288.
21. **Guého, E., M. T. Smith, G. S. de Hoog, G. Billon-Grand, R. Christen, and W. H. Batenburg-van der Vegte.** 1992. Contributions to a revision of the genus *Trichosporon*. *Antonie van Leeuwenhoek* **61:**289–316.
22. **Head, E. S., J. C. Henry, and E. M. MacDonald.** 1984. The cotton swab; technic for the culture of dermatophyte infections: its efficacy and merit. *J. Am. Acad. Dermatol.* **11:**797–801.
23. **Hemashettar, B. M., C. S. Patil, V. V. Yenni, P. R. Malur, and C. K. Campbell.** 1993. Isolation of *Trichophyton yaoundei* in India. *J. Med. Vet. Mycol.* **31:**333–336.
24. **Ingham, E., and A. C. Cunningham.** 1993. *Malassezia furfur*. *J. Med. Vet. Mycol.* **31:**265–288.
25. **Ivanova, L. G., and I. D. Polyakov.** 1983. *Trichophyton sarsikovii*, Ivanova et Polyakov sp. nov.—a new species of pathogenic fungus causing a dermatomycosis of camels. *Mikol. Fitopatol.* **17:**363–367.
26. **Kalter, D. C., J. A. Tschen, P. L. Cernoch, M. E. McBrid, J. Sperber, S. Bruce, and J. E. Wolf, Jr.** 1986. Genital white piedra: epidemiology, microbiology, and therapy. *J. Am. Acad. Dermatol.* **14:**982–993.
27. **Kaminski, G. W.** 1985. The routine use of modified Borelli's lactrimel (MBLA). *Mycopathologia* **91:**57–59.
28. **Kane, J.** 1977. *Trichophyton fischeri* sp. nov.: a saprophyte resembling *Trichophyton rubrum*. *Sabouraudia* **15:**231–241.
29. **Kane, J., and J. B. Fischer.** 1975. The effect of sodium chloride on the growth and morphology of dermatophytes and some other keratolytic fungi. *Can. J. Microbiol.* **21:**742–749.
30. **Kane, J., and J. B. Fischer.** 1976. The differentiation of *Trichophyton rubrum* from *T. mentagrophytes* by use of Christensen's urea broth. *Can. J. Microbiol.* **17:**911–913.
31. **Kane, J., A. A. Padhye, and L. Ajello.** 1982. *Microsporum equinum* in North America. *J. Clin. Microbiol.* **16:**943–947.
32. **Kane, J., I. F. Salkin, I. Weitzman, and C. Smitka.** 1981. *Trichophyton raubitschekii*, sp. nov. *Mycotaxon* **13:**259–266.
33. **Kane, J., J. A. Scott, R. C. Summerbell, and B. Diena.** 1992. *Trichophyton krajdenii*, sp. nov.: an anthropophilic dermatophyte. *Mycotaxon* **40:**307–316.

34. **Kane, J., L. Sigler, and R. C. Summerbell.** 1987. Improved procedures for differentiating *Microsporum persicolor* from *Trichophyton mentagrophytes*. *J. Clin. Microbiol.* **25:**2449–2452.

35. **Kane, J., and C. Smitka.** 1978. Early detection and identification of *Trichophyton verrucosum*. *J. Clin. Microbiol.* **8:**740–747.

36. **Kane, J., R. C. Summerbell, L. Sigler, S. Krajden, and G. Land.** *Laboratory Handbook of Dermatophytes and Other Filamentous Fungi from Skin, Hair and Nails.* Star Publishing Co., Belmont, Calif., in press.

37. **King, R. D., H. A. Khan, J. C. Foye, J. H. Greenberg, and H. E. Jones.** 1975. Transferrin, iron and dermatophytes. Serum dermatophyte inhibitory component definitively identified as unsaturated transferrin. *J. Lab. Clin. Med.* **86:**204–212.

38. **Kwon Chung, K. J., and J. E. Bennett.** 1992. *Medical Mycology*, p. 105–161. Lea & Febiger, Philadelphia.

39. **Lawson, G. T. N., and W. J. McLeod.** 1957. *Microsporum canis*—an intensive outbreak. *Br. Med. J.* **2:**1159–1160.

40. **Mackenzie, D. W. R.** 1963. "Hairbrush diagnosis" in detection and eradication of nonfluorescent scalp ringworm. *Br. Med. J.* **2:**363–365.

41. **Marcon, M. J., and D. A. Powell.** 1992. Human infections due to *Malassezia* spp. *Clin. Microbiol. Rev.* **5:**101–119.

42. **McGinnis, M. R., W. A. Schell, and J. Carson.** 1985. Phaeoannellomyces and the Phaeococcomycetaceae, new dematiaceous blastomycete taxa. *Sabouraudia* **23:**179–188.

43. **Merz, W. G., C. L. Berger, and M. Silva-Hutner.** 1970. Media with pH indicators for the isolation of dermatophytes. *Arch. Dermatol.* **102:**545–547.

44. **Okuda, C., M. Ito, and T. Sato.** 1991. *Trichophyton rubrum* invasion of human hair apparatus in tinea capitis and tinea barbae: light and electron microscope study. *Arch. Dermatol. Res.* **283:**233–239.

45. **Padhye, A. A., F. Blank, P. J. Koblenzer, S. Spatz, and L. Ajello.** 1973. *Microsporum persicolor* infection in the United States. *Arch. Dermatol.* **108:**561–562.

46. **Padhye, A. A., and J. Carmichael.** 1971. The genus Arthroderma Berkeley. *Can. J. Bot.* **49:**1525–1540.

47. **Padhye, A. A., A. S. Sekhon, and J. W. Carmichael.** 1973. Ascocarp production by *Arthroderma* and *Nannizzia* species on keratinous and non-keratinous media. *Sabouraudia* **11:**109–114.

48. **Padhye, A. A., I. Weitzman, and E. Domenech.** 1994. An unusual variant of *Trichophyton tonsurans* var. *sulfureum*. *J. Med. Vet. Mycol.* **32:**147–150.

49. **Padhye, A. A., C. N. Young, and L. Ajello.** 1980. Hair perforation as a diagnostic criterion in the identification of *Epidermophyton*, *Microsporum* and *Trichophyton* species, p. 115–120. *In Superficial Cutaneous and Subcutaneous Infections.* Scientific publication no. 396. Pan American Health Organization, Washington, D.C.

50. **Rippon, J. W.** 1985. The changing epidemiology and emerging patterns of dermatophyte species, p. 209–234. *In* M. R. McGinnis (ed.), *Current Topics in Medical Mycology.* Springer-Verlag, New York.

51. **Rippon, J. W.** 1988. *Medical Mycology: the Pathogenic Fungi and the Pathogenic Actinomycetes*, 3rd ed., p. 154–168. The W. B. Saunders Co., Philadelphia.

52. **Rippon, J. W.** 1988. *Medical Mycology: the Pathogenic Fungi and the Pathogenic Actinomycetes*, 3rd ed., p. 169–275. The W. B. Saunders Co., Philadelphia.

53. **Robinson, B. E., and A. A. Padhye.** 1988. Collection, transport and processing of clinical specimens, p. 11–32. *In* B. B. Wentworth (ed.), *Diagnostic Procedures for Mycotic and Parasitic Infections*, 7th ed. American Public Health Association, Inc., Washington, D.C.

54. **Rosenthal, S. A.** 1974. The epidemiology of tinea pedis, p. 515–526. *In* H. M. Robinson, Jr. (ed.), *The Diagnosis and Treatment of Fungal Infections.* Charles C Thomas, Publisher, Springfield, Ill.

55. **Rosenthal, S. A., and H. Sokolsky.** 1965. Enzymatic studies with pathogenic fungi. *Dermatol. Int.* **4:**72–79.

56. **Salkin, I. F.** 1973. Dermatophyte test medium: evaluation with nondermatophytic pathogens. *Appl. Microbiol.* **26:**134–137.

57. **Sberna, F., V. Farella, V. Geti, F. Taviti, G. Agostini, P. Vannini, B. Knöpfel, and E. M. Difonzo.** 1993. Epidemiology of the dermatophytoses in the Florence area of Italy: 1985–1990. *Mycopathologia* **122:**153–162.

58. **Sequeira, H., J. Cabrita, C. DeVroey, and C. Wuytack-Raes.** 1991. Contributions to our knowledge of *Trichophyton megninii*. *J. Med. Vet. Mycol.* **29:**417–418.

59. **Smith, J. M. B., R. D. Jolly, L. K. George, and M. D. Connole.** 1968. *Trichophyton equinum* var. *autrophicum*: its characteristics and geographic distribution. *Sabouraudia* **6:**296–304.

60. **Stenderup, A., H. Schonheyder, P. Ebbesen, and M. Melbye.** 1986. White piedra and *Trichosporon beigelii* carriage in homosexual men. *J. Med. Vet. Mycol.* **24:**401–406.

61. **Stiller, M. J., S. A. Rosenthal, R. C. Summerbell, J. Pollack, and A. Chan.** 1992. Onychomycosis of the toenails caused by *Chaetomium globosum*. *J. Am. Acad. Dermatol.* **26:**775–776.

62. **Summerbell, R. C.** 1987. *Trichophyton kanei*, sp. nov. a new anthropophilic dermatophyte. *Mycotaxon* **28:**509–523.

63. **Summerbell, R. C., J. Kane, and S. Krajden.** 1989. Onychomycosis, tinea pedis and tinea manuum caused by nondermatophytic filamentous fungi. *Mycoses* **32:**609–619.

64. **Summerbell, R. C., S. A. Rosenthal, and J. Kane.** 1988. Rapid method for differentiation of *Trichophyton rubrum*, *Trichophyton mentagrophytes*, and related dermatophyte species. *J. Clin. Microbiol.* **26:**2279–2282.

65. **Taplin, D.** 1965. The use of gentamicin in mycology. *J. Invest. Dermatol.* **45:**549–550.

66. **Taplin, D., N. Zaias, G. Rebell, and H. Blank.** 1969. Isolation and recognition of dermatophytes on a new medium (DTM). *Arch. Dermatol.* **99:**203–209.

67. **Vanbreuseghem, R., and R. Zaman.** 1963. Contributions á l'identification du *Trichophyton (Langeronia) soudanense* et du *Trichophyton ferrugineum*. *Ann. Soc. Belge Med. Trop.* **3:**259–270.

68. **Van Velsor, H., and H. Singletary.** 1964. Tinea nigra palmaris. *Arch. Dermatol.* **90:**59–61.

69. **Weeks, R. J., and A. A. Padhye.** 1982. A mounting medium for permanent preparations of micro-fungi. *Mykosen* **25:**702–704.

70. **Weitzman, I., and J. Kane.** 1991. Dermatophytes and agents of superficial mycoses, p. 601–616. *In* A. Balows, W. J. Hausler, Jr., K. L. Herrmann, H. D. Isenberg, and H. J. Shadomy (ed.), *Manual of Clinical Microbiology*, 5th ed. American Society for Microbiology, Washington, D.C.

71. **Weitzman, I., M. R. McGinnis, A. A. Padhye, and L. Ajello.** 1986. The genus *Arthroderma* and its later synonym *Nannizzia*. *Mycotaxon* **25:**505–518.

72. **Weitzman, I., and S. McMillen.** 1980. Isolation in the United States of a culture resembling M. *praecox*. *Mycopathologia* **70:**181–186.

73. **Weitzman, I., and S. A. Rosenthal.** 1981/1984. Studies in the differentiation between *Microsporum ferrugineum* Ota and *Trichophyton soudanense* Joyeaux. *Mycopathologia* **84:**95–101.

74. **Weitzman, I., S. A. Rosenthal, and M. Silva-Hutner.** 1988. Superficial and cutaneous infections caused by molds: dermatomycoses, p. 33–97. *In* B. B. Wentworth (ed.), *Diagnostic Procedures for Mycotic and Parasitic Infections*, 7th ed. American Public Health Association, Inc., Washington, D.C.

75. **Weitzman, I., and M. Silver-Hutner.** 1967. Non-keratinous agar media as substrates for the ascigerous state in certain members of the Gymnoascaceae pathogenic for man and animals. *Sabouraudia* **5:**335–339.

76. **West, B. C., and K. J. Kwon-Chung.** 1980. Mycetoma caused by *Microsporum audouinii*. *Am. J. Clin. Pathol.* **73:**447–454.

77. **Young, C. N.** 1972. Range of variation among isolates of *Trichophyton rubrum*. *Sabouraudia* **10:**164–170.

Rhizopus, Rhizomucor, Absidia, and Other Agents of Systemic and Subcutaneous Zygomycoses

MALCOLM D. RICHARDSON AND GILLIAN S. SHANKLAND

66

TAXONOMY

The term zygomycosis embraces deep or subcutaneous mycoses caused by fungi belonging to various genera of the class *Zygomycetes*, namely, mucormycosis and the entomophthoramycoses, the latter being subdivided into basidiobolomycosis and conidiobolomycosis (8).

MUCORMYCOSIS

Definition

The term mucormycosis is used to refer to infections due to moulds belonging to the order *Mucorales*. These organisms can cause rhinocerebral, pulmonary, gastrointestinal, cutaneous, or disseminated infection in predisposed individuals, the different clinical forms often being associated with particular underlying conditions. In most cases, therefore, mucormycosis occurs in patients already receiving treatment because of another disease. For this reason, the infections may not be immediately recognized.

Mucormycosis is an uncommon, frequently fatal fungal infection that rarely arises in otherwise healthy people. An underlying disease, for example, diabetes mellitus, is almost always present. Mucormycosis appears sterotypically in different anatomic sites (paranasal, rhinoorbital, rhinocerebral, cerebral, pulmonary, and gastrointestinal regions) and in the soft tissue of the extremities. It can also appear as disseminated disease.

Geographical Distribution

These infections are worldwide in distribution.

Causal Organisms and Their Habitats

Many different organisms have been implicated, but the most common causes of human infection, listed in order of apparent incidence, are *Rhizopus oryzae* and *Rhizopus microsporus* var. *rhizopodiformis*. Other less frequent etiologic agents, but for which a major pathogenic role in humans has been established, include *Absidia corymbifera*, *Apophysomyces elegans*, *Cunninghamella bertholletiae*, *Mucor* species, *Rhizomucor pusillus*, and *Saksenaea vasiformis*. These moulds are ubiquitous and thermotolerant and can be isolated in large numbers from soil or decomposing organic matter such as fruit and bread.

Nosocomial outbreaks of mucormycosis are not as common as hospital-related aspergillus infections, but they have been reported with leukemic patients. Nosocomial cutaneous infections with *Rhizopus microsporus* var. *rhizopodiformis* have been traced to contaminated dressings.

The major risk factors predisposing individuals to mucormycosis include uncontrolled diabetes mellitus, other forms of metabolic acidosis, burns, and malignant hematological disorders. Treatment is seldom of benefit unless these underlying conditions can be corrected.

Pathogenesis

The pathogenesis of mucormycosis is unclear. Though the infection is undoubtedly exogenous, possible sources have only occasionally been suggested, e.g., adhesive dressings, air conditioning filter units, and food. Cutaneous trauma may be an underestimated event in the initiation of many cases of mucormycosis. However, most infections follow inhalation of spores that have been released into the air, and the lungs and nasal sinuses are common sites of infection.

The infectious propagule responsible for the establishment of zygomycoses is undetermined. Many studies have suggested that sporangiospores are responsible; others have indicated that small hyphal fragments initiate disease. Since sporangiospores of mucoraceous fungi are both uncommon in air and readily contained by phagocytic cells normally found in pulmonary tissues, it is possible that in some cases, especially when trauma is the associated factor, hyphal fragments are the infectious propagules.

Usually, mucormycosis develops first in the nasal mucosa or the sinuses and then spreads to involve the soft parts of the orbit and the brain. However, the extent of disease appears to be host dependent, and not all patients with paranasal mucormycosis ultimately develop rhiocerebral mucormycosis. The infection does not normally progress in patients with disease limited to the paranasal sinuses who have not had extensive surgery or treatment with amphotericin B. Thus, host factors may be as important as treatment in determining the outcome of paranasal mucormycosis. In other cases, the pulmonary vessels are affected, resulting in infarction and parenchymatous necrosis. Less often, infection follows ingestion or traumatic inoculation of organisms into the skin. Invasion of the intestinal blood vessels can cause extensive necrosis, and if the fungus enters

809

the cutaneous blood vessels, there may be ulceration involving all the layers of the skin.

Clinical Manifestations

Detailed reviews of the clinical types of mucormycosis are given in references 6, 9, 10, and 11. Mucormycosis is an opportunistic infection and is seldom seen in healthy persons. Various forms are recognized, each of which is associated with particular underlying conditions. Like the etiologic agents of aspergillosis, the causal organisms of mucormycosis have a predilection for vascular invasion, causing thrombosis, infarction, and necrosis of tissue. The clinical hallmark of mucormycosis is the rapid onset of necrosis and fever. In most cases, progress is rapid, and death follows unless treatment is initiated. The anatomical distribution of lesions appears to correlate to a certain degree with defined predisposing conditions, e.g., craniofacial involvement in individuals with diabetic acidosis, pulmonary and disseminated infections in patients with acute leukemia, and gastrointestinal and cutaneous lesions following local trauma. Indeed, cutaneous infections have possibly replaced craniofacial and pulmonary disease as the prevalent clinical manifestations of mucormycosis, a change mirrored by the emergence of *R. microsporus* var. *rhizopodiformis* (as *R. rhizopodiformis* in many cases) as a significant pathogen.

Rhinocerebral Mucormycosis

The terms rhinocerebral and craniofacial mucormycoses are used to describe infections that begin in the paranasal sinuses and then spread to involve the orbit, face, palate, or brain. The terms should be used only when there is documented brain involvement in addition to involvement of sinuses alone or of both sinuses and orbit, respectively. It is not clear whether paranasal and rhinoorbital mucormycoses are simply rhinocerebral mucormycosis in evolution or whether the extent to which mucormycosis progresses beyond the paranasal sinuses to involve the orbit and brain is dependent on the host response. Rhinocerebral mucormycosis is most commonly seen in acidotic individuals, particularly those with uncontrolled diabetes mellitus, but it also occurs in leukemic patients and organ transplant recipients. It is the most common clinical form of mucormycosis and is often fatal within a week of onset if left untreated.

The initial symptoms include unilateral headache, nasal or sinus congestion or pain, and serosanguinous nasal discharge. Fever is also common. Most patients are not seen during the initial stage of local nasal and sinus infection. Two-thirds or more are lethargic or comatose by the time of their examination.

The primary site of infection is the nasal turbinate, resulting in a florid acute inflammatory reaction and extensive necrotizing sinusitis. Cavernous sinus thrombosis and pituitary infarction are commonly seen in fatal cases. If the infection has spread into the orbit, periorbital or perinasal swelling will progress to induration and discoloration. Ptosis, proptosis, dilatation and fixation of pupil, and loss of vision may occur. Drainage of black pus from the eye is a useful diagnostic sign. From the orbit, infection may spread into the brain, leading to frontal lobe necrosis and abscess formation. These features result from invasion of the fungus through the cribiform plate.

If mucormycosis arises in association with chronic maxillary or frontal sinusitis, there will be unilateral manifestations such as nasal secretion of blood-tinged mucus, facial and orbital pain, palpebral ptosis, and involvement of the hard palate, which becomes gray or blackish and ulcerates. X ray reveals opacification of the sinus in which the infection is located.

If the infection spreads into the palate, a black necrotic lesion, an important diagnostic sign, is often found. Necrotic lesions may also be found on the nasal mucosa. Nasal septum or palatal perforation is frequent.

The cerebrospinal fluid (CSF) findings are nonspecific. The protein concentration may be slightly raised, but the glucose concentration is usually normal. There may be a modest mononuclear pleocytosis. CSF cultures are sterile.

Radiologic findings are nonspecific but are useful in delineating the extent of the infection. Diffuse craniofacial bone destruction is typical. Computed tomography (CT) and magnetic resonance (MR) scans are helpful in defining the extent of bone and soft tissue destruction but are more useful in planning surgical intervention than in establishing a diagnosis. CT scans of the head often reveal sinus opacification, but other changes are minimal, even when there is massive orbital infection. MR scanning may be preferred for the diabetic patient, for whom CT contrast agents may be contraindicated.

Pulmonary Mucormycosis

Pulmonary mucormycosis is seldom diagnosed during life. Mucormycosis may develop in the lungs as a result of aspiration or inhalation of infectious material, or from hematogenous or lymphatic spread during dissemination. Most cases occur in leukemic patients undergoing remission induction treatment. Pulmonary infiltrates develop, with infarction of focal areas of lung. The clinical signs are those of bronchitis, pneumonia, and thrombosis. There is necrosis of the parenchyma, leading to cavitation; the bronchi may be perforated, resulting in hemoptysis. If the infection is left untreated, hematogenous dissemination to other organs, particularly the brain, often occurs. The infection is fatal within 2 to 3 weeks.

The most common presentation is unremitting fever and the development or progression of lung infiltrates despite broad-spectrum antibacterial treatment. Hemoptysis and pleuritic chest pain are uncommon but when present are helpful in suggesting a fungal infection. However, there are no characteristic symptoms or signs to distinguish mucormycosis from aspergillosis.

Chest radiographic findings are nonspecific, but the most common finding is focal or diffuse infiltrates that progress to consolidation, cavitation, or wedge-shaped peripheral lesions representing hemorrhagic infarction. Pleural effusion is uncommon.

Gastrointestinal Mucormycosis

Mucormycosis limited to the gastrointestinal tract is rare. It is an uncommon condition that has usually been encountered in malnourished infants or children. Lesions are most common in the stomach, colon, and ileum. The infection is seldom diagnosed during life. Frequently associated diseases include amebic colitis, pellagra, malnutrition, and kwashiorkor.

The symptoms are varied and depend on the site affected. Nonspecific abdominal pain and hematemesis are typical. Necrotic ulcers develop, and peritonitis follows if intestinal perforation occurs. Intestinal mucormycosis is a fulminant illness ending in death within several weeks due to bowel infarction, sepsis, or hemorrhagic shock.

Cutaneous Mucormycosis

Cutaneous mucormycosis is a particular problem in infected patients with burns, in whom spread to underlying tissue is common. The initial signs include fever, swelling, and changes in the appearance of the burn wound. The development of severe underlying necrosis and infarction in a burn should suggest the diagnosis.

Mucormycotic gangrenous cellulitis can follow other forms of trauma to the skin. In diabetic or immunosuppressed patients, cutaneous lesions may arise at an insulin injection site or a catheter insertion site. More massive trauma, such as open fractures or crash injuries, has been present in other patients. Necrotizing cutaneous mucormycosis has occurred in patients who have had contaminated surgical dressings applied to their skin.

Cutaneous lesions resembling ecthyma gangrenosum may develop following hematogenous dissemination of the fungus in immunosuppressed patients. The lesions begin as erythematous, indurated painful cellulitis and then evolve into ulcers covered with a black eschar.

Disseminated Mucormycosis

Disseminated mucormycosis may follow any of the four forms of mucormycosis described so far, but it is usually seen in neutropenic patients with pulmonary infection. Less commonly, dissemination is from the gastrointestinal tract, burns, or other cutaneous lesions. The most common site of spread is the brain, but metastatic necrotic lesions have also been found in the spleen, heart, and other organs. Disseminated mucormycosis is usually diagnosed after the patient has died of the infection. In occasional patients, metastatic cutaneous lesions permit an earlier diagnosis.

Cerebral infection following hematogenous dissemination results in abscess formation and infarction and is distinct from the rhinocerebral form of mucormycosis. Patients present with sudden onset of focal neurological deficits or coma. Investigation of the CSF is unhelpful: protein, glucose, and cell abnormalities are nonspecific, and cultures are sterile. CT and MR scans are useful in locating the lesions.

Other Forms of Mucormycosis

Isolated mucormycotic brain lesions have been reported in a number of individuals, particularly parenteral drug abusers. The infection presents as a rapid deterioration in neurological status.

Other unusual focal forms of mucormycosis include endocarditis, osteomyelitis, and pyelonephritis.

Mucormycosis in Association with Desferrioxamine Administration

There is a clear association between the use of desferrioxamine (used in the treatment of iron overload) and the development of mucormycosis. Most of the reported cases have involved rhinocerebral disease (reviewed by Skahan et al. [9]). Desferrioxamine is a siderophore derived from *Streptomyces pilosus*. Siderophores are high-affinity iron chelators used by microorganisms to concentrate iron for bioavailability. A number of fungi require iron for growth, and it appears that the fungi responsible for mucormycosis can utilize the iron bound to desferrioxamine. The fact that desferrioxamine is removed by renal clearance means that its circulating time will be prolonged in patients on dialysis, and this may increase the risk for these patients.

Differential Diagnosis

Rhinocerebral mucormycosis is a dramatic and distinctive condition, but it can be confused with cavernous sinus thrombosis, bacterial orbital cellulitis, or rhinocerebral aspergillosis or pseudallescheriosis.

The clinical manifestations of pulmonary mucormycosis cannot be distinguished from those of gram-negative bacterial pneumonia, aspergillosis, or pseudallescheriosis.

Essential Investigations and Their Interpretation

Because mucormycosis is such an aggressive infection, early diagnosis is essential for successful treatment. It is, however, often difficult to obtain suitable and timely specimens that have been obtained correctly.

Collection and Transport of Specimens

If there is any suspicion of mucormycosis, a diligent search for clues that would help establish a specific microbiologic diagnosis should be initiated without delay. The most useful clue is the presence of a black necrotic eschar either in the nasopharynx or on the palate. This eschar or material sloughing from it may be mistaken for clotted blood. Material from the eschar and underlying tissue should be obtained by biopsy or curette and transported to the laboratory as quickly as possible. Regions upstream from the lesion or hemorrhage will most likely yield the fungus.

Direct Microscopy

A definitive diagnosis of zygomycosis is best achieved by direct histopathologic examination and culture. Biopsies of pulmonary, rhinocerebral, and cutaneous lesions and scrapings of necrotic mucocutaneous lesions are appropriate specimens.

Gently break the necrotic material apart, and mount it in a few drops of 20% (wt/vol) potassium hydroxide (KOH). Typical wide, nonseptate hyphae with right-angled branching may be seen. Demonstration of mycelial elements in the correct clinical setting constitutes strong evidence of mucormycosis. If the KOH preparation is unrevealing, surgical biopsy or debridement of necrotic material should be done. Examine any material in KOH, and stain with methenamine silver (Grocott's modification [GMS]) and hematoxylin and eosin (H&E) (see below). If fungal elements are seen, obtain further biopsy specimens. The rapid progression seen in untreated mucormycosis allows no delays in diagnostic efforts.

The microscopic demonstration of *Mucorales* in clinical material taken from necrotic lesions is more significant than their isolation in culture. However, different species cannot be distinguished from each other morphologically. These organisms can be distinguished from other moulds, such as *Aspergillus* spp., by their characteristic broad, nonseptate filaments with right-angled branching. Unlike *Aspergillus* spp., zygomycetes do not have parallel walls with 45° angle branching and do not radiate from a single point in tissues. Zygomycetes are also larger than *Aspergillus* spp. Furthermore, agents of mucormycosis stain poorly with periodic acid-Schiff (PAS) stain but stain well with H&E. Negative microscopic examinations do not eliminate the possibility of mucormycosis even if a portion of necrotic tissue has been examined carefully. Central nervous system involvement in rhinocerebral mucormycosis may suggest the presence of the etiologic agent in CSF. However, agents of mucormycosis have not been isolated from CSF. In cases of suspected pulmonary mucormycosis, examine biopsy ma-

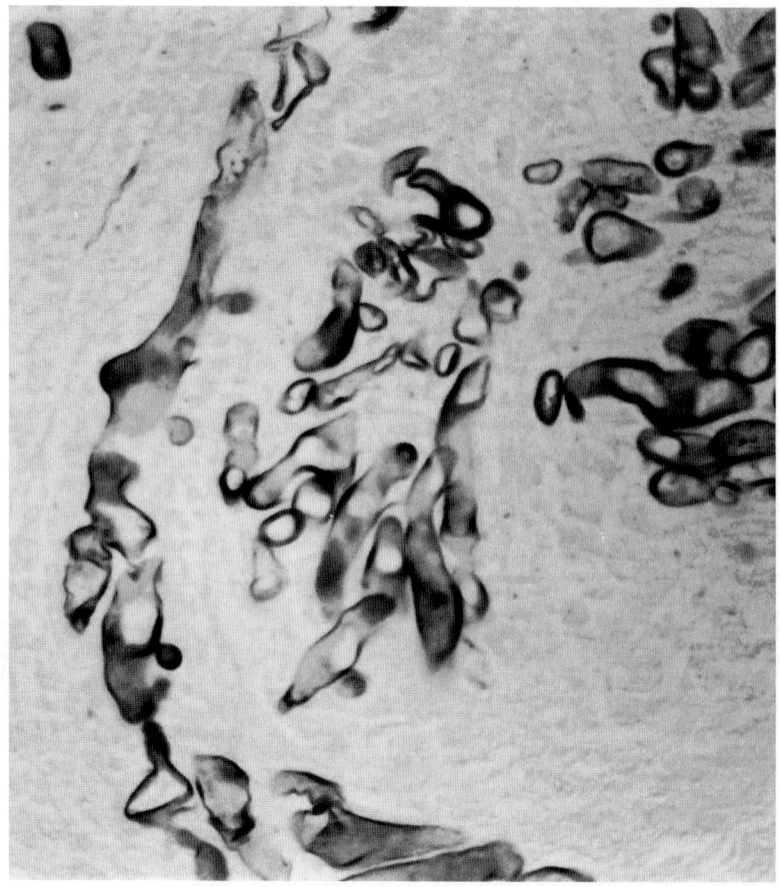

FIGURE 1 Hyphae of *R. oryzae* in H&E-stained section of soft palate from a patient with rhinocerebral mucormycosis. Note the irregular shapes and diameters of the hyphae. Magnification, ×400.

terial or material from fine-needle percutaneous aspiration. Sputum culture is rarely helpful. Metastatic skin lesions, if present, may allow a diagnosis to be made without a lung biopsy being done. In isolated cases, fungal elements have been seen in bronchoalveolar lavage fluid and isolated in culture. Intestinal mucormycosis is usually a rapidly progressive disease ending in death in a few days. In almost all cases, the diagnosis is not suspected until histologic examination of gangrenous tissue obtained at surgery or autopsy reveals the fungus. Antemortem diagnosis is difficult. In suspected cases, the demonstration of fungi in gastric aspirates or feces may be useful.

Histopathology

Biopsy specimens or scrapings from oral lesions are most satisfactory for histologic examination and culture. Demonstrating the fungus in tissue is diagnostic of disease (culture is not), but cultures must be set up to determine the specific etiologic agent. Mucoraceous fungi are readily seen when H&E stain and GMS are used, but their detection rate when stained with the conventional specialized fungal stains of Gridleys and PAS is uneven and barely adequate. One useful stain is cresyl fast violet, which stains zygomycete walls brick red and other fungi blue or purple.

Hyphal elements of mucoraceous fungi consist of broad, twisted, ribbon-like hyphae that are variable in width (up to 50 μm) and often devoid of cytoplasm and have col-

lapsed walls (Fig. 1). True septa are not present, although wrinkles and folds in the cell wall may appear to be septa unless the microscope is focused up and down on the apparent boundary. The hyphae are irregularly and haphazardly branched, and lateral mycelium often grows at nearly 90° (Fig. 1). Although hyphae are broad, their widths may vary greatly, and the hyphal walls also vary in thickness. Branches are often angled 90° from the main hyphal trunk (Fig. 2). The side branches may be much narrower than the main hyphal axis. The mucoraceous fungi can be mistaken for sclerosed capillaries or nematodes, and they may also be confused with empty spherules of *Coccidioides immitis* in regions where coccidioidomycosis is endemic. The small amount of fungus present in tissue sections and the oblique angles at which they are cut make their identification by histologic morphology alone difficult. In some tissue areas, hyphae and hyphal fragments may appear to be septate and constricted at prominent septations. Often, hyphae have terminal bulbous swellings and lateral oval to spherical structures.

Pathology does not vary with the site of infection. Gangrene involvement is common. Surfaces show massive hemorrhage with recent emboli, and in older lesions, the appearance is putty-like. The tissue reaction is not consistent. In some cases, no cellular response is seen, but usually there are various degrees of edema and necrosis with neutrophil accumulations, plasma cells, and some giant cells. The

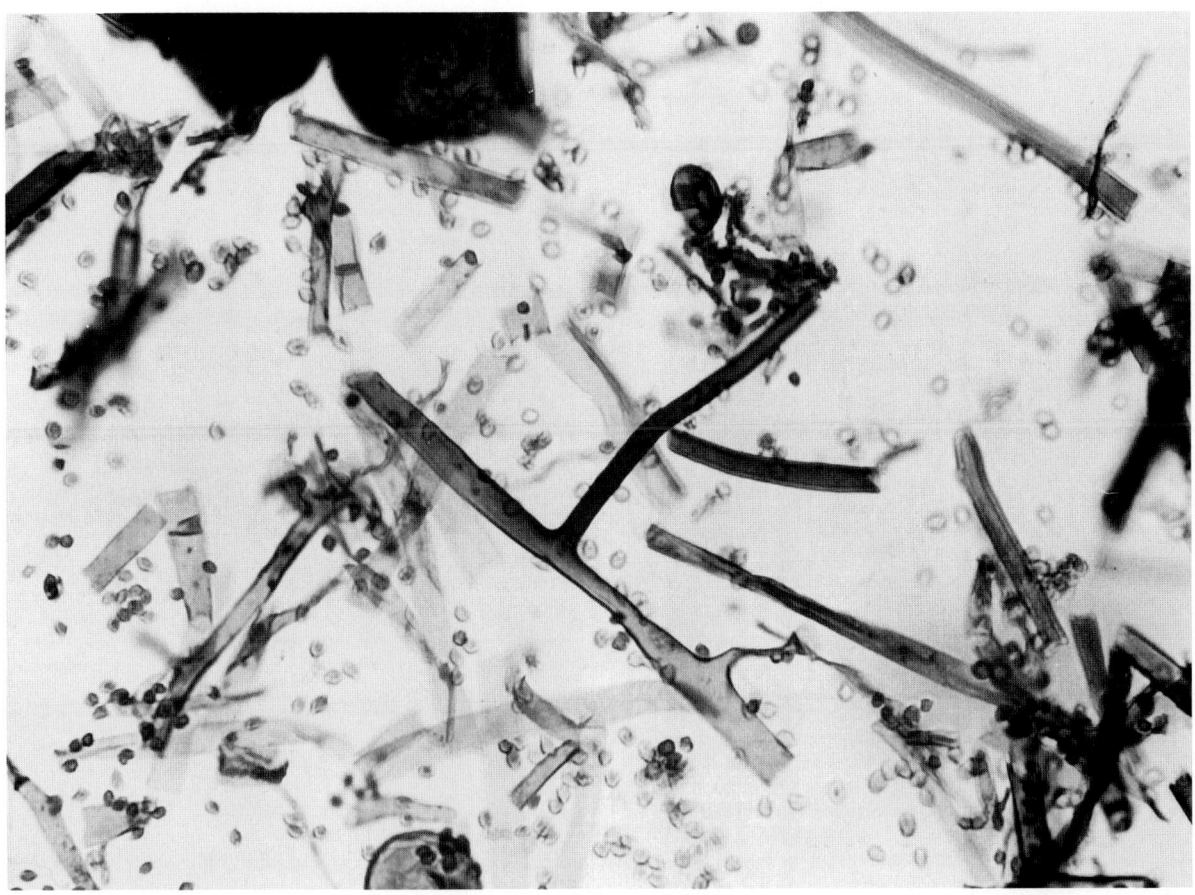

FIGURE 2 Hyphae of *R. oryzae* in a PAS-stained section of an agar block of a glucose-peptone culture. Note the difference in widths of the hyphal fragments and a narrow lateral hypha growing at a 90° angle from the main hyphal trunk. Magnification, ×400.

acute reaction shows the diffuse infiltration of polymorpho-nuclear leukocytes. Eosinophils are not often seen. Tissues often exhibit suppurative reactions but may show granulo-matous changes. The characteristic feature is the involve-ment of blood vessels, with mucoraceous fungi invading the walls of arteries and causing thrombosis and necrosis of the adjacent tissue. The lumens and walls of blood vessels often contain hyphae, chlamydoconidia, fibrin, and inflammatory cells.

Another feature to look for is spherical to ovoid, 15- to 30-μm-diameter, thick-walled, hyaline chlamydospores. Chlamydospores of mucoraceous fungi are PAS and H&E positive but stain either poorly or not at all with GMS. Because of fixation and processing of tissues, the chla-mydospore cytoplasm can retract from the plasma mem-brane, creating an artificial clear space between the con-densed cytoplasm and the thick, rigid cell walls. Some chlamydospores stain entirely; others appear as empty rings. Chlamydospores are seen in the vicinity of mucoraceous hyphae. Detached chlamydoconidia can resemble the yeast form cells of *Blastomyces dermatitidis* or the immature spher-ules and arthroconidia of *Coccidioides immitis*. In pulmonary mucormycosis, chlamydospores are most often observed in alveolar spaces, the lumens of bronchioles, and the visceral pleura. In cutaneous mucormycosis, chlamydoconidia may be seen within necrotic tissue on or near a wound surface.

Immunohistochemical methods of staining can be useful in the diagnosis of mucoraceous infections and their differ-entiation from other fungal diseases. Tissue sections can be deparaffinized, treated with a 1% trypsin solution (pH 8.0) for 1 h at 37°C, and examined by direct immunofluores-cence using specific fluorescent-antibody conjugates to the major species, or a permanent preparation can be made by staining tissue sections by immunoperoxidase techniques with monospecific antibodies to major species. The speci-ficity and activity of these antibodies can be assessed by staining sections of agar blocks cut from seeded cultures in which the fungus grows three-dimensionally throughout the agar. However, the supply of these antibodies is cur-rently limited to specialized or reference laboratories.

Culture

Nasal, palatal, and sputum cultures are seldom helpful, but specimens should be sent for culture if the clinician suspects mucormycosis. Because the mucorales are common con-taminants, isolation of these organisms from sputum, aspi-rated material from sinuses, or bronchial washings must be interpreted with caution. However, if the patient is diabetic or immunosuppressed, the isolation should not be ignored. Isolation of the causal agents has been achieved in a small proportion of cases.

The agents of mucormycosis can be grown on most laboratory media, but glucose-peptone agar (4% glucose, 1% mycologic peptone, 2% agar) or 4% malt extract agar

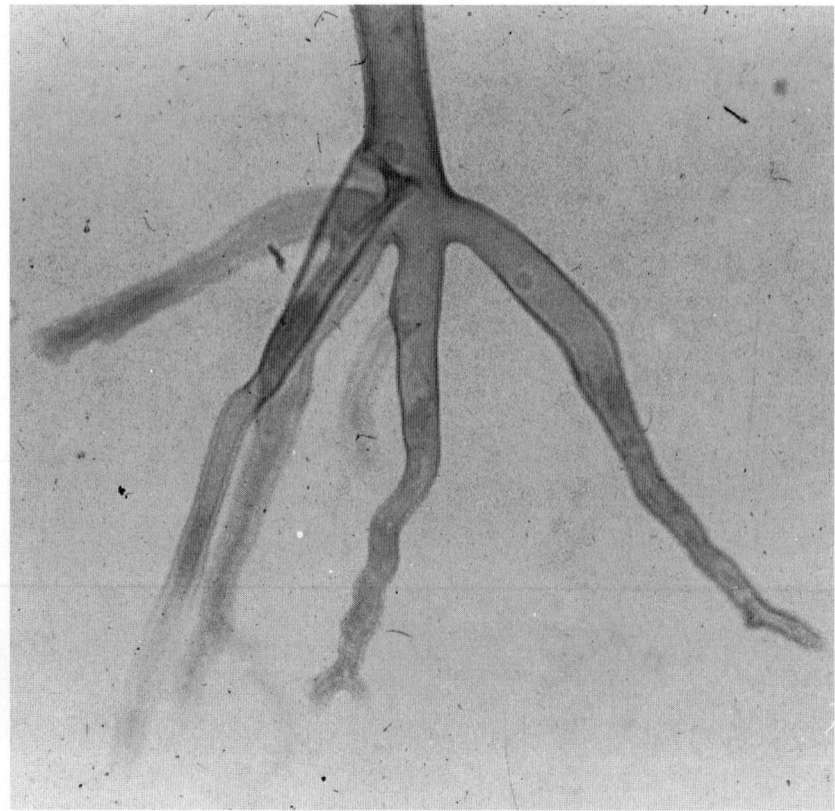

FIGURE 3 Typical formation of rhizoids that are characteristic of *Rhizopus* and *Rhizomucor* spp. and to some extent of *Absidia* spp. also. Magnification, ×400.

(4% malt extract, 2% agar) is recommended, because typical morphology can be demonstrated. Media containing cycloheximide should not be used, because this drug inhibits the growth of these fungi. Point inoculate small portions of biopsy material onto glucose-peptone agar. Swab cultures of the nose or palate are usually negative and are not recommended. Incubate cultures at room temperature and 37°C. Do not homogenize tissue material, as homogenization kills viable hyphal elements of mucoraceous fungi. Refrigeration may also reduce the viability of some of these agents.

Serologic Tests

No serologic tests for mucormycosis are commercially available at present. However, agar gel double-diffusion and enzyme-linked immunosorbent assay formats have been developed and evaluated with a few patients. It is unlikely that these tests are going to be appropriate with fulminating infections. The development of infection-specific antigens and monoclonal antibodies together with oligonucleotide primers and probes may facilitate the early diagnosis of mucormycosis in the future. For a contemporary review of the role and utility of serologic tests in mucormycosis, see the review by Hopwood and Evans (3).

General Description of Fungi Most Likely To Be Isolated and Basis of Identification

The descriptions given below contain morphologic terms specific to mucoraceous fungi. Refer to the glossary of terms in the appendix to this chapter.

Identification of most zygomycetes is based primarily on the morphology of the sporangia, i.e., arrangement and number of sporangiospores, shape, color, presence or absence of columellae and apophyses, arrangement of sporangiophores, and presence or absence of rhizoids. Growth temperature studies can also be especially helpful in identifying and differentiating members of the genera *Rhizopus*, *Rhizomucor*, and *Absidia*. The *Mucorales* are characterized by the production of nonseptate hyphae. However, occasional septa are formed in older cultures. Some genera produce rhizoids (root-like anchors [Fig. 3]) that are connected by hyphae called stolons. The *Mucorales* reproduce asexually by producing spores in sporangia, which are borne on sporangiophores. The sporangiophores of some genera swell where they meet the sporangium, forming the structure called an apophysis (Fig. 4). Typically, a sporangium is a globose swelling at the apex of a sporangiophore. The sporangium contains a central, variously shaped columella that is separated from the spore-containing sac by a membrane. The cytoplasm of the sporangium is cleaved into hundreds of sporangiospores during maturation (Fig. 4). In addition, some genera of *Mucorales* produce small, several-spored sporangia called sporangioles. Finally, a few genera, for example, *Cunninghamella*, produce sporangia that contain a single spore, called conidia. Sexual reproduction in the *Mucorales* is characterized by the formation of zygospores. However, determining the presence of zygospores is not part of a routine laboratory identification scheme.

Several excellent monographs on the taxonomy and detailed identification of zygomycete fungi have been pub-

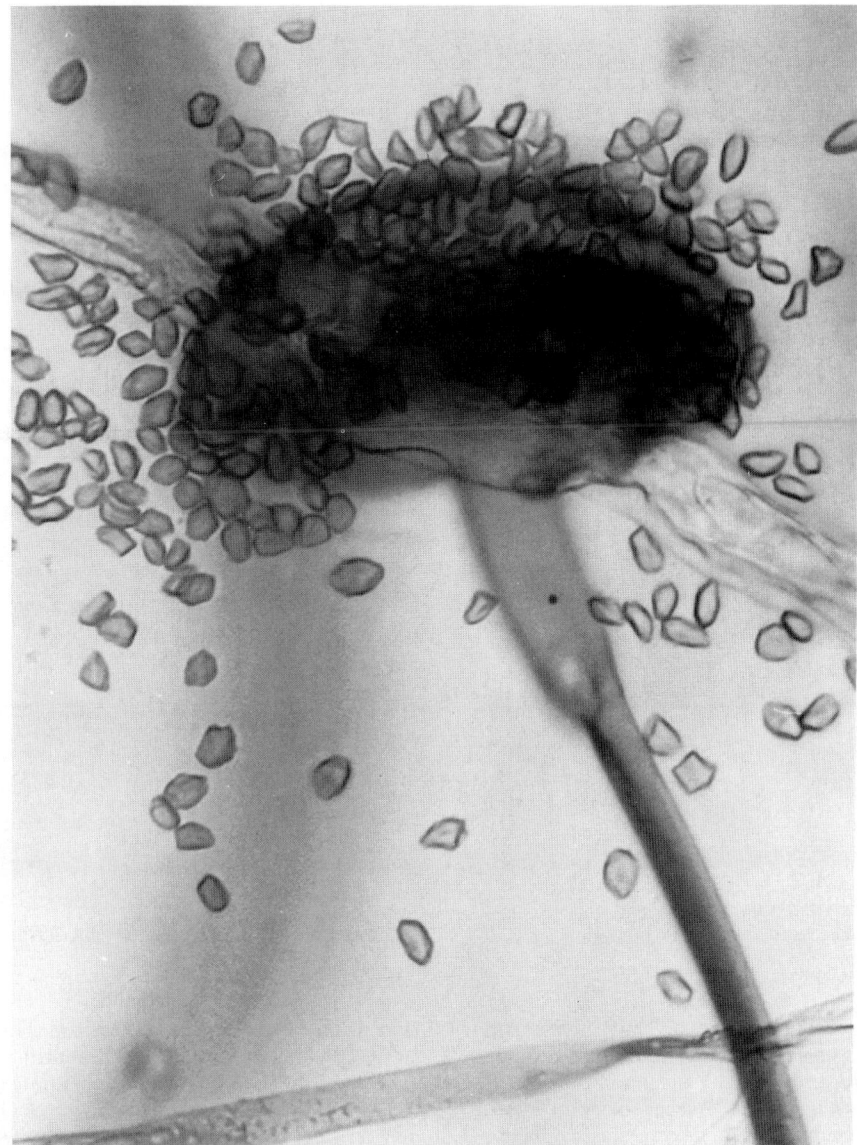

FIGURE 4 Sporangium, apophysis, and sporangiospores of *R. oryzae* (NCPF 2634). Note the rhomboidal and lemon-shaped sporangiospores with striae (wrinkles) on the surface. Magnification, ×400.

lished (2, 4, 5). These sources complement the features highlighted here and give information on isolates not corresponding to the descriptions below.

Description of Causal Agents

The descriptions of the causal agents of zygomycoses given below are based almost entirely on their asexual characteristics, as follows:

1. Type and numbers of sporangiospore.
2. Structure supporting or containing the sporangiospores.
3. Structure supporting the sporangiophore.
4. Presence or absence of branching in the sporangiophore.
5. Shape of the columella.

6. Presence or absence of rhizoids and their locations in relation to the sporophore or stolon.

Spore containers are named according to the numbers of reproductive elements they hold, e.g.,

1. A sporangium contains many spores.
2. A merosporangium contains moderate numbers of spores.
3. A sporangiola contains one to a few spores.

The key morphologic characteristics of the hyphae that typify the zygomycetes are the internodes and the rhizoids. An internode is a runner (or stolon) that runs between contact points on the substrate; rhizoids are root-like hyphae growing within the substrate. Further classification of the zygomycetes can be based on the presence of branched

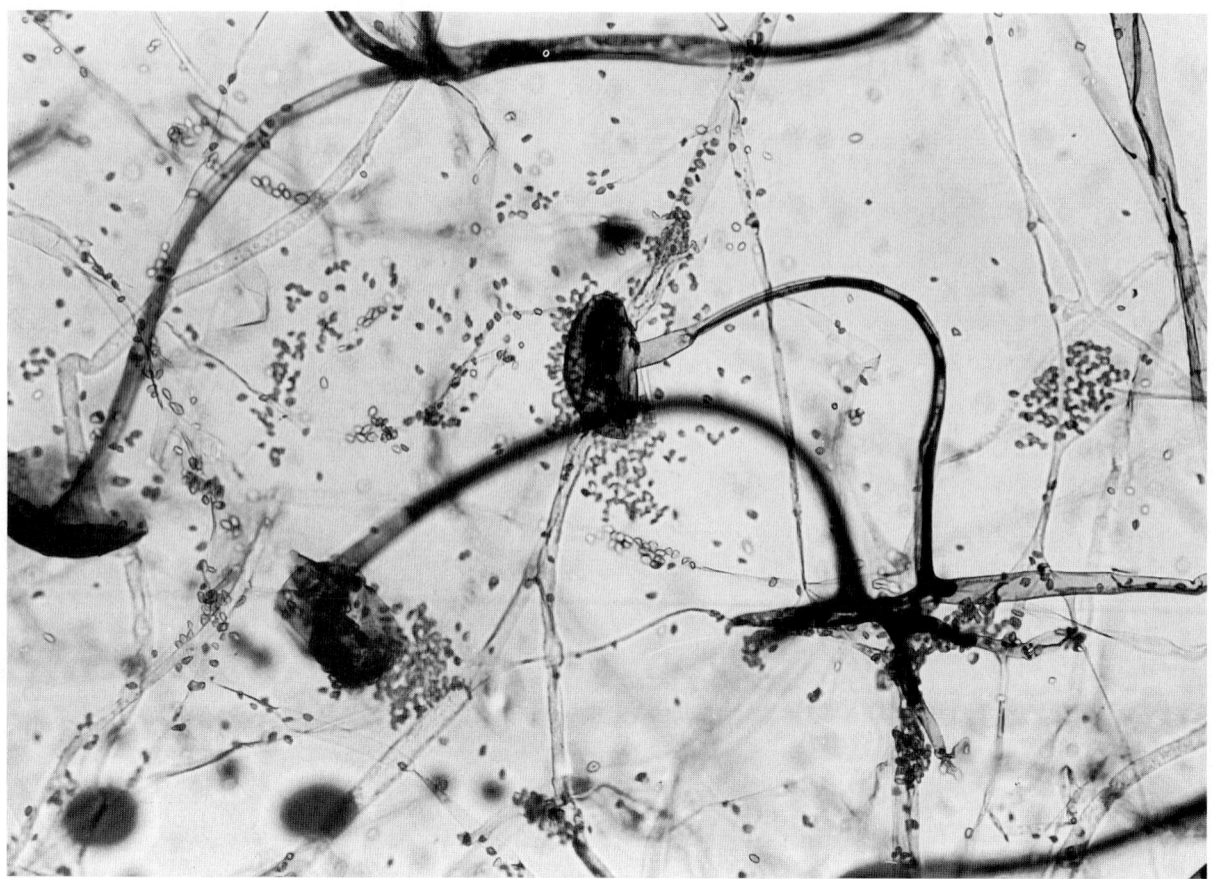

FIGURE 5 *R. oryzae* (NCPF 2634). Note the simple sporangiophores borne singly opposite rhizoids. Some sporangia have collapsed, and the sporangial wall has dissolved. Magnification, ×400.

or unbranched sporangiophores at or between the internodes, the presence or absence of rhizoids, and the maximum temperature for growth. The differentiation of these fungi to species level is difficult, but a scheme of identification based on a combination of asexual morphology and biochemical characteristics has made identification more feasible for the clinical laboratory. The descriptions given below are based on the colony appearances and microscopic morphologies of individual isolates (some deposited in the National Culture Collection of Pathogenic Fungi, Mycology Reference Laboratory, Public Health Laboratory, Bristol, United Kingdom) cultured on 4% malt agar as a primary isolation medium and subcultured on glucose-peptone agar without the addition of cycloheximide. The maximum incubation temperature is indicated, assuming that primary isolation plates of clinical specimens from suspected cases of fungal infection are routinely incubated at a range of temperatures, e.g., 28, 37, and 45°C.

Rhizopus oryzae (*Rhizopus arrhizus*)

Rhizopus oryzae is the most prevalent agent of mucormycosis. Approximately 60% of all culturally proven cases of human mucormycosis and nearly 90% of rhinocerebral cases are caused by *R. oryzae*.

Colony Morphology

R. oryzae forms a fast-growing, white, cottony colony at 28°C, filling a 9-cm-diameter petri dish within 3 to 4 days and reaching about 8 mm in height. The fungus will grow at temperatures up to 45°C if routine primary isolation plates are incubated at this temperature. Colonies tend to collapse, becoming brownish gray to blackish gray, depending on the amount of sporulation. Growth is inhibited by cycloheximide.

Microscopic Morphology

The single most useful characteristic in identifying members of the genus *Rhizopus* is the presence of simple spoangiophores borne singly or in groups opposite rhizoids (Fig. 5). In *R. oryzae*, broad, nonseptate hyphae are seen. Sporangiophores are up to 1,500 μm long and 18 μm wide, smooth walled, nonseptate, and simple with a single axis or branched, and they arise from stolons opposite brownish rhizoids, usually in groups of three or more. Grayish black sporangia are globose, often with a flattened base, powdery in appearance, and up to 180 μm in diameter. Columellae are ellipsoidal on a truncate base, mouse gray, and up to 130 μm long. Sporangiospores vary in shape and size, having a rhomboidal lemon shape with ridges on the surface.

Rhizopus microsporus var. *rhizopodiformis*

The second most common species associated with mucormycosis, causing nearly 15% of all culturally proven cases of the disease in humans, is *R. microsporus* var *rhizopodiformis*. This organism is isolated mainly from cutaneous lesions and

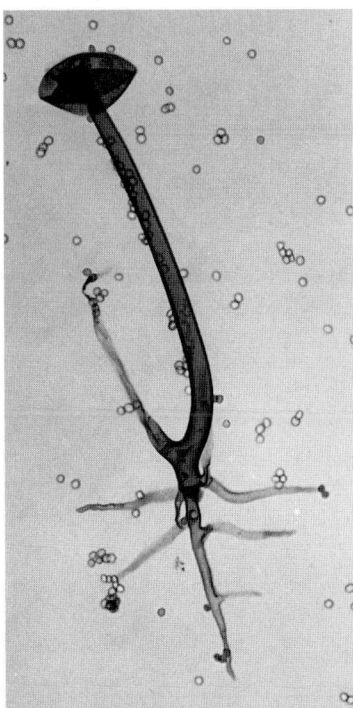

FIGURE 6 *R. microsporus* var. *rhizopodiformis* (NCPF 2955). Note the short sporangiophore compared with that of *R. oryzae* and the branched rhizoids. Magnification, ×400.

only rarely from rhinocerebral lesions. It is associated with necrotizing cutaneous infections following the use of Elastoplast bandages to cover surgical wounds (reviewed in reference 4).

Colony Morphology
Colonies range from gray to dark grayish brown.

Microscopic Morphology
Rhizoids are hyaline to subhyaline. Sporangiophores are shorter (maximum length, 1,000 μm) than those of *R. oryzae* and are produced singly or in groups of up to four (Fig. 6). Sporangia average 100 μm in diameter. Columellae are pyriform or somewhat elongated, with distinct angular apophysis (Fig. 7). Sporangiospores (4 to 6 μm) are subglobose to slightly rhomboidal, with smooth or finely spinulose surfaces. The fungus is thermotolerant, growing well at up to 50°C.

Rhizomucor pusillus
Rhizomucor pusillus has been isolated from patients with pulmonary, disseminated, and cutaneous infections and humans with endocarditis. This fungus can also cause mycotic abortion in cow, horses, and pigs.

Colony Morphology
The organism grows rapidly and is white at first, turning dark smoke gray or brown as sporulation occurs. Colony texture is floccose to wooly. Growth is inhibited by cycloheximide. The maximum growth temperature of 54 to 58°C is a useful distinguishing feature.

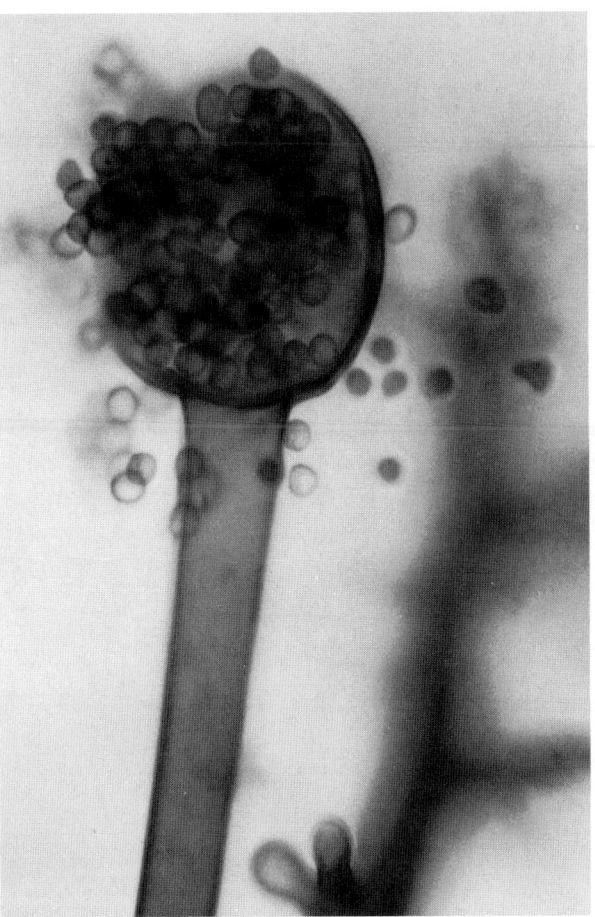

FIGURE 7 *R. microsporus* var. *rhizopodiformis* (NCPF 2955). Note the partially dissolved sporangial wall, smooth-walled globose to subglobose sporangiospores, and angulated, truncated apophysis. Magnification, ×400.

Microscopic Morphology
Rhizomucor pusillus forms broad, nonseptate hyphae. Sporangiophores are mainly monopodially branched and are anchored to the substratum by short rhizoids (Fig. 8). Sporangia are globuse, dark brown, and up to 100 μm in diameter. *Rhizomucor* spp. differ from *Rhizopus* spp. by forming sporangia without apophyses on branched, pale gray-brown sporangiophores. Columellae are subglobose to pyriform. The sporangium wall dissolves to leave a collar around the sporangiophore. Globose to subglobose hyaline sporangiospores 3 to 5 μm in diameter are released.

Absidia corymbifera (Absidia ramosa)
Absidia corymbifera is an infrequent agent of mucormycosis in humans. Documented cases include a case of meningitis in an immunocompetent patient that followed a perforating head injury; a case of cutaneous infection in an immunocompetent patient; and renal infection in an AIDS patient. Infection is most probably acquired by inhalation of sporangiospores. *A. corymbifera* is the only human pathogenic species of this genus.

Colony Morphology
A. corymbifera forms rapidly growing colonies that quickly fill the petri dish. It grows well at 25 to 45°C, and

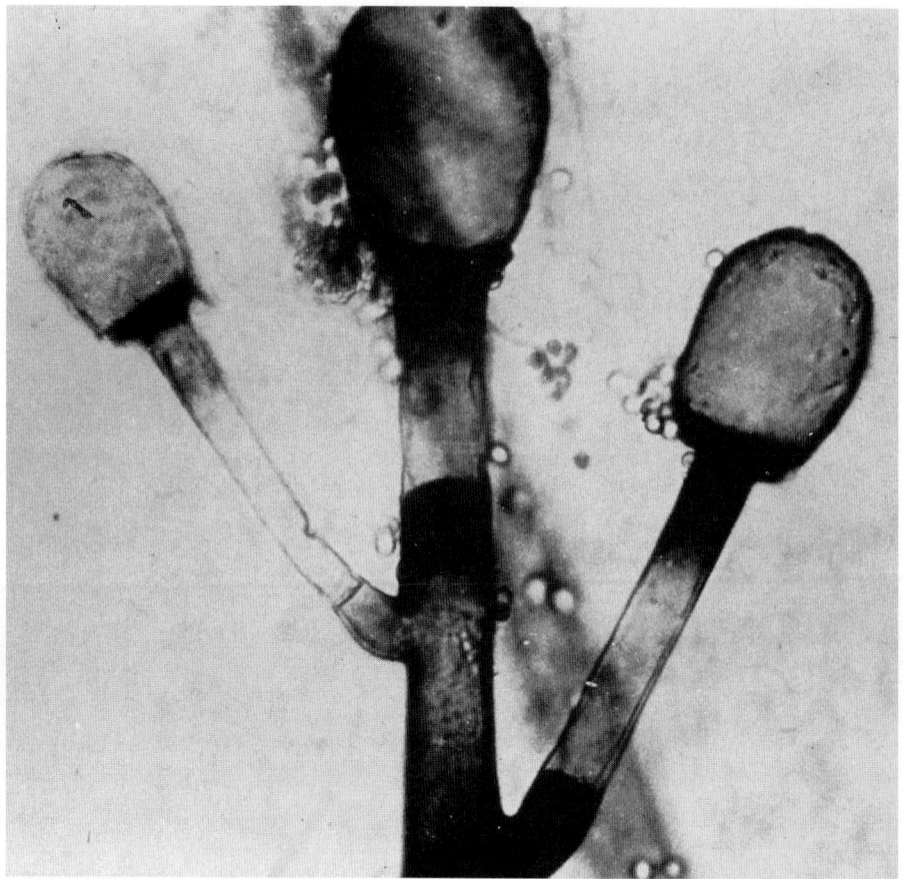

FIGURE 8 *Rhizomucor pusillus.* Note the monopodially branched sporangiophores, pyriform columellae, and small sporangiospores. A collarette can be seen around the sporangiophore. Magnification, ×400. (Reprinted with permission of Kaminski's Teaching Slides on Medical Mycology, Adelaide Children's Hospital, Adelaide, Australia.)

some isolates can grow at 50°C. The floccose colonies are light gray to grayish brown. The colony may originate from stolons between rhizoids or from finely branched aerial hyphae. *A. corymbifera* does not grow in the presence of cycloheximide.

Microscopic Morphology

Like other agents of mucormycosis, *A. corymbifera* forms nonseptate hyphae. Sporangiophores are hyaline to light gray and branched simply or in a corymbiform pattern. Sporangia are pyriform, 100 to 150 μm in diameter, and hyaline when young but become grayish brown to greenish beige (Fig. 9). A swollen apophysis is seen at the point where the sporangiophore merges with the sporangium. Columellae are hemispherical to short ovoidal. Sporangiospores have a greenish yellow tinge and are variable in shape and size, with dimensions varying between 3.2 to 3.9 by 3.1 to 3.5 μm and a long, ellipsoidal shape of 4.1 to 4.9 by 2.3 to 2.9 μm.

Cunninghamella bertholletiae (Synonyms: *Cunninghamella elegans, Cunninghamella echinulata*)

In the genus *Cunninghamella*, only the species *C. bertholletiae* has been verified as a human pathogen (for a recent review, see reference 4). Infection is acquired by inhalation of conidia. Mating reactions have been very important in delineating the species of *Cunninghamella*.

Colony Morphology

C. bertholletiae is very fast growing, filling a petri dish within a week. Colonies are white at first, becoming dark gray and powdery with development. Primary isolates will grow at temperatures up to 42°C. It is sensitive to cycloheximide.

Microscopic Morphology

The sporangiophores of *C. bertholletiae* are up to 20 μm wide and straight, with verticillate or single branches. Terminal vesicles are subglobose to pyriform, with the terminal vesicles up to 40 μm in diameter and the lateral ones 10 to 30 μm in diameter. Sporangiola are globose (7 to 11 μm in diameter) or ellipsoidal (9 to 13 by 6 to 10 μm), verrucose or short echinulate, hyaline when single, but brownish in mass (Fig. 10). *C. bertholletiae* grows optimally between 25 and 30°C; the maximum temperature tolerated is 50°C. The fungus is occasionally associated with infection in humans but, unlike some of the other agents of mucormycosis, can infect immunocompetent individuals.

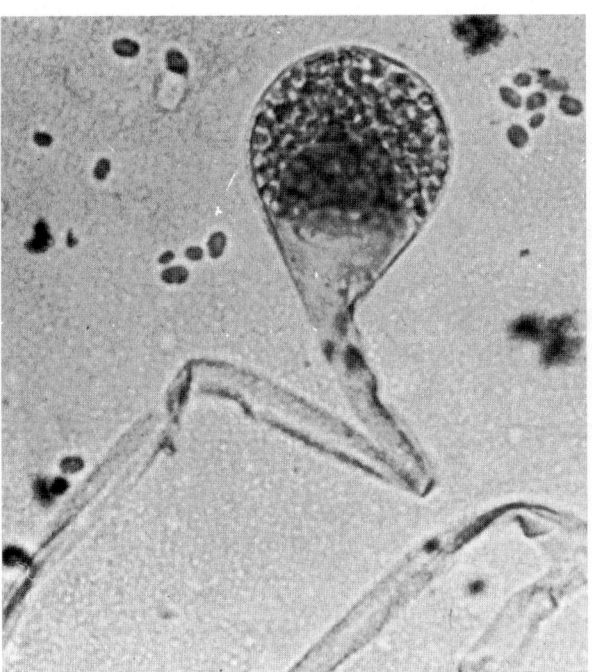

FIGURE 9 *A. corymbifera.* Note the relatively fine sporangiophore, pyriform sporangium, hemispherical columella, and long conical apophysis. Sporangiospores vary in morphology from globose to long ellipsoidal. Magnification, ×400.

Saksenaea vasiformis

Saksenaea vasiformis is widespread in tropical soils and is a known agent of subcutaneous zygomycosis. Infection is acquired by traumatic implantation of the fungus. At least 14 cases of invasive infection have been reported.

Colony Morphology
The colonies of *S. vasiformis* are fast growing, downy, and white with no reverse pigment.

Microscopic Morphology
S. vasiformis forms broad, nonseptate hyphae and flask-shaped sporangia (Fig. 11), each with a distinct spherical venter (that portion of a phialide that lies between the base of a phialide and the conidiogenous locus) and a long neck, that arise singly or in pairs from dichotomously branched, darkly pigmented rhizoids. The columellae are prominent and dome shaped. Small, oblong sporangiospores 1 to 2 by 3 to 4 μm in size are discharged through the neck of the sporangia following the dissolution of an apical mucilaginous plug.

Laboratory identification may be difficult or delayed because of the mould's failure to sporulate on primary isolation media or after subsequent subculture onto potato glucose agar. Sporulation may be stimulated by the use of nutrient-deficient media, such as cornmeal-glucose-sucrose-yeast extract agar. See reference 7 for sporulation procedure.

Mucor Species

Mucor species are infrequently the causative agent of mucormycosis. The genus contains 49 recognized species, many of which occur widely and are of considerable importance. However, few species have been recovered from patients with well-documented cases of mucormycosis, and infections due to members of this genus are rare.

Colony Morphology
Mucor species grow rapidly. Colonies are white and cottony at first but become gray to brown as sporulation occurs. *Mucor* species are inhibited by the concentration of cycloheximide usually incorporated in routine isolation media. Most species grow poorly, if at all, at 37°C.

Microscopic Morphology
Members of the genus *Mucor* can be differentiated from *Absidia*, *Rhizopus*, and *Rhizomucor* spp. by the absence of stolons and rhizoids. Species of *Mucor* produce simple or sympodially branching sporangiophores from nonseptate hyphae. There is no swelling where the sporangiophore and sporangium merge (Fig. 12). The gray-brown sporangia themselves are globose with columellae and contain globose to ellipsoidal sporangiospores. Rhizoids are not formed.

Apophysomyces elegans

Apophysomyces elegans was originally found in soil samples from a mango orchard in India. Subsequently, eight definite cases of mucormycosis caused by this fungus that resulted from traumatic implantation or contamination of burns have been described. *Apophysomyces elegans* closely resembles A. *corymbifera* in its morphology.

Colony Morphology
Apophysomyces elegans is fast growing. Colonies are white initially, becoming creamy white to buff with age and downy. They have no reverse pigment. The organism grows well at temperatures up to 42°C. The growth of *Apophysomyces elegans* is not inhibited by cycloheximide.

Microscopic Morphology
Apophysomyces elegans does not sporulate on routine mycologic media; it forms only broad, nonseptate hyphae. See reference 7 for sporulation procedure. Characteristic identifying features include sporangiophores arising singly from aerial hyphal segments. Thin-walled, hyaline, light grayish brown rhizoids are produced opposite to sporangiophores. Sporangia are pyriform (20 to 50 μm in diameter), with prominent funnel- or wine glass-shaped apophyses and hemispherical columellae produced singly at the tip of each sporangiophore. *Apophysomyces elegans* produces sporangiospores that are smooth, mostly oblong (5.4 to 8.0 by 4.0 to 5.7 μm), occasionally globose, and light brown.

Rare Causes of Mucormycosis
Rare causes of mucormycosis include *Cokeromyces recurvatus*, *Syncephalastrum racemosum*, and *Mortierella* species.

Treatment and Antifungal Susceptibility
Amphotericin B is the drug of choice for mucormycosis. The necessity for prompt administration of amphotericin B does not allow antifungal susceptibility tests to be performed, and they are not necessary. Other antifungal drugs (fluconazole, itraconazole) have no role in the management of mucormycosis.

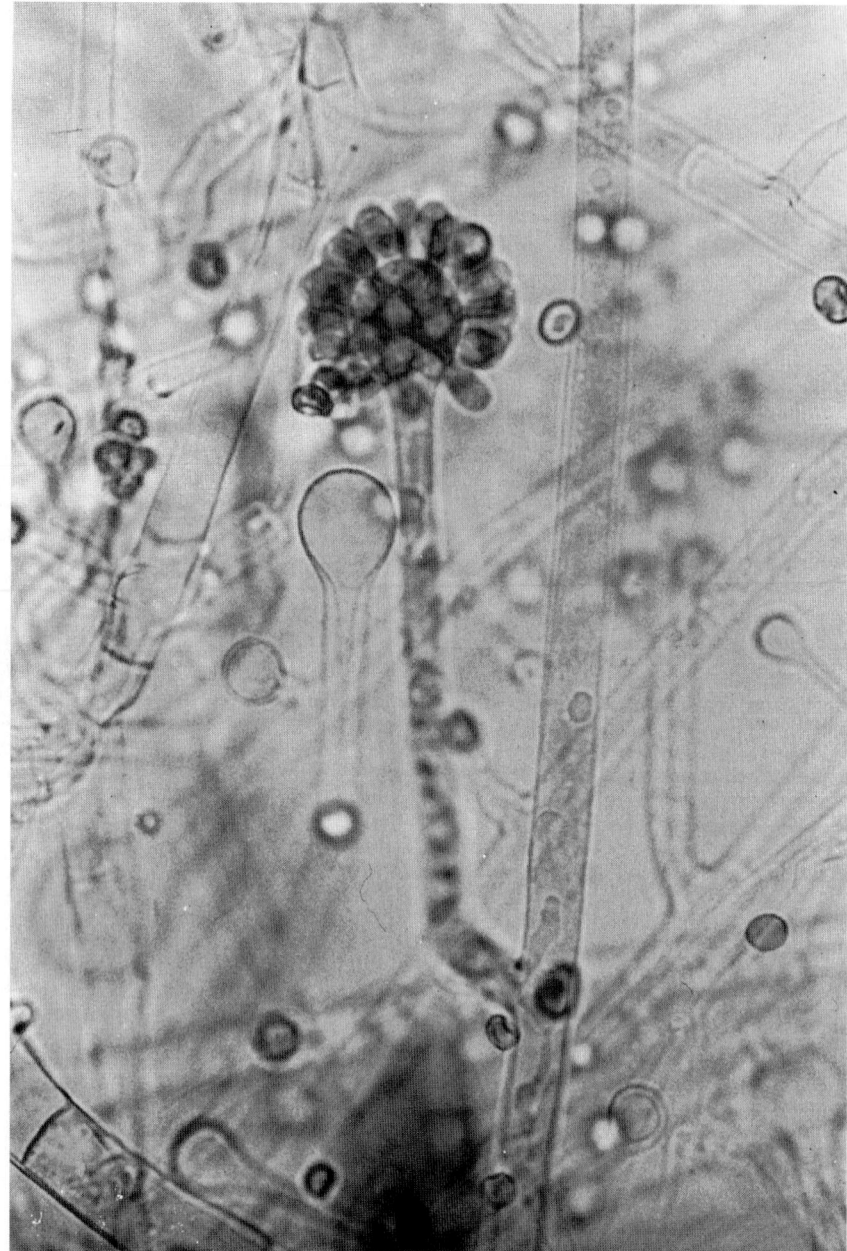

FIGURE 10 *C. bertholletiae*. Note the single sporangiophore branching off a main hyphal trunk and the subglobose to pyriform vesicle bearing ellipsoidal sporangiola. Magnification, ×400. (Reprinted with permission of Kaminski's Teaching Slides on Medical Mycology, Adelaide Children's Hospital, Adelaide, Australia.)

ENTOMOPHTHORAMYCOSES (SUBCUTANEOUS PHYCOMYCOSIS, SUBCUTANEOUS ZYGOMYCOSIS)

Conidiobolomycosis

Definition

Conidiobolomycosis is a chronic mycosis affecting the subcutaneous tissues. It originates in the nasal sinuses and spreads to the adjacent subcutaneous tissue of the face, causing disfigurement.

Geographical Distribution

The disease occurs mainly in the tropical rain forests of Africa, South and Central America, and Southeast Asia.

Causative Organism and Its Habitat

Conidiobolus coronatus (*Entomophthora coronata*) lives as a saphrophyte in soil humus and on decomposing plant matter in moist, warm climates. It can also parasitize certain insects. The disease is most common among adult males, particularly those living or working in tropical rain forests.

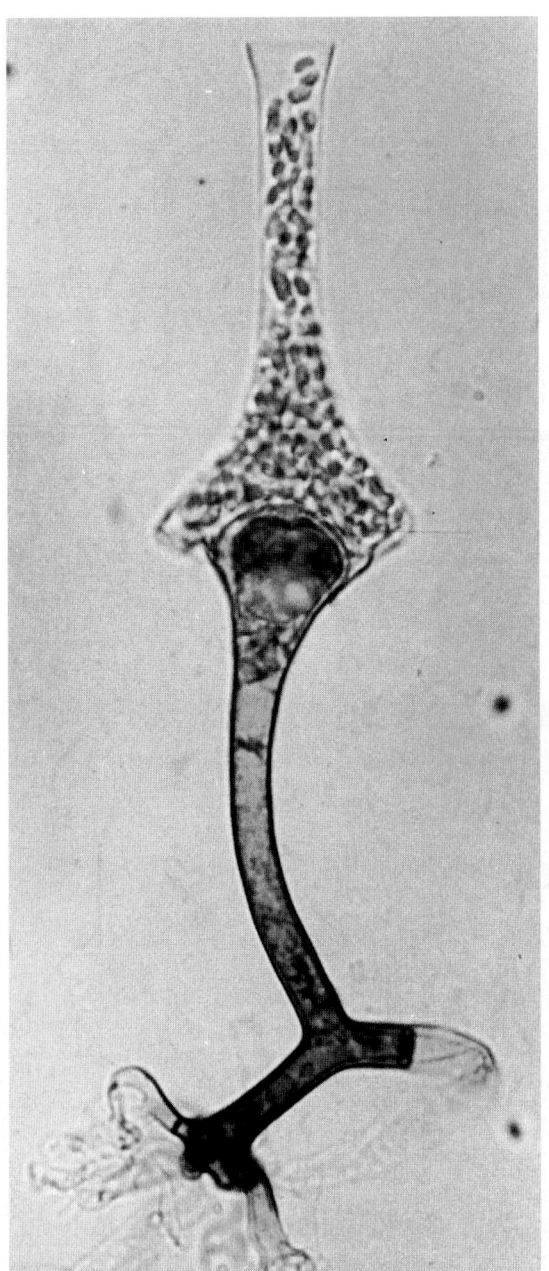

FIGURE 11 *S. vasiformis.* Note the flask-shaped sporangium with a distinct, globose columella and small, smooth, elongated sporangiospores. Magnification, ×400. (Reprinted with permission of Kaminski's Teaching Slides on Medical Mycology, Adelaide Children's Hospital, Adelaide, Australia.)

Infection is acquired through inhalation of spores or through their introduction into the nasal cavities by soiled hands.

Clinical Manifestations

Conidiobolus infection generally begins with unilateral involvement of the nasal mucosa. The most common nasal symptom is obstruction, but frequent nosebleeds can occur and are evidence of the development of a nasal polyp in the anterior region of the inferior turbinate. Subcutaneous nod-ules then develop in the nasal and perinasal regions and may be associated with epidermal lesions.

The spread of the infection is slow but relentless. The infection is usually confined to the face, and the development of gross facial swelling involving the forehead, periorbital region, and upper lip is very distinctive. As a rule, the lesions are firmly attached to the underlying tissue, although the bone is spared. The skin remains intact. Spread to the lymph nodes has been reported.

Differential Diagnosis

Even if, in advanced cases, the diagnosis is obvious from the clinical appearance, mycologic and histologic examinations are essential for confirmation.

Essential Investigations and Their Interpretation

Microscopy

Microscopic examination of smears or tissue from the nasal mucosa will reveal broad, nonseptate, thin-walled mycelial filaments.

Culture

Tissue obtained by biopsy or curettage should be cultured. Cultures are difficult to obtain, and the pathological material must be inoculated onto the largest possible number of media. These media should be incubated at between 25 and 35°C to enhance the growth of *Conidiobolus coronatus.*

Colony morphology. Colonies of *Conidiobolus coronatus* grow rapidly; are flat, cream colored, and glabrous; and become radially folded and covered by a fine, powdery, white surface mycelium and conidiophores. The lid of the petri dish soon becomes covered with conidia, which are forcibly discharged by the conidiophores. The colony may become tan to brown with age.

Microscopic morphology. Conidiophores are simple, forming solitary, terminal conidia that are spherical, 10 to 25 μm in diameter, and single celled and have a prominent papilla. Conidia may also produce hair-like appendages called villae. Conidia germinate to produce either (i) single or multiple hyphal tubes that may also become conidiophores bearing secondary conidia or (ii) multiple short conidiophores, each bearing a small secondary conidium. See reference 1 for additional detail.

Histopathology

Biopsy specimens show fibroblastic proliferation and an inflammatory reaction with lymphocytes, plasma cells, histiocytes, eosinophils, and giant cells. Broad, thin-walled hyphae with occasional septa branched at right angles are seen. The Splendore-Hoepplei phenomenon may be seen.

Treatment and Antifungal Susceptibility

Methods of treatment employed so far have been disappointing. Surgical resection of infected tissue is seldom successful; it may hasten the spread of infection. The condition can be treated with saturated potassium iodide solution (up to 10 ml three times daily as tolerated) or amphotericin B. Susceptibility testing against potassium iodide or amphotericin B is not appropriate.

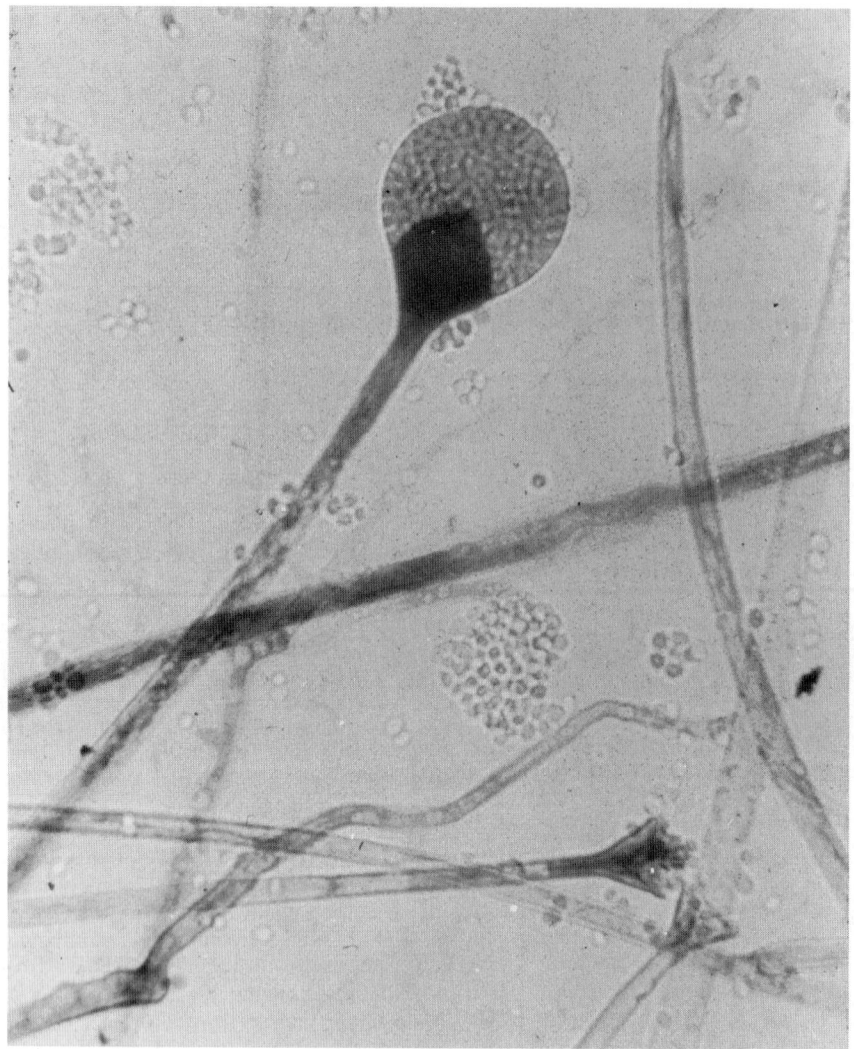

FIGURE 12 *Mucor* species. Note the globose sporangium, elongate columella, and small sporangiospores. Magnification, ×400.

Basidiobolomycosis

Definition
Basidiobolomycosis is a chronic subcutaneous infection of the trunk and limbs.

Geographical Distribution
The disease is encountered chiefly in the tropical regions of East and West Africa, Indonesia, and India.

Causative Organism and Its Habitat
The most widely held view is that *Basidiobolus ranarum* is the sole agent causing the disease and that *Basidiobolus meristosporus* and *Basidiobolus haptosporus* are only synonyms of the former; however, not all authors are of this opinion.

B. ranarum has been isolated from the guts of frogs, toads, and lizards that had apparently swallowed infected insects. It has also been isolated from ants and decaying plant matter.

Still uncertain are how the disease is acquired and what the length of incubation is. Inoculation through a thorn prick or an insect bite has been suggested, as has contamination of a wound or other abrasion. The infection is most common in children.

Clinical Manifestations
The subcutaneous swelling that characterizes this disease is usually localized to the back of the shoulders and to the arms, but it may also be found on the buttocks and thighs.

The initial swelling may be rapid or slow in onset and is hard and painless. The spread is slow but relentless, and a large mass that is attached to the skin but not the underlying tissue (unlike *Conidiobolus* infection) is formed. This is a disfiguring infection, but the skin covering the lesions does not ulcerate. Lymphatic obstruction may occur and can result in massive lymphoedema.

There is no functional impairment as long as the joints are not blocked by the volume of the swellings. The underlying bone and joints are not affected by the disease.

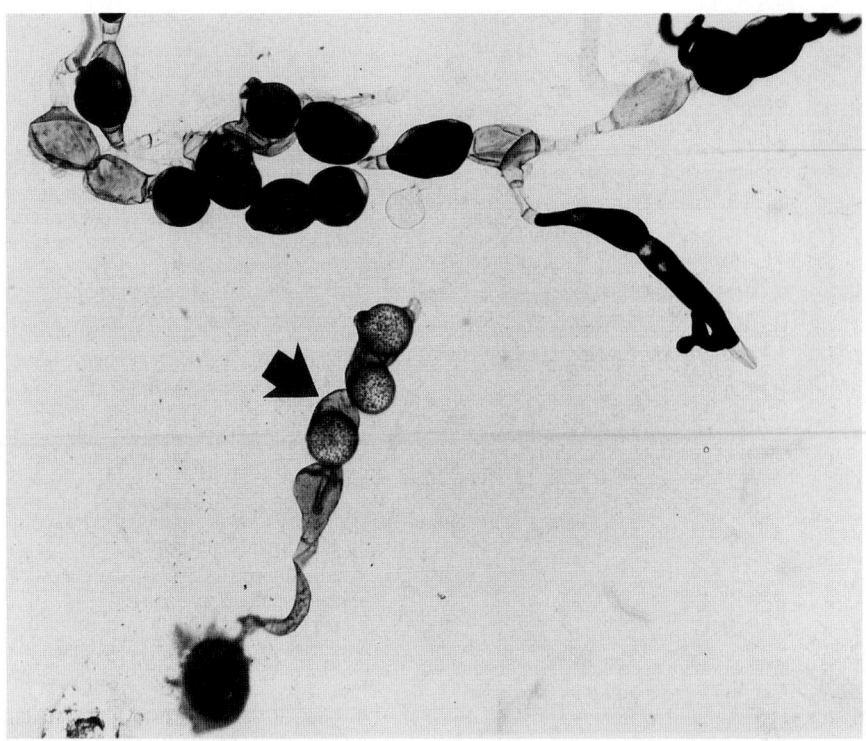

FIGURE 13 Zygospores of B. *ranarum*, showing beak-like appendages (arrow). Magnification, ×400.

Differential Diagnosis

The disease can be diagnosed with confidence on the basis of appearance and the results of mycologic and particularly histologic examinations. Basidiobolomycosis closely resembles soft tissue sarcoma. Mycetoma has the same degree of induration as basidiobolomycosis, but the presence of draining sinuses, fixation to underlying structures, and extension to bone of mycetoma makes the distinction relatively easy. Bacterial cellulitis is much more acute and painful than in zygomycosis. All these conditions can be reliably distinguished from zygomycosis by biopsy and culture of the lesion.

Essential Investigations and Their Interpretation

Specimens must be taken from the subcutaneous tissue, where the infection is seen as nodules that develop into massive, firm, indurated swellings that are freely movable over the underlying muscle but are attached to the skin, which is hyperpigmented but not ulcerated.

Microscopy

Direct microscopy reveals wide, irregular hyphae or fragments thereof with few septa.

Culture

Specimens should be cultured on Sabouraud agar at 30°C. Identifiable colonies should be obtained in less than 1 week.

Colony morphology. Colonies are moderately fast growing at 30°C, flat, yellowish gray, glabrous, becoming radially folded, and covered by a fine, powdery, white surface mycelium. Note that satellite colonies are often formed by germinating conidia ejected from the primary colony.

Microscopic morphology. Large vegetative hyphae (8 to 20 μm in diameter) forming round, smooth, thick-walled zygospores (20 to 50 μm in diameter) that have two closely appressed beak-like appendages are characteristic (Fig. 13). Two types of asexual spores are formed: primary and secondary. Primary spores are globose, one celled, and solitary and are forcibly discharged from a sporophore. The sporophore has a distinct swollen area just below the spore that actively participates in the discharge of the spore. Secondary spores are clavate and one celled and are passively released from a sporophore. These sporophores are not swollen at their bases. The apex of the passively released spore has a knob-like adhesive tip. These spores may function as sporangia, producing several sporangiospores. See reference 1 for additional detail.

Histopathology

The histopathologic features of basidiobolomycosis are very similar to those given above for conidiobolomycosis.

Therapy and Antifungal Susceptibility

The therapy of choice appears to be saturated potassium iodide solution or ketoconazole. Susceptibility testing is not appropriate.

APPENDIX

Glossary of Terms

This glossary includes terms used to describe morphologic structures of the agents of zygomycosis (for detailed mycologic glossaries and definitions, see references 4, 5, and 8).

Apophysis: a swelling. The term is applied to the swelling of a sporangiophore immediately below the columella.

Collar: an annulus of sporangial wall remnants at the tip of a sporangiophore

Collarette: a small collar

Columella: a dome-like structure at the tip of a sporangiophore

Pyriform: pear shaped

Rhizoid: a short branching hypha that resembles a root

Sporangiolum: a small sporangium producing a small number of sporangiospores

Sporangiophore: a specialized hypha upon which a sporangium develops

Sporangiospore: an asexual spore produced within a sporangium

Sporangium: a sac-like structure whose entire internal contents are cleaved into asexual spores

Sporophore: any structure that bears spores

Stolon: a runner or horizontal hypha from which new hyphae, rhizoids, sporangiophores, or any combination of these structures arises

Zygospore: a resting spore resulting from the fusion of two gametangia

REFERENCES

1. **de Vroey, C.** 1989. Identification of agents of subcutaneous mycoses, p. 111–139. *In* E. G. V. Evans and M. D. Richardson (ed.), *Medical Mycology: a Practical Approach*. IRL Press, Oxford, England.
2. **Goodman, N. L., and M. G. Rinaldi.** 1991. Agents of zygomycosis, p. 674–692. *In* A. Balows, W. J. Hausler, Jr., K. L. Herrmann, H. D. Isenberg, and H. J. Shadomy (ed.), *Manual of Clinical Microbiology*, 5th ed. American Society for Microbiology, Washington, D.C.
3. **Hopwood, V., and E. G. V. Evans.** 1991. Serological tests in the diagnosis and prognosis of fungal infection in the compromised patient, p. 312–353. *In* D. W. Warnock and M. D. Richardson (ed.), *Fungal Infection in the Compromised Patient*, 2nd ed. John Wiley & Sons, Chichester, England.
4. **Kwon-Chung, K. J., and J. E. Bennett.** 1992. Mucormycosis, p. 524–559. *In Medical Mycology*. Lea & Febiger, Philadelphia.
5. **McGinnis, M. R.** 1980. *Laboratory Handbook of Medical Mycology*. Academic Press, Inc., New York.
6. **Parfrey, N. A.** 1986. Improved diagnosis and prognosis of mucormycosis. A clinicopathologic study of 33 cases. *Medicine* **65**:113–123.
7. **Rhodes, J. C., and K. J. Kwong-Chung.** 1989. Identification of agents of systemic mycoses, p. 141–170. *In* E. G. V. Evans and M. D. Richardson (ed.), *Medical Mycology: a Practical Approach*. IRL Press, Oxford, England.
8. **Scholer, H. J., E. Muller, and M. A. A. Schipper.** 1983. Mucorales, p. 3–59. *In* D. H. Howard (ed.), *Fungi Pathogenic for Humans and Animals, Part A. Biology*. Marcel Dekker, Inc., New York.
9. **Skahan, K. J., B. Wong, and D. Armstrong.** 1991. Clinical manifestations and management of mucormycosis in the compromised patient, p. 153–190. *In* D. W. Warnock and M. D. Richardson (ed.), *Fungal Infection in the Compromised Patient*, 2nd ed. John Wiley & Sons, Chichester, England.
10. **Smith, J. M. B.** 1989. The Phycomyces, p. 115–169. *In Opportunistic Mycoses of Man and Other Animals*. C.A.B. International Mycological Institute, Wallingford, England.
11. **Walsh, T. J., M. R. Rinaldi, and P. A. Pizzo.** 1993. Zygomycosis of the respiratory tract, p. 149–170. *In* G. A. Sarosi and S. F. Davies (ed.), *Fungal Diseases of the Lung*, 2nd ed. Raven Press, New York.

Bipolaris, Exophiala, Scedosporium, Sporothrix, and Other Dematiaceous Fungi

WILEY A. SCHELL, LESTER PASARELL, IRA F. SALKIN,
AND MICHAEL R. McGINNIS

67

TAXONOMY, DESCRIPTION, AND NATURAL HABITATS

Dematiaceous fungi are characterized by the presence of a brown to black color in the cell walls of their vegetative cells, conidia, or both that results in colonies ranging from olive or gray to black. The dark pigmentation of the majority of medically important fungi is caused by the deposition of dihydroxynaphthalene melanin (92) formed via pentaketide metabolism. This differs from the dihydroxyphenylalanine melanin associated with *Cryptococcus neoformans* formed by the oxidation of ortho- and paradiphenols. These ubiquitous and cosmopolitan pathogens normally are associated with soil and plants, but occasionally they cause infections in humans and animals (25, 50, 51, 53, 77). In medical mycology, dematiaceous fungi often are thought of as being exclusively hyphomycetes, but some ascomycetes, basidiomycetes, coelomycetes, and zygomycetes (see chapter 59 of this Manual) are also dematiaceous.

CLINICAL SIGNIFICANCE

Mycotic infections caused by dematiaceous fungi include mycetoma (see chapter 68 of this Manual), chromoblastomycosis, phaeohyphomycosis, and sporotrichosis and have been reported in healthy and compromised hosts. Typically, they are initiated by the inoculation of the etiologic agent through a penetrating injury. However, in some cases, pulmonary or disseminated infections have been caused by inhalation of conidia from the environment. Chromoblastomycosis is a chronic, localized, cutaneous to subcutaneous opportunistic infection characterized by the presence of muriform (sclerotic) bodies in infected tissue (see Fig. 25 in chapter 60 of this Manual). By definition, mature muriform bodies have intersecting cross walls, though many cells represent younger stages of development and have no or one septum. Muriform bodies result as vegetative growth occurs without the elongation seen in hyphae; they divide by fission along the septa. The presence of muriform bodies in cutaneous or subcutaneous tissue is pathognomonic of chromoblastomycosis. In some cases, dematiaceous hyphae may be seen as well. Because the appearance of the muriform bodies formed by all agents of chromoblastomycosis is similar, the fungus cannot be identified on the basis of tissue morphology but must be cultured.

In contrast to chromoblastomycosis, phaeohyphomycosis is characterized by the presence in tissue of dematiaceous yeastlike cells, pseudohyphae, hyphae, or any combination of these forms. The name phaeohyphomycosis is not restricted to hyphomycetes; it encompasses all fungi having dematiaceous cells in infected tissue, regardless of the taxonomic classification of the etiologic agent. The hyphae observed in clinical specimens may be regular and uniform in diameter or irregular in shape with many swollen cells and may be short or very long. Though most agents of phaeohyphomycosis form dark cells in infected tissue, it should be noted that certain fungi, particularly species of the genera *Alternaria*, *Bipolaris*, and *Curvularia*, often appear to be hyaline in tissue owing to meager formation of melanin yet are considered agents of phaeohyphomycosis because they are manifestly dematiaceous when grown in culture. Furthermore, use of the Masson-Fontana stain (41) on histologic tissue sections reveals the presence of melanin in the cell walls of these fungi. Phaeohyphomycosis (1, 25, 50) encompasses a spectrum of opportunistic entities that ranges from purely cosmetic conditions to fatal cerebral infections. As with chromoblastomycosis, the identities of etiologic agents of phaeohyphomycosis cannot be determined from microscopic examination of clinical specimens. These fungi must be grown on laboratory culture media before they can be identified. The most common phaeohyphomycosis syndromes include allergic fungal sinusitis, keratitis, and subcutaneous infection. Subcutaneous infections commonly present as either a cyst or a diffuse lesion, are chronic, and usually remain localized. Sporotrichosis, caused by *Sporothrix schenckii*, is typically a cutaneous to subcutaneous chronic infection that may undergo lymphatic spread. Musculoskeletal involvement and disseminated infection may occur rarely, and pulmonary infections following inhalation of the conidia have been documented (40). Sporotrichosis is included in this chapter because the etiologic agent is dematiaceous in culture and because the presence of melanin can be demonstrated within the yeast cell walls in tissue by using the Masson-Fontana stain. Certain dematiaceous fungi have also been documented as etiologic agents of mycetoma (see chapter 68 of this Manual).

COLLECTION, TRANSPORT, PROCESSING, AND EXAMINATION OF SPECIMENS

General guidelines for the collection, processing, and examination of specimens are provided in chapter 60 of this Manual. The specimens most frequently submitted for the recovery of dematiaceous fungi include aspirates, biopsy samples, skin scrapings, and surgical tissue. Swabs are not an effective means of specimen collection and should be avoided when possible. Transport media should not be used unless specimens can be easily retrieved in their entirety. Portions of suspicious necrotic, purulent, or caseous specimens should be examined microscopically and inoculated onto isolation media. Tissue specimens should be minced with scalpels into 0.5- to 1.0-mm pieces and inoculated directly to culture media. Tissue homogenizers should not be used, because some moulds do not have regularly septate (compartmentalized) hyphae and thus can be easily killed by homogenization. When fungal cells are found during microscopic examination of clinical specimens, they must be examined for the presence of melanin by using bright-field illumination. The amount of melanin present may be slight; thus, it is imperative that the microscope illumination be correctly adjusted according to the Köhler technique (59). In laboratories that use fluorescence microscopy, bright-field examination of a positive field can be accomplished without moving the slide by blocking off the UV lamp, sliding the UV filter out of the light path, and turning on the incandescent lamp. In addition, it must be noted that heavily melanized fungal cells such as muriform bodies and granules may not be reliably detected by calcofluor white. Therefore, when chromoblastomycosis or mycetoma is suspected, examination of clinical specimens by bright-field microscopy is required. An alternative method for demonstrating the presence of melanin is use of the Masson-Fontana stain (41). In sporotrichosis, the number of yeast cells in clinical specimens is typically very limited, and their sizes and shapes are not distinctive. As a result, most mycologists consider it fruitless to microscopically examine specimens for *Sporothrix schenckii*. However, the fungus can usually be seen in tissue sections treated with diastase and then stained by either the Gomori or the periodic acid-Schiff technique. Fluorescent-antibody-specific conjugates for *Sporothrix schenckii* have been developed, but they are available only at certain reference laboratories (35).

ISOLATION PROCEDURES

To ensure that an isolate is in pure culture, the fungus may be isolated by transferring hyphal tips or individual germinating conidia or by streaking the organism onto plated culture medium containing antibacterial agents and subsequently isolating individual colonies for subculture to a similar medium (49). Ascertaining the purity of the isolate is important, because many of the opportunistic dematiaceous pathogens are pleomorphic, meaning that a single isolate can produce conidia of more than one kind or by more than one mechanism. For an accurate identification, it must be known whether the various types of conidia present in a culture were formed by a single pleomorphic fungus or are due to the presence of more than one fungus.

IDENTIFICATION

The identification of dematiaceous fungi (5, 11, 16, 22, 23, 49, 90) ultimately rests upon their microscopic morphologies and to a lesser extent upon their colony morphologies (Table 1). Morphologic evaluations are normally based on cultures that have been grown on a medium such as potato dextrose agar or cornmeal dextrose agar at 25 to 30°C for approximately 2 weeks. It is essential to study the microscopic and gross characteristics of young as well as fully mature colonies. Many microscopic fields must be examined to observe variant as well as predominant features. The importance of conidium development in defining the genera of dematiaceous fungi makes it essential to determine how a particular fungus forms its conidia. Slide culture preparations (49) using potato dextrose agar or cornmeal dextrose agar are ideal for determining conidiogenesis. These nutritionally minimal media usually stimulate the formation of conidia. If a suspected pathogen does not produce conidia under such conditions, exposure to a naked incandescent light bulb for several days in a 12-h-light, 12-h-dark cycle while the pathogen is growing on a medium such as 2% water agar, V-8 juice agar, potato dextrose agar, cornmeal dextrose agar, or moistened sterile wooden sticks or moistened filter paper may stimulate it to form conidia. Other techniques that may trigger conidiation include exposing the isolate to UV light (310 to 410 nm); growing the isolate at both high and low temperatures; growing it on hay infusion agar, soil extract agar, and cereal agar; or lyophilizing and then regrowing it. For the identification of certain fungi, it may be helpful to demonstrate physiologic phenomena such as thermotolerance and resistance to cycloheximide. When such studies are used, it is important to include controls that will demonstrate viability of the inoculum.

A modern nomenclature that addresses the complex reproductive cycle of fungi and facilitates discussion of their identification has gained acceptance during the past 20 years (10). The term anamorph is used to characterize an asexual reproductive structure or form produced by a fungus. Some fungi seen in the clinical laboratory produce more than one anamorph. For example, the pathogen *Scedosporium apiospermum*, which forms a distinctive anamorph (the *Scedosporium* anamorph) consisting of solitary annellated condiogenous cells (Fig. 8a), may exhibit an additional anamorph typical of the genus *Graphium* (Fig. 8c). When a single fungus produces more than one anamorph, the term synanamorph is used to designate any of the concurrently existing forms. In the present example, the *Graphium* form is a synanamorph of *Scedosporium apiospermum* (and vice versa). In addition, some isolates of *Scedosporium apiospermum* have the ability to produce a sexual form that is characterized by the formation of ascocarps, within which the sexually recombinant ascospores arise. The sexual form of a fungus is referred to as the teleomorph. The teleomorph formed by most clinical strains of *Scedosporium apiospermum* is *Pseudallescheria boydii* (Fig. 8d). Finally, the term holomorph is used to encompass the whole fungus, including all of its forms. In this example, the whole fungus (holomorph) consists of the *Pseudallescheria boydii* teleomorph and the *Scedosporium* and *Graphium* synanamorphs. Because sexual structures are considered taxonomically more significant than asexual structures, the classification of a fungus must be based upon its teleomorph when the teleomorph is known. As a result, the name used for the teleomorph is also used for the holomorph. Problems may arise in certain anamorph-teleomorph connections when a teleomorph is known to have more than one anamorph or when a single anamorph is associated with several different teleomorphs. For example, *Scedosporium*

TABLE 1 Diagnostic features of some medically important dematiaceous fungi

Genus	Diagnostic characteristics	Compare to	Selected references
Alternaria	Colonies rapidly growing, cottony, gray to black. Conidiophores erect, dark, septate, simple or branched. Conidia muriform, obclavate with beak (tapering apex), darkly pigmented, smooth or rough, in simple or branched acropetal chains. Fig. 1a.	*Bipolaris, Curvularia, Exserohilum, Helminthosporium*	22, 23, 49
Aureobasidium	Colonies smooth, moist; yellow, white, cream, light pink, or light brown, finally becoming black from development of arthroconidia (in *Scytalidium* synanamorph). Conidiogenous cells undifferentiated from hyphae; intercalary, terminal, or arising as short lateral branches. Conidia hyaline, smooth, ellipsoidal, variable in shape and size, one celled, often producing secondary blastoconidia. Large, dark, one- or two-celled, thick-walled arthroconidia (i.e., *Scytalidium* synanamorph) usually present. Differs from *Hormonema* spp. by conidia arising in synchrony. Differs from *Phaeococcomyces* spp. by lack of dematiaceous yeast cells. Fig. 4a.	*Hormonema, Scytalidium*	14, 22, 23, 49
Bipolaris	Colonies rapidly growing, woolly, gray to black. Conidiophores dark, erect, simple or branched, septate, geniculate owing to sympodial development. Conidia multiseptate, cylindrical to oblong, dark; septal walls thickened; hilum slightly protruding. Fig. 1c.	*Curvularia, Drechslera, Exserohilum, Helminthosporium*	2, 55, 65
Chaetomium	Colonies rapidly growing, fluffy, initially hyaline, becoming olive-green to brown. Perithecia form on surface of substratum; dark brown to black, globose to flask shaped, ostiolate; with distinctive elongate, hairlike hyphae arising from exterior. Asci clavate, dissolving upon maturity. *C. globosum*, 8 ascospores per ascus, lemon shaped, one celled, brown. Fig. 9a and b.	*Coniothyrium, Lasiodiplodia, Phoma*	20, 90
Cladosporium	Colonies rapidly growing, velvety or cottony, usually some shade of olive-gray to olive-brown or black. Conidiophores dark, erect, long, often septate and branching. Conidia one celled (several celled in some species), smooth or rough, with dark prominent hila; occurring in long, fragile, profusely branched acropetal chains. Conidia at branch points of chains usually shield shaped. *C. carrionii* distinguished by growth up to 37°C; infrequently branched, long chains of conidia; conidia ca. 5 μm long, elliptical, bilaterally symmetrical, mainly uniform in size and shape. Fig. 3a–c.	*Fonsecaea, Xylohypha*	22, 23, 29, 49, 90
Coniothyrium	Colonies of *C. fuckelii* rapidly growing, floccose, hyaline becoming mid to dark brown. Pycnidia globose, averaging 200–300 μm, brown 3–5 cell layers thick, unilocular, with ostiole. Conidia one celled, elliptical, guttulate, smooth, brown at maturity, 1.5–2.5 by 3–4.5 μm. Phialides ampulliform, 4–6 by 3–4.5 μm. Distinguished from *Phoma* spp. by brown conidia. Fig. 9c and d.	*Chaetomium, Phoma, Lasiodiplodia*	37, 87
Curvularia	Colonies rapidly growing, woolly, gray to grayish black or brown. Conidiophores dark, erect, geniculate owing to sympodial development. Conidia multiseptate, usually curved, usually with central cell larger and darker than end cells; thicknesses of septa and outer cell wall approximately the same; hilum dark. Fig. 1b.	*Alternaria, Bipolaris, Drechslera, Exserohilum*	22, 23, 49

(Continued on next page)

TABLE 1 Diagnostic features of some medically important dematiaceous fungi (*Continued*)

Genus	Diagnostic characteristics	Compare to	Selected references
Dactylaria	Colonies flat, olivaceous gray to brown, usually with Bordeaux red diffusible pigment on Emmons' modified Sabouraud glucose agar. Conidiophores hyaline, erect, occasionally geniculate owing to sympodial development. Conidia two celled, cylindrical to oblong, dark, released by rhexolytic dehiscence. One-celled, globose phialoconidia may be present in young colonies. *D. constricta* var. *gallopava* grows at 45°C; inhibited on media containing cycloheximide; gelatin positive in ≥21 days; pathogenic for humans and fowl. *D. constricta* var. *constricta* does not grow at 45°C; grows on media containing cycloheximide; rapidly gelatin positive; not known to be pathogenic. Fig. 2a.	*Scolecobasidium*	17, 73
Exophiala	Colonies usually yeastlike when young (from presence of *Phaeoannellomyces* synanamorph), becoming mouldlike with age; pale brown to black. Conidiophores hyaline to subhyaline, hyphalike or distinct. Conidiogenous cells annellides. Conidia one celled (up to three celled in *E. salmonis*), hyaline to pale brown, accumulating in balls at apices of annellides. *E. castellanii* characterized by poorly differentiated conidiogenous cells having inconspicuous annellations; no growth at 40°C. *E. jeanselmei* has annellides tapering to narrow apices; grows at up to ca. 37°C, no growth at 40°C; assimilates KNO₃. Two varieties of *E. jeanselmei* exist: *E. jeanselmei* var. *jeanselmei* bears well-developed, erect, lageniform to cylindrical annellides; *E. jeanselmei* var. *lecanii-corni* bears annellations arising directly from hyphae. *E. moniliae* exhibits swollen annellides terminating in very long annellated apices; assimilates KNO₃. *E. spinifera* forms distinct spinelike, multicellular conidiophores bearing annellides that are lageniform to cylindrical, tapering to long, annellated apices; assimilates KNO₃. Fig. 5a–g.	*Phaeoannellomyces, Phaeococcomyces, Rhinocladiella, Wangiella*	14, 16, 24, 34, 46, 47–49, 60
Exserohilum	Colonies rapidly growing, woolly, gray to black. Conidiophores dark, erect, geniculate owing to sympodial development. Conidia multiseptate, cylindrical to oblong, dark, with strongly protruding hila. Fig. 1d.	*Bipolaris, Drechslera, Helminthosporium*	2, 55, 65
Fonsecaea	Colonies slow growing, velvety, olivaceous black. Conidiophores pale brown to midbrown, usually erect, apically swollen owing to compact sympodial development. Conidia one celled, pale brown to midbrown; primary conidia function as sympodial conidiogenous cells to produce secondary conidia. Tertiary conidia form similarly. *Cladosporium, Phialophora,* or *Rhinocladiella* synanamorphs may be present. *F. compacta* is differentiated from *F. pedrosoi* by conidia that are subglobose, broadly attached, and arranged in compact conidial heads. Both species grow at 37°C. Fig. 3d.	*Cladosporium, Rhinocladiella*	49, 56
Hormonema	Colonies smooth, moist; yellow, white, cream, light pink, or light brown, finally becoming black from development of arthroconidia. Conidiogenous cells undifferentiated from hyphae; intercalary, terminal, or arising as short lateral branches. Conidia hyaline, smooth, ellipsoidal, variable in shape and size, one celled, often producing secondary blastoconidia. Large, dark one- to two-celled, thick-walled arthroconidia (i.e., *Scytalidium* anamorph) usually present. Differs from *Aureobasidium* spp. by conidia arising successively. Differs from *Phaeococcomyces* spp. by lack of dematiaceous yeast cells. Fig. 4b.	*Aureobasidium, Scytalidium*	14, 22, 23, 49

Genus	Description		References
Lasiodiplodia	Colonies rapidly growing, cottony with abundant aerial mycelium, gray to black. Pycnidia variable in shape, large, up to 5,000 μm wide, 5–7 layers thick, stromatic with ostiole, sometimes setose, forming within 1–3 wk. Conidia initially one celled, elliptical, hyaline; mature conidia two celled, dark brown, longitudinally striate, 20–30 by 10–15 μm. Annellides cylindrical, hyaline, sometimes branched. Fig. 10a–c.	*Coniothyrium, Nattrassia, Phoma*	67
Lecythophora	Colonies rapidly growing, waxy to slimy, pink to salmon or red, often with central brown area in isolates that form chlamydoconidia. Phialides intercalary, with distinct collarette, periclinal wall thickenings, and conically tapering tip >1.2 μm in diameter. Conidia one celled, smooth walled, hyaline, obovate to allantoid, accumulating in slimy balls. Fig. 7a and b.	*Acremonium, Hyphozyma, Phialemonium, Phialophora*	28, 38
Nattrassia	Colonies of *N. mangiferae* rapidly growing, hyaline initially, becoming brownish black. *S. dimidiatum* synanamorph almost always present. Pycnidia develop within 1–2 mo, are variable in shape, several layers thick, up to 500 μm wide, stromatic, ostiolate. Conidia initially hyaline, one celled; mature conidia two to three celled, centrally dark, 12–13 by 4.5–5 μm. Phialides are lageniform and hyaline. Fig. 10d and e.	*Coniothyrium, Lasiodiplodia, Phoma*	88
Phaeoannellomyces	Colonies mucoid, slow growing, smooth, yeastlike, pale brown to black. Yeast cells are annellides, one celled in *P. elegans*, subhyaline to pale brown; pseudohyphae may be formed; hyphae and conidiophores absent but may be produced by associated synanamorphs. *P. elegans* usually occurs as synanamorph associated with *Exophiala* spp. *P. werneckii* differs by large two-celled (and one-celled) yeast annellides, with prominent accumulation of broad annellations (scars). Fig. 4e and f.	*Exophiala, Phaeococcomyces, Wangiella*	48, 57
Phaeococcomyces	Colonies mucoid, slow growing, smooth, yeastlike, pale brown to black. Yeasts one celled, subhyaline to pale brown, formed as multilateral holoblastic conidia; annelloconidia are not formed. Pseudohyphae may be formed; hyphae and conidiophores absent but may be produced by associated synanamorphs. Often produces synanamorphs when grown on cornmeal agar or potato dextrose agar. *P. exophialae* usually occurs as synanamorph associated with *Wangiella dermatitidis*. Fig. 4d.	*Exophiala, Phaeoannellomyces, Wangiella*	12, 14, 49, 57
Phialemonium	Colonies rapidly growing, flat, moist, white, becoming yellowish, with yellow-green pigment diffusing into medium with *P. obovatum*. Conidiophores hyaline, hyphalike. Intercalary and distinct phialides without obvious collarettes arising from hyphae at agar surface or occasionally from short aerial hyphae. Conidia one celled, hyaline, smooth walled, aggregated into slimy balls, obovate to allantoid. In contrast to *Lecythophora* spp, conidiiferous pegs arising from hyphae are cylindrical, ≦1 μm in diameter at tip, without periclinal wall thickening, and usually lacking collarette. Very inconspicuous cylindrical collarette may be formed. Fig. 7c.	*Acremonium, Hyphozyma, Lecythophora, Phialophora*	28, 38

(Continued on next page)

TABLE 1 Diagnostic features of some medically important dematiaceous fungi (*Continued*)

Genus	Diagnostic characteristics	Compare to	Selected references
Phialophora	Colonies rapidly growing, cottony to velvety, olive-gray to black. Conidiophores (if present) usually short, pale brown. Conidiogenous cells phialides with distinct collarettes. Conidia one celled, hyaline to pale brown, accumulating as balls at apices of phialides. *P. parasitica* produces phialides of variable lengths, some isolates forming extremely long phialides swollen near their bases, with prominent cell wall encrustations. Conidia elliptical to cylindrical, often curved. *P. repens* produces intercalary phialides without basal septa or phialides cylindrical to slightly lageniform with delicate collarettes; conidia cylindrical, curved. *P. richardsiae* produces phialides of variable sizes and shapes: some phialides long with flaring, flattened, saucer-shaped collarettes. Conidia of two shapes: globose conidia from phialides with flattened, saucer-shaped collarettes; cylindrical, often curved conidia from phialides of various lengths with inconspicuous collarettes. *P. verrucosa* produces flask-shaped phialides with cup-shaped, dark, often deep collarettes; conidia elliptical. Fig. 6.	*Lecythophora, Phialemonium*	16, 22, 23, 28, 32, 49, 90
Phoma	Colonies slow growing, dark gray to olive-brown. Pycnidia globose, subglobose, or pyriform; brown to black; superficial or immersed; ostiolate. Conidia hyaline, one celled, ovoid to elongate, arising from phialides. Fig. 9e and f.	*Coniothyrium, Nattrassia, Pyrenochaeta*	20
Rhinocladiella	Colonies rapidly growing, velvety, olive-black. Conidiophores pale brown to midbrown, erect, sympodial, usually bearing distinct crowded scars. Conidia one celled, fusiform to obovate, pale brown, with flat basal scar. Conidia occur along conidiophore. Fig. 2b.	*Exophiala, Fonsecaea, Ramichloridium*	22, 23, 49, 78
Scedosporium	Colonies very rapidly growing, cottony, smoky gray to dark brown. Conidiophores hyaline, short or long. Predominant conidiogenous cells are hyaline annellides with or without swollen bases; characterized by prominent bulging annellations. Conidia one celled, obovate, truncate, subhyaline to light black, single or in balls. *S. prolificans* differs from *S. apiospermum* by having conidiogenous cells with prominent swollen bases and by most isolates failing to grow on media containing cycloheximide. Fig. 8a and b.	*Pseudallescheria*	21, 31, 42, 49, 54, 75
Scytalidium	Colonies rapidly growing; hyaline initially, becoming brownish black (colonies of *S. hyalinum* remain hyaline). Hyphae form one- or two-celled arthroconidia that are hyaline or pale brown to brown, subglobose to ellipsoidal. May occur as synanamorphs with *Aureobasidium* spp., *Hormonema* spp., and *Nattrassia magniferae*. Many mycologists consider *S. lignicola* to be synonym of *S. dimidiatum*; others consider *S. lignicola* to be distinct based on its production of both hyaline and dematiaceous arthroconidia, with *S. dimidiatum* producing only dark arthroconidia. Fig. 4c.	*Aureobasidium, Hormonema*	49, 81, 83, 88, 90
Sporothrix	Colonies of typical clinical isolates of *S. schenckii* initially cream colored, smooth, moist, yeastlike; gradually turning brown in irregular patches; becoming velvety to lanose as aerial hyphae develop. Conidiophores elongate, compactly sympodial, with swollen apices; bearing one-celled, hyaline, elliptical to obovate conidia in rosettes. Dissimilar conidia arise directly from vegetative hyphae and are dematiaceous, thick walled, oval (rarely triangular), one celled. Grows as yeast form at 37°C. Fig. 7d–g.	*Rhinocladiella*	13, 18, 49, 82

Genus	Description	Synanamorphs	References
Wangiella	Colonies slow growing, initially yeastlike (from presence of *Phaeococcomyces* synanamorph), smooth, viscous, pale brown to black, becoming filamentous to velvety. Conidiophores hyphalike, subhyaline to pale brown. Conidiogenous cells phialides without distinct collarettes, intercalary or lateral from hyphae. Phialides cylindrical with rounded apices. Conidia one celled, subglobose to elliptical or obovoid, subhyaline to pale brown, accumulating in balls or slipping down conidiogenous cells. Annellidic synanamorph (*Exophiala*) often present. Grows at 40°C; does not assimilate KNO₃. Fig. 5h and i.	*Exophiala, Phaeoannellomyces, Phaeococcomyces, Phialophora*	24, 46, 62
Xylohypha	Colonies slow growing, olivaceous gray to black; velvety. Conidiophores hyphalike, pale brown. Blastoconidia one celled, without evident darkly pigmented hila, occurring in long, sparsely branching chains that are poorly differentiated from conidiophore and vegetative hyphae. *X. bantiana* grows at 42–43°C; conidia measure 2–2.5 by 4–7 μm. *X. emmonsii* differs by conidia that are asymmetric to sigmoid, sometimes two celled, occurring in shorter chains; poor growth at 37°C, no growth at 40°C. Fig. 2c and d.	*Cladosporium*	52, 63

apiospermum is one anamorph that is produced by several species of the teleomorphic genera *Pseudallescheria* and *Petriella*. For that reason, it would be incorrect to use the name *Pseudallescheria boydii* for an isolate of *Scedosporium apiospermum* that failed to develop the *Pseudallescheria boydii* teleomorph in culture. Laboratorians are cautioned that as they strive for mycologic accuracy, they must accept responsibility for ensuring that medical reports remain explicitly clear when new or less familiar names are used.

Several dematiaceous pathogens have no known teleomorphs but do have multiple synanamorphs. Because each of the synanamorphs of a pleomorphic fungus may be referred to by a separate name, the identification process for such fungi can be confusing. A widely accepted approach to this problem is to base the name of an isolate upon the single anamorph that is judged to be the most distinctive, conspicuous, and stable. An accompanying synanamorph may then be referred to, if needed, by using a genus-level name. In the case of the pleomorphic pathogen *Fonsecaea pedrosoi* (Fig. 3d), for example, the synanamorph upon which the identification is based is a compactly sympodial conidiophore that apically produces a series of conidia that in turn give rise to a second series of conidia; similarly, a third series may be formed. This form is the *Fonsecaea pedrosoi* synanamorph. In addition, a *Rhinocladiella* (sympodial) synanamorph, a *Phialophora* (phialide) synanamorph, and a *Cladosporium* synanamorph may also be seen, but these are not essential to the identification of an isolate as *Fonsecaea pedrosoi* (56).

The black yeasts are at times extremely difficult and frustrating to identify. The genera *Phaeococcomyces* and *Phaeoannellomyces* were established (12, 14, 57) to accommodate synanamorphs that exhibit dematiaceous budding yeasts and occasional short pseudohyphae or toruloid hyphal elements. Black yeasts typically represent one synanamorph of pleomorphic fungi. They are recognized by their brown or black, usually mucoid, yeastlike colonies. One of the most frequently isolated black yeasts in the clinical laboratory is the *Phaeoannellomyces* synanamorph associated with *Exophiala jeanselmei*. When the yeastlike isolates of this fungus are transferred from initial isolation media to potato dextrose agar or cornmeal agar, the typical conidiogenous cells and conidia of *Exophiala jeanselmei* rapidly become evident. The assumption that a black yeast, regardless of whether the colony is initially dematiaceous or not, should be identified as *Aureobasidium pullulans* is incorrect.

Sterile isolates represent a second group of medically important fungi that are especially difficult to identify. They commonly are referred to as members of the form order Mycelia Sterilia. These fungi have been shown to cause phaeohyphomycosis and mycetoma. When sterile dematiaceous fungi are isolated, the induction of conidia or fruiting bodies should be attempted as previously discussed. The cultures should be kept for several weeks before being discarded. With time, some of these fungi may develop, either in the agar or at the colony surface, structures such as ascocarps, pycnidia, or synnemata that produce spores or conidia. When fruiting bodies (structures within which propagules are formed) are present, the first step is to determine whether they are ascocarps (sexual reproduction) or pycnidia (asexual reproduction). Ascomycetes, such as *Chaetomium* spp. and *Leptosphaeria* spp., are recognized by the presence of ascocarps, within which asci and ascospores are formed (asci often dissolve at maturity and

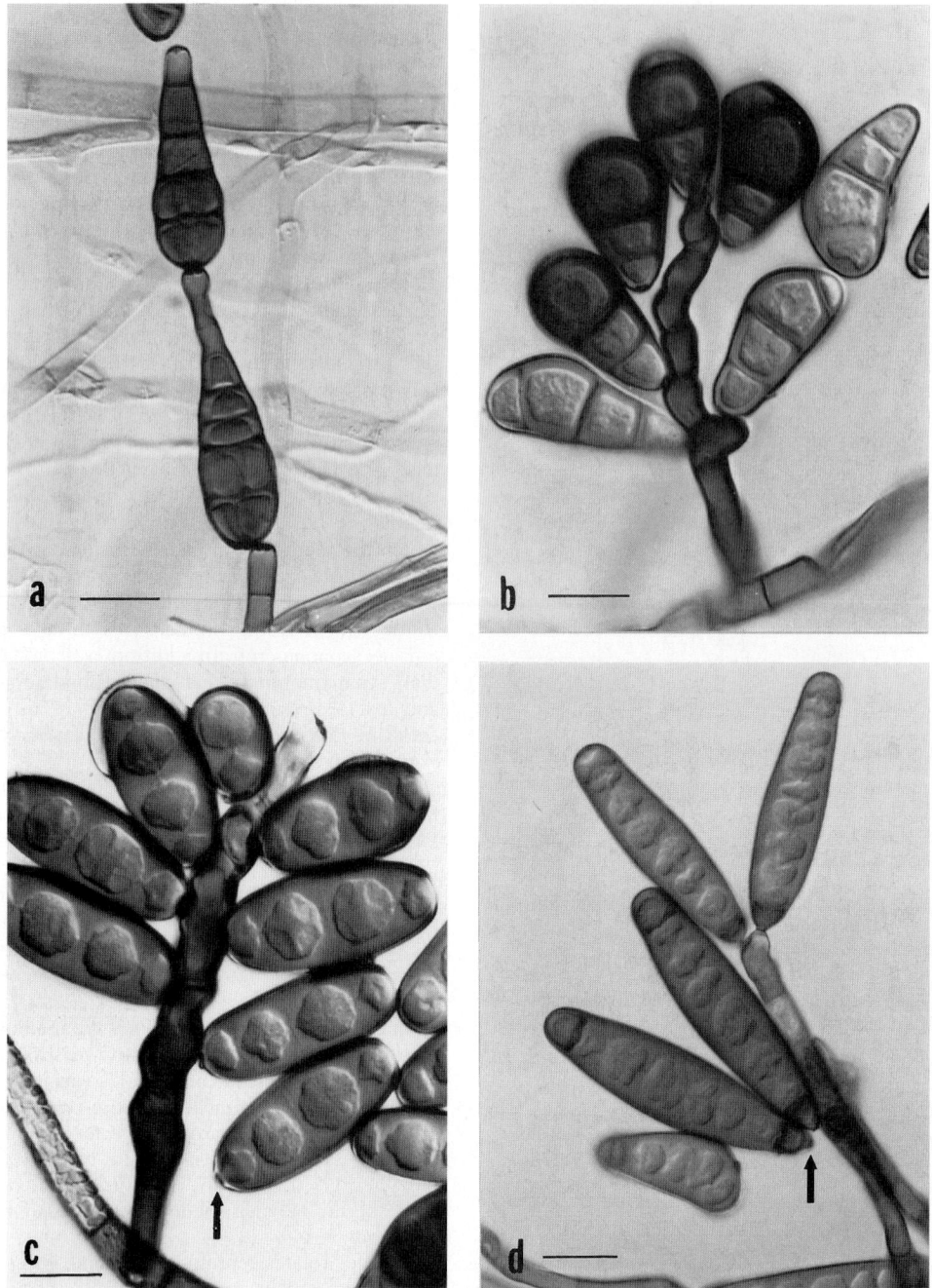

FIGURE 1 Bars equal 10 μm. (a) *Alternaria* sp. Muriform conidia with tapering apices (beaks) arise in chains from the conidiophore. (b) *Curvularia* sp. Curved conidia are developing from a geniculate conidiophore. (c) *Bipolaris spicifera*. The hilum (arrow) of each conidium protrudes only slightly. (d) *Exserohilum rostratum*. Each conidium has a strongly protruding hilum (arrow).

may be visible only in immature ascocarps). If the fruiting bodies are pycnidial, they contain conidia rather than ascospores, and the isolate is then likely to be a member of the sphaeropsidales. Several agents of phaeohyphomycosis, such as *Coniothyrium fuckelii*, *Lasiodiplodia theobromae*, and *Phoma* spp., are pycnidial.

Alternaria spp.

Species of *Alternaria* (Fig. 1a) have been documented in phaeohyphomycotic infections of bone, cutaneous tissue, ears, eyes, paranasal sinuses, and the urinary tract (89, 93). An *Alternaria* sp. and *Alternaria alternata* (synonym, *Alter-*

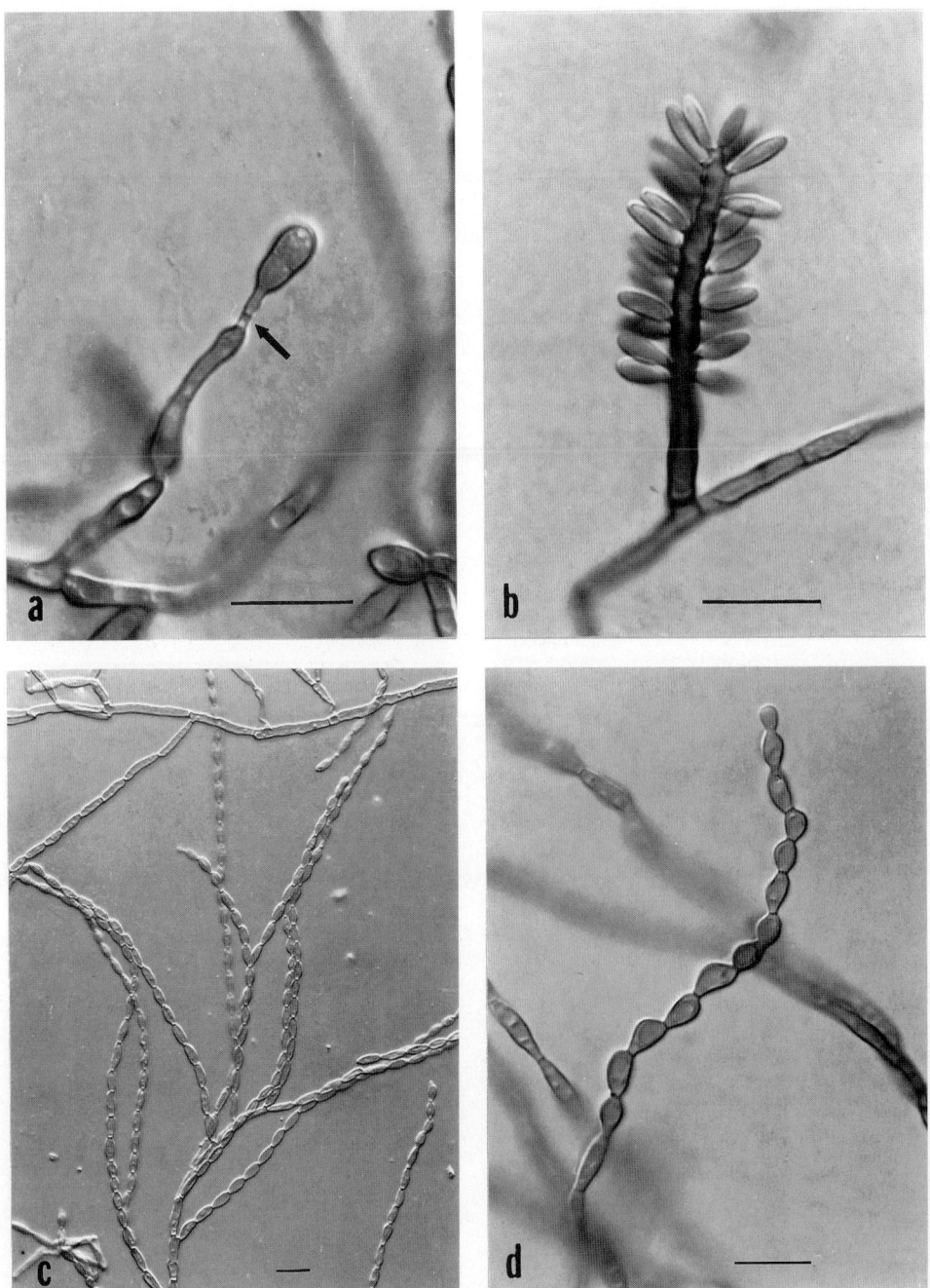

FIGURE 2   Bars equal 10 μm. (a) *Dactylaria constricta.* Two-celled conidium is attached to the conidiophore by a narrow denticle (arrow). (b) *Rhinocladiella aquaspersa.* Numerous one-celled conidia arise from elongate sympodial conidiophores. (c) *Xylohypha bantiana.* Long, sparsely branching chains of blastoconidia arise from nondifferentiated hyphae. (d) *Xylohypha emmonsii.* Conidia are markedly asymmetric and dolliform to sigmoid.

naria tenuis) (30) are the only well-documented human pathogens in this genus. The *Alternaria* anamorph of *Pleospora infectoria* has been reported to be a pathogen of humans but has not been documented convincingly. *Alternaria* isolates are difficult to identify beyond the genus level. If an isolate must be identified to species level, it should be sent to a specialist.

Aureobasidium spp.

Aureobasidium pullulans (Fig. 4a) has been implicated as an agent of phaeohyphomycosis in humans and other animals (74), having been reported from skin, nail, subcutaneous, and deeper tissues. *Aureobasidium* species are often confused with *Hormonema* species. In *Hormonema* species, conidia

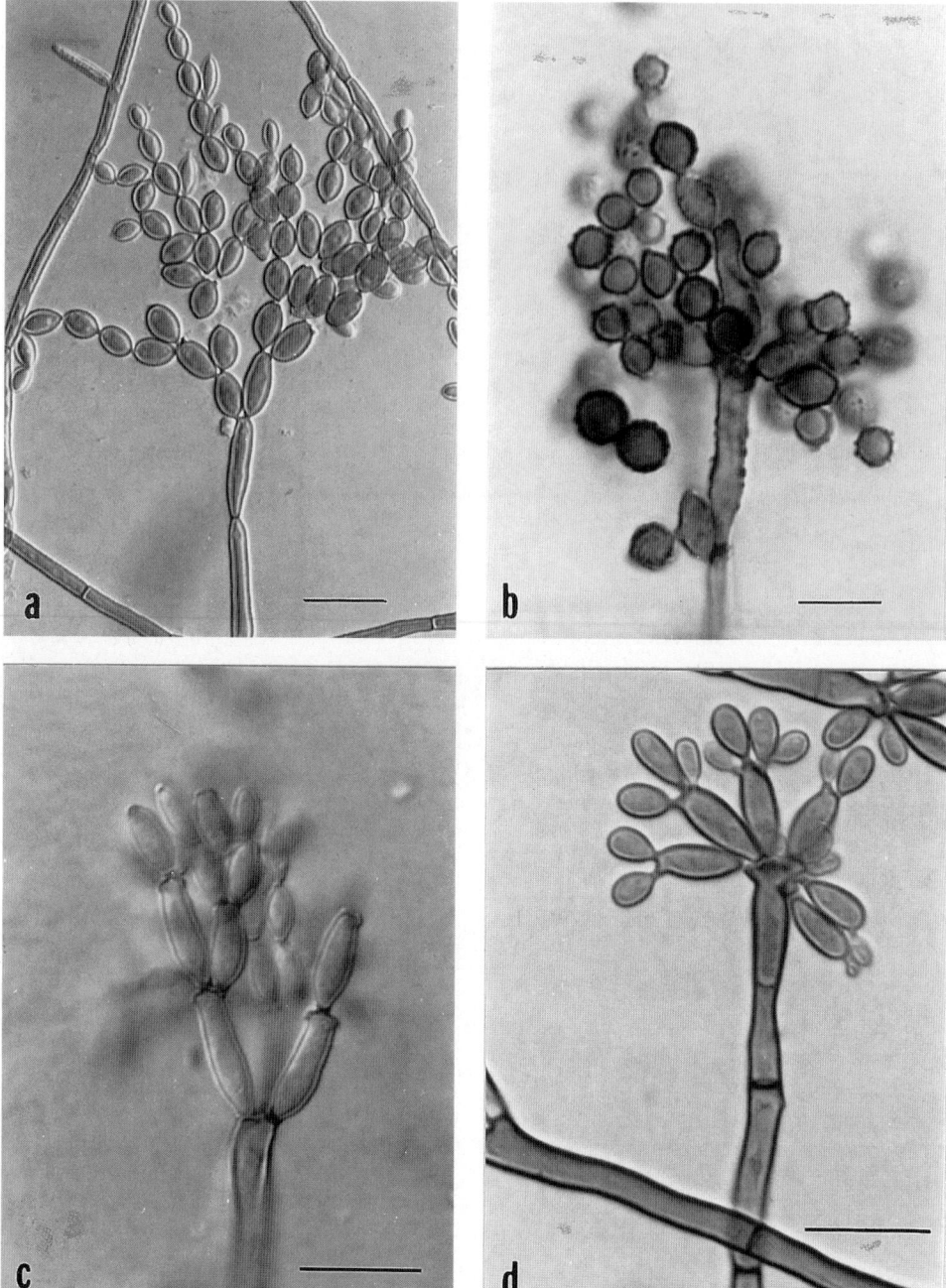

FIGURE 3 Bars equal 10 μm. (a) *Cladosporium carrionii*. Sparsely branched long chains of blastoconidia develop from an erect conidiophore. (b) *Cladosporium sphaerospermum*. Conidia are mostly spherical to subglobose and in very fragile, highly branched chains. (c) *Cladosporium cladosporioides*. Conidia show prominent hila (scars) and are in highly branched fragile chains. Shield-shaped cells are numerous. (d) *Fonsecaea pedrosoi*. Two to four series of conidia arise at the apex of a sympodial conidiophore.

arise in a basipetal succession from either hyaline or dematiaceous, hyphalike conidiogenous cells (Fig. 4b). In contrast, *A. pullulans* produces conidia in a synchronous manner (14). Because several authors have illustrated *Hormonema* spp. under the name *Aureobasidium*, it may be that some cases of infection ascribed to *A. pullulans* were actually caused by misidentified isolates of *Hormonema* spp.

Bipolaris spp.

Several species of *Bipolaris* (Fig. 1c), including *Bipolaris australiensis*, *Bipolaris hawaiiensis*, and *Bipolaris spicifera*, have caused meningitis, paranasal sinusitis, and subcutaneous, eye, and pulmonary infections (55). At least one species of *Drechslera*, *Drechslera biseptata*, has been shown to be an

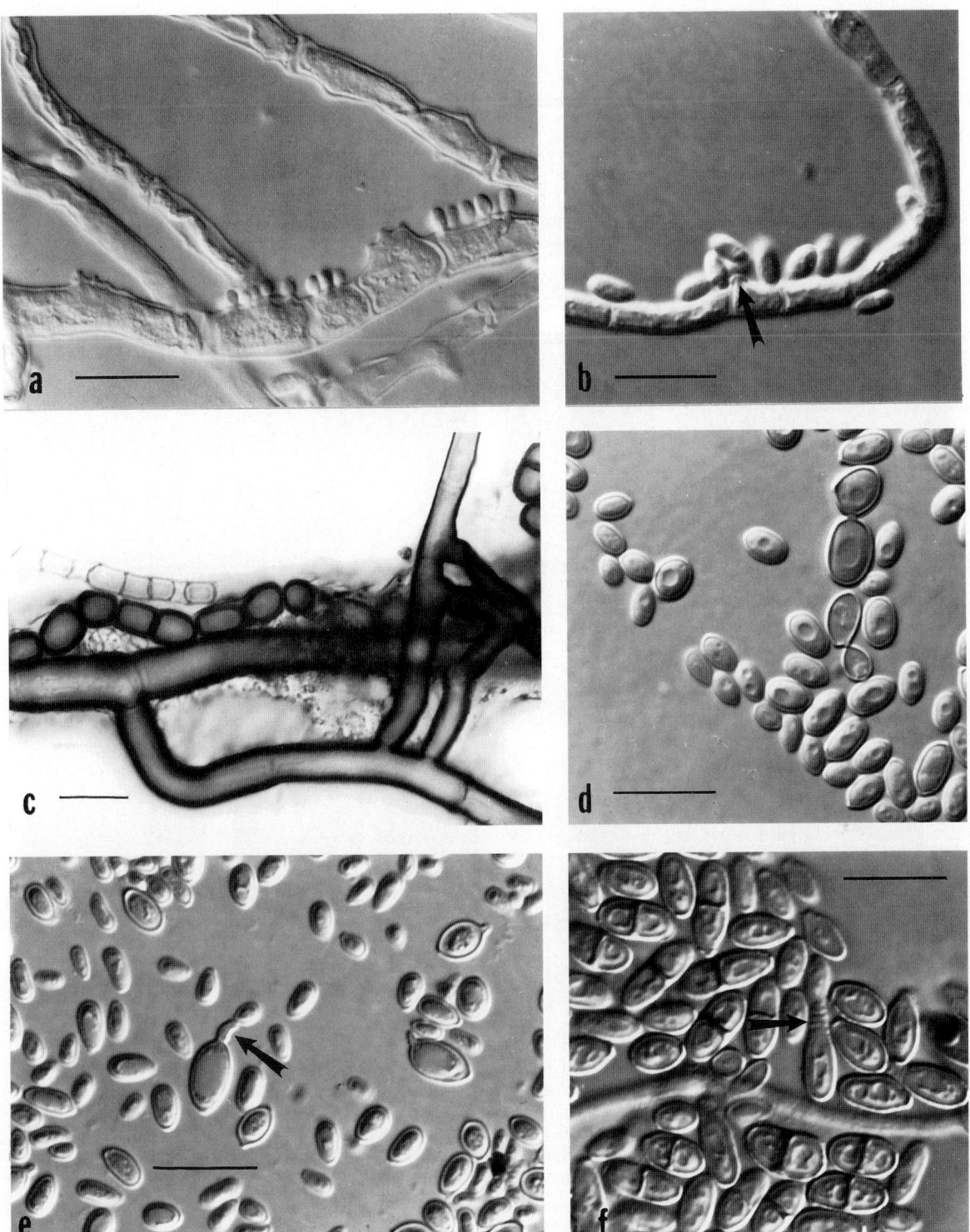

FIGURE 4 Bars equal 10 μm. (a) *Aureobasidium pullulans*. Conidia are developing in a synchronous manner. (b) *Hormonema* sp. Conidia are developing in a sequential manner from a single locus (arrow). (c) *Scytalidium dimidiatum*. Chains consist of one- or two-celled arthroconidia. (d) *Phaeococcomyces* sp. Blastoconidia arise singly from various sites along cell wall. (e) *Phaeoannellomyces elegans*. Yeasts are one-celled annellides with long, narrow zones of annellations (arrow). (f) *Phaeoannellomyces werneckii*. Yeasts are one- to two-celled annellides with broad, prominent annellations (arrow).

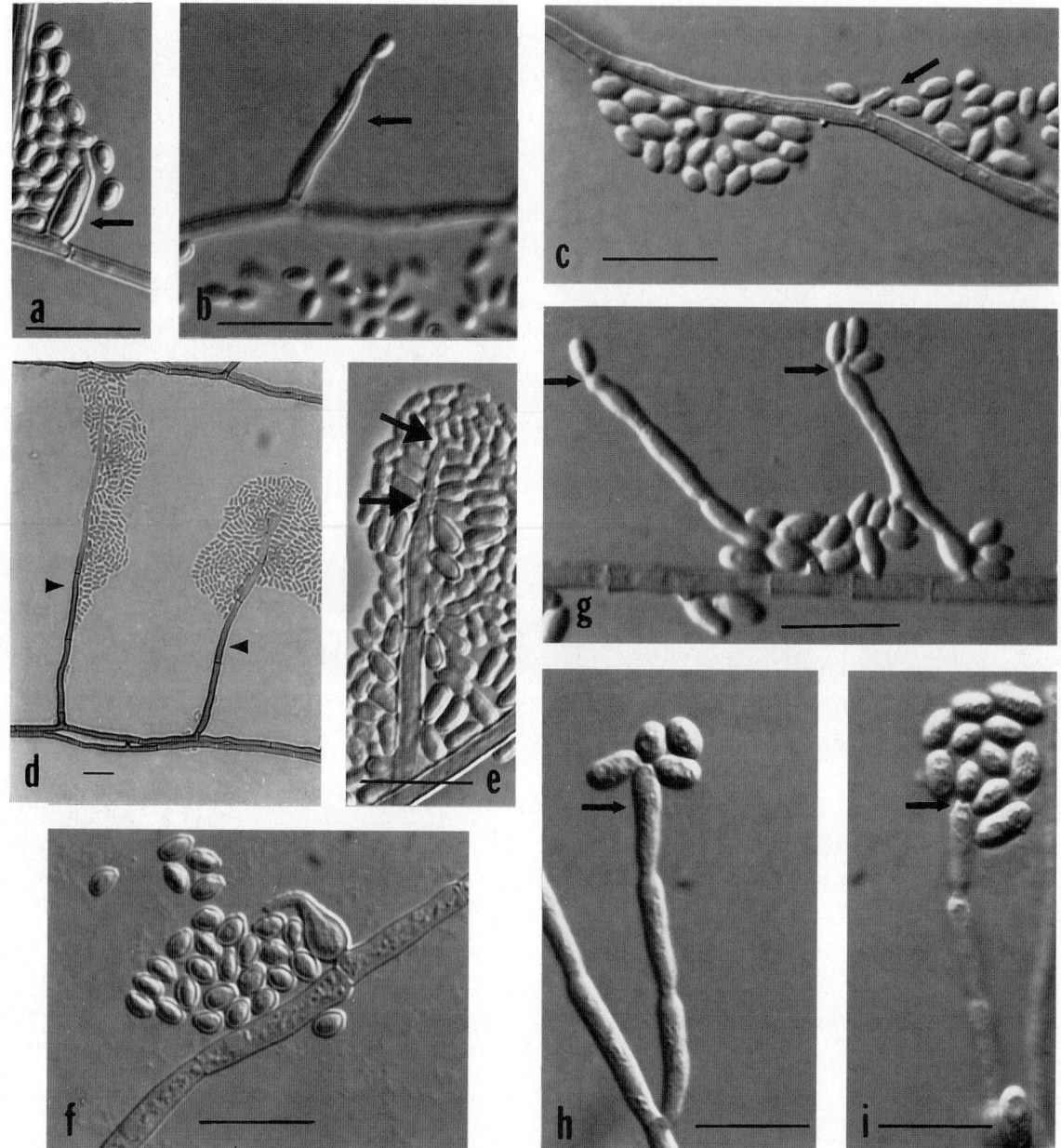

FIGURE 5 Bars equal 10 μm. (a, b) *Exophiala jeanselmei* var. *jeanselmei*. Annellides (arrows) are well developed and cylindrical. (c) *Exophiala jeanselmei* var. *lecanii-corni*. Annellations (arrow) arise directly from hyphae. (d) *Exophiala spinifera*. Conidiophores (arrows) are multicellular and spine-like. (e) *Exophiala spinifera*. Annellated region of annellides (arrows) becomes very long. (f) *Exophiala moniliae*. Annellides are swollen at the base. (g) *Exophiala castellanii*. Annellides have inconspicuous annellations (arrows). (h, i) *Wangiella dermatitidis*. One-celled conidia arise from phialides (arrows) that lack collarettes.

opportunistic pathogen (91). Some medical microbiologists have previously confused species of *Drechslera*, *Exserohilum*, and *Helminthosporium* with *Bipolaris* spp., but recent taxonomic studies (2, 55) have established useful criteria for separating species of these genera.

Chaetomium spp.

Chaetomium species (Fig. 9a and b) have been associated with infections of blood, brain, skin, and nails in both healthy and immunocompromised patients (3, 84). Colonization of bone marrow transplant patients has also been reported. In some cases, *Chaetomium globosum* was specifically identified. When species identification is needed, isolates should be sent to a reference laboratory.

Cladosporium spp.

Cladosporium carrionii (Fig. 3a) is the leading agent of chromoblastomycosis in Australia and Africa (40). Other spe-

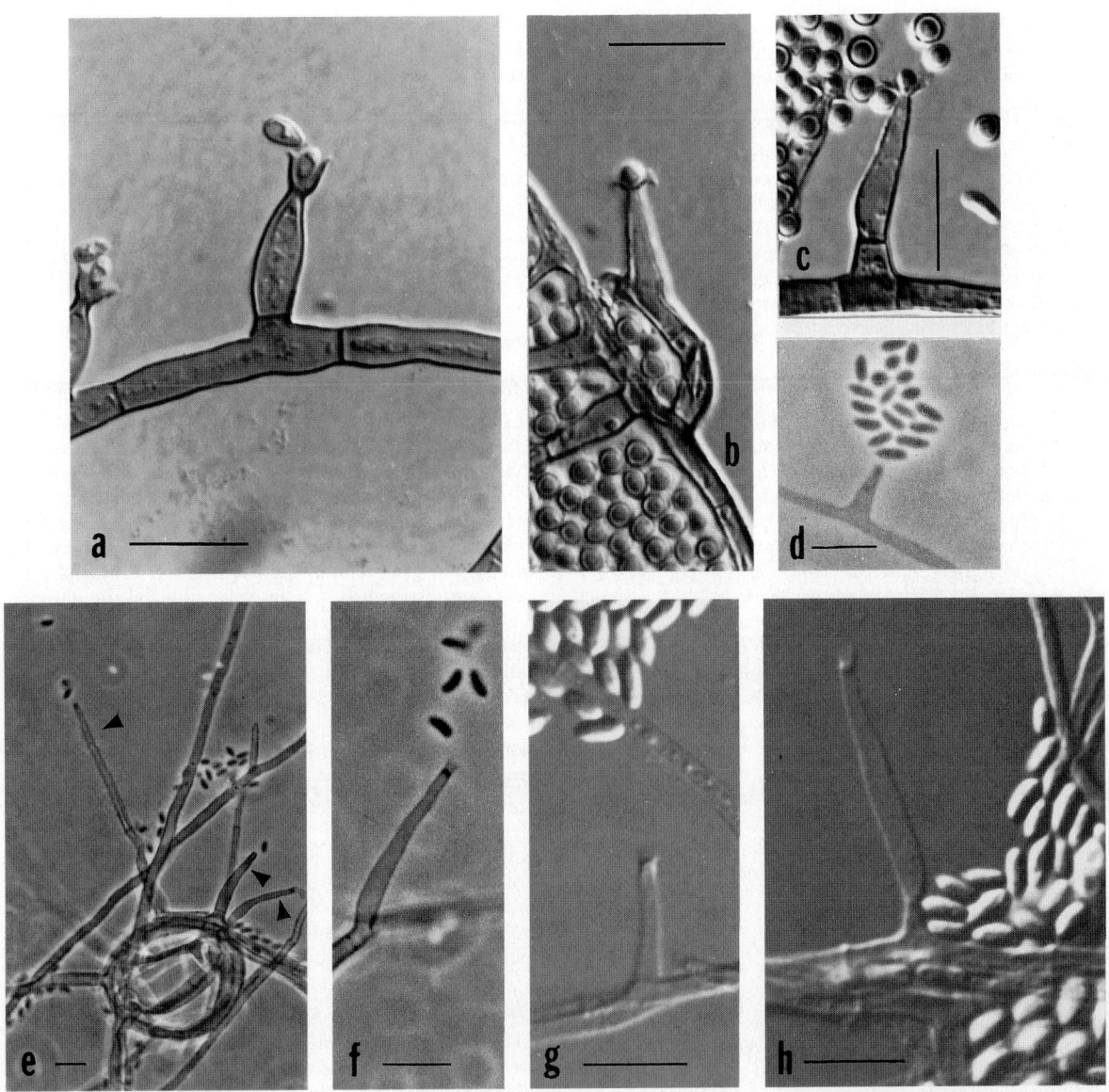

FIGURE 6 Bars equal 10 μm. (a) *Phialophora verrucosa*. Note the presence of a deep collarette at the apex of the phialide. (b, c) *Phialophora richardsiae*. The species is characterized by phialides with flattened collarettes bearing spherical conidia. (d) *Phialophora richardsiae*. Other types of phialides with inconspicuous collarettes form cylindrical conidia. (e, f) *Phialophora parasitica*. Phialides (arrows) are short to extremely long, collarettes are inconspicuous and not flaring, and conidia are curved and cylindrical. (g, h) *Phialophora repens*. Curved, cylindrical conidia arise from cylindrical or lageniform phialides.

cies of *Cladosporium* have been reported from cutaneous, subcutaneous, eye, and nail infections (29, 76). Species of *Cladosporium*, especially *Cladosporium sphaerospermum* and *Cladosporium cladosporioides* (Fig. 3b and c), are among the most common dematiaceous mould contaminants in the clinical laboratory.

Coniothyrium spp.

Coniothyrium fuckelii (Fig. 9c and d) has been recovered from a healthy host with cutaneous phaeohyphomycosis (79), a leukemic patient with liver infection (37), and a heart transplant patient with cutaneous infection. This species is very similar to *Microsphaeropsis olivacea*; taxo-

nomic problems between these taxa need to be addressed (86).

Curvularia spp.

As a group, *Curvularia* species (Fig. 1b) are leading agents of fungal sinusitis and keratitis and may cause endocarditis, mycetoma, pulmonary infection, and subcutaneous phaeohyphomycosis as well (49, 72). *Curvularia geniculata*, *Curvularia lunata*, *Curvularia pallescens*, *Curvularia senegalensis*, and *Curvularia verruculosa* have been specifically identified from infections. Work by Ellis (22, 23) should be consulted if a *Curvularia* isolate must be identified to species.

Dactylaria spp.

Dactylaria constricta var. gallopava (Fig. 2a) has been documented as an etiologic agent of disseminated infections in immunocompromised humans as well as the cause of epizootic encephalitis in flocks of turkeys and chickens (43, 80). This name was proposed (17) as a new combination for the two previously described dematiaceous hyphomycetes Dactylaria gallopava and Scolecobasidium constrictum. The two varieties of Dactylaria constricta are differentiated by physiologic characteristics and pathogenicity in experimental animals (73). Some workers have suggested recognizing the two varieties as distinct species as well as placing them in the genera Ochroconis and Scolecobasidium.

Exophiala spp.

Exophiala jeanselmei (Fig. 5a through c) is a leading etiologic agent of subcutaneous phaeohyphomycosis and may also cause mycetoma (85). Exophiala moniliae (Fig. 5f) and Exophiala spinifera (Fig. 5d and e) have also been reported as agents of phaeohyphomycosis (26, 44). Exophiala werneckii has been renamed Phaeoannellomyces werneckii (57) and is discussed under the genus Phaeoannellomyces. The colonial morphology of Exophiala species is varied. Most isolates initially grow in a yeast form (Phaeoannellomyces synanamorph) that is succeeded by a hyphal Exophiala synanamorph. As a result, colonies are moist and yeastlike at first, becoming velvety to woolly with age. Some isolates that are predominantly the Phaeoannellomyces synanamorph of Exophiala jeanselmei may remain yeastlike. Conidiophores of Exophiala spp. are dematiaceous, simple, or hyphalike. The conidia are one celled in most species and accumulate in balls at the apices of the annellides. With careful study using the oil immersion objective, annellations (rings) usually can be seen at the apices of the annellides.

Exserohilum spp.

Phaeohyphomycosis of skin, subcutaneous tissue, and nasal sinuses has been documented (33, 55, 61). The genus Exserohilum (Fig. 1d) contains three recognized pathogens: Exserohilum longirostratum, Exserohilum mcginnisii, and Exserohilum rostratum. Previously, members of the genus Exserohilum have been confused with species of Bipolaris, Drechslera, and Helminthosporium, but recent taxonomic studies have resolved the confusion (2, 55).

Fonsecaea spp.

The genus Fonsecaea contains two species, Fonsecaea compacta and Fonsecaea pedrosoi, both of which are agents of chromoblastomycosis. Fonsecaea isolates are pleomorphic. They are characterized by the development of one-celled primary conidia that form on erect, dark, compactly sympodial conidiophores (56). These primary conidia in turn become conidiogenous cells and form secondary one-celled conidia. This process usually results in only two or three levels of conidia (in contrast to Cladosporium spp.) and constitutes the distinctive morphology of the genus (Fig. 3d). A second kind of conidial development (incorrectly called Acrotheca-like) results in structures similar to the form seen in the genus Rhinocladiella (a Rhinocladiella anamorph). In yet another synanamorph, the conidia occur as branching chains similar to those found in the genus Cladosporium (a Cladosporium synanamorph). Fonsecaea spp. also may produce phialides with collarettes bearing balls of one-celled conidia that are typical of the genus Phialophora (a Phialophora synanamorph). Because of this pleomorphic

nature, earlier studies of Fonsecaea pedrosoi and Fonsecaea compacta had placed these species in the genera Phialophora and Rhinocladiella. Recently, a Phaeoannellomyces synanamorph has been demonstrated when Fonsecaea pedrosoi is grown at low pH.

Lasiodiplodia theobromae

Lasiodiplodia theobromae (synonym, Botryodiplodia theobromae) (Fig. 10a through c) is a well-documented cause of keratitis following laceration of the eye (70, 76). Rare cases of subcutaneous infection have also been reported.

Lecythophora spp.

Lecythophora hoffmannii (Fig. 7a and b) and Lecythophora mutabilis have been reported from patients with subcutaneous phaeohyphomycosis, endocarditis, and peritonitis (71). The genus Lecythophora was reintroduced for two species that had previously been described in the genus Phialophora. The microscopic morphology of Lecythophora spp. may be very similar to that of Phialemonium spp. Lecythophora spp. differ, in part, by colonies that are pink to salmon or red when young. Lecythophora hoffmannii and Lecythophora mutabilis are distinguished from each other by the absence or presence of brown chlamydoconidia, respectively. Detailed descriptions (28, 38) should be consulted for the identification of isolates, and the use of a reference laboratory is recommended.

Nattrassia mangiferae

Nattrassia mangiferae (Fig. 10d and e) is a pycnidial fungus, previously known as Hendersonula toruloidea, that can cause phaeohyphomycosis of skin and nail (58). The associated synanamorph Scytalidium dimidiatum is the form that is first isolated in the clinical laboratory, and if a special medium, such as sterile banana slices, is used, the Nattrassia mangiferae anamorph may be formed. Pycnidia are produced after incubation for 2 months.

Phaeoannellomyces spp.

Phaeoannellomyces werneckii (synonyms, Cladosporium werneckii, Exophiala werneckii) and Phaeoannellomyces elegans are known agents of phaeohyphomycosis (1, 25, 50). The genus Phaeoannellomyces was created to accommodate those black yeasts that are characterized by the development of yeast cells that function as annellides (57). Phaeoannellomyces werneckii is classified in the blastomycete genus Phaeoannellomyces because the yeast anamorph is stable, distinctive, unique in appearance, and well represented in isolates of the species. Phaeoannellomyces elegans (Fig. 4e), which is a synanamorph associated with Exophiala jeanselmei, is distinguished from Phaeoannellomyces werneckii (Fig. 4f) by its formation of one-celled yeast forms.

Phaeococcomyces spp.

The genus Phaeococcomyces (Fig. 4d) contains black yeasts that form holoblastic conidia from various points on the parent cell (multilateral) in contrast to the annellidic process seen in Phaeoannellomyces spp. Most clinical isolates are Phaeococcomyces exophialae, which usually give rise gradually to the accompanying hyphomycete synanamorph Wangiella dermatitidis. The genus Phaeococcomyces, which was originally named Phaeococcus (12, 14), contains additional species that are distinguished from each other primarily by morphologic criteria.

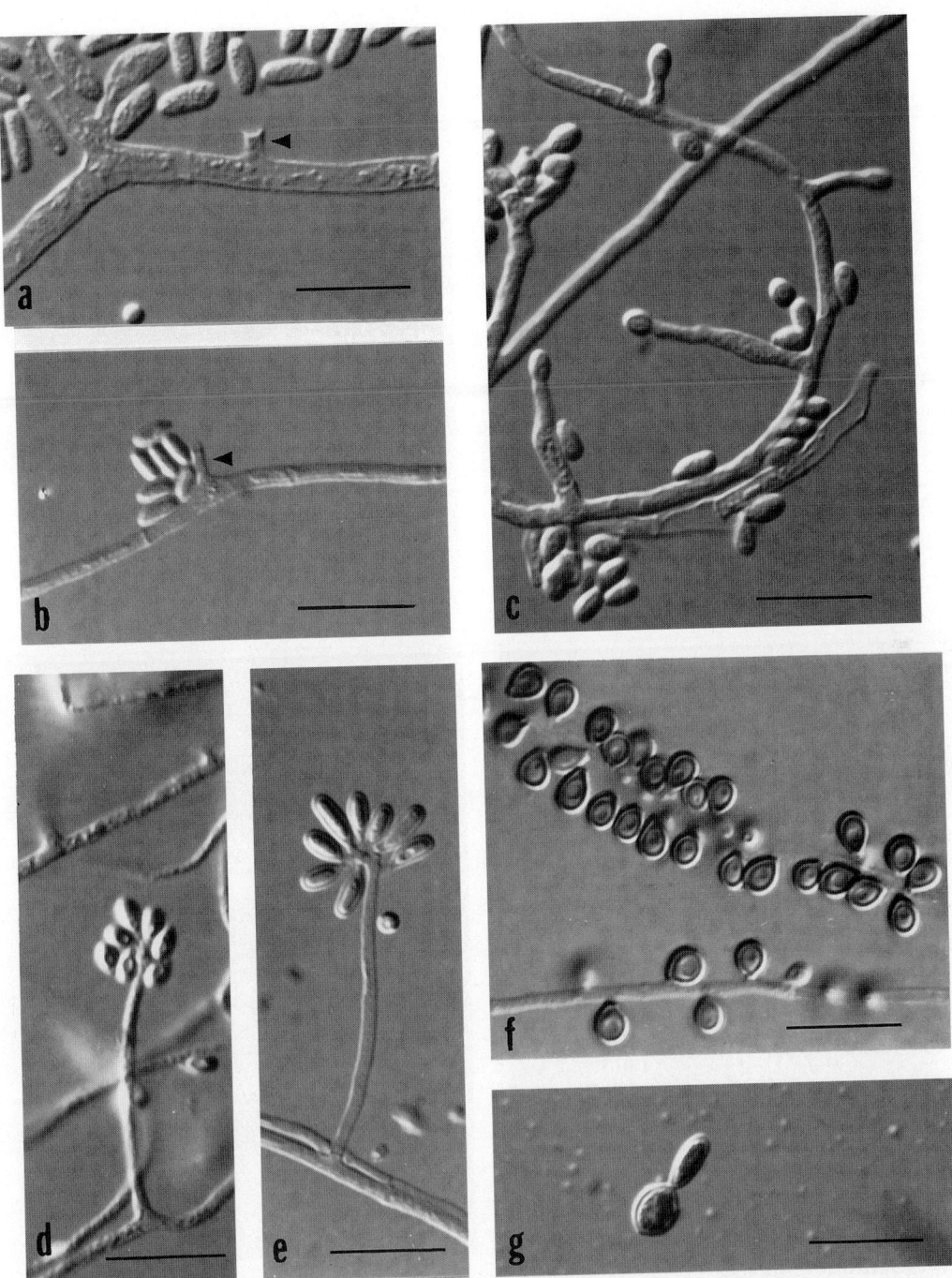

FIGURE 7 Bars equal 10 μm. (a, b) *Lecythophora hoffmannii.* Phialides (arrows) are very short, often conically tapering, and with collarettes; basal septum is usually absent. (c) *Phialemonium* species. Phialides are cylindrical to lageniform and without collarettes. (d, e) *Sporothrix schenckii.* Solitary conidia on denticles arise from sympodial conidiophores (f) *Sporothrix schenckii.* Thick-walled dematiaceous conidia arise along vegetative hyphae. (g) *Sporothrix schenckii.* Yeast synanamorph.

Phialemonium spp.

The genus *Phialemonium* contains three species, two of which (*Phialemonium curvatum* and *Phialemonium obovatum*) are known to cause phaeohyphomycosis. Reported cases include subcutaneous cyst, disseminated infection, and peritonitis (28, 38). The microscopic morphology of *Phialemonium* (Fig. 7c) is intermediate between those of the genera *Acremonium* and *Phialophora* and very similar to that of the genus *Lecythophora*. A diffusible yellow-green pigment is present in some strains and is a useful characteristic

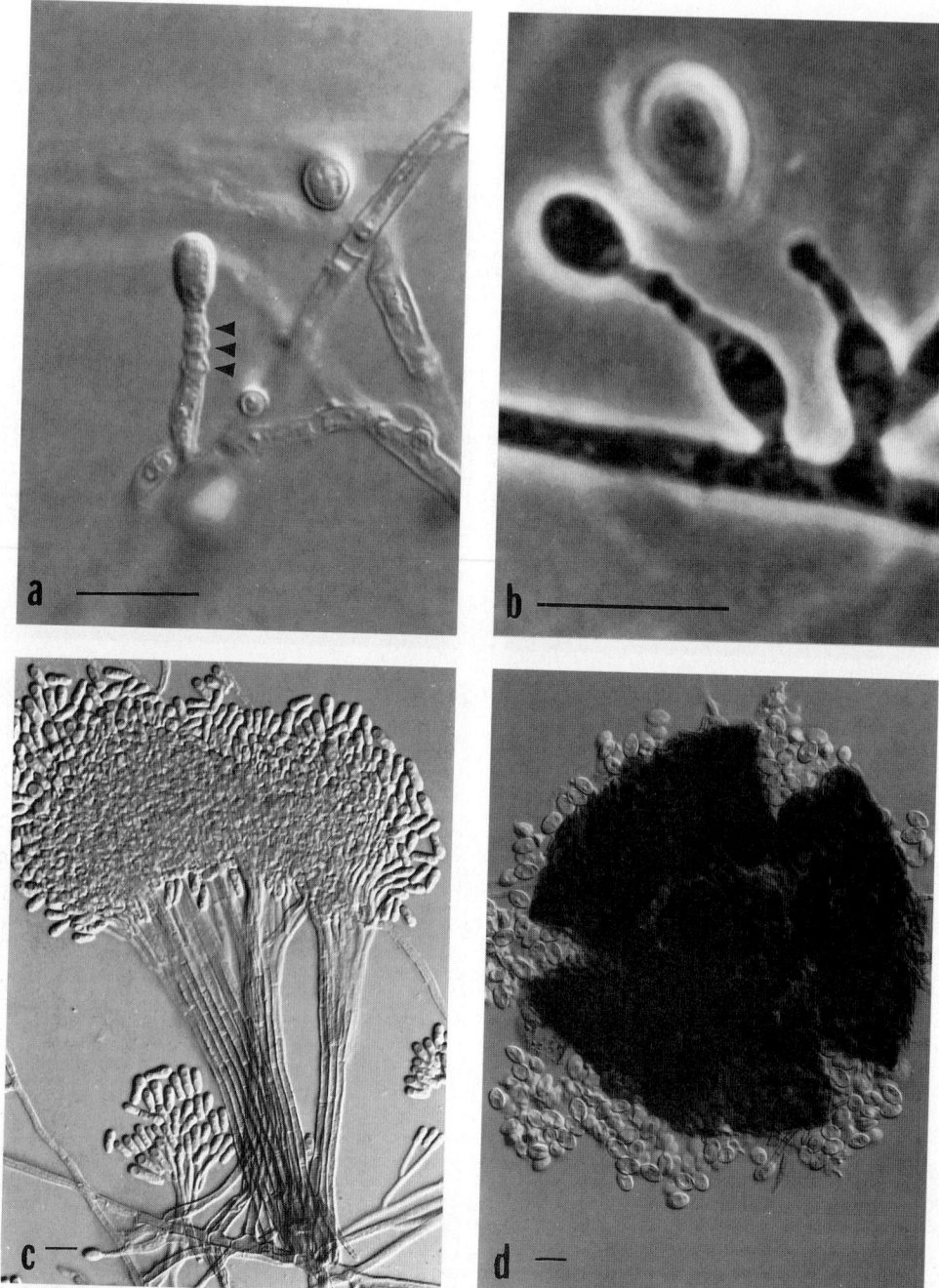

FIGURE 8 Bars equal 10 μm. (a) *Scedosporium apiospermum*. Annellides are characterized by accentuated annellations (arrows). (b) *Scedosporium prolificans*. Annellides are characterized by swollen bases. (c) *Graphium* species. Conidiophores consist of fused hyphae bearing terminal annellides. (d) *Pseudallescheria boydii*. Globose ascocarp ruptures to release football-shaped ascospores.

in identifying these isolates. Detailed descriptions (28, 38) should be consulted for the identification of isolates, and the use of a reference laboratory is recommended.

Phialophora spp.

Phialophora verrucosa (Fig. 6a) is a leading cause of chromoblastomycosis in North America. It is also an etiologic agent of phaeohyphomycosis, as are *Phialophora bubakii*, *Phialophora parasitica* (Fig. 6e and f), *Phialophora repens* (Fig.

6g and h), and *Phialophora richardsiae* (Fig. 6b through d) (27, 66). In addition to causing cutaneous and subcutaneous infections, some species have been reported to cause endocarditis and mycotic keratitis. Two additional agents of phaeohyphomycosis that were previously described as *Phialophora* species have been reclassified to the genus *Lecythophora* (*Lecythophora hoffmannii* and *Lecythophora mutabilis* [28]). This reclassification was necessary because these fungi produce intercalary phialides with short, lateral, conically

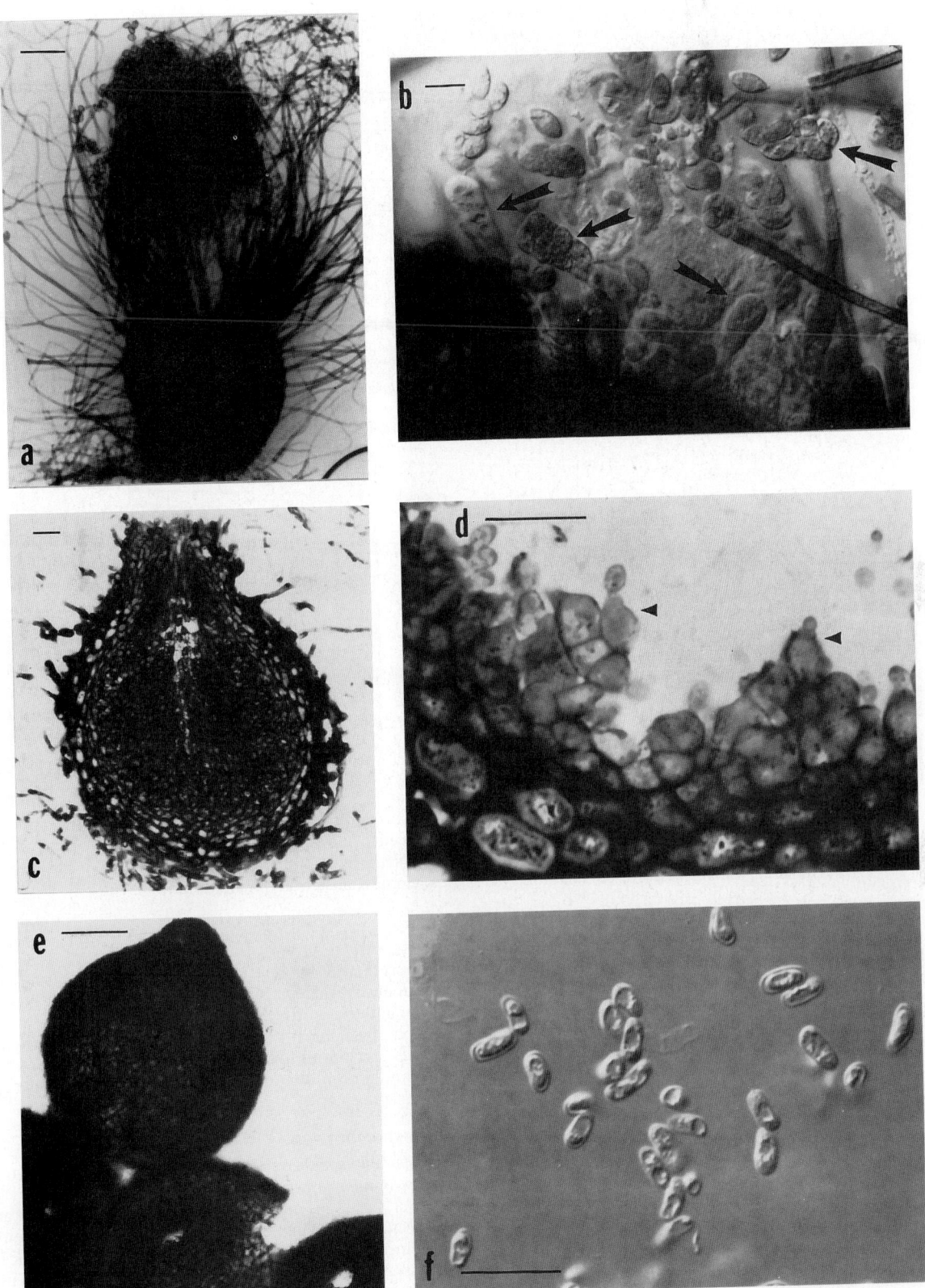

FIGURE 9 (a) *Chaetomium globosum*. Ascocarps are characterized by flexuous, upwardly oriented hairlike hyphae (bar equals 50 μm). (b) Lemon-shaped dark ascospores from within asci (arrows) (bar equals 10 μm). (c) *Coniothyrium fuckelii* (vertical sections) (bar equals 10 μm). Pycnidium is globose and ostiolate. (d) *Coniothyrium fuckelii*. Phialides (arrows) give rise to conidia that are brown at maturity. (e) *Phoma* species. Pycnidia are subglobose and ostiolate (bar equals 50 μm). (f) *Phoma* species. One-celled phialoconidia are hyaline (bar equals 10 μm).

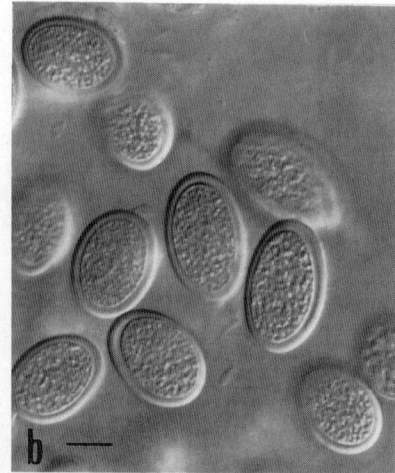

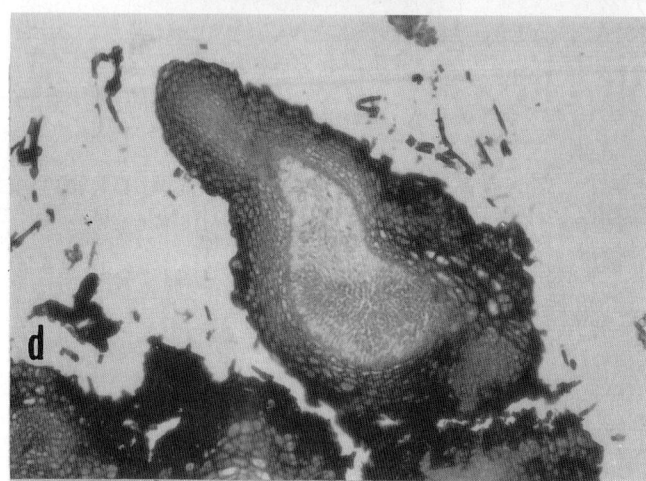
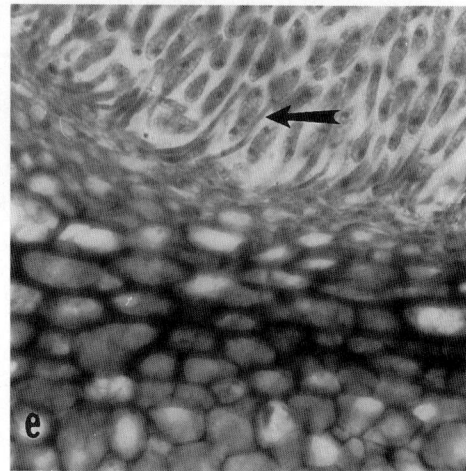

FIGURE 10 (a) *Lasiodiplodia theobromae*. Pycnidia are ostiolate and typically pyriform (bar equals 300 μm). (b, c) *Lasiodiplodia theobromae*. Conidia are one celled and hyaline, becoming two celled, dark, and striate at maturity (bar equals 10 μm). (d) *Nattrassia mangiferae* (vertical sections). Pycnidia are variably shaped. (e) *Nattrassia mangiferae*. Conidia arise from phialides (arrow).

tapering tips exceeding 1.2 μm with periclinal wall thickenings and phialides, collarettes, and hyphae that are hyaline.

Phoma spp.

Phoma eupyrena, *Phoma minutella*, *Phoma minutispora*, *Phoma oculo-hominis*, *Phoma sorghina*, and *Phoma* sp. have been documented as etiologic agents of phaeohyphomycotic infections involving cutaneous and subcutaneous tissues, cornea, and sinuses (4, 69). *Phoma* species (Fig. 9e and f) may also cause hypersensitivity pneumonitis. Because species identification of members of this genus is extremely difficult, isolates should be referred to reference laboratories if necessary.

Rhinocladiella spp.

Rhinocladiella aquaspersa (Fig. 2b), previously known as *Acrotheca aquaspersa*, is rarely an etiologic agent of chromoblastomycosis and has been described in case reports from Brazil and Mexico (78). *Rhinocladiella atrovirens* has caused cerebral phaeophyphomycosis in an AIDS patient (15). Cerebral infections (9) have been also been attributed to *Ramichloridium mackenziei* (a dematiaceous hy-

phomycete that is very similar in appearance to *Ramichloridium obovoideum* and may eventually be proven to be congeneric with *Rhinocladiella*). In addition to the sympodial anamorph that characterizes the genus *Rhinocladiella*, an annellidic anamorph (as seen in *Exophiala* spp.) and a phialidic anamorph (as seen in *Wangiella* sp.) may be present.

Scedosporium spp.

Scedosporium apiospermum (Fig. 8a), previously known as *Monosporium apiospermum*, is an anamorph of *Pseudallescheria boydii* (a fungus once classified as *Petriellidium boydii* and *Allescheria boydii*) (Fig. 8d) (54). Several species of ascomycetes in addition to *Pseudallescheria boydii* may produce a *Scedosporium apiospermum* anamorph. Historically, this fungus has been known as the main cause of fungal mycetoma in North America, but in developed countries, it is now seen more often as an agent of phaeohyphomycosis, infecting paranasal sinuses, eyes, joints, subcutaneous tissue, lung, and brain (40). In such cases, it grows in the form of hyphae that look like those produced by *Aspergillus* spp. and *Fusarium* spp. *Scedosporium prolificans* (previously known as *Sce-*

dosporium inflatum) (Fig. 8b) is a newly emerged pathogen that has been reported numerous times from bone and soft tissue infections (42, 94, 95). The conidiogenous cells of *Scedosporium prolificans* may appear to proliferate sympodially; this, however, is an artifact (21).

Scytalidium spp.
Scytalidium dimidiatum (Fig. 4c), a synanamorph associated with the pycnidial fungus *Nattrassia mangiferae*, is a documented pathogen of nail and skin (58). *Scytalidium hyalinum* has caused subcutaneous infection (96), but because it is not dematiaceous, it should perhaps not be classified in the genus *Scytalidium*. *Scytalidium lignicola* has been reported as a cause of phaeohyphomycosis; its taxonomy, however, is contested. Several mycologists consider *Scytalidium lignicola* to be a synonym of *Scytalidium dimidiatum* (88). Others consider the two to be distinct species but point out that infections previously attributed to *Scytalidium lignicola* were instead caused by misidentified isolates of *Scytalidium dimidiatum* (83). In either case, it is agreed that only *Scytalidium hyalinum* and *Scytalidium dimidiatum* are known to be pathogenic. Several studies should be consulted for the identification of *Scytalidium* spp. (81, 83, 88, 90). The reader is also referred to chapter 64 of this Manual.

Sporothrix spp.
The etiologic agent of sporotrichosis was originally described as *Sporothrix schenckii*. The fungus was later erroneously transferred to the genus *Sporotrichum*. *Sporotrichum* species are basidiomycetes, characterized by the formation of large hyphae with clamp connections and large, one-celled, thick-walled, golden conidia (49). *Sporotrichum* species are neither dimorphic nor pathogenic for humans and other animals. More than one dozen species of *Sporothrix* have been described. *Sporothrix schenckii* (Fig. 7d through g) is the only documented pathogen. *Sporothrix cyanescens*, a recently described species (13) that is distinctive for its purple-blue diffusible pigment, can be recovered from clinical specimens, but animal studies have shown that it is not pathogenic (82). In order to identify a mould recovered from clinical material as *Sporothrix schenckii*, one must demonstrate its dimorphic capability, i.e., that it can grow as both a mould and a stable yeast. To do so, isolates must be subcultured on an enriched medium such as brain heart infusion agar or brain heart infusion broth containing 0.1% agar and then grown under 5% CO_2 at 35 to 37°C. In some isolates, conversion may be limited to small portions of the colony. Occasional isolates can be difficult to convert and may require multiple subcultures and extended incubation. Environmental as well as clinical isolates of *Sporothrix schenckii* form the yeast phase when grown on appropriate media at 37°C. Isolates of *Sporothrix* species that have been recovered from environmental sources such as sphagnum moss and that lack the dematiaceous thick-walled conidia are nonvirulent in mice (18). Although an ascomycete teleomorph has been connected to environmental isolates of *Sporothrix* species, no teleomorph has been found for isolates of *Sporothrix schenckii* recovered from humans with infections.

Wangiella spp.
Wangiella dermatitidis (Fig. 5h and i) is an agent of phaeohyphomycosis that typically causes infections of cutaneous and subcutaneous tissues. Infections of the eye and the brain have also been documented (45). Occasionally, isolates are recovered from sputum and stool specimens, but they are considered clinically insignificant. *Wangiella dermatitidis* is pleomorphic, in that most isolates initially grow in a yeast form (*Phaeococcomyces* synanamorph) with development of toruloid hyphae that is succeeded by the hyphal *Wangiella dermatitidis* synanamorph. In addition to the typical phialides, some conidiogenous cells possess a group of slightly raised, truncate denticles at their apices that occur as a result of polyphialidic development. Furthermore, rare annellides of the kind seen in *Exophiala* spp. may also be produced. The genus *Wangiella* was established because the phialides without collarettes that are typical of *Wangiella dermatitidis* could not be accommodated in any existing genus (46).

Xylohypha spp.
The genus *Xylohypha* contains two medically important species: *Xylohypha bantiana* (Fig. 2c) and *Xylohypha emmonsii* (Fig. 2d). *Xylohypha bantiana* is a neurotropic mould that has caused dozens of cases of cerebral phaeohyphomycosis (19), while *Xylohypha emmonsii* has been reported as the etiologic agent of subcutaneous phaeohyphomycosis (63). *Xylohypha bantiana*, previously known as *Cladosporium trichoides* and *Cladosporium bantianum* (6, 52), was transferred to the genus *Xylohypha* because it lacks the erect conidiophores and prominent conidial scars associated with *Cladosporium* spp. Evidence suggests that *Xylohypha bantiana* infections can have a pulmonary origin (7). Therefore, isolates should be handled with extreme care, employing a biological safety cabinet.

SEROLOGIC AND MOLECULAR-BASED TESTS
Few serologically based tests are applicable to the dematiaceous fungi. A latex agglutination test is commercially available for detecting antibodies to *Sporothrix schenckii*, but despite one study (8) that reported the sensitivity and specificity to be 90 to 94 and 95 to 100%, respectively, the test has not been widely used. The remaining serologic tests consist of fluorescent-antibody conjugates for differential staining of histologic sections (35) and the exoantigen test for identification of cultures (24, 36, 65). None of these reagents are commercially available, and all are maintained only at certain reference laboratories. Molecular probes are being developed for the dematiaceous fungi, but their application is currently limited to taxonomic applications; none are commercially available.

ANTIBIOTIC SUSCEPTIBILITIES
Antifungal susceptibility testing is not routinely performed for dematiaceous fungi. Sporotrichosis can be effectively treated with oral fluconazole (64) or a saturated solution of potassium iodide. Infections with other dematiaceous fungi have historically been difficult to manage, leaving surgical excision, when feasible, as the most effective present treatment option. Chemotherapeutic modalities have achieved only rare successes, but currently, itraconazole appears promising and may become the drug of choice for chromoblastomycosis and phaeohyphomycosis (39, 68).

EVALUATION, INTERPRETATION, AND REPORTING OF RESULTS
Determining whether a particular dematiaceous fungus is involved in a disease process can sometimes be difficult,

because most of these fungi are occasionally recovered from clinical specimens as contaminants. Documentation of a dematiaceous fungus as the etiologic agent of a mycotic infection includes sound evidence that the infection is compatible with a mycosis, that the suspected etiologic agent is seen in clinical specimens, that the morphology of the fungus in the clinical specimens is compatible with that of the suspected etiologic agent, and that the recovered fungus is identified properly. The repeated recovery of a suspected etiologic agent, especially from more than one type of clinical specimen and from more than one culture vessel, is highly significant. The ability of the suspected etiologic agent to grow on media at 37°C may serve as corroborative evidence of its potential to cause an infection. Isolation of a dematiaceous fungus from body sites that normally are sterile is suspicious and should not be quickly dismissed as contamination. Infections attributed to dematiaceous fungi are being diagnosed increasingly among healthy as well as compromised patients. Concurrently, the expanding diversity of etiology within this group of fungi is becoming apparent. Without a doubt, our understanding of infections caused by dematiaceous fungi has increased dramatically in the last 15 years.

REFERENCES

1. **Ajello, L.** 1975. Phaeohyphomycosis: definition and etiology. *Pan Am. Health Organ. Sci. Publ.* **304:**126–133.
2. **Alcorn, J. L.** 1983. Generic concepts in *Drechslera, Bipolaris* and *Exserohilum. Mycotaxon* **17:**1–86.
3. **Anandi, V., T. J. John, A. Walter, J. C. M. Shastry, M. K. Lalitha, A. A. Padhye, L. Ajello, and F. W. Chandler.** 1989. Cerebral phaeohyphomycosis caused by *Chaetomium globosum* in a renal transplant recipient. *J. Clin. Microbiol.* **27:**2226–2229.
4. **Baker, J. G., I. F. Salkin, P. Forgacs, J. H. Haines, and M. E. Kemna.** 1987. First report of subcutaneous phaeohyphomycosis of the foot caused by *Phoma minutella. J. Clin. Microbiol.* **25:**2395–2397.
5. **Barron, G. L.** 1968. *The Genera of Hyphomycetes from Soil.* The Williams & Wilkins Co., Baltimore.
6. **Borelli, D.** 1960. *Torula bantiana,* agente di un granuloma cerebrale. *Riv. Anat. Patol. Oncol.* **17:**615–622.
7. **Borges, M. C., Jr., S. Warren, W. White, and E. V. Pellettiere.** 1991. Pulmonary phaeohyphomycosis due to *Xylohypha bantiana. Arch. Pathol. Lab. Med.* **115:**627–629.
8. **Bulmer, S. O., L. Kaufman, W. Kaplan, D. W. McLaughlin, and D. E. Kraft.** 1973. Comparative evaluation of five serological methods for the diagnosis of sporotrichosis. *Appl. Microbiol.* **26:**4–8.
9. **Campbell, C. K., and S. S. A. Al-Hedaithy.** 1993. Phaeohyphomycosis of the brain caused by *Ramichloridium mackenziei* sp. nov. in Middle Eastern countries. *J. Med. Vet. Mycol.* **31:**325–332.
10. **Carmichael, J. W.** 1979. Cross-reference names for pleomorphic fungi, p. 31–41. *In* B. Kendrick (ed.), *The Whole Fungus.* National Museums of Canada, Ottawa.
11. **Carmichael, J. W., B. Kendrick, I. L. Conners, and L. Sigler.** 1980. *Genera of Hyphomycetes.* University of Alberta Press, Edmonton, Canada.
12. **de Hoog, G. S.** 1979. Nomenclatural notes on some black yeast-like hyphomycetes. *Taxon* **28:**347–348.
13. **de Hoog, G. S., and G. A. de Vries.** 1973. Two new species of *Sporothrix* and their relations to *Blastobotrys nivea. Antonie van Leeuwenhoek J. Microbiol. Serol.* **39:**515–520.
14. **de Hoog, G. S., and E. J. Hermanides-Nijhof.** 1977. *The Black Yeasts and Allied Hyphomycetes. Studies in Mycology* no. 15. Centraalbureau voor Schimmelcultures, Baarn, The Netherlands.
15. **Del Palacio, H. A., M. K. Moore, C. K. Campbell, P. M. Del Palacio, and C. R. Castillo.** 1989. Infection of the central nervous system by *Rhinocladiella atrovirens* in a patient with acquired immunodeficiency syndrome. *J. Med. Vet. Mycol.* **27:**127–130.
16. **Dixon, D. M., and A. Pollack-Wyss.** 1991. The medically important dematiaceous fungi and their identification. *Mycoses* **34:**1–18.
17. **Dixon, D. M., and I. F. Salkin.** 1986. Morphologic and physiologic studies of three dematiaceous pathogens. *J. Clin. Microbiol.* **24:**12–15.
18. **Dixon, D. M., I. F. Salkin, R. A. Duncan, N. J. Hurd, J. H. Haines, M. E. Kemma, and F. B. Coles.** 1991. Isolation and characterization of *Sporothrix schenckii* from clinical and environmental sources associated with the largest U.S. epidemic of sporotrichosis. *J. Clin. Microbiol.* **29:**1106–1113.
19. **Dixon, D. M., T. J. Walsh, W. G. Merz, and M. R. McGinnis.** 1989. Infections due to *Xylohypha bantiana (Cladosporium trichoides). Rev. Infect. Dis.* **11:**515–525.
20. **Domsch, K. H., W. Gams, and T. H. Anderson.** 1980. *Compendium of Soil Fungi,* vol. 1 and 2. Academic Press, Inc., New York.
21. **Dykstra, M. J., I. F. Salkin, and M. R. McGinnis.** 1990. An ultrastructural comparison of conidiogenesis in *Scedosporium apiospermum, Scedosporium inflatum,* and *Scopulariopsis brumptii. Mycologia* **81:**896–904.
22. **Ellis, M. B.** 1971. *Dematiaceous Hyphomycetes.* Commonwealth Mycological Institute, Kew, England.
23. **Ellis, M. B.** 1976. *More Dematiaceous Hyphomycetes.* Commonwealth Mycological Institute, Kew, England.
24. **Espinel-Ingroff, A., P. R. Goldson, M. R. McGinnis, and T. M. Kerkering.** 1988. Evaluation of proteolytic activity to differentiate some dematiaceous fungi. *J. Clin. Microbiol.* **26:**301–307.
25. **Fader, R. C., and M. R. McGinnis.** 1988. Infections caused by dematiaceous fungi: chromoblastomycosis and phaeohyphomycosis. *Infect. Dis. Clin. N. Am.* **2:**925–938.
26. **Fernando Barba-Gómez, J., J. Mayorga, M. R. McGinnis, and A. González-Mendoza.** 1992. Chromoblastomycosis caused by *Exophiala spinifera. J. Am. Acad. Dermatol.* **26:**367–370.
27. **Fincher, R. M. E., J. F. Fisher, A. A. Padhye, L. Ajello, and J. C. H. Steele, Jr.** 1988. Subcutaneous phaeohyphomycotic abscess caused by *Phialophora parasitica* in a renal allograft recipient. *J. Med. Vet. Mycol.* **26:**311–314.
28. **Gams, W., and M. R. McGinnis.** 1983. *Phialemonium,* a new anamorph genus intermediate between *Phialophora* and *Acremonium. Mycologia* **75:**977–987.
29. **Gonzalez, M. S., B. Alfonso, D. Seckinger, A. A. Padhye, and L. Ajello.** 1984. Subcutaneous phaeohyphomycosis caused by *Cladosporium devriesii,* sp. nov. *Sabouraudia J. Med. Vet. Mycol.* **22:**427–432.
30. **Goodpasture, H. C., T. Carlson, B. Ellis, and G. Randall.** 1983. *Alternaria* osteomyelitis. Evidence of specific immunologic tolerance. *Arch. Pathol. Lab. Med.* **107:**528–530.
31. **Gueho, E., and G. S. de Hoog.** 1991. Taxonomy of the medical species of *Pseudallescheria* and *Scedosporium. J. Med. Mycol.* **118:**3–9.
32. **Hironaga, M., K. Nakano, I. Yokoyama, and J. Kitajima.** 1989. *Phialophora repens,* an emerging agent of subcutaneous phaeohyphomycosis in humans. *J. Clin. Microbiol.* **27:**394–399.
33. **Hsu, M. M., and J. Yu-yun Lee.** 1993. Cutaneous and subcutaneous phaeohyphomycosis caused by *Exserohilum rostratum. J. Am. Acad. Dermatol.* **28:**340–344.
34. **Iwatsu, T., K. Sishimura, and M. Makoto.** 1984. *Exophiala castellanii* sp. nov. *Mycotaxon* **20:**307–314.
35. **Kaplan, W., and A. G. Ochoa.** 1963. Application of the fluorescent antibody technique to the rapid diagnosis of sporotrichosis. *J. Lab. Clin. Med.* **62:**835–841.
36. **Kaufman, L., P. Standard, and A. Padhye.** 1980. Serologic relationships among isolates of *Exophiala jeanselmei (Phialophora jeanselmei, P. gouferotii)* and *Wangiella dermatitidis. Pan Am. Health Organ. Sci. Publ.* **396:**252–258.

37. **Kiehn, T. E., B. Polsky, E. Punithalingham, F. F. Edwards, A. E. Brown, and D. Armstrong.** 1987. Liver infection caused by *Coniothyrium fuckelii* in a patient with acute myelogenous leukemia. *J. Clin. Microbiol.* **25:**2410–2412.

38. **King, D., L. Pasarell, D. M. Dixon, M. R. McGinnis, and W. G. Merz.** 1993. A phaeohyphomycotic cyst and peritonitis caused by *Phialemonium* species and a reevaluation of its taxonomy. *J. Clin. Microbiol.* **31:**1804–1810.

39. **Kumar K., I. Kaur, A. Chakrabarti, and V. K. Sharma.** 1991. Treatment of deep mycoses with itraconazole. *Mycopathologia* **115:**169–174.

40. **Kwon-Chung, K. J., and J. E. Bennett.** 1992. *Medical Mycology*. Lea & Febiger, Philadelphia.

41. **Kwon-Chung, K. J., W. B. Hill, and J. E. Bennett.** 1981. New, special stain for histopathological diagnosis of cryptococcosis. *J. Clin. Microbiol.* **13:**383–387.

42. **Malloch, D., and I. F. Salkin.** 1984. A new species of *Scedosporium* associated with osteomyelitis in humans. *Mycotaxon* **21:**247–255.

43. **Mancini, M. C., and M. R. McGinnis.** 1992. *Dactylaria* infections of a human being: pulmonary disease in a heart transplant patient. *J. Heart Lung Transplant.* **11:**827–830.

44. **Matsumoto, T., K. Nishimoto, K. Kimura, A. A. Padhye, L. Ajello, and M. R. McGinnis.** 1984. Phaeohyphomycosis caused by *Exophiala moniliae*. *Sabouraudia J. Med. Vet. Mycol.* **22:**17–26.

45. **Matsumoto, T. T., T. Matsuda, M. R. McGinnis, and L. Ajello.** 1993. Clinical and mycological spectrum of *Wangiella dermatitidis* infections. *Mycoses* **36:**145–155.

46. **McGinnis, M. R.** 1978. Human pathogenic species of *Exophiala, Phialophora*, and *Wangiella. Pan Am. Health Organ. Sci. Publ.* **356:**35–59.

47. **McGinnis, M. R.** 1979. Taxonomy of *Exophiala jeanselmei* (Langeron) McGinnis and Padhye. *Mycopathologia* **65:**79–87.

48. **McGinnis, M. R.** 1979. Taxonomy of *Exophiala werneckii* and its relationship to *Microsporum mansonii. Sabouraudia* **17:**145–154.

49. **McGinnis, M. R.** 1980. *Laboratory Handbook of Medical Mycology*. Academic Press, Inc., New York.

50. **McGinnis, M. R.** 1983. Chromoblastomycosis and phaeohyphomycosis: new concepts, diagnosis, and mycology. *J. Am. Acad. Dermatol.* **8:**1–16.

51. **McGinnis, M. R.** 1992. Black fungi: a model for understanding tropical mycoses, p. 129–149 *In* D. H. Walker (ed.), *Global Infectious Diseases Prevention, Control, and Eradication*. Springer Verlag, New York.

52. **McGinnis, M. R., D. Borelli, A. A. Padhye, and L. Ajello.** 1986. Reclassification of *Cladosporium bantianum* in the genus *Xylohypha. J. Clin. Microbiol.* **23:**1148–1151.

53. **McGinnis, M. R., and A. L. Carrión.** 1993. Chromoblastomycosis, p. 1–13. *In* D. J. Demis (ed.), *Clinical Dermatology*, vol. 3. J. B. Lippincott Co., Philadelphia.

54. **McGinnis, M. R., A. A. Padhye, and L. Ajello.** 1982. *Pseudallescheria* Negroni et Fischer, 1943, and its later synonym *Petriellidium* Malloch, 1970. *Mycotaxon* **14:**94–102.

55. **McGinnis, M. R., M. G. Rinaldi, and R. E. Winn.** 1986. Emerging agents of phaeohyphomycosis: pathogenic species of *Bipolaris* and *Exserohilum. J. Clin. Microbiol.* **24:**250–259.

56. **McGinnis, M. R., and W. A. Schell.** 1980. The genus *Fonsecaea* and its relationship to the genera *Cladosporium, Phialophora, Ramichloridium*, and *Rhinocladiella. Pan Am. Health Organ. Sci. Publ.* **396:**215–234.

57. **McGinnis, M. R., W. A. Schell, and J. Carson.** 1985. *Phaeoannellomyces* and the Phaeococcomycetaceae, new dematiaceous blastomycete taxa. *Sabouraudia J. Med. Vet. Mycol.* **23:**179–188.

58. **Moore, M. K.** 1988. Morphological and physiological studies of isolates of *Hendersonula toruloidea* Nattrass cultured from human skin and nail samples. *J. Med. Vet. Mycol.* **26:**25–39.

59. **Oldham, A. D.** 1988. Care and use of the microscope, p. 574–578. *In* B. B. Wentworth (ed.), *Diagnostic Procedures for Mycotic and Parasitic Infections*, 7th ed. American Public Health Association, Washington, D.C.

60. **Padhye, A. A.** 1978. Comparative study of *Phialophora jeanselmei* and *P. gougerotii* by morphological, biochemical, and immunological methods. *Pan Am. Health Organ. Sci. Publ.* **356:**60–65.

61. **Padhye, A. A., L. Ajello, M. A. Wieden, and K. K. Steinbronn.** 1986. Phaeohyphomycosis of the nasal sinuses caused by a new species of *Exserohilum. J. Clin. Microbiol.* **24:**245–249.

62. **Padhye, A. A., M. R. McGinnis, and L. Ajello.** 1978. Thermotolerance of *Wangiella dermatitidis. J. Clin. Microbiol.* **8:**424–426.

63. **Padhye, A. A., M. R. McGinnis, L. Ajello, and F. W. Chandler.** 1988. *Xylohypha emmonsii* sp. nov., a new agent of phaeohyphomycosis. *J. Clin. Microbiol.* **26:**702–708.

64. **Pappas, P. G., C. A. Kaufman, J. R. Perfect, R. Greenfield, B. MacKenzie, J. Fisher, A. Feldman, and W. E. Dismukes.** 1992. Fluconazole (Flu) for the treatment of sporotrichosis, abstr. 629, p. 215. *Program Abstr. 32nd Intersci. Conf. Antimicrob. Agents Chemother.*

65. **Pasarell, L., M. R. McGinnis, and P. G. Standard.** 1990. Differentiation of selected species of *Bipolaris* and *Exserohilum* with exoantigens. *J. Clin. Microbiol.* **28:**1655–1657.

66. **Pitrak, D. L., E. W. Koneman, R. C. Estupinan, and J. Jackson.** 1988. *Phialophora richardsiae* in humans. *Rev. Infect. Dis.* **10:**1195–1203.

67. **Punithalingham, E.** 1976. *Botryodiplodia theobromae*. CMI Descriptions of Pathogenic Fungi and Bacteria, no. 519. Commonwealth Mycological Institute, Kew, England.

68. **Queiroz-Telles, F., K. S. Purim, and J. N. Fillus.** 1992. Itraconazole in the treatment of chromoblastomycosis due to *Fonsecaea pedrosoi. Int. J. Dermatol.* **31:**805–812.

69. **Rai, M. K.** 1989. *Phoma sorghina* infections in human being. *Mycopathology* **105:**167–170.

70. **Rebell, G., and R. K. Forster.** 1976. *Lasiodiplodia theobromae* as a cause of keratomycosis. *Sabouraudia* **14:**155–170.

71. **Rinaldi, M. G., E. L. McCoy, and D. F. Winn.** 1982. Gluteal abscess caused by *Phialophora hoffmannii* and review of the role of this organism in human mycoses. *J. Clin. Microbiol.* **16:**181–185.

72. **Rinaldi, M. G., P. Phillips, J. G. Schwartz, R. E. Winn, G. R. Holt, F. W. Shagets, J. Elrod, G. Nishioka, and T. B. Aufdemorte.** 1987. Human *Curvularia* infections; report of five cases and review of the literature. *Diagn. Microbiol. Infect. Dis.* **6:**27–39.

73. **Salkin, I. F., and D. M. Dixon.** 1987. *Dactylaria constricta*: description of two varieties. *Mycotaxon* **29:**377–381.

74. **Salkin, I. F., J. A. Martinez, and M. E. Kemna.** 1986. Opportunistic infection of the spleen caused by *Aureobasidium pullulans. J. Clin. Microbiol.* **23:**828–831.

75. **Salkin, I. F., M. R. McGinnis, M. J. Dykstra, and M. G. Rinaldi.** 1988. *Scedosporium inflatum*, an emerging pathogen. *J. Clin. Microbiol.* **26:**498–503.

76. **Schell, W. A.** 1986. Oculomycoses caused by dematiaceous fungi. *Pan Am. Health Organ. Sci. Publ.* **479:**105–109.

77. **Schell, W. A., and M. R. McGinnis.** 1988. Molds involved in subcutaneous infections, p. 99–171. *In* B. B. Wentworth (ed.), *Diagnostic Procedures for Mycotic and Parasitic Infections*, 7th ed. American Public Health Association, Washington, D.C.

78. **Schell, W. A., M. R. McGinnis, and D. Borelli.** 1983. *Rhinocladiella aquaspersa*, a new combination for *Acrotheca aquaspersa. Mycotaxon* **17:**341–348.

79. **Schell, W. A., and J. R. Perfect.** 1993. *Coniothyrium fuckelii*, a species in need of a genus: second case of human infection, abstr. F-34, p. 533. *Abstr. 93rd Gen. Meet. Am. Soc. Microbiol. 1993.*

80. **Sides, E. H., III, J. D. Benson, and A. A. Padhye.** 1991. Phaeohyphomycotic brain abscess due to *Ochroconis gallopavum* in a patient with malignant lymphoma of a large cell type. *J. Med. Vet. Mycol.* **29:**317–322.

81. **Sigler, L., and J. W. Carmichael.** 1976. Taxonomy of *Malbranchea* and some other Hyphomycetes with arthroconidia. *Mycotaxon* **4:**349–488.

82. **Sigler, L., J. L. Harris, D. M. Dixon, A. L. Flis, I. F. Salkin,**

M. Kemna, and R. A. Duncan. 1990. Microbiology and potential virulence of *Sporothrix cyanescens*, a fungus rarely isolated from blood and skin. *J. Clin. Microbiol.* **28:**1009–1015.

83. Sigler, L., and C. J. K. Wang. 1990. *Scytalidium circinatum* sp. nov., a hyphomycete from utility poles. *Mycologia* **82:**399–404.

84. Stiller, M. J., S. Rosenthal, R. C. Summerbell, J. Pollack, and A. Chan. 1992. Onychomycosis of the toenails caused by *Chaetomium globosum*. *J. Am. Acad. Dermatol.* **26:**775–776.

85. Sudduth, E. J., A. J. Crumbley III, and E. W. Farrar. 1992. Phaeohyphomycosis due to *Exophiala* species: clinical spectrum of disease in humans. *Clin. Infect. Dis.* **15:**639–644.

86. Sutton, B. C. 1971. Coelomycetes. IV. The genus *Harknessia*, and similar fungi on *Eucalyptus*. *Mycol. Pap.* **123:**32–34.

87. Sutton, B. C. 1980. *The Coelomycetes: Fungi Imperfecti with Pycnidia Acervuli and Stromata.* Robert MacLehose and Co., Ltd., Glasgow.

88. Sutton, B. C., and B. J. Dyko. 1989. Revision of *Hendersonula*. *Mycol. Res.* **93:**466–488.

89. Viviani, M. A., A. M. Tortorano, G. Laria, A. Giannetti, and G. Bignotti. 1988. Two new cases of cutaneous alternariosis with a review of the literature. *Mycopathologia* **43:**3–12.

90. Wang, C. J. K., and R. A. Zabel. 1990. *Identification Manual for Fungi from Utility Poles in the Eastern United States.* Allen Press, Inc., Lawrence, Kans.

91. Washburn, R. G., D. W. Kennedy, M. G. Begley, D. K. Henderson, and J. E. Bennett. 1988. Chronic fungal sinusitis in apparently normal hosts. *Medicine* **67:**231–247.

92. Wheeler, M. H., and A. A. Bell. 1987. Melanins and their importance in pathogenic fungi, p. 338–387. *In* M. R. McGinnis (ed.), *Current Topics in Medical Mycology*, vol. 2. Springer-Verlag, New York.

93. Wiest, P. M., K. Wiese, M. R. Jacobs, A. B. Morrisse, T. I. Abelson, W. Witt, and M. M. Lederman. 1987. *Alternaria* infection in a patients with acquired immune deficiency syndrome: case report and review of invasive *Alternaria* infections. *Rev. Infect. Dis.* **9:**799–803.

94. Wilson, C. M., E. O'Rourke, M. R. McGinnis, and I. F. Salkin. 1990. *Scedosporium inflatum*: clinical spectrum of a newly recognized pathogen. *J. Infect. Dis.* **161:**102–107.

95. Wood, G. M., J. G. McCormack, D. B. Muir, D. H. Ellis, M. F. Ridley, R. Pritchard, and M. Harrison. 1992. Clinical features of human infection with *Scedosporium inflatum*. *Clin. Infect. Dis.* **14:**1027–1033.

96. Zaatari, G. S., R. Reed, and R. Morewessel. 1984. Subcutaneous hyphomycosis caused by *Scytalidium hyalinum*. *Am. J. Clin. Pathol.* **82:**252–256.

Fungi Causing Eumycotic Mycetoma

ARVIND A. PADHYE

68

DEFINITION OF THE DISEASE

A mycetoma (plural, mycetomata) (33) is a localized, chronic, noncontagious, granulomatous infection involving cutaneous and subcutaneous tissues and eventually, in some cases, the bones. Mycetomata are generally confined to the feet or hands but occasionally affect other parts of the body such as the back, shoulders, buttocks, and, rarely, scalp. The lesions contain granulomas and abscesses that suppurate and drain through sinus tracts. The pus contains granules (sclerotia) (32) that vary in size from microscopic to 2 mm in diameter. Size, color, shape, and texture of the granules vary with the species and sometimes suggest the specific etiology. Both filamentous fungi and actinomycetes are known to cause mycetomata. Mycetomata caused by actinomycetes are called actinomycotic mycetomata, and those incited by species of filamentous fungi are referred to as eumycotic mycetomata. Mycetoma is distributed worldwide and is more commonly seen in humans than in lower animals. Only a few authentic cases of mycetoma involving such animals as dogs, horses, and goats have been described in the literature (2, 10, 23, 28).

OCCURRENCE OF CAUSAL AGENTS IN NATURE

The causal agents of eumycotic mycetoma are saprophytes that live on organic debris in soil. The various causal agents have been isolated from either soil or plant material (1, 4, 18, 25). Segretain (46) and Segretain and Mariat (49), using specific media and techniques, found *Leptosphaeria senegalensis* and *Leptosphaeria tompkinsii* on about 50% of the dry thorns of *Acacia* trees that were examined, particularly thorns that had been stained by mud during the rainy season. *Neotestudina rosatii* was isolated from sandy ground (29), and *Madurella mycetomatis* was recovered from soil and anthills (45, 53). *Acremonium* spp., *Curvularia* spp., *Aspergillus nidulans*, and *Pseudallescheria boydii* have frequently been isolated from soil. Borelli (6) isolated *Madurella grisea* from soil in Venezuela.

MODE OF INFECTION AND SYMPTOMS

A mycetoma develops after a traumatic injury by contaminated thorns, splinters from plants, fish scales or fins, snake bites, insect bites, farm implements, knives, etc. The development of mycetoma after an accidental implantation of the etiologic agent during surgery (41) and in a renal transplant recipient (19) has also been reported. The initial lesions are characterized by a feeling of discomfort and pain at the point of inoculation. Weeks or months later, the subcutaneous tissue at the site of inoculation becomes indurated, abscesses develop, and sinuses may drain to the surface. The lesions are characterized by swelling, suppurating abscesses, granulomas, and sinuses from which serosanguinous fluid containing granules or sclerotia is discharged. The mycetoma develops slowly beneath thick fibrosclerous tissue. The subsequent phase of proliferation involves the invasion of muscles and intramuscular layers. The granulomatous lesions can extend as deep as bony tissue, causing severe bone destruction, formation of small cavities, and complete remodeling. Early osteolytic damage includes loss of the cortical margin and external erosion of the bone. As the disease progresses, blood, lymphatic vessels, and nerves may be damaged. Frequently, secondary bacterial infections and osteomyelitis producing total bone destruction occur. The pain of the disease is often associated with the development of multiple fistulae. Regardless of the etiologic agent involved, the causal organisms develop in the form of soft or hard compact mycelial masses (sclerotia), known as granules or grains, within the infected tissue. The hallmark of a mycetoma is the granule composed of nonconidiating mycelium that may or may not be embedded in a cementlike matrix. Weathered et al. (54) described the cement in M. *mycetomatis* as being an amorphous electron-dense material with areas containing variably sized membrane-bound vesicular inclusions. Some of the host material, especially at the periphery of the granule, provided a protective barrier for the fungus against antifungal agents and antibody.

Mycetoma develops mainly among people such as field workers, farmers, sugarcane workers, and fishermen, who are in contact with contaminated materials. Even though, because of increased exposure, the prevalence of mycetomata is much higher in males than in females, women and children who walk barefoot are also vulnerable to infection.

Fungus balls, or fungomata, caused by *Aspergillus* species, *Coccidioides immitis*, or *P. boydii* in preformed lung cavities, are sometimes mistakenly called mycetomas (20). In the absence of well-organized granules, they should be referred to appropriately as fungus balls, aspergillomas, or coccidioidomas (30). Similarly, mycelial aggregates formed by dermatophytes in deep tissues differ in many respects from the granules of the mycetoma with regard to a general absence of granule ontogeny, a distinct Splendore-Hoeppli reaction surrounding the mycelial aggregates, and the entry of the fungus from the dermis to deeper tissue following rupture of the follicular epithelium. Such infections caused by dermatophytes are best referred to as pseudomycetomata rather than mycetomata (3).

GEOGRAPHIC DISTRIBUTION OF EUMYCOTIC MYCETOMA

Eumycotic mycetomata occur primarily in the tropical and hot temperate zones of the world. They frequently are reported from countries near the Tropic of Cancer, but they also occur beyond this area. Numerous cases have been described from Africa, Asia, and South and Central America. Mycetomata are not seen as commonly in the United States as in tropical countries. A survey of the literature from 1979 to 1992 revealed that not more than 10 cases of eumycotic mycetoma had been reported in the United States.

CLIMATIC CONDITIONS

Climate has a definite influence on the prevalence and distribution of mycetomata. Rivers that flood each year during the wet season in many countries of Africa and Asia influence the distribution of the causal agents. Rainfall also favors the spread of the etiologic agents on organic matter (14).

COLLECTION AND EXAMINATION OF GRANULES

Since all of the agents of eumycotic mycetoma are soil saprophytes and since some are encountered as contaminants of clinical specimens, their etiologic role in mycetomas must be established carefully. A definitive diagnosis must be based on the demonstration of granules in tissue and repeated isolation of the causal fungus from granules aspirated from preferably unopened sinuses.

Pus, exudate, or biopsy material should be examined for the presence of granules, which vary in size from 0.2 to ≥2 mm and are usually detectable with the naked eye. Their color, texture, size, and shape give a fair indication of the identity of the etiologic agent. Actinomycotic and eumycotic mycetomata are differentiated by the examination of crushed granules in Gram-stained preparations. Actinomycotic granules are composed of gram-positive, interwoven, thin filaments 0.5 to 1.0 μm in diameter as well as of coccoid and bacillary forms. Granules of the eumycotic agents, on the other hand, are composed of broad, interwoven, septate hyphae 2 to 5 μm in diameter with many bizarrely shaped swollen cells up to 15 μm in diameter, especially at the periphery of the granules. In many species, the granules also contain a cementlike material.

To maximize the chances of obtaining pure cultures of the etiologic agents, granules from the eumycotic mycetomata should be washed several times with saline containing antibacterial antibiotics such as penicillin and streptomycin. The granules thus freed from the surface bacteria are

TABLE 1 Etiologic agents of eumycotic mycetoma

Etiologic agent	Reference(s)
Ascomycota	
Emericella nidulans (*Aspergillus nidulans*)	27
Leptosphaeria senegalensis	29
Leptosphaeria tompkinisii	16
Neotestudina rosatii	47
Pseudallescheria boydii	35
Deuteromycota	
Acremonium falciforme	21, 36
Acremonium kiliense	21
Acremonium recifei	21, 28
Curvularia geniculata	10
Curvularia lunata	27, 28
Corynespora cassicola	28
Cylindrocarpon destructans	15, 55
Exophiala jeanselmei	34, 40
Fusarium moniliforme	28, 29
Fusarium solani var. *coeruleum*	29, 33, 52
Fusarium solani var. *minus*	33
Madurella grisea	6, 24, 26, 27, 49
Madurella mycetomatis	6, 24, 29
Plenodomus avramii	8
Polycytella hominis	12
Pseudochaetosphaeronema larense	9, 42
Pyrenochaeta mackinnonii	7, 42
Pyrenochaeta romeroi	5, 42, 48

cultured on several petri dishes of Sabouraud dextrose agar (SDA; Difco Laboratories, Detroit, Mich.) containing chloramphenicol (50 mg/liter) and Sabouraud agar containing chloramphenicol (50 mg/liter) and cycloheximide (500 mg/liter). Several of these plates should be incubated at 25°C, and others should be incubated at 37°C. All should be observed at 48-h intervals. Since many of the fungi that cause eumycotic mycetoma grow slowly, culture plates should be incubated for 6 weeks before being discarded as negative. Identification of the isolated fungus, when possible, is based on gross colony morphology and pigmentation, morphology of the conidiophores and conidia, and mechanism of conidiogenesis. Since certain species (*M. grisea, M. mycetomatis, N. rosatii, Pyrenochaeta mackinnonii,* and *Pyrenochaeta romeroi*) do not sporulate readily, sporulation media and physiologic tests such as those for carbohydrate and nitrate utilization must be used for identification.

ETIOLOGIC AGENTS AND THEIR CLASSIFICATION

The 22 species known to cause eumycotic mycetomata are discussed below. Of these species, 5 belong to the division Ascomycota, and 17 are classified under the Deuteromycota (Fungi Imperfecti) (Table 1).

CHARACTERISTICS OF GRANULES IN TISSUE

The granules of eumycotic mycetomata are composed of septate mycelial filaments that are at least 2 to 4 μm in diameter. The mycelium may be distorted and bizarre in form and size, and the cell walls of the fungi, especially toward the periphery of the granules, are thickened.

TABLE 2 Gross characteristics of granules of eumycotic mycetoma agents

Kind of granule in tissue and species	Texture	Size range (mm)	Cementlike matrix
White grain			
Acremonium falciforme	Soft	0.2–0.5	Absent
Acremonium kiliense	Soft	0.2–0.5	Absent
Acremonium recifei	Soft	0.2–0.5	Absent
Aspergillus nidulans	Soft	1–2	Absent
Cylindrocarpon destructans	Soft	0.2–0.5	Absent
Fusarium moniliforme	Soft	0.2–0.5	Absent
Fusarium solani	Soft	0.2–0.6	Absent
Neotestudina rosatii	Soft	0.5–1	Present, peripheral
Polycytella hominis	Soft	0.5–1	Absent
Pseudallescheria boydii	Soft	0.5–1	Absent
Black grain			
Corynespora cassicola	Hard	0.2–0.5	Absent
Curvularia geniculata	Hard	0.5–1	Present, peripheral
Curvularia lunata	Hard	0.5–1	Present, peripheral
Exophiala jeanselmei	Soft	0.2–0.3	Absent
Leptosphaeria senegalensis	Hard	0.5–2	Present, peripheral
Leptosphaeria tompkinsii	Hard	0.5–2	Present, peripheral
Madurella grisea	Soft	0.3–0.6	Present, peripheral
Madurella mycetomatis	Hard	0.5–5	Present, homogeneous
Plenodomus avramii	Soft	0.5–0.8	Absent
Pseudochaetosphaeronema larense	Soft	0.2–0.5	Absent
Pyrenochaeta mackinnonii	Soft	0.2–0.5	Absent
Pyrenochaeta romeroi	Soft	0.2–0.6	Absent

Chlamydospores are frequently present, especially at the periphery of the granule. The mycelium of the granules may or may not be embedded in a cementlike substance, depending on the species involved. The granules often elicit an immunologic response, the so-called Splendore-Hoeppli reaction (antigen-antibody) (13, 33). This reaction is seen histologically in the form of a deposit of an eosinophilic material around the granule.

Depending on the etiologic agent involved, the granules are white to yellow-brown to black and range from 200 μm to 5 mm in diameter. The color gives some clue to the species involved. Most textbooks on medical mycology mention a definite relationship between the color of the granules and the causal species.

The causal agents that produce hyaline mycelium and conidia generally produce white granules in tissue, whereas M. mycetomatis and several other fungi form black granules in tissue. It is believed that melanin, host protein, and dark debris cause the dark color of these granules. Melanin is a low-molecular-weight compound that is anchored to extracellular proteins. The precursor for the extracellular production of melanin may be a compound such as 1,8-dihydroxynaphthalene (32, 33). Generally, fungi that produce pigmented mycelium and/or conidia have dark granules in tissues, but there are several exceptions to this rule. For example, hyaline fungi such as *Acremonium kiliense* and *Fusarium solani* var. *coeruleum* have been described in the literature as producers of black granules (21, 52). In a reverse manner, dematiaceous etiologic agents such as *P. boydii* produce white granules in tissue (25, 35). Although the gross and microscopic characteristics of the granules provide insight into the identity of the etiologic agent or the particular group to which it belongs, definitive identi-fication of the etiologic agent should be based on repeated isolation of the same fungus from several granules, colonial features, microscopic characteristics of conidia, conidiophores, and mechanism of conidiogenesis. The gross characteristics of the granules formed by each of the 22 species discussed are summarized in Table 2.

DESCRIPTION OF THE MORE PREVALENT ETIOLOGIC AGENTS

The 11 etiologic agents described below are the species most commonly isolated from human or lower-animal mycetomata. The physiologic characteristics of the five species most commonly isolated from eumycotic mycetomata are summarized in Table 3. In the United States, cases of eumycotic mycetomata are not common. Among the cases studied, the following species, in diminishing order of frequency, were isolated: *P. boydii*, *Acremonium falciforme*, *M. mycetomatis*, *M. grisea*, and *Exophiala jeanselmei*.

A. falciforme (Carrion) Gams 1971

Three *Acremonium* species (*A. falciforme*, *A. kiliense*, and *A. recifei*) are known to cause mycetomata. *A. falciforme* is the second most common cause of mycetoma in the United States. It was found to be an agent of eumycotic mycetoma in the San Francisco Bay area of California (21). It was also reported as the cause of mycetoma in a renal transplant recipient (19) and as an etiologic agent of mycetoma in a patient from New York (36).

The granules of *A. falciforme* are white to pale yellow, soft, and 0.2 to 0.5 mm in diameter. They are composed of slender, polymorphic, septate hyphae 1.5 to 2.0 μm in

TABLE 3 Physiologic characteristics of some causal agents of eumycotic mycetoma

Etiologic agent	Utilization of:								Starch hydrolysis	Protease activity	Optimum growth temp (°C)
	Galactose	Glucose	Lactose	Maltose	Sucrose	Asparagine	KNO₃	(NH₄)₂SO₄			
A. falciforme	+	+	−	+	+	+	+	+	−	+	30
E. jeanselmei	+	+	−	+	+	+	+	+	+	−	30
M. grisea	+	+	+	+	+	+	+	+	+	+	25
M. mycetomatis	+	+	−	−	+	+	+	+	−	+	37
P. boydii	+	+	−	+	+	+	+	+	−	+	30

diameter with irregular bulbous swellings and peripheral cementing material (Fig. 1). In tissue sections, the granules resemble those produced by *P. boydii*. The diagnosis, therefore, should be based not on morphology of the granule in tissue alone but also on the isolation and identification of the causal agent.

Colonies of *A. falciforme* on SDA are slow growing, reaching 60 to 65 mm in diameter in 2 weeks. They are downy and gray-brown, becoming gray-violet. The reverse of the colony develops a violet-purple pigment. The hyphae are hyaline, septate, smooth, branched, and 1.5 to 2.5 μm in diameter. They bear erect, undifferentiated, unbranched, repeatedly septate conidiophores. The conidia are borne at the tip of the phialidic conidiogenous cells in mucoid clusters. The conidia are sausage shaped, slightly curved, and nonseptate to monoseptate. They measure 7 to 8.5 by 2.7 to 3.2 μm (Fig. 2). The intercalary or, rarely, terminal chlamydospores are smooth, thick walled, and 5 to 8 μm in diameter.

The other two species, *A. kiliense* and *A. recifei*, are not known to cause mycetoma as frequently as does *A. falciforme*. Their granules are similar in morphology to those of *A. falciforme* and cannot be differentiated from each other or from those produced by *P. boydii*. When isolated, *A. kiliense* develops pinkish orange glabrous colonies with coremia. Conidia are broadly elliptical to cylindrical and nonseptate, occurring in gleoid masses at the tips of lateral, long, tapering phialides with vestigial collarettes. Chlamy-

dospores are present. Colonies of *A. recifei* are white to pinkish buff and moist to glabrous. Their conidiophores are branched, bearing long, tapering phialides. Conidia are claviform and nonseptate to monoseptate.

Curvularia geniculata (Tracy and Earl) Boedijin, 1933

C. geniculata is the etiologic agent of mycetoma in dogs in the United States (10). Its granules are black to dark brown, firm, and 0.5 to 1.0 mm or more in size. In tissue sections, the granules are spherical, ovoid, or irregularly shaped and are often surrounded by a zone of epithelioid cells. The periphery of the granule is a dense, interwoven mass of dematiaceous mycelium and thick-walled, chlamydosporelike cells embedded in a cementlike substance. The interior of the granules is vacuolar and consists of a loose network of septate, hyphal filaments.

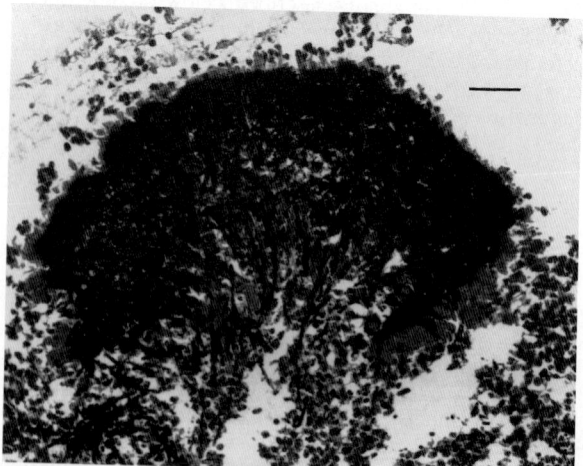

FIGURE 1 White soft granule of *A. falciforme* composed of septate hyphae and peripheral zone of radiating eosinophilic, hyaline, club-shaped material. Gomori methenamine silver stain. Bar = 10 μm.

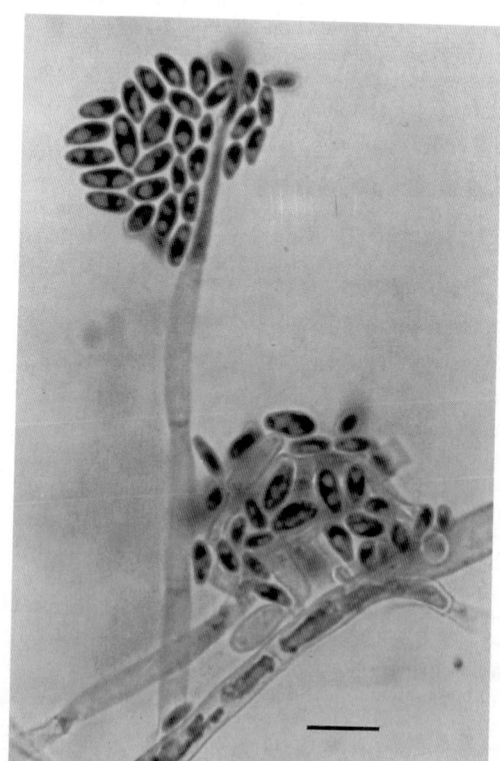

FIGURE 2 *A. falciforme*. Slide culture on SDA showing erect septate conidiophore, phialidic conidiogenous cell, and slightly curved conidia. Bar = 10 μm.

In culture, C. geniculata develops a cottony to downy, olive-gray to black colony. Microscopically, the dematiaceous, septate hyphae bear solitary, geniculate conidiogenous cells. The conidia are without protruding hila. They are smooth walled, predominantly five celled, and curved, with the median cell pale to dark brown.

Curvularia lunata (Wakker) Boedijin 1933

C. lunata has been described as an etiologic agent of mycetoma among humans in Senegal and Sudan (27, 49). Its granules resemble those of C. geniculata in their morphologic characteristics. In culture, C. lunata produces geniculate conidiophores producing terminal, sympodial conidiogenous cells. Conidia are without protruding hila. They are smooth and predominantly four celled, with the penultimate cell pale to dark brown.

E. jeanselmei (Langeron) McGinnis and Padhye 1977

E. jeanselmei is known as an agent of eumycotic mycetoma in India (50, 51), Malaya, Thailand (43), Argentina (39), and the United States (40). E. jeanselmei produces dark granules in host tissue. They are brown to black, irregular in shape, and fragile. Detached portions or fragments of the granules often are found within giant cells. When extruded through open sinuses, the granules often look like worm cases (vermiform) because of their elongated shapes and irregular surfaces. In tissue sections, they appear as hollow spheres or as sinuous bands that are vermiform in appearance. The external surface is composed of brown, thick-walled hyphae and thick-walled chlamydosporelike cells. The granules are cement free. Within the hollow granules, smaller, degenerated hyphal fragments with leukocytes and giant cells may be seen.

Initially, the colonies of E. jeanselmei may be yeastlike and black, gradually spreading, raised or dome shaped, and 20 to 25 mm in diameter. After 2 weeks on SDA, the colonies are covered with short aerial hyphae. At this stage, the colony is mousy gray to olive-gray with an olive-black reverse. The septate mycelium is sometimes toruloid, branched, and pale brown. The conidiophores are partially differentiated from the vegetative hyphae, solitary, branched or unbranched, smooth walled, and pale brown. The conidiogenous cells produce conidia from one or more areas of their cells. The tips of the conidiogenous cells are closely annellated, cylindrical, obclavate, smooth walled, pale brown to black, and elongate as a result of successive conidial formation. The conidia, which aggregate in masses at the tips of conidiophores, tend to slide down the conidiophore or along hyphae. The smooth conidia are exogenous, nonseptate, subglobose, and ellipsoidal to cylindrical, measuring 1.5 to 2.8 μm.

L. senegalensis Baylet, Camain et Segretain, 1959, and L. tompkinsii El Ani, 1966

L. senegalensis and L. tompkinsii are known to cause mycetomata in the northern tropical zone of West Africa, especially in Senegal and Mauritania, and in India. The granules of the two species are indistinguishable from each other. They are black, 0.5 to 2 mm in size, and firm to hard. In tissue sections, the granules are round to polylobulated, with large vesicles. At the periphery, the mycelium is embedded in a black, cementlike substance. The central portion of each granule consists of a loose network of hyphae.

In culture, L. senegalensis and L. tompkinsii grow rapidly and produce gray-brown colonies. On cornmeal agar, both species produce perithecia that are nonstiolate, scattered, immersed or superficial, globose to subglobose, black, and covered with brown, flexuous hyphae. The asci are numerous, eight spored, clavate to cylindrical, and double walled. The major differences between the two species are found in the ascospores, which differ in size, shape, and septation as well as in the nature of the gelatinous sheath that surrounds them (16, 17).

M. grisea Mackinnon, Ferrada and Montemayer, 1949

M. grisea occurs as an etiologic agent of black grain mycetoma in South America, India, Africa, and North and Central America (26, 28). In the United States, five cases caused by M. grisea have been described (11, 37).

The granules are black, 0.3 to 0.6 mm in diameter, and soft to firm. In tissue sections, the granules are oval, lobulated or kidney shaped (reniform), and sometimes vermiform. They are composed of a dense network of hyphae weakly pigmented in the center and brown to blackish brown in the peripheral region as the result of the presence of a brown, cementlike interstitial material.

In culture, M. grisea forms slow-growing velvety colonies that are cerebriform, radially furrowed or smooth, and dark gray to olive-brown to black. The reverses of the colonies are black. In old cultures, a red-brown diffusible pigment is produced by many isolates. Microscopically, the hyphae are septate, dematiaceous, 1 to 3 μm in diameter, and nonsporulating. Chlamydospores rarely are observed. Large moniliform hyphae 3 to 5 μm in diameter often are seen. Some isolates of M. grisea have been described as producing abortive or fertile pycnidia (31, 48). Such isolates are indistinguishable from Pyrenochaeta mackinnonii (7, 44).

M. mycetomatis (Laveran) Brumpt, 1905

The granules produced by M. mycetomatis are reddish brown to black. They may reach 5 mm or more in diameter and are firm to hard. In tissue sections, the granules are compact, variable in size and shape, and frequently multilobulated. They are composed of hyphae 1 to 5 μm in diameter that terminate in enlarged hyphal cells at the periphery of the granules which measure 12 to 15 μm in diameter. The cell wall pigment is minimal, but hyphal cells contain brown particles. The hyphae are embedded in a conspicuous brown matrix that is characteristic of M. mycetomatis. Some granules are vesicular and more regular in size and shape. The vesicles are predominantly visible in the peripheral zone in a dense, brown, cementlike matrix (Fig. 3).

In culture, M. mycetomatis shows wide variation. Colonies are slow growing and white at first, becoming olivaceous, yellow, or brown; flat or dome shaped; and velvety to glabrous, with a characteristic brown diffusible pigment. On nutritionally deficient media, sclerotia 750 μm in diameter develop. These are black and made of undifferentiated polygonal cells. On SDA, the mycelium is sterile. On nutritionally poor media, such as soil extract or hay infusion agar, about 50% of the isolates produce round to pyriform conidia 3 to 4 μm in diameter from the tips of phialides. The phialides are long and tapering, often with a collarette. They range from 3 to 15 μm in length. Occasionally, two or three phialides may arise from a lateral branch (Fig. 4).

M. mycetomatis grows better at 37 than at 30°C, whereas

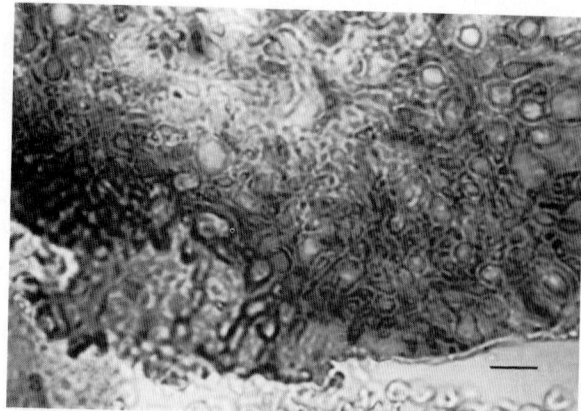

FIGURE 3 Black vesicular granule of M. *mycetomatis*, showing conspicuous cementlike matrix. Hematoxylin and eosin stain. Bar = 10 μm.

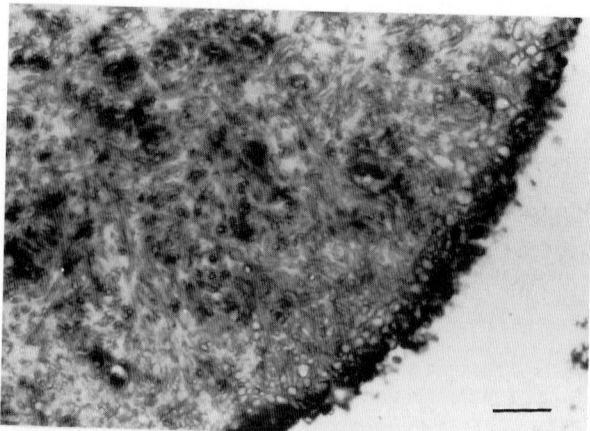

FIGURE 5 White, soft granule of P. *boydii*, showing central area of the granule consisting of loosely interwoven hyaline, septate hyphae and lack of cementing material at the periphery. Bar = 10 μm.

M. *grisea* grows better at 30 than at 37°C. The ability of M. *mycetomatis* isolates to grow at temperatures as high as 40°C serves as a useful diagnostic criterion for distinguishing them from M. *grisea*. M. *mycetomatis* is slowly proteolytic and utilizes glucose, maltose, and galactose but not sucrose. It utilizes potassium nitrate, ammonium sulfate, asparagine, and urea, and it hydrolyzes starch. M. *grisea*, on the other hand, is weakly proteolytic and assimilates glucose, maltose, and sucrose but not lactose (Table 3).

N. rosatii Segretain and Destombes, 1961

The granules of N. *rosatii* are white to brownish white, 0.5 to 1.0 mm in diameter, and soft. In tissue sections, granules appear to be polyhedral to subregular and consist of hyphae which are embedded in the peripheral cementing material. The granules manifest an eosinophilic border. The central portion of each granule consists of more or less disintegrated mycelium and chlamydospores. Mycetomata caused by N. *rosatii* have been described in Australia, Cameroon, Guinea, Senegal, and Somalia (29, 47).

In culture, colonies of N. *rosatii* are slow growing, attaining diameters of 25 to 28 mm in 2 weeks, and have aerial mycelia that are grayish black to brownish black. On potato-carrot or cornmeal agar incubated at 30°C, most of the cleistothecia are submerged. The cleistothecial walls are smooth and are surrounded by interwoven brown to hyaline hyphae. The eight-spored asci, 12 to 35 by 10 to 25 μm, are

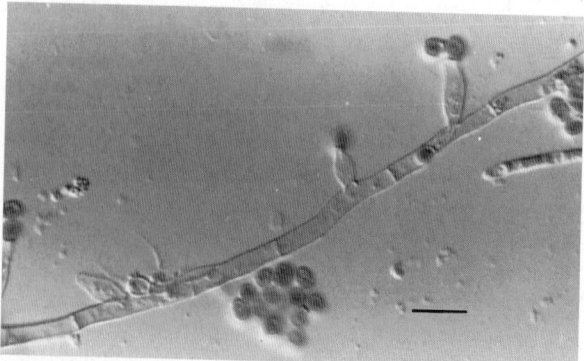

FIGURE 4 M. *mycetomatis*. Slide culture on soil extract agar, showing phialides and conidia. Bar = 10 μm.

scattered in the central part of the cleistothecium and are globose to subglobose, thick walled, and bitunicate, becoming evanescent as ascospores mature. Ascospores vary in size (9 to 12.5 by 4.5 to 8.0 μm) and shape, ranging from ellipsoidal to bicampanulate, asymmetrical, or slightly curved; are constricted at the median transverse septum; and have brown, smooth walls.

P. boydii (Shear) McGinnis, Padhye and Ajello, 1982 (Anamorph: Scedosporium apiospermum Sacc. ex Castellani and Chalmers, 1919)

The granules produced by P. *boydii* are white to yellowish white and soft to firm; they vary from spherical to subspherical to lobulated and are 0.2 to 2.0 mm in diameter. In tissue sections, the granules are composed of hyaline hyphae 1.5 to 5.0 mm in diameter that radiate from the center into terminal thick-walled cells 15 to 20 μm in diameter at the peripheries of the granules. The granules of P. *boydii* lack a cementlike matrix. The central portion of each granule consists of loosely interwoven hyaline mycelium (Fig. 5).

Colonies grow rapidly and are floccose and white at first, becoming gray as conidia are produced. With age, the colonies become dark brown. The anamorphic state, S. *apiospermum*, produces conidia that are egg shaped to clavate, truncate at the tapered end, and subhyaline, becoming pale gray to pale brown in mass. Conidia are produced singly. They remain attached at the tips of conidiogenous cells or multiply from annellides (Fig. 6). Annellations can be detected at the tips of the conidiogenous cells as swollen rings. Occasionally, some isolates also produce a *Graphium* synanamorph, which is characterized by ropelike bundles of hyphae with annellidic conidiogenesis. The hyphae are fused into long stalks known as synnemata. The conidia produced are hyaline, cylindric to clavate, and truncate at the base (Fig. 7).

The ascocarps of the teleomorphic state may be produced when isolates are grown on cornmeal agar. The cleistothecia develop submerged in the agar and appear as black dots. They are spherical, nonostiolate, 140 to 200 μm in diameter, and often covered with brown, thick-walled, septate hyphae 2 to 3 μm wide. They have a wall 4 to 6 μm thick composed of two to three layers of interwoven, flat-

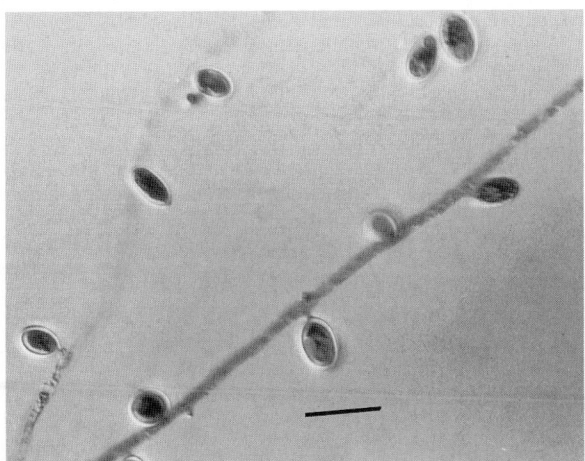

FIGURE 6 *S. apiospermum* anamorph of *P. boydii*. Slide culture on cornmeal agar showing conidia. Bar = 10 μm.

tened, dark brown cells, each 2 to 6 μm wide. The cleistothecia open at maturity by an irregular rupture of the wall. The eight-spored asci are ellipsoidal to nearly spherical. They measure 12 to 18 by 9 to 13 μm. The ascospores are ellipsoidal to oblate (football shaped), symmetrical or slightly flattened, and straw colored. They have two germ pores and measure 6 to 7 by 3.5 to 4 μm.

Even though *P. boydii* is homothallic, many clinical and soil isolates do not form ascocarps. Their identification, then, is based on the morphology of the conidial state alone. *P. boydii* is the commonest agent of mycetoma in humans as well as lower animals in the United States. It has been isolated frequently from manure, soil, and water. It is often encountered as a contaminant of respiratory specimens from patients with chronic lung disease. It is also isolated frequently from open, dirty wounds. Its isolation from such specimens, however, does not always implicate *P. boydii* as an etiologic agent. On the other hand, *P. boydii* is known to cause necrotizing pneumonia and disseminated infections in immunocompromised patients. The signifi-

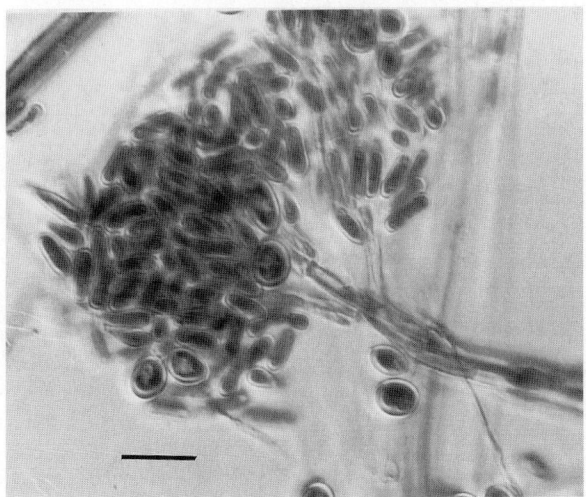

FIGURE 7 *Graphium* anamorph of *P. boydii*. Slide culture showing synnemata and cylindrical to clavate conidia. Bar = 10 μm.

cance of its isolation always should be determined after communication with the attending physician and pathologist.

Pyrenochaeta romeroi Borelli, 1959

Mycetomata caused by *Pyrenochaeta romeroi* have been reported from Somalia, India, and South America (22, 29). *Pyrenochaeta romeroi* produces soft to firm black granules that are oval, lobulated, and sometimes vermiform. They resemble those of *M. grisea* (48).

In culture, colonies of *Pyrenochaeta romeroi* are fast growing and floccose to velvety, with a gray surface and a whitish margin. The reverse of the colony is black, with no diffusible pigment. The hyphae are hyaline to brown, septate, and branched. On nutritionally poor media, *Pyrenochaeta romeroi* produces ostiolate, isolated or aggregated pycnidia 50 to 160 by 40 to 100 μm bearing subhyaline, elliptical pycnidiospores 0.8 to 1.0 by 1.5 to 1.0 μm.

The close resemblance of the granules produced by *Pyrenochaeta romeroi* to those produced by *M. grisea* and the morphologic similarity of the pycnidia formed by some isolates of *M. grisea* and by *Pyrenochaeta romeroi* suggest that these two species may be closely related. However, according to Murray and Buckley (38), the two species show serologic differences that allow differentiation between them. Recently, Romero and Mackenzie (44) studied culture filtrate and cellular antigens prepared from 14 agents of black grain mycetoma by double diffusion and immunoelectrophoresis. They found up to 90% cross-reactivity among *M. grisea* isolates and between *M. grisea* and *Pyrenochaeta mackinnonii* but not between *M. grisea* and *Pyrenochaeta romeroi*. Their results suggested that the isolates of *M. grisea* were antigenically similar or identical to those of *Pyrenochaeta mackinnonii*.

REFERENCES

1. **Ajello, L.** 1962. Epidemiology of human fungus infections, p. 69–83. *In* G. Dalldorf (ed.), *Fungi and Fungous Diseases.* Charles C Thomas, Publisher, Springfield, Ill.
2. **Ajello, L.** 1978. Animal mycetomas—review, p. 270–275. *In Proceedings, Primer Simposio Inetrnacional de Micetomas.* Universidad Centro Occidental, Barquisimeto, Venezuela.
3. **Ajello, L., W. Kaplan, and F. W. Chandler.** 1980. Dermatophyte mycetomas: fact or fiction?, p. 135–140. *In Superficial, Cutaneous and Subcutaneous Infections.* Scientific publication no. 396. Pan American Health Organization, Washington, D.C.
4. **Baylet, R., R. Camain, and M. Rey.** 1961. Champignons de mycetomes isolés des épineux au Senegal. *Bull. Soc. Med. Afr. Noire Lang. Fr.* **6:**317–319.
5. **Borelli, D.** 1959. *Pyrenochaeta romeroi* n. sp. *Rev. Dermatol. Venez.* **1:**325–327.
6. **Borelli, D.** 1962. *Madurella mycetomi* y *Madurella grisea.* *Arch. Venez. Med. Trop. Parasitol. Med.* **4:**195–211.
7. **Borelli, D.** 1976. *Pyrenochaeta mackinnonii* nova species agente de micetoma. *Castellania* **4:**227–234.
8. **Borelli, D.** 1978. *Plenodomus avramii* nova species agente de micetoma, p. 116–126. *In Proceedings, Primer Simposio Internacional de Micetomas.* Universidad Centro Occidental, Barquisimeto, Venezuela.
9. **Borelli, D., R. Zamora, and G. Senabre.** 1976. *Chaetosphaeronema larense* nova specie agente de micetoma. *Gac. Med. Caracas* **84:**307–318.
10. **Brodey, R. S., H. S. Schryver, M. J. Deubler, W. Kaplan, and L. Ajello.** 1967. Mycetoma in a dog. *J. Am. Vet. Med. Assoc.* **151:**442–451.
11. **Butz, W. C., and L. Ajello.** 1971. Black grain mycetoma. *Arch. Dermatol.* **104:**197–201.

12. **Campbell, C. K.** 1987. *Polycytella hominis* gen. et sp. nov., a cause of human pale grain mycetoma. *J. Med. Vet. Mycol.* **25**:301–305.

13. **Chandler, F. W., W. Kaplan, and L. Ajello.** 1980. *A Colour Atlas and Textbook of the Histopathology of Mycotic Diseases.* Wolfe Medical Publications Ltd., London.

14. **Destombes, P., A. Poirier, and O. Nazimoff.** 1970. Mycoses profundes reconnues en 9 ans de pratique histopathologie à l'Institut Pasteur du Cameroun. *Bull. Soc. Pathol. Exot.* **63**:310–315.

15. **De Vries, G. A., G. S. De Hoog, and H. P. De Bruyn.** 1984. *Phialophora cyanescens* sp. nov., with *Phaeosclera*-like synanamorph, causing white-grain mycetoma in man. *Antonie van Leeuwenhoek J. Microbiol. Serol.* **50**:149–153.

16. **El-Ani, A. S.** 1966. A new species of *Leptosphaeria*, an etiologic agent of mycetoma. *Mycologia* **58**:406–411.

17. **El-Ani, A. S., and M. A. Gordon.** 1965. The ascospore sheath and taxonomy of *Leptosphaeria senegalensis. Mycologia* **57**:275–278.

18. **Emmons, C. W.** 1962. Soil reservoirs of pathogenic fungi. *J. Wash. Acad. Sci.* **52**:3–9.

19. **Etta, L. L., V. L. R. Peterson, and D. Gerding.** 1983. *Acremonium falciforme* (*Cephalosporium falciforme*) mycetoma in a renal transplant patient. *Arch. Dermatol.* **119**:707–708.

20. **Fahey, P. J., M. J. Utell, and R. W. Hyde.** 1981. Spontaneous lysis of mycetomas after acute cavitating lung disease. *Am. J. Respir. Dis.* **123**:336–339.

21. **Halde, C., A. A. Padhye, L. D. Haley, M. G. Rinaldi, D. Kay, and R. Leeper.** 1976. *Acremonium falciforme* as a cause of mycetoma in California. *Sabouraudia* **14**:319–326.

22. **Klokke, A. H., G. Swamidasan, R. Anguli, and A. Verghese.** 1968. The causal agents of mycetoma in South India. *Trans. R. Soc. Trop. Med. Hyg.* **62**:509–516.

23. **Lambrechts, N., M. G. Collett, and M. Henton.** 1991. Black grain eumycetoma (*Madurella mycetomatis*) in the abdominal cavity of a dog. *J. Med. Vet. Mycol.* **29**:211–214.

24. **Mackinnon, J. E.** 1954. A contribution to the study of the causal organisms of maduromycosis. *Trans. R. Soc. Trop. Med. Hyg.* **48**:470–480.

25. **Mackinnon, J. E., I. A. Conti-Diaz, E. Gezuele, and E. Civila.** 1971. Datos sobre ecologia de *Allescheria boydii*, Shear. *Rev. Urug. Pathol. Clin. Microbiol.* **9**:37–43.

26. **Mackinnon, J. E., L. V. Ferrada, and L. Montemayor.** 1949. *Madurella grisea* n. sp., a new species of fungus producing black variety of maduromycosis in South America. *Mycopathol. Mycol. Appl.* **4**:385–392.

27. **Mahgoub, E. S.** 1973. Mycetomas caused by *Curvularia lunata, Madurella grisea, Aspergillus nidulans,* and *Nocardia brasiliensis* in Sudan. *Sabouraudia* **11**:179–182.

28. **Mahgoub, E. S., and I. Murray.** 1973. *Mycetoma.* William Heinemann Medical Books Ltd., London.

29. **Mariat, F., P. Destombes, and G. Segretain.** 1977. The mycetomas: clinical features, pathology, etiology and epidemiology. *Contrib. Microbiol. Immunol.* **4**:1–39.

30. **Matsumoto, T., and L. Ajello.** 1986. No granules, no mycetomas. *Chest* **90**:151–152.

31. **Mayorga, R., and J. E. Close De Leon.** 1966. Sur une souche de *Madurella grisea* sporiferé isolée d'un mycetome Guatemalteque à grains noirs. *Sabouraudia* **4**:210–214.

32. **McGinnis, M. R.** 1992. Black fungi: a model for understanding tropical mycosis, p. 129–149. *In* D. H. Walker (ed.), *Global Infectious Diseases, Prevention, Control, and Eradication.* Springer Verlag, New York.

33. **McGinnis, M. R., and R. C. Fader.** 1988. Mycetoma: a contemporary concept. *Infect. Dis. Clin. N. Am.* **2**:939–954.

34. **McGinnis, M. R., and A. A. Padhye.** 1977. *Exophiala jeanselmei*, a new combination for *Phialophora jeanselmei. Mycotaxon* **5**:341–352.

35. **McGinnis, M. R., A. A. Padhye, and L. Ajello.** 1982. *Pseudallescheria boydii* Negroni et Fischer, 1943, and its later synonym *Petriellidium* Malloch, 1970. *Mycotaxon* **14**:94–102.

36. **Milburn, P. B., D. M. Papayanopulos, and B. M. Pomerantz.** 1988. Mycetoma due to *Acremonium falciforme. Int. J. Dermatol.* **27**:408–410.

37. **Montes, L. F., R. G. Freeman, and W. McClarin.** 1969. Maduromycosis due to *Madurella grisea. Arch. Dermatol.* **99**:74–79.

38. **Murray, I., and H. R. Buckley.** 1969. Serological differences between *Pyrenochaeta romeroi* and *Madurella grisea. Sabouraudia* **7**:62–63.

39. **Negroni, R.** 1970. Estudio micologico del primer caso de micetoma por *Phialophora jeanselmei* observado en la Argentina. *Med. Cutan.* **5**:625–630.

40. **Neilson, H. S., N. F. Conant, T. Weinberg, and J. F. Reback.** 1968. Report of a mycetoma due to *Phialophora jeanselmei* and undescribed characteristics of the fungus. *Sabouraudia* **6**:330–333.

41. **Pankovich, A. M., B. J. Auerbach, W. I. Metzer, and T. Barreta.** 1981. Development of maduromycosis (*Madurella mycetomi*) after nailing of a closed tibial fracture. A case report. *Clin. Orthoped. Relat. Res.* **154**:220–222.

42. **Punithalingam, E.** 1979. Sphaeropsidales in culture from humans. *Nova Hedwigia* **37**:119–158.

43. **Pupaibul, K., W. Sindhuphak, and A. Chindaporn.** 1982. Mycetoma of the hand caused by *Phialophora jeanselmei. Mykosen* **25**:321–330.

44. **Romero, H., and D. W. R. Mackenzie.** 1989. Studies on antigens from agents causing black grain eumycetoma. *J. Med. Vet. Mycol.* **27**:303–311.

45. **Segretain, G.** 1964. Recherches sur l'ecologie de *Madurella mycetomi* au Senegal. *Bull. Soc. Fr. Mycol. Med.* **3**:121–124.

46. **Segretain, G.** 1972. Epidemiologie des mycetomes. *Ann. Soc. Belge Med. Trop.* **52**:277–286.

47. **Segretain, G., and P. Destombes.** 1961. Description d'un nouvel agent de maduromycose *Neotestudina rosatii,* n. gen., n. sp. isolé en Afrique. *C. R. Acad. Sci.* **253**:2577–2579.

48. **Segretain, G., and P. Destombes.** 1969. Recherches sur mycetomes à *Madurella grisea* et *Pyrenochaeta romeroi. Sabouraudia* **7**:51–61.

49. **Segretain, G., and F. Mariat.** 1968. Recherches sur la presence d'agents de mycetomes dans le sol et sur les epineux du Senegal et de la Mauritanie. *Bull. Soc. Pathol. Exot.* **61**:194–202.

50. **Talwar, P., and S. C. Sehgal.** 1979. Mycetomas of North India. *Sabouraudia* **17**:287–291.

51. **Thammayya, A., and M. Sanyal.** 1980. *Exophiala jeanselmei* causing mycetoma pedis in India. *Sabouraudia* **18**:91–95.

52. **Thianprasit, M., and A. Sivayathorn.** 1984. Black dot mycetoma. *Mykosen* **27**:219–226.

53. **Thirumalachar, M. J., and A. A. Padhye.** 1968. Isolation of *Madurella mycetomi* from soil in India. *Hind. Antibiot. Bull.* **10**:314–318.

54. **Weathered, D. B., M. A. Markey, R. J. Hay, E. S. Mahgoub, and S. A. Gumaa.** 1986. Ultrastructural and immunologic changes in the formation of mycetoma grains. *J. Med. Vet. Mycol.* **25**:39–46.

55. **Zoutman, D. E., and L. Sigler.** 1991. Mycetoma of the foot caused by *Cylindrocarpon destructans. J. Clin. Microbiol.* **29**:1855–1859.

VIROLOGY

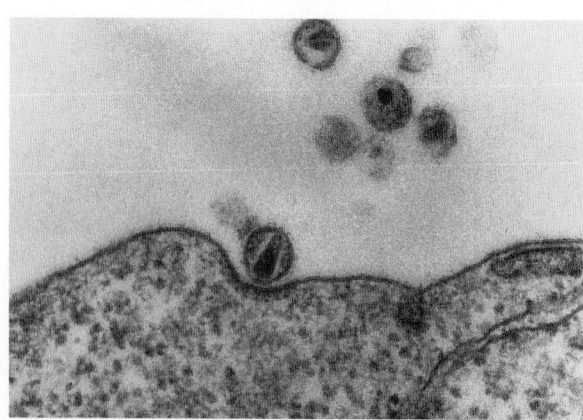

HIV near cell membrane (*courtesy of Visuals Unlimited/© H. Gelderblom*).

Taxonomy of Viruses

JOSEPH L. MELNICK

69

Viruses are separated into families on the basis of the type and form of the nucleic acid genome and the size, shape, substructure, and mode of replication of the virus particle. Within each family, classifications of genera and species depend on antigenicity in addition to other properties, particularly the base sequence homology.

Significant developments in classification and nomenclature of viruses are documented in the reports of the International Committee on Taxonomy of Viruses (ICTV), formerly the International Committee on Nomenclature of Viruses. These reports, published in 1971 (9), 1976 (3), 1979 (6), and 1982 (7), have dealt with viruses of humans, lower animals, insects, plants, bacteria, and fungi and have included summaries of the properties of those groups of viruses as related to their taxonomic placements. Recent decisions of the ICTV were summarized in an interim report published in 1989 (2) and a more definitive report published in 1991 (5).

It seems probable that most of the major groups of viruses, particularly those infecting humans and the vertebrate animals of direct importance to humans, have been recognized. Many of the viruses have now been officially placed in families, genera, and species; within some families, subfamilies or subgenera or both have also been established. Although the focus of this chapter is on these families, progress also has been made with respect to the taxonomy of the viruses of other host groups. For a more detailed discussion of virus taxonomy, references 1 through 9 are recommended to the reader.

The ICTV has approved 73 families and groups of viruses. The kind and strandedness of the viral nucleic acid and the presence or absence of a lipoprotein envelope are the three key properties upon which the 73 families and groups are classified. They have been assigned to seven clusters, six of which infect vertebrates as shown in Fig. 1. Within these clusters, families are further classified on the basis of virion morphology and strategy of genomic replication.

In Table 1, properties of the major families of RNA-containing viruses of humans and other vertebrate animals are summarized; in Table 2, vertebrate viruses with a DNA genome are similarly treated. In Table 3, all of these families are listed along with the genera and individual viruses that may be of special concern for the viral diagnostic laboratory. Subfamilies are also included if they represent agents having direct or indirect bearing on human diseases. Thus, the family *Herpesviridae* includes three subfamilies. Members recently added to this family include human herpesviruses 6 and 7, which have not yet been placed in a genus. Since most of these viral agents are dealt with more fully in the other chapters of this section, the text that follows is confined to explanation of and commentary on some examples from the tables, together with some notes about recent developments.

PICORNAVIRIDAE

Picornaviridae, the smallest of the vertebrate viruses with RNA genomes, have been classed in five genera and several hundred species.

More than 70 members of the genus *Enterovirus* are known to infect humans; these include polioviruses, coxsackieviruses of the A and B groups, echoviruses, and enterovirus serotypes that have been assigned sequential numbers rather than being placed in the echovirus or coxsackievirus subgroups (since the distinctions among these groups have been found to be less sharp than was recognized when these subdivisions were initially established).

More than 100 viruses infecting humans belong to the genus *Rhinovirus*. A third genus in the family is *Cardiovirus*, typified by encephalomyocarditis virus of mice, which may also (rarely) infect humans. A fourth genus, *Aphthovirus*, includes the economically important foot-and-mouth disease viruses of cattle.

After decades of investigation, hepatitis A virus has been classified as a picornavirus. Although it resembles enteroviruses in many respects, it differs in amino acid sequences and also in its greater resistance to thermal inactivation. It has been placed in a separate genus, *Hepatovirus* or *Heparnavirus*.

The picornavirus genome is one piece of linear, single-stranded positive-sense RNA of low molecular weight (about 2.5×10^6). The RNA is infectious and serves as its own messenger for protein translation. The virion is about 27 nm in diameter; it contains 60 copies of each of the four major polypeptides (molecular weights of 33,000, 27,000, 23,000, and 60,000). The enteroviruses and cardioviruses are acid stable and have a buoyant density in CsCl of about 1.34 g/cm^3; in contrast, the rhinoviruses and aphthoviruses

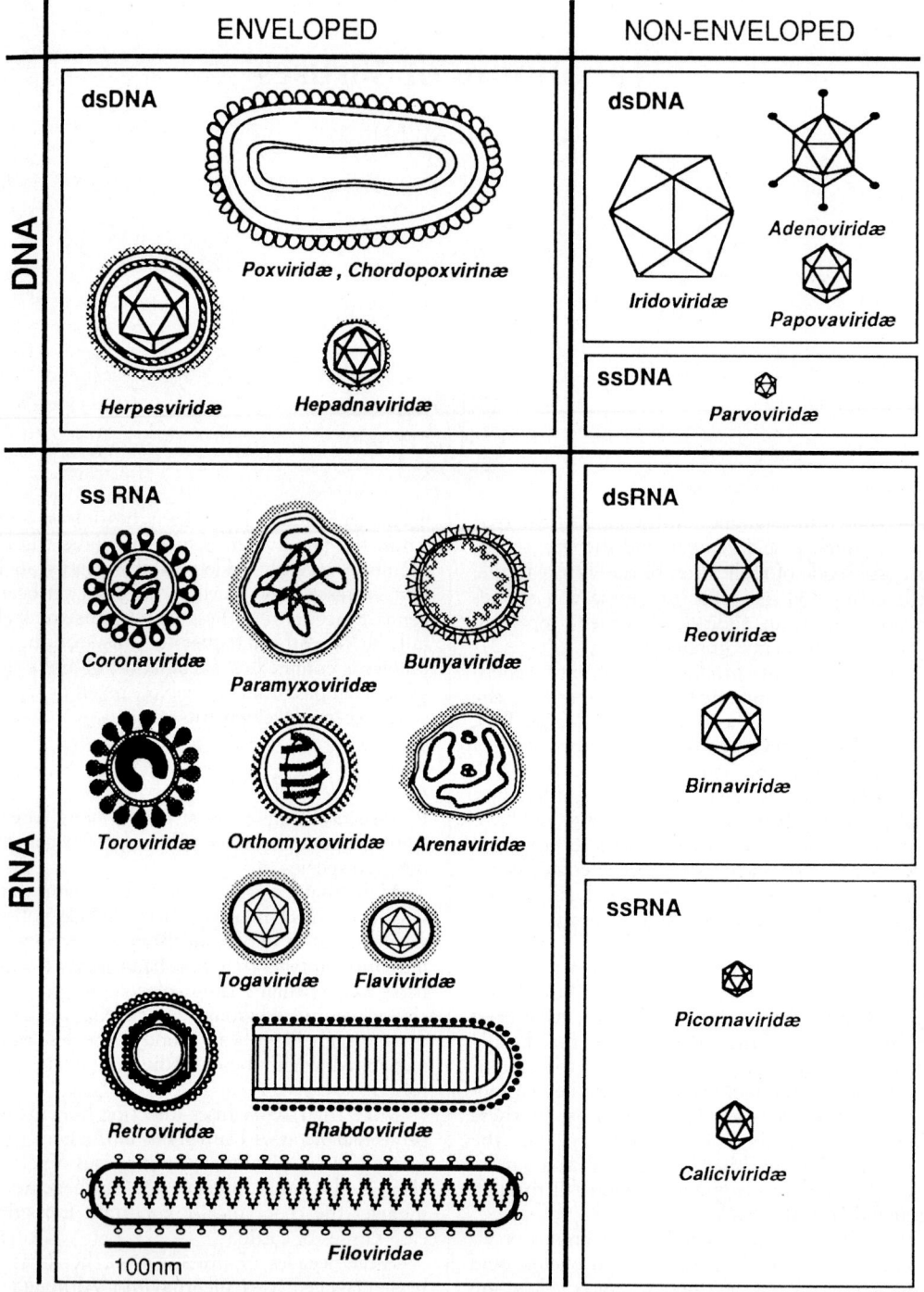

FIGURE 1 ICTV drawings of viruses in the common virus families of vertebrates. All the diagrams have been drawn similarly: vertical lines separate enveloped and nonenveloped viruses, and horizontal lines separate DNA and RNA viruses. Within each of the resulting four sections, single-stranded (ss) and double-stranded (ds) genomes are indicated. As stated by the ICTV (5), "All the diagrams have been drawn approximately to the same scale to provide an indication of the relative sizes of the viruses; but this cannot be taken as definitive for the following reasons. (*i*) Different viruses within a family or group may vary somewhat in size and shape. In general, the size and shape were taken from the type member of the taxon. (*ii*) Dimensions of some viruses are difficult to determine or only approximately known. (*iii*) Some viruses, particularly the larger enveloped ones, are pleomorphic. Only the outlines of most of the smallest viruses are given, with an indication of the icosahedral structure whenever appropriate. The large viruses are given schematically in surface outline, in section, or both, as seems most appropriate to display major morphological characteristics."

are acid labile and have higher buoyant densities of about 1.4 g/cm^3. The base sequences of representative viral genomes have been determined, and the three-dimensional structure of the virion has been established.

The diseases caused by picornaviruses range from paralytic poliomyelitis to aseptic meningitis, hepatitis, pleurodynia, myocarditis, skin rashes, and common colds; inapparent infection is very common. Different viruses may produce the same syndrome; on the other hand, a single picornavirus may cause several different syndromes. Strain differentiation among the polioviruses is of direct concern in public health and medical virology. In the years since 1951 (when the existence of three poliovirus serotypes became known), genomic differences within serotypes have been demonstrated, but the old established strains used in the vaccines continue to confer excellent immunity against current wild strains.

REOVIRIDAE

The RNA genome is single stranded in all of the virus families listed in Table 1 except the family *Reoviridae*, whose RNA is double stranded. The genus *Reovirus* differs somewhat from the other genera in the possession of an outer protein shell and the larger molecular weight of its genome (15 × 10^6 versus 12 × 10^6). Three serotypes within the genus *Reovirus* infect humans, monkeys, dogs, and cattle; in addition, at least five avian reoviruses are known. In the genus *Orbivirus*, important human pathogens include Colorado tick fever virus and the Kemerovo viruses; other important orbiviruses are the bluetongue viruses of sheep and the viruses of African horse sickness. The most recently established genus is *Rotavirus*, which includes several viruses that infect humans as well as viruses of many other mammalian species, typified by simian virus SA-11 and Nebraska calf diarrhea virus. The human rotaviruses are increasingly being recognized as the cause of a large share of the serious episodes of nonbacterial infantile diarrhea. Rotavirus gastroenteritis is one of the most common childhood illnesses throughout the world and is a leading cause of infant deaths in developing countries. These viruses also infect adults, particularly those in close contact with infants and children, but infected adults usually experience no symptoms or may have only minor illness.

CALICIVIRIDAE

Other recent additions to the taxonomic roll of RNA-containing viruses are the members of the families *Caliciviridae* and *Bunyaviridae*. *Caliciviridae* includes a number of viruses of pigs, cats, and sea lions and may include agents that infect humans. Caliciviruslike particles have been observed in human feces in association with gastroenteritis. Norwalk virus, a widespread human agent causing acute epidemic gastroenteritis, has a virion protein structure similar to that of the caliciviruses; it also resembles caliciviruses in several other characteristics. These caliciviruslike viruses have not yet been successfully adapted to tissue culture, but the Norwalk virus has been cloned and the capsid protein has been expressed in a baculovirus system to yield an abundance of nucleic acid-free shells that are similar in size and shape to those of Norwalk virus. These empty-virus-like particles are proving useful in diagnosis and in epidemiologic studies. Other morphologically similar viruses associated with gastroenteritis and named according to the location of the outbreak include Hawaii, Montgomery, and Snow Mountain viruses, which seem to be separate serotypes of the caliciviruslike agents. The genomic organization of the Norwalk virus is similar to that of feline and rabbit caliciviruses, suggesting that this important human pathogen is also a member of this virus family.

BUNYAVIRIDAE

Members of the *Bunyaviridae* form a family of more than 220 viruses, at least 145 of them belonging to the Bunyamwera supergroup of serologically interrelated arboviruses. With the taxonomic placement of this large group, the vast majority of the viruses of the classic arbovirus groupings, initially based on ecological properties and subdivided by serologic interrelationships, have been assigned to families on the basis of biophysical and biochemical characteristics. Human illness caused by Hantaan virus has been recognized in the Far East as Korean hemorrhagic fever, and a variant is known in Scandinavian and eastern European countries as epidemic nephropathy. The illness has been variously named "hemorrhagic fever with renal syndrome" or "muroid virus nephropathy." These agents are now established as a genus, *Hantavirus*. The virus has a labile membrane and a tripartite single-stranded RNA genome. The most common natural hosts are mice (in Korea) and voles (in Europe). There have been several instances of infection of staff members handling laboratory rats infected with the virus, both in the Far East and more recently in Europe. Cases of naturally occurring infection were reported from the southwestern United States in 1993.

RETROVIRIDAE

The family *Retroviridae* includes seven genera, not only the RNA tumor viruses (oncornaviruses and leukoviruses) but also the slowly growing lentiviruses of the maedi-visna group and the foamy virus group of agents that form syncytia in cell cultures. The human immunodeficiency viruses (HIV) associated with AIDS are lentiviruses.

Retroviruses characteristically have a reverse transcriptase (RNA-dependent DNA polymerase) within the virion. The genome is an inverted dimer of linear, single-stranded positive RNA that dissociates readily into two or three pieces. Replication of the viral RNA involves a DNA provirus that is integrated into host cellular DNA.

Endogenous members may be part of the germ line of vertebrate hosts, being inherited as Mendelian genes. The oncovirus genes may not be expressed, but they can be activated by physical and chemical agents, by superinfection with other oncoviruses, and even by herpesviruses. It was through studies of oncoviruses that the cellular oncogenes were recognized.

With some exceptions, oncoviruses fall into host-species-specific groups of agents, including either leukemias or sarcomas, that is, leukemia-sarcoma complexes of avian, murine, feline, and hamster oncoviruses. Other groups are murine mammary tumor virus and primate oncoviruses.

The oncoviruses have been divided into types according to morphologic, antigenic, and enzymatic differences. The type B oncovirus group includes the mouse mammary tumor virus and probably similar viruses from the guinea pig and

TABLE 1 Current classification of RNA-containing viruses of vertebrates

Family	Nucleic acid core	Capsid symmetry	Virion: naked or enveloped	Site of capsid assembly	Site of nucleocapsid envelopment	Reaction to ether treatment	No. of capsomeres	Ribonucleoprotein helix diam (nm)	Virion diam (nm)	Mol wt of nucleic acid in virion (10^6)
Retroviridae	RNA	Icosahedral	Enveloped	Cytoplasm	Surface membrane	Sensitive			About 100	6–7
Bunyaviridae	RNA	Helical	Enveloped	Cytoplasm	Intracytoplasmic membranes	Sensitive		10–12	80–110	6–7
Coronaviridae	RNA	Helical	Enveloped	Cytoplasm	Intracytoplasmic membranes	Sensitive		11–13	80–130	5–6
Rhabdoviridae	RNA	Helical	Enveloped	Cytoplasm	Surface membrane	Sensitive		18	60 × 180	3.5–4.6
Arenaviridae	RNA	Helical	Enveloped	Cytoplasm	Surface membrane	Sensitive			50–300	3–5
Paramyxoviridae	RNA	Helical	Enveloped	Cytoplasm	Surface membrane	Sensitive		18	150–300	5–8
Orthomyxoviridae	RNA	Helical	Enveloped	Cytoplasm	Surface membrane	Sensitive		9–15	80–120	4–5
Flaviviridae	RNA	Icosahedral	Enveloped	Cytoplasm	Intracytoplasmic membranes	Sensitive		?	40–50	4
Togaviridae	RNA	Icosahedral	Enveloped	Cytoplasm	Surface membrane	Sensitive		?	70	3–4
Reoviridae	RNA	Icosahedral	Naked	Cytoplasm		Resistant	32		60–80	12–15
Caliciviridae	RNA	Icosahedral	Naked	Cytoplasm		Resistant	32		35–39	2.6
Picornaviridae	RNA	Icosahedral	Naked	Cytoplasm		Resistant	32		24–30	2.3–2.8

TABLE 2 Current classification of DNA-containing viruses of vertebrates

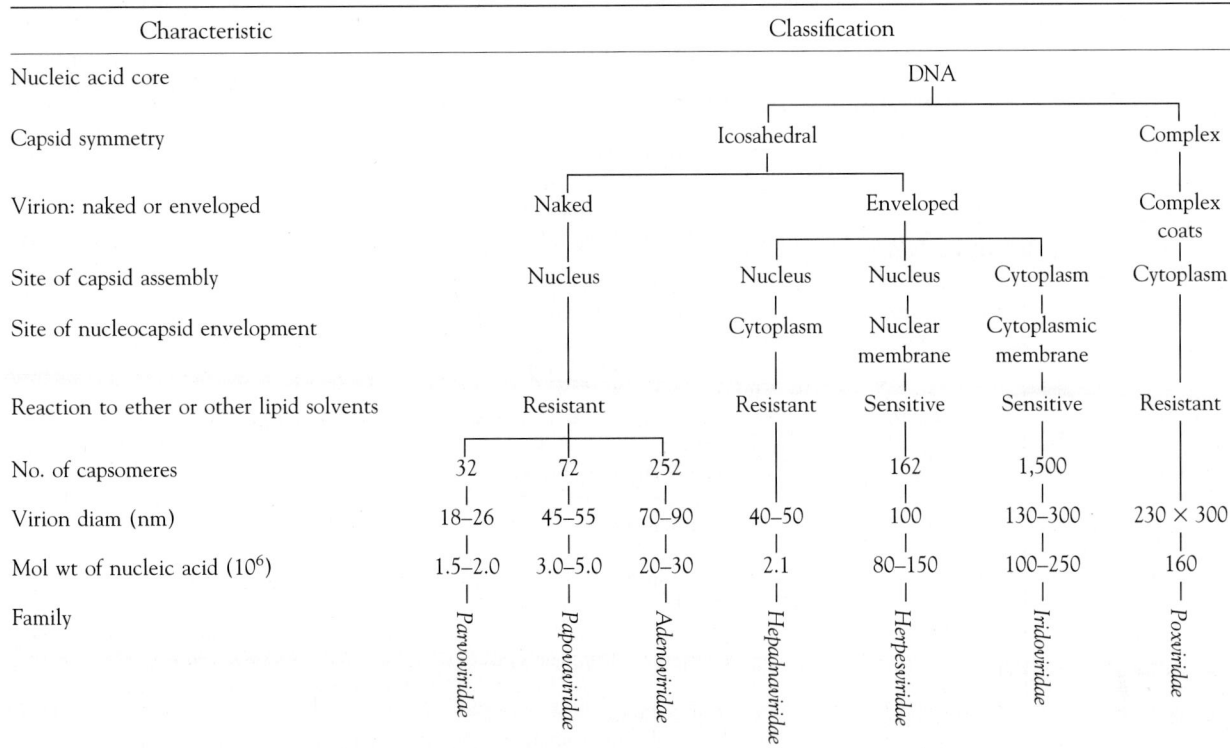

Characteristic	Classification						
Nucleic acid core	DNA						
Capsid symmetry		Icosahedral					Complex
Virion: naked or enveloped		Naked		Enveloped			Complex coats
Site of capsid assembly		Nucleus		Nucleus	Nucleus	Cytoplasm	Cytoplasm
Site of nucleocapsid envelopment				Cytoplasm	Nuclear membrane	Cytoplasmic membrane	
Reaction to ether or other lipid solvents		Resistant		Resistant	Sensitive	Sensitive	Resistant
No. of capsomeres		32	72	252	162	1,500	
Virion diam (nm)		18–26	45–55	70–90	40–50	100	130–300	230 × 300
Mol wt of nucleic acid (10^6)		1.5–2.0	3.0–5.0	20–30	2.1	80–150	100–250	160
Family		*Parvoviridae*	*Papovaviridae*	*Adenoviridae*	*Hepadnaviridae*	*Herpesviridae*	*Iridoviridae*	*Poxviridae*

perhaps other species. The type C retrovirus genus includes mammalian, avian, and reptilian viruses. Members of the HTLV-BLV genus include human T-cell lymphotropic virus types 1 and 2, simian T-cell lymphotropic virus, and bovine leukemia virus. The type D genus includes other viruses of monkeys.

Members of the *Spumavirus* genus, foamy viruses, do not induce tumors or cellular transformation but cause persistent asymptomatic infections in natural and experimental host animals. Foamy or syncytial viruses are known for a number of mammalian species, including humans.

The slow viruses of the maedi-visna group, which have been placed in the genus *Lentivirus*, are morphologically and chemically like other members of the family *Retroviridae* but do not induce tumors. Natural infections are known only in sheep. Visna virus causes panleukoencephalitis. Serologically related viruses (variously designated in different countries as maedi or progressive pneumonia viruses) cause interstitial pneumonitis.

Two types of the human lentivirus HIV have been recognized; type 1 is more widespread and more virulent. HIV is a completely exogenous virus, in contrast to the transforming retroviruses. However, after an individual is exposed to and infected by HIV, proviral DNA is integrated into the cellular DNA of infected cells. The many different isolates of HIV exhibit considerable divergence, particularly in the gene that codes for viral envelope proteins. Visna virus, the prototype of the genus *Lentivirus*, is also known to undergo progressive antigenic variation in reaction to the host's immune response during persistent infection.

PARVOVIRIDAE

All of the virus families shown in Table 2 have their DNA genomes in double-stranded form except for the *Parvoviridae*, whose DNA is single stranded within the virion. Members of the family *Parvoviridae* are very small (Table 2). The molecular weight of the nucleic acid in the virion is relatively very low (1.5×10^6 to 2.0×10^6) compared, for example, with the weight of the DNA of poxviruses (160×10^6). Some members display resistance to high temperatures (60°C, 30 min).

The family *Parvoviridae* encompasses viruses of numerous species of vertebrates, including humans. Two members of the genus *Parvovirus*, the members of which are able to replicate independently, are associated with disease in human beings. Parvovirus B19 causes a transient shutdown of erythrocyte production by killing the late erythroid progenitor cells. This shutdown presents particular problems for individuals already suffering from hemolytic anemias such as sickle cell anemia, causing aplastic crises.

A host range mutant of feline panleukopenia parvovirus, known as canine parvovirus, induces acute enteritis with leukopenia in young and adult dogs as well as myocarditis in puppies. Infections with this virus have reached enzootic proportions around the world.

Several serotypes of adeno-associated viruses, which belong to the genus *Dependovirus*, are known to infect humans, but they have not been shown to be associated with any human disease. Members of this genus cannot multiply in the absence of a replicating adenovirus that serves as a helper virus. The single-stranded DNA is present within the virion as either plus or minus complementary strands in separate particles. Upon extraction, the

TABLE 3 Members of virus families, with emphasis on viruses that infect humans

Family[a]	Genus	Common species	No. of members
Picornaviridae	Enterovirus	Polioviruses	3
		Coxsackieviruses, group A	23
		Coxsackieviruses, group B	6
		Echoviruses	31
		Enteroviruses 68 through 71	4
		Viruses of other vertebrates	>34
	Heparnavirus	Hepatitis A virus	1
		Hepatitis A virus of monkeys	>2
	Cardiovirus	Encephalomyocarditis virus and mengovirus; mouse encephalomyelitis virus	3
	Rhinovirus	Virus types infecting humans	>115
		Viruses of cattle	2
	Aphthovirus	Foot-and-mouth disease viruses of cattle and other cloven-hoofed animals	7
Caliciviridae	Calicivirus	Vesicular exanthema of swine virus	13
		Viruses of cats and sea lions (possible member: Norwalk gastroenteritis virus of humans)	Many
			>5
Reoviridae	Reovirus	Viruses of humans, monkeys, and lower vertebrates	3
		Viruses of birds	>5
	Orbivirus	17 subgroups, including Colorado tick fever and Kemerovo viruses of humans; also bluetongue virus of sheep and African horse sickness virus	>90
	Rotavirus	Human rotaviruses	>6
		Rotaviruses of many mammals, including SA-11 virus of monkeys and Nebraska calf diarrhea virus	Many
Birnaviridae	Birnavirus	Infectious pancreatic necrosis virus of fish; infectious bursal disease virus of chickens	2
Togaviridae	Alphavirus	Sindbis virus and many other mosquito-borne viruses, including viruses of eastern equine, Venezuelan, and western equine encephalitis and Semliki Forest virus	23
	Rubivirus	Rubella virus	1
	Pestivirus	Viruses of cattle and pigs	>3
Flaviviridae	Flavivirus	Yellow fever virus and other mosquito-borne viruses, including viruses of dengue; of Japanese, Murray Valley, and St. Louis encephalitis; and of West Nile fever	26
		Tick-borne viruses, including viruses of Kyasanur Forest disease, Omsk hemorrhagic fever, European and Far Eastern tick-borne encephalitis of humans, and louping ill of sheep	11
		Viruses whose vectors are unknown	17
		Hepatitis C virus	1
Orthomyxoviridae	Influenzavirus	Influenza virus type A	Many
		Influenza virus type B	Several
		Influenza virus type C	1
Paramyxoviridae	Paramyxovirus	Human parainfluenza viruses, including Sendai virus	4
		Mumps virus	1
		Newcastle disease virus of fowl; viruses of other diseases of birds and mammals	>6
	Morbillivirus	Measles virus	1
		Rinderpest virus of cattle	1
		Distemper virus of dogs	1
		Peste-des-petits-ruminants virus of sheep and goats	1
	Pneumovirus	Human respiratory syncytial virus	1
		Respiratory disease viruses of cattle and mice	?
Rhabdoviridae	Vesiculovirus	Vesicular stomatitis virus of horses, cattle, and pigs	Several
	Lyssavirus	Rabies virus	1
		Lagos bat virus and others	>5
Filoviridae	Filovirus	Marburg virus	1
		Ebola virus	2

(Continued on next page)

TABLE 3 *(Continued)*

Family[a]	Genus	Common species	No. of members
Coronaviridae	*Coronavirus*	Human coronavirus	2
		Mouse hepatitis virus, infectious bronchitis virus of fowl, and other agents infecting pigs and other vertebrates	>4
Bunyaviridae	*Bunyavirus*	Bunyamwera virus	>145
		California encephalitis viruses	
		LaCrosse virus, other serologically cross-related groups, and several ungrouped viruses	
	Phlebovirus	Sandfly fever viruses	>30
		Other viruses of humans and lower animals, including Rift Valley fever virus of sheep and other ruminants, which may cause human disease	
	Nairovirus	Crimean-Congo hemorrhagic fever virus	>27
		Viruses of 5 other serogroups, including the virus of Nairobi sheep disease	
	Uukuvirus	Uukuniemi virus and six other agents, all belonging to same serogroup (infect rodents and ticks)	7
	Hantavirus	Hantaan virus of hemorrhagic fever with renal syndrome	Many
Retroviridae			
Oncovirinae	Type C oncovirus group	Sarcoma and leukemia viruses of mice, cats, cattle, birds, snakes, and primates	>15
	HTLV-BLV group	Human T-cell lymphotropic virus types 1 and 2	4
	Type B oncovirus group	Mammary tumor virus of mice (and humans?)	?
	Type D oncovirus group	Monkey (mammary tumor?) virus (Mason-Pfizer monkey virus)	?
Spumavirinae	*Spumavirus*	Syncytial and foamy viruses of humans, monkeys, cattle, and cats	>4
Lentivirinae	*Lentivirus*	Human immunodeficiency virus	2
		Visna, maedi, and progressive pneumonia viruses of sheep	?
Arenaviridae	*Arenavirus*	Lymphocytic choriomeningitis virus of mice	1
		Lassa fever virus	1
		Viruses of Tacaribe complex, including Junin and Machupo viruses of South American hemorrhagic fevers	>8
Parvoviridae	*Parvovirus*	Human parvovirus B19	>1
		Aleutian mink disease virus; viruses of rodents, pigs, cattle, cats, and dogs	Many
	Dependovirus	Adeno-associated virus (adeno-satellite virus): human (types 1–5); monkey (type 4); also of cattle, dogs, and birds	>8
Papovaviridae	*Papillomavirus*	Human papillomaviruses (warts)	Many
		Rabbit (Shope) papillomavirus	1
		Papillomaviruses of other mammals	Many
	Polyomavirus	Polyomavirus of mice	1
		JC and BK viruses of humans	2
		Simian virus 40 of rhesus monkeys	1
		Lymphotropic virus of African green monkeys	1
		Viruses of mice, rabbits, and baboons	>5
Adenoviridae	*Mastadenovirus*	Human adenoviruses	>36
		Viruses of other mammals	>45
	Aviadenovirus	Viruses of birds	>13
Hepadnaviridae	*Hepadnavirus*	Human hepatitis B virus	1
		Hepatitis B viruses of woodchucks, ground squirrels, and ducks	>3
Herpesviridae			
Alphaherpesvirinae	*Simplexvirus*	Human herpes simplex virus types 1 and 2	2
		Bovine mammillitis virus	1
	Varicellovirus	Varicella-zoster virus (herpesvirus 3)	1
		Herpes B virus of monkeys	1
		Pseudorabies virus	1
		Equine rhinopneumonitis virus	1
Betaherpesvirinae	*Cytomegalovirus*	Human cytomegalovirus (herpesvirus 5)	1
	Muromegalovirus	Mouse cytomegalovirus	1
Gammaherpesvirinae	*Lymphocryptovirus*	Epstein-Barr virus (herpesvirus 4)	1
	Thetalymphocryptovirus	Marek's disease herpesvirus of fowl	1
	Rhadinovirus	Herpesvirus saimiri and others	>2

(Continued on next page)

TABLE 3 Members of virus families, with emphasis on viruses that infect humans (*Continued*)

Family[a]	Genus	Common species	No. of members
Iridoviridae	*Iridovirus*	Iridescent insect viruses	Several
		African swine fever virus (?)	1
	Ranavirus	Frog viruses	>30
	Piscinivirus	Fish viruses	Several
Poxviridae			
Chordopoxvirinae (poxviruses of vertebrates)	*Orthopoxvirus*	Vaccinia virus	1
		Smallpox virus (variola)	1
		Poxviruses of lower animals	>6
	Parapoxvirus	Orf virus and other viruses of ungulates	?
		Virus of milker's nodule	1
	Avipoxvirus	Fowlpox virus and other viruses of birds	8
	Capripoxvirus	Viruses of sheep and goats	3
	Leporipoxvirus	Myxoma virus of hares	4
		Fibroma viruses of rabbits and squirrels	
	Suipoxvirus	Swinepox virus	1
	Yatapoxvirus	Yabapox and tanapox viruses	2
	Molluscipoxvirus	Molluscum contagiosum virus	1
Entomopoxvirinae		Poxviruses of insects	>24

[a] Where subfamilies have been designated, they are listed in this column, indented below the family name.

plus and minus DNA strands unite to form a double-stranded helix.

HEPADNAVIRIDAE

The name *Hepadnaviridae* reflects the DNA-containing genomes of its members and their replication within hepatocytes. Each of these viruses has a circular DNA genome that is double stranded except for a region of variable length that is single stranded. In the presence of appropriate substrates, DNA polymerase within the virion can complete the single-stranded region to its full length of 3,200 nucleotides.

Hepatitis B virus of humans and three similar viruses found in woodchucks, Beechey ground squirrels, and Pekin ducks share many basic features. All members of the family share antigens as well as similar morphology and behavior in the infected host. Large amounts of excess viral coat protein are produced in the form of small 22-nm spherical and tubular particles (in the human virus, the antigen is known as hepatitis B surface antigen). The viruses replicate in the liver and are associated with acute and chronic hepatitis. More than 300 million persons are persistent carriers of the human virus and are at very high risk of developing chronic liver diseases, including cancer. The woodchuck hepatitis B virus also causes liver cancer in its natural host. Fragments of viral DNA may be found in the liver cancer cells of both species.

RECENT DEVELOPMENTS

The following groups of viruses of vertebrates were established at the meeting of the ICTV held in Glasgow in 1993. Table 3 summarizes the previously established virus groups but does not include the new groups listed below (8a, 8b).

Birnaviruses

1. The genus *Aquabirnavirus* in the family *Birnaviridae*
2. The genus *Avibirnavirus* in the family *Birnaviridae*
3. The genus *Entomobirnavirus* in the family *Birnaviridae*

Hepatitis Viruses

1. The genus *Deltavirus*
2. Hepatitis delta virus as the type virus in the genus *Deltavirus*

Herpesviruses

1. The genus *Roseolovirus* in the subfamily *Betaherpesvirinae* of the family *Herpesviridae*
2. Human herpesvirus 6 as the type species of the genus *Roseolovirus*

Parvoviruses

1. The subfamilies *Parvovirinae* and *Densovirinae* in the family *Parvoviridae*
2. The genus *Erythrovirus* in addition to the genera *Parvovirus* and *Dependovirus* in the subfamily *Parvovirinae*
3. The genera *Densovirus*, *Iteravirus*, and *Contravirus* in the subfamily *Densovirinae*

Rhabdoviruses

1. The genus *Cytorhabdovirus* in the family *Rhabdoviridae*

REFERENCES

1. **Andrewes, C. H., H. G. Pereira, and P. Wildy.** 1978. *Viruses of Vertebrates*, 4th ed. Macmillan Publishing Co., New York.
2. **Brown, F.** 1989. The classification and nomenclature of vi-

ruses: summary of the results of meetings of the International Committee on Taxonomy of Viruses, Edmonton, Canada, 1987. *Intervirology* **30:**181–186.

3. **Fenner, F.** 1976. Classification and nomenclature of viruses: second report of the International Committee on Taxonomy of Viruses. *Intervirology* **7:**1–16.

4. **Fields, B. N., D. M. Knipe, R. M. Chanock, J. L. Melnick, B. Roizman, and T. P. Monath (ed.).** 1990. *Virology*, 2nd ed. Raven Press, New York.

5. **Francki, R. I. B., C. M. Fauquet, D. L. Knudson, and F. Brown.** 1991. Classification and nomenclature of viruses: fifth report of the International Committee on Taxonomy of Viruses. *Arch. Virol.* Suppl. 2, p. 1–450.

6. **Matthews, R. E. F.** 1979. Classification and nomenclature of viruses: third report of the International Committee on Taxonomy of Viruses. *Intervirology* **12:**129–296.

7. **Matthews, R. E. F.** 1982. Classification and nomenclature of viruses: fourth report of the International Committee on Taxonomy of Viruses. *Intervirology* **17:**1–199.

8. **Melnick, J. L.** 1966–1982. Summaries on viral taxonomy (published annually). *Prog. Med. Virol.*, vol 6 through 28.

8a.**Murphy, F. A.** Personal communication.

8b.**Pringle, C. R.** Personal communication.

9. **Wildy, P.** 1971. Classification and nomenclature of viruses: first report of the International Committee on Nomenclature of Viruses. *Monogr. Virol.* **5:**1–81.

Collection and Preparation of Specimens for Virological Examination

DAVID A. LENNETTE

70

GENERAL CONSIDERATIONS

The laboratory diagnosis of viral infections is based on three general approaches: (i) direct detection of viral components either in cells derived from infected tissues or free in fluid specimens; (ii) isolation of viruses, usually in cell cultures, followed by identification; and (iii) demonstration of a significant increase in serum levels of antibodies to an etiologically plausible virus during the course of an illness or the presence of antibodies of a class (immunoglobulin M [IgM]) believed to occur only transiently after infection. Recent applications of PCR methods and similar schemes of nucleic acid amplification are most analogous to approach ii above and use the same specimens. Each laboratory adopts specimen collection and preparation methods to optimize the detection or recovery of viruses with the methods in use at that facility. This chapter provides guidelines that can be modified locally as necessary.

Specimens for virus isolation and direct detection as well as acute-phase blood samples must be collected within the first few days of an illness if adequate sensitivity of testing is expected. A brief clinical resumé that accompanies the initial specimen from each patient and gives date of onset, major clinical findings, and any agents suspected will assist the laboratory staff in selecting the appropriate tests. Specimens to be collected for various syndromes are shown in Table 1.

Storage Temperature

Specimens should not be held at ambient temperatures, since periods of up to several days may elapse before some specimens are finally inoculated into isolation systems. Most specimens can be held at 4 to 8°C for several days before there is an excessive loss of infectivity. Freezing specimens to −70°C or below will preserve the residual viral infectivities of specimens almost indefinitely, although significant losses of infectivity may occur with each freeze-thaw cycle. Many viruses lose infectivity rapidly when stored at −15 to −20°C, the temperature range found in most standard refrigerator-freezers; the temperature cycling associated with frost-free freezers causes even more rapid loss of infectivity.

Samples for virus isolation attempts should be kept at 4 to 6°C (but not frozen) until just before they are inoculated into the appropriate host systems. If residual specimen material is to be kept for possible retrieval at a later date, it is usually best to freeze it to −70°C (or below). If a virus suspected to be present is labile to freezing, the sample should be diluted with (at least) an equal volume of 2SP (0.2 M sucrose in 0.02 M sodium phosphate buffer; pH 7.2). Many laboratories find that this commonly used chlamydial transport medium (or a 2× medium, 4SP) is effective in preserving infectivities of viruses that are labile to freezing, especially members of the herpesvirus group (6); the use of older cryopreservatives, such as sorbitol syrup or glycerin, is no longer recommended.

Specimen Inspection

All specimens should be inspected when received to determine that they are properly labeled with adequate patient identification by name or code number. The specimen site and collection date should be given on the label; submitters who do not provide this information should be requested to do so. Any discrepancies between specimen label information and the accompanying submission form should be noted on the form, and appropriate follow-up actions should be taken. Unlabeled specimens are not acceptable and should be rejected. Leaking specimen containers should be identified as soon as possible; they should be decontaminated with a general laboratory disinfectant and kept isolated in a closed zipper lock plastic bag. Advise submitters of leaking specimens on how to avoid the problem in the future. If the received specimen volume is less than the minimum that is normally processed (usually, 1 to 2 ml for fluid samples), this fact should be noted and reported to the submitter. Initial specimen processing is completed when the laboratory accession number is affixed to the specimen container and the specimen data and time of receipt are recorded either on paper or in a data processing system.

SPECIMENS FOR VIRUS ISOLATION

Specimens fall into two broad categories: those that are normally sterile and those that are from body fluids that are microbially contaminated. Normally sterile specimens may be inoculated for isolation as soon as the indicated processing steps are completed. Process contaminated specimens to obtain a fluid volume of 2 to 3 ml, and then treat this volume with antimicrobial agents in order to substantially reduce overgrowth of microorganisms in the host cell cul-

TABLE 1 Specimens for virus isolation and direct detection

Source or clinical symptoms and common etiologic agent[a]	Specimen source for virus isolation and direct detection	
	Clinical	Biopsy tissue
Upper respiratory 　Rhinovirus 　Influenza virus 　Parainfluenza virus 　Adenovirus 　Enterovirus 　Cytomegalovirus 　Epstein-Barr virus[b] 　Reovirus	Nasopharyngeal swab, nasal wash (throat swab)	Tonsil, lymph node
Lower respiratory 　Influenza virus 　Respiratory syncytial virus 　Parainfluenza virus 　Cytomegalovirus 　*Chlamydia pneumoniae* 　*Mycoplasma pneumoniae*	Endotracheal aspirate, bronchial wash, bronchoalveolar lavage (sputum)	Lung, bronchus, trachea
Vesicular lesions 　Herpes simplex virus 　Varicella-zoster virus 　Enterovirus	Lesion swab, lesion fluid	Multiple organs
Exanthems 　Herpesvirus 6 　Parvovirus (B19)[b] 　Enterovirus 　Rubeola (measles) virus 　Rubella virus 　*Rickettsia* spp.[b]	Nasopharyngeal swab (throat swab) (stool)	
Central nervous system 　Enterovirus 　Herpesvirus family 　Arboviruses	CSF, nasopharyngeal swab	Brain
Posttransplantation syndromes 　Herpes simplex virus 　Cytomegalovirus 　Epstein-Barr virus 　Herpesvirus 6	Throat swab (urine)	Transplanted organ
Congenital anomalies 　Cytomegalovirus 　Herpes simplex virus 　Rubella virus 　Varicella-zoster virus	Nasopharyngeal swab, urine	Affected organ(s)
Enteritis and diarrhea 　Rotavirus 　Enteric adenovirus 　Astrovirus 　Calicivirus (Norwalk agents)	Stool	Colon

[a]Specimens in parentheses are secondary choices.
[b]Diagnosis by serology, rarely by isolation or direct detection.

tures during virus isolation attempts. Add antimicrobial agents to the cell culture maintenance medium used during isolation attempts also. Although it seems redundant to treat both the extracted specimens and the inoculated cell cultures, no published studies show any benefits in omitting either step. A number of antimicrobial agents for use with cell cultures are commercially available, usually packaged at 100 times the intended final concentration. Some commonly used stock reagents include penicillin-streptomycin-neomycin (5, 5, and 10 mg/ml, respectively), nystatin suspension (10,000 U/ml), amphotericin B (Fungizone; 0.25 mg of amphotericin B and 0.25 mg of sodium deoxycholate per ml), and gentamicin reagent solution (10 mg of gentamicin sulfate per ml). A modification of a mixture previously reported (3), vancomycin-gentamicin-amphotericin B (1.25 mg of gentamicin, 0.25 mg of vancomycin, and 0.25 mg of amphotericin B per ml of stock solution) is the preferred combination in this laboratory.

Virus isolation in cell cultures makes it possible to recover many different agents from appropriately selected specimens and should be used when many viruses are plausible etiologic agents. Collect specimens promptly, preferably within 3 days and not longer than 7 days after the onset of illness. Using aseptic techniques, collect postmortem specimens as soon as possible after death. Label each specimen container with the patient name, site of the specimen, and date and time of collection. Specimens that must be held for long intervals before testing should be promptly frozen to −70°C or below. Otherwise, refrigerate specimens promptly after collection, as indicated above. Fluid specimens (urine, cerebrospinal fluid [CSF], washes, etc.) do not usually require any transport medium and should not be diluted. Suitable holding media, usually packaged in combination with swabs, are available in commercial formulations (10), and formulas for two generic media (Hanks' balanced salt solution with gelatin and buffered tryptose phosphate broth with gelatin) are given in earlier editions of this Manual (10a). Although almost any type of swab may be used satisfactorily with most specimens, calcium alginate fiber tips may inactivate herpes simplex virus and chlamydiae and should be avoided. Charcoal-impregnated swabs also inactivate virus infectivity and should be avoided. Most laboratories now prefer swabs with metal or plastic shafts over those with wooden shafts, which cause cell culture toxicity problems. A comparative discussion of some transport systems is available elsewhere (15).

Nasal and Pharyngeal Swabs

Use a dry swab (either cotton or synthetic fiber) to swab each nostril, and allow the swab to remain in place for a few seconds to absorb secretions. Collect throat swabs by rubbing the tonsils and posterior pharynx with a cotton or synthetic fiber swab, either dry or wetted with viral transport medium.

The swabs may be supplied in a suitable viral transport and storage medium, e.g., Culturette system swabs; if they are not, they should be transferred immediately to a vial containing 2 or 3 ml of such a medium. Do not allow the swabs to dry; most common respiratory viruses lose infectivity rapidly on a dry swab.

Washings

Nasal washings can be obtained by instilling several milliliters of sterile saline into each nostril while the patient's head is tilted back slightly; the head is then brought forward, and the saline is allowed to flow into a small container held beneath the nose. In infants, a small catheter with a suction trap may be employed. Gelatin or bovine serum albumin (1%) may be added to the washing to stabilize any virus that may be recovered. To collect throat washings, adult patients should gargle with the smallest convenient volume (10 ml) of cell culture medium or general purpose bacteriologic broth and then expectorate into a cup. The cup contents are then poured into a screw-cap vial. Specimens may be collected from pediatric patients in the same manner if they are able to cooperate; otherwise, throat swabs will suffice. Throat washings may offer a somewhat higher yield of virus than swabs, but they are not as convenient to collect. Washes from the lower respiratory tract are collected by specialized (invasive) procedures, usually with hospitalized patients; these bronchial and bronchoalveolar washes provide the best sources for recovery of many major respiratory pathogens.

Oral Swabs

Rub a dry cotton swab over any oral lesions, and transfer the swab immediately to a vial of virus transport medium.

Eye Swabs

If any exudate or pus is present in the eye, first remove it with a sterile swab. Then use a second swab, moistened with transport medium or saline, to rub the affected conjunctiva. Immediately clip off the swab tip into a vial of transport medium to retain any cells trapped in the fibers. Corneal specimens should be collected by an ophthalmologist or other adequately trained person, using a spatula.

Cervical Swabs

If more than one swab is used to obtain a cervical specimen, more infected cells will be recovered and better results may be obtained. First, use one swab to clean mucus from the cervix; discard this swab, and then use another swab or Cytobrush to obtain cells from the cervical os. If any lesions are seen, swab them also. Remove the swab or brush, and transfer it to a vial of transport medium. One study in which four swabs were collected sequentially showed improved recovery (of Chlamydia trachomatis) from the last swabs collected. Some physicians have resorted to using scrapers to obtain increased numbers of cells, but this procedure has not been demonstrated to be necessary to obtain satisfactory results.

Vesicle Fluids and Skin Scrapings

Collect specimens of vesicle fluids and cellular material from the bases of lesions before crusting and healing have begun, since the recovery rate from specimens collected later drops sharply. In the case of primary infections with herpes simplex virus, however, the virus may be recovered for up to 7 to 10 days after onset. Disinfection of the site with, e.g., alcohol or iodophors may inactivate viruses; if possible, perform local disinfection after specimens have been collected. Aspirate vesicle fluids with a 26- or 27-gauge needle attached to a tuberculin syringe or with a capillary pipette. The fluids obtained with either method should be rinsed promptly into a small volume of transport medium to prevent loss of the specimen by clotting. Swab open lesions to obtain both fluid and cells from the lesion base. Immediately clip off the swab tip into a vial of transport medium to recover any cells trapped in the fibers.

Stools and Rectal Swabs

Obtain a suitable stool sample by transferring a small (2- to 5-g) portion of stool (either formed or liquid) into a small leakproof container, such as a 1-oz (28.350-g) screw-cap jar. Larger quantities are often submitted, but they are not needed for most procedures. Cardboard or waxed containers are unsuitable, as they are not leakproof and allow desiccation of the sample. No transport medium is required. Rectal swabs are inadequate specimens for the detection of rotavirus or the cytotoxin produced by *Clostridium difficile* (often assayed in the virus laboratory with cell cultures). Therefore, a rectal swab should not be regarded as an expedient substitute for a stool specimen but rather as a specimen appropriate for the recovery of the agents that cause proctitis. Insert a dry swab 3 to 4 cm past the anal sphincter, rotate the swab, and withdraw it. Place the swab in a vial of transport medium, and refrigerate it.

Urine

Clean-voided specimens collected in conventional containers are quite satisfactory for isolation of viruses; special collection methods are not required. Provided that the specimen is refrigerated soon after collection, even viruses often regarded as labile, e.g., cytomegalovirus (CMV), may be recovered up to as much as a week after collection. Adding antibiotics to the specimen may be useful in suppressing bacterial overgrowth but should not be required if the specimen is kept cold. Urine specimens are not adequate substitutes for urethral swabs for the culture of chlamydiae, although nonculture detection methods can be used. Detection of CMV infection is improved by processing several specimens when possible, as virus shedding may be intermittent.

Semen

Semen, usually collected in a wide-mouth container, should be diluted with an equal volume of 2SP transport medium and transferred to a smaller screw-cap container for transport.

CSF

Because the concentration of infectious virus is seldom high in cerebrospinal fluid (CSF), it is important to obtain an adequate sample volume. For virological work, it is desirable to collect 2 ml in a sterile screw-cap tube or vial. When possible, obtain 1-ml samples from infants; volumes of less than 0.5 ml are often all that are available from neonatal infants, but these volumes are of less value, considering the lower recovery rate to be expected. Do not dilute the specimen in any manner; refrigerate it as soon as possible. If the specimen cannot be processed within 24 to 48 h, it may be frozen to below −70°C; if it is to be used for serologic testing only, freezing is not needed.

Serum and Blood

Serum is rarely used for the isolation of viruses; it is, however, a suitable specimen for isolation of enteroviruses from perinatally infected infants. Human parvovirus can also be detected in serum by nonculture methods. Blood clots are used for the recovery of Colorado tick fever virus (found in erythrocytes); this virus may also be detected by immunofluorescent staining of blood smears. The buffy coat cells (leukocytes) are used for detection of viremia, primarily for patients with CMV infections. The high density of cells found in whole blood, particularly erythrocytes, inter-

feres in the examination of inoculated cell cultures; therefore, before inoculation, the specimens must be fractionated by processing at least 5 ml of anticoagulated whole blood with a commercially available separatory medium. The fewest problems occur when blood collected with citrate anticoagulant is used; blood obtained with heparin gives the same rate of CMV isolation as that collected with EDTA.

One widely used procedure combines Histopaque-1077 (Sigma Chemical Co., Inc.) or Ficoll-Hypaque (Pharmacia, Inc.) centrifugation with Macrodex (Pharmacia) sedimentation to recover both mononuclear (MN) cells and polymorphonuclear leukocytes (PMNs). All steps are carried out at room temperature. The recovery of total (unfractionated) leukocytes can be accomplished by a simple procedure using centrifugation with Histopaque-1119 (14).

1. Mix equal (5-ml) volumes of whole blood and sterile phosphate-buffered saline (PBS). Then gently layer the mixture over 3 ml of Histopaque-1077 in a 15-ml conical centrifuge tube.

2. Centrifuge the tube at 400 × g for 30 min. Remove the diluted plasma, and discard it; remove and save the band of cells visible above the Histopaque interface.

3. Wash this MN-cell-containing fraction twice, each time with 10 ml of cell culture medium, and use 800 × g centrifugation for 15 min to collect the cells from each wash cycle. The fractionated MN-cell sample is ready for direct inoculation of cell cultures.

4. Recover PMNs from the erythrocyte-rich pellet at the bottom of the tube. Pipette the pellet into a tube containing 2.5 ml of Macrodex solution. Wash the remaining cells out of the original centrifuge tube with an equal volume of PBS, and transfer the cells to the Macrodex tube.

5. Let the tube with the suspension stand for 2 h, and then collect the supernatant PMN-enriched fraction and discard the sedimented erythrocytes. Wash the Macrodex-PMN fraction in the same manner as the MN-cell fraction; these cells may be combined with the MN-cell fraction for inoculation of cell cultures.

Bone Marrow

Bone marrow is aspirated into a tube or syringe containing citrate or heparin anticoagulant and submitted without other additives or diluents. Mix the marrow thoroughly with the anticoagulant.

Autopsy and Biopsy Specimens

For autopsy or biopsy specimens, collect fresh tissue from any affected site or obvious lesion, using separate sterile instruments for each site sampled. Autopsy samples need not be larger than 1 or 2 g, and biopsy specimens are often much smaller, e.g., obtained by needle. Each specimen should be placed in a separate sterile container and clearly labeled. Frequently sampled tissues for cases of suspected viral etiology include brain, lung, heart muscle, lymph node, and kidney. Liver tissue, sometimes toxic to cell cultures, is useful for direct staining procedures; tracheal and bronchial tissues are often overlooked but are often superior to lung tissue for recovery of respiratory viruses. Samples should be kept refrigerated in a small volume of viral transport medium or saline but should not be placed in any fixative solution, which renders them useless for virus isolation and often for immunofluorescent staining tests as well. If the specimens cannot be processed within 1 or 2 days, it may be preferable to freeze them to −70°C or below.

SPECIMENS FOR SEROLOGIC TESTS

Blood specimens are collected to obtain serum for tests to measure antibodies in order to determine immune status or detect a response to a recent infection. Acute- and convalescent-phase sera must be tested together to determine that antibodies have appeared or increased in titer during the course of an illness. Collect an acute-phase specimen as soon as possible, not later than 5 to 7 days after onset of the illness. Convalescent-phase specimens usually have been collected 10 to 14 days after acute-phase specimens, an interval that was probably needed for the serologic response determined by the complement fixation test. More sensitive tests in current use usually show significant antibody changes with intervals of as little as 3 to 5 days between serum samples. Useful results may sometimes be obtained by testing a single serum specimen; IgM-class antibodies are often detectable within 2 to 3 days after onset of disease. Unfortunately, some individuals do not have detectable serum IgM until as much as a week later, or 9 to 10 days after onset, and IgM testing of sera collected from newborn infants is less reliable than testing sera from older patients. Although IgM-class antibodies are not normally detectable for more than a month or two after infection, persistence for much longer has been encountered with some individuals. Circumstances for different agents vary according to the test methods available, so the performing laboratory should be consulted for advice and information on limitations of such testing.

Blood specimens should be collected without anticoagulants or preservatives, which may affect the results of serologic tests. The usual volume of blood collected is 8 to 10 ml, although 3- to 4-ml specimens (normally collected from pediatric patients) usually provide enough serum to complete all necessary tests. Allow the specimen to clot at room temperature, and then separate the serum by centrifugation, and remove it to a separate vial. Serum should not be shipped to a remote laboratory in its collection tube, as the clot tends to disintegrate and hemolyse in transit. Many laboratories are willing to do limited testing with blood samples obtained by capillary collection techniques, which yield less than 0.5 ml of whole blood. These procedures are very helpful with newborn infants. The serum may be stored at 4 to 6°C for up to several weeks, pending the completion of tests. For longer storage, serum is usually frozen to −20°C or below. Never freeze whole blood; this causes complete hemolysis and may render the specimen unusable for serologic testing. Either submit each specimen when collected, carefully identifying it as "acute" or "convalescent," or store the acute-phase specimen to be submitted for testing with the convalescent-phase specimen. If the date of onset of a patient's illness is obtained, it will usually be evident whether a serum is acute or convalescent phase, even if not otherwise indicated. Paired acute- and convalescent-phase sera from a patient should always be tested simultaneously in one laboratory, since results obtained from two laboratories cannot be accurately compared for changes in antibody titer. If the specimen is a random sample for determination of immunity, it should be identified as "for immunity status."

CSF collected from patients with neurologic disease is sometimes submitted for serologic testing. Such specimens need to be tested in parallel with serum specimens collected at (approximately) the same time in order for the results of testing to be easily interpretable. CSF contaminated with blood is not usable, as it is not possible to tell what proportion of antibodies detected is from the blood and what is produced locally in the central nervous system. CSF is normally well isolated from blood, so that CSF titers of antibodies derived from blood are less than 2 to 3% of those of antibodies in serum of the same person and are usually not detectable (4). CSF antibody titers that are higher than 5% of serum titers are often regarded as evidence for central nervous system infection, leading to local antibody production.

Other body fluids are rarely tested for viral antibodies, so there is little information regarding the significance of detecting antibodies in most of these miscellaneous fluids. Antibodies detected in vitreous humor, however, do provide evidence of eye infections with agents such as varicella-zoster virus (17) and herpes simplex virus (13).

PROCESSING OF SPECIMENS FOR VIRUS ISOLATION

Throat, Nasal, Oral, Rectal, and Genital Swabs

If frozen, specimens may be thawed in a beaker of cold tap water. Dislodge any cells that are trapped in swab fibers by agitating the specimen vial on a Vortex mixer for a few seconds or by shaking it vigorously. The swab(s) may be left in the specimen vial or tube. Add antimicrobial agents to the specimen fluid, and mix thoroughly. After the specimen has been used for inoculating hosts, store the remaining material at or below −70°C for use in any later studies.

Nasal, Throat, Bronchial, and Other Washes

If particulate matter or mucus is present, transfer the specimen to a centrifuge tube containing a few glass beads, and shake or mix it vigorously. Then centrifuge the tube; 15 to 20 min at 3,000 × g is usually sufficient. If a high-speed centrifuge is available, centrifugation times may be reduced in proportion to the force used. Remove as much of the supernatant fluid for use as is feasible without disturbing the pellet and beads; add antibiotics to the fluid, and use it to inoculate hosts. Heavily contaminated fluids may need to be filtered through 200-nm-porosity membrane filters. The pelleted cells may be used to prepare smears for immunofluorescent staining or other direct tests.

Stools

Stool specimens are preferable to rectal swabs, as the swabs are inadequate for some tests commonly performed with stools, such as detection of rotaviruses or *Clostridium difficile* cytotoxin. Because stools vary in consistency from watery to well formed, dilution of specimens to obtain an extract for processing requires some adjustment of the volume of added fluid. Add enough Hanks' balanced salt solution containing antibiotics to dilute liquid stool at least 1:2, but do not dilute a formed stool more than about 1:10. Use only enough specimen to obtain a volume that is easily processed; 3 to 5 ml is usually ample. Centrifuge the diluted stool to obtain a clarified extract; a small, high-speed angle-rotor centrifuge that can run tubes 12 by 75 mm at >3,000 × g is ideal for this purpose. Centrifugation for 10 min at 3,000 × g usually results in a well-clarified supernatant fluid extract; mucoid stools may require pretreatment with a mucolytic agent such as 1.0% N-acetylcysteine or dithiothreitol to obtain satisfactory results. Because of their consistency and very high microbial content, stool specimens result in cell culture contamination problems more often than do other specimens; addition of antibiotics does not alleviate this problem, which can be dealt with, if necessary, by filtering specimen extracts through 200-nm-porosity membrane filters.

Other Body Fluids

Normally sterile body fluids, such as CSF and pleural fluid, need not be treated with antibiotics before inoculation unless bacterial contamination of the specimen is suspected. In that case, the specimen should be treated as described above for swabs.

Urine

Routinely collected urine specimens contain bacteria; add antibiotics to urine specimens not collected by catheter. Urine specimens are commonly cytotoxic and often acidic. Neutralize excessively acidic urine samples by adding a few drops of sodium bicarbonate solution (7.5%), or dilute a portion of the specimen with 2SP. Urine specimens submitted to this laboratory are sterilized by filtration through 200-nm-porosity membrane filters and then diluted 1:5 in 2SP before inoculation of cell cultures.

Semen

Semen submitted for isolation of CMV is often toxic to cell cultures. Recovery of other viruses from semen is uncommon. Special processing is required to reduce cytotoxicity and enhance CMV recovery. Good results have been obtained with the following method (7).

1. Transfer semen, diluted in 2SP, to 1.5-ml microcentrifuge tubes (Eppendorf style), and centrifuge it for 5 min at high speed in a compatible microcentrifuge system.
2. Remove and discard the supernatant fraction. Suspend the pellet in 1 ml of maintenance medium, and disperse the contents thoroughly by vortexing.
3. Inoculate the sample in 0.2-ml amounts into shell vial cultures or conventional tube cultures, as for other specimens.
4. Examine the cultures as usual for evidence of CMV infection.

Autopsy and Biopsy Tissues

Most soft tissue samples are received as small fragments in a small volume of viral transport medium; these can be dispersed in a sonication bath. Larger samples (and harder tissues) are triturated by using a sterile mortar-pestle set. Grinding in a mortar should be done with very small volumes and little added fluid; adding a pinch of sterile sand or alumina may speed the processing of hard or tough tissue samples. Extract the ground-up sample material into enough Hanks' balanced salt solution to form an approximately 10% (wt/vol) suspension, which is then centrifuged as for a wash specimen. Remove the supernatant extract, and dispense it into sterile vials for inoculation of host systems and storage at −70°C.

If viable cells for explantation or cocultivation are needed, use a scalpel or a set of dissecting scissors to finely mince the sample in a petri dish to fragments <0.5 mm wide. Using a minimum volume of fluid, i.e., enough cell culture medium to prevent the specimen from drying, may make this procedure easier.

ENHANCEMENT OF VIRAL INFECTIVITY

An extended discussion of enhancing viral infectivity is available in a recent review (9).

Postinoculation Centrifugation

Although the mechanisms of enhancement are not yet understood, viral infectivity is often increased by low-speed centrifugation of inoculated cell cultures, a phenomenon that has been recognized for more than 25 years (8, 12). Often referred to as shell vial cultures, these cell cultures (usually grown in 1-dram [3.697-ml] shell vials) are widely used for the recovery of herpes group viruses (herpes simplex virus, CMV, and varicella-zoster virus) and increasingly have been adapted for the rapid detection of other viruses.

First, the cell culture medium is removed from sterile shell vials containing monolayer cell cultures grown on 12-mm circular coverslips. The shell vials are then inoculated with specimens for the isolation of herpes simplex virus, CMV, or varicella-zoster virus (1, 5, 12). The inoculated vials are then centrifuged in a swinging-bucket rotor for 1 h at approximately 1,000 × g at ambient temperature (up to 35°C in an uncooled centrifuge). The inoculum is then removed from the vials and replaced with 1 to 2 ml of maintenance medium. The centrifugation technique is used together with examination by immunofluorescent staining after 1 to 3 days of incubation.

Biochemical Enhancement of Infectivity

Reoviruses and rotaviruses, which are tedious or difficult to isolate in cell culture from clinical specimens, may have their infectivities greatly enhanced by treatment with proteolytic enzymes, notably trypsin (2, 16). Similarly, the growth of influenza viruses in some cell cultures is greatly enhanced by trypsin (18). Laboratories that wish to isolate these agents should adopt two simple procedures.

1. Add trypsin (0.01 mg/ml) to processed stool extracts and other specimens of interest.
2. Substitute serum-free cell culture maintenance medium containing 0.5% bovine serum albumin and 5 mg of trypsin per liter for use with cell cultures inoculated with the trypsin-treated specimens.

SPECIMENS FOR DIRECT STAINING OF VIRUS AND VIRAL ANTIGENS

Many common viruses may be detected by immunofluorescent (or immunoenzyme) staining of an appropriate specimen. Immunofluorescence procedures are rapid and reliable when good-quality reagents are available; their use should be considered if the number of agents of interest is small and if a suitable specimen (one containing infected cells) is available. Selection of an appropriate specimen is important, since only a limited amount of a sample can be examined. Immunofluorescence tests are most useful for the detection of respiratory infections and infections causing lesions of the skin or mucous membranes, but rapid direct detection of some other infections, e.g., Epstein-Barr virus (see chapter 74 of this Manual), is sometimes practical.

Specimens for immunofluorescent staining usually consist of cells deposited on some type of microscope slide; plain, precleaned 1-mm-thick glass slides are suitable for any type of specimen. For examination of cell suspensions, some laboratories prefer slides that have printed overlays with wells for the cell spots.

Sections may be prepared from tissue obtained at biopsy or autopsy for immunofluorescent staining. Thin (<7-μm-thick) frozen sections are excellent for almost any tissue; although these are not routinely prepared in many laboratories, the results obtained justify the inconvenience. Some viruses may be detected (after deparaffinization) in sections cut from tissue embedded in paraffin blocks. However, many virus antigens are adversely affected by the fixation and embedding used to prepare these blocks. Do not assume that any given virus will be detectable in such sections without prior evaluation of positive control material that has been prepared in the same way as the test specimen. If cut sections are unavailable, impression smears may be adequate, although they do not allow the examination of as many cells. Touch preparations are made with a freshly cut piece of tissue. Partially dry the wet surface by gently blotting it against the dry surface of a petri dish. Then press the cut surface firmly, with a sliding motion, against a microscope slide.

Cells recovered from the processing of wash specimens (see above) can be used to prepare smears for immunofluorescent staining. Disperse any pellet available from the wash centrifugation in the smallest possible volume (0.01 to 0.05 ml) of PBS containing 0.1% bovine serum albumin, and spot it onto slides as required.

All slides should be air dried as rapidly as possible (the intake grid of a laminar-flow biological safety cabinet is excellent for this purpose) and then fixed briefly. The standard fixation for most purposes is 5 min in acetone at room temperature (cold acetone absorbs water from the air and soon loses efficacy). Fixation by other means, e.g., methanol, should be undertaken only after evaluation to ensure that the reactivity of the antigens to be detected is not destroyed or diminished. Tissues that have been fixed in formalin are generally unsuitable for immunofluorescent staining because of denaturation of viral antigens. Some success has been obtained with such material by careful digestion with trypsin in order to reexpose viral antigen epitopes, but this procedure is not routine for most virology laboratories.

Vesicular Lesion Smears

Exudate or vesicular fluids should first be removed by gentle blotting of open lesions; closed lesions must first be incised around the edge with a suitable instrument. Cells are obtained by rubbing the base of the lesion with the tip of a swab, which is then rolled firmly over a small area on one or more microscope slides. Rubbing motions are ineffective in transferring cells to the slide and should be avoided. Alternatively, scrapings may be obtained from accessible lesions by using a spatula or similar instrument held at a right angle to the plane of the lesion. Good collection technique avoids causing the lesion to bleed. Cells obtained with a spatula should be thinly spread over one or two 5- to 10-mm-diameter areas on a slide, which is then allowed to air dry.

Nasopharyngeal Smears

For detection of viruses such as respiratory syncytial virus, influenza viruses, parainfluenza viruses, and adenoviruses, cells from the nasopharynx are most satisfactory and are readily obtained. Cells may be recovered from nasal washes by centrifugation of the wash sample; cells are more conveniently obtained, however, by use of nasopharyngeal swabs. Good smears are prepared by using two swabs, the first to remove excess mucous secretions and the second to

collect the cells needed for examination. Make smears by rolling the swab tip firmly on the slide surface, as described above for vesicular lesions; if properly made, these smears are entirely satisfactory. Prepare at least one smear for each agent to be tested. Additional smears are often needed for controls.

Smears from Cell Sediments

Cells can be obtained by centrifugation of specimens such as urine, amniotic fluid, and peritoneal fluid. In some cases, it may be worthwhile to attempt rescuing cells from swabs collected for isolation attempts, as follows. Agitate the specimen (swab in viral transport medium) thoroughly with a Vortex mixer to dislodge cells from the swab fibers, and then remove swabs. (This procedure is not effective with calcium alginate swabs, which tend to dissolve partially and gel the transport medium; if these must be processed, acidify the medium to pH 4 to change the gel to a sol.) Centrifuge the remaining specimen for a few minutes at >1,000 × g, after which the supernatant fluid can be used to inoculate cell cultures. Any pellet obtained is dispersed in a small amount (0.025 ml) of buffered saline containing 0.1% bovine serum albumin and spotted onto slides for immunofluorescent examination. This technique may allow the detection of virus in specimens that have been too long in transit for recovery of any virus by isolation.

REFERENCES

1. Brinker, J. P., and G. V. Doern. 1993. Comparison of MRC-5 and A-549 cells in conventional culture tubes and shell vial assays for the detection of varicella-zoster virus. Diagn. Microbiol. Infect. Dis. 17:75–77.
2. Clark, S. M., J. R. Roth, M. L. Clark, B. B. Barnett, and R. S. Spendlove. 1981. Trypsin enhancement of rotavirus infectivity: mechanisms of enhancement. J. Virol. 39:816–822.
3. Forrer, C. B., A. L. Blahy, A. L. Malatico, J. M. Campos, and H. M. Friedman. 1982. Comparison of vancomycin and penicillin for virus isolation. J. Clin. Microbiol. 16:295–298.
4. Gershon, A., S. Steinberg, S. Greenberg, and L. Taber. 1980. Varicella-zoster-associated encephalitis: detection of specific antibody in cerebrospinal fluid. J. Clin. Microbiol. 12:764–767.
5. Gleaves, C. A., D. J. Wilson, A. J. Wold, and T. F. Smith. 1985. Detection and serotyping of herpes simplex virus in MRC-5 cells by use of centrifugation and monoclonal antibodies. J. Clin. Microbiol. 21:29–32.
6. Howell, C. L., and M. J. Miller. 1983. Effect of sucrose-phosphate and sorbitol on infectivity of enveloped viruses during storage. J. Clin. Microbiol. 18:658–662.
7. Howell, C. L., M. J. Miller, and D. A. Bruckner. 1986. Elimination of toxicity and enhanced cytomegalovirus detection in cell cultures inoculated with semen from patients with acquired immunodeficiency syndrome. J. Clin. Microbiol. 24:657–660.
8. Hudson, J. B., V. Misra, and T. R. Mosmann. 1976. Cytomegalovirus infectivity: analysis of the phenomenon of centrifugal enhancement of infectivity. Virology 72:235–243.
9. Hughes, J. H. 1993. Physical and chemical methods for enhancing rapid detection of viruses and other agents. Clin. Microbiol. Rev. 6:150–175.
10. Huntoon, C., R. House, and T. Smith. 1981. Recovery of viruses from three transport media incorporated into Culturettes. Arch. Pathol. Lab. Med. 105:436–437.
10a.Lennette, D. A. 1985. Collection and preparation of specimens for virological examination, p. 687–693. In E. H. Lennette, A. Balows, W. J. Hausler, Jr., and H. J. Shadomy (ed.), Manual of Clinical Microbiology, 4th ed. American Society for Microbiology, Washington, D.C.

11. **Lipson, S. M., P. Costello, S. Forlenza, B. Agins, and K. Szabo.** 1990. Enhanced detection of cytomegalovirus in shell vial cell culture monolayers by preinoculation treatment of urine with low-speed centrifugation. *Curr. Microbiol.* **20:**39–42.

12. **Osborn, J. E., and D. L. Walker.** 1968. Enhancement of infectivity of murine cytomegalovirus in vitro by centrifugal inoculation. *J. Virol.* **2:**853–858.

13. **Sarkies, N., Z. Gregor, T. Forsey, and S. Darougar.** 1986. Antibodies to herpes simplex virus type 1 in intraocular fluids of patients with acute retinal necrosis. *Br. J. Ophthalmol.* **70:**81–84.

14. **Slifkin, M., and R. Cumbie.** 1992. Comparison of the Histopaque-1119 method with the Plasmagel method for separation of blood leukocytes for cytomegalovirus isolation. *J. Clin. Microbiol.* **30:**2722–2724.

15. **Smith, T. F.** 1992. Specimen collection, p. 31–33. *In* S. Specter and G. Lancz (ed.), *Clinical Virology Manual*, 2nd ed. Elsevier, New York.

16. **Spendlove, R. S., M. E. McClain, and E. H. Lennette.** 1970. Enhancement of reovirus infectivity by extracellular removal or alteration of the virus capsid by proteolytic enzymes. *J. Gen. Virol.* **8:**83–94.

17. **Suttorp-Schulten, M. S. A., M. J. W. Zaal, L. Luyendijk, P. J. M. Bos, A. Kijlstra, and A. Rothova.** 1989. Aqueous chamber tap and serology in acute retinal necrosis. *Am. J. Ophthalmol.* **108:**327–328.

18. **Tobita, K., A. Sugiura, C. Enomoto, and M. Furuyama.** 1975. Plaque assay and primary isolation of influenza A viruses in an established line of canine kidney cells (MDCK) in the presence of trypsin. *Med. Microbiol. Immunol.* **162:**9–14.

Herpes Simplex Viruses

ANN M. ARVIN AND CHARLES G. PROBER

71

CLINICAL BACKGROUND

The common clinical manifestations of herpes simplex virus (HSV) infection have been recognized for centuries. With the development of methods for the laboratory isolation and characterization of HSV, two biologically distinct serotypes, HSV type 1 (HSV-1) and HSV-2, were identified. Furthermore, with the development and refinement of a variety of serologic techniques able to identify HSV infections, a better understanding of the epidemiology and pathogenesis of these infections in the human host has been achieved (55, 72). The prevalence of HSV-1 infections increases gradually from childhood, reaching 70 to 80% in later adult years, whereas HSV-2 infection is typically acquired as a sexually transmitted disease, so that its incidence begins to increase in adolescence (10). The prevalence rates for HSV-2 infection range from about 15% to more than 50% in adults, depending on a variety of demographic variables (33, 68). As with the other members of the herpesvirus group (varicella-zoster virus, Epstein-Barr virus, and cytomegalovirus), initial infection with HSV results in the establishment of viral latency, with the potential for subsequent viral reactivation. The initial mucocutaneous infection caused by HSV-1 or HSV-2 is followed by latent infection of neuronal cells in the dorsal root ganglia (6, 30, 67). Subsequent viral reactivation is accompanied by viral excretion from the original mucocutaneous sites of infection, with or without the concomitant appearance of clinical signs and symptoms. HSV transmission can result from direct contact with infected secretions from a symptomatic or an asymptomatic host (13). Although previous infection with HSV-1 does not prevent infection upon exposure to HSV-2, preexisting HSV-1 immunity may modify the severity of HSV-2 infection, rendering it clinically mild or asymptomatic.

The classic presentation of primary HSV-1 infection is herpes gingivostomatitis, an infection of the oral mucosa resulting in extensive, painful vesicular lesions associated with a high temperature and marked submandibular lymphadenopathy. Recurrences of orolabial infections are referred to as fever blisters or cold sores. Other clinical manifestations of HSV-1 infection are conjunctivitis, keratitis, and herpetic whitlow. The most serious infection caused by HSV-1 is sporadic encephalitis, which occurs in older children and adults (74). This infection has an untreated mortality rate of approximately 70%.

The classic presentation of a primary HSV-2 infection is herpes genitalis, an infection characterized by the appearance of extensive, bilaterally distributed lesions in the genital area accompanied by fever, inguinal lymphadenopathy, and dysuria (12). Symptomatic primary genital HSV is caused by HSV-2 in approximately 85% of cases, with the remaining cases caused by HSV-1. Because genital HSV-1 infection is much less likely to produce recurrences, 99% of recurrent genital herpes is due to HSV-2 (26). The most serious consequence of genital HSV infection is neonatal herpes. This infection usually results from exposure of neonates to virus being excreted by mothers at the time of vaginal delivery (14). Unfortunately, the majority (70%) of mothers who infect their neonates are experiencing asymptomatic genital infections at delivery (78). If the neonate is exposed at delivery to a mother with a recurrent infection at delivery, the attack rate is quite low, probably less than 5% (52). However, if the mother is experiencing a primary infection at delivery, the attack rate is probably greater than 50% (75). Neonates may present with infection localized to the skin, eyes, and mucosa or the central nervous system, or the infant may present with a disseminated infection. The mortality rate for untreated infants who develop disseminated infection exceeds 70% (78). The early institution of therapy can substantially reduce the morbidity and mortality of mucocutaneous infections and the mortality rates of disseminated and central nervous system infections (75, 77).

Many individuals with primary HSV-1 or HSV-2 infection do not manifest characteristic clinical disease. In fact, primary infections are often entirely asymptomatic. Similarly, despite the apparently universal establishment of latency, most individuals with past HSV infections do not experience symptomatic recurrences. Nevertheless, individuals who have had HSV-1 or HSV-2 infection are subject to asymptomatic reactivations associated with the isolation of infectious virus from oral or genital sites. Therefore, prevention of the transmission of HSV-1 and HSV-2 in the population is very difficult.

Because of the high prevalence of past HSV infections in the general population, many patients who develop malignancy, an immunodeficiency such as AIDS, or other diseases that require immunosuppressive therapy may experience HSV-1 or HSV-2 infection. These infections, which may be primary or may arise from reactivation of a past infection, can be severe (63). They can be locally invasive, causing considerable mucocutaneous necrosis, or they can spread to contiguous organs, causing infections such as esophagitis or proctitis. Herpes infections in these immunocompromised hosts can also produce viremia with dissemination to multiple organs, causing meningoencephalitis, pneumonitis, hepatitis, and coagulopathy. In addition to the susceptibility of classically immunosuppressed patients, some individuals with chronic skin diseases, particularly eczema, can experience severe primary HSV-1 infection, referred to as Kaposi's varicelliform eruption.

While a presumptive diagnosis of HSV infection can often be made on the basis of clinical findings, a definitive diagnosis is important in many circumstances. Appropriate laboratory testing allows the proper evaluation of mucocutaneous lesions in high-risk patients who are likely to benefit from specific antiviral therapy if the lesion is herpetic (66). The laboratory documentation of HSV infection also provides critical information for managing patients with disseminated disease or encephalitis. In addition, laboratory diagnosis is valuable for many patients with HSV infections that are not life threatening. For example, proving that a genital infection is herpes facilitates counseling regarding the advisability of acyclovir therapy, the risk of recurrences, and the measures to reduce HSV transmission to contacts.

DESCRIPTION OF THE AGENT

HSV-1 and HSV-2, along with varicella-zoster virus, are the human herpesviruses that are classified in the alphaherpesvirus subfamily of herpesviruses (58). Important characteristics of this subfamily include a short replication cycle, production of lytic infection in tissue culture, and establishment of latency in neural ganglia. Like other herpesviruses, HSV-1 and HSV-2 have icosahedral capsids consisting of 162 capsomeres surrounding a core that contains the viral DNA. The phospholipid-rich viral envelope is acquired when the virion buds through regions of the nuclear membrane that have been modified by insertion of viral proteins. The complete virion has a diameter of between 110 and 120 nm. Tegument proteins have been identified between the capsid and the viral envelope. HSV DNA is linear, double stranded, and relatively G+C rich, with a molecular weight of 96×10^6. The viral genome consists of a long unique and a short unique region, each of which is flanked by inverted-repeat regions. Extensive sequence homologies, involving about 40% of the genome, can be demonstrated between HSV-1 and HSV-2 DNAs, to which the antigenic cross-reactivity of the two serotypes and other biological similarities are attributable. The overall pattern of nucleotide sequences composing HSV DNA is conserved; however, restriction endonuclease analysis demonstrates sufficient variability of cleavage sites among clinically unrelated isolates to permit the investigation of HSV transmission by molecular epidemiologic methods.

The HSV genome codes for more than 50 polypeptides, including at least nine glycoproteins (designated glycoprotein A [gA] through gI), at least six capsid proteins, viral protein kinase, DNA polymerase, other enzymes, and DNA-binding proteins that are involved in viral replication (14, 58). The immediate-early proteins of the virus, such as ICP0 and ICP4, constitute products of the alpha genes of HSV that are involved in initiating the cascade by which the expression of HSV genes is regulated. With respect to the glycoproteins, gB and gD carry major epitopes to which neutralizing antibodies bind, and they appear to be involved in viral attachment to the target cell. gC binds to the cell surface complement receptor, C3b; gE and gI are involved in constituting the receptor for immunoglobulin that appears on infected-cell membranes. gB, gD, and gH apparently affect the movement of adsorbed virus from the surface into the cytoplasm of the cell. Although these and other functions of some HSV proteins have been defined or suggested, the understanding of specific effects on virus entry and virus-induced changes in mammalian cells remains quite limited. Because the host cell-derived viral envelope is phospholipid rich, the virus is readily inactivated by lipid solvents; exposure to a pH of <4 and a temperature of 256°C maintained for 20.5 h also eliminates infectivity.

COLLECTION AND STORAGE OF SPECIMENS

HSV-1 and HSV-2 can be recovered from clinical specimens obtained by swabbing mucocutaneous lesions or previously involved mucocutaneous sites in patients with asymptomatic infection. Cotton swabs are preferred for taking specimens for HSV culture; calcium alginate swabs can reduce viral recovery (16). Fresh vesicles, which contain a high concentration of virus, may be aspirated with a small-gauge (e.g., 25-gauge) needle attached to a tuberculin syringe. After aspiration, the surface of the vesicle is removed, and a premoistened swab is used to absorb any remaining fluid. The base of the lesion is then swabbed vigorously to recover infected epithelial cells. Ideally, specimens should be immediately inoculated into tubes of cell cultures. If direct inoculation is not feasible, place the swab specimens into 1 to 2 ml of viral transport medium (discussed in chapter 70 of this Manual). Clinical specimens that might contain HSV can be effectively shipped to reference laboratories at ambient temperatures in Virocult transport tubes (Medical Wire and Equipment Co., Cleveland, Ohio). HSV can survive in these tubes for about 2.75 days at 22°C, and the tubes are compact, enclosed, and resistant to breakage (32).

HSV can be isolated directly from cerebrospinal fluid obtained from patients with meningitis or from peripheral blood leukocytes in those with disseminated infection. Heparin can interfere with viral isolation and should not be used as an anticoagulant. HSV can also be isolated from homogenized sterile tissue specimens, such as brain tissue from patients with encephalitis; see chapter 70 for details. Trypsinizing the tissue may enhance the efficiency of virus isolation from such samples (5). HSV-2 can sometimes be isolated from urine samples obtained from patients with genital herpes complicated by urethritis or cystitis. Rectal swab specimens rarely yield HSV.

The highest isolation rates of HSV are likely if specimens are inoculated on the day they are taken; otherwise, careful attention must be given to conditions of transport and storage, which can affect the recovery of HSV. Specimens may be maintained at 4°C during transport and for up to 48 h (7, 81) but must not be frozen at −20°C. See chapter 70 for additional information.

ISOLATION OF VIRUS

Viral culture is the most sensitive method for the laboratory diagnosis of HSV, and it also allows typing of the viral isolate. HSV-1 and HSV-2 cause typical cytopathic effects (CPE) in a wide variety of cell culture systems (21). Primary human embryonic cells and human diploid cell lines, such as MRC-5, are commonly used because they are commercially available and because other viruses can be isolated in these cell types. Primary rabbit kidney, mink lung, and rhabdomyosarcoma are particularly sensitive cell lines, especially for HSV isolation (35). Continuous human or primate cell lines, such as HEp-2 and Vero, are somewhat less sensitive. The inoculation of two different cell lines in parallel can minimize the periodic variations in sensitivity of different cell lines that are often difficult to avoid (9). The difference in the rates of recovery of HSV in two acceptable cell lines is less than 5% and is apparent with clinical specimens that contain low concentrations of virus. If primary cells are not used, the recovery of HSV is optimal when the cell lines are used at low passage numbers. Most procedures for HSV culture incorporate a l-h adsorption at 36 to 37°C before incubation with a standard tissue culture medium such as Eagle's minimal essential medium supplemented with 2% fetal calf serum and antibiotics.

The presence of HSV is detected by observing foci of enlarged, refractile cells in the monolayer; some clinical isolates also induce the formation of syncytia with multinucleated giant cells. Although the CPE induced by HSV-1 tends to be more diffuse throughout the monolayer than that induced by HSV-2, this observation is not sufficiently specific to be of differentiating value. The incubation time required to observe CPE depends on the concentration of virus in the sample and the condition of the tissue culture cells. With careful maintenance of tissue culture cells, samples with high concentrations of virus produce CPE within 18 to 24 h, and samples containing low concentrations of virus should be identifiable as positive within 4 to 5 days. Cultures from oral or genital lesions can be expected to be positive within 3 days (9). In one study, CPE was observed within 4 days in more than 99% of genital tract specimens from women with asymptomatic HSV-2 reactivation (80). Adding dexamethasone to the culture medium may hasten the appearance of viral CPE in standard tube cultures (79) or may increase the number of infected-cell foci that are visible (73).

The interval to virus identification can be shortened somewhat by preparing the tissue culture cells on removable glass coverslips and inoculating multiple cultures with each specimen. At prescribed intervals after inoculation, the coverslip is removed and the monolayer is stained by using HSV antibodies labeled with fluorescein, biotin-avidin complexes, staphylococcal protein A, or immunoperoxidase reagents (53, 61). Immunologic staining can reveal foci of virus-infected cells before typical CPE is apparent. This immunologic approach has also been applied to regular tissue culture systems and to shell vial systems incorporating centrifugation of the specimen onto the monolayer (26). The yield for earlier viral diagnosis with these techniques depends on the specific tissue culture cell line used, the quality of the HSV-specific antibody reagents, and the viral inoculum in the clinical specimen (19, 31, 46, 50, 82). Typically, results with early immunologic staining at 24 h are not as sensitive as those for standard tube cell cultures that are maintained for 5 to 7 days (20, 60, 79). In an alternative approach, the cells from one of the replicate cultures are lysed with detergent at periodic intervals after inoculation, and the supernatant is tested by an enzyme immunoassay method for HSV antigen detection. Another procedure that may shorten the interval to viral detection

is to infect cells in suspension and then immunostain them (41). Earlier detection of HSV has also been accomplished by using HSV-specific DNA probes to demonstrate the presence of virus (23). The cumulative experience with early detection methods used at 48 h compared with standard tube cultures indicates that the sensitivity with which HSV is identified in clinical specimens is maintained but not enhanced by the former. For clinical purposes, shortening the interval to virus isolation by 24 to 48 h is usually not critical. However, the fact that some of these procedures lend themselves to the semiautomated processing of specimens (Vitek Immuno Diagnostic Assay System; Vitek Systems Inc., Hazelwood, Mo.) may be useful for large diagnostic laboratories (34).

In the case of mucocutaneous HSV infection, the success with which the laboratory can isolate the etiologic agent from clinical specimens depends substantially on the type of lesion being evaluated. For example, in one study, HSV was recovered from 94% of genital herpes lesions cultured during the vesicular stage, from 87% cultured during the pustular phase, and from 70% cultured during the ulcer stage but from only 27% cultured during the crusted stage (44). The specific site of the infection also influences whether the virus is recovered. Examples include the observations that HSV-1 is rarely isolated from the cerebrospinal fluid of patients with herpes encephalitis and that HSV-2 is more likely to be recovered from the usual site of lesion recurrences than from the cervix in women with asymptomatic reactivation of genital HSV-2 infection (2).

IDENTIFICATION OF THE VIRUS

It is important to confirm that CPE is due to HSV and not to varicella-zoster virus, cytomegalovirus, or nonspecific toxic changes. It is also often desirable to distinguish HSV serotypes. Although HSV serotypes have been differentiated in the past by biochemical and biological differences, the availability of HSV-specific monoclonal antibody reagents has facilitated using a simplified immunologic means of verifying and typing HSV isolates (5, 27, 42). Identification of viral isolates with monoclonal antibodies has replaced inoculation of chicken embryo chorioallantoic membrane and guinea pig cells as well as neutralization endpoint and other immunologic procedures for differentiating HSV-1 and HSV-2 isolates. Although in theory a specific monoclonal antibody might not react with epitopes of the relevant viral protein as produced by all HSV strains, in practice it has been possible to develop reagents that are not significantly affected by this potential for antigenic variation. HSV typing with monoclonal antibody reagents is significantly more accurate than typing with polyclonal type-specific rabbit antisera. In a study performed in 1987, three pairs of commercially available monoclonal antibodies (Electro-Nucleonics, Inc., Syva, Inc.; Kallestad Laboratories, Inc.) tested in parallel performed satisfactorily (40). The specificity of the results corresponded closely with typing of isolates by restriction enzyme analysis, which defines the HSV type at the molecular level (5). Use of both HSV-1 and HSV-2 monoclonal antibody reagents makes strain differentiation part of the confirmation procedure.

Another approach for virus identification involves DNA-DNA hybridization in a dot blot system with cloned or synthetic DNA probes that are specific for unique nucleotide sequences of HSV-1 and HSV-2 (19, 49). Such methods are comparable to restriction endonuclease analysis in the specificity with which HSV types are distinguished and may be more widely used when better nonradioactive labeled probes are developed. The fact that HSV-1 is inhibited by antiviral agents such as bromovinyldeoxyuridine, to which HSV-2 strains are resistant, has also been exploited to differentiate HSV types.

DIRECT EXAMINATION

When mucocutaneous lesions are present, direct detection methods can provide a rapid diagnosis of HSV infection. The most common method used for the direct detection of HSV in clinical specimens is direct immunofluorescence or immunoperoxidase staining of cells taken from mucocutaneous lesions (5, 14). These samples are best obtained by exposing the base of the lesion, removing cells with the blunt end of a cotton applicator stick, and immediately streaking the sample onto a glass slide. Since a negative result is reliable only if intact cells are transferred to the slide, the laboratory should confirm the adequacy of the specimen before processing it. With use of fluorescein-conjugated monoclonal antibodies to HSV to stain cytologic preparations of lesion scrapings, sensitivities compared with that of tissue culture isolation are as high as 78 to 88% with relatively few false-positive reactions (27, 51, 64, 70). However, a sample for viral culture should be obtained when the lesion scraping is made to allow confirmation of the direct detection result, since both false-positive and false-negative results can occur with the immunofluorescence and immunoperoxidase staining methods. The Papanicolaou (Pap) stain or Tzanck test can be used to demonstrate cytologic changes in specimens obtained from suspected HSV lesions. The cytologic changes being sought include syncytial giant cells, "ballooning" cytoplasm, and Cowdry type A intranuclear inclusions. These pathologic examinations can be useful, inexpensive methods for evaluating patients with non-life-threatening illness. However, these methods are not specific for HSV and are much less sensitive than direct detection with immunologic reagents (14). Therefore, a negative Tzanck or Pap smear cannot be relied on for the diagnosis of HSV in critical situations such as infections in newborns, pregnant women at term, patients with encephalitis, or immunosuppressed patients.

Direct detection of infected cells with immunologic reagents has been established as reliable for identifying virus-infected cells in brain tissue from patients with HSV encephalitis, but its sensitivity in this setting results from the fact that many cells harbor HSV. This method should not be extended to the analysis of other samples, such as testing cells from cerebrospinal fluid of patients with encephalitis, attempting to detect HSV-infected cells in genital tract specimens from asymptomatic pregnant women, or examining cells in tracheal aspirate or bronchoalveolar lavage samples from immunocompromised patients with pneumonia.

Monoclonal antibodies capable of binding HSV proteins have also been used to enhance the detection of HSV antigens in clinical specimens by using enzyme-linked immunosorbent assays, (ELISAs), immunoperoxidase assays, hemagglutination assays, or avidin-biotin enzyme conjugate assays (1, 24, 43, 47, 56, 64, 65). Most of these methods are applied to the detection of viral antigen in samples collected as described above for viral culture. Antigen detection may be enhanced by concentrating the sample by membrane filtration or by collecting the original sample in

a special transport system (17, 57). The sensitivities of these methods compared with that of viral isolation range from 70 to 95%, with specificities of 65 to 90%. However, none of the antigen detection techniques has been proved to be sensitive enough to detect the asymptomatic shedding of HSV (47, 69, 71). The sensitivities of commercial ELISAs for the detection of HSV antigens are also variable when these assays are compared to viral culture (28).

Whereas methods that identify HSV proteins in the clinical sample are not likely to be improved sufficiently to permit the detection of asymptomatic HSV infection, the sensitivity of the PCR method appears to be adequate for this purpose (11, 29). Viral culture methods effectively amplify HSV by allowing viral replication. Similarly, the PCR method can amplify the "signal" of HSV DNA up to 10^6-fold and should be much more sensitive than methods to detect viral proteins. In genital herpes, PCR detects viral genome or genome fragments for several days after lesions become negative for infectious virus (11). The clinical significance of the prolonged detection of HSV DNA is not known. Correlating PCR positivity and the risk of viral transmission to susceptible contacts and determining when PCR positivity is an indication for antiviral therapy is a subject of continuing clinical investigations. HSV PCR has potential value for testing cerebrospinal fluid in patients with suspected herpes encephalitis, but definitive studies using samples from patients with virologically proved disease have not been completed at present (54, 59).

Direct viral detection by DNA hybridization using radiolabeled or biotinylated probes has also been evaluated for identifying HSV in pathologic specimens (22, 38). These methods, as well as the use of electron microscopy, are sensitive for demonstrating infected cells in tissue sections but are not widely used in clinical laboratories.

SEROLOGIC DIAGNOSIS

Assessing HSV immune status to document whether an individual has had past infection with HSV can be done by many serologic methods. Most methods are generally quite sensitive for detecting HSV immunoglobulin G (IgG) antibodies in individuals with past HSV infection regardless of whether the patient has had any recent signs of HSV disease. For most laboratories, HSV serologic testing is accomplished most efficiently by using commercial kits based on ELISA or latex agglutination procedures (18, 25). However, the extensive cross-reactivity between HSV-1 and HSV-2 makes it impossible to differentiate past HSV-1 from past HSV-2 infection with any of these methods (68). Since physicians are not often aware of this problem, serologic reports that list HSV-1 and HSV-2 antibody titers separately are likely to be misleading. As documented by Ashley et al. (3), currently licensed assays do not discriminate between infections with HSV-1 and HSV-2.

Recurrent HSV infections are not always accompanied by a significant rise in antibody titer. Therefore, serologic tests should not be used to diagnose recurrent HSV infections. Primary HSV infection can be documented by using any of the standard methods to show seroconversion with paired sera. However, just as cross-reactivity prevents the distinction of past HSV-1 from past HSV-2 infection, the serotype of HSV responsible for the seroconversion cannot be determined with commercially available serologic tests.

Testing for IgM antibodies does not improve the specificity of the serologic diagnosis in patients with clinical signs of HSV infection. HSV IgM assays cannot be used to distinguish primary from recurrent infections, because the host response to reactivation can also include IgM antibody production. In addition, the maintenance of quality control for HSV IgM antibody assays is very difficult and is complicated particularly by false-positive results, even when efforts are made to fractionate serum IgG and IgM before testing.

Testing for local production of HSV IgG in cerebrospinal fluid samples can be used to document HSV encephalitis in some patients, but a 2- to 4-week interval may be required before positive results can be demonstrated (45). Therefore, serologic testing is not helpful in providing an early diagnosis to guide the use of antiviral therapy. There is no known diagnostic value in testing cerebrospinal fluid for HSV IgM antibodies.

Recently, serologic methods that circumvent the problem of cross-reactivity between HSV-1 and HSV-2 antibodies have been developed. Serum samples can be tested against HSV-1 and HSV-2 antigens by Western blot (immunoblot) to demonstrate reactivity with type-specific viral proteins (2). The fact that the gG of HSV-1 differs significantly from the HSV-2 homolog has also permitted the development of type-specific serologic assays (2, 4, 8, 39, 68). Monoclonal antibody to HSV-2 gG is used either to capture the protein from an HSV-infected cell sonic extract in a solid-phase ELISA or to prepare immunoaffinity-purified HSV-2 gG for use as antigen in a dot blot assay. With use of the capture ELISA method, antibodies to HSV-2 gG were detected in 96 to 98% of persons with previous episodes of culture-proved HSV-2 infections, with very few false-positive results (48, 62, 68). Western blot analysis or assays detecting antibodies to HSV-2 gG can also be used to document primary HSV-2 infections, even in patients who have had previous infections caused by HSV-1 (4). Unfortunately, none of these HSV type-specific serologic tests have been developed commercially, although new sources of gG protein are now available.

EVALUATION, INTERPRETATION, AND REPORTING OF RESULTS

Rapid Diagnostic Test
Encephalitis caused by HSV and HSV infections in neonates has untreated mortality rates approximating 70% (4, 78). In recent years, effective antiviral therapy has been developed for these life-threatening infections, resulting in the potential to reduce their mortality rates by at least 50% (74, 77). In addition, the course of severe mucocutaneous and disseminated HSV infections in immunosuppressed hosts can be favorably altered by antiviral chemotherapy (66). Furthermore, genital HSV infections, which are the cause of substantial morbidity, can now be effectively managed with acyclovir therapy; the course of clinical attacks can be abrogated, and the frequency of recurrences can be significantly reduced (68). Thus, optimal patient management demands the availability of accurate diagnostic tests for HSV.

For non-life-threatening HSV infections, the availability of a diagnostic laboratory able to perform viral cultures is adequate. Since these infections can usually be recognized clinically, it is generally sufficient for a clinician to obtain virologic confirmation within several days of specimen submission. This can be accomplished readily with viral cultures. However, for life-threatening infections that mandate prompt antiviral therapy, an accurate and rapid

diagnostic test for HSV must be employed. The best currently available test is immunofluorescent staining of infected tissues. Patients with HSV encephalitis are more likely to have a favorable outcome if their antiviral therapy is administered early in the infection (74). Since HSV encephalitis is a difficult infection to diagnose clinically, brain biopsies are recommended for these patients (74). Although antiviral therapy can be started before a definitive diagnosis is made, a positive rapid diagnostic test assures the clinician that optimal therapy has been initiated. A rapid diagnostic test for HSV also must be available to physicians involved with the treatment of newborn infants, because prompt therapy of newborns presenting with skin lesions as the only sign of an HSV infection usually has an excellent outcome. However, if the diagnosis and treatment of this infection are delayed, dissemination beyond the skin may soon follow, and the outcome will be significantly worse (76).

Rapid diagnostic tests for HSV are not sensitive enough to permit the diagnosis of oral or genital HSV infections not associated with lesions, and the clinical use of these tests in these circumstances should be discouraged (55). As noted above, in clinical practice, HSV PCR results must be interpreted very carefully.

Isolate Typing
Although the serotype of HSV responsible for the clinical infection is important epidemiologically, there are only a few situations in which a clinician needs this information. However, if typing of isolates is not done routinely, it is prudent to save the HSV isolate, at least until the clinician has been contacted, to be sure that typing is not desired. One situation in which it is mandatory to type HSV isolates is when the isolate is from the genital tract of a young child. A type 2 isolate recovered from the genital tract of a child should raise concern about the possibility of sexual abuse, whereas a type 1 isolate might be explained on the basis of autoinoculation from the oropharynx.

The type of HSV responsible for the infection can also have prognostic implications. For example, genital infections caused by HSV-1 are less likely to recur than genital infections caused by HSV-2 (37), and HSV encephalitis occurring in newborn infants is likely to be less severe if caused by HSV-1 (15). Although these observations are interesting, they do not demand the immediate availability of viral typing.

Serology
Routinely available serologic tests for HSV cannot reliably differentiate type 1 and type 2 antibodies. This inability underscores the futility of attempting to differentiate past infections with either of these viruses serologically by using commercially available tests. Since most adults have had HSV-1 infection, often without any primary or recurrent symptoms, the serologic diagnosis of acute or past HSV-2 infection is not possible with available methods. Serologic tests can be used to diagnose a true primary HSV infection only if there is no HSV antibody in the acute-phase serum. In this circumstance, if antibody to HSV is present in the convalescent-phase serum sample, a seroconversion can be diagnosed. However, because of the cross-reactivity between HSV-1 and HSV-2, the specific serotype of HSV responsible for the acute infection cannot be established without the use of the type-specific tests described above. It is important to recognize that recurrent infections with HSV cannot be diagnosed serologically. Although a rise in

antibody titer might be evident, such rises can also be nonspecific, occurring, for example, in response to recent infections with other herpesviruses (e.g., varicella-zoster virus).

Considering these limitations, laboratories using commercially available serologic kits should not attempt to report any type-specific HSV antibody titers. Laboratories should report only that serologic evidence of a prior or recent HSV infection is evident. In addition, because the assays detect cross-reacting antibodies, the practice of testing serum samples against both HSV-1 and HSV-2 antigens is not cost-effective for the patient or the laboratory. Until accurate methods are available, genital herpes will continue to be underdiagnosed (36).

REFERENCES

1. Adler-Storthz, K., C. Kendall, R. C. Kennedy, R. D. Henkel, and G. R. Dreesman. 1983. Biotin-avidin amplified enzyme immunoassay for detection of herpes simplex virus antigen in clinical specimens. *J. Clin. Microbiol.* **18:**1329–1334.
2. Arvin, A. M., P. A. Henslelgh, C. G. Prober, D. S. Au, L. L. Yasukawa, A. E. Wittek, P. E. Palumbo, S. G. Paryanl, and A. S. Yeager. 1986. Failure of antepartum maternal cultures to predict the infant's risk of exposure to herpes simplex virus at delivery. *N. Engl. J. Med.* **315:**796–800.
3. Ashley, R., A. Cent, V. Maggs, A. Nahmias, and L. Corey. 1991. Inability of enzyme immunoassays to discriminate between infections with herpes simplex virus types 1 and 2. *Ann. Intern. Med.* **115:**520–526.
4. Ashley, R. L., J. Militoni, F. Lee, A. Nahmias, and L. Corey. 1988. Comparison of Western blot (immunoblot) and glycoprotein G-specific immunodot enzyme assay for detecting antibodies to herpes simplex virus types I and 2 in human sera. *J. Clin. Microbiol.* **26:**662–667.
5. Balachandran, N., B. Franme, M. Chernesky, E. Kraiselburd, Y. Kouri, D. Garcia, C. Lavery, and W. E. Rawls. 1982. Identification of typing of herpes simplex viruses with monoclonal antibodies. *J. Clin. Microbiol.* **16:**205–208.
6. Baringer, J. R., and P. Swoveland. 1973. Recovery of herpes simplex virus from trigeminal ganglions. *N. Engl. J. Med.* **288:**648–650.
7. Bernard, D. L., K. Farnes, D. F. Richards, G. F. Croft, and F. B. Johnson. 1986. Suitability of new chlamydia transport medium for transport of herpes simplex virus. *J. Clin. Microbiol.* **24:**692–695.
8. Boucher, F. D., L. Y. Yasukawa, K. Kerns, M. Kastelein, A. M. Arvin, and C. G. Prober. 1993. Detection of antibodies to herpes simplex virus type 2 with a mammalian cell line expressing glycoprotein gG-2. *Clin. Diagn. Virol.* **1:**29–38.
9. Callihan, D. R., and M. Menegus. 1984. Rapid detection of herpes simplex virus in clinical specimens with human embryonic lung fibroblast and primary rabbit kidney cell cultures. *J. Clin. Microbiol.* **19:**563–565.
10. Coleman, R. M., L. Pereira, P. D. Bailey, D. Dondero, C. Wickliffe, and A. J. Nahmias. 1983. Determination of herpes simplex virus type-specific antibodies by enzyme-linked immunosorbent assay. *J. Clin. Microbiol.* **18:**287–291.
11. Cone, R. W., A. C. Hobson, J. Palmer, M. Remington, and L. Corey. 1991. Extended duration of herpes simplex virus DNA in genital lesions detected by the polymerase chain reaction. *J. Infect. Dis.* **164:**757–760.
12. Corey, L., H. G. Adams, Z. A. Brown, and K. K. Holmes. 1983. Genital herpes simplex virus infections: clinical manifestations, course, and complications. *Ann. Intern. Med.* **98:**958–972.
13. Corey, L., and K. K. Holmes. 1983. Genital herpes simplex virus infections: current concepts in diagnosis, therapy, and prevention. *Ann. Intern. Med.* **98:**973–983.
14. Corey, L., and P. G. Spear. 1986. Infections with herpes simplex viruses. *N. Engl. J. Med.* **314:**686–691, 749–757.
15. Corey, L., E. F. Stone, R. J. Whitley, and K. Mohan. 1988. Difference between herpes simplex virus type 1 and type 2 neonatal encephalitis in neurological outcome. *Lancet* **i:**1–4.
16. Crane, L. R., P. A. Gutterman, T. Chapel, and A. M. Lerner. 1980. Incubation of swab materials with herpes simplex virus. *J. Infect. Dis.* **141:**531.
17. Dascal, A., J. Chan-Thim, M. Morahan, J. Portnoy, and J. Mendelson. 1989. Diagnosis of herpes simplex virus infection in a clinical setting by a direct antigen detection enzyme immunoassay kit. *J. Clin. Microbiol.* **27:**700–704.
18. DeGirolami, P. C., J. Dakos, K. Eichelberger, and S. Biano. 1988. Evaluation of a new latex agglutination method for detection of antibody to herpes simplex virus. *J. Clin. Microbiol.* **26:**1024–1025.
19. Espy, M. J., and T. F. Smith. 1988. Detection of herpes simplex virus in conventional tube cell cultures and in shell vials with a DNA probe kit and monoclonal antibodies. *J. Clin. Microbiol.* **26:**22–24.
20. Espy, M. J., A. D. Wold, D. J. Jespersen, M. F. Jones, and T. F. Smith. 1991. Comparison of shell vials and conventional tubes seeded with rhabdomyosarcoma and MRC-5 cells for the rapid detection of herpes simplex virus. *J. Clin. Microbiol.* **29:**2701–2703.
21. Fayram, L., S. L. Aarnaes, E. M. Peterson, and L. M. de la Maza. 1986. Evaluation of five cell types for the isolation of herpes simplex virus. *Diagn. Microbiol. Infect. Dis.* **5:**127–133.
22. Forghani, B., K. W. Dupuis, and N. J. Schmidt. 1985. Rapid detection of herpes simplex virus DNA in human brain tissue by in situ hybridization. *J. Clin. Microbiol.* **22:**656–658.
23. Forman, M. S., C. S. Merz, and P. Charache. 1992. Detection of herpes simplex virus by a nonradiometric spin-amplified in situ hybridization assay. *J. Clin. Microbiol.* **30:**581–584.
24. Fung, J. C., J. Shanley, and R. C. Tilton. 1985. Comparison of the detection of herpes simplex virus in direct clinical specimens with herpes simplex virus-specific DNA probes and monoclonal antibodies. *J. Clin. Microbiol.* **22:**748–753.
25. Gleaves, C. A., and J. D. Meyers. 1988. Determination of patient herpes simplex virus immune status by latex agglutination. *J. Clin. Microbiol.* **26:**1402–1403.
26. Gleaves, C. A., D. J. Wilson, A. D. Wold, and T. F. Smith. 1985. Detection and serotyping of herpes simplex virus in MRC-5 cells by use of centrifugation and monoclonal antibodies 16 h postinoculation. *J. Clin. Microbiol.* **21:**29–32.
27. Goldstein, L. C., L. Corey, J. K. McDougall, E. Tolentiono, and R. C. Nowinski. 1983. Monoclonal antibodies to herpes simplex viruses: use in antigenic typing and rapid diagnosis. *J. Infect. Dis.* **147:**829–837.
28. Gonik, B., M. Seibel, A. Berkowitz, M. B. Woodin, and K. Mills. 1991. Comparison of two enzyme-linked immunosorbent assays for detection of herpes simplex virus antigen. *J. Clin. Microbiol.* **29:**436–438.
29. Hardy, D. A., A. M. Arvin, L. L. Yasukawa, D. M. Lewinsohn, P. A. Hensleigh, and C. G. Prober. 1990. The successful identification of asymptomatic genital herpes simplex infection at delivery using the polymerase chain reaction. *J. Infect. Dis.* **162:**1031–1035.
30. Hill, T. J. 1985. Herpes simplex virus latency, p. 201–206. In B. Roizman (ed.), *The Herpesviruses*, vol. 3. Plenum Publishing Corp., New York.
31. Hughes, J. H., D. R. Mann, and V. V. Hamparian. 1986. Viral isolation versus immune staining of infected cell cultures for the laboratory diagnosis of herpes simplex virus infections. *J. Clin. Microbiol.* **24:**487–489.
32. Johnson, F. B., R. W. Levitt, and D. F. Richards. 1984. Evaluation of the Virocult transport tube for isolation of herpes simplex virus from clinical specimens. *J. Clin. Microbiol.* **20:**120–122.
33. Johnson, R. E., A. J. Nahmias, L. S. Magder, F. K. Lee, C. A. Brooks, and C. B. Snowden. 1989. A seroepidemiologic survey of the prevalence of herpes simplex virus type 2 infection in the United States. *N. Engl. J. Med.* **321:**7–12.

34. Johnston, S. L. G., S. Hamilton, R. Bindra, D. A. Hursh, and C. A. Gleaves. 1992. Evaluation of an automated immunodiagnostic assay system for direct detection of herpes simplex virus antigen in clinical specimens. *J. Clin. Microbiol.* **30:**1042–1044.

35. Johnston, S. L. G., K. Wellens, and C. S. Siegel. 1990. Rapid isolation of herpes simplex virus by using mink lung and rhabdomyosarcoma cell cultures. *J. Clin. Microbiol.* **28:**2806–2807.

36. Koutsky, L. A., C. E. Stevens, K. K. Holmes, R. L. Ashley, N. B. Kiviat, C. W. Critchlow, and L. Corey. 1992. Underdiagnosis of genital herpes by current clinical and viral-isolation procedures. *N. Engl. J. Med.* **326:**1533–1539.

37. Lafferty, W. E., R. W. Coombs, J. Benedetti, C. Critchlow, and L. Corey. 1987. Recurrences after oral and genital herpes simplex virus infection. Influence of site of infection and viral type. *N. Engl. J. Med.* **316:**1444–1449.

38. Langenberg, A., R. Zbanysek, J. Dragavon, R. Ashley, and L. Corey. 1988. Detection of herpes simplex virus DNA from genital lesions by in situ hybridization. *J. Clin. Microbiol.* **26:**933–937.

39. Lee, F. K., R. M. Coleman, L. Pereira, P. D. Bailey, M. Tatsumo, and A. J. Nahmias. 1985. Detection of herpes simplex virus type 2-specific antibody with glycoprotein G. *J. Clin. Microbiol.* **22:**641–644.

40. Lipson, S. M., T. E. Schutzbank, and K. Szabo. 1987. Evaluation of three immunofluorescence assays for culture confirmation and typing of herpes simplex virus. *J. Clin. Microbiol.* **25:**391–394.

41. Luker, G., C. Chow, D. F. Richards, and F. B. Johnson. 1991. Suitability of infection of cells in suspension for detection of herpes simplex virus. *J. Clin. Microbiol.* **29:**1554–1557.

42. Miller, M. J., and C. L. Howell. 1983. Rapid detection and identification of herpes simplex virus in cell culture by a direct immunoperoxidase staining procedure. *J. Clin. Microbiol.* **18:**550–553.

43. Miranda, Q. R., G. D. Bailey, A. S. Fraser, and H. J. Tenoso. 1977. Solid-phase enzyme immunoassay for herpes simplex virus. *J. Infect. Dis.* **136**(Suppl.)**:**S304–S310.

44. Moseley, R. C., L. Corey, D. Benjamin, C. Winter, and M. L. Remington. 1981. Comparison of viral isolation, direct immunofluorescence, and indirect immunoperoxidase techniques for detection of genital herpes simplex virus infection. *J. Clin. Microbiol.* **13:**913–918.

45. Nahmias, A. J., R. J. Whitley, A. N. Visintine, Y. Takei, C. A. Alford, and the Collaborative Antiviral Study Group. 1982. Herpes simplex virus encephalitis: laboratory evaluations and their diagnostic significance. *J. Infect. Dis.* **145:**829–836.

46. Nerurkar, L. S., A. J. Jacob, D. L. Madden, and J. L. Sever. 1983. Detection of genital herpes simplex infections by a tissue culture–fluorescent-antibody technique with biotin-avidin. *J. Clin. Microbiol.* **17:**149–154.

47. Nerurkar, L. S., M. Namba, C. Brashears, A. J. Jacob, Y. S. Lee, and J. L. Sever. 1984. Rapid detection of herpes simplex virus in clinical specimens using a capture biotin-streptavidin enzyme-linked immunosorbent assay. *J. Clin. Microbiol.* **20:**109–114.

48. Parkes, D. L., C. M. Smith, J. M. Rose, J. Brandis, and S. R. Coates. 1991. Seroreactive recombinant herpes simplex virus type 2-specific glycoprotein G. *J. Clin. Microbiol.* **29:**778–781.

49. Peterson, E. M., S. L. Aarnaes, R. N. Bryan, J. L. Ruth, and L. M. de la Maza. 1986. Typing of herpes simplex virus with synthetic DNA probes. *J. Infect. Dis.* **153:**757–762.

50. Peterson, E. M., B. L. Hughes, S. L. Aarnaes, and L. M. de la Maza. 1988. Comparison of primary rabbit kidney and MRC-5 cells and two stain procedures for herpes simplex virus detection by a shell vial centrifugation method. *J. Clin. Microbiol.* **26:**222–224.

51. Pouletty, P., J. J. Chomel, D. Thouvenot, F. Catalan, V. Rabillon, and J. Kadouche. 1987. Detection of herpes simplex virus in direct specimens by immunofluorescence assay using a monoclonal antibody. *J. Clin. Microbiol.* **25:**958–959.

52. Prober, C. C., W. M. Sullender, L. L. Yasukawa, D. S. Au, A. S. Yeager, and A. M. Arvin. 1987. Low risk of herpes simplex virus infections in neonates exposed to the virus at the time of vaginal delivery to mothers with recurrent herpes simplex virus infections. *N. Engl. J. Med.* **316:**240–244.

53. Pruneda, R. C., and I. Almanza. 1987. Centrifugation-shell vial technique for rapid detection of herpes simplex virus cytopathic effect in Vero cells. *J. Clin. Microbiol.* **25:**423–424.

54. Puchhammer-Stoeckl, E., F. X. Heinz, M. Kundi, T. Popow-Kraupp, G. Grimm, M. M. Millner, and C. Kunz. 1993. Evaluation of the polymerase chain reaction for diagnosis of herpes simplex virus encephalitis. *J. Clin. Microbiol.* **31:**146–148.

55. Rawls, W. E. 1985. Herpes simplex virus, p. 527–561. *In* B. Fields (ed.), *Virology.* Raven Press, Inc., New York.

56. Redfield, D. C., D. D. Richman, S. Albanil, M. N. Oxman, and C. M. Wahl. 1983. Detection of herpes simplex virus in clinical specimens by DNA hybridization. *Diagn. Microbiol. Infect. Dis.* **1:**117–128.

57. Richman, D. D., P. H. Cleveland, D. C. Redfield, M. N. Oxman, and C. M. Wahl. 1984. Rapid viral diagnosis. *J. Infect. Dis.* **149:**298–310.

58. Roizman, B., and W. Batterson. 1985. Herpes viruses and their replication, p. 497–517. *In* B. Fields (ed.), *Virology.* Raven Press, Inc., New York.

59. Rowley, A. H., R. J. Whitley, F. D. Lakeman, and S. M. Wolinsky. 1990. Rapid detection of herpes simplex virus DNA in cerebrospinal fluid of patients with herpes simplex encephalitis. *Lancet* **335:**440–441.

60. Rubin, S. J., and S. Rogers. 1984. Comparison of culture set and primary rabbit kidney cell culture for the detection of herpes simplex virus. *J. Clin. Microbiol.* **19:**920–922.

61. Salmon, V. C., R. B. Turner, M. J. Speranza, and J. C. Overall. 1986. Rapid detection of herpes simplex virus in clinical specimens by centrifugation and immunoperoxidase staining. *J. Clin. Microbiol.* **23:**683–686.

62. Sanchez-Martinez, D., S. Schmid, W. Whittington, D. Brown, W. C. Reeves, S. Chatterjee, R. J. Whitley, and P. E. Pellett. 1991. Evaluation of a test based on baculovirus-expressed glycoprotein G for detection of herpes simplex type-specific antibodies. *J. Infect. Dis.* **164:**1196–1199.

63. Saral, R. 1988. Management of mucocutaneous herpes simplex virus infections in immunocompromised patients. *Am. J. Med.* **85**(2A)**:**57–60.

64. Schmidt, N. J., J. Dennis, V. Devlin, D. Callo, and J. Mills. 1983. Comparison of direct immunofluorescence and direct immunoperoxidase procedures for detection of herpes simplex virus antigen in lesion specimens. *J. Clin. Microbiol.* **18:**445–448.

65. Sewell, D. L. L., and S. A. Horn. 1985. Evaluation of a commercial enzyme-linked immunosorbent assay for the detection of herpes simplex virus. *J. Clin. Microbiol.* **21:**457–458.

66. Shepp, D. H., B. A. Newton, P. S. Dandliker, N. Flournoy, and J. D. Meyers. 1985. Oral acyclovir therapy for mucocutaneous herpes simplex virus infections in immunocompromised marrow transplant recipients. *Ann. Intern. Med.* **102:**783–785.

67. Straus, S. E., J. F. Rooney, J. L. Sever, M. Seidlin, S. Nusinoff-Lehrman, and K. Cremer. 1985. Herpes simplex virus infection: biology, treatment, and prevention. *Ann. Intern. Med.* **103:**404–419.

68. Sullender, W. M., L. L. Yasukawa, M. Schwartz, L. Periera, P. A. Hensleigh, C. C. Prober, and A. M. Arvin. 1988. Type-specific antibodies to herpes simplex virus type 2 (HSV-2) glycoprotein G in pregnant women, infants exposed to maternal HSV-2 infections at delivery, and infants with neonatal herpes. *J. Infect. Dis.* **157:**164–171.

69. Verano, L., and F. J. Michalski. 1990. Herpes simplex virus antigen direct detection in standard virus transport medium by Du Pont Herpchek enzyme-linked immunosorbent assay. *J. Clin. Microbiol.* **28:**2555–2558.

70. Volpi, A., A. D. Lakeman, L. Pereira, and S. Stagno. 1983.

Monoclonal antibodies for rapid diagnosis and typing of genital herpes infections during pregnancy. *Am. J. Obstet. Gynecol.* **146:**813–815.

71. **Warford, A. L., R. A. Levy, K. A. Rekrut, and E. Steinberg.** 1986. Herpes simplex virus testing of an obstetric population with an antigen enzyme-linked immunosorbent assay. *Am. J. Obstet. Gynecol.* **154:**21–28.

72. **Wentworth, B. B., and E. R. Alexander.** 1971. Seroepidemiology of infections due to members of the herpesvirus group. *Am. J. Epidemiol.* **94:**496–507.

73. **West, P. C., B. Aldrich, R. Hartwig, and C. J. Haller.** 1989. Increased detection of herpes simplex virus in MRC5 cells treated with dimethyl sulfoxide and dexamethasone. *J. Clin. Microbiol.* **27:**770–772.

74. **Whitley, R. J.** 1988. Herpes simplex virus infections of the central nervous system. A review. *Am. J. Med.* **85(2A):**61–67.

75. **Whitley, R. J.** 1990. Herpes simplex, p. 282–305. *In* J. O. Klein and J. S. Remington (ed.), *Infectious Diseases of the Fetus and Newborn Infants*, 3rd ed. The W. B. Saunders Co., Philadelphia.

76. **Whitley, R. J., L. Corey, A. Arvin, F. D. Lakeman, C. V. Sumaya, P. F. Wright, L. M. Dunkle, R. W. Steele, S.-J. Soong, A. J. Nahmias, C. A. Alford, D. A. Powell, V. S. San Joaquin, and the NIAID Collaborative Antiviral Study Group.** 1988. Changing presentation of herpes simplex virus infection in neonates. *J. Infect. Dis.* **158:**109–116.

77. **Whitley, R. J., A. J. Nahmias, S.-J. Soong, C. C. Calassco, C. L. Fleming, and C. A. Alford.** 1980. Vidarabine therapy of neonatal herpes simplex virus infections. *Pediatrics* **66:**495–501.

78. **Whitley, R. J., A. J. Nahmias, A. M. Visintine, C. L. Fleming, and C. A. Alford.** 1980. The natural history of herpes simplex virus infection of mother and newborn. *Pediatrics* **66:**489–494.

79. **Woods, C. L., and R. D. Mills.** 1988. Effect of dexamethasone on detection of herpes simplex virus in clinical specimens by conventional cell culture and rapid 24-well plate centrifugation. *J. Clin. Microbiol.* **26:**1233–1235.

80. **Yeager, A. S., A. M. Arvin, and P. A. Hensleigh.** 1982. The validity of reporting results for herpes simplex virus after four days. *J. Reprod. Med.* **27:**447–448.

81. **Yeager, A. S., J. E. Morris, and C. C. Prober.** 1979. Storage and transport of cultures for herpes simplex virus, type 2. *Am. J. Clin. Pathol.* **72:**977–979.

82. **Zhao, L., M. L. Landry, E. S. Balkovic, and C. D. Hsiung.** 1987. Impact of cell culture sensitivity and virus concentration on rapid detection of herpes simplex virus by cytopathic effects and immunoperoxidase staining. *J. Clin. Microbiol.* **25:**1401–1405.

Human Cytomegalovirus

RICHARD L. HODINKA AND HARVEY M. FRIEDMAN

72

CLINICAL BACKGROUND

Cytomegalovirus (CMV) infections are common and usually asymptomatic; however, the incidence and spectrum of disease in newborns and in immunocompromised hosts establish this virus as an important human pathogen. CMV infections can be classified as being acquired before birth (congenital), at the time of delivery (perinatal), or later in life (postnatal). Similar to infections with other herpesviruses, primary infection with CMV results in the establishment of a persistent or latent infection. Reactivation of the virus can occur in response to different stimuli.

CMV infection has been detected in 0.5 to 2.5% of newborn infants and is the most common identified cause of congenital infection. Fewer than 5% of congenitally infected infants develop symptoms during the newborn period; possible manifestations range from severe disease with intrauterine growth retardation, jaundice, hepatosplenomegaly, petechiae, central nervous system abnormalities, and chorioretinitis to more limited involvement. Symptomatic infants may die of complications within the first months of life; more commonly, they survive but are neurologically damaged. It is now recognized that even congenitally infected infants who are asymptomatic early in life may develop hearing defects or learning disabilities.

Newborns can also acquire infection at the time of delivery by contact with virus in the birth canal. Such infants begin to excrete virus at 3 to 12 weeks of age but usually remain asymptomatic. Thus far, it appears that such perinatally infected infants do not develop late neurologic sequelae of infection.

Most postnatal infections are acquired by close contact with individuals who are shedding virus. Since CMV has been detected in several body fluids, including saliva, urine, breast milk, tears, stool, vaginal or cervical secretions, blood, and semen, transmission can occur in a variety of ways. Prolonged shedding of virus after congenital or acquired CMV infection contributes to the ease of virus spread. In addition, CMV can be transmitted by blood transfusion and organ transplantation.

The vast majority of children and adults who acquire CMV infection postnatally remain asymptomatic. In high-risk premature newborns infected as a result of blood transfusions, morbidity and mortality can be significant, and symptoms of hepatosplenomegaly, thrombocytopenia, atyp-

ical lymphocytosis, and hemolytic anemia have been described. Of children attending day-care centers, 20 to 70% of those who enter as toddlers acquire CMV infection over a 1- to 2-year period. Infection is usually asymptomatic, but the children may transmit CMV to their parents, posing a risk to an unborn fetus if the mother is pregnant at the time. In adults, sexual transmission of CMV may occur. Symptoms in young adults include fever, lethargy, and atypical lymphocytosis that can mimic the symptoms caused by Epstein-Barr virus.

CMV infections are frequent and occasionally severe in children or adults with congenital or acquired defects of cellular immunity, such as patients with AIDS, cancer patients (particularly those with leukemia and lymphoma), and recipients of organ transplants. Infections in these patients may be due to reactivation of latent virus or infection with exogenous virus, which may be introduced by blood transfusions or by the grafted organ. Symptoms tend to be most severe after primary infection; however, reactivation infection in a severely immunocompromised host may also cause serious illness. The frequency and severity of CMV infection in organ transplant patients are variable and depend on the type of transplant, the source of the donated organ, the immune status of the recipient, and the duration of the immunosuppressive therapy. Symptoms in these patients include fever, leukopenia, thrombocytopenia, pneumonitis, hepatitis, retinitis, and encephalitis. Death may occur as a result of various complications, including bacterial and fungal superinfections. CMV infection, particularly when associated with pneumonitis, is an important cause of morbidity and mortality after bone marrow transplantation. In patients infected with human immunodeficiency virus, CMV is an important cause of fever, sight-threatening retinitis, encephalitis, and gastrointestinal infections including esophagitis, gastritis, and ulcerative colitis.

DESCRIPTION OF THE AGENT

CMV is a member of the family *Herpesviridae*, which includes Epstein-Barr virus, herpes simplex virus types 1 and 2, varicella-zoster virus, and human herpes virus types 6 and 7. Complete CMV particles have a diameter of 120 to 200 nm and consist of a core containing double-stranded DNA, an icosahedral capsid, and a surrounding envelope. Electron microscopic features of CMV include virions morphologically indistinguishable from those of other herpesviruses, a high ratio of defective viral particles, and the presence of spherical particles called dense bodies. Viral replication occurs in the nucleus of the host cell and involves the expression of immediate-early, early, and late classes of genes. The viral envelope is formed as assembled nucleocapsids bud from the inner surface of the nuclear membrane.

Molecular virologic techniques have been used to study variation among CMV strains. By DNA-DNA reassociation kinetics analysis, various CMV strains have been shown to share considerable homology with AD-169, a standard laboratory strain; by restriction endonuclease analysis, DNAs from various strains have been shown to have similar but distinctive fragment migration patterns (30). These studies suggest that CMV strains are closely related to each other, more so than are herpes simplex virus types 1 and 2. Antigenic heterogeneity among CMV strains has been detected in cross-neutralization and other serologic assays, but evidence for distinct serotypes is limited (70).

CMV is inactivated by a number of physical and chemical treatments, including heat (56°C for 30 min), low pH, ether, UV light, and cycles of freezing and thawing.

COLLECTION AND STORAGE OF SPECIMENS

Specimens for Virus Isolation

CMV can be isolated from a variety of body fluids; however, urine, throat washings, saliva, and blood (buffy coats) are most common for diagnostic purposes. Urine specimens should be clean-voided specimens. Because excretion of CMV in urine is intermittent, increased recovery of the virus is possible by processing more than one specimen. In the evaluation of immunocompromised patients, buffy coat cultures are particularly useful. Detection of CMV in leukocytes is often a better indicator of symptomatic CMV infection than is shedding of virus in urine or throat. Bronchial washings and biopsy and autopsy specimens, particularly of lung, kidney, spleen, liver, brain, and retina, can also be processed for virus isolation. The details of specimen collection and processing are given in chapter 70 of this Manual. Since CMV loses infectivity when subjected to freezing and thawing, specimens should be kept at 4°C in an ice-water bath or refrigerator until they can be used to inoculate cultures, preferably within a few hours after collection. When prolonged transport times are unavoidable, infectivity is reasonably well preserved for at least 48 h at 4°C. If storage in the frozen state is necessary, an equal volume of 0.4 M sucrose-phosphate added to the specimen helps preserve viral infectivity (27). All frozen specimens should be stored at −60 to −80°C or in liquid nitrogen. A complete loss of virus infectivity occurs if specimens are stored at −20°C.

Specimens for Serology

Single-serum specimens are useful in screening for evidence of past infection with CMV and in identifying individuals at risk for CMV infection. This approach is especially helpful in testing sera from organ transplant donors and from donors of blood products that are to be administered to premature infants or bone marrow transplant patients. For the diagnosis of recent CMV infection, paired sera should be obtained at least 2 weeks apart when immunoglobulin G (IgG) antibody is tested for. If congenital infec-

tion is suspected, both maternal and infant sera should be submitted. Detection of IgM in a single serum specimen may be beneficial, but laboratories performing these tests should be aware of the problems encountered with such methods.

DIRECT EXAMINATION OF SPECIMENS

Histopathology

Characteristic large cells (cytomegalic) with basophilic intranuclear and, on occasion, eosinophilic cytoplasmic inclusions can be seen in routine sections of biopsy or autopsy material (Fig. 1). The nuclear inclusion has the appearance of an "owl's eye" because it is typically surrounded by a clear halo that extends to the nuclear membrane. Wright-Giemsa-, hematoxylin-eosin-, or Papanicolaou-stained lung or other biopsy specimens may demonstrate such cells. Although the presence of characteristic cytologic changes suggests CMV infection, virologic or serologic confirmation is suggested. Since CMV can infect tissues without producing morphologic changes, failure to find typical cytomegalic cells does not exclude the possibility of CMV infection.

Exfoliative Cytology

With exfoliative cytologic techniques, a presumptive diagnosis of CMV can be made in 25 to 50% of cases of symptomatic congenital infection. Specimens from older infected individuals are rarely positive. Several fresh urine specimens should be submitted, since exfoliated cells disintegrate rapidly and may be shed only intermittently. Characteristic enlarged cells with prominent inclusions are seen in positive preparations. Exfoliative cytology is most useful when virus isolation techniques are not available. This technique can be applied to other specimens, such as bronchoalveolar lavage fluid (72) and cervical secretions.

Immunofluorescence

Monoclonal antibodies to CMV can be used for the direct detection of CMV antigens by immunofluorescence tests on tissues obtained by biopsy or autopsy (32, 68). Cytospin preparations of bronchoalveolar lavage specimens (42, 72), blood leukocytes (66, 67), and urine sediments (38) have also been examined in this manner. The sensitivity of the assay is improved when mixtures of monoclonal antibodies are used (68). An advantage of immunofluorescence staining is its rapidity; results are available within several hours after tissue is obtained. For most specimens, this method is less sensitive than the shell vial culture technique described below, which has largely replaced tissue immunofluorescence as a rapid test for CMV diagnosis.

Recently, considerable attention has been given to the direct detection of CMV from blood leukocytes (63, 64, 66, 67). A new procedure for the determination of CMV viremia has been developed, and it has been compared to conventional and shell vial culture isolation of CMV from blood. The CMV antigenemia assay has been shown to be a sensitive, specific, and rapid method for the early diagnosis of CMV infection, and it can be used for routine monitoring of patients at high risk for severe CMV disease. The test is relatively simple to perform and is based on the immunocytochemical detection of the 65-kDa lower-matrix phosphoprotein (pp65) in the nuclei of peripheral blood polymorphonuclear leukocytes (Fig. 2). By using this assay, CMV can be detected before the onset of symptoms, and the viral load can be quantified to assist in predicting

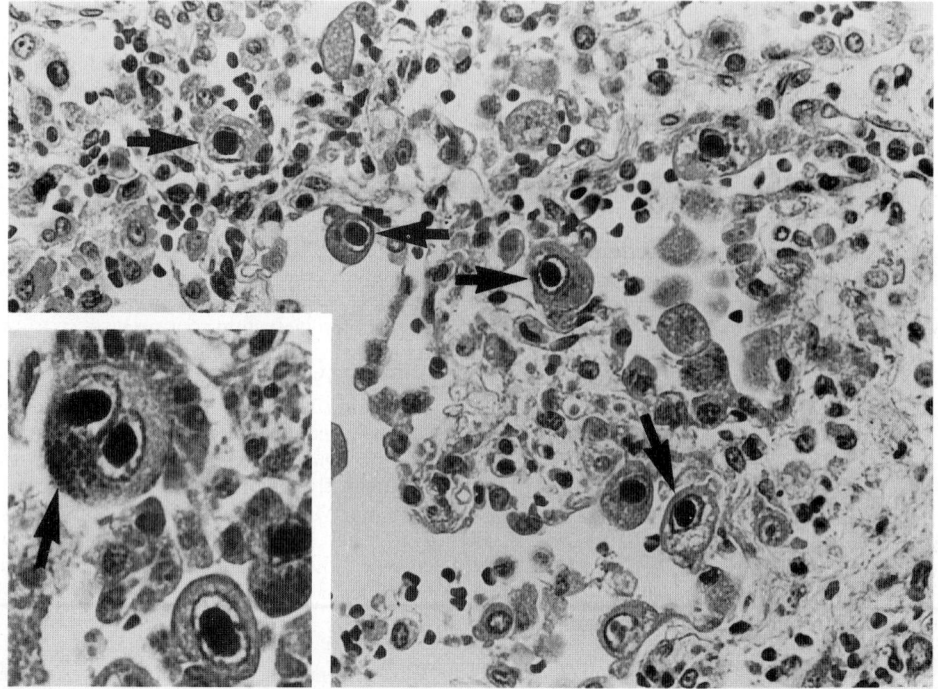

FIGURE 1 Fixed hematoxylin-eosin-stained lung tissue from a patient with interstitial pneumonia. Note the numerous giant cells that possess large intranuclear inclusions surrounded by characteristically clear halos (arrows). Less pronounced granular inclusions may also be present in the cytoplasm (inset). (Courtesy of Eduardo Ruchelli.)

and differentiating CMV disease from asymptomatic infection (18, 36). The procedure has also been utilized to effectively evaluate the efficacy of antiviral therapy (20, 65) and should prove valuable in the management of patients with CMV.

In the antigenemia assay, polymorphonuclear leukocytes are separated from freshly collected whole blood, and the cells are spotted or cytocentrifuged onto microscope slides. The cells are fixed and then permeabilized and stained with suitable monoclonal antibodies directed against CMV pp65. These processes are followed by incubation of the cells either with a horseradish peroxidase-labeled secondary antibody and a chromagen or with a fluorescein isothiocyanate-labeled conjugate diluted in a counterstain. Slides are read by microscopy at ×200 to ×400 magnification. Positive results are viewed as either homogeneous yellow-green fluorescence or a dark brown to red-brown color within the nuclei of infected cells. The procedure has the disadvantages of being labor-intensive and time-consuming; considerable time and effort are spent in reading stained slides, counting cells, and adjusting the cell concentration to 10^6/ml prior to preparing the slides for staining. Also, specimens must be processed without delay for accurate and reliable quantitation of virus load. Commercial preparations of monoclonal antibodies and other reagents needed to perform the antigenemia assay are available.

Electron Microscopy

The pseudoreplica method of electron microscopy can be used to detect CMV in urine and oral specimens of congenitally infected infants (40). Positive results can be obtained with almost all specimens that have infectivity titers of $\geq 10^4$ PFU/ml. An advantage of electron microscopy is

its rapidity; also, stored or contaminated specimens that are unsuitable for isolation attempts can be examined. The main disadvantages of the technique are its relative insensitivity and the need for experienced personnel and expensive equipment.

Hybridization

Molecular dot blot hybridization techniques have been described for detection of the CMV genome in urine (4, 6, 59) and peripheral blood leukocytes (41, 59). Recently, in situ hybridization has been used for the direct detection of CMV in formalin-fixed lung tissue (32, 46), Kaposi's sarcoma tissue (26), and cytospin preparations of bronchoalveolar lavage fluid (23, 72) and peripheral blood leukocytes (10, 62). Biotinylated or horseradish peroxidase-labeled CMV-specific DNA probe kits are commercially available. In situ hybridization has the advantage of being rapid and easy to read by light microscopy. However, current methods lack the sensitivity needed for routine use in clinical laboratories.

PCR DNA Amplification

The PCR is a newly introduced, highly sensitive and rapid method for amplifying and detecting small quantities of specific nucleic acid in clinical specimens. The sensitivity and specificity of PCR for diagnosis of active CMV infection have been evaluated (11, 29, 48, 56, 75). In several studies, the sensitivity of the assay was increased by amplifying gene fragments from both the immediate-early and the late CMV genes or by using nested primers to a single gene fragment. Use of both gene fragments enabled detection of a variety of clinical isolates, indicating that strain variability is not a limiting factor for PCR diagnosis of

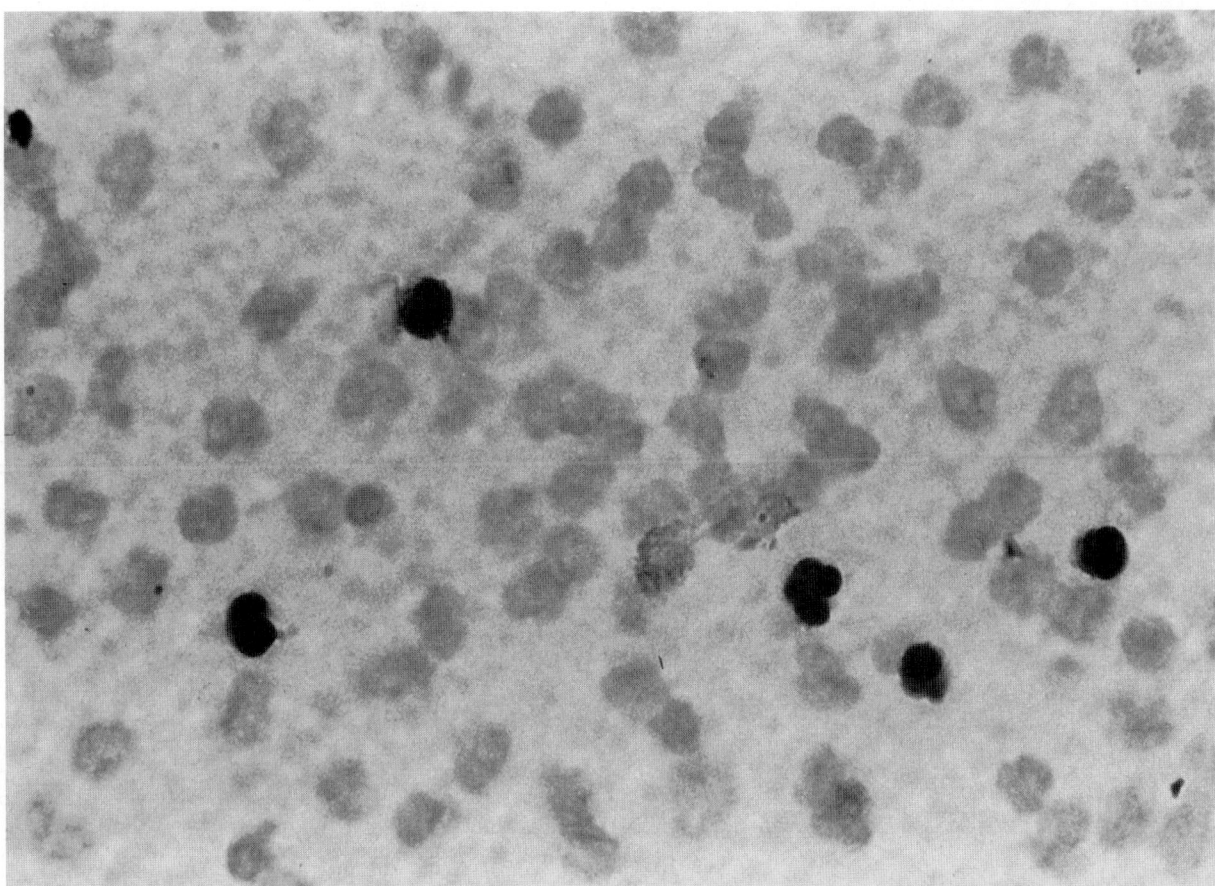

FIGURE 2 CMV antigen-positive polymorphonuclear leukocytes. Note the nuclear staining when a monoclonal antibody directed against the 65-kDa CMV lower-matrix phosphoprotein is used. (Courtesy of Dave Kolsar.)

CMV. PCR has been used successfully to detect CMV DNA in a variety of clinical specimens from organ transplant recipients, patients with AIDS, and infants with congenital infection. The utility of PCR for the continued surveillance of immunocompromised patients and for evaluating the therapeutic efficacy of antiviral drugs has also been demonstrated (19, 20, 64).

An unresolved issue with PCR for CMV diagnosis is whether the test can distinguish between active disease and asymptomatic infection or latency. A study by Stanier et al. (61) detected CMV in peripheral blood mononuclear cells of seropositive healthy blood donors, suggesting that a positive PCR result may correlate better with seropositivity than with active infection. Other investigators, however, have shown that the detection of CMV DNA from blood, cerebrospinal fluid, or tissue specimens of immunocompromised individuals may provide a high predictive value for the accurate diagnosis of clinically significant infection (see reference 64 for a review). A quantitative PCR method has been introduced to show that patients with active CMV disease have higher levels of CMV DNA and that a rapid rise in CMV genomic copy number correlates with symptoms (17). Furthermore, it has been demonstrated that amplification of specific CMV transcripts (mRNA) that are expressed only during active infection may make it possible to identify those patients at greatest risk of developing

symptomatic infection (2, 25). At present, the clinical significance of results provided by PCR is unclear, but it seems likely that the test will prove useful in certain settings. These include diagnosis of less common forms of CMV infection, such as detection of CMV in the cerebrospinal fluids of patients with encephalitis, and diagnosis of CMV in patients at high risk of severe infection, such as CMV donor-positive, recipient-negative transplant patients, so that antiviral therapy can be started early in the course of infection.

Other Direct Detection Methods

An enzyme-linked immunosorbent assay and a dot immunoperoxidase assay using antisera to viral proteins have also been used for the direct detection of CMV from clinical specimens.

DETECTION OF CMV IN CELL CULTURE

Processing of Specimens

Adjustment of urine specimens to pH 7.0 with 0.1 N NaOH or 0.1 N HCl is recommended to reduce toxicity to cell cultures. Centrifuging urine specimens to obtain sediment-enriched samples has been advocated but is usually unnecessary. Sediment-enriched urine samples from renal

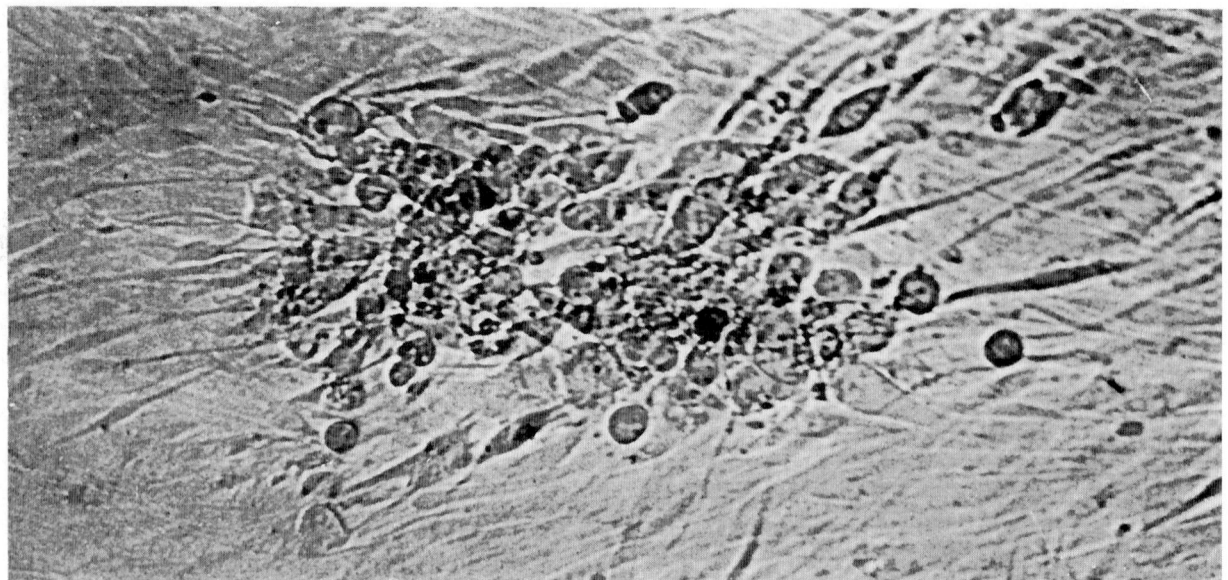

FIGURE 3 CPE produced by a CMV isolate in human skin fibroblasts 10 days postinoculation. Unstained preparation; magnification, ×100. (Courtesy of Sergio Stagno.)

transplant recipients may produce toxicity more frequently than do uncentrifuged urine samples (37).

A number of procedures for obtaining buffy coat cells have been described. We have found density gradient centrifugation using either Ficoll-Hypaque (15, 28) or a mixture of sodium metrizoate and dextran 500 (PMN; Robbins Scientific, Sunnyvale, Calif.) to be most suitable for a clinical laboratory. Fresh blood collected in the presence of heparin, sodium citrate, or EDTA may be used. Both mononuclear cells and granulocytes are efficiently separated from erythrocytes in a single step. The procedure is rapid and easy to perform, and the reagents are commercially available. When compared with traditional sedimentation methods, the technique results in a greater number of virus isolates and an increased yield of infectious foci or plaques (28).

Throat, biopsy, autopsy, and other specimens are processed as described in chapter 70. All specimens are treated with antibiotics before inoculation of cell cultures. Portions of specimens not used to inoculate cell cultures should be mixed with equal volumes of 0.4 M sucrose-phosphate and frozen at −70°C or in liquid nitrogen; some infectivity may be preserved if further isolation attempts are necessary.

Cell Cultures

Human fibroblast cells best support the growth of CMV and therefore are used for diagnostic purposes. Acceptable fibroblast cultures include those prepared from human embryonic tissues or foreskins and serially passaged diploid human fetal lung strains such as WI-38, MRC-5, or IMR-90. Diploid fibroblast cells should be used at a low passage number, since they may become less susceptible to CMV infection with increasing cell generations. Several of these fibroblast cell lines are commercially available. CMV can infect and replicate in certain human epithelial cells, but growth is limited. CMV infection of nonhuman cells can occur but leads to abortive replication with expression of only immediate-early genes and proteins.

Isolation of Virus

Specimens to be tested are added in a volume of 0.2 ml to duplicate tubes of confluent fibroblasts maintained on Eagle minimum essential medium with 2% fetal bovine serum. Alternatively, the tubes are drained of medium, the inocula are absorbed for either 1 h in a stationary position or by centrifugation at 700 × g for 45 min at 30 to 33°C (47), and then fresh medium is added. After inoculation, the tubes can be rolled or kept stationary at 37°C. Twenty-four hours later, the medium is changed for tubes inoculated with urine or buffy coat specimens. Thereafter, and for other types of specimens, medium is changed once a week or more frequently as the pH of the culture medium changes or if toxicity appears. When toxicity necessitates passage of the culture, cells rather than culture medium should be passaged, since CMV remains cell associated. Cells are removed by addition of 0.25% trypsin–0.1% EDTA to the monolayers and incubation at 37°C for 5 or 10 min. When the cells detach, minimum essential medium with 2% fetal bovine serum is added, and the cells are used to inoculate fresh tubes. Tubes are examined for cytopathic effect (CPE) daily for the first 5 days and then twice a week for at least 4 weeks for most specimens and for 6 weeks for buffy coat specimens. Control, uninoculated cultures are handled in the same manner as those inoculated with clinical specimens.

The time of appearance and the extent of CPE depend on the amounts of virus present in specimens. In cultures inoculated with urine from a congenitally infected newborn, CPE may develop by 24 h and progress rapidly to involve most of the monolayer if the virus titer of the urine is extremely high. More commonly, foci of CPE, consisting of enlarged, rounded, refractile cells, appear during the first week, and progression of CPE to surrounding cells proceeds slowly (Fig. 3). In cultures inoculated with urine or throat specimens from older individuals, CPE usually appears within 2 weeks. Buffy coat cultures may not become positive until after 3 to 6 weeks. The usual slow progression of

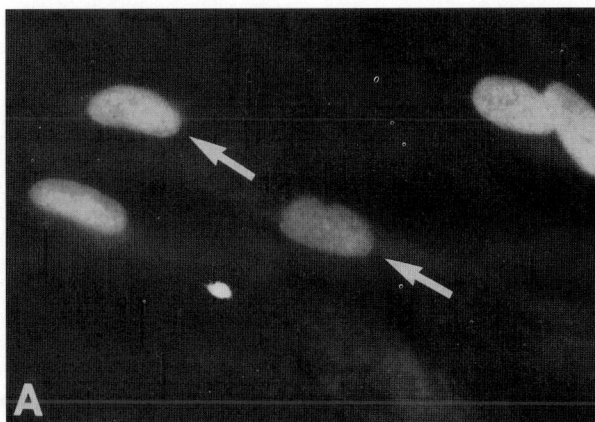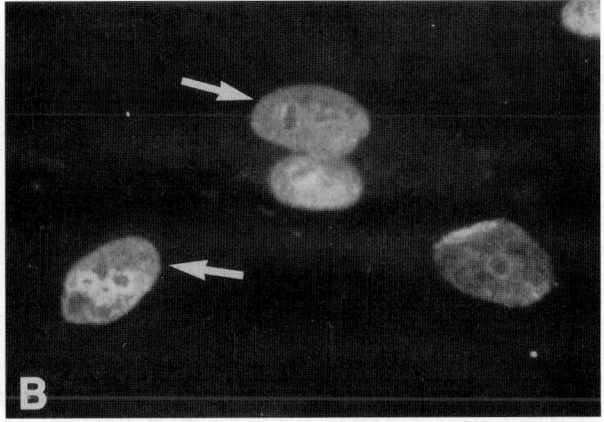

FIGURE 4 Demonstration of CMV early antigens in the nuclei (arrows) of infected MRC-5 cells following shell vial culture and IFA staining. Magnification, ×400.

CPE in cultures inoculated with clinical specimens is due, at least in part, to limited release of virus into extracellular fluid. With strains of CMV that have been serially passaged, including laboratory-adapted strains, greater amounts of extracellular virus are released, and CPE progresses more rapidly.

For storage of fresh isolates, monolayers exhibiting CPE are treated with trypsin-EDTA, and the cells obtained are suspended in Eagle minimum essential medium with 10% fetal bovine serum and 10% dimethyl sulfoxide and then frozen at −70°C. Infectivity can be better maintained for long periods by storage in liquid nitrogen.

Identification of Isolates

In many laboratories, CMV isolates are identified solely on the basis of characteristic cytopathology and host cell range. However, viruses such as adenovirus and varicella-zoster virus may occasionally produce CPE indistinguishable from that of CMV. Suspected CMV isolates can be confirmed by indirect immunofluorescence (IFA) with commercially available monoclonal or polyclonal antibodies. The appearance of typical nuclear fluorescence of infected cells indicates the presence of CMV.

Spin-Amplification Shell Vial Assay

The spin-amplification shell vial assay as described by Gleaves et al. (24) has gained wide acceptance as a rapid method for the detection of CMV in clinical specimens. The technique is based on the amplification of virus in cell cultures after low-speed centrifugation and detects viral antigens produced early in the replication of CMV before the development of CPE. Even low titers of virus present in specimens are easily amplified and rapidly detected within 24 h. Monoclonal antibodies are commercially available and have been used for detection of CMV early antigens. In situ hybridization using commercially available DNA probes to CMV has also been employed (22, 58).

MRC-5 fibroblast cells are grown to confluency on 12-mm round coverslips in 1-dram (3.7-ml) shell vials and inoculated with 0.2 ml of specimen. Shell vials of MRC-5 cells can be obtained commercially or prepared in the laboratory. Monolayers should be inoculated within 1 week after preparation, since older monolayers demonstrate decreased sensitivity to CMV and increased toxicity (14). Pretreatment of the monolayers with dexamethasone, di-

methyl sulfoxide, or these agents plus calcium may increase the sensitivity of the cells to CMV infection (73, 74), although other investigators have failed to confirm these results (13). Two vials should be inoculated for urine, tissue, and bronchoalveolar lavage specimens, and three should be inoculated for blood specimens (50). Alternatively, disruption of purified leukocytes by sonication prior to their use in the shell vial assay may increase the sensitivity of CMV detection from blood (76). After inoculation, vials are centrifuged at 700 × g for 40 min at 25°C, and then 2.0 ml of Eagle minimum essential medium containing 2% fetal bovine serum and antibiotics is added. The cultures are incubated at 37°C for 16 to 24 h, fixed with acetone, and stained. A longer incubation time may be used, but the time should be determined by each laboratory on the basis of individual experience, reagents and staining technique employed, and whether monolayers are purchased or prepared in the laboratory. Uninfected and CMV-infected monolayers are included as negative and positive controls, respectively. Recently, mink lung (ML) cells, a nonhuman continuous cell line, have been shown to be comparable to MRC-5 fibroblasts for the detection of CMV in clinical specimens by shell vial culture (21, 39). A distinct advantage in using ML cells is that this cell line can be propagated and passaged in the laboratory for a long time without a decrease in susceptibility to CMV. Significantly less toxicity and an increase in CMV-positive nuclei were also observed with ML cells. Our laboratory utilizes an IFA to detect the immediate-early antigen of CMV, using the E-13 mouse anti-CMV monoclonal antibody as the primary antiserum followed by a biotin-labeled goat anti-mouse antibody and avidin conjugated with fluorescein isothiocyanate. The monolayers are counterstained with Evans blue. Coverslips are scanned at ×200 to ×250 magnification, and specific staining is confirmed at ×400 to ×630. Positive cells contain apple green fluorescent nuclei against a red cytoplasmic background. Staining of immediate-early antigen appears as an even matte green fluorescence with specks of brighter green (Fig. 4A). Viral inclusions (owl's eyes) may be visible in the nuclei (Fig. 4B).

The spin-amplification shell vial assay is a valuable adjunct to conventional virus isolation. It has the important features of being rapid, sensitive, and specific. However, skilled technical personnel and attention to the quality of specimens, monolayers, and reagents are required for optimum performance. A modification of the shell vial assay to quantify CMV in blood and bronchoalveolar lavage specimens of transplant patients has recently been developed (5, 57). The quantitative method has been used to predict the outcome of CMV infection and to assess the effectiveness of antiviral treatment.

SEROLOGIC DIAGNOSIS

A variety of tests are available for serodiagnosis of CMV infection. In deciding which test to perform, such factors as the patient population, cost, turnaround time, equipment needs, and ease of performance should be considered. Which method is chosen depends on which best meets the needs of individual laboratories. Overall, the determination of a positive seroconversion remains a reliable means for diagnosing primary CMV infection, and the screening of blood and organ donors and recipients plays an important role in preventing the transmission of latent CMV to patients at high risk for severe CMV disease.

EIA

Over the last several years, enzyme immunoassays (EIA) have largely replaced other traditional methods for detecting antibodies to CMV. The main advantages of the EIA are that it is rapid, sensitive, and specific (43, 44, 51). In addition, multiple specimens can be handled daily. The EIA can be used effectively to determine the immune status of a patient (3, 12) and to detect significant rises in antibody titers (43). Kits that detect CMV IgG are available from a number of commercial sources. The kits are easy to use, and the manufacturers provide detailed instructions. All of the materials necessary to perform the assay are included, and the reagents are stable with time. Some companies also provide a spectrophotometer and automated plate washer, which otherwise must be purchased separately at considerable expense. Over the last decade, the development of robotics technology has led to the commercial availability of both fully automated and semi-automated EIA instruments, including sample dispensers, diluters, washers, and spectrophotometers with complete computer programming and generation of written reports.

Passive Latex Agglutination Test

The passive latex agglutination test provides a simple and rapid means of detecting antibodies to CMV in human sera and plasma. The method is highly sensitive and specific (1, 31, 43, 44), and it may serve to determine the immune statuses of patient or blood donor populations (1, 31) or to satisfactorily detect significant antibody rises in paired sera (43, 44). Commercial kits are available. The procedure detects both IgG and IgM antibodies but does not differentiate between the two classes of immunoglobulins. The assay can be completed in 10 to 15 min and does not require extensive washes, long incubation times, or expensive equipment. The main disadvantages are that agglutination patterns can be difficult to discern and readings are subjective.

CF Test

The complement fixation (CF) test has been used for many years in clinical laboratories and is suitable for detecting rises in antibody titers. Reagents for the CF test are commercially available, and detailed procedures have been described elsewhere (8, 35, 49). Choice of antigen preparation for CF should in part depend on the purpose of the test. The CF test with glycine-extracted antigen is more sensitive for distinguishing seropositive from seronegative specimens (35), whereas the freeze-thawed antigen is slightly better for detection of fourfold or greater rises in titer (8). The method is technically demanding, requires rigid standardization, and has a long turnaround time. The CF test is also less sensitive than other methods in detecting low levels of antibody. Therefore, it may be desirable to adopt an alternative system, such as the previously described EIA or latex agglutination assay, for routine diagnostic testing.

ACIF Test

The anticomplement immunofluorescence (ACIF) test is an immunofluorescence assay that detects CMV antibody (34, 54). The method is performed by incubating heat-inactivated serum with CMV-infected fibroblasts fixed to glass slides. CMV-specific antibody bound to the infected cells is detected by using complement and fluorescein-conjugated anticomplement antibodies. An important advantage of this assay compared with the CF test is its rapidity, since results can be available within 2 to 3 h. In addition, sera that are anticomplementary by the CF procedure can be tested by the ACIF test, and nonspecific cytoplasmic staining caused by binding of antibody to Fc receptors is avoided. Disadvantages of the ACIF test are that fewer specimens can be handled daily, the test requires cell culture facilities and an immunofluorescence microscope, and examination of slides requires considerable time and experience.

IFA

As an alternative to the ACIF test, the IFA can be performed (54). This test is slightly faster than the ACIF test but has the disadvantage that the assay should be performed on monolayers of infected cells grown on glass slides; these are more difficult to prepare than the cell suspensions used for making substrate slides in the ACIF test. In addition, specific nuclear fluorescence, which indicates a positive result, is sometimes difficult to distinguish from nonspecific cytoplasmic fluorescence. The latter occurs because CMV infection of fibroblasts induces a cytoplasmic Fc receptor that binds viral and nonviral IgG antibodies (33).

Several immunofluorescence assay systems for CMV antibody detection are commercially available. These include a solid-phase fluorescence immunoassay (trademarked as FIAX) in which CMV antigen and control antigen are absorbed onto solid surfaces. A single dilution of patient serum is applied to the solid surfaces, an anti-human IgG fluorescein conjugate is added, and the intensity of fluorescence is measured in a fluorometer (16). CMV-infected culture cells on glass slides and other IFA kits for measuring CMV antibodies are also available commercially.

CMV IgM Antibodies

Methods for measuring CMV IgM antibodies include IFA, EIA, and radioimmunoassays. The procedures are essentially the same as those employed for detecting IgG antibodies except that anti-human IgM antibodies labeled with suitable markers are used. A recognized pitfall of CMV IgM assays is the occurrence of false-positive and false-negative reactions. False-positive reactions occur when sera contain unusually high levels of rheumatoid factor in the presence of specific CMV IgG (9). Rheumatoid factor is an immunoglobulin, usually of the IgM class, that reacts with IgG. It is produced in some rheumatologic, vasculitic, and viral diseases, including CMV infection. IgM rheumatoid factor forms a complex with IgG that may contain CMV-specific IgG. The CMV IgG binds to CMV antigen, carrying non-viral IgM with it. A test designed to detect IgM will produce a false-positive result. Therefore, testing for rheumatoid factor and removing it if present is important when one is measuring IgM antibodies. False-negative reactions occur if high levels of specific IgG antibodies competitively block the binding of IgM to CMV antigen. Separation of IgG and IgM fractions before testing decreases the incidence of both false-positive and false-negative results.

Rapid and simple methods for the removal of interfering rheumatoid factor and IgG molecules from serum have been developed. These include gel filtration, affinity chromatography, selective absorption of IgM to a solid phase, and removal of IgG by using hyperimmune anti-human IgG, staphylococcal protein A, or recombinant protein G from group G streptococci. Serum pretreatment methods are now incorporated within the procedures of commercially available immunofluorescence and EIA kits, which has resulted in more reliable IgM tests. Recently, reverse capture solid-phase IgM assays have been used as an alternative approach to avoiding false-positive or -negative results. This method uses a solid phase coated with an anti-human IgM antibody to capture the IgM from a serum specimen, after which competing IgG antibody and immune complexes are removed by washing. The bound IgM antibody is then exposed to specific CMV antigen, and then an enzyme-conjugated second antibody and substrate are added.

Although the detection of CMV-specific IgM may be useful in the determination of recent or active infection, results should be interpreted with caution. Because IgM does not cross the placenta, a positive result from a single serum specimen from an infected newborn is diagnostic. However, there may be a lack of or delay in production of IgM in the newborn. IgM antibody can appear in both primary and reactivated CMV infections and can persist for extended periods after a primary infection. This complicates the interpretation of test results, especially if a mother is positive for IgM during pregnancy. Also, patients with Epstein-Barr virus-induced infectious mononucleosis may produce heterotypic IgM responses, causing false-positive IgM results.

Other Serologic Tests

Immune adherence hemagglutination, indirect hemagglutination, and the neutralization test are other serologic tests that are used to measure CMV antibody.

INTERPRETATION OF RESULTS

Recovery of CMV from urine, throat, or other body fluids within the first 2 weeks of life is the most sensitive and specific means of confirming the diagnosis of congenital infection. Urine is the preferred specimen because it contains greater amounts of virus, and the virus will therefore grow quickly in culture. Attempts to isolate virus from blood and tissue of the infected newborn and from amniotic fluid and fetal tissue can also be made. Serologic tests are less useful because of transplacental passage of maternal antibody and technical problems with detection of CMV-specific IgM. A history of seroconversion from a negative to a positive IgG antibody response to CMV is diagnostic of primary infection and may be beneficial in evaluating the pregnant mother with symptoms of viral disease. Most symptomatic congenitally infected infants have relatively high initial CF antibody titers, which persist at stable levels

for many months (60). Congenital infection should therefore be suspected in infants with typical symptoms whose CMV titers persist at levels comparable to or sometimes higher than those of the mothers. Negative IgG titers in both mother and child exclude congenital CMV infection. Infants not previously tested but found to be excreting virus after 2 weeks of age may have either congenital or acquired infection. Standard serologic tests do not differentiate between these possibilities.

Interpretation of serologic tests performed on sera of patients 6 months of age or older is facilitated by the absence of passively acquired maternal antibody. If the initial serum is negative for CMV antibodies and the second serum is positive, a diagnosis of primary infection can be made. If a serum sample from early in an illness contains CMV antibodies and a second sample taken several weeks later demonstrates a fourfold or greater rise in titer, a diagnosis of recent infection due to reactivation or reinfection can be made. Fourfold falls in titer are seldom observed early enough to be useful for laboratory diagnosis, since antibody levels tend to decline slowly over several months after infection. If acute- and convalescent-phase sera are both positive for CMV antibody but the antibody titer is unchanged, the result is interpreted as CMV infection at some time in the past. If the titers of the positive sera are high and the first specimen was obtained late in the illness, the results may indicate active infection, and appropriate specimens for viral isolation should be obtained. Whenever possible, serologic diagnoses of CMV infection should be confirmed by virus isolation, particularly since fourfold fluctuations in CMV antibody titers have been noted in some apparently healthy seropositive individuals (71) and since immunocompromised hosts may not mount a normal immune response.

CMV serologies also play a significant role in the evaluation of an organ donor and recipient before transplantation. A seronegative recipient who receives an organ or blood products from a CMV-seropositive donor is at increased risk for developing primary CMV infection and serious disease. Knowing the serostatus of the donor and recipient is therefore important in determining the treatment or prophylaxis to be used and in considering the type of donor and blood products to be given.

Virologic or serologic detection of CMV indicates active infection but does not establish whether such infection is responsible for symptomatic illness. To implicate CMV as the cause of an illness, laboratory confirmation of active infection in an appropriate clinical setting is required. Routine surveillance for CMV, especially from urine, blood, and respiratory specimens, can identify transplant recipients and patients with HIV infection at high risk for CMV disease (45, 52, 53, 55, 69). The isolation of CMV from urine and respiratory specimens indicates active infection and the need to closely observe a patient so that appropriate medical management can be undertaken if disease develops. When CMV is isolated from leukocytes, the likelihood that the infection is symptomatic increases, but asymptomatic viremia has also been described (7) and is common in patients with AIDS. The use of quantitative detection methods to predict which patients will progress to severe CMV disease may prove beneficial for patient care. Isolation of CMV from liver and particularly lung biopsy specimens must be interpreted with caution, since other pathogens (Chlamydia trachomatis, Pneumocystis carinii, and other bacteria or fungi) may also be present; when dual infection is documented, the relative importance of each pathogen in producing clinical illness may be difficult to determine.

The application of virologic methods, including conventional and shell vial culture, rapid direct detection assays, and serology, should be combined with clinical assessment of the patient to provide an accurate, reliable diagnosis of CMV infection and disease and to allow subsequent prompt, appropriate management.

REFERENCES

1. Beckwith, D. G., D. C. Halstead, K. Alpaugh, A. Schweder, D. A. Blount-Fronefield, and K. Toth. 1985. Comparison of a latex agglutination test with five other methods for determining the presence of antibody against cytomegalovirus. *J. Clin. Microbiol.* **21:**328–331.

2. Bitsch, A., H. Kirchner, R. Dupke, and G. Bein. 1992. Cytomegalovirus transcripts in peripheral blood leukocytes of actively infected transplant patients detected by reverse transcription-polymerase chain reaction. *J. Infect. Dis.* **167:**740–743.

3. Booth, J. C., G. Hannington, T. M. F. Bakir, H. Stern, H. Kangro, P. D. Griffiths, and R. B. Heath. 1982. Comparison of enzyme-linked immunosorbent assay, radioimmunoassay, complement fixation, anticomplement immunofluorescence and passive haemagglutination techniques for detecting cytomegalovirus IgG antibody. *J. Clin. Pathol.* **35:**1345–1348.

4. Buffone, G. J., G. J. Demmler, C. M. Schimbor, and M. D. Yow. 1988. DNA hybridization assay for congenital cytomegalovirus infection. *J. Clin. Microbiol.* **26:**2184–2186.

5. Buller, R. S., T. C. Bailey, N. A. Ettinger, M. Keener, T. Langlois, J. P. Miller, and G. A. Storch. 1992. Use of a modified shell vial technique to quantitate cytomegalovirus viremia in a population of solid-organ transplant recipients. *J. Clin. Microbiol.* **30:**2620–2624.

6. Chou, S., and T. C. Merigan. 1983. Rapid detection and quantitation of human cytomegalovirus in urine through DNA hybridization. *N. Engl. J. Med.* **308:**921–925.

7. Cox, F., and W. T. Hughes. 1975. Cytomegaloviremia in children with acute lymphatic leukemia. *J. Pediatr.* **87:**190–194.

8. Cremer, N. E., M. Hoffman, and E. H. Lennette. 1978. Analysis of antibody assay methods and classes of viral antibodies in serodiagnosis of cytomegalovirus infection. *J. Clin. Microbiol.* **8:**152–159.

9. Cremer, N. E., M. Hoffman, and E. H. Lennette. 1978. Role of rheumatoid factor in complement fixation and indirect hemagglutination tests for immunoglobulin M antibody to cytomegalovirus. *J. Clin. Microbiol.* **8:**160–165.

10. Dankner, W. M., J. A. McCutchan, D. D. Richman, K. Hirata, and S. A. Spector. 1990. Localization of human cytomegalovirus in peripheral blood leukocytes by in situ hybridization. *J. Infect. Dis.* **161:**31–36.

11. Demmler, G. J., G. J. Buffone, C. M. Schimbor, and R. A. May. 1988. Detection of cytomegalovirus in urine from newborns by using polymerase chain reaction DNA amplification. *J. Infect. Dis.* **158:**1177–1184.

12. Dylewski, J. S., L. Rasmussen, J. Mills, and T. C. Merigan. 1984. Large-scale serological screening for cytomegalovirus antibodies in homosexual males by enzyme-linked immunosorbent assay. *J. Clin. Microbiol.* **19:**200–203.

13. Espy, M. J., A. D. Wold, D. M. Ilstrup, and T. F. Smith. 1988. Effect of treatment of shell vial cell cultures with dimethyl sulfoxide and dexamethasone for detection of cytomegalovirus. *J. Clin. Microbiol.* **26:**1091–1093.

14. Fedorko, D. P., D. M. Ilstrup, and T. F. Smith. 1989. Effect of age of shell vial monolayers on detection of cytomegalovirus from urine specimens. *J. Clin. Microbiol.* **27:**2107–2109.

15. Ferrante, A., and Y. H. Thong. 1980. Optimal conditions for simultaneous purification of mononuclear and polymorphonuclear leukocytes from human peripheral blood by the Hypaque-Ficoll method. *J. Immunol. Methods* **36:**109–117.

16. Friedman, H. M., N. B. Tustin, M. M. Hitchings, and S. A. Plotkin. 1981. Comparison of complement fixation and fluorescent immunoassay (FIAX) for measuring antibodies to cytomegalovirus and herpes simplex virus. *Am. J. Clin. Pathol.* **76:**305–307.

17. Gerdes, J. C., E. K. Spees, K. Fitting, J. Hiraki, M. Sheehan, D. Duda, T. Jarvi, C. Roehl, and A. D. Robertson. 1993. Prospective study utilizing a quantitative polymerase chain reaction for detection of cytomegalovirus DNA in the blood of renal transplant patients. *Transplant. Proc.* **25:**1411–1413.

18. Gerna, G., M. G. Revello, E. Percivalle, M. Zavattoni, M. Parea, and M. Battaglia. 1990. Quantitation of human cytomegalovirus viremia by using monoclonal antibodies to different viral proteins. *J. Clin. Microbiol.* **28:**2681–2688.

19. Gerna, G., D. Zipeto, M. Parea, E. Percivalle, M. Zavattoni, A. Gaballo, and G. Milanesi. 1991. Early virus isolation, early structural antigen detection and DNA amplification by the polymerase chain reaction in polymorphonuclear leukocytes from AIDS patients with human cytomegalovirus viraemia. *Mol. Cell. Probes* **5:**365–374.

20. Gerna, G., D. Zipeto, M. Parea, M. G. Revello, E. Silini, E. Percivalle, M. Zavattoni, P. Grossi, and G. Milanesi. 1991. Monitoring of human cytomegalovirus infections and ganciclovir treatment in heart transplant recipients by determination of viremia, antigenemia, and DNAemia. *J. Infect. Dis.* **164:**488–498.

21. Gleaves, C. A., D. A. Hursh, and J. D. Meyers. 1992. Detection of human cytomegalovirus in clinical specimens by centrifugation culture with a nonhuman cell line. *J. Clin. Microbiol.* **30:**1045–1048.

22. Gleaves, C. A., D. A. Hursh, D. H. Rice, and J. D. Meyers. 1989. Detection of cytomegalovirus from clinical specimens in centrifugation culture by in situ DNA hybridization and monoclonal antibody staining. *J. Clin. Microbiol.* **27:**21–23.

23. Gleaves, C. A., D. Myerson, R. A. Bowden, R. C. Hackman, and J. D. Meyers. 1989. Direct detection of cytomegalovirus from bronchoalveolar lavage samples by using a rapid in situ DNA hybridization assay. *J. Clin. Microbiol.* **27:**2429–2432.

24. Gleaves, C. A., T. F. Smith, E. A. Shuster, and G. R. Pearson. 1984. Rapid detection of cytomegalovirus in MRC-5 cells inoculated with urine specimens by using low-speed centrifugation and monoclonal antibody to an early antigen. *J. Clin. Microbiol.* **19:**917–919.

25. Gozlan, J., J.-M. Saloed, C. Chouaid, C. Duvivier, O. Picard, M.-C. Meyohas, and J.-C. Petit. 1993. Human cytomegalovirus (HCMV) late-mRNA detection in peripheral blood of AIDS patients: diagnostic value for HCMV disease compared with those of viral culture and HCMV DNA detection. *J. Clin. Microbiol.* **31:**1943–1945.

26. Grody, W. W., K. J. Lewin, and F. Naeim. 1988. Detection of cytomegalovirus DNA in classic and epidemic Kaposi's sarcoma by in situ hybridization. *Hum. Pathol.* **19:**524–528.

27. Howell, C. L., and M. J. Miller. 1983. Effect of sucrose phosphate and sorbitol on infectivity of enveloped viruses during storage. *J. Clin. Microbiol.* **18:**658–662.

28. Howell, C. L., M. J. Miller, and W. J. Martin. 1979. Comparison of rates of virus isolation from leukocyte populations separated from blood by conventional and Ficoll-Paque/Macrodex methods. *J. Clin. Microbiol.* **10:**533–537.

29. Hsia, K., D. H. Spector, J. Lawrie, and S. A. Spector. 1989. Enzymatic amplification of human cytomegalovirus sequences by polymerase chain reaction. *J. Clin. Microbiol.* **27:**1802–1809.

30. Huang, E.-S., H. A. Kilpatrick, Y.-T. Huang, and J. S. Pagano. 1976. Detection of human cytomegalovirus and analysis of strain variation. *Yale J. Biol. Med.* **49:**29–43.

31. Hursh, D. A., A. D. Abbot, R. Sun, J. P. Iltis, D. H. Rice, and C. A. Gleaves. 1989. Evaluation of a latex particle agglutination assay for the detection of cytomegalovirus antibody in patient serum. *J. Clin. Microbiol.* **27:**2878–2879.

32. Jiwa, N. M., A. K. Raap, F. M. van de Rijke, A. Mulder, J. J. Weening, F. E. Zwaan, T. H. The, and M. van der Ploeg.

1989. Detection of cytomegalovirus antigens and DNA in tissues fixed in formaldehyde. *J. Clin. Pathol.* **42**:749–754.

33. **Keller, R., R. Peitchel, J. N. Goldman, and M. Goldman.** 1976. An IgG-Fc receptor induced in cytomegalovirus-infected human fibroblasts. *J. Immunol.* **116**:772–777.

34. **Kettering, J. D., N. J. Schmidt, D. Gallo, and E. H. Lennette.** 1977. Anticomplement immunofluorescence test for antibodies to human cytomegalovirus. *J. Clin. Microbiol.* **6**:627–632.

35. **Kettering, J. D., N. J. Schmidt, and E. H. Lennette.** 1977. Improved glycine-extracted complement-fixing antigen for human cytomegalovirus. *J. Clin. Microbiol.* **6**:647–649.

36. **Landry, M. L., and D. Ferguson.** 1993. Comparison of quantitative cytomegalovirus antigenemia assay with culture methods and correlation with clinical disease. *J. Clin. Microbiol.* **31**:2851–2856.

37. **Lee, S. L., and H. H. Balfour.** 1977. Optimal method for recovery of cytomegalovirus from urine of renal transplant patients. *Transplantation* **24**:228–230.

38. **Lucas, G., J. M. Seigneurin, J. Tamalet, S. Michelson, M. Baccard, J. F. Delagneau, and P. Deletoille.** 1989. Rapid diagnosis of cytomegalovirus by indirect immunofluorescence assay with monoclonal antibody F6b in a commercially available kit. *J. Clin. Microbiol.* **27**:367–369.

39. **MacKenzie, D., and L. C. McLaren.** 1989. Increased sensitivity for rapid detection of cytomegalovirus by shell vial centrifugation assay using mink lung cell cultures. *J. Virol. Methods* **26**:183–188.

40. **Macris, M. P., A. J. Nahmias, P. D. Bailey, F. K. Lee, A. M. Visintine, and A. W. Braun.** 1981. Electron microscopy in the routine screening of newborns with congenital cytomegalovirus infection. *J. Virol. Methods* **2**:315–320.

41. **Martin, D. C., D. A. Katzenstein, G. S. M. Yu, and M. C. Jordan.** 1984. Cytomegalovirus viremia detected by molecular hybridization and electron microscopy. *Ann. Intern. Med.* **100**:222–225.

42. **Martin, W. J., II, and T. F. Smith.** 1986. Rapid detection of cytomegalovirus in bronchoalveolar lavage specimens by a monoclonal antibody method. *J. Clin. Microbiol.* **23**:1006–1008.

43. **Mayo, D. R., T. Brennan, S. P. Sirpenski, and C. Seymour.** 1985. Cytomegalovirus antibody detection by three commercially available assays and complement fixation. *Diagn. Microbiol. Infect. Dis.* **3**:455–459.

44. **McHugh, T. M., C. H. Casavant, J. C. Wilber, and D. P. Stites.** 1985. Comparison of six methods for the detection of antibody to cytomegalovirus. *J. Clin. Microbiol.* **22**:1014–1019.

45. **Meyers, J. D., P. Lijungman, and L. D. Fisher.** 1990. Cytomegalovirus excretion as a predictor of cytomegalovirus disease after marrow transplantation: importance of cytomegalovirus viremia. *J. Infect. Dis.* **162**:373–380.

46. **Myerson, D., R. C. Hackman, and J. D. Meyers.** 1984. Diagnosis of cytomegaloviral pneumonia by in situ hybridization. *J. Infect. Dis.* **150**:272–277.

47. **Oefinger, P. E., R. M. Shawar, S. H. Loo, L. T. Tsai, and J. K. Arnett.** 1990. Enhanced recovery of cytomegalovirus in conventional tube cultures with a spin-amplified adsorption. *J. Clin. Microbiol.* **28**:965–969.

48. **Olive, D. M., M. Simsek, and S. Al-Mufti.** 1989. Polymerase chain reaction assay for detection of human cytomegalovirus. *J. Clin. Microbiol.* **27**:1238–1242.

49. **Palmer, D. F., L. Kaufman, W. Kaplan, and J. J. Cavallaro.** 1977. *Serodiagnosis of Mycotic Diseases.* Charles C Thomas, Publisher, Springfield, Ill.

50. **Paya, C. V., A. D. Wold, D. M. Ilstrup, and T. F. Smith.** 1988. Evaluation of number of shell vial cell cultures per clinical specimen for rapid diagnosis of cytomegalovirus infection. *J. Clin. Microbiol.* **26**:198–200.

51. **Phipps, P. H., L. Gregoire, E. Rossier, and E. Perry.** 1983. Comparison of five methods of cytomegalovirus antibody screening of blood donors. *J. Clin. Microbiol.* **18**:1296–1300.

52. **Pillay, D., A. A. Ali, S. F. Liu, E. Kops, P. Sweny, and P. D.**

Griffiths. 1993. The prognostic significance of positive CMV cultures during surveillance of renal transplant recipients. *Transplantation* **56**:103–108.

53. **Pillay, D., H. Charman, A. K. Burroughs, M. Smith, K. Rolles, and P. D. Griffiths.** 1992. Surveillance for CMV infection in orthotopic liver transplant recipients. *Transplantation* **53**:1261–1265.

54. **Rao, N., D. T. Waruszewski, J. A. Armstrong, R. W. Atchison, and M. Ho.** 1977. Evaluation of anti-complementary immunofluorescence test in cytomegalovirus infection. *J. Clin. Microbiol.* **6**:633–638.

55. **Salomn, D., F. Lacassin, M. Harzic, C. Leport, C. Perronne, F. Bricaire, F. Brun-Vezinet, and J.-L. Vilde.** 1990. Predictive value of cytomegalovirus viraemia for the occurrence of CMV organ involvement in AIDS. *J. Med. Virol.* **32**:160–163.

56. **Shibata, D., W. J. Martin, M. D. Appelman, D. M. Causey, J. M. Leedom, and N. Arnheim.** 1988. Detection of cytomegalovirus DNA in peripheral blood of patients with human immunodeficiency virus. *J. Infect. Dis.* **158**:1185–1192.

57. **Slavin, M. A., C. A. Gleaves, H. G. Schoch, and R. A. Bowden.** 1992. Quantification of cytomegalovirus in bronchoalveolar lavage fluid after allogeneic marrow transplantation by centrifugation culture. *J. Clin. Microbiol.* **30**:2776–2779.

58. **Sorbello, A. F., S. L. Elmendorf, J. J. McSharry, R. A. Venezia, and R. M. Echols.** 1988. Rapid detection of cytomegalovirus by fluorescent monoclonal antibody staining and in situ DNA hybridization in a dram vial cell culture system. *J. Clin. Microbiol.* **26**:1111–1114.

59. **Spector, S. A., J. A. Rua, D. H. Spector, and R. McMillan.** 1984. Detection of human cytomegalovirus in clinical specimens by DNA-DNA hybridization. *J. Infect. Dis.* **150**:121–126.

60. **Stagno, S., D. W. Reynolds, A. Tsiantos, D. A. Fuccillo, W. Long, and C. A. Alford.** 1975. Comparative serial virologic and serologic studies of symptomatic and sub-clinical congenitally and natally acquired cytomegalovirus infections. *J. Infect. Dis.* **132**:568–577.

61. **Stanier, P., D. L. Taylor, A. D. Kitchen, N. Wales, Y. Tryhorn, and A. S. Tyms.** 1989. Persistence of cytomegalovirus in mononuclear cells in peripheral blood from blood donors. *Br. Med. J.* **299**:897–898.

62. **Stockl, E., T. Popow-Kraupp, F. X. Heinz, F. Muhlbacher, P. Balcke, and C. Kunz.** 1988. Potential of in situ hybridization for early diagnosis of productive cytomegalovirus infection. *J. Clin. Microbiol.* **26**:2536–2540.

63. **The, T. H., W. van der Bij, A. P. van den Berg, M. van der Giessen, J. Weits, H. G. Sprenger, and W. J. van Son.** 1990. Cytomegalovirus antigenemia. *Rev. Infect. Dis.* **12**(Suppl. 7):S737–S744.

64. **The, T. H., M. van der Ploeg, A. P. van den Berg, A. M. Vlieger, M. van der Giessen, and W. J. van Son.** 1992. Direct detection of cytomegalovirus in peripheral blood leukocytes—a review of the antigenemia assay and polymerase chain reaction. *Transplantation* **54**:193–198.

65. **van den Berg, A. P., A. M. Tegzess, A. Scholten-Sampson, J. Schirm, M. van der Giessen, T. H. The, and W. J. van Son.** 1992. Monitoring antigenemia is useful in guiding treatment of severe cytomegalovirus disease after organ transplantation. *Transplant Int.* **5**:101–107.

66. **van der Bij, W., J. Schirm, R. Torensma, W. J. van Son, A. M. Tegzess, and T. H. The.** 1988. Comparison between viremia and antigenemia for detection of cytomegalovirus in blood. *J. Clin. Microbiol.* **26**:2531–2535.

67. **van der Bij, W., R. Torensma, W. J. van Son, J. Anema, J. Schirm, A. M. Tegzess, and T. H. The.** 1988. Rapid immunodiagnosis of active cytomegalovirus infection by monoclonal antibody staining of blood leukocytes. *J. Med. Virol.* **25**:179–188.

68. **Volpi, A., R. J. Whitley, R. Ceballos, S. Stagno, and L.**

Pereira. 1983. Rapid diagnosis of pneumonia due to cytomegalovirus with specific monoclonal antibodies. *J. Infect. Dis.* **147:**1119–1120.

69. **Walker, R. C.** 1991. The role of the clinical microbiology laboratory in transplantation. *Arch. Pathol. Lab. Med.* **115:** 299–305.

70. **Waner, J. L., and T. H. Weller.** 1978. Analysis of antigenic diversity among human cytomegalovirus by kinetic neutralization tests with high-titered rabbit antisera. *Infect. Immun.* **21:**151–157.

71. **Waner, J. L., T. H. Weller, and S. V. Kevy.** 1973. Patterns of cytomegaloviral complement-fixing antibody activity: a longitudinal study of blood donors. *J. Infect. Dis.* **127:**538–543.

72. **Weiss, R. L., G. W. Snow, G. B. Schumann, and M. E. Hammond.** 1990. Diagnosis of cytomegalovirus pneumonitis on bronchoalveolar lavage fluid: comparison of cytology, im-munofluorescence, and in situ hybridization with viral isolation. *Diagn. Cytopathol.* **7:**243–247.

73. **West, P. G., B. Aldrich, R. Hartwig, and G. J. Haller.** 1988. Enhanced detection of cytomegalovirus in confluent MRC-5 cells treated with dexamethasone and dimethyl sulfoxide. *J. Clin. Microbiol.* **26:**2510–2514.

74. **West, P. G., and W. W. Baker.** 1990. Enhancement by calcium of the detection of cytomegalovirus in cells treated with dexamethasone and dimethyl sulfoxide. *J. Clin. Microbiol.* **28:**1708–1710.

75. **Wolf, D. G., and S. A. Spector.** 1992. Diagnosis of human cytomegalovirus central nervous system disease in AIDS patients by DNA amplification from cerebrospinal fluid. *J. Infect. Dis.* **166:**1412–1415.

76. **Wunderli, W., M. K. Kagi, E. Gruter, and J. D. Auracher.** 1989. Detection of cytomegalovirus in peripheral leukocytes by different methods. *J. Clin. Microbiol.* **27:**1916–1917.

Varicella-Zoster Virus

ANNE A. GERSHON, PHILIP LaRUSSA, AND SHARON P. STEINBERG

73

CLINICAL BACKGROUND

Varicella (chickenpox) and zoster represent different clinical manifestations of infection with the same agent, varicella-zoster virus (VZV). Varicella occurs most frequently in children and is characterized by a generalized vesicular exanthem often accompanied by fever. Zoster (shingles) usually occurs in adults or immunocompromised patients (including those with AIDS) and consists of a painful, circumscribed eruption of vesicular lesions with accompanying inflammation of associated dorsal root or cranial nerve sensory ganglia. Varicella is the primary infection with VZV, whereas zoster is a secondary infection due to reactivation of latent VZV in sensory ganglia. That zoster results from reactivation of latent virus rather than reintroduction of virus into the host is supported by the fact that zoster does not exhibit the seasonal prevalence seen with varicella (late winter and spring), nor does zoster frequently occur in young parents, who are often exposed to their own children with chickenpox. Latent VZV has been detected in sensory ganglia of autopsy specimens by PCR (39) and in situ hybridization (7). Molecular studies of VZV isolates from patients with zoster have revealed that the viruses causing the primary infection and the subsequent zoster are the same (28, 59, 62). Studies indicate that reinfection and reactivation of VZV may occur in the absence of clinical symptoms (1, 25–27, 38, 64).

There are several situations in which providing a specific laboratory diagnosis of VZV infection is crucial. VZV infection may cause severe or fatal disease in individuals who are receiving immunosuppressive therapy or who have abnormalities in their cell-mediated immune responses. Progressive, generalized varicella occurs in as many as 30% of children who acquire chickenpox while receiving chemotherapy and radiotherapy for cancer, and mortality in these cases has ranged from 7 to 28% (16, 17). In immunodeficient patients who have had varicella, there is an increased risk of disseminated zoster, and in some studies, mortality rates for these patients have been similar to those for varicella (38, 53). Providing a specific diagnosis of VZV infection in immunosuppressed patients or their contacts may guide the administration of antiviral agents or passive immunization with varicella-zoster immune globulin. Determining the immunity status (presence or absence of antibody) in high-risk immunocompromised individuals

and adults (who are also at some risk to develop severe varicella) exposed to VZV infection also guides the management of these individuals. Live attenuated varicella vaccine is expected to be licensed soon in the United States, and determination of immunity status will also be useful, particularly for adults with no history of varicella, since only about 20% of such individuals are actually susceptible to chickenpox. It is important to provide a specific diagnosis of some of the less common manifestations of VZV infection, such as varicella pneumonia and encephalitic complications, particularly since antiviral drugs are now available. It is sometimes necessary to differentiate VZV infection from other vesicular eruptions such as those caused by herpes simplex virus (HSV).

DESCRIPTION OF THE AGENT

VZV has the morphology typical of members of the family *Herpesviridae*. It has a DNA genome consisting of approximately 125,000 bp. The linear double-stranded genome ($[80 \pm 3] \times 10^6$ Da) consists of a 58×10^6-Da long unique DNA segment U_L flanked by terminal and internal repeats and a 3.4×10^6-Da short unique sequence U_S flanked by internal and terminal repeats (13, 14, 57). The genome exists in four isomers (two major isomers [95%] and two minor isomers [5%]) (8). The virion has a diameter of 150 to 200 nm and consists of an inner icosahedral capsid composed of 162 capsomeres surrounded by a tegument and an envelope composed of two or more membranes. The envelope contains essential lipid, and infectivity of the virus is therefore inactivated by lipid solvents (60). VZV produces approximately 30 structural and nonstructural proteins, including at least five glycoproteins (gps), designated I through V. These are analogous to gps E, B, H, G, and C, respectively, of HSV. The gps play roles in viral attachment and penetration of cells, and their recognition by the immune system of the host results in humoral and cellular immunity to VZV. Immunity to nonstructural antigens may also play a role in host defense. Antigenic variation of VZV isolates has not been demonstrated, although strain differences can be demonstrated by restriction enzyme analysis of the viral genome (58). Restriction enzyme profiles of VZV isolates reveal the following: (i) there are no detectable differences between varicella and zoster isolates, (ii) the agent is remarkably stable after passage in vitro and in vivo, and (iii) nonepidemiologically related strains may be differentiated with only one or two restriction enzymes. Multiple VZV strains are believed to be able to coexist in latent form in the same host. VZV DNA shares partial homology with HSV, Epstein-Barr virus, equine herpesviruses, and pseudorabies viruses (24). VZV shares antigens with herpesviruses that cause varicellalike disease in simian species (12) and with HSV (48, 54).

COLLECTION, SHIPMENT, AND STORAGE OF SPECIMENS

Ideally, to make a laboratory diagnosis of VZV infection, the presence of the virus should be demonstrated in the patient by either isolation in tissue culture or immunofluorescence (IF) staining, and acute- and convalescent-phase antibody titers should also be obtained to document a significant rise in antibody titer. Since VZV is not shed by asymptomatic patients, demonstration of its presence in affected tissues is diagnostic.

Specimens usually examined for VZV are smears of skin scrapings, vesicular fluid, and tissues obtained at autopsy. Smears of cellular material from vesicular lesions should be collected for IF staining wherever feasible, since this procedure is often more sensitive and always more rapid than virus isolation in cell culture (11, 49). Lung is the autopsy tissue from which VZV is most frequently recovered. An acute-phase blood specimen should be taken as soon as possible after the onset of symptoms; it is for testing in parallel with a convalescent-phase specimen to demonstrate a significant increase in VZV antibody titer or the presence of specific immunoglobulin M (IgM). VZV immunity status is determined by testing a single blood specimen.

Smears of Cellular Material from Lesions

Making smears of cellular material from lesions is a very useful method for distinguishing rapidly between VZV and HSV. Smears collected from the base of fresh vesicular lesions may be used for IF staining for demonstration of VZV antigen in the infected epithelial cells. It is essential that the smears contain a large number of infected cells, and ideally, three smears should be made to permit staining with conjugates to VZV and HSV. The direct method of staining, using fluorescein-conjugated mouse monoclonal antibodies to VZV or HSV gps, is the most specific procedure. Direct IF staining is described below.

If insufficient material is available for IF examination and culture, what is available may be submitted for IF or virus isolation only. Material should be placed in holding medium (see chapter 70 of this Manual) to be used for virus isolation. VZV is a very labile agent, so the time that the specimen is kept in holding medium prior to culture should be as brief as possible.

Vesicular Fluids

Vesicular fluids collected into capillary pipettes or syringes can be used for virus isolation attempts and electron microscopy (EM), but fluids are unsatisfactory for IF staining. Fluids collected onto swabs and placed in holding medium may also be used for virus isolation, but it is preferable to culture vesicular fluids without dilution. VZV in vesicular fluids is more stable than that on swabs or diluted into medium. Lesions present on skin for more than 4 days rarely yield infectious virus, but the fluids may give positive results by other techniques such as EM, countercurrent immuno-

electrophoresis (21), molecular hybridization with radioactive or biotinylated DNA and/or RNA probes (20, 52), enzyme immunofiltration (6, 66), or PCR (4, 10, 22, 29, 34–36, 40, 41, 46, 55).

PCR is expected to eventually become a standard method for demonstrating VZV infection, but at present it remains a research tool. Its particular value with regard to VZV is that the infectious virus is much more labile than its DNA, so that it is much easier to demonstrate VZV DNA by PCR than by isolation techniques. Another advantage for VZV is that it is possible to distinguish between vaccine-type and wild-type viruses by PCR, avoiding the step of virus isolation (36). Disadvantages of VZV PCR include false-positive reactions due to contamination of specimens with previously amplified VZV DNA (a problem common to PCR in general) and the unavailability of commercially marketed primers. Laboratories with potential VZV specimens amenable to PCR are encouraged to contact research laboratories where this technique is being done until this assay is more generally available.

Crusts from Lesions

Crusts do not yield infectious virus and are not suitable for IF staining or other specific antigen detection methods. Crusts may be used for EM or PCR.

Lung, Liver, or Skin Tissue

Smears of autopsy tissue for IF staining are prepared by excising pieces of autopsy tissue about 10 mm square, holding the tissue with forceps, and gently pressing the cut surface to the clean area of a slide. Three or four slides should be prepared from each specimen. The slides are allowed to dry at room temperature for 15 to 20 min and then are fixed in acetone for 10 min at room temperature and stained as described below. Suspensions of tissue are prepared for virus isolation as described in chapter 70 of this Manual.

IF staining of frozen sections of punch biopsies from prevesicular skin lesions has also been reported for the rapid and early diagnosis of VZV infections (42). This method is of limited usefulness for routine diagnosis, however, since it is obviously invasive.

CSF and Blood (Buffy Coat)

Virus has been isolated from buffy coat cells (2, 15, 43, 44) and rarely from cerebrospinal fluid (CSF) (30). These materials should be promptly submitted to the laboratory as whole blood in anticoagulant other than heparin or as undiluted CSF.

Cytological Examination of Smears from Vesicular Lesions

Smears of cellular material collected from the bases of vesicular lesions are prepared as described above. They are fixed with methanol and stained with buffered Giemsa stain at pH 7.0 to 7.2. Microscopic examination reveals multinucleated, giant epithelial cells with altered chromatin patterns (Fig. 1); these are characteristic of infection with either VZV or HSV.

EM

Vesicular fluids can be examined by EM. Crusts from lesions are ground in 1 or 2 drops of distilled water. If transportation of specimens presents a problem, heavy smears of fluids and crusts may be prepared on glass slides and air dried;

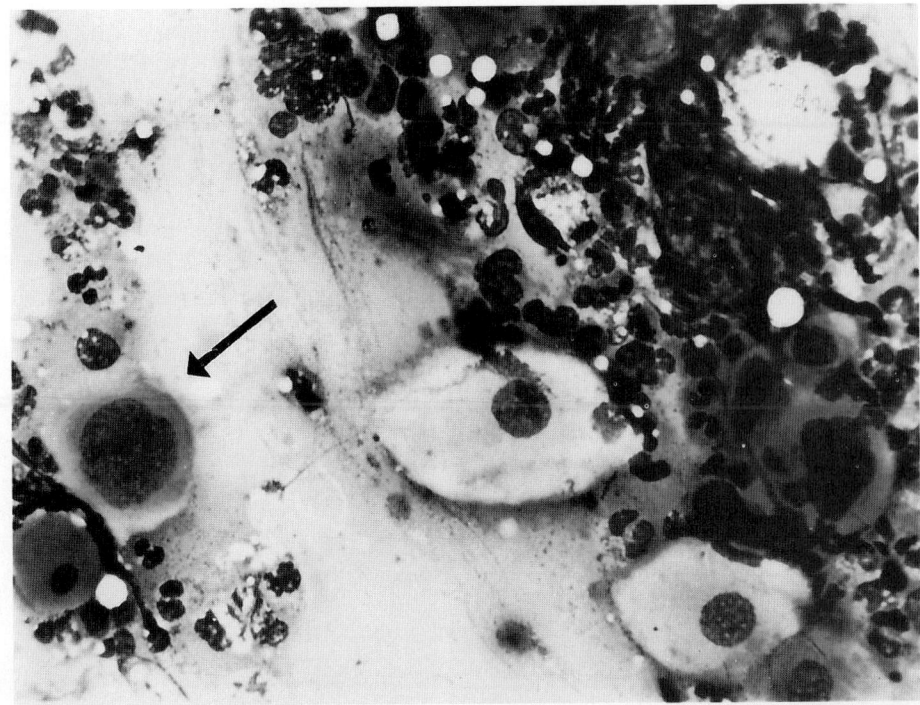

FIGURE 1 Giemsa-stained preparation of material from the base of a vesicular lesion. Magnification, ×250. Arrow indicates a giant cell with a folded nucleus characteristic of VZV or HSV infection.

material from these smears is reconstituted in a drop of water for examination by EM.

A drop of the specimen is placed on a grid and blotted with filter paper. A drop of 3% phosphotungstic acid (prepared in distilled water buffered to neutrality with 1 N KOH) is then added and blotted. Demonstration by EM of virus with typical morphology of herpesviruses (Fig. 2) identifies the etiologic agent as a member of this group and

distinguishes it from the vaccinia virus group, but it does not provide a specific diagnosis of VZV infection.

Storage and Transport of Specimens

Specimens for virus isolation attempts should be inoculated into suitable cell cultures as soon as possible after collection. If inoculation must be delayed (realistically for no longer than 12 h), specimens may be transported or held on

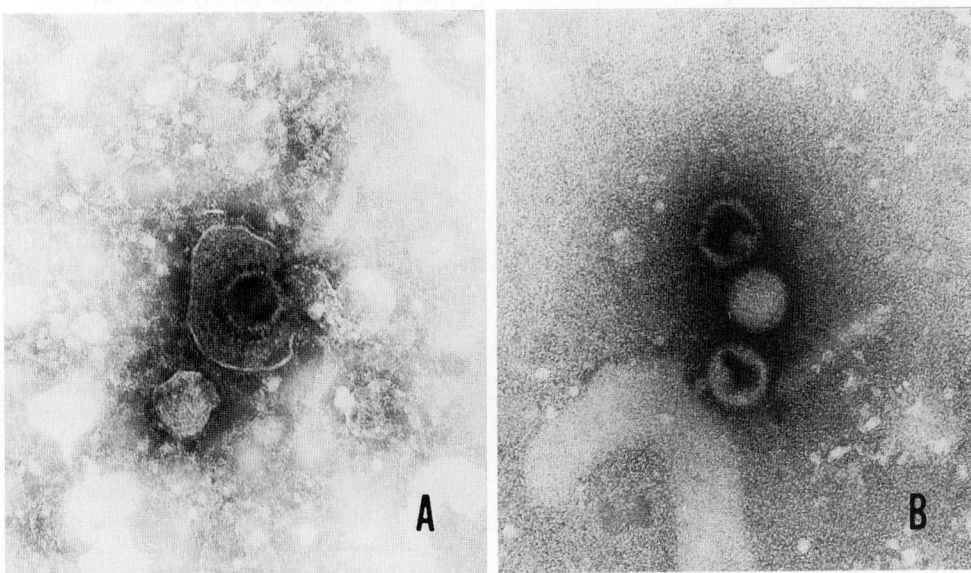

FIGURE 2 Electron micrograph of VZV in a preparation from a lesion swab. Magnification, ×80,000. (A) Enveloped virus; (B) naked nucleocapsids.

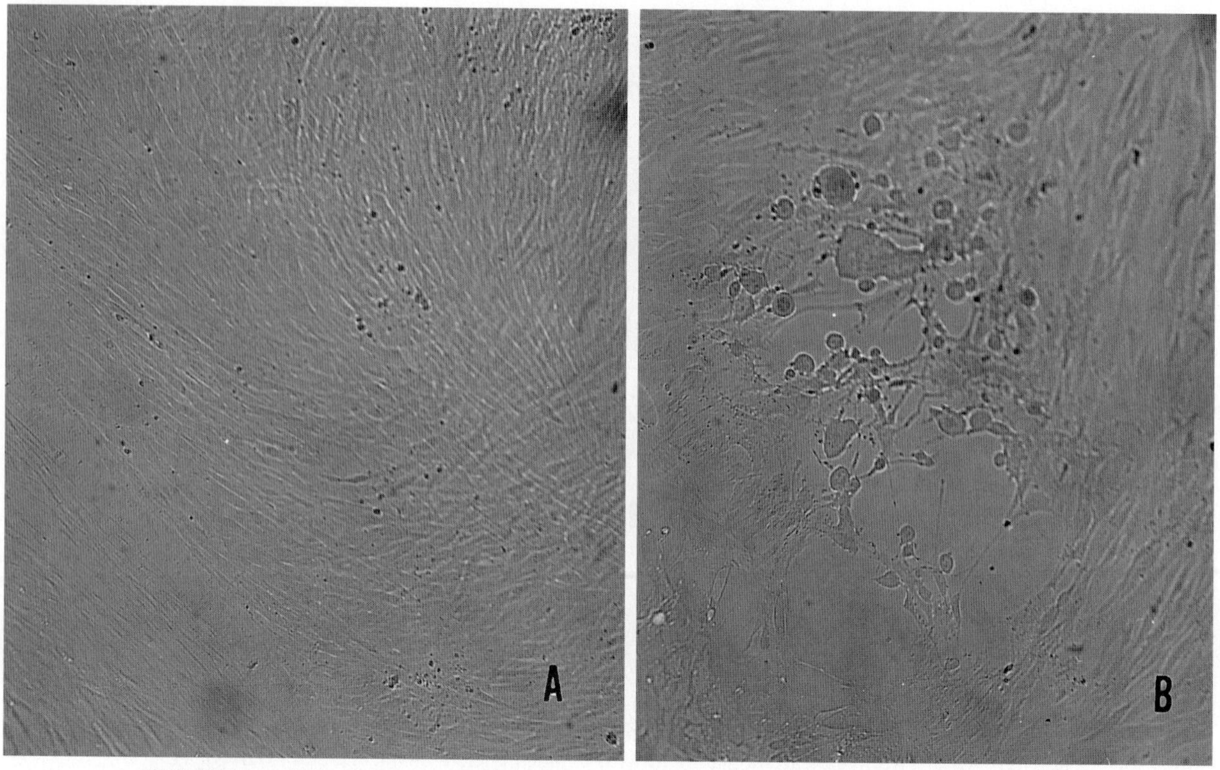

FIGURE 3 CPE of VZV in HFDK cells. Magnification, ×100. (A) Uninfected cells; (B) virus-infected cells.

wet ice or in a refrigerator. For longer holding periods, dry-ice temperatures are required. Refrigeration is not necessary for smears for viral antigen, intranuclear inclusions, or PCR. Specimens from vesicular lesions may be infectious; precautions should be taken in packing to protect postal and laboratory personnel.

ISOLATION AND IDENTIFICATION OF VZV

Host Systems

Diploid human cell lines or primary human cell cultures are the most sensitive host system for isolation of VZV from clinical materials. Human fetal diploid kidney (HFDK) or human fetal diploid lung (HFDL) cells are highly satisfactory for primary isolation of VZV; these cells are available commercially. Some laboratories have used human foreskin fibroblasts for isolation of VZV, but their sensitivity compared to that of HFDL cells is uncertain. Primary monkey kidney cells and embryonic guinea pig cells support the growth of VZV but are less sensitive than HFDL cells. Cell cultures are initiated with Eagle minimal essential medium prepared in a Hanks or Earle balanced salt solution base and supplemented with 10% heat-inactivated (56°C for 30 min) fetal bovine serum. Inoculated cell cultures are maintained in Eagle minimal essential medium prepared in Earle balanced salt solution supplemented with 2% inactivated fetal bovine serum. Although this medium will maintain the cultures for 14 days without a fluid change, a period usually sufficient for the development of viral cytopathic effect (CPE), a change of medium at least once a week is preferable.

Studies have shown virus isolation, even in optimal cell

culture systems, to be less sensitive than IF staining or PCR on lesion materials for diagnosis of VZV infection (3, 11, 21, 36, 49). The lability of viral infectivity in vesicular lesions and the strongly cell-associated nature of the virus probably both contribute to the difficulty of virus isolation.

VZV cannot be isolated in mouse or embryonated egg host cell systems.

Evidence of Infection in Cell Cultures

Figure 3 shows the characteristic CPE of VZV in HFDK cells. Initially, the CPE consists of small, discrete foci of rounded and swollen refractile cells; in HFDK and HFDL cells, these foci may appear from 2 to 14 days after inoculation of the cultures with clinical materials, but in most instances, CPE is first apparent at 3 to 7 days. The foci of infected cells enlarge and may slowly involve much of the monolayer. The spread of infection can be accelerated by dispersing cells in the infected culture with trypsin (see below) and replanting them in growth medium in the same culture vessel.

Subpassage and Storage of Virus

Infectious VZV remains in close association with the host cell, and therefore it is necessary to use virus-infected cells, rather than culture fluids, as an inoculum for serial propagation of the virus. Trypsin-dispersed infected cells are most suitable for subpassage of VZV. The medium is removed from cell cultures in which CPE involves approximately 50% of the cell sheet, and 1 ml of 0.25% trypsin solution is added to each tube culture. After 60 s at room temperature, most of the trypsin is removed, and the small amount of residual trypsin is distributed over the monolayer; as soon as the cells detach from the glass surface, they are dispersed in the original volume of growth medium and inoculated into fresh cell cultures.

Virus in infected cells can be stored in the frozen state if the viability of the cells is maintained through the use of a cryoprotective agent such as dimethyl sulfoxide in the freezing medium. Infected cells are dispersed with trypsin as described above and then suspended in Eagle minimal essential medium containing 10% heat-inactivated fetal bovine serum and 10% dimethyl sulfoxide. Infectious virus can be recovered after 18 months or more of storage at −70°C, but infectivity is preserved more effectively by storage of the infected cells in liquid nitrogen. High-titer cell-free virus prepared by sonic treatment of infected cells is stable at −70°C in minimal essential medium with 10% sorbitol and 10% fetal bovine serum for at least 5 years (51). In our experience, cell-free virus is stable when similarly frozen in Hanks solution alone for as long as 1 year.

Identification of VZV Isolates

Presumptive Identification

Presumptive identification of VZV may be made on the basis of a typical CPE, which is more focal in nature and progresses more slowly than that of HSV. VZV may also be differentiated from HSV by its failure to produce a CPE in rabbit or hamster kidney cell cultures, whereas HSV rapidly produces a CPE. The source of the specimen and the clinical manifestations of the illness in the patient should usually prevent confusion between VZV and human cytomegalovirus (CMV). However, if isolations are made from tissue specimens, there may be a need to distinguish between these herpesviruses. VZV grows well in epithelial cells, whereas CMV generally produces a CPE only in human fibroblast cells. Production of CPE by CMV is generally slower than production by VZV, and the CPE is more focal in nature than that produced by VZV. Furthermore, human CMV strains fail to produce CPE in primary monkey kidney cells, but after initial isolation, VZV strains generally do so. The preferred method for differentiation, however, is virus identification using specific antiserum.

Specific Identification

Specific identification of VZV isolates is made by demonstrating their ability to react with a known positive antiserum. Monoclonal antibodies to VZV are available commercially, and these have overcome problems previously encountered with immune VZV serum produced in animals. In the absence of monoclonal antibodies to VZV, acute- and convalescent-phase sera from patients with a known case of varicella may be employed. Identification is based on the demonstration of a greater degree of IF reactivity of the isolate with the convalescent-phase serum than with the acute-phase serum. However, the use of human sera of uncertain antibody content for identification of viral isolates is not as reliable as the use of monoclonal antibodies. Human sera employed for identification of VZV should be free from antibodies to other human herpesviruses from which it is important to distinguish VZV, namely, HSV and CMV.

Specific identification of VZV isolates is accomplished most readily by IF staining by the direct method. An alternative method is indirect IF staining (61).

Antisera to VZV

Hybridoma technology has been used to produce monoclonal antibodies of high potency and specificity (19, 61). Commercial monoclonal antibodies may be obtained as fluorescein-labeled immunoglobulins for use in direct IF staining or as unlabeled antibodies for use in indirect IF staining.

IF STAINING FOR VZV

Preparation of VZV Slides for IF Staining

Smears of VZV-infected cells are required for standardization of IF reagents, as positive controls when IF staining is used for virus detection in clinical materials or identification of viral isolates, and as an antigen substrate for antibody assays.

To increase infectivity, the stock virus is subpassaged two or three times by inoculating trypsinized infected cells from one culture into two cultures of uninfected cells. When cultures show a 3+ CPE at 24 h after inoculation, they are used for slide preparation. In addition, uninfected cell controls are prepared from trypsinized uninfected cells of the same lot as those infected with VZV. The dispersed cells are suspended in 0.1 M phosphate-buffered saline (PBS) with 2% fetal bovine serum and sedimented by centrifugation at 1,000 × g for 5 min. The supernatant fluids are removed, and the packed cells are suspended in PBS with 2% fetal bovine serum, using a volume of 0.01 ml per tube culture or 0.4 ml per 32-oz (ca 0.95-liter) flask culture. Smears approximately 5 mm in diameter are made from this cell suspension by placing small drops (approximately 0.005 ml each) on microscope slides or by using a cytospin apparatus. The smears are dried at room temperature, fixed with cold acetone for 10 min, and then dried at room temperature. Smears can be stored at −70°C until they are used.

Standardization of Immune Reagents

Fluorescein-conjugated VZV immune conjugates (preferably monoclonal antibodies to VZV) should be titrated against smears of VZV-infected cells to determine the appropriate working dilution for use in virus identification. Ideally, this dilution should be determined in the user's laboratory, although commercially available labeled monoclonal antibodies are packaged in convenient plastic dropper bottles at working dilutions. If unlabeled VZV antibodies are to be used in indirect IF staining, optimal working dilutions of both the viral immune reagent and the conjugated antispecies immune reagent must be determined by preliminary block titrations.

IF staining is done by either the direct or the indirect method (see below), using dilutions of the reagent against smears of VZV-infected and uninfected cells. The working dilution of the reagent is the highest dilution giving specific bright apple green staining and showing no reactivity with uninfected control cells.

Direct IF Staining for VZV Detection and Identification

Working dilutions of the immune conjugates in PBS are added to specimen smears. Positive control slides for VZV and HSV are included as a control on the reactivities and specificities of conjugates. After addition of the conjugates, tests are incubated for 30 min in a warm humidified atmosphere prior to rinsing and mounting.

Examination with a fluorescence microscope should reveal specific bright apple green staining associated with both the cytoplasm and the nucleus of the epithelial cells stained with the VZV conjugate and little or no staining in cells treated with the HSV conjugate. Vesicular-lesion specimens containing too few epithelial cells for definitive examination are reported as unsatisfactory rather than negative.

Identification of VZV isolated in cell cultures is performed by direct IF staining when the inoculated cultures show a 2+ viral CPE. Trypsin-dispersed cells from an infected tube culture are suspended in 3 ml of PBS with 2% fetal bovine serum, and cells are sedimented by centrifugation at 1,000 × g for 5 min. Smears for IF staining are prepared on microscope slides from these cells as described above. Enough smears of each suspension are made to permit staining with both VZV and HSV conjugates, and smears from uninfected cultures are prepared similarly. The smears are fixed and stained as described above. Positive results are based upon staining of strong intensity with the VZV conjugate, little or no staining with the HSV conjugate, and no staining with the uninfected control cells. If commercial reagents are used, there may be a background control stain such as Evans blue, which will stain the uninfected cells a color easily distinguished from yellow-green fluorescence.

Indirect IF Staining for VZV Detection and Identification

Unconjugated monoclonal antibodies can be used for indirect detection of VZV in clinical specimens or for identification of virus isolated in cell culture (61), although direct detection as described above is simpler and preferable. It is essential when using the indirect method to use a conjugate against mouse immunoglobulins that is free from nonspecific reactivity with host tissues and host serum components in clinical specimens. This lack of reactivity is determined by testing dilutions of the conjugate against infected and uninfected cell smears in the absence of intermediate mouse antibodies. The appropriate working dilutions of the VZV reagent and the anti-mouse globulin conjugate are determined by preliminary block titration.

Additional details concerning indirect IF methods in general are included in chapter 11 of this Manual.

Identification of VZV Isolates by Using Shell Vials

The shell vial technique is a combination technique involving virus isolation and early specific identification of virus with monoclonal antibody prior to development of obvious CPE (47). It is a very convenient means for identification of VZV in clinical specimens. Its deficiency is that after a virus has been identified, it is no longer viable, which is not the case with standard virus isolation techniques. Therefore, some clinical laboratories inoculate both shell vials and standard cultures for virus isolation at the same time. Commercially available shell vials containing coverslips on which HFDL cells are growing in a monolayer are used. Swabs or vesicular fluids are diluted in approximately 1 ml of medium, and 0.2 ml is inoculated directly into each of four shell vials from which medium has been removed. The vials are centrifuged for 40 min at 700 × g at room temperature. After centrifugation, 2 ml of maintenance medium is added to each vial. Two uninfected vials

are used as controls. Three vials (two inoculated and one control) are incubated at 35 to 37°C for 24 h, and three are incubated similarly for 72 h. At the appointed time, the vials are rinsed twice with 1 ml of PBS each time, with aspiration of PBS between washings. Monolayers are fixed by removing the medium, washing gently with PBS two times, and adding 2 ml of cold acetone for 10 min. The coverslips are either stored at −70°C or washed by adding 1 ml of PBS for immediate use. For staining, one vial is incubated with 1 or 2 drops of commercial fluorescein-conjugated monoclonal antibody to VZV (already diluted to a working dilution as sold in the bottle) and another vial is incubated with 1 drop of fluorescein-conjugated monoclonal antibody to HSV, both for 30 min at room temperature or 37°C. Ideally, the third control vial is stained with the conjugate that gives a positive reaction with the inoculated culture, either the VZV or the HSV conjugate. The vials are again washed twice with PBS and rinsed once with distilled water, and the coverslips are removed from the vials, air dried, and placed cell side down on microscope slides in mounting medium (phosphate-buffered glycerol, pH 7.2). A coverslip is placed over the specimen and sealed with nail polish, and the slides are examined by fluorescence microscopy as described above. Positive results are based upon staining of 3+ to 4+ intensity with the VZV conjugate, little or no staining with the HSV conjugate, and no staining with the uninfected control cells. If all vials are negative after 24 h of incubation, then the vials incubated for 72 h should be examined. Suggested controls to be included in each run include an uninfected culture with conjugate and a known infected culture with conjugate.

SEROLOGIC METHODS

Serologic procedures are used for laboratory diagnosis of varicella or zoster infections and also for determining immunity status to varicella (presence or absence of VZV antibody) in certain individuals. Serodiagnosis is particularly useful for adults who have no prior history of varicella, since over 75% will have detectable VZV antibody and presumably will therefore have already experienced an episode of varicella (usually subclinical or without symptoms).

Serodiagnosis

The value of serologic procedures for diagnosis of acute VZV infection is limited to some extent by the fact that heterotypic antibody titer rises in response to VZV may occur in certain patients with HSV infection who have experienced a prior infection with VZV. This would appear to be due to a heterotypic antibody response to common antigens in the two viruses (48, 54).

In the past, the complement fixation test was the most widely used procedure for serodiagnosis of VZV infections. However, viral diagnostic laboratories are increasingly adopting enzyme immunoassays (EIAs) as their principal serologic tool for viral diagnosis.

VZV antigens for EIA are available from commercial sources, or they can be prepared from infected HFDL or HFDK cultures. To prepare VZV antigen, cell cultures are inoculated with a high concentration of trypsin-dispersed infected cells and harvested several days later, when they show advanced CPE. Glycine-buffered saline (0.043 M glycine, 0.15 M NaCl, pH 9) is used to rinse infected monolayers, which are then scraped off the glass or plastic with a rubber policeman into approximately 1/20 of the original culture volume of fresh glycine-buffered saline and disrupted by sonication for 30 s at 2.2 to 2.6 A. The antigen is clarified by centrifugation at 1,000 × g for 5 min at room temperature. Uninfected control antigen is prepared in the same manner. Antigens may be stored at −70°C.

For serodiagnosis of infection, acute- and convalescent-phase sera must be analyzed in the same test run. A fourfold or greater increase in antibody titer between acute- and convalescent-phase sera is considered diagnostically significant. In varicella, antibodies are usually demonstrable within several days after onset of rash, and they reach a peak 2 to 3 weeks later. The antibody increase is usually more rapid in zoster, and it is common for antibodies to be detectable at the time of onset.

EIA

The technique for the EIA is described in detail in references 18, 25, and 52; however, a brief description of one EIA for VZV antibody follows (18). Tests are performed in EIA microtiter plates. Optimal dilutions of VZV and control antigens that clearly distinguish known positive and known negative control sera are determined. Ideally, optical densities for positive control sera should be >3 standard deviations higher than the mean optical density of negative control sera. Wells are coated with an optimal dilution of VZV antigen or control antigen in 0.2 ml of 0.06 M bicarbonate buffer (pH 9.5). After overnight incubation at 4°C, the unadsorbed antigen is removed, the wells are washed with 0.01 M PBS (pH 7.3) containing 0.05% Tween 20, and protein adsorption sites are saturated by adding 0.35 ml of a 5% bovine albumin solution in PBS and incubating for 4 h at room temperature. The wells are washed with PBS. Serial twofold dilutions of serum are prepared with 0.05-ml microdilutions in wells coated with VZV antigen or control antigen. (Alternatively, a single dilution of serum such as 1:100 may be tested, and the optical density can be compared with those of a panel of known negative, low, moderate, and high serum VZV antibody levels.) A 0.05-ml volume of PBS is added to each well, and tests are incubated overnight at room temperature. (Alternatively, a shorter incubation period of 1 to 5 h may be tried.) Controls include known positive and negative sera and wells incubated with diluent instead of serum. After the incubation, wells are washed three times with PBS. A 0.1-ml volume of the optimal dilution of alkaline phosphatase-labeled antibodies to human IgG prepared in PBS with 5% bovine albumin is added, and tests are incubated for 2 h at room temperature. The conjugate is then aspirated from the wells, and the wells are washed three times with PBS. A 0.1-ml volume of the enzyme substrate, p-nitrophenylphosphate (1 mg/ml) in diethanolamine (10%) buffer (pH 9.8) with 10^{-3} M MgCl$_2$, is added, and after incubation at room temperature for 30 min, the reaction is stopped by the addition of 0.05 ml of 3 N NaOH per well. Results should be read with an EIA reader; the EIA may be used to identify IgG, IgA, or IgM antibodies against VZV.

A number of commercially marketed tests for measurement of VZV antibody by EIA are available and have been evaluated (5, 9, 37). In general, these tests are fairly specific; however, since they may lack sensitivity, some adults with immunity to varicella may be misidentified as susceptible to chickenpox when tested by this method. EIA may also fail to identify some individuals who have been immunized with live attenuated varicella vaccine (5, 56).

Alternative methods that are applicable to the routine serodiagnosis of VZV infections are anticomplement IF (ACIF) staining (23, 45), latex agglutination (LA) (56), indirect IF with fluorescent-antibody staining of membrane antigen in VZV-infected cells (FAMA) (63), and the immune adherence hemagglutination test (18). Neutralizing antibody to VZV can be assayed by plaque reduction techniques with cell-free virus obtained by sonic treatment of infected human diploid cells (50); however, this is primarily a research rather than a diagnostic tool.

Determination of Immunity Status

A rapid and sensitive serologic method is useful for determining past infection and presumed immunity to varicella. The complement fixation test is not adequately sensitive for this purpose. EIAs are more sensitive than complement fixation, and demonstration of VZV antibody by EIA is good evidence of prior infection with VZV (varicella). In approximately 10 to 15% of sera in which VZV antibody cannot be detected by EIA, however, antibody is detectable by methods such as FAMA and LA. Both of these tests have been used for determining immunity status, and results have correlated well with resistance or susceptibility to infection in clinical settings. The FAMA method is somewhat cumbersome, requiring use of live, unfixed VZV-infected cells; however, its sensitivity and specificity make it ideal for determining immunity status.

FAMA Test

The FAMA test is performed as follows. Doubling dilutions of serum in PBS, beginning at 1:2, are made in a 96-well round-bottom microtiter plate. Heat inactivation of serum is not required and may lead to nonspecific results. Cells infected with VZV are harvested when they exhibit 50 to 75% CPE. Medium is poured off the cultures, and the cells are scraped off the glass into any remaining medium and centrifuged at 700 × g for 2 min. The cells are taken up in PBS and dispersed in 0.025-ml aliquots into microtiter plates containing 0.025-ml dilutions of serum. Cells are then incubated for 30 min at room temperature in a moist chamber. PBS is added to fill the wells, and the plate is centrifuged at 1,000 × g at 4°C for 10 min, after which the PBS is removed by rapidly flipping the plate upside down. The washing steps are repeated two more times. A working dilution of fluorescein-labeled antihuman immunoglobulin (0.025 ml) is added to each well, and incubation for 30 min at room temperature is again carried out. (This assay may also be used to measure specific VZV IgM antibody, in which case a fluorescein-labeled anti-human IgM conjugate is used.) After two washes, a drop of buffered glycerol (9 parts glycerol to 1 part PBS) is added to each well, and the cells are removed with an aspirator and placed on microscope slides. The cells are covered with a glass coverslip sealed with nail polish. The slides are then examined by fluorescence microscopy. The antibody titer is the highest dilution of serum yielding 3+ or 4+ fluorescence.

The sensitivity of this assay derives from the live cells, which preserve the conformational structure of VZV gp antigens, although a modified FAMA assay in which cells are briefly fixed in dilute glutaraldehyde and stored at −70°C has been described (65). The FAMA assay may be used to identify IgG, IgA, or IgM antibodies against VZV. The advantage of the FAMA test is its sensitivity; its disadvantages are that live virus-infected cells are required, it is technically rather difficult, and it cannot be automated.

ACIF Test

A method that yields results comparable to FAMA and may be more convenient to perform is the ACIF test (23, 45). This method avoids the nonspecific reactivity that may be encountered in indirect IF staining and permits examination of serum at dilutions as low as 1:2 and 1:4 (as does the FAMA test). Smears of VZV-infected and uninfected cells are prepared as for IF staining, with fixation in cold acetone for 10 min. Smears may be stored frozen at −20°C for several months. Optimal dilutions of complement and conjugate are determined by box titrations of various dilutions of each reagent (diluted in Veronal-buffered saline, pH 7.2, 0.1 ionic strength) against VZV-infected cells treated with appropriate dilutions of known positive and negative sera. Smears of VZV-infected and uninfected cells are treated with dilutions of heat-inactivated (56°C, 30 min) test serum diluted from 1:2 to 1:1,024 in 0.01 M PBS (pH 7.2) and incubated for 30 min at 37°C in a humidified chamber. After a 5-min rinse of the smear in PBS, an optimal dilution of cold guinea pig complement is added, and slides are incubated for 45 min at 37°C in a humidified chamber. After a 5-min rinse of the smear in PBS, an optimal dilution of fluorescein-conjugated anti-guinea pig complement is added, with incubation for 30 min at 37°C in a moist chamber. After a 5-min rinse, the slides are air dried, mounted in 50% glycerol in PBS, and examined under UV illumination. As controls, known positive and negative sera should be included in each test run. Serum specimens are tested against uninfected cells to control for the presence of antinuclear antibodies. The antibody titer is the highest dilution of serum yielding positive fluorescence on infected cells but not uninfected cells. The advantage of the ACIF test is its specificity; disadvantages include that it cannot be automated, that a high degree of training is required to perform it, and that it cannot distinguish between VZV IgG and IgM.

LA Assay

The LA test makes use of commercially marketed emulsified latex particles coated with gps of VZV. It is performed as follows. Doubling dilutions of sera are made in 1-cm-diameter flat circles on a wax-coated card supplied by the manufacturer. A drop of latex emulsion is added to each circle directly from the dropper bottle. The card is rotated mechanically for 10 min at 90 to 110 rpm in a moist chamber. Circles are examined for agglutination (obvious clumping of the latex particles) under a high-intensity desktop lamp. For screening sera for immunity status, two dilutions (1:4 and 1:50) are tested. A positive reaction in one or both wells is highly correlated with immunity to varicella. If possible, sera (particularly from adults) that are negative at screening dilutions should be retested at dilutions of 1:4 to 1:64 to avoid false-negative results due to prozone phenomena. False-positive reactions are rare (56). For testing acute- and convalescent-phase sera in order to diagnose VZV infection, doubling dilutions beginning at 1:4 can be used. Care must be taken with convalescent-phase specimens to make certain that high antibody titers such as those greater than 1:4,096 are not missed because of a prozone effect. Thus, if a negative result is seen in a convalescent-phase serum specimen, the serum should be retested with higher dilutions of serum. The chief advantages of the LA test are its simplicity, its low cost, and the rapidity with which it can be performed (15 to 30 min). Its disadvantages are that it cannot be automated or used to demonstrate VZV IgM and that prozone reactions occur in 1 to 5% of sera, particularly those with high VZV antibody titers.

EVALUATION AND INTERPRETATION OF RESULTS

Clinical symptoms such as rash may be diagnostically deceiving; in one study, 13% of patients diagnosed as having zoster were found on laboratory testing to have HSV infection (33). Demonstration of VZV, specific antigen, or other viral products such as DNA in skin lesion material or autopsy tissue is diagnostic of active infection, since asymptomatic shedding such as occurs with some of the other herpesviruses such as HSV and CMV does not occur. Demonstration of VZV IgM in serum is highly suggestive of acute VZV infection.

Isolation of VZV is a relatively slow method and is often less sensitive than IF staining of lesion material or EM, since infectious virus persists for a shorter time in vesicles and is more labile than viral particles and antigens. The combination of virus isolation and early IF staining before obvious CPE in shell vials is a practical alternative and provides significant information within 24 to 48 h. Other means for presumptive identification of isolates as VZV include failure of VZV to multiply on rabbit or hamster kidney cells (in contrast to HSV) and ability of VZV to propagate in primary monkey kidney cells and human epithelial cells, which distinguishes it from human CMV. With the increasing availability of VZV-specific immune reagents from commercial sources, specific identification of isolates can readily be made by IF staining or other methods for antigen detection such as EIA. Pitfalls in the use of human sera for specific identification have been indicated.

A fourfold or greater increase in antibody titer to VZV antigen in the absence of a similar rise to HSV antigen is diagnostic of a current VZV infection. However, a high proportion (up to one-third) of individuals with primary HSV infections who have experienced a prior VZV infection show a heterotypic antibody response to VZV antigen, making a differential diagnosis between VZV and HSV infection difficult in the absence of clear-cut clinical findings. Thus, owing to the characteristic clinical picture, it is more reliable to diagnose varicella than zoster by serologic means.

Although negative results by FAMA, ACIF, LA, and EIA are reliable indicators of susceptibility to varicella, the occurrence of clinical varicella in a few individuals with low titers of VZV antibody demonstrated by FAMA and ACIF indicates that low levels of antibody demonstrable by these sensitive methods are not always a reliable indication of protection against clinical illness. This is particularly true for immunocompromised patients. Occasionally, second attacks of varicella can occur (26, 31, 32). Detection of serum antibody to VZV by FAMA in healthy individuals has correlated with protection in up to 96% of persons in one study (26). More recently, similar accuracy with the LA assay has been demonstrated in healthy subjects (56). In general, for determining immunity status to varicella, testing sera from immunocompromised persons for antibodies to VZV may be less reliable than simply asking the patient whether he or she has had varicella. In contrast, testing serum from healthy individuals, particularly adults with no history of varicella, is a useful and reliable means of identifying persons susceptible or immune to varicella.

This chapter is based on the one originally written by Nathalie J. Schmidt (deceased) for the fourth edition of this Manual. The work was supported in part by grant AI 24021.

REFERENCES

1. **Arvin, A., C. M. Koropchak, and A. E. Wittek.** 1983. Immunologic evidence of reinfection with varicella-zoster virus. *J. Infect. Dis.* **148:**200–205.

2. **Asano, Y., N. Itakura, Y. Hiroishi, S. Hirose, T. Nagai, T. Ozaki, T. Yazaki, Y. Yamanishi, and M. Takahashi.** 1985. Viremia is present in incubation period in nonimmunocompromised children with varicella. *J. Pediatr.* **106:**69–71.

3. **Asano, Y., N. Itakura, Y. Hiroishi, S. Hirose, T. Ozaki, K. Okuno, T. Nagai, T. Yazaki, T. Yazaki, K. Yamanishi, and M. Takahashi.** 1985. Viral replication and immunologic responses in children naturally infected with varicella-zoster virus and in varicella vaccine. *J. Infect. Dis.* **152:**863–868.

4. **Baird, R. E., P. Daly, and M. Sawyer.** 1991. Varicella arthritis diagnosed by polymerase chain reaction. *Pediatr. Infect. Dis. J.* **12:**950–951.

5. **Balfour, H. H., C. K. Edelman, C. L. Dirksen, D. R. Palermo, C. S. Suarez, J. T. Kenala, and D. D. Crane.** 1988. Laboratory studies of acute varicella and varicella immune status. *Diagn. Microbiol. Infect. Dis.* **10:**149–158.

6. **Cleveland, P. H., and D. D. Richman.** 1987. Enzyme immunofiltration staining assay for immediate diagnosis of herpes simplex virus and varicella-zoster virus directly from clinical specimens. *J. Clin. Microbiol.* **25:**416–420.

7. **Croen, K. D., J. M. Ostrove, L. Y. Dragovic, and S. E. Straus.** 1988. Patterns of gene expression and sites of latency in human ganglia are different for varicella-zoster and herpes simplex viruses. *Proc. Natl. Acad. Sci. USA* **85:**9773–9777.

8. **Davison, A. J.** 1991. Varicella-zoster virus. The Fourteenth Fleming Lecture. *J. Gen. Virol.* **72:**475–486.

9. **Demmler, G., S. Steinberg, G. Blum, and A. Gershon.** 1988. Rapid enzyme-linked immunosorbent assay for detecting antibody to varicella-zoster virus. *J. Infect. Dis.* **157:**211–212.

10. **Dlugosch, D., A. M. Eis-Hubinger, J. P. Kleim, R. Kaiser, E. Bierhoff, and K. E. Schneweis.** 1992. Diagnosis of acute and latent varicella-zoster virus infections using the polymerase chain reaction. *J. Med. Virol.* **35:**136–141.

11. **Drew, W. L., and L. Mintz.** 1980. Rapid diagnosis of varicella-zoster virus infection by direct immunofluorescence. *Am. J. Clin. Pathol.* **73:**699–701.

12. **Dueland, A. N., J. R. Martin, M. Devlin, M. Wellish, R. Mahlingham, R. Cohrs, K. F. Soike, and D. Gilden.** 1992. Acute simian varicella infection. *Lab. Invest.* **66:**762–773.

13. **Dumas, A. H., J. L. M. C. Geelen, M. W. Weststrate, P. Wertheim, and J. van der Noordaa.** 1981. XbaI, PstI, and BglII restriction enzyme maps of the two orientations of the varicella-zoster virus genome. *J. Virol.* **39:**390–400.

14. **Ecker, J. R., and R. W. Hyman.** 1982. Varicella zoster virus DNA exists as two isomers. *Proc. Natl. Acad. Sci. USA* **79:**156–160.

15. **Feldman, S., and E. Epp.** 1979. Detection of viremia during incubation period of varicella. *J. Pediatr.* **94:**746–748.

16. **Feldman, S., W. Hughes, and C. Daniel.** 1975. Varicella in children with cancer: 77 cases. *Pediatrics* **80:**388–397.

17. **Feldman, S., and L. Lott.** 1987. Varicella in children with cancer: impact of antiviral therapy and prophylaxis. *Pediatrics* **80:**465–472.

18. **Forghani, B., N. Schmidt, and J. Dennis.** 1978. Antibody assays for varicella-zoster virus: comparison of enzyme immunoassay with neutralization, immune adherence hemagglutination, and complement fixation. *J. Clin. Microbiol.* **8:**545–552.

19. **Forghani, B., N. Schmidt, C. K. Myoraku, and D. Gallo.** 1982. Serological reactivity of some monoclonal antibodies to varicella-zoster virus. *Arch. Virol.* **73:**311–317.

20. **Forghani, B., G. Yu, and J. Hurst.** 1991. Comparison of biotinylated DNA and RNA probes for rapid detection of varicella-zoster virus genome by in situ hybridization. *J. Clin. Microbiol.* **29:**583–591.

21. **Frey, H., S. Steinberg, and A. Gershon.** 1981. Varicella-zoster infections: rapid diagnosis by countercurrent immunoelectrophoresis. *J. Infect. Dis.* **143:**274–280.

22. **Furuta, Y., T. Takasu, K. Fukuda, C. Sato-Matsumura, Y. Inuyama, R. Hondo, and K. Nagashima.** 1992. Detection of varicella-zoster virus DNA in human geniculate ganglia by polymerase chain reaction. *J. Infect. Dis.* **166:**1157–1159.

23. **Gallo, D., and N. J. Schmidt.** 1981. Comparison of anticomplement immunofluorescence and fluorescent antibody to membrane antigen test for determination of immunity status to varicella-zoster virus and for serodifferentiation of varicella-zoster and herpes simplex virus infections. *J. Clin. Microbiol.* **14:**539–543.

24. **Gelb, L.** 1990. Varicella zoster virus, p. 2011–2054. *In* B. N. Fields (ed.), *Virology.* Raven Press, New York.

25. **Gershon, A. A., P. LaRussa, and S. Steinberg.** 1991. Live attenuated varicella vaccine: current status and future uses. *Semin. Pediatr. Infect. Dis.* **2:**171–178.

26. **Gershon, A. A., S. Steinberg, L. Gelb, and the NIAID Collaborative Varicella Vaccine Study Group.** 1984. Clinical reinfection with varicella-zoster virus. *J. Infect. Dis.* **149:**137–142.

27. **Gershon, A. A., S. Steinberg, and the NIAID Collaborative Varicella Vaccine Study Group.** 1989. Persistence of immunity to varicella in children with leukemia immunized with live attenuated varicella vaccine. *N. Engl. J. Med.* **320:**892–897.

28. **Hayakawa, Y., T. Yamamoto, K. Yamanishi, and M. Takahashi.** 1986. Analysis of varicella zoster virus (VZV) DNAs of clinical isolates by endonuclease HpaI. *J. Gen. Virol.* **67:**1817–1829.

29. **Isada, N. B., D. P. Paar, M. Johnson, M. Evans, W. Holzgreve, F. Qureshi, and S. Straus.** 1991. In utero diagnosis of congenital varicella zoster infection by chorionic villus sampling and polymerase chain reaction. *Am. J. Obstet. Gynecol.* **165:**1727–1730.

30. **Jemsek, J., S. B. Greenberg, L. Taber, D. Harvey, A. Gershon, and R. Couch.** 1983. Herpes zoster-associated encephalitis: clinicopathologic report of 12 cases and review of the literature. *Medicine* **62:**81–97.

31. **Junker, A. K., E. Angus, and E. Thomas.** 1991. Recurrent varicella-zoster virus infections in apparently immunocompetent children. *Pediatr. Infect. Dis. J.* **10:**569–575.

32. **Junker, K., C. Avnstorp, C. Nielsen, and N. Hansen.** 1989. Reinfection with varicella-zoster virus in immunocompromised patients. *Curr. Prob. Dermatol.* **18:**152–157.

33. **Kalman, C. M., and O. L. Laskin.** 1986. Herpes zoster and zosteriform herpes simplex virus infections in immunocompetent adults. *Am. J. Med.* **81:**775–778.

34. **Kido, S., T. Ozaki, H. Asada, K. Higashi, K. Kondo, Y. Hayakawa, T. Morishima, M. Takahashi, and K. Yamanishi.** 1991. Detection of varicella-zoster virus (VZV) DNA in clinical samples from patients with VZV by the polymerase chain reaction. *J. Clin. Microbiol.* **29:**76–79.

35. **Koropchak, C., G. Graham, J. Palmer, M. Winsberg, S. Ting, M. Wallace, C. Prober, and A. Arvin.** 1991. Investigation of varicella-zoster virus infection by polymerase chain reaction in the immunocompetent host with acute varicella. *J. Infect. Dis.* **163:**1016–1022.

36. **LaRussa, P., O. Lungu, I. Hardy, A. Gershon, S. Steinberg, and S. Silverstein.** 1992. Restriction fragment length polymorphism of polymerase chain reaction products from vaccine and wild-type varicella-zoster virus isolates. *J. Virol.* **66:**1016–1020.

37. **LaRussa, P., S. Steinberg, E. Waithe, B. Hanna, and R. Holzman.** 1987. Comparison of five assays for antibody to varicella-zoster virus and the fluorescent-antibody-to-membrane-antigen test. *J. Clin. Microbiol.* **25:**2059–2062.

38. **Locksley, R. M., N. Flournoy, K. M. Sullivan, and J. Meyers.** 1985. Infection with varicella-zoster virus after marrow transplantation. *J. Infect. Dis.* **152:**1172–1181.

39. Mahalingham, R., M. Wellish, W. Wolf, A. N. Dueland, R. Cohrs, A. Vafai, and D. Gilden. 1990. Latent varicella-zoster viral DNA in human trigeminal and thoracic ganglia. *N. Engl. J. Med.* **323:**627–631.

40. Nahass, G. T., B. A. Goldstein, W. Zhu, U. Serfling, N. S. Penneys, and C. L. Leonardi. 1992. Comparison of Tzanck, viral culture, and DNA diagnostic methods in detection of herpes simplex and varicella-zoster infection (PCR). *JAMA* **268:**2541–2544.

41. Nishi, M., R. Hanashiro, S. Mori, K. Masuda, M. Mochizuki, and R. Hondo. 1992. Polymerase chain reaction for the detection of the varicella-zoster genome in ocular samples from patients with acute retinal necrosis. *Am. J. Ophthalmol.* **114:**603–609.

42. Olding-Stenkvist, E., and M. Grandien. 1976. Early diagnosis of virus-caused vesicular rashes by immunofluorescence on skin biopsies. I. Varicella, zoster, and herpes simplex. *Scand. J. Infect. Dis.* **8:**27–35.

43. Ozaki, T., T. Ichikawa, Y. Matsui, T. Kondo, T. Nagai, Y. Asano, K. Yamanishi, and M. Takahashi. 1986. Lymphocyte-associated viremia in varicella. *J. Med. Virol.* **19:**249–253.

44. Ozaki, T., T. Ichikawa, Y. Matsui, T. Nagai, Y. Asano, K. Yamanishi, and M. Takahashi. 1984. Viremic phase in non-immunocompromised children with varicella. *J. Pediatr.* **104:** 85–87.

45. Presissner, C., S. Steinberg, A. Gershon, and T. F. Smith. 1982. Evaluation of the anticomplement immunofluorescence test for detection of antibody to varicella-zoster virus. *J. Clin. Microbiol.* **16:**373–376.

46. Puchhammer-Stockl, E., T. Popow-Kraupp, F. Heinz, C. Mandl, and C. Kunz. 1991. Detection of varicella-zoster virus DNA by polymerase chain reaction in the cerebrospinal fluid of patients suffering from neurological complications associated with chicken pox or herpes zoster. *J. Clin. Microbiol.* **29:**1513–1516.

47. Schirm, J., J. Meulenberg, G. Pastoor, P. C. Vader, and P. Schroder. 1989. Rapid detection of varicella-zoster virus in clinical specimens using monoclonal antibodies on shell vials and smears. *J. Med. Virol.* **28:**1–6.

48. Schmidt, N. J. 1982. Further evidence for common antigens in herpes simplex and varicella-zoster virus. *J. Med. Virol.* **9:**27–36.

49. Schmidt, N. J., D. Gallo, V. Devlin, J. D. Woodie, and R. W. Emmons. 1980. Direct immunofluorescence staining for detection of herpes simplex and varicella-zoster virus antigens in vesicular lesions and certain tissue specimens. *J. Clin. Microbiol.* **12:**651–655.

50. Schmidt, N. J., and E. H. Lennette. 1975. Neutralizing antibody responses to varicella-zoster virus. *Infect. Immun.* **12:** 606–613.

51. Schmidt, N. J., and E. H. Lennette. 1976. Improved yields of cell-free varicella-zoster virus. *Infect. Immun.* **14:**709–715.

52. Seidlin, M., H. E. Takiff, H. A. Smith, J. Hay, and S. Straus.

1984. Detection of varicella-zoster virus by dot-blot hybridization using a molecularly cloned viral probe. *J. Med. Virol.* **13:**53–61.

53. Shepp, D. H., P. S. Dandliker, and J. D. Meyers. 1986. Treatment of varicella-zoster virus infection in severely immunocompromised patients: a randomized comparison of acyclovir and vidarabine. *N. Engl. J. Med.* **314:**208–212.

54. Shiraki, K., T. Okuno, K. Yamanishi, and M. Takahashi. 1982. Polypeptides of varicella-zoster virus (VZV) and immunological relationship of VZV and herpes simplex virus (HSV). *J. Gen. Virol.* **61:**255–269.

55. Shoji, H., Y. Honda, I. Murai, Y. Sato, K. Oizumi, and R. Hondo. 1992. Detection of varicella-zoster virus DNA by polymerase chain reaction in cerebrospinal fluid of patients with herpes zoster meningitis. *J. Neurol.* **239:**69–70.

56. Steinberg, S., and A. Gershon. 1991. Measurement of antibodies to varicella-zoster virus by using a latex agglutination test. *J. Clin. Microbiol.* **29:**1527–1529.

57. Straus, S. E., H. S. Aulakh, W. T. Ruyechan, J. Hay, et al. 1981. Structure of varicella-zoster virus DNA. *J. Virol.* **40:** 516–526.

58. Straus, S. E., J. Hay, H. Smith, and J. Owens. 1983. Genome differences among varicella-zoster isolates. *J. Gen. Virol.* **64:** 1031–1041.

59. Straus, S. E., W. Reinhold, H. A. Smith, W. Ruyechan, D. Henderson, R. M. Blaese, and J. Hay. 1984. Endonuclease analysis of viral DNA from varicella and subsequent zoster infections in the same patient. *N. Engl. J. Med.* **311:**1362–1364.

60. Takahashi, M. 1983. Chickenpox virus. *Adv. Virus. Res.* **28:**285–356.

61. Weigle, K., and C. Grose. 1983. Common expression of varicella-zoster viral glycoprotein antigens in vitro and in chickenpox and zoster vesicles. *J. Infect. Dis.* **148:**630–638.

62. Williams, D. L., A. Gershon, L. D. Gelb, M. K. Spraker, S. Steinberg, and A. H. Ragab. 1985. Herpes zoster following varicella vaccine in a child with acute lymphocytic leukemia. *J. Pediatr.* **106:**259–261.

63. Williams, V., A. Gershon, and P. Brunell. 1974. Serologic response to varicella-zoster membrane antigens measured by indirect immunofluorescence. *J. Infect. Dis.* **130:**669–672.

64. Wilson, A., M. Sharp, C. Koropchak, S. Ting, and A. Arvin. 1991. Subclinical varicella-zoster virus viremia, herpes zoster, and T lymphocyte immunity to varicella-zoster viral antigens after bone marrow transplantation. *J. Infect. Dis.* **165:**119–126.

65. Zaia, J., and M. Oxman. 1977. Antibody to varicella-zoster virus-induced membrane antigen: immunofluorescence assay using monodisperse glutaraldehyde-fixed target cells. *J. Infect. Dis.* **136:**519–530.

66. Zeigler, T. 1984. Detection of varicella-zoster viral antigens in clinical specimens by solid-phase enzyme immunoassay. *J. Infect. Dis.* **150:**149–154.

Epstein-Barr Virus

EVELYNE T. LENNETTE

74

CLINICAL BACKGROUND

Epstein-Barr virus (EBV), the etiologic agent of infectious mononucleosis (IM), has a worldwide distribution, with 80 to 90% of all adults having been infected. Primary infections occur during the first decade of life in areas with crowded living conditions and poor hygiene. Childhood infections, which are mostly asymptomatic, may infrequently be associated with classic IM. In contrast, 50 to 75% of young adults undergo primary EBV infections, with illness ranging from mild to severe (11).

In most cases of IM, clinical diagnosis can be made from the characteristic triad of fever, pharyngitis and cervical lymphadenopathy lasting 1 to 4 weeks. Normally a self-limiting illness, IM may be complicated by splenomegaly, hepatitis, pericarditis, or central nervous system involvement. Rare fatal primary infections occur in patients with histiocytic hemophagocytic syndrome or in immunologically incompetent patients, such as transplant recipients and those with a genetic X-linked lymphoproliferative syndrome (9, 34). Hematologic features of IM include lymphocytosis with prominent atypical lymphocytes. In 85 to 90% of IM patients, Paul-Bunnell heterophile tests are positive; false positives may occur in 2 to 3% of patients and can be excluded only by EBV-specific serology. Specific laboratory diagnosis is also needed to differentiate the 10 to 15% of heterophile-negative EBV infections from mononucleoses induced by other agents such as cytomegalovirus, adenovirus, and *Toxoplasma gondii*. In primary infections of adults with clinically atypical diseases and of children with negative heterophile response, EBV-specific laboratory diagnosis may also be helpful.

Transmission of EBV requires salivary contact; airborne and blood-borne transmissions are not important routes of infection (11, 15). Familial transmission of EBV infections is common, as demonstrated by intrafamilial sharing of EBV strains (7). As with other herpesviruses, EBV causes a persistent latent infection with intermittent reactivations. Infectious virus can be recovered from the oropharynges of the majority of seropositive, asymptomatic individuals as well as from IM patients. The degree of shedding varies from individual to individual but remains constant within the same individual (36). Salivary EBV shedding in healthy individuals constitutes the primary reservoir for person-to-person transmission. With immunosuppression, infectious

virus can be recovered from the patients with greater frequency and at higher titers (35). Owing to the ubiquity of the virus in seropositive individuals, it has been difficult to ascertain the degree of morbidity attributable to EBV reactivations. While most reactivations are asymptomatic (14), there is evidence that they can be associated with severe, chronic diseases in rare instances (29).

EBV is unique among the herpesviruses in its ability to transform and immortalize human B lymphocytes. In vitro, EBV induces polyclonal immunoglobulin production and leads to establishment of permanent lymphoblastoid lines. In vivo, EBV-infected lymphoid cells are associated with several lymphoproliferative conditions varying from hyperplasia to neoplasia (10).

EBV has long been suspected of having a contributory role in the etiology of Burkitt's lymphoma (BL) and nasopharyngeal carcinoma (NPC) (13). BL is primarily a tumor of children in Africa and New Guinea. Elsewhere in the world, EBV-associated BL has been reported mostly in adults and, more recently, among severely immunosuppressed male homosexuals (37). EBV has also been consistently associated with undifferentiated squamous carcinoma of the nasopharynx, with a particularly high incidence among southern Chinese. In both tumors, viral antigens and genomes can be detected in malignant tissues (19, 23, 25, 39). Unusually high titers of antibodies to several antigens can be correlated with the patient's tumor burden. Serology can be helpful in the management of patients with these two malignancies and in monitoring the effectiveness of their therapies (13).

With the advent of aggressive and immunosuppressive therapies during organ transplantation, EBV lymphoproliferative disorders are seen with increasing frequency. They range from benign, polyclonal hyperplasias with no cytogenetic abnormalities to oligoclonal and monoclonal malignant lymphomas. As in BL and NPC, viral antigens and genomes can both be readily demonstrated in the tumors (10). In addition, EBV is detectable in B-cell lymphomas of patients with acquired and inherited immunodeficiencies (3, 37), in rare T-cell lymphomas of allograft recipients (20), in Reed-Sternberg cells of Hodgkin's lymphoma (33), in gastric carcinomas (31), and in other carcinomas of the head and neck.

DESCRIPTION OF THE AGENT

As a member of the family *Herpesviridae*, EBV has the characteristic herpetic 120-nm enveloped morphology, with 162 capsomeres in icosahedral arrangement. Its double-stranded 172-kbp DNA exists both in linear form in the mature virion and in circular episomal form in latently infected cells. Its linear structure consists of a series of unique sequences alternating with internal repeat sequences, all sandwiched between two terminal-repeat elements that are joined during circularization (4).

In vivo, the virus infects both lymphoid and epithelium-derived tissues of the nasopharynx. In vitro, EBV has been propagated only in B lymphocytes from human and subhuman primates. Although it shares antigenic determinants with other EBV-like subhuman primate viruses, EBV is antigenically distinct from other human herpesviruses. Infection of lymphocytes by EBV leads to their transformation into lymphoblastoid cell lines capable of continuous growth in culture. Infected cells rarely produce infectious virus in vitro. In most transformed cell lines, the viral genome is maintained latently at a constant copy number per cell, with genome expression restricted to the nuclear antigen complex (EBNA) and latent membrane proteins (LMP 1 and LMP 2) (21).

On the molecular level, EBV strains may now be classified into types A and B (sometimes referred to as types 1 and 2) on the basis of organization of their EBNA genes (1). Recent serologic and PCR investigations indicate that both type A and type B have worldwide distribution, although type A appears to be more prevalent than type B (30). Dual infections with both types are not uncommon (32). Biologically, type A strains transform lymphocytes more efficiently than type B viruses (28).

COLLECTION AND STORAGE OF SPECIMENS

Generally, only a single acute-phase serum sample (1 to 5 ml) is needed for diagnosis by serologic testing. Convalescent-phase serum collected 1 to 2 months after onset is occasionally needed for confirmation and interpretations. If collected aseptically, serum can be stored at 5°C for several months. For longer storage, freezing at −20°C is recommended.

For isolation of excreted virus, 5 to 10 ml of throat gargle collected in serum-free tissue culture medium or Hanks' balanced salts solution is satisfactory. Fetal bovine serum (2 to 5%) for use as a stabilizer and antibiotics for suppression of microbial growth can then be added. With serum additive, specimens can be held for 2 to 3 days with prompt refrigeration or frozen at −70°C for longer storage.

Tissues to be examined for viral antigens are collected aseptically and refrigerated in tissue culture medium. Suitable for this purpose are lymph nodes, spleen, liver, and biopsy samples of tumors. Thin (5-μm-thick) cryosections but not formalin-fixed tissues are also suitable. Fresh biopsy samples may be examined for the presence of virus either by selection of EBV-transformed cells or by direct examination of EBV antigens by specific immunostaining. For the detection of EBV nucleic acid, any infected tissue, either fresh, frozen, or paraffin embedded, would be a suitable specimen.

Cerebrospinal fluid is not useful for documentation of EBV-associated central nervous system disease. Neither infectious virus nor the corresponding antibodies have been detected in cerebrospinal fluid, suggesting an immunologic rather than a virologic etiology of the central nervous system disease.

For cultivation of EBV-infected peripheral blood lymphocytes, 10 ml of heparinized (5 to 10 U/ml) blood from IM patients is sufficient. Blood specimens should be processed as soon as possible, although refrigeration is adequate for up to 24 h.

DIRECT EXAMINATION

Direct detection of EB virions by electron microscopy of lymphoid tissues is generally not feasible, as infected cells are usually latent viral carriers. Mature viral particles are rarely seen in infected lymphocytes, even during IM. One exception to this rule is EBV infection of the tongue epithelium in patients with oral leukoplakia. A high concentration of mature virions can be readily demonstrated by electron microscopy of tongue lesion biopsy samples from these patients (8).

Generally, the presence of EBV can be demonstrated by the detection of induced proteins or viral genome. The presence of EBV-induced early antigens (EA) (associated with lytic infection) and the nuclear antigen complex (EBNA) (latent infection) can be readily shown in touch preparations or cryosections of affected tissues, using the anticomplement indirect immunofluorescence (ACIF) staining technique (27) (Fig. 1). With a larger quantity of tissue, EBNA can also be detected by Western blotting (immunoblotting). Alternatively, nucleic acid hybridization techniques such as Southern blotting or PCR can be used to demonstrate the presence of EBV genomes in tissues or cell scrapings (2, 32). More recently, assays for the *EBER-1* gene, coding for a small RNA expressed at a high (10^6 to 10^7) copy number, were found to be highly sensitive in detecting latently infected tissues and tumor (26).

ISOLATION OF VIRUS

Fetal B lymphocytes are the preferred indicator cells for EBV isolation because of their good susceptibility to infection. Lymphoid tissues and leukocytes from seronegative adults are, for practical purposes, not useful for the cultivation of EBV because of suppressive interference from T cells usually found in partially fractionated lymphocyte preparations.

Preparation of Indicator Cells

Human umbilical cord blood (10 ml) collected aseptically with 5 to 10 U of preservative-free heparin per ml should be fractionated through polysaccharide density gradients (Hypaque-Ficoll, Histopaque, etc.) according to the supplier's recommendations. During centrifugation, aggregated erythrocytes and granulocytes sediment to the bottom. Lymphocytes can be recovered at the plasma-gradient interphase, washed once with saline, and resuspended in RPMI 1640 medium supplemented with 10% fetal bovine serum to a density of 5×10^6 cells per ml. After overnight incubation at 37°C in 5% CO_2 to check for sterility, the lymphocytes are ready for inoculation and should be used as soon as possible.

Specimen Inoculation

Throat gargles must be centrifuged at $1,500 \times g$ for 10 min. The supernatant fluid is filtered through a 0.45-μm-pore-size filter to remove remaining cell debris and microorganisms. When possible, prompt inoculation is recommended,

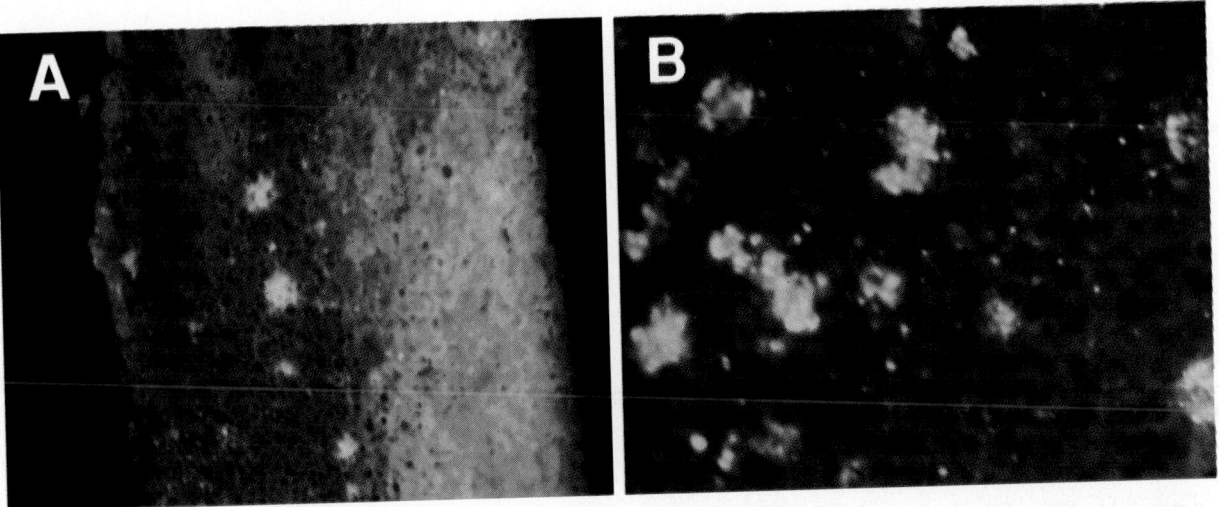

FIGURE 1   (A) EBV EA/D-positive liver cryosection stained by ACIF; (B) EBNA-positive lymph node stained by ACIF.

although the specimen can be frozen at −70°C at this stage for future testing. Just before inoculation, 1 ml of the fractionated leukocyte suspension is centrifuged and suspended in 0.5 ml of growth medium. A 0.5-ml sample of the filtered throat specimen is added to the leukocyte suspension. After an adsorption period of 1 to 2 h at 37°C, 1 ml of culture medium with 10% fetal bovine serum is added. An uninfected cell control and an EBV-infected control should be included in parallel, as rare cord blood lymphocyte preparations have shown spontaneous transformation. All cultures are incubated for 4 weeks with weekly replacements of growth medium. Necrosis of uninfected cultures is usually observable after 2 weeks, whereas virus-positive cultures should contain clusters of large proliferating lymphoblastoid cells. The identification of the transforming agent can then be made by the detection of EBNA in these cells by the ACIF procedure.

IDENTIFICATION OF VIRUS

Viral expression in transformed cell lines varies widely, from almost silent with only a few detectable viral antigens to a fully productive cycle with late products such as viral capsid antigen (VCA) and infectious virions. Depending on the exact "block" on the viral genome expression, various spectra of EBV antigens are detectable in the transformed cells. The only antigen complex present in all EBV-infected cells, however, is EBNA. It can be reliably used as an EBV marker, detectable by ACIF assay within a few hours. No commercial hyperimmune animal or mouse monoclonal anti-EBNA sera are currently available; hence, human sera still provide the only source of reagents suitable for EBNA detection. For reference reagents, sera from EBV-infected and susceptible individuals should be used in parallel. Care should be taken in the selection of negative sera to exclude those with nonspecific antinuclear antibodies.

To detect EBNA in transformed cells in culture, the centrifuged cell pellet is suspended in saline containing 0.5% bovine serum albumin. The cell density is adjusted to 5×10^6 cells per ml, and the suspension is dropped on glass slides. The air-dried slides are fixed for 1 min with an acetone-methanol (1:1) mixture. Fixed smears can be held

at −20°C until testing. Several slides should be prepared in this manner to allow for necessary controls.

Identification of individual viral isolates is now possible with the discovery that the sizes of many of the EBNA family proteins (EBNA 1 through EBNA 6) induced are strain dependent. Each isolate induces proteins with distinctive patterns in Western blotting, thus allowing the fingerprinting of each virus (7).

SEROLOGIC DIAGNOSIS

EB viral isolation is diagnostically not useful owing to the narrow in vitro host range of EBV and the long period needed for isolation of cell-free virus. The ubiquity of EBV in healthy individuals should be taken into consideration in interpreting results obtained either by culture or by PCR.

Serologic testing is the method of choice for the diagnosis of primary infections (16). In patients with symptoms compatible with IM, a positive Paul-Bunnell heterophile-antibody result is diagnostic, and no further testing is necessary. Rapid qualitative agglutination or enzyme-linked immunoassay test kits for Paul-Bunnell heterophile antibodies are widely available and are effective for 80 to 85% of IM patients. Quantitative heterophile measurements, now seldom performed, showed that moderate to high levels of heterophile antibodies are seen during the first month of illness and decrease rapidly after week 4. The false-positive rate for the Paul-Bunnell antibodies is approximately 3%, mostly from individuals who maintain a low but persistent level of these antibodies long after their primary illness (17). The false-negative rate is 10 to 15%; it is more frequent among children than adults. For these patients, EBV-specific serologic testing is needed.

Humoral responses to primary EBV infections appear to be quite rapid. Eighty percent of patients usually have peak titers by the time they consult their physicians. Hence, testing of paired sera is not useful in demonstrating significant antibody changes in the majority of the cases. Effective laboratory diagnosis can be made, on the other hand, from a single acute-phase serum by the testing of antibodies to several EB-associated antigens measured simultaneously. The level and spectrum of antibodies are sufficiently dis-

TABLE 1 Serologic profiles of EBV-associated syndromes

Antibody-antigen	Antibody presence[a]						
	Nonimmune	Infection			Reactive	BL	NPC
		Current primary	Recent primary	Past			
IgM-VCA	−	+	−	−	−	−	−
IgG-VCA	−	+	+	+	+	++	++
IgA-VCA	−	+ or −	−	−	+	++	++
IgG-EA/D	−	+	+	−	+ or −	−	++
IgA-EA/D	−	−	−	−	Not known	−	++
IgG-EA/R	−	+ or −	+ or −	−	+ or −	++	+ or −
Anti-EBNA	−	−	Low	+	+	+	++

[a] −, Negative (<1:10); +, positive (≥1:10).

tinct in 90 to 95% of the cases to allow classification as to whether the patient (i) is still susceptible, (ii) has a current primary infection, (iii) has had a recent primary infection (within 2 to 3 months), (iv) had an infection in the past, or (v) may be having reactivated EBV infection (Table 1).

Antibodies to four antigen complexes, including VCA, EA-diffuse component (EA/D), EA-restricted component (EA/R), and EBNA, may be measured. In addition, differentiation of the immunoglobulin G (IgG), IgM, and IgA subclasses of antibodies to VCA can often be helpful for confirmatory purposes.

Anti-VCA

During the acute phase, both IgG-VCA and IgM-VCA are detectable. Whereas IgM-VCA disappears after 4 weeks, IgG-VCA declines but persists for life. Neutralizing antibodies can also be detected early after onset and persist for life. In practice, the complex neutralizing assays are rarely used to test immunity. Since all patients with IgG-VCA also have neutralizing antibodies, anti-VCA titers are accurate indicators of immunity. In patients with BL and NPC, IgG-VCA is maintained at very high levels, usually 8 to 10 times the geometric mean titers of healthy adults. NPC patients have an additional high IgA-VCA titer as an outstanding EBV serologic feature.

Anti-EA/D and EA/R

In the majority of patients, antibodies to EA/D show a transient rise during the acute phase but are generally undetectable 3 to 6 months after onset. Anti-EA/R antibodies follow the disappearance of anti-EA/D and can be transiently detectable for up to 2 years after onset (18). Antibodies to either or both EA components at moderate titers can reappear during EBV reactivation. With BL patients, anti-EA/R is present at a moderate to high level, whereas with NPC patients, high anti-EA/D titers are common. In the latter patients, both IgG and IGA subclasses of anti-EA/D are present at high titers.

Anti-EBNA

Antibodies to the EBNA complex, as measured by standard ACIF assay with Raji cells, are rarely present in acute-phase serum. A gradual increase in EBNA antibodies occurs during convalescence, and near-peak titers are maintained for life. With severe immunosuppression, anti-EBNA titers may gradually decrease but rarely disappear.

Anti-EBNA 1 through 6

Using transfected cell cultures, it is now possible to measure antibody responses to individual EBNA components in human sera (5). Following IM in healthy individuals, anti-EBNA 2 and -EBNA 6 antibodies increase in titer within the first 3 months after onset. As anti-EBNA 2 wanes and even disappears, anti-EBNA 1 emerges to reach a peak titer between 6 and 12 months. In contrast, many patients with chronic, ill-defined illnesses or a complicated course of EBV primary infections have abnormal EBNA 1/EBNA 2 titer ratios of ≤1, primarily because of the persistence of EBNA 2. Unfortunately, no clear associations can be made between the anti-EBNA 1/anti-EBNA 2 ratio and duration or nature of illness (19). Anti-EBNA 1 is the dominant antibody component in serum as measured by the conventional ACIF with Raji cells (22). The significance of antibodies to other EBNA components is still under investigation.

The levels of each of the mentioned antibodies are usually lower in young patients. However, the profile does not differ with age. The exact titers to each antigen and the time needed to develop a full spectrum of antibodies vary widely with individuals. Also, many individuals maintain EBV antibodies at high levels with or without reactivations (36). For these reasons, diagnosis based on "screening" titers is not feasible.

The optimal combination for EBV serology consists of measurement of IgG-VCA, IgG-EA, and EBNA. This combination allows accurate classification of 90 to 95% of patients with a single specimen (see below). The inclusion of EBNA extends the time during which primary infections can be reliably diagnosed to 2 to 3 months. Diagnosis of primary infection should not rely on the detection of IgM-VCA. Both false-positive and false-negative results occur in IgM-VCA testing. The former are due to the presence of rheumatoid factor (12), while the latter result from late collection of serum samples.

FLUORESCENT-ANTIBODY TESTS

Listed below are the cell lines most commonly used in combination to measure antibodies to VCA, EA/D, EA/R, and EBNA (17). All cell lines grow well in RPMI 1640 supplemented with 10% fetal bovine serum.

1. P3-HR1 is a virus producer cell line. Ten to 15% of cells in the culture express VCA at any given time. This degree of viral expression can be maintained indefinitely.

To prepare smears, centrifuged cells from a 1-week-old culture are suspended in PBS containing bovine serum albumin to a final density of 10^7 cells per ml. The cell suspension (5 to 10 μl) is dropped on glass slides with a Pasteur pipette. Air-dried slides are fixed in acetone for 3 min. Fixed slides can be stored at $-20°C$ indefinitely.

2. The Raji line, in contrast, is positive only for the EBNA complex. VCA is not produced, and EAs are rarely detectable if the cells are grown under normal culture conditions. Hence, Raji cells are suitable for the detection of anti-EBNA. Air-dried smears from 3-day-old cultures are prepared as described for P3-HR1 cells. Fixation, however, is with an mixture of equal volumes of acetone and methanol for 1 to 2 min.

3. EA-positive cells can be prepared several ways. The conventional method involves superinfection of Raji cells with infectious virus concentrated from P3-HR1 culture with centrifugation of culture fluid for 1 h at 10,000 × g. EA smears of consistent quality can be prepared with pre-titered virus stock frozen at $-70°C$. The advantage of this method is its ability to control the density of EA-positive cells. The alternative method involves EA induction of Raji cells by treatment with various chemicals, including tumor-promoting agents (38), iododeoxyuridine (6), and sodium butyrate (24). By controlling the concentration and length of treatment with these chemicals, various degrees of EA expression are achievable.

4. An EBV-negative cell line is necessary as a negative control to exclude nonspecific antinuclear antibodies in some sera. The most appropriate cell line is BJAB, a B-lymphocyte EBV-negative line, although MOLT-4, a T-lymphoid line, is sometime used as an alternative.

For the indirect immunofluorescent-antibody assay, four-fold dilutions of the patient's serum are incubated with the fixed cells and then with fluorescein isothiocyanate (FITC)-conjugated antiserum to the appropriate anti-human Ig subclasses. Incubations are for 30 min at 37°C in moist chambers. Acetone-fixed P3-HR1 cells can be used for measurements of IgG, IgM, and IgA antibodies to VCA. IgM-VCA testing is sometime prolonged to 3 h in the first incubation. This may be necessary to enhance the intensity of the staining.

Acetone-fixed EA smears are suitable for indirect immunofluorescent-antibody assays of antibodies to both EA/D and EA/R, as both antigens are present in the EA-producing cells. The corresponding antigens can be differentiated by their characteristic staining morphologies. EA/D-positive cells show diffuse, speckled staining in the cytoplasm. Frequently, these cells have haloes of fine granular staining where the antigens have leaked out during fixation. EA/R, in contrast, is "restricted" to the cytoplasmic regions within the cells. Methanol fixation of the EA smears preferentially removes EA/R only and hence can be used to prepare smears for differentiation of the two antigens.

The concentration of EBNA in EBV-transformed cells is usually low, and the more sensitive ACIF assay must be used to detect anti-EBNA. Raji smears are incubated with the patient's serum. This incubation is followed by successive 30-min incubations with EBV-negative complement (either pretested human or guinea pig) and then appropriately FITC-conjugated anti-C3 antiserum. If human complement is used, FITC–anti-β_1C/β_1A is preferable. The staining pattern is nuclear and should be present in all the cells.

Commercial smears are available for all of the above antigens. Enzyme-linked immunoassays are now available and are being evaluated in the clinical setting.

INTERPRETATIONS

From the titers and profile of antibodies to VCA, EA, and EBNA in the acute-phase serum, the patient can be classified as (i) susceptible if anti-VCA is absent (<1:10); (ii) having primary EBV infection if anti-VCA is present and anti-EBNA is absent; or (iii) immune, with past infection, if both anti-VCA and EBNA are present. Eighty percent of patients with active EBV infections produce anti-EA/D titers. These antibodies can be very useful as indicators of current or reactivated infections. In the absence of anti-EBNA, anti-EA confirms a primary infection. In the presence of anti-EBNA, anti-EA suggest a reactivated past infection.

IgM-VCA is present in approximately 85 to 90% of sera from IM patients submitted to our laboratory for EBV testing. Most of the 10 to 15% IgM-VCA-negative sera had low or undetectable levels of anti-EBNA, indicative of a primary infection within 6 to 8 weeks. A second serum tested 4 to 6 weeks later would show a significant rise in anti-EBNA titers in all of the patients with primary disease. Hence, reliance on anti-EBNA instead of IgM-VCA effectively extends the acute phase of the illness. IgM-VCA titers found in the presence of elevated EBNA titers (e.g., >1:20) are almost without exception false results due to rheumatoid factors, which can be completely removed by absorption (12).

As with all serologic testing, significant changes in EBV antibodies can be demonstrated only with parallel assays. In general, antibody titers to the various antigens are remarkably stable throughout an individual's lifetime. Significant changes in titers are seldom observed except during the early acute phase of the infection. Antibody level changes seen in EBV-associated tumors occur slowly and are seldom detectable at intervals of less than 3 to 4 months. In patients with chronic conditions other than malignancies, profiles of EBV antibodies compatible with reactivation patterns are already established by the time most patients seek clinical diagnosis. Their EBV profiles do not undergo observable changes, and serial EBV serologies on these patients are not required.

REFERENCES

1. **Addlinger, H. K., H. Delius, U. K. Freese, J. Clark, and G. W. Bornkamm.** 1985. A putative transforming gene of Jijoye virus differs from that of Epstein-Barr prototypes. *Virology* **141:**221–228.
2. **Andiman, W., L. Gradoville, L. Heston, R. Neydorff, M. E. Savage, G. Kitchingman, D. Shedd, and G. Miller.** 1983. Use of cloned probes to detect Epstein-Barr viral DNA in tissues of patients with neoplastic and lymphoproliferative diseases. *J. Infect. Dis.* **148:**967–977.
3. **Cleary, M. L., and R. F. Dorfman.** 1986. Failure in immunological control of the virus infection: post transplant lymphomas, p. 163–181. *In* M. A. Epstein and B. G. Achong (ed.), *The Epstein-Barr Virus: Recent Advances.* William Heinemann Medical Books, London.
4. **Dambaugh, T., K. Hennessy, S. Fennewald, and E. Kieff.** 1986. The virus genome and its expression in latent infection, p. 13–45. *In* M. A. Epstein and B. G. Achong (ed.), *The Epstein-Barr Viruses: Recent Advances.* William Heinemann Medical Books, London.

5. **Dillner, J., and B. Kallin.** 1988. The Epstein-Barr virus encoded proteins. *Adv. Cancer Res.* **50:**95–151.

6. **Gerber, P., and S. Lucas.** 1972. Epstein-Barr virus associated antigens activated in human cells by 5-bromodeoxyuridine. *Proc. Soc. Exp. Biol. Med.* **141:**431–435.

7. **Gratama, J. W., M. A. P. Oosterveer, G. Klein, and I. Ernberg.** 1990. EBNA size polymorphism can be used to trace Epstein-Barr virus spread within families. *J. Virol.* **64:**4703–4708.

8. **Greenspan, J. S., D. Greenspan, E. T. Lennette, D. I. Abrams, M. A. Conant, V. Petersen, and U. K. Freese.** 1985. Replication of Epstein-Barr virus within the epithelial cells of oral "hairy" leukoplakia, an AIDS-associated lesion. *N. Engl. J. Med.* **313:**1564–1571.

9. **Grierson, H., and D. T. Purtillo.** 1987. Epstein-Barr virus infections in males with the X-linked lymphoproliferative syndrome. *Ann. Intern. Med.* **106:**538–545.

10. **Hanto, D. W., G. Frizzera, K. J. Gajl-Peczalska, and R. L. Simmons.** 1985. Epstein-Barr virus, immunodeficiency and B cell lymphoproliferation. *Transplantation* **30:**461–472.

11. **Henle, G., and W. Henle.** 1979. The virus as the etiologic agent of infectious mononucleosis, p. 197–307. *In* M. A. Epstein and B. G. Achong (ed.), *The Epstein-Barr Virus.* Springer-Verlag KG, Berlin.

12. **Henle, G., E. T. Lennette, M. A. Alspaugh, and W. Henle.** 1979. Rheumatoid factor as a cause of positive reactions in tests for Epstein-Barr virus-specific IgM antibodies. *Clin. Exp. Immunol.* **36:**415–422.

13. **Henle, W., and G. Henle.** 1974. Epstein-Barr virus and human malignancies. *Cancer* **34:**1368–1374.

14. **Henle, W., and G. Henle.** 1980. Consequences of persistent Epstein-Barr virus infections, p. 3–9. *In* M. Essex, G. Todaro, and H. zur Hauzen (ed.), *Viruses in Naturally Occurring Cancers. Cold Spring Harbor Conferences on Cell Proliferation.* Cold Spring Harbor Laboratory, Cold Spring Harbor, N.Y.

15. **Henle, W., and G. Henle.** 1985. Epstein-Barr virus and blood transfusions, p. 201–209. *In* R. Y. Dodd and L. F. Baker (ed.), *Infection, Immunity and Blood Transfusion.* Alan R. Liss, Inc., New York.

16. **Henle, W., G. Henle, and C. A. Horwitz.** 1974. Epstein-Barr virus specific diagnostic tests in infectious mononucleosis. *Hum. Pathol.* **5:**551–565.

17. **Henle, W., G. Henle, and C. A. Horwitz.** 1979. Infectious mononucleosis and Epstein-Barr virus-associated malignancies, p. 441–470. *In* E. H. Lennette and N. J. Schmidt (ed.), *Diagnostic Procedures for Viral, Rickettsial and Chlamydial Infections,* 5th ed. American Public Health Association, Washington, D.C.

18. **Horwitz, C. A., W. Henle, G. Henle, H. Rudnick, and E. Latts.** 1985. Long term serological follow-up of patients for Epstein-Barr virus after recovery from infectious mononucleosis. *J. Infect. Dis.* **151:**1150–1153.

19. **Huang, D. P., J. H. C. Ho, W. Henle, and G. Henle.** 1974. Demonstration of Epstein-Barr virus-associated nuclear antigen in nasopharyngeal carcinoma cells from fresh biopsies. *Int. J. Cancer* **14:**580–588.

20. **Jones, J. F., S. Shurin, C. Abramowsky, R. Tubbs, C. G. Sciotto, R. Wahl, J. Sands, D. Gottman, B. Z. Katz, and J. Sklar.** 1988. T-cell lymphomas containing Epstein-Barr viral DNA in patients with chronic Epstein-Barr virus infections. *N. Engl. J. Med.* **318:**733–740.

21. **Kieff, E., and D. Liebowitz.** 1990. Epstein-Barr virus and its replication, p. 1889–1920. *In* B. N. Fields and D. M. Knipe (ed.), *Virology.* Raven Press, New York.

22. **Lennette, E. T., L. Rymo, M. Yadav, G. Masucci, K. Merk, L. Timar, and G. Klein.** 1993. Disease-related differences in antibody patterns against EBV-encoded nuclear antigens EBNA-1, EBNA-2 and EBNA-6. *Eur. J. Cancer* **29:**1584–1589.

23. **Lindahl, T., G. Klein, B. Johansson, and S. Singh.** 1974. Relationship between Epstein-Barr virus (EBV) DNA and the determined nuclear antigen (EBNA) in Burkitt's lymphoma biopsies and other lymphoproliferative diseases. *Int. J. Cancer* **13:**764–772.

24. **Luka, J., B. Kallin, and G. Klein.** 1979. Induction of the Epstein-Barr virus (EBV) cycle in latently infected cells by n-butyrate. *Virology* **94:**228–231.

25. **Nonoyama, M., C. H. Huang, J. S. Pagano, G. Klein, and S. Singh.** 1973. DNA of Epstein-Barr virus detected in tissue of Burkitt's lymphoma and nasopharyngeal carcinoma. *Proc. Natl. Acad. Sci. USA* **70:**3265–3268.

26. **Randhawa, P. S., R. Jaffe, A. J. Demitris, M. Nalesnik, T. E. Starzl, Y. Y. Chen, and L. M. Weiss.** 1992. Expression of Epstein-Barr virus-encoded small RNA (by the EBER-1 gene) in liver specimens from transplant recipients with post-transplantation lymphoproliferative disease. *N. Engl. J. Med.* **24:**1710–1714.

27. **Reedman, B. M., and G. Klein.** 1973. Cellular localization of an Epstein-Barr virus (EBV)-associated complement-fixing antigen in producer and non-producer lymphoblastoid cell lines. *Int. J. Cancer* **11:**499–520.

28. **Rickinson, A. B., L. S. Young, and M. Rowe.** 1987. Influence of the Epstein-Barr nuclear antigen EBNA 2 on the growth phenotype of virus-transformed B cells. *J. Virol.* **61:**1310–1317.

29. **Schooley, R. T., R. W. Carey, G. Miller, W. Henle, R. Eastman, E. J. Mark, K. Kenyon, E. O. Wheeler, and R. Rubin.** 1986. Chronic Epstein-Barr virus infection associated with fever and interstitial pneumonitis. Clinical and serologic features and response to antiviral chemotherapy. *Ann. Intern. Med.* **104:**636–643.

30. **Sculley, T. B., S. M. Cross, P. Borrow, and D. A. Cooper.** 1988. Prevalence of antibodies to Epstein-Barr virus nuclear antigen 2B in persons infected with the human immunodeficiency virus. *J. Infect. Dis.* **158:**186–192.

31. **Shibata, D., M. Tokunaga, Y. Uemura, E. Sato, S. Tanaka, and L. M. Weiss.** 1991. Association of Epstein-Barr virus with undifferentiated gastric carcinomas with intense lymphoid infiltration. *Am. J. Pathol.* **139:**469–474.

32. **Sixbey, J. W., P. Shirley, P. J. Chesney, D. M. Buntin, and L. Resnick.** 1989. Detection of a second widespread strain of Epstein-Barr virus. *Lancet* **ii:**761–765.

33. **Weiss, L. M., L. A. Movahed, R. A. Warnke, and J. Sklar.** 1989. Detection of Epstein-Barr viral genomes in Reed-Sternberg cells of Hodgkin's disease. *N. Engl. J. Med.* **320:**502–506.

34. **Wilson, E. R., A. Malluh, S. Stagno, and W. M. Crist.** 1981. Fatal Epstein-Barr virus-associated hemophagocytic syndrome. *J. Pediatr.* **98:**260–262.

35. **Yao, Q. Y., A. B. Rickinson, and M. A. Epstein.** 1985. In vitro analysis of the Epstein-Barr virus-host balance in long term renal allograft recipients. *Int. J. Cancer* **35:**43–49.

36. **Yao, Q. Y., A. B. Rickinson, and M. A. Epstein.** 1985. A re-examination of the Epstein-Barr virus carrier state in healthy seropositive individuals. *Int. J. Cancer* **35:**35–42.

37. **Ziegler, J. L., W. L. Drew, R. C. Miner, L. Mintz, E. Rosenbaum, J. Gershow, E. T. Lennette, J. Greenspan, E. Shillitoe, J. Beckstead, C. Casavant, and K. Yamamoto.** 1982. Outbreak of Burkitt's-like lymphoma in homosexual men. *Lancet* **ii:**631–633.

38. **zur Hausen, H., F. J. O'Neill, U. K. Freese, and E. Hecker.** 1978. Persisting oncogenic herpesvirus induced by the tumour promoter TPA. *Nature* (London) **272:**373–375.

39. **zur Hauzen, H., H. Schulte-Holthausen, G. Klein, W. Henle, G. Henle, P. Clifford, and L. Santesson.** 1970. EBV DNA in biopsies of Burkitt's tumours and anaplastic carcinomas of the nasopharynx. *Nature* (London) **228:**1056–1058.

Human Herpesvirus 6 and Other Herpesviruses

JOHN A. STEWART AND JOANNE L. PATTON

75

HUMAN HERPESVIRUS 6

Clinical Background

Human Herpesvirus 6 (HHV-6) and Childhood Disease

Roseola (exanthem subitum) is a common infectious disease of childhood, usually occurring from 6 months to 3 years of age. The illness is characterized by an abrupt rise in temperature to as high as 40°C followed in 2 to 4 days by a rapid drop in temperature that coincides with the appearance of an erythematous maculopapular rash that persists for 1 to 3 days. Other major symptoms are listlessness, irritability, and drowsiness. Despite the fever, children often continue to eat and play normally and do not seem nearly as sick as their temperature might indicate. A complication that may occur during the febrile episode is convulsions (25, 38), but other neurologic symptoms and encephalitis occur more rarely. Other physical findings that may occur before the rash appears are palpebral edema and suboccipital, postauricular, and cervical lymphadenopathy. The rash may be either a patchy maculopapular exanthem similar to that seen with rubella or a more confluent eruption like that seen with measles. The skin lesions are never vesicular. They appear first on the neck, behind the ears, and on the back and may spread quickly to involve the scalp, chest, abdomen, and thighs. The face and distal extremities are usually spared. The total illness may last from 2 to 7 days and usually resolves without complications.

Yamanishi et al. (48) were the first to isolate HHV-6 from peripheral blood lymphocytes (PBL) of four children with roseola (exanthem subitum) who had an accompanying rise in HHV-6 antibody. These results verified the long-held belief, based on leukopenia in patients, failure to isolate bacteria, relatively long incubation period (9), and positive transmission (from blood and throat washings) studies in humans (reviewed in reference 31), that this "viral exanthemlike" illness was caused by a virus. Other investigations have used virus isolation (38, 39), seroconversion (20, 39), and PCR detection of HHV-6 DNA in the cerebrospinal fluid (CSF) (25, 38) and PBL of children with roseola to confirm unquestionably that HHV-6 is the causative agent of roseola.

Seroepidemiologic studies of many populations indicate that HHV-6 infection is very common, affecting more than 90% of children before 2 years of age, and has an epidemiology consistent with that of roseola. Infants younger than 5 to 6 months of age appear to be protected by maternal antibodies, and infection is most commonly acquired between 6 and 18 months of age. Primary infection with HHV-6 takes the form of roseola (9) in 30% of children. Various other illnesses have been described, indicating that primary infection can result in a rash without fever and fever without rash. In a study of 243 febrile children under 3 years of age presenting at a pediatric emergency department, HHV-6 was isolated from peripheral blood cells of 34 of the children (14%) (33). The children with viremia had higher temperatures (mean, 39.7°C), lower leukocyte counts, and more frequent findings of malaise and irritability than did the HHV-6-negative group. Serologic studies supported a diagnosis of primary HHV-6 infection in most children. A rash was noted in the emergency department or within a few days of the initial visit in 12 (35%) of the 34 HHV-6-positive children.

A new role for HHV-6 as an invader of the central nervous system in neurologic complications of roseola is emerging. In a series of 21 roseola patients with central nervous system problems during their illness, 15 had generalized seizures, and 6 had focal seizures. HHV-6 DNA was amplified from the CSF of 6 of the 11 infants tested, including 3 with encephalitis-encephalopathy (38). In another series of 10 patients with neurologic symptoms during the febrile phase of roseola, HHV-6 DNA was detected in the CSF of 9 (25). HHV-6 DNA was also detected in all CSF samples from eight children who had three or more febrile convulsions and were studied during their most recent seizure (25). These findings suggest that recurrent febrile convulsions may be associated with reactivation of HHV-6.

HHV-6 Infection and Adult Disease

Active infection with HHV-6 in adults (20, 37, 41) has been associated with a heterophile-negative, mononucleosislike illness and non-A, non-B hepatitis (20, 37). Steeper et al. (37) looked at patients with non-Epstein-Barr virus (non-EBV), noncytomegalovirus (non-CMV) mononucleosislike illnesses and found active HHV-6 infection in 8 of the 27 heterophile-negative cases. The clinical pictures for these patients varied widely and included a short febrile illness, mild cervical lymphadenopathy, liver enzyme elevations suggestive of active viral hepatitis in two patients, and a prolonged febrile illness in a human immunodeficiency virus-positive patient. Three similar patients with mononucleosislike and/or hepatitis illnesses and acute HHV-6 infection have been described (20). Another well-documented case of HHV-6 infection with clinical illness was noted in a 21-year-old HHV-6-seronegative patient after liver transplantation from a HHV-6-seropositive donor (41). Findings included fever, grand mal convulsions, and hepatitis. Primary infection with HHV-6 is a rare event in adults because of the high rate of infection in childhood, but when it occurs, it may have some of the features of a mild mononucleosislike illness and/or mild hepatitis.

HHV-6 activity has also been detected in organ recipients (most of whom are HHV-6 seropositive) following renal, cardiac, and bone marrow transplantation. After renal transplantation, about 50% of patients have an increase in antibody titer, viremia can be detected in a small number, and viral antigen has been detected in some of the rejected kidneys. No consistent association between HHV-6 activity and rejection or immunosuppressive treatment has been found. Symptoms associated with marrow transplantation after HHV-6 reactivation or reinfection are malaise, fever, leukopenia, interstitial pneumonitis, and poor marrow function (10).

Initially, a number of studies reported a higher prevalence of immunofluorescence assay (IFA) antibody to HHV-6 in patients with connective tissue diseases (e.g., sarcoidosis), lymphomas, lymphoid hyperplasia, and immune suppression or deficiency (2). Because many of the IFAs were relatively insensitive, the actual seropositive rate for the healthy population was later shown to exceed 90% (24, 30, 35). Thus, the studies showing an increased seroprevalence rate for patients with certain diseases probably indicate that the mean antibody titers to HHV-6 present in these patients are higher than those in controls. However, mean antibody titers to a number of viruses are higher in patients with connective tissue diseases, leukemia, lymphoid hyperplasia, human immunodeficiency infection, and even chronic fatigue syndrome. This observation may reflect a polyclonal-B-cell stimulation or increased reactivation of latent viruses, such as HHV-6, in patients with these diseases.

Description of the Agent

General Properties of HHV-6
After its discovery in 1986, HHV-6 was quickly recognized as a unique member of the family *Herpesviridae* that is serologically (34) and genetically (22, 28) distinct from the other human herpesviruses. HHV-6 DNA is clearly distinct from the DNAs of the other human herpesviruses (herpes simplex virus [HSV] type 1, HSV type 2, varicella-zoster virus [VZV], CMV, and EBV) by both restriction endonuclease digestion and nucleic acid hybridization (28). The virus envelope encloses an icosahedral capsid of 162 capsomeres and a core containing a double-stranded DNA genome of 161 to 168 kb (27). HHV-6 is readily inactivated by ether and lipid solvents. Cell-free virus does not survive cycles of freezing and thawing unless stored in a protein-rich environment, such as skim milk medium. More than 20 polypeptides ranging in size from 30 to 220 kDa are found in solubilized purified virions (47). One of these, a 101-kDa polypeptide, is strongly reactive on Western immunoblot analysis with both HHV-6-positive human sera and a murine monoclonal antibody and is valuable in diagnostic testing because it is a specific marker for HHV-6 infection.

HHV-6 Strain Groups
HHV-6 (GS) was first isolated from a patient with a lymphoproliferative disorder (34). Additional isolates of importance came from AIDS patients in Uganda (strain U1102) and Zaire (strain Z29). As subsequent isolates were obtained, a number of lines of evidence indicated that HHV-6 isolates form at least two groups that differ in molecular and biologic properties (1, 31). Restriction site polymorphisms that extend across the genome can be used to differentiate all isolates into two groups. However, the nucleotide sequence identity for several regions of the viral genome is very high, ranging from 94 to 96% between the variant groups. Thus, the two groups are very closely related, and the consensus of a group of more than 50 scientists attending the 17th International Herpesvirus Workshop in 1992 was to identify GS- and U1102-like viruses as HHV-6 variant A and Z29-like viruses as HHV-6 variant B (1). The biologic differences in the two variants extend to the clinical level as well, with all HHV-6 isolates from roseola and related febrile illnesses found to be variant B (HHV-6B), with the exception of cultures from two patients from whom both variants were obtained (15). Both variants have been isolated from immunocompromised individuals. Variant A (HHV-6A) has not been independently isolated during a defined primary infection, and no specific disease has been associated with it. No serologic method to distinguish the immune response to the variants has been described.

Collection and Storage of Specimens
Most isolates of HHV-6 have been obtained from PBL samples. The success rate of isolation from the PBL of roseola patients is highest in the febrile stage before rash appears and nears zero after 4 days of rash. Heparinized blood samples should be collected aseptically, and the mononuclear cells should be purified on Ficoll gradients. The cells are usually cocultivated with phytohemagglutinin (PHA)-stimulated cord blood lymphocytes (CBL), but primary culture in RPMI 1640 medium may be attempted. For best results, cultures should be set up within 24 h of blood collection. Freezing mononuclear cells for later isolation is not recommended, because the virus is highly labile and survives freeze-thaw cycles poorly. Although the DNA of HHV-6 is frequently found in PBL and saliva of normal adults, the virus is rarely isolated from PBL of these adults. Isolates identified as HHV-6 have been obtained from saliva samples cultured with CBL and PBL (26, 32).

Direct Examination
Cloned fragments of HHV-6 DNA that can be used to detect the genome in in situ or Southern blot assays have been described elsewhere (31). A number of primer and probe sequences have been developed to perform diagnostic PCR assays. The PCR has been successfully used to detect HHV-6 DNA in (i) the blood, saliva, and CSF of roseola patients (25, 38); (ii) the blood of patients with AIDS and lymphoproliferative disorders; (iii) blood and tissue specimens from immunosuppressed patients (10, 14); and (iv) the blood and saliva of healthy seropositive adults and children. Some of the primer sequences described are useful for discrimination of HHV-6A and HHV-6B and do not react with the closely related HHV-7.

The rate of DNA detection reported from healthy seropositive controls varies widely from study to study, reflecting differences in numbers of PBL sampled, primer pairs used, and number of cycles performed in the assays. Because positive PCR findings can result from the presence of only a few copies of HHV-6 DNA in circulating lymphocytes or tissue samples, quantifying the amount of DNA detected may be needed to correctly assess the role of HHV-6 infection in the disease states being investigated (14). More baseline studies of healthy individuals in the general population are needed to help evaluate positive results.

Virus Isolation

HHV-6 has been isolated from the peripheral blood by primary culture of mononuclear cells and by cocultivation of these cells with either stimulated CBL or adult PBL. The best results are usually obtained with CBL. After isolation, some strains of virus are able to grow in continuous cell lines. For cocultivation studies, mononuclear cells are purified from fresh human cord blood on Ficoll gradients and stimulated to blast formation by culture for 1 to 3 days with 0.002% PHA in growth medium (RPMI 1640 with 10% fetal calf serum, 0.01 mg of hydrocortisone per ml, and antibiotics); 32 U of interleukin-2 per ml is added to the medium. The cord cells are cocultivated with an equal volume of mononuclear cells from patients, incubated at 37°C in 5% CO_2, observed twice a week for cytopathic effects, and checked for the appearance of HHV-6 antigen by the anticomplement immunofluorescence (ACIF) test or IFA after 7 to 10 days. All cell cultures are passaged every 7 to 10 days into CBL freshly stimulated with PHA in growth medium.

Identification of HHV-6

Large balloonlike cells may appear after cocultivation with CBL, suggesting that infection is present. However, such cells may also appear in the absence of infection, and immunologic tests are therefore needed to confirm infection. The IFA and the ACIF assay, using well-characterized human sera, have been used to document HHV-6 infection but will not differentiate HHV-6 from HHV-7 (see below). Commercially available monoclonal antibodies (MAbs) to HHV-6 have been developed, and these reagents or in situ hybridization with HHV-6-specific probe sequences are needed to exclude HHV-7 and give precise, HHV-6 variant-specific, identification. MAbs to the other human herpesviruses should not react with HHV-6-infected cells.

Serologic Diagnosis

Determining an individual's or group's immune status regarding HHV-6 can be done by a variety of methods, including IFA, ACIF, enzyme immunoassay (EIA), immunoblot, immunoprecipitation, and neutralization. All these assays are sensitive and specific enough to evaluate samples from clinical studies in which a change in antibody titer might be expected. However, they are not as useful as virus isolation or DNA detection methods for establishing the diagnosis of acute HHV-6 infection, because immunoglobulin M (IgM) antibody does not appear until 5 to 7 days after onset, and IgG does not appear until 7 to 10 days after onset. The serologic diagnosis of recent HHV-6 infection is based on finding a fourfold or greater increase in IFA or ACIF titer or a significant increase in EIA values between acute- and convalescent-phase sera. Seroconversion in paired serum specimens usually indicates a primary infection with HHV-6. However, when the IFA or the ACIF test, which are somewhat less sensitive, is used, seroconver-

sion may be difficult to distinguish from a rising titer in a person with virus reactivation. Serologic studies of childhood illnesses have clearly established that primary infection with HHV-6 occurs in children with roseola (48). In addition, however, they have established that primary infections are associated with febrile illnesses without rash (33). Serologic studies are of the greatest use in epidemiologic studies and prospective studies of the role of HHV-6 in various diseases. Future studies of this type must take into account the partial cross-reaction between HHV-6 and HHV-7 (46) and the difficulty in distinguishing serologic responses to the two HHV-6 variants. Unfortunately, no commercial antibody kits are available for HHV-6, and assays must be constructed in individual laboratories.

The effects of reactivation upon antibody production are not clear, since the clinical picture in reactivation is not yet defined. Rising or elevated HHV-6 IgG titers have been found in patients actively infected with CMV and EBV (16, 21), but the relative role of each virus in the illness was not clear. Complete removal of CMV IgG reactivity from five serum samples (from patients that had a primary CMV infection with a rise in HHV-6 IgG) had no effect on the HHV-6 titers of four samples, but the fifth had an eightfold drop in HHV-6 IgG titer (21). Two adults with acute febrile illnesses had IgG present in baseline samples and still developed HHV-6 IgM (16). Thus, IgM is produced in both primary and reactivated HHV-6 infections, and major rises in IgG titer may be the result of primary CMV or EBV infections.

IFA

One of the most widely used serologic procedures for HHV-6 is the IFA. As IFAs have become more sensitive, the estimation of antibody prevalence in the general population has risen dramatically to at least 90% (24, 35). With the discovery that HHV-6 (GS) would grow well in several T-cell lines, such as HSB-2, the IFA became easier to perform and more reproducible.

ACIF Test

To avoid the problems with strong nonspecific fluorescent staining of CBL frequently observed with the IFA, an ACIF test that greatly reduced nonspecific fluorescence, gave a stronger signal with positive cells, and was highly specific for HHV-6 without cross-reactivity to the other human herpesviruses was adopted (28). The ACIF test has been effectively used in detecting the antibody responses of roseola patients (48) and in studying seroprevalence.

The ACIF test is performed by adding heat-inactivated serum specimens to slide wells containing acetone-fixed HHV-6 (GS) or HHV-6 (Z29)-infected cells. After incubation and wash steps, a human serum that contains complement but does not have detectable antibodies to HHV-6 at the working dilution is added. After further incubation and wash steps, complement fixation is detected by addition of fluorescein-labeled goat anti-human C3 reagent. Characteristically, cytoplasmic fluorescence is somewhat granular, whereas nuclear fluorescence is solid and bright.

EIA

Several EIA methods that use antigen prepared from purified virions from infected cell culture medium (35) or HHV-6-infected whole-cell lysates (3, 30) have been described. To perform an EIA for antibody to HHV-6, Saxinger et al. (35) prepared antigen by disruption of virions or infected HSB-2 cells in a Tris (pH 8)–Triton X-100–0.6 M

NaCl buffer. The clarified lysate was diluted to a concentration of 0.5 μg/100 μl and used to coat microtitration plates. Other methods of preparing EIA antigen include harvesting infected cells in buffers from pH 7.4 to 9.6, disrupting the cells by freeze-thaw cycles and/or sonication, and coating microtitration plates, usually with a sodium carbonate buffer (pH 9.6). After another wash cycle, antigen-coated plates are incubated with serum dilutions and thoroughly washed with buffer containing detergent, and the bound antibody-antigen complex is detected by enzyme-conjugated anti-human IgG reagents and their appropriate substrates. The specificity of the EIA reaction has been shown by preincubation of sera with soluble HHV-6, HSB-2 control, CMV, EBV, HSV, and VZV antigens. The other herpesvirus antigens showed no competitive cross-reactivity with IgG antibodies to HHV-6 (35), whereas the HHV-6-adsorbing antigens reduced binding to HHV-6 by more than 90% and failed to significantly reduce the titer to any of the other five human herpesviruses (30, 35). In addition to its specificity, the EIA appears to be highly sensitive, with 94 to 97% of population groups above the age of 1 year seropositive to HHV-6.

Other Serologic Procedures

Other tests that have been used to detect HHV-6 antibodies include Western immunoblots, immunoprecipitation, and neutralization. The Western immunoblot and immunoprecipitation assays (4, 47) are key procedures in the detailed analysis of the individual polypeptides and proteins involved in the immune response to HHV-6. The neutralization test has been less widely used.

Interpretation of Results

Recovery of virus from the peripheral blood by quantitative methods is the best indicator of active HHV-6 infection but does not distinguish primary infection from reactivation or latent infection. The results of serologic tests are more difficult to interpret. Demonstration of a fourfold rise in IFA or ACIF antibody titer in a young child with a febrile rash illness confirms the diagnosis of HHV-6 infection with roseola. However, in older children and adults, a concurrent rise in antibody to CMV or EBV has been reported quite frequently in connection with a rise in HHV-6 antibody. When serologic evidence of dual infection is obtained, the relative importance of each pathogen in producing illness may be difficult to ascertain. Confirmation by virus isolation would help resolve such problems. Some caution is also needed in interpreting seroprevalence studies. A positive result in adequately controlled tests indicates infection at some undetermined time in the past, but a negative result can be obtained with relatively insensitive procedures and may not indicate absence of infection. In addition, the potentially confounding role of HHV-7 in producing cross-reactive antibodies must be taken into consideration.

HHV-7

Clinical Background

The role of HHV-7 in disease processes in humans is largely unknown. The only clinical association has been with a roseolalike illness in young children from whom HHV-7 was isolated but HHV-6 was not (40). The finding that HHV-7 can be isolated from the saliva of 75% of healthy adults (7, 45) suggests that it may be transmitted by contact with oral secretions.

Description of the Agent

HHV-7 (RK) was first isolated in 1989 by Frenkel et al. (17) from the peripheral blood of a 25-year-old healthy individual after activation of his CD4$^+$ T cells. Lymphocytes infected with the RK strain exhibited a cytopathic effect that consisted of enlarged ballooning cells and syncytium formation. No virus was recovered from unactivated cells. Typical icosahedral herpesvirus particles were observed in infected cultures by electron microscopy. Each particle had a distinct tegument layer around the capsid similar to that of HHV-6. The DNA of HHV-7 (RK) did not hybridize with large probes from HSV, EBV, VZV, or human CMV, but some HHV-6 DNA probes did cross-hybridize. Thus, the RK strain was more closely related to HHV-6A and HHV-6B than to any of the other human herpesviruses, but it was shown to be clearly distinct by restriction endonuclease analysis (17).

HHV-7 also appears to be immunologically distinct from HHV-6, as shown by Western immunoblot assays using human serum specimens and MAbs (46). Additional studies with some MAbs derived from HHV-6A also indicated limited antigenic similarity with HHV-7. MAbs to two proteins of HHV-6A reacted with many or all of the HHV-7 isolates tested, but four other HHV-6 MAbs failed to react with any of the HHV-7 strains (5, 7, 46). With IFAs (46), it was shown that HHV-7 is a highly prevalent virus that infects children somewhat later than HHV-6, or after 24 months of age. Seroconversion to HHV-6 and HHV-7 were observed as distinct and separate events. Subsequent studies have reported that HHV-7 can be isolated from the saliva of 75% of normal adults (7, 45).

Collection and Storage of Specimens

Heparinized blood samples should be collected aseptically, and the mononuclear cells should be purified on Ficoll gradients. The cells are usually cocultivated with PHA-stimulated CBL. For best results, cultures should be set up within 24 h of blood collection.

Direct Examination

An HHV-7-specific DNA clone has been derived from a saliva isolate that does not cross-react with either variant of HHV-6. This clone can be used in an HHV-7-specific tissue assay by in situ hybridization methods (7). HHV-7-specific PCR primers that amplify HHV-7 DNA and do not amplify the DNAs from any of the other known human herpesviruses, including 12 HHV-6 strains, have been developed from the sequence of a DNA fragment of HHV-7 (JI) (5).

Virus Isolation

HHV-7 can be isolated from saliva and PBL and propagated in PHA-stimulated CBL and PBL by essentially the same procedures used for HHV-6 (7). Because HHV-7 may reactivate HHV-6 latent in PBL, CBL are the preferred cells. Saliva specimens are diluted 1:3 with lymphocyte growth medium (RPMI 1640, 10% fetal calf serum, 32 U of interleukin-2 per ml, and antibiotics) and centrifuged at 2,000 $\times$ g. The supernatant is passed through a 0.45-μm-pore-size filter and then inoculated into two or three individual PHA-stimulated CBL cultures by centrifugal enhancement. Ficoll-purified PBL are cocultivated with at least two individual PHA-stimulated CBL cultures. All cell cultures are incubated at 37°C in 5% CO_2, observed twice a week for cytopathic effects, and passaged weekly into CBL freshly stimulated with PHA in growth medium.

Identification of HHV-7

When a cytopathic effect appears in culture, HHV-7 infection must be confirmed, as this effect may also be induced in CBL by HHV-6A or HHV-6B. Specific reagents such as PCR primers (5), cloned fragments of HHV-7 (7), or restriction endonuclease cleavage patterns of the DNA (7, 17) must be used. Specific MAbs for HHV-7 are not readily available.

Serologic Diagnosis

The human immune response to HHV-7 can be measured by IFA, EIA, and immunoblot assays, with HHV-7-infected cells used for the antigen source (5, 7, 46). Wyatt et al. (46) differentiated the immune response to HHV-7 from that to HHV-6 by studying sequential serum specimens obtained from young children. They noted that seroconversion to HHV-6 and HHV-7, detected by IFA, occurred independently and that high antibody titers to HHV-6 were not associated with high titers to HHV-7. High titers to HHV-6 offered no protection from HHV-7 infection. Paired acute- and convalescent-phase serum specimens from four roseola patients showed strong seroconversion to HHV-6, but three of the pairs also showed some increase in HHV-7 as well (titers increased from <20 to 40, <20 to 80, and <20 to 320). This finding indicates that there may be a limited immune response to cross-reactive epitopes of these viruses.

This antigenic cross-reactivity has also been examined by antigen adsorption of homologous and heterologous antibody before the residual antibody of each specimen was tested by HHV-6B and HHV-7 EIAs. Most cord blood and adult serum specimens contain antibody to both viruses. HHV-6B and HHV-7 antigens adsorbed 10 to 30% of the antibody activity of the heterologous virus but adsorbed far more activity of the homologous virus (6). With this additional evidence that cross-reactive epitopes are present with these viruses, an adsorption step appears necessary to determine specific titers to each virus when infected cells are used as antigens.

Interpretation of Results

The isolation of HHV-7 from or the detection of HHV-7 DNA in saliva, PBL, and other tissue sites indicates that infection is present but does not indicate whether a given illness is due to the infection. Many oral infections are asymptomatic and persistent, and we do not know what type of disease, if any, primary infection may cause. However, isolation of HHV-7 from a diseased tissue (particularly at a high titer) and good epidemiologic correlation of that illness with HHV-7 provides evidence that HHV-7 might be the etiologic agent.

The results of serologic tests are also difficult to interpret. Demonstration of a significant rise in HHV-7 antibody titer in a young child or adult with stable or negative HHV-6 titers (and no evidence that another pathogen is present) indicates that active infection is present. However, serologic findings must be combined with virus isolation results to determine the clinical significance of the serologic results. When serologic evidence of dual infection is obtained, the relative importance of each pathogen in producing illness may be difficult to ascertain. Caution is also required in interpreting seroprevalence studies until the extent of the cross-reaction with HHV-6 is more completely defined and HHV-7-specific tests can be performed.

B VIRUS

Clinical Background

B virus, or cercopithecine herpesvirus 1 (formerly called *Herpesvirus simiae*), is a member of the herpes group of viruses that is indigenous to Old World monkeys, such as rhesus (*Macaca mulatta*), cynomolgus (*Macaca fascicularis*), and other Asiatic monkeys of the genus *Macaca*. As the simian counterpart of HSV, B virus causes subclinical infections as well as dermal, oral, eye, and genital lesions in monkeys. Whitley (44) and Weigler (42) have recently reviewed the biology of B virus and the importance of the virus to humans in its ability to cause life-threatening CNS infections. The virus can be transmitted by monkey bites, by direct or indirect contact with saliva, and even by contact with monkey cell cultures that are widely used in the virus laboratory (43).

After an incubation period of a few days to 6 weeks, an infected person may show localized redness and vesicles and may have pain at the site of virus inoculation. Vesicles on the mucous membranes, pneumonia, diarrhea, abdominal pain, pharyngitis, and lymphocytic pleocytosis have been reported. In almost all untreated symptomatic cases, an ascending myelitis and encephalopathy occur that are frequently fatal; most who survive are left with severe brain damage (29). Although the risk of disease with B virus appears to be low (on the basis of the small number of known cases compared with the thousands of macaque contacts yearly), human infection is a potential hazard of working with Old World monkeys, and precautionary measures to protect workers from infection are of critical importance. The identification of four human cases in Florida in 1987 (11) and two cases in Michigan in 1989 (13) after a 13-year period with no reported cases emphasizes this need.

Description of Agent

B virus is similar to HSV in morphologic features and size and is inhibited by antiviral agents that suppress DNA synthesis. An enveloped herpesvirus, it is sensitive to lipid solvents, acid pH, and detergent solutions. There is an antigenic relationship between HSV and B virus such that antiserum to B virus neutralizes both viruses equally well, whereas antiserum to HSV neutralizes homologous virus at much higher titers than B virus (8, 18). The extent of the cross-reaction is enhanced by complement. B virus appears hardier than HSV; it may survive for 7 days at 37°C and for weeks at 4°C and is very stable when stored at −70°C. The virus can be isolated in a number of cell lines, including primary rabbit kidney and vervet monkey cells or established lines, such as HeLa or BSC1. Rhesus monkey kidney cells readily support the growth of B virus, whereas HSV isolates grow poorly in these cells. The cytopathic effect produced by B virus in many cell lines is quite similar to that produced by HSV, and specific serologic or molecular procedures are needed to identify B virus. The virus can be successfully grown in a number of animal systems as well.

Safety

Because of the potential hazards of B-virus infections, guidelines for preventing infection in monkey handlers have been established (12), and laboratory workers should exercise caution in processing specimens suspected of containing B virus. Laboratory assistance in investigating such specimens can be obtained from the Southwest Foundation for Biomedical Research, San Antonio, Tex., phone (512)

674-1410, or from the Virus Reference Laboratory Inc., San Antonio, Tex., phone (512) 696-5510.

Collection and Storage of Specimens

Vesicular fluid from lesions; swabs of the oropharynx, conjunctiva, and skin lesions; or skin biopsy specimens from affected areas can be used for the diagnosis of B-virus infection. The cotton tips of the applicators used to swab the lesions should be broken off into screw-cap vials containing 2 ml of tryptose phosphate broth with 0.5% gelatin or another viral transport medium that contains protein (skim milk or veal infusion broth). Specimens for virus isolation should be kept cold (4 to 8°C) at all times (but not frozen) and promptly shipped to the appropriate laboratory. Inquiries can be directed to the Division of Viral and Rickettsial Diseases of the Centers for Disease Control and Prevention, phone (404) 639-3091.

Virus Isolation and Serologic Procedures

Since B-virus propagation should be handled under biosafety level 4 conditions, laboratory investigation of specimens from patients or monkeys suspected of B-virus infection is best done by experienced laboratories equipped to handle B-virus cultures. Rapid diagnosis of infection has been greatly facilitated by the development of a PCR assay to detect B-virus DNA in clinical samples (36). The serologic diagnosis of recent B-virus infection is complicated by the extensive cross-reactivity between HSV and B virus. Adequate testing requires the demonstration of at least a fourfold rise in "specific antibody titer" against B virus in paired serum specimens. Specific antibody assays such as the immunoblot (19) or an EIA for specific antibodies that uses adsorption procedures (23) are currently available in only a few reference laboratories equipped to handle B virus. In B-virus-infected patients, a rise in HSV titer may frequently be seen along with an increase in titer to B-virus antigens. An increase in antibody is usually observed 10 to 14 days after the onset of illness but may be greatly delayed if the patient is treated with an antiviral medication such as acyclovir.

REFERENCES

1. Ablashi, D. V., H. Agut, Z. Berneman, G. Campadelli-Fiume, D. Carrigan, L. Ceccerini-Nelli, B. Chandran, S. Chou, H. Collandre, R. Cone, T. Danbaugh, S. Dewhurst, D. DiLuca, L. Foa-Tomasi, B. Fleckenstein, N. Frenkel, R. Gallo, U. Gompels, C. Hall, M. Jones, G. Lawrence, M. Martin, L. Montagnier, F. Neipel, J. Nicholas, P. Pellett, A. Razzaque, G. Torrelli, B. Thomson, S. Salahuddin, L. Wyatt, and K. Yamanishi. 1993. Human herpesvirus-6 strain groups: a nomenclature. Arch. Virol. 129:363–366.
2. Ablashi, D. V., S. F. Josephs, A. Buchbinder, K. Hellman, S. Nakamura, T. Llana, P. Lusso, M. Kaplan, J. Dahlberg, S. Memon, F. Imam, K. L. Ablashi, P. D. Markham, B. Kramarsky, G. R. F. Krueger, P. Biberfeld, F. Wong-Staal, S. Z. Salahuddin, and R. C. Gallo. 1988. Human B-lymphotropic virus (human herpesvirus-6). J. Virol. Methods 21:29–48.
3. Asano, Y., T. Yoshikawa, S. Suga, T. Ozaki, Y. Saito, Y. Hatano, and M. Takahashi. 1990. Enzyme-linked immunosorbent assay for detection of IgG antibody to human herpesvirus 6. J. Med. Virol. 32:119–123.
4. Balachandran, N., R. E. Amelse, W. W. Zhou, and C. K. Chang. 1989. Identification of proteins specific for human herpesvirus 6-infected human T cells. J. Virol. 63:2835–2840.
5. Berneman, Z. N., D. V. Ablashi, G. Li, M. Eger-Fletcher, M. S. Reitz, Jr., C.-H. Hung, I. Brus, A. L. Komaroff, and R. Gallo. 1992. Human herpesvirus 7 is a T-lymphotropic virus and is related to, but significantly different from, human herpesvirus 6 and human cytomegalovirus. Proc. Natl. Acad. Sci. USA 89:10552–10556.
6. Black, J., J. Patton, P. Pellett, and J. Stewart. Unpublished data.
7. Black, J. B., N. Inoue, K. Kite-Powell, S. Zaki, and P. E. Pellett. 1993. Frequent isolation of human herpesvirus 7 from saliva. Virus Res. 29:91–98.
8. Boulter, E. A., S. S. Kalter, R. L. Heberling, J. E. Gualardo, and T. L. Lester. 1982. A comparison of neutralization tests for the detection of antibodies to herpsvirus simiae (monkey B virus). Lab. Anim. Sci. 32:150–152.
9. Breese, B. B. 1941. Roseola infantum (exanthema subitum). N.Y. State J. Med. 41:1854–1859.
10. Carrigan, D. R., W. R. Drobyski, S. K. Russler, M. A. Tapper, K. K. Knox, and R. C. Ash. 1991. Interstitial pneumonitis associated with human herpesvirus 6 infection after marrow transplantation. Lancet 338:147–149.
11. Centers for Disease Control. 1987. B virus infection in humans—Pensacola, Florida. Morbid. Mortal. Weekly Rep. 36:289–290, 295–296.
12. Centers for Disease Control. 1987. Guidelines for prevention of Herpesvirus Simiae (B virus) infection in monkey handlers. Morbid. Mortal. Weekly Rep. 36:680–682, 687–689.
13. Centers for Disease Control. 1989. B virus infections in humans—Michigan. Morbid. Mortal. Weekly Rep. 38:453–454.
14. Cone, R. W., R. C. Hackman, M.-L. W. Huang, R. A. Bowden, J. D. Meyers, M. Metcalf, J. Zeh, R. Ashley, and L. Corey. 1993. Human herpesvirus 6 in lung tissue from patients with pneumonitis after bone marrow transplantation. N. Engl. J. Med. 329:156–161.
15. Dewhurst, S., B. Chandran, K. McIntyre, K. Schnabel, and C. B. Hall. 1992. Phenotypic and genetic polymorphisms among human herpesvirus-6 isolates from North American infants. Virology 190:490–493.
16. Fox, J. D., P. Ward, M. Briggs, W. Irving, T. G. Stammers, and R. S. Tedder. 1990. Production of IgM antibody to HHV6 in reactivation and primary infection. Epidemiol. Infect. 104:289–296.
17. Frenkel, N., E. C. Schirmer, L. S. Wyatt, G. Katsafanas, E. Roffman, R. M. Danovich, and C. H. June. 1990. Isolation of a new herpesvirus from human CD4+ T cells. Proc. Natl. Acad. Sci. USA 87:748–752.
18. Gary, G. W., Jr., and E. L. Palmer. 1977. Comparative complement fixation and serum neutralization antibody titers to herpes simplex virus type 1 and Herpesvirus simiae in Macaca mulatta and humans. J. Clin. Microbiol. 5:465–470.
19. Heberling, R. L., and S. S. Kalter. 1987. A dot-immunobinding assay on nitrocellulose with psoralen inactivated Herpesvirus simiae (B virus). Lab. Anim. Sci. 37:304–308.
20. Irving, W. L., J. Chang, D. R. Raymond, R. Dunstan, P. Grattan-Smith, and A. L. Cunningham. 1990. Roseola infantum and other syndromes associated with acute HHV-6 infection. Arch. Dis. Child. 65:1297–1300.
21. Irving, W. L., V. M. Ratnamohan, L. C. Hueston, J. R. Chapman, and A. L. Cunningham. 1990. Dual antibody rises to cytomegalovirus and human herpesvirus type 6: frequency of occurrence in CMV infections and evidence for genuine reactivity to both viruses. J. Infect. Dis. 161:910–916.
22. Josephs, S. F., D. V. Ablashi, S. Z. Salahuddin, B. Kramarsky, R. B. Franza, P. Pellett, A. Buchbinder, S. Memon, F. Wong-Staal, and R. C. Gallo. 1988. Molecular studies of HHV-6. J. Virol. Methods 21:179–190.
23. Katz, D., J. K. Hilliard, R. Eberle, and S. L. Liper. 1986. ELISA for detection of group-common and virus-specific antibodies in human and simian viruses. J. Virol. Methods 14:99–109.
24. Knowles, W. A., and S. D. Gardner. 1988. High prevalence of antibody to human herpesvirus-6 and seroconversion associated with rash in two infants. Lancet ii:912–913.
25. Kondo, K., H. Nagafuji, A. Hata, C. Tomomori, and

K. Yamanishi. 1993. Association of human herpesvirus 6 infection of the central nervous system with recurrence of febrile convulsions. *J. Infect. Dis.* **167**:1197–2000.

26. Levy, J. A., F. Ferro, D. Greenspan, and E. T. Lennette. 1990. Frequent isolation of HHV-6 from saliva and high seroprevalence of the virus in the population. *Lancet* **335**:1047–1050.

27. Lindquester, G. J., and P. E. Pellett. 1991. Properties of the human herpesvirus 6 strain Z29 genome: G+C content, length, and presence of variable-length directly repeated terminal sequence elements. *Virology* **182**:102–110.

28. Lopez, C., P. Pellett, J. Stewart, C. Goldsmith, K. Sanderlin, J. Black, D. Warfield, and P. Feorino. 1988. Characteristics of human herpesvirus-6. *J. Infect. Dis.* **157**:1271–1273.

29. Palmer, A. E. 1987. B virus, *Herpesvirus simiae*: historical perspectives. *J. Med. Primatol.* **16**:99–130.

30. Parker, C. A., and J. M. Weber. 1993. An enzyme-linked immunosorbent assay for the detection of IgG and IgM antibodies to human herpesvirus type 6. *J. Virol. Methods* **41**:265–276.

31. Pellett, P. E., J. B. Black, and M. Yamamoto. 1992. Human herpesvirus 6: the virus and the search for its role as a human pathogen. *Adv. Virus Res.* **41**:1–52.

32. Pietroboni, G. R., G. B. Harnett, M. R. Bucens, and R. W. Honess. 1988. Antibody to human herpesvirus 6 in saliva. *Lancet* **i**:1059.

33. Pruksananonda, P., C. B. Hall, R. A. Insel, K. McIntyre, P. E. Pellet, C. E. Long, K. C. Schnabel, P. H. Pincus, F. R. Stamey, T. R. Danbaugh, and J. A. Stewart. 1992. Primary human herpesvirus 6 infection in young children. *N. Engl. J. Med.* **326**:1445–1450.

34. Salahuddin, S. Z., D. V. Ablashi, P. D. Markham, S. F. Josephs, S. Sturzenegger, M. Kaplan, G. Halligan, P. Biberfeld, F. Wong-Staal, B. Kramarsky, and R. C. Gallo. 1986. Isolation of a new virus, HBLV, in patients with lymphoproliferative disorders. *Science* **234**:596–601.

35. Saxinger, C., H. Polesky, N. Eby, S. Grufferman, R. Murphy, G. Tegtmeier, V. Parekh, S. Memon, and C. Hung. 1988. Antibody reactivity with HBLV (HHV-6) in U.S. populations. *J. Virol. Methods* **21**:199–208.

36. Scinicariello, F., R. Eberle, and J. K. Hilliard. 1993. Rapid detection of B virus (herpesvirus simiae) DNA by polymerase chain reaction. *J. Infect. Dis.* **168**:747–750.

37. Steeper, T. A., C. A. Horwitz, D. V. Ablashi, S. Z. Salahuddin, C. Saxinger, R. Saltzman, and B. Schwartz. 1990. The spectrum of clinical and laboratory findings resulting from human herpesvirus-6 in patients with mononucleosis-like illnesses not resulting from Epstein-Barr virus or cytomegalovirus. *Am. J. Clin. Pathol.* **93**:776–783.

38. Suga, S., T. Yoshikawa, Y. Asano, T. Kozawa, T. Nakashima, I. Kobayashi, T. Yazaki, H. Yamamoto, Y. Kajita, T. Ozaki, Y. Nishimura, T. Yamanaka, A. Yamada, and J. Imanishi. 1993. Clinical and virological analyses of 21 infants with exanthem subitum (roseola infantum) and central nervous system complications. *Ann. Neurol.* **33**:597–603.

39. Takahashi, K., S. Sonoda, K. Kawakami, K. Miyata, T. Oki, T. Nagata, T. Okuno, and K. Yamanishi. 1988. Human herpesvirus 6 and exanthem subitum. *Lancet* **i**:1463.

40. Tanaka, K., T. Kondo, S. Torigoe, S. Okeda, T. Mukai, and K. Yamanishi. 1994. Human herpesvirus 7: another causal agent for roseola (exanthem subitum). *J. Pediatr.* **125**:1–5.

41. Ward, K. N., J. J. Gray, and S. Efstathiou. 1989. Brief report: primary human herpesvirus 6 infection in a patient following liver transplantation from a seropositive donor. *J. Med. Virol.* **28**:69–72.

42. Weigler, B. J. 1992. Biology of B virus in macaque and human hosts: a review. *Clin. Infect. Dis.* **14**:555–567.

43. Wells, D. L., S. L. Lipper, J. K. Hilliard, J. A. Stewart, G. P. Holmes, K. L. Herrmann, M. P. Kiley, and L. B. Schonberger. 1989. *Herpesvirus simiae* contamination of primary rhesus monkey cell cultures: CDC recommendations to minimize risks to laboratory personnel. *Diagn. Microbiol. Infect. Dis.* **12**:333–336.

44. Whitley, R. J. 1993. The biology of B virus (cercopithecine virus 1), p. 317–328. *In* B. Roizman, R. J. Whitley, and C. Lopez (ed.), *The Human Herpesviruses*. Raven Press, New York.

45. Wyatt, L. S., and N. Frenkel. 1992. Human herpesvirus 7 is a constitutive inhabitant of adult human saliva. *J. Virol.* **66**:3206–3209.

46. Wyatt, L. S., W. J. Rodriguez, N. Balachandran, and N. Frenkel. 1991. Human herpesvirus 7: antigenic properties and prevalence in children and adults. *J. Virol.* **65**:6260–6265.

47. Yamamato, M., J. B. Black, J. A. Stewart, C. Lopez, and P. Pellett. 1990. Identification of a nucleocapsid protein as a specific serological marker of human herpesvirus 6 infection. *J. Clin. Microbiol.* **28**:1957–1962.

48. Yamanishi, K., T. Okuno, K. Shiraki, M. Takahashi, T. Kondo, Y. Asano, and T. Kurata. 1988. Identification of human herpesvirus 6 as a causal agent for exanthem subitum. *Lancet* **i**:1065–1067.

Influenza Viruses

THEDI ZIEGLER AND NANCY J. COX

76

CLINICAL BACKGROUND

Influenza is a highly contagious, acute respiratory disease with symptoms that typically include sudden onset of fever, sore throat, cough, myalgia, and malaise. Rapid spread of the disease through populations, high attack rates for all age groups, and epidemics or pandemics of various severities have also defined influenza through the centuries. On the basis of antigenic differences in two of the major structural proteins, the nucleoprotein (NP) and the matrix (M) protein, influenza viruses are classified as types A, B, and C. Influenza A viruses are further grouped into subtypes according to the characteristics of their membrane glycoproteins, hemagglutinin (HA) and neuraminidase (NA). Variation in these membrane glycoproteins allows influenza viruses to cause repeated infections in the same individual. The influenza A subtypes that have circulated among humans include only three HA (H1, H2, and H3) and two NA (N1 and N2) subtypes. Since 1977, influenza A (H1N1), A (H3N2), and B viruses have been circulating together among humans. In recent years, epidemics caused primarily by type A viruses have alternated with those caused primarily by type B viruses. Epidemics caused by influenza A (H3N2) viruses have most often been associated with an increase in mortality, especially in elderly populations. Influenza C viruses typically cause localized outbreaks of mild upper respiratory tract infection in children and young adults. During an influenza epidemic in the community, there may be a sudden increase in visits to clinics, physicians' offices, or hospital emergency rooms due to acute respiratory illness. High attack rates for schools or residential institutions are common. Epidemics in a community generally last 3 to 6 weeks, but influenza viruses can be present in the community for a variable period before and after the epidemic.

Influenza viruses replicate in the columnar epithelium of the respiratory tract. They are transmitted primarily in small-particle aerosols generated by sneezing, coughing, and speaking but may also be transmitted by direct contact. Transmission is most likely at the onset of symptoms, when infected individuals shed large amounts of infectious virus in their respiratory secretions. Influenza viruses remain infectious in small droplets, particularly in the cold and in low humidity, which may partly explain the winter occurrence of influenza epidemics. The incubation period for influenza lasts between 1 and 4 days. The spectrum of clinical symptoms ranges from asymptomatic or mild respiratory infection to primary viral pneumonia with fatal outcome. The clinical picture of an uncomplicated infection in an adult is characterized by a sudden onset of fever in connection with sometimes severe headaches, myalgias, malaise, and anorexia. The fever can be as high as 41°C and, along with other symptoms, is usually of 3 to 4 days' duration. Respiratory symptoms such as tracheobronchitis, sore throat, nasal congestion, rhinorrhea, and coughing may be present at the beginning of the illness, or they may appear as the fever resolves. In its mildest form, influenza can present as rhinitis or pharyngitis. Higher fever and gastrointestinal symptoms, such as vomiting and abdominal pain, are more frequently seen in children than in adults. Children with influenza often develop otitis media. Lower respiratory tract complications can present as primary viral pneumonia, combined viral and bacterial pneumonia, or bacterial pneumonia after an influenza virus infection.

The antiviral drugs amantadine and rimantadine have been licensed in the United States for prophylactic and therapeutic use against influenza A infections, but annual vaccination of persons who are at high risk for complications remains the most effective way of reducing the impact of influenza. Continuous worldwide surveillance of influenza activity and circulating viruses helps determine which strains are most likely to be circulating during the next season. The composition of the vaccine is updated every year to ensure inclusion of the most recent epidemic strains. More detailed information about influenza is available in references 7, 20, and 27.

CHARACTERISTICS OF THE VIRUSES

Influenza A and B viruses are enveloped viruses with five internal nonglycosylated proteins (NP, M protein, and three polymerase proteins) and three integral membrane proteins (HA, NA, and M2 or NB). Influenza A and B viruses cannot be distinguished morphologically. Virus particles are pleomorphic, usually roughly spherical, and 80 to 120 nm in diameter after multiple passages. Elongated filamentous forms have been noted after a single passage. The most striking morphologic feature of influenza A and B viruses is a surface layer of spikelike projections 10 to 14 nm in length and 4 to 6 nm in diameter formed by the HA and the NA. Viral nucleocapsids have helical symmetry and a length of 30 to 110 nm. Influenza C viruses often have reticular structures on their surfaces.

Influenza A and B viruses contain eight single-stranded RNA segments that are complementary to mRNA (i.e., negative sense) and that encode at least 10 polypeptides, of which 8 are structural proteins found in the virus (see above) and 2 are found in infected cells (NS1 and NS2).

The three largest RNA segments encode the polymerase proteins PA, PB1, and PB2. These proteins are responsible for RNA synthesis and are found in association with the NP and viral RNA as a replication complex. PB1 and PB2 are involved in initiation and elongation functions during transcription of the mRNA. It appears likely that the PA protein is involved in replication of the viral RNA.

The HA protein, which is encoded by RNA segment 4, was named for its role in the agglutination of erythrocytes. It is responsible for virus attachment, virus penetration, and membrane fusion. HA is synthesized as a single polypeptide of approximately 550 amino acids that is subsequently cleaved into two polypeptide chains, HA1 and HA2, by host cell protease. Viruses with uncleaved HA can bind to host cells but cannot cause fusion and are not infectious. The functional HA exists as a homotrimer. X-ray crystallographic studies have revealed that the HA trimer has a globular head domain and a fibrous stem domain. The variable HA1 polypeptide makes up the globular head region and is folded into a β structure and loops that surround the largely conserved receptor-binding pocket. Five overlapping antigenic regions in the globular head domain have been identified from amino acid sequence changes in antigenic variants. The fibrous stem region, which is proximal to the viral membrane, is composed primarily of residues from HA2 but also contains residues from HA1. The HA is anchored in the membrane at its C-terminal end. Neutralizing antibodies against the HA play a crucial role in immunity to influenza infection, and variation in this molecule allows the virus to evade preexisting immunity obtained through infection or immunization.

The NP is encoded by RNA segment 5. NP is found in association with the polymerase proteins in the replication complex (see above). This protein plays a structural role in viral ribonucleoprotein formation, is involved in replication, and in its phosphorylated form has been shown to possess kinase activity. Although NP is targeted by cytotoxic T lymphocytes that recognize conserved or cross-reacting epitopes, immunization with NP provides only limited protection.

The glycoprotein NA is encoded by RNA segment 6 and is present in patches on the surface of the virus. NA cleaves terminal sialic acid residues from oligosaccharide chains. It is thought that NA activity allows penetration of the virus through the mucous layer to the target epithelial cells in the respiratory tract. NA also removes sialic acid from virus and infected cells to prevent self-aggregation and to allow release of virus from infected cells. NA is synthesized and is present in virions as a single polypeptide chain anchored in the membrane at the N-terminal end. Functional NA exists as a homotetramer and appears as a mushroom-shaped spike after release from virus particles by detergent. The three-dimensional structure of the molecule reveals a tetrameric globular head and a long filamentous tail. The globular head consists of mainly β structure containing the sialidase catalytic site surrounded by highly variable antigenic loops. Antibodies to NA do not completely neutralize virus infectivity except at very high concentrations, but they restrict the level of virus replication and attenuate illness.

RNA segment 7 encodes the viral membrane proteins M1 and M2. The highly conserved, type-specific M1 membrane (or M) protein is the most abundant protein in virus particles and is located under the lipid envelope of the virus. This protein has a structural function, is involved in the control of viral RNA polymerase activity, and participates in virus assembly by interacting with both HA and NP. Although antibodies to M1 protein are found after infection or vaccination with influenza virus and although an epitope for recognition of M1 protein by cytotoxic T cells has been demonstrated, the significance of M protein in immunity to influenza virus is uncertain. The M2 protein of influenza A viruses is encoded by a second open reading frame in RNA segment 7. This 97-amino-acid integral membrane protein is encoded by a spliced mRNA transcript and shares 9 amino acids at its N-terminal end with the M1 protein. It is found in large amounts in infected cells but only in small amounts in the mature virus. The M2 protein is a proton channel that performs essential functions in viral uncoating and maturation. Single amino acid changes in the transmembrane domain of the M2 protein confer resistance to the antiviral drugs amantadine and rimantadine. Antibodies to the M2 protein are found in postinfection sera of humans and ferrets, but the role of the M2 protein in immunity to influenza is unknown. Differences exist between influenza A and B viruses with regard to proteins encoded by RNA segments 6 and 7. The protein of influenza B viruses that corresponds to the M2 protein of influenza A viruses is the NB protein encoded by the NA gene. NB is a glycosylated dimeric integral membrane protein that probably also acts as an ion channel protein during virus replication.

Segment 8 codes for the two nonstructural proteins, NS1 and NS2, that are translated from an uninterrupted and a spliced mRNA transcript, respectively. These proteins also share nine amino acids at their N termini, and both are phosphorylated in infected cells. NS1 is synthesized in large amounts early in infection, while NS2 appears later; both proteins accumulate in the nucleus, but their functions remain uncertain.

Although influenza C viruses are similar to influenza A and B viruses in many ways, they contain only seven RNA segments and possess only one surface glycoprotein, which contains hemagglutinating, esterase, and fusion activities. Influenza C viruses have no NA activity, and their receptor-binding specificity differs from that of influenza A and B viruses.

Influenza virus nomenclature includes the type of influenza, the host of origin for nonhuman strains, the geographic origin, the strain number, and the year of isolation. An antigenic description of the HA and the NA is given in parentheses after the year of isolation. For example, A/Hong Kong/1/68 (H3N2) and A/USSR/90/77 (H1N1) are prototype strains for the A (H3N2) and A (H1N1) subtypes.

Two distinct types of antigenic variation that occur in influenza viruses allow them to evade preexisting immunity. The first is antigenic shift, which is the appearance of a new subtype of influenza A virus containing a novel HA and/or NA that is immunologically different from that of viruses circulating previously. This abrupt change in antigenicity may be associated with pandemics, as occurred in 1957 with the emergence of the Asian (H2N2) virus and again in

1968 with the emergence of the Hong Kong (H3N2) strain. The second type of antigenic variation, called antigenic drift, is more gradual and occurs through the accumulation of point mutations in the HA and/or NA within a subtype, resulting in the inability of antibody to previous strains to neutralize the mutant virus. This form of antigenic variation is responsible for periodic epidemics. The considerable antigenic heterogeneity that exists within a subtype may be discerned by hemagglutination inhibition (HAI) and NA inhibition tests as well as by other serologic techniques. This antigenic heterogeneity is reflected in genetic changes that can be detected by sequencing the genes encoding the HA and NA. For more detailed information on the molecular biology of influenza viruses, see references 20, 22, and 27.

LABORATORY DIAGNOSIS

The laboratory diagnosis of influenza virus infections has a number of important functions. Early during the course of the disease, a reliable laboratory result can assist with decisions concerning optimal treatment (e.g., whether the patient would benefit from antiviral therapy with amantadine or rimantadine). In hospitals and nursing homes, prompt recognition of an influenza outbreak can help prevent spread of the virus throughout the institution. Monitoring drug susceptibilities of viruses isolated during antiviral therapy is important in preventing the spread of drug-resistant strains. Isolation and characterization of influenza viruses are necessary for worldwide epidemiologic surveillance and for monitoring of influenza vaccine effectiveness. Antigenic and genetic analyses of currently circulating viruses provide information for the optimal composition of the influenza vaccine.

Many techniques have been developed for the diagnosis of influenza virus infections. Widely used conventional methods include virus isolation in cultured cells or in fertilized chicken eggs, characterization of the isolated virus by HAI testing, and demonstration of a fourfold or greater rise in specific antibodies between acute- and convalescent-phase sera. While these methods are highly sensitive and specific, the time required to obtain a positive result can vary from a few days to 3 weeks. Detection of viral antigens by immunologic methods, visualization of the virus by electron microscopy, and detection of viral nucleic acids by hybridization or by the PCR provide results within hours to 1 or 2 days. Although some of these tests may be less sensitive than virus isolation or serology, they have a direct impact on the patient's care because of the rapid availability of the test result. During recent years, rapid culture assays (i.e., detection of viral antigens in cell cultures by immunologic methods) have found wide application in the identification of many viruses, including influenza virus. This approach is almost as sensitive as conventional virus isolation, but the result is available within 1 or 2 days.

Collection, Transport, and Storage of Specimens

Specimens should be taken as early as possible during the course of disease, preferably within the first 72 h after onset of symptoms. Nasal washings, nasopharyngeal aspirates (NPAs), nasopharyngeal swabs (NPSs), and nasal and throat swabs are routine specimens for influenza virus diagnosis. Bronchoalveolar lavage specimens or transtracheal aspirates may also be considered if clinically warranted. Although taking an NPA causes the patient, particularly small children, some discomfort, it provides an excellent

specimen for virus isolation as well as for detection of viral antigens or nucleic acids. To obtain an optimal NPA, a catheter with an appropriate diameter fitted to a specimen trap is inserted through the nostrils into the nasopharynx. When the catheter is in place, vacuum is applied and the catheter is slowly withdrawn. To increase the volume of mucus, both nostrils should be aspirated, and the catheter can be flushed with 1 to 2 ml of saline or transport medium to collect mucus remaining in the catheter. NPS and nasal and throat swabs, either alone or combined, can be more easily taken from patients of all ages and are suitable specimens if carefully collected (9). Nasal washes taken from older children and adults provide excellent specimens for several diagnostic techniques.

For successful virus isolation, specimens should be placed in transport medium to stabilize the virus. Cell culture medium, tryptose phosphate broth, and veal infusion broth supplemented with 0.5% bovine serum albumin or 0.1% gelatin, antibiotics, and antimycotics are commonly used (1, 15). Although influenza viruses maintain their infectivity rather well when stored in transport media at 4°C for as long as 4 days, specimens should be transported to the laboratory without delay and kept at 4°C until inoculated. If they cannot be inoculated within 4 days, they should be frozen at −70°C immediately after collection and transported on dry ice. For long-term storage at −70°C, cryoprotectives such as dimethyl sulfoxide or extra protein can be added to preserve the virus (15). Such samples can also be employed for the detection of viral RNA by hybridization and PCR.

For the detection of infected nasopharyngeal epithelial cells by immunofluorescence (IF) staining, cells obtained by swabbing the nasopharynx are directly applied to microscope slides. If such staining is to be done on an NPA or on a swab collected into a transport medium, the specimen should be sent refrigerated to the laboratory and processed immediately after arrival to prevent lysis of cells. Mucus collected by aspiration is diluted with phosphate-buffered saline (PBS) and carefully homogenized by gentle pipetting. The cells are pelleted by centrifugation at 400 × g, washed several times with PBS until all mucus is removed, diluted appropriately in PBS, and applied to microscope slides. Mucus-free cell preparations can also be obtained by centrifugation through Percoll in the presence of a mucolytic reagent (40). Slides are air dried and fixed in cold acetone for 10 min.

For the detection of viral antigens in NPAs by enzyme immunoassay (EIA) or similar methods, the specimen can be transported to the laboratory without cooling. Before testing, the mucus is diluted 1:5 in PBS containing fetal bovine serum and detergent and homogenized by sonication (34).

Paired serum specimens are required for serodiagnosis of influenza virus infections. The first serum specimen should be drawn early during the acute phase, and a convalescent-phase serum should be collected about 2 to 3 weeks later. Serum specimens can be stored at 4°C until tested or for extended periods at −20°C.

Isolation of Influenza Viruses in Cell Culture

The Madin-Darby canine kidney cell line (MDCK; American Type Culture Collection, Rockville, Md.; CCL 34) supports the growth of influenza A, B, and C viruses (24). Trypsin at a concentration of 2 μg/ml must be added to the cell culture medium to sustain multiple cycles of virus replication in MDCK cells. 1-Tosylamide-2-phenylethyl

TABLE 1 Hemagglutination inhibition test using World Health Organization influenza reagents from the 1993 to 1994 collection

Reference antigen	Sheep antiserum[a]		
	A/Taiwan/1/86	A/Beijing/32/92	B/Panama/45/90
A/Taiwan/1/86 (N1N1)	1,280	<10	<10
A/Beijing/32/92 (H3N2)	20	1,280	10
B/Panama/45/90	<10	<10	160
B/Panama/45/90 (ether treated)	<10	<10	1,280

[a]Sheep sera are treated with receptor-destroying enzyme. Values are reciprocal of highest serum dilution that inhibits hemagglutination.

chloromethyl ketone-treated trypsin (TPCK-trypsin) is more stable at physiologic temperatures and should preferably be used. Primary cynomolgus or rhesus monkey kidney cells (PMKC) are susceptible to a variety of enteric and respiratory viruses, including influenza viruses; however, the high cost of these cells and the occasional presence of simian viruses in them have limited their use.

Standard virus isolation is done in cell culture tubes. The tubes are seeded with approximately 10^5 MDCK cells in 1 ml of growth medium containing 5 to 10% fetal bovine serum and incubated at 37°C. When the cells are almost confluent, the growth medium is aspirated, the cell sheet is washed with Hanks' buffer or serum-free medium to remove nonspecific inhibitors that may be present in serum, and maintenance medium is added; the cultures should be used within approximately 7 days. The medium is removed before 100 to 300 μl of a clinical specimen is added to each of two or three tubes. The cultures are incubated in a stationary rack at 34°C for at least 2 h, and then the inoculum is replaced by maintenance medium containing TPCK-trypsin. The tubes are incubated at 34°C either in stationary racks or in a roller machine running at a speed of approximately 12 revolutions per hour. Alternatively, 100 to 300 μl of clinical specimen is pipetted directly into the maintenance medium containing TPCK-trypsin, and the medium is changed the following day.

Some influenza viruses cause a distinct cytopathic effect in MDCK cells several days after inoculation. Negative cultures should be checked by a hemadsorption or hemagglutination test at 2- to 3-day intervals (19). The hemagglutination test requires fewer manipulations of the cell culture tube, and the risk of microbial or viral contamination is thus reduced. If no virus is detected after 10 to 14 days, a blind passage may be performed.

Isolation of Influenza Viruses in Embryonated Chicken Eggs

Influenza viruses can be isolated in fertilized chicken eggs. Ten- to 11-day-old embryos are usually used for the isolation of influenza virus A and B viruses, while for influenza virus type C, 7- to 8-day-old embryos are optimal. Both allantoic and amniotic cavities should routinely be inoculated with fresh clinical material (19). Eggs are incubated at 33 to 34°C for 2 to 3 days before the virus is harvested. For influenza C virus, the incubation period should be extended to 5 days. After one or two passages, most influenza A and B viruses will grow efficiently in the allantoic cavity, but influenza C virus grows only in the amniotic cavity. Presence of influenza viruses is verified by the hemagglutination test.

When harvesting egg or cell culture material, care must be taken to avoid cross-contamination between cultures

inoculated with different clinical specimens. If possible, inoculation and harvesting of cell cultures and eggs should be done in different safety cabinets or in different areas of the laboratory. If only one cabinet is available, inoculations with clinical specimens should be done before positive cultures are processed. Influenza viruses are rapidly inactivated by organic solvents such as 70% ethanol and by autoclaving.

Identification and Characterization of Influenza Viruses

Human type O, guinea pig, or chicken erythrocytes can be used in hemagglutination assays for the verification of virus growth in eggs or cell cultures. Guinea pig cells are not agglutinated by influenza C virus, however, so for optimal sensitivity, both chicken and guinea pig erythrocytes should be used. Since influenza C virus and some newly isolated, low-passage-number influenza A viruses may rapidly elute from erythrocytes at room temperature, the settling of erythrocytes should preferably take place at 4°C for 1 to 2 h (19). Because mumps and parainfluenza viruses also can grow in eggs and in cell cultures, any hemagglutinating agent has to be identified by the HAI test using specific sera. HAI is used for typing influenza viruses, for subtyping influenza A viruses, and for antigenically characterizing the isolate (19). Results of a reciprocal HAI test with World Health Organization influenza reagents are shown in Table 1. Viruses grown in cell culture can be typed and subtyped by IF or immunoperoxidase staining with specific antibodies.

Rapid Culture Assays

With the availability of amantadine and rimantadine for the prevention and treatment of influenza A infections, there is an increased need for more rapid diagnostic methods. Several techniques allow detection of viruses in cell cultures before cytopathic effect is visible.

Hemadsorption and hemagglutination have often been used to demonstrate the growth of influenza and certain other viruses in the absence of microscopically visible cytopathic effect. In a recent study, these two methods were applied in parallel to 476 PMKC cultures inoculated with respiratory specimens (17). Testing of the cultures was done at 3- to 5-day intervals for 2 weeks after inoculation. A blind passage was performed on cultures negative after 7 days of incubation. The results from these two methods were in complete agreement on each day of testing. The viruses in this study were identified by IF with commercially available monoclonal antibodies (MAbs) to influenza A and B viruses as well as to parainfluenza viruses 1, 2, and 3. In another rapid culture study, hemadsorption was performed on PMKC daily for 5 days after inoculation; all 73 influenza A viruses and 16 of 17 influenza B viruses were detected by day 3 (26).

With the availability of well-characterized MAbs, the detection of viruses in cultured cells by immunologic methods 1 to 3 days after inoculation has found wide application and is extensively used for a more rapid diagnosis of influenza and other virus infections. Standard cell culture tubes, shell vials, chamber slides, or cell culture clusters may be used for this procedure. Before fixation, cells must be enzymatically or mechanically removed from standard cell culture tubes and transferred to microscope slides. Viral antigen in infected cells can then be detected by IF or immunoperoxidase staining. As an alternative, virus antigens can be extracted from infected cells and detected by immunoassays.

Centrifuged inoculation of specimens increases the sensitivity of virus isolation in cell cultures. Two to three times the number of plaques were detected in a standard plaque assay from the same material when the inoculation step included a centrifugation at $700 \times g$ for 15 to 60 min (36). Furthermore, the number of fluorescent foci after a 24-h incubation increased when specimens were inoculated by centrifugation (8), making the identification of positive specimens easier.

Most commonly, MDCK cells are grown in shell vials or in 24-well cell culture plates. When the cultures are confluent, the growth medium is replaced with serum-free maintenance medium containing trypsin. Clinical specimens are inoculated in duplicate, and the cultures are centrifuged at $700 \times g$ for 30 to 60 min and incubated at 35°C in 5% CO_2 for 18 to 72 h. The medium is then removed, and the cells are washed with PBS and fixed. Acetone or methanol is a suitable fixative, but with plastic culture dishes, acetone should be used at a concentration of 80% in PBS. The fixed cells are incubated with type-specific antibodies followed by an antispecies conjugate, and the cultures are checked for specific staining with a fluorescence or light microscope. The sensitivity of rapid culture assays ranges from 56 to 100% compared with standard isolation (8, 25, 39). The incubation time for the rapid assay is between 18 and 40 h, whereas an average of 4.5 days is needed to obtain a positive result from standard isolation.

Recently, an immunoperoxidase staining assay that allows typing of influenza A and B viruses and simultaneous subtyping of influenza A viruses from clinical specimens within 24 h was developed (48). This test was evaluated on 263 specimens previously found positive by a rapid culture assay (8). The specimens were kept frozen at $-70°C$ for as long as 4 years before testing. Each specimen was inoculated into quadruplicate wells on 24-well cell culture plates. The plates were centrifuged at $700 \times g$ for 45 min and then incubated at 36°C overnight before they were fixed and stained. Two pools of MAbs specific for either influenza A or B virus (42) were used for typing, and two MAbs against conserved epitopes on the denatured HA1 and HA2 subunits of the HA of A/Victoria/3/75 (33) were used for subtyping influenza A viruses. A few specimens contained so little virus that although they were identified as influenza A virus, the subtype could not be determined. However, 94% of the influenza A virus-positive samples were easily subtyped.

Detection of Infected Cells in Clinical Specimens by IF

Demonstration of infected epithelial cells in NPA or NPS is a sensitive method for the diagnosis of respiratory virus infections, and a result can be obtained within a few hours after collection of the specimen. Exfoliated cells in NPA or NPS are prepared, applied to microscopic slides, and fixed as described above. An adequate number of epithelial cells on the slide and the use of well-characterized, specific antibodies are essential to obtaining a high level of sensitivity and specificity by this technique. Polyclonal sera and MAbs have been widely used as immunoreagents in the staining process. Polyclonal sera often show nonspecific fluorescence with cellular debris and with bacteria adherent to epithelial cells, making interpretation of the test result more difficult (37). MAbs against various proteins of influenza virus A and B have been developed (42), and some are commercially available (23, 38). When compared with virus isolation, detection of infected cells by IF is 50 to 90% sensitive. Multiple slides should be prepared from each specimen and stored frozen so that in case of an unclear finding, a second slide is available for restaining. Another advantage of this technique is that the slides can be prepared and fixed in a hospital laboratory or in a doctor's office and then safely transported without losing reactivity.

Detection of Influenza Virus Antigens by Immunoassays

A variety of tests, such as radioimmunoassay, EIA, and fluoroimmunoassay, have been developed for the detection of influenza virus antigens either directly in clinical specimens or after amplification in cell culture (6, 34, 43). In one study, clinical specimens were grown in cell culture for 1 or more days, and after given intervals, the cells were scraped into the supernatant and the material was passively adsorbed to microtiter wells before detection by EIA (43). More commonly, however, antigens in clinical specimens or in cell culture material are captured by specific antibodies that have previously been passively adsorbed to the solid phase. After incubation, unbound material is washed, and the bound antigen is reacted with a secondary antibody and then incubated further with a labeled antispecies antibody. When purified viral antigens are used as antigen, the sensitivities of these assays range from a few picograms to a few nanograms per assay volume. Comparison of radioimmunoassay and EIA with IF using NPAs obtained from patients between 1 month and 90 years of age revealed a high degree of correlation among the three methods (34). Most assays have been evaluated against standard virus isolation in MDCK cells or PMKC, and commonly, the antigen detection assays give a positive result with 50 to 80% of the isolation-positive specimens (6). It is not uncommon to find isolation-negative specimens that give a positive result when tested by immunoassay, particularly if specimens are taken late during the course of infection. Blocking assays have been used to show that many of these "false-positive" results are in fact specific positive findings. To reach optimal sensitivity with immunoassays, incubation of the specimen with the capture antibody usually lasts from one to several hours, and the total time required to perform the tests ranges from 4 to 20 h.

Results can be obtained more rapidly by a fluoroimmunoassay with MAbs against the NP. Microtiter strips are coated with capture antibodies. Diluted specimens are then applied to the wells, and immediately thereafter, a europium-labeled MAb is added and the strips are incubated at 37°C for 60 min. After the strips are washed, an enhancement solution is added, and after a 15-min incubation, fluorescent bound europium is measured with a fluorometer (43). In spite of the short reaction time, this assay detected more positive specimens than an EIA based on polyclonal antibodies.

A commercially available EIA for the detection of influenza A virus antigen in human and animal specimens can be completed in as little as 15 min (16, 32, 44). Nasopharyngeal washes and NPAs work best for this assay, but NPSs and pharyngeal swabs can also be used. Specimens are diluted with a buffer containing detergent and mucolytic agents and then applied to a filter membrane. Influenza virus antigens are nonspecifically adsorbed to the membrane and then detected with enzyme-labeled MAbs to the NP, washed several times, and incubated with two consecutive substrates and the chromogen. A visible purple triangle on the filter membrane indicates a positive result. Only 20 infected cultured cells but more than 1.5×10^3 infectious virus particles in cell-free supernatant are required to give a positive test result. When compared with virus isolation in PMKC or MDCK cells or with direct detection of virus-infected cells by IF, the test shows a sensitivity between 60 and 100%. Weak positive staining of the filter has to be interpreted cautiously, because with some specimens, the test may give false-positive results.

Detection of Viral RNA by Molecular Hybridization

Molecular hybridization has been applied for the detection of influenza A virus RNA in clinical samples (12). NPSs were treated with proteinase K, and nucleic acids were precipitated, denatured, spotted onto nylon filters, and dried. Hybridization with ^{32}P-labeled influenza virus-specific cloned DNA probes was done at 37°C for 66 h. Bound probe was detected by autoradiography. Approximately 1 pg of influenza virus RNA could be detected with this assay, and with optimally stored specimens, 75% of isolation-positive specimens yield a positive result.

Reverse Transcriptase PCR

The reverse transcriptase PCR has been applied to detect influenza virus in clinical specimens and in cell culture- or egg-grown material. The viral RNA is first extracted and then reverse transcribed into DNA, which is then amplified in a second step that uses suitable primers and DNA polymerase. Amplified DNA is electrophoresed in an agarose gel, and DNA is detected by staining with ethidium bromide. Molecular weight markers are included to identify amplified DNA of the appropriate size. DNA can also be detected by molecular hybridization with specific labeled probes either in Southern blots or by spot hybridization. Appropriate selection of primers permits type-specific identification of influenza viruses A, B, and C or subtype-specific identification of influenza A viruses (3, 46, 47). A subtype-specific assay has been evaluated with 26 gargle fluids obtained from patients with influenzalike symptoms. Nucleic acid was amplified by 25 cycles, and if no band was detected in an agarose gel, a fraction of the amplified material was subjected to another 25 amplification cycles. Two blind passages were performed on negative cultures. All 17 isolation-positive specimens were identified as positive by PCR, three of them after 25 cycles of amplification

and 14 after 50 cycles. Six specimens were negative by both assays (46).

Because cross-contamination of specimens or contamination of reagents is a major concern for laboratories using PCR for clinical diagnosis, it is essential to include appropriate controls to detect contamination. In a 6-month prospective study of 434 respiratory specimens obtained from pediatric patients, however, only 2 specimens from which influenza virus could not be isolated were positive by PCR. Two specimens were isolation positive but negative by PCR. Twenty-three specimens were positive and 407 were negative by both methods (4).

Routine sequencing of PCR products of the HA and other genes is now done in a number of laboratories. This combination of methods allows rapid analysis of the molecular changes in circulating influenza viruses (45) and contributes to the worldwide epidemiologic surveillance of influenza and to the selection of optimal vaccine strains (5).

Immune Electron Microscopy

Immune electron microscopy offers a rapid method for the identification of virus particles, including influenza virus, in clinical specimens (29). Nasopharyngeal washings or nasal and pharyngeal swabs collected into transport medium are mixed with appropriately prediluted hyperimmune sera. After incubation at 37°C for 1 h, the mixture is centrifuged at $27,000 \times g$ for 1 h. The supernatant is carefully removed, and the sediment is suspended in a small amount of distilled water. Grids are prepared and stained with phosphotungstic acid. With a limited number of clinical specimens, immune electron microscopy had a sensitivity of slightly more than 50% compared with virus isolation in cell culture.

Serology

Although serologic methods seldom yield a result early enough to influence the patient's treatment, they often establish the diagnosis of influenza virus infection when virus is not detected by any other method (21). When the sensitivities of newly developed diagnostic tests are being evaluated, serology should be considered instead of or in combination with virus isolation as the "gold standard." HAI, the neutralization test (NT), and the complement fixation test (CF) are traditional methods in serodiagnosis and seroepidemiologic studies of influenza, but during recent years, EIA has found wide application.

CF measures antibodies against the NP and thus allows type-specific detection of antibodies to influenza A and B viruses. This test is relatively insensitive in detecting titer rises in serum pairs from individuals with a recent infection. However, since the NP is conserved, this method can be used when antigenic variants emerge for which specific antigens are not yet widely available. HAI and NT are more sensitive and measure antibodies against subtype- and strain-specific antigens. Some human and animal sera contain inhibitors of hemagglutination that may cause false-positive results, but methods to remove these inhibitors from serum are available; alternatively, inhibitor-resistant influenza virus strains can be used as antigens (19). The specificity of EIAs can be designed by choosing a particular protein of the virus as antigen and by using conjugates to different isotypes of immunoglobulins. In some instances, detection of influenza virus-specific immunoglobulin M- or A-class antibodies in serum samples collected soon after the onset of symptoms can help establish a diagnosis (21, 41).

The sensitivities and diagnostic performances of EIAs and HAI tests are influenced by the nature of the antigen.

Cell culture-derived antigens more frequently detect significant titer increases in HAI tests than do egg-grown antigens of the same virus strain (30, 35). With purified virus as antigen, EIA detected slightly more diagnostic titer increases than did the HAI test but significantly more than the CF (18). The NP of influenza viruses has been expressed in *Escherichia coli* (10) and insect cells (31). These recombinant proteins have been used in EIAs, which may replace the CF for the serodiagnosis of influenza virus infections.

The NT most accurately predicts whether an individual has protective immunity against currently circulating virus strains. Two new assay formats that make the NT less time and labor intensive and more applicable for the testing of larger numbers of sera have been introduced. In the former, heat-inactivated sera are mixed with the virus in flat-bottom microtiter plates. After incubation at 37°C for 60 min, a suspension of freshly trypsinized MDCK cells is added, and the plates are further incubated at 37°C overnight. The cells are then fixed, and viral antigen is detected by EIA with MAbs and a peroxidase-labeled anti-mouse conjugate (11). The latter is a focus reduction NT in which after a 1-h incubation at 37°C, the serum-virus mixture is transferred to preformed monolayers of MDCK cells in microtiter plates and allowed to adsorb for 30 min. The virus inoculum is then removed, and the cells are covered with a semisolid overlay and incubated for 24 h. In the final step, cells are fixed, and infected-cell foci are stained by the peroxidase-antiperoxidase technique and counted (28).

Drug Susceptibility Tests

Influenza A viruses resistant to amantadine and rimantadine can frequently be isolated from individuals receiving these antiviral drugs for therapy or prophylaxis (13). The plaque inhibition assay is the conventional assay for drug sensitivity testing (14). Monolayer cultures of MDCK cells are fed with double-strength medium containing trypsin and the test drug. An equal volume of virus suspension containing 5 to 15 PFU/cm^2 of culture surface is added, and the cultures are incubated for 60 min. Duplicate cultures without the drug are infected in the same manner. An agarose overlay containing the drug under study at the appropriate concentration is added, and the cultures are incubated for 2 days before the plaques are counted. This test is very sensitive but requires a careful pretitration of the virus to determine the optimal dilution. Results can be obtained more rapidly with an EIA of infected cells (2). In this assay, confluent MDCK cells in 96-well microtiter plates are inoculated with serial 2- to 10-fold dilutions of the virus, and the appropriate concentration of drug is added. Control wells without drug are inoculated in the same way. After an overnight incubation, the cells are washed and fixed, and virus antigen is detected with a suitable antibody by using standard EIA techniques. Less than 50% inhibition of virus growth in the presence of the drug compared with that in control wells indicates that the virus is to some degree resistant. Resistance should be verified by sequence analysis of the transmembrane region of the M2 gene.

CONCLUSIONS

Laboratory tests for the diagnosis of influenza virus infections should fulfill a number of criteria. The sensitivity and specificity must be high to minimize the number of false-positive and false-negative results. The use of well-standardized reagents is critical. To have an influence on clinical decisions, results must be available within a few hours or, at most, 1 day after collection of the specimen. Performance of the test should not involve the use of any unusual and expensive equipment, and the hands-on time required for the test should be in a range that allows testing a substantial numbers of specimens. Well-trained laboratory staff should be able to read and interpret the test without extensive training. Detecting infected epithelial cells by IF is widely used in the diagnosis of respiratory infections. This test is rapid, and high-quality reagents are available. Reading and interpreting the test, however, require considerable experience. Although sometimes less sensitive than IF, detection of viral antigens by immunoassays offers a good alternative. These tests can be performed within hours, and the specimen can be thoroughly homogenized before testing, which allows retesting of identical material if required. Detection of viral antigens in cell cultures 1 day after inoculation combines the advantages of virus isolation with the speed of antigen detection. Modification of this technique allows screening for drug susceptibility and detection of neutralizing antibodies in sera. Since antigenic and genetic analyses of circulating epidemic viruses are mandatory for global surveillance and vaccine strain selection, laboratories with cell culture facilities should continue to isolate influenza viruses during each influenza season.

REFERENCES

1. **Baxter, B. D., R. B. Couch, S. B. Greenberg, and J. A. Kasel.** 1977. Maintenance of viability and comparison of identification methods for influenza and other respiratory viruses of humans. *J. Clin. Microbiol.* **6:**19–22.
2. **Belshe, R. B., M. Hall Smith, C. B. Hall, R. Betts, and A. J. Hay.** 1988. Genetic basis of resistance to rimantadine emerging during treatment of influenza virus infection. *J. Virol.* **62:**1508–1512.
3. **Claas, E. C. J., M. J. W. Sprenger, G. E. M. Kleter, R. van Beek, W. G. V. Quint, and N. Masurel.** 1992. Type-specific identification of influenza viruses A, B, and C by the polymerase chain reaction. *J. Virol. Methods* **39:**1–13.
4. **Claas, E. C. J., A. J. van Milaan, M. J. W. Sprenger, M. Ruiten-Stuiver, G. I. Arron, P. H. Rothbarth, and N. Masurel.** 1993. Prospective application of reverse transcriptase polymerase chain reaction for diagnosing influenza infections in respiratory samples from a children's hospital. *J. Clin. Microbiol.* **31:**2218–2221.
5. **Cox, N., X. Xu, C. Bender, A. Kendal, H. Regnery, M. Hemphill, and P. Rota.** 1993. Evolution of hemagglutinin in epidemic variants and selection of vaccine viruses, p. 223–230. *In* C. Hannoun et al. (ed.), *Options for the Control of Influenza II.* Elsevier Science Publishers, Amsterdam.
6. **Döller, G., W. Schuy, K. Y. Tjhen, B. Stekeler, and H.-J. Gerth.** 1992. Direct detection of influenza virus antigen in nasopharyngeal specimens by direct enzyme immunoassay in comparison with quantitating virus shedding. *J. Clin. Microbiol.* **30:**866–869.
7. **Douglas, R. G., Jr., and R. F. Betts.** 1979. Influenza virus, p. 1135–1167. *In* G. L. Mandell, R. G. Douglas, Jr., and J. E. Bennet (ed.), *Principles and Practice of Infectious Diseases.* John Wiley & Sons, Inc., New York.
8. **Espy, M. J., T. F. Smith, M. W. Harmon, and A. P. Kendal.** 1986. Rapid detection of influenza virus by shell vial assay with monoclonal antibodies. *J. Clin. Microbiol.* **24:**677–679.
9. **Frayha, H., S. Castricano, J. Mahony, and M. Chernesky.** 1989. Nasopharyngeal swabs and nasopharyngeal aspirates equally effective for the diagnosis of viral respiratory disease in hospitalized children. *J. Clin. Microbiol.* **27:**1387–1389.
10. **Harmon, M. W., I. Jones, M. Shaw, W. Keitel, C. B. Reimer, P. Halonen, and A. P. Kendal.** 1989. Immunoassay for serologic diagnosis of influenza type A using recombinant

DNA produced nucleoprotein antigen and monoclonal antibody to human IgG. *J. Med. Virol.* **27**:25–30.

11. **Harmon, M. W., P. A. Rota, H. H. Walls, and A. P. Kendal.** 1988. Antibody response in humans to influenza virus type B host-cell-derived variants after vaccination with standard (egg-derived) vaccine or natural infection. *J. Clin. Microbiol.* **26**:333–337.

12. **Havlickova, M., A. Z. Pljusnin, and B. Tumova.** 1990. Influenza virus detection in clinical specimens. *Acta Virol.* **34**:446–456.

13. **Hayden, F. G., R. B. Belshe, R. D. Clover, A. J. Hay, M. G. Oakes, and W. Soo.** 1989. Emergence and apparent transmission of rimantadine-resistant influenza A virus in families. *N. Engl. J. Med.* **321**:1696–1702.

14. **Hayden, F. G., K. M. Cote, and R. G. Douglas, Jr.** 1980. Plaque inhibition assay for drug susceptibility testing of influenza viruses. *Antimicrob. Agents Chemother.* **17**:865–870.

15. **Johnson, F. B.** 1990. Transport of viral specimens. *Clin. Microbiol. Rev.* **3**:120–131.

16. **Johnston, S. L. G., and H. Bloy.** 1993. Evaluation of a rapid enzyme immunoassay for detection of influenza A virus. *J. Clin. Microbiol.* **31**:142–143.

17. **Johnston, S. L. G., K. Wellens, and C. Siegel.** 1992. Comparison of hemagglutination and hemadsorption tests for influenza detection. *Diagn. Microbiol. Infect. Dis.* **15**:363–365.

18. **Julkunen, I., R. Pyhälä, and T. Hovi.** 1985. Enzyme immunoassay, complement fixation and hemagglutination inhibition tests in the diagnosis of influenza A and B virus infections. Purified hemagglutinin in subtype-specific diagnosis. *J. Virol. Methods* **1985**:75–84.

19. **Kendal, A. P., J. J. Skehel, and M. S. Pereira.** 1982. *Concepts and Procedures for Laboratory-Based Influenza Surveillance.* Centers for Diseases Control, Atlanta.

20. **Kilbourne, E. D.** 1987. *Influenza.* Plenum Publishing Corp., New York.

21. **Koskinen, P., T. Vuorinen, and O. Meurman.** 1987. Influenza A and B virus IgG and IgM serology by enzyme immunoassays. *Epidemiol. Infect.* **99**:55–64.

22. **Krug, R. M. (ed.).** 1989. *The Influenza Viruses.* Plenum Publishing Corp., New York.

23. **McDonald, J. C., and P. Quennec.** 1993. Utility of a respiratory virus panel containing a monoclonal antibody pool for screening of respiratory specimens in nonpeak respiratory syncytial virus season. *J. Clin. Microbiol.* **31**:2809–2811.

24. **Meguro, H., J. D. Bryant, A. E. Torrence, and P. F. Wright.** 1979. Canine kidney cell line for isolation of respiratory viruses. *J. Clin. Microbiol.* **9**:175–179.

25. **Mills, R. D., K. J. Cain, and G. L. Woods.** 1989. Detection of influenza virus by centrifugal inoculation of MDCK cells and staining with monoclonal antibodies. *J. Clin. Microbiol.* **27**:2505–2508.

26. **Minnich, L. L., and C. G. Ray.** 1987. Early testing of cell cultures for detection of hemadsorbing viruses. *J. Clin. Microbiol.* **25**:421–422.

27. **Murphy, B. R., and R. G. Webster.** 1990. Orthomyxoviruses, p. 1091–1152. *In* B. N. Fields and D. M. Knipe (ed.), *Virology*, 2nd ed. Raven Press, New York.

28. **Okuno, Y., K. Tanaka, K. Baba, A. Maeda, N. Kunita, and S. Ueda.** 1990. Rapid focus reduction neutralization test of influenza A and B viruses in microtiter system. *J. Clin. Microbiol.* **28**:1308–1313.

29. **Ptakova, M., and B. Tumova.** 1985. Detection of type A and B influenza viruses in clinical materials by immunoelectronmicroscopy. *Acta Virol.* **29**:19–24.

30. **Pyhälä, R., L. Pyhälä, M. Valle, and K. Aho.** 1987. Egg-grown and tissue-culture-grown variants of influenza A (H3N2) virus with special attention to their use as antigens in seroepidemiology. *Epidemiol. Infect.* **99**:745–753.

31. **Rota, P. A., R. A. Black, B. K. De, M. W. Harmon, and A. P. Kendal.** 1990. Expression of influenza A and B virus nucleoprotein antigens in baculovirus. *J. Gen. Virol.* **71**:1545–1554.

32. **Ryan-Poirier, K. A., J. M. Katz, R. G. Webster, and Y. Kawaoka.** 1992. Application of Directigen Flu-A for the detection of influenza A virus in human and nonhuman specimens. *J. Clin. Microbiol.* **30**:1072–1075.

33. **Sánchez-Fauquier, A., N. Villanueva, and J. A. Melero.** 1987. Isolation of cross-reactive, subtype-specific monoclonal antibodies against influenza virus HA1 and HA2 hemagglutinin subunits. *Arch. Virol.* **97**:251–265.

34. **Sarkkinen, H. K., P. E. Halonen, and A. A. Salmi.** 1981. Detection of influenza A virus by radioimmunoassay and enzyme-immunoassay from nasopharyngeal specimens. *J. Med. Virol.* **7**:213–220.

35. **Schild, G. C., J. S. Oxford, J. C. de Jong, and R. G. Webster.** 1983. Evidence for host-cell selection of influenza virus antigenic variants. *Nature* (London) **303**:706–709.

36. **Seno, M., Y. Kanamoto, S. Takao, N. Takei, S. Fukuda, and H. Umisa.** 1990. Enhancing effect of centrifugation on isolation of influenza virus from clinical specimens. *J. Clin. Microbiol.* **28**:1669–1670.

37. **Shalit, I., P. A. McKee, H. Beauchamp, and J. L. Waner.** 1985. Comparison of polyclonal antiserum versus monoclonal antibodies for the rapid diagnosis of influenza A virus infections by immunofluorescence in clinical specimens. *J. Clin. Microbiol.* **22**:877–879.

38. **Spada, B., K. Biehler, P. Chegas, J. Kaye, and M. Riepenhoff-Talty.** 1991. Comparison of rapid immunofluorescence assay to cell culture isolation for the detection of influenza A and B viruses in nasopharyngeal secretions from infants and children. *J. Virol. Methods* **33**:305–310.

39. **Stokes, C. E., J. M. Bernstein, S. A. Kyger, and F. G. Hayden.** 1988. Rapid diagnosis of influenza A and B by 24-h fluorescent focus assays. *J. Clin. Microbiol.* **26**:1263–1266.

40. **Ukkonen, P., and I. Julkunen.** 1987. Preparation of nasopharyngeal secretions for immunofluorescence by one-step centrifugation through Percoll. *J. Virol. Methods* **15**:291–301.

41. **Vikerfors, T., G. Lindegren, M. Grandien, and J. van der Logt.** 1989. Diagnosis of influenza A virus infections by detection of specific immunoglobulins M, A, and G in serum. *J. Clin. Microbiol.* **27**:453–458.

42. **Walls, H. H., M. W. Harmon, J. J. Slagle, C. Stocksdale, and A. P. Kendal.** 1986. Characterization and evaluation of monoclonal antibodies developed for typing influenza A and influenza B viruses. *J. Clin. Microbiol.* **23**:240–245.

43. **Walls, H. H., K. H. Johansson, M. W. Harmon, P. E. Halonen, and A. P. Kendal.** 1986. Time-resolved fluoroimmunoassay with monoclonal antibodies for rapid diagnosis of influenza infections. *J. Clin. Microbiol.* **24**:907–912.

44. **Waner, J. L., S. J. Todd, H. Shalaby, P. Murphy, and L. V. Wall.** 1991. Comparison of Directigen Flu-A with viral isolation and direct immunofluorescence for the rapid detection and identification of influenza A virus. *J. Clin. Microbiol.* **29**:479–482.

45. **Xu, X., E. P. Rocha, H. L. Regenery, A. P. Kendal, and N. J. Cox.** 1993. Genetic and antigenic analyses of influenza A (H1N1) viruses, 1986–1991. *Virus Res.* **28**:37–55.

46. **Yamada, A., J. Imanishi, E. Nakajima, K. Nakajima, and S. Nakajima.** 1991. Detection of influenza viruses in throat swab by using polymerase chain reaction. *Microbiol. Immunol.* **35**:259–265.

47. **Zhang, W., and D. H. Evans.** 1991. Detection and identification of human influenza viruses by the polymerase chain reaction. *J. Virol. Methods* **33**:165–189.

48. **Ziegler, T., H. Hall, A. Sánchez-Fauquier, B. Gamble, and N. J. Cox.** Type- and subtype-specific detection of influenza viruses in clinical specimens by rapid culture assay. *J. Clin. Microbiol.*, in press.

Parainfluenza Viruses

JOSEPH L. WANER

77

CLINICAL BACKGROUND

The four parainfluenza viruses (PIVs) of humans cause upper respiratory disease in children and adults. PIV types 1 and 2 (PIV-1 and -2) are the principal causes of laryngotracheobronchitis (croup), although PIV-3 and other infectious agents can also cause croup. PIV-3 is second only to respiratory syncytial virus in importance as a cause of bronchiolitis and pneumonia in infants (3, 5, 9, 20). Type 4 causes mild upper respiratory disease but is rarely encountered when only isolation in cell culture is used for diagnosis. Types 1 and 2 behave similarly, although type 2 generally produces milder illness. The most severe illness caused by PIV-1 and -2 occurs in children between 2 and 4 years of age; type 3 infection produces the most severe illness in infants less than 1 year old. Reinfections with PIVs are common but are generally less severe clinically than primary infections (3, 10). PIV infections in older children and adults are more likely to be asymptomatic or to result in mild disease resembling the common cold syndrome (26). Aside from respiratory disease, PIV-3 was also isolated from spinal fluids of patients with symptoms of meningeal infection (23) and for prolonged periods from adults without acute symptoms (13). Genetic variations and antigenic subgroups exist among the PIVs, but these have not been shown to have biological implications or to affect the choice of the diagnostic methods that are commonly used.

PIV-1 and -2 occur in alternate-year patterns, tending to produce epidemics in the autumn and early winter (5, 11). Type 3 is predominantly endemic and shows little or no seasonality.

PIVs are transmitted via infected respiratory secretions through close person-to-person contact and aerosols. The viruses remain viable on surfaces for several hours (2) but lose infectivity rapidly on hands (1); PIVs are readily inactivated by heat and detergents. Laboratory-associated infections should not occur if careful laboratory practices are followed. Vaccines are not available.

DESCRIPTION OF THE VIRUS

PIVs belong to the family *Paramyxoviridae* and the genus *Paramyxovirus*. The viruses have helical symmetry and single-stranded, nonsegmented RNA genomes with negative polarity. Virions have envelopes obtained by budding from the host cell cytoplasmic membrane. PIVs have six structural proteins, although coding strategies and the total number of gene products may differ in the paramyxoviruses (4). The HN (hemagglutinin-neuraminidase) and F (fusion) proteins are found in the envelope and are glycosylated. The HN protein is larger than the F protein and is necessary for adsorption to the host cell. After adsorption, the F protein mediates virion entry into the cell. The F protein is also responsible for cell-to-cell spread, manifested morphologically in cell cultures as syncytium. The remaining structural proteins are the large nucleocapsid protein, the nucleoprotein, the phosphoprotein, and the matrix protein.

There are common antigens in the four PIVs, mumps virus, Newcastle disease virus, Sendai virus (PIV-1 of mice), bovine PIV-3, and simian virus 5. Isolates of human PIVs can be distinguished according to type with specific antisera, particularly monoclonal antibodies (MAbs) (25).

COLLECTION AND STORAGE OF SPECIMENS

The object of specimen collection is to obtain secretions from the respiratory tract. Efforts should also be made to obtain cellular material, particularly if direct procedures for viral identification are to be attempted. PIV-3 may be excreted for 3 to 10 days following primary infection but for shorter intervals after secondary infections (6). Nevertheless, the greatest quantity of virus is excreted early in the course of illness, making the prompt collection of specimens critical for efficient isolation of virus.

Collection of specimens is most commonly accomplished with a wash of the nasopharynx. However, respiratory secretions may be aspirated by a suction device without the addition of a wash. A convenient and effective adaptation of the method described by Hall and Douglas (14) is recommended. The nasopharynx is vigorously swabbed with a Dacron swab on a flexible aluminum or plastic shaft to loosen mucus and cellular material. Approximately 2 ml of sterile saline is then introduced into the nasopharynx through small tubing attached to a 5-ml syringe; a convenient source of tubing is a Butterfly-22 infusion set. The wash and swab are placed together in 1.5 ml of transport medium and carried to the laboratory on wet ice.

Adults and older children may resist nasopharyngeal

(NP) washes. Throat or NP swabs may be used alone but are generally not as effective as NP washes. The swab should be placed in the transport medium, and the specimen should be promptly taken on wet ice to the laboratory. Specimens should be held at 4°C and should not be frozen if processing will be accomplished within 48 h of collection. A sample (approximately 0.3 ml) should be stored at −70°C for possible use in reevaluating the specimen or for future reference.

Transport medium should consist of a buffered salt solution (pH 7.0) containing a protein stabilizing agent and antibiotics. In this laboratory, a medium consisting of tryptic soy broth with 0.5% gelatin, gentamicin (50 μg/ml), chloramphenicol (5.0 μg/ml), and amphotericin B (2.0 μg/ml) is used.

ISOLATION IN CELL CULTURE

Primary human embryonic kidney and primary monkey kidney are the most sensitive cell cultures for the isolation of PIVs. NCI-H292 is a continuous line of mucoepidermoid cells that may also be used for isolation of PIVs (16). PIV isolates may be adapted to grow in other continuous cell lines (Vero, HEp-2, LLC-MK2, HeLa), but these are not recommended for isolation of virus from clinical specimens. Other viruses that may be present as pathogens in the respiratory tract replicate in primary monkey kidney cultures, which should be an essential component of the laboratory's isolation protocol; primary rhesus monkey kidney (PRMK) cells are recommended. Tubes of uninfected PRMK cultures are maintained on Eagle minimal essential medium (MEME) with 5% newborn calf serum and held at 35°C. Cultures are refed every 4 to 5 days until used but should not be used more than 10 days after receipt.

Before inoculation of cell cultures, specimen material is vigorously mixed, the collection swab is pressed against the side of the container to express fluid, and the swab is discarded. The maintenance medium is removed from two PRMK cultures, and the cultures are washed twice with Hanks balanced salt solution (HBSS). Each tube is inoculated with 0.2 ml of the specimen, and the cultures are absorbed at 37°C. The inoculum is removed after 1 h, and the cultures are washed once with serum-free medium containing trypsin (SFM) and refed with SFM. Cultures are incubated in a roller drum at 35°C; stationary incubation is also satisfactory. In our experience, medium containing serum may delay hemadsorption (HAd) of PIVs for 1 to 3 days and may also diminish the degree of HAd seen. A suitable SFM consists of 50 ml of medium 199, 46 ml of MEME, 3 ml of 100× vitamins for MEME, 0.5 ml of glucose (50% solution), 1 ml of glutamine (200 mM), 0.1 ml of trypsin (0.025%), 50 μg of gentamicin per ml, and 2 μg of amphotericin B per ml; medium 199 and MEME should be buffered with HEPES (N-2-hydroxyethylpiperazine-N'-2-ethanesulfonic acid). Cultures that are refed every 4 days may be maintained on SFM for 10 to 14 days.

Cultures are observed daily for cytopathic effect (CPE) to detect not only PIVs but also other viruses that may have been in the specimen and may replicate in PRMK cells. In our experience, approximately 50% of PIV isolates show CPE between 4 and 5 days after inoculation. CPE of the PIVs may range from unrecognizable to destructive, depending on the PIV type and the isolate. Typically, PIV-1 is identified by HAd before CPE appears (Fig. 1b); CPE consists of small rounded cells that are often difficult to discern as CPE. CPE of PIV-2 is often syncytial (Fig. 1c),

and PIV-3 may show "bridging" of the infected monolayer (Fig. 1d). Degeneration of the entire monolayer often characterizes the CPE of PIV-4.

The detection of a PIV (or other hemadsorbing virus) in cell culture following the appearance of CPE is confirmed by HAd. However, HAd may be detected before identifiable CPE and may be used as an inexpensive rapid method for the detection of hemadsorbing viruses, including the PIVs. More than 50% of the PIV-1 and PIV-3 isolates hemadsorb by 48 h after inoculation (19). A convenient strategy is to examine cultures for HAd 24 h after inoculation and at 48-h intervals thereafter; cultures should also be examined for HAd at the earliest sighting of CPE. Guinea pig erythrocytes (RBCs) are washed four times with HBSS and prepared as a 4% solution that may be stored at 4°C for 3 to 4 days. The medium from the cultures to be tested is removed, 1 ml of cold HBSS is added to each tube, and then 0.2 ml of a 0.4% solution of RBCs prepared in cold HBSS from the stock solution is added. The tubes are incubated first at 4°C for 30 min, then viewed, and then reincubated at room temperature for 30 min; influenza virus should hemadsorb at 4°C and room temperature, whereas PIVs hemadsorb only at 4°C. This is not an absolute phenomenon, but it is helpful. The temperature differences seen are associated with degrees of HAd and are not absolute reactions. Uninfected culture tubes of the same lot number as those inoculated should be hemadsorbed at the same time to control for the presence of hemadsorbing monkey viruses in the cell cultures. Human PIVs are more likely to hemadsorb in a diffuse pattern (Fig. 1b); monkey viruses often hemadsorb in clusters. Virus proteins may adhere to the glass, particularly on the edge of monolayers, resulting in RBCs adsorbing to glass in the absence of visible cellular material. If HAd is not seen, the RBCs may be poured off, and the cultures can be washed with HBSS, refed with SFM, and incubated until the next HAd attempt.

Alternatively, cells may be scraped from an inoculated tube 48 h or more after inoculation and prepared for immunofluorescence (IF) as described below. The cells are stained with a pool of MAbs to the PIVs or with MAbs to each type. The remaining inoculated cultures should be held in case the IF procedure is false negative or to identify a possible dual infection.

DETECTION OF ANTIGEN

IF

The most commonly used method of detecting PIV antigens directly in patient specimens or in infected cell cultures is direct or indirect IF (Fig. 2) (25, 27). MAbs to all four human PIVs may be purchased as individual reagents or as a PIV pool.

For direct examination of cells obtained from NP washes, specimens should be obtained as described above; the nasopharynx should be swabbed prior to washing to obtain increased numbers of NP cells. Cells in the specimen are washed with phosphate-buffered saline (PBS; pH 7.0) until mucus is virtually removed, which usually requires one to three washes. The cells are resuspended in 0.25 ml of PBS, and 10 to 20 μl of the cell suspension is applied to 8-mm-diameter wells circumscribed on eight-well enamel-coated glass slides (Cell-Line Associates, Newfield, N.J.). Wells are examined at ×100 magnification to assess the number of cells and the morphology of the cells present.

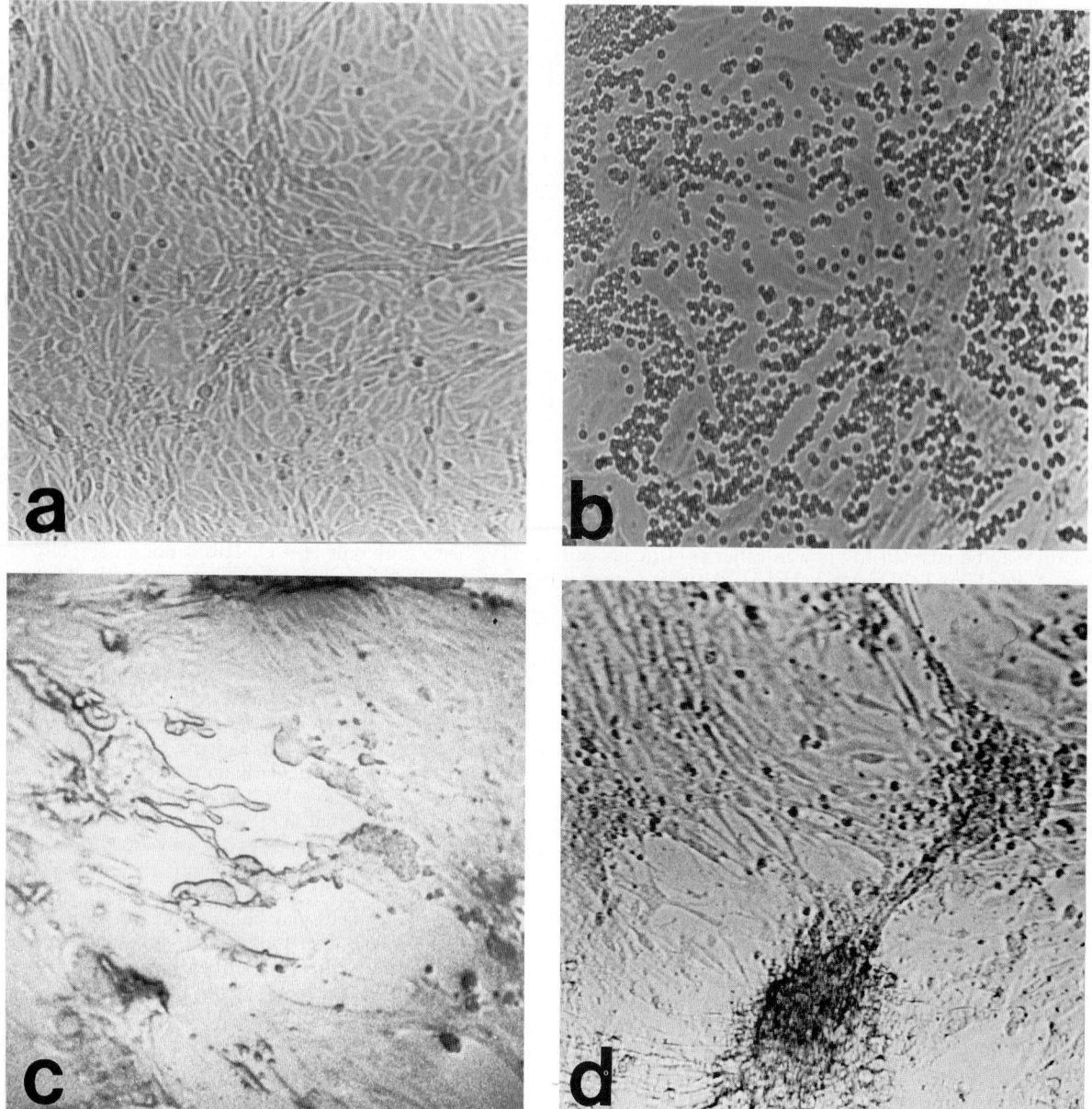

FIGURE 1 HAd of PIV-1 and CPE of PIV-2 and -3 in PRMK cell cultures. (a) Uninfected; (b) HAd of PIV-1; (c) CPE of PIV-2; (d) CPE of PIV-3.

Specimens yielding less than an average of 20 columnar epithelial cells per well should be rejected; squamous epithelial cells are unsatisfactory. The slides are air dried, fixed in cold acetone for 10 min, dipped three times in distilled H$_2$O for rinsing, and air dried before staining. Slides may be stored with a desiccating agent for several months at −70°C.

Cells sufficient to prepare 20 or more wells are usually obtained. The specimen can be comprehensively examined by staining three or four wells each with antibodies to the common respiratory viruses of the season, i.e., respiratory syncytial virus, influenza virus types A and B, adenovirus, and PIVs. MAbs to these viruses are the immune reagents of choice and are commercially available. A pool of MAbs to PIVs may be applied to the specimen; application of the

antibodies to four to six wells provides a good expectation of sensitive and specific results. The diagnosis of a PIV without knowing the type is sufficient for clinical purposes and may be made rapidly with less expense by using the pool; if desired, specific typing may be done later.

Shell Vial Assay

Centrifugation of a specimen onto a monolayer of susceptible cells and subsequent staining of the monolayer for viral antigen(s) provides a rapid and sensitive assay for diagnosis of PIVs (22). Susceptible cell lines growing on coverslips in shell vials may be purchased. Alternatively, cell cultures may be passaged in the laboratory onto coverslips in shell vials. When the monolayer is approximately 90% confluent, 0.2 ml of a specimen is inoculated into

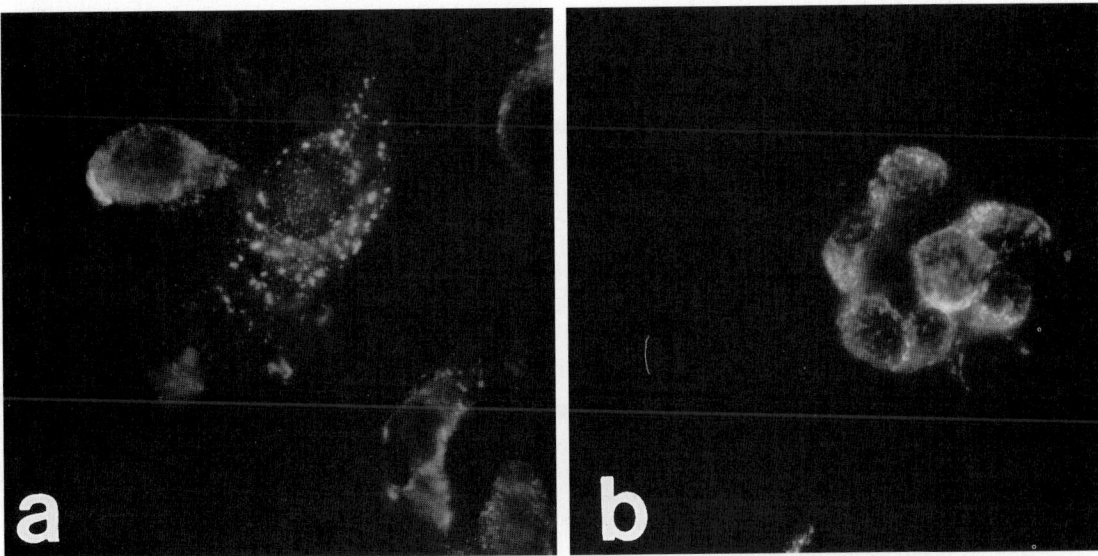

FIGURE 2   Indirect IF reactions utilizing a pool of MAbs to PIV-1, -2, and -3. (a) Infected cell culture; (b) antigen-positive cells from an NP wash.

duplicate vials containing 0.2 ml of medium. Virus and uninfected controls should be included in each centrifugation run. After the vials are centrifuged for 45 min at 700 × g at ambient temperature, the inoculum is aspirated and the cultures are overlaid with medium; SFM is not required. Following 2 days of incubation at 37°C, the coverslips are removed from the vials and prepared for IF as described above; additional cultures may be prepared for staining at 4 days after inoculation. The coverslips may be reacted with a pool of MAbs to the PIVs or with a MAb to one type of PIV. Coverslips are screened at a magnification of ×200, and positive cells are confirmed at a magnification of ×400.

EIA

PIV antigens may be identified in infected cell cultures or clinical specimens by enzyme immunoassay (EIA) (21), which is comparable in specificity to but slightly less sensitive than IF (12). Time-resolved fluoroimmunoassay may be more sensitive than EIA or IF (15). Commercial test kits are not available, but EIAs may be constructed from reagents purchased as single components or prepared locally. PIV antigens and control antigens may be obtained from infected cell cultures. Chapter 11 of this Manual and the publications cited therein should be referred to for preparation of EIAs.

DETECTION OF NUCLEIC ACID

Molecular methods have not been routinely applied to the diagnosis of PIVs but have been developed for research purposes. Diagnosis of PIVs with nucleic acid probes is not likely to be a practical improvement over the available rapid methods. PCR is the most promising molecular method that may be amenable to routine diagnosis. Diversity among isolates of PIV-3 was studied by using reverse transcription to make cDNA and PCR to amplify conserved sequences in the F protein gene; the virus isolates examined were grown in cell culture before the analysis (18). Reverse transcription PCR was also used, however, to detect PIV-3 in clinical specimens (17). RNA was ex-

tracted from respiratory specimens and reverse transcribed to make cDNA; PCR was used to amplify conserved sequences in the HN protein gene. The amplified sequences were identified by EIA using a non-isotope-labeled RNA probe. The method was able to identify PIV-3 in specimens from which virus could not be grown.

SEROLOGY

Antibody to the PIVs can be detected by complement fixation (CF), hemagglutination-inhibition (HI), IF, neutralization, and EIA. Some antigens are shared among the PIVs and between the PIVs and other paramyxoviruses, notably mumps virus. Cross-reactions are therefore common in serologic tests. Heterotypic responses may also occur in patients, and with the possibility of cross-reactions, these responses make a specific serologic diagnosis of a PIV difficult. Immunoglobulin M (IgM) antibodies may be detected by EIA or IF to each of the PIVs and generally show fewer heterologous responses than IgG antibodies (7, 24).

HI and CF tests have been the standard tests of comparison but are being replaced by EIAs, which are recommended for use in most laboratories. The EIA is generally considered more sensitive than HI or CF; the CF test, however, is more specific than HI or EIA. The neutralization test (8) is also very specific but is cumbersome to perform and is not amenable to routine use in most laboratories. Published experience with the IF-antibody test is sparse, but the procedure should be considered by laboratories with expertise in IF.

Blood samples are obtained and allowed to clot, and the sera are collected following centrifugation at low speed. Tests for IgG antibodies should be performed simultaneously on paired sera taken during the acute phase of illness and at least 2 weeks later. If IgM antibody is to be assayed, IgG should be removed from the serum by adsorption onto latex beads coated with anti-IgG or by column chromatography; kits are available commercially.

PIV antigens for use in CF or HI tests may be purchased. Alternatively, antigens may be prepared from infected cell cultures. These antigens may also be used in locally constructed EIAs, as commercial EIA kits are not available. A cell culture infected with a PIV may be prepared as antigen for IF-antibody tests by conventional procedures (7, 25).

EIA

Chapter 11 of this Manual and the references cited therein should be referred to for construction of EIAs. Soluble antigen is prepared in alkaline buffer and incubated overnight at 4°C in wells of 96-well microdilution plates; comparable wells are prepared with control antigen. The wells are washed with PBS containing 0.5% bovine serum albumin (BSA) and 0.05% Tween 20. Sera are serially diluted in the PBS-BSA-Tween 20 buffer and added to PIV- and control antigen-treated wells. Following incubation for 1 h at 37°C, the wells are washed three times and an anti-human IgG-enzyme conjugate (peroxidase is recommended) prepared in the PBS-BSA-Tween 20 buffer is added; controls include positive and negative sera, wells without a test serum but with conjugate, and wells without a test serum or conjugate. The plates are incubated an additional hour at 37°C and washed. An appropriate substrate is added, and the quantity of the developed color is measured spectrophotometrically. Commonly, patient sera are reported as positive when their absorbance values are a minimum of 2 standard deviations above a control value or after the absorbance values are plotted on a dose-response curve obtained with a reference pool of sera. A critical ratio between the absorbance values of an acute-phase serum sample and a convalescent-phase serum sample is sometimes obtained to determine whether the difference is significant.

HI

Antigens for the HI test may be purchased or prepared from infected cell cultures. Antigen must be carefully titrated with guinea pig RBCs for hemagglutination before use in the test. The highest dilution of antigen that shows partial agglutination of the RBCs is the end point. The end point dilution is divided by 8 to arrive at the dilution of antigen required to provide 4 U of hemagglutinin.

Before testing, sera must be treated to remove nonspecific inhibitors and agglutinins. Equal volumes of receptor-destroyer enzyme and serum (0.1 ml is recommended) are incubated together overnight at 37°C. After inactivation at 56°C for 30 min, the serum is incubated with 0.2 ml of 15% guinea pig RBCs for 1 h at 4°C; the RBCs are removed by centrifugation. The serum dilution at this point is 1:4.

Serial twofold dilutions of the treated serum are made in microtiter wells, with PBS as a diluent. Four units of antigen are added to each well and the reactants are mixed. Antigen, antibody, RBCs, and serum controls must be included. In each test, twofold dilutions of the antigen should be made such that 4, 2, 1, 0.5, and 0.25 U of antigen are evaluated for hemagglutination activity. A reference serum of known antibody titer is tested to control for specificity and sensitivity, an RBC control consists of RBCs incubated only with PBS, and serum controls consisting of 1:8 dilutions of each serum tested are incubated only with PBS.

The reaction mixtures are incubated for 1 h at room temperature, and then 0.4% guinea pig RBCs are added to each well. The mixture is incubated again at room temperature until RBC buttons are distinct in the RBC and serum control wells. Hemolysis or agglutination indicates nonspecific reactions associated with the RBCs and/or the serum.

The 4- and 2-U antigen control wells should show complete agglutination, the 1-U well should show partial agglutination, and the 0.5- and 0.25-U wells should not show agglutination. The end point is the highest dilution of a serum that completely inhibits agglutination, as evidenced by a complete RBC button in the well.

EVALUATION OF TEST RESULTS

The isolation of virus or the identification of viral antigen in a clinical specimen is the most valuable information to be conveyed to a clinician. PIVs are rarely present in healthy individuals, and their presence in the context of disease is diagnostic. A policy of immediately reporting isolation or positive antigen detection by telephone is recommended. The quality of a specimen should also be a consideration in reporting results, particularly negative results. A negative result obtained on a poor specimen may have a low negative predictive value. Such instances should be conveyed to the clinician.

The length of time required to identify replication of PIV in cell cultures usually compromises the usefulness of isolation in patient care. Efforts should therefore be made to use rapid techniques to identify PIV antigen directly in clinical specimens or in infected cell cultures; direct IF and the shell vial method are the recommended rapid tests. The identification of a PIV without designation of type is sufficient, as specific therapies are not available.

Antibody determinations are used primarily to make a retrospective diagnosis or to obtain seroepidemiologic data on the PIVs in a community. Since reinfection is common, the presence of antibody per se is of little value in evaluating protective immune status. A fourfold or greater increase in the antibody titer of the convalescent-phase serum over that of the acute-phase serum is diagnostic of recent infection. A fourfold diminution of antibody titer or the presence of IgG antibody in a single serum sample does not have diagnostic significance. The possible presence of maternal antibody and the immature status of the immune system in infants less than 6 months old make serology unreliable for this age group. Heterologous antibody responses in patients make serologic typing of PIVs unreliable. Although IgM antibody shows less heterologous response than IgG antibody, experiences with IgM tests vary considerably and do not support the use of the test for obtaining a reliable diagnosis. Serology, therefore, is of negligible value in obtaining a diagnosis of acute infection.

REFERENCES

1. **Ansari, S. A., S. Springthorpe, S. A. Sattar, S. Rivard, and M. Rahman.** 1991. Potential role of hands in the spread of respiratory viral infections: studies with human parainfluenza virus 3 and rhinovirus 14. *J. Clin. Microbiol.* **29:**2115–2119.
2. **Brady, M. T., J. Evans, and J. Cuartas.** 1990. Survival and disinfection of parainfluenzaviruses on environmental surfaces. *Am. J. Infect. Control* **18:**18–23.
3. **Chanock, R. M., J. A. Bell, and R. H. Parrott.** 1961. Natural history of parainfluenza virus infection, p. 126–137. *In* M. Pollard (ed.), *Perspectives in Virology*, vol. 2. Burgess Publishing Co., Minneapolis.
4. **Chanock, R. M., and K. McIntosh.** 1990. Parainfluenza viruses, p. 963–988. *In* B. N. Fields (ed.), *Virology*. Raven Press, New York.
5. **Chanock, R. M., and P. H. Parrott.** 1965. Acute respiratory disease in infancy and childhood: present understanding and prospects for prevention. *Pediatrics* **36:**21–39.
6. **Chanock, R. M., R. H. Parrott, K. M. Johnson, A. Z.**

Kapikian, and J. A. Bell. 1963. Myxoviruses: parainfluenza. *Am. Rev. Respir. Dis.* **88:**152–166.

7. Fedova, D., J. Novotny, and I. Kubinova. 1992. Serological diagnosis of parainfluenza virus infections: verification of the sensitivity and specificity of the haemagglutination-inhibition (HI), complement fixation (CF), immunofluorescence (IFA) tests and enzyme immunoassay (ELISA). *Acta Virol.* **36:**304–312.

8. Frank, A. L., J. Puck, B. J. Hughes, and T. R. Cate. 1980. Microneutralization test for influenza A and B and parainfluenza 1 and 2 viruses that uses continuous cell lines and fresh serum enhancement. *J. Clin. Microbiol.* **12:**426–432.

9. Glezen, W. P., and F. W. Denny. 1973. Epidemiology of acute lower respiratory disease in children. *N. Engl. J. Med.* **288:**498–505.

10. Glezen, W. P., A. L. Frank, L. H. Taber, and J. A. Kasel. 1984. Parainfluenza virus type 3: seasonality and risk of infection and reinfection in young children. *J. Infect. Dis.* **150:** 851–857.

11. Glezen, W. P., F. A. Loda, and F. W. Denny. 1982. Parainfluenza viruses, p. 441–454. *In* A. S. Evans (ed.), *Viral Infections of Humans. Epidemiology and Control.* Plenum, New York.

12. Grandien, M., C.-A. Petterson, P. S. Gardner, A. Linde, and A. Stanton. 1985. Rapid viral diagnosis of acute respiratory infections: comparison of enzyme-linked immunosorbent assay and the immunofluorescent technique for detection of viral antigens in nasopharyngeal secretions. *J. Clin. Microbiol.* **22:** 757–760.

13. Gross, P. A., R. H. Green, and M. G. M. Curnen. 1973. Persistent infection with parainfluenza type 3 virus in man. *Am. Rev. Respir. Dis.* **108:**894–898.

14. Hall, C. B., and R. G. Douglas, Jr. 1975. Clinically useful method for the isolation of respiratory syncytial virus. *J. Infect. Dis.* **131:**1–5.

15. Hierholzer, J. C., P. G. Bingham, R. A. Coombs, K. H. Johansson, L. J. Anderson, and P. E. Halonen. 1989. Comparison of monoclonal antibody time-resolved fluoroimmunoassay with monoclonal antibody capture-biotinylated detector enzyme immunoassay for respiratory syncytial virus and parainfluenza virus antigen detection. *J. Clin. Microbiol.* **27:**1243–1249.

16. Hierholzer, J. C., E. Castells, G. G. Banks, J. A. Bryan, and C. T. McEwen. 1993. Sensitivity of NCI-H292 human lung mucoepidermoid cells for respiratory and other human viruses. *J. Clin. Microbiol.* **31:**1504–1510.

17. Karron, R. A., J. L. Froehlich, L. Bobo, R. B. Belshe, and R. H. Yolken. 1994. Rapid detection of parainfluenza virus type 3 RNA in respiratory specimens: use of reverse transcription-PCR-enzyme immunoassay. *J. Clin. Microbiol.* **32:**484–488.

18. Karron, R. A., K. L. O'Brien, J. L. Froehlich, and V. A. Brown. 1993. Molecular epidemiology of a parainfluenza type 3 virus outbreak on a pediatric ward. *J. Infect. Dis.* **167:**1441–1445.

19. Minnich, L. L., and C. G. Ray. 1987. Early testing of cell cultures for detection of hemadsorbing viruses. *J. Clin. Microbiol.* **25:**421–422.

20. Parrott, R. H., A. J. Vargosko, H. W. Kim, J. A. Bell, and R. M. Chanock. 1962. Acute respiratory diseases of viral etiology. III. Myxoviruses: parainfluenza. *Am. J. Public Health* **52:**907–917.

21. Sarkkinen, H. K., P. E. Halonen, and A. A. Salmi. 1981. Type-specific detection of parainfluenza viruses by enzyme immunoassay and radioimmunoassay in nasopharyngeal specimens of patients with acute respiratory disease. *J. Gen. Virol.* **56:**49–58.

22. Schirm, J., D. S. Luijt, G. W. Pastoor, J. M. Mandema, and F. P. Schroder. 1992. Rapid detection of respiratory viruses using mixtures of monoclonal antibodies on shell vial cultures. *J. Med. Virol.* **38:**147–151.

23. Vreede, R. W., H. Schellekens, and M. Zuijderwijk. 1992. Isolation of parainfluenza virus type 3 from cerebrospinal fluid. *J. Infect. Dis.* **165:**1166.

24. Vuorinen, T., and O. Meurman. 1989. Enzyme immunoassays for detection of IgG and IgM antibodies to parainfluenza types 1, 2 and 3. *J. Virol. Methods* **23:**63–70.

25. Waner, J. L., N. J. Whitehurst, T. Downs, and D. G. Graves. 1985. Production of monoclonal antibodies against parainfluenza 3 virus and their use in diagnosis by immunofluorescence. *J. Clin. Microbiol.* **22:**535–538.

26. Wenzel, R. P., D. P. McCormick, and W. E. Beam, Jr. 1972. Parainfluenza pneumonia in adults. *JAMA* **221:**294–295.

27. Wong, D. T., R. C. Welliver, K. R. Riddlesberger, M. S. Sun, and P. L. Ogra. 1982. Rapid diagnosis of parainfluenza virus infection in children. *J. Clin. Microbiol.* **16:**164–167.

Respiratory Syncytial Virus

DEBRA A. TRISTRAM AND ROBERT C. WELLIVER

78

CLINICAL BACKGROUND

Epidemiology

Respiratory syncytial virus (RSV) is the most common viral agent causing lower respiratory tract illness in infancy. Outbreaks of RSV infections occur worldwide, with striking epidemics noted in temperate climates in the winter and early spring. In the tropics, epidemics occur during the rainy seasons. Outbreaks usually span approximately 2 to 5 months. Within this period, the majority of cases usually occur within a 1- to 2-month peak (39). Although several investigators have attempted to delineate the climatic conditions responsible for the appearance of RSV, its persistence for several months, and its subsequent disappearance for an approximately equal number of months, no one has been able to clearly establish a link between climate and epidemics (19). While all age groups are susceptible to RSV infections, the majority of hospitalizations for RSV disease occur in 2- to 6-month-old children. Admission rates for bronchiolitis and pneumonia can reach as high as 24/1,000 in young infants experiencing primary RSV infection. Males are slightly more likely to experience severe RSV disease, as are infants from industrialized areas and infants exposed to parental cigarette smoke (43). Both a greater number of siblings and attendance at day-care also increase the risk of infection with RSV (27).

By the first year of life, 50% of all infants experience a primary RSV infection; by 2 years, nearly all have contracted RSV disease (5, 59). Reinfections are common, despite the presence of RSV-specific local and systemic antibodies and neutralizing antibody (21). Nosocomial spread of RSV on pediatric wards is an annual problem encountered by hospital infection control personnel (13).

Disease Manifestations

Common syndromes due to RSV include bronchiolitis and pneumonia. The former is a wheezing-associated illness preceded by 3 to 5 days of upper respiratory tract symptoms. Wheezing and thoracic hyperinflation are prominent. Patients with pneumonia also present after several days of upper respiratory tract symptoms, and crackles are heard on auscultation of the chest. Frequently, the two syndromes overlap. For healthy patients, although illness from RSV disease can be severe, recovery is usual. Among infants and children hospitalized with lower respiratory tract disease

due to RSV, mortality rates are relatively low, i.e., usually less than 1% in otherwise healthy hosts (15). However, among patients with respiratory or cardiac compromise or immune dysfunction, mortality rates as high as 37% have been reported, although 3 to 5% is probably a more accurate estimate (50). Significant morbidity and mortality may occur in RSV infection of the elderly. Outbreaks producing lower respiratory tract disease in nursing homes (16) and institutions for the mentally retarded (17) have been reported.

Another manifestation of RSV disease in infants is apnea, with or without associated respiratory symptomatology. Various studies since the early 1960s have determined the incidence of RSV-associated apnea to be between 16 and 20% (11). Infants with significant risk for apnea include premature infants, young infants (usually <6 weeks old), and infants with a history of apnea of prematurity.

Transmission

Virus transmission has been documented to occur most commonly through contact with infected droplets, most often in fomites. RSV can survive for up to 6 h on environmental surfaces and is most infectious when applied directly to the mucosal surface of the eye or nose by infected hands or objects. Airborne transmission, except by close, direct contact with an infected person, is uncommon. Schools or day-care provide ideal settings for the spread of virus to susceptible individuals.

DESCRIPTION OF THE VIRUS

Morphology

RSV has been assigned to the family *Paramyxoviridae*, although morphologic and antigenic differences distinguish RSV from other members of the paramyxovirus family. RSV has a smaller nucleocapsid (13- to 14-nm diameter) and lacks neuraminidase and a hemagglutinin (53). Thus, RSV has been assigned a separate genus, *Pneumovirus*, within the family *Paramyxoviridae*. It shares this genus with the pneumonia virus of mice and with bovine, ovine, and caprine RSV. Although there are no further members of this genus as yet, antibody studies of other species have demonstrated naturally occurring RSV antibody. This suggests that there may be other viruses related to human RSV and capable of producing disease in other species that have not yet been recovered and identified.

RSV is a small, single-stranded, negative-sense RNA virus. Its genome is linear, with a molecular mass of about 5×10^6 Da (44). Single RSV particles are pleomorphic, ranging in size from 100 to 350 nm (6). Each particle is bounded by a cellular membrane that has 12- to 15-nm spikes composed of F and G glycoproteins on the outer surface at regular intervals. The spherical virion is composed of a lipid bilayer membrane containing an RNA nucleocapsid. The RNA is a nonsegmented helical structure with a periodicity of 65 to 70 nm in a herringbone pattern and appears to be similar to the nucleoproteins of other paramyxoviruses when viewed by negative staining electron microscopy (12).

Biochemistry

Analysis of the nucleotide sequence of RSV has allowed the accurate identification of RSV proteins encoded by the genome. At least 10 specific protein products have been identified thus far (12, 30). The structural proteins are N, P, and L in the nucleocapsid; F and G on the surface of the viral envelope; and M and M2 in the viral envelope. SH (formerly 1A) may be a third integral membrane protein, but it is also expressed on the surfaces of RSV-infected cells. Two nonstructural proteins, formerly designated 1B and 1C, are now called NS1 and NS2.

The first two protein products encoded by the RSV genome reading from the 3' to the 5' end are NS1 and NS2. NS1 is slightly acidic, while NS2 is basic (65).

The N protein is intermediate in size and serves as the major structural protein for the nucleocapsid, while P and L are most likely involved in transcription and replication (44). Monoclonal antibodies raised against N stain intracytoplasmic inclusion bodies in an immunofluorescence assay (68). P is heavily phosphorylated, unlike the other RSV proteins, and is relatively acidic (57). In an immunofluorescence assay, monoclonal antibodies to P also stain intracytoplasmic inclusion bodies (68). The L protein is very large (250 kDa) and relatively hydrophobic. Because of its size and its location in the nucleocapsid, L is believed to be the viral RNA-dependent RNA polymerase (44).

The viral envelope contains several proteins: matrix (M), a 22-kDa protein (M2), and the surface glycoproteins F and G. The M protein is present in detergent-solubilized cores but not in nucleocapsids, suggesting that it is similar to other matrix proteins found in enveloped RNA viruses (44). It is most likely hydrophobic and basic, as suggested by its deduced amino acid sequence (9, 57). The exact location and function of the second putative matrix protein, M2, are not known. M2 is highly basic and hydrophilic and is expressed on the surfaces of infected cells, as determined by immunofluorescence studies (55). F and G are two large glycosylated proteins found on the surface of the viral envelope; they are important immunogens during RSV infection.

The G glycoprotein is highly glycosylated; the protein component is relatively small (32 kDa) compared to the mature, fully processed glycoprotein (85 kDa) (72). It has an unusually high content of serine and threonine, to which the O-linked carbohydrate side chains are attached (72). The function of G protein seems to be host cell adsorption, analogous to the hemagglutinin protein of parainfluenza viruses.

RSV strains can be separated antigenically into two subgroups, A and B. The G glycoprotein seems to have the greatest variation in amino acid sequences in the two RSV subgroups. Comparisons between the G ectodomains of two subgroup A viruses showed a 94% homology, while subgroup B (strain 18537) showed 47% divergence from the subgroup A reference strain (33). The central region, containing cysteine residues and flanking amino acid sequences, was the only region conserved among the strains. In contrast, the F glycoprotein is highly conserved, and antibody to F protein cross-reacts with both A- and B-group F protein (32).

The F glycoprotein has been identified as the fusion protein by studies with monoclonal antibodies (68). F glycoprotein is one of the major components of the projections on the outer surface of the virion and one of the major immunogens of RSV (22). After glycosylation, the F protein is proteolytically cleaved in the infected cell prior to virion assembly. The two subunits F1 and F2 are linked by a disulfide bridge (67). Although direct evidence is not available, it is possible that, similar to what occurs in other paramyxoviruses, posttranslational enzymatic cleavage is required for the formation of infectious viral particles (14).

COLLECTION AND STORAGE OF SPECIMENS

Samples of nasopharyngeal secretions recovered by nasal washing or direct aspiration are two or three times more likely to yield RSV in cell culture than specimens obtained by swabs (45). Nasopharyngeal aspirates apparently result in greater recovery of epithelial cells than washes or swabs, so aspirates should be more effective for immunofluorescence assays, in which antigen is detected on the surface of exfoliated cells. Aspirates are also superior to swabs in enzyme immunoassays (EIAs) (1).

Nasal wash specimens are obtained by placing a 1-oz (28.350-g) tapered rubber bulb syringe containing 5 to 7 ml of sterile saline solution into and occluding the nostril and then rapidly squeezing and releasing the bulb. Fluid recovered in the bulb is emptied into a container containing transport medium. Nasopharyngeal aspirates are best obtained by attaching a size 5 or 8 French polyethylene pediatric feeding tube to a DeLee collection trap, passing the tube into the nasopharynx, and applying suction (approximately 15 to 25 lb/in^2) by wall suction or with a manual suction pump. Optimal results are achieved by suctioning through both nostrils. Recovery of secretions is greater if transport medium is rinsed through the feeding tube into the specimen trap (1, 36). Swab specimens should be collected by carefully and thoroughly swabbing the nasopharynx and placing the swab into a vial containing transport medium.

Specimens need not be placed directly into cell culture at the bedside but should be placed on wet ice before transport to the laboratory to maintain optimal recovery of RSV in cell culture. The transport medium should include a physiologic salt solution and a protein stabilizer such as 0.5% bovine serum albumin or 0.5% gelatin. Since the use of Hanks' balanced salt solution has been reported to decrease the absorbance in an RSV EIA, this solution may not be optimal (25), but several others in common use are acceptable.

RSV is rapidly destroyed at 55°C, and as much as 90% of cell culture infectivity is lost when virus is kept at 37°C for 24 h. Mailing specimens to the laboratory or transporting them at room temperatures can be expected to result in substantial loss of cell culture infectivity. Therefore, all specimens to be analyzed by cell culture or immunofluorescence (in which cellular viability must be maintained) should be transported at refrigerator temperature or on wet ice. Samples for EIAs may be transported at ambient temperatures without loss of reactivity. Approximately 50% of infectivity is lost after a single freeze-thaw cycle. Therefore, freezing should be avoided, but if samples must be stored at refrigerator temperatures for more than 1 day, rapid freezing and holding the virus at −70°C will maintain the titer of virus for at least several months (24). The addition of sucrose or glycerin to the storage medium may enhance viral survival while the virus is frozen. RSV is denatured rapidly at low pH and survives optimally at pH 7.5 (24).

DIRECT EXAMINATION

Immunofluorescence

Immunofluorescence techniques to detect RSV antigens were first applied to cells maintained in vitro and infected with the virus. The characteristic fluorescence in cytoplasm was observed as early as the second day following inoculation of cell cultures. More recently, indirect immunofluorescence techniques have been applied to exfoliated nasopharyngeal epithelial cells present in samples of secretions obtained from patients with RSV infection (20). Several acceptable commercial kits that use both direct immunofluorescence and indirect immunofluorescence techniques for the detection of RSV antigen in respiratory secretions have been developed (8, 10). One review (38) discussed 12 studies in which direct fluorescent-antibody tests were compared with culture and 19 studies in which indirect fluorescent-antibody tests were compared with culture for the detection of RSV infection. With either technique, the sensitivity of the immunofluorescence assay in comparison to culture was usually above 80% and was as high as 97%. Specificity was generally greater than 90%. It should be noted that for all antigen detection techniques (including the EIAs described below), the actual sensitivity is probably greater than that of cell culture. Specimens exposed to temperatures that render RSV unrecoverable from cell culture may still be positive by antigen detection techniques. Similarly, samples obtained later in the course of clinical illness may not contain recoverable virus but may still be positive by antigen detection techniques (36). Indirect immunofluorescence techniques may have a slightly greater sensitivity for detecting RSV (usually ≥95%) than direct fluorescent-antibody tests do (38). Specificities are also generally somewhat higher with indirect fluorescent-antibody tests. Numerous variables can affect the results of either assay, including method of collection of secretions, time during the course of illness when samples were obtained, and use of fresh versus frozen samples. Therefore, comparing separate studies that use different methods is difficult. In any case, reliable results with immunofluorescence tests are available within a few hours of obtaining specimens, and many laboratories have replaced cell culture with antigen detection techniques for establishing a diagnosis of RSV infection.

The preparation of samples for immunofluorescence testing is critical for obtaining optimal results. Samples of respiratory secretions obtained either by washing or by direct aspiration are suspended in growth medium. Excess mucus is removed by 1 to 2 min of vortexing. Particularly tenacious mucus can be broken up by the addition of an equal volume of dimercaptoethanol or N-acetylcysteine before vortexing (47). Several cycles of centrifugation at 300 to 500 × g may be useful in removing mucus. The sediment containing exfoliated cells is then spotted onto microscope slides, air dried, and fixed for 10 min in cold acetone. Teflon-coated slides containing 13-mm-diameter wells have been specifically produced for this purpose, although standard microscope slides are sufficient. Anti-RSV antibodies and fluorescein-conjugated anti-bovine immunoglobulins are available either in commercially available kits or individually. The first antibody should be incubated with cells on slides for 15 to 30 min and then adequately washed. A similar incubation period should be used for fluorescein-conjugated antibodies, again with careful rinsing. Slides are then examined under ×400 magnification with a fluorescence microscope (8, 10, 61).

A satisfactory specimen should have more than 100 cells on the slide. Specific RSV fluorescence is within the cell and is characterized by granular fluorescence on the cell surface or larger inclusionlike fluorescence within the cytoplasm. Most newer commercial kits involve monoclonal antibodies against surface glycoproteins and the nucleocapsid proteins, which are responsible for membrane and cytoplasmic fluorescence, respectively. Polyclonal sera have also been used in commercial kits.

EIAs

Over the past few years, EIAs have probably become the most frequently used method of antigen detection for RSV. A review of 20 studies using EIAs to detect RSV antigen showed sensitivities of ≥90% in 8 of the 20 studies and specificities of ≥90% in 13 of the studies (38). While some investigators have found immunofluorescence assays to be slightly more accurate than EIAs (56), the observed differences are generally fairly minor, and EIAs have other advantages in comparison to immunofluorescence techniques. First, intact respiratory tract cells are not required for EIA, and EIAs are often positive when suboptimal handling of specimens has resulted in the inactivation of culturable virus (18, 45). Therefore, specimen handling requirements are not as strict as those associated with culture or immunofluorescence assays. Other advantages include the technical simplicity of the EIAs, objective end points, the potential for automation, and assay times as short as 15 to 20 min (23).

EIAs generally use plastic microtiter trays as the solid phase. More recently, EIAs that use a membrane filter to absorb the viral antigen present in clinical specimens have been employed. In these assays, a colored reaction product appears bound to the membrane. These tests have internal controls that must be visible for the test to be valid (54, 61, 63). In our experience and in the experience of at least some others, these membrane immunoassays have reduced accuracy in comparison to the immunofluorescence assay and EIAs noted above (54). Freezing and thawing of specimens appear to increase the sensitivity of some of the membrane EIAs. These procedures, however, are time-consuming and tend to override the advantage of a rapid test.

As with immunofluorescence assays, adequate preparation of clinical specimens is essential for optimum results in an EIA. Mucus should be broken up or removed by repeated pipetting, sonication, and vortexing of specimens or by the addition of n-acetylcysteine to specimens. Sonication and treatment with detergents have been reported to increase signals in EIA, presumably by releasing antigens from the cells (26, 56). Since some antigen is probably present in the fluid phase of secretions, whole secretions can be expected to give higher EIA signals than either washed cells alone or secretions from which the cells have been removed by centrifugation.

False-positive results are a particular problem in EIAs. These results should be excluded by using appropriate blocking antibodies if several clinical specimens are positive by EIA but negative by cell culture (8, 42, 60). Both immunofluorescence assays and EIAs can be applied to cell cultures, often revealing the presence of viral antigens before cytopathic effects (CPE) are present (8).

CULTIVATION OF THE VIRUS

Recovery of Cell Culture

Standard cell culture infectivity is a useful method for recovering RSV from clinical specimens. Specimens obtained by direct aspiration into polyethylene catheters or recovered by nasal washes are superior to swabs and, interestingly, to aspirates of tracheal secretions (62). RSV grows best in human epithelial cell lines such as HEp-2 and HeLa cells. The A549 human epithelial cell line seems somewhat less sensitive for recovery of RSV. A variety of primary monkey kidney and human fibroblast cell lines are also useful for recovery of RSV but are not as sensitive as human epithelial lines. Nevertheless, it has been estimated that up to 20% of RSV strains will not be recovered if human epithelial cell lines alone are used (4), although this is not our experience. A combination of human epithelial cell lines, primary monkey kidney cell lines, and human fibroblasts seems optimal for virus recovery. There is considerable variation in sensitivity among lots of the same cell line and a decrease in sensitivity of cell lines as they are maintained by multiple passages (4, 62). Sensitivity may also differ for the same cells grown in different media (35).

Several different culture media are adequate for the growth of RSV. The addition of glutamine appears to be necessary for optimal growth and development of CPE (41). Growth of RSV is enhanced if monolayers of cells are inoculated when they are approximately 50 to 75% confluent (51). Growth at 36°C is equivalent to growth at 33°C (4). CPE usually appear between days 3 and 7 in culture. Large syncytia composed of fused cells with eosinophilic granules within the cytoplasm are characteristic (Fig. 1). However, growth may occur without the development of syncytia, and as many as 26% of isolates of RSV will be recovered only by repeated passage of previously inoculated cells. This is best accomplished by scraping monolayers and adsorbing the recovered cells on a fresh monolayer for 1 h before adding maintenance medium. Subtype B strains of RSV (see below) may develop CPE more slowly than subtype A strains.

Continued passage of large inocula of RSV or passage of temperature-sensitive mutants at nonpermissive temperatures may result in persistent infection of cell lines. CPE is not obvious, and the amount of cell-free virus is diminished (52).

The shell vial technique has been used with several viruses, including RSV (58). In Smith et al.'s study (58), 0.2-ml aliquots of clinical specimens were added directly to shell vials and then centrifuged at 700 × g and 32°C for 60 min. Overall, the recovery rates of RSV in shell vials are similar to those of conventional tube cultures. However, of 46 specimens that were eventually positive for CPE, 43 were already positive at 16 h and the other 3 were positive at 40 h by the shell vial assay. In conventional tube cul-

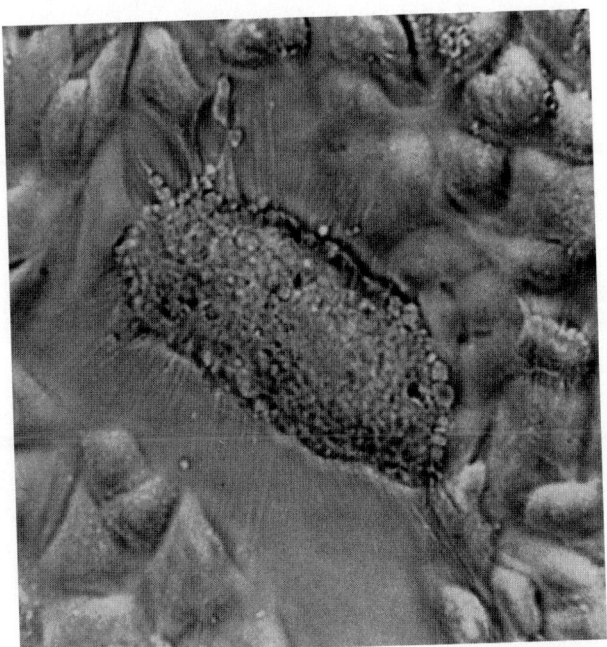

FIGURE 1 Syncytium formation by RSV in human lung epithelial cells.

tures, CPE was first observable at day 2 at the earliest, with a median of 4.5 days. Of 147 clinical specimens that were originally negative for RSV by direct immunofluorescence, 6 were positive after 16 h of incubation in shell vials. The shell vial technique identified RSV in 16 of 17 specimens that were positive by EIA and in an additional 3 specimens that were EIA negative.

Use of shell vials appears to slightly increase the sensitivity of culture methods for detection of RSV and greatly increases the speed at which isolates are detected. However, the increased time and effort required for centrifugation of specimens preclude the use of shell vials for screening large numbers of clinical specimens for RSV.

It is arguable whether antigen detection techniques such as immunofluorescence and EIA should replace cell culture as the standard against which other methods of detection should be tested. For the single purpose of comparison with methods that will be developed in the future, the immunofluorescence assay, EIA, and culture are reasonably similar in sensitivity and specificity, while antigen detection techniques have a considerable advantage in being less time-consuming. However, laboratories interested in detecting any viral pathogen in a given specimen (as opposed to simply ruling in or ruling out RSV) should continue to use cell culture at least as an adjunct to RSV, since viruses other than RSV may be recovered from 5 to 25% of specimens submitted for diagnosis of RSV infection (7).

Subtyping of Viruses

In the past, all RSV strains have been considered to be of a single serotype in that human neutralizing antisera recognize all strains with relatively equal efficiency. Nevertheless, different strains may have detectable differences in the antigenic makeups of individual viral proteins. Application of panels of monoclonal antibodies to different RSV strains allows classification of these strains into two subtypes, based largely on differences in the G glycoprotein (3). Strains of

RSV can be broadly divided into subgroup A (characterized by the Long and A2 strains) and subgroup B (characterized by strain 18537). Nucleotide and amino acid sequencing analyses indicate that the F glycoproteins of each subgroup are 53% antigenically related and have 89% amino acid sequence homology (32). In contrast, the G glycoproteins of group A and group B strains are 5% antigenically related and have 53% amino acid sequence homology (34). Retrospective analyses have indicated that these strains have existed for decades and that they circulate concurrently during RSV epidemics in all areas of the world. Infections due to one subtype or the other can cluster geographically or temporally.

The significance of this antigenic variation remains unclear. Infection with a subgroup A virus tends to induce good cross-reactive antibody responses against subgroup B viruses, while primary infection with a subgroup B virus induces lesser cross-reactive responses to subtype A strains. However, once children are primed by a group A infection, reinfection with either virus induces heterologous rises in neutralizing antibody (48). While no definitive study has been completed, there is currently no indication that reinfection with a heterologous strain is accompanied by any greater severity of illness than reinfection with a homologous strain (49). While epidemiologically important, the identification of subtypes of RSV has not yet entirely explained the frequency with which RSV reinfections occur.

IDENTIFICATION OF ISOLATES

The appearance of characteristic syncytia in cell culture (Fig. 1) is, in many laboratories, considered adequate to establish the presence of RSV. Parainfluenza or measles virus would be expected to produce syncytia on the HEp-2, A549, or Vero cell lines used for RSV only very rarely. However, while ours may not be a general practice, our virology laboratory routinely performs immunofluorescence assays on all cell cultures showing CPE (an educated guess is made as to which virus is present) to confirm the identity of isolates.

SEROLOGIC METHODS

Neutralization Tests for RSV Antibody

Standard neutralization assays are useful in the diagnosis of RSV infection. Neutralization assays performed in test tubes are somewhat more laborious and less precise than plaque reduction or microneutralization assays performed in plastic microtiter plates. A microneutralization assay performed in 96-well microtiter trays is probably the most widely used neutralization assay currently (2). It may be necessary to carry out neutralization assays with both subtype A and subtype B strains, since the use of subtype B strains may give slightly lower neutralizing titers than subtype A strains with the same antisera.

Neutralization antibody titers usually increase by 10 days after the onset of acute illness, reach a peak at 20 to 30 days, and decline substantially over the following year. Reinfections result in a more rapid increase in titer and a higher peak titer. However, up to 40% of subjects with documented RSV infection, particularly infants, do not develop increases in neutralizing titer following infection (16). The addition of complement to assay wells may enhance the observed titers (37).

Immunofluorescence Antibody Testing

Immunofluorescence techniques using commercially available antisera are useful for the diagnosis of RSV infection in all age groups (66, 69). Responses in the immunoglobulin M (IgM), IgG, and IgA isotypes are readily detectable, both in serum and in nasopharyngeal secretions.

In primary infection, IgM responses are detectable 5 to 9 days after the onset of symptoms and persist for several weeks. With secondary infection in childhood, responses are of a somewhat greater magnitude but do not appear to occur earlier then 4 days after the onset of symptoms. In adults, who probably have had multiple infections, IgM is first detectable at about the same time and also persists for several weeks. Significant increases in RSV IgG antibody occur 2 to 4 weeks after the onset of symptoms due to primary infection but are greatly accelerated following secondary infection in infancy. Levels of serum IgA antibody remain relatively low following primary infection but are also enhanced following secondary infection. In one small study, significant increases in RSV IgM antibody were noted in 13% of eight infants less than 3 months of age with RSV infection and in 50 and 71% of infants 3 to 6 months and 6 to 12 months of age, respectively, at the time of RSV infection (69).

False-positive IgM responses may be present in the first month of life as a result of antibody formation against maternal RSV-specific IgG present in the serum of the infant. These complexes bind to RSV antigen used as targets in either EIA or immunofluorescence assays, but they also react with anti-IgM antibodies.

RSV infection can also be documented by significant increases in RSV-specific IgG antibody in immunofluorescence assays. Fourfold or greater increases may occur more commonly with primary infection in infants more than 6 months of age (69).

Titers obtained by immunofluorescence assays for each of the isotypes correlate well with neutralizing-antibody responses. The peak in neutralizing-antibody response correlates most closely with the IgG response detected by immunofluorescence but is attained earlier. Neutralizing titers, especially those measured with the addition of complement, are appreciably higher than those measured by immunofluorescence (37). Titers detected by immunofluorescence also correlate well with complement fixation titers but appear earlier and also have the advantage of making diagnosis of infection with a single serum specimen possible by identification of an IgM-specific response (46).

Antibody EIAs

Several investigators have used EIAs to detect antibody responses to RSV in various isotypes. In primary infection, IgM responses generally first appear in serum on days 5 to 8 following the onset of illness and persist for several weeks, similar to IgM responses determined by immunofluorescence. Hornsleth and colleagues (29) were able to detect RSV-specific IgM responses by EIA in 21 of 26 infants and young children judged to have primary infection. Meurman et al. (46) found that five (63%) of eight patients aged <6 months at the time of infection produced detectable IgM antibodies, whereas all patients older than 12 months had detectable IgM responses. RSV-specific IgM responses can also be used to identify infection in elderly individuals who are presumably undergoing recurrent RSV infection (66).

EIAs can also be used to document significant increases in RSV-specific IgG antibody titers. Meurman and colleagues (46) found significant IgG antibody increases in 24 of 26 infants and young children with RSV infection. The only two patients who failed to develop positive responses were <4 months of age at the time of infection. Other investigators have found diagnostic serologic responses in approximately 50% of infants 1 to 3 months of age and in 85 to 95% of older infants with primary RSV infection. In various studies, titers measured by EIAs correlate well with those measured by immunofluorescence (46), complement fixation (64), and neutralizing antibody (16). Immunofluorescence assays are slightly more sensitive than EIA (46, 66), while both are more sensitive than complement fixation assays (46, 64, 66).

EIAs have also been used to detect RSV-specific IgA antibodies in serum and secretions (46, 70) and RSV-specific IgE responses (70, 71).

EIAs that use a variety of different antigen sources have been developed. These sources include RSV-infected cells grown directly in microtiter plates and fixed with acetone or ethanol; infected cells grown in tissue culture flasks, collected, and then adsorbed to microtiter plates by desiccation (64); virus grown in cell culture and recovered by freezing and thawing of cells before being adsorbed to plates by overnight incubation at 4°C; and virus partially purified by centrifugation over sucrose gradients, diluted in carbonate buffer, and adsorbed to plates overnight at 4°C (71). Plates can also be incubated first with capture antibodies adsorbed overnight at 4°C before incubation with viral antigen (66). In our experience, use of purified virus is necessary to reduce the background absorbance readings sufficiently to allow detection of low titers of antibody. The use of desiccated infected cells as antigen has been associated with the development of a prozone phenomenon, in which negative responses are found at lower serum dilutions and positive responses occur at higher dilutions (64).

Although the immunofluorescence assay is probably somewhat more sensitive than EIA for detecting antibody, the EIA techniques are nevertheless accurate, somewhat easier to perform, and amenable to automation. Therefore, they are probably the method of choice for measuring RSV antibody responses. EIAs have also been used to detect RSV-specific antibody responses in IgG subclasses (28) as well as antibody to specific viral proteins (31). Target proteins are purified by affinity chromatography.

Immunoblot Assays

Antibody responses to individual RSV proteins can be measured by immunoblotting (31, 40). Virus should be partially purified on sucrose gradients and separated by sodium dodecyl sulfate-polyacrylamide gel electrophoresis under reducing conditions on 9% polyacrylamide gels. Viral proteins are then electroblotted onto nitrocellulose membranes.

Fusion Inhibition Antibody

RSV-immune sera contain high titers of antibody to the fusion protein. However, most of this antibody is not functional, in that it does not inhibit fusion of membranes by the virus. Specific antifusion antibody reactivity can be detected and may be of some importance in studying natural immunity as well as in studies of the development of effective vaccines (40).

Summary

Neutralizing antibody (and perhaps fusion-inhibiting antibody) is the only antibody that correlates with protection against reinfection with RSV. In addition, neutralization assays are as sensitive as other assays for detecting seroconversion to RSV. Therefore, they will remain irreplaceable as investigative tools. Nevertheless, immunofluorescence and, particularly, EIA techniques are much more rapidly performed, are approximately as accurate as neutralization assays for detecting increases in antibody titer, and afford the capability of rapid serologic diagnosis by demonstrating the presence of IgM-specific antibody. Therefore, immunofluorescence assays and EIAs may eventually replace neutralization assays in the serologic diagnosis of RSV infection. However, serologic diagnostic techniques are less sensitive than cell culture isolation and rapid diagnostic techniques generally for the diagnosis of RSV infection, and no serologic assay has been standardized or has achieved wide acceptance as a useful diagnostic method for detection of RSV infection. Immunoblot assays and fusion inhibition antibody assays are basically research techniques at this point.

EVALUATION AND REPORTING

With proper obtaining and handling of specimens, the rapid diagnostic tests have reasonable sensitivity, often greater than that of cell culture, for the detection of RSV. Nevertheless, false-negative results by rapid diagnostic methods or cell culture can be expected in a small percentage of cases, and negative tests or cultures should never be interpreted as excluding the possibility of RSV infection. In particular, seriously ill children with highly characteristic illnesses in the peak period of RSV epidemics should not necessarily have antiviral therapy withheld on the basis of a negative test result.

When properly performed, rapid diagnostic tests have high specificity for RSV. The specificity of cell culture is probably even higher. Nevertheless, it seems wise for any laboratory to confirm rapid diagnostic test results at least occasionally by cell culture or by another antigen detection assay to avoid false-positive reports.

Prompt reporting of positive test results may assist clinicians in discontinuing unnecessary forms of therapy or in isolating infected individuals, thereby preventing nosocomial spread of infection.

REFERENCES

1. Ahluwalia, G., J. Embree, P. McNicol, B. Law, and G. W. Hammond. 1987. Comparison of nasopharyngeal aspirate and nasopharyngeal swab specimens for respiratory syncytial virus diagnosis by cell culture, indirect immunofluorescence assay, and enzyme-linked immunosorbent assay. *J. Clin. Microbiol.* **25:**763–767.
2. Anderson, L. J., J. C. Hierholzer, P. G. Bingham, and Y. O. Stone. 1985. Microneutralization test for respiratory syncytial virus based on an enzyme immunoassay. *J. Clin. Microbiol.* **22:**1050–1052.
3. Anderson, L. J., J. C. Hierholzer, Z. Tsou, et al. 1985. Antigenic characterization of respiratory syncytial virus strains with monoclonal antibodies. *J. Infect. Dis.* **151:**626–633.
4. Arens, M. Q., Z. M. Swierkosz, R. R. Schmitt, P. Armstrong, and K. A. Rivetna. 1986. Enhanced isolation of respiratory syncytial virus in cell culture. *J. Clin. Microbiol.* **23:**800–802.
5. Beem, M. 1987. Repeated infections with respiratory syncytial virus. *J. Immunol.* **98:**1115–1122.
6. Bertaiume, L., J. Joncas, and V. Pavilanis. 1974. Comparative structure, morphogenesis and biological characteristics of

the respiratory syncytial (RS) virus and the pneumonia virus of mice (PVM). *Arch. Gesamte Virusforsch.* **45:**39–51.

7. **Blanding, J. G., M. G. Hoshiko, and H. R. Stutman.** 1989. Routine viral culture for pediatric respiratory syncytial specimens submitted for direct immunofluorescence testing. *J. Clin. Microbiol.* **27:**1438–1440.

8. **Bromberg, K., G. Tannis, and B. Daidone.** 1991. Early use of indirect immunofluorescence for the detection of respiratory syncytial virus in HEp-2 cell culture. *Am. J. Clin. Pathol.* **96:**127–129.

9. **Cash, P., C. R. Pringle, and C. M. Preston.** 1979. The polypeptides of human respiratory syncytial virus: products of cell-free protein synthesis and post-translational modifications. *Virology* **92:**375–384.

10. **Cheeseman, S. H., L. T. Pierik, D. Leombruno, K. E. Spinos, and K. McIntosh.** 1986. Evaluation of a commercially available direct immunofluorescence staining reagent for the detection of respiratory syncytial virus in respiratory secretions. *J. Clin. Microbiol.* **24:**155–156.

11. **Church, N. R., N. G. Anas, C. B. Hall, and J. G. Brooks.** 1984. Respiratory syncytial virus-related apnea in infants. *Am. J. Dis. Child.* **138:**247–250.

12. **Collins, P. L., Y. T. Huang, and G. W. Wertz.** 1984. Identification of a tenth mRNA of respiratory syncytial virus and assignment of polypeptides to the 10 viral genes. *J. Virol.* **49:**572–578.

13. **Ditchburn, R. K., J. McQuillin, P. S. Gardner, and S. D. M. Court.** 1971. Respiratory syncytial virus in hospital cross-infection. *Br. Med. J.* **3:**671–733.

14. **Dubovi, E. J., J. D. Geratz, and R. R. Tidwell.** 1983. Enhancement of respiratory syncytial virus-induced cytopathology by trypsin, thrombin, and plasmin. *Infect. Immun.* **40:**351–358.

15. **Eriksson, M., M. Forsgren, S. Sjoberg, M. von Sydow, and S. Wolontis.** 1983. Respiratory syncytial virus identification in young hospitalized children: identification of risk patients and prevention of nosocomial spread by rapid diagnosis. *Acta Paediatr. Scand.* **72:**47–51.

16. **Falsey, A., and E. E. Walsh.** 1992. Humoral immunity to respiratory syncytial virus in the elderly. *J. Med. Virol.* **36:**39–43.

17. **Finger, F., L. J. Anderson, R. C. Dicker, B. Harrison, R. Doan, A. Downing, and L. Corey.** 1987. Epidemic respiratory syncytial virus infection in institutionalized young adults. *J. Infect. Dis.* **155:**1335–1339.

18. **Flander, R. T., P. D. Lindsay, R. Chairez, T. A. Brawner, M. L. Kumar, P. D. Swenson, and K. Bromberg.** 1986. The evaluation of clinical specimens for the presence of respiratory syncytial virus antigen using an enzyme immunoassay. *J. Med. Virol.* **19:**1–9.

19. **Florman, A. L., and L. C. McLaren.** 1988. The effect of altitude and weather on the occurrence of outbreaks of respiratory syncytial virus infections. *J. Infect. Dis.* **158:**1401–1402.

20. **Gardner, P. S., and J. McQuillin.** 1968. Application of immunofluorescent antibody technique in rapid diagnosis of respiratory syncytial virus infection. *Br. Med. J.* **3:**340–343.

21. **Glezen, W. P., L. H. Taber, A. L. Frank, and J. A. Kasel.** 1986. Risk of primary infection and reinfection with respiratory syncytial virus. *Am. J. Dis. Child.* **140:**543–546.

22. **Gruber, C., and S. Levine.** 1983. Respiratory syncytial virus polypeptides. III. The envelope-associated proteins. *J. Gen. Virol.* **64:**825–832.

23. **Halstead, D. C., S. Todd, and G. Fritch.** 1990. Evaluation of five methods for respiratory syncytial virus detection. *J. Clin. Microbiol.* **28:**1021–1025.

24. **Hambling, M. H.** 1964. Survival of the respiratory syncytial virus during storage under various conditions. *Br. J. Exp. Pathol.* **45:**647–655.

25. **Hendry, R. M., and K. McIntosh.** 1982. Enzyme-linked immunosorbent assay for detection of respiratory syncytial virus infection: development and description. *J. Clin. Microbiol.* **16:**324–328.

26. **Hendry, R. M., L. T. Pierik, and K. McIntosh.** 1986. Comparison of washed nasopharyngeal cells and whole nasal secretions for detection of respiratory syncytial virus antigens by enzyme-linked immunosorbent assay. *J. Clin. Microbiol.* **23:**383–384.

27. **Holberg, C. J., A. L. Wright, F. D. Martinez, C. G. Ray, L. M. Taussig, and M. D. Lebowitz.** 1991. Risk factors for respiratory syncytial virus-associated lower respiratory illnesses in the first year of life. *Am. J. Epidemiol.* **133:**1135–1151.

28. **Hornsleth, A., N. Bech-Thomsen, and B. Friis.** 1985. Detection of RS-virus IgG-subclass-specific antibodies: variation according to age in infants and small children and the diagnostic value in RS-virus-infected small infants. *J. Med. Virol.* **16:**329–335.

29. **Hornsleth, A., B. Friis, P. C. Grauballe, and P. A. Krasilnikof.** 1984. Detection by ELISA of IgA and IgM antibodies in secretion and IgM antibodies in serum in primary lower respiratory syncytial virus infection. *J. Med. Virol.* **13:**149–161.

30. **Huang, Y. T., P. L. Collins, and G. W. Wertz.** 1982. The genome of respiratory syncytial virus is a negative-stranded RNA that codes for at least seven mRNA species. *J. Virol.* **43:**150–157.

31. **Jimenez, H. B., H. M. Keir, and P. Cash.** 1987. Immunoblot analysis of human antibody response to respiratory syncytial virus infection. *J. Gen. Virol.* **68:**1267–1275.

32. **Johnson, P. R., and P. L. Collins.** 1988. The fusion glycoproteins of human respiratory syncytial virus of subgroups A and B: sequence conservation provides a structural basis for antigen relatedness. *J. Gen. Virol.* **69:**2623–2628.

33. **Johnson, P. R., R. A. Olmsted, G. A. Prince, B. R. Murphy, D. W. Alling, E. E. Walsh, and P. L. Collins.** 1987. Antigenic relatedness between glycoproteins of human respiratory syncytial virus subgroups A and B: evaluation of the contributions of F and G glycoproteins to immunity. *J. Virol.* **61:**3163–3166.

34. **Johnson, P. R., M. K. Spriggs, R. A. Olmsted, and P. L. Collins.** 1987. The G glycoprotein of human respiratory syncytial virus of subgroups A and B: extensive sequence diversions between antigenically related proteins. *Proc. Natl. Acad. Sci. USA* **84:**5625–5629.

35. **Jordan, W. S., Jr.** 1962. Growth characteristics of respiratory syncytial virus. *J. Immunol.* **88:**581–590.

36. **Kaul, A., R. Scott, M. Gallagher, M. Scott, J. Clement, and P. L. Ogra.** 1978. Respiratory syncytial virus infection: rapid diagnosis in children by use of indirect immunofluorescence. *Am. J. Dis. Child.* **132:**1088–1090.

37. **Kaul, T. N., R. C. Welliver, and P. L. Ogra.** 1981. Comparison of fluorescent-antibody, neutralizing antibody and complement-enhanced neutralizing antibody assays for detection of serum antibody to respiratory syncytial virus. *J. Clin. Microbiol.* **13:**957–962.

38. **Kellog, J. A.** 1991. Culture vs. direct antigen assays for detection of microbial pathogens from lower respiratory tract specimens suspected of containing the respiratory syncytial virus. *Arch. Pathol. Lab. Med.* **115:**451–458.

39. **Kim, H. W., J. O. Arrobio, C. D. Brandt, B. C. Jeffries, G. Pyles, J. L. Reid, R. M. Chanock, and R. H. Parrott.** 1973. Epidemiology of respiratory syncytial virus infection in Washington, DC. I. Importance of the virus in different respiratory tract disease syndromes and temporal distribution of infection. *Am. J. Epidemiol.* **98:**216–225.

40. **Levine, S., A. Dajani, and R. Klaiber-Franco.** 1988. The response of infants with bronchiolitis to the proteins of respiratory syncytial virus. *J. Gen. Virol.* **69:**1229–1239.

41. **Marquez, A., and G. D. Hsiung.** 1967. Influence of glutamine on multiplication and cytopathic effect of respiratory syncytial virus. *Proc. Soc. Exp. Biol. Med.* **124:**95–99.

42. **Masters, H. B., B. J. Bate, C. Wren, and B. A. Lauer.** 1988. Detection of respiratory syncytial virus antigen in nasopharyngeal secretions by Abbott Diagnostics enzyme immunoassay. *J. Clin. Microbiol.* **26:**1103–1105.

43. **McConnochie, K. M., and K. J. Roghmann.** 1989. Wheezing

at 8 and 13 years: changing importance of bronchiolitis and passive smoking. *Pediatr. Pulmonol.* **6:**138–146.

44. **McIntosh, K., and R. M. Chanock.** 1990. Respiratory syncytial virus, p. 963–1074. *In* B. N. Fields and D. M. Knipe (ed.), *Field's Virology.* Raven Press, New York.

45. **McIntosh, K., R. M. Hendry, M. L. Fahnestock, and L. T. Pierik.** 1982. Enzyme-linked immunosorbent assay for the detection of respiratory syncytial virus infection: application to clinical samples. *J. Clin. Microbiol.* **16:**329–333.

46. **Meurman, O., O. Ruuskanen, H. Sarkkinen, P. Hanninen, and P. E. Halonen.** 1984. Immunoglobulin class specific antibody response in respiratory syncytial virus infection measured by enzyme immunoassay. *J. Med. Virol.* **14:**67–72.

47. **Millar, H. R., P. H. Phipps, and E. Rossier.** 1986. Reduction of nonspecific fluorescence in respiratory specimens by pretreatment with *n*-acetylcysteine. *J. Clin. Microbiol.* **24:**470–471.

48. **Muelenear, P. M., F. W. Henderson, V. G. Heming, E. E. Walsh, L. J. Anderson, J. A. Prince, and B. R. Murphy.** 1991. Group-specific serum antibody responses in children with primary and recurrent respiratory syncytial virus infections. *J. Infect. Dis.* **164:**15–21.

49. **Mufson, M. A., R. B. Belshe, C. Orvell, and E. Norrby.** 1987. Subgroup characteristics of respiratory syncytial virus strains recovered from children with two consecutive infections. *J. Clin. Microbiol.* **25:**1535–1539.

50. **Navas, L., E. Wang, V. De Carvalho, and J. Robinson.** 1992. Improved outcome of respiratory syncytial virus infection in a high-risk hospitalized population of Canadian children. *J. Pediatr.* **121:**348–354.

51. **Pons, M. W., A. L. Lambert, D. M. Lambert, and O. M. Rochovansky.** 1983. Improvement of respiratory syncytial virus replication in actively growing HEp-2 cells. *J. Virol. Methods* **7:**217–221.

52. **Pringle, C. R., P. V. Shirodaria, and P. E. Cash.** 1978. Initiation and maintenance of persistent infection by respiratory syncytial virus. *J. Virol.* **28:**199–211.

53. **Richman, A. V., F. A. Pedreia, and N. M. Tauraso.** 1971. Attempts to demonstrate hemagglutination and hemadsorption by respiratory syncytial virus. *Appl. Microbiol.* **21:**1099–1100.

54. **Rothbarth, P. H., M.-C. Hermus, and P. Schrijnemakers.** 1991. Reliability of two new test kits for rapid diagnosis of respiratory syncytial virus infection. *J. Clin. Microbiol.* **29:**824–826.

55. **Routledge, E. G., J. McQuillan, A. C. R. Samson, and G. L. Toms.** 1985. The development of monoclonal antibodies to respiratory syncytial virus and their use in diagnosis by indirect immunofluorescence. *J. Med. Virol.* **15:**305–320.

56. **Sarkkinen, H. K., P. E. Halonen, P. P. Arestila, and A. A. Salmi.** 1981. Detection of respiratory syncytial, parainfluenza type 2 and adenovirus antigens by radioimmunoassay and enzyme immunoassay on nasopharyngeal specimens from children with acute respiratory disease. *J. Clin. Microbiol.* **13:**258–265.

57. **Satake, M., S. Elango, and S. Venkatesan.** 1984. Sequence analysis of the respiratory syncytial virus phosphoprotein gene. *J. Virol.* **52:**991–994.

58. **Smith, M. C., C. Creutz, and Y. T. Huang.** 1991. Detection of respiratory syncytial virus in nasopharyngeal secretions by shell vial technique. *J. Clin. Microbiol.* **29:**463–465.

59. **Suto, T., N. Yano, M. Ikeda, M. Miyamoto, S. Takai, S. Shigeta, Y. Hinuma, and N. Ishida.** 1965. Respiratory syncytial virus infection and its serologic epidemiology. *Am. J. Epidemiol.* **82:**211–224.

60. **Swierkosz, E. M., R. Flander, L. Melvin, J. D. Miller, and M. W. Kline.** 1989. Evaluation of the Abbott TEST PACK RSV enzyme immunoassay for detection of respiratory syncytial virus in nasopharyngeal swab specimens. *J. Clin. Microbiol.* **27:**1151–1154.

61. **Thomas, E. E., and L. E. Book.** 1991. Comparison of two rapid methods for detection of respiratory syncytial virus (Test Pack RSV and Ortho Elisa) with direct immunofluorescence and virus isolation for the diagnosis of pediatric RSV infection. *J. Clin. Microbiol.* **29:**632–635.

62. **Treuhaft, M. W., J. M. Soukup, and B. J. Sullivan.** 1985. Practical recommendations for the detection of pediatric respiratory syncytial virus infections. *J. Clin. Microbiol.* **22:**270–273.

63. **Van Beers, D., M. DeFoor, L. DiCesare, and C. Vandenveld.** 1991. Evaluation of a commercial enzyme and immunomembrane filter assay for detection of respiratory syncytial virus in clinical specimens. *Eur. J. Clin. Microbiol. Infect. Dis.* **10:**1073–1076.

64. **Vaur, L., H. Agut, A. Jarbarg-Cheon, G. Prud'homme, J. C. Nicolas, and F. Bricout.** 1986. Simplified enzyme-linked immunosorbent assay for specific antibodies to respiratory syncytial virus. *J. Clin. Microbiol.* **24:**596–599.

65. **Venkatesan, S., N. Elango, M. Satake, E. Camargo, and R. M. Chanock.** 1984. Organization and expression of respiratory syncytial virus genome, p. 31–44. *In* R. Lerner and R. M. Chanock (ed.), *Modern Approaches to Vaccines.* Cold Spring Harbor Laboratory, Cold Spring Harbor, N.Y.

66. **Vikerfors, T., M. Grandian, M. Johansson, and C.-A. Pettersson.** 1988. Detection of an immunoglobulin M response in the elderly for early diagnosis of respiratory syncytial virus infection. *J. Clin. Microbiol.* **26:**808–811.

67. **Walsh, E. E., M. W. Brandriss, and J. J. Schlesinger.** 1985. Purification and characterization of the respiratory syncytial virus fusion protein. *J. Gen. Virol.* **66:**409–415.

68. **Walsh, E. E., and J. Hrushka.** 1983. Monoclonal antibodies to respiratory syncytial virus proteins: identification of the fusion protein. *J. Virol.* **47:**171–177.

69. **Welliver, R. C., T. N. Kaul, T. I. Putnam, M. Sun, K. Riddlesberger, and P. L. Ogra.** 1980. The antibody response to primary and secondary infection with respiratory syncytial virus: kinetics of class-specific responses. *J. Pediatr.* **96:**808–813.

70. **Welliver, R. C., M. Sun, D. Rinaldo, and P. L. Ogra.** 1985. Respiratory syncytial virus-specific IgE responses following infection: evidence for a predominantly mucosal response. *Pediatr. Res.* **19:**420–424.

71. **Welliver, R. C., D. T. Wong, M. Sun, E. Middleton, Jr., R. S. Vaughan, and P. L. Ogra.** 1981. The development of respiratory syncytial virus-specific IgE and the release of histamine in nasopharyngeal secretions after infection. *N. Engl. J. Med.* **305:**841–846.

72. **Wertz, G. W., P. L. Collins, Y. T. Huang, C. Gruber, S. Levine, and L. A. Ball.** 1985. Nucleotide sequence of the G protein gene of human respiratory syncytial virus reveals an unusual type of viral membrane protein. *Proc. Natl. Acad. Sci. USA* **82:**4075–4079.

Rhinoviruses

MARIE L. LANDRY

79

CLINICAL BACKGROUND

The rhinovirus group derives its name from the predominant site of its replication and symptomatology, the nose. The rhinoviruses constitute the major virus group associated with the acute respiratory illness known as the common cold. The average person suffers two to five colds per year, and one-third to one-half of these colds are due to rhinoviruses (17). Although the common cold is a trivial illness, it is acutely disabling. The cost of the common cold in days lost from work, cold remedies, and analgesics is in the billions of dollars.

In 1930, it was first recognized that the common cold was caused by a "filtrable agent" (22). With the development of cell culture techniques in the 1950s, rhinovirus was isolated (6, 55, 57). With the development of the more sensitive human embryonic lung cell culture and the use of growth conditions that mimicked those of the nose, a number of different serologic types were isolated in the 1960s. In 1967, serotypes 1A through 55 were designated; in 1971, serotypes 56 through 89 were designated; and in 1987, serotypes 90 through 100 were designated (33).

In temperate zones, rhinovirus infections peak in September and show a second peak in spring, but they continue to cause colds throughout the summer. Multiple types appear to circulate in a given area at any time (31, 51). Earlier studies suggested that serotypes with lower numbers were being gradually replaced by higher-numbered serotypes or strains that could not be typed. It was postulated that this change represented either recirculation of a large number of antigenically stable types or antigenic drift (31, 62). More recent studies of rhinovirus strains have found that new serotypes are probably not evolving and that most isolates have already been identified (33, 51).

Rhinovirus infections are spread from person to person by virus-contaminated respiratory secretions. Although transmission of rhinovirus colds among adult volunteers can be interrupted by the use of virucidal tissues (20), the exact mode of transmission of rhinoviruses has been the subject of some controversy. Studies of volunteers have shown that inoculation of virus into the nose or conjunctiva is the most efficient way to initiate infection (23). Virus is present in highest titers in the nose, and the hands of infected persons are commonly contaminated. Some investigators have found that the virus can be transmitted to others via hand-to-hand contact (7, 32) and can be self-inoculated into the nose or conjunctivae (36). Nevertheless, a recent study found that rhinovirus transmission, at least in adults, occurred primarily by the aerosol route (21).

Studies with volunteers show that the incubation period of rhinoviruses after inoculation is 2 to 3 days (23). The peak of virus shedding coincides with acute rhinitis. Virus may become undetectable by 4 to 5 days or may be present in low titers for up to 2 weeks. Nasal epithelial cells are infected and are shed into nasal mucus. However, the symptoms of rhinovirus infection are produced in part by chemical mediators of inflammation, such as bradykinin, lysylbradykinin, and prostaglandins (30, 52, 60). Symptoms persist for 7 days on average and include profuse watery discharge, nasal congestion, sneezing, headache, mild sore throat, cough, and little or no fever. Psychological stress appears to increase both susceptibility to rhinovirus infection and development of clinical symptoms (14, 61).

Infection of the lower respiratory tract may also occur (42–45, 59), and the clinical manifestations of rhinovirus infection in hospitalized children are indistinguishable from those caused by respiratory syncytial virus (43). Rhinovirus infections exacerbate chronic bronchitis and asthma (40, 50), especially in children, and a fatal case in an asthmatic infant has been reported, with isolation of virus from lung and blood postmortem (46). Rhinoviruses have also been associated with otitis media and sinusitis (63). Immunity is type specific and correlates best with local production of immunoglobulin A (IgA) (9).

DESCRIPTION OF THE AGENT

Rhinoviruses are members of the *Picornaviridae* family, which includes the enteroviruses. The rhinovirus virion consists of a nonenveloped icosahedral nucleocapsid that is 20 to 27 nm in diameter and has 60 protomeric units consisting of four protein subunits: VP1, VP2, VP3, and VP4. These structural polypeptides are obtained by posttranslational cleavage of a large polypeptide precursor (3, 48). VP1, VP2, and VP3 reside on the exterior of the virus and make up its protein coat. VP4 resides inside the virion's protein shell. The virion nucleic acid is single-stranded plus-strand RNA approximately 7,200 nucleotides long and having a molecular weight of 2.4×10^6 (12, 31). Rhinovirus genomes have been sequenced and found to share 45 to 62% homology with poliovirus (12). Rhinovirus replicates in the cytoplasms of infected cells, producing infectious virions that sediment at a buoyant density of 1.40 g/liter in cesium chloride. Empty capsids and particles lacking one or more structural polypeptides are also produced. Infectious and noninfectious particles are immunologically distinct and have been referred to as D (dense or native virions), which are fully infectious, and C (coreless virion antigens), which lack VP4 (48, 49).

As they have no lipid envelope, rhinoviruses are resistant to inactivation by organic solvents such as ether, chloroform, ethanol, and 5% phenol. With heating at 50 to 56°C, infectivity progressively decreases as virions become empty capsids with C-type antigenicity. Rhinoviruses are differentiated from enteroviruses by their loss of infectivity upon exposure to pH 3 for 3 h at room temperature. After mild-acid inactivation (pH 5), two types of particles are formed: A particles, which lack VP4, and B particles, which lack both VP4 and RNA (31).

At present, 100 rhinovirus serotypes have been numbered on the basis of neutralization tests. There is no group antigen, but by using high-titer hyperimmune rabbit and guinea pig antisera, cross-reactions between antigenic types have been detected (16, 17, 31). The native D antigenicity of rhinoviruses can be changed to C antigenicity by treatment with low pH, heat, or 2 M urea, producing virus particles that react with heterologous antisera (49).

X-ray crystallography and cryo-electron microscopy studies show that the large depressions or "canyons" found on each of the 60 protomeric units are sites for cell receptor binding (54, 58). Each canyon is too small to admit the combining site of an antibody but is large enough to admit the virus binding site (29, 34, 54). Conformational changes in the canyon floor that inhibit virus attachment to cells can be produced by certain antiviral agents (56).

Rhinoviruses can be divided into two groups based on binding to cell receptors (1, 65). Intercellular adhesion molecule 1 (ICAM-1) is the cell receptor for the majority (major group) of rhinoviruses, or approximately 90% of serotypes. The minor-group receptor has been tentatively identified as a 120-kDa cell surface protein of unknown function. Rhinovirus type 87 appears to utilize neither the major nor the minor group receptor.

Elucidating the role of the major rhinovirus group receptor in viral entry into the host cell and the factors influencing virus-receptor binding (29, 39) is important in devising strategies for disease prevention and therapy.

COLLECTION AND STORAGE OF SPECIMENS

In natural infections, rhinovirus can be isolated from 1 day before to 6 days after onset of cold symptoms but is shed at the highest concentration on day 1 or 2 of illness. Since rhinovirus is excreted in highest titers from the nose, nasal rather than throat specimens should be obtained for virus isolation. Likewise, sputum is a low-yield specimen for the isolation of rhinovirus. A comparison of nasal wash, nose swab, throat gargle, and throat swab specimens for the isolation of rhinovirus from clinical specimens revealed nasal wash to be the best (3, 13). Nasal wash specimens are obtained as follows. Tilt the patient's head backward, and instill 1 ml of sterile phosphate-buffered saline (PBS) into one of the nostrils. Ask the patient to lean forward, and allow the washing to drip into a sterile petri dish or other collection container. Repeat with the other nostril, and continue the procedure until each nostril has been washed with 5 ml of PBS. Transfer the washings into a sterile container with an equal volume of viral transport medium (VTM) containing antibiotics. Including phenol red in the VTM allows the detection of acid pH, which would adversely affect the isolation of rhinovirus.

If a nasal wash cannot be obtained, a combination of a nasal or nasopharyngeal swab and a throat swab should be collected (31). Immerse the swabs in a vial containing 2 ml of VTM.

Specimens should be transported to the laboratory promptly. Best results are obtained with rapid inoculation of cell cultures; however, specimens can be held up to 24 h at 4°C in VTM with neutral pH. If longer delays are necessary, specimens should be frozen at −70°C and thawed just prior to inoculation.

DIRECT EXAMINATION

Because of the great number of rhinovirus serotypes and the lack of a group antigen, it has been difficult to develop tests for direct detection of viral antigen in clinical specimens. Studies in the 1970s that used immunofluorescence to detect rhinovirus antigens in nasopharyngeal specimens revealed a poor correlation between immunofluorescence and virus isolation results (26). Enzyme immunoassays (EIAs) with polyclonal antisera to two different human rhinoviruses have been developed for detection of rhinovirus directly in clinical specimens and after overnight amplification in cell culture (18). A number of stock viruses of different serotypes were cross-reactive to various degrees in these EIA systems. However, when clinical specimens from volunteers infected with homologous virus were tested, correlation with virus isolation was a disappointing 89% for culture-amplified techniques and 67% for direct EIA techniques. Subsequently, an EIA-based system utilizing biotinylated soluble ICAM-1 to detect the major subgroup of rhinoviruses was developed; tests using stock viruses revealed a detection limit of 10^3 50% tissue culture infective doses ($TCID_{50}$) (47).

Since a high degree of homology exists in the 5' noncoding regions of picornavirus genomes, recent emphasis has been on nucleic acid detection methods. Because of genome homology between enteroviruses and rhinoviruses, however, cross-reactions occur. Hybridization using cDNA probes did not detect all rhinoviruses with equal efficiency and was relatively insensitive (2). Oligonucleotide probes detected rhinovirus in nasal washings from volunteers infected with homologous virus (11); however, nonspecific binding was problematic when nasal aspirates were tested (41). In situ hybridization has been used to localize rhinovirus replication in vitro (19). The PCR, using an initial reverse transcriptase step to generate cDNA, has been applied to rhinovirus detection in clinical material (27), and a sensitivity equivalent to (53) or greater than (41) that of culture has been reported. Furthermore, since rhinoviruses have a shorter 5' noncoding region than enteroviruses, rhinoviruses can be differentiated from enteroviruses on the basis of the sizes of their respective PCR products if appropriate primers are used (5, 8, 53). However, in one report, the primers that were able to distinguish rhinovirus from enterovirus were not the most sensitive for detection of virus in clinical specimens and failed to identify many of the positive samples (41). These techniques, though promising, remain research tools at present.

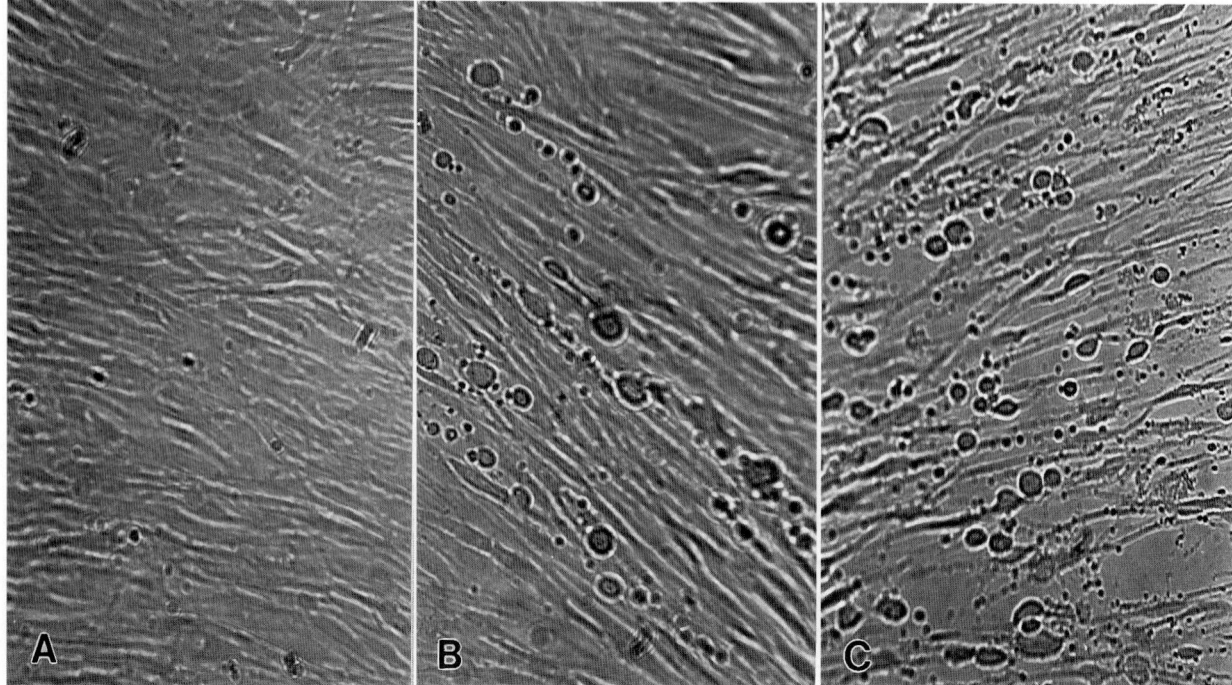

FIGURE 1 Rhinovirus CPE in human fibroblast cells. (A) Uninfected cells; (B) early focus of rhinovirus CPE; (C) More advanced rhinovirus CPE. Magnification, ×100.

ISOLATION OF VIRUS

Cell Culture

Rhinoviruses are best isolated in sensitive cell culture systems and grow only in cells of human or monkey origin. Although the original isolation of rhinoviruses was in primary monkey kidney cells, these cells have not been consistent in yielding a broad range of isolates. The use of sensitive human diploid fibroblasts (HDF) has provided superior results. Rhinoviruses were originally separated into M and H strains on the basis of their abilities to replicate in monkey and human cells (M strains replicate in both monkey and human cells; H strains replicate only in human cells). However, this distinction is no longer used, since H strains can be adapted to grow in monkey cells.

The most commonly used cells in clinical laboratories are HDF cells, such as human embryonic lung strains WI-38 and MRC-5. Human embryonic kidney (HEK) cells can also support rhinovirus replication. HeLa cells, such as HeLa M, Ohio HeLa, and HeLa R-19 cells, support the replication of rhinoviruses to high titers and have been used in research studies and in the preparation of rhinovirus antigens (1, 3, 15). Passage may be necessary for some isolates before cytopathic effects (CPE) are apparent. In one study, a human fetal tonsil cell line was found to be more sensitive than MRC-5 and HeLa cells for rhinovirus isolation (28). Unfortunately, different lots of normally sensitive cell lines have been found to vary over 100-fold in sensitivity to rhinovirus (10); the reasons for this variation are not known. Therefore, for optimal results, simultaneous use of several sensitive systems is recommended.

After inoculation, cultures are incubated in standard cell culture medium such as Eagle minimum essential medium with 2% fetal calf serum and antibiotics at a neutral pH. It is important that culture conditions mimic those of the nose; cultures should be incubated at 33°C with continuous rotation in a roller drum to provide aeration of the monolayer. CPE can be observed as early as 24 to 48 h after inoculation and are often detected by day 4. Cellular changes are easier to read in fibroblasts and are therefore often detected earlier in fibroblasts than in epithelial cell lines. Both large and small rounded, refractile cells with pyknotic nuclei are observed in foci that also contain cellular debris (Fig. 1). Rhinovirus CPE is similar to enterovirus CPE but may sometimes be confused with nonspecific changes. The CPE progresses over a 2- to 3-day period, with the degree of cellular change depending on the serotype and the inoculum dose. Note that rhinovirus CPE can regress, and virus may inactivate if left too long. Therefore, cultures should be promptly passaged. Passage is also necessary for increasing viral titers before identification tests are done.

HDF cultures should be observed for 14 days. HeLa cell cultures can be observed for only up to 8 days, when passage becomes necessary because of nonspecific cell degeneration and rounding.

Organ Culture

Organ cultures of human fetal nasal epithelium or trachea have been used to isolate rhinoviruses not grown in standard cell cultures. However, studies comparing organ cultures with standard cell culture have shown that both systems are necessary for optimal recovery (37). Owing to the limited supply of fetal material, this method is not widely used. Details of the procedure have been reviewed elsewhere (3, 38). In brief, pieces of nasal epithelium or trachea are incubated in culture tubes at 33°C in a roller drum, examined daily for ciliary activity, and monitored for pH changes. If rhinovirus is replicating, ciliary activity ceases within 5 to 7 days. Virus can be harvested in the

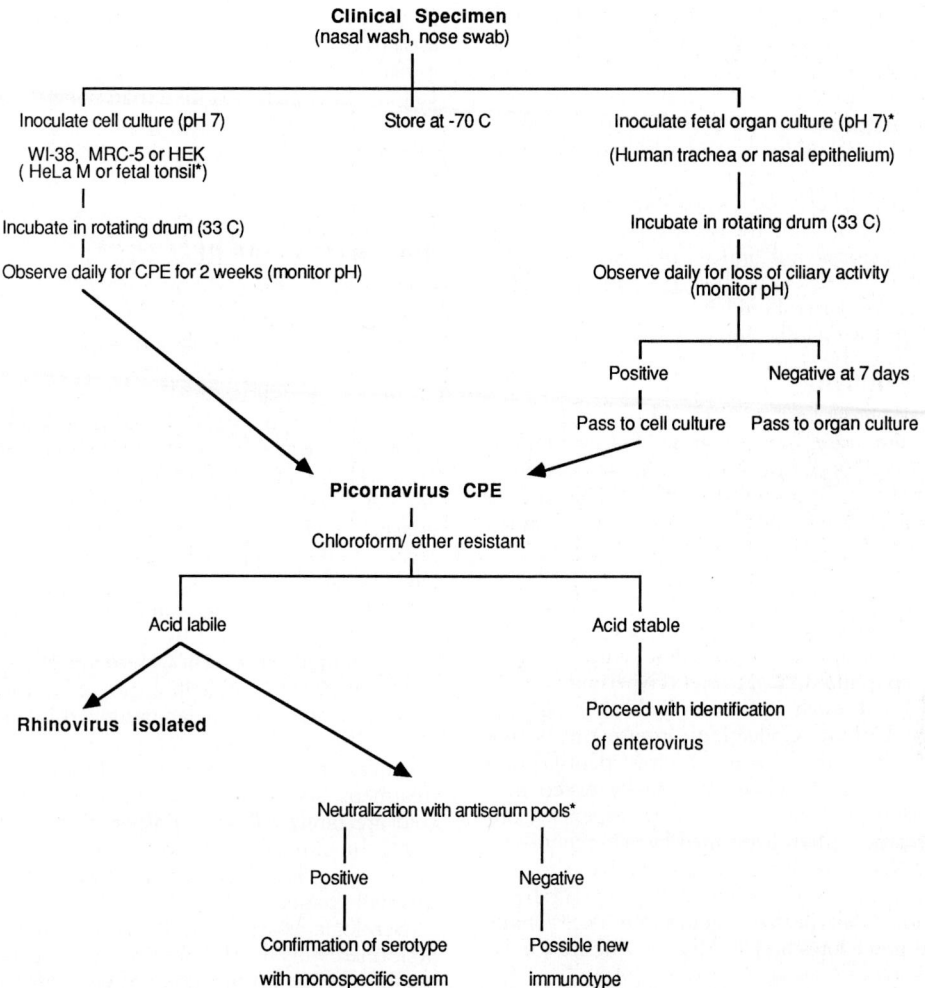

FIGURE 2 Flow scheme for the isolation and identification of rhinovirus. Techniques reserved for the research laboratory are marked by an asterisk.

supernatant fluids and passaged to fresh organ cultures or cell culture.

IDENTIFICATION OF VIRUS

A presumptive diagnosis of a rhinovirus isolate is made by the appearance and progression of characteristic CPE in the appropriate clinical setting. Further identification of a virus isolate requires confirmation of the properties of rhinovirus such as sensitivity to temperature and acid pH and resistance to lipid solvents. For routine viral diagnostic laboratories, these tests are sufficient (Fig. 2). Since 100 rhinovirus serotypes have been identified and no group antigen has been found, specific serotype identification by neutralization is very time-consuming and costly. Only for isolates from unusual clinical situations or in research settings is further identification of rhinovirus serotypes by neutralization warranted. Prior to the performance of these tests, passage of isolates will be necessary to obtain a minimum titer of 10^3 TCID$_{50}$/ml.

Acid pH Stability

Rhinoviruses are sensitive to low pH, whereas enteroviruses are stable at low pH. Thus, a reduction in virus titer by 2 to

3 log$_{10}$ TCID$_{50}$ can be expected upon exposure of rhinovirus to low pH. First, prepare two solutions of a buffer such as HEPES (N-2-hydroxyethylpiperazine-N'-2-ethanesulfonic acid), one at pH 3.0 and one at pH 7.0. Add 0.2 ml of unknown virus suspension to 1.8 ml of HEPES at pH 3.0 and 0.2 ml of virus to 1.8 ml of HEPES at pH 7.0. Keep the mixtures at room temperature for 3 h, adjust the pH to 7.0, make serial dilutions of the mixtures, and inoculate the dilutions into cell culture. If the unknown virus is a rhinovirus, a minimum 2-log$_{10}$ reduction in viral titer should be evident in the acid-treated sample. A known rhinovirus and a known enterovirus should be treated in a similar fashion as controls.

Resistance to Lipid Solvents

Since rhinoviruses lack a lipid envelope, they are resistant to treatment with lipid solvents such as chloroform or ether, and infectivity titers should not be affected by such treatment. Add 0.1 ml of chloroform to 1 ml of isolate, shake vigorously for 10 min, and then centrifuge at 1,000 × g for 5 to 10 min. Remove the top or aqueous phase, prepare serial dilutions, and inoculate the dilutions into cell culture.

Alternatively, add 0.2 ml of diethyl ether to 0.8 ml of isolate in a stoppered tube, and shake vigorously. Allow the mixture to react at room temperature for 1 to 2 h; shake the tube intermittently. Then transfer the mixture into an open glass petri dish in a fume hood, and allow the ether to evaporate. Prepare serial dilutions, and inoculate them into cell cultures.

An equal volume of unknown virus should be handled in a similar manner but not treated with chloroform or ether. A known picornavirus and a chloroform-sensitive virus such as herpes simplex virus should also be included as controls. The picornavirus should be unaffected, and the herpes simplex virus should be completely inactivated.

Temperature Sensitivity

Since rhinoviruses grow best at 33°C, they can be distinguished from enteroviruses by inoculation of serial dilutions of the unknown virus and incubation of one set of cultures at 33°C and a replicate set of cultures at 37°C. The onset of CPE should be more rapid and the titer of virus obtained should be higher at the lower temperature.

Neutralization Tests

Specific identification of rhinoviruses by serotyping is an expensive and labor-intensive procedure currently performed only by specialized laboratories. Hyperimmune antisera for types 1A through 100 are available through the American Type Culture Collection. Intersecting serum pools similar to those used for enterovirus identification have been prepared, and isolates are usually tested in a microneutralization procedure to conserve reagents. Confirmation of serotype is then performed by using monospecific antiserum. Neutralization generally involves the neutralization of 30 to 100 $TCID_{50}$ of virus-induced CPE by 20 U of antiserum. Nontypeable isolates are occasionally found. Detailed procedures are described elsewhere (3, 31, 48).

Other Tests

EIAs that can detect rhinovirus antigens in cell culture have been developed (18). Nucleic acid hybridization techniques can also be used for identification of rhinovirus isolates in cell culture (2); however, cross-reactions with enteroviruses occur. PCR with selected primers can distinguish rhinoviruses from enteroviruses and, when used to identify cell culture isolates, is equivalent to acid sensitivity testing in accuracy and cheaper in reagent costs (5).

SEROLOGIC DIAGNOSIS

Although detection of a rise in antibodies to rhinovirus in natural infections may indicate many infections not diagnosed by virus isolation (35), the number of possible serotypes makes blind serologic testing impractical at present. However, if a virus isolate has been recovered, the patient's serum can be tested against that isolate.

The standard serologic assay to measure antibody in serum or nasal wash specimens is the neutralization test (24, 31). For maximum sensitivity, a low challenge dose (3 to 30 $TCID_{50}$) should be used. Plaque reduction neutralization has also been described. Complement fixation detects heterotypic antibody responses to rhinoviruses and enteroviruses and does not parallel neutralization results (31). Hemagglutination inhibition has also been used but is not as sensitive as neutralization, and not all rhinoviruses react with erythrocytes.

EIAs using HRV-EL and HVR-2 antigens have been used to detect serum and nasal IgG and IgA in volunteers inoculated with these two viruses (9). EIA was 100 to 10,000 times more sensitive than neutralization. The problem of detecting antibodies to multiple serotypes in natural infection was not addressed.

EVALUATION, INTERPRETATION, AND REPORTING OF RESULTS

Laboratory diagnosis of rhinovirus infection is currently based on isolation of virus with characteristic CPE in cell systems sensitive for isolation, such as human fibroblasts, HEK cells, and HeLa cells. Incubation at 33°C and rotation of cultures provide optimal conditions for virus replication. Differentiation of rhinoviruses from enteroviruses, with whose CPE the rhinovirus CPE can be confused, is based primarily on acid stability testing of isolates. Rhinoviruses are inactivated by pH 3 to 5, whereas enteroviruses are unaffected. Sensitivity to temperature and resistance to organic solvents are additional features of rhinovirus isolates.

Owing to the large number of rhinovirus serotypes, identification of viruses by neutralization tests is reserved for specialized laboratories. Serotype identification is primarily useful in epidemiologic or research studies. At present in the clinical laboratory, rhinoviruses are rarely specifically sought, and optimal conditions for rhinovirus isolation are not generally employed. Since there is no treatment, specimens from outpatients with the common cold are rarely cultured. Rather, rhinoviruses are isolated when the clinical diagnosis is influenza or respiratory syncytial virus. However, if specific therapy becomes available, this will change.

Serologic assays for antibody to rhinovirus are also not performed outside the research setting but can provide useful information in studies of viral pathogenesis and immunity.

Because of the tremendous economic cost of rhinovirus infections, much work is being done to understand the pathogenesis of these infections and to develop a treatment for them (4, 25, 30, 60, 64). The development of rapid and accurate diagnostic tests in the research setting should prove useful to the clinical laboratory in the near future.

REFERENCES

1. **Abraham, G., and R. J. Colonno.** 1984. Many rhinovirus serotypes share the same cellular receptor. *J. Virol.* **51:**340–345.
2. **Al-Nakib, W., G. Stanway, M. Forsyth, P. J. Hughes, J. W. Almond, and D. A. Tyrrell.** 1986. Detection of human rhinoviruses and their molecular relationships using cDNA probes. *J. Med. Virol.* **20:**289–296.
3. **Al-Nakib, W., and D. A. J. Tyrrell.** 1988. Picornaviridae: rhinoviruses—common cold viruses, p. 723–742. *In* E. H. Lennette, P. Halonen, and F. A. Murphy (ed.), *Laboratory Diagnosis of Infectious Diseases: Principles and Practice,* vol. 2. Springer-Verlag, New York.
4. **Al-Nakib, W., and D. A. J. Tyrrell.** 1992. Drugs against rhinoviruses. *J. Antimicrob. Chemother.* **30:**115–117.
5. **Altmar, R. L., and P. R. Georghiou.** 1993. Classification of respiratory tract picornavirus isolates as enteroviruses or rhinoviruses by using reverse transcription polymerase chain reaction. *J. Clin. Microbiol.* **31:**2544–2546.
6. **Andrewes, C. H., D. M. Chaproneiri, A. E. H. Gompels, H. G. Pereira, and A. T. Roden.** 1953. Propagation of common cold virus in tissue cultures. *Lancet* **i:**546–547.

7. Ansari, S. A., V. S. Springthorpe, S. A. Sattar, S. Rivard, and M. Rahman. 1991. Potential role of hands in the spread of respiratory viral infections: studies with human parainfluenza virus 3 and rhinovirus 14. *J. Clin. Microbiol.* 29:2115–2119.

8. Balfour-Lynn, I. M., H. B. Valman, G. Stanway, and M. Khan. 1992. Use of polymerase chain reaction to detect rhinovirus in wheezy infants. *Arch. Dis. Child.* 67:760.

9. Barclay, W. S., and W. Al-Nakib. 1987. An ELISA for the detection of rhinovirus specific antibody in serum and nasal secretion. *J. Virol. Methods* 15:53–64.

10. Brown, P. K., and D. A. J. Tyrrell. 1964. Experiments on the sensitivity of strains of human fibroblasts to infection with rhinovirus. *Br. J. Exp. Pathol.* 45:571–578.

11. Bruce, C. B., W. Al-Nakib, J. W. Almond, and D. A. J. Tyrrell. 1989. Use of synthetic oligonucleotide probes to detect rhinovirus RNA. *Arch. Virol.* 105:179–187.

12. Callahan, P. L., S. Mizutani, and R. J. Colonno. 1985. Molecular cloning and complete sequence determination of RNA genome of human rhinovirus type 14. *Proc. Natl. Acad. Sci. USA* 82:732–736.

13. Cate, T. R., R. B. Couch, and K. M. Johnson. 1964. Studies with rhinovirus in volunteers; production of illness, effect of naturally acquired antibody, and demonstration of a protective effect not associated with serum antibody. *J. Clin. Invest.* 43:56–67.

14. Cohen, S., and D. A. J. Tyrrell. 1991. Psychological stress and susceptibility to the common cold. *N. Engl. J. Med.* 325:606–612.

15. Conant, R. M., N. L. Somerson, and V. V. Hamparian. 1968. Rhinovirus: basis for a numbering system. I. HeLa cell for propagation and serologic procedures. *J. Immunol.* 100:107–113.

16. Cooney, M. K., J. P. Fox, and G. E. Kenny. 1982. Antigenic groupings of 90 rhinovirus serotypes. *Infect. Immun.* 37:642–647.

17. Couch, R. B. 1990. Rhinoviruses, p. 607–629. In B. N. Fields and D. M. Knipe (ed.), *Field's Virology*, 2nd ed. Raven Press, New York.

18. Dearden, C. J., and W. Al-Nakib. 1987. Direct detection of rhinoviruses by an enzyme-linked immunosorbent assay. *J. Med. Virol.* 23:179–189.

19. de Arruda, E., III, T. E. Mifflin, J. M. Gwaltney, Jr., B. Winther, and F. G. Hayden. 1991. Localization of rhinovirus replication in vitro with in situ hybridization. *J. Med. Virol.* 34:38–44.

20. Dick, E. C., S. U. Hossain, K. A. Mink, C. K. Meschievitz, S. B. Schultz, W. J. Raynor, and S. L. Inhorn. 1986. Interruption of transmission of rhinovirus colds among human volunteers using virucidal paper handkerchiefs. *J. Infect. Dis.* 153:352–356.

21. Dick, E. C., L. C. Jennings, K. A. Mink, C. D. Wartgow, and S. L. Inhorn. 1987. Aerosol transmission of rhinovirus colds. *J. Infect. Dis.* 156:442–448.

22. Dochez, A. R., G. S. Shibley, and K. C. Mills. 1930. Studies of the common cold. IV. Experimental transmission of the common cold to anthropoid apes and human beings by means of a filtrable agent. *J. Exp. Med.* 52:701–716.

23. Douglas, R. G., Jr. 1970. Pathogenesis of rhinovirus common colds in human volunteers. *Ann. Otol. Rhinol. Laryngol.* 79:563–571.

24. Douglas, R. G., Jr., W. F. Fleet, T. R. Cate, and R. B. Couch. 1968. Antibody to rhinovirus in human sera. I. Standardization of a neutralization test. *Proc. Soc. Exp. Biol. Med.* 127:497–502.

25. Douglas, R. M., B. W. Moore, H. B. Miles, L. M. Davis, N. M. H. Graham, P. Ryan, D. A. Warswick, and J. K. Albrecht. 1986. Prophylactic efficacy of intranasal alpha-2 interferon against rhinovirus infections in the family setting. *N. Engl. J. Med.* 314:65–70.

26. Dreizin, R. S., N. M. Borovkova, T. I. Ponomareva, E. M. Vikhnovich, A. A. Kheinitis, and T. V. Leichinskaya. 1975.

Diagnosis of rhinovirus infections by virological and immunofluorescent methods. *Acta Virol.* 19:413–418.

27. Gama, R. E., P. R. Horsnell, P. J. Hughes, C. Northm, C. B. Bruce, W. Al-Nakib, and G. Stanway. 1989. Amplification of rhinovirus specific nucleic acids from clinical material using the polymerase chain reaction. *J. Med. Virol.* 28:73–77.

28. Geist, F. C., and F. G. Hayden. 1985. Comparative susceptibilities of strain MRC-5 human embryonic lung fibroblast cells and the Cooney strain of human fetal tonsil cells for isolation of rhinoviruses from clinical specimens. *J. Clin. Microbiol.* 22:455–456.

29. Greve, J. M., C. P. Forte, C. W. Marlor, A. M. Meyer, H. Hoover-Litty, D. Wunderlich, and A. McClelland. 1991. Mechanisms of receptor-mediated rhinovirus neutralization defined by two soluble forms of ICAM-1. *J. Virol.* 65:6015–6023.

30. Gwaltney, J. M., Jr. 1992. Combined antiviral and antimediator treatment of rhinovirus colds. *J. Infect. Dis.* 166:776–782.

31. Gwaltney, J. M., Jr., R. J. Colonno, V. V. Hamparian, and R. B. Turner. 1989. Rhinoviruses, p. 579–614. In N. J. Schmidt and R. W. Emmons (ed.), *Diagnostic Procedures for Viral, Rickettsial and Chlamydial Infections*, 6th ed. American Public Health Association, Washington, D.C.

32. Gwaltney, J. M., Jr., P. B. Moskalski, and J. O. Hendley. 1978. Hand-to-hand transmission of rhinovirus colds. *Ann. Intern. Med.* 88:463–467.

33. Hamparian, V. V., R. J. Colonno, M. K. Cooney, E. C. Dick, J. M. Gwaltney, Jr., J. H. Hughes, W. S. Jordan, Jr., A. Z. Kapikian, W. J. Mogabgab, A. Monto, C. A. Phillips, R. R. Rueckert, J. H. Schieble, E. J. Stott, and D. A. J. Tyrrell. 1987. A collaborative report: rhinoviruses—extension of the numbering system from 89 to 100. *Virology* 159:191–192.

34. Harrison, S. C. 1993. Common cold virus and its receptor. *Proc. Natl. Acad. Sci. USA* 90:783.

35. Hendley, J. O., J. M. Gwaltney, Jr., and W. S. Jordan, Jr. 1969. Rhinovirus infections in an industrial population. IV. Infections within families of employees during two fall peaks of respiratory illness. *Am. J. Epidemiol.* 89:184–196.

36. Hendley, J. O., R. P. Wenzel, and J. M. Gwaltney, Jr. 1973. Transmission of rhinovirus colds by self-inoculation. *N. Engl. J. Med.* 288:1361–1364.

37. Higgins, P. G., E. M. Ellis, and D. A. Woolley. 1969. A comparative study of standard methods and organ culture for the isolation of respiratory viruses. *J. Med. Microsc.* 2:109.

38. Hoorn, B. 1966. Organ cultures of ciliated epithelium for the study of respiratory viruses. *Acta Pathol. Microbiol. Scand.* 66(Suppl. 183):1–37.

39. Hoover-Litty, H., and J. M. Greve. 1993. Formation of rhinovirus-soluble ICAM-1 complexes and conformational changes in the virion. *J. Virol.* 67:390–397.

40. Horn, M. E. C., and I. Gregg. 1973. Role of viral infection and host factors in acute episodes of asthma and chronic bronchitis. *Chest* 64:44–48.

41. Johnston, S. L., G. Sanderson, P. K. Pattemore, S. Smith, P. G. Vardin, C. B. Bruce, P. R. Lambden, D. A. J. Tyrrell, and S. T. Holgate. 1993. Use of polymerase chain reaction for diagnosis of picornavirus infection in subjects with and without respiratory symptoms. *J. Clin. Microbiol.* 31:111–117.

42. Kellner, G., T. Popow-Kraupp, C. Binder, I. Goedl, M. Kundi, and C. Kunz. 1991. Respiratory tract infections due to different rhinovirus serotypes and the influence of maternal antibodies on the clinical expression of the disease in infants. *J. Med. Virol.* 35:267–272.

43. Kellner, G., T. Popow-Kraupp, M. Kundi, C. Binder, and C. Kunz. 1989. Clinical manifestations of respiratory tract infections due to respiratory syncytial virus and rhinoviruses in hospitalized children. *Acta Paediatr. Scand.* 78:390–394.

44. Kellner, G., T. Popow-Kraupp, M. Kundi, C. Binder, H. Wallner, and C. Kunz. 1988. Contribution of rhinoviruses to respiratory viral infections in childhood: a prospective study in

a mainly hospitalized infant population. *J. Med. Virol.* **25:**455–469.

45. **Krilov, L., L. Pierik, E. Keller, K. Mahan, D. Watson, M. Hirsch, V. Hamparian, and K. McIntosh.** 1986. The association of rhinoviruses with lower respiratory tract disease in hospitalized patients. *J. Med. Virol.* **19:**345–352.

46. **Las Heras, J., and V. L. Swanson.** 1983. Sudden death of an infant with rhinovirus infection complicating bronchial asthma: case report. *Pediatr. Pathol.* **1:**319–323.

47. **Last-Barney, K., S. D. Marlin, E. J. McNally, C. Cahill, D. Jeanfavre, R. B. Faanes, and V. J. Merluzzi.** 1991. Detection of major group rhinoviruses by soluble intercellular adhesion molecule-1 (sICAM-1). *J. Virol. Methods* **35:**255–264.

48. **Levandowski, R. A.** 1991. Rhinoviruses, p. 412–426. *In* R. B. Belshe (ed.), *Human Virology.* Mosby Year Book, St. Louis.

49. **Longberg-Holm, K., and F. H. Yin.** 1973. Antigenic determinants of infective and inactivated human rhinovirus type 2. *J. Virol.* **12:**114–123.

50. **Minor, T. E., E. C. Dick, A. N. DeMeo, J. J. Ouellette, M. Cohen, and C. E. Reed.** 1974. Viruses as precipitants of asthmatic attacks in children. *JAMA* **227:**292–298.

51. **Monto, A. S., E. R. Bryan, and S. Ohmit.** 1987. Rhinovirus infections in Tecumseh, Michigan: frequency of illness and number of serotypes. *J. Infect. Dis.* **156:**43–49.

52. **Naclerio, R. M., D. Proud, L. M. Lichtenstein, A. Kagey-Sobotka, J. O. Hendley, and J. M. Gwaltney, Jr.** 1988. Kinins are generated during experimental rhinovirus colds. *J. Infect. Dis.* **157:**133–142.

53. **Olive, D. M., S. Al-Mufti, W. Al-Mulla, M. A. Khan, A. Pasca, G. Stanway, and W. Al-Nakib.** 1990. Detection and differentiation of picornaviruses in clinical samples following genomic amplification. *J. Gen. Virol.* **71:**2141–2147.

54. **Olson, N. H., P. R. Kolatkar, M. A. Oliveira, R. H. Cheng, J. M. Greve, A. McClelland, T. S. Baker, and M. G. Rossman.** 1993. Structure of a human rhinovirus complexed with its receptor molecule. *Proc. Natl. Acad. Sci. USA* **90:**507–511.

55. **Pelon, W., W. J. Mogabgab, I. A. Phillips, and W. E. Pierce.** 1957. A cytopathogenic agent isolated from naval recruits with mild respiratory illness. *Proc. Soc. Exp. Biol. Med.* **94:**262–267.

56. **Pevear, D. C., M. J. Fancher, P. J. Felock, M. G. Rossmann, M. S. Miller, G. Diana, A. M. Treasurywala, M. A. McKinlay, and F. J. Dutko.** 1989. Conformational change in the floor of the human rhinovirus canyon blocks adsorption to HeLa cell receptors. *J. Virol.* **63:**2002–2007.

57. **Price, W. H.** 1956. The isolation of a new virus associated with respiratory clinical disease in humans. *Proc. Natl. Acad. Sci. USA* **42:**892–896.

58. **Rossman, M. G., E. Arnold, J. W. Erickson, E. A. Frankenberger, J. P. Griffith, H. J. Hecht, J. E. Johnson, G. Kramer, M. Luo, A. G. Mosser, R. R. Rueckert, B. Sherry, and G. Vriend.** 1985. Structure of a human common cold virus and functional relationship to other picornaviruses. *Nature* (London) **317:**145–153.

59. **Schmidt, H. J., and R. J. Fink.** 1991. Rhinovirus as a lower respiratory tract pathogen in infants. *Pediatr. Infect. Dis. J.* **10:**700–702.

60. **Sperber, S. J., J. O. Hendley, F. G. Hayden, D. K. Riker, J. V. Sorrentino, and J. M. Gwaltney, Jr.** 1992. Effects of naproxen on experimental rhinovirus colds. *Ann. Intern. Med.* **117:**37–41.

61. **Stone, A. A., D. H. Bovbjerg, J. M. Neale, A. Napoli, H. Valdimarsdottir, D. Cox, F. G. Hayden, and J. M. Gwaltney, Jr.** 1992. Development of common cold symptoms following experimental rhinovirus infection is related to prior stressful life events. *Behav. Med.* **18:**115–120.

62. **Stott, E. J., and M. Walker.** 1969. Antigenic variation among strains of rhinovirus type 51. *Nature* (London) **224:**1311–1312.

63. **Turner, B. W., W. S. Cail, J. O. Hendley, F. G. Hayden, W. J. Doyle, J. V. Sorrentino, and J. M. Gwaltney, Jr.** 1992. Physiologic abnormalities in the paranasal sinuses during experimental rhinovirus colds. *J. Allergy Clin. Immunol.* **90:**474–478.

64. **Tyrrell, D. A. J.** 1992. A view from the Common Cold Unit. *Antiviral Res.* **18:**105–125.

65. **Uncapher, C. R., C. M. DeWitt, and R. J. Colonno.** 1991. The major and minor group receptor families contain all but one human rhinovirus serotype. *Virology* **180:**814–817.

Adenoviruses

JOHN C. HIERHOLZER

80

CLINICAL BACKGROUND

Adenoviruses have been recovered from virtually every organ system of humans and have been associated with many clinical syndromes (Table 1). Adenovirus illnesses are endemic throughout the year and occur in all age groups, although they are most common among school-age children, for whom approximately 50% of the infections are asymptomatic. Adenoviruses cause localized outbreaks of respiratory disease in the winter and spring, outbreaks of swimming pool-associated pharyngoconjunctival fever (PCF) in the summer, and epidemics of keratoconjunctivitis associated with industrial eye trauma or ophthalmologic procedures at any time of the year. Adenoviruses cause 5 to 15% of cases of gastroenteritis in infants and preschool children (4, 34). Epidemics of acute respiratory disease that can occur whenever new military recruits are housed together are now preventable by active immunization with the appropriate serotypes in enteric-coated capsules (32). Of the 47 serotypes presently described, the most common are types 1 through 8, 11, 21, 35, 37, and 40 (1–4, 6, 10, 14, 15, 19, 27–29, 34, 36, 38–40).

Upper respiratory illness caused by adenoviruses can be in the form of common colds, pharyngitis, or tonsillitis and occurs chiefly in infants and young children. It is associated primarily with types 1 through 7 (6, 28, 29). Findings include coryza, fever, cough, exudate on the pharyngeal walls, a granular appearance of the mucosa, and tender enlarged cervical nodes. A notable feature of infection with types 1, 2, 5, and 6 is the persistence of virus in a latent state in adenoidal and tonsillar tissues in about 50% of infected children. Another epidemiologically important feature is the excretion of virus in the stool for many months without recurrence of symptoms.

Lower respiratory illness, including bronchitis, bronchiolitis, and pneumonia, often complicates adenovirus infection. Severe pneumonia, sometimes fatal, occurs in infants and children (rarely in adults) and is mostly caused by types 3, 4, 7, and 21 (1, 6, 27–31). Extrapulmonary signs such as kidney and liver involvement and encephalomeningitis can be seen, especially in infants and immunocompromised patients, and in such patients, the generalized disease usually has a fatal outcome. Some children who recover from type 3, 7, or 21 pneumonia have residual lung disease. A cough syndrome similar to whooping cough has been reported in children following type 5 and rarely type 1 through 3 infections (15).

PCF is characterized by fever, pharyngitis, conjunctivitis, and cervical lymphadenopathy, often with headache, diarrhea, and rash. Typical illness with follicular conjunctivitis is generally evident 5 to 6 days after exposure, with infected individuals shedding virus for about 10 days. Types 1 through 7 are usually incriminated in these infections (28, 29).

Epidemic keratoconjunctivitis (EKC) as sporadic cases is caused by many adenoviruses but in epidemic form is caused by type 8 or 37 (14, 37, 40). This severe eye disease becomes apparent after an 8- to 10-day silent incubation period. The initial onset of follicular conjunctivitis with preauricular lymphadenopathy can be accompanied by headache, malaise, and, depending on the serotype, mild upper respiratory illness. After 1 to 2 weeks of these symptoms, subepithelial corneal keratitis develops and may persist for an extended period. Contaminated eye instruments, ophthalmic wash solutions, hands of medical personnel, and towels are the vehicles by which ocular infections are spread to susceptible persons; thus, most cases of EKC in nonindustrial communities are spread iatrogenically. Spread of adenoviruses in eye clinics can be eliminated, however, by triaging patients, separating patients and staff into cohorts, washing hands thoroughly, using unit dosage eyedrops, avoiding invasive procedures on patients with EKC, and properly disinfecting equipment and surfaces. In industrial settings with high levels of airborne particulates, EKC outbreaks have resulted from eye-fomite-eye transmission. Trauma to the corneal epithelium is required for initiating virus infection, particularly with type 8. In type 19- or 37-associated EKC, infection can spread to the eye from respiratory or genital sites (13, 15, 36).

Acute hemorrhagic conjunctivitis and acute hemorrhagic cystitis are both sometimes caused by type 11 and rarely caused by type 21, and the virus is readily recovered from eye secretions or urine as appropriate (5, 15).

Immunocompromised host disease in patients with immune deficiencies; patients with kidney, liver, and marrow transplants or undergoing cancer chemotherapy; and those with AIDS is a major problem, because these infections are often fatal (14, 21). Adenoviruses cause generalized illness in perhaps 10% of such patients; virus has been recovered

TABLE 1 Adenovirus infections and the serotypes involved[a]

Syndrome	Signs and symptoms	Adenovirus serotype(s) involved	
		Frequently	Infrequently
URI	Coryza, pharyngitis, fever, tonsillitis, diarrhea	1–3, 5, 7	4, 6, 11, 18, 21, 29, 31
LRI	Bronchitis, pneumonia, fever, coryza, cough	3, 4, 7, 21	1, 2, 5, 35
Pertussis syndrome	Paroxysmal cough, vomiting, fever, URI	5	1, 2, 3
ARD	Tracheobronchitis, fever, myalgia, coryza, pneumonia	4, 7	3, 14, 21, 35
PCF	Pharyngitis, conjunctivitis, fever, coryza, headache, diarrhea, rash, nodes	3, 4, 7	1, 11, 14, 16, 19, 37
EKC	Keratitis, headache, preauricular nodes, coryza, pharyngitis, diarrhea	8, 37	3, 4, 7, 10, 11, 19, 21
AHC	Chemosis, follicles, subconjunctival hemorrhage, preauricular nodes, fever	11	2–8, 14, 15, 19, 37
Cystitis	Cystitis (usually hemorrhagic), fever, pharyngitis	11	7, 21, 34, 35
Immunocompromised host disease	Diarrhea, rash, URI, pneumonia, hepatitis, cystitis, otitis media	1, 2, 5, 11, 34, 35	7, 21, 29–31, 37–39, 43, 45
Gastroenteritis (infant)	Diarrhea, fever, nausea, vomiting, mild URI	31, 40, 41	1, 2, 12–17, 21, 25, 26, 29
CNS disease	Meningitis, encephalitis, Reye's syndrome	7	3, 32
Venereal disease	Ulcerative genital lesions, urethritis, cervicitis	2, 37	1, 5, 7, 11, 18, 19, 31

[a]URI, upper respiratory illness; LRI, lower respiratory illness; ARD, acute respiratory disease; AHC, acute hemorrhagic conjunctivitis; CNS, central nervous system.

from brain, throat, leukocytes, lung, urine, stool, cerebrospinal fluid, kidney, liver, and pancreas (5, 14, 15, 19, 38).

Gastroenteritis and related syndromes have been associated with adenovirus infection in infants and children (15, 29). Types 40 and 41 and to a lesser extent types 2 and 31 cause acute gastroenteritis (4, 7, 11, 12, 20–22, 34).

Neurologic disease is sometimes caused by adenoviruses, usually as part of multisystem disease, with type 7 being particularly severe. Reye's syndrome is infrequently associated with the childhood serotypes, especially type 3. Types 1 through 3, 6, 7, 12, and 32 have been isolated from cerebrospinal fluid and brain tissue (15, 29).

Serotypes 2, 19, and 37 have been recovered from herpesviruslike genital lesions, often associated with orchitis, cervicitis, or urethritis. Other syndromes have also been described (13, 15, 29). Finally, adenoviruses are often incriminated in nosocomial transmission. This is not surprising, because they cause 10% of the pneumonia cases in hospitalized children, intensive supportive therapy is required for these patients, adenoviruses are shed for long periods in the stool and are transmissible both by fomites and aerosolized droplets, and the viruses can readily infect respiratory and ocular tissues of other patients and of susceptible hospital personnel (4, 6, 10, 28, 29, 31).

DESCRIPTION OF THE AGENT

Human adenoviruses are nonenveloped, double-stranded DNA viruses of the family *Adenoviridae*, genus *Mastadenovirus*. They are 70 to 90 nm in diameter and icosahedral in shape, comprising 10 structural proteins with molecular weights of 5,000 to 120,000. They have buoyant density in cesium chloride of 1.33 to 1.34 g/cm^3 and a molecular weight by sedimentation coefficient of 170×10^6 to 175×10^6.

Adenoviruses replicate in the cell nucleus and tend to be host species specific. The viruses produce characteristic cytopathic effects (CPE) that are accompanied by accumulation of multiple antigenic components and organic acids in the host cell culture fluids. Adenoviruses do not hemadsorb erythrocytes (RBCs) or replicate in embryonated chicken eggs; all possess genus-specific antigenic determinants on the hexon capsomeres. The capsid proteins of adenoviruses are arranged in an icosahedron having 20 triangular faces and 12 vertices. In each virion are 240 hexons and 12 pentons. The hexons are dispersed on the triangular faces and edges, and the pentons are located in the vertices of the icosahedron. Each penton consists of a base (or vertex capsomere) and a fiber that is a rodlike outward projection of variable length with a terminal knob. Inside the capsid is a single molecule of linear, double-stranded DNA of 20×10^6 to 24×10^6 molecular weight. The G+C base compositions of the genomes in different human adenoviruses range from 47 to 60% (26, 35, 36).

Crude suspensions of most adenoviruses are stable for prolonged periods at -20 to $-100°C$ at a pH of 6 to 9. Infectious virus is rapidly inactivated at 56°C and by exposure to 0.25% sodium dodecyl sulfate, free chlorine at 0.5 µg/ml, UV irradiation, or a 1:400 to 1:4,000 dilution of formalin. Viral infectivity is not affected by treatment with ether or chloroform, and the virus can be readily lyophilized without special precautions. Lyophilized viruses appear to retain their infectivity indefinitely at 4 or $-10°C$ or colder.

A classification scheme based on hemagglutination (HA) properties allows a useful separation of the human adenoviruses into six subgroups (18) (Table 2). The scheme is based primarily on complete agglutination of monkey or rat RBCs, partial agglutination of rat RBCs, and level (titer) of agglutination and secondarily on complete agglutination of human, chicken, and other RBCs. Other viral properties such as antigenic relationships, oncogenicity, fiber length, percent DNA homology within and between subgroups, percent G+C content of the DNA, number of cleavage fragments after digestion with SmaI endonuclease, and molecular weights of certain internal proteins tend to coincide with this classification (3, 15, 18, 20, 26, 35, 36).

TABLE 2 Subdivision of human adenoviruses by HA properties

Subgroup[a]	Serotype(s)	Oncogenicity	HA subgroup	Sensitive RBCs
A	12, 18, 31	High	3B[b]	Rat (incomplete)
B	3, 7, 11, 16, 34, 35	Weak	1A	Monkey
	14, 21	Weak	1B[c]	Monkey
C	1, 2, 5, 6	Negative	3A	Rat (incomplete)
D	8, 9, 37	Negative	2A	Rat, mouse, human, guinea pig, dog
	10, 19, 26, 27, 36, 38, 39	Negative	2B	Rat, mouse, human
	13, 43	Negative	2C	Rat, mouse, human, monkey
	15, 22, 23, 30, 44, 45, 46, 47	Negative	2D	Rat, mouse, monkey
	17, 24, 32, 33, 42	Negative	2E	Rat, mouse
	20, 25, 28, 29	Negative	2F[c]	Rat (atypical), monkey
E	4	Negative	3A	Rat (incomplete)
F	40, 41	Negative	3A[c]	Rat (atypical)

[a]Adapted from references 15 and 18.
[b]Very low HA titers.
[c]Moderate-range HA titers.

The replication of adenoviruses is accompanied by the excess formation of antigenic components liberated into the culture fluids as soluble antigens. These proteins are complex structures that carry many antigenic determinants; in particular, the hexons, vertex capsomeres, and fibers possess distinct type- and subgroup-specific antigens. The type-specific determinants on the hexon and fiber are exposed on the surface of the virion and give rise to serum-neutralizing (SN) antibodies. In addition, the fiber is a strong hemagglutinin and thus elicits HA inhibition (HI) antibodies. The genus-specific antigen is the principal determinant on the hexon, but it resides on the internal part of the capsid and thus is not exposed externally and does not elicit protective antibodies. The vertex capsomere is the toxic factor seen in overinoculated cultures. Additional determinants on these components evoke heterotypic SN and HI antibody responses. These determinants are mostly subgroup specific, although intersubgroup relationships may exist (26, 36).

The complete agglutination of monkey and rat RBCs by adenoviruses of subgroups B and D is associated with intact virus particles and three forms of soluble hemagglutinins. The complete soluble forms consist of dodecons (groups of 12 pentons), dimers of pentons, and dimers of fibers. The partial agglutination of rat RBCs seen with subgroup C viruses is due to the relative excess of monomeric pentons and fibers in cell culture fluids. These soluble antigens, however, produce complete agglutination of rat RBCs in the presence of a heterotypic subgroup C antiserum in the test diluent (18, 26, 29, 36). Thus, the adenovirus components have specific biologic properties and serologic reactivities, which form the basis of many laboratory tests.

COLLECTION AND STORAGE OF SPECIMENS

Adenoviruses are stable viruses and are readily recovered from obvious sites of infection. Nasopharyngeal swabs or aspirates, nasal swabs or aspirates, swabs of eye exudates, stool or rectal swabs, urine, urethral or cervical swabs, and biopsy or autopsy tissues are all adequate specimens as dictated by the patient's symptoms. Viral detection rates are greatly enhanced by proper collection of specimens early in the course of disease and by prompt shipment of cold or frozen specimens to the laboratory. Details of the collection

of respiratory, conjunctival, rectal, and other specimens are given in chapter 3 of this Manual.

Paired blood samples are needed to establish or confirm a diagnosis by serologic methods. The first specimen should be collected as soon as possible after the onset of symptoms, and the second should be collected 2 to 4 weeks later. After clotting, serum is separated under sterile conditions and stored at −10 to −20°C.

DIRECT EXAMINATION

Many methods have been applied to the direct detection of adenoviruses in clinical materials, and research in this area is still keen because of the ongoing need to quickly and accurately diagnose a viral infection. Electron microscopy and immunoelectron microscopy have been the principal means of identifying adenoviruses in clinical materials. In one procedure, tissue sections obtained at biopsy or autopsy are fixed, stained, and examined for inclusions or "crystalline" arrays of mature virions in the cell nucleus. In the other method, specimens of respiratory secretions, urine, or stool (as 10 to 20% extracts) are clarified and observed directly by pseudoreplica negative-stain electron microscopy. Sensitivity is increased by concentrating the specimen (in the case of throat or nasal wash, bronchial aspirate, and urine) by ultracentrifugation or membrane ultrafiltration. Sensitivity is further increased by immunoelectron microscopy, in which the specimen (whether concentrated or not) is incubated with a hyperimmune antiserum or a human convalescent-phase serum before grids are prepared for electron microscopy; aggregates of virus particles then indicate both the viral morphology and a specific reaction with the serum used. The first method has proven useful for diagnosis by tissue samples and for pathologic studies. The latter method has been highly useful for diagnosis of adenovirus gastroenteritis (7, 22).

For immunofluorescence assay (IFA) examination, cells in unfrozen specimens are washed, suspended in phosphate-buffered saline (PBS), fixed on slides in acetone, and reacted with antihexon serum and antiglobulin conjugate (see chapter 4 of this Manual). Dense nuclear and stippled cytoplasmic staining are positive reactions for adenoviruses. A sensitive extension of IFA is the time-resolved fluoroimmunoassay (TR-FIA), in which a purified adenovirus monoclonal antibody (MAb) is used as capture antibody in plastic wells, the specimen is added to the wells, and a europium-labeled adenovirus MAb is used as detector antibody. The specific fluorescence of the sample is measured by a fluorometer after a time delay to allow the autofluorescence to decay (15, 17).

For antigen detection by enzyme immunoassay (EIA), a polyclonal capture antibody is adsorbed to the plates, and then antigens or specimens, a second polyclonal antibody (detector) from a different animal species, enzyme-conjugated antispecies antibody, and the substrate-color system are added in that order (see chapter 11 of this Manual). Sensitivity is increased by using MAbs at the detector antibody position and is further increased by using MAbs in both positions in the assay (15, 17, 22).

Radioimmunoassay is equal in sensitivity and ease of performance to EIA but carries the problem of disposal of the radioactive wastes produced. Radioimmunoassays have been used to detect adenoviruses in respiratory secretions and stool specimens (24). Counterimmunoelectrophoresis is a convenient test for detecting adenoviruses in toxic and environmental samples but is not as sensitive as EIA. Counterimmunoelectrophoresis has identified adenovirus antigens in swimming pool water, serum specimens, ophthalmic drops used in eye clinics, and stool specimens (15).

Restriction enzyme analysis (REA) has found limited application in direct detection, being useful only for finding types 40 and 41 in stool specimens (4, 21, 34). The most recent concept for direct detection of adenoviruses is DNA probe technology (see chapter 13 of this Manual). Labeled probes have been most successful with the enteric adenoviruses (11, 15, 25).

The most recent, but not yet generally available, direct test for adenoviruses is DNA amplification by PCR (15, 16, 23). Utilizing primer pairs from highly conserved sequences in the hexon gene, PCR, followed by detection of the amplified gene products by ethidium bromide staining and agarose gel electrophoresis, can detect adenovirus of any serotype in various types of clinical specimens (23). In a further refinement, the products can be reacted in a liquid-phase hybridization system with biotin- and europium-labeled probes; this allows for increased sensitivity when the probes are reacted in streptavidin-coated wells and quantitated by time-resolved fluorometry (16).

ISOLATION OF VIRUS

All of the human adenoviruses except types 40 and 41 replicate and produce CPE in continuous human cell lines of epithelial origin such as HeLa, KB, A549, or HEp-2; primary human embryonic kidney (HEK) cells; and, after adaptation, human embryonic lung fibroblasts (HELF, WI38, MRC-5) and other embryonic fibroblast cells. Types 40 and 41 replicate poorly in the aforementioned cells but grow to consistent (but low) titers in tertiary cynomolgus monkey kidney cells, the Graham-293 adenovirus type 5-transformed secondary HEK cell line, and HEp-2 cells under special conditions (15, 18). For cultivating adenoviruses, Eagle minimal essential medium supplemented with 2% heat-inactivated fetal calf serum and antibiotics is satisfactory as a maintenance medium. Infected cells become enlarged, rounded, and very refractile, and they aggregate into irregular clusters typical of adenovirus CPE (4, 6, 7, 10, 12, 20).

Preparation of the specimen for isolation is important. Respiratory and urine specimens are treated with antibiot-

ics and clarified by low-speed centrifugation for 3 min to remove cells, fibers, and other debris. Stool specimens are brought to 10 to 20% suspensions with PBS or maintenance medium, shaken with glass beads if necessary, and clarified by centrifugation for 30 min. Biopsy or autopsy tissue specimens are minced into small fragments with sterile surgical scissors and then homogenized in a glass tissue grinder or a sterile mortar with a sand abrasive and 5 ml of PBS or maintenance medium. A 10 to 20% suspension is made in this manner and then clarified by low-speed centrifugation. Blood monocytes are prepared for culture by applying 3 ml of heparinized blood cells on a Ficoll-Hypaque gradient. The cells are washed and suspended in RPMI medium with 20% fetal calf serum to a final concentration of 1.5×10^6 cells per ml. Portions (0.5 ml) of the cell suspensions are then placed on human embryonic fibroblast feeder monolayers for cocultivation.

Specimens are inoculated onto cell cultures as soon as practical after they are received. Undiluted test material (0.4 to 0.5 ml) is inoculated into one or two cell culture tubes, adsorbed for 1 h at ambient temperature, overlaid with 1 ml of maintenance medium, and incubated at 35 to 36°C either stationary or on a roller drum. Several uninoculated tubes of each cell type used should be incubated and observed for evidence of nonspecific cellular degeneration and for eventual use as negative controls in identity tests. Sometimes, inoculation of a clinical specimen, especially stool, blood, or urine, leads to transient or irreparable toxic effects on the cell culture. Such toxicity is apparent within 24 h but can be minimized by washing the cell cultures or at least decanting the inoculum after the 1-h adsorption period. All cultures should be observed for CPE twice a week, fed with fresh maintenance medium as required, and subpassaged once after 2 weeks (for HEK and fibroblast cells) or three times at weekly intervals (for continuous epithelial cells), so that all cultures are held and read for at least 28 days. This regimen yields the highest isolation rates. For identification, degeneration should progress until all the cells are affected (4+ CPE) in order to obtain sufficient yields of soluble antigens, hemagglutinins, and infectious virus for identifying the isolate. Soluble-antigen yields are highest in HEp-2 or KB cells; infectious virus yields are highest in HEK cells for types 1 through 39 and 42 through 47 and in Graham-293 cells for types 40 and 41.

A highly practical improvement in adenovirus detection is the shell vial assay (9, 33). In this procedure, the cell monolayer is first established on coverslips, the clinical specimen is added, and the culture is incubated for 1 or 2 days; IFAs on the coverslip with an antihexon MAb or other antisera determine the presence or absence of adenovirus. Sensitivity can be increased by a brief centrifugation of the inoculated cultures (9).

IDENTIFICATION OF VIRUS

Grouping Tests

All of the tests used for direct detection are also used for identifying an isolate as adenovirus. Immunoelectron microscopy, IFA, RIA, TR-FIA, and PCR are solely genus-specific tests except when specialized reagents such as MAbs are employed. Counterimmunoelectrophoresis, EIA, and DNA hybridization assays are genus specific but can be made type specific if selected antisera, MAbs, or DNA fragments are used, as appropriate. REA is type specific if the electrophoretic pattern matches that of a prototype strain (which is unlikely); no antisera are used. It is apparent, therefore, that the test procedure selected is determined by the level of typing desired. In general, IFA, EIA, TR-FIA, or complement fixation (CF) is employed to classify an isolate as adenovirus; agglutination with monkey, human, and rat RBCs is used to place the virus into a subgroup, and then HI and SN tests with selected antisera are utilized to serotype the virus.

The genus-specific IFA, EIA, TR-FIA, and CF test are based on reactions of soluble hexons in the culture fluids and antihexon serum or hexon-specific MAb (15). Procedures for the IFA and EIA are described elsewhere (see chapters 4 and 11 in this Manual), and the CF test is described below under Serological Diagnosis. Which test system to use depends on convenience; all will easily identify hexons in cell cultures.

Typing Tests

Adenovirus serotyping is traditionally based on neutralization; however, HI is more convenient in many laboratory situations (15, 18, 29, 36). In laboratories unable to procure fresh RBCs for HA-HI tests, the only typing method available is the SN test. In this case, antisera to types 1 through 8, 11, 21, 35, and 37 should be routinely included unless epidemiologic or clinical data for the isolate suggest a smaller set of sera (Table 1).

HI Test

Microtiter HA and HI tests are performed in either flexible or hard plastic U-plates. Serial twofold dilutions (1:2 to 1:4,096) of the isolate in 0.05-ml volumes are prepared in plain PBS diluent, in PBS with 1% type 4 antiserum (for subgroup C strains), and in PBS with 1% type 6 antiserum (for subgroup E). (Type 4 and 6 animal antisera are required for adenoviruses giving incomplete HA patterns. The sera should have titers of at least 1:320, be heat inactivated at 56°C for 30 min, and be absorbed with rat RBCs. The absorption is performed by adding 0.1 ml of RBCs for every 1.0 ml of a 1:10 dilution of serum, incubating the mixture overnight at 4°C, and pelleting the cells by centrifugation at 1,800 × g for 5 min at 4°C.)

RBCs are collected and stored in Alsever solution. Monkey and rat RBCs must be handled gently and used within 4 days of collection; immediately before use, the RBCs are washed and adjusted to 0.4% in PBS. An equal volume of 0.4% rhesus or vervet RBCs is now added to one set of duplicate PBS diluent dilutions, and rat RBCs are added to the other set. The same volume of each RBC suspension is added to two separate wells containing 0.05 ml of each type of diluent to serve as cell controls. The plates are then covered with tape, shaken gently, and incubated in a 37°C water bath for 1 h. After this time, the RBC control wells should exhibit complete sedimentation and a teardrop pattern when the plates are tilted slightly. An HA titer is defined as the reciprocal of the highest dilution of virus that shows complete HA; the dilution end point is then considered to contain 1 HA unit (HAU) per 0.05 ml. The patterns of HA titers with monkey, human, and rat RBCs in the PBS diluent and with rat RBCs in the heterotypic serum diluent thus define the subgroup to which the isolate belongs (Table 2) (15, 18).

For HI, the typing antisera are heat inactivated and absorbed with the appropriate RBCs as described above for type 4 and 6 immune sera. The antisera are diluted serially in 0.025-ml amounts in PBS diluent, and the virus is added at 4 HAU/0.025 ml (i.e., the HA titer divided by 8) per well. For a subgroup C or E isolate, antiserum dilutions are prepared in the appropriate heterotypic serum diluent as described above. The mixtures are agitated briefly and incubated at ambient temperature for 1 h before the appropriate RBC suspension is added (at 0.05 ml per well). A backtitration, i.e., serial twofold dilutions of the test antigen dose (0.05 ml in 0.05 ml of PBS or heterotypic serum diluent as appropriate), and cell controls are included in each test. The plates are incubated in the 37°C water bath for 1 h. After this incubation, the first four serial dilutions of the backtitration should exhibit a complete HA pattern, thus indicating that the proper antigen dose was used in the test. The isolate is identified by the typing serum that completely inhibits HA to or near the known homologous titer of the antiserum. Cross-reactions may be observed within antisera of the same subgroup; these are discussed below under Interpretation. Further characterization of the isolate may be obtained by SN tests.

SN Tests

SN tests are readable with definitive results and are the closest approximation the laboratory has to the human serum response following infection, but they are not as convenient to run as HA-HI tests. Three types of SN test can be carried out: the 7-day test in epithelial cell cultures, the 3-day test in MK cells, and microneutralization tests (6, 10, 15, 20, 29).

Conventional 7-day SN tests may be carried out in HEp-2, A549, HeLa, HEK, or Graham-293 cells. The virus isolate must be titrated for infectivity in the cells to be used in the test by preparing serial 10-fold dilutions of virus in Hanks balanced salt solution and inoculating the tubes in triplicate with 0.1 ml of dilution. After 7 days of incubation at 36°C, the end point (or titer) is read by the CPE and is considered to be 1 50% tissue culture infection dose ($TCID_{50}$) of virus per 0.1 ml. Since the slow and inconsistent development of CPE at high dilutions of an adenovirus culture tends to obscure the end point, a working dilution of virus with a titer of 100 $TCID_{50}$/0.1 ml is used. Type-specific antisera are then selected for the SN test, heat inactivated at 56°C for 30 min, and diluted in Hanks balanced salt solution in a twofold dilution series from 1:10 to 1:1,280 (or slightly beyond the homologous titer of each antiserum). For the SN step, 0.3 ml of virus dilution is mixed with 0.3 ml of each antiserum dilution, and the mixture is incubated at ambient temperature for 1 h and inoculated onto fresh cell cultures in triplicate at 0.2 ml per tube. Uninoculated cell controls and a backtitration consisting of a 10-fold dilution series from the working dilution used in the test are included in the test. The tube contents are incubated, fed, and read as usual, with the final reading taken at 7 days. The backtitration should show CPE ending around the second dilution. As in HI, the virus type is indicated by inhibition (of CPE) to or near the homologous titer of an antiserum. The antibody end point is thus defined as the reciprocal of the highest dilution of serum that completely neutralizes CPE.

The most widely used SN procedure is the 3-day test in primary rhesus MK cell cultures. This test does not measure inhibition of infectivity but rather inhibition of the viral toxicity induced by the vertex capsomere. If MK cells are not available, the test can be done in established MK lines, such as Vero, LLC-MK2, or BSC-1. To determine the challenge dose of virus to be used in the test, 0.1 ml of each serial twofold dilution (1:1 to 1:128) prepared in maintenance medium is inoculated into three cell cultures, each of which contains 1 ml of maintenance medium. The titration mixture is incubated at 35 to 36°C, and the degree of CPE is read at 3 days. The highest dilution of virus producing 1+ to 2+ CPE (i.e., 25 to 50% of the cell sheet is affected) in the cells at 3 days is considered the end point, and the virus dose to be used in the SN test is 1 dilution back from this end point. As described above, the selected antisera are heat inactivated, diluted serially to encompass the known homologous antibody titer of each, mixed with an equal volume (0.3 ml) of the isolate dilution, and incubated at ambient temperature for 1 h. Then, 0.2 ml of each virus-serum mixture is inoculated onto fresh cells in triplicate. Uninoculated cell control tubes and three virus control tubes, each containing 0.1 ml of the working dilution used in the test, are included in the test. The tube contents are incubated, fed, and read as usual, with the final reading taken at 3 days. At this time, the virus controls should show 1+ to 2+ CPE; the virus type is indicated by inhibition of CPE to or near the homologous titer of an antiserum.

Microneutralization tests can be done in secondary rhesus MK cells, Vero cells, and HEp-2 cells. In the Vero test, the virus titration is performed in flat-bottom 96-well plates in replicas of six, with the virus being serially diluted in growth medium with an automatic diluter with flame-sterilized loops, manually with a multichannel pipettor, or automatically with a programmable diluter and disposable tips. Then, 0.05 ml of a Vero cell suspension at 175,000 cells per ml is dropped into each well, including the cell control wells. The plate is gently shaken, sealed with sterile tape or a lid, incubated at 35 to 36°C for 5 days, and read by staining with a crystal violet-formaldehyde solution. The end point is the highest dilution of virus showing 0 to 1+ staining (=4+ to 3+ CPE) in 5 days. SN tests then utilize serial dilutions of heat-inactivated antisera made with sterile diluting equipment as described above, 2 U of virus (1 dilution back from the end point) per well, and the cell suspension. Plates are sealed, incubated, and stained as described above. Cell controls and a virus backtitration should be included. The test is read visually for uninfected (stained) cells, and the antibody end point is defined as described above. Types 40 and 41 have peculiar features that separate them from the other adenoviruses in terms of growth and identification in cell cultures; recent reviews concerning the enteric adenoviruses should be consulted for details of the specialized procedures used with them (4, 15, 20, 34, 36).

SEROLOGIC DIAGNOSIS

Serodiagnosis is accomplished with an acute- and convalescent-phase serum pair collected at an interval of 2 to 4 weeks. The timing for collection of the convalescent-phase serum is not critical provided the acute-phase serum was obtained early in the illness. Because of the ubiquity of the adenoviruses and the numerous cross-reactions between related serotypes, seroconversion involving a fourfold or greater rise in antibody titer is necessary to document infection. Adenovirus infection can be documented by a genus-reactive test such as the CF test or EIA, either of which can include the adenovirus antigen as part of a battery of antigens. For illnesses in which adenoviruses play an important role, such as acute respiratory disease, EKC, or PCF, genus-reactive tests can be followed by type-specific tests to pinpoint the serotype involved, or the type-specific

test alone can be carried out for the serotypes usually involved (10).

CF Test

The CF test is the most widely used serologic test in adenovirus infection and the best standardized. The test antigen is commercially available or is easily made by propagating a common serotype in HEp-2 or KB cells and harvesting by three freeze-thaw cycles 2 or 3 days after 4+ CPE occurs. The supernatant fluid after clarification serves as the CF antigen, which is tested for potency by block titrations against a reference human or animal antiserum. Alternatively, the sensitivity of the CF test can be improved by using a semipurified hexon preparation. The adenovirus CF antigen is stable and group specific, enabling it to detect all 47 human serotypes plus intermediate and atypical strains with equal reliability. Further, it can be part of a battery of a dozen or more viral antigens tested simultaneously, thereby making the test cost-effective and efficient in terms of labor and materials. The main disadvantage of the CF test is its low sensitivity: although CF titers are highly reproducible between tests and between laboratories, they detect infection in only 50 to 70% of cases in general and in <50% of cases in young children and adenovirus-immunized military recruits (15, 24, 30, 32).

EIA

EIA is much more sensitive than the CF test and is being used increasingly in diagnostic laboratories (24). Good reagents and automated EIA readers are now commercially available. The test can be carried out in either of two ways. In the first, a common adenovirus (e.g., type 2, 4, or 7) is cultured to produce a high titer of soluble antigens, as for the CF antigen. After clarification and dialysis against an alkaline buffer, the supernatant fluid is added to each well of a 96-well microtiter plate and incubated overnight at 4°C to prepare the solid-phase antigen. A purified or semipurified hexon preparation is preferable for this step, because cellular and medium proteins as well as the type-specific soluble components (fiber, dodecon, etc.) would be removed. After three washes with PBS containing 0.5% gelatin and 0.05% Tween 20, serial dilutions (in PBS-gelatin-Tween) of the patient's paired sera (1:200 to 1:25,600) are added to parallel rows, and the reaction mixture is incubated at 37°C for 1.5 h. After another three washings, a commercially available anti-human immunoglobulin G-peroxidase conjugate at about a 1:1,000 dilution in PBS-gelatin-Tween is added. The plates are placed at 37°C for 1 h, washed, and developed for color (see chapter 11 in this Manual).

In the second procedure, ammonium sulfate-precipitated immunoglobulin G from an adenovirus antiserum in alkaline buffer is attached to each well of a plate for the solid phase to serve as the capture antibody. After three washes as above, antigen (either crude or purified as described above) is added and incubated at 37°C for 1.5 h. The plates are again washed, and serial dilutions of the patient's paired sera (the detector antibody) are added. The patient's sera must be diluted (1:200 to 1:25,600) in PBS-gelatin-Tween solution containing 1.5% heat-inactivated normal serum of the capture antibody species; this procedure greatly reduces the nonspecific background color. After another incubation at 37°C for 1.5 h and a subsequent wash, the anti-human immunoglobulin G-peroxidase conjugate is added and incubated at 37°C for 1 h. The plates are then washed, and the substrate is added as described above.

Both procedures are highly sensitive and can be made quite reproducible if adequate care is taken throughout the test. Controls should include negative virus (cell controls), reagent controls, and positive and negative human sera. The background color in a well-devised test should be negligible, and in most EIA readers, it is automatically subtracted from the test serum readings.

HI Test

HI and SN tests are more sensitive than the CF test, and they measure type-specific antibody rises. Although some reagents are commercially available, most laboratories prepare their own stocks of the common adenovirus serotypes for use as HA antigens or infectious virus in HI and SN tests (14, 18, 30, 36). Human serum specimens are treated for HI tests by heat inactivation (56°C, 30 min) and removal of nonspecific agglutinins by absorption with 50% suspensions of rat or rhesus RBCs as appropriate. Sera are then diluted 1:8 to 1:1,024 in the microtiter plate, leaving 0.025 ml in each well. A virus suspension diluted to contain 4 HAU/0.025 ml is added to each well, mixed gently, and incubated at ambient temperature for 1 h. Then, 0.05 ml of RBC suspension is added per well, and the plate is agitated, sealed with tape, and incubated in a 37°C water bath for 1 h. A virus backtitration, serum controls, and cell controls should be included in the test. The end point (serum titer) is the highest dilution of serum completely inhibiting agglutination by 4 HAU of virus.

SN Tests

SN tests are performed as described above. Acute- and convalescent-phase serum specimens are inactivated at 56°C for 30 min and serially diluted (1:4 to 1:512) in maintenance medium in sterile tubes, leaving 0.3 ml of each dilution in the tubes. To each tube is added 0.3 ml of the dilution of seed virus. The tubes are shaken and incubated at ambient temperature for 1 h, and 0.2 ml of each virus-serum mixture is then added to each of two cell cultures. A virus control is made with a 1:2 dilution of the working virus dilution used in the test and is inoculated at 0.2 ml per tube; a backtitration is made with four serial 10-fold dilutions of the working virus dilution, which is inoculated at 0.1 ml per tube. Uninoculated or sham-inoculated tubes serve as cell controls. Occasionally, known positive and negative sera should be included in the test as a check on the entire system. The cultures are read daily, and the test is concluded when 32 to 320 TCID$_{50}$/0.1 ml is apparent in the virus backtitration. The highest serum dilution in which both tubes show no adenovirus CPE is considered the end point.

Alternatively, the 3-day test in MK cells or a micro-SN test such as in Vero cells may be carried out by the same principles. In both tests, cell controls and a virus backtitration should be included. The micro-SN test is ideal for serum surveys, because a large number of tests can be carried out in a single run, with little cost in cells, media, and supplies. Also, the test is read visually for uninfected (stained) cells. A fourfold or more increase in titer between acute- and convalescent-phase sera indicates a type-specific seroconversion (15, 30, 36).

EVALUATION, INTERPRETATION, AND REPORTING OF RESULTS

IFA, the CF test, EIA, and TR-FIA for antigen identify all human adenoviruses by the group-specific hexon-antihexon reaction, while PCR recognizes adenoviruses by common nucleotide sequences in the hexon gene (16, 23). All five tests perform this task well, but each is subject to its own limitations. HI and SN tests, being type specific, are subject to interpretative problems arising from the multitude of antigenic cross-reactions that occur among the human adenoviruses (8, 14, 15, 18, 20, 21, 26, 29, 30, 36, 39). For instance, all three viruses in subgroup A (Table 2) cross-react in SN tests. Types 12 and 18 exhibit a bilateral cross-reaction between themselves, and type 31 antiserum cross-reacts with type 12 and 18 viruses. Among subgroup B viruses, the most significant reciprocal cross-reactions observed in both HI and SN tests are among types 7, 11, and 14 and between types 3 and 7. Types 34 and 35 are not readily discernible by HI but are by SN. The subgroup C viruses are virtually devoid of cross-reactions.

The major HI and SN cross-reactions in subgroup D do not always coincide, so this subgroup presents the most difficulties in serotyping. Types 8 and 9; types 10, 19, and 37; types 13, 38, and 39; types 15, 22, and 42; types 20 and 47; types 24, 32, 33, and 46; and types 29 and 45 are not easily distinguishable by HI but are readily typed by the SN test. Conversely, types 15, 23, and 29 are clearly identified by HI but are closely related by the SN test. Other crosses in either test are unilateral and low titer.

A notable reciprocal intergroup cross-reaction, seen only in the SN test, occurs between type 4 (in subgroup E) and type 16 (in subgroup B). Type 4 also cross-reacts with types 40 and 41 of subgroup F. Type 40 is difficult to distinguish from type 41 by HI but can be distinguished by carefully constructed SN tests, EIA with MAbs, DNA hybridization probes, and REA (34). Similarly, isolates with restriction enzyme patterns different from those of the prototype are the rule rather than the exception, and these DNA variants appear to be common for all serotypes (2, 3, 15, 21, 35, 36). Thus, restriction enzyme patterns with selected enzymes are often part of the laboratory study of clinical isolates and have epidemiologic value, but they can give misleading results when applied to serotyping (8, 12, 19, 38).

For serologic tests, a significant rise in antibody titer (fourfold or more) between acute- and convalescent-phase sera is required for diagnosis or confirmation of infection. Consequently, the test cannot be performed until late in the convalescent period, when results are of less clinical importance than in acute illness. Nevertheless, the results may be of epidemiologic importance or of clinical interest in establishing an association between an unusual illness and adenovirus infection. Since the CF test and EIA detect antibody stimulated by infection with any human adenovirus, they are valuable only as screening tests for adenovirus infection; they do not yield information on the infecting serotype. Also, the CF test is less sensitive than the HI and SN tests, especially in younger children and in superficial infections such as conjunctivitis, and the EIA test is not yet standardized. For both tests, reagent controls (complement, hemolysin, etc., for CF; diluent, enzyme, etc., for EIA), antigen controls and backtitration, and positive and negative human serum controls should be included in the test for proper interpretation of results.

The HI and SN tests are more sensitive provided the infecting serotype is included as an antigen in these tests. Rises in antibody titers in response to heterotypic adenoviruses may occur with both HI and SN tests in as many as 25% of adult infections, generally between types of the same HA subgroup (15). For these reasons, serology on individual patients cannot be considered to establish the infecting serotype in the absence of virus isolation or epidemiologic factors.

Whether a patient's illness is due to adenovirus infection is a more difficult question. Many infections are asymptomatic and persistent. In infants, antibody may not develop until months after the onset of infection. Isolation of an adenovirus and even serologic evidence of infection may be coincidental to a disease caused by infection with a different agent. On the other hand, isolation of an adenovirus (especially at a high titer) from a diseased organ or its secretions or previous epidemiologic association of the patient's syndrome with a particular adenovirus is considered valid evidence that the adenovirus is the etiologic agent of the illness.

REFERENCES

1. Abzug, M. J., and M. J. Levin. 1991. Neonatal adenovirus infection: four patients and review of the literature. *Pediatrics* **87:**890–896.
2. Adrian, T., M. Becker, J. C. Hierholzer, and R. Wigand. 1989. Molecular epidemiology and restriction site mapping of adenovirus 7 genome types. *Arch. Virol.* **106:**73–84.
3. Adrian, T., G. Wadell, J. C. Hierholzer, and R. Wigand. 1986. DNA restriction analysis of adenovirus prototypes 1 to 41. *Arch. Virol.* **91:**277–290.
4. Albert, M. J. 1986. Enteric adenoviruses. *Arch. Virol.* **88:**1–17.
5. Ambinder, R. F., W. Burns, M. Forman, P. Charache, R. Arthur, W. Beschorner, G. Santos, and R. Saral. 1986. Hemorrhagic cystitis associated with adenovirus infection in bone marrow transplantation. *Arch. Intern. Med.* **146:**1400–1401.
6. Brandt, C. D., H. W. Kim, A. J. Vargosko, B. C. Jeffries, J. O. Arrobio, B. Rindge, R. H. Parrott, and R. M. Chanock. 1969. Infections in 18,000 infants and children in a controlled study of respiratory tract disease. I. Adenovirus pathogenicity in relation to serologic type and illness syndrome. *Am. J. Epidemiol.* **90:**484–500.
7. Brown, M., M. Petric, and P. J. Middleton. 1984. Diagnosis of fastidious enteric adenoviruses 40 and 41 in stool specimens. *J. Clin. Microbiol.* **20:**334–338.
8. de Jong, J. C., R. Wigand, G. Wadell, D. Keller, C. J. Muzerie, A. G. Wermenbol, and G. J. P. Schaap. 1981. Adenovirus 37: identification and characterization of a medically important new adenovirus type of subgroup D. *J. Med. Virol.* **7:**105–118.
9. Espy, M. J., J. C. Hierholzer, and T. F. Smith. 1987. The effect of centrifugation on the rapid detection of adenovirus in shell vials. *Am. J. Clin. Pathol.* **88:**358–360.
10. Fox, J. P., C. E. Hall, and M. K. Cooney. 1977. The Seattle virus watch. VII. Observations of adenovirus infections. *Am. J. Epidemiol.* **105:**362–386.
11. Hammond, G., C. Hannan, T. Yeh, K. Fischer, G. Mauthe, and S. E. Straus. 1987. DNA hybridization for diagnosis of enteric adenovirus infection from directly spotted human fecal specimens. *J. Clin. Microbiol.* **25:**1881–1885.
12. Hammond, G. W., G. Mauthe, J. Joshua, and C. K. Hannan. 1985. Examination of uncommon clinical isolates of human adenoviruses by restriction endonuclease analysis. *J. Clin. Microbiol.* **21:**611–616.
13. Harnett, G. B., P. A. Phillips, and M. M. Gollow. 1984. Association of genital adenovirus infection with urethritis in men. *Med. J. Aust.* **141:**337–338.
14. Hierholzer, J. C. 1992. Adenoviruses in the immunocompromised host. *Clin. Microbiol. Rev.* **5:**262–274.
15. Hierholzer, J. C. Adenoviruses. *In* E. H. Lennette, D. A. Lennette, and E. T. Lennette (ed.), *Diagnostic Procedures for Viral, Rickettsial and Chlamydial Infections,* 7th ed., in press. American Public Health Association, Washington, D.C.
16. Hierholzer, J. C., P. E. Halonen, P. O. Dahlen, P. G. Bingham, and M. M. McDonough. 1993. Detection of adenovirus in clinical specimens by polymerase chain reaction and liquid-phase hybridization quantitated by time-resolved fluorometry. *J. Clin. Microbiol.* **31:**1886–1891.
17. Hierholzer, J. C., K. H. Johansson, L. J. Anderson, C. J. Tsou, and P. E. Halonen. 1987. Comparison of monoclonal time-resolved fluoroimmunoassay with monoclonal capture-biotinylated detector enzyme immunoassay for adenovirus antigen detection. *J. Clin. Microbiol.* **25:**1662–1667.
18. Hierholzer, J. C., Y. O. Stone, and J. R. Broderson. 1991. Antigenic relationships among the 47 human adenoviruses determined in reference horse antisera. *Arch. Virol.* **121:**179–197.
19. Hierholzer, J. C., R. Wigand, L. J. Anderson, T. Adrian, and J. W. Gold. 1988. Adenoviruses from patients with AIDS: a plethora of serotypes and a description of 5 new serotypes of subgenus D (types 43-47). *J. Infect. Dis.* **158:**804–813.
20. Hierholzer, J. C., R. Wigand, and J. C. de Jong. 1988. Evaluation of human adenoviruses 38, 39, 40, and 41 as new serotypes. *Intervirology* **29:**1–10.
21. Johansson, M. E., M. Brown, J. C. Hierholzer, A. Thorner, H. Ushijima, and G. Wadell. 1991. Genome analysis of adenovirus type 31 strains from immunocompromised and immunocompetent patients. *J. Infect. Dis.* **163:**293–299.
22. Leite, J. P., H. G. Pereira, R. S. Azeredo, and H. G. Schatzmayr. 1985. Adenoviruses in faeces of children with acute gastroenteritis in Rio de Janeiro, Brazil. *J. Med. Virol.* **15:**203–209.
23. McDonough, M., O. Kew, and J. Hierholzer. 1993. PCR detection of human adenoviruses, p. 389-393. *In* D. H. Persing, T. F. Smith, F. C. Tenover, and T. J. White (ed.), *Diagnostic Molecular Microbiology: Principles and Applications.* American Society for Microbiology, Washington, D.C.
24. Meurman, O., O. Ruuskanen, and H. Sarkkinen. 1983. Immunoassay diagnosis of adenovirus infections in children. *J. Clin. Microbiol.* **18:**1190–1195.
25. Niel, C., S. A. Gomes, J. P. Leite, and H. G. Pereira. 1986. Direct detection and differentiation of fastidious and nonfastidious adenoviruses in stools by using a specific nonradioactive probe. *J. Clin. Microbiol.* **24:**785–789.
26. Norrby, E. 1969. The structural and functional diversity of adenovirus capsid components. *J. Gen. Virol.* **5:**221–236.
27. Pinto, A., R. Beck, and T. Jadavji. 1992. Fatal neonatal pneumonia caused by adenovirus type 35. *Arch. Pathol. Lab. Med.* **116:**95–99.
28. Schmitz, H., R. Wigand, and W. Heinrich. 1983. Worldwide epidemiology of human adenovirus infections. *Am. J. Epidemiol.* **117:**455–466.
29. Sohier, R., Y. Chardonnet, and M. Prunieras. 1965. Adenoviruses: status of current knowledge. *Prog. Med. Virol.* **7:**253–325.
30. Stalder, H., J. C. Hierholzer, and M. N. Oxman. 1977. New human adenovirus (candidate adenovirus type 35) causing fatal disseminated infection in a renal transplant recipient. *J. Clin. Microbiol.* **6:**257–265.
31. Straube, R. C., M. A. Thompson, R. B. van Dyke, et al. 1983. Adenovirus type-7b in a childrens hospital. *J. Infect. Dis.* **147:**814–819.
32. Takafuji, E. T., J. C. Gaydos, R. G. Allen, and F. H. Top. 1979. Simultaneous administration of live, enteric-coated adenovirus types 4, 7, and 21 vaccines: safety and immunogenicity. *J. Infect. Dis.* **140:**48–53.
33. Trabelsi, A., B. Pozzetto, A. D. Mbida, F. Grattard, A. Ros, and O. G. Gaudin. 1992. Evaluation of four methods for rapid detection of adenovirus. *Eur. J. Clin. Microbiol. Infect. Dis.* **11:**535–539.
34. van der Avoort, H. G., A. G. Wermenbol, T. P. Zomerdijk, J. A. Kleijne, J. A. van Asten, P. Jensma, A. D. Osterhaus, A. H. Kidd, and J. C. de Jong. 1989. Characterization of fastidious adenovirus types 40 and 41 by DNA restriction

enzyme analysis and by neutralizing monoclonal antibodies. *Virus Res.* **12:**139–158.

35. **Wadell, G.** 1984. Molecular epidemiology of human adenoviruses. *Curr. Top. Microbiol. Immunol.* **110:**191–220.

36. **Wadell, G.** 1988. Adenoviridae: the adenoviruses, p. 284–300. *In* E. H. Lennette, P. Halonen, and F. A. Murphy (ed.), *Laboratory Diagnosis of Infectious Diseases: Principles and Practice,* vol. 2. *Viral, Rickettsial, and Chlamydial Diseases.* Springer-Verlag, New York.

37. **Warren, D., K. E. Nelson, J. A. Farrar, E. Hurwitz, J. Hierholzer, E. Ford, and L. J. Anderson.** 1989. A large outbreak of epidemic keratoconjunctivitis: problems in controlling nosocomial spread. *J. Infect. Dis.* **160:**938–943.

38. **Webb, D. H., A. F. Shields, and K. H. Fife.** 1987. Genomic variation of adenovirus type 5 isolates recovered from bone marrow transplant recipients. *J. Clin. Microbiol.* **25:**305–308.

39. **Wigand, R., N. Sehn, J. C. Hierholzer, J. C. de Jong, and T. Adrian.** 1985. Immunological and biochemical characterization of human adenoviruses from subgenus B. I. Antigenic relationships. *Arch. Virol.* **84:**63–78.

40. **Wishart, P. K., C. James, M. S. Wishart, and S. Darougar.** 1984. Prevalence of acute conjunctivitis caused by chlamydia, adenovirus, and herpes simplex virus in a ophthalmic casualty department. *Br. J. Ophthalmol.* **68:**653–655.

Measles Virus

AIMO A. SALMI

81

CLINICAL BACKGROUND

Measles virus (MV) causes an acute, generalized infection that until recently has been one of the most common viral diseases of childhood worldwide (10, 15). The typical epidemiologic pattern of rapidly spreading outbreaks is most often observed in springtime (3). Epidemic peaks coincide with the school year in developed countries. Measles epidemics at 2- to 5-year intervals were common in densely populated countries before the licensure of measles vaccine. Now the goal of the expanded global program for immunization is to eliminate measles entirely. This goal has not yet been achieved. In developing countries in particular, vaccination programs have been only partially successful. Small epidemics of measles have been observed in vaccinated populations even in developed countries in recent years (2), and the eradication of measles seems to be a more distant goal than estimated a few years ago.

Measles is rarely fatal in North America and Europe, but mortality after MV infection can be as high as 20% in developing countries. Poor nutrition and suboptimal hygienic conditions are contributing factors. Measles is traditionally a disease of early childhood, but vaccination programs may change the epidemiologic pattern in the future. Occasional cases or even small epidemics have recently been observed in schoolchildren and young adults (11).

Measles is spread from an MV-infected person to a susceptible person, mainly via aerosol. Virus is excreted late during the 9- to 12-day incubation period, and the patient continues to be infectious until early convalescence. The primary site of infection is the mucosal cells of the respiratory tract. The infection spreads to local lymph nodes, and viremia is established. Circulating virus is found in T cells, B cells, and monocytes. Virus spreads to other susceptible organs during the viremic phase. Virus has been found in lung, gut, bile duct, bladder, skin, and lymphoid organs (9).

The prodromal symptoms of measles are similar to those of other upper respiratory tract infections: running nose, sneezing, cough, and fever. Redness of eyes and photophobia are also early symptoms. Koplik's spots on the mouth epithelium, which usually appear after the prodromal symptoms, are the hallmark of the clinical signs of measles. At the time of the viremia, a typical macular or maculopapular exanthem appears. High fever at the time of rash onset is characteristic. Patients are infectious a few days before the onset of rash and continue to excrete virus until the rash has disappeared. If no complicating sequelae appear, the disease disappears in 1 to 2 weeks after the onset of rash. If the infection is severe, total recovery may take weeks.

Changes in electroencephalograms are seen in about 50% of patients with uncomplicated measles, indicating central nervous system (CNS) involvement. The number of immune cells in the cerebrospinal fluid (CSF) is also frequently increased in measles patients. Although MV can replicate in brain cells, changes observed in the CNS are likely due to autoimmune reactions.

MV rarely causes an acute encephalitis, but postinfectious encephalomyelitis has been reported to occur in about 0.1% of patients. MV readily causes persistent infection in vitro and may also establish persistence in vivo, especially in the CNS. A rare complication of measles is subacute sclerosing panencephalitis (SSPE), which occurs at a frequency of about 1/1,000,000. Patients with SSPE, most often boys, usually have a history of an uneventful MV infection early in life. After a latent period of 1 to 10 years, a progressive neurologic disease develops that is lethal within a few years. Virologically, a persistent, defective infection progresses in the CNS of these patients.

Persistent MV infection has also been suspected in other chronic autoimmune diseases. Patients with multiple sclerosis as well as those with systemic lupus erythematosus have consistently been found to have elevated levels of antibodies to MV. This finding does not necessarily indicate any role of the virus in the etiology or pathogenesis of these diseases but may reflect polyclonal activation of B cells known to occur during the course of the diseases.

Variants of clinical measles have been observed. Atypical measles may occur in persons previously immunized if the immunity after vaccination was only partial. In such patients, the rash may be hemorrhagic or vesicular and develops first in extremities. Enlargement of lymph nodes and spleen is frequently seen. Such atypical cases may be difficult to diagnose clinically. Diarrhea may be a prominent sign of measles patients in developing countries. Measles may be unusually severe in patients in whom T-cell functions are impaired. These patients frequently develop giant-cell pneumonia and severe involvement of other organ systems.

A clinical diagnosis of measles is usually obvious from the clinical symptoms, especially in the Caucasian population. However, recognizing measles rash on the background of a dark skin may be difficult. Laboratory diagnosis by specific virologic tests is necessary to confirm the clinical diagnosis in all atypical cases. Even typical measles patients may not be properly diagnosed clinically by physicians who have not previously seen typical measles. Since small measles epidemics are still found even in vaccinated populations in Europe and North America, specific laboratory confirmation is an important addition to the clinical evaluation of patients suspected of having MV infection.

DESCRIPTION OF AGENT

MV belongs to the family *Paramyxoviridae* and is a member of the genus *Morbillivirus* (15). Other well-characterized members of this genus are rinderpest virus of cattle and distemper virus of dogs. Members of this genus have recently been isolated from different carnivores, including seals (4). The only natural hosts of MV are humans and monkeys, but the virus can be adapted to grow in rodents as well. MV is monotypic, but small biological antigenic variations at the epitope level have been described (21). The variation is based on genetic variability in MV genes (17). Such variations have no effect on protective immunity, since MV infection still provides a lifelong immunity against reinfection.

The main biochemical and structural features of MV are similar to those of other members of the family *Paramyxoviridae*. MV genetic information is carried by a single-stranded linear RNA that sediments at 50S to 52S and has a molecular weight of about 5×10^6. The enveloped measles virions are pleomorphic and 100 to 250 nm in diameter. The envelope is composed of a cell-derived lipid bilayer from which two types of viral glycoproteins protrude. They are seen by electron microscopy as surface projections of 9 to 15 nm.

The MV is composed of six major structural polypeptides, three of which are associated with the envelope. The two protruding glycoproteins, the hemagglutinin (H) and fusion (F) proteins, are the most important from the standpoint of protection by antibodies. Most of the neutralizing antibodies are against the H protein; antibodies against the F protein inhibit the spread of MV in tissue. The third virus protein associated with the envelope and lining its inner surface is the hydrophobic matrix (M) protein. The helical nucleocapsid enclosed in the envelope is composed of the RNA genome surrounded by the major internal protein of MV, the nucleoprotein (NP). Two minor proteins associated with the nucleocapsid core are the polymerase (phospho) (P) protein and the large (L) protein, both of which are involved with the RNA polymerase function of the virion. T-cell immunity to these internal proteins of MV also plays a role in protection.

MV is a labile virus and can be readily inactivated with chemicals affecting the lipid envelope. It is also sensitive to heat inactivation; its infectivity can be maintained by storage at −70°C but not at −20°C. In contrast, MV antigens are relatively stable; hemagglutinating antigens and antigens to be used for enzyme immunoassay (EIA) can be stored at 4°C for months without significant loss of antigenic activity.

COLLECTION AND STORAGE OF SPECIMENS

The success rate for isolation of MV from clinical specimens is low. It has been isolated from nasopharyngeal and conjunctival specimens during the prodrome and early rash illness and from stool and urine specimens during late illness. Although the CNS is frequently involved in the pathogenetic process of measles, virus isolation from CSF is only rarely successful. Specimens for virus isolation are taken by using standard procedures, refrigerated (but not frozen), and transported to the virus laboratory as soon as possible. Since the infectivity of MV is labile, the transport medium should contain protein to stabilize MV infectivity. Media with bovine serum albumin or fetal calf serum are good for this purpose.

Direct demonstration of MV antigen in infected cells is successful more often than virus isolation, since large amounts of virus antigens accumulate in infected cells. Therefore, a specimen for antigen detection should have a high number of infected cells. The best specimen for this purpose is nasopharyngeal secretion collected by a mucus extractor with a mucus trap of the type commonly used in infectious disease wards. The collected mucus is refrigerated and sent immediately to the laboratory for direct immunofluorescence or antigen detection by immunoassay. Virus antigen may also be detected in circulating peripheral blood cells. Heparinized blood or isolated lymphoid cells are sent refrigerated to the virus laboratory for this examination.

Serologic diagnosis is generally based on demonstration of a significant increase in antibody titer in paired serum specimens. The first specimen should be taken as soon as possible after the onset of the disease. The second specimen can be taken as early as 7 days but preferably 10 to 14 days after the first specimen.

Testing of paired acute- and convalescent-phase serum specimens is important for the diagnosis of patients who have been immunized against measles but who subsequently become infected. Such patients usually have a record of measles immunization and often have detectable but low titers of immunoglobulin G (IgG) antibodies to MV in their acute-phase serum specimens. The acute-phase sera do not, however, have detectable IgM antibodies to MV. Acute- and convalescent-phase paired sera show diagnostic rises in IgG MV-specific antibody titers.

A single serum specimen is sufficient for the demonstration of MV-specific IgM antibodies, provided that it has been taken within 4 to 5 weeks of the onset of measles. Intrathecal MV antibody synthesis is pathognomonic for SSPE. Such synthesis can be demonstrated by titrating antibodies in both serum and CSF. For that purpose, paired serum and CSF specimens are taken on the same day and sent to the virus laboratory. Immunoglobulins are more labile in CSF than in serum and require refrigerated transport of paired specimens to the laboratory.

DIRECT EXAMINATION

Typical paramyxoviruslike viruses may be seen in clinical specimens from measles patients by using the electron microscope. However, since all members of the family *Paramyxoviridae* are similar morphologically, no specific viral diagnosis can be made by such an examination. MV-infected cells are typically multinucleated and are formed by fusion of a large number of infected cells. Such cells are pathognomonic for measles, and their presence in throat washings or peripheral blood mononuclear cells can be used as an indication of MV infection. Eosinophilic inclusions are found in the cytoplasms or nuclei of infected cells. The inclusions represent accumulated viral nucleocapsid material and are considered diagnostic in SSPE patients.

Although histologic examination may aid in the diagnosis of measles, specific diagnosis requires a direct demonstration of virus antigens in the cells. This can be done by immunofluorescence staining of tissue cells obtained from patients (6). The most useful substrate for staining is cells from nasopharyngeal secretions, peripheral blood, and urine sediment. Virus antigen is also found in skin biopsy samples obtained during the rash. For immunofluorescence, the specimen is spread on a glass slide, air dried, and fixed in ice-cold acetone for 10 min. Fixed slides can be stored frozen, preferably at −70°C, for a few weeks.

The fixed cells are stained for MV antigens by hyperimmune measles antiserum and fluoresceinated secondary antibodies specific for the immunoglobulins of the hyperimmune serum. Standard staining conditions commonly used in immunofluorescence staining should be employed. The stained samples are examined for a specific staining pattern that is typically cytoplasmic; late during the infectious cycle and in chronic measles such as SSPE, it may also be intranuclear. MV-infected cells, especially late during the infectious cycle, contain a large mass of virus antigens. Especially prominent are the NP and P antigens, which can be seen in small spots or as large fluorescing bodies in the cytoplasm. For reference purposes, cells infected with a laboratory strain of MV should be included as a positive control.

Direct demonstration of viral antigens by radioimmunoassay or EIA has been used for rapid diagnosis of respiratory virus infections in recent years. Nasopharyngeal specimens without fixation can be directly used in such tests. A few nanograms of MV antigen per milliliter can be detected by EIA (19). There is as yet no information on whether such a test is useful for the laboratory diagnosis of MV infection.

Recently introduced hybridization tests for MV RNA in tissues have been used in research laboratories (5) but have not yet been commonly used in diagnostic laboratories.

ISOLATION OF VIRUS

MV can be isolated from patients during the prodromal state, at the time of the rash, and a few days afterwards. Since the success rate for virus isolation is low (18), this approach cannot be used as the primary means for the laboratory diagnosis of MV infection. Virus isolation is more important for the characterization of patients with poor or delayed antibody responses, which occur in many patients with atypical measles.

Although laboratory strains of MV grow readily in a number of continuous cell lines of human or primate origin, the growth of strains directly from measles patients can be difficult to achieve. The best cells for virus isolation are primary human fetal kidney cells. Since such cells are not generally available, primary monkey kidney cells are the second choice. These cells can be prepared from monkey kidney tissue, if available in the virus laboratory, or may be obtained from commercial sources. Primary human amniotic cells have also been used to isolate MV from patients, but the success rate is lower than with primary kidney cells.

The cytopathic effect (CPE) appears late after the inoculation of an MV-containing sample. The inoculated cultures should be observed for typical CPE for at least 2 weeks. Two types of CPE may be observed in the cultures: (i) spindle-shaped single cells that increase in number when the infection advances and (ii) multinucleated giant cells (syncytia) that increase in size during the culture time.

The type of CPE depends on the multiplicity of infection, the virus strain, and the cells used to propagate the virus.

In chronic MV infections (i.e., in SSPE patients), the virus is in a defective, cell-associated form and cannot be isolated by ordinary techniques (8). Instead, cocultivation of patient brain or peripheral blood cells with susceptible monkey cell lines is required. The procedure may take a few weeks and requires many passages of the cell mixture before the CPE appears or antigens can be demonstrated in the coculture. MV isolated by the cocultivation technique may remain cell associated.

IDENTIFICATION OF VIRUS

Methods

The finding that the CPE caused by the isolated virus is syncytial suggests that the isolated virus has a fusion protein. This observation combined with the clinical information justifies the suspicion that the isolated virus may be MV. A preliminary test that can easily be done is hemadsorption with monkey erythrocytes. However, this test is not specific for MV. Proper identification of the virus causing the CPE requires use of specific antisera.

A hyperimmune animal antiserum specific for MV, monoclonal antibodies specific for the internal polypeptides (NP, P, and M), or a mixture of different monoclonal antibodies can be used to identify the isolated virus. The most convenient identification test is direct immunofluorescence with specific antisera. Cells with CPE are scraped off, spread on a glass slide, and fixed in ice-cold acetone for 10 min. Staining with MV-specific antibodies and fluoresceinated secondary antibodies is done by standard procedures. Fluorescence of typical intracellular inclusions stained with the specific antiserum identifies the isolated virus as MV. Staining with a serum specimen taken from a hyperimmunized animal before the immunization should be used, if possible, as a negative control. Virus antigen in infected cells can also be identified by a specific EIA for MV antigens (19).

Preparation of Hyperimmune Sera

Hyperimmune polyclonal antibodies are available from commercial sources but can also be prepared in different species such as sheep, rabbits, or guinea pigs. The following protocol has been successfully used to prepare a high-titer measles antiserum in rabbits. Concentrated virus purified from supernatants of MV-infected cells as described elsewhere in this chapter is a good immunizing antigen. A total of 0.5 to 1 mg of virus in 0.2 ml of a physiologic buffer is added to an equal volume of Freund incomplete adjuvant and mixed thoroughly. The antigen is injected intradermally on multiple sites on the shaved back of a rabbit. After 3 weeks, a booster injection of 0.1 to 0.2 mg of MV in an acceptable adjuvant is given subcutaneously. The boosting is repeated after 3 weeks. Two weeks later, a serum sample is taken and tested by immunofluorescence assay and EIA for antibodies binding to noninfected and MV-infected cells. A specific titer of 1/1,000 or higher by immunofluorescence and 1/100,000 or higher by EIA is normally reached in about 2 months of total immunization time. If the titers have reached that level and the binding of antibodies to noninfected cells is less than 1/100 of the specific binding, the antiserum is acceptable, and a larger amount of the serum can be collected. If binding of the antiserum to noninfected cells is too high, the antiserum might still be useful after absorption with noninfected cells. In such cases the absorbed antiserum should be retested by using the procedure described above.

SEROLOGIC DIAGNOSIS

The presence of specific antibodies in a single serum specimen indicates past MV infection or vaccination. Demonstration of a significant increase of specific antibody titers in a serum pair taken at a 7- to 14-day interval is the basis of the diagnosis of an acute infection. Alternatively, demonstration of MV-specific IgM antibodies in only one specimen gives a specific diagnosis. Since the presence of virus-specific IgM can be determined in less than 24 h, its detection can be used as a rapid diagnostic test.

In all serologic tests, known positive and negative controls should be included in the series. This practice is important, because test conditions affect the level of sensitivity and the degree of nonspecific reactivity in all of these tests.

HI

The hemagglutination inhibition (HI) test is based on the ability of MV H protein to bind to monkey erythrocytes. If the H protein is in a dimeric form or in a larger particle such as the measles virion, erythrocytes are aggregated and can easily be seen. Specific antibodies inhibit this aggregation, the phenomenon on which the HI antibody test is based.

Preparation of the HI Antigen

The antigen for the HI test is available commercially but can also be prepared in a virus laboratory. Continuous cell lines known to give a high yield of MV, such as Vero, CV-1, or HeLa, should be used. Since MV is homotypic, any virus strain can be used for hemagglutinin preparation. When the CPE is spread throughout the cell layer, the culture is harvested. Because a large amount of H protein is found both in the cellular membranes and in released virus, whole cultures can be harvested and used for antigen preparation. The material is frozen and thawed five times and centrifuged at $10,000 \times g$ for 20 min; the supernatant is divided into smaller portions and stored at $-20°C$. The hemagglutinating ability of the antigen can be increased a minimum of fourfold by treatment with Tween 80 and ether (12).

The hemagglutination is titrated on round-bottom microtiter plates using erythrocytes from Old World monkeys such as African green or rhesus monkeys. Add 0.05 ml of 0.5% monkey erythrocytes to 0.025-ml volumes of twofold antigen dilutions, and incubate the plates for 1 h at 37°C. Read the pattern of hemagglutination; the last dilution giving a clear agglutination contains 1 hemagglutinating unit (HAU) of measles antigen. All reagents are diluted in phosphate-buffered saline (PBS). If the erythrocytes aggregate spontaneously, then the dilutions are made in PBS supplemented with 0.2 to 1% bovine serum albumin.

HI Test

Human sera frequently contain nonspecific inhibitors of hemagglutination. Such inhibitors can be removed by adding an equal volume of 25% kaolin in PBS and incubating it for 20 min at room temperature. The mixture is then centrifuged, and the supernatant, which represents a 1:2 dilution of the original serum, can be stored at $-20°C$. Some human sera agglutinate monkey erythrocytes and

must be absorbed with an equal volume of 50% monkey erythrocytes for at least 4 h at 4°C.

For the HI test, twofold dilutions of serum specimens starting with a 1:2 or 1:4 dilution are made on microtiter plates, using a volume of 0.025 ml, and 4 HAU of measles antigen is added in a volume of 0.025 ml. After 1 h of incubation at room temperature, 0.05 ml of 0.5% monkey erythrocytes is added; the plates are then shaken and incubated at 37°C for 1 h. The agglutination pattern is read, and the HI titer is determined as the last dilution inhibiting agglutination. As a control, the hemagglutinating MV antigen is backtitrated to make sure that the amount of hemagglutinin is 4 HAU in the test.

EIA

Measles antibodies can also be measured by EIA. EIA test kits for MV antibodies are commercially available, but such tests can also be set up in any virus laboratory. MV antigen is available from commercial sources and can also be prepared by using standard laboratory virus strains and continuous cell lines supporting MV growth. The cells are maintained in Eagle minimal essential medium without or with 2% or less fetal bovine serum and are infected with a high multiplicity of infectious virus. When the CPE has spread throughout the cell monolayer, the supernatant and the infected cells are harvested separately and processed for antigens.

Infected cells contain a large amount of NP, P, and H proteins and lesser amounts of M, F, and L proteins. For practical purposes, a crude antigen prepared from infected cells is satisfactory for measuring an increase in the level of IgG antibodies and determining the immunity status of the patient. If it is important to measure antibodies to all viral structural components or if the antigen is used in the IgM assays, purified virus should be used.

Preparation of EIA Antigen

For preparation of the crude cellular antigen, the infected cells are washed three times with cold PBS and suspended in 10-fold-diluted PBS. The cells are disrupted with powerful ultrasonic equipment or in a tightly fitting Dounce homogenizer. Cell debris is removed by low-speed centrifugation in an ordinary tabletop refrigerated centrifuge at $500 \times g$ for 15 min. The particulate material in the supernatant is pelleted in an ultracentrifuge at $80,000 \times g$ for 2 h. The pellet is resuspended in a small volume of PBS. The resuspension might require a vigorous homogenization. The amount of protein in the antigen is determined, and the antigen is divided into smaller portions and stored frozen, preferably at $-70°C$. Storage at $-20°C$ is acceptable, but the antigen tends to aggregate, especially if the storage temperature occasionally rises above $-20°C$.

Purification of MV

MV for use as EIA antigen is purified from supernatants of infected cells collected at the time when the CPE has spread to 75 to 100% of the cells. After the supernatant has been clarified by centrifugation at $10,000 \times g$ for 20 min, it can be stored for a few days at 4°C before further purification. The supernatant is then concentrated 10- to 20-fold in a hollow fiber or membrane concentration apparatus. To the concentrated supernatant, NaCl and α-D-methylmannoside are added to final concentrations of 2 mol/liter and 4%, respectively. These additives reduce the nonspecific binding of cellular debris to the virions. After 30 min on ice, the supernatant is layered onto a step gradient consisting of 10 ml of 18% potassium tartrate and 5 ml of 36% potassium tartrate in a buffer consisting of 0.2 M glycine, 0.2 M NaCl, 20 mM Tris, and 2 mM EDTA (pH 7.8) (GNTE buffer). The gradients are centrifuged at 80,000 × g for 90 min. The virus material at the 18 to 36% interface is collected, diluted with GNTE buffer to a tartrate concentration of less than 18%, and layered on a 24-ml linear 18 to 36% potassium tartrate gradient prepared in the GNTE buffer. The gradients are centrifuged at 80,000 × g for 3 h or overnight. The virus band is collected, diluted in GNTE, and pelleted at 80,000 × g for 30 min. The virus is suspended in PBS and stored at −70°C.

EIA for MV Antibodies

Flat-bottom microtiter plates are first coated with crude antigen or purified virus material. The plates should be of a special EIA type or of reliable tissue culture-treated plastic. For coating, the antigen is diluted in PBS or 50 mM carbonate buffer (pH 9.6) to a concentration giving the best binding of specific antibodies. This concentration should be determined experimentally by checkerboard titration, but a concentration of 5 to 50 μg/ml for a crude lysate antigen or 0.5 to 5 μg/ml for purified virions is a good guess for a proper coating concentration. A 0.1-ml volume of the diluted antigen is added to each well in the microtiter plate, which is then incubated overnight at room temperature or 4°C. Blocking of all remaining protein-binding sites on the plastic plate is done by incubating the wells for 1 to 2 h with 0.1 ml of EIA incubation buffer, consisting of PBS supplemented with 0.5% bovine serum albumin and 0.5% Tween 20.

Serum dilutions are made in two- or fourfold series in EIA incubation buffer, starting from a dilution of 1:50 to 1:200. Dilutions can be made with a multichannel pipette directly on the antigen-coated plates. Incubation with the serum dilutions is for 1 h at 37°C or for 2 h at room temperature, after which the plates are washed three times with washing buffer (PBS supplemented with 0.1% Tween 20).

The next step is addition of enzyme-labeled antibodies that recognize human immunoglobulins. The antibodies can be specific for human IgG, which is the major immunoglobulin class of antibodies during measles convalescence and in immune individuals. For demonstration of a titer increase in a specimen pair or of the immunity status of an individual, the antibody can also be specific for all human immunoglobulins. The use of heavy-chain (μ-chain)-specific labeled antibodies is necessary only if IgM antibodies are sought. Incubation with the labeled antibodies is for 1 h at 37°C or for 2 h at room temperature, after which the plates are washed as described above. The final step in the assay is incubation of the plates with the substrate solution. The composition of the substrate depends on the enzyme coupled to the antihuman antibodies. For example, a reliable substrate for horseradish peroxidase is hydrogen peroxide (6 mmol/liter) and o-phenylenediamine (17 mmol/liter) in phosphate (50 mmol/liter)-citrate (20 mmol/liter) buffer (pH. 5.5).

Titers of the serum specimens can be determined from the absorbance values of the different serum dilutions by criteria established in the laboratory. As in other serologic tests, a fourfold increase in antibody titer can be considered proof of a recent virus infection. Since there are no official standards for EIAs, the series should always include known positive and negative serum specimens.

IgM Determinations

The presence of IgM antibodies in a serum specimen indicates an early phase of an immune response. Nearly 100% of serum specimens taken during acute MV infection or in early convalescence are positive for MV-specific IgM antibodies (16, 23). Since an IgM determination can normally be made in less than 24 h in the laboratory, this test can be used for rapid virologic diagnosis of an acute MV infection. The main pitfall of the IgM tests is the possibility of nonspecific positive reactions due to rheumatoid factor interference (22, 23).

IgM can be separated from other classes of human immunoglobulins by methods such as ultracentrifugation in sucrose gradients and exclusion chromatography in columns. The fraction containing the 19S immunoglobulins (IgM) is then tested for antibody activity by any available method. Physical separation is, however, laborious and time-consuming. It is easier to use specific antibodies that recognize the μ chain of human IgM bound to MV antigen in a test.

IgM determination by fluorescent-antibody techniques requires MV-infected fixed cells, fluoresceinated antibodies specific to human IgM, and a microscope equipped with a UV light source and proper fluorescence filters. Any laboratory strain of MV can be used to infect a continuous cell line. Cells left uninfected should also be prepared and used as negative controls. Infected and control cultures are washed three times with PBS, fixed in ice-cold acetone for 10 min, and air dried. Fixed cells can be stored for a couple of months at −20°C without a significant loss of antigenicity. Slides taken from cold storage should first be brought to room temperature and then rinsed in PBS. The serum specimen to be tested is incubated on antigen slides at 37°C in a moist chamber for 30 min, and the slide is washed three times in PBS. A predetermined dilution of fluorescein-labeled antibodies to human IgM (μ chain specific) is added to the slide and incubated for 30 min at 37°C. After three more washings with PBS, the slides are sealed with coverslips and examined under a fluorescence microscope.

Other immunoassays offer more objective means of IgM determination. Although radioimmunoassay (1) is still used, EIA has replaced it because of the hazards associated with the use of radioactive material. MV-specific IgM antibodies can be measured by the direct EIA as described above for IgG antibodies except that specific antibodies for human μ chain, tagged with an enzyme, are used for detection of IgM antibodies bound to measles antigen on the solid phase. Testing of only one serum dilution is sufficient for IgM determination. The best dilution depends on the technical requirements of the system used, but a dilution of 1:50 to 1:200 is generally optimal.

Since many human serum specimens have IgM antibodies capable of binding to cellular proteins, it is necessary to use either antigens prepared from infected and noninfected cells or highly purified virions. Another pitfall in the direct IgM EIA is the nonspecific reaction due to IgM-class rheumatoid factor, which binds to IgG. Such binding of the indicator antibodies to IgM-class rheumatoid factor that is attached to MV-specific IgG gives a false-positive reaction. Therefore, if the level of rheumatoid factor is elevated in the serum specimen, the IgM test results are not reliable. Such specimens should be absorbed with heat-aggregated human immunoglobulins or latex particles coated with human IgG prior to testing (23).

The problems with rheumatoid factor can be avoided if

the order of reagents in the IgM test is reversed (22). In this system (often referred to as "reverse-capture" EIA), the first layer coated to the solid phase consists of antibodies to human IgM (μ specific; commercially available). The second step is incubation of the test specimens on the solid phase, which results in the binding of IgM to the solid phase. The presence of IgM antibodies specific for MV can be detected in two ways. In one approach, specific purified MV is added to the solid phase, and the bound virus is detected by specific antibodies directly bound to an enzyme marker. If the antibodies are used unconjugated, an extra layer of conjugated antibodies specific for the animal species of the MV-specific antiserum is required (22). The second approach is use of purified MV directly labeled with an enzyme (20).

It is easy to recognize negative and strongly positive serum specimens. Difficulties arise when the absorbance reading is only slightly elevated. The only way to decide the cutoff level for positivity in such cases is to test a large number of randomly selected serum specimens likely to be IgM negative and determine the mean and the standard deviation, which can then be used to determine the 95% probability level of true IgM positivity.

Other Serologic Tests

A number of other tests have been used to detect MV antibodies in a serum specimen or an increase in the titer of a serum pair. Indirect immunofluorescence tests for MV IgG and IgM antibodies are still practical alternatives. These tests should be done according to the protocol used for immunofluorescence tests for other viral antibodies. Research laboratories have measured antibodies to different polypeptides of MV by immunoprecipitation of radiolabeled MV proteins and Western immunoblotting (7, 14), but such tests are better suited for qualitative assessment of the immune response to each of the viral polypeptides than for quantitative measurement of the immune response. The neutralization and hemolysis inhibition (HLI) tests can be used for diagnostic tests, since they measure antibodies of direct biological significance. Information about immunity obtained with these tests may be important if epidemics of MV infection appear in a vaccinated population.

Neutralization

The neutralization test generally correlates with immunity, since it measures antibodies capable of inhibiting virus growth in vitro. It is more cumbersome than most of the other antibody tests, since it is based on the inhibition of MV growth in cell culture. The test requires no special equipment beyond the ordinary cell culture facility. Stock MV is diluted to contain 100 50% tissue culture infective doses per 0.1 ml. Equal amounts of serially diluted serum specimen and the diluted virus are incubated together at 4°C for 60 min. A 0.2-ml volume of this mixture is added to each of three to six tissue culture tubes and incubated at 37°C until CPE appears in half of the control tubes with 1 50% tissue culture infective dose of virus alone. The titer is determined to be the dilution at which the CPE is inhibited in more than half of the tubes.

HLI Test

The HLI test measures the presence of antibodies inhibiting hemolysis caused by the F protein of MV. However, since the expression of hemolytic activity requires that virus particles be anchored to erythrocytes via the hemagglutinin, antibodies to the H protein also interfere with the hemolysis reaction. Because the HLI antibodies are effective in inhibiting virus spread in tissues, their measurement is relevant, especially if there is a measles outbreak in a vaccinated population (13).

Antigen used in the HLI test should have a large amount of MV F protein. Since the level of F protein can vary from one preparation to another, the best cell-virus combination for preparing such an antigen should be experimentally determined. Infected cells in a 10% suspension or 10- to 20-times-concentrated supernatant of MV-infected cells are the starting material for preparing the hemolyzing antigen. The material is disrupted by five cycles of freezing and thawing or by vigorous sonication. The hemolyzing activity of the antigen is determined by adding 0.1 ml of washed 10% green monkey erythrocytes to a twofold dilution series of 0.4 ml of hemolyzing antigen in PBS. After 3 h of incubation at 37°C, the erythrocytes are centrifuged at 1,000 × g for 10 min, and the A_{540} of the supernatant is measured. An antigen dilution giving an absorbance of about 0.5 is selected as the working dilution.

For the HLI test, add 0.2 ml of the working dilution of the hemolyzing antigen to equal volumes of serial twofold dilutions of the serum specimen. After 1 h of incubation at room temperature, add 0.1 ml of a 10% suspension of monkey erythrocytes. After incubating the mixture for 3 h at 37°C, centrifuge the erythrocytes, and read the optical density of the supernatant. The HLI titer of the serum is the dilution giving at least 50% reduction of hemolysis as calculated from the control hemolysis with added PBS but without serum specimen.

EVALUATION AND INTERPRETATION OF RESULTS

Isolation of MV indicates active virus replication in the patient and is the basis for a specific virologic diagnosis. Negative results for virus isolation do not exclude MV infection, since the sensitivity of the isolation method is low. Careful timing of specimen collection and testing of more than one specimen are important for successful virus isolation. These are especially important in patients with abnormal immunity and atypical measles, since their antibody responses may be delayed or impaired. Cocultivation of cells from SSPE patients results in virus isolation in only a small percentage of cases. Therefore, negative virus isolation results for these patients do not exclude the diagnosis of SSPE.

HI is one of the best methods for determining the immunity status of an individual, since it is easy to perform and the presence of HI antibodies is known to correlate with protective immunity. HI titers of 1/10 or higher are considered protective against secondary infection, but the lower limit of protection clearly depends on the sensitivity of the test system used. The presence of neutralizing antibodies also correlates well with protective immunity, but the neutralization test is more cumbersome and less often done. EIA measures antibodies to all viral components and can be used as a sensitive test to determine whether an individual has been infected with MV. There are rare individuals, however, who are strongly positive in EIAs but have only low levels of HI or neutralizing antibodies. It is not clear whether such individuals are protected against reinfection by MV.

All of the serologic tests described above can be used to demonstrate a significant increase of antibody level in a serum pair. A fourfold or greater increase indicates an ongoing MV infection. If the first serum specimen is taken during the first 2 days of the rash and the second one is taken a minimum of 7 days later, a significant increase in antibody titer can usually be demonstrated. This is true for patients with ordinary measles, but patients with impaired immunity or atypical measles may synthesize only a small amount of antibodies, and thus no significant increase would be demonstrated in a specimen pair taken at an interval of a few days. An increase may still be seen if the serum specimens are taken a few weeks apart.

A diagnostic serologic test for SSPE patients is the demonstration of intrathecal MV-specific antibody synthesis. This can be shown by calculating the CSF/serum titer ratio. The normal ratio is about 1/200 to 1/500, but in SSPE patients, it is from 1/5 to 1/50. Such a low ratio indicates intrathecal antibody synthesis, provided that ratios of other viral antibodies measured as controls are normal.

Demonstration of MV-specific IgM antibodies is the best rapid diagnostic test. All patients with ordinary measles have IgM antibodies in specimens taken late during the rash and until 4 to 5 weeks later. Delayed elevation of MV-specific IgM antibodies is not found in patients late after infection, and only low levels of IgM antibodies are seen in some SSPE patients early after the onset of the disease (24). The only pitfall of the measles IgM tests is nonspecific positive results due to rheumatoid factor and occasional binding of IgM to nonviral cellular antigens. If these possibilities are excluded, the assays for MV-specific IgM antibodies can be considered very reliable and rapid diagnostic tests.

REFERENCES

1. **Arstila, P., T. Vuorimaa, K. Kalimo, P. Halonen, M. Viljanen, K. Granfors, and P. Toivanen.** 1977. A solid phase radioimmunoassay for IgG and IgM antibodies against measles virus. *J. Gen. Virol.* **34:**167–176.
2. **Atkinson, W. L., W. A. Orenstein, and S. Krugman.** 1992. The resurgence of measles in the United States. *Annu. Rev. Med.* **43:**451–463.
3. **Black, F. L.** 1989. Measles, p. 451–469. *In* A. S. Evans (ed.), *Viral Infections of Humans: Epidemiology and Control*, 3rd ed. Plenum Publishing Corp., New York.
4. **Blixenkorne-Möller, M., V. Svansson, M. Appel, J. Krogsrud, P. Have, and C. Örvell.** 1992. Antigenic relationship between field isolates of morbilliviruses from different carnivores. *Arch. Virol.* **123:**279–294.
5. **Fournier, J. G., M. Tardieu, P. Lebon, O. Robain, G. Ponsot, S. Rozenblatt, and M. Bouteille.** 1985. Detection of measles virus RNA in lymphocytes from peripheral blood and brain perivascular infiltrates of patients with subacute sclerosing panencephalitis. *N. Engl. J. Med.* **313:**910–915.
6. **Fulton, R. E., and P. J. Middleton.** 1975. Immunofluorescence in diagnosis of measles infections in children. *J. Pediatr.* **86:**17–22.
7. **Hankins, R. W., and F. L. Black.** 1986. Western blot analysis of measles virus antibody in normal persons and in patients with multiple sclerosis, subacute sclerosing panencephalitis, or atypical measles. *J. Clin. Microbiol.* **24:**324–329.
8. **Katz, M., and H. Koprowski.** 1973. The significance of failure to isolate infected viruses in cases of SSPE. *Arch. Gesamte Virusforsch.* **41:**390–393.
9. **Moench, T. R., D. E. Griffin, C. R. Obriecht, A. J. Vaisberg, and R. T. Johnson.** 1988. Acute measles in patients with and without neurological involvement: distribution of measles virus antigen and RNA. *J. Infect. Dis.* **158:**433–442.
10. **Morgan, E. M., and F. Rapp.** 1977. Measles virus and its associated diseases. *Bacteriol. Rev.* **41:**636–666.
11. **Mouallem, M., E. Friedman, R. Pauzner, and Z. Farfel.** 1987. Measles epidemic in young adults: clinical manifestations and laboratory analysis in 40 patients. *Arch. Intern. Med.* **147:**1111–1113.
12. **Norrby, E.** 1962. Hemagglutination by measles virus. IV. A simple procedure for production of high potency antigen for hemagglutination-inhibition (HI) tests. *Proc. Soc. Exp. Biol. Med.* **111:**814–818.
13. **Norrby, E., G. Enders-Ruckle, and V. ter Meulen.** 1975. Differences in the appearance of antibodies to structural components of measles virus after immunization with inactivated and live virus. *J. Infect. Dis.* **132:**262–269.
14. **Norrby, E., C. Örvell, B. Vandvik, and D. J. Cherry.** 1981. Antibodies against measles virus polypeptides in different disease conditions. *Infect. Immun.* **34:**718–724.
15. **Norrby, E., and M. N. Oxman.** 1990. Measles virus, p. 1013–1044. *In* B. N. Fields and D. M. Knipe (ed.), *Virology*, 2nd ed. Raven Press, New York.
16. **Pedersen, I. R., A. Antonsdottir, T. Evald, and C. H. Mordhorst.** 1982. Detection of measles IgM antibodies by enzyme linked immunosorbent assay (ELISA). *Acta Pathol. Microbiol. Scand.* **90:**153–160.
17. **Rota, J. S., K. B. Hummel, P. A. Rota, and W. J. Bellini.** 1992. Genetic variability of the glycoprotein genes of current wild-type measles isolates. *Virology* **188:**135–142.
18. **Sakaguchi, M., Y. Yoshikawa, K. Yamanouchi, K. Takeda, and T. Sato.** 1986. Characteristics of fresh isolates of wild measles virus. *Jpn. J. Exp. Med.* **56:**61–67.
19. **Salmi, A., and G. Lund.** 1984. Immunoassays for measles virus nucleocapsid antigen: effect of antigen-antibody complexes. *J. Gen. Virol.* **65:**1655–1663.
20. **Salonen, J., R. Vainionpää, and P. Halonen.** 1986. Assay of measles virus IgM and IgG class antibodies by use of peroxidase labelled viral antigens. *Arch. Virol.* **91:**93–106.
21. **Sheshberadaran, H., S.-N. Chen, and E. Norrby.** 1983. Monoclonal antibodies against five structural components of measles virus. I. Characterization of antigenic determinants on nine strains of measles virus. *Virology* **128:**341–353.
22. **Tuokko, H.** 1984. Comparison of non-specific reactivity in indirect and reverse immunoassays for measles and mumps immunoglobulin M antibodies. *J. Clin. Microbiol.* **20:**972–979.
23. **Tuokko, H., and A. Salmi.** 1983. Detection of IgM antibodies to measles virus by enzyme immunoassay. *Med. Microbiol. Immunol.* **171:**187–198.
24. **Ziola, B., P. Halonen, and G. Enders.** 1986. Synthesis of measles virus-specific IgM antibodies and IgM-class rheumatoid factor in relation to clinical onset of subacute sclerosing panencephalitis. *J. Med. Virol.* **18:**51–59.

Mumps Virus

ELLA M. SWIERKOSZ

82

CLINICAL BACKGROUND

Mumps is an acute, usually self-limited, systemic illness characterized most commonly by bilateral or unilateral parotitis accompanied by high fever and fatigue, although approximately one-third of infections may be asymptomatic. Extra-salivary-gland manifestations include meningitis, encephalitis, orchitis, oophoritis, polyarthritis, and pancreatitis. Nephritis, thyroiditis, mastitis, prostatitis, hepatitis, thrombocytopenia, and deafness have been reported as rare manifestations of mumps infection. Humans are the only natural host of mumps virus. Infection confers long-lasting immunity to clinical disease (2). Most cases of mumps occur in children and teenagers aged 5 to 19 years. Approximately 4,000 cases of mumps were reported in the United States in 1991 (4).

The incubation period of the virus ranges from 2 to 4 weeks, with an average of 16 to 18 days. Transmission is by droplet, with primary viral replication occurring in the upper respiratory mucosal epithelium (13). A live, attenuated mumps vaccine in a trivalent form combined with live, attenuated vaccines to rubella and measles is administered to children at 15 months and at 4 to 6 years (1). Although vaccine administration has dramatically reduced the incidence of mumps, outbreaks continue to occur in vaccinated and unvaccinated populations (14, 18, 37).

DESCRIPTION OF AGENT

Mumps virus is a member of the genus *Paramyxovirus* in the family *Paramyxoviridae*. In addition to mumps virus, the paramyxoviruses include Newcastle disease virus and the parainfluenza viruses. Like other paramyxoviruses, mumps virus is a pleomorphic, enveloped virus between 150 and 200 nm in diameter that contains a helical nucleocapsid composed of negative-sense, single-stranded RNA and three nucleocapsid-associated proteins (L, NP, and P, the last of which is presumed to be RNA-dependent RNA polymerase [44]). Two surface glycoproteins, the F (fusion) and HN (hemagglutinin-neuraminidase) proteins, project from the lipid envelope and are visualized as spikes by electron microscopy. The HN protein, which has both hemagglutinating and neuraminidase activities, mediates adsorption of the virus to host cells, while the F glycoprotein mediates fusion of lipid membranes, allowing penetra-

tion of the nucleocapsid into the cell (44). The HN protein is also responsible for mediating hemolysis of erythrocytes (8). Antibody against the HN protein neutralizes viral infectivity (8). A third membrane-associated protein, called the M, or matrix, protein, is believed to mediate assembly of the virion along the inner surface of the virus-modified host cell membrane (44). Only one antigenic type of mumps has been detected by hemagglutination inhibition (HI) or neutralization assays. Antigenic differences among mumps virus strains have been demonstrated with HN-specific monoclonal antibody (35). Moreover, strain-dependent differences in cytopathology and neurovirulence have been described (24). Studies with monoclonal antibodies show that the paramyxo-group viruses have epitopes in common (30).

Mumps virus infectivity is destroyed by organic solvents, detergents, heating to 56°C, UV irradiation, and formalin treatment (8, 29). Mumps virus, like other paramyxoviruses, is relatively unstable and loses infectivity when stored for longer than 4 h in protein-free medium (8).

The first in vitro cultivation of mumps virus was in chick embryos (22), and the virus was subsequently successfully cultivated in tissue culture derived from chick embryos and in mammalian cell cultures (11).

SPECIMEN SELECTION

Laboratory diagnosis of mumps is achieved by viral isolation or serologic testing (20). Virus can be recovered from saliva from 9 days before and up to 8 days after onset of symptoms (13, 42), from urine for up to 2 weeks postonset (43), from cerebrospinal fluid (CSF) of patients with meningitis (42), and from swabs of the area around Stensen's duct (42). Viremia has rarely been detected. Specimens should be obtained early in the course of illness, when virus titer is highest.

For serologic diagnosis, acute- and convalescent-phase blood samples should be submitted. Acute-phase blood should be collected as soon as possible after onset of symptoms, while convalescent-phase blood should be collected 14 or more days postonset. Mumps virus-specific immunoglobulin M (IgM) can be measured in a single serum specimen. In cases of meningitis, antibody titers of CSF and of a serum specimen collected at the same time as the CSF should be determined (7, 26, 36, 40). A single serum specimen is used to assess immune status.

DETECTION OF VIRAL ANTIGEN AND VIRAL NUCLEIC ACID

Hierholzer et al. (15) found that a time-resolved fluoroimmunoassay is more sensitive than an enzyme-linked immunosorbent assay (ELISA) for detection of mumps virus antigen in clinical specimens. Direct immunofluoresence (IF) of exfoliated cells of the upper respiratory tract has been successfully used for rapid detection of mumps virus, parainfluenza viruses, and respiratory syncytial virus (25a); however, there are relatively few data on the application of this method to mumps diagnosis. The PCR has recently been applied for rapid detection of mumps virus (3).

VIRUS ISOLATION

Mumps virus can be cultured in a variety of primary and continuous cell lines as well as in embryonated chicken eggs. In the diagnostic virology laboratory, commercially available primary rhesus monkey kidney (RhMK) and human embryonic kidney (HEK) cells are commonly used (27). Virus can be isolated from saliva from 9 days before and up to 8 days after the onset of parotitis (13, 42). Viruria can be detected up to 2 weeks after onset of symptoms (43). Virus can also be recovered from the CSF during meningitis (42).

Cell culture tubes are incubated at 35 to 37°C. Placing the tubes on a roller drum apparatus may hasten the appearance of virus-specific cytopathic effect (CPE) and the production of hemagglutinins. Cell cultures should be examined by low-power microscopy (×40 to ×100) on a daily basis for the first week of incubation and every third or fourth day thereafter for a total of 14 days. The appearance of multinucleated giant cells suggests a paramyxovirus, but CPE may not appear or may be very subtle. Primary monkey kidney cells may contain endogenous viruses that produce syncytia. For detection of viral growth in the absence of CPE, the hemadsorption test is performed with guinea pig erythrocytes (0.5%) at days 5 to 7 and day 14 postinoculation (5). Uninfected control tubes are tested in parallel with the inoculated tubes to rule out nonspecific adsorption of erythrocytes and to detect endogenous simian paramyxoviruses. Hemadsorption-positive cultures can be typed by IF staining (see below).

IDENTIFICATION OF VIRUS

Identification of virus by IF is a rapid and technically simple procedure. Titers of polyclonal antisera against parainfluenza viruses should be determined in order to determine the degree of cross-reactivity among these viruses. Monoclonal antibodies obviate the need for titration against parainfluenza viruses. Cells from hemadsorption-positive tubes can be stained directly, or a second passage culture tube showing maximal CPE or hemadsorption can be used. Make a suspension of cells in a small volume of phosphate-buffered saline, apply it to wells of a Teflon-coated microscope slide, and then air dry the slide. Fix the slide in acetone for 10 min, and air dry it. Stain the wells with mumps virus-specific antiserum conjugated to fluorescein isothiocyanate (FITC) for use in the direct IF test or unconjugated for use in the indirect IF test (32). Stain cells from infected and uninfected tubes as controls. A indirect IF kit containing mumps virus monoclonal antibody is available from Chemicon International, Inc. (Temecula, Calif.).

Other methods for identification of mumps isolates are complement fixation (CF), HI, hemadsorption inhibition, and neutralization (38). Reagents for the performance of the CF test are commercially available (BioWhittaker, Walkersville, Md.). Crude antigen prepared from infected cell culture tubes is used in a microtiter CF test (34). The HI test is performed by a microtiter procedure (38). The immune serum for the HI test must be pretreated with receptor-destroying enzyme of *Vibrio cholerae* or with potassium periodate to remove nonspecific inhibitors of agglutination (38). Nonspecific hemagglutinins in the serum must also be removed (38). Immune serum should be tested against parainfluenza viruses to rule out heterologous reactions. The hemadsorption inhibition test is less sensitive than the HI test but is technically simpler to perform (38). The neutralization test has largely been replaced by other methods for identification of viral isolates.

SEROLOGIC DIAGNOSIS

The serologic diagnosis of mumps, traditionally based on the demonstration of a significant titer rise between acute- and convalescent-phase sera, is limited by cross-reactions between mumps and parainfluenza viruses and by delays in collection of the acute-phase serum, so that titer rises are missed. ELISAs, though more sensitive than CF and HI tests (25, 31), are hindered by the same problems. Tests for determination of immunity to mumps virus are now commercially available.

The use of a mumps virus skin test to assay immunity to mumps is not reliable. Both false-positive and false-negative results occur, and application of the skin test antigen to an immune individual can stimulate antibody rises in serum, thus causing interpretive errors (2).

HI Test

The HI procedure is identical to that described above for the identification of viral isolates. Commercially available antigens and immune sera are difficult to find. Patient serum is pretreated to remove nonspecific inhibitors and nonspecific agglutinins and then diluted serially in twofold dilutions, and mumps virus antigen is added. The highest dilution of serum exhibiting HI is the endpoint.

CF Test

The CF test has been used for several decades for the serologic diagnosis of mumps. Mumps S and V antigens, which correspond to the NP and HN virus proteins, respectively, are commercially available (16). The CF test is performed as previously described (34). For effective detection of seroconversions, both antigens should be tested (6, 12), as discussed below under Evaluation and Interpretation of Results. The CF test is not sensitive enough to determine immune status.

Neutralization

Neutralization assays can be performed either in RhMK, HEK, or primary chick embryo cell culture tubes or in microtiter plates seeded with Vero cells (19, 29). The microtiter procedure conserves reagents and patient sera and correlates well with the tube test.

TABLE 1 Commercially available reagents for mumps virus diagnosis

Supplier	Test method	Antibody detected	Intended use
BioWhittaker (Walkersville, Md.)	ELISA	IgG	Determination of immune status
	FIAX	IgG	
Clark Laboratories, Inc. (Jamestown, N.Y.)	ELISA	IgG	Determination of immune status; demonstration of seroconversion by using paired sera
		IgM	Demonstration of acute infection
Schiapparelli Biosystems Inc. (Columbia, Md.)	IFA[a]	IgG	Determination of immune status; demonstration of seroconversion by using paired sera
SCIMEDX (Denville, N.J.)	ELISA	IgG	Determination of immune status
		IgM	Demonstration of acute infection

[a]IFA, indirect fluorescent-antibody assay.

IF Test

The IF procedure allows detection of both IgG and IgM antibodies to mumps. Serial dilutions of patient serum are applied to microscope slide wells containing acetone-fixed infected cells. IgG and IgM are detected by using FITC-conjugated anti-human IgG or IgM. Stained slides are observed under a fluorescence microscope. The highest dilution of serum showing intracytoplasmic fluorescence is the endpoint. Slides containing uninfected cells should be prepared as a control for nonspecific fluorescence. An IF kit for determination of mumps virus titers is commercially available (Table 1).

FIAX

A solid-phase immunofluorescence assay, FIAX (BioWhittaker), is available for measurement of IgG to mumps virus. It is intended for determination of immune status from single serum samples. Mumps virus antigen is coated onto one side of a dual-surface "sampler"; the reverse side contains no antigen and serves as a blank. The sampler is immersed in diluted serum and incubated. After a wash step, the sampler is immersed in FITC-labeled anti-human IgG antibody. After a final wash, the sampler is inserted into a fluorometer, which measures the amount of bound, labeled antibody and expresses it as fluorescence signal units (FSU). The FSU of the blank side is measured and subtracted from the FSU of the antigen side. A calibration curve is constructed by plotting ΔFSU versus titer for each calibrator. Patient titers are then obtained by interpolation of the calibration curve.

ELISA

A number of microplate ELISA procedures that measure IgG and IgM antibodies to mumps virus have been described (10, 23, 25, 28, 31, 33, 39). Antigens used include whole virus, sonicated virus, purified HN (V) and NP (S) antigens, and peroxidase-labeled antigen. Commercial ELISA kits are available from a number of suppliers (Table 1).

Hemolysis-in-Gel Test

The hemolysis-in-gel test (9, 29) is based on the principle that mumps virus antigen coated onto erythrocytes is lysed by specific antibody in the presence of complement. Anti-gen-coated erythrocytes are suspended in 1.5% agarose. After solidification in petri plates, 3-mm-wide holes are punched into the gel. The holes are filled with heat-inactivated patient serum, the plates are incubated for 24 h at 4°C, and guinea pig complement is added as an overlay. The plates are reincubated at 37°C for 2 h, after which the zone of hemolysis is measured to the nearest 0.1 mm. Plates containing uncoated erythrocytes are tested in parallel as a control for nonspecific lysis. The diameter of the hemolytic zone is plotted against the serum dilution, and a regression analysis is done. The diameter of the zone is directly proportional to the antibody titer of the serum.

CSF Antibody

IgG and IgM antibody titers of CSF can be measured in patients with meningitis. CSF titers have been measured by ELISA and by HI, CF, hemolysis inhibition, and mixed hemadsorption tests (7, 26, 36, 40).

EVALUATION AND INTERPRETATION OF RESULTS

The isolation of mumps virus from CSF, respiratory specimens, or urine establishes a diagnosis of mumps. Serologic diagnosis of mumps is problematic because of the cross-reactions between mumps virus and paramyxoviruses (17, 21, 28). Paired sera exhibiting fourfold or greater rises in titer against mumps virus should be tested against the parainfluenza viruses. If a rise in titer against one or more parainfluenza viruses is observed, it is not possible to make a definitive diagnosis of mumps without a compatible clinical picture.

It has been accepted for many years that serologic testing by CF should include measurement of antibody to both V and S antigens of mumps. Moreover, early work describing the CF test demonstrated that antibody to the S antigen arises earlier and more rapidly in developing cases of mumps and that low levels of or absent S antibody in the presence of V antibody is a sign of long-past infection (12). More recent work casts doubt on this interpretation (6). In 40 mumps cases, V antibody appeared earlier in 14 cases, while S antibody appeared earlier in only 2 cases. This study confirms the necessity of testing both V and S for demonstration of seroconversions and extends this recommendation to include the HI test for detection of the maximum number of mumps cases.

Assays for IgM are useful, since cross-reactions between mumps and parainfluenza viruses have not been observed for IgM antibody (10, 25, 41). However, IgM assays false positive because of rheumatoid factor and IgM assays false negative because of high levels of IgG occur (33). IgM capture ELISA (10, 23, 33) is very sensitive and specific and can be used to diagnose a current mumps virus infection. IgM antibody, measured by IgM capture, was present in all patients with mumps by the fifth day of illness, and peak IgM titers were reached at 1 week of illness and persisted for at least 6 weeks (10, 33).

The ELISA has a high sensitivity compared to neutralization and is suited for determination of immune status (23), although cross-reactions with paramyxoviruses also occur. ELISA is more sensitive than the CF and HI tests for detection of low levels of antibodies (31). For diagnosis of acute mumps virus infections, significant titer rises may be more difficult to detect by ELISA than by CF because of the earlier appearance of antibody by ELISA (25).

In cases of mumps meningitis, mumps virus-specific IgG is found in CSF and confirms the diagnosis. The CF and HI tests are relatively insensitive for CSF antibody detection compared to the ELISA (7, 26, 36, 40). IgG is present in CSF within a few days after onset of disease and peaks at 7 to 10 days (40).

REFERENCES

1. **ACIP.** 1989. Mumps prevention: recommendations of the Immunization Practices Advisory Committee (ACIP). *Morbid. Mortal. Weekly Rep.* **38**(Suppl. 9):6–7.
2. **Baum, S. G., and N. Litman.** 1990. Mumps virus, p. 1260–1265. *In* G. L. Mandell, R. G. Douglas, Jr., and J. E. Bennett (ed.), *Principles and Practice of Infectious Diseases*, 3rd ed. Churchill Livingstone, New York.
3. **Boriskin, Y. S., J. C. Booth, and A. Yamada.** 1993. Rapid detection of mumps virus by the polymerase chain reaction. *J. Virol. Methods* **42**:23–32.
4. **Centers for Disease Control.** 1991. Summary of notifiable diseases, United States, 1991. *Morbid. Mortal. Weekly Rep.* **40**:10.
5. **Chanock, R. M., K. M. Johnson, M. K. Cook, D. C. Wong, and A. Vargosko.** 1961. The hemadsorption technique, with special reference to the problem of naturally occurring simian parainfluenza virus. *Am. Rev. Respir. Dis.* **83**:125–129.
6. **Freeman, R., and M. H. Hambling.** 1980. Serological studies on 40 cases of mumps virus infection. *J. Clin. Pathol.* **33**:28–32.
7. **Fryden, A., H. Link, and E. Norrby.** 1978. Cerebrospinal fluid and serum immunoglobulin and antibody titers in mumps meningitis and aseptic meningitis of other etiology. *Infect. Immun.* **21**:852–861.
8. **Ginsberg, H. S.** 1988. Paramyxoviruses, p. 239–259. *In* R. Dulbecco and H. S. Ginsberg (ed.), *Virology*, 2nd ed. J. B. Lippincott Co., Philadelphia.
9. **Grillner, L., and J. Blomberg.** 1976. Hemolysis-in-gel and neutralization tests for determination of antibodies to mumps virus. *J. Clin. Microbiol.* **4**:11–15.
10. **Gut, J. P., C. Spiess, S. Schmitt, and A. Kirn.** 1985. Rapid diagnosis of acute mumps infection by a direct immunoglobulin M antibody capture enzyme immunoassay with labeled antigen. *J. Clin. Microbiol.* **21**:346–352.
11. **Henle, G., and F. Deinhardt.** 1955. Propagation and primary isolation of mumps virus in tissue culture. *Proc. Soc. Exp. Biol. Med.* **89**:556–560.
12. **Henle, G., S. Harris, and W. Henle.** 1948. The reactivity of various human sera with mumps complement fixation antigens. *J. Exp. Med.* **88**:133–147.
13. **Henle, G., W. Henle, K. K. Wendell, and P. Rosenberg.** 1948. Isolation of mumps virus from human being with induced apparent or inapparent infections. *J. Exp. Med.* **88**:223–232.
14. **Hersh, B. S., P. E. M. Fine, W. K. Kent, S. L. Cochi, L. H. Kahn, E. R. Zell, P. L. Hays, and C. L. Wood.** 1991. Mumps outbreak in a highly vaccinated population. *J. Pediatr.* **119**:187–193.
15. **Hierholzer, J. C., P. G. Bingham, E. Castells, and R. A. Coombs.** 1993. Time-resolved fluoroimmunoassays with monoclonal antibodies for rapid identification of parainfluenza type 4 and mumps virus. *Arch. Virol.* **130**:335–352.
16. **Jensik, S. C., and S. Silver.** 1976. Polypeptides of mumps virus. *J. Virol.* **17**:363–373.
17. **Julkunen, I.** 1984. Serological diagnosis of parainfluenza virus infections by enzyme immunoassay with special emphasis on purity of viral antigens. *J. Med. Virol.* **14**:177–187.
18. **Kaplan, K. M., D. C. Marder, S. L. Cochi, and S. R. Preblud.** 1988. Mumps in the workplace. *JAMA* **260**:1434–1438.
19. **Kenny, M. T., K. L. Albright, and R. P. Sanderson.** 1970. Microneutralization test for the determination of mumps antibody in Vero cells. *Appl. Microbiol.* **20**:371–373.
20. **Lennette, D. A.** 1991. Preparation of specimens for virological examination, p. 818–821. *In* A. Balows, W. J. Hausler, Jr., K. L. Herrmann, H. D. Isenberg, and H. J. Shadomy (ed.), *Manual of Clinical Microbiology*, 5th ed. American Society for Microbiology, Washington, D.C.
21. **Lennette, E. H., F. W. Jensen, R. W. Guenther, and R. L. Magoffin.** 1963. Serologic responses to para-influenza viruses in patients with mumps virus infection. *J. Lab. Clin. Med.* **61**:780–788.
22. **Leymaster, G. R., and T. G. Ward.** 1947. Direct isolation of mumps virus in chick embryos. *Proc. Soc. Exp. Biol. Med.* **65**:346–348.
23. **Linde, G. A., M. Granstrom, and C. Orvell.** 1987. Immunoglobulin class and immunoglobulin G subclass enzyme-linked immunosorbent assays compared with microneutralization assay for serodiagnosis of mumps infection and determination of immunity. *J. Clin. Microbiol.* **25**:1653–1658.
24. **Merz, D. C., and J. S. Wolinsky.** 1981. Biochemical features of mumps virus neuraminidases and their relationship with pathogenicity. *Virology* **114**:218–227.
25. **Meurman, O., P. Hanninen, R. V. Krishna, and T. Ziegler.** 1982. Determination of IgG- and IgM-class antibodies to mumps virus by solid-phase enzyme immunoassay. *J. Virol. Methods* **4**:249–257.
25a.**Minnich, L., and C. G. Ray.** 1980. Comparison of direct immunofluorescent staining of clinical specimens for respiratory virus antigens with conventional isolation techniques. *J. Clin. Microbiol.* **12**:391–394.
26. **Morishima, T., M. Miyazu, T. Ozaki, S. Isomura, and S. Suzuki.** 1980. Local immunity in mumps meningitis. *Am. J. Dis. Child.* **134**:1060–1064.
27. **Mufson, M. A.** 1989. Parainfluenza viruses, mumps virus, Newcastle disease virus, p. 669–691. *In* N. J. Schmidt and R. W. Emmons (ed.), *Diagnostic Procedures for Viral, Rickettsial, and Chlamydial Infections*, 6th ed. American Public Health Association, Washington, D.C.
28. **Nicolai-Scholten, M. E., R. Ziegelmaier, F. Behrens, and W. Hopken.** 1980. The enzyme-linked immunosorbent assay (ELISA) for determination of IgG and IgM antibodies after infection with mumps virus. *Med. Microbiol. Immunol.* **168**:81–90.
29. **Norrby, E.** 1985. Mumps, p. 774–778. *In* E. H. Lennette, A. Balows, W. J. Hausler, Jr., and H. J. Shadomy (ed.), *Manual of Clinical Microbiology*, 4th ed. American Society for Microbiology, Washington, D.C.
30. **Orvell, C., R. Rydbeck, and A. Love.** 1986. Immunological relationships between mumps virus and parainfluenza viruses studied with monoclonal antibodies. *J. Gen. Virol.* **67**:1929–1939.
31. **Popow-Kraupp, T.** 1981. Enzyme-linked immunosorbent as-

say (ELISA) for mumps virus antibodies. *J. Med. Virol.* **8:**79–88.

32. **Riggs, J. L.** 1989. Immunofluorescence staining, p. 123–133. *In* N. J. Schmidt and R. W. Emmons (ed.), *Diagnostic Procedures for Viral, Rickettsial, and Chlamydial Infections,* 6th ed. American Public Health Association, Washington, D.C.

33. **Sakata, H., M. Tsurudome, M. Hishiyama, Y. Ito, and A. Sugiura.** 1985. Enzyme-linked immunosorbent assay for mumps IgM antibody: comparison of IgM capture and indirect IgM assay. *J. Virol. Methods* **12:**303–311.

34. **Schmidt, N., and R. W. Emmons.** 1989. General principles of laboratory diagnostic methods for viral, rickettsial and chlamydial infections, p. 21–28. *In* N. J. Schmidt and R. W. Emmons (ed.), *Diagnostic Procedures for Viral, Rickettsial and Chlamydial Infections,* 6th ed. American Public Health Association, Washington, D.C.

35. **Server, A. C., D. C. Merz, M. N. Waxham, and J. S. Wolinsky.** 1982. Differentiation of mumps virus strains with monoclonal antibody to the HN glycoprotein. *Infect. Immun.* **35:**179–186.

36. **Sharief, M. K., and E. J. Thompson.** 1990. A sensitive ELISA system for the rapid detection of virus specific IgM antibodies in the cerebrospinal fluid. *J. Immunol. Methods* **130:**19–24.

37. **Sosin, D. M., S. L. Cochi, R. A. Gunn, C. E. Jennings, and S. R. Preblud.** 1989. Changing epidemiology of mumps and its impact on university campuses. *Pediatrics* **84:**779–784.

38. **Swierkosz, E. M.** 1991. Mumps virus, p. 912–917. *In* A. Balows, W. J. Hausler, Jr., K. L. Herrmann, H. D. Isenberg, and H. J. Shadomy (ed.), *Manual of Clinical Microbiology,* 5th ed. American Society for Microbiology, Washington, D.C.

39. **Ukkonen, P., M. L. Granstrom, and K. Penttinen.** 1981. Mumps-specific immunoglobulin M and G antibodies in natural mumps infection as measured by enzyme-linked immunosorbent assay. *J. Med. Virol.* **8:**131–142.

40. **Ukkonen, P., M. L. Granstrom, J. Rasanen, E. M. Salonen, and K. Penttinen.** 1981. Local production of mumps IgG and IgM antibodies in the cerebrospinal fluid of meningitis patients. *J. Med. Virol.* **8:**257–265.

41. **Ukkonen, P., O. Vaisanen, and K. Penttinen.** 1980. Enzyme-linked immunosorbent assay for mumps and parainfluenza type 1 immunoglobulin G and immunoglobulin M antibodies. *J. Clin. Microbiol.* **11:**319–323.

42. **Utz, J. P., J. A. Kasel, H. G. Cramblett, C. F. Szwed, and R. H. Parrott.** 1957. Clinical and laboratory studies of mumps. I. Laboratory diagnosis by tissue-culture technics. *N. Engl. J. Med.* **257:**497–502.

43. **Utz, J. P., C. F. Szwed, and J. A. Kasel.** 1958. Clinical and laboratory studies of mumps. II. Detection and duration of excretion of virus in urine. *Proc. Soc. Exp. Biol. Med.* **99:**259–261.

44. **Wolinsky, J. S., and A. C. Server.** 1985. Mumps virus, p. 1255–1284. *In* B. N. Fields, D. M. Knipe, R. M. Chanock, J. L. Melnick, B. Roizman, and R. E. Shope (ed.), *Virology.* Raven Press, New York.

Rubella Virus

MAX A. CHERNESKY AND JAMES B. MAHONY

83

DESCRIPTION OF THE GENUS

Rubella virus is classified as a rubivirus and is a member of the family *Togaviridae*, which also contains the arthropod-borne alpha- and flaviviruses as well as members of the genus *Pestivirus*. Classification is based on morphologic, antigenic, and physicochemical properties. Only one type or species of rubella virus has been recognized so far, and it is immunologically distinct from all other known viruses. The virus possesses three major structural proteins: two glycoproteins, E1 (58,000 Da) and E2 (42,000 to 47,000 Da), are associated with the envelope, and nucleocapsid protein C (33,000 Da) is found internally. The virus is spherical, with a diameter of 60 to 70 nm, and contains a dense central core surrounded by a lipid bilayer. Replication of viral RNA and synthesis of protein occur in the cytoplasm of the cell, and the virus matures by budding into cytoplasmic vesicles or from the marginal plasma membrane. The virus is destroyed by lipid solvents or trypsin but not by freezing-thawing or ultrasonification.

NATURAL HISTORY

Rubella virus is found only in human populations and causes rubella (German measles). Postnatal rubella is transmitted chiefly through direct contact with nasopharyngeal secretions or through contact with droplets of these secretions. The peak incidence of infection occurs in the late winter and early spring. Subclinical infection is common. The period of maximum communicability appears to be the few days before and 5 to 7 days after the onset of the rash. Volunteer studies indicate the presence of rubella virus in nasopharyngeal secretions from 7 days before to 14 days after the onset of the rash. Infants with congenital rubella may continue to shed virus in nasopharyngeal secretions and urine for 1 year or more and may transmit infection to susceptible contacts. Virus can be isolated from the nasopharynx at 6 months of age in approximately 10 to 20% of these patients.

Before the widespread use of rubella vaccine, rubella was an epidemic disease with 6- to 9-year cycles; the majority of cases occurred in children. Currently, the incidence of rubella in North America has declined by more than 99% from the incidence in the prevaccine era. Although the risk of acquiring rubella has declined sharply in all age groups, including adolescents and young adults, a greater percentage of the cases are in young, unvaccinated adults in settings in which young adults are grouped together. Recent serologic surveys have indicated that 10 to 20% of young adults are susceptible to rubella. This degree of susceptibility in young adults is due predominantly to underutilization of vaccine in this population and not to waning immunity in immunized persons.

The incubation period for postnatal rubella ranges from 10 to 21 days and is usually 16 to 18 days. It is usually a mild disease characterized by an erythematous maculopapular discrete rash, postauricular and suboccipital lymphadenopathy, and slight fever. Some 25 to 50% of infections are asymptomatic. Transient polyarthralgia and polyarthritis occur occasionally in children but are extremely common in older individuals. Encephalitis and thrombocytopenia are rare complications.

The most commonly described anomalies associated with congenital rubella are ophthalmologic (cataracts, microphthalmia, glaucoma, and chorioretinitis), cardiac (patent ductus arteriosus, peripheral pulmonary artery stenosis, and atrial or ventricular septal defects), auditory (sensorineural deafness), and neurologic (microcephaly, meningoencephalitis, and mental retardation). Infants with congenital rubella frequently are growth retarded and have radiolucent bone disease, hepatosplenomegaly, thrombocytopenia, jaundice, and purpuralike skin lesions ("blueberry muffin" appearance).

Antibodies to the virus appear as the rash fades (Fig. 1), and initially, both immunoglobulin G (IgG) and IgM antibodies can be detected. Antibodies of the IgM class generally do not persist beyond 4 to 5 weeks after onset of illness, but IgG antibodies usually persist throughout life. Reinfection with the virus can occur, but it is almost always asymptomatic and can be detected by a rise in IgG antibodies. The risk of fetal damage resulting from rubella reinfection during pregnancy is negligible. The attenuated virus vaccines induce a production of IgM and IgG antibodies similar to that observed with natural infections except that the titers are somewhat lower. Reinfection rates with wild virus are greater among vaccinees than among persons previously infected under natural conditions.

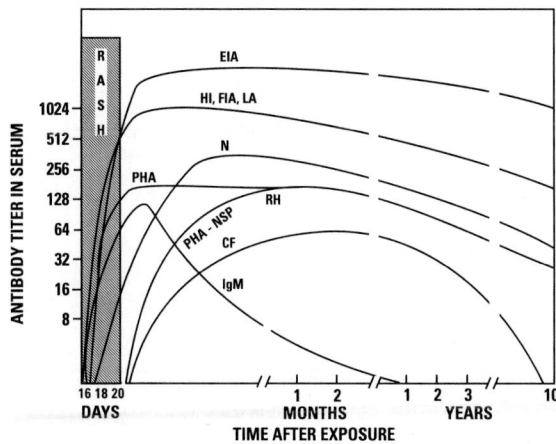

FIGURE 1 Antibody response after rubella virus infection. FIA, fluorescence immunoassay; N, neutralization; NSP, nonstructural protein.

COLLECTION AND STORAGE OF SPECIMENS

For details of procedures for the collection and storage of serum, urine, nasopharyngeal, throat, or cerebrospinal specimens, see chapter 3 of this Manual.

LABORATORY INVESTIGATION

Reliable laboratory technology has been developed for (i) determination of immunity to rubella, (ii) diagnosis of postnatal rubella, and (iii) diagnosis of congenital rubella. Because antibody responses are rapid and specific and virus isolation procedures are slow and expensive, serologic procedures are usually performed for disease diagnosis (the exception is virus isolation for a congenitally infected newborn).

VIRUS ISOLATION

A wide variety of cell types are susceptible to infection by rubella virus. For primary isolation of rubella virus from clinical specimens, however, primary African green monkey kidney (AGMK), Vero, or RK-13 cell cultures are recommended. Isolation of rubella virus in primary cultures of AGMK cells has been considered the standard method (15). Rubella virus is detected in this cell type by interference with the cytopathic effects (CPE) of a challenge virus. For virus isolation in AGMK, four tubes containing cell monolayers are drained of medium; 0.2 ml of the specimen is inoculated into each tube and allowed to adsorb for 1 h at 35 to 36°C. A 1.5-ml amount of maintenance medium (Eagle basal medium with 2% inactivated fetal bovine serum) is then added to each tube, and the tubes are incubated in stationary racks at 35 to 36°C for 10 days. Uninoculated AGMK cell cultures serve as controls. At the end of the incubation period, the tubes are examined for CPE. If none is found, two inoculated and two control tubes are challenged with 100 to 1,000 50% tissue culture infective doses of a challenge virus. The viruses most commonly used as the challenge agent for rubella isolation in AGMK cells are echovirus type 11 or coxsackievirus A9. Challenged tubes are read 3 to 4 days after challenge. The presence of rubella virus is indicated by complete destruc-

tion of the control cells but little or no CPE in the inoculated tubes. Absolute identification of the isolate requires specific neutralization of the interference with rubella antibody.

Specimens with low concentrations of rubella virus may produce little or partial interference in the culture tubes initially inoculated. The culture fluids may have to be passaged to demonstrate the virus. Fluid from the cultures that were not challenged should be inoculated into an additional four tubes of AGMK cells. Control tubes are also included. These tubes should be incubated for an additional 7 to 8 days. Two inoculated tubes and two control tubes are then challenged with echovirus type 11 and observed for CPE. The absence of interference with echovirus cytopathogenicity on passage of fluid from tubes that did not initially demonstrate interference confirms the absence of an interfering agent. If virus is not found after such a passage, it is rarely found by further passages.

Some laboratories use RK-13 or Vero cells for isolation of rubella virus. In these cell systems, rubella virus produces CPE; however, the CPE is not always clear on primary isolation, and cell culture fluids may need to be passaged several times for full detection of virus. These cell systems, however, do offer the advantage of direct neutralization for identification of an isolate. Furthermore, an indirect immunofluorescence staining method has been shown to be specific and sensitive for identifying rubella virus isolates in these cells (14).

Fresh unfrozen tissue specimens may be of particular value in attempts to isolate virus from fetal tissues and organs. A convenient method is to explant a minced tissue fragment with growth medium and allow sufficient time for the outgrowth of cells. When the cells have formed monolayers, the extracellular fluids can be harvested and tested for the presence of an interfering agent as described above. This method of rubella virus isolation is more sensitive than the method in which tissue extracts or homogenates of ground tissue are used.

Rubella virus can be specifically neutralized with rubella antiserum prepared in rabbits. Such antisera are available from several commercial sources. Immune rabbit serum is diluted to contain 4 U of neutralizing antibody. A healthy preimmune (rubella antibody-free) rabbit serum is diluted similarly for the control titration. The media from the two companion, unchallenged cultures containing the interfering agent are pooled, and serial 10-fold dilutions (undiluted to 10^{-6}) are made in maintenance medium; 0.1-ml samples of each dilution are inoculated into three culture tubes. The 10^{-1}, 10^{-2}, and 10^{-3} dilutions are also combined with equal volumes of the prediluted rubella antiserum and of the prediluted healthy rabbit serum. After 1 h of incubation at 35°C, 0.2 ml of each mixture is inoculated into each of three AGMK tubes. The tubes are incubated at 35 to 36°C for 7 to 8 days, challenged with echovirus type 11, and observed for the development of enterovirus CPE. Destruction of the AGMK monolayers inoculated with the isolate dilution containing between 10 and 100 50% tissue culture infective doses plus the immune rabbit serum but not of those inoculated with the healthy rabbit serum indicates that the isolate is rubella virus.

PCR has been successfully performed on RNA extracted from chorionic villi biopsied from the fetus in utero (8). The technique uses a rubella virus-specific oligonucleotide primer to reverse transcribe a cDNA, which is then amplified by PCR and detected by Southern blotting. The technique is limited to research laboratories.

SEROLOGIC TESTING

Serologic techniques for the detection of antibodies to rubella virus provide the approach of choice for laboratory diagnosis of acute and congenital rubella infections and for the determination of rubella immune status. Methods currently available include hemagglutination inhibition (HI), passive hemagglutination (PHA), radial hemolysis (RH) in gel, latex agglutination (LA), enzyme immunoassay (EIA), complement fixation (CF), and a variety of rubella-specific IgM antibody assays. It is not yet fully understood which of this myriad of serologic techniques for measuring rubella IgG antibodies best approximates immunity to reinfection or disease in the fetus.

Investigation of the pregnant patient who has been in contact with another person with rubella presents a challenge for a rapid and accurate serologic diagnosis. If the patient develops clinical signs, a serum specimen should be collected at that time and paired with a second serum collected 5 or more days later. Both are investigated in parallel (same test, same day). A fourfold or greater rise in HI or CF antibodies together with clinical symptoms is diagnostic for recent infection. Alternatively, a significant change in optical densities or binding ratios of the sera in an EIA would be diagnostic. A seroconversion in any patient is conclusive for recent rubella virus infection. If EIA is available for rubella IgM testing, a single serum sample collected between days 3 and 21 after onset of rash usually yields a positive diagnosis (4). Patients without clinical symptoms but with rising titers to rubella virus (the first serum containing antibodies) pose a special problem. They may have a primary infection or reinfection with an anamnestic antibody boost. To confirm these cases, a rubella IgM determination may be performed. Measurement of rubella IgG avidity (7) by enzyme immunoassay may help determine primary infection or reinfection. A third type of problem case involves patients whose sera are collected several days after the infection, when all serologic indicators are at a plateau (Fig. 1); testing for IgM may be helpful. Serologic examination of the suspected contacts may also be helpful in these situations.

Serologic investigation of a suspected case of congenital rubella could involve two approaches. One approach is to make serial determinations of rubella HI or solid-phase immunoassay (SPIA) antibodies during the first 6 months of life; a persistence of titer in the infant during this time (Fig. 2) is highly suggestive of congenital rubella. As a second approach, the demonstration of rubella virus-specific IgM antibody in infant serum during the neonatal period would be diagnostic of congenital rubella.

TEST PROCEDURES

PHA

PHA represents a rapid, inexpensive way of screening for rubella immunity. PHA to nonstructural proteins (Fig. 1) parallels CF and neutralization responses after infection, whereas PHA structural protein responses are more closely aligned with HI and SPIA. Both antibodies remain measurable for years after infection or immunization as an index of immunity. PHA employs erythrocytes stabilized with formaldehyde-pyruvic aldehyde that have been sensitized with rubella virus antigen. The erythrocytes agglutinate in the presence of specific rubella antibody. In performing the test, phosphate buffer is added to V-bottom microwells. Specimens, as well as positive and negative controls, are

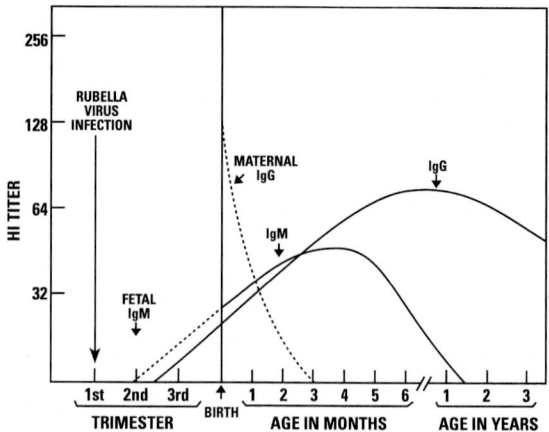

FIGURE 2 Antibody responses in an infant congenitally infected with rubella virus.

added and then mixed before the sensitized cells are added. A row of erythrocytes without rubella antigen is used as a control, which allows more objective scoring of results. A button of erythrocytes signifies the absence of antibody (susceptibility to rubella), whereas a dispersed settling of erythrocytes indicates a positive reaction (immunity to rubella).

RH

RH is popular in Europe and can be used to screen large numbers of serum samples by preparing plates in advance and storing them at 4°C. We have successfully used the method of Russell et al. (12). Freshly drawn sheep erythrocytes are washed with glucose-gelatin-Veronal buffer, treated with 2.5 mg of trypsin (Difco Laboratories, Detroit, Mich.) per ml in glucose-gelatin-Veronal buffer for 1 h at room temperature, and sensitized with rubella hemagglutinating antigen (Flow Laboratories, McLean, Va.; 240 hemagglutination units [HAU] per ml) in HEPES (*N*-2-hydroxyethylpiperazine-*N'*-2-ethanesulfonic acid)-saline-albumin-gelatin buffer (HSAG) (pH 6.2) for 1 h at 4°C. Sensitized erythrocytes (0.15 ml of a 50% suspension) are mixed with 0.4 ml of guinea pig complement (Behringwerke, Marburg, Germany) and then added to 10 ml of 0.8% agarose preheated to 43°C and poured into a square petri dish (100 by 15 mm). Plates can be stored at 4°C and used for up to 14 days. Prepoured RH plates are available from Orion Diagnostics, Helsinki, Finland. Sera are inactivated at 56°C for 30 min, pipetted into 3-mm-diameter wells punched in the agarose, and allowed to diffuse overnight at 4°C in a humidified atmosphere. Plates are then incubated for 2 h at 37°C. Plates with incomplete hemolysis are flooded with 4 ml of guinea pig complement (diluted 1:3) and reincubated. All sera are tested in control plates containing unsensitized erythrocytes to monitor nonspecific hemolysis. Zones of hemolysis in control plates may range from 3.5 to 5 mm in diameter. Zone diameters of >5 mm are taken to indicate immunity.

LA Test

Rapid LA tests are predominantly used for immunity screening. The tests incorporate antigen-coated latex particles and may be performed on a card or in a tray. Agglutination lends itself to processing small numbers of specimens.

HI Test

Although different laboratories use modifications of the standard rubella HI test, the following is a description of the test that we have consistently found to give good results. The test is conveniently performed in disposable plastic or vinyl V-bottom microtiter plates. The rubella hemagglutinating antigen is titrated each time the test is performed. Serial twofold dilutions of the antigen are made in 0.025 ml of HEPES diluent containing 0.025 M HEPES, 0.14 M NaCl, 0.001 M $CaCl_2$, 1% bovine albumin, and 0.00025% gelatin (HSAG) at pH 6.2, to which 0.05 ml of a 0.25% washed suspension of pigeon erythrocytes is added (a broad variety of erythrocytes can be used; the most popular other than pigeon are baby chick or trypsinized human O cells). Control cups containing no antigen are included. The plates are sealed and placed at 4°C for 1 h, after which they are placed at room temperature for 15 min before being read. The highest dilution is considered 1 HAU; 4 HAU is used in the HI test. Before the HI test is performed, nonspecific inhibitors of hemagglutination and nonspecific agglutinins must be removed from sera. Test serum (0.1 ml) is added to 0.1 ml of HSAG and 0.6 ml of a 25% suspension of kaolin. The suspension is mixed and allowed to sit at room temperature for 20 min with frequent agitation. The kaolin is pelleted in a clinical centrifuge, and the supernatant fluid is transferred to a clean tube containing 0.05 ml of a 50% suspension of pigeon erythrocytes. After 60 min on incubation at 4°C, the erythrocytes are centrifuged; the supernatant fluid is then removed and heated at 56°C for 30 min. This final sample, which represents a dilution of 1:8, is now ready to be incorporated into the test. Alternatively, nonspecific inhibitors may be removed by precipitation with heparin and manganous chloride. The serum sample is diluted 1:4 with 0.15 M NaCl. To each 0.8 ml of diluted serum are added 0.03 ml of sodium heparin (200 U) and 0.04 ml of 1 M manganous chloride. The sample is held at 4°C for 20 min, and the precipitate that forms is pelleted by centrifugation. The supernatant fluid is then adsorbed with pigeon erythrocytes as described above.

Known positive and negative control sera are treated similarly. Further serial twofold dilutions of each serum are made in 0.025-ml amounts of HSAG. An area of the plate is reserved for duplication of the first three dilutions of each serum. These dilutions receive HSAG in place of antigen and serve as serum controls. To all other dilutions of serum is added 4 HAU of antigen in a volume of 0.025 ml of HSAG. The antigen is backtitrated in a separate section of the plate by doubling dilutions in 0.025 ml of HSAG to represent 4, 2, 1, and 0.5 HAU. The plates are incubated for 1 h at room temperature, after which 0.05 ml of a 0.25% suspension of pigeon erythrocytes is added to each well and the plate contents are mixed on a plate vibrator. The hemagglutination pattern is read after 1 h at 4°C. The highest dilution of serum that completely inhibits hemagglutination is taken as the endpoint or rubella titer.

Since the original description of the HI test for rubella (15), several modifications have been introduced. These include the choice of indicator erythrocytes (5), optimal pH of reagents (6), methods for removal of nonspecific inhibitors from sera (9), methods for preparation of the antigen (13), and duration of incubation of antigen with the sera (11).

To prepare hemagglutinating antigen, we infect monolayers of BHK-21 cells grown in 32-oz (960-ml) bottles with 5 to 10 ml of rubella stock containing 10^4 or more 50% tissue culture infective doses per ml. After the virus has adsorbed for 2 h at 37°C, the monolayers are covered with Eagle medium containing 2% fetal calf serum that has previously been adsorbed with kaolin. The medium is changed after 24 h and then harvested for hemagglutinating antigen after days 5 and 7 of incubation. High-titer antigen can be extracted from the monolayers by extracting the cell-associated antigen with alkaline buffers. The cell-associated antigen preparations have CF activity and can be used as the antigen source in the CF test. We have found that good-quality CF and hemagglutinating antigens in lyophilized form can be purchased from commercial sources.

Whole blood is collected by drawing a sample from a pigeon wing vein into modified Alsever solution. The erythrocytes are washed three times in HSAG, and the packed cells are suspended in an equal volume of HSAG to make a 50% suspension; part of this suspension is used to absorb nonspecific agglutinins from the test sera. A 10% working suspension is made in HSAG, from which the 0.25% suspension to be used in the test is prepared.

Acid-washed kaolin powder can be purchased from most scientific supply companies. A 25-g amount of kaolin is washed with Tris buffer until a pH of 7.0 or greater is achieved. The Tris buffer is made by mixing 12.1 g of Trizma base (Sigma Chemical Co., St. Louis, Mo.), 80 ml of 1 N HCl, and 0.85 g of NaCl and bringing the volume to 1 liter. This solution is then further diluted 1:10 in distilled water for washing the kaolin. After the final wash in Tris buffer, the kaolin pellet is suspended in 100 ml of Tris-bovine albumin buffer. This buffer is made by adding to 96.67 ml of Tris buffer the following ingredients: 0.33 ml of a 35% sterile solution of bovine serum albumin (BSA; Nutritional Biochemicals Corp., Cleveland, Ohio), 1 ml of a 0.5% $MgCl_2 \cdot 6H_2O$ solution, 1 ml of an 8% NaN_3 solution, and 1 ml of a 0.5% $CaCl_2$ solution.

HSAG is made from three different stock solutions: (i) 5× HEPES-saline is made by adding 29.8 g of HEPES powder, 40.95 g of NaCl, and 0.74 g of $CaCl_2 \cdot 2H_2O$ to 1 liter of water and adjusting the pH to 6.2. BSA is made by adding 20 g of BSA powder to 1 liter of water. Gelatin is made by adding 25.0 mg of gelatin to 2 liters of water. All solutions are filtered. A working solution of HSAG is made by adding 200 ml of HEPES-saline to 500 ml of BSA solution and 100 ml of gelatin stock solution. The volume is made up to 1 liter by adding 200 ml of sterile distilled water. The pH of the working solution should be 6.25. The solution can be stored for 2 months if it remains sterile. HSAG may be purchased from commercial sources such as Flow Laboratories, GIBCO Laboratories (Grand Island, N.Y.), or M.A. Bioproducts.

CF Test

The CF test is performed by a standard technique (16).

EIA

EIA kits are commercially available from a number of North American and European suppliers and have been developed for rubella IgG and IgM.

All of the commercially available EIAs are solid-phase capture assays using microplate wells, beads, or filters and measure antibodies to envelope and/or capsid proteins. Most are performed in a couple of hours and are low in cost. Automation is built into some EIAs, whereas others are manual. Most of the tests are quantitative or semiquantitative and incorporate several positive and negative controls. At present, we are using the novel microparticle EIA

(IMX), which is automated for detecting IgG or IgM on the solid phase. The antibodies are then detected with antihuman IgG or IgM antibody coupled to alkaline phosphatase enzyme followed by methylumbelliferyl phosphate substrate (1). The assay is performed in a high-maintenance machine with reaction cells in a carousel.

IFA Test

An indirect fluorescent-antibody (IFA) test for rubella antibody using chronically infected LLC-MK$_2$ cell cultures as the solid-phase antigen was first described in 1964 (3). Other acutely infected cell systems or even purified rubella antigens have since been used for the rubella IFA test. For the classic IFA test, acutely infected cells are grown either on Leighton tube coverslips or in culture flasks, trypsinized, and deposited on slides to form smears. The cytoplasms of cells infected with rubella virus contain rubella antigens. These antigens are used to detect specific antibodies by the indirect method, in which an anti-human globulin conjugated with fluorescein isothiocyanate is employed. The technique is rapid and relatively inexpensive and allows quantitation of IgG and IgM antibodies; however, IFA results are read visually with a fluorescence microscope and are open to subjective interpretation. Alternatively, a soluble rubella antigen immobilized on an opaque plastic surface has been used in a IFA called FIAX. The antigen-sensitized surface is allowed to react in a two-step procedure with the serum and the fluorescein-labeled conjugate, and the resulting fluorescence signal is measured objectively with a fluorometer. The intensity of the fluorescence signal correlates with the tier of rubella antibody.

Other assays for detecting rubella antibody include mixed hemadsorption, time-resolved fluoroimmunoassay, and silver-enhanced, gold-labeled immunosorbent assay. These tests have their own characteristics and have not yet gained widespread usage.

Specific Rubella IgM Assays

Detection of rubella-specific IgM antibodies can be achieved by sucrose density gradient fractionation followed by HI or SPIA. Density gradient centrifugation entails diluting the test serum 1:3 in phosphate-buffered saline and adsorbing it with pigeon erythrocytes. After removal of the erythrocytes, the serum is placed on a gradient that is constructed by layering the various sucrose solutions in phosphate-buffered saline in a 5-ml cellulose nitrate tube (37, 33, 28, 24, 18, and 12% [wt/vol], as determined by refractometer readings). Before the test serum is layered, the gradient is allowed to equilibrate by overnight diffusions at 4°C. The specimen is carefully laid on top of the gradient and then centrifuged for 16 to 18 min at 150,000 × g. The bottom of the tube is carefully punctured with a needle-type fraction collector, and 10 to 12 fractions are collected (0.3 ml [about 20 drops] per fraction). Fractions 2 to 4 usually contain IgM, fractions 6 to 9 contain IgG, and the top two fractions contain nonspecific inhibitors of hemagglutination. Alternatively, other immunoglobulin separation techniques such as gel filtration, staphylococcal protein A absorption, quaternary aminoethyl-Sephadex chromatography, and 2-mercaptoethanol destruction have been used with variable success. Commercial EIAs for testing whole sera for rubella IgM are available. These EIAs usually employ a pretreatment step or use anti-IgM capture antibody on the solid phase and Fab antibody fragments in the indicator reagent to eliminate rheumatoid factor false-positive results. They possess high levels of sensitivity and specificity (2, 4).

Confirmation of clinical rubella requires the demonstration of a fourfold rise in antibody titers in paired sera or of the presence of rubella virus-specific IgM. There is considerable variability in the antibody titers maintained during life, and a firm diagnosis cannot be made on the basis of the absolute titer of a single serum sample. Because of day-to-day variations in the results of tests, paired sera must be tested in parallel. Differences in titers of paired sera tested on different days may reflect test-to-test variation and not a true change in antibody concentration. This consideration is especially important when paired sera are collected from pregnant women who have not experienced clinical illness. Judgments regarding therapy should be withheld until the sera have been reexamined in parallel. Sera positive for rubella IgM in EIA should be examined for rheumatoid factor to rule out a false-positive result.

Costs are an important consideration for any laboratory performing rubella serology. As expected, the cost per test for all methods decreases as the number of sera tested in a single run increases. The individual cost of testing a small number of serum samples (<10) is highest for RH and HI, since these tests are labor intensive and a large proportion of the cost is for the technician's salary. PHA is the least expensive test for testing 10 or fewer serum samples. As larger numbers of sera are tested, the large difference in the cost per test decreases; the costs are similar for PHA, RH, HI, and EIA, with HI being the least expensive of the four.

Lymphocyte Transformation Assay

A rubella virus-specific lymphocyte transformation assay that uses cryopreserved mononuclear cells has been developed (10). A negative rubella virus-specific lymphocyte transformation response in a seropositive child 3 years of age or younger is highly suggestive of congenitally acquired rubella.

INTERPRETATION

The presence of antibody in patient serum indicates that the patient has been previously infected with the virus and is probably immune to rubella. It is still not certain whether the minute amounts of antibody detected by the highly sensitive EIA are protective. Whenever possible, EIA responses should be converted to HI antibody equivalence.

When tests such as PHA for nonstructural protein or RH are used to screen for immunity in sera collected within 2 to 3 weeks after asymptomatic primary infection with rubella, a second screening employing any of the tests measuring earlier arising antibodies (EIA, HI, etc.) to confirm negativity can indicate recent infection when the second test is positive (Fig. 1). If this occurs, IgM testing of the serum is inevitably positive.

Patients without detectable antibodies are usually susceptible to infection by rubella virus; however, a small percentage of adults may not have detectable antibodies and yet be immune. Neutralizing antibodies at low dilutions can usually be detected in the sera of these patients. The lack of serologic responses to rubella virus vaccine in women who do not have detectable antibodies is often due to low levels of neutralizing antibodies. With the HI test described above, the first dilution in the test is 1:8. A patient serum sample positive only at 1:8 should be retitrated to rule out a possible false-positive reaction.

REFERENCES

1. **Abbott, G. G., J. W. Safford, R. G. MacDonald, M. C. Craine, and R. R. Applegren.** 1990. Development of automated immunoassays for immune status screening and serodiagnosis of rubella virus infection. *J. Virol. Methods* **27:**227–240.
2. **Best, J., S. Palmer, P. Morgan-Capner, and J. Hodgson.** 1984. A comparison of Rubazyme-M and MACRIA for the detection of rubella-specific IgM. *J. Virol. Methods* **8:**99–109.
3. **Brown, G. C., H. F. Maassab, J. A. Veronelli, and T. J. Francis.** 1964. Rubella antibodies in human serum: detection by the indirect fluorescent-antibody technique. *Science* **145:**943–945.
4. **Chernesky, M. A., L. Wyman, J. B. Mahony, S. Castriciano, J. T. Unger, J. W. Safford, and P. S. Metzel.** 1984. Clinical evaluation of the sensitivity and specificity of a commercially available enzyme immunoassay for detection of rubella virus-specific immunoglobulin M. *J. Clin. Microbiol.* **20:**400–404.
5. **Gupta, J. D., and J. D. Harley.** 1970. Use of formalinized sheep erythrocytes in the rubella hemagglutination-inhibition test. *Appl. Microbiol.* **20:**843–844.
6. **Gupta, J. D., and V. J. Peterson.** 1971. Use of a new buffer system with formalinized sheep erythrocytes in the rubella hemagglutination-inhibition test. *Appl. Microbiol.* **21:**749–750.
7. **Hedman, K., and S. A. Rousseau.** 1989. Measurement of avidity of specific IgG for verification of recent primary rubella. *J. Med. Virol.* **27:**288–292.
8. **Ho-Terry, L., G. M. Terry, and P. Londesborough.** 1990. Diagnosis of foetal rubella virus infection by polymerase chain reaction. *J. Gen. Virol.* **71:**1607–1611.
9. **Liebhaber, H.** 1970. Measurement of rubella antibody by hemagglutination inhibition. II. Characteristics of an improved HAI test employing a new method for removal of nonimmunoglobulin HA inhibitors from serum. *J. Immunol.* **104:**826–834.
10. **O'Shea, S., J. Best, and J. E. Banatvala.** 1992. A lymphocyte transformation assay for the diagnosis of congenital rubella. *J. Virol. Methods* **37:**139–148.
11. **Pattison, J. R., D. S. Dane, and J. E. Mace.** 1975. Persistence of specific IgM after natural infection with rubella virus. *Lancet* **i:**185–187.
12. **Russell, S. M., S. R. Benjamin, M. Briggs, M. Jenkins, P. P. Mortimer, and S. B. Payne.** 1978. Evaluation of the single radial hemolysis (SRH) technique for rubella antibody measurement. *J. Clin. Pathol.* **31:**521–526.
13. **Schmidt, N. J., and E. H. Lennette.** 1966. Rubella complement fixing antigens derived from the fluid and cellular phases of infected BHK-21 cells: extraction of cell-associated antigen with alkaline buffers. *J. Immunol.* **97:**815–821.
14. **Schmidt, N. J., E. H. Lennette, J. D. Woodie, and H. H. Ho.** 1966. Identification of rubella virus isolates by immunofluorescent staining, and a comparison of the sensitivity of three cell culture systems for recovery of virus. *J. Lab. Clin. Med.* **68:**502–509.
15. **Stewart, G. L., P. D. Parkman, H. E. Hopps, R. D. Douglas, J. P. Hamilton, and H. M. Meyer, Jr.** 1967. Rubella virus hemagglutination-inhibition test. *N. Engl. J. Med.* **276:**554–557.
16. **U.S. Public Health Service.** 1965. *Standardized Diagnostic Complement Fixation Method and Adaptation to Micro Test.* U.S. Public Health Service Public Health monograph 74. U.S. Government Printing Office, Washington, D.C.

Human Parvoviruses

J. R. PATTISON

84

There are three genera in the family *Parvoviridae*. The densoviruses infect members of the order Insecta but do not infect humans or other vertebrates. The dependoviruses infect a number of animal species, and serologic examination shows that adeno-associated viruses 1 to 4 are common human infectors. However, to date, the adeno-associated viruses are not associated with any human disease, and there is no requirement to test for them in diagnostic laboratories.

The autonomous parvoviruses are widespread in nature and frequently cause disease in their natural hosts. Some of the animal parvoviruses such as feline panleukopenia virus and canine parvovirus are of particular concern to veterinarians, and immunization against these infections is a routine practice.

With respect to humans, one autonomous parvovirus, designated B19 (43), is of considerable importance, and the remainder of this chapter is concerned with this virus. It is quite possible that some of the small round viruses seen in human feces (32) are also autonomous parvoviruses, but at present there are no data on the exact nature of these viruses or any proof of their pathogenicity.

PROPERTIES OF THE VIRUS

B19 virions are relatively uniform, isometric, unenveloped particles ranging in diameter from 20 to 25 nm (mean, 23 nm) (14). The mean buoyant density in $CsCl_2$ is 1.43 g/ml (range, 1.41 to 1.45 g/ml), with a minor peak at 1.39 g/ml that possibly represents empty capsids. There are two capsid proteins of approximately 83 and 60 kDa, of which the latter constitutes about 80% of the total protein mass (48). B19 is resistant to lipid solvents, as would be expected of a parvovirus, but appears to be relatively sensitive to acid and alkali and to heat denaturation. However, in the absence of a cell culture system for the easy replication of the virus, it is not possible to perform definitive studies.

The genome of B19 is a single-stranded DNA 5.5 kb long (44). The virus packages plus and minus DNA strands into separate virions in approximately equal proportions (44). Parvovirus DNA is organized as a linear coding region bounded at each end by terminal palindromic sequences that fold into hairpin duplexes. In B19, these hairpins are relatively long (330 nucleotides versus 100 to 250 nucleotides in animal parvoviruses) inverted terminal repeats (41). This last property, like the packaging of both plus and minus strands, is shared by B19 and the dependoviruses.

The coding regions of B19 are confined to the plus strand and comprise two open reading frames, which code for nonstructural proteins at the 3′ end and for structural proteins at the 5′ end. Restriction endonuclease analysis of multiple isolates of B19 obtained between 1972 and 1984 indicates that the genome of B19 is relatively stable (26).

There appears to be a single, stable antigenic type of B19. Infection is followed by lifelong immunity, indicating a single neutralizable type. Limited studies using immunodiffusion have revealed antigenically identical viruses from different clinical situations. The occasional minor variation in reactivity with mouse monoclonal antibodies has been noted but has so far not proved to be of epidemiologic or diagnostic significance (12).

CLINICAL SIGNIFICANCE

The spectrum of disease caused by parvovirus B19 infection ranges from the wholly asymptomatic to a serious, potentially fatal disease in a minority of patients with particular underlying abnormalities.

Nonspecific Illness

Parvovirus B19 was first recognized in the sera of asymptomatic blood donors (14). The first report of associated clinical illness involved patients with a mild febrile illness and lymphopenia; volunteer studies indicated that the viremic phase of the infection may be associated with fever, headache, chills, myalgia, and malaise (3, 42). A study in a residential school for boys (19) has more closely defined the occurrence of asymptomatic infection and nonspecific illness caused by B19 infection. Among a group of boys 11 to 15 years old who were seronegative before the outbreak began, 72% seroconverted. Approximately two-thirds of these patients were asymptomatic. One quarter of the infections were associated with an influenza-like illness indistinguishable clinically from the disease caused by influenza A and B viruses, which circulated in the boys at approximately the same time as B19 infection. Only 10% had an erythematous rash that was sufficiently typical (see below) to suggest that laboratory investigations for B19 infection were indicated. Thus, it appears that the majority of infections in childhood are asymptomatic or are associated with a nonspecific illness.

Rash Illness

The most common illness caused by B19 virus consists of an erythematous maculopapular rash that in its most clinically distinct form is called erythema infectiosum (EI) (1). EI is common in children between 4 and 11 years of age and is sometimes called fifth disease, since it was the fifth of six erythematous-rash illnesses of childhood in an old classification. Classically, it starts with an intense erythema of the cheeks (hence another of its names, slapped-cheek disease). The rash then proceeds to involve the trunk and limbs and lasts only 1 or 2 days, although transient recrudescences may occur when the individual is hot (as a result of exercise, bathing, or sunlight). The rash on the limbs tends to have a lacy or reticular appearance. There may be associated lymphadenopathy and joint symptoms (see below).

The association between EI and B19 infection was first described in relation to an outbreak in London (7). Cases in Europe, Scandinavia, North America, Australia, and Japan were later shown to be caused by B19, but it is now clear that the illnesses are not always diagnosed clinically as EI. B19 causes an erythematous-rash illness very similar to that caused by rubella virus. Erythema of the cheeks is not always prominent, and the rash often does not have a lacy appearance. The rash sometimes occurs on the palms and soles. In the absence of laboratory tests, EI is most frequently misdiagnosed as rubella, allergy, or viral illness unless there is an associated outbreak of typical EI with red cheeks in young children. With such epidemiologic support, the diagnosis of EI can often be made correctly.

In a few cases of B19 infection, the rash is purpuric (24). In most cases, the platelet count is normal, but thrombocytopenia occasionally occurs (28). The purpura is transient in these patients, and there is no evidence that B19 causes idiopathic thrombocytopenic purpura.

Joint Disease

Symptoms and signs of joint involvement occur frequently in B19 infection, and they sometimes occur in the absence of a rash (1, 7). Approximately 80% of adult females report joint symptoms, although the figure is only approximately 10% in childhood cases. The arthropathy of B19 infection is very similar to that seen with rubella. The most common presentation is a symmetrical arthralgia or arthritis in the small joints of the hands, with wrists, knees, and ankles affected in some cases. The arthropathy tends to be more severe in children. The symptoms and signs usually resolve within 2 weeks, but in a few cases, they persist for months, and very occasionally, they persist for years (37, 46). Some of these patients may be classified clinically as having early benign rheumatoid arthritis, but they are rheumatoid factor negative. B19 virus infection does not seem to be related to rheumatoid arthritis in any way; the frequency of B19 antibody is no greater in patients with seropositive rheumatoid arthritis than in controls, and B19 seroconversion occurs in patients who have had rheumatoid arthritis for years (25).

Aplastic Crisis

An aplastic crisis is a transient, acute event that complicates chronic hemolytic anemia. Hemoglobin levels fall from steady-state values, reticulocytes disappear from the peripheral blood, and erythrocyte precursors are virtually absent from the bone marrow at the beginning of the crisis.

The cessation of erythropoiesis lasts 5 to 7 days, and patients present with symptoms of worsening anemia. The situation is serious in most patients and occasionally fatal. Blood transfusion is required in the acute phase, but after a week or so, the bone marrow recovers rapidly, there is a reticulocytosis, and the hemoglobin concentration returns to steady-state values.

The association between B19 infection and aplastic crisis was first noted in 1981 in patients with sickle cell anemia (31); even with the insensitive diagnostic criteria of the time, B19 infection was shown to be responsible for 90% of cases (40). It is now clear that B19 is the principal cause of aplastic crises worldwide. B19-associated cases have occurred in the West Indies, Europe, and North America, and evidence suggests that B19 has been the principal cause of aplastic crisis in sickle cell anemia patients for at least the last 20 years (38, 40).

Aplastic crisis occurs in hemolytic states other than sickle cell anemia. Investigation of aplastic crises in patients with hereditary spherocytosis, pyruvate kinase deficiency, transfusion-independent beta-thalassemia intermedia, and the dyserythropoietic anemia called HEMPAS has shown that B19 infection is the cause (16, 20, 36, 45).

B19 Infection in Immunosuppressed Patients

Cases of persistent B19 infection have been described in patients with underlying immunodeficiency states (Nezelof's syndrome-affected, acute lymphatic leukemic, and human immunodeficiency virus-positive individuals) (22, 23). The illness is characterized by either persistent anemia or a remitting and relapsing anemia. Viremia occurs and recurs in periods of anemia, and only a weak humoral immune response can be detected. The bone marrow is typical of that seen in aplastic crisis complicating hemolytic anemia. In one patient with acute lymphocytic leukemia, administration of B19 antibodies resulted in a transient fall in virus titer, reticulocytosis, and symptoms of EI (fever, rash, and arthralgia).

B19 Infection in Pregnancy

The intense viremia that is a consistent feature of B19 infection gives ample opportunity for the virus to reach the placenta and the fetus should infection occur during pregnancy, and animal parvoviruses are associated with fetal loss in hamsters, pigs, and cattle.

Early clinical studies showed no increase in birth defects during the year after large outbreaks of EI in the United States and the United Kingdom. When B19 diagnostic tests became available, no viral antigen or specific immunoglobulin M (IgM) antibody could be found in sera taken during the first month of life from infants with birth defects (27). However, this diagnostic approach depends on an analogy with rubella virus and cytomegalovirus infections. Both of these viruses cause persistent infection of the fetus, and it may be that B19 causes an acute infection that is rapidly cleared. Nevertheless, there is as yet no evidence that B19 causes birth defects.

Early studies indicated that B19 infection does not occur with the expected frequency in pregnant women (6, 27), possibly because B19 infection leads to early fetal loss before pregnancy is confirmed. Moreover, in studies of B19 infection in women known to be pregnant, the spontaneous abortion rate appeared to be high. The early studies were small, but later studies in Germany and the United Kingdom (17, 35) showed that fetal loss occurred in 8 to 9% of affected pregnancies. In the United Kingdom study, approximately 1 in 10 of the pregnancies complicated by B19 infection ended in spontaneous abortion during the second trimester, an incidence approximately 10 times that in controls. The pregnancy was lost, on average, 4 to 6 weeks after the onset of symptoms of EI in the mother.

Most pregnancies complicated by B19 continue to full-term delivery of healthy infants. However, damage sometimes occurs as a consequence of second- or third-trimester infection, and in these cases, fetal hydrops appears to be a consistent feature. Maternal B19 infection appears to occur 2 to 12 weeks before the diagnosis of hydrops fetalis (5, 8, 10, 18). B19 infection of the fetus is suggested by cells with eosinophilic intranuclear inclusions, and presence of the virus can be confirmed by in situ or filter hybridization. The percentage of fetal hydrops accounted for by B19 infection is not clear at present. One study reviewed 50 cases of nonimmunologic hydrops fetalis that were seen in a single pathology department in the United Kingdom between 1974 and 1983. Cardiovascular anomalies and chromosomal abnormalities were the most common causes, but 13 remained unexplained or had features of unidentified virus infections. Of these 13, 4 had evidence of B19 infection by in situ hybridization (33). Thus, it is clearly worth investigating B19 as a cause of nonimmunologic hydrops fetalis by detecting virus in fetal blood samples, detecting the B19 genome in extracted tissue, or using specific in situ hybridization. It is also notable that maternal α-fetoprotein levels are raised when fetal infection is due to B19 infection (11).

Studies of the possibility that intrauterine B19 infection causes a low incidence of anatomical and/or functional abnormalities continue. Unpublished observations indicate that B19 infection of the fetus may result in congenital erythrocyte aplasia (47) or a variety of disorders of the central nervous system (1a). Anderson et al. (1a) studied more than 90 cases of intrauterine B19 infection, of which the majority ended in fetal loss. If fetal loss occurs in only 10% of infected pregnancies, then the incidence of central nervous system abnormalities is less than 1%; a study of more than 9,000 infected pregnancies would be required to prove this beyond all doubt.

Pathogenesis of B19 Infection

An understanding of the pathogenesis of B19 virus infection provides the key to understanding the diagnostic approaches used in investigating clinical diseases that might be a consequence of infection. Knowledge of the events of pathogenesis has been accumulated from both volunteer studies and cell culture experiments (3, 34, 48). It is assumed that the natural mode of infection is via the respiratory route, since virus is infectious when given as nasal drops and can be found in nasal secretions at the time of the viremia.

The various events that follow intranasal inoculation of the virus are illustrated in Fig. 1. One week later, an intense viremia (10^{11} particles per ml at peak) develops. In most individuals, this viremia lasts for a matter of days and is followed by a specific antibody response initially of the IgM class but, after a day or so, also of the IgG class.

At the time of the viremia, hematologic changes take place. No erythroid precursors are present in the bone marrow of healthy individuals 10 days after inoculation, and the expected disappearance of reticulocytes from the peripheral blood and a small fall in the hemoglobin level take place during week 2 after inoculation. Levels of lymphocytes, neutrophils, and platelets also show transient drops, but these drops are not due to lack of precursors in

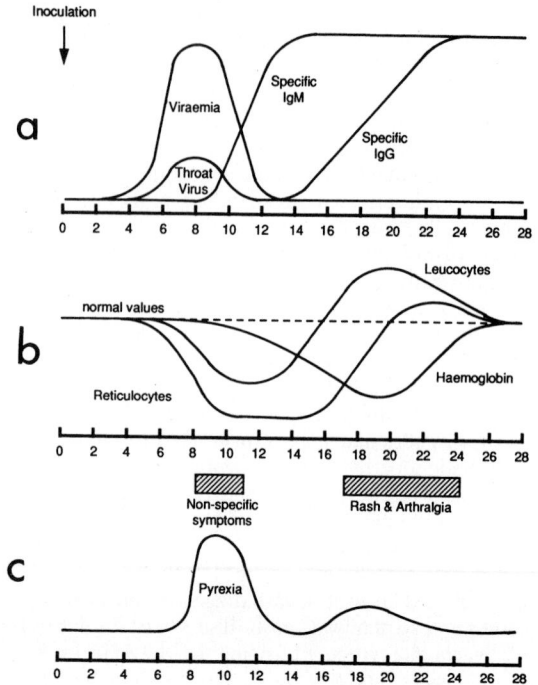

FIGURE 1 Events that follow intranasal inoculation of virus. At the time of inoculation, virus was administered as nasal drops to susceptible volunteers. The virologic (a), hematologic (b), and clinical (c) events occurring during the subsequent 28 days are illustrated schematically. (Reproduced with permission from *Principles and Practice of Clinical Virology*, John Wiley & Sons Ltd., Chichester, England, 1987.)

the bone marrow. Studies with cultured bone marrow cells confirm the in vivo observations. B19 selectively inhibits erythroid colony formation but has no effect on the cells of the myeloid series (29). Secondary cultures of erythroid cells are inhibited by virus, indicating a direct effect on committed erythroid cells, and there is morphologic and antigenic evidence of intracellular virus.

A second phase of illness occurs in infected volunteers during week 3 after inoculation. The characteristic features are rash and arthralgia. As yet, there are no studies of the pathology of either of these features, but since they follow the disappearance of the viremia and occur at a time when there is an easily detectable immune response, the rash and arthralgia may be immune mediated.

The viremia that is characteristic of B19 infection gives ample opportunity for infection of the placenta and fetus if it occurs during pregnancy. In infected fetuses, there appears to be a persistent infection, with damage to hematopoietic cells that leads to (it is assumed) anemia, heart failure, and hydrops fetalis.

Recent work (9) has identified the erythrocyte P antigen as the cellular receptor for B19. P antigen is an erythrocyte antigen and explains the tropism of B19 parvovirus for erythroid cells. P antigen is also found on vascular endothelial cells, and this may partly explain the erythematous-rash illnesses and transplacental infection due to B19 virus. Finally, P antigen is present on fetal myocytes; thus, B19 might cause fetal myocarditis, which could contribute to the pathogenesis of hydrops fetalis.

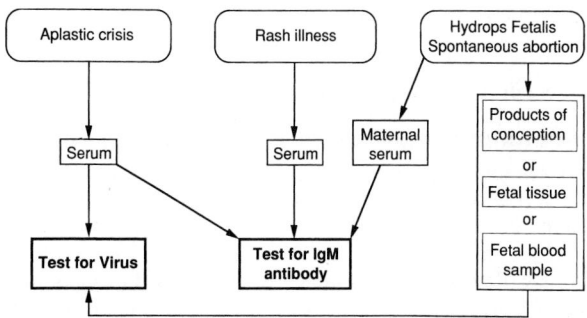

FIGURE 2 Schematic diagram of clinical conditions (upper rank), specimens required (middle rank), and tests done (bold-face type) for the diagnosis of B19 infection.

COLLECTION AND STORAGE OF SPECIMENS

Serum is the principal specimen used for the laboratory diagnosis of B19 infection (Fig. 2). This specimen is suitable for virus detection in cases of aplastic crisis, persistent infection in immunosuppressed patients, and persistent fetal infection. The detection of specific IgM antibody in serum is the cornerstone of the diagnosis of rash illness and arthropathy. Standard blood specimens are drawn at the time of presentation, and serum is removed aseptically from the blood clots in the usual way. These specimens can be stored at 4°C before testing unless there is a long delay. Subsequent long-term storage is generally at −20°C for antibody-positive specimens; storage at −70°C is preferred for virus-positive specimens.

B19 virus in formalin-fixed, paraffin-embedded tissue can be detected by in situ hybridization, so standard pathology protocols can be observed for fetal tissue.

DETECTION OF VIRUS

A variety of techniques for the detection of B19 virus have been described. Laboratories choose from the list of possible tests on the basis of their overall preferences and whether they are using the techniques in other contexts.

CIE

Counterimmunoelectrophoresis (CIE) is one of the simplest methods. Although insensitive, CIE can be undertaken by most laboratories and gives results within 1 h (40). Agarose gel (10% in barbitone acetate buffer) is poured to a depth of at least 1 mm on a glass slide or plate resting on a level surface. Two rows of wells (3 mm in diameter and 5 mm from center to center) are punched into the gel. The wells are filled with test or control sera in one row and B19 antibody-positive sera in the other row. The gel is placed so that the wells filled with known antibody-positive sera are on the anode side, and electrophoresis is carried out for 1 h at 120 V (17 mA). Indirect back illumination is used to examine the gels for lines of precipitate at the end of electrophoresis and after the sample has been held at 4°C for 1 h.

EM

Electron microscopy (EM) is a technique in routine use in many laboratories. It is most frequently used for detection of virus in serum by using a negative staining technique but can also be used for examination of thin sections of tissue.

For detection of viremia, immuno-EM is often used. The specimen is mixed with an equal volume of high-titer B19 antibody-positive serum and held at 37°C for 30 min. The sample is diluted 1:10 in phosphate-buffered saline and centrifuged at 40,000 × g for 1 h. The pellet is then suspended in a small volume of distilled water, mixed with phosphotungstic acid (pH 6.3) to a final concentration of 3%, and touched onto a Formvar-coated electron microscope grid. A preparation containing an antibody-negative serum should be processed in parallel as a control. The grids should be scanned at a magnification of ×30,000, and suggestive structure should be examined at ×100,000.

Immunoassays

In any laboratory providing B19 diagnostic services, immunoassays should be in routine use for the detection of B19 antibody. Such techniques can also be adapted to detect viral antigens. A solid phase is coated with anti-μ-chain antibody and reacted with a serum containing B19-specific IgM antibody. The test material is then reacted with this coated solid phase, and any bound virus is detected by sequential incubation with monoclonal B19 antibody and a labeled anti-mouse immunoglobulin.

The specimen most commonly tested for virus is serum, and a satisfactory working dilution for test serum is 1:10. Other specimens such as nasal or throat washes should be diluted 1:2. A number of variations of the basic test described above are in use, and each requires a different set of detailed steps. The reader is referred to original or recent descriptions for details (2, 12, 13). In general, it must be remembered that appropriate controls must always be included and that if any single reagent or step is changed, the test may need to be reoptimized by making other changes.

Viral-Genome Detection

A variety of techniques have been described for the detection of B19 DNA in serum, body fluids, and tissues. Again, the details of the techniques vary according to whether dot blot hybridization with a radioactive label (4) or a nonradioactive label (15) is being described and whether B19 DNA in tissue is detected after extraction or in situ (5, 33). Most recently, techniques involving amplification of B19 DNA by using the PCR have been described and evaluated (30, 39).

A variety of DNA probes and primers for PCR are available from laboratories that have described their use. Laboratories also require some experience in the technique of hybridization and an ability to perform gel electrophoresis and endonuclease restriction to validate the results of hybridization.

Interpretation of Results

The relative sensitivities of the methods described above are different. CIE detects approximately 10^8 to 10^9 particles per ml. EM, especially if used after antibody aggregation of virus, detects 10^5 to 10^6 particles per ml. In the diagnostic setting, this difference in sensitivity does not matter as much as might be expected. The viremia of a primary infection is intense and short-lived, and with acute-phase specimens taken at presentation in aplastic crisis, CIE can be expected to be positive in about 30% of cases. With use of dot blot hybridization, this figure rises to about 60% (4), and with nested PCR, the figure is 70% (30). In addition, using a nested PCR, serum samples may be positive for months after infection, and bone marrow samples may be positive for >1 year. If an answer is urgently required, CIE and EM should be performed; if the results are negative, radioimmunoassay or DNA-DNA hybridization must be done. The latter is occasionally positive when the former is negative, possibly because of the presence of host antibody coating the virus at the time the viremia is being cleared. If the only positive result on a specimen is by DNA-DNA hybridization, confirmation must include Southern blotting of the putative B19 DNA to ensure that it is of the correct size. If tissue samples such as placenta are to be tested for B19 DNA, confirmation by Southern blotting with and without endonuclease restriction is mandatory, since weak, false-positive dot blot signals do occur with such samples. Needless to say, when PCR is being used, great care must be taken not to contaminate the testing laboratory in any way with B19 DNA.

DETECTION OF ANTIBODY

Solid-phase radiolabeled or enzyme-labeled immunoassays are in routine use in virus diagnostic laboratories. These are the preferred methods for the serologic diagnosis of B19 infection, although CIE and immuno-EM can be used as relatively insensitive tests of seroconversion. A number of commercial kits for B19 antibody testing are now available. Some are more reliable and sensitive than others, and laboratories are advised to evaluate the kits extensively with pedigree sera before putting them into routine use.

Specific IgM Antibody Detection

Procedure

The assay most commonly used for specific IgM antibody detection is based on the antibody capture principle (13). Polystyrene beads are often used as the solid phase, but microdilution wells in strips or plates are also suitable. The solid phase is coated with anti-IgM, and the test serum is reacted with this preparation. Any specific IgM is then detected with the addition of B19 antigen followed by a mouse monoclonal anti-B19 antibody and then by either a radiolabeled or enzyme-labeled detector antibody. The antigen most commonly used is native virus from viremic individuals, but a variety of synthetic antigens are being evaluated.

The amount of specific IgM in a serum sample is quantified by comparing results with those for a set of local standards. The quantification is based on designating a highly reactive serum as containing 100 arbitrary units and including a set of standards in each test (13).

Interpretation of Results

Sera found to contain 10 or more arbitrary units of B19 IgM antibody are unequivocally associated with recent infection. Such antibody concentrations usually appear within 3 to 4 days of the onset of symptoms but occasionally (particularly in cases of aplastic crisis) may not appear until 7 to 10 days after onset. A B19 IgM-negative serum sample taken within 10 days of the onset of illness should also be tested for viral DNA, or a second specimen taken 10 to 14 days later should be requested.

Specific IgM is detectable for 2 to 3 months after acute infection. Interpretation is difficult when equivocal low concentrations of specific IgM are found in sera taken after this time. Testing the sera for IgG may be helpful. Sera containing high concentrations of rubella virus-specific IgM may give low false-positive results when tested for parvovirus-specific IgM, and this fact must be borne in mind when sera taken within 12 weeks of a rubelliform illness are tested (21).

Specific IgG Antibody Detection

Procedure

Assays for detecting anti-B19 IgG analogous to those for detecting specific antibody of the IgM class but differing in some details have been described. Again, there are radio-isotope and enzyme-linked versions of the test, and the choice depends on local preferences (2, 12). The amount of specific IgG antibody present should be quantitated by comparison of the test serum reading with a standard curve as described above for specific IgM.

Interpretation of Results

The test can be used on paired sera to detect seroconversion for the diagnosis of B19 infection. Patients have relatively high concentrations (>50 arbitrary units) of specific IgG for 12 months after acute infection, and such values support the diagnosis of infection during that time. However, individual variation in absolute amounts is very large, and the finding of only low values (<20 arbitrary units) does not exclude recent infection.

Past exposure to B19 virus (and therefore immunity in most individuals) is indicated by the detection of >1.0 arbitrary unit of anti-B19 IgG provided that the test serum/negative serum ratio is greater than 2. Because of the relative insensitivity of the test, a negative result is not synonymous with susceptibility. A positive signal depends on a minimum proportion of the IgG having B19 specificity, and sera with low concentrations may thus not bind sufficient antigen to be detected by the indicator antibody.

Immunocompromised patients with chronic parvovirus infection show poor humoral immune responses, and only low concentrations of anti-B19 IgG (and IgM) can be expected in these patients. None of the detectable antibody neutralizes virus infectivity, and this lack is taken to be an essential component of the pathogenesis of the chronic infection.

REFERENCES

1. **Ager, E. A., T. D. Y. Chin, and J. P. Poland.** 1966. Epidemic erythema infectiosum. *N. Engl. J. Med.* **275:**1326–1331.
1a. **Anderson, L. J., et al. (Centers for Disease Control and Prevention).** Unpublished data.
2. **Anderson, L. J., C. Tsou, R. A. Parker, T. L. Chorba, H. Wulff, P. Tattersall, and P. P. Mortimer.** 1986. Detection of antibodies and antigens of human parvovirus B19 by enzyme-linked immunosorbent assay. *J. Clin. Microbiol.* **24:**522–526.
3. **Anderson, M. J., P. G. Higgins, L. R. Davis, J. S. Williams, S. E. Jones, I. M. Kidd, J. R. Pattison, and D. A. J. Tyrrell.** 1985. Experimental parvoviral infection in humans. *J. Infect. Dis.* **152:**257–265.
4. **Anderson, M. J., S. E. Jones, and A. C. Minson.** 1985. Diagnosis of human parvovirus infection by dot-blot hybridisation using cloned viral DNA. *J. Med. Virol.* **15:**163.
5. **Anderson, M. J., M. N. Khousam, D. J. Maxwell, S. J. Gould, L. C. Happerfield, and W. J. Smith.** 1988. Human parvovirus B19 and hydrops fetalis. *Lancet* **i:**535.
6. **Anderson, M. J., I. M. Kidd, and P. Morgan-Capner.** 1985. Human parvovirus and rubella-like illness. *Lancet* **ii:**663.
7. **Anderson, M. J., E. Lewis, I. M. Kidd, S. M. Hall, and B. J. Cohen.** 1984. An outbreak of erythema infectiosum associated with human parvovirus infection. *J. Hyg.* **92:**85–93.
8. **Bond P. R., E. O. Caul, J. Usher, B. J. Cohen, J. P. Clewley, and A. M. Field.** 1986. Intrauterine infection with human parvovirus. *Lancet* **ii:**448.

9. Brown, K. E., S. M. Anderson, and N. S. Young. 1993. Erythrocyte P antigen: cellular receptor for B19 parvovirus. *Science* 262:114–117.

10. Brown, T., A. Anand, L. D. Ritchie, J. P. Clewley, and A. M. Field. 1986. Intrauterine infection with human parvovirus. *Lancet* i:48.

11. Carrington, D., M. J. Whittle, A. A. M. Gibson, T. Brown, A. M. Field, D. H. Gilmore, D. Aitken, W. J. A. Patrick, E. O. Caul, J. P. Clewley, and B. J. Cohen. 1987. Maternal serum α-fetoprotein—a marker of fetal aplastic crises during intrauterine human parvovirus infection. *Lancet* i:433–435.

12. Cohen, B. J. 1988. Laboratory tests for the diagnosis of infection with B19 virus, p. 69–83. *In* J. R. Pattison (ed.), *Parvoviruses and Human Disease*. CRC Press, Inc., Boca Raton, Fla.

13. Cohen, B. J., P. P. Mortimer, and M. S. Pereira. 1983. Diagnostic assays with monoclonal antibodies for the serum parvovirus-like virus (SLPV). *J. Hyg.* 91:7113.

14. Cossart, Y. E., B. Cant, A. M. Field, and D. Widdows. 1975. Parvovirus-like particles in human sera. *Lancet* i:72.

15. Cunningham, D., J. R. Pattison, and R. D. Craig. 1988. Detection of parvovirus DNA in human serum using biotinylated RNA hybridisation probes. *J. Virol. Methods* 19:279–288.

16. Duncan, J. R., M. D. Capellini, M. J. Anderson, C. G. Potter, J. B. Kurtz, and D. J. Weatherall. 1983. Aplastic crisis due to parvovirus infection in pyruvate kinase deficiency. *Lancet* ii:14–16.

17. Enders, G., and M. Biber. 1990. Parvovirus B19 infections in pregnancy. *Behring Inst. Mitt.* 85:74–78.

18. Gray, E. S., A. Anand, and T. Brown. 1986. Parvovirus infections in pregnancy. *Lancet* i:208.

19. Grilli, E. A., M. J. Anderson, and T. W. Hoskins. 1989. Concurrent outbreaks of influenza and parvovirus B19 in a boys' boarding school. *Epidemiol. Infect.* 103:359–370.

20. Kelleher, J. F., P. P. Mortimer, and T. Kanimura. 1983. Human serum 'parvovirus', a specific cause of aplastic crisis in hereditary spherocytosis. *J. Pediatr.* 102:720.

21. Kurtz, J. B., and M. J. Anderson. 1985. Cross-reactions in rubella and parvovirus-specific IgM tests. *Lancet* ii:1356.

22. Kurtzman, G. J., B. Cohen, P. Meyers, A. Amunullah, and N. S. Young. 1988. Persistent B19 parvovirus infection as a cause of severe chronic anaemia in children with acute lymphocytic leukaemia. *Lancet* i:1159–1162.

23. Kurtzman, G. J., K. Ozawa, B. Cohen, G. Hanson, R. Oseas, and N. S. Young. 1987. Chronic bone marrow failure due to persistent B19 parvovirus infection. *N. Engl. J. Med.* 317:287–294.

24. Lefrere, J.-J., A.-M. Courouce, J.-Y. Muller, M. Clark, and J.-P. Soulier. 1985. Human parvovirus and purpura. *Lancet* ii:730.

25. Lefrere, J.-J., O. Meyer, C.-J. Menkes, M.-J. Beaulier, and A.-M. Courouce. 1985. Human parvovirus and rheumatoid arthritis. *Lancet* i:982.

26. Morinet, F., J.-D. Tratschin, Y. Perol, and G. Siegl. 1986. Comparison of 17 isolates of the human parvovirus B19 by restriction enzyme analysis. Bret report. *Arch. Virol.* 90:165–172.

27. Mortimer, P. P., B. J. Cohen, M. M. Buckley, J. E. Cradock-Watson, M. K. S. Ridehalgh, F. Burkhardt, and U. Schilt. 1985. Human parvovirus and the fetus. *Lancet* ii:1012.

28. Mortimer, P. P., B. J. Cohen, M. A. Rossiter, S. M. Fairhead, and A. F. M. S. Rahman. 1985. Human parvovirus and purpura. *Lancet* ii:730.

29. Mortimer, P. P., R. K. Humphries, J. G. Moore, R. H. Purcell, and N. S. Young. 1983. A human parvovirus-like virus inhibits haemopoietic colony formation in vitro. *Nature* (London) 302:426–429.

30. Patou, G., D. Pillay, S. Myint, and J. R. Pattison. 1993. Characterization of a nested polymerase chain reaction assay for detection of parvovirus B19. *J. Clin. Microbiol.* 31:540–546.

31. Pattison, J. R., S. E. Jones, J. Hodgson, L. R. Davis, J. M. White, C. E. Stroud, and L. Murtuza. 1981. Parvovirus infections and hypoplastic crises in sickle cell anaemia. *Lancet* i:664.

32. Paver, W. K., E. O. Caul, C. R. Ashley, and S. K. R. Clarke. 1973. A small virus in human faeces. *Lancet* i:237.

33. Porter, H. J., T. Y. Khong, M. F. Evans, V. T.-W. Chan, and K. A. Fleming. 1988. Parvovirus as a cause of hydrops fetalis: detection by in situ DNA hybridisation. *J. Clin. Pathol.* 41:381–383.

34. Potter, C. G., A. C. Potter, C. S. R. Hatton, H. M. Chapel, M. J. Anderson, J. R. Pattison, D. A. J. Tyrell, P. G. Higgins, J. S. Willman, H. F. Parry, and P. M. Cotes. 1987. Variation of erythroid and myeloid precursors in the marrow and peripheral blood of volunteer subjects infected with human parvovirus (B19). *J. Clin. Invest.* 79:1486.

35. Public Health Laboratory Service Working Party on 'Fifth' Disease. 1990. Prospective study of human parvovirus (B19) infection in pregnancy. *Br. Med. J.* 300:1166–1170.

36. Rao, K. R. P., A. R. Patel, M. J. Anderson, J. Hodgson, S. E. Jones, and J. R. Pattison. 1983. Infection with a parvovirus-like virus and aplastic crisis in chronic haemolytic anaemia. *Ann. Intern. Med.* 98:930–932.

37. Reid, D. M., T. M. S. Reid, T. Brown, J. A. N. Rennie, and C. J. Eastmond. 1985. Human parvovirus-associated arthritis: a clinical and laboratory description. *Lancet* ii:422.

38. Saarinen, U. M., T. L. Chorba, P. Tattersall, N. S. Young, L. J. Anderson, and P. F. Coccia. 1986. Human parvovirus B19 induced epidemic red cell aplasia in patients with hereditary hemolytic anemia. *Blood* 67:1411.

39. Salimans, M. M. M., S. Holsappel, F. M. van de Rijke, N. M. Jiwa, A. K. Raap, and H. T. Weiland. 1989. Rapid detection of human parvovirus B19 DNA by dot-blot hybridization and the polymerase chain reaction. *J. Virol. Methods* 23:19–28.

40. Sergeant, G. R., J. M. Topley, K. Mason, B. E. Serjeant, J. R. Pattison, S. E. Jones, and R. Mohamed. 1981. Outbreak of aplastic crisis in sickle cell anaemia associated with parvovirus-like agent. *Lancet* ii:595–597.

41. Shade, R. O., M. C. Blundell, S. F. Cotmore, P. Tattersall, and C. R. Astell. 1986. Nucleotide sequence and genome organization of human parvovirus B19 isolated from the serum of a child during aplastic crisis. *J. Virol.* 58:921–936.

42. Shneerson, J. M., P. P. Mortimer, and E. M. Vandervelde. 1980. Febrile illness due to a parvovirus. *Br. Med. J.* 2:1580.

43. Siegl, G., R. C. Bates, K. E. Berns, B. J. Carter, D. C. Kelly, E. Kurstak, and P. Tattersall. 1985. Characteristics and taxonomy of Parvoviridae. *Intervirology* 23:61.

44. Summers, J., S. E. Jones, and M. J. Anderson. 1983. Characterisation of the genome of the agent of erythrocyte aplasia permits its classification as a human parvovirus. *J. Gen. Virol.* 64:2567.

45. West, N. C., R. E. Meigh, M. Mackie, and M. J. Anderson. 1986. Parvovirus infection associated with aplastic crisis in a patient with HEMPAS. *J. Clin. Pathol.* 37:1144.

46. White, D. G., P. P. Mortimer, D. R. Blake, A. D. Woolf, B. J. Cohen, and P. A. Bacon. 1985. Human parvovirus arthropathy. *Lancet* i:419.

47. Young, N., et al. (National Institutes of Health). Unpublished data.

48. Young, N., and K. Ozawa. 1988. Studies of B19 virus in bone marrow cell culture, p. 5–42. *In* J. R. Pattison (ed.), *Parvovirus and Human Disease*. CRC Press, Inc., Boca Raton, Fla.

Arboviruses

THEODORE F. TSAI

85

EPIDEMIOLOGIC AND CLINICAL BACKGROUND

The arboviruses (arthropod-borne viruses) are a taxonomically heterogeneous group of more than 500 viruses that share common modes of vector-borne transmission (2, 4, 5, 8). The *International Catalogue of Arboviruses* (4) also includes certain taxonomically related zoonotic viruses that are spread directly from animals without the agency of a vector. Most arboviruses are transmitted between specific arthropod vectors and vertebrate hosts, e.g., birds and small mammals. Infections of these intermediate hosts are typically asymptomatic, in contrast to those occurring in animals to which the virus has not adapted a parasitic relationship, and often result in various degrees of illness or in death.

Approximately 150 of the recognized arboviruses cause illness in humans; however, the approach to laboratory diagnosis is greatly simplified by considering the possibilities in light of the patient's history of travel and/or exposure (2, 5, 8). Although some arboviruses, such as the dengue viruses, are virtually cosmopolitan, others have a more limited geographic distribution in which they have adapted to specific vectors and intermediate animal hosts. For these reasons, an itinerary of places and circumstances under which infection may have been acquired helps narrow the possibilities in a differential diagnosis.

The epidemiologic and clinical characteristics of medically important arboviruses are summarized in Table 1. More comprehensive descriptions can be found elsewhere (2, 4–6, 8). Current information on arboviral outbreaks and transmission patterns can be obtained from state health departments and from the Division of Vector-Borne Infectious Diseases, Centers for Disease Control and Prevention (CDC), Ft. Collins, CO 80522, (303) 221-6400.

The majority of arboviral infections result in simple febrile illnesses that cannot be distinguished clinically from other common viral infections. Clinical features that characterize acute arboviral fevers include a sudden onset of debilitating symptoms; extreme headache, myalgia, and lumbar pain; relatively rapid resolution (e.g., within 24 to 48 h for Oropouche and sandfly fever); and for some infections (e.g., some cases of Colorado tick fever and dengue), a recrudescent course of fever and symptoms. Lymphadenopathy may be prominent in West Nile fever. Several important arboviruses (e.g., dengue and Sindbis viruses) also produce a rash, but these exanthems are not distinctive, and even when cases have occurred in epidemic proportions, dengue outbreaks have been incorrectly characterized as influenza or measles. The rash of Sindbis fever may have a vesicular component resembling varicella. Several alphaviruses, e.g., Sindbis, chikungunya, Mayaro, and Ross River viruses, produce an acute viral syndrome with polyarthritis and exanthem that must be differentiated clinically from rubella, hepatitis, parvoviral, and mycoplasmal infections among other diseases.

Neurotropic arboviruses such as the viruses of St. Louis, Murray Valley, Western, and Japanese encephalitis cause aseptic meningitis or encephalitis in only a minority of infected persons; most infections are asymptomatic or lead to a mild illness, sometimes with headache. Pathologically, global involvement of cortical, subcortical, and brain stem structures and spinal cord myelitis is typical; consequently, various neurologic presentations are possible, with combinations of coma, weakness, cerebellar and extrapyramidal movement disorders, cranial palsies, and other manifestations of bulbar dysfunction and myelitis.

Hantaviruses, cosmopolitan rodent-borne *Bunyaviridae*, are linked with three clinical syndromes: in Asia, Hantaan and Seoul viruses cause hemorrhagic fever with renal syndrome, an acute pantropic infection marked by acute renal insufficiency due to interstitial nephritis, vascular instability, and hemorrhagic manifestations; in Scandinavia and western Europe, Puumula virus causes a similar but milder form of the disease, nephropathia epidemica, typified by a febrile grippe and clinically insignificant renal insufficiency; in central and eastern Europe, Belgrade virus is the etiology of hemorrhagic fever with renal syndrome; and in North America, fatal adult respiratory distress syndrome (hantaviral pulmonary syndrome) has been linked to at least two hantaviruses, Muerto Canyon and Black Creek Canal viruses. The former (also known as Sin Nombre virus), first linked to an outbreak in the Four Corners region of the American Southwest and subsequently to more than 80 epidemic and sporadic cases (56% fatal) in more than 20 states and Canadian provinces, has its reservoir in the deer mouse, *Peromyscus maniculatus*. Specific therapy of the infection with intravenous ribavirin has not been clearly demonstrated. The pathogenesis of the noncardiogenic pulmonary edema is unknown; however, viral antigen and

980

TABLE 1 Arboviral and zoonotic viral infections by geographic area, mode of transmission, and clinical syndrome[a]

Location and virus	Febrile illness			Meningo-encephalitis	Hemorrhagic fever	Other
	Non-descript	With rash	With arthritis			
North America						
Mosquito borne						
Cache Valley	○					
California encephalitis				○		
Dengue 1–4		○				Hepatitis
EEE				○		
Everglades (VEE type II)				○		Respiratory symptoms
Jamestown Canyon				○		
LaCrosse				●		
StLE				●		
Snowshoe hare				○		
Tensaw				○		Respiratory symptoms
Trivittatus				○		
Western equine encephalitis				●		Respiratory illness
Sand fly borne						
Vesicular stomatitis (New Jersey and Indiana)	○			○		
Tick borne						
Colorado tick fever	●			○	○	
Powassan				○	○	
Zoonoses						
Lymphocytic choriomeningitis	○			○		Pneumonia, parotitis, orchitis, arthritis
Modoc				○		
Black Creek Canal						Noncardiogenic pulmonary edema
Muerto Canyon						Noncardiogenic pulmonary edema
Rio Bravo	○			○		Pneumonia, orchitis
Seoul						Interstitial nephritis
Central and South America						
Mosquito borne						
Bussuquara	○					
Cache valley	○					
Catu	○					
Cotia	○					
Dengue 1–4		●		○	●	Hepatitis
EEE				○		
Fort Sherman	○					
Group C viruses (Apeu, Caraparu, Itaqui, Madrid, Marituba, Murutucu, Nepuyo, Oriboca, Ossa, Restan)	○					

(Continued on next page)

TABLE 1 Arboviral and zoonotic viral infections by geographic area, mode of transmission, and clinical syndrome[a] (Continued)

Location and virus	Febrile illness			Meningoencephalitis	Hemorrhagic fever	Other
	Nondescript	With rash	With arthritis			
Guama	○					
Guaroa	○					
Ilheus	○			?		Hepatitis
Mayaro	○	●	●	○		
Mucambo (VEE type III)				○		
Rocio	○			●		
StLE				○		
Tonate (VEE subtype III)				○		
Tucunduba				○		
Tacaiuma	○					Two cases with concurrent malaria fatal
VEE (epizootic subtypes IABC)	●			●		
Western equine encephalitis				●		
Wyeomyia	○					
Xingu	○					Hepatitis?
YF	●				●	Hepatitis
Sand fly borne						
Alenquer	○					
Candiru	○					
Chagres	○					
Changuinola	○					
Morumbi	○					
Oropouche	●	○		○		
Punta Toro	○			○		
Serra Norte	○			○		
Vesicular stomatitis (New Jersey and Indiana)						
Vesicular stomatitis (Alagoas)	○					
Zoonoses						
Guanarito					●	
Junin					●	
Machupo					●	
Rio Bravo				○		Pneumonia, orchitis
Sabia	○				○	
Europe						
Mosquito borne						
Batai	○					
Calovo	○			○		
Inkoo	○			○		
Sindbis	●	●	●			Respiratory illness
Tahyna	●			○		Respiratory illness
West Nile	○	○	○	○		

Sand fly borne
Sandfly fever (Naples)
Sandfly fever (Sicilian)
Toscana
Tick borne
Bhanja
Central European encephalitis — Hepatitis, optic neuritis
Congo-Crimean
Lipovnik
Louping ill
Thogoto
Zoonoses
Encephalomyocarditis — Pneumonia, arthritis, orchitis, parotitis
Lymphocytic choriomeningitis — Interstitial nephritis, pantropic
Belgrade — Interstitial nephritis
Puumula — Interstitial nephritis, pantropic
Seoul

Asia
Mosquito borne
Batai
Chikungunya
Dengue 1–4 — Hepatitis common
JE
Semliki Forest
Sindbis
Snowshoe hare
Tahyna — Respiratory illness
West Nile — Hepatitis
Yunnan
Zika
Sand fly borne
Chandipura
Sandfly fever (Naples)
Sandfly fever (Sicilian)
Tick borne
Alma-Arasan
Banna
Congo-Crimean
Ganjam
Issyk-kul
Karshi

(Continued on next page)

TABLE 1 Arboviral and zoonotic viral infections by geographic area, mode of transmission, and clinical syndrome[a] (*Continued*)

Location and virus	Febrile illness			Meningo-encephalitis	Hemorrhagic fever	Other
	Non-descript	With rash	With arthritis			
Kemerovo				○		
Kyasanur Forest				●	●	
Negishi				○		
Omsk hemorrhagic fever				○	●	
Powassan				●		
Tamdy	○			●		
Wanowrie				○	○	
Russian spring-summer encephalitis		○				
Syr-Darya valley				○		
Zoonoses						
Encephalomyocarditis	○					Pantropic, interstitial nephritis
Hantaan	○				●	Pneumonia, arthritis, orchitis, parotitis
Lymphocytic choriomeningitis	○			○		Pantropic, interstitial nephritis
Seoul					●	
Africa						
Mosquito borne						
Babanki	●	●	●			
Bangui	○	○				
Banzi	○			○		
Bhanja				○		
Bunyamwera		●		○		
Bwamba	●	●		○		
Chikungunya		●	●	○		
Dengue 1–4		●		○	○	
Germiston	●	○		○	●	
Ilesha		●		○	○	
Koutango	○	○				Hepatitis
Lebombo	○					
Nyando	●					
O'nyong-nyong	●	●	●			
Orungo		●	○		●	
Pongola	●					
Rift Valley fever	●					Hepatitis, retinitis
Semliki Forest	○			●	●	
Shokwe	○	●		●		
Shuni						
Sindbis		●	●		○	

Respiratory illness

Hepatitis
Hepatitis
Hepatitis

Hepatitis, optic neuritis

Pantropic
Pneumonia, arthritis, orchitis, parotitis

Pantropic

Pantropic

Hepatitis

Spondweni
Tahyna
Tataguine
Usutu
Wesselsbron
West Nile
YF
Zika
Sand fly borne
Chandipura
Sandfly fever (Naples)
Sandfly fever (Sicilian)
Tick borne
Bhanja
Congo-Crimean
Dugbe
Nairobi sheep disease
Quaranfil
Thogoto
Zoonoses
Dakar bat
Duvenhage
Lassa
Lymphocytic choriomeningitis
Mokola
Monkeypox
Transmission cycle unknown
Ebola
Kasokero
LeDantec
Marburg
Australia and Oceania
Mosquito borne
Barmah Forest
Dengue 1–4
Edge Hill
GanGan
JE
Kokobera
Kunjin
Murray Valley
Ross River
Sepik
Sindbis
Trubanaman

---

a Adapted from reference 8. Key: O, rare, sporadic; ●, frequent, epidemic. Only arboviruses causing illness after natural infection are listed; viruses causing illness after laboratory exposure only are excluded.

genomic sequences are absent in bronchoalveolar fluid, and no laboratory- or human-to-human-transmitted cases have been reported. Other hantaviruses, identified through genomic amplification but not yet isolated, from Louisiana and Brazil suggest the existence of more unidentified novel rodent-associated hantaviruses.

Viscerotropic infection in yellow fever (YF), in some cases of dengue, and in certain other arboviral infections may result in hepatitis and jaundice, accounting for 5 to 10% of clinically diagnosed cases of viral hepatitis in areas of endemicity. Generalized hemorrhage and multisystem organ failure are the end results of fulminant YF and dengue hemorrhagic fever, mimicking malaria, leptospirosis, and the viral hemorrhagic fevers produced by Lassa and other arenaviruses, the filoviruses Ebola and Marburg viruses, and Congo-Crimean hemorrhagic fever virus. Although YF and dengue do not pose a risk for person-to-person nosocomial transmission, the other viruses are proven hazards to hospital and laboratory workers.

Except for the hemorrhagic fevers addressed in chapter 94 of this Manual, specific therapy of arboviral infections is unavailable. Consequently, the rationale for promptly establishing laboratory diagnosis of an arboviral infection is to hasten confirmation of other treatable conditions and to assist public health officials in identifying arboviral infections with epidemic potential.

DESCRIPTION OF THE AGENTS

Viruses adapted to arthropod vectors occur in several taxonomic families, but the majority are RNA viruses in the families *Togaviridae* (formerly group A arboviruses), *Flaviviridae* (formerly group B arboviruses), *Bunyaviridae*, *Reoviridae*, and *Rhabdoviridae* (Table 2) (4). The majority are enveloped viruses and are unstable in the environment. The *Orbiviridae*, *Bunyaviridae*, and *Orthomyxoviridae* contain multiple genomic segments and may undergo reassortment; however, these events probably occur infrequently in nature and for the arboviruses have not been shown to be epidemiologically significant. Complete nucleotide sequences or sequences of certain genomic segments have been determined for a few arboviruses, facilitating production of synthetic antigens for diagnosis and as candidate vaccines. Important antigenic determinants have been elucidated for the most important arboviruses, and viral and group-specific monoclonal antibodies are available for taxonomic and diagnostic purposes (Table 3).

Most arboviruses elaborate hemagglutinins that aggregate goose, chick, and/or human group O erythrocytes. Hemagglutination (HA) occurs optimally within a narrow and specific pH range, a property that may aid in viral identification. In general, serologic cross-relationships are most evident in hemagglutination inhibition (HI) and binding assays, e.g., enzyme-linked immunosorbent assays (ELISAs) and immunofluorescent-antibody (IFA) tests, and occur to a lesser extent in complement fixation (CF) tests. These broad antigenic relationships can be refined in viral neutralization tests that differentiate individual viruses and even more specific subtype relationships through ratios of homologous and heterologous N antibody titers. Genomic sequence divergence among strains of a single virus in separate geographic locations differentiates "viral topotypes," which apparently result from the isolation and evolutionary divergence of these strains in distinct ecologic niches. Topotypic differences for populations of Japanese encephalitis (JE), YF, dengue, West Nile, eastern equine encephalitis (EEE), Snowshoe hare, and LaCrosse viruses have been described, and such genetic changes have been linked to differences in clinical expression of Sindbis, Venezuelan equine encephalomyelitis (VEE), and St. Louis encephalitis (StLE) viruses.

Most arboviruses can be cultivated in a variety of primary and continuous cell lines, including primary duck and chick embryo cells, primary hamster kidney cells, C6/36 *Aedes albopictus* and AP61 *Aedes pseudoscutellaris* mosquito cell lines, and Vero, LLC-MK$_2$, BHK-21, CER, SW13, and PK cell lines. The rapidity and pattern of cytopathic effects (CPE) vary in each virus-cell system; e.g., primary duck and BHK-21 cells are rapidly destroyed by most arboviruses, and replication in monkey kidney cell lines is generally slower. AP61 and C6/36 cells exhibit nearly universal susceptibility to the arboviruses, including the dengue viruses, which grow poorly in continuous monkey kidney cell lines. The mosquito cell lines, generally incubated at 28°C, exhibit little or no CPE, and evidence of viral replication must be sought by IF or other means. Some flaviviruses form cell syncytia in these lines. The majority of arboviruses are lethal for suckling (2- to 3-day-old) mice, which exhibit signs of illness, paralysis, and death within days to 2 weeks after intracerebral inoculation. Some field isolates exhibit virulence for mice only after adaptation through blind passage.

COLLECTION AND STORAGE OF SPECIMENS

Few primary diagnostic laboratories can justify arboviral isolation and identification as a service, and most laboratories should focus on the appropriate collection of specimens for referral to a reference laboratory. Most arboviruses produce a brief, low level of viremia in humans, and it is only by chance that a viral isolate can be recovered from blood in these instances. However, certain arboviruses, notably the viruses of dengue, YF, chikungunya, and VEE and of Ross River, Oropouche, and sandfly fevers, produce a sufficiently high level of viremia (>dex 3.5) and of sufficient duration (ca. 5 to 7 days after onset of fever) that humans serve as hosts for epidemic transmission. For YF, this represents the urban, *Aedes aegypti*-borne form of the disease. From the perspective of the clinical laboratory, these infections can be diagnosed by isolating virus from blood obtained during the acute phase of illness. The virus of Colorado tick fever infects bone marrow erythrocytic precursors, leading to infection of the erythrocyte for its lifetime; virus can be recovered from blood obtained several weeks after clinical recovery, and infection has been transmitted through transfusion.

Viral isolates can be recovered by biopsy or at autopsy from the viscera of patients with acute YF, dengue hemorrhagic fever, or other viscerotropic arboviral infections and from brain or cerebrospinal fluid (CSF) of patients with central nervous system (CNS) infection. Brain samples should be taken from several areas, including the cortex, brain nuclei, cerebellum, and brain stem. Neurotropic arboviruses sometimes can be isolated from CSF obtained by lumbar puncture during the acute stages of encephalitis or aseptic meningitis. Alphaviruses have been isolated from joint fluids of patients with acute polyarthritis and from the upper respiratory tracts of patients with acute VEE. Under certain circumstances, arboviruses have been recovered from urine, milk, semen, and vitreous fluid. Clinical laboratories sometimes receive arthropods for viral isolation. These samples can be referred in the same manner as tissues

TABLE 2 Taxonomic and physical characteristics of arboviruses and certain zoonotic viruses[a]

Family: virus	No. of arboviruses causing human disease/no. of arboviruses recognized	Genome	No. of genomic segments	Virion size (nm)	Morphology
Togaviridae: alphaviruses	14/29	Positive single-stranded RNA	1	60	Spherical, enveloped
Flaviviridae: flaviviruses	34/69	Positive single-stranded RNA	1	40–50	Spherical, pleomorphic, enveloped
Bunyaviridae: bunyaviruses, phleboviruses, nairoviruses, hantaviruses, uukuviruses	54/254	Negative and ambisense single-stranded RNA	3	80–120	Spherical, pleomorphic, enveloped
Reoviridae: orbiviruses, coltiviruses	5/77	Double-stranded RNA	10/12	60–100	Spherical, nonenveloped
Rhabdoviridae: lyssaviruses, vesiculoviruses, and others	6/70	Negative single-stranded RNA	1	60–85 by 180	Bullet shaped, enveloped
Arenaviridae: arenaviruses	6/16	Ambisense single-stranded RNA	2	50–300	Spherical, pleomorphic
Filoviridae: filoviruses	2/2	Negative single-stranded RNA	1	80 by 14,000	Tubular, filamentous
Orthomyxoviridae	1/2	Negative single-stranded RNA	6–7?	90–110, up to 1,000	Pleomorphic, enveloped
Poxviridae	1/3	Double-stranded DNA		200–400	Pleomorphic, enveloped
Others	3/16				

[a] Adapted from reference 8.

to a reference laboratory, where they are ground, often in pools, for viral isolation or antigen detection.

A plan for dividing available tissues or fluids for viral isolation, electron microscopy, and direct detection assays of antigen and genomic sequences should be devised. Tissues should be collected aseptically and rapidly transported to the laboratory in viral transport medium or on a moist sponge. If the aliquot for viral isolation can be processed within 24 to 48 h, it can be maintained at 4°C; otherwise, the sample should be immediately frozen at −70°C in a mechanical freezer or stored on dry ice. Samples for virus isolation should be kept frozen continuously, avoiding freeze-thaw cycles, which inactivate virus. The aliquot for electron microscopy should be minced and placed directly in glutaraldehyde. Autolytic changes occur rapidly, and tissues should be fixed as quickly as possible. To prepare sections for immunohistochemical examination, a portion of the sample should be either fixed in buffered formalin or, preferably, embedded in freeze medium and frozen. Touch preparations are less reliable. Aliquots for antigen detection by other means and for PCR should be frozen at −70°C. Samples should be shipped by express mail in double, sealed containers following Interstate Commerce Commission (ICC) regulations. Frozen samples should be shipped separately on dry ice. International shipments may require import and/or customs permits and clearances.

Serologic studies are best done by comparing antibody titers in serum samples drawn during the first week of illness and 2 to 3 weeks later. A single serum specimen may be sufficient for diagnosis of certain arboviral infections for which immunoglobulin M (IgM) assays are available. Detection of virus-specific IgM in CSF is a sensitive and highly specific approach to diagnosing CNS infection. Both CSF and serum samples should be tested. IgM capture ELISAs for the arboviral encephalitides transmitted in the United States are offered by several reference laboratories. Serum and CSF samples should be collected aseptically, stored either refrigerated or frozen, and transported according to ICC regulations. Freeze-thaw cycles may reduce antibody titers and should be avoided. Blood or serum blotted onto filter paper and dried may be suitable for some serologic assays.

No specific procedures for patient preparation are required. Ribavirin, chloroquine, rimantadine, amantadine, and other weak bases inhibit replication of some arboviruses and may interfere with their isolation. The isolation of certain arboviruses could be inhibited in patients treated with alpha or gamma interferon or other immunomodulators. Some preparations of gamma globulin, including preparations for intravenous use, may contain arboviral antibodies and could interfere with the isolation of arboviruses from patient specimens or with arboviral serologic procedures.

DIRECT EXAMINATION

Clinical application and experience with direct detection methods for arboviruses generally have been limited, and specimens should be evaluated in parallel with conventional viral isolation and serologic procedures. When tissue is available, direct examination by electron microscopy can rapidly provide evidence of an arboviral infection. Experienced observers can identify virions morphologically, often to the level of a viral family, in thin tissue sections. Visualization of togaviruses in brain tissue helped establish a diagnosis of EEE in some reported cases. Virions may be

TABLE 3 Selected arboviral monoclonal antibodies for serologic and antigen detection assays[a]

Antibody	Prepared against:	Reactivity	Use
6B6C-1	StLE virus (MSI-7)	Flavivirus group	As HRP conjugate in antigen capture and IgM capture ELISAs
4G-2	Dengue 2 (New Guinea C)	Flavivirus group	Viral identification by IF or ELISA; ELISA capture antibody
6B4A-10	JE virus (Nakayama)	JE-StLE-WN-MVE complex	Capture antibody in antigen capture ELISA
JE314H52	JE virus (Nakayama)	JE virus specific	Viral identification by IF or ELISA
6B5A-2	StLE virus (MSI-7)	SLE virus specific	Viral identification by IF and neutralization
4A4C-4	StLE virus (MSI-7)	SLE virus specific	Capture antibody in antigen capture ELISA
4B6C-2	MVE virus (original)	MVE virus specific	Viral identification by IF or ELISA
D2-1F1-3	Dengue 1 (Hawaii)	Dengue 1 virus specific	Viral identification by IF or ELISA
3H5-1-21	Dengue 2 (New Guinea C)	Dengue 2 virus specific	Viral identification by IF or ELISA
D6-8A1-12	Dengue 3 (H87)	Dengue 3 virus specific	Viral identification by IF or ELISA
1H10-6-7	Dengue 4 (H-241)	Dengue 4 virus specific	Viral identification by IF or ELISA
5E-3	YF (17D)	Yellow fever virus specific	Capture antibody in antigen capture ELISA
2D12	YF (17D)	Yellow fever virus specific	Viral identification by IF
4D9	TBE virus	Tick-borne flavivirus complex	Viral identification by IF or ELISA
2A2C-3	WEE virus (McMillan)	Alphavirus group	As HRP conjugate in antigen and IgM capture ELISAs
2A3D-5	WEE virus (McMillan)	WEE complex	Capture antibody for WEE and HJ antigen capture ELISA
2B1C-6	WEE virus (McMillan)	WEE virus specific	Viral identification by IF or ELISA; as HRP conjugate in antigen capture ELISA
1B5C-3	EEE virus (NJ 60)	EEE North American virus specific	Viral identification by IF or ELISA; as HRP conjugate in antigen capture ELISA
1A4B-6	EEE virus (NJ 60)	EEE North American virus specific	Capture antibody for antigen capture ELISA
1B1C-4	EEE virus (BeAn 5122)	EEE complex specific	Viral identification by IF or ELISA
2D4-1	Highlands J virus (B230)	Highlands J virus specific	Viral identification by IF or ELISA; as HRP conjugate in antigen capture ELISA
1A2B-10	VEE virus (E2 peptide)	VEE complex (wild type)	Viral identification by IF or ELISA
5B4D-6	VEE virus (TC-83)	TC-83 (vaccine) specific	Viral identification by IF or ELISA
1A3A-5	VEE virus (IC)	Epizootic VEE virus	Viral identification by IF or ELISA
1A1B-9	VEE virus (IE)	Enzootic VEE virus	Viral identification by IF or ELISA
807-18	LaCrosse virus (original)	LaCrosse virus specific	Viral identification by HI, neutralization, or ELISA
10G-4	LaCrosse virus (original)	California serogroup reactive	Viral identification by ELISA

[a]Adapted from reference 7. Antibodies and/or conjugates are available from the CDC (see text for the address). Abbreviations: MVE, Murray Valley encephalitis; WEE, Western equine encephalitis; WN, West Nile; HRP, horseradish peroxidase.

absent, however, from autopsy tissues of patients dying after a prolonged course of illness in whom infection had previously been cleared.

Immunohistochemical staining of peripheral blood mononuclear cells or tissue sections has been successful in detecting arboviral antigens in CSF and brain specimens from encephalitis patients; in the viscera of patients with YF, dengue, and other viscerotropic infections; and in joint fluids of patients with acute arthritis. Dengue virus, JE virus, and other flaviviruses infect and replicate in mononuclear cells, and their detection in the peripheral blood or CSF is a highly sensitive although somewhat laborious approach to rapid and specific diagnosis. In one study, JE virus-infected mononuclear cells were identified in CSF before intrathecal antiviral antibodies were detectable. Infected peripheral mononuclear cells also have been observed in JE patients with recrudescent symptoms after their previous recovery from acute infection, indicating the possibility of latent infection and recurrent viremia. Monoclonal antibodies specific to many medically important arboviruses are available, and for more obscure arboviruses, hyperimmune

mouse antibodies can be employed (Table 3). Antibodies and information on their working dilutions can be obtained from the CDC and other reference laboratories. Although trypsinized, formalin-fixed tissues are suitable for IFA examination, frozen sections are generally preferred because trypsinization procedures have not been standardized for all arboviruses and because success varies with the duration of formalin fixation.

Antigen capture ELISAs have been developed for several arboviruses; however, the formation of immune antibody-virus complexes in dengue and YF has interfered with detection of viral antigens in viremic blood. Although complexes can be chemically dissociated, the assays have not been shown to be clinically useful. The detection threshold of antigen capture ELISAs for StLE and JE viruses (ca. $10^{3.5}$ PFU) has been sufficiently sensitive for their application in the epidemiologic surveillance of infected arthropod vectors (Table 3).

Viral RNA has been detected by in situ hybridization in archived formalin-fixed and paraffin-embedded tissues as well as in fresh tissues of patients with YF and dengue.

Reverse transcription PCR procedures have been described for the detection of dengue, YF, West Nile, Langat, tick-borne encephalitis (TBE), JE, StLE, Dugbe, Congo-Crimean hemorrhagic fever, Ross River, EEE, and Ockelbo viral RNAs. Prospective clinical evaluation of serum and blood samples submitted for dengue diagnosis suggests that reverse transcription PCR is equal in sensitivity to viral isolation when the most sensitive techniques of mosquito and mosquito cell culture inoculation are used. PCR amplification of viral genomic sequences in skin lesions proved to be more sensitive than viral isolation in the diagnosis of acute Ockelbo (Sindbis virus) infection. Inhibitory effects of human serum on the PCR have been overcome, without significantly diminishing test sensitivity, by diluting samples. Contrary to the expectation that viral RNA in clinical specimens would be rapidly degraded by RNases, under experimental conditions, YF viral RNA could be detected in spiked human serum samples kept for 3 weeks at ambient temperature.

VIRAL ISOLATION

Many of the medically important arboviruses have caused infections in laboratory workers, occasionally in outbreaks. The pathogenesis of infections acquired from aerosols created by pipetting and other procedures may differ from that of natural infection, with the possibility of direct CNS invasion through the olfactory epithelium. On the basis of this experience, the joint CDC-National Institutes of Health (NIH) laboratory biosafety manual recommends biosafety level 2 to 4 containment facilities for laboratory work with arboviruses (3). Arboviruses should be handled in certified laminar-flow hoods in rooms whose air is exhausted directly through HEPA filters and is not commingled with air from other laboratory areas (single-pass air). In addition, laboratory staff should be immunized with available arboviral vaccines. Laboratories that process CNS specimens from patients with encephalitis should consider immunizing staff with rabies vaccine as well.

Tissues, fluids, and serum collected in the acute phase of illness should be inoculated into cultures of several cell lines, including C6/36 or AP61 mosquito cells, a monkey kidney cell line such as Vero or LLC-MK$_2$, BHK-21 or CER (hamster kidney) cells, and, if available, primary duck embryo cells and 2- to 4-day-old suckling mice. Intrathoracic inoculation of uninfected *Toxorhynchites* mosquitoes is the most sensitive system for isolating dengue viruses. Ideally, serum specimens should be inoculated neat and in 10^{-1} and 10^{-2} dilutions to avoid prozone effects due to viral auto-interference and viral antibodies present in the specimen. Tissues should be ground with a mortar and pestle and diluted similarly on a weight-to-volume basis with a protein- and antibiotic-containing medium, e.g., phosphate-buffered saline with 0.75% bovine serum albumin and 1% gentamicin. After adsorption to the drained cell monolayer, cell cultures are fed with medium and observed daily or more often for signs of CPE. Small flasks (25 cm^2), tubes, or coverslips in shell vials may be used. Viral growth in C6/36 and AP61 cells does not reliably produce CPE, and these cells must be examined for viral replication by IFA or other means. When medium from infected cultures is harvested for passage, fetal calf serum should be added to a final concentration of 20 to 40% before the medium is frozen at −70°C. The effect of spin amplification on isolation of arboviruses has not been clinically evaluated.

Intracerebrally inoculated mice (one litter of six to eight mice for each dilution) should be observed twice daily for up to 2 weeks for signs of illness and death. Ill and dead mice should be frozen at −70°C until infected organs can be passaged in cell culture. When a bunyaviral etiology is suspected, it is prudent to harvest livers as well, because certain bunyaviruses replicate to higher titers in mouse liver than in brain.

VIRAL IDENTIFICATION

Viral isolates established in cell culture are most easily identified as arboviruses by IFA examination with a series of polyvalent (NIH grouping fluids) or broadly reactive monoclonal antibodies. Reactivity of an unknown virus with one of the 30 arboviral grouping fluids (produced in mice as hyperimmune ascitic fluids against four or more viruses) permits its broad placement within that group of viruses. The unknown is further identified by its reactivity against a panel of monovalent antibodies representing viruses in the group. Alternatively, monoclonal antibodies reactive against antigens common to the flaviviruses, alphaviruses, or other taxonomic groups may be used (Table 3).

Suspensions of infected and uninfected cells are dropped into wells of a Teflon-coated slide. Slides are dried and fixed in acetone and then processed directly or stored at −20°C. Head squashes of intrathoracically inoculated mosquitoes are processed in the same fashion. Indirect IFA staining is done by conventional procedures with progressive twofold dilutions of antiviral antibodies followed by antispecies fluorescein isothiocyanate conjugate. Endpoint titers of the various antibodies against the unknown antigen are compared.

The unknown is identified when the endpoint IF antibody titer of a monovalent antibody is significantly higher (fourfold or greater) than reactions of other monovalent antibodies in the group. Often, however, IFA reactions among antibodies to related viruses overlap, and identification by cross-neutralization, the most specific means of serologic identification, is indicated. IFA reactivity with a virus-specific monoclonal antibody provides definitive identification (Table 3).

Viral identification by cross-neutralization is best done if the unknown virus is first adapted to produce plaques in a cell culture system. Initially, a unidirectional plaque neutralization test is done with reference antisera to viruses most closely related to the unknown against their homologous viruses and the unknown (see below). The ratios of homologous and heterologous titers are examined (with fourfold differences defined as significant) to determine the relationship of the unknown to the other viruses. Definitive identification necessitates production of an immune serum against the unknown and bidirectional tests in which this and the other antisera are tested against their respective homologous and other viruses represented in the test. Viruses that can be differentiated by fourfold differences in neutralization titer in both directions are considered distinct (homologous and heterologous titers of both antibodies differ by more than fourfold). Virus pairs that show a fourfold difference in one direction only (only one of the two antibodies differs by more than fourfold in its homologous and heterologous neutralization titers) are considered subtypes of the same virus. The comparative reactivities of monoclonal antibodies against individual viruses in more convenient binding assays generally have confirmed or supported this classic approach.

When IFA screening with broadly reactive antibodies

fails to place an unknown virus within a group of arboviruses, morphologic examination of the unknown by negative-stain electron microscopy can rapidly categorize the virus within a taxon. Virions generally can be found in cell culture medium or directly in clinical specimens that have infectivity titers of $>10^6$. Although most pathology laboratories do not routinely prepare negatively stained grids, examination of thin sections of infected cell cultures can often be arranged. Cells or mouse brain infected with EEE (Fig. 1), StLE (Fig. 2), and LaCrosse (Fig. 3) viruses are shown in thin sections (Fig. 1) as examples of the morphologic appearance of an alphavirus, a flavivirus, and a bunyavirus, respectively.

PCR-based assays for detecting and identifying arboviruses are in a state of development, and experience is insufficient to recommend a given procedure. Promising and innovative techniques that detect and identify all medically important flaviviruses (e.g., dengue, YF, StLE, JE, and TBE viruses) by amplifying flavivirus genus-conserved sequences have been reported. Novel flaviviruses can be identified by sequencing amplicons produced from the flavivirus genus-conserved amplimer pair.

SEROLOGY

In most circumstances, laboratory confirmation of arboviral infections still relies on serologic procedures. The classic serologic tests, HI and CF, have gradually been replaced in diagnostic laboratories by IgM capture ELISA. The former two still find use in reference laboratories for differentiating certain arboviral infections. Complete descriptions of arboviral serologic procedures are given in the *Manual of Clinical Laboratory Immunology* (7) and other sources (1).

ELISA

The antibody capture format for detecting IgM in serum or CSF is more sensitive and specific than indirect methods of IgM detection. In dengue infections, virus-specific IgM can be detected in the sera of 90 to 95% of patients by the sixth day after onset of illness. Patients with CNS infections due to flaviviruses, alphaviruses, or bunyaviruses generally have detectable IgM in combinations of serum and/or CSF in 40% of cases by day 4 post-onset and in nearly all cases by 10 days post-onset (approximately 10% cumulative positivity with each post-onset day). The presence of virus-specific IgM in CSF reflects intrathecal antibody production and is considered diagnostic of recent CNS infection. In serum, a declining level of virus-specific IgM can be detected for months and up to a year in some cases, reducing the diagnostic specificity of the test. Therefore, a positive IgM result in a single serum sample is considered only presumptive evidence of recent infection. An analogous IgA capture ELISA for Ross River polyarthritis was shown to be better suited for diagnosis of recent infection, because the level of virus-specific IgA declined more rapidly than that of IgM. Arboviral IgG antibodies are detected by an indirect sandwich method.

Heterologous reactions to antigenically related viruses are a vexing problem in areas where several related viruses cocirculate. Cross-reactions between dengue viruses and JE virus or between YF virus and the numerous flaviviruses that circulate in West Africa frequently produce overlapping values that may be uninterpretable. The specificity of IgM capture ELISA is somewhat better than that of HI, but further improvements have been sought by calculating ratios of absorbance to two or more antigens or by examining the areas under the absorbance curves over several serum dilutions. However, the discriminatory power of these approaches requires further confirmation.

Dilutions of serum (beginning at 1:40 to 1:200 in different procedures) are added to plates coated with infected-cell-culture-fluid-derived viral antigen, and bound IgG is detected with anti-γ-chain conjugate. Crude antigens may need to be captured and concentrated onto plates with a nonhuman antiviral antibody. In a dot blot format, antigen is bound to a membrane solid phase, and unbound serum, conjugate, and substrate are washed off or through the membrane by vacuum filtration or into individual absorptive cartridges. The precipitated substrate can be quantified by densitometry or estimated visually.

Numerous schemes have been devised to define the threshold of a positive or significant ELISA absorbance value. Some laboratories define a positive reaction in multiples (units) of the absorbance of a weak positive specimen, an approach that controls plate-to-plate variations in specimen reactivity. An alternative approach is to define the threshold absorbance as a multiple (e.g., 5 standard deviations) above the mean absorbance of samples from a population of uninfected persons. A simple multiple of two times the absorbance of a negative control generally produces a conservative interpretation, exceeding 5 standard deviations above the mean of negatives.

HI

For the arboviruses that hemagglutinate erythrocytes, i.e., *Flaviviridae*, *Bunyaviridae*, alphaviruses, and certain *Rhabdoviridae*, the HI test is a convenient screening serologic procedure. HI antibodies rise rapidly within the first week after onset of illness and are long-lived, persisting at low levels for decades in some instances. These kinetics are consistent with the mixed IgM and IgG isotype distribution of HI antibodies. Gander erythrocytes are preferred, although chick and trypsinized human O erythrocytes have also been used. Inactivated sucrose-acetone extracts of infected mouse brains provide a high-titered source of antigen; for viruses that do not propagate to a high titer in mouse brain, infected-cell-culture antigens are used. For each virus, HA is optimal within a narrow pH range of less than half a pH unit. Before the HI test is performed, a preliminary HA test is completed in which the specific pH optimum and the antigen titer are determined.

Nonspecific inhibitors of HA present in some sera are removed by kaolin or, preferably, acetone extraction. Serum samples are conventionally tested at 1:10 and at further twofold dilutions to endpoint. Four or 8 HA units of antigen are added to each serum dilution, and the plate is incubated overnight at 4°C. Washed erythrocytes at the appropriate pH are added, and the patterns of settled erythrocytes are scored. The last serum dilution inhibiting HA (i.e., showing button formation) is the endpoint. Serum controls should show no HI activity. Fourfold changes between acute- and convalescent-phase samples are diagnostic of recent infection. In some epidemiologic circumstances, a single or stable elevated titer of >1:80 may be interpreted as presumptive evidence of recent infection; a titer of >2,560 is evidence of a recent secondary antibody response to flaviviral infection.

The HI test is highly sensitive in confirming an infection when adequately timed serum pairs are available, but it is considerably less useful when applied to a single serum sample. Furthermore, HI antibodies tend to be broadly reactive, recognizing common epitopes within antigenic groups or complexes. Although a primary infection may lead to a specific antibody response, repeated infections with related viruses produce an increasing dominance of heterologous antibodies that yield undifferentiated elevated titers to a variety of related viruses, rendering the HI test result uninterpretable.

The HI test is often used in epidemiologic surveys because of the broad reactivity and the longevity of HI antibodies and because the procedure is robust. Capillary samples collected on filter paper and dried can be stored for short periods without refrigeration, mailed, and tested after reconstitution. The procedure is not limited by species-specific reagents and can be used for wildlife and animal surveys, which are often a component of field studies.

Indirect Immunofluorescence

Some laboratories find indirect immunofluorescence to be a convenient procedure that also allows differentiation of IgG and IgM responses. Infected cells fixed onto microscope slides with multiple wells are the usual antigen substrate. A multivalent slide with EEE, Western equine encephalitis, StLE, and LaCrosse virus-infected cells and uninfected control cells is commercially available. Positive and negative human control samples also are supplied when slides are purchased in kit form. Anti-human IgG and IgM fluorescein isothiocyanate conjugates are commercially available, but they should be titrated to ascertain their optimal working dilutions. To conserve antigen, serum samples (1:16 dilution) and CSF (neat) can be screened and then tested in further twofold dilutions to endpoint only if positive. Positive IgM reactions should be verified as specific by ruling out the presence of rheumatoid factor (serum IgM antibodies directed against IgG).

Immunofluorescent IgM antibodies become detectable within a few days after onset of illness, and IgG antibodies appear shortly thereafter. Immunofluorescent IgG antibodies are long-lived, paralleling the longevity of HI and neutralizing antibodies. Fourfold changes are diagnostic of recent infection, and in some circumstances, single or stable elevated titers (>1:128) may indicate recent infection. The sensitivity and specificity of the indirect FA and HI tests are similar. Thus, cross-reactions among the flaviviruses, e.g., between StLE and dengue viruses, occasionally present a problem in interpretation.

CF Test

The CF test is moderately specific and is often used to narrow the definition of a heterologous HI antibody response. The half-life of CF antibodies is 2 to 3 years, and the test can be used as an imprecise measure of recent infection. Thus, some laboratories use the HI test to screen samples and the CF test to define the recency and specificity of reactive samples. Some arboviruses, such as Colorado tick fever and the orbiviruses, do not hemagglutinate erythrocytes, and the CF test is then used in primary diagnosis. Because the test is relatively insensitive, it should be used in combination with some other procedure.

Guinea pig complement, dilutions of test serum (heat inactivated to destroy indigenous complement), and pretitrated antigen are incubated together overnight at 4°C. In the presence of antiviral antibodies, antibody-antigen interactions fix the added complement. Sheep erythrocytes sensitized with hemolysin (rabbit anti-sheep erythrocyte antibody) are added as an indicator, and the plate is incubated at 37°C for 30 min so that lysis will occur. The plates are centrifuged at low speed to sediment unlysed erythro-

cytes, and the extent of lysis and the patterns of sedimentation (buttons) are read semiquantitatively. Wells scoring 3 to 4+ CF show <30% lysis and relatively large buttons. Serum and antigen controls must be free of anticomplement activity. Serum samples generally are tested at twofold dilutions beginning at 1:8. The highest serum-antigen dilution exhibiting 3 to 4+ CF is the endpoint. Fourfold changes are diagnostic of recent infection, and in some circumstances, a single or stable elevated titer of 1:32 may be accepted as presumptive evidence of recent infection.

CF antibodies generally rise slowly after infection, often peaking as late as 6 weeks after onset of illness. Convalescent-phase serum samples should be obtained 2 to 3 months after onset to ensure against missing a late-rising CF antibody response. Some individuals fail to produce detectable CF antibodies, and advanced age may be associated with a blunted, delayed, or undetectable response. In one study of mostly elderly StLE patients, a CF antibody response could not be detected in 20% of cases.

Neutralization

The neutralization test is the most specific of the common serologic procedures and is used principally to sort out heterologous reactions observed in other assays. The presence of viral neutralizing antibodies is also considered the best evidence of protective immunity, and response to immunization is usually monitored by following neutralizing antibody levels. Neutralizing antibodies generally become detectable within the first week after onset of illness, peak in the following 2 weeks, and decay slowly, persisting for years and often over a lifetime. Reexposures and infections with related viruses may stimulate an accelerated secondary response. Although the neutralizing antibody response is relatively specific, repeated infections with related viruses produce a progressively broader heterologous immune response, with the possible extension of cross protection in some instances. Often, these extensive heterologous reactions cannot be resolved with cross-neutralization tests.

Infectious virus is used in the neutralization test; consequently, the procedure should be performed only in laboratories capable of working safely at the appropriate biosafety level. Neutralizing antibodies are best quantitated in a plaque reduction assay, in which dilutions of heat-inactivated serum are mixed with a fixed dose of virus and exogenous complement. After a period of incubation (37°C for 1 h or overnight at 4°C) for neutralization to proceed, the mixtures are applied to drained cell culture monolayers grown in 6- or 24-well panels. After incubation at 37°C for 1 h to allow attachment of unneutralized virus to cells, the monolayers are overlaid with nutrient agarose, methylcellulose, or some other solid or semisolid medium. After several days, with the addition of neutral red, a vital dye, infectious centers or plaques can be counted as clear areas of dead cells. The highest serum dilution inhibiting 90% of the infectious virus dose observed in virus control wells is the endpoint. In a typical test using 100 PFU as the viral input dose, the endpoint would be the last well showing 10 or fewer plaques. Serum samples generally are diluted 1:5, and with the addition of an equal volume of virus, the initial test dilution is 1:10. Fourfold changes in titer are considered diagnostic of recent infection, and in some circumstances, a single or stable elevated titer (>1:80) can be interpreted as presumptive evidence of recent infection.

The neutralizing capacity of a serum sample can also be measured by testing a single dilution of the serum against a

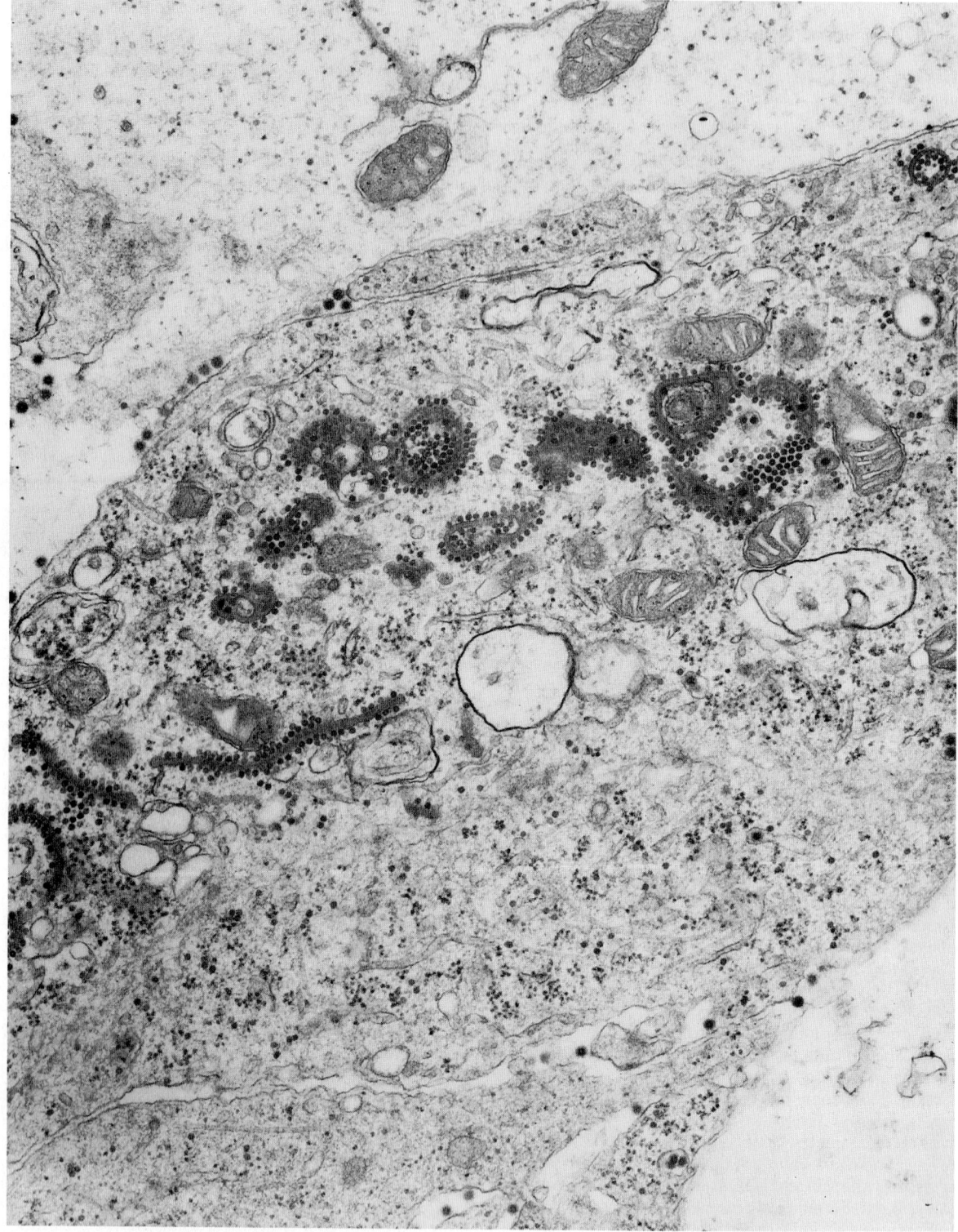

FIGURE 1 *Togaviridae*: alphavirus. EEE virus in mouse brain at 48 h postinfection. Viral nucleocapsids average 28 to 30 nm in diameter and lie within the cytoplasmic matrix and upon electron-dense membranous structures. Enveloped virions are seen passing through the plasma membrane and average 55 to 60 nm in diameter.

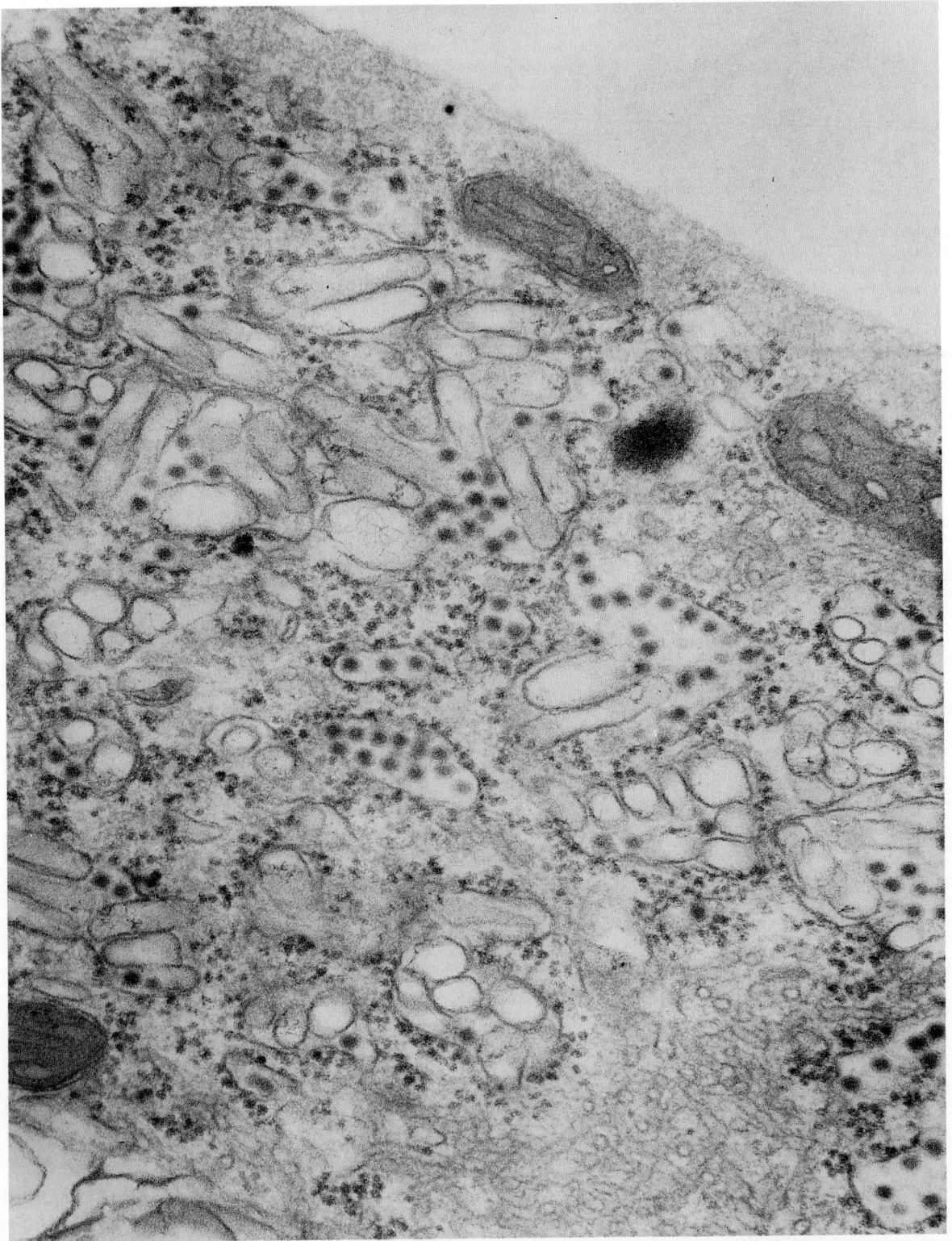

FIGURE 2 *Flaviviridae*: flavivirus. StLE virus-infected cell cultures. Enveloped viral particles accumulate within the cisternae of the endoplasmic reticulum. Cylindrical and round membranous structures are seen within the endoplasmic reticulum at the sites of viral accumulation.

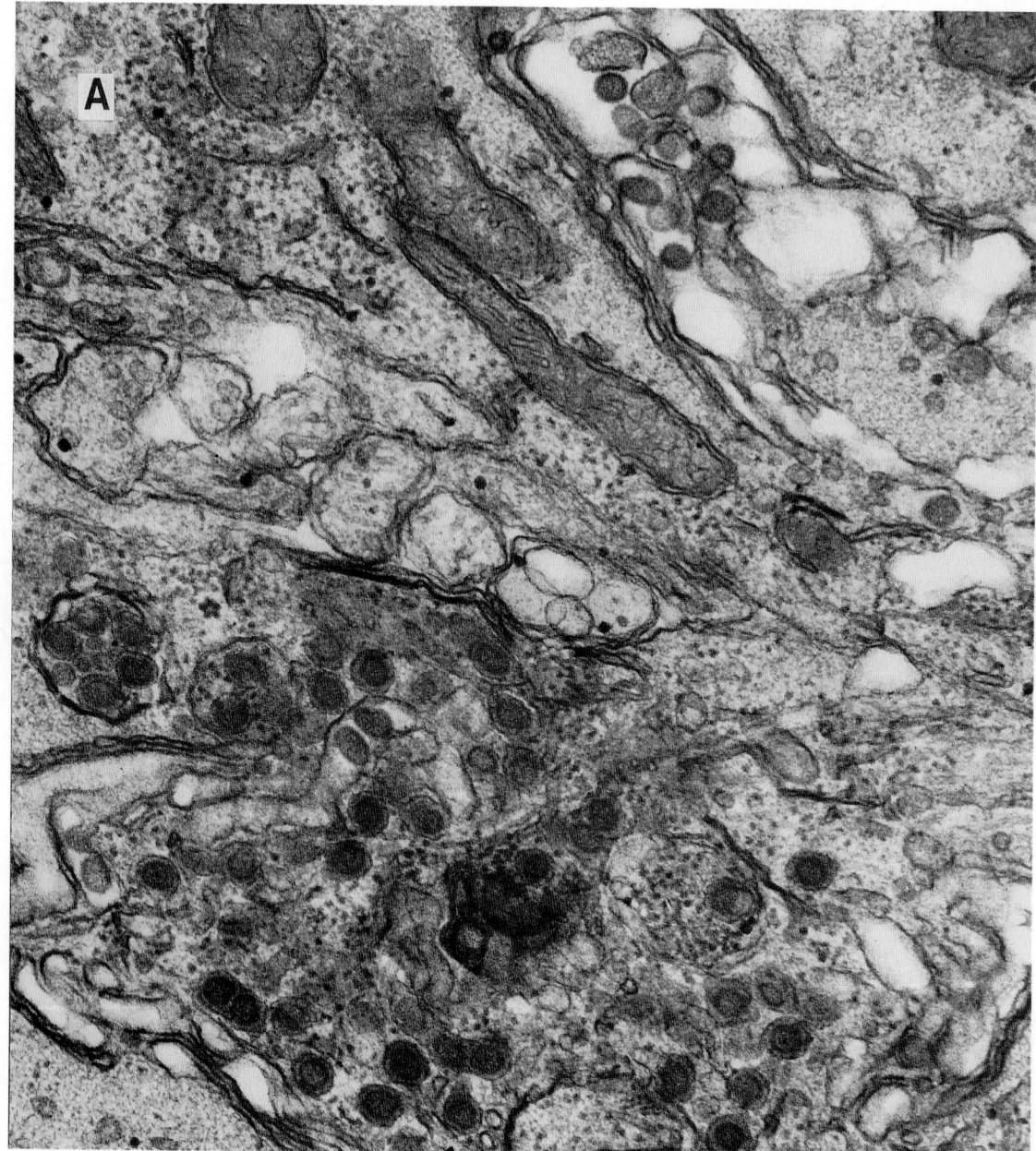

FIGURE 3 *Bunyaviridae*: bunyavirus. LaCrosse virus-infected mouse brain. (A) Virions are seen within cisternae of Golgi complex, endoplasmic reticulum, and vesicles of a neuron. Virus particles measure 90 to 105 nm in diameter. (B) Virions budding into or accumulating in a Golgi complex.

series of virus dilutions. The result is expressed as the serum's log neutralization index.

Other Procedures

As stated previously, heterologous serologic reactions in patients who have had repeated arboviral infections are difficult to sort out even after all available conventional tests have been exhausted. In one attempt to improve specificity, the narrow reactivity provided by viral monoclonal antibodies has been adapted to a serologic test by combining test sample and virus-specific monoclonal anti-

body in a competitive ELISA. Virus-specific antibodies present in the test sample block binding of the monoclonal conjugate to viral antigen, thus reducing absorbance. Synthetic antigens, expressed as polyproteins including areas of the envelope glycoprotein or produced by proteolytic cleavage, have provided some differentiation of flaviviral antibody responses in ELISAs, by immunohistochemical staining of expressed antigen on cell surfaces or by Western blot (immunoblot). Although these techniques may be applicable to population surveys, their discriminatory power for clinical diagnostic purposes requires further confirmation.

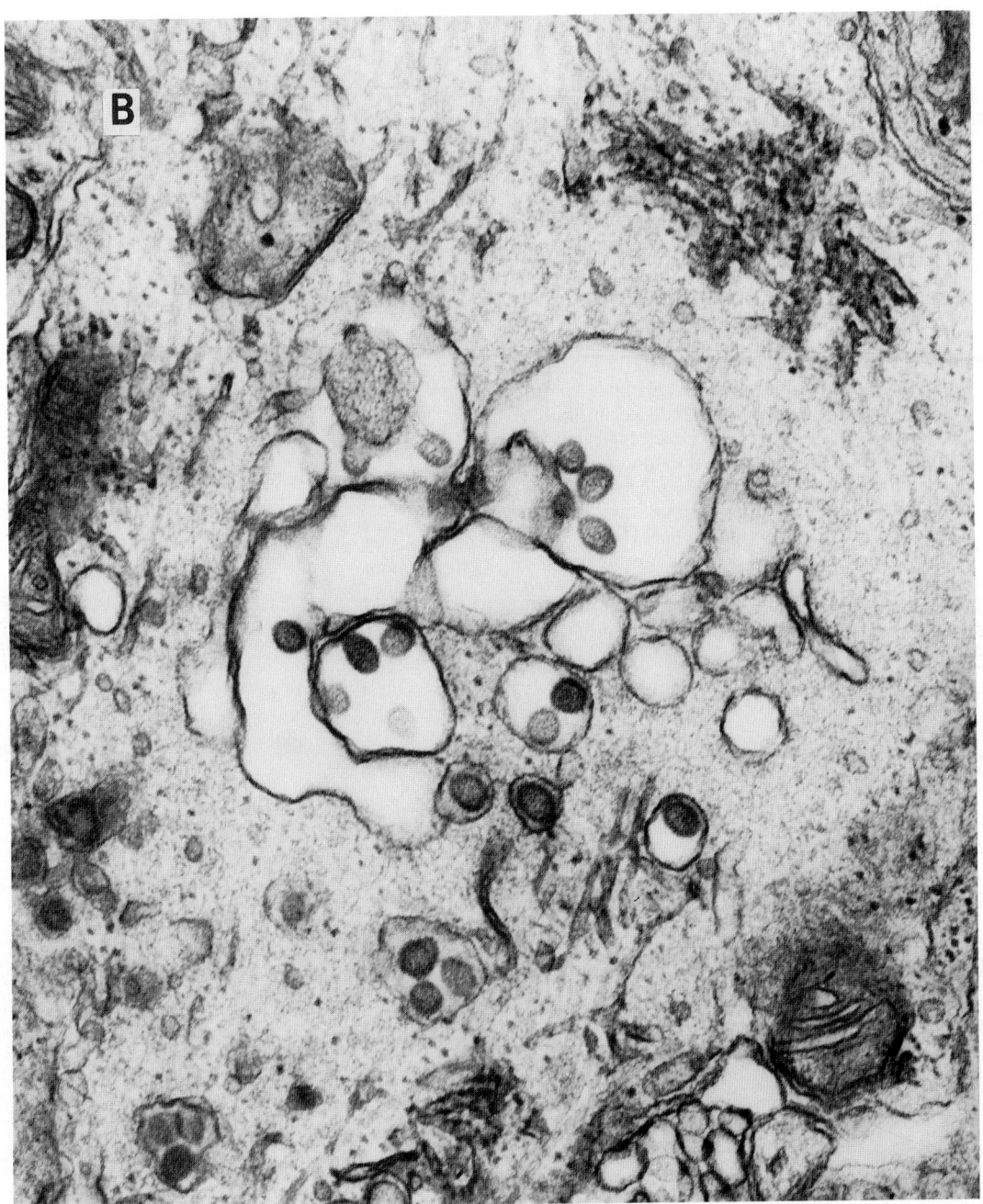

FIGURE 3 *(Continued)*

EVALUATION AND INTERPRETATION OF TEST RESULTS

Unlike herpesviruses, adenoviruses, and other viruses that produce persistent or latent infections, nearly all human arboviral infections are acute and are terminated by viral clearance. Thus, the recovery of an arbovirus from an ill patient is virtually diagnostic. However, rare subacute, progressive, or recrudescent infections and experimental models of persistent infection have been reported for Russian spring-summer encephalitis, West Nile fever, and JE. One study found circulating JE virus-infected peripheral mononuclear cells in patients who had experienced a recrudes- cence of symptoms months after recovery. Persistent Marburg virus infection has also been reported. Many other arboviruses and related zoonotic viruses, e.g., hantaviruses and other *Bunyaviridae*, arenaviruses, *Rhabdoviridae*, and orbiviruses, also produce persistent infections of animals, but it is not known whether infections with these and related viruses (other than rabies) can be established as persistent or latent infections in humans.

The effect of human immunodeficiency virus (HIV) infection and resulting immunodeficiency on the clinical course of arboviral infections is unknown. One unpublished study found that viremia after YF vaccination was sustained for 2 weeks, which is well beyond the usual brief period of several days. Another report suggested that HIV-infected patients may have been at higher risk for acquiring illness in a StLE outbreak. Although AIDS in Africa and Asia is occurring in areas where numerous pathogenic arboviruses are also transmitted, no studies have reported on the occurrence of opportunistic arboviral infections.

Serologic results should be interpreted in view of the epidemiologic history of exposure and the immunization history of the patient. Vaccines for YF, JE, and TBE are commercially available in the United States or abroad, and experimental vaccines for Kayasunar Forest disease, Argentine hemorrhagic fever, hemorrhagic fever with renal syndrome, and Rift Valley fever may be accessible in some circumstances. A vaccination history should be sought, because evidence of immunization might mitigate against certain diagnoses and because vaccine-induced antibodies can interfere with the interpretation of serologic test results. The influence of HIV-associated immunosuppression on the serologic response to arboviral infections is unknown; however, results in immunosuppressed patients should be interpreted with caution. Some HIV-infected patients without clinical evidence of immunosuppression have responded normally to YF vaccine. Hepatitis C virus has recently been identified as a *Flaviviridae*; however, there are no serologic cross-reactions in either direction between hepatitis C virus (in a new-generation ELISA) and arthropod-borne flaviviruses.

REFERENCES

1. **Beaty, B. J., C. H. Calisher, and R. W. Shope.** 1989. Arboviruses, p. 797–856. *In* N. J. Schmidt and R. W. Emmons (ed.), *Diagnostic Procedures for Viral, Rickettsial and Chlamydial Infections*, 6th ed. American Public Health Association, Washington, D.C.

2. **Beran, G. W., and J. H. Steele (ed.).** 1994. *Handbook of Zoonoses*, 2nd ed., section B: Viral. CRC Press, Boca Raton, Fla.

3. **Centers for Disease Control and Prevention and National Institutes of Health.** 1993. *Biosafety in Microbiological and Biomedical Laboratories*, 3rd ed. HHS publication no. (CDC) 93-8395. U.S. Government Printing Office, Washington, D.C.

4. **Karabatsos, N.** 1985. *International Catalogue of Arboviruses Including Certain Other Viruses of Vertebrates*, 3rd ed. American Society of Tropical Medicine and Hygiene, San Antonio, Tex.

5. **Monath, T. P. (ed.).** 1989. *The Arboviruses: Epidemiology and Ecology*, vol. 1–5. CRC Press, Boca Raton, Fla.

6. **Tsai, T. F.** 1991. Arboviral infections in the United States. *Infect. Dis. Clin. N. Am.* **5:**73–102.

7. **Tsai, T. F.** 1992. Arboviruses, p. 606–618. *In* N. R. Rose, E. C. Macario, J. L. Fahey, H. Friedman, and G. M. (ed.), *Manual of Clinical Laboratory Immunology*, 4th ed. American Society for Microbiology, Washington, D.C.

8. **Tsai, T. F.** 1994. Arboviruses and related zoonotic viruses, p. 1266–1288. *In* F. J. Oski (ed.), *Principles and Practice of Pediatrics*, 2nd ed. J. B. Lippincott, Philadelphia.

Rabies Virus

JEAN S. SMITH

86

CLINICAL BACKGROUND

Rabies is a fatal infection of the central nervous system acquired most often through virus transmitted by the bite of a rabid animal. Endemic dog rabies presents the greatest risk to humans, particularly in Africa, Asia, and Latin America, where 50,000 to 60,000 dog rabies cases are reported each year (30). In the United States, rabies control programs have been effective in reducing the number of dog rabies cases to fewer than 200/year; however, a large reservoir of rabies exists in wild animals (more than 5,000 cases per year in a variety of wild carnivore and chiropteran species [16]) and more than 10,000 persons receive antirabies prophylaxis each year in the United States for contact with potentially rabid animals (13).

Rabies virus is present in the saliva of rabid animals and is transmitted to humans through bite wounds or open cuts in skin or mucous membranes. Infection is prevented by the administration of rabies immunoglobulin and a series of five injections of rabies vaccine (1). Since humans exposed to rabies virus ordinarily know when their exposure occurs and can seek treatment promptly, disease prevention is ensured in almost every case. In the United States, human rabies is rare (only 15 cases in the last 10 years). In areas of the world where postexposure prophylaxis is unavailable, human rabies is a common occurrence and probably exceeds 25,000 cases per year (30).

One to two million animal bites per year are treated by physicians in the United States (18). Fortunately, only a small proportion of these bites involves a risk of rabies virus infection. The most important function of the rabies laboratory is to rapidly and accurately identify rabies virus-infected animals so that appropriate rabies preventive measures can be initiated.

DESCRIPTION OF THE AGENT

The rabies virion contains a single-stranded, negative-sense RNA genome of approximately 12,000 nucleotides encoding five proteins from an equal number of monocistronic mRNAs (31). Electron microscopy has shown that the virus particle has the characteristic structure of members of the family *Rhabdoviridae*; it is a rigid, bullet-shaped virus approximately 180 nm in length and 75 nm in diameter. The external surface of the virus consists of a lipid bilayer envelope derived from the host cell plasma membrane and transmembrane glycoprotein spikelike projections encoded by the virus. The surface glycoprotein (or G protein) is thought to bind specifically to cellular receptors and to confer the neurotropism observed in infected animals. On the inner surface of the viral envelope, closely associated with the membrane-bound glycoprotein, is a second membrane or matrix (M) protein thought to play a role in virus budding. The internal, nucleocapsid core of the virus particle is a helix of RNA associated with a phosphorylated nucleoprotein (N protein), a nucleocapsid-associated phosphoprotein (NS protein), and a transcriptase protein (L protein). By analogy with vesicular stomatitis virus, the only rhabdovirus whose biochemistry is known in detail, this RNA-nucleoprotein complex (RNP) is thought to function in the transcription and replication of the virion RNA.

Of the five rabies virus proteins, only the N and G proteins serve important roles in diagnosis and treatment. The G protein is responsible for the induction and binding of neutralizing antibodies. Vaccines containing only the G protein, either as purified viral protein or recombinant DNA-derived protein, confer protective immunity to lethal infection with rabies virus (29). Detection of antibody to this protein is used to assess vaccine efficacy (4).

The internal proteins of rabies virus were thought to contribute little or nothing to protection from rabies virus infection until recent work showed that vaccines containing purified ribonucleoprotein or recombinant DNA-produced N protein would protect some animals from a lethal challenge of rabies virus (6). The immunity conveyed by this protein is not as effective as G-protein-derived immunity, and the mechanism by which the N protein confers protection is not well understood.

Accumulations of viral RNP form the intracytoplasmic inclusions diagnostic for rabies virus infection (21) and can be observed directly by histopathologic means (the Negri body) or indirectly by their reaction with antiserum to the RNP complex or with antibodies specific for the N protein.

COLLECTION AND STORAGE OF SPECIMENS

Safety Recommendations for Specimen Collection

All material collected for rabies diagnosis should be considered infectious, and appropriate handling and shipping precautions should be taken (2). Preexposure rabies immunization is recommended for veterinarians, animal control personnel, and diagnosticians (1). Those assisting in the removal of the brain from a rabies-suspect animal should wear heavy rubber gloves, a face shield, and protective clothing.

Animal Rabies Samples

With the exception of small rodents and lagomorphs, which are not commonly rabid, wild animals that have attacked and bitten a human or a domestic animal are always considered potentially rabid. They should be killed at once, and the brain should be submitted to the laboratory for testing. Stray or unwanted domestic dogs involved in unprovoked biting incidents are also killed and tested for rabies. Healthy domestic dogs or cats that can be confined by their owners and observed for 10 days are not killed; however, any illness in the animal during confinement or before release should be evaluated by a veterinarian and reported to the local health department. If signs suggestive of rabies develop, the animal should be humanely killed, and its head should be removed and shipped to a qualified laboratory for testing (1).

Examination of brain tissue is the only reliable method of rabies diagnosis. Intra vitam diagnosis of rabies in animals, including methods for virus isolation from saliva, viral antigen detection in cutaneous nerves, and serology, may yield false-negative results in some clinically rabid animals (9).

Circumstances may dictate what method of euthanasia is used, but unnecessary damage to the head should be avoided. Large animals should be decapitated, and only the head should be submitted for testing. Specimens should be placed in a suitable watertight container that is tightly sealed. This container, with absorbent material, is then placed in a large, watertight, insulated container. If transit time to state or regional diagnostic laboratories is less than 2 days, specimens may be refrigerated with cold packs. For longer transit times, freezing of the sample on dry ice is recommended. Alternatively, suspension of approximately 1-cm-square pieces of brain samples in 50% glycerine-saline will stabilize the virus in samples shipped on wet ice or cold pack. (Do not freeze samples stored in glycerine-saline.)

Because antirabies treatment is usually delayed pending the results of laboratory tests, the sample should be processed immediately. Brain tissue for confirmatory tests can be stored for a month or longer at −20°C (freezers with autodefrost cycles should not be used for specimen or reagent storage). Rabies virus samples are viable indefinitely at −70°C or below.

The clinical laboratory is occasionally asked to examine specimens that have been handled improperly during transit or that are partially decomposed owing to delay before the animals were found. Positive results on these samples are reliable, but the dependability of negative results varies with the method used for diagnosis (antigen detection is more sensitive than isolation for decomposed samples) and the strain of virus infecting the animal (17). As a general rule, a negative finding is reliable if the condition of the brain tissue is such that intact portions of the brain stem, cerebellum, and hippocampus can be removed from the cranium and if thin, uniform, touch impressions for antigen detection can be made from each section.

Preservation of tissues by fixation in formalin is not recommended if rabies diagnosis is desired. Although a number of procedures for the examination of formalin-fixed brain tissue are available (14, 28), they are inadequate replacements for examination of fresh tissue. Immunofluorescent-antibody tests of formalin-fixed tissue are less sensitive than tests of fresh tissue, and artifacts in the enzyme-mediated detection system are difficult to control for and may result in false-positive tests. The only advantage of formalin fixation is that rabies virus in the tissue is inactivated, thereby reducing the risk of exposure to the specimen during handling; however, this is also the major limitation of the test, since questionable results cannot be confirmed by virus isolation.

Human Rabies Samples

Although human rabies is an extremely rare disease in the United States, rabies should be considered as a possible diagnosis in cases of viral encephalitis of unknown etiology. An early diagnosis, although unlikely to affect patient outcome, can significantly reduce the number of potential exposures to the virus during contact with the patient and permit the identification of persons who are candidates for rabies prophylaxis (11).

Rabies virus antibody does not appear until late in the clinical course; therefore, the most successful diagnostic tests are for viral antigen or nucleic acid. Methods for the intra vitam diagnosis of rabies are based on previous studies of rabies pathogenesis that show that after multiplication in the central nervous system, rabies virus moves centrifugally through the peripheral nervous system. Saliva, cutaneous nerves, or other tissues can be tested for the presence of infectious virus, viral nucleic acid, or viral antigen.

The peripheral nerves most accessible for biopsy are the cutaneous nerves around hair follicles (24). A section of skin 5 to 6 mm in diameter should be taken from the posterior region of the neck at the hairline. The biopsy specimen should include a minimum of 10 hair follicles. The specimens should be placed in gauze moistened with sterile water and enclosed in a sealed container (e.g., test tube) to prevent desiccation. No preservative or fluid should be used. The specimen should be shipped on dry ice.

Saliva can be collected by using a sterile eyedropper pipette. If the patient does not salivate, the oral cavity can be swabbed with a cotton applicator. Tracheal aspirates rarely contain rabies virus and should not be submitted for rabies virus tests. The saliva sample or swab should be enclosed in a sealed container or test tube. No preservative or fluid should be used. The specimen should be shipped on dry ice.

Rabies virus antibody has not been found in the serum or cerebrospinal fluid (CSF) of human patients earlier than days 8 to 10 of clinical illness (which may be several months after the patient's rabies exposure). Specimens collected before day 8 are usually not helpful except as the first of paired samples. If no vaccine or antirabies serum has been given, the presence of antibodies to rabies virus in the serum is diagnostic. Antibody may not appear in spinal fluid for a day to a week (or longer) after antibody is detected in the serum. Rabies virus antibody in the CSF, regardless of the rabies immunization history, suggests a rabies virus infection, since no significant amount of rabies virus antibody has been detected in the CSF of rabies virus-vaccinated subjects. At least 0.5 ml of serum or CSF should be

TABLE 1 Distribution of rabies virus in medullas, cerebella, and hippocampi of animals diagnosed with rabies from 1989 to 1992 by the Texas Department of Health, Austin[a]

Animal	Total no. positive	No. positive for rabies virus antigen							% with focal antigen
		Medulla, cerebellum, and hippocampus	Medulla only	Medulla and cerebellum	Medulla and hippocampus	Cerebellum only	Cerebellum and hippocampus	Hippocampus only	
Bat	196	195	1	0	0	0	0	0	0.5
Cat	100	95	2	2	1	0	0	0	5.0
Coyote	130	128	0	0	2	0	0	0	1.6
Dog	259	250	0	0	7	0	0	2	3.5
Fox	132	131	1	0	0	0	0	0	0.8
Horse	41	35	1	0	3	1	1	0	14.6
Raccoon	35	33	1	0	0	0	0	1	5.7
Skunk	688	687	1	0	0	0	0	0	0.1
Other	103	103	0	0	0	0	0	0	0
Total	1,684	1,657	7	2	13	1	1	3	1.6

[a]Data are from reference 18a.

collected; no preservative should be added. The samples should be shipped on dry ice.

Although a positive test result on any of these samples is an indication of rabies virus infection, no single tissue sample has been positive in every case of human rabies, and therefore, every diagnosis should include tests of a skin biopsy sample, saliva, and serum (CSF should be included if the patient has received antirabies treatment). In some rabies cases, all early samples were negative. If rabies is still suspected despite negative results, additional samples should be taken later in the clinical course.

Corneal epithelium, while positive for rabies virus antigen in some patients, is difficult to sample correctly, especially from comatose patients, and is no longer recommended as a suitable diagnostic sample. Rabies virus has been isolated from the spinal fluid of some patients, but this tends to be possible very late in the clinical course, if at all. Because of the rarity of the disease and the lack of effective treatment, a brain biopsy is not indicated. However, biopsy samples negative for herpes encephalitis should be tested for evidence of rabies virus infection.

Postmortem diagnosis of rabies is made by examination of brain tissue. Portions of the medulla (brain stem), cerebellum, and hippocampus should be frozen and shipped on dry ice to a public health laboratory. Preservation of tissues by fixation in formalin is not recommended if rabies diagnosis is desired.

DIRECT EXAMINATION OF CLINICAL SAMPLES FOR RABIES VIRUS ANTIGEN

Because of its sensitivity and specificity, the speed and ease with which it may be performed, and the economy of its reagents, the direct immunofluorescent-antibody (dIFA) test for rabies virus antigen in brain tissue (5, 26–28) is the preferred test for rabies diagnosis.

dIFA Test

Rabies virus in the brains of infected animals produces intracytoplasmic inclusions of various shapes (dustlike particles <1 μm in diameter and/or large, round to oval masses and strings 2 to 10 μm in diameter). When specifically stained with a fluorescein isothiocyanate (FITC)-labeled

antibody, these inclusions appear smooth with very bright margins and a somewhat less intensely stained central area. Positive tests are graded by the intensity of the staining (+4, a glaring, apple green brilliance; +1, dull but still noticeably apple green) and by the amount of antigen present in a tissue impression (+3 to +4, a massive green infiltration of large inclusions and dustlike particles in every area of the impression; +1 to +2, isolated small inclusions in only a few microscopic fields).

Thin-tissue impressions are made in duplicate by lightly touching cut sections from each of three areas of the brain (medulla, cerebellum, and hippocampus) to clean microscope slides. Before staining, impressions are air dried for 10 to 15 min and fixed in acetone at −20°C for 1 to 4 h or overnight. After removal from the acetone, the slides are allowed to air dry for 10 to 15 min. (Acetone-fixed slides should be handled as containing potentially infectious material.)

Each impression is then reacted with FITC-conjugated anti-rabies virus antibody for 30 min at 37°C in a humidified chamber. The slides are then rinsed and mounted with a phosphate-buffered glycerol mounting medium (pH 8.5). A standard fluorescence microscope with appropriate filters is used to observe fluorescing rabies virus antigen in infected tissues, typically at a magnification of ×400 to ×1,000. Duplicate slides are read independently, preferably by different microscopists, and the results are recorded and compared.

Quality Control in the dIFA Test

Although standardized procedures for the dIFA test exist (5, 27), a recent survey of 103 rabies laboratories in the United States (18b) revealed that several potentially deleterious modifications to the standard procedure are in use.

The modification most likely to compromise a diagnosis is the failure to include in each test slides made from the medulla, the cerebellum, and the hippocampus of a rabies-suspect animal. (Of the 103 surveyed laboratories, 34 do not routinely examine all three areas.) In a small number of cases (Table 1), rabies virus-specific inclusions are found in only one or two areas of the brain. Even when rabies virus antigen is present in all areas of the brain, its distribution is frequently uneven, especially in nonreservoir domestic-animal cases, and a diagnosis can be made with greater confidence if all three areas are examined. For example, a +4 antigen distribution was seen in all areas of the brain of 684 of 688 rabid skunks diagnosed in Texas from 1989 to 1992. In contrast, a uniform antigen distribution was seen in only 12 of 41 horse brains (18a).

A fixation time of 10 min or less was used by 13 of the surveyed laboratories, and 13 laboratories used a fixation temperature of 20 to 25°C. Fixation times of less than 1 h and at temperatures greater than 4°C result in less brilliant and possibly incomplete staining. In emergency situations, a positive diagnosis can be made by examination of slides fixed for shorter times. A second test of slides fixed for the longer times should be used to confirm negative results.

The pH of the mountant was below 8.0 in 18 of the surveyed laboratories. Phosphate-buffered mountants with a pH of >8 should be used; fluorescein is quenched at pH <8, and the instabilities of carbonate-buffered mountants can be particularly deleterious to monoclonal antibody reagents (7). Rinse buffers should have a pH of 7.6 to 7.8.

An objective with a magnification of less than ×40 was used in 15 laboratories. Slides should be examined with 10× oculars and a 40× to 63× objective. Lower magnifications may not be sufficient for all stained rabies virus antigenic forms to be observed.

Virus isolation was used to confirm at least some of the dIFA test results in 40 laboratories. The capability for confirmatory isolation (25), especially of samples for which the dIFA test results are equivocal, is essential to maintaining diagnostic confidence. In areas of high rabies endemicity, dIFA-negative tissue from large animal brains should also be routinely confirmed; the larger sample size afforded by isolation techniques results in a more sensitive assay for rabies virus. Each isolation test should include dIFA-positive samples to verify that culture conditions for virus isolation are appropriate.

Additional aspects of the dIFA test not surveyed but equally important are test controls, conjugate evaluation, and methods for recognizing and avoiding cross-contamination of samples.

Each test should include slides made from known positive and negative animals. The positive control slide serves as a sensitivity control for the anti-rabies virus conjugate, and the negative control slide is used to detect nonspecific staining or adherence of aggregates of labeled antibody that may be mistaken for rabies virus antigen.

Repeat tests should be performed on samples with irregular distributions of antigen (for example, a sample in which only one area of the brain was positive) or in which the inclusions show irregular morphology or stain with less than +3 intensity. On these occasions, slides should be remade from the stored brain tissue, and the test should be repeated. If possible, the repeat test should include a second anti-rabies virus conjugate prepared with antibody from a different source. Additional specificity controls should be included on the repeat test. When these controls are used, two impressions are made on each slide. The specificity of the reaction is ensured by adsorbing the antirabies virus conjugate with either normal brain suspension (on one impression) or rabies virus-infected brain suspension (on the other). When the slide contains rabies virus antigen, rabies virus antibody remaining in the conjugate adsorbed with normal brain tissue should react with the antigen on test slides, and tissue stained with the conjugate adsorbed with rabies virus antigen should not fluoresce. Nonspecific staining should be observed in tissue stained with both adsorption controls. (*Caution:* Adsorption with rabies virus-infected brain tissue does not control for nonspecific adherence of aggregates of FITC-labeled rabies virus antibody and may contribute to the nonrecognition of the problem by specifically removing the antibody aggregates.)

FITC-labeled anti-rabies virus antibody conjugates can be prepared against whole rabies virions or purified RNP or as a mixture of monoclonal antibodies reactive with the N protein. These preparations are available commercially (BBL Microbiology Systems, Cockeysville, Md.; Centocor Inc., Malvern, Pa.; Pasteur Diagnostics, Paris, France). Care should be taken to avoid repeated freeze-thaw cycles of the anti-rabies virus conjugates and long-term storage at +4°C. These practices promote the formation of antibody aggregates that nonspecifically adhere to tissue, especially to thick impressions. Because of the variable morphologic features of rabies virus-specific inclusions and their sometimes sparse distribution, these aggregates are easily mistaken for specific inclusions. If necessary, aggregates can be removed by centrifugation at 10,000 × g for 20 min. The working stock of the conjugate should be stored at +4°C and discarded after 5 days.

Each new lot of anti-rabies virus conjugates should be titrated in twofold dilutions to determine its effective limit. At the optimum dilution, antigen in a positive control slide will stain with a glaring, apple green brilliance (+4 staining intensity), and no background or nonspecific fluorescence will be observed on negative tissue. It is very important that slides used for conjugate evaluation contain the small dustlike rabies virus inclusions along with the larger stringlike inclusions. Detection of the dustlike particles requires a stronger dilution of conjugate and thus is a more stringent evaluation of the reagent. To ensure test sensitivity, the working dilution of the conjugate should be at least twofold more concentrated than the last dilution that stains these small inclusions with +4 intensity. The working dilution should be determined for conjugate diluted in both phosphate-buffered saline (PBS)–0.75% bovine serum albumin (BSA) and mouse or rabbit brain adsorbing suspensions.

One of the most difficult problems to address in rabies diagnosis is cross-contamination of negative samples with small amounts of brain material from a strongly positive sample. A contamination problem is usually not recognized until rabies virus antigen is seen in the brain of an animal in which a diagnosis of rabies is unusual or unexpected (e.g., a vaccinated domestic animal or a rodent). Less often, the contamination is recognized by an increase in the number of samples in which antigen is present in only one area of the brain (Table 1).

The contamination may arise during any of several steps in the diagnostic procedure. All necropsy instruments should be autoclaved, and a separate set should be used for each animal necropsied. Although it may seem primitive, the use of hammers, chisels, and large butcher knives may be preferable to saws for opening the cranium. These instruments are relatively inexpensive and can be maintained in sufficient quantity so that a separate instrument is used for each animal. If used, Stryker saws or bone saws cleaned in a cold sterilant between use should always be considered a possible source of contaminating material. Cold sterilants are effective for surface decontamination only, and while they may inactivate rabies virus, they do not always denature rabies virus antigen in adherent tissue. Deposition of this antigen on subsequent samples may result in false-positive test results. When removing tissue for slide preparation, care should be taken to avoid areas of the brain that the saws may have penetrated. If the entire brain is not saved for confirmatory testing, precautions against contamination should be taken in removing brain sections for storage. Petri dishes or ointment tins are preferable to test tubes for tissue storage so that individual areas of the brain are not in contact with each other.

Rabies virus-infected tissue also accumulates in the vessels used for acetone fixation of slides and can be a source for false-positive results. The acetone should be replaced after each set of slides is prepared, and the jars should be cleaned and autoclaved.

The record of test results for each sample should include the amount of rabies virus antigen variously observed in slides made from the medulla, cerebellum, and hippocampus; the observation of antigen in only one section of the brain (especially the hippocampal section) is unusual (Table 1) and should elicit repeat testing of the sample.

dIFA Test of Skin Biopsy Samples

Serial frozen sections of skin biopsy samples (3 to 6 μm thick) are fixed in cold acetone for 10 min, air dried for 10 min, and stained by dIFA as detailed above. Rabies virus-specific intracytoplasmic inclusions may be seen in the cutaneous nerves surrounding the hair shaft. Serial sections should show antigen throughout the nerve fiber. Only rarely is antigen seen in contiguous tissue.

Other Tests for Rabies Virus Antigen

Histologic stains for Negri bodies (25), although fast, inexpensive, and specific, detect only 50 to 80% of dIFA-positive samples. This lack of sensitivity severely limits the usefulness of these stains except in situations where dIFA is unavailable. Enzyme-linked immunosorbent assays (ELISAs) (25) are better, more economic alternatives to histologic methods for those laboratories without fluorescence microscopes. However, these tests lack the specificity gained from direct immunofluorescence observation of rabies virus intracytoplasmic inclusions, and the relatively high cutoff between negative and positive responses somewhat diminishes the sensitivity of the assay.

ISOLATION OF RABIES VIRUS

Cell Culture

Rabies virus is poorly cytopathic. Evidence of virus growth is by dIFA detection of viral antigen in acetone-fixed cell monolayers. As observed in our laboratory and others, murine neuroblastoma (MNA) cells are the most sensitive cells to street rabies virus strains. These cells are available from BioWhittaker, Walkersville, Md. A similar cell line (CCL131) may be obtained from the American Type Culture Collection. These cells prefer an acidic medium with increased vitamins, which may be prepared from commercially available stocks. Eagle minimum essential medium (Glasgow MEM; Gibco Laboratories Life Technologies Inc., Grand Island, N.Y.) is supplemented with an additional 2× vitamins, 2× glutamine, and 10% fetal bovine serum. Each new lot of the bovine serum should be checked for inhibitors of rabies virus. Penicillin (100 U/ml), streptomycin (100 μg/ml), and amphotericin B (0.25 μg/ml) are also added. The final concentration of sodium bicarbonate should be 0.75 mg/ml. This medium is optimized for cell growth in closed culture. Open-flask cultures or cell culture slides (e.g., for virus isolation) should be incubated in an atmosphere of 0.5% CO_2 at 37°C and 90% humidity.

Higher CO_2 levels require a higher concentration of sodium bicarbonate in the medium.

Brain material for virus isolation is prepared in EMEM as a 20% suspension of medulla, cerebellum, and hippocampus. After centrifugation at 500 × g for 10 min, 0.5 ml of the supernatant is added to a suspension of 2×10^6 MNA cells in 1.5 ml of EMEM. The cells and virus are incubated for 1 h at 37°C. Unless there is obvious evidence of bacterial contamination of the sample, the inoculum may be left on the cells, and the cell suspension may be divided for incubation as flask or slide (or microtiter plate) cultures. Saliva samples for virus isolation should be diluted 1:3 into media with antibiotics, vortexed vigorously, frozen and thawed once, and centrifuged at 500 to 1,000 × g for 20 min. The supernatant (0.5 ml) should be inoculated as described above.

After 40 to 48 h of culture, the slides (or plates) are fixed in cold (−20°C) acetone for 5 min and examined by the dIFA test. Most samples positive by the dIFA test are positive for virus isolation at this time. Flask cultures of samples of expected low initial infectivity or cultures for confirmation of negative results should be given fresh medium after 3 days, and the culture should be continued for an additional 3 to 4 days. At this time, the primary flask culture is trypsinized and used to prepare new slide cultures. There is no need to include supernatant from the primary culture. The secondary slide cultures are incubated for 3 to 4 h (to allow the cells to attach to the slide) or overnight and examined by the dIFA test as before. Samples that are negative upon examination of these secondary cultures are considered negative.

Virus Isolation by Mouse Inoculation

A 20% suspension of approximately 0.3 g of tissue from medulla, cerebellum, and hippocampus is prepared in PBS containing a protein stabilizer (such as 0.75% BSA fraction V). Penicillin (500 U/ml) and streptomycin (2 mg/ml) should also be added to prevent animal death from bacterial contamination of the tissue suspension. Five weanling mice or two families of suckling mice are then inoculated intracerebrally with a 1:2 dilution of centrifuged suspension (500 × g for 5 min) and observed daily for 30 days for signs of rabies (trembling, humping, paralysis, or prostration). A 7- to 20-day incubation period is expected in mice inoculated with street virus preparations, but longer incubation periods are sometimes observed. Virus can be detected by dIFA testing of the brains of inoculated mice well in advance of the appearance of clinical signs or death. An earlier diagnosis can be made if extra mice are inoculated and if their brains are examined at daily intervals postinoculation.

DETECTION OF VIRAL RNA BY MOLECULAR TECHNIQUES

For a variety of reasons, molecular techniques for rabies diagnosis have not achieved widespread use (8, 15, 19). The newer method of nucleic acid detection by reverse transcription of RNA and amplification of cDNA by PCR is as sensitive as immunofluorescence in most instances but much more expensive to perform. No comparative evaluation has been performed, and the presence of numerous genetic variants of rabies (>15% nucleotide difference) in wildlife reservoirs in the United States makes primer selection a problem. Of the primers listed in Table 2, the pair

TABLE 2 Primers for reverse transcription and PCR amplification of rabies RNA

Sequence	Genome position[a]	Use[b]
CTACAATGGATGCCGAC	66	RT-N
GTAGGATGCTATATGGG	1013	RT-N
TTCTTATGAGTCACTCGAATATGTCTTGTTTAG	1426	PCR-N
TTGACGAAGATCTTGCTCAT	1533	PCR-N
ATTTACACGATACCAGACAA	3383	RT-G
CTGAGACGTCTGAAACTCAC	4278	PCR-G
CATCTTGAAGTAAGTCTC	5434	PCR-G

[a]Genome position indicates the 5' nucleotide of the primer in reference to the sequence of SAD B19 (3).
[b]RT, reverse transcription; N, nucleoprotein gene; G, glycoprotein gene.

located at positions 66 and 1533 has the widest range of variant recognition (unpublished observation). Unfortunately, the large size of the amplified product (1,468 bp) reduces the sensitivity of the reaction for samples in which RNA is degraded or limiting. The greatest utility of the technique in rabies diagnosis has come from epidemiologic studies in which precise identification of a rabies variant has provided information about patterns of disease transmission (23).

SEROLOGY

The sensitivity and specificity of the different assay methods for rabies antibody vary greatly (22). The suitability of a particular method is determined by what the results will be used for. Since the surface G protein is the most important protein for conferring immunity to lethal infection with rabies virus, assays of vaccine immunogenicity should measure antibody to this protein. This is most commonly done as a measurement of neutralizing antibody, but G-protein-specific ELISAs are also applicable. Diagnostic tests should be able to detect any antibody present, regardless of the viral protein that induced it, and the ELISA with whole-virus immunosorbents is the most sensitive assay available.

All persons tested 2 to 4 weeks after completion of the recommended schedule for preexposure or postexposure rabies prophylaxis demonstrate an antibody response to rabies virus, and no treatment failures have been documented in the United States (1). Therefore, it is not necessary to document seroconversion to vaccination unless the patient is immunocompromised. Serology may be useful, however, for persons at continued risk who are scheduled to receive booster doses of vaccine 6 months to 2 years after preexposure immunization. A booster dose of vaccine is unnecessary for patients with standing antibody to rabies (1).

Neutralization Test

The standard test for antibody to G protein measures the ability of a serum to neutralize a challenge inoculum, usually indicated by a reduction in the number of fluorescent-antibody-stained microscopic foci of infected cells. There are several modifications of this procedure (22).

First, an appropriate inoculum of the challenge virus standard strain of rabies virus is determined by serial dilution. Tenfold increments are prepared in eight-well Lab-Tek slides, leaving a final volume of 0.1 ml per well. An equal volume of MNA cells (50,000 to 100,000 cells) in EMEM (see Cell Culture above) is added. After a 20-h incubation, infected cells are stained by the fluorescent-antibody technique. At an ocular magnification of ×160,

approximately 50 microscopic fields may be observed in each well of an eight-well slide. One infectious unit of virus is that contained in the virus dilution at which 50% of the observed fields contain one or more infected cells (50% focus-forming dose). In a rabies virus antibody titration, each serum dilution receives 32 to 100 50% focus-forming doses of virus. The antibody titer is the reciprocal of the serum dilution that reduces the challenge virus to one 50% focus-forming dose (a 97 to 99% reduction in virus). The number of international units of rabies antibody in a test serum is determined by comparison with a titration of a rabies reference serum standard included in each test. Serum drawn at 2 to 4 weeks postvaccination should contain 0.5 IU of antibody (approximately equivalent to complete neutralization of the challenge virus with a 1:25 dilution of serum). Booster doses at 6-month or 2-year intervals are required if a 1:5 serum dilution fails to completely neutralize the challenge.

If large serosurveys are anticipated, it is possible to perform all titrations in microtiter plates and detect changes in virus concentration by enzyme immunoassay. This method permits automation of all pipetting, spectrophotometric reading of color development, and titer calculation by computer analysis of data.

ELISA

Although not yet in wide use, several ELISAs for determination of antibodies have been developed. An ELISA based on whole virus as the immunosorbent is one of the most sensitive assays available. Using such an assay, Savy and Atanasiu (20) were able to detect immunoglobulin M antirabies virus antibody very early in the clinical course of three human rabies cases and well before neutralizing antibodies could be detected. If estimates of vaccine immunogenicity are required, the immunosorbent should contain only the G protein. Simple methods for the purification of G protein have been published (12), and materials for a G-protein-specific ELISA are available in kit form from Diagnostics Pasteur. Although extensive comparative trials have not been conducted, data suggest that the ELISA is a reliable and simple alternative to the neutralization test.

REFERENCES

1. **Centers for Disease Control.** 1991. Rabies prevention—United States, 1991: recommendations of the Immunization Practices Advisory Committee (ACIP). *Morbid. Mortal. Weekly Rep.* **40:**1–19.
2. **Centers for Disease Control and National Institutes of Health.** 1988. *Biosafety in Microbiological and Biomedical Laboratories.* U.S. Government Printing Office, Washington, D.C.
3. **Conzelmann, K. K., J. H. Cox, L. G. Schneider, and H. J.**

Thiel. 1990. Molecular cloning and complete nucleotide sequence of the attenuated rabies virus SAD B19. *Virology* **175**:485–499.

4. **Cox, J. H., B. Dietzschold, and L. G. Schneider.** 1977. Rabies virus glycoprotein. II. Biological and serological characterization. *Infect. Immun.* **16**:754–759.

5. **Dean, D. J., and M. K. Abelseth.** 1973. The fluorescent antibody test, p. 73–84. *In* M. M. Kaplan and H. Koprowski (ed.), *Laboratory Techniques in Rabies.* World Health Organization, Geneva.

6. **Dietzschold, B., H. H. Wang, C. E. Rupprecht, E. Cells, M. Tollis, H. Ertl, E. Heber-Katz, and H. Koprowskl.** 1987. Induction of protective immunity against rabies by immunization with rabies virus ribonucleoprotein. *Proc. Natl. Acad. Sci. USA* **84**:9165–9169.

7. **Durham, T. M., J. S. Smith, F. L. Reid, C. T. Hale-Smith, and M. B. Fears.** 1986. Stability of immunofluorescence reactions produced by polyclonal and monoclonal antibody conjugates for rabies virus. *J. Clin. Microbiol.* **24**:301–303.

8. **Ermine, A., N. Tordo, and H. Tsiang.** 1988. Rapid diagnosis of rabies infection by means of a dot hybridization assay. *Mol. Cell. Probes* **2**:75–82.

9. **Fekadu, M., and J. H. Shaddock.** 1984. Peripheral distribution of virus in dogs inoculated with two strains of rabies virus. *Am. J. Vet. Res.* **45**:724–729.

10. **Fischman, H. R., and F. E. Ward III.** 1969. Infectivity of fixed impression smears prepared from rabies virus-infected brain. *Am. J. Vet. Res.* **30**:2205–2208.

11. **Fishbein, D. B.** 1991. Rabies in humans, p. 519–549. *In* G. M. Baer (ed.), *The Natural History of Rabies,* 2nd ed. CRC Press, Inc., Boca Raton, Fla.

12. **Grassi, M., A. I. Wandeler, and E. Peterhans.** 1989. Enzyme-linked immunosorbent assay for determination of antibodies to the envelope glycoprotein of rabies virus. *J. Clin. Microbiol.* **27**:899–902.

13. **Helmick, C. G.** 1983. The epidemiology of human rabies postexposure prophylaxis, 1980–1981. *JAMA* **250**:1990–1996.

14. **Johnson, K. P., P. T. Swoveland, and R. W. Emmons.** 1980. Diagnosis of rabies by immunofluorescence in trypsin-treated histologic sections. *JAMA* **244**:41–43.

15. **Kamolvarin, N., T. Tirawatnpong, R. Rattanasiwamoke, S. Tirawatnpong, T. Panpanich, and T. Hemachudha.** 1993. Diagnosis of rabies by polymerase chain reaction using nested primers. *J. Infect. Dis.* **167**:207–210.

16. **Krebs, J. W., R. C. Holman, U. Hines, T. W. Strine, E. J. Mandel, and J. E. Childs.** 1992. Rabies surveillance in the United States during 1991. *J. Am. Vet. Med. Assoc.* **201**:1836–1848.

17. **Lewis, V. J., and W. L. Thacker.** 1974. Limitations of deteriorated tissue for rabies diagnosis. *Health Lab. Sci.* **11**:8–12.

18. **Moore, R. M., Jr., R. B. Zehmer, J. I. Moulthrop, and R. L. Parker.** 1977. Surveillance of animal-bite cases in the United States, 1971–1972. *Arch. Environ. Health* **32**:267–270.

18a. **Neill, S., and V. Headley.** Unpublished data.

18b. **Powell, J.** Personal communication.

19. **Sacramento, D., H. Bourhy, and N. Tordo.** 1991. PCR technique as an alternative method for diagnosis and molecular epidemiology of rabies virus. *Mol. Cell. Probes* **5**:229–240.

20. **Savy, V., and P. Atanasiu.** 1978. Rapid immunoenzymatic technique for titration of rabies antibodies IgG and IgM. *Dev. Biol. Stand.* **40**:247–253.

21. **Schneider, L. G., B. Dietzschold, R. E. Dierks, W. Matthaeus, P. J. Enzmann, and K. Strohmaler.** 1973. Rabies group-specific ribonucleoprotein antigen and a test system for grouping and typing of rhabdoviruses. *J. Virol.* **11**:748–755.

22. **Smith, J. S.** 1991. Rabies serology, p. 235–252. *In* G. M. Baer (ed.), *The Natural History of Rabies,* 2nd ed. CRC Press, Inc., Boca Raton, Fla.

23. **Smith, J. S., and H. D. Seidel.** 1993. Rabies: a new look at an old disease. *Prog. Med. Virol.* **40**:82–106.

24. **Smith, W. B., D. C. Blenden, T. H. Fuh, and L. Hiler.** 1972. Diagnosis of rabies by immunofluorescent staining of frozen sections of skin. *J. Am. Vet. Med. Assoc.* **161**:1495–1501.

25. **Sureau, P., P. Ravisse, and P. E. Rollin.** 1991. Rabies diagnosis by animal inoculation, identification of negri bodies, or ELISA, p. 203–217. *In* G. M. Baer (ed.), *The Natural History of Rabies,* 2nd ed. CRC Press, Inc., Boca Raton, Fla.

26. **Trimarchi, C. V., and J. Debbie.** 1991. The fluorescent antibody in rabies, p. 219–233. *In* G. M. Baer (ed.), *The Natural History of Rabies,* 2nd ed. CRC Press, Inc., Boca Raton, Fla.

27. **Velleca, W. M., and F. T. Forrester.** 1981. *Laboratory Methods for Detecting Rabies.* U.S. Government Printing Office, Washington, D.C.

28. **Webster, W. A., and G. A. Casey.** 1988. Diagnosis of rabies infection, p. 201–223. *In* J. B. Campbell and K. M. Charlton (ed.), *Rabies.* Kluwer Academic Publishers, Boston.

29. **Wiktor, T. J., R. I. Macfarlan, K. J. Reagan, B. Dietzschold, P. J. Curtis, W. H. Wunner, M. P. Kleny, R. Lathe, J. P. Lecocq, M. Mackett, et al.** 1984. Protection from rabies by a vaccinia virus recombinant containing the rabies virus glycoprotein gene. *Proc. Natl. Acad. Sci. USA* **81**:7194–7198.

30. **World Health Organization.** 1991. *World Survey of Rabies 27 (for Year 1991).* World Health Organization, Geneva.

31. **Wunner, W. H.** 1991. The chemical composition and molecular structure of rabies viruses, p. 31–67. *In* G. M. Baer (ed.), *The Natural History of Rabies,* 2nd ed. CRC Press, Inc., Boca Raton, Fla.

Enteroviruses

HARLEY A. ROTBART

87

CLINICAL BACKGROUND

The enteroviruses (EVs) are among the most common and most important viral pathogens of humans (6, 26). The paralytic potential of the polioviruses, the prototypic EVs, was recognized as early as the 14th century B.C., as illustrated in Egyptian art. Summer epidemics of paralytic poliomyelitis ravaged the United States through the 1950s. Since the introduction of vaccines in the late 1950s and early 1960s, much of the developed world is now virtually free of poliovirus infections. In developing countries, where vaccines are often either unavailable or ineffective, it has been estimated by the World Health Organization that 4 of every 1,000 school-age children contract paralytic disease due to poliovirus infections (2).

Control of poliovirus infections in much of the world has focused attention on the nonpolio EVs (NPEVs), which include the coxsackieviruses, echoviruses, and newer numbered EVs (Table 1). In the United States alone, the NPEVs are estimated to cause 5 to 10 million symptomatic infections annually (51). In temperate climates, these infections occur during the summer and fall months; young children are the most common victims, as both the incidence and severity of NPEV infections vary inversely with the patient's age. In addition to the actual diseases that they cause, the EVs are of great concern because they mimic other pathogens. Distinguishing EV infections from those due to common bacteria and other viruses on clinical grounds alone is often difficult (6). Hence, unnecessary treatment for other infections is frequently instituted during EV infections.

The EVs are responsible for a wide array of clinical diseases affecting many organ systems (Table 2). It is important to note that no disease is uniquely associated with any specific EV serotype and that no serotype is uniquely associated with any one disease (6, 26). This is true even of paralytic poliomyelitis, which has been associated with numerous NPEV serotypes (12, 26). For that reason, it is sufficient in most circumstances to speak of diseases that "EVs cause" and to diagnose "an EV" in the laboratory without necessarily specifying or identifying the particular serotype. Certain clinical syndromes are indeed more likely to be caused by one or a few serotypes (see below), but significant overlap exists among the serotypes and the diseases they cause.

Most EV infections are asymptomatic (6, 26). The most common symptomatic manifestation of EV infection is a nonspecific febrile illness with or without a rash. This so-called viral syndrome is one of the most numerically important causes of fever among children. When accompanied, as it often is, by upper respiratory symptoms, the "summer cold" is indistinguishable from the same illness caused by rhinoviruses (fellow picornaviruses) in the winter months. By far the most vexing clinical EV syndrome that the physician encounters is aseptic meningitis. The EVs are the most common cause of meningitis in the United States. In young infants with the disease, clinical criteria to distinguish EV meningitis from that due to bacteria and herpes simplex virus are unreliable. As a result, even though EV meningitis is generally benign in outcome and no specific therapy is indicated (or currently available), thousands of children annually are hospitalized and treated with unnecessary antibiotics and antiherpes medications because of the fear that a case of meningitis is not due to an EV (7). Additional acute clinical EV syndromes of significance include encephalitis, poliomyelitis (particularly due to the polioviruses), myocarditis (particularly due to the coxsackievirus B group), hemorrhagic conjunctivitis (particularly due to serotypes coxsackievirus A24 and enterovirus type 70), hand-foot-mouth syndrome, Bornholm disease (pleurodynia), and overwhelming neonatal sepsis (particularly due to the echoviruses and type B coxsackieviruses). The last syndrome is thought to be due to perinatal transmission, either transplacentally or during birth, from mother to infant. Despite the name EV, enteric disease is not a prominent manifestation, although diarrhea and vomiting may be significant manifestations of certain outbreaks of "summer flu" due to the EVs. Hepatitis A virus (HAV) was formerly known also as EV 72 but has since been reclassified in its own genus of picornaviruses and is, appropriately, no longer considered an EV (34); typical EV clinical syndromes have not been associated with HAV infections (HAV is discussed separately in chapter 90 of this Manual). Mild hepatitis, almost always in association with other more significant findings, is common during infection with many EV serotypes.

In addition to the well-recognized acute EV diseases, EVs have been implicated in several chronic illnesses including juvenile onset diabetes mellitus, chronic fatigue syndrome, dermatomyositis and polymyositis, congenital

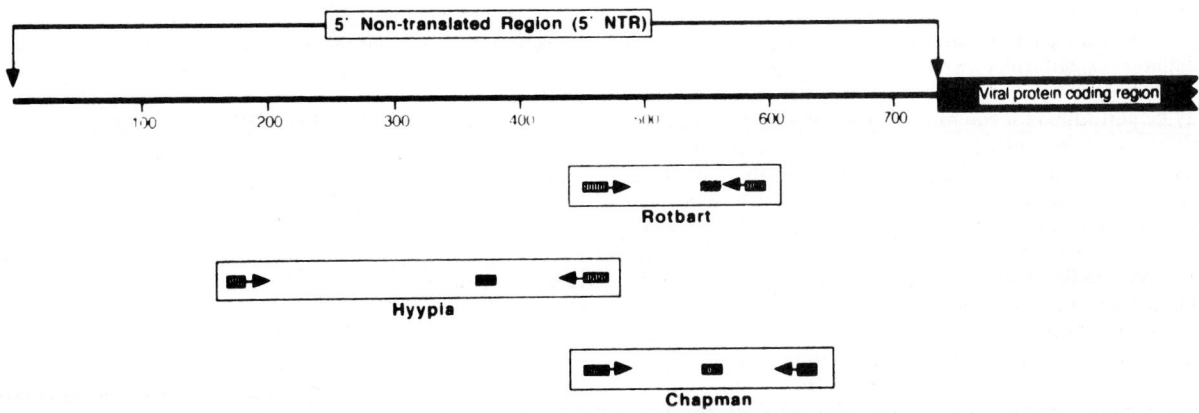

FIGURE 1   Locations, within the 5' nontranslated region of the EV genomes, of the three most completely studied primer and probe combinations for PCR (5, 18, 46).

nation of water and glassware), and reactions are carried out in the presence of RNase inhibitors. Following extraction, reverse transcription of the EV RNA is accomplished with the downstream primer and any of several commercially available reverse transcriptases. PCR itself is performed with *Taq* polymerase in an automated thermal cycler for 30 to 35 cycles of denaturation, annealing, and extension. Detection of the amplified product can be performed by any of several conventional techniques that employ a confirmatory hybridization step using the oligomeric probe molecule, including ethidium bromide-stained agarose gels with confirmatory Southern hybridization, dot blot hybridization, and microtiter plate hybridization assay. The wide array of reported conditions for EV PCR extraction, reverse transcription, amplification, and detection have recently been reviewed (45). The nucleotide dUTP and the enzyme uracil *N*-glycosylase should be used in all PCR reactions (24), and other standardized precautions should also be taken to minimize the risk of carryover contamination from one experiment to another (see chapter 13 of this Manual).

ISOLATION OF VIRUS

The original subclassification of the EVs was based on the ability of individual serotypes to grow in various cell culture and animal systems. Hence, coxsackievirus type A serotypes were characteristically able to replicate in suckling mice, with resultant diffuse myositis and flaccid paralysis. However, these viruses were not readily grown in tissue culture cells derived from monkey or human tissues. In contrast, type B coxsackieviruses grow readily in tissue cultures of both simian and human origins as well as in suckling mice; the pathology in the latter differs from that of coxsackieviruses A in that the myositis with B serotypes is more focal and direct infection of brain, myocardium, pancreas, and liver also occurs. Neither group of coxsackieviruses is pathogenic for monkeys. The echoviruses were defined by their ability to grow only in simian-derived tissue culture cells but not at all in animal systems. The polioviruses grow most prolifically in cells of human and simian origins and in monkeys; they do not grow in murine models. Since these original distinctions were observed, they have been significantly blurred by the identification of new strains and new tissue culture cell lines, resulting in crossover patterns of EV growth. For that reason, recent serotypes have been numbered (EVs 68 to 71) (30).

Isolation of EVs in tissue culture remains the "gold standard" for diagnosis. The commercial availability of increasing numbers and types of continuous cell lines has provided numerous options for routine EV culturing. The general susceptibilities of commonly used cell lines for the EVs are summarized in Table 3; a quick glance at that table shows that no single cell line is optimal for all EV serotypes. Monkey kidney cell lines, the traditional first choice for EV isolation, have good sensitivity for the polioviruses, type B coxsackieviruses, and echoviruses, whereas human diploid fibroblasts like WI-38 and HELF (human embryonic lung fibroblasts) have higher yields for type A coxsackieviruses. RD cells, derived from a human rhabdomyosarcoma, are the most sensitive for detection of coxsackieviruses A but fail with most type B coxsackieviruses; RD cells are notoriously difficult to work with in the laboratory, with rapid overgrowth and cell degeneration, and detection of EV growth frequently requires blind passages. For ease of use, most laboratories use a combination of a continuous monkey kidney cell line, such as CMK (cynomolgus monkey kidney), with a human diploid fibroblast line (such as WI-38, MRC-5, or HELF). Studies have shown improved yield and rapidity of EV detection with the addition of BGM (Bavarian green monkey kidney) and RD cells, albeit associated with greater cost and complexity (8, 9). Duplicate tubes of each cell line do not appear to significantly improve viral isolation with the possible exception of isolation in rhesus monkey kidney cells (8). If enough cell lines are used, the patterns of growth in each by a single EV isolate may accurately predict the subclassification of the isolate (17, 21), although the clinical value of that information is limited. Primary human amnion cells have proven useful for isolation of many EV serotypes including the type A coxsackieviruses (53); the difficulty in preparing and maintaining these cells, however, has resulted in their disuse.

Preparation of clinical samples for inoculation into cell culture has been previously reviewed (22, 31, 32, 40). Feces are suspended (10 to 20% final concentration) in a tube of sterile saline containing glass beads. Vigorous shaking and subsequent settling of the emulsified feces are followed by low-speed centrifugation of the supernatant for about 15 min. Antibiotics are added at concentrations comparable to those in viral transport media (see chapter 70 of this Manual), and a longer low-speed (or shorter high-speed) centrifugation is performed. The resultant supernatant is inoculated directly onto cell cultures. An identical procedure may be performed for the following specimens: rectal swabs in viral transport media, throat swabs in viral transport media, and nasal or pharyngeal washes. Swabs should be squeezed out against the sides of the collection tube and discarded rather than broken off into the media, as swab components may be toxic to cells. Specimens already in transport media containing antibiotics need not be retreated following the initial centrifugation. Cerebrospinal fluid, serum, and urine need no pretreatment and may be inoculated directly onto cell cultures. Serum and urine tend to be more toxic to cells and often require a change in the culture media within 24 h of inoculation. Occasionally, EVs may be antibody bound in serum or cerebrospinal fluid and detectable only following acid dissociation, procedures for which have been presented elsewhere in detail (27, 31). Preparation of body tissues from biopsy or autopsy for EV isolation are as described for viral culture in general in chapter 70 of this Manual.

Although it is the most sensitive method for laboratory diagnosis of coxsackievirus A infection, isolation of EVs in suckling mice is rarely performed any longer because of the difficulty of the technique and of animal maintenance. This method was recently reviewed elsewhere (33).

IDENTIFICATION OF THE AGENTS

As noted above, the determination of the specific serotype of infecting EVs is often unnecessary because the diseases caused by the EVs are not serotype specific. In most circumstances, therefore, it is adequate and useful for the diagnostic laboratory to report the presence of "an EV" without further detail. The most common exception to this principle is in pediatrics, where distinguishing between vaccine strain polioviruses and nonpolioviruses is critical to interpretation of viral culture results. During the first 2 years of life, children are repeatedly immunized with trivalent oral polio vaccine (Sabin strains), whose components, like all EVs, may be shed from the throat for 1 to 2 weeks and in the feces for several weeks to months. Hence, isolates from those two sites must be identified as either nonpolio or polio serotypes, with the latter presumed to be of vaccine origin unless unusual clinical circumstances suggest wild-type infection. Vaccine poliovirus has only rarely been recovered from cerebrospinal fluid or blood (13, 27); further characterization of EV isolates from those sites is thus less important. Distinguishing between polioviruses and NPEVs can be accomplished by two methods, one proven and one experimental. The standard method employs neutralization of the isolate with a pool of antisera directed against the three poliovirus serotypes (17). Successful inhibition of growth by the antisera confirms the poliovirus identity of the isolate, whereas failure to neutralize an isolate with typical cytopathic effect implies the presence of an NPEV. The neutralization method, as previously detailed, utilizes a fixed titer of virus (typically 100 50% tissue culture infectious doses) and pooled antipoliovirus sera (each at a final dilution of 20 U) (17). The experimental method uses a set of PCR primers that is specific for the three poliovirus serotypes; the PCR assay is performed on the culture-passaged isolate and accurately discriminates between polioviruses and NPEVs (1). This latter method has the potential for being simpler and more

rapid than viral neutralization for distinguishing between vaccine strains of poliovirus and disease-causing NPEVs, particularly as PCR assays become more user-friendly (see above).

Further identification of an NPEV to specific serotype is useful under certain circumstances. Indeed, the very principle that most clinical manifestations of EV infections are not serotype specific was historically established by serotyping studies. Epidemiologic studies of patterns of EV infections require knowledge of specific serotypes, as do descriptions of unusual clinical manifestations, such as poliomyelitis due to NPEVs (12) and pandemic hemorrhagic conjunctivitis (52). Only via serotyping will new EVs be discovered, as was the case most recently with the higher-numbered EVs (30). As noted above, an approximation as to subgroup of an isolate can be accomplished by analysis of the cell lines infected by an isolate (21); plaque morphology may give additional clues as to serotype within a subgroup (17). The gold standard for EV serotype determination, however, continues to be the use of intersecting pools of lyophilized antisera, as established by Lim and Benyesh-Melnick (LBM pools) (28, 29). Each isolate can be screened against 16 antisera by using 8 pools or against 25 antisera by using 10 pools. A checkerboard analysis localizes the isolate to a single serotype designation on the basis of the pattern of neutralization with the intersecting serum pools. Microtiter plate neutralization works equally well and saves significant quantities of serum reagents. The LBM pools are available in limited supplies from the World Health Organization.

Broadly reactive and serotype-specific EV monoclonal antibodies have been developed (50, 56, 58) and applied to tissue culture confirmation by immunofluorescence. Preliminary studies have indicated that these reagents, used singly and in pools, may find an important role in serotype identification. Immunofluorescence is a more rapid, simpler procedure than traditional neutralization-based serotyping, and the supply of monoclonal antibodies is unlimited.

SEROLOGIC DIAGNOSIS

Serologic testing, like immunoassays, has had only a limited role in EV diagnosis because of the great diversity of EV serotypes and the lack of a single common antigen. If the specific serotype of an infecting EV is known or suspected, e.g., in community-wide outbreaks, confirmatory immunoglobulin G (IgG) serology can be performed on individual patients to document a rise in antibody titer from the acute to the convalescent phase of infection, thus providing useful epidemiologic information; little actual benefit accrues to the patient. When an EV is recovered from the feces or throat of a patient with unusual clinical manifestations, the etiologic role of the EV may be more firmly established by documenting a fourfold rise in antibody titer to that serotype in paired acute- and convalescent-phase sera. Serosurveys of populations for past exposure to specific EVs have been performed by testing for antibody in single serum specimens; such studies may be useful for retrospective disease associations, such as with EVs and diabetes mellitus (3) or dilated cardiomyopathy (38). Those kinds of studies are subject to many types of sampling bias and therefore are of limited value. In the usual scenario, when a patient presents with meningitis or other acute manifestations of illness and an EV is suspected, serology is not a practical option.

Three traditional types of antibody determinations have

been applied to the EVs: neutralization, complement fixation, and hemagglutination inhibition. Neutralizing antibodies arise early in infection with the EVs and persist for many years or for life. They may be assessed in standard tube dilutions or by microtiter plate; in either format, a fixed titer of a single serotype of EV is inoculated onto cells in the presence of serial dilutions of the patient's serum (22, 31, 32, 40); ideally, acute- and convalescent-phase sera are tested in parallel simultaneously. The rapid appearance of serum neutralizing antibodies following EV infection may make it difficult to document a fourfold rise in titer with convalescence. Complement fixation antibody assays are of much less utility, as these antibodies are transient (weeks to months) and broadly cross-reactive among EV serotypes, i.e., nonspecific. Historically, this assay was successfully applied only in testing for infection with the polioviruses. Only about a quarter to a third of all EV serotypes agglutinate erythrocytes, making hemagglutination inhibition of minimal usefulness; this type of antibody is also cross-reactive among EV serotypes and hence nonspecific.

More recently, coxsackievirus B IgM assays, which take advantage of the shared antigen among the six coxsackievirus B serotypes and the early appearance of the IgM class of antibodies, have appeared promising (4, 11, 41). Although the IgM response measured in these assays may be nonspecific, i.e., a response to a non-coxsackievirus B EV may result in a positive assay, the nonspecificity may actually make this test more broadly reflective of recent exposure to any of numerous EV serotypes; cross-reactivity with non-EV pathogens causing infection has not been thoroughly studied. Many patients in the reported studies had positive IgM assays but were negative for EV infection by culture, making the IgM assay either more sensitive or less specific than traditional culture diagnosis. These IgM tests are not yet widely available, and clinical experience with them remains limited. Attempts to use more broadly reactive EV antigens in IgM assays have met with technical shortcomings (41, 42).

EVALUATION AND INTERPRETATION OF RESULTS

The concept of permissive versus nonpermissive sites of infection is critical to the interpretation of EV assays. The nasopharynx and the gastrointestinal tract are permissive sites of infection; i.e., EVs have ready access to these sites and may remain as "colonizers" for weeks to months. Detection of EVs by virus isolation or PCR at these sites must be interpreted cautiously, because their presence alone does not establish causality of the illness in question. The EV in the feces of a patient today may be leftover shedding from an infection of weeks ago and have nothing to do with the meningitis the patient presents with currently. Indeed, virtually 100% of patients with EV aseptic meningitis have detectable EV in feces (36), but most persons shedding EV in the feces at any particular time are asymptomatic. Feces are thus the most sensitive and least specific site for detecting true EV-associated illness. Since the shedding period in the nasopharynx after EV infection is shorter than that in the feces, the specificity of an EV isolate from the nasopharynx for true causation of current symptoms is better than that with feces but far short of a definitive association. Further complicating the evaluation of results from these two body sites in young children is the frequent administration of live attenuated oral poliovirus vaccine in the first

years of life. Almost every EV isolate from feces and nasopharynges of young children that is encountered by the diagnostic virology laboratory is, in fact, a vaccine strain poliovirus. Reporting an EV isolate in this setting without specifying poliovirus versus NPEV can lead the physician to wrongly discontinue antibiotics or antiherpes therapy in the belief that an EV etiology has been established.

In contrast, the central nervous system, bloodstream, and genitourinary tract are nonpermissive sites of EV infection; i.e., detection of virus in specimens from these sites implies true invasive infection and a high likelihood of association with current illness. Rare reports of coinfections of the cerebrospinal fluid by bacteria and EVs have appeared (55). In these patients, the bacterium-associated clinical sequelae dominated; i.e., the patients were clinically suspected of having bacterial meningitis, and the virus was isolated incidentally; the patients were sick enough that identification of a virus before identification of the bacterium would have been unlikely to dissuade the clinician from continued use of antibiotics. In the much more common situation, where the clinical presentation is typical of viral meningitis, coinfection with a clinically "silent" bacterium would be extraordinarily unlikely. Hence, identification of an EV from a nonpermissive site in a patient with a clinically compatible illness is sufficient evidence for establishing EV causality. The distinction between polioviruses and NPEVs in specimens from nonpermissive sites is less important, since vaccine strains of poliovirus rarely have been reported in such specimens and, in those rare instances, may actually be causing the illness in question (13, 27).

The utilities and shortcomings of serologic assays are described above. In general, results of a single serologic assay are uninterpretable except perhaps for IgM tests, which remain to be standardized and more widely tested. Commercially available serologic or immunoassay panels for the EVs are, in general, derived from only limited numbers of serotypes and lack standardization or published quality controls.

REFERENCES

1. **Abraham, R., T. Chonmaitree, J. McCombs, B. Prabhakar, P. T. Lo Verde, and P. L. Ogra.** 1993. Rapid detection of poliovirus by reverse transcription and polymerase chain amplification: application for differentiation between poliovirus and nonpoliovirus enteroviruses. *J. Clin. Microbiol.* **31:**395–399.
2. **Assaad, F., and K. Ljungars-Esteves.** 1984. World overview of poliomyelitis: regional patterns and trends. *Rev. Infect. Dis.* **6:**S302–S307.
3. **Barrett-Conner, E.** 1985. Is insulin-dependent diabetes mellitus caused by coxsackievirus B infection? A review of the epidemiologic evidence. *Rev. Infect. Dis.* **7:**207–215.
4. **Bell, E. J., R. A. McCartney, D. Basquill, and A. K. R. Chaudhuri.** 1986. μ-Antibody capture ELISA for the rapid diagnosis of enterovirus infections in patients with aseptic meningitis. *J. Med. Virol.* **19:**213–217.
5. **Chapman, N. M., S. Tracy, C. J. Gauntt, and U. Fortmueller.** 1990. Molecular detection and identification of enteroviruses using enzymatic amplification and nucleic acid hybridization. *J. Clin. Microbiol.* **28:**843–850.
6. **Cherry, J. D.** 1987. Enteroviruses: polioviruses (poliomyelitis), coxsackieviruses, echoviruses, and enteroviruses, p. 1729–1841. *In* R. D. Feigin and J. D. Cherry (ed.), *Textbook of Pediatric Infectious Diseases*, 2nd ed. The W. B. Saunders Co., Philadelphia.
7. **Chonmaitree, T., C. Ford, C. Sanders, and H. L. Lucia.**

1988. Comparison of cell cultures for rapid isolation of enteroviruses. *J. Clin. Microbiol.* **26:**2576–2580.

8. **Chonmaitree, T., M. A. Menegus, and K. R. Powell.** 1982. The clinical relevance of CSF viral culture. A two-year experience with aseptic meningitis in Rochester, New York. *JAMA* **247:**1843–1847.

9. **Dagan, R., and M. A. Menegus.** 1986. A combination of four cell types for rapid detection of enteroviruses in clinical specimens. *J. Med. Virol.* **19:**219–228.

10. **Dalakas, M. C., J. L. Sever, D. L. Madden, N. M. Papadopoulos, I. C. Shekarchi, P. Albrecht, and A. Krezlewicz.** 1984. Late postpoliomyelitis muscular atrophy: clinical, virologic, and immunologic studies. *Rev. Infect. Dis.* **6:**S562–S567.

11. **Dorries, R., and V. Ter Meulen.** 1983. Specificity of IgM antibodies in acute human coxsackievirus B infections, analyzed by indirect solid phase enzyme immunoassay and immunoblot technique. *J. Gen. Virol.* **64:**159–167.

12. **Grist, N. R., and E. J. Bell.** 1984. Paralytic poliomyelitis and nonpolio enteroviruses: studies in Scotland. *Rev. Infect. Dis.* **6:**S385–S386.

13. **Gutierrez, K. M., and M. J. Abzug.** 1990. Vaccine-associated poliovirus meningitis in children with ventriculoperitoneal shunts. *J. Pediatr.* **117:**424–427.

14. **Herrmann, E. C., Jr., D. A. Person, and T. F. Smith.** 1972. Experience in laboratory diagnosis of enterovirus infections in routine medical practice. *Mayo Clin. Proc.* **47:**577–586.

15. **Herrmann, J. E., R. M. Hendry, and M. F. Collins.** 1979. Factors involved in enzyme-linked immunoassay of viruses and evaluation of the method for identification of enteroviruses. *J. Clin. Microbiol.* **10:**210–217.

16. **Hogle, J. M., M. Chow, and D. J. Filman.** 1985. Three-dimensional structure of poliovirus at 2.9 A resolution. *Science* **229:**1358–1365.

17. **Hsiung, G. D.** 1973. Enteroviruses, p. 54–67. *In* G. D. Hsiung (ed.), *Diagnostic Virology.* Yale University Press, New Haven, Conn.

18. **Hyypiä, T., P. Auvinen, and M. Maaronen.** 1989. Polymerase chain reaction for human picornaviruses. *J. Gen. Virol.* **70:**3261–3268.

19. **Hyypiä, T., C. Horsnell, M. Maaronen, M. Khan, N. Kalkkinen, P. Auvinen, L. Kinnunen, and G. Stanway.** 1992. A distinct picornavirus group identified by sequence analysis. *Proc. Natl. Acad. Sci. USA* **89:**8847–8851.

20. **Jarvis, W. R., and G. Tucker.** 1981. Echovirus type 7 meningitis in young children. *Am. J. Dis. Child.* **135:**1009–1012.

21. **Johnston, S. L. G., and C. S. Siegel.** 1990. Presumptive identification of enteroviruses with RD, HEp-2 and RMK cell lines. *J. Clin. Microbiol.* **28:**1049–1050.

22. **Kapsenberg, J. G.** 1988. Picornaviridae: the enteroviruses (polioviruses, coxsackieviruses, echoviruses), p. 692–722. *In* E. H. Lennette, P. Halonen, and F. A. Murphy (ed.), *Laboratory Diagnosis of Infectious Diseases: Principles and Practice.* Springer-Verlag, New York.

23. **Lipson, S. M., R. Walderman, P. Costello, and K. Szabo.** 1988. Sensitivity of rhabdomyosarcoma and guinea pig embryo cell cultures to field isolates of difficult-to-cultivate group A coxsackieviruses. *J. Clin. Microbiol.* **26:**1298–1303.

24. **Longo, M. C., M. S. Berninger, and J. L. Hartley.** 1990. Use of uracil DNA glycosylase to control carry-over contamination in polymerase chain reactions. *Gene* **93:**125–128.

25. **McKinney, R. E., S. L. Katz, and C. M. Wilfert.** 1987. Chronic enteroviral meningoencephalitis in agammaglobulinemic patients. *Rev. Infect. Dis.* **9:**334–356.

26. **Melnick, J. L.** 1990. Enteroviruses: polioviruses, coxsackieviruses, echoviruses, and newer enteroviruses, p. 549–605. *In* B. N. Fields and D. M. Knipe (ed.), *Virology.* Raven Press, New York.

27. **Melnick, J. L., R. O. Proctor, A. R. Ocampo, A. R. Diwan, and E. Ben-Porath.** 1966. Free and bound virus in serum after administration of oral poliovirus vaccine. *Am. J. Epidemiol.* **84:**329–342.

28. **Melnick, J. L., V. Rennick, B. Hampil, N. J. Schmidt, and H. H. Ho.** 1973. Lyophilized combination pools of enterovirus equine antisera: preparation and test procedures for the identification of field strains of 42 enteroviruses. *Bull. W.H.O.* **48:**263–268.

29. **Melnick, J. L., N. J. Schmidt, B. Hampil, and H. H. Ho.** 1977. Lyophilized combination pools of enterovirus equine antisera: preparation and test procedures for the identification of field strains of 19 group A coxsackievirus serotypes. *Intervirology* **8:**172–181.

30. **Melnick, J. L., I. Tagaya, and H. Von Magnus.** 1974. Enteroviruses 69, 70, and 71. *Intervirology* **4:**369–370.

31. **Melnick, J. L., H. A. Wenner, and C. A. Phillips.** 1979. Enteroviruses, p. 471–534. *In* E. H. Lennette and N. J. Schmidt (ed.), *Diagnostic Procedures for Viral, Rickettsial and Chlamydial Infections,* 5th ed. American Public Health Association, Washington, D.C.

32. **Melnick, J. L., H. A. Wenner, and L. Rosen.** 1964. The enteroviruses, p. 194–242. *In* E. H. Lennette and N. J. Schmidt (ed.), *Diagnostic Procedures for Viral and Rickettsial Diseases,* 3rd ed. American Public Health Association, Inc., Washington, D.C.

33. **Menegus, M. A.** Enteroviruses, p. 943–947. *In* A. Balows, W. J. Hausler, Jr., K. L. Herrmann, H. D. Isenberg, and H. J. Shadomy (ed.), *Manual of Clinical Microbiology,* 5th ed. American Society for Microbiology, Washington, D.C.

34. **Miller, M. J.** 1993. Viral taxonomy. *Clin. Infect. Dis.* **16:**612–613.

35. **Minnich, L., E. Brown, R. Ashley, C. Andreou, D. Hirsch, E. Yanek, S. Oliver, W. Giles, and S. Maxwell.** 1993. *Abstr. 9th Annu. Clin. Virol. Symp.,* p. 50.

36. **Mintz, L., and W. L. Drew.** 1980. Relation of culture site to the recovery of nonpolio enteroviruses. *Am. J. Clin. Pathol.* **74:**324–326.

37. **Muir, P., F. Nicholson, M. Jhetam, S. Neogi, and J. E. Banatvala.** 1993. Rapid diagnosis of enterovirus infection by magnetic bead extraction and polymerase chain reaction detection of enterovirus RNA in clinical specimens. *J. Clin. Microbiol.* **31:**31–38.

38. **Muir, P., A. J. Tilzey, T. A. H. English, F. Nicholson, M. Signy, and J. E. Banatvala.** 1989. Chronic relapsing pericarditis and dilated cardiomyopathy: serological evidence of persistent enterovirus infection. *Lancet* **i:**804–807.

39. **Olive, D. M., S. Al-Mufti, W. Al-Mulla, M. A. Khan, A. Pasca, G. Stanway, and W. Al-Nakib.** 1990. Detection and differentiation of picornaviruses in clinical samples following genomic amplification. *J. Gen. Virol.* **71:**2141–2147.

40. **Phillips, C. A.** 1980. Enteroviruses and reoviruses, p. 823–828. *In* E. H. Lennette, A. Balows, W. J. Hausler, Jr., and J. P. Truant (ed.), *Manual of Clinical Microbiology,* 3rd ed. American Society for Microbiology, Washington, D.C.

41. **Pozzetto, B., O. G. Gaudin, M. Aouni, and A. Ros.** 1989. Comparative evaluation of immunoglobulin M neutralizing antibody response in acute-phase sera and virus isolation for the routine diagnosis of enterovirus infection. *J. Clin. Microbiol.* **27:**705–708.

42. **Reigel, F., F. Burkhardt, and U. Schilt.** 1985. Cross-reactions of immunoglobulin M and G antibodies with enterovirus-specific viral structural proteins. *J. Hyg.* **95:**469–481.

43. **Romero, J., J. R. Putnak, and E. Wimmer.** 1986. The use of poliovirus proteins VP3 and 2C as group antigens for the detection of enteroviral infections by indirect immunofluorescence, abstr. 967. *Pediatr. Res.* **20:**319.

44. **Romero, J. R., and H. A. Rotbart.** 1993. PCR detection of the human enteroviruses, p. 401–406. *In* D. H. Persing, T. F. Smith, F. C. Tenover, and T. J. White (ed.), *Diagnostic Molecular Microbiology: Principles and Applications.* American Society for Microbiology, Washington, D.C.

45. **Romero, J. R., and H. A. Rotbart.** 1994. PCR-based strategies for the detection of human enteroviruses, p. 341–374. *In* G. D. Ehrlich and S. J. Greenberg (ed.), *PCR-Based Diagnostics in Infectious Disease.* Blackwell Scientific Publications, Boston.

46. **Rotbart, H. A.** 1990. Enzymatic RNA amplification of the enteroviruses. *J. Clin. Microbiol.* **28:**438–442.

47. **Rotbart, H. A.** 1991. Nucleic acid detection systems for enteroviruses. *Clin. Microbiol. Rev.* **4:**156–168.

48. **Rueckert, R. R.** 1990. Picornaviruses and their replication, p. 507–548. *In* B. N. Fields and D. M. Knipe (ed.), *Virology.* Raven Press, New York.

49. **Sawyer, M. H., D. Holland, N. Aintablian, J. D. Connor, E. F. Keyser, and N. J. Waecker, Jr.** Diagnosis of enteroviral central nervous system infection by polymerase chain reaction during a large community outbreak. *Pediatr. Infect. Dis. J.,* in press.

50. **Schnurr, D., S. Yagi, and V. Devlin.** 1993. *Abstr. 9th Annu. Clin. Virol. Symp.,* p. 66.

51. **Strikas, R. A., L. J. Anderson, and R. A. Parker.** 1986. Temporal and geographic patterns of isolates of nonpolio enterovirus in the United States, 1970–1983. *J. Infect. Dis.* **153:**346–351.

52. **Wadia, N. H., S. M. Katrak, V. P. Misra, P. N. Wadia, K. Miyamura, K. Hashimoto, T. Ogino, T. Hikiju, and R. Kono.** 1983. Polio-like motor paralysis associated with acute hemorrhagic conjunctivitis in an outbreak in 1981 in Bombay, India: clinical and serologic studies. *J. Infect. Dis.* **147:**660–668.

53. **Wenner, H. A., and M. F. Lenahan.** 1961. Propagation of group A coxsackie viruses in tissue cultures. *Yale J. Biol. Med.* **34:**421–438.

54. **Wilfert, C. M., and J. Zeller.** 1985. Enterovirus diagnosis, p. 85–107. *In* L. M. de la Maza and E. M. Peterson (ed.), *Medical Virology IV.* Lawrence Erlbaum Associates, Publishers, Hillsdale, N.J.

55. **Wright, H. T., R. M. McAllister, and R. Ward.** 1962. "Mixed" meningitis: reports of a case with isolation of Haemophilus influenzae type b and echovirus type 9 from the cerebrospinal fluid. *N. Engl. J. Med.* **267:**142–144.

56. **Yagi, S., D. Schnurr, and J. Lin.** 1992. Spectrum of monoclonal antibodies to coxsackievirus B-3 includes type- and group-specific antibodies. *J. Clin. Microbiol.* **30:**2498–2501.

57. **Yolken, R. H., and V. M. Torsch.** 1980. Enzyme-linked immunosorbent assay for detection and identification of coxsackie B antigen in tissue cultures and clinical specimens. *J. Med. Virol.* **6:**45–52.

58. **Young, S. A., B. Strong, and R. Radloff.** 1993. *Abstr. 9th Annu. Clin. Virol. Symp.,* p. 67.

59. **Yousef, G. E., I. N. Brown, and J. F. Mowbray.** 1987. Derivation and biochemical characterization of an enterovirus group-specific monoclonal antibody. *Intervirology* **28:**163–170.

60. **Zoll, G. J., W. J. G. Melchers, H. Kopecka, G. Jambroes, H. J. A. Van Der Poel, and J. M. D. Galama.** 1992. General primer-mediated polymerase chain reaction for detection of enteroviruses: application for diagnostic routine and persistent infections. *J. Clin. Microbiol.* **30:**160–165.

Rotaviruses

MARY L. CHRISTENSEN

88

CLINICAL SIGNIFICANCE

Viral gastroenteritis is the second most common clinical disease occurring in developed countries, exceeded only by viral upper respiratory tract illness. Rotaviruses are the major cause of viral gastroenteritis in infants and young children. They can also occasionally cause gastroenteritis in the elderly. Other causes of viral gastroenteritis include enteric adenoviruses, caliciviruses, astroviruses, coronaviruses, and Norwalk and Norwalk-like viruses (see chapters 80 and 89 of this Manual).

Gastroenteritis caused by group A rotaviruses is ubiquitous, occurring in all parts of the world. In countries that have temperate climates, group A epidemics occur yearly during the cooler months of the year, although endemic or sporadic cases may occur during the other months. In tropical countries, group A infection usually occurs endemically throughout the year.

Group B rotaviruses have occurred primarily in China, where they have caused major epidemics, especially infecting adults (3). However, children and neonates can also develop symptomatic group B rotavirus infection. Group B viruses have rarely been seen outside of mainland China. Group C rotaviruses cause gastroenteritis in both children and adults. Although these organisms have occurred worldwide, they have been seen only sporadically or in small outbreaks (11).

Rotavirus infection is by the fecal-oral route. The virus primarily infects the epithelial cells lining the small intestine. After an incubation period of about 1 to 2 days, the onset of rotavirus gastroenteritis is sudden, with vomiting, diarrhea, fever, occasionally abdominal pain, and even respiratory symptoms (3). Loss of fluids and electrolytes in rotavirus gastroenteritis can lead to severe dehydration, hospitalization, and even death. Treatment consists primarily of fluid and electrolyte replacement, either orally or intravenously. Asymptomatic rotaviral shedding can occur in all age groups. Nosocomial infections are frequent and can account for up to 50% of rotavirus cases occurring in hospitals. Transmission of the virus may be difficult to control, since $\geq 10^{10}$ virions per g of stool may be shed by patients, while fewer than 10 focus-forming units of virus can initiate an infection (3).

Rotavirus infections appear to be repetitive, and reinfection usually involves different serotypes. Antibody that develops after infection may persist for as long as 6 months; antibodies can either protect individuals from reinfection or reduce the severity of reinfection (3).

Various experimental vaccines for rotavirus are being developed and evaluated. These are live attenuated vaccines administered by the oral route.

DESCRIPTION OF ROTAVIRUSES

Rotaviruses are major causes of gastroenteritis in the young of many animal species, including humans. Rotaviruses were discovered in the 1970s by the electron microscopic (EM) examination of patients' specimens and were called "rotaviruses" because of their wheel-like appearance. Rotaviruses are in the family *Reoviridae*, each member of which possesses a double layer of icosahedral shells approximately 70 nm in diameter. Each contains a core of double-stranded RNA. The rotaviral double-stranded RNA genome consists of 11 gene segments, which can be separated by polyacrylamide gel electrophoresis (3).

Three of the gene segments that code for the three major rotaviral antigens are of particular interest (3). Gene segment 6 codes for VP6, the major inner-core structural protein (7). VP6 is responsible for the group specificity of rotaviruses, which are divided into groups A through E. Human rotaviruses belong to group A, B, or C. Group A is the most common and most extensively studied rotavirus that infects humans and several animal species. Human group A rotaviruses are subdivided into subgroups I and II. The subgroup specificity is also determined by VP6.

Gene segment 8 or 9, depending on the strain of rotavirus, codes for VP7. VP7 is the major outer capsid protein that is glycosylated. It is responsible for the G serotype specificity of group A viruses. Four major G serotypes, 1 through 4, that infect humans are well recognized and characterized. Other G serotypes that infect humans include types 8, 9, and 12 (8, 16). G serotypes 2 and 8 usually have a subgroup I specificity, whereas G serotypes 1, 3, 4, and 9 usually have a subgroup II specificity. Serotype 1 is the most prevalent G serotype infecting humans worldwide, although serotypes 2, 3, and 4 may predominate in some countries during some seasons (1, 15, 21).

Gene segment 4 codes for VP4. VP4 is a protease-sensitive outer capsid protein. It is responsible for the P serotype specificity of group A viruses. The P serotypes

TABLE 1 General characteristics of commercial rotavirus tests

Test	Kit name (manufacturer)	Type of antibody[a]	Negative control beads, tubes, or wells	Form for final reading	Total incubation time (min)
EIA	Pathfinder (Kallestad)	Polyclonal-monoclonal	No	Tube, microtiter	75
	Rotaclone (Cambridge BioScience)	Monoclonal-monoclonal	No	Microtiter	70
	Rotazyme II (Abbott Laboratories)	Polyclonal-polyclonal	No	Tube	150
	Rotavirus EIA (Isolab)	Polyclonal-polyclonal	Yes	Microtiter	105
	IDEIA Rotavirus (Dako Corp.)	Polyclonal-polyclonal	No	Microtiter	70
EIA, rapid	TestPack Rotavirus (Abbott Laboratories)	Polyclonal-monoclonal/ polyclonal mixture	No	Reaction disk	7
LA	Meritec-Rotavirus (Meridian Diagnostics)	Polyclonal	Yes	Latex beads	5
	Murex Rotavirus Latex (Murex Diagnostics)	Polyclonal	Yes	Latex beads	17
	Rota-Stat (Isolab)	Polyclonal	Yes	Latex beads	13–14
	Rotalex (Orion Diagnostica)	Polyclonal	Yes	Latex beads	2
	Slidex Rota-kit 2 (BioMérieux)	Polyclonal	Yes	Latex beads	7–12
	Virogen Rotatest (Wampole Laboratories)	Polyclonal	Yes	Latex beads	17

[a]For EIA, capture antibody-detector antibody; for LA test, antibody on latex.

infecting humans include P1A, P1B, P2, and P3 (8, 13, 17). P serotypes are independent of G serotypes.

COLLECTION, TRANSPORT, AND STORAGE OF SPECIMENS

For the direct detection of rotavirus, stool specimens should be collected during the acute phase of the illness, preferably during the first 3 to 5 days of illness. Specimens should be placed directly into clean plastic or glass screw-cap jars or cups for submission to the laboratory. The specimen container should not contain any preservatives, detergents, or metal ions. In addition, the container should not contain viral transport medium or tissue culture medium. These media may contain fetal bovine serum or other sera that may contain inhibitors or antibodies to rotavirus. All of these substances could interfere with enzyme immunoassays (EIAs) and latex agglutination (LA) tests. In addition, adding specimens to viral transport medium could dilute the specimen beyond the limits of detection of viral particles.

Very liquid stool specimens from pediatric patients in diapers can be obtained in several ways. Liquid diarrheal stool can be prevented from being absorbed into a diaper either by placing a disposable diaper inside out on an infant or by placing a layer of plastic wrap inside a clean diaper. Also, a clean pediatric urine bag can be placed over the anal area to collect very watery stool. In addition, liquid stool soaked in a diaper can be removed by pressing a cotton swab into the diaper until the swab is well saturated with stool. A saturated swab will take up 0.1 to 0.15 ml of fluid. If this method is used, it is recommended that two or three swabs be submitted at one time, so that sufficient stool material can be eluted from the swabs into the EIA or LA buffer used in the procedure's initial stool processing step. However, rectal swabs per se should not be submitted, since the virions on the rectal swab may not be numerous enough for detection.

Specimens should be transported and stored at 4°C. For long-term storage, specimens should be frozen undiluted between −20 and −60°C and not repeatedly frozen and thawed. Specimens should not be stored in a self-defrosting freezer. Testing fresh specimens stored at 4°C is preferred.

DETECTION OF ROTAVIRUSES

Since rotaviruses are difficult to propagate in cell culture, other detection methods have been developed and used. There are three diagnostic methods that can be easily carried out in diagnostic laboratories on large numbers of specimens and often without specialized equipment. These are the EIA, the rapid membrane EIA, and the LA test. These three tests are antibody-based tests that utilize antigen-antibody reactions. A number of commercial EIA, membrane EIA, and LA test kits that detect group A rotaviruses are available (Table 1).

EIAs

Commercially available EIAs are three-layer double-antibody sandwich assays. They may utilize one of several modifications, and they vary in several respects. First, either polyclonal or monoclonal antibodies may be used for the capture and detector antibody systems, although some of the assays use polyclonal antibodies for both. Second, one commercial kit incorporates a negative control well that is coated with nonimmune serum or globulin, although most of the EIA kits do not include this control. Third, microtiter wells, larger wells, or test tubes are used for the reaction mixtures. A standard laboratory spectrophotometer can be used for reading the tubes, whereas a microtiter reader is needed for the microtiter wells. Companies may furnish computerized spectrophotometers for reading tubes or microtiter plates. One company also provides a specific color chart for visual readings. Some manufacturers suggest that visual readings can be carried out. However, with low-level positive or borderline readings, spectrophotometric readings and mathematical calculations of cutoff values based on the manufacturer's directions are recommended. EIA readings may be considered semiquantitative, since positive readings range from high-level to low-level readings. Thus, more information on the amount of viral antigen being excreted can be obtained from an EIA than from a membrane EIA or an LA test, which are read qualitatively as "positive" or "negative." EIAs are also useful when large numbers of specimens are to be run at one time. Most EIAs are run at ambient temperature, although one commercial EIA requires a 37°C water bath. One company (Cambridge BioScience) manufactures a rotavirus group EIA, an adenovirus types 40 and 41 EIA, and an adenovirus group EIA, all of which have the same format and can be run concomitantly on diarrheal stools from gastroenteritis patients.

Membrane EIA

Simple, fast, easy-to-read qualitative membrane EIAs are also commercially available (TestPack Rotavirus, Abbott Laboratories) and require very little equipment. In these tests, a stool sample and the necessary reagents are added to a small reaction disk. Results are easily read visually after 7 min of total incubation time at ambient temperature. A plus sign appears if the specimen is positive and the test was carried out correctly. If a negative test is carried out correctly, a minus sign appears. No special equipment or expertise is needed, and these EIAs are sensitive and specific (2). Membrane EIAs can be run in less than 30 min and are useful for running one or several rotavirus tests at one time. This type of test can be used in small-volume hospital laboratories, as statim tests in large-volume hospital laboratories, and in physicians' offices.

LA Tests

LA tests have the advantages of requiring very little equipment and being very rapid, usually being completed within 30 min at ambient temperature. All of the LA tests have both test latex particles coated with antirotavirus serum and control latex particles coated with nonimmune serum or globulin. Most LA tests have a high level of specificity (3, 6, 10, 18), owing in part to incorporation into the test procedure of negative control latex particles. LA tests are usually not as sensitive as EIAs (3, 6, 10, 14, 18). The sensitivity of LA tests is adequate with specimens taken early in infection, when large amounts of virus are being excreted. However, the sensitivity of LA tests tends to be lower than that of EIAs with specimens taken late in infection, when significantly fewer virions are being excreted. Like the membrane EIA, the LA test is useful for running one or several rotavirus tests at one time. They are good for laboratories with low-volume rotavirus testing or for statim testing in laboratories that normally use EIA testing for their high-volume rotavirus testing.

Additional EIA, Membrane EIA, and LA Test Criteria

There are additional criteria to take into consideration when choosing among the standard EIAs, the membrane EIA, and the LA tests. Total incubation times of the standard EIAs can range from 75 min to 2.5 h, which is significantly more than is needed to report results of LA tests. However, actual hands-on times for EIAs are significantly less than total incubation times. EIAs are usually more sensitive than LA tests (3, 6, 10, 14, 18), although sensitivity can vary from manufacturer to manufacturer.

Monoclonal antibody-based kits appear to be more sensitive than polyclonal antibody-based kits (3, 6, 10, 14, 18). Kits that contain negative control wells or beads tend to be more specific, since they minimize false-positive reactions. However, neither LA tests nor EIAs detect group B or C rotaviruses, and they may not detect certain group A serotypes (14). Some EIAs give false-positive results on a small percentage of stool specimens from asymptomatic neonates (3, 5, 9).

All of the commercially available standard EIAs, membrane EIAs, and LA tests are carried out as specified by the manufacturers. Quality control standards specified in the kits should be adhered to. Positive and negative controls are provided in the kits and should be used as instructed by the manufacturers.

PCR

Several PCR techniques for detecting and typing rotaviruses have been developed in research laboratories but are not available for routine use. First, a highly sensitive PCR procedure for detecting group A rotaviruses in stools has been developed (20). In this procedure, viral RNA is extracted from stools, and inhibitory substances from the stools are removed from the RNA. The resultant RNA is used in a reverse transcriptase reaction to produce cDNA, which is then amplified by PCR. The same set of primers is used for both the reverse transcriptase step and the PCR step, using directed primers to rotavirus gene 6, which codes for the group-specific antigen. These primers consist of (i) the first 20 bases of gene 6 for primer 1 and (ii) a 25-base segment between bases 234 and 259 for primer 2. As few as 500 genomic copies of purified rotaviral RNA can be detected in this manner. Rotaviral RNA can be detected experimentally in fecal samples at dilutions 1,000- to 10,000-fold beyond the limits of detection by EIA.

A PCR technique has also been developed for typing six of the group A rotaviruses by using type-specific primers derived from six variable regions on gene 9 (12). In addition, a PCR technique for identifying gene 4 (VP4) types (8) and detecting group B and C rotaviruses has been developed.

EM

Before EIAs and LA tests became available, rotaviruses were detected by EM, and this method is still being used by some laboratories. The EM method has several advantages. It is relatively rapid, especially when a small number of samples are evaluated. EM can detect both group A and non-group A rotaviruses as well as other enteric viruses. However, the technique has several disadvantages as well. It requires the use of expensive equipment, and it may be less sensitive than EIA. When a large number of specimens are run at once, EM can be slower and more tedious than other methods now available. The EM procedure for detecting rotaviruses and other enteric viruses in stools has been described previously (4), and more information on detecting diarrheic viruses is available in chapter 89 of this Manual.

Culture

Human rotaviruses are usually not cultured in diagnostic laboratories, since the virus is found in large quantities in stool specimens and can be rapidly detected by antigen detection tests. Only a minority of the virions in a specimen are infectious to cell cultures. However, some research laboratories have cultivated rotaviruses by using various manipulations, including pretreating specimens with trypsin, adding low levels of trypsin to cell culture maintenance medium, and rolling the cell cultures (3). With these methods, the virus can be propagated in primary kidney cells or lines of kidney cells from various species of monkeys. Cytopathic effect can usually be observed after several passages and may consist of cell fusion, cell rounding, granularity of cells, cell lysis, and sloughing of cells. Viruses from patient specimens may grow better in primary cells but can then be passaged in cell lines (3, 19).

SEROLOGY

Detection of virus in stools is the major approach used for diagnosing rotaviral gastroenteritis in the routine diagnostic laboratory. Blood or serum is usually not submitted, since serology for diagnosing rotaviral gastroenteritis is not usually done in diagnostic laboratories. Serologic studies using EIA, radioimmunoassay, and other methods have been carried out primarily in research laboratories for (i) typing or grouping of new isolates or (ii) epidemiologic purposes (3).

EVALUATION, INTERPRETATION, AND REPORTING OF RESULTS

EIAs

Certain standard EIA results can be read visually. With one kit, patient results and control results can be compared to a color chart provided with the kit. With several other kits, patient results can be compared to the color development of the positive and negative controls. All EIAs can also be read spectrophotometrically, with specimens being considered positive if they are greater than a cutoff value specified by the manufacturer. One company's kit has a gray zone in which specimen spectrophotometric readings within 10% of the cutoff value are considered "suspect." Any "suspect" test should be repeated, preferably with a new specimen from the patient. With all kits, all controls should fall within the criteria specified by the manufacturer, and if they do not, the test should be repeated. Spectrophotometric readings are recommended for accuracy, especially with low-level or borderline specimens.

Owing to the wide variation in the composition of the stool specimens being tested, some specimens may contain substances that bind nonspecifically in the EIA tests, giving low-level false-positive test results near the cutoff value. Low-level positive results could also be the result of (i) excretion of low levels of rotaviral antigen or (ii) submission of insufficient, minute amounts of stool specimen on a swab or in a container. If the quantity of stool submitted is less than that specified by the manufacturer of the kit being used, it is recommended that this insufficiency be recorded as additional information if a low-level positive, "suspect," or negative result is obtained. It is recommended that the spectrophotometric cutoff value and the gray zone, if applicable, be reported along with the spectrophotometric reading of the patient's specimen. This information, along with the quantity of specimen submitted, if inadequate, is useful for the physician in deciding whether to send an additional specimen for testing.

The limit of sensitivity of the EIA is about 10^6 virions per ml of stool. Negative results would be obtained with stool specimens containing fewer virions. The rotavirus EIAs, the membrane EIA, and the LA test detect inactive virus particles and viral antigen as well as live rotavirus particles.

Membrane EIA

The membrane EIA is read visually. A plus sign indicates a positive reaction and that the test was carried out correctly. A minus sign indicates a negative reaction and that the test was carried out correctly. These results can be reported as "positive for rotavirus" or "negative for rotavirus," respectively. Tests that give neither a plus nor a minus sign may indicate improper addition of the reagents to the test or deterioration of the reagents being used. These tests should be repeated.

Like the EIAs, the membrane EIA may give a negative result if low levels of rotavirus antigen are below the limits of detection, which is 2×10^7 to 4×10^7 virions per ml of stool. It is suggested that when results are reported, the limits of sensitivity of the test be included with the test results, especially with negative results. Negative results may also be obtained if inadequate amounts of specimen were submitted or if the specimen was diluted prior to the initial test procedure. Inadequate amounts of stool sample or dilution of stool prior to its reception in the laboratory should be recorded along with a negative result so that this information can be given to the physician along with a request for a new specimen.

LA Tests

LA tests are also read visually. The positive rotavirus control reagent should cause agglutination of the test latex particles coated with antirotavirus serum. A patient specimen is positive if it causes agglutination of the test latex but does not agglutinate the negative control latex. A patient specimen is negative if the test latex particles (and the negative control latex particles) do not agglutinate. These results can be reported as "positive for rotavirus" or "negative for rotavirus," respectively.

A patient's specimen agglutinating both the test latex and the negative control latex contains nonspecific agglutinins. The specimen may be a true negative, or the nonspecific agglutinins may be masking any agglutination of the test latex caused by rotavirus particles in the specimen. When nonspecific agglutination occurs, the test should be repeated on a new dilution of the original specimen or, preferably, a new specimen. If nonspecific agglutination persists, the specimen should be tested by an EIA or a membrane EIA.

Owing to incorporation of the negative control latex, LA tests are highly specific, usually approaching 100% specificity in most studies. The limits of specificity of the LA tests are approximately 10^7 virions per gram of stool. It is suggested that the limits of sensitivity of the test be included with the test results, especially when negative results are reported.

REFERENCES

1. **Beards, G. M., U. Desselberger, and T. H. Flewett.** 1989. Temporal and geographical distributions of human rotavirus serotypes, 1983 to 1988. *J. Clin. Microbiol.* **27:**2827–2833.
2. **Chernesky, M., S. Castriciano, J. Mahony, M. Spiewak, and L. Schaeffer.** 1988. Ability of TESTPACK ROTAVIRUS enzyme immunoassay to diagnose rotavirus gastroenteritis. *J. Clin. Microbiol.* **26:**2459–2461.
3. **Christensen, M. L.** 1989. Human viral gastroenteritis. *Clin. Microbiol. Rev.* **2:**51–89.
4. **Christensen, M. L., and C. Howard.** 1991. Viruses causing

gastroenteritis, p. 950–958. *In* A. Balows, W. J. Hausler, Jr., K. L. Herrmann, H. D. Isenberg, and H. J. Shadomy (ed.), *Manual of Clinical Microbiology*, 5th ed. American Society for Microbiology, Washington, D.C.

5. **Cromien, J. L., C. A. Himmelreich, R. I. Glass, and G. A. Storch.** 1987. Evaluation of new commercial enzyme immunoassay for rotavirus detection. *J. Clin. Microbiol.* **25:**2359–2362.

6. **Dennehy, P. H., D. R. Gauntlett, and W. E. Tente.** 1988. Comparison of nine commercial immunoassays for the detection of rotavirus in fecal specimens. *J. Clin. Microbiol.* **26:**1630–1634.

7. **Estes, M. K., B. B. Mason, S. Crawford, and J. Cohen.** 1984. Cloning and nucleotide sequence of the simian rotavirus gene 6 that codes for the major inner capsid protein. *Nucleic Acids Res.* **12:**1875–1887.

8. **Gentsch, J. R., R. I. Glass, P. Woods, V. Gouvea, M. Gorziglia, J. Flores, B. K. Das, and M. K. Bhan.** 1992. Identification of group A rotavirus gene 4 types by polymerase chain reaction. *J. Clin. Microbiol.* **30:**1365–1373.

9. **Giaquinto, C., G. Errico, E. Ruga, I. Naso, and R. D'Elia.** 1986. Evaluation of ELISA test for rotavirus diagnosis in neonates. *J. Pediatr.* **109:**565–566.

10. **Gilchrist, M. J. R., T. S. Bretl, K. Moultney, D. R. Knowlton, and R. L. Ward.** 1987. Comparison of seven kits for detection of rotavirus in fecal specimens with a sensitive, specific enzyme immunoassay. *Diagn. Microbiol. Infect. Dis.* **8:**221–228.

11. **Gouvea, V., J. R. Allen, R. I. Glass, Z.-Y. Fang, M. Bremont, J. Cohen, M. A. McCrae, L. J. Saif, P. Sinarachatanant, and E. O. Caul.** 1991. Detection of group B and C rotaviruses by polymerase chain reaction. *J. Clin. Microbiol.* **29:**519–523.

12. **Gouvea, V., R. I. Glass, P. Woods, K. Taniguchi, H. F. Clark, B. Forrester, and Z.-Y. Fang.** 1990. Polymerase chain reaction amplification and typing of rotavirus nucleic acid from stool specimens. *J. Clin. Microbiol.* **28:**276–282.

13. **Gunasena, S., O. Nakagomi, Y. Isegawa, E. Kaga, T. Nakagomi, A. D. Steele, J. Flores, and S. Ueda.** 1993. Relative frequency of VP4 gene alleles among human rotaviruses recovered over a 10-year period (1982–1991) from Japanese children with diarrhea. *J. Clin. Microbiol.* **31:**2195–2197.

14. **Mathewson, J. J., D. K. Winsor, Jr., H. L. DuPont, and S. L. Secor.** 1989. Evaluation of assay systems for the detection of rotavirus in stool specimens. *Diagn. Microbiol. Infect. Dis.* **12:**139–141.

15. **Matson, D. O., M. K. Estes, J. W. Burns, H. B. Greenberg, K. Taniguchi, and S. Urasawa.** 1990. Serotype variation of human group A rotaviruses in two regions of the USA. *J. Infect. Dis.* **162:**605–614.

16. **Midthun, K. J., Valdesuso, A. Z. Kapikian, Y. Hoshino, and K. Y. Green.** 1989. Identification of serotype 9 human rotavirus by enzyme-linked immunosorbent assay with monoclonal antibodies. *J. Clin. Microbiol.* **27:**2112–2114.

17. **Steele, A. D., D. Garcia, J. Sears, G. Gerna, O. Nakagomi, and J. Flores.** 1993. Distribution of VP4 gene alleles in human rotaviruses by using probes to the hyperdivergent region of the VP4 gene. *J. Clin. Microbiol.* **31:**1735–1740.

18. **Thomas, E. E., M. L. Puterman, E. Kawano, and M. Curran.** 1988. Evaluation of seven immunoassays for detection of rotavirus in pediatric stool samples. *J. Clin. Microbiol.* **26:**1189–1193.

19. **Ward, R. L., D. R. Knowlton, and M. J. Pierce.** 1984. Efficiency of human rotavirus propagation in cell culture. *J. Clin. Microbiol.* **19:**748–753.

20. **Wilde, J., J. Eiden, and R. Yolken.** 1990. Removal of inhibitory substances from human fecal specimens for detection of group A rotaviruses by reverse transcriptase and polymerase chain reaction. *J. Clin. Microbiol.* **28:**1300–1307.

21. **Woods, P. A., J. Gentsch, V. Gouvea, L. Mata, A. Simhon, M. Santosham, Z.-S. Bai, S. Urasawa, and R. I. Glass.** 1992. Distribution of serotypes of human rotavirus in different populations. *J. Clin. Microbiol.* **30:**781–785.

Caliciviruses, Astroviruses, and Other Diarrheic Viruses

MARTIN PETRIC

89

CLINICAL BACKGROUND

Gastroenteritis viruses have been recognized as major human pathogens only over the past 2 decades. This group of viruses includes rotaviruses, adenoviruses, Norwalk and Norwalk-like viruses, caliciviruses, astroviruses, and viruses having a coronaviruslike morphology. While virtually all of these agents were recognized in the 1970s, until recently the major advances have been limited to rotaviruses, which are not only the most commonly recognized agents but also the most amenable to investigation (3). This chapter addresses all gastroenteritis viruses except the rotaviruses, which are discussed elsewhere.

Among the above-named agents, the clinical features of gastroenteritis caused by enteric adenoviruses, Norwalk virus, and astroviruses are well documented. Less is known about calicivirus, Norwalk-like virus, and coronaviruslike infections (3).

Caliciviruses

In addition to the human caliciviruses, this family includes the Norwalk virus. Norwalk virus was tentatively designated a calicivirus on the basis of serologic studies and the characteristics of its capsid protein, notwithstanding the limited morphologic similarity between these agents (8). The recent cloning and sequencing of the genomes of Norwalk virus and the Norwalk-like Southampton virus confirm that these two viruses are indeed members of the family *Caliciviridae* (22, 29).

Human Calicivirus

First documented in 1976 in stool specimens of symptomatic patients, human caliciviruses received their name because of their morphologic similarity to feline calicivirus (7, 31). The five human calicivirus strains identified to date have been designated UK 1 through UK 4 and Japan (8). Evidence that these strains are pathogens came from studies of gastroenteritis outbreaks associated with the excretion of calicivirus in stools followed by documentation of seroconversion. Clinical features of calicivirus-associated gastroenteritis are diarrhea and vomiting, generally without fever. The illness has been reported to last 1 to 11 days and to have an incubation period of 48 to 72 h (7). Symptomatic infections occur in infants and young children, but geriatric outbreaks have also been noted. Infections occur through-

out the year, with increased incidence in the winter months. The virus is transmitted in institutional settings and by contaminated food and water (3). The virus occurs worldwide, and by the age of 4 years, virtually the entire population is seropositive (7).

Norwalk Virus

By applying immunoelectron microscopy (IEM) to fecal specimens of volunteers infected with the stool filtrates from a gastroenteritis outbreak in Norwalk, Ohio, 27-nm-diameter virus particles were visualized in 1972 and designated Norwalk virus. Studies of outbreaks and of volunteers showed that the clinical features of the virus infection include nausea, vomiting, diarrhea, abdominal cramps, headache, and fever (23). The illness has a mean incubation period of 24 h and a duration of 12 to 48 h. It may be severe in debilitated and elderly patients, although asymptomatic infections are documented. Outbreaks of this virus in institutions, cruise ships, and families have been reported (24). The source of infection in outbreaks, which occur throughout the year, is likely to be contaminated food or water, and secondary spread does occur. Antibody response is detectable after infection and is protective for a limited period. According to serologic studies, the virus has a worldwide distribution (23).

Norwalk-Like Viruses

A number of viruses morphologically similar to Norwalk virus have been reported. These are variously called Norwalk-like viruses or small round structured viruses. Currently, at least four serotypes of these viruses are detectable by IEM. These viruses have been classified as Norwalk, Hawaii, Snow Mountain, and Taunton and have been designated small round structured virus serotypes UK 1 through UK 4, respectively (29). The association of these agents with gastroenteritis has been established in outbreak and volunteer studies (23). These viruses cause an illness clinically similar to that caused by Norwalk virus and have epidemiologic features similar to those of Norwalk virus.

Astroviruses

Astroviruses, detected in stool specimens by electron microscopy (EM), were first associated with gastroenteritis in 1975 (28). Astroviruses not antigenically related to the human pathogens have been detected in a number of animal species (28). There are at least five distinct serotypes of human astrovirus. The association of astrovirus with gastroenteritis was established through investigations of outbreaks of mild gastroenteritis, prospective studies, and studies of infection in volunteers (17, 28). Clinical symptoms, which generally last up to 3 days, consist of vomiting, abdominal pain, fever, and diarrhea. The diarrhea has been known to continue for 7 to 14 days and to be accompanied by virus excretion (28). The disease is most common in infants and young children, although asymptomatic infections have been noted. Symptomatic illness occurs more rarely and with diminished symptoms in adults (6). The incubation period is estimated at 3 to 4 days, and secondary spread within families is common (28).

Enteric Adenoviruses

Although adenoviruses have been cultured from stool specimens of children with gastroenteritis, it was only when abundant shedding was documented that these viruses came to be considered etiologic agents of this disease (6, 12). A subset of these agents that could not be isolated in conventional cell cultures was referred to as fastidious enteric adenoviruses (10). They were shown to belong to serotypes 40 and 41, which can be propagated in Graham 293 cells or Chang conjunctival cells. While there were several reports of adenovirus-associated gastroenteritis, the role of enteric adenoviruses in gastroenteritis was definitively demonstrated in prospective studies, in which these viruses were found to account for 59% of the adenoviruses detected in stools and were the only detectable agent associated with 7.25% of patients with diarrhea (10). The distinguishing clinical symptoms of infection with these viruses were prolonged diarrhea and low-grade fever (with or without vomiting) that could exceed 14 days in duration. Secondary spread among children is common. The virus may be second only to rotavirus as the most common cause of pediatric gastroenteritis (6, 10). Infection by enteric adenoviruses is most common in infants under 2 years of age, although asymptomatic shedding has been reported (3, 27). Enteric adenovirus infections occur year-round, with an increased incidence in warmer times of the year (6).

Coronaviruses

Although they are major gastrointestinal pathogens of animals, human coronaviruslike agents have been reported only sporadically in cases of gastrointestinal illness (6, 36). The detection of human toroviruslike agents in patients with gastroenteritis was first reported in 1984, but these agents have been investigated to only a limited extent (2). As these viruses have not been reproducibly isolated in cell culture, their detection is generally limited to EM examination of patient specimens.

Small Round Viruses and Other Agents

Viruses ranging from 25 to 30 nm in diameter have been found in stool specimens from patients with gastroenteritis. These viruses have been termed small round viruses or "picorna-parvo-like agents" because of their morphology. They could represent enteroviruses or phages, and their role in gastroenteritis has not been well established. Picobirnavirus, another small virus reported in an outbreak of diarrhea in Brazil, has a defined double-stranded RNA genome of two segments. Such small round agents are currently under investigation and may well prove to be novel agents of gastroenteritis (3, 6).

DESCRIPTION OF THE AGENTS

Morphology and Structure

Most reports on the structures of gastroenteritis viruses are based on negative-contrast EM. A short description of the characteristic morphological features of each of these agents is therefore presented here.

1. *Caliciviridae* display considerable variability in morphology. The 35-nm-diameter caliciviruses (Fig. 1A) have well-defined scalloped edges and 32 cup-shaped depressions over the surface of the virions, which make for easy unambiguous identification (31). In contrast, Norwalk and Norwalk-like viruses, while similar to caliciviruses in diameter or smaller, have less defined, fuzzier edges, which are particularly distinct if the particles are penetrated by stain (Fig. 1B). The Norwalk virus, measuring 27 nm in diameter, is reported to have a somewhat indistinct, rough outer edge (23).

2. Astroviruses, shown in Fig. 1C, have smooth edges with the characteristic five- or six-pointed star in the center of the 28- to 30-nm-diameter particle. The star shape, which is believed to be due to shallow hollows in the viral surface where stain is differentially deposited, is not necessarily present in all of the particles (31).

3. Adenoviruses (Fig. 1E) are icosahedral structures 70 to 75 nm in diameter that can be readily recognized by EM and differentiated from rotaviruses, which are similar in size (32).

4. Coronaviruses and toroviruses are unique among the human gastroenteritis agents in that they are enveloped viruses with helical symmetry (Fig. 1F). This has made the detection of these agents by EM problematic. The distinct peplomeres of the coronaviruses distinguish them from the less distinct fringes of the toroviruses shown in Fig. 1G (20, 26). The latter also tend to assume a kidney shape. These viruses are 100 to 150 nm in diameter.

5. Small round viruses, shown in Fig. 1D, lack any of the structural features noted for the agents described above. They range between 25 and 30 nm in diameter and have smooth outlines.

Physical-Chemical Characteristics

Recent advances have provided new insight into the physical-chemical nature of these viruses. The salient features of the nucleic acid and protein composition of these agents are summarized in Table 1.

Antigenic Characteristics

Antigenic homology exists among the viruses of each group. There is documented cross-reactivity between the caliciviruses, Norwalk virus, and the Norwalk-like viruses (8). Enteric adenovirus types 40 and 41 can be specifically recognized by monoclonal antibodies and differentiated from other adenoviruses, with whom they share the common adenovirus antigen (38). Similarly, a group-specific monoclonal antibody to the five serotypes of human astroviruses has been developed (16). Cross-reactivity of the human toroviruslike particles with the calf torovirus Breda has been reported (26). The homology between the viruses of each group is important in the design of diagnostic tests for the diagnosis of viruses and for serology.

Genetics

With the cloning and sequencing of the Norwalk virus RNA, its genomic organization has been established (22).

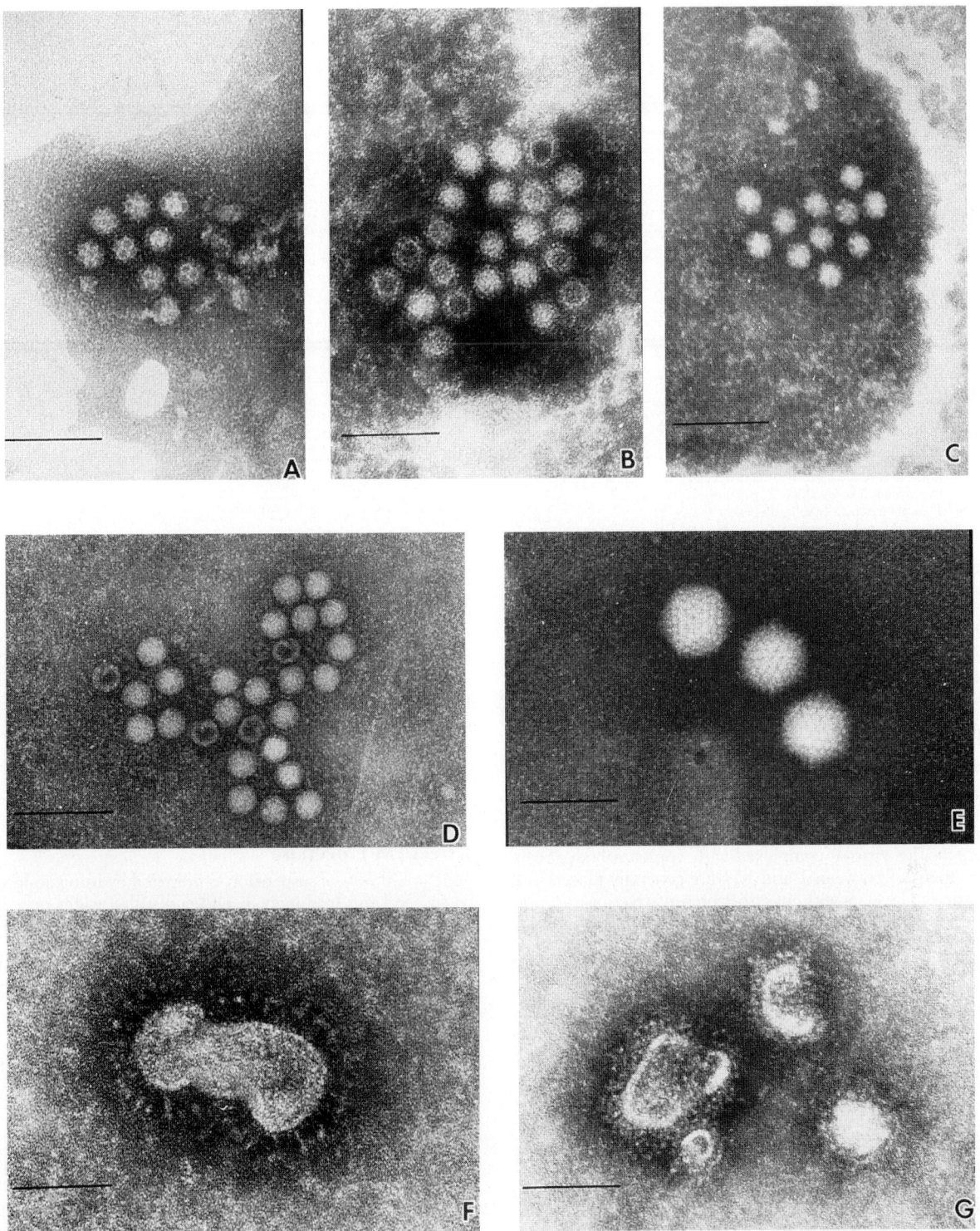

FIGURE 1 Gastroenteritis viruses. (A) Calicivirus; (B) Norwalk-like virus; (C) astrovirus; (D) small round virus; (E) adenovirus; (F) coronavirus; (G) toroviruslike particle. Bars = 100 nm.

The single-stranded RNA has a polymerase open reading frame (ORF) at the 5′ end and a capsid ORF in the 3′ half of the genome. An ORF at the extreme 3′ end has no known function. The polymerase sequence appears to have substantial homology to sequences in the Norwalk-like viruses (29). A subgenomic RNA corresponding to the 3′ end of the astrovirus has been cloned and sequenced (34). This RNA codes for the three capsid proteins. The genomes of the enteric adenoviruses have not yet been completely sequenced, but the defined sequences specific for enteric

TABLE 1 Nucleic acid and protein compositions of gastroenteritis viruses

Virus	Nucleic acid			Major structural protein(s)	
	Type[a]	Genome size (kb)	No. of ORFs	Function[b]	Size(s) (kDa)
Calicivirus	ssRNA	8	3	Capsid	59–65
Norwalk	ssRNA	8	3	Capsid	56.6
Norwalk-like	ssRNA	8	3	Capsid	59
Astrovirus	ssRNA	7	4	Capsid	30, 29, 21
Adenovirus	dsDNA	30	30[c]	Hexon	120
				Penton	80
				Fiber	40
Coronavirus[d]	ssRNA	30	5	N	50–60
				E1	20–30
				E2	180–200
				E3	120–140
Torovirus[e]	ssRNA	30	5	N	18
				E	20
				P	75–90

[a]ss, single stranded; ds, double stranded.
[b]N, nucleocapsid; E, envelope; P, peplomere.
[c]mRNAs are extensively spliced (19).
[d]Based on the animal coronavirus genome (18).
[e]Based on the Berne virus genome (20).

adenoviruses have been reported (1). There is also evidence of homology between the genomes of the human torovirus-like particles and Berne virus (25).

SPECIMEN COLLECTION AND STORAGE

Specimens should be collected as soon as the gastroenteritis is diagnosed, ideally within the first 48 h of illness. Specimens collected after 1 week of illness may yield negative results either because the virus shedding has decreased or because the virus is complexed with coproantibody. Stool specimens of between 1 and 10 g are generally placed into plastic or glass jars without either preservatives or transport media. Very liquid specimens can be collected by scraping them from the diaper with a wooden tongue depressor. Alternatively, the diaper can be lined with plastic wrap to capture a portion of the watery stool. Rectal swabs are discouraged, since they do not contain a sufficient quantity of virus for direct observation. Ileostomy contents are acceptable specimens, as are postmortem segments of bowel tied off at both ends.

Gastroenteritis viruses are relatively stable and preserve well at 4°C for up to a week. For prolonged storage, specimens can be frozen at −70°C, although this may be deleterious for detection of virus by EM. Serologic diagnosis of gastroenteritis virus is generally performed in a research setting. Acute- and convalescent-phase sera are, however, valuable in the performance of IEM when working up an outbreak (30).

DIRECT EXAMINATION BY EM

EM lends itself to gastroenteritis virus diagnosis for a number of reasons. The viruses are generally easily recognizable and are present at concentrations above the threshold for detection by EM. The specimens can thus be examined with a minimum of processing. Moreover, the quantity of virus ($>10^6$/g) generally corresponds to active infection and not asymptomatic shedding. Since the infections caused by gastroenteritis viruses have relatively similar symptomatologies, the broad-spectrum approach offered by EM is best suited for their diagnosis. The virus in stool specimens is stable and can readily be transported. The EM approach is best for examining up to 30 specimens per day. For larger numbers, such as would occur in investigations of outbreaks or research-oriented efforts, EM may prove suboptimal.

Direct EM Procedure

A 10 to 20% stool suspension is prepared by using a pipette to suspend the specimen in either distilled water or a 1% ammonium acetate solution. The specimen is then applied to a copper EM grid, 300 to 400 mesh size, which has been coated with Formvar (polyvinyl formol) and reinforced by carbon shadowing. This application to the grid is best performed over a 50-mm petri dish containing a 42.5-mm-diameter Whatman filter. After approximately 1 min, the specimen is removed from the grid with a fragment of filter paper. If the specimen is particularly viscous, the grid may be washed with a drop of the ammonium acetate solution or distilled water. A drop of 2% phosphotungstic acid (pH 7) is then applied to the grid for 10 to 30 s and removed by blotting with a fragment of filter paper. The dried grid is placed under a short UV light at 900 W/cm² for 10 min to inactivate any viruses and is then examined by EM (11, 32).

This method covers the general approach, and a number of variations are acceptable. The grid may be touched to a virus suspension made on a Parafilm square. The times of exposure of the grid to the specimen may be varied considerably. The grid may be washed at any time prior to staining to remove nonviral debris.

Concentration of Specimens

While stool specimens are generally processed directly, concentration should be considered in suspected Norwalk virus outbreaks or when the specimen has been collected outside the period of maximum excretion. Concentration can be achieved by ultracentrifugation or agar diffusion. By using a Beckman Airfuge with an EM-90 rotor, the specimen can be deposited directly on the EM grid (11). The clarified stool suspension prepared as outlined above is deposited into the sector well of the rotor, and a Formvar-carbon-coated grid is placed in the peripheral depression. After centrifugation at 90,000 rpm for 30 min, the grids are removed and processed as outlined above. If larger quantities of virus are to be concentrated, a conventional ultracentrifuge with a swinging-bucket rotor may be employed. Centrifugation at $100,000 \times g$ for 1 h at 4°C will deposit virtually all viruses (9). The supernatant is discarded, and the well-drained pellet is then suspended in distilled water and processed as outlined above. When ultracentrifugation is employed, it is important to clarify the stool suspension by either filtration or low-speed centrifugation to avoid excessive quantities of debris that may obscure the virus particles.

The resuspended stool specimen can also be subjected to agar diffusion to concentrate the virus particles (11). The specimen is placed on a small block of 1% agar or agarose made up in distilled water, and the grid is floated on the surface of the drop. The preparation is kept at room temperature until the drop is absorbed into the agar, at which time the grid is removed and processed for EM as described above. This procedure not only concentrates virus particles but also removes excessive quantities of salts that diffuse into the agarose. Such salts in the specimen may lead to crystal or precipitate formation on the grid and interfere with the detection of viruses.

IEM

IEM, which employs the principle that antibodies resulting from infection will clump virus particles, played a pivotal role in the establishment of gastroenteritis viruses as pathogens (23). Antisera can be acute- or convalescent-phase patient sera, pooled human serum globulin, or hyperimmune animal antisera specifically prepared with purified virus antigens. The IEM procedure involves incubation of the stool suspension with an equal volume of reference antiserum for 1 h at 37°C (11). The immune complexes, collected by centrifugation for 30 min at $15,000 \times g$, are subjected to negative-contrast EM as outlined above. A simplified version of this technique, called solid-phase IEM, involves coating the EM grid with antibody to the virus (11). This antibody can be either a reference antibody or immune serum globulin. The virus particles in the specimen applied to this grid will be preferentially captured by the immobilized antibody. Virus can also be specifically concentrated by incorporating reference antibody into the agar block described above. In this case, the EM grid is placed on the agar block, and the stool specimen drop is placed on the grid. The antibody diffuses through the grid and forms immune complexes with virus on the grid surface. Alternatives to these approaches involve reacting the virus suspension with antibody labeled with colloidal gold. This facilitates the detection of virus and subviral particles that would not have been apparent on the preliminary examination of the specimen by EM.

In addition to enhancing the sensitivity with which these agents can be detected in patient specimens, IEM has also been applied to the typing of these viruses, as in the cases of the caliciviruses and the Norwalk-like viruses (8).

Examination of Grids by EM

The stained, disinfected grids are examined by EM using EM voltages of 60 to 100 kV. The grid can be examined first at a low scanning magnification of ×1,000 to ×3,000 to determine if it is acceptable for further examination. Grids that are opacified by excess debris or on which no stained material can be detected are inappropriate, and new grids with appropriately altered concentrations of the specimen should be prepared. Grid fields that show a modest amount of granular staining are then examined at ×40,000 to ×50,000 magnification. Under these conditions, the larger viruses are easily seen, and even the smallest viruses are approximately 1 to 2 mm in diameter. To be effective, the EM must be set to maximum resolving power at the high magnification. The fine-structure features used to identify Norwalk-like viruses and especially astroviruses require a resolution of approximately 1 nm.

Virus particles present in the grid field are most readily detected in areas of moderate background staining, where they appear as lighter structures surrounded by a dark contrasting background, as shown in Fig. 1. The most definitive of the spherical structures are the adenoviruses, whose icosahedral capsids are readily discerned, and even the capsomeres are resolved in well-prepared specimens. Among the smaller viruses, the caliciviruses, with their scalloped edges and well-defined cuplike depressions, are readily identified. As shown in Fig. 1A, they may exhibit several distinct orientations. The astroviruses have a comparatively less defined morphology, but in at least a subset of particles, a distinct star is evident, as shown in Fig. 1C. In contrast to the caliciviruses, the astroviruses have smooth edges. The Norwalk-like viruses also have a defined morphology. The capsomeres appear larger and are grouped around a central core. Accordingly, such viruses have been called at times "mini-reo" or "mini-rota," since particles with stain-penetrated centers appear to have capsomeres arranged around a central core. Norwalk virus is generally smaller, measuring 27 nm in diameter. Finally, nonstructured particles described as small round viruses, whose significance is still in question, are also visualized, as shown in Fig. 1D.

Among the nonspherical particles, the coronaviruses form pleomorphic structures with well-delineated dumbbell-shaped peplomeres or spikes. Generally, the peplomeres form a dark halo around the enveloped capsid. Toroviruslike particles have a less well defined peplomere fringe. A proportion of these are kidney shaped, while others have well-established darker-staining regions in the center of the particle, as seen in Fig. 1G. The detection of these particles by EM is less definitive than detection of the spherical structured viruses, and readily available alternative, immunospecific diagnostic tests to confirm the diagnosis are needed. All particles lacking well-defined features, such as adenoviruses or caliciviruses, can pose diagnostic dilemmas if seen as individual structures. When such particles are visualized as clumps or clusters, confidence in identification is increased.

In addition to the above-described well-recognized virus particles, several other structures are visualized in stool specimens. These include the obvious flagella, pili, and bacteria as well as cell wall components, cell membrane blebs, and bacteriophages. The bacteriophages may pose problems, because they may be confused with viruses, especially if their tails are sheared off. Experienced electron microscopists can readily discriminate between these structures and accepted viruses. For those not familiar with EM detection of viruses, a set of grids from a reference laboratory is often useful, especially since each microscope has unique features of resolution.

IMMUNOSPECIFIC ASSAYS

Because of the antigenicity and ease of purification of rotavirus, most immunospecific assays for gastroenteritis viruses have been developed for rotavirus. A limited number of immunospecific assays are available for adenoviruses, and developments in this area are expected with the availability of reference monoclonal antibodies (38). A commercial assay kit containing positive and negative controls and protocols has been marketed under the name Adenoclone Type 40/41 (Cambridge Bioscience, Worchester, Mass.). This kit, which is also available to detect adenovirus group antigen, can be procured and should be considered, since adenoviruses other than types 40 and 41 are involved in the etiology of gastroenteritis (6). There have been reports that the enzyme-linked immunosorbent assay (ELISA) format does not faithfully detect all enteric adenoviruses (37). An alternative to ELISA is the commercially available latex agglutination assay Adenolex (Orion Diagnostica, Helsinki, Finland), with its obvious advantages of speed and simplicity.

Immunospecific tests of the ELISA or radioimmunoassay format, utilizing convalescent-phase human serum, have been described for most of the gastroenteritis viruses. With limited availability of reagents, these tests are applicable only in research settings or reference laboratories (15). The recent cloning and expression of Norwalk virus genes makes possible the production of reference antigen and antibody reagents and thus the development of well-standardized ELISAs for direct diagnosis of the virus and for serology (3). Immunoassays for the detection of Norwalk-like viruses have been described, but they pose the problem that the source of antigen is limited to stool collected in outbreaks (15). An ELISA for the detection of calicivirus antigen and antibody has also been described (35). Monoclonal antibodies to astroviruses have been developed and utilized in the ELISA format (16). Finally, by using the Breda virus hyperimmune antiserum, human torovirus can be detected by ELISA (26). With progress in the cloning and expression of viral genomes, defined recombinant antigens are becoming available. These will allow the commercial production of comprehensive panels of immunodiagnostic reagents for standardized detection of these viruses.

ISOLATION IN CELL CULTURE

The fastidious enteric adenoviruses were first identified because they failed to grow in conventional cell cultures (6, 38). Specific cell lines, such as Graham 293 lung fibroblasts stably transformed with the 5′ end of the adenovirus genome, do promote the growth of these agents and are the cell lines of choice. All astrovirus serotypes grow in HEK or LLC-MK$_2$ cell cultures in the presence of 10 μg of trypsin per ml (28). Neither the calicivirus group nor the human torovirus-coronavirus agents have as yet been reproducibly propagated in cell cultures (6). While growth in culture has a major impact in terms of investigation of the virus, it generally takes time and is not considered sufficiently rapid to contribute to the meaningful management of disease. It is, however, valuable in the preparation of diagnostic reagents and in the characterization of virus pathogens.

ASSAYS BASED ON THE VIRAL GENOME

Identification of Fastidious Enteric Adenovirus Types 40 and 41 by Restriction Endonuclease Digestion of Genomic DNA

In early studies on the characterization of enteric adenoviruses, these agents were found to have restriction sites that on digestion generated uniquely sized fragments that could be visualized by polyacrylamide gel electrophoresis (39). Such an approach was found to be more faithful in characterizing the enteric adenovirus after one passage in cell culture than serum neutralization was after repeat passages (4). In the latter approach, a much smaller quantity of conventional adenovirus that might be present owing to a presumed earlier infection could quickly outgrow the enteric adenovirus and result in inaccurate typing of the agent causing the illness. Among the restriction endonucleases, *Sma*I proved effective in identifying adenovirus types 40 and 41. DNA can also be extracted from adenoviruses present in stool specimens. After clarification of the stool suspension, the virus is precipitated with polyethylene glycol, and the DNA is extracted and digested with *Sma*I (5).

Dot Blot Assays

Using the cloned restriction fragment Ad 41 *Bgl*II-D as a probe, adenovirus types 40 and 41 can be detected directly from patient specimens in a dot blot hybridization assay. The stool specimen, clarified by centrifugation, is spotted onto a nitrocellulose membrane, which is then treated with 1 M NaOH to disrupt the virus and denature the DNA (14). Alternatively, the DNA may be extracted from the clarified specimen and spotted onto the nitrocellulose membrane.

Astroviruses have been detected by dot blot hybridization with a cDNA probe from an internal region of astrovirus type 1. This method proved more sensitive than EM diagnosis and was in fact able to detect astrovirus among specimens considered to be small round viruses (40). A comparison of a dot blot procedure using an RNA probe with an enzyme immunoassay demonstrated that while the former could detect virus at greater dilutions, it was in practice less sensitive than the latter method (33).

Using cDNA probes specific for the 3′ end of the Berne virus genome, torovirus in human stool specimens was successfully detected by dot blot hybridization (25). This was an important finding, as toroviruslike particles are difficult to detect by EM with the same certainty as icosahedral viruses.

PCR

The application of PCR technology to the detection of gastroenteritis viruses was established with the rotavirus system. Among the remaining viruses, applications of this method to the Norwalk virus and, subsequently, Norwalk-like viruses have recently been reported (13, 21).

The virus was extracted from the stool suspension with Freon and concentrated by polyethylene glycol precipitation. The RNA was extracted from the virus pellet with phenol-chloroform after digestion with proteinase K in the presence of 1.25% cetyltrimethylammonium bromide. A PCR product of 456 bases was obtained with the following primers specific for Norwalk virus: 5′-ATTGAGAGCCT CCGCGTG-3′ and 5′-GGTGGCGAAGCGCCCTC-3′. The reverse transcription PCR (RT-PCR) approach, considered 100 times more sensitive than dot blot hybridization of the genomic RNA, was able to detect virus in stool specimens to a dilution of 10^{-4}. An RT-PCR method that uses primers from the polymerase region of Norwalk virus for the detection of Norwalk-like viruses has been described (13). The advantages of RT-PCR are evident in that the amplicon can be sequenced and the degree of genetic identity can be determined in this as yet poorly characterized group of viruses.

PCR has also been applied to the diagnosis of adenoviruses in stool specimens (1). Group-specific primers corresponded to sequences of the adenovirus type 2 hexon gene. For diagnosis of enteric adenoviruses alone, primers with sequences for the E1B gene of adenovirus type 41 produced amplicons of 2,187 bases. It is important to perform PCR with both sets of primers, as there may be considerable asymptomatic shedding of conventional adenoviruses in stools.

INTERPRETATION OF RESULTS

In all studies, the presence of virus in the stool specimen correlates statistically with a gastrointestinal illness. However, it must be borne in mind that a variable number of gastroenteritis virus infections are asymptomatic, as reported, for example, in a recent study of astrovirus gastroenteritis (17). To some extent, the less sensitive EM plays an important role in that it requires a higher virus concentration in the stool specimen, which is more likely to be consistent with the disease process (12). However, EM is by the same logic a suboptimal method for monitoring the shedding of the virus and thus may have an adverse impact on containment of virus infections. Cell culture is relatively sensitive but is under most circumstances too slow to provide meaningful results in what are usually self-limiting infections that spread rapidly. Culture does, however, provide a valuable standard with which to compare other methods. Immunospecific assays provide rapid results with adequate sensitivity and can, in certain instances (Adenolex), be available on a statim basis. These assays have excellent potential in that they are relatively simple to perform and, at present, relatively free of problems. Because of insufficient experience, we cannot yet formulate firm interpretations of nucleic acid probe and PCR technology. While this approach appears to be the most sensitive, the implications of asymptomatic gastrointestinal virus shedding have yet to be established. On the other hand, this methodology has great potential for the detection of viruses in environmental specimens and may in the long run have a major impact on the burden of viral gastroenteritis originating from food and water sources.

REFERENCES

1. **Allard, A., R. Girones, P. Juto, and G. Waddell.** 1990. Polymerase chain reaction for detection of adenoviruses in stool specimens. *J. Clin. Microbiol.* **28:**2659–2667.
2. **Beards, G. M., J. Green, C. Hall, and T. H. Flewett.** 1984. An enveloped virus in stools of children and adults with gastroenteritis that resembles the Breda virus of calves. *Lancet* **i:**1050–1051.
3. **Blacklow, N. R., and H. B. Greenberg.** 1991. Viral gastroenteritis. *N. Engl. J. Med.* **325:**252–264.
4. **Brown, M., M. Petric, and P. J. Middleton.** 1984. Diagnosis of fastidious enteric adenoviruses 40 and 41 in stool specimens. *J. Clin. Microbiol.* **20:**334–338.
5. **Buitenwerf, J., J. J. Lowrens, and J. C. de Jong.** 1985. A simple and rapid method for typing adenoviruses 40 and 41 without cultivation. *J. Virol. Methods* **10:**39–44.
6. **Christiansen, M. L.** 1989. Human viral gastroenteritis. *Clin. Microbiol. Rev.* **2:**51–89.
7. **Cubitt, W. D.** 1987. The candidate caliciviruses. *CIBA Found. Symp.* **128:**127–143.
8. **Cubitt, W. D., N. R. Blacklow, J. E. Herrmann, N. A. Nowak, S. Nakata, and S. Chiba.** 1987. Antigenic relationships between human caliciviruses and Norwalk virus. *J. Infect. Dis.* **156:**806–814.
9. **Davies, H. A.** 1982. Electrin microscopy and immune electron microscopy for detection of gastroenteritis viruses, p. 37–49. *In* D. A. J. Tyrrell and A. Z. Kapikian (ed.), *Virus Infections of the Gastrointestinal Tract.* Marcel Dekker, Inc., New York.
10. **de Jong, J. C., R. Weigand, A. H. Kidd, G. Waddell, J. G. Kapsenberg, C. J. Muzerie, A. G. Wermenbol, and R. G. Firtzlaff.** 1983. Candidate adenoviruses 40 and 41: fastidious adenoviruses from human infant stool. *J. Med. Virol.* **11:**215–231.
11. **Doane, F., and N. Anderson.** 1987. Methods for preparing specimens for electron microscopy, p. 14–31. *In Electron Microscopy in Diagnostic Virology.* Cambridge University Press, New York.
12. **Flewett, T. H.** 1976. Diagnosis of enteritis viruses. *Proc. R. Soc. Med.* **69:**693–696.
13. **Green, J., J. P. Norcott, D. Lewis, C. Arnold, and D. W. G. Brown.** 1993. Norwalk-like viruses: demonstration of genomic diversity by polymerase chain reaction. *J. Clin. Microbiol.* **31:**3007–3012.
14. **Hammond, G., C. Hannan, T. Yeh, K. Fischer, G. Mauthe, and S. E. Straus.** 1987. DNA hybridization for diagnosis of enteric adenovirus infection from directly spotted human fecal specimens. *J. Clin. Microbiol.* **25:**1881–1885.
15. **Hedberg, C. W., and M. T. Osterholm.** 1993. Outbreaks of food-borne and waterborne viral gastroenteritis. *Clin. Microbiol. Rev.* **6:**199–210.
16. **Herrmann, J. E., R. W. Hudson, D. M. Perron-Henry, J. B. Kurtz, and N. R. Blacklow.** 1988. Antigenic characteristics of astrovirus serotypes and development of astrovirus specific monoclonal antibodies. *J. Infect. Dis.* **158:**182–185.
17. **Herrmann, J. E., D. N. Taylor, P. Echeverria, and N. R. Blacklow.** 1991. Astroviruses as a cause of gastroenteritis in children. *N. Engl. J. Med.* **324:**1757–1760.
18. **Holmes, K. V.** 1990. Coronaviridae and their replication, p. 841–856. *In* B. N. Fields, D. M. Knipe, R. M. Chanock, M. S. Hirsch, J. L. Melnick, T. P. Monath, and B. Roizman (ed.), *Virology,* 2nd ed. Raven Press, New York.
19. **Horowitz, M. S.** 1990. Adenoviruses and their replication, p. 1679–1721. *In* B. N. Fields, D. M. Knipe, R. M. Chanock, M. S. Hirsch, J. L. Melnick, T. P. Monath, and B. Roizman (ed.), *Virology,* 2nd ed. Raven Press, New York.
20. **Horzinek, M. C., M. Weiss, and J. Ederveen.** 1987. Toroviridae: a proposed new family of enveloped RNA viruses. *CIBA Found. Symp.* **128:**162–191.
21. **Jiang, X., J. Wang, D. Graham, and M. Estes.** 1992. Detection of Norwalk virus in stool by polymerase chain reaction. *J. Clin. Microbiol.* **30:**2529–2534.
22. **Jiang, X., M. Wang, K. Wang, and M. Estes.** 1993. Sequence and genomic organization of Norwalk virus. *Virology* **195:**51–61.
23. **Kapikian, A. Z., and R. M. Chanock.** 1990. Norwalk group viruses, p. 671–693. *In* B. N. Fields, D. M. Knipe, R. M. Chanock, M. S. Hirsch, J. L. Melnick, T. P. Monath, and B. Roizman (ed.), *Virology,* 2nd ed. Raven Press, New York.
24. **Kaplan, J. E., G. W. Gary, R. C. Baron, N. Singh, L. B. Schonberger, R. Feldman, and H. B. Greenberg.** 1982. Epidemiology of Norwalk gastroenteritis and the role of Norwalk virus in outbreaks of acute non-bacterial gastroenteritis. *Ann. Intern. Med.* **96:**756–761.
25. **Koopmans, M., A. Herrewegh, and M. Horzinek.** 1991. Diagnosis of torovirus infection. *Lancet* **337:**859.
26. **Koopmans, M., M. Petric, R. L. Glass, and S. S. Munroe.**

1993. Enzyme-linked immunosorbent assay reactivity of toro-virus-like particles in fecal specimens from humans with diarrhea. *J. Clin. Microbiol.* **31:**2738–2744.

27. **Kotloff, K. L., G. A. Losonsky, J. G. Morris, S. S. Wasserman, N. Singh-Naz, and M. M. Levine.** 1989. Enteric adenovirus infection and childhood diarrhea: an epidemiologic study in three clinical settings. *Pediatrics* **84:**219–225.

28. **Kurtz, J. B., and T. W. Lee.** 1987. Astroviruses: human and animal. *CIBA Found. Symp.* **128:**92–107.

29. **Lambden, P. R., E. O. Caul, C. R. Ashley, and I. Clarke.** 1993. Sequence and genome organization of a human small round structured (Norwalk-like) virus. *Science* **259:**516–518.

30. **Lew, J., C. W. Lebaron, R. I. Glass, T. Torok, P. M. Griffin, J. G. Wells, D. D. Juranek, and S. P. Wahlquist.** 1990. Recommendations for collection of laboratory specimens associated with outbreaks of gastroenteritis. *Morbid. Mortal. Weekly Rep.* **39:**RR14.

31. **Madeley, C. R.** 1979. Comparison of the features of astroviruses and caliciviruses by electron microscopy. *J. Infect. Dis.* **139:**519–523.

32. **Middleton, P. J., M. Szymanski, and M. Petric.** 1977. Viruses associated with acute gastroenteritis in young children. *Am. J. Dis. Child.* **131:**733–737.

33. **Moe, C. L., J. R. Allen, S. S. Monroe, H. E. Gary, C. D. Humphrey, J. E. Herrmann, N. R. Blacklow, C. Caracamo, M. Koch, K. H. Kim, and R. Glass.** 1991. Detection of astrovirus in pediatric stool specimens by immunoassay and RNA probe. *J. Clin. Microbiol.* **29:**2390–2395.

34. **Monroe, S. S., B. Jiang, S. E. Stine, M. Koopmans, and R. I. Glass.** 1993. Subgenomic RNA sequence of the human astrovirus supports classification of *Astroviridae* as a new family of RNA viruses. *J. Virol.* **67:**3611–3614.

35. **Nakata, S., M. E. Estes, and S. Chiba.** 1988. Detection of human calicivirus antigen and antibody by enzyme-linked immunosorbent assays. *J. Clin. Microbiol.* **26:**2001–2005.

36. **Payne, C. M., C. G. Ray, V. Borduin, L. L. Minich, and M. D. Lebowitz.** 1987. An 8 year study of viral agents of acute gastroenteritis in humans: ultrastructural observations and seasonal distribution with a major emphasis on corona-like particles. *Diagn. Microbiol. Infect. Dis.* **5:**39–54.

37. **Scott-Taylor, T., G. Ahluwalia, B. Klisko, and G. Hammond.** 1990. Prevalent enteric adenovirus variant not detected by commercial monoclonal antibody enzyme immunoassay. *J. Clin. Microbiol.* **28:**2797–2801.

38. **Singh-Naz, N., W. J. Rodriguez, A. H. Kidd, and C. D. Brandt.** 1988. Monoclonal antibody enzyme-linked immunosorbent assay for specific identification and typing of subgroup F adenovirus. *J. Clin. Microbiol.* **26:**297–300.

39. **Wadell, G., A. Allard, M. Johansson, M. Svensson, and I. Unhoo.** 1987. Enteric adenoviruses. *CIBA Found. Symp.* **128:**63–101.

40. **Willcocks, M. M., M. J. Carter, J. G. Silcock, and C. R. Madeley.** 1991. A dot-blot hybridization procedure for the detection of astrovirus in stool samples. *Epidemiol. Infect.* **107:**405–410.

Hepatitis A Virus

JEAN B. CEDERNA AND JACK T. STAPLETON

90

CLINICAL BACKGROUND

Hepatitis A, which has also been termed infectious hepatitis, is caused by the hepatitis A virus (HAV). The Centers for Disease Control and Prevention estimates that 143,000 cases of acute HAV infection occur in the United States each year, accounting for approximately 80 deaths and an estimated $200 million in medical care and lost productivity. Worldwide, the incidence of HAV infection exceeds 1.4 million cases, with a cost of $1.5 to 3 billion annually (35). HAV is transmitted by the fecal-oral route, usually following contact with an index case or during travel to an underdeveloped area where the virus is endemic. Preschool day-care centers are another important source of HAV transmission. In a study conducted within the United States, 11% of all patients with hepatitis A had had contact with children attending a preschool day-care center (36). Sexual transmission of HAV has been described, and an increased frequency of hepatitis A has been observed in homosexual men attending a sexually transmitted disease clinic (10). Common-source outbreaks related to food handlers and transmission via bivalve shellfish have led to large outbreaks of HAV infection; however, these infections account for a small percentage of total cases of hepatitis A (21). Transmission of HAV by blood transfusion or in association with intravenous drug use is rare but well documented (1, 39).

The clinical course of HAV infection can range from a mild, anicteric illness to severe, prolonged, icteric hepatitis. Inapparent or anicteric hepatitis occurs in over 90% of infected children under the age of 5, compared with 25 to 50% of adults. In developing countries, HAV is endemic, and virtually all individuals are infected early in life. However, as developing regions of the world improve their overall levels of sanitation, HAV is acquired at an older age, when symptomatic infection is more likely. Consequently, though the overall rate of HAV infection decreases, the incidence of clinically apparent disease paradoxically increases.

The incubation period of HAV infection (the time between exposure to HAV and the onset of clinical symptoms or biochemical evidence of liver disease) ranges from 15 to 50 days, with a mean of 28 days (34). It appears that the incubation period between exposure and development of hepatitis is inversely related to the size of the inoculum

of virus (38). A short prodomal or preicteric phase, varying from several days to more than a week, often precedes the onset of jaundice. This phase may be characterized by fever, fatigue, malaise, myalgia, anorexia, nausea, and vomiting. Right upper quadrant pain may occur secondary to hepatomegaly. Splenomegaly is present in 10 to 15% of patients. Diarrhea is uncommon and may be due to another enteric bacterial pathogen acquired from the same feces-contaminated source as the HAV (22, 81). The icteric phase of HAV infection may begin with the appearance of dark, golden brown urine due to bilirubinemia and continue one to several days later with pale stools and yellowish discoloration of the mucous membranes, conjunctivae, sclera, and skin. Palmar erythema and spider angiomata may be observed (38). The duration of symptoms resulting from HAV infection is variable, but clinical and biochemical indicators of hepatitis usually resolve within 4 to 8 weeks of the onset of illness (22).

HAV infection does not cause chronic liver disease, although several complications can accompany HAV infection. A cholestatic form of hepatitis characterized by fever, pruritis, prolonged jaundice, and weight loss has been described, and symptoms may last for more than 6 months (30). Skin rashes including urticaria (13), cryoglobulinemia (41), meningoencephalitis (6), Guillain-Barré syndrome, renal failure (9), hematologic complications (33), and cardiovascular complications (29) have all been reported to accompany HAV infection. HAV infection may relapse, and up to 10% of hospitalized patients in Costa Rica suffered a second episode of hepatitis. Increasingly, this biphasic course of hepatitis A is recognized (28). Fulminant HAV infection is uncommon but is estimated to account for 5 to 20% of all cases of fulminant viral hepatitis (54). Fulminant hepatitis is characterized by increasing severity of jaundice, deterioration in liver function, drowsiness, fetor hepaticus, and eventually encephalopathy and coma (64). HAV infection has an overall fatality rate of approximately 0.15 deaths per 1,000 cases; however, a recent study reported 381 deaths out of 115,000 cases, for a case fatality rate of 3.3/1,000. Over 70% of the deaths occurred in adults over 49 years old; thus, the case fatality rate was 27/1,000 (2.7%) in this age group (35).

Laboratory findings in HAV infection are similar to those in other types of viral hepatitis and unfortunately cannot be used to distinguish one form of viral hepatitis from another. The levels of serum aspartate and alanine aminotransferases rise abruptly, with peak levels ranging from less than 300 to more than 3,000 IU/liter. In most cases, though, these levels are higher than 500 IU/liter (22). Serum aminotransferase levels do not correlate well with the degree of liver cell damage. Serum total bilirubin levels typically peak between 5 and 15 mg/100 ml, equally divided between the conjugated and unconjugated fractions (34). Liver enzymes typically normalize within 4 to 6 months and are invariably normal by 12 months. Thrombocytopenia, mild leukopenia and anemia, aplastic anemia, hypoalbuminemia, and prolonged prothrombin and partial thromboplastin times have all been seen in association with HAV infection but are rare (33).

Three forms of prevention for HAV infection are available. First, hygienic measures, including the sanitary disposal of human waste and the maintenance of adequate standards for the purity of drinking water, are critical for preventing diseases. Infections from contaminated drinking water or contaminated shellfish may be prevented by chemical or heat inactivation of the virus (34, 43). Second, passive immunization with immune serum globulin (Ig) can protect individuals after known exposure to HAV or those traveling to high-risk regions of the world. Passive prophylaxis with Ig within the first 2 weeks following exposure to HAV may reduce icteric illness by more than 80% but does not necessarily prevent infection (92). This indicates that antibody to HAV is sufficient for the prevention of HAV infection, independent of cellular immunity. A growing concern is that commercial lots of Ig contain variable concentrations of antibody to HAV (75, 92). In developed countries of the world, the proportion of blood donors who are immune to HAV has decreased to the point that Ig effectiveness may decline (73). For this reason, some countries (e.g., Switzerland) have begun to import lots of Ig to be used for HAV infection prophylaxis.

Finally, active immunization is a recently realized strategy for prevention of hepatitis A. An inactivated HAV vaccine is available in several European countries, and other inactivated, attenuated, or recombinant HAV vaccines are under development (72, 91). Field trials of two formalin-inactivated, cell culture-derived HAV vaccine preparations have demonstrated a high degree of efficacy in preventing disease and in eliciting HAV neutralizing antibodies (42, 87). These inactivated vaccine preparations are very well tolerated, with the major side effect being pain at the injection site (31, 37, 45, 87). It is anticipated that these two vaccine preparations will be licensed in the United States in the next few years. Active immunization in the United States has targeted individuals at the highest risk of HAV infection, such as day-care center workers, travelers to less-developed regions, food handlers, intravenous-drug abusers, and homosexual men. However, it is likely that only universal childhood vaccination will have a major impact on the overall rate of HAV infection (53). Presently, the high cost of production of the inactivated HAV vaccines and the moderate level of morbidity associated with the disease preclude serious consideration of universal childhood vaccination.

DESCRIPTION OF THE AGENT

HAV is a 27-nm nonenveloped icosahedral virus that is a member of the *Picornaviridae* family. HAV was originally classified as an enterovirus; however, because of the size of its proteins, its growth requirements in cell culture, its

resistance to extreme temperatures and pH, and differences in proteolytic processing, HAV has been reclassified as a unique genus called hepatovirus (Table 1). Like those of other picornaviruses, the HAV genome is a positive-sense RNA molecule (11), and viral particles appear to assemble as pentameric precursors that sediment at 14S in sucrose gradients (78). Twelve pentamers assemble in a concentration-dependent manner into empty capsids (approximately 59S to 76S) (68, 78). Following encapsidation of viral RNA, particles sediment at 156S and are infectious. Infectious hepatitis A virions have a buoyant density of 1.32 to 1.34 g/ml in cesium chloride.

The HAV genome is 7,478 nucleotides long and contains a single long open reading frame that begins 735 bases from the 5' end. The 5' noncoding region directs translation by using an internal ribosomal entry site that constitutes 10% of the total genome. A short 3' noncoding region has been poorly characterized. The single open reading frame encodes a polyprotein that undergoes cotranslational, viral-protease-mediated cleavage into nonstructural and structural proteins (Fig. 1). The structural protein precursor (P1 in the nomenclature of Rueckert and Wimmer [69]) is a 98,000-Da protein (5, 90) that appears to be cleaved from the remainder of the polyprotein by the viral 3C protease region (79). This protein is cleaved by the 3C protease region into four proteins named VP0 (1AB), VP3 (1C), VP1 (1D), and PX (1D precursor) (8). HAV structural proteins range in size from 27,000 to 42,000 Da, and the PX protein can be removed from virions by treatment with exogenous proteinases (8, 79, 84). In other picornaviruses, a fourth polypeptide (VP4 [1A]) results from cleavage of VP0 during encapsidation by an uncharacterized mechanism. Although VP4 has not been conclusively demonstrated in HAV, two groups have demonstrated that HAV 1B immunoreactive protein found in empty capsids migrates more slowly (by an estimated 2,500 Da) than the 1B protein found in infectious virions (2, 48), supporting the presence of a small VP4 (1A) molecule that is cleaved from 1B during encapsidation of viral RNA.

HAV has only one serotype and does not react with an enterovirus-specific monoclonal antibody or cDNA probes (67). HAV antigenic structure has been partially characterized by evaluating neutralizing monoclonal antibodies for their abilities to bind virions and subviral particles and by characterizing spontaneously occurring mutants that escape neutralization by these monoclonal antibodies. HAV contains a major immunodominant neutralization site (77) that involves amino acid residues on viral structural proteins 1 and 3 (VP1 and VP3) (57, 60, 61). Although several of these neutralization epitopes are present on 14S pentamer subviral particles, at least one neutralization epitope is formed only when the 14S particles assemble into empty capsids (78). These antibody-binding sites appear to be conformationally dependent, since disruption of the virus by detergents largely destroys antigenicity, and individual capsid proteins do not react with neutralizing monoclonal antibodies. In addition to the antigenic regions on VP1 and VP3, distinct conformationally dependent epitopes involving VP1 and VP2 and a nonconformational neutralization epitope involving VP1 have also been identified (55, 61).

A specific cell receptor for HAV has yet to be characterized, although calcium-dependent HAV binding to permissive tissue culture cells occurs (32, 74), and attachment does not appear to be altered by mutations conferring neutralization resistance to monoclonal antibodies (74). Specific binding of HAV to nonhepatic cells has been

TABLE 1 *Picornaviridae* family

Characteristic	Heparnavirus	Enterovirus	Rhinovirus	Cardiovirus	Aphthovirus
Major members	Human, primate HAVs	Poliovirus, echovirus, coxsackievirus	Human and animal rhinoviruses	Encephalomyocarditis virus	Foot-and-mouth disease viruses
No. of serotypes	1	>70	>110	1	7
Hosts	Humans, other primates	Humans, other vertebrates	Humans, other vertebrates	Humans, other vertebrates, rodents	Most cloven-hoofed animals
Optimal growth temp (°C)	36–37	36–37	33–34	36–37	36–37
Stability:					
60°C	Stable	Labile	Labile	Labile	Labile
pH 3	Stable	Stable	Labile	Stable	Labile
pH 6	Stable	Stable	Becoming stable	Stable	Labile
Structural protein sizes (kDa)					
VPI (1D)	31.9	33.5	31.7	23.3	32.4
VP2 (1B)	24.8	30	29	24.7	28.5
VP3 (1C)	27.8	26.4	25.1	24.3	26.2
VP4 (1A)	2.5	7.4	7.2	8.5	7.2
PX	42	NP[a]	NP	NP	NP
RNA[b]					
G+C content (%)	38	46	39	49	52
5′ ntr[c]	735	741	678	205	1,200
3′ ntr[c]	64	72	47	120	93

[a]NP, not present.
[b]RNAs are from representative strains from each genus.
[c]Number of nucleotides (nt) in addition to those in the poly(A) tail. ntr, nontranslated region.

demonstrated; thus, hepatocyte tropism may not be strictly related to the cell attachment process. In vitro translation of HAV RNA appears to proceed much more efficiently in the presence of hepatocyte cell extracts, suggesting the possibility that preferential replication in the hepatocyte accounts for the hepatic tropism of HAV (27).

The stability of HAV appears to exceed that of all known picornaviruses. HAV is partially resistant to heat and is still infectious after 10 to 12 h at 60°C (63). Inactivation of HAV suspended in buffered saline occurs after 4 min at 70°C and immediately at 85°C. The virus is stable under extremes of pH (34). HAV is generally stable at pH 3 (70). Although variable results have been reported, HAV is generally inactivated by a pH greater than 10 (Table 1). HAV is resistant to treatment with ethyl ether and chloroform, presumably because it lacks a lipid-containing envelope. Hypochlorous acid at a concentration of 1 mg/liter is generally sufficient to inactivate HAV within 30 min. Infectivity can also be abolished by sodium hypochlorite, iodine, and potassium permanganate, although chloramine T and peracetic acid do not inactivate the virus (34).

NATURAL HOSTS

HAV produces disease in humans, chimpanzees, macaque (*Macaca*) monkeys, owl monkeys, stump-tailed monkeys, and several species of South American marmoset monkeys. Disease in nonhuman primates is usually milder, though HAV or viral antigen can generally be detected in serum, liver, bile, and feces. Other primates do not develop clinical infection with HAV. On occasion, monkeys appear to have

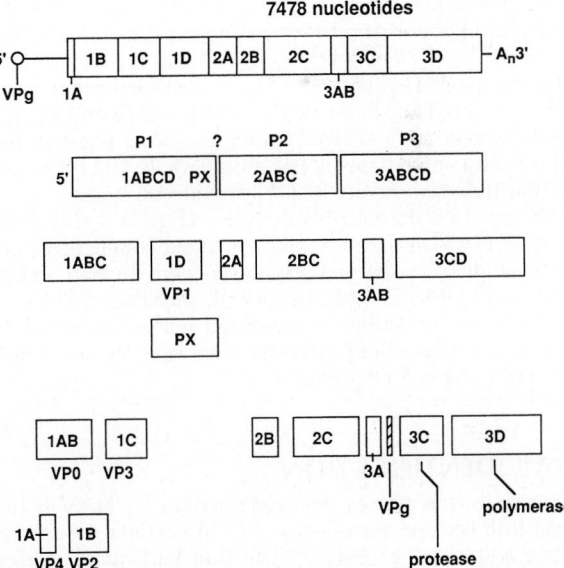

FIGURE 1 Proposed proteolytic cascade for HAV. The HAV genomic RNA, with its 5′ VPg-linked protein, 5′ nontranslated region, long single open reading frame, 3′ nontranslated region, and polyadenylated 3′ end, is depicted in the top line. The single long open reading frame is cleaved into structural proteins (VP0, VP3, VP1, PX) and nonstructural proteins. The nonstructural proteins with identified functions are labeled (protease, polymerase, VPg).

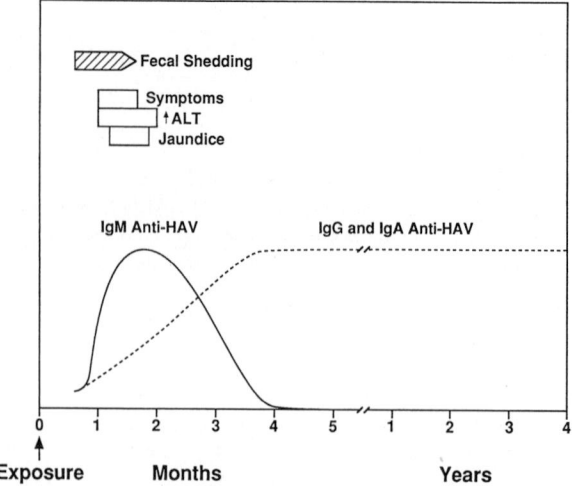

FIGURE 2 Summary of fecal shedding; IgM, IgG, and IgA anti-HAV antibody levels in relationship to exposure; and clinical findings following infection with HAV. ↑ ALT, elevated level of alanine aminotransferase.

anti-HAV antibodies shortly after they are captured, suggesting a nonhuman reservoir for HAV (7, 34, 51).

A great deal of information regarding the virologic and immunologic aspects of HAV infection has been learned from experimental hepatitis A in nonhuman primates. Several virus strains have been used in these primate models. Initial studies done at the National Institutes of Health and the Centers for Disease Control and Prevention used the MS-1 and Phoenix strains of HAV to experimentally infect marmosets (34). The animals developed enzyme elevations after an incubation period of 3 to 4 weeks following intravenous or oral inoculation with the virus. HAV antigen was detected in feces and livers during the acute phase of the illness, and animals developed anti-HAV by the time aminotransferases were elevated. High concentrations of antibody persisted during convalescence (Fig. 2). Fecal shedding of HAV in experimentally infected chimpanzees has been studied in several laboratories with human strains MS-1 and HM-175 (34). In general, shedding of HAV in chimpanzees is neither as prolonged nor as intense as in marmosets; thus, chimpanzees have not been a major source of viral antigen for other studies.

HAV IDENTIFICATION

Direct electron microscopic examination for HAV is impractical, because particles are shed in very low concentrations and may be obscured by other particulate matter. However, immune electron microscopy (IEM) can overcome these obstacles and was the first technique used to identify HAV (16). IEM can distinguish particulate antigen from morphologically similar particles by depositing antibody on the surface of the particle, which may be visualized or identified by labeled second antibodies, and observing for the formation of immune aggregates. Even though this technique is useful, it is time-consuming and expensive and requires an experienced microscopist. New techniques have replaced this one as a diagnostic tool, although a combina-

tion of conventional IEM and solid-phase IEM can be used to identify HAV in stool and cell culture extracts (40).

Although HAV from clinical specimens can be propagated in cell culture (see Isolation Procedures below), growth is slow and unpredictable, and several months of blind passages are usually required before viral antigen can be detected. Consequently, isolation of virus is not a practical tool for identifying HAV from clinical specimens.

Because HAV is highly antigenic, a variety of serologic methods have been developed for identifying HAV antigen. Radioimmunoassay (RIA) techniques were among the first methods established to identify HAV antigen (34, 71). Unlabeled anti-HAV antibody preparations (usually sera obtained from early-convalescent-phase patients) is bound to the plastic surface of a microtiter well, test samples are applied, and ^{125}I-labeled IgG anti-HAV antibody is added to detect bound HAV. The test is considered positive for HAV if the ratio of counts per minute detected by gamma scintillation counting is greater than or equal to 2.1 times the mean counts per minutes for negative control samples. Enzyme-linked immunosorbent assay or enzyme immunoassay (EIA) techniques have largely replaced RIAs for the detection of HAV antigen. EIA substitutes an enzyme-conjugated antibody probe for the radionuclide-conjugated antibody probe of RIA. Horseradish peroxidase or alkaline phosphatase may be used to label the second antibody. Hydrolysis of the substrate occurs when HAV is present and can be measured visually or spectrophotometrically. Antigen detection can be enhanced by using a nitrocellulose membrane as a high-capacity solid-phase material for the nonspecific capture of antigen (56). The nitrocellulose-EIA method can be used to detect as little as 1 ng of HAV protein, which corresponds to 1.5×10^4 virus particles (56).

HAV infection may also be detected by molecular hybridization of HAV RNA with cloned HAV cDNA (82, 83). Hybridization is approximately 10-fold more sensitive than IEM or RIA, and specificity can be enhanced by purifying HAV from clinical specimens in an immunoaffinity step with monoclonal anti-HAV antibodies (43). HAV-specific single-stranded RNA probes have been used in dot blot hybridization tests to detect HAV in environmental samples (94). Radioactive and nonradioactive RNA probes are generated by in vitro transcription of HAV templates inserted in a plasmid vector. HAV RNA is purified from samples by proteinase K digestion followed by phenol-chloroform extraction and ethanol precipitation (44). Hybridization with RNA probes is generally performed at higher stringency than that with cDNA probes.

The PCR has proven to be a useful tool in identifying HAV in clinical specimens. By utilizing the immunoaffinity step described for cDNA-RNA hybridization, specificity is enhanced to the point that the method can be used on fecal specimens (44). RNA may be extracted from immunoaffinity-purified virus with proteinase K digestion, phenol-chloroform extraction, and ethanol precipitation. Alternatively, heating the samples to 95°C following immunoaffinity purification releases viral RNA from the capsids sufficiently for reverse transcription. HAV RNA is converted to cDNA by reverse transcription, and cDNA is amplified by PCR with a thermostable DNA polymerase. By this method, products can be detected by ethidium bromide staining in agarose gels and by Southern transfer or dot blotting using oligonucleotide probes to detect HAV-specific sequences (44).

Strains have been compared by utilizing sequence analysis of cDNA amplified by the PCR. Genotypic differences are detected by comparing the sequences of the PCR products from feces-derived or cell culture-adapted HAV isolates. Sequence analysis of geographically diverse isolates has categorized virus strains into seven subgroups based on their RNA sequence similarities (44, 65). These subgroups of HAV are indistinguishable by EIA and RIA methods but vary by 15 to 25% in their RNA sequences. The conservation of HAV antigenicity suggests that the antigenic site plays a critical role in the life cycle of the virus.

COLLECTION, TRANSPORT, AND STORAGE OF SPECIMENS

Standard separation and storage of sera allow accurate serologic results for both IgM anti-HAV and total anti-HAV antibodies. Antibody titers are stable and anti-HAV levels in lots of IG do not decrease significantly when stored at 4°C for more than 3 weeks; however, repetitive freezing and thawing of sera may lead to a diminution in titer (75, 76). Commercial serologic assays require that sera be stored at 4°C for no longer than 5 days prior to testing.

Samples of feces may contain HAV antigen if they were collected during the 2 weeks prior to illness or several days after the onset of symptoms (Fig. 2). Under unusual circumstances, fecal virus shedding may be protracted, especially in infants (66). Twenty percent fecal slurries or 50% saliva preparations are prepared by using phosphate-buffered saline containing 0.02% sodium azide. These specimens can be stored at −70°C for more than 6 months without significant loss of antigenicity or antibody titer (76). Liver biopsy specimens may also be obtained for detection of HAV particles by immunofluorescence or electron microscopy. Saliva and bile may also be collected for antibody testing; however, these specimens are not approved for commercial diagnostic immunoassays. To preserve infectivity, HAV must be stored at 4°C or frozen to −20 to −70°C. When the virus is stored at 4°C, infectivity is reduced only 0.5 $\log_{10}$ after 16 weeks, and titers remain stable at both −20 and −70°C for over 6 months (71).

ISOLATION PROCEDURES

HAV is very difficult to isolate and propagate in vitro. Even when well adapted to cell culture, HAV takes several days to weeks to reach maximal titers. Initial in vitro propagation of HAV was done with freshly extracted primary adult marmoset hepatocyte cultures (62). Hepatocytelike epithelial cell outgrowth appeared at 7 days. It was later discovered that primary African green monkey cells could be used to grow HAV. Many other human and primate cell lines, including fetal rhesus kidney cells (FrhK-4 and -6) (17, 62), human diploid embryonic lung fibroblast (20, 23, 25), hepatocyte-derived cells (PLC/PRF/5 and HepG2) (23), LLC-MK$_2$, Vero, MRC-5, and other transformed cell lines (91), have since been used to propagate HAV. HAV has been isolated in cell culture from human and experimentally infected primate liver, bile, feces, and sera. Viral antigen yields in the most productive systems are no more than 100 ng/ml. Total viral particle counts are estimated to be about 100-fold greater than the infectious titer. However, after adaptation to cell culture, this ratio diminishes considerably (52, 71).

HAV ANTIBODY DETECTION

Commercially available assays for anti-HAV IgM are the methods of choice for establishing the diagnosis of acute type A hepatitis. To identify total anti-HAV antibody, a competition immunoassay that measures the ability of test sera to compete with labeled anti-HAV antibody (either radiolabeled or enzyme linked) for binding to antigen attached to the solid phase is usually employed. To identify acute infection, an IgM-specific antibody assay is required (Fig. 2). The most commonly used method is a solid-phase antibody capture immunoassay. The solid phase is coated with anti-human IgM. The presence of antibody to HAV in the "captured" IgM from the test serum is indicated by the sequential binding of HAV and labeled anti-HAV reagents (83). Results are usually evaluated by comparison with a cutoff value that is determined from positive and negative controls supplied by the manufacturer. Almost all hepatitis A patients are anti-HAV IgM positive at the onset of symptoms and IgM antibody negative 6 months following infection. Sera obtained from recipients of Ig do not contain levels of anti-HAV antibody detectable by older commercial immunoassays (75, 91); however, current assays may identify anti-HAV antibody in individuals who receive intravenous Ig on a regular basis for immunologic disorders (3).

The Abbott IMx is a fully automated immunoassay based on fluorescence polarization assay and microparticle capture EIA (15). This method is comparable to routine assays for anti-HAV IgM but is reportedly more rapid and convenient.

Several methods have been developed to evaluate sera for HAV neutralizing activity. All utilize cell culture-adapted virus that is incubated with complement-depleted sera prior to virus titration in cell culture. Neutralizing activity is determined on the basis of a 50 or 90% reduction in virus foci or by a 50% reduction in the tissue culture infective dose. Because HAV is generally noncytopathic in cell culture, novel approaches to determining viral infectious titers have been developed. Methods using RIA or EIA to detect virus growth in a terminal-dilution microtiter-based assay have been described (18), as have modifications of a more standard plaque assay (50). This "radioimmunofocus assay" requires a 7- to 14-day incubation of HAV-infected cell monolayers overlaid with agarose. After incubation, the agarose is gently removed, the cells are fixed with acetone, and foci of HAV replication are detected by using ^{125}I-labeled anti-HAV IgG. Variations on these methods have also been described (46). Because of the labor and time involved in these assays (generally 3 weeks per assay), these methods are not practical for general use. Quantitation of neutralizing activity in sera is generally determined by one of two methods. Following incubation with a known inoculum of HAV with serial dilution of sera, the dilution of sera that inhibits growth of virus by 50 or 90% is determined (neutralization titer). Alternatively, a single dilution of antibody (or serum) is incubated with serial dilutions of virus inoculum, and the $\log_{10}$ reduction in HAV infectivity is calculated. The former method is more widely employed.

Saliva and urine provide convenient sources of testing material, particularly for children and for epidemiologic studies. However, these specimens are not approved for commercial immunoassays. EIA can reliably detect IgM anti-HAV antibodies in the saliva of patients with acute or recent hepatitis A but not in the saliva of patients without recent HAV infection (58, 59). The IgM anti-HAV antibody persists for 2 to 4 months. The ratio of IgM to IgG anti-HAV antibodies correlates closely with the interval from the onset of infection. IgA and IgG anti-HAV antibodies can also be detected in the saliva of persons with hepatitis A or in those who have recovered from HAV infection (58). The role that IgA anti-HAV antibody plays in preventing infection is unclear, as saliva and fecal antibodies did not exhibit neutralizing activity after experimental and natural HAV infections (76).

Another approach to diagnostic serologic testing utilizes a hemagglutination inhibition assay. Virus-specific hemagglutination of various species of erythrocytes has been demonstrated for HAV. This activity is blocked by IgG and IgM anti-HAV antibodies and is not affected by inactivation of HAV infectivity by 0.03% β-propriolactone (14, 80). Thus, methods other than EIA may be used for the diagnosis of recent HAV infection.

IMMUNOLOGIC ASPECTS OF HEPATITIS A

It is clear that antibody to HAV is critical and sufficient by itself to prevent reinfection (91); however, several lines of evidence indicate a potential role for cellular immunity in the clearance of HAV from hepatocytes and in HAV pathogenesis (19, 85, 86). HAV persistently infects cells in vitro and does not cause a direct cytopathic effect (34), and peak viral shedding in feces and peak viremia in experimental HAV infection occur before the development of symptoms (34, 49). Anti-HAV antibody is almost always present at the time of symptoms, and in the unusual cases where it is not, gamma interferon is detectable prior to seroconversion (93). All of these data indirectly suggest that cell-mediated immune destruction of HAV-infected hepatocytes leads to hepatitis.

Among the potential mediators of cell-mediated immunity, antibody-dependent cellular cytotoxicity apparently does not occur in HAV infection (24). Conflicting reports about the presence (47) or absence of natural killer cell responses to HAV-infected cells exist (22).

Cell-mediated cytotoxicity is important in the immune response to HAV infection (24). Virus-specific cytotoxic T cells develop in the course of HAV infection, as measured by a microcytotoxicity assay of infected peripheral blood lymphocytes in infected patients (85, 86). Clonal analysis of the T lymphocytes in the liver during acute HAV infection (19) identified antigen-specific CD8$^+$ T lymphocytes from liver biopsy samples obtained during acute HAV infection, implicating these cells as responsible for the destruction of infected hepatocytes (19). Nearly 50% of liver-derived T-cell clones displayed HAV-specific cytotoxicity compared with <1% of peripheral blood lymphocyte clones, further supporting the role of cytolytic T cells in the pathogenesis of acute hepatitis A.

ANTIVIRAL SUSCEPTIBILITIES

A variety of compounds with antiviral activity against picornaviruses or other classes of viruses have been evaluated for activity against HAV. The antipicornavirus drug Disoxaril (WIN 51711) inhibits uncoating of a number of human picornaviruses but has no effect on the H141 strain of HAV and only a modest inhibitory effect on three other strains tested at high concentrations (88). Highly charged antiviral compounds such as protamine and carrageenan demonstrate specific inhibitory effects on HAV replication

in cell culture, although the mechanism of action has not been characterized (4, 26). Finally, ribavirin and amantadine have demonstrated in vitro antiviral activities at concentrations achievable in vivo (12, 89). No clinical trials of these antiviral trials have been reported.

INTERPRETATIONS

Evaluation of IgM and total Ig to HAV by EIA provides the diagnosis of acute or past infection with hepatitis A. Anti-HAV IgM is almost always present in serum at the time of symptoms from hepatitis A and may be present for several months after the acute illness (Fig. 2). Therefore, IgM anti-HAV antibody can be used to make a diagnosis of recent HAV infection (22, 34). Although it is often called the IgG anti-HAV test, the total-anti-HAV assays also measure IgM and IgA antibodies to HAV. The total-anti-HAV-Ig assay is positive in acute hepatitis A and remains positive indefinitely (34, 38). The pattern of positive total anti-HAV antibody and negative IgM anti-HAV antibody indicates that the patient has had an infection with HAV and is protected against reinfection. However, if there is an exogenous source of anti-HAV in the patient, the total-anti-HAV-antibody assay may be positive. For example, patients who have had recent transfusions, newborn infants (in the first 6 months of life), and frequent recipients of intravenous Ig may have received enough circulating anti-HAV antibody for it to be detectable by the EIA (3, 38).

REFERENCES

1. **Akriviadis, E. A., and A. G. Redeker.** 1989. Fulminant hepatitis A in intravenous drug users with chronic liver disease. *Ann. Intern. Med.* **110:**838–839.
2. **Anderson, D. A., and B. C. Ross.** 1990. Morphogenesis of hepatitis A virus: isolation and characterization of subviral particles. *J. Virol.* **64:**5284–5289.
3. **Ballas, Z. K., and J. A. Hudson.** 1993. Hepatitis A, B, and C seroconversion in patients receiving intravenous immunoglobulin (IVIG). *J. Allergy Clin. Immunol.* **91:**148.
4. **Biziagos, E., J. M. Crance, J. Passagot, and R. Deloince.** 1987. Effect of antiviral substances on hepatitis A virus replication in vitro. *J. Med. Virol.* **22:**57–66.
5. **Borovec, S. V., and D. A. Anderson.** 1993. Synthesis and assembly of hepatitis A virus-specific proteins in BS-C-1 cells. *J. Virol.* **67:**3095–3102.
6. **Bromberg, K., D. N. Newhall, and G. Peter.** 1982. Hepatitis A and meningoencephalitis. *JAMA* **247:**815.
7. **Brown, E. A., R. W. Jansen, and S. M. Lemon.** 1989. Characterization of a simian hepatitis A virus (HAV): antigenic and genetic comparison with human HAV. *J. Virol.* **63:**4932–4937.
8. **Cederna, J. B., D. Klinzman, J. McLinden, P. L. Winokur, and J. T. Stapleton.** Unpublished data.
9. **Chio, F., Jr., and A. A. Bakir.** 1992. Acute renal failure in hepatitis A. *Int. J. Artif. Organs* **15:**413–416.
10. **Corey, L., and K. K. Holmes.** 1980. Sexual transmission of hepatitis A in homosexual men. *N. Engl. J. Med.* **302:**435–446.
11. **Coulepis, A. G., S. A. Locarnini, E. G. Westaway, G. A. Tannock, and I. D. Gust.** 1982. Biophysical and biochemical characterization of hepatitis A virus. *Intervirology* **18:**107–127.
12. **Crance, J. M., E. Biziagos, J. Passagot, H. V. Cuyck-Gandre, and R. Deloince.** 1990. Inhibition of hepatitis A virus replication in vitro by antiviral compounds. *J. Med. Virol.* **31:**155–160.
13. **Dollberg, S., Y. Berkun, and E. Gross-Kieselstein.** 1991. Urticaria in patients with hepatitis A virus infection. *Pediatr. Infect. Dis. J.* **10:**702–703.
14. **Dubois, D. R., L. N. Binn, P. L. Summers, R. L. Timchak,**

D. A. Barvir, R. H. Marchwicki, and K. H. Eckels. 1990. Preparation of noninfectious hepatitis A virus hemagglutination for detecting hemagglutination inhibition antibodies. *J. Virol. Methods* **28:**299–304.

15. Fayol, V., and G. Ville. 1991. Evaluation of automated enzyme immunoassays for several markers for hepatitis A and B using the Abbott IMx analyser. *J. Clin. Chem. Clin. Biochem.* **29:**67–70.

16. Feinstone, S. M., A. Z. Kapikian, and R. H. Purcell. 1973. Hepatitis A: detection by immune electron microscopy of a virus like antigen associated with acute illness. *Science* **182:**1026–1028.

17. Flehmig, B. 1981. Hepatitis A virus in cell culture. II. Growth characteristics of hepatitis A virus in Frhk-4/R cells. *Med. Microbiol. Immunol.* **170:**73–81.

18. Flehmig, B., M. Ranke, T. Berthold, and H. J. Gerth. 1979. A solid phase radioimmunoassay for detection of IgM antibodies to hepatitis A virus. *J. Infect. Dis.* **140:**169–175.

19. Fleischer, B., S. Fleischer, K. Maier, K. H. Wiedmann, and A. Vallbracht. 1990. Clonal analysis of infiltrating T lymphocytes in liver tissue in viral hepatitis A. *Immunology* **69:**14–19.

20. Fleming, B., A. Vallbracht, and G. Wurster. 1981. Hepatitis A virus in cell culture. III. Propagation of hepatitis A virus in human embryo kidney and human embryo fibroblasts. *Med. Microbiol. Immunol.* **170:**83–89.

21. Francis, D. P., S. C. Hadler, T. J. Prendergast, E. Peterson, M. M. Ginsberg, C. Lookabaugh, J. R. Holmes, and J. E. Maynard. 1984. Occurrence of hepatitis A, B, and non-A/non-B in the United States. CDC Sentinel County Hepatitis Study 1. *Am. J. Med.* **76:**69–74.

22. Friedman, L. S., and J. L. Dienstag. 1984. The disease and its pathogenesis, p. 55–79. *In* R. J. Gerety (ed.), *Hepatitis A.* Academic Press, London.

23. Frosner, G. G., F. Deinhardt, R. Scheid, V. Gauss-Muller, N. Holmes, V. Messelberger, G. Siegl, and J. J. Alexander. 1979. Propagation of human hepatitis A virus in a hepatoma cell line. *Infection* **7:**303–305.

24. Gabriel, P., A. Vallbracht, and B. Flehmig. 1986. Lack of complement-dependent cytolytic antibodies in hepatitis A virus infection. *J. Med. Virol.* **20:**23–31.

25. Gauss-Muller, V., G. G. Frosner, and F. Deinhardt. 1981. Propagation of hepatitis A virus in human embryo fibroblasts. *J. Med. Virol.* **7:**233–239.

26. Girond, S., J. M. Crance, H. Van Cuyck-Gandre, J. Renaudet, and R. Deloince. 1991. Antiviral activity of carrageenan on hepatitis A virus replication in cell culture. *Res. Virol.* **142:**261–270.

27. Glass, M. J., and D. F. Summers. 1993. Identification of a trans-acting activity from liver that stimulates hepatitis A virus translation *in vitro*. *Virology* **193:**1047–1050.

28. Glikson, M., E. Galun, R. Oren, R. Tur-Kaspa, and D. Shouval. 1992. Relapsing hepatitis A. Review of 14 cases and literature survey. *Medicine* (Baltimore) **71:**14–23.

29. Gordon, S. C., A. S. Patel, R. J. Veneri, K. A. Keskey, and S. M. Korotkin. 1989. Case report: acute type A hepatitis presenting with hypotension, bradycardia, and sinus arrest. *J. Med. Virol.* **28:**219–222.

30. Gordon, S. C., K. R. Reddy, L. Schiff, and E. R. Schiff. 1984. Prolonged intrahepatic cholestasis secondary to acute hepatitis A. *Ann. Intern. Med.* **101:**635–637.

31. Goubau, P., V. Van Gerven, A. Safary, A. Delem, J. Knops, E. D'Hondt, F. E. André, and J. Desmyter. 1992. Effect of virus strain and antigen dose on immunogenicity and reactogenicity of an inactivated hepatitis A vaccine. *Vaccine* **10**(Suppl. 1):S114–S118.

32. Grace, K., E. Amphlett, S. Day, S. Lemon, D. Sangar, D. J. Rowlands, and B. E. Clarke. 1991. *In vitro* translation of hepatitis A virus subgenomic RNA transcripts. *J. Gen. Virol.* **72:**1081–1086.

33. Gundersen, S. G., A. Bjoerneklett, and J. N. Bruun. 1989. Severe erythroblastopenia and hemolytic anemia during a hepatitis A infection. *Scand. J. Infect. Dis.* **21:**225–228.

34. Gust, I. D., and S. M. Feinstone. 1988. Hepatitis A. CRC Press, Inc., Boca Raton, Fla.

35. Hadler, S. C. 1991. *Viral Hepatitis and Liver Disease*, p. 14–20. The Williams & Wilkins Co., Baltimore.

36. Hadler, S. C., H. M. Webster, J. J. Erben, J. E. Swanson, and J. E. Maynard. 1980. Hepatitis A in day-care centers. A community-wide assessment. *N. Engl. J. Med.* **302:**1222–1227.

37. Hoke, C. H., Jr., L. N. Binn, J. E. Egan, R. F. DeFraites, P. O. MacArthy, B. L. Innis, K. H. Eckels, D. Dubois, E. D'Hondt, M. H. Sjorgren, R. Rice, J. C. Sadoff, and W. H. Bancroft. 1992. Hepatitis A in the US Army: epidemiology and vaccine development. *Vaccine* **10**(Suppl. 1):S75–S79.

38. Hollinger, F. B., and A. P. Glombicki. 1990. Hepatitis A virus, p. 1383–1399. *In* G. L. Mandell, Jr., R. G. Douglas, and J. E. Bennett (ed.), *Principles and Practices of Infectious Diseases.* Churchill Livingstone, New York.

39. Hollinger, F. B., N. C. Khan, and P. E. Oefinger. 1983. Posttransfusion hepatitis type A. *JAMA* **250:**2313–2317.

40. Humphrey, C. D., E. H. Cook, Jr., and D. W. Bradley. 1990. Identification of enterically transmitted hepatitis virus particles by solid phase immune electron microscopy. *J. Virol. Methods* **29:**177–188.

41. Inman, R. D., M. Hodge, M. E. A. Johnston, J. Wright, and J. Heathcote. 1986. Arthritis, vasculitis, and cryoglobulinemia associated with relapsing hepatitis A virus infection. *Ann. Intern. Med.* **105:**700–703.

42. Innis, B. L., R. Snitbhan, P. Kunasol, T. Laorakpongse, W. Poopatanakool, S. Suntayakorn, T. Suknantapong, A. Safary, and J. W. Boslego. 1992. Field efficacy trial of inactivated hepatitis A vaccine among children in Thailand (an extended abstract). *Vaccine* **10**(Suppl. 1):S159.

43. Jansen, R. W., J. E. Newbold, and S. M. Lemon. 1985. Combined immunoaffinity cDNA-RNA hybridization assay for detection of hepatitis A virus in clinical specimens. *J. Clin. Microbiol.* **22:**984–989.

44. Jansen, R. W., G. Siegl, and S. M. Lemon. 1990. Molecular epidemiology of human hepatitis A virus defined by an antigen-capture polymerase chain reaction method. *Proc. Natl. Acad. Sci. USA* **87:**2867–2871.

45. Just, M., and R. Berger. 1992. Reactogenicity and immunogenicity of inactivated hepatitis A vaccines. *Vaccine* **10**(Suppl. 1):S110–S113.

46. Krah, D. L., R. D. Amin, D. R. Nalin, and P. J. Provost. 1991. A simple antigen-reduction assay for the measurement of neutralizing antibodies to hepatitis A virus. *J. Infect. Dis.* **163:**634–637.

47. Kurane, I., L. N. Binn, W. H. Bancroft, and F. A. Ennis. 1985. Human lymphocyte responses to hepatitis A virus-infected cells: interferon production and lysis of infected cells. *J. Immunol.* **135:**2140–2144.

48. Kusov, Y. Y., Y. A. Kazachkov, G. K. Dzagurov, G. A. Khozinskaya, M. S. Balayan, and V. Gauss-Müller. 1992. Identification of precursors of structural proteins VP1 and VP2 of hepatitis A virus. *J. Med. Virol.* **37:**220–227.

49. Lemon, S. M., L. N. Binn, R. Marchwicki, P. C. Murphy, L. H. Ping, R. W. Jansen, L. V. S. Asher, J. T. Stapleton, D. W. Taylor, and J. W. LeDuc. 1990. In vivo replication and reversion to a wild type of a neutralization-resistant antigenic variant of hepatitis A virus. *J. Infect. Dis.* **161:**7–13.

50. Lemon, S. M., L. N. Binn, and R. H. Marchwicki. 1983. Radioimmunofocus assay for quantitation of hepatitis A virus in cell culture. *J. Clin. Microbiol.* **17:**834–839.

51. Lemon, S. M., S. F. Chao, R. W. Jansen, L. N. Binn, and J. W. LeDuc. 1987. Genomic heterogeneity among human and nonhuman strains of hepatitis A virus. *J. Virol.* **61:**735–742.

52. Lemon, S. M., R. W. Jansen, and J. E. Newbold. 1985. Infectious hepatitis A particles produced in cell culture consist of three distinct types with different buoyant densities in CsCl. *J. Virol.* **54:**78–85.

53. Lemon, S. M., and J. T. Stapleton. 1993. Prevention of

hepatitis A, p. 61–79. *In* A. J. Zuckerman and H. C. Thomas (ed.), *Viral Hepatitis*. Churchill Livingstone, London.

54. **Ludmerer, K. M., and J. M. Kissane.** 1984. Fulminant hepatic failure in a 21-year old man. *Am. J. Med.* **76:**718–724.

55. **Murphy, P., L. H. Ping, and S. M. Lemon.** 1993. Peptide scanning demonstrates the presence of a linear epitope located near the carboxy terminus of VP1 of hepatitis A virus, abstr. 014, p. 102. *8th Int. Symp. Viral Hepatitis Liver Dis.*

56. **Nadala, E. C. B., and P. C. Loh.** 1990. A nitrocellulose-enzyme immunoassay method for the detection of hepatitis A virus. *J. Virol. Methods* **28:**155–164.

57. **Nainan, O. V., M. A. Brinton, and H. S. Margolis.** 1992. Identification of amino acids located in the antibody binding sites of human hepatitis A virus. *Virology* **191:**984–987.

58. **Parry, J. V., K. R. Perry, S. Panday, and P. P. Mortimer.** 1989. Diagnosis of hepatitis A and B by testing saliva. *J. Med. Virol.* **28:**255–260.

59. **Perry, K. R., J. V. Parry, E. M. Vandervelde, and P. P. Mortimer.** 1992. The detection in urine specimens of IgG and IgM antibodies to hepatitis A and hepatitis B core antigens. *J. Med. Virol.* **38:**265–270.

60. **Ping, L.-H., R. W. Jansen, J. T. Stapleton, J. I. Cohen, and S. M. Lemon.** 1988. Identification of an immunodominant antigenic site involving the capsid protein VP3 of hepatitis A virus. *Proc. Natl. Acad. Sci. USA* **85:**8281–8285.

61. **Ping, L.-H., and S. M. Lemon.** 1992. Antigenic structure of human hepatitis A virus defined by analysis of escape mutants selected against murine monoclonal antibodies. *J. Virol.* **66:** 2208–2216.

62. **Provost, P. J., and M. R. Hilleman.** 1979. Propagation of human hepatitis A virus in cell culture *in vitro. Proc. Soc. Exp. Biol. Med.* **160:**213–221.

63. **Provost, P. J., B. S. Wolanski, W. J. Miller, O. L. Ittensohn, W. J. McAleer, and M. R. Hilleman.** 1975. Physical, chemical and morphological dimensions of human hepatitis A virus strain CR326. *Proc. Soc. Exp. Biol. Med.* **148:**532–539.

64. **Rakela, J., S. M. Lange, J. Ludwig, and W. P. Baldus.** 1985. Fulminant hepatitis: Mayo Clinic experience with 34 cases. *Mayo Clin. Proc.* **60:**289–292.

65. **Robertson, B. H., R. W. Jansen, B. Khanna, A. Totsuka, O. V. Nainan, G. Siegl, A. Widell, H. S. Margolis, S. Isomura, K. Ito, T. Ishizu, Y. Moritsugu, and S. M. Lemon.** 1992. Genetic relatedness of hepatitis A virus strains recovered from different geographical regions. *J. Gen. Virol.* **73:** 1365–1377.

66. **Rosenblum, L. S., M. E. Villarino, O. V. Nainan, M. E. Melish, S. C. Hadler, P. P. Pinsky, W. R. Jarvis, C. E. Ott, and H. S. Margolis.** 1991. Hepatitis A outbreak in a neonatal intensive care unit: risk factors for transmission and evidence of prolonged viral excretion among preterm infants. *J. Infect. Dis.* **164:**476–482.

67. **Rotbart, H. A.** 1989. Human enterovirus infections: molecular approaches to diagnosis and pathogenesis, p. 243–264. *In* B. L. Semler and E. Ehrenfeld (ed.), *Molecular Aspects of Picornavirus Infection and Detection.* American Society for Microbiology, Washington, D.C.

68. **Ruchti, F., G. Siegl, and M. Weitz.** 1991. Identification and characterization of incomplete hepatitis A virus particles. *J. Gen. Virol.* **72:**2159–2166.

69. **Rueckert, R. R., and E. Wimmer.** 1984. Systematic nomenclature of picornavirus proteins. *J. Virol.* **50:**957–959.

70. **Scholz, E., U. Heinricy, and B. Flehmig.** 1989. Acid stability of hepatitis A virus. *J. Gen. Virol.* **70:**2481–2485.

71. **Siegl, G.** 1984. The biochemistry of hepatitis A virus, p. 9–32. *In* R. J. Gerety (ed.), *Hepatitis A.* Academic Press, London.

72. **Siegl, G., and S. M. Lemon.** 1990. Recent advances in hepatitis A vaccine development. *Virus Res.* **17:**75–92.

73. **Stapleton, J. T.** 1992. Passive immunization against hepatitis A. *Vaccine* **10**(Suppl. 1):S45–S47.

74. **Stapleton, J. T., J. Frederick, and B. Meyer.** 1991. Hepatitis A virus attachment to cultured cell lines. *J. Infect. Dis.* **164:** 1098–1103.

75. **Stapleton, J. T., R. Jansen, and S. M. Lemon.** 1985. Neu-

tralizing antibody to hepatitis A virus in immune serum globulin and in the sera of human recipients of immune serum globulin. *Gastroenterology* **89:**637–642.

76. **Stapleton, J. T., D. K. Lange, J. W. LeDuc, L. N. Binn, R. W. Jansen, and S. M. Lemon.** 1991. The role of secretory immunity in hepatitis A virus infection. *J. Infect. Dis.* **163:** 7–11.

77. **Stapleton, J. T., and S. M. Lemon.** 1987. Neutralization escape mutants define a dominant immunogenic neutralization site on hepatitis A virus. *J. Virol.* **61:**491–498.

78. **Stapleton, J. T., V. Raina, P. L. Winokur, K. Walters, D. Klinzman, E. Rosen, and J. H. McLinden.** 1993. Antigenic and immunogenic properties of recombinant hepatitis A virus 14S and 70S subviral particles. *J. Virol.* **67:**1080–1085.

79. **Stapleton, J. T., E. Rosen, D. Klinzman, and J. McLinden.** 1993. Protease digestion of hepatitis A virus proteins present in recombinant subviral particles, abstr. 003, p. 99. *8th Int. Symp. Viral Hepatitis Liver Dis.*

80. **Summers, P. L., D. R. Dubois, W. Houston Cohen, P. O. MacArthy, L. N. Binn, M. H. Sjogren, R. Snitbhan, B. L. Innis, and K. H. Eckels.** 1993. Solid-phase antibody capture hemadsorption assay for detection of hepatitis A virus immunoglobulin M antibodies. *J. Clin. Microbiol.* **31:**1299–1302.

81. **Tabor, E.** 1984. Clinical presentation of hepatitis A, p. 47–53. *In* R. J. Gerety (ed.), *Hepatitis A.* Academic Press, London.

82. **Ticehurst, J., and R. H. Purcell.** 1989. Hepatitis A virus and diagnostic techniques for hepatitis A, p. 957–1065. *In* N. J. Schmidt and R. Emmons (ed.), *Diagnostic Procedures for Viral, Rickettsial, and Chlamydial Infections.* American Public Health Association, Washington, D.C.

83. **Ticehurst, J. R., S. M. Feinstone, T. Chestnut, N. C. Tassopoulos, H. Popper, and R. H. Purcell.** 1987. Detection of hepatitis A virus by extraction of viral RNA and molecular hybridization. *J. Clin. Microbiol.* **25:**1822–1829.

84. **Totsuka, A., H. Yamamoto, and Y. Moritsugu.** 1993. Proteolytic cleavage sites in hepatitis A virus polyprotein, abstr. 004, p. 99. *8th Int. Symp. Viral Hepatitis Liver Dis.*

85. **Vallbracht, A., P. Gabriel, K. Maier, F. Hartmann, and B. Flehmig.** 1986. Cell-mediated cytotoxicity in hepatitis A virus infection. *Hepatology* **6:**1308–1314.

86. **Vallbracht, A., K. Maier, Y. D. Stierhof, K. H. Wiedmann, B. Flehmig, and B. Fleischer.** 1989. Liver-derived cytotoxic T cells in hepatitis A virus infection. *J. Infect. Dis.* **160:**209–217.

87. **Werzberger, A., B. Mensch, B. Kuter, L. Brown, J. Lewis, R. Sitrin, W. Miller, D. Shouval, B. Wiens, G. Calandra, J. Ryan, P. Provost, and D. Nalin.** 1992. A controlled trial of a Formalin-inactivated hepatitis A vaccine in healthy children. *N. Engl. J. Med.* **327:**453–457.

88. **Widell, A.** 1988. Hepatitis A virus virological studies. Ph.D. dissertation. University of Lund, Lund, Sweden.

89. **Widell, A., B. G. Hansson, B. Oberg, and E. Nordenfelt.** 1986. Influence of twenty potentially antiviral substances on *in vitro* multiplication of hepatitis A virus. *Antiviral Res.* **6:**103–112.

90. **Winokur, P. L., J. H. McLinden, and J. T. Stapleton.** 1991. Primary processing of the hepatitis A virus polyprotein expressed by a recombinant vaccinia virus. *Clin. Res.* **39:**215A.

91. **Winokur, P. L., J. H. McLinden, and J. T. Stapleton.** 1991. The hepatitis A virus polyprotein expressed by a recombinant vaccinia virus undergoes proteolytic processing and assembly into virus-like particles. *J. Virol.* **65:**5029–5036.

92. **Winokur, P. L., and J. T. Stapleton.** 1992. Immunoglobulin prophylaxis for hepatitis A. *Clin. Infect. Dis.* **14:**580–586.

93. **Zachoval, R., M. Kroener, M. Brommer, and F. Deinhardt.** 1990. Serology and interferon production during the early phase of acute hepatitis A. *J. Infect. Dis.* **161:**353–354.

94. **Zhou, Y.-J., M. K. Estes, X. Jiang, and T. G. Metcalf.** 1991. Concentration and detection of hepatitis A virus and rotavirus from shellfish by hybridization tests. *Appl. Environ. Microbiol.* **57:**2963–2968.

Hepatitis B and D Viruses

F. BLAINE HOLLINGER AND JULES L. DIENSTAG

91

CLINICAL AND EPIDEMIOLOGIC FEATURES

Viral hepatitis is primarily a disease of the liver that is caused by a number of well-characterized viruses, including hepatitis B virus (HBV) and HDV. The clinical features of hepatitis B are extremely variable. In most acutely infected patients, inapparent (subclinical) hepatitis develops without symptoms or jaundice. Other patients acquire symptoms without jaundice (anicteric hepatitis), while jaundice with symptoms (icteric hepatitis) occurs in approximately 25 to 35% of those infected with HBV. Symptoms may range from mild and transient to severe and prolonged. Patients may recover completely, be overwhelmed by fulminant hepatitis and die (<1.5% of hospitalized patients), or become chronically infected.

Hepatitis B

The incubation period of acute hepatitis B ranges from 6 to 16 weeks in most patients. The incubation period may be shortened by a large inoculum and infection by the percutaneous route; it may be prolonged by physicochemical alteration of HBV (UV irradiation, heat, or storage at room temperature), coinfection with HCV, or administration of high-titer anti-hepatitis B surface antigen (HBsAg) antibody (anti-HBs) to HBV-infected persons shortly after exposure. Variations in host response can also lead to differences in the incubation period.

The prodromal or preicteric phase of hepatitis B tends to be more insidious and prolonged than that of hepatitis A, lasting from 1 to 2 weeks in 80% of patients. It is frequently characterized by malaise, low-grade fever (<39.5°C), fatigability, myalgia, blunting of olfactory and gustatory senses, anorexia, nausea, and vomiting. Weight loss of 1 to 5 kg is common, as is right upper quadrant discomfort or pain associated with hepatomegaly. Physical examination reveals a tender, enlarged liver.

The icteric phase of acute viral hepatitis is often accompanied by the appearance of dark urine followed shortly thereafter by the development of clinical jaundice. With the emergence of jaundice, symptoms tend to improve, and fever usually subsides. The icteric phase lasts, on average, about 1 month. In 10 to 20% of patients infected with HBV, extrahepatic manifestations of the disease develop, most of them believed to be mediated by virus-specific immune complex injury. These include a serum sickness-like syndrome, acute necrotizing vasculitis (polyarteritis nodosa), and membranous glomerulonephritis.

Biochemical tests of liver disease, such as an elevated alanine aminotransferase (ALT) level, are essential for the diagnosis of acute hepatitis B. Other liver enzymes are not as helpful. The total bilirubin is usually less than 10 mg/dl but may occasionally exceed 20 mg/dl. A prolongation of the prothrombin time may reflect early and potentially serious derangements of liver function and may portend a poor outcome.

The factors that determine outcome or chronicity remain largely unknown. In general, the frequency of clinically overt forms of hepatitis B increases with age, whereas the likelihood of becoming chronically infected decreases. Infection in the perinatal period leads to chronicity without symptoms in about 85% of neonates born to HBsAg carrier mothers whose sera contain high concentrations of virus, as manifested by the presence of hepatitis Be antigen (HBeAg) or HBV DNA. In contrast, the rates of transmission and progression to chronicity are dramatically reduced in infants born to mothers whose virus burdens are low. In infants 1 year of age, the incidence of development of a persistent infection following acute disease falls to 26%, and by 3 years of age, the incidence is as low as 8%. The potential for chronicity in immunocompetent adults ranges from 1 to 8%, but symptoms are more common during the acute infection.

HLA class I- and II-restricted cytotoxic T lymphocytes (CTLs) appear to play a major role in facilitating viral clearance during acute HBV infection and in the pathogenesis of hepatocellular injury by recognizing HBV nucleocapsid epitopes expressed on the surfaces of infected hepatocytes. Patients with chronic hepatitis B usually are devoid of or deficient in HLA class I- and II-restricted CTLs directed against these epitopes (19). In transgenic mice that accumulate high concentrations of HBV large-envelope polypeptide within the endoplasmic reticulum (see Description of the Agent below), liver cell injury leads to chronic active hepatitis and hepatocellular carcinoma (15), but a role for HBsAg-directed CTLs in humans with hepatitis B has not been demonstrated.

Pathologic features of acute hepatitis B include both degenerative and regenerative parenchymal changes that lead to lobular disarray. During recovery, regenerative changes predominate; however, in some patients, confluent necrosis, which is potentially progressive and fatal, may occur. Histologic examination is useful for distinguishing among acute viral hepatitis, chronic hepatitis, and cirrhosis. Chronic hepatitis B is associated with hepatocellular carcinoma, especially in males living in Asia and Africa. The incidence of death from liver cancer and cirrhosis increases as a direct function of age; however, cirrhosis is not always a requirement for the development of hepatocellular cancer.

Hepatitis B is transmitted by percutaneous exposure to blood, blood products, and blood-contaminated instruments. Intimate contact, especially sexual contact, and perinatal spread from mother to newborn are the two other common modes of HBV transmission. Approximately one-third of cases of acute hepatitis B in the United States occur in the absence of a recognized mode of transmission. Interferon therapy is effective in converting chronic replicative to nonreplicative HBV infection in approximately 40% of cases. After liver transplantation, HBV reinfection of the new liver is universal, resulting in HBV-induced liver injury in a number of these patients.

Hepatitis D

HDV, previously called the delta agent, is a defective RNA virus that requires concomitant HBV infection for its own replication and expression. In the United States, most cases of hepatitis D occur primarily in populations frequently exposed to blood-contaminated needles and blood products, such as intravenous drug users and hemophiliacs; secondary cases have been recorded in sexual partners of drug users, and such cases can amplify the spread of HDV within population groups other than those recognized as being at high risk for HDV infection. In Mediterranean countries, person-to-person spread, including household contact, is an important mode of transmission. Sustained epidemics of clinically severe acute hepatitis D have been observed in South American populations with high HBV carrier rates (27, 59).

Because HDV infection can become established only in persons with HBV infection, acquisition of acute hepatitis D can occur simultaneously with acute HBV infection (coinfection), or HDV infection can become superimposed upon an already established chronic HBV infection (superinfection). Acute coinfection with both HBV and HDV usually causes an acute illness indistinguishable from acute HBV infection alone except that the infection tends to be more severe, and the likelihood of fulminant hepatitis can be as high as 5%. A biphasic course is sometimes observed during coinfection. HDV infection does not increase the rate of chronicity of acute hepatitis B. The importance of HDV infection lies in its ability to convert an asymptomatic or mild, chronic HBV infection into fulminant or more severe and rapidly progressive disease. In HBV carriers who experience HDV superinfection, chronic HDV infection is the rule unless the patient succumbs to fulminant hepatitis (27, 59). Limited success has been achieved in treating chronic hepatitis D patients with alpha interferon. High-dose, long-term therapy is required, but relapses are common after therapy is stopped (17). Patients with hepatitis D who have hepatic decompensation are candidates for liver transplantation, the outcome of which is better than that for patients with chronic hepatitis B alone (63).

PROCESSING OF SPECIMENS FOR HBV AND HDV AND OCCUPATIONAL SAFETY GUIDELINES

Except for HBV DNA, HDV RNA, and hepatitis delta antigen (HDAg), the stability of serologic markers that develop following HBV or HDV infection eliminates the need for extraordinary collection and storage procedures if bacterial contamination is minimized. Serum or plasma should be separated from blood samples within 24 h and stored at 2 to 8°C if testing is to take place within 5 days. Should longer delays be anticipated, the processed specimens must be frozen. The addition of a bacteriostatic agent to the samples is rarely indicated. Should this become necessary, a final concentration of 0.01% thimerosal, 0.1% sodium azide, or 25 to 50 μg of gentamicin sulfate per ml is preferred. Because heparinized or severely hemolyzed blood may occasionally cause false-positive enzyme immunoassay (EIA) responses, such samples should be avoided.

For nucleic acid analysis, samples should be processed within 6 h and either tested within 24 h or stored at -70°C or lower. Serum is preferred for PCR testing, although plasma recovered from blood collected in anticoagulant citrate dextrose (ACD) or EDTA tubes is suitable. Conversely, heparinized plasma is considered unacceptable, because heparin binds to DNA (2) and interferes with *Taq* polymerase-mediated PCR (33), causing false-negative results. Heparin also inhibits reverse transcription (31, 33, 80). Should heparinized samples be the only material available for HBV DNA analysis, the addition of 1 to 3 U of heparinase I per μg of DNA (in 5 mM Tris [pH 7.5] and 1 mM $CaCl_2$ for 2 h at 25°C) may rescue the sample and allow PCR amplification to be successful (2, 33). For HDV RNA, 40 U of RNasin is included in the heparinase solution.

Processed samples or unseparated blood can be shipped at ambient temperatures by overnight carrier if delivery is expected to be made within 24 h. However, sensitivity and safety are best achieved by shipping these samples in Styrofoam boxes containing at least 5 kg of dry ice, which should maintain samples in a frozen state for at least 72 h when the ambient temperature is <98°F (ca. 37°C). In accordance with the provisions cited in the Interstate Shipment of Etiologic Agents (*Code of Federal Regulations*, Title 42, Part 72), etiologic samples must be shipped in volumes of less than 50 ml, wrapped in sufficient absorbent material to retain all the sample should breakage occur, and placed in the inner container of a double mailing canister. Each container should be sealed with vinyl plastic tape, placed in an insulated Styrofoam mailing carton, and covered with dry ice. Excess space is filled with bubble wrap, Styrofoam peanuts, or other filler. Although HBV and HDV serologic markers are stable for years at -70°C (-30°C for antibodies), repetitive freezing and thawing may lead to substantial losses in concentration.

Laboratory personnel should regard all specimens collected from hepatitis B patients as potentially dangerous. The Occupational Safety and Health Administration (OSHA) standards for occupational exposure to blood-borne pathogens (29 *Code of Federal Regulations* 1910.1030) are designed to provide protection to employees exposed to blood and other potentially infectious materials. Compliance is mandatory. At the core of the standard is the concept of universal precautions, which mandate that all human blood and certain human body fluids be treated as if they are known to be infectious for HIV, HBV, or other blood-borne pathogens. OSHA mandates that all employees who have an occupational exposure be offered hepatitis B vaccine at no cost. OSHA standards and additional safety recommendations can be found in the literature (6, 49, 82).

Ideally, the use of needles or other sharps should be avoided whenever possible; when used, they should be placed in puncture-resistant sharps containers and incinerated or autoclaved before being discarded. All other materials should be placed in discard pans and autoclaved at 121°C for 20 min. A 1 to 10% commercial bleach solution (0.05 to 0.5% sodium hypochlorite), prepared fresh each week to avoid deterioration, is effective against most, if not all, human viruses and is the treatment of choice for disinfection. An exposure time of 15 min is recommended, although the presence of organic material (e.g., serum) may impair the response. Thus, large volumes of disinfectant relative to the volumes of virus-containing materials should be used to inactivate viruses effectively. Because hypochlorites may be corrosive to many metals, the solutions should not be allowed to remain in contact with such materials for more than 10 min. It is safe to autoclave bleach at a final dilution of 1:100. Alternative recommended means of inactivating HBV include 2.0% aqueous alkalinized glutaraldehyde (pH 8.4) and 1% Wescodyne.

HBV

Description of the Agent

HBV is the prototype agent for a new family of DNA viruses designated *Hepadnaviridae*. Related viruses are found in woodchucks, ground squirrels, Peking ducks, and tree squirrels. Features shared by these viruses include virion size and ultrastructure; distinctive polypeptide and antigenic compositions; comparable DNA sizes, structures, and genetic organizations; and a unique mechanism of viral DNA replication. Recently, separate genera for the mammalian and avian hepadnaviruses have been proposed because of differences in their molecular biology.

Ultrastructural examination of particles observed in the sera of patients with HBV infection reveals three distinct morphologic entities in various proportions (Fig. 1). The more numerous forms (by a factor of 10^3 to 10^4) are the small, pleomorphic, spherical, noninfectious particles measuring 17 to 25 nm in diameter (mean of 20 nm). Particle counts of 10^{13} or higher have been detected in some sera. Tubular or filamentous forms of various lengths, but with diameters similar to those of the small particles, also are observed. A third particle, the hepatitis B virion, is a complex double-shelled particle with a diameter of approximately 42 to 47 nm. Originally designated the Dane particle, it consists of an outer shell or lipid-containing envelope, approximately 7 nm in thickness, that surrounds an electron-dense inner core having a diameter of 27 nm. The sera of infected patients may contain as many as 10^{10} infectious virions per ml. The terminology used to describe the various antigens and antibodies associated with HBV is given in Table 1. The complex antigen found on the surface of HBV and comprising the 20-nm-diameter spherical particles and tubular forms is called HBsAg. Previous designations included the Australia or Au antigen and the hepatitis-associated antigen. The envelope or HBsAg proteins are represented by three HBV-encoded polypeptides sometimes designated the small (or major), middle, and large proteins of HBsAg. The major or small protein is encoded by the S region of the S gene and consists of a 226-amino-acid polypeptide (24,000 Da) and its glycosylated form

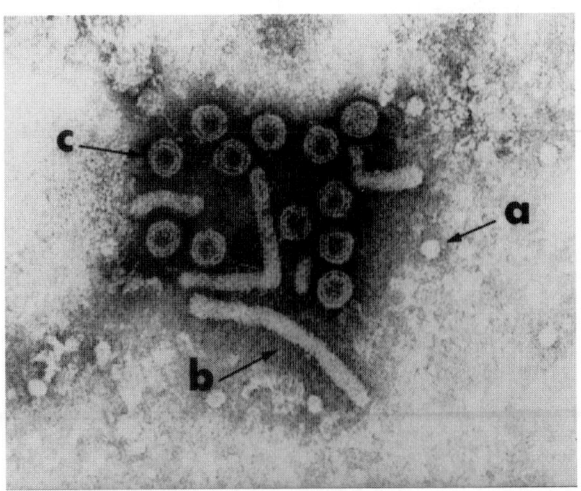

FIGURE 1 Electron micrograph of serum showing the presence of three distinct morphologic entities: a, 17- to 25-nm-diameter pleomorphic, spherical particles; b, tubular or filamentous forms with diameters similar to those of the small particles; c, 42- to 47-nm-diameter double-shelled spherical particles representing the hepatitis B virion (Dane particle). Magnification, ×105,600. Purified preparations of the pleomorphic, spherical HBsAg particles (a), plasma-derived or genetically engineered, are used in human hepatitis B vaccines.

(27,000 Da). The middle protein is encoded by the pre-S2 and S regions of the S gene and contains an additional glycosylation site in the pre-S2 region. The large polypeptide consists of 389 to 400 amino acids encoded by the pre-S1, pre-S2, and S regions of the S gene of the HBV genome but generally lacks the pre-S2 glycosylation site. Analysis has revealed that one antigenic specificity, designated *a*, is common to all HBsAg preparations. In addition, there are two sets of allelic determinants, *d* or *y*, and *w* or *r*. Thus, there are four principal subtypes of HBsAg: *adw*, *ayw*, *adr*, and *ayr*. Because of antigenic heterogeneity in the *w* determinant, there are 10 major serotypes of HBV. The predominant subtype found in North America is *adw*; *ayw* is the second most common. HBV subtypes do not change after infection; therefore, subtype determination can be helpful in tracing infection from one source to another.

The 20-nm-diameter particles primarily contain only S-region-encoded sequences and have average buoyant densities of 1.18 g/cm^3 in CsCl and 1.16 g/cm^3 in sucrose, a molecular weight of 3.7×10^6 to 4.6×10^6, and a sedimentation coefficient that ranges from 39S to 54S. The low buoyant densities of these particles reflect the presence of lipids, which are presumably of cell membrane origin (20). The tubular forms contain mostly small and middle-sized polypeptides, while the lipid-containing envelope of HBV is replete with all S gene products.

The complete virion has a buoyant density of about 1.22 g/cm^3 in CsCl and consists of an envelope of HBsAg, an internal core or nucleocapsid that contains core proteins with hepatitis B core antigen (HBcAg) and HBeAg specificities, and a small (3,200-bp), circular, partially double-stranded DNA molecule with a molecular weight of 1.6×10^6 (72). The HBV genome is partially double stranded because of incomplete extension of the positive-sense strand by the endogenous DNA polymerase enzyme incorporated within the virion. Because some HBV particles may

TABLE 1 Nomenclature for viral hepatitis B

Term	Abbreviation	Description
Hepatitis B virus	HBV	42-nm-diam double-shelled particle that consists of 7-nm-diam outer shell and 27-nm-diam inner core. Core contains small, circular, partially double-stranded DNA molecule and DNA polymerase activity. Originally called the Dane particle. This is the prototype for family *Hepadnaviridae*.
Hepatitis B surface antigen	HBsAg	Complex antigen found on surface of HBV and on 20-nm-diam particles and tubular forms. Formerly designated Australia (Au) antigen or hepatitis-associated antigen.
Hepatitis B core antigen	HBcAg	Antigen associated with 27-nm-diam core of HBV.
Hepatitis B e antigen	HBeAg	Antigen closely associated with nucleocapsid of HBV. Also found as soluble protein in serum.
Antibodies to HBsAg, HBcAg, and HBeAg	Anti-HBs, anti-HBc, and anti-HBe, respectively	Specific antibodies produced in response to their respective antigens.

contain no DNA while others contain genomes with single-stranded gaps of different lengths, the buoyant densities of the nucleocapsids in CsCl can range from 1.30 to 1.36 g/cm^3 (35). Antibodies to HBsAg (anti-HBs), HBcAg (anti-HBc), and HBeAg (anti-HBe) are produced in response to their respective antigens.

HBcAg and HBeAg have different antigenic specificities. The core protein is a polypeptide with a molecular size of 22,000 Da that is encoded by the C gene of HBV. The precore sequence within the C gene is important for expression and secretion of a polypeptide with HBeAg specificity. Proteolytic cleavage of the core protein results in truncated forms of HBeAg that have molecular sizes of approximately 16,000 Da and appear to lack the viral DNA-binding domain. A mutation at the end of the precore region leading to a translational stop codon yields a variant that is unable to produce HBeAg; such precore mutants may contribute to the pathogenesis of chronic hepatitis B and fulminant hepatitis. HBeAg either circulates freely in the blood as a soluble protein or is bound to albumin, α-antitrypsin, or immunoglobulin. It is a reliable marker for the presence of intact virions and thus for infectivity.

The stability of HBV does not always coincide with that of HBsAg. Immunogenicity and antigenicity, but not infectivity, are retained after exposure to ether, acid (pH 2.4 for at least 6 h), and heat (98°C for 1 min, 60°C for 10 h); however, inactivation may be incomplete under these conditions if the concentration of virus is excessively high. Exposure of HBsAg to 0.25% sodium hypochlorite for 3 min destroys antigenicity and infectivity. Infectivity in serum is lost soon after direct boiling for 2 min, autoclaving at 121°C for 20 min, or dry heat at 160°C for 1 h. More recent studies have shown that HBV is inactivated by exposure to sodium hypochlorite (500 mg of free chlorine per liter) for 10 min, 0.1 to 2% aqueous glutaraldehyde, Sporicidin (pH 7.9), 70% isopropyl alcohol, 80% ethyl alcohol at 11°C for 2 min, Wescodyne diluted 1:213, or combined β-propiolactone and UV irradiation (5, 36, 55). HBV retains infectivity when stored at 30 to 32°C for at least 6 months and when frozen at −20°C for 15 years.

Direct Examination

Although not applicable to rapid, large-scale screening of HBV infections by clinical laboratories, immunofluorescence, in situ hybridization, immunohistochemistry, and thin-section electron microscopy have been used extensively to examine pathologic specimens for the presence of HBV-associated antigens or particles (3, 4, 9, 21, 30, 58). Indeed, nonisotopic in situ detection methods for the simultaneous analysis of HBV-related nucleic acids and antigens in paraffin-embedded liver tissue are providing important insights into the relationship between HBV DNA replication and gene expression at the single-cell level (24, 25). Within different hepatocytes, HBsAg is localized exclusively in the cytoplasm, whereas HBcAg is usually detected in the nucleus. Cytoplasmic HBcAg can also be demonstrated, especially during active replication of the virus. Detection of complete virions in the liver is uncommon; however, HBcAg-containing capsids are occasionally observed in the nuclei of hepatocytes.

The techniques used for immunoperoxidase or immunofluorescence are well known. As a general rule, immunofluorescence staining of cryostat sections is superior to the use of paraffin sections. However, cryostat sections are not always available, and cellular outlines are better resolved in uniformly cut paraffin sections than in frozen sections. Reduction of nonspecific background staining of formalin-fixed sections can be achieved by treating tissue with 0.1% trypsin for 2 to 6 h at 37°C before staining. In situations in which autologous antibody might bind to cell-associated antigen, cryostat sections can be treated for 5 min with 0.1 M glycine-HCl buffer (pH 1.2) before specific antibody is added.

Recently, biotin-streptavidin systems have been introduced for detecting viral antigens in cells or in tissues. This methodology offers a number of advantages over other immunodiagnostic clinical techniques: (i) the binding of streptavidin to biotin is extremely rapid and essentially irreversible because of the high affinity constant ($>10^{15}$ M^{-1}) that exists between these two proteins (affinity constants of antibody for most antigens are generally 10^6-fold lower); (ii) sensitivity is enhanced, even when highly diluted primary antibody is used, because the four binding sites on avidin permit amplification of the response; (iii) formalin-fixed, paraffin-embedded histologic sections, smears, and frozen sections can be examined by conventional light microscopy; and (iv) background stain is greatly reduced.

The avidin-biotin-labeled horseradish peroxidase complex (ABC) procedure of Hsu and his associates (28, 29) is highly sensitive and specific. Briefly, formalin-fixed tissue sections are incubated at 37°C for 12 to 16 h, deparaffinized, and hydrated through xylene and a graded alcohol series. The fixation process should utilize a buffered formalin, not exceeding 4% formaldehyde, that is sufficient to maintain integrity of the tissue without destroying the antigenic determinants being evaluated. The sections are exposed sequentially to a primary antibody, a biotinylated anti-species immunoglobulin G (IgG), and Vectastain ABC reagent (Vector Laboratories, Burlingame, Calif.). Next, the tissue antigen is localized by incubating the sections in freshly prepared peroxidase substrate solution. After each step of the staining procedure, slides are rinsed with 10 mM phosphate-buffered saline (PBS; pH 7.6). Sections should not be allowed to dry out during the staining procedure; therefore, a humidified chamber is recommended for incubations. If antigen concentration is low, the substrate incubation step may be lengthened to achieve maximal staining. If nonspecific staining occurs, it can be reduced by incubating the sections for 20 min with diluted normal serum prepared from the species in which the secondary antibody is made. Many mammalian tissues contain endogenous peroxidase. If this problem exists, sections may be incubated for 30 min in 0.3% hydrogen peroxide in methanol either before primary antibody is added or, for those situations in which the antigenic determinant may be destroyed by hydrogen peroxide, before the ABC reagent step.

We have detected HBsAg and HBcAg in liver tissue by an enhanced biotin-streptavidin method (25) that uses a mouse anti-HBs monoclonal antibody and a rabbit anti-HBc polyclonal antibody, respectively, and a commercially available immunodetection kit (StrAviGen Super Sensitive Alkaline Phosphatase Kit; BioGenex Laboratory, Ramon, Calif.). This kit employs an alkaline phosphatase-conjugated streptavidin and a chromogenic substrate solution (naphthol phosphate substrate and fast red chromogen) for localization of tissue antigens.

Paraffin-embedded liver tissue can also be evaluated by in situ hybridization using a digoxigenin-labeled DNA probe and a digoxigenin detection system (24, 25). Digoxigenin is preferred over biotin for several reasons. Endogenous biotin (but not digoxigenin) is present in liver tissue and often results in high levels of background staining (51, 81). In addition, biotin has a high binding affinity for HBsAg (78). Finally, digoxigenin-labeled probes are comparable in sensitivity to radiolabeled probes (23). Probes representing the complete 3.2-kb genomic DNA of HBV and a suitable negative-control DNA preparation are labeled with digoxigenin-dUTP by random priming (18) with a DNA labeling and detection kit (Boehringer Mannheim Biochemical Co., Indianapolis, Ind.). The probes are chemically stable and can be used for extended periods. Before hybridization, deparaffinized and rehydrated sections are rendered permeable by sequential treatment with acid, heat, and proteinase K. Washing the sections at room temperature in PBS containing glycine will stop the proteinase K digestion, and RNA is removed by digestion with RNase A. The sections are prehybridized and then overlaid with the appropriate digoxigenin-labeled probe. A coverslip is placed over each section, and the probe and tissue DNAs are denatured in situ by heating them at 95°C for 4 min. After quenching on ice, the hybridization reaction is carried out at 42°C for 12 to 16 h. Following removal of the coverslips, the digoxigenin-labeled hybrids are detected with an antidigoxigenin alkaline phosphatase conjugate and an enzyme substrate-chromogen solution. The color reaction is stopped with Tris-EDTA buffer (pH 8.0). For the

TABLE 2 U.S. licensed and unlicensed manufacturers of hepatitis B assays

Assay	Manufacturer[a]	
	RIA	EIA
HBsAg	1, 5	1, 2, 3, 4, 5[b]
Anti-HBs	1, 5	1, 5[b]
Anti-HBc	1, 5	1, 4, 5
IgM Anti-HBc	1, 5	1, 5
HBeAg/Anti-HBe	1, 5	1, 5
HBV DNA	1	

[a]1, Abbott Laboratories; 2, Genetic Systems Corp.; 3, Organon Teknika Corp.; 4, Ortho Diagnostic Systems; 5, Sorin/Incstar Corp.
[b]Unlicensed assay: for research use only; not for use in diagnostic procedures.

simultaneous detection of HBV DNA and antigens, in situ hybridization with two different chromogens (red and blue-purple) is carried out after completion of the immunohistochemical procedure. Neither method interferes with or significantly reduces the sensitivity or specificity of the other method; however, too strong an opposing signal could mask or reduce the other signal.

Cell Cultures and Animal Models

Despite recent evidence for the growth of HBV in primary cultures of healthy adult or fetal human hepatocytes (22, 52), serial propagation over a prolonged period has not been accomplished. Still, the use of stably transfected cell lines provides a model of HBV replication, even though the life cycle is incomplete and only low concentrations of virus are produced. Chimpanzees and other high-order primates are highly susceptible to experimental induction of hepatitis B but are not routinely available (1). The pattern of infection is similar to that observed in humans except that the disease is clinically milder. Nevertheless, the chimpanzee model has played an essential role in studies of viral inactivation, vaccine safety and efficacy, disinfection kinetics, infectivity determinations, and immunopathology and in seroepidemiologic investigations. Because of its limited role as a model for studying viral replication or evaluating antiviral agents, it is being replaced by other animal systems such as the woodchuck, Peking duck, and transgenic mouse (46).

Serologic Identification of Hepatitis B Antigens, Antibodies, and Virus

A number of serologic techniques with different degrees of sensitivity and specificity have been developed for the detection of HBV antigens and their specific antibodies (Table 2). These assays rely on the sandwich principle or are competitive binding immunoassays (Table 3). Readers interested in detailed information concerning the purification of hepatitis B-associated antigens, preparation of HBV-specific antiserum, and discontinued assays of historical interest such as agarose gel diffusion, discontinuous counterimmunoelectrophoresis, rheophoresis, complement fixation, reverse passive latex agglutination, and reverse passive hemagglutination are referred to the third edition of this Manual. The procedure for HBV-associated DNA polymerase is detailed in the fourth edition of this Manual.

RIA

Although EIA has replaced radioimmunoassay (RIA) as the most popular method for detecting the various serologic

TABLE 3 Principles of procedures used in various hepatitis B assays

Serologic assay	Principle of procedure (RIA or EIA)	Type of support system	Adsorbed reagent	Label or conjugate
HBsAg	Sandwich	Bead, well	Anti-HBs	Anti-HBs
HBeAg	Sandwich	Bead, well	Anti-HBe	Anti-HBe
Anti-HBs	Sandwich	Bead, well	HBsAg	HBsAg
IgM anti-HBc	Sandwich (modified)[a]	Bead, well	Anti-IgM	Anti-HBc
Anti-HBc	Competitive binding[b]	Bead, well	HBcAg	Anti-HBc[b]
Anti-HBe	Competitive binding[c]	Bead, well	Anti-HBe	Anti-HBe

[a]IgM-specific anti-HBc that is immunologically bound to an anti-IgM-coated (μ-chain-specific) bead is detected by adding recombinant DNA-derived HBcAg and then the appropriate anti-HBc label or conjugate.

[b]The anti-HBc EIA from Ortho Diagnostic Systems is a sandwich assay in which the label or conjugate is murine monoclonal antibodies specific for human IgG and IgM.

[c]Anti-HBe in the test sample and adsorbed anti-HBe compete for a standardized amount of HBeAg added to the reaction well before the addition of anti-HBe label or conjugate.

markers of hepatitis B, the RIA technique continues to be the standard to which all other assays are compared. RIA kits for HBV (Table 2) are currently offered by Abbott Laboratories (Abbott Park, Ill.) and Sorin/Incstar Corp. (Stillwater, Minn.). Detailed procedures for performing these assays are included with each kit.

The HBsAg assay is capable of detecting HBsAg at a level of approximately 0.25 to 0.5 ng/ml. Briefly, serum or plasma is added to a polystyrene bead to which anti-HBs is adsorbed. During incubation, HBsAg forms an immune complex with the anti-HBs at the liquid-surface interface. After the beads are washed, ^{125}I-labeled anti-HBs is added, and incubation is continued. The labeled antibody binds to any HBsAg on the support system, creating an antibody–antigen–^{125}I-tagged antibody sandwich. The washed beads are counted in a gamma counter. By dividing counts per minute (cpm) of the test sample (S) by mean cpm of the negative control samples (N), an S/N ratio is calculated. Samples with values of 2.1 and above are generally considered reactive (positive). An alternative method of determining positivity is to divide S by the cutoff value (C), e.g., 2.1 times the mean cpm of the negative control. Thus, any specimen with an S/C ratio of ≥1.0 is considered reactive for HBsAg. Within limits, there is a direct correlation between the magnitude of the S/C ratio and the concentration of HBsAg in the specimen; however, sizes of the HBsAg-containing particles and surface area of the support system restrict the working range to between 0.25 ng/ml and 1 μg/ml, above which level, saturation (a plateau) is reached. Because sera from many carriers have HBsAg concentrations above 1 μg/ml, the sample must be diluted until a value that falls within the working range of the assay is obtained, in case quantitation of HBsAg becomes necessary.

Repeatably reactive HBsAg results that cannot be confirmed as positive are highly unusual. Nevertheless, the seriousness of the diagnosis, with its attendant medical, social, and economic repercussions, mandates that a confirmatory test be attempted on all repeatably positive specimens. This can be approached in one of two ways: (i) specimens may be tested with another licensed HBsAg assay, or (ii) a blocking or inhibition assay can be performed to see whether unlabeled anti-HBs will specifically inhibit the reaction. The detection of anti-HBc in the absence of anti-HBs also corroborates a positive HBsAg result. Low-level reactivity, e.g., S/C ratios of <1.5, must always be

viewed with suspicion. Partially heparinized plasma is especially prone to cause low-level false-positive HBsAg reactions, which often can be corrected by the addition of thrombin to the sample or by storing the sample at 2 to 8°C for 48 h before retesting. Heparin may appear in the blood of patients on hemodialysis, who are given heparin prior to dialysis; bedridden patients being given heparin to prevent thrombophlebitis; and patients with heparin locks.

Erroneous interpretations also can be avoided by ensuring that the negative control samples are comparable to the test specimen. Thus, tests for HBsAg in fluids of low density, such as cerebrospinal fluid, should employ "normal" (healthy) cerebrospinal fluid (or comparable low-density material) as the control. In general, protein-deficient specimens and recalcified plasma result in higher background levels.

In an emergency, the solid-phase RIA may be modified to provide an answer within 45 min. Although sensitivity is approximately fivefold less than that of the regular assay, the number of positive specimens likely to be missed should be relatively small. To further reduce the number of false-negative values, at the expense of increasing the number of false-positive reactions, the cutoff value can be reduced to 1.5 times the negative control (or an S/C ratio of 0.75). Ultimately, verification of a positive reaction by the regular procedure will be necessary.

Solid-phase RIA for the detection of anti-HBs is similar in principle to HBsAg detection, except that the specimen is incubated with polystyrene beads coated with human HBsAg, subtypes *ad* and *ay*. Radiolabeled HBsAg is used to detect anti-HBs bound to the fixed HBsAg. Specimens with a count rate ≥2.1 times the negative control mean cpm (or an S/C ratio of ≥1.0) are reactive. Unfortunately, inter- and intralot variations in commercial anti-HBs kits and diminishing sensitivity as the kits approach their expiration date limit the usefulness of the S/N and S/C ratios for quantitative comparisons.

With the advent of hepatitis B (HBsAg) vaccine, the need for precision and accuracy in quantifying anti-HBs has become more important. To permit results that can be expressed in milli-international units per milliliter, Hollinger et al. (26) have modified the RIA. Anti-HBs concentrations are based on the First International World Health Organization (WHO) Reference Preparation for hepatitis B immune globulin (HBIG) (lot 26.1.77) provided by the International Laboratory for Biological Standards,

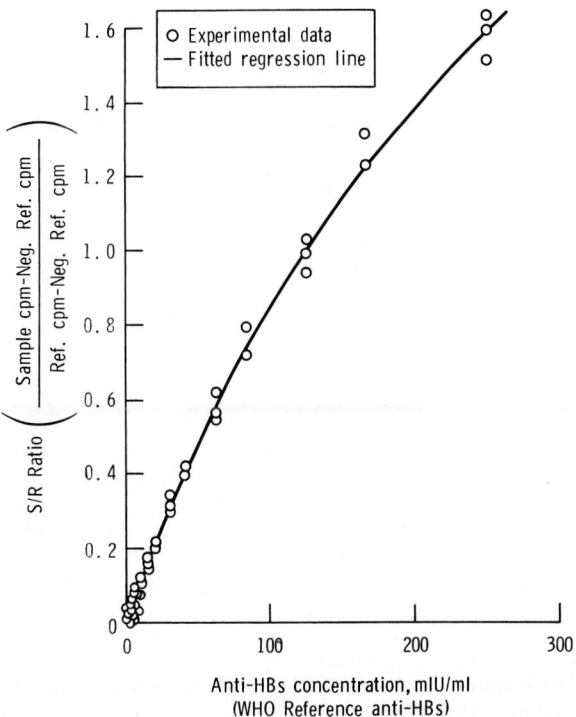

FIGURE 2 RIA standard curve for anti-HBs showing regression of the *S/R* ratio on the anti-HBs concentration. (From reference 26.) For the regression equation, see text.

Central Laboratory of The Netherlands Red Cross Transfusion Services. An arbitrary value of 50 IU of anti-HBs has been assigned to this product. The WHO anti-HBs reference preparation is reconstituted in 0.5 ml of sterile, pyrogen-free water and allowed to equilibrate overnight at 2 to 8°C. It is then diluted in normal recalcified human plasma to contain 125 mIU/ml. To determine the number of milli-international units per milliliter of sample, an *S/R* ratio is computed as follows: (sample cpm − negative-control mean cpm)/(reference cpm − negative-control mean cpm). Determining the regression of the *S/R* ratio (Fig. 2) on the anti-HBs concentration for the WHO Reference Preparation (0.1 to 500 mIU/ml) by using the computer nonlinear regression program BMDP3R (14) yielded the following formula: mIU/ml = $130.75(e^{0.66765(S/R)} - 1)$(reciprocal of dilution). The lower limit of detection for anti-HBs is 1.0 mIU/ml when this procedure is used. Samples whose anti-HBs concentrations exceed 250 mIU/ml are diluted 10-fold in normal recalcified human plasma and retested. To conserve the WHO Reference Preparation, the laboratory should procure a large batch of human serum or recalcified plasma containing polyclonal anti-HBs and determine its relationship to the WHO Reference Preparation (diluted to 125 mIU/ml) by using the *S/R* ratio. The exponent in the formula must be adjusted proportionally to conform with the alternative laboratory reference standard. For example, if the *S/R* relationship between the laboratory standard and the WHO Reference Preparation is 0.500, the exponent in the formula (0.66765) must be reduced by a factor of 0.5. Conversely, an *S/R* ratio of 1.5 would increase the exponent by a factor of 1.5. Once this adjustment is made, the laboratory reference standard can be substituted for the WHO Reference Preparation. The WHO reference reagent

(stored at −70°C) should be periodically reevaluated to verify the accuracy of the laboratory standard, which can be aliquoted and stored at −30°C.

More recently, Abbott Laboratories has released a licensed anti-HBs quantitation panel for monitoring vaccinees. The standards are assayed with the EIA or RIA anti-HBs kits. The panel consists of five vials containing 0, 15, 40, 75, and 150 mIU of human polyclonal anti-HBs per ml based on the WHO anti-HBs Reference Preparation. Specimens are tested in the same assay, and the results are compared to a standard curve generated from the quantitation panel data. Test samples whose values exceed the highest standard are diluted with a specimen dilution buffer and retested. Although results are usually equivalent to those obtained using the WHO Reference Preparation, individual specimens may vary. The lower limit of detection for anti-HBs when this quantitative panel is used is 1 mIU/ml. Sorin/Incstar Corp. also has a panel of six standards that can be used for research purposes only, i.e., not for diagnostic procedures.

RIA kits for anti-HBc are competitive binding assays in which anti-HBc in the test sample competes with a constant amount of human ^{125}I-tagged anti-HBc for a limited number of binding sites on beads coated with recombinant DNA-derived HBcAg. Within limits, the proportion of radioactive anti-HBc bound to the bead is inversely proportional to the concentration of anti-HBc in the test specimen. Incubation is carried out at room temperature for 18 to 22 h. A specimen is considered reactive for anti-HBc if the cpm value of the test sample is less than the cutoff value. An alternative method for reporting results in competitive binding assays is to compute a cutoff-to-sample (or *C/S*) ratio. In this way, any result that is ≥1.0 is considered positive, as it is for direct-binding "sandwich" assays in which an *S/C* ratio is calculated. (As an aid to remembering the correct order of the letters, think of S for sandwich and C for competitive).

Other licensed RIAs available from Abbott Laboratories and Sorin/Incstar Corp. include those for IgM-specific anti-HBc, HBeAg, and anti-HBe. In the IgM anti-HBc assay, anti-HBc of the IgM class is removed or "captured" from the serum by anti-human IgM (μ-chain specific) bound to polystyrene beads. The subsequent addition of recombinant DNA-derived HBcAg results in the attachment of the antigen to any IgM anti-HBc that has combined with the solid-phase anti-human IgM. The sandwich is completed following the addition of radiolabeled anti-HBc. Specifically, 10 μl of a non-heat-inactivated specimen is diluted to 1:1,071 (Abbott Laboratories) or 1:4,200 (Sorin/Incstar). An anti-human-IgM-coated bead is added to each well, and the trays are covered with an adhesive-backed cardboard. Following an incubation period of 1 h at 45°C, the fluid is aspirated from the wells, and the beads are washed with distilled or deionized water. Next, recombinant DNA-derived HBcAg reagent is added to each well, and then ^{125}I-labeled human anti-HBc is added. A cover seal is applied, and the tray is incubated for 18 to 22 h at 15 to 30°C (or for 3 h at 45°C), after which the beads are washed again. Finally, the beads are transferred to tubes and counted in a gamma counter. A cutoff value based on the positive and negative control samples is calculated as described in the manufacturer's instructions. Samples with net counts (minus gamma counter background) repeatedly greater than or equal to the cutoff value are considered positive; those with counts within 10% of the cutoff value should be retested to confirm the original results. An *S/C*

TABLE 4 Comparison of commercial EIAs for detection of HBsAg[a]

Manufacturer	Support system	Absorbed anti-HBs	Anti-HBs conjugate	Sample vol (μl)
Abbott Laboratories	Beads	MM	MM-HRP	200
Genetic Systems Corp.	Wells	MM	MM-HRP	200
Organon Teknika Corp.	Wells	GP	Chimpanzee-AP	200
Ortho Diagnostic Systems	Wells	MM	MM-HRP	200
Sorin/Incstar Corp.[b]	Wells	MM	Sheep-HRP	100

[a]MM, Murine monoclonal antibody; GP, guinea pig; HRP, horseradish peroxidase; AP, alkaline phosphatase.
[b]Unlicensed product: for research use only; not for use in diagnostic procedures.

ratio is computed for this modified direct sandwich assay, and values of ≥1.0 are considered to indicate reactivity.

The commercial tests for HBeAg and anti-HBe are combined into one kit but employ different assay principles (Table 3). The HBeAg test uses the sandwich principle to measure HBeAg in serum or plasma. Polystyrene beads coated with human anti-HBe are added to the test samples, and incubation proceeds for 18 to 22 h at 15 to 30°C. After the beads are washed, anti-HBe tagged with ^{125}I is added, and incubation continues at 45°C for 3 h. Count rates are increased when HBeAg is present in the serum. A positive reaction is one with a value equal to or greater than the cutoff value or with an S/C ratio of ≥1.0. The test for anti-HBe is a modified competitive binding assay in which anti-HBe in 100 μl of the test serum competes with anti-HBe-coated beads for a standardized amount of recombinant DNA HBeAg. Trays containing the beads are incubated for 18 to 22 h at 15 to 30°C, and then the beads are washed. Next, ^{125}I-anti-HBe is added, and incubation continues for 3 h at 45°C. If anti-HBe is present in the test sample, less HBeAg will be available to bind to the bead, and therefore, less ^{125}I-labeled anti-HBe will be immunologically coupled. Within limits, the larger the amount of anti-HBe in the specimen, the lower the count rate. Specimens with count rates equal to or less than the cutoff value are considered reactive for anti-HBe, as in the other competitive binding assays. For this assay, a C/S ratio is computed so that values of ≥1.0 will signify the presence of anti-HBe in the test sample.

EIA

The EIA technique is at least as sensitive as RIA. Improvements in the test have reduced the number of repeatably false-positive reactions, and specificity is almost comparable to that of RIA. Table 2 lists those commercial companies that manufacture qualitative third-generation EIAs (licensed and unlicensed) for the detection of HBV serologic markers, and Table 3 gives the principles of the procedures. EIA kits have longer shelf lives than RIA kits (6 to 15 months versus 2 months). Specific details of the assays can be obtained from the package insert that accompanies each kit, and it is imperative that these instructions be followed explicitly.

Diagnostic EIA products licensed to detect HBsAg (Table 4) follow a generic protocol with a number of minor variations that do not appear to alter the sensitivity or specificity of the assay significantly (except for those procedures with an abbreviated incubation period). Nuances in the various EIA procedures include sample incubation, either simultaneously (one step) with conjugated anti-HBs or consecutively (two step); variation in incubation or color development ranging from 30 min to overnight; use of

polystyrene beads or microwells; use of monoclonal versus polyclonal anti-HBs; and selection of alkaline phosphatase instead of horseradish peroxidase for the enzyme-conjugated antibody. In one scenario, HBsAg is detected by mixing human test serum or plasma with enzyme-conjugated anti-HBs in the presence of a polystyrene bead or microplate well coated with anti-HBs. In each of the assays, appropriate positive and negative controls must be included to ensure validity of the assay. Trays are sealed with an adhesive cover, and incubation is allowed to proceed overnight at room temperature. During the incubation period, HBsAg binds to the antibody-coated solid phase and also combines with the enzyme-labeled antibody to form an anti-HBs–antigen–enzyme-conjugated anti-HBs sandwich. Contents of the wells are aspirated, and the reaction trays are washed. A chromogenic substrate is added, resulting in the development of a measurable colored reaction product. After a 30-min incubation, the enzyme reaction is arrested with either sulfuric acid or sodium hydroxide, depending on the substrate being used, and the sample is read in a spectrophotometer at the appropriate optical density. Specimens giving absorbance values equal to or greater than the calculated cutoff value are considered reactive for HBsAg. Alternatively, an S/C ratio of ≥1.0 for this sandwich assay is evidence for the presence of HBsAg in the test sample (see RIA above for discussion of S/C and C/S ratios). As with RIA results, all repeatably reactive samples must be confirmed with a licensed confirmatory assay. Each run must be validated on the basis of conditions set forth by the manufacturer.

Depending on the assay, final readings should be made within 30 to 120 min after the addition of acid or base to the reaction tray. The control blank used to calibrate the instrument should be reused whenever prolonged interruptions occur or if color develops in the blank. Care should be taken to avoid splashing specimens or reagents outside of wells or high up on the rim of the reaction tray, for such splashes may not be removed in subsequent washing and could cause false-positive test results. Sodium azide poisons the enzyme substrate; therefore, this reagent should not be present in the wash solution. As with the RIA procedure, the wash step is a critical part of the assay. If any delivery port is clogged or dispenses wash solution erratically, reproducibility and precision will suffer. Application of trace amounts of silicone grease to the tube dispensers used to wash beads keeps the water from adhering to the metal and facilitates the washing process. Substitutions for recommended wash solutions should be avoided. For example, phosphate inhibits alkaline phosphatase and could lower the optical density, whereas Tris buffer may increase alkaline phosphatase activity (74).

Solid-phase EIA for the detection of anti-HBs is similar

in principle to HBsAg detection except that the specimen is incubated with polystyrene beads (Abbott Laboratories; Ausab EIA) or in microtiter wells (Sorin/Incstar; for research use only) that have been coated with HBsAg subtypes *ad* and *ay*. Anti-HBs bound to the fixed HBsAg is detected with a complex consisting either of biotin-labeled human HBsAg and horseradish peroxidase-conjugated rabbit antibiotin or of horseradish peroxidase-conjugated human HBsAg. An absorbance value that is equal to or exceeds the cutoff value or an *S/C* ratio of ≥1.0 indicates the presence of anti-HBs in the sample. As with the RIAs, quantitative determination of anti-HBs can be done by using a panel of standards (see RIA above).

Three commercial anti-HBc EIA kits are licensed in the United States (Table 2). Two of the kits are competitive binding EIAs (Abbott Laboratories and Sorin/Incstar) in which anti-HBc from the test sample competes with a constant amount of horseradish peroxidase-conjugated human anti-HBc for a limited number of binding sites on beads (Abbott) or in wells (Sorin) that have been coated with recombinant DNA-derived HBcAg. Within limits, the amount of anti-HBc present in the sample is inversely proportional to the amount of color development. Results are compared to a calculated cutoff value. As with all competitive binding assays, a *C/S* ratio is computed in order for a positive sample to have a value of ≥1.0. The other commercial EIA kit for anti-HBc (Ortho Diagnostic Systems) uses the sandwich principle, in which the specimen is incubated in microtiter wells coated with HBcAg. Bound anti-HBc is detected with horseradish peroxidase-conjugated murine monoclonal antibodies specific for human IgG and IgM. In this direct reacting assay, higher optical densities are recorded in reactive samples; thus, an *S/C* ratio is calculated to achieve positive results of ≥1.0.

Two licensed EIA kits for the detection of IgM-specific anti-HBc, HBeAg, and anti-HBe are available (Table 2). The tests for HBeAg and anti-HBe are performed with the same kit but employ different assay principles (sandwich principle for the HBeAg test and competitive binding for the anti-HBe procedure). The reader is directed to the RIA section above for details of these assays; the concepts of the procedures are identical. To measure IgM-specific anti-HBc in serum or plasma, the EIA IgM anti-HBc test uses a modified sandwich technique similar to that described in the RIA section above. In this assay, 10 μl of test serum is diluted 1:1,071 (Abbott Laboratories) or 1:4,000 (Sorin/Incstar) and then incubated for 1 h at 40°C with a polystyrene bead coated with antibody specific for human IgM (μ-chain specific). During this interval, IgM anti-HBc is captured by the beads. After the liquid is removed and the beads are washed, HBcAg and horseradish peroxidase-conjugated human anti-HBc are added to the bead to which IgM-specific anti-HBc may be immunologically bound. After another incubation period (room temperature for 18 to 22 h or 45°C for 3 h), the beads are rewashed and transferred to tubes, in which color development occurs in the presence of the enzyme substrate. Next, the reaction is stopped, and the intensity of the color is measured with a spectrophotometer. The absorbance is proportional to the quantity of IgM-specific anti-HBc present in the sample. A cutoff value is determined; specimens giving absorbance values equal to or greater than the calculated cutoff value or having *S/C* ratios of ≥1.0 are considered positive for IgM antibodies to HBcAg.

Dot or Southern Blot Hybridization Techniques

Although hybridization procedures are beyond the scope of most clinical laboratories (42), an awareness of their availability is desirable. Human serum and tissue can be analyzed for HBV DNA sequences by a dot hybridization techniques (47, 64–66, 83). In this technique, HBV DNA in the sample is immobilized on a nitrocellulose or nylon membrane and hybridized to a radioactive probe. Typically, sample preparation consists of mixing the serum with alkali to lyse the HBV and separate the DNA strands. After addition of the alkaline serum to the manifold in which the filter membrane is held, vacuum is applied. The DNA strands (and protein) bind to the membrane forming the dot. In order to fix the DNA to the membrane, nitrocellulose filters are baked, and nylon filters are exposed to UV light. The hybridization mixture consists of a buffer solution with the ^{32}P-labeled probe (typically, a recombinant cloned HBV DNA labeled to a specific activity of 2×10^8 to 4×10^8 cpm/μg of DNA). To reduce nonspecific binding of the probe to the membrane, Denhardt solution (0.1% bovine serum albumin, 0.1% Ficoll, 0.1% polyvinylpyrrolidone) and an excess of sheared, denatured, nonhomologous DNA are included in the hybridization mixture. The specificity of the hybridization reaction depends on the temperatures used for hybridization and washing and on the composition of the reagents used in these steps. After hybridization, the filter is washed, dried, and autoradiographed. This method is capable of visually detecting 0.1 to 1.0 pg (28,000 to 280,000 genomic equivalents) of HBV DNA sequences within 24 h, with a 2- to 10-fold increase after 5 days of autoradiography. Similar hybridization studies can be performed by using DNA extracted from small portions of frozen biopsy specimens.

A commercial liquid-phase molecular hybridization assay for the detection and quantitation of HBV DNA in serum (37) has recently become available for research purposes (Abbott Hepatitis B Viral DNA; Abbott Laboratories). The assay can be performed only on serum, and lipemic serum should be clarified by centrifugation prior to use. Specimens may be stored for up to 14 days at 2 to 8°C; otherwise, they should be frozen. Multiple freeze-thawing should be avoided. Nucleases of microbial or human origin may degrade the sample. Thus, extra care must be taken to avoid contamination with nucleases on the skin or on surfaces handled by humans, and pipette tips should be autoclaved before use to destroy nuclease activity. The serum specimen being tested is solubilized to expose the HBV genome and then incubated with an ^{125}I-labeled HBV DNA probe at 65°C for 18 h for hybridization to HBV DNA in the test specimen. The mixture is loaded onto a disposable plastic column for elution of hybrid molecules from the unhybridized ^{125}I-HBV probe. The eluate is counted for 10 min in a gamma counter, and the counts are reported as total counts. The cutoff value is derived by multiplying the mean of the positive control values by 0.015 and adding this number to the mean of the negative control values. Samples with net counts within 10% of the cutoff value should be retested. The assay can detect as few as 4×10^4 HBV molecules, or about 0.15 pg of HBV DNA, a sensitivity that is equivalent to that of filter hybridization (54, 84).

Branched-DNA Amplification Assay

The Chiron Amplex-HBV assay (Fig. 3) is a sandwich nucleic acid hybridization method for the quantitative de-

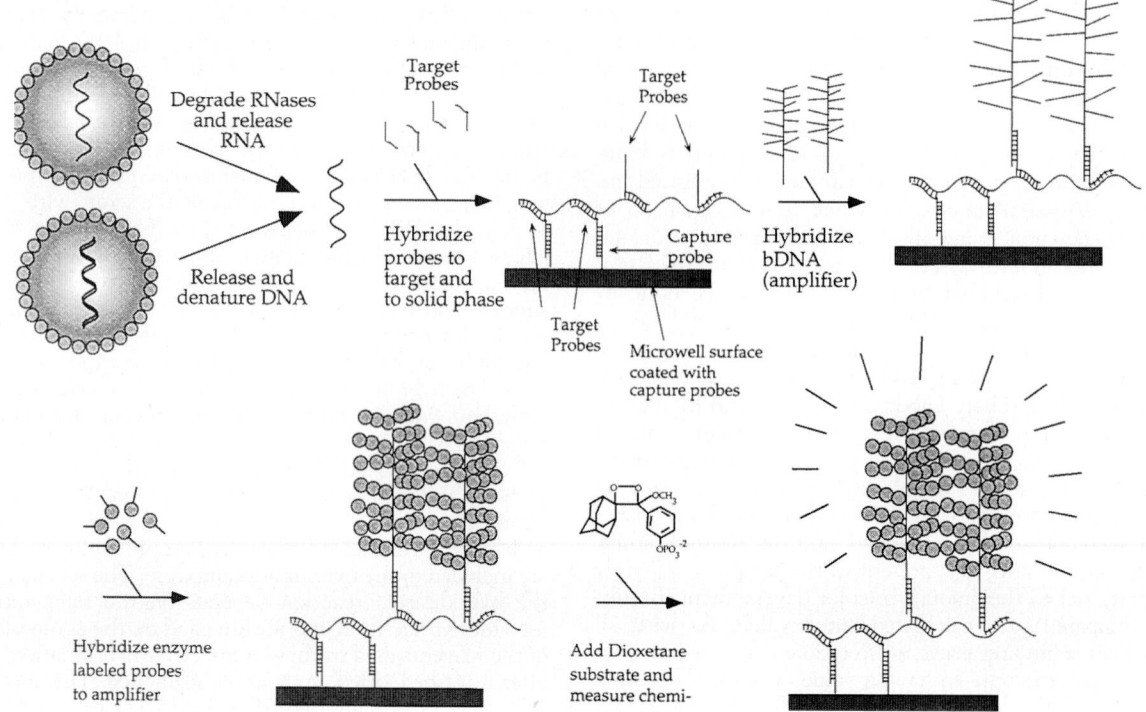

FIGURE 3 Chiron branched DNA (bDNA) signal amplification assay. (© 1993 Chiron Corporation.)

tection of HBV DNA in human serum (76, 77). It is available for research use only (not for use in diagnostic procedures). In this assay, synthetic oligodeoxyribonucleotides containing unique primary segments designed to hybridize to the minus-sense strand of HBV DNA are covalently attached to a secondary fragment through branch points. These target probes hybridize to HBV DNA and to complementary synthetic fragments bound to a solid phase. On another level, the secondary fragments of the target probe direct the binding of multiple copies of synthetic branched-DNA amplifier molecules that contain repeated nucleotide sequences that hybridize with enzyme-modified alkaline phosphatase-labeled probes. The detection scheme relies on alkaline phosphatase-catalyzed chemiluminescence emission from a dioxetane substrate, and the results are recorded as luminescence counts on a plate luminometer (Fig. 3). Light emissions are proportional to the amount of HBV DNA present in each specimen, and HBV DNA is quantitated by comparing light emission values in the test specimen against those obtained on standards. As few as 70,000 HBV DNA equivalents per ml can be detected with this assay.

PCR Assay

Recently, the PCR assay, which makes possible detection of as few as 10 HBV DNA molecules per ml, has been used for amplification of HBV DNA in serum (75). This assay is 100 to 1,000 times more sensitive than the dot hybridization technique for detecting HBV DNA. Synthetic oligonucleotide primers that are complementary to a conserved sequence on the genome that will be amplified are prepared. The procedure allows for denaturing the DNA to separate the strands of the DNA duplex that contains the target region, annealing the primers to their complementary regions, and extending the primers. The discovery of a thermostable enzyme isolated from *Thermus aquaticus*, called *Taq* DNA polymerase, has increased the specificity of the reaction by allowing extension to take place at an elevated temperature. Denaturation, annealing, and extension constitute a cycle and are accomplished by varying the temperature of the reaction. Products of the previous cycle act as templates for succeeding cycles, resulting in an exponential doubling of the DNA. Thus, for every 20 cycles, the target region is amplified about a millionfold, and a second round of 20 cycles with a different pair of primers results in $>10^{12}$ amplification. Alternatively, the second round of amplification may be replaced by a sensitive detection method such as hybridization to a labeled probe.

A number of PCR procedures for detecting HBV DNA in serum have been described (40, 75). Viral DNA can be purified from serum by proteinase K digestion, phenol and chloroform extraction, and precipitation with ethanol, following which the DNA is redissolved in water. Amplification is carried out in a DNA thermal cycler with primers that anneal to highly conserved sequences of the HBV genome. Descriptions of the PCR procedure and guidelines to guard against contamination can be found in references 16, 32, 38, 43, 48, and 70.

Interpretation of Test Results in Hepatitis B Disease

Successful detection of HBV serologic markers depends not only on the relative sensitivity of the test procedures but also on the availability of experienced personnel who comprehend the idiosyncrasies of the procedure and are metic-

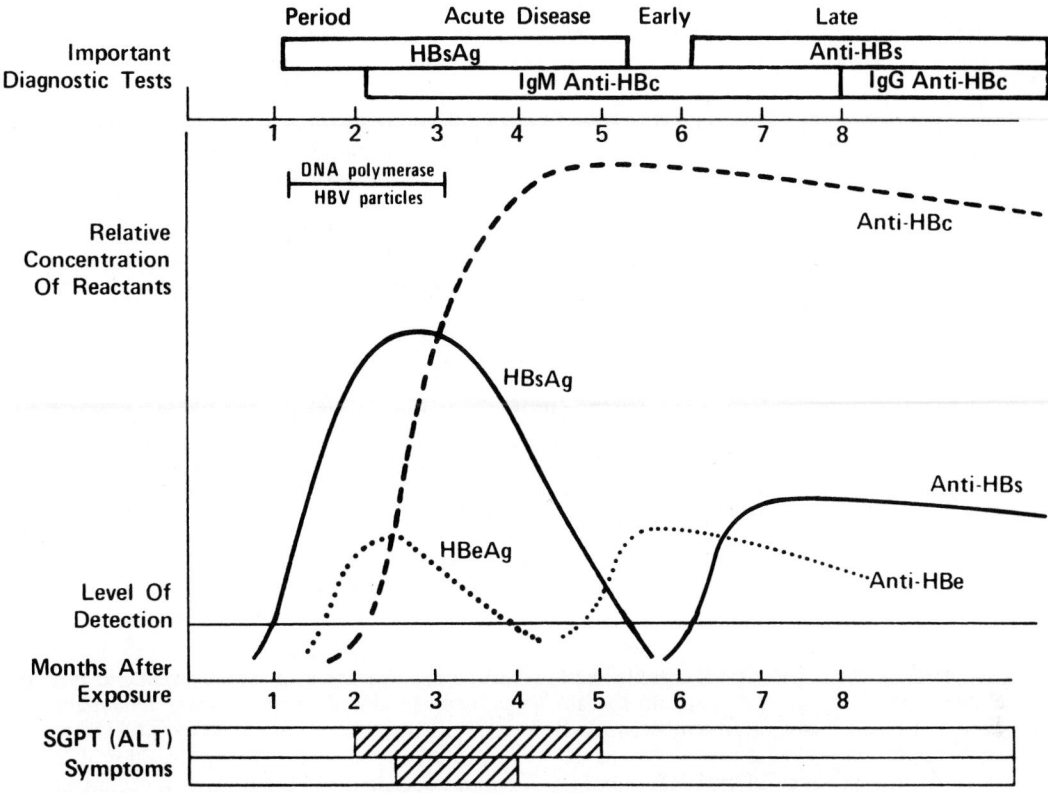

FIGURE 4 Serologic and clinical patterns observed during acute HBV infection. SGPT, serum glutamic pyruvic transaminase.

ulous in their performance of the test. For example, most nonrepeatably reactive specimens result from improper washing of beads or wells (see RIA above), contamination of counting tubes or holders with radioactive material, or cross-contamination of nonreactive specimens with droplets from highly positive samples.

Provided that all procedural details are observed, the final evaluation and interpretation of any test result will be determined by the specificity of the reagents used. It is important for the diagnostic virologist to appreciate the difficulties encountered by commercial companies in preparing quality reagents and automating immunoassays. The preparation of specific polyclonal antibody or the use of monoclonal antibodies has dramatically improved the sensitivity and specificity of these assays, as has the production of HBV antigens by recombinant DNA technology. Confirmatory testing of specimens by blocking assays or specificity testing with a reference antigen or antiserum also assists in the validation of a laboratory result.

The marked increase in sensitivity of the PCR assay for HBV DNA compared with that of the RIA or EIA for HBsAg should not imply that an equivalent increase in the number of HBV-infected persons will be realized. Experience indicates that practically all patients with acute and chronic hepatitis B have HBsAg concentrations that are detectable by conventional assays; thus, methods for detecting HBV DNA are usually unnecessary and should be reserved for specific indications such as monitoring treatment by interferon. It is noteworthy, however, that HBV DNA has been detected in the sera and/or livers of some

HBsAg-negative patients with or without other HBV serologic markers (34, 39, 41, 53, 73). It has also been reported that some hepatitis B vaccine nonresponders may have low-level HBV infection (71), which may account for their nonresponsiveness. Finally, PCR is so sensitive that HBV DNA can be detected even in some patients who have recovered completely from their infections.

As shown in Fig. 4 and summarized in Table 5, the presence of HBsAg in a serum indicates active HBV infection, either acute or chronic. In a typical HBV infection, HBsAg will be detected 2 to 4 weeks before the ALT level becomes abnormal and 3 to 5 weeks before symptoms or jaundice develop. HBV DNA precedes the appearance of HBsAg in the serum. Anti-HBc, primarily of the IgM class, usually appears when the ALT level begins to increase. The presence of IgM-specific anti-HBc at a relatively high titer (>1:1,000) is evidence of an acute infection. These elevated titers decline regardless of whether the disease resolves or becomes chronic. It should be recognized that sera from 5 to 15% of chronic hepatitis B patients may be positive in the IgM anti-HBc assay, especially during reactivation of their disease.

Because the total-anti-HBc test is invariably positive when HBsAg is present in a clinically ill patient (see Fig. 4 and Table 5 for rare exceptions to this statement), this test can be used to validate the HBsAg reaction. Tests with discordant results should always be repeated. In perhaps 5 to 10% of patients with acute hepatitis B (especially those with fulminant disease) and more frequently during early convalescence, serum HBsAg may be undetectable. Exam-

TABLE 5 Interpretation of HBV serologic markers in patients with hepatitis

HBsAg	Assay result Anti-HBs	Anti-HBc	Interpretation
Positive	Negative	Negative	Early acute HBV infection. Confirmation is required to exclude nonrepeatable or nonspecific reactivity (see text).
Positive	(±)[a]	Positive	HBV infection, either acute or chronic. Differentiate with IgM anti-HBc (see text for exceptions). Determine level of replicative activity (infectivity) with HBeAg or HBV DNA.
Negative	Positive	Positive	Indicates previous HBV infection and immunity to hepatitis B.
Negative	Negative	Positive	Possibilities include HBV infection in remote past, "low-level" HBV carrier, "window" between disappearance of HBsAg and appearance of anti-HBs, or false-positive reaction. Investigate with IgM anti-HBc and/or challenge with HBsAg vaccine (see text). When present, anti-HBe helps validate anti-HBc reactivity.
Negative	Negative	Negative	Another infectious agent, toxic injury to liver, disorder of immunity, hereditary disease of liver, or disease of biliary tract.
Negative	Positive	Negative	Vaccine-type response.

[a] ±, anti-HBs is usually absent in this situation, but may occasionally be present.

ination of these sera for IgM-specific anti-HBc may help in establishing the correct diagnosis (see caveat above concerning detection of IgM-specific anti-HBc in chronic hepatitis B patients during reactivation of disease). In the absence of anti-HBc and HBsAg, past or present hepatitis B disease can almost always be excluded (see above for exceptions). In contrast, the presence of anti-HBc alone is plausible evidence for an active HBV infection. This relationship is not infallible, however, as some patients who have recovered from hepatitis B with the development of anti-HBs and anti-HBc may eventually lose their anti-HBs, resulting in a solitary anti-HBc response.

Chronic hepatitis B patients may be tested for HBeAg, anti-HBe, or HBV DNA to determine their levels of HBV replication and their potentials for enhanced infectivity. HBeAg-positive specimens contain high concentrations of HBV DNA in contrast to anti-HBe-positive samples, in which the number of hepatitis B virions is markedly reduced. As anticipated, HBeAg-positive patients are more likely to transmit hepatitis B sexually, percutaneously, or perinatally. HBeAg seroconversion, with loss of HBV DNA, is a desirable objective in the treatment of hepatitis B with antiviral agents such as interferon.

Antibody to HBsAg becomes detectable usually immediately after, but sometimes several weeks after, the disappearance of HBsAg. It is believed that anti-HBs is produced much earlier but is not observed as a result of the formation of immune complexes in the presence of excess HBsAg. Antibody to HBsAg, with or without anti-HBc, specifies immunity against reinfection. Anti-HBs without anti-HBc develops in persons who receive hepatitis B vaccine (which contains only HBsAg), and levels of ≥10 mIU/ml are considered protective. It has been proposed that anti-HBc, not anti-HBs, be used as a screening test to determine susceptibility to HBV for the purpose of recommending hepatitis B vaccination. Depending on the circumstances, passive transfer of anti-HBs or anti-HBc may be observed in patients receiving clotting factors, after immunoglobulin administration, or in neonates of mothers with recent or past hepatitis B. Since blood donations are pretested for HBsAg and total anti-HBc, passive transfer of antibodies following transfusions of blood is unlikely unless the donor has been vaccinated (donors are not tested for anti-HBs).

Recognition of these possibilities will avoid an erroneous diagnosis of HBV exposure or infection, because passive antibodies disappear gradually over 3 to 6 months, while actively produced antibodies are remarkably stable over many years.

Subtyping of specimens by the clinical laboratory provides additional information to the clinician or hospital epidemiologist, because the mutually exclusive subdeterminants are virus specific and not host determined. This fact can be helpful in determining the source of infection or can provide epidemiologic evidence for relatedness among cases. Sequencing a segment of the HBV genome also can be used to determine the source of an infection, but this method is available only as a research tool.

HDV

Description of Agent

HDV is an unclassified 35- to 37-nm-diameter virus that lacks a well-defined nucleocapsid. The structureless core is composed of HDAg and is encapsidated by HBsAg. HDV assumes the HBsAg subtype of the HBV present in the host. Although intracellular HDV replication has been accomplished without HBV in in vitro systems, assembly of intact HDV virions and pathogenicity in vivo require the helper function of HBV to support the replication of HDV (59, 61).

The RNA genome of HDV present in serum is a 1.7-kb single-stranded covalently closed circle of single polarity with viroidlike protein-coding regions (56, 79). Extensive base pairing yields substantial internal complementarity, resulting in an unbranched rodlike structure that is reminiscent of viroids or plant RNA satellite viruses. In tissue, RNA molecules of both polarities, sometimes existing as multimers of the 1.7-kb RNA, are detected. The various forms are consistent with a rolling-circle model of replication. The genome is nonhomologous with HBV DNA and has several sites that are capable of self-cleavage and ligation. HDAg, a phosphoprotein that is nonglycosylated and may bind to the HDV genome, is the product of the only open reading frame (ORF 5) on the antigenomic strand of HDV. It consists of two proteins, a 195-amino-acid 24-kDa protein and a larger 214-amino-acid 27-kDa protein (10). The virus particle has a buoyant density in CsCl of 1.25 g/cm^3 and a sedimentation coefficient that is intermediate between those of HBsAg and intact HBV particles. It is inactivated by formalin under conditions similar to those that inactivate HBV.

Direct Examination

Because HDAg is rarely detectable in serum, even when the serum is known to be infectious, and because antibodies to HDV (anti-HDV) are present indefinitely in patients or experimental animals with chronic HDV infection, liver biopsy with immunohistochemical staining for intrahepatic HDAg is a reliable way to demonstrate ongoing HDV replication. However, these methodologies are restricted to specialized pathology laboratories and are not practical for adoption as rapid screening tests by clinical laboratories. The anti-HDV probe, to which fluorescein or peroxidase is conjugated, is usually derived from serum or plasma that contains a high titer of anti-HDV and a low level of anti-HBc. Thus, the anti-HBc reactivity can be reduced substantially or eliminated entirely by dilution. However, an anti-HDV-negative, anti-HBc-positive control probe should be used in parallel with the anti-HDV probe to validate the anti-HDV specificity of staining. Controls for other sources of nonspecificity, including autoantibodies to nuclear antigens, should be incorporated into procedures for immunohistochemical staining. With these techniques, investigators have shown that the HDAg localizes primarily to the liver cell nucleus and rarely can be detected in the cytoplasm. Although frozen cryostat sections are preferable, HDAg is stable to formalin fixation and paraffin embedding. Therefore, HDAg can be studied in stored paraffin-embedded tissue after digestion of the section with trypsin or pronase (12).

Serologic Identification of Hepatitis D Antigens, Antibodies, and Virus

Anti-HDV has been identified and quantitated primarily by RIA or EIA (12, 60). Tests for anti-HDV require a source of HDAg. Previously, the HDAg had been obtained from HDAg-positive human (postmortem) or chimpanzee liver or serum. Recently, a more reliable source has become available in the form of HDAg-positive woodchuck liver. HDAg can be extracted from human liver only with strong dissociating agents such as 6 M guanidine hydrochloride or 8 M urea. When liver tissue is obtained from experimentally infected chimpanzees or woodchucks at the peak of intrahepatic HDAg expression, i.e., before the appearance of anti-HDV, antigen can be harvested by simple aqueous extraction. Although HDAg is rarely detectable in the sera of patients with acute HDV infection, occasionally, the amount of antigen present is sufficient to serve as a source of antigen for diagnostic testing.

Serum containing anti-HDV can be obtained from patients or experimental animals with acute or chronic HDV infection. As mentioned above, anti-HBc reactivity, invariably present in anti-HDV-positive serum, is relatively low in serum samples with a high level of anti-HDV activity. Thus, residual anti-HBc activity can be diluted out, and HBsAg is removed by ultracentrifugation.

RIA

A commercial RIA for anti-HDV is available for diagnostic purposes (Abbott Anti-Delta; Abbott Laboratories). The test is a competitive-binding RIA in which anti-HDV in the test serum competes with ^{125}I-labeled human anti-HDV for woodchuck liver-derived HDAg coating a solid-phase polystyrene bead. In this assay, 100 μl of ^{125}I-labeled anti-HDV and 100 μl of serum or plasma to be tested are delivered to a reaction well, and an HDAg-coated bead is added. Appropriate negative and positive control samples are incorporated into each test run, and the trays are incubated for 18 to 22 h at room temperature. After incubation, liquid is aspirated from the wells, and the beads are washed with distilled or deionized water, transferred to tubes, and assayed in a gamma counter. Samples with a count rate equal to or below the cutoff value are considered reactive for anti-HDV; counts above the cutoff range are considered negative. A C/S ratio of ≥1.0 is also considered positive. The manufacturer provides specific guidelines to determine whether a test is valid.

HDAg can be detected by solid-phase sandwich RIA as described by Rizzetto et al. (60). The test is based on binding HDAg in serum to anti-HDV adherent to a solid phase followed by incubation with an ^{125}I-labeled IgG anti-HDV probe. Because HDAg always circulates within an HBsAg-encapsidated particle, detection of HDAg requires detergent disruption of the virion (0.5% Tween 80, Nonidet P-40, or deoxycholate) to expose the internal, otherwise sequestered antigen. Adding to the difficulty of HDAg detection is the transient nature of HDV antigenemia occurring during early infection. This simple solid-phase RIA for HDAg can be modified for detection of anti-HDV in a competitive binding RIA. A standardized quantity of HDAg is added to beads coated with anti-HDV. Anti-HDV from the test serum or plasma competes with a constant amount of ^{125}I-labeled anti-HDV for the HDAg immunologically bound to the beads. A ≥50% reduction in the net cpm compared to the cpm in a negative control is evidence for the presence of anti-HDV in the sample. Measurement of anti-HDV titers is achieved by diluting the serum and determining the dilution that inhibits binding of ^{125}I-anti-HDV by 50%.

Acute HDV infection is accompanied by an early, transient anti-HDV response predominantly of the IgM class (68). An antibody class capture, solid-phase RIA for IgM anti-HDV is available outside the United States. The methodology for this technique is analogous to that for the detection of IgM-specific anti-HBc, and interested readers are referred to the hepatitis B RIA section above, in which details of this assay are provided. Since IgM anti-HDV persists in chronic infection and may exist at a high titer in patients with severe chronic hepatitis D (68), the presence of IgM anti-HDV does not distinguish between acute and chronic HDV infection.

EIA

A commercial EIA for anti-HDV based on the configuration described above for the commercial RIA (11) is available for diagnostic purposes (Abbott Anti-Delta EIA; Abbott Laboratories). Similarly, RIA techniques for the detection of HDAg have been applied to the detection of HDAg by EIA, and the same limitation (transient nature of the antigenemia during early acute infection) applies (12).

Western Blot

HDAg can be detected by Western blot (immunoblot). More sensitive than RIA and EIA, blotting is reactive in over 70% of patients with chronic hepatitis D (7). Because of its technical sophistication, Western blotting has not been adapted for routine clinical diagnosis; instead, it remains a research tool confined to a small number of research laboratories.

Hybridization Techniques

A noninvasive approach for detecting HDV RNA in serum has been made possible with the recent availability of cloned cDNA probes (13, 57, 62, 69). The cDNA probe for HDV-associated RNA is incubated with serum, and HDV RNA is identified by dot blot hybridization analysis. This technique remains limited to a small number of research laboratories but holds promise as a simple, noninvasive test for the diagnosis of chronic HDV infection. Techniques for detection of HDV RNA in hepatic tissue by in situ hybridization with ^{125}I and nonradioactive probes have been developed as well (44, 50). Like dot hybridization techniques for the detection of HDV RNA in serum, these techniques remain the domain of specialized research laboratories. Finally, HDV RNA also can be detected by riboprobe hybridization; this is one of the most sensitive assays for HDV infection, reactive in >80% of infected patients (67).

PCR Assay

The most sensitive method for the detection of HDV RNA relies on amplification by PCR. The basic approach is the same as that outlined above for the detection of HBV DNA except that a reverse transcriptase step is required to obtain DNA complementary to HDV RNA in serum or liver tissue (45, 85). A nonradioisotopic PCR assay has been described as well (8). These approaches are not routinely available but are confined to specialized research laboratories.

Interpretation of Test Results

Infection with HDV can occur in the presence of acute or chronic HBV infection. The duration of HBV infection determines the duration of HDV infection. The presence of HDV infection can be identified by demonstrating intrahepatic HDAg or, more practically, anti-HDV seroconversion (a rise in titer of anti-HDV or de novo appearance of IgM-specific anti-HDV, which is detectable only briefly, if at all). Because IgM-specific anti-HDV is transient in acute HDV infection and IgG anti-HDV is often undetectable once HBsAg disappears, retrospective serodiagnosis of acute, self-limited, simultaneous HBV and HDV infections is difficult.

In contrast to patients with acute HBV infection, patients with chronic HBV infection can support HDV replication indefinitely. This can happen when acute HDV infection occurs in the presence of a nonresolving acute HBV infection. More commonly, acute HDV infection becomes chronic when it is superimposed on an underlying chronic HBV infection (Fig. 5). In such cases, the HDV superinfection appears as a clinical exacerbation or as an episode resembling acute viral hepatitis in someone already chronically infected with HBV. In the past, events resembling acute hepatitis in an HBV carrier or a patient with chronic hepatitis B were attributed to superimposed non-A, non-B hepatitis or to the natural history of the disease. A portion of such episodes, however, represents acute superinfection with HDV.

When a patient presents with acute hepatitis and has HBsAg and anti-HDV in the serum, determination of the immunoglobulin class of anti-HBc is helpful in differentiating coinfection from superinfection. Although IgM-specific anti-HBc does not distinguish absolutely between

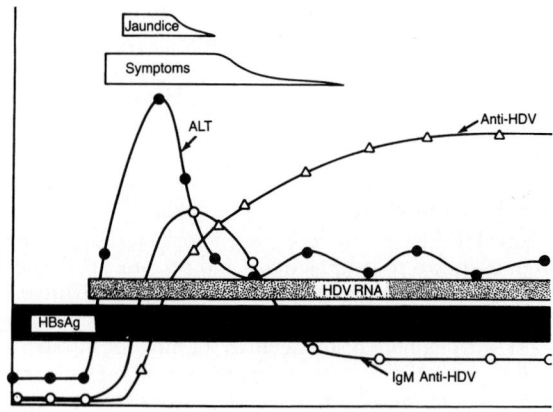

FIGURE 5 Typical serologic course of acute HDV superinfection in an HBsAg carrier. (From reference 27.)

acute and chronic HBV infections, its presence is a presumptive indicator of recent infection, and its absence is a reliable indicator of infection in the remote past. Thus, in simultaneous acute HBV and HDV infections, IgM-specific anti-HBc is detectable, whereas in acute HDV infection superimposed upon chronic HBV infection, anti-HBc is primarily of the IgG class.

As noted above, cDNA tests for the presence of HDV-associated RNA will be useful in the future for determining the presence of ongoing HDV replication and relative infectivity. Currently, probes for this marker are restricted to a limited number of research laboratories.

This effort was supported in part by the Eugene B. Casey Foundation.

REFERENCES

1. **Barker, L. F., J. E. Maynard, R. H. Purcell, J. H. Hoffnagle, K. R. Berquist, and W. T. London.** 1975. Viral hepatitis, type B, in experimental animals. *Am. J. Med. Sci.* **270:**189–195.

2. **Beutler, E., T. Gelbart, and W. Kuhl.** 1990. Interference of heparin with the polymerase chain reaction. *BioTechniques* **9:**166.

3. **Blum, H., A. T. Haase, and G. Vyas.** 1984. Molecular pathogenesis of hepatitis B virus infection: simultaneous detection of viral DNA and antigens in paraffin-embedded liver sections. *Lancet* **ii:**771–774.

4. **Blum, H. E., A. Figus, A. T. Haase, and G. N. Vyas.** 1985. Laboratory diagnosis of hepatitis B virus infection by nucleic acid hybridization analyses and immunohistologic detection of gene products. *Dev. Biol. Stand.* **59:**125–139.

5. **Bond, W. W., M. S. Favero, N. J. Petersen, and J. W. Ebert.** 1983. Inactivation of hepatitis B virus by intermediate-to-high-level disinfectant chemicals. *J. Clin. Microbiol.* **18:**535–538.

6. **Bond, W. W., N. J. Petersen, and M. S. Favero.** 1977. Viral hepatitis B: aspects of environmental control. *Health Lab. Sci.* **14:**235–252.

7. **Buti, M., R. Esteban, R. Jardi, F. Rodriguez-Frias, J. Casacuberta, J. I. Esteban, E. Allende, and J. Guardia.** 1989. Chronic delta hepatitis: detection of hepatitis delta virus in serum by immunoblot and correlation with other markers of delta viral replication. *Hepatology* **10:**907–910.

8. **Cariani, E., A. Ravaggi, M. Pouti, G. Mantero, A. Albertini, and D. Primi.** 1992. Evaluation of hepatitis delta virus RNA

levels during interferon therapy by analysis of polymerase chain reaction products with a nonradioisotopic hybridization assay. *Hepatology* **15**:685–689.

9. Chan, V. T. W., and J. O. McGee. 1990. Non-radioactive probes; preparation, characterization and detection, p. 59–70. *In* J. M. Polak and J. O. McGee (ed.), *In Situ Hybridization: Principles and Practice*. Oxford University Press, Oxford.

10. Chao, M., S.-Y. Hsieh, and J. Taylor. 1990. Role of two forms of the hepatitis delta virus antigen: evidence for a mechanism of self-limiting genome replication. *J. Virol.* **64**:5066–5069.

11. Crivelli, O., M. Rizzetto, C. Lavarini, A. Smedile, and J. L. Gerin. 1981. Enzyme-linked immunosorbent assay for detection of antibody to the hepatitis B surface antigen-associated delta antigen. *J. Clin. Microbiol.* **14**:173–177.

12. Crivelli, O., J. W. K. Shih, and M. Rizzetto. 1983. Methods for detection of the delta antigen and antibody in liver and serum, p. 121–126. *In* G. Verme, F. Bonino, and M. Rizzetto (ed.), *Viral Hepatitis and Delta Infection*. Alan R. Liss, Inc., New York.

13. Denniston, K. J., B. H. Hoyer, A. Smedile, F. V. Wells, J. Nelson, and J. L. Gerin. 1986. Cloned fragment of the hepatitis delta virus RNA genome: sequence and diagnostic application. *Science* **232**:873–875.

14. Dixon, W. J., and M. B. Brown. 1979. *BMDP-79*. University of California Press, Berkeley.

15. Dunsford, H. A., S. Sell, and F. V. Chisari. 1990. Hepatocarcinogenesis due to chronic liver cell injury in hepatitis B virus transgenic mice. *Cancer Res.* **50**:3400–3407.

16. Erlich, H. A. 1989. *PCR Technology: Principles and Applications for DNA Amplification*. Stockton Press, New York.

17. Farci, P., A. Mandas, A. Coiana, M. E. Lai, V. Desmet, P. Van Eyken, Y. Gibo, L. Caruso, S. Scaccabarozzi, D. Criscuolo, J.-C. Ryff, and A. Balestrieri. 1994. Treatment of chronic hepatitis D with interferon alfa-2a. *N. Engl. J. Med.* **330**:88–94.

18. Feinberg, A. P., and B. Vogelstein. 1983. A technique for radiolabeling DNA restriction endonuclease fragments to high specific activity. *Anal. Biochem.* **132**:6–13.

19. Fiaccadori, F., A. Bertoletti, A. Penna, and C. Ferrari. 1992. The immunopathogenesis of hepatitis B. *Ann. Ital. Med. Int.* **7**:153–159.

20. Gavilanes, F., J. M. Gonzalez-Ros, and D. L. Peterson. 1982. Structure of hepatitis B surface antigen. *J. Biol. Chem.* **257**:7770–7777.

21. Gowans, E. J., A. R. Burrel, B. P. Jilbert, and B. P. Marmion. 1985. Cytoplasmic (but not nuclear) hepatitis B virus core antigen reflects HBV DNA synthesis at the level of the infected hepatocyte. *Intervirology* **24**:220–225.

22. Gripon, P., C. Diot, and C. Guguen-Guillouzo. 1993. Reproducible high level infection of cultured adult human hepatocytes by hepatitis B virus: effect of polyethylene glycol on adsorption and penetration. *Virology* **192**:534–540.

23. Guo, K. J., and D. S. Bowden. 1991. Digoxigenin-labeled probes for the detection of hepatitis B virus DNA in serum. *J. Clin. Microbiol.* **29**:506–509.

24. Han, K. H., F. B. Hollinger, C. A. Noonan, H. Solomon, G. B. G. Klintmalm, R. M. Genta, and B. Yoffe. 1993. Southern-blot analysis and simultaneous *in situ* detection of hepatitis B virus-associated DNA and antigens in patients with end-stage liver disease. *Hepatology* **18**:1032–1038.

25. Han, K. H., F. B. Hollinger, C. A. Noonan, and B. Yoffe. 1992. Simultaneous detection of HBV-specific antigens and DNA in paraffin-embedded liver tissue by immunohistochemistry and in situ hybridization using a digoxigenin-labeled probe. *J. Virol. Methods* **37**:89–97.

26. Hollinger, F. B., E. Adam, D. Heiberg, and J. L. Melnick. 1981. Response to hepatitis B vaccine in a young adult population, p. 451–466. *In* W. Szmuness, H. J. Alter, and J. E. Maynard (ed.), *Viral Hepatitis, 1981 International Symposium*. The Franklin Institute Press, Philadelphia.

27. Hoofnagle, J. H. 1989. Type D (delta) hepatitis. *JAMA* **261**:1321–1325.

28. Hsu, S.-M., L. Raine, and H. Fanger. 1981. Use of avidin-biotin-peroxidase complex (ABC) in immunoperoxidase techniques: a comparison between ABC and unlabeled antibody (PAP) procedures. *J. Histochem. Cytochem.* **29**:577–580.

29. Hsu, S.-M., L. Raine, and H. Fanger. 1981. A comparative study of the peroxidase-antiperoxidase method and an avidin-biotin complex method for studying polypeptide hormones with radioimmunoassay antibodies. *Am. J. Clin. Pathol.* **75**:734–738.

30. Huang, S., H. Minassian, and J. D. More. 1976. Application of immunofluorescent staining on paraffin sections improved by trypsin digestion. *Lab. Invest.* **35**:383–390.

31. Imai, H., O. Yamada, S. Morita, S. Suehiro, and T. Kurimura. 1992. Detection of HIV-1 RNA in heparinized plasma of HIV-1 seropositive individuals. *J. Virol. Methods* **36**:181–184.

32. Innis, M. A., D. H. Gelfand, J. J. Sninsky, and T. H. White. 1990. *PCR Protocols: a Guide to Methods and Applications*. Academic Press, Inc., San Diego.

33. Izraeli, S., C. Pfleiderer, and T. Lion. 1991. Detection of gene expression by PCR amplification of RNA derived from frozen heparinized whole blood. *Nucleic Acids Res.* **19**:6051.

34. Kaneko, S., R. H. Miller, S. M. Feinstone, M. Unoura, K. Kobayashi, N. Hattori, and R. H. Purcell. 1989. Detection of serum hepatitis B DNA in patients with chronic hepatitis using the polymerase chain reaction assay. *Proc. Natl. Acad. Sci. USA* **86**:312–316.

35. Kaplan, P. M., E. C. Ford, R. H. Purcell, and J. L. Gerin. 1976. Demonstration of subpopulations of Dane particles. *J. Virol.* **17**:885–893.

36. Kobayashi, H., M. Tsuzuki, K. Koshimizu, H. Toyama, N. Yoshihara, T. Shikata, K. Abe, K. Mizuno, N. Otomo, and T. Oda. 1984. Susceptibility of hepatitis B virus to disinfectants or heat. *J. Clin. Microbiol.* **20**:214–216.

37. Kuhns, M., V. Thiers, A. Courouce, J. Scotto, P. Tiollais, and C. Brechot. 1984. Quantitative detection of HBV DNA in human serum, p. 665–666. *In* G. N. Vyas, J. L. Deinstag, and J. H. Hoofnagle (ed.), *Viral Hepatitis and Liver Disease*. Grune and Stratton, Orlando, Fla.

38. Kwok, S., and R. Higuchi. 1989. Avoiding false positives with PCR. *Nature* (London) **339**:237–238.

39. Lai, M. E., P. Farci, A. Figus, A. Balestrieri, M. Arnone, and G. N. Vyas. 1989. Hepatitis B virus DNA in the serum of Sardinian blood donors negative for the hepatitis B surface antigen. *Blood* **73**:17–19.

40. Larzul, D., F. Guigue, J. J. Sninsky, D. H. Mack, C. Brechot, and J. L. Guesdon. 1988. Detection of hepatitis B virus sequences in serum by using *in vitro* enzymatic amplification. *J. Virol. Methods* **20**:227–237.

41. Liang, T. J., H. E. Blum, K. Hasegawa, H. Takahashi, E. Galun, and J. R. Wands. 1991. Detection and transmission of low-level hepatitis B-related virus in HBsAg-negative patients, p. 684–687. *In* F. B. Hollinger, S. M. Lemon, and H. S. Margolis (ed.), *Viral Hepatitis and Liver Disease*. The Williams & Wilkins Co., Baltimore.

42. Lin, H. J. 1989. Biochemical detection of hepatitis B virus constituents. *Adv. Clin. Chem.* **27**:143–199.

43. Lin, H. J., N. Shi, M. Mizokami, and F. B. Hollinger. 1992. Polymerase chain reaction assay for hepatitis C virus RNA using a single tube for reverse transcription and serial rounds of amplification with nested primer pairs. *J. Med. Virol.* **38**:220–224.

44. Lopez-Talavera, J., M. Buti, J. Casacuberta, H. Allende, R. Jardi, R. Esteban, and J. Guardia. 1993. Detection of hepatitis delta virus RNA in human liver tissue by non-radioactive in situ hybridization. *J. Hepatol.* **17**:199–203.

45. Madejón, A., I. Castillo, J. Bartolomé, M. L. Campillo, J. C. Porres, A. Moreno, and V. Carreño. 1990. Detection of HDV-RNA by PCR in serum of patients with chronic HDV infection. *Hepatology* **11**:381–384.

46. Marion, P. L., C. Trepo, K. Matsubara, and P. M. Price. 1990. Experimental models in hepadnavirus research: report of a workshop, p. 866–874. *In* F. B. Hollinger, S. M. Lemon,

and H. S. Margolis (ed.), *Viral Hepatitis and Liver Disease*. The Williams & Wilkins Co., Baltimore.

47. **Morace, G., K. von der Helm, W. Jilg, and F. Deinhardt.** 1985. Detection of hepatitis B virus DNA in serum by a rapid filtration-hybridization assay. *J. Virol. Methods* **12:**235–242.

48. **Mullis, K. B., and F. A. Faloona.** 1987. Specific synthesis of DNA in vitro via a polymerase catalyzed chain reaction. *Methods Enzymol.* **155:**335–350.

49. **National Committee for Clinical Laboratory Standards.** 1989. *Protection of Laboratory Workers from Infectious Disease Transmitted by Blood, Body Fluids, and Tissue: Tentative Guideline*, document M29-T. National Committee for Clinical Laboratory Standards, Villanova, Pa.

50. **Negro, F., F. Bonino, A. Di Bisceglie, J. Hoofnagle, and J. Gerin.** 1989. Intrahepatic markers of hepatitis delta virus infection: a study by *in situ* hybridization. *Hepatology* **10:**916–920.

51. **Niedobitek, G., T. Finn, H. Herbst, and H. Stein.** 1989. Detection of viral genomes in the liver by *in situ* hybridization using ³⁵S-bromodeoxyuridine- and biotin-labeled probes. *Am. J. Pathol.* **134:**633–639.

52. **Ochiya, T., T. Tsurimoto, K. Ueda, K. Okubo, M. Shiozawa, and K. Matsubara.** 1989. An *in vitro* system for infection with hepatitis B virus that uses primary human fetal hepatocytes. *Proc. Natl. Acad. Sci. USA* **86:**1875–1879.

53. **Paterlini, P., G. Gerken, F. Khemeny, D. Franco, A. D'Errico, W. Grigioni, J. Wands, M. Kew, E. Pisi, P. Tiollais, and C. Brechot.** 1991. Primary liver cancer in HBsAg-negative patients: a study of HBV genome using the polymerase chain reaction, p. 605–610. *In* F. B. Hollinger, S. M. Lemon, and H. S. Margolis (ed.), *Viral Hepatitis and Liver Disease*. The Williams & Wilkins Co., Baltimore.

54. **Pontisso, P., M. G. Ruvaletto, G. Fattovich, L. Chemello, G. Morsica, L. Brollo, V. Matteotti, and A. Alberti.** 1989. Serum HBV-DNA in anti-HBe positive patients detected by filter and liquid phase hybridization assays. *Mol. Cell. Probes* **3:**245–249.

55. **Prince, A. M., W. Stephan, and B. Brotman.** 1983. β-Propiolactone/ultraviolet irradiation: a review of its effectiveness for inactivation of virus in blood derivatives. *Rev. Infect. Dis.* **5:**92–107.

56. **Purcell, R. H., and J. L. Gerin.** 1990. Hepatitis delta virus, p. 2275–2287. *In* B. N. Fields and D. M. Knipe (ed.), *Virology*. Raven Press, New York.

57. **Rashoffer, R., M. Buti, R. Esteban, R. Jardi, and M. Roggendorf.** 1988. Demonstration of hepatitis D virus RNA in patients with chronic hepatitis. *J. Infect. Dis.* **157:**191–195.

58. **Ray, M. B.** 1979. *Hepatitis B Virus Antigens in Tissue*. University Park Press, Baltimore.

59. **Rizzetto, M.** 1983. The delta agent. *Hepatology* **3:**729–737.

60. **Rizzetto, M., J. W. Shih, and J. L. Gerin.** 1980. The hepatitis B virus-associated δ antigen: isolation from liver, development of solid-phase radioimmunoassays for δ antigen and anti-δ and partial characterization of δ antigen. *J. Immunol.* **125:**318–324.

61. **Ryu, W.-S., M. Bayer, and J. Taylor.** 1992. Assembly of hepatitis delta virus particles. *J. Virol.* **66:**2310–2315.

62. **Saldanha, J., F. di Blasi, C. Blas, J. Velosa, F. M. Ramalho, V. di Marco, I. Mora, M. C. de Moura, V. Carreno, A. Craxi, H. C. Thomas, and J. Monjardino.** 1989. Detection of hepatitis delta virus RNA in chronic liver disease. *J. Med. Virol.* **9:**23–28.

63. **Samuel, D., R. Muller, G. Alexander, L. Fassati, B. Ducot, J.-P. Benhamou, H. Bismuth, and Investigators of the European Concerted Action on Viral Hepatitis Study.** 1993. Liver transplantation in European patients with the hepatitis B surface antigen. *N. Engl. J. Med.* **329:**1842–1847.

64. **Scotto, J., M. Hadchouel, C. Hery, J. Yvart, P. Tiollais, and C. Brechot.** 1983. Detection of hepatitis B virus DNA in serum by a simple spot hybridization technique: comparison with results for other viral markers. *Hepatology* **3:**279–284.

65. **Shafritz, D. A., and S. J. Hadziyannis.** 1984. Hepatitis B virus DNA in liver and serum, viral antigens and antibodies,

66. **Shimizu, Y., S. Ida, T. Matsukura, and T. Yuasa.** 1984. Determination of hepatitis B virus DNA in serum by molecular hybridization. *Microbiol. Immunol.* **28:**1117–1123.

67. **Smedile, A., K. F. Bergmann, B. M. Baroudy, F. V. Wells, R. H. Purcell, F. Bonino, M. Rizzetto, and J. L. Gerin.** 1990. Riboprobe assay for HDV RNA: a sensitive method for the detection of the HDV genome in clinical serum samples. *J. Med. Virol.* **30:**20–24.

68. **Smedile, A., C. Lavarini, O. Crivelli, G. Raimondo, M. Fassone, and M. Rizzetto.** 1982. Radioimmunoassay detection of IgM antibodies to the HBV-associated delta (δ) antigen: clinical significance in δ infection. *J. Med. Virol.* **9:**131–138.

69. **Smedile, A., M. Rizzetto, K. Denniston, F. Bonino, F. Wells, G. Verme, F. Consolo, B. Hoyer, R. H. Purcell, and J. L. Gerin.** 1986. Type D hepatitis: the clinical significance of hepatitis D virus RNA in serum as detected by a hybridization-based assay. *Hepatology* **6:**1297–1302.

70. **Sninsky, J. J.** 1991. Application of the polymerase chain reaction to the detection of viruses, p. 799–805. *In* F. B. Hollinger, S. M. Lemon, and H. S. Margolis (ed.), *Viral Hepatitis and Liver Disease*. The Williams & Wilkins Co., Baltimore.

71. **Takahashi, H., J. T. Liang, H. E. Blum, M. Zeniya, K. Fujise, H. Kameda, and J. R. Wands.** 1991. Identification of low-level hepatitis B viral genome in hepatitis B vaccine nonresponders in Japan, p. 779–781. *In* F. B. Hollinger, S. M. Lemon, and H. S. Margolis (ed.), *Viral Hepatitis and Liver Disease*. The Williams & Wilkins Co., Baltimore.

72. **Takahashi, T., S. Nakagawa, T. Hashimoto, K. Takahashi, M. Imai, Y. Miyakawa, and M. Mayumi.** 1976. Large-scale isolation of Dane particles from plasma containing hepatitis B antigen and demonstration of a circular double-stranded DNA molecule extruding directly from their cores. *J. Immunol.* **117:**1392–1397.

73. **Thiers, V., E. Nakajima, D. Kremsdorf, D. Mack, H. Schellekens, F. Driss, A. Goudeau, J. Wands, J. Sninsky, P. Tiollais, and C. Brechot.** 1988. Transmission of hepatitis B from hepatitis B-seronegative subjects. *Lancet* **ii:**1273–1276.

74. **Tijssen, P.** 1985. *Practice and Theory of Enzyme Immunoassays*. Elsevier Science Publishers, Amsterdam.

75. **Ulrich, P. P., R. A. Bhat, B. Seto, D. Mack, J. Sninsky, and G. N. Vyas.** 1989. Enzymatic amplification of hepatitis B virus DNA in serum compared with infectivity testing in chimpanzees. *J. Infect. Dis.* **160:**37–43.

76. **Urdea, M. S.** 1989. Synthesis and characterization of branched DNA (bDNA) for the direct and quantitative detection of CMV, HBV, HCV, and HIV. *Clin. Chem.* **39:**725–726.

77. **Urdea, M. S., J. A. Running, T. Horn, J. Clyne, L. L. Ku, and B. D. Warner.** 1987. A novel method for the rapid detection of specific nucleotide sequences in crude biological samples without blotting or radioactivity; application to the analysis of hepatitis B virus in human serum. *Gene* **61:**253–264.

78. **Van den Oord, J. J., F. Facchetti, C. De Wolf-Peeters, and V. J. Desmet.** 1989. Binding of biotin to hepatitis B surface antigen: a possible pitfall in immunohistochemistry. *J. Histochem. Cytochem.* **37:**551–554.

79. **Wang, K.-S., W.-L. Choo, A. J. Weiner, J.-H. Ou, R. C. Najarian, R. M. Thayer, G. T. Mullenbach, K. J. Denniston, J. L. Gerin, and M. Houghton.** 1986. Structure, sequence and expression of the hepatitis delta (δ) viral genome. *Nature* (London) **323:**508–513.

80. **Willems, M., H. Moshage, F. Nevens, J. Fevery, and S. H. Yap.** 1993. Plasma collected from heparinized blood is not suitable for HCV-RNA detection by conventional RT-PCR assay. *J. Virol. Methods* **42:**127–130.

81. **Wood, G. S., and R. Warnke.** 1981. Suppression of endogenous avidin-binding activity in tissues and its relevance to

biotin-avidin detection systems. *J. Histochem. Cytochem.* **29:** 1196–1204.

82. **World Health Organization.** 1983. *Laboratory Biosafety Manual.* World Health Organization, Geneva.

83. **Yoffe, B., D. K. Burns, H. S. Bhatt, and B. Combes.** 1990. Extrahepatic hepatitis B virus DNA sequences in patients with acute hepatitis B infection. *Hepatology* **12:**187–192.

84. **Zarski, J. P., M. Kuhns, L. Berck, F. Degos, S. W. Schalm,** P. Tiollais, and C. Brechot. 1989. Comparison of a quantitative standardized HBV-DNA assay and a classical spot hybridization test in chronic active hepatitis B patients undergoing antiviral therapy. *Res. Virol.* **140:**283–291.

85. **Zignego, A. L., P. Dény, C. Féray, A. Ponzetto, P. Gentilini, P. Tiollais, and C. Bréchot.** 1990. Amplification of hepatitis delta virus RNA by polymerase chain reaction: a tool for viral detection and cloning. *Mol. Cell. Probes* **4:**43–51.

Hepatitis C Virus

JUDITH C. WILBER

92

CLINICAL BACKGROUND

Hepatitis C virus (HCV) is now known to be the major cause of parenterally transmitted non-A, non-B hepatitis. Until the virus was characterized, diagnosis was made by exclusion of all other known causes of hepatitis. Antibody to HCV is found in over 80% of patients with well-documented non-A, non-B hepatitis. The worldwide prevalence of HCV is 0.2 to 2% in blood donors and up to 80% in intravenous-drug users. In a large percentage of HCV cases, transmission is by transfusion and other parenteral means such as sharing of needles, occupational exposure to blood, and hemodialysis. However, in close to half of HCV infections, the route of transmission is unknown (3).

HCV establishes a chronic infection in 50 to 80% of cases. Chronic infection is often asymptomatic even in the presence of liver damage discernible on biopsy (7). Chronic HCV is characterized by fluctuating alanine aminotransferase levels (20) and recognizable changes in liver histology. Chronic infection can lead to cirrhosis and hepatocellular carcinoma.

Therapy with alpha interferon has been approved by the U.S. Food and Drug Administration (FDA) and can be effective in eliminating the virus (16). However, fewer than half of HCV-infected individuals respond to treatment, and relapse is common (22).

DESCRIPTION OF THE AGENT

HCV is a small, enveloped, positive-sense, single-stranded RNA virus. The genome has approximately 9,400 nucleotides that contain a single large open reading frame spanning almost the entire genome, which could code for a polyprotein of 3,011 amino acids (26). The virus is distantly related to both the animal pestiviruses and the human flaviviruses, which have similar genetic and polypeptide organization. Although there is little homology between HCV and any other known viral sequences, there is a 45 to 49% homology with pestivirus sequences in the 5' terminal region, while the hydropathicity of the polyprotein is closer to that of the flaviviruses (14). The putative organization of the virus is shown in Fig. 1.

HCV shows significant sequence diversity throughout the genome. Comparison of sequences has led to the description of a number of viral types, which may differ by as much as 33% throughout the genome. The 5' untranslated region is the most conserved, while the envelope region is the most variable. Several genotypes of HCV have been described (10, 11, 18, 32–35, 37). Currently, differentiation of genotypes has been done by sequencing, differential PCR, PCR with type-specific probes, restriction fragment length polymorphism analysis, and serotyping. These techniques have been used in the 5' untranslated region (10), the core (33), and NS5 (18, 34). Sequence comparisons of NS5 distinguish all of the known types. Nomenclature has been a confusing issue, since new types were named by each investigator. Recently, a naming scheme that accounts for similarities in subtypes and includes a naming convention for new isolates was proposed (Table 1) (34). It remains to be determined whether genotyping of HCV is important clinically. There have been numerous reports that genotype 1 infection may be typified by high viral load and may be less responsive to interferon treatment than are other viral genotypes.

DISCOVERY OF THE AGENT

Once serologic testing for hepatitis A and B was developed and hepatitis B testing was instituted for screening blood donors in the early 1970s, it was noted that at least one other agent was responsible for transfusion-associated hepatitis (21, 27). The sera of implicated donors were capable of transmitting disease to chimpanzees, and further studies with chimpanzees revealed that the agent was a small enveloped virus probably related to the flaviviruses (4). Despite extensive research for over a decade, conventional methods failed to isolate the virus in cell culture, visualize it by electron microscopy, or find specific antibodies or antigens for use in diagnostic tests.

Cloning of the viral genome led to an antigen with which the first diagnostic tests could be developed and to the viral nucleic acid sequence, from which numerous antigens, nucleic acid probes, and phylogenetic relationships could be derived (13, 26). A large volume of plasma from a chimpanzee infected by contaminated factor XIII concentrate was ultracentrifuged, and all of the nucleic acid was extracted from the pellet. Random primers were used to copy the nucleic acid into cDNA fragments, which were cloned into bacteriophage λgt11. The encoded polypeptides were expressed in *Escherichia coli*. More than a million

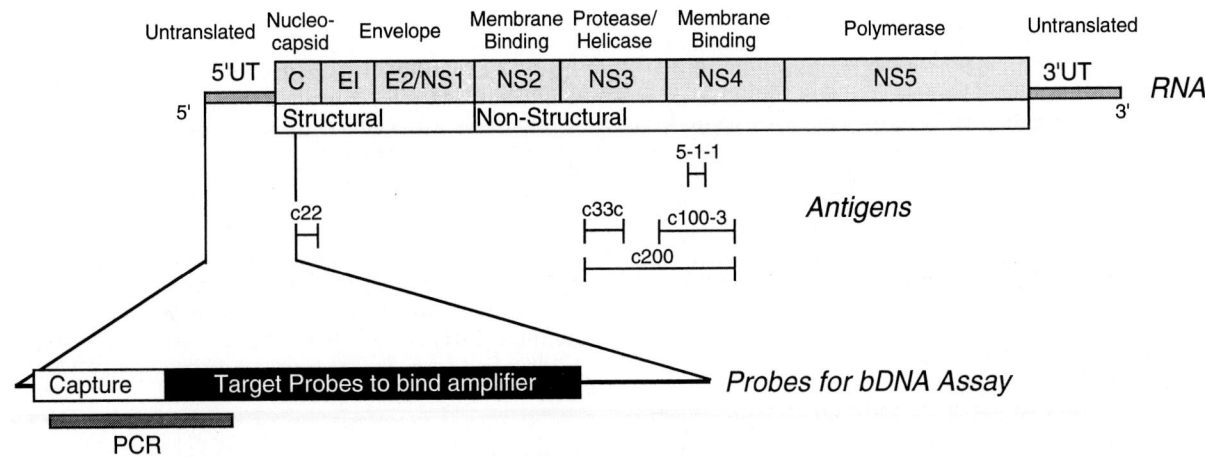

FIGURE 1 Map of the putative functional regions of the HCV genome and derivations of the recombinant HCV antigens. Also shown are the recombinant antigens used in serologic assays, the region used most commonly for RT-PCR primers, and the areas in which target probes for bDNA signal amplification are located.

polypeptides were screened before clone 5-1-1 was found to produce a polypeptide reactive with antibody from patients with non-A, non-B hepatitis. By using 5-1-1 as a hybridization probe, overlapping clones were found and ligated together to form the gene fragment C100. Antigen was expressed in recombinant yeast cells by fusing the human gene for superoxide dismutase (SOD), which facilitates efficient expression of foreign proteins in yeast cells, with the C100 portion of the HCV genome. The resulting recombinant antigen, c100-3, was used to develop the first screening test for transfusion-associated non-A, non-B hepatitis (28). Numerous antigens have now been identified and incorporated into serologic assays.

SEROLOGIC TESTS

Cloning the viral genome made it possible to develop serologic assays that use recombinant antigens, even though the virus had not been visualized by electron microscopy or grown in cell culture. The c100-3 antigen was first used in a radioimmunoassay to determine whether this newly identified antigen would indeed detect antibody specific for the agent responsible for transfusion-transmitted non-A, non-B hepatitis (28). This assay was used to test a coded panel of sera from the National Institutes of Health that had proved all previous putative non-A, non-B hepatitis agents incorrect (1). All but one of the non-A, non-B hepatitis sera were detected, and the one sample that was missed was obtained during the acute phase of the disease. There were no false positives and no discordant results on duplicate samples included in the panel. The c100-3 antigen was then used by Ortho Diagnostic Systems and Chiron to develop a microwell enzyme immunoassay (EIA), which was approved by the FDA in May 1990. HCV antibody testing was immediately instituted in all voluntary blood collection agencies

TABLE 1 Comparison of nomenclature for HCV types[a]

Proposed name	Published example	Nomenclature of:				
		Cha et al. (10)	Chan et al. (11) and Simmonds et al. (35)	Enomoto et al. (18)	Mori et al. (32) and Okamoto et al. (33)	Tsukiyama-Kohara et al. (37)
1a	HCV-1, HCV-H	I	1a	K-PT	I	NC
1b	HCV-J, HCV-BK	II	1b	K-1	II	I
1c		NC	NC	NC	NC	NC
2a	HC-J6	III	2a	K-2a	III	II
2b	HC-J8	III	2b	K-2b	IV	II
2c		III	NC	NC	NC	NC
3a	Ta, E-b1	IV	3	NC	V	NC
3b	Tb	IV	NC	NC	VI	NC
4a		NC	4	NC	NC	NC
5a		V	NC	NC	NC	NC
6a		NC	NC	NC	NC	NC

[a] NC, Sequences not classified by originating authors. Adapted from *Journal of General Virology* (34) with permission of the publisher.

in the United States as a part of routine donor screening. A bead-based EIA using the same antigen was manufactured by Abbott Laboratories, and it was approved in July 1990.

The first-generation HCV EIAs were reactive in 80 to 90% of patients who developed chronic non-A, non-B hepatitis following transfusion. However, antibody is not detected in the earliest stages of infection (2). The window of seronegativity is 10 to 15 weeks in some cases, but antibody may not appear for 6 to 12 months in other cases. Antibody to c100-3 generally disappears in patients whose infections resolve, although this may take years. Anti-HCV immunoglobulin M (IgM) is detectable intermittently in acute and chronic HCV infection (6).

As with other immunoassays, screening low-risk populations can yield false-positive reactions. A strip immunoassay was developed (RIBA; Chiron Corp.) to help distinguish specific HCV antibody from nonspecific reactivity. The earliest version available had bands on nitrocellulose strips that corresponded to the original 5-1-1-encoded antigen produced in *E. coli*, the c100-3 antigen produced in yeast cells, an SOD band, and moderate- and low-IgG control bands. A specimen was considered reactive if it reacted with both HCV bands with an intensity equal to or greater than that of the low-IgG control and not with the SOD band, which represents the carrier protein. Even though the antigens on the strip were from the same region of the virus as the antigen in the EIA, reactivity by RIBA gave an additional indication that the reactivity by EIA was specific for HCV antibody and correlated with infectivity when donors implicated in the transmission of non-A, non-B hepatitis were tested (17, 39). The Abbott Laboratories reference laboratory employs a neutralization or blocking assay.

New serologic tests were developed when additional reactive recombinant antigens that increase the sensitivity of the HCV antibody tests were found. Multiple-antigen EIAs (EIA-2) were approved by the FDA in 1992. The new antigens incorporated into the assays are c22-3 from the nucleocapsid (core) region and c33c from the nonstructural region NS3. c33c was combined with c100-3 to form a single expression protein from NS3 and NS4 called c200. The addition of c22-3 and c33c to HCV serologic tests has resulted in earlier detection of antibody to HCV, and a greater number of acute and chronic non-A, non-B hepatitis patients are identified than with the first-generation assays (12, 31, 39). It is estimated that 10 to 30% more HCV antibody-positive individuals are detected by using second-generation assays. Anti-c22-3 and anti-c33c appear to be present in a significant number of HCV-infected individuals who lack detectable antibody to c100-3.

The four-antigen RIBA Hepatitis C Virus 2.0 Strip Immunoassay (RIBA-2) was approved by the FDA in June 1993. The RIBA-2 assay incorporates antigens 5-1-1 from *E. coli*, c100-3 and c22-3 from yeast cells, c33c from *E. coli*, and SOD. Other nitrocellulose strip-based assays are under development.

Figure 2 shows examples of interpretation of the RIBA-2 strip immunoassay. A specimen is considered positive if there are reactions with two or more bands representing at least two different gene regions with intensities equal to or greater than that of the weak IgG control ($\geq 1+$) and no reaction with the SOD band. If the specimen reacts with bands from only one gene region, such as 5-1-1 and c100-3 (both from the NS4 region), it is considered indeterminate (8). A specimen is also considered indeterminate if it reacts

with the SOD band in addition to HCV antigens with $\geq 1+$ reactivity. The absence of HCV antigen bands with $\geq 1+$ reactivity or a strip with an SOD band alone is considered negative.

Many blood specimens that are reactive on multiple-antigen EIAs have indeterminate reactions by RIBA-2. A large proportion of these exhibit immunoreactivity to c22-3 only (36). A new strip immunoassay (RIBA 3.0) has been developed to resolve the status of RIBA-2 indeterminates; this assay is in use in Europe (41). In it, the c100-3 and c22-3 bands are synthetic peptides, while c33c and a new band representing NS5 are recombinant proteins (5-1-1 is not included). The synthetic peptides have been designed to avoid nonspecific cross-reactivity and to provide increased sensitivity.

DIRECT TESTS

Since no routine culture methods are available to clinical laboratories, detection of viral RNA in serum is used as a marker for the virus itself. Quantitation of viral RNA may be useful in monitoring the course of disease and the effectiveness of antiviral treatment. Many of the early studies to determine the specificity of HCV antibody tests relied on retrospective analysis of transmission of HCV infection through infected units of blood. It was shown that 80 to 90% of donors who were RIBA positive transmitted non-A, non-B hepatitis to recipients (19). Some individuals with true HCV antibody may have recovered from the infection. Detection of HCV RNA is accomplished by amplification methods such as reverse transcriptase PCR (RT-PCR) (5, 23, 25, 30, 40) or branched-DNA (bDNA) signal amplification (Quantiplex HCV-RNA Assay; Chiron Corp.) (29, 38). Over 90% of RIBA-positive specimens are also positive for HCV RNA, corresponding to the infectivity noted in transfusion studies.

RT-PCR transcribes a portion of extracted viral RNA into DNA and amplifies the number of DNA molecules to a detectable level through the use of primers, enzymes, and temperature cycling. Most laboratories use primers from the 5′ untranslated region of the HCV genome because this region is the most conserved. Many different RT-PCR methods with numerous detection methods for amplified product are being used in research laboratories. For example, many researchers use two-stage "nested" PCR to increase the sensitivity enough that only ethidium bromide staining is necessary to detect bands on gel, thus avoiding the use of radioactive probes. Nested PCR amplifies product from the first PCR reaction, using an internal set of primers. However, the second round of amplification is not necessary to achieve adequate sensitivity, and it may lead to contamination (25, 43). Using a labeled HCV-specific probe to detect the PCR product increases both the sensitivity and the specificity, and nonradioactive labels are now available. Numerous precautions should be used to avoid contamination of future reactions with PCR products and carryover from one patient specimen to another.

Detection and quantitation of HCV RNA can also be achieved by using a technique based on bDNA signal amplification. In this technique, 50 μl of serum is placed directly into two wells of a 96-well plate. Proteinase K and detergent are added in a reagent that also contains target oligonucleotide probes. These probes include a set of 5 target probes that each hybridize to a portion of the HCV RNA and to capture probes in the well and another set of 18 target probes that mediate hybridization of the RNA to

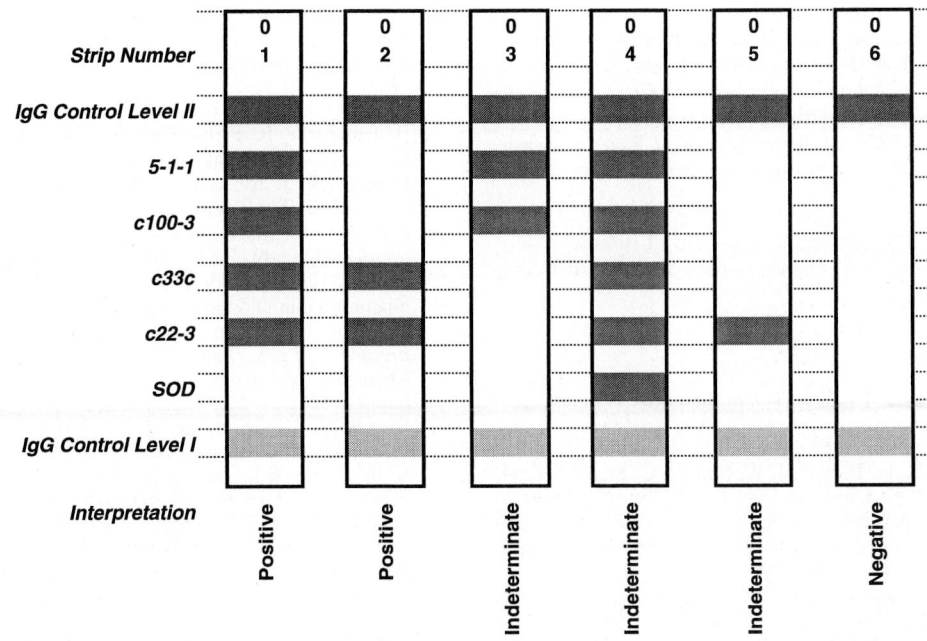

FIGURE 2 Examples of results using the RIBA HCV 2.0 strip immunoblot assay, which utilizes four HCV-encoded antigens immobilized on nitrocellulose test strips. The strips are incubated with patient serum or plasma and run as an EIA; a specific antigen-antibody reaction results in a colored band. Strip 01: positive; reaction with three different gene regions (5-1-1 and c100-3 are from the same region); no SOD band (carrier protein). Strip 02: positive; reaction with two different gene regions; no SOD band. Strips 03 and 05: indeterminate; reaction with one gene region (5-1-1 and c100-3 are from the same region). Strip 04: indeterminate; reaction with the SOD carrier protein. Strip 06: negative; the visible bands show reaction with the moderate- and low-level IgG controls. Reproduced from the *Journal of Clinical Immunoassay* (42) with permission.

synthetic bDNA amplifier molecules. All of the probes are located in the 5′ untranslated region and the core region of the HCV genome (Fig. 1). Each bDNA molecule then hybridizes 45 alkaline phosphatase-labeled probes. The complex is incubated with a chemiluminescent substrate, dioxetane, and the light emitted is measured in a luminometer and compared to a standard curve to determine the level of RNA in the sample.

COLLECTION OF SAMPLES

HCV antibody assays use either serum or plasma. For HCV RNA detection and quantitation, the specimen of choice is aseptically collected serum, although plasma may also be used. Plasma may be collected in EDTA, acid citrate dextrose, sodium citrate, or other anticoagulants, although heparin should be avoided if PCR studies are planned, as it can inhibit enzymatic reactions. Viral RNA is vulnerable to degradation by high levels of RNase in blood, although it appears to be somewhat protected by the viral coat (9). It is very important to separate the serum or plasma from the cellular blood components as soon as possible to avoid viral degradation by granulocytes (15). The separated serum or plasma should be refrigerated or frozen within 4 to 6 h of collection. Specimens should be kept cold after thawing, and repeated freezing and thawing should be avoided.

SELECTION OF TESTS AND INTERPRETATION

HCV diagnosis is currently achieved by detecting specific antibody by EIA and strip immunoblot assays that use antigens derived from the cloning of the HCV genome. If both antibody tests are positive, there is a high likelihood that the patient is infected (24). Liver enzyme levels should be monitored, and biopsy should be considered.

Tests for HCV RNA may be used to help diagnose cases before seroconversion or if RIBA is indeterminate. However, even with the most sensitive techniques, HCV RNA is not detectable in serum during all stages of infection. A single negative HCV RNA test does not indicate that an individual is uninfected or has responded to treatment. Reasons for using HCV RNA detection and quantitation in evaluating patients are as follows (5).

1. Diagnosis of acute infection. High levels of HCV RNA are detectable before seroconversion.
2. Decision to treat patients with interferon
   a. Evaluation of patients with chronic hepatitis who are negative for antibody to HCV or have indeterminate results
   b. Evaluation of patients with chronic hepatitis with autoantibodies who may have false-positive HCV antibody tests
3. Evaluation of HCV infection after liver transplantation, when ALT levels may be normal
4. Monitoring antiviral therapy

a. Patients with sustained responses to alpha interferon therapy have lower pretreatment levels of HCV RNA than do nonresponders (29).

b. HCV RNA levels during treatment may indicate the need to alter the course of therapy.

c. Sustained loss of detectable HCV RNA in serum may indicate response to treatment.

Further clinical studies will clarify the interpretation of levels of HCV RNA during the course of disease. Algorithms for diagnosis and prognosis of HCV are being developed.

REFERENCES

1. **Alter, H. J.** 1991. Descartes before the horse: I clone, therefore I am: the hepatitis C virus in current perspective. *Ann. Intern. Med.* **115:**644–649.
2. **Alter, H. J., R. H. Purcell, J. W. Shih, J. C. Melpolder, M. Houghton, Q.-L. Choo, and G. Kuo.** 1989. Detection of antibody to hepatitis C virus in prospectively followed transfusion recipients with acute and chronic non-A non-B hepatitis. *N. Engl. J. Med.* **321:**1494–1500.
3. **Alter, M. J., S. C. Hadler, F. N. Judson, A. Mares, W. J. Alexander, P. Y. Hu, and J. K. Miller.** 1990. Risk factors for acute non-A non-B hepatitis in the United States and association with hepatitis C infection. *JAMA* **264:**2231–2235.
4. **Bradley, D. W., K. A. McCaustland, E. H. Cook, C. A. Schable, J. W. Ebert, and J. E. Maynard.** 1985. Posttransfusion non-A non-B hepatitis in chimpanzees: physicochemical evidence that the tubule-forming agent is a small, enveloped virus. *Gastroenterology* **88:**773–779.
5. **Bréchot, C.** 1993. Polymerase chain reaction for the diagnosis of hepatitis B and C viral hepatitis. *J. Hepatol.* **17**(Suppl. 3)**:**S35–S41.
6. **Brillanti, S., M. Foli, P. Perini, C. Masci, M. Miglioli, and L. Barbara.** 1991. Long-term persistence of IgM antibodies to HCV in chronic hepatitis C. *Hepatology* **14:**969–974.
7. **Brillanti, S., S. Gaiani, M. Miglioli, M. Foli, C. Masci, and L. Barbara.** 1993. Persistent hepatitis C viraemia without liver disease. *Lancet* **341:**464–465.
8. **Busch, M. P., L. Tobler, S. Quan, J. C. Wilber, P. Johnson, A. Polito, E. Steane, A. Zola, C. Bahl, M. Nelles, and S. R. Lee.** 1993. A pattern of 5-1-1 and c100-3 only on hepatitis C virus (HCV). *Transfusion* **33:**84–88.
9. **Busch, M. P., J. C. Wilber, P. Johnson, L. Tobler, and C. S. Evans.** 1992. Impact of specimen handling and storage on detection of hepatitis C virus RNA. *Transfusion* **32:**420–425.
10. **Cha, T. A., E. Beall, B. Irvine, J. Kolberg, D. Chien, G. Kuo, and M. S. Urdea.** 1992. At least five related, but distinct hepatitis C viral genotypes exist. *Proc. Natl. Acad. Sci. USA* **89:**7144–7148.
11. **Chan, S.-W., F. McOmish, E. C. Holmes, B. Dow, J. F. Peutherer, E. Follett, P. L. Yap, and P. Simmonds.** 1992. Analysis of a new hepatitis C virus type and its phylogenetic relationship to existing variants. *J. Gen. Virol.* **73:**1131–1141.
12. **Chemello, L., D. Cavalletto, P. Pontisso, F. Bortolotti, C. Donada, V. Donadon, M. Frezza, P. Casarin, and A. Alberti.** 1993. Patterns of antibodies to hepatitis C virus in patients with chronic non-A, non-B hepatitis and their relationship to viral replication and liver disease. *Hepatology* **17:**179–182.
13. **Choo, Q.-L., G. Kuo, A. J. Weiner, L. R. Overby, D. W. Bradley, and M. Houghton.** 1989. Isolation of a cDNA clone derived from a blood-borne non-A, non-B viral hepatitis genome. *Science* **244:**359–364.
14. **Choo, Q.-L., K. H. Richman, J. H. Han, K. Berger, C. Lee, C. Dong, C. Gallegos, D. Coit, A. Medina-Selby, P. J. Barr, A. J. Weiner, D. W. Bradley, G. Kuo, and M. Houghton.** 1991. Genetic organization and diversity of the hepatitis C virus. *Proc. Natl. Acad. Sci. USA* **88:**2451–2455.
15. **Cuypers, H. T. M., D. Bresters, N. Winkel, H. W. Reesink, A. J. Weiner, M. Houghton, C. L. van der Poel, and P. N. Lelie.** 1992. Storage conditions of blood samples and primer selection affect the yield of cDNA polymerase chain reaction products of hepatitis C virus. *J. Clin. Microbiol.* **30:**3220–3224.
16. **Davis, G. L., L. A. Balart, E. R. Schiff, K. Lindsay, H. C. Bodenheimer, R. P. Perrillo, W. Carey, I. M. Jacobson, J. Payne, J. L. Dienstag, D. H. VanThiel, C. Tamburro, J. Lefkowitch, J. Albrecht, C. Meschievitz, T. J. Ortego, A. Gibas, and The Hepatitis Interferon Treatment Group.** 1989. Treatment of chronic hepatitis C with recombinant interferon alfa. *N. Engl. J. Med.* **321:**1501–1505.
17. **Ebeling, F., R. Naukkarinen, and J. Leikola.** 1990. Recombinant immunoblot assay for hepatitis C virus antibody as predictor of infectivity. *Lancet* **335:**982–983.
18. **Enomoto, N., A. Takada, T. Nakao, and T. Date.** 1990. There are two major types of hepatitis C virus in Japan. *Biochem. Biophys. Res. Commun.* **170:**1021–1025.
19. **Esteban, J. I., A. Gonzales, J. M. Hernandez, L. Viladomiu, C. Sanchez, J. C. Lopez-Talavera, G. Martin-Vega, X. Vidal, R. Esteban, and J. Guardia.** 1991. Evaluation of antibodies to hepatitis C virus in a contemporary study of transfusion-associated hepatitis. *N. Engl. J. Med.* **232:**1107–1112.
20. **Farci, P., H. J. Alter, D. Wong, R. H. Miller, J. W. Shih, B. Jett, and R. H. Purcell.** 1991. A long-term study of hepatitis C virus replication in non-A, non-B hepatitis. *N. Engl. J. Med.* **325:**98–104.
21. **Feinstone, S. M., A. Z. Kapikian, R. H. Purcell, H. J. Alter, and P. V. Holland.** 1975. Transfusion-associated hepatitis not due to viral hepatitis type A or B. *N. Engl. J. Med.* **292:**767–770.
22. **Garson, J. A., S. Brillanti, C. Ring, P. Perini, M. Miglioli, and L. Barbara.** 1992. Hepatitis C viraemia rebound after "successful" interferon therapy in patients with chronic non-A, non-B hepatitis. *J. Med. Virol.* **37:**210–214.
23. **Garson, J. A., R. S. Tedder, M. Briggs, P. Tuke, J. A. Glazebrook, A. Trute, D. Parker, J. A. J. Barbara, M. Contreras, and S. Aloysius.** 1990. Detection of hepatitis C viral sequences in blood donations by "nested" polymerase chain reaction and prediction of infectivity. *Lancet* **335:**1419–1422.
24. **Gretch, D., W. Lee, and L. Corey.** 1992. Use of aminotransferase, hepatitis C antibody, and hepatitis C polymerase chain reaction RNA assays to establish the diagnosis of hepatitis C virus infection in a diagnostic virology laboratory. *J. Clin. Microbiol.* **30:**2145–2149.
25. **Gretch, D. R., J. J. Wilson, R. L. Carithers, C. dela Rosa, J. H. Han, and L. Corey.** 1993. Detection of hepatitis C virus RNA: comparison of one-stage polymerase chain reaction (PCR) with nested-set PCR. *J. Clin. Microbiol.* **31:**289–291.
26. **Houghton, M., A. Weiner, J. Han, G. Kuo, and Q.-L. Choo.** 1991. Molecular biology of the hepatitis C viruses: implications for diagnosis, development and control of viral disease. *Hepatology* **14:**381–388.
27. **Koretz, R. L., S. Ritman, and G. L. Gitnick.** 1973. Posttransfusion hepatitis in recipients of blood screened by newer assays. *Lancet* **ii:**694–696.
28. **Kuo, G., Q.-L. Choo, H. J. Alter, G. L. Gitnick, A. G. Redeker, R. H. Purcell, T. Miyamura, J. L. Dienstag, M. J. Alter, C. E. Stevens, G. E. Tegtmeier, F. Bonino, M. Colombo, W.-S. Lee, C. Kuo, K. Berger, J. R. Shuster, L. R. Overby, D. W. Bradley, and M. Houghton.** 1989. An assay for circulating antibodies to a major etiologic virus of human non-A, non-B hepatitis. *Science* **244:**362–364.
29. **Lau, J. Y. N., G. L. Davis, J. Kniffen, K.-P. Qian, M. S. Urdea, C. S. Chan, M. Mizokami, P. D. Neuwald, and J. C. Wilber.** 1993. Significance of serum hepatitis C virus RNA levels in chronic hepatitis C. *Lancet* **341:**1501–1504.
30. **Lok, A. S. F., R. Cheung, R. Chan, and V. Liu.** 1992. Hepatitis C viremia in patients with hepatitis C virus infection. *Hepatology* **15:**1007–1012.
31. **McHutchison, J. G., J. L. Person, S. Govindarajan, B. Valinluck, T. Gore, S. R. Lee, M. Nelles, A. Polito, D. Chien, R. DiNello, S. Quan, G. Kuo, and A. G. Redeker.**

1992. Improved detection of hepatitis C virus antibodies in high-risk populations. *Hepatology* **15:**19–25.

32. **Mori, S., N. Kato, A. Yagyu, T. Tanaka, Y. Ikeda, B. Petchclai, P. Chiewsilp, T. Kurimura, and K. Shimotohno.** 1992. A new type of hepatitis C virus in patients in Thailand. *Biochem. Biophys. Res. Commun.* **183:**334–342.

33. **Okamoto, H., Y. Sugiyama, S. Okada, K. Kurai, Y. Akahane, Y. Sugai, T. Tanaka, K. Sato, F. Tsuda, Y. Miyakawa, and M. Mayumi.** 1992. Typing hepatitis C virus by polymerase chain reaction with type-specific primers: application to clinical surveys and tracing infectious sources. *J. Gen. Virol.* **73:**673–679.

34. **Simmonds, P., E. C. Holmes, T.-A. Cha, S.-W. Chan, F. McOmish, B. Irvine, E. Beall, P. L. Yap, J. Kolberg, and M. S. Urdea.** 1993. Classification of hepatitis C virus into six major genotypes and a series of subtypes by phylogenetic analysis of the NS-5 region. *J. Gen. Virol.* **74:**2391–2399.

35. **Simmonds, P., F. McOmish, P. L. Yap, S.-W. Chan, C. K. Lin, G. Dusheiko, A. A. Saeed, and E. C. Holmes.** 1993. Sequence variability in the 5′ non-coding region of hepatitis C virus: identification of a new virus type and restrictions on sequence diversity. *J. Gen. Virol.* **74:**661–668.

36. **Tobler, L. H., M. P. Busch, J. Wilber, R. DiNello, S. Quan, A. Polito, R. Kochesky, C. Bahl, M. Nelles, and S. R. Lee.** 1994. Evaluation of indeterminate c22-3 reactivity in volunteer blood donors. *Transfusion* **34:**130–134.

37. **Tsukiyama-Kohara, K., M. Kohara, K. Yamaguchi, N. Maki, A. Toyoshima, K. Miki, S. Tanaka, N. Hattori, and A. Nomoto.** 1991. A second group of hepatitis C viruses. *Virus Genes* **5:**243–254.

38. **Urdea, M. S.** 1993. Synthesis and characterization of branched DNA (bDNA) for the direct and quantitation detection of CMV, HBV, HCV, HIV. *Clin. Chem.* **39:**725–726.

39. **van der Poel, C. L., H. T. M. Cuypers, H. W. Reesink, A. J. Weiner, S. Quan, R. Di Nello, J. J. P. van Boven, I. Winkel, D. Mulder-Folkerts, P. J. Exel-Oehlers, W. Schaasberg, A. Leentvaar-Kuypers, A. Polito, M. Houghton, and P. N. Lelie.** 1991. Confirmation of hepatitis C virus infection by new four-antigen recombinant immunoblot assay. *Lancet* **337:**317–319.

40. **Weiner, A. J., G. Kuo, D. W. Bradley, F. Bonino, G. Saracco, C. Lee, J. Rosenblatt, Q.-L. Choo, and M. Houghton.** 1990. Detection of hepatitis C viral sequences in non-A, non-B hepatitis. *Lancet* **335:**1–3.

41. **Wicki, A. N., and H. Joller-Jemelka.** 1993. Indeterminate hepatitis C. *Lancet* **341:**1534.

42. **Wilber, J. C.** 1993. Development and use of laboratory tests for hepatitis C infection: a review. *J. Clin. Immunoassay* **16:** 204–207.

43. **Wilber, J. C., P. J. Johnson, and M. S. Urdea.** 1993. Reverse-transcriptase PCR for hepatitis C virus RNA, p. 327–331. *In* D. H. Persing, T. F. Smith, F. C. Tenover, and T. J. White (ed.), *Diagnostic Molecular Microbiology: Principles and Applications.* American Society for Microbiology, Washington, D.C.

Hepatitis E Virus

JOHN TICEHURST

93

Hepatitis E virus (HEV) is the only known agent of enterically transmitted non-A, non-B (NANB) hepatitis. Like all hepatitis viruses, HEV appears to be a unique and important agent of acute disease around the world. In the United States, however, typical cases of hepatitis E have been recognized only as a result of travel in known areas of endemicity, including Africa, Asia, and Mexico. Hepatitis E should be suspected after such travel, especially if there is no evidence for another cause of acute hepatitis. Although there is no practical diagnostic technique for acute hepatitis E, an appropriate assay is likely to be available within the next few years.

DESCRIPTION OF THE AGENT

After assays for infection with hepatitis A virus (HAV) and hepatitis B virus (HBV) were developed during the 1970s, it became apparent that there were additional agents of viral hepatitis. The diseases associated with these undiscovered agents were known as NANB hepatitis. In industrialized countries, NANB hepatitis was usually associated with blood transfusion and other forms of parenteral transmission. Thus, most references to NANB hepatitis in the American literature more precisely described parenterally transmitted NANB hepatitis. The vast majority of parenterally transmitted NANB hepatitis has been associated with hepatitis C virus (HCV) since its identification in 1989 (8). Hepatitis D virus (HDV; also known as delta agent) is another NANB hepatitis agent by definition and is usually transmitted by the parenteral route; infection always requires active HBV replication (39).

It was also recognized that enterically transmitted epidemics of hepatitis occurred in southern Asia and other developing regions. The first such epidemic involved 29,000 cases in Delhi, India, during 1955 and 1956 and resulted from contamination of a major water supply with raw sewage (2, 9, 40). An etiologic agent was not isolated, so it was thought that HAV caused such outbreaks. However, differences from hepatitis A (Table 1) were clues that another agent was involved. In 1980, Khuroo (30) and Wong et al. (cited in references 2 and 40, among others) demonstrated that the Delhi epidemic and others were associated with at least one unidentified agent. The disease was referred to as enterically transmitted (ET) NANB hepatitis.

In 1983, Balayan et al. (5) identified a viral agent of ET-NANB hepatitis, subsequently named HEV (8, 40, 44). By 1988, confirmatory data identified HEV as a worldwide cause of ET-NANB hepatitis since the Delhi epidemic, if not earlier, because hepatitis E-like illness occurred in Europe during the 19th century (3, 9, 40). The identity of HEV was independently established when its genome was molecularly cloned by Reyes et al. in 1990 (44). There is only one known serotype of HEV, but strains from different epidemics and sporadic cases have distinguishing properties; it is useful to designate strains by nation of origin and year of collection. Because HEV may not be the only cause of ET-NANB hepatitis (2) (see Evidence for Other Agents below), hepatitis E should refer only to disease caused by HEV.

HEV is a small, round, nonenveloped, positive-strand RNA virus. Classification of HEV as a calicivirus has been suggested (25), but characterization is not yet sufficient for assignment to a family or genus (4). Analysis of amino acid sequences suggested that HEV, rubella virus, and beet necrotic yellow vein virus could represent a distinct category among the "supergroup" of alpha-like viruses (31).

HEV particles in feces or bile are nearly spherical and approximately 29 nm in diameter, which is slightly larger than HAV and other picornaviruses but smaller than typical caliciviruses (Fig. 1) (3, 5, 8, 9, 25, 32, 48–50). HEV sedimented at 183S and had a buoyant density of 1.29 g/cm^3 in potassium tartrate-glycerol (8), which is similar to certain caliciviruses (25). Little is known about the resistance of HEV, and there are no published descriptions of its inactivation. Virus particles are morphologically resistant to trichlorotrifluoroethane, a lipid solvent, and morphologically intact when prepared from lyophilized feces (48–50). HEV may be labile in dense hyperosmolar solutions (sucrose or CsCl) or during storage between -120 and $8°C$ (although it may more be stable at the lower temperature) (9). Successful efforts to control HEV outbreaks have included chlorination of water supplies. Most laboratories assume that HEV is destroyed by iodinated disinfectants or autoclaving.

HEV has a small, positive-strand RNA genome with distinctive characteristics (7, 22, 44, 46). The prototype nucleotide sequence, representing HEV strain Myanmar/82, is markedly dissimilar from those of other viruses (46) (Fig. 2). Judging from amino acid similarities, possible re-

TABLE 1 Comparison of hepatitis E and hepatitis A[a]

Similarities
    Worldwide distribution with epidemic and sporadic cases
    Fecal-oral transmission
    Low potential for transmission by transfusion or other
      parenteral modes
    Chronic disease not recognized

Differences
    Distribution not uniform: typical, endemic hepatitis E not
      recognized in United States, northern Europe, Japan, and
      Australia
    Slightly longer incubation period (average of 40 days) for
      hepatitis E
    Cholestasis more prominent in hepatitis E
    High mortality (10–20%) with hepatitis E during pregnancy
    Clinically apparent secondary cases of hepatitis E unusual
    Endemic hepatitis E predominantly disease of young adults
    Levels of antibody to HEV highest during acute phase and
      may wane during convalescence
    Anti-HEV IgG detectable in nearly all patients when they
      present with hepatitis

[a]For details, see reference 40 and other reviews.

lationships to other viruses appear to be very distant (22, 25, 31, 40, 51, 55).

HEV proteins are poorly characterized. The only data for native proteins are visual, based on immunologic reactivity observed during immune electron microscopy (IEM) of HEV particles and immunofluorescence (IF) microscopy of viral antigen (HEVAg) in liver tissue (5, 34). Most putative functions (Fig. 2) are based on very limited amino acid similarities with sequences of known function. When a complete ORF2 was fused into baculovirus DNA and then expressed in eukaryotic cells, 75- and 55-kDa polypeptides were produced (52, 53). These infrequently formed spherical particles with diameters of 18 or 30 nm, and the latter resembled native HEV capsids. Immunoreactive polypeptides (Table 2) were first identified by using postinfection sera to screen recombinant bacteriophage (55); synthetic peptides (Table 3) were similarly identified.

In addition to sequence variation (Fig. 2), HEV strains may have differences in virulence and host range (8, 48, 52). However, it is generally thought that only one serotype of HEV has been identified: with nearly all immunoassays, reactivity has not been affected by the geographic origin of the antigen or antibody (17, 32–34, 42, 48, 56). The concept of a single serotype is also supported by cross-protection: previously infected animals were protected from a second HEV strain, and they appeared to develop anamnestic anti-HEV responses (8, 9, 48, 52, 54). The only evidence for serologic variation between HEV strains resulted from assaying for anti-HEV with short (≤21-amino-acid) ORF3 peptides: sera from Kenya and Turkmenistan were less reactive with peptides representing Mexico/86 than with those of Myanmar/82, whereas Mexican sera reacted similarly with peptides of both strains (29). An unknown agent, which could represent a new serotype of HEV, appears to have caused an epidemic of ET-NANB hepatitis in India (2) (see Evidence for Other Agents below).

Intensive efforts to isolate HEV in cell culture date to the Delhi epidemic of 1955 and 1956 (48). Among several

reports of in vitro propagation, the most convincing data resulted from cocultivation of kidney or liver cells from HEV-infected cynomolgus monkeys (*Macaca fascicularis*) with cells from continuous lines (4). Even if these reports were confirmed, it appears that concentrations of progeny virus were too low to be useful.

NATURAL AND EXPERIMENTAL HOSTS

Humans are the only definite host for HEV. Although there is no conclusive evidence for a nonhuman reservoir, anti-HEV antibody has been detected in sera from wild-caught cynomolgus monkeys and rodents (6, 26, 50). These data suggest that such animals were previously infected with HEV or an antigenically similar agent. Experimentally infected animals have yielded valuable data for understanding the pathogenesis of hepatitis E and have been essential sources of anti-HEV antibody, virus, and HEVAg. Most such infections have been in primates, especially cynomolgus monkeys and chimpanzees (*Pan troglodytes*) (3, 5, 6, 8, 32, 34, 36, 40, 48, 50–54). Other studies have used domestic piglets (*Sus scrofa domestica*) and noninbred white rats (4, 26).

HEPATITIS E

Clinical Significance

HEV is responsible for much of the viral hepatitis in areas of endemicity. Individuals can be affected in large epidemics (reported nearly every year), in focal outbreaks (a recognized threat for military personnel), or as sporadic cases (3, 4, 9, 13, 33, 40, 49, 54). ET-NANB hepatitis was estimated to account for 60 to 70% of the sporadic illness in India (3). Fulminant hepatitis E is unusual except in pregnant women: the mortality rate has been 10 to 20% and is highest during the third trimester (9, 12, 16, 30, 40). All Americans should be considered at risk for infection with HEV (1).

Epidemiology

Outbreaks and epidemics are often associated with fecal contamination of drinking water (Table 1) (9, 40, 49). Although children are affected (19, 24), hepatitis E is predominantly a disease of young adults, especially males (8, 9, 12, 30, 37, 40). It is not known whether the high frequency of fulminant hepatitis E in pregnant women is due to epidemiologic or pathogenic factors (see Pathogenesis below).

Known areas of endemicity are found in most of Africa, central and southern Asia (including Indonesia), and North America (Mexico) (4, 18, 29, 34, 40, 48). South America and southern Europe may also have endemic hepatitis E but much less frequently (4, 42). Among 7- to 46-year-olds in the Kathmandu Valley of Nepal, an area of endemicity with summer-long epidemics every few years, only 10.2% had anti-HEV antibodies by polypeptide enzyme immunoassay (EIA) (Table 4) (37). Prevalence was 33.8% in those with a history of jaundice. Among Turkish individuals, 5.9% had anti-HEV antibodies (47). The prevalence of anti-HEV antibodies increased during the third decade of life in both regions. Although more-sensitive assays for anti-HEV antibody (see below) might detect additional past infections, many natives of regions of endemicity are probably susceptible to HEV. Several factors could explain why the prevalence of anti-HEV antibody is

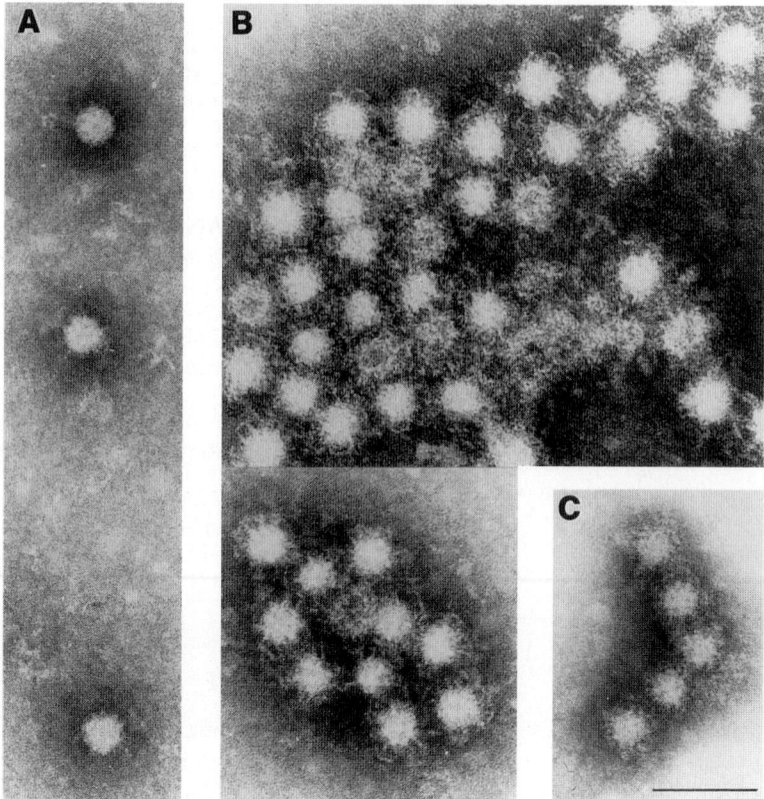

FIGURE 1 HEV particles detected in bile by IEM. Bile was collected from a cynomolgus monkey 30 days after inoculation with HEV Mexico/86. In the original experiments, IEM was performed to screen for viruslike particles in serum from a Mexican patient who was convalescent from hepatitis E (34) (C) and then to identify particles as HEV by using reagent-paired sera (panel A, preinoculation; panel B, acute phase) from another monkey that had been infected with HEV Myanmar/82. Antibody was rated on a semiquantitative scale of 0 (no antibody coating) to 4+ (heavy coating that nearly obscured the HEV particles). (A) Three virus particles; antibody rating was 0 to 1+ (interpreted as no antibody). (B) Top, portion of an immune complex that contained approximately 60 particles; bottom, complex of 11 particles. Antibody rating was 4+. (C) Complex of five virus particles; antibody rating was 4+. This bile was subsequently shown to contain HEV RNA (38) and infectious HEV (36, 50). It was also used as a reagent source of HEV particles for IEM. The assay could demonstrate the absence (A) or presence (B and C) of anti-HEV antibody. Bar = 100 nm. (Reprinted from reference 50 with permission of the publisher.)

low and why hepatitis E predominantly affects young adults in regions where other enteric viruses infect nearly all young children. HEV is usually excreted at very low concentrations (see Pathogenesis below), and the virus may be labile (see Description of the Agent above), so environmental circulation of HEV may be lower than that of other viruses. The proportion of subclinical HEV infections is probably highest in children. In addition, anti-HEV antibody sometimes declines to undetectable levels (see Pathogenesis below). It is not known whether anyone is susceptible to reinfection, but second cases have not been observed, and individuals with anti-HEV antibody did not develop hepatitis during an outbreak (11).

No epidemics or typical cases of hepatitis E originating in the United States, northern Europe, Japan, or Australia have been described. Hepatitis E and ET-NANB hepatitis have been recognized in people who traveled to or emigrated from a region of endemicity (1, 16, 48). Anti-HEV antibody has not been detected in American patients with sporadic, community-acquired NANB hepatitis (33). However, possible evidence of past infection in healthy individuals from regions thought not to be areas of endemicity has been detected by using anti-HEV antibody assays that are based on recombinant DNA (rDNA) technology (15, 56). For example, 1.2 to 3.4% of U.S. blood donors were reactive, but the specificity of such reactivity has not been determined (see Assays for Anti-HEV Antibody below). There is disagreement about a possible role for HEV in fulminant hepatitis in the United States and the United Kingdom (35).

Pathogenesis: Replication, Excretion, and Immune Response

How HEV travels from the gastrointestinal tract to the liver, the only known site of HEV replication, is not known. Hepatic viral replication is usually the first detectable event during experimental infections, occurring before

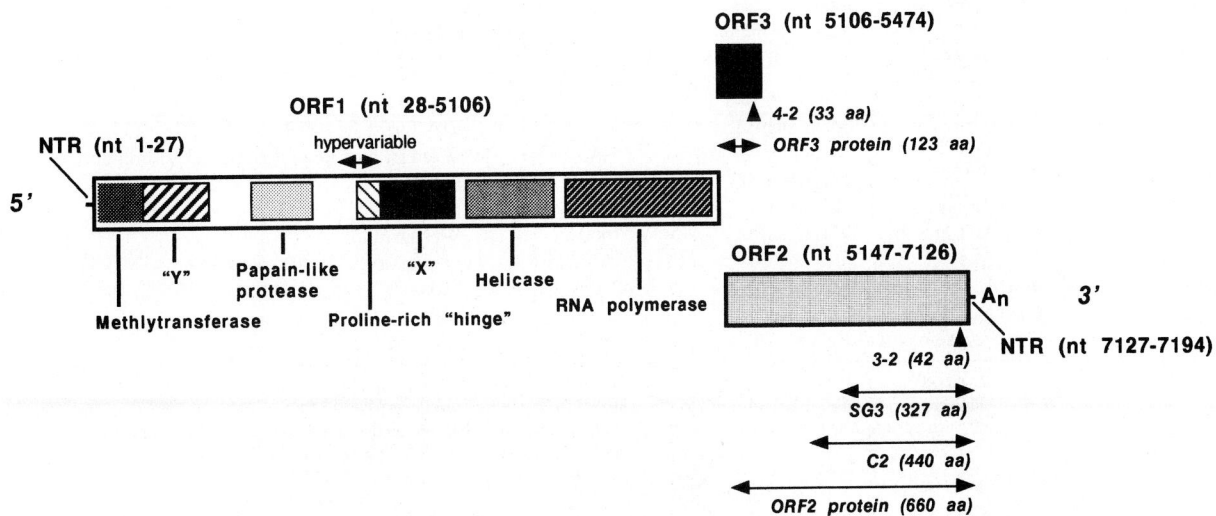

FIGURE 2 Schematic diagram of the HEV genome, a single strand of positive-sense RNA. Large rectangles demarcate the three open reading frames (ORFs) known to encode proteins of putative function and immunoreactive peptides and polypeptides. Numerals refer to nucleotide positions in the HEV Myanmar/82 genome, which is 7,194 nt in length before a 3' poly(A) tail of several hundred nucleotides (A$_n$) (46). There are short nontranslated regions (NTR) at each end. ORF1 and ORF3 overlap for 1 nt, ORF1 and ORF2 are separated by 41 nt, and ORF2 and ORF3 overlap by 28 nt. Among characterized HEV strains, only Mexico/86 was collected outside of Asia. The Mexico/86 genome has an additional 9 nt in the 3' NTR and ORFs with slightly different lengths, and it may lack the 5'-terminal 24 nt (22). When sequences of Mexican and Asian HEV strains are compared, nucleotide identity is approximately 75% for ORF1, 81% for ORF2, and 91% for ORF3; amino acid (aa) identities are 84, 93, and 87%, respectively. In contrast, sequences of Myanmar/82 and Pakistan/87 are highly similar (93 to 99% nucleotide identity and 99 to 100% amino acid identity for the three ORFs) (51). In the hypervariable region (nt 2011 to 2325, above the ORF1 rectangle), nucleotide and amino acid identities are approximately 90% among geographically distinct Asian strains; Mexican and Asian nucleotide sequences are approximately 56% identical, resulting in extensive amino acid changes (7, 22, 51). However, this region is much less variable among strains collected in Pakistan and western China during the late 1980s, suggesting that there was little mutational pressure in that geographic area (57). ORF1 is thought to encode nonstructural proteins (44). Domains of amino acid similarity are represented by shaded and hatched portions of the ORF1 rectangle, and putative functions are indicated below, according to Koonin et al. (31). Several domains of unknown function were identified by similarity to rubella virus ("Y," proline-rich "hinge," and "X"), beet necrotic yellow vein virus ("Y"), and alphaviruses ("X"). Koonin et al. (31) concluded that HEV, rubella virus, and beet necrotic yellow vein virus might represent a distinct group among the supergroup of alpha-like viruses. Although HEV ORF1 and ORF2 resemble those of caliciviruses with regard to organization, size, and genomic location, HEV does not have an ORF near the 3' terminus like that of typical caliciviruses. A small ORF somewhat like HEV ORF3 has been recognized only in the genome of a Norwalk-like candidate calicivirus (25). Synthetic peptides of the ORF1-encoded amino acid sequence have reacted with postinfection sera, especially peptides representing the C terminus of the proposed RNA-dependent RNA polymerase (not shown) (27). ORF2 is thought to encode the major capsid protein of HEV. Polypeptides that react with postinfection sera are shown below the ORF2 and ORF3 rectangles; double-headed arrows and italics indicate the coding regions, designations, and amino acid lengths for larger polypeptides SG3 (56), C2 (17, 41, 42), ORF2 protein (20, 52, 53), and ORF3 protein (20). ORF2 protein includes the complete ORF2 protein sequence of 660 amino acids and can form capsidlike structures (52, 53). Short polypeptides 3-2 and 4-2 are indicated by arrowheads immediately below the 3' ends of ORF2 and ORF3, respectively; these were the first immunoreactive polypeptides to be identified (55). Postinfection sera have also reacted with peptides corresponding to the C termini and other regions of the ORF2 and ORF3 proteins (not shown) (14, 28, 29). This figure was adapted from similar figures in references 7, 31, and 56, among others.

hepatocellular changes and long before inflammatory changes and biochemical evidence of hepatitis (Fig. 3) (32, 34, 36, 50). Progeny virus are released from hepatocytes to the gallbladder by an unknown mechanism and then ex-creted into the feces. Viremia has been detected, approximately coinciding with fecal HEV excretion (51). Recent cell culture studies suggested that HEV was also present in kidneys (4). However, the origin, duration, and diagnostic

TABLE 2 Immunoreactive HEV polypeptides produced by rDNA techniques[a]

Genome region	Designation	HEV strain	Length (amino acids)	Uses
ORF2	3-2(M)	Mexico/86	42	EIA
	3-2(B)	Myanmar/82	42	EIA formerly; will be replaced by SG3
	C2	Myanmar/82	440	Western blot assay for anti-HEV; immunogen
	SG3	Myanmar/82	327	EIA
	ORF2(P) protein	Pakistan/87	660	EIA
	ORF2(B) protein	Myanmar/82	660	Western blot
ORF3	4-2(M)	Mexico/86	33	EIA
	4-2(B)	Myanmar/82	33	EIA
	ORF3 protein	Myanmar/82	123	Western blot

[a]Polypeptides 3-2(M), 3-2(B), 4-2(M), and 4-2(B) were originally reported in reference 55; their use in EIAs was reported in references 19 and 45. The construction of C2 and its use in a Western blot were first described in reference 42; its use as an immunogen is described in reference 41. SG3 was originally reported in reference 56. ORF2 and ORF3 proteins are synthesized in baculovirus-infected insect cells (20, 52), and the others are produced as fusion proteins in *Escherichia coli*.

significance of nonenterohepatic HEV are not known. In general, in vivo concentrations of HEV appear to be very low. HEV is predominantly excreted during the incubation period and the early acute phase (5, 12, 38, 43, 49, 50, 58). Fecal HEV concentrations are usually very low: only 1 of 13 HEV-positive specimens had as many as 10^6 detectable virions per g of feces (58).

In experimentally infected animals, development of anti-HEV antibody usually coincides with the end of viral excretion and precedes the most severe hepatitis (36, 38, 50–52). Anti-HEV antibody (total or immunoglobulin G [IgG]) is almost always detectable when patients present with acute hepatitis E. Antibody levels are usually highest during the acute phase and may persist, decline, or become undetectable but rarely rise during convalescence (3, 11, 13, 17, 40, 48, 49, 56). Acute-phase anti-HEV antibody has also included IgM and IgA; these antibodies were usually no longer detectable after several months (13). Specific cellular immune responses are not yet defined but are implied by involvement of CD8 lymphocytes and other mononuclear cells in hepatic lesions (10, 16) and by the rapid, probably anamnestic, response of previously infected animals to a

new inoculum of HEV (see Description of the Agent above). Hepatitis and anti-HEV antibody appear to develop concurrently. Thus, it is thought that hepatitis E is an immunopathologic disease: viral damage to hepatocytes may be minimal, but host responses to infected cells appear to result in hepatitis (8, 10, 32, 36, 48, 50).

The pathogenesis of fulminant hepatitis E in pregnant women is not understood. The late Hans Popper (cited in reference 40) proposed that HEV-induced injury to Kupffer and other sinusoidal cells could diminish their abilities to protect hepatocytes from the endotoxins of intestinal bacteria. Direct (endotoxin-mediated) and secondary (eicosanoid-mediated) damage would then lead to infiltration of neutrophils, edema, and cholestasis. This possible mechanism, a modification of the Shwartzman phenomenon, is consistent with the known sensitivity of pregnant women to endotoxins.

Symptoms and Signs

Most patients have a moderately severe but self-limited illness (Fig. 3) that cannot be distinguished from other forms of acute hepatitis. A prodrome may occur (11, 39,

TABLE 3 Synthetic peptides for detecting anti-HEV antibody by EIA[a]

ORF	Amino acid position	Peptide No.	Peptide Sequence	Corresponding polypeptide[b]	Concn (μg/ml) in mixture[c] Unconjugated	Individually conjugated	Mixture conjugated
2	319–340	11	RVSRYSSTARHRLRRGADGTAE	C2	5		
	422–437	12	DKGIAIPHDIDLGESR	SG3	1		
	442–460	13	DYDNQHEQDRPTPSPAPSR	SG3		4	4
	631–648	22	RPLGLQGCAFQSTVAELQ	3-2(B)		4	4
	641–660	23	QSTVAELQRLKMKVGKTREL	3-2(B)	10	20	20
3	91–110	5	ANPPDHSAPLGVTRPSAPPLA	4-2(B)	10	8	8
		28	ANQPGHLAPLGEIRPSAPPLA[d]	4-2(M)	2		
	105–123	6	PSAPPLPHVVDLPQLGPRR	4-2(B)	20	20	14
		29	PSAPPLPPVADLPQPGLRR[d]	4-2(M)	2		

[a]From peptide numbers, methods, data, and conclusions in reference 18.
[b]The polypeptide (Fig. 2, Table 2) is the smallest that corresponds to the peptide listed here.
[c]Attached to wells of a microtiter plate as (i) a mixture of unconjugated peptides, (ii) a mixture of peptides individually conjugated to bovine serum albumin, or (iii) bovine serum albumin conjugated to a mixture of peptides (mixture conjugated). Results from the last method were most similar to those obtained by Western blot (18).
[d]Sequence derived from HEV strain Mexico/86; all other sequences are from Myanmar/82.

TABLE 4 Detection of anti-HEV antibody in serum

Method[a]	Detection of anti-HEV	Reagent antigen(s)	Specificity[b]	Sensitivity[c]	Practicality
Based on rDNA technology					
Polypeptide EIA	Indirect (IgG, IgM, or IgA)	HEV polypeptide fused with bacterial proteins	High?	Moderate to high?	High
ORF2 protein EIA	Indirect (IgG or IgM)	ORF2(P) protein from baculovirus-infected cells	High?	High?	High
Peptide EIA	Indirect (IgG or IgM)	Synthetic peptides	High?	Moderate to high?	High
Western blot	Indirect (IgG or IgM)	C2 (HEV-TrpE fusion protein) in *E. coli* lysate[d]	High?	Moderate to high?	Moderate
Based on native antigens[e]					
IEM	Direct (total or IgM-like)	Source of HEV particles	High	High	Low
IF microscopy	Blocking (total)	Fluorescein-conjugated anti-HEV IgG and frozen liver containing HEVAg	High	Moderate?	Moderate

[a] All are solid-phase immunoassays except traditional IEM, in which anti-HEV antibody and virus bind in suspension to form immune complexes. For EIAs, the solid phase is a well of a microtiter plate coated with reagent antigen. IEM can be performed as a solid-phase assay if grids coated with protein A or anti-Ig are used to capture immune complexes (23).
[b] None of these assays has detected reactivity during infections with other hepatitis viruses or during other liver diseases. However, the specificities of rDNA-based assays are not completely understood: reactivity has been detected in a few percent of healthy individuals from regions thought not to be areas of endemicity and where typical cases of hepatitis E have not been identified. The specificity of IEM was questioned when reactivity was detected in sera from a monkey that was subsequently infected and developed hepatitis (50); however, the IEM result was corroborated by ORF2 protein EIA (52), and others have detected anti-HEV reactivity in wild-caught monkeys and rodents (6, 26).
[c] Few sera have been tested in multiple assay formats, but IEM appears to be as sensitive as if not more so than other formats (52). ORF2 protein EIA was more sensitive than a polypeptide EIA for assaying sera from experimentally infected monkeys (53). Otherwise, it is assumed that assays with reagent ORF2 protein (including IEM) are likely to be more sensitive than IF microscopy for assaying sera from six owl monkeys (50). EIA development continues, so it is likely that the sensitivity and specificity of this format will improve.
[d] Preliminary data were recently reported for Western blots with proteins from baculovirus-infected cells; these data suggested that an assay with reagent ORF2(B) protein is more sensitive than one with ORF3 protein (20).
[e] Technical capability and reagent antigens are currently available at only a few research institutions. Suspensions of feces, bile, or liver have been sources of HEV particles for IEM. A few convalescent-phase serum samples have been identified as good sources of anti-HEV IgG for IF microscopy. HEVAg-containing liver tissue has usually been collected from experimentally infected cynomolgus monkeys.

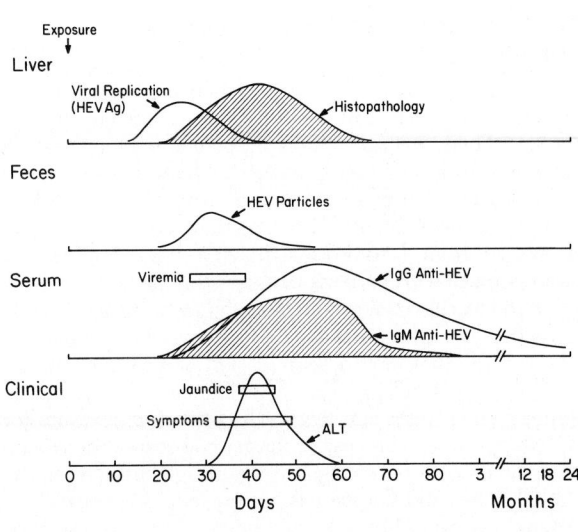

FIGURE 3 Diagram of virologic, histopathologic, serologic, and clinical events in a typical human case of hepatitis E. These events are somewhat speculative, because detailed knowledge of natural history is limited, especially for humans. Somewhat similar patterns have been described for cynomolgus monkeys and other experimentally infected animals (5, 6, 8, 36, 50–54). However, the incubation period is longer in humans (in this putative course, 37 days from exposure to development of jaundice), and disease is more severe (among nonhumans, only swine develop overt signs such as jaundice). The earliest recognized evidence of infection is viral replication in the liver, detected as HEVAg by IF microscopy, HEV particles by IEM, or HEV RNA by PCR (Table 5) (33, 34, 36, 50). The liver is the only known site of HEV replication; release of progeny virions from hepatocytes into bile leads to HEV excretion (HEV particles in feces, detected by PCR, IEM, or inoculation of susceptible primates) (12, 38, 43, 49, 50, 58). Viremia (detected by PCR) is also thought to reflect HEV released from the liver, but few serum samples have been assayed, and HEV concentration has not been determined (51). As HEV replication wanes, histopathologic changes, antibody to HEV, and signs and symptoms develop. However, HEV may still be detectable during clinical illness (12, 43, 49, 58). Curves for antibody (IgG anti-HEV and IgM anti-HEV) appear to be similar for different types of assay (Table 4) (11, 13, 19, 33, 34, 50, 52, 53). Note that anti-HEV IgG develops rapidly after anti-HEV IgM, is present at high levels when jaundice appears, and usually reaches peak levels before convalescence. Recent data indicate that serum anti-HEV IgA may develop and diminish in a pattern similar to that of anti-HEV IgM (13). ALT, the serum concentration of alanine aminotransferase, is the usual marker for biochemical evidence of hepatitis. The figure was revised from that in reference 39.

49). Acute-phase symptoms can include jaundice, anorexia, abdominal pain, nausea, vomiting, fever, and pruritus (5, 8, 32, 39, 40). Jaundice, the cardinal sign of acute viral hepatitis, can be manifested as yellow skin, scleral icterus, dark urine, or clay-colored feces. Nonspecific signs include fever, hepatomegaly, and abdominal tenderness. HEV infections occur subclinically or without jaundice, but the proportions of such disease patterns are not known, especially as they relate to age (but milder infections are probably more frequent in children) (10, 16, 30, 37, 39, 40).

Patient Management, Infection Control, and Prevention

As with other forms of acute viral hepatitis, there are no specific modes of therapy, and patient care is largely supportive. There are no established guidelines for isolation (4), but feces from a newly ill patient should be considered biohazardous until proven otherwise. Health care workers have been infected with HEV. Pregnant women should be prohibited from laboratories studying HEV and should avoid contact with patients or their specimens.

Outbreaks have been controlled by instituting or restoring public sanitation measures. These typically include coincident chlorination and elimination of fecal contamination from water supplies. Avoidance of potentially contaminated water and raw foods is the only measure for protecting individuals against infection, because there is no effective immunoprophylaxis. Immune globulin from the United States and Europe has not prevented hepatitis E in certain travelers (1), and the evidence for protection by using immune globulin prepared in regions of endemicity is inconclusive (12). Because propagation of HEV in cell culture has not been practical, an effort to develop an HEV vaccine used rDNA technology: C2 protein (Table 2, Fig. 2) induced a protective effect in cynomolgus monkeys (41).

COLLECTION AND TRANSPORT OF SPECIMENS

Samples from individuals with suspected hepatitis E should be collected as early in the course of illness as possible. Storage should be as cold as possible, with avoidance of freezing and thawing. If contacts of patients were also exposed to HEV, preillness specimens are ideal. Several considerations explain these approaches: (i) levels of anti-HEV antibody are usually highest during the acute phase, (ii) HEV may be labile and is predominantly excreted at very low concentrations before or during the first week of jaundice, and (iii) clinical specimens are needed for validating research methods. For transportation, solid CO_2 (dry ice; $-70°C$) is usually the most practical freezing medium. The vapor phase of liquid N_2 ($-120°C$) is preferred for specimens suspected to contain HEV. Biohazard identification labels should be used on packages thought to contain HEV (feces and preacute- or acute-phase sera) or other agents of disease. Shipments of HEV into the United States need an importation permit, obtained from the Centers for Disease Control and Prevention.

Acute-phase serum is often the only specimen available. Preexposure serum is essential for demonstrating a serologic response to HEV; such serum is rarely stored but might be part of an institutional collection (3) or could be obtained before travel. Serum collected during convalescence, i.e., several weeks or more after the hepatitis, can be useful for detecting anti-HEV IgG as a marker of past infection. Storage at $\leq -20°C$ is acceptable for preserving anti-HEV antibody, but $\leq -70°C$ should be used for viremic specimens.

Feces should be collected early in the course of disease, at least during the first week of jaundice. Specimens should be stored at $\leq -70°C$, and $-120°C$ is preferred for specimens known to contain HEV. Lyophilization preserves HEV particles (49), but its effects on particle concentration and infectivity are not known.

ASSAYS FOR ANTI-HEV ANTIBODY

If diagnostic tests become available, they will probably be based on immunoassay approaches for detecting anti-HEV IgM in acute-phase serum. Currently, nondiagnostic EIAs for anti-HEV IgG are sold outside of the United States, and several formats are being used in research studies of anti-HEV IgM, IgG, and IgA (Table 4). However, the specificities of rDNA-based assays have not been completely determined (see Epidemiology above and discussion immediately below). Older methods that use native HEV antigens are useful but cumbersome and are available at only a few research institutions.

Assays Based on rDNA Technology

Recent efforts have yielded EIAs in which binding of anti-HEV antibody to reagent antigen is detected by enzyme-labeled anti-human Ig. Most work has been done with antigens that are encoded near the 3′ ends of ORF2 and ORF3, but larger polypeptides and smaller peptides are now being evaluated (Tables 2 through 4, Fig. 2). Solid phases include antigen-coated wells of microtiter plates and Western blot (immunoblot) strips. Typically, test serum (diluted 1:25 to 1:100) is incubated with the solid phase. Bound antibody is detected by labeled anti-γ (IgG heavy chain), anti-μ (IgM heavy chain), or anti-IgA (13). Detailed procedures are not described below because commercially available methods come with instructions, and research methods are changing during development; all are based on standard EIA techniques. These EIAs have detected IgG in up to 3.4% of healthy people from countries thought not to contain regions of endemicity for HEV (15, 18, 56). It is not known whether this low-prevalence reactivity indicates infection acquired in a region of high endemicity, low-level endemicity of HEV, infection with antigenically related virus (nonpathogenic HEV or an undiscovered agent), or nonspecificity in the assays from sources other than viruses (32). Although specificity and sensitivity have not been fully established, these assays are rapidly yielding results that are being used to develop seroepidemiologic concepts. More comparisons between promising rDNA-based methods and impractical native-antigen techniques are needed.

Polypeptide EIA

One type of EIA uses reagent antigens of HEV polypeptide fused into carrier protein and produced by bacteria. A test for anti-HEV IgG is sold by Genelabs Diagnostics in Asia and uses HEV polypeptides 3-2(M), 4-2(M), and 4-2(B) (Table 2, Fig. 2). Future versions may incorporate SG3 antigen, because it appears to yield superior sensitivity (56). A similar assay is marketed in Europe by Abbott Laboratories. Anti-HEV IgA and IgM have been detected in research studies: levels usually peaked within 1 month of acute hepatitis and then became undetectable within several months (13, 19, 45, 56). Both were less frequently detected than anti-HEV IgG, which usually persisted but at lower concentrations than during the acute phase. Thus, a diagnostic format will need to detect an acute-phase-specific response.

ORF2 Protein EIA

Another EIA detects anti-HEV IgM and IgG by using reagent antigen synthesized from complete ORF2(P) in baculovirus-infected cells (52). When sera from experimentally infected primates were tested, this assay appeared to correlate well with IEM and was more sensitive than polypeptide EIAs (52, 53). Temporal patterns of anti-HEV responses during outbreaks were similar to those detected by polypeptide EIAs (2, 11).

Peptide EIA

Several groups have prepared synthetic peptides that react with sera from patients with HEV infections (14, 18, 27–29, 54). Many such peptides have sequences that are included in the 3-2 and 4-2 antigens, but others correspond to putative nonstructural proteins and to central portions of the ORF2 protein (27, 28) (Table 3, Fig. 2). Sequence differences in the amino terminus of ORF3 protein might permit discrimination between infections with HEV Mexico/86 and those with Asian strains (29). Peptide EIAs have identified antibodies, predominantly IgG, in sera from patients with geographically diverse HEV infections (14, 18, 27, 54). Results have correlated with those from Western blot and IF microscopy. One such assay also had an unusually low frequency (0.5%) of reactivity with sera from healthy individuals in regions thought not to be regions of endemicity (18) (however, different control sera have been used for different EIAs). Another potential advantage is that any laboratory with resources for peptide synthesis could set up such assays (Table 3).

Western Blot

C2 (Table 2, Fig. 2) has been used in an immunoblot format for detecting IgG and IgM responses to epidemic, sporadic, and experimental HEV infections (17, 24, 42). Reactivity was detected only in postinoculation sera during experimental infections (42). IgG was detectable in 7 of 39 controls among Sudanese children (24); it is not known if all such reactivity represents past HEV infections. IgM reactivity in humans was no longer detectable after 8 months of convalescence (17, 42). A recent report suggested that in Western blots, complete ORF2(B) protein more effectively detected anti-HEV antibody than did complete ORF3 protein (20).

Assays with Native HEV Antigens as Reagent

Available quantities of HEV have been insufficient for rapid, practical, reproducible assays (48), but development efforts continue (4). IEM and IF microscopy are "gold standard" methods that were crucial to the early understanding of HEV infections and are highly specific when performed under blind code (Table 4). However, each requires reagents that are difficult to obtain, a special microscope, training, and patience (analysis of each specimen takes 15 min or much longer).

IEM

Test serum (a typical final dilution is 1:50) is mixed with reagent HEV to form immune complexes of antibody-coated virus particles (Fig. 1). Anti-HEV antibody is semi-quantitatively rated, but concentration (titer) can be determined by serial 10-fold dilution of test serum (50). "Staplelike" molecules are suggestive of anti-HEV IgM (3, 10). Antibody class can be identified by solid-phase IEM, a more rapid method that uses EM grids coated with anti-IgM or anti-IgG to capture immune complexes (23). IEM for anti-HEV antibody appears to be at least as sensitive and specific as other methods, but it should probably be used only for confirming certain rDNA-based EIA results, because reagent HEV is scarce.

IF Microscopy

Anti-HEV antibody is usually detected in a blocking format: antibody in a test specimen (typical final dilutions are 1:4 or 1:8) prevents binding of fluorescein-conjugated anti-HEV IgG to HEVAg in reagent tissue (33, 34). Titer can be determined by serial twofold dilution. Results appear to be very specific and are much more rapidly obtained than by IEM. Like IEM, IF microscopy has been valuable for studying epidemiology and experimental infections (32–34, 50) and should be useful for confirming results from rDNA-based EIAs.

ASSAYS FOR HEV

There are no routine methods for detecting HEV in clinical materials. Research techniques are fairly cumbersome. In addition, HEV is present in feces predominantly before and during the first week after the onset of jaundice. Limited data suggest that the temporal pattern of viremia is similar (Fig. 3). HEV concentrations in most clinical specimens are too low for detection by immunoassay or direct nucleic acid hybridization. Thus, it is not likely that practical assays will be developed for detecting HEV in clinical laboratories.

Detection of HEV Particles or Macromolecules

Reverse Transcription PCR

Because HEV nucleotide sequences are unique among those that are known, PCR is thought to be specific for detecting HEV. PCR also appears to be sensitive (Table 5). Many researchers have used chemical (phenol or guanidinium isothiocyanate-phenol) methods to extract total RNA from the specimen. Other approaches to RNA purification have used glass powder to adsorb total RNA or solid-phase anti-IgM to capture reagent anti-HEV antibody and HEV in the specimen (7, 36, 38, 58). The latter, termed affinity-capture PCR (36, 38), has been especially useful because it eliminates labor-intensive steps and potential losses during chemical RNA purification and yields template that is primarily, if not entirely, HEV RNA. In one version (7), microcentrifuge tubes are coated with anti-human IgM, washed, incubated with reagent serum, washed, incubated with 100 μl of 5% (wt/vol) fecal suspension, and then washed before PCR. A similar method, in which immune complexes are formed and then captured, yielded results that were similar to those obtained with chemically extracted RNA (38).

Many different oligonucleotides have been successfully used for amplifying HEV RNA, but primers for detecting all known strains have not yet been identified. It has been suggested that high sequence identity within ORF2 (nucleotide [nt] 6227 to 6603) might yield such primers (7). Oligonucleotides spanning a conserved ORF1 sequence (nt 3880 to 4930) were more sensitive for amplifying Chinese strains than sets representing ORF2 or the "hypervariable" region of ORF1 (57). However, analogous HEV Mexico/86 primers (nt 4272 to 4942) did not amplify HEV Myanmar/82 or Pakistan/87 (38). The hypervariable region should be avoided except for studying sequence variability (7).

By means of PCR, HEV has been detected in feces, bile, serum, and liver (7, 36, 38, 43, 51, 57, 58). When serial fecal specimens from an outbreak were assayed, HEV was detected in 29% of 39 specimens (representing 5 of 11 patients), most often within the first 7 days of icterus but as late as 16 days (58). One patient from the outbreak ex-

TABLE 5 Detection of HEV

Method	Detects[a]	Specimen(s)	HEV reagents	Sensitivity[b]	Practicality[c]
PCR	HEV RNA	Feces, bile, liver serum	Oligonucleotides	High	Low
Affinity-capture PCR	RNA in HEV particles	Feces, bile, liver	Oligonucleotides and serum containing anti-HEV antibody	High	Low to moderate
IEM	HEV particles	Feces, bile, liver	Paired sera: one with anti-HEV antibody, one from before inoculation[d]	Low	Very low
IF microscopy	HEVAg	Liver[e]	Fluorescein-conjugated anti-HEV IgG	Low?	Very low
Inoculation of animals[f]	Infectious HEV	Feces, bile, liver[g]	Assay for anti-HEV antibody or HEV	High	Very low

[a]All methods appear to be specific for HEV.

[b]It was recently demonstrated that $10^{6.7}$ genomes of HEV Pakistan/87, detected by PCR, corresponded to 10^6 cynomolgus monkey 50% infectious doses (53). As few as 20 molecules of cloned HEV Mexico/86 cDNA were detected in another study (38). The sensitivity of IEM is approximately 10^6 particles per ml; direct comparison has shown IEM to be less sensitive than affinity-capture PCR for HEV (36, 38, 58). Rating for IF microscopy reflects only the assumption that a method based on detecting antigen is less sensitive: comparison with PCR (36) or other methods has been insufficient for an accurate assessment of relative sensitivity.

[c]For clinical specimens, low or very low, because HEV replication ceases or wanes to undetectable levels in most patients before jaundice develops.

[d]Preinoculation serum is essential as an anti-HEV-negative reagent because acute-phase serum nearly always has high levels of anti-HEV.

[e]In frozen-tissue embedding medium (34, 36, 50).

[f]Inoculation of cell culture is a potential method, but even if preliminary reports (4) are confirmed, it will be impractical unless nearly all clinical isolates propagate and unless rapid methods are developed for detecting replication.

[g]Potential inocula include serum or kidney tissue. Serum has induced hepatitis in chimpanzees and tamarins, but these animals have not been assayed for evidence of HEV replication (Purcell et al., cited in reference 48). HEV was reportedly isolated in cell culture of monkey kidney (4).

creted HEV but did not have anti-HEV IgM in acute-phase serum; thus, PCR has the potential for identifying HEV infections in seronegative patients.

IEM

IEM makes visual identification of HEV possible but requires reagent sera and a trained observer with great patience (often requires >4 h per specimen) (3, 36, 49, 50). The method is highly specific when paired sera are used under code (Fig. 1, Table 5) and the observer knows how to recognize HEV particles in immune complexes. A solid-phase method, which uses grids coated with protein A or anti-Ig to capture immune complexes, appeared to be more sensitive than traditional IEM (23).

IF Microscopy

HEVAg in liver can be detected with reagent-labeled anti-HEV IgG. Semiquantitative ratings of 0 to 3+ (34) or 0 to 4+ (36) have been used to estimate the concentration of HEVAg. The antigens designated HEVAg have not been characterized, but their presence coincided with HEV that was detectable in liver and bile by IEM and PCR (36). In addition, fluorescence was blocked by absorbing the reagent IgG with ORF2 protein (32).

Detection of Infectious HEV

There are no practical methods for isolating infectious HEV. Most research laboratories isolate and propagate virus by inoculating animals. Reports of HEV propagation in cell culture (4, 48) suggest that replication is low level and without cytopathologic effect; to be at all useful, culture methods need to be enhanced.

INTERPRETATION OF TEST RESULTS

Precise conclusions about a particular patient may not be possible with current assays. However, the likelihood of

identifying an acute, recent, or past infection may be enhanced if several generalizations are considered. First, the presence of HEV in a clinical specimen indicates acute infection (predominantly late incubation or first week of jaundice). Because the virus is not detectable in many patients, however, absence of HEV does not rule out acute infection. Positive PCR results should be interpreted with caution for patients with low risk of HEV infection because of the possibility of contamination during assay. Second, anti-HEV IgM identifies acute or recent infection (within several months of exposure); anti-HEV IgA appears to be a similar marker. However, because many patients in outbreaks have had negative results with current polypeptide EIAs for anti-HEV IgM or IgA, absence of these antibodies does not rule out acute infection. Third, total anti-HEV antibody and anti-HEV IgG are the only HEV-specific markers that are currently detectable in nearly all acutely infected patients, but they do not define when the infection occurred. Anti-HEV antibody concentrations are almost always highest during acute hepatitis, so "seroconversion" or a rise in antibody levels between acute and convalescent phases rarely occurs. In addition, the absence of detectable anti-HEV antibody does not rule out past infection. Fourth, the specificity of rDNA-based assays for anti-HEV is not completely understood. Thus, detection of anti-HEV antibody in a low-risk individual, such as a healthy American with no history of travel to regions of endemicity, should be interpreted with caution. Some of these concepts may change as data accumulate.

With current assays, testing for both anti-HEV antibody and HEV provides the most information (Table 6). A more practical approach is to test first for anti-HEV antibody and then, if anti-HEV IgM is not detected, for HEV. Other data and historical information can also be helpful. Clinical laboratories can exclude other causes of hepatitis, including microorganisms (such as HAV, HBV, HCV, HDV, Epstein-Barr virus, cytomegalovirus, dengue viruses, yellow

TABLE 6 Hepatitis E and other forms of HEV infection: interpretation based on recent data

Diagnosis	Jaundice	↑ ALT[a]	Virologic findings
Icteric hepatitis E	+	+	Anti-HEV IgM, acute phase (majority of patients),[b] or HEV in feces, serum, liver, or bile (unusual finding), or anti-HEV antibody, higher concentration during convalescence (rare finding), or anti-HEV antibody, acute phase or convalescence and not detected preexposure (rare specimen)
Anicteric hepatitis E[c]	−	+	Same as icteric
Acute subclinical infection[c]	−	−	HEV in feces or serum ± anti-HEV IgM
Recent infection[c]	−	−	Anti-HEV IgM
Possible acute hepatitis E	+	+	Anti-HEV antibody (IgG, total, or IgM-like by IEM), acute phase (diagnosis more likely if first exposure and no evidence for other causes of hepatitis)
Past infection[d]	−	−	Anti-HEV antibody (IgG or total)

[a] ↑ ALT, elevated level of alanine aminotransferase in serum or other biochemical evidence of hepatitis.

[b] Detected in assay specifically for anti-HEV IgM or by testing the IgM fraction in an assay for total anti-HEV antibody (3); recent data suggest that the serum IgA response to HEV may be similar to that of IgM (13).

[c] Could represent prodrome and progress to icteric hepatitis E.

[d] Indeterminate but probably >1 month before serum was collected. Because anti-HEV may wane to undetectable levels, the absence of anti-HEV IgG does not mean that an individual was never infected.

fever virus, leptospira, or *Coxiella burnetii*), drugs (such as ethanol, anesthetics, or antimicrobial agents), tumor, autoimmunity, or gallbladder disease. For patients in the United States, hepatitis E is suggested by two travel scenarios: a visit to a region of endemicity when previous exposure to HEV is unlikely (especially the first such trip for a native-born American) or recent emigration from a country where HEV is endemic. Hepatitis E may be more likely if a hepatitis outbreak was occurring in the region of travel.

When a patient has jaundice, biochemical evidence of hepatitis, and detectable HEV and anti-HEV IgM, interpretation is not difficult: the diagnosis is icteric hepatitis E (Table 6). Similar data without jaundice could represent anicteric hepatitis E or subclinical infection, depending on the presence of other symptoms and signs. A patient with only anti-HEV IgM most likely has acute hepatitis E but could have had a recent infection with persistent IgM response followed by a second liver disease. Possible acute hepatitis E is represented only by anti-HEV antibody (total, IgG, or IgM-like by IEM) with a low probability of past exposure and no evidence for another agent; this diagnosis is less likely for patients from areas of endemicity. In a healthy individual, detection only of total anti-HEV or anti-HEV IgG probably represents past infection (at an indeterminate time, >1 month before serum collection); however, this result could also represent an incubating HEV infection. Finally, acute and recent infections may not have detectable HEV or anti-HEV IgM (but are almost certain to have anti-HEV antibody, total or IgG), and past infections may not have any HEV markers.

EVIDENCE FOR OTHER AGENTS OF ET-NANB HEPATITIS

There are patients who have hepatitis but no evidence for acute infection with known viruses, including HEV or HAV, in settings where enteric transmission is common. Examples have been reported as sporadic disease in Egypt, Ethiopia, India, Tajikistan, and Uzbekistan (17) and as a possible outbreak in Belize (21). Some such cases might be identified as hepatitis E by using more sensitive assays.

However, it was recently determined that an epidemic of non-A, non-E hepatitis occurred in the Andaman Islands, India, during 1987 (2). These data will probably lead to discovery of at least one more agent of ET-NANB hepatitis or a new serotype of HEV or HAV.

I thank Joe Bryan, George Dawson, Howard Fields, Charles Longer, Sergei Tsarev, and Patrice Yarbough for access to their unpublished manuscripts. Although nearly all of the HEV literature is new, it has been necessary to limit the number of references. I acknowledge many other contributions, most of which are cited per reviews or newer work among the references, and apologize for not directly recognizing them. Please contact me for additional references. I also appreciate helpful discussions with Patricia Charache, William Merz, and David Thomas and with my former colleagues at the National Institute of Allergy and Infectious Diseases and the Walter Reed Army Institute of Research.

Preparation of this chapter was supported by a fellowship from Baxter Diagnostics.

REFERENCES

1. **Anonymous.** 1993. Hepatitis E among U.S. travelers, 1989–1992. *Morbid. Mortal. Weekly Rep.* **42:**1–4.
2. **Arankalle, V. A., M. S. Chadha, S. A. Tsarev, S. U. Emerson, A. R. Risbud, K. Banerjee, and R. H. Purcell.** 1994. Seroepidemiology of water-borne hepatitis in India and evidence for a third enterically-transmitted hepatitis agent. *Proc. Natl. Acad. Sci. USA* **91:**3428–3432.
3. **Arankalle, V. A., J. Ticehurst, M. A. Sreenivasan, A. Z. Kapikian, H. Popper, K. M. Pavri, and R. H. Purcell.** 1988. Aetiological association of a virus-like particle with enterically transmitted non-A, non-B hepatitis. *Lancet* **i:**550–554.
4. **Balayan, M. S.** 1993. Hepatitis E virus infection in Europe: regional situation regarding laboratory diagnosis and epidemiology. *Clin. Diagn. Virol.* **1:**1–9.
5. **Balayan, M. S., A. G. Andzhaparidze, S. S. Savinskaya, E. S. Ketiladze, D. M. Braginsky, A. P. Savinov, and V. F. Poleschuk.** 1983. Evidence for a virus in non-A, non-B hepatitis transmitted via the fecal-oral route. *Intervirology* **20:**23–31.
6. **Balayan, M. S., N. A. Zamyatina, T. A. Ivannikova, V. I. Khaustov, A. G. Andjaparidze, and V. F. Poleschuk.** 1991. Laboratory models in the studies of enterically transmitted non-A, non-B hepatitis, p. 213–220. *In* T. Shikata, R. H.

Purcell, and T. Uchida (ed.), *Viral Hepatitis C, D and E.* Elsevier Science Publishers B.V., Amsterdam.

7. **Bi, S. L., M. A. Purdy, K. A. McCaustland, H. S. Margolis, and D. W. Bradley.** 1993. The sequence of hepatitis E virus isolated directly from a single source during an outbreak in China. *Virus Res.* **28:**233–247.

8. **Bradley, D. W.** 1990. Hepatitis non-A, non-B viruses become identified as hepatitis C and E viruses. *Prog. Med. Virol.* **37:**101–135.

9. **Bradley, D. W., K. Krawczynski, E. H. Cook, Jr., K. A. McCaustland, B. H. Robertson, C. D. Humphrey, M. A. Kane, J. Spelbring, I. Weisfuse, A. Andjaparidze, M. Balayan, H. Stetler, and O. Velazquez.** 1988. Enterically transmitted non-A, non-B hepatitis: etiology of disease and laboratory studies in nonhuman primates, p. 138–147. *In* A. J. Zuckerman (ed.), *Viral Hepatitis and Liver Disease.* Alan R. Liss, Inc., New York.

10. **Brown, E. A., J. Ticehurst, and S. M. Lemon.** 1994. Immunopathology of hepatitis A and hepatitis E virus infections, p. 11–37. *In* H. C. Thomas and J. Waters (ed.), *Immunology of Liver Disease.* Kluwer Academic Publishers, Lancaster, United Kingdom.

11. **Bryan, J. P., S. A. Tsarev, M. Iqbal, J. Ticehurst, S. Emerson, A. Ahmed, J. Duncan, A. R. Rafiqui, I. A. Malik, R. H. Purcell, and L. J. Legters.** 1994. Epidemic hepatitis E in Pakistan: patterns of serologic response and evidence that antibody to hepatitis E virus protects against disease. *J. Infect. Dis.* **170:**517–521.

12. **Cao, X.-Y., X.-Z. Ma, Y.-Z. Liu, X.-M. Jin, Q. Gao, H.-J. Dong, H. Zhuang, C.-B. Liu, and G.-M. Wang.** 1991. Epidemiological and etiological studies on enterically transmitted non-A, non-B hepatitis in the south part of Xinjiang, p. 297–312. *In* T. Shikata, R. H. Purcell, and T. Uchida (ed.), *Viral Hepatitis C, D and E.* Elsevier Science Publishers B.V., Amsterdam.

13. **Chau, K. H., G. J. Dawson, K. M. Bile, L. O. Magnius, M. H. Sjogren, and I. K. Mushahwar.** 1993. Detection of IgA class antibody to hepatitis E virus in serum samples from patients with hepatitis E virus infection. *J. Med. Virol.* **40:** 334–338.

14. **Coursaget, P., Y. Buisson, N. Depril, P. Le Cann, M. Chabaud, C. Molinié, and R. Roué.** 1993. Mapping of linear B cell epitopes on open reading frames 2- and 3-encoded proteins of hepatitis E virus using synthetic peptides. *FEMS Microbiol. Lett.* **109:**251–255.

15. **Dawson, G. J., K. H. Chau, C. M. Cabal, P. O. Yarbough, G. R. Reyes, and I. K. Mushahwar.** 1992. Solid-phase enzyme-linked immunosorbent assay for hepatitis E virus IgG and IgM antibodies utilizing recombinant antigens and synthetic peptides. *J. Virol. Methods* **38:**175–186.

16. **Dienes, H. P., T. Hütteroth, L. Bianchi, M. Grun, and W. Thoenes.** 1986. Hepatitis A-like non-A, non-B hepatitis: light and electron microscopic observations of three cases. *Virchows Arch.* (*Pathol. Anat.*) **409:**657–667.

17. **Favorov, M. O., H. A. Fields, M. A. Purdy, T. L. Yashina, A. G. Aleksandrov, M. J. Alter, D. M. Yarasheva, D. W. Bradley, and H. S. Margolis.** 1992. Serologic identification of hepatitis E virus infections in epidemic and endemic settings. *J. Med. Virol.* **36:**246–250.

18. **Favorov, M. O., Y. E. Khudyakov, H. A. Fields, N. S. Khudyakova, N. Padhye, M. J. Alter, E. Mast, L. Polish, T. L. Yashina, D. M. Yarasheva, G. G. Onischenko, and H. S. Margolis.** 1994. Enzyme immunoassay for the detection of antibody to hepatitis E virus (anti-HEV) based on synthetic peptides. *J. Virol. Methods* **46:**237–250.

19. **Goldsmith, R., P. O. Yarbough, G. R. Reyes, K. E. Fry, K. A. Gabor, M. Kamel, S. Zakaria, S. Amer, and Y. Gaffar.** 1992. Enzyme-linked immunosorbent assay for diagnosis of acute sporadic hepatitis E in Egyptian children. *Lancet* **339:** 328–331.

20. **He, J., A. W. Tam, P. O. Yarbough, G. R. Reyes, and M. Carl.** 1993. Expression and diagnostic utility of hepatitis E

virus putative structural proteins expressed in insect cells. *J. Clin. Microbiol.* **31:**2167–2173.

21. **Hoffman, K. J., J. C. Gaydos, R. E. Krieg, J. F. Duncan, Jr., P. O. Macarthy, J. R. Ticehurst, R. Jaramillo, L. G. Reyes, M. H. Sjogren, and L. J. Legters.** 1993. Initial report of a hepatitis investigation in rural Belize. *Proc. R. Soc. Trop. Med. Hyg.* **87:**259–262.

22. **Huang, C.-C., D. Nguyen, J. Fernandez, K. Y. Yun, K. E. Fry, D. W. Bradley, A. W. Tam, and G. R. Reyes.** 1992. Molecular cloning and sequencing of the Mexico isolate of hepatitis E virus (HEV). *Virology* **191:**550–558.

23. **Humphrey, C. D., E. H. Cook, Jr., and D. W. Bradley.** 1990. Identification of enterically transmitted hepatitis virus particles by solid phase immune electron microscopy. *J. Virol. Methods* **29:**177–188.

24. **Hyams, K. C., M. A. Purdy, M. Kaur, M. C. McCarthy, M. A. Hussain, A. El-Tigani, K. Krawczynski, D. W. Bradley, and M. Carl.** 1992. Acute sporadic hepatitis E in Sudanese children: analysis based on a new Western blot assay. *J. Infect. Dis.* **165:**1001–1005.

25. **Kapikian, A. Z. (ed.).** 1994. *Viral Infections of the Gastrointestinal Tract.* Marcel Dekker, Inc., New York.

26. **Karetnyi, Y. V., D. I. Dzhumalieva, R. K. Usmanov, I. P. Titova, Y. I. Litvak, and M. S. Balayan.** 1993. Probable involvement of rodents in the spread of hepatitis E. *J. Microbiol. Epidemiol. Immunol.* **4:**52–56.

27. **Kaur, M., K. C. Hyams, M. A. Purdy, K. Krawczynski, W. M. Ching, K. E. Fry, G. R. Reyes, D. W. Bradley, and M. Carl.** 1992. Human linear B-cell epitopes encoded by the hepatitis E virus include determinants in the RNA-dependent RNA polymerase. *Proc. Natl. Acad. Sci. USA* **89:**3855–3858.

28. **Khudyakov, Y. E., M. O. Favorov, D. L. Jue, T. K. Hine, and H. A. Fields.** 1994. Immunodominant antigenic regions in a structural protein of the hepatitis E virus. *Virology* **198:** 390–393.

29. **Khudyakov, Y. E., N. S. Khudyakova, D. L. Jue, T. W. Wells, N. Padhya, and H. A. Fields.** 1994. Comparative characterization of antigenic epitopes in the immunodominant region of the protein encoded by open reading frame 3 in Burmese and Mexican strains of hepatitis E virus. *J. Gen. Virol.* **75:**641–646.

30. **Khuroo, S. M.** 1980. Study of an epidemic of non-A, non-B hepatitis. Possibility of another human hepatitis virus distinct from post-transfusion non-A, non-B type. *Am. J. Med.* **68:** 818–824.

31. **Koonin, E. V., A. E. Gorbalenya, M. A. Purdy, M. N. Rozanov, G. R. Reyes, and D. W. Bradley.** 1992. Computer-assisted assignment of functional domains in the nonstructural polyprotein of hepatitis E virus: delineation of an additional group of positive-strand RNA plant and animal viruses. *Proc. Natl. Acad. Sci. USA* **89:**8259–8263.

32. **Krawczynski, K.** 1993. Hepatitis E. *Hepatology* **17:**932–941.

33. **Krawczynski, K., D. Bradley, Ajdukiewicz, M. Alter, F. Caredda, J. Dilawari, F. Hlady, B. Innis, M. Kane, A. Nasidi, A. Redeker, H. Stetler, A. Toukan, and B. Yoffee.** 1991. Virus-associated antigen and antibody of epidemic non-A, non-B hepatitis: serology of outbreaks and sporadic cases, p. 229–236. *In* T. Shikata, R. H. Purcell, and T. Uchida (ed.), *Viral Hepatitis C, D and E.* Elsevier Science Publishers B.V., Amsterdam.

34. **Krawczynski, K., and D. W. Bradley.** 1989. Enterically transmitted non-A, non-B hepatitis: identification of virus-associated antigen in experimentally infected cynomolgus macaques. *J. Infect. Dis.* **159:**1042–1049.

35. **Kuwada, S. K., V. M. Patel, F. B. Hollinger, H. J. Lin, P. O. Yarbough, R. H. Wiesner, D. Kaese, and J. Rakela.** 1994. Non-A, non-B fulminant hepatitis is also non-E and non-C. *Am. J. Gastroenterol.* **89:**57–61.

36. **Longer, C. F., S. L. Denny, J. D. Caudill, T. A. Miele, L. V. S. Asher, S. A. Myint, C.-C. Huang, W. F. Engler, J. W. LeDuc, L. N. Binn, and J. R. Ticehurst.** 1993. Experimental hepatitis E: pathogenesis in cynomolgus macaques (*Macaca fascicularis*). *J. Infect. Dis.* **168:**602–609.

37. **Longer, C. F., M. P. Shrestha, P. O. Macarthy, K. S. A. Myint, C. H. Hoke, Jr., J. Ticehurst, and B. L. Innis.** 1994. Epidemiology of hepatitis E virus (HEV): a cohort study in Kathmandu, Nepal, p. 409–411. In K. Nishioka, H. Suzuki, S. Mishiro, and T. Oda (ed.), *Viral Hepatitis and Liver Disease.* Springer-Verlag, Tokyo.

38. **Miele, T. A., J. D. Caudill, R. W. Jansen, G. R. Reyes, H. Y. Zhang, S. M. Lemon, D. W. Bradley, and J. Ticehurst.** Detection of hepatitis E virus (HEV) by an affinity-capture/polymerase chain reaction method (AC/PCR), abstr. 696. In K. Nishioka (ed.), *Program Abstr. 1993 Int. Symp. Viral Hepatitis Liver Dis., Tokyo.* Manuscript in preparation.

39. **Purcell, R. H., J. H. Hoofnagle, J. Ticehurst, and J. L. Gerin.** 1989. Hepatitis viruses, p. 957–1065. In N. J. Schmidt and R. W. Emmons (ed.), *Diagnostic Procedures for Viral, Rickettsial, and Chlamydial Infections.* American Public Health Association, Washington, D.C.

40. **Purcell, R. H., and J. R. Ticehurst.** 1988. Enterically transmitted non-A, non-B hepatitis: epidemiology and clinical characteristics, p. 131–137. In A. J. Zuckerman (ed.), *Viral Hepatitis and Liver Disease.* Alan R. Liss, Inc., New York.

41. **Purdy, M. A., K. A. McCaustland, K. Krawczynski, J. Spelbring, G. R. Reyes, and D. W. Bradley.** 1993. Preliminary evidence that a *trpE*-HEV fusion protein protects cynomolgus macaques against challenge with wild-type hepatitis E virus (HEV). *J. Med. Virol.* **41:**90–94.

42. **Purdy, M. A., K. A. McCaustland, K. Krawczynski, A. Tam, M. J. Beach, N. C. Tassopoulos, G. R. Reyes, and D. W. Bradley.** 1992. Expression of a hepatitis E virus (HEV)-*trpE* fusion protein containing epitopes recognized by antibodies in sera from human cases and experimentally infected primates. *Arch. Virol.* **123:**335–349.

43. **Ray, R., R. Aggarwal, P. N. Salunke, N. N. Mehrotra, G. P. Talwar, and S. R. Naik.** 1991. Hepatitis E virus genome in stools of hepatitis patients during large epidemic in north India. *Lancet* **338:**783–784.

44. **Reyes, G. R., M. A. Purdy, J. P. Kim, K.-C. Luk, L. M. Young, K. E. Fry, and D. W. Bradley.** 1990. Isolation of a cDNA from the virus responsible for enterically transmitted non-A, non-B hepatitis. *Science* **247:**1335–1339.

45. **Skidmore, S. J., P. O. Yarbough, K. A. Gabor, and G. R. Reyes.** 1992. Hepatitis E virus: the cause of a waterbourne hepatitis outbreak. *J. Med. Virol.* **37:**58–60.

46. **Tam, A. W., M. M. Smith, M. E. Guerra, C.-C. Huang, D. W. Bradley, K. E. Fry, and G. R. Reyes.** 1991. Hepatitis E virus (HEV): molecular cloning and sequencing of the full-length viral genome. *Virology* **185:**120–131.

47. **Thomas, D. L., R. W. Mahley, S. Badur, K. E. Palaoglu, and T. C. Quinn.** 1993. Epidemiology of hepatitis E virus infection in Turkey. *Lancet* **341:**1561–1562.

48. **Ticehurst, J.** 1991. Identification and characterization of hepatitis E virus, p. 501–513. In F. B. Hollinger, S. M. Lemon, and H. S. Margolis (ed.), *Viral Hepatitis and Liver Disease.* The Williams & Wilkins Co., Baltimore.

49. **Ticehurst, J., T. J. Popkin, J. P. Bryan, B. L. Innis, J. F. Duncan, A. Ahmed, M. Iqbal, I. Malik, A. Z. Kapikian, L. J. Legters, and R. H. Purcell.** 1992. Association of hepatitis E virus with an outbreak of hepatitis in Pakistan: serologic responses and pattern of virus excretion. *J. Med. Virol.* **36:**84–92.

50. **Ticehurst, J., L. L. Rhodes, Jr., K. Krawczynski, L. V. Asher, W. F. Engler, T. L. Mensing, J. D. Caudill, M. H. Sjogren, C. H. Hoke, Jr., J. W. LeDuc, D. W. Bradley, and L. N. Binn.** 1992. Infection of owl monkeys (*Aotus trivirgatus*) and cynomolgus monkeys (*Macaca fascicularis*) with hepatitis E virus from Mexico. *J. Infect. Dis.* **165:**835–845.

51. **Tsarev, S. A., S. U. Emerson, G. R. Reyes, T. S. Tsareva, L. J. Legters, I. A. Malik, M. Iqbal, and R. H. Purcell.** 1992. Characterization of a prototype strain of hepatitis E virus. *Proc. Natl. Acad. Sci. USA* **89:**559–563.

52. **Tsarev, S. A., T. S. Tsareva, S. U. Emerson, A. Z. Kapikian, J. Ticehurst, W. London, and R. H. Purcell.** 1993. ELISA for antibody to hepatitis E virus (HEV) based on complete open-reading frame-2 protein expressed in insect cells: identification of HEV infection in primates. *J. Infect. Dis.* **168:**369–378.

53. **Tsarev, S. A., T. S. Tsareva, S. U. Emerson, P. O. Yarbough, L. J. Legters, T. Moskal, and R. H. Purcell.** 1994. Infectivity titration of a prototype strain of hepatitis E virus (HEV) in cynomolgus monkeys. *J. Med. Virol.* **43:**135–142.

54. **Uchida, T., T. T. Aye, X. Ma, F. Iida, T. Shikata, M. Ichikawa, T. Rikihisa, and K. M. Win.** 1993. An epidemic outbreak of hepatitis E in Yangon of Myanmar: antibody assay and animal transmission of the virus. *Acta Pathol. Jpn.* **43:**94–98.

55. **Yarbough, P. O., A. W. Tam, K. E. Fry, K. Krawczynski, K. A. McCaustland, D. W. Bradley, and G. R. Reyes.** 1991. Hepatitis E virus: identification of type-common epitopes. *J. Virol.* **65:**5790–5797.

56. **Yarbough, P. O., A. W. Tam, K. Gabor, E. Garza, R. A. Moeckli, I. Palings, C. Simonsen, and G. Reyes.** 1994. Assay development of diagnostic tests for hepatitis E, p. 367–370. In K. Nishioka, H. Suzuki, S. Mishiro, and T. Oda (ed.), *Viral Hepatitis and Liver Disease.* Springer-Verlag, Tokyo.

57. **Yin, S., S. A. Tsarev, R. H. Purcell, and S. U. Emerson.** 1993. Partial sequence comparison of eight new Chinese strains of hepatitis E virus suggests the genome sequence is relatively stable. *J. Med. Virol.* **41:**230–241.

58. **Zhang, H. Y., J. D. Caudill, T. J. Popkin, J. F. Duncan, I. Hussain, A. Ahmed, M. Iqbal, I. A. Malik, L. J. Legters, and J. Ticehurst.** 1993. Patterns of hepatitis E virus (HEV) excretion from selected patients in an outbreak, abstr. 707. In K. Nishioka (ed.), *Program Abstr. 1993 Int. Symp. Viral Hepatitis Liver Dis., Tokyo.* Manuscript in preparation.

Filoviruses and Arenaviruses

PETER B. JAHRLING

94

CLINICAL BACKGROUND

Viruses from two families, *Arenaviridae* (19) and *Filoviridae* (24), are important human pathogens that cause severe, frequently fatal viral hemorrhagic fevers. Although exotic to North America, these viruses are important public health problems in Africa and South America and may be introduced to North America by travelers returning from these areas. The arenavirus and filovirus pathogens have been associated with severe nosocomial outbreaks involving health care workers and laboratory personnel. Continuing isolation of new arenavirus and filovirus subtypes from new geographic locations categorizes these agents as emerging infections whose potential importance to public health has only recently been recognized (27). The similarity in initial isolation, clinical management, and viral diagnostic procedures for patients with suspected arenavirus or filovirus infections is the rationale for grouping these taxonomically distinct viruses together in this chapter. Patients with these infections frequently present with similar, nonspecific clinical signs resembling those of malaria, typhoid, and pharyngitis. A detailed travel history coupled with a high index of suspicion and the availability of definitive, virologic tools should facilitate rapid viral diagnosis and the timely implementation of appropriate patient isolation and clinical management procedures.

Arenaviruses are maintained in nature by association with specific rodent hosts (Table 1), in which they produce chronic viremia and/or viruria. Naturally occurring human disease can usually be traced to direct or indirect contact with infected rodents. Aerosol infectivity is also thought to be an important natural route of infection as well. Attempts to implicate arthropod vectors have been negative, but ectoparasites taken from viremic mammalian hosts have occasionally yielded arenavirus isolates.

Of the 16 recognized members of the family *Arenaviridae* (19), 6 are human pathogens. Lassa virus is the etiologic agent of Lassa fever, a West African disease. The original, isolated outbreaks described were hospital associated, occurring first in Nigeria in 1969 and later in Liberia and Sierra Leone. Case-fatality rates ranging from 20 to 40% were reported. Subsequently, extensive clinical disease and high seroprevalence were well documented in Nigeria, Sierra Leone, and Liberia. Other countries with serologic evidence of Lassa fever include Guinea, Ivory Coast, Ghana, Senegal, Upper Volta, The Gambia, and Mali. The true incidence of human Lassa fever cases is believed to be thousands to tens of thousands annually. Related viruses, apparently not pathogenic for humans, have been isolated in Mozambique, Zimbabwe, and Central African Republic (Table 1). Lassa virus is now recognized to persist in foci of endemicity in Liberia and Sierra Leone, although current epidemiologic data are sparse, owing to civil unrest in that area. In hospitals located near foci of endemicity in Sierra Leone, 30% of all medical deaths were attributed to Lassa fever, and mortality among hospitalized patients with Lassa fever is still estimated to be 15 to 20%. Overall mortality, including subclinical cases identified serologically, is substantially less, perhaps less than 1%. Deaths occur year-round, with some increases during the dry season.

Junin virus, which causes Argentine hemorrhagic fever (AHF), was first isolated in 1959, although the disease was recognized as early as 1943 in Argentina. Junin virus infections occur primarily among young, male agricultural workers. Annual numbers of AHF cases from 1958 to the present have fluctuated from 100 to 3,500, with several hundred cases occurring in most years. Peak seasonal incidence is from March to June, coinciding with both the season for corn harvesting and the expansion of wild rodent populations, that of including *Calomys musculinis*. The reported mortality rate for patients with laboratory-confirmed AHF was 14 to 17% before the routine use of immune plasma therapy was implemented.

Machupo virus, the etiologic agent of Bolivian hemorrhagic fever (BHF), was first isolated in 1963, although reports of sporadic outbreaks began in 1959 and a series of devastating epidemics occurred from 1962 through 1964. These epidemics involved more than 1,000 patients with an 18% mortality rate. *Calomys callosus* was implicated as the rodent reservoir, and the incidence of human cases was correlated with the presence of infected rodents in households. Control of *C. callosus* has eliminated epidemic BHF and reduced the annual incidence of cases to fewer than 10. Another severe outbreak of BHF occurred in Cochabamba, Bolivia, in 1971. This outbreak was associated entirely with nosocomial spread from a single index patient who had returned from the region of endemicity. This outbreak, plus the explosive human-to-human chain of lethal transmission for Lassa fever and anecdotal reports of nosocomial spread for Junin virus infection, highlights the necessity for

TABLE 1 Currently recognized arenaviruses and associated human diseases

Virus, date isolated	Natural host	Geographic distribution	Naturally occurring human disease	Human laboratory infections
Old World				
LCM, 1933	*Mus musculus* (house mouse)	Americas, Europe	Undifferentiated febrile illness, aseptic meningitis; rarely serious	Common; usually mild, but 5 were fatal
Lassa, 1969	*Mastomys* sp. (multimammate rat)	West Africa	Lassa fever; mild to severe and fatal disease	Common; often severe
Ippy, 1984	*Arvicanthus* sp.	Central African Republic	Unknown	None
Mopeia, 1977	*Mastomys natalensis*	Mozambique, Zimbabwe	Unknown	None; little experience
Mobala, 1983	*Praomys* sp.	Central African Republic	Unknown	None; little experience
New World				
Junin, 1958	*Calomys musculinus*	Argentina	AHF	Common; often severe
Machupo, 1963	*Calomys callosus*	Bolivia	BHF	Common; often severe
Tacaribe, 1956	*Artibeus* bats	Trinidad, West Indies	None detected	One suspected; moderately symptomatic
Amapari, 1966	*Oryzomys gaeldi*, *Neacomys guianae*	Brazil	None detected	None detected
Parana, 1970	*Oryzomys buccinatus*	Paraguay	None detected	None detected
Tamiami, 1970	*Sigmodon hispidus* (cotton rat)	Florida	Antibodies detected	None detected
Pichinde, 1971	*Oryzomys albigularis*	Colombia	None detected	Occasional, mild to asymptomatic
Latino, 1973	*Calomys callosus*	Bolivia	None detected	None detected
Flexal, 1977	*Oryzomys* spp.	Brazil	None detected; moderately severe	One recognized
Guanarito, 1989	*Sigmodon alstoni*	Venezuela	Venezuelan hemorrhagic fever	None detected
Sabia, 1993	Unknown	Brazil	Viral hemorrhagic fever	One recognized (severe)

biocontainment precautions when one is dealing with arenavirus-infected patients and materials.

Guanarito virus is one of the truly emerging arenaviruses. It was first recognized in association with an outbreak of hemorrhagic fever thought initially to be dengue fever in Venezuela in 1989 (38). During 3 years of epidemiologic surveillance, from September 1989 to December 1991, 88 cases of viral hemorrhagic fever with a fatality rate of 34% were documented. Since then, only sporadic cases have emerged, with 105 cases and 26 deaths reported by the end of 1992. The patients affected ranged from 6 to 54 years old; the highest rate was in males over 15 years old. Most viral hemorrhagic fever cases have been reported from the municipality of Guanarito and neighboring Barinas State. These areas are predominantly rural, with small farms used for agriculture and cattle production; thus, most homes are near fields. Field studies have shown that the cotton rat, *Sigmodon alstoni*, is the major host and natural reservoir of Guanarito virus (42). In addition, the cane rat, *Zygdontomys breviocauda*, may serve as a transiently infected host with importance in maintenance of the viral cycle. As for the other arenaviruses, transmission from rodents to humans is thought to relate primarily to contact with food, water, or air contaminated with rodent urine. Preliminary epidemiologic information, including reports of low antibody rates among family contacts of cases and hospital personnel caring for infected patients, suggests that person-to-person transmission is uncommon. More extensive epidemiologic studies are in progress.

Yet another new arenavirus pathogen, Sabia virus, has

been described (11). It was isolated in Brazil in 1993 from a patient with a fatal case of hemorrhagic fever thought initially to be yellow fever. During characterization of this virus, a laboratory technician became infected, probably via the aerosol route; he eventually recovered. The emergence of both Guanarito and Sabia virus infections and the initial confusion of these infections with other viral fevers including dengue and yellow fever illustrate the importance of developing specific, definitive diagnostic tools for these agents.

The relatively limited geographic distributions of Lassa, Machupo, and Junin viruses probably relate to the limited distributions of the chronically infected rodent hosts. In contrast, lymphocytic choriomeningitis (LCM) virus is associated with the house mouse, *Mus musculus*, and the virus, like its rodent reservoir, is much more widespread. The presence of LCM virus is well documented in the Americas and Europe but not in Scandinavia. LCM virus has also not been isolated from sources in Australia or Africa. Infected mouse colonies have long been associated with transmission to humans; more recently, the role of infected hamster colonies in dissemination, particularly to laboratory workers and pet owners, has become evident, as has the occupational hazard of manipulating certain human tumor cell lines propagated in rodents, including hamsters and nude mice (22). The absence of LCM from Africa is surprising; some of the Lassa-like viruses (e.g., Mopeia) isolated from that continent are phylogenetically as closely related to LCM virus as they are to Lassa virus according to sequence analysis of structural proteins (10).

Marburg (MBG) and Ebola hemorrhagic fevers are caused by taxonomically distinct viruses that form two groups within one genus, *Filovirus*, within the family *Filoviridae* (24). MBG virus was first recognized in 1967, when 25 persons in Germany and Yugoslavia became infected following contact with monkey kidneys or primary tissue cultures derived from monkeys imported from Uganda. Seven of these patients died (30). In addition, six secondary human cases occurred. Since then, sporadic, virologically confirmed MBG virus cases have occurred in Zimbabwe, South Africa, and Kenya. Ebola virus first emerged in two major disease outbreaks that occurred almost simultaneously in Zaire and Sudan in 1976. Over 500 cases were reported, with case-fatality rates of 88% in Zaire and 53% in Sudan. Despite the simultaneous emergence of Ebola virus strains in Zaire (EBO-Z) and Sudan (EBO-S), these Ebola virus isolates are serologically distinguishable and differ in host range, virulence potential, and biochemical properties. Serologic studies to date suggest that infections with Ebola virus or related viruses have occurred in Zaire, Sudan, Central African Republic, Gabon, Nigeria, Ivory Coast, Liberia, Cameroon, and Kenya. The geographic range of Ebola viruses may extend to other African countries, for which adequate serosurveys are lacking. Serologic studies suggest that subclinical infections may occur frequently, although the validity of the methods used for population-based serosurveys of filovirus activity has recently been questioned (36). The natural reservoirs for human infection with MBG and Ebola viruses have not been identified. Recently, another virus serologically and phylogenetically related to Ebola virus was isolated from cynomolgus monkeys. These animals originated in the Philippines and were being held in a quarantine facility in Reston, Va., in 1989 when an explosive epizootic occurred; most of the monkeys died, and a filovirus, Ebola-Reston (EBO-R), was isolated (20). EBO-R may be less virulent for humans than the African filovirus strains, since four workers at the quarantine facility seroconverted to EBO-R without disease. EBO-R shares a high degree of sequence homology with EBO-S and EBO-Z in the region of the glycoprotein gene (39) and is as closely related to the African strains as they are to each other. The epidemiology and significance of this newly emerging virus strain are the subjects of intensive investigation.

CLINICAL PRESENTATION

Among the arenavirus pathogens, LCM virus produces the least severe infection (28). A modest proportion of LCM virus infections are thought to be subclinical. A typical LCM virus case is usually heralded by fever, myalgia, retro-orbital headache, weakness, and anorexia. Especially during the first week, prominent symptoms include sore throat, chills, vomiting, cough, retrosternal pain, and arthralgia. Rash occurs, but infrequently. In about one-third of patients, fever recurs, coinciding with the onset of frank neurologic involvement, usually aseptic meningitis, or, less frequently, meningoencephalitis. Complete recovery is almost always the rule. Thus, LCM virus infections are temporarily debilitating but rarely fatal, even when neurologic complications arise.

Lassa, Junin, and Machupo virus infections are more severe. For Lassa virus infection, fever and myalgia develop insidiously between 3 and 16 days after exposure and increase in severity during the following week. There are few specific clinical manifestations. Lassa fever patients usually come to the hospital within 5 to 7 days of onset with sore throat, severe lower back pain, and conjunctivitis. These symptoms usually increase in severity during the following week and are accompanied by nausea, vomiting, diarrhea, chest and abdominal pain, headache, cough, dizziness, and tinnitus. Later, pneumonitis and pleural and pericardial effusions with friction rub frequently occur. A maculopapular rash may develop, but frank hemorrhage is seen in only some of the more severe cases. Oozing from puncture sites and mucous membranes and melena are more common. Approximately 15 to 20% of hospitalized patients die. Death relates to sudden cardiovascular collapse resulting from hepatic, pulmonary, and myocardial damage. Few Lassa fever patients develop central nervous system signs, although tinnitus or deafness may develop as recovery begins. Lassa fever is a particularly severe disease among pregnant women, for whom mortality rates are somewhat higher. In children, the disease course is similar to that in adults, but in infants, a condition described as "swollen-baby syndrome," characterized by anascara, abdominal distention, and bleeding, has been described (32). Clinical laboratory studies are not usually helpful for Lassa fever; specific virologic testing is required, especially in a setting where Lassa fever is less common or when mild, atypical cases are occurring.

The clinical pictures for AHF (Junin virus) and BHF (Machupo virus) are well characterized; for viral hemorrhagic fever (Guanarito virus and Sabia virus), less information is available, yet all these infections are sufficiently similar to each other to be discussed as a single entity, referred to as South American hemorrhagic fevers. Incubation periods range from 7 to 14 days, and very few subclinical cases are thought to occur. Following a gradual onset of fever, anorexia, and malaise over several days, constitutional signs involving the gastrointestinal, cardiovascular, and central nervous systems become apparent by the time patients present to the hospital. On initial examination, AHF and BHF patients are febrile, acutely ill, and mildly hypotensive. They frequently complain of back pain, epigastric pain, headache, retro-orbital pain, photophobia, dizziness, constipation or diarrhea, and coughing. Vascular phenomena, including flushing of the face, neck, and chest, and bleeding from the gums are common. Exanthem is almost invariably present; petechiae or tiny vesicles spread over the erythematous palate and fauces. Neurologic involvement, ranging from mild irritability and lethargy to abnormalities in gait, tremors of the upper extremities, and, in severely ill patients, coma, delirium, and convulsions, occurs in more than half of the patients. During the second week of illness, clinical improvement may begin, or complications may develop. Complications include extensive petechial hemorrhages, oozing from puncture wounds, melena, and hematemesis. These manifestations of capillary damage and thrombocytopenia do not result in life-threatening blood loss. However, hypotension and shock may develop, often in combination with serious neurologic signs among the 15% of patients who die. Survivors begin to show improvement by the third week. Recovery is slow; weakness, fatigue, and mental difficulties may last for weeks, and a significant proportion of patients relapse with a late neurologic syndrome. In contrast to the situation with Lassa fever, clinical laboratory studies are frequently useful. Total leukocyte counts usually fall to 1,000 to 2,000 cells per mm³, although the differential remains normal. Platelet counts fall precipitously, usually to 25,000 to 100,000/mm³ but occasionally lower. Routine clotting parameters are usually normal or slightly deranged, but in severe cases, evidence of disseminated intravascular coagulation is apparent.

MBG and Ebola virus infections are clinically similar, although the frequencies of reported signs and symptoms vary. Following incubation periods of 4 to 16 days, onset is sudden, marked by fever, chills, headache, anorexia, and myalgia. These signs are soon followed by nausea, vomiting, sore throat, abdominal pain, and diarrhea. When first examined, patients are usually overtly ill, dehydrated, apathetic, and disoriented. Pharyngeal and conjunctival injection are usual. Within several days, a characteristic maculopapular rash over the trunk, petechiae, and mucous membrane hemorrhages appears. Gastrointestinal bleeding accompanied by intense epigastric pain is common, as are petechiae and bleeding from puncture wounds and mucous membranes. Shock develops shortly before death, often 6 to 16 days after onset of illness. Abnormalities in coagulation parameters include fibrin split products and prolonged prothrombin and partial thromboplastin times, suggesting that disseminated intravascular coagulation is a terminal event. Clinical laboratory studies usually reveal profound leukopenia early, followed by a rapid rise in association with secondary bacteremia. Platelet counts decline to 50,000 to 100,000/mm^3 during the hemorrhagic phase.

DESCRIPTION OF THE AGENTS

Arenaviridae

The family *Arenaviridae* comprises 16 named viruses that share unique morphologic and physiochemical characteristics. In nature, arenaviruses are associated with unique mammalian hosts, usually rodents (Table 1). Antigenic relationships are established mainly on the basis of broadly reactive antibody binding assays, i.e., historically, the complement fixation (CF) test and more recently, the indirect fluorescent-antibody (IFA) test and enzyme-linked immunosorbent assays (ELISAs). Finer discriminations among virus strains have been made by using neutralization tests; more recently, monoclonal antibodies have been used for fine discriminations in IFA and ELISA formats (4). Molecular phylogeny relationships based on protein sequences generally validate the serologic groupings (10). Two taxonomic complexes are generally accepted (6, 47). The LCM, or Old World, complex includes the LCM and Lassa viruses plus a number of Lassa virus-like strains from Mozambique, Zimbabwe, and Central African Republic. These may be unique viruses or substrains of Lassa virus. All have been isolated from rodents of the family *Muridae*. The Tacaribe, or New World, complex includes Tacaribe, Junin, Machupo, Amapari, Parana, Latino, Pichinde, Tamiami, and Flexal viruses. All New World complex viruses have been isolated from rodents of the family *Cricetidae*; the single exception is Tacaribe virus, which is from bats. Old and New World complex viruses are distantly related, and cross-reactions by CF or IFA test using crude antigens are obtained only when high-titer antisera are used. Cross-reactivity is much broader when synthetic peptides are used for conserved antigenic epitopes.

The morphology of arenaviruses is distinctive in thin-section electron microscopy (EM) (34) and was the basis for associating LCM virus first with Machupo virus and ultimately with all the viruses in the family. Individual virions are pleomorphic and range in size from 60 to 280 nm (mean, 110 to 130 nm). A unit membrane envelops the

structure and is covered with club-shaped, 10-nm projections. No symmetry has been discerned. The most prominent and distinctive feature of these virions is the presence of various numbers of electron-dense particles (usually 2 to 10), which may be connected by fine filaments. These particles, each 20 to 25 nm in diameter, are identical with host cell ribosomes according to biochemical and oligonucleotide analyses. Three major virion structural proteins are usually found (5). Two are glycosylated: G1 (50,000 to 72,000 Da) and G2 (31,000 to 41,000 Da) constitute the virion envelope and spikes. Both G1 and G2 serve as highly type-specific neutralization targets. The N protein (63,000 to 72,000 Da) is clearly associated with the virion RNA and is considered the nucleocapsid protein. The N protein in intact cells or virions is not accessible to antibody but can readily be detected in acetone-fixed cells by immunofluorescence. Nucleocapsids can be isolated by treatment of intact virions with detergent. Liberated nucleocapsids are 10 nm in diameter and range to 450 nm in length. Two size classes of closed or circular nucleocapsid have been identified: 640 and 1,300 nm. Small, 3- to 4-nm beaded strands can also be resolved. Some arenaviruses elaborate a soluble protein antigenically related to the N protein into supernatant fluids of infected cell cultures. The group-reactive arenavirus antigen measured in both the CF and the IFA tests is associated with the nucleocapsid and is probably N protein. No hemagglutinins have been found.

Four RNA species with distinct oligonucleotide fingerprints can be isolated from intact arenavirus virions. Two species are virus specific: the small (S) RNA (22S) codes for N and GPC (the precursor of G1 and G2), and the large (L) RNA (31S) codes for a 200-kDa protein believed to be an RNA-dependent RNA polymerase. The 28S and 18S species, isolated in various proportions, are ribosomal. The sizes of L and S viral RNA have been estimated at 2.6 × 10^6 and 1.4 × 10^6 daltons. The coding strategy for arenaviruses is unique and has been termed "ambisense," since the 3' half of the S RNA codes for N in the viral complementary sense and is separated by an intragenic hairpin structure from the 5' region, which codes for GPC in the viral sense. Through this mechanism, GPC and N gene expressions are independently regulated. Viral RNA must be translated before GPC can be expressed, and this regulation may be fundamental to the maintenance of persistent infections in chronically infected hosts and cells in culture.

Arenaviruses mature by budding at the cytoplasmic membrane, and host proteins are incorporated into the virion envelope. Vero cells infected with each of the viruses contain distinctive intracytoplasmic inclusion bodies. When immunohistochemistry and monoclonal antibodies are used, these intracytoplasmic inclusions are demonstrably immunoreactive with antinucleocapsid but not G1 or G2 antibodies. Buoyant densities of intact arenavirus particles in cesium chloride range from 1.17 to 1.18. All arenaviruses are readily inactivated by ethyl ether, chloroform, sodium deoxycholate, and acid media (pH less than 5). β-Propiolactone (44) and gamma irradiation (14) are both reported to inactivate arenavirus infectivity while preserving reactivity in standard serologic tests.

Filoviridae

The filoviruses, MBG and Ebola viruses, have a common morphology, similar densities, similar polyacrylamide gel electrophoresis (PAGE) profiles, and some RNA sequence homology. MBG and Ebola virion RNAs are nonsegmented, negative, and single stranded (NNS RNA); 19.1

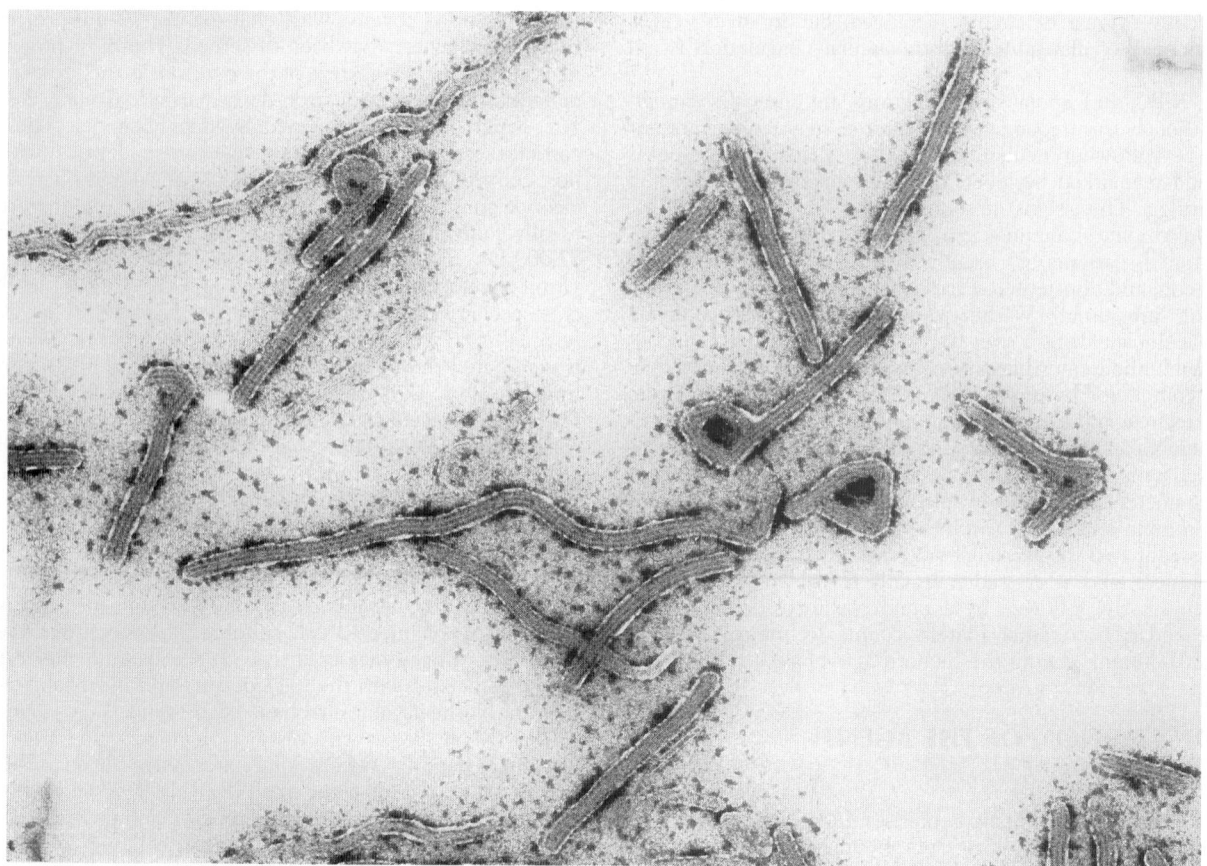

FIGURE 1 Electron micrograph of EBO-R. Filamentous particles are 80 nm in diameter with 50-nm cores and various lengths, some being >10,000 nm long. Negative stain (sodium phosphotungstate) of supernatant fluids from MA-104 cells. Magnification, ×32,550. (Figure courtesy of T. W. Geisbert.)

kb long; and 4.0×10^6 to 4.5×10^6 Da. The genomes of MBG and Ebola virus are linearly arranged in a manner consistent with other NNS-RNA viruses. A substantial degree of similarity to the paramyxoviruses has been noted, especially in the nucleocapsid and polymerase genes. However, comparison with other filoviral protein sequences confirms that filoviruses are distinct. Further, filoviruses are sufficiently distinct by ultrastructural and serologic criteria to warrant separate taxonomic status as *Filoviridae* (24).

MBG and Ebola virions mature by the budding of cytoplasmic tubular structures, which contain the nucleocapsids, through the cell membrane. The resulting viral particles are very large, typically 790 to 970 nm in length and consistently 80 nm in diameter. Bizarre structures of widely varying lengths, sometimes exceeding 14,000 nm, as well as branching, circular, or "6" shapes, probably resulting from coenvelopment of multiple nucleocapsids during budding, are frequently found in negatively stained preparations (33) (Fig. 1). When infected tissue culture fluids are purified by rate zonal centrifugation, the infectivity is associated with uniformly sized bacilliform particles 790 nm in length for MBG virus and 970 nm for Ebola virus with sedimentation coefficients of 1,400S (25). Each infectious particle contains a nucleocapsid with a 20- to 30-nm central axis surrounded by a helical capsid 40 to 50 nm in diameter with cross-striations at 3- to 5-nm intervals. The nucleocapsid is

surrounded by a host cell-derived envelope bearing 7- to 10-nm projections in a regular array.

Both viruses have at least seven virus-specific structural proteins expressed from seven genes (39). For Ebola virus, the ribonucleoprotein complex contains L (180 kDa), N (104 kDa), VP30 (30 kDa), and VP35 (35 kDa) in loose association. L is an RNA-dependent RNA polymerase, and VP35 may have a role similar to that of the P proteins of paramyxoviruses and rhabdoviruses. GP (125,000) is the major spike protein; VP40 (a matrix protein) and VP24 make up the remaining protein content of the multilayered envelope (25, 39). The GP of Ebola and MBG viruses can be differentiated by the presence or absence of N- and O-linked glycans. When grown in Vero or MA-104 cells, the GP of MBG virus is totally devoid of terminal sialic acid, while the GP of Ebola virus has abundant (2-3)-linked sialic acid. A region of the GP of both Ebola and MBG viruses has a high degree of homology with the immunosuppressive protein of oncogenic retroviruses. The structural proteins of MBG and Ebola viruses are very similar in sodium dodecyl sulfate (SDS)-PAGE profiles, but there is no evidence that they are antigenically related. Phylogenetic analysis of GP genes from filoviruses clearly separates MBG virus from Ebola virus. The nucleotide sequence differences among three Ebola virus strains is 47%, with alignments containing two or three gaps, while MBG virus has about 72% divergence and six to eight gaps (39).

Despite their unusual morphologic properties, MBG and Ebola viruses resemble the arenaviruses and other lipid-enveloped viruses in being susceptible to lipid solvents, β-propiolactone (44), formaldehyde, UV light, and gamma radiation (14). These viruses are stable at room temperature for several hours but are inactivated by incubation at 60°C for 1 h.

COLLECTION AND STORAGE OF SPECIMENS

For virus isolation, serum, heparinized plasma, or, less ideally, whole blood should be collected during the acute, febrile stages of these illnesses and frozen on dry ice or in liquid nitrogen vapor. Throat wash and urine specimens should also be collected and mixed with an equal volume of buffered diluent containing serum proteins to stabilize viral infectivity prior to freezing. Storage at higher temperatures (over −40°C) may be unavoidable but will lead to rapid losses in infectivity. Lassa virus can be isolated from most acute-phase sera and throat washings obtained within several weeks of onset with great success; it is isolated less frequently from urine. Junin virus is usually recoverable from serum for 3 to 10 days after onset and often from throat washings for a similar period but rarely from urine. Machupo virus is recovered from only one in five acute-phase sera and even less frequently from throat washings or urine. LCM virus may be recovered from acute-phase sera obtained within the first week after onset but rarely if ever from throat washings or urine specimens. LCM virus may also be isolated from cerebrospinal fluid during the period of meningeal involvement and from the brain at autopsy. MBG and Ebola viruses are usually recoverable from acute-phase sera; various specimens, including throat washings, urine, soft tissue effusates, semen, and anterior eye fluid, have yielded these viruses, even when the specimens were obtained late in convalescence. Lassa, Machupo, Junin, MBG, and Ebola viruses are all readily isolated from spleen, lymph nodes, liver, and kidney obtained at autopsy but rarely if ever from brain or other central nervous system tissues. Notably, Lassa virus is usually isolated from the placentas of infected pregnant women. Specimens collected for viral isolation are also suitable for testing by antigen capture ELISA; maintenance at −20°C for several weeks is adequate for antigen preservation. Impression smears of infected tissues, prepared as described in the next section, may be fixed by immersion in cold acetone and stored frozen for viral antigen staining by IFA. Formaldehyde-fixed tissues and paraffin-embedded blocks are also suitable for immunohistochemical identification of viral antigens.

Blood obtained in early convalescence for serodiagnosis may be infectious despite the presence of antibodies and should be handled accordingly (8, 9). Maintenance of samples at −20°C or below is sufficient to preserve antibody titers and antigenicity, but lower temperatures are required to preserve infectivity. Certain anticoagulants should be avoided: citrate interferes with the IFA test, both citrate and oxalate cause nonspecific cytopathic effect in Vero and MA-104 cells used for virus isolation, and EDTA can interfere with some ELISA techniques. Heparinized plasma or serum samples are best.

Manipulation of these specimens and tissues, including sera obtained from convalescent patients, may pose a serious biohazard, and their handling outside a maximum containment laboratory should be minimized. Current recom- mendations in the United States are that such samples be manipulated only at the biosafety level 4 (BL-4) containment level (9). At a minimum, barrier nursing procedures should be implemented, and personnel caring for the patient and handling diagnostic specimens should wear disposable caps, gowns, shoe covers, surgical gloves, and face masks (preferably full-face respirators equipped with high-efficiency particulate air [HEPA] filters) (8). Gloves should be disinfected immediately if they come in direct contact with infected blood or secretions. All disposable and reusable equipment should be placed directly in a suitable disinfectant solution, such as sodium hypochlorite, phenolic detergent, or a solution of quaternary ammonium compounds. Reusable equipment can then be sterilized. To minimize the risk of autoinoculation, needle sheaths should not be replaced on used needles; needles should be placed immediately in disinfectant. Glass equipment (e.g., microscope slides, microhematocrit tubes, syringes) poses a significant hazard also; substitutes should be found for all glass equipment except Vacutainer tubes for blood collection. Use of Vacutainers is considered safer than use of syringes and needles, which must be disassembled before tube contents are transferred to another tube. Procedures that generate aerosols (e.g., trituration, centrifugation) should be minimized and done only if additional protective equipment, such as a flexible plastic film isolator capable of maintaining negative pressure and HEPA-filtered exhaust, is available. For specialized procedures, samples may be inactivated by the addition of β-propiolactone and may then be safely tested by serologic or antigen capture assays in the field. Heating to 60°C for 1 h renders diagnostic specimens noninfectious and is feasible for measurement of heat-stable substances such as electrolytes, blood urea nitrogen, and creatinine.

For all testing of infectious material, samples should be packaged in accordance with current recommendations (7) and forwarded, after consultation, to one of the following laboratories that maintain BL-4 facilities and a diagnostic capability for these agents:

1. Centers for Disease Control and Prevention
   Special Pathogens Branch
   Division of Viral and Rickettsial Diseases
   Center for Infectious Diseases
   Atlanta, GA 30333
   Tel: 404-639-1115 FAX: 404-639-1118
2. U.S. Army Medical Research Institute of Infectious Diseases
   Headquarters, Fort Detrick
   Frederick, MD 21702-5011
   Tel: 301-619-2833 FAX: 301-610-4625
3. Center for Applied Microbiology and Research
   Special Pathogens Unit
   Salisbury
   Porton Down, Wiltshire, SP4 OJG, England
   Tel: 0980-610391 FAX: 0980-611093
4. National Institute for Virology
   Private Bag X4
   Sandringham, Johannesburg
   South Africa
   Tel: 27-11-882-9910 FAX: 27-11-882-0596

DIRECT EXAMINATION

MBG and Ebola viruses have been successfully visualized directly by EM of both heparinized blood and urine ob-

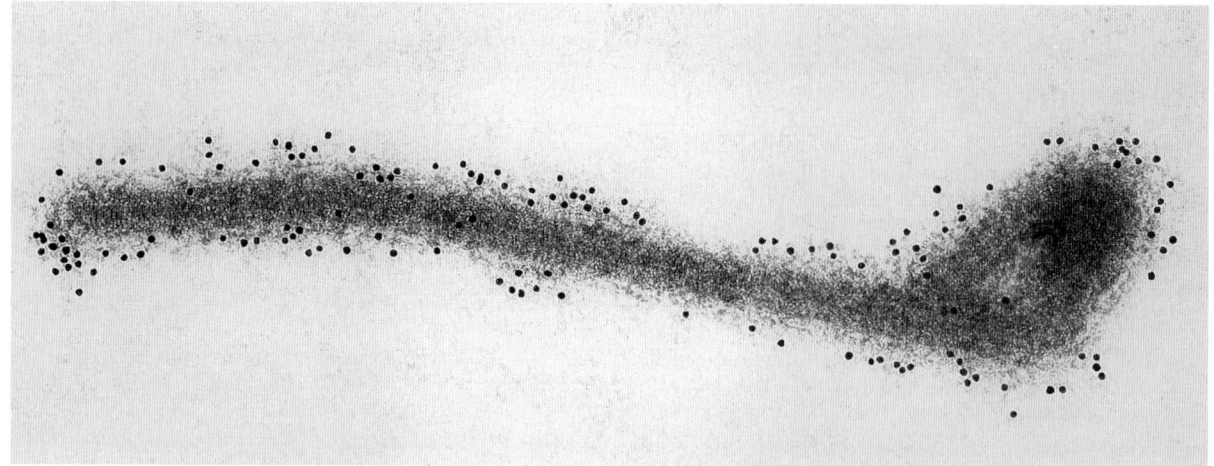

FIGURE 2 Immunoelectron microscopic staining of EBO-R incubated with polyclonal guinea pig anti-Ebola virus serum followed by anti-guinea pig IgG labeled with gold spheres. Magnification, ×119,000. (Figure courtesy of T. W. Geisbert.)

tained during the febrile period as well as of tissue culture supernatant fluids. These materials are processed by immediate fixation with 0.5% glutaraldehyde followed by low-speed centrifugation. Virions are then sedimented at 12,000 × g for 15 min, resuspended in 1/100 of the original sample volumes, and placed on Formvar-carbon-coated EM grids, which are then negatively stained with 1% phosphotungstic acid for 30 s and examined. The combination of size and shape of the virions is sufficiently characteristic to allow a morphologic diagnosis of filovirus (33), even when the isolate is a new entity, such as EBO-R (20). Differentiation of these virions and identification as MBG or Ebola virus are accomplished by immunoelectron microscope techniques (15, 16). Murine monoclonal antibodies and guinea pig polyclonal antisera have been successfully applied to recent filovirus isolates (Fig. 2). These immunoelectron microscope techniques also work well for diagnosis of arenavirus infections, although the morphology of the virions is less striking. Retrospective examination of ultrathin sections of formalin-fixed tissues obtained from patients at autopsy have occasionally revealed typical MBG virus, Ebola virus, or arenavirus particles, most often in the liver, spleen, or kidney but only when infectious virus concentrations are very high and fixation is good. EBO-R particles were easily demonstrated by thin-section EM of tissues of monkeys dying in the recent epizootic (20) (Fig. 3).

Direct immunofluorescence staining (DFA) and IFA staining of impression smears or air-dried suspensions of liver, spleen, or kidney have been used successfully to detect cytoplasmic inclusion bodies associated with MBG infection; clumps of MBG virus antigen have also been observed by DFA examination of infected, dried, citrated blood smears. This approach was successfully adapted to diagnosis of EBO-R in impression smears of blood, tissues, and nasal turbinates, as it was to detect Junin virus-infected cells in peripheral blood and urinary sediment. Development of immunohistochemical techniques for detection of filovirus and arenavirus antigens in formalin-fixed tissues has recently advanced to the point that results are far more satisfactory than for IFA examination of frozen, acetone-fixed sections (12, 20). Obtaining frozen sections for diagnosis is rarely worth the biohazard incurred, especially since

the threshold sensitivity for detection exceeds 6 $\log_{10}$ PFU/g. For filoviruses, paraffin blocks of tissues are sectioned and mounted on silane-coated slides. They are deparaffinized, hydrated, digested with protease, and stained for the presence of viral antigens with cocktails of murine monoclonal antibodies (20). Biotinylated horse anti-mouse antiserum is then reacted, and the product is developed with a streptavidin-alkaline phosphatase system. Substitution of other chromogens can further increase sensitivity while reducing background (12). These techniques are expected to prove especially useful for the retrospective diagnosis of viral infections in archived tissues and paraffin blocks. However, systematic examinations of archived materials have not yet been reported.

Development of in situ nucleic acid hybridization techniques has progressed significantly, especially for LCM virus (41), but application of these methods in the clinical setting has not been reported.

Diagnostic tools based on use of reverse transcriptase PCR (RT-PCR) to detect viral RNA in clinical materials have been evaluated in a small series of Lassa fever patient sera from West Africa (43). The sensitivity of PCR relative to conventional viral isolation was 0.82; the specificity was 0.68. PCR was positive longer during the disease course than isolation, owing perhaps to the greater sensitivity of the former in the presence of the early antibodies associated with acute Lassa fever. These results predict good success for detection of Lassa virus in tissue samples obtained at autopsy, but this has never been field tested. Likewise, various RT-PCR strategies for detecting Junin virus RNA in clinical materials have been devised (3, 29). Some of these strategies are reputed to be far more sensitive than conventional isolation procedures, especially in the presence of antibody; however, none of these procedures has yet been tested against conventional techniques for Junin virus infection diagnosis. RT-PCR strategies also exist for Guanarito virus, but they remain to be field tested. The Junin virus PCR is reported to detect less than 1 PFU of virus; alternatively, it can detect one infected cell in a population of 10^4 (29). The initial processing step entails guanidinium thiocyanate, which inactivates the virus. Sensitivity is enhanced by inclusion of an acid phenol-isoamyl

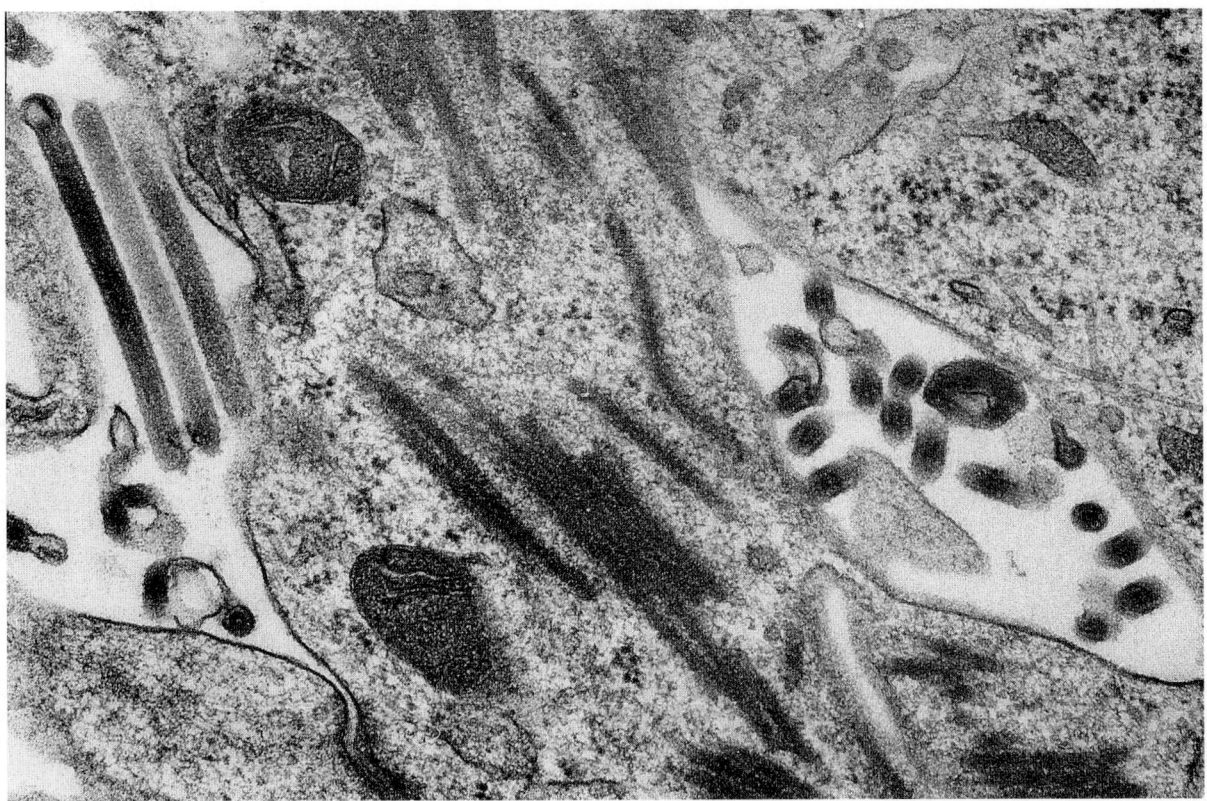

FIGURE 3 Thin section of infected MA-104 cells, with Ebola virions budding into intracytoplasmic vacuoles in longitudinal and cross-section and intracytoplasmic inclusions formed by nucleocapsids. Magnification, ×54,000. (Figure courtesy of T. W. Geisbert.)

alcohol extraction step that removes most of the hemoglobin derivatives; subsequent n-butanol extraction yields a clear, colorless pellet lacking inhibitors for *Taq* polymerase and reverse transcriptase. For Ebola virus, RT-PCR was applied to the detection of EBO-R in tissues and sera of infected primates during the recent outbreak with excellent sensitivity and correlation with conventional isolation techniques. Any method that promises to be a reliable substitute for procedures entailing manipulation of infectious BL-4 virus deserves attention.

ISOLATION OF VIRUS

The best general method currently available for isolation of MBG virus, Ebola virus, and the pathogenic arenaviruses is inoculation of appropriate cell cultures (usually Vero cells) followed by IFA or other immunologically specific testing of the inoculated cells for the presence of viral antigens at intervals. Especially for EBO-R and EBO-S, MA-104 cells (a fetal rhesus monkey kidney cell line) are more sensitive than Vero cells. Supernatant fluids should be collected for backtitration and confirmatory testing as detailed below. Primary isolations in cell cultures have been routinely employed for years to obtain Lassa, Machupo, Junin, and MGB viruses and EBO-Z from field-collected materials. Vero cell inoculation should also work for LCM virus, but intracranial (i.c.) inoculation of weanling mice is still regarded as the most sensitive established indicator of LCM (18). EBO-S has also been isolated by Vero cell inoculation but less reliably, since several blind passages are usually required

(31); SW-13 and MA-104 cells are more sensitive. Other cell lines, including human diploid lung (MRC-5) and BHK-21 cells, also work. Although Machupo and Junin viruses were historically isolated by i.c. inoculation of newborn hamsters and mice, respectively, Vero cells are approximately as sensitive and are far less cumbersome to manage in BL-4 containment. Furthermore, Vero cells permit isolation and identification, usually within 1 to 5 days, a significant advantage over animals, which require 7 to 20 days of incubation for illness to develop.

Clinical specimens and clarified tissue homogenates (usually 10%, wt/vol) are diluted in a suitable maintenance medium such as Eagle's minimal essential medium with Earle's salts and 2% heat-inactivated calf serum and adsorbed in small volumes to cell monolayers grown in suitable vessels such as tissue culture T-25 flasks or 60-mm petri dishes. It is important to test higher dilutions of these specimens as well as more-concentrated material, since autointerference (probably from defective interfering particles present in lower dilutions) may totally inhibit viral replication as measured by IFA testing, ELISA, or plaque formation. Following adsorption, sufficient maintenance medium is added to maintain the cells for 7 days. Inoculation of replicate vessels permits destructive testing of cells at frequent intervals. Cells inoculated with high-titer samples contain antigen within 1 to 2 days of inoculation. Cells inoculated with low-titer material may require up to 7 days to accumulate detectable antigen. If no antigen is detected by 7 days, the sample is considered negative, but supernatant fluids should be blind passaged to confirm the absence of virus. When EBO-S is suspected, the requirement for blind passage is anticipated and supernatant fluids are harvested and passed at 3- to 5-day intervals; cells are usually stained "real time" to accelerate the identification process. To confirm the presence and identity of the virus, supernatant fluids are tested for evidence of viral replication by plaquing or by antigen capture ELISA techniques. Cocultivation of Hypaque-Ficoll-separated peripheral blood leukocytes with susceptible cells has increased the isolation frequency of Junin virus (1). Cocultivation of lymphocytes from spleens of experimentally infected animals has yielded Lassa virus late in convalescence, even after neutralizing antibody has appeared. The technique merits systematic development for the remaining arenavirus and filovirus pathogens.

Although adequate cell culture systems exist for the isolation of LCM virus, most isolations to date have been obtained in mice. Weanling mice 3 to 4 weeks old are inoculated i.c. with undiluted samples. Care must be taken to use mice from a colony known to be free of LCM virus. Since the "high-dose" phenomenon of viral interference is occasionally a problem, a dilution of each sample (perhaps 1:100) is usually also inoculated into a second group of mice. Many LCM virus isolates produce a characteristic convulsive disease within 5 to 7 days that is nearly pathognomonic. Brains from dead mice may be used to prepare CF or ELISA antigens or may be stained by IFA to obtain presumptive identification. Clarified mouse brain may also be used as antigen for confirmatory testing by neutralization.

Animal inoculations are recommended for primary isolation of the other arenaviruses and of MBG and Ebola viruses only when cell cultures are not available and when adequate biocontainment facilities exist to maintain the animals. Newborn mice (1 to 3 days old) are highly susceptible to Junin virus inoculated i.c.; newborn hamsters are believed to be more sensitive to Machuco virus, although newborn mice have been used with success. Newborn hamsters and mice die 7 to 20 days after i.c. inoculation, usually exhibiting a characteristic convulsion after being spun by the tail. For the South American hemorrhagic fever viruses, particularly Junin virus, young adult guinea pigs inoculated either i.c. or peripherally have been used. Guinea pigs die 7 to 18 days after Junin virus inoculation. Most LCM virus strains are lethal for guinea pigs also. For Lassa virus, strain 13 guinea pigs are exquisitely sensitive to most Lassa virus strains and uniformly die 12 to 18 days after inoculation; outbred Hartley strain guinea pigs are somewhat less susceptible. The pathogenicity of virulent Lassa virus strains for outbred Swiss albino mice inoculated i.c. seems to vary with different sources; mice should not be seriously considered for Lassa virus isolations. MBG virus, EBO-Z, and EBO-S produce febrile responses in guinea pigs 4 to 10 days after inoculation; however, none of these viruses kills guinea pigs consistently on primary inoculation, and only EBO-Z has been adapted to uniform lethality by sequential guinea pig passages. EBO-Z is usually pathogenic for newborn mice inoculated i.c., but EBO-S, EBO-R, and MBG virus are not.

IDENTIFICATION OF VIRUS

Typing Antisera

Detection of viral antigens in infected tissue culture cells (usually Vero or MA-104) merits a presumptive diagnosis, provided the serologic reagents have been tested against all the reference arena- and filoviruses expected in a given laboratory, thus permitting an interpretation of viral cross-reactions. Virus isolates in cell culture supernatant fluids or tissue homogenates are presumptively or specifically identified by their reactivities with diagnostic antisera in various serologic tests (described below). Specific polyclonal antisera are prepared in adult guinea pigs, hamsters, rats, or mice inoculated intraperitoneally with infectious virus. Rhesus and cynomolgus monkeys that are convalescent from experimental infections are also reasonable sources for larger quantities of immune sera. Diagnostic antisera produced by a single injection of infectious virus are less cross-reactive and usually higher titered than those produced by multiple injections of inactivated antigens. To further reduce the induction of extraneous antibodies, inoculum virus should be derived from tissues or cells homologous to the species being immunized; likewise, the virus suspension should be stabilized with homologous serum or serum proteins. Sera produced for use in the CF, DFA, and IFA tests and in the ELISA should be collected 30 to 60 days after inoculation; sera for neutralization tests should be collected later. All sera must be rigorously tested for the presence of live virus before being removed from a maximum containment system.

Production and use of specific murine monoclonal antibodies with fine specificities for N and GP epitopes of LCM virus, Junin virus (40), and other arenaviruses have been reported, and monoclonal antibodies for Lassa virus, MBG virus, EBO-Z, EBO-S, and EBO-R with similar potentials exist. These have proven useful for defining epitopes and taxonomic relationships among these virus strains. One antibody, raised to a conserved sequence in the G2 region of LCM virus, reacts by ELISA and the IFA test with all arenavirus strains tested. This antibody might find utility as a panarenavirus detector in IFA and antigen capture ELISA procedures (45). Reference reagents for Lassa, Machuco, MBG, and Ebola viruses are not generally available; hyperimmune mouse ascitic fluids for LCM, Junin, and Tacaribe viruses are available from the National Institutes of Health, Research Resources Branch.

Immunofluorescence Procedures

To process infected cells for DFA examination and presumptive identification, inoculated cell monolayers are dispersed by trypsinization (0.05% trypsin with 0.02% EDTA for 10 min at 37°C), diluted in phosphate-buffered saline (PBS) containing 10% calf serum, and centrifuged at 400 × g for 10 min. The cell pellet is washed by resuspension in PBS, centrifugation, and resuspension in PBS to a final concentration of 10^6 cells per ml. Small drops (10 to 20 μl each) of cell suspensions are placed onto circular areas of specially prepared epoxy-coated slides that have been cleaned by immersion in ethanol and polished to remove residual oily deposits. These "spot slides" are air dried, fixed in acetone at room temperature for 10 min, and either stained immediately or stored frozen at −70°C. Although acetone fixation greatly reduces the number of infectious intracellular virus, spot slides prepared in this manner should still be considered infectious and handled accordingly. Recently, spot slides have been rendered noninfectious by gamma radiation (10) with no diminution in fluorescent-antigen intensity. Alternatively, infected cells may be biologically inactivated with β-propiolactone (35). Gamma radiation is recommended if the appropriate equipment is available.

For DFA tests, specific immune globulin prepared by ethanol or ammonium sulfate precipitation of immune sera followed by conjugation with fluorescein is diluted to a working concentration predetermined by box titration and flooded onto the infected cells, which are incubated at room temperature for 30 min in a moist chamber. Evans blue (0.005%) is sometimes added to the conjugate as a counterstain. After incubation, the slides are washed by immersion in PBS for two 10-min periods, dipped in water to remove salts, dried by evaporation, and mounted under coverslips in PBS-glycerol (pH 7.8). Specific viral fluorescence is characterized as intense, punctate to granular aggregates confined to the cytoplasms of infected cells. Specific MBG and Ebola virus fluorescence may include large, bizarrely shaped aggregates up to 10 μm across. Nonspecific fluorescence is rarely a problem in DFA procedures for these viruses. Detection of MBG, Ebola, Lassa, and LCM virus antigens by DFA testing is usually considered sufficient for a definitive diagnosis, although Lassa and LCM viruses cross-react at low levels in this test. Detection of Junin or Machupo virus antigens by DFA testing constitutes a presumptive diagnosis, since these viruses can be reliably distinguished from each other only by neutralization tests. IFA formats for viral detection are more cumbersome and cross-reactive, but they can be used if direct conjugates are unavailable. Immunohistochemical techniques for detecting arenaviruses (12) and filoviruses (20) by using a variety of chromogens can also be applied.

CF Test

The CF test was routinely used in early investigations for detection and presumptive identification of the arenaviruses and MBG virus. However, the CF test is rarely used now, since the development of reliable, simplified, and more sensitive immunofluorescence procedures described above. Detailed instructions for performing the CF test for arenaviruses are available (6), and the method can be applied to MBG and Ebola viruses as well. In brief, CF antigens are prepared from infected Vero cell culture supernatant fluids or from suckling mouse or suckling hamster brains extracted by sucrose-acetone. For reasons of safety, CF antigens are frequently inactivated by the addition of β-propiolactone and may be stored frozen at $-70°C$ indefinitely or preserved by lyophilization.

Antigen Capture ELISA

The recent development of an antigen capture ELISA for quantitative detection of arenavirus and filovirus antigens in viremic sera and tissue culture supernatants has facilitated early detection and identification of these agents (26, 35). These tests reliably detect antigens in samples inactivated by either β-propiolactone or irradiation; thus, they can be conducted safely without elaborate containment facilities. The threshold sensitivities for these assays are approximately 2.1 to 2.5 $\log_{10}$ PFU/ml, so they are sufficiently sensitive to detect antigen in most acute-phase viral hemorrhagic fevers viremias plus virus concentrations in throat washings and urine samples as well. The antigen capture ELISA developed for Ebola virus serves as a generic model for antigen capture ELISAs for both the arenavirus and the filovirus groups (26). The assay is a double-sandwich capture ELISA. Plates are coated overnight with a mixture of monoclonal antibodies and washed, and unknowns are added at fourfold dilutions in SerDil (5% [wt/vol] Bacto skim milk in 0.01 M PBS [pH 7.4] plus 0.1% Tween 20). Following the incubation and wash steps, poly-

clonal rabbit anti-Ebola virus serum is added, incubated, and washed, and then anti-rabbit immunoglobulin G (IgG) is added. Color development, by the horseradish peroxidase-ABTS (2,2'-azino-di-[3-ethylbenzthiazoline sulfonate] [6]) system, is proportional to the quantity of Ebola virus antigen captured on the plate. Samples are considered positive if the optical density at 410 nm exceeds the mean plus 3 standard deviations for the normal controls. In the EBO-R epizootic setting, very close concordance was obtained between conventional isolation and antigen capture ELISA techniques. The suitability of this test for clinical materials from human patients infected with EBO-Z or EBO-S is predicted to be good but awaits systematic testing of field-collected materials. Similar formats have been devised and tested for Lassa (35), Junin, and LCM viruses. For Lassa virus, good correlation with viral isolation was found for serum samples collected in West Africa, provided the infectious virus titers exceeded 2.1 $\log_{10}$ PFU/ml. Substitution of a monoclonal antibody of high avidity and the appropriate specificities for polyclonal sera generally increases the sensitivities and specificities of these antigen capture ELISAs. However, use of monoclonal antibodies with broad cross-reactivities (45) is potentially useful for development of a general panarenavirus detector system.

Neutralization Tests

Neutralization tests for these viruses take many forms; depending on the virus, neutralization tests range from extremely sensitive and reliable (e.g., those for Junin and Machupo viruses) and moderately insensitive but reliable (those for Lassa and LCM viruses) to totally unreliable (those for MBG and Ebola viruses). The common denominator in all neutralization tests is measurement of inhibition of viral replication by reaction with immune serum. Thus, the form of the test is determined in part by the availability of tools to measure infectious virus or viral antigens. The reasons for the poor neutralizing-antibody responses elicited by the filoviruses are thought to relate to a high degree of glycosylation and to the presence of a possibly immunosuppressive epitope in the GP (membrane) protein.

For the New World arenaviruses (Junin and Machupo viruses), the most generally applied neutralization test is a plaque reduction test that uses Vero cells and the serum dilution-constant virus format. The serum dilution calculated (by probit analysis) to reduce the control number of plaques by 50% is usually taken as the endpoint, although in some laboratories the highest serum dilution producing 80% reduction is used. The 80% reduction test is commonly used to distinguish Junin from Machupo virus. For Lassa and LCM viruses, neutralizing-antibody activity is rapidly lost on dilution; for this reason, the constant serum-varying virus dilution format is preferred. Neutralization of Old World arenaviruses is also markedly enhanced by addition of complement; thus, 10% fresh guinea pig serum is routinely added to the diluent. Plaque reduction tests in various formats have largely replaced animal protection tests, which were notoriously imprecise for measurement of arenavirus neutralizing antibodies.

SEROLOGIC DIAGNOSIS

IFA Test

The IFA test is clearly the method of choice for documenting recent infections with MBG and Ebola viruses and the arenaviruses. Preparation of spot slides with infected Vero cells is identical to the procedure described above. Some refinements to enhance reproducibility and quality between spot slide lots have been suggested (13). Refinements include gamma irradiation of infected cell suspensions prior to distribution in 1-ml aliquots adjusted to a concentration of 12×10^6 cells per ml and frozen. To prepare new slides, an aliquot is thawed and diluted with 12 ml of saline to net 75 to 100 slides. Uninfected cells may also be admixed with the virus-infected cells to aid discrimination between specific and nonspecific fluorescence. Although monovalent spot slides are usually desired and are prepared with cells optimally infected with a single virus, polyvalent spot slides can also be prepared by mixing cells infected with other viruses selected from these or other taxonomic groups that have similar geographic distributions (23). Test sera are diluted serially, usually in twofold or fourfold increments starting at 1:10 or 1:16. Infected cells (and uninfected control cells) are incubated with serum dilutions, washed, reincubated with appropriate fluorescein-conjugated antiglobulin (or specific anti-IgM or -IgG), washed, mounted, and observed. Endpoint determination is very subjective. Most experienced observers consider the endpoint to be the highest dilution producing typical cytoplasmic fluorescence that is clearly positive relative to that in uninfected cells. Although it is possible to obtain reproducible endpoints within individual laboratories, discrepancies in titers determined by different laboratories are common and probably relate to variations in interpretation, epi-illumination intensity, filtration systems, and fluorescein conjugates. There is general agreement that use of oil immersion objectives increases background, leading to lower apparent endpoints. For Ebola virus antibodies, backgrounds may be sufficiently high that 1:64 is adopted as the cutoff titer (13). Antibodies measured by the IFA test are usually the first to appear, often becoming detectable within the first few days of hospitalization for Lassa virus, within 10 days of onset for MBG and Ebola viruses, and somewhat later for Junin and Machupo viruses. The presence of specific IgM antibodies or a rising IFA titer constitutes a presumptive diagnosis of acute infection (46). IgM antibodies measured by IFA testing decline to undetectable titers within several months, while IgG antibodies, which are thought to compete with IgM in IFA tests, persist for at least several years.

CF Test

The CF test was used initially to classify the arenaviruses and to detect seroconversions (6). However, the CF test is rarely used now, because it is inferior to the IFA test for detecting both arenavirus and filovirus infections. CF antibody titers evolve more slowly (3 to 4 weeks after onset) and recede rapidly to undetectable titers (usually within 1 year). The CF test is less specific than the IFA test, and anticomplementary sera are frequently encountered.

Neutralization Tests

As described above, reliable tests for measuring neutralizing antibody to MBG and Ebola viruses are not available. For the arenaviruses, plaque reduction tests with Vero cells are generally used. For measuring neutralizing antibody to Lassa and LCM viruses, which are difficult to neutralize and are poor inducers of this antibody, test sera are diluted, usually 1:10, in medium containing 10% guinea pig serum as a complement source and mixed with serial dilutions of challenge virus. Titers are expressed as a $\log_{10}$ neutralization index, defined as ($\log_{10}$ PFU in control) − ($\log_{10}$ PFU in

test serum). For Junin and Machupo viruses the more conventional serum dilution-constant virus format is usually used, although the constant serum-virus dilution format is equally useful for distinguishing among strains. Neutralizing-antibody responses require weeks to months to evolve but persist for years. Performance of these tests is restricted to laboratories equipped to handle infectious viruses.

ELISA for Detection of IgG and IgM Antibodies

ELISA procedures for detecting Lassa virus-specific IgG and IgM have been developed (35) and successfully employed with field-collected human sera. When this ELISA is used in combination with the Lassa virus antigen capture ELISA described above, virtually all Lassa fever patients can be specifically diagnosed within hours of hospital admission. As is true of the antigen capture ELISA, success of the antibody ELISA is critically dependent on highly avid, purified capture antibodies or globulins. A simplification of this procedure has been developed for Ebola virus. It entails use of infected Vero cell lysate diluted in SerDil and adsorbed directly to the microtiter plate wells as antigen. Test sera are initially serially diluted 1:16 and incubated with antigen in a format analogous to that of the antigen capture ELISA described above. Following incubation, washing, and addition of rabbit anti-human serum, color development is measured by the horseradish peroxidase-ABTS system. Samples are considered positive if the optical density at 410 nm exceeds the mean plus 3 standard deviations for normal serum controls. This procedure can be further modified to detect virus-specific IgM by coating the plates with anti-human IgM followed by test serum dilutions and cell lysate antigens and then following the antigen capture protocol. These developmental procedures have worked well with animals experimentally infected with Lassa and Ebola viruses, but for Junin virus, the IgM ELISA has been less successful. All of these promising assays are sufficiently developed to warrant field testing. If successful, IgG and IgM ELISAs will probably replace the more subjective IFA tests as the serologic tests of choice.

WB (Immunoblot)

Western blots (WB) are feasible for demonstrating antibodies to arenaviruses and filoviruses. However, WB has never been applied systematically or routinely to diagnosis. Recently, WB was proposed as a confirmatory test to supplement the IFA test for filovirus antibodies (13). Detection of the nucleocapsid (N) band plus either VP30 or VP24 was taken as diagnostic. The WB procedure was further refined by miniaturization, using the Phast system SDS-PAGE and transblot apparatus (Phast Western Blot), or PWB system. The 10 to 15% gradient gels provided the best results, and the optimum amount of protein loaded per lane was 50 ng. A dilution of 1:200 ensured detection of all positives and an absence of background staining. In the PWB system, VP30 and VP24 were rarely seen; thus, a sole band (N) was taken as positive. When skim milk (0.5%) was added to the serum diluent to reduce background, specific reaction with N alone correlated closely with IFA test positivity. All Ebola virus-positive sera tested from earlier outbreaks and experimentally infected primates were positive in this test. There remain problematic sera that test positive in the IFA test (at 1:64 or greater) and negative by PWB, raising the suspicion that other, serologically related but distinct filoviruses may be circulating.

Other Serologic Tests

Other serologic tests have been applied to diagnosis. Gel diffusion tests have been employed for arenaviruses, but those tests detect antibodies directed primarily against nucleocapsid and are less sensitive than the IFA test or even the CF test; thus, gel diffusion tests have little role in modern diagnosis. Another test developed for Lassa virus (and antibodies) is a reversed passive hemagglutination (and inhibition) test employing Lassa virus antibody-coated erythrocytes, which agglutinate in the presence of viral antigen (17). The test has not gained widespread acceptance, perhaps because meticulous care is required to obtain satisfactory antibody-erythrocyte conjugates. For Ebola virus, a radioimmunoassay that uses ^{125}I-labeled staphylococcal protein A was successfully used on many human and animal sera; it discriminates EBO-Z from EBO-S strains (37). However, it is considered too labor-intensive for general use in the diagnostic laboratory. The PWB described above holds promise as a confirmatory test for the filovirus IFA test and may be adapted to the arenaviruses as well.

EVALUATION AND INTERPRETATION OF RESULTS

Early diagnosis of arenavirus and filovirus infections is desirable, since specific immune plasma and appropriately selected antiviral drugs are often effective when treatment is initiated soon after onset. Early recognition of these infections should also trigger strict isolation procedures to prevent spread of the disease to patient contacts. In areas where specific viruses are endemic, the index of suspicion is often high, and experienced clinicians may be remarkably accurate in rendering a diagnosis of fully developed cases on clinical grounds alone. Yet even in these areas, specific virologic and serologic tests are required to confirm clinical impressions, since many other diseases, including malaria, typhoid, rickettsiosis, idiopathic thrombocytopenia, and viral hepatitis, may masquerade as an arena- or filovirus infection. While the availability of inactivated antigen spot slides for IFA testing in field hospitals has facilitated diagnosis based on seroconversion determined by IFA testing, timely diagnosis requires a means of detecting infectious virus or antigen in the field. The antigen capture ELISA holds promise of detecting clinically relevant concentrations of virus in β-propiolactone-inactivated sera, body fluids, and tissues, as is convincingly demonstrated in EBO-R-infected primates (20, 26).

PCR assays may eventually eliminate almost all need to isolate infectious virus to establish definitive diagnosis and thus will reduce the need for BL-4 biocontainment. In the interim, however, detection of viral antigens by DFA testing with Vero cells inoculated with patient specimens will continue to be important in establishing a definitive diagnosis. Inoculation of tissue cultures for DFA examination and isolation of these viruses should not be done outside a maximum containment laboratory except with LCM virus, which may be handled at a lower containment level (9). Identification by DFA testing yields a definitive diagnosis except for Junin and Machupo viruses, which are reliably discriminated from each other only by the neutralization test or the more sophisticated tools of molecular virology. Quantitative cross-testing by DFA can discriminate among other virus strains that cross-react in the DFA test such as Lassa virus with LCM virus and EBO-Z with EBO-S. For

virus isolates originating from areas where the geographic distributions of related viruses overlap, it is essential that quantitative DFA testing be done and desirable that identification be confirmed by neutralization tests when available.

While interpretation of serologic data is usually facilitated by the generally restricted geographic ranges of these viruses, ranges do overlap, and IFA and CF data are occasionally ambiguous. Investigations spanning several decades (2) have shown the simultaneous development of CF antibodies against Junin and LCM virus antigens in patients in the zone of AHF endemicity clinically diagnosed as cases of AHF, including some from whose acute-phase blood Junin virus was isolated. This fact creates serious difficulties in reaching a definite diagnosis not heretofore encountered with other arenavirus infections of humans; as a result, epidemiologic evaluation of data may be faulty. The geographic distributions of Junin and Machupo viruses certainly overlap that of LCM virus in South America. In Africa, the distribution of Lassa virus may overlap those of newly isolated virus strains from rodents in Zimbabwe, Mozambique, and Central African Republic, which cross-react strongly by IFA with Lassa virus and, to a lesser extent, with LCM virus. While these strains are not known to be associated with human disease, their presence may confuse the interpretation of serologic data from African surveys. The extent to which heterologous arenavirus infection and/or reinfection broadens antibody specificity has not been systematically evaluated for any of the available serologic tests. For Ebola virus, IFA tests have been sensitive and specific in detecting seroconversions and acutely convalescent individuals in the midst of an outbreak. However, the IFA test yielded misleading results when applied to population-based serosurveys; a significant proportion of human and primate sera are reactive by IFA even when there is no clinical or epidemiologic evidence of previous infection with known filoviruses. Despite these potential problems, the experience to date has been that in the midst of outbreaks caused by MBG or Ebola virus or the arenaviruses, identifications of the etiologic agents by DFA and IFA tests have been clear and unambiguous, especially when the diagnoses were confirmed by neutralization tests.

Because of the biohazard, virus isolation data for these viruses are usually available only retrospectively. MBG virus and EBO-Z are usually isolated from acute-phase sera, while EBO-S is isolated less often, perhaps because of the need for blind passage. In the past, opportunities to isolate some of these viruses may have been missed because newborn mice were used as the primary detection system. Lassa virus is usually recovered from acute-phase serum samples taken from hospitalized patients soon after admission, and it frequently occurs in the presence of specific IgM antibody. Junin, Machupo, and LCM viruses are recovered less frequently, and diagnosis is usually based on seroconversion. The IFA and ELISA responses are the earliest for all these viruses, being detectable 7 to 10 days after onset for Lassa and LCM viruses, 10 to 14 days for MBG and Ebola viruses, 12 to 17 days for Junin virus, and 17 to 30 days for Machupo virus. The CF antibodies evolve several days after the IFA response. The presence of specific IgM antibodies detected by IFA testing is indicative of recent infection, since IgM IFA or ELISA titers persist for less than 3 months. The presence of specific IgM IFA titers in the cerebrospinal fluid of an LCM patient constitutes a definitive diagnosis. For all the arenavirus and filovirus pathogens, a rising IgM or IgG titer constitutes a strong presumptive diagnosis. Since IgM titers, like CF titers, do not persist long, a decreasing titer suggests a recent infection that occurred perhaps several months previously.

For Lassa fever patients, a detectable IFA or ELISA response does not necessarily signal imminent recovery; viremia frequently persists, and patients die after an IFA response. For Junin, LCM, MBG, and Ebola virus infections, the appearance of antibodies detectable by IFA testing coincides with the disappearance of viremia and with recovery. In Machupo virus infection, IFA titers appear even later, 1 week or more after the crisis has passed. For the arenaviruses, neutralizing antibodies appear much later in convalescence than IFA or CF antibodies. Reliable neutralizing-antibody data for MBG and Ebola virus infections are not currently available. Neutralizing antibodies against arenaviruses persist for long periods, perhaps for life, and thus provide the most reliable basis for determining the minimum resistance of a population to reinfection. The role of neutralizing antibody in acute recovery is less clear. The protective efficacy of passively administered immune plasma is believed to be a function of neutralizing-antibody titers, and selection of plasma should be on this basis, especially for Lassa fever, since protective efficacy is predicted by neutralizing antibody titers and not by IFA testing (21). A serologic test to predict the efficacy of MBG and Ebola virus-immune plasmas is urgently needed. Among the group-specific antibody tests for arenaviruses, the IFA test gained practical acceptance over the CF test more than a decade ago. Now, however, ELISAs are widely regarded as providing a more objective method than fluorescence. IgM capture ELISA and antigen capture ELISA are rapidly becoming routine.

The highest priority for future development is refinement of the available diagnostic tools to permit definitive virus identifications in the field. PCR-based assays will add another dimension to the capability of field laboratories to diagnose acute disease almost in real time. Proper tailoring of primers should permit the design of tests with the proper degree of specificity. The recent emergence of two new arenaviral pathogens, Guanarito and Sabia viruses, as well as of EBO-R serves as a reminder that broadly reactive grouping reagents are still required to augment the newly evolving tools of PCR and capture ELISAs based on extremely specific monoclonal antibodies and gene sequences. An investment in rapid diagnosis should result in more timely intervention with effective treatment regimens and, through implementation of appropriate public health measures, may reduce dissemination of these highly virulent viral pathogens.

REFERENCES

1. **Ambrosio, A. M., D. A. Enria, and J. I. Maiztegui.** 1986. Junin virus isolation from lympho-mononuclear cells of patients with Argentine hemorrhagic fever. *Intervirology* **25:**97–102.
2. **Ambrosio, A. M., M. R. Feuillade, G. S. Gamboa, and J. I. Maiztegui.** 1994. Prevalence of lymphocytic choriomeningitis virus infection in a human population in Argentina. *Am. J. Trop. Med. Hyg.* **50:**381–386.
3. **Bochstahler, F. E., P. G. Carney, G. Bushar, and J. L. Sagripanti.** 1992. Detection of Junin virus by the polymerase chain reaction, *J. Virol. Methods* **39:**231–235.
4. **Buchmeier, M. J., H. A. Lewicki, O. Tomori, and M. B. A. Oldstone.** 1981. Monoclonal antibodies to lymphocytic choriomeningitis and Pichinde viruses: generation, characterization, and cross-reactivity with other arenaviruses. *Virology* **113:**73–85.
5. **Buchmeier, M. J., and B. S. Parekh.** 1987. Protein structure and expression among arenaviruses. *Curr. Top. Microbiol. Immunol.* **133:**41–58.
6. **Casals, J.** 1977. Serologic reactions with arenaviruses. *Medicina* (Buenos Aires) **37**(Suppl. 3):59–68.
7. **Centers for Disease Control.** 1980. Interstate shipment of etiologic agents. *Fed. Regist.* **45:**48626–48629.
8. **Centers for Disease Control.** 1988. Management of patients with suspected viral hemorrhagic fever. *Morbid. Mortal. Weekly Rep.* **37**(Suppl. 3):1–16.
9. **Centers for Disease Control and National Institutes of Health.** 1993. *Biosafety in Microbiology and Biomedical Laboratories.* HHS publication no. (CDC) 93-8395. U.S. Government Printing Office, Washington, D.C.
10. **Clegg, J. C. S.** 1993. Molecular phylogeny of the arenaviruses and guide to published sequence data, p. 175–187. *In* M. S. Salvato (ed.), *The Arenaviridae.* Plenum Press, New York.
11. **Coimbra, T. L. M., E. S. Nassar, M. N. Burattini, L. T. M. De Souza, I. B. Ferreira, I. M. Rocco, A. De Rosa, P. Vasconcelos, F. P. Pinheiro, J. W. LeDuc, R. Rico-Hesse, J. Gonzalez, P. B. Jahrling, and R. B. Tesh.** 1994. New arenavirus isolated in Brazil. *Lancet* **343:**391–392.
12. **Connolly, B. M., A. B. Jenson, C. J. Peters, S. J. Geyer, J. F. Barth, and R. A. McPherson.** 1993. Pathogenesis of Pichinde virus infection in strain 13 guinea pigs: an immunocytochemical, virologic, and clinical chemistry study. *Am. J. Trop. Med. Hyg.* **49:**10–23.
13. **Elliott, L. H., S. P. Bauer, G. Perez-Oronoz, and E. S. Lloyd.** 1993. Improved specificity of testing methods for filovirus antibodies. *J. Virol. Methods* **43:**85–100.
14. **Elliott, L. H., J. B. McCormick, and K. M. Johnson.** 1982. Inactivation of Lassa, Marburg, and Ebola viruses by gamma irradiation. *J. Clin. Microbiol.* **16:**704–708.
15. **Geisbert, T. W., and P. B. Jahrling.** 1990. Use of immunoelectron microscopy to show Ebola virus during the 1989 United States epizootic. *J. Clin. Pathol.* **43:**813–816.
16. **Geisbert, T. W., P. B. Jahrling, M. A. Hanes, and P. M. Zack.** 1992. Association of Ebola-related Reston virus particles and antigen with tissue lesions of monkeys imported to the United States. *J. Comp. Pathol.* **106:**137–152.
17. **Goldwasser, R. A., L. H. Elliott, and K. M. Johnson.** 1980. Preparation and use of erythrocyte-globulin conjugates to Lassa virus in reversed passive hemagglutination and inhibition. *J. Clin. Microbiol.* **11:**593–599.
18. **Hotchin, J., and E. Sikora.** 1975. Laboratory diagnosis of lymphocytic choriomeningitis. *Bull. W.H.O.* **52:**555–558.
19. **International Committee on Taxonomy of Viruses.** 1991. Classification and nomenclature of viruses. Fifth report of the International Committee on Taxomy of Viruses. *Arch. Virol.* Suppl. 2:247–249.
20. **Jahrling, P. B., T. W. Geisbert, D. W. Dalgard, E. D. Johnson, T. G. Ksiazek, W. C. Hall, and C. J. Peters.** 1990. Preliminary report: isolation of Ebola virus from monkeys imported to USA. *Lancet* **335:**502–505.
21. **Jahrling, P. B., and C. J. Peters.** 1984. Passive antibody therapy of Lassa fever in cynomolgus monkeys. *Infect. Immun.* **44:**528–533.
22. **Jahrling, P. B., and C. J. Peters.** 1992. Lymphocytic choriomeningitis virus: a neglected pathogen of man. *Arch. Pathol. Lab. Med.* **116:**486–488.
23. **Johnson, K. M., L. H. Elliott, and D. L. Heymann.** 1981. Preparation of polyvalent viral immunofluorescent intracellular antigens and use in human serosurveys. *J. Clin. Microbiol.* **14:**527–529.
24. **Kiley, M. P., E. T. W. Bowen, G. A. Eddy, M. Isaacson, K. M. Johnson, J. B. McCormick, F. A. Murphy, S. R. Pattyn, D. Peters, O. W. Prozesky, R. L. Regnery, D. I. H. Simpson, W. Slenczka, P. Sureau, G. Van der Groen, P. A. Webb, and H. Wulff.** 1982. Filoviridae: taxonomic home for Marburg and Ebola viruses? *Intervirology* **18:**24–32.
25. **Kiley, M. P., N. J. Cox, L. H. Elliott, A. Sanchez, R. Defries, M. J. Buchmeier, D. D. Richman, and J. B. McCormick.** 1988. Physiochemical properties of Marburg virus:

evidence for three distinct virus strains and their relationship to Ebola virus. *J. Gen. Virol.* **69:**1957–1967.

26. **Ksiazek, T. G., P. E. Rollin, P. B. Jahrling, E. Johnson, D. W. Dalgard, and C. J. Peters.** 1992. Enzyme immunosorbent assay for Ebola virus antigens in tissues of infected primates. *J. Clin. Microbiol.* **30:**947–950.

27. **Lederberg, J., R. E. Shope, and S. C. Oals, Jr. (ed.).** 1992. *Emerging Infections. Microbial Threats to Health in the United States.* National Academy Press, Washington, D.C.

28. **Lehmann-Grube, F.** 1971. Lymphocytic choriomeningitis virus. *Virol. Monogr.* **10:**1–173.

29. **Lozana, M. E., P. D. Ghiringhelli, V. Romanowski, and O. Grau.** 1993. A simple nucleic acid amplification assay for the rapid detection of Junin virus in whole blood samples. *Virus Res.* **27:**37–53.

30. **Martini, G. A.** 1971. Marburg virus disease. Clinical syndrome, p. 1–9. *In* G. A. Martini and R. Siegert (ed.), *Marburg Virus Disease.* Springer-Verlag, New York.

31. **McCormick, J. B., S. P. Bauer, L. H. Elliott, P. A. Webb, and K. M. Johnson.** 1983. Biologic differences between strains of Ebola virus for Zaire and Sudan. *J. Infect. Dis.* **147:**264–267.

32. **Monson, M. H., A. K. Cole, J. D. Frame, J. R. Serwint, S. Alexander, and P. B. Jahrling.** 1987. Pediatric Lassa fever: a review of 33 Liberian cases. *Am. J. Trop. Med. Hyg.* **36:**408–415.

33. **Murphy, F. A., G. Van der Groen, S. G. Whitfield, and J. V. Lange.** 1978. Ebola and Marburg virus morphology and taxonomy, p. 61–84. *In* S. R. Pattyn (ed.), *Ebola Virus Haemorrhagic Fever.* Elsevier/North Holland Biomedical Press, Amsterdam.

34. **Murphy, F. A., and S. G. Whitfield.** 1975. Morphology and morphogenesis of arenaviruses. *Bull. W.H.O.* **52:**409–419.

35. **Niklasson, B. S., P. B. Jahrling, and C. J. Peters.** 1984. Detection of Lassa virus antigens and Lassa-specific IgG and IgM by enzyme-linked immunosorbent assay. *J. Clin. Microbiol.* **20:**239–244.

36. **Peters, C. J., A. Sanchez, H. Feldmann, P. E. Rollin, S. Nichol, and T. G. Ksiazek.** 1994. Filoviruses as emerging pathogens. *Semin. Virol.* **5:**147–154.

37. **Richman, D. D., P. H. Cleveland, J. B. McCormick, and K. M. Johnson.** 1983. Antigenic analysis of strains of Ebola virus: identification of two Ebola virus serotypes. *J. Infect. Dis.* **147:**268–271.

38. **Salas, R., N. De Manzione, R. B. Tesh, R. Rico-Hesse, R. E. Shope, A. Betancourt, O. Godoy, R. Bruzual, M. E. Pacheco, B. Ramos, M. E. Taibo, J. G. Tamayo, R. Jaimes, C. Vasquez, F. Araoz, and J. Querales.** 1991. Venezuelan haemorrhagic fever. *Lancet* **338:**1033–1036.

39. **Sanchez, A., M. P. Kiley, B. P. Holloway, and D. D. Auperin.** 1993. Sequence analysis of the Ebola virua genome: organization, genetic elements, and comparison with the genome of Marburg virus. *Virus Res.* **29:**215–240.

40. **Sanchez, A., D. Y. Pifat, R. H. Kenyon, C. J. Peters, J. B. McCormick, and M. P. Kiley.** 1989. Junin virus monoclonal antibodies: characterization and cross-reactivity with other arenaviruses. *J. Gen. Virol.* **70:**1125–1132.

41. **Southern, P. J., and M. B. A. Oldstone.** 1986. Molecular anatomy of viral infection: study of viral nucleic acid sequences and proteins in whole body sections, p. 147–156. *In* M. B. A. Oldstone (ed.), *Concepts in Viral Pathogenesis II.* Springer-Verlag, New York.

42. **Tesh, R. B., M. L. Wilson, R. Salas, N. M. C. De Manzione, D. Tovar, T. G. Ksiazek, and C. J. Peters.** 1993. Field studies on the epidemiology of Venezuelan hemorrhagic fever. *Am. J. Trop. Med. Hyg.* **49:**227–235.

43. **Trappier, S. G., A. L. Conaty, B. B. Farrar, D. D. Auperin, D. D., J. B. McCormick, and S. P. Fisher-Hoch.** 1993. Evaluation of the polymerase chain reaction for diagnosis of Lassa virus infection. *Am. J. Trop. Med. Hyg.* **49:**214–221.

44. **Van der Groen, G., and L. H. Elliott.** 1982. Use of beta-propiolactone-inactivated Ebola, Marburg, and Lassa intracellular antigens in immunofluorescent antibody assay. *Ann. Soc. Belg. Med. Trop.* **62:**49–54.

45. **Weber, E. L., and M. J. Buchmeier.** 1988. Fine mapping of a peptide sequence containing an antigenic site conserved among arenaviruses. *Virology* **164:**30–38.

46. **Wulff, H., and K. M. Johnson.** 1979. Immunoglobulin M and G responses measured by immunofluorescence in patients with Lassa or Marburg virus infections. *Bull. W.H.O.* **57:**631–635.

47. **Wulff, H., J. V. Lange, and P. A. Webb.** 1978. Interrelationships among arenaviruses measured by indirect immunofluorescence. *Intervirology* **9:**344–350.

Human Papillomavirus

NANCY B. KIVIAT AND LAURA A. KOUTSKY

95

STRUCTURE AND NOMENCLATURE

Papillomaviruses are classified in the family *Papovaviridae*. These viruses have naked icosahedral capsids composed of 72 capsomeres 45 to 55 nm in diameter with a double-stranded supercoiled circular DNA genome of approximately 8,000 bp (10). Within the genus *Papillomavirus*, several species-specific papillomaviruses have been identified, including among others those that infect humans (human papillomaviruses [HPV]), cows, deer, dogs, monkeys, rabbits, and horses (5, 23). Furthermore, over the last decade, it has become clear that there are many different types of HPV. Two strains of HPV are classified as being the same or different types depending on the amount of DNA homology they share as demonstrated by reassociation kinetics. If two HPV strains demonstrate less than 50% DNA homology, they are classified as distinct types. If DNA homology with another virus is more than 50%, but less than 100%, then the two viruses are considered subtypes or variants of the same HPV type (10). Over 70 different types of HPV have been identified from clinical specimens collected throughout the world (35) (Table 1). As more and more clinical samples are analyzed, it is becoming apparent that many subtypes also exist for each type of HPV (8, 65, 67).

Detailed descriptions of the organization of the viral genome have appeared elsewhere (5). Briefly, the papillomavirus genome consists of an upstream regulatory region; an early region, which encodes protein necessary for viral DNA replication and transformation; and a late region, which encodes viral structural proteins, including the major capsid proteins necessary for productive viral replication. The early regions include the open reading frames E6 and E7, which are important in transformation of healthy cells to malignant ones. The E6 and E7 open reading frames encode proteins that appear to interfere with the activity of normal cellular tumor suppressor proteins. E6 binds to and increases the degradation of p53, while E7 binds to and antagonizes the function of the retinoblastoma tumor suppressor protein (pRB) (5, 19, 42). In nonmalignant lesions, HPV is episomal, while it appears to be integrated into the host cell genome in most invasive cancers associated with this viral infection (35).

PATHOGENESIS, NATURAL HISTORY, AND TRANSMISSION

Replication of HPV is tightly linked to squamous epithelial cell differentiation, with viral capsids being produced only in terminally differentiated squamous cells. For this reason, traditional techniques for culturing viruses cannot be used for culturing HPV. Difficulties in propagating the virus have hindered the development of a sensitive and type-specific serologic test for detecting infection by specific types of HPV. Because of the problem of establishing current or past infection with HPV, we know relatively little about the natural history and transmission of these infections.

Transmission of cutaneous types of HPV is by direct contact or by fomites, and transmission of genital types is thought to be primarily through sexual contact (31). Several investigators have reported that approximately two-thirds of the partners of women with clinically diagnosed warts also have or develop such pathology within several months (46). Transmission from mothers to children during childbirth rarely results in laryngeal papillomatosis (12).

HPV infects stratified squamous, metaplastic squamous, and columnar epithelial cells. The virus appears to gain access to the lower portion of the epithelium in areas of local trauma and then infects basal epithelial cells (26, 59, 68). Local trauma is also associated with development of new lesions. For example, laryngeal warts frequently develop near tracheotomy sites, and genital warts develop at the edges of areas that have undergone laser ablation (50, 54). Infection frequently results in focal areas of hyperplasia in all layers of the epithelium. The majority of HPV infections are self-limited, but it is now clear that malignant tumors develop in a subset of infections that involve oncogenic viruses (43). It is well established that specific types of HPV play a central role in the pathogenesis of most cervical cancers. Furthermore, a number of cofactors (such as smoking or birth control methods) for genital tract cancer have been identified, but the exact relationship between HPV infection, specific cofactors, and cancer is not well understood (4).

CLINICAL PRESENTATION AND ANATOMIC DISTRIBUTION

Specific types of HPV tend to infect different types of epithelia and produce different clinical and pathologic

TABLE 1 Common types of HPV, sites of infection, clinical manifestations, and associations with invasive cancers

HPV type(s)	Site(s) infected	Clinical manifestation	Association with malignancy
Cutaneous			
1	Soles of feet	Deep plantar warts	None
2–4	Hands, arms	Common warts	None
7	Hands, arms	Butcher's warts	None
5, 8, 9, 12, 14, 15, 17, 19–25, 36–38	Forehead, arms, trunk	Flat macular warts and neoplastic lesions, primarily in patients with epidermodysplasia verruciformis	Over 30% of epidermodysplasia verruciformis patients with types 5, 8, 14, 17, and 20 develop malignancy.
26–29, 34	Forehead, arms, trunk	Common flat warts	Frequent, especially in immunosuppressed patients; Bowen's disease
Mucosal			
6, 11	Anogenital epithelium, larynx, upper respiratory tract, conjunctiva, oral cavity	Papillomatosis primarily laryngeal, also in upper respiratory tract; condylomata; intraepithelial neoplasia	Rare; Buschke-Löwenstein tumors
42–44	Anogenital epithelium	Condylomata; intraepithelial neoplasia	Rare
31, 33, 35, 51, 52	Anogenital epithelium	Condylomata; intraepithelial neoplasia	Occasionally
16, 18, 45, 56	Anogenital epithelium	Condylomata; intraepithelial neoplasia	Most frequently of all types

manifestations (Table 1). This tissue tropism is not completely site specific; for example, HPV types 6 and 11 infect both the genital tract and the larynx. Depending on the type of epithelium infected, HPV types are often referred to as cutaneous or mucosal. In general, cutaneous types infect keratinizing epithelium, while mucosal HPVs infect non-keratinizing epithelium. Some types of HPV that infect the nongenital or keratinizing epithelium are found almost exclusively among patients with epidermodysplasia verruciformis or with other forms of immunosuppression (35).

Cutaneous HPV Infection

Infection with HPV types 1 through 4 is thought to occur almost universally during childhood or adolescence and is generally self-limited without consequence. In patients with epidermodysplasia verruciformis, a rare autosomal dominant genetic disorder of cellular immunity, or in patients with iatrogenic immunosuppression, cutaneous infection with many HPV types is commonly associated with malignancy. At least 30% of patients with epidermodysplasia verruciformis and many patients with renal transplants have developed HPV-related cancers of the skin, especially in areas exposed to the sun (23, 52, 68). An unusual form of cutaneous HPV infection usually associated with HPV type 7 also occurs in meat handlers (40).

Mucosal HPV Infection

Mucosal HPV primarily infects the anogenital tract epithelium, but these HPV types can also be found in the oral mucosa, conjunctiva, and respiratory tract. Genital tract HPV infection is thought to be the most common sexually transmitted viral infection in the United States (31). Some infections can be easily identified by the clinician, others require microscopy, while others can be detected only through the use of molecular hybridization assays. The prevalence of HPV infection, whether it is diagnosed clinically by microscopy or by molecular assays, varies significantly with age, gender, and sexual behavior of the population (13).

Genital warts occur on the vulva, vagina, cervix, penis, and anus. Descriptions of genital warts can be found even in ancient Greek writings (46); however, only recently has it become clear that several different sexually transmitted viruses are responsible for these lesions. In addition, over the last 10 years, we have learned that the flat lesions seen by gynecologists with the aid of a colposcope after application of acetic acid are related to the typical genital warts seen on the vulva and penis. Women and men with clinically visible papillary warts frequently also have such "flat warts" present. As was seen with several other sexually transmitted diseases (STDs), there appears to have been a marked increase in the age-adjusted incidence of genital warts between the 1950s and 1980s. Some suggest that the incidence of genital warts increased 2.5- to 8-fold during this period (9). In the early 1990s, it was estimated that about 1% of men and women in the United States between 15 and 49 years of age had clinically evident genital warts (diagnosed with the unaided eye) (13).

Association with Malignancy

In addition to being a common STD, genital HPV infection is of considerable importance because specific types of genital HPVs play a major role in the pathogenesis of epithelial cancers of the female and male genital tracts (31, 35, 55). Over the last decade, several case control and cohort studies have convincingly demonstrated that specific types of HPV are the causal agents of at least 90% of cervical cancers (44). The genital HPVs are commonly referred to as high, intermediate, or low risk depending on the frequency with which they are present in cancers. The most common high-risk types include HPV types 16, 18, 45, and 56, while HPV types 31, 33, and 35 are referred to as intermediate-risk types, and HPV types 6, 11, 42, 43, and 44 are considered low risk because they are almost never associated with malignancy (37). HPV has also been implicated in head and neck squamous cell cancers (62). The lack of a serologic assay has made it difficult to estimate the risk of malignancy conferred by infection with oncogenic types of HPV. Several cohort studies examining the risk of development of cervical intraepithelial neoplasias (CIN) in relationship to HPV type have recently appeared (7, 32). In one recent study of initially cytologically negative women attending an STD clinic, the cumulative incidence of CIN 2 or 3 at 2 years was 38% among those who had oncogenic HPV type 16 or 18 (detected by Southern transfer hybridization) at some time during the study. After HPV was adjusted for, factors that were independently associated with development of CIN included serologic evidence of *Chlamydia trachomatis* and a positive culture for *Neisseria gonorrhoeae* (32). The probability that the risk of developing CIN 2–3, the lesion thought to be the precursor of invasive cancer, will be similar in other populations is unknown. Laboratory-based studies have begun to provide an explanation for the association between infection with high-risk types of HPV and malignancy. As mentioned above, high-risk but not low-risk types of genital HPV produce oncoproteins (designated E6 and E7) that bind normal tumor suppressor proteins p53 and pRB. This interaction is thought to be central to carcinogenesis (42).

PREVALENCE OF HPV INFECTION

Lacking a serologic assay, we know little about the lifetime risk of infection with genital types of HPV. The prevalence of HPV infection varies with the diagnostic method used, as well as the age and sexual behavior of the population assayed. As with other STDs, genital warts and cytologic evidence of cervical HPV infection are most frequently detected among sexually active young women between 15 and 25 years of age. Manifestations of productive HPV infections such as warts and HPV-associated squamous intraepithelial lesions of the cervix, which are common shortly after women become sexually active, are only rarely detected in menopausal and postmenopausal women, perhaps reflecting development of immunity (13).

The prevalence of HPV infection detected cytologically appears to be about 2% and to range from 0.7 to 3% (16, 39, 53). Among women attending STD clinics, 8 to 13% of Pap smears show signs of HPV infection (18). In contrast to the apparent rise in the prevalence of clinically diagnosed warts over the last 20 years mentioned above, Pap smears have not shown a similar increase in the prevalence of HPV-associated changes. One possible explanation of this apparent discrepancy is that clinicians have simply become increasingly aware of the disease. Alternatively, the absence of an increase in koilocytosis might reflect changes in the cytologic classification schemes that have occurred over the last 15 years.

Recent studies that have used very sensitive PCR-based methods suggest that most adults who have HPV DNA detected in genital tract samples do not have clinical or morphologic evidence of HPV infection (2, 11, 15, 20, 21, 30, 41, 44, 49, 56, 61, 63, 64). Detection of HPV DNA in women under 40 years of age and with normal cytology ranges from 3.8 to 50%, with most studies reporting prevalences over 20%. In many cases, most of the HPV detected cannot be classified as one of the known HPV types. The prevalence of the high-risk HPV types in normal Pap smears is considerably lower (2.9 to 30%). Until it is possible to accurately sample the epithelium and to dem-

onstrate the topographic relationship between virus and tissue, it will remain unclear whether this prevalence in Pap smears represents infection of that area of the cervix where most cancers occur (the transformation zone) or contamination from adjacent epithelium known to be less likely to manifest HPV infection with morphologic changes and subsequent development of malignancy.

DIAGNOSIS OF HPV INFECTION

Because HPV cannot be cultured and because a reliable serologic assay for either acute or past infection with these agents is not available, several other methods have been used to diagnose infection.

Clinical and Colposcopic Examination

The use of clinical examination for detection of genital HPV is convenient but extremely insensitive. Only the larger, primarily papillary lesions can be identified. The majority of HPV infections, especially those of the cervix, are flat and grossly invisible (47). Many such cervical lesions become visible on colposcopic examination, especially if colposcopy is performed after application of acetic acid to the cervical epithelium. However, among women who have not been preselected on the basis of cytology and who are at high risk for STDs (including HPV), colposcopy lacks specificity, and compared to molecular tests, colposcopy is insensitive (48).

Microscopic Examination of Exfoliated Cell Samples (Pap Smears) or Tissue Biopsy Specimens

Squamous cells infected with HPV frequently show a variety of changes, the most characteristic being perinuclear clearing, with an increase in the density of the surrounding rim of cytoplasm (29, 58). Cells exhibiting such changes are referred to as koilocytes and can be seen in both exfoliated-cell samples and tissue biopsy specimens. Other changes that are frequently present in HPV-infected cells include changes in the shape and size of both the cytoplasm and the nucleus and in the amount and distribution of the nuclear chromatin. Abnormalities of keratinization are frequently present but often subtle. While most pathologists agree on the diagnosis of HPV infection in biopsy specimens or smears when all such changes are clearly present, several studies have shown poor intra- and interobserver reproducibilities for the diagnosis of HPV infection (28). Furthermore, as mentioned above, since most subjects who have HPV DNA detected by molecular assays do not have microscopic evidence of HPV (2, 11, 15, 20, 21, 30, 41, 44, 49, 56, 61, 63, 64), microscopic diagnosis appears to be insensitive.

Immunocytochemistry

Late structural antigens of HPV in tissue biopsy specimens can be detected by using polyclonal antisera directed to a prototype HPV. Detection of late structural antigens by using polyclonal antibodies raised against Shope papillomavirus allows confirmation of the presence of HPV in characteristic tissue biopsy specimens. Comparisons of this assay with HPV DNA probes have shown that such polyclonal antibodies are at least 90% specific but not more than 50% sensitive, most likely because in many cases, only small amounts of late structural proteins are present (29, 60).

Electron Microscopy

Although electron microscopic identification of these viruses in either cellular scrapes or tissue is possible, this approach is costly, time-consuming, and of low sensitivity. Furthermore, as with clinical or microscopic examination, one is unable to determine the specific type of HPV present.

Detection of Specific Types of HPV DNA by DNA Hybridization

At present, the "gold standard" for diagnosis of HPV is detection of HPV DNA by hybridization technology with either DNA or RNA probes directed against specific types of HPV DNA. Several different hybridization techniques have been used, including Southern blot, dot blot, sandwich, filter in situ, or hybridization in solution. All these techniques can be used on fresh, frozen, or fixed tissue or exfoliated-cell samples from which DNA can be extracted. In situ hybridization potentially offers the advantages of hybridization while conserving morphology and could allow one to determine whether the HPV is integrated or episomal in nature (51, 66). In situ hybridization with PCR has been reported; however, this technique has not been widely adapted by most laboratories because it remains less sensitive than other available assays. Recently, hybridization with RNA probes directed to HPV DNA has been performed in conjunction with the PCR, resulting in a significant increase in the detection of HPV nucleic acid (38, 45). Each of these procedures (hybridization and PCR) has advantages and disadvantages, and the choice of method depends on the information desired, the expertise of the laboratory performing the test, and the turnaround time required for a result.

Hybridization without Prior Amplification of HPV DNA

Prior to the development of PCR-based assays, Southern transfer hybridization was the assay of choice. This assay is performed by extracting all DNA from the clinical sample and then cutting the DNA with one or more restriction enzymes chosen to best demonstrate characteristic banding patterns (for example, PstI). The DNA fragments are then separated electrophoretically in an agarose gel, denatured, and transferred and fixed to a nylon membrane for hybridization with one or more type-specific labeled HPV probes (36). To achieve maximum sensitivity for detection of HPV, multiple DNA (RNA) probes are used, frequently under conditions of varying stringency, to detect as many HPVs of interest as possible. Nylon membranes are generally stripped and then reprobed with a different set of type-specific HPV probes. Although nonradiolabeled probes can be used, radiolabeled DNA probes increase the sensitivity approximately 10-fold over that obtained with nonradiolabeled probes (14). From 0.1 to 0.01 HPV genome copy per cell is generally detectable if radiolabeled probes are used (17).

The advantages of Southern transfer hybridization are that it is specific and relatively sensitive and that it allows visualization of specific banding patterns associated with specific types of viruses. In addition, by lowering the stringency of the hybridization, it is possible to screen for uncharacterized HPV types. This assay was widely used in early studies of the prevalence of HPV, but it has several major drawbacks. The assay is very time-consuming and labor intensive, and performing the technique in a standardized manner requires highly trained personnel. Processing, hybridizing, and probing techniques for this assay vary considerably from laboratory to laboratory, and since even minor changes in protocol can alter test results, reproducibility is less than optimal. Brandsma et al. (3) showed that interlaboratory variation was substantial for the performance of Southern blots, with interlaboratory agreement varying between 67 and 97% for HPV positivity and between 50 and 92% for HPV type specificity (3). Factors that have the potential of affecting the test outcome include the way in which the genital epithelium is sampled (biopsy versus smear, direct scrape versus vaginal wash), the amount of DNA used in the assay, the number and types of specific probes used, the stringency conditions, and the number of rounds of hybridization performed. Furthermore, interpretation of Southern transfer hybridization blots is often extremely difficult and not highly reproducible.

Dot hybridization, another commonly used membrane-based hybridization assay, is in many ways similar to Southern transfer hybridization but considerably less difficult to perform, especially if one of the commercially available type-specific dot tests is used. In these assays, purified cellular DNA without prior electrophoresis is fixed in the form of a dot onto the membrane (27). Membrane filters are then probed, frequently with RNA probes, to detect specific HPV DNA sequences. Many of the commercially available assays have extremely good specificities, with sensitivities similar to that of Southern transfer hybridization (27). These assays offer several important advantages over Southern transfer hybridization. In addition to being relatively easy to perform, they are very reproducible, and several specimens can be processed at one time with a turnaround time considerably shorter than that of Southern transfer hybridization (17, 24, 35, 57).

HPV DNA Detection Assays Combining Amplification of HPV DNA and Hybridization

Amplifying HPV DNA by PCR technology prior to hybridization allows for a significant increase in sensitivity (24). Although a Food and Drug Administration-approved commercially available kit for detection of specific types of HPV has not yet become available, DNA-typing results obtained by PCR are in good agreement with those obtained by Southern blot. Not surprisingly, PCR-based assays frequently detect additional HPV types. Only rarely (<5%) are specimens positive by dot or Southern blot and negative by PCR (1, 24, 34, 57). In one study of 893 anal samples tested by PCR, dot filter hybridization, and Southern transfer hybridization, 44% were HPV positive by all three tests, 12% were negative by all three tests, 31% were positive by PCR only, and 0.4% were positive by dot filter or Southern transfer hybridization but negative by PCR (34). The specific HPV types detected by Southern transfer hybridization were also detected by PCR in 90% of specimens. Reproducibility of PCR was assessed by analysis of (i) paired same-day specimens (89% agreement), (ii) intralaboratory repeat analysis (96% agreement), and (iii) interlaboratory analysis (88% agreement). In addition, control specimens prepared from HPV-negative cell lines and samples obtained from other dermal surfaces of patients were never positive for genital HPV types, further supporting the validity of the PCR results (34).

Several different primers and procedures for PCR have been described. The sensitivity of this technique for detection of HPV and specific types of HPV varies not only with the specific primers and probes used but also with the specific procedures employed. One approach is to first amplify HPV DNA by using consensus primers designed to facilitate amplification of a broad spectrum of HPV types by targeting the well-conserved L1 or E1 open reading frame of the HPV genome (17, 24). PCR products generated from such primers are then applied to a membrane (as a dot or in a fashion similar to the Southern blot procedure) and hybridized with labeled DNA or RNA type-specific probes. While a detailed discussion of the different primer sets that have been described is beyond the scope of this chapter, the choice of primers to be used is of considerable importance. For example, if the target DNA is damaged, as is frequently the case in older paraffin-embedded biopsy specimens or on Pap smears, primers designed to produce smaller amplification products offer the greatest sensitivity. If one is searching for uncharacterized HPV types, increasing the degeneracy of the primers will broaden the spectrum of detectable HPVs, although the likelihood of amplifying non-HPV DNA sequences will also increase (24). Well-characterized primers can currently be purchased, and it is anticipated that a PCR detection assay for HPV will be commercially available in the near future. PCR-based detection of HPV DNA offers greatly increased sensitivity, but several potential pitfalls have been identified (24), the most important being the possibility for contamination from either the clinic or the laboratory. Contamination from HPV plasmids, PCR products, or simply sample-to-sample transfer has been reported. Some of the anticontamination precautions that are common in laboratories that process large numbers of specimens by PCR include preparation of a master mix of reaction reagents in a sterile biosafety hood; use of positive displacement pipettes or plugged pipette tips for all amplification reactions; and routine inclusion of negative control samples, consisting of a known number of HPV-negative K562 cells, through all of the sample preparation steps. This last procedure allows one to assess laboratory technique and check for interspecimen contamination. In addition, sample preparation, PCR amplification, and PCR product analyses should all be done in separate areas with equipment and glassware that are specifically dedicated to those tasks. If plasmids are present in the laboratory, contamination with plasmid DNA can be reduced by using anticontamination primers. These primers flank the cloning side of the plasmid so that in case of contamination, no PCR product is produced under standard conditions (2, 63). Ten percent of randomly selected positive specimens and 5% of randomly selected negative specimens should be rerun. We have also found that frequent interlaboratory comparisons with other laboratories that regularly test large numbers of specimens are useful.

There is no question that PCR allows increased detection of a large number of HPV infections. However, the association between cervical or anal intraepithelial neoplasia and the detection of higher levels of HPV DNA that is possible without amplification is stronger than that between such pathology and the presence of only low levels of HPV detected by PCR alone. Furthermore, studies examining the risk of development of malignancy in relationship to infection by specific types of HPV have thus far detected HPV without using PCR technology. The relationship between risk of malignancy and infection with specific types of HPV as detected by PCR is not known. Therefore, the clinical relevance of detecting viral DNA at extremely low levels remains uncertain.

COLLECTION OF SPECIMENS

Several collection media for HPV specimens have been described elsewhere and are commercially available (35). The ideal collection medium allows one to proceed to hybridization without extensive manipulation of the clinical specimen. This is of considerable importance in avoiding specimen contamination, especially when DNA is being amplified prior to hybridization. Care must therefore be taken to collect specimens in a medium that is free of reagents that inhibit amplification. It is preferable that a collection medium also have a relatively long shelf life and be stable at room temperature.

There has been much debate as to how to best collect samples from the female genital tract for detection of HPV DNA by hybridization. When Southern transfer hybridization was widely used, some researchers advocated performing vaginal washes rather than directly sampling the cervix (6). Such washes were thought to increase the amount of HPV collected for hybridization. While vaginal washes did increase the amount of virus collected, the additional HPV DNA collected primarily reflected vaginal infection rather than cervical disease. Since the cervix is the primary site of interest, the usefulness of collecting vaginal HPV DNA is unclear. With the widespread addition of DNA amplification to hybridization assays, the question of whether an adequate amount of HPV DNA can be obtained from direct cervical sampling has become moot. On the other hand, even with PCR technology, it is difficult to obtain an adequate number of cells from the more keratinized epithelium of the external genitalia of men and women (25).

SEROLOGIC DETECTION OF HPV

PCR may amplify the DNA of HPV that exists as either a productive or a latent infection. We currently have no accurate test to assess past infection with this virus. It is now known that antibodies to HPV are produced; however, it is not known whether infection with HPV produces long-lasting antibodies at levels that will be readily detectable by serologic assays. It is also hoped that specific serologic markers that will allow us to identify women at risk for development of invasive disease can be developed. Current serologic assays for specific types of HPV use various sources for antigens, including bacterially expressed fusion proteins, synthetic peptides, or HPV propagated in a mouse xenograft system. Both Western blot (immunoblot) assays and enzyme-linked immunosorbent assays using these antigens have been developed. Results from interlaboratory comparisons have indicated poor to moderate agreement, with good agreement generally obtained when a large proportion of sera tests negative for HPV (22). More promising are the capsid proteins that are produced by cloning HPV in vaccinia virus or baculovirus, with expression of capsids containing conformationally correct L1 and L2 epitopes (22).

REPORTING AND INTERPRETATION OF RESULTS

At present, with our very limited understanding of the natural history, transmission, and risk of development of malignancy associated with HPV infection, the relevance of detecting HPV DNA is unclear. Similarly, the clinical implications of a negative test are for the moment uncertain. As discussed above, many sexually active young men and women have detectable oncogenic types of genital tract

HPV DNA, but probably only a subset of these people will ever develop HPV-related cancer. The meaning of a negative test result is difficult to assess, since we know little about the likelihood of misclassification due to intermittent shedding of virus or to problems with sampling or reproducibility of even the most sensitive assays, such as PCR. At present, treatment of genital tract pathology is based on morphology rather than on detection of specific types of HPV. The clinical impact of detection of HPV DNA is unclear. Because of the central role of specific types of HPV in the pathogenesis of cancer, detection of HPV DNA may play a role in cervical cancer control in the future, either as an adjunct to cytology in assessing which women need further workup or close follow-up, or perhaps in certain populations as a primary screening method.

REFERENCES

1. **Bauer, H. M., Y. Ting, C. E. Greer, J. C. Chambers, C. J. Tashiro, J. Chimera, A. Reingold, and M. M. Manos.** 1991. Genital human papillomavirus infection in female university students as determined by a PCR-based method. *JAMA* **265:** 472–477.
2. **Beyer-Finkley, E., H. Pfister, and F. Girardi.** 1990. Anticontamination primers to improve specificity of polymerase chain reaction in human papillomavirus screening. *Lancet* **i:**1289–1290.
3. **Brandsma, J., R. D. Burk, W. D. Lancaster, H. Pfister, and M. H. Schiffman.** 1989. Inter-laboratory variation as an explanation for varying prevalence estimates of human papillomavirus infection. *Int. J. Cancer* **43:**260–262.
4. **Briton, L. A.** 1992. Epidemiology of cervical cancer—overview, p. 3–24. *In* N. Munoz, F. X. Bosch, K. V. Shah, and A. Meheus (ed.), *The Epidemiology of Cervical Cancer and Human Papillomavirus.* JARC scientific publication no. 119. International Agency for Research on Cancer, Lyon, France.
5. **Broker, T. R., and M. Botchan.** 1986. Papillomaviruses: retrospectives and prospectives. *Cancer Cells* **4:**17–36.
6. **Burk, R. D., A. S. Kadish, S. Calderin, and S. L. Romney.** 1986. Human papillomavirus infection of the cervix detected by cervicovaginal lavage and molecular hybridization: correlation with biopsy results and Papanicolaou smear. *Am. J. Obstet. Gynecol.* **154:**982–989.
7. **Campion, M. J., D. J. McCance, J. Cuzick, and A. Singer.** 1986. Progressive potential of mild cervical atypia: prospective cytological, colposcopic, and virological study. *Lancet* **ii:**237–240.
8. **Chan, S. Y., H. U. Bernard, C. K. Ong, S. P. Chan, B. Hofmann, and H. Delius.** 1992. Phylogenetic analysis of 48 papillomavirus types and 28 subtypes and variants: a showcase for the molecular evolution of DNA viruses. *J. Virol.* **66:** 5714–5725.
9. **Chuang, T. Y., H. O. Perry, L. T. Kurland, and D. M. Ilstrup.** 1984. Condyloma acuminatum in Rochester, Minnesota, 1950–1978. I and II. *Arch. Dermatol.* **120:**469–483.
10. **Coggin, J. R., and H. zur Hausen.** 1979. Workshop on papillomaviruses and cancer. *Cancer Res.* **39:**545–546.
11. **Coll-Seck, A., M. A. Faye, C. W. Critchlow, A. D. Mbaye, J. Kuypers, G. Woto-Gaye, C. Langley, E. B. De, K. K. Holmes, and N. B. Kiviat.** Cervical intraepithelial neoplasia and human papillomavirus infection among Senegalese women seropositive for HIV-1 or HIV-2 or seronegative for HIV. Submitted for publication.
12. **Corbitt, G., A. P. Zarod, J. R. Arrand, M. Longson, and W. T. Farrington.** 1988. Human papillomavirus (HPV) genotypes associated with laryngeal papilloma. *J. Clin. Pathol.* **41:**284–288.
13. **Critchlow, C. W., and L. A. Koutsky.** *Epidemiology of Human Papillomavirus Infection,* in press.
14. **Crum, C. P., G. Nuovo, D. Friedman, and S. V. Silverstein.** 1988. A comparison of biotin and isotope-labeled ribonucleic acid probes for in situ detection of HPV 16 ribonucleic acid in genital precancers. *Lab. Invest.* **58:**354–359.
15. **Czegledy, J., K. O. Rogo, M. Evander, and G. Wadell.** 1992. High-risk human papillomavirus types in cytologically normal cervical scrapes from Kenya. *Med. Microbiol. Immunol.* **180:** 321–326.
16. **de Brux, J., G. Orth, O. Croissant, B. Cochard, and M. Ionesco.** 1983. Condylomatous lesions of the cervix uteri: development in 2466 patients. *Bull. Cancer* **70:**410–422.
17. **de Villiers, E.-M.** 1992. Hybridization methods other than PCR: an update, p. 111–120. *In* N. Munoz, F. X. Bosch, K. V. Shah, and A. Meheus (ed.), *The Epidemiology of Cervical Cancer and Human Papillomavirus.* IARC scientific publication no. 119. International Agency for Research on Cancer, Lyon, France.
18. **Drake, M., G. Medley, and H. Mitchell.** 1987. Cytological detection of human papillomavirus infection. *Obstet. Gynecol. Clin. N. Am.* **14:**431–450.
19. **Dyson, N., P. M. Howley, K. Munger, and E. Harlow.** 1989. The human papillomavirus-16 E7 oncoprotein is able to bind to the retinoblastoma gene product. *Science* **243:**934–937.
20. **Engels, H., A. Nyongo, M. Temmerman, W. G. V. Quint, E. van Marck, and W. J. Eylenbosch.** 1992. Cervical cancer screening and detection of genital HPV-infection and chlamydial infection by PCR in different groups of Kenyan women. *Ann. Soc. Belg. Med. Trop.* **72:**53–62.
21. **Fairley, C. K., S. Chen, S. N. Tabrizi, K. Leeton, M. A. Quinn, and S. M. Garland.** 1992. The absence of genital human papillomavirus DNA in virginal women. *Int. J. Sex. Transm. Dis. AIDS* **3:**414–417.
22. **Galloway, D. A.** 1992. Serological assays for the detection of HPV antibodies, p. 147–161. *In* N. Munoz, F. X. Bosch, K. V. Shah, and A. Meheus (ed.) *The Epidemiology of Human Papillomavirus and Cervical Cancer.* IARC scientific publication no. 119. International Agency for Research on Cancer, Lyon, France.
23. **Gissmann, L.** 1984. Papillomaviruses and their association with cancer in animals and in man. *Cancer Surv.* **3:**162–181.
24. **Gravitt, P. E., and M. M. Manos.** 1992. Polymerase chain reaction-based methods for the detection of human papillomavirus DNA, p. 121–133. *In* N. Munoz, F. X. Bosch, K. V. Shah, and A. Meheus (ed.), *The Epidemiology of Human Papillomavirus and Cervical Cancer.* IARC scientific publication no. 119. International Agency for Research on Cancer, Lyon, France.
25. **Grussendorf-Conen, E. I., E.-M. de Villiers, and L. Gissman.** 1986. Human papillomavirus genomes in penile smears of healthy men. *Lancet* **ii:**1092.
26. **Jenson, A. B., R. J. Kurman, and W. D. Lancaster.** 1987. Tissue effects of and host response to human papillomavirus infection. *Obstet. Gynecol. Clin. N. Am.* **14:**397–406.
27. **Kiviat, N., L. Koutsky, C. Critchlow, D. Galloway, D. Vernon, M. Peterson, P. McElhose, S. Pendras, C. Stevens, and K. Holmes.** 1990. Comparison of Southern transfer hybridization and dot filter hybridization for detection of cervical human papillomavirus infection with types 6, 11, 16, 18, 31, 33, and 35. *Am. J. Clin. Pathol.* **94:**561–565.
28. **Kiviat, N. B., C. W. Critchlow, and R. J. Kurman.** Reassessment of the morphological continuum of cervical intraepithelial lesions: does it reflect different stages in the progression to cervical carcinoma?, p. 59–66. *In* N. Munoz, F. X. Bosch, K. V. Shah, and A. Meheus (ed.), *The Epidemiology of Cervical Cancer and Human Papillomavirus.* IARC scientific publication no. 119. International Agency for Research on Cancer, Lyon, France.
29. **Kiviat, N. B., L. A. Koutsky, C. W. Critchlow, A. T. Lorinca, A. P. Cullen, J. Brockway, and K. K. Holmes.** 1992. Prevalence and cytologic manifestations of HPV types 6, 11, 16, 18, 31, 33, 35, 42, 43, 44, 45, 51, 52, and 56 among 500 consecutive women. *Int. J. Gynecol. Pathol.* **11:**197–203.
30. **Kjaer, S. K., and E. Lynge.** 1989. Incidence, prevalence and time trends of genital HPV infection determined by clinical examination and cytology, p. 113–124. *In* N. Munoz, F. X.

Bosch, and O. M. Jensen (ed.), *Human Papillomavirus and Cervical Cancer*. IARC scientific publication no. 94. International Agency for Research on Cancer, Lyon, France.

31. **Koutsky, L. A., D. A. Galloway, and K. K. Holmes.** 1988. Epidemiology of genital human papillomavirus infection. *Epidemiol. Rev.* **10:**122–163.

32. **Koutsky, L. A., K. K. Holmes, C. W. Critchlow, C. E. Stevens, J. Paavonen, A. M. Beckmann, T. A. DeRouen, D. A. Galloway, D. Vernon, and N. B. Kiviat.** 1992. A cohort study of the risk of cervical intraepithelial neoplasia grade 2 or 3 in relation to human papillomavirus infection. *N. Engl. J. Med.* **327:**1272–1278.

33. **Kreider, J. W.** 1987. Susceptibility of various human tissues to transformation in vivo with human papillomavirus type 11. *Int. J. Cancer* **39:**459.

34. **Kuypers, J. M., C. W. Critchlow, P. E. Gravitt, D. A. Vernon, J. B. Sayer, M. M. Manos, and N. B. Kiviat.** 1993. Comparison of dot filter hybridization, Southern transfer hybridization, and polymerase chain reaction amplification for diagnosis of anal human papillomavirus infection. *J. Clin. Microbiol.* **31:**1003–1006.

35. **Lorincz, A.** Human papillomaviruses. *In* N. J. Schmidt and R. W. Emmons (ed.), *Diagnostic Procedures for Viral, Rickettsial, and Chlamydial Diseases*, 7th ed., in press. American Public Health Association, Washington, D.C.

36. **Lorincz, A. T.** 1990. Human papillomavirus detection tests, p. 953–959. *In* K. K. Holmes, P.-A. Mardh, P. F. Sparling, P. J. Wiesner, W. Cates, S. M. Lemon, and W. E. Stamm (ed.), *Sexually Transmitted Diseases*, 2nd ed. McGraw-Hill Book Co., New York.

37. **Lorincz, A. T., R. Reid, A. B. Jenson, M. D. Greenberg, W. Lancaster, and R. J. Kurman.** 1992. Human papillomavirus infection of the cervix: relative risk associations of fifteen common anogenital types. *Obstet. Gynecol.* **79:**328–337.

38. **Manos, M. M., Y. Ting, D. K. Wright, A. J. Lewis, T. R. Broker, and S. M. Wolinsky.** 1989. The use of polymerase chain reaction amplification for the detection of genital human papillomavirus. *Cancer Cells* **7:**209–214.

39. **Meisels, A., and C. Morin.** 1986. Flat condyloma of the cervix: two variants with different prognosis, p. 115–119. *In* R. Peto and H. zur Hausen (ed.), *Viral Etiology of Cervical Cancer*. Banbury report no. 21. Cold Spring Harbor Laboratory, Cold Spring Harbor, N.Y.

40. **Melchers, W., S. de Mare, E. Kuitert, J. Galama, J. Walboomers, and A. J. van den Brule.** 1993. Human papillomavirus and cutaneous warts in meat handlers. *J. Clin. Microbiol.* **31:**2547–2549.

41. **Melkert, P. W. J., E. Hopman, A. van den Brule, E. K. J. Risse, P. J. van Diest, O. P. Bleker, T. Helmerhorst, M. I. Schipper, C. J. L. M. Meijer, and J. M. M. Walboomers.** 1993. Prevalence of HPV in cytomorphologically normal cervical smears, as determined by the polymerase chain reaction, is age-dependent. *Int. J. Cancer* **53:**919–923.

42. **Munger, K., M. Scheffner, J. M. Huibregtse, and P. M. Howley.** 1992. Interactions of HPV E6 and E7 oncoproteins with tumor suppressor gene products, p. 197–217. *In* A. J. Legine (ed.), *Tumor Suppressor Genes, the Cell Cycle and Cancer*. Cold Spring Harbor Laboratory, Cold Spring Harbor, N.Y.

43. **Munoz, N., and F. X. Bosch.** 1992. HPV and cervical cancer: review of case-control and cohort studies, p. 251–262. *In* N. Munoz, F. X. Bosch, K. V. Shah, and A. Meheus (ed.), *The Epidemiology of Cervical Cancer and Human Papillomavirus*. IARC scientific publication no. 119. International Agency for Research on Cancer, Lyon, France.

44. **Munoz, N., F. X. Bosch, S. de Sanjose, L. Fafur, I. Izarzugaza, M. Gili, P. Viadiu, C. Navarro, C. Martos, N. Ascunce, L. C. Gonzalez, J. M. Kaldor, E. Guerrero, A. Lorincz, M. Samtamaria, P. Alonso de Ruiz, N. Aristizabal, and K. Shah.** 1992. The causal link between human papillomavirus and invasive cervical cancer: a population-based case-control study in Colombia and Spain. *Int. J. Cancer* **52:**743–749.

45. **Nuovo, G. J., P. MacConnell, A. Forde, and P. Delvenne.** 1991. Detection of human papillomavirus DNA in formalin-fixed tissues by in situ hybridization after amplification by polymerase chain reaction. *Am. J. Pathol.* **139:**847–854.

46. **Oriel, J. D.** 1971. Natural history of genital warts. *Br. J. Vener. Dis.* **47:**1–13.

47. **Paavonen, J., L. A. Koutsky, and N. Kiviat.** 1990. Cervical neoplasia and other STD-related genital and anal neoplasias, p. 561–592. *In* K. K. Holmes, P.-A. Mardh, P. F. Sparling, P. J. Wiesner, W. Cates, S. M. Lemon, and W. E. Stamm (ed.), *Sexually Transmitted Diseases*, 2nd ed. McGraw-Hill Book Co., New York.

48. **Paavonen, J., C. E. Stevens, C. W. Critchlow, N. Kiviat, P. Wolner-Hanssen, L. A. Koutsky, T. DeRouen, and K. K. Holmes.** 1988. Colposcopic correlates of cervical human papillomavirus infection. *Obstet. Gynecol. Surv.* **43:**323.

49. **Pao, C. C., C.-Y. Lin, J.-S. Maa, C.-H. Lai, S.-Y. Wu, and Y.-K. Soong.** 1990. Detection of human papillomaviruses in cervicovaginal cells using polymerase chain reaction. *J. Infect. Dis.* **161:**113–115.

50. **Papay, F., B. Wood, and M. Coulson.** 1988. Squamous cell papilloma at the tracheoesophageal puncture stoma. *Arch. Otolaryngol. Head Neck Surg.* **144:**564–568.

51. **Park, R. S., R. J. Kurman, T. D. Keissis, and K. V. Shah.** 1991. Comparison of peroxidase-labeled DNA probes with radioactive RNA probes for detection of human papillomaviruses by in situ hybridization in paraffin sections. *Mod. Pathol.* **4:**81–85.

52. **Pfister, H., A. Gasseenmairer, F. Nurnberger, and I. Stuttgen.** 1983. HPV 5 in a carcinoma of an epidermodysplasia verruciformis patient infected with various human papillomavirus types. *Cancer Res.* **43:**1436–1441.

53. **Rakoczy, P., G. Sterret, J. Kulski, D. Whitaker, L. Hutchison, J. McKenzie, and E. Pixley.** 1990. Time trends in the prevalence of human papillomavirus infections in archival Papanicolaou smears: analysis by cytology, DNA hybridization, and polymerase chain reaction. *J. Med. Virol.* **32:**10–17.

54. **Reid, R., and M. D. Greenberg.** 1991. Human papillomavirus related diseases of the vulva. *Clin. Obstet. Gynecol.* **34:**630–650.

55. **Roman, A., and K. H. Fife.** 1989. Human papillomaviruses: are we ready to type? *Clin. Microbiol. Rev.* **2:**166–90.

56. **Schiffman, M. H., H. M. Bauer, R. N. Hoover, A. G. Glass, D. M. Cadell, B. B. Rush, D. R. Scott, M. E. Serman, R. J. Kurman, S. Wacholder, C. K. Stanton, and M. M. Manos.** 1993. Epidemiologic evidence showing that human papillomavirus causes most cervical intraepithelial neoplasia. *J. Natl. Cancer Inst.* **85:**958–964.

57. **Schiffman, M. H., H. M. Bauer, A. T. Lorincz, M. M. Manos, J. C. Byrne, A. G. Glass, D. M. Cadell, and P. M. Howley.** 1991. A comparison of Southern blot hybridization and polymerase chain reaction methods for the detection of human papillomavirus DNA. *J. Clin. Microbiol.* **29:**573–577.

58. **Schneider, A., and L. Koutsky.** 1992. Natural history and epidemiological features of genital HPV infection, p. 25–52. *In* N. Munoz, F. X. Bosch, K. V. Shah, and A. Meheus (ed.), *The Epidemiology of Cervical Cancer and Human Papillomavirus*. IARC scientific publication no. 119. International Agency for Research on Cancer, Lyon, France.

59. **Schneider, A., T. Oltersdorf, and V. Schneider.** 1987. Distribution pattern of human papilloma virus 16 genome in cervical neoplasia by molecular in situ hybridization of tissue sections. *Int. J. Cancer* **39:**717–721.

60. **Shah, K. V.** 1990. Biology of human genital tract papillomaviruses, p. 425–431. *In* K. K. Holmes, P.-A. Mardh, P. F. Sparling, and P. J. Wiesner (ed.), *Sexually Transmitted Diseases*. McGraw Hill Book Co., New York.

61. **ter-Muelen, J., H. C. Eberhardt, J. Luande, H. N. Mgaya, J. Chang-Clause, H. Mtiro, M. Mhina, P. Kashaija, S. Ockert, X. Yu, G. Meinhardt, L. Gissmann, and M. Pawlita.** 1992. Human papilloma (HPV) infection, HIV infection and cervical cancer in Tanzania, East Africa. *Int. J. Cancer* **51:**515–521.

62. **Tyan, Y. S., S. T. Liu, W. R. Ong, M. L. Chen, C. H. Shu, and Y. S. Chang.** 1993. Detection of Epstein-Barr virus and human papillomavirus in head and neck tumors. *J. Clin. Microbiol.* **31:**53–56.

63. **van den Brule, A. J., E. C. Claas, M. du Maine, W. J. Melchers, T. Helmerhorst, W. G. Quint, J. Lindeman, C. J. Meijer, and J. M. Walboomers.** 1989. Use of anticontamination primers in the polymerase chain reaction for the detection of human papilloma virus genotypes in cervical scrapes and biopsies. *J. Med. Virol.* **29:**20–27.

64. **van Doornum, G. J. J., C. Hooykaas, L. H. J. Juffermans, S. M. G. A. van der Lans, and M. M. D. van der Linden.** 1992. Prevalence of human papillomavirus infections among heterosexual men and women with multiple sexual partners. *J. Med. Virol.* **37:**13–21.

65. **Van Ranst, M., J. B. Kaplan, and R. D. Burk.** 1992. Phylogenetic classification of human papillomaviruses: correlation with clinical manifestations. *J. Gen. Virol.* **73:**2653–2660.

66. **Wolber, R. A., and P. B. Clement.** 1991. In situ DNA hybridization of cervical small cell carcinoma and adenocarcinoma using biotin-labeled human papillomavirus probes. *Mod. Pathol.* **4:**96–100.

67. **Xi, L. F., G. W. Demers, N. B. Kiviat, J. Kuypers, A. M. Beckmann, and D. A. Galloway.** 1993. Sequence variation in non-coding region of HPV type 16 detected by single strand conformation polymorphism analysis. *J. Infect. Dis.* **168:**610–617.

68. **Zur Hausen, H.** 1989. Papillomaviruses as carcinoma viruses. *Adv. Viral Oncol.* **8:**1–26.

Polyomaviruses

EUGENE O. MAJOR

CLINICAL BACKGROUND

Diseases associated with infection with the human polyomaviruses JC virus (JCV) and BK virus (BKV) occur mostly in immunocompromised adults, indicating a reactivation of a latent infection. JCV was isolated from brain tissue of a Hodgkin's lymphoma patient with the demyelinating disease progressive multifocal leukoencephalopathy (PML). Although virus particles resembling polyomaviruses have been associated with lesions in the white matter of PML patients, it was not until JCV was consistently isolated from numerous other PML patients' brain tissues that JCV was considered that etiologic agent (52, 56, 67). The virus lytically infects the oligodendrocyte in the white matter of the brain, which results in focal or multifocal plaques of demyelination. Such lesions can be demonstrated by magnetic resonance imaging or computerized tomography scans as seen in Fig. 1, which also demonstrates the essential elements of the histopathology of PML. The clinical outcome most frequently seen in PML patients is hemiparesis or motor dysfunction, visual deficits such as cortical blindness or hemianopsia, and progressive dementia or cognitive impairment (8, 44, 56). Although JCV is tumorigenic in hamsters and owl and squirrel monkeys (42, 68), it has not been associated with human neoplasms. BKV has been associated with kidney infections, particularly in renal transplant patients such as the original patient, who suffered ureteral stenosis. BKV infection has also been associated with renal infections in bone marrow transplant patients. However, BKV has not been specifically identified as the etiologic agent in these infections.

There have also been reports of pancreatitis in a limited number of patients (5, 16). BKV DNA has been found in a human adenoma, pancreatic islet cells, and an insulinoma. Genomic sequences have also been found in gliomas and other tumors in the brain (22). It is not known whether BKV plays any role in the etiology of these tumors. BKV has not been associated with central nervous system infections such as encephalitis or demyelination.

On the basis of seroepidemiologic studies, the population in general appears to be susceptible to human polyomavirus infections. Serum conversion occurs at ages 5 to 8 in almost 80% of the population worldwide (2, 3, 13, 27, 39, 53). Many individuals maintain antibody titers to one or the other of the viruses, sometimes both, throughout their lives. It is not clear in all cases whether these titers represent recurrent infection or reactivation of a latent infection. The nature of the primary infection of BKV or JCV is also not known. Since both viruses can be excreted in the urine, however, even under conditions of no apparent disease, it is assumed that the kidney can be infected at some early stage following viral contact. For JCV, however, lymphoid tissues such as bone marrow and spleen have also been identified as sites for initial and/or latent infection (31, 36, 43).

BKV infection has been associated with kidney infections in renal allograft recipients and bone marrow transplant patients. BKV was detected in urine by the PCR in an average of 10 to 25% of this population. A similar percentage of the healthy population also excretes virus in the urine (4, 6, 33). There is also a report that a number of human immunodeficiency virus type 1 (HIV-1)-seropositive individuals shed BKV in the urine (47). Analysis of antibody of BKV from graft recipients revealed a higher incidence of infection in this population, i.e., 25% with primary infection and 21% with viral reactivation (2). These studies also revealed that primary infection could arise from the donor organ. Presence of virus, however, does not always predict poor graft survival or illness in the host.

JCV has been associated with human brain infection of the oligodendrocyte, the myelin-producing cell in the white matter, leading to PML. JCV is not currently associated with any other pathological condition, although no exhaustive study of its association with neoplastic diseases has been done. Viral antigen or viral DNA is always present in lesions of the white matter normally identified from biopsy or autopsy tissues. Use of immunocytochemistry or in situ DNA hybridization confirms the presence of JCV in brain tissue and therefore confirms the diagnosis of PML (1, 15, 21, 61). These assays also provided evidence that JCV was present in B cells in the bone marrow and spleen (36). JCV DNA has also been identified in peripheral blood lymphocytes (PBLs) and cerebrospinal fluid (CSF) by using PCR analysis (29, 32, 63–65). Although the exact nature of primary infection is not known, it appears that JCV can establish latency in lymphoid organs and upon immune suppression can be reactivated in lymphocytes and travel to the brain. JCV-infected B lymphocytes in PBLs and brain tissue have been described (43, 64). The recognition of JCV-induced PML in transplant and cancer patients has

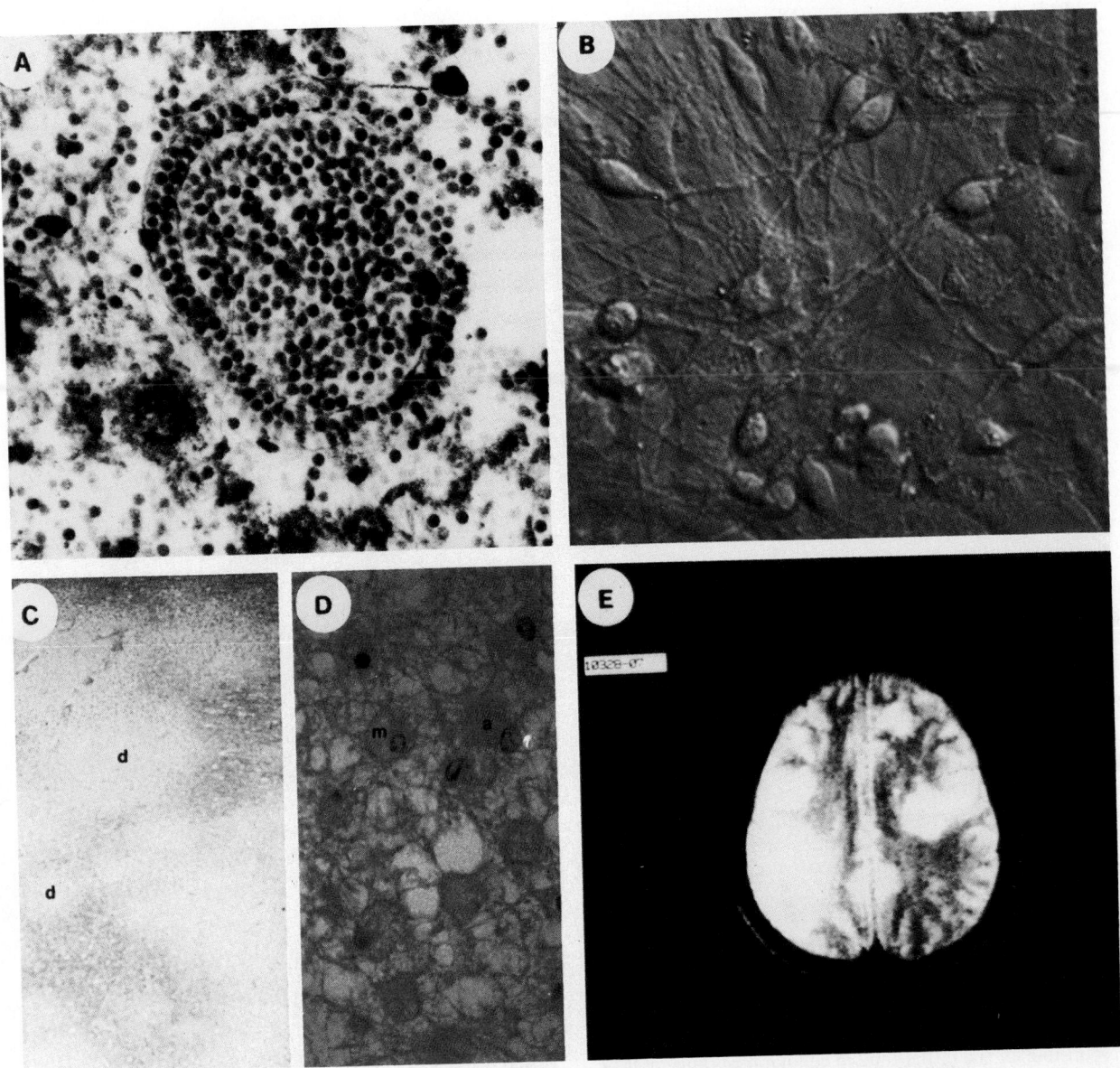

FIGURE 1 Histopathology of JCV-induced PML. (A) Electron micrograph of brain tissue from a PML patient, showing assembly of JCV particles in the nucleus of an infected oligodendrocyte. (B) Human fetal neuroglial cells in culture, demonstrating morphologically heterogeneous population of cells. (C) Luxol fast blue stain of white matter from a PML patient with demyelinated plaque lesion; d, area of demyelination. (D) Hematoxylin stain of lesion showing a macrophage (m) and an astrocyte (a). (E) Magnetic resonance image of PML-affected brain, with lesions of the subcortical white matter. (Reprinted with permission from reference 44.)

been rare in the past, accounting for only several hundred cases reported each year in the United States. However, PML occurs in approximately 5% of all AIDS patients, which represents thousands of reported cases worldwide each year (10, 66). The increased awareness and diagnosis of PML in this patient group have alerted laboratories and clinicians to consider PML in non-AIDS cases as well.

With both JCV and BKV, the majority of recorded infections occur in immunosuppressed or immunocompromised individuals. This observation suggests that impairment of the immune system through HIV-1 infection, neoplastic disease, or drug treatment for graft protection can lead to reactivation of these viruses.

DESCRIPTION OF THE AGENTS

Both viruses were described in 1971. They were isolated from a cell culture made from inoculations of urine from one patient and brain tissue from another patient into human fetal kidney or brain cell cultures, respectively (28, 54). These viruses were classified in the genus *Polyomavirus* of the family *Papovaviridae* because of their architecture (icosahedral) and size (40-nm diameter) and because of the physical and genetic makeups of their double-stranded-DNA genomes (44, 52). Both viruses have the ability to hemagglutinate human type O erythrocytes, unlike simian polyomavirus 40. Virion particles are assembled in the

TABLE 1 Laboratory selection and appropriate assays for virus identification

Specimen	Virus to identify				
	Virus isolation[a]	Hemagglutination	PCR[b]	Antigen detection[c]	DNA hybrid[d]
Body fluids					
Urine	BKV, JCV	BKV, JCV	BKV, JCV		BKV, JCV[e]
Blood (lymphocytes)		JCV			
CSF		JCV	BKV, JCV		
Tissues[f]					
Brain	JCV		BKV, JCV	JCV	JCV
Kidney	BKV, JCV	BKV, JCV	BKV, JCV	BKV, JCV	BKV, JCV
Marrow		JCV	JCV	JCV	JCV

[a] Most permissive cell cultures are derived from human fetal kidney for BKV and fetal brain for JCV. An established cell line of fetal astrocytes, SVG (45), is equally efficient for JCV isolation.

[b] Primer pair selection for genome amplification is made from conserved regions around the coding and noncoding sequences.

[c] Antibody to either the viral T protein or the capsid protein VP1 is used in immunofluorescence and immunocytochemistry assays (30, 41, 48, 61).

[d] Radioactive or biotin-labeled probes are used for the in situ DNA hybridization test. Biotin-labeled probes distinguish between JCV and BKV and detect active viral replication.

[e] Pelleted uroepithelial cells are targets for hybridization.

[f] Tissue sections may routinely be formalin fixed and paraffin embedded. Frozen sections are also useful for in situ DNA hybridization (1, 35). Either embedded or frozen sections can be extracted for DNA templates for PCR test.

nucleus from capsid proteins VP1, VP2, and VP3. VP1 is the external protein used for attachment to a presumed glycoprotein cell receptor and is responsible for the virion's ability to agglutinate human type O erythrocytes. Treatment of cells with neuraminidase eliminates or greatly reduces virus infectivity. Antibodies to VP1 neutralize infection and prevent hemagglutination. VP2 and VP3 are internal to the virion particle and presumably assist in assembly by interactions with the viral genome.

DNA sequence analysis of BKV and JCV showed an approximately 75% homology between the two at the DNA level and 68% homology at the amino acid level (25). The most significant differences in the viral genomes were in the noncoding nucleotide sequences responsible for viral DNA replication and transcription. These nucleotides exist as tandem repeats located directly between the regions coding for the multifunctional nonstructural T proteins and the structural capsid proteins. These regulatory sequences are responsible for the rather limited cellular host range of JCV (primarily human glial cells) and the less restricted host range of BKV (epithelial cells of human and nonhuman-primate origin as well as other cell types) (9, 23, 26, 44, 46). It appears that during replication, these sequences are subject to substantial rearrangements both in vivo and in vitro (20, 25, 50, 51).

SPECIMENS: BODY FLUIDS AND TISSUES

Relevant body fluids for analysis include urine, blood, and CSF. BKV is most frequently detected in urine. Whether the sample is to be used for future PCR assays, virus isolation, or other tests, the urine can be stored at −80°C for long periods. Processing of urine involves centrifugation at approximately 1,000 × g to collect cells and then at high speeds (100,000 × g) to pellet virus from the urine. Cell or high-speed pellets can then be extracted for nucleic acid analysis or suspended in neutral pH buffer for virus isolation. These samples can be used for hemagglutination assays of human type O erythrocytes (18, 19) or for inoculation of susceptible cell cultures. Direct immunofluorescence exam-

ination of cells pelleted from urine has also been described (34).

Blood can be collected in either heparin or citrate buffers. Buffy coats of PBLs can be prepared by using a Ficoll-Hypaque or comparable separation medium. CSF can be collected and stored at −80°C. Blood should be kept at room temperature until processing, which should be done within 24 h. CSF and PBLs can be processed for viral DNA by standard phenol-chloroform extraction procedures (44, 65). Nucleic acid samples are quantitated by UV spectroscopy and used as templates for PCR analysis. They can be kept at 4°C for long periods in a Tris buffer with 0.01 M EDTA. DNA can be frozen as well, but continual freeze-thaw cycles may damage the DNA sufficiently that it does not give consistent results in PCR. A summary of the most frequently collected clinical samples and virus assays is presented in Table 1.

JCV is most frequently detected in situ in infected brain tissues from suspected PML patients. Samples are often brain biopsy or autopsy specimens that are either formalin fixed and paraffin embedded or frozen. Tissue sections can also be processed for antigen detection by using antibodies to the virion capsid VP1 protein. Several mouse monoclonal antibodies to the nonstructural viral T protein have been used for detecting JCV or BKV infection (30, 41, 48, 60). Tissue sections from bone marrow or kidney biopsies can be processed in a similar manner. Bone marrow aspirates should be centrifuged prior to plating onto gelatin coated slides, air dried, and fixed in 4% paraformaldehyde prior to hybridization with a JCV probe.

DIAGNOSIS OF POLYOMAVIRUS INFECTIONS

Hemagglutination inhibition assays form the basis of routine serologic testing for the presence of antibody against both BKV and JCV. Also, an enzyme-linked immunosorbent assay for the detection of immunoglobulin M antibody has been developed for these polyomaviruses (40, 41). Serology is generally not helpful in the diagnosis of acute infections, since the majority of the population demonstrates antibodies from previous exposure. There is little evidence for rise in antibody titers as an indication of acute infection. Serology, however, does play an important role in establishing the epidemiology of spread of human polyomaviruses and has contributed information on the worldwide distribution of both JCV and BKV (42).

VIRUS ISOLATION

With the availability of antibody and DNA probes for BKV and JCV, virus isolations are not routinely done. The most susceptible cell types that support virus multiplication are primary cultures derived from human embryonic tissues. BKV grows well in human embryonic kidney cells and epithelial cells and also forms plaques on these cells. Viral cytopathic effect is evidenced by the enlargement of nuclei and gradual but continual peeling of cells in the monolayer. Some virus is shed into the cell culture medium, while a portion of the virus remains cell associated. Freeze-thaw cycles assist in releasing virus that can be detected by hemagglutination assays.

JCV grows best in human fetal brain cultures and multiplies in the astrocyte and the precursor oligodendroglial cell. JCV does not infect neurons in culture or in the brain. Even under optimum conditions of sufficient viral inoculum, JCV growth takes weeks, and multiple passages are sometimes required before progeny virus can be detected. Viral cytopathic effect is not that prominent but is present as translucent cells with large nuclei that begin to shed from the culture layer. Virus is not shed into the medium, and therefore, treatment of cell lysates with a nonionic detergent such as sodium deoxycholate is necessary for sample preparation for the hemagglutination assay. As an alternative to primary cultures from human fetal brain, an established human fetal cell line, SVG, is susceptible to JCV productive infection and grows readily in the laboratory (45).

NUCLEIC ACID TECHNIQUES

Tissue sections have also been analyzed successfully by using in situ DNA hybridization, as demonstrated in Fig. 2A. There are several reports on the advantages in specificity and sensitivity of DNA hybridization over antigen detection (1, 4, 17, 44, 57). In one of these assays, the DNA probe used was biotin labeled. Although not as sensitive in detecting very low copy numbers of JCV, biotin-labeled probes detect as few as 100 copies, which is sufficient for diagnostic use in clinical tissues. This hybridization assay measures DNA replication, an earlier event in the life cycle of viral multiplication than virion capsid formation. It is possible that some cells are infected but do not produce sufficient antigen for immunocytochemical tests. DNA hybridization can also be employed to identify these cells. Use of gelatin-coated or sialinate-treated slides may improve the adherence of the tissues for immunocytochemistry or hybridization. Biopsy of other tissues such as bone marrow and bone marrow aspirates has been reported (36, 43). Figure 2B shows cells in a bone marrow aspirate that give a positive signal upon hybridization with a JCV biotin-labeled probe.

Application of nucleic acids from extracted clinical samples to membrane filters is also in common use for viral detection. Extracted nucleic acid samples can be either directly blotted to a filter (dot blot procedures) or electro-phoresed through agarose gels and then blotted to a filter (Southern procedure). Either viral DNA can then be identified by using radiolabeled viral genomic probes. These methods are routinely used in laboratories but require several days to complete and use labeled probes of relatively high specific activities, e.g., 10^8 dpm/μg of [^{32}P]DNA. Under these conditions, it can be difficult to differentiate BKV and JCV owing to their nucleotide sequence homologies.

PCR of Urine Samples

The use of PCR and other nucleic acid amplification techniques results in increased sensitivity. However, there are difficulties with the interpretation of a positive result. Since both BKV and JCV are excreted in urine and appear in kidney tissue under normal as well as pathological conditions, PCR has been used to detect viral genomes in urine in a number of studies (7, 24, 38, 47, 49). Essentially, both viruses can be found in 10 to 27% of samples from patients with renal tumors, pregnant women, bone marrow and renal transplant recipients, older individuals, and individuals who are infected with HIV-1. The high rate of shedding of viral particles in the urine negates the utility of these assays for the diagnosis of acute infections.

PCR Analysis of Blood and CSF

PCR analysis of other fluids or tissues for JCV DNA presents a setting different from that of BKV infection. The pathogenesis of JCV infection leading to demyelination in the brain suggests viral latency in kidney and lymphoid cells in bone marrow, spleen, or lymph nodes. Upon reactivation due to weakened immune surveillance, JCV becomes associated with PBLs, most notably the B cells. Evidence for the presence of JCV DNA in PBLs came from examination of a large number of PML patients, of whom >90% had circulating JCV in their blood (64), and of leukemia patients with bone marrow transplants (58). JCV DNA has also been detected in a large number of CSF samples from PML patients with or without AIDS but not in non-PML patients (12, 65). There have been several reports of PCR detection of JCV DNA in brain tissue of PML patients, as expected, and one report of JCV in B cells in brains of PML patients (32, 62).

EVALUATION AND INTERPRETATION OF RESULTS

The number of immune-suppressed individuals will continue to increase owing to neoplastic diseases, AIDS, and therapies requiring immune suppression for renal, bone marrow, heart, and other graft recipients. Since serology is not a principal diagnostic tool for human polyomaviruses, detection of viral proteins or nucleic acids is essential. Also, virus isolation for either BKV or JCV is very time-consuming and prone to false negatives. Since BKV is associated with renal infections, identification of virus in urine by nucleic acid hybridization or immunocytochemistry for viral proteins is useful. Brain tissue is still needed to diagnose JCV-induced PML. In situ DNA hybridization is the most reliable assay for detecting the presence of JCV in brain tissue and has proven to be more specific and more sensitive than immunocytochemistry. The list of samples routinely collected for these viruses and the most appropriate assays using these samples are shown in Table 1.

In cases of suspected PML when brain biopsy is not appropriate or not possible, the presence of JCV in PBLs

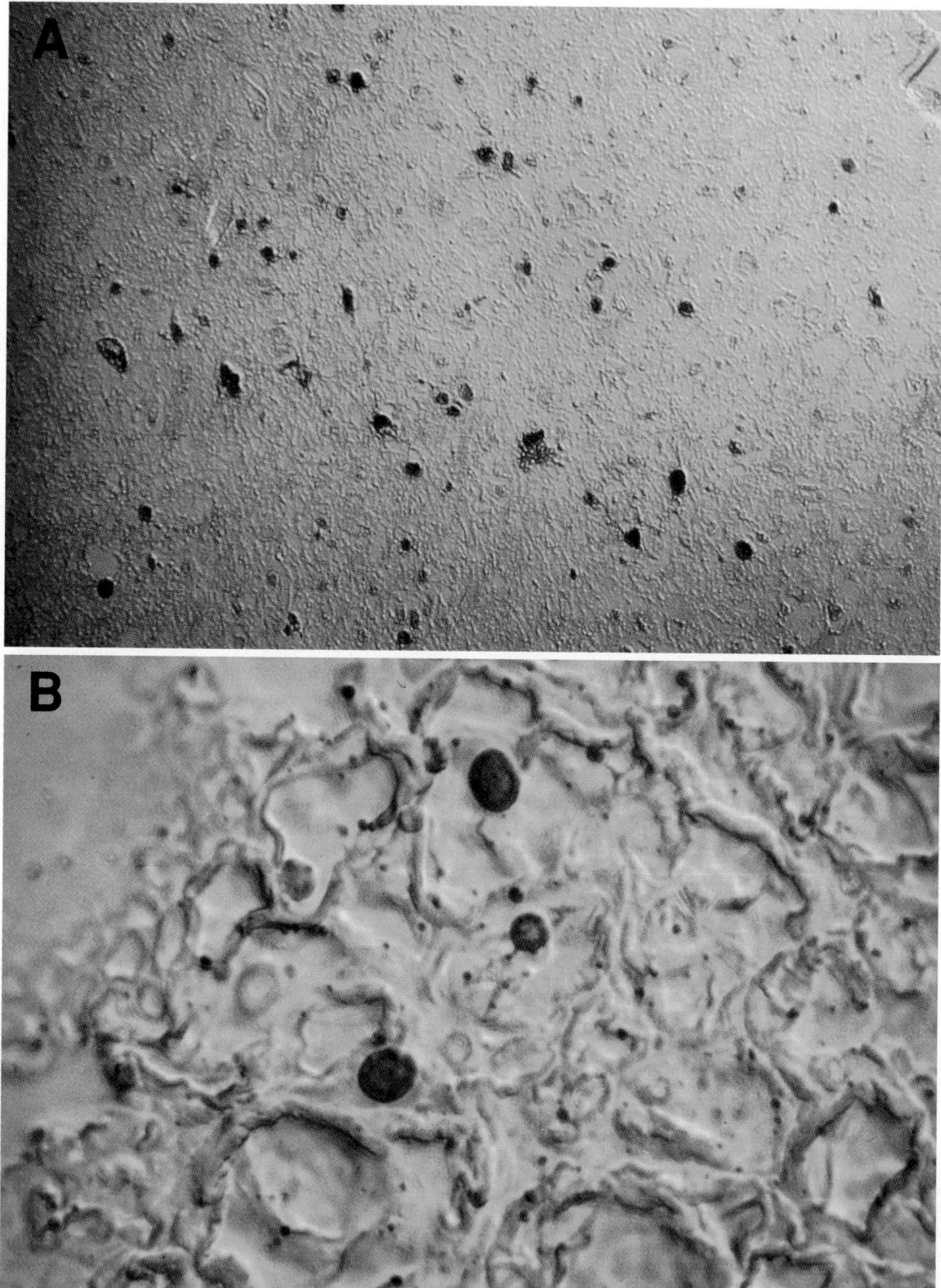

FIGURE 2 (A) Biopsy section of PML-affected brain tissue assayed by in situ DNA hybridization to a biotin-labeled JCV DNA probe. Positive hybridization signal is a dark precipitate in the infected cell nuclei generated by the oxidation of diaminobenzidene in the presence of streptavidin-horseradish peroxidase and H_2O_2. (B) Bone marrow aspirate from PML patient fixed onto gelatin-coated slide and hybridized to a biotin-labeled JCV DNA probe as described for panel A.

and CSF along with radiographic findings in an immuno-deficient patient may all but confirm the diagnosis of PML. The use of noninvasive procedures to collect samples for diagnosis of viral infections in the nervous system will be helpful.

THERAPY STRATEGIES FOR POLYOMAVIRUS INFECTION

There is currently no accepted treatment for JCV and BKV infections and only anecdotal accounts of therapies for JCV-induced PML. A recent review catalogs and describes in detail these attempts at therapy (44). The use of nucle-oside analogs, particularly cytosine arabinoside (ara-C), has been tried on the largest number of PML patients. Reports have not shown a consistent positive response, but for a few patients, treatment seems to have produced some improve-ment in cognitive ability and motor function (14, 55). One recent report on treatment with combined cytarabine and alpha interferon described a marked improvement in the clinical course of disease (59). However, use of nucleoside analogs in patients with impaired immunity is difficult to manage. A clinical trial protocol using ara-C in PML pa-tients with AIDS has been initiated through the AIDS Clinical Trial Groups. Laboratory data show that ara-C blocks JCV replication without toxicity to human fetal glial cells in culture. In the AIDS Clinical Trial Group ara-C will be administered intrathecally or intravenously for 6 months only to biopsy-proven cases of PML. The presence and concentration of virus in peripheral blood and CSF will be monitored by PCR. Clinical improvement will be the goal of the study. This trial, which will enroll 90 patients, will be the first controlled treatment trial for PML. There are also data suggesting that Camptothecin, a drug that interferes with viral replication by inhibition of topoi-somerase, may be effective in cell culture experiments (37). This drug may also have activity against BKV.

A number of cases of remission of PML have been assessed both clinically and neuroradiologically, essentially without therapeutic intervention (11). The mechanisms that protect individuals who have JCV in their peripheral circulation from developing PML or who are long-term survivors of or in remission from the clinical course of PML are currently under investigation.

REFERENCES

1. Aksamit, A., J. L. Sever, and E. O. Major. 1986. Progressive multifocal leukoencephalopathy: JC virus detection by in situ hybridization compared with immunocytochemistry. Neurology 36:499–504.
2. Andrews, C., K. Shah, R. Daniel, M. Hirsch, and R. Rubin. 1988. A serologic investigation of BK virus and JC virus infections in recipients of renal allografts. J. Infect. Dis. 158:176–181.
3. Andrews, C. A., R. Daniel, and K. Shah. 1983. Serologic studies of papovavirus infections in pregnant women and renal transplant recipients, p. 133–141. In J. L. Sever and D. L. Madden (ed.), Polyomaviruses and Human Neurological Disease. Alan R. Liss, Inc., New York.
4. Arthur, R., A. Beckmann, C. L. Chou, R. Saral, and K. Shah. 1985. Direct detection of the human papovavirus BK in urine of bone marrow transplant recipients: comparison of DNA hybridization with ELISA. J. Med. Virol. 16:29–36.
5. Arthur, R., K. Shah, S. Baust, G. Santos, and R. Saral. 1986. Association of BK viruria with hemorrhagic cystitis in recipients of bone marrow transplants. N. Engl. J. Med. 315:230–234.
6. Arthur, R. A., K. V. Shah, P. Charache, and R. Saral. 1988. BK and JC virus infections in recipients of bone marrow transplants. J. Infect. Dis. 158:563–569.
7. Arthur, R. A., S. Dagostin, and K. Shah. 1989. Detection of BK virus and JC virus in urine and brain tissue by the poly-merase chain reaction. J. Clin. Microbiol. 27:1174–1179.
8. Åstrom, K.-E., E. L. Mancall, and E. P. Richardson, Jr. 1958. Progressive multifocal leukoencephalopathy. Brain 81:93–127.
9. Beckman, A., and K. V. Shah. 1983. Propagation and pri-mary isolation of JCV and BKV in urinary epithelial cell cultures, p. 3–14. In J. L. Sever (ed.), Polyomaviruses and Human Neurological Diseases. Alan R. Liss, Inc., New York.
10. Berger, J. R., B. Kaszovitz, M. J. Post, and G. Dickinson. 1987. Progressive multifocal leukoencephalopathy associated with human immunodeficiency virus infection. A review of the literature with a report of sixteen cases. Ann. Intern. Med. 107:78–87.
11. Berger, J. R., and L. Mucke. 1988. Prolonged survival and partial recovery in AIDS-associated progressive multifocal leukoencephalopathy. Neurology 38:1060–1065.
12. Brouqui, P., C. Bollet, J. Delmont, and A. Bourgeade. 1992. Diagnosis of progressive multifocal leukoencephalopathy by PCR detection of JC virus from CSF. Lancet 339:1182.
13. Brown, P., T. Tsai, and D. C. Gajdusek. 1975. Seroepide-miology of human papovaviruses: discovery of virgin popula-tions and some unusual patterns of antibody prevalence among remote peoples of the world. Am. J. Epidemiol. 102:331–340.
14. Buckman, R., and E. Wiltshaw. 1976. Progressive multifocal leukoencephalopathy successfully treated with cytosine arabi-noside. Br. J. Haematol. 34:153–154.
15. Budka, H., and K. Shah. 1983. Papovavirus antigens in paraffin section of PML brains, p. 299–309. In J. L. Sever and D. L. Madden (ed.), Polyomaviruses and Human Neurological Diseases. Alan R. Liss, Inc., New York.
16. Cheeseman, S., P. Black, R. Rubin, K. Cantell, and M. Hirsch. 1980. Interferon and BK papovavirus-clinical and laboratory studies. J. Infect. Dis. 141:157–161.
17. Chesters, P. M., J. Heritage, and D. J. McCance. 1983. Persistence of DNA sequences of BK virus and JC virus in normal human tissues and in diseased tissues. J. Infect. Dis. 147:676–682.
18. Coleman, D. V., S. D. Gardner, and A. M. Field. 1973. Human polyomavirus infection in renal allograft recipients. B. Med. J. 3:371–375.
19. Coleman, D. V., M. R. Wolfendale, R. A. Daniel, N. K. Dhanjal, S. D. Gardner, P. E. Gibson, and A. M. Field. 1980. A prospective study of human polyomavirus infection in preg-nancy. J. Infect. Dis. 142:1–8.
20. Dörries, K. 1984. Progressive multifocal leukoencephalopa-thy: analysis of JC virus DNA from brain and kidney tissue. Virus Res. 1:25–38.
21. Dorries, K., R. T. Johnson, and V. Ter Meulen. 1979. Detection of polyoma virus DNA in PML-brain tissue by (in situ) hybridization. J. Gen. Virol. 42:49–57.
22. Dörries, K., G. Loeber, and J. Meixensbarger. 1987. Associ-ation of polyomaviruses JC, SV40, and BK with human brain tumors. Virology 160:268–270.
23. Feigenbaum, L., K. Khalili, E. O. Major, and G. Khoury. 1987. Regulation of the host range of human papovavirus JCV. Prog. Natl. Acad. Sci. USA 84:3695–3698.
24. Flægstad, T., A. Sundsfjord, R. R. Arthur, M. Pedersen, T. Traavik, and S. Subramani. 1991. Amplification and se-quencing of the control regions of BK and JC virus from human urine by polymerase chain reaction. Virology 180:553–560.
25. Frisque, R., G. Bream, and M. Cannella. 1984. Human polyomavirus JC virus genome. J. Virol. 51:458–469.
26. Gardner, S., E. Mackenzie, C. Smith, and A. Porter. 1984. Prospective study of the human polyomaviruses BK and JC and cytomegalovirus in renal transplant recipients. J. Clin. Pathol. 37:578–586.

27. **Gardner, S. D.** 1973. Prevalence in England of antibody to human polyomavirus (BK). *B. Med. J.* **1:**77–78.

28. **Gardner, S. D., A. M. Field, D. V. Coleman, and B. Hulme.** 1971. New human papovavirus (BK) isolated from urine after renal transplantation. *Lancet* **i:**1253–1257.

29. **Gibson, P., W. Knowles, J. Hand, and D. Brown.** 1993. Detection of JC virus DNA in cerebrospinal fluid of patients with progressive multifocal leukoencephalopathy. *J. Med. Virol.* **39:**278–281.

30. **Greenlee, J., and P. M. Keeney.** 1986. Immunoenzymatic labelling of JC papovavirus T antigen in brains of patients with progressive multifocal leukoencephalopathy. *Acta Neuropathol.* **71:**150–153.

31. **Grinnel, B. W., B. L. Padgett, and D. L. Walker.** 1983. Distribution of nonintegrated DNA from JC papovavirus in organs of patients with progressive multifocal leukoencephalopathy. *J. Infect. Dis.* **147:**669–675.

32. **Henson, J., M. Rosenblum, D. Armstrong, and H. Furneaux.** 1991. Amplification of JC virus DNA from brain and cerebrospinal fluid of patients with progressive multifocal leukoencephalopathy. *Neurology* **41:**1967–1971.

33. **Hogan, F., E. Borden, J. McBain, B. L. Padgett, and D. Walker.** 1980. Human polyomavirus infections with JC virus and BK virus in renal transplant patients. *Ann. Intern. Med.* **92:**373–378.

34. **Hogan, T. F., B. L. Padgett, D. L. Walker, E. C. Borden, and J. A. McBain.** 1980. Rapid detection and identification of JC virus and BK virus in human urine by using immunofluorescence microscopy. *J. Clin. Microbiol.* **11:**178–183.

35. **Houff, S. A., D. Katz, C. Kufta, and E. O. Major.** 1989. A rapid method for in situ hybridization for viral DNA in brain biopsies from patients with acquired immunodeficiency syndrome (AIDS). *AIDS* **3:**843–845.

36. **Houff, S. A., E. O. Major, D. Katz, C. Kufta, J. Sever, S. Pittaluga, J. Roberts, J. Gitt, N. Saini, and W. Lux.** 1988. Involvement of JC virus-infected mononuclear cells from the bone marrow and spleen in the pathogenesis of progressive multifocal leukoencephalopathy. *N. Engl. J. Med.* **318:**301–305.

37. **Kerr, D., C. Chang, J. Gordon, M. Bjornsti, and K. Khalili.** 1993. Inhibition of human neurotropic virus (JCV) DNA replication in glial cells by Camptothecin. *Virology* **196:**612–618.

38. **Kitamura, T., Y. Aso, N. Kuniyoshi, K. Hara, and Y. Yogo.** 1990. High incidence of urinary JC virus infection in nonimmunocompromised older patients. *J. Infect. Dis.* **161:**1128–1133.

39. **Knight, R. S., N. M. Hyman, S. D. Gardner, P. E. Gibson, M. M. Esiri, and C. P. Warlow.** 1988. Progressive multifocal leukoencephalopathy and viral antibody titres. *J. Neurol.* **235:**458–461.

40. **Knowles, W., P. Gibson, J. Hand, and D. Brown.** 1992. An M-antibody capture radioimmunoassay for detection of JC virus-specific IgM. *J. Virol. Methods* **40:**95–106.

41. **Knowles, W., I. Sharp, L. Efstratiou, J. Hand, and S. Gardner.** 1991. Preparation of monoclonal antibodies to JC virus and their use in the diagnosis of progressive multifocal leukoencephalopathy. *J. Med. Virol.* **34:**127–131.

42. **London, W. T., S. A. Houff, D. L. Madden, D. A. Fuccillo, M. Gravell, W. C. Wallen, A. E. Palmer, J. L. Sever, B. L. Padgett, D. L. Walker, G. M. ZuRhein, and T. Ohashi.** 1978. Brain tumors in owl monkeys inoculated with a human polyomavirus (JC virus). *Science* **201:**1246–1249.

43. **Major, E. O., K. Amemiya, G. Elder, and S. A. Houff.** 1990. Glial cells of the human developing brain and B cells of the immune system share a common DNA binding factor for recognition of the regulatory sequences of the human polyomavirus, JCV. *J. Neurosci. Res.* **27:**461–471.

44. **Major, E. O., K. Amemiya, C. S. Tornatore, S. A. Houff, and J. R. Berger.** 1992. Pathogenesis and molecular biology of progressive multifocal leukoencephalopathy, the JC virus-induced demyelinating disease of the human brain. *Clin. Microbiol. Rev.* **5:**49–73.

45. **Major, E. O., A. E. Miller, P. Mourrain, R. Traub, E. De**

Widt, and J. L. Sever. 1985. Establishment of a line of human glial cells that supports JC virus multiplication. *Proc. Natl. Acad. Sci. USA* **82:**1257–1261.

46. **Major, E. O., and D. A. Vacante.** 1989. Human fetal astrocytes in culture support the growth of the neurotropic human polyomavirus, JCV. *J. Neuropathol. Exp. Neurol.* **48:**425–436.

47. **Markowtiz, B., H. Thompson, J. Mueller, J. Cohen, and W. Dynan.** 1993. Incidence of BK virus and JC virus viruria in human immunodeficiency virus infected and uninfected subjects. *J. Infect. Dis.* **167:**13–20.

48. **Marshall, J., A. Smith, and S. Cheng.** 1991. Monoclonal antibody specific for BK virus large T antigen allows discrimination among the different papovaviral large T antigens. *Oncogene* **6:**1673–1676.

49. **Marshall, W., A. Telenti, J. Proper, A. Aksamit, and T. Smith.** 1991. Survey of urine from transplant recipients for polyomaviruses JC and BK using the polymerase chain reaction. *Mol. Cell. Probes* **5:**125–128.

50. **Martin, J. D., D. M. King, J. M. Slauch, and R. J. Frisque.** 1985. Differences in regulatory sequences of naturally occurring JC virus variants. *J. Virol.* **53:**306–311.

51. **Miyamura, T., A. Furuno, and K. Yoshiike.** 1985. DNA rearrangement in the control region for early transcription in a human polyomavirus JC host range mutant capable of growing in human embryonic kidney cells. *J. Virol.* **54:**750–756.

52. **Padgett, B. L., C. M. Rogers, and D. L. Walker.** 1977. JC virus, a human polyomavirus associated with progressive multifocal leukoencephalopathy: additional biological characteristics and antigenic relationships. *Infect. Immun.* **15:**656–662.

53. **Padgett, B. L., and D. L. Walker.** 1973. Prevalence of antibodies in human sera against JC virus, an isolate from a case of progressive multifocal leukoencephalopathy. *J. Infect. Dis.* **127:**467–470.

54. **Padgett, B. L., G. ZuRhein, D. Walker, R. Echroade, and B. Dessel.** 1971. Cultivation of papova-like virus from human brain with progressive multifocal leukoencephalopathy. *Lancet* **i:**1257–1260.

55. **Portegies, P., P. R. Algra, C. E. M. Hollar, J. M. Prins, P. Reiss, J. Valk, and J. Lange.** 1991. Response to cytarabine in progressive multifocal leukoencephalopathy in AIDS. *Lancet* **337:**680–681.

56. **Richardson, E. P., Jr.** 1961. Progressive multifocal leukoencephalopathy. *N. Engl. J. Med.* **265:**815–823.

57. **Schmidbauer, M., H. Budka, and K. V. Shah.** 1990. Progressive multifocal leukoencephalopathy (PML) in AIDS and in the pre-AIDS era. A neuropathological comparison using immunocytochemistry and in situ DNA hybridization for virus detection. *Acta Neuropathol.* **80:**375–380.

58. **Schneider, E., and K. Dorries.** 1993. High frequency of polyomavirus infection in lymphoid cell preparations after allogeneic bone marrow transplantation. *Transplant. Proc.* **25:**1271–1273.

59. **Steiger, M., G. Tarnsby, S. Gabe, J. McLaughlin, and A. Schapira.** 1993. Successful outcome of progressive multifocal leukoencephalopathy with cytarabine and interferon. *Ann. Neurol.* **33:**407–411.

60. **Stoner, G. L., C. F. Ryschkewitsch, D. F. L. Walker, and H. D. Webster.** 1986. JC papovavirus large tumor (T)-antigen expression in brain tissue of acquired immune deficiency syndrome (AIDS) and non-AIDS patients with progressive multifocal leukoencephalpathy. *Proc. Natl. Acad. Sci. USA* **83:**2271–2275.

61. **Stoner, G. L., D. Soffer, C. F. Ryschkewitsch, D. L. Walker, and H. D. Webster.** 1988. A double-label method detects both early (T-antigen) and late (capsid) proteins of JC virus in progressive multifocal leukoencephalopathy brain tissue from AIDS and non-AIDS patients. *J. Neuroimmunol.* **19:**223–236.

62. **Telenti, A., A. J. Aksamit, J. Proper, and T. F. Smith.** 1990. Detection of JC virus DNA by polymerase chain reaction in patients with progressive multifocal leukoencephalopathy. *J. Infect. Dis.* **162:**858–861.

63. **Telenti, A., W. Marshall, A. Aksamit, J. Smilack, and T. Smith.** 1992. Detection of JC virus by polymerase chain

reaction in cerebrospinal fluid from two patients with progressive multifocal leukoencephalopathy. *Eur. J. Clin. Microbiol. Infect. Dis.* **11:**253–254.

64. **Tornatore, C., J. Berger, S. Houff, B. Curfman, K. Meyers, D. Winfield, and E. Major.** 1992. Detection of JC virus DNA in peripheral lymphocytes from patients with and without progressive multifocal leukoencephalopathy. *Ann. Neurol.* **31:**454–462.

65. **Weber, T., R. Turner, S. Frye, B. Ruf, J. Haas, E. Schielke, W. Luke, W. Luer, K. Felgenhauer, and G. Hunsmann.** 1994. Specific diagnosis of progressive multifocal leukoencephalopathy by polymerase chain reaction. *J. Infect. Dis.* **169:**1138–1141.

66. **Wiley, C. A., M. Grafe, C. Kennedy, and J. A. Nelson.** 1988. Human immunodeficiency virus (HIV) and JC virus in acquired immune deficiency syndrome (AIDS) patients with progressive multifocal leukoencephalopathy. *Acta Neuropathol.* **76:**338–346.

67. **ZuRhein, G. M.** 1969. Association of papovavirions with a human demyelinating disease (progressive multifocal leukoencephalopathy). *Prog. Med. Virol.* **11:**185–247.

68. **ZuRhein, G. M.** 1983. Studies of JC virus-induced nervous system tumors in the Syrian hamster: a review, p. 205–221. *In* J. L. Sever and D. L. Madden (ed.), *Polyomavirus and Human Neurological Diseases.* Alan R. Liss, Inc., New York.

Human Immunodeficiency Viruses

EDWARD BARKER AND SUSAN W. BARNETT

97

In the early 1980s, reports of the unusual occurrence of *Pneumocystis carinii* pneumonia and Kaposi's sarcoma in previously healthy young men in the United States led to the recognition of AIDS as a new disease syndrome (33, 43). Further studies indicated that this disease was also found in Haiti and Africa. Initial epidemiologic evidence strongly suggested that an infectious agent was responsible for the disease and that the agent was transmitted by blood or sexual contact.

By early 1983, scientists at the Pasteur Institute had isolated a novel retrovirus from a young homosexual man with lymphadenopathy (3). This virus, called lymphadenopathy-associated virus, was distinct from other known human retroviruses, such as the human T-cell leukemia virus type I (HTLV-I) and a rarer virus also isolated from leukocytes, HTLV-II. Soon afterward, biologically and molecularly similar viruses were isolated from AIDS patients in the United States and termed HTLV-III (28) and AIDS-associated retrovirus (53). These AIDS viruses differ from HTLV-I and HTLV-II in that they replicate quickly and to high titers in T-lymphocyte cultures, causing distinct cytopathic effects; they are nononcogenic; and they demonstrate unique antigenic and molecular properties.

In 1986, a subcommittee of the International Committee on Taxonomy of Viruses recommended that this new subfamily of viruses variously referred to as lymphadenopathy-associated virus, HTLV-III, and AIDS-associated retrovirus be named human immunodeficiency virus (HIV) (17). It was also suggested that the various strains of the virus be designated by a code with geographically informative letters and sequential numbers placed in brackets or as a subscript (for example, HIV-1$_{SF33}$ for San Francisco isolate 33).

In 1986, a new subtype of HIV, HIV type 2 (HIV-2), was isolated from West African AIDS patients and from seropositive asymptomatic individuals (16). HIV-2 has since become increasingly prevalent in Europe (81) and Brazil (19) and more recently in India (2). Several cases have also been reported in the United States (9, 90). While HIV-2 isolates exhibit several biologic characteristics in common with HIV-1, HIV-2 is 3 times less efficient at being transmitted sexually and 10 times less efficient at being transmitted vertically. Although HIV-2 can induce immunosuppression and AIDS (15, 90), the viral load in individuals infected with this virus is much lower than that

found in HIV-1-infected people. In addition, HIV-2 can be serologically and molecularly distinguished from HIV-1. Antigenic cross-reactivity between these subtypes is generally limited to the viral core and polymerase proteins, but most HIV-2-positive sera can be identified by using HIV-1-based serologic tests (22). Nevertheless, the spread of HIV-2 and the failure to consistently detect HIV-2-infected specimens by using HIV-1-specific reagents highlight the need to employ HIV-2-specific tests when such an infection is suspected.

DESCRIPTION OF THE AGENT

HIVs are classified in the family *Retroviridae*, genus *Lentivirinae*, by virtue of their physicochemical, molecular, and biologic characteristics. The human *Lentivirinae* are distinct from other genera of human retroviruses, such as the *Spumavirinae* (human foamy virus), HTLV-I, and HTLV-II.

Electron microscopy of HIV reveals a sphere with a cone-shaped core (Fig. 1). Virions measure 80 to 130 nm in diameter and have a unique three-layered structure (Fig. 2 and Table 1). Innermost is the genome-nucleocapsid complex, which is associated with reverse transcriptase (RT) molecules. This complex is enclosed within a capsid, which is surrounded by a host cell membrane-derived envelope from which project viral-envelope glycoprotein "spikes." The virion genome is a single-stranded positive-sense RNA molecule more than 9 kb long. Each virion is diploid, containing two identical copies of its RNA genome.

The HIV-1 and HIV-2 genomes have similar genomic structures, although subtype-specific genes have been identified (Fig. 3). These genomes contain the three genes common to all retroviruses: the *gag* gene (group-specific antigen) encodes the virion core proteins; the *pol* region encodes an RT (or RNA-dependent DNA polymerase), a protease, and an endonuclease (or integrase); and the *env* gene encodes the two major envelope glycoproteins, gp120 and gp41.

The 53-kDa precursor core polypeptide (p53 or p55) encoded by the *gag* gene is cleaved by an HIV protease (p10) into four major components. p25, the most abundant and the only phosphorylated protein of the *gag* proteins, noncovalently associates with the viral genome to form a nucleoid shell. Other *gag*-encoded core proteins are the p7 and p9 proteins, which are derived from a p15 precursor

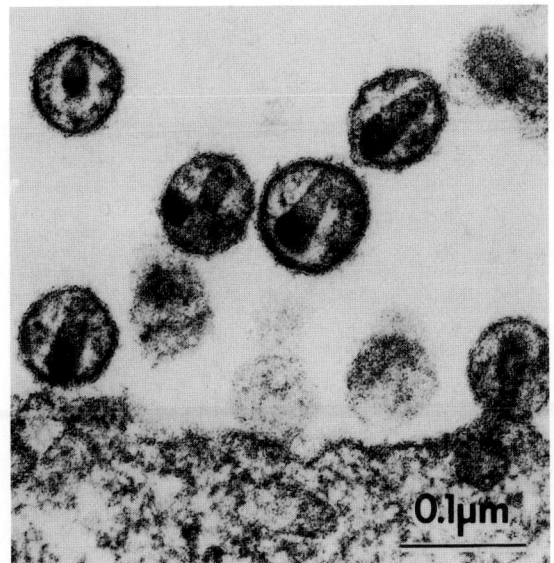

FIGURE 1 Transmission electron micrograph of HIV replicating in a T cell. Reprinted with permission from reference 52.

cleaved from the carboxy terminus of p53, and p17, which is myristylated and comes from the amino terminus of p53. p17 is found outside the viral nucleoid and forms the matrix of the virion (Fig. 2).

The *pol* polyprotein (p160) is translated by a frameshifting mechanism as ribosomes read through full-length HIV RNA transcripts. This *pol* precursor is cleaved into the self-cleaving viral protease p10; the RT, which is active in two forms, p66 and p51; and the endonuclease p32. The RT also possesses an RNase H activity required for viral replication. In the course of viral replication, these enzymes catalyze the synthesis of a double-stranded DNA provirus on the viral RNA template. Proviral cDNA is integrated into the host cell genome in reactions involving the viral endonuclease.

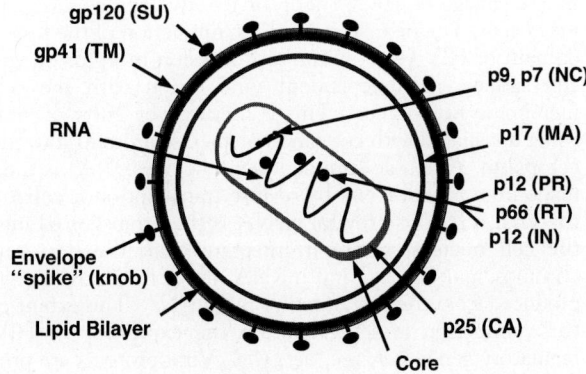

FIGURE 2 HIV virion with structural and other virion proteins identified. Abbreviated viral protein designations are listed in Table 3. Reprinted with permission from reference 52. SU, surface; TM, transmembrane; NC, nucleocapsid; MA, matrix; PR, protease; IN, integrase; CA, capsid.

TABLE 1 HIV proteins and their functions[a]

Protein	Designation(s)	Function
Group-specific antigens (Gag)	p25 (p24)	Capsid structural protein
	p17	Matrix protein, myristoylated
	p9	? RNA-binding protein
	p6	? RNA-binding protein; helps in virus budding
Polymerase (Pol)	p55, p63	RT; RNase H inside core
Protease (PR)	p15	Posttranslational processing of viral proteins
Integrase (IN)	p11	Viral cDNA integration
Envelope	gp120	Envelope surface protein
	gp41 (gp36)	Envelope transmembrane protein
Tat[+]	p14	Transactivation
Rev[+]	p19	Regulates viral mRNA expression
Nef[+]	p27	Pleiotropic, including virus suppression; myristoylated
Vif	p23	Increases virus infectivity and cell-cell transmission; ? role in virion assembly
Vpr[+]	p18	Role in virus replication; ? transactivation
Vpu[b]	p15	Role in virus release
Vpx[c]	p15	Role in infectivity
Tev[d]	p26	Tat and Rev activator

[a]See Fig. 2 for locations of viral genes on the HIV genome. Table is reprinted with permission from reference 52.
[b]Present only with HIV-1. Expression appears to be regulated by Vpr.
[c]Present only with HIV-2. May be a duplication of Vpr.
[d]Not found in association with the virion; Vpx, not certain.

The *env* gene encodes a glycosylated precursor polypeptide, gp160. This precursor is processed by the host's cellular proteases into gp120, the external envelope glycoprotein that forms the spikes on the virion, and the transmembrane protein gp41. During infection, gp120 binds to the CD4 receptor on the cell surface before penetration. The envelope proteins, especially gp120, can differ substantially among HIV-1 and HIV-2 strains (58). In addition to substantial genetic variability in *env* among these strains, some HIV-2 subtypes possess an external glycoprotein of a slightly higher molecular weight, gp140. The gp120 (or gp140) external glycoprotein is attached to the virion surface by noncovalent interactions with the viral gp41 protein, whose hydrophobic carboxy terminus serves as a transmembrane anchor. The amino-terminal residues of gp41 resemble the fusogenic domains of paramyxoviruses (32), and there is evidence that gp41 plays an important role in the cytopathic changes (e.g., syncytium formation) observed in HIV-infected cells (86) as well as in fusion of HIV with the cell surface.

In addition to these three characteristic retroviral genes, the HIV genomes encode at least four regulatory proteins. Two major genes, *tat* and *rev*, code for proteins p14 and p19, respectively, which are essential positive regulators of viral replication. *tat* is believed to act at both the transcriptional and the posttranscriptional levels to increase viral gene

HIV-1

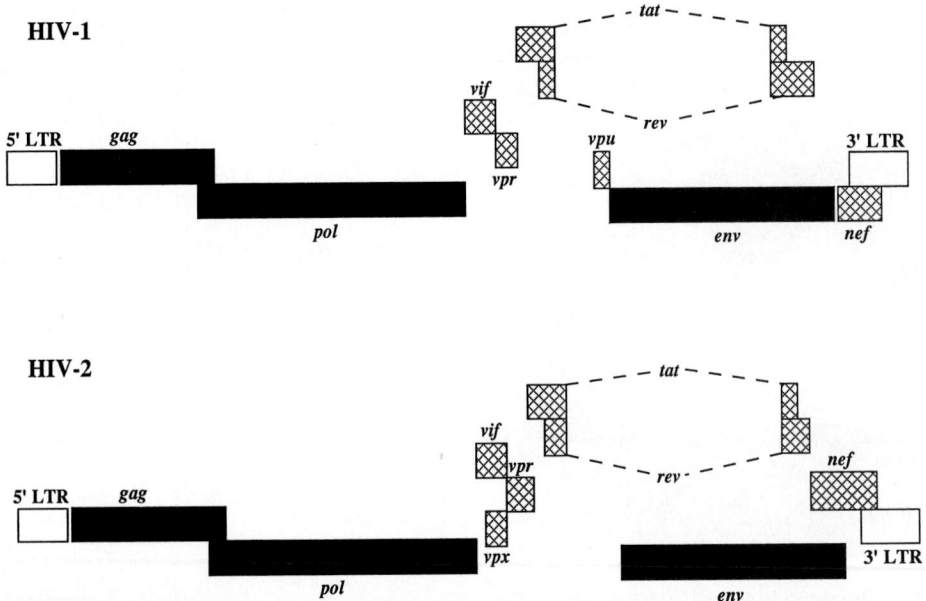

FIGURE 3   Genetic organization of HIV-1 and HIV-2. Shown are regions encoding precursors for virus proteins (solid bars [*gag*, *pol*, *env*]) and regulatory proteins (cross-hatched bars [*tat*, *rev*, *nef*, *vif*, *vpr*, *vpu*, *vpx*]). *tat* and *rev* are translated from overlapping reading frames in spliced RNA that join two open reading frames. The integrated virus contains identical long terminal repeats (open bars; LTR) at both ends of the viral coding sequence. Reprinted with permission from reference 51.

expression. *rev* is required for the transport of full-length mRNA molecules (which encode the viral structural proteins) out of the nucleus. The *nef* gene encodes a 27-kDa myristylated protein that regulates viral replication in some cells and may be important in viral latency (11). It has been shown in some studies that the *nef* protein is required for the maintenance of high viral load during infection. *vif* is a 23-kDa protein that appears to be important in later stages of virion maturation in that it allows for the processing and transport of internalized HIV core protein. *vif* appears to be required for infection of some cell types (82). Other viral genes include *vpu*, which is HIV-1 specific; *vpx*, which is HIV-2 specific; and *vpr*, which is found in both subtypes. The *vpu* gene encodes a 16-kDa protein that functions in the release of virus particles from infected cells (29) and is involved in downmodulation of the CD4 molecule on the surface of the infected cell by degrading the C-terminal end of the CD4 molecule in the endoplasmic reticulum. The *vpr* gene encodes a 15-kDa protein that increases the rate of replication of the virus and accelerates the cytopathic effect of the virus in T cells by increasing the level of viral protein expression (18). Transcriptional regulation of the viral genome appears to be mediated through indirect viral and direct cellular protein interactions with DNA sequences within the viral 5′ long terminal repeat in proviral DNA. Several binding sites for known *trans*-acting regulatory factors have been mapped in these sequences (91).

VIRAL INFECTION

HIV infects a wide variety of tissues in humans (Table 2), including bone marrow, lymph node, blood, brain, skin, and bowel. HIV infection can induce cytopathic effects such as syncytium formation (57) or apoptosis (49, 50) in some instances and latent or persistent infections in others

(27). The outcome of this infection depends at least in part on the particular strain of virus (52). In addition, susceptibility and viral production can vary depending on the target cell. For example, macrophages, which are usually susceptible to infection, produce relatively low levels of most strains of the virus, while cultured dendritic cells can produce high titers of the same virus strains (59).

HIV infection of cells is believed to involve several steps, which are shown in Fig. 4. The first step involves attachment of viral envelope protein gp120 to the cell surface receptor CD4 (20) or an alternative receptor (78). It is thought that a conformation change in gp120 and the receptor then occurs and that this change is followed by displacement of gp120. As a result of this process, proteolytic cleavage of the V3 loop or its envelope counterpart may occur. The next step involves interaction of the fusion domain of HIV (e.g., gp41) with a fusion receptor on the cell surface. pH-independent viral fusion with the cell membrane then occurs. This is followed by entry of viral RNA associated with core and polymerase proteins into the cytoplasm. A double-stranded DNA copy (cDNA) is then made from a viral RNA by reverse transcription involving the viral RT. This proviral cDNA is then transported into the cell nucleus, where it integrates into the host cell chromosomal DNA. Viral mRNA and genomic RNA are produced from the integrated proviral DNA. The extent of this production is dependent on the expression of HIV regulatory genes (*tat*, *rev*, *nef*, *vpr*). Viral proteins are produced from full-length and spliced mRNAs, and viral genomic RNA is incorporated into the capsid forming at the cell membrane. Gag and Gag-Pol polyproteins are processed via protease at the cell surface or in budding virions. Finally, the viral capsid buds through the cell membrane, incorporating both the processed viral-envelope glycopro-

TABLE 2 Human cells susceptible to HIV[a]

Hematopoietic	Brain	Skin	Bowel	Other
T lymphocytes	Capillary endothelial cells	Langerhans cells (?)	Columnar and goblet cells	Myocardium
B lymphocytes	Astrocytes	Fibroblasts	Enterochromaffin cells	Renal tubular cells
Macrophages	Macrophages (microglia)		Colon carcinoma cells	Synovial membrane
NK cells	Oligodendrocytes			Hepatic sinusoid epithelium
Megakaryocytes	Ganglia cells			Hepatic carcinoma cells
Dendritic cells	Neuroblastoma cells			Kupffer cells
Promyelocytes	Glioma cell lines			Pulmonary fibroblasts
Stem cells	Neurons (?)			Fetal adrenal cells
Thymic epithelium				Adrenal carcinoma cells
Follicular dendritic cells				Retina
				Cervix (? epithelium)
				Prostate
				Testes
				Osteosarcoma cells
				Rhabdomyosarcoma cells
				Fetal chorionic villi
				Trophoblast cells

[a]Susceptibility to HIV is determined by in vitro or in vivo studies. Table is reprinted with permission from reference 52.

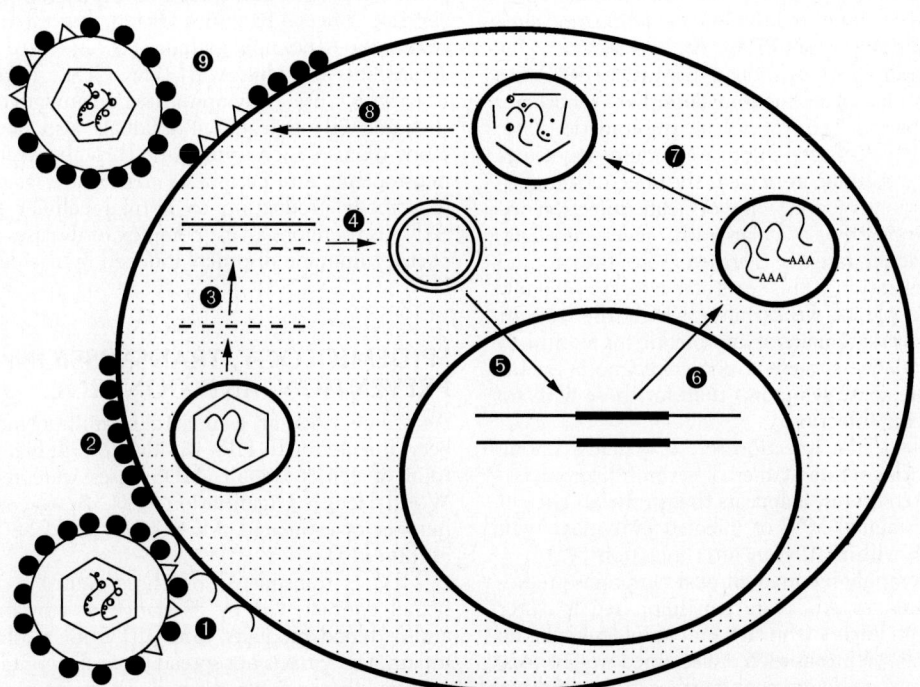

FIGURE 4 HIV infection cycle. In steps 3 through 5, viral core proteins are associated with the viral genome. – – –, RNA; ——, DNA. Double-stranded circular forms can be covalently or noncovalently bound. The latter are the forms that integrate into the cell chromosome. Step 1, attachment and fusion; step 2, uncoating and nucleocapsid entry; step 3, reverse transcription; step 4, cDNA formation; step 5, integration; step 6, mRNA and viral RNA transcription; step 7, translation; step 8, core particle assembly; step 9, final assembly and budding. Symbols: ●, gp120; △, cellular proteins; ∪, viral receptor (i.e., CD4). Reprinted with permission from reference 52.

TABLE 3 Lentiviruses[a]

Virus	Host infected	Primary cell type infected	Clinical disorder
Equine infectious anemia virus	Horse	Macrophages	Cyclical infection in first year: hemolytic anemia and sometimes encephalopathy
Visna-maedi virus	Sheep	Macrophages	Encephalopathy
Caprine arthritis-encephalitis virus	Goat	Macrophages	Immune deficiency, encephalopathy
Bovine immunodeficiency virus	Cow	Macrophages	Lymphadenopathy; lymphocytosis; ? central nervous system disease
Feline immunodeficiency virus	Cat	T lymphocytes	Immune deficiency
Simian immunodeficiency virus	Primate	T lymphocytes	Immune deficiency, encephalopathy
HIV	Human	T lymphocytes	Immune deficiency, encephalopathy

[a]Table is reprinted with permission from reference 52.

teins and some host cell proteins present on the cell surface (1). This final stage may involve the function of the viral *vif* gene product (75).

CLINICAL SIGNIFICANCE OF HIV INFECTION

Lentivirus infections in general are associated with a number of diseases that have long incubation periods and involve the hematopoietic and central nervous systems. In most instances, lentivirus infections can lead to immune suppression, arthritis, and autoimmune diseases. Lentiviruses that have been identified in other species include visna-maedi virus of sheep, equine infectious anemia virus, caprine-encephalitis virus, and the bovine, feline, and simian immunodeficiency viruses (Table 3).

HIV infection in humans can lead to a variety of disease states, including an acute mononucleosislike syndrome, prolonged asymptomatic infection, a symptomatic state, and AIDS. In a large proportion of cases, adults may experience the acute syndrome at or near the time of seroconversion. The distinctive symptoms of acute infection include lymphadenopathy, macular rash, fever, myalgia, arthralgia, headache, fatigue, diarrhea, sore throat, and neurologic manifestations. This syndrome may last for a few days up to a couple of weeks, after which the patient generally recovers and remains asymptomatic for months to years. Recent evidence suggests a poorer prognosis for individuals with the acute symptoms than for those with no symptoms following infection.

Progression from HIV infection to AIDS differs among different populations. From studies of several different cohorts in the United States, it appears that in the absence of treatment, an estimated 50% of infected individuals will progress to AIDS within 10 years after infection (45).

The most commonly reported clinical symptoms predictive for progression to AIDS are persistent herpes zoster infections, oral candidiasis (thrush), oral hairy leukoplakia, and constitutional symptoms such as sustained weight loss, fatigue, night sweats, and persistent diarrhea. Some of these symptoms characterize what historically has been referred to as AIDS-related complex. Herpes zoster is an early sign of progression, candidiasis and hairy leukoplakia occur later, and the persistent constitutional symptoms are highly indicative of progression to AIDS. Persistent generalized lymphadenopathy, which often appears during infection, was originally believed to predict progression to AIDS in HIV-infected homosexual men but now appears not to be a good predictor. Most recently, the Centers for Disease Control and Prevention (CDC) has expanded the AIDS-predicting symptoms to include those observed in HIV-infected women progressing to AIDS. These include vulvovaginal candidiasis, moderate or severe cervical dysplasia, and pelvic inflammatory disease.

The AIDS-defining diseases and symptoms include Kaposi's sarcoma, *P. carinii* pneumonia, chronic diarrhea often caused by cryptosporidia, cryptococcal meningitis, toxoplasmosis, encephalopathies and dementia, cytomegalovirus retinitis, esophageal candidiasis, anal-rectal carcinomas, and B-lymphocytic lymphomas. Since 1993, this list has been extended to include three other clinical conditions: pulmonary tuberculosis, recurrent pneumonia, and invasive cervical cancer (10). In addition, the CDC classification system for AIDS now includes HIV-infected asymptomatic adolescents and adults with <200 CD4$^+$ T lymphocytes per μl or with CD4$^+$ T-lymphocyte percentages of less than 14.

Some HIV-infected individuals also develop autoimmune diseases such as immune thrombocytopenic purpura, neutropenia, and peripheral neuropathy as a result of autoantibody production to normal cellular proteins (44). AIDS is therefore a vast complex of diseases resulting from primary immune disorders induced by the causative agent, HIV.

EPIDEMIOLOGY, TRANSMISSION, PREVENTION, AND CONTROL

As of late 1993, an estimated 14 million individuals have been infected with HIV worldwide (94; Fig. 5). In 1992, a total of 484,163 adult AIDS cases were reported to the World Health Organization (12). By the year 2000, the number of estimated AIDS cases could be as high as 24 million (13).

HIV is transmitted by three routes: (i) by intimate sexual contact, (ii) by exchange of contaminated blood and/or blood products, and (iii) from mother to fetus or infant. The virus is not spread by casual contact or by insect vectors. Transmission most likely occurs via virus-infected cells, since relatively little free virus is found in plasma or body fluids compared to that in infected cells (Table 4). The cell-free virus titer in saliva, tears, urine, and milk is at least 10-fold less than that in plasma (52). Plasma usually contains between 1 and 5,000 particles per ml, a number that differs depending on the stage of infection. This amount in plasma is much less than the 100 million to 1 billion particles of cell-free infectious hepatitis B virus

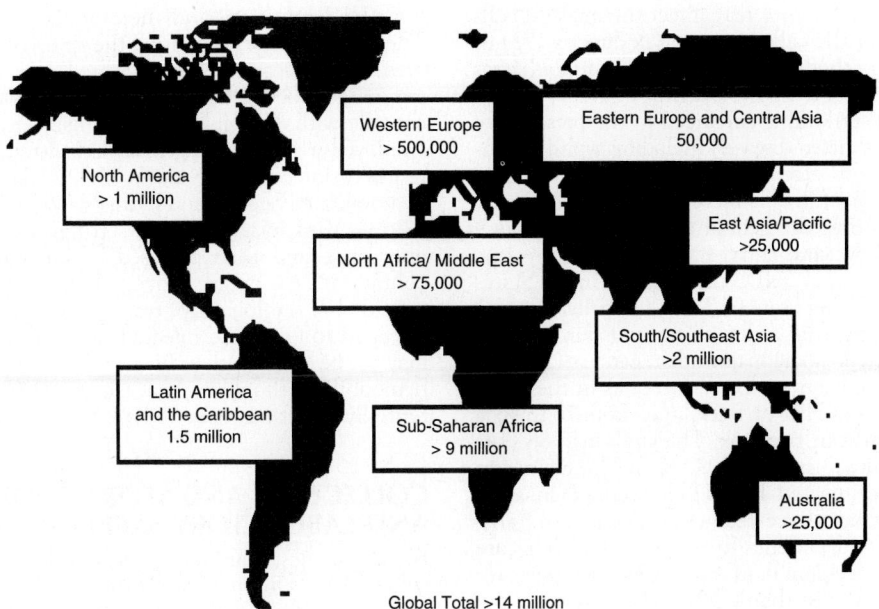

FIGURE 5 Global estimates of cases of HIV infection in 1993. Reprinted with permission from references 64 and 94.

particles found in 1 ml of blood from some hepatitis B virus-infected individuals (see chapter 91 of this Manual).

The largest group of individuals infected with HIV were exposed to the virus through unprotected sexual contact. Heterosexual intercourse is the predominant mode of transmission globally (89). On a regional scale, however, there is a difference in the incidence of infection depending on sexual preference. For example, in a recent study in Ger-

many, the incidence of infection by male heterosexual contacts is five times lower than that by male homosexual contacts (84). In contrast, in sub-Saharan Africa, the incidence of infection by heterosexual contacts is much higher than that by homosexual contacts (41). Even though this situation exists in Africa, heterosexual intercourse is a less efficient mechanism of transmission than homosexual intercourse. In either case, the receptive part-

TABLE 4 Isolation of HIV from body fluids[a]

Source	Virus isolation (no. positive/no. tested)	Estimated quantity of HIV (no. of infectious particles/ml)	% of cells infected
Cell-free fluid			
Plasma	33/33	1–5,000[b]	
Tears	2/5	<1	
Ear secretions	1/8	5–10	
Saliva	3/55	<1	
Sweat	0/2	NK	
Feces	0/2	NK	
Urine	1/5	<1	
Vaginal, cervical	5/16	<1	
Semen	5/15	10–50	
Milk	1/5	<1	
Cerebrospinal fluid	21/40	10–10,000	
Infected cells			
PBMCs	89/92		0.001–1[b]
Saliva	4/11		<0.01
Bronchial fluid	3/24		NK
Vaginal and cervical fluids	7/16		NK
Semen	11/28		0.01–5

[a]Table is reprinted with permission from reference 52. NK, not known.
[b]High levels are associated with symptoms and advanced disease.

ner is more susceptible to infection than the insertive or active partner (24). The virus may infect the mucosal cells of the bowel (66) or the cells of the cervix directly (72) or may gain access to other susceptible cells via injured areas in the mucosal lining of the bowel, vagina, or uterus. Sexual transmission of either kind is enhanced in the presence of other sexually transmitted diseases, including syphilis, gonorrhea, and especially ulcerating diseases such as genital herpes and chancroid (48). These conditions bring many HIV-infected cells into the genital fluids as well as cause ulcerations that allow viral entry.

In 1990, a quarter of all AIDS cases in the United States were in intravenous-drug users (IVDU). Globally, the rate of seroconversion by this route of transmission has increased dramatically in some major urban centers (23). For example, there have been explosive increases in the numbers of transmissions of HIV by IVDU, i.e., from 16 to 46% in a matter of months in Bangkok, Thailand, in 1988 (14) and >50% in Manipur, India, in 1990 (76). This change in percentage of HIV-infected IVDU has declined in areas where needle exchanges have been used. The rate of transmission from infected patients to uninfected health care workers following accidental needle sticks, however, remains extremely low (less than 0.5%). The reason for the difference between accidental needle sticks and IVDU is in the volume of blood that is transferred. In accidental needle stick simulations, a mean volume of 0.034 μl of blood is transferred by hollow-bore needles (40). In contrast, IVDU needle-sharing practices result in the transfer of volumes approximately 1,000 times larger. Even though the incidence of infection by needle sticks is low (0.5%), widespread infections have occurred in areas where it is common practice to reuse syringes and needles because of inadequate or nonexistent sterilization equipment.

Another group of HIV-infected individuals that has emerged much later in the course of the epidemic are children of infected mothers. The estimated frequency of transmission by this route is between 15 and 40% (25). Transmission occurs in three different periods: in utero, intrapartum, and postpartum. Transplacental infection contributes to the major means of transmission (in utero transmission), usually during the third trimester, although there have been cases in which this infection occurred in the first trimester (25, 74). HIV-seropositive mothers are more infectious during the clinical disease, in which the second peak of viremia appears. At delivery, infants can become infected through the birth canal. First-born twins experience more trauma during delivery and are more often infected than second-born twins. Infected mothers do not usually transmit HIV postpartum unless they experience viremia (i.e., after being acutely infected) while breastfeeding. Breast-feeding is not, however, a major route of transmission. In fact, in developing countries, the advantages of breast-feeding by infected mothers in decreasing childhood mortality far outweigh the slight chances of infecting the child. In addition, breast-feeding greatly delays the onset of AIDS in children who are already infected with the virus (85).

The other groups at risk for HIV infection are hemophiliacs and blood transfusion recipients. Transmission by contaminated blood products is the most efficient way of acquiring HIV infection. Fortunately, transmission in these groups has been almost eliminated as the result of serologic screening of blood donors and heat treatment of clotting factor preparations (54) in countries where these programs have been possible.

At present, there is no vaccine to protect an individual from HIV infection, and there are no curative therapies. The best means of preventing the spread of the infection at present is education. Simply stated, one must avoid contaminated blood and blood products and/or intimate sexual contact with HIV-infected persons. Condoms should be employed when engaging in sexual activities involving potentially infected individuals. IVDU must avoid sharing contaminated needles and syringes. When hypodermic needles are used in the medical setting, they should be used only once and then disposed of immediately in proper containers. As noted above, blood and blood products should be serologically tested and donors should be screened to eliminate infected individuals, some of whom may yet be seronegative. Blood factor concentrates used by hemophiliacs should be heat treated to inactivate any potentially infectious virus (54).

COLLECTION AND STORAGE OF SPECIMENS AND LABORATORY SAFETY

Blood samples, routinely collected by venipuncture, are generally used for viral detection and serologic analyses. To isolate virus from peripheral blood mononuclear cells (PBMCs), heparinized or EDTA-treated blood samples must be collected. HIV can also be detected in and isolated from other body fluids such as plasma, serum, cerebrospinal fluid, saliva, tears, milk, urine, and genital secretions and from biopsy specimens of infected tissues such as the bowel. Even though the amount of virus present in some biopsy specimens may be low, precautions should be taken when handling any potentially HIV-infected clinical specimens. Gloves should be worn, and needles should be handled by safe procedures after collection of blood.

Plasma, serum, or other body fluids to be tested serologically should be stored at -20 or $-70°C$. For later detection of viral antigens, clinical specimens should preferably be stored at $-70°C$. For best results, virus isolations should be performed immediately following specimen collection (see below). Nevertheless, PBMCs and other infected cells may be stored in liquid nitrogen in culture medium plus 10% dimethyl sulfoxide with reasonable recovery of virus from the PBMCs in most cases.

Laboratories involved in HIV antibody testing should use biosafety level 2 standards. Clinical specimens from all individuals should be considered infected so as to avoid the confusion of having certain specimens handled differently. A class II type A biosafety cabinet with HEPA-filtered air should be employed for viral isolation work; gloves (two pairs) and a surgical gown with sleeve cuffs that can be tucked into gloves should be worn. Because the levels of free virus in body fluids including blood are very low, transmission via aerosol from primary specimens is unlikely. However, when large quantities of virus are being grown, when concentrated viral preparations are being used, and when procedures that may produce droplets or aerosols are being performed, biosafety level 2 and 3 standards should be employed. In our laboratory, we take the following precautions.

1. All specimens are stored in well-constructed containers with secure lids that prevent leakage during transport. When glass containers are used, a secondary container that prevents breakage is desirable. Care is taken to avoid contaminating the outsides of the containers. These containers should be labeled to indicate that they contain HIV, and

they should be stored in areas with appropriate biohazard warning labels.

2. Persons processing primary specimens or culturing virus wear gloves, gowns, surgical masks, and protective eyewear. Two pairs of gloves are suggested. To avoid contamination of the laboratory, at least the outer pair of gloves should be removed and disposed of in an autoclavable biohazard bag after completion of tasks. Gloves should also be removed whenever they are visibly contaminated.

3. The generation of droplets, aerosols, and spills should be avoided. A certified biologic safety cabinet is employed for any work that could produce aerosols and for all manipulations of infected cell cultures. Safety cabinets are not required for routine procedures that involve already-fixed infected cells or inactivated virus.

4. Mechanical pipetting devices are used for manipulating all fluids in the laboratory. All pipettes should be plugged with cotton to avoid cross-contamination of specimens and contamination of pipetting devices.

5. The use of needles and glassware is restricted to situations for which there is no alternative. Needles and glass are disposed of in closed puncture-proof containers specifically designed for this purpose. All plasticware is decontaminated with either bleach or Wescodyne and then disposed of in autoclavable biohazard bags.

6. Laboratory work surfaces are decontaminated with an appropriate germicide after a spill and when work is completed. Contaminated materials are decontaminated and disposed of according to institutional policies on infectious waste.

7. Hand washing with soap and water immediately after infectious materials have been handled and after work is completed, even when gloves have been worn, should be a routine practice.

HIV can be readily inactivated under clinical laboratory conditions by several methods (73, 83). Although relatively stable in a dry or lyophilized state (55), the virus is very sensitive to many detergents (for example, Triton-X and Nonidet P-40 but not Tween 20), including soap, and can be eliminated rapidly by bleach (0.5% sodium hypochlorite). Alcohol and acetone-alcohol mixtures (>70%) can also inactivate virus efficiently at room temperature. HIV strains are also sensitive to iodophores such as organic iodine compounds, pH extremes (pH 2, pH 12), UV and X irradiation, and heating in liquid solution at 56°C. For inactivation of virus in serum prior to serologic testing, specimens should be routinely heated at 56°C for 30 min.

SEROLOGIC ASSAYS TO DETECT HIV INFECTION

Detection of HIV antibodies is still the most efficient and most common way to determine whether an individual has been exposed to HIV and to screen blood and blood products for this infectious agent. A test is considered positive when assays such as the enzyme-linked immunosorbent assay (ELISA), indirect immunofluorescence assay (IFA), and Western blots (immunoblots) are considered reactive. A positive test result indicates exposure and, outside of the perinatal and neonatal periods, is presumed to indicate infection by the virus.

ELISA

The ELISA is the assay most commonly used to screen for HIV because of its relatively low cost, standardized proce-

dure, high reliability, and ability to give rapid results. The sensitivity of the test has ranged from 93 to 100%. Under optimal laboratory conditions, the sensitivity and specificity in most cases are >99% (87). Commercial ELISA kits are available and are licensed by the Food and Drug Administration.

Generally, the test procedure involves a noncompetitive indirect staining procedure using immobilized HIV antigen to bind anti-HIV antibodies. Bound HIV antibodies are detected after being complexed with enzyme-labeled antihuman immunoglobulin G (IgG), which catalyzes a colorless substrate to a colored product. The color change, read spectrophotometrically as optical density, is directly proportional to the concentration of HIV antibodies in the test sample. The result is evaluated on the basis of the cutoff optical density values for positive and negative controls determined in each test.

Most assays use viral antigens obtained from HIV-infected T-lymphocyte cell lines. These preparations are usually rich in p24, p17, gp160, gp120, and gp41. False-positive results from this test can arise from contamination of the cell culture preparation with culture material to which non-HIV human antibodies respond. The antigen that most commonly produces false-positive results is HLA-DR (77). Because of this, many commercial ELISA kits use recombinant and synthetic peptides as antigens.

False-negative results can occur when the tests are done before seroconversion, when the patient is immunosuppressed, and sometimes when the individual is late in the course of AIDS. In addition, some ELISA kits used to detect HIV-1 may not detect antibodies to HIV-2. Other problematic clinical states that lead to false results include hemodialysis, autoimmune disorders, multiple myeloma, and hemophilia (7). Human error and variability in test kits may also contribute to some false results. Because of the possibility of false results, it is important to perform repetitive ELISA testing and to confirm ELISA results with Western blots and IFA as described below (Fig. 6).

Rapid Agglutination Screening Assays

Rapid latex agglutination screening is most effective when ELISA cannot be performed properly because of inadequate testing facilities.

The latex agglutination test uses latex beads coated with recombinant or synthetic antigen. Small amounts of whole blood or serum are mixed with the latex beads and rotated on a card. After several minutes, the tests are evaluated by eye in bright light for level of agglutination relative to that in the negative control.

The advantages of this method are the ease of performance and the ability to perform tests without electrical equipment. The main disadvantages are the method's variable sensitivity, which ranges from 71 to 99%, and its somewhat variable specificity, which ranges from 93 to 99% (87). In addition, one study demonstrated that positive and negative results scored by different observers varied by 20% (61), a difference related to the difficulty of assessing weak agglutination reactions. Thus, the need for confirmation following this test is much greater than that for ELISA.

Western Blot

The Western blot assay is the test most commonly used for confirming the presence of HIV-specific antibodies. However, Western blots are more expensive and more time-consuming than ELISAs, and they require more technical expertise owing to subjective interpretations. Specificity and sensitivity are high but are somewhat dependent on the interpretive criteria employed (see below; Table 5).

Western blots are prepared with partially purified whole virus, infected cell lysates, or recombinant viral proteins electrophoresed in denaturing gels. The advantage of using cell lysates rather than whole virus is the presence of relatively abundant viral proteins not found in large amounts in the virus. The viral proteins separated on the gel are transferred onto nitrocellulose paper. The nitrocellulose is treated with a buffer containing nonspecific proteins, and the antigen-impregnated nitrocellulose is then exposed to the test serum. Afterward, the nitrocellulose is exposed to an enzyme-linked secondary antibody. The profile of antibodies bound to the various viral proteins is identified by adding substrate. The color change that occurs can be visualized as bands on the nitrocellulose that correspond to the migrations of the different viral proteins (Fig. 7). Positive and negative controls along with molecular weight markers are run simultaneously to allow identification of the antibodies reactive with specific viral proteins.

Positive results are interpreted as the presence of two or three bands. Which bands should be present for the specimen to be considered positive depends on the health organization that sets the criteria (Table 5). The absence of all bands on the nitrocellulose is interpreted as a negative

TABLE 5 Different interpretive criteria in the United States for positive Western blot tests[a]

Health organization(s)	Criteria
Food and Drug Administration (Bethesda, Md.), Consortium for Retrovirus Serology Standardization (Davis, Calif.)	Bands for p24 or p31 and for gp41 or gp160/120
American Red Cross (Washington, D.C.)	Three or more bands, with one band each from *gag*, *pol*, and *env*
CDC (Atlanta, Ga.), Association of State and Territorial Public Health Laboratory Directors (Iowa City, Iowa)	Two of three bands: p24, gp41, or gp160/gp120

[a]Table is reprinted with permission from reference 7.

result. If any band that does not meet the criteria for a positive result is present, it is regarded as indeterminate. When indeterminate results are seen, the test subject's serum should be tested again 6 months later to see whether the same band appears. For reasons similar to those described above for ELISA, false-positive results can occur with Western blot assays. If they do, other tests, such as viral culture or PCR, should be used to confirm whether results are correct (see below).

RIPA

Antibodies to HIV can also be detected by radioimmunoprecipitation assay (RIPA), which combines immunoprecipitation of radioactively labeled HIV proteins with sodium dodecyl sulfate-polyacrylamide electrophoresis (SDS-PAGE). Radiolabeling can be performed either by incorporation of [^{35}S]cysteine (or [^{35}S]methionine),

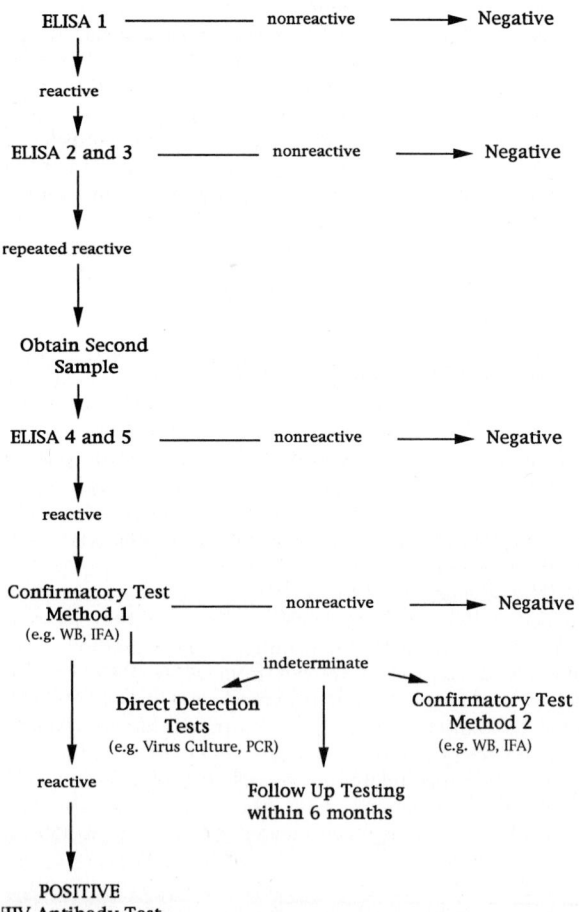

FIGURE 6 Algorithm for testing HIV antibodies in serum. WB, Western blot. Reprinted with permission from reference 7.

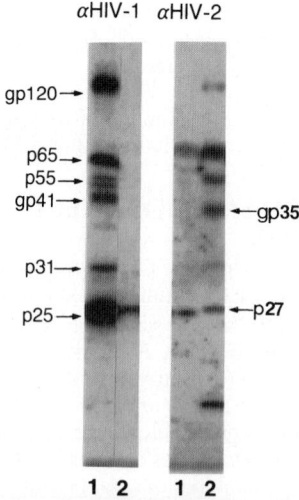

FIGURE 7 Proteins from purified HIV-1 (lanes 1) and HIV-2 (lanes 2) isolates were reacted with serum from an HIV-1- or HIV-2-infected individual. Reprinted with permission from reference 26.

[³H]leucine (or [³H]isoleucine), or [¹⁴C]glucosamine into culture or by direct iodination of purified recombinant HIV proteins. After labeling, the samples are washed extensively (cells) or separated from excess label by gel filtration chromatography (proteins). Lysates or protein preparations are preabsorbed with healthy serum and clarified with the addition of *Staphylococcus* protein A conjugated to Sepharose beads. Subsequently, immunoprecipitation is performed by overnight incubation of the radiolabeled antigen preparation with the test serum. Antibody-antigen complexes are precipitated by the addition of protein A-Sepharose, and precipitates are washed extensively with buffer. Precipitated proteins are eluted by boiling in a sample buffer and are analyzed by SDS-PAGE followed by autoradiography. The main advantage of this method over the Western blot is that antibodies can react with antigens in native conditions before undergoing electrophoresis. RIPA can also be employed to detect viral antigens with a known positive antiserum. The RIPA is useful in resolving the antibody status of individuals infected with HIV-1 and HIV-2.

Indirect IFA

The indirect IFA, like the Western blot, is an excellent test for confirming results obtained in the ELISA. However, IFA produces results much faster and at a fraction of the cost of a Western blot, provided a fluorescence microscope is available.

This test uses cells from chronically infected T-cell lines that are fixed on a slide with acetone, air dried, and then incubated with patient serum. After incubation, the cells are washed and incubated with fluorescein-conjugated anti-human IgG. After the second incubation, the cells are washed again and stained with Evan's blue. Negative and positive control sera are tested simultaneously. The slides are evaluated microscopically for fluorescent staining patterns and intensity of fluorescence. Fluorescences of infected and uninfected cell lines are compared alone with the negative and positive controls to eliminate background fluorescence and nonspecific binding. Indeterminate results are usually followed up by Western blots. In some cases, false results arise as described above for ELISAs, and these results should be reconfirmed with other tests.

ISOLATION AND CULTIVATION OF VIRUS

While current serologic tests (described above) determine whether an individual has been exposed to HIV, a definitive test for an active infection is the recovery of HIV from cells or cell-free fluids of the infected individual. In fact, the initial isolation of HIV was made by cultivating PBMCs from a person with lymphadenopathy syndrome (3). Patient specimens can be cocultured with cell lines permissive for HIV to promote the replication of the virus. Preferably, patient specimens can be cocultured with phytohemagglutinin-stimulated PBMCs obtained from seronegative individuals in a medium containing interleukin-2.

Primary viral isolation from patient material depends on the level of HIV being expressed. This level usually ranges from 1 infected CD4⁺ cell in 100 to 1 in 10,000 (5). Thus, an adequate amount of heparinized blood is needed (no less than 2 ml). For best results, the blood specimen should be cultured within a few hours after venipuncture. Nevertheless, whole blood can be held up to 36 h at room temperature without significant reduction of virus recovery from the cells. Any blood sample to be kept for more than 36 h should be stored at room temperature for 24 h and held at 4°C thereafter. PBMCs should be recovered from the blood by Ficoll-Hypaque gradient centrifugation. Once purified, PBMCs can then be maintained in serum-containing medium with interleukin-2 for 7 to 10 days before being cultured for virus. The coculture method is also applicable for detecting virus in body fluids; however, fluids should be stored at −70°C within 3 h following their removal from an individual.

Usually, 3×10^6 to 6×10^6 PBMCs from the infected individual are mixed with an equal number of uninfected PBMCs (which were previously stimulated with phytohemagglutinin for 3 days and washed) to initiate the culture. In cases of low numbers of PBMCs (e.g., infants), as few as 10^6 cells can be used if the culture vessel is kept small (8). Cultures are kept at 37°C in a 5% CO_2 atmosphere. The cultures are monitored for the presence of virus in the medium when the medium is replaced every 3 to 4 days. For the detection of virus in culture fluid, the fluid is first filtered (0.45-μm-pore-size filter) or centrifuged at 2,000 × g for 15 min to remove cell debris and then subjected to the RT assay and/or the p24 antigen ELISA described below. Culture supernatants possessing positive RT activity or p24 antigen ELISA reactivity can be saved and stored at −70°C for further studies. For most HIV-infected individuals, RT activity or detectable p24 levels become apparent within the first 2 weeks of culture; by 30 days, more than 90% of those infected cultures yield positive results (8). In some cases, when levels of infectious virus are very low, these cultures may take longer, sometimes up to 60 days. In some instances, CD8⁺ cells within the PBMC group should be removed to enhance the recovery of virus, since these cells suppress the ability of HIV to replicate in cells (88). These methods can also be adapted and employed to detect and isolate HIV in other tissue specimens or cell-free body fluids, provided the specimen is collected and used immediately or is stored at −70°C within 3 h (69).

DETECTION OF VIRUS AND VIRUS-INFECTED CELLS

Detection of HIV Antigens by ELISA

Virus can be detected in serum, plasma, or cell culture supernatant by using the solid-phase capture technique, or "sandwich" ELISA. This assay utilizes anti-HIV polyclonal or monoclonal antibodies that are coated onto wells of plastic plates to capture viral antigen. After being washed, the attached antigens are reacted with another anti-HIV antibody in which an enzyme is bound. A colorless substrate that is converted by the enzyme to give a color is added, and results are measured spectophotometrically. The absorbance is proportional to the amount of HIV antigen present. One of the antigens most commonly measured is the p24 protein. Some commercially available kits can detect picogram amounts of HIV p24 protein per milliliter of test sample.

RT Assay

The RT assay can be used to monitor retrovirus replication during virus isolation procedures. Since the RT assay is not specific for HIV, it must be used in conjunction with other clinical, serologic, and/or HIV-specific antigen data to confirm HIV infection. The RT assay for HIV detection, well described elsewhere (39), employs magnesium instead of manganese as the divalent cation in the reaction mixture. This mixture also contains a poly(A) RNA template, an oligo(dT) primer, a detergent to disrupt virions, appropriate buffers, unlabeled deoxynucleoside triphosphates (dNTPs), and radioactively labeled dTTP (usually, [³H]dTTP). RT assays are routinely performed on virus pelleted from 1 ml of culture fluid (centrifuged for 2 h at 16,000 × *g* and 4°C). Reaction products are precipitated onto filters by trichloroacetic acid precipitation, washed with ethanol, and counted by scintillation spectroscopy. This RT method can be adapted to automated procedures suitable for running large numbers of samples. RT assays can also be performed on smaller aliquots of culture fluids (without centrifuging the virus) by using ³²P-labeled dNTPs. Reaction products are spotted onto paper, and autoradiography is used as a means of detecting RT activity (93).

Western Blot Assay

Western blots, more commonly employed to detect HIV antibodies in serum (see above), also can be used to analyze infected cell lysates or concentrates of culture supernatants for the presence of HIV proteins. The advantage of this method over ELISA and IFA is that the different HIV gene products can be identified and distinguished by their sizes. For antigen detection, a high-titer control serum with activity against all or most HIV proteins is preferable. When low antigen levels are to be detected, the use of a radiolabeled probe such as ¹²⁵I-protein A in the final step is preferred over the use of enzyme conjugates because of the sensitivity of the probe. Reactive bands are revealed by autoradiography, and their molecular weights are determined by comparison with molecular weight markers run on the same gel. The Western blot is very useful in distinguishing HIV-1 and HIV-2 isolates following virus culture. In most cases, HIV-1- and HIV-2-specific sera can be employed to unambiguously identify which HIV subtype is present. In addition, the sizes of the envelope glycoproteins generally are distinct (gp120 for HIV-1, gp140 for HIV-2); differences in the core protein may also be observed.

Indirect IFA

The indirect IFA is also useful as a means of detecting virus in infected cells. The presence of virus in PBMC cultures or in other cells can be confirmed by the detection of viral antigens with HIV-specific antiserum. For this analysis, cells are washed three times in an isotonic solution at neutral pH and suspended at an appropriate concentration before being spotted onto a pretreated IFA slide. After air drying, samples are fixed in acetone, air dried, and reacted with a known anti-HIV-antibody-containing serum. Uninfected cells and serum from HIV-negative individuals should always be run in parallel with test cells to rule out antibodies made to cellular proteins. Monoclonal antibodies to specific HIV antigens can also be employed. Following incubation with antibody, the cells are washed, incubated with fluorescein-labeled anti-human IgG, and washed. Slides are then mounted and observed under a fluorescence microscope. If immunofluorescence microscopes are not available, horseradish peroxidase enzyme-conjugated secondary antibodies can be substituted for the fluorescent antibody. Following the addition of appropriate substrates, these specimens can be visualized by using conventional microscopes.

IFA is a rapid, sensitive method in which as little as a single HIV-expressing cell can be detected in a population of hundreds or thousands of cells examined. However, the success of this assay is dependent on the experience of the reader in interpreting fluorescence patterns.

Antibodies conjugated with fluorescent molecules can also be used in conjunction with flow cytometric analysis to test the level of viral antigen in cells and the frequency of cells expressing these antigens. Using a primary fixation step in 90% methanol allows the detection of intracellular antigens by this technique. The level of p24 detected in PBMCs by flow cytometry is inversely related to the CD4⁺ count of the tested individual (62). Thus, this technique can be used to determine the viral burden of the individual (see below).

PCR

Presently, the most sensitive method of detecting HIV infection is the PCR assay (42). The assays described above are incapable of detecting the proviral form of the virus. The molecular targets of the PCR procedure are most commonly integrated and/or free episomal HIV proviral DNA sequences, but the assay can also be modified to permit detection of HIV RNA (37). The PCR assay for HIV DNA can be performed using only 1 μg of cellular DNA, which is equivalent to only 21 μl of whole blood (47). This DNA amplification procedure uses two oligonucleotide primers that are complementary to the plus and minus strands of target DNA (Fig. 8). In order to ensure detection of many different HIV strains, the sequence of the primers should correspond to regions of the HIV genome that are highly conserved among different HIV isolates. Such regions have been found in *gag*, *env*, *pol*, and the long terminal repeat. Repeated denaturation, renaturation, and elongation of the primers with a thermostable DNA polymerase is achieved by repeated temperature cycling. This process results in an exponential increase in the number of copies of the DNA region that is flanked by the primers. This DNA is then denatured and hybridized in solution to a ³²P-labeled HIV probe that contains sequences complementary to those amplified. When HIV sequences are present, a target-probe complex is formed that can be visualized by autoradiography following PAGE. For some studies, the use of multiple primer pairs (and their respective probes) is recommended to further guarantee detection of HIV variants (68).

The sensitivity of detection of HIV sequences by PCR with a *gag* primer pair and a *gag* probe has been reported to be 18 copies of viral genome in DNA from 150,000 cells (47). In a study employing 57 infected individuals, the virus could be recovered from the PBMCs in at least 89% of cases and from plasma in 75% of cases. In contrast, by PCR methods, proviral HIV DNA was detected in PBMCs from all individuals tested. HIV DNA was not detected in any of the HIV-seronegative controls. Thus, PCR appears to be the most sensitive method of detecting virus in infected individuals.

While low levels of HIV can often be detected after in vitro propagation of infected cells in culture, the PCR method possesses the unique advantage of being able to detect nonreplicating (latent) viral genomes. Another advantage of the PCR procedure is that it can be applied to DNA isolated directly from PBMCs of fresh blood, thus eliminating the need for cell culturing. The main disadvantage of this method is that owing to the extreme sensitivity of the assay (97 to 100%), it is highly prone to cross-contamination with nucleic acids, which cause many false-positive results. If measures are taken to minimize nucleic acid contamination (46), the specificity can be as high as 100%. Primer pairs should be carefully selected in order to

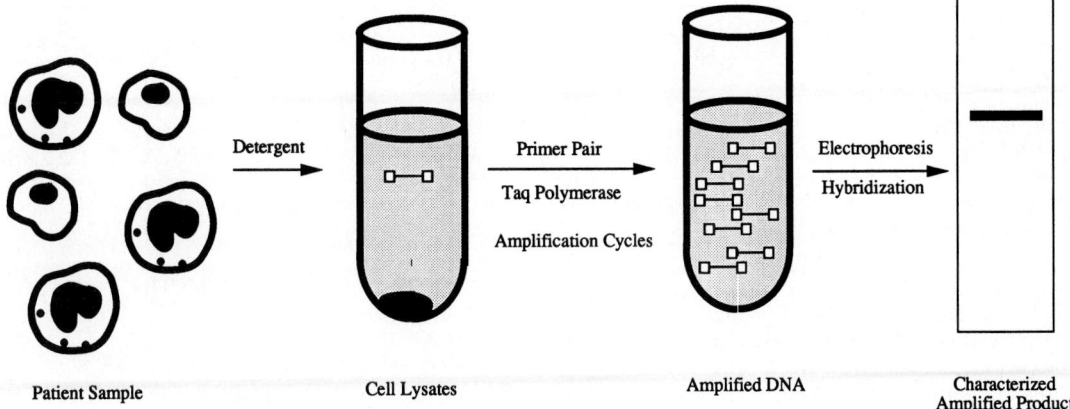

FIGURE 8 PCR procedure for amplification of DNA or RNA at low copy number. Reprinted with permission from reference 21.

avoid detection of noninfectious endogenous retroviruses (47).

In Situ Hybridization

In situ hybridization is a histomolecular method with which to detect HIV nucleic acids in different cells and tissues. Slides containing fixed specimens are probed for the presence of intracellular HIV nucleic acids. The technique, described in detail elsewhere (36, 92), involves fixation in paraformaldehyde followed by rehydration, digestion with proteinase K, acetylation, and treatment with either RNase or DNase. Cells are then refixed with paraformaldehyde and dehydrated. To detect both double- and single-stranded nucleic acids, samples are denatured in 95% formamide at 65°C for 15 min, and the reaction is stopped on ice prior to dehydration. Hybridization is performed by applying a denatured ^{35}S-labeled HIV DNA probe (approximate specific activity, 2×10^8 dpm/μg) to the cytologic or histologic sample. After a long incubation, slides are washed at high stringency, dehydrated, coated with Kodak NTB-2 photographic emulsion, and allowed to expose for several days at 4°C. Slides are then developed, stained with Wright-Giemsa stain (or other cell type-specific reagents), and microscopically examined for autoradiographic grains over cells, indicating intracellular HIV. Positive cells are defined as those with a number of grains five times that of background.

Several studies have shown that under optimal conditions, each grain above background represents about five genome equivalents of HIV RNA per cell (36). One study reports that some infected cells are positive by in situ hybridization 1 to 2 days prior to the detection of cell-associated viral antigens by immunocytochemistry (IFA or immunoprecipitation assay) and 2 to 5 days before RT activity or viral antigen could be measured in culture supernatants (6). The in situ method has successfully identified HIV infections of various cell types, including those of the lymphatic system (36), the central nervous system and brain (92), and the bowel (66). Thus, in situ hybridization is a very sensitive method that also allows direct correlation of a specific hybridization signal with the morphologic and immunophenotypic properties of the cells and tissues studied.

PROGNOSTIC INDICATORS

At this time, there are no direct measures that determine the consequence of infection other than the use of surrogate markers. These markers are not only important in predicting the outcome of HIV infection but also useful in determining when treatment should be administered. For example, surrogate markers can be used to monitor the course of treatment of an individual receiving antiretroviral therapy or to determine when to administer prophylaxis for *P. carinii* pneumonia.

CD4$^+$ T-Cell Counts

The best predictor of disease progression thus far is the absolute CD4$^+$ count. An overt decline in the CD4$^+$ T-cell count usually precedes clinical disease. CD4 is a cell surface molecule found mostly on T lymphocytes that is important for the activation of the lymphocytes when they are stimulated by antigens. Since CD4 is a major receptor for viral surface glycoprotein gp120, T cells expressing this surface marker are primary targets for infection. Cells expressing these markers are important in regulating the immune responses necessary for controlling pathogens and neoplasms. Thus, loss of these cells leads to an immunocompromised state that eventually leads to disease. Decline in response to certain stimuli by these cells leads to disease (35, 80).

Although in asymptomatic individuals an average of 1/10,000 CD4$^+$ cells is actually infected, the number of CD4$^+$ cells lost greatly exceeds this amount (5). Recently, Groux et al. demonstrated that this loss of CD4$^+$ cells is partially attributable to programmed cell death, or apoptosis, which occurs in infected individuals but not in seronegative donors (34).

In general, the lower the CD4$^+$ T-cell count, the greater the chance of entry to disease. In a study at San Francisco General Hospital (65), 87% of individuals having CD4$^+$ T-cell counts of less than 200 cells per mm^3 at entry to the study progressed to AIDS within 3 years, while only 46% of those whose CD4$^+$ counts were between 200 and 400 progressed to AIDS. Only 16% of those individuals with >500 CD4$^+$ cells per mm^3 progressed to AIDS within 3 years. Findings of studies carried out at other centers were similar (60). These trends occur in all risk groups analyzed,

TABLE 6 Viral titers of plasma and PBMCs recovered from HIV-infected individuals[a]

Clinical status	Mean CD4 count (cells/mm³)	Plasma titer (infectious dose/μl)	PBMC titer (infectious dose/cell)
Asymptomatic (n = 16)	462	0.03	2×10^{-5}
Symptomatic (n = 18)	147	3.33	2.7×10^{-3}
AIDS (n = 20)	88	3.33	2.2×10^{-3}

[a]Patients did not receive antiviral therapy for at least 4 weeks prior to viral titration. Table is reprinted with permission from reference 38.

including hemophiliacs and IVDU. Moreover, the rate of decline of CD4$^+$ cells is also a good predictor of AIDS. In a study of HIV-infected hemophiliacs over an 11-year period, those who remained AIDS free in that period had an average loss of approximately 60 CD4$^+$ cells per mm³ per year compared to 125/mm³/year for those who advanced to AIDS (70).

Measurement of Viral Loads

Direct measurements of viral burdens can also be good predictors of the outcome of infection. These measurements include plasma p24 antigen levels, titration of infectious virus in plasma or PBMCs, and quantitation of proviral DNA and RNA by PCR. All procedures are discussed in detail in the section above.

P24 Antigen in Plasma

p24 antigen levels determined by the ELISA described above are used to measure virus levels in the plasma. Although the ELISA is very specific, the sensitivity is usually too low for screening viral antigen in the plasma of asymptomatic individuals. Since the levels of p24 antigen in the plasma increase during the course of the disease, this method is a useful prognostic tool in monitoring individuals who are symptomatic or have AIDS (60). The sensitivity of this assay can be increased by acid treatment of the plasma to be tested to break down any antigen-antibody complexes that may block the ability to detect p24 antigens (67).

A study of 16 HIV-infected individuals revealed that about half of the individuals who acquired AIDS in a 4-year period had serum p24 levels of >50 pg/ml (4). In contrast, only 5% of the individuals who acquired AIDS in the same period had p24 levels of <50 pg/ml.

Titration of Infectious Virus in Plasma and Cells

Virus in plasma or cells is titrated by culturing various dilutions of cells or plasma with phytohemagglutin-stimulated PBMCs or other susceptible cells (see above). The cultures should be set up within 3 h of acquiring the specimen from the test subject (69). The level of viral infection is measured either by determining RT activity or by measuring p24 levels as described above in Detection of

Virus and Virus-Infected Cells. The results of one such study are shown in Table 6.

Titration of Viral DNA or RNA by PCR

The PCR method described above has enhanced our ability to detect 1 HIV-infected cell in 100,000. This technique has enabled researchers to demonstrate that the viral burden was substantially higher than was believed and that the burden changed with a change in the clinical state of the individual (Table 7). In this regard, viral culture isolation showed 1 provirus-containing CD4$^+$ T cell per 4,000 to 150,000 CD4$^+$ cells, while PCR demonstrated 1 per 2,500 to 26,000 cells (5).

Recently, PCR has been used to quantify virus in the plasma by using a technique called quantitative competitive PCR (QC-PCR). This method was used to measure RNA in plasma by use of a competitive RNA template matched to the target sequence but differing from it slightly by virtue of an introduced internal deletion. Increasing amounts of the competitive template are added to replicate portions of the test specimen, and quantitation is based on determination of relative, not absolute, amounts of the target sequences and the competitor after electrophoretic separation. QC-PCR can be used to detect virus in 0.5 ml of plasma; however, larger volumes increase the sensitivity of assay. In one study, QC-PCR-determined HIV RNA levels in plasma differed between clinical stages, with means of 78,200 copies per ml of plasma in asymptomatic patients, 352,000 copies per ml of plasma in symptomatic patients, and 2,448,000 copies per ml of plasma in AIDS patients (71).

Immune Activation Markers

Infection by HIV induces activation of the immune system, which in turn increases the pool of HIV-susceptible cells available and results in increased virus production. Thus, measurement of immune activation markers is an indirect means of measuring virus production in an individual. An indirect way to measure immune activation is to measure the amount of β_2-microglobulin in plasma. This protein is part of the class I major histocompatibility complex and is present in almost all nucleated cells. This marker correlates

TABLE 7 Frequency of HIV-infected cells in a group of HIV-infected individuals in a 14-month period[a]

Clinical status at end of study	CD4 count (cells/μl)[b]		Frequency of HIV-infected CD4$^+$ cells[c]	
	Beginning of study	End of study	Beginning of study	End of study
Asymptomatic	527	592	0.001–0.0001	0.001–0.0001
Symptomatic or AIDS	569	219	0.001–0.0001	0.1–0.01

[a]Table is reprinted with permission from reference 79.
[b]Mean CD4 count of test subjects.
[c]Frequency among PBMCs was detected by PCR.

with the degree of lymphocyte activation and also with progression to disease in HIV-infected individuals. For example, analysis of HIV-infected individuals in the San Francisco Men's Health Study showed that in a 3-year period, of those individuals who had β_2-microglobulin levels of <246 nmol/liter, 7.3% developed AIDS, while of individuals with levels of >322 nmol/liter, 34.2% developed AIDS (56).

In addition to β_2-microglobulin, neopterin, a metabolite of GTP, is produced largely by lymphocytes and macrophages-monocytes during activation. Unlike most of the tests discussed above, the test for neopterin commonly measures levels in the urine and is usually correlated with the level of creatinine present, although serum neopterin levels can also be measured. In a Los Angeles multicenter AIDS cohort study, the risks of developing AIDS over a 3-year period were 12% in those with a serum neoptrin level of 12 nmol/liter and 37% in those with levels of >20 nmol/liter (63).

Although the immune activation markers mentioned above are commonly used to measure disease progression, other activation markers have also been shown to associate with disease progression. For example CD38 expression on $CD8^+$ cells showed a direct correlation to development of AIDS (30). In addition, soluble-cytokine receptors have been detected in the sera of infected individuals and have been shown to be predictors of disease progression (31).

INTERPRETATION OF TEST RESULTS

Since the discovery of HIV as the causative agent of AIDS, the screening of individuals as well as the blood supply for HIV infection has primarily relied on the use of anti-HIV antibody assays (7). Positive antibody status is considered equivalent to HIV infection, and the epidemiologic evidence appears to support this assumption. However, the absence of antibodies to HIV is sometimes an unreliable indicator of the absence of infection, especially in the early stages of infection and in certain populations at high risk for infection. HIV has been identified in individuals known to be at risk who have remained seronegative for up to 3 years following infection (42). Thus, risk status should be considered when evaluating an individual for HIV infection, and tests should be chosen accordingly. In addition, self-deferral of high-risk individuals as blood donors is essential.

A general scheme for the determination of HIV-infection is outlined in Fig. 6. Most individuals are initially tested for serum antibodies to HIV by ELISA. HIV antibodies can also be detected in other body fluids from infected individuals. If the patient has been recently infected, antibodies to HIV can develop within 2 to 6 weeks but may take longer. After initial screening, confirmation of the presence of HIV antibodies is necessary. This is particularly important after ELISA screening because of the occurrence of nonspecific reactions that yield false-positive results. In general, sera giving a positive ELISA are retested by a second ELISA and in some cases by a third. If these results are also positive, the possibility of HIV antibodies in the sample is examined further by IFA or Western blot.

When serologic tests for HIV antibodies are inconclusive, viral detection methods may be desirable, especially in high-risk individuals who, for treatment purposes, could benefit from early diagnosis. Although time-consuming and technically involved, virus cultivation and PCR assays are the most reliable tests for the determination of HIV infec-

tion in these individuals. Nevertheless, the currently predominant HIV assay for clinical investigation is the p24 antigen ELISA. This assay can sometimes detect an initial viremic stage (in both blood and cerebrospinal fluid samples) occurring immediately after HIV infection (60). However, this viremia is usually transient, and a negative antigen status is generally found a few weeks afterward and corresponds to a rise in anti-HIV antibody levels. This rise may reflect the formation of antibody-antigen complexes that mask p24. To circumvent the problem with complexes, it may be necessary to treat the serum at low pH to remove these antibody-antigen complexes before testing (67).

If HIV infection is highly suspected and HIV-1-specific test results are inconclusive, serologic tests for HIV-2 that employ HIV-2-specific reagents should be used. In addition, viral cultivation may be desirable to further identify possible HIV-2 by Western blot methods employing viral lysates and HIV subtype-specific sera.

In conclusion, HIV-1 and HIV-2 are members of the *Lentivirus* genus of human retroviruses. These viruses are the causative agents of a complex array of diseases that characterize various clinical states, including what is referred to as AIDS. As described here, numerous serologic and viral detection assays that can be used to identify HIV infection are available. While awaiting the development of more-effective antiviral therapies, early testing for HIV in potentially infected individuals is recommended to allow prompt medical treatment and surveillance and to prevent the spread of the virus to other individuals. At present, medical treatment is limited to the use of antiviral drug zidovudine, ddI, and ddC and to treatment of the many opportunistic infections and conditions that afflict HIV-infected individuals. Thus, until a vaccine for HIV is developed, continual education on how to prevent HIV transmission must be given major emphasis.

REFERENCES

1. Arthur, L. O., J. W. Bess, Jr., R. C. Sowder II, R. E. Benveniste, D. L. Mann, J. C. Chermann, and L. E. Henderson. 1992. Cellular proteins bound to immunodeficiency viruses: implications for pathogenesis and vaccines. *Science* 258:1935–1938.
2. Babu, P. G., N. K. Saraswathi, F. Devapriya, and T. J. John. 1993. The detection of HIV-2 in southern India. *Indian J. Med. Res.* 97:49–52.
3. Barre-Sinoussi, F., J.-C. Chermann, F. Rey, M. T. Nugeyre, S. Chamaret, J. Gruest, C. Dauguet, C. Axler-Blin, F. Vezinet-Brun, C. Rouzioux, W. Rozenbaum, and L. Montagnier. 1983. Isolation of a T-lymphotropic retrovirus from a patient at risk for acquired immune deficiency syndrome (AIDS). *Science* 220:868–871.
4. Bollinger, R. C., Jr., R. L. Kline, H. L. Francis, M. W. Moss, J. G. Bartlett, and T. C. Quinn. 1992. Acid dissociation increases the sensitivity of p24 antigen detection for the evaluation of antiviral therapy and disease progression in asymptomatic human immunodeficiency virus-infected persons. *J. Infect. Dis.* 165:913–916.
5. Brinchmann, J. E., J. Albert, and F. Vartdal. 1991. Few infected $CD4^+$ T cells but a high proportion of replication-competent provirus copies in asymptomatic human immunodeficiency virus type 1 infection. *J. Virol.* 65:2019–2023.
6. Busch, M. P., J. H. Beckstead, H. Hollander, and G. N. Vyas. 1988. A histomolecular approach to the detection and study of human immunodeficiency virus infection, p. 209–242. In P. A. Luciw and K. S. Steimer (ed.), *HIV Detection by Genetic Engineering Methods*. Marcel Dekker, Inc., New York.
7. Bylund, D. J., U. H. M. Ziegner, and D. G. Hooper. 1992.

Review of testing for human immunodeficiency virus. *Clin. Lab. Med.* **12**:305–333.

8. **Castro, B. A., C. D. Weiss, L. D. Wiviott, and J. A. Levy.** 1988. Optimal conditions for recovery of the human immunodeficiency virus from peripheral blood mononuclear cells. *J. Clin. Microbiol.* **26**:2371–2376.

9. **Centers for Disease Control.** 1989. Update: HIV-2 infection—United States. *Morbid. Mortal. Weekly Rep.* **38**:572–580.

10. **Centers for Disease Control.** 1993. Update: acquired immunodeficiency syndrome—United States, 1992. *JAMA* **270**:930.

11. **Cheng-Mayer, C., P. Ianello, K. Shaw, P. A. Luciw, and J. A. Levy.** 1989. Differential effects of *nef* on HIV replication: implications for viral pathogenesis in the host. *Science* **246**:1629–1632.

12. **Chin, J., and S. K. Lwanga.** 1991. Estimation and projection of adult AIDS cases: a simple epidemiological model. *Bull. W.H.O.* **69**:399–406.

13. **Chin, J., P. A. Sato, and J. M. Mann.** 1990. Projection of HIV infection and AIDS cases to the year 2000. *Bull. W.H.O.* **68**:1–11.

14. **Choopanya, K., S. Vanichseni, D. C. Des Jarlais, K. Plangsringarm, W. Sonchai, M. Carballo, P. Friedmann, and S. R. Friedman.** 1991. Risk factors and HIV seropositivity among injecting drug users in Bangkok. *AIDS* **5**:1509–1513.

15. **Clavel, F.** 1987. HIV-2, the West African AIDS virus. *AIDS* **1**:135–140.

16. **Clavel, F., K. Mansinho, S. Chamaret, D. Guetard, V. Favier, J. Nina, M.-O. Santos-Ferreira, J.-L. Champalimaud, and L. Montagnier.** 1987. Human immunodeficiency virus type 2 infection associated with AIDS in West Africa. *N. Engl. J. Med.* **316**:1180–1185.

17. **Coffin, J., A. Haase, J. A. Levy, L. Montagnier, S. Oroszlan, N. Teich, H. Temin, K. Toyoshima, H. Varmus, P. Vogt, and R. Weiss.** 1986. Human immunodeficiency viruses. *Science* **232**: 697. (Letter.)

18. **Cohen, E. A., E. F. Terwilliger, Y. Jalinoos, J. Proulx, J. G. Sodroski, and W. A. Haseltine.** 1990. Identification of HIV-1 vpr product and function. *J. Acquired Immune Defic. Syndr.* **3**:11–18.

19. **Cortes, E., R. Detels, D. Aboulafia, X. Ling Li, T. Moudgil, M. Alam, C. Bonecker, A. Gonzag, L. Oyafuso, M. Tondo, C. Boite, N. Hammershalk, C. Capitani, D. J. Slamon, and D. Ho.** 1989. HIV-1, HIV-2, and HTLV-1 infection in high-risk group in Brazil. *N. Engl. J. Med.* **320**:953–958.

20. **Dalgleish, A. G., P. C. Beverley, P. R. Clapham, D. H. Crawford, M. F. Greaves, and R. A. Weiss.** 1984. The CD4 (T4) antigen is an essential component of the receptor for the AIDS retrovirus. *Nature* (London) **312**:763–767.

21. **Davey, R. T., and H. C. Lane.** 1990. Laboratory methods in the diagnosis and prognostic staging of infection with human immunodeficiency virus type 1. *Rev. Infect. Dis.* **12**:912–930.

22. **Denis, F., G. Leonard, A. Sangare, G. Gershy-Damet, J.-L. Rey, B. Soro, D. Schmidt, M. Mounier, M. Verdier, A. Baillou, and F. Barin.** 1988. Comparison of 10 enzyme immunoassays for detection of antibody to human immunodeficiency virus type 2 in West Africa. *J. Clin. Microbiol.* **26**:1000–1004.

23. **Des Jarlais, D. C., S. R. Friedman, K. Choopanya, S. Vanichseni, and T. P. Ward.** 1992. International epidemiology of HIV and AIDS among injecting drug users. *AIDS* **6**:1053–1068.

24. **Detels, R., P. English, B. R. Visscher, L. Jacobson, L. Kingsley, J. S. Chmiel, J. P. Dudley, L. J. Eldred, and H. M. Ginzburg.** 1989. Seroconversion, sexual activity, and condom use among 2915 HIV seronegative men followed for up to 2 years. *J. Acquired Immune Defic. Syndr.* **2**:77–83.

25. **Ehrnst, A., S. Lingren, M. Dictor, B. Johansson, A. Sonnerborg, J. Czajkowski, G. Sudin, and A.-B. Bohlin.** 1991. HIV in pregnant women and their offspring evidence for late transmission. *Lancet* **338**:203–207.

26. **Evans, L. A., J. Moreau, K. Odehouri, H. Legg, A. Barboza,** C. Cheng-Mayer, and J. A. Levy. 1988. Characterization of a noncytopathic HIV-2 strain with unusual effects on CD4 expression. *Science* **240**:1522–1525.

27. **Folks, T. M., and D. P. Bednarik.** 1992. Mechanisms of HIV-1 latency. *AIDS* **6**:3–16.

28. **Gallo, R. C., S. Z. Salahuddin, M. Popovic, G. M. Shearer, M. Kaplan, B. F. Haynes, T. J. Palker, R. Redfield, J. Oleske, and B. Safai.** 1984. Frequent detection and isolation of cytopathic retroviruses (HTLV-III) from patients with AIDS and at risk for AIDS. *Science* **224**:500–503.

29. **Geraghty, R. J., and A. T. Panganiban.** 1993. Human immunodeficiency virus type 1 vpu has a CD4$^-$ and an envelope glycoprotein-independent function. *J. Virol.* **67**:4190–4194.

30. **Giorgi, J. V., Z. Liu, L. E. Hultin, W. G. Cumberland, K. Hennessey, and R. Detels.** 1993. Elevated levels of CD38$^+$ CD8$^+$ T cells in HIV infection add to the prognostic value of low CD4$^+$ T cell levels: results of 6 years of follow-up. *J. Acquired Immune Defic. Syndr.* **6**:904–912.

31. **Godfried, M. H., T. van der Poll, J. Jansen, J. A. Romijn, J. K. M. E. Schattekerk, E. Endert, S. J. H. van Deventer, and H. P. Sauerwein.** 1993. Soluble receptors for tumour necrosis factor: a putative marker of disease progression in HIV infection. *AIDS* **7**:33–36.

32. **Gonzalez-Scarano, F., M. N. Waxham, A. M. Ross, and J. A. Hoxie.** 1987. Sequence similarities between human immunodeficiency virus gp41 and paramyxovirus fusion proteins. *AIDS Res. Hum. Retroviruses* **3**:245–252.

33. **Gottlieb, M. D., R. Schroff, H. M. Schanker, J. D. Weisman, P. T. Fan, R. A. Wolf, and A. Saxon.** 1981. Pneumocystis carinii pneumonia and mucosal candidiasis in previously healthy homosexual men. *N. Engl. J. Med.* **305**:1425–1431.

34. **Groux, H., G. Torpier, D. Monte, Y. Mouton, A. Capron, and J. C. Ameisen.** 1992. Activation-induced death by apoptosis in CD4$^+$ T cells from human immunodeficiency virus-infected asymptomatic individuals. *J. Exp. Med.* **175**:331–340.

35. **Gruters, R. A., F. G. Terpstra, R. De Jong, C. J. M. Van Noesel, R. A. W. Van Lier, and F. Miedema.** 1990. Selective loss of T cell function in different stages of HIV infection. Early loss of anti-CD3-induced T cell proliferation followed by decreased anti-CD3-induced cytotoxic T lymphocyte generation in AIDS-related complex and AIDS. *Eur. J. Immunol.* **20**:1039–1044.

36. **Harper, M. E., L. M. Marselle, R. C. Gallo, and F. Wong-Staal.** 1986. Detection of lymphocytes expressing human T-lymphotropic virus type III in lymph nodes and peripheral blood from infected individuals by *in situ* hybridization. *Proc. Natl. Acad. Sci. USA* **83**:772–776.

37. **Hart, C., T. Spira, and J. Moore.** 1988. Direct detection of HIV RNA expression in seropositive subjects. *Lancet* **ii**:596–599.

38. **Ho, D. D., T. Moudgil, and M. Alam.** 1989. Quantitation of human immunodeficiency virus type 1 in the blood of infected persons. *N. Engl. J. Med.* **321**:1621–1625.

39. **Hoffman, A. D., B. Banapour, and J. A. Levy.** 1985. Characterization of the AIDS-associated retrovirus reverse transcriptase and optimal conditions for its detection in virions. *Virology* **147**:326–335.

40. **Hoffman, P. N., D. P. Larkin, and D. Samuel.** 1989. Needlestick and needleshare—the difference. *J. Infect. Dis.* **160**:545.

41. **Hunter, D. J.** 1993. AIDS in sub-Saharan Africa: the epidemiology of heterosexual transmission and the prospects for prevention. *Epidemiology* **4**:63–72.

42. **Imagawa, D. T., M. H. Lee, S. M. Wolinsky, K. Sano, F. Morales, S. Kwok, J. J. Sninsky, P. G. Nishanian, J. Giorgi, J. L. Fahey, J. Dudley, B. R. Visscher, and R. Detels.** 1989. Human immunodeficiency virus type 1 infection in homosexual men who remain seronegative for prolonged periods. *N. Engl. J. Med.* **320**:1458–1462.

43. **Jaffe, H. W., D. J. Bregman, and R. M. Selik.** 1983. Acquired immune deficiency syndrome in the United States: the first 1,000 cases. *J. Infect. Dis.* **148**:339–345.

44. **Katz, D. H.** 1993. AIDS: primarily a viral or an autoimmune disease. *AIDS Res. Hum. Retroviruses* **9:**489–493.

45. **Kuo, J. M., J. M. Taylor, and R. Detels.** 1991. Estimating the AIDS incubation period from a prevalent cohort. *Am. J. Epidemiol.* **133:**1050–1057.

46. **Kwok, S., and R. Higuchi.** 1989. Avoiding false positives with PCR. *Nature* (London) **339:**237.

47. **Kwok, S., D. H. Mack, J. J. Sninsky, G. D. Ehrlich, B. J. Poiescz, N. L. Dock, H. J. Alter, D. Mildvan, and M. H. Grieco.** 1988. Diagnosis of human immunodeficiency virus in seropositive individual: enzymatic amplification of HIV viral sequences in peripheral blood mononuclear cells, p. 243–255. *In* P. A. Luciw and K. S. Steimer (ed.), *HIV Detection by Genetic Engineering Methods.* Marcel Dekker, Inc., New York.

48. **Laga, M., N. Nzilambi, and J. Goeman.** 1991. The interrelationship of sexually transmitted diseases and HIV infection: implications for the control of both epidemics in Africa. *AIDS* **5:**S55–S63.

49. **Laurent-Crawford, A. G., B. Krust, S. Muller, Y. Riviere, M.-A. Rey-Cuille, J.-M. Bechet, L. Montagnier, and A. G. Hovanessian.** 1991. The cytopathic effect of HIV is associated with apoptosis. *Virology* **185:**829–839.

50. **Laurent-Crawford, A. G., B. Krust, Y. Riviere, C. Desgranges, S. Muller, M. P. Kieny, C. Dauguet, and A. G. Hovanessian.** 1993. Membrane expression of HIV envelope glycoproteins triggers apoptosis in CD4 cells. *AIDS Res. Hum. Retroviruses* **9:**761–773.

51. **Levy, J. A.** 1989. Human immunodeficiency viruses and the pathogenesis of AIDS. *JAMA* **261:**2997–3006.

52. **Levy, J. A.** 1993. Pathogenesis of human immunodeficiency virus infection. *Microbiol. Rev.* **57:**183–289.

53. **Levy, J. A., A. D. Hoffman, S. M. Kramer, J. A. Landis, J. M. Shimabukuro, and L. S. Oshiro.** 1984. Isolation of lymphocytopathic retroviruses from San Francisco patients with AIDS. *Science* **225:**840–842.

54. **Levy, J. A., G. Mitra, and M. M. Mozen.** 1984. Recovery and inactivation of infectious retroviruses added to factor VIII concentrates. *Lancet* **ii:**722–723.

55. **Levy, J. A., G. A. Mitra, M. F. Wong, and M. Mozen.** 1985. Survival of AIDS associated retrovirus during factor VIII purification from plasma inactivation by wet and dry heat procedures. *Lancet* **i:**1456–1457.

56. **Lifson, A. R., N. A. Hessol, S. P. Buchbinder, P. M. O'Malley, L. Barnhart, M. Segal, M. H. Katz, and S. D. Holmberg.** 1992. Serum beta$_2$-microglobulin and prediction of progression to AIDS in HIV infection. *Lancet* **339:**1436–1440.

57. **Lifson, J. D., G. R. Reyes, M. S. McGrath, B. S. Stein, and E. G. Engleman.** 1986. AIDS retrovirus induced cytopathology: giant cell formation and involvement of CD4 antigen. *Science* **232:**1123–1127.

58. **Looney, D. J., S. Hayashi, M. Nicklas, R. R. Redfield, S. Broder, F. Wong-Staal, and H. Mitsuya.** 1990. Differences in the interaction of HIV-1 and HIV-2 with CD4. *J. Acquired Immune Defic. Syndr.* **3:**649–657.

59. **Macatonia, S. E., R. Lau, S. Patterson, J. Pinching, and S. C. Knight.** 1990. Dendritic cell infection, depletion and dysfunction in HIV-infected individuals. *Immunology* **71:**38–45.

60. **MacDonell, K. B., J. S. Chimel, L. Poggensee, S. Wu, and J. P. Phair.** 1990. Predicting progression to AIDS: combined usefulness of CD4 lymphocyte counts and p24 antigenemia. *Am. J. Med.* **89:**706–712.

61. **Malone, J. D., E. S. Smith, J. Sheffield, D. Bigelow, K. C. Hyams, S. G. Beardsley, R. S. Lewis, and C. R. Roberts.** 1993. Comparative evaluation of six rapid serological tests for HIV-1 antibody. *J. Acquired Immune Defic. Syndr.* **6:**115–119.

62. **McSharry, J. J., R. Constantino, E. Robbiano, R. Echols, R. Stevens, and J. M. Lehman.** 1990. Detection and quantitation of human immunodeficiency virus-infected peripheral blood mononuclear cells by flow cytometry. *J. Clin. Microbiol.* **28:**724–733.

63. **Melmed, R. N., J. M. G. Taylor, R. Detels, M. Bozorgmehri, and J. L. Fahey.** 1989. Serum neopterin changes in HIV-infected subjects: indicator of significant pathology, CD4 T-cell changes, and development of AIDS. *J. Acquired Immune Defic. Syndr.* **2:**70–76.

64. **Merson, M. H.** 1993. Slowing the spread of HIV: agenda for the 1990's. *Science* **260:**1266–1272.

65. **Moss, A. R., P. Bacchetti, D. Osmond, W. Krampf, R. E. Chaisson, D. Stites, J. Wilber, J.-P. Allain, and J. Carlson.** 1988. Seropositivity for HIV and the development of AIDS or AIDS related condition: three year follow up of the San Francisco General Hospital cohort. *Br. Med. J.* **296:**745–750.

66. **Nelson, J. A., C. A. Wiley, C. Reynolds-Kohler, C. E. Reese, W. Margaretten, and J. A. Levy.** 1988. Human immunodeficiency virus detected in bowel epithelium from patients with gastrointestinal symptoms. *Lancet* **i:**259–262.

67. **Nishanian, P., K. R. Huskins, S. Stehn, R. Detels, and J. L. Fahey.** 1990. A simple method for improving assay demonstrates that HIV p24 antigen is present as immune complexes in most sera from HIV-infected individuals. *J. Infect. Dis.* **162:**21–28.

68. **Ou, C.-Y., S. Kwok, S. W. Mitchell, D. H. Mack, J. J. Sninsky, J. W. Krebs, P. Feorino, D. Warfield, and G. Schochetman.** 1988. DNA amplification for direct detection of HIV-1 in DNA of peripheral blood mononuclear cells. *Science* **239:**295–297.

69. **Pan, L.-Z., A. Werner, and J. A. Levy.** 1993. Detection of plasma viremia in HIV-infected individuals at all clinical stages. *J. Clin. Microbiol.* **31:**283–288.

70. **Phillips, A. N., C. A. Lee, J. Elford, G. Janossy, A. Timms, M. Bofill, and P. B. A. Kernoff.** 1991. Serial CD4 lymphocyte counts and development of AIDS. *Lancet* **337:**389–392.

71. **Piatak, M., Jr., S. Saag, L. C. Yang, J. C. Kappes, K.-C. Luk, B. H. Hahn, G. M. Shaw, and J. D. Lifson.** 1993. High levels of HIV-1 in plasma during all stages of infection determined by competitive PCR. *Science* **259:**1749–1754.

72. **Pomerantz, R. J., S. M. de la Monte, S. P. Donegan, T. R. Rota, M. W. Vogt, D. E. Craven, and M. S. Hirsch.** 1988. Human immunodeficiency virus (HIV) infection of the uterine cervix. *Ann. Intern. Med.* **108:**321–327.

73. **Resnick, L., K. Veren, S. Z. Salahuddin, S. Tondreai, and P. D. Markham.** 1984. Stability and inactivation of HTLV III/LAV under clinical and laboratory environment. *JAMA* **255:**1887–1891.

74. **Ryder, R. W., and M. Termmerman.** 1991. The effect of HIV-1 infection during pregnancy and the perinatal period on maternal and child health in Africa. *AIDS* **5:**S75–S85.

75. **Sakai, H., R. Shibata, J.-I. Sakuragi, S. Sakuragi, M. Kawamura, and A. Adachi.** 1993. Cell-dependent requirement of human immunodeficiency virus type 1 Vif protein for maturation of virus particles. *J. Virol.* **67:**1663–1666.

76. **Sarkar, S., P. Mookerjee, A. Roy, T. N. Naik, J. K. Singh, A. R. Sharma, Y. I. Singh, P. K. Singh, S. P. Tripathy, and S. C. Pal.** 1991. Descriptive epidemiology of intravenous heroin users—a new risk group for transmission of HIV in India. *J. Infect.* **23:**201–207.

77. **Sayers, M. H., P. G. Beatty, and J. A. Hansen.** 1986. HLA antibodies as a cause of false-positive reactions in screening enzyme immunoassays for antibodies to human T-lymphotropic virus type III. *Transfusion* **26:**113–115.

78. **Schneider-Schaulies, J., S. Schneider-Schaulies, R. Brinkman, P. Tas, M. Halbrugge, U. Walter, H. C. Holmes, and V. Ter Meulen.** 1992. HIV-1 gp120 receptor on CD4-negative brain cells activates a tyrosine kinase. *Virology* **191:**765–772.

79. **Schnittman, S. M., J. J. Greenhouse, M. C. Psallidopoulos, M. Baseler, N. P. Salzman, A. S. Fauci, and H. C. Lane.** 1990. Increasing viral burden in CD4$^+$ T cells from patients with human immunodeficiency virus (HIV) infection reflects rapidly progressive immunosuppression and clinical disease. *Ann. Intern. Med.* **113:**438–443.

80. **Shearer, G. M., S. M. Payne, S. M. Joseph, and W. E. Biddison.** 1984. Functional T lymphocyte immune deficiency in a population of homosexual men who do not exhibit

symptoms of acquired immune deficiency syndrome. *J. Clin. Invest.* **74**:496–506.

81. **Smallman-Raynor, M., and A. Cliff.** 1991. The spread of human immunodeficiency virus type 2 into Europe: a geographical analysis. *Int. J. Epidemiol.* **20**:480–489.

82. **Sova, P., and D. J. Volsky.** 1993. Efficiency of viral DNA synthesis during infection of permissive and nonpermissive cells with *vif*-negative human immunodeficiency virus type 1. *J. Virol.* **67**:6322–6326.

83. **Spire, B., F. Barre-Sinoussi, and D. Dormont.** 1985. Inactivation of lymphadenopathy-associated virus by heat, gamma rays, and ultraviolet light. *Lancet* **i**:188–189.

84. **Steiger, M. T., M. Peters, B. Fienbork, W. Stille, and H. W. Doerr.** 1993. Results of studies from AIDS counseling center of the Frankfurt/M city health office January 1990 to December 1991. *Gesundheitswesen* **55**:68–73.

85. **Tozzi, A. E., P. Pezzotti, and D. Greco.** 1990. Does breast-feeding delay progression to AIDS in HIV-infected children. *AIDS* **4**:1293–1304.

86. **Vanini, S., R. Longhi, A. Lazzarin, E. Vigo, A. G. Siccardi, and G. Viale.** 1993. Discrete regions of HIV-1 gp41 defined by syncytia-inhibiting affinity-purified human antibodies. *AIDS* **7**:167–174.

87. **Van Kerckhoven, I., G. Vercauteren, P. Piot, and G. van der Groen.** 1991. Comparative evaluation of 36 commercial assays for detecting antibodies to HIV. *Bull. W.H.O.* **69**:753–760.

88. **Walker, C. M., D. J. Moody, D. P. Stites, and J. A. Levy.** 1986. CD8$^+$ lymphocytes can control HIV infection *in vitro* by suppressing virus replication. *Science* **234**:1563–1566.

89. **Weber, T., G. Hunsmann, W. Stevens, and A. F. Fleming.** 1992. Human retroviruses. *Bailliere's Clin. Haematol.* **5**:273–314.

90. **Weiss, S. H., J. Lombardo, J. Michaels, L. R. Sharer, M. Rayyarah, J. Leonard, A. Mangia, P. Kloser, S. Sathe, R. Kapila, N. M. Williams, R. Altman, J. French, and W. E. Parkin.** 1988. AIDS due to HIV-2 infection—New Jersey. *Morbid. Mortal. Weekly Rep.* **259**:969–972.

91. **West, M., J. Mikovits, G. Princler, Y. L. Liu, F. W. Ruscetti, H. F. Kung, and H. Raziuddin.** 1992. Characterization and purification of a novel transcriptional repressor from HeLa cell nuclear extracts recognizing the negative regulatory element region of human immunodeficiency virus-1 long terminal repeat. *J. Biol. Chem.* **267**:24948–24952.

92. **Wiley, C. A., R. D. Schrier, J. A. Nelson, P. W. Lambert, and M. B. A. Oldstone.** 1986. Cellular localization of human immunodeficiency virus infection within the brains of acquired immune deficiency syndrome patients. *Proc. Natl. Acad. Sci. USA* **83**:7089–7093.

93. **Willey, R. L., D. H. Smith, L. A. Lasky, T. S. Theodore, P. L. Earl, B. Moss, D. J. Capon, and M. A. Martin.** 1988. In vitro mutagenesis identifies a region within the envelope gene of the human immunodeficiency virus that is critical for infectivity. *J. Virol.* **62**:139–147.

94. **World Health Organization.** 1993. *WHO World Program on AIDS.* World Health Organization, Geneva.

Human T-Cell Lymphotropic Virus Types I and II

HELEN LEE, JOHN D. BURCZAK, AND JESSIE SHIH

98

CLINICAL BACKGROUND

Human T-lymphotropic virus types I (HTLV-I) and II (HTLV-II) are distributed worldwide in discrete foci of endemic disease. HTLV-I, the first retrovirus associated with human diseases, is the etiologic agent for adult T-cell leukemia-lymphoma (ATL) and a neurologic disorder known as tropical spastic paraparesis or HTLV-I-associated myelopathy (HAM/TSP) (5, 17). HTLV-I infections are endemic in southwestern Japan, the Caribbean basin, Taiwan, parts of Central and South America, and sub-Saharan Africa. In areas of Japan where the disease is endemic, the HTLV-I infection rate increases with age, with seroprevalence ranging from 0.3 to 8% of the adult population (13). In the United States, HTLV-I prevalence ranges from 0.025% in asymptomatic blood donors to 7 to 49% among intravenous-drug users and prostitutes (12). HTLV-II infections are detected mainly in intravenous-drug users in the United States and, more recently, in Native Americans in New Mexico and native Panamanians.

The routes of transmission for HTLV-I are similar to those of another human retrovirus, human immunodeficiency virus (HIV). However, HTLV-I transmission is almost exclusively cell associated. In blood transfusion studies, recipients of HTLV-I-infected units containing cellular blood products seroconverted, while recipients of fresh frozen plasma from infected individuals did not (14). This result is in contrast to that of HIV, which is transmitted by both cellular and cell-free blood components. Transmission of HTLV-I may be vertical (from mother to child), sexual (heterosexual or homosexual), or parenteral (transfusion or sharing of contaminated needles and syringes). Vertical HTLV-I transmission is principally postnatal by breast-feeding of mother's milk containing infected lymphocytes (15). Seropositivity ranging from 20 to 90% has been reported in breast-fed children of seropositive mothers (1). In certain parts of Japan, where HTLV-I is highly endemic, pregnant women are screened for antibody against HTLV-I. Those who are seropositive are counseled not to breast-feed.

Most individuals infected with HTLV-I are asymptomatic carriers. Integrated proviral genomes are detectable in their lymphocytic DNA. Asymptomatic carriers are seropositive and capable of transmitting the virus. Approximately 4 to 5% of these carriers progress to ATL over their lifetimes; the disease occurs at a mean age of approximately 40 years.

ATL

Four transitionally related categories of ATL have been described: aymptomatic, preleukemic, smoldering-chronic, and acute. Preleukemic patients are usually asymptomatic and HTLV-I seropositive. Pre-ATL is usually discovered by incidental detection of leukocytosis or morphologically abnormal ATL lymphocytes. These ATL cells frequently have lobulated or flower-shaped nuclei, although the morphologic appearance may vary. The HTLV-I provirus can be integrated into the abnormal T lymphocytes in a monoclonal or oligoclonal pattern, suggestive of a clonal development of cells, some of which may progress to malignancy. Approximately half of pre-ATL patients have spontaneous regression of abnormal T lymphocytes; in the remaining half, lymphocytosis persists, with some patients progressing to ATL.

Patients with smoldering ATL have characteristic skin lesions with modest bone marrow involvement, whereas chronic-ATL patients have leukocytosis with substantial numbers of circulating ATL cells. Patients in the smoldering-chronic stage may progress, after months or years, into acute ATL.

Acute-ATL patients have elevated leukocyte counts, often accompanied by granulocytosis and eosinophilia. Common presentations include lymphadenopathy without mediastinal tumors, hepatosplenomegaly, infiltration of leukemic cells (resulting in skin lesions), sometimes interstitial pneumonitis, elevation of serum lactate dehydrogenase, hyperbilirubinemia, and hypercalcemia with bone resorption. Remission following chemotherapy is rare; the mean survival time of patients with acute ATL is less than 1 year. Prognosis for acute ATL is poor if ascites are present. In acute ATL, HTLV-I provirus is monoclonally integrated into the malignant T cell's DNA. Virtually all leukemic cells express the CD4 antigen and, with rare exceptions, the CD8 antigen. A high level of *Tac* antigen, a subunit of the interleukin-2 receptor, is characteristic of ATL tumor cells. Chromosomal abnormalities are frequently found in acute-ATL patients but are less frequent in smoldering-chronic-ATL patients.

Little is known about the role of HTLV-I in the development of ATL, in part because of the long latency period

1115

(30 to 40 years between infection and the onset of disease) and in part because of the insensitivity of classic molecular techniques in detecting proviral sequences in the small numbers of infected cells in asymptomatic carriers, pre-ATL patients, and even some subacute-ATL patients. By the time HTLV-I-infected cells are present at detectable levels, the viral genomes are integrated either oligoclonally or monoclonally. Occasionally, some healthy asymptomatic carriers and smoldering-ATL patients have polyclonally integrated HTLV-I provirus. These observations are similar to those made for bovine leukemia virus, a related bovine retrovirus, in which polyclonal integration is evident in the early phases of a disease known as persistent lymphocytosis. The observation suggests that initial infection by HTLV-I may result in proviral integration and transformation of several cells. Subsequent selection processes may result in the development of a dominant HTLV-I-infected clone that develops into a tumor. During acute ATL, there are many malignant cells, and the provirus can generally be detected at a unique monoclonal integration site in the patient's lymphocytes.

A distinguishing feature of HTLV-I in acute ATL as well as in earlier stages of ATL is a lack of detectable expression of viral genes even though the HTLV-I provirus is present in all tumor cells in vivo. The lack of in vivo viral expression is not due to defective provirus, as expression can be induced when ATL cells are placed in culture. The lack of expression may be due to a latent state of the virus or to immunoselection against tumor cells that express viral antigens on their surfaces. HTLV-I appears to be expressed in some cells in vivo, since an antibody response against viral proteins can be detected.

Differential diagnoses of ATL include non-Hodgkin's lymphoma, mycosis fungoides or Sezary's syndrome, Lennert's lymphoma, and T-cell chronic lymphocytic leukemia. Differentiating characteristics include seropositivity to HTLV-I, hypercalcemia, elevated lactate dehydrogenase level, negative histologic staining for terminal deoxynucleotide transferase, and detection of CD4 and *Tac* antigen on ATL cells. Definitive evidence for the disease is the demonstration of HTLV-I proviral sequences in the leukemic ATL cells.

HAM/TSP

The neurologic disease HAM/TSP is reported in all parts of the world where HTLV-I is endemic (16, 19), with a frequency of 0.3% of HTLV-I carriers. It is mainly a disease of adults and usually begins with an insidious weakness of the legs that slowly progresses to a spastic paraparesis with associated sphincter disturbance causing incontinence. The clinical symptoms of HAM/TSP are always slowly progressive without the temporary remissions often observed with multiple sclerosis. Neurologic findings include weakness and spasticity of the extremities, hyperreflexia, Babinski sign, urinary or fecal incontinence, and mild peripheral sensory loss. HAM/TSP patients generally have normal lymphocyte numbers, with occasional morphologically atypical lymphocytes resembling ATL cells in the peripheral blood or cerebrospinal fluid. The cerebrospinal fluid may contain antibodies against HTLV-I and exhibit lymphocytic pleocytosis and an elevation in protein. In most HAM/TSP patients, magnetic resonance imaging shows lesions in both the white matter and the paraventricular regions of the brain. HAM/TSP has also been described in patients coinfected with HTLV-I and HIV type 1; treatment with corticosteroids has been of some clinical benefit.

As in ATL, the role of HTLV-I in the development of HAM/TSP is poorly understood. However, in contrast to ATL, HAM/TSP may develop in some patients within a few years following infection by transfusion of infected blood (6). The presence of intrathecal antibodies against HTLV-I indicates that the virus infects the central nervous system in addition to the blood. Polyclonal, rather than oligoclonal or monoclonal, integration of the virus occurs in peripheral blood and cerebrospinal fluid lymphocytes of HAM/TSP patients, similar to that seen in asymptomatic carriers and smoldering-ATL patients; thus, the development of HAM/TSP is not the consequence of a malignant clone. HAM/TSP patients generally have higher titers of antibodies against HTLV-I than ATL patients or asymptomatic carriers (5). Additionally, there is an association between HAM/TSP, but not ATL, and certain HLA haplotypes (20). Characterization of HTLV-I isolates from HAM/TSP and ATL patients does not reveal any distinguishing features.

HTLV-I has been found in rare cases of B-cell chronic lymphocytic leukemia (B-CLL). In Jamaica, a region where HTLV-I is endemic, the rate of seropositivity to HTLV-I was higher in B-CLL patients than in the general population. However, no HTLV-I provirus could be demonstrated in the B-CLL leukemic cells of one such patient.

HTLV-II, a virus closely related to and showing 60% DNA homology with HTLV-I, was isolated from a patient with hairy T-cell leukemia (8). A second HTLV-II isolate was recovered from a patient with a biclonal disease with malignant B hairy cells and atypical T lymphocytosis. Molecular analysis of the HTLV-II from the second patient showed two copies of the provirus in malignant CD8 cell clones without detectable proviral RNA. This HTLV-II clonal integration pattern and lack of viral expression are analogous to those seen for HTLV-I in ATL.

The association of HTLV-II with disease is uncertain, in large part because of the limited number of patients with specific diseases who carry HTLV-II. HTLV-II infection has been demonstrated in a significant portion of intravenous-drug abusers in the United States (11). Most HTLV-II-infected intravenous-drug abusers are asymptomatic and have no overt hematologic or immunologic abnormalities, although some manifest benign lymphocytosis.

HTLV-II shows extensive serologic cross-reactivity with and nucleic acid homology to HTLV-I. Like HTLV-I, HTLV-II is present in T lymphocytes of infected individuals and can transform T lymphocytes in vitro (3).

DESCRIPTION OF THE VIRUSES

HTLV-I and HTLV-II are closely related members of the retrovirus family. Previously known as RNA tumor viruses, retroviruses are characterized by their ability to cause connective tissue tumors (sarcomas), reticuloendothelial and hematopoietic system cancers (leukosis and leukemias), and mammary carcinomas. Although retroviruses were initially suspected to be etiologic agents for these malignancies, it is now clear that not all retroviruses cause cancer and that some are associated with other diseases such as AIDS. Every retrovirus has an RNA genome. Retroviral replication requires a virus-determined reverse transcriptase to produce a proviral DNA intermediate, which is integrated into the host cell.

HTLV-I and HTLV-II are in the oncovirus subfamily, whose members are generally leukemogenic and cause cell proliferation. However, the biological properties and genomic structures of HTLV-I and HTLV-II, as well as of bovine leukemia virus and simian T-cell leukemia virus, are distinct from those of the other oncoviruses. While these four viruses, sometimes known as *trans*-regulating oncoviruses, are oncogenic and cause transformation, they do not carry oncogene sequences; their replication is regulated by at least two proteins from the regulatory gene *tax-rex*. Thus, it may be more appropriate to group these viruses into a separate subfamily.

According to electron microscopy, HTLV-I and HTLV-II are C-type viruses. In infected cells, there is usually no intracellular viral structure until budding from the host cell has started. Extracellularly, HTLV-I and HTLV-II virions (densities of 1.16 to 1.18 g/ml in sucrose) are approximately 100 nm in diameter, are enveloped, and have electron-dense cores. They are sensitive to organic solvents, detergents, and heat inactivation but resistant to UV light and X irradiation. Like all retroviruses, HTLV-I and HTLV-II contain two identical copies of a single-stranded RNA genome; each RNA strand is approximately 9 kb long and is positive sense (protein can be translated directly from the sequence). The strands are discontinuous, and it has been suggested that during reverse transcription, the enzyme reverse transcriptase is able to move between strands around the breaks.

The HTLV-I and HTLV-II genomes carry the typical retroviral genes plus a *trans*-regulating gene. From the 5' to the 3' direction, the genome carries *gag* (coding for structural proteins that constitute the viral core), *pol* (coding for enzymes necessary for viral replication), and *env* (coding for envelope proteins required for infection and involved in mitogenic stimulation) and ends with the *trans*-regulating gene *tax-rex* (coding for proteins involved in the regulation of viral expression) at the 3' end. The *tax* protein activates expression of viral and a number of cellular genes, including the lymphokine interleukin-2 and the interleukin-2 receptor. In the case of HTLV-I, this autocrine stimulation may cause cell proliferation, the first step in a multistep pathway for oncogenic transformation. During infection, a double-stranded DNA provirus is integrated into the host cell DNA. Both the 5' and the 3' ends of the provirus contains identical regions, termed the long terminal repeats, that are involved in viral expression.

LABORATORY DIAGNOSIS

Direct Detection

Southern blot restriction mapping is the classic method for demonstrating HTLV-I proviral sequences in lymphocytes from acute-ATL patients, who have high levels of HTLV-I-infected cells. This technique, however, is not sufficiently sensitive to detect the low number of infected peripheral blood lymphocytes, which may contain only one or two copies of provirus, of asymptomatic HTLV-I–HTLV-II carriers.

HTLV-I and HTLV-II in peripheral blood lymphocytes can be detected with PCR using primers (commercially available SK43/44) directed against a conserved region in *tax-rex*. The amplified sequences may be differentiated by restriction enzyme digestion (10, 11) or detected by hybridization using HTLV-I- or HTLV-II-specific probes (probe SK45). The product of the *tax-rex* primers is 159 nucleotides for either HTLV-I or HTLV-II. Alternatively, *pol* primers (SK110, SK111), which specifically bind and amplify either the HTLV-I or the HTLV-II polymerase region, may

be used and probed with SK112 (polymerase I) or SK188 (polymerase II).

Virus Isolation

For virus isolation of HTLV-I or HTLV-II, 20 to 50 ml of heparinized blood from the infected patient is promptly fractionated through gradient material (Ficoll-Hypaque, Histopaque, etc.) to separate the mononuclear cells. The recovered leukocytes are grown in RPMI 1640 medium containing phytohemagglutinin P and interleukin-2 to activate the T cells. Cocultivation of the infected leukocytes with similarly fractionated and activated uninfected adult or cord blood lymphocytes increases the sensitivity of the cultures. Alternatively, the infected cells may be irradiated prior to cocultivation. Isolates are generated by transfection of the virus, which is highly radiation resistant, from the dying infected cells to the uninfected growth-stimulated cells (7). Thus, infected cell lines obtained in this manner are not the same clone as the original infected cell. The transformed infected cell lines are continuous lines, with persistent production of reverse transcriptase and other viral proteins.

Immunofluorescence assays (IFA) are effective for identification of HTLV-I- and HTLV-II-transformed cell cultures derived from patient lymphocytes. Acetone-fixed cell smears prepared from cell cultures are stained with reference human polyclonal or mouse monoclonal antibodies against specific viral proteins (e.g., *gag* or *env* gene proteins). For differentiation between HTLV-I and HTLV-II, monoclonal antibodies against HTLV type-specific epitopes are used. Cross-reactivity between the two types is often found, and endpoint titrations are used to determine type. Positive reactions with the primary sera can then be detected with fluorescein-labeled antibody directed against the intermediate antibody used. Characterization of the virus in the cell lines can also be made by electron microscopy, detection of reverse transcriptase, and genomic analysis. Genomic analysis of HTLV-I and HTLV-II provirus can be accomplished either by classic techniques such as Southern blot restriction mapping or by amplified probe techniques.

Serologic Diagnosis

Demonstration of serum antibodies against HTLV-I or HTLV-II indicates infection by these viruses. Owing to the homology between the two viruses, current serologic procedures used to screen blood donors cannot differentiate between them. The cross-reacting procedures include IFA, latex agglutination, and enzyme immunoassays; they detect both HTLV-I and HTLV-II infection.

Current U.S. Food and Drug Administration-licensed serologic screening enzyme immunoassays use purified viral lysates coupled to polystyrene solid phases (18). Their sensitivities (ranging from 97.3 to 100%) and specificities (99.3 to 99.9%) are insufficient to permit unequivocal serodiagnosis in low-prevalence populations. Similar to the algorithm recommended for use with HIV testing, if a sample is repeatedly reactive to HTLV-I in the screening enzyme immunoassay, the results must be confirmed by more specific methods, which include immunofluorescence staining, immunoblotting (IB), and radioimmunoprecipitation assay (RIPA). These supplemental tests, which are not yet approved by the Food and Drug Administration, are available as research procedures.

Typically, confirmation of the positive screening test is by IB followed by RIPA. IB is more sensitive for the *gag*

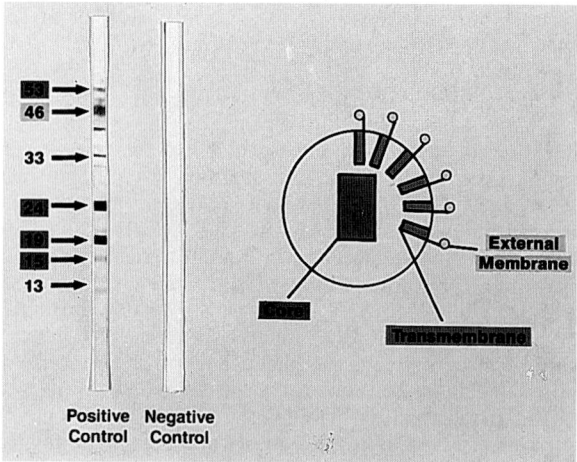

FIGURE 1 Typical HTLV-I pattern in Western blot.

proteins p19 and p24 than for the *env* protein gp46. The standard viral bands detected by Western blot (IB) are *gag* p19, p24, and p33 and *env* gp46 (Fig. 1). Additional bands such as p26, p28, p32, p36, and p53 may be present depending on the HTLV-I cell line used to generate the blots. The criteria established by the U.S. Public Health Service for confirmation of HTLV-I seropositivity include antibodies against both *env* (gp46 and/or gp61/68) and *gag* (p24) gene products by Western blot or RIPA alone or in combination. A sample is considered indeterminate if antibodies against HTLV-I-specific proteins are detected in a combination other than those given above, e.g., p19 only, p19 and p28, or p19 and *env*. Sera with no reactivity to any HTLV-I proteins are considered negative or false positive (2, 18). Indeterminate IB samples are rarely positive by

PCR; however, sera with anti-*gag* p19 and *env* gp61/68 without *gag* p24 reactivity can be PCR positive for HTLV-I (4).

An indeterminate IB is usually followed by RIPA, which has a higher sensitivity for *env* gp61/68. The standard bands detected by RIPA (using the HUT 102.B2 cell line) are *gag* p24, and p55, *tax* p40x, and *env* gp61 (Fig. 2). Again, different banding patterns are seen with different cell lines. The results of the RIPA can be negative (no bands), indeterminate (incomplete banding pattern), or positive (p24, gp61 present). A combination of IB and RIPA results can lead to indeterminate or positive identification of samples (Fig. 3).

Currently, confirmatory assays based on a full complement of HTLV-II proteins are performed by only a few research laboratories. Confirmatory assays under development include an improved differential HTLV IB that incorporates virus-specific recombinant *env* gp46 peptides of HTLV-I and HTLV-II.

IFA, used in some laboratories as a confirmatory assay, is technically less demanding and less expensive than IB or RIPA. The major disadvantages of IFA are the requirements for a fluorescence microscope and highly trained technicians, subjective reading, and lack of automation and a permanent record. False-positive and false-negative results can occur. When performed with adequate internal controls, the rate of indeterminate results is lower for IFA than for IB.

In general, different HTLV-I-infected groups display different anti-HTLV-I titers as well as different IB patterns. Sera from HAM/TSP patients have higher antibody titers against most HTLV-I proteins than do sera from ATL patients and asymptomatic blood donors in areas where HTLV-I is endemic. Sera from both blood donors and intravenous-drug abusers in areas where the virus is not endemic have antibodies predominantly against p24 *gag*

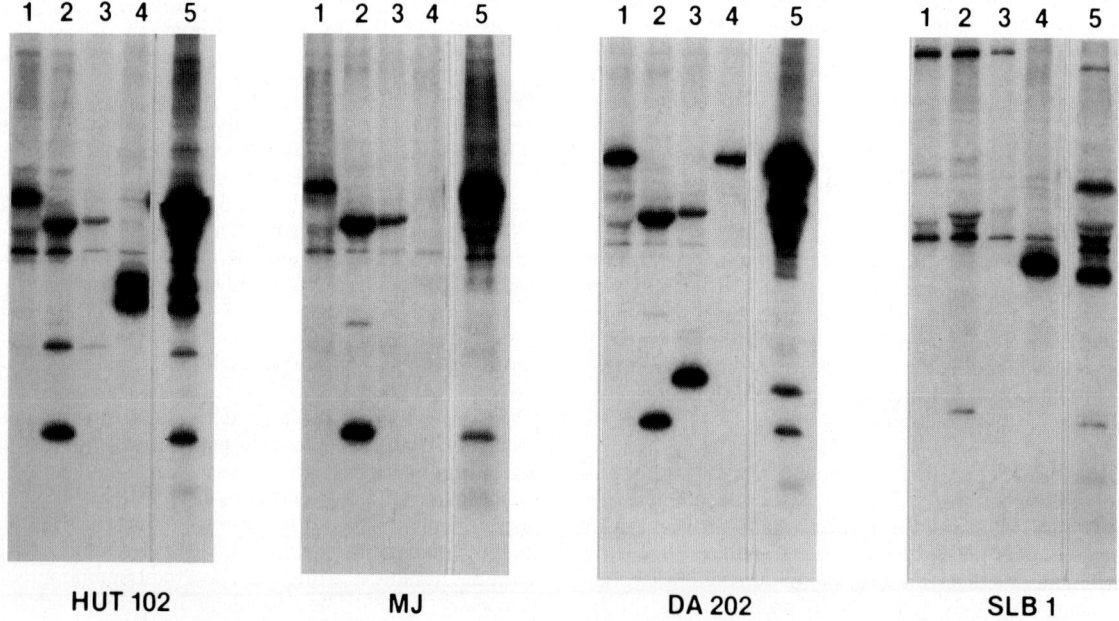

FIGURE 2 RIPA reactivities of various anti-HTLV-I antibodies with different cell lines. Lanes: 1, mouse monoclonal anti-gp46 antibody; 2, mouse monoclonal anti-p24 antibody; 3, mouse monoclonal anti-p19 antibody; 4, mouse monoclonal anti-p40x antibody; 5, positive control.

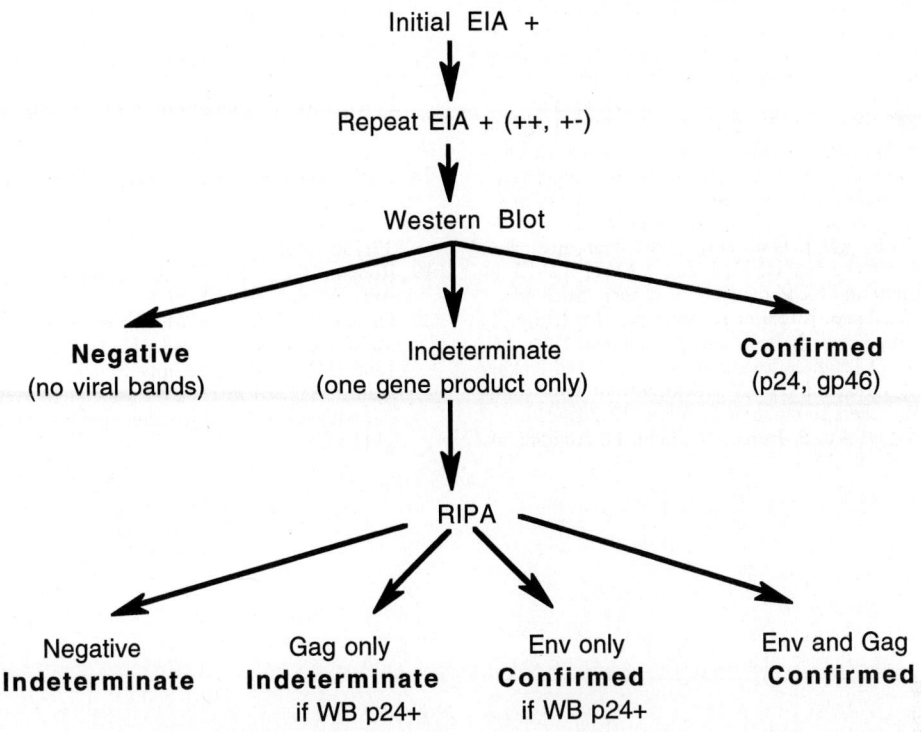

FIGURE 3 HTLV confirmatory algorithm. WB, Western blot.

and to a lesser level against p19 *gag*. These sera generally require a combination of IB and RIPA for confirmation. HTLV-II, a virus previously thought to be rare, appears to be more prevalent than HTLV-I, particularly among intravenous-drug abusers and some native North and South American groups.

The use of HTLV-I-infected cell lysate in RIPA allows for the detection of antibodies against p40 *tax*. The prevalence of anti-p40 *tax* was reported to be significantly higher among HAM/TSP patients than among asymptomatic carriers, smoldering-chronic-ATL patients, or acute-ATL patients (9). Children from mothers seropositive for both HTLV-I structural proteins and p40 *tax* show a higher rate of HTLV-I infection than those from mothers seropositive for HTLV-I but negative for p40 *tax*. Thus, antibodies against p40 *tax* may prove useful as a prognostic marker for HAM/TSP and for the transmission of HTLV-I from mother to child.

REFERENCES

1. Ando, Y., S. Nakano, K. Saito, I. Shimamoto, M. Ichijo, T. Toyama, and Y. Hinuma. 1987. Transmission of adult T-cell leukemia retrovirus (HTLV-I) from mother to child: comparison of bottle with breast fed babies. *Jpn. J. Cancer Res.* **78:**322–324.
2. Centers for Disease Control. 1993. Recommendations for counseling persons infected with human T-lymphotropic virus, types I and II. *Morbid. Mortal. Weekly Rep.* **42:**1–13.
3. Chen, I. S. Y., S. G. Quan, and D. W. Golde. 1983. Human T-cell leukemia virus type II transforms normal human lymphocytes. *Proc. Natl. Acad. Sci. USA* **80:**7006–7009.
4. Donegan, E., P. Pell, H. Lee, G. M. Shaw, J. W. Moseley, and The Transfusion Safety Study Group. 1992. Transmission of human T-lymphotropic virus type I by blood components

5. from a donor lacking anti-p24: a case report. *Transfusion* **32:**68–71.
6. Gessain, A., J. C. Vernant, L. Maurs, F. Barin, O. Gout, A. Calender, and G. de-The. 1985. Antibodies to human T-lymphotropic virus type I in patients with tropical spastic paraparesis. *Lancet* **ii:**407–409.
7. Gout, O., M. Baulac, A. Gessain, F. Semah, F. Saal, J. Peries, C. Cabrol, C. Foucault-Fretz, D. Laplane, F. Sigaux, and G. de The. 1990. Rapid development of myelopathy after HTLV-I infection acquired by transfusion during cardiac transplantation. *N. Engl. J. Med.* **322:**383–388.
8. Graziano, S. L., B. M. Lehr, S. A. Merl, G. D. Ehrlich, J. L. Moore, E. J. Hallinan, C. Hubbell, F. R. Davey, J. Vournakis, and B. J. Poiesz. 1987. Quantitative assay of human T-cell leukemia/lymphoma virus transformation. *Cancer Res.* **47:**2468–2473.
9. Kalyanaraman, V. S., M. G. Sarngadharan, M. Robert-Guroff, I. Miyoshi, D. Blayney, D. Golde, and R. C. Gallo. 1982. A new subtype of human T-cell leukemia virus (HTLV-II) associated with a T-cell variant of hairy cell leukemia. *Science* **218:**571–573.
10. Kamihira, S., K. Toriya, T. Amagasaki, S. Momita, S. Ikeda, Y. Yamada, M. Tomonaga, M. Ichimaru, K. Kinoshita, and T. Sawada. 1989. Antibodies against p40 tax gene product of human T-lymphotropic virus type-I (HTLV-I) under various conditions of HTLV-I infection. *Jpn. J. Cancer Res.* **80:**1066–1071.
11. Lee, H., P. Swanson, J. D. Rosenblatt, I. S. Y. Chen, W. C. Sherwood, D. E. Smith, G. E. Tegtmeier, L. P. Fernando, C. T. Fang, M. Osame, and S. H. Kleinman. 1991. Relative prevalence and risk factors of HTLV-I and HTLV-II infection in US blood donors. *Lancet* **337:**1435–1439.
12. Lee, H., P. Swanson, V. S. Shorty, J. A. Zack, J. D. Rosenblatt, and I. S. Y. Chen. 1989. High rate of HTLV-II infection in seropositive IV drug abusers in New Orleans. *Science* **244:**471–475.
13. Lee, H. H., S. H. Weiss, L. S. Brown, D. Mildvan,

V. Shorty, L. Saravolatz, A. Chu, H. M. Ginzburg, N. Markowitz, D. C. Des Jarlais, W. A. Blattner, and J. P. Allain. 1990. Patterns of HIV-1 and HTLV-I/II in intravenous drug abusers from the Middle Atlantic and central regions of the United States. *J. Infect. Dis.* **162**:347–352.

13. Maeda, Y., M. Furukawa, Y. Takehara, K. Yoshimura, K. Miyamoto, T. Matsuura, Y. Morishima, K. Tajima, K. Okochi, and Y. Hinuma. 1984. Prevalence of possible adult T-cell leukemia virus-carriers among volunteer blood donors in Japan: a nationwide study. *Int. J. Cancer* **33**:717–721.

14. Melief, C. J. M., and J. Goudsmit. 1986. Transmission of lymphotropic retroviruses HTLV-I and LAV/HTLV-III by blood transfusion and blood products. *Vox Sang.* **50**:1–11.

15. Nakano, S., Y. Ando, K. Saito, I. Moriyama, T. Ichijo, T. Toyama, K. Sugamura, J. Imai, and Y. Hinuma. 1986. Primary infection of Japanese infants with adult T-cell leukaemia-associated retrovirus (ATLV): evidence for viral transmission from mothers to children. *J. Infect.* **12**:205–212.

16. Osame, M., K. Usuku, S. Izumo, N. Ijichi, H. Amitani, A. Igata, M. Matsumoto, and M. Tara. 1986. HTLV-I associated myelopathy, a new clinical entity. *Lancet* **i**:1031–1032.

17. Poiesz, B. J., F. W. Ruscetti, A. F. Gazdar, P. A. Bunn, J. D. Minna, and R. C. Gallo. 1980. Detection and isolation of type C retrovirus particles from fresh and cultured lymphocytes of a patient with cutaneous T-cell lymphoma. *Proc. Natl. Acad. Sci. USA* **77**:7415–7419.

18. Public Health Service Working Group (FDA/CDC/NIH). 1988. Licensure of screening tests for antibody to human T-lymphotropic virus type I. *Morbid. Mortal. Weekly Rep.* **37**:736–747.

19. Roman, G. C. 1987. Retrovirus-associated myelopathies. *Arch. Neurol.* **44**:659–663.

20. Usuka, K., S. Sonoda, M. Osame, S. Yashiki, K. Takahashi, M. Matsumoto, T. Sawada, K. Tsuji, M. Tara, and A. Igata. 1988. HLA haplotype-linked high immune responsiveness against HTLV-I in HTLV-I associated myelopathy: comparison with adult T-cell leukemia/lymphoma. *Ann. Neurol.* **23**:S143–S150.

Spongiform Encephalopathies

DAVID M. ASHER

99

The spongiform encephalopathies comprise a group of two or three infectious diseases of humans and at least four of animals (Table 1) with roughly similar clinical courses and histopathologies (54, 55). All are elicited by similar unique pathogenic agents. The diseases are often called "slow" infections, a term coined by Sigurdsson (140) to describe infectious diseases with long incubation periods, protracted clinical courses (months or years), fatal outcomes, pathologies restricted to a single organ system, and limited ranges of susceptible hosts. The pathogenic agents themselves have been termed prions (125), infectious amyloids (55), or unconventional viruses (134). Since the structure of the pathogens remains in dispute, here they are simply called agents.

EPIDEMIOLOGY

Kuru was the most common cause of death in women of the Fore language group of Papua New Guinea in the 1950s and early 1960s; it also affected many children of both sexes over the age of 4 years and a few adult males (57). The disease has now almost completely disappeared among the Fore and has not affected anyone born after 1960 (88). This peculiar epidemiologic pattern suggests that the infection was naturally transmitted only by cannibalism, which ended in the late 1950s, and implies that the incubation period may be as short as 4 years or longer than 35 years. Creutzfeldt-Jakob disease (CJD) (86) is an uncommon cause of dementia, affecting an estimated $1/10^6$/year or fewer in most populations studied, although frequencies in several isolated areas are more than 30 times higher (98). Males and females are affected in approximately equal numbers (20). The natural mechanisms of transmission are not yet understood, although a growing number of iatrogenic cases have been recognized (23, 55): contaminated surgical instruments (113, 149), cortical electrodes (9), corneal transplants (52), dural grafts (36-38, 51, 95, 96, 99, 114, 143a, 144), and human pituitary hormones (19, 23, 42, 75) have been implicated as sources of iatrogenic CJD in young people. (By experimentally transmitting CJD to animals, we confirmed that a batch of growth hormone and an electrode remained contaminated long after their last use in humans [60, 61].) After iatrogenic transmission, incubation periods for CJD have been as short as 14 months and longer than 20 years.

TABLE 1 Slow infections of the nervous system caused by unconventional agents (spongiform encephalopathies, prion diseases)

Disease	Naturally infected host(s)
CJD, GSS	Humans
Kuru	Humans
Bovine spongiform encephalopathy ("mad cow" disease)	Cattle, captive exotic ungulates, domestic cats
Chronic wasting disease	American mule deer, elk
Scrapie	Sheep, goats
Transmissible mink encephalopathy	Mink

Two instances of married couples with CJD (80, 100), two cases in neuropathology technicians (107, 141), and single cases in a neurosurgeon (56), a pathologist (70), and an oral surgeon (8) have been described. However, the empirical risk for medical personnel and other contacts of CJD patients appears to be small.

Scrapie is an important disease of sheep in Great Britain and on the European continent and has become increasingly common in American sheep in recent years (71). No epidemiologic evidence implicates scrapie as a source of human CJD. However, an outbreak of bovine spongiform encephalopathy ("mad cow" disease) in England, apparently introduced by accidental incorporation into cattle feed of contaminated meat and bone meal prepared from renderings of scrapie-infected sheep and later cattle (82, 148), raised fears that the scrapie agent, having crossed one "species barrier" from sheep into cattle and later affecting other ungulates and domestic cats, may have broadened its range of susceptible hosts to include humans. This possibility remains under intense scrutiny. Transmissible mink encephalopathy in the United States is postulated to have resulted from feeding contaminated beef to mink (94). Nothing is known about the origin and occurrence in the wild of the "wasting disease" of American elk and mule deer (150) that closely resembles scrapie.

CLINICAL PICTURE AND COURSE OF ILLNESS

Kuru (57) is characterized by progressive loss of coordination, cerebellar ataxia of extremities and trunk, and a variety of signs of brain stem degeneration. There is little clinical evidence of cerebral cortical involvement and no frank dementia. Most patients have the shivering tremors that gave kuru its name in the Fore language. Convulsive seizures and myoclonic jerks have not been observed. Patients become progressively incapacitated and usually die within a year from pneumonia, infected decubitus ulcers with septicemia, or accidental burns. The disease has not been recognized outside Papua New Guinea (55).

CJD, first described almost 70 years ago (86), typically affects middle-aged adults, though iatrogenically infected young adults and older adolescents have been described (15). The disease usually begins insidiously with vague complaints of sensory disturbance, particularly visual, or confusion. Patients become progressively demented, sometimes quite rapidly over only a few weeks. Most patients eventually have repetitive myoclonic jerking movements of the extremities and may have generalized "startle" myoclonus as well. Rigidity and pathologic cortical-release reflexes appear as the cerebral gray matter degenerates. A variety of other neurologic abnormalities, including convulsions and an asymmetrical weakness and spasticity that may suggest a space-occupying lesion, are less common. Most patients die in less than a year after onset, though a few patients survive much longer (20). Death usually results from pneumonia or one of the other complications to which patients with terminal neurologic diseases are subject. Remission of CJD has been claimed but never well documented. In many series of CJD, about 10% of cases have a family history of presenile dementia. The Gerstmann-Sträussler-Scheinker syndrome (GSS) is an extremely rare familial spongiform encephalopathy with progressively severe cerebellar ataxia and relatively later onset of dementia (97). It may be considered a variant of familial CJD (FCJD). Scrapie, transmissible mink encephalopathy, wasting disease of mule deer and elk, and bovine spongiform encephalopathy are degenerative brain diseases that have been recognized in a variety of animals and resemble the human diseases to some degree in clinical picture and course and strikingly in histopathologic findings (55).

Fatal familial insomnia (FFI) is a newly recognized syndrome (92, 104) described below. Although patients with FFI have the same mutation as do some patients with FCJD, there are substantial differences in pathology between the two syndromes, and no infectious agent has been implicated in FFI yet.

PATHOLOGY

The spongiform encephalopathies are all characterized by progressive relentless degeneration of the central nervous system (5–7) that is most severe in the gray matter, though there may be secondary loss of myelin and white-matter involvement as well. The histopathology consists of neuronal vacuolation and ultimate loss of neurons; in humans with CJD and in some animals with scrapie or experimental CJD, the vacuolation may be severe enough to cause status spongiosus of the cerebral gray matter, giving the group of diseases its name (62). Neuronal loss is accompanied by proliferation and hypertrophy of fibrous astrocytes. The cerebellum is prominently involved in kuru, with marked loss of Purkinje and granule cells; cerebellar degeneration also occurs in CJD, though it is generally not so severe as in kuru. Amyloid plaques are universally present in brains of patients with GSS, very common in kuru patients (at least 70% of cases), and occasionally found in CJD patients (about 15% of cases); they are most often seen in the cerebellum but occur elsewhere as well. Notably lacking in the brain in spongiform encephalopathies are inflammatory cells and changes of primary demyelination.

There is an accumulation in the brain of an abnormal protein that is relatively resistant to protease digestion. This protein was first recognized morphologically by negative-stain electron microscopy as scrapie-associated fibrils (105) and was later recognized as a unique band of protein with an apparent molecular mass of 27 to 30 kDa on sodium dodecyl sulfate-polyacrylamide gel electrophoresis (13). The protein, designated PrP27-30 (101), is a sialoglycoprotein (14) consisting of 55 amino acids (109) with attached carbohydrates, a neuraminic acid (14), and an inositol (34, 142). Particularly in disease of long duration, this protein (now most often called the prion protein, or PrP) may form amyloid plaques visible by light microscopy as noted above. These plaques are amorphous accumulations of protein that stain with Congo red dye and are birefringent under polarized light. The amyloid plaques of spongiform encephalopathies resemble those of Alzheimer's disease, but antibodies prepared against PrP do not react with Alzheimer amyloid and vice versa (11, 123). Even in brains without plaques, immunostaining may reveal extracellular accumulations of PrP in subependymal and periventricular areas (46) and in the Golgi apparatus or endoplasmic reticula of neurons (123).

The prion proteins of various species are very similar in amino acid sequences and antigenicity but are not identical (12, 89). These proteins are derived from a larger precursor protein of 253 amino acids that was originally designated PrP33-35 (106) and was later referred to as PrP^C in healthy individuals and as PrP^{Sc} or PrP^{CJD} in scrapie or CJD patients, respectively. PrP^{Sc} and PrP^{CJD}, extracted from brain tissue afflicted with spongiform encephalopathy, are, like PrP27-30, relatively resistant to protease digestion, while PrP^C from healthy brains is protease sensitive. Since the amino acid sequences of the two forms of protein are identical (116), the change in protease resistance is attributed to some posttranslational alteration whose nature is not known. The conversion of PrP^C to PrP^{Sc} and PrP^{CJD} is accompanied by a shift from an alpha-helical structure to a beta-sheet conformation (47, 122); the relationship between those changes and the infectious agents of the spongiform encephalopathies is under investigation (136). PrP^C, secreted at the surface membranes of a variety of healthy cells and rapidly catabolized by cells (33, 34), is apparently not essential for life (at least in transgenic and other experimental animals) (126–128). Its normal functions remain unknown.

PATHOGENESIS

Most studies investigating the pathogenesis of the spongiform encephalopathies have been conducted with scrapie, both natural and experimental (1, 72). The agent appears to enter the body through the integument; in sheep, the intact intestinal tract may be a portal of entry as well. The agent appears to replicate first in local lymph nodes draining the portal of entry. Later, there is replication in the spleen and other lymphoid organs throughout the body.

TABLE 2 Infectivity of tissues, body fluids, and excretions from patients with CJD

Tissue, fluid, or excreta containing infectious agent				
Consistently (≥50% of attempts)	Sometimes (4–50% of attempts)	Occasionally[a]	Never[b]	
Brain	CSF	Blood	Adrenal	Nasal mucus
Spinal cord	Kidney	Urine	Feces	Nerve
Eye	Liver		Heart	Saliva
	Lung		Marrow	Sputum
	Lymph node		Muscle	Tears
	Spleen			

[a] Transmission of CJD from blood and urine of patients to animals was not confirmed in our laboratory (blood or blood components, 12 attempts; urine, 11 attempts).

[b] Three or more attempts were negative. A smaller number of attempts were also negative for intestine, thyroid, fat, gingiva, prostate, testis, placenta or amnion, semen, vaginal secretions, and breast milk.

There appears to be "viremia" in some animals and possibly in humans as well. In mice with scrapie, the agent may ascend peripheral nerves to enter the spinal cord (85). In any event, the agents appear in the nervous system only relatively late in the course of experimental infections initiated by the subcutaneous route, which is presumed to mimic the route of infection with natural disease. Tissues and body fluids demonstrated to contain the infectious agent in patients dying of CJD are summarized in Table 2. Scrapie has not been transmitted from affected rodents to offspring in utero, nor have pregnant women with kuru or CJD transmitted disease to their children (55). However, sheep can apparently be infected with scrapie in utero as well as by contact with affected animals after birth. In none of the spongiform encephalopathies have antibodies or a cell-mediated immune response to the infectious agents been detected, nor has interferon been found. The possible role of cytokines in the spongiform encephalopathies has been investigated recently (29, 91).

HOST GENETICS AND SPONGIFORM ENCEPHALOPATHIES

CJD has long been recognized in families (3); in many series, five or 10% of CJD patients have a family history of presenile dementia. Most pedigrees of FCJD suggest an autosomal dominant mode of inheritance: the disease occurs without skipping generations and affects approximately half the siblings of propositi, both males and females in equal numbers. GSS also shows an autosomal dominant pattern of inheritance (76). The genetic basis for FCJD and GSS appears to reside in a series of mutations in the gene coding for the PrP precursor protein, currently designated the PRNP gene, on the short arm of human chromosome 20. PRNP has an open reading frame of 759 nucleotides (253 codons) in which 10 point mutations and a variety of insertions have been linked to FCJD or GSS.

The importance of the analogous gene in spongiform encephalopathies of animals was earlier noted for experimental scrapie (and later for CJD) in mice (30, 31), in which the sinc gene, long known to control incubation periods of scrapie in mice (79), was found to be closely linked if not identical to the PrP gene; a similarly close linkage was found between PrP and the scrapie incubation period genes of sheep (78). In mice of some genotypes infected with certain strains of scrapie agent, the incubation periods are so long that they exceed the life span of the

animal, producing a phenotype of resistance to disease though not to inapparent infection (50).

The important role of the PrP gene in the incubation, susceptibility, and pathogenesis of the spongiform encephalopathies has been further clarified by a series of elegant studies with transgenic mice. Mice are ordinarily resistant to hamster-adapted strains of scrapie; however, when a portion of the PrP gene sequence from hamsters was engineered into mice, the mice acquired susceptibility to those strains, and the agent recovered from mice had hamster-adapted biological properties (129, 138). Increased expression in mice of an engineered PrP gene bearing a mutation similar to one found in most patients with GSS caused a disease resembling scrapie (77), although replicating infectious scrapie agent has apparently not been recovered from those mice; a similar complex neurologic disease was also observed in mice expressing abnormal amounts of wild-type PrP (147). Animals not expressing PrP^C (called PrP knockout mice because a disruption was engineered into their PrP genes) developed and behaved normally, and they have thus far been resistant to scrapie disease, though possibly not to infection (27, 128).

Owen and coworkers discovered an abnormal restriction endonuclease pattern in the PRNP genes of affected members in a family with typical autosomal dominant FCJD (119); Goldgaber and coworkers found that patients from other families with FCJD lacked that abnormality (69), suggesting that other mutations might also be involved. Owen and others soon learned that their mutation was a large insertion of about 150 bases (118) in a region of five normal octapeptide tandem repeats between codons 51 and 91 of the PRNP gene (coding for six extra repeats there). Several other insertions in the 51-91 octapeptide repeat region (coding for five to eight extra octapeptides) besides the six-octapeptide insertion first discovered have been associated with FCJD, and one (coding for nine extra octapeptides) is associated with GSS (55, 66).

Hsiao and colleagues found that a single nucleotide change in codon 102 of the PRNP gene (Leu-102; the encoded amino acid was changed from proline to leucine) was linked to GSS (76). Goldgaber and coworkers found another point mutation in PRNP at codon 200 (Lys-200) that cosegregated with FCJD (68). Leu-102 and Lys-200 are the two most common point mutations associated with familial spongiform encephalopathies, but other point mutations (summarized in Table 3 and Fig. 1) have been found; in addition to the most common Lys-200 mutation,

TABLE 3 Point mutations in prion protein gene associated with familial spongiform encephalopathies

Codon no.	Amino acid		Spongiform encephalopathy
	Normal	Mutant	
178	Asp	Asn	CJD
180	Val	Ile	CJD
200[a]	Glu	Lys	CJD
210	Val	Ile	CJD
102[a]	Pro	Leu	GSS
105	Pro	Leu	GSS
117	Ala	Val	GSS
145	Tyr	Stop[b]	GSS
198	Phe	Ser	GSS
217	Glu	Arg	GSS
129	Met or Val	Met or Val	CJD/Asn-178 + Val-129 FFI/Asn-178 + Met-129
117	Ala	Ala	?None[c]
124	Gly	Gly	?None
232	Met	Arg	?None

[a] Largest numbers of familial spongiform encephalopathies were associated with these mutations.
[b] Stop, stop codon, amber mutant.
[c] ?None, normal polymorphism.

FCJD has been linked to each of three other point mutations (Asn-178, Ile-180, and Ile-210), and GSS has been linked to point mutations at five other codons (Leu-105, Val-117, Stop-145, Ser-198, and Arg-217) as well as to Leu-102. Other mutations in the *PRNP* gene associated with FCJD and GSS probably await discovery.

Recently, FFI, an inherited syndrome with an autosomal dominant pattern of occurrence that is characterized by progressive severe insomnia and dysautonomia with selective atrophy of two thalamic nuclei, was described first in a large northern Italian kindred and later in several others (92, 102, 104). Patients with FFI have ataxia, myoclonus, and other signs resembling those of CJD and GSS, and a few affected patients had spongiform changes in the cerebral cortex. Those findings prompted the hypothesis that FFI might be a new prion disease; indeed, protease-resistant PrP was detected in brains of patients with FFI, although it apparently differed in size from the PrP found in CJD and GSS patients (104, 108). In patients with FFI, a mutation in the *PRNP* gene at codon 178 (Asn-178) was identical to that found in some kindreds with FCJD. However, the two groups of patients differed in *PRNP* sequences at another codon on the abnormal allele, 129 (a codon encoding methionine in some normal subjects and valine in others); FFI patients had Val-129 linked to Asn-178, while FCJD patients had Met-129 (67, 103). It remains unknown how the same point mutation in one codon of a gene might interact with otherwise normal polymorphic nucleotides in another codon of the same gene to produce different clinical illnesses. FFI may also differ from FCJD in another important way: to date, no infectious agent has been demonstrated in FFI patients, though the number of attempts to transmit encephalopathy to animals from their brain tissues has been small. It is possible that FFI is caused by an abnormality in the *PRNP* gene itself without involvement of the infectious agent of spongiform encephalopathy, while patients with FCJD have both an abnormal gene product and an acquired infection. This is an area of active research (21, 67, 103).

Patients with iatrogenic and sporadic CJD were much more often homozygous at codon 129 of the *PRNP* gene (>90%) than were subjects in the general population (about 50%) (17, 49, 121), suggesting that homozygotes are more susceptible to CJD infection than are heterozygotes. In addition to the methionine-valine polymorphism at codon 129 and a methionine-arginine polymorphism at codon 232 in healthy subjects, other normal polymorphisms in the *PRNP* gene occur at codons 117 and 124 (silent differences, where both variants encode the same amino acid) and in the 51-91 octapeptide repeat region, where both deletions and short insertions of fewer than five extra repeats have been detected in healthy subjects (55). It is not known whether these polymorphisms play any role in neurologic diseases.

Not all subjects with point mutations in the *PRNP* gene express disease, even in affected families. While penetrance appears to be quite high for GSS in family members with the Leu-102 mutation and for FCJD in those with the Asn-178 mutation, fewer than 60% of subjects with the Lys-200 and Ile-210 mutations have gotten CJD (124), at least by the usual expected age. It is not known whether unaffected family members bearing those mutations in the *PRNP* gene have inapparent infections, perhaps owing to exceptionally long incubation periods, or have escaped infection. Several individuals with the Asn-178 mutation in families with FFI have also survived past the age of 60 without showing signs of illness (58). Whatever the explanation, variations in expression must be kept in mind when dealing with unaffected family members.

INFECTIOUS AGENTS

The agents of the spongiform encephalopathies are unique in their physical properties. They are all experimentally transmissible to a variety of animals including monkeys and rodents (55). Although cell cultures have been infected (4, 28, 32–35, 117, 130, 131, 137), the cells display no cytopathic effects, and the agents generally propagate to much

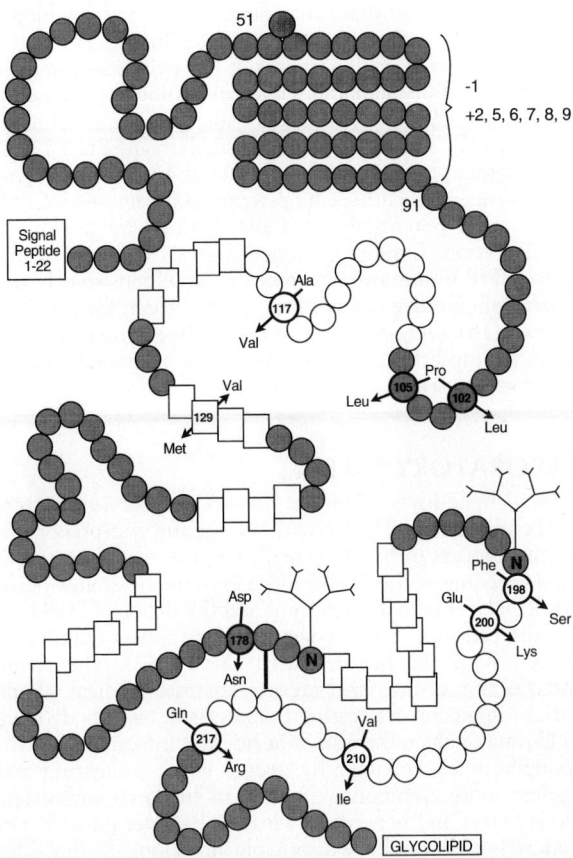

FIGURE 1 Schematic diagram of PrP, the amyloid or prion protein found in brains of humans and animals with spongiform encephalopathies (reprinted with the kind permission of P. Brown [16] and Humana Press). The 230-amino-acid backbone of PrP is depicted, with random turns and coils indicated by gray circles, the alpha helix indicated by open circles, and the beta sheet indicated by open rectangles. Cysteine residues 179 and 214 are linked by a disulfide bond, and sugar groups are attached to asparagines at positions 181 and 197. Important posttranslational modifications include N-terminal cleavage of a 22-amino-acid signal peptide and replacement of a C-terminal 23-residue sequence by a glycophosphatidylinositol by which the protein is anchored to the cell membrane. The point mutations, insertions, and deletions mentioned in the text, both those associated with disease and those found in healthy subjects, are indicated by arrows and a bracket.

lower titers in culture than in brain tissue. Most studies characterizing the properties of the spongiform encephalopathy agents have used strains of scrapie agent adapted to mice and hamsters; these strains have incubation periods as short as 60 days and high infectivity titers (84).

The infectious agents display extreme resistance to physical inactivation by heat and by ionizing and UV radiation as well as to chemical inactivation by a wide array of substances that destroy most viruses. This behavior, plus the fact that the agents show significant decreases in titers of infectivity after exposure to proteases (41) and other treatments that denature or degrade proteins (125), has convinced some authorities that the infectious agents are most probably replicating proteins devoid of nucleic acid

(59, 90). Prusiner suggested that these novel agents be called prions in recognition of their probable protein component and in anticipation that they would be found to lack nucleic acids. He and others preferring the all-protein hypothesis have emphasized that the aberrant physical behavior of the agents is inconsistent with properties of nucleic acids and that infectivity has seldom if ever been separated from the prion or amyloid protein, which they propose to be the infectious agent itself (126).

For the following reasons, several other authorities remain unconvinced by the all-protein hypothesis (81) and skeptical of the evidence marshaled by its proponents (39, 40, 93, 134). (i) Reinterpretation of irradiation inactivation kinetic studies suggests that the possibility of a small nucleic acid genome has not been excluded (133). (ii) No confirmed studies have demonstrated the infectivity-bearing agent to be physically smaller than a small virus (63). (iii) No studies have convincingly demonstrated that a purified protein uncontaminated with nucleic acid from an infected host contains the self-replicating agent of a spongiform encephalopathy. (iv) In several experimental systems, the kinetics of PrP formation failed to correlate with kinetics of infectivity (43, 44, 111, 153). (v) Tissues of PrP knockout mice that have been genetically manipulated so that they cannot synthesize the PrP precursor protein (PrP^C), although not susceptible to overt scrapie disease, nonetheless appear to have supported replication of the infectious scrapie agent as assayed in wild-type mice, albeit to a much lower titer and possibly in a form more susceptible to inactivation by heat (27). (vi) Finally, the all-protein hypothesis has not yet satisfactorily explained the phenomenon of strainlike behavior by the spongiform encephalopathy agents. Various "strains" of the scrapie agent have properties (like incubation period, distribution of histopathologic lesions, and presence or absence of visible amyloid-plaque formation) that breed true on passage in inbred mice (25, 26, 81), and those properties occasionally undergo sudden changes resembling the mutations that occur in genomes of conventional pathogens (83). (Similar "strains" of CJD agent have also been proposed [48, 64].) The molecular basis for such strainlike biological behavior is easily understood if the agents have small nucleic acid genomes but not if they consist entirely of a ubiquitous host-encoded protein like PrP that is posttranslationally modified but without changes in normal primary amino acid structure.

Proponents of the all-protein hypothesis have put forth a variety of ingenious additional hypotheses to explain how proteins might transmit self-replicating information not encoded in their primary amino acid sequences (55, 126, 145, 146, 151, 152). Other authorities maintain that the agents of spongiform encephalopathies more probably contain small nucleic acid genomes (134). However, no unique nonhost nucleic acids have been identified in infectious materials from the spongiform encephalopathies. Viruslike particles have often been noted in sections of tissues (45, 112), and recently, smaller particles were found in extracts of tissues from infected animals (120). However, such particles have not been convincingly associated with the infectious agents. Thus, in spite of more than 20 years of controversy, the nature of the infectious agents of the spongiform encephalopathies, although unique in microbiology, remains disputed.

LABORATORY DIAGNOSIS

The spongiform encephalopathies are generally diagnosed during life by typical constellations of clinical findings. CJD is suspected in a middle-aged patient (or in a younger person with a history suggestive of iatrogenic transmission) having progressive dementia and myoclonic jerks. Later in disease, the electroencephalogram frequently shows periodic high-voltage complexes on a slow and disorganized background (65). Hematologic studies and clinical chemistries are unrevealing. Some patients with CJD have abnormal liver functions suggestive of parenchymal disease (135); the significance of that is not known. The cerebrospinal fluid (CSF) often contains moderately increased amounts of protein, which is not diagnostic; cell content of the CSF is normal. A research technique involving two-dimensional electrophoresis and isoelectric focusing of CSF in gels has proven useful in the diagnosis of CJD; silver staining reveals the presence of two abnormal polypeptides (designated 130 and 131) in the CSF of most patients with CJD but not in those with Alzheimer's disease (10, 74, 154). The polypeptides have not been identified or fully sequenced. They are apparently not related to PrP or to other known proteins. The 130 and 131 peptides in CSF are not specific to CJD; they are also found in about half of the CSF specimens from patients with acute herpes encephalitis. However, in practice, the usual diagnostic problem is to differentiate between CJD and Alzheimer's disease, and the presence of the 130 and 131 peptides in CSF strongly militates against the latter. This test is available only in a few research laboratories.

Definitive laboratory diagnosis of spongiform encephalopathies can be made only from brain tissue. Histopathologic examination by a skilled and experienced neuropathologist remains the standard by which other tests must be judged. In addition to evaluation of neuronal vacuolation in sections stained with hematoxylin and eosin, stains for astrocytes (gold stains if available, or immunostaining for glial fibrillary acidic protein) and for amyloid help in diagnosis. Immunostaining of amyloid may be enhanced by pretreating sections with formic acid (87) or by autoclaving them (73, 143), although there remains substantial disagreement about the optimal conditions for these techniques. A recent report suggested that detection of PrP^{Sc} in lymphoid tissues may provide early diagnosis of scrapie in sheep (110); similar methods of early diagnosis not requiring brain biopsy would also be useful for the human spongiform encephalopathies. Although intracerebral inoculation of brain tissue suspensions into susceptible animals with induction of typical spongiform encephalopathy is diagnostic, it is extremely time-consuming and expensive; for example, squirrel monkeys, perhaps the animals most consistently susceptible to experimental CJD agent among commercially available animals, must be observed for at least 3 years before a transmission attempt is considered tentatively negative. Although laboratory rodents are susceptible to some strains of CJD agent, in our experience, fewer than 20% of attempts at primary transmission to rodents from well-documented cases have been successful. Attempts at primary transmission to animals have sometimes been confusing owing to problems with contamination in laboratories where the agents of spongiform encephalopathies are under study. The demonstration of PrP in brains of patients with progressive dementia is useful in confirming the diagnosis of spongiform encephalopathy (18). Until recently, the preparation of the protein required more brain tissue than is generally obtained by biopsy; recent reports suggest that modified techniques permitting detection of very small amounts of PrP by immunostaining on nitrocellulose membranes after guanidine isothiocyanate treatment (139) or in embedded tissue sections as described above may be suitable for studying small fragments of biopsy tissue. Since there is still no effective therapy for the spongiform encephalopathies, biopsy can be recommended only when other potentially treatable diseases are also under consideration. Limited evidence suggests that demonstration of PrP in autopsy specimens is more consistently successful when tissue has not been stored frozen for very long periods (18). Unfortunately, at the moment, the techniques for detecting PrP remain available only by special arrangement with research laboratories.

LABORATORY SAFETY

Although, as discussed above, the risk to laboratory personnel of accidental infection with spongiform encephalopathy agents appears to be very small, because of the uniformly fatal outcome of the diseases, the extreme resistance of the etiologic agents to disinfection, and the difficulty of detecting the presence of these agents, potentially contaminated tissues should be handled with caution (2). The recent introduction of universal precautions into medical laboratories responsible for testing human tissues and body fluids (115) may reduce the risk of accidental infection with the spongiform encephalopathy agents while protecting staff against more common exposures to human immunodeficiency virus and hepatitis B virus. Whenever possible, the materials used should be disposable and should be discarded carefully by laboratories handling potentially contaminated specimens; if possible, such materials should be incinerated. Although the spongiform encephalopathy agents may not be completely destroyed even at very high temperatures (22), titers of infectivity are reduced markedly by heating (132); 2 h in the stream autoclave at 132°C is currently recommended to decontaminate objects that withstand such treatment. The infectivity titers of spongiform encephalopathy agents are also reduced substantially by exposure to 5.25% sodium hypochlorite (undiluted household chlorine bleach) and sodium hydroxide (2 M or stronger recommended). The same formic acid treatment reported to improve immunostaining of amyloid in tissue sections may inactivate the spongiform encephalopathy agents as well (24). Recently, a commercial phenolic disinfectant was also found to be very effective in inactivating the scrapie agent (53).

REFERENCES

1. **Asher, D. M., C. J. Gibbs, Jr., and D. C. Gajdusek.** 1976. Pathogenesis of spongiform encephalopathies. *Ann. Clin. Lab. Sci.* **6:**84–103.
2. **Asher, D. M., C. J. Gibbs, Jr., and D. C. Gajdusek.** 1986. Slow viral infections: safe handling of the agents of the subacute spongiform encephalopathies, p. 59–71. *In* B. M. Miller, D. H. M. Gröschel, J. H. Richardson, D. Vesley, J. R. Songer, R. D. Housewright, and W. E. Barkley (ed.), *Laboratory Safety: Principles and Practice.* American Society for Microbiology, Washington, D.C.
3. **Asher, D. M., C. L. Masters, D. C. Gajdusek, and C. J. Gibbs, Jr.** 1983. Familial spongiform encephalopathies, p. 273–291. *In* S. Kety, L. Rowland, R. Sidman, and S. Matthysse (ed.), *Genetics of Neurological and Psychiatric Disorders.* Raven Press, New York.

4. Asher, D. M., R. T. Yanagihara, D. C. Gajdusek, and C. J. Gibbs, Jr. 1979. Studies of spongiform encephalopathies in cell culture, p. 235–242. *In* S. Prusiner and W. Hadlow (ed.), *Slow Transmissible Agents of the Nervous System*, vol. 2. Academic Press, Inc., New York.

5. Beck, E., P. M. Daniel, A. J. Davey, D. C. Gajdusek, and C. J. Gibbs, Jr. 1982. The pathogenesis of transmissible spongiform encephalopathy: an ultrastructural study. *Brain* **105:** 755–786.

6. Beck, E., P. M. Daniel, W. B. Matthews, D. L. Stevens, M. P. Alpers, D. M. Asher, D. C. Gajdusek, and C. J. Gibbs, Jr. 1969. Creutzfeldt-Jakob disease: the neuropathology of a transmission experiment. *Brain* **92:**699–716.

7. Bell, J. E., and J. W. Ironside. 1993. Neuropathology of spongiform encephalopathies in humans. *Br. Med. Bull.* **49:** 738–777.

8. Berger, J. R., and N. J. David. 1993. Creutzfeldt-Jakob disease in a physician—a review of the disorder in health care workers. *Neurology* **43:**205–206.

9. Bernoulli, C., J. Siegfried, G. Baumgartner, F. Regli, T. Rabinowics, D. C. Gajdusek, and C. J. Gibbs, Jr. 1977. Danger of accidental person-to-person transmission of Creutzfeldt-Jakob disease by surgery. *Lancet* **i:**478–479.

10. Blisard, K., L. Davis, M. Harrington, J. Lovell, M. Kornfeld, and M. Berger. 1990. Pre-mortem diagnosis of Creutzfeldt-Jakob disease by detection of abnormal cerebrospinal fluid proteins. *J. Neurol. Sci.* **99:**75–81.

11. Bobin, S. A., J. R. Currie, P. A. Merz, D. L. Miller, J. Styles, W. A. Walker, G. Y. Wen, and H. M. Wisniewski. 1987. The comparative immunoreactivities of brain amyloids in Alzheimer's disease and scrapie. *Acta Neuropathol.* (Berlin) **74:**313–323.

12. Bode, L., M. Pocchiari, H. Gelderblom, and H. Diringer. 1985. Characterization of antisera against scrapie-associated fibrils (SAF) from affected hamster and cross-reactivity with SAF from scrapie-affected mice and from patients with Creutzfeldt-Jakob disease. *J. Gen. Virol.* **66:**2471–2478.

13. Bolton, D. C., M. P. McKinley, and S. B. Prusiner. 1982. Identification of a protein that purifies with the scrapie prion. *Science* **218:**1309–1311.

14. Bolton, D. C., R. K. Meyer, and S. B. Prusiner. 1985. Scrapie PrP 27-30 is a sialoglycoprotein. *J. Virol.* **53:**596–606.

15. Brown, P. 1990. Iatrogenic Creutzfeldt-Jakob disease. *Aust. N.Z. J. Med.* **20:**633–635.

16. Brown, P. The 'brave new world' of transmissible spongiform encephalopathy (infectious cerebral amyloidosis). *Mol. Neurobiol.* **8:**79–87.

17. Brown, P., L. Cervenakova, L. G. Goldfarb, M. R. Mccombie, R. Rubenstein, R. G. Will, M. Pocchiari, J. F. Martinez Lage, C. Scalici, C. Masullo, G. Graupera, J. Ligan, and D. C. Gajdusek. 1994. Iatrogenic Creutzfeldt-Jakob disease—an example of the interplay between ancient genes and modern medicine. *Neurology* **44:**291–293.

18. Brown, P., M. Coker-Vann, K. Pomeroy, M. Franko, D. M. Asher, C. J. Gibbs, Jr., and D. C. Gajdusek. 1986. Diagnosis of Creutzfeldt-Jakob disease by Western blot identification of marker protein in human brain tissue. *N. Engl. J. Med.* **314:** 547–551.

19. Brown, P., D. C. Gajdusek, C. J. Gibbs, Jr., and D. M. Asher. 1985. Potential epidemic of Creutzfeldt-Jakob disease from human growth hormone therapy. *N. Engl. J. Med.* **313:** 728–731.

20. Brown, P., C. Gibbs, Jr., P. Rodgers-Johnson, D. Asher, M. Sulima, A. Bacote, L. Goldfarb, and D. Gajdusek. 1994. Human spongiform encephalopathy: the NIH series of 300 cases of experimentally transmitted disease. *Ann. Neurol.* **35:** 513–529.

21. Brown, P., P. Kaur, M. P. Sulima, L. G. Goldfarb, C. J. Gibbs, Jr., and D. C. Gajdusek. 1993. Real and imagined clinicopathological limits of prion dementia. *Lancet* **341:**127–129.

22. Brown, P., P. P. Liberski, A. Wolff, and D. C. Gajdusek. 1990. Resistance of scrapie infectivity to steam autoclaving after formaldehyde fixation and limited survival after ashing at 360°C: practical and theoretical implications. *J. Infect. Dis.* **161:**467–472.

23. Brown, P., M. A. Preece, and R. G. Will. 1992. Friendly fire in medicine—hormones, homografts, and Creutzfeldt-Jakob disease. *Lancet* **340:**24–27.

24. Brown, P., A. Wolff, and D. C. Gajdusek. 1990. A simple and effective method for inactivating virus infectivity in formalin-fixed tissue samples from patients with Creutzfeldt-Jakob disease. *Neurology* **40:**887–890.

25. Bruce, M. E. Scrapie strain variation and mutation. *Br. Med. Bull.* **49:**822–838.

26. Bruce, M. E., and A. G. Dickinson. 1987. Biological evidence that scrapie agent has an independent genome. *J. Gen. Virol.* **68:**79–89.

27. Büeler, H., A. Aguzzi, A. Sailer, R. A. Greiner, P. Autenried, M. Aguet, and C. Weissmann. 1993. Mice devoid of PrP are resistant to scrapie. *Cell* **73:**1339–1347.

28. Butler, D. A., M. R. Scott, J. M. Bockman, D. R. Borchelt, A. Taraboulos, K. K. Hsiao, D. T. Kingsbury, and S. B. Prusiner. 1988. Scrapie-infected murine neuroblastoma cells produce protease-resistant prion proteins. *J. Virol.* **62:**1558–1564.

29. Campbell, I. L., M. Eddleston, P. Kemper, M. B. A. Oldstone, and M. V. Hobbs. 1994. Activation of cerebral cytokine gene expression and its correlation with onset of reactive astrocyte and acute-phase response gene expression in scrapie. *J. Virol.* **68:**2383–2387.

30. Carlson, G. A., D. T. Kingsbury, P. A. Goodman, S. Coleman, S. T. Marshall, S. DeArmond, D. Westaway, and S. B. Prusiner. 1986. Linkage of prion protein and scrapie incubation time genes. *Cell* **46:**503–511.

31. Carlson, G. A., D. Westaway, P. A. Goodman, M. Peterson, S. T. Marshall, and S. B. Prusiner. 1988. Genetic control of prion incubation period in mice. *CIBA Found. Symp.* **135:** 84–99.

32. Caughey, B. 1993. Scrapie associated PrP accumulation and its prevention—insights from cell culture. *Br. Med. Bull.* **49:**860–872.

33. Caughey, B., R. E. Race, and B. Chesebro. 1988. Detection of prion protein mRNA in normal and scrapie-infected tissues and cell lines. *J. Gen. Virol.* **69:**711–716.

34. Caughey, B., R. E. Race, D. Ernst, M. J. Buchmeier, and B. Chesebro. 1989. Prion protein biosynthesis in scrapie-infected and uninfected neuroblastoma cells. *J. Virol.* **63:**175–181.

35. Caughey, B., and G. J. Raymond. 1993. Sulfated polyanion inhibition of scrapie-associated PrP accumulation in cultured cells. *J. Virol.* **67:**643–650.

36. Centers for Disease Control. 1987. Rapidly progressive dementia in a patient who received a cadaveric dura mater graft. *Morbid. Mortal. Weekly Rep.* **36:**49–50, 55.

37. Centers for Disease Control. 1987. Update: Creutzfeldt-Jakob disease in a patient receiving a cadaveric dura mater graft. *Morbid. Mortal. Weekly Rep.* **36:**324–325.

38. Centers for Disease Control. 1989. Update: Creutzfeldt-Jakob disease in a second patient who received a cadaveric dura mater graft. *Morbid. Mortal. Weekly Rep.* **38:**37–38, 43.

39. Chesebro, B. 1992. Spongiform encephalopathies—PrP and the scrapie agent. *Nature* (London) **356:**560.

40. Chesebro, B., and B. Caughey. 1993. Scrapie replication without the prion protein? *Curr. Biol.* **3:**696–698.

41. Cho, H. 1980. Requirement of a protein component for scrapie infectivity. *Intervirology* **14:**213–216.

42. Cochius, J. I., N. Hyman, and M. M. Esiri. 1992. Creutzfeldt-Jakob disease in a recipient of human pituitary-derived gonadotrophin—a second case. *J. Neurol. Neurosurg. Psychiatry* **55:**1094–1095.

43. Czub, M., H. R. Braig, and H. Diringer. 1986. Pathogenesis of scrapie: study of the temporal development of clinical symptoms, of infectivity titres and scrapie-associated fibrils in brains of hamsters infected intraperitoneally. *J. Gen. Virol.* **67:**2005–2009.

44. Czub, M., H. R. Braig, and H. Diringer. 1988. Replication of

the scrapie agent in hamsters infected intracerebrally confirms the pathogenesis of an amyloid-inducing virosis. *J. Gen. Virol.* **69:**1753–1756.

45. **David-Ferreira, J. F., K. L. David-Ferreira, C. J. Gibbs, Jr., and J. A. Morris.** 1968. Scrapie in mice: ultrastructural observations in the cerebral cortex. *Proc. Soc. Exp. Biol. Med.* **28:**313–320.

46. **DeArmond, S. J., M. P. McKinley, R. A. Barry, M. B. Braunfeld, J. R. McColloch, and S. B. Prusiner.** 1985. Identification of prion amyloid filaments in scrapie-infected brain. *Cell* **41:**221–235.

47. **Degioia, L., C. Selvaggini, E. Ghibaudi, L. Diomede, O. Bugiani, G. Forloni, F. Tagliavini, and M. Salmona.** 1994. Conformational polymorphism of the amyloidogenic and neurotoxic peptide homologous to residues 106-126 of the prion protein. *J. Biol. Chem.* **269:**7859–7862.

48. **Deslys, J.-P., C. Lasmézas, and D. Dormont.** 1994. Selection of specific strains in iatrogenic Creutzfeldt-Jakob disease. *Lancet* **343:**848–849.

49. **Deslys, J. P., D. Marce, and D. Dormont.** 1994. Similar genetic susceptibility in iatrogenic and sporadic Creutzfeldt-Jakob disease. *J. Gen. Virol.* **75:**23–27.

50. **Dickinson, A. G., H. Fraser, and G. W. Outram.** 1979. Scrapie incubation time can exceed natural lifespan. *Nature* (London) **256:**732–733.

51. **Diringer, H., and H. R. Braig.** 1989. Infectivity of unconventional viruses in dura mater. *Lancet* **i:**439–440.

52. **Duffy, P., G. Collins, A. G. Devoe, B. Streeten, and D. Cohen.** 1974. Possible person-to-person transmission of Creutzfeldt-Jakob disease. *N. Engl. J. Med.* **290:**693.

53. **Ernst, D. R., and R. E. Race.** 1993. Comparative analysis of scrapie agent inactivation methods. *J. Virol. Methods* **41:**193–201.

54. **Gajdusek, D. C.** 1990. Subacute spongiform encephalopathies: transmissible cerebral amyloidoses caused by unconventional viruses, p. 2289–2324. *In* B. N. Fields and D. M. Knipe (ed.), *Virology*, 2nd ed., vol. 2. Raven Press, New York.

55. **Gajdusek, D. C.** Infectious amyloids. Subacute spongiform encephalopathies as transmissible cerebral amyloidoses. *In* B. N. Fields and D. M. Knipe (ed.), *Virology*, 3rd ed., vol. 2, in press. Raven Press, New York.

56. **Gajdusek, D. C., C. J. Gibbs, Jr., K. Earle, G. J. Dammin, M. C. Schoene, and H. R. Tyler.** 1974. Transmission of subacute spongiform encephalopathy to the chimpanzee and squirrel monkey from a patient with papulosis maligna of Köhlmeyer-Degos. *Excerpta Med. Int. Congr. Ser.* **319:**390–392.

57. **Gajdusek, D. C., and V. Zigas.** 1957. Degenerative disease of the central nervous system in New Guinea: epidemic occurrence of "kuru" in the native population. *N. Engl. J. Med.* **257:**974–978.

58. **Gambetti, P., R. Petersen, L. Monari, M. Tabaton, L. Autilio-Gambetti, P. Cortelli, P. Montagna, and E. Laugaresi.** 1993. Fatal familial insomnia and the widening spectrum of prion diseases. *Br. Med. Bull.* **49:**980–994.

59. **Gibbons, R. A., and G. D. Hunter.** 1967. Nature of the scrapie agent. *Nature* (London) **215:**1041–1043.

60. **Gibbs, C. J., Jr., D. M. Asher, P. W. Brown, J. E. Fradkin, and D. C. Gajdusek.** 1993. Creutzfeldt-Jakob disease infectivity of growth hormone derived from human pituitary glands. *N. Engl. J. Med.* **328:**358–359.

61. **Gibbs, C. J., Jr., D. M. Asher, A. Kobrine, H. L. Amyx, and D. C. Gajdusek.** Transmission of Creutzfeldt-Jakob disease to a chimpanzee by electrodes contaminated during neurosurgery. *J. Neurol. Neurosurg. Psychiatry* **57:**757–758.

62. **Gibbs, C. J., Jr., and D. C. Gajdusek.** 1969. Infection as the etiology of spongiform encephalopathy (Creutzfeldt-Jakob disease). *Science* **165:**1023–1025.

63. **Gibbs, C. J., Jr., and D. C. Gajdusek.** 1971. Transmission and characterization of the agents of spongiform virus encephalopathies: kuru, Creutzfeldt-Jakob disease, scrapie and mink encephalopathy. *Res. Publ. Assoc. Res. Nerv. Ment. Dis.* **49:**383–410.

64. **Gibbs, C. J., Jr., D. C. Gajdusek, and H. Amyx.** 1979. Strain variation in the viruses of Creutzfeldt-Jakob disease and kuru, p. 87–110. *In* S. B. Prusiner and W. J. Hadlow (ed.), *Slow Transmissible Diseases of the Nervous System*, vol. 1. Academic Press, Inc., New York.

65. **Gloor, P., O. Kalabay, and N. Giard.** 1968. The electroencephalogram in diffuse encephalopathies: electroencephalographic correlates of grey and white matter lesions. *Brain* **91:**779–801.

66. **Goldfarb, L. G., P. Brown, and D. C. Gajdusek.** 1992. The molecular genetics of human transmissible spongiform encephalopathy, p. 139–153. *In* S. B. Prusiner, J. Collinge, J. Powell, and B. Anderton (ed.), *Prion Diseases of Humans and Animals.* Ellis Horwood Ltd., New York.

67. **Goldfarb, L. G., R. B. Petersen, M. Tabaton, P. Brown, A. C. LeBlanc, P. Montagna, P. Cortelli, J. Julien, C. Vital, W. W. Pendelbury, M. Haltia, P. R. Wills, J. J. Hauw, P. E. McKeever, L. Monari, B. Schrank, G. D. Swergold, L. Autilio-Gambetti, D. C. Gajdusek, E. Lugaresi, and P. Gambetti.** 1992. Fatal familial insomnia and familial Creutzfeldt-Jakob disease: disease phenotype determined by a DNA polymorphism. *Science* **258:**806–808.

68. **Goldgaber, D., L. G. Goldfarb, P. Brown, D. M. Asher, W. T. Brown, W. S. Linn, J. W. Tenner, S. M. Feinstone, R. Rubenstein, R. J. Kascsak, J. W. Boellard, and D. C. Gajdusek.** 1989. Mutations in familial Creutzfeldt-Jakob disease and Gerstmann-Sträussler-Scheinker's syndrome. *Exp. Neurol.* **106:**204–206.

69. **Goldgaber, D., J. W. Teener, L. G. Goldfarb, D. M. Asher, P. W. Brown, S. Feinstone, and D. C. Gajdusek.** 1988. No Msp 1 polymorphism in the open reading frame of the PrP gene in patients with familial Creutzfeldt-Jakob disease. *Alzheimer Dis. Assoc. Disord.* **2:**311.

70. **Gorman, D. G., D. F. Benson, D. G. Vogel, and H. V. Vinters.** 1992. Creutzfeldt-Jakob disease in a pathologist. *Neurology* **42:**463.

71. **Hadlow, W. J.** 1990. An overview of scrapie in the United States. *J. Am. Vet. Med. Assoc.* **196:**1676–1677.

72. **Hadlow, W. J., R. C. Kennedy, and R. E. Race.** 1982. Natural infection of Suffolk sheep with scrapie virus. *J. Infect. Dis.* **146:**657–664.

73. **Haritani, M., Y. I. Spencer, and G. A. H. Wells.** 1994. Hydrated autoclave pretreatment enhancement of prion protein immunoreactivity in formalin-fixed bovine spongiform encephalopathy-affected brain. *Acta Neuropathol.* (Berlin) **87:**86–90.

74. **Harrington, M. G., C. R. Merril, D. M. Asher, and D. C. Gajdusek.** 1986. Abnormal proteins in the cerebrospinal fluid of patients with Creutzfeldt-Jakob disease. *N. Engl. J. Med.* **315:**279–283.

75. **Hintz, R., M. MacGillivray, A. Joy, and R. Tintner.** 1985. Fatal degenerative neurological disease in patients who received pituitary-derived human growth hormone. *Morbid. Mortal. Weekly Rep.* **34:**359–366.

76. **Hsiao, K., H. F. Baker, T. J. Crow, M. Poulter, F. Owen, J. D. Terwilliger, D. Westaway, J. Ott, and S. B. Prusiner.** 1989. Linkage of a prion protein missense variant to Gerstmann-Sträussler syndrome. *Nature* (London) **338:**342–345.

77. **Hsiao, K. K., M. Scott, D. Foster, D. F. Groth, S. J. DeArmond, and S. B. Prusiner.** 1990. Spontaneous neurodegeneration in transgenic mice with mutant prion protein. *Science* **250:**1587–1590.

78. **Hunter, N., J. D. Foster, A. G. Dickinson, and J. Hope.** 1989. Linkage of the gene for the scrapie-associated fibril protein (PrP) to the Sip gene in Cheviot sheep. *Vet. Rec.* **124:**364–366.

79. **Hunter, N., J. Hope, I. McConnell, and A. G. Dickinson.** 1987. Linkage of the scrapie-associated fibril protein (PrP) gene and Sinc using congenic mice and restriction fragment length polymorphism analysis. *J. Gen. Virol.* **68:**2711–2716.

80. **Jellinger, K., F. Seitelberger, W.-D. Heiss, and W. Holczabek.** 1972. Konjugale Form der subakuten spongiöse Enzepha-

lopathie (Jakob-Creutzfeldt Erkrankung). *Wien. Klin. Wochenschr.* **84:**245–249.

81. **Kimberlin, R. H.** 1986. Scrapie: how much do we really understand? *Neuropathol. Appl. Neurobiol.* **12:**131–147.

82. **Kimberlin, R. H.** 1993. Bovine spongiform encephalopathy—an appraisal of the current epidemic in the United Kingdom. *Intervirology* **35:**208–218.

83. **Kimberlin, R. H., S. Cole, and C. A. Walker.** 1987. Temporary and permanent modifications to a single strain of mouse scrapie on transmission to rats and hamsters. *J. Gen. Virol.* **68:**1875–1881.

84. **Kimberlin, R. H., and C. A. Walker.** 1977. Characteristics of a short incubation model of scrapie in the golden hamster. *J. Gen. Virol.* **34:**295–304.

85. **Kimberlin, R. H., and C. A. Walker.** 1988. Pathogenesis of experimental scrapie. *CIBA Found. Symp.* **135:**37–62.

86. **Kirschbaum, W. R.** 1968. *Jakob-Creutzfeldt Disease.* American Elsevier, New York.

87. **Kitamoto, T., K. Ogomori, J. Tateishi, and S. B. Prusiner.** 1987. Formic acid pretreatment enhances immunostaining of cerebral and systemic amyloids. *Lab. Invest.* **57:**230–236.

88. **Klitzman, R. L., M. P. Alpers, and D. C. Gajdusek.** 1984. The natural incubation period of kuru and the episodes of transmission in three clusters of patients. *Neuroepidemiology* **3:**3–20.

89. **Kretzschmar, H. A., L. E. Stowring, D. Westaway, W. H. Stubblebine, S. B. Prusiner, and S. J. DeArmond.** 1986. Molecular cloning of a human prion protein cDNA. *DNA* **5:**315–324.

90. **Lewin, P.** 1972. Scrapie: an infective peptide? *Lancet* **i:**748.

91. **Liberski, P. P., R. Yanagihara, V. R. Nerurkar, and D. C. Gajdusek.** 1993. Tumour necrosis factor-alpha produces Creutzfeldt-Jakob disease-like lesions in vivo. *Neurodegeneration* **2:**215–225.

92. **Manetto, V., R. Medori, P. Cortelli, P. Montagna, P. Tinuper, A. Baruzzi, G. Rancurel, J. J. Hauw, J. J. Vanderhaeghen, P. Mailleux, O. Bugiani, F. Tagliavini, C. Bouras, N. Rizzuto, E. Lugaresi, and P. Gambetti.** 1992. Fatal familial insomnia—clinical and pathologic study of five new cases. *Neurology* **42:**312–319.

93. **Manuelidis, L., T. Sklaviadis, and E. E. Manuelidis.** 1987. Evidence suggesting that PrP is not the infectious agent in Creutzfeldt-Jakob disease. *EMBO J.* **6:**341–347.

94. **Marsh, R. F.** 1990. Bovine spongiform encephalopathy in the United States. *J. Am. Vet. Med. Assoc.* **196:**1677.

95. **Martinez Lage, J. F., M. Poza, and J. G. Tortosa.** 1993. Creutzfeldt-Jakob disease in patients who received a cadaveric dura mater graft—Spain, 1985–1992. *Morbid. Mortal. Weekly Rep.* **42:**560–563.

96. **Martinez Lage, J. F., J. Sola, M. Poza, and J. A. Esteban.** 1993. Pediatric Creutzfeldt-Jakob disease—probable transmission by a dural graft. *Childs Nerv. Syst.* **9:**239–242.

97. **Masters, C. L., D. C. Gajdusek, and C. J. Gibbs, Jr.** 1981. Creutzfeldt-Jakob disease virus isolation from the Gerstmann-Sträussler syndrome, with an analysis of the various forms of amyloid deposition in the virus-induced spongiform encephalopathies. *Brain* **104:**559–588.

98. **Masters, C. L., J. O. Harris, D. C. Gajdusek, C. J. Gibbs, Jr., C. Bernoulli, and D. M. Asher.** 1979. Creutzfeldt-Jakob disease: patterns of world wide occurrence and the significance of familial and sporadic clustering. *Ann. Neurol.* **5:**177–188.

99. **Masullo, C., M. Pocchiari, G. Macchi, G. Alema, G. Piazza, and M. A. Panzera.** 1989. Transmission of Creutzfeldt-Jakob disease by dural cadaveric graft. *J. Neurosurg.* **71:**954–955.

100. **Matthews, W. B.** 1975. Epidemiology of Creutzfeldt-Jakob disease in England and Wales. *J. Neurol. Neurosurg. Psychiatry* **38:**210–213.

101. **McKinley, M. P., D. C. Bolton, and S. B. Prusiner.** 1983. A protease-resistant protein is a structural component of the scrapie prion. *Cell* **35:**57–62.

102. **Medori, R., P. Montagna, H. J. Tritschler, A. LeBlanc, P. Cortelli, P. Tinuper, E. Lugaresi, and P. Gambetti.** 1992. Fatal familial insomnia—a second kindred with mutation of prion protein gene at codon-178. *Neurology* **42:**669–670.

103. **Medori, R., and H. J. Tritschler.** 1993. Prion protein gene analysis in 3 kindreds with fatal familial insomnia (FFI)—codon-178 mutation and codon-129 polymorphism. *Am. J. Hum. Genet.* **53:**822–827.

104. **Medori, R., H. J. Tritschler, A. LeBlanc, F. Villare, V. Manetto, H. Y. Chen, R. Xue, S. Leal, P. Montagna, P. Cortelli, P. Tinuper, P. Avoni, M. Mochi, A. Baruzzi, J. J. Hauw, J. Ott, E. Lugaresi, L. Autilio-Gambetti, and P. Gambetti.** 1992. Fatal familial insomnia, a prion disease with a mutation at codon-178 of the prion protein gene. *N. Engl. J. Med.* **326:**444–449.

105. **Merz, P. A., R. A. Somerville, H. M. Wisniewski, and K. Iqbal.** 1981. Abnormal fibrils from scrapie-infected brain. *Acta Neuropathol.* (Berlin) **54:**63–74.

106. **Meyer, R. K., M. P. McKinley, K. A. Bowman, M. B. Braunfeld, R. A. Barry, and S. B. Prusiner.** 1986. Separation and properties of cellular and scrapie prion proteins. *Proc. Natl. Acad. Sci. USA* **83:**2310–2314.

107. **Miller, D.** 1988. Creutzfeldt-Jakob disease in histopathology technicians. *N. Engl. J. Med.* **318:**853–854.

108. **Monari, L., S. G. Chen, P. Brown, P. Parchi, R. B. Petersen, J. Mikol, F. Gray, P. Cortelli, P. Montagna, B. Ghetti, L. G. Goldfarb, D. C. Gajdusek, E. Lugaresi, P. Gambetti, and L. Autilio Gambetti.** 1994. Fatal familial insomnia and familial Creutzfeldt-Jakob disease—different prion proteins determined by a DNA polymorphism. *Proc. Natl. Acad. Sci. USA* **91:**2839–2842.

109. **Multhaup, G., H. Diringer, H. Hilmert, H. Prinz, J. Heukeshoven, and K. Beyreuther.** 1985. The protein component of scrapie-associated fibrils is a glycosylated low molecular weight protein. *EMBO J.* **4:**1495–1501.

110. **Muramatsu, Y., A. Onodera, M. Horiuchi, N. Ishiguro, and M. Shinagawa.** 1994. Detection of PRP^{Sc} in sheep at the preclinical stage of scrapie and its significance for diagnosis of insidious infection. *Arch. Virol.* **134:**427–432.

111. **Muramoto, T., T. Kitamoto, J. Tateishi, and I. Goto.** 1993. Accumulation of abnormal prion protein in mice infected with Creutzfeldt-Jakob disease via intraperitoneal route: a sequential study. *Am. J. Pathol.* **143:**1470–1479.

112. **Narang, H. K., D. M. Asher, K. L. Pomeroy, and D. C. Gajdusek.** 1987. Abnormal tubulovesicular particles in brains of hamsters with scrapie. *Proc. Soc. Exp. Biol. Med.* **184:**504–509.

113. **Nevin, S., W. H. McMenemy, D. Behrman, and D. P. Jones.** 1960. Subacute spongiform encephalopathy: a subacute form of encephalopathy attributed to vascular dysfunction (spongiform cerebral atrophy). *Brain* **83:**519–564.

114. **Nisbet, T. J., I. MacDonaldson, and S. N. Bishara.** 1989. Creutzfeldt-Jakob disease in a second patient who received a cadaveric dura mater graft. *JAMA* **261:**1118.

115. **Occupational Safety and Health Administration, U.S. Department of Labor.** 1991. Occupational exposure to bloodborne pathogens; final rule. (29 CFR Part 1910.1030). *Fed. Regist.* **56:**64175–64182.

116. **Oesch, B., D. Westaway, M. Wälchli, M. P. McKinley, S. H. B. Kent, R. Aebersold, R. A. Barry, P. Tempst, D. B. Teplow, L. E. Hood, S. B. Prusiner, and C. Weissmann.** 1985. A cellular gene encodes scrapie PrP 27-30 protein. *Cell* **40:**735–746.

117. **Oleszak, E. L., G. Murdoch, L. Manuelidis, and E. E. Manuelidis.** 1988. Growth factor production by Creutzfeldt-Jakob disease cell lines. *J. Virol.* **62:**3103–3108.

118. **Owen, F., M. Poulter, R. Lofthouse, J. Collinge, T. J. Crow, D. Risby, H. F. Baker, R. M. Ridley, K. Hsiao, and S. B. Prusiner.** 1989. Insertion in prion protein gene in familial Creutzfeldt-Jakob disease. *Lancet* **i:**51–52.

119. **Owen, F., M. Poulter, R. Lofthouse, T. J. Crow, D. Risby, H. F. Baker, and R. M. Ridley.** 1988. A rare Msp1 polymorphism in the human prion gene in a family with a history of early onset dementia. *Neurosci. Lett. Suppl.* **32:**S53.

120. **Özel, M., and H. Diringer.** 1994. Small virus-like structure

in fractions from scrapie hamster brain. *Lancet* **343**:894–895.

121. **Palmer, M. S., A. J. Dryden, J. T. Hughes, and J. Collinge.** 1991. Homozygous prion protein genotype predisposes to sporadic Creutzfeldt-Jakob disease. *Nature* (London) **352**:340–342.

122. **Pan, K. M., M. Baldwin, J. Nguyen, M. Gasset, A. Serban, D. Groth, I. Mehlhorn, Z. W. Huang, R. J. Fletterick, F. E. Cohen, and S. B. Prusiner.** 1993. Conversion of alpha-helices into beta-sheets features in the formation of the scrapie prion proteins. *Proc. Natl. Acad. Sci. USA* **90**:10962–10966.

123. **Piccardo, P., J. Safar, M. Ceroni, D. C. Gajdusek, and C. J. Gibbs, Jr.** 1990. Immunohistochemical localization of prion protein in spongiform encephalopathies and normal tissue. *Neurology* **40**:518–522.

124. **Pocchiari, M., M. Salvatore, F. Cutruzzola, M. Genuardi, C. T. Allocatelli, C. Masullo, G. Macchi, G. Alema, S. Galgani, Y. G. Xi, R. Petraroli, M. C. Silvestrini, and M. Brunori.** 1993. A new point mutation of the prion protein gene in Creutzfeldt-Jakob disease. *Ann. Neurol.* **34**:802–807.

125. **Prusiner, S. B.** 1982. Novel proteinaceous infectious particles cause scrapie. *Science* **216**:136–144.

126. **Prusiner, S. B.** 1989. Scrapie prions. *Annu. Rev. Microbiol.* **43**:345–374.

127. **Prusiner, S. B.** 1993. Genetic and infectious prion diseases. *Arch. Neurol.* **50**:1129–1153.

128. **Prusiner, S. B., D. Groth, A. Serban, R. Koehler, D. Foster, M. Torchia, D. Burton, S. L. Yang, and S. J. DeArmond.** 1993. Ablation of the prion protein (PrP) gene in mice prevents scrapie and facilitates production of anti-PrP antibodies. *Proc. Natl. Acad. Sci. USA* **90**:10608–10612.

129. **Prusiner, S. B., M. Scott, D. Foster, K. M. Pan, D. Groth, C. Mirenda, M. Torchia, S. L. Yang, D. Serban, G. A. Carlson, et al.** 1990. Transgenetic studies implicate interactions between homologous PrP isoforms in scrapie prion replication. *Cell* **63**:673–686.

130. **Race, R. E., B. Caughey, K. Graham, D. Ernst, and B. Chesebro.** 1988. Analyses of frequency of infection, specific infectivity, and prion protein biosynthesis in scrapie-infected neuroblastoma cell clones. *J. Virol.* **62**:2845–2849.

131. **Race, R. E., L. H. Fadness, and B. Chesebro.** 1987. Characterization of scrapie infection in mouse neuroblastoma cells. *J. Gen. Virol.* **68**:1391–1399.

132. **Rohwer, R. G.** 1984. Virus-like sensitivity of scrapie agent to heat inactivation. *Science* **223**:600–602.

133. **Rohwer, R. G.** 1986. Estimation of scrapie nucleic acid MW from standard curves for virus sensitivity to ionizing radiation. *Nature* (London) **320**:381.

134. **Rohwer, R. G.** 1991. The scrapie agent: "a virus by any other name." *Curr. Top. Microbiol. Immunol.* **172**:195–232.

135. **Roos, R., D. C. Gajdusek, and C. J. Gibbs, Jr.** 1973. The clinical characteristics of transmissible Creutzfeldt-Jakob disease. *Brain* **96**:1–20.

136. **Safar, J., P. P. Roller, D. C. Gajdusek, and C. J. Gibbs, Jr.** 1993. Thermal stability and conformational transitions of scrapie amyloid (prion) protein correlate with infectivity. *Protein Sci.* **2**:2206–2216.

137. **Scott, M. R., D. A. Butler, D. E. Bredesen, M. Wälchli,** K. K. Hsiao, and S. B. Prusiner. 1988. Prion protein gene expression in cultured cells. *Protein Eng.* **2**:69–76.

138. **Scott, M. R., D. Foster, C. Mirenda, D. Serban, F. Coufal, M. Wälchli, M. Torchia, D. Groth, G. Carlson, S. J. DeArmond, et al.** 1989. Transgenic mice expressing hamster prion protein produce species-specific scrapie infectivity and amyloid plaques. *Cell* **59**:847–857.

139. **Serban, D., A. Taraboulos, S. J. DeArmond, and S. B. Prusiner.** 1990. Rapid detection of Creutzfeldt-Jakob disease and scrapie prion proteins. *Neurology* **40**:110–117.

140. **Sigurdsson, B.** 1954. Observations on three slow infections of sheep. *Br. Med. J.* **110**:255–270, 307–322, 341–354.

141. **Sitwell, L., B. Lach, E. Atack, and D. Atack.** 1988. Creutzfeldt-Jakob disease in histopathology technicians. *N. Engl. J. Med.* **318**:854.

142. **Stahl, N., D. R. Borchelt, K. Hsiao, and S. B. Prusiner.** 1987. Scrapie prion protein contains a phosphatidylinositol glycolipid. *Cell* **51**:229–240.

143. **Tateishi, J., and T. Kitamoto.** 1993. Developments in diagnosis for prion diseases. *Br. Med. Bull.* **49**:971–979.

143a. **Thadani, V., P. L. Penar, J. Partington, R. Kalb, R. Janssen, L. B. Schonberger, C. S. Rabkin, and J. W. Prichard.** 1988. Creutzfeldt-Jakob disease probably acquired from a cadaveric dura mater graft. Case report. *J. Neurosurg.* **69**:766–769.

144. **Weber, T., H. Tumani, B. Holdorff, J. Collinge, M. Palmer, H. A. Kretzschmar, and K. Felgenhauer.** 1993. Transmission of Creutzfeldt-Jakob disease by handling of dura mater. *Lancet* **341**:123–124.

145. **Weissmann, C.** 1991. A 'unified theory' of prion propagation. *Nature* (London) **352**:679–683.

146. **Weissmann, C., H. Büeler, M. Fischer, and M. Aguet.** 1993. Role of the PrP gene in transmissible spongiform encephalopathies. *Intervirology* **35**:164–175.

147. **Westaway, D., S. J. DeArmond, J. Cayetano-Canlas, D. Groth, D. Foster, S.-L. Yang, M. Torchia, G. A. Carlson, S. B. Prusiner, and C. This.** 1994. Degeneration of skeletal muscle, peripheral nerves, and the central nervous system in transgenic mice overexpressing wild-type prion proteins. *Cell* **76**:117–129.

148. **Wilesmith, J. W.** 1993. Epidemiology of bovine spongiform encephalopathy and related diseases. *Arch. Virol. Suppl.* **7**:245–254.

149. **Will, R. G., and W. B. Matthews.** 1982. Evidence for case-to-case transmission of Creutzfeldt-Jakob disease. *J. Neurol. Neurosurg. Psychiatry* **45**:235–238.

150. **Williams, E. S., and S. Young.** 1980. Chronic wasting disease of mule deer: a spongiform encephalopathy. *J. Wildl. Dis.* **16**:89–98.

151. **Wills, P. R.** 1992. Potential pseudoknots in the PrP-encoding messenger RNA. *J. Theor. Biol.* **159**:523–527.

152. **Wills, P. R.** 1993. Self-organization of genetic coding. *J. Theor. Biol.* **162**:267–287.

153. **Xi, Y. G., L. Ingrosso, A. Ladogana, C. Masullo, and M. Pocchiari.** 1992. Amphotericin-B treatment dissociates in vivo replication of the scrapie agent from PrP accumulation. *Nature* (London) **356**:598–601.

154. **Yun, M., W. Wu, L. Hood, and M. Harrington.** 1992. Human cerebrospinal fluid protein database—edition 1992. *Electrophoresis* **13**:1002–1013.

Poxviruses Infecting Humans

JOSEPH J. ESPOSITO AND ROBERT F. MASSUNG

100

CLINICAL EPIDEMIOLOGIC PERSPECTIVE

This chapter focuses on human poxvirus infections. An extensive description of the taxonomic structure of the family *Poxviridae* (9) and several comprehensive collections of information on poxvirus characteristics have recently been published (2–4, 6, 12, 19, 22, 29, 40, 41). Several different laboratory protocols for the diagnosis of poxvirus have been described in earlier editions of this Manual and elsewhere (13, 14, 21, 30–33). Poxviruses that currently infect humans by natural transmission are zoonoses, except for molluscum contagiosum virus (34, 35), the only entity in the genus *Molluscipoxvirus*, which is transmitted strictly between humans. The zoonotic poxviruses generally are of low human-to-human transmissibility. They include members of the genus *Orthopoxvirus* (monkeypox [24, 25], cowpox viruses [1], and the vaccinia virus subspecies buffalopox virus [8, 28]); the genus *Parapoxvirus* (Orf, milker's nodule, and papulosa stomatitis viruses [34, 37]; and the genus *Yatapoxvirus* (tanapox and Yaba monkey tumor viruses [YMTVs]) (26, 39). Orf and molluscum contagiosum virus infections are the most commonly occurring poxvirus infections worldwide; they usually resolve within weeks or months without serious sequelae. There has been relatively little need for full-scale laboratory virologic and serologic confirmation of Orf and molluscum contagiosum infections, since the lesions usually can be identified clinically and histopathologically (19, 34, 35, 37). The most studied poxvirus is the orthopoxvirus vaccinia virus, which was used worldwide to eradicate smallpox, the disease caused by variola virus (7, 19, 20, 22, 29). Vaccinia and certain other poxviruses are now being developed as recombinant viruses for expressing a variety of foreign proteins, including vaccine antigens (2, 12, 29, 37, 41).

GENERAL DESCRIPTION OF AGENTS

Morphology

All poxviruses described in this chapter belong to the family *Poxviridae*, subfamily *Chordopoxvirinae* (9). By microscopy, virions are large and either brick shaped, if orthopoxvirus, yatapoxvirus, or molluscipoxvirus, or ovoid, if parapoxvirus. Virions range in length from 220 to 450 nm and in width from 140 to 260 nm. They all contain a nucleoprotein core, which in conventional negatively stained thin sections appears dumbbell shaped and surrounded by a complex series of membranes. The concavities between the core membrane and the outer membrane are usually occupied by lateral bodies of undefined function. Depending on cell type and virus strain, as many as ~20% of virions may be released by exocytosis through the Golgi, producing extracellular enveloped virions (EEVs).

EEVs are mature infectious virions that have distinct surface antigenic properties; unlike EEVs, the bulk of progeny virions, which are also infectious, stay resident in the cytoplasm until cell lysis occurs. The EEV outermost tegument, the envelope, covers the outer membrane and derives from the Golgi cisternal membranes. The envelope is a single lipid bilayer interspersed with virus proteins. The cell-associated virus particles, i.e., the intracellular naked virions (INVs), lack the EEV envelope. INVs of the brick-shaped viruses have a lipoprotein outer membrane studded with globular surface tubules; the ovoid parapoxviruses have what appears to be a single filamentous tubule spiraling around the particle surface. It has long been suspected that the outer membrane is synthesized de novo; however, recent data have suggested that it is derived from cell membranes of the intermediate compartment that lies between the endoplasmic reticulum and the Golgi membrane stacks. As virion morphogenesis begins, the viral outer membrane is first seen as a spicule-coated lipid bilayer. The membrane then elongates into a crescent shape interspersed with virus proteins. Subsequently, the membrane encloses the entire nucleoprotein to form an immature virus particle. One layer of the membrane constitutes the outer membrane, and the other constitutes the core membrane. (*Note:* In the literature, the surface outer membrane of the INV has sometimes been referred to as the envelope; however, the International Committee on Taxonomy of Viruses currently accepts the following definitions: the envelope is the outer tegument of the EEV, and the outer membrane is the surface tegument of the INV.)

Virus particles contain about 100 proteins, including structural proteins and a virtually complete system for primary transcription (29). The genome, which is in the nucleoprotein complex (synonym: nucleosome), consists of a single linear molecule of double-stranded DNA comprising ~130 to ~375 kbp that are covalently closed at each end to form hairpinlike telomeres. Complete genome DNA sequences have been reported for the Copenhagen strain of vaccinia virus (23) and the highly virulent Bangladesh 1975 strain of variola virus (27).

Differentiating Features

Although DNA restriction endonuclease assay and sequencing are the most precise methods for poxvirus identification, histopathology, electron microscopy, determination of virus growth features, and antigenic testing have also been used for differentiating poxviruses (4, 7, 8, 11, 15–17, 21, 24, 26, 28, 42). Poxviruses grow solely in the cytoplasm, producing basophilic or B-type inclusions; the inclusions are also known as "virus factories" and represent the sites of virus replication. Certain species, such as cowpox virus, also produce acidophilic A-type inclusions, so-called ATI particles, which are proteinaceous deposits detectable histopathologically with stains such as Giemsa and hematoxylin-eosin. Depending on virus strain, ATI particles may (V^+) or may not (V^-) contain virions. When samples are visualized by negative-stain electron microscopy, the human orthopoxviruses mentioned above cannot be differentiated by virion morphology, but very experienced microscopists have claimed that tanapox and molluscum contagiosum viruses can be distinguished by microscopy, especially if good clinical histories accompany the specimens. Parapoxviruses are virtually indistinguishable from each other morphologically, but their distinctive ovoid shape differentiates them from the brick-shaped viruses.

A cross-reactive antigen (nucleoprotein antigen) common to poxviruses can be prepared by acid extraction and can be resolved by various immunologic tests. Poxvirus genera are antigenically distinguishable by the neutralization test (NT) with hyperimmune reference sera (4, 13, 14, 30–33). The orthopoxviruses are very closely related antigenically. Current routine serologic tests with patients' serum samples are not useful in routinely determining the infecting orthopoxvirus species; virus specimens are most useful for this purpose. *Orthopoxvirus* appears to be the only genus whose members produce a hemagglutinin (HA), which is mainly found in the infected cell membrane. HA and hemadsorption tests with properly prepared sensitive chicken erythrocytes are useful for visualizing a functional HA.

Orthopoxviruses are the only human poxviruses that produce pocks (42) on fertile chicken egg chorioallantoic membranes (CAMs). Parapoxviruses, tanapox virus, and molluscum contagiosum virus do not form pocks on the CAM; however, members of other genera that infect animals (e.g., avipoxviruses, leporipoxviruses, capripoxviruses) form pocks.

DNA analysis has been most reliable for differentiating poxviruses (8, 11, 17, 18, 26, 35). Orthopoxviruses are ~33% G+C, yatapoxviruses are ~32%, molluscum contagiosum is ~60%, and parapoxviruses are ~63%. Restriction endonuclease assay of virion DNA gives genome structural data that are extremely powerful for determining differences between genera, species, strains, and variants. The appendix to this chapter provides methods for purifying virions and viral DNA that make detailed examination of poxvirus samples possible. The following sections provide overviews and specific details on infections caused by poxviruses that infect humans.

Orthopoxviruses

Overview

In addition to the zoonotic orthopoxviruses, the genus *Orthopoxvirus* contains the etiologic agent of smallpox, variola virus, which was highly transmissible and had a strict human host range and no animal reservoir. As mentioned above, vaccinia virus was used for mass vaccination against smallpox. In December 1979, a global commission of the World Health Organization (WHO) declared smallpox eradicated, and this declaration was sanctioned by the World Health Assembly in May 1980 (20). In conjunction with eradication, the so-called "whitepox" viruses, which had been implicated as naturally occurring variolalike mutants of monkeypox virus, were concluded to have arisen by cross-contamination in laboratories (18, 20).

Monkeypox virus, which causes a smallpoxlike disease in people living in the African rain forest, is now regarded as the most serious poxvirus infection (25). Vaccinia virus infections are not generally regarded as naturally occurring, although vaccinee-to-animal (cattle)-to-human transmissions occurred during smallpox vaccination campaigns. Vaccinia is discussed here mainly because its use is associated with rare postvaccinal side effects and accidental infections, because it appears to have a very close relationship to zoonotic buffalopox virus (8), and because recombinant vaccinia viruses are now in human clinical trials, including of candidate AIDS vaccines, and in field trials of a wildlife-baited vaccine against rabies (2, 12). Routine smallpox vaccination has now ceased worldwide; however, in the United States, vaccination is recommended by the Advisory Committee on Immunization Practices (ACIP) for protecting research and health care workers at high risk of infection from handling human orthopoxviruses.

Buffalopox virus is an orthopoxvirus of unresolved natural origin that has caused outbreaks involving milking buffalo, dairy cattle, and humans in parts of Asia and Africa. DNA analyses of a few isolates, including isolates from an outbreak in India in 1985, after cessation of routine vaccinia vaccination, indicated that buffalopox virus is a vaccinia virus subspecies. Its origin is suggested to be somehow rooted in the widespread use of vaccinia for immunization against smallpox (8, 22). Cowpox virus, which appears mainly in Europe and neighboring parts of Asia, is generally believed to be transmitted via a rodent reservoir to zoo, wild, and domestic animals and through these sources to humans (1).

Ectromelia mousepox virus is an orthopoxvirus that has been implicated in various human infections such as hepatitis and so-called erythromelalgia syndrome. However, ectromelia virus is historically a notorious contaminant of laboratory mice and diagnostic cultures, and contamination must be rigorously ruled out if ectromelia virus is implicated in human disease. Ectromelia virus will not be discussed here, because data on ectromelia virus causing human diseases are inconclusive.

Smallpox

Natural variola virus infections were transmitted between humans usually by aerosol or by direct or indirect (fomite) physical contact (20). Variola major strains produced severe prodrome, fever, prostration, and rash; there was a toxemia or other form of systemic shock that led to case fatality rates ranging to 40%. Variola minor strains (synonyms: alastrim, amass, or kaffir virus) produced less severe systemic complications; case fatality rates ranged to 1%.

DNA and biologic data have suggested that European isolates and South American isolates of variola minor virus (true alastrim viruses) are genetically the same but distinct from certain so-called African variola minor virus isolates, which seem to be attenuated variants of true variola major virus (7).

Naturally acquired ordinary smallpox progressed from an oropharyngeal enanthem to a dermal exanthem, a skin rash that developed from macule to papule to vesicle to pustule (20). A noticeably higher density of the lesions appeared on the face and extremities than on the trunk, so that lesions appeared centrifugally distributed. Severe chickenpox rash caused by varicella-zoster virus has often been confused with and misdiagnosed as smallpox rash. Smallpox rarely developed without rash, but it sometimes was misdiagnosed as acute leukemia, meningococcemia, or idiopathic thrombocytopenic purpura.

Variola major smallpox was differentiated into four main clinical types. (i) Ordinary smallpox (90% of cases) produced viremia, fever, prostration, and rash. The severe viremia and inflammatory reaction led to a variety of problems, including ocular (keratoconjunctivitis, blindness), osteopathic (arthritis), respiratory (bronchitis, pneumonitis), gastrointestinal (diarrhea), genitourinary (unilateral orchitis), and central nervous system (encephalitis) difficulties. These sequelae and secondary bacterial infections produced complications that contributed to the general toxemia and systemic shock leading to fatality rates as high as 40%. (ii) Modified smallpox (5% of cases) produced mild prodrome with few skin lesions in previously vaccinated people. (iii) Flat smallpox (5% of all cases) produced slow-developing focal lesions with generalized infection and a ~50% fatality rate. (iv) Hemorrhagic smallpox (<1% of cases) induced bleeding into the skin and mucous membranes; it was invariably fatal within 1 week of onset. A discrete type of the ordinary form resulted from variola minor infections (20).

The last naturally occurring smallpox infection was in Somalia in October 1977 (African variola minor), although in August 1978, a fatal laboratory-associated infection with variola major virus occurred at the University of Birmingham in England. With the achievement of eradication, storage of variola virus stocks and research with the infectious virus is now restricted to the two remaining WHO Collaborating Centers for Smallpox and Other Poxvirus Infections at the Centers for Disease Control and Prevention (CDC) in Atlanta, Ga., and the Research Institute for Viral Preparations in Moscow, Russia. Recently, the complete nucleotide sequence (186,103 bp) of the DNA of the Bangladesh 1975 strain (27), which caused the last case of variola major, and parts of the DNAs of eight other strains have been determined under the aegis of the WHO. It has been recommended by the WHO Committee on Orthopoxvirus Infections that all remaining samples of variola virus be destroyed. The final decision on whether or not to destroy the samples is planned to be made at the 1995 World Health Assembly.

Human Monkeypox

A review describing human monkeypox is available (25). The clinical appearance of human monkeypox is much like that of smallpox, with a centrifugally distributed vesiculopustular rash predominating; however, lymphadenopathy is more prominent in human monkeypox than in smallpox, and the epidemiology is different. The disease is a rare zoonosis, with virus transmission between humans occurring extremely infrequently. It is today the most severe naturally occurring poxvirus disease of humans. Laboratory differentiation of human monkeypox and smallpox generally depends on isolation and identification of monkeypox virus, although with the sequences of variola DNA determined, PCR-based, type-specific DNA identification methods are being developed. Differentiation of monkeypox by using patients' sera is rather difficult, as is serodifferentiation of all orthopoxvirus infections, because of the extremely close antigenic relatedness of the viruses within this genus (15, 16, 22).

Like variola virus, monkeypox virus appears to enter mainly through the mucosa of the upper respiratory tract or through skin abrasions; it then migrates to regional lymph nodes. The virus moves via a primary viremia to internal organs, and then a secondary viremia provides general dissemination of virus throughout the body and causes a rash. During prodrome, lymphadenopathy with fever and generalized headache are common symptoms. Usually, the rash appears first on the face, and then eruptions occur on the rest of the body. The lesions, like smallpox rash, develop through stages of macules, papules, vesicles, and pustules. Sequelae involve scarring and pitting at the sites of the lesions.

Monkeypox was first recognized in 1958 as an exanthem of primates; hence, the virus was named monkeypox virus. Later, the disease was seen in other captured animals, and the first human case was reported in Zaire (Congo) in 1970. Serosurveys and virologic investigations of wild animals in the 1980s have suggested that primates are sporadically infected, as are human beings. The role of wild primates in sustaining virus transmission or as a source of human infection has been difficult to define, but occurrence of the disease in humans does correlate with contact with infected animals. Serosurvey data and virus isolation have strongly implicated African arboreal squirrels (*Funisciurus* and *Heliosciurus* spp.) as the potential reservoir. From 1970 to the present, human monkeypox has been observed only in the rain forest of Africa, mostly in Zaire but also in Liberia, Ivory Coast, Sierra Leone, Nigeria, Benin, Cameroon, and Gabon. The virus seems to present little risk of being introduced widely in the human populace; 209 cases in a population of about 8 million were observed during the last active surveillance in Zaire from 1980 to 1984. The fatality rate of human monkeypox is about 15%; unvaccinated children often contract the most severe form of the smallpoxlike syndrome. Malaria and other immunodeficiency cofactors, such as AIDS, which is occurring in epidemic proportions in these countries, may influence prognoses, but current epidemiologic information is lacking.

Vaccinia

As mentioned above, except for laboratory and health care workers at risk of contracting human orthopoxvirus infection, including infection with recombinant vaccinia viruses, routine smallpox vaccination has been stopped worldwide. The ACIP recommends as a safeguard that workers at high risk of infection be vaccinated within 10 years of potential exposure. The vaccine is contraindicated in persons with eczema and other immunocompromising conditions and as a treatment for herpesvirus and certain other infections, for which smallpox vaccine sometimes had been used illegitimately. In the United States, CDC provides vaccine for health care and laboratory workers at risk of contracting human orthopoxvirus infections. CDC also dispenses vaccinia immunoglobulin should it be needed to treat postvaccinal side effects or a human orthopoxvirus infection. Administering physicians should contact the CDC Drug Service and the CDC National Immunization Program for details on obtaining smallpox vaccine and immunoglobulin and for advice on clinical questions.

Serious complications rarely occur when generally healthy individuals are vaccinated with the vaccinia vaccine strains recommended by the WHO and authorized in the country of use (e.g., the New York Board of Health strain, the component in Wyeth Dryvax vaccine, is used in the United States; the Lister strain [same as Padwadanger strain in India] is used in many countries outside the United States, and the Tian Tan strain is used in China). Side effects may include postvaccinal encephalitis and encephalopathy, severe skin reactions including eczema vaccinatum, progressive vaccinia (vaccinia necrosum), generalized vaccinia, and certain autoinoculation complications, including ocular involvement (20–22, 34). The vesiculopustular vaccinial lesions have on rare occasion been the source of infection of family contacts. Vaccinia infections may be life threatening in persons with compromised immune systems.

Buffalopox

Sporadic outbreaks of buffalopox virus that involve transmission between milking buffalo, cattle, and people have been reported, mainly in India but also in Egypt, Bangladesh, Pakistan, and Indonesia. Vaccinialike lesions have been observed on the animals' teats and the milkers' hands; milk is infectious. Biologic data and limited DNA analyses of isolates from an outbreak in India in 1985 suggest that buffalopox virus may be derived from vaccinia vaccine strains that were transmitted from humans to livestock during the smallpox vaccination era (8, 22). However, the natural histories of all the buffalopox viruses have not been determined. It is not known whether buffalopox virus is rodent borne, as seems to be the case with cowpox virus.

Cowpox

Cowpox is a rare occupational infection of humans that is acquired through contact with infected cows or other animals (infected rats, cats, and zoo and circus elephants have been sources of the disease). Cowpox virus has been isolated from humans and animals in Europe and in adjoining regions of Asia (1). The lesions in humans occur mainly on the fingers, with reddening and swelling; autoinoculation of other parts of the body may occur. The lesions are likened to those of a primary vaccinia virus vaccination; the site becomes papular, in 4 to 5 days a vesicle develops, and healing takes about 3 weeks (19, 21, 22).

Yatapoxviruses

Overview

The three genus members, tanapox virus, Yaba-like disease virus (YLDV) of monkeys, and YMTV, are serologically related (39). DNA maps of tanapox virus and YLDV are extremely similar, suggesting that they are likely the same agent; however, these DNA maps are markedly different from YMTV DNA maps, although the three DNAs cross-hybridize extensively (26). The epidemiology and natural history of yatapoxviruses are very poorly defined. YMTV and YLDV infections have occurred in animal handlers; however, tanapox virus is the prime naturally occurring human pathogen in the genus (19, 21).

Tanapox

Tanapox is an endemic zoonosis of equatorial Africa that is thought to be transmitted to humans mechanically by biting insects, especially during the rainy season. Tanapox virus was first isolated from human skin biopsy specimens taken during outbreaks in 1957 and 1962 that occurred in the Tana River Valley, Kenya. Tanapox virus was isolated from infected persons during surveillance for human monkeypox in Zaire in the late 1970s (24). Six isolates from different years and locations showed DNA profiles identical to that of Kenya tanapox virus of the 1950s, suggesting that very little virus variation occurs in nature (26).

YLDV and YMTV

YLDV was first recognized during epizootics in 1965 and 1966 in primate centers in California, Oregon, and Texas. The origin of the epizootics was traced to a primate-importing company. During these outbreaks, animal handlers contracted YLDV infections that were clinically identical to classic tanapox.

Tanapox virus and YLDV in humans produce a brief fever followed by development of firm, elevated, round, necrotic maculopapular nodules that are distinct from the vesiculopustular lesions of orthopoxviruses. Generally, very few lesions develop, and these are primarily on the skin of the upper arms, face, neck, or trunk (21). Symptoms include a low-grade fever, backache, and headache prior to appearance of lesions. On recovery, lesions umbilicate without pustulation, usually healing in 2 to 4 weeks.

YMTV, isolated in 1958 during an outbreak in a colony of rhesus monkeys in Yaba near Lagos, Nigeria, produces epidermal histiocytomas, which are tumorlike masses of histiocytic polygonal mononuclear cell infiltrates that advance to suppurative inflammatory reactions. Animal handlers have been accidentally infected with YMTV (39). There have been no reports of monkey or human YMTV or YLDV infection for over 2 decades.

Parapoxviruses

Overview

Three different parapoxvirus diseases occur in humans, generally as occupational infections: milker's nodule (in dairy cattle, the disease is termed pseudocowpox or paravaccinia), Orf (in sheep and goats, the disease has been referred to as Orf, contagious ecthyma, contagious pustular dermatitis, contagious pustular stomatitis, or sore mouth), and papulosa stomatitis (in calves and beef cattle, the disease is termed bovine papular stomatitis) (21, 37).

Parapoxvirus infections are transmitted to humans by direct contact with infected livestock through abraded skin on the hands and fingers, and ocular autoinoculation sometimes occurs (21, 34). Milker's nodule produces a reddened hemispheric papule that matures to a purplish, smooth, firm nodule varying in size up to 2 cm; the lesions usually are not painful and can persist for about 6 weeks. Human Orf virus infection is usually found on the fingers, hands, and arms but may also be found on the face and neck; fever and swelling of draining lymph nodes may be present, and the lesions often ulcerate and are painful. Autoinoculation of the eye may lead to serious sequelae (34). Wildlife (skinning animals such as deer and reindeer) have also been sources of Orf virus infection.

Genus *Molluscipoxvirus*

Overview

Molluscum contagiosum virus is the sole member of the genus *Molluscipoxvirus*. This organism appears to have a human-restricted host range.

Molluscum contagiosum

Molluscum contagiosum occurs worldwide as a contact disease in children or as a sexually transmitted disease in adults. In children and teenagers, the lesions generally appear on the trunk, limbs (except palms and soles), and face, where there may be ocular involvement (19, 21, 34). Infection is usually transmitted by direct or indirect skin contact, such as among wrestlers or in public baths and swimming pools. As a sexually transmitted disease in teenagers and adults, the lesions are mostly on the lower abdominal wall, pubis, inner thighs, and genitalia. Lesions are pearly, flesh-colored, raised, firm, umbilicated nodules about 4 cm in diameter. Lesions tend to disseminate by autoinoculation. Severe dissemination has been noted in AIDS patients; multiple lesions may become confluent, forming a single, often disfiguring plaque (34). Two main DNA restriction patterns have been observed for a limited number of samples analyzed, but no correlation of DNA type, disease syndrome, or geographic distribution has been confirmed (34, 35).

COLLECTION, HANDLING, AND STORAGE OF SPECIMENS

In the United States and its territories, a suspected case of smallpox must be immediately reported to the respective state or territorial health department. After the health department reviews the case and if it still appears to be a suspected case, it should be immediately reported to the WHO Collaborating Center for Smallpox and Other Poxvirus Infections, Poxvirus Section, CDC, Atlanta, GA 30333. The WHO Collaborating Center at the CDC also serves internationally. The CDC Poxvirus Section functions primarily as a research and reference laboratory and includes appropriate containment facilities for working with variola, monkeypox, and other exotic human poxviruses. In the United States and territories, other routine poxvirus diagnostic specimens, including suspected Orf and molluscum contagiosum samples, generally should be examined through private diagnostic laboratories or local governmental health laboratories.

Basic procedures for collecting clinical specimens can be found elsewhere in this Manual. It is most important to collect an amount of specimen sufficient to permit effective testing. Use of an inadequate amount of a specimen decreases the dependability of laboratory tests for diagnosing diseases. Suitable specimens for virologic tests of most suspected poxvirus infections are two to four biopsy specimens, scabs, and vesicular fluids, including the cells at the base of the vesicles. Preferably, collect vesicular fluid in capillary tubes or on glass microscope slides as a thick droplet without spreading the smear, or collect the fluid on dry swabs. Suitable autopsy specimens include sections of skin, liver, spleen, lung, or kidney. After contacting the appropriate laboratory, send the collected specimens in a dry condition in a Parafilm-sealed container (e.g., screw-cap glass or plastic vial or bottle, small slide holder in plastic bag). Do not add so-called transport fluid or glycerol to poxvirus speci-

mens. Specimens collected from patients with poxvirus infections can be stored at 4°C for a short time, but −20 to −70°C should be used for long-term storage. Changes in pH of specimens, such as may occur when dry ice vapors enter containers, should be avoided. A sealed metal, plastic, or paperboard outer container should be considered. International, U.S., and local packing and shipping regulations must be followed.

LABORATORY DIAGNOSTIC METHODS

Virologic Methods

The Poxvirus Section at CDC mainly uses a combination of methods, including electron microscopy, virus growth in cell culture and/or on the CAMs of 12-day-old chick embryos, and DNA restriction assay for characterizing poxviruses. A PCR protocol is under development. The assays are generally accurate, dependable, and relatively rapid. Various laboratory protocols for poxvirus diagnostics, including our recent methods for DNA restriction assay, have been detailed in previous editions of this Manual and elsewhere (13, 14, 21, 30–33). The most powerful and definitive virus diagnostic methods involve DNA assays. These methods are becoming more readily available to diagnostic facilities at all levels.

Serologic Methods

When virus specimens are not available (for example, if the patient is seen too late in the course of infection), antibody assessment may be the only course for establishing the etiology of the disease; in this case, serum specimens are useful. There is no one immunologic test that defines a person's degree of protection against a poxvirus infection. Protection is genetically defined and requires a concert of cell-mediated and humoral immune responses; immunity depends on the infecting virus, the dose, and the portal of entry. For example, it is documented that 5% of persons vaccinated against smallpox developed a modified, albeit less severe, disease. Thus, subsequent infection with members of the same poxvirus genus and species can occur.

The primary focus of serologic methods used at the CDC is the assay of antibodies from human orthopoxviruses and yatapoxviruses, in which case the virus NT and the enzyme-linked immunosorbent assay (ELISA) are mainly used. Molluscum contagiosum, for which there is no test because the virus cannot be grown in culture, is not tested for and serology for parapoxviruses is not performed at CDC; such infections are readily diagnosed clinically, often with the aid of histopathology. The techniques for the NT and the ELISA, plus methods for other poxvirus serologic tests that have been used at CDC, have been detailed elsewhere (4, 5, 13, 14, 21, 30–33, 38). More recently, the ELISA has been done with sucrose gradient-purified orthopoxvirions in an attempt to have a more standardized antigen than the infected-cell homogenates used previously, which tended to give high backgrounds with certain serum samples. An ELISA based on a single type-specific orthopoxvirus protein is currently under development. Usually, a fourfold rise in titer between serum drawn at the acute and convalescent phases is considered diagnostic. Often, only a single serum specimen taken at one phase or the other is available. In such situations, it is generally very difficult to interpret serologic test results properly; thus, serologic testing at the CDC is discretionary and on a case-by-case basis, depending on the epidemiologic circumstances. Because orthopoxviruses are very closely related, serum cross-absorption tests, such as those performed with immunofluorescent-antibody or immunodiffusion methods, have been used with variable success with patient and animal sera. More research, especially on antigen standardization and development of type-specific antigens based on sequence determinations, is needed to further optimize such tests. Generally, sera from patients do not have the high antibody levels and specificities necessary for performing cross-absorption tests.

The NT

The antigen for routine poxvirus antibody NT can be lysates of infected cell cultures clarified by low-speed centrifugation or various purified or partly purified virus preparations that may contain INVs, EEVs, or both. Purified virus preparations are more defined and, depending on the particle/PFU ratio, may be more suitable than lysates, which can bind neutralizing antibodies via virions as well as by other intracellular antigens. The appendix to this chapter describes a general method for purifying INVs from poxvirus-infected cell cultures; the method is adaptable for use with infected CAMs or lesion material. Neutralizing antibodies for variola, monkeypox, cowpox, and vaccinia viruses are often detectable as early as 6 days after infection (vaccination). The half-life of NT antibodies is variable, but they have been detected more than 20 years after smallpox vaccination. When only one serum specimen is drawn from an infected patient, especially after onset of rash, and when the sample shows a moderate titer (<500 recipocal dilution), a diagnosis is difficult, particularly if the patient has been vaccinated before the infection, such as in cases of cowpox and monkeypox in previously vaccinated persons. A vaccination scar (most countries have not routinely vaccinated civilians for >15 years) and a titer of >1,000 are presumptive evidence for the diagnosis of a superinfection by a second orthopoxvirus.

ELISA

Original versions of the ELISA method for poxviruses used 96-well microtitration plates coated with a crude concentrate of orthopoxvirus-infected cell culture lysate. Experience with the test showed that crude lysates give variable titers with different antigen batches; thus, the question of antigen standardization has now come into focus. The antigen currently being evaluated at CDC for use in ELISAs for orthopoxvirus antibodies is partly purified or sucrose gradient-purified INVs (see the appendix to this chapter). Reasons for investigating the use of INVs in poxvirus serology have been that (i) antibodies against virion structural proteins on INVs may be longer lasting than those against other virus antigens; (ii) compared with most infected-cell lysate antigens whose quality and quantity could be affected by culture conditions (poxviruses produce more than 200 proteins), fewer virion surface antigens are present on INVs; and (iii) INVs can be semi-quantified by spectrophotometry. With the advent of more rapid, large-scale DNA sequencing methods, the use of more specific, possibly genetically engineered reagents is likely to be forthcoming.

Serology for Milker's Nodule, Orf, Tanapox, and Molluscum Contagiosum

The serologic method used at CDC for examining samples (accepted on a case-by-case basis) in order to determine antibodies for confirming parapoxvirus infection has been an ELISA that uses infected cell culture concentrate as antigen. Tanapox virus serologic testing has been done by ELISA with infected cell culture concentrate antigen, by an indirect immunofluorescent-antibody test, and by the NT with moderate effectiveness. As in orthopoxvirus serology, there is hope that better-defined antigens will lead to improved serologic methods. Generally, serum should be collected at 3 to 5 weeks or later after a presumed onset date; however, compared with our understanding of vaccinia and variola virus infections in humans, our understanding of the immunologic response to parapoxvirus infections is poor. As mentioned above, no practical serodiagnostic test for routine or large-scale studies of molluscum contagiosum infections is yet available, because the virus cannot be propagated outside of the human host.

Miscellaneous Comments

Immunoprecipitation in agar, immunofluorescence, and complement fixation tests are relatively less sensitive and less informative than NT and ELISA. Immunoprecipitation in agar is often difficult to achieve, because precipitating antibodies are not seen in many clinically proven cases of variola, monkeypox, or vaccinia virus infection; however, the test has been useful in research with hyperimmune animal sera. Immunodiffusion results are rarely positive in individuals recently vaccinated or revaccinated against smallpox. In the complement fixation test, antibodies from smallpox patients may not appear until the second week after clinical diagnosis. A substantial proportion of sera from unvaccinated smallpox patients tested for such antibodies, even after the eighth day after onset, have been negative. After primary vaccination or revaccination for smallpox, complement-fixing antibodies may not be detectable. For Orf and milker's nodule infections, antibodies have not always been detected in virologically confirmed cases. For these diseases, negative serologic results present a problem in interpretation. Experience suggests that the complement fixation test has rather limited value in poxvirus diagnostics.

EVALUATION OF VIROLOGIC DIAGNOSTIC METHODS

Experience at CDC has been that orthopoxvirus infections, especially smallpox and human monkeypox, can be confidently diagnosed in the laboratory by a combination of three methods: electron microscopy, virus growth on CAMs, and DNA restriction assay. Human and animal cowpox virus infections do not occur naturally in the United States; however, we have correctly diagnosed the disease for the few human cowpox samples sent from foreign laboratories. It cannot be overemphasized that appropriate specimens in adequate amounts greatly improve the probability of a correct laboratory diagnosis. We note that a recurring problem is making a correct negative diagnosis. Generally, Orf, milker's nodule, and tanapox infections are more difficult to correctly diagnose in the laboratory than are orthopoxvirus infections, mainly because the viruses causing the former diseases are more difficult to propagate on primary cultivation of the samples. Although visualization of enveloped virus particles by electron microscopy in suspected tanapox specimens has led to a high degree of successful laboratory diagnoses, it has been more difficult to correctly confirm clinical diagnoses of Orf or milker's nodule infection. Many of the samples we receive are collected incorrectly; thus, the tasks connected with laboratory diagnosis become time-consuming and costly, and the odds of correct identification decrease.

APPENDIX

Diagnostic reagents and antigen preparations for several poxviruses are not generally available commercially. Prior editions of this Manual contain detailed methods for preparing orthopoxvirus-infected CAM homogenates, HA antigen, and chicken erythrocytes for the HA and HA inhibition tests (13, 32). A periodate treatment method for removal of nonspecific inhibitors from sera to be used in the HA inhibition test and a procedure for preparing vaccinia virus hyperimmune antisera are available (13). The HA inhibition test may still be useful for large-scale screening of orthopoxvirus infections, but the ELISA method seems more promising, as many diagnostic laboratories continue to develop automated ELISA protocols. Below is a protocol for purification of orthopoxvirus INVs for those laboratories desiring to perform a poxvirus ELISA with virus particles as antigen. We have also included a protocol adopted at CDC for relatively rapid small-scale extraction of poxvirus DNA from infected tissue culture (36) to complement a large-scale cytoplasmic DNA extraction procedure published elsewhere (10). INVs can also be used for preparing DNA. We have found that DNA prepared by the procedure below is suitable for restriction assay, PCR, and cloning into plasmids.

Purification of Virions from Infected Cell Monolayers

1. Inoculate 30 cell culture 150-cm^2 monolayers (e.g., Fogh and Lund [FL] human amnion, RK-13 rabbit kidney, BSC, LLC MK$_2$, or CV-1 monkey kidney) with 2 ml each of vaccinia virus or other poxvirus (titer, ca. 10^7 PFU/ml). Absorb the inoculum for 1 h at 37°C (use 34°C for yatapoxviruses) in a CO$_2$ incubator, and then add 50 ml of prewarmed Dulbecco high-glucose or other reinforced minimal essential medium containing 2% fetal calf serum.

2. Harvest infected cells in about 2 to 4 days when they are at 90 to 100% cytopathic effect by scraping cells into the medium (cooling cultures to 4°C also dislodges cells into the medium).

3. Centrifuge cells at 2,500 rpm for 10 min (Beckman JA-10 rotor or equivalent).

4. Combine and resuspend pellets in 300 ml of cold hypotonic reticulocyte swelling buffer (10 mM Tris, 10 mM NaCl, 1.5 mM MgCl$_2$; pH 7.8).

5. Pellet cells as in step 4, resuspend cells in 40 ml of reticulocyte swelling buffer, and incubate the suspension for 2 h to overnight at 4°C to swell cells.

6. Rupture the cells by 20 to 30 strokes of a Dounce homogenizer; observe the release of nuclei microscopically.

7. Centrifuge at 600 rpm (100 × g; Beckman JA-10 rotor) for 15 min to pellet the nuclei. Save the supernatant fluid (cytoplasmic material).

8. Recycle the nucleus-containing pellet as described in steps 5 through 7 if intact cells are seen microscopically.

9. Pool the supernatant fluids of steps 7 and 8, proceed to the next step, or *safely* perform sonication or Potter-Elvehjem homogenization of supernatant fluid, using an appropriate aerosol and sound wave containment to increase the number of particles recovered (virions are more dispersed by either of these methods).

10. Using several SW-28 Beckman rotor tubes or the equivalent, each containing a 15- to 20-ml cushion of 40% (wt/wt) sucrose (40 g of sucrose in 60 ml of 10 mM Tris [pH 8.5] buffer), gently layer on an aliquot of the supernatant fluid. With a plastic 1-ml pipette, underlay a 0.1-ml pad of 100% glycerol beneath the sucrose cushion to prevent formation of compacted pellets. Centrifuge at 20,000 rpm (60,000 × g) for 90 min at 4°C.

11. Aspirate the supernatant fluid and most of the sucrose. Vortex the virus "soft" pellet in the glycerol pad and the bottom 0.5 ml of sucrose.

12. Dilute the pellet about 1:3 with 10 mM Tris (pH 8.5) buffer or enough to reduce the density so that the virus suspension can be layered gently onto a prechilled sucrose gradient of 20% (wt/wt) to 40% (wt/wt) sucrose in 10 mM Tris (pH 8.5) buffer (load the gradients with <5 ml of sample). Before loading the gradients, sonicate or Potter-Elvehjem homogenize the pellet to disperse the virions. Cast the gradients at room temperature in Beckman SW-28 rotor tubes or the equivalent, and then chill.

13. Centrifugation of the gradient at 13,000 rpm (25,000 × g) at 4°C will sediment virions to the middle of the gradient (the speed may be increased or decreased for optimized banding of specific poxviruses; poxviruses have a sedimentation coefficient of about 5,000 Svedberg units). Aspirate the sucrose above the virus band. Harvest the virus band, and dispense it in 0.5-ml aliquots. Store frozen at −20 to −70°C. Dilute (1:20 to 1:30) a small sample of the preparation to determine the optical density at 260 nm (use disposable UV cuvettes). One unit of optical density at 260 nm is equivalent to 1.2 × 10^{10} particles (vaccinia) per ml. Titrate on monkey kidney cells (e.g., CV-1, Vero, BSC-40, Rat-2) to obtain the ratio of particles to PFU.

14. Antigen for ELISA is prepared by diluting virus to 10^8 particles per ml in standard carbonate-bicarbonate ELISA buffer. Coat the solid-phase surface (e.g., 96-well microtiter plates, plastic rods) with virus suspension. For concentrated stock antigen for the NT, dilute the virus in sucrose with an equal volume of sterile 100% glycerol, and keep the mixture at −20°C. To prepare for the test, further dilute virus in cell culture media to 100 PFU per appropriate volume to be mixed with an equal volume of patient's diluted serum.

Small-Scale Preparation of Poxvirus DNA

DNA can be prepared on a large scale from infected cells or from purified or partially purified virion preparations as described by Esposito et al. (10). For clinical applications, small-scale preparations of DNA are suitable for PCR or restriction endonuclease assay. A procedure that we adopted and found most useful has been described by Rice et al. (36). The protocol is as follows.

1. One 75-cm^2 tissue culture monolayer at ~95% confluency is infected with virus and incubated until severe cytopathology is evident. Infected cells are scraped from the surface (chilling the culture for a few hours will also loosen attached cells) and collected by low-speed centrifugation. Cells are washed once in phosphate-buffered saline (PBS) and then suspended in 600 μl of PBS, transferred to a 1.5-ml microcentrifuge tube, and thrice frozen (at −70°C or in dry ice) and thawed (at room temperature).

2. Adjust with concentrated high-grade reagents to 0.5% Triton X-100, 35 mM 2-mercaptoethanol (stock is 14.3 M), and 20 mM disodium EDTA (stock is 200 mM, pH 8), and mix gently by inverting the tube a few times.

3. Centrifuge for 2.5 min at 3,000 rpm (500 × g) at room temperature (Tomy microcentrifuge model RC-15A or equivalent) to pellet nuclei and large debris.

4. Transfer the supernatant to a fresh tube, and pellet viral cores (10 min at 15,000 rpm [15,000 × g]) at room temperature.

5. Aspirate the supernatant, and then suspend the pelleted viral cores in 100 μl of 10 mM Tris-HCl (pH 8) buffer containing 1 mM disodium EDTA, 5 mM 2-mercaptoethanol, 150 μg of proteinase K per ml, 200 mM NaCl, and 1% sodium dodecyl (lauryl) sulfate. Incubate at 56°C for 30 to 45 min.

6. Cool the mixture to room temperature, and extract it twice with an equal volume of a mixture of TE buffer (10 mM Tris-HCl [pH 7.5], 1 mM EDTA) and saturated phenol-chloroform-isoamyl alcohol (50:48:2) (genetic-technology-grade phenol is washed thrice with TE buffer and then combined with chloroform and isoamyl alcohol). Extraction phases are separated at 10,000 rpm (8,000 × g) for 1 min.

7. The final phenol (bottom) phase is aspirated completely, and the aqueous phase (ca. 0.5 ml) is extracted twice with an equal volume of chloroform-isoamyl alcohol mixture (98:2).

8. The aqueous phase is transferred to a fresh tube and precipitated with 2 volumes of cold ethanol. The precipitate is collected by centrifugation at 15,000 rpm (15,000 × g) for 5 min at 4°C, the supernatant fluid is aspirated completely, and the pellet is washed with 70% ethanol, repelleted as described above, and air dried for a few minutes at room temperature.

9. The DNA (ca. 10 to 20 μg per flask) is soaked in 25 μl of TE buffer for about 1 h or until it enters solution.

REFERENCES

1. Bennett, M., C. J. Gaskill, D. Baxby, R. M. Gaskill, D. F. Kelly, and J. Naidoo. 1990. Feline cowpox virus infection. A review. *J. Small Anim. Pract.* **14:**167–173.

2. **Binns, M. M., and G. L. Smith (ed.).** 1992. *Recombinant Poxviruses.* CRC Press, Inc., Boca Raton, Fla.

3. **Buller, R. M. L., and G. J. Palumbo.** 1991. Poxvirus pathogenesis. *Microbiol. Rev.* **55:**80–122.

4. **Cole, G. A., and R. V. Blanden.** 1982. Immunology of poxviruses, p. 1–19. *In* A. J. Nahmias and R. J. O'Reilly (ed.), *Comprehensive Immunology,* vol. 9. *Immunology of Human Infection,* part II. *Viruses and Parasites: Immunodiagnosis and Prevention of Infectious Diseases.* Plenum Press, New York.

5. **Conroy, J. M., R. W. Stevens, and K. E. Hechemy.** 1991. Enzyme immunoassay, p. 87–92. *In* A. Balows, W. J. Hausler, Jr., K. L. Herrmann, H. D. Isenberg, and H. J. Shadomy (ed.), *Manual of Clinical Microbiology,* 5th ed. American Society for Microbiology, Washington, D.C.

6. **Dumbell, K. R.** 1989. Poxviruses, p. 395–427. *In* J. Porterfield (ed.), *Andrews' Viruses of Vertebrates,* 5th ed. Bailliere Tindall, London.

7. **Dumbell, K. R., and F. Huq.** 1986. The virology of variola minor. Correlation of laboratory tests with the geographic distribution and human virulence of variola isolates. *Am. J. Epidemiol.* **123:**403–415.

8. **Dumbell, K. R., and M. Richardson.** 1993. Virological investigations of specimens from buffaloes affected by buffalopox in Maharashtra State, India, between 1985 and 1987. *Arch. Virol.* **128:**257–267.

9. **Esposito, J. J.** *Poxviridae. In* F. A. Murphy, C. Fauquet, D. H. L. Bishop, S. A. Gabrial, A. W. Jarvis, G. P. Martelli, M. Mayo, and M. D. Summers (ed.), *The Classification and Nomenclature of Viruses: 6th Report of the International Committee on Taxonomy of Viruses,* in press. Springer-Verlag, Vienna.

10. **Esposito, J. J., R. Condit, and J. F. Obijeski.** 1981. The preparation of orthopoxvirus DNA. *J. Virol. Methods* **2:**175–179.

11. **Esposito, J. J., and J. C. Knight.** 1985. Orthopoxvirus DNA: a comparison of restriction profiles and maps. *Virology* **143:**230–251.

12. **Esposito, J. J., and F. A. Murphy.** 1989. Infectious recombinant vectored virus vaccines. *Adv. Vet. Sci. Comp. Med.* **33:**195–247.

13. **Esposito, J. J., and J. H. Nakano.** 1991. Poxvirus infections in humans, p. 858–867. *In* A. Balows, W. J. Hausler, Jr., K. L. Herrmann, H. D. Isenberg, and H. J. Shadomy (ed.), *Manual of Clinical Microbiology,* 5th ed. American Society for Microbiology, Washington, D.C.

14. **Esposito, J. J., and J. H. Nakano.** 1992. Human poxviruses, p. 643–668. *In* E. H. Lennette (ed.), *Laboratory Diagnosis of Viral Infections,* 2nd ed. Marcel Dekker, Inc., New York.

15. **Esposito, J. J., J. F. Obijeski, and J. H. Nakano.** 1977. Serological relatedness of monkeypox, variola, and vaccinia viruses. *J. Med. Virol.* **1:**35–47.

16. **Esposito, J. J., J. F. Obijeski, and J. H. Nakano.** 1977. The virion and soluble antigen proteins of variola, monkeypox, and vaccinia viruses. *J. Med. Virol.* **1:**95–110.

17. **Esposito, J. J., J. F. Obijeski, and J. H. Nakano.** 1978. Orthopoxvirus DNA: strain differentiation by electrophoresis of restriction endonuclease fragmented virion DNA. *Virology* **89:**53–66.

18. **Esposito, J. J., J. F. Obijeski, and J. H. Nakano.** 1985. Can variola viruses be derived from monkeypox virus? An investigation based on DNA mapping. *Bull. W.H.O.* **63:**695–703.

19. **Fenner, F.** Poxviruses. *In* B. N. Fields, D. Knipe, and P. Howley (ed.), *Fields' Virology,* 3rd ed., in press. Raven Press, Inc., New York.

20. **Fenner, F., D. A. Henderson, I. Arita, Z. Jezek, and I. Ladnyi.** 1988. *Smallpox and Its Eradication.* World Health Organization, Geneva.

21. **Fenner, F., and J. H. Nakano.** 1988. *Poxviridae:* the poxviruses, p. 177–207. *In* E. H. Lennette, P. Halonen, and F. A. Murphy (ed.), *Laboratory Diagnosis of Infectious Diseases,* vol. 2. *Viral, Rickettsial, and Chlamydial Diseases.* Springer-Verlag, New York.

22. **Fenner, F., R. Wittek, and K. R. Dumbell.** 1989. *The Orthopoxviruses.* Academic Press, Inc., New York.

23. **Goebel, S. J., G. P. Johnson, M. E. Perkus, S. W. Davis, J. P. Winslow, and E. Paoletti.** 1990. The complete DNA sequence of vaccinia virus. *Virology* **179:**247–266.

24. **Jezek, Z., I. Arita, M. Szczeniowski, K. M. Paluku, R. Kalisa, and J. H. Nakano.** 1985. Human tanapox in Zaire: clinical and epidemiological observations on cases confirmed by laboratory studies. *Bull. W.H.O.* **63:**1027–1035.

25. **Jezek, Z., and F. Fenner.** 1988. Human monkeypox. *Monogr. Virol.* **17:**1–140.

26. **Knight, J. C., F. J. Novembre, D. R. Brown, C. S. Goldsmith, and J. J. Esposito.** 1989. Studies on Tanapox virus. *Virology* **172:**116–124.

27. **Massung, R. F., J. J. Esposito, L. Liu, Q. Jin, T. R. Utterback, J. C. Knight, L. Aubin, T. E. Yuran, J. M. Parsons, V. N. Loparev, N. A. Selivanov, K. F. Cavallaro, A. R. Kerlavage, B. W. J. Mahy, and J. C. Venter.** 1993. Potential virulence determinants in terminal regions of variola smallpox virus. *Nature* (London) **366:**748–751.

28. **Mathew, T.** 1987. *Advances in Medical and Veterinary Virology, Immunology and Epidemiology: Cultivation and Immunological Studies on Pox Group of Viruses with Special Reference to Buffalo Pox Virus.* Thajema Publishers, New Delhi, India.

29. **Moss, B.** *Poxviridae* and their replication. *In* B. N. Fields, D. Knipe, and P. Howley (ed.), *Fields' Virology,* 3rd ed., in press. Raven Press, Inc., New York.

30. **Nakano, J. H.** 1978. Comparative diagnosis of poxvirus diseases, p. 267–339. *In* E. Kurstak and C. Kurstak (ed.), *Comparative Diagnosis of Viral Diseases,* vol. 1. Academic Press, New York.

31. **Nakano, J. H.** 1979. Poxviruses, p. 257–308. *In* E. H. Lennette and N. J. Schmidt (ed.), *Diagnostic Procedures for Viral, Rickettsial and Chlamydial Infections,* 5th ed. American Public Health Association, Inc., Washington, D.C.

32. **Nakano, J. H.** 1985. Poxviruses, p. 733–741. *In* E. H. Lennette, A. Balows, W. J. Hausler, Jr., and H. J. Shadomy (ed.), *Manual of Clinical Microbiology,* 4th ed. American Society for Microbiology, Washington, D.C.

33. **Nakano, J. H., and J. J. Esposito.** 1989. Poxviruses, p. 224–265. *In* N. J. Schmidt and R. W. Emmons (ed.), *Diagnostic Procedures for Viral, Rickettsial, and Chlamydial Infections,* 6th ed. American Public Health Association, Inc., Washington, D.C.

34. **Pepose, J. S., and J. J. Esposito.** *Molluscum contagiosum,* Orf, and vaccinia virus ocular infections in humans. *In* G. N. Holland and D. Battaglia (ed.), *Ocular Infection and Immunity,* in press. C. V. Mosby, Inc., Philadelphia.

35. **Porter, C. D., N. W. Blake, J. J. Cream, and L. C. Archard.** 1992. *Molluscum contagiosum* virus, p. 233–257. *In* D. Wright and L. C. Archard (ed.), *Molecular and Cell Biology of Sexually Transmitted Diseases.* Chapman and Hall, Ltd., London.

36. **Rice, C. M., C. A. Franke, J. H. Strauss, and D. E. Hruby.** 1985. Expression of Sindbis virus structural proteins via recombinant vaccinia virus: synthesis, processing, and incorporation into mature Sindbis virus. *J. Virol.* **56:**227–239.

37. **Robinson, A. J., and D. J. Lyttle.** 1992. Parapoxviruses: their biology and potential as recombinant vaccine vectors, p. 285–327. *In* M. M. Binns and G. L. Smith (ed.), *Recombinant Poxviruses.* CRC Press, Inc., Boca Raton, Fla.

38. **Rosebrock, J. A.** 1991. Labeled-antibody techniques: fluorescent, radioisotopic, immunochemical, p. 79–86. *In* A. Balows, W. J. Hausler, Jr., K. L. Herrmann, H. D. Isenberg, and H. D. Shadomy (ed.), *Manual of Clinical Microbiology,* 5th ed. American Society for Microbiology, Washington, D.C.

39. **Rouhandeh, H.** 1988. Yaba virus, p. 1–15. *In* G. Darai (ed.), *Virus Diseases in Laboratory and Captive Animals.* Martinus Nijhoff, Boston.

40. **Smith, G. L.** 1993. Vaccinia virus glycoproteins and immune evasion. *J. Gen. Virol.* **74:**1725–1740.

41. **Turner, P. C., and R. W. Moyer (ed.).** 1990. Poxviruses. *Curr. Top. Microbiol. Immunol.* **163:**1–211.

42. **Westwood, J. C. N., P. H. Phipps, and E. A. Boulter.** 1957. The titration of vaccinia virus on the chorioallantoic membrane of the developing chick embryo. *J. Hyg.* **52:**123–139.

PARASITOLOGY

IX

VOLUME EDITOR
MICHAEL A. PFALLER

SECTION EDITOR
THOMAS R. FRITSCHE

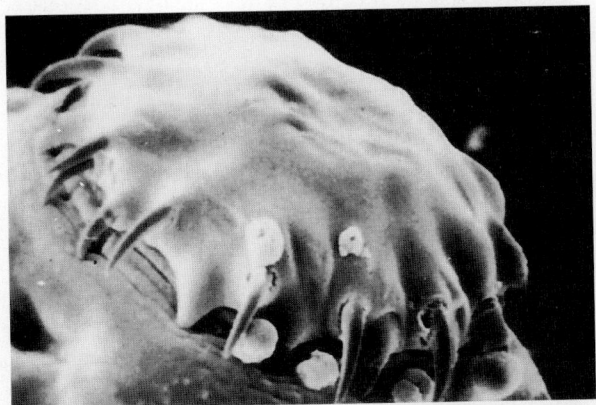

Scanning electron micrograph of *Taenia solium* scolex from cysticercus (*courtesy of Marietta Voge; from E. K. Markell, M. Voge, and D. T. John, Medical Parasitology, 7th ed., The W. B. Saunders Co., Philadelphia, 1992*).

Introduction to Diagnostic Parasitology: Biologic, Clinical, and Laboratory Considerations

THOMAS R. FRITSCHE AND JAMES W. SMITH

101

Parasitic diseases are important causes of morbidity and mortality in many parts of the world and are especially problematic when they occur in individuals weakened by concurrent nutritional and other diseases. While malaria has long been eradicated from the United States and the occurrence of indigenous nematode infections such as ascariasis, trichuriasis, and hookworm infection has markedly decreased, other parasitic diseases, especially those caused by intestinal protozoa, have come into prominence. Giardiasis and cryptosporidiosis are frequent public health problems, with outbreaks related to contaminated water supplies and day-care centers for children. Immunocompromised hosts, especially those with AIDS, are well known for being at risk from *Cryptosporidium parvum*, *Toxoplasma gondii*, various species of microsporidia, and *Strongyloides stercoralis*, all of which (except for *S. stercoralis*) are endemic worldwide.

In addition to infections indigenous to the United States, a wide variety of other parasites may be seen in citizens who have traveled to or worked in foreign countries or in foreign nationals who are visiting or now residing in the United States. Such individuals, including those with malaria or leishmaniasis, may be asymptomatic for months before disease develops or relapses occur. Occasionally, latent malaria is recognized only following blood donation or delivery of a congenitally infected infant. Other diseases, such as schistosomiasis, strongyloidiasis, or echinococcosis, may become clinically evident only years latter. Because immigrants and travelers are at particular risk from various parasitic infections, a thorough geographic history should be an important component of the overall medical history.

The incidence of parasitic infections in the U.S. population is difficult to determine, as most such infections are not reported to public health authorities (malaria being an exception). The actual prevalence for most parasites is undoubtedly much lower than available incidence figures suggest, because these figures are based on specimens specifically submitted for parasite examination (Table 1). Numbers of parasitic infections worldwide and associated mortality are also difficult to estimate, but the World Health Organization and other authorities have published their best estimates (Table 2). Furthermore, the World Health Organization has targeted seven of these parasitic diseases as significant impediments to improving the health of the world's population and is placing special emphasis in research and training on helping control these infections, which include malaria, schistosomiasis, lymphatic filariasis, onchocerciasis, leishmaniasis, Chagas' disease, and African trypanosomiasis.

CLASSIFICATION, LIFE CYCLES, AND APPROACHES FOR CONTROL

Parasitic organisms known to infect humans have arisen from a variety of taxonomic groups within the protozoal, helminthic, and arthropod phyla. The arthropods comprise a diverse group of true parasites, parasite vectors, and pestiferous species that can be especially problematic both for humans and for domestic animals.

A knowledge of parasite life cycles is crucial to the understanding of disease transmission, pathogenesis, and control. Humans may be infected by obligate human parasites, zoonotic parasites, or opportunistic free-living organisms. Certain species with narrow host specificities (i.e., intestinal nematodes, malarial parasites) rely exclusively on humans as the reservoir, whereas others (i.e., *Leishmania* spp., *Trichinella* spp., hermaphroditic flukes) infect numerous animal species that consequently serve as the reservoir for human infection. In some cases (i.e., certain microsporidia), the reservoir for a particular parasite has not been determined. Classification of the more commonly found parasites affecting humans is presented in Table 3.

While control of the obligate human parasites may theoretically be achievable, we can exert relatively little control over the zoonotic parasites without threatening extinction of their host species. Some species of parasites are actually free-living organisms that may infect humans in an opportunistic fashion, causing serious or fatal disease. Amebae of the genera *Naegleria*, *Acanthamoeba*, and *Balamuthia* are common inhabitants of water and soil and are known to cause encephalitis, keratitis, and systemic infections (depending on the genus), often in individuals weakened by preexisting disease. Control of such organisms is not possible because of their free-living life cycle and ubiquity in the environment.

Parasites are transmitted by several mechanisms: ingestion of an infective stage, direct penetration of skin by infective larvae, and inoculation via an arthropod vector. In the simplest life cycle, a parasite stage from humans is immediately infectious to other humans following inges-

TABLE 1 Incidence of most commonly identified intestinal parasites in fecal specimens examined by state health department laboratories, 1987[a]

Parasite	% of specimens positive
Giardia lamblia	7.2
Blastocystis hominis	2.6
Hookworms	1.5
Trichuris trichiura	1.2
Entamoeba histolytica	0.9
Ascaris lumbricoides	0.8
Clonorchis and Opisthorchis spp.	0.6
Dientamoeba fragilis	0.5
Strongyloides stercoralis	0.4
Enterobius vermicularis	0.4
Hymenolepis nana	0.4
Cryptosporidium parvum	0.2
Commensal protozoa	10.8

[a]Adapted from reference 12.

tion; this is true of certain of the intestinal parasites, including *Entamoeba histolytica*, *Giardia lamblia*, *Cryptosporidium parvum*, *Enterobius vermicularis*, and *Hymenolepis nana*. The increased prevalence of these infections in children reflects the ease with which transmission occurs directly from person to person. In infections such as ascariasis or trichuriasis, an additional maturation period outside of the host is required before the parasite egg is infective. Children, however, are also at high risk of acquiring these parasites by handling and ingesting fecally contaminated soil.

Ingestion of infective stages may also occur coincidentally with ingestion of parasite vectors (*Dipylidium*, *Dracunculus*, and *Hymenolepis* spp.), plants (*Fasiolopsis* and *Fasciola* spp.), or animal products (*Toxoplasma*, *Taenia*, *Diphyllobothrium*, *Clonorchis*, *Heterophyes*, *Metagonimus*, *Paragonimus*, *Nanophyetus*, and *Trichinella* spp.).

In other parasitic diseases, the parasite directly penetrates the skin. Hookworm disease, strongyloidiasis, and schistosomiasis are examples of diseases in which larval stages actively penetrate the skin and migrate through various body tissues before they mature and initiate egg laying. Diseases actively or passively transmitted by arthropod vectors include malaria, babesiosis, leishmaniasis, trypanosomiasis, and filariasis.

Measures to eradicate parasitic infections vary tremendously depending on the species involved and must include efforts at prevention and control as well as treatment. Person-to-person spread, above all else, requires attention to personal hygiene, especially adequate hand washing and cleaning of fingernails. Improved hygiene in food production and preparation, including thorough cooking, eliminates most food-associated parasites. Proper sewage treatment or use of latrines along with widespread availability of potable water is mandatory in order to dramatically reduce parasitic diseases transmitted in feces or urine. Elimination of insect vectors and use of residual insecticides continues to play an important role in control of the arthropod-transmitted diseases. For the traveler, however, personal protection by using insect repellents, netting, and appropriate clothing will best limit the opportunities for exposure to blood- and tissue-feeding arthropods.

Much of the earlier enthusiasm surrounding the possible eradication of malaria has been tempered by the realization that *Plasmodium falciparum* and, more recently, *Plasmodium*

TABLE 2 Prevalence of parasitic infections worldwide[a]

Source and disease	Estimated population involved	Annual no. of deaths
Protozoan		
Amebiasis	10% of world population	40,000–110,000
African trypanosomiasis	100,000 new cases/yr	5,000
American trypanosomiasis	24 million	60,000
Giardiasis	200 million	
Leishmaniasis	1.2 million	
Malaria	400–490 million	2.2–2.5 million
Helminthic		
Ascariasis	1 billion	1,550
Cestodiases	65 million	
Clonorchiasis-opisthorchiasis	19 million	
Dracunculiasis	10 million	
Fasciolopsiasis	10 million	
Filariasis	85–110 million	
Hookworm	900 million	
Onchocerciasis	>30 million	
Paragonimiasis	3.2 million	
Schistosomiasis	>200 million	500,000–1 million
Strongyloidiasis	35 million	
Trichostrongyliasis	5.5 million	
Trichuriasis	500–800 million	

[a]Adapted from reference 10.

TABLE 3 Summary of more commonly found human parasites and their primary sites of infection[a]

Subkingdom *Protozoa*	Subkingdom *Metazoa*
Phylum *Sarcomastigophora*	Phylum *Platyhelminthes* (flatworms)
Subphylum *Sarcodina* (amebae)	Class *Cestoidea* (tapeworms)
Entamoeba histolytica (I)[b]	*Diphyllobothrium* spp. (I)[b]
Entamoeba hartmanni (I)	*Dipylidium caninum* (I)[b]
Entamoeba coli (I)	*Echinococcus granulosus* (H)[b]
Iodamoeba butschlii (I)	*Echinococcus multilocularis* (H)[b]
Endolimax nana (I)	*Hymenolepis nana* (I)[b]
Naegleria fowleri (T, C)[b]	*Hymenolepis diminuta* (I)[b]
Acanthamoeba spp. (T, C, E)[b]	*Taenia saginata* (I)[b]
Balamuthia mandrillaris (T, C)[b]	*Taenia solium* (I, T)[b]
Blastocystis hominis (I)	Class *Trematoda* (flukes)
Subphylum *Mastigophora* (flagellates)	*Clonorchis sinensis* (H)
Giardia lamblia (I)[b]	*Fasciola hepatica* (H)[b]
Dientamoeba fragilis (I)[b]	*Fasciolopsis buski* (I)
Chilomastix mesnili (I)	*Heterophyes heterophyes* (I)
Retortamonas intestinalis (I)	*Metagonimus yokogawai* (I)
Enteromonas hominis (I)	*Nanophyetus salmincola* (I)[b]
Trichomonas hominis (I)	*Opisthorchis viverrini* (H)
Trichomonas vaginalis (V)[b]	*Paragonimus* spp. (L)[b]
Leishmania tropica (T)	*Schistosoma haematobium* (B)
Leishmania major (T)	*Schistosoma japonicum* (B)
Leishmania aethiopica (T)	*Schistosoma mekongi* (B)
Leishmania mexicana (T)[b]	*Schistosoma mansoni* (B)
Leishmania braziliensis (T)	
Leishmania donovani (H, T)	Phylum *Nematoda* (roundworms)
Trypanosoma brucei gambiense (B, C)	Class *Adenophorea* (Aphasmidia)
Trypanosoma brucei rhodesiense (B, C)	*Trichinella spiralis* (T, I)[b]
Trypanosoma cruzi (B, T)[b]	*Trichuris trichiura* (I)[b]
	Capillaria philippinensis (I)
Phylum *Ciliophora* (ciliates)	Class *Secernentia* (Phasmidia)
Balantidium coli (I)[b]	*Enterobius vermicularis* (I)[b]
	Ascaris lumbricoides (I)[b]
Phylum *Apicomplexa* (apicomplexans)	*Ancylostoma duodenale* (I)
Class *Sporozoea*	*Necator americanus* (I)[b]
Subclass *Piroplasmea*	*Strongyloides stercoralis* (I)
Babesia spp. (B)[b]	*Trichostrongylus* spp. (I)[b]
Subclass *Coccidia*	*Anisakis* spp. (I)[b]
Plasmodium falciparum (B)	*Wuchereria bancrofti* (T, B)
Plasmodium malariae (B)	*Brugia malayi* (T, B)
Plasmodium ovale (B)	*Loa loa* (T, B)
Plasmodium vivax (B)	*Onchocerca volvulus* (T)
Cryptosporidium parvum (I)[b]	*Mansonella perstans* (T, B)
Cyclospora cayetanensis (I)[b]	*Mansonella ozzardi* (T, B)
Isospora belli (I)[b]	*Mansonella streptocerca* (T)
Toxoplasma gondii (T)[b]	*Dracunculus medinensis* (T)
	Angiostrongylus cantonensis (T)
Phylum *Microspora* (microsporidia)	*Angiostrongylus costaricensis* (T)
Encephalitozoon spp. (E, H, T)[b]	
Enterocytozoon bieneusi (I)[b]	
Septata intestinalis (I, T)[b]	

[a] Abbreviations: B, blood; C, spinal fluid; E, eye; H, liver; I, intestine; L, lung; T, tissue; V, vagina.
[b] Pathogenic parasite that occurs naturally in the United States.

vivax display resistance to commonly used antimalarial agents and that some of the anopheline mosquito vectors have become resistant to the commonly used insecticides. Also, schistosomiasis appears to be spreading in some areas as a result of attempts to build dams and improve irrigation systems to feed burgeoning populations. Recent efforts to develop vaccines by using recombinant DNA technologies offer the hope that these diseases will someday be controlled. While advances are being made and field trials are under way for several vaccines, progress has been slow, partly because of the biochemical and immunological complexity of working with these eukaryotic organisms.

LABORATORY APPROACHES FOR DIAGNOSIS OF PARASITES

The diagnosis of parasitic infections almost always depends on laboratory methods, and close communication between laboratory and physician is necessary to ensure that the laboratory is aware of the physician's expectations and that the physician is aware of test limitations. For intestinal and blood parasites, morphologic demonstration of a diagnostic stage(s) is the principal means of diagnosis, whereas for tissue infections, immunodiagnostic techniques are generally more useful.

Timing of specimen acquisition is an important consideration, because symptoms in a patient may occur before the appearance of diagnostic forms (prepatent period). Both ascariasis and schistosomiasis may be responsible for symptomatic diseases, i.e., Löffler's pneumonia and Katayama fever, respectively, before eggs appear in clinical specimens. In such patients, the physician may suspect parasitic infection, but the tentative diagnosis must be based on clinical and historical impressions or on immunodiagnostic tests. Definitive diagnosis must await detection of diagnostic stages, which means that a certain amount of time must pass. Many parasites, both in stool and in blood, appear continuously during an infection, and specific timing for specimen collection is not necessary. Others, however, have predictable variations in their appearance (malaria, microfilariae) or appear sporadically at best (*Giardia* cysts, *Taenia* eggs). Specific details are presented in chapters 102, 104, and 106 of this Manual.

In establishing a diagnosis, the clinician places a great deal of trust in the laboratory. This trust can be misplaced if laboratory personnel are not fully trained to identify or exclude parasites. At a minimum, laboratories involved in the diagnosis of parasites should be able to properly perform complete stool and blood examinations. The literature clearly documents outbreaks that have been overlooked because of incompetent laboratory diagnosis and pseudo-outbreaks that have occurred because of misidentification of inflammatory cells or artifacts as parasites (8, 11). The results of proficiency testing programs also suggest that some laboratories have difficulty with the identification of certain parasites, especially the intestinal protozoa (15, 17). A well-functioning quality assurance program, both internal and external, is crucial to high-quality parasitology testing.

A large variety of immunodiagnostic tests have been described and are available for the detection of either antibodies or antigens to specific parasites. Many of these tests are commercially available and provide additional information to augment the routine parasite examinations. Application of such tests, however, requires careful consideration as to their usefulness in clinical decision making. A variety of recently developed gene amplification and DNA probe assays for the detection of parasitic infections show promise of enhanced sensitivity and specificity over existing methods for certain infections. Until these methodologies and costs are within the capabilities of most clinical laboratories, they will remain as research tools. Immunodiagnostic and molecular methods are more thoroughly discussed in chapter 103 of this Manual.

The following chapters describe widely used methods for the identification and diagnosis of parasitic diseases. Those wishing more detailed information about clinical parasitology are referred to a number of well-known texts (3–6, 9, 10, 20). A variety of other publications are available to further assist laboratorians in the evaluation of clinical specimens and the identification of possible parasites (1, 2,

5, 13, 14, 18, 19). Descriptive morphology of parasites as seen in histologic sections is summarized in several other recent books (7, 16, 21, 22).

REFERENCES

1. **Ash, L. R., and T. C. Orihel.** 1987. *Parasites: a Guide to Laboratory Procedures and Indentification.* ASCP Press, American Society of Clinical Pathologists, Chicago.
2. **Ash, L. R., and T. C. Orihel.** 1990. *Atlas of Human Parasitology,* 3rd ed. American Society of Clinical Pathologists Press, Chicago.
3. **Beaver, P. C., R. C. Jung, and E. W. Cupp.** 1984. *Clinical Parasitology,* 9th ed. Lea & Febiger, Philadelphia.
4. **Binford, C. H., and D. H. Connor.** 1976. *Pathology of Tropical and Extraordinary Diseases.* Armed Forces Institute of Pathology, Washington, D.C.
5. **Garcia, L. S., and D. A. Bruckner.** 1993. *Diagnostic Medical Parasitology.* American Society for Microbiology, Washington, D.C.
6. **Goldsmith, R., and D. Heyneman (ed.).** 1989. *Tropical Medicine and Parasitology.* Appleton and Lange, Norwalk, Conn.
7. **Gutierrez, Y.** 1989. *Diagnostic Pathology of Parasitic Infections, with Clinical Correlations.* Lea & Febiger, Philadelphia.
8. **Krogstad, D. J., H. C. Spencer, Jr., G. R. Healy, N. N. Gleason, D. J. Sexton, and C. A. Herron.** 1978. Amebiasis: epidemiologic studies in the United States, 1971–1974. *Ann. Intern. Med.* **88:**89–97.
9. **Manson-Bahr, P. E. C., and D. R. Bell (ed.).** 1987. *Manson's Tropical Diseases,* 19th ed. The W. B. Saunders Co., Philadelphia.
10. **Markell, E. K., and M. Voge.** 1992. *Medical Parasitology,* 7th ed. The W. B. Saunders Co., Philadelphia.
11. **Morbidity and Mortality Weekly Report.** 1985. Pseudo-outbreak of intestinal amebiasis—California. *Morbid. Mortal. Weekly Rep.* **34:**125–126.
12. **Morbidity and Mortality Weekly Report.** 1991. Results of testing for intestinal parasites by state diagnostics laboratories, United States, 1987. *Morbid. Mortal. Weekly Rep.* **40:**25–45.
13. **National Committee for Clinical Laboratory Standards.** 1992. *Slide Preparation and Staining of Blood Films for the Laboratory Diagnosis of Parasitic Diseases; Tentative Guideline.* NCCLS document M15-T. National Committee for Clinical Laboratory Standards, Villanova, Pa.
14. **National Committee for Clinical Laboratory Standards.** 1993. *Procedures for the Recovery and Identification of Parasites from the Intestinal Tract; Proposed Guideline.* NCCLS document M28-P. National Committee for Clinical Laboratory Standards, Villanova, Pa.
15. **National Research Council, Office of International Affairs, Board on Sciences and Technology for International Development, and the Institute of Medicine, National Academy of Sciences.** 1987. *The U.S. Capacity to Address Tropical Infectious Disease Questions,* p. 93–96. National Academy Press, Washington, D.C.
16. **Orihel, T. C., and L. R. Ash.** 1994. *Parasites in Human Tissues.* ASCP Press, Chicago.
17. **Smith, J. W.** 1979. Identification of fecal parasites in the special parasitology survey of the College of American Pathologists. *Am. J. Clin. Pathol.* **72**(Suppl.)**:**371–373.
18. **Smith, J. W., and Y. Gutierrez.** 1991. Medical parasitology, p. 1158–1217. *In* J. B. Henry (ed.), *Clinical Diagnosis and Management by Laboratory Methods,* 18th ed. The W. B. Saunders Co., Philadelphia.
19. **Spencer, F. M., and L. S. Monroe.** 1982. *The Color Atlas of Intestinal Parasites,* 2nd ed. Charles C Thomas, Publisher, Springfield, Ill.
20. **Strickland, G. T. (ed.).** 1991. *Tropical Medicine,* 7th ed. The W. B. Saunders Co., Philadelphia.
21. **Von Lichtenberg, F.** 1991. *Pathology of Infectious Diseases.* Raven Press, New York.
22. **Woods, G. L., Y. Gutierrez, D. H. Walker, D. T. Purtilo, and J. D. Shanley.** 1993. *Diagnostic Pathology of Infectious Diseases.* Lea & Febiger, Philadelphia.

Diagnosis of Parasitic Infections: Collection, Processing, and Examination of Specimens

LYNNE S. GARCIA, SANDRA BULLOCK-IACULLO, JOSEPHINE PALMER, AND
ROBYN Y. SHIMIZU

102

Laboratory results based on parasite recovery and identification depend on proper specimen collection, fixation, and processing. With cost containment and clinical relevance becoming more important issues, specimen rejection criteria have become more strict. As a part of any overall total quality management or continuous quality improvement program for the laboratory, the generation of test results must begin with stringent criteria for specimen collection, fixation, and processing; results based on improperly handled specimens may lead to wasted resources and misleading reports (14, 18, 28). Specific quality control recommendations can be found in the *Clinical Microbiology Procedures Handbook* (18). Universal precautions as outlined by the Occupational Safety and Health Act should be used at all times; diagnostic parasitology procedures require the same safety guidelines as procedures in any other area of microbiology (8, 27).

FECAL SPECIMENS

Collection of Specimens

Certain substances and medications can interfere with the detection of intestinal protozoa: mineral oil, barium, bismuth, antibiotics, antimalarial agents, and nonabsorbable antidiarrheal preparations. Parasites may not be recovered for 1 to several weeks after administration of any of these compounds. After barium and/or antibiotics are administered, specimen collection should be delayed for 5 to 10 days and for at least 2 weeks, respectively (14, 24).

Fecal specimens should be collected in clean, dry, wide-mouth containers with tight-fitting lids. The specimens should not be contaminated with water or urine, because water may contain free-living organisms that can be mistaken for human parasites, and urine may destroy motile organisms. For safety reasons, stool specimen containers should be placed in plastic bags for transport to the laboratory for testing. *A series of three specimens is considered a minimum for an adequate examination* (14, 29). Since numbers of organisms and identities of species may vary from day to day, it is recommended that the three specimens be collected on separate days (every other day if possible) or that the series of three specimens be collected within no more than 10 days. If a patient has severe watery diarrhea, multiple specimens can be accepted on a single day; any

organisms present might be missed owing to a dilution factor related to fluid loss (14). One or more specimens per day taken over several days may be appropriate, and more than three specimens may be needed to rule out parasitic infection. Eliminating ova and parasite (O and P) examinations for patients hospitalized for more than 3 days may become an accepted procedure (26). Three posttherapy specimens should be collected on alternate days once therapy is completed. Specimens collected during therapy may give false-negative results and should not be considered indicative of a cure.

Fresh Stool Specimens

Freshly passed specimens are mandatory for the recovery of motile trophozoites (amebae, flagellates, or ciliates), which are normally found in cases of diarrhea; the gastrointestinal tract contents are moving through the system too rapidly for cysts to form. Once the stool specimen is passed from the body, trophozoites do not encyst but will disintegrate if not examined or preserved within a short time. Liquid specimens should be examined *within 30 min of passage*, not 30 min from the time they reach the laboratory, or they should be preserved. Soft (semiformed) specimens may contain both protozoan trophozoites and cysts and should be examined within 1 h of passage; again, if this time frame is not possible, then preservatives should be used. Formed stools usually contain few trophozoites, so immediate examination is not as critical, but rapid fixation of the specimen immediately after passage is recommended for recovery and identification of intestinal protozoa (29). The importance of preserving organisms before distortion or disintegration occurs outweighs the limited motility information that might be gained by examining fresh specimens.

Macroscopic Examination

Stool consistency (formed, soft, loose, or watery) may indicate the types of organisms present. Trophozoites are usually found in loose or watery specimens, both trophozoites and cysts are usually found in a soft specimen, and cysts are generally found in formed specimens. Helminth eggs, coccidian oocysts, and microsporidian spores can be found in any type of fecal specimen; in the case of coccidian oocysts, the more liquid the stool, the more oocysts are present. Tapeworm proglottids can migrate and may be found on or beneath the stool or anywhere in the collection

TABLE 1 Preservatives

Preservative[a]	Advantages	Disadvantages
5 or 10% formalin (buffered or not buffered)	Good stool fixative for concentrates; easy to prepare; long shelf life. Concentrated sediments can be used with new monoclonal detection kits and for acid-fast stains.	Does not preserve trophozoites well; does not preserve organism morphology for acceptable permanent stained smear
MIF	Components both fix and stain organisms for wet mounts; easy to prepare; long shelf life; contains no mercury compounds; useful for field surveys	Organism morphology on permanent stained smears generally not as good as that seen with Schaudinn's fluid or PVA (mercuric chloride base)
SAF	Used for concentration and permanent stained smears; contains no mercury compounds; easy to prepare; long shelf life. Concentrated sediment can be used with new monoclonal detection kits or for acid-fast stains.	Albumin-glycerin required for adhesion of specimen to slide; organism morphology on stained smears better with iron hematoxylin stains than with trichrome
Schaudinn's fixative	Used for smears prepared from fresh stools or specimens from intestinal mucosal surfaces; excellent preservation of protozoan trophozoites and cysts	Not recommended for use in concentration procedures; contains mercuric chloride; poor adhesive qualities with liquid or mucoid specimens
PVA	Excellent preservation of protozoan trophozoites and cysts for permanent stained smears, which can be prepared in ~1 h or wk or some mo later; good adhesion to slide; specimen can be concentrated but not as well as with formalin-preserved specimens; long shelf life	Contains mercuric chloride (Schaudinn's fluid); difficult to prepare in laboratory; not as good as formalin for concentration procedures; cannot use specimens for monoclonal detection kits or acid-fast stains
Modified PVA (copper, zinc bases)	Can prepare permanent stained smears and perform concentration techniques; does not contain mercury compounds	Protozoan morphology poor when preserved in copper sulfate-based PVA; morphology better with zinc sulfate-based PVA; organisms may be difficult to see, but with experience, can be identified

[a]MIF, merthiolate-iodine-formalin; SAF, sodium acetate-acetic acid-formalin; PVA, Schaudinn's fixative plus PVA, a synthetic resin that helps stool adhere to the slide; modified PVA, PVA fixative prepared with modified Schaudinn's fluid (either zinc sulfate or copper sulfate is substituted for mercuric chloride).

container, and adult pinworms and *Ascaris* spp. may occasionally be found on the surface or within the stool.

In certain parasitic infections, blood and mucus may be present in the stool; areas of the specimen that contain blood and mucus should be carefully examined for the presence of amebic trophozoites. Diarrheic specimens and those containing obvious amounts of mucus should not be filtered during processing; parasites may be trapped in the mucus.

Dark stools may indicate bleeding high in the gastrointestinal tract, whereas fresh (bright red) blood usually suggests bleeding in the colon. Occult blood in the stool may or may not be related to a parasitic infection. Various compounds may also give color to the stool (iron and bismuth, black; barium, light tan to white).

Stool Specimen Preservatives

If there is a delay between the time the specimen is passed and the time it is received in the laboratory or will be processed, collection kits with stool preservatives are recommended (29). Collection kits should be convenient and easy to use and should have easily understood instructions, preferably in several languages and with diagrams. Improper collection results in inappropriate specimens and wasted

laboratory resources. Table 1 lists the most commonly used preservatives, and Table 2 summarizes their usefulness for the different laboratory procedures.

Microscopic Examination

The O and P examination comprises three techniques: direct wet mount, concentration, and permanent stained smear. Each technique is designed for a particular purpose and forms an integral part of the total examination, although changes have recently been recommended (14, 15, 24, 25, 29). Many laboratories receive all fecal specimens in preservatives. This eliminates the delay between specimen passage and fixation, thus providing overall better organism morphology. Since organisms in preservative do not exhibit motility, the direct wet mount is no longer a mandatory part of the routine O and P examination. However, if fresh fecal specimens are received, motile organisms may be seen in the direct wet mount, particularly if the stool is loose or watery.

Direct Wet Mount

The direct wet mount (Table 3) is used primarily to detect motile protozoan trophozoites; detecting any parasite depends on the number of organisms present and the tech-

TABLE 2 Stool preservatives and procedures

Specimen[a]	Direct wet mount	Concn	Permanent stained smear[b]
Fresh stool	Yes	Yes	Yes (IH, T)
Schaudinn's[c]	No	No	Yes (IH, T)
5 or 10% formalin	No (no motility)	Yes	No
MIF	No (no motility)	Yes	No[d]
SAF	No (no motility)	Yes	Yes (IH, T)
PVA	No	Yes[e]	Yes (IH, T)
Modified PVA	No	Yes[e]	Yes

[a]Abbreviations are as defined in the footnote to Table 1.

[b]IH, iron hematoxylin; T, trichrome.

[c]Schaudinn's fixative is used for fresh stool smeared onto the slide and as the base fixative for PVA and provides the best definition of internal morphology for protozoa.

[d]This fixative can be used for concentration; permanent stained smears are sometimes prepared with polychrome IV stain.

[e]Although specimens preserved in PVA can be concentrated, some organisms do not concentrate well (G. lamblia, Trichuris trichiura), exhibit poor morphology (S. stercoralis), or may not be visible at all (I. belli).

niques used. These organisms are very pale and transparent, two characteristics that make the use of low-intensity light a necessity. The direct wet mount is prepared by mixing a small amount of stool (a low cone on the end of a wooden applicator stick; about 2 mg) with a drop of 0.85% NaCl for examination under a coverslip (22 by 22 mm). Any blood or mucus should be examined as a direct wet mount. The coverslip should be systematically examined under the low-power (10×) objective with low-intensity light; any suspicious objects may then be examined under the high dry objective (40×). Even if nothing suspicious is seen on low power, at least one-third of the coverslip area should be examined under the 40× (high dry) objective. Use of the oil immersion objective (100×) on mounts of this kind is not recommended. Protozoan organisms in a saline preparation are usually refractile; if suspicious objects are seen with high dry power, allow at least 15 s for detecting motility of slow-moving protozoa. Heat applied by placing

a hot penny on the edge of a slide may enhance the motility of trophic protozoa.

After checking for trophic amebae, add a drop of iodine at the edge of the coverslip, or prepare a new wet mount with iodine alone. Lugol's and D'Antoni's iodine are recommended; Gram's iodine is not recommended for staining parasites. Once iodine is added to the preparation, the trophozoites will be killed, and motility will be lost. The purpose of the iodine is to enhance the visibility of nuclei and glycogen vacuoles in any cysts that are present.

If preserved specimens are received, it is more cost-effective to omit the direct wet mount and begin the stool examination with the concentration procedure, particularly since motility ceases with the addition of preservative. Results obtained from wet mount examinations should be confirmed by permanent stained smears. Findings from the direct wet mount examination can be reported as preliminary, and the final report can then be submitted after the concentration and permanent stain procedures are completed.

Concentration (Sedimentation, Flotation)

Fecal concentration (Table 4) is a part of the complete O and P examination and allows the detection of small numbers of organisms that may be missed when only a direct wet smear is used (13, 24, 29, 34). There are two types of concentration procedures: sedimentation and flotation. Both are designed to separate protozoan cysts and oocysts, microsporidian spores, and helminth eggs and larvae from fecal debris by centrifugation and/or differences in specific gravity (29).

Biphasic sedimentation methods result in removal of most of the fecal debris by use of a solvent, most commonly ethyl acetate, while protozoan cysts, oocysts, eggs, and larvae are sedimented during centrifugation. The sedimentation procedure is recommended as being the easiest to perform, least subject to technical error, and suitable for both small- and large-volume laboratories. In order to concentrate the smaller organisms, i.e., coccidia and microsporidia, the centrifuge should reach the recommended speed before monitoring of the centrifugation time begins (500 ×

TABLE 3 Direct wet mount: a review

Topic	Comments
Principle	To assess worm burden of patient, provide quick diagnosis of heavily infected specimen, and detect motile protozoa
Specimen	Any fresh stool specimen that has not been refrigerated
Reagents	0.85% NaCl, Lugol's or D'Antoni's iodine
Examination	Low power (100×) for entire coverslip (22 by 22 mm) preparation (both saline and iodine); high dry power (400×) for at least 1/3 of coverslip area (both saline and iodine)
Results and laboratory reports	Results should be considered presumptive; however, some organisms could be definitively identified (G. lamblia cysts and Entamoeba coli cysts, helminth eggs and larvae, I. belli oocysts). Results should be categorized as preliminary; final report would be available after concentration and permanent stained smear were examined.
Procedure notes and limitations	Once iodine is added to preparation, organisms are killed and motility is lost, but nuclei of cysts may be enhanced. Direct wet mount examination is not necessary for specimens received in preservative; examination of concentration and permanent stained smear is usually sufficient. Direct wet mounts are normally examined at low power (100×) and high dry power (400×).

TABLE 4 Concentration: a review

Topic	Comments
Principle	Concentrates parasites through sedimentation or flotation; designed to allow recovery of protozoan cysts and oocysts, microsporidian spores, and helminth eggs and larvae
Specimen	Any stool specimen that is fresh or is preserved in formalin, SAF, or MIF; PVA-fixed specimens can also be concentrated by sedimentation but are not optimal.
Reagents	5 or 10% formalin (buffered or nonbuffered), ethyl acetate, zinc sulfate (specific gravity of 1.18 for fresh stool or 1.20 for preserved stool), 0.85% NaCl, Lugol's or D'Antoni's iodine
Examination	Low power (100×) for entire coverslip (22 by 22 mm) preparation (iodine optional); high dry power (400×) for at least 1/3 of coverslip area (both saline and iodine)
Results and laboratory reports	Concentration results should be considered presumptive; however, some organisms could be definitively identified (*G. lamblia* cysts and *Entamoeba coli* cysts, helminth eggs and larvae, *I. belli* oocysts). Reports are categorized as preliminary; final report would be available after concentration and permanent stained smear were examined.
Procedure notes and limitations	Formalin-ethyl acetate sedimentation concentration is commonly used. Zinc sulfate flotation will not detect operculated or heavy eggs; both surface film and sediment should be examined before negative result is reported. Smears prepared from concentrated stool are normally examined at low power (100×) and high dry power (400×). Too much iodine may obscure helminth eggs (will mimic debris).

g for 10 min) (14, 18, 29). If the centrifugation time at the proper speed is reduced for any of the centrifugation steps, these parasites may not sediment properly and may be discarded with the fecal debris.

Flotation concentrations permit the separation of protozoan cysts, coccidian oocysts, and certain helminth eggs and larvae by use of a liquid (zinc sulfate) with a specific gravity higher than that of the organisms. Parasites float to the surface, while debris remains in the bottom of the tube. Although flotation methods yield a cleaner preparation than sedimentation, some helminth eggs (operculated eggs and/or very dense eggs, e.g., unfertilized *Ascaris* eggs) do not float and would be missed. Flotation preparations must be examined immediately after centrifugation; to ensure detection of all organisms in the flotation procedure, both the surface film and the sediment should be examined. Concentrated specimens are examined as wet mounts with or without iodine in the same way as direct wet smears.

Permanent Stained Smears (Trichrome, Iron Hematoxylin)

Detection and identification of intestinal protozoa frequently depend on examination of a permanent stained smear under the oil immersion lens (100× objective) (Table 5). These slides provide a permanent record of organisms identified and may be used for consultations and teaching. Many identifications are tentative until confirmed by the permanent stained slide. Protozoa may be seen on the stained smear when they are totally missed by the direct smear and concentration methods. *The permanent stain is recommended for every stool sample submitted for a routine parasite examination.* For routine diagnostic work, most laboratories select the trichrome method or one of the modified methods with iron hematoxylin (12, 20, 25, 29, 30).

Staining problems occur because the specimen is too old when placed in fixative or stained, the smears are too dense, and/or the specimen is not fixed properly. Fresh specimens should not be allowed to dry before fixation, whereas poly-

vinyl alcohol (PVA)-fixed specimens must dry thoroughly before staining. Parasite variation also causes problems; immature cysts fix more easily than old cysts, and *Entamoeba coli* cysts require a longer fixation time than those of other species. Fixation for 1 h is usually adequate.

Trichrome

The trichrome technique of Wheatly (29) for fecal specimens is a modification of Gomori's original tissue stain (29). The trichrome stain is a rapid, simple procedure that produces uniformly well stained smears of intestinal protozoa, human cells, yeast cells, and artifact material in approximately 45 min or less. Specimens consist of fresh stool fixed in Schaudinn's fixative or PVA-preserved stool; trichrome is generally not as good with sodium acetate-acetic acid-formalin (SAF)- and merthiolate-iodine-formalin (MIF)-preserved specimens.

Iron Hematoxylin

Iron hematoxylin stain was used for most of the original morphologic descriptions of intestinal protozoa (29). A new method now available incorporates a carbol fuchsin step in the procedure; when the smears are stained, acid-fast organisms (*Cryptosporidium parvum*, *Cyclospora cayetanensis*, and *Isospora belli*) as well as the more traditionally staining protozoa can be seen (14, 30). Iron hematoxylin stains can be used on SAF-, Schaudinn-, or PVA-fixed specimens (38).

Modified Acid-Fast Smears

Cryptosporidium parvum, *I. belli*, and *Cyclospora cayetanensis* can cause severe diarrhea in both immunocompromised and immunocompetent hosts (10, 14, 21, 22). Oocysts may be difficult to detect without special staining (14). Fluorescent stains such as auramine-rhodamine, auramine-carbol fuchsin, and acridine orange can be used as screening techniques (10). However, modified acid-fast stains are recom-

TABLE 5 Permanent stained smears: a review

Topic	Comments
Principle	Provides contrasting colors for background debris and parasites; allows examination and recognition of detailed organism morphology under oil immersion (1,000×); allows recovery and identification of intestinal protozoa
Specimen	Stool specimens (fresh or preserved with Schaudinn's, PVA, SAF, or MIF). Formalin-fixed specimens cannot be used for permanent staining.
Reagents	Trichrome, iron hematoxylin, modified iron hematoxylin, or polychrome IV stains and associated solutions; dehydrating solutions (alcohols and xylenes); mounting fluid optional; immersion oil
Examination	Oil immersion (1,000×) for at least 300 fields. Additional fields may be required if suspect organisms have been seen in wet preparations from concentrated specimen. Screening can be done with 50× or 60× oil immersion objective, but identifications should be confirmed with 100× oil immersion objective.
Results and laboratory reports	Most suspect protozoa and/or human cells can be identified or confirmed with permanent stained smear. These reports should be categorized as final (direct wet smear and concentration examination provide preliminary results).
Procedure notes and limitations	Most common stains include trichrome and iron hematoxylin. Permanent stained smears are normally examined under oil immersion (100× objective). Identification of intestinal trophozoites and cysts is primary purpose of this technique. Unfortunately, helminth eggs and larvae take up too much stain, may be distorted, and are more easily identified in direct or concentrated wet mounts.

mended to confirm these organisms (Table 6), and a hot method (Ziehl-Neelsen modified acid-fast stain) (14, 18) or a cold method (Kinyoun's modified acid-fast stain) (14, 18) can be used. Although modified acid-fast stains are useful, the newer monoclonal reagents for *Cryptosporidium* spp. are more sensitive (1, 10, 13, 33). The sizes of microsporidial spores (1 to 2 μm) make their identification very difficult without special stains or the use of monoclonal or poly-

TABLE 6 Modified acid-fast stains: a review

Topic	Comments
Principle	Provides contrasting colors for background debris and parasites; allows recognition of acid-fast characteristics of organisms such as *Cryptosporidium*, *Cyclospora*, and *Isospora* spp. under high dry examination (40× objective). Internal morphology (sporozoites) is seen in some *Cryptosporidium* oocysts under oil immersion (100× objective).
Specimen	Any stool specimen that is fresh or is preserved in formalin or SAF
Reagents	Kinyoun's acid-fast stain, modified Ziehl-Neelsen stain, and associated solutions; dehydrating solutions (alcohols and xylenes); mounting fluid optional. Decolorizing agents are less intense than routine acid-alcohol used in routine acid-fast staining (this fact makes these procedures "modified" acid-fast procedures).
Examination	High dry for at least 300 fields; additional fields may be required if suspect organisms have been seen but are not clearly acid fast. Sporozoites may be seen under oil immersion objective (100×).
Results and laboratory reports	Identification of *Cryptosporidium* and *Isospora* oocysts should be possible; *Cyclospora* oocysts are twice the size of *Cryptosporidium* oocysts and should be visible but tend to be more acid-fast variable. Quality control slides of known positive specimens should be stained and compared with patient specimens.
Procedure notes and limitations	Both cold and hot modified acid-fast methods are excellent for staining of coccidian oocysts. Procedure limitations are related to specimen handling (proper filtration of specimen, use of recommended centrifugation speeds and time). Also, there is some controversy concerning whether or not organisms lose ability to take up acid-fast stains after long-term storage in 10% formalin. Organisms are more difficult to find in specimens from patients who do not have typical watery diarrhea (more formed stool = more artifact material and fewer oocysts).

TABLE 7 Modified trichrome (chromotrope 2R)-stained smears: a review

Topic	Comments
Principle	Provides contrasting colors for background debris and parasites; allows examination and recognition of organism morphology under oil immersion (1,000×); allows recovery and identification of microsporidial spores. Internal morphology (horizontal or diagonal "stripes") may be seen in some spores under oil immersion (100× objective).
Specimen	Any stool specimen that is fresh or is preserved in formalin or SAF
Reagents	Modified trichrome stain (using high-dye-content chromotrope 2R) and associated solutions, dehydrating solutions (alcohols and xylenes), mounting fluid optional
Examination	Oil immersion for at least 300 fields; additional fields may be required if suspicious objects are seen but not clearly identified
Results and laboratory reports	Identification of microsporidial spores possible, but their small size makes recognition difficult. Final results depend on appearance of quality control slides compared with that of patient specimens.
Procedure notes and limitations	Because of difficulty in getting dye to penetrate spore wall, this staining approach can be very helpful. Limitations are related to specimen handling (proper filtration of specimens, use of recommended centrifugation speeds and time) and complete understanding of difficulties in recognizing microsporidial spores because of their small size (1–1.5 μm).

clonal antibody reagents; acid-fast methods are not routinely used for these spores (16, 39).

Concentrated sediment of fresh, formalin-preserved, or SAF-preserved stool may be used. Other clinical specimens such as duodenal fluid, bile, and pulmonary specimens (induced sputum, bronchial washings, biopsy samples) may also be stained.

Modified Trichrome Stain for Microsporidia

Although microsporidia have been found in most body sites, the current laboratory emphasis is on intestinal specimens (6, 11, 20, 32). The diagnosis of intestinal microsporidiosis (*Enterocytozoon bieneusi, Septata intestinalis*) often depends on invasive procedures and subsequent examination of biopsy specimens by electron microscopy (4). The need for a practical method for the routine clinical laboratory has led to the development of newer methods. Slides prepared from fresh or formalin-fixed stool specimens can be stained by using new chromotrope-based techniques (Table 7) and can be examined with light microscopy. These staining methods are based on the fact that stain penetration of the microsporidial spore is very difficult; thus, the dye content of the chromotrope 2R is higher (10-fold) than that of the trichrome routinely used for stools, and the staining time is generally much longer (90 min) (14, 19a, 31, 36).

ADDITIONAL TECHNIQUES FOR STOOL EXAMINATION

In addition to the routine O and P examination, other diagnostic techniques for the recovery and identification of parasitic organisms are available. Most laboratories do not routinely offer all of these techniques, but many are relatively simple and inexpensive to perform. Both the clinician and the laboratorian should be aware of the possibilities and the clinical relevance of information obtained by such techniques.

Culture of Larval-Stage Nematodes

Nematode larval stages that hatch in soil or tissues may be diagnosed by using certain fecal culture methods to concentrate the larvae (Table 8). *Strongyloides stercoralis* larvae are the most common larvae in stool specimens. Depending on the fecal transit time through the intestine and the patient's condition, rhabditiform and occasionally filariform larvae may be present. Also, if examination of the stool is delayed, embryonated eggs and larvae of hookworm may be present. Cultures are used to (i) reveal the presence of larvae when their numbers are very low, (ii) determine whether the infection is due to *S. stercoralis* or hookworm on the basis of rhabditiform larval morphology (by allowing hookworm eggs to hatch and release first-stage larvae), and (iii) allow development of larvae into the filariform stage for further differentiation.

Certain fecal culture methods (sometimes referred to as coproculture) are especially helpful in detecting light infections of hookworm, *S. stercoralis*, and *Trichostrongylus* spp. and specifically identifying parasites. The examination of infective-stage nematode larvae also helps in the specific diagnosis of hookworm and trichostrongyle infections, because the eggs of many of these species are identical, and specific identifications are based on larval morphology. Careful handling of these cultures is essential, as the nematode larvae are at the infective stage and could result in laboratory-acquired infections through skin penetration.

Egg Studies

Estimation of Worm Burdens

The only human parasites that reasonably correlate egg production with adult worm burdens are *Ascaris lumbricoides*, *Trichuris trichiura*, and the hookworms (*Necator americanus* and *Ancylostoma duodenale*). Although not commonly used, information on approximate worm burdens is useful when determining the intensity of infection, deciding on possible chemotherapy, and evaluating the efficacy

TABLE 8 Culture of larval-stage nematodes: a review

Topic	Comments
Principle	Culture of feces for larvae is useful to (i) reveal their presence when they are too scanty to be detected by concentration methods; (ii) distinguish whether infection is due to *S. stercoralis* or hookworm on basis of rhabditiform larval morphology by allowing hookworm eggs to hatch, thus releasing first-stage larvae; and (iii) allow development of larvae into filariform stage for further differentiation. Cultures include Harada-Mori filter paper strip (14, 18, 29), filter paper-slant petri dish (14, 18, 29), charcoal (14), and Baermann funnel apparatus (14, 18, 29).
Specimen	Any stool specimen that is fresh and has not been refrigerated
Examination	Daily checking of fluid for presence of larvae. Hold cultures for 10 days prior to final report.
Results and laboratory reports	Failure to recover larvae does not completely rule out possibility of infection; however, infection is less likely when results are negative.
Procedure notes and limitations	Because recovery of infective larvae is always possible, gloves must be worn at all times when these procedures are used. Make sure culture systems are kept hydrated; certain amt of water evaporates and is lost owing to culture equilibration, particularly during first couple of days.

of the drugs administered (Table 9). With current therapies, the need for monitoring therapy through egg counts is rarely relevant. Egg counts obtained by the direct smear method of Beaver (29) or the Stoll dilution method (29) are estimates only, and variations in counts may occur, regardless of how carefully the procedure is followed. If two or more fecal specimens are to be compared, it is best to have the same individual use the technique on both samples and do multiple counts.

Hatching of Schistosome Eggs

When schistosome eggs are recovered from either urine or stool, they should be carefully examined to determine viability. The presence of living miracidia within the eggs indicates an active infection that generally requires therapy. The viability of the miracidium can be determined in two ways: (i) the cilia of the flame cells (primitive excretory cells) may be seen on a wet smear under high dry power and are usually actively moving, and (ii) the miracidia may be released from the eggs if a hatching procedure is used (24, 29). The eggs will usually hatch within several hours when placed in 10 volumes of dechlorinated or spring water (hatching may begin soon after contact with the water). Use of a sidearm flask covered with foil (except for the side arm) has been recommended, but an Erlenmeyer flask may be an acceptable substitute. The eggs that are recovered in

TABLE 9 Egg studies: a review

Topic	Comments
Principle	Information on approx worm burdens can be useful when determining intensity of infection, deciding on possible chemotherapy, and evaluating efficacy of drugs administered. With current therapy, monitoring egg counts may not be relevant. Hatching of schistosome eggs can provide very important information for physician; if eggs are viable, then treatment is recommended. If nothing but egg shells or eggs containing distorted larvae are seen, therapy may not be required (no living adult worms are present).
Specimen	Any stool specimen that is fresh and has not been refrigerated. For hatching test or microscopic review for miracidium viability, stool specimen must be processed (rinsed, centrifuged) with 0.85% NaCl rather than water, which would stimulate eggs to hatch prematurely.
Examination	Perform multiple egg counts on same specimen; it is also recommended that clinical decisions be based on series of egg counts over time, particularly when monitoring therapy. Schistosome eggs recovered (but not hatched) should be examined under microscope for cilium activity (flame cells) within miracidia.
Results and laboratory reports	Failure to recover nematode eggs does not completely rule out infection; however, probability of infection is very low when results are negative, particularly with *Trichuris* and *Ascaris* spp. and hookworm. When schistosome eggs are reported, viability of eggs (miracidium larva within egg shell) must be indicated. If nothing but egg shells or eggs containing distorted larvae are seen, nonviability of these eggs should be indicated.
Procedure notes and limitations	Egg counts are always subject to error; multiple counts are recommended. Stool consistency must be taken into account when calculating total no. of eggs. Failure to find schistosome eggs or to see evidence of egg viability does not rule out possibility of schistosomiasis, particularly when long-standing infection is suspected. Multiple counts are recommended.

the urine (24-h specimen collected with *no preservatives*) are easily obtained from the sediment and can be examined under the microscope to determine viability.

OTHER SPECIMENS FROM THE INTESTINAL TRACT AND UROGENITAL SYSTEM

Examination for Pinworm

Enterobius vermicularis (pinworm) is found worldwide, often in children. The adult female worm migrates from the anus, usually at night, and deposits her eggs on the perianal area. The adult female (8 to 13 mm long) may be found on the surface of the stool or on the perianal skin. Diagnosis is usually based on the recovery of typical eggs, which are thick shelled and football shaped with one slightly flattened side. Each egg often contains a fully developed embryo and is infective within a few hours of being deposited. Eggs are not commonly found in feces and must be detected by some technique other than the routine fecal examination. The most widely used procedure is the cellulose tape (adhesive cellophane tape) method (24, 29). Commercial collection kits containing sticky plastic paddles are also available. Specimens should be obtained in the morning before the patient bathes or goes to the bathroom. At least four consecutive negative slides should be observed before the patient is considered free of infection (12).

Sigmoidoscopy Material

In addition to the O and P examination, sigmoidoscopy specimens can be helpful in the diagnosis of amebiasis; however, sigmoidoscopy is not a substitute for the series of three routine stool examinations. Material from the mucosal surface should be aspirated or scraped, not obtained with cotton-tipped swabs. A minimum of six representative areas of the mucosa should be sampled and examined. The specimen must be processed or fixed immediately. There are three methods of examination: direct wet mount or trichrome- or iron hematoxylin-stained permanent smears; monoclonal reagent-based fluorescent antibody (FA); or enzyme immunoassay (EIA). Depending on the availability of trained personnel, proper fixation fluids, the amount of specimen obtained, or the suspect diagnosis, only one or two procedures may be required. If the amount of material limits the examination to one procedure, the use of PVA fixative, from which a permanent stained smear can be prepared, is highly recommended.

If the material is going to be examined by any of the new FA or EIA monoclonal detection kits (*Cryptosporidium parvum*, *Giardia lamblia*), then 5 or 10% formalin or SAF fixative is recommended. Physicians performing sigmoidoscopy procedures often do not realize the importance of selecting the proper fixative for material to be examined for parasites. A parasitology specimen tray (containing saline [direct wet mount]; Schaudinn's fixative, PVA, and SAF [permanent stained smears]; and 5 or 10% formalin or SAF [FA or EIA monoclonal reagents]) can be provided, or a trained technologist can be available at the time of sigmoidoscopy to prepare the slides. False-negative results may occur if the specimen has been poorly prepared.

Duodenal Drainage

Routine stool examinations may not reveal infections with *G. lamblia* or *S. stercoralis*. Examination of duodenal drainage material may reveal these or other parasites inhabiting the duodenum or bile duct. The specimen should be sub-mitted to the laboratory immediately in a tube containing no preservative; the amount may vary from <0.5 ml to several milliliters of fluid. The specimen may be centrifuged (10 min at 500 × g) and should be examined *immediately* as a direct wet mount for motile organisms (iodine may be added later to facilitate identification of any protozoa present). If the specimen cannot be completely examined within 2 h of collection, any remaining material should be preserved in 5 or 10% buffered formalin or SAF. The "falling leaf" motility often described for *Giardia* trophozoites is rarely seen in these preparations. The organisms are frequently caught in mucous strands, and movement of the flagella on the *Giardia* trophozoites may be the only subtle motility seen for these flagellates. *Strongyloides* larvae will usually be very motile.

Often the duodenal fluid contains mucus, which is where the organisms tend to be found. Therefore, centrifugation of the specimen is important, and the sedimented mucus should be examined. Monoclonal FA or EIA detection kits (*Cryptosporidium parvum*, *G. lamblia*) could also be used with fresh, formalinized, or SAF-preserved material.

If a presumptive diagnosis of giardiasis is obtained on the basis of the direct wet mount examination, the coverslip can be removed, and the specimen can be fixed with either Schaudinn's fluid or PVA for subsequent staining with either trichrome or iron hematoxylin. If the amount of duodenal material submitted is very small, then permanent stains can be prepared rather than using any of the specimen for a direct wet mount. This approach provides a permanent record; potential problems with unstained organisms, very minimal motility, and a lower-power examination can be avoided by using oil immersion examination (1,000×) of the stained specimen.

Duodenal Capsule or String Test Technique (Entero-Test)

One method of sampling duodenal contents eliminates the need for intestinal intubation (14). The device used consists of a length of nylon yarn coiled inside a gelatin capsule. The yarn protrudes through one end of the capsule; this end of the line is taped to the side of the patient's face. The capsule is then swallowed, the gelatin dissolves in the stomach, and the weighted string is carried by peristalsis into the duodenum. The yarn is attached to the weight by a slipping mechanism; the weight is released and passes out in the stool when the line is retrieved after 4 h. Bile-stained mucus clinging to the yarn is then scraped off (mucus can also be removed by pulling the yarn between gloved thumb and finger) and collected in a small petri dish. Usually, 4 or 5 drops of material are obtained.

The specimen should be examined *immediately* as a direct wet mount for motile organisms (iodine may be added later). Motility appears as described above under Duodenal Drainage. If the specimen cannot be completely examined immediately after the yarn has been removed, the material should be preserved in 5 or 10% formalin, PVA, or SAF. Smears should be prepared from any mucus present.

Note: The pH of the terminal end of the yarn should be checked to ensure adequate passage into the duodenum (a very low pH means it never left the stomach). The terminal end of the yarn should be yellow-green, indicating that it was in the duodenum (the bile duct drains into the intestine at this point).

Urogenital Specimens

The identification of *Trichomonas vaginalis* is usually based on the examination of multiple wet preparations of vaginal and urethral discharges and prostatic secretions or urine sediment. Specimens are diluted with a drop of saline and examined under low power and reduced illumination for the presence of actively motile organisms; as the jerky motility begins to diminish, it may be possible to observe the undulating membrane, particularly under high dry power. Observation of motile organisms from the direct wet mount or from appropriate culture media (14, 18) or direct detection by using monoclonal antibodies (7) is more reliable than use of stained smears.

Urinary sediment may be examined for certain filarial infections, especially in areas of endemicity. The membrane filtration technique can be used with urine for the recovery of microfilariae (14).

SPUTUM, ASPIRATES, AND BIOPSY MATERIAL

Expectorated Sputum

Organisms that can be detected in sputum and may cause pneumonia, pneumonitis, or Loeffler's syndrome include the migrating larval stages of *Ascaris lumbricoides*, *S. stercoralis*, and hookworm; the eggs of *Paragonimus westermani*; the protoscolices or hooklets of *Echinococcus granulosus*; *Pneumocystis carinii* (now classified with the fungi); trophozoites of *Entamoeba histolytica*, *Entamoeba gingivalis*, and *Trichomonas tenax*; oocysts of *Cryptosporidium parvum*; and possibly microsporidia (6, 32). In a *Paragonimus* infection, the sputum may be viscous, streaked with blood, and tinged with brownish flecks, which are clusters of eggs ("iron filings"). Sputum is generally examined as a wet mount and is usually not concentrated. If the sputum is thick, an equal amount of 3% sodium hydroxide (NaOH) (or undiluted chlorine bleach) can be added. The specimen is then thoroughly mixed, and the mixture is centrifuged. If there is a delay, the sputum should be fixed in 10% formalin to preserve helminth eggs or larvae and in PVA or SAF. This will preserve any eggs or larvae and allow later staining for protozoa. If *Cryptosporidium* sp. is suspected, then acid-fast or monoclonal antibody techniques normally used for stool specimens could be used (1, 13, 16, 29, 33, 39).

Induced Sputum

Stained smears of induced sputa can be used to detect *P. carinii* and to differentiate trophozoite and cyst forms from other possible causes of pneumonia, particularly in the AIDS patient (5). Organisms must be differentiated from other fungi such as *Candida* spp. and *Histoplasma capsulatum*. Although induced sputum specimens have been used successfully in the diagnosis of *P. carinii*, the yield varies. This variation may be due in part to the lack of adherence to specimen rejection criteria. If clinical evaluation of a patient suggests *P. carinii* pneumonia and the induced sputum specimen is negative, a bronchoalveolar lavage specimen should be evaluated by various methods, including use of the newer monoclonal reagents (2, 3) and calcofluor white (19, 35). Patients may also have to be evaluated for extrapulmonary *P. carinii* infections (9).

Aspirates

The examination of aspirated material may be valuable, particularly when routine testing methods have failed to demonstrate the organisms. Specimens should be transported to the laboratory immediately after collection. Techniques and potential parasites are reviewed in Table 10.

Biopsy Material

In certain cases, a biopsy may be the only means of confirming a suspected parasitic infection. Specimens examined as fresh material rather than as tissue sections should be kept moist in saline and submitted to the laboratory immediately. Detection of parasites in tissue depends in part on specimen collection and quantity. Multiple tissue samples often improve diagnostic results, and the total specimen must be examined. If the tissue specimen is too small for all the diagnostic procedures requested, prioritize the procedures after consultation with the physician. In addition to standard histologic preparations of biopsy specimens, impression smears and teased and pressed preparations of biopsy tissue from skin, muscle, cornea, intestine, liver, lung, and brain can be used. Tissue to be examined by permanent sections or electron microscopy should be fixed and processed as specified.

Tissue can be submitted in a sterile container on a sterile sponge dampened with saline. Impression smears, teased preparations, pressed preparations, and skin scrapings can be prepared for examination. Remaining specimen may be used for protozoan cultures after direct examination or after permanent stains have been prepared. Most protozoa will be found on the permanent stained smears (impression smears, touch or pressed preparations, teased preparations) or in wet preparations made with calcofluor white (17, 37). When culture is used, permanent stained smears of the culture medium or sediment may also reveal some of the protozoa. Skin snips can be processed as teased preparations; pressed preparations can be used for the examination of muscle for *Trichinella spiralis*. Filarial infections may be confirmed by the recovery and identification of microfilariae in skin scrapings and/or biopsy samples. Skin scrapings are recommended for *Sarcoptes scabei*.

Skin

Onchocerca volvulus and *Mansonella streptocerca*

Skin snips are recommended for the diagnosis of human filarial infections with *O. volvulus* and *M. streptocerca* (14). A razor blade can be used to cut a small slice from a skin fold held between thumb and forefinger, or a slice may be taken from a small cone of skin pulled up by a needle. The skin snip should be thick enough to include the outer part of the dermal papillae but so thin that significant bleeding does not occur, just a slight oozing of fluid. In cases of African onchocerciasis, it is preferable to take skin snips from the buttock region (above the iliac crest); in cases of Central American onchocerciasis, the preferred skin snip sites are the shoulders (over the scapula). Skin snips are placed immediately in a drop of normal saline or distilled water and covered; teasing the specimen with dissecting needles may facilitate release of the microfilariae; usually they emerge within 30 min to 1 h and can be examined with low-intensity light and the 10× objective of the microscope. To see morphologic details, allow the teased snip preparation to dry, fix it in absolute methyl alcohol, and stain it with Giemsa.

TABLE 10 Aspirates and biopsy tissues: a review

Body site	Possible parasites (most common)	Comments (diagnostic methods)
Lung	*Toxoplasma gondii*	Giemsa stain, tissue culture, immunospecific reagents
	Amebae	Trichrome stain
	Cryptosporidium parvum	Modified acid-fast stain, immunospecific reagents
	Microsporidia	Modified trichrome (chromotrope 2R) stain; electron microscopy
Liver	*Toxoplasma gondii, Leishmania donovani*	Giemsa stain
	Cryptosporidium parvum	Modified acid-fast stain, immunospecific reagent
	Entamoeba histolytica	Giemsa or trichrome stain (may require use of proteolytic enzymes)
	Microsporidia	Modified trichrome (chromotrope 2R) stain; electron microscopy
	Echinococcus spp.	Centrifugation (hydatid sand)
Brain	*Naegleria* sp., *Acanthamoeba* spp., *Balamuthia* spp.	Giemsa or trichrome stain; culture
	Entamoeba histolytica	Giemsa or trichrome stain
	Toxoplasma gondii	Giemsa stain, immunospecific reagent
	Microsporidia	Acid-fast, Giemsa, or modified trichrome (chromotrope 2R) stain; electron microscopy
Skin	*Leishmania* spp.	Giemsa stain; culture
	Onchocerca volvulus, Mansonella streptocerca	Giemsa or hematoxylin and eosin stain
Intestine	*Cryptosporidium parvum*	Modified acid-fast stain, immunospecific reagents
	Cyclospora cayetanensis	Modified acid-fast stain
	Enterocytozoon bieneusi, Septata intestinalis	Acid-fast, Giemsa, or modified trichrome (chromotrope 2R) stain; electron microscopy
	Giardia lamblia	Giemsa or trichrome stain
	Entamoeba histolytica	Giemsa or trichrome stain
Eye	Various genera of microsporidia	Acid-fast, Giemsa, or modified trichrome (chromotrope 2R) stain; electron microscopy
	Acanthamoeba sp.	Giemsa or trichrome stain, calcofluor for cysts, culture
Muscle	*Trichinella spiralis*	Wet preparation, pressed preparation
	Trypanosoma cruzi	Giemsa stain, culture
Lymph nodes	*Trypanosoma brucei* subsp. *gambiense* or *T. brucei* subsp. *rhodesiense, Leishmania donovani, Trypanosoma cruzi*	Giemsa stain, culture
Bone marrow	*Leishmania donovani*	Giemsa stain, culture

Cutaneous Amebiasis and Cutaneous Leishmaniasis

Skin biopsy samples for the diagnosis of cutaneous amebiasis and cutaneous leishmaniasis should be processed for impression smears and tissue sectioning and subsequently stained by the appropriate technique.

Lymph Nodes: Trypanosomiasis, Leishmaniasis, Chagas' Disease, and Toxoplasmosis

Lymph node material should be processed for tissue sectioning and impression smears, which should be processed like thin blood films and stained with Giemsa. Culture media can also be inoculated, provided the specimen has been collected under sterile conditions (14, 18).

Muscle: Trichinosis

Depending on the numbers present, encapsulated *Trichinella spiralis* larvae may be seen in fresh muscle if small pieces of muscle are pressed between two slides and exam-ined under the microscope. Larvae are usually most abundant in the diaphragm, masseter muscle, or tongue, although the gastrocnemius or deltoid muscles are usually sampled. Routine histologic sections can also be prepared. In human infections, serologic tests are more often used for diagnosis (see chapter 103 of this Manual).

Rectal or Bladder Biopsy: Schistosomiasis

In an old, chronic, or light infection with *Schistosoma mansoni* or *Schistosoma japonicum*, stool examinations may be negative; however, an examination of the rectal mucosa may reveal eggs. Fresh tissue should be compressed between two microscope slides and examined with the 10× objective and low-intensity light. Because treatment may depend on the viability of the eggs, they should be examined carefully to determine the presence of living miracidia. Bladder wall mucosa may contain eggs of *Schistosoma haematobium* in cases where the urine is negative. Eggs

TABLE 11 Blood parasites: a review

Procedure	Specimen	Comments
Thick and thin blood films	Whole or EDTA-treated blood	*Plasmodium* spp., *Babesia* spp., *Leishmania donovani*, *Trypanosoma* spp., microfilariae
Buffy coat preparation	Buffy coat cells	*Leishmania donovani*, *Trypanosoma* spp., microfilariae
QBC malaria tube (14, 18, 23)	EDTA-treated blood	*Plasmodium* spp., *Trypanosoma* spp., microfilariae
Knott's concentration (14, 18)	EDTA-treated blood	Microfilariae
Membrane filtration (14, 18)	EDTA-treated blood	Microfilariae
Gradient centrifugation (14)	EDTA-treated blood	Microfilariae
Triple centrifugation (14, 18)	EDTA-treated blood	*Trypanosoma* spp.
Delafield's stain (14, 18)	Whole or EDTA-treated blood	Microfilarial sheath

found in the bladder wall should be checked for viability (18, 29).

Other Tissues: Cestode Larval Stages

In addition to *Echinococcus granulosus* (hydatid disease) and the larval stage of *Taenia solium* (cysticercosis), other larval cestodes occasionally cause human disease. The larval stage of *Multiceps* spp., a parasite of dogs and wild canids, may cause human coenurosis, which resembles a cysticercus but is larger and has multiple scolices developing from the germinal membrane surrounding the fluid-filled bladder. These larvae occur in extraintestinal locations, including the eye, central nervous system, and muscle. Human sparganosis is caused by the larval stages of *Spirometra* spp., which are parasites of various canine and feline hosts and are closely related to the members of the genus *Diphyllobothrium*. Spargana are elongated, ribbonlike larvae without a bladder and with a slightly expanded anterior end lacking suckers. The larvae may be found in superficial tissues or nodules, and they can also cause ocular sparganosis, a more serious disease. Recognition of prominent calcareous corpuscles occurring in tapeworm tissue can be helpful in making a general identification; specific identification usually depends on referral to specialists.

CULTURE

Very few clinical laboratories routinely use culture techniques for parasites. Culture methods are often complex and not practical for the routine diagnostic laboratory. In larger institutions, some techniques may be available, particularly where consultative and research services are provided. With the development of newer molecular methods, other diagnostic methods may become less relevant. Culture procedures have been developed for *Entamoeba histolytica*, *Naegleria fowleri*, *Acanthamoeba* spp., *Trichomonas vaginalis*, *Toxoplasma gondii*, *Trypanosoma cruzi*, and the leishmanias. These procedures are usually available on special request and after consultation with the laboratory. Proper specimen collection is essential for optimum results. For those interested in trying these techniques, there are some excellent references for review (14, 18, 29).

BLOOD

Parasites, including *Plasmodium* spp., *Babesia* spp., *Trypanosoma* spp., *Leishmania donovani*, and microfilariae, may be recovered in whole blood, buffy coat preparations, or various types of concentrations (Table 11) (14, 23, 24, 28).

Identification to species level will usually be based on the examination of permanent stained thick and thin blood films. Films can be prepared from fresh whole blood collected with no anticoagulants, anticoagulated blood, or sediment from the various concentration procedures. Giemsa stain is recommended; however, Wright's stain or a Wright-Giemsa combination stain can also be used. Delafield's hematoxylin stain is useful in that it stains both microfilariae and the microfilarial sheath; in some cases, Giemsa does not stain the sheath.

Preparation of Thick and Thin Blood Films

Specific identification of most blood parasites usually requires a permanent stain. Thick films allow a larger amount of blood to be examined; however, the morphologic characteristics of blood parasites that allow species identification are best seen in thin films. The accurate examination of thick and thin blood films and the identification of parasites depend on the use of absolutely clean, grease-free slides for preparation of all blood films. Subsequent to a finger stick, the blood should flow freely; blood diluted with tissue fluids will show a decreased number of parasites per field. Unless you are certain that you will receive well-prepared slides, request a tube of fresh blood (EDTA anticoagulant is preferred), and prepare the films yourself. To detect stippling, the films must be prepared within 1 h after the blood is drawn. After that time, stippling may not be visible on stained films and the organism morphology may not be as good; however, identification is still frequently possible.

The specimen collection time should be clearly indicated on the tube of blood and also on the final report. This information will support correlation of the results with any fever pattern or other symptoms. There should also be some indication on the final report that one negative specimen does not rule out the possibility of a parasitic infection. Blood films should be prepared when the patient is first seen or admitted; when malaria is a possible diagnosis, then after the first set of negative smears, samples should be taken at intervals of 6 to 8 h for at least three successive days.

Staining Blood Films

It is better to select one staining method that will provide reproducible results than to do several on a hit-or-miss basis. Blood films should be stained as soon as possible. Delay of more than 3 days may result in failure to demonstrate typical staining characteristics for individual species.

The most common stains are of two types. In Wright's stain, the fixative is combined with the staining solution;

both fixation and staining occur at the same time, so the thick film must be lysed before staining. Giemsa stain represents the other type of staining solution, in which the fixative and the stain are separate; thus, the thin film must be fixed with absolute methanol before staining. Thick films will be lysed during the staining process.

When slides are removed from either type of staining solution, they should be dried in a vertical position. After being air dried, they may be examined under oil immersion by placing the oil directly on the uncovered blood film. If films are to be kept for a permanent record, they should be protected with a cover glass by mounting them in a medium such as Permount. Any oil must be removed from the film(s) by using xylene before mounting with a cover glass.

Giemsa Stain

Giemsa stain is sold as a concentrated stock solution or as a powder for those who prefer to make up their own stain. Each new lot should be tested for optimal staining times before being used on patient specimens. If the blood cells appear to be adequately stained, the timing and stain dilution should be appropriate to demonstrate the presence of malarial and other parasites. The commercial liquid stain or the stock solution prepared from powder should be diluted to prepare the working stain solution. Stock Giemsa stain is diluted with buffer 1:20 for thin blood films (stained for 20 min) or 1:50 for thick films (stained for 50 min). Some people prefer to stain both thick and thin films for a longer time in a more dilute solution. Phosphate buffer used to dilute the stock stain should be neutral or slightly alkaline (pH 7.0 to 7.2), although some workers recommend using pH 6.8 to emphasize Schüffner's dots. A good general rule for stain dilution versus staining time is as follows: if dilution is 1:20, stain for 20 min; if dilution is 1:30, stain for 30 min; etc. However, a series of stain dilutions and staining times should be tried to determine the best dilution and time for each batch of stock stain.

Stain Results

Giemsa stain colors the components of blood as follows: erythrocytes (RBCs), pale red; nuclei of leukocytes, purple with pale purple cytoplasm; eosinophilic granules, bright purple-red; and neutrophilic granules, deep pink-purple. If malaria parasites are present, the cytoplasm stains blue and the nuclear material stains red to purple-red. Schüffner's dots and other inclusions in the RBCs stain red to purple-red. Nuclear and cytoplasmic staining characteristics of other blood parasites such as *Babesia* spp., trypanosomes, and leishmanias stain like those in the malaria parasites. Although the sheath of a microfilaria may not always stain with Giemsa, the nuclei within the microfilaria itself stains blue to purple.

Proper Examination of Thick and Thin Blood Films

Thick Blood Films

If thick films are properly prepared, the greatest concentration of blood cells will be in the center of the film, but this area also often contains the most-distorted organisms. Parasitic organisms should initially be looked for at low magnification (10× objective) to detect microfilariae more readily. Malarial organisms, *Babesia* spp., and trypanosomes are best looked for under oil immersion (100× objective). Intact RBCs are frequently seen at the very periphery of the thick film; such cells, if infected, may prove useful in malaria diagnosis, since they may demonstrate the charac-

teristic morphology necessary to identify the organisms to species level. The presence of clustered brown pigment granules may also be a clue to the presence of malarial parasites. Examination of a thick film usually requires 5 to 10 min (approximately 100 oil immersion fields). Parasitic organisms may be distorted in thick films, and care must be taken not to confuse stain deposits or platelets with malarial parasites. Confirmation and identification to species should be made with the thin film.

Thin Blood Films

In any examination of thin blood films for parasitic organisms, the initial screen should be carried out with the low-power objective (10×). Microfilariae may be missed if the entire thin film is not examined. Microfilariae are rarely present in large numbers, and frequently, only a few organisms occur in each thin film preparation. Microfilariae are commonly found at the edges of the thin film or at the feathered end of the film, because they are carried to these sites during spreading of the blood. The feathered end of the film where the RBCs are drawn out into one single, distinctive layer of cells should be examined for the presence of malarial organisms, *Babesia* spp., and trypanosomes. In these areas, the morphologies and sizes of the infected RBCs are most clearly seen. Examination for these parasites requires the oil immersion objective (100×).

Examination of the thin film by an experienced microscopist usually takes 15 to 20 min (200 to 300 oil immersion fields). Because people tend to scan blood films at different rates, it is important to examine a minimum number of fields, regardless of the time it takes to perform this procedure. Although some people use a 50× or 60× oil immersion objective to screen stained blood films, there is some concern that small parasites such as malarial organisms, *Babesia* spp., or *L. donovani* may be missed at this total magnification (×500 or ×600), which is smaller than the ×1,000 total magnification obtained with the more traditional 100× oil immersion objective. If something suspicious is seen in the thick film, the number of fields examined on the thin film may often be considerably more than 200 to 300. The request for blood film examination should always be considered a STAT request, with all reports (negative as well as positive) being telephoned to the physician as soon as possible. Appropriate governmental agencies (local, state, and federal) should be notified within the specified time frame per guidelines and/or law.

Diagnostic problems with automated differential instruments have been reported (14, 28). Malaria and *Babesia* infections can be missed and therapy can be delayed when these instruments are used. They are not designed to detect intracellular blood parasites, and the inability of the automated systems to discriminate between uninfected RBCs and those infected with parasites may pose serious diagnostic problems.

Buffy Coat Smears

Smears prepared from the buffy coat layer (just below the layer of leukocytes after centrifugation) and stained with Giemsa are recommended when trypanosomes, *L. donovani*, or microfilariae are suspected (28).

Concentration for Microfilariae

Whole blood containing an anticoagulant can be concentrated by the Knott technique (28). Blood is lysed with 2% formalin (in a 1:5 ratio) and centrifuged, and the sediment is examined as a wet preparation. The wet mount can also be allowed to air dry and then can be stained with Giemsa stain or hematoxylin. Giemsa-stained smears can be restained with hematoxylin to demonstrate the sheath if microfilariae are found.

REFERENCES

1. **Arrowood, M. J., and C. R. Sterling.** 1989. Comparison of conventional staining methods and monoclonal antibody-based methods for *Cryptosporidium* oocyst detection. *J. Clin. Microbiol.* **27:**1490–1495.
2. **Baselski, V. S., M. K. Robinson, L. W. Pifer, and D. R. Woods.** 1990. Rapid detection of *Pneumocystis carinii* in bronchoalveolar lavage samples by using cellufluor staining. *J. Clin. Microbiol.* **28:**393–394.
3. **Baughman, R. P., S. S. Strohofer, S. A. Clinton, A. D. Nichol, and P. T. Frome.** 1989. The use of an indirect fluorescent antibody test for detecting *Pneumocystis carinii*. *Arch. Pathol. Lab. Med.* **113:**1062–1065.
4. **Blanshard, C., W. S. Hollister, C. S. Peacock, D. G. Tovey, D. S. Ellis, E. U. Canning, and B. G. Gazzard.** 1992. Simultaneous infection with two types of intestinal microsporidia in a patient with AIDS. *Gut* **33:**418–420.
5. **Blumenfeld, W., and J. A. Kovacs.** 1988. Use of a monoclonal antibody to detect *Pneumocystis carinii* in induced sputum and bronchoalveolar lavage fluid by immunoperoxidase staining. *Arch. Pathol. Lab. Med.* **112:**1233–1236.
6. **Bryan, R. T.** 1990. Microsporidia, p. 2130–2134. *In* G. L. Mandell, R. G. Douglas, and J. E. Bennett (ed.), *Principles and Practice of Infectious Diseases*, 3rd ed. Churchill Livingstone, New York.
7. **Chang, T. H., S. Y. Tsing, and S. Tzang.** 1986. Monoclonal antibodies against *Trichomonas vaginalis*. *Hybridoma* **5:**43–51.
8. **Code of Federal Regulations.** 1991. Occupational exposure to bloodborne pathogens; final rule. 29CFR1910.1030. *Fed. Regist.* Dec. 6, 1991.
9. **Cohen, O. J., and M. Y. Stoeckle.** 1991. Extrapulmonary *Pneumocystis carinii* infections in the acquired immunodeficiency syndrome. *Arch. Intern. Med.* **151:**1205–1214.
10. **Current, W. L., and L. S. Garcia.** 1991. Cryptosporidiosis. *Clin. Microbiol. Rev.* **3:**325–358.
11. **Friedberg, D. N., S. M. Stenson, J. M. Orenstein, P. M. Tierno, and N. C. Charles.** 1990. Microsporidial keratoconjunctivitis in acquired immunodeficiency syndrome. *Arch. Ophthalmol.* **108:**504–508.
12. **Garcia, L. S.** 1990. Laboratory methods for diagnosis of parasitic infections, p. 776–861. *In* E. J. Baron, L. R. Peterson, and S. M. Finegold (ed.), *Bailey & Scott's Diagnostic Microbiology*, 9th ed. The C. V. Mosby Co., St. Louis.
13. **Garcia, L. S., T. C. Brewer, and D. A. Bruckner.** 1989. Incidence of *Cryptosporidium* in all patients submitting stool specimens for ova and parasite examination: monoclonal antibody-IFA method. *Diagn. Microbiol. Infect. Dis.* **11:**25–27.
14. **Garcia, L. S., and D. A. Bruckner.** 1993. *Diagnostic Medical Parasitology*, 2nd ed. American Society for Microbiology, Washington, D.C.
15. **Garcia, L. S., R. Y. Shimizu, A. Shum, and D. A. Bruckner.** 1993. Evaluation of intestinal protozoan morphology in polyvinyl alcohol preservative: comparison of zinc sulfate- and mercuric chloride-based compounds for use in Schaudinn's fixative. *J. Clin. Microbiol.* **31:**307–310.
16. **Garcia, L. S., R. Y. Shimizu, A. C. Shum, and C. H. Zierdt.** 1992. Diagnosis of *Enterocytozoon bieneusi* microsporidiosis by detection of spores in human fecal specimens, abstr. 801, p. 244. *Program Abstr. Intersci. Conf. Antimicrob. Agents Chemother.*
17. **Griffin, J. L.** 1978. Pathogenic free-living amoebae, p. 507–549. *In* J. P. Kreier (ed.), *Parasitic Protozoa*, vol. 15. Academic Press, Inc., New York.
18. **Isenberg, H. D. (ed.).** 1992. *Clinical Microbiology Procedures Handbook*, vol. 1 and 2. American Society for Microbiology, Washington, D.C.
19. **Kim, Y. K., S. Parulekar, P. K. W. Yu, R. J. Pisani, T. F. Smith, and J. P. Anhalt.** 1990. Evaluation of calcofluor white stain for detection of *Pneumocystis carinii*. *Diagn. Microbiol. Infect. Dis.* **13:**307–310.
19a.**Kokoskin, E., T. W. Gyorkos, A. Camus, L. Cedilotte, T. Purtill, and B. Ward.** 1994. Modified technique for efficient detection of microsporidia. *J. Clin. Microbiol.* **32:**1074–1075.
20. **Lacey, C. J. N., A. M. T. Clarke, P. Fraser, T. Metcalfe, G. Bonsor, and A. Curry.** 1992. Chronic microsporidian infection of the nasal mucosae, sinuses and conjunctivae in HIV disease. *Genitourin. Med.* **68:**179–181.
21. **Long, E. G., A. Ebrahimzadeh, E. H. White, B. Swisher, and C. S. Callaway.** 1990. Alga associated with diarrhea in patients with acquired immunodeficiency syndrome and in travelers. *J. Clin. Microbiol.* **28:**1101–1104.
22. **Long, E. G., E. H. White, W. W. Carmichael, P. M. Quinlish, R. Raja, B. L. Swisher, H. Daugharty, and M. T. Cohen.** 1991. Morphologic and staining characteristics of a *Cyanobacterium*-like organism associated with diarrhea. *J. Infect. Dis.* **164:**199–202.
23. **Long, G. W., L. S. Rickman, and J. H. Cross.** 1990. Rapid diagnosis of *Brugia malayi* and *Wuchereria bancrofti* filariasis by an acridine orange/microhematocrit tube technique. *J. Parasitol.* **76:**278–280.
24. **Markell, E. K., M. Voge, and D. T. John.** 1994. *Medical Parasitology*, 8th ed. The W. B. Saunders Co., Philadelphia.
25. **Melvin, D. M., and M. M. Brooke.** 1985. *Laboratory Procedures for the Diagnosis of Intestinal Parasites*, p. 163–189. U.S. Department of Health, Education and Welfare publication no. (CDC) 85-8282. U.S. Government Printing Office, Washington, D.C.
26. **Morris, A. J., M. L. Wilson, and L. B. Reller.** 1992. Application of rejection criteria for stool ovum and parasite examinations. *J. Clin. Microbiol.* **30:**3213–3216.
27. **National Committee for Clinical Laboratory Standards.** 1991. *Protection of Laboratory Workers from Infectious Disease Transmitted by Blood, Body Fluids, and Tissue; Proposed Guideline.* Document M29-T2. National Committee for Clinical Laboratory Standards, Villanova, Pa.
28. **National Committee for Clinical Laboratory Standards.** 1992. *Slide Preparation and Staining of Blood Films for the Laboratory Diagnosis of Parasitic Diseases, Tentative Guideline.* Document M15-T. National Committee for Clinical Laboratory Standards, Villanova, Pa.
29. **National Committee for Clinical Laboratory Standards.** 1993. *Procedures for the Recovery and Identification of Parasites from the Intestinal Tract; Proposed Guideline.* Document M28-P. National Committee for Clinical Laboratory Standards, Villanova, Pa.
30. **Palmer, J.** 1991. Modified iron hematoxylin/Kinyoun stain. *Clin. Microbiol. Newsl.* **13:**39–40. (Letter to the editor.)
31. **Ryan, R. J., G. Sutherland, K. Coughlan, M. Globan, J. Doultree, J. Marshall, R. W. Baird, J. Pedersen, and B. Dwyer.** 1993. A new trichrome-blue stain for detection of microsporidial species in urine, stool, and nasopharyngeal specimens. *J. Clin. Microbiol.* **31:**3264–3269.
32. **Schwartz, D. A., R. T. Bryan, K. O. Hewanlowe, G. S. Visvesvara, R. Weber, A. Call, and P. Angritt.** 1992. Disseminated microsporidiosis (*Encephalitozoon hellem*) and acquired immunodeficiency syndrome—autopsy evidence for respiratory acquisition. *Arch. Pathol. Lab. Med.* **116:**660–668.
33. **Siddons, C. A., P. A. Chapman, and B. A. Rush.** 1992. Evaluation of an enzyme immunoassay kit for detecting *Cryptosporidium* in faeces and environmental samples. *J. Clin. Pathol.* **45:**479–482.
34. **Smith, J. W., and M. S. Bartlett.** 1985. Diagnostic parasitology: introduction and methods, p. 595–611. *In* E. H. Lennette, A. Balows, W. J. Hausler, Jr., and H. J. Shadomy (ed.), *Manual of Clinical Microbiology*, 4th ed. The American Society for Microbiology, Washington, D.C.

35. **Stratton, N., J. Hryniewicki, S. L. Aarnaes, G. Tan, L. M. de la Maza, and E. M. Peterson.** 1991. Comparison of monoclonal antibody and calcofluor white stains for the detection of *Pneumocystis carinii* from respiratory specimens. *J. Clin. Microbiol.* **29:**645–647.

36. **Weber, R., R. T. Bryan, R. L. Owen, C. M. Wilcox, L. Gorelkin, G. S. Visvesvara, and The Enteric Opportunistic Infections Working Group.** 1992. Improved light-microscopical detection of microsporidia spores in stool and duodenal aspirates. *N. Engl. J. Med.* **326:**161–166.

37. **Wilhelmus, K. R., M. S. Osato, R. L. Font, N. M. Robinson,** and **D. B. Jones.** 1986. Rapid diagnosis of *Acanthamoeba* keratitis using calcofluor white. *Arch. Ophthalmol.* **104:**1309–1312.

38. **Yang, J., and T. H. Scholten.** 1977. A fixative for intestinal parasites permitting the use of concentration and permanent staining procedures. *Am. J. Clin. Pathol.* **67:**300–304.

39. **Zierdt, C. H., V. J. Gill, and W. S. Zierdt.** 1993. Detection of microsporan spores in clinical samples by indirect fluorescent-antibody assay using whole-cell antisera to *Encephalitozoon cuniculi* and *Encephalitozoon hellum. J. Clin. Microbiol.* **31:**3071–3074.

Diagnosis of Parasitic Infections: Immunologic and Molecular Methods

MARIANNA WILSON, PETER SCHANTZ, AND NORMAN PIENIAZEK

103

INTRODUCTION

Diagnosis of parasitic infections is definitively made by identification of parasites in host tissue or excreta. Such identification is not generally possible in diseases such as toxoplasmosis or toxocariasis, in which parasites are located deep in tissue, and is not initially recommended in diseases such as cysticercosis or echinococcosis, in which invasive techniques with some risk to the patient are necessary to obtain material. Detection of antibodies can be very useful as an indicator that an individual has been infected with a specific parasite. A positive result in a person with no exposure to the parasite prior to recent travel in an area where the disease is endemic can be interpreted as indicating recent infection. However, detection of specific antibodies in a person native to an area where the parasite is endemic may reflect only a past infection unrelated to current clinical status. In general, detection of antibodies to parasitic diseases indicates only infection at some indeterminate time and not necessarily acute or current infection. Levels of antibodies to parasites slowly decline after the patient is cured of the infection but generally last for at least 6 months to many years, depending on the infecting parasite, and thus are not generally reliable indicators of successful cure.

The detection of specific immunoglobulin M (IgM) and IgA antibodies may be of value in determining the approximate time of initial infection with *Toxoplasma gondii* but is not recommended for any other parasitic disease. If infection with a parasite is suspected and stool or urine examinations are either not indicated or negative, then the appropriate serology test for specific IgG antibodies should be requested. If the parasite-specific IgG is negative, a positive IgM, IgA, or IgE result (generally a false-positive reaction) may be misinterpreted as a true reaction, resulting in unnecessary and inappropriate treatment of the patient.

The past decade has seen several changes in the availability of parasitic immunodiagnostic procedures (Table 1). Commercial kits for parasitic diseases other than toxoplasmosis and amebiasis have recently become available in the United States. Several private laboratories offer tests that were previously available only at the Centers for Disease Control and Prevention (CDC). Duplication of a published procedure does not necessarily mean that the results of several laboratories are identical without comparable evaluation and, ideally, exchange of reagents and sera. Therefore, the interpretation derived from the original results should not be applied to a duplicated test procedure without evaluations to establish test sensitivity, specificity, and interpretation for the new test.

Toxoplasma antibody detection tests are now performed with commercially available kits by a large number of laboratories. Unfortunately, it is virtually impossible to compare results obtained with kits from different companies other than as just positive or negative because of the lack of standardization of expression of results (147). The College of American Pathologists now offers *Toxoplasma* antibody proficiency testing as part of the Virology Antibody Detection survey. Participation of all laboratories that perform *Toxoplasma* antibody testing should be encouraged. The summary of data obtained in this program is an indication not only of individual laboratory performance but also of kit performance. Laboratories with aberrant results should question the accuracy of the kit they use as well as their own technical performance.

The diagnosis of human intestinal protozoa depends on microscopic detection of the various parasite stages in feces, duodenal fluid, or small intestine biopsy specimens. Since fecal examination is very labor intensive and requires a skilled microscopist, both antibody tests and antigen detection tests have been investigated as alternatives. Antibody tests are of limited value except for invasive amebiasis. Detection of parasite antigens should be more indicative of current infection and could be quickly performed by laboratory personnel other than an experienced morphologist. Much work has been accomplished during the last 10 years on the development of antigen detection systems, resulting in commercially available reagents for *Cryptosporidium parvum*, *Entamoeba histolytica*, *Giardia lamblia*, and *Trichomonas vaginalis*.

Molecular biology shows promise for improved diagnosis of a variety of parasitic infections. Nucleic acid probes may allow detection of parasite nucleic acid in material from patients with various infections, including malaria (71, 150) and leishmaniasis (148). Probes allow improved strain differentiation, which is useful for both epidemiology and taxonomy. Molecular biology may also allow development of more specific antigens for serologic tests, as has been described for Chagas' disease (39), leishmaniasis (24), and echinococcosis (57, 67). Even though these probes may

TABLE 1 Antibody and antigen tests for parasitic diseases

Disease	Available antibody test(s)	Commercial antigen detection test(s)
Amebiasis	EIA, IHA	EIA
Babesiosis	IFA	
Chagas' disease	EIA, IFA, CF	
Cryptosporidiosis		EIA, DFA, IFA
Cysticercosis	EIA, IB	
Echinococcosis	EIA, IHA, IB	
Fascioliasis	EIA	
Filariasis	EIA	
Giardiasis		EIA, DFA, IFA
Leishmaniasis	IFA, CF	
Malaria	IFA	
Paragonimiasis	EIA, IB	
Schistosomiasis	EIA, IB	
Strongyloidiasis	EIA	
Toxocariasis	EIA	
Toxoplasmosis	EIA, IFA	
Trichinellosis	BF, EIA	
Trichomonas		DFA, EIA, DNA probe
Trypanosoma cruzi (see Chagas' disease)		

show improved accuracy and ultimately be less labor intensive, they are currently not practical for routine use in most laboratories.

ANTIBODY DETECTION

African Trypanosomiasis

Immunodiagnostic tests for the detection of *Trypanosoma brucei rhodensiense* or *T. brucei gambiense* infections are not available without prior consultation with CDC. If African trypanosomiasis is suspected, please call the Division of Parasitic Diseases, CDC, at (404) 488-7760 for information.

Amebiasis

Several clinical syndromes of infection with *Entamoeba histolytica* are recognized: asymptomatic infections, involving only luminal colonization of the gut, account for the vast majority of human infections; symptomatic infections may be limited to the gastrointestinal tract (amoebic colitis) or may be extraintestinal (amoebic liver abscess). Serologic testing is most useful in patients with extraintestinal disease in which organisms are not generally found on stool examination.

The indirect hemagglutination (IHA) test continues to be used at CDC for routine serodiagnosis of amebiasis. The IHA test detects antibody specific for *E. histolytica* at titers of ≥1:256 in approximately 95% of patients with extraintestinal amebiasis, 70% of patients with active intestinal infection, and 10% of asymptomatic persons who are passing cysts of *E. histolytica*. Positive titers may persist for months or years even after successful treatment, so the presence of a titer does not necessarily indicate acute or

current infection. Specificity is excellent: false-positive reactions rarely occur at titers of ≥1:256 (65).

Several kits, including IHA, enzyme immunoassays (EIAs) (88), and immunodiffusion, are commercially available in the United States. Although the immunodiffusion test is specific, it is slightly less sensitive than the IHA test and requires a minimum of 24 h to obtain a result, in contrast to 2 h required for the IHA test or EIA. However, the simplicity of the procedure makes it ideal for the laboratory that has only an occasional specimen to test. The IHA test and EIA are more suitable for laboratories that have frequent requests for amebiasis serology. EIAs are as sensitive and specific as the IHA test (80). Although detection of IgM antibodies specific for *E. histolytica* has been reported, sensitivity is only about 64% in patients with current invasive disease (119).

Babesiosis

Natural transmission of *Babesia microti* occurs in the coastal areas of the northeastern United States and, rarely, in Minnesota and Wisconsin. Recently, human infection on the United States West Coast with a *Babesia* species that is antigenically and genotypically distinct from *B. microti* was reported (114). Because the tick vector also transmits the causative agent of Lyme disease, dual infections of *B. microti* and *Borrelia burgdorferi* have been reported. Diagnosis of *Babesia* infection should be made by observation of parasites in patients' blood films. However, antibody detection tests are useful for detecting infected individuals with very low levels of parasitemia (such as asymptomatic blood donors in transfusion-associated cases), for posttherapy diagnosis after parasitemia is no longer detectable, and for discrimination between *Plasmodium falciparum* and *Babesia* infection in patients whose blood film examinations are inconclusive and whose travel histories cannot exclude either parasite (55).

The indirect immunofluorescence test (IFA) is quite sensitive for *B. microti* infection (17, 81). Patients' titers generally rise to ≥1:1,024 during the first weeks of illness and fall gradually over 6 months to titers of 1:16 to 1:256, remaining at that level typically for about 13 months and sometimes longer. Specificity is 100% in patients with other tick-borne diseases or persons not exposed to the parasite. Cross-reactions may occur in serum specimens from patients with malaria infections, but titers are generally highest with the homologous antigen (18). The extent of cross-reactivity between *Babesia* species is variable; a negative result with *B. microti* antigen for a patient exposed on the West Coast may be a false-negative reaction for *Babesia* infection (114).

Chagas' Disease

Infections with *Trypanosoma cruzi* are common in Central and South America (78). Many immigrants from areas where Chagas' disease is endemic currently reside in the United States and are potential sources of parasite transmission via infected blood (127). During the acute stage of illness, blood film examination generally reveals the presence of trypomastigotes. During the chronic stage of infection, parasites are rare or absent from the circulation; immunodiagnosis is the method of choice for determining whether the patient is infected.

The IFA, complement fixation (CF) test, and EIA are employed at CDC. Although IFA is very sensitive, cross-reactivity occurs with sera from patients with leishmaniasis, a protozoan disease that occurs in the same geographical areas as *T. cruzi*. The CF test is less sensitive but more specific than the IFA (77). Sensitivities and specificities of EIAs that use crude antigens are similar to those of the IFA. Two EIA kits for antibody detection (one of which is approved by the Food and Drug Administration) are available in the United States (109). Although differentiating between acute and chronic infection is very important in determining therapy, serology cannot be used to do so. A positive titer indicates only infection at some unknown time and not acute infection.

Cysticercosis (Larval *Taenia solium*)

The CDC immunoblot assay with purified *Taenia solium* antigens (135) is the immunodiagnostic test of choice for confirming a clinical and radiologic presumptive diagnosis of neurocysticercosis. It is 100% specific and has a sensitivity superior to that of any other test yet evaluated. Serum specimens from 97% of 108 parasitologically confirmed cases of cysticercosis had detectable antibodies. No serum samples from 376 patients with other microbial infections reacted with any of the *Cysticercus*-specific antigens. The most important factors identified as determining positive immunoblot reactions are the numbers and stages of development of cysticerci. Cumulative clinical experience has confirmed that in patients with multiple (more than two) lesions, the test has more than 90% sensitivity (31, 34, 135, 145). Seropositivity in biopsied patients with single, enhancing parenchymal cysticercal cysts was <50%, but in nonbiopsied clinically defined patients with a single cyst, it was 70% (145). Seropositivity in serum and cerebrospinal fluid (CSF) of patients with multiple but only calcified cysts was 82 and 77%, respectively. In all patients, regardless of their clinical presentation, the immunoblot assay is slightly more sensitive in serum than in CSF; consequently, there is no need to obtain CSF solely for use in the immunoblot assay (145).

The CDC glycoprotein immunoblot assay is both more specific and more sensitive than the EIA systems with which it has been compared (31, 34). Nonspecificity has been a major problem in most EIAs because of cross-reacting components in crude antigens derived from cysticerci; these components react with antibodies specific for other helminth infections, especially hydatidosis and filariasis (124). Most partially purified fractions evaluated in an EIA appear to have lower sensitivity than crude antigens and do not necessarily achieve higher specificity. Currently available antibody detection tests for cysticercosis do not distinguish between active and inactive infections and thus have not been useful in evaluating the outcomes and prognoses of medically treated patients.

Echinococcosis

Cystic Hydatid Disease (*Echinococcus granulosus*)

Since hydatid cysts are located in deep-seated organs and since closed biopsy could induce anaphylaxis or cause dissemination of protoscolices, immunodiagnostic tests can be very helpful in confirming or ruling out a presumptive diagnosis of hydatid disease. However, the clinician must have some knowledge of the characteristics of the available tests and the patient and parasite factors associated with false results (122). False-positive reactions may occur in persons with other helminthic infections, cancer, and chronic immune disorders. Negative test results do not rule out echinococcosis, because some cyst carriers do not have detectable antibodies. Detectable immune responses have been associated with the location, integrity, and vitality of the larval cyst. Cysts in the liver are more likely to elicit antibody response than cysts in the lungs, and regardless of localization, antibody detection tests are least sensitive in patients with intact hyaline cysts. Cysts in the lungs, brain, and spleen are associated with lowered serodiagnostic reactivity, whereas those in bone appear to more regularly stimulate detectable antibody. Fissuration or rupture of a cyst is followed by an abrupt stimulation of antibodies. Often, a patient with senescent, calcified, or dead cysts is seronegative. Antibody binding with antigen in circulating immune complexes may also cause false-negative test results (22).

IHA, IFA, and EIA are sensitive tests for detecting antibodies in serum of patients with cystic hydatid disease (CHD); sensitivity rates vary from 60 to 90% depending on the characteristics of the cases (122). Recent research has resulted in identification and characterization of many of the diagnostically important antigens; it is hoped that these antigens will be used in the next generation of tests to increase specificity without sacrificing sensitivity (57, 129). At present, however, the best available serologic diagnosis is obtained by using combinations of tests. EIA or IHA is used to screen all specimens; a positive reaction is confirmed by immunoblot (IB) assay or one of the variety of gel diffusion assays that demonstrate the echinococcal "arc 5" (87, 92, 99, 140). Although these confirmatory assays give false-positive reactions with sera of 5 to 25% of persons with neurocysticercosis, the clinical presentation of the latter disease is rarely confused with that of cystic echinococcosis (99, 122).

Circulating antigens have been detected in the sera of only about half of patients with cystic echinococcosis; however, these antigens may be detectable in the sera of many patients with negative or borderline antibody titers (22). Antigen assays might be useful as additional tests for those patients with no detectable *Echinococcus* antibodies, but these assays are not yet available in the United States.

Antibody responses have also been monitored as a way of evaluating the results of treatment, but with mixed results. Following successful radical surgery, antibody titers decline and sometimes disappear; titers rise again if secondary hydatid cysts develop. Tests for arc 5 or IgE antibodies appear to reflect antibody decline during the first 24 months postsurgery, whereas the IHA and other tests remain positive for 4 years or more (8). Chemotherapy has not been followed by consistent declines in antibody titers, but a recent study concluded that tests based on specific IgE antibody detection showed a strong association with chemotherapeutic cure (37).

AHD (*Echinococcus multilocularis*)

Most persons infected with *Echinococcus multilocularis* have a detectable immune response, and most patients with alveolar hydatid disease (AHD) are positive in serologic tests done with heterologous *Echinococcus granulosus* antigens. With crude *Echinococcus* antigens, nonspecific reactions create the same difficulties as described above for diagnosis of CHD; however, immunoaffinity-purified *Echinococcus multilocularis* antigens (Em2) used in the EIA give positive serologic reactions in more than 95% of AHD cases (56). Comparing the serologic reactivity to Em2 antigen with that to antigens containing components of both *Echinococcus multilocularis* and *Echinococcus granulosus* permits discrimination of patients with AHD from those with CHD (59, 84). Combining two purified *Echinococcus mul-*

tilocularis antigens (Em2 and recombinant antigen II/3-10) in a single immunoassay optimizes sensitivity and specificity (58). As in cystic echinococcosis, serologic tests are more useful for postoperative follow-up than for determining the outcome of chemotherapy (60).

Fascioliasis

The acute manifestations of human fascioliasis may precede the appearance of eggs in the stool by several weeks; immunodiagnostic tests may be useful for early indication of *Fasciola* infection as well as for confirmation of chronic fascioliasis when egg production is low or sporadic and for ruling out "pseudofascioliasis" associated with ingestion of parasite eggs in sheep or calves' liver (68).

The current test of choice for immunodiagnosis of human *Fasciola hepatica* infection is EIA with excretory-secretory (ES) antigens (32, 69, 70). Specific antibodies to *F. hepatica* may be detectable within 2 to 4 weeks after infection, which is 5 to 7 weeks before eggs appear in stool. Sensitivity for the Falcon assay screening test–enzyme-linked immunosorbent assay (FAST-ELISA) was reported to be 95%, while sensitivity for the immunoblot with 12-, 17-, and 63-kDa antigens appeared to be 100% (70). However, some cross-reactivity may occur with serum specimens of patients with schistosomiasis. Antibody levels decrease shortly after chemotherapeutic cure and can be used to predict the success of therapy (33).

Filariasis

Lymphatic Filariasis (*Wuchereria bancrofti* and *Brugia* spp.)

Immunodiagnostic tests for human filariasis may be useful, because the detection of filariae and microfilariae is often difficult owing to low numbers and parasite sequestration. In the past, antigens have been crude and fractionated extracts of filarial worms of other mammals, such as *Dirofilaria immitis*, the dog heartworm (77). Such antigens are usually reactive with serum samples of patients with various forms of filariasis but are completely lacking in specificity. ES or surface antigens of homologous *Wuchereria* or *Brugia* parasites are more specific than whole-worm somatic extracts (94), but supplies of such material are limited because of the lack of animal models and the difficulties of in vitro cultivation. EIAs that measure only IgG4 or IgE antibodies to *Brugia* antigens greatly increase specificity, because humans do not produce IgG4 antibodies to phosphocholine, an immunodominant molecule present on a wide range of organisms, including filariae (83).

Antibody detection methods cannot distinguish past exposure from current infection, nor is there any correlation of antibodies with worm burden. Detection of parasite material (antigen or other products) in blood or urine offers potential advantages for assessing the state of an individual infection. Although circulating antigens can be detected in patients with bancroftian filariasis, positive results were initially confined almost exclusively to serum specimens from persons with microfilaremia (7, 141, 142). However, many of the lymphatic filarial syndromes are not associated with microfilaremia. Initial studies with an assay based on the Og4C3 monoclonal antibody suggest that this assay detects antigen in asymptomatic *W. bancrofti* microfilaremic patients with 100% sensitivity (138). Further studies are necessary to determine whether this assay will aid in the diagnosis of patients with clinical manifestations of lymphatic filariasis.

At present, positive serum antibody tests are of little diagnostic value except in patients not native to areas of filarial endemicity. Most residents of regions where filaria is endemic have been sensitized to filarial antigens through years of exposure to infected mosquitoes; also, filarial antigens may cross-react with those of other nematode parasites found in these areas. Thus, a positive immunodiagnostic test is a necessary but not sufficient procedure for the diagnosis of brugian or bancroftian filariasis (108). Conversely, a negative test result in an American or European whose symptoms suggest lymphatic filariasis after travel in an area where filariasis is endemic strongly suggests no infection.

Onchocerciasis

Current diagnostic methods (detection of parasites in the skin or eye or the presence of suggestive clinical changes) are crude and often laborious and insensitive. Accurate immunodiagnostic tests are needed to identify persons requiring treatment, to evaluate the success of treatment, and to perform epidemiologic surveys.

IgG and IgE antibodies that react with onchocercal antigens are readily detected in the sera of infected persons; however, they are not specific for the presence or extent of the infection, nor do they allow discrimination between onchocerciasis and other filarial infections. Tests based on IgE detection are more specific than those based on IgG (143). EIA restricted to detection of IgG4 antibodies and employing purified, surface-derived, low-molecular-weight onchocercal antigens is highly sensitive and specific, although tests still cross-react with serum specimens of some patients with *W. bancrofti* infection (14). Antibodies directed against a 16-kDa *Onchocerca volvulus* antigen have been reported to be species specific and to indicate active infection (89). Antigen detection methods are feasible, but these and antibody assays require refinement of the reagents to control specificity (29).

Leishmaniasis

Visceral leishmaniasis is the most difficult clinical presentation of the three types of leishmaniasis (visceral, mucocutaneous, and cutaneous) to diagnose. Biopsy of liver, spleen, or bone marrow is necessary to demonstrate the parasite in tissue. Antibody detection is a useful adjunct for diagnosing visceral leishmaniasis but is much less reliable for cutaneous leishmaniasis. If leishmaniasis is suspected, consultation on diagnosis and therapy may be obtained by calling the Division of Parasitic Diseases, CDC, at (404) 488-7760.

The IFA and CF tests are used at CDC for detection of antibodies (110). These procedures can differentiate leishmaniasis from other clinically similar conditions but cannot determine the species of *Leishmania*. Cross-reactions occur in patients with antibodies to *T. cruzi* (Chagas' disease), which is endemic in some of the same Central and South American countries as leishmaniasis. Only the IFA test is performed in cases of suspected cutaneous leishmaniasis. The sensitivities of all antibody tests for cutaneous leishmaniasis are poor: most patients do not have a detectable circulating-antibody response to the parasite.

The immunology and classification of *Leishmania* species have been very active areas of research. Biochemical, immunologic, and molecular biology techniques have been used extensively, resulting in more specific antigens and probes for the detection and identification of parasites (9, 12, 63, 148).

Malaria

The serodiagnosis of malaria is not recommended except for (i) screening blood donors involved in cases of transfusion-induced malaria when the donor's parasitemia may be below the detectable level of blood film examination and (ii) testing a patient with a febrile illness who is suspected of having malaria and from whom repeated blood smears are negative. IFA tests with antigens of the four human *Plasmodium* species for malarial-antibody detection are very sensitive and specific, but the presence of antibodies indicates that infection occurred at some time in the past and does not necessarily indicate current infection (134).

Paragonimiasis

Pulmonary paragonimiasis is the most common presentation of patients infected with *Paragonimus* spp., although extrapulmonary (cerebral, abdominal) paragonimiasis may occur (152). Detection of eggs in sputum or feces of patients with paragonimiasis is often very difficult; therefore, serodiagnosis may be very helpful in confirming infections and monitoring the results of individual chemotherapy (126). In the United States, detection of antibodies to *Paragonimus westermani* has helped physicians differentiate paragonimiasis from tuberculosis in Indochinese immigrants (75).

The CF test has been the standard test for pargonimiasis; it is highly sensitive for diagnosis and for assessing cure after therapy. Because of the technical difficulties of the CF test, EIAs were developed as a replacement (19, 79). The IB assay performed with a crude antigen extract of *Paragonimus westermani* has been in use at CDC since 1988 (130). Positive reactions, based on demonstration of an 8-kDa antigen-antibody band, were obtained with serum samples of 96% of patients with parasitologically confirmed *Paragonimus westermani* infection. Specificity was ≥99%; of 210 serum specimens from patients with other parasitic and nonparasitic infections, only 1 serum sample from a patient with *Schistosoma haematobium* reacted. Antibody levels detected by EIA and IB do decline after chemotherapuptic cure but not as rapidly as those detected by the CF test (20, 96).

Most published literature deals with pulmonary paragonimiasis due to *Paragonimus westermani*, although in some geographic areas, other *Paragonimus* species cause similar or distinct clinical manifestations in human infections. Cross-reactivity between species does occur but at different levels for different species (79).

Schistosomiasis

Detection of antibodies to schistosomes can be useful in indicating infections when eggs cannot be demonstrated in fecal or urine specimens. Test sensitivity and specificity vary widely among the many tests currently available for the serologic diagnosis of schistosomiasis and are dependent on both the type of antigen preparations used (crude, purified, adult worm, egg, cercarial) and the test procedure (21, 91).

At CDC, all serum specimens are initially tested by FAST-ELISA with *Schistosoma mansoni* adult microsomal antigen (62). A positive reaction (>8 U/μl of serum) indicates infection with *Schistosoma* species. The specificity of this assay for detecting schistosome infection is 99%. The sensitivities are 99% for *S. mansoni* infection, 95% for *S. haematobium* infection, and ≥50% for *Schistosoma japonicum* infection. Because the EIA with *S. mansoni* adult microsomal antigen is not as sensitive for detection of the other species as for detection of *S. mansoni*, immunoblots of all three species are used (based on the patient's travel history) to ensure detection of *S. haematobium* and *S. japonicum* infections. Species-specific diagnosis is achieved by using adult worm antigens on IBs (136, 137). The presence of antibody is indicative only of schistosome infection at some time and cannot be correlated with worm burden, egg production, clinical status, or prognosis.

Strongyloidiasis

Diagnosis and treatment of chronic strongyloidiasis is critical for preventing dissemination of the infection, which may prove lethal in immunosuppressed patients (30, 48). Immunodiagnostic tests for strongyloidiasis are indicated when the infection is suspected and the organism cannot be demonstrated by duodenal aspiration, string tests, or repeated examinations of stool. Antibody detection tests should use antigens derived from *Strongyloides stercoralis* filariform larvae for the highest sensitivity and specificity (103). Although IFA and IHA tests have been used, EIA is currently recommended because of its greater sensitivity (84 to 92%) (41, 47, 121). Immunocompromised persons with disseminated strongyloidiasis usually have detectable IgG antibodies despite their immunodepression (49). Cross-reactions in patients with filariasis and some other nematode infections may occur (41, 47, 102). Important test limitations are that (i) 8 to 16% of *Strongyloides* carriers are seronegative and (ii) antibody test results cannot be used to differentiate between past and current infections. A positive test warrants continuing efforts to establish a parasitologic diagnosis followed by anthelminthic treatment. Serologic monitoring may be of use in the follow-up of immunocompetent treated patients. In a limited number of patients, antibody titers decrease markedly, and in some cases disappear, after successful thiabendazole therapy (102, 111). The potential use of IBs for the immunodiagnosis of strongyloidiasis has been discussed but requires additional evaluation (107, 120).

Toxocariasis (Larva Migrans)

Antibody detection tests are usually the only means of confirmation of a presumptive clinical diagnosis of toxocaral visceral larva migrans (VLM) or ocular larva migrans (OLM), the two most common clinical syndromes associated with *Toxocara* infections. The currently recommended serologic test for toxocariasis is EIA with larval-stage antigens. Such antigens can be extracted from embryonated eggs or obtained when they are released in vitro by cultured infective larvae (28). The latter, *Toxocara* ES (TES) antigens, are preferable to larval extracts because they are convenient to produce and because an absorption-purification step is not required for obtaining maximum specificity (51). When serum specimens from patients with high antibody titers to other helminth antigens are tested, positive reactions for TES antigens are only rarely observed (54); the few samples that did react with TES antigens showed patterns of banding by IB identical to those of antigens from patients with toxocaral larva migrans, thus suggesting concurrent infection. Cross-reactions with anti-A and anti-B blood group antibodies (51) or C-reactive protein were not observed, although such cross-reactivity with *Toxocara* antigens has been reported by others (131, 132).

Evaluation of the true sensitivities and specificities of serologic tests for toxocariasis in human populations is not possible because of the lack of parasitologic methods for detecting *Toxocara* parasites. These inherent problems result in underestimations of sensitivity and specificity. Evaluation of the *Toxocara* EIA in groups of patients with presumptive diagnoses of VLM indicated a sensitivity and specificity of 78 and 92%, respectively, at a titer of ≥1:32 (53). The sensitivity of EIA for the diagnosis of OLM is less than that for VLM (113, 123); in our experience, sensitivity and specificity were 73 and 95%, respectively, at a titer of ≥1:32. When the cutoff titer was lowered to 1:8, sensitivity increased to 90% (113). Further confirmation of the specificity of the serologic diagnosis of OLM can be obtained by examining aqueous or vitreous humor samples for antibodies (11, 50).

When interpreting the serologic findings, clinicians must be aware that a measurable titer does not necessarily indicate current clinical *Toxocara canis* infection. In most human populations, a small number of those tested have positive EIA titers that apparently reflect the prevalence of asymptomatic toxocariasis. In the United States, 2.8% (cutoff titer, 1:32) of nearly 9,000 persons tested were positive, but the percentage of reactors varied significantly according to age, race, and socioeconomic status (52).

Work in progress promises new techniques that may yield much more clinically useful information than the positive or negative result provided by current antibody assays. Studies include the detection of circulating TES antigens (3); with further evaluation, such tests may be suitable for clinical use. Monoclonal-antibody-based tests reportedly distinguish between *Toxocara canis* and *Toxocara cati* and distinguish recent from chronic infections (95).

Toxoplasmosis

Because of the incidence of *Toxoplasma* infection (40) and the availability of commercial kits, antibody detection of *Toxoplasma* infection constitutes the largest number of serologic test requests for parasitic diseases in the United States. Comparisons of kits indicated that most perform comparably in detecting *Toxoplasma*-specific IgG antibodies (139, 147) but vary markedly in detecting *Toxoplasma*-specific IgM antibodies (61, 146). The College of American Pathologists offers *Toxoplasma* antibody proficiency testing as a component of the Virology Antibody Detection survey.

The IFA test and EIA for IgG and IgM antibodies are the tests most commonly used today. Persons should be tested initially for the presence of *Toxoplasma*-specific IgG antibodies in order to determine immune status. A positive IgG titer indicates infection with the organism at some time. If more precise knowledge of the time of infection is necessary, then an IgG-positive person should have an IgM test performed by a procedure with minimal nonspecific reactions, such as IgM-capture EIA (38, 61, 101). Newborn infants suspected of congenital toxoplasmosis should be tested by both an IgM- and an IgA-capture EIA. Detection of *Toxoplasma*-specific IgA antibodies is more sensitive than IgM detection in congenitally infected babies (25–27, 86, 133). Although a commercial kit is available in Europe (27), none is available yet in the United States. *Toxoplasma*-specific IgM antibodies may be detected by EIA for as long as 18 months after acute acquired infection (100, 146), but high levels may indicate infection within the past 3 to 4 months.

Another new method for determining acute infection measures the antigen-binding avidities of IgG antibodies; low-affinity antibodies are produced during the early stages of infection, while high-affinity antibodies appear to reflect past immunity (66, 76, 85). A commercial kit is available in Europe but not in the United States.

Serologic determination of active central nervous sys-

tem toxoplasmosis in immunocompromised patients is not possible at this time (90). *Toxoplasma*-specific IgG antibody levels in AIDS patients are often low to moderate, but occasionally no specific IgG antibodies can be detected. Tests for IgM antibodies are generally negative. Detection of circulating antigen in AIDS patients has been evaluated, but the procedure lacks sensitivity.

Trichinellosis

Human trichinellosis patients rarely develop detectable antibody until 3 to 5 weeks postinfection, well after the onset of acute-stage illness. Antibody development is also affected by the infecting dose of larvae: the larger the infecting dose, the faster the patient's antibody response will develop. Multiple serum specimens should be drawn several weeks apart to demonstrate seroconversion in patients whose initial specimens were negative.

Immunodiagnostic tests currently available in the United States include EIA and bentonite flocculation (BF) (106). Antigen preparations may be crude antigens prepared from homogenates of *Trichinella spiralis* muscle larvae or ES products produced by cultured larvae (42, 43). Positive reactions are detectable in serum samples of 80 to 100% of patients with clinically symptomatic trichinellosis. Antibody levels are often not detectable in the first month postinfection, peak in the second or third month postinfection, and then decline slowly for several years (35, 98). IgG, IgM, and IgE antibodies are detectable in many patients; however, tests based on IgG antibodies are most sensitive (36, 98).

In our experience at CDC, EIA with ES antigen detects antibodies earlier than BF in 25% of serum specimens from patients with acute infection but is less specific than BF. Antibodies are detectable longer in EIA than in BF. The EIA is used for routine screening; all EIA-positive specimens are then tested by BF for confirmation. A positive result in both tests indicates infection with *Trichinella spiralis* within the last several years.

IB tests have revealed parasite-specific antigens that when isolated from the parent antigens could perhaps be useful in the development of more sensitive and specific tests (72, 93). Recent studies have demonstrated circulating antigens in patients with trichinellosis; however, tests for antibodies were more sensitive diagnostic indicators (73, 105).

ANTIGEN DETECTION

Commercial availability of IFA, direct fluorescent-antibody (DFA), and EIA tests has provided the clinical laboratory with the means of identifying or confirming the presence of parasites or their antigens in clinical specimens (144). Table 2 lists which kits are available for which parasite and summarizes test parameters and references.

Cryptosporidium parvum

Most *Cryptosporidium* infections result in symptomatic illness and commonly occur in neonates and immunocompromised persons. Immunodiagnostic methods that detect either whole organisms or soluble antigens provide increased sensitivity over microscopy without loss of specificity.

Table 2 summarizes the features of four commercial products (one DFA and three EIAs) available in the United States for the immunodiagnosis of cryptosporidial infections. The most sensitive and specific method is reported to

TABLE 2 Summary of commercially available kits for immunodetection of parasitic organisms or antigens

Organism and kit name	Manufacturer and/or distributor[a]	Type of test	Sensitivity (%)[b]	Specificity (%)[b]	Reference(s)
Cryptosporidium parvum					
ProSpecT	Alexon	EIA	97	98	
IDEIA	Dako	EIA	100	100	128
MeriFluor	Meridian	DFA, IFA	100	100	46
Color Vue	Seradyn	EIA	93	93	116
Entamoeba histolytica					
ProSpecT	Alexon	EIA	87	99	
	TechLab	EIA			
Giardia lamblia					
ProSpecT	Alexon	EIA	96	100	2, 118, 125
GiardEIA	Antibodies	EIA	91	96	16
MeriFluor	Meridian	DFA, IFA	100	100	46, 151
Color Vue	Seradyn	EIA	97	94	117
DD System	Trend	EIA	96	97	
Trichomonas vaginalis					
MeriFluor	Meridian	DFA, EIA	86	99	82
Affirm VP$_{III}$	MicroProbe	DNA probe	80	100	13

[a]Alexon, Inc., 1190 Borregas Ave., Sunnyvale, CA 94089; Antibodies, Inc., P.O. Box 1560, Davis, CA 95617; Dako Corp., 6392 Via Real, Carpinteria, CA 93013; Meridian Diagnostics, Inc., 3471 River Hills Dr., Cincinnati, OH 45244; MicroProbe Corp., 1725 220th St. NE, Bothell, WA 98021; Seradyn, P.O. Box 1210, Indianapolis, IN 46206; TechLab, VPI Research Park, 1861 Pratt Dr., Blacksburg, VA 24060; Trend Scientific, Inc., P.O. Box 12266, St. Paul, MN 55112.

[b]Compared to those of conventional methods.

be the MeriFluor IFA test (Meridian), which identifies oocysts in concentrated or unconcentrated fecal samples by using a fluorescein isothiocyanate (FITC)-labeled monoclonal antibody (5, 6, 15, 44, 46). A combined DFA test for the simultaneous detection of cryptosporidial oocysts and *Giardia* cysts is also available. Fluorescence microscopy provides a significantly improved signal-to-noise ratio compared to that of the conventional microscopic stool examination (46). Three commercial EIAs (ProSpecT/ Cryptosporidium, Alexon, Inc.; IDEIA, Dako Corp.; and Color Vue, Seradyn, Inc.) have been introduced for the detection of cryptosporidial antigen in fresh or frozen stool samples and also in stool specimens treated with the fixatives 10% formalin, merthiolate-iodine-formalin, or sodium acetate-acetic acid-formalin. Concentrated or polyvinyl alcohol-treated samples are unsuitable for testing with these EIA kits. The kits are reportedly superior to microscopy (especially acid-fast staining) and show good correlation with the monoclonal-antibody-based immunofluorescence assay. Kit sensitivities and specificities ranged from 93 to 100% when they were used in a clinical setting (104, 116, 128). When the Color Vue kit was used for an epidemiologic survey of Brazilians, the authors of the study concluded that the kit was good for diarrheal specimens but had greatly decreased sensitivity for nondiarrheal asymptomic infections (104).

E. histolytica

Initial attempts at immunodetection of *E. histolytica* antigens in fecal specimens indicated that the sensitivities and specificities of these methods were not as good as those of routine stool examination. However, more recent articles in the literature have indicated improved sensitivity and specificity with the use of monoclonal antibodies (97, 149)

and differentiation of pathogenic from nonpathogenic *E. histolytica* infections by detection of galactose-inhibitable adherence protein (1, 64). Commercial kits are available from Alexon and TechLab.

Giardia lamblia

G. lamblia is the most common human intestinal protozoan pathogen in the United States and is an important cause of diarrhea in children and adults. Outbreaks of disease are common within day-care centers as a result of person-to-person contact and may also occur in communities as a result of drinking water contaminated by infected human or animal feces. Infected persons often excrete cysts intermittently, so multiple stools collected over at least several days must sometimes be examined to detect the parasite.

Table 2 summarizes the features of five commercial products (one DFA and four EIAs) available in the United States for the immunodiagnosis of giardiasis. The ProSpecT/Giardia assay (Alexon) detects GSA 65 cyst antigen in eluates of fresh or fixed stools with a sensitivity of 96% and a specificity of 100%, compared to 74% sensitivity and 100% specificity for stool examination (118). Concentrated or polyvinyl alcohol-treated samples are not suitable for testing with EIA kits. Minor modification in the application of this product increased the sensitivity from 93.9 to 98% while maintaining specificity at 100% (2). Although this fecal antigen assay was initially reported not to detect trophozoite antigens, more recent data (138a) indicate that it does. The GiardEIA kit (Antibodies, Inc.) requires a fresh stool specimen but detects multiple cyst and trophozoite antigens. The manufacturer's evaluation found a kit sensitivity of 91% and a specificity of 96%. The Color Vue Giardia kit (Seradyn, Inc.) is also reported to detect multiple cyst and trophozoite antigens but can be used to test fresh, frozen, or fixed specimens. Test sensitivity was 97% and specificity was 96% (117).

A DFA assay that employs FITC-labeled monoclonal antibody for detection of *Giardia* cysts in stools is available. This assay is also available in a combined DFA kit for the simultaneous detection of *Giardia* cysts and cryptosporidial oocysts. The sensitivity and specificity of the kit were both 100% compared to those of microscopy (46). This kit can be used for quantitation of cysts and oocysts and thus is potentially useful for epidemiologic and control studies (151).

Microsporidia

Microsporidia are obligate intracellular parasites that have a variety of animal hosts. A number of infections have been reported in AIDS patients. Five genera of microsporidia have been recognized in humans: *Encephalitozoon*, *Nosema*, *Pleistophora*, *Enterocytozoon*, and *Microsporidium*, a genus comprising all those organisms not yet classified (45). Polyclonal antisera have been reported to be useful in IFA detection of *Enterocytozoon bieneusi* in clinical samples (duodenal fluid of biopsy sample, colonic fluid, feces) from AIDS patients (153). Monoclonal antibody reagents for *Encephalitozoon hellem* as well as polyclonal murine antisera raised against *Encephalitozoon cuniculi* and *Encephalitozoon hellem* used in an IFA test detected *Encephalitozoon cuniculi*, *Encephalitozoon hellem*, *Enterocytozoon bieneusi*, and *Septata intestinalis* (4). Monoclonal antibody reagents developed at the National Institutes of Health are being evaluated at several institutions for efficacy of microsporidium detection in clinical specimens. No reagents for immunodetection of microsporidial parasites or soluble parasite antigens are currently available commercially.

Trichomonas vaginalis

Trichomoniasis, an infection caused by *Trichomonas vaginalis*, is a common sexually transmitted disease. Diagnosis is made by detection of trophozoites in vaginal secretions or urethral specimens by wet mount microscopic examination, DFA staining of specimens, or culture. Sensitivities of the assays were reported as 60% for wet mounts and 86% for DFA compared to those of cultures (82). A kit that employs FITC- or enzyme-labeled monoclonal antibodies for use in a DFA or EIA procedure is available for detection of whole parasites in fluids. The Affirm VP Microbial Identification Test system was designed for use in clinics, physicians' offices, and clinical laboratories. The system uses synthetic oligonucleotide probes for the simultaneous detection of *Trichomonas vaginalis* and *Gardnerella vaginalis* from a single vaginal swab. Sensitivity for detection of *Trichomonas vaginalis* infection was 100% compared with that of wet mount examination but only 80% compared with that of culture (13).

MOLECULAR DIAGNOSTICS

Molecular diagnostic methods are gaining acceptance in clinical microbiology laboratories. Development of the PCR technique initiated a large number of reports on the use of DNA-based molecular methodologies. These methods are useful for microbial detection, strain typing, and epidemiology. A number of commercial diagnostic kits for viral and bacterial detection by molecular methods are available, but only one kit is available for detection of a parasite (13).

To date, most applications of molecular diagnostic techniques for parasites are better suited to research laboratories than to diagnostic laboratories (10). In some cases, molecular methods can provide important diagnostic information that is unavailable through traditional techniques. A good example is application of PCR and molecular probes for strain differentiation of *P. falciparum* (150) and detection of pyrimethamine resistance in the same organism (23). In a similar application, PCR and nonradioactive DNA probes were used for differentiation of pathogenic and nonpathogenic *E. histolytica* in stool samples (115). Other important PCR applications are assays for detection of parasites in environmental samples (74).

Apart from these special applications, simple PCR and DNA probe hybridization tests have been reported for most species of parasites. Recent examples include detection of *Trichomonas*, *Plasmodium*, *Trypanosoma*, *Echinococcus*, *Taenia*, *Leishmania*, *Brugia*, *Toxoplasma*, and *Onchocerca* spp. All PCR applications suffer from various limitations, principally at the stage of detection of the PCR product (112). To overcome these limitations, some commercial companies are offering solutions that transform the test tube or agarose gel format of detection into a 96-microwell format. Even so, the complexity of preparation (purification) of diagnostic samples for amplification and the cost of thermocyclers used for amplification will be prohibitive for many diagnostic laboratories in the near future. In principle, compared to serologic assays, PCR offers a tremendous advantage, but expanding the usefulness of this method, especially for a routine parasitic diagnostic laboratory, will take some time.

REFERENCES

1. **Abd-Alla, M. D., T. F. H. G. Jackson, V. Gathiram, A. M. El-Hawey, and J. I. Ravdin.** 1993. Differentiation of pathogenic *Entamoeba histolytica* infections from nonpathogenic infections by detection of galactose-inhibitable adherence protein antigen in sera and feces. *J. Clin. Microbiol.* **31:** 2845–2850.
2. **Addiss, D. G., H. M. Mathews, J. M. Stewart, S. P. Wahlquist, R. M. Williams, R. J. Finton, H. C. Spencer, and D. D. Juranek.** 1991. Evaluation of a commercially available enzyme-linked immunosorbent assay for *Giardia lamblia* antigen in stool. *J. Clin. Microbiol.* **29:**1137–1142.
3. **Aguilar, C., C. Cuellar, S. Fenou, and J. L. Guillen.** 1987. Comparative study of assays detecting circulating immune complexes and specific antibodies in patients infected with *Toxocara canis*. *J. Helminthol.* **61:**196–202.
4. **Aldras, A. M., J. M. Orenstein, D. P. Kotler, J. A. Shadduck, and E. S. Didier.** 1994. Detection of microsporidia by indirect immunofluorescence antibody test using polyclonal and monoclonal antibodies. *J. Clin. Microbiol.* **32:**608–612.
5. **Anusz, K. Z., P. H. Mason, M. W. Riggs, and L. E. Perryman.** 1990. Detection of *Cryptosporidium parvum* oocysts in bovine feces by monoclonal antibody capture enzyme-linked immunosorbent assay. *J. Clin. Microbiol.* **28:** 2770–2774.
6. **Arrowood, M. J., and C. R. Sterling.** 1989. Comparison of conventional staining methods and monoclonal antibody-based methods for *Cryptosporidium* oocyst detection. *J. Clin. Microbiol.* **27:**1490–1495.
7. **Au, A. C. S., D. A. Denham, M. W. Steward, C. C. Draper, M. D. Ismail, C. K. Rao, and J. W. Mak.** 1981. Detection of circulating antigens and immune complexes in feline and human lymphatic filariasis. *S.E. Asian J. Trop. Med. Public Health* **12:**492–498.
8. **Baldelli, F., R. Papili, D. Francisci, C. Tassi, G. Stagni, and S. Pauluzzi.** 1992. Post operative surveillance of human hydatidosis: evaluation of immunodiagnostic tests. *Pathology* **24:**75–79.

9. **Barker, D. C.** 1987. DNA diagnosis of human leishmaniasis. *Parasitol. Today* **3:**177–184.

10. **Bendall, R. P., and P. L. Chiodini.** 1993. New diagnostic methods for parasitic infections. *Curr. Opin. Infect. Dis.* **6:**318–322.

11. **Biglan, A. W., L. T. Glickman, and L. A. Lobes.** 1979. Serum and vitreous *Toxocara* antibody in nematode endophthalmitis. *Am. J. Ophthalmol.* **88:**898–901.

12. **Blaxter, M. L., M. A. Miles, and J. M. Kelly.** 1988. Specific serodiagnosis of visceral leishmaniasis using a *Leishmania donovani* antigen identified by expression cloning. *Mol. Biochem. Parasitol.* **30:**454–459.

13. **Briselden, A. M., and S. L. Hillier.** 1994. Evaluation of Affirm VP Microbial Identification Test for *Gardnerella vaginalis* and *Trichomonas vaginalis*. *J. Clin. Microbiol.* **32:**148–152.

14. **Cabrera, Z., R. M. E. Parkhouse, K. Forsyth, A. Gomez Priego, R. Pabon, and L. Yarzabol.** 1989. Specific detection of human antibodies to *Onchocerca volvulus*. *Trop. Med. Parasitol.* **40:**454–459.

15. **Chapman, P., B. Rush, and J. McLauchlin.** 1990. An enzyme immunoassay for detecting *Cryptosporidium* in faecal and environmental samples. *J. Med. Microbiol.* **32:**233–237.

16. **Chappell, C. L., and C. C. Matson.** 1992. *Giardia* antigen detection in patients with chronic gastrointestinal disturbances. *J. Fam. Pract.* **35:**49–53.

17. **Chisholm, E. S., T. K. Ruebush, A. J. Sulzer, and G. R. Healy.** 1978. *Babesia microti* infection in man: evaluation of an indirect immunofluorescent antibody test. *Am. J. Trop. Med. Hyg.* **27:**14–19.

18. **Chisholm, E. S., A. J. Sulzer, and T. K. Ruebush.** 1986. Indirect immunofluorescence test for human *Babesia microti* infection: antigenic specificity. *Am. J. Trop. Med. Hyg.* **35:**921–925.

19. **Cho, S. Y., S. T. Hong, Y. H. Rho, S. Choi, and Y. C. Han.** 1981. Application of micro-ELISA in serodiagnosis of human paragonimiasis. *Korean J. Parasitol.* **19:**151–156.

20. **Cho, S. Y., S. I. Kim, S. Y. Kang, Y. Kong, S. K. Han, Y. S. Shim, and Y. C. Han.** 1989. Antibody changes in paragonimiasis patients after praziquantel treatment as observed by ELISA and immunoblot. *Korean J. Parasitol.* **27:**15–21.

21. **Correa-Oliveira, R., L. M. S. Dusse, I. R. C. Viana, D. G. Colley, O. S. Carvalho, and G. Gazzinelli.** 1988. Human antibody responses against schistosomal antigens. I. Antibodies from patients with *Ancylostoma*, *Ascaris lumbricoides* or *Schistosoma mansoni* infections react with schistosome antigens. *Am. J. Trop. Med. Hyg.* **38:**348–355.

22. **Craig, P. S.** 1986. Detection of specific circulating antigen, immune complexes and antibodies in human hydatidosis from Trukana (Kenya) and Great Britain by enzyme-immunoassay. *Parasite Immunol.* **8:**171–188.

23. **David, J. R., and D. A. Harn.** 1993. Immunology and molecular biology of tropical infectious diseases. *Parasitol. Today* **9:**349–350.

24. **de Andrade, C. R., L. V. Kirchhoff, J. E. Donelson, and K. Otsu.** 1992. Recombinant *Leishmania* Hsp90 and Hsp70 are recognized by sera from visceral leishmaniasis patients but not Chagas' disease patients. *J. Clin. Microbiol.* **30:**330–335.

25. **Decoster, A., F. Darcy, A. Caron, and A. Capron.** 1988. IgA antibodies against P30 as markers of congenital and acute toxoplasmosis. *Lancet* **ii:**1104–1107.

26. **Decoster, A., F. Darcy, A. Caron, D. Vinatier, D. Houze de l'Aulnoit, G. Vittu, G. Niel, F. Heyer, B. Lecolier, M. Delcroix, J. C. Monnier, M. Duhamel, and A. Capron.** 1992. Anti-P30 IgA antibodies as prenatal markers of congenital toxoplasma infection. *Clin. Exp. Immunol.* **87:**310–315.

27. **Decoster, A., B. Slizewicz, J. Simon, C. Bazin, F. Darcy, G. Vittu, C. Boulanger, Y. Champeau, J. L. Demory, M. Duhamed, and A. Capron.** 1991. Platelia-Toxo IgA, a new kit for early diagnosis of congenital toxoplasmosis by detec-

tion of anti-P30 immunoglobulin A antibodies. *J. Clin. Microbiol.* **29:**2291–2295.

28. **de Savigny, D. H.** 1975. In vitro maintenance of *Toxocara canis* larvae and a simple method for the production of *Toxocara* ES antigen for use in serodiagnositic tests for visceral larva migrans. *J. Parasitol.* **61:**781–782.

29. **des Moutis, I., A. Ouaissi, J. M. Grzych, L. Yarzabal, A. Haque, and A. Capron.** 1983. *Onchocerca volvulus*: detection of circulating antigen by monoclonal antibodies in human onchocerciasis. *Am. J. Trop. Med. Hyg.* **32:**533–542.

30. **DeVault, G. A. J., J. W. King, M. S. Rohr, M. D. Landreneau, S. T. Brown, and J. C. McDonald.** 1990. Opportunistic infections with *Strongyloides stercoralis* in renal transplantation. *Rev. Infect. Dis.* **12:**653–671.

31. **Diaz, J. F., M. Verastegui, R. H. Gilman, V. C. Tsang, J. B. Pilcher, C. Gallo, H. H. Garcia, P. Torres, T. Montenegro, E. Miranda, and The Cysticercosis Working Group in Peru.** 1992. Immunodiagnosis of human cysticercosis (Taenia solium): a field comparison of an antibody-enzyme-linked immunosorbent assay (ELISA), an antigen-ELISA, and an enzyme-linked immunoelectrotransfer blot (EITB) assay in Peru. *Am. J. Trop. Med. Hyg.* **46:**610–615.

32. **Espino, A. M., B. E. Dumenigo, R. Fernandez, and C. M. Finlay.** 1987. Immunodiagnosis of human fascioliasis by enzyme-linked immunosorbent assay using excretory-secretory products. *Am. J. Trop. Med. Hyg.* **37:**605–608.

33. **Espino, A. M., J. C. Millan, and C. M. Finlay.** 1992. Detection of antibodies and circulating excretory-secretory antigens for assessing cure in patients with fascioliasis. *Trans. R. Soc. Trop. Med. Hyg.* **86:**649.

34. **Feldman, M., A. Plancarte, M. Sandoval, M. Wilson, and A. Flisser.** 1990. Comparison of two assays (EIA and EITB) and two samples (saliva and serum) for the diagnosis of neurocysticercosis. *Trans. R. Soc. Trop. Med. Hyg.* **84:**559–562.

35. **Feldmeier, H., U. Bienzle, R. Jansen-Rosseck, P. G. Kremsner, H. Wieland, G. Dobos, S. Schroeder, D. Fengler-Dopp, and H. H. Peter.** 1991. Sequelae after infection with *Trichinella spiralis*: a prospective cohort study. *Wien. Klin. Wochenschr.* **103:**111–116.

36. **Feldmeier, H., H. Fischer, and G. Blaumeiser.** 1987. Kinetics of humoral response during the acute and the convalescent phase of human trichinosis. *Zentralbl. Bakteriol. Hyg. A* **264:**221–234.

37. **Force, L., J. M. Torres, A. Carrillo, and J. Busca.** 1992. Evaluation of eight serological tests in the diagnosis of human echinococcosis and follow-up. *Clin. Infect. Dis.* **15:**473–480.

38. **Franco, E. L., K. W. Walls, and A. J. Sulzer.** 1981. Reverse enzyme immunoassay for detection of specific anti-*Toxoplasma* immunoglobulin M antibodies. *J. Clin. Microbiol.* **13:**859–864.

39. **Frasch, A. C. C., and M. B. Reyes.** 1990. Diagnosis of Chagas disease using recombinant DNA technology. *Parasitol. Today* **6:**137–139.

40. **Frenkel, J. K.** 1990. Toxoplasmosis in human beings. *J. Am. Vet. Med. Assoc.* **196:**240–248.

41. **Gam, A. A., F. A. Neva, and W. A. Krotoski.** 1987. Comparative sensitivity and specificity of ELISA and IHA for serodiagnosis of strongyloidiasis with larval antigens. *Am. J. Trop. Med. Hyg.* **37:**157–161.

42. **Gamble, H. R., and C. E. Graham.** 1984. Comparison of monoclonal antibody-based competitive and indirect enzyme-linked immunosorbent assays for the diagnosis of swine trichinosis. *Vet. Immunol. Immunopathol.* **6:**379–389.

43. **Gamble, H. R., D. Rapic, A. Marinculic, and K. D. Murrell.** 1988. Evaluation of excretory-secretory antigens for the serodiagnosis of swine trichinellosis. *Vet. Parasitol.* **30:**131–137.

44. **Garcia, L. S., T. C. Brewer, and D. A. Bruckner.** 1987. Fluorescent detection of *Cryptosporidium* oocysts in human fecal specimens by using monoclonal antibodies. *J. Clin. Microbiol.* **25:**119–121.

45. **Garcia, L. S., and R. Y. Shimizu.** 1993. Diagnostic parasitology: parasitic infections and the compromised host. *Lab. Med.* **24:**205–215.

46. **Garcia, L. S., A. C. Shum, and D. A. Bruckner.** 1992. Evaluation of a new monoclonal antibody combination reagent for the direct fluorescent detection of *Giardia* cysts and *Cryptosporidium* oocysts in human fecal specimens. *J. Clin. Microbiol.* **30:**3255–3257.

47. **Genta, R. M.** 1988. Predictive value of an enzyme-linked immunosorbent assay (ELISA) for the serodiagnosis of strongyloidiasis. *Am. J. Clin. Pathol.* **89:**391–394.

48. **Genta, R. M.** 1989. Global prevalence of strongyloidiasis: critical review with epidemiologic insights into the prevention of disseminated disease. *Rev. Infect. Dis.* **11:**755–767.

49. **Genta, R. M., R. W. Douce, and P. D. Walzer.** 1986. Diagnostic implications of parasite-specific immune responses in immunocompromised patients with strongyloidiasis. *J. Clin. Microbiol.* **23:**1099–1103.

50. **Glickman, L. T., R. Dypess, D. Hiles, and T. Gessner.** 1979. *Toxocara* specific antibody in the serum and aqueous humor of a patient with presumed ocular and visceral toxocariasis. *Am. J. Trop. Med. Hyg.* **28:**29–35.

51. **Glickman, L. T., R. B. Grieve, S. S. Lauria, and D. L. Jones.** 1985. Serodiagnosis of ocular toxocariasis: a comparison of two antigens. *J. Clin. Pathol.* **38:**103–107.

52. **Glickman, L. T., and P. M. Schantz.** 1981. Epidemiology and pathogenesis of zoonotic toxocariasis. *Epidemiol. Rev.* **3:**230–250.

53. **Glickman, L. T., P. M. Schantz, R. Dombroske, and R. Cypess.** 1978. Evaluation of serodiagnostic test for visceral larva migrans. *Am. J. Trop. Med. Hyg.* **27:**492–498.

54. **Glickman, L. T., P. M. Schantz, and R. B. Grieve.** 1986. Toxocariasis, p. 201–231. *In* K. W. Walls and P. M. Schantz (ed.), *Immunodiagnosis of Parasitic Diseases*, vol. 1. *Helminthic Diseases.* Academic Press, Inc., New York.

55. **Golightly, L. M., L. R. Hirschhorn, and P. F. Weller.** 1989. Fever and headache in a splenectomized woman. *Rev. Infect. Dis.* **11:**629–637.

56. **Gottstein, B.** 1985. Purification and characterization of a specific antigen from *Echinococcus multilocularis. Parasite Immunol.* **7:**202–212.

57. **Gottstein, B.** 1992. Molecular and immunological diagnosis of echinococcosis. *Clin. Microbiol. Rev.* **5:**248–261.

58. **Gottstein, B., P. Jacquier, S. Bresson-Hadni, and J. Eckert.** 1993. Improved primary immunodiagnosis of alveolar echinococcosis in humans by an enzyme-linked immunosorbent assay using the Em2plus antigen. *J. Clin. Microbiol.* **31:**373–376.

59. **Gottstein, B., P. M. Schantz, T. Todorov, A. G. Saimot, and P. Jacquier.** 1986. An international study on the serological differential diagnosis of human cystic and alveolar echinococcosis. *Bull. W.H.O.* **64:**101–105.

60. **Gottstein, B., K. Tschudi, J. Eckert, and R. Ammann.** 1989. Em2-ELISA for the follow-up of alveolar echinococcosis after complete surgical resection of liver lesions. *Trans. R. Soc. Trop. Med. Hyg.* **83:**389–393.

61. **Gretch, D. R., J. J. Warren, R. M. Bacina, E. D. Stefansson, and T. R. Fritsche.** 1992. Performance characteristics of a commercial antibody-capture enzyme immunoassay for detection of *Toxoplasma*-specific IgM antibodies. *Diagn. Microbiol. Infect. Dis.* **15:**587–593.

62. **Hancock, K., and V. C. W. Tsang.** 1986. Development and optimization of the FAST-ELISA for detecting antibodies to *Schistosoma mansoni. J. Immunol. Methods* **92:**167–176.

63. **Handman, E., G. F. Mitchell, and J. W. Goding.** 1987. *Leishmania major*: a very sensitive dot-blot ELISA for detection of parasites in cutaneous lesions. *Mol. Biol. Med.* **4:**377–383.

64. **Haque, R., K. Kress, S. Wood, T. F. Jackson, D. Lyerly, T. Wilkins, and W. A. J. Petri.** 1993. Diagnosis of pathogenic *Entamoeba histolytica* infection using a stool ELISA based on monoclonal antibodies to the galactose-specific adhesin. *J. Infect. Dis.* **167:**247–249.

65. **Healy, G. R.** 1988. Serology, p. 650–719. *In* J. I. Ravdin (ed.), *Amebiasis: Human Infection by Entamoeba histolytica.* John Wiley & Sons, Inc., New York.

66. **Hedman, K., M. Lappalainen, I. Seppaia, and O. Makela.** 1989. Recent primary toxoplasma infection indicated by a low avidity of specific IgG. *J. Infect. Dis.* **159:**736–740.

67. **Helbig, M., P. Frosch, P. Kern, and M. Frosch.** 1993. Serological differentiation between cystic and alveolar echinococcosis by use of recombinant larval antigens. *J. Clin. Microbiol.* **31:**3211–3215.

68. **Hillyer, G. V.** 1986. Fascioliasis, paragonimiasis, clonorchiasis, and opisthorchiasis, p. 39–68. *In* K. W. Walls and P. M. Schantz (ed.), *Immunodiagnosis of Parasitic Diseases*, vol. 1. *Helminthic Diseases.* Academic Press, Inc., New York.

69. **Hillyer, G. V., and M. Soler de Galanes.** 1991. Initial feasibility studies of the fast-ELISA for the immunodiagnosis of fascioliasis. *J. Parasitol.* **77:**362–365.

70. **Hillyer, G. V., M. Soler de Galanes, J. Rodriguez-Perez, J. Bjorland, M. Silva de Lagrava, S. Ramirez Guzman, and R. T. Bryan.** 1992. Use of the Falcon assay screening test–enzyme-linked immunosorbent assay (FAST-ELISA) and the enzyme-linked immunoelectrotransfer blot (EITB) to determine the prevalence of human fascioliasis in the Bolivian Altiplano. *Am. J. Trop. Med. Hyg.* **46:**603–609.

71. **Holmberg, M., A. B. Vaidya, F. C. Shenton, R. W. Snow, B. M. Greenwood, H. Wigzell, and U. Pettersson.** 1990. A comparison of two DNA probes, one specific for *Plasmodium falciparum* and one with wider reactivity, in the diagnosis of malaria. *Trans. R. Soc. Trop. Med. Hyg.* **84:**202–205.

72. **Homan, W. L., A. C. Derksen, and F. van Knapen.** 1992. Identification of diagnostic antigens from *Trichinella spiralis. Parasitol. Res.* **78:**112–119.

73. **Ivanoska, D., K. Cuperlovic, H. R. Gamble, and K. D. Murrell.** 1989. Comparative efficacy of antigen and antibody detection tests for human trichinellosis. *J. Parasitol.* **75:**38–41.

74. **Johnson, D. W., N. J. Pieniazek, and J. B. Rose.** 1993. DNA probe hybridization and PCR detection on *Cryptosporidium* compared to immunofluorescense assay. *Water Sci. Technol.* **27:**77–84.

75. **Johnson, R. J., and J. R. Johnson.** 1983. Paragonimiasis in Indochinese refugees—roentgenographic findings with clinical correlations. *Am. Rev. Respir. Dis.* **128:**534–538.

76. **Joynson, D. H., R. A. Payne, and B. K. Rawal.** 1990. Potential role of IgG avidity for diagnosing toxoplasmosis. *J. Clin. Pathol.* **43:**1032–1033.

77. **Kagan, I. G.** 1980. Serodiagnosis of parasitic diseases, p. 724–750. *In* E. H. Lennette, A. Balows, W. J. Hausler, Jr., and J. P. Truant (ed.), *Manual of Clinical Microbiology.* American Society for Microbiology, Washington, D.C.

78. **Kirchhoff, L. V.** 1990. *Trypanosoma* species, p. 2077–2084. *In* G. L. Mandell, R. G. Douglas, and J. E. Bennett (ed.), *Principles and Practice of Infectious Diseases.* Churchill Livingstone, New York.

79. **Knobloch, J., and I. Lederer.** 1983. Immunodiagnosis of human paragonimiasis by an enzyme immunoassay. *Trop. Med. Parasitol.* **34:**21–23.

80. **Knobloch, J., and E. Mannweiler.** 1983. Development and persistence of antibodies to *Entamoeba histolytica* in patients with amebic liver abscess. *Am. J. Trop. Med. Hyg.* **32:**727–732.

81. **Krause, P. J., S. Telford, R. Ryan, P. A. Conrad, M. Wilson, J. W. Thomford, and A. Speilman.** 1994. Antibody testing for babesiosis: evaluation of a serologic test for the diagnosis of human *Babesia microti* infection. *J. Infect. Dis.* **169:**923–926.

82. **Krieger, J. N., M. R. Tam, C. E. Stevens, I. O. Nielsen, J. Hale, N. B. Kiviat, and K. K. Holmes.** 1988. Diagnosis of trichomoniasis. Comparison of conventional wet-mount examination with cytologic studies, cultures, and monoclonal antibody staining of direct specimens. *JAMA* **259:**1223–1227.

83. **Lal, R. B., and E. A. Ottesen.** 1988. Enhanced diagnostic

specificity in human filariasis by IgG$_4$ antibody assessment. *J. Infect. Dis.* **158:**1034–1037.

84. **Lanier, A. P., D. E. Trujillo, P. M. Schantz, J. F. Wilson, and B. Gottstein.** 1987. Comparison of serologic tests for diagnosis and followup of alveolar hydatid disease. *Am. J. Trop. Med. Hyg.* **37:**609–615.

85. **Lappalainen, M., P. Koskela, M. Koskiniemi, P. Ammala, V. Hiilesmaa, K. Teramo, K. O. Raivio, J. S. Remington, and K. Hedman.** 1993. Toxoplasmosis acquired during pregnancy: improved serodiagnosis based on avidity of IgG. *J. Infect. Dis.* **167:**691–697.

86. **Le Fichoux, Y., P. Marty, and H. Chan.** 1987. Les IgA seriques specifiques dans le diagnostic de la toxoplasmose. *Ann. Pediatr.* (Paris) **34:**375–379.

87. **Leggatt, G. R., W. Yang, and D. P. McManus.** 1992. Serological evaluation of the 12 kDa subunit of antigen B in *Echinococcus granulosus* cyst fluid by immunoblot analysis. *Trans. R. Soc. Trop. Med. Hyg.* **86:**189–192.

88. **Lin, T. M., C. M. Schubert, D. J. Kiefer, S. Troy, R. Cort, J. L. Giegel, J. H. Lin, and L. Neill.** 1989. Two rapid and simple enzyme immunoassays for human antibodies to *Entamoeba histolytica. J. Immunoassay* **10:**301–313.

89. **Lobos, F., N. Weiss, M. Karam, H. R. Taylor, E. A. Ottesen, and T. B. Nutman.** 1991. An immunogenic *Onchocerca volvulus* antigen: a specific and early marker of infection. *Science* **251:**1603–1605.

90. **Luft, B. J., and J. S. Remington.** 1992. Toxoplasmic encephalitis in AIDS. *Clin. Infect. Dis.* **15:**211–222.

91. **Maddison, S. E.** 1987. The present status of serodiagnosis and seroepidemiology of schistosomiasis. *Diagn. Microbiol. Infect. Dis.* **7:**93–105.

92. **Maddison, S. E., S. B. Slemenda, P. M. Schantz, J. A. Fried, M. Wilson, and V. C. W. Tsang.** 1989. A specific diagnostic antigen of *Echinococcus granulosus* with an apparent molecular weight of 8 kDa. *Am. J. Trop. Med. Hyg.* **40:**377–383.

93. **Mahannop, P., W. Chaicumpa, P. Setasuban, N. Morakote, and P. Tapchaisri.** 1992. Immunodiagnosis of human trichinellosis using excretory-secretory (ES) antigen. *J. Helminthol.* **66:**297–304.

94. **Maizels, R. M., M. Philipp, and B. M. Ogilivie.** 1982. Molecules on the surface of parasitic nematodes as probes of the immune response in infection. *Immunol. Rev.* **61:**109–136.

95. **Maizels, R. M., and B. D. Robertson.** 1989. *Toxocara canis:* secreted glycoconjugate antigens in immunobiology and immunodiagnosis, p. 1–9. *In* M. W. Kennedy (ed.), *Parasitic Nematodes—Antigens, Membranes and Genes.* Taylor and Francis, Ltd., New York.

96. **Maleewong, W., C. Wongkham, S. Pariyanonda, and P. Intapan.** 1992. Analysis of antibody levels before and after praziquantel treatment in human paragonimiasis heterotremus. *Asian Pac. J. Allergy Immunol.* **10:**69–72.

97. **Merino, E., W. Glender, R. del Muro, and L. Ortiz-Ortiz.** 1990. Evaluation of the ELISA test for detection of *Entamoeba histolytica* in feces. *J. Clin. Lab. Anal.* **4:**39–42.

98. **Morakote, N., K. Sukhavat, C. Khamboonruang, V. Siriprasert, S. Suphawitayanukul, and W. Thamasonthi.** 1992. Persistence of IgG, IgM, and IgE antibodies in human trichinosis. *Trop. Med. Parasitol.* **43:**167–169.

99. **Moro, P. L., R. H. Gilman, M. Wilson, P. M. Schantz, M. Verastegui, H. H. Garcia, and E. Miranda.** 1992. Immunoblot (Western blot) and double diffusion (DD5) tests for hydatid disease cross-react with sera from patients with cysticercosis. *Trans. R. Soc. Trop. Med. Hyg.* **86:**422–423.

100. **Naot, Y., D. R. Guptill, and J. S. Remington.** 1982. Duration of IgM antibodies to *Toxoplasma gondii* after acute acquired toxoplasmosis. *J. Infect. Dis.* **145:**770.

101. **Naot, Y., and J. S. Remington.** 1980. An enzyme-linked immunosorbent assay for detection of IgM antibodies to *Toxoplasma gondii:* use for diagnosis of acute acquired toxoplasmosis. *J. Infect. Dis.* **142:**757–766.

102. **Neva, F. A.** 1986. Biology and immunology of human strongyloidiasis. *J. Infect. Dis.* **153:**397–405.

103. **Neva, F. A., A. A. Gam, and J. Burke.** 1981. Comparison of larval antigens in an enzyme-linked immunosorbent assay for strongyloidiasis in humans. *J. Infect. Dis.* **144:**427–432.

104. **Newman, R. D., K. L. Jaeger, T. Wuhib, A. A. M. Lima, R. L. Guerrant, and C. L. Sears.** 1993. Evaluation of an antigen capture enzyme-linked immunosorbent assay for detection of *Cryptosporidium* oocysts. *J. Clin. Microbiol.* **31:**2080–2084.

105. **Nishiyama, T., T. Araki, N. Mizuno, T. Wada, T. Ide, and T. Yamaguchi.** 1992. Detection of circulating antigens in human trichinellosis. *Trans. R. Soc. Trop. Med. Hyg.* **86:**292–293.

106. **Norman, L., and I. Kagan.** 1975. An evaluation of crude and fractionated trichina antigens in the diagnosis of trichinosis. *Bol. Chil. Parasitol.* **30:**58–64.

107. **Northern, C., and D. I. Grove.** 1990. *Strongyloides stercoralis:* antigenic analysis of infective larvae and adult worms. *Int. J. Parasitol.* **20:**381–387.

108. **Ottesen, E. A.** 1984. Filariasis and tropical eosinophilia, p. 390–411. *In* K. S. Warren and A. F. Mahmoud (ed.), *Tropical and Geographical Medicine.* McGraw-Hill Book Co., New York.

109. **Pan, A. A., G. B. Rosenberg, M. K. Hurley, G. J. Schock, V. P. Chu, and A. Aiyappa.** 1992. Clinical evaluation of an EIA for the sensitive and specific detection of serum antibody to *Trypanosoma cruzi* (Chagas' disease). *J. Infect. Dis.* **165:**585–588.

110. **Pappas, M. G., P. B. McGreevy, R. Hajkowski, L. D. Hendricks, C. N. Oster, and W. T. Hockmeyer.** 1983. Evaluation of promastigote and amastigote antigens in the indirect fluorescent antibody test for American cutaneous leishmaniasis. *Am. J. Trop. Med. Hyg.* **32:**1260–1267.

111. **Pelletier, L. L. J., C. B. Baker, A. A. Gam, T. B. Nutman, and F. A. Neva.** 1988. Diagnosis and evaluation of treatment of chronic strongyloidiasis in ex-prisoners of war. *J. Infect. Dis.* **157:**573–576.

112. **Persing, D. H.** 1991. Polymerase chain reaction: trenches to benches. *J. Clin. Microbiol.* **29:**1281–1285.

113. **Pollard, Z. F., W. H. Jarrett, W. S. Hagler, D. S. Allain, and P. M. Schantz.** 1979. ELISA for diagnosis of ocular toxocariasis. *Ophthalmology* **86:**750–752.

114. **Quick, R. E., B. L. Herwaldt, J. W. Thomford, M. E. Garnett, M. L. Eberhard, M. Wilson, D. H. Spach, J. W. Dickerson, S. R. Telford, K. R. Steingart, R. Pollock, D. H. Persing, J. M. Kobayashi, D. D. Juranek, and P. A. Conrad.** 1993. Babesiosis in Washington state: a new species of *Babesia? Ann. Intern. Med.* **119:**284–290.

115. **Romero, J. L., S. Descoteaux, S. Reed, E. Orozco, J. Santos, and J. Samuelson.** 1992. Use of polymerase chain reaction and nonradioactive DNA probes to diagnose *Entamoeba histolytica* in clinical samples. *Arch. Med. Res.* **23:**277–279.

116. **Rosenblatt, J. E., and L. M. Sloan.** 1993. Evaluation of an enzyme-linked immunosorbent assay for detection of *Cryptosporidium* spp. in stool specimens. *J. Clin. Microbiol.* **31:**1468–1471.

117. **Rosenblatt, J. E., L. M. Sloan, and S. K. Schneider.** 1993. Evaluation of an enzyme-linked immunosorbent assay for the detection of *Giardia lamblia* in stool specimens. *Diagn. Microbiol. Infect. Dis.* **16:**337–341.

118. **Rosoff, J. D., C. A. Sanders, S. S. Sonnad, P. R. De Lay, W. K. Hadley, F. F. Vincenzi, D. M. Yajko, and P. D. O'Hanley.** 1989. Stool diagnosis of giardiasis using a commercially available enzyme immunoassay to detect *Giardia*-specific antigen 65 (GSA 65). *J. Clin. Microbiol.* **27:**1997–2002.

119. **Sathar, M. A., A. E. Simjee, J. Delarey Nel, B. L. F. Bredenkamp, V. Gathiram, and T. F. H. G. Jackson.** 1988. Evaluation of an enzyme-linked immunosorbent assay in the serodiagnosis of amoebic liver abscess. *S. Afr. Med. J.* **74:**625–628.

120. **Sato, Y., F. Inoue, R. Matsuyama, and Y. Shiroma.** 1990.

Immunoblot analysis of antibodies in human strongyloidiasis. *Trans. R. Soc. Trop. Med. Hyg.* **84:**403–406.

121. **Sato, Y., M. Takara, and M. Otsuru.** 1985. Detection of antibodies in strongyloidiasis by enzyme-linked immunosorbent assay (ELISA). *Trans. R. Soc. Trop. Med. Hyg.* **79:**51–55.

122. **Schantz, P. M., and B. Gottstein.** 1986. Echinococcosis (hydatidosis), p. 69–107. In K. W. Walls and P. M. Schantz (ed.), *Immunodiagnosis of Parasitic Diseases*, vol. 1. *Helminthic Diseases.* Academic Press, Inc., New York.

123. **Schantz, P. M., D. Myer, and L. T. Glickman.** 1979. Clinical, serologic, and epidemiologic characteristics of ocular toxocariasis. *Am. J. Trop. Med. Hyg.* **28:**24–28.

124. **Schantz, P. M., D. Shanks, and M. Wilson.** 1980. Serological cross-reactions with sera from patients with echinococcosis and cysticercosis. *Am. J. Trop. Med. Hyg.* **29:**609–612.

125. **Schieven, B. D., and Z. Hussain.** 1990. Evaluation of an enzyme-immunoassay test kit for diagnosing infections with *Giardia lamblia. Serodiagn. Immunother. Infect. Dis.* **4:**109–113.

126. **Shim, Y. S., S. Y. Cho, and Y. C. Han.** 1991. Pulmonary paragonimiasis: a Korean perspective. *Semin. Respir. Med.* **12:**35–45.

127. **Shulman, I. A., and M. D. Appleman.** 1991. Transmission of parasitic and bacterial infections through blood transfusion within the U.S. *Crit. Rev. Clin. Lab. Sci.* **28:**447–459.

128. **Siddons, C. A., P. A. Chapman, and B. A. Rush.** 1992. Evaluation of an enzyme immunoassay kit for detecting *Cryptosporidium* in faeces and environmental samples. *J. Clin. Pathol.* **45:**479–482.

129. **Siracusano, A., S. Ioppolo, S. Notargiacomo, E. Ortona, R. Rigano, A. Teggi, F. De Rosa, and G. Vicari.** 1991. Detection of antibodies against *Echinococcus granulosus* major antigens and their subunits by immunoblotting. *Trans. R. Soc. Trop. Med. Hyg.* **85:**239–243.

130. **Slemenda, S. B., S. E. Maddison, E. C. Jong, and D. D. Moore.** 1988. Diagnosis of paragonimiasis by immunoblot. *Am. J. Trop. Med. Hyg.* **39:**469–471.

131. **Smith, H. V., H. R. Hinson, and R. W. A. Girdwood.** 1985. Serodiagnosis of human toxocariasis. *Trans. R. Soc. Trop. Med. Hyg.* **81:**516.

132. **Smith, H. V., J. R. Kusel, and R. W. A. Girdwood.** 1985. The production of human A and B blood group-like substances by in vitro maintained second stage *Toxocara caris* larvae: their presence on the outer larval surfaces and in their excretions/secretions. *Clin. Exp. Immunol.* **54:**625–633.

133. **Stepick-Biek, P., P. Thulliez, F. G. Araujo, and J. S. Remington.** 1990. IgA antibodies for diagnosis of acute congenital and acquired toxoplasmosis. *J. Infect. Dis.* **162:**270–273.

134. **Sulzer, A. J., and M. Wilson.** 1971. The indirect fluorescent antibody test for the detection of occult malaria in blood donors. *Bull. W.H.O.* **45:**375–379.

135. **Tsang, V. C., J. A. Brand, and A. E. Boyer.** 1989. An enzyme-linked immunoelectrotransfer blot assay and glycoprotein antigens for diagnosing human cysticercosis (*Taenia solium*). *J. Infect. Dis.* **159:**50–59.

136. **Tsang, V. C. W., K. Hancock, S. E. Maddison, A. L. Beatty, and D. M. Moss.** 1984. Demonstration of species-specific and cross-reactive components of the adult microsomal antigens from *Schistosoma mansoni* and *S. japonicum* (MAMA and JAMA). *J. Immunol.* **132:**2607–2613.

137. **Tsang, V. C. W., and P. P. Wilkins.** 1991. Immunodiagnosis of schistosomiasis: screen with FAST-ELISA and confirm with immunoblot. *Clin. Lab. Med.* **11:**1029–1039.

138. **Turner, P., B. Copeman, D. Gerisi, and R. Speare.** 1993. A comparison of the Og4C3 antigen capture ELISA, the Knott test, and IgG$_4$ assay and clinical signs in the diagnosis of Bancroftian filariasis. *Trop. Med. Parasitol.* **44:**45–48.

138a. **Turner, S.** Personal communication.

139. **Van Enk, R. A., K. K. James, and K. D. Thompson.** 1991. Evaluation of three commercial enzyme immunoassays for toxoplasma and cytomegalovirus antibodies. *Am. J. Clin. Pathol.* **95:**428–434.

140. **Verastegui, M., P. Moro, A. Guevara, T. Rodriguez, E. Miranda, and R. H. Gilman.** 1992. Enzyme-linked immunoelectrotransfer blot test for diagnosis of human hydatid disease. *J. Clin. Microbiol.* **30:**1557–1561.

141. **Weil, G. J., D. C. Jain, S. Santhanam, A. Malhotra, H. Kumar, K. V. P. Sethumadhavan, F. Liftis, and T. K. Ghosh.** 1987. A monoclonal antibody-based enzyme immunoassay for detecting parasite antigenemia in Bancroftian filariasis. *J. Infect. Dis.* **156:**350–355.

142. **Weil, G. J., K. V. P. Sethomadhavan, S. Santhanam, D. C. Jain, and T. K. Ghosh.** 1988. Persistence of parasite antigenemia following diethylcarbamazine therapy of Bancroftian filariasis. *Am. J. Trop. Med. Hyg.* **38:**589–595.

143. **Weiss, N., R. Hussain, and E. A. Ottesen.** 1982. IgE antibodies are more species-specific than IgG antibodies in human onchocerciasis and lymphatic filariasis. *Immunology* **45:**129–137.

144. **Wilson, M., and M. J. Arrowood.** 1993. Diagnostic parasitology: direct detection methods and serodiagnosis. *Lab. Med.* **24:**145–149.

145. **Wilson, M., R. T. Bryan, J. A. Fried, D. A. Ware, P. M. Schantz, J. B. Pilcher, and V. C. Tsang.** 1991. Clinical evaluation of the cysticercosis enzyme-linked immunoelectrotransfer blot in patients with neurocysticercosis. *J. Infect. Dis.* **164:**1007–1009.

146. **Wilson, M., D. A. Ware, and D. D. Juranek.** 1990. Serologic aspects of toxoplasmosis. *J. Am. Vet. Med. Assoc.* **196:**277–281.

147. **Wilson, M., D. A. Ware, and K. W. Walls.** 1987. Evaluation of commercial serodiagnostic kits for toxoplasmosis. *J. Clin. Microbiol.* **25:**2262–2265.

148. **Wirth, D. F., W. D. Rogers, H. Dourado, and B. Albuquerque.** 1989. Leishmaniasis: DNA probes for diagnosis, p. 249–258. In F. C. Tenover (ed.), *DNA Probes for Infectious Diseases.* CRC Press, Inc., Boca Raton, Fla.

149. **Wonsit, R., N. Thammapalerd, S. Tharavanij, P. Radomyos, and D. Bunnag.** 1992. Enzyme-linked immunosorbent assay based on monoclonal and polyclonal antibodies for the detection of *Entamoeba histolytica* antigens in faecal specimens. *Trans. R. Soc. Trop. Med. Hyg.* **86:**166–169.

150. **Wooden, J., S. Kyes, and C. H. Sibley.** 1993. PCR and strain identification in *Plasmodium falciparum. Parasitol. Today* **9:**303–305.

151. **Xiao, L., and R. P. Herd.** 1993. Quantitation of *Giardia* cysts and *Cryptosporidium* oocysts in fecal samples by direct immunofluorescence assay. *J. Clin. Microbiol.* **31:**2944–2946.

152. **Yokogawa, M.** 1982. Paragonimiasis, p. 123–142. In J. H. Steel (ed.), *CRC Handbook Series in Zoonoses: Parasitic Zoonoses.* CRC Press, Inc., Boca Raton, Fla.

153. **Zierdt, C. H., V. J. Gill, and W. S. Zierdt.** 1993. Detection of microsporidian spores in clinical samples by indirect fluorescent-antibody assay using whole-cell antisera to *Encephalitozoon cuniculi* and *Encephalitozoon hellem. J. Clin. Microbiol.* **31:**3071–3074.

Blood and Tissue Protozoa

LYNNE S. GARCIA, ALEXANDER J. SULZER, GEORGE R. HEALY,
KATHARINE K. GRADY, AND DAVID A. BRUCKNER

104

MALARIA

Malaria has probably had a greater impact on world history than any other infectious disease. It is still a common disease in many parts of the world, particularly tropical and subtropical areas; each year there are 200 to 300 million malaria cases with 2 to 3 million deaths, many of the victims being children. Cases in countries where malaria is not endemic may present some difficult diagnostic dilemmas, especially for infected nonimmune travelers, who are often misdiagnosed.

Of the four most common species that infect humans, *Plasmodium vivax* and *Plasmodium falciparum* account for ~95% of infections, with *P. vivax* accounting for ~80%. Evidence indicates that there may be a fifth species of human malaria (68). Means of transmission other than vectors include blood transfusions, use of contaminated hypodermic needles, and congenital infection.

No therapy is universally effective, and vaccines are not yet in general use. Although in the United States over a thousand cases of malaria are reported to the Centers for Disease Control and Prevention each year, the actual number of cases may be much higher. There has also been a definite increase over recent years in the number of cases of *P. falciparum* malaria, possibly because of increased chloroquine resistance.

Life Cycle and Morphology of the Parasite

When taking a blood meal, the female anopheline mosquito injects infective sporozoites from her salivary glands (Fig. 1). These sporozoites are carried by the bloodstream to the liver, where they invade parenchymal cells, initiating the preerythrocytic or primary exoerythrocytic cycle. The sporozoite increases in size and finally develops into a tissue schizont (meront) containing several thousand merozoites. Distended by the parasitic meront, the cell ruptures, releasing the merozoites into the bloodstream, where they invade erythrocytes (RBCs). Within the RBC, the merozoite (or young trophozoite) is vacuolated, ring shaped, more or less ameboid, and uninucleate. The excess proteins, an iron porphyrin and hematin left over from the metabolism of hemoglobin, combine and form what is called "malarial pigment." In the RBC, most of the merozoites develop into trophozoites and finally into erythrocytic schizonts, each containing merozoites in numbers characteristic of the par-

asite species. The schizont ruptures, releasing the merozoites, which again invade RBCs and continue the cycle. After a few erythrocytic cycles, some trophozoites develop into macrogametocytes or microgametocytes. Whether development is predetermined genetically or as a response to some specific stimulus is unknown. Asexual and sexual forms of three species of *Plasmodium* circulate in the bloodstream. However, in *P. falciparum* infections, as the parasite continues to grow, the RBC membrane becomes sticky, and the cells tend to adhere to the endothelial linings of the capillaries of the internal organs; only the ring forms and the gametocytes (occasionally mature schizonts) normally appear in the peripheral blood.

When gametocytes are taken up by an anopheline mosquito with its blood meal, they either become macrogametes or, in the case of microgametes, send forth flagella and divide into six to eight microgametes. A macrogamete fuses with a microgamete and becomes a zygote. The zygote develops into an ameboid ookinete that penetrates the gut wall and encysts on the outside of the gut within the hemocoel, becoming an oocyst. On rupture of the oocyst, sporozoites escape into the body cavity of the mosquito. Some of them make their way to the salivary glands, from which they may infect another vertebrate host when the next blood meal is taken by the mosquito.

If the circulating parasitemia is not eliminated by the immune system or therapy and if the numbers of infected RBCs, after an initial decrease, begin to increase, a recrudescence has occurred. Although all species may exhibit recrudescence, *Plasmodium malariae* may be in very low numbers and remain undetected for many years until a blood transfusion from an infected donor leads to symptoms in a susceptible host. If the erythrocytic infection is eliminated but then restored later because of a new invasion of the RBCs from liver merozoites, a recurrence or true relapse has occurred. Such a situation theoretically occurs only with *P. vivax* and *Plasmodium ovale*. These dormant liver forms may remain quiescent for a year or more and are called hypnozoites (46). Some of the morphologic features of the four different malarias are compared in Table 1 and shown in Fig. 2 through 6.

Clinical Disease

Symptoms of clinical disease include anemia, splenomegaly, and the classic paroxysm, with its chills, fever, and sweats.

1171

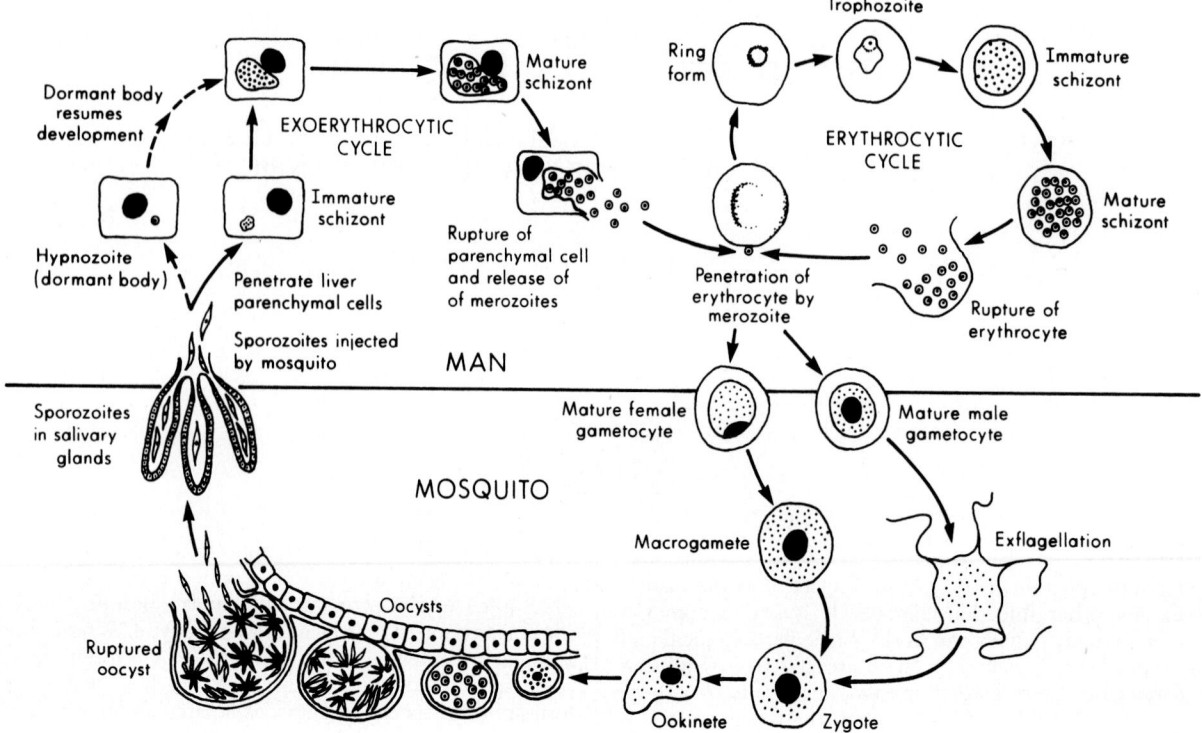

FIGURE 1 Life cycle of the malaria parasite. (Reproduced with permission from D. J. Krogstad and M. A. Pfaller, Prophylaxis and treatment of malaria, *Curr. Clin. Top. Infect. Dis.* **3:**56–73, 1983.)

Although the febrile paroxysms strongly suggest infection, many patients seen in medical facilities, particularly patients in the early stages, do not exhibit the typical fever pattern. They may have a steady low-grade fever or several small, random peaks each day. The typical paroxysm begins with a cold stage and rigors lasting for 1 to 2 h. During the next few hours, the patient spikes a high fever and feels very hot, and the skin is warm and dry. Marked sweating and a drop in body temperature to normal or subnormal occur in the last few hours of the paroxysm.

Malaria can mimic other diseases, and symptoms may include lethargy, anorexia, nausea, vomiting, diarrhea, and headache. Leukopenia, eosinophilia, and thrombocytopenia may also occur.

P. vivax

The primary clinical attack of *P. vivax* usually occurs from 7 to 10 days after infection. Once the typical paroxysms begin and after a period of irregular periodicity, a regular 48-h cycle is established. Since *P. vivax* infects only reticulocytes, parasitemia is usually limited to approximately 2.5% of the available RBCs. Splenomegaly occurs during the first few weeks (Fig. 2).

P. ovale

Although *P. ovale* and *P. vivax* infections are clinically similar, *P. ovale* malaria is less severe, tends to relapse less frequently, and usually ends with spontaneous recovery, often after no more than 6 to 10 paroxysms. *P. ovale* (like *P. vivax*) infects only reticulocytes (Fig. 3).

P. malariae

P. malariae primarily invades older RBCs, so the number of infected cells is limited. The incubation period may be much longer than that seen with *P. vivax* or *P. ovale* malaria, ranging from about 27 to 40 days. Regular periodicity of 72 h is seen from the beginning, and the paroxysm is more severe, including a longer period of chills and more severe symptoms during the febrile stage (Fig. 4).

Proteinuria is common in *P. malariae* infections and in children may be associated with clinical signs of the nephrotic syndrome. Kidney problems may result from deposition within the glomeruli of circulating antigen-antibody complexes in an antigen-excess situation seen with chronic infection. Infection may end with spontaneous recovery, or there may be a recrudescence or a series of recrudescences over many years.

P. falciparum

P. falciparum invades all ages of RBCs; parasitemia may exceed 50% (Fig. 5). Schizogony (merogony) occurs in the internal organs (spleen, liver, bone marrow, etc.) rather than in the circulating blood. Onset of a *P. falciparum* malaria attack occurs 8 to 12 days after infection and is preceded by 3 to 4 days of vague symptoms such as aches, pains, headache, anorexia, or nausea. Cycle periodicity will not be established during the early stages, and presumptive diagnosis may be totally unrelated to a possible malaria infection. If the fever does develop a synchronous cycle, the cycle is usually somewhat less than 48 h. An untreated

TABLE 1 Information on plasmodia and parasite morphology in Giemsa-stained thin blood smears[a]

Characteristic	P. vivax	P. malariae	P. falciparum	P. ovale
Persistence of exoerythrocytic cycle	Yes	No	No	Yes
Relapses	Yes	No, but long-term recrudescences are recognized	No long-term relapses	Possible, but usually spontaneous recovery
Time of cycle	44–48 h	72 h	36–48 h	48 h
Appearance of parasitized RBCs (size and shape)	1.5–2 times larger than normal; oval to normal; may be normal size until ring fills half of cell; RBC often ameboid	Normal shape; may be normal size or slightly smaller	Normal	60% of cells larger than normal and oval; 20% have irregular frayed edges; often compressed to oval
Schüffner's dots (James stippling in P. ovale; eosinophilic stippling)	Usually present in all cells except early ring forms	None	None; occasionally commalike red dots (Maurer's dots) are present	Present in all stages including early ring forms; dots may be larger and darker than in P. vivax
Color of cytoplasm	Decolorized, pale	Normal	Normal, bluish tinge at times	Decolorized, pale
Multiple rings/cell	Occasionally	Rarely	Commonly	Occasionally
Developmental stages present in peripheral blood	All	Ring forms few, as ring stage brief; mostly growing and mature trophozoites and schizonts	Young ring forms and no older stages; few gametocytes	All
Appearance of parasite Young trophozoite (early ring form)	Ring is 1/3 diameter of cell with cytoplasmic circle around vacuole; heavy chromatin dot	Ring is often smaller than in P. vivax, occupying 1/8 of cell; heavy chromatin dot; vacuole at times "filled in"; pigment forms early	Delicate, small ring with small chromatin dot (frequently 2); scanty cytoplasm around small vacuoles; sometimes at edge of RBC (appliqué form) or filamentous slender form; may have multiple rings/cell	Ring is smaller and more compact than in P. vivax; otherwise similar to that of P. vivax
Growing trophozoite	Multishaped irregular ameboid parasite; streamers of cytoplasm close to large chromatin dot; vacuole retained until close to maturity; increasing amounts of brown pigment	Compact rounded or band-shaped solid forms; chromatin may be hidden by coarse dark brown pigment	Heavy ring forms; fine pigment grains	Ring shape maintained until late in development; nonameboid compared to P. vivax

[a]Table is from information in reference 28.

(Continued on next page)

TABLE 1 Information on plasmodia and parasite morphology in Giemsa-stained thin blood smears[a] (Continued)

Characteristic	P. vivax	P. malariae	P. falciparum	P. ovale
Mature trophozoite	Irregular ameboid mass; ≥1 small vacuoles retained until schizont stage; fills almost entire cell; fine golden-brown pigment	Vacuoles disappear early; cytoplasm compact, oval, band shaped, or nearly round, almost filling cell; chromatin may be hidden by peripheral coarse, dark brown pigment	Not seen in peripheral blood (except in severe infections); all phases following ring form develop in capillaries of viscera	Compact; vacuoles disappear; pigment dark brown but less so than in P. malariae
Schizont (presegmenter, meront)	Progressive chromatin division; cytoplasmic bands containing clumps of brown pigment	Similar to P. vivax but smaller; darker, larger pigment granules peripheral or central	Not seen in peripheral blood (see above)	Smaller; more compact than P. vivax
Mature schizont (meront)	16 (12–24) merozoites, each with chromatin and cytoplasm, filling entire RBC, which can hardly be seen; small, with irregular size and shape of nuclei	8 (6–12) merozoites in rosettes or irregular clusters filling normal-size cells, which can hardly be seen; central arrangement of brown-green pigment	Not seen in peripheral blood	3/4 of cells occupied by 8 (8–12) merozoites in rosettes or irregular clusters; large, regular size and shape of nuclei
Gametocyte	Rounded or oval homogeneous cytoplasm; diffuse delicate golden to light brown pigment throughout parasite; eccentric compact chromatin mass may vary in size	Similar to P. vivax but fewer, pigment darker and coarser	Sex differentiation difficult; crescent or sausage shapes characteristic; may appear in "showers"; heavy black pigment near chromatin dot, which is often central	Smaller than P. vivax
Main criteria	Large pale RBC; trophozoite irregular; pigment usually present; fine Schüffner's dots not always present; several phases of growth seen in one smear; gametocytes appear as early as day 3	RBC normal in size and color; trophozoites compact, stain usually intense, band forms not always seen; coarse pigment; no stippling of RBCs; merozoites in schizont (meront) often in rosette around centrally located pigment; gametocytes appear after few wk	Development following ring stage takes place in blood vessels of internal organs; delicate ring forms and crescent-shaped gametocytes are only forms normally seen in peripheral blood; gametocytes appear after 7–10 days	RBC enlarged, oval, with fimbriated edges; coarse Schüffner's (James) dots in all stages; merozoites in schizont (meront) often in rosette around centrally located pigment; gametocytes appear after 4 days or as late as 18 days

primary attack of *P. falciparum* malaria usually ends within 2 to 3 weeks. True relapses from the liver do not occur, and recrudescences after a year are rare.

Cerebral malaria is most often seen with *P. falciparum*; the patient may become disoriented or violent, may develop severe headaches and lapse into coma, or may suddenly become comatose with few or no prior symptoms. There is no correlation between the severity of the symp-

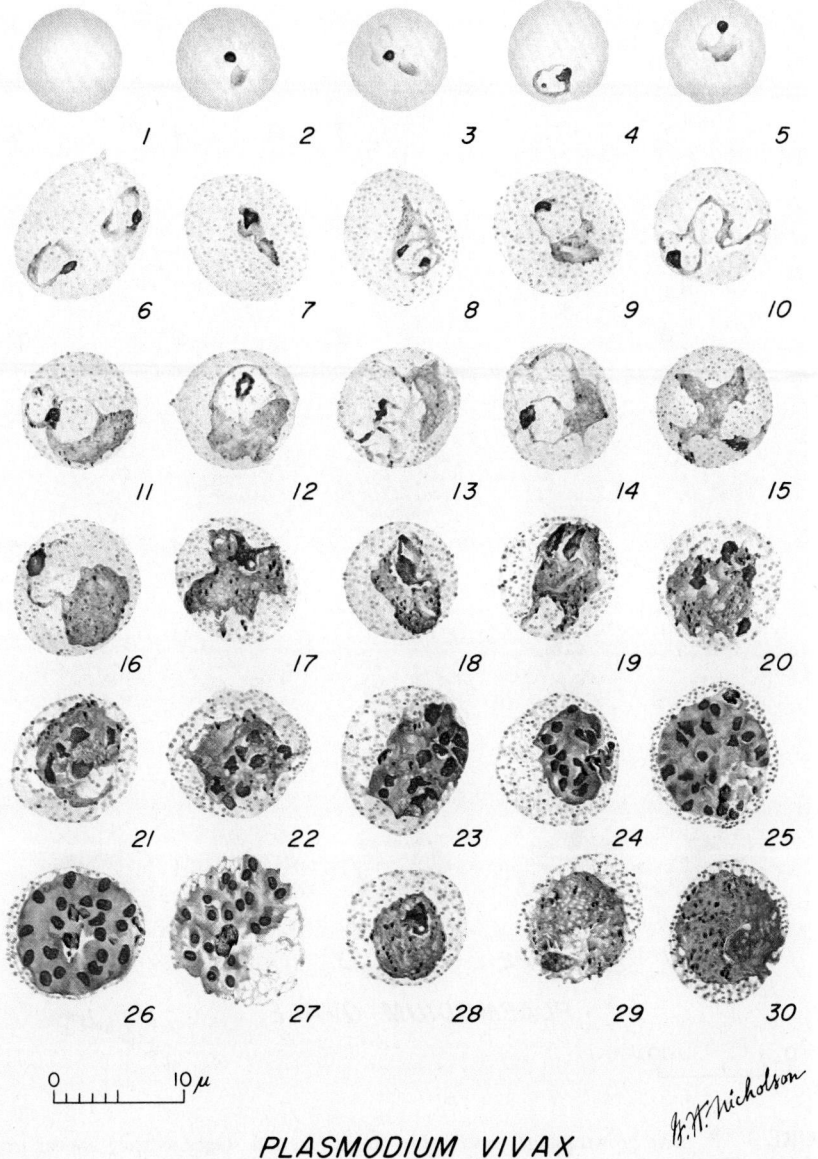

PLASMODIUM VIVAX

FIGURE 2 *P. vivax* growth stages: successive developmental stages as they appear in stained blood films. The cytoplasm of the developing trophozoites and immature schizonts is normally much paler than shown in this figure; color intensity has been increased for purposes of illustration. Drawings: 1, normal RBC; 2 to 5, young trophozoites (rings); 6 to 16, growing trophozoites; 17 and 18, mature trophozoites; 19 to 21, young (early) immature schizonts; 22 and 23, older immature schizonts; 24 to 27, nearly mature and mature schizonts; 28 and 29, nearly mature and mature macrogametocytes; 30, mature microgametocyte. (Reproduced with permission from G. R. Coatney, W. E. Collins, M. Warren, and P. G. Contacos, *The Primate Malarias*, U.S. Department of Health, Education and Welfare, Wasington, D.C., 1971.)

toms and the peripheral blood parasitemia. Patients with cerebral malaria are infected with RBC-rosette-forming *P. falciparum*, and plasma from these patients generally has no antirosetting activity; both of these situations may lead to plugging of the blood vessels. In contrast, *P. falciparum* parasites from patients with mild malaria lack the rosetting phenotype or have a much lower rosetting rate (21). Extreme fevers of 107°F (ca. 42°C) or

higher may develop in an uncomplicated attack or in cerebral malaria. Without vigorous therapy, the patient usually dies.

Although infrequent, blackwater fever is most often associated with *P. falciparum* malaria. There is usually a history of previous malarial attacks. Sudden intravascular hemolysis results in a dramatic color change in the urine that is due to the high methemoglobin content.

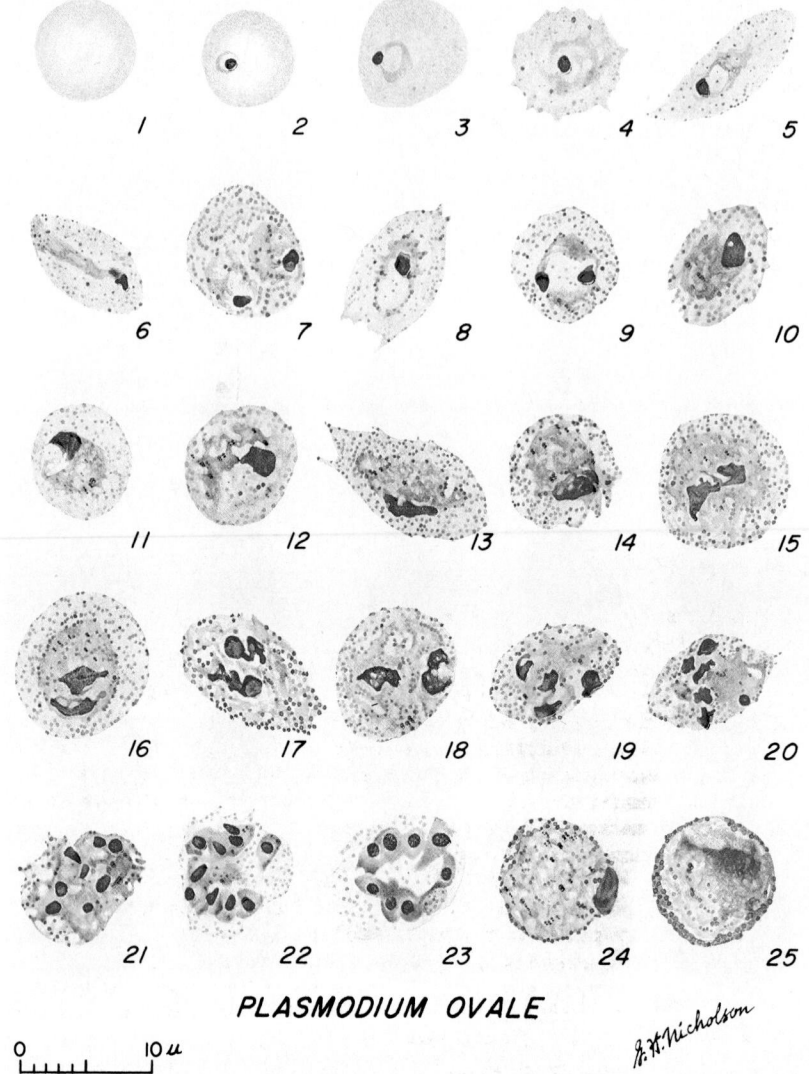

PLASMODIUM OVALE

F. H. Nicholson

0   10 µ

FIGURE 3 *P. ovale* growth stages: successive developmental stages as they appear in stained blood films. The Schüffner's dots (stippling) are normally larger than this figure shows. Drawings: 1, normal RBC; 2 to 5, young trophozoites (rings); 6 to 12, growing trophozoites; 13 to 15, mature trophozoites; 16 to 22, immature schizonts; 23, mature schizont; 24, mature macrogametocyte; 25, mature microgametocyte. (Reproduced with permission from G. R. Coatney, W. E. Collins, M. Warren, and P. G. Contacos, *The Primate Malarias*, U.S. Department of Health, Education and Welfare, Washington, D.C., 1971.)

Diagnosis

Malaria is life threatening, and laboratory requests for blood smear examination and organism identification should be treated as STAT requests. It is also important to obtain some specific patient history information. Malaria is usually associated with patients having a history of travel within an area of endemicity, although other routes of infection are well documented. The smears should be examined at length under oil immersion. Even with a low parasitemia on blood smears, the patient may still be faced with a serious, life-threatening disease.

Organism recovery and subsequent identification may be more difficult than textbooks imply. It is very important that this fact be recognized, particularly when one is dealing with a possible fatal infection with *P. falciparum*. When requests for malaria smears are received, patient history should be available to the laboratory and should include the following (28, 77).

1. Where has the patient been, and what was the date of return to the United States? Where does the patient live (possible "airport" malaria) (39)?

2. Has malaria ever been diagnosed in the patient before? If so, what species was identified?

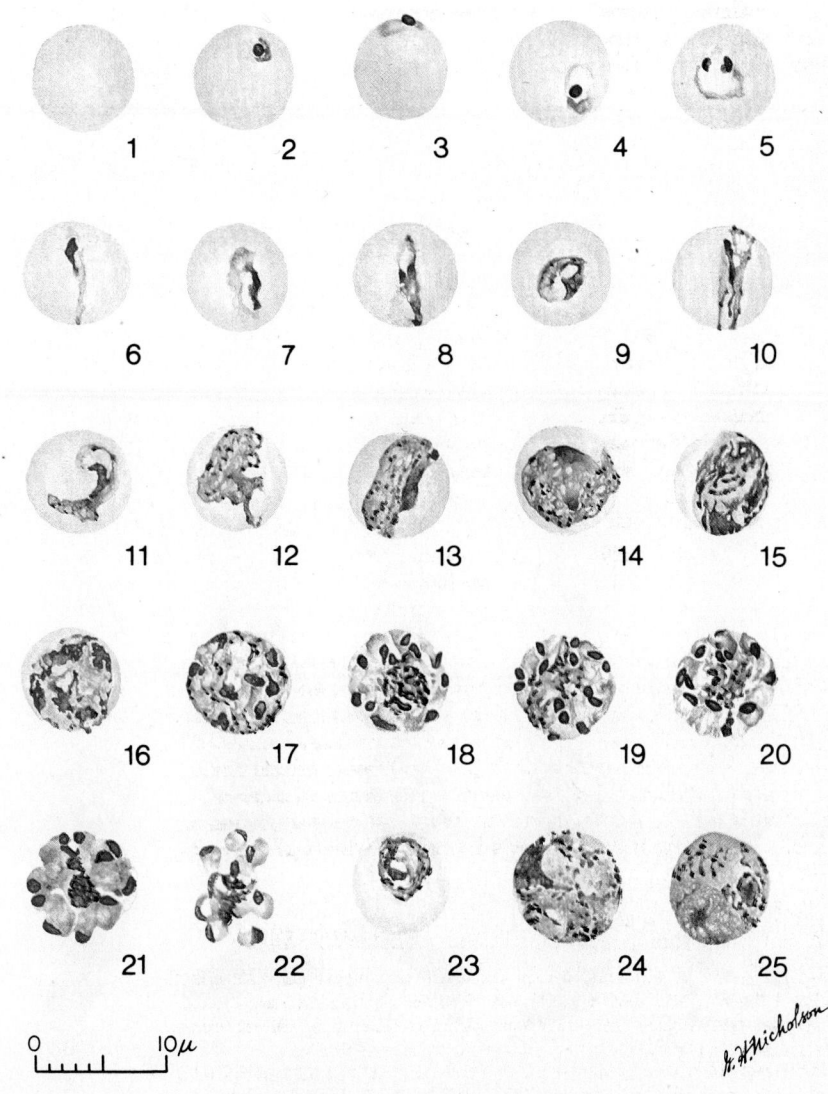

PLASMODIUM MALARIAE

FIGURE 4 *P. malariae* growth stages: successive developmental stages as they appear in stained blood films. Drawings: 1, normal RBC; 2 to 5, young trophozoites (rings); 6 to 11, growing trophozoites; 12 and 13, nearly mature and mature trophozoites; 14 to 20, immature schizonts; 21 and 22, mature schizont; 23, developing gametocyte; 24, mature macrogametocyte; 25, mature microgametocyte. (Reproduced with permission from G. R. Coatney, W. E. Collins, M. Warren, and P. G. Contacos, *The Primate Malarias*, U.S. Department of Health, Education and Welfare, Washington, D.C., 1971.)

3. What medication (prophylaxis or otherwise) has the patient received, and how often? When was the last dose taken?

4. Has the patient ever received a blood transfusion? Is there a possibility of other needle transmission (drug user)?

5. When was the blood specimen drawn, and was the patient symptomatic at the time? Is there any evidence of a fever periodicity?

One set of negative blood smears will not rule out malaria, especially when partial prophylactic medication or therapy has been used and the smear contains few organisms. Although the parasitemia may be low, serious disease may be present. Patients with a relapse or early primary disease may also have a low parasitemia. Additional blood specimens should be examined over a 36-h time frame.

Both thick and thin blood films should be prepared on patient admission, and at least 200 to 300 oil immersion fields of the thin film should be examined before a negative report is given (19, 28, 62). A thick blood film contains 10 to 15 times the volume used for a thin blood film. The sensitivity of a thick blood film is also due to the fact that

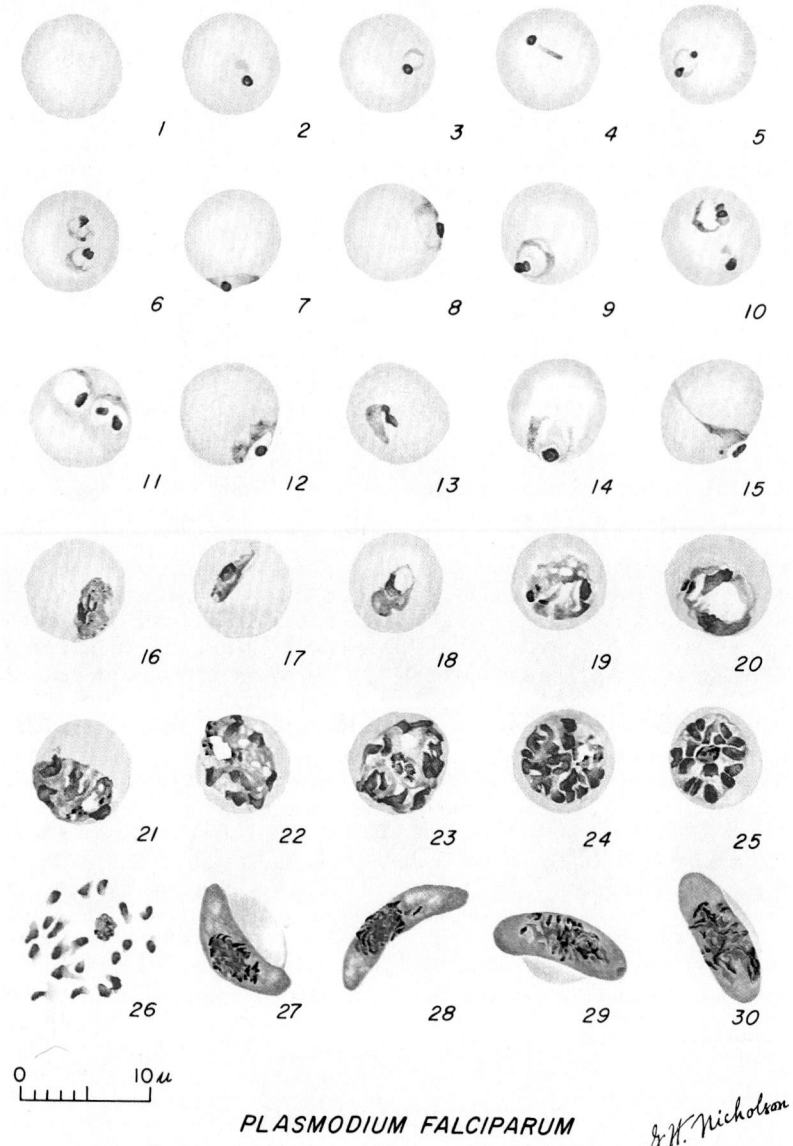

1 2 3 4 5

6 7 8 9 10

11 12 13 14 15

16 17 18 19 20

21 22 23 24 25

26 27 28 29 30

0 10 μ

PLASMODIUM FALCIPARUM

G. H. Nicholson

FIGURE 5 *P. falciparum* growth stages: successive developmental stages as they appear in stained blood films. The cytoplasm of the rings is normally much paler than shown in this figure; color intensity has been increased for purposes of illustration. Rings and gametocytes will be found in peripheral blood films; other stages occur in RBCs but are sequestered in the capillaries of internal organs and are rarely found in peripheral blood films. Drawings: 1, normal RBC; 2 to 11, young trophozoites (rings); 12 to 15, growing trophozoites; 16 to 18, mature trophozoites; 19 to 22, immature schizonts; 23 to 26, nearly mature and mature schizonts; 27 and 28, mature macrogametocytes; 29 and 30, mature microgametocytes. (Reproduced with permission from G. R. Coatney, W. E. Collins, M. Warren, and P. G. Contacos, *The Primate Malarias*, U.S. Department of Health, Education and Welfare, Washington, D.C., 1971.)

RBCs are lysed during staining; remaining structures include parasites, leukocytes, and platelets (Fig. 6 through 9). Although there are fewer parasites per unit area in the thin film, methanol fixation preserves RBC morphology; the relationship between the sizes of infected and uninfected RBCs will be obvious, as will other characteristics of the parasite itself. Although organisms can be seen by using other blood stains such as Wright's stain, Giemsa stain is recommended for all parasitic blood work. Blood collected using EDTA anticoagulant is acceptable; however, if the blood remains in the tube for longer than 1 h, true stippling may not be visible within the infected RBCs, and other deceptive changes in parasite morphology may occur. The proper ratio between blood and anticoagulant is necessary for good organism morphology. Heparin can also be used, but EDTA is preferred. Finger stick blood is recommended, particularly when the volume of blood required is minimal (no other hematologic procedures ordered). The blood should be free flowing when taken for smear preparation and should not be contaminated with alcohol used to clean the finger prior to the stick.

Species identification is essential to good patient management, since this information will influence which drug(s) will be used. Many patients with *P. falciparum* infections have few or no crescent-shaped gametocytes visible in the blood. Low parasitemias with the delicate ring forms may be missed; consequently, oil immersion examination at ×1,000 magnification is mandatory. If the majority of organisms are young rings, it may be impossible to differentiate *P. falciparum* from other species; mixed infections (5 to 7% of malaria infections) present another difficulty.

Microhematocrit centrifugation with the QBC tube (glass capillary tube and closely fitting plastic insert; QBC capillary blood tubes; Becton Dickinson, Franklin Lakes, N.J.) has been used for the detection of blood parasites (6, 49). Although a malaria infection could be detected by this method (which may be more sensitive than the thick or thin blood smear), appropriate thick and thin blood films would still need to be examined to accurately identify the organisms. When the QBC system is used, *P. vivax* and *P. ovale* look alike; in a mixed infection, the less common species, which may be *P. falciparum*, may be missed.

Diagnostic problems with automated differential instruments have been reported (28). Cases of malaria as well as of *Babesia* infection have been missed with these methods. Although the instruments are not designed to detect intracellular blood parasites, the inability of the automated systems to discriminate between uninfected RBCs and those infected with parasites may pose serious diagnostic problems.

Although antigen detection and other methods such as PCR have been developed, they are not yet commonly used (7, 8, 65, 83, 88).

Serologic Diagnosis

The time required for development of antibody and for appropriate testing prevents serologic testing from being practical, particularly in a severely ill patient. Situations in which serologic results might be helpful include blood donor testing and epidemiologic studies. Specific testing for antibodies to *P. falciparum*, *P. vivax*, *P. malariae*, and *P. ovale* is currently performed at the Centers for Disease Control and Prevention. There is usually some cross-reactivity between species (see chapter 102 of this Manual for additional details).

Treatment

Antimalarial drugs are often referred to as tissue schizonticides (kill tissue schizonts), blood schizonticides (kill blood schizonts), gametocytocides (kill gametocytes), and sporonticides (prevent formation of sporozoites within the mosquito). Although chloroquine-resistant *P. falciparum* is well recognized, chloroquine has generally been effective for both chemoprophylaxis and treatment of *P. vivax*, *P. ovale*, and *P. malariae* infections. Suppressive therapy is used to prevent clinical symptoms in individuals who are going into areas where the parasites are endemic. Antimalarial drugs are effective against erythrocytic forms but do not prevent sporozoites from entering the liver and beginning the preerythrocytic cycle. The radical-cure approach eradicates both the liver and the RBC stages, including gametocytes.

Epidemiology and Prevention

Malaria is primarily a rural disease and depends on transmission by female anopheline mosquitoes. There are great variations in vector efficiency, which is also related to parasite strain differences. Even when the vector is present in an area, an average number of bites per person per day must be sustained, or the infection gradually dies out. Once an area is clear of the infection, there may also be a drop in population immunity, a situation that may lead to a severe epidemic if and when parasites are reintroduced.

Immunity

Genetic alterations in RBCs may confer natural immunity to malaria. Duffy antigen-negative RBCs lack surface receptors for *P. vivax* invasion. Many West Africans and some American blacks are Duffy antigen negative, which may explain the low incidence of *P. vivax* in West Africa. In other areas of Africa, *P. vivax* is much more prevalent. Partial resistance to *P. falciparum* is seen in individuals with the sickle cell trait and in those with sickle cell anemia. Resistance is related to sickling of hemoglobin S (HbS) containing RBCs. Other resistance factors include formation of deoxyhemoglobin aggregates within the cells, loss of potassium from sickled cells, and the fact that parasites actually cause HbS cells to sickle. Resistance is also seen in glucose-6-phosphate dehydrogenase-deficient cells. Partial immunity to malarial infection is seen in areas of endemicity when HbC, HbE, β-thalassemia, and phenylketonuria deficiencies exist (58).

During the first year of life, infants are relatively immune to malarial infections because they have a large percentage of HbF, passive immunity from maternal antibodies, and diets deficient in *para*-aminobenzoic acid (37).

Vaccines

Techniques developed for in vitro culture of *P. falciparum* have enhanced studies of parasite metabolism, resistance, and mechanisms of attachment and invasion. With the use of genetic engineering and other newer techniques, progress toward vaccine production may lead to effective protection against malarial infections (36, 63, 78).

BABESIOSIS

Recognition of the spectrum of human disease caused by organisms in the genus *Babesia* has recently resulted in an increase in the number of cases reported. The infection can be transmitted via transfusion, can cause serious illness (particularly in immunocompromised patients), and is difficult to treat (52).

Babesiosis has been documented worldwide, and several outbreaks in humans have occurred in the northeastern United States, particularly in Long Island, Cape Cod, and the islands off the East Coast, and in the Midwest (3, 43, 71–73). *Babesia microti* is the cause of most infections in nonsplenectomized patients in the United States. Other species involved in human infection following splenectomy include *Babesia bovis* and *Babesia divergens*. A new species, as yet unnamed, from Washington State has been identified (69). A case in a 24-year-old male who had been splenectomized in 1985 for an immune-mediated thrombocytopenia was reported from California (20). This is the third reported case of presumed indigenous transmission of babe-

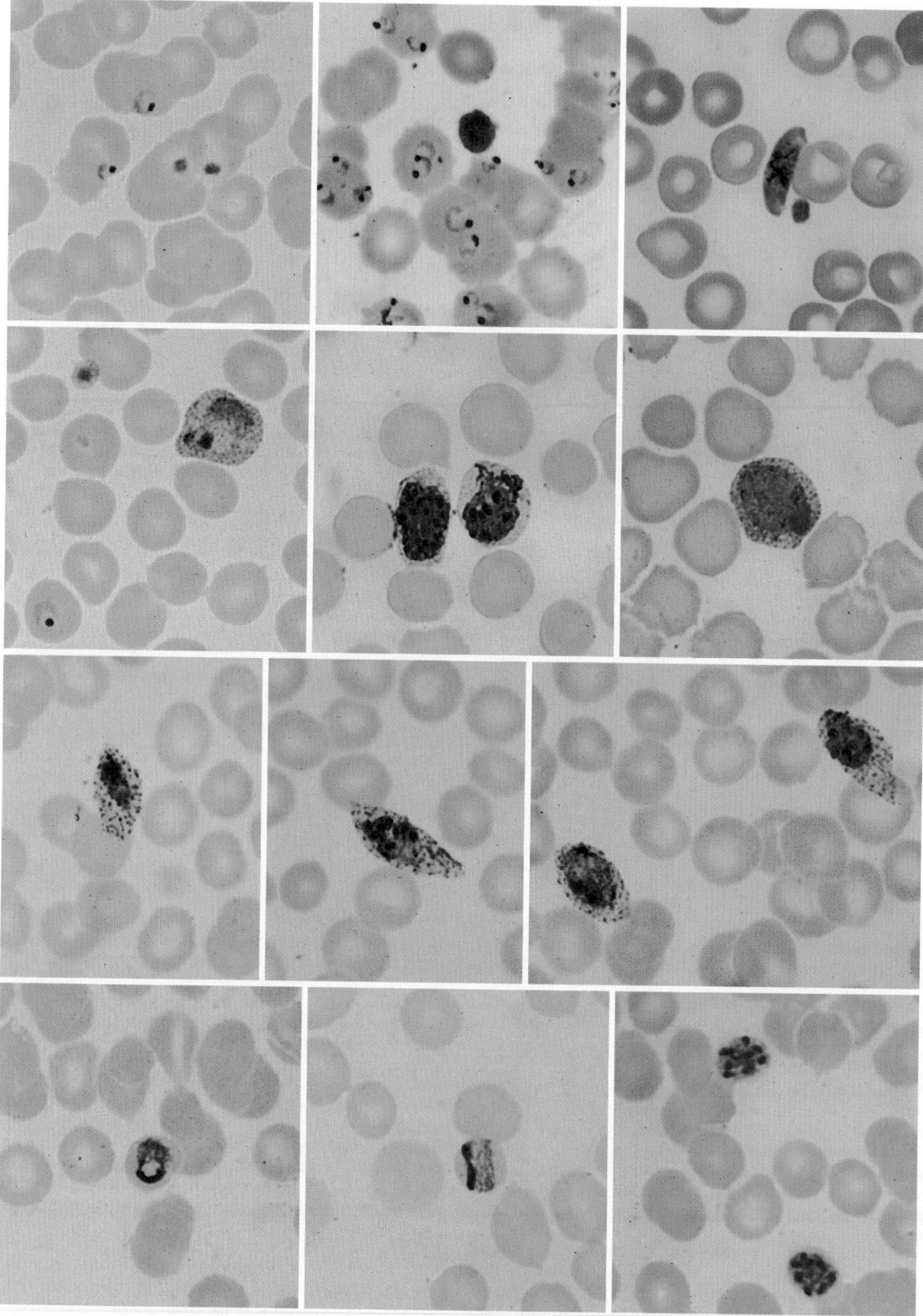

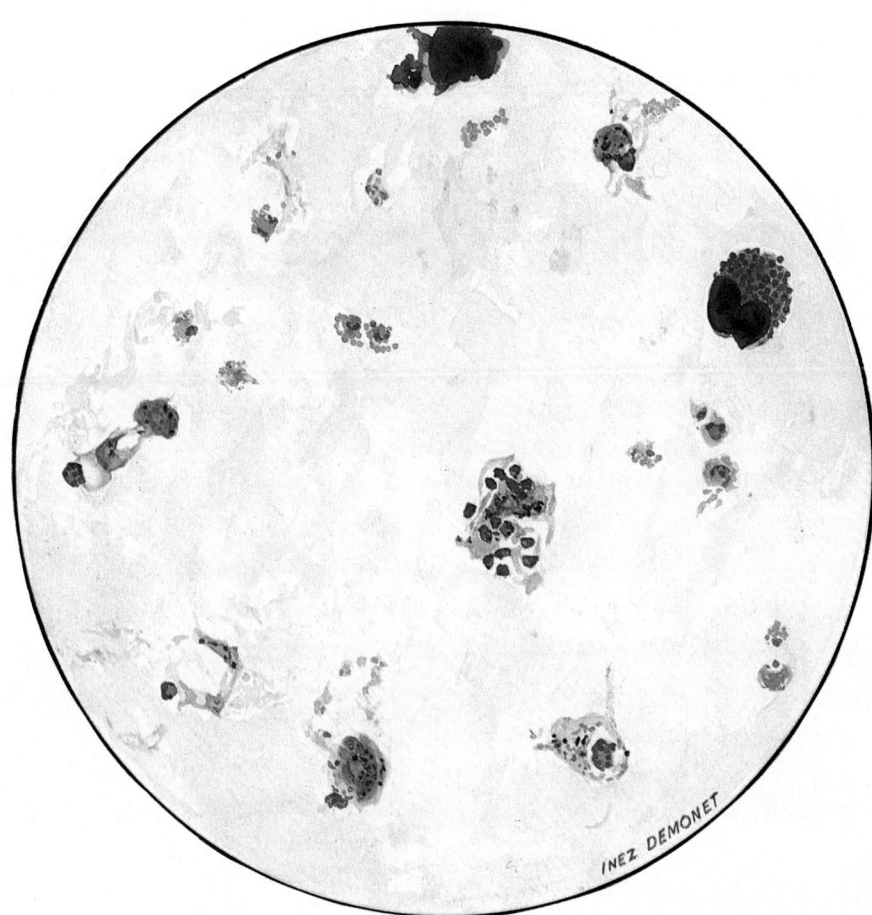

INEZ DEMONET

FIGURE 7 Morphology of *P. vivax* in a thick blood film. There are ameboid trophozoites at 2, 7, and 9 o'clock; an immature schizont with two chromatin dots (nuclear material) at 6 o'clock; a macrogametocyte at 5 o'clock; and a mature schizont near the center. *Note:* Outside of areas of endemicity, the parasites on a thick film are normally fewer than in this figure. (Reproduced from A. Wilcox, *Manual for the Microscopical Diagnosis of Malaria in Man*, U.S. Department of Health, Education and Welfare, Washington, D.C., 1960.)

siosis in California, all in patients who had been splenectomized.

Life Cycle and Morphology of the Parasite

Infection is transmitted by several species of ticks in which the sexual multiplication cycle occurs. When a tick takes a blood meal, infective forms are introduced into the human host. Organisms infect the RBCs, in which they appear as pleomorphic, ringlike structures (Fig. 10). They resemble the early trophozoite (ring) forms of malarial parasites, particularly *P. falciparum*. The organisms measure 1.0 to 5.0 μm, the RBCs are not enlarged or pale, the cells do not contain stippling, and malarial pigment is never seen. The early form contains very little cytoplasm and a very small nucleus. In mature forms, two or more chromatin dots may be seen. Occasionally, a tetrad formation, referred to as a Maltese cross, is seen (28).

FIGURE 6 Photomicrographs of *Plasmodium* species in thin blood films (Giemsa stain; magnification, ×1,000.) (First row) *P. falciparum.* (Left) Two RBCs containing trophozoites (rings). Also note the two platelets present. (Center) Numerous rings in RBCs of a patient with severe *P. falciparum* malaria. (Right) Sausage-shaped microgametocyte. (Second row) *P. vivax.* (Left) Growing trophozoite in enlarged RBC with Schüffner's dots (stippling). A ring-form trophozoite is present in another RBC. (Center) Immature and mature schizonts. (Right) Macrogametocyte. (Third row) *P. ovale.* (Left) Growing trophozoite, which is compact, in an enlarged, elongated RBC with fimbriated edges and Schüffner's dots. (Center) Mature schizont. (Right) Schizont and macrogametocyte. (Fourth row) *P. malariae.* (Left) Normal-size RBC containing growing trophozoite. (Center) Band form trophozoite. (Right) Two mature schizonts. (Photomicrographs from Lawrence Ash; used with permission.)

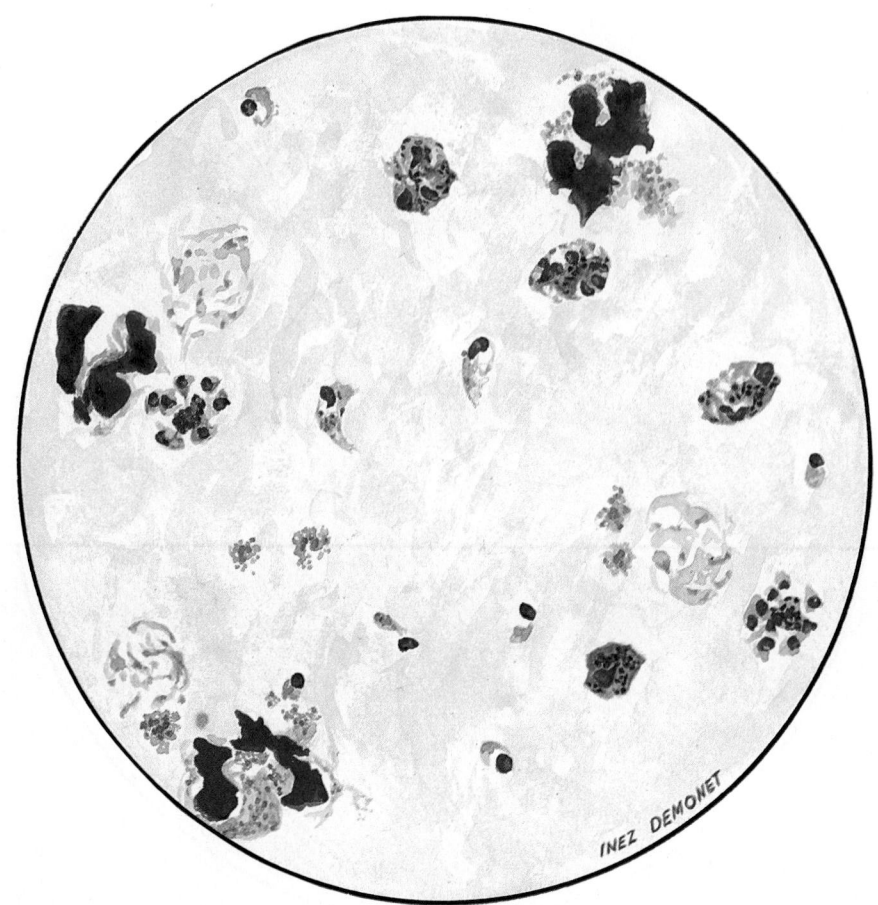

FIGURE 8 Morphology of *P. malariae* in a thick blood film. There are small trophozoites toward 6 o'clock; growing trophozoites slightly above the center; mature trophozoites at 5 o'clock; immature schizonts at 12, 1, and 3 o'clock; and mature schizonts at 4 and 9 o'clock. *Note*: Outside of areas of endemicity, the parasites on a thick film are normally fewer than in this figure. (Reproduced from A. Wilcox, *Manual for the Microscopical Diagnosis of Malaria in Man*, U.S. Department of Health, Education and Welfare, Washington, D.C., 1960.)

Clinical Disease

In patients infected by *B. microti* from the northeastern United States who were not splenectomized, symptoms appeared from 10 to 24 days after a tick bite and continued for several weeks. First, there was general malaise and then fever, shaking chills, profuse sweating, arthralgias, myalgias, headache, fatigue, weakness, and dark urine. In some cases, hepatosplenomegaly is present, and serum bilirubin and transaminase levels may be elevated owing to hemolytic anemia (73). Although it is not always possible to detect organisms in the blood smears, the parasitemia may reach 80 to 90%, particularly in splenectomized patients.

By the end of 1991, 13 cases of babesiosis had been reported in Connecticut, the largest number of human cases reported on the mainland United States. Twelve of these infections were probably acquired via tick bites, and one was acquired from a blood transfusion (3). Ages ranged from 61 to 95 for those with tick-acquired infection. Two patients died with active infections, and one died from chronic obstructive pulmonary disease.

Babesia infections described in California and other parts of the world are different from those seen in the northeastern United States, where the infection is most often subclinical. Those in California and Europe present as a fulminant, febrile, hemolytic disease affecting splenectomized or immunosuppressed individuals. Work suggests that species other than *B. microti* or the bovine parasites were involved in these acute infections (20, 69).

A case of transfusion-induced babesiosis accompanied by disseminated intravascular coagulopathy in an elderly patient suffering fever, chills, nausea, arthralgias, and lethargy was reported in 1982. The symptoms began after transfusion with 2 U of packed RBCs during surgery. One of the donors had a high indirect-fluorescent-antibody titer against *B. microti*. Blood films were negative for *B. microti* infections, and the infection was confirmed only by hamster inoculation (52).

Diagnosis

Most documented human cases of babesiosis have had a low parasitemia, thus the need for both thick and thin blood films stained with Giemsa stain. When the organisms can-

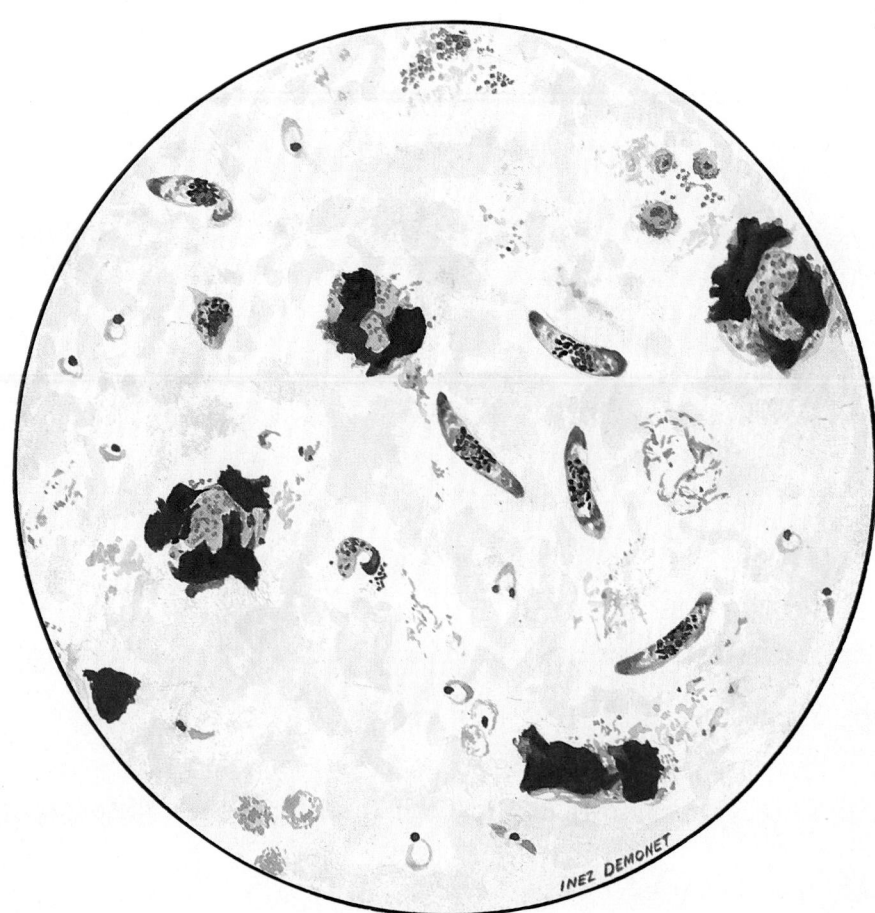

FIGURE 9 Morphology of *P. falciparum* in thick blood film. There are multiple small tropho-zoites (rings) and sausage (crescent)-shaped gametocytes. Also, there are two gametocytes with different shapes toward 10 o'clock. *Note:* Outside of areas of endemicity, the parasites on a thick film are normally fewer than in this figure. (Reproduced from A. Wilcox, *Manual for the Microscopical Diagnosis of Malaria in Man,* U.S. Department of Health, Education and Welfare, Washington, D.C., 1960.)

not be demonstrated, blood can be inoculated intraperito-neally into a hamster or gerbil; if the infecting species is *B. microti*, organisms with the same morphology as that seen in humans can usually be found in the blood within a few days. If the QBC system is used, *Babesia* rings could easily be confused with ring forms of malaria species.

In differentiating *Babesia* spp. from *Plasmodium* spp. in blood slides, the occasional presence of tetrads (Maltese crosses) is diagnostic for *Babesia* spp., and *Babesia* spp. are never accompanied by pigment. *B. microti* is much more variable in size than *P. falciparum* rings, and extracellular parasites are much more common with *B. microti* than with *P. falciparum* (Fig. 10).

Serologic studies have been reported; however, cross-reactivity has been seen among different species of *Babesia* and several species of *Plasmodium* (72). In one study, a titer of ≥1:1,024 indicated a recent infection, and a titer of ≤1:64 indicated past infection (72).

Treatment

B. microti babesiosis can often be effectively managed with supportive care. There is no specific drug of choice; how-

ever, the recommended therapy is clindamycin and quinine or intravenous quinidine.

Epidemiology and Prevention

B. microti organisms are transmitted from mice to humans primarily by the deer tick, *Ixodes dammini*, in the East and Midwest and by closely related species elsewhere. Infected persons may be asymptomatic. Prevention relies on avoid-ing ticks or promptly removing them once they have been detected, since there are apparently no fully effective tick repellents. If symptoms do appear 1 to 2 weeks after a tick bite, a physician should be consulted.

The duration of parasitemia following asymptomatic or symptomatic infection is not known. In areas of endemic-ity, exclusion of blood donors who are febrile, are known to have had babesiosis, or have an anti-*Babesia* titer of 1:16 or higher may reduce the risk. However, until a practical, effective, sensitive screening method is developed, more cases of transfusion-induced babesiosis will probably occur (52).

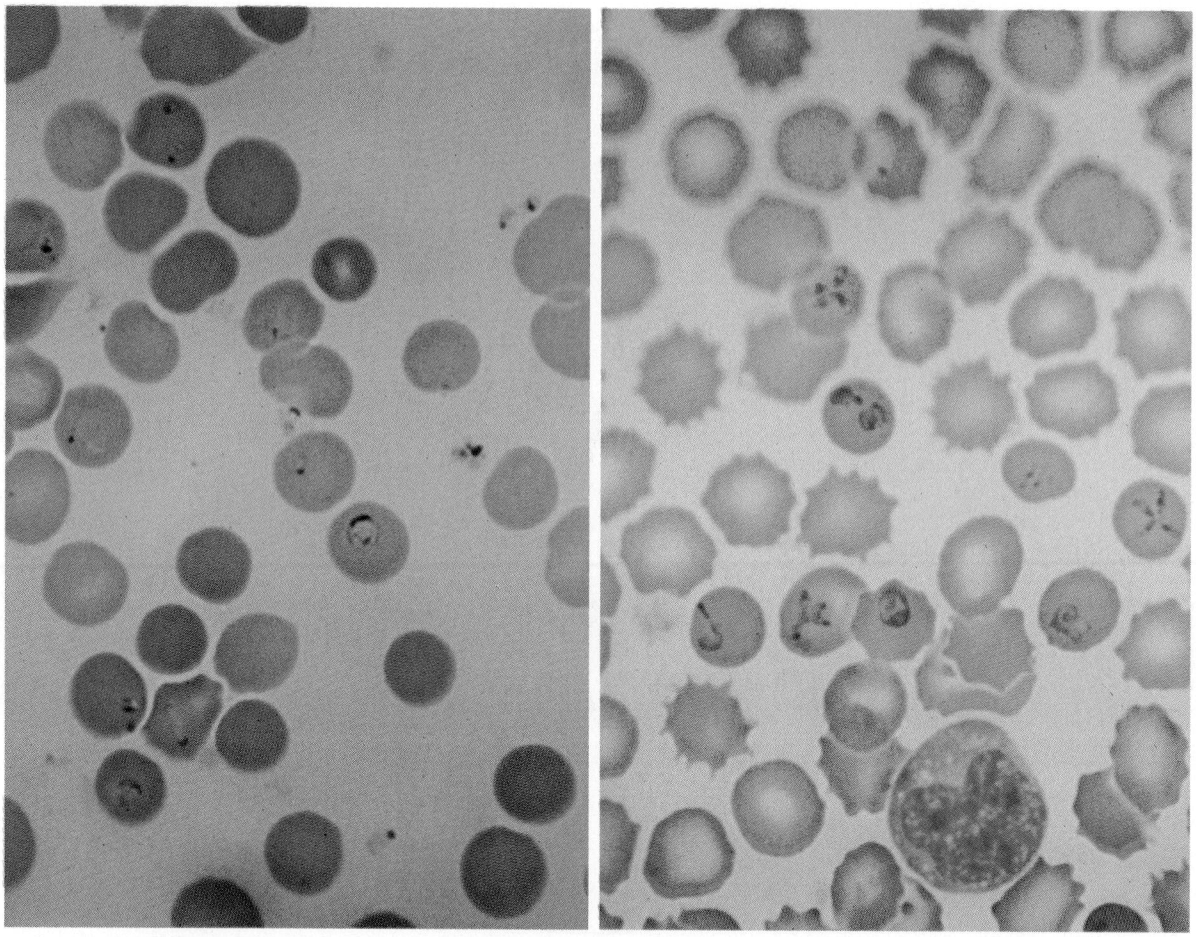

FIGURE 10 *Babesia* spp. in thin blood films. (Left) *B. microti* in human blood: parasites in RBCs and extracellular parasites; (right) *Babesia* sp. in classic Maltese cross (tetrad) configuration, a form diagnostic for babesiosis but rare in slides from humans infected with *B. microti*.

LEISHMANIASIS

Leishmania spp. are obligate intracellular parasites transmitted by bites of infected sand flies (Fig. 11). Leishmaniasis is mainly a zoonosis, although in certain areas, humans serve as the reservoir. More than 400,000 new cases are reported annually, and estimates indicate that as many as 350 million people are at risk, with approximately 12 million currently infected (4). Depending on the species, infection with *Leishmania* spp. can result in cutaneous, mucocutaneous, or visceral disease.

In cutaneous leishmaniasis, parasites are found in macrophages and are morphologically identical, although different species produce different clinical syndromes. The outcome also depends on age, previous exposure, and immune status of the host (2). Taxonomy is controversial, and disease categories are undergoing extensive review (12), particularly since clinical evidence suggests that the parasites may not be limited to specific tissues but may be hematogenously spread throughout the body (53).

Several species cause classic as well as aberrant forms of Old World cutaneous leishmaniasis (*Leishmania tropica*, *Leishmania major*, *Leishmania aethiopica*). *L. tropica* (*L. tropica* minor) infections, the cause of human or urban cutaneous leishmaniasis, are seen in urban Middle Eastern settings, Afghanistan, India, the former U.S.S.R., and Tur-

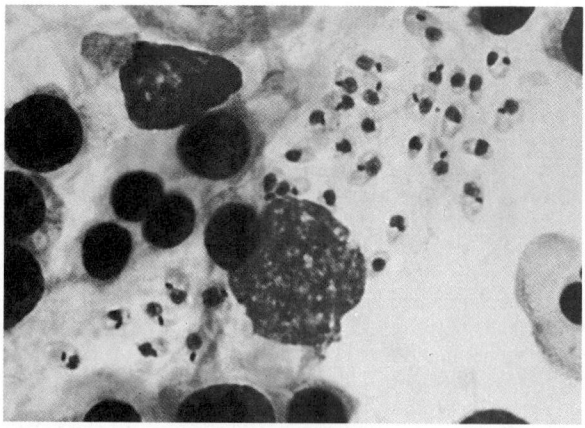

FIGURE 11 Macrophages containing intracellular amastigotes (Leishman-Donovan or LD bodies) of *L. donovani*. (Reproduced with permission from J. W. Smith, D. M. Melvin, T. C. Orihel, L. R. Ash, R. M. McQuay, and J. H. Thompson, Jr., *Atlas of Diagnostic Medical Parasitology: Blood and Tissue Parasites*, American Society of Clinical Pathologists, Chicago, 1976.)

key. *L. major* (*L. tropica* major) infections, the cause of zoonotic or rural cutaneous leishmaniasis, are seen in the Middle East, Afghanistan, Africa, and the former U.S.S.R. *L. aethiopica* infections are seen in Ethiopia, Kenya, possibly Uganda, the former U.S.S.R., and southern Yemen.

New World cutaneous leishmaniasis is found from southern Texas to Latin and South America; has been found in at least 24 countries, including the Caribbean islands; and has been confirmed in 59 United States travelers (34). Both classic and aberrant forms are seen, and there is no absolute separation of species based on clinical signs and symptoms (32). Originally, identification was based on clinical symptoms and geographic and epidemiologic information. Genetic, biochemical, and immunologic methods provide more precise taxonomic characteristics of the parasite itself (32, 44, 47, 85, 87). Human infections are primarily contracted through forest-related activities in zoonotic areas. Organisms causing New World leishmaniases are *Leishmania mexicana* (syn. *Leishmania pifanoi*) (causes chiclero's ulcer or bay sore), *Leishmania amazonensis* (syn. *Leishmania garnhami*), *Leishmania venezuelensis*, *Leishmania lainsoni*, *Leishmania peruviana* (causes uta), *Leishmania colombiensis*, and *Leishmania guyanensis* (causes pian bois or forest yaws). *L. amazonensis* and *L. mexicana* cause diffuse cutaneous leishmaniasis, while infections with *Leishmania panamensis* (causes ulcera de Bejuco) often metastasize and cause mucocutaneous lesions (32).

The *Leishmania braziliensis* complex is the main cause of mucocutaneous leishmaniasis in the New World. *L. braziliensis* (causes espundia) is the primary species. Mucocutaneous lesions have also been seen in *L. panamensis* infections.

Visceral leishmaniasis (causes kala azar) is caused by organisms in the *Leishmania donovani* complex (*L. donovani*, *Leishmania infantum*, and *Leishmania chagasi*) (85). Kala azar is characterized by darkening of the skin on the forehead and hands. *L. donovani* complex organisms parasitize reticuloendothelial cells throughout the body, not just mucous membranes or subcutaneous tissues.

Life Cycle and Morphology of the Parasite

The organism is engulfed by reticuloendothelial cells of the mammalian host and exists as the amastigote form (Leishman-Donovan body or LD body) within the phagocytic cell (Fig. 11). The amastigote forms are small, round or oval bodies of 3 to 5 μm. The large nucleus and small kinetoplast in the amastigote can be demonstrated with Giemsa or Wright's stain. The amastigote multiplies by binary fission until the host cell is destroyed, and the liberated organisms are then phagocytized by other reticuloendothelial cells. In visceral leishmaniasis, the parasitized macrophages enter the bloodstream and are carried to bone marrow, liver, and spleen, where rapid multiplication occurs.

As the sand fly (*Phlebotomus* spp. or *Lutzomyia* spp.) takes a blood meal, the amastigote is ingested and transforms into the promastigote stage, a motile, slender organism (10 to 15 μm) with a single anterior flagellum. Promastigotes multiply by longitudinal fission in the gut of the insect. The promastigotes then migrate to the hypostome of the sand fly and are released when the next blood meal is taken, thus initiating the infection. Depending on the parasite species, the duration of the life cycle in the sand fly is from 4 to 18 days.

Old World Cutaneous Leishmaniasis

Clinical Disease

The incubation period of Old World cutaneous leishmaniasis may vary from 2 weeks (*L. major*) to several months to 3 years (*L. tropica* and *L. aethiopica*). The first sign of cutaneous disease is a firm, painless papule at the bite site, usually on an exposed part of the body. Papules may be intensely pruritic, will grow to 2 cm or more in diameter, and may resemble a pimple. Destruction of the papule by necrosis of the epidermis leads to ulceration. Secondary bacterial and fungal infections are frequent complications. Patients with acute cutaneous leishmaniasis fail to produce interleukin or gamma interferon, and the parasite interferes with the killing mechanism of activated macrophages (14, 51, 74).

L. tropica (*L. tropica* minor) produces chronic dry-type lesions that rarely lead to visceral leishmaniasis. However, military personnel who were in the Middle East for Operation Desert Storm contracted infections with *L. tropica*; the infections were viscerotropic and were diagnosed by examination of bone marrow aspirates (22).

Papules of *L. major* ulcerate rapidly and are covered with a serous exudate (moist, wet sore). Lesions of *L. major* are frequently multiple, particularly in nonimmune patients; they may become confluent and secondarily infected; and they heal slowly.

Infections with *L. aethiopica* may cause classic oriental sore, mucocutaneous leishmaniasis, or diffuse cutaneous leishmaniasis, the last being a condition with thickening of the skin in plaques, papules, or multiple nodules, particularly on the face and limbs, that resembles lepromatous leprosy.

Epidemiology and Prevention

Female *Phlebotomus* sand flies transmit Old World cutaneous leishmaniasis. Infection may also be transmitted by direct contact with infected lesions and mechanically through bites by the stable or dog fly (*Stomoxys* spp.).

In areas of endemicity, vaccination is performed by inoculating exudate from naturally acquired lesions into an inconspicuous body area of a nonimmune person. The infection is usually limited to single lesions, mimics the natural disease, and leads to lifelong protection against the homologous strain of the species. About 2% of those vaccinated do not self-cure and require therapy. Thus far, prepared vaccines have not been effective (57, 61).

Humans are the primary reservoir and source of infection for *L. tropica* in urban areas of endemicity. Active case detection and treatment help reduce transmission. Unlike *L. tropica*, *L. major* (the cause of rural oriental sore) is a zoonosis, with human infections occurring sporadically. Gerbils and other rodents are the principal reservoir hosts; recently, naturally infected dogs have been found. *L. aethiopica* is also a zoonosis, with the rock hyrax as the main reservoir.

Prevention includes the use of residual insecticides and insect repellents. Reservoir control has generally been unsuccessful. Individuals with lesions should be warned to protect the lesion from insect bites with some type of covering. Patients should also be educated about the possibility of autoinoculation or infection.

Treatment

Lesions generally heal spontaneously over a period of a year or so and confer immunity on the individual within the

TABLE 2 Clinical manifestations of leishmania in the New World[a]

Disease	Comments
American visceral leishmaniasis, *L. chagasi*	Most common cause of visceral leishmaniasis in New World
American cutaneous leishmaniasis[b]	
Cutaneous, all New World species	May heal spontaneously
Hyperergic form (ML), *L. braziliensis* complex, *L. mexicana* complex	Progressive ulcerative infection with hypersensitivity and autoimmunity
Anergic form (DCL), *L. mexicana* complex	Diffuse cutaneous, defective cellular immunity

[a]Depending on the immune status of the host and possibly other factors, most New World *Leishmania* spp. can produce many disease manifestations. Data are from reference 32.

[b]ML, mucosal leishmaniasis; DCL, diffuse cutaneous leishmaniasis.

immediate area of endemicity. It may therefore be beneficial not to treat the lesions unless there is the possibility of a disfiguring scar.

Treatment has included cryotherapy, heat, surgical excision of lesions, and chemotherapy with antimonial compounds. Therapeutic failure may be due to developing resistance. To ensure that treatment has been effective, follow-up smears and cultures should be done 1 to 2 weeks posttherapy. Amphotericin B and ketoconazole are successful alternatives to antimonial compounds.

New World Cutaneous Leishmaniasis

Clinical Disease

L. mexicana lesions are usually single, self-limiting cutaneous papules, nodules, or ulcers on the face and ears (~60%); they are painless and usually heal within a few months. Ulcerative lesions on the ears (chiclero's ulcer) may cause destruction of ear cartilage. *L. colombiensis, L. amazonensis, L. lainsoni, L. peruviana,* and *L. venezuelensis* infections are usually associated with single cutaneous lesions, while infections caused by *L. guyanensis* and *L. panamensis* are characterized by lymphatic dissemination. *L. peruviana* primarily affects children, who have a single or a few painless lesions that heal spontaneously within several months. Individuals infected with *L. mexicana* and approximately 30% of those infected with *L. amazonensis* are more likely to develop diffuse cutaneous leishmaniasis. Parasites capable of producing a spectrum of disease in human hosts are listed in Table 2.

Epidemiology and Prevention

Reservoir hosts include rodents, opossums, anteaters, sloths, and dogs. Human infections occur primarily in forest habitat. However, *L. peruviana* is found in villages rather than forests, and the probable reservoir is the domestic dog. At least 20 cases of *L. mexicana* infection have been reported in south central Texas (26), where reservoirs include opossums, cotton rats, and house rats. There is strong evidence that *L. mexicana* is endemic in south central Texas, and the majority of infected patients have presented with ulcerating lesions.

Prevention has been discussed above in the section on Old World cutaneous leishmaniasis.

Treatment

L. mexicana lesions are self-limiting and usually require no treatment; an exception is lesions on the pinna of the ear, which should be treated. Treatment regimens are similar to those used for Old World cutaneous leishmaniasis. Follow-up smears and cultures should be made 1 to 2 weeks posttherapy.

Mucocutaneous Leishmaniasis

Clinical Disease

The primary lesions of mucocutaneous leishmaniasis are similar to those of cutaneous leishmaniasis; if left untreated, serious mucocutaneous disease may develop in up to 80% of the cases (33). Involvement of the nasal or oral mucosa may occur concurrently with the active primary lesion or many years later, after the primary lesion has healed. Metastasis may be due to hematogenous spread of organisms (53). Ulceration and erosion of soft tissue and cartilage can be progressive, leading to loss of the lips, soft parts of the nose, and soft palate. In the nonulcerative form, the patient may develop a "tapir nose" due to local edema and hypertrophy of the upper lip and nose. In acute and chronic forms of *L. braziliensis*, detection of amastigotes in tissues is very difficult, but the patient will have a high antibody titer. Mucosal lesions do not heal spontaneously and may also be complicated by secondary bacterial infections.

A few mucocutaneous leishmaniasis cases have been reported in the Sudan, with the causative agent being *L. donovani*, and in Ethiopia, where the agent is *L. aethiopica*.

Epidemiology and Prevention

L. braziliensis infections are found in Argentina, Bolivia, Belize, Brazil, Colombia, Costa Rica, French Guiana, Guatemala, Honduras, Mexico, Panama, Paraguay, Peru, and Venezuela. Sand fly vectors belong to the genus *Lutzomyia*. Natural reservoir hosts are rodents and dogs.

L. peruviana infections are principally found in Peru, and dogs and rodents are natural reservoirs. Personal protection against sand fly bites is the most effective means of prevention.

Treatment

Most individuals with infections caused by *L. braziliensis* complex respond to treatment with antimonial compounds. Severe mucosal lesions are more refractory to treatment, and individuals may relapse or suffer recrudescence after therapy has stopped. Longer courses of therapy are recommended for mucocutaneous leishmaniasis than for other types of leishmaniasis. Amphotericin B has been used as an alternative drug. Follow-up smears and cultures should be made 1 to 2 weeks posttherapy.

Visceral Leishmaniasis

Clinical Disease

The clinical features of visceral leishmaniasis vary from asymptomatic, self-resolving infection to acute visceral disease, with an incubation period ranging from 10 days to 2 years but normally 2 to 4 months. Symptoms may be nonspecific or may mimic those of typhoid fever or malaria; they include fever, anorexia, malaise, weight loss, and, frequently, diarrhea. A double (dromedary) or triple fever peak may be seen daily, and symptoms may include nontender hepatomegaly and splenomegaly, lymphadenopathy, and occasional acute abdominal pain. Darkening of facial, hand, foot, and abdominal skin (kala azar) is seen in light-colored persons in India. As the disease progresses, anemia, cachexia, and marked enlargement of liver and spleen are seen. Death may ensue after a few weeks or after 2 to 3 years. Most individuals are asymptomatic or have minor symptoms that resolve without therapy. In many areas, visceral leishmaniasis is a disease of children. Only a few of those infected develop the classic signs. The infection has been reported in human immunodeficiency virus (HIV)-positive persons, in whom there may be an absence of splenomegaly (30, 31). American visceral leishmaniasis has always been a disease of young, malnourished children. However, as the prevalence of AIDS increases in tropical America, American visceral leishmaniasis will probably appear more often as an opportunistic infection, similar to the pattern of L. infantum infection in southern Europe (30). Congenital infections leading to abortion can occur (24). Death frequently results from complications such as bacillary or amebic dysentery, septicemia, or pneumonia (2, 27).

Postdermal leishmaniasis lesions may be hypopigmented or may appear as erythematous macules that may become nodular later. A butterfly rash resembling that of systematic lupus erythematosus may appear on the face and may resemble lepromatous leprosy or disseminated cutaneous leishmaniasis. Erythematous and nodular lesions contain many organisms (1).

Hyperplasia of reticuloendothelial cells occurs in the liver, spleen, bone marrow, lymph nodes, small intestinal mucosa, and other lymphoid tissues. Hematopoiesis is initially normal; however, reduction in life spans of RBCs and leukocytes leads to anemia and granulocytopenia. Extensive hypertrophy of the liver and spleen may occur, but both organs may return to normal size after successful treatment. Decreases in T-cell blastogenesis and production of interleukin-1 and -2, gamma interferon, and tumor necrosis factor return to normal after therapy (35).

Suppression of cell-mediated immunity leads to parasite dissemination and uncontrolled multiplication, with high levels of immunoglobulin G (IgG) and IgM. Circulating immune complexes involving immunoglobulin classes A, G, and M can be found in the serum. Delayed-hypersensitivity reactions to skin test antigens are suppressed or absent; however, with resolution of infection, the reaction returns, and patients recovering from infection are thought to be immune (67).

Epidemiology and Prevention

Visceral leishmaniasis is a zoonosis except in India and parts of China and Kenya, where humans serve as the reservoir host. Individuals with post-kala azar dermal leishmaniasis may be very important reservoirs for maintaining the infection (1). Protection against sand fly bites is the most effective method of prevention. Reducing the numbers of reservoir hosts such as rodents and dogs and treatment of infected individuals have also been successful in reducing transmission.

L. donovani occurs in Africa and Asia, with Phlebotomus sand flies as the vector. L. infantum is found in Africa, Europe, the Mediterranean area, and southwestern Asia. Infections occur primarily in children and are endemic in nature. Phlebotomus sand flies are the vector.

L. chagasi is found in Central and South America and primarily affects children. The disease is a zoonosis, involving dogs and cats as natural reservoirs. The sand fly vector in the New World is Lutzomyia spp. In Oklahoma, visceral leishmaniasis has been described in dogs that had never been in areas noted for leishmaniasis.

Treatment

Sodium stibogluconate (Pentostam) is used for the treatment of visceral leishmaniasis, and pentamidine isethionate, an aromatic diamidine, is recommended for cases that are resistant to antimonial compounds. Combination therapy with daily pentavalent antimony plus gamma interferon every other day may accelerate an early response. Amphotericin B can be used when the disease does not respond to antimonial or pentamidine therapy.

Follow-up smears and cultures should be made 1 to 2 weeks posttherapy, and the patient should be monitored clinically.

Diagnosis

Cutaneous leishmaniasis may go unrecognized and may have to be differentiated from tropical ulcers, syphilis, yaws, leprosy, South American blastomycosis, sporotrichosis, pyogenic nodules or abscesses, furuncles, insect bites, and tuberculosis of the skin. Detection of amastigotes in clinical specimens or promastigotes in culture confirms the infection (19).

After lesions have been cleaned with 70% alcohol, specimens are collected from the most recent or most active lesions. Material collected from the center of a necrotic ulcer may reveal only contaminating bacterial or fungal organisms. Specimens are collected from the margins by aspiration, scraping, or biopsy; local anesthesia may be recommended for the procedure. Multiple slides should be prepared.

A punch biopsy specimen can be used for routine histology and for making multiple imprints or touch preparations. Recognition of amastigotes in tissues is more difficult than recognition in smears or imprints, because the organisms appear smaller and are cut at various angles. Although fine-needle aspiration can be performed, a biopsy specimen is recommended.

Slides should be air dried and fixed with 100% methanol prior to Giemsa staining. Amastigote stages will be found within macrophages or close to disrupted cells. Extensive examination may be necessary to find organisms. Monoclonal antibodies and DNA probes have also been used to detect infections (7, 11, 86, 95).

Specimens for culture should be collected aseptically to prevent contamination of the culture media, usually Novy-MacNeal-Nicolle's medium or Schneider's Drosophila medium supplemented with 30% fetal bovine serum (90). Cultures should be examined periodically for 4 weeks. Promastigote stages can be detected microscopically in wet mounts and then stained with Giemsa stain to visualize their morphology.

Intranasal inoculation of patient material into the golden hamster has been useful for diagnosis but is not practical for most laboratories. Also, it may take several months for the animal to become positive. A combination of tissue smears and cultures is recommended.

The leishmanin (Montenegro) test is a delayed-hypersensitivity reaction to a suspension of killed promastigotes that has been administered intradermally and can be used for epidemiologic surveys of a population to identify groups at risk of infection (93). The test is not species specific; positive reactions are usually seen with cutaneous and mucocutaneous leishmaniases but not with active visceral and diffuse cutaneous leishmaniases.

Studies now indicate that in visceral leishmaniasis, organisms can be found in nasal secretions, tonsillopharyngeal mucosa, and centrifuged urine (56). Splenic biopsy tissue yields the highest rate of positive specimens. Other specimens include lymph node aspirates, liver biopsy samples, sternal or iliac crest bone marrow, and buffy coat preparations. Multiple tissue imprints or smears should be examined, and material should be sent for routine histology. Both microscopic examination and culture should be used. DNA probes and Western blot (immunoblot) analysis have also been suggested (15, 38). Hamsters can be inoculated intraperitoneally. The animal should be killed after 2 to 3 months, and the liver and spleen should be examined for parasites by microscopic examination of histologic sections and culture of macerated tissue.

Gamma globulins may make up 60 to 70% of the serum proteins. Both IgG and IgM levels are elevated, the basis for the aldehyde or formol-gel screening test used in areas of endemicity. Patients with active infections generally have positive test results. In a test tube, a drop of formaldehyde is added to 1 ml of the patient's serum; a positive test results in an opaque, stiff-jelly consistency within 3 min to 24 h.

In active visceral leishmaniasis, the delayed-hypersensitivity (Montenegro) test is negative. However, within several months to a year after recovery, individuals exhibit positive responses.

While a number of serologies have been developed for diagnostic purposes, commercial reagents and standardization are lacking (48). Except in areas of endemicity, serologic testing is generally not available.

TRYPANOSOMIASIS

Trypanosomes are hemoflagellates that live in human blood and tissues. Vector transmission occurs in two different ways. African trypanosomes (*Trypanosoma brucei gambiense*, *T. brucei rhodesiense*) are transmitted via vector bite from salivary secretions, while American trypanosomes (*T. cruzi*) are transmitted through the bite wound contaminated by the vector's feces. African trypanosomiasis (sleeping sickness) is found on the African continent, and *T. cruzi* (Chagas' disease) is limited to the New World.

East African (*T. brucei rhodesiense*) trypanosomiasis has a much higher parasitemia and a more rapid disease progression than the West African form (*T. brucei gambiense*); American trypanosomiasis (*T. cruzi*) infections can also be very serious, with high morbidity and mortality. In African trypanosomiasis, the diagnostic stage is the trypomastigote. In American trypanosomiasis, both amastigote and trypomastigote stages are diagnostic.

African Trypanosomiasis

Life Cycle and Morphology of the Parasite

Trypomastigotes multiply by longitudinal binary fission in fluid aspirated from the trypanosomal chancre, blood, cerebrospinal fluid (CSF), and lymph nodes. They range from a long, slender-bodied organism with a long flagellum to a short, fat, stumpy form without a free flagellum, which is the infective stage for the tsetse fly (Fig. 12).

At the organism's posterior end is the kinetoplast, from which the flagellum and undulating membrane arise. When a blood meal is taken, organisms are ingested by the tsetse fly (*Glossina* spp.). They multiply in the gut of the fly, and after 2 weeks, the organisms migrate back to the salivary glands, where the metacyclic (infective) forms develop

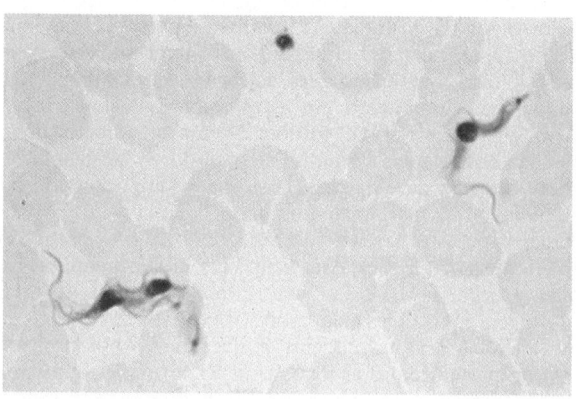

FIGURE 12 Trypomastigotes of *T. brucei rhodesiense* in mouse (thin film; Giemsa stain). One trypomastigote is in the process of cell division. (Reproduced with permission from J. W. Smith, D. M. Melvin, T. C. Orihel, L. R. Ash, R. M. McQuay, and J. H. Thompson, Jr., *Atlas of Diagnostic Medical Parasitology: Blood and Tissue Parasites*, American Society of Clinical Pathologists, Chicago, 1976.)

from the epimastigotes in a few days. The tsetse fly, which is now infective, remains infective for life and can introduce trypanosomes into the puncture wound at each subsequent blood meal.

Clinical Disease

Infective forms in the skin multiply and produce a local inflammatory reaction; a nodule or chancre may develop, but resolution usually occurs within 1 to 2 weeks. Once trypomastigotes enter the bloodstream, a symptom-free low-grade parasitemia may continue for months, and self-cure may occur with no symptoms. Also, lymph node invasion can occur, but the blood films will be negative.

Lymph node invasion is followed by remittent, irregular fevers with night sweats, headaches, malaise, and anorexia. Febrile and afebrile periods alternate, with more organisms in the circulating blood during fevers. Enlarged lymph nodes are soft, painless, and nontender. Posterior cervical lymph nodes are most often involved (Winterbottom's sign). The spleen and liver become enlarged. With Gambian trypanosomiasis, the blood-lymphatic stage may last for years before sleeping sickness syndrome occurs.

Patients may have delayed sensation to pain (Kerandel's sign), an irregular erythematous rash and anemia, granulocytopenia, increased sedimentation rate, and marked increases in serum IgM. African trypanosomes are able to change the surface coats of the outer membranes, a feat called antigenic variation. This change is responsible for successive waves of parasitemia every 7 to 14 days and allows the parasite to evade the host's humoral immune response, resulting in a sustained high IgM level (9, 98). In an immunocompetent host, the absence of elevated serum IgM rules out trypanosomiasis. Sleeping sickness is initiated when trypomastigotes invade the central nervous system (CNS).

T. brucei rhodesiense produces a more rapid, fulminating disease with a short incubation period, and many trypomastigotes appear earlier in the blood. Early stages mimic *T. brucei gambiense* infections; however, death may occur before extensive CNS involvement, even though CNS invasion occurs early.

Diagnosis

Symptoms include irregular fever, posterior cervical lymph node enlargement, delayed sensation to pain, and erythematous skin rashes. Finding trypomastigotes in blood, lymph node and sternum bone marrow aspirates, and CSF confirms the diagnosis.

Health care personnel must adhere to blood-borne-pathogen precautions when handling blood, CSF, or aspirates. Blood can be collected by finger stick or venipuncture (with EDTA). Multiple thick and thin blood films and buffy coat films should be examined before trypanosomiasis is ruled out. More trypomastigotes will be seen in the blood during the febrile period. If CSF is examined, at least 5 ml should be collected. If organisms are not found in the blood, lymph node aspirates can be examined. Blood and CSF specimens should be examined during, 1 to 2 months after, and 2 to 3 years after completion of therapy for CNS involvement. The enzyme-linked immunosorbent assay (ELISA) method has been used to detect antigen in serum and CSF (41, 59), for confirmation of presence or absence of CNS infection, and for follow-up.

Serologic techniques include indirect fluorescent antibody, ELISA, indirect hemagglutination, and the card agglutination trypanosomiasis test. However, false-positive results due to exposure to animal trypanosomes or other flagellates that are noninfectious to humans can be misleading. Although culture is not practical for most diagnostic laboratories, it is more appropriate than animal inoculation.

Epidemiology and Prevention

T. brucei gambiense is transmitted from person to person by the bites of male and female tsetse flies (*Glossina palpalis* and *Glossina tachinoides*); asymptomatic individuals are probably the residual disease reservoir. Tsetse fly vectors of Rhodesian trypanosomiasis (*Glossina pallidipes*, *Glossina morsitans*) are game feeders that may transmit the disease from human to human or animal to human. Because infection results in acute disease, asymptomatic carriers are not a source of transmission as they are in Gambian trypanosomiasis. Mechanical and congenital transmissions also occur in humans.

Treatment

Drugs used in the therapy of African trypanosomiasis are toxic and require prolonged administration. Different drugs may be required for the blood stages before CNS involvement, when sleeping sickness is suspected, for patients who have exhibited both early and late stages of sleeping sickness, or in cases refractory to treatment.

Individuals should be monitored for 2 to 3 years after therapy. About 2% of the patients treated for CNS disease experience relapse. Antitrypanosomal drugs are not routinely available in the United States; the Centers for Disease Control and Prevention serves as a resource.

American Trypanosomiasis

American trypanosomiasis (Chagas' disease) is a zoonosis caused by *T. cruzi* that causes acute or chronic disease and invades the cells of many organs, e.g., heart, esophagus, and colon. Chagas' disease is a major health problem in Latin America, where 100 million persons are at risk of infection, and 16 to 18 million are infected (94). In certain areas, 10% of all adult deaths are due to Chagas' disease.

Life Cycle and Morphology

Humans are infected when metacyclic trypomastigotes are released with the feces while the reduviid bug (triatomids, kissing bugs, or conenose bugs) is taking a blood meal; allergic reaction to the insect's saliva stimulates scratching, and the feces are scratched into the bite wound or onto mucosal surfaces. After entering the wound, metacyclic forms invade local tissues, transform to the amastigote stage, and begin to multiply. In humans, *T. cruzi* can be found in two forms, as amastigotes and trypomastigotes (Fig. 13). Trypomastigotes do not divide in the blood but carry the infection to all parts of the body. Amastigotes multiply in reticuloendothelial system cells, cardiac muscle, skeletal muscle, smooth muscle, and neuroglia cells.

Trypomastigotes are approximately 20 μm long and generally assume a C or U shape in stained blood films. Trypomastigotes occur in the blood in two forms, a long slender form and a short stubby one. Morphology is similar to that of African trypomastigotes. *Trypanosoma rangeli* (31-μm average length, small kinetoplast) trypomastigotes may have to be differentiated from those of *T. cruzi* (21-μm maximum length, large kinetoplast). Human infections with *T. rangeli* are apparently asymptomatic, and trypomastigotes have been noted in the blood for longer than a year. No evidence of pathogenicity has been noted in human volunteers.

Within the cell, the trypomastigote loses its flagellum and undulating membrane and divides by binary fission to form an amastigote, which continues to divide and eventually fills and destroys the infected cell, releasing the organisms. The amastigote is indistinguishable from those in leishmanial infections.

Clinical Disease

Acute Chagas' disease mimics brucellosis, endocarditis, salmonellosis, schistosomiasis, toxoplasmosis, tuberculosis, connective tissue diseases, and leukemia and should be suspected in any individual from an area of endemicity who develops an acute febrile illness with lymphadenopathy and myocarditis. A chagoma or Romaña's sign may be diagnostic; however, allergic reactions to insect bites may produce similar lesions. Diagnosis depends on finding trypomastigotes in the blood, amastigote stages in tissues, or positive serologic reactions. Blood examination is similar to that used for African trypanosomiasis. Trypomastigotes may be detected in acute cases; however, in chronic disease, this stage is rare or absent except during febrile exacerbations. Chronic Chagas' disease with cardiomyopathy may be confused with endocarditis, ischemic heart disease, or rheumatic heart disease.

The severity of Chagas' disease varies with geographic area and may be related to strain differences in *T. cruzi* (18, 97). In children under the age of 5 years, the disease is generally acute. A mild, localized inflammatory reaction may occur at the infection site; a painful, erythematous subcutaneous nodule (chagoma) may form and may take 2 to 3 months to subside. In early stages of infection, amastigotes or trypomastigotes may be aspirated from the chagoma. If the route of inoculation is the ocular mucosa, edema of the eyelids and conjunctivitis may occur (Romaña's sign). Regional lymph nodes may become enlarged, hard, and tender. Trypomastigotes appear in the blood about 10 days after infection, remain through the acute phase, and become rare or absent during the chronic phase. In older children and adults, the disease is subacute or

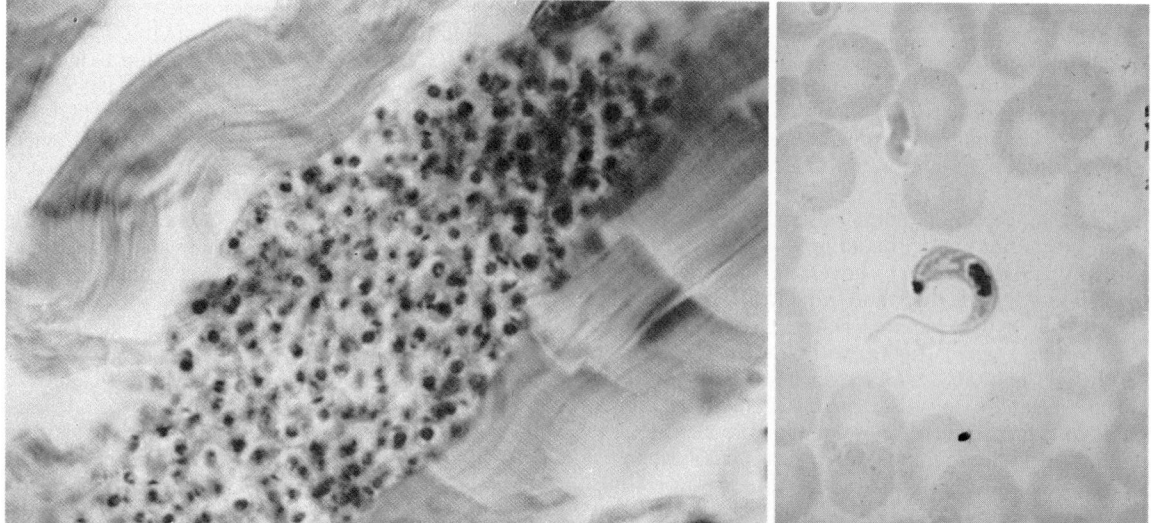

FIGURE 13 *T. cruzi* amastigotes in skeletal muscle (hematoxylin and eosin stain) and trypomastigote in peripheral blood (Giemsa stain). (Reproduced with permission from J. W. Smith, D. M. Melvin, T. C. Orihel, L. R. Ash, R. M. McQuay, and J. H. Thompson, Jr., *Atlas of Diagnostic Medical Parasitology: Blood and Tissue Parasites*, American Society of Clinical Pathologists, Chicago, 1976.)

chronic. The chronic phase may be initially asymptomatic; transmission by blood transfusion is a serious problem in areas of endemicity (84).

Autoantibodies cross-reacting with parasites and host tissues cause the megasyndrome (megaesophagus or megacolon). The most frequent clinical sign of chronic Chagas' disease is cardiomyopathy with cardiomegaly and conduction changes leading to heart failure or to a slow, continuing loss of cardiac function, with possible ventricular rupture and thromboembolus formation. Recently, in an effort to explain pathogenesis, the potential roles of CD8+ T cells and possible autoimmune mechanisms have been emphasized (16, 40, 70). Although this symptom is generally less common, patients from certain areas are more likely to have dilation of the digestive tract with or without cardiomyopathy.

Congenital transmission can occur in both acute and chronic disease, with stillbirth, low birth weight, myocarditis, neurologic alterations, and death shortly after birth. Infants of seropositive mothers should be monitored with blood examinations and serologic tests for up to a year after birth. Chagas' disease has been reported in HIV-positive patients, and neurologic sequelae due to *T. cruzi* have been a major finding in these patients (25, 64). Because of the immunodeficiency associated with HIV infections, a concomitant infection with *T. cruzi* may be difficult to recognize, particularly in individuals who have moved to areas where the trypanosomes are not endemic (29).

The high IgM levels present in patients with African trypanosomiasis are not found in patients with Chagas' disease. Antigenic variation is less common in *T. cruzi*, but despite humoral and cellular immunity, *T. cruzi* is able to persist in the host.

Diagnosis

Universal precautions should be used when patient specimens are handled: trypomastigotes are very infectious. Cha-

gas' disease should be considered when there is a history of consistent exposure to reduviid bug bites, residence or travel in areas of endemicity, laboratory accident, or recent blood transfusion in the area of endemicity or where immigrants from areas of endemicity are donating blood.

Thick and thin blood films and buffy coat films are recommended. Aspirates from chagomas and enlarged lymph nodes and biopsy samples can be examined for amastigotes and trypomastigotes. Aspirates, blood, and tissues can also be cultured, which is valuable in detecting low-grade parasitemias. The medium of choice is Novy-MacNeal-Nicolle's medium (28). Cultures should be incubated at 25°C and observed for epimastigote stages for up to 30 days before they are considered negative. If practical, laboratory animals (rats, mice) could be inoculated, and the blood should be observed for trypomastigotes.

In areas where reduviid bugs are available, xenodiagnosis can be used to detect light infections with few trypomastigotes in the blood (28). However, individuals previously bitten by triatomids may be sensitized to their salivary secretions and may develop an anaphylactic reaction.

Serologic tests used for the diagnosis of Chagas' disease include complement fixation (Guerreiro-Machado test), indirect fluorescent antibody, indirect hemagglutination, and ELISA (45, 50, 54, 66, 89). Cross-reactions have been noted in patients with *T. rangeli* and *T. brucei* infections, leishmaniasis, syphilis, toxoplasmosis, hepatitis, leprosy, schistosomiasis, infectious mononucleosis, systemic lupus erythematosus, and rheumatoid arthritis.

In chronic Chagas' disease, trypomastigotes are rare or absent in the peripheral blood except during febrile episodes; diagnosis depends on culture, xenodiagnosis, PCR (5), or serologic tests, although some individuals with chronic Chagas' disease may have negative serologic results. Follow-up blood specimens should be reexamined 1 to 2 months after therapy. If available, xenodiagnosis may be

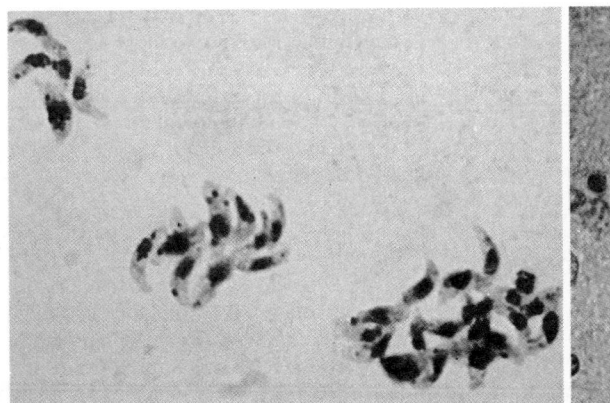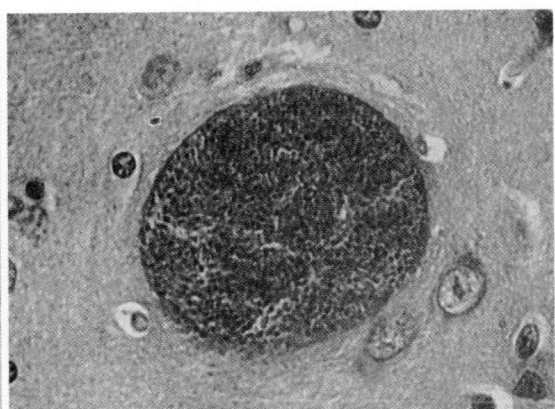

FIGURE 14 *T. gondii.* (Left) Tachyzoites (trophozoites); (right) tissue cyst in brain containing numerous bradyzoites. (Reproduced with permission from J. W. Smith, D. M. Melvin, T. C. Orihel, L. R. Ash, R. M. McQuay, and J. H. Thompson, Jr., *Atlas of Diagnostic Medical Parasitology: Blood and Tissue Parasites*, American Society of Clinical Pathologists, Chicago, 1976.)

done. The patient may also be monitored with serologies and electrocardiograms.

Epidemiology and Prevention

Chagas' disease is a zoonosis involving reduviid bugs living in close association with human reservoirs (dogs, cats, armadillos, opossums, raccoons, and rodents). Human infections occur mainly in rural areas with poor sanitary and socioeconomic conditions (79). Reduviids in the United States have not adapted themselves to household habitation.

Transmission to humans depends on the defecation habits of the insect vector. In the United States and other areas where reduviid bugs do not ordinarily defecate while feeding, human infections are rare. Only three autochthonous cases have been reported in the United States: two from Texas and one from California (82).

Transmission can also occur through organ transplants and blood transfusions (84). Some countries in areas of endemicity have mandatory laws for serologic testing of blood donors.

Treatment

Various drugs have shown promise in acute and early chronic disease; however, they must be administered over a prolonged period, and there are side effects. Surgery is required in advanced cases of megaesophagus and megacolon. Administration of corticosteroids should be avoided because of exacerbation of the infection.

TOXOPLASMOSIS

Toxoplasma gondii is distributed worldwide, is closely related to other coccidia, and has certain similarities to malarial parasites (also in the phylum Apicomplexa). Although serologic evidence indicates a high rate of human exposure to the organisms, the disease is relatively rare. The organisms can infect many vertebrates besides humans, but the definitive host is the house cat and other members of the Felidae family.

Life Cycle and Morphology of the Parasite

Humans can acquire the infection in several ways, either by the accidental ingestion of oocysts shed in cat feces, by the ingestion of rare or raw meats, in utero, or by transfusion. Although ingestion of rare or raw meats may be a source of infection through ingestion of tissue forms, two of the largest outbreaks of human toxoplasmosis were due to oocyst infection (10, 81), and exposure to oocysts may be a common source of individual isolated cases.

Obligate intracellular parasites are found in two forms in humans. Proliferating tachyzoites are usually seen in the early, more acute infection (Fig. 14). Resting forms or cysts contain the more slowly growing trophozoites or bradyzoites and are found primarily in muscle and brain, probably as a result of the host's immune response (Fig. 14). Bradyzoites encyst 8 to 10 days after entry into the host, are more resistant to pepsin, have a slower generation time, and are the only stage to initiate the enteroepithelial cycle and transform into oocysts in the feline intestine (55).

Trophozoites (tachyzoites) are banana shaped (2 to 3 μm wide and 4 to 8 μm long) and have one end rounder than the other. Cysts are formed during chronic infections, and organisms within the cyst wall are strongly periodic acid-Schiff positive. During the acute phase, there may be groups of tachyzoites that appear to be cysts; however, they are not strongly periodic acid-Schiff positive and have been termed "pseudocysts."

The cycle in cats includes asexual multiplication and sexual reproduction in the small intestinal epithelial cells, with the final stage being the oocyst, which is passed out in the feces. This cycle takes approximately 20 to 24 days after infection with the oocysts but only 3 to 5 days after the ingestion of meat infected with cysts, e.g., an infected mouse.

Clinical Disease

Large numbers of people with positive serologies to *T. gondii* suggest that most infections are benign, with few or no symptoms (75). Severe symptoms are seen with congenital (transplacental) infections or those in the compromised patient. Recent advances in our understanding of the immunology of toxoplasmosis are reviewed by Subauste and Remington (80).

Tachyzoites rupture from infected cells and invade adjacent cells, resulting in expanding focal lesions. Once cysts are formed, the process becomes quiescent. In the immunocompromised or immunodeficient patient, a cyst rupture or primary exposure to the organism often leads to lesions. Organisms can be disseminated via the lymphatics and the bloodstream to other tissues.

Congenital infections may be severe if the mother acquires the infection during the first or second trimester of pregnancy. At birth or soon thereafter, symptoms in these infants may include retinochoroiditis, cerebral calcification, and occasionally hydrocephalus or microcephaly. Symptoms of congenital CNS involvement may not appear until several years later.

Postnatal infections can be placed in four groups: (i) lymphadenitis, fever, headache, and myalgia, with a possibility of splenomegaly and a brief erythematous rash; (ii) typhuslike exanthematous form with myocarditis, meningoencephalitis, atypical pneumonia, and possibly death; (iii) CNS involvement, which is usually fatal; and (iv) retinochoroiditis, which may be severe, requiring enucleation. The most common manifestation in adults is lymphadenopathy, local or generalized, and the nodes most commonly involved are those of the neck.

Compromised host infections can lead to severe complications; underlying conditions include Hodgkin's disease, other non-Hodgkin's lymphomas, leukemias, solid tumors, collagen vascular disease, organ transplant, and AIDS. There may be diffuse encephalopathy, meningoencephalitis, or cerebral-mass lesions. More than 50% of these patients will show altered mental status, motor impairment, seizures, abnormal reflexes, and other neurologic sequelae. Even in these patients, studies show that 80% receiving chemotherapy for toxoplasmosis improve significantly or have complete remission. However, in those with AIDS, therapy has to be continued for long periods to maintain a clinical response (91). *Toxoplasma* encephalitis has been reported as a life-threatening opportunistic infection among patients with AIDS (23, 96).

Respiratory disease due to *T. gondii* has been recognized only rarely in immunocompromised patients, and the few cases that have been reported in HIV-positive patients have been in association with CNS disease. A report of six cases of pulmonary toxoplasmosis in HIV-positive patients in the absence of neurologic findings emphasizes the importance of considering this organism as the cause of pulmonary disease in these patients with respiratory symptoms (76).

Diagnosis

Diagnostic procedures include examining biopsy specimens, buffy coat cells, or spinal fluid or isolating the organism in tissue culture or laboratory animals. It is difficult to demonstrate the parasite in lymph node biopsy samples, and since many individuals have cysts within the tissues, recovery of organisms from tissue culture or animal inoculation may be misleading, since the organisms may not be the etiologic agent of disease (92). However, two representative situations in which finding organisms may be very significant include tachyzoite-positive smears and/or tissue cultures inoculated from CSF and, in cases of acute pulmonary disease, the demonstration of tachyzoites in Giemsa-stained smears of bronchoalveolar lavage fluid (17).

Serologic tests are recommended in all cases of suspected toxoplasmosis. Newborns from mothers who are antibody positive have passively transferred maternal IgG. The detection of IgM antibodies provides a more accurate indication of infection in the newborn. IgM studies giving false-negative (IgG saturation of antigen receptors) and false-positive (presence of rheumatoid factor and antinuclear antibody) results are possible. Presence of IgM antibody titers in single serum specimens cannot be used to indicate recent exposure (13, 42, 60, 81). Additional information can be found in chapter 103 of this Manual.

Treatment

Treatment is recommended for clinically active disease, diagnosed congenital toxoplasmosis, and symptomatic compromised patients. Therapy for pregnant patients and newborns with *Toxoplasma* antibody is somewhat controversial. Prophylactic therapy is recommended for the newborn until it can be demonstrated that IgM antibody is not present.

Epidemiology and Prevention

Hand washing is recommended after potential exposure to oocysts. Meat should be cooked so that the internal temperature reaches 150°F (66°C). Indoor cats fed on dry, canned, or boiled food are unlikely to be infected. Cats who have access to infected prey animals (birds or mice) may shed oocysts in their feces, so other preventive measures have been recommended, including changing the kitty litter box daily and disinfecting the pan with boiling water. Also, the feces should not be placed onto the soil but should be disposed of either in the toilet or within bags.

Cellular probes and antigens that could be useful for future vaccine development have been identified and characterized.

REFERENCES

1. **Addy, M., and A. Nandy.** 1992. Ten years of kala-azar in West Bengal. I. Did post-kala azar dermal leishmaniasis initiate the outbreak in 24-Parganas. *Bull. W.H.O.* **70:** 341–346.
2. **Altes, J., A. Salas, M. Riera, M. Udina, A. Galmes, J. Balanzat, A. Ballesteras, J. Buades, F. Salva, and C. Villalonga.** 1990. Visceral leishmaniasis: another HIV-associated opportunistic infection? Report of eight cases and review of the literature. *AIDS* **58:**201–207.
3. **Anderson, J. F., E. D. Mintz, J. J. Gadbaw, and L. A. Magnarelli.** 1991. *Babesia microti,* human babesiosis, and *Borrelia burgdorferi* in Connecticut. *J. Clin. Microbiol.* **29:**2779–2783.
4. **Ashford, R. W., P. Desjeux, and P. deRaadt.** 1992. Estimation of population at risk of infection and number of cases of leishmaniasis. *Parasitol. Today* **8:**104–105.
5. **Avila, H. A., J. B. Pereira, O. Thiemann, E. Depaiva, W. Degrave, C. M. Morel, and L. Simpson.** 1993. Detection of *Trypanosoma cruzi* in blood specimens of chronic chagasic patients by polymerase chain reaction amplification of kinetoplast minicircle DNA: comparison with serology and xenodiagnosis. *J. Clin. Microbiol.* **31:**2421–2426.
6. **Baird, J. K., Purnomo, and T. R. Jones.** 1992. Diagnosis of malaria in the field by fluorescence microscopy of QBC capillary tubes. *Trans. R. Soc. Trop. Med. Hyg.* **86:**3–5.
7. **Barker, R. H.** 1990. DNA probe diagnosis in parasitic infections. *Exp. Parasitol.* **70:**494–499.
8. **Barker, R. H., T. Banchongaksorn, J. M. Courval, W. Suwonkerd, K. Rimwungtragoon, and D. F. Wirth.** 1992. A simple method to detect *Plasmodium falciparum* directly from blood samples using the polymerase chain reaction. *Am. J. Trop. Med. Hyg.* **46:**416–426.
9. **Barry, J. D., and C. M. R. Turner.** 1991. The dynamics of

antigenic variation and growth of African trypanosomiasis. *Parasitol. Today* **7:**207–211.

10. **Beneson, M. W., E. T. Takafuji, S. M. Lemon, R. L. Greenup, and A. J. Sulzer.** 1982. Oocyst-transmitted toxoplasmosis with ingestion of contaminated water. *N. Engl. J. Med.* **307:**666–669.

11. **Beverly, S. M.** 1991. Gene amplification in *Leishmania. Annu. Rev. Microbiol.* **45:**417–444.

12. **Blackwell, J. M.** 1992. Leishmaniasis epidemiology: all down to the DNA. *Parasitology* **104:**S19–S34.

13. **Bobic, B., D. Sibalic, and O. Djurkovic-Djakovic.** 1991. High levels of IgM antibodies specific for *Toxoplasma gondii* in pregnancy 12 years after primary *Toxoplasma* infection. *Gynecol. Obstet. Invest.* **31:**182–184.

14. **Bogdan, C., M. Röllunghoff, and W. Solbach.** 1990. Evasion strategies of *Leishmania* parasites. *Parasitol. Today* **6:**183–187.

15. **Bogdan, C., N. Stosiek, H. Fuchs, M. Röllinghoff, and W. Solbach.** 1990. Detection of potentially diagnostic leishmanial antigens by Western blot analysis of sera from patients with kala azar or multilesional cutaneous leishmaniases. *J. Infect. Dis.* **162:**1417–1418.

16. **Bonfa, E., V. S. T. Viana, A. C. P. Barreto, N. H. Yoshinari, and W. Cossermelli.** 1993. Autoantibodies in Chagas' disease—an antibody cross reactive with human and *Trypanosoma cruzi* ribosomal proteins. *J. Immunol.* **150:**3917–3923.

17. **Bottone, E. J.** 1991. Diagnosis of acute pulmonary toxoplasmosis by visualization of invasive and intracellular tachyzoites in Giemsa-stained smears of bronchoalveolar lavage fluid. *J. Clin. Microbiol.* **29:**2626–2627.

18. **Breniere, S. F., P. Braquemond, A. Solari, J. F. Agnese, and M. Tibayrenc.** 1991. An isoenzyme study of naturally occurring clones of *Trypanosoma cruzi* isolated from both sides of the West Andes highland. *Trans. R. Soc. Trop. Med. Hyg.* **85:**62–66.

19. **Bullock-Iacullo, S.** 1992. Detection of blood parasites, p. 7.8.1.1–7.8.10.3. *In* H. D. Isenberg (ed.), *Clinical Microbiology Procedures Handbook*, vol. 2. American Society for Microbiology, Washington, D.C.

20. **California Morbidity.** 1992. *Babesiosis in California.* Issue 3/4, January 24. California Division of Communicable Disease Control, Berkeley.

21. **Carlson, J., H. Helmby, A. V. S. Hill, D. Brewster, B. M. Greenwood, and M. Wahlgren.** 1990. Human cerebral malaria: association with erythrocyte rosetting and lack of anti-rosetting antibodies. *Lancet* **336:**1457–1460.

22. **Centers for Disease Control.** 1992. Viscerotropic leishmaniasis in persons returning from Operation Desert Storm—1990–1991. *Morbid. Mortal. Weekly Rep.* **40:**131–134.

23. **Dannemann, B. R., D. M. Israelski, G. S. Leoung, T. McGraw, J. Mills, and J. S. Remington.** 1991. *Toxoplasma* serology, parasitemia and antigenemia in patients at risk for toxoplasmic encephalitis. *AIDS* **5:**1363–1365.

24. **Eltoun, I. A., E. E. Zulstra, M. S. Ali, H. W. Ghalib, M. M. H. Satti, B. Eltoum, and A. E. El-Hassan.** 1992. Congenital kala azar and leishmaniasis in the placenta. *Am. J. Trop. Med. Hyg.* **46:**57–62.

25. **Ferreira, M. S., S. D. Nishioka, A. Rocha, A. M. Silva, R. G. Ferreira, W. Oliver, and S. Tostes, Jr.** 1991. Acute fatal *Trypanosoma cruzi* meningoencephalitis in a human immunodeficiency virus-positive hemophiliac patient. *Am. J. Trop. Med. Hyg.* **45:**723–727.

26. **Furner, B. B.** 1990. Cutaneous leishmaniasis in Texas: report of a case and review of the literature. *J. Am. Acad. Dermatol.* **23:**368–371.

27. **Garces, J. M., S. Tomas, J. Rubies-Prat, J. L. Gimeno, and L. Drobnic.** 1990. Bacterial infection as a presenting manifestation of visceral leishmaniasis. *Rev. Infect. Dis.* **12:**518–519.

28. **Garcia, L. S., and D. A. Bruckner.** 1993. *Diagnostic Medical Parasitology*, 2nd ed. American Society for Microbiology, Washington, D.C.

29. **Gluckstein, D., F. Ciferri, and J. Ruskin.** 1992. Chagas' disease: another cause of cerebral mass in the acquired immunodeficiency syndrome. *Am. J. Med.* **92:**429–432.

30. **Gradon, J. D., J. G. Timpone, and S. M. Schnittman.** 1992. Emergence of unusual opportunistic pathogens in AIDS: a review. *Clin. Infect. Dis.* **15:**134–157.

31. **Gradoni, L., A. Scalone, and M. Gramiccia.** 1993. HIV-*Leishmania* co-infections in Italy—serological data as an indication of the sequence of acquisition of the 2 infections. *Trans. R. Soc. Trop. Med. Hyg.* **87:**94–96.

32. **Grimaldi, G., Jr., and R. B. Tesh.** 1993. Leishmaniases of the New World: current concepts and implications for future research. *Clin. Microbiol. Rev.* **6:**230–250.

33. **Gutierrez, Y., G. H. Salinas, G. Palma, L. B. Valderrama, C. V. Santrech, and N. G. Saravia.** 1991. Correlation between histopathology, immune response, clinical presentation and evaluation in *Leishmania braziliensis* infection. *Am. J. Trop. Med. Hyg.* **45:**281–289.

34. **Herwaldt, B. L., S. L. Stokes, and D. D. Juranek.** 1993. American cutaneous leishmaniasis in United States travelers. *Ann. Intern. Med.* **118:**779–784.

35. **Ho, J. L., R. Badaro, A. Schwartz, C. A. Dinarello, J. A. Gelfand, J. Sobel, A. Barral, M. B. Netto, E. M. Carvalho, S. G. Reed, and W. D. Johnson, Jr.** 1992. Diminished in vitro production of interleukin 1 and tumor necrosis factor α during acute visceral leishmaniasis and recovery after therapy. *J. Infect. Dis.* **165:**1094–1102.

36. **Hoffman, S. L., V. Nussenzweig, J. C. Sadoff, and R. A. Nussenzweig.** 1991. Progress toward malaria preerythrocytic vaccines. *Science* **252:**520–521.

37. **Hommel, M.** 1981. Malaria: immunity and prospects for vaccination. *West. J. Med.* **135:**285–299.

38. **Howard, M. K., J. M. Kelly, R. P. Lane, and M. A. Miles.** 1991. A sensitive repetitive DNA probe that is specific to the *Leishmania donovani* complex and its use as an epidemic logical and diagnostic reagent. *Mol. Biochem. Parasitol.* **44:**63–72.

39. **Isaacson, M.** 1989. Airport malaria: a review. *Bull. W.H.O.* **67:**737–743.

40. **Jones, E. M., D. G. Colley, S. Tostes, E. R. Lopes, C. L. Vnencakjones, and T. L. Mccurley.** 1993. Amplification of a *Trypanosoma cruzi* DNA sequence from inflammatory lesions in human chagasic cardiomyopathy. *Am. J. Trop. Med. Hyg.* **48:**348–357.

41. **Komba, E., M. Odiit, D. B. Mbulamberi, E. C. Chimfwembe, and V. M. Nantulya.** 1992. Multicentre evaluation of an antigen-detection ELISA for the diagnosis of *Trypanosoma brucei rhodesiense* sleeping sickness. *Bull. W.H.O.* **70:**57–61.

42. **Konishi, E.** 1991. Naturally occurring immunoglobulin M antibodies to *Toxoplasma gondii* in Japanese populations. *Parasitology* **102:**157–162.

43. **Krause, P. J., S. R. Telford III, R. Ryan, A. B. Hurta, I. Kwasnik, S. Luger, J. Niederman, M. Gerber, and A. Spielman.** 1991. Geographical and temporal distribution of babesial infection in Connecticut. *J. Clin. Microbiol.* **29:**1–4.

44. **Kreutzer, R. D., A. Corredor, G. Grimaldi, Jr., M. Grogl, E. D. Rowton, D. G. Young, A. Morales, D. McMahon-Pratt, H. Guzman, and R. B. Tesh.** 1991. Characterization of *Leishmania colombiensis* sp. N (Kinetoplastida: Trypanosomatidae), a new parasite infecting humans, animals and phlebotomine sandflies in Colombia and Panama. *Am. J. Trop. Med. Hyg.* **44:**662–675.

45. **Krieger, M. A., E. Almeida, W. Oelemann, J. S. Lafaille, J. B. Perreira, H. Krieger, M. R. Carvalho, and S. Goldenberg.** 1992. Use of recombinant antigens for the accurate immunodiagnosis of Chagas' disease. *Am. J. Trop. Med. Hyg.* **46:**427–434.

46. **Krotoski, W. A., P. C. C. Garnham, R. S. Bray, D. M. Krotoski, R. Killick-Kendrick, C. C. Draper, G. A. T. Targett, and M. W. Guy.** 1982. Observations on early and late post sporozoite tissue stages in primate malaria. 1. Discovery of a new latent form of *Plasmodium cynomolgi* (the hypnozoite), and failure to detect hepatic forms within the first 24 hours after infection. *Am. J. Trop. Med. Hyg.* **31:**24–35.

47. **Laskay, T., R. Kiessling, T. F. R. DeWit, and D. F. Wirth.** 1991. Generation of species-specific DNA probes for *Leishmania aethiopica. Mol. Biochem. Parasitol.* **44:**279–286.

48. **Locksley, R. M., and J. A. Louis.** 1992. Immunology of leishmaniasis. *Curr. Opin. Immunol.* **4:**413–418.

49. **Long, G. W., T. R. Jones, L. S. Rickman, R. Trimmer, and S. L. Hoffman.** 1991. Acridine orange detection of *Plasmodium falciparum* malaria: relationship between sensitivity and optical configuration. *Am. J. Trop. Med. Hyg.* **44:**402–405.

50. **Lorca, M., A. Gonzalez, C. Veloso, V. Reyes, and U. Vergara.** 1992. Immunodetection of antibodies in sera from symptomatic and asymptomatic Chilean Chagas' disease patients with *Trypanosoma cruzi* recombinant antigens. *Am. J. Trop. Med. Hyg.* **46:**44–49.

51. **Magill, A. J., M. Grogl, R. A. Gasser, W. Sun, and C. N. Oster.** 1993. Visceral infection caused by *Leishmania tropica* in veterans of Operation Desert Storm. *N. Engl. J. Med.* **328:**1384–1387.

52. **Marcus, L. C., J. M. Valigorsky, W. L. Fanning, T. Joseph, and B. Glick.** 1982. A case report of transfusion-induced babesiosis. *JAMA* **248:**465–467.

53. **Martinez, J. E., A. L. Arias, M. A. Escobar, and N. G. Saravia.** 1992. Haemoculture of *Leishmania* (*Vianna*) braziliensis from two cases of mucosal leishmaniasis: re-examination of haematogenous dissemination. *Trans. R. Soc. Trop. Med. Hyg.* **86:**392–394.

54. **Matsumoto, T. K., S. Hoshinoshimizu, P. M. Nakamura, H. F. Andrade, and E. S. Umezawa.** 1993. High resolution of *Trypanosoma cruzi* amastigote antigen in serodiagnosis of different clinical forms of Chagas' disease. *J. Clin. Microbiol.* **31:**1486–1492.

55. **McLeod, R., D. Mack, and C. Brown.** 1991. *Toxoplasma gondii:* new advances in cellular and molecular biology. *Exp. Parasitol.* **72:**109–121.

56. **Mebrahtu, Y. B., L. D. Hendricks, C. N. Oster, P. G. Lawyer, P. V. Perkins, H. Pamba, D. Koech, and C. R. Roberts.** 1993. *Leishmania donovani* parasites in the nasal secretions, tonsillopharyngeal mucosa, and urine centrifugates of visceral leishmaniasis patients in Kenya. *Am. J. Trop. Med. Hyg.* **48:**530–535.

57. **Modabber, F.** 1990. Development of vaccines against leishmaniasis. *Scand. J. Infect. Dis. Suppl.* **76:**72–78.

58. **Nagel, R. L.** 1990. Innate resistance to malaria: the intraerythrocytic cycle. *Blood Cells* **16:**321–339.

59. **Nantulya, V. M., F. Doua, and S. Molisho.** 1992. Diagnosis of *Trypanosoma brucei gambiense* sleeping sickness using an antigen detection enzyme-linked immunosorbent assay. *Trans. R. Soc. Trop. Med.* **86:**42–45.

60. **Naot, Y., D. R. Guptill, and J. S. Remington.** 1982. Duration of IgM antibodies of *Toxoplasma gondii* after acute acquired toxoplasmosis. *J. Infect. Dis.* **145:**770.

61. **Nascimento, E., W. Mayrink, C. A. da Costa, M. S. M. Michalick, M. N. Melo, G. C. Barros, M. Dias, C. M. F. Antunes, M. S. Lima, D. C. Taboada, and T. Y. Liu.** 1990. Vaccination of humans against cutaneous leishmaniasis: cellular and humoral immune responses. *Infect. Immun.* **58:**2198–2203.

62. **National Committee for Clinical Laboratory Standards.** 1993. *Slide Preparation and Staining of Blood Films for the Laboratory Diagnosis of Parasitic Diseases.* Tentative guideline M15-T. National Committee for Clinical Laboratory Standards, Villanova, Pa.

63. **Nussenzweig, V., and R. S. Nussenzweig.** 1990. Progress toward a malaria vaccine. *Hosp. Pract.* **15:**45–57.

64. **Oddo, D., M. Casanova, G. Acuna, J. Ballesteros, and B. Morales.** 1992. Acute Chagas' disease (*trypanosomiasis americana*) in acquired immunodeficiency syndrome. *Hum. Pathol.* **23:**41–44.

65. **Oliveira, D. A., B. P. Holloway, E. L. Durigon, W. E. Collins, and A. A. Lal.** Polymerase chain reaction and a liquid phase non-isotopic hybridization for detection of *Plasmodium falciparum* infection. *Am. J. Trop. Med. Hyg.*, in press.

66. **Pan, A. A., G. B. Rosenberg, M. K. Hurley, G. J. H.** Schock, V. P. Chu, and A. Aiyappa. 1992. Clinical evaluation of an EIA for the sensitive and specific detection of serum antibody to *Trypanosoma cruzi* (Chagas' disease). *J. Infect. Dis.* **165:**585–588.

67. **Pearson, R. D., G. Cox, S. M. B. Jeronimo, J. Castracane, J. S. Drew, T. Evans, and J. E. de Alencar.** 1992. Visceral leishmaniasis: a model for infection-induced cachexia. *Am. J. Trop. Med. Hyg.* **47:**8–15.

68. **Qari, S. H., Y. P. Shi, I. F. Goldman, V. Udhayakumar, W. E. Collins, and A. A. Lal.** 1993. Identification of *Plasmodium vivax*-like human malaria parasite. *Lancet* **341:**780–783.

69. **Quick, R. E., B. L. Herwaldt, J. W. Thomford, M. E. Garnett, M. L. Eberhard, M. Wilson, D. H. Spach, J. W. Dickerson, S. R. Telford, K. R. Steingart, R. Pollock, D. H. Pershing, J. M. Kobayashi, D. D. Juranek, and P. A. Conrad.** 1993. Babesiosis in Washington State—a new species of Babesia. *Ann. Intern. Med.* **119:**284–290.

70. **Reiss, D. D., E. M. Jones, S. Tostes, E. R. Lopes, G. Gazzinelli, D. G. Colley, and T. L. McCurley.** 1993. Characterization of inflammatory infiltrates in chronic chagasic myocardial lesions—presence of tumor necrosis factor-alpha+ cells and dominance of granzyme A+, CD8+ lymphocytes. *Am. J. Trop. Med. Hyg.* **48:**637–644.

71. **Ruebush, T. K., II, P. B. Cassaday, H. J. Marsh, et al.** 1977. Human babesiosis on Nantucket Island: clinical features. *Ann. Intern. Med.* **86:**6–9.

72. **Ruebush, T. K., II, E. S. Chisholm, A. J. Sulzer, et al.** 1981. Development and persistence of antibody in persons infected with *Babesia microti. Am. J. Trop. Med. Hyg.* **30:**291–292.

73. **Ruebush, T. K., II, D. D. Juranek, F. S. Chisholm, et al.** 1977. Human babesiosis on Nantucket Island: evidence for self limited and subclinical infections. *N. Engl. J. Med.* **297:**825–827.

74. **Russo, D. M., S. J. Turco, J. M. Burns, Jr., and S. G. Reed.** 1992. Stimulation of human T-lymphocytes by *Leishmania* lipophosphoglycan-associated proteins. *J. Immunol.* **148:**202–207.

75. **Saavedra, R., and P. Herion.** 1991. Human T-cell clones against *Toxoplasma gondii:* production of interferon-γ, interleukin-2, and strain cross-reactivity. *Parasitol. Res.* **77:**379–385.

76. **Schnapp, L. M., S. M. Geaghan, A. Campagna, J. Fahy, D. Steiger, V. Ng, W. K. Hadley, P. C. Hopewell, and J. D. Stansell.** 1992. *Toxoplasma gondii* pneumonitis in patients infected with the human immunodeficiency virus. *Arch. Intern. Med.* **152:**1073–1077.

77. **Schwartz, I. K., E. M. Kackritz, and L. C. Patchen.** 1991. Chloroquine-resistant *Plasmodium vivax* from Indonesia. *N. Engl. J. Med.* **324:**927.

78. **Siddiqui, W. A.** 1991. Where are we in the quest for vaccines for malaria? *Drugs* **41:**1–10.

79. **Starr, M. D., J. C. Rohas, R. Zeledon, D. W. Hird, and T. E. Carpenter.** 1991. Chagas' disease: risk factors for house infestation by *Triatoma dimidiata,* the major vector of *Trypanosoma cruzi* in Costa Rica. *Am. J. Epidemiol.* **133:**740–747.

80. **Subauste, C. S., and J. S. Remington.** 1993. Immunity to *Toxoplasma gondii. Curr. Opin. Immunol.* **5:**532–537.

81. **Sulzer, A. J., E. L. Franco, E. Takafuji, M. Benenson, K. W. Walls, and R. L. Greenup.** 1986. An oocyst-transmitted outbreak of toxoplasmosis: patterns of immunoglobulin G and M over one year. *Am. J. Trop. Med. Hyg.* **35:**290–296.

82. **Tanowitz, H. B., L. V. Kirchoff, D. Simon, S. A. Morris, L. M. Weiss, and M. Wittner.** 1992. Chagas' disease. *Clin. Microbiol. Rev.* **5:**400–419.

83. **Taylor, D. W., and A. Voller.** 1993. The development and validation of a simple antigen detection ELISA for *Plasmodium falciparum* malaria. *Trans. R. Soc. Trop. Med. Hyg.* **87:**29–31.

84. **Theis, J. H.** 1990. Latin American immigrants—blood donation and *Trypanosoma cruzi* transmission. *Am. Heart J.* **120:**1483–1484.

85. **Uliana, S. R. B., M. H. T. Affonso, E. P. Camago, and L. M. Floeter-Winter.** 1991. *Leishmania:* genus identification based

on specific sequence of the 18S ribosomal RNA sequence. *Exp. Parasitol.* **72:**157–163.

86. **Van Eys, G. J. J. M., L. Guizani, G. S. Lighthart, and K. Dellagi.** 1991. A nuclear DNA probe for the identification of strains within the *Leishmania donovani* complex. *Exp. Parasitol.* **72:**459–463.

87. **Van Eys, G. J. J. M., G. J. Schoone, N. C. M. Kroon, and S. B. Ebeling.** 1992. Sequence analysis of small subunit ribosomal genes and its use for detection and identification of *Leishmania* parasites. *Mol. Biochem. Parasitol.* **51:**133–142.

88. **Vanvianen, P. H., A. Vanengen, S. Thaithong, M. Vanderkeur, H. J. Tanke, H. J. Vanderkaay, B. Mons, and C. J. Janse.** 1993. Flow cytometric screening of blood samples for malaria parasites. *Cytometry* **14:**276–280.

89. **Vergara, U., C. Veloso, A. Gonzalez, and M. Lorca.** 1992. Evaluation of an enzyme-linked immunosorbent assay for the diagnosis of Chagas' disease using synthetic peptides. *Am. J. Trop. Med. Hyg.* **46:**39–43.

90. **Visvesvara, G. S.** 1992. Parasite culture: *Leishmania* spp. and *Trypanosoma cruzi,* p. 7.9.4.1–7.9.4.7. *In* H. D. Isenberg (ed.), *Clinical Microbiology Procedures Handbook,* vol. 2. American Society for Microbiology, Washington, D.C.

91. **Wanke, C., C. U. Tuazon, A. Kovacs, T. Dina, D. O. David, N. Barton, D. Katz, M. Lunde, C. Levy, F. K. Conley, H. C. Lane, A. S. Fauci, and H. Masur.** 1987. *Toxoplasma* encephalitis in patients with acquired immune deficiency syndrome: diagnosis and response to therapy. *Am. J. Trop. Med. Hyg.* **36:**509–516.

92. **Wasylenko, M., L. Racela, C. J. Papasian, and I. Watanabe.** 1991. *Toxoplasma gondii* in a cervical smear. *Acta Cytol.* **35:**145–146.

93. **Weigle, K. A., L. Valderrama, A. L. Arias, C. Santrich, and N. G. Saravia.** 1991. Leishmanin skin test standardization and evaluation of safety, dose storage, longevity of reaction, and sensitization. *Am. J. Trop. Med. Hyg.* **44:**260–271.

94. **WHO Expert Committee.** 1991. *Control of Chagas' Disease.* World Health Organization technical report, series 811. World Health Organization, Geneva.

95. **Wilson, S. M., R. McNerney, M. B. Moreno, I. Frame, and M. A. Miles.** 1992. Adaptation of a radioactive *L. donovani* complex DNA probe to a chemiluminescent detection system gives enhanced sensitivity for diagnostic and epidemiologic applications. *Parasitology* **104:**421–426.

96. **Zangerle, R., F. Allerberger, P. Pohl, P. Fritsch, and M. P. Dierich.** 1991. High risk of developing toxoplasmic encephalitis in AIDS patients seropositive to *Toxoplasma gondii. Med. Microbiol. Immunol.* **180:**59–66.

97. **Zavola-Castro, J. E., O. Velasco-Castrejon, and R. Hernandez.** 1992. Molecular characterization of Mexican stocks of *Trypanosoma cruzi* using total DNA. *Am. J. Trop. Med. Hyg.* **47:**201–209.

98. **Ziegelbauer, K., and P. Overath.** 1992. Identification of invariant surface glycoproteins in the bloodstream stage of *Trypanosoma brucei. J. Biol. Chem.* **267:**10791–10796.

Pathogenic and Opportunistic Free-Living Amebae

GOVINDA S. VISVESVARA

105

BACKGROUND

Small free-living amebae belonging to the genera *Naegleria*, *Hartmannella*, and *Acanthamoeba* are commonly found in soil, fresh water, and even sewage and sludge (4, 10, 13, 16, 23, 30). Several species of *Acanthamoeba* have also been isolated from brackish water and seawater (4, 13, 21) and from ear discharge, pulmonary secretions, nasopharyngeal swabs, maxillary sinus samples, mandibular autografts, and stool samples (4, 10, 13, 16, 30). *Acanthamoeba* spp. have been also known to harbor *Legionella* spp. and mycobacteria, which signifies the public health importance of these organisms (13, 30). Excellent reviews (2, 10, 13) and books (16, 23) have been published on the biology, disease potential, pathology, pathogenicity, and epidemiology of these amebae.

Naegleria and *Acanthamoeba* spp. normally feed on bacteria and multiply in their environmental niche as free-living organisms. However, when given the right conditions and opportunity, they have also been known to cause fatal disease of the central nervous system (CNS) in humans (2, 5, 9, 10, 13, 14–16, 18, 30, 31) and *Acanthamoeba* keratitis.

The concept that these small free-living amebae could become human pathogens was proposed by Culbertson and his colleagues, who isolated *Acanthamoeba* sp. strain A-1 (now designated *Acanthamoeba culbertsoni*) from tissue culture medium thought to contain an unknown virus (3). They also demonstrated amebae in brain lesions of mice and monkeys that died within a week after intracerebral inoculation with these amebae. Culbertson hypothesized that similar infection might exist in nature in humans. in *Fowler and Carter* in 1965 were the first to describe a fatal infection due to free-living amebae, which occurred in the brain of an Australian patient (see reference 2). The infection is now believed to have been due to a *Naegleria* sp. (2).

NAEGLERIA MENINGOENCEPHALITIS

Only one species of *Naegleria*, *Naegleri fowleri*, is known to cause human infection. In 1966, Butt et al. (see references 2 and 13) described the first case of free-living-ameba infection in the United States and coined the term primary amebic meningoencephalitis (PAM). PAM due to *N. fowleri* is an acute fulminating disease with an abrupt onset

that occurs generally in previously healthy children and young adults who have had contact with fresh water about 7 to 10 days before the onset of symptoms. It is characterized by severe headache, spiking fever, stiff neck, photophobia, and coma leading to death within 3 to 10 days after the onset of symptoms (2, 13, 16). The portal of entry of these amebae is the nasal passages. When people swim in lakes and other bodies of water that harbor these amebae, the amebae may enter the nostrils of the swimmers, make their way into the olfactory lobes via the cribriform plate, and cause acute hemorrhagic necrosis leading to destruction of the olfactory bulbs and the cerebral cortex. Large numbers of amebic trophozoites, many with ingested erythrocytes and brain tissue, are usually seen interspersed with brain tissue (Fig. 1). Cysts of *N. fowleri* are not usually seen in the brain tissue. More than 160 cases of PAM have been reported worldwide, and only a few patients have survived. It is believed that 70 cases of PAM have occurred in the United States as of January 1994.

ACANTHAMOEBA ENCEPHALITIS

Several species of *Acanthamoeba* (*A. culbertsoni*, *A. castellanii*, *A. polyphaga*, *A. astronyxis*, *A. healyi*, *A. divionensis*) have been known to cause a chronic granulomatous amebic encephalitis (GAE), primarily in immunosuppressed (either iatrogenically or because of human immunodeficiency virus infection or AIDS), chronically ill, or otherwise debilitated persons with no previous history of exposure to fresh water (5, 9, 13–16, 31). GAE has an insidious onset and is usually chronic, lasting for more than a week and sometimes even for months (14–16). GAE is characterized by headache, confusion, dizziness, drowsiness, seizures, and sometimes hemiparesis. Cerebral hemispheres are usually the most affected CNS tissue. They are often edematous, with extensive hemorrhagic necrosis involving the temporal, parietal, and occipital lobes. Amebic trophozoites and cysts are usually scattered throughout the tissue. Many blood vessels are thrombotic with fibrinoid necrosis and cuffed by polymorphonuclear leukocytes, amebic trophozoites, and cysts (Fig. 2). Multinucleated giant cells forming granulomas may be seen in immunocompetent patients. Some patients, especially those with human immunodeficiency virus or AIDS, develop chronic, ulcerative skin lesions, abscesses, or erythematous nodules (5, 7–9, 16, 31).

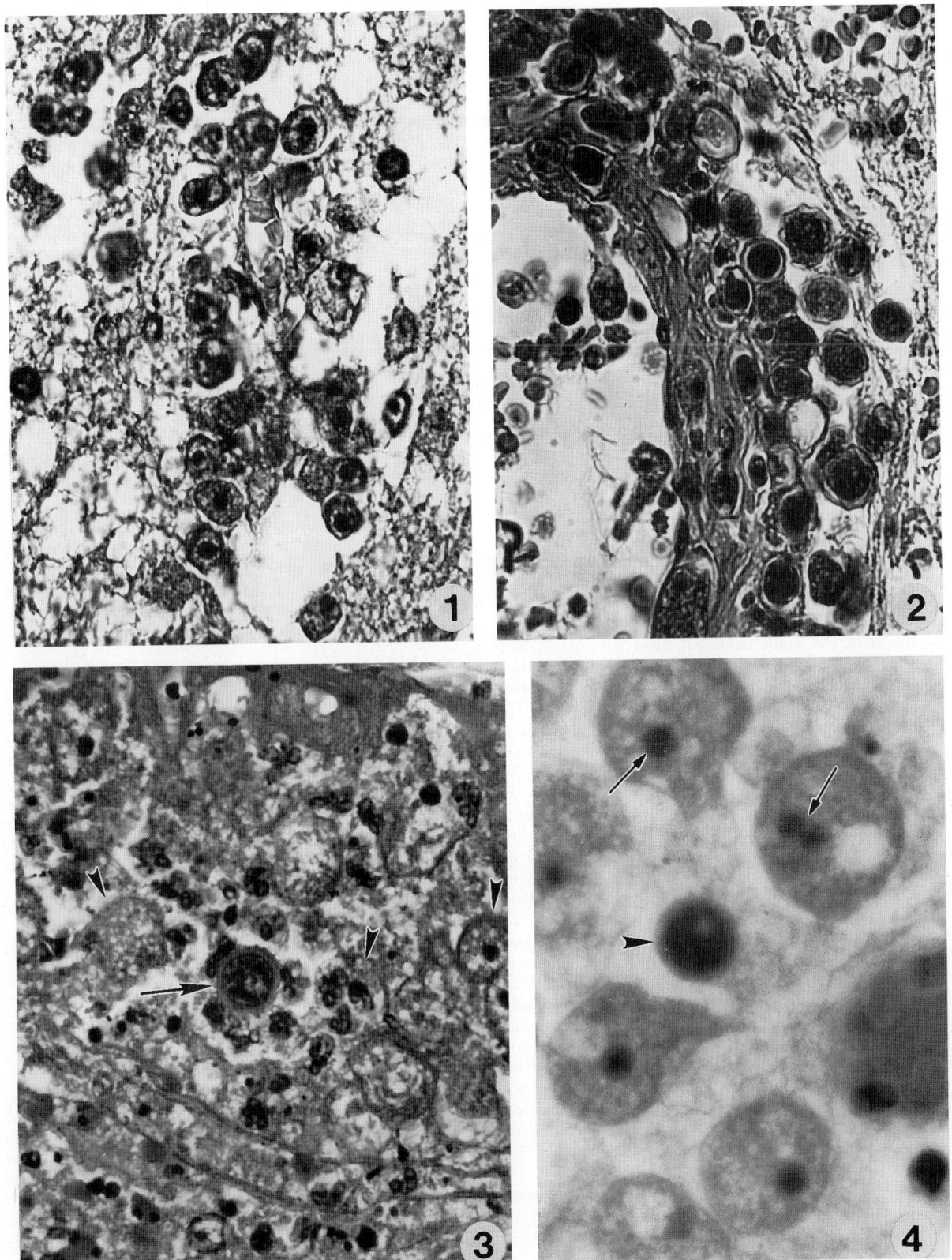

FIGURE 1 Large numbers of *N. fowleri* trophozoites in a section of CNS, showing extensive necrosis and destruction of brain tissue. Magnification, ×750.

FIGURE 2 *A. castellanii* trophozoites and cysts around a blood vessel in a section of CNS from a GAE patient. Magnification, ×650.

FIGURE 3 *Balamuthia* trophozoites (arrowheads) and a cyst (arrow) in the brain section of a GAE patient. Magnification, ×550.

FIGURE 4 Trophozoites and a darkly staining cyst (arrowhead) in a brain section of a GAE patient. Note the double nucleolar elements (arrow) within the nucleus of the trophozoites. Magnification, ×1,000.

It is believed that the route of invasion and penetration into the CNS is hematogenous, probably from a primary focus in the lower respiratory tract or the skin (14–16). Since *Acanthamoeba* cysts have been isolated from dust in air (4, 13, 30), it is quite likely that some persons, especially those who are immunologically compromised, inhale cyst forms of these amebae when passing through areas where soil has been freshly turned. The amebae may subsequently excyst into trophozoites and invade the nasal mucosa. More than 110 cases of GAE have been recorded worldwide, and ~64 of these cases (30 in AIDS patients) have occurred in the United States as of January 1994.

Because of the confusion that existed in the earlier literature with regard to the nomenclature of *Acanthamoeba* and *Hartmannella* spp., some workers in the field referred to these amebae as belonging to the *Hartmannella-Acanthamoeba* group or simply as "*H-A* amebae." Since no true *Hartmannella* species has been found to be pathogenic to humans, all references in the literature to *Hartmannella* spp. in human tissues should be corrected to read *Acanthamoeba* or *Balamuthia* spp.

BALAMUTHIA (LEPTOMYXID) ENCEPHALITIS

Until recently, it was believed that all cases of GAE were caused by *Acanthamoeba* spp. Though a number of these cases were confirmed by serologic techniques as being caused by *Acanthamoeba* spp., the causative organisms in a few cases could not be definitively identified. Since cysts were found in the brain tissues of these patients, it was believed that the infections were caused by some other species of *Acanthamoeba* that did not cross-react with the anti-*Acanthamoeba* sera used in the serologic test (14–16, 30). It was only recently that *Balamuthia mandrillaris* (leptomyxid ameba) was definitively identified by an immunofluorescence test as the causal agent of these cases (9, 28–30).

The pathology and pathogenesis are similar to those of *Acanthamoeba* GAE. Both trophozoites and cysts are found in CNS tissue (Fig. 3), and their sizes overlap those of *Acanthamoeba* (1, 13, 27, 31, 32). Hence, it is difficult to differentiate *Balamuthia* from *Acanthamoeba* spp. in tissue sections under the light microscope. In some cases, *Balamuthia* trophozoites in tissue sections appear to have more than one nucleolus in the nucleus (Fig. 4). In such cases, it may be possible to distinguish *Balamuthia* amebae from *Acanthamoeba* organisms on the basis of nuclear morphology, since *Acanthamoeba* trophozoites have only one nucleolus. In a majority of cases, electron microscopy, immunohistochemical technique, or both are necessary to identify *Balamuthia* organisms. Ultrastructurally, the cysts are characterized by three layers in the cyst wall: an outer wrinkled ectocyst, a middle structureless mesocyst, and an inner thin endocyst (28, 29). *Balamuthia* amebas are antigenically distinct from *Acanthamoeba* organisms; they can easily be distinguished by immunofluorescence or other immunochemical assays (1, 9, 28, 29).

ACANTHAMOEBA KERATITIS

In addition to GAE, *Acanthamoeba* spp. also cause a painful vision-threatening disease of the human cornea, *Acanthamoeba* keratitis. If the infection is not treated promptly, it may lead to ulceration of the cornea, loss of visual acuity, and eventually blindness and enucleation (6, 10–13, 32). The first case of *Acanthamoeba* keratitis in the United States was reported in 1973 in a south Texas rancher with a history of trauma to his right eye (11). Both trophozoites and cyst stages of *A. polyphaga* were shown in corneal sections and repeatedly cultured from corneal scrapings and biopsy specimens. It is estimated that as of February 1994, more than 500 cases of *Acanthamoeba* keratitis have occurred in the United States. Between 1973 and July 1986, 208 cases were diagnosed and reported to the Centers for Disease Control and Prevention (24, 25, 30). The numbers of cases gradually increased between 1973 and 1984, with a dramatic increase beginning in 1985. An in-depth epidemiologic and case-control study (24, 25) revealed that a major risk factor was the use of contact lenses, predominantly daily wear or extended-wear soft lenses, and that patients with *Acanthamoeba* keratitis were significantly more likely than controls to use homemade saline instead of commercially prepared saline (78 versus 30%, respectively), disinfect their lenses less frequently than recommended by the lens manufacturers (72 versus 32%), and wear their lenses while swimming (63 versus 30%). *Acanthamoeba* keratitis is characterized by severe ocular pain, a 360° or partial paracentral stromal ring infiltrate, recurrent corneal epithelial breakdown, and a corneal lesion refractory to commonly used ophthalmic antibacterial medications. *Acanthamoeba* keratitis in the early stages is frequently misdiagnosed as herpes simplex virus keratitis because of the irregular epithelial lesions, stromal infiltrative keratitis, and edema that are commonly seen in herpes simplex virus keratitis (6, 10–13, 24, 25, 30, 32).

MORPHOLOGY

N. fowleri in its trophic form is a small limaxlike ameba measuring 10 to 35 μm that exhibits an eruptive locomotion by producing smooth hemispherical bulges. The posterior end, termed the uroid, appears to be sticky and often has several trailing filaments. During its life cycle, this ameba produces a transient pear-shaped biflagellate stage, resulting from altered environmental conditions, and smooth-walled cysts (Fig. 5–8). The flagellates do not have cytostomes. Cysts are usually spherical and measure 7 to 15 μm, and the cyst wall may have one or more pores plugged with a mucoid material (20).

An *Acanthamoeba* organism is a slightly larger ameba (15 to 45 μm) that produces from the surface of its body fine, tapering, hyaline projections called acanthopodia (Fig. 9 and 10). It has no flagellate stage but produces a double-walled cyst (10 to 25 μm) with a wrinkled outer wall, an ectocyst, and a stellate, polygonal, or even round inner wall, the endocyst (20).

Balamuthia trophozoites are in general irregular in shape; a few, however, may be limacine. Actively feeding amebae may measure from 12 to 60 μm in length, with a mean of 30 μm (Fig. 11). The trophic forms, while feeding on tissue culture cells, produce broad pseudopodia without any clearly discernible movement. However, when tissue culture cells are destroyed, the trophozoites resort to a spider-like walking movement by producing fingerlike determinate pseudopodia (29). Like *Acanthamoeba* spp., *Balamuthia* spp. do not have a flagellate stage. Cysts are generally spherical and measure from 6 to 30 μm in diameter (Fig. 12). Under the light microscope, each cyst appears to have an irregular and slightly wavy outer wall and a round inner wall. A layer of refractile granules immediately below the inner cyst wall is often seen in mature cysts. In the electron microscope, the cyst wall can be seen to consist of three walls: an outer

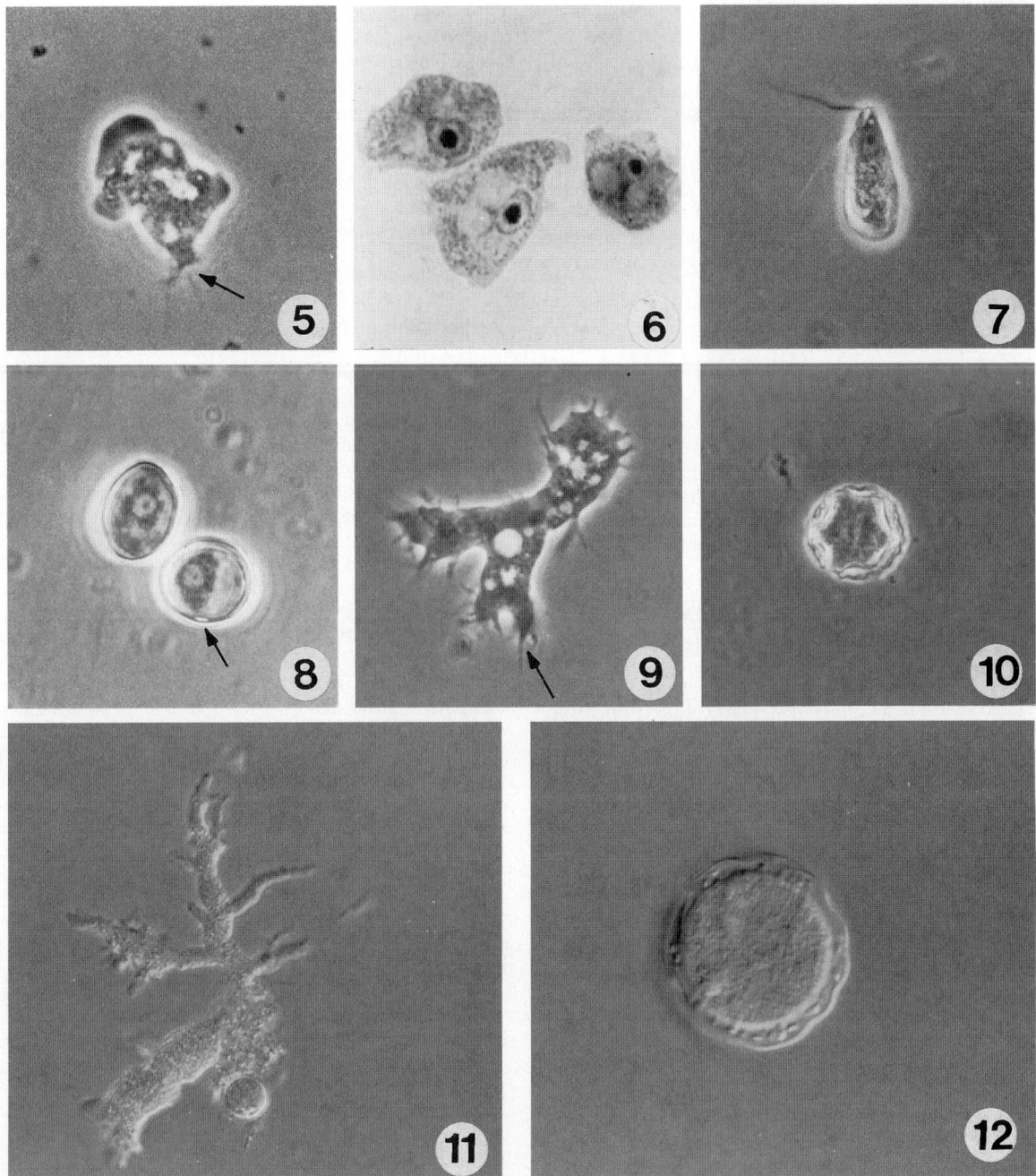

FIGURES 5–8 *N. fowleri.* (FIGURE 5) Trophozoite, phase-contrast (note the uroid and filaments at arrow); (FIGURE 6) trophozoite, trichrome stain; (FIGURE 7) biflagellate, phase-contrast; (FIGURE 8) smooth-walled cyst, phase contrast (note the pore at the arrow). All magnifications are ×1,100.
FIGURES 9 and 10 *A. castellanii.* (FIGURE 9) Trophozoite, phase-contrast (note the acanthopodia at the arrow); (FIGURE 10) double-walled cyst, phase-contrast. Both magnifications are ×1,100.
FIGURES 11 and 12 *B. mandrillaris.* (FIGURE 11) Trophozoite, phase-contrast; (FIGURE 12) cyst, phase-contrast. Both magnifications are ×1,500.

thin, irregular ectocyst; a thick, electron-dense inner endocyst; and a middle amorphous layer, the mesocyst (28, 29).

Acanthamoeba, Balamuthia, and *Naegleria* spp. are predominantly uninucleate, although binucleate forms are occasionally seen. The nucleus is characterized by a large, dense, centrally located nucleolus. *Naegleria* amebae exhibit a promitotic pattern of cell division wherein the nucleolus and the nuclear membrane persist during nuclear division. *Acanthamoeba* spp., however, divide by conventional mitosis, in which the nucleolus and the nuclear membrane disappear during cell division. The pattern of nuclear division in *Balamuthia* spp. is termed metamitosis: the nuclear membrane breaks down, and the nucleolus eventually disappears (29).

COLLECTION, HANDLING, AND STORAGE OF SPECIMENS

For isolating the etiologic agent, cerebrospinal fluid, small pieces of tissue (brain, lungs, corneal biopsy material), or corneal scrapings from the affected area must be obtained aseptically. The specimens should be kept at room temperature (24 to 28°C) and should never be frozen. The specimens may be kept at 4°C for short periods but never for more than 24 h. Personnel handling the specimens must take appropriate precautions, such as wearing surgical masks and gloves and working in a biological safety cabinet. Remaining tissues must be preserved in 10% neutral buffered formalin so that they can be examined histologically for amebae.

METHODS OF EXAMINATION

Direct Examination

Direct examination of the sample as a wet mount preparation is of paramount importance in the diagnosis of PAM and other diseases caused by these amebae. Since the amebae tend to attach to the surface of the container, the container should be shaken gently; then a small drop of fluid is placed on a clean microscope slide and covered with a no. 1 coverslip. The cerebrospinal fluid may have to be centrifuged at 150 × g for 5 min to concentrate the amebae. After the specimen has been centrifuged, most of the supernatant is carefully aspirated, and the sediment is gently suspended in the remaining fluid. A drop of this suspension is prepared as described above for microscopic observation.

The slide preparation should be examined under a compound microscope with 10× and 40× objectives. Phase-contrast optics are preferable. If regular bright-field illumination is used, the slide should be examined under diminished light. A step in which the slide is warmed to 35°C is optional. Amebae, especially *N. fowleri*, if present, can easily be detected by their active directional movements.

Permanently Stained Preparations

A small drop of the sedimented cerebrospinal fluid or other sample is placed in the middle of a slide and allowed to stand in a moist chamber for 5 to 10 min at 37°C. This will allow any amebae to attach to the surface of the slide. Several drops of warm (37°C) Schaudinn's fixative are dropped directly onto the sample and allowed to stand for 1 min. The slide is then transferred to a Coplin jar containing the fixative and fixed for 1 h. It may be stained in Wheatley's trichrome or Heidenhain's iron hematoxylin stain. Corneal scrapings smeared on microscope slides may be fixed with methanol and stained with Hemacolor stain (Harleco, a division of EM Industries, Inc.) (13).

ISOLATION

The recommended procedure for isolating free-living pathogenic amebae from biological specimens is as follows.

Materials

1. Page's ameba saline (13) (see Appendix 1)
2. Nonnutrient agar plates with ameba saline (see Appendix 1)
3. 18- to 24-h-old culture of *Escherichia coli* or *Enterobacter aerogenes*
4. Sterile distilled water
5. Bacteriologic loop
6. Fine spatula made of nichrome wire
7. Sterile Pasteur pipettes, rubber bulbs
8. Sterile 1-ml serologic pipettes
9. Sterile screw-cap test tubes, 100 by 13 or 125 by 16 mm
10. Moist chamber
11. Vaspar (1:1 petrolatum-paraffin), melted
12. Microscope slides, 3 by 1 or 3 by 2 in. (7.6 by 2.5 or 7.6 by 5 cm)
13. Coverslips, 22 mm square, no. 1 thickness

Preparation of Agar Plates

1. Remove plates from refrigerator, and place them in a 37°C incubator for 30 min.
2. Add 0.5 ml of ameba saline to a slant culture of *E. coli* or *Enterobacter aerogenes*. Gently scrape the surface of the slant with a sterile bacteriologic loop (do not break the agar surface). Using a sterile Pasteur pipette, gently and uniformly suspend the bacteria. Add 2 or 3 drops of this suspension to the middle of a warmed (37°C) agar plate, and spread the bacteria over the surface of the agar with a bacteriologic loop. The plate is then ready for inoculation.

Inoculation of Plates with Specimens

1. For cerebrospinal fluid samples, centrifuge the cerebrospinal fluid at 250 × g for 5 to 8 min. With a sterile serologic pipette, carefully transfer all but 0.5 ml of the supernatant to a sterile tube, and store the tube at 4°C for possible future use. Mix the sediment with the remaining fluid. With a sterile Pasteur pipette, place 2 or 3 drops in the center of the agar plate precoated with bacteria, and incubate in room air at 37°C.
2. For tissue samples, triturate a small piece of the tissue in a small amount of ameba saline. With a sterile Pasteur pipette, place 2 or 3 drops of the mixture in the center of the agar plate. Incubate the plate in room air at 37°C for CNS tissues and at 30°C for tissues from other sites.
3. Water and soil samples are handled in the same manner as cerebrospinal fluid and tissue specimens, respectively.
4. Control cultures are recommended for comparative purposes, although care should be exercised to prevent cross-contamination of patient cultures.

Examination of Plates

1. Using the low-power (10×) objective of a microscope, observe the plates daily for 7 days for amebae.
2. If you see amebae anywhere, circle that area with a wax pencil. With a fine spatula, cut a small piece of agar from the circled area, and place it face down on the surface of a fresh agar plate precoated with bacteria; incubate as described above. Both *Naegleria* and *Acanthamoeba* spp. can easily be cultivated in this way and, with periodic transfers, maintained indefinitely. When the plate is examined under a microscope, the amebae will look like small blotches, and if they are observed carefully, their movement can be discerned. After 2 to 3 days of incubation, the amebae will start to encyst. If a plate is examined after 4 to 5 days of incubation, trophozoites as well as cysts will be visible. *Balamuthia* spp., however, will not grow on agar plates seeded with bacteria. While *Balamuthia* spp. can be grown

on monkey kidney or lung fibroblast cell lines, such techniques are not routinely available.

Identification and Culture

Identification of living organisms to the generic level is based on characteristic patterns of locomotion, morphologic features of the trophic and cyst forms, and enflagellation experiments. Immunofluorescence or immunoperoxidase tests using monoclonal or polyclonal antibodies (available from the Centers for Disease Control and Prevention) will be helpful in differentiating *Acanthamoeba* spp. from *Balamuthia* spp. in the fixed tissue and in identifying the species, especially *Acanthamoeba* spp., in fixed tissue.

Enflagellation Experiment

1. Mix a drop of the sedimented cerebrospinal fluid containing amebae with about 1 ml of sterile distilled water in a sterile tube, or, with a bacteriologic loop, scrape the surface of a plate that is positive for amebae, transferring a loopful of scraping to a sterile tube that contains approximately 1 ml of distilled water.

2. Gently shake the tube, and transfer a drop of this suspension to the center of a coverslip whose edges have been coated thinly with petroleum jelly. Place a microscope slide over the coverslip, and invert the slide. Seal the edges of the coverslip with Vaspar. Place the slide in a moist chamber, and incubate as before for 2 to 3 h. In addition, incubate the tube as described above.

3. Periodically examine the tube and the slide preparation microscopically for free-swimming flagellates. *Naegleria* spp. have a flagellated stage; *Acanthamoeba* spp. do not. If the sample contains *N. fowleri*, about 30 to 50% of the amebae will have undergone transformation into pear-shaped biflagellated organisms.

Other Culture Methods

Axenic Culture

Acanthamoeba spp. can easily be cultivated axenically, without the addition of serum or host tissue, in many different types of nutrient media, e.g., Proteose Peptone-yeast extract-glucose medium (Appendix 2), Trypticase soy broth medium, or chemically defined medium (10, 13). *N. fowleri*, however, requires fetal calf serum or brain extract in the medium, e.g., Nelson's medium (Appendix 3). A chemically defined medium has only recently been developed for *N. fowleri* (19). *Balamuthia* spp. cannot be cultivated on agar plates with bacteria or in an axenic medium. They can, however, be cultivated on mammalian cell lines (28, 29).

Axenic cultures of *Acanthamoeba* and *Naegleria* spp. can be established as follows. Actively growing 24- to 36-h-old amebae are scraped from the surface of the plate, suspended in 50 ml of ameba saline, and centrifuged at 150 × *g* for 5 min. The supernatant is aspirated, and the sediment is inoculated into Proteose Peptone-yeast extract-glucose medium or Nelson's medium, depending on the ameba isolate, and incubated at 37°C. Penicillin and streptomycin to final concentrations of 400 U/ml and 400 µg/ml, respectively, are added aseptically to the medium before the amebae are inoculated. Three subcultures into the antibiotic-containing medium at weekly intervals are usually sufficient to eliminate the associated bacteria (*E. coli* or *Enterobacter aerogenes*).

Cell Culture

Pathogenic *Acanthamoeba*, *Balamuthia*, and *Naegleria* spp. can also be inoculated onto many types of mammalian cell cultures. The amebae grow vigorously in these cell cultures and produce cytopathic effects somewhat similar to those caused by viruses. Because of such cytopathic effects, these amebae were mistaken for transformed cell types presumed to contain viruses and were erroneously termed lipovirus and Ryan virus (13, 30).

Animal Inoculation

Two-week-old Swiss Webster mice weighing 12 to 15 g each can be infected with these amebae. Mice are anesthetized with ether, and a drop of ameba suspension is instilled into their nostrils. Mice infected with *N. fowleri* die within 5 to 7 days after developing characteristic signs such as ruffled fur, aimless wandering, partial paralysis, and, finally, coma and death. Mice infected with *Acanthamoeba* and *Balamuthia* spp. may die of acute disease within 5 to 7 days or of chronic disease after several weeks. In all cases, amebae can be demonstrated in the mouse brain either by culture or histologic examination.

Antigenic Characteristics

Pathogenic *N. fowleri* is morphologically indistinguishable from nonpathogenic *N. gruberi* at the trophic stage. Differences between these amebae, however, have been demonstrated antigenically by the gel diffusion, immunoelectrophoretic, and indirect immunofluorescence techniques (10, 13, 30). Antigenic differences have also been shown among various species of *Acanthamoeba* (10, 13, 16, 18). Some recent studies indicate that *N. fowleri* can be distinguished from other species of *Naegleria* on the basis of isoenzyme patterns as well as restriction fragment length polymorphisms of genomic DNA (10, 13, 16, 17). Analyses of isoenzyme pattern, mitochondrial DNA, and small-subunit rRNA have also been carried out on various *Acanthamoeba* species and isolates in order to understand inter- and intraspecific diversity and phylogenetic relationships (10, 13, 16, 18).

Serology

The serologic techniques discussed here have been developed as research tools and are not routinely available to clinical laboratories.

Complement fixation antibody to *Acanthamoeba* spp. has been shown in the serum of patients suffering from upper respiratory tract distress and in those with optic neuritis and macular disease (10, 13, 16, 26, 30). Kenney (as discussed in reference 13) demonstrated increasing complement fixation antibody titers to *Acanthamoeba* in three successive samples of serum collected over a 2-month period from each of two patients. Precipitin antibody has also been shown in the serum of a patient suffering from *A. polyphaga* keratitis (11, 13, 30). Recently, an increase in ameba immobilization antibody titer over a 16-month period was demonstrated in the serum of a Nigerian patient, who made a partial recovery from an *A. rhysodes*-induced CNS disease (see reference 13).

An antibody response to *N. fowleri*, however, has not yet been defined. Most of the patients with *Naegleria* PAM died within a very short time (5 to 10 days), before they had time to produce detectable levels of antibody. In one case, however, in which the patient survived PAM due to *Naegleria* sp., an indirect immunofluorescence titer of

1:4,096 against *N. fowleri* was detected in the serum after 42 days of hospitalization (13, 22).

TREATMENT

Pathogenic *N. fowleri* is exquisitely susceptible to amphotericin B in vitro (2, 10, 13, 16, 22). At least three patients with PAM were reported to have recovered after receiving intrathecal and intravenous injections of this drug alone or in combination with miconazole (10, 13, 16, 22). Culbertson found sulfadiazine to be active against experimental *Acanthamoeba* infections in mice (10, 13, 16). Jones et al. (11) found that paromomycin, clotrimazole, and hydroxystilbamidine isethionate were active against *A. polyphaga* in vitro. In one study, 0.1% propamidine isethionate eye drops (Brolene) and 0.15% dibromopropamidine ointment together with topical neomycin sulfate were successful in the management of *Acanthamoeba* keratitis (32). In another study, 1% clotrimazole used in combination with Brolene and Neosporin on four patients gave excellent results (6). Recent studies indicate that medical cure has been achieved with topical administration of polyhexamethylene biguanide (12).

APPENDIX 1
Nonnutrient Agar Plates

Agar	1.5 g
Page's ameba saline	100.0 ml

Dissolve agar in the saline with heat, and sterilize at 15 lb/in^2 for 15 min. Cool to 60°C, and aseptically pour into plastic petri dishes. Use 20 ml for 100- by 15-mm dishes or 5 ml for 16- by 15-mm dishes. After the agar gels, store the plates in canisters at 4°C (refrigerator). Plates may be kept in the refrigerator for 3 months.

Prepare Page's ameba saline as follows.

Sodium chloride (NaCl)	120 mg
Magnesium sulfate (MgSO$_4$ · 7H$_2$O)	4 mg
Calcium chloride (CaCl$_2$ · 2H$_2$O)	4 mg
Disodium hydrogen phosphate (Na$_2$HPO$_4$)	142 mg
Potassium dihydrogen phosphate (KH$_2$PO$_4$)	136 mg
Distilled water, to	1 liter

Dissolve the chemicals in water, and sterilize the mixture by autoclaving it at 15 lb/in^2 for 15 min. The solution may be stored in the refrigerator for up to 6 months.

APPENDIX 2
Peptone-Yeast Extract-Glucose Medium for *Acanthamoeba* spp.

Proteose Peptone (Difco)	20.0 g
Yeast extract (Difco)	2.0 g
MgSO$_4$ · 7H$_2$O	0.980 g
CaCl$_2$	0.059 g
Sodium citrate · 2H$_2$O	1.0 g
Fe(NH$_4$)$_2$(SO$_4$)$_2$ · 6H$_2$O	0.02 g
KH$_2$PO$_4$	0.34 g
Na$_2$HPO$_4$ · 7H$_2$O	0.355 g
Glucose	18.0 g
Distilled water, to	1.0 liter

Final pH, 6.5 + 0.2
Dissolve all ingredients except CaCl$_2$ in about 900.0 ml of distilled water. Add CaCl$_2$ while stirring. Bring the volume to

1,000 ml. Dispense 5 ml per tube into screw-cap tubes (16 by 125 mm). Autoclave for 15 min (15 lb/in^2).

APPENDIX 3
Nelson's Medium for *N. fowleri*

Liver digest (Oxoid)	1.0 g
Glucose	1.0 g
Ameba saline, to	1.0 liter

Dissolve the ingredients in ameba saline (see Appendix 1). Dispense 10 ml per tube into screw-cap tubes (16 by 125 mm). Autoclave for 15 min. Add 0.2 ml of heat-inactivated fetal calf serum to each tube before inoculating the medium with amebae.

REFERENCES

1. **Anzil, A. P., C. Rao, M. A. Wrozlek, G. S. Visvesvara, J. H. Sher, and P. B. Kozlowski.** 1991. Amebic meningoencephalitis in a patient with AIDS caused by a newly recognized opportunistic pathogen: leptomyxid ameba. *Arch. Pathol. Lab. Med.* **115:**21–25.
2. **Carter, R. F.** 1972. Primary amoebic meningo-encephalitis. An appraisal of present knowledge. *Trans. R. Soc. Trop. Med. Hyg.* **66:**193–208.
3. **Culbertson, C. G., J. W. Smith, and J. R. Minner.** 1958. *Acanthamoeba:* observations on animal pathogenicity. *Science* **127:**1506.
4. **DeJonckheere, J. F.** 1987. Epidemiology, p. 127–147. *In* E. G. Rondanelli (ed.), *Amphizoic Amoebae: Human Pathology.* Piccin Nuova Libraria, Padua, Italy.
5. **Di Gregorio, C., F. Rivasi, N. Mongiardo, B. De Rienzo, and G. S. Visvesvara.** 1991. *Acanthamoeba* meningoencephalitis in an AIDS patient: first report from Europe. *Arch. Pathol. Lab. Med.* **116:**1363–1365.
6. **Driebe, W. T., G. A. Stern, R. J. Epstein, G. S. Visvesvara, M. Adi, and T. Komadina.** 1988. *Acanthamoeba* keratitis: potential role for topical clotrimazole in combination chemotherapy. *Arch. Ophthalmol.* **106:**1196–1201.
7. **Friedland, L. R., S. A. Raphael, E. S. Deutsch, J. Johal, L. J. Martyn, G. S. Visvesvara, and H. Lischner.** 1992. Disseminated *Acanthamoeba* infection in a child with symptomatic human immunodeficiency virus infection. *Pediatr. Infect. Dis. J.* **11:**404–407.
8. **Gonzalez, M. M., E. Gould, G. Dickinson, A. J. Martinez, G. S. Visvesvara, T. J. Cleary, and G. T. Hensley.** 1986. Acquired immunodeficiency syndrome associated with *Acanthamoeba* infection and other opportunistic organisms. *Arch. Pathol. Lab. Med.* **110:**749–751.
9. **Gordon, S. M., J. P. Steinberg, M. DuPuis, P. Kozarsky, J. F. Nickerson, and G. S. Visvesvara.** 1992. Culture isolation of *Acanthamoeba* species and leptomyxid amebas from patients with amebic meningoencephalitis, including two patients with AIDS. *Clin. Infect. Dis.* **15:**1024–1030.
10. **John, D. T.** 1993. Opportunistically pathogenic free-living amebae, p. 143–246. *In* J. P. Kreier and J. R. Baker (ed.), *Parasitic Protozoa,* vol. 3. Academic Press, Inc., New York.
11. **Jones, D. B., G. S. Visvesvara, and N. M. Robinson.** 1975. *Acanthamoeba polyphaga* keratitis and *Acanthamoeba* uveitis associated with fatal meningoencephalitis. *Trans. Ophthalmol. Soc. U.K.* **95:**221–232.
12. **Larkin, D. F. P., S. Kilvington, and J. K. G. Dart.** 1992. Treatment of *Acanthamoeba* keratitis with polyhexamethylene biguanide. *Ophthalmology* **99:**185–191.
13. **Ma, P., G. S. Visvesvara, A. J. Martinez, F. H. Theodore, P.-M. Daggett, and T. K. Sawyer.** 1990. Naegleria and Acanthamoeba infections: review. *Rev. Infect. Dis.* **12:**490–513.
14. **Martinez, A. J.** 1980. Is acanthamoebic encephalitis an opportunistic infection? *Neurology* **30:**567–574.
15. **Martinez, A. J.** 1982. Acanthamoebiasis and immunosuppression. Case report. *J. Neuropathol. Exp. Neurol.* **41:**548–557.

16. **Martinez, A. J.** 1985. *Free-Living Amebas: Natural History, Prevention, Diagnosis, Pathology, and Treatment of the Disease.* CRC Press, Inc., Boca Raton, Fla.

17. **McLaughlin, G. L., F. H. Brandt, and G. S. Visvesvara.** 1988. Restriction fragment length polymorphisms of the DNA of selected *Naegleria* and *Acanthamoeba* amoebae. *J. Clin. Microbiol.* **26:**1655–1658.

18. **Moura, H., S. Wallace, and G. S. Visvesvara.** 1992. *Acanthamoeba healyi* n. sp. and the isoenzyme and immunoblot profiles of *Acanthamoeba* spp., groups 1 and 3. *J. Protozool.* **39:**573–583.

19. **Nerad, T. A., G. S. Visvesvara, and P.-M. Daggett.** 1983. Chemically defined media for the cultivation of *Naegleria*: pathogenic and high temperature tolerant species. *J. Protozool.* **30:**383–387.

20. **Page, F. C.** 1985. *A New Key to Fresh Water and Soil Gymnamoebae.* Fresh Water Biological Association, Cumbria, England.

21. **Sawyer, T. K., G. S. Visvesvara, and B. A. Harke.** 1976. Pathogenic amebas from brackish and ocean sediments with a description of *Acanthamoeba hatchetti*, n. sp. *Science* **196:**1324–1325.

22. **Seidel, J. S., P. Harmatz, G. S. Visvesvara, A. Cohen, J. Edwards, and J. Turner.** 1982. Successful treatment of primary amebic meningoencephalitis. *N. Engl. J. Med.* **306:**346–348.

23. **Singh, B. N.** 1975. *Pathogenic and Non-pathogenic Amebae.* John Wiley & Sons, Inc., New York.

24. **Stehr-Green, J. K., T. M. Bailey, F. H. Brandt, J. H. Carr, W. W. Bond, and G. S. Visvesvara.** 1987. *Acanthamoeba* keratitis in soft contact lens wearers: a case-control study. *JAMA* **258:**57–60.

25. **Stehr-Green, J. K., T. M. Bailey, and G. S. Visvesvara.** 1990. The epidemiology of *Acanthamoeba* keratitis in the United States. *Am. J. Ophthalmol.* **107:**331–336.

26. **Visvesvara, G. S.** 1985. Laboratory diagnosis, p. 193–215. *In* E. G. Rondanelli (ed.), *Amphizoic Amoebae: Human Pathology.* Piccin Nuova Libraria, Padua, Italy.

27. **Visvesvara, G. S.** 1988. Acanthamoebiasis and naegleriosis, p. 723–730. *In* A. Balows, W. J. Hausler, Jr., M. Ohashi, and A. Turano (ed.), *Laboratory Diagnosis of Infectious Diseases: Principles and Practice*, vol. 1. Springer-Verlag, New York.

28. **Visvesvara, G. S., A. J. Martinez, F. L. Schuster, G. J. Leitch, S. V. Wallace, T. K. Sawyer, and M. Anderson.** 1990. Leptomyxid ameba, a new agent of amebic meningoencephalitis in humans and animals. *J. Clin. Microbiol.* **28:** 2750–2756.

29. **Visvesvara, G. S., F. L. Schuster, and A. J. Martinez.** 1993. *Balamuthia mandrillaris*, new genus, new species, agent of amebic meningoencephalitis in humans and animals. *J. Eukaryot. Microbiol.* **40:**504–514.

30. **Visvesvara, G. S., and J. K. Stehr-Green.** 1990. Epidemiology of free-living ameba infections. *J. Protozool.* **37:**25S–33S.

31. **Wiley, C. A., R. E. Safrin, C. E. Davis, P. W. Lampert, A. I. Braude, A. J. Martinez, and G. S. Visvesvara.** 1987. *Acanthamoeba* meningoencephalitis in a patient with AIDS. *J. Infect. Dis.* **155:**130–133.

32. **Wright, P., D. Warhurst, and B. R. Jones.** 1985. *Acanthamoeba* keratitis successfully treated medically. *Br. J. Ophthalmol.* **69:**778–782.

Intestinal and Urogenital Protozoa

GEORGE R. HEALY AND LYNNE S. GARCIA

106

The protozoa that parasitize the intestinal and urogenital systems of humans belong to five groups: amebae, flagellates, ciliates, coccidia, and microsporidia (Table 1). With the exception of *Trichomonas vaginalis* and microsporidia of the genera *Pleistophora*, *Nosema*, and *Encephalitozoon*, all of these organisms live in the intestinal tract.

The species of intestinal protozoa vary in prevalence and pathogenicity (Table 1). Some species are rarely encountered in patients in the United States but may be found in Americans who travel to areas in which the organisms are endemic and in persons from those areas who visit or emigrate to the United States. Therefore, clinicians and laboratory personnel should be aware of both common and uncommon parasite species that might be found in their patients.

In addition to the protozoan species generally considered human parasites, some species parasitic in animals may also infect humans. For example, *Cryptosporidium* species, long recognized as pathogens in calves, lambs, turkeys, and other animals, is now well recognized as a human pathogen and has caused severe infections in patients with AIDS.

During the past few years, several outbreaks of diarrhea have been associated with a spherical organism measuring 8 to 10 μm in diameter; the distribution is worldwide. These are acid-fast-variable organisms that have been found in the feces of immunocompetent travelers to developing countries, those with no travel history, and patients with AIDS. Patients reported symptoms of a flulike illness with nausea, vomiting, anorexia, weight loss, and explosive diarrhea lasting 1 to 3 weeks. The organisms were subsequently thought to be a new pathogen, possibly an oocyst, a flagellate, an unsporulated coccidian, a large *Cryptosporidium* sp., a blue-green alga (cyanobacteriumlike body) (18, 19), or a coccidianlike body. An epidemiologic report of one outbreak implicated exposure to a contaminated water source (18). These organisms have now been classified as coccidia in the genus *Cyclospora* (24).

The microsporidia are obligate intracellular parasites that have also been recognized in a variety of animals, particularly invertebrates. Those organisms found in humans tend to be quite small, ranging from 1.5 to 2 μm. Until recently, awareness and understanding of human infections have been marginal; only with increased understanding of AIDS within the immunosuppressed population has attention been focused on these organisms (9). Limited availability of electron microscope capability has also played a role in our inability to recognize and diagnose these infections. However, with the introduction of newer diagnostic methods, our ability to identify these parasites has definitely improved. To date, six genera have been recognized in humans: *Encephalitozoon*, *Nosema*, *Pleistophora*, *Enterocytozoon*, *Septata*, and "*Microsporidium*," a catchall genus for those organisms not yet classified (Table 1).

Most of the intestinal protozoa (except *Sarcocystis hominis* and *Sarcocystis suihominis*, which are acquired by ingestion of the infective stages in raw or improperly cooked beef and pork, respectively) are transmitted through fecally contaminated food, water, or other materials. Prevalence of intestinal protozoa is correlated with socioeconomic conditions, and higher rates of infection occur in people who have poor personal hygiene or who live in areas with poor sanitation. Contaminated water supplies are a particular problem, because cysts are not killed by usual levels of chlorination. Filtration is required. Endemic and epidemic diseases have been traced to water supplies that use surface water that either is not filtered or is filtered through improperly functioning filters.

Some of the intestinal protozoa are commensals or nonpathogenic organisms that produce no evidence of disease; however, microscopists must be able to distinguish pathogenic from nonpathogenic species. In addition, the presence of nonpathogenic species indicates that the person has been exposed to fecal contamination. Several species are capable of causing mild to severe gastrointestinal symptoms. *Entamoeba histolytica* may produce extraintestinal lesions in various areas of the body, and microsporidia have been diagnosed in tissues other than the intestine, such as muscle, liver, brain, kidney, and cornea. However, pathogenic or potentially pathogenic protozoa do not always produce symptoms in infected people or may remain after symptoms have resolved. Such asymptomatic persons may serve as reservoirs for the infection. In addition, detection of a potentially pathogenic protozoan does not necessarily prove that the organism is causing the illness. Patients may have diarrhea caused by other organisms such as *Salmonella* spp., *Shigella* spp., *Escherichia coli*, or rotavirus. The pathogenicity of some species (*Blastocystis hominis*, for example) has been questioned. *T. vaginalis*, a urogenital protozoan, is also

TABLE 1 Intestinal and urogenital protozoa that may be found in specimens from patients in the United States

Type and species	Relative prevalence[a]	Pathogenicity[b]
Amebae		
Entamoeba histolytica	+	+
Entamoeba hartmanni	+	−
Entamoeba coli	++	−
Entamoeba polecki	R	−
Endolimax nana	++	−
Iodamoeba bütschlii	+	−
Blastocystis hominis	++	±
Ciliate		
Balantidium coli	R	+
Flagellates		
Dientamoeba fragilis	++	+
Giardia lamblia	++	+
Trichomonas vaginalis	++	+
Trichomonas hominis	+	−
Chilomastix mesnili	+	−
Enteromonas hominis	R	−
Retortamonas intestinalis	R	−
Coccidia		
Isospora belli	R	+
Sarcocystis spp.	R	?
Cryptosporidium parvum	+	+
Cyclospora cayetanensis	+	+
Microsporidia		
Pleistophora spp.	R	+
Nosema spp.	R	+
Encephalitozoon spp.	R	+
Enterocytozoon bieneusi	R	+
Septata intestinalis	R	+

[a] +, common; ++, very common; R, rare.
[b] +, pathogenic; −, nonpathogenic; ±, no consensus on pathogenicity.

considered pathogenic and may cause mild to severe vaginitis and other urogenital problems.

This chapter covers information on the morphologic identification of organisms (presented in tabular form and diagrams), recommended procedures for laboratory diagnosis, and clinical aspects of important pathogens.

In the descriptions of diseases, the clinical manifestations noted refer to findings in patients with symptomatic disease and do not necessarily refer to findings in every person infected with the parasite species.

LABORATORY DIAGNOSIS

Because the symptoms produced by pathogenic intestinal protozoa are usually nondiagnostic, diagnosis requires laboratory detection of the parasite by the microscopic examination of feces or other body material. Immunodiagnostic methods are useful for the diagnosis of extraintestinal amebiasis, but they are of limited usefulness for intestinal diseases.

Although not all intestinal protozoa are pathogenic, microscopists must be capable of identifying both pathogenic and nonpathogenic species, with the possible excep-

tion of species that are rarely found in patients in the United States. Morphology, especially that of amebae, varies, and species characteristics often overlap such that individual nonpathogenic organisms may have characteristics resembling those of pathogens and vice versa. Thus, for reliable identification, microscopists must be able to differentiate all species regardless of their potential for causing disease. Special attention will be given to the recognition and identification of the clinically significant pathogens, especially *E. histolytica*, *Giardia lamblia*, *Dientamoeba fragilis*, *Cryptosporidium parvum*, and the microsporidia, particularly *Enterocytozoon bieneusi* and *Septata intestinalis*.

The identification of protozoan species is based on the morphology of the diagnostic stages (13, 29). The particular features or characteristics used for identification vary with the group of organisms (for example, amebae or flagellates), the species, and the stage(s) of parasite present. The diagnostic stages are trophozoites or cysts for the amebae, flagellates, and ciliates; oocysts or sporocysts for the coccidia; vacuolated forms for *B. hominis*; and spores for the microsporidia.

The type of material to be examined depends on the parasite and its location in the body. For intestinal protozoa, feces are commonly submitted for examination, although other materials such as duodenal or sigmoidoscopic aspirates or biopsy samples are occasionally obtained.

Four types of procedures are used to recover and demonstrate intestinal protozoa: direct wet mount examinations, concentration techniques, permanently stained preparations, and cultivation. All of these methods may not be needed in every case. The selection of appropriate techniques depends on the species of parasite suspected and the stage(s) of parasite likely to be found in the specimen. For example, trophozoite stages of amebae and flagellates are more likely to be present in soft or diarrheic stools, and cysts are more likely in formed feces. Thus, the techniques used to examine diarrheic fecal specimens may differ from those used for formed specimens.

For accurate and reliable identification, specimens must be properly collected and handled before examination (4, 13). Protozoa, especially trophozoites, may develop atypical morphology in old or poorly collected specimens (22). Ideally, fecal specimens should be placed in fixative immediately after being evacuated or should reach the laboratory within 1 to 2 h after passage; other materials such as urine, aspirate, or biopsy samples from duodenal or colonic lesions should be sent to the laboratory immediately after collection. If transportation is delayed, specimens should be appropriately preserved to maintain the diagnostic characteristics of organisms that might be present.

AMEBAE

Seven species of intestinal amebae may live in the ceca and colons of humans: *E. histolytica*, *Entamoeba hartmanni*, *Entamoeba polecki*, *Entamoeba coli*, *Endolimax nana*, *Iodamoeba bütschlii*, and *B. hominis* (Fig. 1 to 5). Infection is acquired by the ingestion of cysts, which excyst in the intestine. The cysts are quite hardy and can survive for days or weeks in water or soil. Infection is usually diagnosed by the identification of organisms in feces, although other materials may be examined in symptomatic cases. Immunodiagnostic tests are useful for the diagnosis of extraintestinal amebiasis and may assist in diagnosis of invasive intestinal disease (see chapter 103 of this Manual).

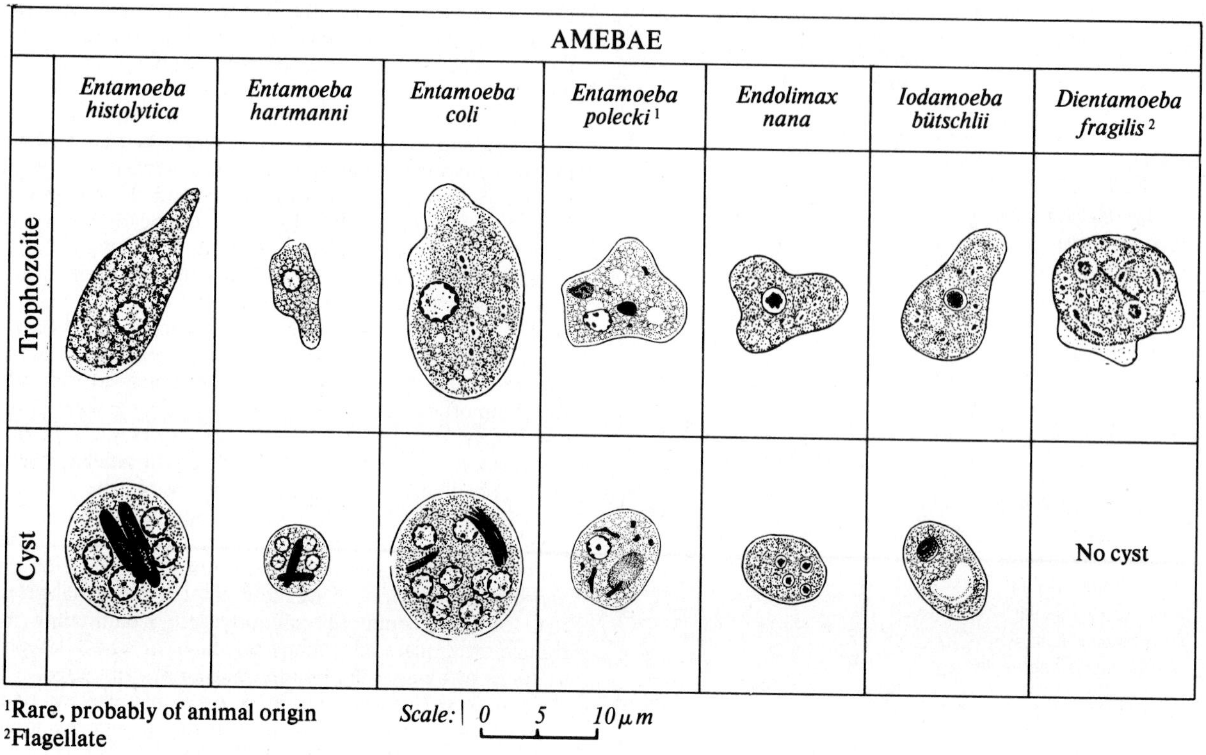

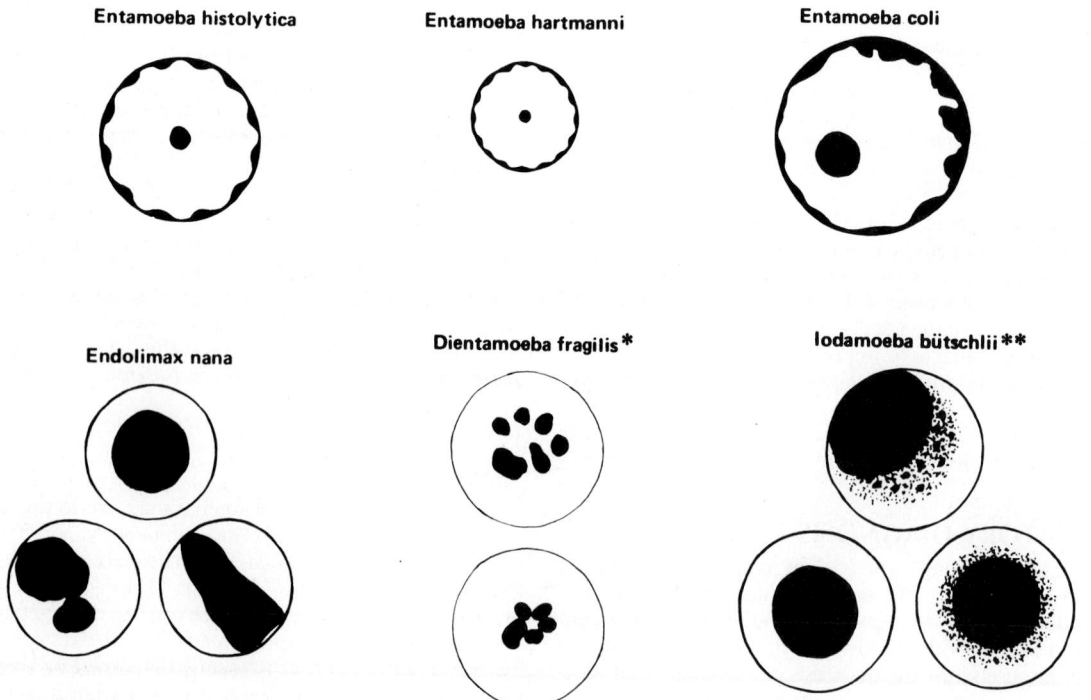

AMEBAE						
	Entamoeba histolytica	*Entamoeba hartmanni*	*Entamoeba coli*	*Entamoeba polecki* [1]	*Endolimax nana*	*Iodamoeba bütschlii*	*Dientamoeba fragilis* [2]

(Trophozoite row and Cyst row; *Dientamoeba fragilis*: No cyst)

[1]Rare, probably of animal origin
[2]Flagellate

Scale: 0 5 10 μm

FIGURE 1 Amebae and flagellate (*Dientamoeba fragilis*) found in human stool specimens. (From reference 4.)

Entamoeba histolytica   Entamoeba hartmanni   Entamoeba coli

Endolimax nana   Dientamoeba fragilis *   Iodamoeba bütschlii **

FIGURE 2 Nuclei of amebae. *, flagellate. **, *Iodamoeba* cysts, which may have eccentric karyosomes against the nuclear membrane, as seen in the upper *Iodamoeba* nucleus; trophozoite nuclei are usually not against the nuclear membrane. Achromatic granules are not always visible (lower left) in *Iodamoeba* nuclei. (From reference 29 with permission.)

E. histolytica

E. histolytica causes amebiasis and is the only ameba pathogenic for humans (5). Infections with *E. histolytica* are classified as amebiasis irrespective of whether the person exhibits symptoms. A number of outbreaks, usually caused by contaminated food or water, have occurred in the United States. An unusual outbreak was caused by inadequate disinfection of a colonic irrigation apparatus (17). Strains isolated from patients with clinical disease have been shown by isoenzyme analysis to differ from commensal strains (5), which suggests that only strains with certain isoenzyme patterns are capable of causing invasive disease. In the future, the name *E. histolytica* may be used to designate pathogenic strains, while another species of *Entamoeba* (*Entamoeba dispar*) would include those that are nonpathogenic (11a). The incubation period is variable, from as short as a few days to as long as weeks or even months (5). Clinical amebiasis, i.e., infection with symptoms produced presumably by an amebic invasion of colonic tissue, may present with various manifestations, including dysentery, colitis, and ameboma. Amebic dysentery is an acute diarrhea with ulcerations of the colonic mucosa. Symptoms include crampy, lower abdominal pain with bloody mucoid diarrhea in severe cases, and the infection is occasionally complicated by intestinal perforation. In some people, an increased frequency of bowel movements with or without blood and mucus may occur. A chronic form, amebic colitis, produces symptoms similar to those of ulcerative colitis or other forms of inflammatory bowel disease, with diarrhea, sometimes bloody, occurring over a long period, sometimes alternating with periods of constipation or normal bowel function. Some patients with amebiasis have been misdiagnosed as having ulcerative colitis (17). Another, less common form of intestinal disease, ameboma, is produced by the growth of granulomatous tissue in response to the infecting amebae, resulting in a large local lesion of the bowel that radiologically resembles a carcinoma.

Other than trophozoites containing ingested erythrocytes, we have no way of differentiating pathogenic from nonpathogenic organisms on the basis of morphologic differences, so our approach to reporting and therapy may not change in the near future. Any *E. histolytica* organisms should be reported to the physician; the decision to treat or not treat must be a clinical one. There are DNA probes that can differentiate various strains, but they are not yet available for clinical use.

Infections with *E. histolytica*, with or without a history of gastrointestinal symptoms, may result in hematogenous spread of the organisms to the liver via the portal system, resulting in amebic abscesses of the liver. This occurs in up to 5% of patients with symptomatic intestinal amebiasis. Approximately 40% of patients with amebic liver abscess do not have a history of bowel symptoms, and in some patients, *E. histolytica* may not be present in stool at the time liver disease becomes manifest. Amebic abscesses also occasionally occur in the lung, brain, or other organs.

Intestinal infection is usually diagnosed by the microscopic identification of organisms in feces or in sigmoidoscopic material from ulcerations. Only trophozoites are found in tissue lesions, but both trophozoites and cysts may be found in the intestinal lumen (Tables 2 and 3). Some patients with invasive disease may have only trophozoites in fecal specimens, and direct wet mount of fresh material or the more sensitive permanent stain may be required to establish the diagnosis. Trophozoites are difficult to recog-

nize in wet mounts of fixed or concentration material. Depending on the type of fecal specimen, morphologic examination by direct wet mount, concentration, and permanent stain may be useful. Purged stool specimens occasionally show parasites when they are not detected in normally passed stools. Cultures for amebae may be helpful and are essential if isoenzyme patterns or other techniques are to be employed, but they are not used routinely in most laboratories (13, 21, 22). Suspected amebic abscesses are generally diagnosed by positive serologic tests, although organisms may sometimes be identified microscopically in abscess aspirates or liver tissue biopsy. The last material aspirated is more likely to contain amebae. Cultivation of liver abscess fluid is not routinely attempted. Aspirates of abscesses or intestinal lesions may show amebic trophozoites, sometimes containing ingested erythrocytes in direct wet mounts or with permanent stains such as trichrome. In sections of liver or intestinal lesions, periodic acid-Schiff (PAS) stain counterstained with hematoxylin is particularly helpful. Amebic trophozoite cytoplasm is more PAS positive than most tissue and inflammatory cells, and identification is confirmed by demonstrating the typical *Entamoeba* nucleus (13).

Other species of intestinal amebae are not pathogenic but must be differentiated from *E. histolytica* (see below). *E. polecki* (Tables 4 and 5) is seen occasionally in refugees from Southeast Asia and may be confused with *E. histolytica* (4).

Entamoeba gingivalis is a common inhabitant of the oral cavity, particularly in patients with poor oral hygiene (4). It resembles *E. histolytica* but has no known cyst stage. As a result, trophozoites of *E. gingivalis* (Tables 4 and 5) may cause confusion in diagnosing amebic lung abscess by morphologic examination of pulmonary material, especially sputum.

B. hominis

B. hominis inhabits the large intestine, and organisms are passed in feces (Fig. 6 and 7). Three morphologic forms have been described: amebic, granular, and vacuolated (34). The vacuolated form is most commonly seen in fecal specimens (34).

Although *B. hominis* may be found in up to 25% of stool specimens examined, only occasional patients have clinical symptoms, and there is much controversy about whether *B. hominis* is pathogenic or nonpathogenic (20, 34). *B. hominis* may be suspected when the complete battery of parasitologic, bacterial, and viral tests on stools has failed to disclose any other agent and *Blastocystis* organisms are numerous in the specimen. The predominant and virtually only symptom has been persistent, mild diarrhea.

Infection is diagnosed by finding the familiar spherical or ovoid form with a large central vacuole and nuclei and various other organelles arranged around the periphery. The organism is described in Table 6 and shown in Fig. 6 and 7. *Blastocystis* organisms can be demonstrated by any of the methods usually used for the diagnosis of intestinal parasite infections, although exposing unfixed feces to water in the performance of concentration procedures causes lysis of *Blastocystis* organisms.

Although we normally do not quantitate protozoa on the report slip, *B. hominis* should be reported and quantitated (rare, few, moderate, many) in the examination of the permanent stained smear. In some cases when the organism has been thought to be the causative agent, large numbers of organisms have been present in the stool. The percentage

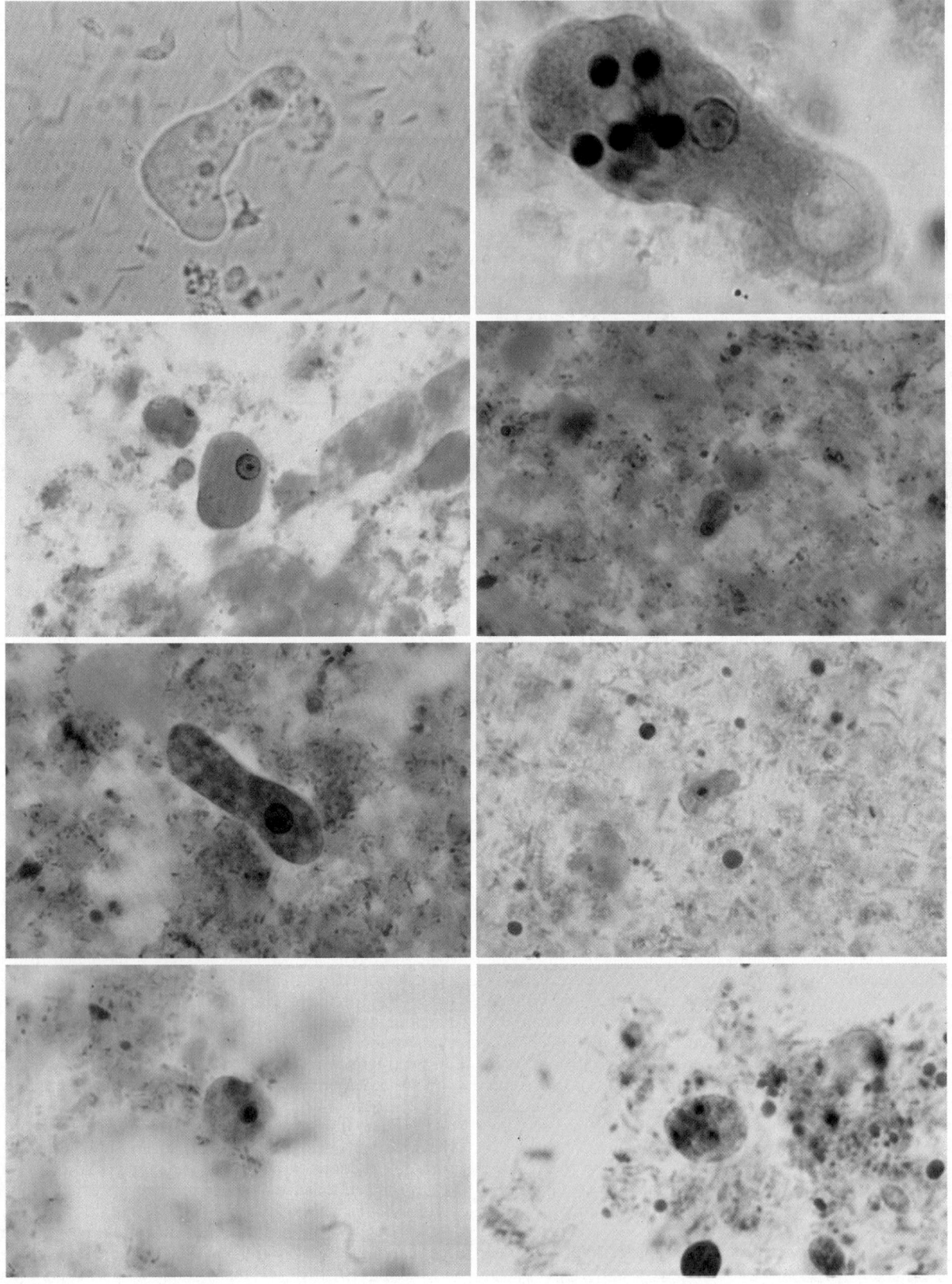

of people who harbor this parasite can range from about 10% to as high as 20% in some areas of the country.

Morphologic Identification of Amebae in Fecal Specimens

Both trophozoites and cysts are diagnostic stages of the amebae, and either or both stages can be detected in feces. Microscopists must be familiar with the morphologic characteristics used for the differentiation of species and must be able to distinguish trophozoites from epithelial cells, polymorphonuclear leukocytes, and macrophages as well as from pus cells, yeasts, pollen, molds, and other objects that may be present in feces.

The typical characteristics of trophozoites and cysts of the ameba species are listed in Tables 2, 3, 4, and 5; diagrams of amebic cysts and trophozoites are presented in Fig. 1. Diagrams of nuclei are shown in Fig. 2; photomicrographs are presented in Fig. 3 through 5. Although *D. fragilis* is a flagellate, it is included in these figures because it resembles and must be differentiated from the amebae.

Trophozoites

Motility of trophozoites is progressive or nonprogressive, may be difficult to see, and is most evident in saline wet mounts of fresh feces. However, many laboratories receive specimens in preservatives, so direct wet mounts are not prepared. Other than motility, morphologic characteristics of the trophozoites are best seen on permanent stained smears. The cytoplasm may contain fine or coarse granules, erythrocytes, yeast cells, molds, or bacteria. The arrangement, size, and uniformity of nuclear chromatin are important characteristics, as are the size and position of the karyosome.

Cysts

The numbers and positions of the nuclei, absence or distribution of peripheral chromatin, size and position of the karyosome, and presence of cytoplasmic inclusions such as chromatoidal bodies or glycogen are important characteristics for the identification of amebic cysts.

Not all of the characteristics discussed can be seen in a single type of preparation; stained and unstained wet mounts are helpful, but permanent stained smears are necessary to demonstrate all of the features. Unstained wet mounts may reveal trophozoites and cysts. The motility of

trophozoites in physiologic saline mounts of fresh material and the presence of cytoplasmic inclusions such as erythrocytes in trophozoites and chromatoid bodies in cysts can be observed. However, stained preparations are recommended for reliable species identification. Buffered methylene blue solution (Nair stain) can be used for temporary stains of trophozoites in fresh specimens and will permit the microscopist to distinguish host cells from amebae and trophozoites of *Entamoeba* spp. from those of other genera. Iodine solutions are used for temporary cyst stains of fresh or fixed specimens. Characteristics of cysts are less variable than those of trophozoites, and species of cysts can often be identified in iodine-stained wet mounts. However, examination of permanent stained smears using oil immersion (1,000×) is recommended for definitive identification of both trophozoites and cysts. Size is not a reliable feature for species differentiation of either trophozoites or cysts except in separating *E. histolytica* and *E. hartmanni*.

Morphologic characteristics of species overlap, and some organisms may be atypical, thus making identification difficult. For example, distinguishing trophozoites of *E. histolytica* from those of *E. coli* is often difficult because of morphologic variations. Rarely can species of trophozoites be identified by a single feature, such as karyosome location, or with a single organism. The microscopist must observe both the cytoplasmic and the nuclear characteristics of several organisms before making a species identification. Although cysts are more easily identified than trophozoites, several cysts (particularly if they are immature) should be observed to ensure that the identification is reliable. If two species are identified, there should be distinct populations of each.

Sometimes, although amebic organisms are recognized, species cannot be identified. In these instances, the laboratory should report "unidentified ameba trophozoites (or cysts)"; if the genus can be determined but the species cannot, "unidentified *Entamoeba* trophozoites or cysts" should be reported, and another specimen should be requested.

FLAGELLATES

The flagellates inhabit the intestinal and atrial areas (Fig. 1, 5, 8, and 9). Intestinal infections by flagellates that have cyst stages are acquired by the ingestion of cysts. Infections

FIGURE 3 (Opposite) Ameba trophozoites (trichrome stain except upper left; oil immersion magnification, ×1,000). Row 1: (left) *E. histolytica*, unstained. The clear pseudopod is evident at the bottom of the organism. Several ingested erythrocytes are present in the cytoplasm. The nucleus is not visible. (Right) *E. histolytica*. This large trophozoite contains numerous ingested erythrocytes (RBCs) to the left of the nucleus. The nucleus has a small, dotlike central karyosome and peripheral chromatin that, although granular, is distributed fairly evenly around the nuclear membrane. This large trophozoite is from a patient with amebic dysentery. Row 2: (left) *E. histolytica*. The trophozoite is typical of those seen in clinical specimens and does not contain ingested RBCs (more common than those containing ingested RBCs). The nucleus is typical, with a dotlike central karyosome and evenly distributed peripheral chromatin. (Right) *E. hartmanni*. This trophozoite has characteristics similar to those of *E. histolytica* but is smaller. The nucleus is in the lower portion of the trophozoite. Row 3: (left) *E. coli*. This elongated trophozoite of *E. coli* has a nucleus with a large eccentric karyosome. The chromatin is arranged somewhat unevenly around the nuclear membrane. There are numerous vacuoles within the cytoplasm. (Right) *E. nana*. This trophozoite has a nucleus with a large karyosome and generally no peripheral chromatin on the nuclear membrane. A clear halo surrounds the karyosome and extends to the nuclear membrane; however, there is often tremendous morphologic variation in the nuclei of these organisms, particularly if many are present in a clinical specimen. The cytoplasm is relatively smooth with small vacuoles. Row 4: (left) *I. bütschlii*. This organism has a large karyosome and normally no chromatin on the nuclear membrane. The cytoplasm often contains larger vacuoles with ingested bacteria and more debris than are seen in *E. nana*. (Right) *D. fragilis*. This flagellate is included here because it mimics the amebae. There is no known cyst stage, and the trophozoite contains either one or two nuclei that are usually fragmented into several chromatin granules. There may be a wide size variation, even in a single permanent stained smear. The cytoplasm is vacuolated.

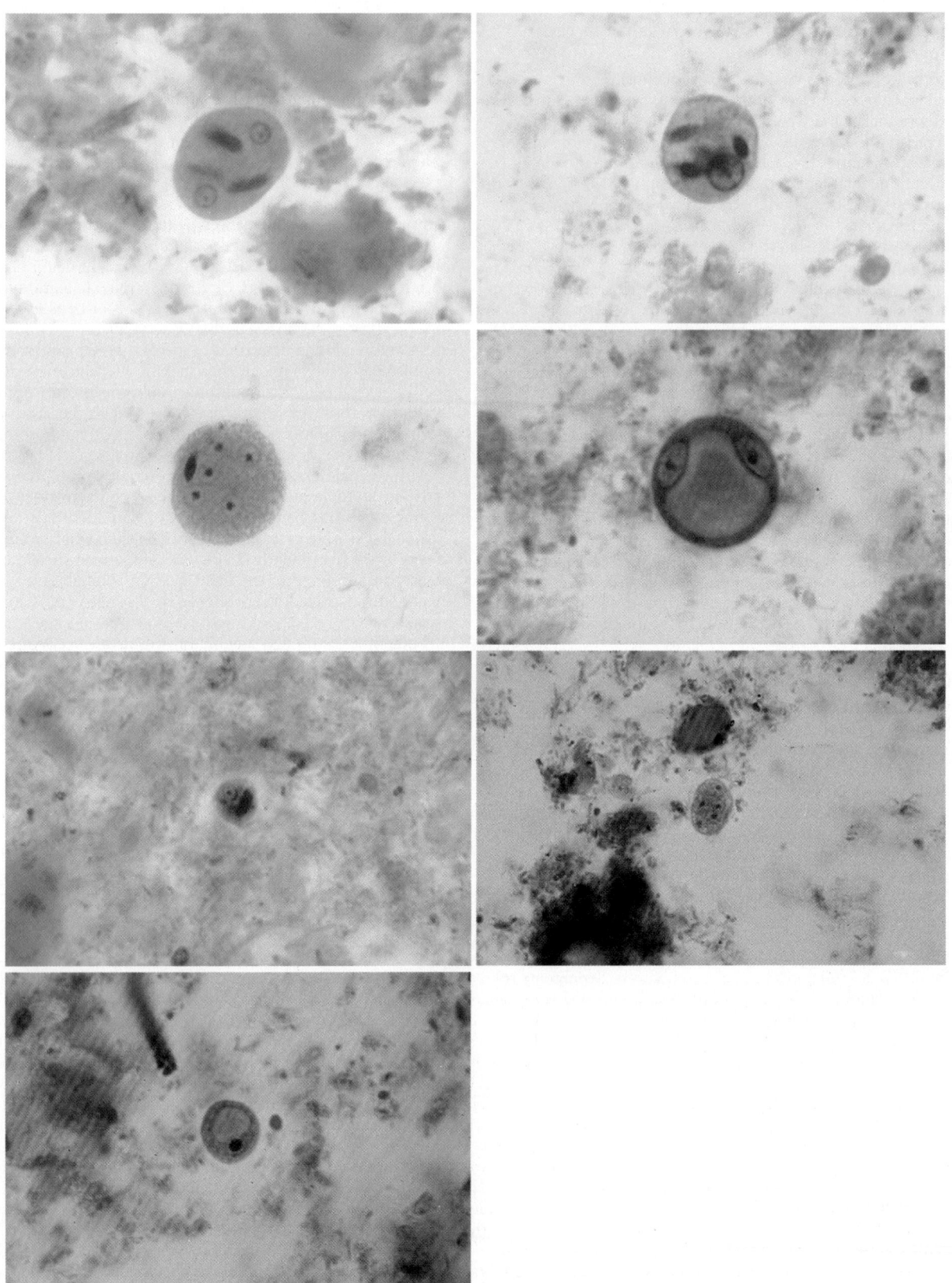

are usually diagnosed by morphologic examinations of feces or other body specimens (Tables 7 and 8). For G. *lamblia*, immunodiagnostic tests for the detection of antibody in serum specimens and parasite antigens in stool specimens have been reported (see chapter 103 of this Manual). Species identifications are generally based on microscopic identification of morphologic features of trophozoites or cysts. Two species, G. *lamblia* and D. *fragilis*, cause clinically significant intestinal disease, and T. *vaginalis* is a frequent cause of vaginitis.

G. lamblia

Giardiasis is acquired by ingestion of the hardy cysts of G. *lamblia*. Although some institutions have changed or are considering a species name change to G. *intestinalis*, we have decided to maintain G. *lamblia* for this edition. Outbreaks related to contaminated water are common in the United States, and infections are frequent in day-care centers and among campers and male homosexuals (13). Various conditions associated with *Giardia* infection in the compromised patient include hypogammaglobulinemia, protein or caloric malnutrition, previous gastrectomy, histocompatibility antigen HLA-B12, gastric achlorhydria, blood group A, differences in mucolytic proteins, immunoglobulin deficiencies, and reduced secretory immunoglobulin A levels in the gut. Trophozoites infect the upper small intestine but do not invade the tissues to produce ulcers. Infection may elicit a variety of symptoms or may be asymptomatic. The incubation period is variable, ranging from a few days to several weeks, with an average of about 9 days (13). In acute giardiasis, symptoms include nausea, upper intestinal cramping or pain, and malaise. There is often explosive, watery diarrhea characterized by foul-smelling stools. These symptoms are accompanied by flatulence and abdominal distention. The acute stage of clinical giardiasis may be followed by a chronic stage, or the chronic type of infection may be the first indication of infection. In such infections, there are flatulence, mushy foul-smelling stools, upper intestinal cramping, and abdominal distention. A number of patients also exhibit belching, nausea, anorexia, vomiting, and symptoms of heartburn. Fever and chills may be present but to a lesser degree. Symptoms may mimic peptic ulcer or gallbladder disease. In some patients, the cysts may be excreted in stools in a variable pattern, although the reasons for this are not clear. This variable shedding of cysts may occur even when there are classic symptoms of disease and numerous trophozoites in the upper small intestine.

Diagnosis is usually established by the demonstration of cysts or, occasionally, trophozoites in feces or of trophozoites in duodenal contents obtained by string test (1), aspiration, or biopsy (Fig. 8 and 9). Because of the variable

shedding of organisms, several stool specimens should be examined before the infection is ruled out. It is best to examine a total of three specimens collected 2 to 3 days apart, although daily specimens for a total of 3 days can be used. A normal series of three stools (or even six stools) can be examined and be negative and yet the patient will still have giardiasis. It is important for both the laboratory and the clinician to recognize this fact. If viable trophozoites are present, they may be identified by the characteristic falling-leaf or tumbling motion in saline mounts of fresh feces. The large, ventral sucking disk can be seen as the organism turns. In a lateral view, the trophozoite appears spoon shaped. Permanent stains are recommended if organisms cannot be identified in wet mounts or by concentration (22). Immunofluorescence and antigen detection methods are discussed in chapter 103 of this Manual. Small plant cells can resemble *Giardia* cysts, so organisms should be carefully examined for the fibrils and nuclei characteristic of *Giardia* species. Permanent stains are recommended.

When *Giardia* organisms are not found in stool specimens, duodenal aspirates, string test mucus, or biopsied mucosal tissue can be examined. The string test (1) is used to collect mucus from the duodenal area, and it may be less traumatic for the patient than other methods. Materials obtained by drainage, aspiration, or the string test can be examined by simple, direct wet mounts. Biopsy tissue may be processed and stained by the usual histopathologic methods; however, before preservation, a fresh imprint smear of the mucosal surface on a slide can be made and stained with trichrome or Giemsa stain.

There are enzyme immunoassay and fluorescent-antibody monoclonal antigen detection systems for this organism (see chapter 103). These may be appropriate in screening situations and may help diagnose infections when the stools are repeatedly negative. In-service training for the medical staff is mandatory to ensure that tests are appropriately ordered and that the limits of the information generated (test results limited to absence or presence of G. *lamblia*) are completely understood. Recently, a great deal of interest has been shown in these reagents by industrial companies and municipalities interested in water testing. This is particularly relevant when the water sources are used for drinking and/or for recreational purposes. One of the commercially available reagents (fluorescent antibody) has monoclonal antibodies for both *Giardia* cysts and *Cryptosporidium* oocysts and has now been used in a number of situations for water testing (15).

Chilomastix mesnili

C. *mesnili* is a nonpathogenic flagellate that inhabits the cecum and colon. The diagnostic stages (trophozoites and cysts) are passed in feces (Tables 7 and 8). The trophozoites

FIGURE 4 (Opposite) Ameba cysts (trichrome stain; oil immersion magnification, ×1,000). Row 1: (left) E. *histolytica*. Two of the four nuclei are seen in this plane of focus. Three chromatoid bodies are evident; they stain dark blue and have rounded ends. (Right) E. *histolytica*. This uninucleate cyst has numerous rounded chromatoid bodies that in this organism stain red. The palely staining areas represent glycogen masses. Row 2: (left) E. *coli*. Five nuclei are evident in this plane of focus, and there is a red chromatoid body to the left. Nuclear karyosomes are large and central. Although karyosomes are typically eccentric, they may be centered, as in the nuclei of this cyst. Cytoplasm is granular. (Right) E. *coli*. This binucleate cyst of E. *coli* contains a large glycogen vacuole. Immature cysts such as this are typical of E. *coli*. Row 3: (left) E. *hartmanni*. This small cyst has one nucleus in this plane of focus; these cysts frequently contain numerous chromatoid bodies and two nuclei instead of the mature four. A large chromatoid body is present on the right. (Right) E. *nana*. All four dotlike nuclei are evident at this focal plane. Halos are evident around some of the nuclei. Row 4: I. *bütschlii*. This cyst has a large glycogen vacuole, with the nucleus below it. Achromatic granules are not visible (they make up part of the "basket" nucleus), and the karyosome is large and rounded.

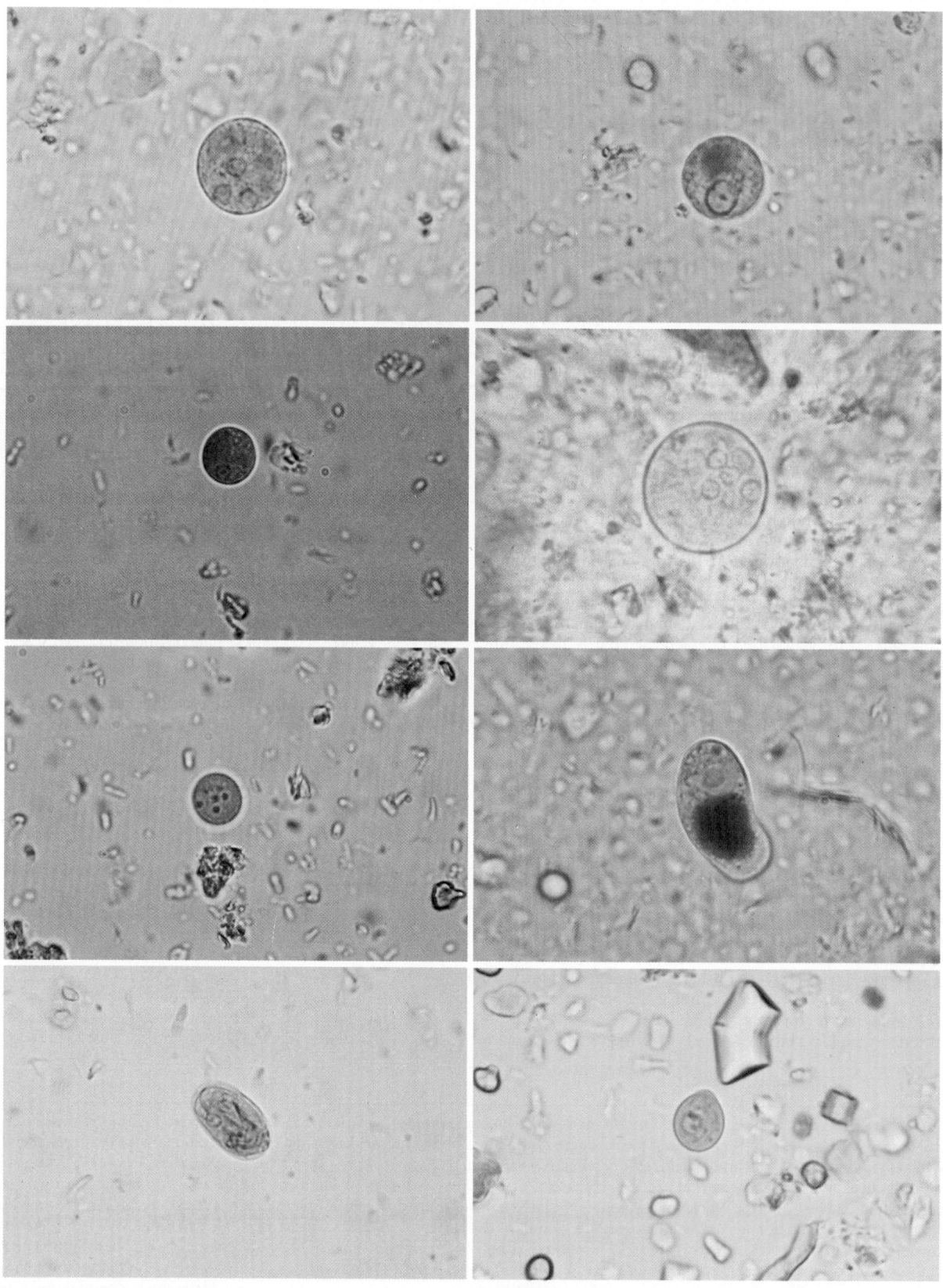

have a characteristic rotating, wobbling motion that may be recognized in saline mounts of fresh material. The spiral groove, which extends along the body, is sometimes visible as the organism turns. In permanent stained preparations, trophozoites are usually lightly stained and sometimes distorted, and they may be overlooked. The most prominent feature is the long cytostome, which extends about one-third to one-half the length of the body. The nucleus also may have a collection of chromatin along one side, giving it a lopsided look (Fig. 8 and 9).

The presence of the cytostome and spiral groove, the location of the nucleus at one end with tapering of the opposite end, and nuclear characteristics aid in the differentiation of C. mesnili from other intestinal protozoa.

C. mesnili cysts are often identified by their usual lemon shape, single large nucleus, and fibrils (Fig. 8). Cysts are not always lemon shaped; they may be rounded or, if viewed on end, may appear perfectly round.

D. fragilis

D. fragilis is a flagellate with no external flagellum; thus, the disease is discussed in the flagellate section, but the organism is included in Fig. 1 to 3, which concern amebae, because it must be differentiated from the amebae (Table 7). The organism does not have a cyst stage, and the exact means of spread is not clear. D. fragilis infection is commonly associated with enterobiasis, and it has been suggested that D. fragilis may infect Enterobius eggs and thus bypass gastric acidity (13). Although clinical infections with D. fragilis occur, they are infrequently reported, probably because of the self-limited nature of the infection such that stool examination is not requested and because of the difficulty in detecting and identifying the trophozoites in stool specimens from individuals with suggestive symptoms. The incubation period for clinical disease is not known with certainty. Symptoms have been reported more frequently in children than in adults and are predominantly diarrhea and abdominal distention. Outbreaks in day-care centers have been described. Nausea, vomiting, and weight loss have been recorded in from one-third to one-fifth of the cases reported in the literature.

Permanent stains are required to diagnose this infection, and multiple specimens may be required, since shedding varies from day to day (22). The delicately staining trophozoites are usually (60 to 80%) binucleate, though the nuclei may be in different planes of focus. The nuclei tend to be fragmented and can often be seen in a "tetrad" formation. They must be differentiated from the trophozoites of E. nana, I. bütschlii, and E. hartmanni. Nuclear characteristics, the presence of binucleate forms, tremendous size variation, and the absence of cysts aid in identification of this organism.

It is strongly recommended that the permanent stained smear be examined for every stool specimen submitted to the laboratory for an ova and parasite examination. If this approach is not used, then many infections with D. fragilis can be missed. If a laboratory from time to time finds and identifies Giardia sp. but never sees D. fragilis, then collection and diagnostic methods should be reviewed.

Trichomonas hominis

T. hominis, a nonpathogenic protozoan that inhabits the colon, has only a trophozoite stage (Table 7). The motile trophozoites in saline wet mounts display nervous, jerky motions. Each possesses an undulating membrane that extends most of its length (Fig. 8 and 9) and often can be seen in wet mounts, especially if the organisms have slowed down or are trapped in fecal debris. Iodine stains are of little value for the identification of T. hominis, because the organisms tend to become distorted. Permanent stains are recommended; with this stain organisms can be seen, even though they are occasionally distorted and difficult to recognize. The axostyle may be visible beyond the end of the teardrop-shaped trophozoite.

T. vaginalis

T. vaginalis inhabits the urogenital systems of both males and females and is considered a pathogen. The trophozoites (the only stage) are found in the urine of both sexes, in material from the vagina, and in prostatic secretions (Table 9). It is estimated that approximately 5 million women in the United States have trichomoniasis and that roughly 1 million men harbor the parasite. Infection is usually but not always acquired by sexual contact. The infection in males is generally asymptomatic, but 25 to 50% of infected women exhibit symptoms (13), which include dysuria, vaginal itching and burning, and, in severe infections, a foamy, yellowish green discharge with a foul odor. In many women, the infection becomes symptomatic and chronic, with periods of relief in response to therapy. Recurrences of infection and disease may be caused by reinfection from an asymptomatic sexual partner, in the true sense of a sexually transmitted disease, or by failure of the drug metronidazole to completely eliminate the parasite. Symptomatic infections in males are rarely reported but include prostatitis, urethritis, epididymitis, and urethral stricture. Rarely, T. vaginalis infections occur in ectopic sites, and parasites may be recovered from areas of the body other than the urogenital system.

Infections with T. vaginalis are usually detected by finding the motile trophozoites in wet mounts of vaginal fluid, prostatic fluid, or sediments of freshly passed urine. In wet mounts, the trophozoite moves with a nervous, jerky motion and possesses an undulating membrane, which extends only half the length of the organism (Table 9). In old urine specimens, the organisms may be dead or badly distorted

---

FIGURE 5 (Opposite) Ameba and flagellate cysts (iodine-stained wet mounts; oil immersion magnification, ×1,000). Although these organisms are photographed with oil immersion (1,000×), the majority of wet preparations will be examined using low power (100×) or high dry power (400×); the permanent stained smear will be examined using oil immersion (1,000×). Row 1: (left) E. histolytica. Three nuclei are evident in this focal plane. (Right) E. histolytica. This immature uninucleate cyst has a reddish-staining glycogen mass above the nucleus. Row 2: (left) E. hartmanni. There is one nucleus in this focal plane. An irregular glycogen mass is evident above the nucleus. (Right) E. coli. In this focal plane, a cluster of six nuclei may be recognized toward the right of the cyst. (Right) I. bütschlii. In this focal plane, the one in the center is out of focus. Row 3: (left) E. nana. Four nuclei are evident, though the one in the center is out of focus. The large, reddish glycogen mass is prominent. Above it is the nucleus, which has a large pale karyosome. Row 4: (left) G. lamblia. Two nuclei are evident toward the upper left, and multiple fibrils are present. (Right) C. mesnili. This small, lemon-shaped cyst has the nucleus on the left and faint fibrils on the right.

TABLE 2 Intestinal protozoa: trophozoites of common amebae[a]

Organism	Size[b] (diam or length)	Motility	Nucleus (no. and visibility)	Appearance of stained:			
				Peripheral chromatin	Karyosome	Cytoplasm	Inclusions
E. histolytica	12–60 μm; usual range, 15–20 μm; invasive forms may be >20 μm	Progressive, with hyaline, fingerlike pseudopodia; may be rapid	1; difficult to see in unstained preparations	Fine granules, uniform in size and usually evenly distributed; may appear beaded	Small, usually compact; centrally located but may also be eccentric	Finely granular, "ground glass"; clear differentiation of ectoplasm and endoplasm; if present, vacuoles are usually small	Noninvasive organism may contain bacteria; erythrocytes present diagnostic
E. hartmanni	5–12 μm; usual range, 8–10 μm	Usually nonprogressive	1; usually not seen in unstained preparations	Nucleus may stain more darkly than that of E. histolytica, although morphology is similar; chromatin may appear as solid ring rather than beaded	Usually small and compact; may be centrally located or eccentric	Finely granular	May contain bacteria; no erythrocytes
E. coli	15–50 μm; usual range, 20–25 μm	Sluggish, nondirectional, with blunt, granular pseudopodia	1; often visible in unstained preparations	May be clumped and unevenly arranged on membrane; may also appear as solid dark ring with no beads or clumps	Large, not compact; may or may not be eccentric; may be diffuse and darkly stained	Granular, with little differentiation into ectoplasm and endoplasm; usually vacuolated	Bacteria, yeast cells, other debris
Endolimax nana	6–12 μm; usual range, 8–10 μm	Sluggish, usually nonprogressive	1; occasionally visible in unstained preparations	Usually no peripheral chromatin; nuclear chromatin may be quite variable	Large, irregularly shaped; may appear "blotlike"; many nuclear variations are common; may mimic E. hartmanni or D. fragilis	Granular, vacuolated	Bacteria
I. bütschlii	8–20 μm; usual range, 12–15 μm	Sluggish, usually nonprogressive	1; usually not visible in unstained preparations	Usually no peripheral chromatin	Large; may be surrounded by refractile granules that are difficult to see ("basket nucleus")	Coarsely granular; may be highly vacuolated	Bacteria, yeast cells, other debris

[a]From Garcia and Bruckner (13).
[b]Wet preparation measurements (in permanent stains, organisms usually measure 1 to 2 μm less).

TABLE 3 Intestinal protozoa: cysts of common amebae[a]

Organism	Size[b] (diam or length)	Shape	Nucleus (no. and visibility)	Appearance of stained: Peripheral chromatin	Karyosome	Cytoplasm, chromatoidal bodies	Glycogen[c]
E. histolytica	10–20 μm; usual range, 12–15 μm	Usually spherical	Mature cyst, 4; immature, 1 or 2; characteristics difficult to see on wet preparation	Fine, uniform granules, evenly distributed; nuclear characteristics may not be as clearly visible as in trophozoite	Small, compact, usually centrally located but occasionally eccentric	May be present; bodies usually elongate with blunt, rounded, smooth edges; may be round or oval	May be diffuse or absent in mature cyst; clumped chromatin mass may be present in early cysts (stains reddish brown with iodine)
E. hartmanni	5–10 μm; usual range, 6–8 μm	Usually spherical	Mature cyst, 4; immature, 1 or 2; nucleated cysts very common	Fine granules evenly distributed on membrane; nuclear characteristics may be difficult to see	Small, compact, usually centrally located	Usually present; bodies usually elongate with blunt, rounded, smooth edges; may be round or oval	May or may not be present, as in E. histolytica
E. coli	10–35 μm; usual range, 15–25 μm	Usually spherical; may be oval, triangular, or other; may be distorted on permanent stained slide owing to inadequate fixative penetration	Mature cyst, 8; occasionally, ≥16; immature cysts with ≥2 nuclei occasionally seen	Coarsely granular; may be clumped and unevenly arranged on membrane; nuclear characteristics not as clearly defined as in trophozoite; may resemble E. histolytica	Large, may or may not be compact and/or eccentric; occasionally centrally located	May be present (less frequently than in E. histolytica); splinter shaped with rough, pointed ends	May be diffuse or absent in mature cyst; clumped mass occasionally seen in mature cysts (stains reddish brown with iodine)
Endolimax nana	5–10 μm; usual range, 6–8 μm	Usually oval; may be round	Mature cyst, 4; immature cysts, 2, very rarely seen and may resemble cysts of Enteromonas hominis	Rarely present; small granules or inclusions are occasionally seen; fine linear chromatoidal bodies may be faintly visible on well-stained smears	Smaller than karyosome seen in trophozoites but generally larger than those of genus Entamoeba	No peripheral chromatin	Usually diffuse if present (stains reddish brown with iodine)
I. bütschlii	5–20 μm; usual range, 10–12 μm	May vary from oval to round; cyst may collapse owing to large glycogen vacuole space	Mature cyst, 1	No peripheral chromatin	Larger, usually eccentric refractile granules may be on one side of karyosome ("basket nucleus")	None; small granules are occasionally present	Large, compact, well-defined mass (stains reddish brown with iodine)

[a]From Garcia and Bruckner (13).
[b]Wet preparation measurements (in permanent stains, organisms usually measure 1 to 2 μm less).
[c]Stained with iodine.

TABLE 4 Intestinal protozoa: trophozoites of less common amebae[a]

Organism	Size (diam or length)	Motility	Nucleus (no. and visibility)	Appearance of stained:			
				Peripheral chromatin	Karyosome	Cytoplasm	Inclusions
E. polecki	10–12 μm	Usually nonprogressive, sluggish (like E. coli)	1; can occasionally be seen on wet preparation; intermediate between E. histolytica and E. coli	Fine granules (may be interspersed with large granules) evenly arranged on membrane; chromatin may also be clumped at one or both edges of membrane	Small, usually centrally located	Finely granular	May contain ingested bacteria
E. gingivalis	5–20 μm (avg, 10–15 μm)	Multiple pseudopodia; vary from long and lobose to short and blunt	1; similar to that of E. histolytica	Fine granules, closely packed	Small, well defined, usually centrally located	Finely granular	Ingested epithelial cells and host leukocytes

[a]From Garcia and Bruckner (13).

TABLE 5 Intestinal protozoa: cysts of less common amebae[a]

Organism	Size (diam or length)	Shape	Nucleus (no. and visibility)	Appearance of stained:		
				Peripheral chromatin	Karyosome	Cytoplasm[b]
E. polecki	5–11 μm	Usually spherical	Mature cyst, 1; may be visible in wet preparations (rarely 2 or 4 nuclei)	Similar to that seen in trophozoite	Similar to that seen in trophozoite	Abundant, angular pointed ends; threadlike chromatoidal bodies may be present; half of cysts contain "inclusion mass" that is spherical or ovoidal
E. gingivalis	No known cyst stage	NA	NA	NA	NA	NA

[a]From Garcia and Bruckner (13). NA, not applicable.
[b]Chromatoidal bodies, glycogen, inclusions (stained).

TABLE 6 Morphologic criteria used to identify *B. hominis*[a]

Characteristic	Comments
Shape and size	Organisms are generally round, approx 6–40 μm in diam, and usually characterized by large central body (looks like large vacuole) surrounded by multiple small nuclei; central body area can stain various colors (trichrome) or remain clear; tremendous size range, even in single specimen
Other features	More amebic form can be seen in diarrheal fluid but is difficult to identify; because of variation in size, may be confused with various yeast cells

[a]From Garcia and Bruckner (13).

and thus cannot be identified or may be confused with host cells. In addition, old urine specimens may be contaminated with airborne, free-living flagellates, especially if the urine collection vessel is open to the air and not sterile.

Vaginal materials commonly used for the diagnosis of *T. vaginalis* infections are vaginal fluid, scrapings, or washings. These samples may be examined morphologically in a saline wet mount or as a stained smear, or the material can be cultured. Although some workers feel that wet mount examinations are as efficient as cultures in revealing infections, current evidence suggests that cultivation methods are superior (1, 13, 22). Immunofluorescence and enzyme-linked immunosorbent assay (ELISA) methods have recently been described (see chapter 103 of this Manual).

Morphologic Identification of Flagellates

The flagellates are a more diverse group than the amebae. The diagnostic features of flagellate trophozoites and cysts are described in Tables 7 and 8, respectively; diagrams are shown in Fig. 8, and photomicrographs are presented in Fig. 5 and 9.

Trophozoites

The type of motility, shape, number of nuclei, and other characteristics such as an undulating membrane, sucking disk, cytostome, spiral groove, and numbers and locations of flagella are important characteristics used to identify flagellate trophozoites.

Cysts

The organism shape and size, numbers and positions of nuclei, and absence or arrangement of fibrils are used to identify flagellate cysts.

Not all of the features listed in Tables 7 and 8 can be seen in a single type of preparation. In some cases, species can be determined by the examination of either direct or concentrated wet mounts. Species of cysts may be identified in iodine-stained mounts. However, permanent stains are always recommended for every stool specimen submitted; organisms identified in wet preparations may not represent all types of organisms present.

CILIATE

Balantidium coli, a pathogenic ciliate inhabiting the colon, is the only ciliate and the largest protozoan parasitizing

humans (Fig. 7). Both trophozoites and cysts may be found in the feces.

Balantidiasis in humans is rarely reported in the United States. The disease is more prevalent where there is a close association of humans with pigs, the natural hosts from which humans contract the infection. The organism also infects nonhuman primates, especially the great apes. The symptoms of infection with *Balantidium coli* are referable to the large bowel and similar to those of amebiasis: lower abdominal pain, nausea, vomiting, and tenesmus. Chronic infections may present with cramps, frequent episodes of watery mucoid diarrhea, and rarely with bloody diarrhea. Chronic infections have been known to last for several months. In tropical areas in which the parasite is endemic, the infection is often severe because patients have other parasitic, bacterial, or viral infections or are undernourished. *Balantidium coli* causes colonic ulcers similar to those caused by *E. histolytica*, but it does not spread to other organs.

In human feces, trophozoites are readily recognized by their large size, shape, and rapid rotating motion. Cysts are less easily identified, but they usually cause few diagnostic problems. The morphology of trophozoites and cysts is described in Table 10; diagrams are shown in Fig. 7.

The examination of direct saline mounts is the most practical method of detecting infections. Cysts can be recovered by concentration, but in human infections, trophozoites are usually more numerous than cysts. Iodine-stained mounts and permanent stains are of little value, because the organisms tend to overstain.

COCCIDIA

The intestinal coccidia that parasitize humans belong to the subphylum *Sporozoa* and are obligatory tissue parasites that inhabit the mucosa of the small intestine (Fig. 6 and 7). Species of four genera (*Isospora*, *Sarcocystis*, *Cryptosporidium*, and *Cyclospora*) parasitize humans. The intestinal phase of toxoplasmosis that occurs in cats is similar to the intestinal infections of *Isospora* and *Sarcocystis* species in humans. The growth stages resemble those of malaria (also a sporozoan) and involve asexual and sexual generations. Therefore, the diagnostic stages, which are passed in feces, are unlike those of other intestinal protozoa. For *I. belli*, *Cryptosporidium parvum*, and *Cyclospora cayetanensis*, oocysts, either unsporulated (immature) as in *I. belli* or *Cyclospora cayetanensis* or sporulated (mature) as in the cryptosporidia, are diagnostic stages. The diagnostic stages for *Sarcocystis* species are free sporocysts and mature oocysts. Oocysts and sporocysts are transparent and are difficult to see in unstained preparations unless the microscope light is reduced and carefully regulated. Descriptions of the diagnostic stages are presented in Table 11, diagrams are shown in Fig. 7, and photomicrographs are presented in Fig. 6.

Isospora and *Sarcocystis* Species

Isosporiasis is caused by *I. belli* and, although infrequently recognized, can produce severe intestinal disease (Fig. 7). The disease occurs in AIDS patients (30). Deaths from overwhelming infections have been reported in immunocompromised patients. Symptoms, apparently caused by the schizogony of the developing forms in the epithelial cells and perhaps by toxins produced, include diarrhea, nausea, fever, steatorrhea, headache, and weight loss. The disease may persist for months or years.

Sarcocystis infection occurs in a variety of hosts, includ-

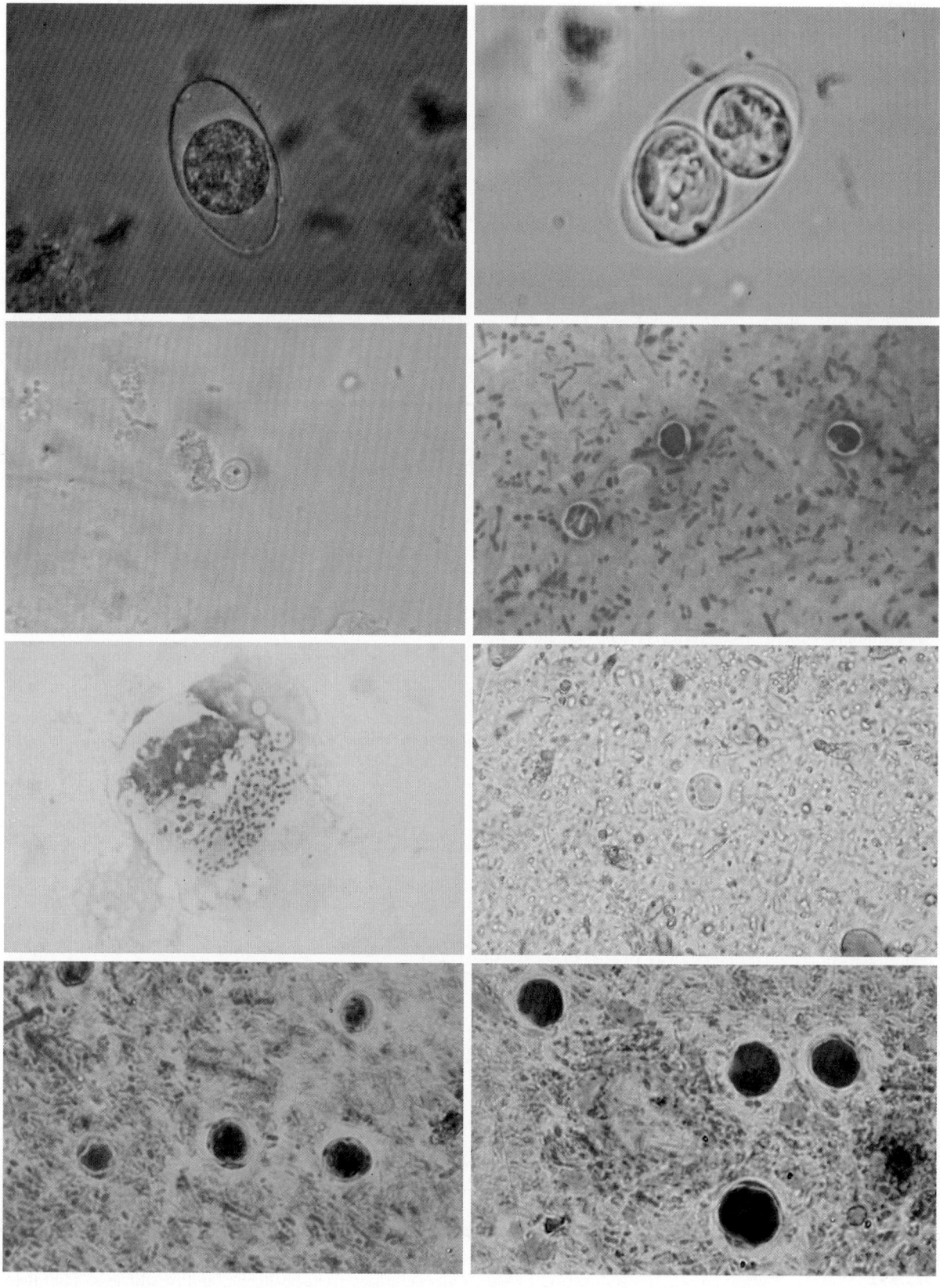

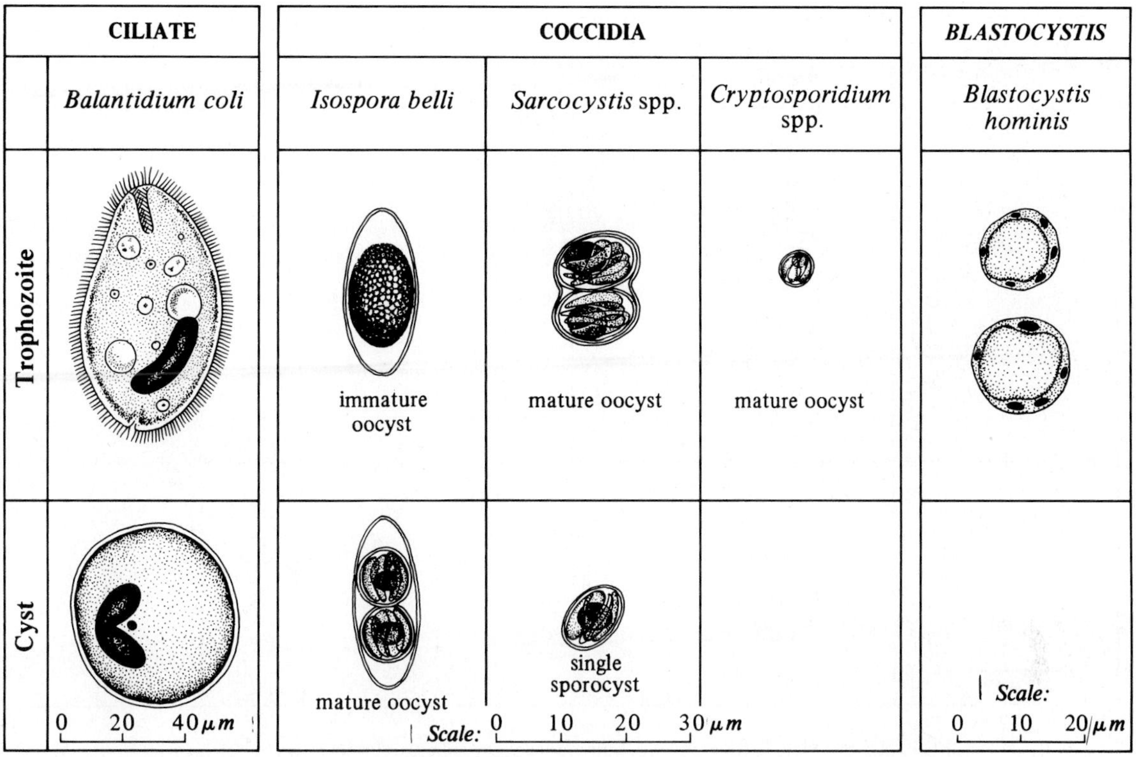

FIGURE 7 Ciliate, coccidia, and *Blastocystis hominis* found in human stool specimens. (From reference 4.)

ing humans. The sexual stage of *Sarcocystis* species, similar to that of *Isospora* species, occurs in the intestine of the carnivorous definitive host, and the asexual stage occurs in the muscles of an intermediate host animal that ingests infective sporozoites from the feces of the carnivore (Table 11). Humans are infected with the sexual stages of *S. hominis* (formerly called *Isospora hominis*), acquired from ingestion of improperly cooked or raw beef, and *S. suihominis*, acquired from ingestion of pork. Infected humans excrete sporulated oocysts that possess a thin oocyst wall (13). Humans are occasionally infected with a number of other species of *Sarcocystis* in which the asexual stages parasitize skeletal and cardiac muscles.

Isospora and *Sarcocystis* intestinal infections are diagnosed by identification of the organisms in direct or concentrated wet mounts of feces. Iodine stains the oocysts and sporocysts and makes them more readily visible. The oocysts usually stain acid fast. Trichrome stains, however, are of little or no value for demonstration of the organisms. In addition to being difficult to detect microscopically, oocysts of *I. belli* are sometimes not passed in feces until the symptoms of the infection have subsided. The duration of oocyst passage varies considerably, from a few days to a few weeks; therefore, in attempting to establish the diagnosis,

several stool specimens should be collected and examined. This pattern of passage of diagnostic stages of coccidia in feces may also be true of *Sarcocystis* species. The diagnostic stage of *I. belli* is an unsporulated or immature oocyst that will mature in several days at room temperature to an oocyst containing two sporocysts, each of which in turn contains four sporozoites. *Sarcocystis* species already have mature sporocysts, sometimes surrounded by a thin-walled oocyst.

The laboratory diagnosis of *I. belli* is accomplished as follows.

1. The oocysts are more easily recovered and identified when wet preparations (direct smear, concentration smear) are examined.

2. Oocysts recovered in a concentrate sediment from polyvinyl alcohol-preserved stool may be very difficult to detect (oocyst wall is very difficult to see).

3. A biopsy can be positive even though no organisms are seen in the stool. This contradiction is *not* necessarily due to poor-quality laboratory work but may reflect normal findings.

4. Although these organisms can be stained with aura-

FIGURE 6 (Opposite) Coccidia, stained and unstained. Magnification, ×1,000, except row 3, left (×2,000). Row 1: (left) *I. belli*, immature oocyst, unstained. (Right) *I. belli* mature oocyst, unstained. Row 2: (left) *Cryptosporidium parvum* oocyst, unstained. (Right) *Cryptosporidium parvum* oocysts, modified acid-fast stain. Row 3: (left) *Enterocytozoon* spores in intestinal cell, Giemsa stain. (Courtesy of Elizabeth Canning; used with permission.) (Right) *B. hominis*, central body forms, iodine stain. Row 4: (left) *B. hominis*, central body forms, trichrome stain. (Right) *B. hominis*, central body forms, trichrome stain.

FLAGELLATES				
Trichomonas hominis	*Chilomastix mesnili*	*Giardia lamblia*	*Enteromonas hominis*	*Retortamonas intestinalis*

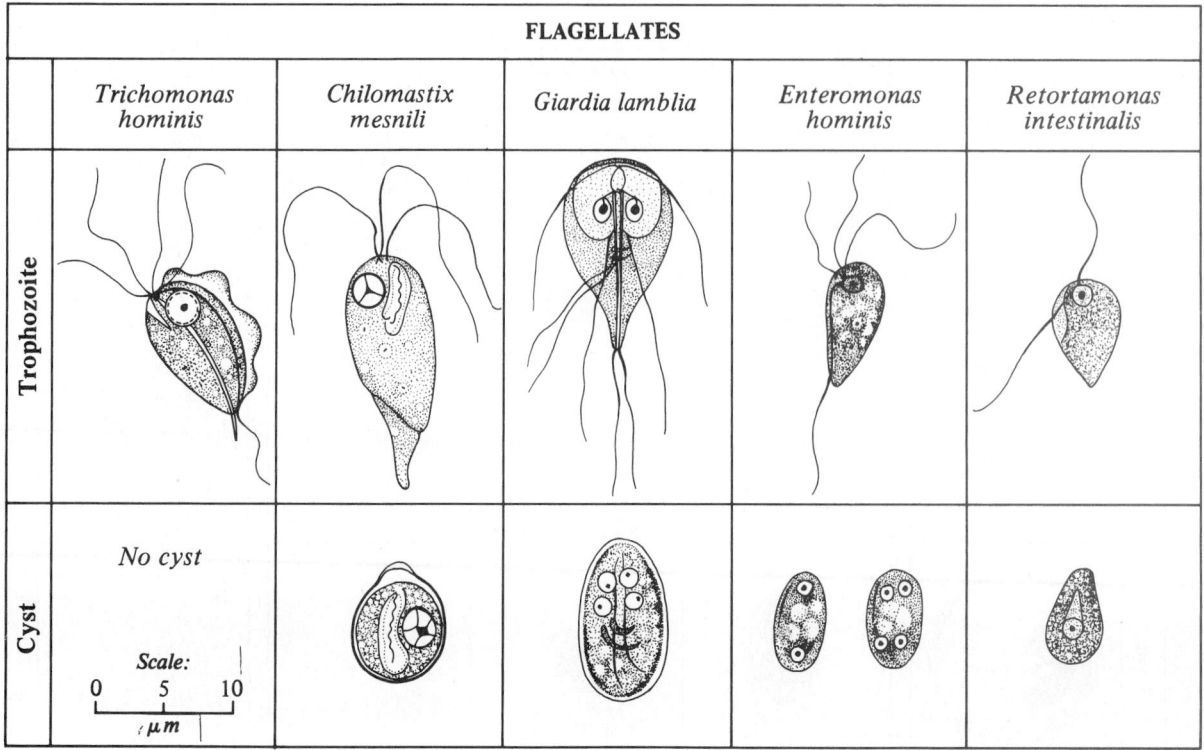

FIGURE 8 Flagellates found in human stool specimens. (From reference 4.)

mine-rhodamine stains, they should be confirmed by wet smears or acid-fast stains.

Cryptosporidium parvum

Cryptosporidium species infect the brush borders of columnar epithelial cells and cause cryptosporidiosis (11). Human infections are caused by *Cryptosporidium parvum*. Dissemination has been reported, and the organisms occasionally infect the cells of other organs in immunocompromised hosts. Clinically apparent infections with *Cryptosporidium* species can be separated into two groups. Patients with intact immune function develop a profuse, watery diarrhea accompanied by mild epigastric cramping pain, nausea, and anorexia, which is generally self-limited, all of which last for 10 to 15 days (11). Immunocompromised patients, such as those having AIDS or receiving immunosuppressive therapy, develop a more severe, long-lasting infection. Symptoms are as noted above, but the disease is prolonged, with profuse, watery diarrhea persisting from several weeks to months or years. There is no effective specific therapy for cryptosporidiosis. Outbreaks have been reported in day-care centers and have been associated with contaminated municipal water supplies (16).

Specimens from patients suspected of having cryptosporidiosis should be preserved in formalin before being sent to the laboratory or immediately upon receipt. *Cryptosporidium* oocysts are very small (4 to 5 μm) and can easily be overlooked in fecal preparations or confused with yeast cells. The infection has likely been common in humans but was undiagnosed because the organisms were not recognized by the usual diagnostic methods. The oocysts contain sporozoites and are immediately infective when passed in the stool, both formed and diarrheic (Table 11; Fig. 6 and

7). Identification is often difficult. Direct wet mounts can be used to examine feces, but oocysts are difficult to differentiate from yeasts, and in light infections, organisms may not be detected. Specimens can also be concentrated by the Sheather sugar flotation method or the formalin-ethyl acetate method (recommended). It is important to remember that the centrifugation speed and time should be 500 × *g* and 10 min for all wash steps and the final spin. Both unstained and iodine-stained mounts as well as slides for acid-fast staining should be prepared. In unstained mounts containing mature oocysts, the refractile residual body can be detected, but the sporozoites may not be distinct. Oocysts do not stain with iodine (unless they are exposed to it for long periods), but yeast cells do stain, thus helping distinguish oocysts from yeast cells. Various acid-fast stains that are more sensitive than wet mounts are the usual methods used to detect oocysts in feces or concentrates or to confirm the identification of organisms seen in other preparations. Oocysts stain intensely acid fast, whereas yeast cells and fecal material do not. There is usually some irregularity in staining of organisms, which aids in differentiating them from homogeneously staining particles sometimes seen in feces. Both fluorochrome and carbol-fuchsin acid-fast stains are useful. Both immunofluorescence and ELISA diagnostic kits are commercially available (15) (Fig. 10) (see chapter 103 of this Manual).

The laboratory diagnosis of *Cryptosporidium parvum* is accomplished as follows.

1. If the specimen represents the typically watery diarrhea, there should be numerous organisms caught up in the mucus.

2. The more normal the stool (semiformed, formed),

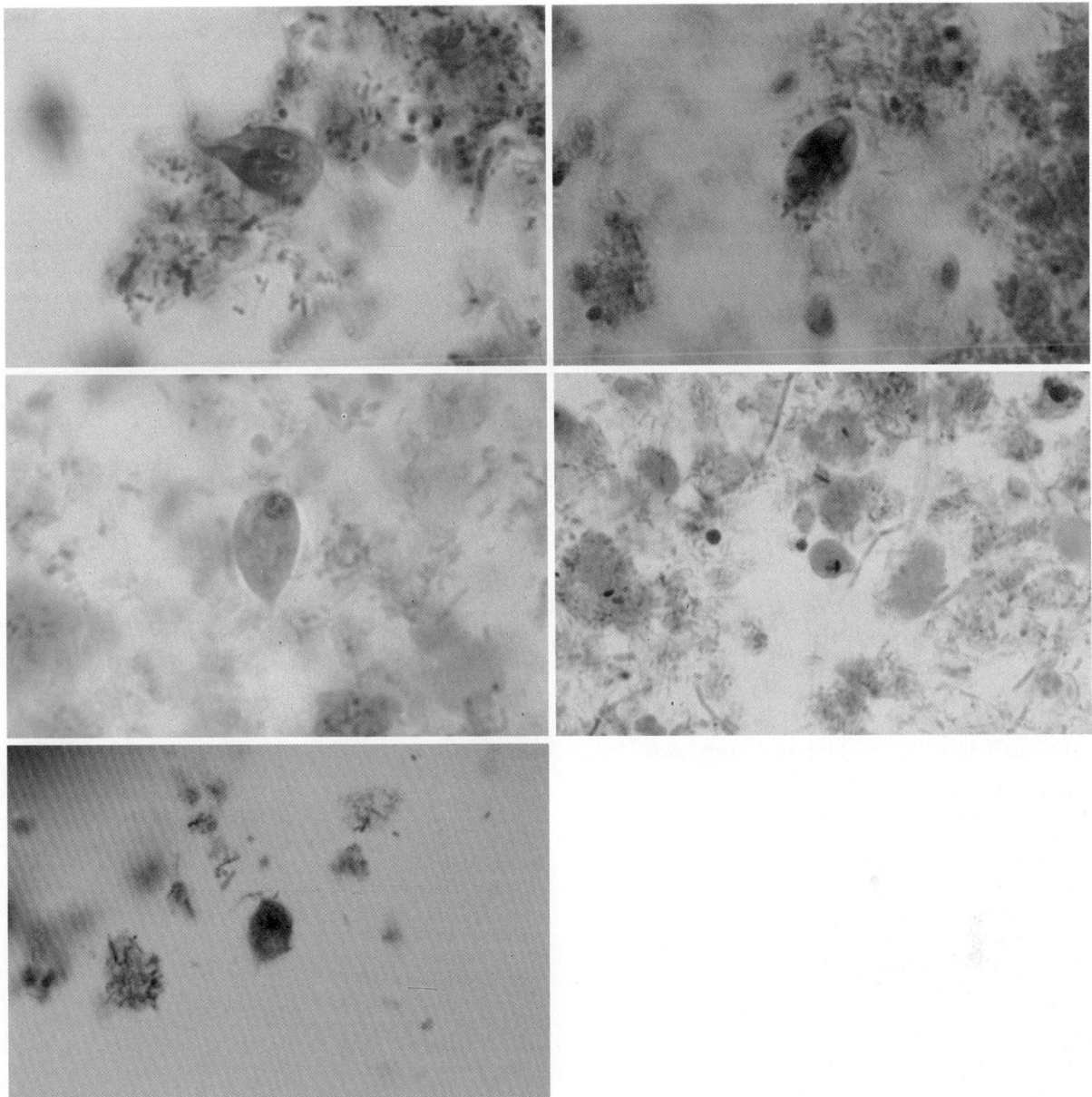

FIGURE 9 Flagellates (trichrome stain; oil immersion magnification, ×1,000). Row 1: (left) G. *lamblia* trophozoite. Two nuclei and a prominent median body are evident in this pyriform organism. (Right) G. *lamblia* cyst. There are two nuclei toward the bottom in this plane of focus. Fibrils are evident in the cytoplasm. Row 2: (left) C. *mesnili* trophozoite. The nucleus is at the upper end. The pale cytostome is evident to the left of the nucleus. (Right) C. *mesnili* cyst. The lemon-shaped cyst is in the center of the field. The nucleus is in the lower portion of the cyst. Fibrils are faintly visible. Row 3: (left) T. *hominis* trophozoite. The nucleus is toward the top of the organism. A portion of the undulating membrane is evident to the right of the nucleus. The axostyle is evident at the bottom of the organism.

the fewer the organisms and the more the artifact material that will be present.

3. Several concentrates and at least five or six acid-fast smears should be examined before a patient is considered negative, especially if the patient has been on experimental medications.

4. The normal permanent stains (trichrome, iron hematoxylin) do not adequately stain *Cryptosporidium* spp.

5. Although specimens can be stained with auramine-rhodamine stains, results should be confirmed by using other acid-fast stains or monoclonal antibody-based reagents (more sensitive and specific). This is particularly true if the stool contains other cells or a lot of artifact material (more normal stool consistency).

6. Sputum specimens should be submitted in 5 or 10% formalin and processed the same way as a stool sample.

TABLE 7 Intestinal protozoa: trophozoites of flagellates[a]

Organism	Shape and size	Motility	Nucleus (no. and visibility)	No. of flagella[b]	Other features
Dientamoeba fragilis	Shaped like amebae; 5–15 μm; usual range, 9–12 μm	Usually nonprogressive; pseudopodia are angular, serrated, or broad lobed and almost transparent	Percentage may vary, but 40% of organisms have 1 nucleus and 60% have 2 nuclei; not visible in unstained preparations; no peripheral chromatin; karysome is cluster of 4–8 granules	No visible flagella	Cytoplasm finely granular and may be vacuolated with ingested bacteria, yeasts, and other debris; may be great variation in size and shape on single smear
Giardia lamblia	Pear shaped; 10–20 μm long; 5–15 μm wide	Falling-leaf motility may be difficult to see if organism is in mucus	2; not visible in unstained mounts	4 lateral, 2 ventral, 2 caudal	Sucking disk occupying 1/2–3/4 of ventral surface; pear shaped from front, spoon shaped from side
Chilomastix mesnili	Pear shaped; 6–24 μm long; usual range, 10–15 μm long; 4–8 μm wide	Stiff, rotary	1; not visible in unstained mounts	3 anterior, 1 in cytostome	Prominent cytostome extending 1/3–1/2 length of body; spiral groove across ventral surface
Trichomonas hominis	Pear shaped; 5–15 μm long; usual range, 7–9 μm long; 7–10 μm wide	Jerky, rapid	1; not visible in unstained mounts	3–5 anterior, 1 posterior	Undulating membrane extends length of body; posterior flagellum extends free beyond end of body
Trichomonas tenax	Pear shaped; 5–12 μm long; usual range, 6.5–7.5 μm; 7–9 μm wide	Jerky, rapid	1; not visible in unstained mounts	4 anterior, 1 posterior	Seen only in preparations from mouth; axostyle (slender rod) protrudes beyond posterior end and may be visible; posterior flagellum extends only halfway down body, and there is no free end
Enteromonas hominis	Oval, 4–10 μm long; usual range, 8–9 μm long; 5–6 μm wide	Jerky	1; not visible in unstained mounts	3 anterior, 1 posterior	One side of body flattened; posterior flagellum extends free posteriorly or laterally
Retortamonas intestinalis	Pear shaped or oval; 4–9 μm long; usual range, 6–7 μm long; 3–4 μm wide	Jerky	1; not visible in unstained mounts	1 anterior, 1 posterior	Prominent cytostome extending approximately 1/2 length of body

[a]From Garcia and Bruckner (13).
[b]Usually difficult to see.

Cyclospora cayetanensis

Organisms in the genus *Cyclospora* were first described in 1881 but were not connected with human infections until recently. The mature oocyst contains two sporocysts, each with two sporozoites. The oocysts recovered from human stool are immature, so the structure of the mature oocyst is rarely seen in those specimens (Table 11). It takes approximately 5 days or more for oocysts to sporulate, according to

TABLE 8 Intestinal protozoa: cysts of flagellates[a]

Species	Size	Shape	No. of nuclei	Other features
Dientamoeba fragilis, Trichomonas hominis, Trichomonas tenax	No cyst stage	NA	NA	NA
Giardia lamblia	8–19 μm long; usual range, 11–14 μm long; 7–10 μm wide	Oval, ellipsoidal, or round	4; not distinct in unstained preparations; usually located at one end	Longitudinal fibers in cysts may be visible in unstained preparations; deeply staining median bodies usually lie across longitudinal fibers; there is often shrinkage, and cytoplasm pulls away from cyst wall; may also be "halo" effect around outside of cyst wall due to shrinkage caused by dehydrating reagents
Chilomastix mesnili	6–10 μm long; usual range, 7–9 μm long; 4–6 μm wide	Lemon shaped with anterior hyaline knob	1; not distinct in unstained preparations	Cytostome with supporting fibrils, usually visible in stained preparation; curved fibril along side of cytostome usually referred to as "shepherd's crook"
Enteromonas hominis	4–10 μm long; usual range, 6–8 μm long; 4–6 μm wide	Elongate or oval	1–4; usually 2 lying at opposite ends of cyst; not visible in unstained mounts	Resembles E. nana cyst; fibrils or flagella usually not seen
Retortamonas intestinalis	4–9 μm long; usual range, 4–7 μm long; 5 μm wide	Pear shaped or slightly lemon shaped	1; not visible in unstained mounts	Resembles Chilomastix cyst; shadow outline of cytostome with supporting fibrils extends above nucleus; bird beak fibril arrangement

[a]From Garcia and Bruckner (13). NA, not applicable.

evidence obtained from recent excystation experiments. Electron microscopy has confirmed the presence of characteristic organelles for coccidian organisms of the phylum Apicomplexa. The name Cyclospora cayetanensis has been proposed for this newly described disease-producing coccidian organism from humans (24).

The symptoms of Cyclospora infection are rather nonspecific. The clinical presentation of patients infected with this organism is similar to the presentation of those infected with Cryptosporidium parvum. There is generally one day of malaise and low-grade fever, with diarrhea of up to seven stools per day. There may also be fatigue, anorexia, vomiting, myalgia, and weight loss, with remission of self-limiting diarrhea in 3 to 4 days followed by relapses lasting up to 4 weeks. Clinical clues include any unexplained prolonged diarrheal illness during the summer and travel in a tropical area (18). The majority of infected individuals have intermittent diarrhea for 2 to 3 weeks, and many complain of intense fatigue as well as anorexia and myalgia during the illness.

When examined in formalin-preserved human stool specimens and with electron microscopy, these organisms exhibit the characteristics of a cyanobacterium. The organisms stain orange with safranin and are acid fast variable, with some organisms staining deep red with a mottled appearance but no internal organization. Preliminary electron microscopy reveals internal structures that resemble those of the cyanobacteria (2). In clean wet mounts, the organisms are seen as nonrefractile spheres that are acid fast variable with the modified acid-fast stain; those that are unstained appear as glassy, wrinkled spheres (19) (Fig. 11). Modified acid-fast stains stain the oocysts from light pink to deep red; some of the oocysts contain granules or appear bubbly. It is very important to be aware of these organisms when using the modified acid-fast stain for Cryptosporidium parvum and when other similar but larger structures (approximately twice the size of Cryptosporidium oocysts) are seen in the stained smear. It will be important for laboratories to measure all acid-fast oocysts, particularly if they appear to be somewhat larger than those of Cryptosporidium parvum. The oocysts autofluoresce (strong green or intense blue) under UV epifluorescence (19).

The laboratory diagnosis of Cyclospora cayetanensis is accomplished as follows.

TABLE 9 Characteristics of *T. vaginalis*[a]

Characteristic	Comment
Shape and size	Pear shaped; 7–23 μm long (avg, 13 μm); 5–15 μm wide
Motility	Jerky, rapid
No. of nuclei and visibility	1; not visible in unstained mounts
No. of flagella	3–5 anterior, 1 posterior; usually difficult to see
Other features	Seen in urine, urethral discharges, and vaginal smears; undulating membrane extends 1/2 length of body; no free posterior flagellum; axostyle easily seen
Infective stage	Trophozoite
Usual location	Vagina (male, urethra)
Striking clinical findings	Leukorrhea, pruritus vulva (thin white urethral discharge in male)
Other sites of infection	Urethra (prostate in male)
Stage usually recovered during clinical phase	Trophozoite only, no cyst

[a]From Garcia and Bruckner (13).

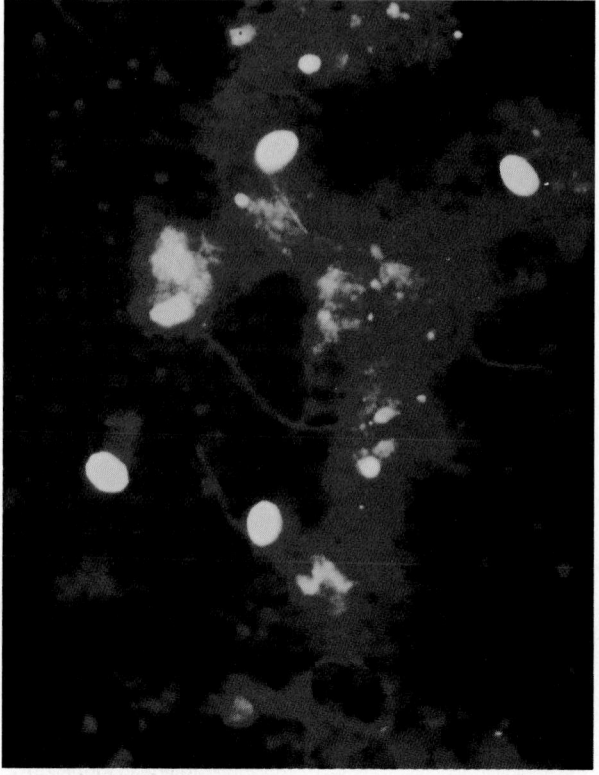

FIGURE 10 *G. lamblia* cysts and *Cryptosporidium parvum* oocysts (monoclonal antibody fluorescent stain combination product). (Courtesy of Meridian Diagnostics, Inc.)

1. Specimens may appear as larger than normal *Cryptosporidium parvum* oocysts; make sure you measure these organisms to confirm sizes. *Cyclospora* oocysts measure from 8 to 10 μm, and *Cryptosporidium* oocysts measure from 4 to 6 μm.

2. On wet smears, the oocysts appear as nonrefractile spheres and autofluoresce from bright green to intense blue with UV epifluorescence.

3. On modified acid-fast-stained smears, the oocysts stain from light pink to deep red; some of them contain granules or appear bubbly. Those that do not stain may have a wrinkled appearance.

4. If they are seen in a regular trichrome-stained smear of stool, the oocysts may appear as clear, round, somewhat wrinkled objects larger than *Cryptosporidium parvum* oocysts.

TABLE 10 Intestinal protozoa: the ciliate *Balantidium coli*

Stage	Shape and size	Motility	No. of nuclei	Other features
Trophozoite	Ovoid with tapering anterior end; 50–100 μm long; 40–70 μm wide; usual width range, 40–50 μm	Rotary, boring; may be rapid	1 large kidney-shaped macronucleus; 1 small round micronucleus, which is difficult to see even in stained smear; macronucleus may be visible in unstained preparation	Body covered with cilia, which tend to be longer near cytostome; cytoplasm may be vacuolated
Cyst	Spherical or oval; 50–70 μm in diam; usual range, 50–55 μm		1 large macronucleus visible in unstained preparation; micronucleus difficult to see	Macronucleus and contractile vacuole are visible in young cysts; in older cysts, internal structure appears granular; cilia difficult to see within cyst wall

[a]From Garcia and Bruckner (13).

TABLE 11 Coccidia[a]

Species	Shape and size	Other features
Cryptosporidium parvum	Oocyst generally round, 4–5 μm in diam; each mature oocyst contains sporozoites, which may or may not be visible	Oocyst is usual diagnostic stage in stool. Various other stages in life cycle can be seen in biopsy specimens taken from gastrointestinal tract (brush border of epithelial cells, intestinal tract) and other tissues (respiratory tract, biliary tract). Disseminated disease and nosocomial transmission documented
Cyclospora cayetanensis	Organisms generally round, 8–9 μm in diam; mimic *Cryptosporidium* spp. (acid fast) but are larger	In wet smears, look like nonrefractile spheres; also autofluoresce with epifluorescence; acid fast variable from no color to light pink to deep red; those that do not stain may appear wrinkled; in trichrome-stained stool smear, appear clear, round, somewhat wrinkled; cause diarrhea in both immunocompetent and immunosuppressed patients
Isospora belli	Ellipsoidal oocyst; usual range, 20–30 μm long, 10–19 μm wide; sporocysts rarely seen broken out of oocysts, but measure 9–11 μm	Mature oocyst contains 2 sporocysts with 4 sporozoites each; usual diagnostic stage in feces of patient with diarrhea is immature oocyst containing spherical mass of protoplasm
Sarcocystis hominis, *Sarcocystis suihominis,* *Sarcocystis bovihominis*	Oocyst thin walled and contains 2 mature sporocysts, each containing 4 sporozoites; thin oocyst wall frequently ruptures; ovoid sporocysts are each 9–16 μm long and 7.5–12 μm wide	Thin-walled oocyst or ovoid sporocysts occur in stool
Sarcocystis "lindemanni"	Considerable variation in skeletal and cardiac muscle sarcocysts	Sarcocysts contain from several hundred to several thousand trophozoites, each 4–9 μm wide and 12–16 μm long. Sarcocysts may also be divided into compartments by septa, which are not seen in *Toxoplasma* cysts (tissue or muscle).

[a]From Garcia and Bruckner (13).

Microsporidia

The microsporidian organisms that parasitize humans belong to the phylum *Microspora*, order *Microsporida*. They are obligate, intracellular parasites of invertebrate animals. Although the five genera that have been implicated in human infections are acquired by ingestion of the small (1- to 2-μm) spores containing a coiled polar tubule, only *Enterocytozoon* and *Septata* sp. remain as parasites of the intestine, although *Septata* sp. does disseminate to other tissues (3, 7, 9). The other three genera (*Nosema*, *Encephalitozoon*, and *Pleistophora*) are found in a variety of tissues (8, 12, 27, 28, 32). Microsporidial keratoconjunctivitis has recently been recognized in patients with AIDS (10).

The parasites have been diagnosed primarily in AIDS patients or other immunocompromised individuals (6, 23). Reports also indicate that approximately 30% of AIDS patients positive with *Cryptosporidium parvum* infection in the gastrointestinal tract also have microsporidiosis, with spores being identified in stool specimens (14, 33) (Fig. 12). Biopsied tissues have usually been diagnosed by electron microscopic studies. Since the spores are small (1 to 2 μm), finding them in stool specimens by light microscopy has been difficult, but techniques are improving (Table 12). Spores have been identified in Giemsa-stained intestinal biopsy samples (25) (Fig. 12) by using oil immersion light microscopy. The modified trichrome stain has been recom-

TABLE 12 Microsporidia[a]

Body site	Diagnosis	Comments
All organs, including eye	Routine histology (fair); acid-fast, PAS stains recommended (spores); animal inoculation not recommended (laboratory animals may carry occult infections); electron microscopy may be necessary; modified trichrome stains helpful; monoclonal reagents under development; nonspecific fluorescent stains helpful but nonspecific	Have been found as insect or other animal parasites; route of infection may be ingestion, inhalation, or direct inoculation (eye); dissemination from GI tract has been documented with *Septata intestinalis*; GI tract infections most common in AIDS patients (*Septata* and *Enterocytozoon* spp.); "*Microsporidium*" used as catchall genus for isolates not yet identified

[a]From Garcia and Bruckner (13). Microsporidia include *Nosema*, *Encephalitozoon*, *Enterocytozoon*, *Septata*, *Pleistophora*, and "*Microsporidium*" spp. GI, gastrointestinal.

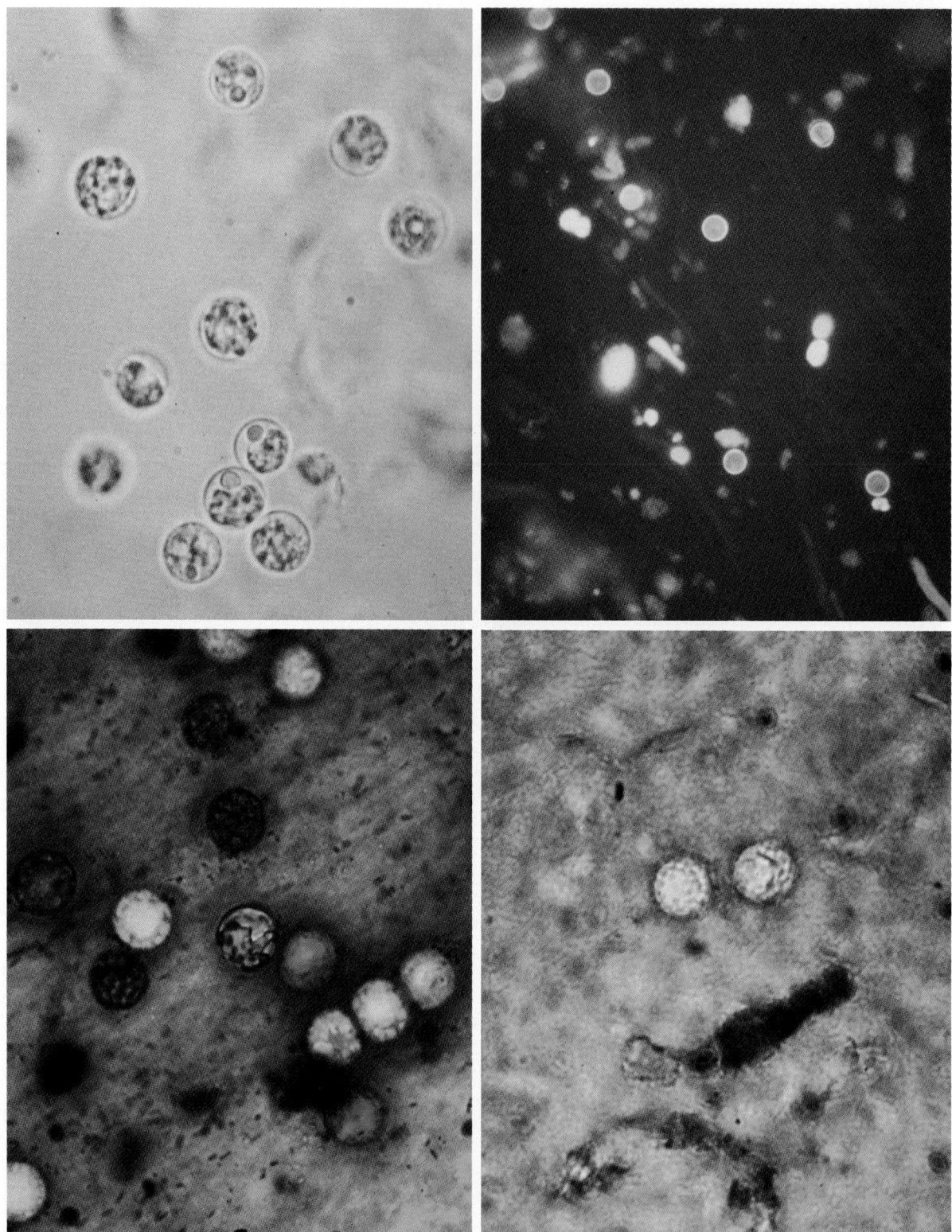

FIGURE 11 *Cyclospora cayetanensis*, stained and unstained. Magnification, ×1,000 except when indicated. Row 1: (left) *Cyclospora cayetanensis* oocysts (saline, wet preparation). These organisms were formerly called CLBs (cyanobacteriumlike or coccidiumlike organisms). (Courtesy of Earl G. Long; used with permission.) (Right) *Cyclospora cayetanensis* oocysts (×400) (autofluorescence). Note the fluorescence around the oocysts (lower magnification than other photographs). (Courtesy of Earl G. Long; used with permission.) Row 2: (left) *Cyclospora cayetanensis* oocysts (trichrome stain). Note that the oocysts do not stain with trichrome; they appear as "ghost cells" and are approximately twice the size of *Cryptosporidium parvum* oocysts. (Courtesy of Earl G. Long; used with permission.) (Right) *Cyclospora cayetanensis* oocysts (modified acid-fast stain). Note the variability in staining; this is common and demonstrates the difference between these organisms and *Cryptosporidium parvum* oocysts, which stain more consistently with this stain. (Courtesy of Earl G. Long; used with permission.)

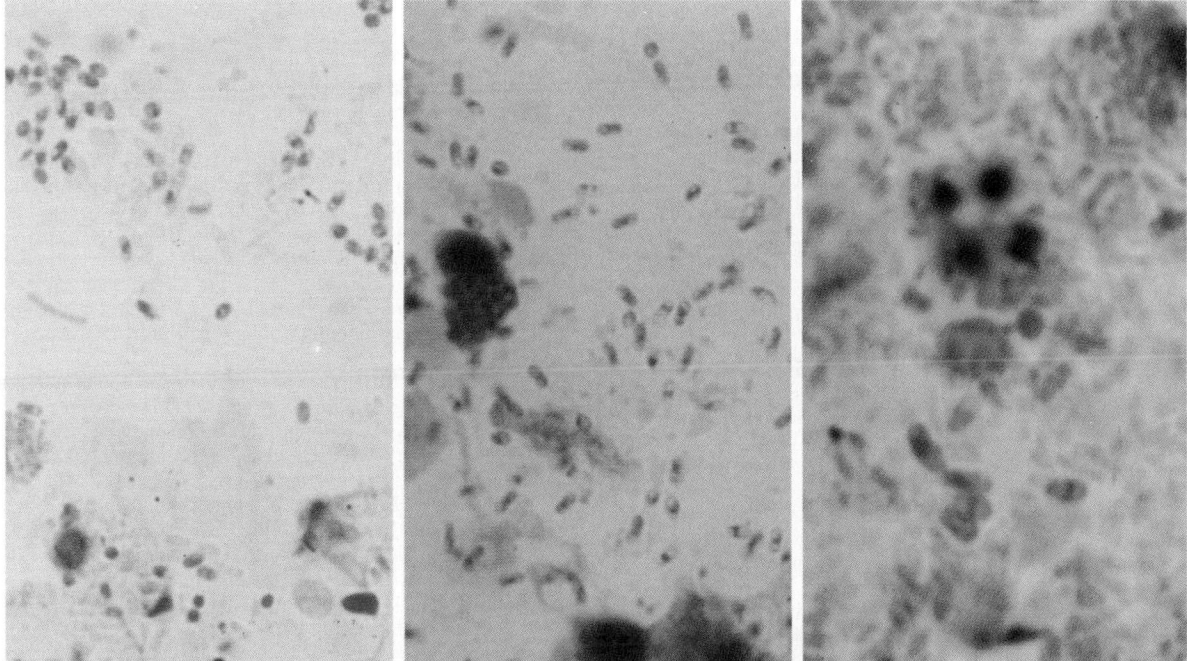

FIGURE 12 Microsporidian spores. Stained with modified trichrome-blue stain; magnification, ×400. (Courtesy of Norbert J. Ryan; used with permission.) (Left) *Enterocytozoon bieneusi* or *Septata intestinalis* in stool. (Center) *Encephalitozoon* or *Septata* species in nasopharyngeal aspirate cytospin preparation. Stained with modified trichrome-green stain; magnification, ×1,000. (Courtesy of Rainer Weber; used with permission.) (Right) *Enterocytozoon bieneusi* or *Septata intestinalis* in stool; higher magnification.

mended for stool specimens; however, the organism size still prevents easy identification (26, 31). In some cases, parts of the polar tubule can be seen as a horizontal or diagonal line within the spore; however, this feature is usually pale and not always easy to see. The use of nonspecific fluorescent stains has also been recommended, although some of the reagents are difficult to obtain. Less costly and less cumbersome diagnostic methods are needed and may include the use of monoclonal antibody reagents, some of which are currently under development.

The laboratory diagnosis of microsporidia is accomplished as follows.

1. It appears from the literature that approximately 30% of AIDS patients and those with cryptosporidiosis may also have infections with microsporidia. In this group of patients with chronic diarrhea, microsporidian infections must be considered (14, 33).

2. The modified trichrome staining procedure for stool may be difficult to interpret without positive controls to review. Make sure that the material on the slides is *very* thin, the smear is stained for the full 90 min, and the smear is examined under oil immersion (total magnification of at least ×1,000) (26, 31).

3. As monoclonal or polyclonal antibody reagents are produced, we may see diagnostic procedures with greater specificity and sensitivity. Reagents are currently being used in a few institutions; these preparations are generally easier to read than the routine stains.

4. Touch preparations of tissue can be methanol fixed and stained with Giemsa.

5. Plastic-embedded tissues stained with PAS, silver,

acid-fast, and routine hematoxylin and eosin stains generally stain better than paraffin-embedded tissues. This difference may be related to the use of formalin as a tissue fixative.

6. Make sure to gain some experience in examining these preparations before sending out patient specimen results. *This work mandates the use of positive control material*, regardless of which technique(s) you are using. Once a laboratory has gained experience with these methods, it is then appropriate to begin accepting clinical specimens.

7. Electron microscopy is the "gold standard" when confirming infections and when attempting to classify the organisms seen in tissues.

REFERENCES

1. **Beal, C., R. Goldsmith, M. Kotby, M. Sherit, A. El-Tagi, A. Fariud, S. Zakaria, and J. Eapen.** 1992. The plastic envelope method, a simplified technique for culture diagnosis of trichomoniasis. *J. Clin. Microbiol.* **30:**2265–2268.

2. **Bendall, R. P., S. Lucas, A. Moody, G. Tovey, and P. L. Chiodini.** 1993. Diarrhoea associated with cyanobacterium-like bodies: a new coccidian enteritis of man. *Lancet* **341:** 590–592.

3. **Blanshard, C., W. S. Hollister, C. S. Peacock, D. G. Tovey, D. S. Ellis, E. U. Canning, and B. G. Gazzard.** 1992. Simultaneous infection with two types of intestinal microsporidia in a patient with AIDS. *Gut* **33:**418–420.

4. **Brooke, M., and D. Melvin.** 1984. *Morphology of Diagnostic Stages of Intestinal Parasites of Humans*, 2nd ed. U.S. Department of Health and Human Services publication no. (CDC) 84-8116. Centers for Disease Control, Atlanta.

5. Bruckner, D. A. 1992. Amebiasis. *Clin. Microbiol. Rev.* **5:**356–369.

6. Bryan, R. T., A. Cali, R. L. Owen, and H. C. Spencer. 1991. Microsporidia: opportunistic pathogens in patients with AIDS, p. 1–26. *In* T. Sun (ed.), *Progress in Clinical Parasitology*, vol. 2. Field and Wood Medical Publishers, New York.

7. Cali, A., D. P. Kotler, and J. M. Orenstein. 1993. *Septata intestinalis* N.G., N.Sp., an intestinal microsporidian associated with chronic diarrhea and dissemination in AIDS patients. *J. Eukaryot. Microbiol.* **40:**101–112.

8. Cali, A., D. M. Miesler, I. Rutherford, C. Y. Lowder, J. T. McMahon, D. L. Longworth, and R. T. Bryan. 1991. Corneal microsporidiosis in a patient with AIDS. *Am. J. Trop. Med. Hyg.* **44:**463–468.

9. Cali, A., and R. L. Owen. 1990. Intracellular development of *Enterocytozoon*, a unique microsporidian in the intestine of AIDS patients. *J. Protozool.* **37:**145–155.

10. Centers for Disease Control. 1990. Microsporidia keratoconjunctivitis in patients with AIDS. *Morbid. Mortal. Weekly Rep.* **39:**188–189.

11. Current, W. L., and L. S. Garcia. 1991. Cryptosporidiosis. *Clin. Microbiol. Rev.* **4:**325–358.

11a. Diamond, L. S., and C. G. Clark. 1993. A redescription of *Entamoeba histolytica* Schaudinn, 1903 (emended Walker, 1911) separating it from *Entamoeba dispar* Brumpt, 1925. *J. Eukaryot. Microbiol.* **40:**340–344.

12. Didier, E. S., P. J. Didier, D. N. Friedberg, S. M. Stenson, J. M. Orenstein, R. W. Yee, F. O. Tio, R. M. Davis, C. Vossbrinck, N. Millichamp, and J. A. Shadduck. 1991. Isolation and characterization of a new human microsporidian, *Encephalitozoon hellem* (n. sp.), from three AIDS patients with keratoconjunctivitis. *J. Infect. Dis.* **163:**617–621.

13. Garcia, L. S., and D. A. Bruckner. 1993. *Diagnostic Medical Parasitology*, 2nd ed. American Society for Microbiology, Washington, D.C.

14. Garcia, L. S., R. Y. Shimizu, and D. A. Bruckner. 1994. Detection of microsporidial spores in fecal specimens from patients diagnosed with cryptosporidiosis. *J. Clin. Microbiol.* **32:**1739–1741.

15. Garcia, L. S., A. C. Shum, and D. A. Bruckner. 1992. Evaluation of a new monoclonal antibody combination reagent for direct fluorescence detection of *Giardia* cysts and *Cryptosporidium* oocysts in human fecal samples. *J. Clin. Microbiol.* **30:**3255–3257.

16. Hayes, E., T. Matte, T. O'Brien, T. McKinley, G. Logsdon, J. Rose, B. Ungar, D. Word, P. Pinsky, M. Cummings, M. Wilson, E. Long, E. Hurwitz, and D. Juranek. 1989. Large community outbreak of cryptosporidiosis due to contamination of a filtered public water supply. *N. Engl. J. Med.* **320:**1372–1376.

17. Istre, G. R., K. Kreiss, R. Hopkins, G. Healy, M. Benziger, T. Canfield, P. Dickinson, T. Englert, R. Compton, H. Mathews, and R. Smith. 1982. An outbreak of amebiasis spread by colonic irrigation at a chiropractic clinic. *N. Engl. J. Med.* **307:**339–342.

18. Long, E. G., A. Ebrahimzadeh, E. H. White, B. Swisher, and C. S. Callaway. 1990. Alga associated with diarrhea in patients with acquired immunodeficiency syndrome and in travelers. *J. Clin. Microbiol.* **28:**1101–1104.

19. Long, E. G., E. White, W. E. Carmichael, P. M. Quinlisk, R. Raja, B. L. Swisher, H. Daugharty, and M. T. Cohen. 1991. Morphologic and staining characteristics of a *Cyanobacterium*-like organism associated with diarrhea. *J. Infect. Dis.* **164:**199–202.

20. Markell, E. K., and M. P. Udkow. 1986. *Blastocystis hominis*: pathogen or fellow traveler? *Am. J. Trop. Med. Hyg.* **35:**1023–1026.

21. Melvin, D., and M. Brooke. 1982. *Laboratory Procedures for the Diagnosis of Intestinal Parasites*, 3rd ed. U.S. Department of Health and Human Services publication no. (CDC) 82-8282. Centers for Disease Control, Atlanta.

22. National Committee for Clinical Laboratory Standards. 1993. *Procedures for the Recovery and Identification of Parasites from the Intestinal Tract; Proposed Guideline*. Document M28-P. National Committee for Clinical Laboratory Standards, Villanova, Pa.

23. Orenstein, J. M. 1991. Microsporidiosis in the acquired immunodeficiency syndrome. *J. Parasitol.* **77:**843–864.

24. Ortega, Y., C. R. Sterling, R. H. Gilman, V. A. Cama, and F. Diaz. 1993. *Cyclospora* species: a new protozoan pathogen of humans. *N. Engl. J. Med.* **328:**1308–1312.

25. Rijpstra, A., E. Canning, R. VanKetel, J. Eeftinck-Schattenkerk, and J. Laarman. 1988. Use of light microscopy to diagnose small-intestinal microsporidiosis in patients with AIDS. *J. Infect. Dis.* **157:**1827–1831.

26. Ryan, R. J., G. Sutherland, K. Coughlan, M. Globan, J. Doultree, J. Marshall, R. W. Baird, J. Pedersen, and B. Dwyer. 1993. A new trichrome-blue stain for detection of microsporidial species in urine, stool, and nasopharyngeal specimens. *J. Clin. Microbiol.* **31:**3264–3269.

27. Schwartz, D. A., R. T. Bryan, K. O. Hewanlowe, G. S. Visvesvara, R. Weber, A. Call, and P. Angritt. 1992. Disseminated microsporidiosis (*Encephalitozoon hellem*) and acquired immunodeficiency syndrome—autopsy evidence for respiratory acquisition. *Arch. Pathol. Lab. Med.* **116:**660–668.

28. Shadduck, J., and E. Greeley. 1989. Microsporidia and human infections. *Clin. Microbiol. Rev.* **2:**158–165.

29. Smith, J., R. McQuay, L. Ash, D. Melvin, T. Orihel, and J. Thompson, Jr. 1976. *Atlas of Diagnostic Parasitology*, vol. 2. *Intestinal Protozoa*. American Society for Clinical Pathology, Chicago.

30. Soave, R., and W. Johnson. 1988. AIDS commentary, *Cryptosporidium* and *Isospora belli* infection. *J. Infect. Dis.* **157:**225–229.

31. Weber, R., R. T. Bryan, R. L. Owen, C. M. Wilcox, L. Gorelkin, and G. S. Visvesvara. 1992. Improved light-microscopical detection of microsporidia spores in stool and duodenal aspirates. *N. Engl. J. Med.* **326:**161–166.

32. Weber, R., H. Kuster, G. S. Visvesvara, R. T. Bryan, D. A. Schwartz, and R. Luthy. 1993. Disseminated microsporidiosis due to *Encephalitozoon hellem*—pulmonary colonization, microhematuria, and mild conjunctivitis in a patient with AIDS. *Clin. Infect. Dis.* **17:**415–419.

33. Weber, R., B. Sauer, R. Luthy, and D. Nadal. 1993. Intestinal coinfection with *Enterocytozoon bieneusi* and *Cryptosporidium* in a human immunodeficiency virus-infected child with chronic diarrhea. *Clin. Infect. Dis.* **17:**480–483.

34. Zierdt, C. 1991. *Blastocystis hominis*: past and future. *Clin. Microbiol. Rev.* **4:**61–79.

Intestinal Helminths

LAWRENCE R. ASH AND THOMAS C. ORIHEL

107

Intestinal helminths are usually diagnosed by detection of eggs or larvae in feces. Characteristics used in identifying eggs (Fig. 1) include size, shape, thickness of shell, special structures of the shell (mammillated covering, operculum, knob, spine), and developmental stage of egg contents (undeveloped, developing, embryonated).

Objects that might be confused with helminth eggs include pollen grains, mushroom spores, vegetable material, and the eggs of mites and free-living nematodes (1). Plant hairs may be confused with nematode larvae. Confusing artifacts found in feces typically can be distinguished from parasite eggs and larvae because artifacts usually do not have precise morphologic characteristics and appropriate sizes; however, until individuals have considerable experience in examining feces, there are many opportunities for mistaking artifacts for parasite eggs or larvae.

In the period after infection is acquired but before helminths have matured to produce eggs, i.e., the prepatent period, infections are difficult to diagnose. Larvae occasionally may be found in sputum as a result of their migration through the lungs, but in most instances, diagnosis during the prepatent phase of an infection can be established only on the basis of clinical symptomatology if at all.

NEMATODES

The nematodes, or roundworms, are small to large, elongate, cylindrical parasites living primarily in the intestinal tract as adult worms. Nematodes typically have life cycles that require no intermediate hosts, although some species may utilize one or more intermediate hosts. All of the nematodes of humans, with the exception of *Strongyloides* spp., are dioecious (have two sexes) and characteristically have four larval stages and an adult stage. The infective stage for the human host varies from the first- to the third-stage larva depending on the species of parasite involved. The infective stages of nematodes are usually within eggs, but in a few species (hookworms, trichostrongyles, *Strongyloides* spp.), the larvae hatch from the eggs and become infective in the soil. Diagnosis of most of the human intestinal nematode parasites depends on finding characteristic eggs in feces. An important exception is *Strongyloides stercoralis* infection, in which first-stage larvae are excreted in feces.

The parasites that commonly infect humans are described below.

Enterobius vermicularis

Infection with *E. vermicularis*, the human pinworm parasite, has worldwide distribution, and its true prevalence is probably considerably underestimated. The organism is primarily a parasite of young children; the rapid development of its eggs to the infective stage and their ability to persist for extended periods on fomites lead to rapid dissemination of the infection from child to child and to adults. Infections are especially common in institutional settings.

Adult pinworms are small, females being 8 to 13 mm long and barely visible to the naked eye as white motile worms on the surfaces of stool specimens or on the perianal skin. Males are only 2 to 3 mm long and usually are not seen. The adult female has a characteristic long, pointed tail, whereas in males, the posterior end is blunt. Adult worms typically live in the cecum, colon, appendix, and rectum. Usually, adult females migrate out of the anal orifice at night and lay their eggs in the perianal area, where they adhere to the skin, hair, or bed clothing and bed linen that come in contact with these eggs. Because they live so far posterior in the intestinal tract, females often lay their eggs on the surface of the fecal mass; as a result, the eggs are not well mixed within the feces and usually will not be detected in routine stool examinations.

Enterobius eggs are elongate and flattened on one side, with a thick, colorless shell. They are 50 to 60 μm long by 20 to 40 μm wide and are partially embryonated when laid. The eggs develop rapidly and become infective within 4 to 6 h, at which time the egg contains a tadpolelike larva (Fig. 2A). Adherence of these eggs to fingers and fomites results in their ready transfer to the mouth and thus further infection.

The infection is especially troublesome in young children; large worm burdens may cause extreme pruritus, loss of sleep, and irritability. Adult females migrating out of the anus may occasionally enter the vagina and subsequently the uterus or fallopian tubes, where they die. Disintegration of the dead worms and liberation of the eggs contained in utero results in an inflammatory response and granuloma formation about the eggs in these sites.

Infection is best diagnosed by the use of a cellulose tape technique (2). In this procedure, a strip of cellulose tape

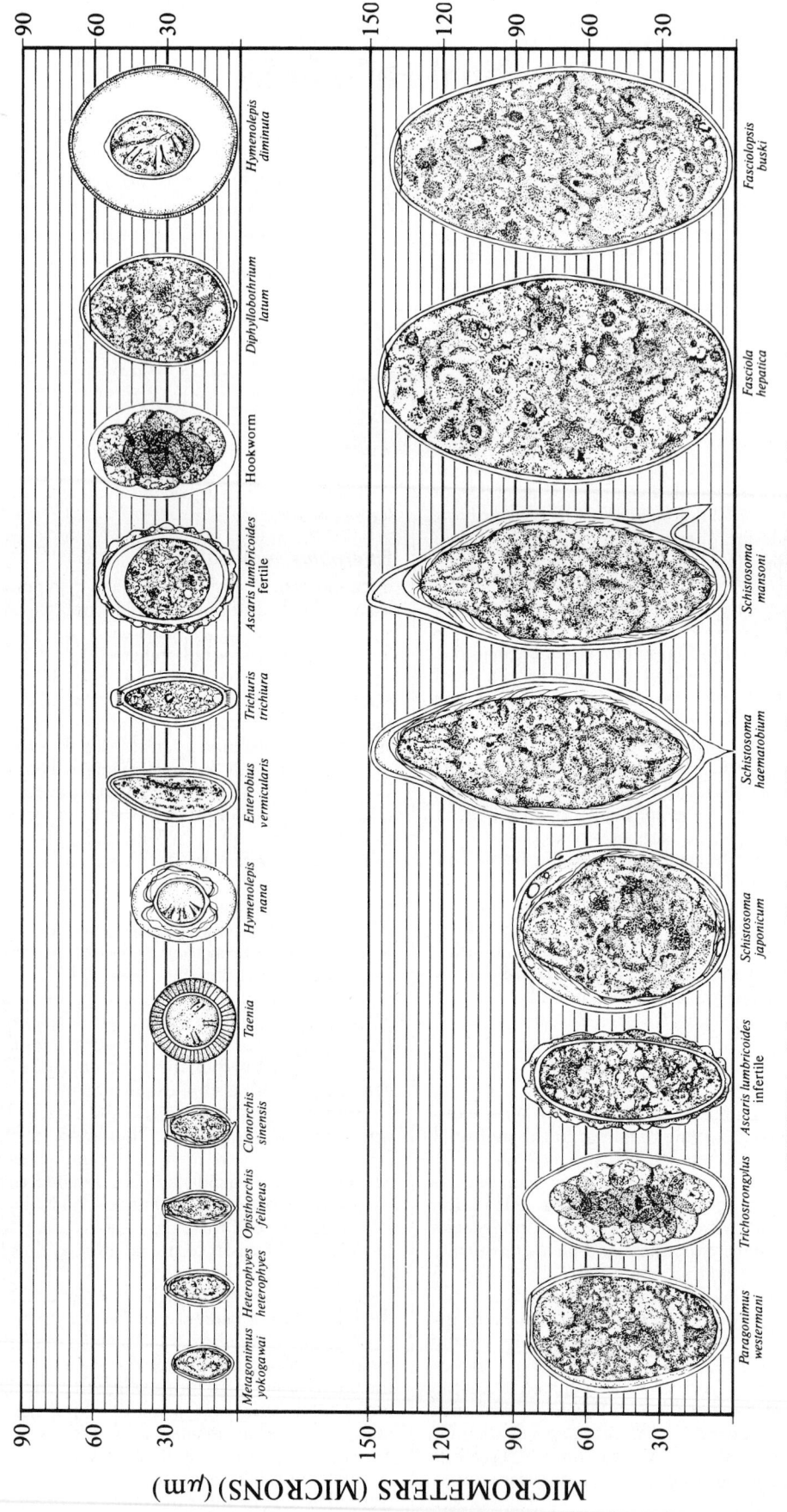

FIGURE 1 Relative sizes of helminth eggs (from Centers for Disease Control and Prevention). *Schistosoma mekongi* and *Schistosoma intercalatum* have been omitted.

MICROMETERS (MICRONS) (μm)

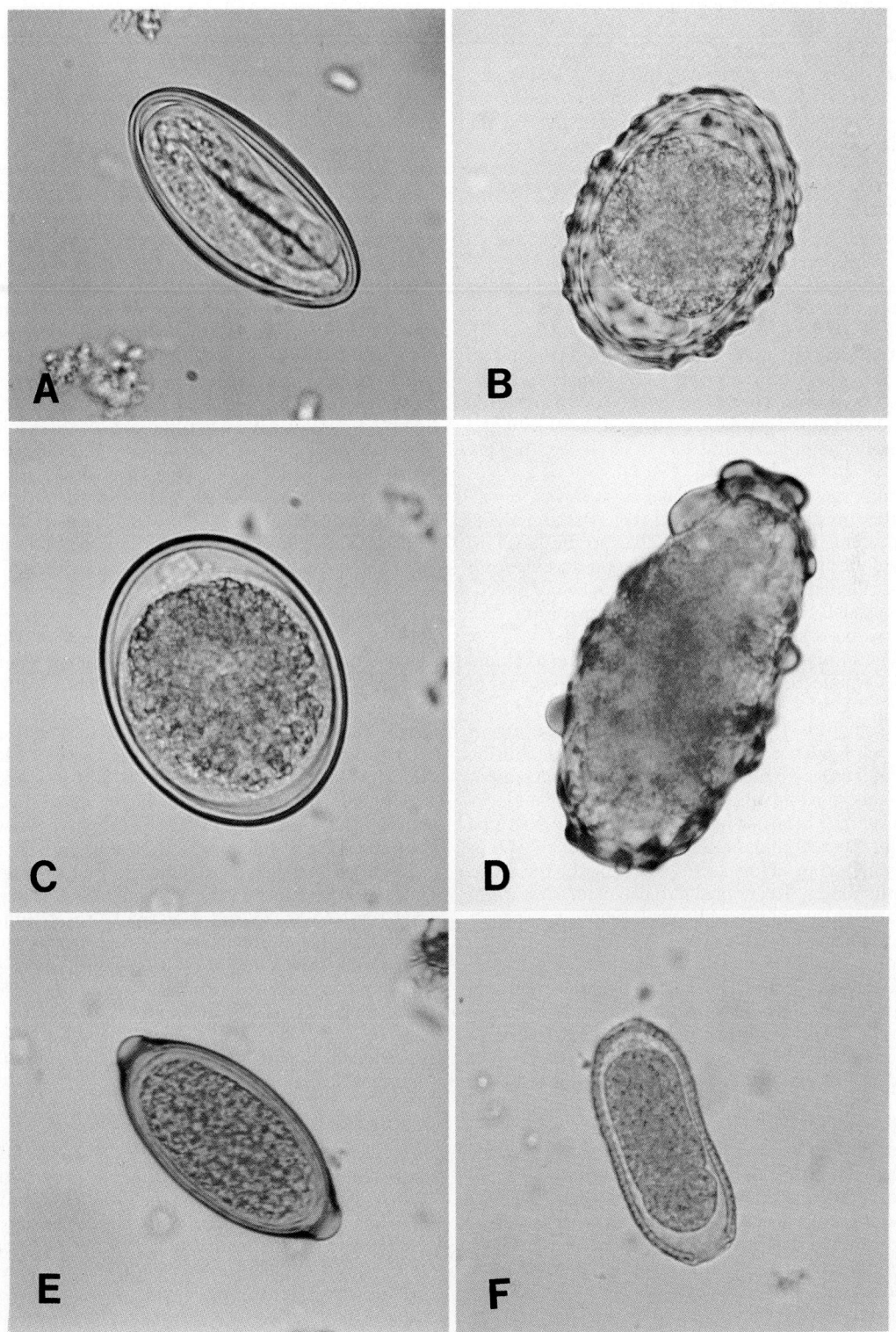

FIGURE 2 Eggs of intestinal nematode parasites (magnification, ×850). (A) Embryonated, infective egg of *E. vermicularis*; (B) fertile egg of *A. lumbricoides*; (C) decorticated fertile egg of *A. lumbricoides*; (D) infertile egg of *A. lumbricoides*; (E) *T. trichiura*; (F) *C. philippinensis*. (Panels A, C, and D are from reference 1; used with permission.)

held adhesive side outward on a microscope slide is pressed firmly against the right and left perianal folds. The tape is then spread back over the slide with the adhesive side down and is examined directly under the microscope. Examinations on multiple days may be required to diagnose infection. Other methods (in particular, the use of anal swabs) have been used to demonstrate this infection; however, there is no clear advantage to swab techniques, and the cellulose tape method remains the most widely used procedure.

Ascaris lumbricoides

The largest and probably the most prevalent of the human intestinal roundworms, A. lumbricoides, has worldwide distribution. The flesh-colored adult worms are large; females usually range from 20 to 35 cm long by 3 to 6 mm wide, and males range from 15 to 31 cm by 2 to 4 mm. Males, with their ventrally curved tails, can be readily distinguished from females, which have straight tails. Females produce large numbers of eggs, perhaps up to 200,000 per female worm per day. These eggs must undergo a developmental period in soil of approximately 2 to 3 weeks before they are infective.

Following ingestion by the human host, infective eggs hatch in the intestine, and the third-stage larvae undergo an obligatory migration through the liver to the lungs. In the course of their migration, they grow and develop for 8 or 9 days to a length of approximately 1 mm before returning to the small intestine to complete maturity. The prepatent period is approximately 2 months.

Pathology in humans can be caused by both larval and adult stages. In infections with large numbers of eggs and in repeated infections, the larval migration phase may result in Ascaris pneumonitis (Loeffler's syndrome), consisting of dyspnea, cough, rales, eosinophilia, and transient, shifting lung infiltrates as seen by X-ray examination. When present in large numbers, adult worms may cause intestinal blockage. However, the presence of small numbers or even one adult worm is potentially dangerous because of their tendency to migrate to ectopic sites, particularly the liver, during febrile illness. The normal life span of an adult worm is approximately 1 year, although female worms may persist for 16 to 20 months.

Infection is usually diagnosed by demonstration of typical fertile eggs in feces. The egg is ovoid, contains a single-celled ovum, and is 55 to 75 μm long by 35 to 50 μm wide. The egg is yellow-brown and has a thick, transparent shell that is covered by a mammillated, albuminoid outer layer (Fig. 2B). Occasionally, the outer mammillated layer is absent; in this circumstance, the eggs are called decorticated eggs (Fig. 2C). Female worms that have never been fertilized or have exhausted their supply of sperm produce infertile eggs. These are elongate (85 to 90 μm long by 43 to 47 μm wide) and have a thin shell that may lack mammillations entirely or may have grossly irregular mammillations scattered over the surface of the shell. These eggs contain a mass of disorganized, highly refractive granules and fat globules of various sizes (Fig. 2D). Because of the large numbers of eggs produced by the female worm, the eggs can usually be found in direct fecal smears and are readily detected by either flotation or sedimentation concentration procedures. Infertile eggs do not float in the standard zinc sulfate solution (specific gravity, 1.18), and they may be missed if only this flotation concentration is used. Adult worms may be spontaneously passed in feces or may emerge from the anus, mouth, or nares; young devel-

oping worms, especially in heavy infections, may be found in feces. The characteristic, prominent three lips at the anterior end of the adult worm aid in identification of the immature or adult parasites.

Trichuris trichiura

Trichuriasis, also known as whipworm infection, is found worldwide in the warmer, moist regions. Adult worms live attached to the wall of the cecum and, less commonly, to the walls of the large intestine, appendix, and lower part of the ileum. Males and females are of similar sizes, ranging from 30 to 50 mm long, and have long, attenuated anterior portions and thicker, short posterior ends. The long, slender anterior end is threaded into the mucosal epithelium, and the posterior portion hangs free in the lumen. Adult worms are long-lived, commonly living for up to 10 years and often longer. Though frequently found in association with A. lumbricoides infection because soil requirements for the development of the two species' infective eggs are similar, Trichuris infections are more frequently seen in older children because of the longer life span of the adult parasites. Egg production by Trichuris females probably does not exceed several thousand eggs per day.

Eggs are barrel shaped and yellow-brown, have thick shells, and usually measure 50 to 55 μm long by 22 to 24 μm wide. At both ends of the egg are prominent, clear, mucoid "plugs" (Fig. 2E). The egg contains an unsegmented ovum when passed in feces, and once the eggs are in the soil, it takes 2 to 3 weeks for the infective, first-stage larva to develop. When infected eggs are ingested, the prepatent period in humans is approximately 3 months.

Light infections with T. trichiura are usually not troublesome, but when large numbers of parasites are present, there may be diarrhea or even dysentery with abdominal cramping; occasionally, rectal prolapse occurs. Heavy infections may result in dehydration, weight loss, and anemia.

Diagnosis, in particular of heavy infections, is readily made by finding the characteristic eggs in direct wet mounts of feces. In light infections, when eggs are few, concentration procedures may be required to find the eggs. In some instances, for reasons not well understood, larger than normal eggs (65 to 83 μm long by 27 to 36 μm wide) are produced (10). In addition, the dog whipworm (Trichuris vulpis) can reach maturity in humans; the eggs of this parasite are larger than those in typical T. trichiura infections, being 72 to 90 μm long by 32 to 40 μm wide. The eggs of T. vulpis, though similar in length to the large eggs occasionally seen in T. trichiura infections, are usually wider and more barrel shaped.

Capillaria philippinensis

C. philippinensis has been recognized since the mid-1960s as causing human infection. Although its geographic distribution was initially restricted to the Philippines and Thailand, occasional cases in recent years have been reported from various parts of the world, including Japan, Taiwan, Egypt, Iran, and Colombia. C. philippinensis is normally a parasite of fish-eating birds, and various fish serve as obligatory intermediate hosts. In areas where the disease is endemic, human infection is acquired by the ingestion of raw or poorly cooked fish harboring infective larvae in their tissues (4). Patent infections develop in approximately 1 month. Female worms may contain thick-shelled unembryonated eggs, thin-shelled embryonated eggs, or first-stage larvae in their uteri; this bizarre variation appears to be dependent on the age of the infection and possibly other factors. Internal autoinfection from the larvae is a normal feature of the life cycle in mammalian hosts, resulting in large worm burdens that cause diarrhea, wasting, dehydration, and, if untreated, death of the host.

Diagnosis of the infection depends on finding the characteristic thick-shelled, unembryonated eggs in feces. The egg is 36 to 45 μm long by 21 μm wide, has a moderately thick striated shell, and possesses inconspicuous mucoid "plugs" at both ends (Fig. 2F). It is not uncommon in individuals with chronic diarrhea for eggs, larvae, and even adult worms to be passed simultaneously in feces.

Hookworm Infections

The two principal human hookworm parasites are *Necator americanus* and *Ancylostoma duodenale*, but other species of the genus *Ancylostoma* may also produce such infections in various parts of the world. *N. americanus* is found in the United States as well as other areas of the world, but *A. duodenale* does not occur in the United States, although its geographic distribution elsewhere frequently overlaps that of *N. americanus*. Hookworm infection is widely distributed in the tropics and subtropics and also extends into moist, temperate climates.

Adult hookworms are characterized by an anterior end modified into a buccal capsule that contains either teeth or cutting plates with which they can anchor themselves to the wall of the small intestine. Male hookworms are further characterized by having posterior ends modified to form an umbrellalike structure referred to as a bursa; this structure is not present in female worms, which have straight, pointed tails. *Necator* adults are 7 to 11 mm long by 0.3 mm wide, and they have buccal capsules provided with cutting plates. *A. duodenale* adults are somewhat larger (8 to 13 mm long by 0.4 mm wide), and they have buccal capsules containing two pairs of teeth.

The thin-shelled eggs of hookworms are partially embryonated when passed in feces, and they are essentially indistinguishable from each other (Fig. 3A): they range from 55 to 75 μm long by 36 to 40 μm wide. In the soil, the eggs embryonate and hatch within 1 to 2 days as first-stage, rhabditoid larvae measuring 250 to 350 μm long by 17 μm wide.

In the hookworm life cycle, the first-stage larvae that hatch in the soil develop into infective, third-stage, filariform larvae in approximately 1 week, and these larvae initiate human infection by direct penetration of the skin. *A. duodenale* infective larvae can also infect via the mouth, but *Necator* spp. cannot; in addition, *Necator* spp. require an obligatory lung migration, whereas if *A. duodenale* infection is acquired orally, there is direct maturation in the intestine to the adult stage. Patent infections develop in 5 to 6 weeks, and the life spans of the adults of both species are usually only 1 to 2 years but may be as long as 10 years or more.

The pathogenesis of hookworm infection is directly related to the worm burden. Light infections are well tolerated and cause few symptoms. Acute, heavy infections may result in fatigue, weakness, abdominal pain, and diarrhea with blood loss; the blood loss is more severe in *A. duodenale* infections. Chronic hookworm infection results in iron deficiency anemia, listlessness, pallor, and general retardation of development in afflicted children.

Diagnosis is accomplished by demonstration of characteristic eggs in feces. Counts of fewer than five eggs per coverslip in wet mounts indicate light infections unlikely to cause anemia. Counts of more than 25 eggs per coverslip

suggest heavy infection. If there has been prolonged delay in examination of the feces (usually more than a day), larvae may develop and hatch, and it is then necessary to differentiate hookworm first-stage larvae from those of *S. stercoralis*, the stage typically passed in the feces in humans with strongyloidiasis. Hookworm rhabditoid larvae have long, narrow buccal chambers (Fig. 3E) and inconspicuous genital primordia. Accurate identification of the hookworm species causing infection depends on recovering and examining adult worms or culturing larval stages to the infective filariform stage, with subsequent morphologic study of these larvae to distinguish between *Necator* and *Ancylostoma* species (2).

Trichostrongylus Species

Human infection with species of the genus *Trichostrongylus* is found throughout the world, in particular in rural areas where herbivorous animals are raised. Adult trichostrongyles typically live in the digestive tracts of sheep, cattle, goats, and other herbivores; the principal species occurring in humans are *Trichostrongylus colubriformis* and *Trichostrongylus orientalis*. Although these infections are rarely troublesome in terms of human disease, they do present occasional diagnostic difficulties, since trichostrongyle eggs resemble hookworm eggs in shape but tend to be much larger.

Adult worms are small and slender, usually measuring less than 1 cm in length; males have a prominent bursa. Trichostrongyle eggs are 75 to 95 μm long by 40 to 50 μm wide. They have colorless, thin shells and are tapered slightly at one end (Fig. 3B). The inner vitelline membrane around the ovum is frequently wrinkled at the tapered end of the egg. The germinal mass does not fill the shell. Eggs passed in feces develop in the soil; third-stage larvae hatch in the external environment, and typically these infective larvae are lying on grass or other vegetation when ingested by the definitive hosts. Infection is acquired by the oral route only; these larvae are not capable of invading skin.

S. stercoralis

Human strongyloidiasis is widely distributed in the tropics and subtropics, but it also extends into moist, temperate regions. Even in areas where the disease is endemic, its distribution is extremely focal, because the existence of the parasite is dependent on a high groundwater table. Adult parasitic females are parthenogenetic, and parasitic males do not occur. The minute females, only 2 to 3 mm long by 30 to 40 μm in diameter, live within the mucosal epithelium of the small intestine, where they produce thin-shelled eggs that embryonate and hatch in this location. The first-stage rhabditoid larvae migrate into the intestinal lumen, enter the fecal stream, and are the usual stage passed in feces.

In the soil, first-stage larvae follow a direct or indirect course of development. In the direct cycle, the larvae rapidly develop into filariform, third-stage infective larvae that can initiate human infection by direct penetration of the skin. Alternatively, in the indirect cycle, the first-stage larvae develop into a free-living generation of adult male and female worms. When these free-living adult worms mate, the female lays eggs that embryonate and hatch in the soil as first-stage larvae; these larvae develop into third-stage filariform larvae, which may then initiate human infection by skin penetration. Although multiple free-living generations in the soil have been suggested to occur, it appears that this is not common.

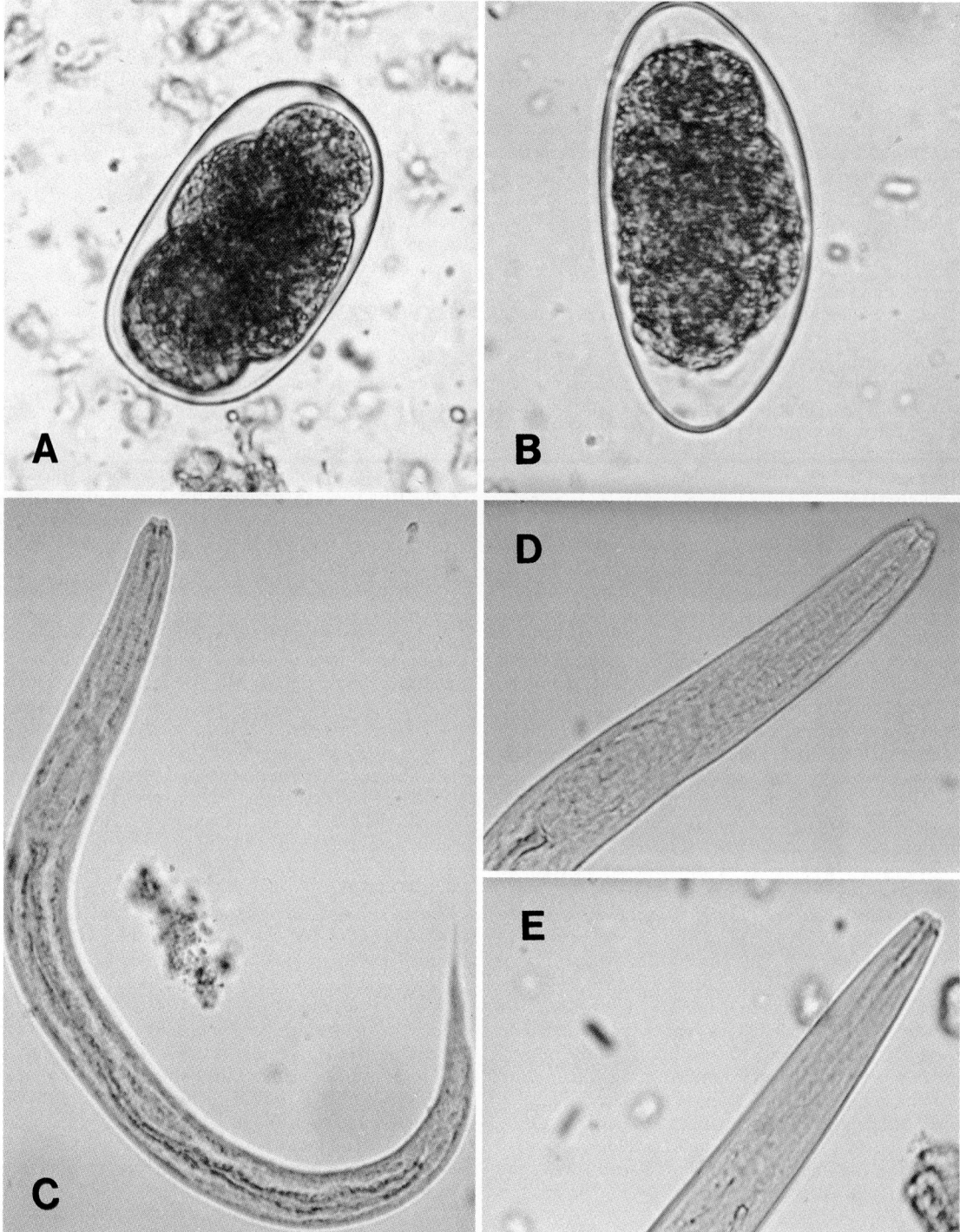

FIGURE 3 Eggs and larvae of intestinal nematode parasites (magnification, ×850). (A) Hookworm egg; (B) *Trichostrongylus* sp.; (C) first-stage larva of *S. stercoralis*; (D) anterior end of first-stage larva of *S. stercoralis* showing short buccal cavity; (E) anterior end of first-stage hookworm larva showing long buccal cavity. (Panels A, C, and E are from reference 1; used with permission.)

Internal autoinfection (hyperinfection) is a common and dangerous sequela in individuals with latent infections who are immunocompromised, receiving immunosuppressive therapy, alcoholic, or malnourished (7). Latent infec-

tions may persist for decades, as evidenced by former World War II prisoners of war who acquired infections in prison camps in Southeast Asia; such individuals continue to experience recurrent bouts of serpigenous urticarial rashes on the trunk, buttocks, and groin that are due to larval migration (8). Hyperinfection results in rapid multiplication of these parasites within the intestinal tract, with subsequent reinvasion of the bowel wall or perianal skin by filariform larvae. In addition to the rapid development of new adult females by this process, the extensive migration of these third-stage larvae into virtually all tissues and organs can result in overwhelming infection characterized by extensive hemorrhage and death if not treated rapidly. Interestingly, although latent strongyloidiasis may be a serious complication in individuals who are immunocompromised, this hyperinfection syndrome does not appear to be an important complication in individuals with AIDS, for reasons not well understood.

Diagnosis may be difficult, especially for individuals with long-standing chronic infections. For patients with no symptoms and few parasites, direct wet mount and standard fecal concentration procedures may fail to reveal the first-stage larvae. Individuals who are candidates for immuno-suppressive therapy (e.g., organ transplantation, treatment for malignancies) and who are from geographic areas where strongyloidiasis is known to be endemic (e.g., many parts of Latin America and Southeast Asia) must be carefully screened prior to treatment. In latent infections, it is common for only small numbers of larvae to be passed in feces, and in many instances, the larvae are passed on an irregular basis. Thus, multiple stool samples taken several days apart should be examined. In addition to normal concentration procedures, examination of the whole fecal specimen by the Baermann procedure is recommended (2). In this procedure, based on the active migration of first-stage larvae out of fecal material into surrounding water, the fecal specimen is placed on top of wire mesh in a large funnel filled with water; the water in the funnel is retained by means of a clamp on rubber tubing at the stem end of the funnel; the feces are allowed to stand for several hours before aliquots of water are drawn off and examined for the presence of larvae. Multiple examinations performed in this manner are likely to detect light infections, if the larvae are present. Duodenal aspiration techniques, including the Enterotest, have also resulted in improved diagnosis of this infection.

First-stage *Strongyloides* larvae are the diagnostic stage found in feces; they measure 180 to 380 μm long by 14 to 20 μm wide. They have a short buccal chamber and a prominent cluster of cells, the genital primordium, that is located at midbody between the intestine and the ventral body wall (Fig. 3C and D). If stool specimens have been allowed to sit in a warm room for 24 h or longer before examination, it may be necessary to distinguish between *Strongyloides* first-stage larvae and the first-stage larvae of hookworms that may have embryonated and hatched from eggs present in the feces. Hookworm first-stage larvae are the same size as *Strongyloides* larvae, but the former have long buccal chambers (Fig. 3E) and their genital primordia are inconspicuous and usually cannot be seen.

Occasionally, filariform *Strongyloides* larvae may be seen in feces or sputum, particularly in cases of hyperinfection. These larvae are approximately 500 to 600 μm long by 16 μm wide; their ratio of esophagus length to intestinal length is 1:1, and the tail of the larva is notched and not pointed as in filariform hookworm larvae.

TREMATODES

Adult trematode parasites of humans live in the intestine, liver, lung, or blood vessels. The flukes, as they are fre-quently called, all have complex life cycles that always involve snails as first intermediate hosts. In addition, many must utilize a second intermediate host in which the infective stage for humans and other definitive hosts develops. The life cycles involve specific freshwater molluscs that are used as first intermediate hosts by each species of trematode. The molluscs are infected by a ciliated larva, the miracidium, which emerges from the trematode egg. Within the tissues of the snails, a complex process of reproduction involving several different parasite stages results in the production of tailed, free-swimming larvae called cercariae. Cercariae are released into water, and some (such as the schistosomes) may infect humans directly; however, in most species, they invade the tissues of a second interme-diate host (e.g., fish, crabs) and develop into the infective, encysted metacercarial stage. Some cercariae attach to various types of aquatic vegetation, and the metacercarial stages become encysted on the plant material. Human infections with the flukes are usually acquired by ingestion of infective stages, although schistosomiasis results from direct skin penetration by cercariae. Trematode infections usually are diagnosed by identification of eggs in feces or, more rarely, in sputum or urine. The eggs of all human trematodes, except the schistosomes, have an operculum through which the miracidium escapes. Small trematode eggs, typically those less than 50 μm long, usually contain a fully developed miracidium when passed in feces, as do the schistosome eggs. Larger trematode eggs are usually undeveloped when excreted and must undergo a period of development in water for several weeks before the miracidium is produced.

Intestinal Flukes

Fasciolopsis buski

F. buski is the largest and most pathogenic of the human intestinal flukes. It occurs in many parts of Asia, including China, India, Indonesia, Taiwan, Thailand, and Vietnam. Pigs and humans are the primary hosts for this parasite, and infection is acquired by ingestion of metacercariae encysted externally on various types of aquatic vegetation (e.g., water chestnuts and the water caltrop). It takes approximately 3 months from ingestion of metacercariae until eggs are found in feces of the human host. The eggs are large, broadly ellipsoidal, and thin shelled and measure 130 to 140 μm long by 80 to 85 μm wide; they have inconspicuous opercula and are unembryonated when passed in feces (Fig. 4A). Although *Fasciolopsis* eggs are similar in size and morphology to those of *Fasciola hepatica*, the sheep, cattle, and human liver fluke, the shell of a *Fasciolopsis* egg at the abopercular end is usually smooth and not roughened as it is in *Fasciola hepatica*. Since *F. buski* is restricted to the Orient, whereas *Fasciola hepatica* is distributed worldwide, it is important to consider the geographic history and clinical symptomatology of the patient in order to establish the correct diagnosis based on eggs found in feces.

Heterophyid Infections

A large number of genera and species of minute intestinal flukes parasitize humans in many parts of the world. For the most part, these infections are of minor medical significance, but the presence of the eggs in feces requires that the eggs be identified and distinguished from the morphologically similar eggs of more pathogenic flukes (e.g., *Clonorchis* and *Opisthorchis* spp.). The most common human hetero-phyid-type parasites are *Heterophyes heterophyes* and *Metag-*

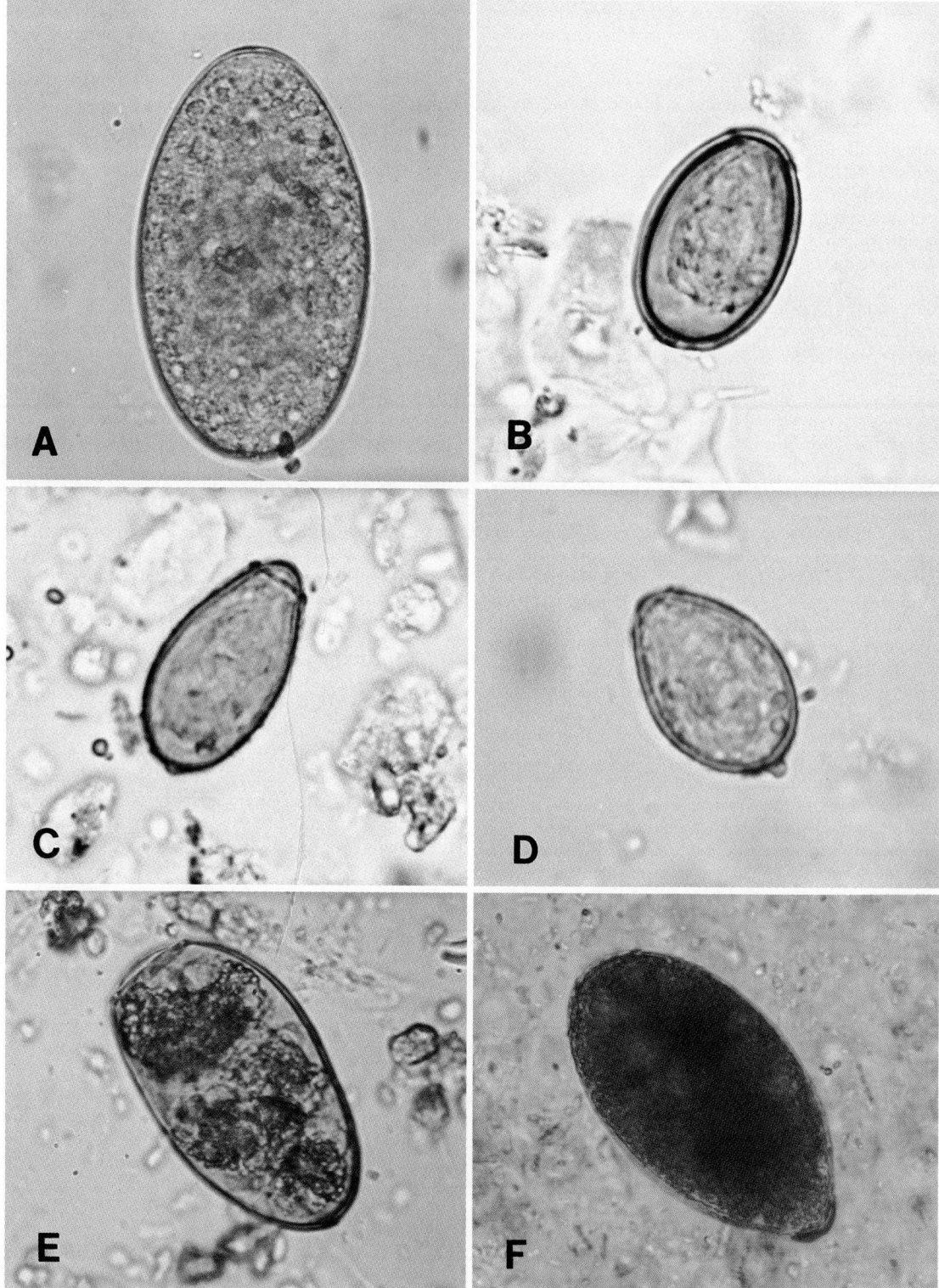

FIGURE 4 Eggs of trematode parasites. (A) *F. buski* (magnification, ×500); (B) *H. heterophyes* (magnification, ×1,500); (C) *Clonorchis sinensis* (magnification, ×1,500); (D) *O. viverrini* (magnification, ×1,500); (E) *P. westermani* (magnification, ×600); (F) *Nanophyetus salmincola* (magnification, ×750). (Panels B through D are from reference 1; used with permission.)

onimus yokogawai. The former occurs in the Orient and the Nile Delta, whereas *Metagonimus* spp. are found not only in the Far East but also in Turkey, the Balkans, and other parts of Europe. However, heterophyid and related heterophyid-like species lack host specificity and will mature in a wide range of mammals and birds that serve as reservoir hosts for human infections. As a consequence, in any particular geographic area where fish are inadequately cooked or eaten raw, the minute intestinal flukes that are found in humans reflect the particular parasite species endemic in such local animals as dogs, cats, rodents, and birds.

Adult flukes are only a few millimeters long and live in the crypts or the superficial mucosal epithelium of the small intestine; in general, they have a life span of several months or less. Patent infections develop within 2 to 3 weeks of ingestion of metacercariae in the flesh of fish. When large numbers of worms are present, they may cause a mild diarrhea and abdominal cramping, but in most cases when small numbers of adults are present, little or no symptomatology is associated with the infection.

The eggs of all the heterophyid and heterophyidlike species are typically small (17 to 30 μm long by 13 to 18 μm wide), have more or less inconspicuous opercula, and are embryonated when discharged in the feces (Fig. 4B). Though somewhat similar in size and morphology to the eggs of *Clonorchis* and *Opisthorchis* spp., heterophyid eggs usually lack the seated opercula seen in the eggs of these liver flukes (5). In addition to *Heterophyes* and *Metagonimus* spp., other genera commonly found in people include *Pygidiopsis* (Korea), *Haplorchis* (Thailand, Vietnam), and *Stellantchasmus* (Hawaii) spp. and numerous others in other areas of the world. Morel mushroom spores, which can be found in human feces, are occasionally misidentified as heterophyid or heterophyidlike eggs; however, these spores lack opercula and do not contain miracidia as do the fluke eggs.

Another small-fluke infection that has been reported from humans in the northwestern United States who ingest raw, incompletely cooked, or smoked salmon is caused by *Nanophyetus salmincola* (6). Although the adult worms are similar in size to the heterophyids, the eggs they produce are much larger (64 to 97 μm long by 43 to 55 μm wide) and are readily distinguishable from those of heterophyids (Fig. 4F).

Liver Flukes

Fasciola hepatica

Fasciola hepatica occurs in sheep- and cattle-raising areas worldwide, and it causes occasional to extensive human infection. Infection is acquired by ingestion of aquatic vegetation, such as watercress in salads, on which metacercariae have encysted. Metacercariae migrate from the intestine to the liver by passing through the intestinal wall into the abdominal cavity, entering the liver by penetration of Glisson's capsule, and migrating through the parenchyma to the bile ducts, where they reach maturity and live as adult worms. The prepatent period is approximately 2 months. As a consequence of the need for this extraintestinal migration, migrating worms also may end up in ectopic locations (e.g., body wall, cutaneous tissues, lungs), where they may cause abscesses or fibrotic lesions. Adult worms are large, fleshy flukes that may cause severe liver damage, particularly when present in large numbers.

Eggs of *Fasciola hepatica* are large, broadly ellipsoid, and 130 to 150 μm long by 63 to 90 μm wide. The operculum

is inconspicuous, the shell appears roughened at the abopercular end, and the eggs are unembryonated when passed in feces. The eggs are similar in size and morphology to those of *F. buski*, but the roughening of the shell at the abopercular end is absent in *F. buski* eggs. Since the eating of parasitized cattle or sheep liver results in the passage of *Fasciola* eggs in feces, it is necessary to rule out spurious infections by examination of the feces several days after individuals have stopped eating liver.

Clonorchis sinensis

The so-called Oriental liver fluke is commonly seen in the United States, particularly in immigrants from Southeast Asia. *Clonorchis sinensis* and the closely related species, *Opisthorchis viverrini*, live as adults in the bile ducts of humans and reservoir host animals, including cats and dogs. Infections are acquired by ingestion of metacercariae encysted under the scales of fish that have been insufficiently cooked. In the human or animal host, metacercariae migrate to the bile ducts of the liver by direct migration via the common bile duct. Adult worms reach maturity and begin to lay eggs approximately 1 month following infection. Infections may persist for 20 years or longer. Pickled fish imported from the Orient have been occasional sources of human infection in the United States. *O. viverrini* is a common infection in Thailand.

Diagnosis of infection depends on finding the characteristic eggs in feces. *Clonorchis* eggs are ovoid, thick shelled, yellowish brown, and 27 to 35 μm long by 12 to 19 μm wide; they are embryonated when passed in feces (Fig. 4C). The eggs have prominent, seated opercula; the abopercular end of the shell usually has a prominent knob or short, commalike protuberance. *O. viverrini* eggs are morphologically similar to those of *Clonorchis sinensis* but tend to be somewhat broader and have less conspicuous shoulders (Fig. 4D). The eggs of these liver flukes are somewhat similar in size and appearance to heterophyid eggs except that the latter are usually smaller and lack a seated operculum and the abopercular knob.

Lung Flukes: Paragonimus westermani

Human lung fluke infections are caused by a number of species of *Paragonimus* in various parts of the world, including Asia, Africa, and Latin America. *P. westermani* is widespread in Asia and is perhaps the most important species causing human disease. Dogs, cats, and other wild animals serve as important reservoirs of human infection for all species. Adult flukes usually live in pairs in fibrous capsules in the lung parenchyma of their hosts. The eggs that are produced pass up the bronchial tree and may be found in sputum or feces. Crabs and crayfish serve as second intermediate hosts, and human infections usually derive from eating raw or poorly cooked infected crustaceans. The practice of marinating raw crab in brine, wine, or vinegar is usually ineffectual in killing metacercariae.

In the human host, metacercariae migrate from the intestine into the body cavity, move through the diaphragm into the thoracic cavity, and then invade the lungs, where the worms mature and begin to lay eggs in 5 to 6 weeks. These infections may be long-lived, frequently 1 to 2 decades or longer. In the course of larval migration from the intestine, the flukes may end up in ectopic locations, such as the rib cage, body wall, and brain. Infection in the brain is frequently fatal. Histologic sections of these ectopic locations frequently demonstrate only eggs and not the adult worms.

Diagnosis depends on finding eggs in feces or, less frequently, sputum. Eggs are broadly ovoid, thick shelled, yellow-brown, and 80 to 120 μm long by 45 to 70 μm wide (Fig. 4E). They have distinct opercula, the shell is distinctly thickened at the abopercular end of the shell, and they are unembryonated when passed in feces. Although the operculate eggs of the fish tapeworm *Diphyllobothrium latum* are sometimes misidentified as those of *P. westermani*, the former are considerably smaller and have a knoblike structure at the abopercular end of the shell.

Several other species of *Paragonimus* have been described as causing human infection: in Africa, *Paragonimus uterobilateralis* and *Paragonimus africanus*, and in the Western Hemisphere, *Paragonimus mexicanus*, *Paragonimus caliensis*, and *Paragonimus ecuadoriensis*. Other species also have been reported from China and Southeast Asian countries. The eggs of all these species differ somewhat in size and morphology from those of *P. westermani*, but they are sufficiently similar that a generic diagnosis usually can be made, with specific identification depending to a certain extent on where the infection was found.

Blood Flukes (Schistosomes)

Schistosomiasis (bilharziasis) afflicts more than 250 million people in the world and therefore is, along with malaria, one of the most important of all human parasitic diseases. The etiologic agents are markedly different from the other human trematodes. The schistosomes have separate sexes and live in blood vessels of the abdominal cavity. The three most important human species are *Schistosoma mansoni*, *Schistosoma japonicum*, and *Schistosoma haematobium*; other species of lesser importance include *Schistosoma mekongi* (Asia) and *Schistosoma intercalatum* (Africa). Though each of the schistosomes utilizes specific and different snail intermediate hosts, their life cycles are similar. Each produces thin-shelled eggs that lack opercula and contain miracidia when excreted in feces or urine. The egg shell of each species typically has a lateral or terminal spine that aids in identification. Miracidia hatch from the eggs and penetrate appropriate snails to establish infection. Infected snails produce fork-tailed cercariae that are liberated into water and directly penetrate the skin to establish infections in human and animal hosts. In the human host, the larval blood flukes migrate through the lungs and become established in venous blood vessels of the mesenteries or bladder, where they mature and mate and the females deposit eggs. The eggs make their way through the wall of the intestine or bladder and are excreted in feces or urine. Pathology due to schistosome infection is primarily due to egg deposition in tissues, with resulting granuloma formation.

Schistosoma mansoni

Schistosoma mansoni has the widest geographic distribution of the schistosomes; it is found in Africa, the Arabian peninsula, Brazil, Puerto Rico, and some islands in the Caribbean. Adult worms live in the portal system of the liver and the small venules of the lower ileum and colon. Eggs are laid in the blood vessels, make their way through the wall of the intestine, and are passed in feces. A *Schistosoma mansoni* egg is 114 to 175 μm long by 45 to 70 μm wide; it contains a miracidium, and the shell has a prominent lateral spine (Fig. 5A and B). The prepatent period for the infection is approximately 6 weeks. In acute schistosomiasis, blood and mucus appear in feces along with the lateral-spined eggs. In chronic schistosomiasis, eggs accumulate in the walls of the intestine, rectum and liver;

correspondingly, fewer eggs are found in feces, and concentration procedures are required to diagnose the infection reliably. Rectal biopsies of the mucosa may be useful in chronic infections (2).

Schistosoma japonicum

Schistosoma japonicum is found in China, the Philippines, and countries of Southeast Asia. A zoophilic strain of the parasite infects animals but not humans in Taiwan, and the infection has been virtually eliminated from Japan, where occasional infections may be found in cattle. Animal reservoirs of infection are important and include water buffaloes, pigs, dogs, cats, and wild rodents. Adult worms live in mesenteric veins, and the prepatent period is 5 to 6 weeks. The embryonated eggs found in feces are round to ovoid, lack opercula, and are 70 to 100 μm long by 55 to 65 μm wide. The thin shell has a small, inconspicuous spine that frequently is difficult to see (Fig. 5C). In addition, the surface of the egg often has fecal debris adhering to it, and this debris can obscure the egg and make it difficult to recognize. Rectal biopsy is an important diagnostic tool when fecal examinations are negative. Female worms produce larger numbers of eggs than the other schistosome species, resulting in extensive pathology in the wall of the intestine and liver. Because of the size of the eggs, they may be disseminated widely in the body via the vascular system. The brain and spinal cord are often involved, and in the Philippines, severe epileptic seizures are a frequent feature of this infection.

Schistosoma haematobium

Schistosoma haematobium occurs in Africa, Lebanon, Syria, Iran, the Arabian peninsula, and Malagasy, where it causes urinary schistosomiasis. Adult worms reside in the venous plexuses of the bladder, and eggs that are laid move through the wall of the bladder and are passed in urine. In chronic infections, accumulation of eggs in the bladder wall can lead to bladder and ureter pathology. Symptomatology includes hematuria, difficulty with micturition, and development of hydroureter and hydronephrosis; renal complications and renal failure are common sequelae of infection. *Schistosoma haematobium* infection is believed to be a predisposing factor for squamous cell carcinoma of the bladder (9). The elongated, thin-shelled egg is 112 to 170 μm long by 40 to 70 μm wide and contains a miracidum; there is a terminal spine (Fig. 5E). Diagnosis typically is made by examination of urine. Eggs can sometimes be found in feces and in the wall of the rectum as well as in the bladder wall.

Schistosoma intercalatum

The human schistosome *Schistosoma intercalatum* occurs in Zaire, Gabon, Cameroon, and the Central African Republic. The egg of this species has a terminal spine, is 140 to 240 μm long by 50 to 85 μm wide, and is found in feces (Fig. 5F). Because the eggs of *Schistosoma haematobium* also sometimes occur in feces, it is necessary to differentiate *Schistosoma intercalatum* from *Schistosoma haematobium*; the egg of *Schistosoma intercalatum* not only is larger but also frequently has an equatorial bulge, and the terminal spine is more pointed, slightly curved, and longer than that of *Schistosoma haematobium*.

Schistosoma mekongi

Closely related to *Schistosoma japonicum*, *Schistosoma mekongi* is a parasite of humans and dogs in countries bordering on the Mekong River, especially Laos and Cam-

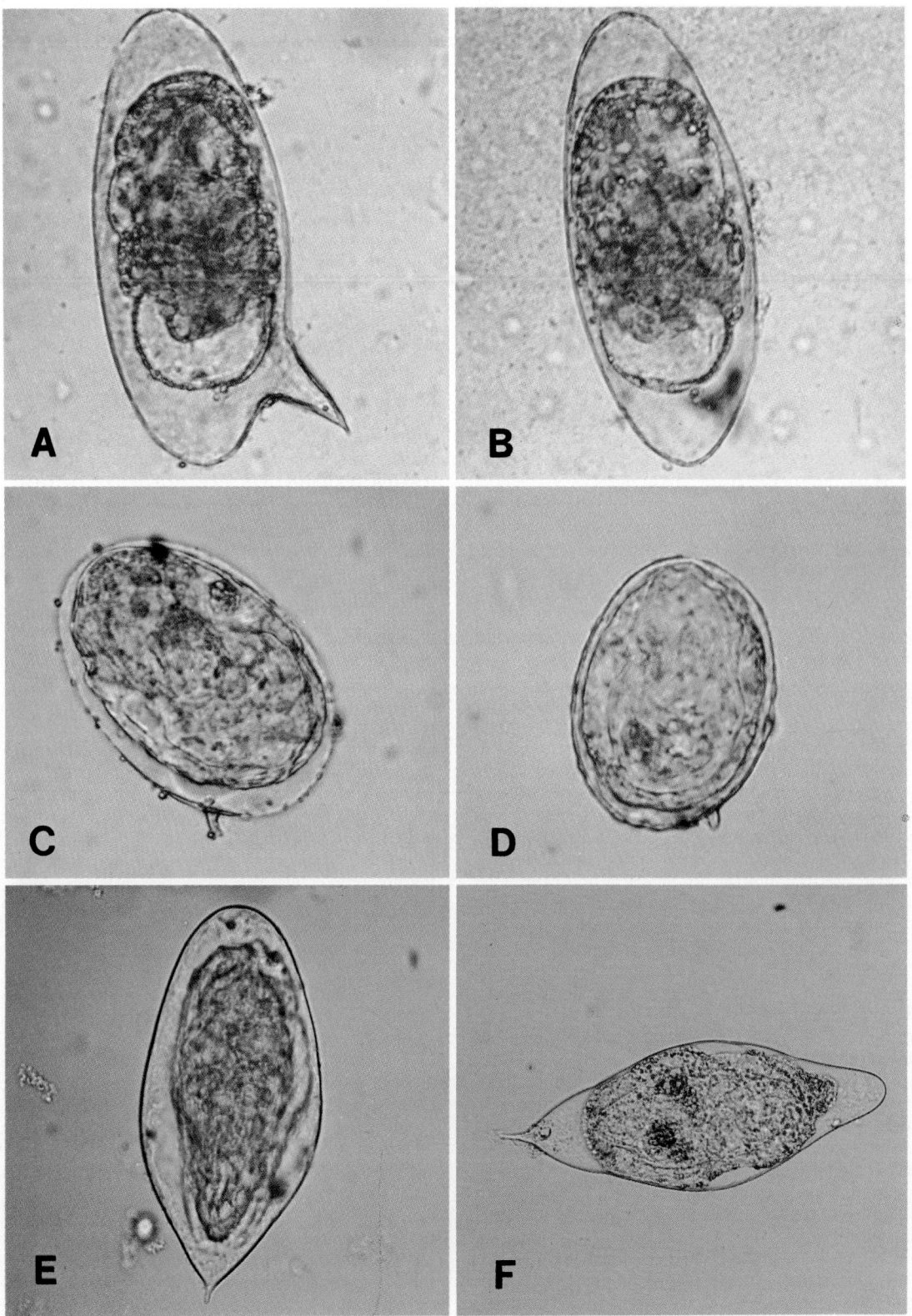

FIGURE 5 Eggs of schistosome species (magnification, ×600). (A) *Schistosoma mansoni*; (B) *Schistosoma mansoni* egg with typical lateral spine not in view; (C) *Schistosoma japonicum*; (D) *Schistosoma mekongi*; (E) *Schistosoma haematobium*; (F) *Schistosoma intercalatum*. (Panels C through E are from reference 1; used with permission.)

bodia. The egg is similar in morphology to that of *Schistosoma japonicum* but is smaller, ranging from 51 to 78 μm long by 39 to 66 μm wide (Fig. 5D). Eggs from dogs are usually smaller than those from humans. The knoblike spine on the egg may be difficult to see.

CESTODES

The four most common adult tapeworm parasites of the human small intestine are *D. latum*, *Taenia saginata*, *Taenia solium*, and *Hymenolepis nana*. Two other tapeworms, primarily animal parasites, reach maturity in humans and may occasionally cause infections; they are *Hymenolepis diminuta* and *Dipylidium caninum*. In addition to the adult tapeworms, a number of larval cestodes can produce serious human disease, including cysticercosis (*Taenia solium*), hydatid disease (*Echinococcus* species), and coenurosis (*Taenia multiceps*); these larval tapeworms occurring in human tissues are discussed in chapter 108 of this Manual. The large adult tapeworms, *D. latum*, *Taenia saginata*, and *Taenia solium*, usually occur singly in the small intestine and may live 20 years or longer. Diagnosis of tapeworm infections is achieved by finding characteristic eggs, proglottids, or both, depending on the species involved, in feces.

All of the adult tapeworms parasitizing humans require an intermediate host with the exception of *Hymenolepis nana*, which may infect humans directly by eggs or indirectly by using beetle intermediate hosts. Morphologically, adult tapeworms have a scolex at the anterior end and a short, undifferentiated neck region that gives rise to the proglottids; the main body (strobila) of the tapeworm adult consists of the neck and immature, mature, and gravid proglottids. As tapeworms increase in age and size, the most posterior gravid proglottids may break off or disintegrate and pass in feces. Some tapeworms (*D. latum*, *Hymenolepis nana*, and *Hymenolepis diminuta*) lay eggs that are passed in feces; for others (*Taenia saginata* and *Taenia solium*), their gravid proglottids typically break off and are passed in feces or actively migrate out of the anus. Sometimes, *Taenia* proglottids rupture in the intestine, and eggs then appear in feces. The eggs of all species of *Taenia* and the related genus *Echinococcus* are morphologically identical and cannot be distinguished from one another. The eggs of *Taenia solium*, *Taenia multiceps*, and *Echinococcus* species can infect humans directly and cause disease by their larval stages; the eggs of *Taenia saginata* are not directly infective to humans.

Diphyllobothrium latum

Known as the fish tapeworm, *D. latum* differs from other adult tapeworms infecting humans in its morphology, biology, and epidemiology. Its geographic distribution includes areas with cold, clear lakes such as Scandinavia, other areas of northern Europe, the former USSR, northern Japan, and North America, principally the upper Midwest, Alaska, and Canada. The adult parasite can attain a length of 10 to 15 m, is ivory in color, and has a scolex that is provided with shallow grooves (bothria) rather than suckers on its dorsal and ventral aspects.

In its life cycle, unembryonated operculate eggs resembling those of trematodes are passed in feces and must undergo embryonation in water for several weeks. Ciliated, six-hooked embryos (coracidia) hatch from these eggs and must be ingested by appropriate species of freshwater copepods. Within the copepod, a solid-bodied larval stage, the procercoid, develops and becomes infective to the second intermediate host, fish. In fish, the procercoids migrate into

the flesh and develop to the plerocercoid (sparganum) stage, which is then infective to human or animal hosts. After ingestion of the sparganum, it takes 3 to 5 weeks for the adult tapeworm to attain maturity and begin to lay eggs. The parasite may produce no clinical symptoms in some people, but when it becomes large, it may cause mechanical obstruction of the bowel, may cause diarrhea and abdominal pain, and in some individuals, particularly in northern European countries, may be responsible for a vitamin B_{12} deficiency that results in pernicious anemia.

Diagnosis is made by finding the characteristic operculate eggs in feces. They are 58 to 75 μm long by 44 to 50 μm wide. The abopercular end frequently has a small knoblike protrusion (Fig. 6F). Individuals with long-standing infection are likely to have large numbers of eggs in feces. As infections grow older, one or a small chain of proglottids may break off and be passed in feces. Proglottids passed in feces are wider than long (3 by 1 mm), and the genital pore is situated on the midventral surface rather than laterally as in the other human tapeworms. In freshly passed proglottids, the coiled uterus in the center of the proglottid is yellow-brown.

Taenia saginata

Taenia saginata is known as the beef tapeworm and is distributed worldwide, but it is especially prevalent in Mexico, South America, eastern and western Asia, and many countries in Europe. Cattle serve as the intermediate host, and ingestion of eggs from contaminated pasturelands by grazing cattle results in development in cattle tissue of the infective cysticercus stage (*Cysticercus bovis*).

The cysticercus is 4 to 6 mm long by 7 to 10 mm wide and has a pearllike appearance in tissues; the unarmed scolex is invaginated into a fluid-filled bladder. After ingestion of the cysticercus in raw or poorly cooked beef, it takes approximately 2 to 3 months for the infection to become patent in the human host. An adult tapeworm attains a length of 4 to 8 m and has a scolex provided with four suckers and an unarmed rostellum; gravid proglottids are longer than they are wide (18 to 20 by 5 to 7 mm). Each proglottid has a genital pore at the midlateral margin. In mature proglottids, the ovary has only two lobes, and a vaginal sphincter muscle is present. Gravid proglottids, which are highly muscular and active, break off from the strobila and can actively migrate out of the anus. Although patients may exhibit no symptomatology with this infection, the mature worm may cause abdominal discomfort, diarrhea, and occasionally intestinal obstruction as a result of its large size.

Diagnosis of species usually is made by identification of gravid proglottids that have been passed in feces or have actively migrated out of the anus. Identification of the proglottids is based on morphology of the uterus, which can be demonstrated after injection with India ink or staining with carmine or hematoxylin stains (2). In *Taenia saginata*, there are 15 to 20 lateral branches on each side of the central uterine stem. If the proglottids rupture in the intestine, the typical *Taenia* eggs can be seen in feces. A taeniid egg is spherical and has a thick, yellow-brown, prismatic shell (Fig. 6A). Within the egg is a six-hooked embryo, the oncosphere. Occasionally, especially when eggs are liberated directly from proglottids, there is a thin outer membrane around the egg (Fig. 6B). Eggs are 31 to 43 μm in diameter.

It is generally accepted that the eggs of *Taenia saginata* are not directly infective to humans, but caution should be

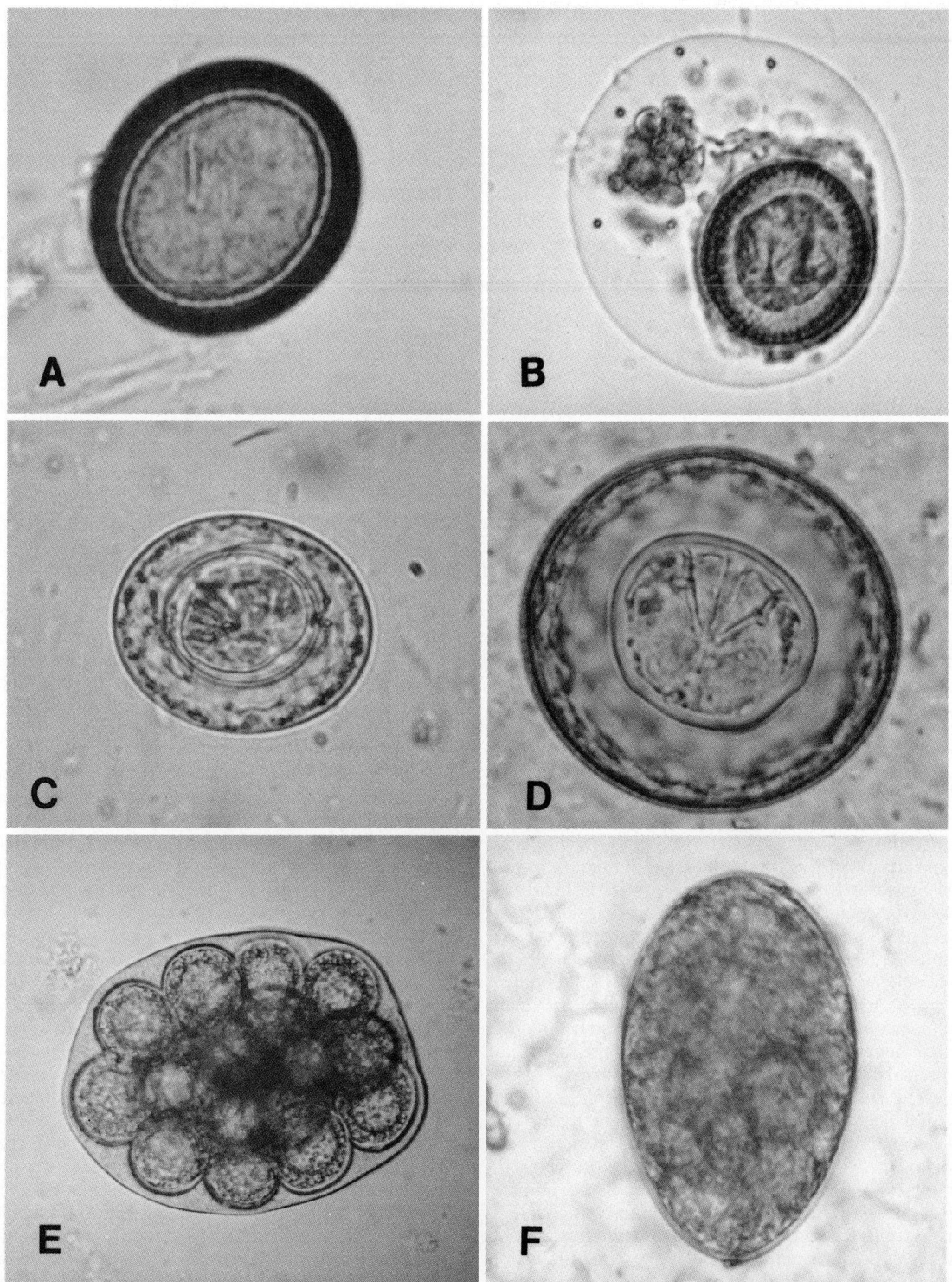

FIGURE 6 Eggs of cestodes. (A) *Taenia* species. Eggs of all species of *Taenia* and *Echinococcus* are identical (magnification, ×1,000). (B) *Taenia* egg surrounded by the primary membrane frequently seen around eggs directly liberated from gravid proglottids (magnification, ×800); (C) *Hymenolepis nana* (magnification, ×900); (D) *Hymenolepis diminuta* (magnification, ×900); (E) egg packet of *Dipylidium caninum*; (F) *D. latum* (magnification, ×800). (Panels B through F are from reference 1; used with permission.)

exercised in the handling of all proglottids and taeniid eggs, since the eggs of *Taenia solium*, *Taenia multiceps*, and *Echinococcus* spp. are directly infective to humans and can cause cysticercosis, coenurosis, and hydatidosis, respectively. In Taiwan, the Philippines, Korea, and perhaps parts of Southeast Asia, there is a human tapeworm morphologically identical as an adult worm to *Taenia saginata* (3). However, with this Asian species, the cysticercus stage is limited to the livers of pigs and less frequently of cattle, and unlike the typical *Taenia saginata* cysticercus, the scolex has hooklets. The adult tapeworm is not distinguishable from typical *Taenia saginata*; it even lacks hooks on the scolex. It has been suggested that this tapeworm may represent a new species or subspecies, and studies to clarify the status of this parasite are in progress.

Taenia solium

Known as the pork tapeworm, *Taenia solium* has an extensive geographic distribution throughout Europe, Mexico, Central and South America, China, and India. It is no longer commonly found in the United States, although recent immigrants from Mexico and Latin America, in particular, may commonly harbor the parasite. Human infection is acquired by ingestion of infective cysticerci (*Cysticercus cellulosae*) in poorly cooked pork or pork products. The adult worm may reach lengths of 2 to 7 m; it has a scolex with four suckers and a rostellum armed with two rows of hooklets. In mature proglottids, the ovary has two lobes and an accessory lobe (the accessory lobe is lacking in *Taenia saginata*), and a vaginal sphincter muscle is lacking (it is present in *Taenia saginata*). Gravid proglottids have 7 to 13 lateral branches off the central uterine stem. Since the eggs of *Taenia solium* are infective to humans and can cause cysticercosis, extreme caution in the handling of these proglottids is recommended.

Hymenolepis nana

Hymenolepis nana is the smallest of the adult tapeworms infecting humans, attaining lengths of 2.5 to 4.0 mm, and is the most common tapeworm infection of humans in the United States. It is normally a parasite of mice, in which the life cycle characteristically involves various beetles as intermediate hosts, and transmission to humans is usually accomplished by direct ingestion of infective eggs containing oncospheres. When eggs are ingested, a solid-bodied larva, a cysticercoid, first develops in the wall of the small intestine; subsequently, the larva migrates back into the intestinal lumen, where it reaches maturity as an adult tapeworm in 2 to 3 weeks. In beetles that ingest eggs of *Hymenolepis nana*, the cysticercoids develop in the body cavity and have thick protective walls about them. Although humans may acquire infection by accidental ingestion of infected beetles (often occurring in dry cereals), direct infection is far more common and is the primary reason *Hymenolepis nana* usually occurs in institutional and familial settings where hygiene is substandard. A feature of human *Hymenolepis nana* infection is the opportunity for internal autoinfection with the parasite, which may result in large worm burdens. Autoinfection occurs when eggs discharged by adult tapeworms in the lumen of the small intestine hatch rapidly and invade the wall of the intestine; here, cysticercoids are formed, and they subsequently reenter the intestine to mature to adult worms.

Diagnosis of the infection rests on finding the spherical to subspherical embryonated eggs in feces. The egg is 30 to 47 μm in diameter and thin shelled, and it contains a six-hooked oncosphere that lies in the center of the egg and is separated from the outer shell by considerable space (Fig. 6C). The oncosphere is surrounded by a membrane that has two polar thickenings from which arise four to eight filaments extending into the space between the embryo and the outer shell. Proglottids are rarely seen, since gravid proglottids do not ordinarily break off from the main strobila of the adult worm.

Hymenolepis diminuta

An occasional human parasite, *Hymenolepis diminuta* is primarily a parasite of rats. Beetles and other arthropods serve as obligatory intermediate hosts, with humans generally acquiring infection accidentally by ingestion of infected meal beetles present in various grains and cereals. Cysticercoids, the infective stage, develop in the hemocoels of beetles after ingestion of eggs. As an adult, *Hymenolepis diminuta* may be 20 to 60 cm long, and numerous tapeworms may be present in the same host. In humans, hyperinfection or direct infection by ingestion of eggs, as occurs with *Hymenolepis nana*, has not been recorded.

Diagnosis of infection is by demonstration of eggs in feces. The egg is spherical and large (70 to 85 by 60 to 80 μm) and has a yellow-brown, moderately thick shell (Fig. 6D). The six-hooked embryo in the egg is central and is considerably separated from the outer membrane; however, there are no polar thickenings or filaments such as those in the eggs of *Hymenolepis nana*. As in *Hymenolepis nana* infection, proglottids do not usually break away from the strobila of the adult worm and are not usually a diagnostic consideration as in human large-tapeworm infections.

Dipylidium caninum

Dipylidium caninum is the most common and widespread adult tapeworm of dogs and cats. Human infection has been reported in many parts of the world. Children are more frequently infected than adults as a result of their more intimate contact with dogs and their fleas, which serve as obligatory intermediate hosts. Because the infection is not a troublesome one and is self-limiting, there are probably many more cases than are reported in the literature.

Adult tapeworms may be present in considerable numbers and may vary in length from 10 to 70 cm. The scolex is conical, with four prominent suckers and a small, retractile rostellum that bears multiple rows of small spines. A gravid proglottid is elongate (23 by 8 mm), has a genital pore on both lateral margins (hence the name double-pored dog tapeworm), and is divided into small compartments, each of which contains 8 to 15 oncospheres that are enclosed in a thin, embryonic membrane. In dogs, the white proglottids are frequently passed in feces or may be seen dangling from the anus as small chains; on carpeting and floors, these segments undergo dehydration and resemble grains of rice. In the life cycle, wormlike larval fleas in soil or carpeting ingest the eggs in the proglottids, and cysticercoids develop in the hemocoel. When the larval fleas metamorphose into adult fleas, the cysticercoids remain viable, and animal or human infections usually are acquired by ingestion of the adult fleas. Adult tapeworms reach maturity in the small intestine in approximately 1 month.

Diagnosis of infection is usually made by finding the typical double-pored, compartmented proglottids in feces or by finding egg packets containing multiple oncospheres liberated by disintegration of the proglottids (Fig. 6E).

REFERENCES

1. **Ash, L. R., and T. C. Orihel.** 1990. *Atlas of Human Parasitology*, 3rd ed. ASCP Press, Chicago.
2. **Ash, L. R., and T. C. Orihel.** 1991. *Parasites: a Guide to Laboratory Procedures and Identification.* ASCP Press, Chicago.
3. **Bowles, J., and D. P. McManus.** 1994. Genetic characterization of the Asian *Taenia*, a newly described taeniid cestode of humans. *Am. J. Trop. Med. Hyg.* **50:**33–44.
4. **Cross, J. H.** 1992. Intestinal capillariasis. *Clin. Microbiol. Rev.* **5:**120–129.
5. **Ditrich, O., M. Giboda, T. Scholz, and S. A. Beer.** 1992. Comparative morphology of eggs of the Haplorchiinae (Trematoda: Heterophyidae) and some other medically important heterophyid and opisthorchiid flukes. *Folia Parasitol.* **39:**123–132.
6. **Eastburn, R. L., T. R. Fritsche, and C. A. Terhune, Jr.** 1987. Human intestinal infection with *Nanophyetus salmincola* from salmonid fishes. *Am. J. Trop. Med. Hyg.* **36:**586–591.
7. **Genta, R. M.** 1992. Dysregulation of strongyloidiasis: a new hypothesis. *Clin. Microbiol. Rev.* **5:**345–355.
8. **Pelletier, L. L.** 1984. Chronic strongyloidiasis in World War II Far East ex-prisoners of war. *Am. J. Trop. Med. Hyg.* **33:**55–61.
9. **Smith, J. H., and J. D. Christie.** 1986. The pathobiology of *Schistosoma haematobium* infection in humans. *Hum. Pathol.* **17:**333–345.
10. **Yoshikawa, H., M. Yamada, Y. Matsumoto, and Y. Yoshida.** 1989. Variations in egg size of *Trichuris trichiura. Parasitol. Res.* **75:**649–654.

Tissue Helminths

THOMAS C. ORIHEL AND LAWRENCE R. ASH

108

A large number of helminth parasites, including nematodes, flukes, and tapeworms, live in human tissues as adults or larvae. Some are natural parasites of humans, while a larger number are zoonotic species. Diagnosis of patent infections usually depends on identification of the parasite's reproductive products discharged in blood, feces, or other body fluids or, in the case of larval parasites, on recovery of the parasite itself from the tissues. The nematodes are represented by a diversity of species, many of which are not well known. The trematodes are equally diverse in both structure and tissue habitats. These organisms have already been covered in large part in the preceding chapter on intestinal parasites, because their eggs are passed in feces and urine. The cestodes are represented by larval tapeworms, which cause cysticercosis, coenurosis, sparganosis, and hydatid disease.

Although these parasites are relatively easily identified when they are intact, recognition and identification of them in histopathologic material is often challenging.

NEMATODES

Filariae

The filarial worms are arthropod-transmitted parasites of the lymphatic, subcutaneous, and cutaneous tissues of humans. All share a unique characteristic: the adult female worm produces a primitive larva called a microfilaria, which is found in the peripheral blood or the skin. Certain species of microfilariae circulate in the blood with a well-defined circadian rhythm or "periodicity" that may be nocturnal or diurnal; other species lack periodicity and are found in the peripheral blood at all hours of the day and night. When absent from the peripheral blood, microfilariae are to be found in the deeper visceral capillaries. Because the adult worms are typically sequestered in the tissues, diagnosis of infection depends on finding microfilariae in blood or skin, depending on the species.

The microfilaria is relatively simple in its organization and structure. It is vermiform and in stained preparations appears to be composed of a column of nuclei that is interrupted along its length by spaces and special cells, which are the precursors of body organs or organelles. Some species of microfilariae are enveloped in a sheath, whereas in others, the sheath is absent (1).

All of the filariae are transmitted by species of bloodsucking arthropods such as mosquitoes, midges, blackflies, and tabanid flies, in which the microfilaria develops to the infective stage. Subsequent development of the infective larva to the gravid, adult stage in the vertebrate host requires several months, in some cases as long as a year or more. Although these parasites are not endemic in humans in the United States, they are often seen in immigrants or in individuals who have resided or traveled in areas where the parasites are endemic. Several species of filariae infect humans.

Wuchereria bancrofti, which produces an infection frequently referred to as bancroftian filariasis, is the most common species of filaria infecting humans. It is widely distributed throughout tropical and subtropical areas of the world. The adult worms live in the lymphatic system (see Fig. 3A) and produce lymphangitis, lymphadenitis, and obstructive fibrosis, which restricts the flow of lymph, with resultant lymphedema. Long-term chronic infection may result in elephantiasis of the extremities and genitalia and, in males, hydrocele. The microfilaria circulates in the peripheral blood with a nocturnal periodicity in most regions of the world; however, in the South Pacific, it is essentially without any periodicity. The microfilaria is sheathed, lies in smooth curves in stained smears, and measures about 298 μm in length by 7.5 to 10.0 μm in diameter. The column nuclei are dispersed; there is a short headspace, and the pointed tail is devoid of nuclei (Fig. 1A). In Giemsa stain, the sheath stains faintly or not at all. The microfilaria must be distinguished from other sheathed microfilariae. This is accomplished most easily on the basis of the arrangement of nuclei, particularly in the tail (Fig. 2). Since microfilariae may be present in the blood only in small numbers, sensitive procedures such as thick blood films, saponin lysis, Knott concentration, and membrane filter concentration are used routinely to detect infections (2).

Brugia malayi is another mosquito-borne filaria that inhabits the lymphatic system of humans. Its geographical distribution is restricted to Asia and the Indian subcontinent. In some regions, it is coendemic with *W. bancrofti*. Lymphatic pathology similar to that produced in chronic bancroftian filariasis occurs in chronic infections with this parasite; however, hydrocele is not a common complication in males. The microfilaria, which circulates in the blood, may be periodic or subperiodic, usually the former. It is

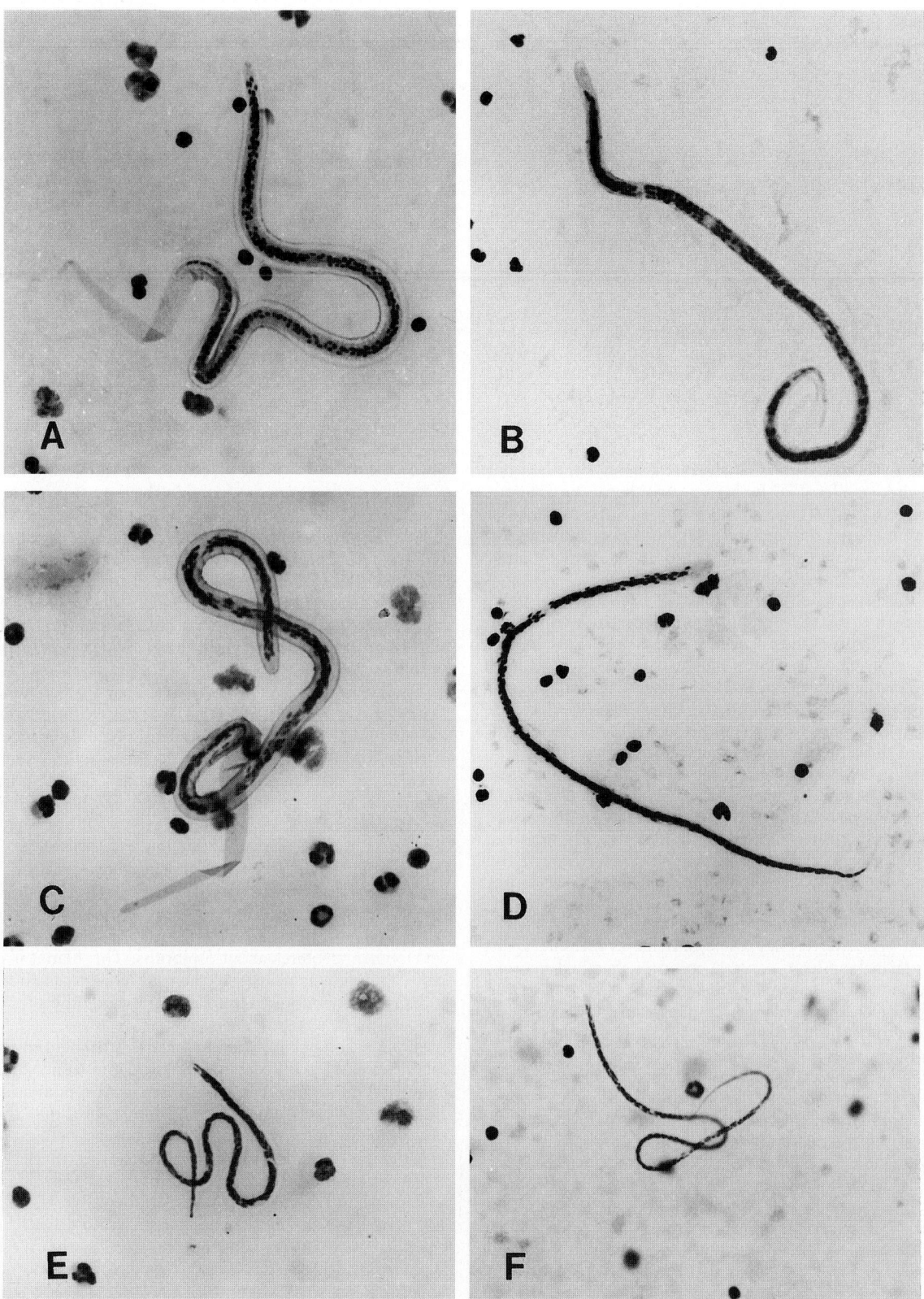

FIGURE 1 Common microfilariae found in humans. Hematoxylin stain, Magnification, ×400. (A) *W. bancrofti*; (B) *B. malayi*; (C) *L. loa*; (D) *O. volvulus*; (E) *M. perstans*; (F) *M. ozzardi*.

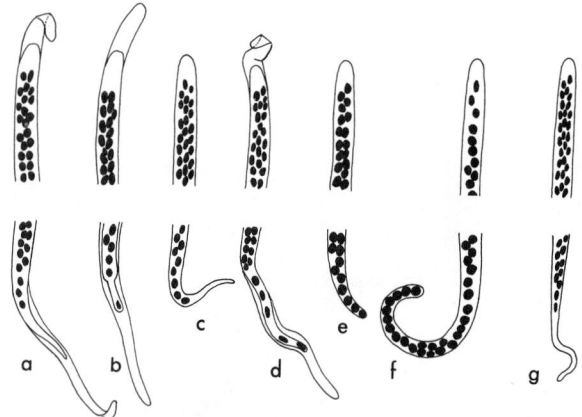

FIGURE 2 Diagrammatic representation of the anterior and posterior extremities of the common microfilariae found in humans. (a) *W. bancrofti*; (b) *B. malayi*; (c) *O. volvulus*; (d) *L. loa*; (e) *M. perstans*; (f) *M. streptocerca*; (g) *M. ozzardi*.

similar in morphology to *W. bancrofti*, being sheathed but somewhat smaller (270 by 5 to 6 μm). It can be differentiated from the *W. bancrofti* microfilaria by the presence of subterminal and terminal nuclei in the tail (Fig. 1B and 2b). With Giemsa stain, the sheath stains a bright pink, whereas that of *W. bancrofti* does not (1).

A third species of lymphatic filaria, *Brugia timori*, infects humans in the eastern end of the Indonesian archipelago, particularly the islands of Timor and Flores. This species produces a microfilaria very similar to that of *B. malayi*, with conspicuous subterminal and terminal nuclei. The two can be most easily differentiated on the basis of size. *B. timori* is larger, measuring more than 300 μm in length; also, its sheath tends not to stain with Giemsa stain.

Loa loa, a common filarial parasite of humans, is endemic only in West and Central Africa. It is often referred to as the "eye worm," because the adult worms, which live in the subcutaneous tissues, often migrate into the orbit and the conjunctivae. The adult worms move freely through the tissues, often producing transient inflammatory reactions referred to as "Calabar swellings." The microfilariae circulate in the blood, often in very large numbers, with a diurnal periodicity. They are sheathed and measure up to 300 μm in length. In contrast to other sheathed microfilariae we have discussed, nuclei extend to the end of the tail; however, they are somewhat irregularly arranged along the length of the tail (Fig. 1C and 2d). The sheath of the microfilaria does not stain with Giemsa stain. When adult worms enter the conjunctivae, they can be extracted surgically. Diagnosis of infection depends on identification of the microfilaria in daytime blood films or removal of adult worms from the conjunctivae.

Onchocerca volvulus is an important human filarial parasite in both hemispheres. It is endemic across Central Africa, in a small area in the Middle East (Yemen), in Central America (Mexico, Guatemala), and in northern South America. The adult worms are embedded in fibrous nodules in the subcutaneous tissues and frequently in deeper tissues. These nodules, or "onchocercomata," may be found on the head, trunk, and extremities; their anatomical location frequently correlates with the geographical strain of the parasite. The microfilaria lives in the skin. It lacks a sheath and measures approximately 309 μm in

length by 5 to 9 μm in diameter. The tail is tapered, usually bent or flexed, and without nuclei (Fig. 1D and 2c). The microfilaria is sometimes found in the blood and urine, typically after treatment with diethylcarbamazine. Diagnosis is made by finding the typical microfilaria in skin snips teased in water or saline solution, fluids expressed from scarified skin, or aspirates from nodules (1). Also, adult worms may be demonstrated in excised nodules that have been sectioned and stained (Fig. 3B). Adult worms may be freed from the fibrous tissues of the nodule by using digestive enzymes such as collagenase. Microfilariae may also be seen in the cornea and in the anterior chamber of the eye viewed with the aid of a slit lamp.

Skin snips are best obtained by biopsy punch. Samples may be taken from the scapular region, the iliac crest, and the calf; the first two are the preferred sites. Teasing the skin snips in saline or tap water tends to liberate the microfilariae from the tissues.

Mansonella streptocerca is another skin-dwelling filaria that infects humans in the rain forest belt of Africa. This parasite, previously allocated to various genera (*Dipetalonema*, *Acanthocheilonema*, and *Tetrapetalonema*) at different times, has recently been placed in the genus *Mansonella* (14). The adult worms are found in the dermal layers of the skin, as are the microfilariae. The microfilaria has no sheath, is long and slender, and measures approximately 210 by 5 to 6 μm. Its most characteristic feature is its "crooked" tail (Fig. 2f); in addition, the column of nuclei extends to the end of the tail. In areas where the distribution of this species overlaps that of *O. volvulus*, great care must be given to proper identification of microfilariae found in skin snips.

Two other species of *Mansonella* are parasites of humans. One of these, *Mansonella ozzardi*, is restricted in geographical distribution to the Western Hemisphere. It is endemic in parts of Mexico, Panama, and northern South America, especially in the Amazon Basin and as far south as northern Argentina. It is also found in several Caribbean islands, including Hispaniola. The adult worms inhabit subcutaneous tissues, and the microfilariae circulate in the blood. The microfilaria is small, measuring about 224 by 4 to 5 μm, and has a long attenuated tail devoid of nuclei (Fig. 1F and 2g). It circulates in the blood at all hours of the day and night. According to studies of experimental animals, prepatent development requires about 5 months. This filaria, like *L. loa*, readily infects visitors to areas in which it is endemic and is often encountered in missionaries and others residing temporarily in endemic areas.

Mansonella perstans, a filaria widely distributed in tropical Africa and less so in South America, is also encountered frequently in individuals who have lived temporarily in areas where it is endemic. The location of the adult worms in the human host is less well known, but it is believed that they inhabit the abdominal cavity and the mesenteries. The microfilaria has no sheath, is small (approximately 203 by 4 to 5 μm), and circulates in the peripheral blood without any periodicity. Its most characteristic feature is its blunt tail filled with nuclei (Fig. 1E and 2e). In the Western Hemisphere, *M. perstans* is often found in association with *M. ozzardi*. In areas where the two species coexist, special care must be taken to make accurate identifications.

In general, *Mansonella* species are regarded as innocuous filariae that produce little or no pathology in the human host.

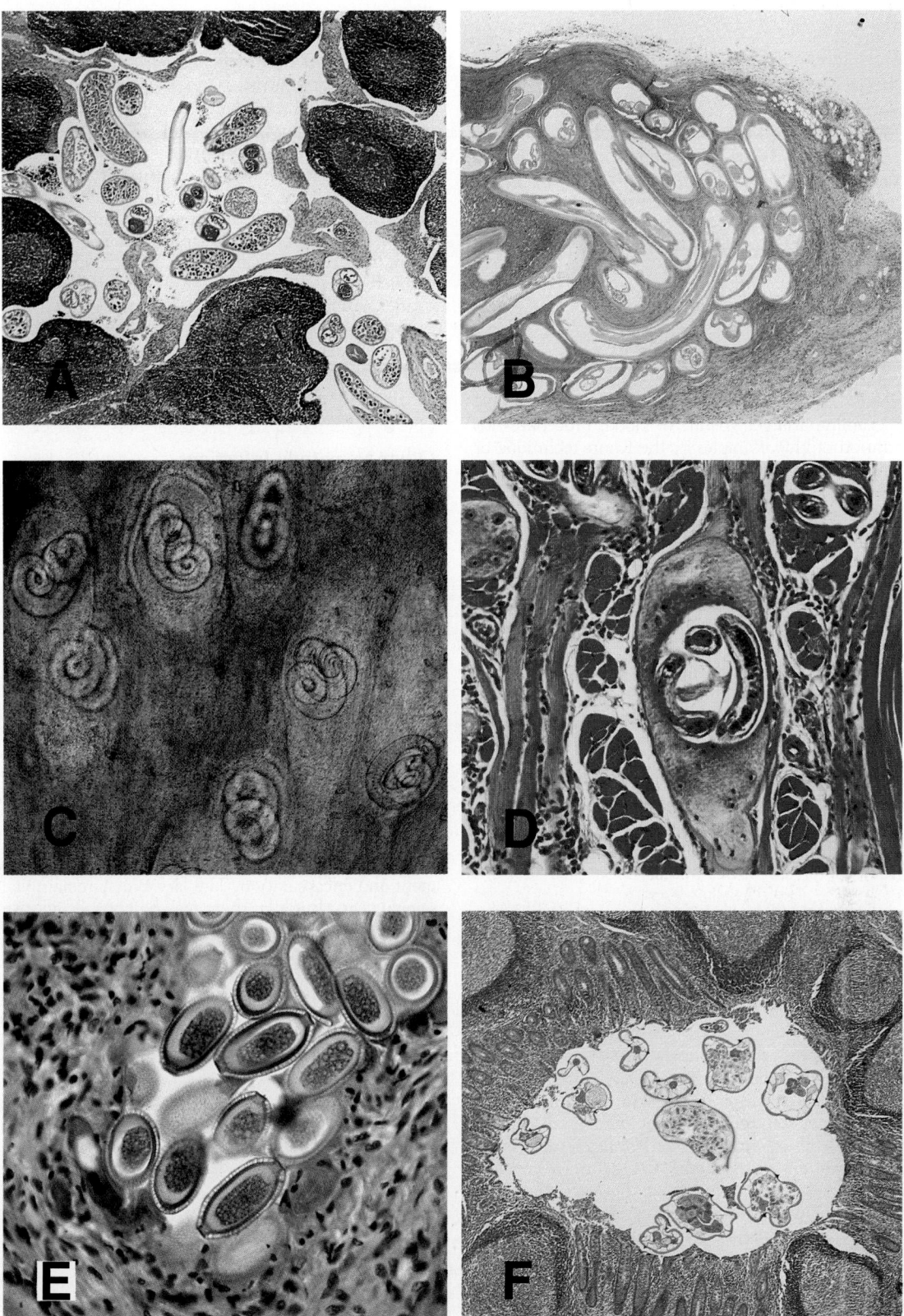

FIGURE 3 (A) Adult *W. bancrofti* in a sinus of a lymph node removed from an infected individual. Hematoxylin and eosin stain; magnification, ×50. (B) Adult *O. volvulus* enmeshed in a fibrous nodule removed from the subcutaneous tissues of an individual in West Africa. Hematoxylin and eosin stain; magnification, ×20. (C) *T. spiralis* infective-stage larvae. In this press preparation of diaphragm muscle, one can see several encapsulated larvae. Magnification, ×100. (D) *T. spiralis* infective-stage larvae in section of rat tongue. Hematoxylin and eosin stain; magnification, ×200. (E) *C. hepatica* eggs in liver. Characteristic morphologic features of eggs are evident even in tissue sections. Hematoxylin and eosin stain; magnification, ×290. (F) *Enterobius vermicularis* adult worms in lumen of human appendix. Hematoxylin and eosin stain; magnification, ×50.

1247

Laboratory Diagnosis

Microfilariae can be detected in samples of blood by a variety of techniques. In thick, wet blood films prepared during either the day or the night, microfilariae may be quickly recognized by their size and rapid movement among the blood cells. However, species are not likely to be identified in this manner, so stained blood films are required. Thin blood films generally are inadequate because of the small amount of blood involved. On the other hand, blood films of 20-μl volume are large enough to detect even scanty numbers of microfilariae, and when these are stained with Giemsa or hematoxylin stains, the important diagnostic morphologic features of each species generally can be seen.

Sometimes, because the numbers of microfilariae in the thick blood films may be very scanty, the microscopist may wish to examine a larger volume of blood. Procedures for concentration of blood samples by the Knott technique or membrane filtration are found in most laboratory guides (2). Species of microfilariae are readily identified on the basis of size, presence or absence of a sheath, and the structure of the tail and its staining characteristics with Giemsa stain.

Skin snips must be examined in order to find microfilariae of *O. volvulus* and *M. streptocerca*. Care must be taken to obtain bloodless snips in order not to contaminate the sample with other species of microfilariae that may be present in the blood. One may simply abrade a small area of skin and collect the exuded tissue juices on a slide; these may be examined immediately for living microfilariae or allowed to dry and then stained and examined. Serologic tests have poor sensitivity and specificity and generally are not used for diagnosis. Two conceptually different techniques, i.e., antigen detection assays and antibody detection assays, for the diagnosis of filariasis, especially lymphatic filariasis, are currently being evaluated in research laboratories (see chapter 103 of this Manual).

Guinea Worm

The nematode parasite *Dracunculus medinensis* dates back to ancient Egyptian literature (15th century B.C.). It is still widely distributed in India, Pakistan, and at least 17 countries in Africa. The adult worms live in subcutaneous tissues, and infections become clinically evident when the female migrates to the body surface, usually on the feet and ankles, and produces a blister on the skin. On contact with water, the blister ruptures, and the female discharges swarms of motile larvae into the water. Only a small portion of the female worm is extruded from the lesion. However, it can be and usually is removed from the body by gentle traction. If the female fails to reach the surface of the body, it will die in the tissues and often calcify in situ. The male worms, which are very small, are rarely seen and also die in the tissues. Secondary bacterial infections of the lesions can disable the individual and complicate recovery.

The life cycle involves transmission of the parasite by a biological vector, a copepod, which ingests larvae liberated by the female into the water. Larvae develop to infectivity in the copepod and are transmitted when the infected copepod is ingested by humans in drinking water. Diagnosis depends on the appearance of the female at the surface of the skin. Guinea worm infection can be controlled easily by provision of piped water or covered wells for drinking water.

Enterobius vermicularis

Although the pinworm is an intestinal parasite and is treated as such in this Manual, the propensity of the female worm for migration from the anus and perianal tissues into other organs warrants mention here. The adult female worm may migrate into the reproductive tract of a human female and become encapsulated in the uterus or Fallopian tubes and may wander even further into the peritoneal cavity, often becoming encapsulated in the mesenteries. Adult female worms also have been found in the liver parenchyma, pulmonary nodules, spleen, lymph nodes, and ovaries. Adult worms are seen with the greatest frequency in the lumen of the appendix, where there may be little or no evidence of inflammation and no clinical manifestations of their presence (Fig. 3F). This parasite is easily recognized on the basis of its anatomical features as seen in tissue sections (12).

Other Nematode Infections

Several species of nematode parasites that are natural parasites of lower animals may gain entry into the human host and undergo partial development. The degree of development achieved by the parasite and the attendant pathology that is manifested vary with the parasite and may be significant and even life-threatening. Only the more common parasite species will be mentioned here.

Trichinella spiralis

T. spiralis, the common agent of human trichinosis, is a parasite of carnivores that shows little evidence of host specificity. Infections in humans result from the ingestion of insufficiently cooked or raw pork or of pork products containing the encysted larvae. Bear meat is also a known source of human infections. Adult worms live in the mucosa of the small intestine. The female produces and discharges larvae that enter the bloodstream and invade the skeletal musculature, where they undergo further development and encapsulation. The larvae may remain viable for several years. Initially, there may be nonspecific gastroenteritis, fever, eosinophilia, myositis, and circumorbital edema. The adult worms survive in the intestine for up to 6 to 8 weeks and then are expelled by the host.

The definitive diagnosis is made by demonstration of encapsulated larvae in biopsy samples of skeletal muscle, particularly deltoid and gastrocnemius muscles (Fig. 3C and D).

The digestion of muscle tissue in artificial gastric juice followed by examination of the sediment for larvae is a more sensitive test. Serologic tests are widely used with good results. A highly antigenic parasite, *T. spiralis* stimulates a very strong antibody response that can be measured by a variety of serologic procedures.

Capillaria hepatica

Though normally a parasite of rodents, *C.* (=*Calodium*) *hepatica* can produce infections in humans. There are well-documented human cases from all over the world. Most infections are found in children of dirt-eating ages, who present with a clinical picture of visceral larva migrans. Human and animal infections are acquired by ingestion of infective eggs in soil; adult worms mature and deposit eggs directly into the liver parenchyma, where they remain in an undeveloped state until the liver is eaten and the eggs are digested free of the tissues to pass in the feces of the predator animal. If, as in the human host, the liver is not eaten, the eggs are never liberated into the external environment. Diagnosis is established by liver biopsy or at necropsy. In genuine human cases, eggs are never passed in the feces (3). Eggs of C. *hepatica* passed by humans indicate a spurious infection and represent an instance in which the livers of infected animals have been eaten. Eggs are passed usually for several days and then disappear from the feces.

The eggs of C. *hepatica* must be distinguished from those of *Trichuris* spp. or other human *Capillaria* spp. Eggs are 51 to 67 μm long by 30 to 35 μm wide, have thick striated walls and inconspicuous "plugs" at both ends, and are unembryonated when seen in the feces. Eggs in liver biopsy specimens are readily recognized on the basis of their characteristic morphologic features (Fig. 3E).

Anisakis Species

The growing popularity of raw-fish dishes—sushi, sashimi, ceviche, and others—has led to an increased frequency of human infections with various larval ascaridoid nematodes (11). Species in a number of genera, including *Anisakis* and *Pseudoterranova*, are parasites, as adults, of marine mammals, utilizing shrimplike crustaceans as first intermediate hosts and fish and squid as second intermediate hosts in their life cycles. When they are ingested, the larval stages encysted in the mesenteries or flesh of fish have the ability to pass from one fish to another without maturing to adult worms. When humans accidentally ingest the immature stages of these nematodes, the larvae may partially penetrate the wall of the stomach or the intestine and produce an eosinophilic granuloma. Many human infections present with acute abdomen or other signs suggestive of intestinal obstruction. In some instances, the long, whitish worms, up to several centimeters in length, may be coughed up by the patient or be removed from the throat.

Diagnosis of the nematodes producing anisakiasis can be accomplished by study of the distinctive morphologic features of the intact worms or by examination of the microanatomical features of the parasites in histologic sections (Fig. 4E).

Eustrongylides Species

Nematode larvae of the genus *Eustrongylides*, parasites belonging to the family Dioctophymoidea, have in recent years been incriminated as agents of human infections. These parasites normally occur in the adult stage in the alimentary tracts of fish-eating birds and utilize fish, amphibians, and reptiles as intermediate hosts. Although five of the cases occurred in individuals who had eaten live minnows while fishing, one case occurred as a result of eating home-prepared sushi (17). In most instances, these large, bright red larvae invade the abdominal cavity of a patient and require surgical removal. These parasites, along with the anisakine nematodes, further illustrate the potential for acquisition of parasitic infections by eating raw infected fish.

Angiostrongylus cantonensis and *Angiostrongylus costaricensis*

Metastrongylid nematodes of the genus *Angiostrongylus* have become important human parasites (3). A. *cantonensis*, a lungworm that lives in the pulmonary arteries of rats (*Rattus* spp.), is a cause of human eosinophilic meningitis or eosinophilic meningoencephalitis. Human infections have been reported in many Pacific Islands, including Hawaii, and in Thailand, Indonesia, other parts of Southeast Asia,

and Cuba. Although the parasite has been found in rats in Puerto Rico and New Orleans, human cases have not been reported. In the natural life cycle, rats excrete first-stage larvae in the feces, and these larvae penetrate directly into or are ingested by many species of terrestrial snails or slugs. Larvae reach the infective third stage in the molluscs, and when the latter are eaten by rats, the larvae make a prolonged month-long migration through the brain before they mature to adults in the pulmonary artery. Here, the worms mate, and the females lay eggs that will develop and hatch in the lung tissue; larvae will be found in the feces 7 weeks after rats ingest infected molluscs.

Infective larvae in terrestrial molluscs may be eaten by and remain in the tissues of a wide range of vertebrate and invertebrate animals, including planarians, shrimp, crabs, fish, amphibians, and reptiles. Human infections are rarely derived from eating ordinary terrestrial molluscs; instead, infection is acquired by eating raw or poorly cooked infected shrimp, crabs, or large edible snails such as *Pila* spp. Accidental infection also may result from eating planarians or small slugs on improperly washed vegetables, such as lettuce, or on fruits, such as strawberries.

In the human host, larvae migrate to the brain, spinal cord, or eye, where they become immature adults. Rarely can the worms complete their migration to the pulmonary artery; instead, they usually die in the meninges or the parenchyma of the brain, giving rise to meningeal symptoms. Diagnosis is usually based on a history of residence in an area where the parasite occurs combined with the development of appropriate clinical symptoms such as severe and prolonged headaches, nerve involvement, and the presence of eosinophils in cerebrospinal fluid. Occasionally, immature worms may be found in aspirated cerebrospinal fluid. Serologic diagnosis has not been reliable.

A closely related parasite, A. *costaricensis*, causes human abdominal angiostrongyliasis in most countries of Central and South America. The adult parasite normally lives in small arteries and arterioles of the ileocecal region of the intestines of various rodents. Eggs are produced and develop in the intestinal wall, and first-stage larvae then migrate into the intestinal lumen to pass in the feces. Various terrestrial slugs serve as intermediate hosts; as is the case for A. *cantonensis*, it appears that most snails and slugs support development of the parasite to the infective third stage. Human infection is acquired by accidental ingestion of slugs or other hosts; transport hosts probably play an important role as well. Most human infections occur in children, but all age groups may be infected. The distal small intestine is the usual habitat of the adults and is the site where eggs are discharged and where they may stimulate granulomatous inflammation leading to symptoms of acute abdomen. Surgery is often performed, and the affected segment of the intestine may be resected; infection is subsequently diagnosed by finding eggs in the tissues. Larvae have not been found in human feces. Though serologic tests have been described, their reliability is still in question.

Gnathostoma Species

The gnathostomes are spiruroid nematodes that are natural parasites of a variety of mammals, including dogs, cats, swine, and wild animals such as raccoons, otters, and opossums, that feed on fish and other cold-blooded vertebrates. Accidental human infections with *Gnathostoma* species typically involve extraintestinal migration of larval stages that produces a visceral larva migrans syndrome (16). Human infections are especially common in Southeast Asia and

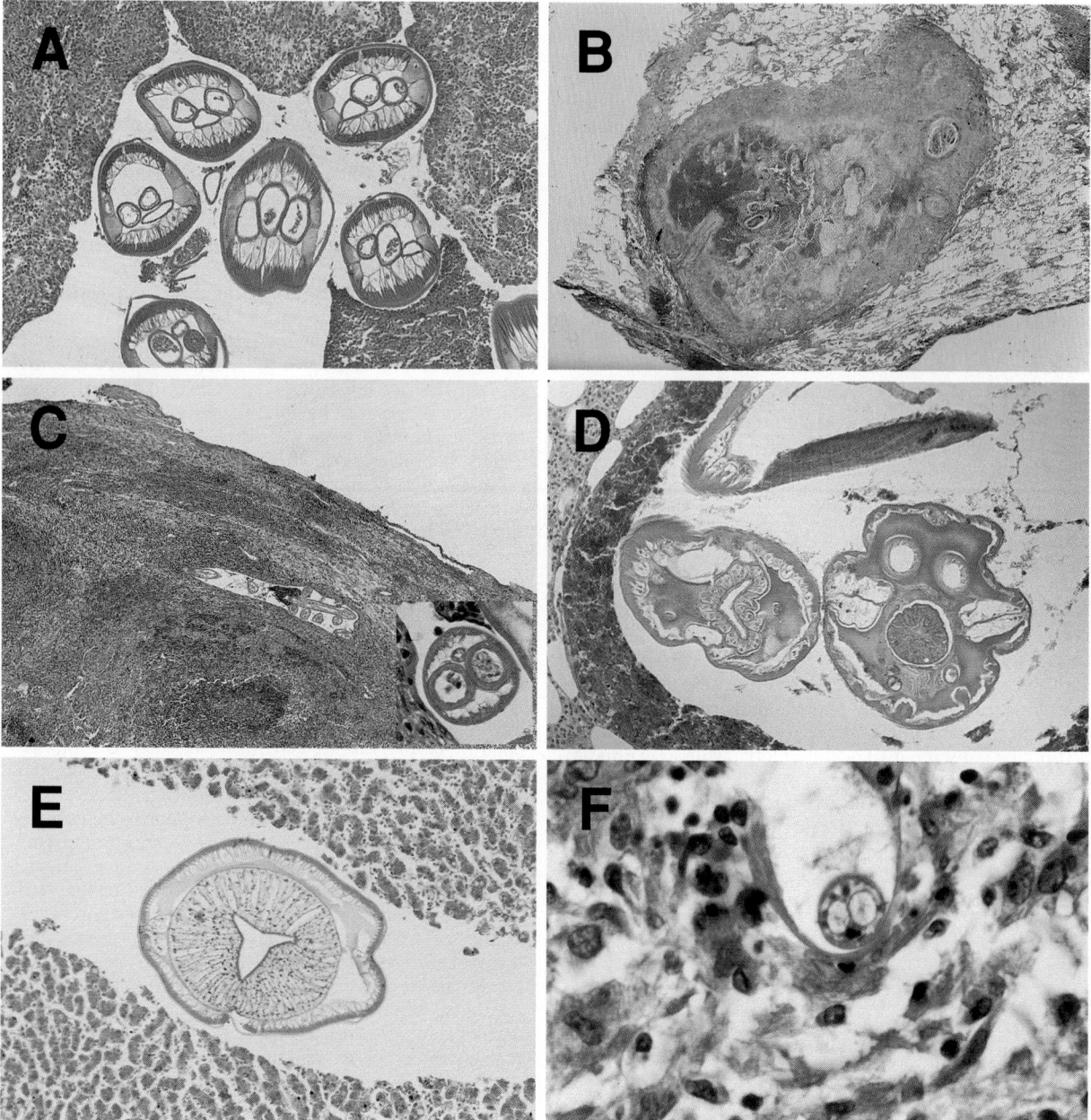

FIGURE 4 Helminth parasites in tissues; hematoxylin and eosin stain. (A) *Dirofilaria tenuis*. Sections of an immature adult female worm in a subcutaneous nodule of a resident of Florida. The typical morphologic features of the body wall are evident here even though the worm is dead. Magnification, ×100. (B) *Dirofilaria immitis* in a granulomatous nodule in human lung. This immature worm is trapped in a small pulmonary artery. Magnification, ×7. (C) *Brugia* species of zoonotic origin in a subcapsular vessel of a lymph node from a human in Ohio. Magnification, ×35. Inset shows a transverse section of the female worm. Its morphologic features immediately identify it as a filaria of the *Brugia* type. Magnification, ×250. (D) *Gnathostoma* species. Transverse sections through an advanced third-stage larva removed from the subcutaneous tissue of an individual who acquired the infection in Mexico. The morphology of the body wall including its spines, the presence of ballonets in the body cavity, and the structure of the intestine indicate that the worm is a larval gnathostome. Magnification, ×30. (E) *Anisakis* species. Transverse section through the body showing its unusual lateral chords and structure of the intestine. Magnification, ×120. (F) *Toxocara canis* larva. Transverse section in an inflammatory granulomatous lesion in the liver of a small child. Magnification, ×1,000.

have been reported in the Western Hemisphere as well, particularly in Mexico and Ecuador.

The adult worms in their natural hosts are found, typically, embedded in the gastric mucosa; female worms pro-

duce a characteristic egg that passes in the fecal stream in an unembryonated condition. The life cycle involves a copepod as a first intermediate host in which the parasite reaches the infective stage. Other animals such as fish and amphibians serve as second intermediate hosts, and reptiles, birds, and mammals may be paratenic hosts. Humans become infected by ingesting infective larvae encysted in the flesh of intermediate and/or paratenic hosts that have been inadequately cooked prior to consumption. In humans, the advanced third-stage larva migrates through the superficial subcutaneous tissues (Fig. 4D); frequently, larvae may invade the orbit, the eye itself, and the central nervous system. During its migratory phase in the subcutaneous tissues, which usually begins 2 to 4 weeks after infection, there may be periodic migratory subcutaneous swellings, or the larva may become stationary and provoke an intense inflammatory reaction. Invasion of the central nervous system is extremely serious and often fatal. Larvae and attendant symptoms may persist for several months.

Diagnosis is usually based on clinical presentation, i.e., history of eating raw fish in geographical areas where infections are common combined with subsequent migratory lesions. Removal of the worm(s) from the tissues is therapeutic; it also permits specific identification of the parasite.

Zoonotic Filariae

A wide variety of filarial worms that are natural parasites of wild mammals have been recovered from the tissues of people in many parts of the world. Infections have been recorded with greatest frequency in the United States and the Mediterranean region, especially Italy. These filariae are transmitted by blood-sucking arthropods that in some cases feed on both animals and humans. Their feeding behavior, together with the overlapping environments of certain animals and humans, makes it possible for the infective stages of these filariae to accidentally infect humans.

Species of *Dirofilaria*, among them *Dirofilaria tenuis*, *Dirofilaria repens*, and *Dirofilaria ursi*, which have been found in the subcutaneous tissues of their animal hosts and are transmitted by mosquitoes, are the filariae most frequently encountered in human infections (3, 9, 12). These infections are characteristically cryptic; i.e., there is no microfilaremia. Female worms may be sexually mature but are usually infertile. The worms have a propensity for migration to the face, orbit, and conjunctivae but may be found, dead or alive, in subcutaneous nodules on any part of the body (Fig. 4A). Diagnosis is based on recovery of the intact worm from the tissues or on identification of the filaria in histologic sections of the affected tissues (Fig. 4A). Usually, only a single worm is recovered from human tissues, although two or more have been found on rare occasions.

Dirofilaria immitis, the heartworm of dogs and other canids, has been reported from humans wherever the filaria is found in dogs. Typically, immature stages of the parasite lodge in and obstruct small pulmonary arteries, producing an infarct and eventually a granulomatous nodule (Fig. 4B) that appears as a "coin lesion" in the lung on X rays (3, 12, 15). At present, diagnosis depends on microscopic examination of the excised lesion or of a needle biopsy specimen (8, 12). There are no serologic or immunologic tests that will differentiate the parasite lesion from neoplastic tumors. On rare occasions, adult worms have been found in the right side of the heart or in the great vessels.

Lymphatic-dwelling *Brugia* species are natural parasites of animals in many parts of the world. Human infections of zoonotic origin have been reported most frequently from the northeastern United States and northern South America (13). The worms are invariably located in lymph nodes or lymphoid tissues (Fig. 4C). As stated for the dirofilarias, patent infections are extremely rare but have been reported

(13); most often, the female worms found in the tissues are immature and infertile. The animal reservoirs of infection in the natural environment have not been established precisely, but raccoons, rabbits, wild felines, and possibly other animals may be involved.

Larva Migrans

Several species of hookworms, ascarids as well as *Strongyloides* species, infect wild and domestic animals and may gain entry into the human host and undergo partial development. The severity of subsequent disease may vary from mild to severe, with involvement of various tissues and organs.

Cutaneous larva migrans, also known as creeping eruption or ground itch, refers to the production of serpiginous, inflamed trails in the skin resulting in intense pruritis and may be caused by various species of *Strongyloides*, *Ancylostoma*, and other hookworms. It occurs as the result of skin contact with sandy-loam types of soil that contain filariform larvae of these species that have developed to the infective stage from eggs or larvae discharged in the feces of infected animals. In the southeastern part of the United States, this may be an occupational hazard of electricians, plumbers, and construction workers, who inadvertently come in contact with infected soil when working under or near houses where animals have defecated. In addition, species of *Strongyloides* parasitizing wild animals, such as nutria and raccoons, have been incriminated in causing dermatitis in oil field workers and trappers in Louisiana. *Ancylostoma braziliense*, a hookworm of dogs and cats, is the species usually associated with classic creeping eruption acquired along sandy beaches, but other species of the genus may be involved as well. Invasion of human skin by the dog hookworm, *Ancylostoma caninum*, can produce creeping eruption; these larvae may also invade other tissues and organs of the body, including the eye, where they may produce granulomatous lesions.

The diagnosis of cutaneous larva migrans is based almost entirely on clinical findings and demonstration of the classical serpiginous trails (5).

Visceral larva migrans is a syndrome originally associated with larval migration in humans, especially children, of dog and cat ascarid species of the genus *Toxocara*. Although many other helminths of animals may infect humans and migrate through organs and deep tissues of the body, *Toxocara canis* is still the classic example of this type of infection.

The dog ascarid, *Toxocara canis*, is more important as a human parasite than is *Toxocara cati*, the species occurring in felines. Toxocariasis results from the ingestion of infective eggs of the parasite from soil and is characterized by hypereosinophilia, hepatomegaly, fever, pneumonitis, and sometimes death. The visceral larva migrans syndrome may persist for many years and may result in severe complications involving the eye and the central nervous system. Ocular larval migrans may mimic retinoblastoma, a malignant tumor of the eye.

Diagnosis of visceral larva migrans is difficult and is generally based on clinical findings and immunodiagnostic procedures. Occasionally, diagnosis is established by finding focal inflammatory lesions containing larvae in tissues such as liver, brain, and other organs (Fig. 4F). Sections of nematode larvae can sometimes be identified in human tissues obtained by biopsy or at necropsy, and these can be identified by specialists expert in helminth microanatomy.

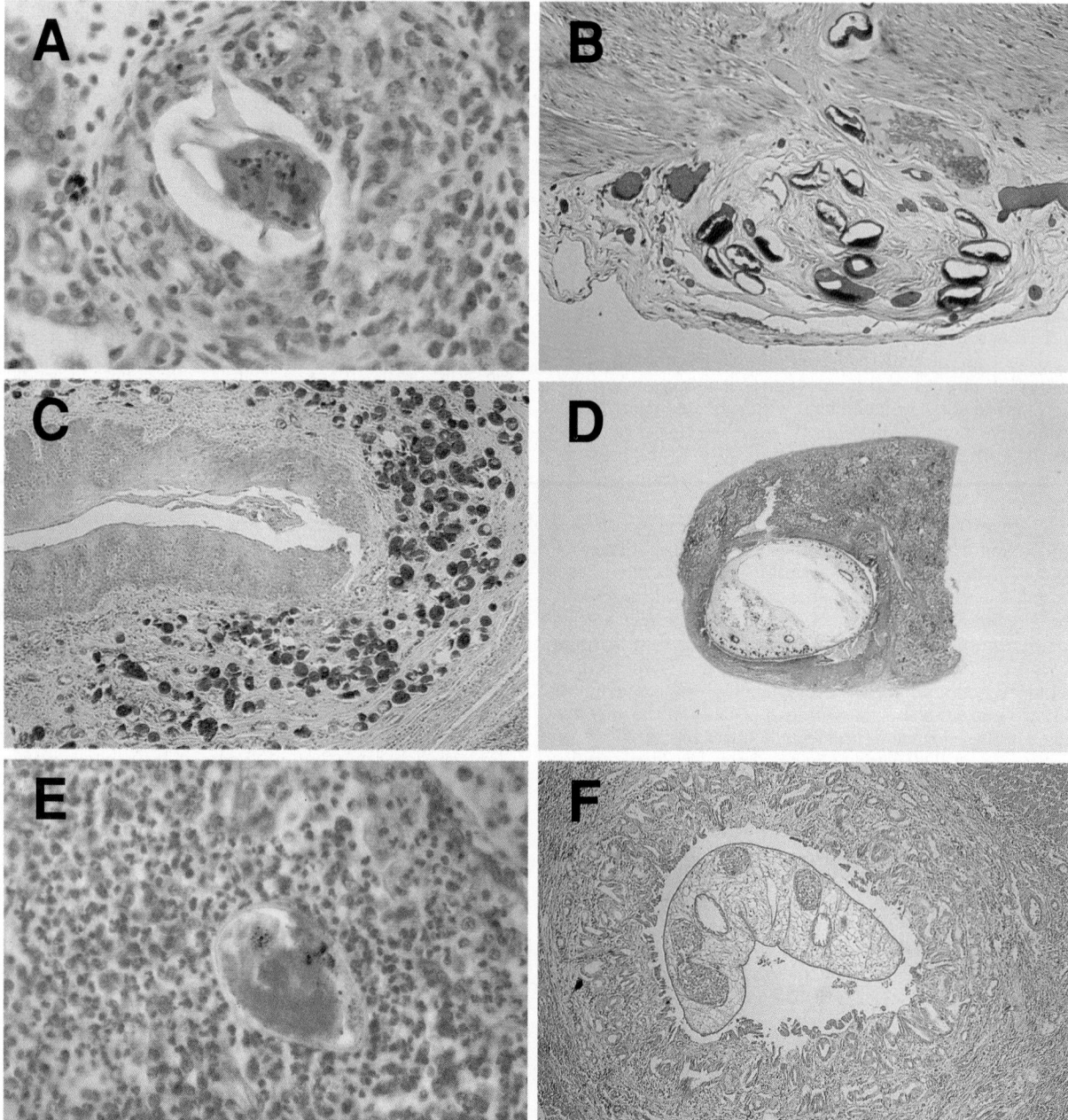

FIGURE 5 Trematode parasites in tissues. (A) Egg of *S. mansoni* in early granuloma in liver; the lateral spine is visible. Hematoxylin and eosin stain; magnification, ×400. (B) Cluster of *S. japonicum* eggs in a periodic acid-Schiff-stained section of human liver. Magnification, ×200. (C) Large number of *S. haematobium* eggs in human bladder. Many eggs (black) are undergoing calcification. Magnification, ×50. (D) Section through an adult *Paragonimus* worm in the lung of a cat. Trichrome stain; magnification, ×14. (E) Egg of *P. westermani* in a section of human lung. Hematoxylin and eosin stain; magnification ×350. (F) Adult *C. sinensis* in human bile duct. Hematoxylin and eosin stain; magnification, ×40.

TREMATODES

The important human trematode parasites are described in chapter 107 of this Manual inasmuch as their eggs are passed in feces or, more rarely, sputum and urine. However, adult trematodes and/or their eggs may be found in the wall of the alimentary tract, the liver, lungs, mesenteries, bladder, brain, and subcutaneous and muscular tissues. The trematodes commonly found in human tissues include the liver flukes (*Fasciola hepatica*, *Clonorchis sinensis*, and *Opis-*

thorchis species), the lung flukes (*Paragonimus* species), and the blood flukes (*Schistosoma* species).

Liver Flukes

F. hepatica, *Clonorchis sinensis*, and *Opisthorchis viverrini* live in the bile ducts of their human and animal hosts and may be found in sections of the livers of infected individuals who live in areas where the parasites are endemic or in others who have emigrated from endemic regions (Fig. 5F).

In sections through adult parasites in liver, such as *Clonorchis* and *Opisthorchis*, it is not uncommon to see typical eggs clustered in the uterine branches of these worms, which is a significant aid in the identification of the parasite.

The normal migratory path of larval *F. hepatica* to the bile ducts of the human host entails passage from the intestine into the abdominal cavity and entrance into the liver by penetration of Glisson's capsule. However, not infrequently, aberrant migration by these larval stages results in their presence in ectopic sites, including the abdominal wall, subcutaneous tissues, lungs, and other organs and tissues. Abscesses are produced in these locations, and degenerating adult worms may be found in histologic sections; in some instances, the worms are so degenerate or decomposed that they are unrecognizable; however, their eggs, liberated from the uteri, may still be identified. Size and morphology of the eggs, cut in various aspects, can aid in the correct identification of this parasitic infection.

Lung Flukes

Paragonimus westermani, as well as related species found in humans, make a larval migration from the intestine to the lungs (Fig. 5D) via the abdominal cavity, the diaphragm, and the thoracic cavity. As with *F. hepatica*, aberrant migration by larval *P. westermani* may result in its presence in the abdominal wall, rib cage, subcutaneous tissues, brain, and other sites where abscesses typically should be found. Degenerate adult worms or remnants of eggs may be found in histologic sections (Fig. 5E). Size and morphology of the egg remnants often allow specific identification of this infection in these ectopic sites. Abscesses in the brain have frequently been described for patients from Korea.

Blood Flukes (Schistosomes)

Each of the three major human schistosome parasites lives in some part of the venous side of the vascular system. *Schistosoma mansoni* adults reside typically in the hemorrhoidal venous plexus draining the lower ileum and colon; they also may be found in intrahepatic blood vessels of the liver. *Schistosoma japonicum* adults are usually in radicles of the superior mesenteric vein draining the small intestine, but some adult worms may also live in the portal system of the liver. Adults of *Schistosoma haematobium* usually are found in the vesical plexuses of the venous system. Granulomas produced by the eggs of *S. mansoni* and *S. japonicum* are usually found in the wall of the intestinal tract or in the liver, but they may also disseminate via the bloodstream to distant sites, including the lungs, heart, spleen, kidneys, and spinal cord (Fig. 5A and B). The smaller size of the eggs of *S. japonicum* aids in their extensive dissemination via the bloodstream. In the Philippines, eggs are not uncommonly found in the brain, where they provoke epileptiform seizures. Eggs of *S. haematobium* are commonly found in the wall of the bladder, but they are also seen in the kidney, ureters, wall of the rectum, and other tissues (Fig. 5C). Identification of schistosome eggs in tissues is somewhat dependent on the tissues in which they are found, the size of the eggs, and whether or not lateral or terminal spines can be visualized on whole or partial eggshell remnants. Adult schistosomes are not frequently seen in histologic sections.

CESTODES

A number of tapeworms live as adults in the small intestines of humans (or animals), and their larval stages often occur in human tissues. These larvae have significant morphologic features that usually allow them to be identified when they are examined grossly or studied in tissue sections. The most significant diseases caused by larval tapeworms include sparganosis, cysticercosis, coenurosis, and several forms of hydatid disease.

Sparganosis

Human sparganosis is caused by larval cestodes of the genus *Spirometra* that as adult worms usually parasitize cats, dogs, and various wild canids and felids. Species of this genus are closely related biologically and morphologically to *Diphyllobothrium* spp. and have similar life cycles involving copepods as first intermediate hosts and fish as second intermediate hosts. *Spirometra mansoni* is found in many parts of Asia, in particular China, Japan, Korea, Vietnam, and other areas, and *Spirometra mansonoides* is the common species found in the United States.

Characteristically in sparganosis, after ingestion of the plerocercoid larva (also called a sparganum) in poorly cooked fish, accidental ingestion of infected copepods containing procercoid larvae, or use of infected hosts such as frogs and snakes as poultices, the larva migrates in the tissues of the human host. In the United States, typical cases of sparganosis present as subcutaneous swellings that may be migratory and sometimes tender; usually, there is little or no peripheral eosinophilia. In Asia, many cases of sparganosis in the brain have been reported (10). Diagnosis is usually made after surgical removal of the intact worm or histopathologic study of the tissue mass. Spargana are typically ribbonlike and ivory colored and may range from several millimeters to 30 cm long by approximately 3 to 4 mm wide. The anterior portion of the worm is flattened and has a ventral groove; the body is transversely ridged, but this is not true segmentation. In histologic sections, the tegument typically has deep folds, and prominent bundles of longitudinal muscle fibers and excretory channels lie within the parenchyma; calcareous corpuscles are present and prominent (Fig. 6F).

Cysticercosis

Human cysticercosis is produced by ingestion of the eggs of the pork tapeworm *Taenia solium*. It occurs throughout the world, although its greatest prevalence is in Mexico; other areas with significant prevalence include Latin America, India, China, Africa, and Europe (7). Acquisition of cysticercosis in the United States has been uncommon in the past, but immigrants from areas where the parasite is endemic are frequently diagnosed with the infection, and reports of locally acquired cysticercosis are increasing in number in the United States.

The cysticercus of *Taenia solium* is often referred to as *Cysticercus cellulosae*; it is round to oval, translucent, and about 5 mm or more in diameter (Fig. 6A). It has a scolex bearing four suckers and a rostellum with a circle of hooks that is invaginated into the fluid-filled bladder. The cysticercus develops in any organ or tissue of the body; it is most serious when it occurs in the central nervous system and the eye (6). Diagnosis of the infection may be difficult. It may be established by the recovery of whole cysticerci or by the histologic demonstration of cysticerci in surgically removed tissues (Fig. 6B and C). Diagnosis may also be made by

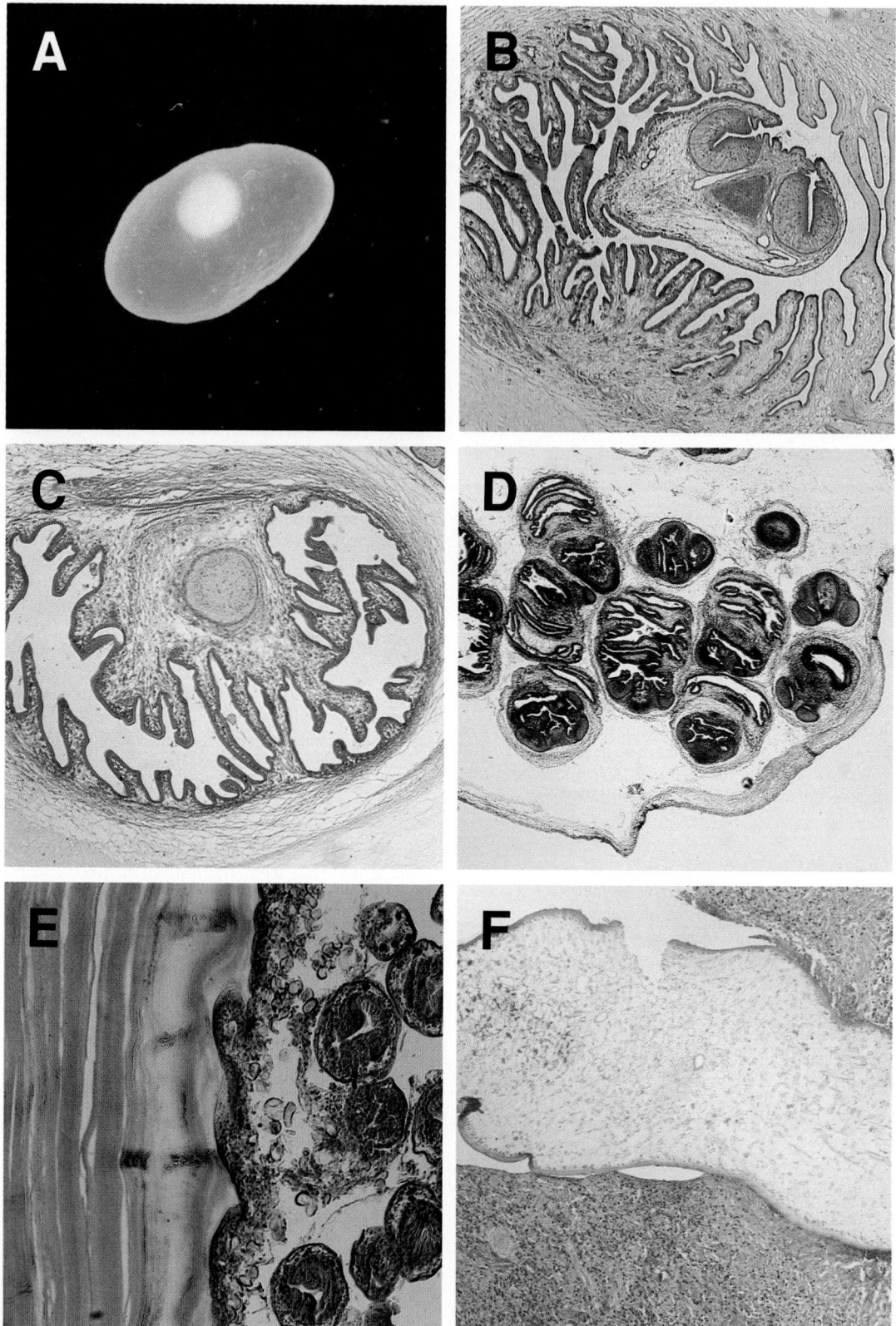

FIGURE 6 Cestode parasites in tissues. (A) Cysticercus of *Taenia solium* (also called *Cysticercus cellulosae*). Denser opaque area represents invaginated neck and scolex in fluid-filled bladder. Magnification, ×8. (B) Section through cysticercus in tissue showing scolex, neck, and bladder wall. Hematoxylin and eosin stain; magnification, ×300. (C) Section through neck region of cysticercus; a portion of a sucker is seen in the center. Hematoxylin and eosin stain; magnification, ×30. (D) Section through a coenurus of *Taenia multiceps*, showing multiple scoleces. Hematoxylin and eosin stain; magnification, ×30. (E) *E. granulosus*. Hydatid cyst in liver. Note laminated membrane, germinal layer, and individual protoscoleces. Hematoxylin and eosin stain; magnification, ×125. (F) Section through a sparganum (*Spirometra* species) in human tissue. Magnification, ×50.

detection of calcifying cysticerci by X ray and the use of computerized tomography.

Coenurosis

Coenurosis is produced by the larval stages of tapeworms of the genus *Taenia* that mature in the intestines of various canids and felines. The two most important species are *Taenia multiceps* (formerly *Multiceps multiceps*) and *Taenia serialis* (formerly *Multiceps serialis*). The eggs of these species are typical taeniid eggs and are indistinguishable from the eggs of other species of *Taenia* and *Echinococcus*. Sheep (for *Taenia multiceps*) and rodents (for *Taenia serialis*) are the usual intermediate hosts, but humans who ingest eggs of these species develop coenurus infection. As with cysticerci, a coenurus may develop in subcutaneous tissues, muscle, the eye, or the central nervous system. Typically, it is from several millimeters to several centimeters in diameter and has multiple scoleces invaginated into a fluid-filled bladder. Each scolex has four suckers and a rostellum armed with hooklets. Diagnosis is usually made by recovering the whole coenurus or by demonstrating in histologic section the typical bladder with multiple invaginated scoleces (Fig. 6D).

Unilocular Hydatid Infection

Although various forms of hydatid disease are caused by a number of species of *Echinococcus*, *Echinococcus granulosus* is the most important species producing human hydatidosis. *E. granulosus* is a minute tapeworm that as an adult is to 6 mm long and lives in the small intestines of domestic and wild canids. The parasite is distributed extensively in sheep- and cattle-raising areas of the world, where these herbivorous animals serve as intermediate hosts for the infection. Humans acquire infection by ingestion of typical taeniid eggs, which are excreted in the feces of infected dogs. When eggs are ingested by the intermediate hosts or by humans, the oncospheres liberated from the eggs migrate via the bloodstream to the liver, lungs, and other tissues and organs to develop into hydatid cysts. Within the cysts, brood capsules and protoscoleces develop and proliferate from an inner germinal membrane, so that ultimately the cysts may become very large and contain many hundreds, and even thousands, of protoscoleces. Each protoscolex is a potentially infective organism.

Cysts grow slowly, and their presence frequently goes undetected for many years until they reach sufficient size to cause clinical symptoms. Infections of the central nervous system usually become apparent much earlier than those of other organs.

Diagnosis of infection may be difficult but can be accomplished by the use of X ray, ultrasonic scanning, and computerized tomography to detect cysts in tissues. When the presence of cysts is combined with an appropriate residential or travel history or an occupation such as sheep raising in areas where the parasites are endemic, there are reasonable grounds for suspicion of infection. Many immunodiagnostic tests have been utilized, but there is wide variation in sensitivity and specificity. Aspiration of material from a hydatid cyst in situ is dangerous and is not a recommended procedure, since accidental spillage of cyst contents may result in dissemination of liberated scoleces and/or an anaphylactic reaction. Fluid in hydatid cysts typically consists of disintegrating brood capsules and scoleces, hooklets, and calcareous corpuscles; this material is referred to as "hydatid sand." In histologic section, the wall of the unilocular cyst comprises three layers: an outer fibrous layer of host tissue, an acellular laminated layer, and an inner thin layer of germinal epithelium (Fig. 6E). Brood capsules arising from the germinal layer are stalked and project into the lumen of the cyst; each of these contains several invaginated protoscoleces with four suckers and a crown of hooklets.

Multilocular Hydatid (Alveolar Hydatid) Infection

A second form of human hydatid disease is alveolar hydatidosis, caused by the larval stage of *Echinococcus multilocularis*. The life cycle of this parasite involves foxes, wolves, and dogs as definitive hosts and small rodents and voles as intermediate hosts. The adult worm is even smaller than *E. granulosus*, measuring only 1.2 to 3.7 mm in length, and has a strobila of only three proglottids. In the livers of rodents, the invasive larval stage has a multicompartmented appearance, each compartment containing many protoscoleces proliferating from the germinal membrane. Ingestion of the eggs of *E. multilocularis* by humans results in development of invasive cysts in the liver. However, protoscoleces are rarely seen in human alveolar hydatid disease, and instead, the germinal layer proliferates exogenously and endogenously, resulting in folded and collapsed membranes scattered throughout the tissue. The pathologic picture of the liver resembles a hepatic carcinoma. The geographic distribution of the infection includes northern Europe, Japan, China, India, and North America; the infection is widespread in Alaska, Canada, and the northern tier of states in the midwestern United States. Diagnosis is difficult and frequently is accomplished only by histologic examination of tissues removed surgically.

Polycystic Hydatid Infection

A third type of hydatid infection, caused by *Echinococcus vogeli*, has been reported in Latin America (4). The normal life cycle of the parasite involves bush dogs and large rodents (pacas). A polycystic hydatid develops primarily in the liver, where, in humans, it has an invasive behavior. These cysts develop brood capsules and protoscoleces in human infections, which distinguishes them from the cysts of human alveolar hydatid infection.

REFERENCES

1. **Ash, L. R., and T. C. Orihel.** 1990. *Atlas of Human Parasitology,* 3rd ed. American Society of Clinical Pathologists, Chicago.
2. **Ash, L. R., and T. C. Orihel.** 1991. *Parasites: a Guide to Laboratory Procedures and Identification.* American Society of Clinical Pathologists, Chicago.
3. **Beaver, P. C., R. C. Jung, and E. W. Cupp.** 1984. *Clinical Parasitology,* 9th ed. Lea & Febiger, Philadelphia.
4. **D'Alessandro, A., R. L. Rausch, C. Cuello, and N. Aristizabal.** 1979. *Echinococcus vogeli* in man with a review of polycystic hydatid disease in Colombia and neighboring countries. *Am. J. Trop. Med. Hyg.* **28:**303–317.
5. **Davies, H. D., P. Sakuls, and J. S. Keystone.** 1993. Creeping eruption. A review of clinical presentation and management of 60 cases presenting to a tropical disease unit. *Arch. Dermatol.* **129:**588–591.
6. **Del Britto, O. H., and J. Sotelo.** 1988. Neurocysticercosis: an update. *Rev. Infect. Dis.* **10:**1075–1087.
7. **Flisser, A., K. Willms, J. P. Laclette, C. Larralde, C. Ridaura, and F. Beltran (ed.).** 1982. *Cysticercosis: Present State of Knowledge and Perspectives.* Academic Press, Inc., New York.
8. **Gutierrez, Y.** 1984. Diagnostic features of zoonotic filariae in tissue sections. *Hum. Pathol.* **15:**514–525.

9. **Gutierrez, Y.** 1990. *Diagnostic Pathology of Parasitic Infections with Clinical Correlations*. Lea & Febiger, Philadelphia.

10. **Holodniy, M., J. Almenoff, J. Loutit, and G. K. Steinberg.** 1991. Cerebral sparganosis: case report and review. *Rev. Infect. Dis.* **13**:155–159.

11. **Ishikura, H., K. Kikuchi, K. Nagasawa, T. Ooiwa, H. Takamiya, N. Sato, and K. Sugane.** 1993. Anisakidae and anisakidosis. *Prog. Clin. Parasitol.* **3**:43–102.

12. **Orihel, T. C., and L. R. Ash.** 1995. *Parasites in Human Tissues*. American Society of Clinical Pathologists, Chicago.

13. **Orihel, T. C., and P. C. Beaver.** 1989. Zoonotic *Brugia* infections in North and South America. *Am. J. Trop. Med. Hyg.* **40**:638–647.

14. **Orihel, T. C., and M. L. Eberhard.** 1982. *Mansonella ozzardi*: a redescription with comments on its taxonomic relationships. *Am. J. Trop. Med. Hyg.* **31**:1142–1147.

15. **Ro, J. Y., P. J. Tsakalakis, V. A. White, M. A. Luna, E. G. Chang-Tung, L. Green, L. Cribbett, and A. C. Ayala.** 1989. Pulmonary dirofilariasis: the great imitator of primary or metastastic lung tumor. A clinicopathologic analysis of seven cases and a review. *Hum. Pathol.* **20**:69–76.

16. **Rusnak, J. M., and D. R. Lucy.** 1993. Clinical gnathostomiasis: case report and review of English language literature. *Clin. Infect. Dis.* **16**:33–50.

17. **Wittner, M., J. W. Turner, G. Jacquette, L. R. Ash, M. P. Salgo, and H. B. Tanowitz.** 1989. Eustrongylidiasis—a parasitic infection acquired by eating sushi. *N. Engl. J. Med.* **320**:1124–1126.

Arthropods of Medical Importance

THOMAS R. FRITSCHE AND MICHAEL A. PFALLER

109

Arthropods make up one of the largest of the animal phyla, a phylum consisting of more than a million described species, with estimates of the total number of species ranging up to 30 million. An exceedingly small number of this total directly or indirectly affect human health, but those that do are responsible for significant morbidity and mortality.

The mechanisms of injury by which arthropods pose a threat to humans and their domestic animals include direct tissue invasion, envenomation, vesication, blood loss, mechanical and biological transmission of infectious agents, and allergic diatheses. Humans are also susceptible to exaggerated fears of arthropods (entomophobia) and delusions of infection (delusory parasitosis). Species directly or indirectly responsible for human disease include representatives of all major arthropod classes, including the Diplopoda (millipedes), Chilopoda (centipedes), Crustacea (crustaceans), Insecta (bugs, lice, fleas, dipterans, hymenopterans, and others), Pentastomida (tongue worms), and Arachnida (scorpions, spiders, ticks, and mites) (Table 1).

Clinical laboratories are often the initial contact for both clinicians and patients with questions concerning arthropod-related diseases or with actual specimens submitted for identification. Such specimens may take the form of intact organisms or parts of organisms, skin scrapings or other tissue preparations, articles of clothing or bedding, and foodstuffs, among others. Arthropods recovered from the toilet bowl following a bowel movement or urination are often disconcerting to patients, who think they may have an intestinal or bladder infection, but such recovery is not uncommon. It is usually coincidental and not related to infection.

Clinical microbiologists should have a general understanding of the more commonly encountered medically important arthropods (especially ectoparasites) and should be able to perform limited identifications, at least of the higher taxonomic levels. It is equally important for laboratorians to recognize the clinical situations in which outside expertise should be sought. Owing to the complex nature of medical entomology, this chapter can present only a limited overview of the field. Also, descriptions of the groups covered here refer primarily to adult stages, though many arthropods, especially insects, are clinically problematic in their larval stages as well. Differentiation of larvae is often quite difficult and requires the use of specialized literature (3, 10, 26, 28) or the help of a specialist. A variety of general and medical entomology texts and guides that provide more detail about all of the groups described here are available (1, 2, 10, 12–14, 23, 26, 28, 38).

MECHANISMS OF INJURY

Direct Tissue Invasion

A variety of arthropods, including scabies mites, chigoe fleas, pentastomids, and dipteran larvae (maggots), may be responsible for invasion of superficial tissues (by the first two groups; referred to as infestation) or of deeper body tissues and cavities (by the last two groups; referred to as infection). Tissue invasion of any sort by maggots is referred to as myiasis and may occur in either living or devitalized tissues (1, 30, 38).

Envenomation

Every day, large numbers of people are subjected to the stings and bites of many different arthropods. Reactions to these attacks vary greatly and depend on the species involved; the type of saliva or venom introduced into the body; the location of the wound; and the individual's physiologic response, which may vary depending on whether the person has been previously exposed. For most people, arthropod bites and stings cause only local tissue reactions, but for some, serious systemic, life-threatening reactions may occur.

Arthropods capable of stinging come from many different groups, with the greatest offenders including hymenopterans (ants, bees, and wasps) and scorpions. Systemic hypersensitivity reactions are more frequently seen following sensitization to the venom of hymenopterans. These venoms are mixtures of biogenic amines, polypeptide toxins, and enzymes, including phospholipases and hyaluronidase, which are responsible for both local and systemic effects. The venom of scorpions is composed of peptide toxins that act primarily on the nervous system and have few local effects (13, 38).

Many arthropods are capable of biting, either in the process of feeding or in self-defense, and are liable to introduce either saliva or more potent toxins intended to paralyze their intended prey. While arthropod bites generally cause far fewer serious reactions than stings, some individuals may develop systemic reactions to saliva. The

TABLE 1 Classification of arthropods of medical importance

Class Diplopoda (millipedes)

Class Chilopoda (centipedes)

Class Crustacea (crustaceans)
   Order Copepoda (copepods)
   Order Decapoda (crabs, crayfish)

Class Insecta (insects)
   Order Hemiptera (bedbugs, kissing bugs)
   Order Siphonaptera (fleas)
   Order Anopleura (sucking lice)
   Order Dictyoptera (cockroaches)
   Order Hymenoptera (ants, wasps, bees)
   Order Coleoptera (beetles)
   Order Diptera (flies, mosquitos, midges)
   Order Lepidoptera (moths, butterflies, caterpillars)

Class Pentastomida (tongue worms)

Class Arachnida (arachnids)
   Subclass Scorpiones (scorpions)
   Subclass Araneae (spiders)
   Subclass Acari (ticks, mites, chiggers)

---

groups usually capable of introducing saliva include the centipedes (Chilopoda); mosquitoes, flies, and biting midges (Diptera); bedbugs, kissing bugs, and assassin bugs (Hemiptera); sucking lice (Anopleura); fleas (Siphonaptera); ticks and mites (Acari); and spiders (Araneae). While almost all spiders are venomous, only a few groups pose a distinct danger to humans; these include the widow spiders, violin spiders, and certain tarantulas (1, 13, 26).

Envenomation may also result from exposure to certain urticating caterpillars (larval insects of the order Lepidoptera). These larval insects may contain numerous tiny hairs that are both irritating and capable of injecting a venom to which some individuals may become sensitized. The barbed hairs may also be responsible for dermatitis and serious eye disease (1, 13, 31).

Vesication

Some millipedes (Diplopoda) are capable of spraying an irritating and vesicating (blister-causing) chemical for defensive purposes from glands located on each body segment. While most of the millipedes seen in the United States are small and innocuous, some of the larger tropical species are capable of squirting their fluids for some distance and causing irritating skin burns. Blister beetles (Coleoptera) are so named because of the presence of the vesicating compound cantharidin in their body fluids. When these attractive beetles are handled, cantharidin may be released, resulting in the appearance of fluid-filled skin blisters several hours later (1, 13, 26, 38).

Blood Loss

Blood loss from arthropod infestation is not an insignificant problem among domestic animals, although in humans it rarely becomes life-threatening. Blood-sucking arthropods include bedbugs and kissing bugs; lice; fleas; flies, mosquitoes, and biting midges; and ticks and mites. More commonly, these arthropods are involved in the transmission of infectious agents.

Transmission of Infectious Agents

Many species of arthropods are capable of either mechanical or biological transmission of agents responsible for serious infectious disease. Filth flies, which include the common housefly *Musca domestica*, are easily incriminated in the mechanical transmission of the agents of bacillary dysentery, cholera, typhoid, viral diarrhea, amebic dysentery, giardiasis, pinworms, and tapeworms.

Arthropod vectors responsible for the biological transmission of infectious agents may serve either as simple amplification vehicles for the organism or in a more complex role involving the changing life cycle stages of certain parasites (1, 26, 38). Major groups of arthropods involved in the transmission of biological agents are summarized in Table 2.

Hypersensitivity Reactions

Allergic hypersensitivity to arthropod bites and stings may develop in any individual upon repeated exposure. Because of the frequency with which hymenopteran stings occur, severe systemic reactions including anaphylaxis are well-described clinical entities. Such serious reactions may also develop in response to bites and stings of many other arthropods as well.

Repeated exposure to the saliva, excrement, and/or body parts of many arthropods, including mites, ticks, lice, bedbugs, certain caterpillars, moths, and butterflies, is also well known to exacerbate allergic conditions. Because of their widespread occurrence in the environment, a large variety of house, dust, and animal mites may also be responsible for the development of allergic manifestations such as asthma and hay fever in predisposed individuals. Certain of these problems can be controlled or minimized through the use of skin desensitization programs (1, 9, 26, 30, 38).

Psychologic Manifestations

Unreasonable or excessive fear of seeing or touching arthropods is termed entomophobia. Delusory parasitosis is a potentially more serious emotional disorder in which an individual is convinced that arthropod or worm parasites are present on or in the body despite repeated examinations that fail to reveal their presence. In such circumstances, the person may relate that the "infestation" has ruined his or her life, resulting often in job loss, divorce, movement from domicile to domicile, and repeated use of pesticides and extermination services. The victim often describes numerous, unsatisfying visits to physicians or other health care providers. These problems may originate in the home or the workplace and may be transferred from one to the other. Sometimes the delusions can be so convincing that others come to believe in the existence of the infestation or become "infested" themselves.

Because domiciles or the workplace can be infested with many different insects and mites, it is imperative that thorough inspections be performed as part of the overall evaluation of an affected individual. A mysterious onset of irritation and itching may be due to bites from unrecognized mites following abandonment of a rodent or bird nest in a human dwelling or may result from exposure to a variety of chemical, mineral, or other noxious stimuli. Infestations with scabies mites, lice, fleas, or bedbugs may also result in unexplained irritation unless their presence is specifically looked for (1, 13, 38).

TABLE 2 Summary of major arthropod genera involved in biological transmission of infectious diseases

Arthropod	Etiologic agent	Disease
Arachnida		
Acari (ticks)		
Ixodes spp.	*Borrelia burgdorferi*	Lyme disease
	Babesia spp.	Babesiosis
Ornithodoros spp.	*Borrelia* spp.	Relapsing fever
Dermacentor spp., *Amblyomma* spp.	*Francisella tularensis*	Tularemia
	Ehrlichia spp.	Ehrlichiosis
	Rickettsia rickettsii	Rocky Mountain spotted fever
Dermacentor spp.	Coltivirus	Colorado tick fever
Acari (mites)		
Leptotrombidium spp.	*Rickettsia tsutsugamushi*	Scrub typhus
Liponyssoides spp.	*Rickettsia akari*	Rickettsialpox
Crustacea		
Decapods	*Paragonimus* spp.	Paragonimiasis
Copepods	*Diphyllobothrium* spp.	Diphyllobothriasis
	Dracunculus medinensis	Guinea worm disease
	Gnathostoma spinigerum	Gnathostomiasis
Insecta		
Anopleura		
Pediculus spp.	*Rickettsia prowazekii*	Epidemic typhus
	Rochalimaea quintana	Trench fever
	Borrelia recurrentis	Epidemic relapsing fever
Diptera		
Aedes spp.	Flaviviruses	Dengue, yellow fever
	Other arboviruses	Encephalitis
Anopheles spp.	*Plasmodium* spp.	Malaria
	Brugia malayi	Filariasis
	Arboviruses	Encephalitis
Culex spp.	*Wuchereria* and *Brugia* spp.	Filariasis
	Arboviruses	Encephalitis
Culicoides spp.	*Mansonella* spp.	Filariasis
Glossina spp.	*Trypanosoma brucei*	African sleeping sickness
Chrysops spp.	*Loa loa*	Loiasis
	Francisella tularensis.	Tularemia
Simulium spp.	*Onchocerca volvulus*	Onchocerciasis
	Mansonella ozzardi	Filariasis
Phlebotomus spp., *Lutzomyia* spp.	*Leishmania* spp.	Leishmaniasis
	Bartonella bacilliformis	Bartonellosis
	Phlebovirus	Sandfly fever
Hemiptera		
Panstrongylus, *Rhodnius*, and *Triatoma* spp.	*Trypanosoma cruzi*	Chagas' disease
Siphonaptera		
Xenopsylla spp.	*Yersinia pestis*	Plague
	Rickettsia typhi	Murine typhus
Nosopsyllus spp.	*Rickettsia typhi*	Murine typhus
Ctenocephalides spp.	*Dipylidium caninum*	Dog tapeworm disease

BIOLOGICAL CHARACTERISTICS OF ARTHROPODS

Arthropods are bilaterally symmetrical invertebrate animals characterized by a segmented body, a rigid chitinous exoskeleton that is molted periodically during growth, and several pairs of jointed appendages. Development proceeds from egg to adult form through the process of metamorphosis, which may be either gradual or complete.

The class Insecta demonstrates both types of metamorphosis depending on the group. Bedbugs, kissing bugs, lice, and cockroaches display gradual metamorphosis, in which the organism passes through three life cycle stages: egg,

nymph, and adult. Nymphs closely resemble adults but are sexually immature and grow gradually through a series of molts.

Insects that display complete metamorphosis pass through four different stages: egg, larva, pupa, and adult. This type of growth is typical of butterflies and moths; medically important arthropods that display it include flies and mosquitoes; fleas; ants, bees, and wasps; and beetles. Larval forms differ significantly from adults and are often wormlike; they include (for example) fly maggots, mosquito wrigglers, beetle grubs, and caterpillars. They often live in different ecological niches from the adults and have different nutritional needs. True metamorphosis occurs when the larvae stop feeding and begin to pupate. During this time, the pupa undergoes a major reorganization of both internal and external organs and is often susceptible to predation. Emergence of the adult signifies the onset of sexual maturity and the initiation of reproductive activities (2, 26).

Spiders and scorpions undergo developmental changes most similar to the process of gradual metamorphosis as seen in insects. Life cycles of ticks and mites typically include egg, larva, nymph, and adult. Larvae of ticks and mites can be differentiated from other stages by determining the number of legs present: larvae have six legs, whereas nymphs and adults have eight (14, 23, 26, 32).

SPECIMEN HANDLING AND EXAMINATION

Invariably, arthropods brought to the attention of clinicians are submitted to the clinical laboratory. These specimens most commonly originate from human sources (skin surface, tissue, stool, sputum, and other anatomic sites) and a variety of nonhuman sources (foodstuffs, water, clothing, bedding, carpeting, and animals, among others).

Depending on the laboratorian's interest and comfort in examining arthropods, specimens may be retained for study and identification or forwarded to someone trained in the field of entomology. Identification of many arthropods is not difficult and can be attempted provided the individual has some background in the speciality and has access to the various texts and other aids, including dichotomous keys, available. *Depending on the seriousness of the problem, however, expert help should be sought when significant clinical decisions regarding therapy and prognosis are being made.* Usually, state public health laboratories have the expertise available or know of individuals trained in medical aspects of entomology at regional educational institutions, museums, or other state or federal agencies, including the Centers for Disease Control and Prevention.

Methods of killing and preserving arthropods should be dictated by the type of arthropod being collected and the characteristics it is necessary to preserve for identification purposes. Many specimens deteriorate beyond recognition if left to die without preservation. Exposure to the fumes of standard killing agents, including ethyl acetate, chloroform, and ether, is often appropriate for long-term storage and identification. Standard techniques used subsequent to killing include dry preservation, immersion in preservative fluids, and slide mounting.

Adult mosquitoes, midges, and flies are best preserved dry to retain key taxonomic characteristics such as body and wing scales and are usually pinned for future examination. Most beetles, true bugs, and cockroaches are also best killed and maintained as pinned specimens. Arthropods that are pinned and dried need to be stored in tight-fitting boxes with naphthalene or dichlorobenzene to protect them from pests (2, 37).

For many other medically important arthropods, including larval forms (maggots, grubs, caterpillars, etc.), ectoparasites (lice, fleas, ticks, and mites), spiders, and scorpions, immersion in 70 to 80% ethyl alcohol provides adequate preservation. It is often preferable to kill large maggots and other larvae in hot water to extend their bodies and prevent contraction when they are placed in alcohol. Also, when such larvae are received with attached tissue or in a fecal specimen, it is best to gently remove or wash away attached debris prior to preservation (26).

Certain of the smaller medically important arthropods

(mites, small ticks, fleas, sand flies, and midges) are best killed in alcohol and subsequently prepared as permanent slide mounts. Prior to dehydration and mounting, the darker specimens can be cleared by placing them in 10% potassium hydroxide for several days. Mounting media include water-based types (modified Berlese's medium) and organic-solvent-based types (Canada balsam, Permount, and isobutyl methacrylate, among others) (1, 12, 37).

Many different kinds of specialized collecting equipment have been described and are in use by entomologists: nets, tubes, traps (baited and suction), and aspirators, the last being especially useful for the collection of mosquitoes and sand flies in epidemiologic investigations. A variety of entomology and parasitology books and laboratory guides provide more detailed information about the collection, preservation, and preparation of arthropod specimens for examination (1, 2, 10, 12, 26, 37).

CLASSIFICATION, CLINICAL SIGNIFICANCE, AND IDENTIFYING CHARACTERISTICS

The classification scheme followed here is traditional in its approach to the arthropods. The recognized classes, subclasses, and orders are presented in Table 1. Arthropods of medical importance are differentiated into two subphyla determined by the presence of either articulated mandibles (Mandibulata) or pincherlike chelicerae (Chelicerata). Classes of mandibulate arthropods include the myriapods (millepedes and centipedes), crustaceans, and insects; the chelicerate arthropods include the arachnids. The affinities of the pentostomes have yet to be determined (1, 2, 26). Differentiation at the ordinal level is only somewhat more difficult but does require some biological training and familiarity with arthropod anatomy. Distinction of various groups and species beyond this level may require detailed reference sources and use of often complex dichotomous keys. Despite these difficulties, the list of medically important groups of arthropods is not overwhelming and should be within the grasp of most laboratorians committed to learning about them. A simplified key to the classes, subclasses, and orders of adult arthropods of medical importance is given in Table 3, although the reader may need to consult illustrated texts as well (2, 10, 13, 26, 28). The groups listed in Table 3 but not discussed here (termites, order Isoptera; earwigs, order Dermaptera; silverfish and firebrats, order Thysanura; and chewing lice, order Mallophaga) are discussed elsewhere (2, 14, 23, 28).

The following sections include a discussion of each group of medically important arthropods that indicates their major medical importance and in some instances provides assistance to their identification. Information on specific infectious agents transmitted by arthropods may be found in the appropriate sections of this Manual.

CLASS DIPLOPODA (MILLIPEDES)

Millipedes are wormlike arthropods with numerous apparent body segments, each of which has two pairs of legs. While somewhat similar in appearance to centipedes, millipedes are more rounded and have mouthparts that point downward. Most of the more than 8,000 described species feed nocturnally on plants or decaying vegetation and do not have mouthparts capable of piercing bites. Many do produce secretions for defensive purposes from glands located on the sides of each body segment. Secreted fluids are

TABLE 3 Key to adult stages of common arthropod classes, subclasses, and orders of medical importance[a]

1. Three or four pairs of legs [2]
   Five or more pairs of legs [22]

2. Three pairs of legs with antennae (**insects class insecta**) [3]
   Four pairs of legs without antennae (**spiders, ticks, mites, scorpions: class Arachnida**) [20]

3. Wings present, well developed [4]
   Wings absent or rudimentary [12]

4. One pair of wings (**flies, mosquitoes, midges: order Diptera**) [5]
   Two pairs of wings [6]

5. Wings with scales (**mosquitoes: order Diptera**)
   Wings without scales (**other flies: order Diptera**)

6. Mouthparts adapted for sucking, with elongate proboscis [7]
   Mouthparts adapted for chewing, without elongate proboscis [8]

7. Wings densely covered with scales, proboscis coiled (**butterflies and moths: order Lepidoptera**)
   Wings not covered with scales; proboscis not coiled but directed backward (**bedbugs and kissing bugs: Hemiptera**)

8. Both pair of wings membranous, similar in structure, although size may vary [9]
   Front pair of wings leathery or shell-like, serving as covers for the second pair [10]

9. Both pairs of wings similar in size (**termites: order Isoptera**)
   Hind wing much smaller than front wing (**wasps, hornets, and bees: order Hymenoptera**)

10. Front wings horny or leathery without distinct veins, meeting in a straight line down the middle [11]
    Front wings leathery or paperlike with distinct veins, usually overlapping in the middle (**cockroaches: order Dictyoptera**)

11. Abdomen with prominent cerci or forceps; wings shorter than abdomen (**earwigs: order Dermaptera**)
    Abdomen without prominent cerci or forceps; wings covering abdomen (**beetles: order Coleoptera**)

12. Abdomen with three long terminal tails (**silverfish and firebrats: order Thysanura**)
    Abdomen without three long terminal tails [13]

13. Abdomen with narrow waist (**ants: order Hymenoptera**)
    Abdomen without narrow waist [14]

14. Abdomen with prominent pair of cerci or forceps (**earwigs: order Dermaptera**)
    Abdomen without cerci or forceps [15]

15. Body flattened laterally; antennae small, fitting into grooves in side of head (**fleas: order Siphonaptera**)
    Body flattened dorsoventrally; antennae projecting from side of head, not fitting into grooves [16]

16. Antennae with nine or more segments [17]
    Antennae with three to five segments [18]

17. Pronotum covering head (**cockroaches: order Dictyoptera**)
    Pronotum not covering head (**termites: order Isoptera**)

18. Mouthparts consisting of tubular jointed beak; three- to five-segmented tarsi (**bedbugs: order Hemiptera**).
    Mouthparts retracted into head or of the chewing type; one- or two-segmented tarsi [19]

19. Mouthparts retracted into the head, adapted for sucking blood (**sucking lice: order Anopleura**)
    Mouthparts of the chewing type (**chewing lice: order Mallophaga**)

20. Body oval, consisting of a single saclike region (**ticks and mites: subclass Acari**)
    Body divided into two distinct regions, a cephalothorax and an abdomen [21]

21. Abdomen joined to the cephalothorax by a slender waist; abdomen with segmentation indistinct or absent; stinger absent (**spiders: subclass Araneae**)
    Abdomen broadly joined to the cephalothorax; abdomen distinctly segmented, ending with a stinger (**scorpions: subclass Scorpiones**)

22. Five to nine pairs of legs or swimmerets; one or two pairs of antennae; principally aquatic organisms (**copepods, crabs, and crayfish: class Crustacea**)
    Ten or more pairs of legs; swimmerets absent; one pair of antennae; terrestrial organisms [23]

23. Only one pair of legs per body segment (**centipedes: class Chilopoda**)
    Two pairs of legs per body segment (**millipedes: class Diplopoda**)

[a]Adapted from references 13 and 28.

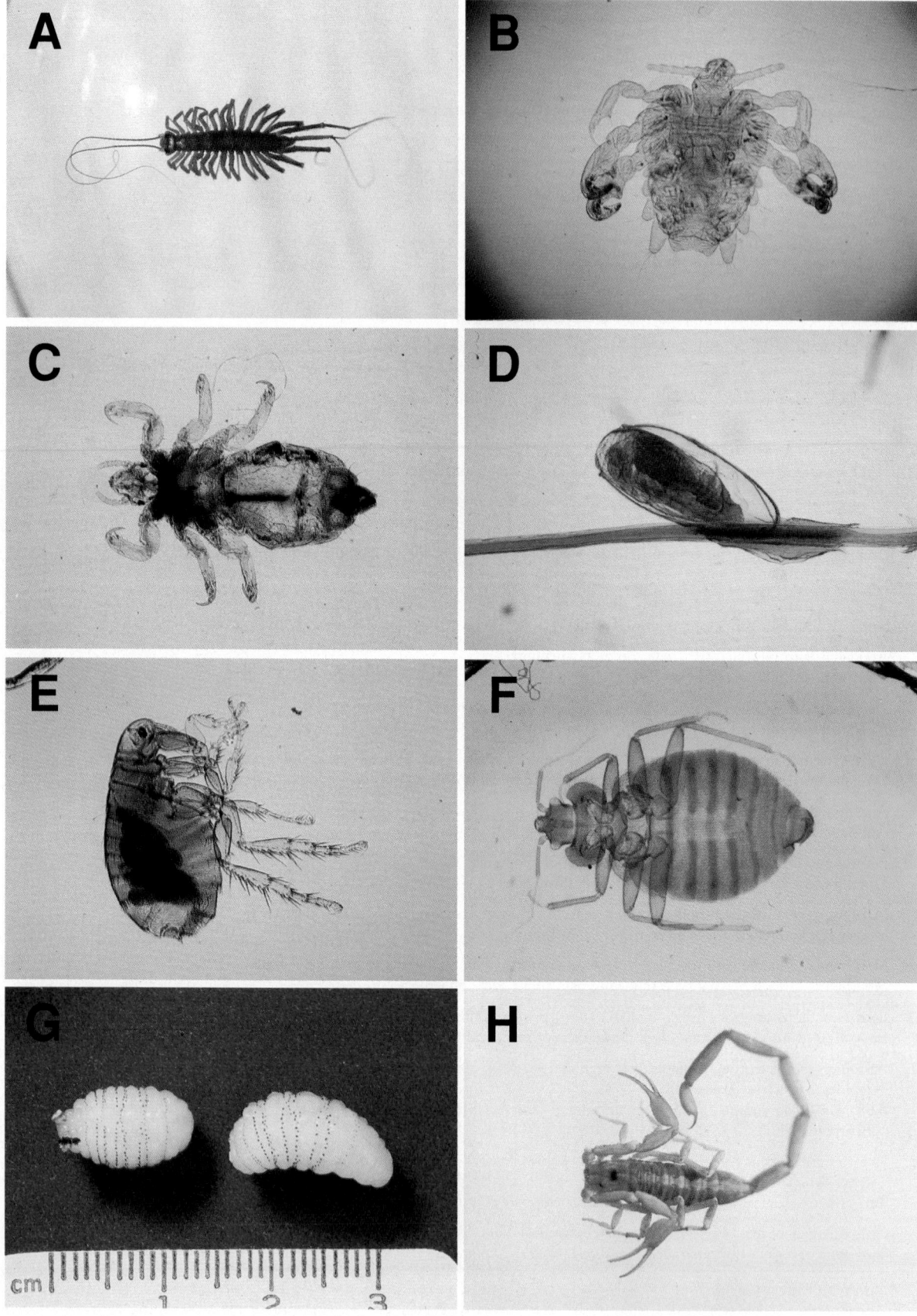

irritating to a potential enemy and in some of the larger (up to 30-cm) tropical species can be forcefully discharged over several centimeters. Exposure of human skin to these fluids can result in a burning sensation, discoloration, and blister formation. Active or passive eye exposure can result in severe conjunctivitis and occasionally corneal ulceration (1, 2, 13, 26, 28).

CLASS CHILOPODA (CENTIPEDES)

Centipedes are flattened, multisegmented, wormlike arthropods with one pair of legs per segment and long antennae (Fig. 1). Approximately 2,800 species from both temperate and tropical climates have been described. Unlike millipedes, centipedes are fast-moving and can inflict a painful bite from forward-facing "pincers," which are modified from the first pair of legs. Bites usually occur when someone steps on, picks up, or rolls onto a centipede while sleeping on the ground. Most species are nocturnal and feed on other arthropods after paralyzing them with poison from the pincers, although larger species are also thought to feed on small vertebrates. The larger (26- to 45-cm) venomous species found in the southern United States and in tropical climates are able to penetrate the skin of humans, giving a painful, burning bite with local reactions including edema, blistering, and tissue necrosis. Systemic reactions may occur, especially in sensitized individuals, but fatalities are rare (1, 2, 13, 26, 28).

CLASS CRUSTACEA (CRUSTACEANS)

Crustaceans of medical importance are few, and their mechanisms of injury are related to their role as vectors of certain helminths. They are primarily aquatic organisms that use gills for breathing or absorb oxygen directly through the cuticle. Crustacea possess at least five pair of legs, which may be specialized for particular functions (i.e., swimmerets), and two pairs of antennae. Decapod crustacea include the prawns, shrimps, lobsters, crayfish, and crabs, which are characterized by the presence of a cephalothorax (fused head and thorax) covered by a protective shield known as a carapace. Certain crabs and crayfish serve as the second intermediate host for the various species of the lung fluke *Paragonimus* sp. found around the world. When infected crabs or crayfish are eaten raw, the encysted metacercariae are transmitted to humans, and adult worms eventually come to reside in the lungs (1, 2, 28).

Copepods are microscopic zooplankton that lack a carapace, and certain freshwater genera (*Cyclops, Diaptomus*) are capable of serving as the first intermediate host for several different human helminths, including the nematodes *Dracunculus medinensis* and *Gnathostoma spinigerum* and cestodes of the genera *Diphyllobothrium* and *Spirometra* (1, 2, 26, 28). Copepods are occasionally found in stool specimens but undoubtedly result from ingestion of unfiltered water.

CLASS INSECTA (INSECTS)

The insects, or hexapods, are the most important of the arthropod classes, constituting more than 90% of all described species in the phylum, and are the only class in which flight has developed. Insects are easily distinguished from other arthropods by a body that is divided into three parts (head, thorax, and abdomen); one pair of antennae and three pairs of legs; and one, two, or no pairs of wings. Insects may be involved in all mechanisms of injury reviewed above, although their impact on the transmission of infectious agents, including viral, bacterial, protozoan, and metazoan pathogens, is especially noteworthy. Of the more than 30 described orders of insects, only those of major medical importance are touched on here (Table 1).

Sucking Lice: Order Anopleura (Phthiraptera)

Sucking lice are small, wingless ectoparasites that are dorsoventrally flattened and have a specialized claw on the end of each leg that allows them to cling to hair shafts or clothing fibers. The three species of human lice include the head louse, *Pediculus capitis*; the body louse, *Pediculus humanus*; and the pubic or crab louse, *Phthirus pubis* (28). Although named for their primary sites of infection, lice are not always confined to a certain part of the body. All of these species suck blood intermittently and produce eggs known as nits, which are deposited either on hair shafts (*P. capitis* and *P. pubis*) or clothing (*P. humanus*). Metamorphosis is gradual, and larval forms live and feed side by side with adults. In addition to their blood-sucking habits, body lice are capable of transmitting the rickettsial agents of typhus and trench fever and the spirochetes responsible for relapsing fever (Table 2) (1, 2, 13).

Head and body lice are very similar morphologically, and their distinction as two species has been controversial (Fig. 1). Anatomic and ecologic differences have been detected, however, and support the concept of two species (24, 26). In addition, only the body louse is known to transmit pathogens between humans. Both species grow to about 3 mm in length and are longer than they are wide. The male is usually smaller than the female and has a more rounded posterior end (13, 24, 26, 28).

Head lice are found predominantly on the head and neck and behind the ears, with the nits being glued to hair shafts very close to the skin surface in these areas. Because incubation takes 7 to 10 days, hatched eggs appear farther up (ca. 0.5 in. [ca. 1.27 cm]) on the hair shaft. Embyonating eggs have intact opercula, whereas hatched eggs do not. Because a variety of objects (hair casts, dander, hair spray, and fungal hair infection) may mimic nits, which are typically 1 mm in length, it is important to differentiate them carefully (Fig. 1). Schoolchildren are at particular risk for acquiring these parasites through the sharing of infested caps, scarves, or combs; from exposure to infested carpeting, beds, or furniture; or from close personal contact. Parents and other family members may subsequently become in-

FIGURE 1 Medically important arthropods. Row 1: (left) centipede; (right) *Phthirus pubis* (crab louse). Row 2: (left) *Pediculus capitis* (head louse); (right) nit of head louse attached to hair shaft. Row 3: (left) *Pulex irritans* (human flea); (right) *Cimex lectularius* (bedbug). Row 4: (left) *Dermatobia hominis* (human botfly) third-stage larvae (photograph courtesy of J. Brad Thomas, David L. Bergeron, and James J. Plorde); (right) scorpion.

fested when children bring lice home from school or day-care centers (1, 13, 18, 26, 30).

Adult and immature body lice are found predominantly on the hairy body regions below the neck when they are feeding, and they move readily between the body and clothing when they are not feeding. Unlike head lice, body lice tend to lay clusters of eggs on clothing, especially on seams or waistbands. Body lice are readily spread among individuals held in institutional settings or other close quarters. Sharing of infested clothing and exposure to infested bedding are common routes of transmission (1, 13, 26, 30).

Pubic or crab lice are much rounder than the other lice, measuring approximately 2 mm in diameter, and their first pair of legs are significantly smaller and more slender than the other pairs (Fig. 1). The abdomen is more crablike and displays hairy tufts on the lateral margins. Adults, nymphs, and nits are usually found on the pubic hairs but may be found on chest, armpit, and facial hair, including eyebrows and eyelashes. Transmission occurs through close personal contact, especially during sexual intercourse (1, 13, 26, 30).

Fleas: Order Siphonaptera

Fleas are small wingless ectoparasites capable of sucking or "siphoning" blood, hence the origin of the ordinal name. Unlike lice, fleas have bodies that are laterally compressed and long, muscular legs that are adapted for jumping (Fig. 1). They undergo complete metamorphosis during their life cycle, which includes four stages: egg, larva, pupa, and adult. Of the approximately 2,500 described species and subspecies, 94% are ectoparasites of mammals, with the remainder occurring on birds (26).

Only a small number directly affect humans. These include blood-sucking pests; tissue-penetrating jiggers; intermediate hosts for certain cestode parasites; and vectors for the agents of plague, murine typhus, and perhaps others. Identification of fleas is difficult for the clinical microbiologist because of the specialized nature of the terminology used to describe flea morphology, the difficulty in following complex dichotomus keys, and the large number of described species (8, 15, 16, 26, 28). Fortunately, the number of flea species commonly associated with humans is small, and the infestation is often related to contact with domestic animals and pets.

The Oriental rat flea, *Xenopsylla cheopis*, is the most important species affecting human health because of its ability to transmit the bacterial agents of plague and murine typhus. This species normally parasitizes Norway and roof rats but will readily bite humans should the rodent host die. The northern rat flea, *Nosopsyllus fasciatus*, also parasitizes Norway and roof rats and may be involved in the transmission of murine typhus to humans. The sticktight flea, *Echidnophaga gallinacea*, is a common pest of poultry but will readily attack humans. The name originates from the flea's habit of attaching firmly to the skin for extended periods during feeding (1, 13, 26).

The human flea, *Pulex irritans*, normally parasitizes a variety of other mammals, including domestic and wild pigs, canids, mustelids, and deer (Fig. 1). In certain settings, such as farms, they may be a particular nuisance in causing flea bite hypersensitivity reactions. *Pulex* fleas may play a role in maintaining the sylvatic cycle of plague. Throughout North America, the cat flea, *Ctenocephalides felis*, and to a lesser extent the dog flea, *Ctenocephalides canis*, are the most pestiferous flea species encountered. Both species move easily between cats and dogs and readily attack hu-

mans. In addition to causing typical delayed hypersensitivity skin reactions from their bites, cat and dog fleas are the usual intermediate hosts for the cysticercoid stage of the tapeworm *Dipylidium caninum* and less frequently for *Hymenolepis diminuta* and *Hymenolepis nana* (1, 13, 26, 30).

The jigger or chigoe flea, *Tunga penetrans*, is the cause of tungiasis in Central and South America and parts of tropical Africa. Its mechanism of injury is unusual because of the habit of the female flea of embedding itself in the skin during its reproductive life. Usual sites of infestation include the soft skin between the toes and under the toenails, but other sites, including the hands, arms, elbows, and genital region, may also be affected. As the flea engorges with blood and begins to produce eggs, it swells to approximately 1,000 times its original size (about the size of a small pea). Several thousand eggs are produced, after which the flea dies. Resulting inflammatory responses and secondary bacterial infections may result in loss of digits, septicemia, or tetanus. Detection of infestation is made by recognizing the dark portion of the chigoe flea's abdomen, which displays the respiratory spiracles, as it protrudes from the skin surface in the center of an enlarging, inflamed lesion (1, 13, 38).

Cockroaches: Order Dictyoptera

Cockroaches are dorsoventrally flattened insects with long multisegmented antennae, biting mouthparts, two pairs of wings, two posterior abdominal projections known as cerci, and legs adapted for running. Metamorphosis is gradual, and the nymphs appear similar to the adults, although the nymphs are smaller and have no wings. Cockroaches are omnivorous nocturnal feeders and prefer secluded areas. Several species have closely adapted themselves to human habitation, where they share our food, water, shelter, and warmth (2, 14).

Mostly considered nuisance pests, cockroaches impart an unpleasant odor to living areas by fouling their environment with feces, disgorging partially digested food at intervals, and discharging secretions from abdominal glands (14). They are also potential carriers of fecal pathogens because of their ease of movement from sewers and drains to areas of habitation, especially food preparation areas. In addition to harboring many enteric bacteria, cockroaches can also harbor hepatitis and polio viruses, *Entamoeba histolytica*, and several species of the most common enteric nematodes. They also may occur in such high densities as to cause allergy and asthma in some individuals following exposure to their excreta, cast skins, or body parts (1, 13, 26).

The most important pest species include the German cockroach, *Blattella germanica*; the Oriental cockroach, *Blatta orientalis*; the brown-banded cockroach, *Supella longipalpa*; and several species in the genus *Periplaneta*, including the American cockroach, *Periplaneta americana* (14, 28). The different species have somewhat different habits. The German roach prefers the warm, moist conditions commonly found in kitchens, restaurants, and bathrooms. It is considered the number one pest in buildings in the United States. The Oriental cockroach is found in cooler locations such as cellars and basements and the cooler parts of kitchens as well as in outdoor drainage areas and rubbish heaps. The brown-banded cockroach is a common pest in most parts of the tropics and has spread to North America. It may be found in many different parts of a building and is known to cause damage to book bindings. The American cockroach is the most widespread and frequent cause of domiciliary infestation in both the tropics and the subtropics. In temperate climates, it seeks out conditions similar to those preferred by the German cockroach (13, 26).

Bedbugs and Kissing Bugs: Order Hemiptera

Bedbugs (family Cimicidae) and kissing and assassin bugs (family Reduviidae) are blood-sucking insects characterized by a long proboscis that is folded ventrally under the body when not in use. Members of both families undergo gradual metamorphosis and may be found in areas of human habitation.

The common bedbug, *Cimex lectularius*, and the tropical bedbug, *Cimex hemipterus*, are reddish brown, oval, dorsoventrally flattened insects approximately 5 mm in length. Adults have small wing pads that are nonfunctional (Fig. 1). Kissing bugs are black or brown with elongate, cone-shaped heads, narrowed necks, and four-segmented antennae that insert laterally into the head. They average 1 to 3 cm in length. The abdomen is widened in the middle and may display orange and black markings. They have wings and are good fliers (2, 13, 26, 28).

Both bedbugs (adult males and females and nymphs) and kissing bugs feed nocturnally, will attack almost any mammal, and are cosmopolitan in distribution. Bedbugs hide in bedsteads, mattresses, and under loose wallpaper, whereas kissing bugs live in cracks and crevices of walls and in thatched roofs. Kissing bugs feed exclusively on vertebrate blood and are named from their habit of biting around the face, often painlessly. Assassin bugs predominantly attack and feed on other insects but will attack humans, giving a painful bite.

The bites of both groups of bugs produce lesions that range from small red inconspicuous marks to hemorrhagic bullae, depending on an individual's sensitivity to their saliva. Periorbital edema, known as Romana's sign, is often seen following a bite by kissing bugs to the face. Bedbugs produce a peculiar odor in heavily infested homes, and their repeated attacks may make sleep impossible. While bedbugs are not known to transmit disease, reduviids are well known to transmit the agent of Chagas' disease, *Trypanosoma cruzi*, in Mexico, Central America, and South America (see chapter 104 of this Manual). The infective trypomastigotes are passed in the feces of the reduviid during feeding and are usually self-inoculated by the host secondary to scratching (1, 13, 26, 30).

Bees, Wasps, and Ants: Order Hymenoptera

Hymenopterans are characterized by two pairs of membranous wings, with the front wings much larger than the hind wings, and by biting-chewing mouthparts. Ant queens and males display wings only during certain times of the year. Many hymenopterans are social insects that attack and sting any intruder disturbing their nests. The stinging apparatus is actually the modified ovipositor of the female, which is capable of injecting venom and is used either for the capture of prey or for defense (2, 14, 23, 26, 28).

The stings of honeybees and bumblebees (family Apidae) and of wasps, hornets, and yellow jackets (family Vespidae) cause only transient pain and swelling in most individuals but may be responsible for severe systemic reactions, including anaphylaxis, in sensitized persons. In the United States, an estimated 50 to 100 people die each year from such reactions. The introduction of the Africanized honeybee to Brazil in 1956 has resulted in the spread of these bees to much of South and Central America and more recently to the southern United States. These bees are

almost identical to domesticated strains but are known to be more readily provoked in defense of the hive and exhibit massive stinging behavior (1, 13, 26, 28).

Depending on their size, almost all ants (family Formicidae) are capable of piercing human skin with their bite, and some species, such as harvester ants and fire ants, may sting as well. The imported fire ant *Solenopsis invicta* has become particularly troublesome in the southeastern United States because it is easily disturbed and capable of both biting with strong mandibles and injecting venom with a sting (1, 13, 26, 28).

Beetles: Order Coleoptera

Beetles constitute the largest group of insects, with over a quarter of a million described species, and are characterized by front wings modified into leathery coverings that meet in a straight line down the middle of the abdomen (2, 26, 28). While beetles are considered of minor medical importance, many can bite, and others, especially the blister beetles (family Meloidae), may exude irritating and vesicating fluids that cause dermatitis or blister formation. Blister beetles are found in many parts of the world and are widespread in the United States. Cantharidin is the vesicating substance found in blister beetles, and the chief component of the aphrodisiac Spanish fly (1, 13, 14, 26).

Larvae of some larder beetles (Dermestidae) have urticating hairs that may penetrate the skin and cause dermatitis. If ingested, these hairs may irritate the intestinal tract. Dermestid beetles and mealworms (Tenebrionidae) serve as intermediate hosts for the rodent tapeworms *H. diminuta* and *H. nana*, which are known to infect humans. Reports of invasion of living tissues by adult or larval beetles are not uncommon in the literature, but such invasions are usually considered accidental occurrences. Pseudoparasitism of the intestinal tract may result from eating infested foodstuffs and passing intact larvae or adults in the stool (1, 13, 26, 28).

Moths and Butterflies: Order Lepidoptera

Medically important moths and butterflies are primarily those having caterpillars that possess urticating hairs or spines capable of secreting venom when the creature is handled. Exposure to urticating hairs usually results in no more than localized dermatitis, whereas exposure to stinging spines may result in local manifestations of stabbing pain, a burning sensation, and inflammation. In some individuals and with certain species, the effects may become systemic and lead to shock and paralysis (1, 9, 26, 28).

Urticating scales and hairs of certain adult moths, especially those of the tussock moths, gypsy moths, and *Hylesia* spp., may also be problematic, causing dermatitis, upper respiratory tract irritation, and eye irritation. Tussock and gypsy moths are known to cause problems seasonally for forestry workers and others in the United States, whereas *Hylesia* moths are common in South America and may emerge in large numbers and congregate around bright lights in populated areas (9, 13, 31).

Flies, Mosquitoes, and Midges: Order Diptera

Dipterans include the insects of greatest medical importance, although most species are harmless to humans. They are characterized by a single pair of functional membranous wings, and they undergo complete metamorphosis (i.e., egg, larva, pupa, and adult) in which the larvae develop in environments completely different from the habitats of the adults (2, 26, 28).

Human disease and injury from these insects are engendered primarily by three mechanisms: blood sucking, biological and mechanical transmission of infectious agents, and myiasis or infection of the body by larval dipterans. Blood sucking may be done by dipterans equipped with puncturing-sucking mouthparts (mosquitoes), piercing-sucking mouthparts (stable flies, tsetse flies), or bladelike mouthparts (deerflies, horseflies, and blackflies). Bites from the last two groups are usually much more painful than those from the first group. Saliva from most dipteran bites causes local irritation and, in some individuals, systemic reactions, including fever, urticaria, and respiratory distress. Continuous or repeated exposure to blood-sucking diptera can often be a traumatic experience both physically and psychologically (1, 13, 26, 30).

Blood-sucking diptera are also responsible for the transmission of major human pathogens, including malaria, filariasis, and arboviruses by mosquitoes; African trypanosomiasis by tsetse flies; leishmaniasis by sand flies; and onchocerciasis by blackflies. Nonbiting synanthropic flies such as houseflies, flesh flies, and blowflies are well known for mechanically transmitting certain viral, bacterial, protozoal, and helminthic agents responsible for intestinal diseases (1, 14, 38). These flies readily feed on human food as well as excrement and easily move between the two. Identification of adult diptera is beyond the scope of this chapter and requires the use of specialized literature.

Certain dipteran larvae (maggots) are capable of surviving or developing in living human tissues and body cavities in an obligatory, facultative, or accidental fashion. Maggots are often found feeding on dead tissue in wounds following removal of dressings or casts. In some cases, they may invade adjacent viable tissues in a facultative fashion. Flies producing accidental myiasis have no requirement for development in human or animal tissues and include the housefly *Musca domestica*. A smaller number of fly species have an obligate need to invade living tissues and may cause serious, even life-threatening infection. While infections may occur in any anatomic site, most myiases are reported as being gastrointestinal, dermal or cutaneous, auricular, ocular, nasopharyngeal, or genitourinary (1, 13, 26, 38). Gastrointestinal infections are most often accidental, occurring following ingestion of contaminated foodstuffs, or represent pseudomyiasis, i.e., a stool sample that has been secondarily contaminated following improper collection or storage.

Obligatory myiasis in humans is caused by relatively few species, all of which are zoonotic in origin. *Dermatobia hominis*, the human botfly, may appear in boil-like lesions of individuals who have spent time in certain parts of Africa, Central America and South America. As the larvae mature, the posterior end, which bears the spiracles, is seen at the surface of the dermal lesion. Mature larvae emerge from these lesions and fall to the ground, where they pupate (Fig. 1). Another arthropod, such as a mosquito or tick, upon which the female botfly has deposited her eggs, is the actual vector of this species. The tumbu fly (*Cordylobia anthropophagia*) is one of the more common myiasis-producing flies affecting humans in sub-Saharan Africa. Eggs are usually laid on dry ground that has been fecally contaminated or on hanging laundry. Upon contact with human or animal skin, larvae rapidly penetrate and produce a furuncular type of myiasis. In some areas, clothing, diapers, and bedding that have been air dried must be routinely ironed to kill the eggs of this species (1, 13, 26, 38).

The Old World screwworm, *Chrysomya bezziana*, and the New World screwworm, *Cochliomyia hominivorax*, deposit their eggs directly on wounds or near the nostrils. Larvae actively feed and move through tissues, causing extensive destruction. The New World screwworm has been eradicated from the United States through the release of sterilized male flies but continues to exist in parts of Central and South America. *Wohlfahrtia magnifica* and *Wohlfahrtia vigil* are sarcophagid flies responsible for obligatory myiasis in parts of Europe, Asia, and North Africa and in North America, respectively. *W. magnifica* can produce a traumatic myiasis similar to that of screwworm, whereas *W. vigil* produces furuncular myiasis (1, 14, 26, 38).

Facultative myiasis is most often caused by blowflies (family Calliphoridae), which ordinarily feed on dead and decaying tissues such as in wounds but may move into adjacent viable tissues opportunistically. Flesh flies (family Sarcophagidae) and houseflies (family Muscidae) occasionally demonstrate similar habits. Examples of some myiasis-causing flies and a pictoral key for identifying their spiracles (breathing pores) are shown in Fig. 2 (1, 17, 26, 28).

CLASS PENTASTOMIDA (PENTASTOMES OR TONGUE WORMS)

Classification of the pentastomes or tongue worms remains uncertain because of a lack of morphologic characteristics from which to determine their affinities. While the usual arthropodlike features are lacking, pentastome larvae do resemble those of mites. The body of the adult is wormlike with pseudosegmentation and appears to be divided into a head and abdomen. A pair of sclerotized hooks or claws is present on each side of the mouth (1, 26).

These organisms live primarily as adults in the nasal and respiratory passages of reptiles, birds, and mammals, where they attain lengths of 1 to 10 cm. Intermediate hosts include rodents, large herbivores, and freshwater fish. Humans may be infected with either the larval or the nymphal stage, depending on whether exposure is to the eggs or the larval forms, respectively, of the parasites. Only rarely have adult pentastomes been recovered from humans. Lesions produced by larval pentastomes usually occur in the liver and lungs and are reported to be caused by *Armillifer moniliformis* in parts of Asia and *Armillifer armillatus* in tropical Africa. In parts of the Middle East and Africa, immature (nymphal) stages of *Linguatula serrata* are known to lodge in the human nasopharynx, producing an irritative and obstructive condition known as halzoun or marrara (1, 12, 38).

CLASS ARACHNIDA (SCORPIONS, SPIDERS, TICKS, AND MITES)

Arachnids of medical importance are found in three subclasses, namely, Scorpiones (scorpions), Araneae (spiders), and Acari (ticks and mites). Scorpions and spiders are characterized by two body regions, the cephalothorax and abdomen, whereas in ticks and mites, these two have been reduced to one. A pair of prehensile, pincherlike chelicerae characterize all members of the class and reflect the predaceous nature of the group. Of the three medically important subclasses, only the Acari are not predaceous. All lack antennae, mandibles, and wings. Metamorphosis is gradual, and molting occurs repeatedly to permit growth. While all arachnid nymphs and adults have four pairs of legs, tick and mite larvae have only three pairs of legs. Arachnids are

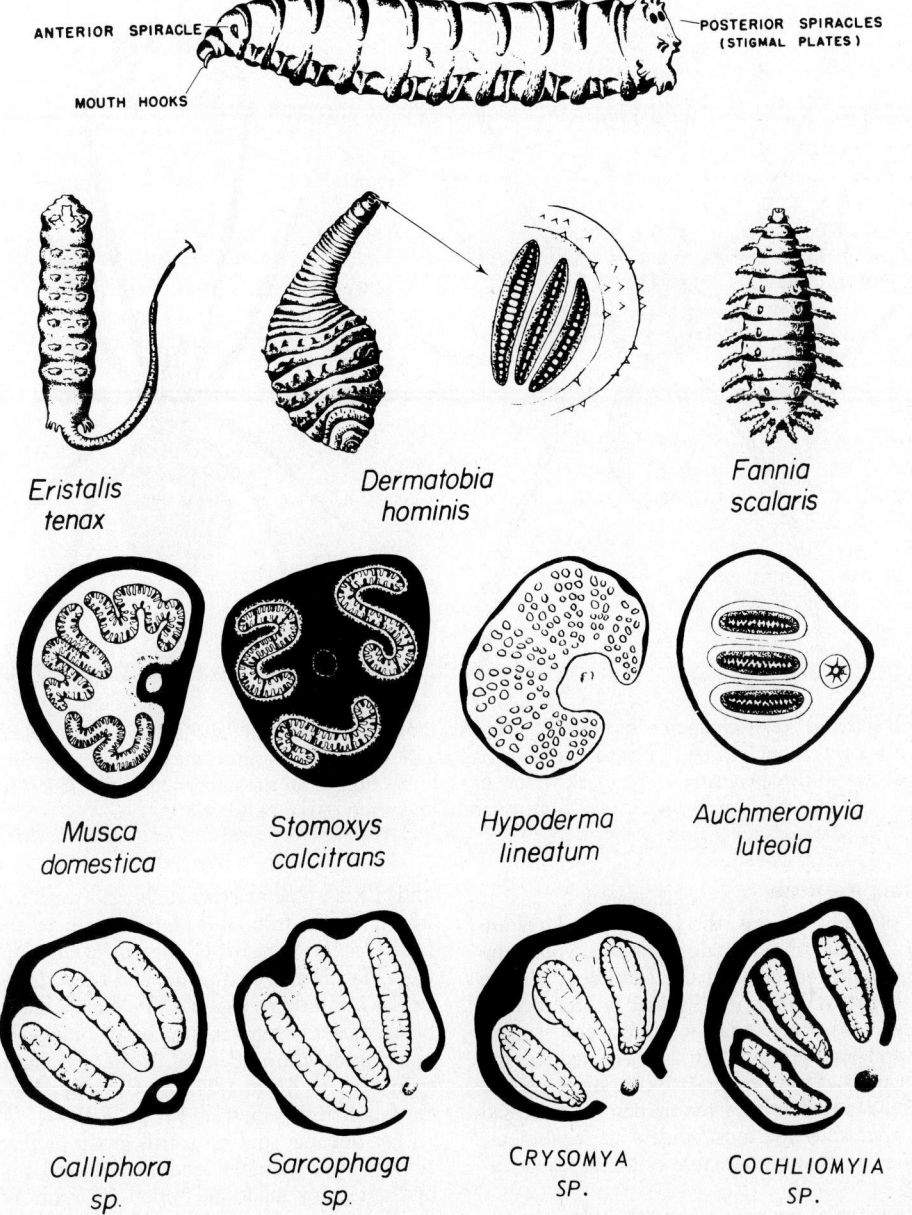

ANTERIOR SPIRACLE

POSTERIOR SPIRACLES
(STIGMAL PLATES)

MOUTH HOOKS

Eristalis tenax

Dermatobia hominis

Fannia scalaris

Musca domestica

Stomoxys calcitrans

Hypoderma lineatum

Auchmeromyia luteola

Calliphora sp.

Sarcophaga sp.

Crysomya SP.

Cochliomyia SP.

FIGURE 2 Key characters of myiasis-producing fly larvae (Centers for Disease Control and Prevention, Atlanta, Ga.). (Top) Mature larva of a muscoid fly (from R. Hegner et al., copyright 1938 by D. Appleton-Century Co., Inc., New York).

perhaps best recognized by their abilities to inject poisonous venoms (scorpions and spiders) and to transmit viral, bacterial, and protozoal pathogens (ticks and mites) (Table 2).

Scorpions: Subclass Scorpiones

The body of a scorpion is divided into the cephalothorax and the segmented abdomen, to the latter of which is attached a five-segmented tail that ends in a bulbous stinging apparatus. In addition to having four pairs of legs, a scorpion has a pair of forward-directed pincer claws or pedipalps that gives it a crablike appearance (Fig. 1). Length varies from 2 to 10 cm, depending on the species (1, 26, 28).

Scorpions are nocturnal feeders that paralyze their in-

tended victim with venom. When the scorpion is disturbed, the stinger may be used for defense. Depending on the species, the venom varies in its neurologic and hemolytic toxicity for humans, with the stings of many species eliciting no more reaction than those of bees and wasps. The stings of some species, however, may be deadly, accounting for over 1,000 deaths annually. Symptoms encountered during systemic poisoning include, among others, spreading paralysis, hypertension, convulsions, and respiratory arrest. While several relatively harmless species are routinely found in the southwestern United States, *Centruroides sculpturatus* is potentially deadly, especially to children. Other poisonous species are found in Europe, Africa, and the Middle East (1, 13, 26, 38).

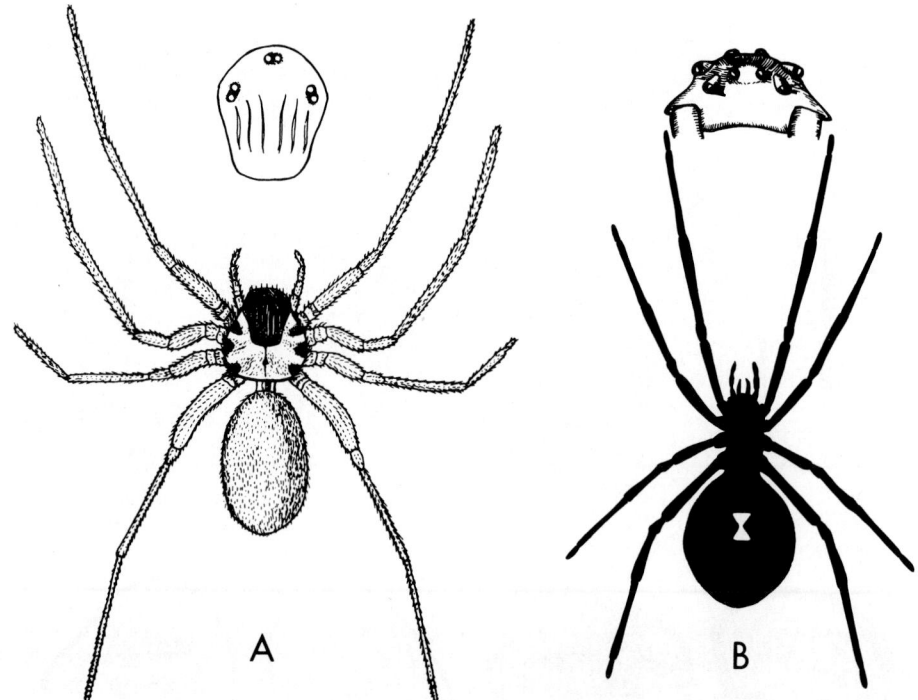

FIGURE 3 Key characters of *Loxosceles reclusa* (brown recluse spider) and *Lactrodectus mactans* (black widow spider). (A) The brown recluse has a fiddle-shaped marking on the cephalothorax and six eyes in three pairs. (B) The black widow has a red hourglass on the underside of the abdomen and eight eyes (Centers for Disease Control and Prevention, Atlanta, Ga.).

Spiders: Subclass Araneae

Like the bodies of the scorpions, the bodies of spiders are divided into two regions: the cephalothorax, to which the four pairs of legs are attached, and the abdomen. Unlike that of scorpions, the abdomen of a spider is superficially unsegmented and is missing the tail with attached stinger. In place of the tail are spinnerets, which connect to the abdominal silk glands. The mouthparts contain a pair of fanglike chelicerae through which venom can be expressed from associated glands. While most spiders are venomous, few possess chelicerae that are capable of penetrating human skin (1, 2).

Spiders are terrestrial animals that use silk webs to ensnare their prey and venom to immobilize it. The majority of spiders able to bite humans produce little more than transitory irritation and pain. The urticating hairs on some species of tarantulas may produce more discomfort than their bite does. Of the 30,000 or more species that have been described, only 20 to 30 are known to cause more serious disease (20, 26, 28).

Systemic arachnidism may be caused by several groups of spiders, of which the widow spiders (genus *Lactrodectus*) are best known. Other groups include certain tarantulas (several genera), funnel web spiders (genus *Atrax*), wolf spiders (genus *Lycosa*), and wandering spiders (genus *Phoneutria*). Widow spiders are cosmopolitan in distribution, with five closely related species occurring in the United States: *Lactrodectus mactans* (black widow), which occurs from southern New England to Florida and west to California and Oregon; *Lactrodectus variolus* (northern black widow), which occurs from the northeastern United States and adjacent areas in Canada to Florida and west to Oklahoma

and Texas; *Lactrodectus hesperus* (western black widow), which occurs from the Plains States to the Pacific coast; *Lactrodectus geometricus* (brown widow spider), reported from Florida and California; and *Lactrodectus bishopi*, which occurs in southern Florida. The female black widow is shiny black, has a characteristic red or orange hourglass-shaped marking on the underside of the abdomen, and has a leg span of 3 to 4 cm (Fig. 3). Considerable color variation exists among species (1, 13, 26, 28).

For nesting, spiders in this group prefer protected locations with a suitable food source, and in past years, the outdoor privy made an opportune spot. When provoked, they bite in self-defense, injecting a potent nonhemolytic neurotoxin that acts on the central nervous system, producing weakness, myalgia, and muscle spasm that may progress to paralysis, convulsions, and occasionally death. Mortality has been estimated at <1 to 6% (1, 13, 26, 30, 38).

Necrotic arachnidism or loxoscelism occurs following envenomation by members of the genus *Loxosceles*, known as violin spiders. In the United States, *Loxosceles reclusa*, the brown recluse or fiddleback spider, is the most commonly implicated species. It occurs in the eastern half of the country and westward into Wyoming and Arizona. Other species implicated in the United States include *Loxosceles deserta*, *Loxosceles arizonica*, *Loxosceles rufescens*, and *Loxosceles devia*. *Loxosceles laeta* and *Loxosceles rufipes* are commonly encountered in parts of South America and produce a more severe form of cutaneous necrosis. The brown recluse is a tan to dark brown spider 1 to 2 cm long that is characterized by a darkened violin-shaped marking oriented base forward on the dorsum of the cephalothorax and extending from the eyes to the base of the abdomen (1, 13,

26). Six eyes are grouped in a semicircle of three pairs on top of the head (Fig. 3).

In areas of human habitation, the reclusive habits of these spiders often place them in closets among clothes, in basements, under porches, or in other areas that are normally little disturbed. When provoked, the spiders may give a painless bite and inject a dermonecrotic and hemolytic toxin. Several hours later, the area will become red, swollen, and painful. Cutaneous necrosis and sloughing then occur over a matter of days, leaving a deep lesion that is slow to heal and prone to scarring. Depending on the amount and potency of the toxin injected, complications may remain superficial or occasionally may become systemic, with a high mortality rate (1, 13, 26, 30, 38).

Reports of necrotic arachnidism also occur in areas where *Loxosceles* spiders are not found. In the U.S. Pacific Northwest, the hobo spider, *Tegenaria agrestis*, has been implicated as a cause of cutaneous necrosis, with 40% of victims having some systemic symptoms as well (7).

Ticks and Mites: Subclass Acari

Ticks

Ticks are obligate blood-sucking ectoparasites of many reptiles, birds, and mammals, including humans. Although they constitute a small group of only about 800 described species, they are of extreme importance to human and animal health because of their impact in transmitting a variety of viral, bacterial, and protozoal pathogens and in causing additional loss to livestock and wildlife because of their feeding activities (26).

Ticks and mites may be differentiated from other arachnids by their fused cephalothorax and abdomen and by their larvae, which have three pairs of legs. Many mites are microscopic, whereas ticks may be easily seen with the naked eye. In addition, a characteristic part of the tick feeding apparatus is the toothed hypostome, which allows the mouthparts to remain solidly imbedded (1, 2, 14).

Two of the three families of ticks, the hard ticks (Ixodidae) and the soft ticks (Argasidae), contain species known to transmit pathogens to humans. Hard ticks may be recognized by the presence of a dorsal sclerotized plate known as the scutum and by mouthparts that project anteriorly when viewed from the top. The scutum covers the entire dorsal surface of the male but only the anterior dorsal surface of the female (Fig. 4). This allows for greater abdominal engorgement with blood by females during feeding, ensuring a greater supply of the nutrients needed for egg production. While males do ingest blood, the amount is much less.

Argasids, or soft ticks, differ from hard ticks by having a leathery integument, downwardly directed mouthparts (not visible from the top), and no scutum (Fig. 4). Life cycles vary considerably between the two groups, with soft ticks often taking multiple blood meals of short duration (usually less than 30 min) during each stage, having several nymphal stages, and being long-lived. They may feed repeatedly on the same or different hosts, and females may lay eggs several times during their lives. Hard ticks have a single nymphal stage, take only one blood meal on a host at each growth stage, and remain attached to the host for hours or days while feeding. The life cycle of most hard ticks requires 2 years for completion and may involve from one to three hosts, depending on the tick species. Female hard ticks die after ovipositing (13, 14, 23, 32–34).

Ticks found crawling on or imbedded in the skin of humans are most often hard ticks. Soft ticks that feed on humans tend to be most active at night, feeding only briefly and then often painlessly, making their discovery uncommon. Care must be taken not to misidentify an engorged hard tick for a soft tick, which it superficially resembles. Identification of ticks found in the United States to the genus level is only moderately difficult for the nonspecialist (Fig. 5) (1, 14, 26–28). Identification of specimens to the species level (4–6, 11, 21, 22, 27, 28) is more problematic for the clinical microbiologist and should generally be referred to an outside specialist (entomologist, parasitologist, university extension service, state health laboratory, Centers for Disease Control and Prevention, etc.). This is especially true when the submitting clinician is contemplating postexposure prophylaxis for such tick-vectored diseases as Lyme borreliosis or Rocky Mountain spotted fever.

Several important species of hard ticks found in North America include *Dermacentor variabilis* (the American dog tick), *Dermacentor andersoni* (the Rocky Mountain wood tick), *Ixodes scapularis* (black-legged tick, northern and southern forms), *Ixodes pacificus* (western black-legged tick), *Amblyomma americanum* (the Lone Star tick), and *Rhipicephalus sanguineus* (the brown dog tick). *Dermacentor* and *Amblyomma* ticks are referred to as ornate because of the easily recognizable whitish markings on the scutum. The other genera of North American ticks lack this ornamentation and are known as inornate ticks (Fig. 4).

The American dog tick, *D. variabilis*, is found throughout the United States exclusive of the Rocky Mountain region and is the primary vector for Rocky Mountain spotted fever in the eastern states (Fig. 4). The Rocky Mountain wood tick, *D. andersoni*, occurs throughout the Rocky Mountain states and into Canada, where it serves as the primary vector for Rocky Mountain spotted fever. Both species have been implicated in cases of tularemia, tick paralysis (see below), and, rarely, Q fever. In addition, *D. andersoni* is the chief vector for transmission of the coltivirus responsible for Colorado tick fever (1, 34, 38).

The genus *Ixodes* consists of inornate ticks that differ from other hard ticks by having the anal groove surround the anus anteriorly (Fig. 4 and 5). This feature may be difficult to appreciate on engorged ticks. While there are 36 described species in the United States that parasitize a variety of small and large mammals (6, 21), primarily 2 species are involved in the transmission of the Lyme disease spirochete to humans (34, 36). *I. scapularis*, the black-legged tick, is the common vector for this spirochete along the eastern seaboard, in the southern states, and in the north central states. The northern form of this tick was renamed *Ixodes dammini* in 1979 because certain structural differences from the southern form were detected, but recent studies on mating compatibility and genetic similarities have shown the two to be conspecific (29, 35). *I. pacificus*, the western black-legged tick, is the most common vector for transmission of Lyme disease in the western states. These two species are also known to transmit the protozoal agent of babesiosis, *Babesia microti*, and mixed infections are being described with increasing frequency. The vector for Lyme disease in Europe is the related species *Ixodes ricinus*. In other parts of the world, *Ixodes* spp. are involved in the transmission of encephalitis-causing arboviruses. Tick paralysis may also be caused by members of this genus. Identification of *Ixodes* ticks to the species level is a difficult task and requires the assistance of a specialist (6, 11, 21).

A. americanum, the Lone Star tick, is so named because

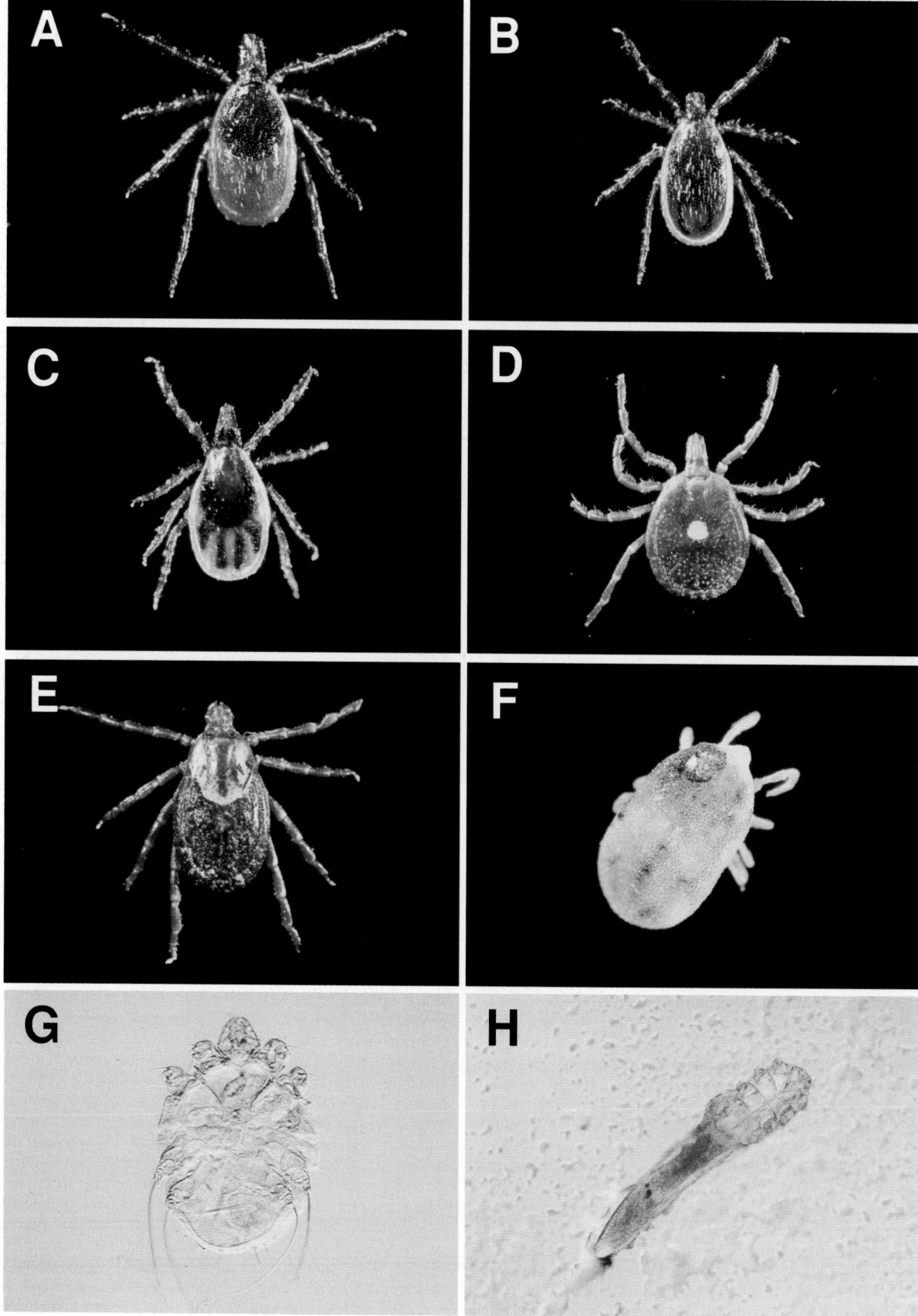

of a single white marking at the posterior end of the scutum in the female (Fig. 4) and its occurrence in the Lone Star State, Texas (1, 13, 26, 28). It is one of the more pestiferous species, and its range extends from Texas to the Atlantic coast and north to New York. All stages attack humans and are able to inflict painful bites because of their long mouthparts. This species is capable of transmitting Rocky Mountain spotted fever as well as tularemia and possibly Lyme disease. It is also capable of causing tick paralysis (34).

R. sanguineus, the brown dog tick, is an inornate tick distributed worldwide that is commonly found on dogs and in the homes in which they reside. All stages feed on the same canine host and occasionally bite humans. This species has been implicated in the transmission of the agents of Rocky Mountain spotted fever and ehrlichiosis in North America and of boutonneuse fever in the Mediterranean basin (1, 13, 26, 28).

Soft ticks of the genus Ornithodoros are important vectors worldwide for the spirochetal agents of relapsing fever (Borrelia spp.). In the United States, at least four tick species are involved in transmission. These include Ornithodoros hermsi (Rocky Mountain and Pacific coast states and British Columbia) (Fig. 4), Ornithodoros parkeri (western United States), Ornithodoros turicata (western United States and northern Mexico), and Ornithodoros talaje (southern United States). A variety of rodents serve as the usual tick hosts and reservoir for the spirochetes. Human exposure to these ticks usually occurs outdoors, while camping near rodent burrows or in rodent-infested buildings, especially older vacation cabins (1, 13, 14, 26). Species identifications require use of specialized keys (5, 27, 28).

Tick paralysis is another important and potentially lethal disease caused in North America by ticks of the genera Dermacentor, Amblyomma, and Ixodes. A neurotoxin produced in the tick's salivary glands produces an ascending flaccid paralysis with generalized toxemia within several days after initiation of feeding by a female tick. The disorder may readily be differentiated from other causes of paralytic disease, including poliomyelitis, by the finding of an attached, engorged tick in an area usually hidden from view, such as the scalp. Careful removal of the tick and its mouthparts will generally cause resolution of symptoms within several hours or days (1, 34, 38).

Mites

A mite is a tiny arachnid with a fused cephalothorax and abdomen that has eight legs as a nymph and adult and six legs as a larva. The study of acarology is largely a matter for the specialist, as keys are complex and not always helpful for the relatively few medically important species (1, 13, 25–28, 39). Identification of the obligate human mites, including the itch mite (Scarcoptes scabei) and the follicle mite (Demodex folliculorum), should be within the abilities of most clinical laboratories, however.

Mites are medically important pests for several reasons: they may attack humans directly either as obligate or facultative parasites, certain species may act as reservoirs and vectors for infectious diseases, and many have been impli-

cated in cases of dermatitis and dust allergies. Delusions of mite infestations, one of the forms of delusory parasitosis, are also well-known psychiatric problems that require careful investigation before actual infestations are ruled out.

Sarcoptes scabei, the itch mite and agent of sarcoptic mange, is an obligate human parasite that lives in the upper layers of the epidermis, where it creates serpiginous burrows. Mites live for about 2 months, during which their tunneling activity and deposition of eggs and excreta produce inflammation and intense itching. Most mites occur on the hands and arms, especially in the interdigital folds, but they may involve the wrists, breasts, buttocks, and external genitalia, among other areas. Clinical manifestations of sarcoptic mange may vary considerably depending on the development of hypersensitivity. Immunocompromised individuals may be at risk for developing Norwegian scabies, a generalized dermatitis characterized by the presence of thousands of mites in the epidermis (1, 13, 14, 26, 30, 38).

Diagnosis is made by demonstrating the mite and its eggs in skin scrapings taken preferably from the terminal portion of a burrow. Scrapings may be placed on a microscope slide, cleared with a drop of potassium hydroxide, covered with a coverslip, and examined. Another method recommends using a drop of mineral oil when scrapings are taken and examining this fluid directly. Adult female mites average 330 to 450 μm in length and have oval, saclike bodies in which the first two pairs of legs are separated from the last two pairs. The dorsal body surface has multiple parallel ridges, several toothlike spines, and hairs of various lengths (Fig. 4) (28). Males are somewhat smaller than females.

While humans primarily contract scabies from being in close contact with infested persons (often among families, in institutional settings, or where personal hygiene has been neglected), temporary infestation with animal strains of the parasite may occur. Transmission during sexual contact is also known to occur (1, 19).

The human follicle mites Demodex folliculorum and Demodex brevis are obligate parasites that inhabit hair follicles and sebaceous glands, respectively. They are minute organisms (100 to 400 μm in length, with Demodex brevis being the shorter of the two) well adapted for life in follicles by having a wormlike body, four pairs of stubby legs, and an abdomen with numerous annulations (Fig. 4). Infestations occur primarily on the face, especially on the forehead, eyelids, and nasolabial folds. While the presence of these mites has been associated with various skin conditions, including acne and blepharitis, their high incidence in healthy individuals makes their medical significance hard to assess (1, 13, 26, 28, 30).

A number of free-living animal mite species that are parasitic at one or more stages of their life cycle also attack humans, facultatively causing a variety of dermatologic and allergic conditions. Larval chigger mites of the family Trombiculidae are well known to attack humans and other vertebrates throughout the world; their saliva produces a wheal-and-flare reaction with intense itching. Larvae may appear as tiny red dots attached to the skin, especially in areas where clothing is tight and restricts their movement,

---

FIGURE 4 Medically important ticks and mites. Row 1: (left) I. scapularis (black-legged tick or deer tick, formerly named I. dammini) nonengorged adult female; (right) I. scapularis adult male. Row 2: (left) I. scapularis nonengorged nymph; (right) A. americanum (Lone Star tick) nonengorged adult female. Row 3: (left) D. variabilis (dog tick) nonengorged adult female; (right) Ornithodoros hermsi (soft tick) nonengorged adult. Row 4: (left) Sarcoptes scabei (itch mite) adult; (right) Demodex folliculorum (follicle mite) adult. Photographs in rows 1 through 3 are from Northwest Infectious Disease Consultants and are reproduced with permission.

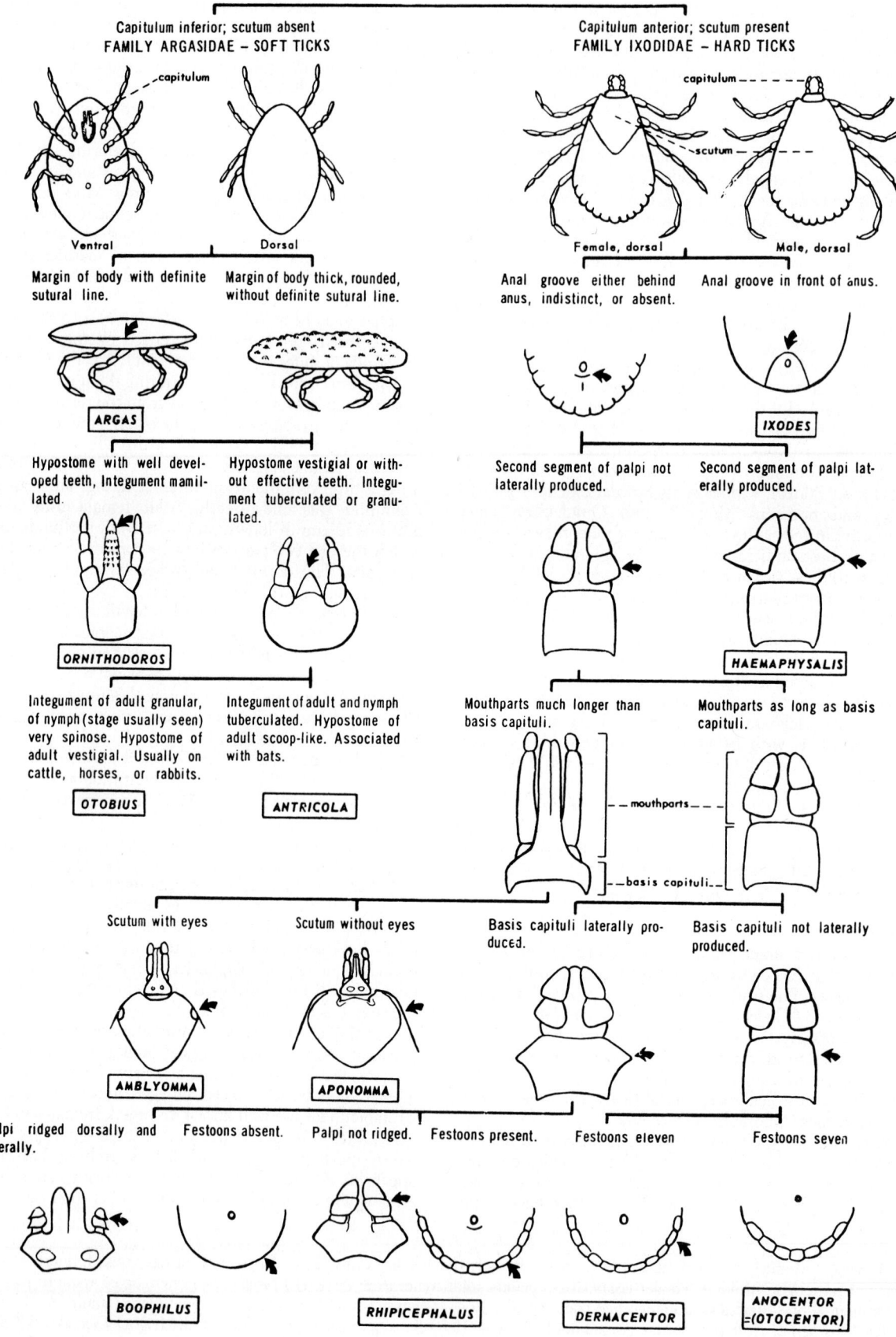

FIGURE 5 Pictorial key to the tick genera found in the United States (Centers for Disease Control and Prevention, Atlanta, Ga.) (28).

such as ankles, groin, waistline, armpits, and wrists. Following feeding, engorged larvae fall to the ground, where they develop into nonparasitic nymphs and adults. *Eutrombicula alfreddugesi* is a common pest in the New World, whereas *Neotrombicula autumnalis* occurs in Europe. Chiggers in the genus *Leptotrombidium* are more problematic in parts of Asia and Australia because they serve as vectors for *Rickettsia tsutsugamushi*, the agent of scrub typhus (1, 26, 38).

A variety of other biting mites normally parasitic on other animals may occasionally be a nuisance to humans, causing dermatitis. Exposure may be occupational (grain, flour, and cheese mites; chicken mite; northern fowl mite; tropical fowl mite) or accidental (straw itch mite; tropical rat mite; spiny rat mite; house mouse mite; and cheyletiellid mites of birds, rodents, cats, dogs, and other mammals). Humans living in older buildings and houses that are inhabited by rodents or birds may be susceptible to mite infestations following the abandonment of a nest and subsequent searching by its mite inhabitants for new hosts. The house mouse mite (*Liponyssoides sanguineus*), in addition to being pestiferous, is an important vector for the transmission of *Rickettsia akari*, the agent of rickettsialpox (1, 13, 14, 23, 28).

Mites in the family Pyroglyphidae are particularly problematic as allergens responsible for house dust allergy. House dust mites (especially *Dermatophagoides farinae* and *Dermatophagoides pteronyssinus*) may occur in great numbers in dust from bedroom mattresses, blankets, pillows, and carpets, where they feed on sloughed human skin, food particles, and other available organic debris. Warm ambient temperatures and high humidity accentuate the problem. The secretions, excreta, and body parts of these mites act as potent allergens responsible for allergic rhinitis and asthma as well as some skin conditions (1, 9, 13, 26, 28, 39).

REFERENCES

1. **Beaver, P. C., R. C. Jung, and E. W. Cupp.** 1984. *Clinical Parasitology*, 9th ed. Lea & Febiger, Philadelphia.
2. **Borror, D. J., N. F. Johnson, and C. A. Triplehorn.** 1989. *An Introduction to the Study of Insects*, 6th ed. Saunders College Publishing Co., Philadelphia.
3. **Chu, H. F.** 1949. *How To Know the Immature Insects.* William C. Brown Co., Dubuque, Iowa.
4. **Cooley, R. A.** 1938. *The Genera Dermacentor and Otocentor (Ixodidae) in the United States with Studies in Variation.* National Institute of Health bulletin no. 171. National Institute of Health, Washington, D.C.
5. **Cooley, R. A., and G. M. Kohls.** 1944. *The Argasidae of North America, Central America, and Cuba.* Am. Midl. Nat. Monogr. 1:1–152.
6. **Cooley, R. A., and G. M. Kohls.** 1945. *The Genus Ixodes in North America.* National Institute of Health bulletin no. 184. National Institute of Health, Washington, D.C.
7. **Fisher, R. G.** 1994. Necrotic arachnidism. *West. J. Med.* 160:570–572.
8. **Fox, I.** 1940. *Fleas of the Eastern United States.* Iowa State College Press, Ames.
9. **Frazier, C. A., and P. A. Brown.** 1980. *Insects and Allergy and What To Do about Them.* University of Oklahoma Press, Norman.
10. **Furman, D. P., and E. P. Catts.** 1970. *Manual of Medical Entomology*, 3rd ed. Mayfield Publishing Co., Palo Alto, Calif.
11. **Furman, D. P., and E. C. Loomis.** 1984. The ticks of California (Acari:Ixodida). *Bull. Calif. Insect Surv.* 25:1–79.
12. **Garcia, L. S., and D. A. Bruckner.** 1993. *Diagnostic Medical Parasitology*, 2nd ed. American Society for Microbiology, Washington, D.C.
13. **Goddard, J.** 1993. *Physician's Guide to Arthropods of Medical Importance.* CRC Press, Inc., Boca Raton, Fla.
14. **Harwood, R. F., and M. T. James.** 1979. *Entomology in Human and Animal Health*, 7th ed. Macmillan, New York.
15. **Holland, G. P.** 1985. The fleas of Canada, Alaska, and Greenland (Siphonaptera). *Mem. Entomol. Soc. Can.* 130:1–630.
16. **Hubbard, C. A.** 1947. *Fleas of Western North America. Their Relation to Public Health.* Iowa State College Press, Ames.
17. **James, M. T.** 1947. *The Flies That Cause Myiasis in Man.* USDA miscellaneous publication no. 631. U.S. Department of Agriculture, Washington, D.C.
18. **Juranek, D. D.** 1985. *Pediculus capitis* in school children: epidemiologic trends, risk factors, and recommendations for control, p. 199–211. *In* M. Orkin and H. I. Maibach (ed.), *Cutaneous Infestations and Insect Bites.* Marcel Dekker, Inc., New York.
19. **Juranek, D. D., R. W. Currier, and L. E. Millikan.** 1985. Scabies control in institutions, p. 139–158. *In* M. Orkin and H. I. Maibach (ed.), *Cutaneous Infestations and Insect Bites.* Marcel Dekker, Inc., New York.
20. **Kaston, B. J.** 1978. *How To Know the Spiders*, 3rd ed. William C. Brown Co., Dubuque, Iowa.
21. **Keirans, J. E., and C. M. Clifford.** 1978. The genus *Ixodes* in the United States: a scanning electron microscope study and key to the adults. *J. Med. Entomol. Suppl.* 2:1–149.
22. **Keirans, J. E., and T. R. Litwak.** 1989. Pictoral key of the adults of hard ticks, family Ixodidae (Ixodida: Ixodoidea) east of the Mississippi river. *J. Med. Entomol.* 26:435–448.
23. **Kettle, D. S.** 1982. *Medical and Veterinary Entomology.* John Wiley & Sons, Inc., New York.
24. **Kim, K. C., H. D. Pratt, and C. J. Stojanovich.** 1986. *The Sucking Lice of North America.* Pennsylvania State University Press, University Park.
25. **Krantz, G. W.** 1978. *A Manual of Acarology*, 2nd ed. Oregon State University, Corvallis.
26. **Lane, R. P., and R. W. Crosskey.** 1993. *Medical Insects and Arthropods.* Chapman & Hall, London.
27. **McDaniel, B.** 1979. *How To Know the Mites and Ticks.* William C. Brown Co., Dubuque, Iowa.
28. **National Communicable Disease Center.** 1969. *Pictorial Keys: Arthropods, Reptiles, Birds, and Mammals of Public Health Significance.* Communicable Disease Center, Atlanta.
29. **Oliver, J. H., Jr., M. R. Owlsly, H. J. Hutcheson, A. M. James, C. Chen, W. S. Irby, E. M. Dotson, and D. K. McLain.** 1993. Conspecificity of the ticks *Ixodes scapularis* and *I. dammini* (Acari: Ixodidae). *J. Med. Entomol.* 30:54–63.
30. **Orkin, M., and H. I. Maibach.** 1985. *Cutaneous Infestations and Insect Bites.* Marcel Dekker, Inc., New York.
31. **Shama, S. K., P. H. Etkind, T. M. Odell, A. T. Canada, A. M. Finn, and N. A. Soter.** 1982. Gypsy-moth-caterpillar dermatitis. *N. Engl. J. Med.* 306:1300–1301.
32. **Sonenshine, D. E.** 1991. *Biology of Ticks*, vol. 1. Oxford University Press, New York.
33. **Sonenshine, D. E.** 1993. *Biology of Ticks*, vol. 2. Oxford University Press, New York.
34. **Spach, D. H., W. C. Liles, G. L. Campbell, R. E. Quick, D. E. Anderson, and T. R. Fritsche.** 1993. Tick-borne diseases in the United States. *N. Engl. J. Med.* 329:936–947.
35. **Spielman, A., C. M. Clifford, J. Piesman, and M. D. Corwin.** 1979. Human babesiosis on Nantucket Island, USA: description of the vector, *Ixodes (Ixodes) dammini*, n. sp. (Acarina: Ixodidae). *J. Med. Entomol.* 15:218–234.
36. **Steere, A. C.** 1989. Lyme disease. *N. Engl. J. Med.* 321:585–596.
37. **Steyskal, G. C., W. L. Murphy, and E. M. Hoover.** 1987. *Insects and Mites: Techniques for Collection and Preservation.* USDA miscellaneous publication no. 1443. U.S. Department of Agriculture, Washington, D.C.
38. **Strickland, G. T. (ed.).** 1991. *Hunter's Tropical Medicine*, 7th ed. The W. B. Saunders Co., Philadelphia.
39. **Woolley, T. A.** 1988. *Acarology: Mites and Human Welfare.* John Wiley & Sons, Inc., New York.

ANTIMICROBIAL AGENTS AND SUSCEPTIBILITY TESTING

X

VOLUME EDITOR
FRED C. TENOVER

SECTION EDITORS
JAMES H. JORGENSEN AND DANIEL F. SAHM

Antimicrobial Susceptibility Testing: General Considerations

JAMES H. JORGENSEN AND DANIEL F. SAHM

110

One of the most important tasks of the clinical microbiology laboratory is the performance of antimicrobial susceptibility tests on significant bacterial isolates. The goal of a susceptibility test is to predict through an in vitro assessment the likelihood of successfully treating a patient's infection with a particular antimicrobial agent. With some bacteria, fungi, viruses, and parasites, the selection of antimicrobial therapy continues to be empiric either because resistance problems have not yet developed or because effective alternative agents do not exist. Susceptibility testing is most important with bacterial species that are not predictably susceptible to drugs of choice for such infections, e.g., staphylococci, enterococci, pneumococcus, members of the family *Enterobacteriaceae*, *Pseudomonas* spp., *Haemophilus influenzae*, *Neisseria gonorrhoeae*, and *Mycobacterium tuberculosis*. Those organisms may be resistant to commonly used antimicrobial agents because of various acquired resistance mechanisms. The most important function of a susceptibility test is the detection of clinically relevant antimicrobial resistance in organisms causing an infection. It is important to emphasize that susceptibility testing should not be performed on normal flora or colonizing organisms and that testing should be performed only if well-standardized methods exist for that species. This section of the Manual focuses on susceptibility testing methods for aerobic and anaerobic bacteria and mycobacteria and provides important updates on progress toward development of standardized methods for fungi, viruses, and protozoa. Conventional, standardized dilution, and agar diffusion susceptibility methods and some automated and nontraditional methodologies are described in the following chapters. A number of choices exist in antibacterial testing with respect to methodology and selection of drugs for routine testing. Fewer choices exist in antifungal, antiviral, and antiprotozoal testing.

SELECTING AN ANTIMICROBIAL SUSCEPTIBILITY TESTING METHOD

Clinical microbiology laboratories can choose from among several conventional or novel methods for performance of routine antibacterial susceptibility testing. These include the disk diffusion test, broth microdilution, agar gradient, and rapid automated instrument methods.

In recent years there has been a trend toward use of either commercial broth microdilution or automated instrument methods instead of the disk diffusion procedure. However, interest in the disk diffusion test may be renewed because of the flexibility in drug selection and the low cost inherent in this test. The availability of numerous antibacterial agents and the diversity in antibiotic formularies in different institutions has made it difficult for manufacturers of commercial test systems to provide standard test panels that fit every facility's needs. Thus, the flexibility of the disk diffusion test, which allows a laboratory to easily custom design its own test batteries, is an undeniable asset. Other advantages of the disk diffusion procedure include the facts that it is one of the most established, best proven of all tests and that it continues to be updated and refined through frequent National Committee for Clinical Laboratory Standards (NCCLS) publications (14). The interpretive category results provided by the disk test (susceptible, intermediate, and resistant) are readily understood by all clinicians.

Advantages of microdilution or agar gradient methods include the generation of a quantitative result (i.e., an MIC) rather than a category result, the ability to accurately test some anaerobic or fastidious species that may not be tested accurately by disk diffusion, and the ancillary benefits of the computer systems that accompany many of the microdilution systems (2, 6, 7, 9). Indeed, the computerized data management systems are extremely useful to laboratories that do not possess a laboratory information system. However, an MIC method should not be chosen on the assumption that MICs are more useful to physicians than susceptibility category results are. There is no clear evidence that the former are more relevant than the latter to appropriate selection of antibacterial therapy.

A laboratory may choose to perform rapid, automated antibacterial susceptibility testing in order to generate test results faster than can be accomplished by manual methods. Provision of susceptibility results 1 day sooner than they can be provided by conventional methods seems a logical advance in patient care. However, there is little objective evidence that rapid susceptibility test results reduce mortality or morbidity (4) unless very aggressive communication measures are initiated by the laboratory to make physicians aware of the results (22). This lack of impact may be because physicians have come to expect antimicrobial susceptibility test results on the second day after a specimen is submitted or because the rapid results are still not available

TABLE 1 Examples of susceptibility testing profiles requiring further evaluation[a]

Organism	Susceptibility profile to investigate
Staphylococci	Vancomycin resistance
Viridans streptococci	Vancomycin resistance
Viridans streptococci	High-level aminoglycoside resistance
Beta-hemolytic streptococci	Penicillin resistance
Neisseria gonorrhoeae	Ceftriaxone resistance
Neisseria meningitidis	Penicillin resistance
Enterobacteriaceae	Imipenem resistance
Klebsiella spp. and *Escherichia coli*	Cefoxitin or cefotetan resistance
Pseudomonas aeruginosa	Amikacin resistance; gentamicin or tobramycin susceptibility
Xanthomonas maltophilia	Imipenem susceptibility or trimethoprim-sulfamethoxazole resistance

[a]Adapted from reference 3. Patterns may indicate technical error or emerging resistance.

soon enough to assist with initial selection of antimicrobial therapy.

A previously cited shortcoming of rapid susceptibility testing methods has been some compromise in detection of some inducible or subtle resistance mechanisms (1, 20–21). The instruments most notorious for such problems are no longer marketed, and the remaining instruments have made significant efforts to correct their earlier problems (11, 23). However, some deficits with the detection of extended-spectrum β-lactamase-mediated resistance and some forms of vancomycin resistance continue to be noted with some current systems (10, 19). Accuracy should not be sacrificed in an effort to generate a rapid susceptibility result.

SELECTION OF ANTIBACTERIAL AGENTS FOR ROUTINE TESTING

The laboratory has the responsibility to test and report on the antimicrobial agents that are most appropriate for the bacterium isolated, the site of infection, and the clinical-practice setting in which the laboratory functions. The battery of antimicrobial agents routinely tested and reported on by the laboratory will depend on the characteristics of patients cared for in the institution and the likelihood of encountering organisms resistant to the first-choice agents for therapy (8). Thus, a laboratory serving a tertiary care medical center that specializes in organ transplantation and care of immunosuppressed patients may need to routinely test agents that are broader in spectrum than those tested by a laboratory that primarily supports an ambulatory outpatient practice, in which antibiotic-resistant organisms are less common.

When determining a laboratory's routine susceptibility testing batteries, several principles should be followed. First, test the antimicrobial agents that are included in the institution's formulary and that physicians prescribe on a daily basis. Second, consider the bacterial species to be tested, which also strongly influence the choice of antimicrobial agents for testing. The NCCLS publishes tables that list the antimicrobial agents appropriate for testing various groups of aerobic, fastidious, and anaerobic bacteria (12–14). Tables 1 and 1A in references 12 through 14 include recommendations regarding which agents are the most important to test routinely and which may be tested only by special request or circumstance, i.e., selectively. The guidelines indicate the drugs that are most appropriate for testing for

each organism group and for some infections at specific anatomic sites (e.g., isolates from cerebrospinal fluid, blood, urine, or feces). The listing also includes a few agents that may be tested as representatives of other agents because of the greater ability of an agent to detect resistance to other closely related drugs (e.g., the use of oxacillin to predict beta-lactam resistance in staphylococci and the use of the oxacillin disk test to predict penicillin resistance in pneumococci). This initial list of agents should then be tailored to an individual institution's specific needs through discussions with infectious disease physicians, pharmacists, and committees concerned with infection control and the institutional formulary.

A third important step in defining routine testing batteries is to ascertain the availability of specific antimicrobial agents for testing by the laboratory's routine testing methodology. Certain methods (i.e., disk diffusion, gradient diffusion, or in-house-prepared broth or agar dilution) allow the greatest flexibility in selecting test batteries. In contrast, some commercial test systems may have less flexibility in matching test batteries to an institution's formulary or in prompt availability of the latest antimicrobial agents available for clinical use. However, practicality limits the maximum number of drugs that can be tested simultaneously by any susceptibility method. For example, a maximum of 12 disks can be placed on a 150-mm Mueller-Hinton agar plate, and only a similar number can be accommodated in a microdilution tray if full concentration ranges of each agent are to be included for routine determination of MICs.

INTERPRETATION AND REPORTING OF SUSCEPTIBILITY TEST RESULTS

Careful periodic review of susceptibility data is important both for patient management and for surveillance of emerging resistance. Although information regarding antimicrobial resistance mechanisms is complex, it is important to be familiar with what antibiotic resistance patterns are likely, unlikely, and yet to be recognized with respect to the organisms encountered within one's own institution and in general (Table 1). Establishing and maintaining effective monitoring programs may be labor intensive; nevertheless, because of the continual emergence of new resistance patterns, such systematic review of susceptibility data is as crucial as selecting reliable and accurate testing methods and equipment. Computerization of results can greatly facilitate the administration of such quality assurance programs. This computerized approach is available with certain automated systems or may be established through certain laboratory information systems. Once flagged, apparently aberrant susceptibility profiles should be promptly reevaluated, and the isolate may need to be retested. If the original findings are confirmed, a new or modified resistance mechanism may be identified.

In two key examples, careful reevaluation of the profiles ultimately led to the characterization of new or uncommon susceptibility profiles. Identification of *Klebsiella pneumoniae* isolates exhibiting the unusual profile of susceptibility to a broad-spectrum cephalosporin (cefoxitin) but resistance to an extended-spectrum cephalosporin (ceftazidime) and aztreonam resulted in the discovery and characterization of a new extended-spectrum β-lactamase (TEM-10) (16). Similarly, careful laboratory analysis of results obtained with an *Enterococcus faecalis* isolate resulted in one of the earliest reports of acquired vancomycin resistance among enterococci in the United States (18). This isolate has since become the prototypical low-level vancomycin resistance (*vanB*) strain (5). The ability to recognize aberrant susceptibility profiles will continue to play a key role in the discovery of the new resistances currently anticipated (Table 1) (3).

Laboratorians must make an effort to ensure that susceptibility data are communicated to clinicians in an understandable manner that supports optimal antimicrobial utilization. For example, methicillin-resistant staphylococci are considered to be cross-resistant to all beta-lactams, even though in vitro results may indicate susceptibility to certain cephalosporins, beta-lactam–β-lactamase inhibitor combinations, or imipenem. Reports indicating susceptibility to these drugs clearly would be misleading (see also chapters 112 and 116 of this Manual). Thus, such strains should be reported as resistant to all beta-lactam drugs.

The emergence of extended-spectrum β-lactamases, especially among isolates of *Klebsiella* spp. and *Escherichia coli*, has complicated our interpretation and reporting of cephalosporin and aztreonam results (see also chapter 112 of this Manual). Depending on the enzyme produced, an isolate may exhibit susceptibility to cefoxitin and ceftriaxone but resistance to ceftazidime and aztreonam. Because any cephalosporin may be of questionable efficacy for an extended-spectrum-β-lactamase-producing strain (17), simply reporting the profile obtained by the laboratory would be inadequate. When such a strain is encountered, the laboratory should contact the physician either directly or through an infectious disease consultation, so that concerns regarding the most appropriate beta-lactam therapy may be adequately addressed.

Susceptibility tests for enterococci are essentially screens for resistance to combination therapy (see also chapters 24 and 116 in this Manual). Therefore, reporting results is slightly different than for most other organism-antimicrobial agent combinations. Reports must avoid giving the false impression that a susceptibility result for any component of combination therapy (i.e., ampicillin, penicillin, vancomycin, gentamicin, or streptomycin) indicates that that drug is an appropriate choice for single-agent therapy of serious infections such as endocarditis. An explanatory note that clearly conveys this information should accompany the susceptibility report.

FUTURE DIRECTIONS AND NEEDS IN ANTIMICROBIAL SUSCEPTIBILITY TESTING

Antimicrobial resistance is becoming widespread among a variety of clinically significant bacterial species (15). Therefore, the microbiology laboratory is critically positioned to play a pivotal role not only in patient management but also in the surveillance and control of antimicrobial resistance by prudent testing and reporting of antimicrobial agents.

To meet the challenges and responsibilities of this unique position effectively, we must continuously assess and, if necessary, redesign our testing strategies. This ongoing process involves addressing several issues. The first priority is to maintain accurate and reliable testing methods, whether they are conventional or molecular (see also chapter 117 in this Manual). This is best accomplished through continued monitoring of performance with well-characterized strains that thoroughly challenge the testing methods and procedures used. This task is more cumbersome as the complexity of the resistance mechanisms emerging among bacteria increases. Therefore, flexibility in the selection and application of testing methods will continue to be an important component of maintaining reliable testing services within the laboratory. Monitoring of results for trends and changes, the other vital component of test accuracy, will need to be enhanced through the wider availability of automated, computer-based systems that allow rapid and accurate review of antimicrobial susceptibility profiles.

Additionally, reporting schemes and formats will have to be periodically reviewed and updated. Basically, clinical microbiology personnel will need to become more proactive in their reporting practices to ensure effective and meaningful communication of significant results to physicians in a manner that supports optimal use of antimicrobial agents.

Finally, there is the broader public health responsibility to monitor resistance trends in order to detect new or uncommon susceptibility profiles. Susceptibility data should be disseminated in a manner that supports effective surveillance systems for local, national, and even international resistance patterns.

REFERENCES

1. **Boyce, J. M., R. L. White, M. C. Bonner, and W. R. Lockwood.** 1982. Reliability of the MS-2 system in detecting methicillin-resistant *Staphylococcus aureus*. *J. Clin. Microbiol.* **15:**220–225.
2. **Citron, D. M., M. I. Ostovari, A. Karlsson, and E. J. C. Goldstein.** 1991. Evaluation of the E test for susceptibility testing of anaerobic bacteria. *J. Clin. Microbiol.* **29:**2197–2203.
3. **Courvalin, P.** 1992. Interpretive reading of antimicrobial susceptibility tests. *ASM News* **58:**368–375.
4. **Doern, G. V., D. R. Scott, and A. L. Rashad.** 1982. Clinical impact of rapid antimicrobial susceptibility testing of blood culture isolates. *Antimicrob. Agents Chemother.* **21:**1023–1024.
5. **Evers, S., D. F. Sahm, and P. Courvalin.** 1993. The *vanB* gene of vancomycin-resistant *Enterococcus faecalis* V583 is structurally related to genes encoding D-ala:D-ala ligases and glycopeptide-resistance proteins Van A and Van C. *Gene* **124:**143–144.
6. **Hill, G. B., and S. Schalkowsky.** 1990. Development and evaluation of the spiral gradient endpoint method for susceptibility testing of anaerobic gram-negative bacilli. *Rev. Infect. Dis.* **12**(Suppl. 2):S200–S209.
7. **Jorgensen, J. H.** 1993. Selection criteria for an antimicrobial susceptibility testing system. *J. Clin. Microbiol.* **31:**2841–2844.

8. **Jorgensen, J. H.** 1993. Selection of antimicrobial agents for routine testing in a clinical microbiology laboratory. *Diagn. Microbiol. Infect. Dis.* **16:**245–249.

9. **Jorgensen, J. H., M. J. Ferraro, M. L. McElmeel, J. Spargo, J. M. Swenson, and F. C. Tenover.** 1994. Detection of penicillin and extended-spectrum cephalosporin resistance among *Streptococcus pneumoniae* clinical isolates by use of the E test. *J. Clin. Microbiol.* **32:**159–163.

10. **Katsanis, G. P., J. Spargo, M. J. Ferraro, L. Sutton, and G. A. Jacoby.** 1994. Detection of *Klebsiella pneumoniae* and *Escherichia coli* strains producing extended-spectrum β-lactamases. *J. Clin. Microbiol.* **32:**691–696.

11. **Nadler, H. L., C. Dolan, L. Mele, and S. R. Kurtz.** 1985. Accuracy and reproducibility of the AutoMicrobic System Gram-Negative General Susceptibility-Plus card for testing selected challenge organisms. *J. Clin. Microbiol.* **22:**355–360.

12. **National Committee for Clinical Laboratory Standards.** 1993. *Methods for Antimicrobial Susceptibility Testing of Anaerobic Bacteria.* Approved standard M11-A3. National Committee for Clinical Laboratory Standards, Villanova, Pa.

13. **National Committee for Clinical Laboratory Standards.** 1993. *Methods for Dilution Antimicrobial Susceptibility Testing for Bacteria That Grow Aerobically.* Approved standard M7-A3. National Committee for Clinical Laboratory Standards, Villanova, Pa.

14. **National Committee for Clinical Laboratory Standards.** 1993. *Performance Standards for Antimicrobial Disk Susceptibility Tests.* Approved standard M2-A5. National Committee for Clinical Laboratory Standards, Villanova, Pa.

15. **Neu, H. C.** 1992. The crisis in antibiotic resistance. *Science* **257:**1064–1073.

16. **Quinn, J. P., D. Miyashiro, D. Sahm, R. Flamm, and K. Bush.** 1989. Novel plasmid-mediated β-lactamase (TEM-10) conferring selective resistance to ceftazidime and aztreonam in clinical isolates of *Klebsiella pneumoniae. Antimicrob. Agents Chemother.* **33:**1451–1456.

17. **Rice, L. B., J. D. C. Yao, K. Klimm, G. M. Eliopoulos, and R. C. Moellering, Jr.** 1991. Efficacy of different β-lactams against an extended-spectrum β-lactamase-producing *Klebsiella pneumoniae* strain in the rat intra-abdominal abscess model. *Antimicrob. Agents Chemother.* **35:**1243–1244.

18. **Sahm, D. F., J. Kissinger, M. S. Gilmore, P. R. Murray, R. Mulder, J. Solliday, and B. Clarke.** 1989. In vitro susceptibility studies of vancomycin-resistant *Enterococcus faecalis. J. Clin. Microbiol.* **33:**1588–1591.

19. **Sahm, D., and L. Olsen.** 1990. In vitro detection of enterococcal vancomycin resistance. *Antimicrob. Agents Chemother.* **34:**1846–1848.

20. **Stone, L. L., and D. L. Jungkind.** 1983. False-susceptible results from the MS-2 system used for testing resistant *Pseudomonas aeruginosa* against two third-generation cephalosporins, moxalactam and cefotaxime. *J. Clin. Microbiol.* **18:**389–394.

21. **Tenover, F. C., J. Tokars, J. Swenson, S. Paul, K. Spitalny, and W. Jarvis.** 1993. Ability of clinical laboratories to detect antimicrobial-agent-resistant enterococci. *J. Clin. Microbiol.* **31:**1695–1699.

22. **Trenholme, G. M., R. L. Kaplan, P. H. Karahusis, T. Stine, J. Fuhrer, W. Landau, and S. Levin.** 1989. Clinical impact of rapid identification and susceptibility testing of bacterial blood culture isolates. *J. Clin. Microbiol.* **27:**1342–1345.

23. **Washington, J. A., C. C. Knapp, and C. C. Sanders.** 1988. Accuracy of microdilution and the AutoMicrobic System in detection of β-lactam resistance in gram-negative bacterial mutants with derepressed β-lactamase. *Rev. Infect. Dis.* **10:**824–829.

Antibacterial Agents

JOSEPH D. C. YAO AND ROBERT C. MOELLERING, JR.

111

Antimicrobial chemotherapy has played a vital role in the treatment of human infectious diseases in the 20th century. Since the discovery of penicillin in the 1920s, literally hundreds of antimicrobial agents have been developed or synthesized, and dozens are currently available for clinical use. While the broad number and variety of agents available provide a great deal of flexibility for the clinician in the use of these agents, the sheer numbers and continuing development of agents available make it difficult for clinicians to keep up with progress in the field. Similarly, this variety presents significant challenges for the clinical microbiologist who must decide which agents are appropriate for inclusion in routine and specialized susceptibility testing.

This chapter provides an overview of the antibacterial agents currently marketed in the United States, with major emphasis on their mechanisms of action, spectra of activity, important pharmacologic parameters, and toxicities. Antibiotics that have fallen into disuse or remain investigational will be mentioned only briefly.

PENICILLINS

The penicillins (Table 1) are a group of natural and semi-synthetic antibiotics containing the chemical nucleus 6-aminopenicillanic acid, which consists of a beta-lactam ring fused to a thiazolidine ring (Fig. 1a). The naturally occurring compounds are produced by a number of *Penicillium* spp. The penicillins differ from one another in substitutions at position 6, where changes in the side chain may modify the pharmacokinetic and antibacterial properties of the drug.

Mechanisms of Action

The major antibacterial action of penicillins is derived from their ability to inhibit a number of bacterial enzymes, namely, penicillin-binding proteins (PBPs), that are essential for peptidoglycan synthesis (199). This ability to inhibit bacterial cell wall enzymes such as the transpeptidases usually confers on the penicillins bactericidal activity against gram-positive bacteria. The bactericidal activity of the penicillins is often related to their ability to trigger membrane-associated autolytic enzymes that destroy the cell wall. Other minor mechanisms of action include inhibition of bacterial endopeptidase and glycosidase, enzymes involved in bacterial cell growth. There is also recent evidence

suggesting that penicillins may inhibit RNA synthesis in some bacteria, causing death without cell lysis, but the significance of these observations remains to be proven (115).

Pharmacology

Oral absorption differs markedly among the penicillins. As a natural congener of penicillin G, penicillin V resists gastric acid inactivation and is better absorbed from the gastrointestinal tract than is penicillin G. Amoxicillin is a semisynthetic analog of ampicillin and has greater gastro-intestinal absorption than ampicillin (95 versus 40% absorption). Bacampicillin is an ampicillin ester that is considerably better absorbed from the gastrointestinal tract than is ampicillin or amoxicillin. This ester is inactive until naturally occurring esterases in the intestinal mucosa and serum hydrolyze it to release the parent compound, ampicillin, into the serum. The isoxazolyl penicillins, such as oxacillin, cloxacillin, and dicloxacillin, as well as nafcillin are acid stable and are also absorbed from the gastrointestinal tract, in contradistinction to certain other antistaphylococcal penicillins, such as methicillin, which are not acid resistant and cannot be given by the oral route.

Repository forms of penicillin G, available in procaine or benzathine, delay absorption from an intramuscular depot. Procaine penicillin G provides detectable levels for 12 to 24 h, suitable for treatment of uncomplicated pneumococcal pneumonia and gonorrhea due to fully susceptible organisms. Benzanthine penicillin G achieves very low levels in blood for prolonged periods (3 to 4 weeks) and is useful for the therapy of syphilis and the prophylaxis of streptococcal pharyngitis and rheumatic fever.

Penicillins are well distributed to many body compartments, including lung, liver, kidney, muscle, bone, and placenta. Penetration into the eye, brain, cerebrospinal fluid (CSF), and prostate is poor in the absence of inflammation. These drugs are metabolized to a small degree and are rapidly excreted, essentially unchanged, via the kidney. With average half-lives of 0.5 to 1.5 h, they are usually administered every 4 to 6 h to maintain effective levels in blood. The renal tubular excretion of penicillins can be blocked by probenecid, thus prolonging their half-lives in serum.

Dosage reduction of most penicillins is necessary only in severe renal insufficiency (creatinine clearance of ≤10 ml/

TABLE 1 Penicillins

Natural
   Benzylpenicillin (penicillin G)
   Phenoxymethyl penicillin (penicillin V)

Semisynthetic
   Penicillinase resistant
      Methicillin
      Nafcillin
      Isoxazolyl penicillins
         Cloxacillin
         Dicloxacillin
         Oxacillin
   Extended spectrum
      Aminopenicillins
         Ampicillin
         Amoxicillin
         Bacampicillin
      Carboxypenicillins
         Carbenicillin
         Ticarcillin
      Ureidopenicillins
         Azlocillin
         Mezlocillin
         Piperacillin

Penicillin + β-lactamase inhibitor combinations
   Ampicillin-sulbactam (Unasyn)
   Amoxicillin-clavulanate (Augmentin)
   Ticarcillin-clavulanate (Timentin)
   Piperacillin-tazobactam (Zosyn)

min). Dosages of all penicillins except nafcillin and the isoxazolyl penicillins are adjusted for hemodialysis. Peritoneal dialysis requires dosage reduction of carbenicillin and ticarcillin.

Spectrum of Activity

The penicillins have antibacterial activity against most gram-positive and many gram-negative and anaerobic organisms. Penicillin G is very effective against penicillin-susceptible *Staphylococcus aureus*, *Streptococcus pneumoniae*, *Streptococcus pyogenes*, viridans streptococci, *Streptococcus bovis*, *Neisseria gonorrhoeae*, *Neisseria meningitidis*, *Pasteurella multocida*, anaerobic cocci, *Clostridium* spp., *Fusobacterium* spp., and many *Bacteroides* spp. except for members of the *Bacteroides fragilis* group. However, the occurrence of penicillin-resistant pneumococci has recently been increasing worldwide (103). Penicillin is the drug of choice for syphilis and *Actinomyces* infections. Penicillin V has a spectrum of activity similar to that of penicillin G except that it is less active against *N. gonorrhoeae*. Penicillinase-resistant penicillins, of which methicillin is the prototype, are primarily effective against penicillinase-producing staphylococci (77). The agents are at least 25 times more active than other penicillins against penicillinase-positive *S. aureus* and *Staphylococcus epidermidis*. Although they are also active against *S. pneumoniae* and *S. pyogenes*, their MICs for these organisms are higher than those of penicillin G. They are not active against enterococci, members of the family *Enterobacteriaceae*, *Pseudomonas* spp., and members of the *B. fragilis* group.

Ampicillin and amoxicillin have spectra of activity sim-

ilar to that of penicillin G, but they are more active against enterococci and *Listeria monocytogenes*. Although they are also more active against *Haemophilus influenzae* and *Haemophilus parainfluenzae*, up to 25% of *H. influenzae* isolates are resistant, usually because of β-lactamase production. *Salmonella* and *Shigella* spp., including *Salmonella typhi*, and many strains of *Escherichia coli* and *Proteus mirabilis* are susceptible to these agents. Ampicillin is more effective against shigellae, whereas amoxicillin is more effective against salmonellae. Both of these agents are degraded by β-lactamase and are inactive against many *Enterobacteriaceae* and *Pseudomonas* spp.

The carboxypenicillins and ureidopenicillins have increased activity against gram-negative bacteria that are resistant to ampicillin. Although these drugs are susceptible to staphylococcal penicillinase, they are more stable against hydrolysis by the β-lactamases of *Enterobacteriaceae* and *Pseudomonas aeruginosa*. Carbenicillin and ticarcillin are relatively active against streptococci and enterococci and are also active against *Haemophilus* spp., *Neisseria* spp., and a variety of anaerobes (67). They inhibit *Enterobacteriaceae* but are inactive against *Klebsiella* spp.

The ureidopenicillins have greater in vitro activity against streptococci and enterococci than do the carboxypenicillins, and they inhibit more than 75% of *Klebsiella* spp. (45, 189). They have excellent activity against many *Enterobacteriaceae* and anaerobic bacteria, including members of the *B. fragilis* group. On a weight basis, their activities against *P. aeruginosa* in decreasing order of potency are as follows: piperacillin, azlocillin > mezlocillin, ticarcillin > carbenicillin (33). These agents also act synergistically with aminoglycosides against *P. aeruginosa*.

Adverse Effects

Common reactions to penicillins include allergic skin rashes, diarrhea, and drug fever. Severe anaphylactic reactions, which can be fatal, may occur in previously sensitized patients rechallenged with penicillins, but fortunately, such reactions are quite rare. At high doses (usually >30 × 10^6 U/day), penicillin G can cause myoclonic twitching and seizures due to central nervous system toxicity. All of the penicillins may cause interstitial nephritis on an allergic basis, but methicillin is more likely than the other penicillins to cause this complication. Hepatitis has been associated with prolonged use of oxacillin. High-dose carbenicillin can result in sodium overload and hypokalemia. Neutropenia may occur with any of the penicillins. Thrombocytopenia and Coombs-positive hemolytic anemia are rare complications of penicillin therapy. Bleeding tendencies due to interference with platelet function can occur with the use of carboxypenicillins and ureidopenicillins (55). Although pseudomembranous colitis has been associated with all the penicillins, it occurs more frequently with ampicillin (11).

CEPHALOSPORINS

Cephalosporins are derivatives of the fermentation products of *Cephalosporium acremonium* (also designated *Acremonium chrysogenum*). They contain a 7-aminocephalosporanic acid nucleus, which consists of a beta-lactam ring fused to a dihydrothiazine ring (Fig. 1b). Various substitutions at positions 3 and 7 alter their antibacterial activities and pharmacokinetic properties. Addition of a methoxy group at position 7 of the beta-lactam ring results in a new

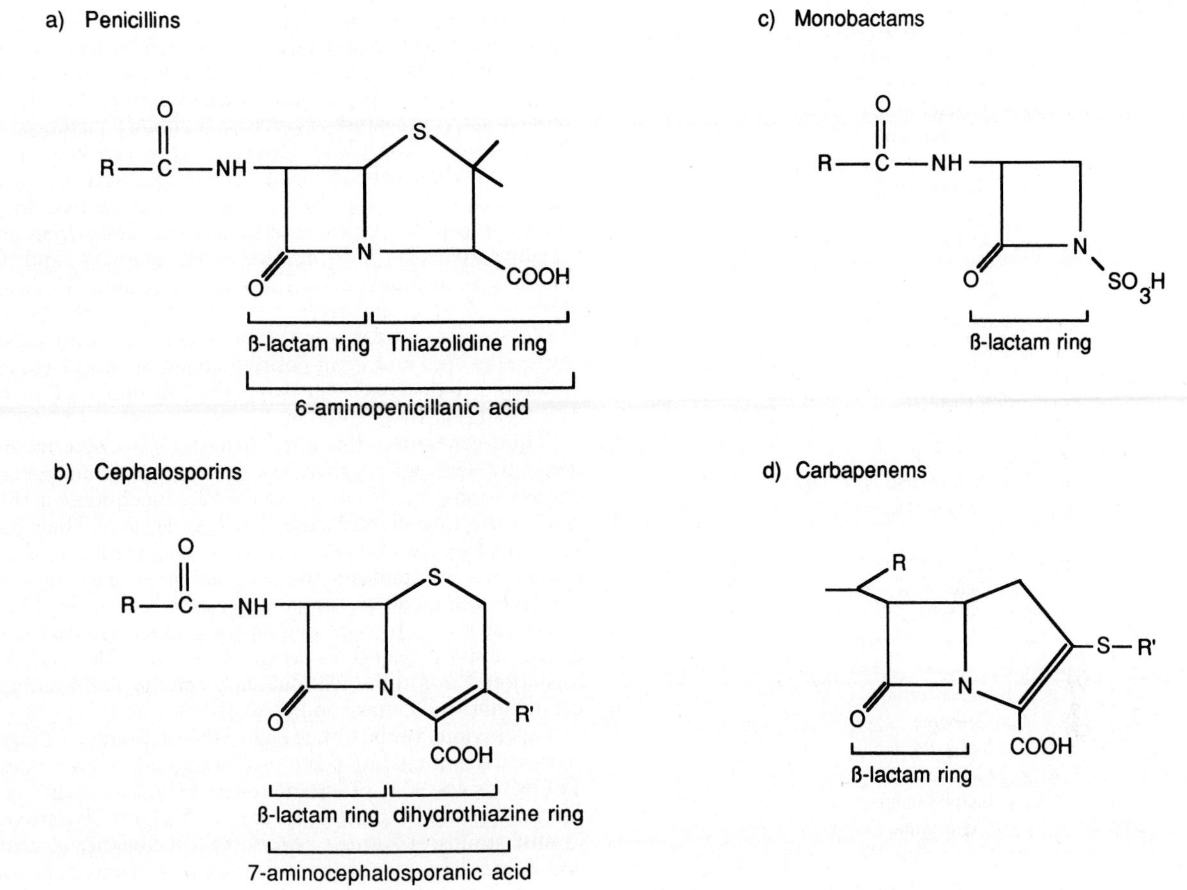

a) Penicillins

β-lactam ring Thiazolidine ring

6-aminopenicillanic acid

b) Cephalosporins

β-lactam ring dihydrothiazine ring

7-aminocephalosporanic acid

c) Monobactams

β-lactam ring

d) Carbapenems

β-lactam ring

FIGURE 1 Chemical structures of beta-lactam antibiotics.

group of compounds called cephamycins, which are highly resistant to a variety of β-lactamases.

Mechanism of Action

Similar to the penicillins, cephalosporins act by binding to PBPs of susceptible organisms, thereby interfering with synthesis of peptidoglycan of the bacterial cell wall. In addition, these beta-lactam agents may produce bactericidal effects by triggering autolytic enzymes in the cell envelope (199).

Pharmacology

Many cephalosporins require parenteral administration, but a growing number are available in oral form. Cephalexin, cephradine, cefadroxil, cefaclor, cefuroxime axetil, cefprozil, loracarbef, cefixime, and cefpodoxime are given orally with good gastrointestinal absorption (>90% of oral dose). Cefuroxime axetil is an acetoxyethyl ester of cefuroxime, and it is deesterified in the intestinal mucosa and absorbed into the bloodstream as cefuroxime. Relatively high concentrations of these agents are attained across the placenta and in synovial, pleural, pericardial, and peritoneal fluids. Levels in bile are usually high, especially with cefoperazone, which is excreted mainly in the bile. Ceftizoxime, cefotaxime, ceftriaxone, cefoperazone, and moxalactam enter the CSF well and are useful for the treatment of meningitis. Cefuroxime penetrates inflamed meninges, but levels in CSF are borderline and are inadequate in providing bactericidal activity against certain susceptible bacteria.

Cephalothin, cephapirin, and cefotaxime are converted to the desacetyl forms before excretion. All cephalosporins except cefoperazone are excreted primarily by the kidney, and for these drugs, dosage adjustments are necessary in patients with renal insufficiency (creatinine clearance of <50 ml/min). Like that of the penicillins, the renal excretion of cephalosporins is impeded by probenecid. In general, these agents are removed by hemodialysis but not by peritoneal dialysis. Of the cephalosporins, cefonicid and ceftriaxone have the longest elimination half-lives, at 4.5 and 8 h, respectively, permitting once- or twice-daily drug administration in the treatment of serious infections.

Spectrum of Activity

Cephalosporins are classified by a well-accepted but somewhat arbitrary scheme of grouping by generations that is based on general features of their antibacterial activities (Table 2). The first-generation (narrow-spectrum) drugs, exemplified by cephalothin and cefazolin, have good grampositive activity and relatively modest gram-negative activity. They are active against penicillin-susceptible and -resistant S. aureus as well as S. pneumoniae, S. pyogenes, and other aerobic and anaerobic streptococci. Methicillin-resistant S. aureus, S. epidermidis, and enterococci are resistant. Some Enterobacteriaceae, including many strains of E. coli,

TABLE 2 Cephalosporins

Narrow spectrum
Cefadroxil
Cefazolin
Cephalexin
Cephaloridine
Cephalothin
Cephapirin
Cephradine

Expanded spectrum
Cefaclor
Cefamandole
Cefonicid
Cefcoranide
Cefprozil
Cefuroxime
Loracarbef

Cefmetazole
Cefotetan
Cefoxitin

Broad spectrum
Cefixime
Cefoperazone
Cefotaxime
Cefpodoxime
Ceftazidime
Ceftizoxime
Ceftriaxone

Klebsiella spp., and *Proteus mirabilis*, are susceptible. *Pseudomonas* spp., including *P. aeruginosa*, many *Proteus* spp., and *Serratia* and *Enterobacter* spp. are resistant. These agents are active against penicillin-susceptible anaerobes except members of the *B. fragilis* group. They have only modest activities against *H. influenzae*.

The second-generation (expanded-spectrum) cephalosporins are stable to certain β-lactamases found in gram-negative bacteria and as a result have increased activities against gram-negative organisms. The agents are more active than narrow-spectrum drugs against *E. coli*, *Klebsiella* spp., and *Proteus* spp. Their activities also extend to cover some *Enterobacter* and *Serratia* strains, and they have good activities against *Haemophilus* spp., *Neisseria* spp., and many anaerobes. Cefaclor, cefuroxime (as well as its axetil ester), cefamandole, cefonicid, and cefprozil are active against ampicillin-resistant *H. influenzae* and *Moraxella catarrhalis*. However, cefamandole exhibits a significant inoculum effect and is not suitable for treating life-threatening infections due to *H. influenzae*. Ceforanide and cefonicid have spectra of antibacterial activities similar to that of cefamandole, but they are less active than cefamandole against gram-positive cocci. Loracarbef belongs to a new class of cephalosporin derivatives known as carbacephems in which the sulfur atom of the dihydrothiazine ring is replaced by a methylene group to form a tetrahydropyridine ring (32). Since this structural modification of the cephalosporin nucleus is minor, loracarbef is considered a cephalosporin. Its spectrum of antibacterial activity is very similar to those of

cefaclor, cefuroxime, and cefprozil. None of the expanded-spectrum agents is active against *Pseudomonas* spp.

Cefoxitin, cefotetan, and cefmetazole belong to a unique group of expanded-spectrum cephalosporins that have marked activity against anaerobes, including members of the *B. fragilis* group (201). Cefotetan is two to four times less active than cefoxitin and cefmetazole against gram-positive cocci, but it is more potent than these two drugs against susceptible *Enterobacteriaceae*. The three drugs are equally active against *H. influenzae*, *M. catarrhalis*, and *N. gonorrhoeae*, including penicillin-resistant strains. While all three drugs are comparable in their activities against *B. fragilis* species, cefoxitin is the most active against other *Bacteroides* spp. and gram-positive anaerobic cocci (201). Cefotetan and cefmetazole have the advantage of more prolonged half-lives in serum.

Third-generation (broad-spectrum) cephalosporins are generally less active than the narrow-spectrum agents against gram-positive cocci, but they are much more active against the *Enterobacteriaceae* and *P. aeruginosa*. Their potent broad spectra of gram-negative activity are due to their stability to β-lactamases and their ability to pass through the outer cell envelopes of gram-negative bacilli (54, 132). There are two subgroups among these agents: those with potent activity against *P. aeruginosa* (ceftazidime and cefoperazone) and those without such activity (ceftizoxime, cefotaxime, and ceftriaxone).

Cefotaxime inhibits more than 90% of strains of *Enterobacteriaceae*, including those resistant to aminoglycosides. The MICs for 90% of strains tested (MIC$_{90}$s) of *E. coli*, *Proteus* spp., and *Klebsiella* spp. are <0.5 μg/ml. Its activity against strains of *Serratia marcescens*, *Enterobacter cloacae*, and *Acinetobacter* spp. is variable, and it is inactive against *P. aeruginosa*. It has moderate activity against anaerobes but is inferior to cefoxitin and cefotetan against most of these isolates.

Ceftizoxime, ceftriaxone, and moxalactam have spectra of activity similar to that of cefotaxime with a few exceptions. Ceftriaxone is the most active agent against penicillinase-positive or -negative strains of *N. gonorrhoeae* (58). It is effective as single-dose therapy for infections caused by these organisms (118). Because of its long half-life in serum (the longest of the currently available cephalosporins), ceftriaxone is used frequently in outpatient antibiotic therapy of serious infections, including Lyme disease (144).

Cefoperazone is less active than cefotaxime against many *Enterobacteriaceae* and gram-positive cocci. However, it has activity against *P. aeruginosa*, with an MIC$_{50}$ of ≤16 μg/ml. Its activity against anaerobes is similar to that of cefotaxime. Ceftazidime is the most potent of the currently available cephalosporins against *P. aeruginosa*, with an MIC$_{90}$ of <8 μg/ml (132). It is more active than the ureidopenicillins against these strains. This agent has activity similar to that of cefotaxime against the *Enterobacteriaceae* but is not as active against gram-positive cocci. It has little activity against gram-negative anaerobes.

Cefixime (7) and cefpodoxime (160) are broad-spectrum oral cephalosporins that are more stable than the narrow- and expanded-spectrum oral cephalosporins against gram-negative bacterial β-lactamases. Compared with these agents, cefixime and cefpodoxime are equally active against streptococci but considerably less active against *S. aureus*. With potent activities similar to that of ceftizoxime against many *Enterobacteriaceae*, *H. influenzae*, *M. catarrhalis*, and *N. gonorrhoeae* (including β-lactamase-producing strains), they are inactive against *Pseudomonas*, *Enterobacter*, *Serra-*

tia, and *Morganella* spp. and anaerobes. None of the currently available cephalosporins is clinically useful against enterococci.

Cefpirome (formerly HR 810) and cefepime (BMY 28142) are broad-spectrum cephalosporins currently undergoing preclinical evaluations for therapeutic use. They have the unique features of reduced affinity for class I β-lactamases (therefore activity against stably derepressed class I β-lactamase mutants of *Enterobacteriaceae* and *P. aeruginosa*) and increased abilities to pass through the bacterial outer membrane compared to the broad-spectrum agents. In addition to enhanced activity against gram-negative bacteria, they remain active against gram-positive cocci, particularly streptococci, with MICs comparable to those of the narrow-spectrum cephalosporins (98, 101). However, they are not active clinically against anaerobes.

Adverse Effects

Cephalosporins are generally very well tolerated. The most common side effects are diarrhea and hypersensitivity reactions such as rash, drug fever, and serum sickness. Cross-reactions with these drugs occur in only 3 to 7% of penicillin-allergic patients (2). Other infrequent side effects include pseudomembranous colitis, elevated serum creatinine and transaminase levels, leukopenia, thrombocytopenia, and Coombs-positive hemolytic anemia. These abnormalities are usually mild and reversible. Prolonged use of ceftriaxone has been associated with formation of gallbladder sludge, which usually resolves after the drug is discontinued (164), and, rarely, cholecystis.

Disulfiramlike reactions have been described in patients receiving cefamandole, cefotetan, and cefoperazone. This reaction is attributed to the *N*-methylthiotetrazole side chains of these antibiotics, which are similar to the chemical structure of disulfiram. Hypoprothrombinemia and bleeding tendencies have been observed with these cephalosporins. Causes of the coagulopathy included (i) alteration of healthy gut flora by the antibiotics, thus inhibiting the synthesis of vitamin K and its precursors, and (ii) the *N*-methythiotetrazole side chain, which inhibits the vitamin K-dependent carboxylase enzyme responsible for converting clotting factors II, VII, IX, and X to their active forms and also prevents regeneration of active vitamin K from its inactive form (162).

OTHER BETA-LACTAM ANTIBIOTICS

Aztreonam

Aztreonam is the only monobactam antibiotic currently in clinical use. The monobactams are beta-lactams with various side chains affixed to a monocyclic nucleus (Fig. 1c).

Mechanism of Action

Azetreonam binds primarily to PBP 3 of gram-negative aerobes, including *P. aeruginosa*, thereby disrupting bacterial cell wall synthesis. It is not hydrolyzed by most commonly occurring plasmid- and chromosomally mediated β-lactamases, and it does not induce the production of these enzymes (18).

Pharmacology

Given intravenously, azetreonam is widely distributed to body tissues and fluids. Average serum drug concentrations exceed the MIC_{90}s of most *Enterobacteriaceae* by four to eight times for 8 h and are inhibitory to *P. aeruginosa* for 4

h. It crosses inflamed meninges in sufficient amounts to be potentially therapeutic for meningitis caused by susceptible organisms. Its half-life in serum is about 1.7 h, and it is excreted mainly unchanged by the kidney. Dosage modification is necessary for patients with renal failure. The drug is removed by both hemodialysis and peritoneal dialysis.

Spectrum of Activity

The antibacterial activity of aztreonam is limited to aerobic gram-negative bacilli, inhibiting most *Enterobacteriaceae*, *Neisseria* spp., and *Haemophilus* spp. with MIC_{90}s of ≤0.5 μg/ml (9, 182). It has significant activity against *Enterobacter* spp. and *Serratia marcescens*, with most strains being inhibited at ≤16 μg/ml. However, many *Acinetobacter* spp., *Pseudomonas cepacia*, and *Xanthomonas maltophilia* are resistant. It shows in vitro synergism when combined with aminoglycosides against 30 to 60% of aztreonam-susceptible organisms, including *P. aeruginosa* and aminoglycoside-resistant gram-negative bacilli (17). Bacterial tolerance and inoculum effect are generally not seen with this agent. Aztreonam is not active against gram-positive bacteria and anaerobes.

Adverse Effects

Aztreonam is generally a safe agent, with a toxicity profile similar to those of other beta-lactam drugs. Nausea, diarrhea, skin rash, eosinophilia, mild elevation of serum transaminase levels, and transiently elevated serum creatinine level have occurred. It has minimal cross-reactivity with other beta-lactams and can be used safely in patients allergic to penicillins or cephalosporins (163). Hematologic abnormalities have not been reported.

Carbapenems

Carbapenems are a unique class of beta-lactam agents with the widest spectrum of antibacterial activity of the currently available antibiotics. Structurally, they differ from other beta-lactams in having a hydroxyethyl side chain in a *trans* configuration at position 6 and lacking a sulfur or oxygen atom in the bicyclic nucleus (Fig. 1d). The unique stereochemistry of the hydroxyethyl side chain confers stability against β-lactamases. Imipenem (*N*-formimidoyl thienamycin), a semisynthetic derivative of thienamycin produced by *Streptomyces* spp., is the first carbapenem antimicrobial agent developed for clinical use (12). Other members of this class currently undergoing preclinical evaluation or clinical trials include meropenem (SM-7339), panipenem (RS-533), and biapenem (L 647 or LJC 10,627).

Mechanism of Action

Carbapenems bind to PBP 1 and PBP 2 of gram-negative and gram-positive bacteria, causing cell elongation and lysis (176). They are stable toward most plasmid- or chromosomally mediated β-lactamases except those produced by *X. maltophilia* and some strains of *B. fragilis* (106).

Pharmacology

After intravenous administration, the carbapenems distribute widely in the body but undergo no significant biliary excretion. Imipenem penetrates inflamed meninges, with levels in CSF of 5 to 11 μg/ml (154). The drug is metabolized and inactivated in the kidneys by a dehydropeptidase-I (DHP-I) enzyme found in the brush borders of proximal renal tubular cells. To achieve adequate concentrations in serum and urine, a DHP inhibitor, cilastatin, was developed; it is combined with imipenem in a 1:1 dosage ratio for clinical use. Cilastatin has no antibacterial activity, nor does it alter the activity of imipenem. It has a renal protective effect by preventing excessive accumulation of potentially toxic imipenem metabolites in the renal tubular cells. Meropenem and biapenem, each having a β-methyl group at position 1 of the bicyclic nucleus, are thought to be resistant to the action of DHP-I without the need of a DHP-I inhibitor.

The pharmacokinetics of imipenem, meropenem, and biapenem are very similar, with elimination half-lives in serum of about 1 h. Peak concentrations in serum are about 25 to 35 μg/ml following a 500-mg dose. Dosage adjustment is necessary for creatinine clearances of $\leq$30 ml/min. These agents, including cilastin, are effectively removed by hemodialysis.

Spectrum of Activity

In general, all the carbapenems have similar antibacterial potencies, with minor differences. They have excellent in vitro activity against aerobic gram-positive species: staphylococci (penicillin-susceptible and -resistant isolates); viridans streptococci; group A, B, C, and G streptococci; *Bacillus* spp.; and *L. monocytogenes*. Imipenem is slightly more active than the other agents against streptococci and staphylococci, but methicillin-resistant staphylococci are usually resistant to all. Most enterococci are inhibited by the carbapenems at $\leq$4 μg/ml, with slightly better activity by imipenem, but *Enterococcus faecium* strains are usually resistant.

More than 90% of *Enterobacteriaceae*, including those resistant to other beta-lactams and aminoglycosides (138), are susceptible to carbapenems, with the following decreasing order of activity: meropenem > biapenem, penipenem $\geq$ imipenem (88, 106, 112). Most *Acinetobacter* spp., *Enterobacter* spp., *Citrobacter* spp., and *Serratia* spp. are inhibited by $\leq$2 μg/ml. Activity against *P. aeruginosa* is equivalent among the carbapenems and is similar to that of ceftazidime, with most strains inhibited at 0.5 to 8 μg/ml. While they inhibit *P. cepacia* and *Pseudomonas stutzeri*, they are inactive against *X. maltophilia* (88). Emergence of resistant *Pseudomonas* spp. has been observed during therapy with imipenem (207). The drug may show in vitro antagonism when combined with broad-spectrum cephalosporins or extended-spectrum penicillins as a result of its ability to induce class I β-lactamase production (133).

Carbapenems are the most potent beta-lactams against anaerobes, with activities comparable to those of clindamycin and metronidazole. The MIC$_{90}$s against *B. fragilis*, *Prevotella melaninogenica*, other *Bacteroides* spp., *Fusobacterium* spp., and anaerobic gram-positive cocci are $\leq$1 μg/ml (112, 170). Most *Clostridium* spp. except some strains of *Clostridium difficile* are susceptible. Imipenem is also active against *Actinomyces* spp., *Nocardia asteroides*, and *Mycobacterium* spp. (13, 35, 44).

Adverse Effects

Side effects of imipenem are similar to those of other beta-lactam antibiotics. Gastrointestinal distress occurs in up to 5% of patients. Allergic reactions such as drug fever, skin rashes, and urticaria are seen in about 3% of patients. Cross-reactivity with other beta-lactam agents is possible but not fully studied. Seizures of unclear etiology have occurred in about 1% of patients, particularly in the elderly age group and in patients with renal insufficiency or underlying neurologic disorders. Reversible elevation of serum transaminase levels, leukopenia, and thrombocytopenia have been described, but no coagulopathy has been reported.

β-LACTAMASE INHIBITORS

Clavulanic Acid

Clavulanic acid is a naturally occurring weak antimicrobial agent found initially in cultures of *Streptomyces clavuligerus* (136). It inhibits β-lactamases from staphylococci and many gram-negative bacteria. This agent acts primarily as a "suicide inhibitor" by forming an irreversible acyl enzyme complex with the β-lactamase, leading to loss of activity of the enzyme.

Clavulanic acid acts synergistically with various penicillins and cephalosporins against β-lactamase-producing staphylococci, klebsiellae, *H. influenzae*, *M. catarrhalis*, *N. gonorrhoeae*, *E. coli*, *Proteus* spp., and members of the *B. fragilis* group (5, 67). Recently discovered plasmid-mediated TEM β-lactamases in ceftazidime-resistant strains of *Klebsiella pneumoniae* and *E. coli* are inactivated by this agent (145, 203). However, the inducible β-lactamases (chromosomal class I) of *Enterobacter*, *Citrobacter*, *Serratia*, and *Pseudomonas* spp. are not inhibited by clavulanic acid. The combination of clavulanic acid with ampicillin, amoxicillin, or ticarcillin is active in vitro against *Mycobacterium tuberculosis*, which is known to produce β-lactamases (36, 214).

In the United States, clavulanic acid is available for clinical use in a 1:2 or 1:4 combination with oral amoxicillin and in a 1:15 or 1:30 parenteral combination with ticarcillin. The pharmacologic parameters of amoxicillin and ticarcillin are not significantly altered when either drug is combined with clavulanic acid. Amoxicillin-clavulanate is moderately well absorbed from the gastrointestinal tract, with a half-life in serum of about 1 h for each component. One-third of a dose is metabolized, and the remainder is excreted unchanged in the urine. The drug is widely distributed to various body tissues and fluids, but it penetrates uninflamed meninges very poorly.

Adverse reactions are similar to those reported for amoxicillin or ticarcillin used alone. Nausea, vomiting, abdominal cramps, and diarrhea occur in 5 to 15% of patients taking amoxicillin-clavulanate. The incidence of allergic skin reactions is similar to that with ampicillin alone.

Sulbactam

Sulbactam is a semisynthetic 6-desaminopenicillin sulfone with weak antibacterial activity (4). It functions as an effective inhibitor of certain plasmid- and chromosomally mediated β-lactamases of *S. aureus*, many *Enterobacteriaceae*, *H. influenzae*, *M. catarrhalis*, *Neisseria* spp., *Legionella* spp., *Bacteroides* spp., and *Mycobacterium* spp. (66, 151). Sulbactam alone is active against *N. gonorrhoeae*, *N. meningitidis*, some *Acinetobacter* spp., and *P. cepacia* (96, 152). It acts synergistically with penicillins and cephalosporins against organisms that are otherwise resistant to the beta-lactam drugs because of the production of β-lactamases. A combination of sulbactam (8 μg/ml) and ampicillin (16 μg/ml) inhibits most strains of staphylococci, *Klebsiella* spp., *E. coli*, *H. influenzae*, *M. catarrhalis*, *Neisseria* spp., and members of the *B. fragilis* group that are ampicillin resistant (152, 202). Like clavulanic acid, sulbactam does not inhibit the β-lactamases of *Enterobacter*, *Citrobacter*, *Providencia*, indole-positive *Proteus*, or *Pseudomonas* spp. or of *X. maltophilia*.

For clinical use, sulbactam is combined with ampicillin as a parenteral preparation in a 1:2 ratio. The pharmacologic properties of the drugs are not affected by each other in this combination. Ampicillin-sulbactam penetrates well into body tissues and fluids, including peritoneal and blister fluids. It enters the CSF in the presence of inflamed meninges. Like ampicillin, sulbactam has a half-life in serum of 1 h, and 85% of the drug is excreted unchanged via the kidneys. Since clearances of both sulbactam and ampicillin are affected similarly in patients with impaired renal function, dosage adjustments are similar for the two drugs.

The most common side effects of the ampicillin-sulbactam combination have been nausea, diarrhea, and skin rash. Transient eosinophilia and transaminasemia have been reported. Adverse reactions attributed to ampicillin may also occur with the use of ampicillin-sulbactam.

Tazobactam

Tazobactam (formerly YTR 830) is a penicillanic acid sulfone derivative structurally related to sulbactam. Like clavulanic acid and sulbactam, tazobactam acts as a suicidal β-lactamase inhibitor and binds to bacterial PBP 1 or PBP 2 (123). Despite having very poor intrinsic antibacterial activity by itself, it is comparable to clavulanate and sulbactam in lowering MICs for many organisms by up to 20-fold when combined with various beta-lactams and used against β-lactamase-producing organisms. Tazobactam actively inhibits the β-lactamases of staphylococci, *H. influenzae*, *N. gonorrhoeae*, *E. coli*, and members of the *B. fragilis* group (3, 82, 107). It also has activity against the class I β-lactamases of *Citrobacter*, *Proteus*, *Providencia*, and *Morganella* spp. but remains inactive against those of *Enterobacter* spp., *Pseudomonas* spp., *X. maltophilia*, and some *Klebsiella* spp. (104). Of the penicillin–β-lactamase inhibitor combinations, piperacillin-tazobactam is the one most active (two- to eightfold-lower MICs) against β-lactamase-producing aerobic and anaerobic gram-negative bacilli (51, 104).

Available as a 1:8 ratio dosage combination with piperacillin, tazobactam is administered parenterally. The two drugs do not affect each other's metabolism or pharmacokinetics. High concentrations of both agents are achieved in the intestinal mucosa, lung, and skin, with relatively poor distribution to muscle, fat, prostate, and CSF (in the absence of inflamed meninges). With a half-life in serum of about 1 h, elimination of tazobactam is mainly via the renal route and is not affected by hepatic failure (173). Major adverse effects of the piperacillin-tazobactam combination are similar to those of piperacillin alone, such as diarrhea, skin rash, and allergic reactions. Mild elevation in serum transaminase levels may be encountered in about 10% of patients.

AMINOGLYCOSIDES AND AMINOCYCLITOLS

Since the first aminoglycoside (aminoglycosidic aminocyclitol), streptomycin, was introduced in 1944, this class of antibiotics has played a vital role in the treatment of serious gram-negative infections. Among the unique features of the aminoglycosides are bactericidal activity against aerobic gram-negative bacilli (including *Pseudomonas* spp.), activity against *M. tuberculosis*, and relatively low incidence of bacterial resistance. The currently available aminoglycosides are derived from *Micromonospora* spp. (gentamicin, sisomicin, and netilmicin) or from *Streptomyces* spp. (streptomycin, neomycin, kanamycin, tobramycin, and paromomycin). The difference in origin of these compounds accounts for the differences of their suffixes, "-micin" versus "-mycin." Streptomycin, neomycin, kanamycin, tobramycin, and gentamicin are naturally occurring aminoglycosides, whereas amikacin and netilmicin are semisynthetic derivatives of kanamycin and sisomicin, respectively. Isepamicin, which is presently undergoing clinical trial in the United States and elsewhere, is a semisynthetic derivative of gentamicin B. Structurally, each of these aminoglycosides contains two or more amino sugars linked by glycosidic bonds to an aminocyclitol ring nucleus.

Spectinomycin is an aminocyclitol antibiotic isolated from *Streptomyces spectabilis*. Although it contains an aminocyclitol nucleus, it is not strictly an aminoglycoside because it does not contain an amino sugar or a glycosidic bond.

Mechanism of Action

Aminoglycosides are bactericidal agents that inhibit bacterial protein synthesis by binding irreversibly to the bacterial 30S ribosomal subunit. The aminoglycoside-bound bacterial ribosomes then become unavailable for translation of mRNA during protein synthesis, and this situation leads to cell death (41). The aminoglycosides also cause misreading of the genetic code, with resultant production of nonsense proteins. To reach the intracellular ribosomal binding targets, an aerobic energy-dependent process is necessary to enable successful penetration of the inner cell membrane by the aminoglycosides. Bacterial uptake of these agents is facilitated by inhibitors of bacterial cell wall synthesis such as beta-lactams and vancomycin. This interaction forms the basis of the antibacterial synergism between aminoglycosides and beta-lactam antibiotics (43, 71).

Spectinomycin acts similarly to the aminoglycosides by binding to the 30S ribosomal subunits and inhibiting protein synthesis. However, it does not cause misreading of the mRNA and is not bactericidal.

Pharmacology

All aminoglycosides have similar pharmacologic properties. Gastrointestinal absorption of these agents is unpredictable and always low. Because of its severe toxicity with systemic administration, neomycin is available only for oral and topical use. After intravenous administration, aminoglycosides are freely distributed in the extracellular space but penetrate poorly into the CSF, vitreous fluid of the eye, biliary tract, prostate, and tracheobronchial secretions, even in the presence of inflammation.

In adults with normal renal function, the aminoglycosides have half-lives in serum of about 2 to 3 h. They are primarily excreted, essentially unchanged, via the kidneys. There is considerable variation in the elimination of aminoglycosides among individuals, especially in patients with impaired renal function. Monitoring of serum aminoglycoside levels in these patients is essential for providing adequate therapy and reducing toxicity. In renal failure, the drugs accumulate, and dosage reductions are necessary. Aminoglycosides are substantially removed by hemodialysis and to a lesser extent by peritoneal dialysis.

Spectrum of Activity

Aminoglycoside antibiotics are active primarily against aerobic gram-negative bacilli and S. aureus. As a group, they are particularly potent against the Enterobacteriaceae, P. aeruginosa, and Acinetobacter spp. Certain differences in antimicrobial spectra among the various aminoglycosides do exist. Kanamycin is limited in its spectrum because of the common resistance of P. aeruginosa and the frequent occurrence of plasmid-mediated inactivating enzymes among other gram-negative bacilli (41). It is now used occasionally as a "second-line" drug in combination with other antibiotics for the therapy of mycobacterial infections (191, 192). Similarly, widespread resistance among the Enterobacteriaceae has limited the usefulness of streptomycin. As a single agent, streptomycin is used in the therapy of infections due to Francisella tularensis (tularemia) and Yersinia pestis (plague). It is often used in conjunction with tetracycline for the treatment of brucellosis. It has the greatest in vitro activity of the aminoglycosides against M. tuberculosis. It may also be used in combination with penicillin or vancomycin for the treatment of infective endocarditis due to viridans streptococci or enterococci, provided that the organisms do not possess high-level ribosomal or enzymatic resistance to streptomycin (15, 197, 206).

Although gentamicin and tobramycin have very similar antibacterial activity profiles, gentamicin is more active in vitro against Serratia spp., whereas tobramycin is more active against P. aeruginosa (131). However, these minor differences have not been correlated with greater efficacy of one agent over the other. For the most part, gentamicin and tobramycin are susceptible to inactivation by the same modifying enzymes produced by resistant bacteria, except that in contrast to gentamicin, tobramycin can be inactivated by 6-acetyltransferase and 4'-adenyltransferase and has a variable susceptibility to 3-acetyltransferase. Netilmicin and amikacin are resistant to many of these aminoglycoside-modifying enzymes and therefore are active against most Enterobacteriaceae that are resistant to gentamicin and tobramycin (126). Netilmicin is intrinsically less active than gentamicin or tobramycin against P. aeruginosa, and most gentamicin-resistant Serratia, Proteus, Providencia, and Pseudomonas isolates are also usually resistant to netilmicin (65). Amikacin is often used as the aminoglycoside of choice when gentamicin and tobramycin resistances are prevalent. In addition, amikacin is active against many Mycobacterium spp. (191, 192). Aminoglycosides are only moderately active against Haemophilus and Neisseria spp.

Although active against staphylococci, aminoglycosides are not recommended as single agents for the treatment of staphylococcal infections. Gentamicin is often combined with a penicillin or vancomycin for synergy in the treatment of serious infections due to staphylococci, enterococci, or viridans streptococci (15, 197, 198). The aminoglycosides are not active against anaerobes.

Paromomycin is an aminoglycoside notable for its amebicidal and antihelminthic effects, and it is used clinically for the treatment of intestinal amebiasis and tapeworm infections (119). It has modest antibacterial activity against gram-positive cocci and Enterobacteriaceae, but P. aeruginosa isolates are generally resistant.

Spectinomycin is used primarily for uncomplicated anogenital infections due to N. gonorrhoeae (195), including β-lactamase-producing strains, and resistant gonococci are rare (58, 219). It is useful in patients with penicillin allergy.

Spectinomycin is ineffective for pharyngeal gonococcal infections, syphilis, or chlamydial infections.

Adverse Effects

Considerable intrinsic toxicity, mainly in the form of nephrotoxicity and auditory or vestibular toxicity, is characteristic of all the aminoglycosides. The nephrotoxic potential varies among the aminoglycosides, with neomycin being the most nephrotoxic and streptomycin the least. This effect is usually reversible when the drug is discontinued. The presence of hypotension, prolonged duration of therapy, preexisting renal insufficiency, and possibly excessive trough serum aminoglycoside concentrations increase the risk of nephrotoxicity.

All aminoglycosides are capable of causing damage to the eighth cranial nerve in humans. Vestibular toxicity is more frequently associated with streptomycin, gentamicin, and tobramycin, whereas auditory toxicity is more typical of kanamycin and amikacin. This frequently irreversible side effect may occur even after discontinuation of the drug and is cumulative with repeated courses of the agent. The ototoxicity is a result of selective destruction of the hair cells in the cochlea. Clinically detectable auditory and vestibular dysfunctions have been reported to occur in 3 to 5% of patients receiving gentamicin, tobramycin, or amikacin who underwent audiometric testing (56).

Neuromuscular paralysis, which is usually reversible, can occur after rapid intravenous infusion of aminoglycosides. This phenomenon occurs particularly in the setting of myasthenia gravis or concurrent use of succinylcholine during anesthesia. Other minor adverse reactions include local pain and allergic skin rashes. No known serious adverse reactions have been reported for spectinomycin.

QUINOLONES

Quinolones belong to a group of potent antibiotics biochemically related to nalidixic acid, which was developed initially as a urinary antiseptic. Nalidixic acid and its early analogs, oxolinic acid and cinoxacin, have limited clinical applications as a result of the widespread emergence of bacterial resistance. Newer quinolones have been synthesized by modifying the original two-ring quinolone (or naphthyridine) nucleus with different side chain substitutions (212). These new agents, also known as fluoroquinolones, each contain a fluorine atom attached to the nucleus at position 6. Cinoxacin, norfloxacin, ciprofloxacin, enoxacin, ofloxacin, and lomefloxacin are currently available for clinical use in the United States. Temafloxacin has been withdrawn from clinical use because of toxicities. Fleroxacin and pefloxacin will likely be licensed for clinical use in the near future, while newer agents such as sparfloxacin, levofloxacin, and clinafloxacin are undergoing preclinical and clinical investigation in the United States.

Mechanism of Action

The primary bacterial target of the quinolones is DNA gyrase, an enzyme essential for DNA replication (93, 212). Bacterial DNA gyrases are related to a series of mammalian topoisomerase enzymes that have a similar function. Their therapeutic index stems from the fact that the clinically useful fluoroquinolones inhibit bacterial DNA gyrase at concentrations far below those required to inhibit mammalian topoisomerases. By inhibiting bacterial DNA synthesis, these agents are bactericidal. However, the antibacterial activity of quinolones is reduced in the presence of low pH, urine, and divalent cations (Mg^{2+} and Ca^{2+}) (213). Bacterial resistance to quinolones results from chromosomal mutations that alter DNA gyrase or decrease drug permeation. Plasmid-encoded resistance to the fluoroquinolones has not been reported.

Pharmacology

Pharmacokinetic parameters are similar among the fluoroquinolones, with minor differences. These drugs are well absorbed from the gastrointestinal tract, with bioavailability varying from 60 to 95% for the various fluoroquinolones (91, 208). After oral administration, concentrations in serum peak after 1 to 2 h. The presence of food does not significantly alter the absorption of these drugs, but coadministration with aluminum- or magnesium-containing antacids substantially reduces the bioavailability and subsequent peak concentrations in serum as a result of chelation between quinolones and antacids. The degree of serum protein binding is generally low, ranging from 8% for ofloxacin to 60% for enoxacin. The long half-lives of elimination for fluoroquinolones in serum, ranging from 3.5 to 4 h for ciprofloxacin to 12 h for fleroxacin, allow twice- or once-daily dosing.

These agents have good penetration into lung, kidney, muscle, bone, intestinal wall, and extravascular body fluids. Concentrations in prostate are about 2 times those in the serum, and concentrations of 25 to 100 times above peak concentrations in serum are achieved in the urine. Concentrations are variable but low in CSF obtained from patients with meningitis (211). Quinolones penetrate phagocytes so well that concentrations within neutrophils are 14 times as high as those in serum (186). This feature may account for the excellent in vivo activity of these agents against such intracellular pathogens as *Brucella, Listeria, Salmonella*, and *Mycobacterium* spp.

Pefloxacin is extensively metabolized in the liver, being converted to norfloxacin in vivo. Ofloxacin exhibits little or no in vivo metabolism and is excreted mainly (90%) by the kidneys. The other quinolones are cleared by both hepatic and renal routes in various proportions, with elimination primarily via the kidneys. This renal elimination is blocked by probenecid. Small amounts are also excreted in the bile.

Hepatic insufficiency prolongs the half-life of pefloxacin, whereas the elimination of other fluoroquinolones is significantly diminished in the presence of renal failure. These drugs are only partially removed by hemodialysis and are minimally affected by peritoneal dialysis because of their marked extravascular penetration, as reflected in their very large volumes of distribution.

Spectrum of Activity

Fluoroquinolones possess excellent activity in vivo against *Enterobacteriaceae, P. aeruginosa, Acinetobacter* spp., β-lactamase-positive and -negative *H. influenzae*, and gram-negative cocci such as *N. gonorrhoeae, N. meningitidis*, and *M. catarrhalis* (14, 30, 50, 212). Potency against *S. aureus* and coagulase-negative staphylococci is good; activity against streptococci and enterococci is lower. However, methicillin-resistant staphylococci are becoming increasingly resistant to these agents. The currently available agents are generally inactive against anaerobic bacteria, with little or no clinically useful activity against members of the *B. fragilis* group and *C. difficile* (28, 73).

Enteropathogenic gram-negative bacilli such as salmonellae, shigellae, *Yersinia enterocolitica, Vibrio* spp., *Aeromo-*

nas spp., *Plesiomonas* spp., *Campylobacter jejuni*, and enteroinvasive and enterotoxigenic *E. coli* are all susceptible to the quinolones (50, 213). Clinical studies have shown these drugs to be effective in the prophylaxis as well as the treatment of infectious diarrheas. However, reduced susceptibility and resistance to quinolones have emerged in clinical isolates of *Shigella* and *Campylobacter* spp. (52, 94). *Legionella* spp. are susceptible to these agents, with MICs of most fluoroquinolones being 0.12 to 1.0 μg/ml for these organisms (48). Fluoroquinolones are the first class of oral agents with outstanding potency against *P. aeruginosa*. Ciprofloxacin is the most active among these drugs against *P. aeruginosa*, with a MIC_{90} of ≤0.5 μg/ml (50). However, *P. cepacia* and *X. maltophilia* are generally resistant to the quinolones.

The fluoroquinolones, especially ciprofloxacin and ofloxacin, are active in vitro against *M. tuberculosis, Mycobacterium fortuitum, Mycobacterium kansasii*, and *Mycobacterium xenopi* (42, 217). Their activity against *Mycobacterium avium-Mycobacterium intracellulare* complex is fair to poor. They also exhibit activity against *Chlamydia trachomatis* and *Mycoplasma hominis* but are less potent against *Ureaplasma urealyticum* (100). Ciprofloxacin and pefloxacin inhibit *Rickettsia conorii, Rickettsia rickettsii*, and *Coxiella burnetii* (149, 150, 216). Although quinolones possess in vitro activity against *Plasmodium falciparum* at achievable concentrations in serum, they are relatively ineffective when used clinically for the treatment of malaria. *Nocardia* spp. are relatively resistant to the quinolones (13, 44).

No significant inoculum effect has been observed in the bacteria susceptible to quinolones. Combinations of quinolones with beta-lactam drugs or aminoglycosides are usually indifferent or additive in their effects against gram-negative and gram-positive bacteria and mycobacteria (213). However, bactericidal activities of quinolones can be antagonized by rifampin or chloramphenicol.

Adverse Effects

Gastrointestinal symptoms, which occur in up to 5% of patients as nausea, vomiting, abdominal discomfort, and diarrhea, are the most common side effects (84). However, *C. difficile* colitis rarely occurs with the use of quinolones. Headaches, fatigue, insomnia, dizziness, agitation, and, rarely, seizures can occur. These adverse neurologic effects are usually associated with high dosages in elderly patients or concurrent use of nonsteroidal anti-inflammatory drugs (especially with enoxacin).

Allergic reactions are uncommon and often manifest as rash, urticaria, or generalized pruritus. Dose-related photosensitivity has also been described. Rare laboratory abnormalities occurring during fluoroquinolone therapy include elevations in serum transaminases, eosinophilia, leukopenia, and thrombocytopenia. Shortly after its release in the United States and Europe, temafloxacin was withdrawn from clinical use because its use was associated with hemolysis, thrombocytopenia, and acute renal failure occasionally resulting in death.

Enoxacin and to a lesser extent ciprofloxacin and pefloxacin increase the levels of concomitant theophylline and other methylxanthines in serum as a result of decreased hepatic clearance (146, 204). Other reported drug interactions include augmentation of the anticoagulant effects of warfarin by ciprofloxacin, norfloxacin, and ofloxacin and an increase in serum cyclosporin levels by ciprofloxacin (146).

Although irreversible cartilage erosions and skeletal abnormalities were observed in studies of quinolone toxicity in animals (29), such effects have not yet been documented clinically. However, quinolones are generally contraindicated for use in patients less than 18 years old and in pregnant or nursing mothers.

TETRACYCLINES

Tetracyclines are broad-spectrum bacteriostatic antibiotics with the hydronaphthacene nucleus, which contains four fused rings. The congeners form three groups based on duration of action. Chlortetracycline, oxytetracycline, and tetracycline are short-acting, demeclocycline and methacycline are intermediate-acting, and doxycycline and minocycline are long-acting compounds.

Mechanism of Action

The tetracyclines act against susceptible microorganisms by inhibiting protein synthesis. They enter bacteria by an energy-dependent process and bind reversibly to the 30S ribosomal subunits of the bacteria (27). This process blocks the access of aminoacyl-tRNA to the RNA-ribosome complex, preventing bacterial polypeptide synthesis. Bacterial resistance to tetracycline occurs as a result of active efflux of the drug from the cell, an altered ribosomal target site that prevents binding of the drug, or production of modifying enzymes that inactivate the drug (175).

Pharmacology

Tetracyclines are incompletely absorbed from the gastrointestinal tract, but their absorption is improved in the fasting state. Ingestion of food, especially dairy products, and other substances such as antacids and iron preparations impairs the absorption of these drugs. Less interference with absorption by foods occurs with doxycycline and minocycline. These long-acting tetracyclines are more readily absorbed, and therefore, lower doses are required. Peak concentrations in serum of 3 to 5 μg/ml are reached in 2 h after standard oral dosages. Intravenous preparations are available, and peak concentrations in serum of 10 to 20 μg/ml are reached in 1 h after intravenous administration. Tetracyclines are usually bacteriostatic at these clinically achievable concentrations in serum.

Tetracyclines are metabolized by the liver and concentrated in the bile. Biliary concentrations of tetracyclines are three to five times higher than concurrent levels in plasma. These drugs accumulate in the blood in patients with hepatic insufficiency or biliary obstruction. They should be avoided or used cautiously in reduced dosages in patients with impaired liver function.

These antibiotics are excreted primarily in the urine except for doxycycline, which is excreted primarily (90%) as an inactive conjugate via the biliary tract in the feces. Renal failure prolongs the half-lives of the tetracyclines except doxycycline. Therefore, doxycycline is considered the tetracycline of choice for extrarenal infections in the presence of renal failure.

Tissue penetration of these drugs is excellent, but levels in CSF are low even in the presence of meningeal inflammation. Tetracyclines cross the placenta and are incorporated into fetal bone and teeth. They are excreted in high concentrations in human milk. Therefore, tetracyclines are not advised for pregnant or lactating women. Minocycline, the most lipophilic tetracycline at physiologic pH, reaches relatively high concentrations in saliva and tears, making it an ideal antibiotic for eradicating the meningococcal carrier state (83, 89).

Spectrum of Activity

All tetracyclines have similar antimicrobial spectra, with activity against many gram-positive and gram-negative bacteria, mycoplasmas, chlamydiae, rickettsiae, and some protozoa. Many gram-positive aerobic cocci, including *S. aureus*, *S. pyogenes*, and *S. pneumoniae*, are susceptible at concentrations achievable in serum. However, emergence of tetracycline-resistant strains of *S. pneumoniae* is frequent (103). Although many *E. coli* isolates are susceptible to tetracyclines, pseudomonads and many *Enterobacteriaceae* are resistant. *Shigella* and *Salmonella* spp. are increasingly resistant to these agents (25). Tetracyclines are used mainly for the treatment of acute, uncomplicated urinary tract infections due to *E. coli* (178) and as effective prophylactic therapy for traveler's diarrhea caused by enterotoxigenic *E. coli* (62). With activity against *Pseudomonas pseudomallei*, *Brucella* spp., *Vibrio* spp., and *Mycobacterium marinum* (192), they have been used successfully in the treatment of infections due to these bacteria. Their efficacy in the therapy of cholera is diminishing owing to the emergence of resistant *Vibrio cholerae* isolates (215). Minocycline is active against *Nocardia* spp. (44). Many anaerobic bacteria, including members of the *B. fragilis* group and *Actinomyces* spp., are susceptible to tetracyclines (34, 127). These drugs are commonly used preoperatively with or without neomycin as oral bowel preparations.

These drugs are useful in the treatment of urethritis and acute pelvic inflammatory diseases caused by *N. gonorrhoeae*, *Chlamydia trachomatis*, *U. urealyticum*, and *Mycoplasma hominis*. Emergence of resistance to tetracyclines among *N. gonorrhoeae* strains is increasing (58). The drugs are effective for the treatment of other chlamydial infections (psittacosis, lymphogranuloma venereum, and trachoma) (118). Other infections responsive to tetracyclines include granuloma inguinale, chancroid, relapsing fever, and tularemia.

Tetracyclines are the drug of choice for treating rickettsial infections (Rocky Mountain spotted fever, endemic and scrub typhus, and Q fever). Many pathogenic spirochetes, including *Treponema pallidum* and *Borrelia burgdorferi*, are susceptible (118, 128). Protozoans such as *Plasmodium falciparum* and *Entamoeba histolytica* are also inhibited by these drugs (118, 140).

Adverse Effects

Tetracyclines have irritative effects on the upper gastrointestinal tract, producing esophageal ulcerations, nausea, vomiting, and epigastric distress. Alterations in the enteric flora occur with the use of tetracyclines, often resulting in diarrhea, and pseudomembranous colitis can develop with prolonged use.

Hypersensitivity reactions are unusual, generally manifesting themselves as urticaria, fixed drug eruptions, morbilliform rashes, and anaphylaxis. Cross-reactivity among tetracyclines is the rule. Photosensitivity reactions consist of an erythematous rash on areas exposed to sunlight and can occur with all analogs, especially demeclocycline (64).

Tetracycline causes depression of bone growth, permanent discoloration of the teeth, and enamel hypoplasia when given during tooth and skeletal development (80). Therefore, these drugs are usually avoided in childhood (<8 years of age) and during pregnancy.

Minocycline has been known to cause vertigo, and benign intracranial hypertension (pseudotumor cerebri) has been described with many of the analogs (193). Tetracycline can aggravate preexisting renal failure by inhibiting protein synthesis, increasing the azotemia from amino acid metabolism.

MACROLIDES

Macrolides have been in use since the early 1950s, with erythromycin as the prototypical antibiotic of this class (196). Their chemical structures consist of a macrocyclic lactone ring attached to two sugar moieties, desosamine and cladinose. They differ from each other in the size (14 to 16 atoms) and substitution pattern of the lactone ring. Erythromycin is a naturally occurring 14-membered macrolide derived from *Streptomyces erythreus*, and other natural analogs include oleandomycin, spiramycin, and josamycin. Clarithromycin and azithromycin are 14- and 15-membered semisynthetic macrolides, respectively, that offer significant advantages over erythromycin because of expanded antimicrobial spectra, improved pharmacokinetic parameters, and less frequent adverse effects and drug interactions. Azithromycin is also known as an azalide because it has a nitrogen atom incorporated into its lactone ring. Dirithromycin, roxithromycin, flurithromycin, and rokitamycin are new macrolides currently under preclinical and clinical evaluation (102).

Mechanism of Action

Macrolides are generally bacteriostatic agents that inhibit bacterial RNA-dependent protein synthesis. They may be bactericidal at high drug concentrations and against a low inoculum of bacteria. They bind reversibly to the 50S ribosomal subunits of susceptible microorganisms, thereby blocking the translocation reaction of polypeptide chain elongation. Presence of rRNA methylases is the primary mechanism of macrolide resistance and confers macrolide-lincosamide-streptogramin B coresistance (47). Other uncommon mechanisms of macrolide resistance are production of macrolide-inactivating enzymes (esterases, phosphotransferases) and active efflux of macrolides.

Pharmacology

Erythromycin is available in various topical, parenteral (lactobionate and gluceptate), and oral (base, stearate, ethylsuccinate, and estolate) preparations, while clarithromycin and azithromycin are available only in oral forms. When administered orally, erythromycin base is rapidly inactivated by gastric acid, whereas the newer macrolides are stable against acid degradation. Intestinal absorption of erythromycin (except for the estolate form) and azithromycin is reduced up to 50% in the presence of food. Peak levels in serum of 2 to 3, 0.5 to 1, and 0.4 μg/ml are reached at 3 h after oral doses of erythromycin (500 mg), clarithromycin (250 mg), and azithromycin (500 mg), respectively. Much higher concentrations of erythromycin are achieved with intravenous infusion. Tissue distributions of macrolides are excellent, with concentrations in various tissues 10- to 100-fold higher than that in serum (205). The high concentrations reached rapidly within neutrophils and macrophages account for the potent activity of these drugs against intracellular pathogens (165). They penetrate poorly into the brain and CSF, but they do cross the placenta and are excreted in breast milk.

Erythromycin and clarithromycin are metabolized by the liver and primarily excreted in the bile. Clarithromycin exhibits first-pass metabolism, which produces a microbio-logically active 14-hydroxy derivative that is two to four times more potent than the parent drug against some organisms. Azithromycin is excreted largely unchanged in the bile. Erythromycin, clarithromycin, clarithromycin's active metabolite, and azithromycin have terminal half-lives in serum of 1.5, 5, 8.5, and >40 h, respectively. Because of its exceptionally high tissue penetration, azithromycin has a half-life in tissue of 2 to 4 days (165). Dosage adjustment of clarithromycin is necessary with moderate to severe renal failure (creatinine clearance of <30 ml/min). Except for clarithromycin, macrolides are removed minimally by hemodialysis or peritoneal dialysis.

Spectrum of Activity

Macrolides are relatively broad-spectrum antibiotics, with activity against gram-positive and some gram-negative bacteria, mycoplasmas, chlamydiae, treponemes, and rickettsiae (85, 196, 205). Erythromycin shows good activity against staphylococci and streptococci, including *S. pneumoniae*, but emergence of resistance among these isolates (especially group A streptococci) is a problem in certain parts of the world (26, 103). Clarithromycin is two- to fourfold more active than erythromycin and azithromycin is less active than erythromycin against most staphylococci and streptococci (8). These drugs are bactericidal against susceptible strains of streptococci but bacteriostatic toward staphylococci and enterococci. Erythromycin-resistant strains display cross-resistance to these drugs, and methicillin-resistant staphylococci and many enterococci are resistant to all macrolides. These drugs are also active against *Corynebacterium* spp., *L. monocytogenes*, and *Actinomyces israelii* (8).

The antibacterial activities of macrolides against gram-negative bacilli are influenced by pH, with increasing potency (lower MICs) as the pH increases to 8.5. *H. influenzae* and *M. catarrhalis* are more susceptible to azithromycin (MIC$_{90}$ of 0.5 μg/ml) than to erythromycin or clarithromycin (8- to 16-fold-higher MIC$_{90}$s) (134). However, additive (and possibly synergistic) activity between clarithromycin and its 14-hydroxy metabolite reduces the MIC of clarithromycin for *H. influenzae* by two- to fourfold (135). Clarithromycin is the most active drug in this class against *Chlamydia pneumoniae* (MIC$_{90}$ of 0.03 μg/ml) and *Legionella* isolates (MIC$_{90}$ of 0.25 μg/ml) (8). All three macrolides are equally potent against *Bordetella pertussis* and *Mycoplasma pneumoniae*, and erythromycin has long been established as the drug of choice for the therapy of infections due to these pathogens and *Legionella* spp. Macrolides are active against *Campylobacter* spp., *Helicobacter pylori*, *Pasteurella multocida*, *N. meningitidis*, and *Borrelia burgdorferi* (8, 117, 188). Unlike other macrolides, azithromycin is also active in vitro against *E. coli*, *Shigella* spp., *Salmonella* spp., and *Y. enterolitica* (134).

Macrolide antibiotics are effective in vitro against many pathogens that cause sexually transmitted diseases. *N. gonorrhoeae*, *Haemophilus ducreyi*, *Chlamydia trachomatis*, and *U. urealyticum* are all susceptible, but only azithromycin is active against *Mycoplasma hominis* (8, 134, 135). Erythromycin may be used for treatment of gonorrhea and syphilis in patients who cannot tolerate penicillin G (118), but data on the new macrolides for these indications are limited. Azithromycin is effective as an alternative to tetracyclines for the treatment of genital chlamydial infections (177).

The macrolides have good activity against anaerobic bacteria such as *B. fragilis*, other *Bacteroides* spp., and anaerobic gram-positive cocci, with MIC$_{90}$s of 1 to 4 μg/ml (8).

Most *Clostridium* spp., especially *Clostridium perfringens*, are inhibited at ≤1 μg/ml. Atypical mycobacteria are more susceptible than *M. tuberculosis* to macrolide antibiotics (134, 135). The MIC$_{90}$s of clarithromycin and azithromycin for *M. avium-M. intracellulare* are in the range of 2 to 4 μg/ml, allowing additive or synergistic killing activity of these organisms within infected macrophages when these drugs are combined with other antimycobacterial drugs (6). Erythromycin is used occasionally to treat infections due to *Mycobacterium scrofulaceum*, *M. kansasii*, and *Mycobacterium chelonae* (122, 191) and in combination with ampicillin against *Nocardia asteroides* (60).

Spiramycin and the new macrolides offer comparable in vitro activity against *Toxoplasma gondii*, and they have been reported to show efficacy in the treatment of toxoplasmosis (119). Although spiramycin has been used to treat cryptosporidiosis, the therapeutic efficacy remains to be proven (39).

Adverse Effects

The incidence of serious side effects related to the use of erythromycin is relatively low. Gastrointestinal irritation, such as abdominal cramps, nausea, vomiting, and diarrhea, is common with oral administration and can occur when the drug is given intravenously. These side effects occur less frequently with clarithromycin and azithromycin. Thrombophlebitis is associated with intravenous infusion, but it can be avoided by dilution of the dose in a large volume of fluid and by a slow infusion rate. Hypersensitivity reactions may include skin rash, fever, and eosinophilia. Cholestatic hepatitis occurring in adults has frequently been associated with the estolate form but has also been reported with other forms of erythromycin (179). For this reason, erythromycin estolate is no longer recommended for use in adults.

Reversible hearing loss may occur with use of large doses and very high concentrations of erythromycin in serum (4 g/day), usually in elderly patients with renal insufficiency (16, 86). Ototoxicity has also been reported with high doses of clarithromycin and azithromycin used to treat M. avium-M. intracellulare infections. Pseudomembranous colitis and superinfection of the gastrointestinal tract or vagina with *Candida* spp. or gram-negative bacilli occur rarely. Concurrent erythromycin therapy increases the levels of theophylline, cyclosporine, and digoxin in serum by interfering with their hepatic metabolism (143). It also increases the anticoagulant effect of warfarin (143). To date, no clinically significant interactions have been observed between these drugs and clarithromycin or azithromycin. However, cardiac arrhythmias have occurred with concurrent use of erythromycin or clarithromycin and terfenadine.

LINCOSAMIDES

The lincosamide antibiotics include lincomycin, which was initially isolated from *Streptomyces lincolnensis*, and clindamycin, which is a chemical modification of lincomycin. The chemical structure of each drug consists of an amino acid linked to an amino sugar. Clindamycin has increased antibacterial activity and improved absorption after oral administration compared with those of lincomycin (116). Both drugs are available for parenteral and oral use, but lincomycin is now used very infrequently in the United States.

Mechanism of Action

Lincosamides bind to the 50S ribosomal subunits of susceptible bacteria and prevent elongation of peptide chains by interfering with peptidyl transfer, thereby suppressing protein synthesis. The ribosomal binding sites are the same as or closely related to those that bind macrolides and chloramphenicol. Clindamycin can be bactericidal or bacteriostatic, depending on the drug concentration, bacterial species, and inoculum of bacteria.

Pharmacology

About 90% of an oral clindamycin dose is absorbed from the gastrointestinal tract, with no interference from the ingestion of food. A single oral dose of 150 mg yields a peak concentration in serum of 2 to 3 μg/ml in 1 h. Peak levels in serum of 10 to 12 μg/ml are obtained at 1 h after an intravenous dose of 600 mg. Therapeutic serum drug levels are maintained for 6 to 9 h after these dosages.

Clindamycin distributes well into bone, lungs, pleural fluid, and bile, but it penetrates poorly into CSF, even with meningitis. It readily crosses the placenta and enters fetal tissues. Clindamycin is actively concentrated in neutrophils and macrophages.

The normal half-life of clindamycin is 2.4 h. Most of the drug is metabolized by the liver and excreted in an inactive form in the urine. Its half-life is prolonged by severe liver dysfunction, necessitating dosage reduction in patients with severe liver disease. Although serum drug levels are increased in patients with severe renal failure, dose modification is not essential. The drug is not removed significantly by hemodialysis or peritoneal dialysis.

Spectrum of Activity

Lincosamides have a broad spectrum of activity against the aerobic gram-positive cocci and anaerobes. Clindamycin is more potent than lincomycin against methicillin-susceptible *Staphylococcus* spp., *S. pneumoniae*, and group A and viridans streptococci (110). The MIC$_{90}$s are in the range of 0.01 to 0.1 μg/ml for these strains. However, resistance to clindamycin has emerged in clinical isolates of these bacteria that are also resistant to erythromycin (26). The prevalence of clindamycin-resistant *S. aureus* may be 15 to 20% in some institutions. Enterococci are uniformly resistant to the lincosamides.

Clindamycin is one of the most active antibiotics available against anaerobes, including members of the *B. fragilis* group and *C. perfringens*, with MIC$_{90}$s of ≤2 μg/ml (10, 180). However, clindamycin resistance (which appears to be increasing) is found in 5 to 10% of *B. fragilis* group species, 10 to 20% of clostridial species, 10% of peptococci, and most *Fusobacterium varium* strains (34, 110, 180). All of the *Enterobacteriaceae* are resistant to clindamycin. Clindamycin has been used successfully as single-agent therapy for actinomycosis (156), babesiosis (119, 209), and malaria (167). It is also effective in combination with pyrimethamine for toxoplasma encephalitis (38, 155) and in combination with primaquine for *Pneumocystis carinii* pneumonia (184).

Adverse Effects

Clindamycin-associated diarrhea occurs in up to 20% of patients, and use of this drug has been associated with pseudomembranous colitis caused by toxin-producing *C. difficile* (11). This complication is not dose related and may occur after oral or parenteral therapy. Prompt cessation of the antibiotic in conjunction with oral vancomycin, metronidazole, or bacitracin therapy is effective in reversing this complication.

Other uncommon side effects include skin rashes, fever, and reversible elevation of serum transaminases. Clindamycin can block neuromuscular transmission and may potentiate the action of neuromuscular blocking agents during anesthesia.

GLYCOPEPTIDES AND LIPOPEPTIDES

Vancomycin, a bactericidal antibiotic obtained from *Streptomyces orientales*, is the only glycopeptide marketed for clinical use in the United States. Initially introduced for its efficacy against penicillin-resistant staphylococci, it has become most useful against methicillin-resistant staphylococci and in patients allergic to penicillins or cephalosporins. Teicoplanin (formerly teichomycin A), a new complex glycopeptide chemically related to vancomycin (172), is still undergoing clinical evaluation in the United States. Daptomycin (LY 146032) and ramoplanin (MDL 62198) are an investigational semisynthetic lipopeptide and lipoglycopeptide, respectively, with spectra of activity similar to those of the glycopeptides. Because of its excessive neuromuscular toxicity, daptomycin will not be developed for clinical use. Systemic toxicity also precludes the use of ramoplanin for anything other than topical application.

Mechanism of Action

Glycopeptides inhibit peptidoglycan synthesis in the bacterial cell wall by complexing with the D-alanyl–D-alanine portion of the cell wall precursor (129). Daptomycin also acts by inhibiting bacterial peptidoglycan synthesis, but this is likely secondary to its interference with membrane transport of precursors (1). The exact mechanism of action of daptomycin has yet to be determined.

Pharmacology

Vancomycin and teicoplanin can be administered orally or parenterally. After oral administration, the drugs are poorly absorbed, and high concentrations in stool are achieved, accounting for the efficacy of these drugs in treating pseudomembranous colitis (59). Desirable peak and trough levels in serum of 20 to 50 and 5 to 15 μg/ml, respectively, are obtained after a 1-g intravenous dose of vancomycin every 12 h in healthy subjects. Similar serum drug concentrations are reached with intravenous teicoplanin, which has the advantage of a longer half-life in serum and can be administered once daily. Therapeutic levels of both drugs are achieved in synovial, ascitic, pericardial, and pleural fluids, with variable penetration into the CSF only in the presence of inflamed meninges (81, 120).

Vancomycin and teicoplanin have half-lives in serum of 6 and 45 h, respectively, in patients with healthy renal function, and they are eliminated from the body by glomerular filtration. In severe renal insufficiency, their excretion is prolonged to about 9 days, and they are not removed by hemodialysis or peritoneal dialysis.

Spectrum of Activity

Glycopeptides and daptomycin are active mainly against aerobic and anaerobic gram-positive organisms, including methicillin-susceptible and -resistant staphylococci, streptococci, enterococci, *Corynebacterium* spp., *Bacillus* spp., *L. monocytogenes*, *Clostridium* spp., and *Actinomyces* spp. The MICs of vancomycin against *S. aureus*, *S. epidermidis*, streptococci, and enterococci are typically in the range of 0.25 to 2 μg/ml. The bactericidal activity varies, with MBCs 20-fold higher than MICs for viridans streptococci. These agents are essentially bacteriostatic against enterococci. Teicoplanin (76, 78, 111) and daptomycin (111) are two to four times more active than vancomycin against these gram-positive cocci. Unlike vancomycin and teicoplanin, daptomycin exhibits moderate bactericidal activity toward this group of organisms. Increasing resistance to vancomycin has emerged among clinical isolates of *Enterococcus faecalis* (121), *Enterococcus faecium* (63, 69, 109), *Enterococcus gallinarum* (99), and coagulase-negative staphylococci (166). Cross-resistance with teicoplanin is variable in these strains, but most are susceptible to daptomycin and ramoplanin (97). Other naturally vancomycin-resistant gram-positive organisms include *Leuconostoc*, *Lactobacillus*, and *Pediococcus* spp., most of which are susceptible to ramoplanin (31, 97, 159).

Vancomycin is useful in the prevention and treatment of endocarditis due to gram-positive bacteria in patients who are allergic to penicillin (15, 37). It is the drug of choice for treating *Corynebacterium jeikeum* infections (70) and is useful for *Flavobacterium meningosepticum* meningitis (81) and antibiotic-associated *C. difficile* colitis (11, 59).

The glycopeptides and lipopeptides are not active against gram-negative organisms, mycobacteria, or fungi. They show no cross-resistance with other unrelated antibiotics. They can be synergistic with aminoglycosides against staphylococci, streptococci, enterococci, and listeriae (185, 198).

Adverse Effects

The most frequent side effects of vancomycin are fever, chills, and phlebitis at the site of infusion. Rapid or bolus infusion of vancomycin causes tingling and flushing of the face, neck, and thorax, known as the red neck or red man syndrome, as a result of histamine release by basophils and mast cells (147). This phenomenon is not due to allergic hypersensitivity. Allergic maculopapular or diffuse erythematous rashes can occur in up to 5% of patients (174). Reversible leukopenia or eosinophilia can rarely develop with glycopeptide use.

Hearing loss due to ototoxicity has been described occasionally in patients in whom serum vancomycin concentrations exceeded 50 μg/ml, but it is hard to find unequivocal evidence of vancomycin ototoxicity in humans or animals (16). Vancomycin-induced nephrotoxicity is rare since the recent availability of highly purified vancomycin preparations. However, the risk of nephrotoxicity appears to be increased during combination therapy with vancomycin and aminoglycosides.

Teicoplanin is generally well tolerated and does not produce the red man syndrome or nephrotoxicity. It does cause irritation at the site of intravenous infusion, and ototoxicity has been reported (40).

SULFONAMIDES AND TRIMETHOPRIM

Sulfonamides were the first effective systemic antimicrobial agents used in the United States during the 1930s. They are derived from sulfanilamide, which shares chemical similarities with *para*-aminobenzoic acid, a factor essential for bacterial folic acid synthesis. Various substitutions at the sulfonyl radical attached to the benzene ring nucleus en-

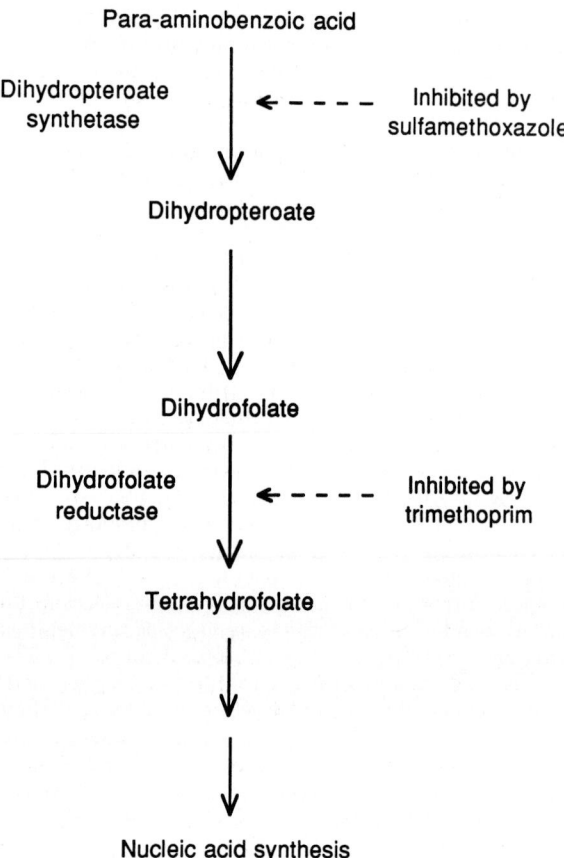

Para-aminobenzoic acid

Dihydropteroate
synthetase ← − − − − Inhibited by
sulfamethoxazole

Dihydropteroate

Dihydrofolate

Dihydrofolate
reductase ← − − − − Inhibited by
trimethoprim

Tetrahydrofolate

Nucleic acid synthesis

FIGURE 2 Mechanism of action of TMP-SMX.

hance the antibacterial activity and also determine pharmacologic properties of the drug.

Trimethoprim (TMP) is a pyrimidine analog that inhibits the enzyme dihydrofolate reductase, thus interfering with folic acid metabolism, subsequent pyrimidine synthesis, and one-carbon fragment metabolism in the bacteria. Since TMP and sulfonamides block the bacterial folic acid metabolic pathway at different sites, they potentiate the antibacterial activity of one another and act additively or synergistically against a wide variety of organisms. Such a fixed combination, TMP-sulfamethoxazole (SMX), also called co-trimoxazole, was introduced clinically in 1968 and has proven to be very effective in the treatment of many infections (158, 161).

Mechanism of Action

Sulfonamides competitively inhibit bacterial modification of *para*-aminobenzoic acid into dihydrofolate, whereas TMP inhibits bacterial dihydrofolate reductase (Fig. 2). This sequential inhibition of folate metabolism ultimately prevents the synthesis of bacterial DNA (87). Since mammalian cells do not synthesize folic acid, human purine synthesis is not affected significantly by sulfonamides or TMP. The antibacterial effect of these agents may be reduced in patients receiving high doses of folinic acid.

Pharmacology

Sulfonamides are usually administered in the oral and topical forms; the intravenous preparations (sulfadiazine and

sulfisoxazole) are rarely used. The sulfonamides vary in their durations of action. Thus, sulfamethizole and sulfisoxazole are short-acting, sulfadiazine and SMX are intermediate-acting, and sulfadoxine is a long-acting compound. Mafenide acetate (Sulfamylon cream) and silver sulfadiazine are applied topically in burn patients and have significant percutaneous absorption. Sulfacetamide is available as an ophthalmic preparation, and various combinations of other sulfonamides are available orally (triple sulfa, or trisulfapyrimidine) or as vaginal creams or suppositories.

The orally administered sulfonamides are absorbed rapidly and completely from the gastrointestinal tract. They are metabolized in the liver by acetylation and glucuronidation and are excreted by the kidney as free drug and inactive metabolites. Sulfonamides compete for bilirubin-binding sites on plasma albumin and increase levels of unconjugated bilirubin in blood. For this reason they should not be given to neonates, in whom increased serum bilirubin levels may cause kernicterus.

Sulfonamides are well distributed throughout the body, with levels in the cerebrospinal, synovial, pleural, and peritoneal fluids about 80% of concentrations in serum. They readily cross the placenta and enter the fetal circulation. Sulfonamides may be used in patients with renal failure, but the drugs may accumulate during prolonged therapy as a result of reduced renal excretion.

TMP is available only for oral use and is absorbed almost completely from the gastrointestinal tract. After the usual 100-mg dose, peak levels in serum reach 1 μg/ml in 1 to 4 h. This drug distributes widely in body tissues, including the kidney, lung, prostate, and body fluids (142). Concentrations in CSF are about 40% of levels in serum. Its half-life in serum is about 10 h in healthy subjects and is prolonged in those with renal insufficiency. Up to 80% of a dose is excreted unchanged in the urine by tubular secretion; the remaining fraction is excreted as inactive metabolites by the kidney or in the bile.

A fixed combination of TMP-SMX in a dose ratio of 1:5 is available for oral and intravenous use. An intravenous dose of 160 mg of TMP with 800 mg of SMX produces average peak levels in serum of 3.4 and 47.3 μg/ml, respectively, in 1 h. Similar peak levels are reached at 2 to 4 h after the same dose is taken orally. Widely distributed in the body, both drugs reach therapeutic levels in the CSF (40% of levels in serum). Excretion is primarily by the kidney; dosage reduction is necessary in patients with creatinine clearances of ≤30 ml/min. Both TMP and SMX are removed by hemodialysis and partially by peritoneal dialysis.

Spectrum of Activity

Sulfonamides are inhibitory to a variety of gram-positive and gram-negative bacteria, actinomycetes, chlamydiae, toxoplasmas, and plasmodia. Their in vitro antimicrobial activities are irregular, being strongly influenced by inoculum size and composition of the test media. Susceptibility testing endpoints are often difficult to determine because of the presence of hazy growth within zones of inhibition in disk diffusion tests and because of the phenomenon of "trailing" in dilution tests. Sulfadiazine and sulfisoxazole are effective for rheumatic fever prophylaxis, but they are not useful in treating established group A beta-hemolytic streptococcal pharyngitis. These drugs may be used for prophylaxis of close contacts of patients with meningitis due to sulfonamide-susceptible *N. meningitidis*. Sulfisoxazole can be used to treat chlamydial urethritis, and sulfacetamide ophthalmic solution is effective for trachoma and inclusion conjunctivitis.

Sulfadiazine in combination with pyrimethamine has been used successfully to treat toxoplasmosis, and sulfadoxine combined with pyrimethamine (Fansidar) is effective in the prophylaxis and therapy of *Plasmodium falciparum* malaria. Sulfonamides are active against *Nocardia asteroides* (44), and they show moderate activity against M. *kansasii*, M. *fortuitum*, M. *marinum*, and M. *scrofulaceum* (153). Other uses of sulfonamides include therapy of meliodosis, dermatitis herpetiformis, lymphogranuloma venereum, and chancroid.

Among the gram-negative bacilli, *E. coli* strains were initially susceptible to the sulfonamides, especially at levels achievable in the urine. Therefore, these drugs have been used primarily in the treatment of first-episode acute urinary tract infections due to *E. coli*. However, increasing bacterial resistance has limited their efficacy in recent years. *Serratia marcescens*, *P. aeruginosa*, enterococci, and anaerobes are usually resistant to the sulfonamides.

TMP is active in vitro against many gram-positive cocci and most gram-negative bacilli. *P. aeruginosa*, most anaerobes, *Mycoplasma pneumoniae*, and mycobacteria are resistant. The MIC varies considerably with the test media used. Like the sulfonamides, TMP is primarily used in the therapy of uncomplicated and recurrent urinary tract infections due to susceptible organisms. However, the prevalence of TMP-resistant *Enterobacteriaceae* is increasing (74).

Combinations of TMP with other agents, such as rifampin, polymyxins, and aminoglycosides, have demonstrated in vitro synergistic antibacterial activity against various gram-negative bacilli. TMP combined with dapsone is effective in the treatment of *Pneumocystis carinii* pneumonia in immunocompromised patients.

Many gram-positive cocci, including staphylococci and streptococci, and most gram-negative bacilli except *P. aeruginosa* are susceptible to TMP-SMX (19). The drug combination has variable bactericidal effects on enterococci in vitro, depending on the test media used for susceptibility testing (130). Unlike many bacteria that can utilize only thymidine for growth, enterococci can use thymidine, thymine, exogenous folinic acid, dihydrofolate, and tetrahydrofolate, resulting in higher MICs (25- to 50-fold increase) on media containing these compounds (125). This fact also explains the ineffectiveness of TMP-SMX against enterococci in vivo.

With excellent activity against *S. pneumoniae*, *M. catarrhalis*, and *H. influenzae*, including β-lactamase-producing strains, TMP-SMX is useful for the therapy of acute otitis media, sinusitis, acute bronchitis, and pneumonia. It has shown excellent results in the prophylaxis and therapy of acute and chronic urinary tract infections (178). It is an effective alternative therapy for uncomplicated urogenital gonorrhea, including cases caused by penicillinase-producing *N. gonorrhoeae* (118). It can be used also for the treatment of chancroid, but resistance to TMP-SMX in *H. ducreyi* is increasing. The drug combination is also useful in treating infections due to salmonellae, shigellae, enteropathogenic *E. coli*, and *Y. enterocolitica* (158). It has been used successfully for prophylaxis and treatment of traveler's diarrhea (53), but resistance to TMP-SMX in *Shigella* spp. and *E. coli* now severely limits its usefulness in many parts of the world.

Other microorganisms susceptible to TMP-SMX include *Brucella* spp., *P. pseudomallei*, *P. cepacia*, X. *maltophilia*, M. *kansasii*, M. *marinum*, and M. *scrofulaceum*. M. *tuberculosis*

and M. *chelonae* are generally resistant. It is a valuable antibiotic for the treatment of *Nocardia asteroides* infections (190), *P. cepacia* and X. *maltophilia* bacteremia, *L. monocytogenes* meningitis, *Isospora belli* gastroenteritis (141), and Whipple's disease. In immunocompromised hosts (e.g., those with leukemia or AIDS, organ transplant recipients), TMP-SMX is effective for the prophylaxis and treatment of *Pneumocystis carinii* pneumonia (23, 218).

Adverse Effects

Sulfonamides are known to cause nausea, vomiting, headache, and fever. Hypersensitivity reactions can occur as rashes, vasculitis, erythema nodosum, erythema multiforme, and Stevens-Johnson syndrome (20). Very high doses of less water soluble sulfonamides such as sulfadiazine may result in crystalluria, with renal tubular deposits of sulfonamide crystals. Bone marrow toxicity with anemia, leukopenia, or thrombocytopenia can occur. Sulfonamides should be avoided in patients with glucose-6-phosphate dehydrogenase deficiency because of associated hemolytic anemia. Sulfonamides may potentiate the effects of warfarin, phenytoin, and oral hypoglycemic agents.

In general, TMP is well tolerated. With prolonged use, megaloblastic anemia, neutropenia, and thrombocytopenia can develop, especially in folate-deficient patients. Adverse reactions to TMP-SMX due to either the TMP or, more commonly, the SMX component can occur. Mild gastrointestinal symptoms and allergic skin rashes occur in about 3% of patients (108). Megaloblastic bone marrow changes with leukopenia, thrombocytopenia, or granulocytopenia may develop, usually in patients with preexisting folate deficiency. Nephrotoxicity usually occurs only in patients with underlying renal dysfunction. Patients with AIDS have a much higher frequency of adverse reactions (as much as 70%) (75).

POLYPEPTIDES

Polymyxins

Polymyxins are a group of related cyclic basic polypeptides originally derived from *Bacillus polymyxa*. They have limited spectra of antimicrobial activity and significant toxicity. Only polymyxins B and E (colistin) are available for therapeutic use in humans.

Mechanism of Action

Acting like detergents or surfactants, members of this group of antibiotics interact with the phospholipids of the bacterial cell membrane, thereby increasing cell permeability and disrupting osmotic integrity. This process results in leakage of intracellular constituents, leading to cell death. The bactericidal action is reduced in the presence of calcium, which interferes with the attachment of drugs to the cell membrane.

Pharmacology

The polymyxins are usually administered by the parenteral, oral, or topical route. They are not significantly absorbed when given orally or topically, and intramuscular injections can be painful. Peak concentrations in serum of 5 μg/ml are obtained with a total daily dose of intravenous polymyxin B at 2.5 mg (25,000 U)/kg of body weight. Colistin sulfate is given orally for local antibacterial effect in the gut, while colistimethate sodium, a sulfomethyl derivative of colistin, is used for intramuscular injections. The half-life of polymyxin B in serum is 6 to 7 h, and that of colistin is 2 to 4 h. The drugs do not penetrate well into pleural fluid, synovial fluid, or CSF even in the presence of inflammation. Excretion is mostly via the kidneys by glomerular filtration. Levels in serum and toxicity are increased in states of renal insufficiency. These drugs are not removed by hemodialysis, but small amounts can be removed by peritoneal dialysis.

Polymyxin is often used topically as 0.1% polymyxin in combination with bacitracin or neomycin for treatment of skin, mucous membrane, eye, and ear infections. It is poorly absorbed from these surfaces. When the drug is used for irrigation of serous or wound cavities, systemic absorption can be significant enough to produce toxicity.

Spectrum of Activity

Polymyxins are active only against gram-negative bacilli, especially *Pseudomonas* spp. The MIC_{90}s for *Pseudomonas* spp., including *P. aeruginosa*, are <8 μg/ml. *Proteus*, *Providencia*, *Serratia*, and *Neisseria* isolates are usually resistant. Emergence of resistance during therapy is rare, and there is no cross-resistance with other antibiotics. Polymyxins B and E have identical antimicrobial spectra and show complete cross-resistance to one another.

The combination of polymyxins with TMP-SMX may be synergistic in the treatment of serious infection due to multiply resistant *Serratia* spp., *P. aeruginosa*, *P. cepacia*, and *X. maltophilia* (157). The polymyxins are usually reserved for serious, life-threatening *Pseudomonas* or gram-negative bacillary infections caused by organisms resistant to all other antibiotics. Aerosolized polymyxins have been used successfully to treat *P. aeruginosa* colonization or respiratory infections in patients with cystic fibrosis or bronchiectasis (57).

Adverse Effects

Neurotoxicity and nephrotoxicity are the two major side effects of polymyxins. Paresthesia with flushing, dizziness, vertigo, ataxia, slurred speech, drowsiness, or mental confusion occurs when levels in serum exceed 1 to 2 μg/ml. Polymyxins also have a curarelike effect and can block neuromuscular transmission. Dose-related renal dysfunction occurs in about 20% of patients receiving appropriate therapeutic dosages. Allergic reactions such as fever and skin rashes are rare, but urticaria and shock after rapid intravenous infusion have occurred.

Bacitracin

Originally isolated from *Bacillus licheniformis* (formerly *Bacillus subtilis*), bacitracin is a peptide antibiotic consisting of peptide-linked amino acids. Although it was introduced initially for the systemic treatment of severe staphylococcal infections, it is now mainly restricted to topical use because of its systemic toxicity.

Mechanism of Action

Bacitracin inhibits dephosphorylation of a lipid pyrophosphate, a step essential for bacterial wall synthesis. It also disrupts the bacterial cytoplastic membrane.

Pharmacology

Bacitracin is often used in various topical preparations, such as creams, ointments, antibiotic sprays and powders, and solutions for wound irrigation or bladder instillation. When used as a topical antibiotic, no significant amount of baci-

tracin is absorbed systemically. Large doses used to irrigate serous cavities may be associated with systemic toxicity.

Spectrum of Activity

The drug is active mainly against gram-positive bacteria, particularly staphylococci and group A beta-hemolytic streptococci. However, group C and G streptococci are less susceptible, and group B streptococci are resistant (61). *Neisseria* spp. are also susceptible, but gram-negative bacilli are resistant. Bacitracin is often combined with neomycin, polymyxin B, or both in topical preparations to provide broad-spectrum antibacterial coverage. Orally administered bacitracin is effective in treating antibiotic-associated *C. difficile* colitis (46).

Adverse Effects

Systemic administration of bacitracin results in significant nephrotoxicity. Side effects are rare when the drug is given orally or applied topically. The drug is nonirritating to skin or mucous membranes. Allergic skin sensitization is rare.

CHLORAMPHENICOL

Chloramphenicol is a unique antibiotic originally derived from *Streptomyces venezuelae*. It contains a nitrobenzene ring. It is a highly effective broad-spectrum antimicrobial agent with specific indications for use in seriously ill patients. Thiamphenicol is an analog of chloramphenicol with a similar spectrum of antimicrobial activity (137). Only chloramphenicol is available for clinical use in the United States.

Mechanism of Action

The drug is a bacteriostatic agent that inhibits protein synthesis by binding reversibly to the peptidyltransferase component of the 50S ribosomal subunit and preventing the transpeptidation process of peptide chain elongation. At therapeutic concentrations achievable in serum, it can be bactericidal against common meningeal pathogens such as *S. pneumoniae*, *N. meningitidis*, and *H. influenzae* (148). Bacterial resistance occurs with plasmid-mediated production of chloramphenicol acetyltransferase, which inactivates the drug (169).

Pharmacology

Chloramphenicol is available for oral and parenteral use. It is rapidly and completely absorbed from the gastrointestinal tract. After an oral or intravenous dose of 1 g, peak concentrations in serum at 2 h can reach 10 to 15 μg/ml. It diffuses well into many tissues and body fluids, including CSF, where levels are generally 30 to 50% of concentrations in serum even without meningeal inflammation (171). The antibiotic readily crosses the placental barrier and is present in human milk.

Chloramphenicol is metabolized and inactivated by glucuronidation in the liver, with a half-life of 4 h in adults. The active drug (5 to 10%) and its inactive metabolites are excreted by the kidneys. Careful monitoring of serum chloramphenicol levels, maintaining peak concentrations in serum in the therapeutic range of 10 to 20 μg/ml, is useful for ensuring therapeutic efficacy and reduced toxicity. Patients with hepatic failure have high levels of active drug in serum owing to prolonged half-life. Dosage modification is not necessary in the presence of renal insufficiency, since the metabolites are not as toxic as the active drug. Levels in serum are not affected by hemodialysis or peritoneal dialysis.

Spectrum of Activity

Chloramphenicol is very active against many gram-positive and gram-negative bacteria, chlamydiae, mycoplasmas, and rickettsiae. $MIC_{90}s$ for most gram-positive aerobic and anaerobic cocci are ≤ 12.5 μg/ml (137). However, the drug is usually inactive against methicillin-resistant *S. aureus* and *S. epidermidis* and is variably active against enterococci. *N. meningitidis*, *H. influenzae* (ampicillin-resistant and -susceptible strains), and most *Enterobacteriaceae* are susceptible. Its activity against *Serratia* and *Enterobacter* isolates is variable, and *Pseudomonas* spp. are usually resistant. Salmonellae, including *Salmonella typhi*, are also susceptible, but resistant isolates are being encountered with increasing frequency (25).

Chloramphenicol has excellent activity against anaerobic bacteria, including members of the *B. fragilis* group. Almost all of these isolates are inhibited at concentrations of ≤ 10 μg/ml (34, 127). It is also active against *Rickettsia* spp. and *Coxiella burnetii*.

Adverse Effects

Bone marrow toxicity is the major complication of chloramphenicol use. This side effect may occur either as a dose-related bone marrow suppression or as an idiosyncratic aplastic anemia. Reversible bone marrow depression with anemia, leukopenia, and thrombocytopenia occurs as a result of a direct pharmacologic effect of the drug on hematopoiesis. High doses (>4 g/day), prolonged therapy, and excessively high levels in serum (>20 μg/ml) predispose patients to develop this type of complication. The second form of bone marrow toxicity is a rare but usually fatal complication that manifests as aplastic anemia. This response is not dose related, and the precise mechanism is unknown. It can occur weeks to months after the use of chloramphenicol, and it can develop after the use of oral, intravenous, or topical preparations.

Gray baby syndrome, characterized by vomiting, abdominal distention, cyanosis, hypothermia, and circulatory collapse, may occur in premature infants and neonates. This toxicity results from the immature hepatic function of neonates, which impairs hepatic inactivation of the drug. Reversible optic neuritis causing decreased visual acuity has been reported in patients receiving prolonged therapy. Chloramphenicol can occasionally cause hypersensitivity reactions, including skin rashes, drug fevers, and anaphylaxis. It potentiates the action of warfarin, phenytoin, and oral hypoglycemic agents by competitive inhibition of hepatic microsomal enzymes.

METRONIDAZOLE

Metronidazole is a nitroimidazole derivative that was first introduced in 1959 for the treatment of *Trichomonas vaginalis* infections. It now has an important therapeutic role in the treatment of infections due to anaerobic bacteria and certain protozoan parasites.

Mechanism of Action

Metronidazole owes its bactericidal activity to the nitro group of its chemical structure. After the drug gains entry into the cells of susceptible organisms, the nitro group is reduced by a nitroreductase enzyme in the cytoplasm, generating certain short-lived, highly cytotoxic intermediate compounds or free radicals that disrupt host DNA (49). Resistance to nitroimidazole drugs may be due to decreased uptake of the drug or the presence of intracellular enzymes that can scavenge the free-radical intermediates.

Pharmacology

Metronidazole is absorbed rapidly and almost completely when given orally. Peak levels in serum of 6 μg/ml are obtained 1 h after an oral dose of 250 mg. Intravenous doses of 7.5 mg/kg result in peak concentrations in serum of 20 to 25 μg/ml. The drug has a half-life in serum of 8 h. Therapeutic levels are achieved in all body tissues and fluids, including abscess cavities and CSF, even without meningeal inflammation. The drug crosses the placenta and is secreted in breast milk. The drug is primarily metabolized by the liver, and 60 to 80% is excreted by the kidneys. With impaired hepatic function, plasma clearance of metronidazole is delayed and dosage adjustments are necessary. The pharmacokinetics are minimally affected by renal insufficiency. Metronidazole and its metabolites are removed completely by dialysis.

Spectrum of Activity

Metronidazole exhibits potent activity against almost all anaerobic bacteria, including the *B. fragilis* group, *Fusobacterium* spp., and *Clostridium* spp. (10). It is the only antimicrobial agent with consistent bactericidal activity against *B. fragilis*. However, the susceptibility of gram-positive anaerobic cocci is somewhat variable, with $MIC_{90}s$ of 16 μg/ml for these organisms. Most strains of the genera *Actinomyces*, *Arachnia*, and *Propionibacterium* are usually resistant. Frequencies of metronidazole-resistant *Bacteroides* isolates (MIC of >16 μg/ml) in the range of 2 to 5% have been reported from various institutions (34, 127). Metronidazole has no activity against aerobic bacteria, including the *Enterobacteriaceae*.

The drug is effective in the treatment of antibiotic-associated colitis caused by *C. difficile* (11, 24), with efficacy equivalent to that of oral vancomycin for this indication (183). It is also useful in combination with an aminoglycoside for treating polymicrobial soft tissue infections and mixed aerobic-anaerobic intra-abdominal and pelvic infections.

Metronidazole is active against the protozoa *Trichomonas vaginalis*, *Giardia lamblia*, and *Entamoeba histolytica*. It is the drug of choice for the treatment of trichomoniasis, giardiasis, and intestinal and invasive amebiasis, including amebic liver abscess (119, 181, 210).

Tinidazole and ornidazole are other nitroimidazole derivatives with antianaerobe activities somewhat more potent than that of metronidazole. At present, however, they are investigational drugs in the United States.

Adverse Effects

The drug is generally well tolerated, and adverse side effects are uncommon. It can cause mild gastrointestinal symptoms such as nausea, abdominal cramps, and diarrhea. An unpleasant, metallic taste may be experienced with oral therapy. Metronidazole can potentiate the effect of warfarin and prolong the prothrombin time.

Although metronidazole is carcinogenic in mice and rats, there is no evidence for increased carcinogenicity in humans. Use of this agent in pregnancy, especially during the first trimester and in nursing mothers, however, should be avoided.

RIFAMPIN

Rifampin, also known as rifampicin, is a semisynthetic antibiotic derived from rifamycin B, which belongs to a group of macrocyclic compounds produced by the mold *Streptomyces mediterranei*. Introduced for clinical use in 1968 as an effective antituberculous drug, this agent has activity against many bacteria. A closely related compound, rifabutin, a derivative of rifamycin S, is another potent antimycobacterial agent, especially against M. *avium*-M. *intracellulare* (139).

Mechanism of Action

Rifampin exerts its bactericidal effect by forming a stable complex with bacterial DNA-dependent RNA polymerase, preventing the chain initiation process of DNA transcription (200). Mammalian RNA synthesis is not affected, because the mammalian enzyme is much less sensitive to the drug. Rifampin-resistant isolates possess an altered RNA polymerase enzyme that arises easily from single-step mutations during monotherapy with rifampin.

Pharmacology

The drug is well absorbed after oral administration, reaching peak concentrations in serum of 5 to 10 μg/ml in 2 to 4 h following a 600-mg dose. A parenteral preparation is also available. Rifampin is deacetylated in the liver to an active metabolite and excreted in the bile, and it undergoes enterohepatic circulation. Normal half-life in serum varies from 1.5 to 5 h. Dosage adjustments are necessary for patients with severe hepatic dysfunction.

Rifampin is well distributed to almost all body tissues and fluids, reaching concentrations equal to or exceeding that in the serum. Levels in the CSF are highest in the presence of inflamed meninges. It is able to enter phagocytes and kill living intracellular organisms (113), and it crosses the placenta. About 30 to 40% of the drug is excreted in the urine, and it does not accumulate in patients with impaired renal function. Hemodialysis and peritoneal dialysis do not eliminate the drug.

Spectrum of Activity

In addition to its well-known antimycobacterial effects (30), rifampin has a wide spectrum of antimicrobial activity. It is bactericidal against gram-positive cocci such as staphylococci (including methicillin-resistant strains), streptococci, and anaerobic cocci, with MICs in the range of 0.01 to 0.5 μg/ml. It remains an important adjunct in the combination therapy of serious and chronic staphylococcal infections (187). However, it is bacteriostatic against enterococci, with usual MICs of <16 μg/ml (124).

N. *gonorrhoeae*, N. *meningitidis*, and H. *influenzae*, including β-lactamase-producing strains, are susceptible to rifampin, which is used frequently in the prophylaxis of meningococcal and H. *influenzae* type b meningitis (83). MICs for Enterobacteriaceae are ≤12 μg/ml, while MICs for *Serratia marcescens* and P. *aeruginosa* are higher (124). Besides quinolones, rifampin is one of the most active agents against *Legionella pneumophila* and other *Legionella* spp., with MICs of ≤0.03 μg/ml. Because of its ability to enter phagocytes, rifampin inhibits the growth of *Brucella* spp. and *Coxiella burnetii* intracellularly (150), and it is used frequently in the combination therapy of infections due to these organisms. Although *Chlamydia* spp. are very susceptible to rifampin in vitro, resistance emerges rapidly when rifampin is used alone.

Adverse Effects

Rifampin has many side effects, including gastrointestinal discomfort and hypersensitivity reactions, such as drug fever, skin rashes, and eosinophilia. It produces a harmless, orange-red coloration of saliva, tears, urine, and sweat. In up to 20% of patients, an influenzalike syndrome with fever, chills, arthralgias, and myalgias may develop after several months of intermittent therapy (79). This immunologic reaction may be associated with hemolytic anemia, thrombocytopenia, and renal failure. Rifampin-induced hepatitis occurs in <1% of patients and is more frequent during concurrent isoniazid therapy for tuberculosis. The drug is known to antagonize the effect of oral contraceptives and diminish the anticoagulant activity of warfarin.

NITROFURANTOIN

Nitrofurantoin belongs to a class of compounds consisting of a primary nitro group joined to a heterocyclic ring. Its role in human therapeutics is limited to treatment of urinary tract infections.

Mechanism of Action

The precise mechanism of action of nitrofurantoin is unknown. The drug is believed to inhibit various bacterial enzymes and to damage DNA (114).

Pharmacology

The drug is available in microcrystalline (Furadantin) and macrocrystalline (Macrodantin) forms. It is administered orally and is well absorbed from the gastrointestinal tract. Very low levels of the drug are achieved in serum and most body tissues after usual oral doses. With a half-life in serum of about 20 min, two-thirds of the drug is rapidly metabolized and inactivated in various tissues. The remaining one-third is excreted unchanged into the urine. An average dose of nitrofurantoin yields a concentration in urine of 50 to 250 μg/ml in patients with healthy renal function. In alkaline urine, more of the drug is dissociated into the ionized form, with lowered antibacterial activity. Nitrofurantoin accumulates in the sera of patients with creatinine clearances of <60 ml/min. The drug is removed by hemodialysis. The risk of systemic toxicity increases in the presence of severe uremia. It is contraindicated in patients with significant renal impairment and hepatic failure.

Spectrum of Activity

Nitrofurantoin has a broad spectrum of antibacterial activity against gram-positive and gram-negative bacteria, particularly the common urinary tract pathogens. It is active against gram-positive cocci, such as S. *aureus*, S. *epidermidis*, *Staphylococcus saprophyticus*, and *Enterococcus faecalis*, with MICS in the range of 4 to 25 μg/ml (90). S. *pneumoniae*, S. *pyogenes*, *Bacillus subtilis*, and *Corynebacterium* spp. are also susceptible, but they rarely cause urinary tract infections. Over 90% of E. *coli* and many coliform bacteria are susceptible to nitrofurantoin at MICs of <32 μg/ml. However, only one-third of *Enterobacter* and *Klebsiella* isolates are susceptible. *Pseudomonas* and most *Proteus* spp. are resistant. Susceptible organisms rarely become resistant to this drug during therapy.

Adverse Effects

Gastrointestinal irritation with anorexia, nausea, and vomiting is the most common side effect. Diarrhea and abdominal cramps may occur. Hypersensitivity reactions, such as drug fever, chills, arthralgia, skin rashes, and a lupuslike syndrome, have been observed (92).

Pulmonary reactions are the most common serious side effects associated with nitrofurantoin use. Acute pneumonitis with fever, cough, dyspnea, eosinophilia, and pulmonary infiltrates present on chest X-ray films can occur after a few days of therapy (92). This immunologically mediated reaction is more common in elderly patients and is rapidly reversible after cessation of therapy. Chronic pulmonary reactions with interstitial pneumonitis leading to irreversible pulmonary fibrosis can occur in patients on continuous therapy for 6 months or more.

Peripheral polyneuropathy is a serious side effect that occurs more often in patients with renal failure (92). Hemolytic anemia, megaloblastic anemia, and bone marrow suppression with leukopenia can occur. Rare hepatotoxic reactions, such as cholestatic jaundice and chronic active hepatitis, have been reported (168).

METHENAMINE

Methenamine is a tertiary amine with properties of a monoacidic base; it is used as a urinary antiseptic. To be activated, it is combined chemically with a poorly metabolized acid and administered as the mandelate (Mandelamine) or hippurate (Hiprex, Urex) salt.

Mechanism of Action

Methenamine has no antibacterial action by itself, but it is metabolized at acid pH to ammonia and formaldehyde, which provides the antiseptic action. This hydrolytic process occurs in the urine, and an effective bacteriostatic concentration of formaldehyde is reached at a urine pH of <5.5. Since the serum is at physiologic pH, formaldehyde is not released while methenamine circulates in the body.

Pharmacology

The agent is well absorbed from the gastrointestinal tract and is rapidly excreted in the urine. The elimination half-life of methenamine is about 4 h. At a urine pH of 5.0, about 20% of the methenamine excreted in the urine is hydrolyzed to formaldehyde and ammonia. Bactericidal levels (>20 μg/ml) of formaldehyde are generated in the bladder urine at 2 h after oral administration and may be maintained for at least 6 h or until the patient voids. The mandelate and hippurate moieties are also rapidly excreted in the urine in active, unchanged forms by glomerular filtration and tubular secretion. The agent is contraindicated in patients with hepatic insufficiency because of the ammonia produced.

Spectrum of Activity

With the liberation of enough formaldehyde into the urine, methenamine is essentially active against all gram-positive and gram-negative bacteria and against fungi (105). However, it is not effective for treating urinary tract infections due to urea-splitting organisms such as *Proteus* spp., which can convert urea to ammonium hydroxide, thereby preventing the hydrolysis of methenamine to formaldehyde. Combination with acetohydroxamic acid, a urease inhibitor, has been suggested for treating these infections by *Proteus* spp. Since bacteria and fungi do not become resistant to formaldehyde, emergence of resistance to methenamine is not a problem.

Methenamine is not useful for acute urinary tract infections. It has been used successfully as prophylactic therapy for recurrent bacteriurias, particularly those caused by highly resistant gram-negative bacilli or yeasts. It is also effective as prolonged suppressive therapy for chronic bacteriuria in the absence of structural abnormalities of the urinary tract.

Adverse Effects

Methenamine and its acid salts are generally well tolerated. Some patients may develop nausea, vomiting, abdominal cramps, and diarrhea. High doses or prolonged administration of the drug can cause urinary tract irritation by the free formaldehyde, resulting in urinary frequency, dysuria, albuminuria, and hematuria. Skin rashes may also occur. To avoid precipitation of urate crystals in the urine, methenamine salts should not be used in patients with gout or hyperuricemia.

MUPIROCIN

Mupirocin, formerly pseudomonic acid A, is a novel topical antibacterial agent derived from the fermentation products of *Pseudomonas fluorescens* (68). Structurally, it contains a unique 9-hydroxynonanoic acid moiety. It was developed for the topical treatment of superficial soft tissue infections, particularly those due to staphylococci.

In the United States, mupirocin is available as a 2% ointment (Bactroban). After topical application, <1% of the drug is absorbed systemically, with no detectable levels in the urine or feces. Penetration into deeper dermal layers of the skin is increased with traumatized skin and occlusive dressings. The drug is highly protein bound (95%), and its activity is lowered in the presence of serum. It is most active at moderately acid pH, with no inoculum effect (194). Mupirocin is slowly metabolized in the skin to inactive monic acid.

Mupirocin inhibits isoleucyl-tRNA synthetase, resulting in cessation of bacterial tRNA and protein synthesis (95). It has excellent in vitro activity, primarily against the gram-positive cocci. *S. aureus*, including methicillin-resistant strains, and coagulase-negative staphylococci are uniformly very susceptible, with MIC$_{90}$s of <0.5 μg/ml (21). Emergence of resistant strains of staphylococci occurs at very low frequency. Most streptococci (including *S. pneumoniae*; beta-hemolytic streptococci of groups A, B, C, and G; and viridans streptococci) are inhibited by concentrations of ≤1 μg/ml. Resistant bacteria include enterococci, *Corynebacterium* spp., *Erysipelothrix* spp., *Propionibacterium acnes*, gram-positive anaerobes, and most gram-negative bacteria. However, *H. influenzae*, *N. gonorrhoeae*, *N. meningitidis*, *M. catarrhalis*, *Bordetella pertussis*, and *Pasteurella multocida* are quite susceptible, with MICs in the range of 0.02 to 0.025 μg/ml. There is no cross-resistance between mupirocin and other major groups of antibiotics.

Clinically, mupirocin is efficacious in the therapy of superficial skin infections, such as impetigo, folliculitis, and burn wound infections, that are due to staphylococci or streptococci (72). It has been used successfully to eradicate nasal carriage of *S. aureus*, including methicillin-resistant strains (22).

No systemic toxic effects have been reported with mupirocin. Local irritation, such as burning, stinging, itch, and rash, which may be due to the polyethylene glycol base in the vehicle ointment, may occur.

APPENDIX
Approximate Concentrations of Antibacterial Agents in Serum

The concentrations of antimicrobial agents listed below are approximations taken from various reports and publications. Several factors can influence the levels of antimicrobial agent in individual patients, including inherent differences in the patients themselves, their physical conditions, drug dosages, and routes of administration. Levels can also be influenced by the assay methods used to determine them. Therefore, the concentrations given here should be used only as approximate values, and clinicians should use their knowledge of the patient and the drugs, the recommendations from U.S. Food and Drug Administration-approved package inserts, or other reputable sources in planning therapeutic regimens.

Antimicrobial agent	Half-life in serum (h)	Unit dose	Avg peak level in serum (μg/ml)[a]		
			p.o.	i.m.	i.v.
Amikacin	2–2.5	7.5 mg/kg		15–20	20–40
Amoxicillin	1	500 mg	6–8		
Amoxicillin-clavulanate	1.3/1.0	500/125 mg	4.4 (Amox) 2.3 (Clav)		
Ampicillin	1.1	500 mg	2.5–5	8–10	
		1 g			40
Ampicillin-sulbactam	1.1/1.0	3 g			120 (Amp) 60 (Sulb)
		1.5 g		18 (Amp) 13 (Sulb)	
Azithromycin	11–48	500 mg	0.4		
Azlocillin	0.9–1.2	2 g			130
Aztreonam	1.7	1 g		45	90–160
Bacampicillin	1.1	800 mg	13		
Carbenicillin	1.1	1 g		20–30	150
Carbenicillin indanyl sodium	1.1	764 mg	10		
Cefaclor	0.6	500 mg	16		
Cefadroxil	1.5	500 mg	10		
Cefamandole	0.5–1	1 g		20–36	90–140
Cefazolin	1.8	1 g		65	185
Cefepime	2	1 g		30	65
Cefixime	3–4	400 mg	3.5		
Cefmetazole	1.2–1.5	1 g			70
Cefonicid	3.5–4.5	1 g		98	220
Cefoperazone	2	1 g		65–75	153
Ceforanide	2.7–3	1 g		70	125
Cefotaxime	1	1 g		20	40–45
Cefotetan	3–4.5	1 g		50–80	160
Cefoxitin	0.7–1	1 g		20–25	55–110
Cefpirome	2	1 g		45	86
Cefpodoxime	2.5	200 mg	2.3		
Cefprozil	1.3	500 mg	10.5		
Ceftazidime	1.9	1 g		40	70
Ceftizoxime	1.4–1.8	1 g		39	80–90
Ceftriaxone	6–9	500 mg		40–45	
		1 g			150
Cefuroxime	1.3	750 mg		27	50
Cefuroxime axetil	1.3	500 mg	9		
Cephalexin	0.9	500 mg	18		
Cephalothin	0.6	1 g			30–60
Cephapirin	0.6	1 g			40–70
Cephradine	0.8	500 mg	16		
		1 g		12	60–80
Chloramphenicol	4	1 g	10–18		10–15
Chlortetracycline	6	500 mg	??		
Cinoxacin	1–1.5	500 mg	15		
Ciprofloxacin	3.5	500 mg	2.5		
		400 mg			4.6
Clarithromycin	5	250 mg	0.5–1		
Clindamycin	2.4–3	300 mg	3	6	
		600 mg			10–12
Cloxacillin	0.5	500 mg	10		

(Continued on next page)

Antimicrobial agent	Half-life in serum (h)	Unit dose	Avg peak level in serum (µg/ml)[a]		
			p.o.	i.m.	i.v.
Colistimethate sodium	2–4.5	150 mg		5–6	
Cyclacillin	0.5–0.75	500 mg	12		
Demeclocycline	12	300 mg	??		
Dirithromycin	44	500 mg	0.45		
Dicloxacillin	0.5–0.7	500 mg	15		
Doxycycline	18–22	100 mg	2.5		4
Enoxacin	4–6	400 mg	3–5		
Erythromycin	1.5	500 mg	2–3		
		1 g			10
Fleroxacin	12	400 mg	5		7–8
Fusidic acid	13–19	500 mg	25–30		50
Gentamicin	2–3	1.5 mg/kg		4–6	4–8
Imipenem	1	500 mg			25–35
Kanamycin	2.2–3	7.5 mg/kg		20–25	
Lomefloxacin	6.5	400 mg	3		
Loracarbef	1	400 mg	14		
Meropenem	1	500 mg			25–35
Metronidazole	8	500 mg	12		20–25
Mezlocillin	1	1 g		15	
		3 g			260
Minocycline	14–16	100 mg	1		
Nafcillin	0.5	500 mg		5–8	
		1 g			20–40
Nalidixic acid	1.5	1 g	20–50		
Netilmicin	2.2–2.5	2 mg/kg		5–7	6–8
Nitrofurantoin	0.3	100 mg	<2		
Norfloxacin	3.3	400 mg	1.5		
Ofloxacin	5	400 mg	4		
Oxacillin	0.5	500 mg	4–6	14–16	
		1 g			40
Oxytetracycline	9	250 mg	3–4		
Pefloxacin	10	400 mg	3		
Penicillin G	0.5	500 mg	1.5–2.5		
Aqueous	0.5	1×10^6 U		8–10	10
Benzathine		1.2×10^6 U		0.1–0.15	
Procaine		1.2×10^6 U		3	
Penicillin V	0.5	500 mg	3–5		
Piperacillin	1.1	2 g		36	
		4 g			240
Piperacillin-tazobactam	1.1/1	2.25 g		38 (Pip) 7 (Tazo)	
		4.5 g			280 (Pip) 35 (Tazo)
Pivampicillin	0.5–1	350 mg	2		
Polymyxin B	6–7	2.5 mg/kg			5
Rifampin	2–5	600 mg	7–9		10
Spectinomycin	1–3	2 g			100
Spiramycin	3.8	2 g	3		
Streptomycin	2–3	1 g		25–50	
Sulfadiazine	17	2 g	100–150		
Sulfadoxine	150–200	1 g	50–75		
Sulfamethizole	4–7	2 g	60		
Sulfamethoxazole	10–12	1 g	40		
Sulfisoxazole	5–7	2 g	170		
Teicoplanin	45	200 mg		7	
		400 mg			20–40
Tetracycline	8	500 mg	4		8
Ticarcillin	1.2	1 g		20–30	
		3 g			190
Ticarcillin-clavulanate	1.2/1.0	3.1 g			330 (Ticar) 8 (Clav)

(Continued on next page)

Antimicrobial agent	Half-life in serum (h)	Unit dose	Avg peak level in serum (μg/ml)[a]		
			p.o.	i.m.	i.v.
Tobramycin	2–2.8	1.5 mg/kg		4–6	4–8
Trimethoprim	10–12	100 mg	1		
Trimethoprim-sulfamethoxazole		160/800 mg	3 (TMP)		9 (TMP)
			46 (SMX)		106 (SMX)
Vancomycin	6	500 mg			20–40

[a] After 30-min intravenous infusion. p.o., oral; i.m., intramuscular; i.v., intravenous.

REFERENCES

1. **Allen, N. E., J. N. Hobbs, Jr., and W. E. Alborn, Jr.** 1987. Inhibition of peptidoglycan biosynthesis in gram-positive bacteria by LY146032. *Antimicrob. Agents Chemother.* **31:** 1093–1099.
2. **Anderson, J. A.** 1986. Cross-sensitivity to cephalosporins in patients allergic to penicillin. *Pediatr. Infect. Dis. J.* **5:**557–561.
3. **Appelbaum, P. C., M. R. Jacobs, S. K. Spangler, and S. Yamabe.** 1986. Comparative activity of β-lactamase inhibitors YTR 830, clavulanate, and sulbactam combined with β-lactams against β-lactamase-producing anaerobes. *Antimicrob. Agents Chemother.* **30:**789–791.
4. **Aswapokee, N., and H. C. Neu.** 1978. A sulfone beta-lactam compound which acts as a beta-lactamase inhibitor. *J. Antibiot.* **31:**1238–1244.
5. **Bansal, M. B., S. K. Chuah, and H. Thadepalli.** 1985. In vitro activity and in vivo evaluation of ticarcillin plus clavulanic acid against aerobic and anaerobic bacteria. *Am. J. Med.* **79**(Suppl. 5B):33–38.
6. **Barradell, L. B., G. L. Plosker, and D. McTavish.** 1993. Clarithromycin: a review of its pharmacological properties and therapeutic use in *Mycobacterium avium-intracellulare* complex infection in patients with acquired immune deficiency syndrome. *Drugs* **46:**289–312.
7. **Barry, A. L., and R. N. Jones.** 1987. Cefixime: spectrum of antibacterial activity against 16,016 clinical isolates. *Pediatr. Infect. Dis. J.* **6:**954–957.
8. **Barry, A. L., R. N. Jones, and C. Thornsberry.** 1988. In vitro activities of azithromycin (CP 62,993), clarithromycin (A-56268, TE-031), erythromycin, roxithromycin, and clindamycin. *Antimicrob. Agents Chemother.* **32:**752–754.
9. **Barry, A. L., C. Thornsberry, R. N. Jones, and T. L. Gavan.** 1985. Aztreonam: antibacterial activity, β-lactamase stability, and interpretive standards and quality control guidelines for disk-diffusion susceptibility test. *Rev. Infect. Dis.* **7**(Suppl. 4):S594–S604.
10. **Bartlett, J. G.** 1982. Anti-anaerobic antibacterial agents. *Lancet* **ii:**478–481.
11. **Bartlett, J. G.** 1992. Antibiotic-associated diarrhea. *Clin. Infect. Dis.* **15:**573–579.
12. **Barza, M.** 1985. Imipenem: first of a new class of β-lactam antibiotics. *Ann. Intern. Med.* **103:**552–560.
13. **Berkey, P., D. Moore, and K. Rolston.** 1988. In vitro susceptibilities of *Nocardia* species to newer antimicrobial agents. *Antimicrob. Agents Chemother.* **32:**1078–1079.
14. **Beskid, G., and B. L. T. Prosser.** 1993. A multicenter study on the comparative in vitro activity of fleroxacin and three other fluoroquinolones: an interim report from 27 centers. *Am. J. Med.* **94**(Suppl. 3A):2S–8S.
15. **Bisno, A. L., W. E. Dismukes, D. T. Durack, E. L. Kaplan, A. W. Karchmer, D. Kaye, S. H. Rahimtoola, M. A. Sande, J. P. Sanford, C. Watanakunakorn, and W. R. Wilson.** 1989. Antimicrobial treatment of infective endocarditis due to viridans streptococci, enterococci, and staphylococci. *JAMA* **261:**1471–1477.
16. **Brummett, R. E., and K. E. Fox.** 1989. Vancomycin- and erythromycin-induced hearing loss in humans. *Antimicrob. Agents Chemother.* **33:**791–796.
17. **Buesing, M. A., and J. H. Jorgensen.** 1984. In vitro activity of aztreonam in combination with newer β-lactams and amikacin against multiply resistant gram-negative bacilli. *Antimicrob. Agents Chemother.* **25:**283–285.
18. **Bush, K., J. S. Freudenberger, and R. B. Sykes.** 1982. Interaction of azthreonam and related monobactams with β-lactamases from gram-negative bacteria. *Antimicrob. Agents Chemother.* **22:**414–420.
19. **Bushby, S. R. M.** 1973. Trimethoprim-sulfamethoxazole: in vitro microbiological aspects. *J. Infect. Dis.* **128**(Suppl.): S442–S462.
20. **Carroll, O. M., P. A. Bryan, and R. J. Robinson.** 1966. Stevens-Johnson syndrome associated with long-acting sulfonamides. *JAMA* **195:**691–693.
21. **Casewell, M. W., and R. L. R. Hill.** 1985. In-vitro activity of mupirocin (pseudomonic acid) against clinical isolates of *Staphylococcus aureus. J. Antimicrob. Chemother.* **15:**523–531.
22. **Casewell, M. W., and R. L. R. Hill.** 1986. Elimination of nasal carriage of *Staphylococcus aureus* with mupirocin (pseudomonic acid): a controlled trial. *J. Antimicrob. Chemother.* **17:**365–372.
23. **Centers for Disease Control.** 1992. Guidelines for prophylaxis against *Pneumocystis carinii* pneumonia for children infected with human immunodeficiency virus. *JAMA* **265:** 1637–1644.
24. **Cherry, R. D., D. Portnoy, M. Jabbari, D. S. Daly, D. G. Kinnear, and C. A. Goresky.** 1982. Metronidazole: an alternate therapy for antibiotic-associated colitis. *Gastroenterology* **82:**849–851.
25. **Cherubin, C. E.** 1981. Antibiotic resistance of *Salmonella* in Europe and the United States. *Rev. Infect. Dis.* **3:**1105–1126.
26. **Cherubin, C. E., and D. B. Azabache.** 1992. While nearly no one was watching: the rise of erythromycin and clindamycin resistance in *Streptococcus pneumoniae* and *Streptococcus pyogenes. Antimicrob. Newsl.* **8:**37–44.
27. **Chopra, I., P. M. Hawkey, and M. Hinton.** 1992. Tetracyclines, molecular and clinical aspects. *J. Antimicrob. Chemother.* **29:**245–277.
28. **Chow, A. W., N. Chang, and K. H. Bartlett.** 1985. In vitro susceptibility of *Clostridium difficile* to new β-lactam and quinolone antibiotics. *Antimicrob. Agents Chemother.* **28:** 842–844.
29. **Christ, W., T. Lehnert, and B. Ulbrich.** 1988. Specific toxicologic aspects of the quinolones. *Rev. Infect. Dis.* **10**(Suppl. 1):S141–S146.
30. **Clark, J., and A. Wallace.** 1967. The susceptibility of mycobacteria to rifamide and rifampin. *Tubercle* **48:**144–148.
31. **Collins, L. A., G. M. Eliopoulos, C. B. Wennersten, M. J. Ferraro, and R. C. Moellering, Jr.** 1993. In vitro activity of ramoplanin against vancomycin-resistant gram-positive organisms. *Antimicrob. Agents Chemother.* **37:**1364–1366.
32. **Cooper, R. D. G.** 1992. The carbacephems: a new beta-lactam antibiotic class. *Am. J. Med.* **92**(Suppl. 6A):2S–6S.
33. **Coppens, L., and J. Klastersky.** 1979. Comparative study of anti-*Pseudomonas* activity of azlocillin, mezlocillin, and ticarcillin. *Antimicrob. Agents Chemother.* **15:**396–399.
34. **Cuchural, G. J., Jr., F. P. Tally, N. V. Jacobus, K. Aldridge, T. Cleary, S. M. Finegold, G. Hill, P. Iannini, J. P. O'Keefe, C. Pierson, D. Crook, T. Russo, and D. Hecht.** 1988. Susceptibility of the *Bacteroides fragilis* group in the

United States: analysis by site of isolation. *Antimicrob. Agents Chemother.* **32:**717–722.

35. **Cynamon, M. H., and G. S. Palmer.** 1982. In vitro susceptibility of *Mycobacterium fortuitum* to N-formimidoyl thienamycin and several cephamycins. *Antimicrob. Agents Chemother.* **22:**1079–1081.

36. **Cynamon, M. H., and G. S. Palmer.** 1983. In vitro activity of amoxicillin in combination with clavulanic acid against *Mycobacterium tuberculosis. Antimicrob. Agents Chemother.* **24:**429–431.

37. **Dajani, A. S., A. L. Bisno, K. J. Chung, D. T. Durack, M. Freed, M. A. Gerber, A. W. Karchmer, H. D. Millard, S. Rahimtoola, S. T. Shulman, C. Watanakunakorn, and K. A. Taubert.** 1990. Prevention of bacterial endocarditis: recommendations of the American Heart Association. *JAMA* **264:**2919–2922.

38. **Dannemann, B., J. A. McCutchan, D. Israelski, D. Antoniskis, C. Leport, B. Luft, J. Nussbaum, N. Clumeck, P. Morlat, J. Chiu, J.-L. Vilde, P. Haseltine, J. Leedom, J. Remington, M. Orellana, D. Feigal, A. Bartok, and the California Collaborative Treatment Group.** 1992. Treatment of toxoplasmic encephalitis in patients with AIDS: a randomized trial comparing pyrimethamine plus clindamycin to pyrimethamine plus sulfadiazine. *Ann. Intern. Med.* **116:**33–43.

39. **Davey, P., J.-C. Pechère, and D. Speller (ed.).** 1988. Spiramycin reassessed. *J. Antimicrob. Chemother.* **22**(Suppl. B):1–210.

40. **Davey, P. G., and A. H. Williams.** 1991. A review of the safety profile of teicoplanin. *J. Antimicrob. Chemother.* **27**(Suppl. B):69–73.

41. **Davies, J. E.** 1983. Resistance to aminoglycosides: mechanisms and frequency. *Rev. Infect. Dis.* **5**(Suppl. 2):S261–S267.

42. **Davies, S., P. D. Sparham, and R. C. Spencer.** 1987. Comparative in-vitro activity of five fluoroquinolones against mycobacteria. *J. Antimicrob. Chemother.* **19:**605–609.

43. **Davis, B. D.** 1982. Bactericidal synergism between β-lactams and aminoglycosides: mechanism and possible therapeutic implications. *Rev. Infect. Dis.* **4:**237–245.

44. **Dewsnup, D. H., and D. N. Wright.** 1984. In vitro susceptibility of *Nocardia asteroides* to 25 antimicrobial agents. *Antimicrob. Agents Chemother.* **25:**165–167.

45. **Drusano, G. L., S. C. Schimpff, and W. L. Hewitt.** 1984. The acylampicillins: mezlocillin, piperacillin, and azlocillin. *Rev. Infect. Dis.* **6:**13–32.

46. **Dudley, M. N., J. C. McLaughlin, G. Carrington, J. Frick, C. H. Nightingale, and R. Quintiliani.** 1986. Oral bacitracin vs vancomycin therapy for *Clostridium difficile*-induced diarrhea: a randomized double-blind trial. *Arch. Intern. Med.* **146:**1101–1104.

47. **Eady, E. A., J. I. Ross, and J. H. Cove.** 1990. Multiple mechanisms of erythromycin resistance. *J. Antimicrob. Chemother.* **26:**461–465.

48. **Edelstein, P. H., E. A. Gaudet, and M. A. C. Edelstein.** 1989. In vitro activity of lomefloxacin (NY-198 or SC 47111), ciprofloxacin, and erythromycin against 100 clinical *Legionella* strains. *Diagn. Microbiol. Infect. Dis.* **12**(Suppl.):93S–95S.

49. **Edwards, D. I.** 1993. Nitroimidazole drugs—action and resistance mechanisms. I. Mechanisms of action. *J. Antimicrob. Chemother.* **31:**9–20.

50. **Eliopoulos, G. M., and C. T. Eliopoulos.** 1993. Activity in vitro of the quinolones, p. 161–193. *In* D. C. Hooper and J. S. Wolfson (ed.), *Quinolone Antimicrobial Agents*, 2nd ed. American Society for Microbiology, Washington, D.C.

51. **Eliopoulos, G. M., K. Klimm, M. J. Ferraro, G. A. Jacoby, and R. C. Moellering, Jr.** 1989. Comparative in vitro activity of piperacillin combined with the beta-lactamase inhibitor tazobactam (YTR 830). *Diagn. Microbiol. Infect. Dis.* **12:**481–488.

52. **Endtz, H. P., G. J. Ruijs, B. van Klingeren, W. H. Jansen, T. van Reyden, and R. P. Mouton.** 1991. Quinolone resis-

53. **Ericsson, C. D., and H. L. DuPont.** 1993. Travelers' diarrhea: approaches to prevention and treatment. *Clin. Infect. Dis.* **16:**616–624.

54. **Fass, R. J.** 1983. Comparative in vitro activities of third-generation cephalosporins. *Arch. Intern. Med.* **143:**1743–1745.

55. **Fass, R. J., E. A. Copelan, J. T. Brandt, M. L. Moeschberger, and J. J. Ashton.** 1987. Platelet-mediated bleeding caused by broad-spectrum penicillins. *J. Infect. Dis.* **155:**1242–1248.

56. **Fee, W. E., Jr.** 1980. Aminoglycoside ototoxicity in the human. *Laryngoscope* **90**(Suppl. 24):1–19.

57. **Feeley, T. W., G. C. DuMoulin, J. Hedley-Whyte, L. S. Bushnell, J. P. Gilbert, and D. S. Feingold.** 1975. Aerosol polymyxin and pneumonia in seriously ill patients. *N. Engl. J. Med.* **293:**471–475.

58. **Fekete, T.** 1993. Antimicrobial susceptibility testing of *Neisseria gonorrhoeae* and implication for epidemiology and therapy. *Clin. Microbiol. Rev.* **6:**22–33.

59. **Fekety, R., J. Silva, B. Buggy, and H. G. Deery.** 1984. Treatment of antibiotic-associated colitis with vancomycin. *J. Antimicrob. Chemother.* **14**(Suppl. D):97–102.

60. **Finland, M., M. C. Bach, C. Garner, and O. Gold.** 1974. Synergistic action of ampicillin and erythromycin against *Nocardia asteroides*: effect of time of incubation. *Antimicrob. Agents Chemother.* **5:**344–353.

61. **Finland, M., C. Garner, C. Wilcox, and L. D. Sabath.** 1976. Susceptibility of beta-hemolytic streptococci to 65 antibacterial agents. *Antimicrob. Agents Chemother.* **9:**11–19.

62. **Freeman, L. D., D. R. Hopper, D. F. Lathen, D. P. Nelson, W. O. Harrison, and D. S. Anderson.** 1983. Brief prophylaxis with doxycycline for the prevention of traveler's diarrhea. *Gastroenterology* **84:**276–280.

63. **Frieden, T. R., S. S. Munsiff, D. E. Low, B. M. Willey, G. Williams, Y. Faur, W. Eisner, S. Warren, and B. Kreiswirth.** 1993. Emergence of vancomycin-resistant enterococci in New York City. *Lancet* **342:**76–79.

64. **Frost, P., G. D. Weinstein, and E. C. Gomez.** 1972. Phototoxic potential of minocycline and doxycycline. *Arch. Dermatol.* **105:**681–683.

65. **Fu, K. P., and H. C. Neu.** 1976. In vitro study of netilmicin compared with other aminoglycosides. *Antimicrob. Agents Chemother.* **10:**526–534.

66. **Fu, K. P., and H. C. Neu.** 1979. Comparative inhibition of β-lactamases by novel β-lactam compounds. *Antimicrob. Agents Chemother.* **15:**171–176.

67. **Fuchs, P. C., A. L. Barry, C. Thornsberry, and R. N. Jones.** 1984. In vitro activity of ticarcillin plus clavulanic acid against 632 clinical isolates. *Antimicrob. Agents Chemother.* **25:**392–394.

68. **Fuller, A. T., G. Mellows, M. Woolford, G. T. Banks, K. D. Barrow, and E. B. Chain.** 1971. Pseudomonic acid: an antibiotic produced by *Pseudomonas fluorescens. Nature* (London) **234:**416–417.

69. **George, R. C., and A. H. C. Uttley.** 1989. Susceptibility of enterococci and epidemiology of enterococcal infection in the 1980s. *Epidemiol. Infect.* **103:**403–413.

70. **Geraci, J. E., and W. R. Wilson.** 1981. Vancomycin therapy for infective endocarditis. *Rev. Infect. Dis.* **3**(Suppl.):S250–S258.

71. **Giamarellou, H.** 1986. Aminoglycosides plus β-lactams against gram-negative organisms: evaluation of in vitro synergy and chemical interactions. *Am. J. Med.* **80**(Suppl. 6B):126–137.

72. **Goldfarb, J., D. Crenshaw, J. O'Horo, E. Lemon, and J. L. Blumer.** 1988. Randomized clinical trial of topical mupirocin versus oral erythromycin for impetigo. *Antimicrob. Agents Chemother.* **32:**1780–1783.

73. **Goldstein, E. J. C., and D. M. Citron.** 1985. Comparative activity of the quinolones against anaerobic bacteria isolated

at community hospitals. *Antimicrob. Agents Chemother.* **27:** 657–659.

74. **Goldstein, F. W., B. Papadopoulou, and J. F. Acar.** 1986. The changing pattern of trimethoprim resistance in Paris, with a review of worldwide experience. *Rev. Infect. Dis.* **8:**725–737.

75. **Gordin, F. M., G. L. Simon, C. B. Wofsy, and J. Mills.** 1984. Adverse reactions to trimethoprim-sulfamethoxazole in patients with acquired immunodeficiency syndrome. *Ann. Intern. Med.* **100:**495–499.

76. **Gorzynski, E. A., D. Amsterdam, T. R. Beam, Jr., and C. Rotstein.** 1989. Comparative in vitro activities of teicoplanin, vancomycin, oxacillin, and other antimicrobial agents against bacteremic isolates of gram-positive cocci. *Antimicrob. Agents Chemother.* **33:**2019–2022.

77. **Gravenkemper, C. F., J. V. Bennett, J. L. Brodie, and W. M. M. Kirby.** 1965. Dicloxacillin: in vitro and pharmacologic comparisons with oxacillin and cloxacillin. *Arch. Intern. Med.* **116:**340–345.

78. **Greenwood, D.** 1988. Microbiological properties of teicoplanin. *J. Antimicrob. Chemother.* **21**(Suppl. A):1–13.

79. **Grosset, J., and S. Leventis.** 1983. Adverse effects of rifampin. *Rev. Infect. Dis.* **5**(Suppl. 3):S440–S446.

80. **Grossman, E. R., A. Walcheck, and H. Freedman.** 1971. Tetracycline and permanent teeth: the relationship between doses and tooth color. *Pediatrics* **47:**567–570.

81. **Gump, D. W.** 1981. Vancomycin for treatment of bacterial meningitis. *Rev. Infect. Dis.* **3**(Suppl.):S289–S292.

82. **Gutmann, L., M. D. Kitzis, S. Yamabe, and J. F. Acar.** 1986. Comparative evaluation of a new β-lactamase inhibitor, YTR 830, combined with different β-lactam antibiotics against bacteria harboring known β-lactamases. *Antimicrob. Agents Chemother.* **29:**955–957.

83. **Guttler, R. B., G. W. Counts, C. K. Avent, and H. N. Beaty.** 1971. Effect of rifampin and minocycline on meningococcal carrier rates. *J. Infect. Dis.* **124:**199–205.

84. **Halkin, H.** 1988. Adverse effects of the fluoroquinolones. *Rev. Infect. Dis.* **10**(Suppl. 1):S258–S261.

85. **Hardy, D. J., D. M. Hensey, J. M. Beyer, C. Vojtko, E. J. McDonald, and P. B. Fernandes.** 1988. Comparative in vitro activities of new 14-, 15-, and 16-membered macrolides. *Antimicrob. Agents Chemother.* **32:**1710–1719.

86. **Haydon, R. C., J. W. Thelin, and W. E. Davis.** 1984. Erythromycin ototoxicity: analysis and conclusions based on 22 case reports. *Otolaryngol. Head Neck Surg.* **92:**678–684.

87. **Hitchings, G. H.** 1973. Mechanism of action of trimethoprim-sulfamethoxazole. I. *J. Infect. Dis.* **128**(Suppl.):S433–S436.

88. **Hoban, D. J., R. N. Jones, N. Yamane, R. Frei, A. Trilla, and A. C. Pignatari.** 1993. In vitro activity of three carbapenem antibiotics: comparative studies with biapenem (L-627), imipenem, and meropenem against aerobic pathogens isolated worldwide. *Diagn. Microbiol. Infect. Dis.* **17:**299–305.

89. **Hoeprich, P. D., and D. M. Warshauer.** 1974. Entry of four tetracyclines into saliva and tears. *Antimicrob. Agents Chemother.* **5:**330–336.

90. **Hof, H., O. Zak, E. Schweizer, and A. Danzler.** 1984. Antibacterial activities of nitrothiazole derivatives. *J. Antimicrob. Chemother.* **14:**31–39.

91. **Hoffken, G., H. Lode, C. Prinzing, K. Borner, and P. Koeppe.** 1985. Pharmacokinetics of ciprofloxacin after oral and parenteral administration. *Antimicrob. Agents Chemother.* **27:**375–379.

92. **Holmberg, L., G. Boman, L. E. Bottiger, B. Eriksson, R. Spross, and A. Wessling.** 1980. Adverse reactions to nitrofurantoin: analysis of 921 reports. *Am. J. Med.* **69:**733–738.

93. **Hooper, D. C., and J. S. Wolfson.** 1993. Mechanisms of quinolone action and bacterial killing, p. 53–75. *In* D. C. Hooper (ed.), *Quinolone Antimicrobial Agents*, 2nd ed. American Society for Microbiology, Washington, D.C.

94. **Horiuchi, S., Y. Inagaki, N. Yamamoto, N. Okamura, Y. Imagawa, and R. Nakaya.** 1993. Reduced susceptibilities of *Shigella sonnei* strains isolated from patients with dysentery to fluoroquinolones. *Antimicrob. Agents Chemother.* **37:**2486–2489.

95. **Hughes, J., and G. Mellows.** 1978. Inhibition of isoleucyl-transfer ribonucleic acid synthetase in *Escherichia coli* by pseudomonic acid. *Biochem. J.* **176:**305–318.

96. **Jacoby, G. A., and L. Sutton.** 1989. *Pseudomonas cepacia* susceptibility to sulbactam. *Antimicrob. Agents Chemother.* **33:**583–584.

97. **Johnson, A. P., A. H. C. Uttley, N. Woodford, and R. C. George.** 1990. Resistance to vancomycin and teicoplanin: an emerging clinical problem. *Clin. Microbiol. Rev.* **3:**280–291.

98. **Jones, R. N., M. A. Pfaller, S. D. Allen, E. H. Gerlach, P. C. Fuchs, and K. E. Aldridge.** 1991. Antimicrobial activity of cefpirome: an update compared to five third-generation cephalosporins against nearly 6000 recent clinical isolates from five medical centers. *Diagn. Microbiol. Infect. Dis.* **14:**361–364.

99. **Kaplan, A. H., P. H. Gilligan, and R. R. Facklam.** 1988. Recovery of resistant enterococci during vancomycin prophylaxis. *J. Clin. Microbiol.* **26:**1216–1218.

100. **Kenny, G. E., T. M. Hooton, M. C. Roberts, F. D. Cartwright, and J. Hoyt.** 1989. Susceptibilities of genital mycoplasmas to the newer quinolones as determined by the agar dilution method. *Antimicrob. Agents Chemother.* **33:**103–107.

101. **King, A., C. Boothman, and I. Phillips.** 1990. Comparative in vitro activity of cefpirome and cefepime, two new cephalosporins. *Eur. J. Clin. Microbiol. Infect. Dis.* **9:**677–685.

102. **Kirst, H. A., and G. D. Sides.** 1989. New directions for macrolide antibiotics: structural modifications and in vitro activity. *Antimicrob. Agents Chemother.* **33:**1413–1418.

103. **Klugman, K. P.** 1990. Pneumococcal resistance to antibiotics. *Clin. Microbiol. Rev.* **3:**171–196.

104. **Knapp, C. C., J. Sierra-Madero, and J. A. Washington.** 1989. Activity of ticarcillin/clavulanate and piperacillin/tazobactam (YTR 830; CL-298,741) against clinical isolates and against mutants derepressed for class I beta-lactamase. *Diagn. Microbiol. Infect. Dis.* **12:**511–515.

105. **Knight, V., J. W. Draper, E. A. Brady, and C. A. Attmore.** 1952. Methenamine mandelate: antimicrobial activity, absorption and excretion. *Antibiot. Chemother.* **2:**615–635.

106. **Kropp, H., L. Gerckens, J. G. Sundelof, and F. M. Kahan.** 1985. Antibacterial activity of imipenem: the first thienamycin antibiotic. *Rev. Infect. Dis.* **7**(Suppl. 3):S389–S410.

107. **Kuck, N. A., N. V. Jacobus, P. J. Petersen, W. J. Weiss, and R. T. Testa.** 1989. Comparative in vitro and in vivo activities of piperacillin combined with the β-lactamase inhibitors tazobactam, clavulanic acid, and sulbactam. *Antimicrob. Agents Chemother.* **33:**1964–1969.

108. **Lawson, D. H., and B. J. Paice.** 1982. Adverse reactions to trimethoprim-sulfamethoxazole. *Rev. Infect. Dis.* **4:**429–433.

109. **Leclercq, R., E. Derlot, J. Duval, and P. Courvalin.** 1988. Plasmid-mediated resistance to vancomycin and teicoplanin in *Enterococcus faecium*. *N. Engl. J. Med.* **319:**157–161.

110. **Leigh, D. A.** 1981. Antibacterial activity and pharmacokinetics of clindamycin. *J. Antimicrob. Chemother.* **7**(Suppl. A):3–9.

111. **Low, D. E., A. McGeer, and R. Poon.** 1989. Activities of daptomycin and teicoplanin against *Staphylococcus haemolyticus* and *Staphylococcus epidermidis*, including evaluation of susceptibility testing recommendations. *Antimicrob. Agents Chemother.* **33:**585–588.

112. **Malanoski, G. J., L. Collins, C. Wennersten, R. C. Moellering, Jr., and G. M. Eliopoulos.** 1993. In vitro activity of biapenem against clinical isolates of gram-positive and gram-negative bacteria. *Antimicrob. Agents Chemother.* **37:**2009–2016.

113. **Mandell, G. L.** 1983. The antimicrobial activity of rifampin: emphasis on the relation to phagocytes. *Rev. Infect. Dis.* **5**(Suppl. 3):S463–S467.

114. **McCalla, D. R.** 1977. Biological effects of nitrofurans. *J. Antimicrob. Chemother.* **3:**517–520.

115. **McDowell, T. D., and K. E. Reed.** 1989. Mechanism of penicillin killing in the absence of bacterial lysis. *Antimicrob. Agents Chemother.* 33:1680–1685.

116. **McGehee, R. F., Jr., C. B. Smith, C. Wilcox, and M. Finland.** 1968. Comparative studies of antibacterial activity in vitro and absorption and excretion of lincomycin and clindamycin. *Am. J. Med. Sci.* 256:279–292.

117. **McNulty, C. A. M., J. Dent, and R. Wise.** 1985. Susceptibility of clinical isolates of *Campylobacter pyloridis* to 11 antimicrobial agents. *Antimicrob. Agents Chemother.* 28:837–838.

118. **Medical Letter on Drugs and Therapeutics.** 1992. Drugs for sexually transmitted diseases. *Med. Lett. Drugs Therapeut.* 33:119–124.

119. **Medical Letter on Drugs and Therapeutics.** 1993. Drugs for parasitic infections. *Med. Lett. Drugs Therapeut.* 35:111–122.

120. **Moellering, R. C., Jr.** 1984. Pharmacokinetics of vancomycin. *J. Antimicrob. Chemother.* 14(Suppl. D):43–52.

121. **Moellering, R. C., Jr.** 1991. The enterococcus: a classic example of the impact of antimicrobial resistance on therapeutic options. *J. Antimicrob. Chemother.* 28:1–12.

122. **Molavi, A., and L. Weinstein.** 1971. In-vitro activity of erythromycin against atypical mycobacteria. *J. Infect. Dis.* 123:216–219.

123. **Moosdeen, F., J. D. Williams, and S. Yamabe.** 1988. Antibacterial characteristics of YTR 830, a sulfone β-lactamase inhibitor, compared with those of clavulanic acid and sulbactam. *Antimicrob. Agents Chemother.* 32:925–927.

124. **Morris, A. B., R. B. Brown, and M. Sands.** 1993. Use of rifampin in nonstaphylococcal, nonmycobacterial disease. *Antimicrob. Agents Chemother.* 37:1–7.

125. **Murray, B. E.** 1990. The life and times of the enterococcus. *Clin. Microbiol. Rev.* 3:46–65.

126. **Muscato, J. J., D. W. Wilbur, J. J. Stout, and R. A. Fahrlender.** 1991. An evaluation of the susceptibility patterns of gram-negative organisms isolated in cancer centres with aminoglycoside usage. *J. Antimicrob. Chemother.* 27(Suppl. C):1–7.

127. **Musial, C. E., and J. E. Rosenblatt.** 1989. Antimicrobial susceptibilities of anaerobic bacteria isolated at the Mayo Clinic during 1982 through 1987: comparison with results from 1977 through 1981. *Mayo Clin. Proc.* 64:392–399.

128. **Nadelman, R. B., S. W. Luger, E. Frank, M. Wisniewski, J. J. Collins, and G. P. Wormser.** 1992. Comparison of cefuroxime axetil and doxycycline in the treatment of Lyme disease. *Ann. Intern. Med.* 117:273–280.

129. **Nagarajan, R.** 1991. Antibacterial activities and modes of action of vancomycin and related glycopeptides. *Antimicrob. Agents Chemother.* 35:605–609.

130. **Najjar, A., and B. E. Murray.** 1987. Failure to demonstrate a consistent in vitro bactericidal effect of trimethoprim-sulfamethoxazole against enterococci. *Antimicrob. Agents Chemother.* 31:808–810.

131. **Neu, H. C.** 1976. Tobramycin: an overview. *J. Infect. Dis.* 134(Suppl.):S3–S19.

132. **Neu, H. C.** 1982. The new beta-lactamase-stable cephalosporins. *Ann. Intern. Med.* 97:408–419.

133. **Neu, H. C.** 1985. Carbapenems: special properties contributing to their activity. *Am. J. Med.* 78(Suppl. 6A):33–40.

134. **Neu, H. C.** 1991. Clinical microbiology of azithromycin. *Am. J. Med.* 91(Suppl. 3A):12S–18S.

135. **Neu, H. C.** 1991. The development of macrolides: clarithromycin in perspective. *J. Antimicrob. Chemother.* 27(Suppl. A):1–9.

136. **Neu, H. C., and K. P. Fu.** 1978. Clavulanic acid, a novel inhibitor of beta-lactamases. *Antimicrob. Agents Chemother.* 14:650–655.

137. **Neu, H. C., and K. P. Fu.** 1980. In vitro activity of chloramphenicol and thiamphenicol analogs. *Antimicrob. Agents Chemother.* 18:311–316.

138. **Neu, H. C., and P. Lubthavikul.** 1982. Comparative in vitro activity of N-formimidoyl thienamycin against gram-positive and gram-negative aerobic and anaerobic species

and its β-lactamase stability. *Antimicrob. Agents Chemother.* 21:180–187.

139. **O'Brien, R. J., M. A. Lyle, and D. E. Snider, Jr.** 1987. Rifabutin (ansamycin LM 427): a new rifamycin-S derivative for the treatment of mycobacterial diseases. *Rev. Infect. Dis.* 9:519–530.

140. **Pang, L. W., N. Limsomwong, E. F. Boudreau, and P. Singharaj.** 1987. Doxycycline prophylaxis for falciparum malaria. *Lancet* i:1161–1164.

141. **Pape, J. W., R. I. Verdier, and W. D. Johnson, Jr.** 1989. Treatment and prophylaxis of *Isospora belli* infection in patients with the acquired immunodeficiency syndrome. *N. Engl. J. Med.* 320:1044–1047.

142. **Patel, R. B., and P. G. Welling.** 1980. Clinical pharmacokinetics of co-trimoxazole (trimethoprim-sulfamethoxazole). *Clin. Pharmacokinet.* 5:405–423.

143. **Periti, P., T. Mazzei, E. Mini, and A. Novelli.** 1992. Pharmacokinetic drug interactions of macrolides. *Clin. Pharmacokinet.* 23:106–131.

144. **Pfister, H. W., V. Preac-Mursic, B. Wilske, E. Schielke, F. Sörgel, and K. M. Einhaupl.** 1991. Randomized comparison of ceftriaxone and cefotaxime in Lyme neuroborreliosis. *J. Infect. Dis.* 163:311–318.

145. **Phillipon, A., R. Labia, and G. Jacoby.** 1989. Extended-spectrum β-lactamases. *Antimicrob. Agents Chemother.* 33:1131–1136.

146. **Polk, R. E.** 1989. Drug-drug interactions with ciprofloxacin and other fluoroquinolones. *Am. J. Med.* 87(Suppl. 5A):76S–81S.

147. **Polk, R. E., D. P. Healy, L. B. Schwartz, D. T. Rock, M. L. Garson, and K. Roller.** 1988. Vancomycin and the red-man syndrome: pharmacodynamics of histamine release. *J. Infect. Dis.* 157:502–507.

148. **Rahal, J. J., Jr., and M. S. Simberkoff.** 1979. Bactericidal and bacteriostatic action of chloramphenicol against meningeal pathogens. *Antimicrob. Agents Chemother.* 16:13–18.

149. **Raoult, D., P. Roussellier, V. Galicher, R. Perez, and J. Tamalet.** 1986. In vitro susceptibility of *Rickettsia conorii* to ciprofloxacin as determined by suppressing lethality in chicken embryos and by plaque assay. *Antimicrob. Agents Chemother.* 29:424–425.

150. **Raoult, D., H. Torres, and M. Drancourt.** 1991. Shell-vial assay evaluation of a new technique for determining antibiotic susceptibility, tested in 13 isolates of *Coxiella burnetii*. *Antimicrob. Agents Chemother.* 35:2070–2077.

151. **Retsema, J. A., A. R. English, and A. E. Girard.** 1980. CP-45,899 in combination with penicillin or ampicillin against penicillin-resistant *Staphylococcus*, *Haemophilus influenzae*, and *Bacteroides*. *Antimicrob. Agents Chemother.* 17:615–622.

152. **Retsema, J. A., A. R. English, A. Girard, J. E. Lynch, M. Anderson, L. Brennan, C. Cimochowski, J. Faiella, W. Norcia, and P. Sawyer.** 1986. Sulbactam/ampicillin: in vitro spectrum potency, and activity in models of acute infection. *Rev. Infect. Dis.* 8(Suppl. 5):S528–S534.

153. **Rodloff, A. C.** 1982. In-vitro susceptibility test of nontuberculous mycobacteria to sulphamethoxazole, trimethoprim, and combinations of both. *J. Antimicrob. Chemother.* 9:195–199.

154. **Rogers, J. D., M. A. P. Meisinger, F. Gerber, G. B. Calandra, J. L. Demetriades, and J. A. Bland.** 1985. Pharmacokinetics of imipenem and cilastatin in volunteers. *Rev. Infect. Dis.* 7(Suppl. 3):S435–S446.

155. **Rolston, K. V. I., and J. Hoy.** 1987. Role of clindamycin in the treatment of central nervous system toxoplasmosis. *Am. J. Med.* 83:551–554.

156. **Rose, H. D., and M. W. Rytel.** 1972. Actinomycosis treated with clindamycin. *JAMA* 221:1052.

157. **Rosenblatt, J. E., and P. R. Stewart.** 1974. Combined activity of sulfamethoxazole, trimethoprim, and polymyxin B against gram-negative bacilli. *Antimicrob. Agents Chemother.* 6:84–92.

158. **Rubin, R. H., and M. N. Swartz.** 1980. Trimethoprim-sulfamethoxazole. *N. Engl. J. Med.* **303:**426–432.

159. **Ruoff, K. L., D. R. Kuritzkes, J. S. Wolfson, and M. J. Ferraro.** 1988. Vancomycin-resistant gram-positive bacteria isolated from human sources. *J. Clin. Microbiol.* **26:**2064–2068.

160. **Sader, H. S., R. N. Jones, J. A. Washington, P. R. Murray, E. H. Gerlach, S. D. Allen, and M. E. Erwin.** 1993. In vitro activity of cefpodoxime compared with other oral cephalosporins tested against 5556 recent clinical isolates from five medical centers. *Diagn. Microbiol. Infect. Dis.* **17:**143–150.

161. **Salter, A. J.** 1982. Trimethoprim-sulfamethoxazole: an assessment of more than 12 years of use. *Rev. Infect. Dis.* **4:**196–236.

162. **Sattler, F. R., M. R. Weitekamp, and J. O. Ballard.** 1986. Potential for bleeding with the new beta-lactam antibiotics. *Ann. Intern. Med.* **105:**924–931.

163. **Saxon, A., A. Hassner, E. A. Swabb, B. Wheeler, and N. F. Adkinson, Jr.** 1984. Lack of cross-reactivity between aztreonam, a monobactam antibiotic, and penicillin in penicillin-allergic subjects. *J. Infect. Dis.* **149:**16–22.

164. **Schaad, U. B., J. Wedgwood-Krucko, and H. Tschaeppeler.** 1988. Reversible ceftriaxone-associated biliary pseudolithiasis in children. *Lancet* **ii:**1411–1413.

165. **Schentag, J. J., and C. H. Ballow.** 1991. Tissue-directed pharmacokinetics. *Am. J. Med.* **91**(Suppl. 3A):5S–11S.

166. **Schwalbe, R. S., J. T. Stappleton, and P. H. Gilligan.** 1987. Emergence of vancomycin resistance in coagulase-negative staphylococci. *N. Engl. J. Med.* **316:**927–931.

167. **Seaberg, L. S., A. R. Parquette, I. Y. Gluzman, G. W. Phillips, Jr., T. F. Brodasky, and D. J. Krogstad.** 1984. Clindamycin activity against chloroquine-resistant *Plasmodium falciparum*. *J. Infect. Dis.* **150:**904–911.

168. **Sharp, J. R., K. G. Ishak, and H. J. Zimmerman.** 1980. Chronic active hepatitis and severe hepatic necrosis associated with nitrofurantoin. *Ann. Intern. Med.* **92:**14–19.

169. **Shaw, W. V.** 1984. Bacterial resistance to chloramphenicol. *Br. Med. Bull.* **40:**36–41.

170. **Sheikh, W., D. H. Pitkin, and H. Nadler.** 1993. Antibacterial activity of meropenem and selected comparative agents against anaerobic bacteria at seven North American Centers. *Clin. Infect. Dis.* **16**(Suppl. 4):S361–S366.

171. **Smith, A. L., and A. Weber.** 1983. Pharmacology of chloramphenicol. *Pediatr. Clin. N. Am.* **30:**209–236.

172. **Somma, S., L. Gastaldo, and A. Corti.** 1984. Teicoplanin, a new antibiotic from *Actinoplanes teichomyceticus* nov. sp. *Antimicrob. Agents Chemother.* **26:**917–923.

173. **Sörgel, F., and M. Kinzig.** 1993. The chemistry, pharmacokinetics and tissue distribution of piperacillin/tazobactam. *J. Antimicrob. Chemother.* **31**(Suppl. A):39–60.

174. **Sorrell, T. C., and P. J. Collignon.** 1985. A prospective study of adverse reactions associated with vancomycin therapy. *J. Antimicrob. Chemother.* **16:**235–241.

175. **Spear, B. S., N. B. Shoemaker, and A. A. Salyers.** 1992. Bacterial resistance to tetracycline: mechanisms, transfer, and clinical significance. *Clin. Microbiol. Rev.* **5:**387–399.

176. **Spratt, B. G., V. Jobanputra, and W. Zimmermann.** 1977. Binding of thienamycin and clavulanic acid to the penicillin-binding proteins of *Escherichia coli* K-12. *Antimicrob. Agents Chemother.* **12:**406–409.

177. **Stamm, W. E.** 1991. Azithromycin in the treatment of uncomplicated genital chlamydial infections. *Am. J. Med.* **91**(Suppl. 3A):19S–22S.

178. **Stamm, W. E., and T. M. Hooton.** 1993. Management of urinary tract infections in adults. *N. Engl. J. Med.* **329:**1328–1334.

179. **Sullivan, D., M. E. Csuka, and B. Blanchard.** 1980. Erythromycin ethylsuccinate hepatoxicity. *JAMA* **243:**1074.

180. **Sutter, V. L.** 1977. In vitro susceptibility of anaerobes: comparison of clindamycin and other antimicrobial agents. *J. Infect. Dis.* **135**(Suppl.):S7–S12.

181. **Swedberg, J., J. F. Steiner, F. Deiss, S. Steiner, and D. A. Driggers.** 1985. Comparison of single-dose vs one-week

182. **Sykes, R. B., and D. P. Bonner.** 1985. Aztreonam: first monobactam. *Am. J. Med.* **78**(Suppl. 2A):2–10.

183. **Teasley, D. G., D. N. Gerding, M. M. Olson, L. R. Peterson, R. L. Gebhard, M. J. Schwartz, and J. T. Lee, Jr.** 1983. Prospective randomized trial of metronidazole versus vancomycin for *Clostridium difficile*-associated diarrhea and colitis. *Lancet* **ii:**1043–1046.

184. **Toma, E., S. Fournier, M. Dumont, P. Bolduc, and H. Deschamps.** 1993. Clindamycin/primaquine versus trimethoprim-sulfamethoxazole as primary therapy for *Pneumocystis carinii* pneumonia in AIDS: a randomized, double-blind pilot trial. *Clin. Infect. Dis.* **17:**178–184.

185. **Tuazon, C. U., and H. Miller.** 1984. Comparative in vitro activities of teichomycin and vancomycin alone and in combination with rifampin and aminoglycosides against staphylococci and enterococci. *Antimicrob. Agents Chemother.* **25:**411–412.

186. **Van der Auwera, P., T. Matsumoto, and M. Husson.** 1988. Intraphagocytic penetration of antibiotics. *J. Antimicrob. Chemother.* **22:**185–192.

187. **Van der Auwera, P., F. Meunier-Carpentier, and J. Kastersky.** 1983. Clinical study of combination therapy with oxacillin and rifampin for staphylococcal infections. *Rev. Infect. Dis.* **5**(Suppl. 3):S515–S522.

188. **Vanhoff, R., B. Gordts, R. Dierickx, H. Coignau, and J. P. Butzler.** 1980. Bacteriostatic and bactericidal activities of 24 antimicrobial agents against *Campylobacter fetus* subsp. *jejuni*. *Antimicrob. Agents Chemother.* **18:**118–121.

189. **Verbist, L.** 1979. Comparison of the activities of the new ureidopenicillins, piperacillin, mezlocillin, azlocillin, and Bay k 4999 against gram-negative organisms. *Antimicrob. Agents Chemother.* **16:**115–119.

190. **Wallace, R. J., Jr., E. J. Septimus, T. W. Williams, Jr., R. H. Conklin, T. K. Satterwhite, M. B. Bushby, and D. C. Hollowell.** 1982. Use of trimethoprim-sulfamethoxazole for treatment of infections due to *Nocardia*. *Rev. Infect. Dis.* **4:**315–325.

191. **Wallace, R. J., Jr., J. M. Swenson, V. A. Silcox, and M. G. Bullen.** 1985. Treatment of non-pulmonary infections due to *Mycobacterium fortuitum* and *Mycobacterium chelonei* on the basis of in vivo susceptibilities. *J. Infect. Dis.* **152:**500–514.

192. **Wallace, R. J., Jr., and K. Wiss.** 1981. Susceptibility of *Mycobacterium marinum* to tetracyclines and aminoglycosides. *Antimicrob. Agents Chemother.* **20:**610–612.

193. **Walters, B. N. J., and S. S. Gubbay.** 1981. Tetracycline and benign intracranial hypertension: report of five cases. *Br. Med. J.* **282:**19–20.

194. **Ward, A., and D. M. Campoli-Richards.** 1986. Mupirocin: a review of its antibacterial activity, pharmacokinetic properties and therapeutic use. *Drugs* **32:**425–444.

195. **Ward, M. E.** 1977. The bactericidal action of spectinomycin on *Neisseria gonorrhoeae*. *J. Antimicrob. Chemother.* **3:**323–329.

196. **Washington, J. A., II, and W. R. Wilson.** 1985. Erythromycin: a microbial and clinical perspective after 30 years of clinical use. *Mayo Clin. Proc.* **60:**189–203, 271–278.

197. **Watanakunakorn, C., and C. Bakie.** 1973. Synergism of vancomycin-gentamicin and vancomycin-streptomycin against enterococci. *Antimicrob. Agents Chemother.* **4:**120–124.

198. **Watanakunakorn, C., and J. C. Tisone.** 1982. Synergism between vancomycin and gentamicin or tobramycin for methicillin-susceptible and methicillin-resistant *Staphylococcus aureus* strains. *Antimicrob. Agents Chemother.* **22:**903–905.

199. **Waxman, D. J., and J. L. Strominger.** 1983. Penicillin-binding proteins and the mechanism of action of beta-lactam antibiotics. *Annu. Rev. Biochem.* **52:**825–869.

200. **Wehrli, W.** 1983. Rifampin: mechanisms of action and resistance. *Rev. Infect. Dis.* **5**(Suppl. 3):S407–S411.

201. **Wexler, H. M., and S. M. Finegold.** 1988. In vitro activity of cefotetan compared with that of other antimicrobial

agents against anaerobic bacteria. *Antimicrob. Agents Chemother.* **32:**601–604.

202. **Wexler, H. M., B. Harris, W. T. Carter, and S. M. Finegold.** 1985. In vitro efficacy of sulbactam combined with ampicillin against anaerobic bacteria. *Antimicrob. Agents Chemother.* **27:**876–878.

203. **Wiedemann, B., C. Kliebe, and M. Kresken.** 1989. The epidemiology of beta-lactamases. *J. Antimicrob. Chemother.* **24**(Suppl. B):1–22.

204. **Wijnands, W. J. A., and T. B. Vree.** 1988. Interaction between the fluoroquinolones and the bronchodilator theophylline. *J. Antimicrob. Chemother.* **22**(Suppl. C):109–114.

205. **Williams, J. D., and A. M. Sefton.** 1993. Comparison of macrolide antibiotics. *J. Antimicrob. Chemother.* **31**(Suppl. C):11–26.

206. **Wilson, W. R., R. L. Thompson, C. J. Wilkowske, J. A. Washington II, E. R. Giuliani, and J. E. Geraci.** 1981. Short-term therapy for streptococcal infective endocarditis: combined intramuscular administration of penicillin and streptomycin. *JAMA* **245:**360–363.

207. **Winston, D. J., M. A. McGrattan, and R. W. Busuttil.** 1984. Imipenem therapy of *Pseudomonas aeruginosa* and other serious bacterial infections. *Antimicrob. Agents Chemother.* **26:**673–677.

208. **Wise, R., D. Lister, C. A. McNulty, D. Griggs, and J. M. Andrews.** 1986. The comparative pharmacokinetics of five quinolones. *J. Antimicrob. Chemother.* **18**(Suppl. D):71–81.

209. **Wittner, M., K. S. Rowin, H. B. Tanowitz, J. F. Hobbs, S. Saltzman, B. Wenz, R. Hirsch, E. Chisholm, and G. R. Healy.** 1982. Successful chemotherapy of transfusion babesiosis. *Ann. Intern. Med.* **96:**601–604.

210. **Wolfe, M.** 1992. Giardiasis. *Clin. Microbiol. Rev.* **5:**92–100.

211. **Wolff, M., L. Boutron, E. Singlas, B. Clair, J. M. Decazes, and B. Regnier.** 1987. Penetration of ciprofloxacin into cerebrospinal fluid of patients with bacterial meningitis. *Antimicrob. Agents Chemother.* **31:**899–902.

212. **Wolfson, J. S., and D. C. Hooper.** 1985. The fluoroquinolones: structures, mechanisms of action and resistance, and spectra of activity in vitro. *Antimicrob. Agents Chemother.* **28:**581–586.

213. **Wolfson, J. S., and D. C. Hooper.** 1989. Fluoroquinolone antimicrobial agents. *Clin. Microbiol. Rev.* **2:**378–424.

214. **Wong, C. S., G. S. Palmer, and M. H. Cynamon.** 1988. In-vitro susceptibility of *Mycobacterium tuberculosis*, *Mycobacterium bovis* and *Mycobacterium kansasii* to amoxycillin and ticarcillin in combination with clavulanic acid. *J. Antimicrob. Chemother.* **22:**863–866.

215. **World Health Organization.** 1993. *Guidelines for Cholera Control.* World Health Organization, Geneva.

216. **Yeaman, M. R., L. A. Mitscher, and O. G. Baca.** 1987. In vitro susceptibility of *Coxiella burnetii* to antibiotics, including several quinolones. *Antimicrob. Agents Chemother.* **31:**1079–1084.

217. **Young, L. S., O. G. W. Berlin, and C. B. Inderlied.** 1987. Activity of ciprofloxacin and other fluorinated quinolones against mycobacteria. *Am. J. Med.* **82**(Suppl. 4A):23–26.

218. **Young, L. S., and J. Hindler.** 1987. Use of trimethoprim-sulfamethoxazole singly and in combination with other antibiotics in immunocompromised patients. *Rev. Infect. Dis.* **9**(Suppl. 2):S177–S181.

219. **Zenilman, J. M., L. J. Nims, M. A. Menegus, F. Nolte, and J. S. Knapp.** 1987. Spectinomycin-resistant gonococcal infections in the United States, 1985–1986. *J. Infect. Dis.* **156:**1002–1004.

Mechanisms of Resistance to Antimicrobial Agents

RICHARD QUINTILIANI, JR., AND PATRICE COURVALIN

112

Antibiotics have reduced the mortality from infectious diseases but not the prevalence of these diseases. Use, and often abuse, of antimicrobial agents encourages the evolution of bacteria toward resistance, resulting often in therapeutic failure. This evolution is due to the emergence of "new" resistance mechanisms and to the spread of well-characterized mechanisms of resistance to the majority of bacterial species. Knowledge of biochemical mechanisms of resistance aids in the in vitro detection of resistance phenotypes, which is useful in guiding therapy and in the elucidation of cross-resistance to antibiotics. It is also the basis for the rational design of drugs that elude resistance or are directed against new targets.

Bacterial resistance can be intrinsic or acquired. Intrinsic resistance is species or genus specific and delineates the spectrum of activity of the antibiotic. Acquired resistance is present in only certain strains of a species or of a genus. The latter results from mutation in a gene located in the host chromosome or a plasmid or from acquisition of new genetic information by a bacterium, mainly by conjugation or transformation.

Bacteria can resist antibiotics by four distinct mechanisms: alteration of the target site, enzymatic detoxification of the antibiotic, decreased drug accumulation, and bypass of an antibiotic-sensitive step. The first three mechanisms can be due to chromosomal mutations or may be plasmid mediated. The last one occurs primarily by acquisition of an antibiotic resistance plasmid or transposon. The biochemical mechanisms and the genetic support of resistance will be considered for the antibiotic families of clinical importance.

RESISTANCE TO β-LACTAMS

Of the various mechanisms of acquired resistance to β-lactam antibiotics, resistance due to β-lactamases is the most prevalent. Alterations in the preexisting penicillin-binding proteins (PBPs), acquisition of a novel PBP insensitive to β-lactams, and changes in the outer membrane proteins of gram-negative organisms that prevent these compounds from reaching their targets can also confer resistance.

Penicillin-Interactive, Active-Site Serine Proteins

All PBPs and most β-lactamases constitute a superfamily of evolutionarily related active-site serine peptidases that interact with β-lactams in a similar fashion. β-Lactamases and PBPs catalytically disrupt the cyclic amide bond of penicillin and other β-lactams via formation of a serine-ester-linked acylenzyme derivative (Fig. 1). Three groups of proteins that mediate this reaction have been distinguished: the β-lactamases, the low-molecular-weight PBPs (DD-carboxypeptidases), and the high-molecular-weight PBPs (transpeptidases-transglycosylases). Although DD-carboxypeptidase and transpeptidase activities are inhibited by β-lactams as described below, DD-carboxypeptidases are dispensable, since their inhibition is not lethal to the cell.

β-Lactams are structural analogs of the peptidyl–D-alanyl–D-alanine termini of peptidoglycan cell wall precursors and are recognized by PBPs and β-lactamases. Following β-lactam binding, these groups of proteins undergo sequential acylation and deacylation reactions (Fig. 1). The β-lactam ring is opened by a nucleophilic attack on the β-lactam amide bond by the hydroxyl group of a serine residue located at the active site of the enzyme. Acylation of the enzyme generates an acylenzyme intermediate characterized by an ester bond between the enzyme and the penicilloyl (or cephalosporyl) moiety.

Differentiation between PBPs and β-lactamases reflects differences in the rate of deacylation, which is dependent on the capacity of water to act as an attacking nucleophile. In the case of β-lactamases, water is an effective attacking nucleophile that efficiently hydrolyzes the ester linkage of the acylenzyme intermediate, thereby releasing the penicilloyl (or cephalosporyl) moiety and regenerating the active enzyme (Fig. 1). The reaction catalyzed by PBPs is identical to that with β-lactamases except that the acylenzyme intermediate is resistant to nucleophilic attack. Thus, the β-lactam remains covalently bound to the PBP, which is itself inactivated. Because the stability of the enzyme-penicillin complex allows easy detection of these enzymes, they are termed penicillin-binding proteins. In general, the lability of the β-lactamase–β-lactam interaction determines resistance, whereas the stability of the high-molecular-weight-PBP–β-lactam interaction determines susceptibility. It should be noted, however, that a distinction between β-lactamases and PBPs based on the stability or lability of the acylenzyme intermediate is somewhat oversimplified, since depending on the particular enzyme adduct, the rates of enzyme turnover can be highly variable. Thus, β-lactamases may sometimes generate stable acylenzyme interme-

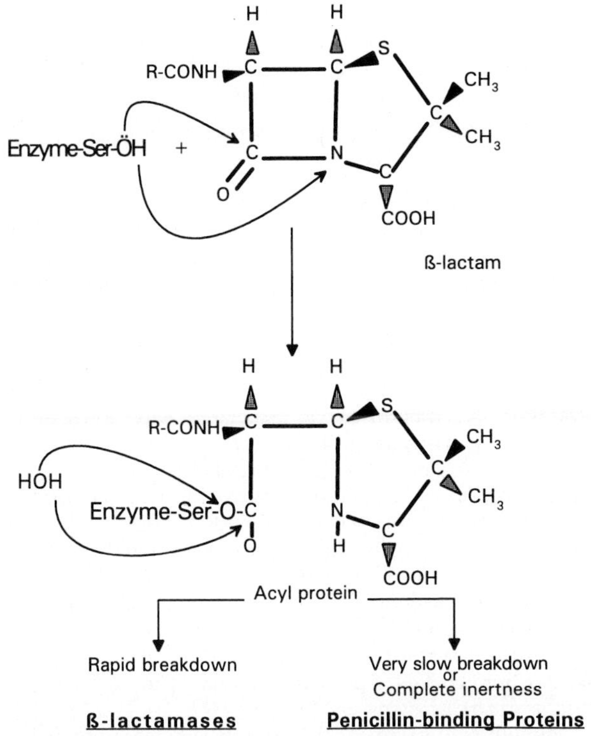

FIGURE 1 Penicillin-interactive, active-site serine peptidases and their reactions with β-lactam carbonyl donors. Modified with permission from reference 61.

diates, while PBPs occasionally exhibit weak β-lactamase activity (133).

Resistance Due to PBP Alteration

Penicillin resistance due to altered high-molecular-weight PBPs is more common in gram-positive than in gram-negative bacteria. Almost all *Eubacteria* spp. examined contain three to eight PBPs on the outer face of the cytoplasmic membrane. The PBPs are named numerically: the higher the molecular weight, the smaller the numeric designation. Note that although the PBPs of different species and genera may be designated by the same number, they are not necessarily related. β-Lactams have multiple potential targets, since two to four PBPs are normally essential for cell survival. Cell death may result not from the binding of the antibiotic per se but from the action of peptidoglycan-degradative enzymes termed autolysins that are triggered by the binding. The concept of "tolerance" refers to the ability of autolysin-defective mutants to remain viable in the presence of cell wall-active bactericidal antibiotics.

Gram-Positive Bacteria

Methicillin-resistant *Staphylococcus aureus* (MRSA) is resistant to all β-lactams, including cephalosporins, carbapenems, and monobactams. Methicillin resistance in *S. aureus* is unique in that it is due to acquisition of DNA of unknown origin that codes for a supernumerary PBP that is not inhibited by β-lactams. *S. aureus* normally contains four PBPs, of which PBPs 1, 2, and 3 are essential (156). The low-affinity PBP in MRSA, termed PBP2a (or PBP2′), is encoded by the chromosomal gene *mecA* (59) and is thought to function as a β-lactam-resistant transpeptidase.

The *mecA* gene is widely distributed among different species of staphylococci and is highly conserved (163, 198).

Although MRSA exhibits high-level cross-resistance to all β-lactams, strains may be homogeneously or, more often, heterogeneously resistant. In the latter case, the majority of cells express low-level methicillin resistance, but small subpopulations of highly resistant cells may be present (77). Phenotypic expression of *mecA* is affected by a number of factors, including pH, temperature, osmolarity, regulatory sequences, and unlinked chromosomal genes (120, 162). Expression of PBP2a can be inducible or constitutive. Regulation of *mecA* in *S. aureus* and *Staphylococcus epidermidis* occurs at the transcriptional level. Three resistance phenotypes (immediately inducible, delayed inducible, and constitutive) are associated with differences in regulation (162). Immediately inducible strains harbor a plasmid containing the β-lactamase (*bla*) operon. The DNA sequence of the *mecA* promoter is similar to that of the promoter of *blaZ*, the β-lactamase structural gene, suggesting that *mecA* is regulated by genes encoded by the penicillinase plasmid (see Resistance Due to β-Lactamases below) (162). Elimination of the penicillinase plasmid converts some of these strains to a constitutive phenotype. Strains of *S. aureus* and *S. epidermidis* with the delayed inducible phenotype contain an upstream regulatory locus consisting of the *mecI* and *mecR1* genes, which are homologous to the penicillinase regulator genes *blaI* and *blaR1*, respectively, and mediate repression of the *mecA* gene (183, 186). These strains are of particular clinical importance, since they are very slowly derepressed, making their detection difficult. Forty-eight hours of induction by methicillin is required for full expression of resistance (162). At the transcriptional level, there appears to be no difference in the abilities of various β-lactams to induce, and thus, differences in inducibility may be strain specific (162).

In addition to *mecA*, *mecR1*, and *mecI*, unlinked determinants also contribute to phenotypic expression of methicillin resistance. These genes, designated *fem* (factors essential for expression of methicillin resistance) or *aux* (auxiliary), are present in the chromosome of MRSA and methicillin-susceptible *S. aureus*. Two such genes, *femA* and *femB*, have no effect on the synthesis of PBP2a but are involved in biosynthesis of the pentaglycine interpeptide bridge of peptidoglycan (117). Disruption of *femA* or *femB* decreases the glycine content of peptidoglycan precursors and confers increased susceptibility to β-lactams (117). The presence of *femA* is also necessary for the increased autolytic activity observed in *mecA* strains.

Borderline homogeneous methicillin resistance in strains of *S. aureus* that do not hybridize with a *mecA* probe has been associated with elevated amounts of PBP4 (192). Decreased amounts of PBP3 confer moderate levels of resistance to oral cephalosporins, including cephalexin, cephradine, cefadroxil, and cefaclor (60).

Penicillin-resistant strains of *Streptococcus pneumoniae* possess modified PBPs that have low affinities for β-lactams (69). β-Lactamases have not been detected in this species. Penicillin-susceptible *Streptococcus pneumoniae* contains six PBPs (1a, 1b, 2a, 2x, 2b, and 3). Selection of mutants in the laboratory indicated that resistance to penicillin can occur in a stepwise fashion. Analysis of the mutants confirmed that resistance was due to changes in PBPs (74, 102). Penicillin resistance in clinical isolates is usually associated with the presence of up to five low-affinity PBPs, including 1a, 1b, 2a, 2b, and 2x (69). However, the emergence of low-β-lactam-affinity PBPs in clinical isolates of *Streptococ-*

cus pneumoniae appears to have evolved not by mutation but by acquisition of foreign DNA.

The chromosomal PBP genes of penicillin-susceptible strains of *Streptococcus pneumoniae* are highly conserved (<1% sequence variation) (44). In contrast, the PBP genes of resistant strains show significant sequence divergence. They are called "mosaic" genes, since they consist of blocks of DNA sequences that are very similar or identical to the corresponding portions of the susceptible genes and of blocks that diverge significantly. The genes encoding resistant PBPs 1a, 2x, and 2b consist of mosaics, while the PBP1b and PBP2a genes have not yet been sequenced (44, 103, 119). Mosaic genes have presumably arisen by homologous recombination following transformation with DNA from a closely related species (44). Certain strains of resistant *Streptococcus pneumoniae* contain mosaic PBP2b genes consisting of blocks of nucleotides that originated from *Streptococcus mitis*; others possess blocks from another unknown species or from both (43, 44). In *Streptococcus pneumoniae* containing *Streptococcus mitis*-derived sequences, the amino acids responsible for the low affinity to penicillin appear to be encoded by at least one of the acquired blocks, even though the donor strains were penicillin susceptible and did not contain a low-affinity PBP2b (43). The PBP genes of the donors consist of mosaics themselves, indicating the extent to which this naturally transformable genus can acquire exogenous DNA. Interestingly, the amino acid substitution Thr-445 to Ala, which is known to dramatically reduce the affinity of PBP2b for penicillin, was present in all resistant strains of *Streptococcus pneumoniae* but not in the susceptible *Streptococcus mitis* donor strains examined (43). It has been suggested that this substitution may be present in the strains of *Streptococcus mitis* responsible for donating PBP2b genes to *Streptococcus pneumoniae*, is present as a polymorphism, or is the result of a spontaneous mutation (43). Importantly, PBP2x and not PBP2b is the primary target of penicillin-susceptible *Streptococcus pneumoniae*. Thus, strains containing a low-affinity PBP2b may or may not be phenotypically resistant to penicillin, depending on which and how many low-affinity PBPs they possess (43).

In *Streptococcus pneumoniae*, high-level resistance to extended-spectrum cephalosporins involves changes in PBP1a and PBP2x, which are encoded by closely linked genes that can be transferred en bloc into a susceptible host (126). Penicillin-resistant *Streptococcus sanguis* and *Streptococcus oralis* contain altered low-affinity PBPs that probably arose

by horizontal transfer of altered PBP2b genes from resistant *Streptococcus pneumoniae* (45).

Enterococcus faecalis and *Enterococcus faecium* are intrinsically resistant to low levels of penicillins. Resistance in non-β-lactamase-producing strains has been associated with overproduction of low-affinity PBP5 (56, 204).

Gram-Negative Bacteria

As mentioned above, resistance due to modifications in PBPs is less common in gram-negative organisms. However, PBP alterations in clinical isolates of *Pseudomonas aeruginosa*, *Haemophilus influenzae*, *Neisseria gonorrhoeae*, *Neisseria meningitidis*, *Acinetobacter calcoaceticus*, and *Bacteroides fragilis* have been reported (177).

Penicillin resistance in non-β-lactamase-producing strains of *N. gonorrhoeae* is associated with low-affinity forms of the essential PBPs 1 and 2 (49). The deduced amino acid sequences of PBP2 encoded by the *penA* gene from susceptible and resistant strains have been compared. Since the *penA* genes were almost identical over the first two-thirds of the gene but extensively altered over the remainder, it appears that resistant *penA* genes are mosaics generated in a fashion similar to what occurs in *Streptococcus pneumoniae* (176). Penicillin resistance in *N. meningitidis* is also associated with hybrid PBP2 genes. In both species, foreign DNA probably originated in *Neisseria flavescens* (178).

Penicillin resistance in non-β-lactamase-producing strains of *H. influenzae* and *P. aeruginosa* has been associated with low-affinity PBPs 3, 4, and 5 (124) and decreased amounts of PBP3 (63), respectively. Cefoxitin resistance in clinical isolates of *B. fragilis* has been associated with alterations in PBPs 1 and 2.

Resistance Due to β-Lactamases

β-Lactamases are considered dispensable enzymes, since their only known function is the hydrolysis of β-lactams. In gram-positive bacteria, β-lactamases are mainly secreted into the growth medium, whereas in gram-negative organisms, they are secreted into the periplasmic space.

Because β-lactamases are a family of tremendous diversity, a number of classification schemes have been suggested. The most recent scheme, proposed by Bush, includes both plasmid- and chromosomally specified enzymes and is based on substrate profile, inhibitor profile, and physical characteristics such as molecular weight and isoelectric point (24) (Table 1). In addition, knowledge of the

TABLE 1 Classification scheme for β-lactamases[a]

Group	Designation[b]	Preferred substrates	Representative enzymes or hosts
1	CEP-N	Cephalosporins	Chromosomal enzymes from gram-negative bacteria
2a	PEN-Y	Penicillins	Penicillinases from gram-positive bacteria
2b	BDS-Y	Cephalosporins, penicillins	TEM-1, TEM-2
2b'	EBS-Y	Cephalosporins, penicillins, cefotaxime	TEM-3, TEM-5
2c	CAR-Y	Penicillins, carbenicillin	PSE-1, PSE-3, PSE-4
2d	CLX-Y[c]	Penicillins, cloxacillin	OXA-1, PSE-2
2e	CEP-Y	Cephalosporins	*Proteus vulgaris*
3	MET-N	Variable	*Bacillus cereus* II, *Pseudomonas* L1
4	PEN-N	Penicillins	*Pseudomonas cepacia*

[a]Modified with permission from reference 24.
[b]Y, β-lactamases that are inhibited by 10 μM clavulanic acid; N, β-lactamases that are not inhibited by 10 μM clavulanic acid.
[c]Inhibition by clavulanic acid may occur at higher concentrations for some members of this group.

amino acid sequences of some β-lactamases allows them to be classified into one of four evolutionary distinct molecular classes (A, B, C, and D). For the most part, β-lactamases of a particular molecular class correlate with a specific group in the Bush classification system.

β-Lactamases of classes A, C, and D operate by a serine-ester-linked acyl enzyme mechanism, as described above. Class A comprises enzymes that preferentially hydrolyze penicillins and includes the β-lactamases of *S. aureus* (Bush group 2a) and many of the plasmid-specified β-lactamases of gram-negative bacteria, such as the TEM- and SHV-type enzymes (groups 2b and 2b′). In gram-negative bacteria, β-lactamases of classes C (group 1) and D (group 2d) include the chromosomal cephalosporinases and the oxacil-lin-hydrolyzing enzymes (OXA-1, OXA-2, and OXA-10 [PSE-2]), respectively. Class B (group 3) comprises the rare metallo-β-lactamases, which require Zn^{2+} as a cofactor (1, 10). These enzymes, which have been identified in *B. fragilis*, *Xanthomonas maltophilia*, *Flavobacterium* spp., and *Legionella* spp., hydrolyze all classes of β-lactams and are not inhibited by penicillinase inhibitors (10).

The widely disseminated plasmid-mediated TEM-1, TEM-2, and SHV-1 β-lactamases of gram-negative bacteria hydrolyze penicillins and narrow-spectrum cephalosporins, but they hydrolyze extended-spectrum cephalosporins such as cefotaxime and ceftazidime as well as aztreonam and imipenem only poorly. Unlike the chromosomal AmpC enzymes (see below), overproduction of these enzymes does not significantly affect resistance to the extended-spectrum antibiotics. However, variants capable of hydrolyzing many of the newer extended-spectrum cephalosporins and monobactams have emerged because of mutations in the structural genes for these enzymes. Naturally selected amino acid substitutions frequently occur adjacent to three of four evolutionary conserved amino acid motifs that form the active sites of serine peptidases (93).

In certain cases, enzymatic kinetic parameters and resistance phenotypes remain unchanged despite amino acid substitutions. For example, TEM-1 and TEM-2, which differ by a single amino acid at position 37, confer similar resistance phenotypes (13). In contrast, in the case of TEM-7 and SHV-2, the Arg-162 and Gly-236 residues of TEM-1 and SHV-1, respectively, have been replaced by a serine (14, 32, 174). In both instances, the single amino acid substitution extends the enzyme's substrate range to include oxime-substituted molecules such as ceftazidime and cefotaxime. However, the various substitutions do not necessarily allow the extended-spectrum β-lactamases to hydrolyze the substrates equally. For instance, TEM-7 hydrolyzes ceftazidime and cefotaxime at approximately the same rate, whereas SHV-2 hydrolyzes cefotaxime about 10 times more rapidly than ceftazidime. In other instances, members of the TEM and SHV families are encoded by genes with multiple mutations that may have been selected sequentially. An additional set of amino acid alterations, involving the substitution of Glu-102 (TEM-2)→Lys in TEM-9 or of Glu-237 (SHV-2)→ Lys in SHV-4 and -5, further increase the hydrolysis rates of ceftazidime (66). Similar amino acid substitutions have been described for other extended-spectrum TEM, SHV, and related enzymes (26). More recently, OXA-11 was shown to be an extend-ed-spectrum variant of the class D oxacillinase OXA-10 (PSE-2) because of two amino acid substitutions (70).

In *S. aureus*, class A β-lactamases appear to be encoded by several closely related transposable elements (161). Un-like the constitutive class A enzymes of gram-negative

bacteria, the enzymes of *S. aureus* are usually inducible, and regulation is thought to be under repressor control (1). Three regulatory genes, *blaI*, *blaR1*, and *blaR2*, have been distinguished. The *blaI* gene encodes a repressor that neg-atively regulates transcription of the β-lactamase structural gene *blaZ*. The BlaR1 protein, encoded by *blaR1*, displays homology with gram-negative chemotactic transducers, with an extracellular domain responsible for signal recep-tion and an intracellular domain responsible for production of an intracellular signal. Thus, the BlaR1 protein functions during the induction process as a receptor for β-lactams and results, directly or indirectly, in the derepression of *blaZ*. The unlinked *blaR2* chromosomal gene downregulates chromosomally and plasmid-specified β-lactamase expres-sion, but its precise role in regulation remains to be determined.

Virtually all enterobacteria studied to date produce a chromosomally encoded cephalosporinase. These enzymes belong to molecular class C and are sufficiently distinct biochemically to be considered species specific (92, 184). Inducible cephalosporinases have been found in many spe-cies, including *Enterobacter cloacae*, *Citrobacter freundii*, *Ser-ratia marcescens*, and *P. aeruginosa* (165). In these species, β-lactamase is normally produced at very low levels but is induced to rates several hundred times higher by the pres-ence of β-lactams. Inducible strains mutate at high frequen-cies to high-level constitutive resistance (112). Such mu-tants are clinically important since they are frequently responsible for the rapid emergence of resistance to a num-ber of the newer β-lactam antibiotics. Practically all enter-obacteria that do not produce inducible cephalosporinases possess a class C cephalosporinase that is expressed consti-tutively at very low levels and does not contribute to β-lactam resistance. However, mutations resulting in con-stitutive high-level expression that confer β-lactam resis-tance do occur. Although rare, the newer cephalosporins have also selected for several plasmid-specified forms of class C β-lactamases (143, 147).

The control mechanisms for class C inducible β-lacta-mases, mostly studied in *E. cloacae* and *C. freundii*, are highly complex and not completely understood. The β-lac-tamases are encoded by the *ampC* gene. Four genes in-volved in the induction of *ampC* expression have been identified: *ampR*, *ampG*, *ampD*, and *ampE*. The *ampR* gene encodes a protein that binds upstream from *ampC* and acts as a transcriptional activator in the presence of β-lactams and as a repressor in their absence (113). AmpR appears to be activated by the binding of specific peptidoglycan frag-ments generated by the action of β-lactams on the cell wall (114). AmpG may be involved in signal transmission, pos-sibly as a transporter of the activating ligands (114). Resis-tance by hyperproduction of β-lactamase is often due to mutations in the unlinked *ampD* gene that derepress ex-pression of *ampC* (111). It is thought that *ampD* is involved in peptidoglycan synthesis and that *ampD* mutants produce, in the absence of β-lactam antibiotics, the putative pep-tidoglycan fragments that activate AmpR (114, 197). The role of *ampE* has yet to be determined, but it may be involved in induction.

The *ampC* gene of *Escherichia coli* is expressed constitu-tively at very low levels owing to an inefficient promoter and the presence of an attenuator structure between the promoter and the structural gene. Hyperproduction of β-lactamase can be caused by mutations in the promoter and attenuator regions or by replacement of the native promoter by a more efficient one from another species, such as *Shigella sonnei* (140). Hyperproducing mutants of AmpC can also result from amplification of the *ampC* gene (138).

The mechanism by which resistance to extended-spectrum β-lactams occurs in gram-negative bacteria that hyperproduce class C β-lactamases remains controversial, since the enzymes exhibit extremely low or undetectable hydrolytic activities against these compounds (131, 164). Because of this discrepancy and the observation that the enzymes have a strong affinity for these "β-lactamase-resistant" substrates, it has been suggested that the mechanism of resistance is "trapping" (191). This model proposes that β-lactamases confer resistance by binding to and trapping β-lactam molecules as they slowly cross the outer membrane and enter the periplasmic space rather than by hydrolyzing them. However, others have suggested that outer membrane impermeability coupled with slow hydrolysis rates adequately explains *ampC*-mediated resistance to the extended-spectrum drugs (80). If trapping were the explanation, it has been calculated that the enzyme would have to be synthesized at a rate of 40% of the dry mass per generation (80).

β-Lactam Resistance Due to Decreased Outer Membrane Permeability and Active Efflux

In contrast to gram-positive organisms, in which β-lactams have unhindered access to their PBP targets, the outer membrane in a gram-negative bacterium is a barrier to these compounds. The balance between antibiotic influx and clearance, whether due to hydrolysis or trapping, determines the susceptibility or resistance of the cell; permeability barriers alone rarely produce significant resistance. In *Escherichia coli* and probably numerous other gram-negative bacteria, β-lactams diffuse across the outer membrane primarily through water-filled channels consisting of a specific class of proteins, termed porins (135). *Escherichia coli* produces at least two porin types, called OmpF and OmpC. Mutations that cause reduced expression or alteration of OmpF and/or OmpC result in decreased susceptibility to many β-lactams (76, 91, 115). The intrinsic resistance of *P. aeruginosa* to a variety of antimicrobial agents, including β-lactams, tetracycline, fluoroquinolones, and chloramphenicol, is partly due to the presence of a low-efficiency porin (OprF). However, the major contribution to the multidrug resistance phenotype of this species appears to be a unique efflux system with wide substrate specificity whose physiologic function is the export of the siderophore pyoverdine (134). The emergence in *P. aeruginosa* of resistance specific to imipenem is associated with the loss of the specific channel OprD (196). It appears that imipenem preferentially uses this channel to cross the outer membrane, which under normal circumstances probably transports basic amino acids (57). In *Escherichia coli*, crossresistance to various classes of antibiotics, including β-lactams, tetracycline, chloramphenicol, and quinolones, may be due to mutations in the *marRAB* operon (see Tetracycline Resistance below).

AMINOGLYCOSIDE RESISTANCE

Aminoglycosides are particularly active against aerobic gram-negative bacilli and gram-positive cocci. Intrinsic resistance most often results from impaired uptake, while acquired resistance is frequently due to plasmid-encoded modifying enzymes. Acquired resistance can also be due to chromosomal mutations that alter the ribosomal target or the antibiotic uptake.

Intrinsic and Acquired Resistance Due to Decreased Uptake

Intrinsic or acquired resistance secondary to decreased antibiotic uptake confers low-level cross-resistance to all aminoglycosides. Aminoglycosides are highly positively charged compounds that must cross the outer membrane (in gram-negative organisms) and the cytoplasmic membrane (in gram-negative and -positive bacteria) before they reach their cytoplasmic ribosomal target. In gram-negative bacteria, the initial step involves ionic binding of aminoglycosides to anionic sites on the outer membrane surface; these sites include lipopolysaccharides (LPS), outer membrane proteins, and the polar heads of phospholipids. In *Escherichia coli*, it has been generally believed that aminoglycosides reach the periplasmic space by diffusion through outer membrane porin proteins. However, the role of porin-mediated uptake has been questioned, since no difference in susceptibility between wild-type and porin-deficient strains of *Escherichia coli* or other gram-negative organisms can be demonstrated (166). Studies of *P. aeruginosa* and *Escherichia coli* indicate that uptake across the outer membrane may instead be due to a "self-promoted uptake" mechanism (73). The cationic aminoglycosides displace divalent cations (e.g., Mg^{2+}) that cross-bridge adjacent LPS molecules, thereby permeabilizing the outer membrane and allowing entry of the antibiotic (95). Disruption of the outer membrane may contribute to the bactericidal activity of these drugs since neither inhibition of protein synthesis nor codon misreading alone can adequately account for this activity (40, 41, 95). Thus, species-related differences in aminoglycoside activity may be at least partly due to differences in the compositions (LPS, outer membrane proteins, porins, phospholipids) of outer membranes. Decreased gentamicin binding was observed in a strain of *Escherichia coli* with a mutation in LPS phosphates (149). Resistance to gentamicin in *P. aeruginosa* mutants has been associated with overproduction of the major outer membrane protein H1, which may protect the bacterium from cationic antibiotics by substituting for Mg^{2+}.

The electrical potential established by the electron transport system provides the driving force for aminoglycoside entry across the cytoplasmic membrane. This accounts for the relative resistance of electron transport-deficient anaerobic organisms and the decreased susceptibility of facultative anaerobes such as enterococci, streptococci, and members of the family *Enterobacteriaceae* when they are grown anaerobically. Acquired resistance to all aminoglycosides in bacteria with a variety of "energy deficiencies" is likely due to decreased uptake (18). These strains frequently form small colonies consistent with impaired growth rates and may therefore be less pathogenic (18).

Acquired Resistance

Aminoglycoside-Modifying Enzymes

The most common mechanism of acquired resistance to aminoglycosides is antibiotic inactivation by plasmid- or transposon-encoded modifying enzymes. The cytoplasmic modifying enzymes are present in quantities only sufficient to inactivate drug that enters the cytoplasm (18). Resistant cells are therefore characterized by growth in the presence of unchanged antibiotic in the growth medium.

There are three classes of aminoglycoside-modifying en-

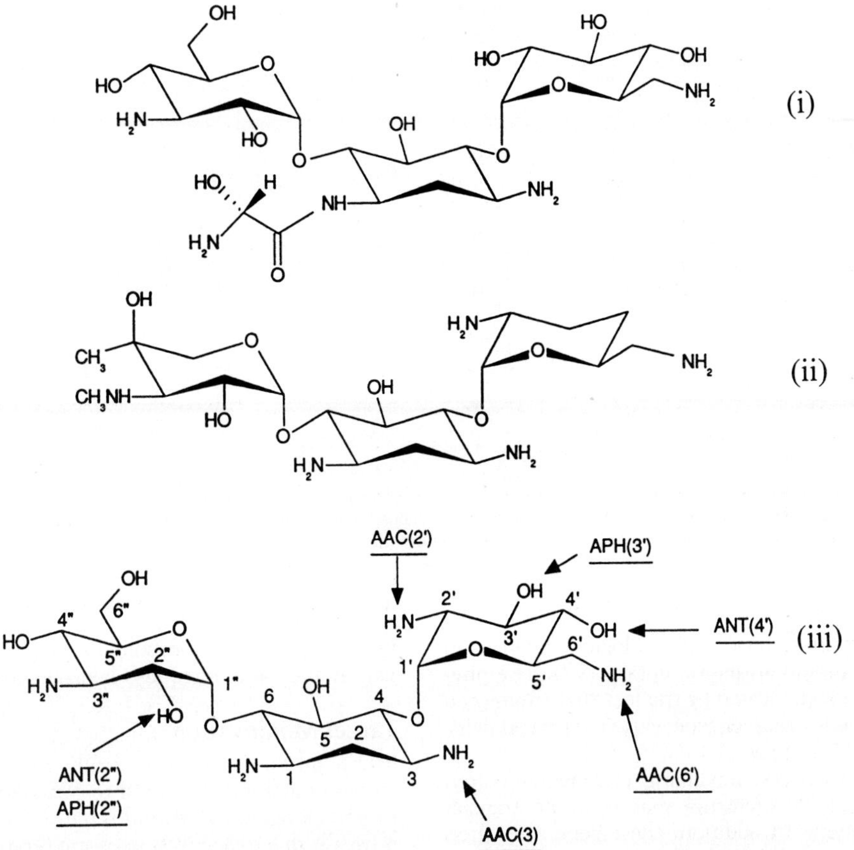

FIGURE 2 Structures of amikacin (i), gentamicin C1a (ii), and kanamycin B (iii). Arrows indicate sites of modification by resistance enzymes (Table 2). Tobramycin is 3'-deoxykanamycin B. Modified with permission from reference 39.

zymes: acetyltransferases (AAC), adenylyltransferases (ANT), and phosphotransferases (APH) (39, 167). The structures of typical aminoglycosides and their sites of enzymatic modification are shown in Fig. 2. ANT and APH adenylylate and phosphorylate hydroxyl groups, respectively, while AAC acetylates amino groups. In general, only phosphorylating enzymes confer very high levels of resistance. Since each class comprises numerous enzymes that can modify different hydroxyl or amino groups, the enzymes are further classified into subclasses. For example, the AACs consist of three subclasses that can acetylate amino groups at the 3, 2', or 6' position. Some aminoglycoside-modifying enzymes have broad substrate ranges and determine resistance to structurally unrelated compounds (Table 2). For instance, ANT(3"/9) can modify both the 3"-hydroxyl of streptomycin and the 9-hydroxyl of spectinomycin. The bifunctional enzyme AAC(6')+APH(2") consists of an amino-terminal portion that catalyzes acetylation of 6'-amino groups and a carboxy terminus that catalyzes phosphorylation of 2"-hydroxyl groups.

Production of an enzyme does not necessarily confer resistance to all aminoglycosides that are substrates in vitro. Enzyme subclasses comprise various enzyme types (designated by roman numerals) that act on the same site but vary in their substrate ranges and thus confer different resistance phenotypes. For example, although purified APH(3')-I is capable of phosphorylating hydroxyl groups at

the 3' position in kanamycin, neomycin, and amikacin, only resistance to kanamycin and neomycin is determined (35, 37). The susceptibility to amikacin is due to a significantly higher K_m of the enzyme for amikacin than for kanamycin or neomycin. Consequently, APH(3')-I modifies amikacin less well than the last two drugs. Similarly, the AAC(3)-I enzyme determines resistance to gentamicin, whereas AAC(3)-II determines resistance to gentamicin, netilmicin, and tobramycin. These examples highlight the importance of an enzyme's catalytic efficiency on the resistance phenotype. In certain situations, however, the resistance phenotype may be misleading. The presence of certain aminoglycoside-modifying enzymes may affect the antibacterial activities of combinations of antibiotics even though the strain appears to be susceptible to the aminoglycoside. For example, in *S. aureus* and *Enterococcus* spp., APH(3')-III does not determine resistance to amikacin, although synergy with β-lactams or vancomycin is abolished (107).

Inference of the enzyme content of a cell from the resistance phenotype is difficult. As mentioned, an enzyme may be produced without conferring resistance. In addition, the level of resistance depends significantly on the host bacterium. Enzymes that determine resistance to gentamicin or tobramycin in *P. aeruginosa* may not confer detectable resistance in *Escherichia coli* (97). Species-related phenotypic differences may be due to differences in gene copy

TABLE 2 Substrate profiles of selected aminoglycoside-modifying enzymes[a]

Aminoglycoside	APH		ANT				AAC	
	3'	2"	2"	4'	(3"/9)	9	3	6'
Gentamicin C1a	−	+	+	−	−	−	+	+
Gentamicin C1	−	+	+	−	−	−	−	−
Gentamicin C2	−	+	+	−	−	−	−	+
Tobramycin	−	(+)	+	+	−	−	+	+
Amikacin	(+)	(+)	−	+	−	−	−	+
Kanamycin A	+	(+)	+	+	−	−	+	−
Kanamycin B	+	(+)	+	+	−	−	+	+
Kanamycin C	+	(+)	+	+	−	−	−	+
Netilmicin	−	+	−	−	−	−	(+)	+
Sisomicin	−	+	+	−	−	−	(+)	+
Isepamicin	(+)	(+)	−	+	−	−	−	+
Streptomycin	−	−	−	−	+	+	−	−
Spectinomycin	−	−	−	−	+	−	−	−

[a]Symbols: +, normal substrate; (+), substrate for some forms of the enzyme; −, nonsubstrate for the enzyme. The fact that an antibiotic is a substrate for an enzyme in vitro does not necessarily imply that a strain producing that enzyme is resistant to the antibiotic.

number or gene expression or to physiologic differences in membrane structure and antibiotic uptake (97). The phenotype can also be complicated by the fact that strains can harbor multiple genes that encode various enzymes with overlapping substrate ranges.

The genes that encode modifying enzymes are often carried on transposable elements that facilitate transfer among diverse genera. In addition, these genes are often located in integrons that assist in their dissemination (see Resistance to Sulfonamides and Trimethoprim below). Briefly, integrons are mobile genetic elements that each encode their own recombination system and contain an insertion site for the integration of various resistance genes. Aminoglycoside resistance genes associated with integrons include ant(2")-Ia, aac(3)-Ia, aac(3)-VIa, ant(3")-Ia, aac(6')-Ia (3), aac(6')-Ib, and aac(6')-IIa. The ability of integrons to acquire multiple resistance determinants may partly explain the increased prevalence of multiply aminoglycoside-resistant strains noted in recent studies. In general, aminoglycoside resistance genes are expressed constitutively, but there are exceptions, such as the chromosomal aac(6')-Ic gene that is present in strains of S. marcescens and the aac(2') gene of Providencia stuartii (155).

Ribosomal Target Modification

In the laboratory strain, Escherichia coli K-12 mutations in the rpsL gene that alter ribosomal protein S12 have been shown to determine high-level resistance to streptomycin (18). Streptomycin resistance due to ribosomal modification has also been reported in clinical isolates of N. gonorrhoeae, S. aureus, P. aeruginosa, and E. faecalis. In streptomycin-resistant strains of Mycobacterium tuberculosis, mutational alterations in 16S and in ribosomal protein S12 have been identified (50).

MACROLIDE, LINCOSAMIDE, AND STREPTOGRAMIN RESISTANCE

Intrinsic resistance to macrolide, lincosamide, and streptogramin B (MLS$_B$) antibiotics in gram-negative bacilli is due to low permeability of the outer membrane to these hydrophobic compounds. Acquired resistance occurs most often

by alteration of the ribosomal target, although drug inactivation and efflux have also been described.

Target Modification

MLS$_B$ antibiotics are chemically distinct but have similar modes of action. They bind to 50S ribosomal subunits and inhibit elongation of peptide chains (148). MLS$_B$ resistance is usually due to acquisition of erm (erythromycin resistance methylase) genes that encode enzymes that N^6-dimethylate a specific adenine residue of 23S rRNA (101). Methylases from numerous gram-positive and -negative bacteria modify an analogous adenine residue that is located in a conserved region of the rRNA (172, 187). Ribosomal methylation confers cross-resistance to MLS$_B$ antibiotics, the so-called MLS$_B$ resistance phenotype. Presumably, methylation leads to a conformational change in the ribosome that results in decreased affinity for all MLS$_B$ antibiotics, because the binding sites for these drugs overlap.

Nucleotide sequencing of various clinically important bacterial species allow distinction of at least nine erm genes (105) (Table 3). Because a number of these genes cross-hybridize, clinical isolates can be assigned to one of four hybridization classes: ermA, ermC, ermAM, and ermF. Despite distinct hybridization classes, the amino acid sequences of these enzymes are highly conserved, which suggests that they evolved from a common ancestor, possibly an antibiotic producer (5).

Expression of MLS$_B$ resistance can be inducible or constitutive. The type of expression is not related to the class of erm determinant but depends on a regulatory region upstream from the methylase structural gene. Regulation by these regions occurs by a translational attenuation mechanism in which mRNA secondary structure influences the level of translation (89). In laboratory mutants and clinical isolates, single nucleotide changes, deletions, or duplications in the regulatory region convert inducibly resistant strains to constitutively resistant ones that are cross-resistant to MLS$_B$ antibiotics (104).

In staphylococci, inducible strains are resistant to 14-membered (erythromycin, roxithromycin) and 15-membered (azithromycin) macrolides only. The 16-membered macrolides (spiramycin, josamycin), lincosamides, and

TABLE 3 Distribution of *erm* genes in clinically important bacterial species[a]

Hybridization class	Gene	Host(s)
ermA	*ermA*	*Staphylococcus aureus*, coagulase-negative staphylococci
ermAM	*ermAM*	*Streptococcus sanguis*, *Streptococcus pneumoniae*, *Streptococcus agalactiae*, *Streptococcus pyogenes*
	ermB	*Staphylococcus aureus*, *Bacillus subtilis*, *Lactobacillus* spp.
	ermB-like	*Enterococcus faecalis*
	ermBC	*Escherichia coli*
	ermP	*Clostridium perfringens*
	ermZ	*Clostridium difficile*
	NI	*Lactobacillus reuteri*
ermC	*ermC*	*Staphylococcus aureus*, coagulase-negative staphylococci
	ermM	*Staphylococcus epidermidis*
ermF	*ermF*	*Bacteroides fragilis*, *Bacteroides ovatus*

[a]Modified from reference 105. NI, gene not identified.

streptogramin B antibiotics remain active in vitro. This so-called dissociated resistance is due to the fact that only 14- and 15-membered macrolides are effective inducers of methylase synthesis. MLS$_B$ resistance in streptococci can also be expressed constitutively or inducibly. However, in the latter case, all MLS$_B$ antibiotics can to various degrees act as inducers. Thus, in streptococci, inducible MLS$_B$ resistance is characterized by cross-resistance to all MLS$_B$ antibiotics.

In *Staphylococcus* spp., MLS$_B$ resistance is often mediated by small (2- to 4-kb) nonconjugative plasmids carrying the *ermC* determinant (203). With the notable exception of *Streptococcus pneumoniae*, MLS$_B$ resistance plasmids have been found in nearly all *Streptococcus* spp. (88). Transfer of MLS$_B$ resistance in the absence of plasmid DNA has been demonstrated in *Streptococcus pneumoniae*, *Streptococcus bovis*, viridans group, and group A, B, F, and G streptococci (25, 90). In *Streptococcus pneumoniae*, chromosomal MLS$_B$ resistance is mediated by the broad-host-range transposon Tn*1545*, a member of a closely related family of conjugative elements that includes Tn*916* (see Tetracycline Resistance below) (36). In *E. faecalis*, inducible erythromycin resistance is often mediated by the nonconjugative transposon Tn*917*, which harbors an *ermAM* determinant (158, 193). Interestingly, the frequency of Tn*917* transposition increases in the presence of erythromycin.

Antibiotic Inactivation

Unlike target modification, which causes resistance to structurally distinct antibiotics, enzymatic inactivation confers resistance only to structurally related drugs.

Members of the family *Enterobacteriaceae* that are resistant to high levels of erythromycin, often isolated from neutropenic patients receiving oral erythromycin for gastrointestinal decontamination, destroy the lactone rings of 14-membered macrolides by production of an erythromycin esterase (12) or by phosphorylation catalyzed by a 2'-phosphotransferase (139). Two types of esterases (I and II) are encoded by the *ereA* and *ereB* genes, respectively (4, 142). The combination of *ereB* and *ermB* (encoding an rRNA methylase) is frequently found in enterobacteria highly resistant to erythromycin, demonstrating the synergism of the two gene products (6). Clinical isolates of *Staphylococcus haemolyticus* and *S. aureus* highly resistant to lincomycin and apparently susceptible to clindamycin have been described; however, the MBCs of clindamycin are greatly increased. Such strains produce a 3-lincomycin 4-clindamycin O-nucleotidyltransferase (17).

Active Efflux

Resistance due to active efflux has been reported in *S. epidermidis*. The resistance gene *msrA* encodes a transport-related protein (160) that confers inducible resistance to 14- and 15-membered macrolides. It remains unknown why induced strains are also cross-resistant to streptogramin type B antibiotics.

TETRACYCLINE RESISTANCE

Tetracycline resistance is the most common antibiotic resistance encountered in nature. Although it can result from chromosomal mutations affecting outer membrane permeability, it more commonly results from acquisition of exogenous DNA that encodes proteins involved in active efflux of tetracycline or in protection of the ribosome.

Altered Permeability

In *Escherichia coli*, chromosomal mutations producing a deficiency in the outer membrane porin OmpF, through which tetracycline normally diffuses, confer low-level resistance to tetracycline as well as to β-lactams, chloramphenicol, and quinolones. Moreover, exposure to tetracycline or chloramphenicol can select mutations in a genetic locus that confer high-level resistance to the selective agent and structurally unrelated antibiotics, including penicillins, cephalosporins, nalidixic acid, and fluoroquinolones (30, 58). This system, designated MAR (multiple antibiotic resistance), involves the *marRAB* operon, in which a reduction in OmpF porin is only part of the resistance mechanism (29). The *marRAB* operon is regulated and derepressed by tetracycline and chloramphenicol (68). Whereas point mutations in the *marRAB* operon confer MAR, deletion of this region prevents *Escherichia coli* from acquiring the MAR phenotype (29, 68). The *marA* gene product, MarA, is related to a family of positive transcriptional activators, which suggests that the *marRAB* operon is part of a larger regulon influencing the expression of various distant genes (29). One such gene, *micF*, when activated produces an antisense RNA that inhibits *ompF* translation and results in a reduction of OmpF (31). However, other *mar*-related changes must be involved, since OmpF mutants are less resistant than their MAR counterparts (39, 40).

Recently, activation of the *mar* operon was also shown to stimulate the active efflux of tetracycline and chloramphenicol (58, 122). Fluoroquinolone resistance in MAR mutants is also only partly attributable to reductions in OmpF. An active efflux mechanism has not been detected, and it is believed that decreased permeability is due to alterations of additional entry proteins that contribute to drug resistance (30). The *mar* locus has also been detected in other members of the family *Enterobacteriaceae*, including *Salmonella* and *Shigella* spp.

Active Efflux

Resistance to tetracycline in gram-negative and gram-positive bacteria often occurs as a result of energy-dependent pumping of the antibiotic from the cell so that the levels of drug are decreased and the ribosome is not inhibited (123). Antibiotic efflux is mediated by the Tet membrane proteins, which use an antiport mechanism of transport involving the exchange of a proton for a tetracycline-cation complex (96, 132). Even though protein synthesis in resistant cells is unimpaired in the presence of tetracycline, the measured residual intracellular levels of the drug appear to be sufficient to inhibit protein synthesis (9, 109). It remains to be determined in what compartment or in what form the residual tetracycline exists. Since tetracycline can exist in several ionic forms, it is possible that the active form is preferentially pumped from the cell (175).

The tetracycline efflux genes of clinically important bacteria have been assigned to eight classes on the basis of DNA-DNA hybridization. Genes *tet*(A) to *tet*(E) have been detected in various gram-negative bacteria. The *tet*(P) determinant is apparently confined to *Clostridium* spp. and consists of a composite determinant encoding an efflux and a ribosomal protection mechanism (see below) (173). The *tet*(K) and *tet*(L) genes have been found only in gram-positive bacteria. With the exception of *tet*(B), the tetracycline efflux-type genes do not handle minocycline (a lipophilic analog of tetracycline) well, providing only 10% or less of the resistance to tetracycline (109). Recently, an efflux mechanism has been shown to act in *Proteus mirabilis*, a species that was previously believed to be intrinsically resistant (110). Despite mechanistic differences in regulation of *tet* genes (81, 99, 195), the presence of tetracycline induces synthesis of resistance proteins that display significant similarity to bacterial transport proteins such as sugar transporters, from which they may have evolved (169).

The intrinsic low-level resistance of *P. aeruginosa* to a wide variety of antimicrobial agents, including tetracycline, appears to be mostly due to a single efflux system with broad substrate specificity (134).

Ribosome Protection

Tetracycline acts by reducing the affinity of the A and P sites of the 30S ribosomal subunit for aminoacyl-tRNA. Tetracycline resistance can result from production of a protein that interacts with the ribosome such that protein synthesis is unaffected by the presence of the antibiotic. The exact mechanism of resistance remains unclear, since experiments with TetM and TetO indicate that these proteins do not prevent tetracycline from binding to the ribosome (20, 118). For TetM, modification of tRNA may be necessary for resistance (21). Regardless of the precise mechanism, TetM permits aminoacyl-tRNAs to bind productively to the ribosome. Ribosomal protection genes share significant homology with elongation factors

G and Tu (27, 118), and like them, TetM has GTPase activity (20).

Four classes (M, O, Q, S) of ribosomal protection genes have been characterized: *tet*(M) is found in a large variety of gram-positive and -negative bacteria; *tet*(O) is detected in *Campylobacter* spp., *Streptococcus* spp., and *Enterococcus* spp. (211); *tet*(Q) exists in *Bacteroides* spp.; and *tet*(S) is found in *Listeria monocytogenes* and *E. faecalis* (27, 28). All classes of ribosomal protection genes also confer resistance to minocycline. Although *tet*(M) is found on plasmids in *Neisseria* spp., *Kingella denitrificans*, *Eikenella corrodens*, and *Haemophilus ducreyi* (100), it is usually found in the chromosome as part of Tn916-related conjugative transposons. These elements are self-transferable, possess an extremely broad host range, and transfer at increased frequency in the presence of tetracycline (15, 171).

The *tet*(M) gene is inducible by tetracycline (20). In *Campylobacter coli*, *tet*(O) may also be regulated, since its upstream region is almost identical to that of *tet*(M). Furthermore, this region is necessary for high-level tetracycline resistance (201). In *Campylobacter jejuni* and *Escherichia coli*, however, *tet*(O) appears to be expressed constitutively (201).

In *Bacteroides* spp., tetracycline resistance is specified by *tet*(Q), which is usually carried on very large (>60-kb) conjugative transposons called tetracycline resistance (Tc^r) elements (136). These elements have the unusual property of mediating in *trans* the excision, circularization, and transfer of a family of noncontiguous 10- to 12-kb segments of chromosomal DNA (170).

RESISTANCE TO SULFONAMIDES AND TRIMETHOPRIM

Biosynthesis of several amino acids and purines depends on the availability of tetrahydrofolate (THF) derivatives required by enzymes that catalyze transfer of one-carbon units. Moreover, THF is a prerequisite cofactor for essential thymidylate synthesis. In contrast to mammalian cells that use dietary folates, most bacteria are unable to take up preformed folic acid derivatives and must synthesize THF de novo. The first of the last three steps in THF biosynthesis in bacteria is the condensation of 2-amino-4-hydroxyl-6-hydroxymethyl-pteridine with *p*-aminobenzoic acid (PABA) by dihydropteroic acid synthase (DHPS). The product, dihydropteroic acid, undergoes a second condensation with glutamic acid, yielding dihydrofolic acid. The final step is reduction of dihydrofolic acid to THF by dihydrofolate reductase (DHFR).

Bacterial THF biosynthesis is inhibited by agents that interfere with DHFR or DHPS. Sulfonamides are structural analogs of PABA that competitively inhibit DHPS. Trimethoprim is a competitive inhibitor of bacterial DHFR.

Intrinsic Resistance

Outer membrane impermeability confers trimethoprim and sulfonamide resistance on *P. aeruginosa* (188). Intrinsic resistance to trimethoprim in a number of species, including *Neisseria* spp., *Clostridium* spp., *Brucella* spp., *Bacteroides* spp., *Moraxella catarrhalis*, and *Nocardia* spp., is due to host DHFR enzymes with low affinities for the drug (190). Folate auxotrophs such as *Enterococcus* spp. and *Lactobacillus* spp. that are able to use exogenous preformed folates exhibit decreased susceptibilities to sulfonamides and trimethoprim (210). Enterococci can also utilize exogenous thymine or thymidine to escape inhibition by folate pathway antagonists (72).

Acquired Resistance to Trimethoprim

Chromosomal Mutations

Chromosomal mutations in the DHFR structural gene *folA* that confer low-level resistance to trimethoprim are easily obtained in the laboratory and occur in clinical isolates. Overproduction of host DHFR confers resistance in clinical isolates of *Escherichia coli* (52), *Streptococcus pneumoniae* (121), and *H. influenzae* (42).

Chromosomal mutations that result in decreased outer membrane permeability and confer cross-resistance to nalidixic acid, trimethoprim, and chloramphenicol have been observed in *Klebsiella pneumoniae*, *Enterobacter aerogenes*, *Enterobacter cloacae*, and *Serratia marcescens* (67). The pleiotropic resistance observed in these strains was associated with a decrease in levels of certain outer membrane proteins (67).

Combined high-level resistance to trimethoprim and sulfonamides can be due to chromosomal mutations that inactivate thymidylate synthetase, an enzyme that converts deoxyuridylate to thymidylate. These so-called *thy* mutants require exogenous thymine or thymidine for DNA synthesis and are thus resistant to folate pathway antagonists.

Plasmid-Mediated Resistance

High-level resistance to trimethoprim in enterobacteria is almost always caused by the acquisition of DNA that specifies a trimethoprim-resistant DHFR with an altered active site (2). At least 11 DHFRs have been characterized in gram-negative organisms on the basis of kinetic data, antisera, primary structure, and DNA-DNA hybridization. Type I DHFR is the most common and confers high-level resistance to trimethoprim (53, 146). The corresponding gene, *dhfrI*, is specified by the transposon Tn*7* (54).

Recently, the *dhfrI* and *aadA1* genes of Tn*7* have been detected in genomic environments other than Tn*7* as part of mobile genetic elements called cassettes (181). A cassette consists of a resistance gene flanked by conserved DNA sequences that promote site-specific insertion into so-called integrons, elements first noted in transposon Tn*21* (64). Integrons consist of an insertion site and an integrase gene that encodes a system for site-specific integration of one or more antibiotic resistance cassettes (33, 34, 71). Thus, integrons function as mobile expression cassettes for various inserted genes. The *dhfrI* gene has been found inserted in DNA sequences similar to the integron of Tn*21* (181). The appearance of the *dhfrI* gene in Tn*21*-like structures not only increases its potential for dissemination but also links trimethoprim and sulfonamide resistance, since the integron contains a conserved sulfonamide resistance determinant (180).

Type II DHFRs are the least sensitive to trimethoprim and have been found on various plasmids (146). Type III enzymes are more closely related to gram-negative chromosomal DHFRs than to the plasmid-specified type I enzyme (53). They confer a moderate level of trimethoprim resistance and have been detected in *Salmonella typhimurium* and *Shigella* spp. (3, 55). Type IV DHFR is unique in its ability to be induced by trimethoprim (208). In *Escherichia coli*, type X DHFR is also encoded by a gene as part of an integron (145).

Transposon-mediated trimethoprim resistance has also been observed in *S. aureus* (209). DNA sequences related to the *dfrA* gene have also been detected in *S. epidermidis* and *Staphylococcus hominis* (185).

Acquired Resistance to Sulfonamides

Chromosomal Mutation

Chromosomal mutations that cause overproduction of PABA have been reported to confer sulfonamide resistance in *Neisseria* spp. and *S. aureus* (47). However, this mechanism has not been studied recently and its significance has yet to be determined. Resistance in *Escherichia coli*, *Streptococcus pneumoniae*, and *Neisseria* spp. is associated with alterations of DHPS that result in a low affinity for sulfonamides (189). As already mentioned, mutation to thymine auxotrophy in many bacterial species can confer resistance to trimethoprim and sulfonamides (47).

Plasmid-Mediated Resistance

Acquired sulfonamide resistance can be due to acquisition of plasmids that encode a drug-resistant DHPS. Two types (I and II) of resistant DHPSs, encoded by the *sulI* and *sulII* genes, respectively, have been identified in gram-negative organisms (153, 180). The *sulI* gene is almost always located in conserved segments of integrons in Tn*21*-like elements carried by large conjugative plasmids (154). The *sulII* gene is frequently linked genetically to a streptomycin resistance gene on IncQ broad-host-range and small nonconjugative plasmids (153).

QUINOLONE RESISTANCE

The bacterial chromosome replicates by a semiconservative mechanism in which each strand of the parental duplex acts as a template for synthesis of a complementary daughter strand. However, since the chromosome is circular, unwinding of the intertwined parental DNA strands during replication would result in positive supercoiling farther along the molecule that would rapidly generate insurmountable resistance to further movement. This problem is overcome by DNA gyrase, which, among other functions, removes positive DNA supercoils from closed circular DNA (182).

DNA gyrase consists of a tetramer of two nonidentical subunits (A and B) in the form A_2B_2. The A and B subunits are encoded by chromosomal genes *gyrA* and *gyrB*, respectively. In all genera studied, DNA gyrases are highly homologous (87). The enzyme introduces a 4-bp staggered double-strand break in one duplex strand such that each A subunit is attached via the Tyr-122 residue to the newly created 5' terminus of each strand. Subsequently, another segment of duplex DNA is passed through the break, and the break is resealed. The A subunit, which is responsible for the breakage-ligation reaction, is the primary target of quinolones (86, 179). Recently, it has been proposed that quinolones bind to the complex of DNA and DNA gyrase and that the quinolone binding site is exposed only after the DNA strands are cleaved. Quinolones are known to inhibit the resealing of cleaved DNA and presumably inhibit gyrase function by trapping the covalent enzyme-DNA intermediate (179).

Alteration of DNA Gyrase

It has been possible to select spontaneous, single-step mutants highly resistant to nalidixic acid but not to the newer fluoroquinolones (85). However, highly resistant strains can be selected by serial passage in the presence of increasing antibiotic concentrations, indicating that multiple mutations are likely to be responsible for high-level fluoroquinolone resistance (85).

Low- and high-level resistance to nalidixic acid and fluoroquinolones can be due to or associated with a mutation(s) in the gyrA gene (129). Studies of Escherichia coli laboratory-derived mutants indicated that resistance was associated with various allelic forms (nalA, nfxA, norA, cfxA, and ofxA) of the gyrA gene. Similar mutations have been found in C. freundii, Serratia marcescens, P. aeruginosa, and H. influenzae.

More recently, mutations in gyrA of clinical isolates have been reported. Examination of eight quinolone-resistant uropathogenic Escherichia coli strains revealed that in all but one, Ser-83→Leu or Trp substitutions were associated with high-level resistance to nalidixic acid (141). The remaining strain, which expressed low-level resistance, had a Ser-87→Val change. trans-Complementation analysis, based on the fact that quinolone-susceptible gyrA alleles are usually dominant over their resistant counterparts, indicated that high-level fluoroquinolone resistance in clinical isolates of Escherichia coli, P. aeruginosa, Providencia stuartii, K. pneumoniae, and A. calcoaceticus was at least partly due to gyrA mutations (79). To distinguish neutral mutations from mutations responsible for resistance, a chimeric gyrA gene that was wild type (susceptible) except for a Ser-83→Trp substitution was constructed. When present in a strain of Escherichia coli harboring a temperature-sensitive gyrA allele, the hybrid gene conferred high-level resistance to nalidixic acid, confirming its direct contribution to quinolone resistance. In similar experiments, a chimeric gyrA gene that included the double mutation Ser-83→Leu and Asp-87→Gly was demonstrated to account for high-level resistance to fluoroquinolones (79). Alterations in GyrA have also been reported in resistant clinical isolates of E. faecalis (130), Campylobacter jejuni (200), and Salmonella typhimurium (150).

In clinical isolates of S. aureus highly resistant to ciprofloxacin, partial sequencing of gyrA revealed Ser-84→Leu, Ser-84→Ala, Ser-84→Leu, and Ser-85→Pro or Glu-88→Lys amino acid substitutions (62). In S. epidermidis, a novel Ser-84→Phe substitution was found. Isolates of S. aureus with a Ser-84→Leu substitution had the highest levels of ciprofloxacin resistance. As in Escherichia coli, multiple gyrA mutations or a gyrA mutation(s) in addition to a mutation(s) outside the gyrA locus is probably necessary for high-level fluoroquinolone resistance.

All gyrA resistance mutations described have been located in a 130-bp sequence termed the quinolone-resistance-determining region (206). The amino acid substitutions are close to the catalytic Tyr-122 residue, which is consistent with the observation that quinolones prevent resealing of cleaved DNA (179). As mentioned above, clinical isolates of Escherichia coli and Staphylococcus spp. highly resistant to quinolones appear to be associated with "mutational hot spots" at positions Ser-83 and Ser-84, respectively. In almost all instances, amino acid substitutions at these positions involve the replacement of a hydroxyl group with a bulky hydrophobic residue. This suggests that mutations in gyrA induce changes in the binding site conformation and/or charge that may be important for quinolone-DNA gyrase interaction (129).

In laboratory strains of Escherichia coli, two amino acid substitutions in GyrB have been shown to determine low-level resistance to quinolones. An Asp-426→Asn substitution conferred resistance to all quinolones, whereas a Lys-447→Glu change led to resistance to nalidixic and oxolinic acids but hypersusceptibility to ciprofloxacin, norfloxacin, ofloxacin, enoxacin, and sparfloxacin (128, 205). Despite the similar frequencies of gyrA and gyrB mutations in vitro, mutations in gyrA appear to be mostly responsible for resistance in clinical isolates (128). Since gyrA mutations are associated with high-level resistance, they may have a selective advantage over gyrB mutations which confer low-level resistance (129).

Decreased Permeability

Decreased drug accumulation associated with changes in the outer membrane confer low-level resistance to quinolones and structurally unrelated drugs. Quinolone resistance is a characteristic of the MAR phenotype (see Tetracycline Resistance above). In Escherichia coli, exposure to tetracycline or chloramphenicol selects for mutations that confer resistance to the selective agent as well as to a wide range of structurally unrelated antibiotics, including β-lactams, nalidixic acid, and the fluoroquinolones. In Escherichia coli, the mutations norB, nfxC, and cfxB are associated with decreased permeability and reductions in OmpF (82, 86). However, the mutations map in or very near the marRAB operon and may be alleles of the same gene(s) (84). In vitro studies of P. aeruginosa have identified various loci at which mutations confer resistance by decreased membrane permeability due to alterations of outer membrane proteins and LPS (83, 125, 134).

A uropathogenic isolate of S. aureus was shown to contain an efflux pump encoded by the norA gene that is intrinsic to this species. Cloned norA confers high-level resistance to hydrophilic quinolones but low or no resistance to hydrophobic ones (207). Resistance does not result from mutations in the structural gene but may be due to a mutation in a regulatory region, since resistance is associated with increased norA transcription (94).

CHLORAMPHENICOL RESISTANCE

Chloramphenicol resistance in gram-positive and gram-negative organisms most commonly results from the acquisition of plasmids encoding chloramphenicol acetyltransferases (CAT), which enzymatically inactivate the drug. In certain gram-negative bacteria, decreased outer membrane permeability can confer resistance to chloramphenicol and structurally unrelated compounds.

Enzymatic Inactivation

Chloramphenicol contains two hydroxyl groups that are acetylated in a reaction catalyzed by CAT in which acetyl coenzyme A is the acyl donor. Initial acetylation occurs at the C-3 hydroxyl group to give 3-acetoxy-chloramphenicol. Following nonenzymatic rearrangement to 1-acetoxy-chloramphenicol and reacetylation, the 1,3-diacetoxy-chloramphenicol product is formed (168). Both the mono- and the diacetylated derivatives are unable to bind the 50S ribosomal subunit to inhibit prokaryotic peptidyltransferase (168).

Knowledge of CAT structure, specificity, and mechanism relies significantly on studies of the enteric type III variant, for which a detailed structure is available. All CAT variants appear to have a trimeric structure composed of identical subunits. Although the DNA sequences encoding these enzymes show little similarity, their deduced amino acid sequences are significantly conserved, particularly in a region that includes a conserved histidyl residue analogous to the catalytic His-195 active-site residue of the type III enzyme (98).

CATs in Gram-Positive Bacteria

In *S. aureus*, five variants of CAT (A, B, C, D, and that encoded by the prototypic plasmid pC194) have been differentiated on the basis of electrophoretic mobility (51). Kinetic studies and N-terminal sequencing indicated that the variants are very closely related (51). The structural genes (*cat*) for the enzymes are generally located on small (3- to 5-kb) multicopy plasmids. Production of CAT is inducible by chloramphenicol by a translational attenuation mechanism (116).

Chloramphenicol-resistant strains of *E. faecalis* and *Streptococcus pneumoniae* produce an inducible CAT that resembles the type D variant of staphylococci (38). In *Clostridium perfringens*, there are two constitutively expressed determinants: *catP*, which is usually found on conjugative plasmids as part of the transposable element Tn*4451*, and chromosomally located *catQ* (11), which is nearly identical to *catD*, the CAT gene found in resistant *Clostridium difficile* (159).

CATs in Gram-Negative Bacteria

In gram-negative bacteria, resistance to chloramphenicol is usually mediated by plasmid- or transposon-mediated genes that are generally expressed constitutively (116). Three types of enzymes (I, II, and III) have been identified. Ubiquitous type I enzymes bind and confer resistance to fusidic acid and are usually encoded by Tn*9* or related elements (168). Interestingly, the Tn*9*-encoded *cat* gene is under cyclic-AMP-mediated catabolite repression, in which glucose inhibits CAT synthesis (108). Various CATs closely related to the type II enteric enzymes have been isolated from *H. influenzae* and *Haemophilus parainfluenzae* (157). A novel CAT gene related to the genes for chromosomally encoded enzymes of *P. aeruginosa* and *Agrobacterium tumefaciens* has been found on Tn*2424* of *Escherichia coli* (144). This transposon is closely related to Tn*21*, and nucleotide sequencing of flanking regions indicated that the *cat* gene is part of an integron (see Trimethoprim Resistance).

Decreased Permeability

In gram-negative bacteria, resistance may also be due to chromosomal mutations that result in decreased outer membrane permeability. In *H. influenzae* (22) and *Salmonella typhi* (194), high-level resistance to chloramphenicol has been associated with a decrease in a major outer membrane protein and a lack of OmpF entry porin, respectively. In *P. aeruginosa*, nonenzymatic chloramphenicol resistance is associated with the presence of the *cmlA* gene on the In4 integron of Tn*1696* (16). The CmlA protein appears to result in reduced expression of two outer membrane porins (OmpA and OmpC) and decreased chloramphenicol uptake (16, 23). However, it is possible that Tn*1696*-mediated chloramphenicol resistance is due to efflux, since the protein sequence of CmlA indicates that it is a transporter belonging to the major facilitator family. In *Escherichia coli*, resistance to chloramphenicol and structurally unrelated antibiotics is also part of the MAR phenotype (see Tetracycline Resistance above).

GLYCOPEPTIDE RESISTANCE

The glycopeptide antibiotics vancomycin and teicoplanin bind to the peptidyl–D-alanyl–D-alanine termini of peptidoglycan precursors and prevent, presumably by steric hindrance, the transglycosylation and transpeptidation steps of cell wall peptidoglycan synthesis (127). Gram-negative organisms are intrinsically resistant because of impermeability of the outer cell membrane to these large, rigid, hydrophobic molecules. Among pathogenic gram-positive bacteria, glycopeptide resistance is restricted to *Enterococcus* spp., *S. haemolyticus*, and intrinsically resistant species such as *Lactobacillus* spp., *Leuconostoc* spp., *Pediococcus* spp., and *Erysipelothrix rhusiopathiae*. The mechanism of resistance has been extensively studied in enterococci (7). Two distinct phenotypes of acquired resistance, VanA and VanB, can be distinguished on the basis of the antibacterial and inducing activities of glycopeptides (7, 152).

Strains with the VanA phenotype include mostly isolates of *E. faecalis* and *E. faecium* with inducible high-level resistance to vancomycin and teicoplanin but not to other cell wall inhibitors (106). Resistance is generally transferable to susceptible enterococci by conjugation (106). The genes necessary and sufficient for expression of the VanA phenotype are carried by the 11-kb transposon Tn*1546* (8). Transposition of Tn*1546* or related elements into self-transferable plasmids and subsequent transfer by conjugation appears to be responsible for the spread of high-level resistance among clinical isolates of enterococci. Tn*1546* encodes nine polypeptides that can be assigned to four functional groups: transposition functions (ORF1 and ORF2), regulation of vancomycin resistance genes (VanR and VanS), resistance to glycopeptides by production of depsipeptides (VanH, VanA, and VanX), and accessory proteins that are not necessary for glycopeptide resistance (VanY and VanZ).

The deduced amino acid sequence of VanA is similar to those of the chromosomal D-alanine–D-alanine ligases of *E. faecalis* (48) and members of the family *Enterobacteriaceae* (46). These enzymes are part of the D-alanine branch of peptidoglycan synthesis and catalyze the conversion of two D-alanines into the dipeptide D-Ala–D-Ala, which is subsequently incorporated into a disaccharide pentapeptide cell wall precursor by the D-Ala–D-Ala-adding enzyme (199). However, unlike the physiologic chromosomal ligases, the VanA ligase preferentially synthesizes the depsipeptide D-Ala–D-lactate (7, 75). The replacement of the NH group of the amide D-Ala–D-Ala linkage by an oxygen in the D-Ala–D-lactate ester linkage eliminates a hydrogen bond between the NH group of the terminal D-alanine and vancomycin and results in a 1,000-fold-decreased affinity for vancomycin (19). Consequently, such strains are resistant, since vancomycin is no longer able to inhibit peptidoglycan synthesis.

Biochemical analysis of the VanH protein indicated that it produces D-2-hydroxyacid substrates, specifically D-lactate from pyruvate for VanA (19). In addition to VanA and VanH, the VanX protein is necessary for expression of glycopeptide resistance (13). VanX is a DD-dipeptidase that contributes to VanA-mediated resistance by hydrolyzing the dipeptide D-Ala–D-Ala, which would otherwise compete with D-Ala–D-lactate for incorporation into pentapeptide cell wall precursors.

The distal portion of Tn*1546* encodes two proteins, VanY and VanZ, that are not necessary for glycopeptide resistance. The *vanY* gene is required for vancomycin-inducible production of a DD-carboxypeptidase that hydrolyzes peptidoglycan precursors terminated by D-Ala or D-lactate (7). Under certain conditions, VanY may contribute to resistance by hydrolyzing peptidoglycan precursors terminated by D-Ala. The normal sensitive D-Ala-terminated precursors could result from the incorporation of D-Ala–D-Ala that escapes hydrolysis by VanX. As yet, no function has been assigned to the deduced product of the *vanZ* gene, which is not significantly related in structure to other proteins.

Amino acid sequence similarities between VanR and response regulators and between VanS and the histidine protein kinases of the so-called two-component regulatory systems were detected (7). In the VanS-VanR regulatory system, VanS undergoes autophosphorylation in response to an unknown signal triggered by the presence of glycopeptides. Subsequently, VanS transfers the phosphate group to its cognate response regulator, VanR, thereby activating a promoter for cotranscription of *vanH*, *vanA*, *vanX*, and possibly *vanY* (7).

High-level glycopeptide resistance has been detected in strains of *E. faecium*, *E. faecalis*, *Enterococcus avium*, *Enterococcus durans*, *Enterococcus raffinosus*, and *Enterococcus mundtii*. There is no barrier to heterospecific expression of the resistance genes in *Bacillus thuringiensis*, *L. monocytogenes*, *S. aureus*, and *Streptococcus* spp., and dissemination of this type of resistance to other human pathogens should be anticipated. Moreover, glycopeptide resistance has been transferred from *E. faecalis* to *S. aureus* by conjugation under laboratory conditions (137).

The VanB phenotype includes strains *E. faecalis* and *E. faecium*, which have variable-level resistance to vancomycin (MICs, 4 to >1,024 μg/ml) and susceptibility to teicoplanin (152). These strains are inducible by vancomycin only, but induced strains are cross-resistant to vancomycin and teicoplanin. Despite the wide range of MICs of vancomycin, all these strains contain *vanB*-related sequences (152). Similar to *vanA*, *vanB* encodes a ligase of altered specificity, since these strains also contain peptidoglycan precursors terminated by D-lactate (48). Additional nucleotide sequencing indicated that *vanB* is bordered by genes that are very similar to *vanH* and *vanX*. In contrast to *vanA*, the *vanB* gene cluster is located on the chromosome as part of very large (90- to 250-kb) conjugative elements (151). VanB-type resistance in certain strains with low- and high-level resistance is transferable to susceptible enterococci by conjugation (151, 152). A variant of a VanB-type strain constitutively resistant to vancomycin and teicoplanin has occurred under therapy, suggesting that the distinction between the VanA and the VanB phenotypes can be attributed, at least in part, to differences in regulation (78).

Three species of enterococci (*Enterococcus gallinarum*, *Enterococcus casseliflavus*, and *Enterococcus flavescens*) are intrinsically resistant to glycopeptides. These species express constitutive resistance to low levels of vancomycin only (VanC phenotype). The *vanC1* gene of *E. gallinarum* encodes protein VanC, which is also related to D-Ala–D-Ala ligases and appears to catalyze the formation of the novel dipeptide D-Ala–D-serine (65). Among other intrinsically resistant species, *Pediococcus* spp., *Lactobacillus* spp., and *Leuconostoc* spp. contain D-lactate-terminated precursors (65).

RESISTANCE TO RIFAMPIN

Rifampin is particularly active against gram-positive bacteria and mycobacteria by inhibition of DNA-dependent RNA polymerase. Chromosomal mutations resulting in single amino acid changes in the β-subunit of RNA polymerase confer resistance to the drug. Most gram-negative

bacteria are intrinsically resistant owing to decreased uptake of the hydrophobic antibiotic. Purified RNA polymerase from such bacteria is sensitive to rifampin (202).

We thank M. Arthur, B. Berger-Bächi, S. Cole, J. Davies, J.-M. Frère, D. Hooper, S. Levy, R. Leclercq, H. Nikaido, O. Sköld, and B. Spratt for critical review of portions of the manuscript.

REFERENCES

1. **Ambler, R. P.** 1980. The structure of β-lactamases. *Phil. Trans. R. Soc. Lond. (Biol.)* **289:**321–331.
2. **Amyes, S. G. B., and J. T. Smith.** 1974. R-factor trimethoprim resistance mechanism: an insusceptible target site. *Biochem. Biophys. Res. Commun.* **58:**412–418.
3. **Amyes, S. G. B., and K. J. Towner.** 1990. Trimethoprim resistance; epidemiology and molecular aspects. *J. Med. Microbiol.* **31:**1–19.
4. **Arthur, M., D. Autissier, and P. Courvalin.** 1986. Analysis of the nucleotide sequence of the *ereB* gene encoding the erythromycin esterase type II. *Nucleic Acids Res.* **14:**4987–4999.
5. **Arthur, M., A. Brisson-Noël, and P. Courvalin.** 1987. Origin and evolution of genes specifying resistance to macrolide, lincosamide, and streptogramin antibiotics: data and hypotheses. *J. Antimicrob. Chemother.* **20:**783–802.
6. **Arthur, M., and P. Courvalin.** 1986. Contribution of two different mechanisms to erythromycin resistance in *Escherichia coli. Antimicrob. Agents Chemother.* **30:**694–700.
7. **Arthur, M., and P. Courvalin.** 1993. Genetics and mechanisms of glycopeptide resistance in enterococci. *Antimicrob. Agents Chemother.* **37:**1563–1571.
8. **Arthur, M., C. Molinas, F. Depardieu, and P. Courvalin.** 1993. Characterization of Tn*1546*, a Tn3-related transposon conferring glycopeptide resistance by synthesis of depsipeptide peptidoglycan precursors in *Enterococcus faecium* BM4147. *J. Bacteriol.* **175:**117–127.
9. **Avtalion, R. R., R. Ziegler-Schlomowitz, M. Perl, A. Wojdani, and D. Sompolinsky.** 1971. Depressed resistance to tetracycline in *Staphylococcus aureus. Microbios.* **3:**165–180.
10. **Bandoh, K., Y. Muto, K. Watanabe, N. Katoh, and K. Ueno.** 1991. Biochemical properties and purification of metallo-β-lactamase from *Bacteroides fragilis. Antimicrob. Agents Chemother.* **35:**371–372.
11. **Bannam, T. L., and J. I. Rood.** 1991. Relationship between the *Clostridium perfringens catQ* gene product and chloramphenicol acetyltransferases from other bacteria. *Antimicrob. Agents Chemother.* **35:**471–476.
12. **Barthélémy, P., D. Autissier, G. Gerbaud, and P. Courvalin.** 1984. Enzymatic hydrolysis of erythromycin by a strain of *Escherichia coli:* a new mechanism of resistance. *J. Antibiot.* **37:**1692–1696.
13. **Barthélémy, M., J. Péduzzi, and R. Labia.** 1985. Distinction entre les structures primaires des β-lactamases TEM-1 et TEM-2. *Ann. Inst. Pasteur Microbiol.* **136A:**311–321.
14. **Barthélémy, M., J. Péduzzi, H. B. Yaghlane, and R. Labia.** 1988. Single amino acid substitution between SHV-1 β-lactamase and cefotaxime-hydrolyzing SHV-2 enzyme. *FEBS Lett.* **231:**217–220.
15. **Bertram, J., M. Strätz, and P. Dürre.** 1991. Natural transfer of conjugative transposon Tn*916* between gram-positive and gram-negative bacteria. *J. Bacteriol.* **173:**443–448.
16. **Bissonnette, L., S. Champetier, J.-P. Buisson, and P. H. Roy.** 1991. Characterization of the nonenzymatic chloramphenicol resistance (*cmlA*) gene of the In4 integron of Tn*1696*: similarity of the product to transmembrane transport proteins. *J. Bacteriol.* **173:**4493–4502.
17. **Brisson-Noël, A., P. Delrieu, D. Samain, and P. Courvalin.** 1988. Inactivation of lincosamide antibiotics in *Staphylococcus. J. Biol. Chem.* **263:**15880–15887.
18. **Bryan, L. E.** 1989. Aminoglycoside resistance, p. 35–57. *In*

L. E. Bryan (ed.), *Microbial Resistance to Drugs.* Springer-Verlag, Berlin.

19. **Bugg, T. D. H., G. D. Wright, S. Dutka-Malen, M. Arthur, P. Courvalin, and C. T. Walsh.** 1991. Molecular basis for vancomycin resistance in *Enterococcus faecium* BM4147: biosynthesis of a depsipeptide peptidoglycan precursor by vancomycin resistance proteins VanH and VanA. *Biochemistry* **30:**10408–10415.

20. **Burdett, V.** 1991. Purification and characterization of Tet(M), a protein that renders ribosomes resistant to tetracycline. *J. Biol. Chem.* **266:**2872–2877.

21. **Burdett, V.** 1993. tRNA modification activity is necessary for Tet(M)-mediated tetracycline resistance. *J. Bacteriol.* **175:**7209–7215.

22. **Burns, J. L., P. M. Mendelman, J. Levy, T. L. Stull, and A. Smith.** 1985. A permeability barrier as a mechanism of chloramphenicol resistance in *Haemophilus influenzae. Antimicrob. Agents Chemother.* **27:**46–54.

23. **Burns, J. L., C. E. Rubens, P. M. Mendelman, and A. L. Smith.** 1986. Cloning and expression in *Escherichia coli* of a gene encoding nonenzymatic chloramphenicol resistance from *Pseudomonas aeruginosa. Antimicrob. Agents Chemother.* **29:**445–450.

24. **Bush, K.** 1989. Characterization of β-lactamases. *Antimicrob. Agents Chemother.* **33:**259–263.

25. **Buu-Hoi, A., and T. C. Horodniceanu.** 1980. Conjugative transfer of multiple antibiotic resistance markers in *Streptococcus pneumoniae. J. Bacteriol.* **142:**313–320.

26. **Chanal, C., M.-C. Poupart, D. Sirot, R. Labia, J. Sirot, and R. Cluzel.** 1992. Nucleotide sequences of CAZ-2, CAZ-6, and CAZ-7 β-lactamase genes. *Antimicrob. Agents Chemother.* **36:**1817–1820.

27. **Charpentier, E., G. Gerbaud, and P. Courvalin.** 1993. Characterization of a new class of tetracycline-resistance gene *tet*(S) in *Listeria monocytogenes* BM4210. *Gene* **131:**27–34.

28. **Charpentier, E., G. Gerbaud, and P. Courvalin.** Unpublished results.

29. **Cohen, S. P., H. Hächler, and S. B. Levy.** 1993. Genetic and functional analysis of the multiple antibiotic resistance (*mar*) locus in *Escherichia coli. J. Bacteriol.* **175:**1484–1492.

30. **Cohen, S. P., L. M. McMurry, D. C. Hooper, J. S. Wolfson, and S. B. Levy.** 1989. Cross-resistance to fluoroquinolones in multiple-antibiotic-resistant (Mar) *Escherichia coli* selected by tetracycline or chloramphenicol: decreased drug accumulation associated with membrane changes in addition to OmpF reduction. *Antimicrob. Agents Chemother.* **33:**1318–1325.

31. **Cohen, S. P., L. M. McMurry, and S. B. Levy.** 1988. *marA* locus causes decreased expression of OmpF porin in multiple-antibiotic-resistant (Mar) mutants of *Escherichia coli. J. Bacteriol.* **170:**5416–5422.

32. **Collatz, E., G. Tran Van Nhieu, D. Billot-Klein, and L. Gutmann.** 1989. Substitution of serine for arginine in position 162 of TEM-type β-lactamases extends the substrate profile of mutant enzymes, TEM-7 and TEM-101, to ceftazidime and aztreonam. *Gene* **78:**349–354.

33. **Collis, C. M., G. Grammaticopoulos, J. Briton, H. W. Stokes, and R. Hall.** Site-specific insertion of gene cassettes into integrons. *Mol. Microbiol.* **9:**41–52.

34. **Collis, C. M., and R. Hall.** 1992. Site-specific deletion and rearrangement of integron insert genes catalyzed by the integron DNA integrase. *J. Bacteriol.* **174:**1574–1585.

35. **Courvalin, P., and C. Carlier.** 1981. Resistance towards aminoglycoside-aminocyclitol antibiotics in bacteria. *Antimicrob. Agents Chemother.* **8**(Suppl. A):57–69.

36. **Courvalin, P., and C. Carlier.** 1986. Transposable multiple antibiotic resistance in *Streptococcus pneumoniae. Mol. Gen. Genet.* **205:**291–297.

37. **Courvalin, P., and J. Davies.** 1977. Plasmid-mediated aminoglycoside phosphotransferase of broad substrate range that phosphorylates amikacin. *Antimicrob. Agents Chemother.* **11:**619–624.

38. **Courvalin, P., W. V. Shaw, and A. E. Jacob.** 1978. Plasmid-mediated mechanisms of resistance to aminoglycoside-aminocyclitol antibiotics and to chloramphenicol in group D streptococci. *Antimicrob. Agents Chemother.* **13:**716–725.

39. **Davies, J.** 1991. Aminoglycoside-aminocyclitol antibiotics and their modifying enzymes, p. 691–713. *In* V. Lorian (ed.), *Antibiotics in Laboratory Medicine.* The Williams & Wilkins Co., Baltimore.

40. **Davis, B. D.** 1987. Mechanism of bactericidal action of aminoglycosides. *Microbiol. Rev.* **51:**341–350.

41. **Davis, B. D.** 1988. The lethal action of aminoglycosides. *J. Antimicrob. Chemother.* **22:**1–3.

42. **DeGroot, R., J. Campos, S. L. Moseley, and A. L. Smith.** 1988. Molecular cloning and mechanism of resistance in *Haemophilus influenzae. Antimicrob. Agents Chemother.* **32:**477–484.

43. **Dowson, C. G., T. J. Coffey, C. Kell, and R. A. Whiley.** 1993. Evolution of penicillin resistance in *Streptococcus pneumoniae*; the role of *Streptococcus mitis* in the formation of a low affinity PBP2B in *S. pneumoniae. Mol. Microbiol.* **9:**635–643.

44. **Dowson, C. G., A. Hutchison, J. A. Brannigan, R. C. George, D. Hansman, J. Liñares, A. Tomasz, J. M. Smith, and B. G. Spratt.** 1989. Horizontal transfer of penicillin-binding protein genes in penicillin-resistant clinical isolates of *Streptococcus pneumoniae. Proc. Natl. Acad. Sci. USA* **86:**8842–8846.

45. **Dowson, C. G., A. Hutchison, N. Woodford, A. P. Johnson, R. C. George, and B. G. Spratt.** 1990. Penicillin-resistant viridans streptococci have obtained altered penicillin-binding protein genes from penicillin-resistant strains of *Streptococcus pneumoniae. Proc. Natl. Acad. Sci. USA* **87:**5858–5862.

46. **Dutka-Malen, S., C. Molinas, M. Arthur, and P. Courvalin.** 1990. The VanA glycopeptide resistance protein is related to D-alanyl-D-alanine ligase cell wall biosynthesis enzymes. *Mol. Gen. Genet.* **224:**364–372.

47. **Elwell, L. P., and M. E. Fling.** 1989. Resistance to trimethoprim, p. 249–290. *In* L. E. Bryan (ed.), *Microbial Resistance to Drugs.* Springer-Verlag, Berlin.

48. **Evers, S., P. E. Reynolds, and P. Courvalin.** 1994. Sequence of the *vanB* and *ddl* genes encoding D-alanine:D-lactate and D-alanine:D-alanine ligases in vancomycin-resistant *Enterococcus faecalis* V583. *Gene* **140:**97–102.

49. **Faruki, H., and P. F. Sparling.** 1986. Genetics of resistance in a non-β-lactamase-producing gonococcus with relatively high-level penicillin resistance. *Antimicrob. Agents Chemother.* **30:**856–860.

50. **Finken, M., P. Kirschner, A. Meier, A. Wrede, and E. C. Böttger.** 1993. Molecular basis of streptomycin resistance in *Mycobacterium tuberculosis*: alterations of the ribosomal protein S12 gene and point mutations within a functional 16S ribosomal RNA pseudoknot. *Mol. Microbiol.* **9:**1239–1246.

51. **Fitton, J. E., and W. V. Shaw.** 1979. Comparison of chloramphenicol acetyltransferase variants in staphylococci. *Biochem. J.* **177:**575–582.

52. **Flensburg, J., and O. Sköld.** 1984. Regulatory changes in the formation of chromosomal dihydrofolate reductase causing resistance to trimethoprim. *J. Bacteriol.* **159:**184–190.

53. **Fling, M. E., J. Kopf, and C. Richards.** 1988. Characterization of plasmid pAZ1 and the type III dihydrofolate reductase gene. *Plasmid* **19:**30–38.

54. **Fling, M. E., and C. Richards.** 1983. The nucleotide sequence of the trimethoprim-resistant dihydrofolate reductase gene harbored by Tn7. *Nucleic Acids Res.* **11:**5147–5158.

55. **Fling, M. E., L. Walton, and L. P. Elwell.** 1982. Monitoring of plasmid-encoded, trimethoprim-resistant dihydrofolate reductase genes: detection of a new resistant enzyme. *Antimicrob. Agents Chemother.* **22:**882–888.

56. **Fontana, R., A. Grossato, L. Rossi, Y. R. Cheng, and G. Satta.** 1985. Transition from resistance to hypersusceptibility to β-lactam antibiotics associated with loss of a low-affinity penicillin-binding protein in a *Streptococcus faecium* mutant

highly resistant to penicillin. *Antimicrob. Agents Chemother.* **28**:678–683.

57. **Fukuoka, T., S. Ohya, T. Narita, M. Katsuta, M. Iijima, N. Masuda, H. Yasuda, J. Trias, and H. Nikaido.** 1993. Activity of the carbapenem panipenem and role of the OprD (D2) protein in its diffusion through the *Pseudomonas aeruginosa* outer membrane. *Antimicrob. Agents Chemother.* **37**:322–327.

58. **George, A. M., and S. B. Levy.** 1983. Amplifiable resistance to tetracycline, chloramphenicol, and other antibiotics in *Escherichia coli*: involvement of a non-plasmid-determined efflux of tetracycline. *J. Bacteriol.* **155**:531–540.

59. **Georgopapadakou, N. H.** 1993. Penicillin-binding proteins and bacterial resistance to β-lactams. *Antimicrob. Agents Chemother.* **37**:2045–2053.

60. **Georgeopapadakou, N. H., S. A. Smith, and D. P. Bonner.** 1982. Penicillin-binding proteins in a *Staphylococcus aureus* strain resistant to specific β-lactam antibiotics. *Antimicrob. Agents Chemother.* **22**:172–175.

61. **Ghuysen, J.-M.** 1991. Serine β-lactamases and penicillin-binding proteins. *Annu. Rev. Microbiol.* **45**:37–67.

62. **Goswitz, J. J., K. E. Willard, C. E. Fasching, and L. E. Peterson.** 1992. Detection of *gyrA* gene mutations associated with ciprofloxacin resistance in methicillin-resistant *Staphylococcus aureus*: analysis by polymerase chain reaction and automated direct DNA sequencing. *Antimicrob. Agents Chemother.* **36**:1166–1169.

63. **Gotoh, H. K., K. Nunomura, and T. Nishino.** 1990. Resistance of *Pseudomonas aeruginosa* to cefsulodin: modification of penicillin-binding protein 3 and mapping of its chromosomal gene. *J. Antimicrob. Chemother.* **25**:513–523.

64. **Grinsted, J., F. de la Cruz, and R. Schmitt.** 1990. The Tn21 subgroup of bacterial transposable elements. *Plasmid* **24**:163–189.

65. **Gutmann, L., D. Billot-Klein, D. Shlaes, and J. Van Heijenoort.** 1993. Analysis of peptidoglycan precursors in naturally vancomycin-resistant gram-positive organisms, abstr. 115. *Program Abstr. 33rd Intersci. Conf. Antimicrob. Agents Chemother.*

66. **Gutmann, L., B. Ferré, F. W. Goldstein, N. Rizk, E. Pinto-Schuster, J. F. Acar, and E. Collatz.** 1989. SHV-5, a novel SHV-type β-lactamase that hydrolyzes broad-spectrum cephalosporins and monobactams. *Antimicrob. Agents Chemother.* **33**:951–956.

67. **Gutmann, L., R. Williamson, N. Moreau, M. Kitzis, E. Collatz, J. F. Acar, and F. Goldstein.** 1985. Cross-resistance to nalidixic acid, trimethoprim, and chloramphenicol associated with alterations in outer membrane proteins of *Klebsiella*, *Enterobacter*, and *Serratia*. *J. Infect. Dis.* **151**:501–507.

68. **Hächler, H., S. P. Cohen, and S. B. Levy.** 1991. *marA*, a regulated locus which controls expression of chromosomal multiple antibiotic resistance in *Escherichia coli*. *J. Bacteriol.* **173**:5532–5538.

69. **Hakenbeck, R., M. Tarplay, and A. Tomasz.** 1980. Multiple changes of penicillin-binding proteins in penicillin-resistant clinical isolates of *Streptococcus pneumoniae*. *Antimicrob. Agents Chemother.* **17**:364–371.

70. **Hall, L. M. C., D. M. Livermore, D. Gur, M. Akova, and H. E. Akalin.** 1993. OXA-11, an extended-spectrum variant of OXA-10 (PSE-2) β-lactamase from *Pseudomonas aeruginosa*. *Antimicrob. Agents Chemother.* **37**:1637–1644.

71. **Hall, R. M., D. E. Brooks, and H. W. Stokes.** 1991. Site-specific insertion of genes into integrons: role of the 59-base element and determination of the recombination cross-over point. *Mol. Microbiol.* **5**:1941–1959.

72. **Hamilton-Miller, J. M. T.** 1988. Reversal of activity of trimethoprim against gram-positive cocci by thymidine, thymine, and folates. *Antimicrob. Agents Chemother.* **22**:35–39.

73. **Hancock, R. E. W., S. W. Farmer, Z. Li, and K. Poole.** 1991. Interaction of aminoglycosides with the outer membranes and purified lipopolysaccharide and OmpF porin of *Escherichia coli*. *Antimicrob. Agents Chemother.* **35**:1309–1314.

74. **Handwerger, S., and A. Tomasz.** 1986. Alterations in penicillin-binding proteins of clinical and laboratory isolates of pathogenic *Streptococcus pneumoniae* with low levels of penicillin resistance. *J. Infect. Dis.* **153**:83–89.

75. **Handwerger, S. M. J., M. J. Pucci, K. J. Vol, J. Liu, and M. S. Lee.** 1992. The cytoplasmic peptidoglycan precursor of vancomycin-resistant *Enterococcus faecalis* terminates in lactate. *J. Bacteriol.* **174**:5982–5984.

76. **Harder, K. J., H. Nikaido, and M. Matsuhashi.** 1981. Mutants of *Escherichia coli* that are resistant to certain beta-lactam compounds lack the *ompF* porin. *Antimicrob. Agents Chemother.* **20**:549–552.

77. **Hartman, B. J., and A. Tomasz.** 1986. Expression of methicillin resistance in heterogeneous strains of *Staphylococcus aureus*. *Antimicrob. Agents Chemother.* **29**:85–92.

78. **Hayden, M. K., G. M. Trenholme, J. E. Schultz, and D. F. Sahm.** 1993. In vivo development of teicoplanin resistance in a VanB *Enterococcus faecium*. *J. Infect. Dis.* **167**:1224–1227.

79. **Heisig, P., and B. Wiedemann.** 1991. Use of a broad-host-range *gyrA* plasmid for genetic characterization of fluoroquinolone-resistant gram-negative bacteria. *Antimicrob. Agents Chemother.* **35**:2031–2036.

80. **Hewinson, R. G., S. J. Cartwright, M. P. E. Slack, R. D. Whipp, M. J. Woodward, and W. W. Nichols.** 1989. Permeability to cefsulodin of the outer membrane of *Pseudomonas aeruginosa* and discrimination between β-lactamase-mediated trapping and hydrolysis as mechanisms of resistance. *Eur. J. Biochem.* **179**:667–675.

81. **Hillen, W., K. Schollmeir, and C. Gatz.** 1984. Control of expression of the Tn10-encoded tetracycline resistance operon. *J. Mol. Biol.* **172**:185–201.

82. **Hirai, K., H. Aoyama, T. Irikura, S. Iyobe, and S. Mitsuhashi.** 1986. Differences in susceptibility to quinolones of outer membrane mutants of *Salmonella typhimurium* and *Escherichia coli*. *Antimicrob. Agents Chemother.* **29**:535–538.

83. **Hirai, K., S. Suzue, T. Irikura, S. Iyobe, and S. Mitsuhashi.** 1987. Mutations producing resistance in *Pseudomonas aeruginosa*. *Antimicrob. Agents Chemother.* **31**:582–586.

84. **Hooper, D. C., J. S. Wolfson, M. A. Bozza, and E. Y. Ng.** 1992. Genetics and regulation of outer membrane protein expression by quinolone resistance loci *nfxB*, *nfxC*, and *cfxB*. *Antimicrob. Agents Chemother.* **36**:1151–1154.

85. **Hooper, D. C., J. S. Wolfson, E. Y. Ng, and M. N. Swartz.** 1987. Mechanisms of action and resistance to ciprofloxacin. *Am. J. Med.* **82**(Suppl. 4A):12–20.

86. **Hooper, D. C., J. S. Wolfson, K. S. Souza, C. Tung, G. L. McHugh, and M. N. Swartz.** 1986. Genetic and biochemical characterization of norfloxacin resistance in *Escherichia coli*. *Antimicrob. Agents Chemother.* **29**:639–644.

87. **Hopewell, R., M. Oram, R. Briesewitz, and L. M. Fisher.** 1990. DNA cloning and organization of the *Staphylococcus aureus gyrA* and *gyrB* genes: close homology among gyrase proteins and implications for 4-quinolone action and resistance. *J. Bacteriol.* **172**:3481–3484.

88. **Horaud, T., C. Le Bouguénec, and K. Pepper.** 1985. Molecular genetics of resistance to macrolides, lincosamides, and streptogramin B (MLS) in streptococci. *J. Antimicrob. Chemother.* **16**:111–135.

89. **Horinouchi, S., and B. Weisblum.** 1980. Post-transcriptional modification of RNA conformation: mechanism that regulates erythromycin-induced resistance. *Proc. Natl. Acad. Sci. USA* **77**:7079–7083.

90. **Horodniceanu, T., C. Le Bouguénec, A. Buu-Hoi, and G. Bieth.** 1982. Conjugative transfer of antibiotic resistance markers in β-hemolytic streptococci in the presence and absence of plasmid DNA, p. 105–108. *In* D. Schlessinger (ed.), *Microbiology—1982*. American Society for Microbiology, Washington, D.C.

91. **Jaffe, A., Y. A. Chabbert, and O. Semonin.** 1982. Role of porin proteins OmpF and OmpC in the permeation of β-lactams. *Antimicrob. Agents Chemother.* **22**:942–948.

92. **Jaurin, B., and T. Grundstrom.** 1981. AmpC cephalospori-

nase of *Escherichia coli* K-12 has a different evolutionary origin from that of β-lactamases of the penicillinase type. *Proc. Natl. Acad. Sci. USA* **78:**4897–4901.

93. **Joris, B., J.-M. Ghuysen, G. Dive, A. Renard, O. Dideberg, P. Charlier, J.-M. Frere, J. A. Kelly, J. C. Boyington, P. C. Moews, and J. R. Knox.** 1988. The active-site serine penicillin-recognizing enzymes as members of the *Streptomyces* R61 D,D-peptidase family. *Biochem. J.* **250:**313–324.

94. **Kaatz, G. W., S. M. Seo, and C. A. Ruble.** 1993. Efflux-mediated fluoroquinolone resistance in *Staphylococcus aureus. Antimicrob. Agents Chemother.* **37:**1086–1094.

95. **Kadurugamuwa, J. L., A. J. Clarke, and T. J. Beveridge.** 1993. Surface action of gentamicin on *Pseudomonas aeruginosa. J. Bacteriol.* **175:**5798–5805.

96. **Kaneko, M., A. Yamaguchi, and T. Sawai.** 1985. Energetics of tetracycline efflux system encoded by Tn*10* in *Escherichia coli. FEBS Lett.* **193:**194–198.

97. **Kato, T., Y. Sato, S. Iyobe, and S. Mitsuhashi.** 1982. Plasmid-mediated gentimicin resistance of *Pseudomonas aeruginosa* and its lack of expression in *Escherichia coli. Antimicrob. Agents Chemother.* **22:**358–363.

98. **Kleanthous, C., P. M. Cullis, and W. V. Shaw.** 1985. 3-(Bromoacetyl)chloramphenicol, an active site directed inhibitor for chloramphenicol acetyltransferase. *Biochemistry* **24:**5307–5313.

99. **Klock, G., B. Unger, C. Gatz, W. Hillen, J. Altenbuchner, K. Schmid, and R. Schmitt.** 1985. Heterologous repressor-operator recognition among four classes of tetracycline resistance determinants. *J. Bacteriol.* **161:**326–332.

100. **Knapp, J., S. R. Johnson, J. M. Zenilman, M. C. Roberts, and S. A. Morse.** 1988. High-level tetracycline resistance resulting from TetM in strains of *Neisseria* spp., *Kingella denitrificans,* and *Eikenella corrodens. Antimicrob. Agents Chemother.* **32:**765–767.

101. **Lai, C. J., and B. Weisblum.** 1971. Altered methylation of ribosomal RNA in an erythromycin-resistant strain of *Staphylococcus aureus. Proc. Natl. Acad. Sci. USA* **68:**856–860.

102. **Laible, G., and R. Hakenbeck.** 1987. Penicillin-binding proteins in β-lactam-resistant laboratory mutants of *Streptococcus pneumoniae. Mol. Microbiol.* **1:**355–363.

103. **Laible, G., B. G. Spratt, and R. Hakenbeck.** 1991. Interspecies recombinational events during the evolution of altered PBP 2× genes in penicillin-resistant clinical isolates of *Streptococcus pneumoniae. Mol. Microbiol.* **5:**1993–2002.

104. **Lampson, B. C., and J. T. Parisi.** 1986. Naturally occurring *Staphylococcus epidermidis* plasmid expressing constitutive macrolide-lincosamide-streptogramin B resistance contains a deleted attenuator. *J. Bacteriol.* **166:**479–483.

105. **Leclercq, R., and P. Courvalin.** 1991. Bacterial resistance to macrolide, lincosamide, and streptogramin antibiotics by target modification. *Antimicrob. Agents Chemother.* **35:**1267–1272.

106. **Leclercq, R., E. Derlot, J. Duval, and P. Courvalin.** 1988. Plasmid-mediated resistance to vancomycin and teicoplanin in *Enterococcus faecium. N. Engl. J. Med.* **319:**157–161.

107. **Leclercq, R., S. Dutka-Malen, A. Brisson-Noël, C. Molinas, E. Derlot, M. Arthur, J. Duval, and P. Courvalin.** 1992. Resistance of enterococci to aminoglycosides and glycopeptides. *Clin. Infect. Dis.* **15:**495–501.

108. **Le Grice, S. F., H. Matzura, R. Marcoli, S. Iida, and T. A. Bickle.** 1982. The catabolite-sensitive promoter for the chloramphenicol acetyl transferase gene is preceded by two binding sites for the catabolite gene activator protein. *J. Bacteriol.* **150:**312–318.

109. **Levy, S. B.** 1984. Resistance to the tetracyclines, p. 191–240. *In* L. E. Bryan (ed.), *Antimicrobial Drug Resistance.* Academic Press, Orlando, Fla.

110. **Levy, S. B.** Personal communication.

111. **Lindberg, F., S. Lindquist, and S. Normark.** 1988. Genetic basis of induction and overproduction of chromosomal class I β-lactamase in nonfastidious gram-negative bacilli. *Rev. Infect. Dis.* **10:**782–785.

112. **Lindberg, F., L. Westman, and S. Normark.** 1985. Regula-

tory components in *Citrobacter freundii ampC* β-lactamase induction. *Proc. Natl. Acad. Sci. USA* **82:**4620–4624.

113. **Lindquist, S., F. Lindberg, and S. Normark.** 1989. Binding of *Citrobacter freundii* AmpR regulator to a single DNA site provides both autoregulation and activation of the inducible *ampC* β-lactamase gene. *J. Bacteriol.* **171:**3746–3753.

114. **Lindquist, S., K. Weston-Hafer, H. Schmidt, C. Pul, G. Korfmann, J. Erickson, C. Sanders, H. H. Martin, and S. Normark.** 1993. AmpG, a signal transducer in chromosomal β-lactamase induction. *Mol. Microbiol.* **9:**703–715.

115. **Livermore, D. M.** 1988. Permeation of β-lactam antibiotics into *Escherichia coli, Pseudomonas aeruginosa,* and other gram-negative bacteria. *Rev. Infect. Dis.* **10:**691–698.

116. **Lovett, P. S.** 1990. Translational attenuation as the regulator of inducible *cat* genes. *J. Bacteriol.* **172:**1–6.

117. **Maidhof, H., B. Reinicke, P. Blümel, B. Berger-Bächi, and H. Labischinski.** 1991. *femA,* which encodes a factor essential for expression of methicillin resistance, affects glycine content of peptidoglycan in methicillin-resistant and methicillin-susceptible *Staphylococcus aureus* strains. *J. Bacteriol.* **173:**3507–3513.

118. **Manavathu, E. K., C. L. Fernandez, B. S. Cooperman, and D. E. Taylor.** 1990. Molecular studies on the mechanism of tetracycline resistance mediated by Tet(O). *Antimicrob. Agents Chemother.* **34:**71–77.

119. **Martin, C., T. Briese, and R. Hakenbeck.** 1992. Nucleotide sequences of genes encoding penicillin-binding proteins from *Streptococcus pneumoniae* and *Streptococcus oralis* with high homology to *Escherichia coli* penicillin-binding proteins 1A and 1B. *J. Bacteriol.* **174:**4517–4523.

120. **Marty, V., V. S. Madiraju, D. P. Brunner, and B. J. Wilkinson.** 1987. Effects of temperature, NaCl, and methicillin on penicillin-binding proteins, growth, peptidoglycan synthesis, and autolysis in methicillin-resistant *Staphylococcus aureus. Antimicrob. Agents Chemother.* **31:**1727–1733.

121. **McCuen, R. W., and F. M. Sirotnak.** 1974. Hyperproduction of dihydrofolate reductase in *Diplococcus pneumoniae* by mutation in the structural gene. *Biochim. Biophys. Acta* **338:**540–544.

122. **McMurray, L. M., A. M. George, and S. B. Levy.** 1994. Active efflux of chloramphenicol in susceptible *Escherichia coli* strains and in multiple-antibiotic-resistant (Mar) mutants. *Antimicrob. Agents Chemother.* **38:**542–546.

123. **McMurray, L. M., R. E. Petrucci, Jr., and S. B. Levy.** 1980. Active efflux of tetracycline encoded by four genetically different tetracycline resistance determinants in *Escherichia coli. Proc. Natl. Acad. Sci. USA* **77:**3974–3977.

124. **Mendelman, P. M., D. O. Chaffin, and G. Kalaitzoglou.** 1990. Penicillin-binding proteins and ampicillin resistance in *Haemophilus influenzae. J. Antimicrob. Chemother.* **25:**525–534.

125. **Michéa-Hamzehpour, M., Y. X. Furet, and J. C. Pechère.** 1991. Role of protein D2 and lipopolysaccharide in diffusion of quinolones through the outer membrane of *Pseudomonas aeruginosa. Antimicrob. Agents Chemother.* **35:**2091–2097.

126. **Muñoz, R., C. G. Dowson, M. Daniels, T. J. Coffey, C. Martin, R. Hakenbeck, and B. G. Spratt.** 1992. Genetics of resistance to third-generation cephalosporins in clinical isolates of *Streptococcus pneumoniae. Mol. Microbiol.* **6:**2461–2465.

127. **Nagarajan, R.** 1991. Antibacterial activities and modes of action of vancomycin and related glycopeptides. *Antimicrob. Agents Chemother.* **35:**605–609.

128. **Nakamura, S., M. Nakamura, T. Kojima, and H. Yoshida.** 1989. *gyrA* and *gyrB* mutations in quinolone-resistant strains of *Escherichia coli. Antimicrob. Agents Chemother.* **33:**254–255.

129. **Nakamura, S., H. Yoshida, M. Bogaki, M. Nakmura, and T. Kojima.** 1993. Quinolone resistance mutations in DNA gyrase, p. 135–143. *In* T. Andoh, H. Ikeda, and M. Oguro (ed.), *Molecular Biology of DNA Topoisomerases and Its Application to Chemotherapy.* CRC Press, London.

130. **Nakanishi, N., S. Yoshida, H. Wakebe, M. Inoue, and S.**

Mitsuhashi. 1991. Mechanisms of clinical resistance to fluoroquinolones in *Enterococcus faecalis*. *Antimicrob. Agents Chemother.* **35:**1053–1059.

131. **Nayler, J. H. C.** 1987. Resistance to β-lactams in Gram-negative bacteria: relative contributions of β-lactamase and permeability limitations. *J. Antimicrob. Chemother.* **19:**713–732.

132. **Neyfakh, A. A.** 1992. The multidrug efflux transporter of *Bacillus subtilis* is a structural and functional homolog of the *Staphylococcus* NorA protein. *Antimicrob. Agents Chemother.* **36:**484–485.

133. **Nicholas, R. A., and J. Strominger.** 1988. Relations between β-lactamases and penicillin-binding proteins: β-lactamase activity of penicillin-binding protein 5 from *Escherichia coli. Rev. Infect. Dis.* **10:**733–738.

134. **Nikaido, H.** 1994. Prevention of drug access to bacterial targets: role of permeability barriers and active efflux. *Science* **264:**382–388.

135. **Nikaido, H., and T. Nakae.** 1979. The outer membrane of gram-negative bacteria. *Adv. Microb. Physiol.* **20:**163–250.

136. **Nikolich, M. P., N. B. Shoemaker, and A. Salyers.** 1992. A *Bacteroides* tetracycline resistance gene represents a new class of ribosome protection tetracycline resistance. *Antimicrob. Agents Chemother.* **36:**1005–1012.

137. **Noble, W. C., Z. Virani, and R. G. A. Cree.** 1992. Cotransfer of vancomycin and other resistance genes from *Enterococcus faecalis* NCTC 12201 to *Staphylococcus aureus. FEMS Microbiol. Lett.* **93:**195–198.

138. **Normark, S., T. Edlund, T. Grundström, S. Bergström, and H. Wolf-Watz.** 1977. *Escherichia coli* K-12 mutants hyperproducing chromosomal beta-lactamase by gene repetitions. *J. Bacteriol.* **132:**912–922.

139. **O'Hara, K., T. Kanda, K. Ohmiya, T. Ebisu, and M. Kono.** 1989. Purification and characterization of macrolide 2'-phosphotransferase from a strain of *Escherichia coli* that is highly resistant to erythromycin. *Antimicrob. Agents Chemother.* **33:**1354–1357.

140. **Olsson, O., S. Bergström, F. P. Lindberg, and S. Normark.** 1983. *ampC* β-lactamase hyperproduction in *Escherichia coli*: natural ampicillin resistance generated by horizontal chromosomal DNA transfer from *Shigella. Proc. Natl. Acad. Sci. USA* **80:**7556–7560.

141. **Oram, M., and M. Fisher.** 1991. 4-Quinolone resistance mutations in the DNA gyrase of *Escherichia coli* clinical isolates identified by using the polymerase chain reaction. *Antimicrob. Agents Chemother.* **35:**387–389.

142. **Ounissi, H., and P. Courvalin.** 1985. Nucleotide sequence of the gene *ereA* encoding the erythromycin esterase in *Escherichia coli. Gene* **35:**271–278.

143. **Papanicolaou, G. A., A. A. Medieros, and G. A. Jacoby.** 1990. Novel plasmid-mediated β-lactamase (MIR-1) conferring resistance to oxyimino- and α-methoxy β-lactams in clinical isolates of *Klebsiella pneumoniae. Antimicrob. Agents Chemother.* **34:**2200–2209.

144. **Parent, R., and P. H. Roy.** 1992. The chloramphenicol acetyltransferase gene of Tn*2424*: a new breed of *cat. J. Bacteriol.* **174:**2891–2897.

145. **Parsons, Y., R. M. Hall, and H. W. Stokes.** 1991. A new trimethoprim resistance gene, *dhfrX*, in the In7 integron of plasmid pDGO100. *Antimicrob. Agents Chemother.* **35:**2436–2439.

146. **Pattishall, K. H., J. Acar, J. J. Burchall, F. W. Goldstein, and R. J. Harvey.** 1977. Two distinct types of trimethoprim-resistant dihydrofolate reductases specified by R plasmids of different compatibility groups. *J. Biol. Chem.* **252:**2319–2323.

147. **Payne, D. J., N. Woodruff, and S. G. B. Amyes.** 1992. Characterization of the plasmid-mediated β-lactamase BIL-1. *J. Antimicrob. Chemother.* **30:**119–127.

148. **Pestka, S.** 1974. Binding of ^{14}C-erythromycin to *Escherichia coli* ribosomes. *Antimicrob. Agents Chemother.* **6:**474–478.

149. **Peterson, A. A., S. W. Fesik, and E. J. McGroaty.** 1987. Decreased binding of antibiotics to lipopolysaccharides from polymyxin-resistant strains of *Escherichia coli* and *Salmonella typhimurium. Antimicrob. Agents Chemother.* **31:**230–237.

150. **Piddock, L. J. V., D. J. Griggs, M. C. Hall, and Y. F. Yin.** 1993. Ciprofloxacin resistance in clinical isolates of *Salmonella typhimurium* obtained from two patients. *Antimicrob. Agents Chemother.* **37:**662–666.

151. **Quintiliani, R., Jr., and P. Courvalin.** 1994. Conjugal transfer of the vancomycin resistance determinant *vanB* between enterococci involves the movement of large genetic elements from chromosome to chromosome. *FEMS Microbiol. Lett.* **119:**359–364.

152. **Quintiliani, R., Jr., S. Evers, and P. Courvalin.** 1993. The *vanB* gene confers various levels of self-transferable resistance to vancomycin in enterococci. *J. Infect. Dis.* **167:**1220–1223.

153. **Radström, P., and G. Swedberg.** 1988. RSF1010 and a conjugative plasmid contain *sulII*, one of two known genes for plasmid-borne sulfonamide resistance dihydropteroate synthase. *Antimicrob. Agents Chemother.* **32:**1684–1692.

154. **Radström, P., G. Swedberg, and O. Sköld.** 1991. Genetic analysis of sulfonamide resistance and its dissemination in gram-negative bacteria illustrate new aspects of R plasmid evolution. *Antimicrob. Agents Chemother.* **35:**1840–1848.

155. **Rather, P. N., E. Orosz, K. J. Shaw, R. Hare, and G. Miller.** 1993. Characterization and transcriptional regulation of the 2'-N-acetyltransferase gene from *Providencia stuartii. J. Bacteriol.* **175:**6492–6498.

156. **Reynolds, P. E.** 1988. The essential nature of staphylococcal penicillin-binding proteins, p. 343–351. *In* P. Actor, L. Daneo-Moore, M. L. Higgins, M. R. J. Salton, and G. D. Shockman (ed.), *Antibiotic Inhibition of Bacterial Cell Surface Assembly and Function.* American Society for Microbiology, Washington, D.C.

157. **Roberts, M., A. Corney, and W. Shaw.** 1982. Molecular characterization of three chloramphenicol acetyltransferases isolated from *Haemophilus influenzae. J. Bacteriol.* **151:**737–741.

158. **Rollins, L. D., L. N. Lee, and D. J. LeBlanc.** 1985. Evidence for a disseminated erythromycin resistance determinant mediated by Tn*917*-like sequences among group D streptococci isolated from pigs, chickens, and humans. *Antimicrob. Agents Chemother.* **27:**439–444.

159. **Rood, J. I., S. Jefferson, T. I. Bannan, J. M. Wilkie, P. Mullany, and B. W. Wren.** 1989. Hybridization analysis of three chloramphenicol resistance determinants from *Clostridium perfringens* and *Clostridium difficile. Antimicrob. Agents Chemother.* **33:**1569–1574.

160. **Ross, J. I., E. A. Eady, J. H. Cove, W. J. Cunliffe, S. Baumberg, and J. C. Wootton.** 1990. Inducible erythromycin resistance in staphylococci is encoded by a member of the ATP-binding transport super-gene family. *Mol. Microbiol.* **4:**1207–1214.

161. **Rowland, S.-J., and K. G. H. Dyke.** 1990. Tn*552*, a novel transposable element from *Staphylococcus aureus. Mol. Microbiol.* **4:**961–975.

162. **Ryffel, C., F. H. Kayser, and B. Berger-Bächi.** 1992. Correlation between regulation of *mecA* transcription and expression of methicillin resistance in staphylococci. *Antimicrob. Agents Chemother.* **36:**25–31.

163. **Ryffel, C., W. Tesch, I. Birch-Machin, P. E. Reynolds, L. Barberis-Maino, F. H. Kayser, and B. Berger-Bächi.** 1990. Sequence comparison of *mecA* genes isolated from methicillin-resistant *Staphylococcus aureus* and *Staphylococcus epidermidis. Gene* **94:**137–138.

164. **Sanders, C. C.** 1987. Chromosomal cephalosporinases responsible for multiple resistance to newer β-lactam antibiotics. *Annu. Rev. Microbiol.* **41:**573–593.

165. **Sanders, C. C.** 1989. The chromosomal beta-lactamases, p. 129–144. *In* L. E. Bryan (ed.), *Microbial Resistance to Drugs.* Springer-Verlag, Berlin.

166. **Sawai, T., R. Hiruma, N. Kawana, M. Kaneko, F. Taniyasu, and A. Inami.** 1982. Outer membrane permeation of β-lactam antibiotics in *Escherichia coli, Proteus mirabilis,* and

Enterobacter cloacae. Antimicrob. Agents Chemother. **22:**585–592.

167. **Shaw, K. J., P. N. Rather, R. S. Hare, and G. H. Miller.** 1993. Molecular genetics of aminoglycoside genes and familial relationships of the aminoglycoside-modifying enzymes. *Microbiol. Rev.* **57:**138–163.

168. **Shaw, W. V.** 1983. Chloramphenicol acetyltransferase: enzymology and molecular biology. *Crit. Rev. Biochem.* **14:**1–46.

169. **Sheridan, R. P., and I. Chopra.** 1991. Origin of tetracycline efflux proteins: conclusions from nucleotide sequence analysis. *Mol. Microbiol.* **5:**895–900.

170. **Shoemaker, N. B., R. D. Barker, and A. Salyers.** 1989. Cloning and characterization of a *Bacteroides* conjugal tetracycline-erythromycin resistance element by using a shuttle cosmid vector. *J. Bacteriol.* **171:**1294–1302.

171. **Showsh, S., and R. Andrews, Jr.** 1992. Tetracycline enhances Tn*916*-mediated conjugal transfer. *Plasmid* **28:**213–224.

172. **Skinner, R., E. Cundliffe, and F. J. Schmidt.** 1983. Site of action of a ribosomal RNA methylase responsible for resistance to erythromycin and other antibiotics. *J. Biol. Chem.* **258:**12702–12706.

173. **Sloan, J., L. M. McMurray, D. Lyras, S. B. Levy, and J. I. Rood.** 1994. The *Clostridium perfringens* TetP determinant comprises two overlapping genes: *tet*A(P) which mediates active tetracycline efflux and *tet*B(P) which is related to the ribosomal protection family of tetracycline resistance determinants. *Mol. Microbiol.* **11:**403–415.

174. **Sougakoff, W., S. Goussard, G. Gerbaud, and P. Courvalin.** 1988. Plasmid-mediated resistance to third-generation cephalosporins caused by point mutations in TEM-type penicillinase genes. *Rev. Infect. Dis.* **10:**879–884.

175. **Speer, B. S., N. B. Shoemaker, and A. A. Salyers.** 1992. Bacterial resistance to tetracycline: mechanisms, transfer, and clinical significance. *Clin. Microbiol. Rev.* **5:**387–399.

176. **Spratt, B. G.** 1988. Hybrid penicillin-binding proteins in penicillin-resistant isolates of *Neisseria gonorrhoeae. Nature* (London) **332:**173–176.

177. **Spratt, B. G.** 1989. Resistance to β-lactam antibiotics mediated by alterations of penicillin-binding proteins, p. 77–97. In L. E. Bryan (ed.), *Microbial Resistance to Drugs.* Springer-Verlag, Berlin.

178. **Spratt, B. G., Q. Zhang, D. M. Jones, A. Hutchinson, J. A. Brannigan, and C. G. Dowson.** 1989. Recruitment of a penicillin-binding protein gene from *Neisseria flavescens* during the emergence of penicillin resistance in *Neisseria meningitidis. Proc. Natl. Acad. Sci. USA* **86:**8988–8992.

179. **Sugino, A., C. L. Peebles, K. N. Kreuzer, and N. R. Cozzarelli.** 1977. Mechanism of action of nalidixic acid: purification of *Escherichia coli nal*A gene product and its relationship to DNA gyrase and a novel nicking-closing enzyme. *Proc. Natl. Acad. Sci. USA* **74:**4767–4771.

180. **Sundström, L., P. Radström, G. Swedberg, and O. Sköld.** 1988. Site-specific recombination promotes linkage between trimethoprim- and sulfonamide-resistance genes. Sequence characterization of *dhfr*V and *sul*I and a recombination active locus of Tn*21. Mol. Gen. Genet.* **213:**191–201.

181. **Sundström, L., and O. Sköld.** 1990. The *dhfr*I trimethoprim resistance gene of Tn*7* can be found at specific sites in other genetic surroundings. *Antimicrob. Agents Chemother.* **34:**642–650.

182. **Sutcliffe, J. A., T. D. Gootz, and J. F. Barrett.** 1989. Biochemical characteristics and physiological significance of major DNA topoisomerases. *Antimicrob. Agents Chemother.* **33:**2027–2033.

183. **Suzuki, E., K. Kuwahara-Arai, J. F. Richardson, and K. Hiramatsu.** 1993. Distribution of *mec* regulator genes in methicillin-resistant *Staphylococcus* clinical isolates. *Antimicrob. Agents Chemother.* **37:**1219–1226.

184. **Sykes, R. B., and M. Matthew.** 1976. The β-lactamases of gram-negative bacteria and their role in resistance to β-lactam antibiotics. *J. Antimicrob. Chemother.* **2:**115–157.

185. **Tennent, J. M., H. Young, B. R. Lyon, S. G. B. Amyes, and R. A. Skurray.** 1988. Trimethoprim resistance determinants encoding a dihydrofolate reductase in clinical isolates of *Staphylococcus aureus* and coagulase-negative staphylococci. *J. Med. Microbiol.* **26:**67–73.

186. **Tesch, W., C. Ryffel, A. Strässle, F. H. Kayser, and B. Berger-Bächi.** 1990. Evidence of a novel staphylococcal *mec*-encoded element (*mecR*) controlling expression of penicillin-binding protein PBP2′. *Antimicrob. Agents Chemother.* **34:**1703–1706.

187. **Thakker-Varia, S., A. C. Ranzini, and D. T. Dubin.** 1985. Ribosomal RNA methylation in *Staphylococcus aureus* and *Escherichia coli:* effect of the MLS (erythromycin resistance) methylase. *Plasmid* **14:**152–161.

188. **Then, R. L.** 1982. Mechanisms of resistance to trimethoprim, the sulfonamides, and trimethoprim-sulfamethoxazole. *Rev. Infect. Dis.* **4:**261–269.

189. **Then, R. L.** 1989. Resistance to sulfonamides, p. 291–312. In L. E. Bryan (ed.), *Microbial Resistance to Drugs.* Springer-Verlag, Berlin.

190. **Then, R. L., and P. Angehrn.** 1979. Low trimethoprim susceptibility of anaerobic bacteria due to insensitive dihydrofolate reductases. *Antimicrob. Agents Chemother.* **15:**1–6.

191. **Then, R. L., and P. Angehrn.** 1982. Trapping of nonhydrolyzable cephalosporins by cephalosporinases in *Enterobacter cloacae* and *Pseudomonas aeruginosa* as a possible resistance mechanism. *Antimicrob. Agents Chemother.* **21:**711–721.

192. **Tomasz, A., H. B. Drugeon, H. M. de Lencastre, D. Jabes, L. McDougall, and J. Bille.** 1989. New mechanism for methicillin resistance in *Staphylococcus aureus:* clinical isolates that lack the PBP2a gene and contain normal penicillin-binding proteins with modified penicillin-binding capacity. *Antimicrob. Agents Chemother.* **33:**1869–1874.

193. **Tomich, P. K., F. Y. An, and D. B. Clewell.** 1980. Properties of erythromycin-inducible transposon Tn*917* in *Streptococcus faecalis. J. Bacteriol.* **141:**1366–1374.

194. **Toro, C. S., S. R. Lobos, I. Calderón, M. Rodríguez, and G. C. Mora.** 1990. Clinical isolate of a porinless *Salmonella typhi* resistant to high levels of chloramphenicol. *Antimicrob. Agents Chemother.* **34:**1715–1719.

195. **Tovar, K., A. Ernest, and W. Hillen.** 1988. Identification and nucleotide sequence of the class E *tet* regulatory elements and operator and inducer binding of the encoded purified Tet repressor. *Mol. Gen. Genet.* **215:**76–80.

196. **Trias, J., J. Dufresne, R. C. Levesque, and H. Nikaido.** 1989. Decreased outer membrane permeability in imipenem-resistant mutants of *Pseudomonas aeruginosa. Antimicrob. Agents Chemother.* **33:**1201–1206.

197. **Tuomanen, E., S. Lindquist, S. Sande, M. Galleni, K. Light, D. Gage, and S. Normark.** 1991. Co-ordinate regulation of β-lactamase induction and peptidoglycan composition by the *amp* operon. *Science* **251:**201–204.

198. **Ubukata, K., R. Nonoguchi, M. D. Song, M. Matsuhashi, and M. Konno.** 1990. Homology of *mec*A gene in methicillin-resistant *Staphylococcus haemolyticus* and *Staphylococcus simulans* to that of *Staphylococcus aureus. Antimicrob. Agents Chemother.* **34:**170–172.

199. **Walsh, C. T.** 1989. Enzymes in the D-alanine branch of bacterial cell wall peptidoglycan assembly. *J. Biol. Chem.* **264:**2393–2396.

200. **Wang, Y., W. M. Huang, and D. E. Taylor.** 1993. Cloning and nucleotide sequence of the *Campylobacter jejuni gyr*A gene and characterization of quinolone resistance mutations. *Antimicrob. Agents Chemother.* **37:**457–463.

201. **Wang, Y., and D. E. Taylor.** 1991. A DNA sequence upstream of the *tet*(O) gene is required for full expression of tetracycline resistance. *Antimicrob. Agents Chemother.* **35:** 2020–2025.

202. **Wehrli, W.** 1983. Rifampin: mechanisms of action and resistance. *Rev. Infect. Dis.* **5:**S407–S411.

203. **Weisblum, B.** 1985. Inducible resistance to macrolides, lincosamides and streptogramin type B antibiotics: the resis-

tance phenotype, its biological diversity, and structural elements that regulate expression—a review. *J. Antimicrob. Chemother.* **16**(Suppl. A):63–90.

204. **Williamson, R., C. Le Bouguénec, L. Gutmann, and T. Horaud.** 1985. One or two low affinity penicillin-binding proteins may be responsible for the range of susceptibility of *Enterococcus faecium* to benzylpenicillin. *J. Gen. Microbiol.* **131**:1933–1940.

205. **Yamagishi, J., H. Yoshida, M. Yamayoshi, and S. Nakamura.** 1986. Nalidixic acid-resistant mutations of the *gyrB* gene of *Escherichia coli*. *Mol. Gen. Genet.* **204**:367–373.

206. **Yoshida, H., M. Bogaki, M. Nakamura, and S. Nakamura.** 1990. Quinolone resistance-determining region in the DNA gyrase *gyrA* gene of *Escherichia coli*. *Antimicrob. Agents Chemother.* **34**:1271–1272.

207. **Yoshida, H., M. Bogaki, S. Nakamura, K. Ubukata, and M. Konno.** 1990. Nucleotide sequence and characterization of the *Staphylococcus aureus norA* gene, which confers resistance to quinolones. *J. Bacteriol.* **172**:6942–6949.

208. **Young, H., and S. G. B. Amyes.** 1986. A new mechanism of plasmid trimethoprim resistance. Characterization of an inducible dihydrofolate reductase. *J. Biol. Chem.* **261**:2503–2505.

209. **Young, H., R. A. Skurray, and S. G. B. Amyes.** 1987. Plasmid-mediated trimethoprim resistance in *Staphylococcus aureus*: characterization of the first gram-positive plasmid dihydrofolate reductase (type S1). *Biochem. J.* **243**:309–312.

210. **Zervos, M. J., and D. R. Schaberg.** 1985. Reversal of the in vitro susceptibility of enterococci to trimethoprim-sulfamethoxazole by folinic acid. *Antimicrob. Agents Chemother.* **28**:446–448.

211. **Zilhao, R., B. Papadopoulou, and P. Courvalin.** 1988. Occurrence of the *Campylobacter* resistance gene *tet*(O) in *Enterococcus* and *Streptococcus* spp. *Antimicrob. Agents Chemother.* **32**:1793–1796.

Antibacterial Susceptibility Tests: Dilution and Disk Diffusion Methods

GAIL L. WOODS AND JOHN A. WASHINGTON

113

Antibacterial susceptibility testing may be performed by either dilution or diffusion methods. The choice of methodology is often based on many factors, including relative ease of performance, flexibility, use of automated or semiautomated devices for both identification and susceptibility testing, and, in some instances, misconceptions regarding the clinical importance of MICs versus the interpretive results that can be provided by the disk diffusion method. These misconceptions are often based on an assumption that the dilution test is the more accurate of the two methods and that physicians prefer a quantitative result. Since there is a direct relationship between the zone of inhibition diameter and the MIC and since MICs and zone of inhibition diameters with reference strains are reproducible on an interlaboratory and intralaboratory basis, there is no evidence to support the impression that one method is more accurate than the other. The fact is that a numerical or quantitative result reported from a conventional dilution test should not be misconstrued as representing greater accuracy, since the actual MIC may be a concentration somewhere between the concentration inhibiting the organism and the next lowest concentration tested (i.e., 1 $\log_2$ dilution lower). For example, an organism may grow in the presence of an antimicrobial agent at a concentration of 4 μg/ml and be inhibited at the next higher concentration tested, 8 μg/ml. The actual MIC could be anywhere between 4.1 and 8 μg/ml. Moreover, few physicians other than those specializing in the field of infectious diseases can interpret MICs, so the laboratory is obliged to provide an interpretation of MICs anyway to avoid any risk of their misinterpretation. Specific indications for performing dilution tests are not well agreed upon, but they may be limited to isolates from patients with endocarditis, osteomyelitis, and meningitis. MICs can be useful for evaluating relative degrees of susceptibility of bacteria to various antimicrobial agents and for comparing the relative activities of drugs against various species. Finally, MICs can be useful in the delineation of relative resistance and the intermediate category. In general, however, the decision as to whether to perform dilution or disk diffusion testing is usually based on logistical reasons, such as the laboratory's selection of a system for providing both identification and susceptibility testing. Many such systems are now commercially available; however, the major challenge posed by such systems is the flexibility available in the panel of antimicrobial agents to

be tested. Limited flexibility of commercially available products has led many laboratories back to the use of disk diffusion testing, which is highly flexible in this regard.

The selection of antibacterial agents is complicated by the sheer number of agents available today. Many of these compounds, however, exhibit similar if not identical activities in vitro, so that one compound representing the particular spectrum class of agents can be selected for testing purposes. It is important in this regard that the microbiologist work with the hospital formulary committee to ensure that the antibacterial agents being tested in the laboratory reflect those in the hospital's formulary. Failure to do so can contribute to antibiotic misuse (29). Guidelines for the selection of antibacterial agents to be tested have been published by the National Committee for Clinical Laboratory Standards (NCCLS) (Table 1) (24, 25). While this listing is sometimes regarded as the standard for selection of those agents to be tested, it should be emphasized that it is a guideline only and that many variables go into the decision as to what agents should be tested in any particular setting. The NCCLS also cautions that the decision as to which agents should be tested and reported on selectively should be made by the clinical microbiologist in conjunction with the infectious disease practitioner, the pharmacy, and/or the infection control committee. (For further discussion of the selection of antimicrobial agents to be tested, see chapter 110 of this Manual.)

DILUTION METHODS

Dilution susceptibility testing methods are used to determine the minimal concentration, usually expressed in micrograms per milliliter, of antimicrobial agents required to inhibit or kill a microorganism. Procedures for determining antimicrobial inhibitory activity are carried out by either agar- or broth-based methods. Antimicrobial agents are usually tested at $\log_2$ (twofold) serial dilutions, and the lowest concentration that inhibits visible growth of an organism is recorded as the MIC. The concentration range used may vary with the drug, organism identification, and site of infection. Ranges should include concentrations that allow determinations of the interpretive categories (i.e., susceptible, intermediate, and resistant) and also the acceptable ranges for quality control reference strains. Other dilution methods include those that test a single or a

TABLE 1 Antimicrobial agents recommended for routine dilution and disk diffusion susceptibility testing[a]

Antimicrobial agent	Group[b] recommended for testing:				
	Enterobacteriaceae	Pseudomonas spp. and other non-Enterobacteriaceae	Staphylococci	Enterococci	Streptococci other than S. pneumoniae
Penicillins					
Penicillin G			A[c]	A[c,d]	A
Ampicillin	A			A[c]	A
Oxacillin or methicillin			A[d]		
Ticarcillin[d]	B	A			
Mezlocillin[d]	B	A			
Piperacillin	B	A			
Azlocillin[d]		A			
Ampicillin-sulbactam	B		B[e]		
Amoxicillin-clavulanic acid	B		B[e]		
Piperacillin-tazobactam	B	B	B[e]		
Ticarcillin-clavulanic acid	B	B[f]	B[e]		
Cephalosporins					
Cephalothin	A[g]		A[g]		B
Cefazolin	A		A		
Cefamandole	B				
Cefonicid	B				
Cefuroxime	B				
Cefmetazole	B				
Cefoperazone	B	B			
Cefoxitin	B				
Cefotetan	B				
Cefotaxime	B			C	
Ceftriaxone	B			C	
Ceftizoxime	B	U			
Ceftazidime	C				
Cefepime	C	C			
Other beta-lactams					
Imipenem	B	B	C[e]		
Aztreonam	B	B			
Aminoglycosides					
Gentamicin	A	A		A[h]	
Netilmicin	C	C			
Tobramycin	C	B			
Amikacin	C	B			
Macrolides					
Azithromycin			B		B
Clarithromycin			B		B
Erythromycin			B		B
Quinolones					
Ciprofloxacin	B	B	C	U	
Lomefloxacin	D	D	D	D	D
Norfloxacin	D	D	D	D	D
Ofloxacin	C	C	C	C	C
Miscellaneous					
Chloramphenicol	C	C	C		
Clindamycin			B		B
Nitrofurantoin	D		D	D	D
Rifampin			C		
Sulfisoxazole	D		D		
Tetracycline	C	C[i]	C[i]	D	C
Trimethoprim-sulfamethoxazole	B	B	B		
Trimethoprim	D		D		
Vancomycin			A	B[d]	B

(Footnotes on next page)

selected few concentrations of antimicrobial agents (i.e., breakpoint susceptibility testing and single-drug-concentration screens).

Dilution methods offer flexibility in the sense that the standard medium used to test routinely encountered microorganisms (e.g., staphylococci, members of the family Enterobacteriaceae, Pseudomonas aeruginosa) may be readily supplemented or even replaced with another medium to allow accurate testing of various fastidious bacteria not reliably tested by disk diffusion. Dilution methods are also adaptable to automated systems. In addition, if plates or panels are prepared in-house, the combination of antibiotics to be included is not limited. Any drug available in powder form may be used.

The flexibility of dilution testing is also evident in the reporting formats that may be used. Quantitative results (MICs in micrograms per milliliter), category results (susceptible, intermediate, or resistant), or both may be used.

DILUTION TESTING: AGAR METHOD

Dilution of Antimicrobial Agents

The solvents and diluents needed to prepare stock solutions of most commonly used antimicrobial agents are presented in the NCCLS document on dilution testing (25).

Preparation and Storage of Antimicrobial Solutions

A twofold dilution scheme for an antimicrobial agent that starts with an antimicrobial stock concentration of 5,120 μg/ml is shown in Table 2 (17, 25). Depending on the number of tests to be prepared, the actual volumes used are proportionally increased. Certain intermediate concentrations are also used as the starting concentrations for the preparation of subsequent dilutions, and this should be taken into account when actual volume needs are being calculated. Although there are other approaches to obtaining a range of doubling dilutions, the scheme outlined in Table 2 is convenient, is economical in pipette use, and avoids some of the cumulative error inherent in traditional serial dilution methods.

Preparation, Supplementation, and Storage of Media

Mueller-Hinton agar is the recommended medium for testing most commonly encountered aerobic and facultatively anaerobic bacteria. The dehydrated agar base is commercially available and should be prepared as described by the manufacturer. Before sterilization, the molten agar is distributed into screw-cap containers in exact aliquots sufficient to dilute the intermediate antimicrobial concentrations 10-fold. Containers, one for each drug concentration to be tested, are sterilized by autoclaving at 121°C for 15 min and allowed to equilibrate to 48 to 50°C in a preheated water bath. Once the containers are equilibrated, the appropriate volume of intermediate antimicrobial concentration is added, and the container contents are mixed by gentle inversion, poured into 100-mm round or square sterile plastic plates set on a level surface, and allowed to solidify. For growth controls, plates containing drug-free agar are also prepared. All plates should be poured to a depth of 3 to 4 mm (20 to 25 ml of agar per plate), and the pH of each batch should be checked to confirm the appropriate pH range of 7.2 to 7.4.

After sterilization and temperature equilibration of the molten agar, any necessary supplements are aseptically added to the Mueller-Hinton agar. For testing streptococci, supplementation with 5% defibrinated sheep or horse blood is recommended (25). However, sheep blood supplementation may antagonize the activities of sulfonamide and trimethoprim (8); therefore, if these drugs are to be tested, 5% lysed horse blood should be used. The presence of blood also affects results with novobiocin and nafcillin as well as the in vitro activities of cephalosporins against enterococci (9, 10, 31, 34); therefore, blood supplementation should not be used unless necessary for bacterial growth. Performance standards of Mueller-Hinton agar have been defined sufficiently that calcium and magnesium supplementation should not be done. The agar should be supplemented with 2% NaCl for testing of methicillin or oxacillin against staphylococci (21).

Prepared plates should be sealed in plastic bags and stored at 4 to 8°C. If the plates are for research or reference studies, they should be used within 5 days of preparation; for routine clinical laboratory purposes, there may be no time limit as long as control strains that are tested routinely give accurate results within acceptable ranges. Before inoculation, plates that have been stored under refrigeration should be allowed to equilibrate to room temperature, and the agar surface should be dry.

Inoculation Procedures

Variations in inoculum size may substantially affect MIC endpoint determinations; therefore, careful preparation is required to obtain accurate results. The recommended final

---

(Footnotes to Table 1)

^{a}Modified from NCCLS data (24, 25) with permission. Current standards and supplements to them may be obtained from NCCLS, 771 East Lancaster Ave., Villanova, PA 19085.

^{b}Group A comprises primary drugs to be tested and reported; group B comprises those to be tested as primary drugs but reported selectively; group C comprises supplemental drugs to be reported selectively; and group D comprises drugs to be tested against urinary isolates only. Groups known to have nearly identical antimicrobial activities are enclosed in boxes. Because the clinical efficacies of the agents within each group should be similar, only one from each group need be tested.

^{c}Results of testing penicillin apply to other penicillins (e.g., ampicillin, amoxicillin, carboxy- and ureidopenicillins) versus β-lactamase-negative enterococci.

^{d}Combination therapy consisting of penicillin, ampicillin, or vancomycin and an aminoglycoside is recommended.

^{e}Staphylococci resistant to the penicillinase-resistant penicillins should also be considered resistant to penicillins, beta-lactam–β-lactamase inactivating combinations, cephalosporins, and imipenem.

^{f}Ticarcillin-clavulanic acid should not be considered as a therapeutic alternative for P. aeruginosa isolates resistant to carboxy- or ureidopenicillins.

^{g}Cephalothin test results may also be used to represent cephapirin, cephradine, cephalexin, cefaclor, cefadroxil, and cefazolin (except against the Enterobacteriaceae). Cefuroxime, cefpodoxime, cefproxil, and loracarbef may be tested separately on a supplemental basis because their activities may be greater against the Enterobacteriaceae than are those of cephalothin or cefazolin.

^{h}For use of aminoglycosides to screen enterococci for synergy resistance, see section on Breakpoint Susceptibility Testing and Single-Drug-Concentration Screens for Resistance.

^{i}Doxycycline or minocycline may be tested on a supplemental basis against some nonfermentative gram-negative bacilli and staphylococci.

TABLE 2 Antibacterial agent dilution schedule for agar and broth susceptibility methods[a]

Step	Starting solution of antimicrobial agent		No. of vol[b] of distilled water	Intermediate concn (μg/ml)[c]	Final concn (μg/ml) after 1:10 dilution in agar or broth
	Concn (μg/ml)	No. of vol[b]			
1	5,120	1	0	**5,120**	512 (256)
2	5,120	1	1	2,560	256 (128)
3	5,120	1	3	**1,280**	128 (64)
4	1,280	1	1	640	64 (32)
5	1,280	1	3	320	32 (16)
6	1,280	1	7	**160**	16 (8)
7	160	1	1	80	8 (4)
8	160	1	3	40	4 (2)
9	160	1	7	**20**	2 (1)
10	20	1	1	10	1 (0.5)
11	20	1	3	5	0.5 (0.25)
12	20	1	7	**2.5**	0.25 (0.125)
13	2.5	1	1	1.25	0.125 (0.0625)

[a]Adapted from references 17 and 25.

[b]Volume proportions remain constant, but the actual number of volumes selected depends on the number of plates, tubes, or trays to be prepared.

[c]Concentrations in boldface type are dilutions used for preparing the corresponding final concentration and as starting concentrations in the preparation of subsequent dilutions.

[d]For agar dilution and broth microdilution methods that use inoculum volumes of 0.001 to 0.005 ml to inoculate 0.1 ml of drug-containing broth, the final concentration is also the actual testing concentration. For macrodilution and microdilution methods in which the inoculum dilutes the final drug concentration 1:2, the actual testing concentrations are given in parentheses (see text for discussion of macrodilution inoculation procedures).

inoculum for agar dilution is 10^4 CFU per spot, and it may be prepared in either of two ways. Four or five colonies are picked from overnight growth on agar-based medium and inoculated into 4 to 5 ml of suitable broth that will support good growth (Mueller-Hinton, brain heart infusion, Trypticase soy [BBL Microbiology Systems, Cockeysville, Md.], or tryptic soy [Difco Laboratories, Detroit, Mich.]). Broths are incubated at 35°C until turbid, and the turbidity is adjusted to match that of a 0.5 McFarland standard (ca. 10^8 CFU/ml). Alternatively, colonies from overnight growth may be suspended in broth to a turbidity that matches the McFarland standard, eliminating the time needed for growing the inoculum in broth (25). By using either sterile broth or normal saline, a 1:10 dilution of the suspension is made to give an adjusted concentration of 10^7 CFU/ml (25).

Once the adjusted inoculum is prepared, inoculation of the antimicrobial agent plates should be accomplished within 30 min, since longer delays may lead to significant changes in inoculum size. By using either a calibrated loop or an inoculum replicating device, 0.001 to 0.002 ml of the 10^7-CFU/ml suspension is delivered to the agar surface, resulting in the final desired inoculum of approximately 10^4 CFU per spot. For convenience, use of a replicator is preferred, because a consistent inoculum volume for up to 36 different isolates is simultaneously delivered. To use this device, an aliquot of the adjusted inoculum for each isolate is pipetted into the appropriate well of the seed plate, and a multiprong inoculator is used to pick up and gently transfer 0.002 ml from the wells to the agar surfaces. The surface of the agar plates must be dry before inoculation, which should begin with the lowest drug concentration. To check for viability of each test isolate and also as an added check for purity, control plates not containing drug are inoculated last. Finally, plates should be clearly marked so that orientations of the different isolates being tested on each plate are known.

Incubation

Inoculated plates are allowed to stand until inocula have been completely absorbed by the medium; then they are inverted and incubated in air at 35°C for 16 to 20 h before being read. To facilitate detection of vancomycin-resistant enterococci and methicillin-resistant staphylococci, plates containing vancomycin and either oxacillin or methicillin, respectively, should be incubated for a full 24 h before being read (see chapter 116 of this Manual). Incubation in the presence of increased CO_2 is not recommended.

Interpretation and Reporting of Results

Before reading and recording the results obtained with clinical isolates, those obtained with quality control strains should be checked to ensure that they are within the acceptable accuracy ranges (see Quality Control below), and drug-free control plates should be examined for isolate viability and purity. Endpoints for each antibiotic are best determined by placing plates on a dark background and observing for the lowest concentration that inhibits visible growth, which is recorded as the MIC. A single colony or a faint haze left by the initial inoculum should not be regarded as growth. If two or more colonies persist at antimicrobial concentrations beyond an otherwise obvious endpoint or if there is no growth at lower concentrations but growth at higher concentrations, the isolate should be subcultured to confirm purity, and the test should be repeated. Substances that may antagonize the antibacterial activities of sulfonamides and trimethoprim may be carried over with the inoculum and cause "trailing," or less definite endpoints (8, 9). Therefore, the MICs for these antimicrobial agents should be interpreted as the endpoint at which an obvious 80 to 90% diminution of growth occurs.

The MIC of each antimicrobial agent is usually recorded in micrograms per milliliter. These quantitative results may be reported with the appropriate corresponding interpretive categories (susceptible, intermediate, or resistant), or the interpretive category may be reported alone. The MIC interpretive standards for these susceptibility categories, as recommended by the NCCLS, are provided in Table 3. For detailed instructions concerning the use of these criteria and categories, the latest NCCLS standards for dilution testing methods should be consulted. Note that the interpretive standards for most of the penicillin-class drugs vary with the organism being tested.

The three interpretive categories are defined as follows. Susceptible indicates that an infection caused by the tested microorganism may be appropriately treated with the usually recommended dose of antibiotic. Intermediate indicates that the isolate may be inhibited by attainable concentrations of certain drugs (e.g., beta-lactams) if higher dosages can be used or if the infection involves a body site in which the drug is physiologically concentrated (e.g., urinary tract). The intermediate category also serves as a buffer zone that prevents technical artifacts from causing major interpretive discrepancies. Resistant isolates are not inhibited by the concentration of antibiotic achievable with the recommended dose and/or yield results that fall within a range in which specific resistance mechanisms are likely to be present (25).

Advantages and Disadvantages

Dilution testing by the agar method is a well-standardized, reliable susceptibility-testing technique that may be used as a reference for evaluating the accuracy of other testing systems. In addition, the simultaneous testing of several isolates is possible, and microbial contamination or heterogeneity is more readily detected than by broth methods. The major disadvantages of the agar method are associated with the time-consuming and labor-intensive tasks of preparing the plates, especially as the number of different antimicrobial agents to be tested against each isolate increases.

DILUTION TESTING: BROTH METHODS

The general approaches for broth methods include macrobroth dilution, in which the broth volume for each antimicrobial concentration is ≥1.0 ml contained in 13- by 100-mm tubes, and microbroth dilution, in which antimicrobial dilutions are in 0.05- to 0.1-ml volumes contained in wells of microtiter trays.

Macrobroth Dilution Methods

Dilution of Antimicrobial Agents

Stock solutions are prepared as discussed in the NCCLS document on dilution testing (25), and a typical dilution scheme that may be used to obtain full-range doubling dilutions is outlined in Table 2. As in the agar method, the actual volumes used for the dilutions would be proportionally increased according to the number of tests being prepared, with a minimum of 1.0 ml needed for each drug concentration. Because addition of the inoculum results in a 1:2 dilution of each concentration, all final drug concentrations must be prepared at twice the actual desired testing concentration (see Inoculation Procedures below). If not, the actual testing concentrations will be those given parenthetically in Table 2.

Preparation, Supplementation, and Storage of Media

Cation-adjusted Mueller-Hinton broth (CAMHB) is recommended for routine testing of commonly encountered nonfastidious organisms (25). Adjustment with the cations Ca^{2+} (20 to 25 mg/liter) and Mg^{2+} (10 to 12.5 mg/liter) is required to ensure acceptable results when $P.\ aeruginosa$ isolates are tested against aminoglycosides and when tetracycline is tested against other bacteria. Many manufacturers provide Mueller-Hinton broth that already has appropriate concentrations of divalent cations, so care must be taken not to supplement such products with additional cations. If cation concentrations are below the recommended levels, adjustment is accomplished by the addition of suitable volumes of filter-sterilized, chilled $CaCl_2$ stock (3.68 g of $CaCl_2 \cdot 2H_2O$ dissolved in 100 ml of deionized water for a concentration of 10 mg of Ca^{2+} per ml) and $MgCl_2$ stock (8.36 g of $MgCl_2 \cdot 6H_2O$ in 100 ml of deionized water for a concentration of 10 mg of Mg^{2+} per ml) to the cooled broth (6, 25). Insufficient cation concentrations result in increased aminoglycoside activity (13, 19, 30, 35), while excess cation content results in decreased aminoglycoside activity against $P.\ aeruginosa$ (5, 13, 19, 30). Reliable detection of staphylococcal resistance to oxacillin, methicillin, or nafcillin requires that the CAMHB used for testing these drugs be supplemented with 2% NaCl (21, 25).

Appropriately adjusted and supplemented Mueller-Hinton broth is used to prepare the final antimicrobial agent concentrations according to the scheme shown in Table 2. To minimize evaporation and antibiotic deterioration, tubes should be tightly capped and stored at 4 to 8°C. For reference studies, the dilutions should be used within 5 days of preparation; for routine use, there may be no time limit as long as quality control accuracy ranges are maintained (see Quality Control below).

Inoculation Procedures

The recommended final inoculum is 5×10^5 CFU/ml. Isolates are inoculated into a broth that will support good growth and incubated until turbid, and the turbidity is adjusted to match that of a 0.5 McFarland standard (approximately 10^8 CFU/ml). Alternatively, four or five colonies from overnight growth on a blood agar plate may be directly suspended in broth to match the turbidity of the McFarland standard (25). This alternative is recommended for testing methicillin, oxacillin, or nafcillin against staphylococci but may also be used with other bacteria (25). A portion of the suspension is diluted 1:100 (10^6 CFU/ml) with broth or 0.85% saline. When 1 ml of this dilution is added to each tube containing 1 ml of the drug diluted in CAMHB, a final inoculum of 5×10^5 CFU/ml is achieved. Broth not containing antimicrobial agent is inoculated as a control for organism viability (growth control). All tubes should be inoculated within 15 to 20 min of inoculum preparation, and an aliquot of the inoculum should be plated to check for purity and inoculum size.

Incubation

Tubes are incubated in air at 35°C for 16 to 20 h before being read. Incubation should be extended to a full 24 h for the detection of vancomycin-resistant enterococci and methicillin-resistant staphylococci (25). Use of increased CO_2 is not recommended.

Interpretation and Reporting of Results

Before MICs for the test strains are read and recorded, the growth controls should be examined for viability, inoculum

TABLE 3 Interpretive standards for dilution and disk diffusion susceptibility testing[a]

Antimicrobial agent and organism	MIC (µg/ml)			Zone diam (mm)		
	Susceptible	Intermediate	Resistant	Susceptible	Intermediate	Resistant
Penicillins						
Penicillin G						
Staphylococci[b]	≤0.12		≥0.25	≥29		≤28
Enterococci[c]	≤8		≥16	≥15		≤14
Listeria monocytogenes	≤2		≥4	≥20		≤19
Streptococci other than *Streptococcus pneumoniae*	≤0.12	0.25–2.0	≥4	≥28	20–27	≤19
Methicillin[d]	≤8		≥16	≥14	10–13	≤9
Oxacillin-nafcillin[d]	≤2		≥4	≥13	11–12	≤10
Ampicillin	≤8		≥16	≥14		≤9
Enterobacteriaceae	≤8	16	≥32	≥17	14–16	≤13
Staphylococci	≤0.25		≥0.5	≥29		≤28
L. monocytogenes	≤2		≥4	≥20		≤19
Enterococci[c]	≤8		≥16	≥17		≤16
Streptococci other than *S. pneumoniae*	≤0.12	0.25–2	≥4	≥30	22–29	≤21
Amoxicillin-clavulanic acid						
Staphylococci	≤4/2		≥8/4	≥20		≤19
Other organisms	≤8/4	16/8	≥32/16	≥18	14–17	≤13
Ampicillin-sulbactam	≤8/4	16/8	≥32/16	≥15	12–14	≤11
Azlocillin	≤64		≥128	≥18		≤17
Pseudomonas aeruginosa						
Carbenicillin						
P. aeruginosa	≤128	256	≥512	≥17	14–16	≤13
Other gram-negative bacilli	≤16	32	≥64	≥23	20–22	≤19
Mezlocillin						
P. aeruginosa	≤64		≥128	≥16		≤15
Other gram-negative bacilli	≤16	32–64	≥128	≥21	18–20	≤17
Piperacillin						
P. aeruginosa	≤64		≥128	≥18		≤17
Other gram-negative bacilli	≤16	32–64	≥128	≥21	18–20	≤17
Piperacillin-tazobactam						
P. aeruginosa	≤64/4		≥128/4	≥18		≤17
Other gram-negative bacilli	≤16/4	32/4–64/4	≥128/4	≥21	18–20	≤17
Staphylococci	≤8/4		≥16/4	≥18		≤17
Ticarcillin						
P. aeruginosa	≤64		≥128	≥15		≤14
Other gram-negative bacilli	≤16	32–64	≥128	≥20	15–19	≤14
Ticarcillin-clavulanic acid						
P. aeruginosa	≤64/2		≥128/2	≥15		≤14
Other gram-negative bacilli	≤16/2	32/2–64/2	≥128/2	≥20	15–19	≤14
Cephalosporins						
Cefaclor	≤8	16	≥32	≥18	15–17	≤14
Cefamandole	≤8	16	≥32	≥18	15–17	≤14
Cefazolin	≤8	16	≥32	≥18	15–17	≤14
Cefepime	≤8	16	≥32	≥18	15–17	≤14
Cefetamet	≤4	8	≥16	≥18	15–17	≤14
Cefixime	≤1	2	≥4	≥19	16–18	≤15
Cefmetazole	≤16	32	≥64	≥16	13–15	≤12
Cefonicid	≤8	16	≥32	≥18	15–17	≤14
Cefoperazone	≤16	32	≥64	≥21	16–20	≤15
Cefotaxime	≤8	16–32	≥64	≥23	15–22	≤14
Cefotetan	≤16	32	≥64	≥16	13–15	≤12
Cefoxitin	≤8	16	≥32	≥18	15–17	≤14
Cefpodoxime	≤2	4	≥8	≥21	18–20	≤17
Cefprozil	≤8	16	≥32	≥18	15–17	≤14
Ceftazidime	≤8	16	≥32	≥18	15–17	≥14
Ceftizoxime	≤8	16–32	≥64	≥20	15–19	≤14
Ceftriaxone	≤8	16–32	≥64	≥21	14–20	≤13
Cefuroxime axetil	≤4	8–16	≥32	≥23	15–22	≤14

(Continued on next page)

TABLE 3 *(Continued)*

Antimicrobial agent and organism	MIC (μg/ml)			Zone diam (mm)		
	Susceptible	Intermediate	Resistant	Susceptible	Intermediate	Resistant
Cefuroxime sodium	≤8	16	≥32	≥18	15–17	≤14
Cephalothin	≤8	16	≥32	≥18	15–17	≤14
Loracarbef	≤8	16	≥32	≥18	15–17	≤14
Moxalactam	≤8	16–32	≥64	≥23	15–22	≤14
Other beta-lactams						
Aztreonam	≤8	16	≥32	≥22	16–21	≤15
Imipenem	≤4	8	≥16	≥16	14–15	≤13
Aminoglycosides						
Amikacin	≤16	32	≥64	≥17	15–16	≤14
Gentamicin	≤4	8	≥16	≥15	13–14	≤12
Enterococci (high-level resistance)	≤500		>500	≥10	7–9	6
Netilmicin	≤8	16	≥32	≥15	13–14	≤12
Tobramycin	≤4	8	≥16	≥15	13–14	≤12
Streptomycin						
Enterococci (high-level resistance)						
Broth microdilution	≤1,000		>1,000			
Agar based	≤2,000		>1,000	≥10	7–9	6
Glycopeptides						
Teicoplanin	≤8	16	≥32	≥14	11–13	≤10
Vancomycin	≤4	8–16	≥32	≥17	15–16	≤14
Enterococci						
Other organisms	≤4	8–16	≥32	≥12	10–11	≤9
Macrolides						
Azithromycin	≤2	4	≥8	≥18	14–17	≤13
Clarithromycin	≤2	4	≥8	≥18	14–17	≤13
Erythromycin	≤0.5	1–2	≥8	≥23	14–22	≤13
Quinolones						
Ciprofloxacin	≤1	2	≥4	≥21	16–20	≤15
Enoxacin	≤2	4	≥8	≥18	15–17	≤14
Fleroxacin	≤2	4	≥8	≥19	16–18	≤15
Lomefloxacin	≤2	4	≥8	≥22	19–21	≤18
Norfloxacin[e]	≤4	8	≥16	≥17	13–16	≤12
Ofloxacin	≤2	4	≥8	≥16	13–15	≥12
Other						
Chloramphenicol	≤8	16	≥32	≥18	13–17	≤12
Clindamycin	≤0.5	1–2	≥4	≥21	15–20	≤14
Nitrofurantoin	≤32	64	≥128	≥17	15–16	≤14
Rifampin	≤1	2	≥4	≥20	17–19	≤16
Sulfonamide	≤256		≥512	≥17	13–16	≤12
Tetracycline	≤4	8	≥16	≥19	15–18	≤14
Trimethoprim[e]	≤8		≥16	≥16	11–15	≤10
Trimethoprim-sulfamethoxazole	≤2/38		≥4/76	≥16	11–15	≤10

[a]Adapted from NCCLS data (25, 26) with permission. The interpretive data are valid only if the methodologies in M2-A5 (25) and M7-A3 (26) are followed. NCCLS frequently updates the interpretive tables through new additions of the standards and supplements to them. Users should refer to the most recent additions. The current standards and supplements to them may be obtained from NCCLS, 771 East Lancaster Ave., Villanova, PA 19085.

[b]Penicillin should be used as the class representative for all penicillins (e.g., ampicillin, amoxicillin, mezlocillin, piperacillin, ticarcillin). Isolates for which MICs are ≤0.03 μg of penicillin per ml generally do not produce β-lactamase, whereas those for which MICs are ≥0.25 μg/ml do and should be regarded as resistant to penicillins. Isolates with penicillin MICs of 0.06 or 0.12 μg/ml should be tested for β-lactamase.

[c]Therapy of serious enterococcal infections requires high doses of penicillin or ampicillin in combination with an aminoglycoside. Vancomycin may be substituted for the penicillin in instances of penicillin hypersensitivity or of penicillin or ampicillin resistance.

[d]Oxacillin or methicillin may be tested; however, oxacillin is preferred because of its greater stability in vitro. The results from testing oxacillin apply also to other penicillinase-resistant penicillins. Oxacillin-resistant staphylococci should be considered resistant to all penicillins, cephalosporins, monobactams, carbacephems, carbapenems, and beta-lactam–β-lactamase inhibitor combinations.

[e]For the treatment of urinary tract infections only.

subcultures should be checked for contamination and appropriate inoculum size, and the accuracy of the MICs obtained with the quality control strains should be confirmed (see Quality Control below). Growth or lack thereof in the antimicrobial agent-containing tubes is best determined by comparison with the growth control. Generally, growth is indicated by turbidity, a single sedimented button ≥2 mm in diameter, or several buttons with smaller diameters. As with the agar method, trailing endpoints may be seen when trimethoprim or sulfonamides are tested, and the concentration at which an obvious 80 to 90% diminution of growth, compared with that of the growth control, occurs should be recorded as the MIC (25). Other interpretation problems include the "skipped tube," in which growth is not observed at one concentration but is observed at lower and higher drug concentrations. When this occurs, the skipped tube should be ignored and the concentration that finally inhibits growth at serially higher concentrations should be recorded as the MIC. If more than one skipped tube occurs or if there is growth at higher antimicrobial concentrations but not at lower, the results should not be reported and the test for that drug should be repeated.

The lowest concentration that completely inhibits visible growth of the organism as detected by the unaided eye is recorded as the MIC. MIC interpretive standards for the susceptibility categories are provided in Table 3. The definitions of and comments concerning these categories that were given for the agar method also pertain to the macrodilution broth method.

Advantages and Disadvantages

The macrodilution broth method is a well-standardized and reliable reference method that is useful for research purposes, but because of the laborious nature of the procedure and the availability of more convenient dilution systems (i.e., microdilution), this procedure is generally not useful for routine susceptibility testing in most clinical microbiology laboratories.

Microbroth Dilution Methods

The convenience afforded by the availability of dilution susceptibility testing in microdilution trays has led to the widespread use of microdilution broth methods. The trays, containing several antimicrobial agents to be tested simultaneously, may be prepared in-house or obtained commercially either frozen or lyophilized. When commercial systems are used, the manufacturers' recommendations concerning storage, inoculation, incubation, and interpretation should be followed. The primary focus of this section will be the in-house preparation and use of microbroth dilution trays. However, many of the principles and practices discussed here are pertinent to the microbroth dilution method in general, regardless of the source of the trays.

Dilution of Antimicrobial Agents

Antimicrobial stock solutions are prepared as outlined in the NCCLS document on dilution testing (25). The dilution scheme used for agar and macrobroth dilution methods is applicable to the antimicrobial dilutions needed for the preparation of microbroth dilution panels (Table 2). Available automated dispensing systems require that at least 10 ml of broth containing each antimicrobial concentration be prepared. From the 10-ml samples, aliquots of either 0.05 or 0.1 ml are simultaneously dispensed to the corresponding wells of each microbroth dilution tray. If 0.05-ml volumes are dispensed, allowances must be made for the 1:2 dilution

of the final drug concentration that will occur when the 0.05 ml of inoculum is added (see Inoculation Procedures below). When 0.1-ml aliquots are dispensed, the volume of inoculum used is sufficiently small (≤0.005 ml) that adjustments in the antimicrobial dilution scheme are not needed. As a general rule, when the inoculum volume is less than 10% of the broth volume in the well, dilution of the antimicrobial concentration by the inoculum does not have to be taken into account (25).

Preparation, Supplementation, and Storage of Media

CAMHB is the recommended medium for routine microbroth dilution testing and should be prepared as discussed above for the macrobroth dilution method. Also, supplementation of the broth with 2% NaCl is required for detection of methicillin-resistant staphylococci (25). After the antimicrobial dilutions have been dispensed to the trays, they are stacked in groups of 5 to 10, with an empty tray placed on top to minimize contamination and evaporation. Each stack is sealed in a plastic bag and frozen immediately at −20°C or, preferably, at −60°C or colder. At −20°C, preservation is ensured for at least 6 weeks, but the shelf life may be extended to months if the trays are stored at −60 to −70°C. Thawed panels must be used or discarded. They should not be refrozen, since freeze-thaw cycles cause substantial deterioration of beta-lactam antibiotics, which is why freezers with self-defrosting units should be avoided.

Inoculation Procedures

The final desired inoculum concentration is 5×10^5 CFU/ml. The isolates may be grown in broth to match the turbidity of a 0.5 McFarland standard (ca. 10^8 CFU/ml), or such a suspension can be made from colonies grown on an agar medium overnight (3), which is the method preferred for methicillin-resistant staphylococci (25). For microbroth dilution procedures that require 0.001- to 0.005-ml volumes to inoculate wells containing 0.1 ml of broth, a portion of the 0.5 McFarland suspension is diluted 1:10 (10^7 CFU/ml) in sterile saline or broth. Multipoint plastic or metal inoculum replicators designed to collect and deliver appropriate volumes are used to transfer the inoculum from the diluted suspension to the wells of the microbroth dilution tray, resulting in further dilutions ranging from 1:20 to 1:100 and final inoculum concentrations of 1×10^5 to 5×10^5 CFU/ml (1×10^4 to 5×10^4 CFU per well). For systems that use an inoculum volume of 0.05 ml to inoculate 0.05 ml of broth, a 1:100 dilution of a 0.5 McFarland suspension (10^6 CFU/ml) is used. When the inoculum is added to the wells, the 1:2 dilution of the 10^6-CFU/ml inoculum results in a final inoculum concentration of 5×10^5 CFU/ml (5×10^4 CFU per well) and also halves the antibiotic concentration in each well. Special care should be taken to confirm the inoculum density on a periodic basis to ensure that the appropriate amount of inoculum is achieved. Insufficient inoculum is a frequent problem in microbroth dilution tests that may not be detected with the use of recommended quality control strains.

Microbroth dilution trays should be inoculated within 15 to 20 min of inoculum preparation; during preparation, an aliquot should be subcultured to check the purity of the isolates, and colony counts should be set up to check the accuracy of the inoculum concentration. Finally, one well not containing an antimicrobial agent should be inoculated and used as a growth control.

Incubation

After inoculation, each tray should be covered with plastic tape, sealed in a plastic bag, or tightly fitted with an empty tray to prevent evaporation during incubation. Trays are incubated in air at 35°C for 16 to 20 h before being read and should not be incubated in stacks of more than five. The incubation chamber should be kept sufficiently humid to avoid evaporation but not so humid that condensation results in contamination problems. A full 24 h of incubation is recommended for the detection of vancomycin-resistant enterococci and methicillin-resistant staphylococci (25). Use of increased CO_2 is not recommended unless it is necessary for growth of a particular isolate.

Interpretation and Reporting of Results

Before MICs for the clinical isolates are read and recorded, the growth control wells should be examined for organism viability. It is advisable to check inoculum purity by subculture and periodically to verify inoculum size by quantitative subculture. Also, the accuracy of the MICs obtained with the quality control strains should be confirmed (see Quality Control below). Various viewing apparatuses are available and should be used to facilitate examination of the microbroth dilution wells for growth. Growth is best determined by comparison with that in the growth control well and generally is indicated by turbidity throughout the well or by buttons, single or multiple, in the well bottom. The occurrence of trailing endpoints when trimethoprim or sulfonamides are tested, a single skipped well, or more than one skipped well should not be reported as they are for the macrobroth dilution test.

The MIC interpretive standards for susceptibility categories are given in Table 3. The definitions of these categories and the comments concerning the use of these standards for agar and macrobroth dilution methods are also applicable to microbroth dilution methods.

Advantages and Disadvantages

The use of microbroth dilution trays prepared in-house provides a reliable standardized reference method for susceptibility testing. Inoculation and reading procedures allow relatively convenient simultaneous testing of several antimicrobial agents against individual organisms. Not all laboratories have the facilities required for preparation of microbroth dilution trays; however, a wide variety of products are commercially available today. Such products provide trays with wells containing prepared antimicrobial dilutions either frozen or lyophilized. The former types of trays must be stored frozen in the laboratory, whereas lyophilized trays can be stored under refrigeration. All such products are accompanied by multipoint inoculating devices. Results of testing may be determined by visual examination or with semiautomated or automated ("walkaway") instrumentation. An alternative approach to the use of trays containing either frozen or lyophilized antimicrobial agents is the Alamar system (Alamar, Sacramento, Calif.). By contrast, each series of wells for any given antimicrobial agent in this system has a paper disk, each containing a specified amount of the antimicrobial agent so that in any series of wells, once the inoculum broth is added, there is a series of concentrations of antimicrobial agent (on a $\log_2$ scale). In other words, after addition of the inoculum broth, the antimicrobial agent elutes from the paper disk into the inoculum broth, providing the desired concentration of antimicrobial agent. Also in contrast to

other microbroth dilution systems, the Alamar system employs an oxidation-reduction color indicator that changes from blue to red when growth occurs (28). The Alamar system therefore combines the simplicity of antimicrobial carriage in a paper disk and a color reaction for differentiating between growth and growth inhibition.

The versatility of antimicrobial components available with commercial microbroth dilution trays is limited compared with the flexibility of preparing panels in-house or using the disk diffusion method. Newer devices that combine short incubations with automated reading of trays can offer speed in test performance and interpretation of results but are problematic in detecting resistance due to certain specific mechanisms (e.g., the *mec* gene in staphylococci, one of the *van* phenotypes in enterococci [32], or beta-lactam resistance due to Bush group I β-lactamase in species of *Citrobacter*, *Enterobacter*, *Proteeae*, and *Serratia* [36]) (see chapter 118 of this Manual).

BREAKPOINT SUSCEPTIBILITY TESTING AND SINGLE-DRUG-CONCENTRATION SCREENS FOR RESISTANCE

Breakpoint Susceptibility Testing

Breakpoint susceptibility testing refers to methods by which antimicrobial agents are tested only at the specific concentrations necessary for differentiating between the interpretive categories of susceptible, intermediate, and resistant rather than in the full range of doubling-dilution concentrations used to determine MICs. When two appropriate drug concentrations are selected, any one of the interpretive categories may be determined. Growth at both concentrations indicates resistance, growth only at the lower concentration signifies an intermediate result, and no growth at either concentration is interpreted as susceptibility. A single concentration may be used for testing antimicrobial agents specifically for urinary tract infection (e.g., nitrofurantoin), with growth interpreted as resistance, and no growth interpreted as susceptibility.

As for full-range dilution testing, breakpoint methods require the use of appropriately adjusted and supplemented Mueller-Hinton broth or agar. In addition, the standard inoculation, incubation, and interpretation procedures recommended for the full-range dilution methods should be followed.

Because breakpoint testing is a direct measure of antimicrobial activity, there is the advantage of avoiding inherent errors associated with extrapolating disk diffusion zone sizes and MIC results (15, 16). Considering expense and convenience, a greater number and variety of antimicrobial agents can be incorporated into a microbroth dilution tray set up for breakpoint testing than in trays designed for full-range dilution testing. However, breakpoint testing provides only interpretive category results and cannot be used to generate the doubling-dilution MIC data required in certain clinical situations. Also, convenient quality control procedures to ensure that appropriate concentrations of each antimicrobial agent are present are lacking. One possible approach is to use one organism whose modal MIC is equal to or no less than 1 doubling dilution less than the lower or lowest concentration tested and a second organism whose modal MIC is equal to or no more than 1 doubling dilution greater than the higher or highest concentration tested (25). One of these two quality control organisms should provide on-scale results (25). Selective testing of antibiotics that are most unstable in vitro (e.g., clavulanic acid combinations, imipenem, cefaclor) is advisable for routine purposes (25).

Resistance Screens

In some circumstances, testing a single drug concentration may be the most reliable and convenient method for detecting antimicrobial resistance. The most clinically useful resistance screens are those for staphylococcal resistance to the penicillinase-resistant penicillins (i.e., methicillin, nafcillin, and oxacillin) and for high-level resistance of *Enterococcus* spp. to gentamicin, streptomycin, or vancomycin, which are described in chapter 116 of this Manual.

E Test

The E Test (PDM epsilometer; AB Biodisk NA, Piscataway, N.J.) is an in vitro method for quantitative antimicrobial susceptibility testing whereby a preformed antimicrobial gradient from a plastic-coated strip diffuses into an agar medium inoculated with the test organism. In this test, the MIC is read directly from a scale on the strip at the point where the ellipse of growth inhibition intercepts the strip. Several strips, each containing a different antimicrobial agent, can be placed radially on the surface of a round Mueller-Hinton agar plate inoculated with a suspension of a bacterial isolate that has been adjusted to match the turbidity of a 0.5 McFarland turbidity standard. This test has been evaluated by, among others, Huang et al. (20) and Sanchez et al. (33). Both groups found a high level of agreement between MICs obtained by the E Test and reference dilution methods when enterococci and gram-negative bacilli with a variety of resistance mechanisms were tested. In addition, Huang et al. (20) and Novak et al. (28) determined a high level of correlation between the E Test and reference methodology for the detection of methicillin-resistant staphylococci. The test, though not inexpensive, combines the simplicity and flexibility of the disk diffusion test with the opportunity to determine MICs of any particular antimicrobial agent for a particular bacterial isolate. The strength of this method may be the testing of fastidious or anaerobic bacteria, since the strips may be placed onto various enriched media (see chapters 114 and 115 of this Manual).

QUALITY CONTROL

Quality control recommendations are designed to effectively evaluate the precision and accuracy of the dilutions and test procedures used, monitor reagent reliability, and evaluate the performance of individuals who are conducting the tests.

Reference Strains

A key to accomplishing the goals of quality control is the careful selection and use of reference bacterial strains that are genetically stable and give MICs that are in the midrange of each antimicrobial agent tested (25). That is, if there are seven dilutions in a series, the reference strain should give an MIC from the third to fifth dilution, preferably the fourth. If there are four or fewer dilutions in a series or if nonconsecutive dilutions are tested (i.e., breakpoint susceptibility testing), quality control for the correct interpretive category rather than an actual MIC or MIC ranges may be more appropriate (see recommendations under Breakpoint Susceptibility Testing above). *Escherichia coli* ATCC 25922, *P. aeruginosa* ATCC 27853, *Enterococcus*

faecalis ATCC 29212, and *Staphylococcus aureus* ATCC 29213 are the recommended reference strains for both agar and broth methods (25). *E. coli* ATCC 35218 is recommended only for beta-lactam–β-lactamase inhibitor combinations. These organisms may be obtained from the American Type Culture Collection or other reliable commercial sources. For proper storage and subculture procedures, the recommendations of either NCCLS (25) or the commercial provider should be followed.

MIC Ranges

The acceptable quality control MIC ranges for the various reference strains are given in the NCCLS document on dilution testing (25). Updates of these MIC ranges are generally published annually. An out-of-control result is defined as an MIC not within the acceptable range. Certain out-of-control results can frequently be directly related to the medium used for testing. High MICs of gentamicin for *P. aeruginosa* ATCC 27853 indicate an inappropriately high cation content of the Mueller-Hinton medium, and low MICs indicate an insufficient cation concentration. Although trimethoprim-sulfamethoxazole is not recommended therapy for *Enterococcus faecalis* infections, results obtained with the ATCC 29212 strain are useful for detecting inappropriate concentrations of substances such as thymidine that interfere with the in vitro activity of antifolate drugs. Trimethoprim-sulfamethoxazole MICs of >0.5/9.5 μg/ml indicate the presence of such interfering substances.

Batch and Lot Quality Control

Representative plates, panels, or trays from each new batch if prepared in-house or from each new shipment lot if obtained from a commercial source are quality controlled for accuracy and sterility. MICs obtained by testing reference quality control strains should be within acceptable accuracy ranges. If such accuracy is not achieved, the batch or lot should be rejected, or results obtained with the antimicrobial agent(s) in question should not be reported (see below). Similarly, if selected uninoculated plates or trays fail the sterility check after incubation, the batch or lot should be rejected. In addition to these formal quality control procedures that use reference strains, careful review of susceptibility results obtained during daily testing of clinical isolates is necessary to identify aberrant or unusual susceptibility patterns that may indicate quality control problems.

Quality Control Frequency

In addition to batch and lot testing, quality control should be performed daily, or at least every day that the plates or trays are being used to test clinical isolates. When quality control is performed, two consecutive out-of-control MICs or more than two nonconsecutive out-of-control values in 20 consecutive tests indicate problems in the dilution testing procedure that must be identified and solved. However, if accuracy can be sufficiently documented as outlined below, daily testing may be replaced by weekly testing (25).

Each drug-reference strain combination is tested for 30 consecutive days to obtain a total of 30 MICs for each combination. If three or fewer MICs per combination are outside the accuracy range, weekly testing may replace daily testing. During weekly testing, a single MIC outside the accuracy range requires that daily testing be performed for 5 consecutive days unless there is an obvious source of error (e.g., contamination, incorrect reference strain used, incorrect antimicrobial agent tested, incorrect atmosphere of incubation). In such a circumstance, the quality control test need only be repeated. If all five MICs for a problem drug-organism combination are within the accuracy range, weekly testing may be resumed. If one or more of the five MICs for the problem drug-organism combination are outside the accuracy range, daily testing must be initiated. Returning to weekly testing requires again documenting 30 consecutive days with three or fewer MICs outside the accuracy range. If more than three MICs per combination are outside the accuracy range, daily quality control testing must be continued.

DISK DIFFUSION TESTING

The disk diffusion method of susceptibility testing allows categorization of bacterial isolates as susceptible, resistant, or intermediate to a variety of antimicrobial agents. To perform the test, commercially prepared filter paper disks impregnated with a specified amount of an antimicrobial agent are applied to the surface of an agar medium that has been inoculated with the test organism. The drug in the disk diffuses through the agar (2). As the distance from the disk increases, the concentration of the antimicrobial agent decreases logarithmically, creating a gradient of drug concentrations in the agar medium surrounding each disk. Concomitant with diffusion of the drug, the bacteria that were inoculated on the surface and are not inhibited by the concentration of antimicrobial agent continue to multiply until a lawn of growth is visible. In areas where the concentration of drug is inhibitory, no growth occurs, forming a zone of inhibition around each disk.

The disk diffusion procedure has been standardized primarily for testing rapidly growing bacteria (7, 18, 24). This method should not be used to evaluate antimicrobial susceptibilities of bacteria that show marked strain-to-strain variability in growth rates. The test, however, has been modified to allow reliable testing of certain fastidious bacteria (discussed in chapters 114 and 115 of this Manual).

The diameter of the zone of inhibition is influenced by the rate of diffusion of the antimicrobial agent through the agar, which may vary among different drugs. The zone size, however, is inversely proportional to the MIC, measured as discussed earlier in this chapter. Criteria currently recommended for interpreting zone diameters and MIC results for commonly used antimicrobial agents are listed in Table 3.

Zone of Inhibition Diameter

Interpretive Criteria

The first step in determining interpretive criteria for the disk diffusion test is selection of MIC breakpoints that define resistance and susceptibility categories for each antimicrobial agent. Zone of inhibition diameters that correspond to these breakpoints are determined by testing a minimum of 300 and preferably 500 bacterial isolates by both dilution and disk diffusion methods and correlating zone of inhibition diameters and MICs for each drug tested (27). Isolates tested should include not only those commonly encountered in clinical laboratories but also those with resistance mechanisms pertinent to the class of antimicrobial agent being tested. Organisms evaluated should be those most likely to be tested against the antimicrobial agent in question. The data are analyzed by preparing a scattergram. By convention, each MIC ($\log_2$ scale) is plotted on the y axis, and the corresponding zone of inhibition

diameter (arithmetic scale) is plotted on the x axis. Regression analysis is then performed, and a straight regression line showing the best fit is drawn. From this line, an approximate MIC can be inferred from any zone of inhibition diameter. For antimicrobial agents to which isolates are either susceptible or resistant and only infrequently intermediate, regression analysis is not valid. In such cases, the data are plotted as a scattergram, and the interpretive standards are selected so as to allow optimal separation of the two populations (23, 27).

Antimicrobial Agent Disks

The concentrations of antimicrobial agent in the disks used for agar diffusion testing are standardized, and in the United States, only a single disk for each drug is recommended. The optimal amount of antimicrobial agent per disk is determined by testing disks with several different drug contents that are then evaluated in scattergrams and regression lines generated from the data. The most desirable amount of drug per disk is that which produces a zone of inhibition diameter of at least 10 mm with all resistant isolates and a zone diameter no larger than 30 mm (rarely, 40 mm) with susceptible isolates.

Commercially prepared antimicrobial disks usually are supplied in separate containers, each with a desiccant. They must not be used beyond the specified expiration date and should be stored under refrigeration (2 to 8°C) or frozen in a non-frost-free freezer at −14°C or colder until needed. Disks containing a beta-lactam agent should always be frozen to ensure that they retain their potency, although a small supply may be stored in the refrigerator for up to 1 week. Unopened disk containers should be removed from the refrigerator or freezer 1 to 2 h before use. This allows the disks to equilibrate to room temperature before the container is opened, thus minimizing the amount of condensation that will occur when warm air contacts the cold disks. If a mechanical disk-dispensing apparatus is used, it should be fitted with a tight cover, supplied with an adequate desiccant, stored in the refrigerator when not in use, and warmed to room temperature before being opened.

Method

Agar Medium

Currently, the recommended medium for disk diffusion testing is Mueller-Hinton agar. This unsupplemented medium has been selected by the NCCLS for several reasons: (i) it demonstrates good batch-to-batch reproducibility for susceptibility testing; (ii) it is low in sulfonamide, trimethoprim, and tetracycline inhibitors; (iii) it supports the growth of most nonfastidious bacterial pathogens; and (iv) many data and much experience regarding its performance have been accrued. Some bacteria, especially streptococci, do not grow satisfactorily on unsupplemented Mueller-Hinton agar but will grow if the medium is supplemented with defibrinated sheep, horse, or other animal blood at a final concentration of 5% (vol/vol). Susceptibility testing of more fastidious bacteria, such as *Haemophilus* species, *Neisseria gonorrhoeae*, and *Streptococcus pneumoniae*, is discussed in chapters 114 and 115 of this Manual.

Plates of Mueller-Hinton agar may be purchased, or the agar may be prepared from a commercially available dehydrated base according to the manufacturer's directions. If the agar is prepared, only formulations that have been tested according to and have met acceptance limits recommended by the NCCLS should be used (26). The prepared medium is autoclaved and immediately placed in a 45 to 50°C water bath. When cool, it is poured into plastic or glass flat-bottomed petri dishes on a level, horizontal surface to give a uniform depth of about 4 mm (60 to 70 ml of medium for 150-mm plates and 25 to 30 ml for 100-mm plates) and allowed to cool to room temperature. Agar deeper than 4 mm may cause false-resistance results, whereas agar less than 4 mm deep may be associated with a false-susceptibility report.

Each batch of Mueller-Hinton agar should be checked when the medium is prepared to ensure that the pH is between 7.2 and 7.4 at room temperature, which means that the pH must be measured after the medium has solidified. This can be done by allowing a small amount of agar to solidify around the tip of a pH electrode in a beaker or a cup, by macerating a sufficient amount of agar in neutral distilled water, or by using a properly calibrated surface electrode. A pH outside the range of 7.2 to 7.4 may adversely affect susceptibility test results. If the pH is too low, drugs such as the aminoglycosides and macrolides will appear to lose potency, whereas others (for example, the penicillins) may appear to have excessive activity. The opposite effects are possible if the pH is too high.

Freshly prepared plates may be used the same day or stored in a refrigerator (2 to 8°C). If plates are not used within 7 days of preparation, they should be wrapped in plastic to minimize evaporation. Just before use, if excess moisture is on the surface, plates should be placed in an incubator (35°C) or, with lids ajar, in a laminar-flow hood at room temperature until the moisture evaporates (usually 10 to 30 min). When the medium is inoculated, no droplets of moisture should be visible on its surface or on the petri dish cover.

Various components of or supplements to Mueller-Hinton medium may affect susceptibility test results; therefore, appropriate quality control procedures (see Quality Control below) must be performed, and zone diameters must be within acceptable limits. For example, media containing excessive amounts of thymidine or thymine can reverse the inhibitory effects of sulfonamides and trimethoprim, causing zones of growth inhibition to be smaller or less distinct. Organisms may therefore appear resistant to these drugs when in fact they are not. Variation in the concentration of divalent cations, primarily calcium and magnesium, affects results of aminoglycoside, tetracycline, and colistin tests with *P. aeruginosa* isolates (5, 6). A cation content that is too high reduces zone sizes, whereas a cation content that is too low has the opposite effect. When Mueller-Hinton agar is supplemented with blood, the zone of inhibition diameters for oxacillin and methicillin may be 2 to 3 mm smaller than those obtained with unsupplemented agar. On the other hand, sheep blood may markedly increase the zone diameters of some newer cephalosporins when they are tested against enterococci (10). Sheep blood may also cause indistinct zones or a film of growth within the zones of inhibition around sulfonamide and trimethoprim disks.

Inoculation Procedure

To ensure reproducibility of disk diffusion susceptibility test results, the inoculum must be standardized (1, 7, 12, 24). This is accomplished by adjusting the density of the inoculum to equal the turbidity of a barium sulfate ($BaSO_4$) 0.5 McFarland turbidity standard. The standard may be purchased or may be prepared as described above under Dilution Methods. The accuracy of the density of a prepared standard should be verified by using a spectrophotometer

with a 1-cm light path; for the 0.5 McFarland standard, the A_{625} should be 0.08 to 0.10.

The inoculum of the test isolate may be prepared by the growth method or directly from colonies on the agar plate, as described above for dilution testing.

When trimethoprim-sulfamethoxazole is tested by the direct inoculum method, colonies from blood agar medium may carry over enough trimethoprim or sulfonamide antagonists to produce a haze of growth inside the zones of inhibition surrounding susceptible isolates.

The Mueller-Hinton agar plate should be inoculated within 15 min after the inoculum suspension has been adjusted. A sterile cotton swab is dipped into the suspension, rotated several times, and pressed firmly on the inside wall of the tube above the fluid level to remove excess inoculum from the swab. The swab is then streaked over the entire surface of the agar plate three times, with the plate rotated approximately 60° each time to ensure even distribution of the inoculum. A final sweep of the swab is made around the agar rim. The lid is left ajar for 3 to 5 min but no longer than 15 min to allow any excess surface moisture to be absorbed before the drug-impregnated disks are applied.

Antimicrobial Disks

Within 15 min after the plates are inoculated, selected antimicrobial agent disks are distributed evenly on the surface, with at least 24 mm (center to center) between them. Disks are placed individually with sterile forceps or with a mechanical dispensing apparatus and then gently pressed down onto the agar. Generally, place no more than 12 disks on one 150-mm plate and no more than 5 disks on a 100-mm plate. Some of the antimicrobial agent in the disk diffuses almost immediately; therefore, once a disk contacts the agar surface, the disk should not be moved.

Incubation

No longer than 15 min after the disks are applied, the plates are inverted and incubated at 35°C in ambient air. A delay of more than 15 min before incubation permits excess prediffusion of the antimicrobial agents. The interpretive standards for nonfastidious bacteria are based on results of tests incubated in ambient air, and the zone of inhibition diameters of some drugs, such as the aminoglycosides, macrolides, and tetracyclines, are significantly altered by CO_2; therefore, plates should not be incubated in increased CO_2. Testing isolates of some fastidious bacteria, however, requires incubation in 5 to 7% CO_2 (see chapters 114 and 115 of this Manual).

Interpretation and Reporting

Each plate is examined after incubation for 16 to 18 h except for isolates of staphylococci and enterococci, which must be incubated a full 24 h to allow detection of resistance to oxacillin and vancomycin, respectively (24). If plates are inoculated correctly, the zone diameters of inhibition are uniformly circular and the lawn of growth is confluent. Growth that consists of individual colonies indicates that the inoculum was too light, and the test must be repeated. The diameters of the zones of complete inhibition, including the diameter of the disk, are measured to the nearest whole millimeter with sliding calipers, a ruler, or a template prepared specifically for the purpose of reading disk diffusion plates (4). When unsupplemented Mueller-Hinton agar is used, the measuring device is held on the back of the inverted petri dish, which is illuminated with reflected light located a few inches above a black, nonreflecting background. If the medium was supplemented with blood, zones are measured at the agar surface with the cover removed.

The zone margin is the area where no obvious growth is visible. Generally, tiny colonies detected only by close scrutiny or by using transmitted light or mechanical enlargers should generally be ignored. However, when isolates of staphylococci or enterococci are tested, any discernible growth within the zone of inhibition around the oxacillin disk (for staphylococci) or vancomycin disk (for enterococci) indicates resistance. For other bacteria, we recommend that discrete colonies growing within a clear zone of inhibition be subcultured, reidentified, and retested. With *Proteus* species, if a thin film of swarming growth is visible in an otherwise obvious zone of inhibition, the margin of heavy growth is measured and the film is disregarded. With trimethoprim, the sulfonamides, and combinations of the two agents, antagonists in the medium may allow minimal growth; therefore, the zone diameter is measured at the obvious margin, and slight growth (20% or less of the lawn of growth) is disregarded. With Mueller-Hinton agar supplemented with blood, the zone of growth inhibition, not a zone of inhibition of hemolysis, is measured.

The zone diameters measured around each disk are interpreted on the basis of guidelines published by the NCCLS, and the organisms are reported as susceptible, intermediate, or resistant to the antimicrobial agents tested (Table 3) (24). A description of the interpretation of the categories of susceptible, intermediate, and resistant has already been provided above under Dilution Methods.

Advantages and Disadvantages

The disk diffusion test has several advantages: (i) it is technically simple to perform and very reproducible, (ii) the reagents are relatively inexpensive, (iii) it does not require any special equipment, (iv) it provides category results that are easily interpreted by clinicians, and (v) it is flexible regarding selection of antimicrobial agents for testing. The primary limitation of the disk diffusion test is the spectrum of organisms for which it has been standardized. Currently, studies are not adequate to develop reproducible, definitive standards for interpretation of test results with bacteria (not listed in the NCCLS document regarding disk diffusion) that may require different media or atmospheres of incubation for adequate growth or that show delayed or variable growth rates. A potential disadvantage of disk diffusion susceptibility testing is that it provides a qualitative result, but a quantitative result indicating the degree of susceptibility may be desirable in some cases. Moreover, the disk diffusion test may not be adequate for the detection of oxacillin-heteroresistant staphylococci (14, 21, 22) or vancomycin-resistant (low level) enterococci (32).

Quality Control

The goals of a quality control program for disk diffusion are to monitor the precision and accuracy of the procedure, the performance of the reagents (medium, disks), and the performance of persons who do the test and read, interpret, and report results. To best achieve these goals, reference strains are selected for their genetic stability and their usefulness in the disk diffusion test.

Reference Strains

Reference strains recommended by the NCCLS to control the precision and accuracy of the disk diffusion procedure when nonfastidious bacteria are tested are *E. coli* ATCC 25922, *P. aeruginosa* ATCC 27853, *S. aureus* ATCC 25923, *Enterococcus faecalis* ATCC 29212 (or 33186), and *E. coli* ATCC 35218. *E. coli* ATCC 35218 is recommended only as a control for β-lactamase inhibitor combinations containing clavulanic acid, sulbactam, or tazobactam. *Enterococcus faecalis* ATCC 29212 (or 33186) is used to monitor the levels of inhibitors of trimethoprim or sulfonamides in Mueller-Hinton agar. *Enterococcus faecalis* ATCC 29212 also is used to control disks containing a high concentration of gentamicin or streptomycin (see chapter 116 of this Manual).

The reference strains listed above should be obtained from a reliable source, and stock cultures should be maintained in such a way that viability is ensured and the opportunity for selection of resistant variants is minimal (11). The procedures for maintaining and storing working stock cultures have already been described above under Dilution Methods. A culture of a reference strain may be used to monitor the disk diffusion test provided there is no significant change in the mean zone of inhibition diameter that cannot be attributed to an error in methodology. If an unexplained result indicates that the inherent susceptibility of the strain has been altered, a fresh culture of that organism should be obtained.

Zone of Inhibition Diameter Ranges

The ranges of zone diameters for reference strains used to monitor performance of the disk diffusion test are updated frequently; therefore, readers should refer to the most recent NCCLS document for this information. Generally, 1 in every 20 tests in a series of tests might be out of the accepted limits. If a second result falls outside the stated limits, corrective action must be taken. The action taken and the results of that action must be documented.

Frequency of Testing

Each new batch or lot of Mueller-Hinton agar must be tested with the reference strains listed above before the medium is released for use with clinical specimens, and quality control must be done before a new lot of antimicrobial disks is introduced. Appropriate reference strains also should be tested each day the disk diffusion test is performed. The frequency of testing, however, may be reduced if satisfactory performance is documented for 30 consecutive days of testing: For each combination of drug and reference strain, no more than 3 of the 30 zone of inhibition diameters may be outside the accepted limits published by the NCCLS (24). When this criterion is fulfilled, each reference strain need be tested only once per week and any time a reagent component of the test is changed. However, if a zone of inhibition diameter falls outside the acceptable control limits, corrective action must be taken. If the problem appears to be caused by an obvious error such as testing the wrong disk or the wrong reference strain, contamination of the reference strain, or incubation in the incorrect atmosphere, repeating the test with the appropriate reference strain is acceptable. However, if a cause of the error is not obvious, quality control must be performed daily for a period that will allow discovery of the source of the aberrant result and documentation of how the problem was resolved. This may be accomplished by the same approach described under Quality Control in Dilution Methods above.

COMMON SOURCES OF ERROR IN ANTIBACTERIAL SUSCEPTIBILITY TESTING

Potential sources of error in antibacterial susceptibility testing may be categorized as those that relate to the test system and its components, those associated with the test procedure, those peculiar to certain organism and drug combinations, and those that relate to reporting. The most common sources of error encountered in clinical microbiology laboratories are reviewed in the following paragraphs.

Various components of the susceptibility test system may be a source of error. First, the system itself may have limitations regarding the organisms that should be tested. For example, the disk diffusion method should be used only to test rapidly growing bacterial pathogens that have consistent growth rates (those for which interpretive criteria have been developed by the NCCLS), and some commercial systems are not recommended for testing certain organisms. Second, the medium used may be a source of error if it fails to conform to recommended guidelines. For agar dilution and disk diffusion, the Mueller-Hinton agar should be 3 to 4 mm deep. Factors common to both agar-based and broth-based systems are the pH of the medium, which for Mueller-Hinton agar or broth should be between 7.2 and 7.4, and its cation content. The concentration of magnesium and calcium in the broth medium should be that recommended by the NCCLS to ensure reliable results when isolates of *P. aeruginosa* are tested against aminoglycosides. For detection of oxacillin-resistant staphylococci, it is essential that the amount of sodium chloride in the medium containing oxacillin be appropriate for the type of system being used. Third, the components of the system (antimicrobial disks, agar plates, and trays) must be stored properly, and they should not be used beyond the stated expiration dates.

Steps in the susceptibility test procedure that may be a source of error if they are not performed correctly include inoculum preparation, incubation, endpoint interpretation, and performance of appropriate quality control. The inoculum must be pure, and it must contain an adequate concentration of bacteria. With rare exceptions, all systems should be incubated in ambient air at 35°C. The incubation time, however, varies. For conventional dilution and disk diffusion systems, incubation for 16 to 20 h and 16 to 18 h, respectively, is recommended except for isolates of staphylococci and enterococci, which must be incubated a full 24 h (24, 25). Some commercial systems have been designed to provide results in less than 16 h, but they may not be reliable for all organism and drug combinations. The endpoints of all susceptibility tests must be measured accurately, following guidelines published by the NCCLS. If endpoints are interpreted by an instrument, the reliability of that instrument must be monitored. Moreover, with all susceptibility test systems, appropriate reference strains must be tested at regular intervals, and any problems that occur must be thoroughly investigated.

Testing some bacteria against certain antimicrobial agents may yield misleading results, because these in vitro results do not necessarily correlate with in vivo activity. Examples include narrow- and expanded-spectrum cephalosporins and aminoglycosides tested against *Salmonella* and *Shigella* spp.; all beta-lactam agents except the penicillinase-resistant penicillins (oxacillin, nafcillin, methicillin) tested against oxacillin-resistant staphylococci; cephalosporins, aminoglycosides (except concentrations used to detect high-level resistance), clindamycin, and trimethoprim-sul-

famethoxazole tested against enterococci; and cephalosporins tested against *Listeria* spp. Therefore, for these combinations of organisms and drugs, results should not be reported. Other potential problems associated with reporting are possible transcriptional errors for laboratories that use a manual recording and reporting system and possible errors in transmission of data for laboratories in which an automated susceptibility test system is interfaced with the laboratory and/or hospital information system.

REFERENCES

1. **Baker, C. N., C. Thornsberry, and R. W. Hawkinson.** 1983. Inoculum standardization in antimicrobial susceptibility tests: evaluation of the overnight agar cultures and the Rapid Inoculum Standardization System. *J. Clin. Microbiol.* **17:**450–457.
2. **Barry, A. L.** 1991. Procedures and theoretical considerations for testing antimicrobial agents in agar media, p. 1–16. *In* V. Lorian (ed.), *Antibiotics in Laboratory Medicine*, 3rd ed. The Williams & Wilkins Co., Baltimore.
3. **Barry, A. L., R. E. Badal, and R. W. Hawkinson.** 1983. Influence of inoculum growth phase on microdilution susceptibility tests. *J. Clin. Microbiol.* **18:**645–651.
4. **Barry, A. L., L. J. Joyce, A. P. Adams, and E. J. Benner.** 1973. Rapid determination of antimicrobial susceptibility for urgent clinical situations. *Am. J. Clin. Pathol.* **59:**693–699.
5. **Barry, A. L., G. H. Miller, C. Thornsberry, R. S. Hare, R. N. Jones, R. R. Lorber, and C. Cramer.** 1987. Influence of cation supplements on activity of netilmicin against *Pseudomonas aeruginosa* in vitro and in vivo. *Antimicrob. Agents Chemother.* **31:**1514–1518.
6. **Barry, A. L., L. B. Reller, G. H. Miller, J. A. Washington, F. D. Schoenknecht, L. R. Peterson, R. S. Hare, and C. Knapp.** 1992. Revision of standards for adjusting the cation content of Mueller-Hinton broth for testing susceptibility of *Pseudomonas aeruginosa* to aminoglycosides. *J. Clin. Microbiol.* **30:**585–589.
7. **Bauer, A. W., W. M. M. Kirby, J. C. Sherris, and M. Turck.** 1966. Antibiotic susceptibility testing by standardized single disk method. *Am. J. Clin. Pathol.* **45:**493–496.
8. **Bauer, A. W., and J. C. Sherris.** 1964. The determination of sulfonamide susceptibility of bacteria. *Chemotherapia* **9:**1–19.
9. **Brenner, V. C., and J. C. Sherris.** 1972. Influence of different media and bloods on the results of diffusion antibiotic susceptibility tests. *Antimicrob. Agents Chemother.* **1:**116–122.
10. **Buschelman, B. J., R. N. Jones, and M. J. Bale.** 1994. Effects of blood medium supplements on activities of newer cephalosporins tested against enterococci. *J. Clin. Microbiol.* **32:**565–567.
11. **Coyle, M. B., M. F. Lampe, C. L. Aitkin, P. Feigl, and J. C. Sherris.** 1976. Reproducibility of control strains for antibiotic susceptibility testing. *Antimicrob. Agents Chemother.* **10:**436–440.
12. **D'Amato, R. F., and L. Hochstein.** 1982. Evaluation of a rapid inoculum preparation method for agar disk diffusion susceptibility testing. *J. Clin. Microbiol.* **15:**282–285.
13. **D'Amato, R. F., C. Thornsberry, C. N. Baker, and L. A. Kirven.** 1975. Effect of calcium and magnesium ions on the susceptibility of *Pseudomonas* species to tetracycline, gentamicin, polymyxin B, and carbenicillin. *Antimicrob. Agents Chemother.* **7:**596–600.
14. **De Lencastre, H., A. M. Sa Figueiredo, C. Urban, J. Rahal, and A. Tomasz.** 1991. Multiple mechanisms of methicillin resistance and improved methods for detection in clinical isolates of *Staphylococcus aureus*. *Antimicrob. Agents Chemother.* **35:**632–639.
15. **Doern, G. V.** 1987. Breakpoint susceptibility testing. *Clin. Microbiol. Newsl.* **9:**81–84.
16. **Doern, G. V., A. Dascal, and M. Keville.** 1985. Susceptibility testing with the Sensititre breakpoint broth microdilution system. *Diagn. Microbiol. Infect. Dis.* **3:**185–191.
17. **Ericsson, H. M., and J. C. Sherris.** 1971. Antibiotic sensi-

tivity testing. Report of an international collaborative study. *Acta Pathol. Microbiol. Scand. Sect. B Suppl.* **217:**1–90.

18. **Federal Register.** 1972. Rules and regulations: antibiotic susceptibility disks. *Fed. Regist.* **37:**20525–20529.

19. **Garrod, L. P., and P. M. Waterworth.** 1969. Effect of medium composition and the apparent sensitivity of *Pseudomonas aeruginosa* to gentamicin. *J. Clin. Pathol.* **22:**534–538.

20. **Huang, M., P. N. Baker, S. Banerjee, and F. C. Tenover.** 1992. Accuracy of the E Test for determining antimicrobial susceptibilities of staphylococci, enterococci, *Campylobacter jejunii,* and gram-negative bacteria resistant to antimicrobial agents. *J. Clin. Microbiol.* **30:**3243–3248.

21. **Huang, M. B., E. T. Gay, C. N. Baker, S. N. Banerjee, and F. C. Tenover.** 1993. Two percent sodium chloride is required for susceptibility testing of staphylococci with oxacillin when using agar-based dilution methods. *J. Clin. Microbiol.* **31:**2683–2688.

22. **Knapp, C. C., M. D. Ludwig, and J. A. Washington.** 1994. Evaluation of differential inoculum disk diffusion method and Vitek GPS-SA card for detection of oxacillin-resistant staphylococci. *J. Clin. Microbiol.* **32:**433–436.

23. **Metzler, C., and R. M. Dettaan.** 1974. Susceptibility tests of anaerobic bacteria: statistical and clinical considerations. *J. Infect. Dis.* **130:**588–594.

24. **National Committee for Clinical Laboratory Standards.** 1993. *Performance Standards for Antimicrobial Disk Susceptibility Tests.* Approved standard M2-A5. National Committee for Clinical Laboratory Standards, Villanova, Pa.

25. **National Committee for Clinical Laboratory Standards.** 1993. *Methods for Dilution Antimicrobial Susceptibility Tests for Bacteria That Grow Aerobically.* Approved standard M7-A3. National Committee for Clinical Laboratory Standards, Villanova, Pa.

26. **National Committee for Clinical Laboratory Standards.** 1993. *Evaluating Production Lots of Dehydrated Mueller-Hinton Agar.* Tenetative standard M6-T. National Committee for Clinical Laboratory Standards, Villanova, Pa.

27. **National Committee for Clinical Laboratory Standards.** 1994. *Development of In Vitro Susceptibility Testing Criteria and Quality Control Parameters.* NCCLS document M23-A. National Committee for Clinical Laboratory Standards, Villanova, Pa.

28. **Novak, S. M., J. Hindler, and D. A. Bruckner.** 1993. Reliability of two novel methods, Alamar and E Test, for detection of methicillin-resistant *Staphylococcus aureus. J. Clin. Microbiol.* **31:**3056–3057.

29. **Pestotnik, S. L., R. S. Evans, J. P. Burke, and R. M. Gardner.** 1990. Therapeutic antibiotic monitoring: surveillance using a computerized expert system. *Am. J. Med.* **88:**43–48.

30. **Reller, L. B., F. D. Schoenknecht, and M. A. Kenny.** 1974. Antibiotic susceptibility testing of *P. aeruginosa:* selection of a control strain and criteria for magnesium and calcium content in media. *J. Infect. Dis.* **130:**454–463.

31. **Sahm, D. F., C. N. Baker, R. N. Jones, and C. Thornsberry.** 1984. Influence of growth medium on the in vitro activities of second- and third-generation cephalosporins against *Streptococcus faecalis. J. Clin. Microbiol.* **20:**561–567.

32. **Sahm, D. F., J. Kissinger, M. S. Gilmore, P. R. Murray, R. Mulder, J. Solliday, and B. Clarke.** 1989. In vitro susceptibility studies of vancomycin-resistant *Enterococcus faecalis. Antimicrob. Agents Chemother.* **33:**1588–1591.

33. **Sanchez, M. L., M. S. Barrett, and R. N. Jones.** 1992. The E-Test applied to susceptibility tests of gonococci, multiply-resistant enterococci, and *Enterobacteriaceae* producing potent β-lactamases. *Diagn. Microbiol. Infect. Dis.* **15:**459–463.

34. **Sherris, J. C., A. L. Rashad, and G. A. Lighthart.** 1967. Laboratory determination of antibiotic susceptibility to ampicillin and cephalothin. *Ann. N.Y. Acad. Sci.* **145:**248–265.

35. **Washington, J. A., II, R. J. Synder, P. C. Kohner, C. G. Wiltsle, D. M. Ilstrup, and J. T. McCall.** 1978. Effect of cation content of agar on the activity of gentamicin, tobramycin, and amikacin against *Pseudomonas aeruginosa. J. Infect. Dis.* **137:**103–111.

36. **York, M. K., G. F. Brooks, and E. H. Fiss.** 1992. Evaluation of the autoSCAN-W/A rapid system for identification and susceptibility testing of gram-negative fermentative bacilli. *J. Clin. Microbiol.* **30:**2903–2910.

Susceptibility Tests of Fastidious Bacteria

GARY V. DOERN

114

The criteria used to interpret the results of in vitro susceptibility tests are predicated in large measure on the specific method used to perform the test. Important features of any susceptibility test are composition of the medium and conditions of incubation. Fastidious organisms frequently require supplemented media and often require increased CO_2 for growth. As a result, the disk diffusion and dilution procedures described in chapter 113 of this Manual that are based on use of unsupplemented Mueller-Hinton media and incubation in ambient atmospheric air are not applicable. Specialized procedures are necessary for susceptibility tests with fastidious organisms.

This chapter provides practical guidelines for susceptibility tests of *Haemophilus influenzae*, *Streptococcus pneumoniae*, *Moraxella catarrhalis*, and *Neisseria gonorrhoeae*. Susceptibility testing of *Neisseria meningitidis* and viridans and beta-hemolytic streptococci is also briefly discussed. Other fastidious organisms such as *Listeria* spp., lipophilic diphtheroids, *Campylobacter* spp., *Helicobacter* spp., *Bordetella* spp., *Brucella* spp., *Francisella* spp., and the HACEK (*Haemophilus*, *Actinobacillus*, *Cardiobacterium*, *Eikenella*, *Kingella*) bacilli are not discussed. Susceptibility tests of proven reliability for these organisms have not been described. Therefore, routine susceptibility testing of these organisms in the clinical microbiology laboratory should not be undertaken. In selected cases, determination of MICs by using specialized media and conditions of incubation in a research setting may be appropriate; however, the clinical utility of the results of such tests is questionable.

In general, four issues are addressed as they pertain to each of the major organism groups: current resistance patterns, preferred testing methods, alternative testing methods, and laboratory strategies for testing organisms recovered from human clinical material. Information regarding patterns of antimicrobial resistance in the patient population served by an individual laboratory is extremely useful in developing rational, cost-effective approaches to in vitro susceptibility testing. For instance, there is little justification for routinely testing antimicrobial agents known to be either uniformly active or inactive against a particular organism. The preferred testing methods described here are largely those that have received the broadest application in the United States, those that have been endorsed by the National Committee for Clinical Laboratory Standards (NCCLS), and those for which interpretive criteria for

results and quality control guidelines have been developed. A variety of alternative methods are also described. The major limitation for use of these alternative procedures is lack of defined quality control criteria by which proper test performance can be judged. Laboratories employing alternative methods for susceptibility testing must develop and utilize their own performance standards, and even then, such methods should be used only with discretion. Finally, a practical approach for laboratories confronted with performing in vitro susceptibility tests with fastidious bacteria is outlined. These recommendations may warrant modification based on an individual laboratory's patient population, known patterns of resistance, technical expertise, and availability of resources. Unless special circumstances exist, it is assumed that susceptibility tests will be performed only on organisms known or suspected to be clinically significant.

H. INFLUENZAE

Current Patterns and Mechanisms of Resistance

Currently, ca. 20 to 40% of *H. influenzae* strains produce a TEM-1-type β-lactamase and as a result are resistant to ampicillin and amoxicillin (4, 14, 26, 28). Penicillin inherently possesses less activity than these two beta-lactams against *Haemophilus* spp. The actual prevalence of β-lactamase production varies with geographic location, patient demographics, and the presence or absence of type b capsular antigen. The last association is both intriguing and relevant. Typically, rates of β-lactamase-mediated ampicillin resistance are two- to fourfold higher among encapsulated type b strains than among non-type b strains. Strains in the former category are usually associated with systemic, life-threatening infections such as meningitis, while strains in the second category cause non-life-threatening, often self-limiting infections of the respiratory tract such as otitis media, sinusitis, and bronchopulmonary infections. Through widespread use of the recently licensed conjugate-*H. influenzae* type b vaccine, systemic *Haemophilus* infections due to encapsulated type b strains have become far less prevalent in the United States. As a result, the form of *H. influenzae* most commonly encountered in the clinical laboratory today, i.e., nontypeable *H. influenzae*, is less likely to be resistant to ampicillin by virtue of production of

TABLE 1 Description of disk diffusion susceptibility test methods for *H. influenzae*, *S. pneumoniae*, and *N. gonorrhoeae*

Organism	Medium[a]	Inoculum[b]	Incubation time (h)[c]	Zone measurement
H. influenzae	HTM agar	20- to 24-h chocolate agar growth	16–18	From bottom of plate
N. gonorrhoeae	GC agar base + supplement	20- to 24-h chocolate agar growth	20–24	From top of plate with lid removed
S. pneumoniae	MH-SBA	20- to 24-h SBA growth	20–24	From top of plate with lid removed

[a]The compositions of HTM and GC agar base with supplement are defined in the text.
[b]Growth for all inocula is at 35°C in 5 to 7% CO_2. Suspension is in MHB or 0.9% NaCl to a 0.5 McFarland standard.
[c]All incubations are at 35°C in 5 to 7% CO_2.

β-lactamase. The β-lactamase of *H. influenzae* is readily detected by a variety of different β-lactamase assays, including acidometric, iodometric, and chromogenic cephalosporin methods that employ either nitrocefin or PADAC as substrates. These β-lactamase assays are described in detail in chapter 116 of this Manual.

Strains of *H. influenzae* that are resistant to ampicillin by mechanisms other than production of a TEM-1-type β-lactamase have been described elsewhere (36, 37). The mechanism of resistance in such strains may be altered targets of action (i.e., penicillin-binding proteins), diminished penetration of antimicrobial agent to its site of action because of decreased permeability of the outer membrane of this gramnegative bacterium, or production of a different β-lactamase (i.e., ROB-1 enzyme). In the first two cases, ampicillin resistance will not be detected with a standard β-lactamase assay. Fortunately, β-lactamase-negative ampicillin-resistant strains (BLNAR) remain extremely uncommon in the United States, with a prevalence of <0.1% (4, 14, 26, 28).

Except for narrow-spectrum cephalosporins such as cephalothin, cefazolin, cephadroxil, and cephalexin, cephalosporin resistance is uncommon in *H. influenzae*. Rates of resistance of 1 to 5% are observed with cefaclor, cefprozil, and loracarbef (a carbacephem). Fewer then 1% of strains are resistant to cefuroxime, cefixime, and cefpodoxime. Resistance to ceftriaxone, cefotaxime, ceftizoxime, and ceftazidime has not been described. Other agents with near-uniform activity against *H. influenzae* include the β-lactamase inhibitor combinations amoxicillin-clavulanate and ampicillin-sulbactam, imipenem, aztreonam, the two new macrolides azithromycin and clarithromycin, and the fluoroquinolone antimicrobial agents. Chloramphenicol resistance also remains uncommon. Rates of resistance of 1 to 5% are observed with the tetracyclines, trimethoprim-sulfamethoxazole (TMP-SMX), and rifampin. Several general references regarding antimicrobial resistance in *H. influenzae* are references 4, 14, 26, and 28.

Preferred Test Methods

Table 1 lists the general features of the optimum disk diffusion susceptibility test method for *H. influenzae*. Table 2 describes the salient characteristics of the preferred dilution procedure for this bacterium. The methods described are those advocated by the NCCLS. A list of the antimicrobial agents for which these methods are applicable when *H. influenzae* is tested together with interpretive criteria for results and quality control ranges can be obtained from the appropriate NCCLS documents, i.e., M2-A5 for disk diffusion tests and M7-A3 for dilution susceptibility tests (40–42). It is important to realize that the breakpoints and quality control guidelines published by the NCCLS apply only to tests performed precisely as described in NCCLS documents. Changing any of the test parameters renders the interpretive criteria for the results and quality control ranges meaningless, since the results of any antimicrobial susceptibility test are completely dependent on the method used.

Haemophilus test medium (HTM) in either its agar or broth form is now advocated for susceptibility tests with *Haemophilus* spp. (32, 40, 41). HTM consists of Mueller-Hinton base, 15 μg of hematin per ml, 15 μg of NAD per ml, and 5 mg of yeast extract per ml (15, 32, 40, 41). The broth version of the medium is cation adjusted and contains 0.2 IU of thymidine phosphorylase per ml. The principal

TABLE 2 Description of dilution susceptibility test methods for *H. influenzae*, *S. pneumoniae*, and *N. gonorrhoeae*

Organism	Medium[a]	Format	Final vol (μl)	Inoculum[b]	Final concn (CFU)	Incubation conditions
H. influenzae	HTM broth	Broth microdilution	100	20- to 24-h chocolate agar growth	5×10^5/ml	35°C for 20–24 h in ambient air
N. gonorrhoeae	GC agar base + supplement	Agar dilution	NA	20- to 24-h chocolate agar growth	10^4/spot	35°C and 5–7% CO_2 for 20–24 h
S. pneumoniae	MHB-LHB	Broth microdilution	100	20- to 24-h SBA growth	5×10^5/ml	35°C for 20–24 h in ambient air

[a]The compositions of HTM and GC agar base with supplement are defined in the text.
[b]See Table 1, footnote *b*.

advantages of HTM are its transparency; its suitability for testing TMP-SMX; its commercial availability, at least in the agar form; and the availability of interpretive criteria for results and quality control guidelines for numerous antimicrobial agents. As with any new medium, problems have been noted by some laboratories that use HTM (12, 22, 38). Most notable among these problems have been poor growth or growth failures with rare strains of *H. influenzae* that have higher-than-normal hemin (X-factor) requirements, indistinct zones of inhibition in disk diffusion tests, and trailing endpoints in broth microdilution tests, especially when relatively β-lactamase-labile cephalosporins such as cefaclor, cefprozil, loracarbef, and cefuroxime are tested versus BLNAR strains. Another problem is the inability of disk diffusion tests to correctly categorize BLNAR strains as being resistant to these cephalosporins and to β-lactamase inhibitor combinations such as amoxicillin-clavulanate.

Two American Type Culture Collection (ATCC) strains of *H. influenzae* are advocated for quality control tests with HTM: ATCC 49766 and ATCC 49247. ATCC 49766 is used for tests of cefaclor, cefuroxime, cefonicid, loracarbef, and cefprozil activities, and ATCC 49247 is used with all of the other agents for which quality control criteria for *H. influenzae* have been developed (42). The reason for having two quality control strains relates to the BLNAR nature of ATCC 49247. This strain was actually first recommended by the NCCLS for all quality control tests on HTM; however, it proved to be too rigorous a challenge for the five cephalosporins noted above (2). As a result, a second strain, ATCC 49766, is now advocated for testing these agents. In addition, laboratories experiencing poor growth or growth failures with HTM can employ a third ATCC strain of *H. influenzae*, ATCC 10211, as a means of assessing the growth-supporting properties of HTM. This strain has a high X-factor growth requirement. All commercial manufacturers of HTM are encouraged to use ATCC 10211 as an internal growth control prior to shipping media.

Alternative Test Methods

Numerous different methods and media have been advocated for susceptibility tests with clinical isolates of *H. influenzae* (43). Disk diffusion tests may be performed using chocolate Mueller-Hinton agar (choc-MHA) and four antimicrobial agents: ampicillin, chloramphenicol, amoxicillin-clavulanate, and ampicillin-sulbactam (39, 53). Disk diffusion susceptibility tests with choc-MHA must be restricted to these four agents because these are the only antimicrobial agents for whose results interpretive criteria based on systematic studies have been developed. Mueller-Hinton broth (MHB) containing 3 to 5% lysed horse blood (LHB) may be used as an alternative to HTM broth for microdilution tests (3, 4, 32, 48). Any antimicrobial agent may be tested, but the MIC interpretive breakpoints established for HTM do not necessarily apply to MHB-LHB, and interpretive criteria based specifically on the use of MHB-LHB have never been developed. Indeed, MICs obtained with *H. influenzae* in HTM broth are typically 0.5 to 2 doubling dilutions lower than those obtained in MHB-LHB (3, 4, 32, 48). One exception is TMP-SMX, for which MICs determined in HTM broth are usually 1 or 2 doubling dilutions higher than those obtained in MHB-LHB. MICs can also be determined for *H. influenzae* by using the Epsilometer or E test on either HTM agar or choc-MHA (30). However, this method must be more extensively evaluated before its routine use can be advocated.

Laboratory Strategies for Susceptibility Testing

A rapid β-lactamase assay should be performed on all isolates as a convenient and reliable means of quickly assessing the activities of ampicillin and related compounds such as amoxicillin. A positive β-lactamase test result indicates ampicillin resistance; a negative result indicates that an isolate is highly likely to be susceptible. As noted above, rare β-lactamase-negative strains of *H. influenzae* are ampicillin resistant. Obviously, a negative β-lactamase assay result with a BLNAR strain would falsely predict ampicillin susceptibility. If such strains were common, all organisms yielding negative β-lactamase test results would have to be confirmed with some direct test of ampicillin activity before being reported as ampicillin susceptible. Fortunately, BLNAR strains of *H. influenzae* remain distinctly uncommon in the United States, and when they do occur, they are nontypeable and associated with non-life-threatening infections (4, 14, 26, 28). For this reason, to all intents and purposes, a negative β-lactamase test result translates into ampicillin susceptibility and can be reported as such without performing a confirmatory test of ampicillin activity. In those rare cases when the activity of ampicillin versus a known β-lactamase-negative organism needs to be determined directly, either disk diffusion or broth microdilution tests should be performed.

Most other antimicrobial agents have very predictable activities or lack of activity against *H. influenzae* and therefore need not be tested routinely. Agents with nearly uniform activity include the extended-spectrum cephalosporins, the carbapenems, the tetracyclines, azithromycin, clarithromycin, and the quinolones. With the exception of BLNAR strains of *H. influenzae*, the β-lactamase inhibitor combinations amoxicillin-clavulanate and ampicillin-sulbactam are also uniformly active. Selected agents may warrant testing because of documented resistance. These agents include cefaclor, loracarbef, cefprozil, and TMP-SMX. Tests with these agents should be performed at the discretion of individual laboratories, and the disk diffusion or broth microdilution tests described above may be used. Lastly, mention should be made of chloramphenicol. Rates of resistance to this agent remain at <0.5% in the United States (4, 14, 26, 28). Furthermore, chloramphenicol is rarely used today in the management of *Haemophilus* infections. For these reasons, there rarely exists a need for testing chloramphenicol. However, in those rare circumstances in which chloramphenicol susceptibility information is necessary, a rapid assay for chloramphenicol acetyltransferase (CAT) activity should be performed. CAT assays are described in chapter 116 of this Manual. When the tests are performed correctly, the results of either the conventional tube CAT assay or a commercially available disk test for CAT production predict the activity of chloramphenicol versus *H. influenzae* with nearly complete accuracy (13).

S. PNEUMONIAE

Current Patterns and Mechanisms of Resistance

Until recently, penicillin was nearly uniformly active against *S. pneumoniae*. During the last decade, however, first in other parts of the world and now in the United States, strains of *S. pneumoniae* with higher-than-normal penicillin MICs have appeared (1). Currently, ca. 20% of clinical isolates in the United States are apparently not susceptible to penicillin (4, 51). Among these, 1 to 3% express high-level resistance (i.e., MICs of ≥2.0 μg/ml), while the remaining 17 to 19% have intermediate penicillin MICs (i.e., MICs of 0.1 to 1.0 μg/ml). Strains in the former category are referred to as resistant and are refractile to penicillin therapy irrespective of disease association. Strains in the latter category were initially termed relatively resistant (45, 52). More recently, they have been called intermediate (41) and appear to be responsive to treatment with penicillin except in patients with meningitis. The mechanism of resistance is alteration of the targets of penicillin action, i.e., penicillin-binding proteins (see chapter 112 of this Manual). As a result, the activities of other beta-lactam antimicrobial agents such as ampicillin, amoxicillin, the β-lactamase inhibitor combinations, cephalosporins, and carbapenems against these strains are also diminished. Use of these agents to treat pneumococcal infections due to penicillin-intermediate or -resistant strains must therefore be guided by suitable susceptibility tests.

With the exception of three cephalosporins with intrinsically limited activities against *S. pneumoniae* (ceftazidime, cefixime, and ceftibuten), all of the beta-lactams noted above are uniformly active against penicillin-susceptible pneumococci. Resistance of *S. pneumoniae* to vancomycin has not been described. In the United States, rates of resistance to the macrolides, erythromycin, azithromycin, and clarithromycin as well as to chloramphenicol and tetracycline have traditionally been very low, i.e., 0.5 to 3% (4, 28). This appears to be changing, at least with selected agents. For instance, TMP-SMX resistance has become more prevalent in *S. pneumoniae* and now occurs at rates of up to 40% in the United States (4). Macrolide resistance, often linked to resistance to the tetracyclines and/or chloramphenicol, occurs most frequently in penicillin-resistant strains of *S. pneumoniae* (20). As penicillin-resistant strains become more prevalent, rates of resistance to the macrolides, tetracyclines, and chloramphenicol are also expected to increase. The quinolone class of antimicrobial agents, at least those that are currently available in the United States, in general have limited pneumococcal activity.

Preferred Test Methods

Disk diffusion susceptibility tests with pneumococci are best performed by using MHA plates supplemented with 5% defibrinated sheep blood (MH-SBA) (41, 50, 53). Table 1 lists the general features of the optimum disk diffusion susceptibility test method for *S. pneumoniae*. Dilution tests should be done in a broth microdilution format with cation-adjusted MHB-LHB used as a medium (50). It is essential that the LHB used as a supplement be antibiotic free and completely devoid of erythrocyte ghosts. Table 2 describes the salient characteristics of the preferred dilution procedure for this bacterium. The methods described are those advocated by the NCCLS. A list of the antimicrobial agents for which these methods are applicable when *S. pneumoniae* is tested, interpretive criteria for results, and quality control ranges can be obtained from the appropriate NCCLS disk diffusion and dilution susceptibility test documents (M2-A5 and M7-A3, respectively) (40–42).

As noted above, strains of *S. pneumoniae* that are not susceptible to penicillin have now become a problem throughout the world. Such strains are most definitively characterized by using the broth microdilution method in MHB-LHB. Organisms for which penicillin MICs are $\leq$0.06 μg/ml should be considered susceptible. MICs of 0.1 to 1.0 μg/ml define an isolate as penicillin intermediate. MICs of $\geq$2.0 μg/ml denote frank resistance.

A disk diffusion screening test has been developed to presumptively identify penicillin-resistant and -intermediate strains of *S. pneumoniae* (10, 24, 41, 50, 52). It is performed on MH-SBA plates with disks that contain 1 μg of oxacillin each. Penicillin disks are not used. A zone of inhibition of $\geq$20 mm surrounding a 1-μg oxacillin disk indicates that the organism is penicillin susceptible (i.e., has a penicillin MIC of $\leq$0.06 μg/ml). An inhibitory zone of $\leq$19 mm implies that the test strain is either intermediate or penicillin resistant. The disk diffusion test, however, does not distinguish between these two groups. An MIC must be determined in order to precisely categorize an isolate of *S. pneumoniae* as being either intermediate or frankly resistant to penicillin. Furthermore, at least a small percentage of susceptible strains will also have zone sizes of $\leq$19 mm in the oxacillin screening test; i.e., they will be falsely categorized as resistant or intermediate. For these reasons, all isolates that yield an oxacillin zone size of $\leq$19 mm should have their penicillin MICs determined. Prior to the availability of the penicillin MIC, such strains may be considered presumptively penicillin intermediate or resistant. Quality control testing with *S. pneumoniae*, both disk diffusion and broth microdilution, is accomplished by using the ATCC strain of *Pneumococcus*, ATCC 49619.

Alternative Test Methods

Some investigators have suggested that HTM can be used with *S. pneumoniae* at least for broth dilution MIC determinations if not for disk diffusion tests (3, 27, 31). This practice should be discouraged, primarily because of high rates of growth failure when fresh clinical isolates of *Pneumococcus* are grown on or in HTM. The E test is, however, a very useful means of determining MICs with a variety of antimicrobial agents versus *S. pneumoniae* (23, 29, 30, 35). Indeed, the E test may currently represent the optimum susceptibility test method for this organism. The medium of choice for E-test MIC determinations with *S. pneumoniae* is MH-SBA. An organism suspension equivalent to a 0.5 McFarland turbidity standard is used to inoculate plates in a manner identical to that used with the standardized disk diffusion procedure. Plastic strips impregnated with a gradient of antimicrobial concentrations (E-test strips, PDM Epsilometer; AB Biodisk, Solna, Sweden) are applied radially to plates such that the lowest concentrations of drug are oriented toward the center. Plates are incubated for 20 to 24 h, and elliptical zones of inhibition are observed. MICs are determined by the line at which the inhibition ellipse intersects the test strip.

Laboratory Strategies for Susceptibility Testing

All isolates of *S. pneumoniae* should be tested by using the oxacillin disk screening procedure for penicillin susceptibility. Organisms yielding zones of inhibition of $\geq$20 mm may be considered penicillin susceptible. Strains with zones of $\leq$19 mm should be reported as presumptively not susceptible to penicillin. A definitive report of penicillin activity on such strains should be predicated on the results of a confirmatory MIC determination. The results of penicillin susceptibility tests also appear to predict the activities of ampicillin, amoxicillin, amoxicillin-clavulanate, and ampicillin-sulbactam.

In most circumstances, only strains that are penicillin intermediate or resistant need be evaluated for susceptibility to relevant cephalosporins and imipenem, since penicillin-susceptible strains of *S. pneumoniae* are also uniformly susceptible to these agents. In selected situations, however, e.g., systemic-disease isolates in a geographic area where extended-spectrum cephalosporin resistance has been noted, it might be prudent to initiate tests with relevant cephalosporins and perform the oxacillin screen for penicillin activity simultaneously. Unfortunately, at present, the only beta-lactam antimicrobial agents other than penicillin for which the NCCLS promulgates interpretive criteria of results are cefotaxime and ceftriaxone, and then only for broth microdilution tests performed in MH-LHB (40, 42).

The tetracyclines, the macrolides, and chloramphenicol probably need not be routinely tested against *S. pneumoniae*. However, as noted above, rates of resistance to these agents may be increasing, and there may come a time when routine testing will be justifiable, at least for pneumococcal isolates from infections that could be treated with these antimicrobial agents. Susceptibility tests with tetracycline versus *S. pneumoniae* predict the activities of other agents in this class. The same is true of erythromycin, whose activity accurately predicts the activities of other macrolides (34). In contrast, on the basis of higher current rates of resistance, a case can be made for routinely testing TMP-SMX by using either a disk diffusion or an MIC method on isolates of *S. pneumoniae* obtained from patients with infections that can be managed with an oral antimicrobial agent. Currently available quinolone-class agents need not be tested against *S. pneumoniae*. In general, these agents lack sufficient activity to warrant use in pneumococcal infections. Testing the quinolones might mistakenly encourage their inappropriate use. Obviously, as patterns of resistance change and as new antimicrobial agents are developed, these recommendations for routine testing of *S. pneumoniae* may warrant modification.

M. CATARRHALIS

Current Patterns and Mechanisms of Resistance

M. catarrhalis, a newly recognized pathogen of the human respiratory tract, has an extremely predictable susceptibility profile. Nearly all clinically significant isolates of *M. catarrhalis* produce a β-lactamase that is chromosomal, constitutively produced in small amounts, and inhibited by clavulanate and sulbactam (11, 56). Only one β-lactamase assay reliably detects the *M. catarrhalis* enzyme: the chromogenic cephalosporin method that utilizes nitrocefin as a substrate (16; see also chapter 116 of this Manual). Two principal β-lactamase types have been described: BRO-1 and BRO-2 (56). These enzymes can be distinguished by their isoelectric profiles, substrate specificities, and relative amounts of enzyme produced (BRO-1 >>> BRO-2) (56). They also differ with respect to prevalence, with BRO-1-producing strains being 10-fold more prevalent than BRO-2-producing strains (11). Perhaps the most important difference between strains that produce these two different types of β-lactamases is the MICs obtained with beta-lactam antimicrobial agents such as penicillin and ampicillin. BRO-1-producing strains have high MICs of these drugs (e.g., ≥4.0 μg/ml) and clearly appear to be resistant, whereas strains that produce the BRO-2 enzyme typically have very low MICs (e.g., ≤0.5 μg/ml), indicating susceptibility (11). The chromogenic cephalosporin β-lactamase assay does not distinguish between these two β-lactamase types.

The activities of most other antimicrobial agents are predictable (11, 56). *M. catarrhalis* is typically susceptible to cephalosporins, amoxicillin-clavulanate, ampicillin-sulbac-

tam, and imipenem. The *M. catarrhalis* β-lactamase does not inactivate these compounds. The tetracyclines and erythromycin also appear to be nearly uniformly active against *M. catarrhalis*, with resistant strains documented only infrequently (6). The same has been true of TMP-SMX; however, my own recent experience suggests that rates of TMP-SMX resistance with *M. catarrhalis* may be increasing.

Preferred Test Methods

Although often considered a fastidious bacterium, *M. catarrhalis* grows readily on unsupplemented Mueller-Hinton medium in ambient atmospheric air at 35 to 37°C and is therefore best tested by using the disk diffusion and dilution procedures recommended by the NCCLS for nonfastidious bacteria (M2-A5 and M7-A3, respectively) (40–42). The utility of these methods for tests with *M. catarrhalis* have been documented for seven antimicrobial agents: ampicillin, amoxicillin-clavulanate, cephalothin, cefaclor, erythromycin, tetracycline, chloramphenicol, and TMP-SMX (17). As noted above, other antimicrobial agents are either uniformly active or inactive against *M. catarrhalis*. Therefore, they need not be tested, at least routinely. However, when such tests are deemed necessary, they can probably also be performed with unsupplemented Mueller-Hinton media.

Laboratory Strategies for Susceptibility Testing

The vast majority of clinical isolates of *M. catarrhalis* need be tested only with the nitrocefin-based chromogenic cephalosporin β-lactamase assay. The results of this assay predict the activities of penicillin, ampicillin, and amoxicillin with near certainty (16). Rare reports of erythromycin resistance have appeared (6); however, the prevalence of such strains remains far less than 0.1%, and as a result, erythromycin need not be tested routinely. With one exception, other antimicrobial agents are either uniformly active or inactive against *M. catarrhalis* and therefore also need not be tested (18). The single exception is TMP-SMX. Increasing rates of resistance to TMP-SMX justify routine testing of this drug combination, at least in clinical situations in which TMP-SMX might be used.

N. GONORRHOEAE

Current Patterns and Mechanisms of Resistance

In the United States, the three most important mechanisms of antibiotic resistance among clinical isolates of *N. gonorrhoeae* are plasmid-associated TEM-1-type β-lactamase-mediated penicillin resistance (in PPNG isolates), penicillin resistance due to mutations in chromosomal genes that result in altered penicillin-binding proteins or diminished outer membrane permeability (in CMRNG isolates), and high-level plasmid-mediated tetracycline resistance (in TRNG isolates). PPNG express high-level resistance to penicillin, ampicillin, and amoxicillin. Since the mechanism of resistance is production of a TEM-1-type β-lactamase, such strains are readily identified by using any of the β-lactamase assays noted in the *Haemophilus* section above. PPNG remain susceptible to most nonpenicillin beta-lactams such as the β-lactamase inhibitor combinations, most cephalosporins, and imipenem. TRNG usually occur in the absence of resistance to other agents. Rare strains of *N. gonorrhoeae*, however, possess the plasmid determinants for both β-lactamase production and high-level tetracycline resistance (33). CMRNG, in addition to being penicillin resistant, usually also express at least a low-level resistance to a variety of other agents, including tetracycline, certain cephalosporins, erythromycin, the aminoglycosides, and TMP-SMX. The actual prevalence of these three forms of antibiotic resistance among strains of *N. gonorrhoeae* in the United States is extremely variable and depends on geographic location, patient demographics, and time. Other types of resistance that have been described with *N. gonorrhoeae* but that occur infrequently include resistance to the quinolone class of agents; expanded-spectrum cephalosporins such as cefoxitin, cefotetan, and cefuroxime; and spectinomycin (21).

Preferred Test Methods

An excellent review of issues surrounding antimicrobial susceptibility testing of *N. gonorrhoeae* was recently provided by Fekety (21). According to the results of a recent multicenter study, there now exist defined methods for both disk diffusion and agar dilution susceptibility tests with *N. gonorrhoeae* (25, 40, 41). Agar dilution remains the method of choice for determining MICs for this organism because of the tendency of the organism to lyse in broth media. Both disk diffusion and agar dilution tests with *N. gonorrhoeae* are best performed using GC agar base to which has been added a 1% defined growth supplement. Supplement containing a much lower concentration of cystine (i.e., 1.1 g/liter) than is usually present is required to avoid inhibition of carbapenem and clavulanate activities in agar dilution MIC tests with *N. gonorrhoeae* (25). Table 1 lists the general features of the optimum disk diffusion susceptibility test method for *N. gonorrhoeae*. Table 2 describes the salient characteristics of the preferred dilution procedure for this bacterium. The methods described are those advocated by the NCCLS. A listing of the antimicrobial agents for which these methods are applicable when *N. gonorrhoeae* is tested, interpretive criteria for results, and quality control ranges can be obtained from the appropriate NCCLS documents (M2-A5 [41, 42] for disk diffusion tests and M7-A3 [40, 42] for dilution susceptibility tests). *N. gonorrhoeae* ATCC 49226 should be used for quality control.

Alternative Test Methods

An alternative to GC agar base with 1% defined supplement for disk diffusion susceptibility tests with *N. gonorrhoeae* is choc-MHA. Interpretive criteria for disk diffusion results have been promulgated for four antimicrobial agents: penicillin, tetracycline, spectinomycin, and cefoxitin (54). As an alternative to using agar dilution, MICs can be determined with the E test with either choc-MHA or GC agar base containing 1% defined supplement as a test medium (47).

Laboratory Strategies for Susceptibility Testing

The currently recommended drug of first choice in the management of all gonococcal infections except pelvic inflammatory disease is ceftriaxone. Although mean MICs of ceftriaxone for *N. gonorrhoeae* appear to be increasing, frank resistance has never been described anywhere with clinical isolates of *N. gonorrhoeae*. Therefore, the drug need not be tested routinely. Other drugs of potential utility in infections due to *N. gonorrhoeae*, such as penicillin, tetracycline, spectinomycin, selected other cephalosporins, and the quinolones, need be tested only when clinically indicated. Examples might include treatment failure isolates; organisms recovered from body sites indicative of pelvic

inflammatory disease, for which cefoxitin is the drug of choice; epidemiologic surveillance of resistance; isolates recovered from patients with ceftriaxone allergy; and circumstances in which local physician bias leads to use of drugs other than ceftriaxone for primary treatment. PPNG are best detected with a β-lactamase assay. CMRNG require use of disk diffusion or dilution susceptibility tests, procedures that may also be applied to the other antimicrobial agents noted above.

N. MENINGITIDIS

Until recently, there has been no need to perform susceptibility tests with clinical isolates of *N. meningitidis*. This organism has remained nearly uniformly susceptible to those antimicrobial agents of utility in the management of meningococcal infections, i.e., penicillin, ampicillin, ceftriaxone, cefotaxime, and chloramphenicol. In some parts of the world, this pattern of susceptibility appears to be changing. For instance, in South Africa, Spain, England, Canada, and the United States, isolates of *N. meningitidis* with elevated penicillin MICs (i.e., ≥0.1 μg/ml) have been described (5, 44, 46, 49, 57). In addition, β-lactamase production by this organism has been reported (9). Finally, resistance to agents such as the sulfonamides and rifampin, drugs used for chemoprophylaxis, is now well documented (20, 55). Because of the changes that are occurring in *N. meningitidis*, susceptibility test methods and interpretive criteria for results that are of proven reliability may become necessary in the future. In general, however, such methods do not yet exist.

One method that has been proposed for screening for penicillin and ampicillin susceptibility is a 1-μg oxacillin disk diffusion test performed with MH-SBA plates (8). Organisms with zones of inhibition of ≤10 mm may be considered relatively resistant to penicillin and ampicillin. Similar conclusions can be derived from strains of *N. meningitidis* with inhibitory zones of ≤26 mm surrounding 2-U penicillin disks (7). This oxacillin disk screening procedure is analogous to that used with *S. pneumoniae*. Further study is required, however, before this procedure can be advocated as a method for routine testing of *N. meningitidis*. Currently, when a need arises to determine the susceptibility of an isolate of *N. meningitidis*, an MIC should probably be determined. This can be accomplished by using a broth microdilution test with MHB-LHB or by performing an E test on either an MH-SBA plate or a choc-MHA plate. In the absence of interpretive criteria of results of proven clinical value for meningococcal infections, the results of these MIC tests should be interpreted with caution.

VIRIDANS AND BETA-HEMOLYTIC STREPTOCOCCI

As with *N. meningitidis*, susceptibility test methods of proven reliability for viridans and beta-hemolytic streptococci have not yet been described. This is problematic, since resistance to penicillin (viridans and beta-hemolytic streptococci) and macrolide antimicrobial agents (beta-hemolytic streptococci) has now been observed and appears to be increasing in prevalence in at least some parts of the world (19). In addition, the beta-hemolytic streptococci in particular have reemerged as important if not reasonably common causes of systemic infections in patients with antecedent localized diseases of the skin or upper respiratory tract. As we await the development of susceptibility test methods suitable for testing these two organism groups, performance of an MIC test is probably prudent in clinical situations in which a susceptibility test result is essential to management decisions. MICs should be determined by using a broth microdilution procedure in MHB-LHB.

REFERENCES

1. **Appelbaum, P. C.** 1992. Antimicrobial resistance in *Streptococcus pneumoniae*: an overview. *Clin. Infect. Dis.* **15:**77–83.
2. **Barry, A. L., and R. R. Packer.** 1992. Performance of *Haemophilus* test media prepared with 12 different lots of Mueller-Hinton agar from four manufacturers. *J. Clin. Microbiol.* **30:** 1145–1147.
3. **Barry, A. L., M. A. Pfaller, and P. C. Fuchs.** 1993. Broth medium for antimicrobial susceptibility tests of *Haemophilus influenzae, Streptococcus pneumoniae, Streptococcus pyogenes,* and *Moraxella catarrhalis.* Haemophilus test medium versus Mueller-Hinton broth with lysed horse blood. *Eur. J. Clin. Microbiol. Infect. Dis.* **12:**548–553.
4. **Barry, A. L., M. A. Pfaller, P. C. Fuchs, and R. R. Packer.** 1994. In vitro activities of twelve orally administered antimicrobial agents against four species of bacterial respiratory pathogens from U.S. medical centers in 1992 and 1993. *Antimicrob. Agents Chemother.* **38:**2419–2425.
5. **Botha, P.** 1988. Penicillin-resistant *Neisseria meningitidis* in southern Africa. *Lancet* **i:**54.
6. **Brown, B. A., R. J. Wallace, Jr., C. W. Flanagan, R. W. Wilson, J. I. Luman, and S. D. Redditt.** 1989. Tetracycline and erythromycin resistance among clinical isolates of *Branhamella catarrhalis. Antimicrob. Agents Chemother.* **33:**1631–1633.
7. **Campos, J., P. M. Mendelman, M. U. Sako, D. O. Chaffin, A. L. Smith, and J. A. Saez-Nieto.** 1987. Detection of relatively penicillin G-resistant *Neisseria meningitidis* by disk susceptibility testing. *Antimicrob. Agents Chemother.* **31:**1478–1482.
8. **Campos, J., G. Trujillo, T. Seuba, and A. Rodriguez.** 1992. Discriminative criteria for *Neisseria meningitidis* isolates that are moderately susceptible to penicillin and ampicillin. *Antimicrob. Agents Chemother.* **36:**1028–1031.
9. **Dillon, J. R.** 1984. β-Lactamase-producing *Neisseria meningitidis. Clin. Microbiol. Newsl.* **6:**165–166.
10. **Dixon, J. M. S., A. E. Lipinski, and M. E. P. Graham.** 1977. Detection and prevalence of pneumococci with increased resistance of penicillin. *Can. Med. Assoc. J.* **117:**1159–1161.
11. **Doern, G. V.** 1986. *Branhamella catarrhalis*—an emerging human pathogen. *Diagn. Microbiol. Infect. Dis.* **4:**191–201.
12. **Doern, G. V.** 1992. In vitro susceptibility testing of *Haemophilus influenzae*: review of new National Committee for Clinical Laboratory Standards recommendations. *J. Clin. Microbiol.* **30:**3035–3038.
13. **Doern, G. V., G. S. Daum, and T. A. Tubert.** 1987. In vitro chloramphenicol susceptibility testing of *Haemophilus influenzae*: disk diffusion procedures and assays for chloramphenicol acetyltransferase. *J. Clin. Microbiol.* **25:**1453–1455.
14. **Doern, G. V., J. H. Jorgensen, C. Thornsberry, D. A. Preston, T. Tubert, J. S. Redding, and L. A. Maher.** 1988. National collaborative study of the prevalence of antimicrobial resistance among clinical isolates of *Haemophilus influenzae. Antimicrob. Agents Chemother.* **32:**180–185.
15. **Doern, G. V., J. H. Jorgensen, C. Thornsberry, and H. Snapper.** 1990. Disk diffusion susceptibility testing of *Haemophilus influenzae* using Haemophilus test medium. *Eur. J. Clin. Microbiol. Infect. Dis.* **9:**329–336.
16. **Doern, G. V., and T. Tubert.** 1987. Detection of β-lactamase activity among clinical isolates of *Branhamella catarrhalis* using six different β-lactamase assays. *J. Clin. Microbiol.* **25:**1380–1383.
17. **Doern, G. V., and T. Tubert.** 1988. Disk diffusion susceptibility testing of *Branhamella catarrhalis* with ampicillin and seven other antimicrobial agents. *Antimicrob. Agents Chemother.* **31:**1519–1523.
18. **Doern, G. V., and T. A. Tubert.** 1988. In vitro activities of 39 antimicrobial agents for *Branhamella catarrhalis* and comparison of results with different quantitative susceptibility test methods. *Antimicrob. Agents Chemother.* **32:**259–261.
19. **Dowson, C. G., A. Hutchison, N. Woodford, A. P. Johnson, R. C. George, and B. G. Spratt.** 1990. Penicillin-resistant viridans streptococci have obtained altered penicillin binding protein genes from penicillin-resistant strains of *Streptococcus pneumoniae. Proc. Natl. Acad. Sci. USA* **87:**5853–5862.
20. **Facinelli, B., and P. E. Varaldo.** 1987. Plasmid-mediated sulfonamide resistance in *Neisseria meningitidis. Antimicrob. Agents Chemother.* **31:**1642–1643.
21. **Fekety, T.** 1993. Antimicrobial susceptibility testing of *Neisseria gonorrhoeae* and implications for epidemiology and therapy. *Clin. Microbiol. Rev.* **6:**22–23.
22. **Heelan, J. S., D. Chesney, and G. Guadagno.** 1992. Investigation of ampicillin-intermediate strains of *Haemophilus influenzae* by using the disk diffusion procedure and current National Committee for Clinical Laboratory Standards guidelines. *J. Clin. Microbiol.* **30:**1674–1677.
23. **Jacobs, M. R., S. Bajahsouzian, P. C. Applebaum, and A. Bolmstrom.** 1992. Evaluation of the E-test for susceptibility testing of pneumococci. *Diagn. Microbiol. Infect. Dis.* **15:**473–478.
24. **Jacobs, M. R., M. N. Jaspar, R. M. Robins-Browne, and H. J. Kooruhof.** 1980. Antimicrobial susceptibility testing of pneumococci. II. Determination of optimum disc diffusion test for detection of penicillin G resistance. *J. Antimicrob. Chemother.* **6:**53–54.
25. **Jones, R. N., T. L. Gavan, C. Thornsberry, P. C. Fuchs, E. H. Gerlach, J. S. Knapp, P. Murray, and J. A. Washington II.** 1989. Standardization of disk diffusion and agar dilution susceptibility tests for *Neisseria gonorrhoeae*: interpretive criteria and quality control guidelines for ceftriaxone, penicillin, spectinomycin, and tetracycline. *J. Clin. Microbiol.* **27:** 2758–2766.
26. **Jorgensen, J. H.** 1992. Update on mechanisms and prevalence of antimicrobial resistance in *Haemophilus influenzae. Clin. Infect. Dis.* **14:**119–123.
27. **Jorgensen, J. H., G. V. Doern, M. J. Ferraro, C. C. Knapp, J. M. Swenson, and J. A. Washington.** 1992. Multicenter evaluation of the use of *Haemophilus* test medium for broth microdilution antimicrobial susceptibility testing of *Streptococcus pneumoniae* and development of quality control limits. *J. Clin. Microbiol.* **30:**961–966.
28. **Jorgensen, J. H., G. V. Doern, L. A. Maher, A. W. Howell, and J. S. Redding.** 1990. Antimicrobial resistance among respiratory isolates of *Haemophilus influenzae, Moraxella catarrhalis,* and *Streptococcus pneumoniae* in the United States. *Antimicrob. Agents Chemother.* **34:**2075–2080.
29. **Jorgensen, J. H., M. J. Ferraro, M. L. McElmeel, J. Spargo, J. M. Swenson, and F. C. Tenover.** 1994. Detection of penicillin and extended-spectrum cephalosporin resistance among *Streptococcus pneumoniae* clinical isolates by use of the E-test. *J. Clin. Microbiol.* **32:**159–163.
30. **Jorgensen, J. H., A. W. Howell, and L. A. Maher.** 1991. Quantitative antimicrobial susceptibility testing of *Haemophilus influenzae* and *Streptococcus pneumoniae. J. Clin. Microbiol.* **29:**109–114.
31. **Jorgensen, J. H., L. A. Maher, and A. W. Howell.** 1990. Use of *Haemophilus* test medium for broth microdilution antimicrobial susceptibility testing of *Streptococcus pneumoniae. J. Clin. Microbiol.* **28:**430–434.
32. **Jorgensen, J. H., J. S. Redding, L. A. Maher, and A. W. Howell.** 1987. Improved medium for antimicrobial susceptibility testing of *Haemophilus influenzae. J. Clin. Microbiol.* **25:**2105–2113.
33. **Knapp, J. S., J. M. Zenilman, J. W. Biddle, G. H. Perkins, W. E. DeWitt, M. L. Thomas, S. R. Johnson, and S. A. Morse.** 1987. Frequency and distribution in the United States

of strains of *Neisseria gonorrhoeae* with plasmid-mediated, high-level resistance to tetracycline. *J. Infect. Dis.* **155:**819–822.

34. **Knudsen, J. D., M. Petersen, R. Poulsen, N. Frimodt-Møller, and F. Espersen.** 1994. Susceptibility testing of pneumococci, susceptible as well as resistant towards penicillin, to six macrolides, abstr. D36, p. 123. *Program Abstr. 34th Intersci. Conf. Antimicrob. Agents Chemother.*

35. **Macias, E. A., E. O. Mason, H. Y. Ocera, and M. T. LaRocco.** 1994. Comparison of E test with standard broth microdilution for determining antibiotic susceptibilities of penicillin-resistant strains of *Streptococcus pneumoniae. J. Clin. Microbiol.* **32:**430–432.

36. **Markowitz, S. M.** 1980. Isolation of an ampicillin-resistant, non-beta lactamase-producing strain of *Haemophilus influenzae. Antimicrob. Agents Chemother.* **17:**80–83.

37. **Mendelman, P. M., D. O. Chaffin, T. L. Stull, C. E. Rubens, K. D. Mack, and A. L. Smith.** 1990. Characterization of non-β-lactamase-mediated ampicillin resistance in *Haemophilus influenzae. Antimicrob. Agents Chemother.* **26:**235–244.

38. **Mendelman, P. M., E. A. Wiley, T. L. Stull, C. Clausen, D. O. Chaffin, and O. Onay.** 1990. Problems with current recommendations for susceptibility testing of *Haemophilus influenzae. Antimicrob. Agents Chemother.* **34:**1480–1484.

39. **National Committee for Clinical Laboratory Standards.** 1985. *Performance Standards for Antimicrobial Disk Susceptibility Tests.* Approved standard M2-A3. National Committee for Clinical Laboratory Standards, Villanova, Pa.

40. **National Committee for Clinical Laboratory Standards.** 1993. *Methods for Dilution Antimicrobial Susceptibility Tests for Bacteria That Grow Aerobically,* 3rd ed. (M7-A3). National Committee for Clinical Laboratory Standards, Villanova, Pa.

41. **National Committee for Clinical Laboratory Standards.** 1993. *Performance Standards for Antimicrobial Disk Susceptibility Tests,* 5th ed. (M2-A5). National Committee for Clinical Laboratory Standards, Villanova, Pa.

42. **National Committee for Clinical Laboratory Standards.** 1994. *Performance Standards for Antimicrobial Susceptibility Testing;* fifth informational supplement (M100-S5). National Committee for Clinical Laboratory Standards, Villanova, Pa.

43. **Needham, C. A.** 1988. *Haemophilus influenzae:* antibiotic susceptibility. *Clin. Microbiol. Rev.* **1:**218–227.

44. **Riley, G., S. Brown, and C. Krishnan.** 1992. Penicillin resistance in *Neisseria meningitidis. Clin. Microbiol. Newsl.* **6:**165–166.

45. **Saah, A. J., J. P. Mallonee, M. Tarpay, C. Thornsberry,** M. A. Roberts, and E. R. Rhoades. 1980. Relative resistance to penicillin in pneumococcus. A prevalence and case-control study. *JAMA* **243:**1824–1827.

46. **Saez-Nieto, J. A., and J. Campos.** 1988. Penicillin-resistant strains of *Neisseria gonorrhoeae* in Spain. *Lancet* **i:**1452–1453.

47. **Sanchez, M. L., M. S. Barrett, and R. N. Jones.** 1992. The E test applied to susceptibility tests of gonococci, multiply-resistant enterococci and *Enterobacteriaceae* producing potent beta-lactamases. *Diagn. Microbiol. Infect. Dis.* **15:**459–463.

48. **Scriver, S. R., D. E. Low, A. E. Simor, B. Toye, A. McGeer, R. Jaeger, and Canadian Haemophilus Study Group.** 1992. Broth microdilution testing of *Haemophilus influenzae* with *Haemophilus* test medium versus lysed horse blood broth. *J. Clin. Microbiol.* **30:**2284–2289.

49. **Sutcliff, E. M., D. M. Jones, S. El-Sheikh, and A. Perdival.** 1988. Penicillin-insensitive meningococci in the UK. *Lancet* **i:**657.

50. **Swenson, J. M., B. C. Hill, and C. Thornsberry.** 1986. Screening pneumococci for penicillin resistance. *J. Clin. Microbiol.* **24:**749–752.

51. **Thornsberry, C., J. K. Marler, and T. J. Rich.** 1992. Increased penicillin resistance in recent U.S. isolates of *Streptococcus pneumoniae,* abstr. C-268, p. 465. *Abstr. 92nd Gen. Meet. Am. Soc. Microbiol.*

52. **Thornsberry, C., and J. M. Swenson.** 1980. Antimicrobial susceptibility tests for *Streptococcus pneumoniae. Lab. Med.* **11:**83–86.

53. **Thornsberry, C., J. M. Swenson, C. N. Baher, L. K. McDougal, S. A. Stocker, and B. C. Hill.** 1988. Methods for determining susceptibility of fastidious and unusual pathogens to selected antimicrobial agents. *Diagn. Microbiol. Infect. Dis.* **9:**135–138.

54. **U.S. Department of Health and Human Services.** 1987. Antibiotic-resistant strains of *Neisseria gonorrhoeae. Morbid. Mortal. Weekly Rep.* **36:**s10–s12.

55. **Vanderlaan, J.** 1991. Meningitis due to a rifampin resistant group C *Neisseria meningitidis. Clin. Microbiol. Newsl.* **13:**47–48.

56. **Wallace, R. J., Jr., R. D. Nash, and V. A. Steingrube.** 1990. Antibiotic susceptibilities and drug resistance in *Moraxella* (*Branhamella*) *catarrhalis. Am. J. Med.* **88:**s46–s53.

57. **Woods, C. B., B. Wasilauskas, and L. Givner.** 1992. Meningitis caused by *Neisseria meningitidis* relatively resistant (NmRRP) in North Carolina. Paper presented at 32nd Interscience Conference on Antimicrobial Agents and Chemotherapy.

Susceptibility Testing of Anaerobic Bacteria

HANNAH M. WEXLER AND GARY V. DOERN

115

INTRODUCTION AND LABORATORY STRATEGIES FOR SUSCEPTIBILITY TESTING

Although anaerobic bacteria are well-documented causes of a variety of common, often serious infections in humans, the appropriateness of routine testing of anaerobes is not universally agreed upon (3, 14, 29). Traditionally, anaerobic infections have been treated empirically without need for in vitro antimicrobial susceptibility testing. This was done for several reasons. Anaerobic infections, especially endogenous infections in which the host's own commensal flora causes disease, are usually polymicrobic. As a result, the clinical meaning of in vitro susceptibility tests has been questioned, since such tests are typically performed on pure cultures with individual organisms. Also, anaerobic infections are often associated with tissue necrosis and abscess formation, leading to impaired delivery of antimicrobial agents in blood to the actual site of infection. This explains why anaerobic infections are often aggressively managed with debridement, aspiration, and/or surgical removal of infected tissue. Indeed, in many cases, antimicrobial therapy is used as an adjunct to primary surgical management. Additionally, methods for recovery of anaerobic bacteria from clinical specimens as well as techniques for identifying isolates have traditionally been very cumbersome and time-consuming, often leading to the availability of isolates for susceptibility testing only many days after clinical specimens were first obtained. Finally, since many anaerobic infections are endogenous (arising in sites contiguous to a mucosal surface that harbors a resident anaerobic flora), distinguishing clinically significant isolates from contaminants is often difficult if not impossible. In view of these considerations, the clinical utility of anaerobic susceptibility testing is frequently considered limited.

On the other hand, evidence of regional variation in susceptibility profiles suggests that hospitals could periodically consider batch testing organisms either in their own facility or by sending them to a reference laboratory (8). Certain circumstances also may justify performance of anaerobic susceptibility tests on individual patient isolates (3, 24). However, even when anaerobic isolates are recovered and characterized in a timely manner and determined to be clinically significant, standardized methods for anaerobic susceptibility testing are not always reliable and often present growth patterns that are difficult to interpret. It is

little wonder that most clinical microbiology laboratories have not felt compelled to develop extensive anaerobic susceptibility testing capabilities.

Because of the technical and interpretive difficulties associated with anaerobic susceptibility testing, presentation of definitive recommendations is difficult. This chapter describes currently available methodologies and their interpretation with the understanding that many uncertainties still surround in vitro susceptibility testing of anaerobic bacteria (28).

CURRENT PATTERNS OF RESISTANCE

A general outline of current resistance patterns for anaerobic bacteria is provided below. For more specific information, a variety of publications may be consulted (16, 19, 28–31). Note that patterns reported in one geographic location do not necessarily correspond to those observed at other locations (5, 15, 27). Reasons for such differences include variations in test methods, lack of agreement between countries on appropriate interpretive breakpoints, and lack of clearly defined breakpoints in published reports. Given the clustering of the MICs for many strains near the breakpoint, changes in interpretive breakpoints of 1 or 2 dilutions may change reports of susceptibility percentages by as much as 50%.

Bacteroides spp. and Related Gram-Negative Bacilli

Penicillin, ampicillin, and the anti-*Pseudomonas* penicillins ticarcillin, mezlocillin, and piperacillin are active against about 70% of *Prevotella* and *Porphyromonas* (formerly *Bacteroides*) spp. but against only 5 to 20% of the *Bacteroides fragilis* group. The principal mechanism of resistance is β-lactamase production (9, 10, 13, 27). Other penicillins, such as oxacillin and dicloxacillin (the isoxazolyl penicillins), are intrinsically much less active against all of these organisms. Not surprisingly, the β-lactam–β-lactamase inhibitor combinations are effective against essentially all strains of the *B. fragilis* group. There have been only rare reports of resistance to these agents. Other antimicrobial agents generally active against strains of the *B. fragilis* group in the United States include the carbapenems (imipenem and meropenem), chloramphenicol, and metronidazole. While the currently available quinolones have poor to fair activities against anaerobes, some newer quinolone agents under development have excellent activities against anaerobes (16, 17, 31).

Cefoxitin, an expanded-spectrum cephalosporin that is often used for treating anaerobic infections, and clindamycin are active against 70 to 80% of the *B. fragilis* group species (11). Among the members of this group, *B. fragilis* appears to be most susceptible. Cefotetan, a cephalosporin similar to cefoxitin, is active against *B. fragilis* but is much less active against the other members of the group (i.e., *Bacteroides distasonis*, *Bacteroides ovatus* and *Bacteroides thetaiotaomicron* [11]). The broad-spectrum cephalosporins vary greatly with respect to activity against the *B. fragilis* group, usually inhibiting ca. 50% of the non-*B. fragilis* members of the group. Ceftizoxime appears to be the most active broad-spectrum cephalosporin. However, its activity is influenced by the test method used; that is, ceftizoxime appears to be more active when tested with a broth microdilution method than when tested with an agar dilution technique (2, 6, 24). Isolated strains of *Bacteroides* resistant to metronidazole have been reported in the United Kingdom (15) and, more commonly, in France (25). A few isolates resistant to amoxicillin-clavulanate and one isolate resistant to chloramphenicol have also been reported (27).

Other Gram-Negative Bacilli

Penicillin resistance has been observed among isolates of the genus *Fusobacterium*. Nineteen percent of isolates were found to be resistant to amoxicillin at 4 μg/ml, with most resistant strains producing a β-lactamase (19). The rate of resistance to extended-spectrum penicillins such as ticarcillin and piperacillin is ≤10%. In general, ≥90% of *Fusobacterium* spp. are susceptible to cephalosporins. Rates of resistance to clindamycin of <5% are seen among *Fusobacterium* spp. (19). The motile anaerobic gram-negative rods such as *Wolinella* spp. vary in susceptibility to β-lactams but are almost always susceptible to chloramphenicol and metronidazole.

Gram-Positive Bacilli and Cocci

Nonsporeforming gram-positive bacilli (*Eubacterium*, *Actinomyces*, *Propionibacterium*, and *Lactobacillus* spp.) are typically susceptible to penicillins and to β-lactam–β-lactamase inhibitor combinations. Organisms in the first three genera are also usually susceptible to the cephalosporins, while lactobacilli may be resistant. All four organism groups are often resistant to metronidazole. *Clostridium difficile*, the organism responsible for pseudomembranous colitis and most cases of antibiotic-associated diarrhea, is often resistant to cephalosporins (though it is quite susceptible to cefotetan) and clindamycin but susceptible to metronidazole, vancomycin, and various penicillins. Note, however, that ampicillin and cefotetan can both lead to overgrowth of *C. difficile* in the colon despite the susceptibility of this organism to these agents in vitro. *Clostridium perfringens* is almost always susceptible to the agents commonly used to treat anaerobic infections. With other *Clostridium* species, there are resistance rates of 15 to 30% to clindamycin and cefoxitin and ≥30% to other cephalosporins. Occasional resistance to penicillin is also seen. β-Lactamase production has been observed in some species, i.e., *Clostridium ramosum*, *Clostridium butyricum*, and *Clostridium clostridioforme*. Chloramphenicol and metronidazole are almost always active against the clostridia. The anaerobic cocci, *Peptococcus* and *Peptostreptococcus* spp., are normally susceptible to β-lactam–β-lactamase inhibitor combinations, penicillins,

and cephalosporins. Also, some strains are resistant to metronidazole, but upon reexamination, these strains are usually shown to be microaerophilic.

DESCRIPTION OF TEST METHODS

As with all susceptibility tests, preparation of inocula, medium composition, test format, conditions of incubation, results interpretation, and quality control are essential determinants of the results obtained with anaerobes.

Inoculum Preparation

The inoculum for anaerobic susceptibility tests may be prepared by suspending colonies taken from a 24- to 72-h blood agar plate directly in any clear broth that does not suppress growth of anaerobes to a density equivalent to a 0.5 McFarland standard. Examples are thioglycolate broth and brucella broth (24). With broth dilution MIC determinations, inocula may be conveniently prepared in aliquots of the growth medium used to perform the test. Alternatively, the initial suspension may be prepared by inoculating five or more colonies into enriched fluid thioglycolate medium and incubating it for 4 to 6 h (or overnight for slow-growing organisms). This suspension is then diluted to a density equivalent to a 0.5 McFarland standard (24, 26).

Media

The preferred medium for agar dilution tests with anaerobes is Wilkins-Chalgren agar (33) or brucella agar base. For broth microdilution tests, several broth media have been used successfully, including Schaedler's, West-Wilkins, brain heart infusion (BHI), and a broth with the same formulation as Wilkins-Chalgren agar but with the agar omitted. The last medium is available commercially (Anaerobe Broth; Difco Laboratories, Detroit, Mich.)

Supplements must be added to these media to support the growth of certain fastidious anaerobes such as *Bacteroides gracilis*, *Bilophila wadsworthia*, many pigmenting *Prevotella* and *Porphyromonas* spp., *Fusobacterium* spp., and anaerobic cocci. A general guideline is that unless growth of the test organism after 48 h is sufficient to permit easy visual discernment of growth versus no growth, an accurate MIC cannot be determined and a supplement must therefore be used. Use of supplements, however, requires that the microbiologist prove that the supplement does not interfere with the test by ensuring that quality control organisms yield the expected results.

Brucella agar, for instance, may be supplemented with vitamin K$_1$ (1 μg/ml) and hemin (5 μg/ml). Vitamin K$_1$ is prepared in a stock solution (10 mg/ml) by mixing 0.2 ml of vitamin K$_1$ (3-phytylmenadione) with 20 ml of 95% ethanol. This stock solution should be stored at 4 to 8°C in dark bottle. A working solution (1 mg/ml stored at 4 to 8°C in a dark bottle for no more than 30 days) is added to the brucella agar base in a 1:1,000 ratio to achieve a final concentration of 1 μg/ml just prior to autoclaving. The hemin supplement is prepared in a stock solution (5 mg/ml) by dissolving 0.5 g of hemin in 10 ml of 5 N NaOH (certified ACS), adding distilled water to 100 ml, and sterilizing the solution at 121°C for 15 min. This stock may be stored at 4 to 8°C for 1 month. One milliliter of stock solution is added to 1 liter of medium just prior to autoclaving. Defibrinated sheep blood or lysed (laked) sheep blood prepared by alternate freezing and thawing may also be used to supplement both Wilkins-Chalgren agar and brucella agar base. Both of these supplements are used at a final concentration of 5%. Other supplements that may be considered for agar dilution tests include 1.0% Tween 80 for gram-positive cocci and rabbit serum or 2 to 3% laked horse blood for pigmented *Prevotella* and *Porphyromonas* spp. and other fastidious anaerobes. The most commonly used supplement for broth microdilution susceptibility tests with anaerobes is 3 to 5% lysed horse blood. Finally, formate-fumarate should routinely be added to media (agar or broth) for testing of *B. gracilis* and other *Bacteroides ureolyticus* group strains.

The National Committee for Clinical Laboratory Standards (NCCLS) recommends that for routine agar dilution testing, plates should not be stored longer than 7 days before being used (24). Extensive experience obtained in the laboratory of one of us (H.M.W.) indicates that optimal performance is obtained when agar dilution plates are prepared within 24 h of use. Plates containing imipenem should never be prepared more than 24 h in advance of the test.

Conditions of Incubation

Anaerobic jars equipped with disposable hydrogen-carbon dioxide generators and palladium-coated catalyst pellets or an anaerobic chamber are the methods recommended for incubating agar dilution plates, broth microdilution trays, and broth macrodilution tubes. The incubation atmosphere should contain 4 to 7% CO_2. Catalyst pellets may be reactivated after use by heating them at 160°C for 1.5 to 2 h. Indicators of anaerobiosis should be included. Incubation in an anaerobic atmosphere should be at 35 to 37°C for 48 h.

The control plates, microdilution trays, or broth macrodilution tubes are divided, and then one is incubated anaerobically (growth control) and one is incubated aerobically (aerobic contaminant control). For agar dilution, an additional plate should be inoculated and subsequently refrigerated to be used as an inoculum control for distinguishing slight growth from dried inoculum. Agar dilution test plates should be left at room temperature until the inoculum is absorbed, and they should then be stacked and incubated upside down (to prevent condensate from falling on the inoculated spots). Microdilution trays should not be stacked more than four high to allow for even incubation temperature. Trays may be covered with plastic lids or perforated tape to reduce evaporation (cellophane tape should not be used).

Agar Dilution Test Methods

Agar dilution MIC determinations with anaerobic bacteria are best accomplished by using the methods described by the NCCLS (24) or in the *Wadsworth Anaerobic Bacteriology Manual* (26). These methods are also discussed in detail in chapter 113 of this Manual. The preferred medium is either Wilkins-Chalgren agar (for the NCCLS reference technique) or supplemented brucella-laked blood agar. The final inoculum should be 10^5 CFU per spot, an inoculum that is 10-fold larger than that advocated for use in agar dilution MIC determinations with aerobic or facultative bacteria.

Broth Microdilution Tests

Microdilution trays may be prepared and frozen or may be purchased commercially either frozen or with antibiotic solutions added to wells in either a freeze-dried or a lyophilized state. Frozen trays are stored in plastic bags at

−70°C or lower for up to 6 months prior to use. Dried trays may be stored at room temperature for the periods indicated by the manufacturer's expiration date. The final volume in each well of a microdilution tray should be at least 100 µl. Well volumes of less than 100 µl should not be used for anaerobes both because of evaporation, which can result in concentration of antimicrobial agents, and because of the marked effect of larger inocula in smaller volumes. Frozen trays should be allowed to equilibrate to room temperature before inoculation. Dried trays may be used directly.

Inocula should be prepared as for agar dilution, with a final inoculum concentration in each well of ca. 10^6 CFU/ml. The actual volume of inoculum transferred to each well of a microdilution tray is generally 0.01 ml (or less than 10% of the volume of broth in wells) with frozen trays or 100 µl with dried trays. The preparation of the inoculum for broth microdilution testing is very important, because variations in the inoculum density significantly affect results. Trays should be inoculated within 15 min after inocula are prepared. The actual inoculation of trays must be standardized and can be accomplished by using a hand-held inoculator or a mechanized dispenser. Prereducing the trays prior to inoculation, i.e., placing them in an anaerobic environment for 2 to 4 h, may enhance the growth of certain fastidious anaerobes. In addition, trays *must* be prereduced if metronidazole is to be tested, since the antimicrobial activity of metronidazole is dependent on the formation of an active intermediate that demands a reduced atmosphere.

Control wells should include a well with broth but no drug (growth control) and an uninoculated well as a sterility check. This well may also be used as a "negative" control for visual comparison with growth in inoculated wells. MBCs cannot be reliably determined by using broth microdilution trays with anaerobic bacteria (29a).

Alternative Test Procedures

The limited agar dilution or breakpoint method combines the advantages of agar dilution with economy and clinical utility (24). Two to four concentrations of each antimicrobial agent are tested in divided plates. As many as 16 organisms may be tested against 12 antimicrobial agents in a single anaerobe jar. At least three concentrations of each antimicrobial agent should be tested: the MIC breakpoint that defines resistance for a particular agent plus 1 dilution higher and 1 dilution lower than the breakpoint concentration. Other concentrations that are midrange for susceptibility may be added.

The broth macrodilution procedure is particularly useful for organisms with swarming growth such as some *Clostridium* spp. or if MBCs are to be determined. Serial twofold dilutions of antimicrobial stock solutions are prepared in 2.5 ml of brucella broth containing 5 µg of hemin, 1 mg of $NaHCO_3$ (26), and 1 µg of vitamin K_1 per ml or in one of the microdilution broths mentioned above. Tubes should be prepared within 3 h of use or frozen at −70°C. The inoculum should be a 1:200 dilution of a 0.5 McFarland standard prepared as described above and made in the same broth used to dilute the drugs, and an inoculum volume of 2.5 ml should be added to the broth containing the drug. The final inoculum will be ca. 2.5×10^5 CFU/ml. An inoculated broth containing no antimicrobial agent is included as a growth control for each strain tested, and a tube of uninoculated broth is included as a sterility control.

The E test (A-B Biodisk, Solna, Sweden) is gaining popularity, and a number of studies have validated its utility and indicate that results obtained correlate well with those obtained using the NCCLS-approved agar dilution method (7, 23). In the E test, a suspension of test organism equivalent to a 0.5 McFarland standard is streaked confluently over the surface of a 150-mm-diameter petri dish containing agar that will support growth of the test organism. Plastic strips impregnated with a gradient of antimicrobial concentrations are applied radially to the surface of the plate such that the lowest concentrations of drug are toward the center. Plates are incubated for 24 to 48 h in an anaerobic atmosphere until growth is readily apparent. Elliptical zones of inhibition are observed, and MICs are defined by the line of intersection between the inhibitory ellipse and the test strip.

The medium of choice for E tests with anaerobes appears to be brucella agar supplemented with vitamin K, hemin, and 5% laked (lysed) sheep blood, although Wilkins-Chalgren agar without added blood may also be used (7). E-test MICs determined on supplemented brucella agar are nearly identical to those determined with the NCCLS reference agar dilution method for cefoxitin, ceftizoxime, imipenem, penicillin, metronidazole, and clindamycin (7, 23).

The advantages of the E test are its flexibility, convenience, and simplicity. It is particularly well suited to laboratories that test individual isolates of anaerobic bacteria infrequently. If additional studies continue to confirm its reliability, the E test may represent the optimal approach to anaerobic susceptibility testing, especially for smaller laboratories or those that do not perform such testing in batches. Its principal drawback is its high cost.

The spiral streak method (Spiral Systems Instruments, Bethesda, Md.), used for the spiral gradient endpoint determination, deposits the antibiotic in a concentration gradient that decreases radially from the center of the plate. Isolates are deposited on the plate in radial streaks, and after incubation for 48 h in an anaerobic atmosphere, the endpoints of growth are marked. Data are entered into a computer software program that determines the concentration of drug at that point based on the radius of growth and the molecular weight (i.e., diffusion characteristics) of the antimicrobial agent. This method correlates well with agar dilution (18, 32) and may be very useful for batch testing of isolates or for research laboratories that perform studies with large numbers of isolates.

Modifications of the agar disk diffusion technique have been used for rapidly growing anaerobes (4). While disk diffusion tests correlated well with an agar dilution technique for fast-growing anaerobes with some antimicrobial agents, there were many instances in which the correlation was poor. At present, disk diffusion tests are not considered appropriate for anaerobic susceptibility testing.

Finally, a broth disk elution technique has commonly been used in the past for determining the antimicrobial susceptibilities of anaerobic bacteria. As originally described, specified numbers of disks containing antimicrobial agents (such as those commonly used for aerobic disk diffusion susceptibility testing) were added to tubes of prereduced, anaerobically sterilized (PRAS) BHI broth and left to sit for 30 min (34). During this time, the antimicrobial agent eluted off the disk into the broth; the number of disks and the disk contents as well as the volume of BHI determined the final concentration of antimicrobial agent in the tube. The tubes were inoculated with test organism, incubated in an anaerobic environment, and then examined for macroscopic evidence of growth after a specified period of incubation. In effect, this method represents a breakpoint

broth macro-tube dilution procedure. Limitations of this technique included the necessity of using a specialized broth and the requirement for anaerobic incubation. The procedure was modified by Kurzynski and colleagues, who substituted the more commonly available thioglycolate broth for PRAS BHI broth, which in turn permitted incubation of tubes in ambient atmospheric air (22). This latter method became very popular, undoubtedly because of its simplicity.

Unfortunately, despite its convenience and the broad application this technique has experienced in the past, the broth disk elution method can no longer be advocated for use in testing anaerobes in routine clinical microbiology laboratories. It has been found to yield unreliable results when compared to a standardized broth microdilution method (1) and the reference agar dilution method of the NCCLS (21, 35), particularly with *Bacteroides* spp., the most commonly encountered anaerobe in most laboratories. Indeed, in the most recent NCCLS guidelines for anaerobic susceptibility testing, the broth disk elution method has been eliminated (24).

Currently, the only β-lactamase assay of any value for testing anaerobic bacteria is the nitrocefin disk assay. This procedure is described in chapter 116 of this Manual. Acidometric or iodimetric assays should not be used because of the high rates of false-negative results obtained with anaerobes. In general, while the nitrocefin disk assay is simple and rapid and may detect the β-lactamases produced by species of *Prevotella*, *Porphyromonas*, and *Bacteroides* as well as selected other anaerobes, even this assay has limited clinical utility owing to a lack of sensitivity. Other newer, investigational reagents may be of some practical value in detecting β-lactamase production among anaerobic isolates in the future (12, 20). A reliable β-lactamase assay for anaerobes would be very useful. β-Lactamase-producing strains are resistant to many penicillins and cephalosporins but susceptible to β-lactam agents combined with β-lactamase inhibitors such as sulbactam, tazobactam, or clavulanic acid. It must be remembered, however, that resistance to β-lactam agents is not always mediated by β-lactamases, as other mechanisms of resistance have been described.

Interpretation of Results

Recently, the NCCLS revised the definition of an anaerobic agar dilution MIC endpoint to emphasize that the endpoint should be read as the concentration at which there is the most marked change from the growth control (24). This change might be manifested as no growth, as tiny colonies, as a haze, or as greatly diminished growth compared to that of the control. Ambiguous endpoints occur most frequently with gram-negative organisms and certain β-lactams, in particular, cephalosporins and β-lactam–β-lactamase inhibitor combinations. The plates should be read against a dark background, thus diminishing the appearance of the "haze." Limited viability data support this approach to endpoint determination. Further study is required to determine whether this approach is optimal for all anaerobes.

For broth microdilution MIC determinations, trays should be examined by indirect transmitted light, preferably with the assistance of a viewing device. The criteria for endpoint determination for the microdilution technique are similar to those for the agar dilution technique: the concentration at which the most significant reduction in growth is observed should be chosen as the endpoint. This reduction in growth may be manifested as complete inhibition of growth or as a tiny, gradually diminishing button of growth. Trailing endpoints may be observed with some antimicrobial agents, certain media, or some antimicrobial agent-organism combinations.

MIC determinations should be interpreted according to the interpretive criteria recommended by the NCCLS (24). These criteria apply to agar dilution, broth microdilution, and broth macrodilution tests. The NCCLS has recommended that an intermediate category be established for anaerobic bacteria because of the difficulty in determining endpoints and the clustering of MICs at breakpoint concentrations. In general, a maximum dosage of an antimicrobial agent is recommended for anaerobic infections to achieve the highest possible drug levels at the site of infection. With these dosage regimens, it is reasoned that organisms in the susceptible or intermediate categories will probably respond to therapy if appropriate surgical intervention is taken.

Quality Control

The use of quality control strains is necessary for monitoring all of the variables that may influence anaerobic susceptibility tests. Appropriate control tests should be performed whenever new media, reagents, or antimicrobial agents are prepared. Quality control must be performed for all of the techniques described above. The recommended quality control strains are *B. fragilis* ATCC 25285, *B. thetaiotaomicron* ATCC 29741, and *Eubacterium lentum* ATCC 43055. Two quality control strains should be used for each test run, preferably including a strain with an MIC in the same concentration range as those of the test strains. It is recognized that this may not always be possible. Expected values for quality control strains are published by the NCCLS (24). As a general rule, MICs for both broth macrodilution and broth microdilution tests are often 1 twofold dilution lower than those obtained with agar dilution tests. Also, values in agar media containing blood may be 1 twofold dilution higher than the NCCLS-approved values. Certain organism-antimicrobial agent combinations are not recommended for quality control because endpoints are difficult to determine or are too variable. In some cases, there is only one recommended quality control strain. Efforts to establish strains that give reliable and consistent quality control readings are being undertaken by the NCCLS. The latest NCCLS guidelines add a new quality control strain, *E. lentum*, and delete the *Clostridium perfringens* quality control strain because of inconsistent results in laboratory tests (24).

REFERENCES

1. **Aldridge, K. E., A. Henderberg, D. D. Schiro, and C. V. Sanders.** 1990. Discordant results between the broth disk elution and broth microdilution susceptibility tests with *Bacteroides fragilis* group isolates. *J. Clin. Microbiol.* **28:**375–378.
2. **Aldridge, K. E., H. M. Wexler, C. V. Sanders, and S. M. Finegold.** 1990. Comparison of in vitro antibiograms of *Bacteroides fragilis* group isolates: differences in resistance rates in two institutions because of differences in susceptibility testing methodology. *Antimicrob. Agents Chemother.* **34:**179–181.
3. **Baron, E. J., D. M. Citron, and H. M. Wexler.** 1990. Son of anaerobic susceptibility testing—revisited. *Clin. Microbiol. Newsl.* **12:**69–70.
4. **Barry, A. L., P. C. Fuchs, E. H. Gerlach, S. D. Allen, J. F. Acar, K. E. Aldridge, A.-M. Bourgault, H. Grimm, G. S. Hall, W. Heizmann, R. N. Jones, J. M. Swenson, C. Thornsberry, H. Wexler, J. D. Williams, and J. Wust.** 1990. Multilaboratory evaluation of an agar diffusion disk suscepti-

bility test for rapidly growing anaerobic bacteria. *Rev. Infect. Dis.* **2:**s210–s217.
5. **Betriu, E., E. Campos, C. Cabronero, C. Rodriguez-Aveil, and J. J. Picaxo.** 1990. Susceptibilities of species of the *Bacteroides fragilis* group to 10 antimicrobial agents. *Antimicrob. Agents Chemother.* **34:**671–673.
6. **Borobio, M. V., A. Pascual, M. C. Dominguez, and E. J. Perea.** 1986. Effect of medium, pH and inoculum size on activity of ceftizoxime and Sch-34343 against anaerobic bacteria. *Antimicrob. Agents Chemother.* **30:**626–627.
7. **Citron, D. M., M. I. Ostavari, A. Karlsson, and E. J. C. Goldstein.** 1991. Evaluation of the Epsilometer (E-test) for susceptibility testing of anaerobic bacteria. *J. Clin. Microbiol.* **29:**2197–2203.
8. **Cornick, N. A., G. J. Cuchural, Jr., D. R. Syndman, N. V. Jacobus, P. Iannini, G. Hill, T. Cleary, J. P. O'Keefe, C. Pierson, and S. M. Finegold.** 1990. The antimicrobial susceptibility patterns of the *Bacteroides fragilis* group in the United States, 1987. *J. Antimicrob. Chemother.* **25:**1011–1019.
9. **Cuchural, G. J., S. Hurlbut, M. H. Malamy, and F. P. Tally.** 1988. Permeability to beta-lactams in *Bacteroides fragilis*. *J. Antimicrob. Chemother.* **22:**785–790.
10. **Cuchural, G. J., F. P. Tally, N. V. Jacobus, P. K. Marsh, and J. W. Mayhew.** 1983. Cefoxitin inactivation by *Bacteroides fragilis*. *Antimicrob. Agents Chemother.* **34:**936–940.
11. **Dias, M. B. S., N. V. Jacobus, F. P. Tally, and S. L. Gorbach.** 1986. Activity of cefotetan against anaerobic bacteria. *Diagn. Microbiol. Infect. Dis.* **4:**359–363.
12. **Doern, G. V., R. N. Jones, J. A. Washington, H. G. Gerlach, A. Brueggemann, M. Irwin, C. Knapp, and D. Biedenbach.** Multi-center clinical laboratory evaluation of S1, a new chromogenic cephalosporin for beta-lactamase detection. Submitted for publication.
13. **Eley, A., and D. Greenwood.** 1986. Beta-lactamases of type culture strains of the *Bacteroides fragilis* group and of strains that hydrolyse cefoxitin, latamoxef and imipenem. *J. Med. Microbiol.* **21:**49–57.
14. **Finegold, S. M.** 1988. Susceptibility testing of anaerobic bacteria. *J. Clin. Microbiol.* **26:**1253–1256.
15. **Fox, A. R., and I. Phillips.** 1987. The antibiotic sensitivity of the *Bacteroides fragilis* group in the United Kingdom. *J. Antimicrob. Chemother.* **20:**477–488.
16. **Goldstein, E. J. C.** 1992. Patterns of susceptibility to fluoroquinolones among anaerobic bacterial isolates in the United States. *Clin. Infect. Dis.* **4:**s377–s381.
17. **Goldstein, E. J. C., and D. M. Citron.** 1992. Comparative activity of ciprofloxacin, ofloxacin, sparfloxacin, temafloxacin, CI-960, CI-990, and WIN 57273 against anaerobic bacteria. *Antimicrob. Agents Chemother.* **36:**1158–1162.
18. **Hill, G. B., and S. Schalkowsky.** 1990. Development and evaluation of the spiral gradient endpoint method for susceptibility testing of anaerobic gram-negative bacilli. *Rev. Infect. Dis.* **2:**s200–s209.
19. **Johnson, C.** 1992. Susceptibility of anaerobic bacteria to beta-lactam antibiotics in the United States. *Clin. Infect. Dis.* **4:**s371–s376.
20. **Jones, R. N., L. D. Sutton, and D. Biedenbach.** Development, characterization and initial evaluation of S1, a new chromogenic cephalosporin for beta-lactamase detection. Submitted for publication.
21. **Jorgensen, J. H., J. S. Redding, and A. W. Howell.** 1986. Evaluation of broth disk elution methods for susceptibility testing of anaerobic bacteria with the newer β-lactam antibiotics. *J. Clin. Microbiol.* **23:**545–550.
22. **Kurzynski, T. A., J. W. Yrios, A. G. Helstad, and C. R. Field.** 1976. Aerobically incubated thioglycollate broth disk method for antibiotic susceptibility testing of anaerobes. *Antimicrob. Agents Chemother.* **10:**727–732.
23. **Nachnani, S., A. Scuteri, M. G. Newman, A. B. Avanessian, and S. L. Lomeli.** 1992. E-test: a new technique for antimicrobial susceptibility testing for periodontal microorganisms. *J. Periodontol.* **63:**576–583.
24. **National Committee for Clinical Laboratory Standards.**

1993. *Methods for Antimicrobial Susceptibility Testing of Anaerobic Bacteria*, 3rd ed. Approved standard M11-A3. National Committee for Clinical Laboratory Standards, Villanova, Pa.

25. **Reysset, G., A. Haggoud, and M. Sebald.** 1992. Genetics of resistance of *Bacteroides* species to 5-nitroimidazole. *Clin. Infect. Dis.* **4:**s401–s403.

26. **Summanen, P., E. J. Baron, D. Citron, C. Stron, H. M. Wexler, and S. M. Finegold.** 1993. *Wadsworth Anaerobic Bacteriology Manual*, 5th ed. Star Publishing Co., Belmont, Calif.

27. **Tuner, K., and C. E. Nord.** 1992. Antibiotic susceptibility of anaerobic bacteria in Europe. *Clin. Infect. Dis.* **4:**s387–s389.

28. **Wexler, H. M.** 1991. Susceptibility testing of anaerobic bacteria: myth, magic, or method? *Clin. Microbiol. Rev.* **4:**470–484.

29. **Wexler, H. M.** 1993. Susceptibility testing of anaerobic bacteria: the state of the art. *Clin. Infect. Dis.* **4:**s328–s333.

29a. **Wexler, H. M.** Unpublished observations.

30. **Wexler, H. M., E. Molitoris, and S. M. Finegold.** 1991. Effect of β-lactamase inhibitors on the activities of various β-lactam agents against anaerobic bacteria. *Antimicrob. Agents Chemother.* **25:**1219–1224.

31. **Wexler, H. M., E. Molitoris, and S. M. Finegold.** 1992. In vitro activities of three of the newer quinolones against anaerobic bacteria. *Antimicrob. Agents Chemother.* **36:**239–243.

32. **Wexler, H. M., E. Molitoris, F. Jashnian, and S. M. Finegold.** 1991. Comparison of spiral gradient with conventional agar dilution for susceptibility testing of anaerobic bacteria. *Antimicrob. Agents Chemother.* **35:**1196–1202.

33. **Wilkins, T. D., and S. Chalgren.** 1976. Medium for use in antibiotic susceptibility testing of anaerobic bacteria. *Antimicrob. Agents Chemother.* **10:**926–928.

34. **Wilkins, T. D., and T. Thiel.** 1973. Modified broth-disk method for testing the antibiotic susceptibility of anaerobic bacteria. *Antimicrob. Agents Chemother.* **3:**350–356.

35. **Zabransky, R. J., R. J. Birk, T. A. Kurzynski, and K. L. Toohey.** 1986. Predicting the susceptibility of anaerobes to cefoperazone, cefotaxime, and cefoxitin with the thioglycolate broth disk procedure. *J. Clin. Microbiol.* **24:**181–185.

Special Tests for Detecting Antibacterial Resistance

JANA M. SWENSON, JANET A. HINDLER, AND LANCE R. PETERSON

116

Special tests for detecting antibacterial resistance range from the rapid and simple spot β-lactamase test to the more time-consuming and complex MBC assays. These tests may either supplement or replace traditional testing methods depending on the organism and the assay. The tests explained in this chapter include tests for detection of high-level aminoglycoside resistance in enterococci, a test for detection of vancomycin resistance in enterococci, tests to detect oxacillin resistance in staphylococci, tests for detection of β-lactamases, and tests for determination of bactericidal activity.

Information concerning quality control of the screening tests described is given in each section; however, guidelines for frequency of quality control testing are not given, because they are not available. A practical approach would be to perform quality control testing each day patient isolates are tested or less frequently (e.g., weekly) once a laboratory has thoroughly documented that less-frequent quality control testing can validate the reliability of the screening procedures. In addition, quality control tests should be performed each time new lots of materials are put into use.

TESTS TO DETECT RESISTANCE IN ENTEROCOCCI

Systemic enterococcal infections such as endocarditis are commonly treated with a combination of an antimicrobial agent whose site of action is the cell wall (a beta-lactam drug or vancomycin) and an aminoglycoside (usually gentamicin or streptomycin). These agents act synergistically to enhance killing (88), but when an enterococcal strain is resistant to the cell wall-active agent or has high-level resistance (HLR) to the aminoglycosides, there is no synergism, and combination therapy will likely be unsuccessful (63). Because of this, it is important to detect resistance to both the aminoglycosides and the cell wall-active agents in order to predict the likelihood of synergy.

Detection of HLR to Aminoglycosides

Because enterococci are inherently resistant to aminoglycosides with MICs normally ranging from 8 to 256 μg/ml, these antimicrobial agents cannot be used as single agents for therapy (34, 63). This intrinsic, moderate-level resistance is due to poor uptake of the aminoglycoside (63). Acquired aminoglycoside resistance in enterococci is due

either to mutations resulting in decreased binding of the agent to the ribosome, which occurs with streptomycin only (called ribosomal resistance), or, more commonly, to the acquisition of genes that encode aminoglycoside-modifying enzymes. Acquired resistance usually corresponds to MICs of >2,000 μg/ml and is designated HLR (63; see also chapters 24 and 112 in this Manual).

Synergy between an aminoglycoside and a cell wall-active agent can be determined directly by performing complex time-kill studies (46) or can be predicted by using less cumbersome screening tests. Gentamicin and streptomycin are the only two agents that need to be tested on a routine basis. All enterococcal isolates that are resistant to gentamicin are also resistant to other aminoglycosides except streptomycin, which elicits a different resistance mechanism. Consequently, streptomycin resistance must be determined independently. Isolates of *Enterococcus faecium* are intrinsically resistant to amikacin, kanamycin, tobramycin, and netilmicin despite in vitro testing results (61). *Enterococcus faecalis* susceptible to gentamicin may be resistant to kanamycin and amikacin. If amikacin is being considered for therapy, in vitro tests with amikacin cannot reliably predict HLR to amikacin in *E. faecalis*, but kanamycin could be tested to predict HLR to amikacin (81).

In 1992, the National Committee for Clinical Laboratory Standards (NCCLS) recognized that confusion existed about correct methods to use for detection of HLR to aminoglycosides in enterococci and has now developed agar dilution, broth microdilution, and disk diffusion methods for detecting HLR (67, 68). These methods are summarized in Table 1.

Agar Dilution Screening Method

Agar plates are prepared using brain heart infusion agar (BHIA) supplemented with 500 μg of gentamicin or 2,000 μg of streptomycin per ml (see chapter 113 of this Manual). The plates are inoculated by spotting 10 μl of a suspension that is equivalent to a 0.5 McFarland standard prepared from growth on an 18- to 24-h agar plate, giving a final inoculum of 10^6 CFU per spot. The plates are incubated for a full 24 h in ambient air. The presence of more than one colony or a haze of growth should be read as resistance. For streptomycin, plates should be reincubated for an additional 24 h if there is no growth at 24 h. Mueller-Hinton agar (MHA), MHA plus 5% sheep blood, or dextrose phosphate

TABLE 1 Screening methods for detecting vancomycin and high-level aminoglycoside resistance in enterococci

Parameter	Screening procedure			
	Vancomycin agar dilution	Aminoglycoside agar dilution	Aminoglycoside broth microdilution	Aminoglycoside disk diffusion
Medium	BHIA	BHIA	BHI	MHA
Inoculum	10^5–10^6 CFU/spot	10^6 CFU/spot	5×10^5 CFU/ml	0.5 McFarland[a]
Incubation (h)	24	24^b	24^b	18–24
Drug concn				
Vancomycin	6 μg/ml			
Gentamicin	NA	500 μg/ml	500 μg/ml	120 μg/disk
Streptomycin	NA	2,000 μg/ml	1,000 μg/ml	300 μg/disk
Endpoint	>1 colony	>1 colony	Any growth	6 mm = resistant, 7–9 mm = inconclusive,[c] ≥10 mm = susceptible

[a]NCCLS disk diffusion standard (68).
[b]If negative for streptomycin at 24 h, reincubate for an additional 24 h.
[c]If the zone is 7 to 9 mm, the test is inconclusive, and an agar dilution or broth microdilution test should be performed to confirm susceptibility or resistance.

agar may be substituted for BHIA, but because growth is better on BHIA, this is the preferred medium. A commercially available agar screen plate (Remel, Lenexa, Kans.) that uses MHA has been available since 1991 and has performed well (77, 78). Kanamycin screening tests have not been as extensively evaluated and are not standardized, but it has been reported that for determining HLR to both amikacin and kanamycin, kanamycin at 2,000 μg/ml in BHIA can be used (81).

Broth Microdilution Screening Method
Broth microdilution plates are prepared with single wells containing BHI broth supplemented with 500 μg of gentamicin or 1,000 μg of streptomycin per ml. The final inoculum is that recommended for routine broth microdilution tests: 5×10^5 CFU/ml. The plates are incubated for 24 h in ambient air. Any growth is interpreted as resistance. For streptomycin, plates should be reincubated for an additional 24 h if there is no growth at 24 h.

The recommended streptomycin concentration for use in the broth microdilution screen is 1,000 μg/ml, which is half of that used in the agar dilution screen. Because this test is often included as a part of a routine gram-positive MIC panel, the inoculum is that commonly used in broth microdilution testing (5×10^5 CFU/ml). The total number of cells tested in the agar dilution screening procedure (10^6 CFU/spot) is 20-fold higher than that normally used in the broth microdilution test (5×10^4 CFU/0.1-ml well). In order to provide a test that uses a low inoculum and at the same time maximizes the detection of HLR to streptomycin, it was necessary to lower the concentration recommended for testing streptomycin from 2,000 to 1,000 μg/ml in the broth microdilution test. Because of the poorer growth and the lower inoculum, Mueller-Hinton broth was inadequate for use by broth microdilution. The performance of other aminoglycosides has not been evaluated.

Disk Diffusion Screening Method
The standard disk diffusion procedure (68) described in chapter 113, which uses unsupplemented MHA, is followed except that special high-content disks (gentamicin, 120 μg; streptomycin, 300 μg) are used (82). Zones are measured after 18 to 24 h of incubation in ambient air at 35°C. Isolates with zone diameters of ≥10 mm are categorized as

susceptible. Absence of a zone of inhibition corresponds to the presence of HLR. Strains with zone diameters in the 7- to 9-mm range usually display HLR, but a few show only moderately elevated MICs (unpublished observations). Therefore, strains giving 7- to 9-mm zones should be tested by either the standard agar or the broth microdilution screening method to determine susceptibility or resistance. High-content gentamicin and streptomycin disks are not yet available commercially but soon may be. In the meantime, methods for preparing high-content disks have been described (52).

Quality Control
E. faecalis ATCC 29212 is used as a susceptible control for both gentamicin and streptomycin; E. faecalis ATCC 51299, resistant to both gentamicin and streptomycin, is the resistant control strain. Only E. faecalis ATCC 29212 is used for control of disk diffusion tests. Expected control limits are 16 to 22 mm for gentamicin 120-μg disks and 14 to 19 mm for streptomycin 300-μg disks.

Detection of Penicillin and Ampicillin Resistance
Compared to streptococci, for which MICs of penicillin are usually ≤0.12 μg/ml, all enterococci are relatively resistant to beta-lactams, with MICs of penicillin usually ≥2 μg/ml (63). Isolates of E. faecium are inherently more resistant to penicillin than are isolates of E. faecalis; the usual MICs of penicillin for E. faecium are 16 to 32 μg/ml, whereas the usual MICs for E. faecalis are 2 to 4 μg/ml (27, 61). MICs of ampicillin are generally 1 dilution lower than MICs of penicillin (28, 63). This intrinsic relative resistance as well as higher levels of resistance to penicillin and ampicillin (MICs of ≥16 μg/ml) has been associated with alterations in penicillin-binding proteins (PBPs) (25, 63). In addition to changes in PBPs, resistance to beta-lactam agents can also be mediated by the production of β-lactamase (64), but this is less common.

No screening tests for the detection of penicillin or ampicillin resistance have been described. Routine antimicrobial susceptibility tests (see chapter 113 in this Manual) will detect resistance due to changes in PBPs, and this resistance will be evident as higher MICs or smaller zones of inhibition. However, routine tests (such as broth microdilution or disk diffusion) will not detect resistance in enterococci that produce β-lactamase. An inoculum 100-fold greater than that routinely recommended (e.g., 10^7 CFU/ml) is necessary in order for resistance to be demonstrated by standard dilution methods (64). The nitrocefin β-lactamase test is recommended for detection of strains that are resistant because of the production of β-lactamase (see β-Lactamase Tests below).

As defined by the NCCLS (67), resistance to penicillin and ampicillin corresponds to MICs of ≥16 μg/ml. The NCCLS recommends that results of penicillin susceptibility tests be used to predict susceptibility to ampicillin, amoxicillin, and beta-lactam–β-lactamase inhibitor combinations (67, 68). The NCCLS does not currently extend that recommendation to include acylureidopenicillins (piperacillin, mezlocillin, and azlocillin). In general, enterococcal strains resistant to penicillin or ampicillin because of altered PBPs (MICs of ≥16 μg/ml) should also be considered resistant to ureidopenicillins and imipenem (34). In the absence of PBP-mediated resistance, β-lactamase-producing enterococci should be considered resistant to penicillin, ampicillin, and the ureidopenicillins but susceptible to imipenem as well as to beta-lactam–β-lactamase inhibitor combinations (64).

Since MICs of ampicillin are generally 1 dilution lower than MICs of penicillin (28), it is possible to encounter strains that appear susceptible to ampicillin but resistant to penicillin (94). The clinical significance of this finding is unknown. In addition, some confusion exists about cross-resistance of piperacillin and penicillin or ampicillin. Herman and Gerding reported that ureidopenicillins have in vitro activities similar to those of penicillin and ampicillin (32). However, others have reported higher MICs of piperacillin than of penicillin for some isolates of enterococci (26, 27, 31), suggesting that penicillin breakpoints probably should not be used to interpret results for ureidopenicillins. Categorizing penicillin- and ampicillin-resistant strains as being resistant to piperacillin is a conservative but appropriate approach.

Although Torres et al. (99) recently recommended that, in the absence of aminoglycoside resistance, strains of *E. faecium* with penicillin MICs of 16 μg/ml should be considered synergy susceptible, additional clinical studies are needed to clarify the level of ampicillin and penicillin resistance that correlates with the absence of synergy. Currently, most commonly used antimicrobial susceptibility systems do not include concentrations above 16 μg/ml, so differentiating between borderline resistance (ampicillin MICs of 16 to 32 μg/ml) and higher-level resistance (MICs of ≥64 μg/ml) is not readily possible. For a β-lactamase-negative strain, testing a single concentration of ampicillin or penicillin at 32 or 64 μg/ml by an agar dilution or broth macrodilution test with Mueller-Hinton medium and an inoculum normally used for MIC testing (67) could help define the level of resistance in such strains.

Detection of Vancomycin Resistance

Definitions of vancomycin resistance in enterococci are undergoing modification as more is learned about the genetics and clinical significance of the resistance. However, there are essentially three phenotypes of resistance: (i) high-level vancomycin resistance (MICs of ≥64 μg/ml) with accompanying teicoplanin resistance (≥16 μg/ml) (VanA phenotype), (ii) low- to high-level resistance (MICs of 16 to 512 μg/ml) without teicoplanin resistance (VanB phenotype), and (iii) intrinsic low-level resistance associated with *Enterococcus gallinarum* and *Enterococcus casselifla-*

vus (MICs of 2 to 32 μg/ml) (VanC phenotype) (49, 50). Both the VanA and the VanB phenotypes are most commonly seen in *E. faecalis* and *E. faecium*, but they have been described in other species (12). As defined by the NCCLS, vancomycin resistance breakpoints are ≤4 μg/ml for susceptibility, 8 to 16 μg/ml for intermediate resistance, and ≥32 μg/ml for resistance.

At this time, many methods commonly employed by clinical laboratories, including disk diffusion and the Vitek and MicroScan systems, may fail to detect low-level vancomycin resistance in enterococci (both VanB and VanC types) (79, 80, 92, 102). New recommendations for disk diffusion testing of vancomycin (including extending incubation to 24 h and examining zones by transmitted light) have improved the accuracy of that test (90). Because of the failure of some systems to detect the vancomycin resistance expressed by certain enterococcal strains, an agar screening test first described by Willey et al. (102) has been adopted by NCCLS (67, 68, 89) (Table 1).

Agar Dilution Screen

Agar plates are prepared (see chapter 113 in this Manual for the general procedure for agar plate preparation) with BHIA supplemented with 6 μg of vancomycin per ml. The plates are inoculated by spotting 1 to 10 μl of a suspension on the agar surface, using growth from an 18- to 24-h agar plate to make a suspension equivalent in turbidity to a 0.5 McFarland standard. Final inoculum is 10^5 to 10^6 CFU per spot. The plates are incubated for a full 24 h in ambient air at 35°C. The presence of more than one colony or a haze of growth should be read as resistance.

Quality Control

For quality control, *E. faecalis* ATCC 29212 (no growth = susceptibility) and *E. faecalis* ATCC 51299 (growth = resistance) should be tested. Plates made with BHIA from certain manufacturers may allow light growth of *E. faecalis* ATCC 29212, especially if the higher inoculum is used or if the plates are held for longer than 24 h. New lots of plates should always be quality controlled before being used for testing clinical strains.

Reporting Resistance in Enterococci

For any serious enterococcal infection, results of the screen for HLR to gentamicin and streptomycin must be reported in concert with results of the test for the cell wall-active agent, because synergy would not be expected if any one of the agents reported is resistant. Helpful suggestions on reporting the results of enterococcal tests are given by Hindler and Sahm in a recent publication (34) and in chapter 24 of this Manual.

OXACILLIN SCREEN TEST FOR DETECTION OF PENICILLIN RESISTANCE IN PNEUMOCOCCI

Soon after an outbreak of *Streptococcus pneumoniae* resistant to multiple antimicrobial agents was described in South Africa (38), a screening test using 1-μg oxacillin disks for detection of penicillin resistance was described (18). Since then, this test has been used extensively and shown to work well (91). Details of the procedure and suggestions about its use can be found in chapter 114 of this Manual.

TABLE 2 Characterization of oxacillin (methicillin) resistance phenotypes in staphylococci[a]

Resistance phenotype[b]	mec gene encoded	Mechanism	Borderline resistance[c]	β-Lactamase inhibitor effect[d]	Beta-lactam cross-resistance	Multiple resistance to non-beta-lactams[e]
Classic						
Homogeneous	+	Supplemental PBP (PBP 2a)	−	−	+	(+)
Heterogeneous	+	Supplemental PBP (PBP 2a)	+/−	−	+	(+)
Inactivation by β-lactamase	−	Increased β-lactamase production	+	+	−	−
MOD-SA	−	Modified preexisting PBPs 1, 2, and 4	+	−	−	−

[a]Table is modified from reference 76 with permission.

[b]β-Lactamase inactivation and MOD-SA resistance have been described only in *S. aureus*.

[c]Borderline resistance phenotype: oxacillin MICs of 2 to 8 µg/ml with unclear endpoints, or disk diffusion zone diameters of 10 to 13 mm with poorly defined zone edges.

[d]Addition of β-lactamase inhibitor lowers MIC $\geq$2 dilutions.

[e]Parentheses indicate that exceptions may occur.

AGAR SCREEN FOR DETECTION OF OXACILLIN (METHICILLIN) RESISTANCE IN STAPHYLOCOCCI

At least three different resistance mechanisms contribute to oxacillin (methicillin) resistance in *Staphylococcus aureus* (see chapter 112 in this Manual). These include (i) production of a supplemental PBP (PBP 2a) that is encoded by a chromosomal *mec* gene, (ii) inactivation by β-lactamase, and (iii) production of modified PBPs (MOD-SA) (10, 17, 29). From a clinical perspective, it is important to differentiate isolates that have *mec*-positive resistance, which is the classic type of oxacillin (methicillin) resistance, from the infrequently encountered isolates that have one of the other types of more subtle or borderline resistance. Characteristics that might help differentiate the three types of oxacillin (methicillin) resistance are outlined in Table 2. Strains that possess the *mec* gene (classic resistance) are either heterogeneous or homogeneous in their expression of resistance. With homogeneous expression, virtually all cells express resistance when tested by standard in vitro tests. However, testing of a heteroresistant isolate results in some cells that appear susceptible and others that appear resistant. Often, only 1 in 10^4 to 1 in 10^8 *mec*-positive cells in the test population expresses resistance (30, 75, 98). Heterogeneous expression occasionally results in MICs that appear to be borderline, i.e., oxacillin MICs of 2 to 8 µg/ml. Isolates that have classic resistance are usually resistant to other agents such as erythromycin, clindamycin, chloramphenicol, tetracycline, trimethoprim-sulfamethoxazole, a quinolone, or an aminoglycoside. Resistance mediated by β-lactamase or the presence of MOD-SAs also results in borderline resistance. β-Lactamase-mediated resistance can usually be distinguished from the classic type (*mec* positive) or MOD-SA resistance by the fact that the addition of a β-lactamase inhibitor (e.g., clavulanic acid) to the oxacillin MIC test lowers the MIC by 2 or more dilutions. Isolates that are resistant by either the β-lactamase or the MOD-SA mechanism usually do not have multiple-drug resistance.

The presence of classic resistance can be detected simply and reliably with the agar screening test (67); β-lactamase or MOD-SA strains are unlikely to grow on the agar screen plate. The *mec* gene has been identified in coagulase-negative staphylococci, and the agar screening test can also detect *mec*-positive resistance in these organisms (29, 59); the other types of resistance have been described only in *Staphylococcus aureus*.

Test Method

MHA supplemented with 4% sodium chloride and 6 µg of oxacillin per ml is used for the agar screen, as recommended by the NCCLS (67). Plates containing 4% NaCl and 10 µg of methicillin per ml have also been described (97) but are currently not recommended. Because oxacillin is more stable and appears to be superior to other penicillinase-resistant penicillins (e.g., methicillin, nafcillin, and the isoxazolyl penicillins cloxacillin, dicloxacillin, and flucloxacillin) in detecting resistance to this group of compounds, it is preferred in the agar screen and other diagnostic tests. Agar screen plates are available from several commercial manufacturers. The procedure for preparing agar dilution plates that is outlined in chapter 113 of this Manual provides a useful guideline for agar screen plate preparation.

In the agar screening test, inoculum suspensions are prepared by the direct inoculum preparation method by selecting colonies from overnight growth on a nonselective agar plate. The colonies are transferred to broth (e.g., tryptic soy broth) or saline to produce a suspension that matches the turbidity of a 0.5 McFarland standard. This suspension is used to inoculate the oxacillin agar screen plate by dipping a cotton swab into the test suspension, expressing the excess liquid from the swab, and touching the swab to a spot on the agar surface or streaking the swab across a small section of the agar surface. Alternatively, 10 µl of a 1:100 dilution of a 0.5 McFarland standardized suspension can be deposited on the agar surface, resulting in 10^4 CFU per spot (53). Test plates are incubated for a full 24 h at 35°C (no higher) in ambient air and examined for any evidence of growth, which indicates resistance.

Quality Control

The NCCLS currently does not make any specific recommendations for quality control of oxacillin agar screen plates. However, *S. aureus* ATCC 29213, which is recommended for dilution MIC tests, can be used as an oxacillin-susceptible control strain. An oxacillin-resistant strain, preferably a heteroresistant isolate, should also be included. *S. aureus* ATCC 38591 has been suggested for this purpose (53).

Ability of Agar Screen Plate To Detect Oxacillin Resistance and Reporting Results

Growth of a staphylococcal isolate on an oxacillin agar screen plate generally means that the isolate is *mec* positive. If performed properly, the agar screen method will detect most *mec*-positive staphylococci. Occasionally, however, a particular heteroresistant *mec*-positive strain may not be detected, perhaps in part because of a low frequency of resistance expression and/or lot-to-lot or manufacturer-to-manufacturer variation in the test medium (33, 35). The oxacillin agar screening test generally does not detect borderline resistant strains. Although MOD-SA isolates, particularly those with MICs of >8 μg/ml, occasionally grow on agar screen plates (88a), isolates with borderline resistance due to β-lactamase usually have oxacillin MICs of ≤6 μg/ml and do not grow on the screen plates. Since both types of borderline resistant isolates are infrequently encountered in clinical specimens, the possibility of their presence affects the utility of the agar screen test only minimally.

The NCCLS recommends that oxacillin (methicillin)-resistant staphylococci be considered resistant to all beta-lactam agents, including penicillins, cephalosporins, beta-lactam–β-lactamase inhibitor combinations, and imipenem. These agents are clinically ineffective against such staphylococci, even though they may demonstrate in vitro activity (67, 68). Consequently, an isolate that grows on the oxacillin agar screen plate should be considered resistant to these agents as well as to all penicillinase-resistant penicillins. Isolates that appear to be oxacillin (methicillin) resistant by an alternative test method but fail to grow on the agar screen plate are probably borderline resistant and lack *mec*. It is likely that these isolates would be clinically susceptible to beta-lactam agents that appear to be active in vitro (9, 10, 57, 95); much less is known about borderline resistance than about *mec*-positive resistance. If oxacillin (methicillin)-resistant *S. aureus* is isolated from a seriously ill patient and the isolate is presumed to be resistant by a mechanism other than the *mec* gene, the laboratory worker should convey this possibility to the patient's clinician.

DETECTION OF ENZYMES MEDIATING RESISTANCE

Detection of antimicrobial agent-modifying enzymes in the clinical laboratory is limited to tests for β-lactamase and chloramphenicol acetyltransferase (CAT). For more detailed information on these and other types of resistance enzymes, refer to chapters 112 and 117 in this Manual.

β-Lactamase Tests

In the clinical laboratory, β-lactamase tests must be used only when they can provide clinically useful information, and the definitions of positive or negative reactions must not be extended beyond their intended meanings. For example, a β-lactamase-positive result for a *Neisseria gonorrhoeae* isolate means that the isolate is resistant to penicillin but does not imply that the isolate is resistant to the cephalosporin group of beta-lactam agents. Similarly, direct β-lactamase tests for members of the family *Enterobacteriaceae* or for *Pseudomonas* spp. (all of which produce a variety of β-lactamases that result in various susceptibilities to beta-lactam agents) have little clinical value and should not be used for these species. The organisms for which β-lactamase tests may be useful are listed in Table 3.

Direct Tests for β-Lactamase Activity

In the direct β-lactamase test, a positive reaction indicates that the isolate is resistant to the beta-lactam agents noted in Table 3, but a negative reaction is inconclusive. For example, most ampicillin-resistant *Haemophilus influenzae* isolates produce β-lactamase, which can be detected by direct β-lactamase tests; however, rare strains may be ampicillin resistant but β-lactamase negative (19). For the latter, conventional disk diffusion or dilution tests are needed to detect the resistance (see chapter 114 in this Manual).

Three direct β-lactamase assays have been widely used. These include the acidometric, iodometric, and chromogenic methods (54). Each of these involves testing bacteria grown on nonselective media, and results are available within 1 to 60 min. The acidometric and iodometric methods use a colorimetric indicator to detect the presence of penicilloic acid in the reaction vessel following β-lactamase hydrolysis of penicillin. In the acidometric method, the substrate is citrate-buffered penicillin with a phenol red indicator. A decreasing pH associated with the presence of penicilloic acid results in a color change from red (negative result) to yellow (positive result) (22). The substrate in the iodometric test is phosphate-buffered penicillin plus a starch-iodine complex. Penicilloic acid, if present, reduces the iodine and prevents it from combining with starch, resulting in a colorless reaction (positive); a bluish purple color corresponds to a negative result (8).

The chromogenic cephalosporin nitrocefin can be utilized in a test tube assay (70) but has been incorporated into a commercial product, the cefinase disk (Becton Dickinson Microbiology Systems, Cockeysville, Md.). This disk method is very easy to use and is the β-lactamase test method used by most clinical laboratories (42). β-Lactamase hydrolysis of the chromogenic cephalosporin molecule causes an electron shift that results in a colored product (70). Although there has been some variability in performance or a lack of experience or both when the acidometric and iodometric methods have been used with some bacteria, the nitrocefin method is reliable in detecting β-lactamases produced by all of the organisms indicated in Table 3 (42, 69). PADAC, another chromogenic β-lactamase assay, is reliable only for testing *H. influenzae*, *N. gonorrhoeae*, and *Moraxella catarrhalis*; PADAC must not be used for staphylococci (1).

The colorimetric β-lactamase tests rely on visualization of a colored product that presumably results from β-lactamase destruction of the substrate beta-lactam molecule. However, these tests are not 100% specific, and other substances may yield colored endpoints. Serum may cause a colored reaction with the nitrocefin test (70), and if reagents are not stored properly, spontaneous degradation of penicillin may produce false-positive acidometric or iodometric β-lactamase reactions. An agar plate disk bioassay for β-lactamase testing has been described (51). Use of this bioassay eliminates the specificity concerns for colorimetric endpoints in direct tests; however, the method is cumbersome and rarely performed in routine clinical laboratories.

While some bacteria (e.g., *H. influenzae*, *N. gonorrhoeae*,

TABLE 3 Bacteria for which β-lactamase tests have been used in the clinical laboratory

Species	Method(s) commonly employed	Predicted resistance[a]
Bacteroides spp. (and other gram-negative anaerobes)	Direct β-lactamase tests[b]	Penicillins[c]
Enterococcus spp.	Direct β-lactamase tests	Penicillins[c]
Haemophilus influenzae	Direct β-lactamase tests	Penicillins[c]
Moraxella catarrhalis	Direct β-lactamase tests (nitrocefin only)	Penicillins[c]
Neisseria gonorrhoeae	Direct β-lactamase test	Penicillins[c]
Staphylococcus spp.	Direct β-lactamase tests with prior induction	Penicillins[c]
Gram-negative bacilli		
Acinetobacter spp.	Disk approximation test for inducible β-lactamase[d]	Extended-spectrum cephalosporins
Citrobacter freundii	Disk approximation test for inducible β-lactamase[d]	Extended-spectrum cephalosporins
Enterobacter spp.	Disk approximation test for inducible β-lactamase[d]	Extended-spectrum cephalosporins
Proteus,[e] Providencia, and Morganella spp.	Disk approximation test for inducible β-lactamase[d]	Extended-spectrum cephalosporins
Pseudomonas aeruginosa	Disk approximation test for inducible β-lactamase[d]	Extended-spectrum cephalosporins
Serratia marcescens	Disk approximation test for inducible β-lactamase[d]	Extended-spectrum cephalosporins
Escherichia coli	Double-disk potentiation test for ESBLs[d]	Cephalosporins and aztreonam
Klebsiella pneumoniae	Double-disk potentiation test for ESBLs[d]	Cephalosporins and aztreonam

[a] A positive result indicates resistance; however, a negative result is inconclusive, since other resistance mechanisms may occur.
[b] Includes chromogenic cephalosporin, acidometric, and iodometric tests.
[c] A positive result indicates resistance to all penicillinase-labile penicillins, including ampicillin, amoxicillin, azlocillin, carbenicillin, mezlocillin, piperacillin, and ticarcillin.
[d] Used only to confirm unusual results from conventional tests for research or academic purposes.
[e] Except Proteus mirabilis.

and enterococci) constitutively produce β-lactamase, others, such as staphylococci, produce detectable amounts of enzyme only after exposure to an inducing agent, which is generally a beta-lactam agent (21). If staphylococci produce a positive β-lactamase result without induction, results can be reported. However, if no β-lactamase is detected, then before a negative result is reported, the test must be performed on cells that have been exposed to an inducing agent. This can be accomplished by testing organisms that have been grown in the presence of subinhibitory concentrations of a beta-lactam agent (e.g., 0.25 μg of cefoxitin per ml) in a broth or agar system. Alternatively, cell paste from around the periphery of the zone surrounding a beta-lactam disk (e.g., a 1-μg oxacillin disk) can be tested. A positive result may take longer to develop in staphylococci than in other organisms, and the test should not be considered negative until it has been allowed to react for at least 60 min.

The S. aureus strains recommended by the NCCLS for quality control of routine disk diffusion and dilution tests (67, 68) can be used for quality control of β-lactamase tests. S. aureus ATCC 25923 is β-lactamase negative, whereas S. aureus ATCC 29213 is β-lactamase positive.

Tests for Inducible β-Lactamases

Virtually all Acinetobacter spp., Citrobacter freundii, Enterobacter spp., Morganella spp., Proteus vulgaris, Providencia spp., Pseudomonas aeruginosa, and Serratia marcescens are capable of producing inducible β-lactamases. Upon exposure to an inducing agent (a beta-lactam agent), these organisms produce β-lactamases that hydrolyze primary and extended-spectrum penicillins and cephalosporins. Generally, the bacterium stops producing the β-lactamase when the inducing agent is removed, although some cells may mutate to a state in which they produce the β-lactamase continuously (83).

With conventional in vitro susceptibility tests, isolates that have mutated to the resistant state should test resistant; sometimes, however, the resistant cells may be present at a low frequency, and resistance may then be difficult to detect. In contrast, isolates that have not mutated to the resistant state appear susceptible in in vitro test systems (101).

Several tests for assessing the ability of an isolate to produce inducible β-lactamases have been described. However, because virtually all isolates of the genera and species mentioned above have the potential to produce inducible β-lactamases, the β-lactamase induction test (14) has limited clinical application. Instead, physicians must understand the organisms that can produce inducible β-lactamases and the potential for therapeutic failure if selected beta-lactams are used for treating infections caused by these organisms.

Tests for ESBLs

The genes generally responsible for β-lactamase-mediated ampicillin resistance in *Escherichia coli* and *Klebsiella* spp. can undergo simple point mutations that result in the production of novel β-lactamases capable of hydrolyzing extended-spectrum cephalosporins (e.g., cefotaxime, ceftriaxone, ceftizoxime, and ceftazidime) and aztreonam as well as older beta-lactam drugs. These enzymes are referred to as extended-spectrum β-lactamases (ESBLs) (73) and are discussed in chapter 112 in this Manual. Many types of ESBLs have been noted in several gram-negative species, and these are associated with a variety of in vitro antimicrobial susceptibility profiles.

The in vitro susceptibility results obtained with an isolate that produces ESBLs often defy typical "hierarchy" rules of beta-lactam (particularly cephalosporin) activity. Sometimes, narrow- or expanded-spectrum cephalosporins appear to be more active than broad-spectrum agents. Some suggest that ESBL-producing isolates should be considered resistant to all extended-spectrum penicillins, cephalosporins, and monobactams, even if they appear to be susceptible to these agents in vitro. The more β-lactamase-stable beta-lactams (e.g., imipenem) are active in vitro (39, 41, 96) and appear to be clinically effective also (39, 60). The genes that code for production of ESBLs are often linked to other resistance genes such that ESBL-producing isolates are often multiply resistant (e.g., resistant to aminoglycosides and trimethoprim-sulfamethoxazole).

Routine disk diffusion and MIC tests may not always identify isolates that produce ESBLs, and even a slight decrease in susceptibility (e.g., a slightly elevated MIC or a smaller zone) to an extended-spectrum cephalosporin or aztreonam should be a clue to the presence of an ESBL (41, 44, 96). Many ESBL-producing clinical isolates have demonstrated frank resistance to ceftazidime and aztreonam, and resistance to these agents can serve as markers to identify ESBL-producing strains (39, 44, 60). Unlike inducible β-lactamases, ESBLs are inhibited by β-lactamase inhibitors, and this property has been applied to in vitro tests to identify them. In the double-disk potentiation procedure (41), test inocula are applied to agar plates as for the standard disk diffusion test (68). Then, a disk containing the substrate agent (e.g., cefotaxime) is strategically placed 30 mm (center to center) from a disk containing a β-lactamase inhibitor such as amoxicillin or clavulanic acid. Following overnight incubation, clearing in the area where both clavulanic acid and cefotaxime molecules have diffused together with decreased susceptibility suggests the presence of an ESBL (41). A more sophisticated three-dimensional disk diffusion type of test has been described by Thomson and Sanders (96). Additionally, cefotaxime or ceftazidime MIC tests performed with and without a β-lactamase inhibitor such as clavulanic acid can be used to identify ESBLs. MICs for ESBL-producing strains are considerably lower when obtained in the presence of the β-lactamase inhibitor. The effectiveness of these methods for detecting ESBL-producing strains may vary with the isolate and the type of enzyme produced. Therefore, there is currently no standardized recommended method.

Tests for CAT

Chloramphenicol resistance is often mediated through the production of chloramphenicol acetyltransferase (CAT). A rapid tube test to detect CAT has been described elsewhere (3) and is also described in the *Clinical Microbiology Proce-dures Handbook* (40). Either a commercial CAT kit (Remel) or the tube method can be used. A positive result indicates chloramphenicol resistance; however, a negative result is inconclusive, since other chloramphenicol resistance mechanisms exist. A conventional disk diffusion or dilution test is needed to identify the latter.

The CAT test has been used for *H. influenzae* (3), *Salmonella* spp. (16), and *S. pneumoniae* (58). However, it is currently used primarily in research settings.

TESTS FOR DETECTION OF BACTERICIDAL ACTIVITIES OF ANTIMICROBIAL AGENTS

As a result of the recent outbreaks of multidrug-resistant pathogens (6, 11, 13, 47, 48, 87), the pressure to perform special tests for bactericidal activity may increase as physicians prescribe various combinations of antimicrobial agents for treatment of infections for which there are no standard regimens or established therapeutic guidelines. In these circumstances, physicians may ask clinical laboratories to perform bactericidal tests for individual patients to assess the adequacy of therapy. The two tests most commonly requested are those for determining the MBC and the serum bactericidal titer (SBT).

Historically, tests of bactericidal activity have been recommended in several clinical situations in order to monitor therapy. These situations include bacterial endocarditis (15, 36, 66, 74), meningitis (66), sepsis in the immunocompromised patient (15, 66, 74), infections in those unable to mount an immune response (22, 36, 45), osteomyelitis (36, 66, 100), chronically infected implants (36), and other types of chronic infections (36). Bactericidal tests have also been used for detecting tolerance (66) and for predicting (or monitoring) therapeutic efficacy (e.g., SBT) in infectious diseases lacking therapeutic guidelines (24). Despite the continued use of these tests, it is crucial for laboratory personnel to realize that the scientific data supporting the performance of bactericidal tests are limited and that clinicians who order these tests often do so because of training or personal experience. This section addresses the pros and cons of performing bactericidal tests in a clinical laboratory setting and suggests methods that might enhance the reliability of bactericidal testing.

Clinical Use of Bactericidal Testing

Two approaches have been used for assessment of bactericidal activity. In one, the lowest concentration of antimicrobial agent needed to kill the organism (MBC) is determined and compared to the levels of drug estimated or measured in serum or tissue at various time points. The goal of this pharmacokinetic approach is to attain during all or part of the dosing interval a drug concentration (in the serum or at the site of infection) that exceeds the MBC of the infecting microbe. This type of testing is especially cumbersome, since procedures to measure drug concentrations in the patient's specimen as well as to perform MBC testing are required. In the other approach, the bactericidal activity or titer (SBT) of serum or body fluid from a patient receiving antimicrobial therapy is assessed (84, 85). The body fluid (usually serum) is serially diluted and inoculated with the patient's own infecting organism. In the SBT test, the result is expressed as a titer indicating the dilution of serum or body fluid that is lethal to the microbe, and a higher number or titer indicates better activity. This test is in contrast to the MBC test, in which a lower number (the

TABLE 4 Methods of bactericidal testing (MBC or SBT)

Test component	NCCLS document[a]	This Manual[b]
Broth medium	Mueller-Hinton with appropriate supplements	Same
pH	7.2–7.4	Same
Serum additives	Optional	Not recommended
Inoculum preparation	Growing cells from 20–30 colonies	Growing cells from 4 or 5 colonies
Inoculum density	5×10^5 CFU/ml	Same
Inoculation	Add below broth surface without agitation	Same
Incubation	35°C for 24 h (agitate macrodilution tubes at 20 h)	No agitation until sampled
Micro- or macrodilution preferred	Microdilution	Either. For microdilution, run test in duplicate. If 48-h subculture is needed, microdilution requires quadruplicate testing.
Subculture	Duplicate 0.01-ml samples for macrodilution; 0.01- to 0.1-ml samples for microdilution	Single 0.1-ml sample. Run test in duplicate; do 48-h subculture if nonagreement.
Endpoint	99.9% reduction of initial inoculum using Poisson-calculated rejection values	99.9% reduction of initial inoculum using $n + 2\sqrt{n}$ for final endpoint cutoff determination
Quality control	Described in documents	Use quality control strains with MBC tests
Interpretation	Proposed	By consultation only. Standard interpretation of results is not established.

[a]See references 65 and 66.
[b]See reference 72.

lethal concentration of drug) is better. However, while the SBT type of testing appears to be much simpler to perform and may be the fastest laboratory method to give any assessment of therapy in a patient infected with a multi-drug-resistant pathogen, technical and biologic variables affecting test performance make interpretation of test results difficult (55, 56, 72).

The MBC is designed to be a direct measure of a given antimicrobial agent's bactericidal or killing effect on a selected bacterial isolate under defined laboratory conditions. However, results in humans demonstrate (i) that a quantitative relationship between the measured MBC and the magnitude of the SBT cannot be easily demonstrated and (ii) that neither test is accurate in directly predicting the clinical response to therapy. The application of bactericidal test results to therapy of infections other than endocarditis has not received rigorous scrutiny but is often accepted as inherently useful (43, 86). In summary, these reports show the need to improve testing methods (72). While the latest tentative guideline from the NCCLS proposes an interpretation scheme for the SBT test (65), it is questionable whether there are sufficient data to support a standardized interpretation at this time (4, 72). Obviously, the greater the titer that can be achieved throughout the treatment dosing interval, the more likely a better outcome of therapy

is. However, the interpretation needs to minimally take into account knowledge regarding the physiology of drug-organism interactions, the seriousness of the infection, the site of the infection, and the potential toxicity (and cost) of alternative treatment regimens.

Recommended Use of Bactericidal Testing

In the clinical laboratory, bactericidal testing (either MBC or SBT) should be offered only with expert microbiologic consultation, and any of the tests for bactericidal activity should be interpreted by someone knowledgeable in both bactericidal testing and infectious diseases. Should testing be performed as an adjunct to patient care, a rigidly standardized procedure must be followed. The method described here is a combination of the NCCLS method and that described by Peterson and Shanholtzer (65, 66, 72). The differences are highlighted in Table 4.

Methodology

There are certain basic requirements for standardization of bactericidal tests. For a detailed description of methods and definitions of terms (such as tolerance and paradoxical effect, etc.), the NCCLS documents or the review by Peterson and Shanholtzer should be consulted (65, 66, 72). The differences between the methods compared in Table 4 require brief comments. Serum additives are not recommended because of the risk of transmission of blood-borne diseases. Use of four or five colonies is recommended to minimize the possibility of introducing contaminants (mixed cultures) that may occur when larger numbers of colonies are used for the inoculum. Agitation is not recommended during the actual procedure, since any agitation presents the potential problem of removing viable bacterial cells from continued exposure to the test antimicrobial agent (37, 93). Taylor et al. demonstrated early agitation to be helpful only when a large initial inoculum volume (1.0 ml) was used and not when a small volume (≤0.1 ml) was used (93). A subculture of 0.1 ml is preferred to give sufficient precision for the endpoint determination used in the evaluation of the MBC or SBT result. Using a subculture of only 0.01 ml will not provide the ≥10 colonies that are required to demonstrate a 99.9% endpoint reduction, unless the initial inoculum is ≥10^6 CFU/ml. Because of the poor reproducibility of bactericidal tests, they should be run in duplicate (72). The endpoint determination is easily made by using the method of Anhalt et al. (2), since the readily available Poisson tables are based on a subculture volume of only 0.01 ml (72).

"Skipped" growth in tubes or wells makes interpretation of results difficult. When tests are performed in duplicate and one of the two sets demonstrates a skip at 24 h, then the other set can be used for determining results. If both sets show skips at 24 h, then 48-h subcultures should be performed on both sets, and one or both sets with results free of skipped tubes at 48 h should be used for determining endpoints. This method, used with the formula $n + 2\sqrt{n}$ to calculate the cutoff endpoint (2), provides consistent, reproducible results for MBC testing (72). Other problems, such as discrepant results (a >1-doubling-dilution difference) between paired tests at both 24 and 48 h, bacterial growth at high drug concentrations with none at lower concentrations, failure of growth in the control well or tube, growth in the sterility well or tube, etc., require that the entire test be repeated. Using this method since mid-1992, the laboratory at Northwestern Memorial Hospital has found that only 0.9% of tests need to be held for 48 h of incubation and another 0.9% need to be completely repeated.

Specific guidelines for interpretation cannot be given at present. In treatment settings where bactericidal therapy is desired, documentation of the in vitro bactericidal activities of specific agents against individual organisms can be provided by MIC and MBC testing. When there is a need to document bactericidal activity of therapy in vivo, the laboratory can perform SBT-type testing.

Other Tests for Assessing Bactericidal Activity

Measuring the rate of bactericidal activity, such as with time-kill analysis, is a potential replacement of or adjunct to MBC testing (20, 46, 62, 71). While procedures for the time-kill test have been proposed by the NCCLS and in the *Clinical Microbiology Procedures Handbook* (46, 66), it is important to remember that the same critical factors affecting the outcome of the MBC and SBT tests apply to the performance of time-kill assays.

Correlation of other tests with more traditional tests of assessment of bactericidal activity and clinical outcome of infectious disease therapy continues to be controversial. Bayer and Morrison have commented on the disparity between results of time-kill tests and the checkerboard test for evaluating synergistic bactericidal interactions when combinations of antimicrobial agents are used (5). Recently, however, Briceland et al. proposed that time-kill tests (particularly those directly involving the assay of a patient's serum sample) may improve the clinical utility of in vitro determination of bactericidal activity (7). We are not aware that any prospective, controlled clinical comparison of these tests has been done. Several other approaches to bactericidal analysis of antimicrobial activity have been suggested and reviewed elsewhere (72). Less is known about the variables that might influence the performance of any of these newer tests than about the more traditional MBC or SBT test, and none of these tests has been evaluated prospectively in any clinical investigation. Until such evaluations are done and interpretative guidelines are established, it is difficult to recommend the use of any tests other than the MBC or SBT test in a clinical laboratory setting.

REFERENCES

1. **Anhalt, J. P., and R. Nelson.** 1982. Failure of PADAC test strips to detect staphylococcal β-lactamase. *Antimicrob. Agents Chemother.* **21:**983–984.
2. **Anhalt, J. P., L. D. Sabath, and A. L. Barry.** 1980. Special tests: bactericidal activity, activity of antimicrobics in combination, and detection of β-lactamase production, p. 478–481. *In* E. H. Lennette, A. Balows, and W. J. Hausler, Jr. (ed.), *Manual of Clinical Microbiology*, 3rd ed. American Society for Microbiology, Washington, D.C.
3. **Azemun, P., T. Stull, M. Roberts, and A. L. Smith.** 1981. Rapid detection of chloramphenicol resistance in *Haemophilus influenzae*. *Antimicrob. Agents Chemother.* **20:**168–170.
4. **Baron, E. J., L. R. Peterson, and S. M. Finegold (ed.).** 1994. *Bailey and Scott's Diagnostic Microbiology*. The C. V. Mosby Co., St. Louis.
5. **Bayer, A. S., and J. O. Morrison.** 1984. Disparity between timed-kill and checkerboard methods for determination of in vitro bactericidal interactions of vancomycin plus rifampin versus methicillin-susceptible and -resistant *Staphylococcus aureus*. *Antimicrob. Agents Chemother.* **26:**220–223.
6. **Boyle, J. F., S. A. Soumakis, A. Rendo, J. A. Herrington, D. G. Gianarkis, B. E. Thurberg, and B. G. Painter.** 1993. Epidemiologic analysis and genotypic characterization of a nosocomial outbreak of vancomycin-resistant enterococci. *J. Clin. Microbiol.* **31:**1280–1285.
7. **Briceland, L. L., M. T. Pasko, and J. M. Mylotte.** 1987. Serum bactericidal rate as a measure of antibiotic interactions. *Antimicrob. Agents Chemother.* **31:**679–685.
8. **Catlin, B. W.** 1975. Iodometric detection of *Haemophilus influenzae* beta-lactamase: rapid, presumptive test for ampicillin resistance. *Antimicrob. Agents Chemother.* **7:**265–270.
9. **Chambers, H. F., G. Archer, and M. Matsuhashi.** 1990. Low-level methicillin resistance in *Staphylococcus aureus*. *Antimicrob. Agents Chemother.* **33:**424–428.
10. **Chambers, H. F., and C. J. Hackbarth.** 1992. Methicillin-resistant *Staphylococcus aureus*: genetics and mechanisms of resistance, p. 21–35. *In* M. T. Cafferkey (ed.), *Methicillin-Resistant Staphylococcus aureus: Clinical Management and Laboratory Aspects*. Marcel Dekker, Inc., New York.
11. **Chow, J. W., A. Kuritza, D. M. Shales, M. Green, D. F. Sahm, and M. J. Zervos.** 1993. Clonal spread of vancomycin-resistant *Enterococcus faecium* between patients in three hospitals in two states. *J. Clin. Microbiol.* **31:**1609–1611.
12. **Clark, N. C., R. C. Cooksey, B. C. Hill, J. M. Swenson, and F. C. Tenover.** 1993. Characterization of glycopeptide-resistant enterococci from U.S. hospitals. *Antimicrob. Agents Chemother.* **37:**2311–2317.
13. **Cohen, M. L.** 1992. Epidemiology of drug resistance: implications for a post-antimicrobial era. *Science* **257:**1050–1055.
14. **Cox, V.** 1992. β-Lactamase induction test for gram-negative bacilli, p. 5.7.1–5.7.6. *In* H. D. Isenberg (ed.), *Clinical Mi-*

crobiology Procedures Handbook, vol. 1. American Society for Microbiology, Washington, D.C.

15. **DeGirolami, P. C., and G. Eliopoulos.** 1987. Antimicrobial susceptibility tests and their role in therapeutic drug monitoring. *Clin. Lab. Med.* **7:**499–513.

16. **de la Maza, H., S. I. L. Miller, and M. J. Ferraro.** 1990. Use of commercially available rapid chloramphenicol acetyltransferase test to detect resistance in *Salmonella* spp. *J. Clin. Microbiol.* **28:**1867–1869.

17. **DeLencaster, H., S. A. Figueiredo, C. Urban, J. Rahal, and A. Tomasz.** 1991. Multiple mechanisms of methicillin resistance and improved methods for detection in clinical isolates of *Staphylococcus aureus*. *Antimicrob. Agents Chemother.* **35:**632–639.

18. **Dixon, J. M. S., A. E. Lipinski, and M. E. P. Graham.** 1977. Detection and prevalence of pneumococci with increased resistance to penicillin. *Can. Med. Assoc. J.* **117:**1159–1161.

19. **Doern, G. V., J. H. Jorgensen, C. Thornsberry, D. A. Preston, T. Tubert, J. S. Redding, and L. A. Maher.** 1988. National collaborative study of the prevalence of antimicrobial resistance among clinical isolates of *Haemophilus influenzae*. *Antimicrob. Agents Chemother.* **32:**180–185.

20. **Drake, T. A., C. J. Hackbarth, and M. A. Sande.** 1983. Value of serum test in combined drug therapy of endocarditis. *Antimicrob. Agents Chemother.* **24:**653–657.

21. **Dyke, K. G. H.** 1979. Beta-lactamases of *Staphylococcus aureus*, p. 291–310. *In* J. M. T. Hamilton-Miller and J. T. Smith (ed.), *Beta-Lactamases*. Academic Press, London.

22. **Eliopoulos, G. M., and R. C. Moellering, Jr.** 1982. Principles of antibiotic therapy. *Med. Clin. N. Am.* **66:**3–15.

23. **Escamilla, J.** 1976. Susceptibility of *Haemophilus influenzae* to ampicillin as determined by use of a modified one-minute beta-lactamase test. *Antimicrob. Agents Chemother.* **9:**196–198.

24. **Fasching, C. E., D. N. Gerding, and L. R. Peterson.** 1987. Treatment of ciprofloxacin- and ceftizoxime-induced resistant gram-negative bacilli. *Am. J. Med.* **82**(Suppl. 4A)**:**80–86.

25. **Fontana, R., P. Canepari, M. M. Lleo, and G. Satta.** 1992. Mechanisms of resistance of enterococci to beta-lactam antibiotics. *Eur. J. Clin. Microbiol. Infect. Dis.* **9:**103–105.

26. **Gootz, T. D., C. C. Sanders, and W. E. Sanders, Jr.** 1979. In vitro activity of furazlocillin (Bay k 4999) compared with those of mezlocillin, piperacillin, and standard beta-lactam antibiotics. *Antimicrob. Agents Chemother.* **15:**783–791.

27. **Gordon, S., J. M. Swenson, B. C. Hill, N. E. Pigott, R. R. Facklam, R. C. Cooksey, C. Thornsberry, Enterococcal Study Group, W. R. Jarvis, and F. C. Tenover.** 1992. Antimicrobial susceptibility patterns of common and unusual species of enterococci causing infections in the United States. *J. Clin. Microbiol.* **30:**2373–2378.

28. **Grayson, M. L., G. M. Eliopoulos, C. B. Wennersten, K. L. Ruoff, P. C. De Girolami, M. J. Ferraro, and R. C. Moellering, Jr.** 1991. Increasing resistance to beta-lactam antibiotics among clinical isolates of *Enterococcus faecium*: a 22-year review at one institution. *Antimicrob. Agents Chemother.* **35:**2180–2184.

29. **Hackbarth, C. J., and H. F. Chambers.** 1989. Methicillin-resistant staphylococci: genetics and mechanisms of resistance. *Antimicrob. Agents Chemother.* **33:**991–994.

30. **Hartman, B. J., and A. Tomasz.** 1986. Expression of methicillin resistance in heterogeneous strains of *Staphylococcus aureus*. *Antimicrob. Agents Chemother.* **29:**85–92.

31. **Hartzen, S. H., N. Frimodt-Möller, and J. J. Andreasen.** 1988. *In vitro* antibacterial activities of eleven antibiotics against *S. faecalis*. APMIS **96:**584–588.

32. **Herman, D. J., and D. N. Gerding.** 1991. Antimicrobial resistance among enterococci. *Antimicrob. Agents Chemother.* **35:**1–4.

33. **Hindler, J. A., and C. B. Inderlied.** 1985. Effect of source of Mueller-Hinton agar and resistance frequency on the detec-

34. **Hindler, J. A., and D. F. Sahm.** 1992. Controversies and confusion regarding antimicrobial susceptibility testing of enterococci. *Antimicrob. Newsl.* **8:**65–74.

35. **Hindler, J. A., and N. L. Warner.** 1987. Effect of source of Mueller-Hinton agar on detection of oxacillin resistance in *Staphylococcus aureus* using a screening methodology. *J. Clin. Microbiol.* **25:**734–735.

36. **Isenberg, H. D.** 1988. Antimicrobial susceptibility testing: a critical evaluation. *J. Antimicrob. Chemother.* **22**(Suppl. A)**:**73–86.

37. **Ishida, K., P. A. Guze, G. M. Kalmanson, K. Albrandt, and L. B. Guze.** 1982. Variables in demonstrating methicillin tolerance in *Staphylococcus aureus* strains. *Antimicrob. Agents Chemother.* **212:**688–690.

38. **Jacobs, M. R., H. J. Koornhof, R. M. Robins-Browne, C. M. Stevenson, Z. A. Vermaak, I. Freiman, G. B. Miller, M. A. Witcomb, M. Isaäcson, J. I. Ward, and R. Austrian.** 1978. Emergence of multiply resistant pneumococci. *N. Engl. J. Med.* **299:**735–740.

39. **Jacoby, G. A., and A. A. Mederios.** 1991. More extended-spectrum β-lactamases. *Antimicrob. Agents Chemother.* **35:**1697–1704.

40. **Jankins, M.** 1992. Chloramphenicol acetyltransferase test, p. 5.8.1–5.8.7. *In* H. D. Isenberg (ed.), *Clinical Microbiology Procedures Handbook*, vol. 1. American Society for Microbiology, Washington, D.C.

41. **Jarlier, V., M. H. Nicolas, G. Fournier, and A. Philippon.** 1988. Extended broad-spectrum β-lactamases conferring transferable resistance to newer β-lactam agents in Enterobacteriaceae: hospital prevalence and susceptibility patterns. *Rev. Infect. Dis.* **10:**867–878.

42. **Jones, R. N., D. C. Edson, and CAP Microbiology Resource Committee of the College of American Pathologists.** 1991. Antimicrobial susceptibility testing trends and accuracy in the United States. *Arch. Pathol. Lab. Med.* **115:**429–436.

43. **Jordan, G. W., and M. M. Kawachi.** 1981. Analysis of serum bactericidal activity in endocarditis, osteomyelitis, and other bacterial infections. *Medicine* **60:**49–61.

44. **Katsanis, G. P., J. Spargo, M. J. Ferraro, L. Sutton, and G. A. Jacoby.** 1994. Detection of *Klebsiella pneumoniae* and *Escherichia coli* strains producing extended-spectrum β-lactamases. *J. Clin. Microbiol.* **32:**691–696.

45. **Kiehn, T. E., P. D. Ellner, and D. Budzko.** 1989. Role of the microbiology laboratory in care of the immunosuppressed patient. *Rev. Infect. Dis.* **11**(Suppl. 7)**:**S1706–S1710.

46. **Knapp, C., and J. A. Moody.** 1992. Tests to assess bactericidal activity, p. 5.16.1–5.16.33. *In* H. D. Isenberg (ed.), *Clinical Microbiology Procedures Handbook*, vol. 1. American Society for Microbiology, Washington, D.C.

47. **Kunin, C. M.** 1993. Resistance to antimicrobial drugs—a worldwide calamity. *Ann. Intern. Med.* **118:**557–561.

48. **Landman, D., N. K. Mobarakai, and J. M. Quale.** 1993. Novel antibiotic regimens against *Enterococcus faecium* resistant to ampicillin, vancomycin, and gentamicin. *Antimicrob. Agents Chemother.* **37:**1904–1908.

49. **Leclercq, R., S. Dutka-Malen, A. Brisson-Noel, C. Molinas, E. Derlot, M. Arthur, J. Duval, and P. Courvalin.** 1992. Resistance of enterococci to aminoglycosides and glycopeptides. *Clin. Infect. Dis.* **15:**495–501.

50. **Leclercq, R., S. Dutka-Malen, J. Duval, and P. Courvalin.** 1992. Vancomycin resistance gene *vanC* is specific to *Enterococcus gallinarum*. *Antimicrob. Agents Chemother.* **36:**2005–2008.

51. **Lee, W. W., and L. Komary.** 1976. New method for detecting in vitro inactivation of penicillins by *Haemophilus influenzae*. *Antimicrob. Agents Chemother.* **10:**654–656.

52. **Leitch, C., and S. Boonlayangoor.** 1992. Tests to detect high-level aminoglycoside resistance in enterococci, p. 5.4.1–5.4.9. *In* H. D. Isenberg (ed.), *Clinical Microbiology*

Procedures Handbook, vol. 1. American Society for Microbiology, Washington, D.C.

53. **Leitch, C., and S. Boonlayangoor.** 1992. Tests to detect oxacillin (methicillin)-resistant staphylococci with an oxacillin screen plate, p. 5.5.1–5.5.7. *In* H. Isenberg (ed.), *Clinical Microbiology Procedures Handbook*, vol. 1. American Society for Microbiology, Washington, D.C.

54. **Leitch, C., and S. Boonlayangoor.** 1992. β-Lactamase tests, p. 5.3.1–5.3.8. *In* H. D. Isenberg (ed.), *Clinical Microbiology Procedures Handbook*, vol. 1. American Society for Microbiology, Washington, D.C.

55. **Mackowiak, P. A., M. Marling-Cason, and R. L. Cohen.** 1982. Effects of temperature on antimicrobial susceptibility of bacteria. *J. Infect. Dis.* **145:**550–553.

56. **MacLowry, J. D.** 1989. Perspective. The serum dilution test. *J. Infect. Dis.* **160:**624–626.

57. **Massanari, R. M., M. A. Pfaller, D. S. Wakesfield, G. T. Hammons, L. A. McNut, R. F. Woolson, and C. M. Helms.** 1988. Implications of acquired oxacillin resistance in management and control of *Staphylococcus aureus* infections. *J. Infect. Dis.* **158:**701–709.

58. **Matthews, H. W., C. N. Baker, and C. Thornsberry.** 1988. Relationship between in vitro susceptibility test results for chloramphenicol and production of chloramphenicol acetyltransferase by *Haemophilus influenzae*, *Streptococcus pneumoniae*, and *Aerococcus* species. *J. Clin. Microbiol.* **26:**2387–2390.

59. **McDougal, L. K., and C. Thornsberry.** 1984. New recommendations for disk diffusion antimicrobial susceptibility tests for methicillin-resistant (heteroresistant) staphylococci. *J. Clin. Microbiol.* **19:**482–488.

60. **Meyer, K. S., O. C. Urban, J. A. Eagan, B. J. Berger, and J. J. Rahal.** 1993. Nosocomial outbreak of *Klebsiella* infection resistant to late-generation cephalosporins. *Ann. Intern. Med.* **119:**353–358.

61. **Moellering, R. C., Jr., O. M. Koraeniowski, M. A. Sande, and C. B. Wennersten.** 1979. Species-specific resistance to antimicrobial synergism in *Streptococcus faecium* and *Streptococcus faecalis*. *J. Infect. Dis.* **140:**203–208.

62. **Moody, J. A., C. E. Fasching, L. R. Peterson, and D. N. Gerding.** 1987. Ceftazidime and amikacin alone and in combination against *Pseudomonas aeruginosa* and *Enterobacteriaceae*. *Diagn. Microbiol. Infect. Dis.* **6:**59–67.

63. **Murray, B. E.** 1990. The life and times of the enterococcus. *Clin. Microbiol. Rev.* **3:**46–65.

64. **Murray, B. E.** 1992. β-Lactamase-producing enterococci. *Antimicrob. Agents Chemother.* **36:**2355–2359.

65. **National Committee for Clinical Laboratory Standards.** 1992. *Methodology for the Serum Bactericidal Test.* Tentative standard, M21-T. National Committee for Clinical Laboratory Standards, Villanova, Pa.

66. **National Committee for Clinical Laboratory Standards.** 1992. *Methods for Determining Bactericidal Activity of Antimicrobial Agents.* Tentative standard, M26-T. National Committee for Clinical Laboratory Standards, Villanova, Pa.

67. **National Committee for Clinical Laboratory Standards.** 1993. *Methods for Dilution Antimicrobial Susceptibility Tests for Bacteria That Grow Aerobically.* Approved standard M7-A3. National Committee for Clinical Laboratory Standards, Villanova, Pa.

68. **National Committee for Clinical Laboratory Standards.** 1993. *Performance Standards for Antimicrobial Disk Susceptibility Tests.* Approved standard M2-A5. National Committee for Clinical Laboratory Standards, Villanova, Pa.

69. **Neumann, M. A., D. F. Sahm, C. Thornsberry, and J. E. McGowan, Jr.** 1991. *Cumitech 6A, New Developments in Antimicrobial Agent Susceptibility Testing: a Practical Guide.* Coordinating ed., J. E. McGowan, Jr. American Society for Microbiology, Washington, D.C.

70. **O'Callaghan, C. H., A. Morris, S. M. Kirby, and A. H. Shingler.** 1972. Novel method for detection of β-lactamase by using a chromogenic cephalosporin substrate. *Antimicrob. Agents Chemother.* **1:**283–288.

71. **Peterson, L. R., J. A. Moody, C. E. Fasching, and D. N. Gerding.** 1987. In vivo and in vitro activity of ciprofloxacin plus azlocillin against 12 streptococcal isolates in a neutropenic-site model. *Diagn. Microbiol. Infect. Dis.* **7:**127–136.

72. **Peterson, L. R., and C. J. Shanholtzer.** 1992. Tests for bactericidal effects of antimicrobial agents: technical performance and clinical relevance. *Clin. Microbiol. Rev.* **5:**420–432.

73. **Philippon, A., R. Labia, and G. Jacoby.** 1989. Extended-spectrum β-lactamases. *Antimicrob. Agents Chemother.* **33:**1131–1136.

74. **Rosenblatt, J. E.** 1987. Laboratory tests used to guide antimicrobial therapy. *Mayo Clin. Proc.* **62:**799–805.

75. **Sabath, L.** 1977. Chemical and physical factors influencing methicillin resistance of *Staphylococcus aureus* and *Staphylococcus epidermidis*. *J. Antimicrob. Chemother.* **3:**47–51.

76. **Sahm, D. F.** 1994. Streptococci and staphylococci: laboratory considerations for in vitro susceptibility testing. *Clin. Microbiol. Newsl.* **16:**9–14.

77. **Sahm, D. F., S. Boonlayangoor, P. C. Iwen, J. L. Baade, and G. L. Woods.** 1991. Factors influencing determination of high-level aminoglycoside resistance in *Enterococcus faecalis*. *J. Clin. Microbiol.* **29:**1934–1939.

78. **Sahm, D. F., S. Boonlayangoor, and J. E. Schulz.** 1991. Detection of high-level aminoglycoside resistance in enterococci other than *Enterococcus faecalis*. *J. Clin. Microbiol.* **29:**2595–2598.

79. **Sahm, D. F., J. Kissinger, M. S. Gilmore, P. R. Murray, R. Mulder, J. Solliday, and B. Clarke.** 1989. In vitro susceptibility studies of vancomycin-resistant *Enterococcus faecalis*. *Antimicrob. Agents Chemother.* **33:**1588–1591.

80. **Sahm, D. F., and L. Olsen.** 1990. In vitro detection of enterococcal vancomycin resistance. *Antimicrob. Agents Chemother.* **34:**1846–1848.

81. **Sahm, D. F., and C. Torres.** 1988. Effects of medium and inoculum variations on screening for high-level aminoglycoside resistance in *Enterococcus faecalis*. *J. Clin. Microbiol.* **26:**250–256.

82. **Sahm, D. F., and C. Torres.** 1988. High-content aminoglycoside disks for determining aminoglycoside-penicillin synergy against *Enterococcus faecalis*. *J. Clin. Microbiol.* **26:**257–260.

83. **Sanders, C. C.** 1992. β-Lactamases of gram-negative bacteria: new challenges for new drugs. *Clin. Infect. Dis.* **14:**1089–1099.

84. **Schlichter, J. G., and H. MacLean.** 1947. A method for determining the effective therapeutic level in the treatment of subacute bacterial endocarditis with penicillin: a preliminary report. *Am. Heart J.* **34:**209–211.

85. **Schlichter, J. G., H. MacLean, and A. Milzer.** 1949. Effective penicillin therapy in subacute bacterial endocarditis and other chronic infections. *Am. J. Med. Sci.* **217:**600–608.

86. **Sculier, J. P., and J. Klastersky.** 1984. Significance of serum bactericidal activity in gram-negative bacillary bacteremia in patients with and without granulocytopenia. *Am. J. Med.* **76:**429–435.

87. **Silver, L. L., and K. A. Bostian.** 1993. Discovery and development of new antibiotics: the problem of antibiotic resistance. *Antimicrob. Agents Chemother.* **37:**377–383.

88. **Standiford, H. D., J. B. deMaine, and W. M. M. Kirby.** 1970. Antibiotic synergism of enterococci. *Arch. Intern. Med.* **126:**255–259.

88a. **Swenson, J. M.** Unpublished observations.

89. **Swenson, J. M., N. C. Clark, M. J. Ferraro, D. F. Sahm, G. Doern, M. A. Pfaller, L. B. Reller, M. P. Weinstein, R. J. Zabransky, and F. C. Tenover.** 1994. Development of a standardized screening method for detection of vancomycin-resistant enterococci. *J. Clin. Microbiol.* **32:**1700–1704.

90. **Swenson, J. M., M. J. Ferraro, D. F. Sahm, P. Charache, The National Committee for Clinical Laboratory Standards Working Group on Enterococci, and F. C. Tenover.** 1992. New vancomycin disk diffusion breakpoints for enterococci. *J. Clin. Microbiol.* **30:**2525–2528.

91. **Swenson, J. M., B. C. Hill, and C. Thornsberry.** 1986. Screening pneumococci for penicillin resistance. *J. Clin. Microbiol.* **24:**749–752.

92. **Swenson, J. M., B. C. Hill, and C. Thornsberry.** 1989. Problems with the disk diffusion test for detection of vancomycin resistance in enterococci. *J. Clin. Microbiol.* **27:**2140–2142. (Erratum, **28:**403, 1990.)

93. **Taylor, P. C., F. D. Schoenknecht, J. C. Sherris, and E. C. Linner.** 1983. Determination of minimum bactericidal concentrations of oxacillin for *Staphylococcus aureus*: influence and significance of technical factors. *Antimicrob. Agents Chemother.* **23:**142–150.

94. **Tenover, F. C., J. Tokars, J. Swenson, S. Paul, K. Spitalny, and W. Jarvis.** 1993. Ability of clinical laboratories to detect antimicrobial agent-resistant enterococci. *J. Clin. Microbiol.* **31:**1695–1699.

95. **Thauvin-Eliopoulos, E., L. B. Rice, G. M. Eliopoulos, and R. C. Moellering, Jr.** 1990. Efficacy of oxacillin and ampicillin-sulbactam combination in experimental endocarditis caused by β-lactamase-hyperproducing *Staphylococcus aureus*. *Antimicrob. Agents Chemother.* **34:**728–732.

96. **Thomson, K. S., and C. C. Sanders.** 1992. Detection of extended-spectrum β-lactamases in members of the family *Enterobacteriaceae*: comparison of the double-disk and three-dimensional tests. *Antimicrob. Agents Chemother.* **36:**1877–1882.

97. **Thornsberry, C., and L. K. McDougal.** 1983. Successful use of broth microdilution in susceptibility tests for methicillin-resistant (heteroresistant) staphylococci. *J. Clin. Microbiol.* **18:**1084–1091.

98. **Tomasz, A., S. Nachman, and H. Leaf.** 1991. Stable classes of phenotypic expression in methicillin-resistant clinical isolates of staphylococci. *Antimicrob. Agents Chemother.* **35:**124–129.

99. **Torres, C., C. Tenorio, M. Lantero, M. Gastañares, and F. Baquero.** 1993. High-level penicillin resistance and penicillin-gentamicin synergy in *Enterococcus faecium*. *Antimicrob. Agents Chemother.* **37:**2427–2431.

100. **Tubbs, R. R.** 1977. Insuring effective antimicrobial therapy: laboratory evaluation. *J. Am. Osteopath. Assoc.* **76:**617–624.

101. **Washington, J. A., II, C. C. Knapp, and C. C. Sanders.** 1988. Accuracy of microdilution and the Automicrobic system in detection of β-lactam resistance in gram-negative bacterial mutants with derepressed β-lactamases. *Rev. Infect. Dis.* **10:**824–829.

102. **Willey, B. M., B. N. Kreiswirth, A. E. Simor, G. Williams, S. R. Scriver, A. Phillips, and D. E. Low.** 1992. Detection of vancomycin resistance in *Enterococcus* species. *J. Clin. Microbiol.* **30:**1621–1624.

Genetic Methods for Detecting Antibacterial Resistance Genes

FRED C. TENOVER, TANJA POPOVIC, AND ØRJAN OLSVIK

117

Resistance to antimicrobial agents in bacteria can be mediated by several mechanisms, including (i) changes in the permeability of the cell envelope that limit the amount of drug that has access to cellular targets, (ii) changes in or elimination of the site of drug action, (iii) provision of alternate enzymatic pathways around those blocked by antibacterial therapy, and (iv) destruction or inactivation of the antimicrobial agent (56, 58; also see chapter 112 of this Manual). While some of these changes are due to random mutations in the bacterial genome, most resistance to antimicrobial agents in bacteria is mediated by acquired genes whose presence in a cell is usually synonymous with a resistance phenotype. Although it is presumed that most resistance genes harbored by bacteria are expressed, exceptions to this rule have been documented (29, 62, 73). Nonetheless, genetic tests aimed at the detection of resistance genes in bacterial isolates by using DNA probes or the PCR are based on the supposition that gene carriage equals resistance. Genetic methods, including DNA probes and PCR-based assays, are now used in many clinical laboratories. Thus, the application of genetic methods to the detection of resistance genes is a natural extension of this technology. However, microbiologists must realize that while 90 to 95% of resistance to a given drug may be due to acquisition of one particular resistance gene, a subset of isolates will be resistant because of either chromosomal mutation or the presence of a novel resistance gene that will not be detected by probes or PCR, although such resistance would be detected by traditional susceptibility testing methods. Thus, no genetic assay is 100% sensitive or specific.

REASONS FOR USING GENETIC TESTS TO DETECT RESISTANCE GENES

There are four reasons to pursue the identification of antimicrobial resistance genes by genetic methods. First, DNA probes or nucleic acid amplifications techniques are helpful for arbitrating MIC results that are at or near the breakpoint for resistance. For example, oxacillin-resistant isolates of *Staphylococcus aureus* with MICs between 2 and 8 μg/ml may contain the *mec* (methicillin) resistance gene determinant or may produce high levels of β-lactamase that slowly hydrolyze oxacillin (13). While vancomycin would be the drug of choice for the former cases, β-lactamase hyperpro-

ducers can be effectively treated with more effective penicillinase-stable beta-lactams or beta-lactam–β-lactamase inhibitor compounds (52). A test showing the absence of the *mec* gene suggests that a physician could use an antimicrobial agent other than vancomycin to treat the infection.

Second, genetic methods can be used to directly detect resistance genes or mutations that result in resistance in organisms from clinical specimens in order to guide therapy early in the course of a patient's disease, long before cultures are positive. For example, PCR assays can detect mutations in the *rpoB* locus associated with rifampin resistance in *Mycobacterium tuberculosis* (85). Such mutations indicate that the strain is at least resistant to rifampin and may be resistant to several drugs. A positive PCR result for mutations in the *rpoB* locus directs the physician to avoid rifampin and use alternative antimycobacterial agents.

Third, genetics-based tests are more accurate than antibiograms for monitoring the epidemiologic spread of a particular resistance gene in a hospital or a patient population. For example, tracking the spread of the *vanA* vancomycin resistance gene in enterococci with PCR assays has helped document the spread of multiresistant enterococci along the eastern seaboard of the United States (15). Antibiograms cannot differentiate between organisms containing the *vanA* gene and those with derepressed *vanB* genes (6).

Fourth, genetics-based tests can be used as the "gold standard" for resistance when the accuracy of new susceptibility testing methods that use clinical isolates or stock cultures with borderline MICs is being evaluated (37).

DESIGN OF GENETIC TESTS FOR RESISTANCE GENES

General Guidelines

The ideal genetic test targets nucleic acid sequences within the open reading frame (or coding region) of the resistance gene and avoids sequences outside of the gene that may contain insertion elements or promoter sequences that may be present in susceptible strains or in strains with other types of resistance genes. Probes that extend beyond the coding regions of the gene have been shown to produce false-positive results (51). Among the probes and primers

TABLE 1 DNA probes for antimicrobial agent resistance genes

Antimicrobial agent(s)	Organism	Target	Fragment or sequence	Reference
Aminoglycosides	*Escherichia coli*(pGH54)	*aph(3′)-I*	274-bp *Ava*I	106
	E. coli(pSAY16)	*aph(3′)-II*	377-bp *Sph*I-*Eag*I	106
	Enterococcus faecalis(pJH1)	*aph(3′)-III*	530-bp *Hpa*II	99
	Acinetobacter baumannii(pIP1841)	*aph(3′)-VI*	370-bp *Eco*RI-*Acc*I	45
	E. coli(pFCT3103)	*ant(2″)-Ia*	310-bp *Ava*I	88
	E. coli(pFT1045)	*ant(3″)-Ia*	483-bp *Rsa*I	86
	E. coli(pSCH2005)	*ant(4′)-Ia*	318-bp *Sac*I-*Bgl*II	74
	E. coli(pMG235)	*ant(4′)-IIa*	204-bp *Eco*RV-*Ssp*I	74
	E. coli(pFCT4392)	*aac(3)-Ia*	307-bp *Ava*I-*Eco*RV	91
	E. coli(pFCT1165)	*aac(6′)-Ia*	405-bp *Ssp*I	87
	E. coli(pAZ505)	*aac(6′)-Ib*	344-bp *Dde*I-*Ava*I	98
	E. coli(pSCH2014)	*aac(6′)-Ic*	359-bp *Bam*HI-*Eco*RV	76
	E. faecalis CIP 54-32	*aac(6′)-Ii*	600-bp *Eco*RI-*Hind*III	17
	E. coli(pC390)	*aac(3)-Va*	514-bp *Sal*I-*Cla*I	67
	E. coli(pSCH4203)	*aac(3)-Vb*	742-bp *Cla*I-*Eco*RV	67
	E. faecalis(pIP800)	*aac(6′)-aph(2″)*	1,317-bp *Hinc*II-*Taq*I	27
	Staphylococcus aureus(pUB110)	*ant(4′)*	473-bp *Hinc*II	11
	E. faecalis(pJH-1)	*ant(6′)-Ia*	470-bp *Hpa*II	62
Beta-lactams	*S. aureus*(pGO164)	*mec*	1.1-kb *Bgl*I-*Xba*I or	3
			400-bp *Hinc*II-*Cla*I	37
	E. coli(pBR322)	*bla*$_{TEM}$	424-bp *Bgl*I-*Hinc*II	47
	E. coli(pMON38)	*bla*$_{SHV}$	780-bp *Pvu*II	38
	Klebsiella pneumoniae	*bla*$_{SHV}$	5′ CTG CAG TGG ATG GTG GAC GAT CGG G 3′	86
	E. coli(pMON401)	*bla*$_{ROB}$	250-bp *Sau*3A	47
	E. coli(pMON301)	*bla*$_{OXA}$	315-bp *Bgl*II	47
Chloramphenicol	*E. coli*(pJIR62)	*catP*	380-bp *Eco*RV-*Hinf*I	72
	E. coli(pJIR260)	*catQ*	350-bp *Dra*I-*Pst*I	72
	Clostridium difficile(pPPM9)	*catD*	270-bp *Eco*RV-*Taq*I	104
	E. coli(pBR325)	*catI*	1.1-kb *Eco*RI-*Hind*III	10
	E. coli(pJHC-A18)	*catII*	1.4-bp *Eco*RI-*Hind*III	69
	E. coli(pJHC-A34)	*catIII*	2.7-kb *Pst*I	69
Macrolides	*E. coli*(pJIR233)	*ermP*	500-bp *Eco*RI-*Hind*III	8
	E. coli(pIP1100)	*ereA*	716-bp *Eco*RI-*Pst*I	5
	E. coli(pIP1527)	*ereB*	838-bp *Eco*RI-*Pst*I	5
	S. aureus::Tn554	*ermA*	417-bp *Mbo*I	22
	E. coli(pIP1527)	*ermB*	349-bp *Rsa*I	5
	S. aureus(pE194)	*ermC*	472-bp *Hae*III-*Hinc*II	22
	Bacteroides fragilis(pVA1476)	*ermF*	230-bp *Taq*I-*Hind*III	30
	S. aureus(pUL5054)	*msrA*	1.9-kb *Hind*III	22
Sulfonamides	*E. coli*(RSF1010)	*sulI*	660-bp *Sac*II-*Bgl*II	66
	E. coli(pGS05)	*sulII*	780-bp *Hinc*II	66
Tetracycline	*E. coli*(pSL107)	*tet*(A)	750-bp *Sma*I	50
	E. coli(pRT11)	*tet*(B)	1.27-kb *Hinc*II	70
	E. coli(pBR322)	*tet*(C)	928-bp *Bst*N1	50
	E. coli(pSL106)	*tet*(D)	3.0-kb *Hind*III-*Pst*I	50
	E. coli(pSL1504)	*tet*(E)	2.5-kb *Cla*I-*Pst*I	81
	Streptococcus agalactiae(pAT102)	*tet*(K)	870-bp *Hinc*II	109
	E. faecalis(pBC16)	*tet*(L)	310-bp *Cla*I-*Hpa*II	109
	E. faecalis(pIP804)	*tet*(M)	850-bp *Hind*III-*Cla*I	51
	Campylobacter coli(pIP1433)	*tet*(O)	1,458-bp *Hind*III-*Nde*I	82
	E. coli(pFKT686)	*tet*(O)	500-bp *Rsa*I-*Hinc*II	90
	E. coli(pJIR39)	*tet*(P)	800-bp *Eco*RI-*Sph*I	1
	B. fragilis(pPH7Δ1)	*tet*(Q)	2.0-kb *Eco*RI-*Cla*I	30
	Listeria monocytogenes(pAT451)	*tet*(S)	300-bp *Eco*RI-*Bgl*I	14

(*Continued on next page*)

TABLE 1 DNA probes for antimicrobial agent resistance genes *(Continued)*

Antimicrobial agent(s)	Organism	Target	Fragment or sequence	Reference
Trimethoprim	*E. coli*(pLKO627)	*dhfrI*	5' TTT AGG CCA GTT TTT ACC AAG ACT TCG C 3'	84
	E. coli(pLKO22)	*dhfrIII*	5' GAG AAA CAT TGC CCA TGG CCT CTA CGC TC 3'	84
	E. coli(pLKO221)	*dhfrVI*	5' AGT GTC GAG GAA AGG AAT TTC AAG CTC A 3'	84
	E. coli(pCJO-01-1)	*dhfrIX*	5' AAA ACA GTA CCA CCC ACC CAG AAC ACT 3'	40
	E. coli(pFE872)	*dhfrI*	500-bp *Hpa*I	31
	E. coli(pLKO627)	*dhfrI*	500-bp *Bam*HI-*Kpn*I	40
	E. coli(pWZ820)	*dhfrII*	275-bp *Sau*3A-*Eco*RI	110
	E. coli(pLKO601)	*dhfrII*	280-bp *Eco*RI-*Sal*I	40
	E. coli(pUN972)	*dhfrIII*	850-bp *Pst*I	93
	E. coli(pUK1148)	*dhfrIV*	1.6-kb *Cla*I	97
	E. coli(pLK09)	*dhfrV*	500-bp *Hinc*II	97
	E. coli(pUN1043)	*dhfrVII*	300-bp *Eco*RV	96
	E. coli(pCJO01-1)	*dhfrIX*	340-bp *Eco*RV-*Hind*III	40
	E. coli(pMAQ41)	*dhfrX*	450-bp *Xho*I-*Bgl*II	63
	E. coli(pBEM155)	*dhfrXII*	500-bp *Bam*HI-*Hind*III	79
Vancomycin	*Enterococcus faecium*(pIP816)	*vanA*	265-bp *Bam*HI-*Pst*I	20
	E. faecalis V583	*vanB*	620-bp cloned fragment	24
	Enterococcus gallinarum BM4174	*vanC*	680-bp *Eco*RI-*Hinc*II	21

that have been described for studying antibacterial resistance are those directed to β-lactamase genes and the genes that encode resistance to aminocyclitols, aminoglycosides, chloramphenicol, glycopeptides, macrolides, mupirocin, quinolones, rifampin, sulfonamides, tetracyclines, and trimethoprim. Examples of DNA probes that can be used to detect resistance genes are shown in Table 1, and examples of PCR primers that target resistance genes or mutations associated with resistance are shown in Table 2. Most of the DNA probes target regions within the open reading frame of the resistance gene; however, some of the probes were developed before the gene sequence data were available and may contain sequences outside of the open reading frame. Therefore, appropriate specificity controls (organisms that have the same resistance pattern but contain resistance genes other than the target gene) should always be included in all reactions using these probes and primers.

Using Commercial DNA Probes To Monitor Growth as an Indicator of Drug Resistance

In 1989, Kawa et al. (42) reported that the Gen-Probe DNA probe assay for *M. tuberculosis* could be used to monitor the growth of mycobacteria in pairs of BACTEC vials with and without various antituberculous agents as a means of rapidly determining those isolates that were likely to be drug resistant. While these early studies looked promising, other investigators have had difficulty in repeating the success of Kawa et al. with the nonradioactive Gen-Probe kits currently available. Similar approaches to the susceptibility testing of *Legionella pneumophila* (68) and *S. aureus* (for methicillin resistance) (105) have been reported. However, additional data are needed to confirm the validity of this approach.

AMINOGLYCOSIDE RESISTANCE GENES

Aminoglycoside resistance genes are common in both gram-positive and gram-negative organisms (75). However, the large number of different types of aminoglycoside resistance genes present in gram-negative organisms (acetyltransferases, adenylyltransferases, and phosphotransferases) and the lack of consensus sequences that would allow detection of multiple types of genes with a single DNA probe or PCR primer set (75) make it difficult to use probes and PCR tests to predict resistance. Rather, genetic methods are better suited for classification of new determinants and for epidemiologic studies (29, 73, 101, 102).

Aminoglycoside resistance in gram-positive organisms, however, is more uniform (62). Thus, probes and PCR primers can be used to identify strains that carry genes encoding high-level aminoglycoside resistance. PCR assays and probes have been used, particularly with enterococci, to identify the ANT(6) streptomycin resistance gene and the gene encoding the AAC(6')-APH(2") bifunctional enzyme responsible for high-level gentamicin resistance (27, 41, 101).

DETECTING GENES ASSOCIATED WITH RESISTANCE TO BETA-LACTAM DRUGS

Methicillin Resistance in Staphylococci

Detection of oxacillin resistance in staphylococci, which is primarily mediated by the *mec* gene determinant, continues to be a problem, particularly with the coagulase-negative strains (13). A *mec* gene probe can differentiate isolates that are borderline resistant to oxacillin because of the production of large quantities of β-lactamase from those that are

TABLE 2 PCR assays for resistance genes

Antimicrobial agent and gene	Primers (5′→3′)[a]	Product size		Use	Reference
Aminoglycosides					
aac(3)-Ia	ACC TAC TCC CAA CAT CAG CC				
	ATA TAG ATC TCA CTA CGC GC	169	bp	Detection	101
aac(3)-IIa	ACT GTG ATG GGA TAC GCG TC				
	CTC CGT CAG CGT TTC AGC TA	237	bp	Detection	101
aac(3)-IIIa	CAC AAG AAC GTG GTC CGC TA				
	AAC AGG TAA GCA TCC GCA TC	185	bp	Detection	101
aac(3)-IVa	CTT CAG GAT GGC AAG TTG GT				
	TCA TCT CGT TCT CCG CTC AT	286	bp	Detection	101
aad(2″)-Ia	ATG TTA CGC AGC AGG GCA GTC G				
	CGT CAG ATC AAT ATC ATC GTG C	188	bp	Detection	102
aad(4′)-Ia	GCA AGG ACC GAC AAC ATT TC				
	TGG CAC AGA TGG TCA TAA CC	165	bp	Detection	101
aac(6′)-aph(2″)	CCA AGA GCA ATA AGG GCA TA				
	CAC TAT CAT AAC CAC TAC CG	220	bp	Detection	101
aph(3′)-IIIa	GCC GAT GTG GAT TGC GAA AA				
	GCT TGA TCC CCA GTA AGT CA	292	bp	Detection	101
aac(6′)-Ie	ATA TGA TTA TGA AAA AGG TGA				
	ATA ATC AAT CTT TAT AAG TCC	1,470	bp	Detection	41
Beta-lactams					
pbp1A[b]	CGG CAT TCG ATT TGA TTC GCT TCT				
	CTG AGA AGA TGT CTT CTC AGG	2.4	kb	Fingerprinting	16
pbp2B[b]	GAT CCT CTA AAT GAT TCT CAG GTG G				
	CAA TTA GCT AGC AAT AGG TGT TG G	1.5	kb	Fingerprinting	19
pbp2X[b]	CGT GGG ACT ATT TAT GAC CGA AAT GG	2.0	kb	Fingerprinting	44
	AAT TCC AGC ACT GAT GG				
	A AAT AAA CAT ATT A				
bla$_{TEM}$[c]	AGT TAT CTA CAC GAC GG				
	GGC GTA CTA TTC ACT CT	761	bp	Detection	78
bla$_{TEM}$	ATA AAA TTC TTG AAG ACG AAA				
	GAC AGT TAC CAA TGC TTA ATC A	1,079	bp	Sequencing	49
bla$_{TEM}$	TTG GGT GCA CGA GTG GGT TA				
	TAA TTG TTG CCG GGA AGC TA	503	bp	Probe	4
bla$_{SHV}$	TCG GGC CGC GTA GGC ATG AT				
	AGC AGG GCG ACA ATC CCG CG	625	bp	Probe	4
bla$_{CARB}$	AAT GGC AAT CAG CGC TTC CC				
	GGG GCT TGA TGC TCA CTC CA	586	bp	Probe	4
mecA	AAA ATC GAT GGT AAA GGT TGG C				
	AGT TCT GCA GTA CCG GAT TTG C	533	bp	Detection	55
mecA	GAA ATG ACT GAA CGT CCG AT				
	GCG ATC AAT GTT ACC GTA GT	150	bp	Detection	100
Macrolides					
erm	GAA/G ATI GGI III GGI AAA/G GGI CA				
	AAC/T TGA/G TTC/T TTI GTA/G AA	530	bp	Detection	7
Mupirocin					
IRS[d]	CCA TGC CTT ACC AGT TGA ATT				
	GGA TCC CCG AGC ACT ATC CGA				
		1.65	kb	Probe	32
Rifampin					
rpoB[e]	TAC GGT CGG CGA GCT GAT CC				
	TAC GGC GTT TCG ATG AAC C	411	bp	Sequencing	85
rpoB[e]	TCG CCG CGA TCA AGG AGT				
	TGC ACG TCG CGG ACC TCC A	157	bp	SSCP assay	85
rpoB[f]	CAG GAC GTC GAG GCG ATC AC				
	AAC GAC GAC GTG GCC AGC GT	710	bp	Detection	35

(Continued on next page)

TABLE 2 PCR assays for resistance genes   *(Continued)*

Antimicrobial agent and gene	Primers (5'→3')[a]	Product size	Use	Reference
Tetracyclines				
tet(M)	GAA CTC GAA CAA GAG GAA AGC ATG GAA GCC CAG AAA GGA T	741 bp	Detection	59
tet(M)	TTA TCA ACG GTT TAT CAG G CGT ATA TAT GCA AGA CG	397 bp	Detection	9
tet(O)	AAC TTA GGC ATT CTG GCT CAC TCC CAC TGT TCC ATA TCG TCA	515 bp	Detection	59
tet(M)-*tet*(O)	GTT TAT CAC GGA AG(TC) GC(AT) A GGA GCC CAG AAA GGA TT(CT) GG	686 bp	Detection	71
Vancomycin				
ddl[g]	GGI GA(AG) GA(TC) GGI (TA)(CG)I (TCA)TI CA(AG)GG TG (AG)AA ICC IGG IA(TAG) IGT (AG)TT	640 bp	Probe	21
ddl[g]	GA(AG) GAT GGI T(CG)C AT(AC) CA(AG) GG(AT) (AC)GT (AG)AA ICC IGG CA(GT) (AG)GT (AG)TT	630 bp	Probe	33
vanA	ATG GCA AGT CAG GTG AAG ATG G TCC ACC TCG CCA ACA ACT AAC G	357 bp	Probe	103
vanA	CAT GAA TAG AAT AAA AGT TGC AAT A CCC CTT TAA CGC TAA TAC GAT CAA	1,030 bp	Detection	15
vanB	GTG ACA AAC CGG AGG CGA GGA CCG CCA TCC TCC TGC AAA AAA	433 bp	Detection	15
vanC	GAA AGA CAA CAG GAA GAC CGC ATC GCA TCA CAA GCA CCA ATC	796 bp	Detection	15

[a]X, any base; I, inosine.
[b]For *S. pneumoniae* only.
[c]Specific for *bla*$_{TEM}$ genes in *N. gonorrhoeae*.
[d]IRS, isoleucyl-tRNA synthetase variants.
[e]For *M. tuberculosis* only.
[f]For *M. leprae* only.
[g]*ddl*, D-alanyl–D-alanine ligase gene.

resistant because of the presence of the *mec* determinant (13, 95). Several PCR assays based on the DNA sequence information reported by Song et al. (80) have also been described (55, 94, 100). The rare strains of *S. aureus* that are resistant to oxacillin by virtue of containing modified penicillin-binding proteins (PBPs) with reduced affinity for oxacillin (the so-called MOD strains) may be misclassified as oxacillin susceptible by the *mec* gene test, since these strains are truly oxacillin resistant but do not contain the *mec* gene (95). Nevertheless, a probe or a PCR assay that can be used on an isolated colony of *S. aureus* or other staphylococcal species to identify organisms that have the *mec* gene could be a valuable tool for the clinical microbiology laboratory, presuming that the caveats to its use are understood. For example, a recent study by Huang et al. reported that three strains of *Staphylococcus simulans* carried the *mec* gene but had MICs of oxacillin of 2 μg/ml, a borderline susceptibility result (37). These isolates would be classified as resistant by the probe assay. Conversely, seven isolates of *S. aureus* that exhibited the MOD phenotype were *mec* probe negative (i.e., susceptible) but resistant to oxacillin by broth microdilution testing (MICs of 4 to 16 μg/ml) (37). Although these strains, along with the *S. simulans* strains, are rare clinical isolates, they do present potential failures of the genetic methods.

Detection of the *mec* gene by PCR with organisms present in blood samples in BACTEC bottles (Becton

Dickinson Microbiology Systems, Cockeysville, Md.) has been reported by Ubukata et al. (100). A minimum of 500 CFU of *S. aureus* were required for detection of the *mec* gene, while the number of CFU of coagulase-negative species of staphylococci required to give a positive PCR test was at least 10-fold higher. Detection was accomplished by using a nonradioactive enzyme-linked colorimetric assay. Although detection of a *mec* gene carried by a skin contaminant is a potential drawback to this approach (a false-positive result), a rapid *mec* gene-negative result could help reduce pharmacy costs by signalling the clinician that a semisynthetic penicillin could be used for therapy instead of vancomycin.

Beta-Lactam Resistance in Pneumococci

Resistance to penicillin and other antimicrobial agents in pneumococci has become a global problem (43). Resistance develops when pneumococcal PBPs are remodeled through the acquisition of chromosomal DNA from other pneumococci or other streptococcal species (16, 19, 44). Because of the apparent random nature of the remodeling process, it has not yet been possible to develop DNA probes to target resistant strains. However, PCR primers have been developed for amplifying regions of the 1a, 2b, and 2x PBP genes to facilitate DNA sequence analysis of the remodeled genes (16, 19, 44). The amplification products can also be cleaved with restriction enzymes, and the fragment patterns can be used as strain markers for epidemiologic studies of pneumococcal resistance (54).

β-Lactamase Genes in Gram-Negative Organisms

Several DNA probes and PCR primer sets have been developed to detect the genes encoding the TEM, SHV, OXA, CARB, and ROB β-lactamases present in gram-negative organisms (4, 47). For example, the bla_{TEM} gene has been detected in many members of the family *Enterobacteriaceae*, in *Haemophilus* spp., and in *Neisseria gonorrhoeae*. Perine and coworkers have used DNA probes to detect the bla_{TEM} gene directly in urethral exudates from males (64), and Carter et al. have used a similar approach to detect ampicillin-resistant organisms directly in urine (12). Recently, PCR was used to detect bla_{TEM} directly in cerebrospinal fluid samples containing *Haemophilus influenzae* (89). Thus, the utility of direct detection has clearly been demonstrated, although this method is rarely, if ever, used in clinical laboratories, primarily because the test is not commercially available.

Several important issues regarding the clinical use of probes and primers are highlighted by these studies. First, at least two different β-lactamase genes can mediate ampicillin resistance in *H. influenzae*: bla_{TEM} and bla_{ROB} (18). In addition, alterations in the affinities of the PBPs for beta-lactam drugs can result in non-β-lactamase-mediated ampicillin resistance in *H. influenzae* (53). Thus, when *H. influenzae* is screened for resistance, a multitude of mechanisms must be considered. However, strains containing the bla_{ROB} gene and strains with PBP-mediated ampicillin resistance are rare. Non-β-lactamase-mediated penicillin resistance also occurs in *N. gonorrhoeae* (25) and could potentially cause misleading results if probes to the bla_{TEM} gene were used to predict ampicillin resistance. Such strains, however, are also rare.

The numbers of nosocomial infections caused by *Klebsiella pneumoniae* and other *Enterobacteriaceae* that produce extended-spectrum β-lactamases capable of hydrolyzing cefotaxime, ceftriaxone, ceftazidime, and aztreonam are increasing in the United States and Europe (39). Better tools are needed for the recognition and classification of these enzymes. A subclassification scheme for bla_{TEM} genes based on hybridization of oligonucleotide probes that differentiate the various genes at critical base pairs involved in resistance to extended-spectrum cephalosporins has been described by Mabilat and Courvalin (48). This method is very helpful for epidemiologic studies of extended-spectrum β-lactamase genes but is tedious to perform, since as many as 20 different hybridization reactions must be completed per strain tested. However, the system is more sensitive than isoelectric focusing for identifying novel β-lactamase genes (48) and is therefore a valuable addition to reference laboratories studying the epidemiology of resistance. As rapid DNA-sequencing equipment becomes more widely accessible, direct sequencing of bla_{TEM} genes may become the standard for epidemiologic studies.

CHLORAMPHENICOL RESISTANCE

Several classes of chloramphenicol acetyltransferases (CAT) can mediate resistance to chloramphenicol (77). DNA probes to three CAT genes commonly found in gram-negative organisms (*catI, catII,* and *catIII*) and genes specific to gram-positive anaerobes including *Clostridium perfringens* and *Clostridium difficile* (*catP, catQ,* and *catD*) have been described (72, 77, 104). Additional genes that

show relatively little DNA sequence homology with those mentioned above are present in staphylococci, streptococci, and aerobic gram-positive bacilli. DNA probes for many of these genes have been described (Table 1), although their clinical utilities have yet to be established. Direct detection of these genes in clinical samples has not been reported but may become important as resistance levels to extended-spectrum beta-lactam agents, currently the drugs of choice for treating bacterial meningitis in children, become more pronounced. The data supporting the use of the DNA probes for detection of the CAT genes in gram-negative organisms (*catI, catII, catIII*), although referred to in reference 84, have apparently never been published. Thus, specificity controls are important when these probes are used.

GLYCOPEPTIDE RESISTANCE

Resistance to vancomycin in enterococci has become a major global issue during the last few years (6, 15, 103). Resistance in enterococci can be mediated by several different genes, including *vanA, vanB,* and *vanC* (6, 15, 20, 21, 24, 33, 46, 103). Both DNA probes and PCR assays that detect these genes have been described. To date, none of the nucleic acid-based assays have been used to directly detect resistance genes in bacteria in clinical samples. However, these assays can be valuable for epidemiologic studies of resistance and for verifying the gene present in borderline resistant strains.

MACROLIDE, LINCOSAMIDE, AND STREPTOGRAMIN RESISTANCE

The genes mediating resistance to erythromycin have been the targets both of DNA probes and of PCR assays. DNA probe assays that detect several erythromycin methylase genes (*erm* genes) that mediate resistance to macrolides, lincosamides (such as clindamycin), and streptogramins, i.e., the MLS resistance phenotype, have been used for epidemiologic studies of resistance (5, 22, 92). In addition, a novel erythromycin gene, *msrA*, which mediates resistance only to macrolides and streptogramins (MS resistance), has recently been described (92). Studies by Eady et al. demonstrate that the *msrA* gene is common in isolates of *S. aureus* and produces a phenotype of erythromycin resistance but clindamycin susceptibility (22). This mimics the action of the *ermA* gene, which produces the same phenotype. Since most strains of staphylococci that are erythromycin resistant are presumed to be resistant to clindamycin as well, these probes may be useful in determining the presence of the *msrA* and *ermA* genes if clindamycin therapy is a critical issue. The PCR primers for the *erm* genes are degenerate in that one of several bases can be substituted at certain positions in the primers and still yield an amplification product. This broadens the types of *erm* genes that can be detected to include those that mediate resistance to erythromycin in *Bacteroides* species (7). Neither the DNA probes nor primer sets have been used directly on clinical samples.

MUPIROCIN RESISTANCE

Mupirocin is an antistaphylococcal agent that is used to reduce carriage of staphylococci among infected patients and hospital personnel. Recently, a PCR assay that can detect high-level mupirocin resistance was described (32).

However, the practical value of the assay has not been assessed in a clinical laboratory setting.

QUINOLONE RESISTANCE

There are two major mechanisms of quinolone resistance: decreased permeability of the cell, which limits access of the drug to the target site, and alteration of the target site itself, which is the cell's DNA gyrase (36). While there is relatively little interest in detecting resistance to nalidixic acid among various bacterial genera, there is significant interest in detecting resistance to the 4-fluoroquinolones. Although probes to detect mutations in staphylococcal *gyrA* genes associated with resistance are available (26), the probes have proved difficult to use, and not all resistant strains demonstrate differential binding of the probes. In lieu of probes, several investigators have reported the use of PCR coupled with direct sequencing of the amplification products to detect changes in the nucleotide sequence of the *gyrA* gene (34, 61). The primers, however, appear to be species specific. Thus, a DNA probe or PCR tests for quinolone resistance are not yet available except for laboratories that have DNA-sequencing capability.

SULFONAMIDE RESISTANCE

There are two major sulfonamide resistance genes, *sul*I and *sul*II. Both have been cloned and sequenced, and probes have been described for each gene (66). Neither probe has been used to directly identify the presence of the gene in bacteria in clinical samples. Interestingly, the *sul* genes are often associated with transposable DNA elements, such as Tn*21* (66), which can shuttle multiple resistance genes from organism to organism. Thus, the *sul* genes can serve as indicators of multiple resistance in gram-negative organisms.

TETRACYCLINE RESISTANCE

Tetracycline resistance can be mediated by at least 15 different genes (83). DNA probes have been used for epidemiologic studies of the *tet*(A), *tet*(B), *tet*(C), *tet*(D), *tet*(E), *tet*(F), *tet*(M), *tet*(N), *tet*(O), and *tet*(Q) determinants (1, 30, 50, 51, 60, 70, 71, 81, 82, 90, 109). Of these, *tet*(M) is the most widespread, having been located in a very diverse group of organisms including staphylococci; streptococci; pneumococci; gram-negative bacilli such as campylobacters, *Gardnerella vaginalis*, and fusobacteria; mycoplasmas; and ureaplasmas (83). A *tet*(G) determinant has been identified and sequenced, but no probe has been described (108). PCR primers for the *tet*(M), *tet*(O), and *tet*(Q) genes have been described (9, 59, 71). These primers may have particular value for directly detecting tetracycline resistance genes in periodontal pathogens (60), since the presence of resistant organisms may indicate the patients who are likely to fail therapy with tetracycline, which is the drug of choice for periodontal disease. Because the *tet*(M) and *tet*(O) genes are highly related, some primer sets produce amplification products with both genes, making it necessary to use specific DNA probes to confirm the identities of the amplification products.

TRIMETHOPRIM RESISTANCE

The number of genes capable of mediating trimethoprim resistance in bacteria continues to grow (2, 79). DNA probes have proven to be powerful tools for detecting and classifying novel trimethoprim resistance genes (called *dhfr*) (2, 40, 65, 79). However, because no consensus sequences common to all the *dhfr* genes have been identified, PCR primers that could simplify the detection of this family of genes have not been developed. The probes continue to be useful as epidemiologic markers for studies of the dissemination of *dhfr* resistance genes.

DETECTING RESISTANCE IN MYCOBACTERIA

Multidrug-resistant strains of M. *tuberculosis* have been recognized in many hospitals in the eastern United States and have become a major public health problem (23). Consequently, the rapid identification of resistant strains has become a critical issue for the laboratory. PCR assays have been developed to detect mutations in the *rpoB* locus of the M. *tuberculosis* genome that are associated with the development of rifampin resistance (85). PCR and a single-strand conformation polymorphism (SSCP) assay (a technique that examines the mobility of PCR products in a polyacrylamide gel) can differentiate rifampin-resistant strains of M. *tuberculosis* from susceptible strains with high confidence (85). Rifampin resistance also serves as a marker for multidrug resistance in M. *tuberculosis*. Rifampin resistance primers have also been developed to detect resistance in *Mycobacterium leprae* (35).

Recently, mutations associated with streptomycin resistance in M. *tuberculosis* have been identified by PCR (28). However, SSCP assays similar to those used for rifampin resistance have not been described. Instead, DNA sequencing is required for identification of the altered sequences.

Although a genetic mechanism for isoniazid resistance has been identified in some resistant strains of M. *tuberculosis* (107), no DNA probe or PCR-based test for detecting resistant strains is available.

The use of molecular methods to detect resistance markers in mycobacteria is an area of great potential benefit to the clinical mycobacteriology laboratory. However, our understanding of resistance to a variety of drugs in mycobacteria is just beginning to unfold, and other mechanisms of resistance, perhaps the predominant mechanism of resistance in some cases, have yet to be described.

GUIDELINES FOR USING GENETIC TESTS

Few clinical laboratories make their own DNA probes to detect resistance genes because of the time and expense required for probe production and quality control. Nevertheless, probes continue to be valuable research tools for studying the epidemiology of resistance gene spread. Laboratories that use PCR for detection of bacteria and viruses should consider the detection of resistance genes as another important application of the technology, particularly for documenting resistance in organisms such as M. *tuberculosis*.

The critical issue with PCR assays is the reliability of results. The need for quality control measures, including use of amplification controls such as simultaneous amplification of sequences of rRNA or of the genes coding for rRNA (rDNA) to ensure the availability of amplifiable nucleic acid and the absence of inhibitory substances in the reaction, cannot be stressed enough. Multiple blanks interspersed between the test vials to ensure specificity and the absence of carryover contamination are also critical. The temperatures used in PCR assays optimized for use with purified DNA or DNA from bacterial isolates obtained in pure culture may not be stringent enough to avoid false-positive results when used with clinical samples, such as blood or cerebrospinal fluid, where considerably more non-specific priming occurs (89). It may be necessary to increase the temperatures of the assays, particularly the annealing temperatures, to avoid this problem.

One should never assume that PCR primers reported in the literature have undergone rigorous testing. Rather, primer sets should be thoroughly tested for specificity, self-complementarity, and dimer formation before use. According to the Clinical Laboratory Improvement Act of 1988, validation of DNA probe and PCR tests by the clinical laboratory in which they are to be used is mandatory before these methods can be used for analysis of clinical specimens. Methods for validation are published by the National Committee for Clinical Laboratory Standards (57).

Finally, now that DNA probes and PCR assays for detecting and differentiating resistance genes are becoming available, more surveys of resistance mechanisms, such as those described by Eady et al. (22), Ounissi et al. (62), and Shaw et al. (73) should be undertaken to determine the reservoirs of resistance genes and the ways resistance genes disseminate in hospitals and community settings. Such studies would also help determine the frequency with which organisms carry resistance genes that are not expressed. Although still considered experimental, many of the probe and PCR methods for detecting resistance genes described here are already having a positive effect on guiding therapy early in the course of infection and are making the treatment of infectious diseases less empiric.

REFERENCES

1. **Abraham, L. J., D. I. Berryman, and J. I. Rood.** 1988. Hybridization analysis of the class P tetracycline resistance determinant from the *Clostridium perfringens* R-plasmid, pCW3. *Plasmid* **19**:113–120.
2. **Amyes, S. G. B., and K. J. Towner.** 1990. Trimethoprim resistance; epidemiology and molecular aspects. *J. Med. Microbiol.* **31**:1–19.
3. **Archer, G. L., and E. Pennell.** 1990. Detection of methicillin resistance in staphylococci by using a DNA probe. *Antimicrob. Agents Chemother.* **34**:1720–1724.
4. **Arlet, G., and A. Philippon.** 1991. Construction by polymerase chain reaction and intragenic DNA hybridization for three main types of transferable beta-lactamases (TEM, SHV, CARB). *FEMS Microbiol. Lett.* **82**:19–26.
5. **Arthur, M., A. Andremont, and P. Courvalin.** 1987. Distribution of erythromycin esterase and rRNA methylase genes in members of the family Enterobacteriaceae highly resistant to erythromycin. *Antimicrob. Agents Chemother.* **31**:404–409.
6. **Arthur, M., and P. Courvalin.** 1993. Genetics and mechanisms of glycopeptide resistance in enterococci. *Antimicrob. Agents Chemother.* **37**:1563–1571.
7. **Arthur, M., C. Molinas, C. Mabilat, and P. Courvalin.** 1990. Detection of erythromycin resistance by the polymerase chain reaction using primers in conserved regions of *erm* rRNA methylase genes. *Antimicrob. Agents Chemother.* **34**:2024–2026.
8. **Berryman, D. I., and J. I. Rood.** 1989. Cloning and hybridization of *ermP*, a macrolide-lincosamide-streptogramin resistance determinant from *Clostridium perfringens*. *Antimicrob. Agents Chemother.* **33**:1346–1353.
9. **Blanchard, A., D. M. Crabb, K. Dybvig, L. B. Duffy, and G. H. Cassell.** 1992. Rapid detection of *tetM* in *Mycoplasma*

hominis and *Ureaplasma urealyticum* by PCR: *tetM* confers resistance to tetracycline but not necessarily to doxycycline. *FEMS Microbiol. Lett.* **95**:277–282.
10. **Bolivar, F., R. L. Rodriguez, P. J. Greene, M. C. Betlach, H. L. Heyneker, H. W. Boyer, J. H. Crosa, and S. Falkow.** 1977. Construction and characterization of new cloning vehicles. II. A multipurpose cloning system. *Gene* **2**:95–113.
11. **Carlier, C., and P. Courvalin.** 1990. Emergence of 4′,4″-aminoglycoside nucleotidyltransferase in enterococci. *Antimicrob. Agents Chemother.* **34**:1565–1569.
12. **Carter, G. I., K. J. Towner, N. J. Pearson, and R. C. B. Slack.** 1989. Use of a non-radioactive hybridization assay for direct detection of gram-negative bacteria carrying TEM beta-lactamase genes in infected urine. *J. Med. Microbiol.* **28**:113–117.
13. **Chambers, H. F.** 1988. Methicillin-resistant staphylococci. *Clin. Microbiol. Rev.* **1**:173–186.
14. **Charpenteir, E., G. Gerbaud, and P. Courvalin.** 1993. Characterization of a new class of tetracycline-resistance gene *tet*(S) in *Listeria monocytogenes* BM4210. *Gene* **131**:27–34.
15. **Clark, N. C., R. C. Cooksey, B. C. Hill, J. M. Swenson, and F. C. Tenover.** 1993. Characterization of glycopeptide resistant enterococci from U.S. hospitals. *Antimicrob. Agents Chemother.* **37**:2311–2317.
16. **Coffey, T. J., C. G. Dowson, M. Daniels, J. Zhou, C. Martin, B. G. Spratt, and J. M. Musser.** 1991. Horizontal transfer of multiple penicillin-binding protein genes, and capsular biosynthetic genes, in natural populations of *Streptococcus pneumoniae*. *Mol. Microbiol.* **5**:2255–2260.
17. **Costa, Y., M. Gaslimand, R. Leclercq, J. Duval, and P. Courvalin.** 1993. Characterization of the chromosomal *aac(6′)-Ii* gene specific for *Enterococcus faecium*. *Antimicrob. Agents Chemother.* **37**:1896–1903.
18. **Daum, R. S., M. Murphy-Corb, E. Shapira, and S. Dipp.** 1988. Epidemiology of Rob β-lactamase among ampicillin-resistant *Haemophilus influenzae* in the United States. *J. Infect. Dis.* **157**:450–455.
19. **Dowson, C. G., A. Hutchison, and B. G. Spratt.** 1989. Extensive remodeling of the transpeptidase domain of penicillin-binding protein 2B of a penicillin-resistant South African isolate of *Streptococcus pneumoniae*. *Mol. Microbiol.* **3**:95–102.
20. **Dutka-Malen, S., C. Molinass, M. Arthur, and P. Courvalin.** 1990. The VANA glycopeptide resistance protein is related to D-alanyl-D-alanine ligase cell wall biosynthesis enzymes. *Mol. Gen. Genet.* **224**:364–372.
21. **Dutka-Malen, S., C. Molinass, M. Arthur, and P. Courvalin.** 1992. Sequence of the *vanC* gene of *Enterococcus gallinarum* BM4174 encoding a D-alanine:D-alanine ligase-related protein necessary for vancomycin resistance. *Gene* **112**:53–58.
22. **Eady, E. A., J. I. Ross, J. L. Tipper, C. E. Walters, J. H. Cove, and W. C. Noble.** 1993. Distribution of genes encoding erythromycin ribosomal methylases and an erythromycin efflux pump in epidemiologically distinct groups of staphylococci. *J. Antimicrob. Chemother.* **31**:211–217.
23. **Edlin, B. R., J. I. Tokars, M. H. Grieco, J. T. Crawford, J. Williams, E. M. Sordillo, K. R. Ong, J. O. Kilburn, S. W. Dooley, K. G. Castro, W. R. Jarvis, and S. D. Holmberg.** 1992. An outbreak of multidrug resistant tuberculosis among hospitalized patients with acquired immunodeficiency syndrome. *N. Engl. J. Med.* **326**:1514–1521.
24. **Evers, S., D. F. Sahm, and P. Courvalin.** 1993. The *vanB* gene of vancomycin-resistant *Enterococcus faecalis* V583 is structurally related to genes encoding D-Ala:D-Ala ligases and glycopeptide-resistance proteins VanA and VanC. *Gene* **124**:143–144.
25. **Faruki, H., R. N. Kohmescher, W. P. McKinney, and P. F. Sparling.** 1985. A community-based outbreak of infection with penicillin-resistant *Neisseria gonorrhoeae* not producing penicillinase (chromosomally mediated resistance). *N. Engl. J. Med.* **313**:607–611.

26. **Fasching, C. E., F. C. Tenover, T. G. Slama, L. M. Fisher, S. Sreedharan, M. Oram, and L. R. Peterson.** 1991. *gyrA* mutations in ciprofloxacin-resistant, methicillin-resistant *Staphylococcus aureus* from Indiana, Minnesota and Tennessee. *J. Infect. Dis.* **164:**976–979.

27. **Ferretti, J. J., K. S. Gilmore, and P. Courvalin.** 1986. Nucleotide sequence analysis of the gene specifying the bifunctional 6′-aminoglycoside acetyltransferase 2″-aminoglycoside phosphotransferase enzyme in *Streptococcus faecalis* and identification and cloning of gene regions specifying the two activities. *J. Bacteriol.* **167:**631–638.

28. **Finken, M., P. Kirschner, A. Meier, A. Wrede, and E. Böttger.** 1993. Molecular basis of streptomycin resistance in *Mycobacterium tuberculosis*: alterations of the ribosomal protein A12 and point mutations within a functional 16S ribosomal RNA pseudoknot. *Mol. Microbiol.* **9:**1239–1246.

29. **Flamm, R. K., K. L. Phillips, F. C. Tenover, and J. J. Plorde.** 1993. A survey of clinical isolates of *Enterobacteriaceae* using a series of DNA probes for aminoglycoside resistance genes. *Mol. Cell. Probes* **7:**139–144.

30. **Fletcher, H. M., and F. L. Macrina.** 1991. Molecular survey of clindamycin and tetracycline resistance determinants in *Bacteroides* species. *Antimicrob. Agents Chemother.* **35:**2415–2418.

31. **Fling, M. E., and C. Richards.** 1983. The nucleotide sequence of the trimethoprim resistance dihydrofolate reductase gene harbored by Tn7. *Nucleic Acids Res.* **11:**5147–5158.

32. **Gilbert, J., C. E. Perry, and B. Slocombe.** 1993. High-level mupirocin resistance in *Staphylococcus aureus*: evidence for two distinct isoleucyl-tRNA synthetases. *Antimicrob. Agents Chemother.* **37:**32–38.

33. **Gold, H. S., S. Unal, E. Cercenado, C. Thauvin-Eliopoulos, G. G. Eliopoulos, C. B. Wennersten, and R. C. Moellering, Jr.** 1993. A gene conferring resistance to vancomycin but not teicoplanin in isolates of *Enterococcus faecalis* and *Enterococcus faecium* demonstrates homology with *vanB*, *vanA*, and *vanC* genes of enterococci. *Antimicrob. Agents Chemother.* **37:**1604–1609.

34. **Goswitz, J. J., K. E. Willard, C. E. Fashing, and L. R. Peterson.** 1992. Detection of *gyrA* mutations associated with ciprofloxacin resistance in methicillin-resistant *Staphylococcus aureus*: analysis by polymerase chain reaction and automated direct DNA sequencing. *Antimicrob. Agents Chemother.* **36:**1166–1169.

35. **Honoré, N., and S. Cole.** 1993. The molecular basis of rifampin resistance in *Mycobacterium leprae*. *Antimicrob. Agents Chemother.* **37:**414–418.

36. **Hooper, D. C., J. S. Wolfson, E. Y. Ng, and M. N. Schwartz.** 1987. Mechanisms of action of and resistance to ciprofloxacin. *Am. J. Med.* **82**(Suppl. 4A):12–20.

37. **Huang, M. B., T. E. Gay, C. N. Baker, S. N. Bannerjee, and F. C. Tenover.** 1993. Two percent sodium chloride is required for susceptibility testing of staphylococci with oxacillin when using agar-based dilution methods. *J. Clin. Microbiol.* **31:**2683–2688.

38. **Huovinen, S., P. Huovinen, and G. A. Jacoby.** 1988. Detection of plasmid-mediated β-lactamases with DNA probes. *Antimicrob. Agents Chemother.* **32:**175–179.

39. **Jacoby, G. A., and A. A. Medeiros.** 1991. More extended-spectrum beta-lactamases. *Antimicrob. Agents Chemother.* **35:**1697–1704.

40. **Jansson, C., A. Franklin, and O. Sköld.** 1992. Spread of newly found trimethoprim resistance gene *dhfrIX* among porcine isolates and human pathogens. *Antimicrob. Agents Chemother.* **36:**2704–2708.

41. **Kaufhold, A., A. Podbielski, T. Horaud, and P. Ferrieri.** 1992. Identical genes confer high-level resistance to gentamicin upon *Enterococcus faecalis*, *Enterococcus faecium*, and *Staphylococcus aureus*. *Antimicrob. Agents Chemother.* **36:**1215–1218.

42. **Kawa, D., D. Pennell, L. Kubista, and R. Schell.** 1989. Development of a rapid method for determining the susceptibility of *Mycobacterium tuberculosis* to isoniazid using the Gen-Probe DNA hybridization system. *Antimicrob. Agents Chemother.* **33:**1000–1005.

43. **Klugman, K. P.** 1990. Pneumococcal resistance to antibiotics. *Clin. Microbiol. Rev.* **3:**171–196.

44. **Laible, G., B. G. Spratt, and R. Hakenbeck.** 1991. Interspecies recombinational events during the evolution of altered PBP 2X genes in penicillin-resistant clinical isolates of *Streptococcus pneumoniae*. *Mol. Microbiol.* **5:**1993–2002.

45. **Lambert, T., G. Gerbaud, P. Bouvet, J.-F. Vieu, and P. Courvalin.** 1990. Dissemination of amikacin resistance gene *aphA6* in *Acinetobacter* spp. *Antimicrob. Agents Chemother.* **34:**1244–1248.

46. **Leclercq, R., S. Dutka-Malen, J. Duval, and P. Courvalin.** 1992. Vancomycin resistance gene *vanC* is specific to *Enterococcus gallinarum*. *Antimicrob. Agents Chemother.* **36:**2005–2008.

47. **Levesque, R. C., A. A. Medeiros, and G. A. Jacoby.** 1987. Molecular cloning and DNA homology of plasmid-mediated β-lactamase genes. *Mol. Gen. Genet.* **206:**252–258.

48. **Mabilat, C., and P. Courvalin.** 1990. Development of "oligotyping" for characterization and molecular epidemiology of TEM beta-lactamases in *Enterobacteriaceae*. *Antimicrob. Agents Chemother.* **34:**2210–2216.

49. **Mabilat, C., and S. Goussard.** 1994. PCR detection and identification of genes for extended-spectrum β-lactamases, p. 553–559. *In* D. H. Persing, T. F. Smith, F. C. Tenover, and T. J. White (ed.), *Diagnostic Molecular Microbiology: Principles and Applications*. American Society for Microbiology, Washington, D.C.

50. **Marshall, B., C. Tachibana, and S. B. Levy.** 1983. Frequency of tetracycline resistance determinant classes among lactose-fermenting coliforms. *Antimicrob. Agents Chemother.* **24:**835–840.

51. **Martin, P., P. Trieu-Cuot, and P. Courvalin.** 1986. Nucleotide sequence of the *tetM* tetracycline resistance determinant of the streptococcal conjugative shuttle transposon Tn1545. *Nucleic Acids Res.* **14:**7047–7058.

52. **Massanari, R. M., M. A. Pfaller, D. S. Wakefield, G. T. Hammons, L.-A. McNutt, R. F. Woolson, and C. H. Helms.** 1988. Implications of acquired oxacillin resistance in the management and control of *Staphylococcus aureus* infections. *J. Infect. Dis.* **158:**702–709.

53. **Mendelman, P., D. O. Chaffin, T. L. Stull, C. E. Rubens, K. D. Mack, and A. L. Smith.** 1984. Characterization of non-β-lactamase-mediated ampicillin resistance in *Haemophilus influenzae*. *Antimicrob. Agents Chemother.* **26:**235–244.

54. **Muñoz, R., T. J. Coffey, M. Daniels, C. G. Dowson, G. Laible, J. Casal, R. Hakenbeck, M. Jacobs, J. M. Musser, B. G. Spratt, and A. Tomasz.** 1991. Intercontinental spread of a multiresistant clone of serotype 23F *Streptococcus pneumoniae*. *J. Infect. Dis.* **164:**302–306.

55. **Murakami, K., W. Minamide, K. Wada, E. Nakamura, H. Teraoka, and S. Watanabe.** 1991. Identification of methicillin-resistant strains of staphylococci by polymerase chain reaction. *J. Clin. Microbiol.* **29:**2240–2244.

56. **Murray, B. M.** 1991. New aspects of antimicrobial resistance and the resulting therapeutic dilemmas. *J. Infect. Dis.* **163:**1185–1194.

57. **National Committee for Clinical Laboratory Standards.** 1994. *Specifications for Molecular Microbiology Methods for Infectious Diseases*. Proposed guideline MM3-P, vol. 1, no. 1. National Committee for Clinical Laboratory Standards, Villanova, Pa.

58. **Neu, H. C.** 1992. The crisis in antibiotic resistance. *Science* **257:**1064–1073.

59. **Olsvik, B., I. Olsen, and F. C. Tenover.** Detection of *tet*(M) using the polymerase chain reaction in bacteria isolated from patients with periodontal pathogens. *Oral Microbiol. Immunol.*, in press.

60. **Olsvik, B., and F. C. Tenover.** 1993. Tetracycline resistance in periodontal pathogens. *Clin. Infect. Dis.* **16**(Suppl. 4):S310–S313.

61. **Oram, M., and L. M. Fisher.** 1991. 4-Quinolone resistance mutations in the DNA gyrase of *Escherichia coli* clinical isolates identified by using the polymerase chain reaction. *Antimicrob. Agents Chemother.* **35**:387–389.

62. **Ounissi, H., E. Derlot, C. Carlier, and P. Courvalin.** 1990. Gene homogeneity for aminoglycoside-modifying enzymes in gram-positive cocci. *Antimicrob. Agents Chemother.* **34:** 2164–2168.

63. **Parsons, Y., R. M. Hall, and H. W. Stokes.** 1991. A new trimethoprim resistance gene, *dhfrX*, in the In7 integron of plasmid pDGO100. *Antimicrob. Agents Chemother.* **35**:2436–2439.

64. **Perine, P. L., P. A. Totten, K. K. Holmes, E. H. Sng, A. V. Ratman, R. Widy-Wersky, H. Nsanze, E. Habte-Gabr, and W. G. Westbrook.** 1985. Evaluation of a DNA hybridization method for detection of African and Asian strains of *Neisseria gonorrhoeae* in men with urethritis. *J. Infect. Dis.* **152:** 59–63.

65. **Pulkkinen, L., P. Houvinen, E. Vuorio, and P. Tiovanen.** 1984. Characterization of trimethoprim resistance by use of probes specific for transposon Tn7. *Antimicrob. Agents Chemother.* **26**:82–86.

66. **Rådström, P., G. Swedberg, and O. Sköld.** 1991. Genetic analyses of sulfonamide resistance and its dissemination in gram-negative bacteria illustrate new aspects of R plasmid evolution. *Antimicrob. Agents Chemother.* **35**:1840–1848.

67. **Rather, P. N., R. Mierzwa, R. S. Hare, G. H. Miller, and K. J. Shaw.** 1992. Cloning and DNA sequence analysis of an *aac(3)-Vb* gene from *Serratia marcescens. Antimicrob. Agents Chemother.* **36**:2222–2227.

68. **Reznicek, M., M. Bale, and M. Pfaller.** 1991. Application of DNA probes to antimicrobial susceptibility testing of *Legionella pneumophila. Diagn. Microbiol. Infect. Dis.* **14**:7–10.

69. **Roberts, M. C., L. A. Actis, and J. H. Crosa.** 1985. Molecular characterization of chloramphenicol-resistant *Haemophilus parainfluenzae* and *Haemophilus ducreyi. Antimicrob. Agents Chemother.* **28**:176–180.

70. **Roberts, M. C., B. A. Brown, V. A. Steingrube, and R. J. Wallace.** 1990. Genetic basis of tetracycline resistance in *Moraxella (Branhamella) catarrhalis. Antimicrob. Agents Chemother.* **34**:1816–1818.

71. **Roberts, M. C., Y. Pang, D. E. Riley, S. L. Hillier, R. C. Berger, and J. N. Krieger.** 1993. Detection of Tet M and Tet O tetracycline resistance genes by polymerase chain reaction. *Mol. Cell. Probes* **7**:387–393.

72. **Rood, J. I., S. Jefferson, T. L. Bannam, J. M. Wilkie, P. Mullany, and B. W. Wren.** 1989. Hybridization analysis of three chloramphenicol resistance determinants from *Clostridium perfringens* and *Clostridium difficile. Antimicrob. Agents Chemother.* **33**:1569–1574.

73. **Shaw, K. J., R. S. Hare, F. J. Sabatelli, M. Rizzo, C. A. Cramer, G. H. Miller, L. Verbist, H. Van Landuyt, Y. Glupczynski, M. Catalano, and M. Woloj.** 1991. Correlation between aminoglycoside resistance profiles and DNA hybridization of clinical isolates. *Antimicrob. Agents Chemother.* **35**:2253–2261.

74. **Shaw, K. J., H. Munayyer, P. N. Rather, R. S. Hare, and G. H. Miller.** 1993. Nucleotide sequence analysis and DNA hybridization studies of the *ant(4')-IIa* gene from *Pseudomonas aeruginosa. Antimicrob. Agents Chemother.* **37**:708–714.

75. **Shaw, K. J., P. N. Rather, R. S. Hare, and G. H. Miller.** 1993. Molecular genetics of aminoglycoside resistance genes and familial relationships of the aminoglycoside-modifying enzymes. *Microbiol. Rev.* **57**:138–163.

76. **Shaw, K. J., P. N. Rather, F. J. Sabatelli, P. Mann, U. Munayyer, R. Mierzwa, G. L. Petrikkos, R. S. Hare, G. H. Miller, P. Bennett, and P. Downey.** 1992. Characterization of the chromosomal *aac(6')-Ic* gene from *Serratia marcescens. Antimicrob. Agents Chemother.* **36**:1447–1455.

77. **Shaw, W. V.** 1983. Chloramphenicol acetyltransferases: enzymology and molecular biology. *Crit. Rev. Biochem.* **14**:1–46.

78. **Simard, J.-L., and P. H. Roy.** 1993. PCR detection of penicillinase-producing *Neisseria gonorrhoeae*, p. 543–546. *In* D. H. Persing, T. F. Smith, F. C. Tenover, and T. J. White (ed.), *Diagnostic Molecular Microbiology: Principles and Applications.* American Society for Microbiology, Washington, D.C.

79. **Singh, K. V., R. R. Reves, L. K. Pickering, and B. E. Murray.** 1992. Identification by DNA sequence analysis of a new plasmid-encoded trimethoprim resistance gene in fecal *Escherichia coli* isolates from children in day-care centers. *Antimicrob. Agents Chemother.* **36**:1720–1726.

80. **Song, M. D., M. Wachi, M. Doi, F. Ishino, and M. Matsuhashi.** 1987. Evolution of an inducible penicillin-target protein in methicillin-resistant *Staphylococcus aureus* by gene fusion. *FEBS Lett.* **221**:167–171.

81. **Sørum, H., M. C. Roberts, and J. H. Crosa.** 1992. Identification and cloning of a tetracycline resistance gene from a fish pathogen *Vibrio salmonicida. Antimicrob. Agents Chemother.* **36**:611–615.

82. **Sougakoff, W., B. Papadopoulou, P. Nordmann, and P. Courvalin.** 1987. Nucleotide sequence and distribution of gene *tetO* encoding tetracycline resistance in *Campylobacter coli. FEMS Microbiol. Lett.* **44**:153–159.

83. **Speer, B. S., N. B. Shoemaker, and A. A. Salyers.** 1992. Bacterial resistance to tetracycline: mechanisms, transfer, and clinical significance. *Clin. Microbiol. Rev.* **5**:387–399.

84. **Sundström, L., G. Swedberg, and O. Sköld.** 1993. Characterization of transposon Tn5086, carrying the site-specifically inserted gene *dhfrVII* mediating trimethoprim resistance. *J. Bacteriol.* **175**:1796–1805.

85. **Telenti, A., P. Imboden, F. Marchesi, T. Schmidheini, and T. Bodmer.** 1993. Direct, automated detection of rifampicin-resistant *Mycobacterium tuberculosis* by polymerase chain reaction and single-strand confirmation polymorphism analysis. *Antimicrob. Agents Chemother.* **37**:2054–2058.

86. **Tenover, F. C.** Unpublished data.

87. **Tenover, F. C., D. Filpula, K. L. Phillips, and J. J. Plorde.** 1988. Cloning and sequencing of a gene encoding a 6'-N-acetyltransferase from an R factor of *Citrobacter diversus. J. Bacteriol.* **170**:471–473.

88. **Tenover, F. C., T. D. Gootz, K. P. Gordon, and J. J. Plorde.** 1984. Development of a DNA probe for the structural gene of the 2''-O-adenyltransferase aminoglycoside modifying enzyme. *J. Infect. Dis.* **159**:678–687.

89. **Tenover, F. C., M. B. Huang, J. K. Rasheed, and D. H. Persing.** 1994. Development of PCR assays to detect ampicillin resistance genes in cerebrospinal fluid samples containing *Haemophilus influenzae. J. Clin. Microbiol.* **32**:2729–2737.

90. **Tenover, F. C., D. J. LeBlanc, and P. Elvrum.** 1987. Cloning and expression of a tetracycline resistance determinant from *Campylobacter jejuni* in *Escherichia coli. Antimicrob. Agents Chemother.* **31**:1301–1306.

91. **Tenover, F. C., K. L. Phillips, T. Gilbert, P. Lockhart, P. J. O'Hara, and J. J. Plorde.** 1989. Development of a DNA probe from the deoxyribonucleotide sequence of a 3-N-aminoglycoside acetyltransferase [AAC(3)-I] resistance gene. *Antimicrob. Agents Chemother.* **33**:551–559.

92. **Thakker-Varia, S., J. W. Jenssen, L. Mon-McDermott, M. P. Weinstein, and D. T. Dubin.** 1987. Molecular epidemiology of macrolides-lincosamides-streptogramin B resistance in *Staphylococcus aureus* and coagulase-negative staphylococci. *Antimicrob. Agents Chemother.* **31**:735–743.

93. **Thomson, C. J., K. J. Towner, H.-K. Young, and S. G. B. Amyes.** 1990. Identification and cloning of the type IIIa plasmid-encoded dihydrofolate reductase gene from trimethoprim-resistant gram-negative bacteria isolated in Britain. *J. Med. Microbiol.* **31**:213–218.

94. **Tokue, Y., S. Shoji, K. Satoh, A. Watanabe, and M. Motomiya.** 1992. Comparison of a polymerase chain reaction assay and a conventional microbiologic method for detection of methicillin-resistant *Staphylococcus aureus. Antimicrob. Agents Chemother.* **36**:6–9.

95. **Tomasz, A., H. B. Drugeon, H. M. de Lencastre, D. Jabes, L. McDougal, and J. Bille.** 1989. New mechanism for me-

thicillin resistance in *Staphylococcus aureus*: clinical isolates that lack the PBP 2a gene and contain normal penicillin-binding capacity. *Antimicrob. Agents Chemother.* **33**:1869–1874.

96. **Towner, K. J., and G. I. Carter.** 1990. Cloning of the type VII trimethoprim resistant dihydrofolate reductase gene and identification of a specific DNA probe. *FEMS Microbiol. Lett.* **70**:19–22.

97. **Towner, K. J., H.-K. Young, and S. G. B. Amyes.** 1988. Biotinylated DNA probes for trimethoprim-resistant dihydrofolate reductases types IV and V. *J. Antimicrob. Chemother.* **22**:285–291.

98. **Tran Van Nhieu, G., and E. Collatz.** 1987. Primary structure of an aminoglycoside 6'-N-acetyltransferase AAC(6')-4, fused in vivo with the signal peptide of the Tn3-encoded β-lactamase. *J. Bacteriol.* **169**:5708–5714.

99. **Trieu-Cuot, P., and P. Courvalin.** 1983. Nucleotide sequence of the *Streptococcus faecalis* plasmid gene encoding 3'5"-aminoglycoside phosphotransferase type III. *Gene* **23**:331–334.

100. **Ubukata, K., S. Nakagami, A. Nitta, A. Yamane, S. Kawakami, M. Sugiura, and A. Konno.** 1992. Rapid detection of the *mecA* gene in methicillin-resistant staphylococci by enzymatic detection of polymerase chain reaction products. *J. Clin. Microbiol.* **30**:1728–1733.

101. **van de Klundert, J. A. M., and J. S. Vliegenthart.** 1993. PCR detection of genes coding for aminoglycoside modifying enzymes, p. 547–552. *In* D. H. Persing, T. F. Smith, F. C. Tenover, and T. J. White (ed.), *Diagnostic Molecular Microbiology: Principles and Applications*. American Society for Microbiology, Washington, D.C.

102. **vanHoof, R., J. Content, E. Van Bossuyt, L. Dewit, and E. Hannecart-Pokorni.** 1992. Identification of the *aadB* gene coding for the aminoglycoside-2"-O-nucleotidyltransferase, ANT(2"), by means of the polymerase chain reaction. *J. Antimicrob. Chemother.* **29**:365–374.

103. **Woodford, N., D. Morrison, A. P. Johnson, V. Briant, R. C. George, and B. Cookson.** 1993. Application of DNA probes for *rRNA* and *vanA* genes to investigation of a nosocomial cluster of vancomycin-resistant enterococci. *J. Clin. Microbiol.* **31**:653–658.

104. **Wren, B. W., P. Mullany, and S. Tabachali.** 1988. Molecular cloning and genetic analysis of a chloramphenicol acetyltransferase determinant from *Clostridium difficile*. *Antimicrob. Agents Chemother.* **32**:1213–1217.

105. **Youmans, G., T. E. Davis, and D. D. Fuller.** 1993. Use of chemiluminescent DNA probes in the rapid detection of oxacillin resistance in clinically isolated strains of *Staphylococcus aureus*. *Diagn. Microbiol. Infect. Dis.* **16**:99–104.

106. **Young, S. A., F. C. Tenover, T. D. Gootz, K. P. Gordon, and J. J. Plorde.** 1985. Development of two DNA probes for differentiating the structural genes of subclasses I and II of the aminoglycoside-modifying enzyme 3'-aminoglycoside phosphotransferase. *Antimicrob. Agents Chemother.* **27**:739–744.

107. **Zhang, Y., B. Heym, B. Allen, D. Young, and S. Cole.** 1992. The catalase-peroxidase gene and isoniazid resistance of *Mycobacterium tuberculosis*. *Science* **358**:591–593.

108. **Zhao, J., and T. Aoki.** 1992. Nucleotide sequence analysis of the class G tetracycline resistance determinant from *Vibrio anguillarum*. *Microbiol. Immunol.* **36**:1051–1060.

109. **Zilhao, R., B. Papadopoulou, and P. Courvalin.** 1988. Occurrence of the *Campylobacter* resistance gene *tetO* in *Enterococcus* and *Streptococcus* spp. *Antimicrob. Agents Chemother.* **32**:1793–1796.

110. **Zolg, J. W., U. J. Hanggi, and H. G. Zachau.** 1978. Isolation of a small DNA fragment carrying the gene for a dihydrofolate reductase from a trimethoprim resistance factor. *Mol. Gen. Genet.* **164**:15–29.

Instrument-Based Antibacterial Susceptibility Testing

MARY JANE FERRARO AND JAMES H. JORGENSEN

Clinical microbiology laboratories can choose from several different manual or instrument-assisted methods for performance of routine antibacterial susceptibility tests. These methods include disk diffusion, agar dilution, broth microdilution, antibiotic gradient, and overnight or short incubation with automated instrument. Until recently, there has been a trend in the United States away from the use of the disk diffusion test in favor of either broth microdilution or automated instrument methods, as evidenced by the methods used by participants in the College of American Pathologists microbiology surveys (13). This chapter focuses on overnight or short-incubation antimicrobial susceptibility test methods available in the United States that offer different levels of instrumentation. Some test systems are computer assisted rather than automated; the computer facilitates data entry and interpretation of visually determined growth endpoints.

The least automated of the instrument-based systems only interpret growth endpoints when microdilution trays are inserted into a reader device; other systems incubate microdilution trays or special cards with microcuvettes and perform serial interpretations of growth patterns in the presence of antimicrobial agents. The instruments vary in their methods of presenting the bacterial test module to the optical detector system. The instruments that offer the highest degree of automation generally do so by incorporating simple internal robotics to manipulate the test modules during the incubation and reading sequences. Current instruments utilize either the principle of turbidimetric detection of bacterial growth in a liquid medium or that of fluorometric detection of fluorescent indicators (34) or of hydrolysis of fluorogenic substrates (2, 10, 26, 38) incorporated in a special liquid medium. The use of suppression of turbidity as evidence of the inhibitory effect of an antimicrobial agent or, conversely, an increase in turbidity in the presence of a drug as an indication of microbial resistance has been firmly established in manual broth dilution susceptibility test methods (42). Inference of susceptibility or resistance based on the activities of specific bacterial enzymes produced during bacterial growth and their hydrolysis of fluorogenic substrates is a more recent method of growth determination. All of the instruments rely heavily on microprocessor-controlled functions, and they utilize microcomputers to provide final printed reports and to aid in data storage and retrieval. Most of the instruments can also be used to perform additional functions, usually to identify gram-negative or gram-positive bacteria and, in some cases, to assemble combined identification-antimicrobial susceptibility reports (37).

COMPUTER-ASSISTED, MANUAL, OVERNIGHT BROTH MICRODILUTION AND AGAR DILUTION TEST SYSTEMS

Virtually every manufacturer of broth microdilution trays for antimicrobial susceptibility testing offers a mechanized device for hydration or inoculation of trays and a device for facilitating manual visualization of results after incubation. Most manufacturers also offer some type of reader device whereby the user indicates the results of manual interpretation of the growth patterns in the tray by the use of a light pen, a video display screen resembling the configuration of the tray, or a touch-sensitive template that overlays the microdilution tray. The Baxter MicroScan Touch-SCAN-SR (Baxter Diagnostics Inc., West Sacramento, Calif.), Becton Dickinson Sceptor (Becton Dickinson Diagnostic Instrument Systems, Sparks, Md.), bioMérieux Vitek UniScept (bioMérieux Vitek, Hazelwood, Mo.), Difco Pasco (Difco Laboratories, Inc., Detroit, Mich.), and Radiometer Sensititre Sensi Touch (Radiometer America, Inc., Westlake, Ohio) are examples of such systems. An autoinoculator is also available for purchase with the UniScept and Sensititre systems. Although growth in the test wells is manually interpreted, the reader devices assist in recording data, applying interpretive criteria, and generating a computer-printed report. The microcomputers that are included with these systems enable the user to store and later retrieve data for the generation of periodic reports, e.g., cumulative susceptibility profiles for various organisms during a defined time period.

Agar dilution tests can be mechanized through the use of a Steers replicator inoculum delivery device (39) or a Cathra Replicator (AutoMed, Inc., Arden Falls, Minn.). Microcomputer-assisted reading of agar dilution plates can be accomplished by the use of the Cathra Replianalyzer II (29), which accepts, stores, interprets, and prints results of manually visualized growth spots on agar dilution plates. An automated reader that feeds, indexes, and reads (using video image capture analysis) agar dilution plates, the AutoMed Cathra AutoReader has been developed and awaits submission to the Food and Drug Administration (FDA).

SEMIAUTOMATED, OVERNIGHT BROTH MICRODILUTION TEST SYSTEMS

Automated tray reader devices that interpret growth patterns in trays represent the next level of instrumentation available from several manufacturers. The Baxter MicroScan AutoSCAN-4, Becton Dickinson AutoSceptor, Radiometer Sensititre AutoReader, and Alamar READar (Alamar, Sacramento, Calif.) are examples of contemporary instruments for automated interpretation of the results of microdilution trays following overnight incubation in a standard incubator. All readers determine growth photometrically except for the Alamar READar, which incorporates an oxidation-reduction indicator in the growth medium that changes color and fluorescence in response to chemical reduction resulting from bacterial growth (34). Fluorescence intensity (indicating reduction) is also determined by a fluorometer in the READar. All of these instruments are configured with microcomputers for report preparation and data storage. With the exceptions of AutoReader and AutoSceptor, which come with an autoinoculator, these instruments automate only the final, reading step involved in the performance of MIC or breakpoint microdilution tests. Evaluations of one instrument have shown that MIC endpoints can be interpreted by such instruments with reasonable accuracy (1). However, instances of haze or small pellets of growth that may be important in the determination of resistance can be missed by automated readers with photometers (21, 50).

AUTOMATED SHORT-INCUBATION SUSCEPTIBILITY TEST SYSTEMS

The most profound change to occur in antimicrobial susceptibility testing during the 1980s was increased number of laboratories using automated short-incubation susceptibility test systems (13). Continuing improvements in microprocessors, robotics, and microcomputers have allowed the development of instruments capable of providing antimicrobial susceptibility results in as short a time as 3.5 h. In order to accomplish this, a variety of strategies that deviate from those of conventional susceptibility test methods are employed. For example, the concentration of bacteria in the test inoculum is generally adjusted upward from that used in standard procedures. Often, the actual concentration of the antimicrobial agent in the test well is manipulated to provide results comparable to those of conventional microdilution tests. The MICs produced by these instruments are algorithmically derived from growth rate comparisons. Sophisticated computer software used with the instruments affords editing and adjustment of results for certain drug-bacterium combinations. The growth media employed in these test systems are not those of traditional methods and often include modifications to promote more rapid growth of certain bacteria.

The first short-incubation susceptibility system, the TAAS system, was developed in the early 1970s by Technicon Instruments Corp. (Tarrytown, N.Y.) but never marketed (12). Subsequent instruments such as the Autobac System developed and marketed by Pfizer Diagnostics in the early 1970s (later called Autobac Series II; Organon Teknika, Durham, N.C.) and the Abbott MS-2 system

(later called the Avantage MICROBIOLOGY CENTER; Abbott Laboratories Diagnostic Division, Irving, Tex.) were based on the photometric growth detection principles used in TAAS. Although these systems enjoy the distinction of being the first commercially available short-incubation susceptibility test systems, they are no longer manufactured and are virtually extinct in laboratories today. Short-incubation systems such as the Vitek System, formerly called the AutoMicrobic System, or AMS (bioMérieux Vitek), and the WalkAway (2, 10) (formerly called the AutoScan W/A; Baxter Diagnostics) are currently the only ones with FDA approval for use in clinical laboratories in the United States. Both the short-incubation Vitek and the WalkAway systems were initially approved by the FDA for testing most common bacteria and antimicrobial agents prior to more stringent 1991 requirements (see below). However, all new antimicrobial agents to be marketed for testing with these systems must undergo the extensive FDA review described below. Other systems, such as the ATB-plus (bioMérieux, France), COBAS-BACT (Roche Diagnostics, Basel, Switzerland) (24), and short-incubation Sensititre ARIS (Radiometer America) (27, 40), are used to various extents throughout the world but are not currently approved by the FDA for use in the United States.

Vitek System

The Vitek System is a by-product of U.S. space exploration efforts in the 1960s. It was designed and originally manufactured by the McDonnell-Douglas Corp. for the National Aeronautics and Space Administration as an onboard test system for detecting and identifying common urinary tract pathogens in specimens from astronauts in spacecraft. Because of its intended use aboard a spacecraft, it was highly automated and relatively compact from its inception. Very small plastic reagent cards were designed to contain microliter quantities of biochemical test and selective growth media for detection and identification of organisms. The Vitek System was modified in the 1970s for clinical laboratory use, principally for screening and identification of urinary organisms and for qualitative antimicrobial susceptibility testing with S, I, or R results for *Escherichia coli* or *Proteus mirabilis* only. In the last decade, MIC cards that provide semiquantitative results for most rapidly growing gram-positive and gram-negative aerobic bacteria in 4 to 10 h have been developed.

Vitek hardware includes a filling-sealer module for inoculation of the cards; an incubator-reader module that incorporates a carousel-like device to hold the test cards, a robotic system to manipulate the cards, and a photometer for measurement of optical density and biochemical reaction color changes in the cards; and a computer module, including video display monitor and printer (37). Vitek also offers an elaborate information management system for storing and retrieving test data and for producing a variety of statistical reports. The Vitek uses turbidimetrically determined kinetic measurements of growth in the presence of antimicrobial agents to provide linear regression analysis and ultimately algorithm-derived MICs. The user may elect to have susceptibility test results reported in either qualitative (S, I, R) or doubling-dilution MIC formats.

The fixed format of the 30-well Vitek cards imposes some limitations in the selection of drugs (9 to 11) for testing; however, a large number of standard panels are available. For laboratories wishing to test a larger number of antimicrobial agents, Vitek offers a combination of two cards (called Flex cards) at a reduced price that are tested together and allow merging of results from both cards into one report containing many different drugs. A 45-well card that will allow testing of 15 to 17 agents may be introduced in early 1995.

WalkAway System

The WalkAway System was developed in the late 1980s and is capable of automating either overnight or short-incubation susceptibility tests. The instrument consists of a large self-contained incubator-reader unit and a microcomputer with video display terminal and printer (37). The WalkAway uses standard-size microdilution trays that are read either photometrically (overnight testing) or fluorometrically (short-incubation testing). Once the microdilution trays have been manually inoculated with a multiprong device, they are placed in one of the incubator positions in the instrument. The instrument then incubates the trays for the appropriate period and robotically positions and aligns the trays under the central photometer-fluorometer to perform the final readings of growth endpoints when the value of a growth index exceeds a predetermined level (10).

The WalkAway offers a choice of MIC or breakpoint testing of gram-positive and gram-negative bacteria after overnight or rapid (3.5- to 5.5-h) readings. Special combination trays that provide susceptibility and organism identification in the same tray are available.

In theory, the detection of resulting fluorescence as a marker of growth is more sensitive than photometric technology, thus allowing more rapid assessment of bacterial growth. However, fluorometric detection of growth is indirect and assumes that all bacteria are capable of metabolizing the fluorogenic substrates, that enzymatic activities of nonmultiplying cells are insignificant in comparison to those of multiplying cells, and that the fluorogenic substrates do not interfere with the activities of any of the antimicrobial agents tested (2, 10).

Sensititre ARIS

The Sensititre ARIS was developed in the 1980s and is currently marketed in the United States only for overnight susceptibility testing, although it is available in other countries for short-incubation testing. The instrument includes an incubator-reader unit with associated microcomputer (including video display terminal and printer). Similar to the WalkAway, the ARIS incorporates the use of standard-size microdilution trays; once the trays are individually inoculated with an autoinoculator, the instrument incubates them, positions the trays by robotics under the fluorometer, and automatically interprets the growth patterns at the conclusion of tests. As with the short-incubation WalkAway, growth is determined following fluorogenic-substrate hydrolysis. The ARIS uses a conventional overnight incubation period for performance of MIC, breakpoint, or combination susceptibility identification tests with gram-positive and gram-negative bacteria.

REGULATORY OVERSIGHT OF SUSCEPTIBILITY TEST INSTRUMENTS AND CRITERIA FOR ACCEPTABLE PERFORMANCE

At present, none of the professional consensus organizations (e.g., National Committee for Clinical Laboratory Standards [NCCLS]) promulgates guidelines regarding the specific quality control or performance of short-incubation susceptibility test methods. Since enactment of the Safe Medical Devices Act of 1990, the short-incubation instruments and their drug panels require premarket approval (PMA) by the FDA for clearance to be marketed in the United States. This PMA has become more difficult to obtain in the 1990s because of new, more stringent FDA controls (9). Extensive studies that provide data on how accurately short-incubation systems determine susceptibility test results compared to a reference method are required for PMA. A large number of organisms, including quality control organisms, a challenge set with known mechanisms of resistance, and fresh or stock clinical isolates representing each antimicrobial agent's spectrum of activity, must be tested. It is expected that very major errors will be ≤1.5%, major errors will be ≤3%, essential agreement (± 1 $\log_2$ dilution) will be ≥90%, and growth failures will be ≤10%. If any microorganism, antimicrobial agent, or combination thereof does not meet these specifications, the labeling (package insert) of the device should include this contraindication and recommend the use of an alternative method for testing. If fewer than 20 strains representing each known mechanism of resistance for that antimicrobial agent have been tested, a statement that the ability to detect resistance to an agent among these species is unknown because resistant strains were not available at the time of comparative testing must be included in the labeling. Instruments that employ incubation times of 16 h or more are subject to less extensive processing by the FDA.

Over the years, a variety of criteria for statistically defining the acceptable accuracy of a new susceptibility test system (23, 24, 35, 41) have been proposed. Most recently (14), it has been suggested that a new susceptibility test method should provide >90% agreement with MICs determined by a reference method. Furthermore, very major errors determined on a large sample ($n \geq 35$) of known resistant isolates should be ≤3% and the combination of major and minor errors attributable to the new test should be ≤7% when determined on a large known susceptible population or a large unselected sample of clinical isolates. Modification of these criteria may be necessary for antibiotics without an intermediate interpretive category or for certain resistance mechanisms and specific groups of bacteria that may be difficult to detect by any reference method (14).

ADVANTAGES OF CURRENT INSTRUMENTS

The antimicrobial susceptibility testing instruments discussed above represent the highest level of automation now available for use in clinical microbiology laboratories. However, automation in clinical microbiology is still in a very early state of development compared with the level of automation that has been achieved in clinical chemistry, hematology, and immunology laboratories. All of the current microbiology instruments offer the potential for improved intra- and interlaboratory reproducibility of antimicrobial susceptibility tests, and in some cases, they significantly reduce the time required to perform the tests (e.g., 3.5 to 10 h versus overnight incubation). Most evaluations have reported that the instruments are mechanically reliable and that their operation is readily mastered by well-trained microbiologists and medical technologists. Some of these instruments offer modest labor savings over manually performed tests (4). Because all of the instruments can perform some of the most common analyses in the microbiology laboratory, i.e., organism identification and antimicrobial susceptibility testing, the potential exists to automate a sizable proportion of a laboratory's tasks by using only one instrument. The potential to establish a link between the laboratory computer and the microcomputer that controls the function of the instrument affords potential savings in both labor and transcription errors associated with manual entry of results. When such instruments are interfaced, however, verification of susceptibility data by supervisory personnel is advisable before the automated transfer of data to a patient file. In addition, data management systems available with these instruments provide neat, legible test reports for the laboratory and can archive and periodically analyze trends in antimicrobial susceptibilities of the microorganisms encountered in a given institution.

One of the advantages, yet to be fully exploited, of instrument-based susceptibility test systems is the potential use of artificial intelligence to create an expert system (6, 11) for automated review and verification of the data generated. Appropriately programmed software can contain early-detection rules similar to those used by knowledgeable microbiologists to detect impossible or rare phenotypes while allowing for correction of technical errors that may have occurred in the testing. Other rules, designed to recognize rare or unlikely resistance markers or rare antibiogram phenotypes, can also be designed. Following detection, the possibility of potential errors or unusual phenotypes would be flagged on the data terminal or laboratory report for technologists or supervisory review. At least one system, the Vitek, has added this enhancement to the instrument software.

It has yet to be definitively demonstrated that rapid susceptibility test results have a major positive impact on patient care. Three studies (7, 44, 45) have shown that providing rapid susceptibility test results is more likely to result in a more timely change to appropriate antimicrobial therapy than if conventional reporting times are employed. One of the studies (44) also documented direct cost savings attributable to changes to less expensive antimicrobial agents based on the rapid susceptibility test reports. Further studies must be conducted if the clinical benefits of routine, same-day susceptibility testing are to be defined. However, it has been aptly pointed out that rapid generation of any test results in the laboratory can have a clinical impact only when the results are linked to a reporting system that permits rapid transmission of these test results to the physician (15, 22, 28), preferably not too late in the working day (3).

DISADVANTAGES OF CURRENT INSTRUMENTS

The instruments discussed in this chapter offer automation of several of the steps involved in performing an antimicrobial susceptibility test, although none of the existing instruments is able to execute the entire process, i.e., standardize an inoculum and continue through the final interpretation of results, without operator intervention. Microbiologists must still isolate bacteria by using conventional culture methods and prepare a bacterial suspension of defined density for testing by the instrument. Thus, the automated aspects of the test do not begin until after the manual inoculum preparation. Indeed, inadequacy of the inoculum is a major variable affecting performance of all of the instruments.

Laboratories using these systems must purchase antibiotic test panels that are specified by the manufacturer unless their volume warrants purchase of a more expensive custom panel. The flexibility to change their test battery or to quickly begin testing a newly marketed antimicrobial agent is limited.

Another major disadvantage is that these instrument-based systems are not applicable for testing all clinically relevant groups of bacteria. For example, testing fastidious bacteria, anaerobes, and certain nonfermentative gram-negative bacilli is often not possible. Therefore, alternative susceptibility test methods must be available in the laboratory.

For short-incubation systems, the range of doubling dilutions or drug concentrations tested for a given antimicrobial agent is not wide. MICs, when produced, are reliable only within the concentrations tested, and MICs higher or lower than the test concentrations can be expressed only as semiquantitative results.

Quality control is dependent on the use of American Type Culture Collection organisms specified by the manufacturer, since the NCCLS does not provide standards for test procedures or quality control of short-incubation susceptibility test systems. Some of the quality control organisms suggested by manufacturers do not produce on-scale quality control values (17, 18). The quality control check suggested by the manufacturer may not detect subtle deteriorations in instruments or reagent performance.

When purchased, a capital investment of $28,000 to $125,000 for hardware is required. The disposables used with the rapid systems may be more expensive than the components needed for manual test methods. In addition to space and electrical requirements, fees for service contracts add to operational costs. Laboratories may wish to obtain use of an instrument system by using a reagent rental or lease agreement to avoid a large initial capital outlay for hardware.

Manufacturers' package inserts often include frequent disclaimers or limitations for organisms that can be tested on the system. If the disclaimed antimicrobial agent is critical to patient care, an alternative procedure must be used to assess resistance, or the antimicrobial agent cannot be reported. In some cases, the accuracy of instrument-generated results has been lower than that of manual reference systems, particularly if the instrument has used a short incubation period or a nontraditional growth detection method (19, 46, 52).

False susceptibility and false resistance results (when results are compared to those of reference methods) have been reported by users of instrument-read or short-incubation susceptibility test systems. In some cases, these findings have been sporadic and not easily explained. In many instances, careful attention to the importance of inoculum preparation from fresh colonies and appropriate inoculum standardization by using a nephelometer rather than visual adjustment will correct spurious results.

An incubation period of only 3.5 to 6 h may not be adequate for expression of all bacterial resistance mechanisms, e.g., inducible β-lactamase-mediated resistance among gram-negative bacilli to some enzyme-labile β-lactam antibiotics (20, 33, 47). This is particularly true for *Citrobacter freundii*, *Enterobacter* spp., *Serratia* spp., *Morganella morganii*, *Providencia* spp., and *Pseudomonas aeruginosa*. Although detection of plasmid-mediated extended-spectrum β-lactamases that confer resistance to ceftazidime, cefotaxime, and other broad-spectrum cephalosporins and monobactams is a problem for reference susceptibility test methods, it may be even more of a problem for short-incubation systems (16). Future inclusion of test wells with low concentrations of broad-spectrum cephalosporins with and without a β-lactamase inhibitor (e.g., sulbactam or clavulanate) may help point out the existence of strains with these enzymes (8, 30), which appear to be more common in *Klebsiella* spp. and *E. coli*.

False resistance to aztreonam, especially for *Proteus* and *Morganella* spp., can occur in the systems that detect growth photometrically (unpublished observations). This phenomenon results because elongation of the cells, which occurs just prior to lysis, is interpreted as growth. Additionally, false resistance to imipenem has been reported for various species by both overnight and short-incubation systems. Appearing sporadically, such false results may be due to degradation of the antimicrobial agent in the specific test system (49; unpublished observations) and/or to the zinc concentration in the medium (5).

Although recently described as a problem with short-incubation susceptibility test systems (36), the detection of oxacillin-resistant staphylococci has been greatly improved (47). The inclusion of increased NaCl in the growth medium, improvements in software algorithms, and the use of software rules to look for resistance to multiple antimicrobial agents have all contributed to the correction of this problem. Detection of the heterogeneously resistant expression class 1 strains (43), however, remains a problem with both reference and short-incubation systems, because the frequency of methicillin-resistant organisms among some heteroresistant strains can be as low as 1 in 10^6 to 1 in 10^8 resistant cells per population.

Low- or moderate-level vancomycin resistance in *Enterococcus* spp., especially that of the Van B type (40), may not be detected by short-incubation susceptibility test systems (32). Since some vancomycin-resistant strains exhibit inducible resistance, failure to detect them may be due in part to the short incubation period. In one study, however, intrinsic, low-level resistance of the Van C type in *Enterococcus gallinarum* appears to be detected accurately (40).

Problems with the detection by overnight and short-incubation susceptibility test systems of high-level gentamicin or high-level streptomycin resistance in *Enterococcus* spp. have also been reported (31, 48). For the short-incubation systems, a decrease in the concentration of aminoglycosides in the test medium promises to correct the problems (27, 51; unpublished observations). Failure of an optical reading device to detect subtle amounts of growth in high-level aminoglycoside wells following overnight incubation has been reported (21). Changes in the growth medium employed in these wells for overnight systems (48), as recently recommended by the NCCLS (25), or extended incubation (51) may correct these problems.

SUMMARY

Instruments for the performance of antimicrobial susceptibility testing should be viewed as some of the best examples of automation in clinical microbiology. The performance of susceptibility tests seems well suited to instrumentation, since an objective measurement of microbial proliferation is the basis for the test. Some instruments provide more rapid results than can be generated by using manual systems, and all have the potential to improve intra- and interlaboratory standardization. Efforts to develop test methods with even shorter analysis times must continue, so that rapid reporting

can occur in a more compressed and thus more relevant time period. Moreover, efforts toward the development of an instrument capable of testing a wide variety of bacteria and accurately detecting clinically relevant resistance mechanisms should continue. Further advances may result from the exploration of more innovative means of detecting the inhibitory effects of antimicrobial agents on microorganisms or from the use of nucleic acid probes to determine the presence of the genetic determinants of resistance.

REFERENCES

1. **Baker, C. N., S. A. Stocker, D. L. Rhoden, and C. Thornsberry.** 1986. Evaluation of the MicroScan antimicrobial susceptibility system with the AutoScan-4 automated reader. *J. Clin. Microbiol.* **19:**744–747.
2. **Bascomb, S., J. H. Godsey, M. Kangas, L. Nea, and K. M. Tomfohrde.** 1991. Rapid antimicrobial susceptibility testing of gram-positive cocci using Baxter Microscan Rapid Fluorogenic Panels and autoSCAN-W/A. *Pathol. Biol.* **39:**466–470.
3. **Cherubin, C. E., R. Eng, and M. Appleman.** 1987. A critique of semiautomated susceptibility systems. *Rev. Infect. Dis.* **9:**655–659.
4. **College of American Pathologists.** 1992. *Workload Recording Method and Personnel Management Manual.* College of American Pathologists, Northfield, Ill.
5. **Cooper, G. L., A. Louie, A. L. Baltch, R. C. Chu, R. P. Smith, W. J. Ritz, and P. Michelsen.** 1993. Influence of zinc on *Pseudomonas aeruginosa* susceptibilities to imipenem. *J. Clin. Microbiol.* **31:**2366–2370.
6. **Courvalin, P.** 1992. Interpretive reading of antimicrobial susceptibility tests. *ASM News* **58:**368–375.
7. **Doern, G. V., D. R. Scott, and A. L. Rashad.** 1982. Clinical impact of rapid antimicrobial susceptibility testing of blood culture isolates. *Antimicrob. Agents Chemother.* **21:**1023.
8. **Ferraro, M. J., G. A. Jacoby, G. Katsanis, J. T. Solliday, J. Spargo, and J. P. Gayral.** 1992. Improving the reliability of extended-spectrum b-lactamase detection by the Vitek System, abstr. 123. *Program Abstr. 32nd Intersci. Conf. Antimicrob. Agents Chemother.*
9. **Food and Drug Administration.** 1991. *Federal Guidelines. Review Criteria for Assessment of Antimicrobial Susceptibility Devices.* Food and Drug Administration, Washington, D.C.
10. **Godsey, J. H., S. Bascomb, T. Bonnette, M. Kangas, K. Link, K. Richards, and K. M. Tomfohrde.** 1991. Rapid antimicrobial susceptibility testing of gram-negative bacilli using Baxter MicroScan Rapid Fluorogenic Panels and autoSCAN-W/A. *Pathol. Biol.* **39:**461–465.
11. **Hirtz, P., C. Recule, P. Le Noc, D. Sirot, and J. Croize.** 1992. Detection des betalactamases e spectre elargi par technique Rapid ATB E. Interet du systeme expert API V2.1.1. *Pathol. Biol.* **40:**551–555.
12. **Isenberg, H. D., A. Reichler, and D. Wiseman.** 1971. Prototype of a fully automated device for determination of bacterial antibiotic susceptibility in the clinical laboratory. *Appl. Microbiol.* **22:**980–986.
13. **Jones, R. N., and D. C. Edson.** 1991. Antimicrobial susceptibility testing trends and accuracy in the United States. *Arch. Pathol. Lab. Med.* **115:**429–436.
14. **Jorgensen, J. H.** 1993. Selection criteria for an antimicrobial susceptibility testing system. *J. Clin. Microbiol.* **31:**2841–2844.
15. **Jorgensen, J. H., and J. M. Matsen.** 1987. Physician acceptance and application of rapid microbiology instrument test results, p. 209–212. *In* J. H. Jorgensen (ed.), *Automation in Clinical Microbiology.* CRC Press, Inc., Boca Raton, Fla.
16. **Katsanis, G. P., J. Spargo, M. J. Ferraro, L. Sutton, and G. A. Jacoby.** 1994. Detection of *Klebsiella pneumoniae* and *Escherichia coli* strains producing extended-spectrum β-lactamases. *J. Clin. Microbiol.* **32:**691–696.
17. **Kellogg, J. A.** 1984. Inability to control selected drugs on commercially-obtained microdilution MIC panels. *Am. J. Clin. Pathol.* **82:**455–458.

18. **Kellogg, J. A.** 1985. Inability to adequately control antimicrobial agents on AutoMicrobic System gram-positive and gram-negative cards. *J. Clin. Microbiol.* **21:**454–456.

19. **Kelly, M. T., and C. Leicester.** 1992. Evaluation of the AutoScan Walkaway System for rapid identification and susceptibility testing of gram-negative bacilli. *J. Clin. Microbiol.* **30:**1568–1571.

20. **Lampe, M. F., C. L. Aitken, P. G. Dennis, P. S. Forsythe, K. E. Patrick, F. D. Schoenknecht, and J. C. Sherris.** 1975. Relationship of early readings of minimal inhibitory concentrations to the results of overnight tests. *Antimicrob. Agents. Chemother.* **8:**429–433.

21. **Louie, M., A. E. Simor, S. Szeto, M. Patel, B. Kreiswirth, and D. E. Low.** 1992. Susceptibility testing of clinical isolates of *Enterococcus faecium* and *Enterococcus faecalis. J. Clin. Microbiol.* **30:**41–45.

22. **Matsen, J. M.** 1985. Means to facilitate physician acceptance and use of rapid test results. *Diagn. Microbiol. Infect. Dis.* **3:**35s–78s.

23. **Metzler, C. M., and R. M. Dehaan.** 1974. Susceptibility tests of anaerobic bacteria: statistical and clinical considerations. *J. Infect. Dis.* **130:**588–594.

24. **Murray, P. R., A. C. Niles, and R. L. Ilceren.** 1987. Comparison of a highly automated 5-h susceptibility testing system, the Cobas-Bact, with two reference methods: Kirby-Bauer disk diffusion and broth microdilution. *J. Clin. Microbiol.* **25:**2372–2377.

25. **National Committee for Clinical Laboratory Standards.** 1993. *Methods for Dilution Antimicrobial Susceptibility Tests for Bacteria That Grow Aerobically,* 3rd ed. Approved standard M7-A3. National Committee for Clinical Laboratory Standards, Villanova, Pa.

26. **Nolte, F. S., K. K. Krisher, L. A. Beltran, N. P. Christianson, and G. E. Sheridan.** 1988. Rapid and overnight microdilution antibiotic susceptibility testing with the Sensititre AutoReader System. *J. Clin. Microbiol.* **261:**1079–1084.

27. **Nolte, F. S., J. M. Williams, K. L. Maher, and B. Metchock.** 1993. Evaluation of modified MicroScan screening tests for high-level aminoglycoside resistance in *Enterococcus faecalis. Am. J. Clin. Pathol.* **99:**286–288.

28. **Pestotnik, S. L., R. S. Evans, J. P. Burke, P. M. Gardner, and D. C. Classen.** 1990. Therapeutic antibiotic monitoring: surveillance using computerized expert system. *Am. J. Med.* **88:**43–48.

29. **Reiber, N. E., M. T. Kelly, J. M. Latimer, D. L. Tison, and R. M. Hysmith.** 1985. Comparison of the Cathra Repliscan II, the AutoMicrobic System Gram-Negative General Susceptibility-Plus card, and the Micro-Media System Fox panel for dilution susceptibility testing of gram-negative bacilli. *J. Clin. Microbiol.* **21:**959–962.

30. **Ronco, E., M. L. Migueres, M. Guenounou, and A. Philippon.** 1991. Detection des betalactamases a spectre elargi avec le systeme ATB CMI. *Pathol. Biol.* **39:**480–485.

31. **Sahm, D. F., S. Boonlayangoor, P. C. Iwen, J. L. Baade, and G. L. Woods.** 1991. Factors influencing determination of high-level aminoglycoside resistance in *Enterococcus faecalis. J. Clin. Microbiol.* **29:**1934–1939.

32. **Sahm, D. F., and L. Olsen.** 1990. In vitro detection of enterococcal vancomycin resistance. *Antimicrob. Agents Chemother.* **34:**1846–1848.

33. **Schadow, K. H., D. K. Giger, and C. C. Sanders.** 1993. Failure of the Vitek AutoMicrobic System to detect beta-lactam resistance in *Aeromonas* species. *Am. J. Clin. Pathol.* **100:**308–310.

34. **Schembra, C., C. Dennett, S. Killian, T. Chavez, and J. Kihara.** 1993. Microwell fluorometer for reading of colorimetric/fluorometric bacterial susceptibility and identification panels, abstr. C306. *Abstr. 93rd Gen. Meet. Am. Soc. Microbiol. 1993.*

35. **Sherris, J. C., and K. J. Ryan.** 1982. Evaluation of automated and rapid methods, p. 105. *In* R. C. Tilton (ed.), *Rapid Methods and Automation in Microbiology.* American Society for Microbiology, Washington, D.C.

36. **Skulnick, M., A. E. Simor, D. Gregson, M. Patel, G. W. Small, B. Kreiswirth, D. Hathoway, and D. E. Low.** 1992. Evaluation of commercial and standard methodology for determination of oxacillin susceptibility in *Staphylococcus aureus. J. Clin. Microbiol.* **30:**1985–1988.

37. **Stager, C. E., and J. R. Davis.** 1992. Automated systems for identification of microorganisms. *Clin. Microbiol. Rev.* **5:**302–327.

38. **Staneck, J. L., S. D. Allen, E. E. Harris, and R. C. Tilton.** 1985. Automated reading of MIC microdilution trays containing fluorogenic enzyme substrates with the Sensititre Autoreader. *J. Clin. Microbiol.* **22:**187–191.

39. **Steers, E., F. Foltz, B. S. Graves, and J. Riden.** 1959. An inocula replicating apparatus for routine testing of bacterial susceptibility to antibiotics. *Antibiot. Chemother.* **9:**307–311.

40. **Tenover, F. C., J. Tokars, J. Swenson, S. Paul, K. Spitalny, and W. J. Jarvis.** 1993. Ability of clinical laboratories to detect antimicrobial agent-resistant enterococci. *J. Clin. Microbiol.* **31:**1695–1699.

41. **Thornsberry, C., J. P. Anhalt, J. A. Washington II, L. R. McCarthy, F. D. Schoenknecht, J. C. Sherris, and H. J. Spencer.** 1980. Clinical laboratory evaluation of the Abbott MS-2 automated antimicrobial susceptibility testing system: report of a collaborative study. *J. Clin. Microbiol.* **12:**375–390.

42. **Thrupp, L. D.** 1986. Susceptibility testing of antibiotics in liquid media, p. 93–150. *In* V. Lorian (ed.), *Antibiotics in Laboratory Medicine,* 2nd ed. The Williams & Wilkins Co., Baltimore.

43. **Tomasz, A., S. Nachman, and H. Leaf.** 1991. Stable classes of phenotypic expression in methicillin-resistant clinical isolates of staphylococci. *Antimicrob. Agents Chemother.* **35:**124–129.

44. **Trenholme, G. M., R. L. Kaplan, P. H. Karakusis, T. Stine, J. Fuhrer, W. Landau, and S. Levin.** 1989. Clinical impact of rapid identification and susceptibility testing of bacterial blood culture isolates. *J. Clin. Microbiol.* **27:**1342–1345.

45. **Vincent, P., D. Izard, T. Lebrun, J. C. Sailly, G. Arbon, A. Hassoun, and H. Leclerc.** 1985. Interet cliniques des resultats rapides de bacteriologie au de l'infection nosocomiales: comparaison avec les methods traditionelles. *Presse Med.* **14:**1697–1700.

46. **Visser, M. R., L. Bogaards, M. Rozenberg-Arska, and J. Verhoef.** 1992. Comparison of the autoSCAN W/A and Vitek Automicrobic systems for identification and susceptibility testing of bacteria. *Eur. J. Clin. Microbiol. Infect. Dis.* **11:**979–984.

47. **Washington, J. A.** 1993. Rapid antimicrobial susceptibility testing: technical and clinical considerations. *Clin. Microbiol. Newsl.* **15:**153–155.

48. **Weissmann, D., J. Spargo, C. Wennersten, and M. J. Ferraro.** 1991. Detection of enterococcal high-level aminoglycoside resistance with MicroScan freeze-dried panels containing newly modified medium and Vitek gram-positive susceptibility cards. *J. Clin. Microbiol.* **29:**1232–1235.

49. **White, R. L., M. B. Kays, L. V. Friedrich, E. W. Brown, and J. R. Koonce.** 1991. Pseudoresistance of *Pseudomonas aeruginosa* resulting from degradation of imipenem in an automated susceptibility testing system with predried panels. *J. Clin. Microbiol.* **29:**398–400.

50. **Willey, B. M., B. N. Kreiswirth, A. E. Simor, G. Williams, S. R. Scriver, A. Phillips, and D. E. Low.** 1992. Detection of vancomycin-resistance in *Enterococcus* spp. *J. Clin. Microbiol.* **30:**1621–1624.

51. **Woods, G. L., B. DiGiovanni, M. Levison, P. Pitsakis, and D. LaTemple.** 1993. Evaluation of MicroScan rapid panels for detection of high-level aminoglycoside resistance in enterococci. *J. Clin. Microbiol.* **31:**2786–2787.

52. **York, M. K., G. F. Brooks, and E. H. Fiss.** 1992. Evaluation of the autoSCAN-W/A Rapid System for identification and susceptibility testing of gram-negative fermentative bacilli. *J. Clin. Microbiol.* **30:**2903–2910.

Antimicrobial Agents and Susceptibility Tests: Mycobacteria

CLARK B. INDERLIED AND MAX SALFINGER

119

In their monograph entitled *Chemotherapy of Tuberculosis*, Russel and Middlebrook (98) summarized the treatment results for nearly 1,000 tuberculosis patients hospitalized at the National Jewish Hospital in Denver, Colo., during the 1950s. The vast majority (97%) of patients were infected with *Mycobacterium tuberculosis*. In the preface, Russel and Middlebrook clearly expected that 40 years later, things would be quite different:

> The principal purpose of this monograph is to present a scientific analysis and practical recommendations concerning the contemporary method of chemotherapy. It will surely contain errors which will amuse those who come after our time and, if we live long enough, will amuse—nay, disappoint—ourselves should we reread this presentation some years hence. . . . With the understanding that science means a method and not a static body of knowledge; and that our knowledge, and therefore our concepts, will change with more investigation, along with the readers', we wish to apologize a priori for the seeming pedantry and the dogmatism which is unintentionally, though inevitably, introduced into such a work as we have undertaken here.

One of the most alarming aspects of the contemporary increase in tuberculosis has been the nosocomial outbreaks of multidrug-resistant (MDR) *M. tuberculosis* (27, 30, 66, 95). The recommendations of the advisory council for the elimination of tuberculosis (16) require in vitro drug susceptibility testing of initial *M. tuberculosis* isolates from all tuberculosis patients and reporting of these results to the local health department. Also, it is recommended that patients receive four-drug therapy until microbiology results are known. More than 40 years later, we have returned to the perspective of Russel and Middlebrook.

The clinical mycobacteriology laboratory continues to play a pivotal role in the control of tuberculosis and an expanded role in the diagnosis and treatment of other types of mycobacterioses. The antimicrobial agents used for the treatment of mycobacterial diseases and the methods of testing the susceptibilities of *M. tuberculosis* complex, *M. avium* complex (MAC), rapidly growing mycobacteria, and other selected clinically significant mycobacteria are described and discussed in this chapter with the intent of providing guidelines and information that will be useful to clinical microbiologists and physicians who need to mea-

sure the in vitro susceptibilities of mycobacteria to antimicrobial agents.

ANTIMICROBIAL AGENTS

Although a variety of antimicrobial agents are available for the treatment of mycobacterial diseases, only for tuberculosis and perhaps leprosy is there a consensus on the best treatment regimens. For other mycobacterioses, the clinician is not infrequently faced with a dilemma in choosing a treatment regimen because of a lack of clinical precedence in the treatment of rare mycobacterioses or unclear efficacy. The situation is further confounded by the need to treat mycobacterial infections with a mixture of agents both to improve efficacy and to prevent resistance or overcome inherent resistance. The antimicrobial agents that are used in the treatment of mycobacterioses are discussed below and summarized in terms of clinical efficacy and candidates for susceptibility testing in Table 1.

INH

Isoniazid (isonicotinic acid hydrazide) (INH), a synthetic antimicrobial agent introduced in 1952 for the treatment of tuberculosis, is a remarkably specific and potent bactericidal agent against tubercle bacilli. INH has comparatively low toxicity and is active against virtually all wild-type strains of *M. tuberculosis*. While the exact mechanism of action of INH is not known, the primary effect is on mycolic acid synthesis, as evidenced by increased fragility of the mycobacterial cell, increased intracellular viscosity, decreased cellular hydrophobicity, and loss of acid fastness (124). Some evidence indicates that INH inhibits a desaturation step in the production of long-chain fatty acids (111) and may also inhibit the elongation of fatty acids and hydroxy lipids (26). In addition to the effect on mycolic acid synthesis, there is evidence that bactericidal free radicals form as a result of the interaction of INH and catalase or peroxidase (23), observations that are consistent with the established correlation between INH resistance and a loss of catalase activity. In addition, INH appears to interfere with NAD metabolism, which would lead to pleiotropic effects on energy metabolism and macromolecular synthesis (29, 125, 126).

Zhang et al. (134) reported on the detection of the *katG* gene in *M. tuberculosis* and correlated the deletion of this

TABLE 1 Antimycobacterial agents ranked by clinical utility and expected in vitro susceptibility[a]

Mycobacterium species	Antimycobacterial agent			
	Primary or 1st choice	Secondary or 2nd choice	Alternative or investigational	Resistance likely
Well documented				
M. tuberculosis, M. africanum, M. bovis[b]	INH, RMP, PZA, SM, EMB	Ciprofloxacin, ofloxacin, ethionamide, cycloserine	Amikacin, rifabutin, sparfloxacin	
M. leprae	Dapsone, RMP, clofazimine	Ethionamide prothionamide	Clarithromycin, minocycline	
M. avium, M. intracellulare	Azithromycin, clarithromycin, EMB	Amikacin, clofazimine, ciprofloxacin, rifabutin	Cycloserine, SM	INH, PZA
M. chelonae	Amikacin, clarithromycin	Kanamycin		Primary antituberculosis agents, ciprofloxacin, other quinolones
M. fortuitum	Amikacin, cefoxitin	Cefmetazole, imipenem, minocycline, capreomycin, kanamycin, ciprofloxacin		Primary antituberculosis agents
M. abscessus	Amikacin, clarithromycin	Kanamycin, cefoxitin		Primary antituberculosis agents, ciprofoxacin, other quinolones
Less well documented				
M. kansasii	INH, RMP, EMB	Amikacin, ethionamide, SM, cycloserine		
M. malmoense	INH, RMP, EMB			
M. ulcerans	RMP, amikacin	EMB, TMP-SMX[c]	SM, clofazimine	
M. marinum	RMP, EMB	TMP-SMX, minocycline	Clarithromycin	
M. simiae	Insufficient information			
M. haemophilum	Ciprofloxacin, cycloserine, RMP, rifabutin	Azithromycin, clarithromycin, amikacin, clofazimine	Ofloxacin, sparfloxacin	
M. scrofulaceum	Excision without chemotherapy?	INH, RMP, SM, cycloserine		
M. szulgai	INH, RMP, EMB	SM, capreomycin, viomycin		
M. xenopi	None	INH, RMP, SM, cycloserine		

[a]For species for which there is well-documented experience, the table lists primary and secondary agents. For species for which choices are more speculative, largely because of the paucity of information available, first- and second-choice agents are given. First-choice agents are expected to be active against wild-type isolates (i.e., from untreated patients). Second-choice agents are less preferred, usually for reasons of toxicity, expense, or unclear efficacy. The information contained in this table was compiled from a variety of sources, many of which provide more specific information on the use of these and other agents for the treatment of various mycobacterial diseases (62, 77, 80, 105).

[b]M. bovis and M. bovis BCG are considered PZA resistant.

[c]TMP-SMX, trimethoprim-sulfamethoxazole.

gene with resistance to INH. The katG gene encodes for both catalase and peroxidase, and they showed that the transfer of this gene into an INH-resistant strain of M. smegmatis conferred susceptibility to INH and that deletion of the gene from clinical isolates of M. tuberculosis correlated with INH resistance. The same group showed that INH resistance in M. aurum and M. smegmatis correlated with the loss or reduction of peroxidase and catalase activities and that both species had gene sequences that were homologous to that of the katG gene in M. tuberculosis (54). The recent sequence analysis of the katG gene should lead to more insight into the functional and regulatory roles of this gene in INH susceptibility and resistance (55), which is likely to assist in the cloning of INH resistance determi-

nants and the identification of new, more potent INH-like drugs (97).

Banerjee et al. (6) recently reported that INH and ethionamide resistance in M. tuberculosis correlates with a missense mutation in the inhA gene. The inhA gene encodes a protein that has sequence similarity (75% similarity and 40% identity) to the Escherichia coli enzyme EnvM, and the authors reported that cell-free assays indicate that this enzyme may be a component of the mycolic acid biosynthetic pathway. The sequence of the resistant inhA results in a Ser-to-Ala substitution that is conserved in the naturally INH-resistant, rapidly growing M. smegmatis.

INH is active only against replicating tubercle bacilli; slowly replicating bacilli in the caseous lesions are not readily killed by INH, and dormant bacilli are unlikely to be affected. INH resistance develops rapidly in patients on monotherapy, and the frequency of INH resistance within a population of tubercle bacilli ranges from 10^{-5} to 10^{-6}. Wild-type isolates of M. tuberculosis are inhibited by INH, with MICs of 0.05 to 0.2 μg/ml (MICs quoted here are not based on standardized methods and are intended only to convey a sense of the relative potencies of certain agents), while other susceptible isolates of slowly growing mycobacteria such as M. kansasii and M. xenopi are inhibited at 1 to 5 μg/ml (28). Most other nontuberculous mycobacteria, including the MAC, M. marinum, M. ulcerans, and all rapidly growing mycobacteria, are resistant to INH.

INH is well absorbed when administered perorally or intramuscularly. The drug is distributed throughout the body, and levels in cerebrospinal fluid (CSF) may equal levels in plasma with meningeal inflammation or 20% of levels in plasma without inflammation (1). INH is metabolized in the liver and intestines, primarily by acetylation by an N-acetyltransferase that can vary significantly from person to person. However, the acetylator phenotype of an individual does not appear to influence either the efficacy of INH or the risk for hepatotoxicity. Adverse drug reactions include infrequent age-related hepatitis and, even less frequently, peripheral neuropathy, hypersensitivity reactions such as fever and rash, and arthralgias.

RMP

Rifampin (RMP), or rifampicin, is 3,4-(methylpiperazinyl-iminomethylidene)-rifamycin SV. It was introduced in 1968 as a potent antituberculosis agent. RMP is active against a wide variety of non-acid-fast bacteria (non-AFB) and several other slowly growing mycobacteria, notably M. leprae, M. kansasii, M. haemophilum, and M. marinum, but is only variably active against the MAC and is inactive against the rapidly growing mycobacteria. RMP inhibits the prokaryotic DNA-dependent RNA polymerase by binding to the β subunit at the presumed catalytic center of the enzyme. Mammalian RNA polymerase is inhibited by rifampin only at significantly higher concentrations. The RNA polymerase of MAC isolates appears to be susceptible to rifampin; therefore, the primary mechanism of intrinsic resistance is most likely impermeability.

Rifampin resistance correlates with changes in the structure of the β subunit of the prokaryotic RNA polymerase. These changes are primarily due to missense mutations within highly conserved regions of the rpoB gene. Telenti et al. (113) showed that RMP resistance correlated with changes, primarily amino acid substitutions, within a conserved region of the rpoB gene that encodes the β subunit of the M. tuberculosis polymerase. They developed a rapid RMP resistance screening test in which mutations within the rpoB gene were detected by PCR amplification of a 157-bp region that contains the majority of mutation sites and subsequent use of single-strand conformational polymorphism (114). The method was successfully applied to the detection of RMP-resistant M. tuberculosis in minimally grown cultures (BACTEC 12B vials with growth indexes [GIs] of ≤100) and smear-positive sputa with >10 organisms per 250× field. The molecular basis of RMP resistance in M. leprae is similar to that in M. tuberculosis, and it appears that the same methodology can be applied to the detection of RMP resistance in M. leprae (58). Williams (123) developed a rapid PCR-based DNA-sequencing protocol that targets a 305-bp region of the rpoB gene and showed that there was a high degree of sequence similarity

(90 to 100%) between the region from M. tuberculosis and those from M. leprae, M. avium, and M. africanum. In her analysis of 110 RMP-resistant strains of M. tuberculosis, 16 mutations were identified. Nine of these 16 were identical to those described by Telenti et al. (113), and 7 were newly identified. In two strains of RMP-resistant M. avium, missense mutations were detected, but in two other strains, no mutations were detected. These results agree with the observations of Guerrero et al. (41), who showed that mutations in the rpoB genes of M. avium and M. intracellulare are rare and that RMP resistance is most likely a reflection of impermeability to the drug.

RMP is well absorbed from the gastrointestinal tract, and peak concentrations of 5 to 10 μg/ml are reached within 1 to 2 h after an oral dose of 600 mg; concentrations in CSF reach 50% of levels in plasma with meningeal inflammation. RMP is available in combination with INH as a single capsule (Rifamate) containing 300 mg of RMP and 150 mg of INH and in combination with INH and pyrazinamide (PZA) also as a single capsule (Rifater) containing 120 mg of RMP, 50 mg of INH, and 300 mg of PZA. RMP concentrations of ≤0.5 μg/ml are bactericidal for wild-type isolates of M. tuberculosis, and the drug affects intracellular, slowly replicating bacilli in caseous lesions as well as actively replicating tubercle bacilli in open pulmonary cavities. Adverse drug reactions include gastrointestinal and hypersensitivity reactions; however, the major effect is hepatotoxicity and a red-orange discoloration of urine, tears, and other body fluids. RMP also induces increased hepatic metabolism of a wide variety of other drugs (1).

PZA

PZA is a synthetic derivative (pyrazine analog) of nicotinamide and in combination with INH is rapidly bactericidal for replicating forms of M. tuberculosis, with an average MIC of 20 μg/ml. PZA is inactive against nonreplicating tubercle bacilli and totally inactive against other species of mycobacteria, including M. bovis, MAC, and the rapidly growing mycobacteria. PZA is active only at a slightly acidic pH; thus, in vitro susceptibility test media must be adjusted to pH 5.5 to 6 in order to accurately measure the activity of the drug. Most likely, PZA is active only in the acidic milieu of the phagolysosome, and depending on the concentration achieved at the site of the infection, PZA may be bacteriostatic or bactericidal. PZA is hydrolyzed in the liver to the active metabolite pyrazinoic acid, and although the mechanisms of action of PZA are unknown, activity depends on this conversion to pyrazinoic acid. M. tuberculosis produces a pyrazinamidase, and most strains of PZA-resistant M. tuberculosis lack this enzyme; however, some PZA-resistant isolates retain pyrazinamidase activity, suggesting that there are other mechanisms of resistance. PZA is well absorbed from the gastrointestinal tract and widely distributed throughout the body, with maximum levels in serum of approximately 45 μg/ml at 1 to 4 h after an oral dose of 1 g (20 to 25 mg/kg of body weight). Hepatotoxicity occurs in a small number of patients, and photosensitivity and rash occur rarely. Gout is an important contraindication because of the hyperuricemia associated with PZA therapy. PZA is usually discontinued after the first 2 months of short-course therapy for tuberculosis, while INH and RMP are continued for an additional 4 months.

EMB

Ethambutol [dextro-2,2′-(ethylenediimino)-di-1-butanol-dihydrochloride] (EMB) is a potent synthetic antituberculosis compound introduced in 1961. The MICs of EMB tested against wild-type isolates of M. tuberculosis range from 1 to 5 μg/ml, but the activity of the drug against other slowly growing Mycobacterium spp. is much more variable. The primary mechanism of action of EMB is a bacteriostatic inhibition of cell wall synthesis, while evidence points to a specific effect on arabinogalactan synthesis (110). The frequency of mutation to EMB resistance in M. tuberculosis is on the order of 10^{-5}. Although most members of the MAC are considered intrinsically resistant to EMB, a variety of studies have showed that combinations of EMB and other agents, notably quinolones and macrolides, are synergistic (56, 69) and that EMB appears to have a permeabilizing effect on the MAC cell wall (57).

Peak serum EMB concentrations of 5 μg/ml are achieved by 2 to 4 h after a dose of 25 mg/kg. The primary adverse effect associated with EMB is a decrease in visual acuity due to optic neuritis that is related to both the dose and the duration of treatment. The effects are generally reversible upon discontinuation of the drug. A variety of other adverse reactions have been reported, but these are infrequent and sometimes difficult to ascribe to EMB, since they may be due to concurrent therapy with other antituberculosis agents.

Rifabutin

Rifabutin is a spiropiperidyl rifamycin with potent in vitro activity against M. tuberculosis (50) and the MAC (52, 99). The mode of action and mechanism of resistance of rifabutin appear to be identical to those of RMP; however, approximately 30% of RMP-resistant M. tuberculosis isolates are susceptible to rifabutin, a fact that may be correlated with certain specific mutations in the rpoB gene (8). Rifabutin decreases the incidence of disseminated MAC disease in human immunodeficiency virus (HIV)-infected patients when it is used as a prophylactic agent, and the drug was recently approved by the U.S. Food and Drug Administration for that indication (78, 90). The role of rifabutin as a therapeutic agent for MAC disease is unclear, but there may be a significant dose effect (63). In addition to being more active than RMP on a weight basis, rifabutin has a long elimination half-life in humans, and it concentrates in tissues, notably in lung tissue, where levels are 10-fold higher than those in serum. This propensity may account for the reported effectiveness of rifabutin in the therapy of MAC pulmonary infections (32). Rifabutin is absorbed from the gastrointestinal tract and reaches peak levels in serum of 0.5 μg/ml in about 4 h after a 300-mg dose. Adverse drug reactions with rifabutin are similar to those observed with RMP, including an effect on the metabolism of other drugs (1). Some unique rifabutin toxicities have been described, including leukopenia, thrombocytopenia, arthralgias, and uveitis when it is coadministered with clarithromycin.

Aminoglycosides

The aminoglycosides that are used for the treatment of tuberculosis and other mycobacterial infections include amikacin, kanamycin, and streptomycin (SM). In addition, capreomycin and viomycin, basic peptide antibiotics with mechanisms of action similar to that of the aminoglycosides, are active against M. tuberculosis and certain other species of mycobacteria. The other aminoglycosides, gentamicin and tobramycin, are inactive against mycobacteria at the usual concentrations in serum. Kanamycin is a glycoside of 2-deoxystreptamine, and amikacin is a derivative of kanamycin; thus, the structures of these antimycobacterial aminoglycosides are similar. The primary mechanism of action of the aminoglycosides is inhibition of the post- to pretranslocation step of protein synthesis by blocking of the binding of aminoacyl-tRNA (e-type binding). Viomycin also blocks aminoacyl-tRNA translocation, and viomycin resistance confers resistance to capreomycin, suggesting that the mechanism of action is the same. Streptomycin MICs for wild-type isolates of M. tuberculosis are usually well below the peak concentration in serum of 25 to 50 μg/ml achieved by 1 to 2 h following a 1-g intramuscular dose. Amikacin is the most potent of the aminoglycosides, with average MICs of 1 μg/ml for M. tuberculosis and 12.5 μg/ml for M. chelonae and M. abscessus. There is comparatively little clinical experience with amikacin in the treatment of tuberculosis because of the expense of the drug, but amikacin in combination with cefoxitin is standard empirical therapy for serious infections suspected to be caused by rapidly growing mycobacteria. Amikacin is also active against MAC, with about 75% of isolates susceptible to 30 μg/ml, a level that approaches the maximum concentration in serum when the drug is administered by intravenous infusion. In an uncontrolled trial, amikacin was shown to be the active component of a multidrug treatment regimen that was associated with a microbiological and clinical response in HIV-infected patients with disseminated MAC disease (18).

The molecular basis of SM resistance in M. tuberculosis was recently investigated in two studies and shown to result from mutations in the gene that encodes ribosomal protein S12 or the 16S rRNA region that is linked to the S12 protein (33, 87). Finken et al. (33) showed that mutations in the S12 protein were present in 20 of 38 SM-resistant strains, and a mutation in the 16S rRNA gene was present in 9 strains. Nair et al. (87) determined the nucleotide sequence of the rpsL gene and showed that in a small number of isolates, SM resistance appeared to be a result of point mutations at codon 43 of the rpsL gene, a site of SM resistance in E. coli. More recently, Meier et al. (81) showed that in two isolates, SM resistance was associated with single-base (C→T) mutations in position 491 or 512 of the 16S rRNA and a single-base (A→G) mutation at position 904 in a third isolate. The latter mutation is equivalent to a mutation in E. coli that correlates with a functional change in the ribosome. Adverse drug reactions associated with aminoglycosides and peptide antibiotics, including hearing loss, tinnitus, loss of balance, and renal failure, are well known.

Cycloserine

D-Cycloserine (4-amino-3-iso-oxazolidinone) is an analog of D-alanine that inhibits the synthesis of D-alanyl–D-alanine, an essential component of the mycobacterial cell wall. Cycloserine is active against all mycobacteria and several other types of bacteria. The average MICs for M. tuberculosis range from 5 to 20 μg/ml, while peak levels in serum of 20 to 40 μg/ml are achieved by 4 h after an oral dose of 250 mg. The drug is widely distributed through the body, including the CSF. Significant adverse drug reactions are associated with cycloserine treatment, notably peripheral neuropathy and central nervous system dysfunction, including seizures and psychotic disturbances.

Ethionamide

Ethionamide (2-ethyl-pyridine-4-carbonic acid thioamide) is a derivative of isonicotinic acid and, like INH, blocks mycolic acid synthesis; however, isolates of M. *tuberculosis* that are resistant to high concentrations of INH are susceptible to ethionamide, suggesting that the site of action may be different from that of INH. The average MIC for M. *tuberculosis* is 0.6 to 2.5 μg/ml, and levels in serum of 2 to 20 μg/ml are achieved by 3 to 4 h after an oral dose of 0.5 to 1 g. Significant side effects are associated with ethionamide, including gastrointestinal irritation with nausea, vomiting, and cramps, and neurologic symptoms may require discontinuation of the drug.

Dapsone

Dapsone (diaminodiphenyl sulfone) is a synthetic compound first shown to be active against M. *leprae* in the early 1940s. Dapsone is an antifolate that, like other inhibitors of folic acid synthesis, exerts primarily a bacteriostatic effect and is only weakly bactericidal. Dapsone is administered orally and is well absorbed and distributed throughout the body. Levels in tissue are approximately 2 μg/ml following a 200-mg dose. The drug has a long half-life in serum of 10 to 50 h, depending on the individual patient. Adverse drug reactions include gastrointestinal intolerance with nausea, vomiting, anorexia, and methemaglobinemia (common). Hematuria, rash, pruritus, and fever can occur. Traditionally, dapsone is used in combination with RMP and clofazimine for the treatment of leprosy. Acedapsone is a diacetylated form of dapsone with an extraordinarily long half-life of 46 days; as a result, the drug is administered infrequently (e.g., five injections per year), and peak concentrations in tissue occur 20 to 35 days after a dose. Acedapsone is relatively inactive against M. *leprae*, but in vivo, it is deacetylated to the parent compound.

Clofazimine

Clofazimine, or lamprene [3-(p-chloroanilino)-10-(p-chlorophenyl)-2,10-dihydro-2-isopropyliminophenazine], is a substituted iminophenazine, bright red dye with potent in vitro activity against MAC (MICs range from 0.1 to 5 μg/ml) but unclear therapeutic efficacy either alone or in combination with other agents. The drug also has potent in vitro activity against M. *tuberculosis*, but there is little or no information on in vivo activity. Clofazimine has weak bactericidal activity against M. *leprae* but is used in combination with RMP and dapsone as a conventional treatment regimen for leprosy. However, it may take up to 50 days of treatment before there is evidence of tissue antimicrobial activity, which may be relevant to the role of this drug in the treatment of both M. *leprae* and MAC. The precise mechanism of action is unknown; however, clofazimine is highly lipophilic and binds preferentially to mycobacterial DNA. The absorption of clofazimine following an oral dose is variable, ranging from 45 to 60%. Average concentrations in serum are 0.7 to 1.0 μg/ml following a dose of 100 to 300 mg. The half-life is extraordinarily long (estimated to be 70 days), and the drug tends to be deposited in fatty tissues and in cells of the reticuloendothelial system. Adverse drug reactions are primarily limited to discoloration of the skin, conjunctiva, cornea, and body fluids and to gastrointestinal intolerance, including pain, diarrhea, nausea, and vomiting.

Azithromycin and Clarithromycin

Azithromycin and clarithromycin are important agents in the treatment of all forms of MAC disease, infections caused by rapidly growing mycobacteria, and leprosy. These new macrolides also appear to be useful in the treatment of diseases caused by M. *marinum*, M. *haemophilum*, and M. *kansasii*. Azithromycin, an azalide (a subclass of macrolides), and clarithromycin are structurally similar to erythromycin but with modifications that improve acid stability and increase potency, drug half-life, achievable concentrations in tissue, and bioavailability without toxicity. These macrolides are bacteriostatic agents and inhibit the growth of microorganisms by binding to the 50S subunit of the prokaryotic ribosome, blocking protein synthesis probably at the peptidyl transferase step. Meier et al. (81) recently showed that clarithromycin mutants of M. *intracellulare* have a single-base mutation at Ade-2058 in the 23S rRNA gene, a mutation that has been associated with macrolide resistance in other bacteria.

The in vitro activity of azithromycin against MAC appears to be quite modest, with MICs 32- to 64-fold above the maximum concentration in serum; however, the remarkable ability of azithromycin to concentrate in tissues probably accounts for the activity of this drug in an animal model of disseminated MAC disease (64). In a small, uncontrolled clinical trial (500-mg dose) (133) and a larger unpublished trial (600- and 1,200-mg doses) that was presented, in part, at a recent conference (72), the treatment of disseminated MAC disease in HIV-infected patients with azithromycin was associated with a positive clinical and microbiological response. Azithromycin is rapidly absorbed from the gastrointestinal tract and widely distributed throughout the body. Peak concentrations in serum following a 500-mg dose are 0.4 to 0.6 μg/ml; however, the drug has a terminal half-life of 68 h, which contributes to the ability of azithromycin to concentrate in tissues to high levels. In a small study in humans, levels of azithromycin in polymorphonuclear neutrophils were nearly 1,000-fold higher than levels in serum (3).

Clarithromycin inhibits 90% of MAC isolates at MICs of 0.25 to 0.5 μg/ml when measured by a radiometric broth macrodilution method at a neutral to slightly alkaline pH (the activities of all macrolides are strongly influenced by pH). The results of recent clinical trials indicate that 500 mg of clarithromycin administered twice a day to HIV-infected patients with disseminated MAC disease results in a clinical and microbiological response (24, 61a). In December 1993, the U.S. Food and Drug Administration approved use of the 500-mg dose of clarithromycin in combination with at least one other agent, usually EMB, for the treatment of MAC disease. Clarithromycin peak levels in serum of 2 to 3 μg/ml are achieved within 5 to 6 h of a 500-mg dose; concentrations of clarithromycin in tissue are 4 to 5 times above levels in serum, and concentrations in macrophages are 20 to 30 times higher than those in serum. The elimination half-life is 5 to 7 h following twice-a-day 500-mg doses.

From the information presently available, it appears that azithromycin and clarithromycin resistance develops relatively quickly (8 to 12 weeks) when either drug is used as monotherapy for MAC disease and perhaps other nontuberculous mycobacterial infections. A U.S. Public Health Service task force recommended that azithromycin or clarithromycin be used only in combination with a second and perhaps a third agent such as EMB or clofazimine (78). The premise of this recommendation is based on experience with M. tuberculosis, and at present, there are no corroborating clinical or laboratory data that a drug with marginal potency or therapeutic efficacy such as EMB or clofazimine will prevent the development of macrolide resistance. Azithromycin and clarithromycin appear to be equally effective in the treatment of MAC disease, and it is unclear whether the differences between these two drugs, including pharmacokinetics and frequency of dosing, are significant in terms of long-term effectiveness or compliance. Adverse drug reactions, including diarrhea, nausea, abnormal taste, dyspepsia, abdominal pain, and headache, appear to occur with equally low frequency (less than 3% of patients) with both azithromycin and clarithromycin.

Ciprofloxacin and Ofloxacin

Ciprofloxacin and ofloxacin are fluorinated carboxyquinolones with moderate in vitro activities against M. tuberculosis and variable activities against MAC and rapidly growing mycobacteria. Levofloxacin (L enantiomer of ofloxacin) and sparfloxacin are investigational fluorinated quinolones with potent in vitro activities and promising in vivo activities against M. tuberculosis, but they will not be discussed further. The mechanism of action of all fluorinated quinolones is inhibition of DNA synthesis as a result of binding to the DNA gyrase (bacterial topoisomerase II). While this is the presumed mechanism of action in mycobacteria, fewer studies have been performed with mycobacteria than with other microorganisms. However, Takiff et al. (112) recently showed that quinolone resistance in M. tuberculosis can be ascribed to mutations in the gyrA and gyrB genes, which encode the DNA gyrase subunits. It will be of interest to correlate the specific mutations with functional and structural changes in order to prove that these changes are the cause of the resistance.

The MICs of ciprofloxacin for susceptible isolates of M. tuberculosis range from 0.25 to 3 μg/ml. While there are no verified interpretive standards for testing ciprofloxacin against mycobacteria, in a recent study of 275 isolates of MDR M. tuberculosis from 142 patients, the MIC that inhibited 90% of strains was ≤1.0 μg/ml when the proportion method was used. These results suggest that the serum breakpoint for ciprofloxacin is a useful tentative interpretive criterion for testing M. tuberculosis (38). Most isolates of M. fortuitum are susceptible to ciprofloxacin, while most isolates of M. chelonae are resistant. The MIC of ciprofloxacin that inhibited 90% of MAC strains is 16 μg/ml, and only 30% of isolates are susceptible to 2 μg/ml (75, 132). Ciprofloxacin is well absorbed from the gastrointestinal tract and is rapidly distributed throughout the body. Maximum concentrations in serum of 2.4 and 4.3 μg/ml are achieved by 1 to 2 h after an oral dose of 500 or 750 mg, respectively. The elimination half-life of ciprofloxacin is 4 h in subjects with healthy renal function. The MICs of ofloxacin for susceptible isolates of M. tuberculosis range from 0.5 to 2.5 μg/ml. Ofloxacin maximum concentrations in serum are achieved by 1 to 2 h after an oral dose. The concentration of ofloxacin in serum is 2.9 μg/ml following a single 400-mg dose or 4.6 μg/ml after a steady-state dose. Ofloxacin has biphasic elimination, with the majority of the drug (95% of the area under the curve) eliminated by 4 to 5 h after a steady-state dose and the remainder eliminated after 20 to 25 h. With accumulation, the half-life extends to 9 h in patients with healthy renal function. The efficacies of ciprofloxacin and ofloxacin in the treatment of pulmonary tuberculosis may relate, in part, to the fact that both of these quinolones concentrate in lung tissue to levels fourfold or more above the concentrations in serum (4). Ciprofloxacin and ofloxacin should be tested as secondary agents or when resistance to other antituberculous agents is suspected or known (115).

PAS

para-Aminosalicylic acid (PAS) is an antifolate that is active against M. tuberculosis but inactive against most other mycobacteria. There is some evidence that PAS may also affect iron transport in M. tuberculosis. The average MIC for susceptible isolates of M. tuberculosis is 1 μg/ml, and peak levels in serum of 7 to 8 μg/ml are achieved by 1 to 2 h following a 4-g dose. PAS is incompletely absorbed in the gastrointestinal tract and is associated with significant gastrointestinal side effects that in combination with the need for large dosages (10 to 12 g/day) lead to frequent compliance problems.

Amithiozone

Amithiozone (Thiacetazone, Tibione, or Panthrone) is a thiosemicarbazole that is active against M. tuberculosis and has an average MIC for wild-type strains of 1 μg/ml. Resistance develops quickly on monotherapy; therefore, the drug is administered with a second agent, usually INH. Peak levels in serum of 1 to 4 μg/ml are achieved following an oral dose of 150 mg. Adverse drug reactions include gastrointestinal irritation and bone marrow suppression; hepatotoxicity can occur in patients receiving concomitant INH. Amithiozone in combination with INH has been successfully used in Africa for the treatment of tuberculosis. However, recent evidence associated Stevens-Johnson syndrome and severe epidermal necrolysis in HIV-infected patients with tuberculosis treated with regimens containing amithiozone (31, 40), and as a result, the World Health Organization recommended that amithiozone not be used to treat HIV-infected patients (91). Amithiozone is not available in the United States and is not used in Europe because of the adverse effects.

SUSCEPTIBILITY TESTING OF THE M. TUBERCULOSIS COMPLEX

Drug Resistance

Subsequent to the publication of Russel and Middlebrook (98), the World Health Organization organized two meetings that led to the description of reliable criteria and techniques for testing mycobacteria for resistance to antituberculosis drugs (14, 15). The critical proportion for resistance on Lowenstein-Jensen slants varied according to the drug; e.g., 1% for INH and RMP and 10% for SM, EMB, PZA, ethionamide, kanamycin, cycloserine, and capreomycin. However, on the basis of the excellent experience of Russel and Middlebrook with 7H10 agar (98), the Centers for Disease Control and Prevention (CDC) recommended Middlebrook 7H10 agar and 1% as the critical proportion for all drugs (74).

Resistance is fundamentally a phenomenon linked to large initial bacterial populations. In lung tuberculosis, the greatest populations are those in cavities, which can contain 10^7 to 10^9 organisms, whereas those in hard caseous foci, the most common type of lesion, do not exceed 10^2 to 10^4 organisms (13). The far greater frequency of resistance during the treatment of cavitary tuberculosis than of noncavitary tuberculosis was shown as early as 1949 (59, 60).

David at CDC (25) demonstrated the probability distribution of drug-resistant mutants and in a fluctuation test showed that M. tuberculosis spontaneously mutated to resistance to INH, SM, EMB, and RMP. The average mutation rates for resistance to INH, SM, EMB, and RMP were calculated to be 3×10^{-8}, 3×10^{-8}, 1×10^{-7}, and 2×10^{-10} mutations per bacterium per generation, respectively. Thus, the mutation rate for resistance to two drugs is less than 10^{-15}.

All studies of the genetic basis of antimicrobial resistance in M. tuberculosis to date indicate that the MDR phenotype is the result of accumulative mutations rather than the acquisition of an MDR transfer factor (54a). Although acquisition of the MDR phenotype as a result of a single genetic event (resistance transfer or mutation) cannot be ruled out, at present it seems unlikely that such a single event is the cause of the extant MDR strains of M. tuberculosis.

Special-Population Hypothesis

According to the generally accepted theory, resistance appearing during drug treatment is due to the selection and multiplication of resistant mutants preexisting in the tubercle bacillus population of the host. Inasmuch as the susceptible bacilli are the predominant part of the population, the initial killing involves a large number of microorganisms, and the consequence is a sharp fall in the population of bacilli during the initial period of treatment. The rise due to multiplication of resistant mutants is revealed later. This "fall-and-rise" phenomenon, as demonstrated in the patient's sputum, was described in the late 1940s (22, 92). In 1979, Mitchison (83) suggested the special-populations hypothesis to explain the action of the major antituberculous drugs against the various subpopulations of tubercle bacilli. The subpopulations include (i) rapidly growing bacilli in the pulmonary lesions, (ii) bacilli that grow in short metabolic spurts, (iii) bacilli that reside in the acidic environment of the caseous lesions, and (iv) dormant, nonreplicating bacilli. Each of the agents of the conventional multiple-drug treatment regimen for tuberculosis is more or less effective in eradicating tubercle bacilli within each of these special populations. Thus, the use of multiple drugs in the treatment of tuberculosis is aimed at both preventing drug resistance and achieving a maximum therapeutic effect.

Critical Concentrations

The criteria for defining drug-resistant M. tuberculosis were established on an empirical basis, i.e., that there is a certain proportion of drug-resistant mutants above which therapeutic success is less likely to be realized. The procedures used to perform drug susceptibility tests and the criteria for interpreting the results take into account two factors: (i) the critical proportion of drug-resistant mutants and (ii) the critical concentration of the drug in the test medium. On the basis of clinical and bacteriologic studies, the significant proportion of bacilli resistant to an antituberculosis drug above which a clinical response is unlikely was generally set at 1% (74, 98). The critical concentration of a drug is the level of drug that inhibits the growth of most cells within the population of a wild-type strain of tubercle bacilli without appreciably affecting the growth of the resistant mutant cells that might be present (82, 96, 98). In other words, if the proportion of tubercle bacilli that are resistant to the critical concentration of a drug exceeds 1%, it is unlikely that the use of that drug will lead to a therapeutic success. It should be noted that this concentration may not

bear a direct relationship to the peak level in serum. The critical concentrations of antituberculosis drugs in different media are given in Table 2 (70, 79).

Extent of Service and Susceptibility Testing

In the United States, mycobacteriology laboratories are classified by the levels or extents of services provided based on criteria proposed by the CDC, American Thoracic Society (ATS), and College of American Pathologists (CAP). In general, the criteria emphasize the importance of the number of specimens processed, the expertise of the laboratory staff, and the cost-effectiveness of the procedures and protocols (2, 46, 74). The CAP recognizes four levels of service, ranging from no mycobacteriological procedures performed (extent 1) to definitive and comprehensive procedures that may or may not include susceptibility testing (extent 4). The ATS recommends three levels of service; however, the ATS levels of service are roughly equivalent to CAP levels 2, 3, and 4. The CAP recommends that susceptibility testing be performed only in an extent 3 or 4 laboratory but that even an extent 4 laboratory might refer isolates to another laboratory with more extensive experience in susceptibility testing of mycobacteria. In general, a CAP extent 4 laboratory should perform susceptibility testing only if the laboratory is capable of identifying the isolate to species and if the laboratory regularly performs susceptibility tests on that type of isolate (e.g., >10 tests per week). Since certain drugs are tested infrequently (e.g., cycloserine or amikacin) or must be prepared from reference powders (e.g., rifabutin or ciprofloxacin) or are problematic (e.g., cycloserine), secondary drug testing should be promptly referred to a laboratory with specific expertise in testing these agents.

When To Perform Susceptibility Tests

Kubica and Dye (74) suggested more than 25 years ago that susceptibility testing of primary drugs be performed in ATS level II laboratories in order to provide clinically meaningful data in the shortest possible time, but routine testing of all new isolates (from patients not previously treated or not suspected to have primary resistance) was not recommended (5). Given the low incidence of drug resistance in the United States at that time and the high probability of successful drug treatment of tuberculosis, it was difficult to justify the cost of more frequent testing. Indeed, in communities with an incidence of primary drug-resistant tuberculosis of less than 5%, there may still be little need to perform routine susceptibility testing. While the primary responsibility for ensuring that susceptibility tests are scheduled appropriately is that of the physician (5), current recommendations are designed to ensure that the laboratory diagnosis and susceptibility testing of M. tuberculosis proceed in an expeditious manner and avoid inadvertent delays caused by lapses in ordering or reporting practices. Whether a laboratory will adopt the current recommendations should be decided in collaboration with infectious disease experts and the hospital infection control department with the guidance of the local tuberculosis control program and public health department.

Current Recommendations

In light of the increasing incidence of tuberculosis in the United States and the outbreaks of MDR M. tuberculosis, the advisory council for the elimination of tuberculosis issued new recommendations in 1993 that recommend in vitro susceptibility testing of initial M. tuberculosis isolates

TABLE 2 Recommended concentrations of antimycobacterial agents to test against M. *tuberculosis* by radiometric or proportion methods

Antimicrobial agent	Avg MIC (μg/ml) of susceptible M. tuberculosis	Concn (μg/ml) in serum[a]	Medium and concn (μg/ml)				
			BACTEC 12B low	BACTEC 12B high	7H10 low	7H10 high	7H11
Primary agents							
INH	0.05–0.2	7	0.1	0.4	0.2	1	0.2
RMP	0.5	10	2		1		1
PZA[b]	20	45	100		25	50	
EMB	1–5	2–5	2.5[c]	7.5	5	10	7.5
SM	8	25–50	2[c]	6	2	10	2
Secondary agents							
Capreomycin	1–50	30	5		10		10
Kanamycin	5	14–29	5		5		6
Cycloserine	5–20	20–40	50		30		30
Ethionamide	0.6–2.5	2–20	5		5		10
PAS	1	7.5	4		2		8
Alternative agents[d]							
Rifabutin[e]	0.06–8	0.2–0.5	1[f]				
Amikacin	1	16–38	1[f]				
Ciprofloxacin	0.25–3	2–4	2[f]				
Ofloxacin	0.5–2.5	3–11	2[f]				

[a]Concentrations 1 to 4 h after usual dosage.
[b]Tested at pH 6 in BACTEC 12B medium and at pH 5.5 in 7H10 medium.
[c]Woodley (127) showed that 2.5 μg of EMB and 2.0 μg of streptomycin per ml tested against susceptible and resistant isolates of M. *tuberculosis* agreed with the 7H10 proportion method 97 to 99 and 100%, respectively.
[d]Critical concentrations have not been defined for these agents. Each should be tested over a range of three to five concentrations.
[e]RMP-susceptible versus RMP-resistant isolates. About 30% of RMP-resistant isolates will be rifabutin susceptible.
[f]Tentative critical concentration.

from all patients and reporting of test results to the local health department (16). Also, the recommendations suggest that patients receive four-drug therapy until microbiology results are known. Despite these recommendations and as recent surveys demonstrated, the most rapid methods are not available in all CAP extent 3 or 4 laboratories (61, 101, 129). Furthermore, the clinician must realize the importance of rapidly submitting specimens to the laboratory, identifying the need for susceptibility testing, and being knowledgeable about the test methods and reporting criteria used by the laboratory to which specimens are submitted. In addition, the clinician must recognize that the responsibilities for interpreting susceptibility results and making the medical decision remain with them (5).

Model Fast-Track Program

Motivated by the equipment and training costs necessary to upgrade mycobacteriology laboratories to CAP extent 3 or 4 service, the New York State Department of Health introduced a fast-track program for tuberculosis testing. The goal is to provide the fastest turnaround times for susceptibility testing results for highly contagious smear-positive patients. This program relies on the use of rapid technologies (radiometric detection, species-specific nucleic acid hybridization identification, and radiometric susceptibility testing), rapid specimen transport by overnight courier to a central laboratory (Wadsworth Center for Laboratories and Research), same-day reporting by fax, and 7-days-per-week service. Preliminary results show that susceptibility test results are available to the clinician in an average of 2.5

weeks after sputum collection (101). While such a program may not be feasible for many laboratories, the results demonstrate that mycobacteriology services can be provided in an expeditious manner that results in timely information that has a meaningful impact on clinical care and public health. The centralization of such services will assist in controlling the costs.

Methods Available for Testing M. tuberculosis

The methods generally accepted for determining the antimicrobial susceptibilities of mycobacteria are based on growth of the microorganisms on or in a solid or liquid medium containing a specified concentration of a single drug. Four methods have been described. These include the proportion method, the radiometric or BACTEC method, the absolute concentration method, and the resistance ratio method. Methods that are commonly used to test rapidly growing aerobic and facultative anaerobic bacteria are, for a variety of reasons, unsuitable for testing mycobacteria. For example, the conventional disk diffusion method is not suitable for testing slowly growing mycobacteria because the drug diffuses throughout the medium before the growth of the mycobacteria is significantly affected. The BACTEC and agar proportion methods are most commonly used in the United States, and a standard procedure was proposed by the National Committee for Clinical Laboratory Standards (NCCLS) (88). The technical details of these procedures have been described by others (43, 62, 70, 106), and only the proportion and BACTEC methods are discussed

TABLE 3 Guidelines for selecting dilution of specimen concentrate prior to inoculation of 7H10 medium for susceptibility testing by direct method

Dilutions to test[a]	No. of AFB observed	
	Carbol fuchsin stain	Fluorochrome stain
Undiluted, 10^{-2}	0	<25
10^{-1}, 10^{-2}	1–10	25–250
10^{-2}, 10^{-3}	>10	>250

[a]Prepare dilutions of concentrated specimen based on the number of bacilli observed in the initial acid-fast smear. Use sterile distilled water to prepare the dilutions, and examine the carbol-fuchsin stain with the oil immersion objective (1,000×) and the fluorochrome stain with the high-dry objective (450×). If the patient is on therapy, it may not be that all bacilli observed in the smear will be viable; therefore, test the undiluted specimen as well as the appropriate dilution based on the microscopic criteria given in this table.

further in this chapter. The absolute concentration and resistance ratio methods are described elsewhere (62).

Source of Inoculum for Susceptibility Testing

The source of the inoculum for a susceptibility test may be either a smear-positive specimen (direct method) or growth from a primary culture or subculture (indirect method). The direct method is used when antimicrobial resistance is known or suspected. The indirect method is considered the standard method for inoculum preparation, and results of the direct method are usually confirmed by subsequent testing by the indirect method. With both the direct and the indirect methods, the inoculum must be a pure culture, and careful attention must be given to avoid over- or underinoculation. For the direct method, the inoculum is either a digested, decontaminated clinical specimen or an untreated normally sterile body fluid in which AFB are seen in stained smears. To ensure adequate but not excessive growth in the direct susceptibility test on solid medium, specimens are diluted according to the number of organisms observed in the stained smear of the clinical specimen (a comparative dilution scheme is shown in Table 3). Theoretically, this type of inoculum is more representative of the population of the tubercle bacilli in a particular lesion in the host. It is prudent to include an undiluted inoculum if the smear-positive specimen is from a patient who is receiving antimicrobial therapy, since a significant proportion of the bacilli seen on the smear may be nonviable.

Use of the direct method may be warranted in situations where there is a high prevalence of drug resistance (35), especially when second-line drugs are included in the initial test panel. However, cost can become a critical factor when there is a high incidence of smear-positive specimens containing nontuberculous mycobacteria. The direct method is not recommended for routine use with the BACTEC procedure, since this application has been the subject of only limited evaluation (76). In addition, the direct method may take 3 to 5 days longer than the indirect method, thus impairing the rapidity of reporting a result.

For the indirect method, the source of the inoculum is a subculture, usually from the primary isolation medium. Careful attention should be given to the selection of colony types so that the final inoculum is representative of all types present in order to ensure a balance of potentially resistant and susceptible bacilli. The source of inoculum for the BACTEC method can be growth on Middlebrook 7H10 or

7H11 agar or an egg-based medium, but the sample should be taken less than 4 weeks after growth is first detected. Turbid growth in a liquid medium such as Middlebrook 7H9 broth or sufficient growth in BACTEC 12B medium is also acceptable; however, one should remember that a mixed culture can result in apparent resistance, especially when the inoculum is derived from a broth culture.

Media Used for Susceptibility Testing

Although a variety of media have been used for drug susceptibility testing of slowly growing mycobacteria, including Middlebrook 7H10 or 7H11 agar and Löwenstein-Jensen egg-based media, there is considerable variability in the results. Egg-based media are unsuitable for susceptibility testing not only because of uncertainty about the potency of drugs after inspissation but also because some drugs are affected by phospholipids, proteins, and certain amino acids present in the medium. In an effort to provide uniformity in the testing of mycobacteria, the CDC (70) and the NCCLS (88) recommend that Middlebrook 7H10 agar supplemented with oleic acid-albumin-dextrose-catalase (OADC) be used as the standard medium for susceptibility testing of slowly growing mycobacteria by the proportion method. A majority of M. tuberculosis clinical isolates grow on this medium, and when a dissecting microscope is used, the transparency of 7H10 agar facilitates the recognition of mixed mycobacterial species or the presence of contaminants. Occasionally, the growth of drug-resistant strains of M. tuberculosis on 7H10 medium is not sufficient (fewer than 50 colonies) for the test to be valid. With these isolates, 7H11 medium may be substituted for 7H10 agar, but it is necessary to use higher concentrations of some drugs, as shown in Table 2 (44, 62, 79). At present, there is no standard formulation for Middlebrook 7H10 agar or OADC (such as there is for Mueller-Hinton medium), and a recent study demonstrated that quality control of the medium, especially the OADC supplement, is critically important (42). The medium used for studies with mycobacteria in the BACTEC instrument is an enriched Middlebrook 7H9 broth containing 4 μCi of ^{14}C-labeled palmitic acid per vial that is referred to as 7H12 or BACTEC 12B medium. Mycobacteria as well as other microorganisms metabolize the fatty acid and produce $^{14}CO_2$; therefore, growth or the inhibition of growth in BACTEC 12B medium is measured radiometrically as changes in the growth-dependent production of carbon dioxide.

Drugs Used for Susceptibility Testing

Antimicrobial agents for susceptibility testing (reference powders) can be obtained directly from the manufacturer or from commercial sources. The reference powder should be accompanied by information on potency (micrograms per milligram), expiration date, and lot number as well as on stability and solubility of the agent. Do not use preparations formulated for therapeutic use in humans or animals. Store unopened vials of powders according to the directions of the manufacturer, and store opened containers in a desiccator at the recommended temperature. Stock solutions of most agents at concentrations of ≥1,000 μg/ml remain stable for at least 6 months at −20°C and for 1 year at −70 to −80°C. Note that cycloserine solutions at neutral or acid pH are unstable at room temperature and must be used immediately; however, if they are alkalinized with Na_2CO_3 to a pH of 10, cycloserine stock solutions of 10,000 μg/ml can be stored for 1 week at 5°C and for 1 month at −20°C without loss of activity (62). Directions provided by the drug manufacturer should be consulted in addition to these general recommendations. Paper disks impregnated with standardized amounts of the primary and secondary drugs are available from commercial sources for use with the disk elution modification of the proportion method. Use of these disks obviates errors in weighing and dilution as well as errors in labeling, since the disks are coded with drug names and concentrations. This technique provides results equivalent to those obtained with solutions prepared from reference powders.

BACTEC Method

After more than 10 years of clinical use and several adjustments, radiometric susceptibility testing has become a reliable and rapid susceptibility test method. The BACTEC 12B broth vial contains ^{14}C-labeled palmitic acid as a growth substrate. Mycobacteria catabolize the substrate as a primary carbon and energy source and release $^{14}CO_2$ as a metabolic end product into the headspace above the liquid medium. The BACTEC 460 instrument quantitatively detects the amount of $^{14}CO_2$ released, expressed in terms of GI, and then automatically replaces the headspace with 5 to 10% unlabeled CO_2 in air, thereby maintaining the recommended CO_2 atmosphere. The rate and amount of $^{14}CO_2$ produced are directly proportional to the rate and amount of growth.

The BACTEC system can be used to test all primary drugs (RMP, INH, PZA, EMB, and SM) as well as secondary drugs, including the newer quinolones and rifamycins. The rapid availability of results when the BACTEC procedure is used may take precedence over cost considerations. If resistance to a primary drug is suspected, secondary drugs can be tested by the BACTEC method; however, in most situations, secondary drugs should be tested by the agar proportion method by a laboratory with experience testing these drugs. In this regard, it is prudent to consider the limitations of the BACTEC method, which in some clinical situations might be best considered a screening test, since the BACTEC method does not allow an estimate of the percentage of resistant bacilli and is vulnerable to major errors (false susceptibility or resistance) caused by mixed populations of mycobacterial species. Indeed, when an MDR isolate of M. tuberculosis is detected for the first time by the BACTEC method, the identity of the isolate should be confirmed and the presence of a contaminant or mixed culture should be ruled out before one proceeds with the testing of secondary agents. While the importance of promptly reporting an MDR M. tuberculosis isolate cannot be overstated, the consequences of a false report of multiple resistance must be recognized as well.

Four of the primary drugs are available as lyophilized powders in the BACTEC SIRE (SM, INH, RMP, and EMB) kit, designed specifically for use with 12B medium and the BACTEC 460 instrument. Alternatively, stock solutions of these agents can be prepared by using reference powders. Lyophilized vials of PZA are also available for the pH 6.0 modified BACTEC PZA test medium. Secondary drugs, including capreomycin, cycloserine, ethionamide, and kanamycin (39), as well as newer agents such as ciprofloxacin, ofloxacin, and rifabutin (critical concentrations have not been established for these newer agents) must be prepared from reference powders. Although not all drugs have been tested, the results of the radiometric and proportion methods correlate well except for cycloserine. Procedures for testing secondary and newer drugs are currently

being evaluated in a multicenter study, and recommendations should be available in the near future.

Inoculum

The source of inoculum for susceptibility testing should be fresh growth of actively growing bacilli. If growth from a primary BACTEC 12B vial is used for the inoculum, the daily GI reading should be at least 500 or, even better, between 800 and 900. The higher the GI, the shorter the period until the control vial reaches >30 and the earlier the test results can be interpreted. If the primary vial is past the GI peak or if the vial was kept at room temperature for more than 2 days, a fresh 12B vial should be inoculated. Other types of broth cultures (Dubos, Middlebrook 7H9, or Septi-Chek AFB) can be used as a source of inoculum; however, one must be sure that the broth culture is uncontaminated or not mixed. After thorough mixing to break up clumps, the turbidity should be adjusted to match a McFarland 0.5 standard rather than a McFarland 1 standard; use a 1:100 dilution of the final inoculum as a control. Use BACTEC diluting fluid or distilled water but not a broth or other substrate-containing medium to prepare the dilutions. Use an undiluted control when testing slowly growing resistant isolates of M. tuberculosis, since an undiluted control can also serve as a source of inoculum if a test must be repeated.

Fresh growth on any solid medium may also serve as an inoculum source. It is important to harvest several (5 to 10) colonies in order to ensure a representative sample of growth (do not pick from only 2 or 3 single colonies), and it is very important to thoroughly homogenize the inoculum in a tube containing 3 to 5 ml of diluting fluid and glass beads. After vortexing for 1 to 2 min, let the tube stand undisturbed for 30 min to allow larger particles to settle, and then transfer the upper one-half of the supernatant into a new tube. Adjust the turbidity to match a McFarland 0.5 standard, and use this suspension to inoculate the drug-containing vials. Use a 1:100 dilution of the final inoculum as a control as described above. Inoculate blood agar and 7H10 plates with a few drops of the inoculum suspension to check for purity.

As safety precautions, use disposable tuberculin syringes with permanently attached needles; wear rubber gloves, a respirator, and a gown; and perform all steps involving preparation of inocula and inoculation of vials in a class II or III biological safety cabinet.

Incubation

Prior to inoculation, BACTEC 12B vials must be prerun in the BACTEC 460 instrument in order to establish a gas phase of 5% CO_2 in air in the headspace of the vial; reject any vial with an initial GI of ≥20. Inoculate each prerun vial with 0.1 ml of the adjusted suspension of the test or quality control isolate, and incubate the vials at 37 ± 1°C (the temperature of incubation is critical) in the dark. Sunlight on BACTEC 12B medium may cause formaldehyde to form, and drugs such as RMP are light sensitive.

Reading, Interpreting, and Reporting

Vials are read on the BACTEC instrument at 24 ± 1-h intervals for a minimum of 4 days and until the control vial reaches a GI of ≥30. For laboratories with limited weekend coverage, an alternative schedule is to inoculate vials on Friday and read them first on Monday (45). The Monday reading is disregarded (the vials must be read in order to replace the gas in the headspace), and the vials are read daily for at least 3 days (5 days total). Batch testing is discouraged unless the susceptibility test can be started during the same week that the isolate is identified to species, i.e., identified as M. *tuberculosis* complex. In those situations, to avoid any delay, the isolate should be sent to a reference laboratory that works 7 days a week.

When the GI of the control is ≥30 after a minimum of 4 days, the difference in GIs from one day to the next, designated ΔGI, should be interpreted as follows:

$$\Delta GI \text{ of control} > \Delta GI \text{ of drug} = \text{susceptible}$$

$$\Delta GI \text{ of control} < \Delta GI \text{ of drug} = \text{resistant}$$

If the GI is ≥500 and then is >500 on the next reading, the isolate should be considered resistant to that drug regardless of the ΔGI (check that the GI of the control was not >30 in less than 3 days to rule out overinoculation). Continued incubation for a few days may allow one to resolve borderline results, but do not extend the incubation beyond 3 days after the control vial is positive. Never report susceptibility testing results without a preliminary or final identification of the isolate as M. *tuberculosis* complex or M. *tuberculosis*, especially when the isolate is resistant to a drug(s). Usually, the BACTEC results are straightforward; however, if the ΔGIs are close (±10%), issue a preliminary report as pending, and repeat the test. In most situations, resistant strains (especially RMP resistant) should be immediately tested against secondary drugs. Use of the term "borderline," as recommended by the BACTEC manufacturer, is not encouraged, because the meaning of this term is unclear, and the term may mislead the clinician to believe that an active drug has no use in the treatment of tuberculosis.

Susceptibility test results should be reported without delay. The report should include the name of the drug, the concentration tested, and the result (susceptible or resistant). It is not possible to report the percentage of resistance when the BACTEC method is used. Drug resistance should be reported by telephone and/or fax to the requesting physician, infection control program, and local tuberculosis control program, and a follow-up hard-copy report should be submitted. It is prudent to confirm the receipt of a fax report in lieu of direct communication with the physician or public health official.

Quality Control

Reference strains of known susceptibility patterns should be tested with each new batch of drug and BACTEC 12B medium. In addition, quality control tests should be performed (i) at least once per week in laboratories that perform tests daily or weekly or (ii) whenever a patient isolate is tested if tests are less frequently performed. The H37Rv strain of M. *tuberculosis* ATCC 27294, which is susceptible to all standard antituberculosis agents at recommended test concentrations, is used in most laboratories. A second strain that is resistant to two or more of the commonly used drugs (e.g., INH and RMP) should also be used to test new batches of drugs or media. Mutants of H37Rv selected for in vitro resistance to single or multiple drugs are also available; however, these strains are resistant to very high concentrations and are not particularly suitable for quality control purposes. In addition, reporting the test results for a patient isolate should not be delayed because of pending test results for a resistant quality control strain. Suspensions of quality control strains may be prepared in BACTEC diluting solution and frozen at −70°C in 1-ml test samples for up to 6 months (106). Thaw a single tube

for quality control testing, and use the same procedure as that used with clinical isolates. At present, there is no published consensus on what quality control results should be available before the test results on a patient isolate are released. However, it seems reasonable that the quality control test results for medium components be available and acceptable as well as the results for the susceptible quality control strain (e.g., ATCC 27294). The procedure manual should clearly state the corrective action to be taken when a quality control test fails, including how the results on patient isolates should be handled.

Testing PZA

Testing PZA is different from testing the other primary drugs because PZA activity must be measured at pH 5.5 rather than pH 6.8, the usual pH of the growth medium. However, most strains of M. *tuberculosis* grow poorly at pH 5.5, and some fail to grow altogether (12). As a compromise between testing at the pH for optimal PZA activity and testing at that for optimal growth, Salfinger et al. (103) recommended that PZA be tested at pH 5.9 ± 0.1. In addition, in order to accurately and reliably test PZA, several other adjustments are made to the standard radiometric method. (i) To prepare the inoculum, each isolate must be subcultured in a fresh BACTEC 12B vial supplemented with BACTEC reconstituting fluid, which is an aqueous solution of polyoxyethylene stearate (POES) (108). Test the isolate daily, and once the GI is 300 to 500, use this culture for susceptibility testing. Alternatively, use a 1:10 dilution of the inoculum (GI = 999 or McFarland 0.5) used for testing the other primary drugs (100), and vigorously homogenize the suspension before use. (ii) Reconstitute the lyophilized PZA with POES or dissolve the PZA reference powder in POES to achieve a final concentration of 100 µg/ml, the critical concentration for the BACTEC method that is equivalent to the 25- to 50-µg/ml concentration used in the proportion method (106). (iii) A special BACTEC 12B medium at pH 6.0 supplemented with POES is used to test PZA. (iv) Inoculate the PZA-containing vial and the control vial with the same suspension; i.e., do not dilute the inoculum for the control vial. (v) Test daily until the control vial reaches a GI of ≥200. If the GI fails to reach ≥200 within 14 days, the test is uninterpretable. (vi) The isolate is considered susceptible to PZA if the GI of the PZA vial is <10% of the GI of the control vial. If the GIs of the test and control vials are very close, the result is considered borderline, and the test must be repeated. (vii) In addition to M. *tuberculosis* H37Rv (PZA susceptible), M. *bovis* BCG (PZA resistant) can be used as a control for PZA resistance. If an isolate tests resistant to PZA, especially if the isolate is resistant to PZA alone, the identity of the isolate should be confirmed, since M. *bovis* and M. *bovis* BCG are PZA resistant, whereas the majority of M. *tuberculosis* isolates are PZA susceptible (104). This is especially important if the laboratory identifies isolates only to the level of the M. *tuberculosis* complex. Some M. *tuberculosis* isolates grow poorly at pH 6.0 (102), and other isolates may be inhibited by the POES supplement (106).

Proportion Method

The agar proportion method for susceptibility testing of slowly growing mycobacteria was developed in the early 1960s and is the method most commonly used in mycobacteriology laboratories in the United States (61). The proportion method, as presently used in the United States, has undergone various modifications (44, 70, 122). A proposed standard method was published by the NCCLS, and this procedure will be updated in an expanded form in 1994 or 1995 (71, 88). The modified proportion method is also described in detail by Hacek (43) and Inderlied (62). The preferred medium for the proportion test is Middlebrook 7H10 agar, and drugs used in the proportion method can be prepared from reference powders or added as drug-impregnated disks as described above. PZA can be tested by the proportion method with a low-pH 7H10 agar; however, 25% of M. *tuberculosis* isolates may fail to grow at pH 5.5 with oleic acid in the medium, and ADC rather than OADC should be used to supplement the medium (12).

Inoculum and Incubation

The inoculum can be prepared by the direct or indirect method. When the direct method is used, examine cultures weekly for 3 weeks, and even though mature colonies may appear on control media in less than 3 weeks, do not submit a report of susceptible until week 3. Resistant strains of M. *tuberculosis* complex may be slow to produce visible growth because of metabolic differences from susceptible wild-type isolates. If cultures are incubated beyond 3 weeks, degradation of the antimicrobial compound may permit the appearance of colonies of organisms that were susceptible to the initial level of the drug. Since the controls are inoculated with a processed specimen, also examine isolated colonies for colonial morphology and pigmentation differences and for mixed species of mycobacteria or the presence of a contaminant. Small colonies of rapid growers as well as the rough, dry colonies of some MAC strains are similar in appearance to M. *tuberculosis* colonies on 7H10 agar. Also, bear in mind that rapidly growing mycobacteria may be slow to develop on primary isolation media and that rapidly growing mycobacteria will appear to be MDR when tested against primary antituberculosis agents. For a test to be valid, the control must show good growth (at least 50 to 150 colonies). Never report susceptibility testing results without a preliminary identification.

When using the indirect method, pick a sufficient number of colonies to make a suspension that is equivalent to a McFarland 1 standard, or if there is insufficient growth, pick sufficient colonies into Dubos Tween-albumin broth, and incubate them at 35 to 37°C until the turbidity matches a McFarland 1 standard. The suspension should not contain clumps of organisms. The actual number of CFU per milliliter is likely to vary from suspension to suspension, requiring that two dilutions (usually 10^{-2} and 10^{-4}) of the suspension be used to inoculate two sets of media. If the source of the inoculum is old or if there is scant growth, use 10^{-1} and 10^{-3} dilutions, or subculture the isolate in broth. Quality control strains should be tested at dilutions of 10^{-2} and 10^{-3}. Quadrant plates are commonly used for the agar proportion method, and 0.1 ml (about 3 drops from a Pasteur pipette) of each dilution of the inoculum is placed into each quadrant, using one dilution per set of plates. Allow the plates to thoroughly dry after inoculation, and place the plates in individual CO_2-permeable polyethylene bags (clear sandwich bags). Incubate at 35 to 37°C in 5 to 10% CO_2 in air, and protect the plates from light during storage and incubation to prevent the formation of formaldehyde from the medium ingredients. To check strains for purity, inoculate blood agar and 7H11 agar plates with 1 or 2 drops of the suspension, and streak for isolation.

Reading, Interpreting, and Reporting

Examine the purity control plates for contamination after 1 to 2 days, and continue to check the 7H10 plate throughout the incubation period and the blood agar plate for 1 week. Once the no-drug control quadrant of the test isolate or control strain shows at least 50 to 150 colonies, the plates can be read and interpreted. If there are fewer than 50 colonies in the no-drug quadrant by 3 weeks, repeat the test. Use of the designations 1+ to 4+ to grade the amount of growth is discouraged. Obvious resistance should be immediately reported. Note the presence of microcolonies in the drug-containing quadrants, and report the number of colonies observed in each quadrant as a percentage of the number of colonies in the no-drug control quadrant. If the percentage of colonies is ≥1%, the isolate should be reported as resistant to that drug at the concentration tested. The percentage of resistant colonies can be used as an indicator of therapeutic efficacy and should be reported to the physician.

Quality Control

M. *tuberculosis* H37Rv (ATCC 27294) is susceptible to all primary and secondary antituberculosis drugs and can be used for quality control. Strains of M. *tuberculosis* that are resistant to INH, RMP, and other drugs are available from the American Type Culture Collection; however, these strains are resistant to high concentrations of the respective drugs and are not ideal for quality control testing (62). Aliquots of suspensions of quality control strains of M. *tuberculosis* adjusted to match a McFarland 1 standard can be stored at −70°C for up to 6 months. Quality control testing should be performed with each new batch of medium or antimicrobial agent, and media should be checked for sterility and shown to support adequate growth. Guthertz et al. (42) demonstrated the importance of controlling for all components of 7H10 agar, including lots of Middlebrook 7H10 agar, glycerol, and especially OADC. The issue of which quality control results should be available before patient test results are released is discussed above.

SUSCEPTIBILITY TESTING OF MAC

Clinical Significance

In many clinical laboratories, and perhaps in all laboratories in large urban areas, the MAC is the most common mycobacterium isolated from clinical specimens. Clearly, the incidence of MAC isolates reflects the fact that disseminated MAC disease is the most common bacterial opportunistic infection afflicting HIV-infected patients, including both adults and children. However, the incidence of MAC lung disease has been steadily increasing in other patient populations, notably patients with a history of chronic pulmonary disease, nonsmoking women, and young adults with cystic fibrosis. The isolation of MAC from blood or other sterile sites is almost always clinically significant but is especially so in patients with profound CD4 T-cell lymphocytopenia, i.e., ≤50 CD4 T lymphocytes per μl. The detection of MAC in respiratory or gastrointestinal tract specimens can be difficult to interpret, especially in patients who are not clearly at risk for MAC disease. Clinical, histopathologic, and/or roentgenographic evidence of MAC disease, repeated isolation of MAC, increasing numbers of bacilli in sequential specimens, and the absence of other identifiable causes of signs and symptoms are important factors to consider in assessing the clinical significance of MAC isolates.

Drug Resistance

Most MAC isolates are intrinsically resistant to the conventional antimycobacterial agents, and many clinicians with extensive experience in treating MAC disease believe that INH and PZA have no role in the treatment of MAC disease (78). MAC isolates are variably susceptible to aminoglycosides (amikacin, kanamycin, and SM) and rifamycins (RMP and rifabutin), and neither of these classes of agents is clearly effective alone or in combination with other agents in the treatment of MAC disease. Rifabutin was approved by the U.S. Food and Drug Administration as a prophylactic agent for disseminated MAC disease; however, the mechanism of action of rifabutin as a prophylactic agent is unknown. The intrinsic antimicrobial resistance is most likely due to the impermeability of the MAC cell wall and membrane to these compounds (93), and cell-free in vitro studies have shown that certain drug targets (e.g., ribosomes, ribosomal subunits, RNA polymerase) in MAC cells bind the corresponding drugs (e.g., macrolides, aminoglycosides, and rifamycins) and that target functions are inhibited. While MAC organisms produce β-lactamase (85), there is otherwise no evidence that they actively degrade or inactivate antimicrobial agents. Although most MAC isolates carry plasmids of various sizes (20, 21, 68, 86) and these plasmids have been associated with virulence (36) and antimicrobial resistance (34, 84), resistance transfer factors have not been identified in MAC organisms, and there is no evidence for the transfer of antimicrobial resistance between strains in nature.

The role of the cell wall as a primary factor in antimicrobial resistance is also supported by the observation that surfactants such as Tween 80 potentiate the activities of antimicrobial agents and that combinations of certain agents (e.g., fluoroquinolones) and EMB are synergistic, probably because EMB partially disrupts the integrity of the MAC cell wall and membrane. The MAC displays at least three colony type variants: a smooth, opaque, domed type; a smooth, transparent, flat type; and a rough type. Colony variant types have been associated with certain phenotypic properties, and the smooth, transparent, flat type is generally more virulent and resistant to antimicrobial agents than the opaque type. Conversion between colony types appears to be a phenotypic rather than a genotypic phenomenon and occurs at a high rate. The transparent-to-opaque transition occurs at a rate of approximately 5×10^{-4}, while the reverse transition occurs at a rate of approximately 10^{-6} per bacterium per generation. The clinical significance of this phenomenon is unknown, and both the transparent and opaque colony types are observed in primary cultures of various clinical specimens. The transparent colony type may not be evident unless the specimen is diluted first or streaked for isolation. The occurrence of colony type variants is problematic for susceptibility testing, since the difference in susceptibility to agents that are now commonly used to treat MAC infections can be significant.

When To Perform Susceptibility Tests

At present, in the vast majority of clinical situations, it is inappropriate and unnecessary to perform in vitro susceptibility tests on initial MAC isolates, whether they involve disseminated or localized disease. Susceptibility testing may be useful if a patient on therapy relapses, especially when the therapy includes a macrolide, or if the infection is intractable and the clinical situation is acute or desperate. However, the clinician and laboratorian must be aware that the interpretation of in vitro results lacks verified criteria. Indeed, at present, the detection of macrolide resistance may be the only test result that is of value in the management of patients with MAC disease.

Methods

There are no standardized methods for in vitro susceptibility testing of MAC isolates, and the methods and interpretive criteria described for testing *M. tuberculosis* isolates should not be used to guide the treatment of patients with MAC disease (78). The critical concentrations of antimycobacterial agents such as INH, RMP, PZA, and EMB are based on the lowest concentrations of these drugs necessary to distinguish wild-type strains of *M. tuberculosis* from treated or resistant strains on certain types of media. Furthermore, the critical concentrations relate to clinical efficacy in the treatment of tuberculosis and the prevention of drug resistance when a threshold of 99% of the test population is used. A number of factors make it inappropriate to apply these interpretive criteria to MAC. First, the susceptibilities of MAC wild-type strains are considerably more variable than those of *M. tuberculosis* wild-type strains (48, 65, 116). Second, no comprehensive, controlled clinical trials have been performed with the intent of establishing interpretive criteria for MAC isolates based on microbiological and clinical efficacy. A small number of trials have provided some information on azithromycin (63, 133) and clarithromycin (17, 24, 53). Third, most MAC isolates are intrinsically resistant to the primary antituberculous drugs (37). Fourth, antimicrobial resistance to drugs that are initially active against MAC isolates is poorly understood but appears to involve both genotypic and phenotypic mechanisms (49, 53, 63).

Elements of a Standard Method

While there is no standard method, there is some consensus on certain aspects of in vitro susceptibility testing of MAC isolates (62, 107). First, a broth medium appears to be more reliable than an agar medium. Second, radiometric (BACTEC 12B medium, pH 6.8), broth macrodilution, and broth microdilution (7H9, 7HSF [130], or Mueller-Hinton medium supplemented with OADC) have yielded consistent and reproducible results in various studies (62, 107). Third, careful attention must be paid to the preparation of the inoculum, especially to avoiding selection of colony type variants during subculture. Only transparent colony types should be tested. Fourth, the inoculum should be between 10^4 and 10^5 for the radiometric broth macrodilution test and approximately 5×10^5 CFU/ml for the broth microdilution test. Fifth, test three to five concentrations of drugs (Table 1) in $\log_2$ increments, and measure the activity as a MIC. Sixth, the period of incubation should not extend beyond 7 days, and the no-drug control should not exceed a GI of 999 in fewer than 4 days in the radiometric test. Seventh, the endpoint for the radiometric test is defined by the GI for the inoculum diluted 1:100. Eighth, Tween 80 or other surfactants should not be used to disperse clumps of bacilli because of the synergistic effect between surfactants and antimicrobial agents. At present, the majority of literature on in vitro susceptibility testing of the MAC favors use of the BACTEC method.

Areas of Controversy

The areas of remaining controversy include preparation of the inoculum, range of drug concentrations to test, reading and interpretation of BACTEC results, and MIC interpretive criteria. Some laboratories advocate the use of "seed" BACTEC vials (subcultures of fresh growth) as sources of inoculum, while others prefer to prepare a suspension of mycobacteria directly from agar plates in a manner similar to the direct inoculum method advocated by the NCCLS for testing fastidious rapidly growing aerobic bacteria (89). Interpretation of the BACTEC GI readings follows the recommendations of the manufacturer (106) or are somewhat more restrictive (107). Alternatively, Inderlied (62) described a method of GI analysis based on the generation of dose-response curves. With all of these methods, the MIC is defined as the lowest concentration of drug that inhibits the growth of the microorganism to less than the growth of the 1:100 control. In the absence of well-established correlations with clinical efficacy and outcome, the choice of MIC interpretive criteria (resistant or susceptible) is problematic. Some workers (51, 107) have advocated the use of criteria based on maximum (peak) concentrations in serum and the highest MICs for wild-type strains of M. tuberculosis, while others suggest no criteria (106). It is important to recognize that disseminated MAC disease is principally an infection of the blood, macrophages, bone marrow, spleen, and other tissues and that clinical effectiveness is likely to relate to both potent activity and the ability of drugs to accumulate in tissues to levels above the MIC for the infecting microorganism.

Other Slowly Growing Mycobacteria

If there is a lack of consensus on the testing of the MAC, there is even less consensus on the testing of slowly growing mycobacteria such as M. gordonae (the isolation of M. gordonae has doubtful clinical significance), M. marinum, M. xenopi, M. ulcerans, M. genavense, and others. Woods and Washington reviewed the nontuberculosis mycobacterioses and concluded that in many cases, no clear recommendations could be made for treatment of these infections and that the need for susceptibility testing was difficult or impossible to assess (128). Certain mycobacterioses are quite rare (e.g., disease caused by M. simiae) or are initially misdiagnosed as tuberculosis, which confused the analysis of response to therapy. Some have argued that susceptibility test methods as applied to M. tuberculosis should not be applied to any of the other slowly growing mycobacteria (94). However, there is somewhat more experience with M. kansasii, and the results of susceptibility testing with the proportion method may provide useful information in guiding antimicrobial therapy. Isolates of M. kansasii from patients not previously treated with RMP are predictably susceptible to 1 μg of RMP per ml, and patients have been successfully treated with RMP, INH, and a third agent, usually EMB; however, INH dosages are increased because of relative resistance (28). The recommendation of the ATS is to treat M. kansasii disease with a combination of INH (300 mg), RMP (600 mg), and EMB (15 mg/kg). M. haemophilum has emerged as a potentially important pathogen in immunocompromised patients, and there is some correlation between susceptibility test results and clinical efficacy, although virtually all treatment regimens examined included combinations of agents (109). Wild-type isolates of M. haemophilum appear to be susceptible to quinolones, rifamycins, clarithromycin, and azithromycin

and resistant to PZA and EMB, and they are likely to be resistant to INH and SM (7, 109). The most prudent approach to the testing of slowly growing mycobacteria other than M. tuberculosis is to restrict such testing to reference laboratories that have extensive experience working with these species. The interpretation of the susceptibility test results will require good communication between the laboratory and clinician. Clinicians without experience in the treatment of these infections should consult with others who have such experience.

SUSCEPTIBILITY TESTING OF RAPIDLY GROWING MYCOBACTERIA

Clinical Significance

Rapidly growing mycobacteria are defined as AFB that form visible colonies from a dilute inoculum on a solid medium within 5 to 7 days. Although there are over 30 species of rapidly growing mycobacteria, disease in humans is primarily caused by only three species: M. fortuitum, M. chelonae (formerly M. chelonae subsp. chelonae), and M. abscessus (formerly M. chelonae subsp. abscessus) (11, 120). These three species are important causes of cutaneous, pulmonary, and nosocomial infections, especially following catheter insertions, augmentation mammaplasty, and cardiac bypass surgery. Disseminated disease is rare and is usually associated with immunodeficiency, including corticosteroid therapy, although not in HIV-infected patients (80). M. smegmatis, M. peregrinum, and M. chelonae-like organisms also have been implicated as rare causes of disease in humans (118, 119), and at least two species of rapidly growing mycobacteria are considered animal pathogens (37).

The importance of distinguishing rapidly growing mycobacteria from slowly growing mycobacteria isolated from clinical specimens must be emphasized, since conventional antimycobacterial agents (except aminoglycosides) are ineffective in the treatment of disease caused by rapid growers. Furthermore, this intrinsic resistance to conventional antimycobacterial agents may not be appreciated by clinicians with limited experience in treating this type of mycobacterial infection. Thus, it is incumbent on the laboratory to assist in assessing the clinical significance of these isolates and to provide information on appropriate antimicrobial therapy. Isolation of the aforementioned species of rapidly growing mycobacteria from wounds is almost always clinically significant, and susceptibility testing should be done with antimicrobial agents with proven therapeutic effectiveness. The isolation of rapidly growing mycobacteria from respiratory specimens is difficult to interpret, but these isolates are frequently not clinically significant. However, the repeated isolation of a rapid grower in pure culture or as the predominant species in respiratory specimens is consistent with true respiratory tract disease, especially in patients with a history of chronic respiratory disease. In the case of wound infections, debridement and excision of the infected tissue are frequently necessary adjuncts to antimicrobial therapy, although 20% of cases of cutaneous infection are likely to spontaneously resolve without surgical intervention or antimicrobial therapy (117).

Although these mycobacteria are considered rapid growers when they are subcultured to blood or chocolate agar, on primary isolation it may take much longer than 5 to 7 days to isolate them. When rapid growers, especially M. chelonae and M. abscessus, are suspected to be the cause of infection, specimens should be inoculated onto Löwen-

TABLE 4 Antimicrobial agents to test against rapidly growing mycobacteria by broth microdilution method[a]

Antimicrobial agent	Concn range (μg/ml)	MIC (μg/ml)			
		Susceptible[b]	Intermediate	Resistant	M. fortuitum ATCC 6841[c]
Amikacin[d]	0.5–64	≤16	32	≥64	≤0.25–1
Cefoxitin[e]	2–256	≤16	32	≥64	16–32
Ciprofloxacin[f]	0.06–8	≤1	2	≥4	≤0.06–0.25
Clarithromycin[g]	0.12–16	≤2	4	≥8	
Doxycycline[h]	0.25–32	≤1	2–8	≥16	≤0.25–4
Imipenem[i]	0.5–32	≤4	8	≥16	1–4
Sulfamethoxazole[j]	0.5–256	≤32		≤64	2–16
Sulfisoxazole[j]	0.5–256	≤32		≤64	
Tobramycin[k]	0.5–32	≤4	8	≥16	8–32

[a]Table and footnotes are adapted from Brown et al. (10) and Inderlied (62).

[b]Breakpoints follow NCCLS standards except as noted.

[c]Expected range for quality control purposes. Potencies of drug preparations should be tested with *Staphylococcus aureus* ATCC 29213, *E. coli* ATCC 25922, and *Pseudomonas aeruginosa* ATCC 27853 and compared with expected ranges as described in NCCLS document M7-A3 (89).

[d]Kanamycin may be more active than amikacin against M. *abscessus* and M. *chelonae*.

[e]Cefoxitin is 1 dilution higher than the conventional NCCLS-suggested breakpoint, and cefoxitin is the only cephalosporin that should be tested.

[f]Not active against most strains of M. *chelonae* and M. *abscessus*.

[g]Interpretive values are as declared by Abbott Laboratories, Abbott Park, Ill.

[h]Breakpoint is 2 dilutions lower than the conventional NCCLS-suggested breakpoint; in general, doxycycline and minocycline are four- to eightfold more active than tetracycline.

[i]Solutions are unstable at ambient temperatures and above.

[j]Sulfamethoxazole and sulfisoxazole breakpoints are closest values for the sulfamethoxazole component of trimethoprim-sulfamethoxazole; e.g., 64 μg of sulfamethoxazole alone per ml corresponds to 76 μg of sulfamethoxazole per ml in combination with trimethoprim.

[k]Likely to be active against only M. *chelonae* and M. *fortuitum*.

stein-Jensen and/or Middlebrook medium and incubated at both 30 and 37°C in order to promote growth of the organisms. Identification of rapidly growing mycobacteria to the level of species is important because of significant differences in the wild-type susceptibility patterns of the three species most commonly associated with disease in humans.

Methods

Four methods of measuring the in vitro susceptibilities of rapidly growing mycobacteria have been described: broth microdilution, agar disk elution, disk diffusion, and the E test. None of these methods has been verified and approved by the NCCLS for testing rapid growers. The disk diffusion method was described by Hawkins et al. (47) and Inderlied (62), but the method has important limitations, notably disk contents for certain drugs that are too low and a lack of verified interpretive criteria. The disk diffusion method is not recommended for testing rapidly growing mycobacteria and will not be described further in this chapter. Use of the E test for rapid growers was recently described elsewhere (73), and although the initial results are promising, the method will not be considered here because of the limited information presently available about the performance of this test.

Broth Microdilution Method

The broth microdilution method for testing rapidly growing mycobacteria is essentially a modification of the NCCLS standard method for non-AFB that grow aerobically (88) and was recently described in detail by Brown et al. (10). This method is most suitable for laboratories that test large numbers of isolates. The antimicrobial agents, concentration ranges, and interpretive criteria that should be considered for testing rapidly growing mycobacteria by this

method are shown in Table 4. Commercially prepared broth microdilution panels can be used if the appropriate drugs are available at the necessary concentrations. Alternatively, broth microdilution panels can be prepared by following the protocol described by Brown et al. (10), which recommends use of a dispensing device such as the Quick Spense II (Dynatech, Inc., Chantilly, Va.). Once prepared, the plates can be sealed in plastic bags and stored at −70°C for up to 6 months.

The inoculum can be either a subculture in broth or a preparation made directly by picking colonies from a plate. In either case, the inoculum should be prepared in Trypticase soy broth or cation-supplemented Mueller-Hinton broth with 0.02% Tween 80 and four or five sterile 0.5-mm-diameter glass beads. Care should be taken to avoid clumping of the mycobacteria. The final inoculum should be 1×10^5 to 5×10^5 CFU/ml or 1×10^4 to 5×10^4 CFU/100 μl per well of a microtiter plate. The purity of the inoculum should be checked for each isolate, and the size of the inoculum should occasionally be verified by quantitative plate culture, especially if there is a problem with clumping.

Incubate the plates for 3 to 5 days at 30°C, but do not incubate them beyond 5 days because of drug instability. If there is not sufficient growth to allow the results to be interpreted after 5 days of incubation, repeat the test. The MIC is defined as the lowest concentration of antimicrobial agent that completely inhibits visible growth. As with other types of bacteria, "trailing" is common when the sulfonamides are tested, and the MICs of these agents should be read at approximately 80% inhibition of growth.

Agar Disk Elution Method

The agar disk elution method can be viewed as an adaptation of the disk elution modification of the proportion

TABLE 5 Antimicrobial agents to test against rapidly growing mycobacteria by agar disk elution method[a]

Antimicrobial agent	Resistance breakpoint (μg/ml)	Disk content (μg)	No. of disks/well	Final concn (μg/ml)	M. fortuitum ATCC 6841[b]
Amikacin	32	30	1	6	S
Amikacin	32	30	5	30	S/R
Cefoxitin	32	30	5	30	S
Ciprofloxacin	2	5	2	2	S
Doxycycline	8	30	1	6	S
Imipenem	8	10	4	8	S
Tobramycin	8	10	4	8	S/R
Trimethoprim-sulfamethoxazole	32	25	6	30	S
RMP[c]	1	5	1	1	R

[a]Table and footnotes are adapted from Brown et al. (9) and Inderlied (62).

[b]S, susceptible, no growth; R, resistant, growth equivalent to no drug control. Expected results for M. fortuitum ATCC 6841 are for quality control purposes. Brown et al. (9) suggest using Pseudomonas aeruginosa ATCC 27853 for additional quality control of the method.

[c]RMP is suggested to provide a quality control test for the detection of resistance and for the testing of M. marinum, which may require incubation for up to 14 days.

method for testing M. tuberculosis (122). The disk elution method is particularly suitable for occasional or infrequent testing of small numbers of isolates. The method was recently described in detail by Brown et al. (9). Commercially prepared disks containing antimicrobial agents are placed into the wells of a standard 35-mm-wide, six-well tissue culture plate and eluted with 0.5 ml of either 10% OADC or Trypticase soy broth (Table 5). One well is reserved for a growth control, and no drug is added. OADC should be used for testing M. chelonae or M. abscessus, while Trypticase soy broth is sufficient for testing M. fortuitum. Molten Mueller-Hinton agar is added to each well of the plate, and the agar is allowed to solidify. Standard quadrant plates can be substituted for the tissue culture plates; however, Brown et al. (9) indicate that Middlebrook 7H10 medium should not be substituted for Mueller-Hinton agar because of discrepancies in MIC results, notably with aminoglycosides. Once prepared, the plates can be stored for 7 days at 4°C except when imipenem is being tested (plates containing imipenem must be inoculated within 24 h). Prepare the inoculum as described above for the broth microdilution procedure, and use a suspension of mycobacteria adjusted to match the turbidity of a McFarland 0.5 standard to prepare a 1:100 dilution. Use sterile deionized water to dilute the suspension, and inoculate 10 μl onto the surface of each well of the plate. The final inoculum is approximately 1.5 × 10^4 CFU per well. The plates should be incubated in air at 30°C for 3 to 5 days. A purity plate should be prepared for each isolate tested and for the quality control strain (M. fortuitum ATCC 6841). The inoculum preparation can be verified by quantitative culture on blood agar or other suitable nonselective medium. The plates should be first examined at 3 days. If the growth control well shows uniformly dispersed, discrete colonies, the test results can be read. The endpoint is no growth at the concentration of drug tested except for sulfonamides, for which the endpoint is considered an 80% reduction in growth compared with that of the control.

One important limitation to this procedure is that the drugs are tested at only one or two (amikacin) concentrations that are at or near the NCCLS breakpoint for susceptible (the breakpoint for cefoxitin is 1 dilution higher, i.e., 32 versus 16 μg/ml); as a result, the tolerance for experimental fluctuations is extremely low. Nevertheless, the test

has proven reliable and useful for guiding the therapy of infections caused by rapidly growing mycobacteria (9). At present, there are no recommended concentrations or interpretive criteria for testing clarithromycin or azithromycin.

Alternative Susceptibility Testing Methods

Jacobs et al. described an innovative approach to the susceptibility testing of mycobacteria based on the use of a luciferase reporter mycobacteriophage (phasmid) (67). The premise of this approach is that viable mycobacteria support the infection and multiplication of mycobacteriophages, while mycobacteria that are inhibited and killed by antimicrobial agents do not. Mycobacteriophage TM4 was engineered to contain a firefly luciferase gene linked to a high-efficiency promoter of the heat shock protein gene (hsp60) from BCG. Luciferin, the substrate for the luciferase, readily crosses the mycobacterial cell wall and membrane, and the luciferase gene is efficiently transcribed and translated by the mycobacterial expression system. Mycobacteria that were susceptible to INH or RMP supported the infection and multiplication of the engineered phage and expressed luciferase. When luciferin was added, light production was proportional to the number of mycobacteria in the assay. When exposed to INH or RMP, the cells were killed, and no light was produced. The success of this assay as a routine method depends on the availability of mycobacteriophages in reagent quantities, the host range of the phage(s), and the efficiency of infection (partly related to the expression of receptors by the test isolates). Perhaps the most immediate application of this assay is in drug development. Cooksey et al. (19) recently showed that the firefly luciferase gene could be introduced into M. tuberculosis H37Ra by electroporation of a plasmid containing the luciferase gene linked to the hsp60 promoter. Furthermore, they showed that the in vitro susceptibility test results of a broth microtiter assay using the luciferase-H37Ra strain was comparable for several drugs to a conventional broth macrodilution assay. However, the bioluminescence assay required lysis of the mycobacteria, and the kinetics of antimicrobial activity significantly influenced the assay. Nevertheless, the potential of this method or a refinement of it as a useful tool in the effort to identify new and more potent antimycobacterial agents is clear.

Finally, other methods either have already been identified or are recognized as having potential for the improved testing of mycobacteria. Alamar Blue, a proprietary oxidation-reduction dye, has been successfully used in a microtiter assay for the testing of non-AFB, and preliminary results indicate that the method can be applied to the testing of M. *tuberculosis* and M. *avium* (131). The epsilometer or E test has also been successfully applied to the testing of mycobacteria (73, 121), and the manufacturer has expressed interest in developing strips that contain conventional antimycobacterial agents.

This chapter benefited from the comments and suggestions of Richard J. Wallace, Jr.

REFERENCES

1. **Alford, R. H.** 1990. Antimycobacterial agents, p. 350–360. *In* G. L. Mandell, R. G. Douglas, Jr., and J. E. Bennett (ed.), *Principles and Practices of Infectious Diseases.* Churchill Livingstone, Inc., New York.
2. **American Thoracic Society.** 1983. Levels of laboratory services for mycobacterial disease: official statement of the American Thoracic Society. *Am. Rev. Respir. Dis.* **128:**213.
3. **Amsden, G. W., and C. H. Ballow.** 1993. PMN and RBC uptake of azithromycin in healthy volunteers during a 5-day dosing regimen, abstr. 725, p. 252. *Program Abstr. 33rd Intersci. Conf. Antimicrob. Agents Chemother.*
4. **Andriole, V. T.** 1990. Quinolones, p. 334–345. *In* G. L. Mandell, R. G. Douglas, Jr., and J. E. Bennett (ed.), *Principles and Practices of Infectious Diseases.* Churchill Livingstone, Inc., New York.
5. **Bailey, W. C., J. B. Bass, J. E. Hawkins, G. P. Kubica, and R. J. Wallace.** 1984. Drug susceptibility testing for mycobacteria. *Am. Thoracic Soc. Newsl.* **10:**9–10.
6. **Banerjee, A., E. Dubnau, A. Quemard, V. Balasubramanian, K. S. Um, T. Wilson, D. Collins, G. de Lisle, and W. R. Jacobs, Jr.** 1994. inhA, a gene encoding a target for isoniazid and ethionamide in *Mycobacterium tuberculosis.* *Science* **263:**227–230.
7. **Bernard, E. M., F. F. Edwards, T. E. Kiehn, S. T. Brown, and D. Armstrong.** 1993. Activities of antimicrobial agents against clinical isolates of *Mycobacterium haemophilum.* *Antimicrob. Agents Chemother.* **37:**2323–2326.
8. **Bodmer, T., C. Bernasconi, G. Zuercher, B. Haeberli, F. Marchesi, P. Imboden, and A. Telenti.** 1993. Molecular basis of rifabutin susceptibility in rifampicin resistant *Mycobacterium tuberculosis*, abstr. 1583, p. 408. *Program Abstr. 33rd Intersci. Conf. Antimicrob. Agents Chemother.*
9. **Brown, B. A., J. M. Swenson, and R. J. Wallace, Jr.** 1992. Agar disk elution test for rapidly growing mycobacteria, p. 5.10.1–5.10.11. *In* H. D. Isenberg (ed.), *Clinical Microbiology Procedures Handbook*, vol. 1. American Society for Microbiology, Washington, D.C.
10. **Brown, B. A., J. M. Swenson, and R. J. Wallace, Jr.** 1992. Broth microdilution test for rapidly growing mycobacteria, p. 5.11.1–5.11.10. *In* H. D. Isenberg (ed.), *Clinical Microbiology Procedures Handbook*, vol. 1. American Society for Microbiology, Washington, D.C.
11. **Bruckner, D. A., and P. Colonna.** 1993. Nomenclature for aerobic and facultative bacteria. *Clin. Infect. Dis.* **16:**598–605.
12. **Butler, W. R., and J. O. Kilburn.** 1982. Improved method for testing susceptibility of *Mycobacterium tuberculosis* to pyrazinamide. *J. Clin. Microbiol.* **16:**1106–1109.
13. **Canetti, G.** 1965. Present aspects of bacterial resistance in tuberculosis. *Am. Rev. Respir. Dis.* **92:**687–702.
14. **Canetti, G., W. Fox, A. Khomenko, H. T. Mahler, N. K. Menon, D. A. Mitchison, N. Rist, and N. A. Smelev.** 1969. Advances in techniques of testing mycobacterial drug sensitivity, and the use of sensitivity tests in tuberculosis control programs. *Bull. W.H.O.* **41:**21–43.
15. **Canetti, G., S. Froman, J. Grosset, P. Hauduroy, M. Lagerova, H. T. Mahler, G. Meissner, D. A. Mitchison, and L. Sula.** 1963. Mycobacteria: laboratory methods for testing drug sensitivity and resistance. *Bull. W.H.O.* **29:**565–578.
16. **Centers for Disease Control.** 1993. Initial therapy for tuberculosis in the era of multidrug resistance. Recommendations of the advisory council for the elimination of tuberculosis. *Morbid. Mortal. Weekly Rep.* **42**(RR-7):1–8.
17. **Chaisson, R. E., C. A. Benson, M. Dube, R. Hafner, M. Dellerson, S. Lichter, T. Smith, and F. R. Sattler.** 1992. Clarithromycin for disseminated *Mycobacterium avium* complex in AIDS patients, abstr. WeB 1052, p. We54. *8th Int. Conf. AIDS/III STD World Congr.*
18. **Chiu, J., J. Nussbaum, S. Bozette, J. G. Tilles, L. S. Young, J. Leedom, P. N. R. Heseltine, and J. A. McCutchan.** 1990. Treatment of disseminated *Mycobacterium avium* complex infection in AIDS with amikacin, ethambutol, rifampin, and ciprofloxacin. *Ann. Intern. Med.* **113:**358–361.
19. **Cooksey, R. C., J. T. Crawford, W. R. Jacobs, Jr., and T. M. Shinnick.** 1993. A rapid method for screening antimicrobial agents for activities against a strain of *Mycobacterium tuberculosis* expressing firefly luciferase. *Antimicrob. Agents Chemother.* **37:**1348–1352.
20. **Crawford, J. T., and J. H. Bates.** 1986. Analysis of plasmids in *Mycobacterium avium-intracellulare* isolates from persons with acquired immunodeficiency syndrome. *Am. Rev. Respir. Dis.* **134:**659–661.
21. **Crawford, J. T., and J. O. Falkinham III.** 1990. Plasmids of the *Mycobacterium avium* complex, p. 97–119. *In* J. J. McFadden (ed.), *Molecular Biology of the Mycobacteria.* Surrey University Press, London.
22. **Crofton, J., and D. A. Mitchison.** 1948. Streptomycin resistance in pulmonary tuberculosis. *Br. Med. J.* **2:**1009–1015.
23. **Dackett, P. S., V. R. Aber, and D. B. Lowrie.** 1980. The susceptibility of strains of *Mycobacterium tuberculosis* to catalase mediated peroxidative killing. *J. Gen. Microbiol.* **121:**381–386.
24. **Dautzenberg, B., C. Truffot, S. Legris, M.-C. Meyohas, H. C. Berlie, A. Mercat, and J. Grosset.** 1991. Activity of clarithromycin against *Mycobacterium avium* infection in patients with the acquired immune deficiency syndrome. *Am. Rev. Respir. Dis.* **144:**564–569.
25. **David, H. L.** 1970. Probability distribution of drug-resistant mutants in unselected populations of *Mycobacterium tuberculosis.* *Appl. Microbiol.* **20:**810–814.
26. **Davidson, L. A., and K. Takayama.** 1979. Isoniazid inhibition of the synthesis of monosaturated long-chain fatty acids in *Mycobacterium tuberculosis* H37Ra. *Antimicrob. Agents Chemother.* **16:**104–105.
27. **Davidson, P. T.** 1987. Drug resistance and the selection of therapy for tuberculosis. *Am. Rev. Respir. Dis.* **136:**255–257.
28. **Davidson, P. T.** 1989. The diagnosis and management of disease caused by M. *avium* complex, M. *kansasii*, and other mycobacteria. *Clin. Chest Med.* **10:**431–443.
29. **Davis, W. B., and M. M. Weber.** 1977. Specificity of isoniazid on growth inhibition and competition for an oxidized nicotiniamide adenine dinucleotide regulatory site on the electron transport pathway in *Mycobacterium phlei.* *Antimicrob. Agents Chemother.* **12:**213–218.
30. **Dooley, S. W., W. R. Jarvis, W. J. Martone, and D. E. Snider.** 1992. Multidrug-resistant tuberculosis. *Ann. Intern. Med.* **117:**257–259.
31. **Dukes, C. S., J. Sugarman, J. P. Cegielski, G. J. Lallinger, and D. H. Mwakyusa.** 1992. Severe cutaneous hypersensitivity reactions during treatment of tuberculosis in patients with HIV infection in Tanzania. *Trop. Geogr. Med.* **44:**308–311.
32. **Farr, B. M., and G. L. Mandell.** 1990. Rifamycins, p. 295–303. *In* G. L. Mandell, R. G. Douglas, Jr., and J. E. Bennett (ed.), *Principles and Practices of Infectious Diseases.* Churchill Livingstone, Inc., New York.
33. **Finken, M., P. Kirschner, A. Meier, A. Wrede, and E. C.**

Böttger. 1993. Molecular basis of streptomycin resistance in *Mycobacterium tuberculosis*: alterations of the ribosomal protein S12 gene and point mutations within a functional 16S ribosomal RNA pseudoknot. *Mol. Microbiol.* **9:**1239–1246.

34. **Franzblau, S. G., T. Takeda, and M. Nahamura.** 1986. Mycobacterial plasmids: screening and possible relationship to antibiotic resistance in *Mycobacterium avium-Mycobacterium intracellulare*. *Microbiol. Immunol.* **30:**903–907.

35. **Frieden, T. R., T. Sterling, A. Pablos-Mendez, J. O. Kilburn, G. M. Cauthen, and S. W. Doolery.** 1993. The emergence of drug-resistant tuberculosis in New York City. *N. Engl. J. Med.* **328:**521–526.

36. **Gangadharam, P. R., V. K. Perumal, J. T. Crawford, and J. H. Bates.** 1988. Association of plasmids and virulence of *Mycobacterium avium* complex. *Am. Rev. Respir. Dis.* **137:** 212–214.

37. **Good, R. C.** 1985. Opportunistic pathogens in the genus *Mycobacterium*. *Annu. Rev. Microbiol.* **39:**347–369.

38. **Gross, W. M., D. A. Bonato, F. S. Vadney, and S. Campbell.** 1993. In vitro activity of ciprofloxacin in multiple drug resistant strains of *Mycobacterium tuberculosis*: 1985–1992, abstr. U-26, p. 173. *Abstr. 93rd Gen. Meet. Am. Soc. Microbiol. 1993.*

39. **Gross, W. M., and J. E. Hawkins.** 1986. Radiometric susceptibility testing of *M. tuberculosis* with secondary drugs, abstr. C-378, p. 391. *Abstr. Annu. Meet. Am. Soc. Microbiol. 1986.*

40. **Grosset, J. H.** 1992. Treatment of tuberculosis in HIV infection. *Tubercle Lung Dis.* **73:**378–383.

41. **Guerrero, C., L. Stockman, F. Marchesi, T. Bodmer, G. D. Roberts, and A. Telenti.** 1994. Evaluation of the *rpoB* gene in rifampicin-susceptible and resistant *Mycobacterium avium* and *Mycobacterium intracellulare*. *J. Antimicrob. Chemother.* **33:**661–663.

42. **Guthertz, L. S., M. E. Griffith, E. G. Ford, J. M. Janda, and T. F. Midura.** 1988. Quality control or individual components used in Middlebrook 7H10 medium for mycobacterial susceptibility testing. *J. Clin. Microbiol.* **26:**2338–2342.

43. **Hacek, D.** 1992. Modified proportion agar dilution test for slowly growing mycobacteria, p. 5.13.1–5.13.15. *In* H. D. Isenberg (ed.), *Clinical Microbiology Procedures Handbook*, vol. 1. American Society for Microbiology, Washington, D.C.

44. **Hawkins, J. E.** 1984. Drug susceptibility testing, p. 177–193. *In* G. P. Kubica and L. G. Wayne (ed.), *The Mycobacteria: a Sourcebook*, part A. Marcel Dekker, Inc., New York.

45. **Hawkins, J. E.** 1986. Non-weekend schedule for BACTEC susceptibility testing of *Mycobacterium tuberculosis*. *J. Clin. Microbiol.* **23:**934–937.

46. **Hawkins, J. E., R. C. Good, G. P. Kubica, P. R. Gangadharam, H. M. Gruft, and K. D. Stottmeier.** 1983. The levels of service concept in mycobacteriology. *Am. Thoracic Soc. News* **9:**19–25.

47. **Hawkins, J. E., R. J. Wallace, Jr., and B. A. Brown.** 1991. Antibacterial susceptibility tests: mycobacteria, p. 1138–1152. *In* A. Balows, W. J. Hausler, Jr., K. L. Herrmann, H. D. Isenberg, and H. J. Shadomy (ed.), *Manual of Clinical Microbiology*, 5th ed. American Society for Microbiology, Washington, D.C.

48. **Heifets, L.** 1988. MIC as a quantitative measurement of the susceptibility of *Mycobacterium avium* strains to seven antituberculosis drugs. *Antimicrob. Agents Chemother.* **32:**1131–1136.

49. **Heifets, L., N. Mor, and J. Vanderkolk.** 1993. *Mycobacterium avium* strains resistant to clarithromycin and azithromycin. *Antimicrob. Agents Chemother.* **37:**2364–2370.

50. **Heifets, L. B., and M. D. Iseman.** 1985. Determination of in vitro susceptibility of mycobacteria to ansamycin. *Am. Rev. Respir. Dis.* **132:**710–711.

51. **Heifets, L. B., and M. D. Iseman.** 1990. Choice of antimicrobial agents for *M. avium* disease based on quantitative tests of drug susceptibility. *N. Engl. J. Med.* **323:**419–420.

52. **Heifets, L. B., M. D. Iseman, P. J. Lindholm-Levy, and W.** **Kanes.** 1985. Determination of ansamycin MICs for *Mycobacterium avium* complex in liquid medium by radiometric and conventional methods. *Antimicrob. Agents Chemother.* **28:**570–575.

53. **Heifets, L. B., P. J. Lindholm-Levy, and R. D. Comstock.** 1992. Clarithromycin minimal inhibitory and bactericidal concentrations against *Mycobacterium avium*. *Am. Rev. Respir. Dis.* **145:**856–858.

54. **Heym, B., and S. T. Cole.** 1992. Isolation and characterization of isoniazid-resistant mutants of *Mycobacterium smegmatis* and *M. aurum*. *Res. Microbiol.* **143:**721–730.

54a. **Heym, B., N. Honoré, C. Truffot-Pernot, A. Banerjee, C. Schurra, W. R. Jacobs, Jr., J. D. A. van Embden, J. H. Grosset, and S. T. Cole.** 1994. Implications of multidrug resistance for the future of short-course chemotherapy of tuberculosis: a molecular study. *Lancet* **344:**293–298.

55. **Heym, B., Y. Zhang, S. Poulet, D. Young, and S. T. Cole.** 1993. Characterization of the *katG* gene encoding a catalase-peroxidase required for isoniazid susceptibility of *Mycobacterium tuberculosis*. *J. Bacteriol.* **175:**4255–4259.

56. **Hoffner, S. E., M. Kratz, B. Olsson-Liljequist, S. B. Svenson, and G. Källenius.** 1989. In-vitro synergistic activity between ethambutol and fluorinated quinolones against *Mycobacterium avium* complex. *J. Antimicrob. Chemother.* **24:** 317–324.

57. **Hoffner, S. E., S. B. Svenson, and A. E. Beezer.** 1990. Microcalorimetric studies of the initial interaction between antimycobacterial drugs and *Mycobacterium avium*. *J. Antimicrob. Chemother.* **25:**353–359.

58. **Honore, N., and S. T. Cole.** 1993. Molecular basis of rifampin resistance in *Mycobacterium leprae*. *Antimicrob. Agents Chemother.* **37:**414–418.

59. **Howard, W. L., F. Maresh, E. E. Mueller, S. A. Yanitelli, and G. F. Woodruff.** 1949. The role of pulmonary cavitation in the development of bacterial resistance to streptomycin. *Am. Rev. Tuberc.* **59:**391–401.

60. **Howlett, H. S., J. B. O'Connor, J. F. Sadusk, J. E. Swift, and F. A. Beardsley.** 1949. Sensitivity of tubercle bacilli to streptomycin: the influence of various factors upon the emergence of resistant strains. *Am. Rev. Tuberc.* **59:**402–414.

61. **Huebner, R. E., R. C. Good, and J. I. Tokars.** 1993. Current practices in mycobacteriology: results of a survey of state public health laboratories. *J. Clin. Microbiol.* **31:**771–775.

61a. **Husson, R. N., L. A. Ross, S. Sandelli, C. B. Inderlied, D. Venzon, L. L. Lewis, L. Woods, P. S. Conville, F. G. Witebsky, and P. A. Pizzo.** 1994. Orally administered clarithromycin for the treatment of systemic *Mycobacterium avium* complex infection in children with acquired immunodeficiency syndrome. *J. Pediatr.* **124:**807–814.

62. **Inderlied, C. B.** 1991. Antimycobacterial agents: in vitro susceptibility testing, spectrums of activity, mechanisms of action and resistance, and assays for activity in biological fluids, p. 134–197. *In* V. Lorian (ed.), *Antibiotics in Laboratory Medicine*. The Williams & Wilkins Co., Baltimore.

63. **Inderlied, C. B., C. A. Kemper, and L. E. M. Bermudez.** 1993. The *Mycobacterium avium* complex. *Clin. Microbiol. Rev.* **6:**266–310.

64. **Inderlied, C. B., P. T. Kolonski, M. Wu, and L. S. Young.** 1989. In vitro and in vivo activity of azithromycin (CP 62,993) against the *Mycobacterium avium* complex. *J. Infect. Dis.* **159:**994–997.

65. **Inderlied, C. B., L. S. Young, and J. K. Yamada.** 1987. Determination of in vitro susceptibility of *Mycobacterium avium* complex isolates to antimicrobial agents by various methods. *Antimicrob. Agents Chemother.* **31:**1697–1702.

66. **Iseman, M. D.** 1993. Treatment of multidrug-resistant tuberculosis. *N. Engl. J. Med.* **329:**784–791.

67. **Jacobs, W. R., Jr., R. G. Barletta, R. Udani, J. Chan, G. Kalkut, G. Sosne, T. Kieser, G. J. Sarkis, G. F. Hatfull, and B. R. Bloom.** 1993. Rapid assessment of drug susceptibilities of *Mycobacterium tuberculosis* by means of luciferase reporter phages. *Science* **260:**819–822.

68. **Jucker, M. T., and J. O. Falkinham III.** 1990. Epidemiology of infection by nontuberculous mycobacteria. IX. Evidence for two DNA homology groups among small plasmids in *Mycobacterium avium, Mycobacterium intracellulare,* and *Mycobacterium scrofulaceum. Am. Rev. Respir. Dis.* **142:**858–862.

69. **Källenius, G., S. G. Svenson, and S. E. Hoffner.** 1989. Ethambutol: a key for *Mycobacterium avium* complex chemotherapy. *Am. Rev. Respir. Dis.* **140:**264.

70. **Kent, P. T., and G. P. Kubica.** 1985. *Public Health Mycobacteriology—a Guide for the Level III Laboratory.* Centers for Disease Control, U.S. Department of Health and Human Services, Atlanta.

71. **Kiehn, T. E.** 1993. Personal communication.

72. **Koletar, S. L., D. J. Williams, and A. Berry.** 1994. Serum level and MIC do not correlate with the efficacy of azithromycin for disseminated *Mycobacterium avium* complex (MAC) infection in patients with AIDS, abstr. 292, p. 65. *Program Abstr. 2nd Int. Conf. Macrolides, Azalides, Streptogramins.*

73. **Koontz, F. P., M. E. Erwin, M. S. Barrett, and R. N. Jones.** 1994. E-test for routine clinical antimicrobial susceptibility testing of rapid growing mycobacteria isolates. *Diagn. Microbiol. Infect. Dis.* **19:**183–186.

74. **Kubica, G. P., and W. E. Dye.** 1967. *Laboratory Methods for Clinical and Public Health Mycobacteriology.* U.S. Government Printing Office, Washington, D.C.

75. **Leysen, D. C., A. Haemers, and S. R. Pattyn.** 1989. Mycobacteria and the new quinolones. *Antimicrob. Agents Chemother.* **33:**1–5.

76. **Libonati, J. P., C. E. Stager, J. R. Davis, and S. H. Siddiqi.** 1988. Direct antimicrobial drug susceptibility testing of *Mycobacterium tuberculosis* by the radiometric method. *Diagn. Microbiol. Infect. Dis.* **10:**41–48.

77. **Mandell, G. L., R. G. Douglas, Jr., and J. E. Bennett.** 1992. *Handbook of Antimicrobial Therapy 1992.* Churchill Livingstone, Inc., New York.

78. **Masur, H.** 1993. Recommendations on prophylaxis and therapy for disseminated *Mycobacterium avium* complex disease in patients infected with the human immunodeficiency virus. *N. Engl. J. Med.* **329:**898–904.

79. **McClatchy, J. K.** 1978. Susceptibility testing of mycobacteria. *Lab. Med.* **9:**47–52.

80. **McFarland, E. J., and D. R. Kuritzkes.** 1993. Clinical features and treatment of infection due to *Mycobacterium fortuitum/chelonae* complex. *Curr. Clin. Top. Infect. Dis.* **13:**188–202.

81. **Meier, A., P. Kirschner, B. Springer, V. A. Steingrube, B. A. Brown, R. J. Wallace, Jr., and E. C. Böttger.** 1994. Identification of mutations in 23S rRNA gene of clarithromycin-resistant *Mycobacterium intracellulare. Antimicrob. Agents Chemother.* **38:**381–384.

82. **Mitchison, D. A.** 1952. Titration of strains of tubercle bacilli against isoniazid. *Lancet* **ii:**858–860.

83. **Mitchison, D. A.** 1979. Basic mechanisms of chemotherapy. *Chest* **76**(Suppl):771–781.

84. **Mizuguchi, Y., M. Fukunaga, and H. Taniguchi.** 1981. Plasmid deoxyribonucleic acid and translucent-to-opaque variation in *Mycobacterium intracellulare* 103. *J. Bacteriol.* **146:**656–659.

85. **Mizuguchi, Y., M. Ogawa, and T. Udou.** 1985. Morphological changes induced by β-lactam antibiotics in *Mycobacterium avium-M. intracellulare* complex. *Antimicrob. Agents Chemother.* **27:**541–547.

86. **Morris, S. L., D. A. Rouse, A. Malik, S. D. Chaparas, and F. G. Witebsky.** 1990. Characterization of plasmids extracted from AIDS-associated *Mycobacterium avium* isolates. *Tubercle* **71:**181–185.

87. **Nair, J., D. A. Rouse, G. H. Bai, and S. L. Morris.** 1993. The *rpsL* gene and streptomycin resistance in single and multiple drug-resistant strains of *Mycobacterium tuberculosis. Mol. Microbiol.* **10:**521–527.

88. **National Committee for Clinical Laboratory Standards.** 1990. *Antimycobacterial Susceptibility Testing.* Proposed standard M24-P. National Committee for Clinical Laboratory Standards, Villanova, Pa.

89. **National Committee for Clinical Laboratory Standards.** 1993. *Methods for Dilution Antimicrobial Susceptibility Tests for Bacteria That Grow Aerobically.* Document M7-A3. National Committee for Clinical Laboratory Standards, Villanova, Pa.

90. **Nightingale, S. D., W. D. Cameron, F. M. Gordin, P. M. Sullam, D. L. Cohn, R. E. Chaisson, L. J. Eron, P. D. Saprti, B. Bihari, D. L. Kaufman, J. J. Stern, D. D. Pearce, W. G. Weinberg, A. LaMarca, and F. P. Siegel.** 1993. Two controlled trials of rifabutin prophylaxis against *Mycobacterium avium* complex infection in AIDS. *N. Engl. J. Med.* **329:**828–833.

91. **Nunn, P., J. Porter, and P. Winstanley.** 1993. Thiacetazone—avoid like poison or use with care. *Trans. R. Soc. Trop. Med. Hyg.* **87:**578–582.

92. **Pyle, M.** 1947. Relative number of resistant tubercle bacilli in sputa of patients before and during treatment with streptomycin. *Proc. Mayo Clin.* **22:**465–473.

93. **Rastogi, N., C. Frehel, A. Ryter, H. Ohayon, M. Lesourd, and H. L. David.** 1981. Multiple drug resistance in *Mycobacterium avium*: is the wall architecture responsible for the exclusion of antimicrobial agents? *Antimicrob. Agents Chemother.* **20:**666–677.

94. **Rastogi, N., K. S. Goh, N. Guillou, and V. Labrousse.** 1992. Spectrum of drugs against atypical mycobacteria: how valid is the current practice of susceptibility testing and the choice of drugs. *Zentralbl. Bakteriol.* **277:**474–484.

95. **Riley, L. W.** 1993. Drug-resistant tuberculosis. *Clin. Infect. Dis.* **17**(Suppl 2):S442–S446.

96. **Rist, N.** 1953. The application to clinical practice of the laboratory experience with combinations of antituberculosis drugs. *Bull. Int. Union Tuberc.* **23:**416–427.

97. **Rosner, J.** 1993. Susceptibilities of *oxyR* regulon mutants of *Escherichia coli* and *Salmonella typhimurium* to isoniazid. *Antimicrob. Agents Chemother.* **37:**2251–2253.

98. **Russel, W. R., and G. Middlebrook.** 1961. *Chemotherapy of Tuberculosis.* Charles C Thomas, Publisher, Springfield, Ill.

99. **Saito, H., K. Sato, and H. Tomioka.** 1988. Comparative in vitro and in vivo activity of rifabutin and rifampicin against *Mycobacterium avium* complex. *Tubercle* **69:**187–192.

100. **Salfinger, M.** 1994. Unpublished observations.

101. **Salfinger, M., G. T. DiFerdinando, and H. W. Taber.** 1993. Availability of a fast track for TB susceptibility testing in New York State, abstr. 332, p. 180. *Program Abstr. 33rd Intersci. Conf. Antimicrob. Agents Chemother.*

102. **Salfinger, M., and L. B. Heifets.** 1988. Determination of pyrazinamide MICs for *Mycobacterium tuberculosis* at different pHs by the radiometric method. *Antimicrob. Agents Chemother.* **32:**1002–1004.

103. **Salfinger, M., L. B. Reller, B. Demchuk, and Z. T. Johnson.** 1989. Rapid radiometric method for pyrazinamide susceptibility testing of *M. tuberculosis. Res. Microbiol.* **140:**301–309.

104. **Salfinger, M., L. B. Reller, and F. M. Kafader.** 1990. Pyrazinamide resistance of *Mycobacterium tuberculosis* complex isolates, abstr. U55, p. 150. *Abstr. 90th Annu. Meet. Am. Soc. Microbiol.*

105. **Sanford, J. P.** 1993. *Guide to Antimicrobial Therapy.* Antimicrobial Therapy, Inc., Dallas.

106. **Siddiqi, S. H.** 1992. Radiometric (BACTEC) tests for slowly growing mycobacteria, p. 5.14.1–5.14.25. *In* H. D. Isenberg (ed.), *Clinical Microbiology Procedures Handbook,* vol. 1. American Society for Microbiology, Washington, D.C.

107. **Siddiqi, S. H., L. B. Heifets, M. H. Cynamon, N. M. Hooper, A. Laszlo, J. P. Libonati, P. J. Lindholm-Levy, and N. Pearson.** 1993. Rapid broth macrodilution method for determination of MICs for *Mycobacterium avium* isolates. *J. Clin. Microbiol.* **31:**2332–2338.

108. **Siddiqi, S. H., J. P. Libonati, M. E. Carter, N. M. Hooper, J. F. Baker, C. C. Hwangbo, and L. E. Warfel.** 1988. Enhancement of mycobacterial growth in Middlebrook

7H12 medium by polyoxyethylene stearate. *Curr. Microbiol.* **17:**105–110.

109. **Straus, W. L., S. M. Ostroff, D. B. Jernigan, T. E. Kiehn, E. M. Sordillo, D. Armstrong, N. Boone, N. Schneider, J. O. Kilburn, V. A. Silcox, V. LaBombardi, and R. C. Good.** 1994. Clinical and epidemiologic characteristics of *Mycobacterium haemophilum*, an emerging pathogen in immunocompromised patients. *Ann. Intern. Med.* **120:**118–125.

110. **Takayama, K., and J. O. Kilburn.** 1989. Inhibition of synthesis of arabinogalactan by ethambutol in *Mycobacterium smegmatis. Antimicrob. Agents Chemother.* **33:**1493–1499.

111. **Takayama, K., and N. Qureshi.** 1984. Structure and synthesis of lipids, p. 315–344. *In* G. P. Kubica and L. G. Wayne (ed.), *The Mycobacteria: a Sourcebook*, part A. Marcel Dekker, Inc., New York.

112. **Takiff, H. E., L. Salazar, C. Guerrero, W. Philipp, W. M. Huang, B. Kreiswirth, S. T. Cole, W. R. Jacobs, and A. Telenti.** 1994. Cloning and nucleotide sequence of *Mycobacterium tuberculosis gyrA* and *gyrB* genes and detection of quinolone resistance mutations. *Antimicrob. Agents Chemother.* **38:**773–780.

113. **Telenti, A., P. Imboden, F. Marchesi, D. Lowrie, S. Cole, M. J. Colston, L. Matter, K. Schopfer, and T. Bodmer.** 1993. Detection of rifampin-resistance mutations in *Mycobacterium tuberculosis. Lancet* **341:**647–650.

114. **Telenti, A., P. Imboden, F. Marchesi, T. Schmidheini, and T. Bodmer.** 1993. Direct, automated detection of rifampin-resistant *Mycobacterium tuberculosis* by polymerase chain reaction and single-strand conformation polymorphism analysis. *Antimicrob. Agents Chemother.* **37:**2054–2058.

115. **Tenover, F. C., J. T. Crawford, R. E. Huebner, L. J. Geiter, C. R. Horsburgh, and R. C. Good.** 1993. The resurgence of tuberculosis: is your laboratory ready? *J. Clin. Microbiol.* **31:**767–770.

116. **Tsukamura, M., and T. Miyachi.** 1989. Correlations among naturally occurring resistances to antituberculosis drugs in *Mycobacterium avium* complex strains. *Am. Rev. Respir. Dis.* **139:**1033–1035.

117. **Wallace, R. J., Jr.** 1989. The clinical presentation, diagnosis, and therapy of cutaneous and pulmonary infections due to the rapidly growing mycobacteria, M. *fortuitum* and M. *chelonae. Clin. Chest Med.* **10:**419–429.

118. **Wallace, R. J., Jr., J. M. Musser, S. I. Hull, V. A. Silcox, L. C. Steele, G. D. Forrester, A. Labidi, and R. K. Selander.** 1989. Diversity and sources of rapidly growing mycobacteria associated with infections following cardiac surgery. *J. Infect. Dis.* **159:**708–716.

119. **Wallace, R. J., Jr., D. R. Nash, M. Tsukamura, Z. M. Blacklock, and V. A. Silcox.** 1988. Human disease due to *Mycobacterium smegmatis. J. Infect. Dis.* **158:**52–59.

120. **Wallace, R. J., Jr., J. M. Swenson, V. A. Silcox, R. C. Good, J. A. Tschen, and M. S. Stone.** 1983. Spectrum of disease due to rapidly growing mycobacteria. *Rev. Infect. Dis.* **4:**326–331.

121. **Wanger, A. R., and K. Mills.** 1994. E test for susceptibility testing of *Mycobacterium tuberculosis* and *Mycobacterium avium-intracellulare. Diagn. Microbiol. Infect. Dis.* **19:**179–181.

122. **Wayne, L. G., and I. Krasnow.** 1966. Preparation of tuberculosis susceptibility testing media by means of impregnated disks. *Am. J. Clin. Pathol.* **45:**769–771.

123. **Williams, D. L., C. Waguespack, K. Eisenach, J. T. Crawford, F. Portaels, M. Salfinger, C. M. Nolan, C. Abe, V. Sticht-Groh, and T. P. Gillis.** 1994. Characterization of rifampin resistance in pathogenic mycobacteria. *Antimicrob. Agents Chemother.* **38:**2380–2386.

124. **Winder, F. G.** 1982. Mode of action of the antimycobacterial agents and associated aspects of the molecular biology of the mycobacteria, p. 353–438. *In* C. Ratledge and J. Stanford (ed.), *The Biology of the Mycobacteria*. Academic Press, Inc., New York.

125. **Winder, F. G., and P. B. Collins.** 1968. The effect of isoniazid on nicotinamide nucleotide levels in *Mycobacterium bovis* strain BCG. *Am. Rev. Respir. Dis.* **97:**719–720.

126. **Winder, F. G., and P. B. Collins.** 1969. The effect of isoniazid on nicotinamide nucleotide concentrations in tubercle bacilli. *Am. Rev. Respir. Dis.* **100:**101–103.

127. **Woodley, C. L.** 1986. Evaluation of streptomycin and ethambutol concentrations for susceptibility testing of *Mycobacterium tuberculosis* by radiometric and conventional procedures. *J. Clin. Microbiol.* **23:**385–386.

128. **Woods, G. L., and J. A. Washington 3rd.** 1987. Mycobacteria other than *Mycobacterium tuberculosis*: review of microbiologic and clinical aspects. *Rev. Infect. Dis.* **9:**275–294.

129. **Woods, G. L., and F. G. Witebsky.** 1993. Current status of mycobacterial testing in clinical laboratories. *Arch. Pathol. Lab. Med.* **117:**876–884.

130. **Yajko, D. M., P. S. Nassos, and W. K. Hadley.** 1987. Broth microdilution testing of susceptibilities to 30 antimicrobial agents of *Mycobacterium avium* strains from patients with acquired immune deficiency syndrome. *Antimicrob. Agents Chemother.* **31:**1579–1584.

131. **Yajko, D. M., C. A. Sanders, P. S. Nassos, J. Madej, M. Lancaster, and W. K. Hadley.** 1993. A colorimetric method for determining the antimicrobial susceptibility of M. *tuberculosis*, M. *avium* complex and rapid growing mycobacteria, abstr. U-20, p. 172. *Abstr. 93rd Gen. Meet. Am. Soc. Microbiol. 1993.*

132. **Young, L. S., O. G. Berlin, and C. B. Inderlied.** 1987. Activity of ciprofloxacin and other fluorinated quinolones against mycobacteria. *Am. J. Med.* **82:**23–26.

133. **Young, L. S., L. Wiviott, M. Wu, P. Kolonoski, R. Bolan, and C. B. Inderlied.** 1991. Azithromycin for treatment of *Mycobacterium avium-intracellulare* complex infection in patients with AIDS. *Lancet* **338:**1107–1109.

134. **Zhang, Y., B. Heym, B. Allen, D. Young, and S. Cole.** 1992. The catalase-peroxidase gene and isoniazid resistance of *Mycobacterium tuberculosis. Nature* (London) **358:**591–593.

Antifungal Agents and Susceptibility Testing

ANA ESPINEL-INGROFF AND MICHAEL A. PFALLER

120

The higher incidence of fungal infections in the last 10 years has been attributed to the increased use of newer and more effective antibacterial agents, the AIDS pandemic, and the rapidly expanding number of chemically induced immunosuppression patients and bone marrow and solid-organ transplantation and oncology patients (2, 7, 55, 63). As a result of improved management protocols, AIDS, cancer, and transplantation populations now survive longer and become highly susceptible to life-threatening fungal infections (7). Fungi are emerging as important nosocomial pathogens that cause severe morbidity and mortality in hospitalized patients. The National Nosocomial Infections Surveillance System reported that the rate of reported fungal infections rose from 2.0 to 3.8 infections per 1,000 discharges between 1980 and 1990, with a total of 30,477 fungal infections reported (4). *Candida* species accounted for 7.7% of all nosocomial bloodstream infections in the United States from 1985 to 1988 and were the fourth most common cause of nosocomial bloodstream infections, following coagulase-negative staphylococci, *Staphylococcus aureus*, and *Enterococcus* spp. (65). The impact of these infections is substantial, with the mortality and excess length of hospital stay attributable to nosocomial candidemia estimated at ≥38% and 30 days (median), respectively (65). Recently, several centers have reported a higher frequency of colonization and a greater incidence of systemic infections with less virulent species such as *Candida krusei*, *Candida parapsilosis*, and *Candida lusitaniae*.

In addition to *Candida* spp., new fungal pathogens such as *Fusarium* spp., *Trichosporon beigelii*, the zygomycetes, dematiaceous fungi, and other usually nonpathogenic fungi are being recognized as significant problems in some patients (2). Among the pathogenic opportunistic fungi, *Histoplasma capsulatum*, *Coccidioides immitis*, and *Cryptococcus neoformans* are frequently found in patients with chemically induced immunosuppression and in patients with AIDS. *Cryptococcus neoformans* is a cause of infection in approximately 6 to 10% of individuals with AIDS in the United States and in up to 30% of AIDS patients in Africa. About 90% of AIDS patients infected with *Cryptococcus neoformans* develop meningitis (47). Histoplasmosis occurs in 5% of individuals with AIDS who come from areas of endemicity and in up to 25% of AIDS patients in selected U.S. cities (12, 66). Paralleling the increase in the incidence of opportunistic fungal infections over the last decade has

been the introduction of new antifungal agents with systemic activities. Although more extensive research in the development of antifungal drugs has been done lately, only six antifungal agents are currently licensed for use against systemic fungal infections. These six systemic antifungal agents include the polyene amphotericin B, the imidazoles miconazole and ketoconazole, the triazoles fluconazole and itraconazole, and the pyrimidine synthesis inhibitor flucytosine (5-FC) (Table 1).

SYSTEMIC ANTIFUNGAL AGENTS

Polyenes

Amphotericin B (1956; E.R. Squibb & Sons, Princeton, N.J.) and liposomal and lipid complex amphotericin B are polyene macrolide antibiotics used primarily in the treatment of systemic and life-threatening fungal infections. Liposomal and lipid complex formulations of amphotericin B are undergoing clinical evaluation. The polyene antifungal agents act by binding to ergosterol in the fungal cell membrane, causing osmotic instability and loss of membrane integrity. The direct membrane toxicity is due in part to oxidative damage and is frequently fungicidal (58, 63). This effect is extended to mammalian cells, in which the drug binds to cholesterol, creating the high toxicity associated with the polyene agents. Amphotericin B is produced by *Streptomyces nodosus* (62) and has been used for many years in the treatment of mycotic diseases. It is unstable when subjected to heat, light, or acid pH. Although resistance to amphotericin B is rare, changes in membrane sterols have been correlated with the development of resistance both in vitro and in vivo (11, 29, 63). Such resistance has assumed clinical importance, particularly with certain species, such as *Candida lusitaniae* (11, 29).

Liposomal and lipid complex formulations of amphotericin B were designed to maximize the delivery of amphotericin B to patients with deep-seated fungal infections such as hepatosplenic candidiasis and invasive pulmonary aspergillosis (32, 63). These preparations have selective toxicity for fungal cells but not for erythrocytes (32) and theoretically promote the delivery of drug to the site of infection while avoiding the toxicity of supramaximal doses of amphotericin B. The limited amount of uncontrolled

TABLE 1 Established antifungal agents with systemic activity

Antifungal agent	Mechanism of action	Route[a]	Comments
Polyenes			
Amphotericin B	Binds to ergosterol, causing direct oxidative membrane damage	i.v.	Established agent; broad spectrum; toxic
Liposomal and lipid complex amphotericin B	Same as amphotericin B	i.v.	Investigational agent with broad-spectrum activity and decreased toxicity; role in therapy not yet established
Azoles			
Miconazole	Inhibition of membrane sterol synthesis	i.v.	Toxic agent with modest anticandidal activity; active against *P. boydii*
Ketoconazole	Same as miconazole	Oral, topical	Modest broad-spectrum activity
Itraconazole	Same as miconazole	Oral	Triazole with broad-spectrum activity
Fluconazole	Same as miconazole	Oral, i.v.	Triazole with broad-spectrum activity; good central nervous system penetration; good in vivo activity
Pyrimidine synthesis inhibitor			
5-FC	Inhibition of DNA and RNA synthesis	Oral	Toxicity and resistance are problems. Used in combination with amphotericin B

[a] i.v., intravenous.

data available suggests that these novel formulations of amphotericin B may be useful in treating infections that are refractory to conventional therapy (32, 63); however, these agents must be further studied in controlled, randomized clinical trials.

Azoles

The azole class of antifungal compounds was introduced in the late 1960s as a group of antifungal therapeutic agents, but only a few of them have demonstrated systemic antifungal activity. Four azoles have been approved for the treatment of systemic fungal diseases: miconazole, ketoconazole, fluconazole, and itraconazole (Table 1). The azoles inhibit fungal cytochrome P-450-dependent enzymes, with resulting impairment of ergosterol synthesis and depletion of ergosterol in the fungal cell membrane. Although resistance to azoles has been demonstrated in *Candida albicans*, the detection of azole resistance in vitro seems to be quite variable and method dependent (52). In general, clinical correlations with in vitro susceptibility test results for azole antifungal agents have been quite poor (15, 23, 38, 45).

Miconazole and ketoconazole are established, broad-spectrum imidazole agents active against a variety of fungal pathogens, including yeasts, dimorphic organisms, dermatophytes, and opportunistic pathogens. Miconazole (1970; Janssen Pharmaceutica, Piscataway, N.J.) was the first azole derivative to be administered intravenously for the therapy of systemic fungal infections. However, owing to its toxicity and high relapse rates, its use is limited to certain cases of pseudallescherosis (33), rare refractory cryptococcal meningitis, and coccidioidal meningitis in children, to whom it is administered intraventricularly in addition to oral ketoconazole (57, 60). Ketoconazole (1977; Janssen Pharmaceutica) is an orally absorbed antifungal agent. It requires a normal intragastric pH for absorption and penetrates poorly into the cerebrospinal fluid (10, 30). Ketoconazole has been established as an effective alternative to amphotericin B for

the treatment of immunocompetent individuals with non-threatening, non-central nervous system, localized or disseminated histoplasmosis, blastomycosis, mucocutaneous candidiasis, paracoccidioidomycosis, and selected forms of coccidioidomycosis (5, 13).

In the last 2 years, itraconazole and fluconazole have been approved as orally active systemic agents with less potential for toxicity than the imidazoles (3, 21, 23, 53) because of their more specific binding to fungal cell cytochromes than to mammalian cytochromes. Fluconazole (Pfizer Pharmaceuticals, New York, N.Y.) is a relatively small molecule that is partially water soluble, is easily absorbed, has a prolonged half-life (up to 25 h in humans), is minimally protein bound, is excreted largely as an active drug in the urine, and penetrates well into the cerebrospinal fluid (3). Fluconazole has an important role in the maintenance therapy of cryptococcal meningitis in patients with AIDS as well as in the treatment of fungal diseases caused by *Candida* spp. (except *Candida krusei* and *Torulopsis glabrata*), and *Coccidioides immitis*. Both fluconazole and itraconazole are alternatives to amphotericin B and ketoconazole in the treatment of histoplasmosis in AIDS patients (12, 53).

Itraconazole (Janssen Pharmaceutica) is a lipophilic compound characterized by good oral absorption, extensive distribution in tissues, and long half-life in serum. Itraconazole is very insoluble in aqueous fluids and is highly protein bound (>90%). This drug binds strongly to plasma proteins and penetrates poorly into cerebrospinal fluid and urine but well into skin and soft tissues. Itraconazole appears to be more active in vitro and in vivo than ketoconazole and fluconazole against *H. capsulatum*, *Blastomyces dermatitidis* (12), *Coccidioides immitis*, and *Sporothrix schenckii*. Itraconazole holds the most promise against *Aspergillus* spp. The efficacy of these triazoles is being evaluated further in several clinical trials by the Mycoses Study Group.

Pyrimidine Synthesis Inhibitor

5-FC (Hoffmann-LaRoche Inc., Nutley, N.J.) is a water-soluble, stable compound used orally in the treatment of systemic infections caused by susceptible pathogenic or opportunistic yeasts and other fungi. It acts as a competitive antimetabolite for uracil in the synthesis of yeast RNA, and it also interferes with thymidylate synthetase (37, 46). Five enzymes are involved in the mode of action of 5-FC. The first step is initiated by the uptake of the drug by a membrane-bound permease. Inside the cell, the drug is deaminated to 5-fluorouracil, which is the main active form of the drug. These activities can be antagonized in vitro by a variety of purine and pyrimidine bases and nucleosides (46, 56). Because of this antagonism, the antifungal activity of 5-FC can be demonstrated in vitro only in synthetic media free of substances such as cytosine, uracil, and other pyrimidines and purines. At least two metabolic sites are responsible for resistance to 5-FC (37, 46); one involves the enzyme cytosine permease, which is responsible for the uptake of 5-FC into fungal cells, and the other involves the enzyme cytosine deaminase, which is responsible for the deamination of 5-FC to 5-fluorouracil, the metabolically active form of the drug.

ANTIFUNGAL SUSCEPTIBILITY TESTING

Rationale

In vitro antifungal susceptibility tests are similar in design to tests with antibacterial agents and are performed for the same reasons. Ideally, in vitro susceptibility tests (i) provide a reliable measure of the relative activities of two or more antifungal agents, (ii) correlate with in vivo activity and predict the likely outcome of therapy, (iii) provide a means with which to monitor the development of resistance among a normally susceptible population of organisms, and (iv) predict the therapeutic potentials of newly discovered investigational agents (26, 38, 56). Unfortunately, there is little evidence to support the clinical correlations of antifungal susceptibility test results with in vivo outcomes (45, 49, 59). Although recent studies by Powderly et al. (48) and Radetsky et al. (49) suggest some correlation between MIC results for amphotericin B and clinical outcome, the general applicability of these results remains confused by the retrospective nature of the studies, the documented variability of the nonstandardized in vitro test methods used (28), and the difficulty in defining fungal diseases and their responses to therapy.

With the use of both established and investigational agents has come the recognition of resistance to one or more antifungal agents in selected isolates (11, 26, 29, 34, 38, 46, 48, 52, 63). As a result, clinical laboratories are now being asked to assume a greater role in the selection and monitoring of antifungal chemotherapy. The methods that have been applied to antifungal susceptibility testing include broth dilution (macro- and microdilution), agar dilution, and disk diffusion. Currently, two commercial antifungal susceptibility tests are being evaluated in this country: the colorimetric microdilution test developed by Alamar Biosciences, Inc. (Sacramento, Calif.) and the agar diffusion antifungal strip E test by AB Biodisk (Solna, Sweden) (Table 2). The National Committee for Clinical Laboratory Standards (NCCLS) Subcommittee on Antifungal Susceptibility Testing has proposed a reference method for broth dilution susceptibility testing of yeast cells (36). This reference method was the result of a series of

TABLE 2 Methods used for antifungal susceptibility testing

Test method	Means of endpoint determination
Broth dilution	
Macrodilution (yeasts)	Visual comparison of turbidity with turbidity of 1:5 growth control dilution, ATP photometry, turbidimetry, colorimetry, radiometry, dry weight
Microdilution (yeasts)	Visual comparison of turbidity with turbidity of growth control, ATP photometry, turbidimetry, colorimetry, radiometry, dry weight
Colorimetric microdilution (yeasts)	Visual observation of color change
Macro- and microdilution (filamentous fungi)	Visual comparison of growth (approx 75% inhibition) with that of growth control, also spectrophotometric reading
Agar dilution	
Macrodilution (standard dishes)	Visual
Agar diffusion	
Disk	Zone diam (visual)
Antifungal strip (E test)	Zone of inhibition (visual)

collaborative studies that focused on some of the testing variables discussed below (18, 24, 40).

Infrequently, it is necessary to perform susceptibility tests with amphotericin B. These occasions include the rare cases of resistance to this drug in some isolates of *Candida* spp., the zygomycetes, and *Pseudallescheria boydii*. The other fungi are usually susceptible to this drug (21, 56). Susceptibility testing of the azoles is of some value against yeast isolates, because yeast susceptibility to these drugs is variable. However, these tests are not warranted with the dimorphic fungi, since these isolates are usually susceptible to them. In vitro testing with 5-FC is more important clinically than testing with either the polyene compounds or the azoles because of the repeated demonstration of de novo resistance to 5-FC as well as the emergence of resistant strains of yeasts and other fungi after therapeutic exposure to the drug (21, 37, 56). Susceptibility tests with 5-FC should be performed on all isolates of pathogenic yeasts isolated from patients destined to receive the drug, some isolates of *Aspergillus* spp., including *Aspergillus fumigatus* (resistance is not uncommon) and certain dematiaceous fungi (pathogenic species, including *Phialophora* spp., have demonstrated variable susceptibilities). Most of the other fungi, including the dimorphic fungi, are resistant to 5-FC. Such tests should also be performed on all isolates recovered during therapy. However, the results of such tests must be interpreted with caution, as the results cannot always be regarded as fully predictive of clinical responses (37, 45, 59).

Variables

In vitro antifungal susceptibility testing is influenced by a number of technical variables, including inoculum size and preparation, medium formulation and pH, duration and temperature of incubation, and the criterion used for MIC endpoint determination (8, 14, 15, 21, 26–28, 38). In addition, antifungal susceptibility testing is complicated by problems unique to fungi, such as slow growth rates (relative to bacteria) and the ability of certain dimorphic fungi to grow either as a unicellular yeast form that produces blastoconidia or as a hyphal or filamentous fungal form that may produce asexual spores, depending on pH, temperature, and medium composition (34, 56). It appears that azoles are more potent in inhibiting hyphal outgrowth than blastoconidium formation in *Candida albicans* (23, 38, 52, 63), which suggests that the morphologic or developmental form of the organism may be another important variable. Finally, the basic properties of the antifungal agents themselves, such as solubility, chemical stability, modes of action, and the tendency to produce partial inhibition of growth over a wide range of concentrations, must be taken into account. These factors pose problems relative to in vitro susceptibility testing. The polyenes are water insoluble and are inactivated by heat, light, and acid. The medium used should be well buffered (pH 7.0), and the test solutions should be protected from light. On the other hand, the azoles, except fluconazole, have relatively good chemical stabilities but also poor solubility in aqueous media.

The determination of MIC endpoints is a critical step in antifungal susceptibility testing, especially with the azoles. The usual partial inhibition or trailing that is observed with 5-FC and the azoles precludes the determination of well-defined endpoints and creates a great deal of variability. Several different methods of MIC endpoint determination have been applied in efforts to develop a test method that is both objective and easy to perform and interpret in the routine clinical laboratory (Table 2). Traditionally, the MIC was considered the lowest concentration of an antifungal agent that inhibited total growth of the fungi, as detected visually. The determination of MIC endpoints with a less stringent criterion (slight turbidity is ignored above the MIC endpoint) has increased interlaboratory reproducibility and produced a shift in the MIC distribution toward lower drug concentrations for *Candida albicans* and *Candida tropicalis*, especially of the azoles (18, 24). The less stringent criterion also permits discrimination between putatively susceptible and resistant isolates when the presence of the trailing turbidity is similar for all the drug concentrations above the MIC. The amount of this trailing or partial inhibition can be estimated by diluting the drug-free growth control 1:5 (0.2 ml of growth control plus 0.8 ml of medium), which approximates an 80% inhibition standard (18, 36). This approach provides a convenient and direct method of establishing a specific turbidity endpoint for each isolate that more precisely reflects an 80% inhibition. For amphotericin B, endpoints are defined easily, because trailing or partial inhibition is not usually observed with this drug.

Standardized Methods for Yeasts

The higher incidence of fungal infections has increased the use of antifungal agents, especially the azoles. It also has increased the interest in clinical laboratory testing of new fungal isolates from patients. The need to develop standardized methods for antifungal susceptibility tests was perceived as a means of providing good interlaboratory agreement among clinical laboratories as well as meaningful communication between clinical laboratories and physicians (8, 25). Standardization can also play an important

role in the needed evaluations of in vitro versus in vivo correlations of drug efficacy. In response to these needs, the NCCLS in 1982 established a subcommittee to coordinate work on antifungal susceptibility tests. The goal of the subcommittee was to develop a reliable reference method for in vitro susceptibility testing of yeasts first and other fungi later with the hope of correlating the results of the reference method with clinical effectiveness.

As a result of several collaborative studies (18, 24, 40), the NCCLS subcommittee in 1992 proposed the reference method (M27-P; described below) (36) for testing yeast fungi such as *Candida* spp., *T. glabrata*, and *Cryptococcus neoformans*. This method has not been used in studies of the yeast forms of dimorphic fungi such as *B. dermatitidis* and *H. capsulatum* (36). The relationship established between the results obtained with the reference method and patient response needs further evaluation and additional studies by the NCCLS subcommittee.

Macrodilution

The broth macrodilution test is the most widely used technique for antifungal susceptibility testing and is the reference method for yeast cells proposed by the NCCLS Subcommittee on Antifungal Susceptibility Testing (8, 36). Broth macrodilution tests are adequate for testing all antifungal agents against any fungal isolate. Macrodilution tests are suitable for small clinical laboratories in which the volume of these tests is low. The proposed NCCLS reference method is outlined in Fig. 1 and described below.

Medium

The test medium should be a completely defined synthetic medium. Broth RPMI 1640 with L-glutamine and a pH indicator and without sodium bicarbonate (04-525Y from BioWhittaker, Walkersville, Md., and American Biorganics, Inc., Niagara Falls, N.Y.; R-6504 from Sigma Chemical Co., St. Louis, Mo.) is the medium recommended by the NCCLS subcommittee. The medium should be buffered to a pH of 7.0 at 25°C. One buffer that has given satisfactory results for antifungal testing is MOPS (morpholinepropanesulfonic acid; final molarity, 0.165 for pH 7.0). This medium is suitable for the testing of amphotericin B, 5-FC, and azoles against *Candida* spp. and several of the filamentous fungi (17, 18, 20, 21, 24, 36); however, RPMI 1640 may not be adequate to support the growth of some strains of *Cryptococcus neoformans* or distinguish susceptible isolates from potential resistant strains to amphotericin B.

Drug Stock Solutions

Antifungal stock solutions should be prepared at concentrations at least 10 times the highest concentration to be tested (e.g., 1,280 μg/ml for fluconazole and 5-FC). A solution of standard 5-FC powder (Hoffmann-LaRoche) is prepared in distilled water and sterilized by filtration. For testing of amphotericin B (Squibb) or other polyenes, sufficient standard drug is weighed to prepare a solution of 1,600 μg/ml. The actual amount to be weighed must be adjusted according to the specific biological activity of the standard. If standard amphotericin B is not available, amphotericin B either as the pharmaceutical preparation (Fungizone for injection; Squibb) or as a laboratory reagent (Fungizone for laboratory use; Squibb) may be substituted. Standard substances of amphotericin B and other polyenes may be solubilized in dimethyl sulfoxide or dimethylformamide. These solutions must be protected from light and

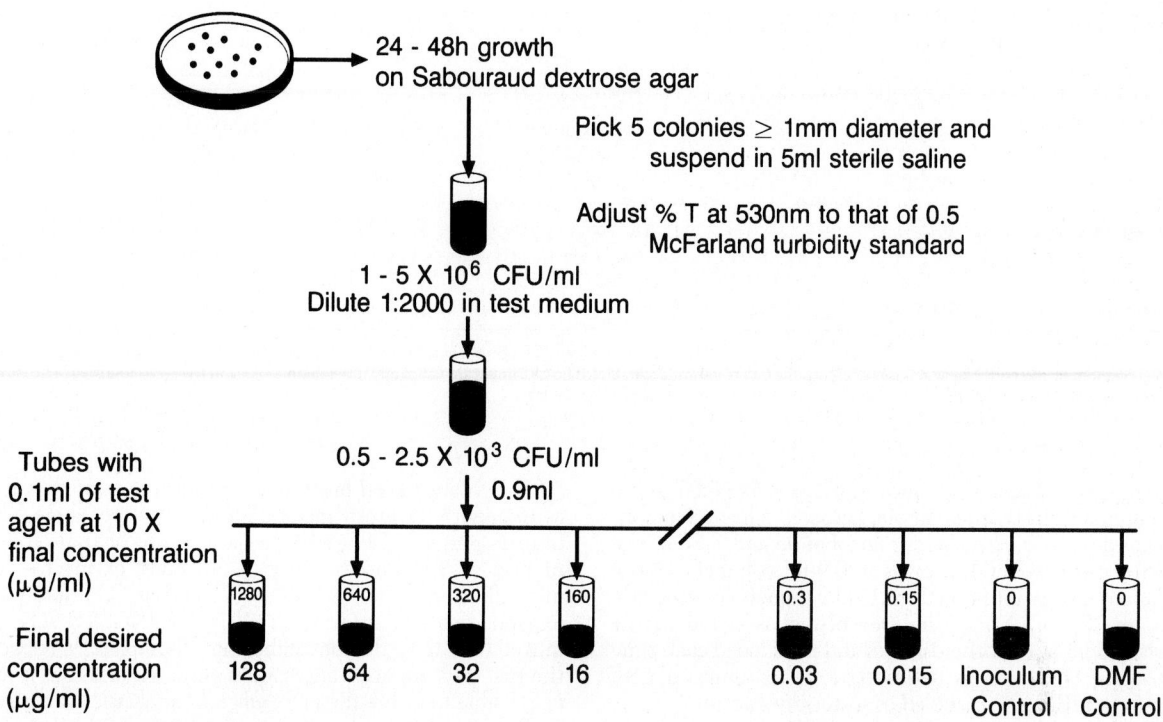

FIGURE 1 Protocol for broth macrodilution susceptibility testing of yeast cells. % T, percent transmission; DMF, dimethylformamide.

should be allowed to stand for 30 min before use to permit autosterilization.

A variety of solvent systems are used with the azoles. These systems include 10% stock solutions prepared in polyethylene glycol, dimethyl sulfoxide, dimethylformamide (36), ethanol, mixtures of ethanol and HCl (22), and sterile distilled water (5-FC and fluconazole only). Some of the antifungal agents, including itraconazole, will not remain completely solubilized upon dilution into aqueous media. Thus, at higher concentrations, a certain amount of turbidity will be encountered, and this may interfere with the interpretation of MIC endpoints. This can be avoided by using the same amount of dimethyl sulfoxide (e.g., 1%) to do the drug dilutions.

The sterile stock solutions may be stored in small volumes in sterile polypropylene or polyethylene vials carefully sealed at −60°C (preferably) or below but never at a temperature greater than −20°C. Vials are removed as needed and used the same day. Any unused drug is to be discarded at the end of the day. Stock solutions of most antifungal agents can be stored at −60°C or below for 6 months or more without significant loss of activity. The use of quality control (QC) strains will help in evaluating drug activity (36).

Preparation of Inocula

Inocula should be prepared by using the spectrophotometric method (40) as outlined in Fig. 1. The test organisms are grown on plates of Sabouraud agar (Sabouraud dextrose agar Emmons, 11589, from BBL; Sabouraud agar modified, 0747, from Difco) at 35°C and subcultured at least twice to ensure purity and viability. The inoculum suspension is prepared by picking five colonies, each at least

1 mm in diameter, from 24-h-old cultures of *Candida* and *Torulopsis* species or 48-h-old cultures of *Cryptococcus neoformans* and suspending the material in 5 ml of sterile 0.85% NaCl. The turbidity of the cell suspension measured at 530 nm is adjusted with sterile saline to match the transmittance produced by a 0.5 McFarland barium sulfate standard. This produces a cell suspension containing 1×10^6 to 5×10^6 organisms per ml, which is then diluted 1:2,000 with RPMI medium to provide a testing inoculum of 0.5×10^3 to 2.5×10^3 organisms per ml.

Drug Dilutions and Performance of Macrodilution Test

The following procedure gives sufficient material to test one isolate and the appropriate QC organism. With the RPMI medium, prepare 5 ml of a solution of the drug or drugs to be tested. For the example outlined in Fig. 1, a 0.5-ml sample of the 1,600 μg/ml stock solution of amphotericin B is added to 4.5 ml of medium to give a solution of 160 μg of amphotericin B per ml. Similar 1:10 dilutions are prepared with the other antifungal compounds. The broth macrodilution test drug dilutions are prepared to be 10 times the strength of the final drug concentration, with medium used as diluent according to the NCCLS standard additive twofold drug dilution schema (36) (e.g., 640 to 1.2 μg/ml for fluconazole and 5-FC and 160 to 0.3 μg/ml for amphotericin B, ketoconazole, and itraconazole).

The 10× drug dilutions are dispensed in 0.1-ml volumes into round-bottom polystyrene, snap-cap, sterile tubes (12 by 75 mm; Falcon 2054; Becton Dickinson Labware, Lincoln Park, N.J.). The day of the test, each tube is inoculated by adding a 0.9-ml volume of the corresponding diluted yeast inoculum suspension. This step brings the drug dilu-

TABLE 3 Expected MIC ranges for six QC isolates[a]

	MIC range (μg/ml)					
Antifungal agent	*Candida albicans* ATCC 90028	*Candida albicans* ATCC 90029	*Candida parapsilosis* ATCC 90018	*Cryptococcus neoformans* ATCC 90112	*Cryptococcus neoformans* ATCC 90113	*Torulopsis glabrata* ATCC 90030
Amphotericin B	0.25–1	0.25–1	0.25–1	0.25–1	0.5–2	0.25–1
5-FC	1–4	≥64	0.25–1	2–8	>64	≤0.12
Ketoconazole	≤0.03–0.12	≤0.03–0.12	≤0.03–0.125	0.12–0.5		0.25–1
Fluconazole	≤0.03–0.5	≤0.12–0.5	≤0.125–0.5	1–4	1–2	8–16

[a]Reproduced from reference 36. Permission to use portions of M27-P (*Reference Method for Broth Dilution Antifungal Susceptibility Testing of Yeasts: Proposed Standard*) has been granted by the NCCLS, which is not responsible for errors or inaccuracies. The QC limits are valid only if the methodology in M27-P is followed. The current M27 edition may be obtained from NCCLS, 771 E. Lancaster Ave., Villanova, PA 19085.

tions to the final test drug concentrations (16 to 0.03 μg/ml for amphotericin B, itraconazole, ketoconazole, and miconazole and 64 to 0.12 μg/ml for fluconazole and 5-FC). The growth control tube(s) receives an 0.9-ml volume(s) of the inoculum suspension(s) and a 0.1-ml volume(s) of drug-free medium. The QC organism (see below) is tested in the same manner as the other isolates and is included each time an isolate is tested. In addition, 1 ml of uninoculated, drug-free medium is included as a sterility control.

The MIC tubes are incubated at 35°C without agitation for 46 to 50 h in ambient air and observed for the presence or absence of visible turbidity or growth. When *Cryptococcus neoformans* is tested, tubes should be incubated for a total of 70 to 74 h. The MIC endpoints are determined by visually grading turbidity, and the MIC is judged to be the lowest concentration of an antifungal agent that substantially inhibits the growth of the organism. For amphotericin B, the MIC is read as the lowest concentration that prevents any discernible growth. For azoles and 5-FC, a less stringent endpoint allowing for slight turbidity above the MIC is recommended (18, 24, 36). In determining the endpoint for the azoles and 5-FC, the turbidities in all tubes are estimated visually, and the MIC is defined as the lowest drug concentration that reduces growth by 80% relative to that of the growth control (MIC_{80}). The MIC_{80} endpoint may be estimated by diluting the drug-free growth control tube 1:5 with test medium.

QC

QC of MIC tests is essential to good laboratory practice. Suitable QC organisms include *Candida albicans* ATCC 90028 and ATCC 90029, *Candida parapsilosis* ATCC 90018, *Cryptococcus neoformans* ATCC 90112 and ATCC 90113, and *T. glabrata* ATCC 90030 (36). Table 3 summarizes the expected MICs of various antifungal agents for these QC isolates. With repeated testing, more than 95% of the MICs should be within the MIC ranges represented in this table. The overall performance of the test system should be monitored by testing an appropriate QC isolate(s) each day a test is performed for each drug. More detailed information can be found in NCCLS reference method M27-P (36). NCCLS (NCCLS document M27-T) is recommending *Candida parapsilosis* ATCC 22019 and *Candida krusei* ATCC 6258 as reference strains.

Microdilution

Broth microdilution antifungal tests are similar to broth macrodilution tests and give comparable results (17, 18).

Although antifungal broth microdilution tests are not used as frequently as broth macrodilution tests are, some laboratories perform them with yeasts because of their convenience. A preliminary comparison study of macro- and microdilution antifungal tests (17) and an additional and expanded similar study reported by the NCCLS subcommittee (18) have demonstrated that discrepancies between the two tests are not statistically significant and that it may be possible to utilize the more efficient and easier to perform broth microdilution method in the clinical laboratory. In addition to low discrepancies between the two tests, the NCCLS study indicated that the microdilution test provides consistent MIC results and that interlaboratory agreement of the microdilution MICs was higher than that of the macrodilution MICs of some drugs (18). Only the steps and testing conditions that are relevant to the microdilution test will be discussed in detail.

The 10× drug dilutions described above should be diluted 1:5 with RPMI to achieve the 2× strength needed for the microdilution test. The stock inoculum suspensions are prepared and adjusted as described above for the broth macrodilution test. The stock yeast suspension is mixed for 15 s with a vortex, diluted 1:50, and further diluted 1:20 with medium to obtain the 2× test inoculum (1×10^3 to 5×10^3 CFU/ml). The 2× inoculum is diluted 1:1 when the wells are inoculated and thus the desired final inoculum size is achieved (0.5×10^3 to 2.5×10^3 CFU/ml).

The broth microdilution test is performed by using sterile, disposable, multiwell microdilution plates (96 U-shaped wells) (Dynatech Laboratories, Inc., Alexandria, Va.). A multichannel pipette is used to dispense 2× drug concentrations into the wells of rows 1 to 10 of the microdilution plates in 100-μl volumes. Row 1 contains the highest drug concentration (either 64 or 16 μg/ml), and row 10 contains the lowest drug concentration (either 0.12 or 0.03 μg/ml). Each well is inoculated on the day of the test with 100 μl of the corresponding 2× diluted inoculum suspension, which brings the drug dilutions and inoculum densities to the final concentrations mentioned above. The growth control wells contain 100 μl of sterile drug-free medium and are inoculated with 100 μl of the corresponding diluted (2×) inoculum suspensions. The QC organism is tested in the same manner and is included each time an isolate is tested. Row 11 of the microdilution plate can be used for the sterile control (drug-free medium only).

The microdilution plates are incubated at 35°C and observed for the presence or absence of visible growth. The microdilution wells are scored with the aid of a reading mirror (Cooke Engineering Co., Alexandria, Va.); the growth in each well is compared with that of the growth control (drug-free) well. A numerical score ranging from 0 to 4 on the following scale is given to each well: 0, optically clear; 1, slightly hazy; 2, prominent decrease in turbidity; 3, slight reduction in turbidity; 4, no reduction in turbidity. The MIC for amphotericin B is defined as the lowest concentration at which a score of 0 (complete absence of growth) is observed, and those for 5-FC and the azoles are described as the lowest concentrations at which a score of 2 is observed.

Expected Results

Most isolates of susceptible yeasts will be inhibited by 8.0 μg or less of 5-FC per ml. Isolates of *Cryptococcus neoformans* with MICs of 16 to 64 μg/ml as well as totally resistant isolates (MICs of ≥128 μg/ml) may be recovered from patients during treatment with 5-FC. Amphotericin B at a concentration of 1.0 μg or less per ml is both inhibitory and fungicidal for most yeasts. An amphotericin B MIC of 2.0 μg/ml (or greater) suggests probable clinical resistance, since this MIC only approximates the concentrations achievable in serum at high doses of amphotericin B (≥1 mg/kg of body weight per day) and exceeds that achievable in cerebrospinal fluid. Studies by both Powderly et al. (48) and Bryce et al. (6) have found that patients with fungemia due to *Candida* spp. with amphotericin B MICs greater than 0.8 μg/ml were significantly more likely to die because of that infection than were patients with infection due to *Candida* spp. with in vitro susceptibilities to amphotericin B at 0.8 μg/ml or less. Although these studies are interesting, additional investigation will be necessary before a clinically significant breakpoint value for amphotericin B susceptibility can be defined.

Of the azoles, miconazole and ketoconazole are fungistatic for most yeast isolates at concentrations of 0.5 to 8 μg/ml. Fluconazole inhibits the majority of *Candida* spp. and *Cryptococcus neoformans* at concentrations of 4 μg/ml or less; however, isolates of *T. glabrata* and *Candida krusei* generally have MICs of 4 to 8 and 16 to 32 μg/ml or greater, respectively. Itraconazole is generally quite active in vitro, with MICs of 0.01 to 0.5 μg/ml or less for most yeast isolates.

Macro- and Microdilution Methods for Filamentous Fungi

Although the number of serious infections caused by the filamentous fungi is lower than the number of yeast infections, antifungal susceptibility testing of these opportunistic pathogens is important in the clinical laboratory. The problems in developing a standardized susceptibility testing method for yeast fungi have been considerable, and such testing for the filamentous fungi represents an even greater challenge. It is hoped that standardization of the various antifungal susceptibility testing steps for yeasts (Fig. 1) has opened the door to developing standards for antifungal susceptibility tests for the filamentous fungi by adopting and tailoring the steps in yeast testing. These efforts are just getting under way. Recent multicenter studies have begun to address the problems of inoculum preparation and of interlaboratory reproducibility of both macro- and microdilution tests (19, 20).

In a recent multicenter study, Espinel-Ingroff et al. (19) evaluated a modification of the spectrophotometric method used to prepare yeast inocula for the preparation of inocu-

TABLE 4 Transmission ranges and sums of mean inoculum sizes for filamentous fungi in six laboratories

Fungus[a] (no. of isolates tested)	% T range[b]	Mean inoculum size (10^6 CFU/ml)
Aspergillus flavus (5)	78–82	1.2
Aspergillus fumigatus (5)	80–82	1.4
Scedosporium apiospermum (*Pseudallescheria boydii*) (5)	68–71	1.2
Rhizopus arrhizus (*Rhizopus orizae*) (5)	68–71	1.2
Sporothrix schenckii (5)	80–82	2.4
QC isolate: *Paecilomyces* sp.[c] (1)	74–76	1.1

[a]Each isolate was tested at least once in each laboratory.
[b]Percent transmission at 30 nm.
[c]Tested three to five times in each laboratory.

lum suspensions of selected filamentous fungi. Conidial suspensions were prepared, and the turbidities were adjusted with a spectrophotometer to 68 to 82% transmission at 530 nm according to the species tested. Inoculum densities of 1.4×10^6 (CFU/ml, mean value) were obtained in 90% of the six participating laboratories (Table 4). These results indicate that suitable inoculum suspensions may be reliably prepared for use in antifungal susceptibility testing of mould fungi.

Collaborative efforts have also been initiated to evaluate the suitability of the macro- and microdilution methods for yeasts as reference methods for in vitro testing of filamentous fungi (20). The in vitro susceptibilities of 25 isolates of *A. fumigatus*, *Aspergillus flavus*, *P. boydii*, *Rhizopus arrhizus*, and *Sporothrix schenckii* to five antifungal agents (amphotericin B, fluconazole, itraconazole, miconazole, and ketoconazole) were determined in six centers by both macro- and microdilution methods (20). All tests were performed in accordance with the NCCLS reference method for yeasts (Table 1). Overall, the agreement between macro- and microdilution methods was quite good: 74 to 100%, depending on the antifungal agent and organism tested. Interlaboratory agreement was similar to that observed for yeast testing and ranged from 80 to 95% for microdilution testing and from 77 to 100% for macrodilution testing. These results are quite promising and suggest that it may be possible to develop a reference method for antifungal testing of filamentous fungi.

Since the reference method for yeasts was described above, only the modified steps and testing conditions for the filamentous fungi are described below. The inoculum for each isolate is prepared by first growing the fungus on potato dextrose agar slants (Remel, Lenexa, Kans.) for 7 days at 35°C. A conidial suspension is prepared by flooding each slant with approximately 2 ml of sterile 0.85% saline. The resulting mixture is withdrawn, and the heavy particles are allowed to settle for 3 to 5 min. The upper homogeneous suspension containing the conidia is mixed for 15 s with a vortex. The turbidity of the mixed suspension is measured by using a spectrophotometer at 530 nm and adjusted to a specific final transmission range (Table 4) for each species tested (19). Only conidial suspensions of approximately 0.5×10^4 to 5×10^4 CFU/ml have been evaluated for antifungal susceptibility testing of mould fungi by this method (20). The QC organisms used for yeast testing (Table 3) plus an isolate of *Paecilomyces variotti* ATCC 22319 (Table 5) may be tested in the same manner

TABLE 5 Tentative QC limits for testing filamentous fungi with reference strain of *P. variotti*[a]

Drug	MIC-1 range[b] (μg/ml)	
	Microdilution test	Macrodilution test
Amphotericin B	0.5–1	0.25–1
Fluconazole	4–64	0.5–4
Itraconazole	0.03–4	0.03–1
Miconazole	0.03–0.25	0.03–0.25
Ketoconazole	0.06–0.5	0.06–0.5

[a]The QC isolate (*P. variotti* ATCC 22319; University of Texas) was tested with each drug and by both tests 12 times in six laboratories.
[b]Encompasses 95% of the values; MIC-1, approximately 75% reduction in growth.

as the other isolates and should be included each time an isolate is evaluated with any antifungal agent.

All tubes and microdilution trays are incubated at 35°C and observed each day for the presence of growth; when growth is visible in the growth control, each tube is vortexed for 10 s immediately prior to being scored, which allows the detection of small amounts of growth. The growth in each tube and well is compared with that of the growth control (drug free) and given a numerical score as follows: 0, optically clear or showing no growth; 1, approximately 75% reduction in growth; 2, approximately 50% reduction in growth; 3, approximately 25% reduction in growth; 4, no reduction in growth.

The MIC endpoint criterion for moulds (by both macro- and microdilution tests) is the lowest drug concentration that is given a growth score of 1 or the drug concentration that inhibits approximately ≥75% of the growth of the fungus being tested compared with growth of the growth control. This less stringent criterion allows for the trailing or partial inhibition frequently seen with the azoles.

Modifications of Reference Method for the Clinical Laboratory

Although a reference method for in vitro susceptibility testing is essential for purposes of standardization and interlaboratory reproducibility, it may not be the best method for testing all organisms or the most convenient method for routine use in the clinical laboratory. Thus, modifications of the reference method are acceptable and expected. With this in mind, several modifications of the macrobroth reference method of antifungal susceptibility testing are currently under investigation. They offer promise as alternative approaches that may better serve practical clinical laboratory needs. These methods include colorimetric endpoint methods for MIC endpoint determination and agar diffusion MIC methods.

Colorimetric Methods for MIC Endpoint Determination

A novel alternative to the standard method of visually grading turbidity is the use of a colorimetric oxidation-reduction indicator to aid in the determination of MIC endpoints (39, 42, 61). This approach is a reliable indicator of the number of viable yeast cells present and allows the determination of a metabolic MIC (61). Recently, the NCCLS-based microdilution assay described above (18) has been adapted to a colorimetric format by the simple addition of a commercially available oxidation-reduction

indicator, Alamar Blue (Alamar) (39, 42). The result is a colorimetric microdilution method that provides clear MIC endpoints that can be determined visually and minimizes the trailing endpoints that are observed frequently with the azoles. In testing fluconazole against *Candida albicans*, the agreement of the colorimetric oxidation-reduction microdilution MIC readings with the standard macrodilution MIC$_{80}$ or microdilution MIC readings of 2 were excellent: 94 and 95%, respectively (42). A similar high level of agreement has been observed with other species of *Candida* and with amphotericin B and 5-FC (39). The method is simple to perform and may provide a more objective and reproducible endpoint than those based on visual estimates of turbidity.

Agar Diffusion Methods

Agar diffusion methods for antimicrobial susceptibility testing are in common use in clinical microbiology laboratories. Worldwide, the most commonly employed technique for antibacterial susceptibility testing is the disk diffusion test, which yields a quantitative result (zones of inhibition) and a qualitative interpretive category (e.g., susceptible or resistant). Recently, the E test (AB Biodisk) has been introduced as a means of producing an accurate, reproducible quantitative MIC result by using an agar diffusion format (54). The E test is based on the diffusion of a continuous concentration gradient of an antimicrobial agent from a plastic strip into an agar medium. This newly developed in vitro test is comparable to the current broth and agar methods for antibacterial testing and is superior to disk test methods (54). The flexibility of the E test makes it applicable to a wide range of antimicrobial susceptibility testing, including antifungal testing.

The application of agar diffusion methods to the testing of antifungal agents has been limited to date. An international collaborative study compared fluconazole disk and MIC tests and found the disk test to be comparable to the MIC test (41, 43). The E test also has had limited application to antifungal testing; however, preliminary evaluations suggest that it correlates well with the NCCLS reference method (16). The E test appears to be most valuable in testing organisms for which there is no routine method of quantitative testing and may be useful in testing antimicrobial agents, such as the azoles, that give trailing endpoints with broth dilution methods.

CLINICAL CORRELATION

In order to be useful clinically, in vitro susceptibility testing should give a reliable prediction of in vivo response to therapy in human infections. Unfortunately, the evidence for clinical correlation with in vitro antifungal susceptibility tests is limited at best (15, 23, 27, 38, 45, 51). This is not surprising, given the problems with in vitro susceptibility testing methods and the difficulties encountered in interpreting the outcome of therapy in the very complex patients involved (27, 31, 44, 51). In addition, therapeutic failures due to resistant organisms appear to be rare. The recent recognition of relative resistance to amphotericin B among isolates of *Candida* spp. from immunocompromised patients (6, 11, 48); the well-known problem of amphotericin B-resistant *Candida lusitaniae*; and the emergence of azole resistance among isolates of *Candida albicans*, *T. glabrata*, and *Candida krusei* (1, 9, 35, 50, 64, 67, 68) suggest that this picture may change. The fact that in vitro susceptibility testing was able to detect these clinically resistant isolates is promising and suggests that prediction of clinical outcome by in vitro susceptibility tests is possible.

One major problem with in vitro antifungal susceptibility tests is the trailing endpoint observed with the azole antifungal agents. Clinical isolates of *Candida* spp. frequently have fluconazole MICs of 32 to ≥ 64 µg/ml yet appear to respond to therapy (43). It is possible that the use of the less stringent MIC_{80} endpoint recommended by the NCCLS will correct this problem; however, this must be validated in clinical trials. We emphasize that a critical next step in establishing the clinical usefulness of the NCCLS reference procedure is to determine the relationship between test results and patient responses to therapy (36). *This has not yet been done.* Studies to define the population distribution of MICs of amphotericin B, 5-FC, and fluconazole for several species of yeast fungi have begun. However, the findings are preliminary and include no clinical correlations. Thus, additional clinical studies will be necessary before meaningful interpretive breakpoint criteria can be developed.

REFERENCES

1. **Akova, M., H. E. Akalin, O. Uzun, and D. Gür.** 1991. Emergence of *Candida krusei* infections after therapy of oropharyngeal candidiasis with fluconazole. *Eur. J. Clin. Microbiol. Infect. Dis.* **10:**598–599.
2. **Anaissie, E. J., G. P. Bodey, and M. G. Rinaldi.** 1989. Emerging fungal pathogens. *Eur. J. Clin. Microbiol. Infect. Dis.* **8:**323–330.
3. **Arndt, C. A. S., T. J. Walsh, C. L. McCully, F. M. Balis, P. A. Pizzo, and D. G. Poplack.** 1988. Fluconazole penetration into cerebrospinal fluid: implications for treating fungal infections of the central nervous system. *J. Infect. Dis.* **157:**178–180.
4. **Beck-Sagué, C. M., W. R. Jarvis, and the National Nosocomial Infections Surveillance System.** 1993. Secular trends in the epidemiology of nosocomial fungal infections in the United States, 1980–1990. *J. Infect. Dis.* **167:**1247–1251.
5. **Bradsher, R. W.** 1992. Blastomycosis. *Clin. Infect. Dis.* **14**(Suppl. 1)**:**S82–S90.
6. **Bryce, E. A., F. J. Roberts, A. S. Sekhon, and A. Coldman.** 1992. Yeast in blood cultures: evaluation of factors influencing outcome. *Diagn. Microbiol. Infect. Dis.* **15:**233–237.
7. **Bullock, W., T. Kozel, S. Scherer, and D. M. Dixon.** 1993. Medical mycology in the 1990s: involvement of NIH and the wider community. *ASM News* **59:**182–185.
8. **Calhoun, D. L., G. D. Roberts, J. N. Galgiani, J. E. Bennett, D. S. Feingold, J. Jorgensen, G. S. Kobayashi, and S. Shadomy.** 1986. Results of a survey of antifungal susceptibility tests in the United States and interlaboratory comparison of broth dilution testing of flucytosine and amphotericin B. *J. Clin. Microbiol.* **23:**298–301.
9. **Casanovas, R. O., D. Caillot, E. Solary, B. Bonotte, P. Chavanet, A. Bonin, P. Camerlynck, and H. Guy.** 1992. Prophylactic fluconazole and *Candida krusei* infections. *N. Engl. J. Med.* **326:**891–892.
10. **Daneshmend, T. K., D. W. Warnock, M. D. Ene, E. M. Johnson, M. R. Potten, M. D. Richardson, and P. J. Williamson.** 1984. Influence of food on the pharmacokinetics of ketoconazole. *Antimicrob. Agents Chemother.* **25:**1–3.
11. **Dick, J. D., W. G. Merz, and R. Saral.** 1980. Incidence of polyene-resistant yeasts recovered from clinical specimens. *Antimicrob. Agents Chemother.* **18:**158–163.
12. **Dismukes, W. E., R. W. Bradsher, G. C. Cloud, C. A. Kauffman, and the NIAID Mycoses Study Group.** 1992. Itraconazole therapy for blastomycosis and histoplasmosis. *Am. J. Med.* **93:**489–497.
13. **Dismukes, W. E., A. M. Stamm, J. R. Graybill, P. C. Craven, D. A. Stevens, R. L. Stiller, G. A. Sarosi, G. Medoff, C. R. Gregg, H. A. Gallis, B. T. Fields, Jr., R. L.**

14. **Marier, T. M. Kerkering, L. G. Kaplowitz, G. Cloud, C. Bowles, and S. Shadomy.** 1983. Treatment of systemic mycoses with ketoconazole: emphasis on toxicity and clinical response in 52 patients. *Ann. Intern. Med.* **98:**13–20.
14. **Doern, G. V., T. A. Tubert, K. Chopin, and M. G. Rinaldi.** 1986. Effect of medium composition on results of macrobroth dilution antifungal susceptibility testing of yeasts. *J. Clin. Microbiol.* **24:**507–511.
15. **Drutz, D. J.** 1987. In vitro antifungal susceptibility testing and measurement of levels of antifungal agents in body fluids. *Rev. Infect. Dis.* **9:**392–397.
16. **Espinel-Ingroff, A.** E test for antifungal susceptibility testing of yeasts. *Diagn. Microbiol. Infect. Dis.* **19,** in press.
17. **Espinel-Ingroff, A., T. M. Kerkering, P. R. Goldson, and S. Shadomy.** 1991. Comparison study of broth macrodilution and microdilution antifungal susceptibility tests. *J. Clin. Microbiol.* **29:**1089–1094.
18. **Espinel-Ingroff, A., C. W. Kish, Jr., T. M. Kerkering, R. A. Fromtling, K. Bartizal, J. N. Galgiani, K. Villareal, M. A. Pfaller, T. Gerarden, M. G. Rinaldi, and A. Fothergill.** 1992. Collaborative comparison of broth macrodilution and microdilution antifungal susceptibility tests. *J. Clin. Microbiol.* **30:**3138–3145.
19. **Espinel-Ingroff, A., M. Pfaller, E. Anaissie, B. Breslin, D. Dixon, A. Fothergill, J. Peter, M. Rinaldi, and T. Walsh.** 1993. Multicenter evaluation of standardized inoculum for antifungal susceptibility testing of filamentous fungi, abstr. F-11, p. 529. *Abstr. 93rd Gen. Meet. Am. Soc. Microbiol. 1993.*
20. **Espinel-Ingroff, A., M. Pfaller, E. Anaissie, B. Breslin, D. Dixon, A. Fothergill, J. Peter, M. Rinaldi, and T. Walsh.** Standardizing antifungal susceptibility testing for filamentous fungi: a comparative and collaborative study. *Antimicrob. Agents Chemother.,* in press.
21. **Espinel-Ingroff, A., and S. Shadomy.** 1989. In vitro and in vivo evaluation of antifungal agents. *Eur. J. Clin. Microbiol. Infect. Dis.* **8:**352–361.
22. **Espinel-Ingroff, A., S. Shadomy, and R. J. Gebhart.** 1984. In vitro studies with R 51,211 (itraconazole). *Antimicrob. Agents Chemother.* **26:**5–9.
23. **Fromtling, R. A.** 1988. Overview of medically important antifungal azole derivatives. *Clin. Microbiol. Rev.* **1:**187–217.
24. **Fromtling, R. A., J. N. Galgiani, M. A. Pfaller, A. Espinel-Ingroff, K. F. Bartizal, M. S. Bartlett, B. A. Body, C. Frey, G. Hall, G. D. Roberts, F. B. Nolte, F. C. Odds, M. G. Rinaldi, A. M. Sugar, and K. Villareal.** 1993. Multicenter evaluation of a macrobroth antifungal susceptibility test for yeasts. *Antimicrob. Agents Chemother.* **37:**39–45.
25. **Galgiani, J. N.** 1984. Why not standardize antifungal susceptibility testing? *Antimicrob. Newsl.* **1:**40.
26. **Galgiani, J. N.** 1987. Antifungal susceptibility tests. *Antimicrob. Agents Chemother.* **31:**1867–1870.
27. **Galgiani, J. N.** 1990. Susceptibility of *Candida albicans* and other yeasts to fluconazole: relation between in vitro and in vivo studies. *Rev. Infect. Dis.* **12:**S272–S275.
28. **Galgiani, J. N., J. Reiser, C. Brass, A. Espinel-Ingroff, M. A. Gordon, and T. M. Kerkering.** 1987. Comparison of relative susceptibilities of *Candida* species to three antifungal agents as determined by unstandardized methods. *Antimicrob. Agents Chemother.* **31:**1343–1347.
29. **Hadfield, T. L., M. B. Smith, R. E. Winn, M. G. Rinaldi, and C. Guerra.** 1987. Mycoses caused by *Candida lusitaniae*. *Rev. Infect. Dis.* **9:**1006–1012.
30. **Huang, Y. C., J. Colaizzi, R. H. Bieman, R. Woestenborghs, and J. Heykants.** 1986. Pharmacokinetics and dose proportionality of ketoconazole in normal volunteers. *Antimicrob. Agents Chemother.* **30:**206–210.
31. **Kobayashi, G., and E. D. Spitzer.** 1989. Testing of organisms for susceptibility to triazoles: is it justified? *Eur. J. Clin. Microbiol. Infect. Dis.* **8:**387–389.
32. **Lopez-Berestein, G., G. P. Bodey, L. S. Frankel, and K. Mehta.** 1987. Treatment of hepatosplenic candidiasis with liposomal amphotericin B. *J. Clin. Oncol.* **5:**310–317.
33. **Lutwick, L. I., M. W. Rytel, J. P. Yanez, J. N. Galgiani, and**

D. A. Stevens. 1979. Deep infections from *Petriellidium boydii* treated with miconazole. *JAMA* **241**:272–273.

34. **McGinnis, M. R., and M. G. Rinaldi.** 1986. Antifungal drugs: mechanisms of action, drug resistance, susceptibility testing, and assays of activity in biological fluids, p. 223–281. *In* V. Lorian (ed.), *Antibiotics in Laboratory Medicine.* The Williams & Wilkins Co., Baltimore.

35. **McIlroy, M. A.** 1991. Failure of fluconazole to suppress fungemia in a patient with fever, neutropenia, and typhlitis. *J. Infect. Dis.* **163**:420–421.

36. **National Committee for Clinical Laboratory Standards.** 1992. *Reference Method for Broth Dilution Susceptibility Testing of Yeast: Proposed Standard.* NCCLS document M27-P (ISBM 1-56238-186-5). National Committee for Clinical Laboratory Standards, Villanova, Pa.

37. **Normark, S., and J. Schönebeck.** 1972. In vitro studies of 5-fluorocytosine resistance in *Candida albicans* and *Torulopsis glabrata. Antimicrob. Agents Chemother.* **2**:114–121.

38. **Odds, F. C.** 1985. Laboratory tests for the activity of imidazole and triazole antifungal agents in vitro. *Semin. Dermatol.* **4**:260–270.

39. **Pfaller, M. A., and A. L. Barry.** 1994. Evaluation of a novel colorimetric broth microdilution method for antifungal susceptibility testing of yeast isolates. *J. Clin. Microbiol.* **32**:1992–1996.

40. **Pfaller, M. A., L. Burmeister, M. S. Bartlett, and M. G. Rinaldi.** 1988. Multicenter evaluation of four methods of yeast inoculum preparation. *J. Clin. Microbiol.* **26**:1437–1441.

41. **Pfaller, M. A., B. DuPont, G. S. Kobayashi, J. Müller, M. G. Rinaldi, A. Espinel-Ingroff, S. Shadomy, P. F. Troke, T. J. Walsh, and D. W. Warnock.** 1992. Standardized susceptibility testing of fluconazole: an international collaborative study. *Antimicrob. Agents Chemother.* **36**:1805–1809.

42. **Pfaller, M., C. Grant, V. Morthland, and J. Rhine-Chalberg.** 1994. Comparative evaluation of alternative methods for broth dilution susceptibility testing of fluconazole against *Candida albicans. J. Clin. Microbiol.* **32**:506–509.

43. **Pfaller, M. A., and M. G. Rinaldi.** 1992. In vitro testing of susceptibility of fluconazole, p. 10–22. *In* W. B. Powderly and J. W. Van't Wout (ed.), *The Antifungal Agents,* vol. 1. *Fluconazole.* Marius Press, Lancashire, United Kingdom.

44. **Pfaller, M. A., and M. G. Rinaldi.** 1993. Antifungal susceptibility testing: current state of technology, limitations, and standardizations. *Infect. Dis. Clin. N. Am.* **7**:435–444.

45. **Polak, A., and D. M. Dixon.** 1987. In vitro/in vivo correlation of antifungal susceptibility testing using 5-fluorocytosine and ketoconazole as examples of two extremes, p. 45–59. *In* R. A. Fromtling (ed.), *Recent Trends in the Discovery, Development and Evaluation of Antifungal Agents.* J. R. Prous Science Publisher, Barcelona, Spain.

46. **Polak, A., and H. J. Scholer.** 1975. Mode of action of 5-fluorocytosine and mechanisms of resistance. *Chemotherapy* **21**:113–130.

47. **Powderly, W. G.** 1992. Therapy for cryptococcal meningitis in patients with AIDS. *Clin. Infect. Dis.* **14**(Suppl. 1):S54–S59.

48. **Powderly, W. G., G. S. Kobayashi, G. P. Herzig, and G. Medoff.** 1988. Amphotericin B-resistant yeast infection in severely immunocompromised patients. *Am. J. Med.* **84**:826–832.

49. **Radetsky, M., R. C. Wheeler, M. H. Roe, and J. K. Todd.** 1986. Microtiter broth dilution method for yeast susceptibility testing with validation by clinical outcome. *J. Clin. Microbiol.* **24**:600–606.

50. **Redding, S., J. Smith, G. Farinacci, M. Rinaldi, A. Fothergill, J. Rhine-Chalberg, and M. Pfaller.** 1994. Development of resistance to fluconazole by *Candida albicans* during treatment of oropharyngeal candidiasis in AIDS: documentation by in vitro susceptibility testing and DNA subtype analysis. *Clin. Infect. Dis.* **18**:240–242.

51. **Rex, J. H., M. A. Pfaller, M. G. Rinaldi, A. Polak, and J. N. Galgiani.** 1993. Antifungal susceptibility testing. *Clin. Microbiol. Rev.* **6**:367–381.

52. **Ryley, J. F., R. G. Wilson, and K. J. Barrett-Bee.** 1984. Azole resistance in *Candida albicans. Sabouraudia* **22**:53–63.

53. **Saag, M. S., and W. E. Dismukes.** 1988. Azole antifungal agents: emphasis on new triazoles. *Antimicrob. Agents Chemother.* **32**:1–8.

54. **Sanchez, M. L., and R. N. Jones.** 1992. E test, an antimicrobial susceptibility testing method with broad clinical and epidemiologic application. *Antimicrob. Newsl.* **8**:1–7.

55. **Sarosi, G. A., and P. C. Johnson.** 1992. Disseminated histoplasmosis in patients infected with human immunodeficiency virus. *Clin. Infect. Dis.* **14**(Suppl. 1):S60–S67.

56. **Shadomy, S., A. Espinel-Ingroff, and R. Y. Cartwright.** 1985. Laboratory studies with antifungal agents: susceptibility tests and bioassays, p. 991–999. *In* E. H. Lennette, A. Balows, W. J. Hausler, Jr., and H. J. Shadomy (ed.), *Manual of Clinical Microbiology,* 4th ed. American Society for Microbiology, Washington, D.C.

57. **Shehab, Z. M., H. Britton, J. H. Dunn.** 1988. Imidazole therapy of coccidioidal meningitis in children. *Pediatr. Infect. Dis. J.* **7**:40–44.

58. **Sokol-Anderson, M. L., J. Brajtburg, and G. Medoff.** 1986. Amphotericin B-induced oxidative damage and killing of *Candida albicans. J. Infect. Dis.* **154**:76–83.

59. **Stiller, R. L., J. E. Bennett, H. J. Scholer, H. J. Wall, A. Polak, and D. A. Stevens.** 1983. Correlation of in vitro susceptibility test results with in vivo response: flucytosine therapy in a systemic candidiasis model. *J. Infect. Dis.* **147**:1070–1076.

60. **Sung, J. P., and J. G. Grendahl.** 1977. Clinical experimental therapy with miconazole for human disseminated coccidioidomycosis, p. 293–309. *In* L. Ajello (ed.), *Coccidioidomycosis, Current Clinical and Diagnostic Status.* Symposia Specialists, Miami.

61. **Tellier, R., M. Krajden, G. A. Grigoriew, and I. Campbell.** 1992. Innovative endpoint determination system for antifungal susceptibility testing of yeasts. *Antimicrob. Agents Chemother.* **36**:1619–1625.

62. **Trejo, W. H., and R. Bennett.** 1963. *Streptomyces nodosus* sp. nov. The amphotericin-producing organism. *J. Bacteriol.* **85**:436–439.

63. **Walsh, T. J., and A. Pizzo.** 1988. Treatment of systemic fungal infections: recent progress and current problems. *Eur. J. Clin. Microbiol. Infect. Dis.* **7**:460–475.

64. **Warnock, D. W., J. Burke, N. J. Cope, E. M. Johnson, N. A. vonFraunhofer, and E. W. Williams.** 1988. Fluconazole resistance in *Candida glabrata. Lancet* **ii**:1310.

65. **Wey, S. B., M. Mori, M. A. Pfaller, R. F. Woolson, and R. P. Wenzel.** 1988. Hospital acquired candidemia: the attributable mortality and excess length of stay. *Arch. Intern. Med.* **148**:2642–2645.

66. **Wheat, L. J.** 1992. Histoplasmosis in Indianapolis. *Clin. Infect. Dis.* **14**(Suppl. 1):S91–S99.

67. **Willocks, L., C. L. S. Leen, R. P. Brettle, D. Urguhart, T. B. Russell, and L. J. R. Milne.** 1991. Fluconazole resistance in AIDS patients. *J. Antimicrob. Chemother.* **28**:937–939.

68. **Wingard, J. R., W. G. Merz, M. G. Rinaldi, T. R. Johnson, J. E. Karp, and R. Saral.** 1991. Increase in *Candida krusei* infection among patients with bone marrow transplantation and neutropenia treated prophylactically with fluconazole. *N. Engl. J. Med.* **325**:1274–1277.

Antiviral Agents and Susceptibility Testing

ELLA M. SWIERKOSZ AND KAREN K. BIRON

121

One of the areas of most rapid advancement in the study of infectious diseases is the treatment of viral infections. Many viral infections can now be successfully managed with an expanding repertoire of antiviral agents. However, long-term use of some antiviral agents has led to emergence of drug-resistant virus strains, particularly in immunocompromised patients. In certain clinical situations, when patients fail to respond to antiviral therapy, diagnostic virology laboratories are being asked to perform antiviral susceptibility testing. This chapter briefly discusses the major antiviral agents in use and the clinical situations whereby resistant virus strains emerge, thus necessitating in vitro susceptibility testing. An overview of the laboratory methods used for detecting antiviral resistance is presented.

The major antiviral agents in use are listed in Table 1 and have been discussed in detail in a review by Bean (1). Alpha interferon and the topical agents idoxuridine and trifluridine are not considered in this chapter but are discussed by Bean (1). An understanding of the mechanisms of drug action is essential for a discussion of antiviral resistance. Most of the agents discussed here are nucleoside analogs with selective action on viral replication.

ANTIVIRAL AGENTS AND MECHANISMS OF ACTION AND RESISTANCE

Herpesviruses

Agents active against the herpes group viruses include vidarabine, acyclovir, ganciclovir, and foscarnet. Vidarabine (9-beta-D-arabinofuranosyladenine; adenine arabinoside; Vira-A), an adenosine nucleoside analog, was the first systemic antiviral agent to be licensed for use in the United States (1). It is active in vitro against herpes simplex virus (HSV), varicella-zoster virus (VZV), human cytomegalovirus (CMV), and vaccinia virus. It revolutionized the treatment of HSV encephalitis and neonatal herpes by dramatically reducing mortality and neurologic sequelae. Vidarabine is phosphorylated by cellular kinases to its triphosphate form, which acts as an inhibitor of viral DNA polymerase. However, cellular DNA polymerase is also inhibited, as are other cellular functions (1). Its toxicity and poor solubility in water contributed to its replacement by acyclovir. Vidarabine-resistant virus can be generated in the laboratory (14) but is not a problem clinically. Despite its in vitro activity against acyclovir-resistant HSV, it has not proven useful for treatment of human immunodeficiency virus (HIV)-infected individuals (65).

Acyclovir [9-(2-hydroxyethoxymethyl)guanine; acycloguanosine; Zovirax], a guanosine analog, has in vitro activity against a number of herpes group viruses. Clinically it is useful only for treatment of HSV and VZV infections. Because it is relatively nontoxic and easier to administer, acyclovir has largely replaced vidarabine as the treatment of choice for HSV and VZV infections. Acyclovir is phosphorylated by a virus-specific enzyme, thymidine kinase (TK), to acyclovir monophosphate, which is subsequently phosphorylated to acyclovir triphosphate by cellular kinases. The triphosphate form of acyclovir competes with the natural substrate dGTP for viral DNA polymerase. Because it has a higher affinity for the viral polymerase than dGTP does, it is preferentially incorporated into newly synthesized viral DNA. This results in termination of DNA synthesis, because acyclovir lacks the 3' hydroxyl group necessary to form phosphodiester linkages with incoming nucleotides (1). Inactivation of the HSV DNA polymerase also occurs as a consequence of the tight binding of the enzyme to the template-primer-acyclovir monophosphate complex that forms during incorporation of acyclovir monophosphate (60). This inactivation provides an additional and unique means of shutting down viral DNA synthesis with acyclovir treatment.

Acyclovir resistance arises as a result of mutations in the viral TK or DNA polymerase genes. Two types of TK mutations have been recognized for HSV. One mutation produces strains deficient in TK, which results in reduced or absent phosphorylation of acyclovir. The second mutation results in altered substrate binding properties for acyclovir (TK^A). Most acyclovir-resistant clinical isolates of HSV and VZV are TK deficient (12). TK mutations render these viruses less pathogenic in animals (29). However, in immunocompromised hosts, such as bone marrow transplant and HIV-infected patients, TK^- virus can cause progressive disease (2, 25). TK^- viruses are, however, also restricted in their ability to reactivate from latency (15, 39). Therefore, subsequent recurrences in the same patient could again be responsive to acyclovir (TK^+) (63, 65).

DNA polymerase mutations of HSV, VZV, and CMV may confer only marginal shifts in 50% inhibitory concentrations (IC_{50}); these modest shifts in drug susceptibility

TABLE 1 Major antiviral agents in use

Agent	Clinical indications for use
Acyclovir	Localized and systemic HSV and VZV infections; prophylaxis of transplant recipients
Amantadine-rimantidine	Prophylaxis and treatment of influenza A
AZT	HIV infection
ddI, d4T, ddC	HIV-1 in patients not responsive to AZT or with AZT-resistant virus
Foscarnet	Acyclovir-resistant HSV and VZV, ganciclovir-resistant CMV, and CMV infections in immunocompromised patients
Ganciclovir	CMV infections in immunocompromised patients; prophylaxis of transplant recipients
Idoxuridine	HSV keratitis
Alpha interferon	Chronic hepatitis B and C and condyloma acuminata
Ribavirin	Severe respiratory syncytial virus infections, Lassa fever
Trifluridine	HSV keratitis
Vidarabine[a]	Herpes encephalitis, neonatal herpes, zoster in immunocompromised host

[a] Acyclovir therapy has largely supplanted vidarabine therapy.

may not be clinically important if levels of drug in plasma exceed the in vitro resistance level (18). However, the virulence and ability to establish and reactivate from latency of these mutants, unlike TK mutants, can be like those of wild-type virus (29).

Acyclovir-resistant HSV has rarely been recovered from patients with normal immunity (48). It is not uncommon to recover resistant HSV from transplant and HIV-infected patients after prolonged treatment with acyclovir (25, 63, 65). Acyclovir-resistant VZV has been recovered from AIDS patients after chronic acyclovir therapy (49, 55, 64). Recently, repeated reactivations of acyclovir-resistant HSV in an immunocompetent individual have been documented. This virus was shown to be TK[A] (44).

Ganciclovir [9-(1,3-dihydroxy-2-propoxy)methylguanine; Cytovene] is another guanosine analog that is similar in structure to acyclovir. However, it is up to 100 times more active than acyclovir against human CMV (1). Its spectrum of activity includes HSV and CMV. Because it is more toxic than acyclovir, ganciclovir is generally used only to treat serious CMV infections. Ganciclovir is activated to its monophosphate form by virus-specific TK in HSV-infected cells and by a CMV-specific phosphotransferase in CMV-infected cells (50). It is subsequently phosphorylated to ganciclovir triphosphate by cellular kinases. Ganciclovir triphosphate acts as a competitive inhibitor of viral DNA polymerase but, unlike acyclovir, does not cause DNA chain termination (1). Resistance of clinical isolates of CMV to ganciclovir has been associated with point mutations in the catalytic domain of the phosphotransferase gene, which results in failure to phosphorylate ganciclovir (13). Clinical isolates of ganciclovir-resistant CMV have been recovered from transplant and AIDS patients (6, 23, 26).

Foscarnet (trisodium phosphonoformate; Foscavir) is a pyrophosphate analog with activity against herpes viruses, hepatitis B virus, and HIV type 1 (HIV-1) (1). Foscarnet serves as a noncompetitive inhibitor of viral DNA polymerases and of reverse transcriptase of HIV-1. It has recently been licensed for treatment of CMV retinitis in AIDS patients. Acyclovir-resistant HSV and VZV have also been successfully treated with foscarnet (63–65). Resistance to foscarnet is due to mutation of the viral DNA polymerase gene (74). Clinical isolates of foscarnet-resistant CMV and HSV have been recovered (43, 66, 74).

Interestingly, foscarnet-resistant HSV is susceptible or borderline susceptible to acyclovir, suggesting that the binding site of acyclovir triphosphate on the viral DNA polymerase may differ from that for foscarnet (66).

Respiratory Viruses

Ribavirin (1-β-D-ribofuranosyl-1,2,4-trizole-3-carboxamide; Virazole) is a guanosine analog with in vitro activity against a wide variety of viruses, including respiratory syncytial virus, influenza A and B viruses, parainfluenza virus, Lassa fever virus, and HIV. Ribavirin is phosphorylated intracellularly; both monophosphate and triphosphate forms are active. The exact mechanisms of action are unclear, but ribavirin inhibits several steps in viral replication, including capping of viral mRNA and elongation of viral mRNAs (1). Clinically, it is used almost exclusively for treatment of respiratory syncytial virus infections in infants. To date, no clinically significant resistance to ribavirin has been encountered.

Amantadine (1-adamantanamine hydrochloride; Symmetrel) is a primary symmetrical amine active against influenza A virus. It prevents viral uncoating and release of viral RNA into the cytoplasm (1). Rimantadine (Flumadine) is a structurally similar compound with an identical spectrum and mechanism of activity. Resistance to these drugs is due to a mutation in the M2 (matrix) protein (4). Clinical isolates of influenza A virus resistant to amantadine and rimantadine have readily been recovered during therapy, which may limit the widespread use of these two drugs (35, 37).

HIV-1

Zidovudine (azidothymidine; 3′-azido-3′-dexoxythymidine; AZT; Retrovir) is a thymidine analog with activity against HIV and other human retroviruses. It is approved for primary therapy of HIV-1 infection. Other agents approved for treatment of HIV in patients refractory to or intolerant of AZT include didanosine (2′,3′-dideoxyinosine; ddI; Videx) and zalcitabine (dideoxycytidine; ddC; Hivid). 2′,3′-Didehydro-2′,3′-dideoxythymidine (d4T, stavudine, Zerit) is currently being evaluated in clinical trials. In HIV-infected cells, 2′,3′-dideoxynucleosides are converted by cellular kinases, nucleotidases, or other enzymes to their respective 5′ triphosphates, which serve as competitive inhibitors of HIV reverse transcriptase. Incorporation of the triphosphate form into the growing DNA chain also leads to chain termination, because these agents lack free 3' hydroxyl groups. Moreover, AZT monophosphate competitively inhibits the cellular enzyme thymidylate kinase, causing a reduction in the levels of thymidine triphosphate (1, 70).

Prolonged monotherapy with AZT results in recovery of AZT-resistant variants of HIV-1. Larder and colleagues (45) measured AZT susceptibility of HIV-1 isolates from HIV-infected patients on therapy for various times by a plaque assay with CD4$^+$ HeLa cells. IC$_{50}$s for isolates from untreated patients were 0.01 to 0.05 μM, and those for isolates obtained after 6 or more months of therapy were 0.04 to 6 μM. Genotypic characterization of these isolates demonstrated five amino acid substitutions in the reverse transcriptase gene at positions 41, 67, 70, 215, and 219 (42, 47). Further analysis revealed that a mutation at either codon 70 or 215 alone (or in combination) caused partial resistance, while four mutations were required for high-level resistance (codons 70, 215, 67, and either 41 or 219) (11, 42, 62). Furthermore, a single patient can harbor mixtures of HIV-1 isolates with different in vitro susceptibilities to AZT (62). Cross-resistance to other 2',3'-dideoxynucleosides was not observed with AZT-resistant isolates (70). Similar characterization of ddI- and ddC-resistant isolates has determined that mutations at codon 74 (Leu-74→Val) and codon 184 (Met-184→Val) resulted in resistance to ddI and cross-resistance to ddC (32, 73). The Leu-74→Val change also resulted in a reversal of the AZT resistance conferred by a Thr-215→Tyr substitution in the reverse transcriptase (73). However, reversal of AZT resistance does not always occur with acquisition of the mutation at codon 74; clinical isolates that contained the mutation at codon 74 in combination with mutations at codons 41 and 215 exhibited some degree of AZT resistance (27). Eron et al. further demonstrated that the AZT resistance mutations at codons 215 and 219 augmented the ddI resistance conferred by the mutation at codon 74 (27). A double mutation at codon 69 resulting in a Thr-69→Asp substitution also confers resistance to ddC (30).

CLINICAL INDICATIONS FOR ANTIVIRAL SUSCEPTIBILITY TESTING

Because drug-resistant mutants do emerge in clinical practice, in vitro antiviral susceptibility testing is warranted in certain clinical situations. Persistent or worsening HSV or VZV infection while on acyclovir may indicate drug resistance. Alternative therapy, i.e., foscarnet, is available; therefore, susceptibility testing can influence patient management decisions. Furthermore, HSV or VZV causing recurrent infection is often TK competent and therefore again susceptible to acyclovir, which has minimal toxicity compared to foscarnet (1). Persistent or worsening CMV retinitis, pneumonitis, or colitis unresponsive to ganciclovir may also indicate drug-resistant virus. Foscarnet can be used as an alternative agent. Influenza A virus isolates resistant to amantadine and rimantadine readily emerged in clinical trials of these drugs; therefore, continuous shedding or transmission of influenza A virus in a population on prophylaxis or being treated with amantadine or rimantadine may be due to drug resistance. Monitoring of HIV isolates is essential to investigating the relationship between the development of disease progression and the emergence of in vitro resistance to nucleoside analog inhibitors of HIV-1 reverse transcriptase, to test for cross-resistance to alterna-

tive HIV-1 antiviral agents, to standardize HIV susceptibility testing, and to develop rapid susceptibility testing assays. The prospect of combination therapy for HIV-1 further underscores the urgency for reliable susceptibility testing.

DEFINITION OF ANTIVIRAL RESISTANCE

Antiviral resistance is a decrease in susceptibility to an antiviral drug that can be clearly established by in vitro testing and can be confirmed by genetic analysis of the virus and biochemical study of the altered enzymes. In vitro drug resistance must be distinguished from clinical resistance, in which the viral infection fails to respond to therapy. Clinical failures may or may not be due to the presence of a drug-resistant virus. Failure to achieve clinical response also hinges on other factors such as the patient's immunologic status and the pharmacokinetics of the drug in that individual patient. These factors, and perhaps others, may contribute to a poor clinical response, even though the antiviral susceptibility testing of the viral isolate denotes in vitro susceptibility (79). In vitro resistance has been correlated with clinical resistance; for example, patients with HIV infection and other immune disorders who have HSV infections refractory to acyclovir frequently shed TK-deficient HSV strains for which IC$_{50}$s are >2 μg/ml (25, 65). In contrast, it is less clear how directly in vitro resistance of HIV-1 to AZT, which develops after 6 or more months of therapy, correlates with clinical deterioration of the patient (10, 45, 72, 77, 78).

VARIABLES OF ANTIVIRAL SUSCEPTIBILITY TESTING

Unlike bacterial susceptibility testing, antiviral susceptibility testing has not yet had any standards established for it. Many variables affect susceptibility testing results. Testing of a single virus isolate may lead to greatly different endpoints depending on the type of cell culture used, the size of the virus inoculum (too large an inoculum can make a susceptible isolate appear resistant; too small an inoculum can make all isolates appear susceptible), and the method used (20, 22, 34, 45, 57, 77). For example, dye uptake (DU) assays for HSV susceptibility testing produce higher IC$_{50}$s than the plaque reduction assay (PRA) (38). The concentration range of the drug tested affects the quality of the dose-response curve and therefore the validity of endpoint calculations. Susceptibility results are usually expressed as IC$_{50}$s because of a greater mathematical precision of the 50% endpoint than of a 90 or 95% endpoint. (Synonyms for IC$_{50}$ are 50% inhibitory dose [ID$_{50}$] and 50% effective dose [ED$_{50}$].) However, debate continues concerning the appropriate endpoint, i.e., 50 versus 90 or 95% inhibition; IC$_{90}$s may correlate better with clinical response, but they are less precise and reproducible than IC$_{50}$ measurements. Moreover, few studies have correlated in vitro results with clinical response (2, 26, 40, 48, 61, 72, 78).

Another critical variable is the heterogeneity within the population of a virus "isolate." A single clinical isolate actually represents a mixture of drug-susceptible and drug-resistant phenotypes (39, 56, 62, 73). A virus population that has never encountered an antiviral agent is predominantly drug susceptible; resistant virus may be present at low levels. The presence of low levels of resistant virus in a population that is predominantly drug susceptible might not be reflected in the IC$_{50}$ but would manifest its presence in higher IC_{90}s or IC_{95}s. At this time, it is unknown whether a small fraction of drug-resistant virus is important to the behavior of the virus in vivo or how such a fraction might affect the response of the infection to therapy in an otherwise healthy host. However, in immunocompromised patients, under the continued selective pressure of antiviral therapy, resistant virus can emerge, and its presence can often be correlated with progressive viral disease (2, 12, 23, 25, 26, 40, 55, 63–65, 69).

The genetic locus at which a mutation occurs also affects susceptibility testing endpoints. For example, DNA polymerase mutations of HSV, VZV, and CMV usually confer relatively smaller increases in in vitro susceptibility values than do TK mutations, which could go undetected in a mixed population of wild-type and mutant virus. Good definition of such mixtures can be obtained only by testing appropriate concentrations of drug and a sufficiently large fraction of the population to detect resistant strains. Since an alternative therapy for acyclovir-resistant HSV and VZV and ganciclovir-resistant CMV, i.e., foscarnet, is currently available and since newer therapies, i.e., protease inhibitors and other reverse transcriptase inhibitors, are being developed for AZT-resistant HIV, it is important to identify resistant populations of virus as they begin to emerge and to develop rapid methods of laboratory characterization of drug-resistant isolates.

TESTING METHODS

Antiviral susceptibility assays in use include PRA, DU, and DNA hybridization assays for the herpes group viruses; enzyme immunoassay (EIA) for HSV, VZV, and influenza A virus; and peripheral blood mononuclear cell (PBMC) cocultivation assay for HIV-1. A plaque autoradiography assay to determine the TK activities of HSV and VZV strains is useful for defining the mechanism of resistance to agents requiring activation by TK. Detailed protocols for these procedures have been published elsewhere and are not presented here (38, 41, 74a, 75). However, the principle of each test and guidelines for interpretation of results are discussed.

Control Strains

Simultaneous testing of control strains is crucial when antiviral susceptibility testing is being done. Reference strains should include genetically and phenotypically well-characterized drug-susceptible and drug-resistant isolates. Drug-resistant strains chosen for reference should include those with drug resistance phenotypes relevant to the mode of action of the drug to be tested. For example, for testing nucleoside analogs, which require phosphorylation by viral TK, i.e., acyclovir versus HSV and VZV, TK-negative or -deficient strains should be included. Susceptibility testing of CMV should include the ganciclovir resistance (UL97 alteration), phosphorylation deficiency phenotype and DNA polymerase variants; DNA polymerase mutants resistant to foscarnet (PFA[r] mutants) should be included for foscarnet testing. The National Institute of Allergy and Infectious Diseases AIDS Research and Reference Reagent Program (Bethesda, Md.) provides upon request a number of reference strains of HSV, VZV, and CMV, including the drug-resistant strains mentioned above and laboratory control strains CMV AD169 and VZV Oka. For AZT susceptibility testing of HIV-1, pairs of AZT-susceptible and AZT-resistant isolates obtained pretherapy and posttherapy, respectively, are also available from the National

Institute of Allergy and Infectious Diseases AIDS Research and Reference Reagent Program.

PRA for CMV, HSV, and VZV

The PRA has classically been the "standard" method of antiviral susceptibility testing to which newer methods are compared (7, 9, 36, 38, 52). The principle of the PRA is the inhibition of viral plaque formation in the presence of antiviral agent. Although the PRA is tedious and consumes more reagents than other methods, for small-scale testing of isolates, it is appropriate. Prior to performing the antiviral susceptibility assay per se, titers of HSV isolates must be determined to ensure an inoculum appropriate for the surface area of the assay wells or plates (i.e., approximately 100 PFU/60-mm-wide tissue culture plate). Because clinical strains of CMV and VZV are cell associated and because low-titer cell-free stocks are less stable during storage, infected cell suspensions (obtained by trypsin treatment of the infected monolayer) can be conveniently used for these viruses. Continued passage of clinical CMV strains in culture usually increases the virus yield. However, low-passage infected cultures are used because they are more likely to be representative of the original mixed population of the clinical isolate than a higher-passage stock would be. Well-characterized drug-susceptible and drug-resistant strains of HSV, CMV AD169, and VZV Oka or Ellen serve as reference strains and are available from the National Institute of Allergy and Infectious Diseases AIDS Research and Reference Reagent Program. Stepwise instructions for performance of the PRA for HSV, VZV, and CMV have been published elsewhere (38, 74a, 75). Proposed susceptibility breakpoints determined by the PRA are listed in Table 2.

DU Assay

The DU assay has been used for many years for susceptibility testing of HSV (38, 52). This assay is based on the preferential uptake of a vital dye (neutral red) by viable cells but not by nonviable cells. The extent of viral lytic activity is determined by the relative amount of dye bound to viable cells after infection with HSV compared with the amount bound to uninfected cells. The dye bound by viable cells is eluted by ethanol and measured colorimetrically. The drug concentration inhibiting viral lytic activity by 50% is considered the IC_{50}.

The DU assay consistently gives IC_{50}s of acyclovir that are three to five times greater than those given by the PRA. This difference is most likely due to the higher inoculum used in the DU assay (500 PFU/ml) and to the use of a liquid overlay. This format does allow drug-resistant virus to "amplify," thus resulting in a more sensitive detection of small amounts of drug-resistant virus. Therefore, the DU assay uses a cutoff IC_{50} of >3 μg/ml to denote acyclovir resistance (Table 2).

Advantages of the DU method include its ability to be semiautomated, allowing for efficient testing of large numbers of isolates, and its ability to detect smaller amounts of resistant virus than the PRA can detect. Disadvantages are the relatively high cost of automated equipment and the technical problems caused by overseeding of cells into the culture wells and precipitation of neutral red onto the monolayer. The stepwise procedure for the DU assay has been published in a previous edition of this Manual (38).

DNA Hybridization

Another approach to antiviral susceptibility testing is measuring the effect of different antiviral compounds on viral

TABLE 2 Proposed guidelines for interpretation of viral susceptibility results

Virus	Antiviral agent	Method	IC$_{50}$ denoting resistance	Reference(s)
HSV	Acyclovir	PRA	>2 μg/ml	17, 29, 52
		DNA hybridization	>2 μg/ml	25, 76
		DU assay	>3 μg/ml	38
	Foscarnet	PRA	>100 μg/ml	14
	Vidarabine	PRA	≥2-fold increase in IC$_{50}$ compared to that of control or pretherapy isolate	14
VZV	Acyclovir	PRA and DNA hybridization	≥3–4-fold increase in IC$_{50}$ compared to that of pretherapy isolate or to control strain	7, 40, 49, 64
CMV	Foscarnet	PRA and DNA hybridization	>100 μg/ml	74
	Ganciclovir	PRA and DNA hybridization	≥3–4-fold increase in IC$_{50}$ compared to that of pretherapy isolate or control strain (≈3 μg/ml)	23, 26, 57, 71
Influenza A	Amantadine-rimantadine	EIA	>0.1 μg/ml	4
HIV	ddI	PRA	≥5-fold increase in IC$_{50}$ compared to that of pretherapy isolate	73
	AZT	PRA	≥30-fold increase in IC$_{50}$ compared to that of pretherapy isolate	42, 45

DNA synthesis. A dot blot hybridization assay (31) and a commercially available hybridization assay (Hybriwix Probe Systems; Diagnostic Hybrids, Inc., Athens, Ohio) are discussed here in detail. A second commercial hybridization assay in development by Chiron Corp. shows promise for susceptibility testing of CMV (67, 68).

Dot Blot Hybridization Assay

The dot blot hybridization assay, developed by Gadler (31), has been successfully used for susceptibility testing of CMV. It is a conventional hybridization assay requiring transfer of viral DNA to nitrocellulose filters, denaturation of DNA, baking of filters, prehybridization of filters, overnight hybridization, and autoradiography. Since the DNA probes are labeled with ^{32}P, probes have short shelf lives. Correlation between the PRA and the dot blot hybridization assay has been excellent (8, 31). The stepwise procedure has been detailed elsewhere (75).

Hybriwix DNA Hybridization Assays

Commercially available DNA-DNA hybridization test kits (Hybriwix Probe Systems) permit quantitation of HSV, CMV, and VZV DNAs in the presence of various concentrations of antiviral agents. Each kit contains cell culture plates, replacement medium, lysing reagent, negatively charged nylon filters ("wicks"), ^{125}I-labeled probe, hybridization solution, wash reagent, and positive and negative control wicks. Infection of tissue culture plates with virus, addition of antiviral agent, and incubation of plates are the same as for the conventional hybridization assay. After appropriate incubation times to allow viral growth, cells are lysed and DNA is denatured by the addition of lysing reagent. Wicks are placed vertically in each well, and lysates are transferred to the wicks by capillary action. Once dry, the wicks are placed in a vial containing single-stranded ^{125}I-labeled probe. After a 2-h hybridization reaction, wicks are washed and dried, and radioactivity is counted in a gamma counter. The concentration of drug

that causes a 50% reduction in counts compared with that of untreated virus controls is the IC$_{50}$. The turnaround time of the Hybriwix hybridization reactions, including lysing, wicking, and washing, is approximately 4 h. These assays have successfully detected both in vitro- and in vivo-derived acyclovir-resistant HSV (25, 63, 65, 76) and VZV (40, 64) and ganciclovir-resistant CMV (19, 71). Drawbacks of the Hybriwix assays include the use of a radioisotope, the short shelf life of the probe (2 months), and the relatively high cost of the kit. It is more cost-effective in a laboratory where large numbers of isolates are analyzed.

Branched-DNA Signal Amplification Assay

Chiron Corp. (Emeryville, Calif.) has in development a new method for detection of various viral nucleic acids, the branched-DNA signal amplification assay, which has been applied to antiviral susceptibility testing of CMV (67, 68). This hybridization assay can be completed in one working day, does not use radioisotopes, and requires fewer days of culturing in the presence of antiviral agents than does the PRA. Although still in development, this technology promises to be very useful for antiviral susceptibility testing.

Assessment of Viral TK Activity by Plaque Autoradiography

The mechanism of acyclovir resistance for most HSV and VZV strains is a mutation in the virus-specific TK gene. To determine whether resistance to acyclovir is due to diminished and/or altered TK, two plaque autoradiography methods are used (51). Incorporation of [^{125}I]iododeoxycytidine (IdC), a pyrimidine analog selectively phosphorylated by VZV- or HSV-specific TK, correlates well with the acyclovir-phosphorylating potential of HSV and VZV isolates. Incorporation of [^{14}C]thymidine (dT) specifically assesses the thymidine-phosphorylating activity of these isolates and is useful for analyzing resistance to pyrimidine nucleoside analogs (acyclovir is a purine nucleoside analog). Most acyclovir-resistant HSV and VZV fail to incorporate both substrates owing to diminished TK activity (TK$^-$); occasionally, strains with altered substrate (TK^a) activity that fail to incorporate IdC but are able to incorporate dT are seen. For IdC incorporation, Vero cells (HSV) and MRC-5 cells (VZV) are used. For dT incorporation, LMTK$^-$ TK$^-$ mouse LM cells (Roswell Park Memorial Institute, Buffalo, N.Y.) are used for HSV; the TK$^-$ cell line 143B is used for VZV dT incorporation. These assays provide both quantitative and qualitative evaluations of the TK status of a mixed population of TK$^+$ and TK$^-$ viruses. The IdC and dT plaque autoradiograph methods have been detailed elsewhere (51, 75).

EIAs

HSV and VZV

EIAs have been developed for susceptibility testing of HSV and VZV (5, 59). The EIA permits quantitative measurement of viral activity by spectrophotometric analysis; IC$_{50}$s are calculated as the concentrations of antiviral agent that reduce the absorbance to 50% of that of the virus control. The EIA is well suited for routine diagnostic laboratories, where these assays are commonly run, and is less labor-intensive and resource expensive than the PRA. Susceptibilities of HSV and VZV determined by this method have correlated well with those obtained by the PRA.

Influenza A Virus

Susceptibility testing of influenza A virus isolates against amantadine and rimantadine can be performed by the PRA (36); however, the EIA is less tedious and more suitable for testing multiple isolates (3). The EIA utilizes antibodies to influenza A virus hemagglutinins (H1 or H3); viral hemagglutinin expression correlates with viral growth. Amantadine and rimantadine activities are measured, therefore, by inhibition of hemagglutinin expression. Amantadine-susceptible and -resistant isolates, whose M2 gene sequences are known, serve as controls and must be tested in parallel with patient isolates. A protocol for the EIA for susceptibility testing of influenza A virus has been published (75).

Susceptibility Testing of HIV

A number of antiviral susceptibility assays have been proposed for testing HIV-1 isolates (45, 47, 73). A serious limitation of these procedures is that not all clinical isolates grow in the cell cultures used in these assays. The AIDS Clinical Trials Group has developed an HIV drug susceptibility assay for testing AZT that is performed in PBMCs (41). This assay allows growth of almost all clinical isolates. Viral activity is quantitated by measurement of p24 antigen. Details of the AIDS Clinical Trial Group protocol have been published elsewhere (41, 75).

DETECTION OF ANTIVIRAL RESISTANCE BY GENE AMPLIFICATION TECHNIQUES

Genetic analysis of HIV-1 isolates resistant to AZT, ddI, and ddC has identified specific *pol* gene mutations that code for specific amino acid substitutions in the HIV reverse transcriptase (11, 27, 30, 32, 42, 47, 62, 73). Because isolation of HIV in PBMCs and subsequent susceptibility testing also performed in PBMC can require weeks, monitoring for drug resistance by conventional methods cannot be completed in a time frame relevant to management of acute cases. Furthermore, analysis of a large number of isolates is necessary to correlate the appearance of resistance with clinical deterioration. In order to determine AZT resistance in a more timely fashion and to efficiently test multiple isolates, the PCR has been employed to directly detect mutations in the reverse transcriptase gene from peripheral blood lymphocytes of HIV-infected individuals receiving AZT therapy (46, 62). PCR assays were developed by using oligonucleotide primers specific for either wild-type or mutant *pol* DNA at codons 67, 70, 215, and 219. These primers were used separately with common upstream primers that annealed with both wild-type and mutant virus DNAs. Amplification occurred only when the 3' end of a primer matched the target sequence. PBMCs obtained at the initiation of AZT therapy were all wild type by PCR analysis. Mixed populations of wild-type and mutant DNAs could be detected in PBMCs obtained during therapy; later specimens from the same individuals revealed only mutant DNA. PCR and self-sustained sequence replication amplification have also been used to detect the *pol* mutations associated with AZT resistance in HIV isolates grown in PBMCs or MT-2 cells. (28, 33, 62). PCR continues to be a powerful tool for characterizing HIV resistance to other nucleoside analogs (27, 30).

Four mutations in the CMV UL97 phosphotransferase gene have recently been identified in seven ganciclovir-resistant clinical isolates of CMV (13). The presence of any one of these four mutations, which all confer a single amino acid change in the putative catalytic domain, resulted in reduced phosphorylation of ganciclovir. Judging from the successful application of PCR to the detection the AZT-resistant HIV, PCR should prove valuable in analyzing suspected ganciclovir-resistant CMV isolates. PCR can be applied to any virus for which the specific mutation(s) causing a decrease in antiviral susceptibility is known.

INTERPRETATION OF ANTIVIRAL SUSCEPTIBILITY RESULTS

The concentration of antiviral agent to which a virus is considered susceptible or resistant has generally been based on median susceptibilities of large numbers of clinical isolates from patients prior to, during, and after antiviral therapy (17, 21). However, in vitro results indicating susceptibility or resistance may not correlate with response of the infection to therapy in vivo. The clinical response of the patient depends on a number of other factors such as immunologic status and pharmacokinetics of the drug in that particular patient (dose or route of administration could be inappropriate). For example, oral administration of an antiviral agent to an AIDS patient with underlying absorption problems due to gastroenteritis may result in a poor clinical response even though the IC$_{50}$ of the isolate denotes in vitro susceptibility. Englund et al. (25) and Safrin et al. (63, 65) have documented HSV infections in patients with HIV infections and other immunologic disorders who failed to respond to therapy despite in vitro IC$_{50}$s indicating susceptibility to vidarabine or acyclovir. Conversely, HSV isolates for which IC$_{50}$s of acyclovir are >2 μg/ml can occasionally be recovered from otherwise healthy hosts who have responded to acyclovir therapy (48). Thus, a high IC$_{50}$ derived by in vitro susceptibility testing may indicate that a viral strain is drug resistant yet not be an absolute predictor of clinical failure of the infection to respond to therapy. Neither can in vitro susceptibility to a drug a priori predict successful clinical outcome. Whenever possible, evidence of genetic alteration of the virus should be considered as well. As drug combination therapies, including immune modulators, become more commonplace, clinical response may be achieved with viruses whose $IC_{50}s$ indicate in vitro resistance to a single agent. Interpretation of antiviral susceptibility results is further complicated by the variability in endpoint due to differences in testing methodologies (34, 57). DU assays for HSV susceptibility testing produce higher $IC_{50}s$ than the PRA (38, 52). Interlaboratory variation has also been observed even when the same assay has been employed (16, 57, 58). One approach to interpreting susceptibility endpoints is to compare the $IC_{50}s$ of an isolate obtained prior to therapy (or of a well-characterized reference control strain) with that of an isolate obtained during therapy; a significant increase in the ratio of such $IC_{50}s$ denotes resistance. However, pretherapy isolates are often unavailable, and the IC_{50} ratio considered clinically significant is unclear. Prospective studies correlating in vivo response with in vitro susceptibility are essential to resolving these issues.

There is controversy concerning the significance of in vitro susceptibility testing results for HIV-1 (10, 61, 77, 78). Despite clear documentation of AZT resistance of HIV-1 isolates from patients on prolonged therapy, the clinical importance of these findings remains problematic. While there is a correlation between AZT resistance and disease progression, there is a much closer association between the presence of the syncytium induction phenotype and disease progression on therapy (53, 72). It is unlikely that any single parameter, such as antiviral resistance, will alone be predictive of disease progression because of the complexity of interactions leading to disease. For example, the immune status of the infected patient, the presence of drug-induced toxicity, the viral load, the virus phenotypes, and the relative proportion of resistant virus in the total virus population are important determinants of clinical response (61). In pediatric patients, there is preliminary evidence that increased resistance to AZT correlates with clinical disease progression (54, 78).

With these caveats in mind, tentative guidelines correlating $IC_{50}s$ with the in vitro drug resistance phenotype are proposed (Table 2). As mentioned above, endpoints are dependent on the test method, so each new method must be correlated with a historic standard that has been used to test large numbers of isolates. Because of the variables that affect antiviral susceptibility results, the absolute IC_{50} may vary from assay to assay and laboratory to laboratory. Large-scale collaborative comparisons of methods with the same viral isolates are necessary to standardize antiviral susceptibility testing and to establish definitive interpretive guidelines.

FUTURE DIRECTIONS

In the development of clinically relevant antiviral susceptibility testing, much work remains to be done. Susceptibility testing must be standardized, and consensus regarding method, cell line, inoculum, endpoint determination criteria, and control strains must be reached. Drew et al. (24) and Japour et al. (41) recently discussed some additional drawbacks of antiviral susceptibility testing. A major problem with culture-based susceptibility testing of viruses other than HSV is that assays may require weeks to complete, a fact that limits their utility in management of acute cases. Second, virus isolates must often be passaged to obtain a

sufficient inoculum for susceptibility testing, and such passage may alter results by selecting for a virus population different from that originally present. Third, patients with CMV and HIV infections can shed multiple strains concurrently; susceptibility testing results for a single isolate may not be representative of the isolate causing disease. Moreover, as resistant virus strains emerge, combination drug therapy will become more prevalent, and it will thus be necessary to evaluate drug combinations for synergy, indifference, or antagonism. Genetic characterization of drug-resistant mutants is essential to permit application of such techniques as PCR to obtain results in a clinically useful time frame.

REFERENCES

1. **Bean, B.** 1992. Antiviral therapy: current concepts and practices. *Clin. Microbiol. Rev.* **5:**146–182.
2. **Bean, B., C. Fletcher, J. Englund, S. N. Lehrman, and M. N. Ellis.** 1987. Progressive mucocutaneous herpes simplex infection due to acyclovir-resistant virus in an immunocompromised patient: correlation of viral susceptibilities and plasma levels with response to therapy. *Diagn. Microbiol. Infect. Dis.* **7:**199–204.
3. **Belshe, R. B., B. Burk, F. Newman, R. L. Cerruti, and I. S. Sim.** 1989. Resistance of influenza A virus to amantadine and rimantadine: results of one decade of surveillance. *J. Infect. Dis.* **159:**430–435.
4. **Belshe, R. B., M. H. Smith, C. B. Hall, R. Betts, and A. J. Hay.** 1988. Genetic basis of resistance to rimantadine emerging during treatment of influenza virus infection. *J. Virol.* **62:**1508–1512.
5. **Berkowitz, F. E., and M. J. Levin.** 1985. Use of an enzyme-linked immunosorbent assay performed directly on fixed infected cell monolayers for evaluating drugs against varicella-zoster virus. *Antimicrob. Agents Chemother.* **28:**207–210.
6. **Biron, K. K.** 1991. Ganciclovir-resistant human cytomegalovirus isolates; resistance mechanisms and in vitro susceptibility to antiviral agents. *Transplant. Proc.* **23**(Suppl. 3):162–167.
7. **Biron, K. K., and G. B. Elion.** 1980. In vitro susceptibility of varicella-zoster virus to acyclovir. *Antimicrob. Agents Chemother.* **18:**443–447.
8. **Biron, K. K., J. A. Fyfe, S. C. Stanat, L. K. Leslie, J. B. Sorrell, C. U. Lambe, and D. M. Coen.** 1986. A human cytomegalovirus mutant resistant to the nucleoside analog 9 - {[2 - hydroxy - 1 - (hydroxymethyl)ethoxy]methyl}guanine (BW B759U) induces reduced levels of BW B759U triphosphate. *Proc. Natl. Acad. Sci. USA* **83:**8769–8773.
9. **Biron, K. K., S. C. Stanat, J. B. Sorrell, J. A. Fyfe, P. M. Keller, C. U. Lambe, and D. J. Nelson.** 1985. Metabolic activation of the nucleoside analog q-[2-hydroxy-1-(hydroxymethyl)ethoxymethyl]guanine in human diploid fibroblasts infected with human cytomegalovirus. *Proc. Natl. Acad. Sci. USA* **82:**2473–2477.
10. **Boucher, C. A. B.** 1992. Clinical significance of zidovudine-resistant human immunodeficiency viruses. *Res. Virol.* **43:**134–136.
11. **Boucher, C. A. B., E. O'Sullivan, J. W. Mulder, C. Ramautarsing, P. Kellam, G. Darby, J. M. A. Lange, J. Goudsmit, and B. A. Larder.** 1992. Ordered appearance of zidovudine resistance mutations during treatment of 18 human immunodeficiency virus-positive subjects. *J. Infect. Dis.* **165:**105–110.
12. **Chatis, P. A., and C. S. Crumpacker.** 1992. Resistance of herpesviruses to antiviral drugs. *Antimicrob. Agents Chemother.* **36:**1589–1595.
13. **Chou, S., C. L. Talarico, S. Stanat, and K. K. Biron.** 1993. Genetic analysis of the Pol and UL97 coding regions of clinical cytomegalovirus (CMV) isolates including those with ganciclovir resistance, abstr. 180, p. 31-A. *Abstr. 1993 Infect. Dis. Soc. Am. Annu. Meet.*

14. **Coen, D. M., H. E. Fleming, Jr., L. K. Leslie, and M. J. Retondo.** 1985. Sensitivity of arabinosyladenine-resistant mutants of herpes simplex virus to other antiviral drugs and mapping of drug hypersensitivity mutations to the DNA polymerase locus. *J. Virol.* **53:**477–488.

15. **Coen, D. M., M. Kosz-Vnenchak, J. G. Jacobson, D. A. Leib, C. L. Bogard, P. A. Schaffer, K. L. Tyler, and D. M. Knipe.** 1989. Thymidine kinase-negative herpes simplex virus mutants establish latency in mouse trigeminal ganglia but do not reactivate. *Proc. Natl. Acad. Sci. USA* **86:**4736–4740.

16. **Cole, N. L., and H. H. Balfour, Jr.** 1987. In vitro susceptibility of cytomegalovirus isolates from immunocompromised patients to acyclovir and ganciclovir. *Diagn. Microbiol. Infect. Dis.* **6:**255–261.

17. **Collins, P., G. Appleyard, and N. M. Oliver.** 1982. Sensitivity of herpes virus isolates from acyclovir clinical trials. *Am. J. Med.* **73**(Suppl.):380–382.

18. **Collins, P., B. A. Larder, N. M. Oliver, S. Kemp, I. W. Smith, and G. Darby.** 1989. Characterization of a DNA polymerase mutant of herpes simplex virus from a severely immunocompromised patient receiving acyclovir. *J. Gen. Virol.* **70:**375–382.

19. **Dankner, W. M., D. Scholl, S. C. Stanat, M. Martin, R. L. Sonke, and S. A. Spector.** 1990. Rapid antiviral DNA-DNA hybridization assay for human cytomegalovirus. *J. Virol. Methods* **28:**293–298.

20. **De Clercq, E.** 1982. Comparative efficacy of antiherpes drugs in different cell lines. *Antimicrob. Agents Chemother.* **21:**661–663.

21. **Dekker, C., M. N. Ellis, C. McLaren, G. Hunter, J. Rogers, and D. W. Barry.** 1983. Virus resistance in clinical practice. *J. Antimicrob. Chemother.* **12:**137–152.

22. **Drew, W. L., and T. R. Matthews.** 1989. Susceptibility testing of herpes viruses. *Clin. Lab. Med.* **9:**279–286.

23. **Drew, W. L., R. C. Miner, D. F. Busch, S. E. Follansbee, J. Gullett, S. G. Mehalko, S. M. Gordon, W. F. Owen, Jr., T. R. Matthews, W. C. Buhles, and B. DeArmond.** 1991. Prevalence of resistance in patients receiving ganciclovir for serious cytomegalovirus infection. *J. Infect. Dis.* **163:**716–719.

24. **Drew, W. L., R. Miner, and E. Saleh.** 1993. Antiviral susceptibility testing of cytomegalovirus: criteria for detecting resistance to antivirals. *Clin. Diagn. Virol.* **1:**179–185.

25. **Englund, J. A., M. E. Zimmerman, E. M. Swierkosz, J. L. Goodman, D. R. Scholl, and H. H. Balfour, Jr.** 1990. Herpes simplex virus resistant to acyclovir. A study in a tertiary care center. *Ann. Intern. Med.* **112:**416–422.

26. **Erice, A., S. Chou, K. K. Biron, S. C. Stanat, H. H. Balfour, Jr., and M. C. Jordan.** 1989. Progressive disease due to ganciclovir-resistant cytomegalovirus in immunocompromised patients. *N. Engl. J. Med.* **320:**289–293.

27. **Eron, J. J., Y.-K. Chow, A. M. Caliendo, J. Videler, K. M. Devore, T. P. Cooley, H. A. Liebman, J. D. Kaplan, M. S. Hirsch, and R. T. D'Aquila.** 1993. *pol* mutations conferring zidovudine and didanosine resistance with different effects in vitro yield multiply resistant human immunodeficiency virus type 1 in vivo. *Antimicrob. Agents Chemother.* **37:**1480–1487.

28. **Eron, J. J., P. Gorczyca, J. C. Kaplan, and R. T. D'Aquila.** 1992. Susceptibility testing by polymerase chain reaction DNA quantitation: a method to measure drug resistance of human immunodeficiency virus type 1 isolates. *Proc. Natl. Acad. Sci. USA* **89:**3241–3245.

29. **Field, H. J., and G. Darby.** 1980. Pathogenicity in mice of strains of herpes simplex virus which are resistant to acyclovir in vitro and in vivo. *Antimicrob. Agents Chemother.* **17:**209–216.

30. **Fitzgibbon, J. E., R. M. Howell, C. A. Haberzettl, S. J. Sperber, D. J. Gocke, and D. T. Dubin.** 1992. Human immunodeficiency virus type 1 *pol* gene mutations which cause decreased susceptibility to 2′-3′-dideoxycytidine. *Antimicrob. Agents Chemother.* **36:**153–157.

31. **Gadler, H.** 1983. Nucleic acid hybridization for measurement of effects of antiviral compounds on human cytomegalovirus DNA replication. *Antimicrob. Agents Chemother.* **24:**370–374.

32. **Gao, Q., Z. Gu, M. A. Parniak, X. Li, and M. A. Wainberg.** 1992. In vitro selection of variants of human immunodeficiency virus type 1 resistant to 3′-azido-3′-deoxythymidine and 2′,3′-dideoxyinosine. *J. Virol.* **66:**12–19.

33. **Gingeras, T. R., P. Prodanovich, T. Latimer, J. C. Guatelli, D. D. Richman, and K. J. Barringer.** 1991. Use of self-sustained sequence replication amplification reaction to analyze and detect mutations in zidovudine-resistant human immunodeficiency virus. *J. Infect. Dis.* **164:**1066–1074.

34. **Harmenberg, J., B. Wahren, and B. Oberg.** 1980. Influence of cells and virus multiplicity on the inhibition of herpesviruses with acycloguanosine. *Intervirology* **14:**239–244.

35. **Hayden, F. G., R. B. Belshe, R. D. Clover, A. J. Hay, M. G. Oakes, and W. Soo.** 1989. Emergence and apparent transmission of rimantadine-resistant influenza A virus in families. *N. Engl. J. Med.* **321:**1696–1702.

36. **Hayden, F. G., K. M. Cote, and G. D. Douglas, Jr.** 1980. Plaque inhibition assay for drug susceptibility testing of influenza viruses. *Antimicrob. Agents Chemother.* **17:**865–870.

37. **Hayden, F. G., S. J. Sperber, R. B. Belshe, R. D. Clover, A. J. Hay, and S. Pyke.** 1991. Recovery of drug-resistant influenza A virus during therapeutic use of rimantadine. *Antimicrob. Agents Chemother.* **35:**1741–1747.

38. **Hill, E. L., M. N. Ellis, and P. Nguyen-Dinh.** 1991. Antiviral and antiparasitic susceptibility testing, p. 1184–1188. *In* A. Balows, W. J. Hausler, Jr., K. L. Herrmann, H. D. Isenberg, and H. J. Shadomy (ed.), *Manual of Clinical Microbiology,* 5th ed. American Society for Microbiology, Washington, D.C.

39. **Hill, E. L., G. A. Hunter, and M. N. Ellis.** 1991. In vitro and in vivo characterization of herpes simplex virus clinical isolates recovered from patients infected with human immunodeficiency virus. *Antimicrob. Agents Chemother.* **35:**2322–2328.

40. **Jacobson, M. A., T. G. Berger, S. Fikrig, P. Cecherer, J. W. Moohr, S. C. Stanat, and K. K. Biron.** 1990. Acyclovir-resistant varicella zoster virus infection after chronic oral acyclovir therapy in patients with the acquired immunodeficiency syndrome (AIDS). *Ann. Intern. Med.* **112:**187–191.

41. **Japour, A. J., D. L. Mayers, V. A. Johnson, D. R. Kuritzkes, L. A. Beckett, J.-M. Arduino, J. Lane, R. J. Black, P. S. Reichelderfer, R. T. D'Aquila, C. S. Crumpacker, the RV-43 Study Group, and the AIDS Clinical Trials Group Virology Committee Resistance Working Group.** 1993. Standarized peripheral blood mononuclear cell culture assay for determination of drug susceptibilities of clinical human immunodeficiency virus type 1 isolates. *Antimicrob. Agents Chemother.* **37:**1095–1101.

42. **Kellam, P., C. A. B. Boucher, and B. A. Larder.** 1992. Fifth mutation in human immunodeficiency virus type 1 reverse transcriptase contributes to the development of high-level resistance to zidovudine. *Proc. Natl. Acad. Sci. USA* **89:**1934–1938.

43. **Knox, K. K., W. R. Drobyski, and D. R. Carrigan.** 1991. Cytomegalovirus isolate resistant to ganciclovir and foscarnet from a marrow transplant patient. *Lancet* **337:**1292–1293.

44. **Kost, R. G., E. L. Hill, M. Tigges, and S. E. Straus.** 1993. Recurrent acyclovir resistant genital herpes in an immunocompetent patient. *N. Engl. J. Med.* **329:**1777–1782.

45. **Larder, B. A., G. Darby, and D. D. Richman.** 1989. HIV with reduced sensitivity to zidovudine (AZT) isolated during prolonged therapy. *Science* **243:**1731–1734.

46. **Larder, B. A., P. Kellam, and S. D. Kemp.** 1991. Zidovudine resistance predicted by direct detection of mutations in DNA from HIV-infected lymphocytes. *AIDS* **5:**137–144.

47. **Larder, B. A., and S. D. Kemp.** 1989. Multiple mutations in HIV-1 reverse transcriptase confer high-level resistance to zidovudine (AZT). *Science* **246:**1155–1158.

48. **Lehrman, S. N., J. M. Douglas, L. Corey, and D. W. Barry.** 1986. Recurrent genital herpes and suppressive oral acyclovir therapy. Relation between clinical outcome and in-vitro drug sensitivity. *Ann. Intern. Med.* **104:**786–790.

49. **Linnemann, C. C., Jr., K. K. Biron, W. G. Hoppenjans, and A. M. Solinger.** 1990. Emergence of acyclovir-resistant varicella zoster virus in an AIDS patient on prolonged acyclovir therapy. *AIDS* **4:**577–579.

50. **Littler, E., A. D. Stuart, and M. S. Chee.** 1992. Human cytomegalovirus UL97 open reading frame encodes a protein that phosphorylates the antiviral nucleoside analogue ganciclovir. *Nature* (London) **358:**160–162.

51. **Martin, J. L., M. N. Ellis, P. M. Keller, K. K. Biron, S. N. Lehrman, D. W. Barry, and P. A. Furman.** 1985. Plaque autoradiography assay for the detection and quantitation of thymidine kinase-deficient and thymidine kinase-altered mutants of herpes simplex virus in clinical isolates. *Antimicrob. Agents Chemother.* **28:**181–187.

52. **McLaren, C., M. N. Ellis, and G. A. Hunter.** 1983. A colorimetric assay for the measurement of the sensitivity of herpes simplex viruses to antiviral agents. *Antiviral Res.* **3:**223–224.

53. **Montaner, J. S. G., J. Singer, M. T. Schecter, J. M. Raboud, C. Tsoukas, M. O'Shaughnessy, J. Ruedy, K. Nagai, H. Salomon, B. Spira, and M. A. Wainberg.** 1993. Clinical correlates of *in vitro* HIV-1 resistance to zidovudine. Results of the multicenter Canadian AZT trial. *AIDS* **7:**189–196.

54. **Ogino, M. T., W. M. Dankner, and S. A. Spector.** 1993. Development and significance of zidovudine resistance in children infected with human immunodeficiency virus. *J. Pediatr.* **123:**1–8.

55. **Pahwa, S., K. Biron, W. Lim, P. Swenson, M. H. Kaplan, N. Sadick, and R. Pahwa.** 1988. Continuous varicella-zoster infection associated with acyclovir resistance in a child with AIDS. *JAMA* **260:**2879–2882.

56. **Parris, D. S., and J. E. Harrington.** 1982. Herpes simplex virus variants resistant to high concentrations of acyclovir exist in clinical isolates. *Antimicrob. Agents Chemother.* **22:**71–77.

57. **Pepin, J.-M., F. Simon, M. C. Dazza, and F. Brun-Vezinet.** 1992. The clinical significance of in vitro cytomegalovirus susceptibility to antiviral drugs. *Res. Virol.* **143:**126–128.

58. **Plotkin, S. A., W. L. Drew, D. Felsenstein, and M. S. Hirsch.** 1985. Sensitivity of clinical isolates of human cytomegalovirus to 9-(1,3-dihydroxy-2-propoxymethyl)guanine. *J. Infect. Dis.* **152:**883–884.

59. **Rabalaiss, G. P., M. J. Levin, and F. E. Berkowitz.** 1987. Rapid herpes simplex virus susceptibility testing using an enzyme-linked immunosorbent assay performed in situ on fixed virus-infected monolayers. *Antimicrob. Agents Chemother.* **31:**946–948.

60. **Reardon, J. E., and T. Spector.** 1989. Herpes simplex virus type 1 DNA polymerase. Mechanism of inhibition by acyclovir triphosphate. *J. Biol. Chem.* **264:**7405–7411.

61. **Richman, D. D.** 1992. The clinical significance of drug-resistant mutants of human immunodeficiency virus. *Res. Virol.* **143:**130–131.

62. **Richman, D. D., J. C. Guatelli, J. Grimes, A. Tsiatis, and T. R. Gingeras.** 1991. Detection of mutations associated with zidovudine resistance in human immunodeficiency virus by utilizing the polymerase chain reaction. *J. Infect. Dis.* **164:**1075–1081.

63. **Safrin, S., T. Assaykeen, S. Follansbee, and J. Mills.** 1990. Foscarnet therapy for acyclovir-resistant mucocutaneous herpes simplex virus infection in 26 AIDS patients: preliminary data. *J. Infect. Dis.* **161:**1078–1084.

64. **Safrin, S., T. G. Berger, I. Gilson, P. R. Wolfe, C. B. Wofsy, J. Mills, and K. K. Biron.** 1991. Foscarnet therapy in five patients with AIDS and acyclovir-resistant varicella-zoster virus infection. *Ann. Intern. Med.* **115:**19–21.

65. **Safrin, S., C. Crumpacker, P. Chatis, R. Davis, R. Hafner, J. Rush, H. A. Kessler, B. Landry, J. Mills, and the AIDS Clinical Trials Group.** 1991. A controlled trial comparing foscarnet with vidarabine for acyclovir-resistant mucocutaneous herpes simplex in the acquired immunodeficiency syndrome. *N. Engl. J. Med.* **325:**551–555.

66. **Safrin, S., and L. Phan.** 1993. In vitro activity of penciclovir against clinical isolates of acyclovir-resistant and foscarnet-resistant herpes simples virus. *Antimicrob. Agents Chemother.* **37:**2241–2243.

67. **Shen, L. P., R. Kokka, D. Besemer, P. Johnson, J. Kolberg, D. Miner, and L. Drew.** 1992. Direct detection of HCMV in clinical specimens using bDNA assay, abstr. 1244, p. 318. *Program Abstr. 32nd Intersci. Conf. Antimicrob. Agents Chemother.*

68. **Shen, L. P., J. A. Kolberg, R. R. Spaete, R. Miner, and W. L. Drew.** 1992. A quantitative method of detection of human cytomegalovirus DNA using a branched DNA enhanced label amplification assay, abstr. S-54, p. 408. *Abstr. 92nd Gen. Meet. Am. Soc. Microbiol. 1992.*

69. **Sibrack, C. D., L. T. Gutman, C. M. Wilfert, C. McLaren, M. H. St. Clair, P. M. Keller, and D. W. Barry.** 1982. Pathogenicity of acyclovir-resistant herpes simplex virus type 1 from an immunodeficient child. *J. Infect. Dis.* **146:**673–682.

70. **Sommadossi, J.-P.** 1993. Nucleoside analogs: similarities and differences. *Clin. Infect. Dis.* **16**(Suppl. 1)**:**S7–S15.

71. **Stanat, S. C., J. E. Reardon, A. Erice, M. C. Jordan, W. L. Drew, and K. K. Biron.** 1991. Ganciclovir-resistant cytomegalovirus clinical isolates: mode of resistance to ganciclovir. *Antimicrob. Agents Chemother.* **35:**2191–2197.

72. **St. Clair, M. H., P. M. Hartigan, J. C. Andrews, C. L. Vavro, M. S. Simberkoff, J. D. Hamilton, and the VA Cooperative Study Group.** 1993. Zidovudine resistance, syncytium-inducing phenotype, and HIV disease progression in a case-control study. *J. Acquired Immune Defic. Syndr.* **6:**891–897.

73. **St. Clair, M. H., J. L. Martin, G. Tudor-Williams, M. C. Bach, C. L. Vavro, D. M. King, P. Kellam, S. D. Kemp, and B. A. Larder.** 1991. Resistance to ddI and sensitivity to AZT induced by a mutation in HIV-1 reverse transcriptase. *Science* **253:**1557–1559.

74. **Sullivan, V., and D. M. Coen.** 1991. Isolation of foscarnet-resistant human cytomegalovirus patterns of resistance and sensitivity to other antiviral drugs. *J. Infect. Dis.* **164:**781–784.

74a.**Swierkosz, E. M., and K. K. Biron.** 1994. Antiviral susceptibility testing, p. 8.26.1–8.26.21. *In* H. D. Isenberg (ed.), *Clinical Microbiology Procedures Handbook,* supplement 1. American Society for Microbiology, Washington, D.C.

75. **Swierkosz, E. M., and K. K. Biron.** Antiviral susceptibility testing. *In* E. H. Lennette, D. A. Lennette, and E. T. Lennette (ed.), *Diagnostic Procedures for Viral, Rickettsial and Chlamydial Infections,* 7th ed., in press. American Public Health Association, Washington, D.C.

76. **Swierkosz, E. M., D. R. Scholl, J. L. Brown, J. D. Jollick, and C. A. Gleaves.** 1987. Improved DNA hybridization method for detection of acyclovir-resistant herpes simplex virus. *Antimicrob. Agents Chemother.* **31:**1465–1469.

77. **Tudor-Williams, G., and V. C. Emery.** 1992. Development of in vitro resistance to zidovudine is likely to be clinically significant. *Rev. Med. Virol.* **2:**123–129.

78. **Tudor-Williams, G., M. H. St. Clair, R. E. McKinney, M. Maha, E. Walter, S. Santacroce, M. Mintz, K. O'Donnell, R. Rudoll, and C. L. Vavro.** 1992. HIV-1 sensitivity to zidovudine and clinical outcome in children. *Lancet* **339:**15–19.

79. **Wade, J. C., C. McLaren, and J. D. Meyers.** 1983. Frequency and significance of acyclovir resistant herpes simplex virus isolated from marrow transplant patients receiving multiple courses of treatment with acyclovir. *J. Infect. Dis.* **148:**1077–1082.

Acquired Drug Resistance and Susceptibility Testing in Parasitic Diseases

SAMUEL L. STANLEY, JR.

122

Protozoan and helminthic diseases cause extensive morbidity and mortality worldwide. Helminths such as ascaris and hookworm infect more than half a billion people worldwide (44), while malaria alone causes disease in more than 120 million people yearly (48). Drug resistance among parasites is an increasingly recognized phenomenon. The development of drug resistance in *Plasmodium falciparum*, the most deadly of the malaria parasites, has become a major threat to public health in much of the world. Yet despite the tremendous importance of parasitic diseases and the growing problem of drug resistance, we know very little about the mechanisms underlying drug resistance in most parasites, and standardized methods for performing susceptibility studies of parasites are few.

This chapter examines mechanisms underlying drug resistance in three clinically important parasites (*P. falciparum*, *Trichomonas vaginalis*, and *Leishmania* spp.) and describes methods that have been developed for susceptibility testing of these parasites. I emphasize that this chapter deals with tests that are largely confined to specialized research laboratories and have not found their way into general practice (12).

P. FALCIPARUM

The malaria parasites *P. falciparum*, *Plasmodium vivax*, *Plasmodium ovale*, and *Plasmodium malaria* cause more than 1 million deaths annually (48). Among these organisms, drug resistance was first recognized in *P. falciparum*. Resistance was first seen with chloroquine, previously the safest, most effective, and most widely used agent for the treatment and prophylaxis of malaria infection. *P. falciparum* resistant to chloroquine is now found in most malarious regions of the world. Clinically important resistance of *P. falciparum* to a number of other previously effective antimalarial agents, including quinine, pyrimethamine-sulfadoxine, and the new agents mefloquine and halofantrine, has also been recognized. The problem of drug resistance is no longer confined to *P. falciparum*, as *P. vivax* resistant to chloroquine has now been described (37).

The mechanism of chloroquine resistance in *P. falciparum* is a subject of controversy. Recent studies have suggested that chloroquine may kill *P. falciparum* by inhibiting a novel heme polymerase located in the digestive vacuole of the parasite (38). Chloroquine-resistant *P. falciparum* appears to accumulate less chloroquine than susceptible parasites (9, 20). Thus, in chloroquine-resistant parasites, chloroquine levels in the digestive vacuole may be too low to inhibit the heme polymerase (42). Whether this reduced accumulation is due to a decrease in the energy-dependent uptake of chloroquine by the parasite (3) or to enhanced efflux of chloroquine by resistant *P. falciparum* (20) remains unclear. Support for increased efflux of chloroquine as the mechanism for chloroquine resistance came from the finding that *P. falciparum* has a P glycoprotein-like molecule similar to the molecules implicated in the multidrug resistance phenotype seen in tumor cells (11, 46). The P glycoprotein molecule (encoded by the *mdr* genes) functions as an ATP-dependent pump that can export cytotoxic drugs from cells. In tumor cells, amplification and overexpression of P glycoprotein is associated with resistance to a variety of drugs. However, studies of chloroquine-resistant and chloroquine-susceptible *P. falciparum* isolates have not demonstrated a consistent correlation between chloroquine resistance and amplification or overexpression of the *pfmdr* genes of *P. falciparum*. In addition, in a genetic cross between chloroquine-resistant and chloroquine-susceptible *P. falciparum*, chloroquine resistance was not linked to the *pfmdr* locus (11, 43). While the role of *pfmdr* genes in chloroquine resistance remains uncertain, there is now evidence to suggest that amplification of these genes may be associated with the development of drug resistance to a newer antimalarial agent, mefloquine (1, 47).

Pyrimethamine-sulfadoxine has been widely used for the treatment of *P. falciparum* and *P. vivax* malaria. All plasmodia require folic acid for the synthesis of pyrimidines, and the parasite dihydrofolate reductase enzyme (DHFR; 5,6,7,8-tetrahydrofolate:NADP$^+$ oxidoreductase; EC 1.5.1.3) is approximately 1,000-fold more sensitive to inhibition by pyrimethamine than the mammalian enzyme is (17). However, *P. falciparum* resistant to pyrimethamine-sulfadoxine is now found in many malarious areas of the world. Analysis of the DHFR gene from pyrimethamine-sulfadoxine-resistant parasites has demonstrated specific point mutations that result in amino acid changes at the pyrimethamine-binding site of the enzyme (32). The alterations caused by these mutations probably act to inhibit DHFR binding of pyrimethamine. Analyses of DHFR genes from *P. falciparum* organisms resistant to another antifolate drug, proguanil, have demonstrated point mutations at different locations in the enzyme (31). This explains the clinical finding of *P. falciparum* isolates resistant to pyrimethamine that remain susceptible to proguanil.

The most widely used tests for drug resistance in *P. falciparum* measure the ability of a given agent to inhibit parasite development in an in vitro culture system (27, 28, 35, 40). While originally developed for use with *P. falciparum* in continuous culture, these tests have now been applied in the field. In the field version of the test available from the World Health Organization, capillary blood from an infected patient is diluted in parasite culture medium (RPMI 1640 supplemented with 25 mM HEPES [N-2-hydroxyethylpiperazine-*N'*-2-ethanesulfonic acid] buffer) and then dispensed into microplate wells that contain known concentrations of the drug to be tested (30). After incubation in a candle jar at 37°C for 24 to 30 h, the microplates are harvested, and thick smears of the contents of each well are stained with Giemsa stain and evaluated for the number of parasites maturing to the schizont form in drug-containing versus control wells (30). Standard values for the test have been described by the World Health Organization. Antimalarial agents tested by this assay system include chloroquine, quinine, mefloquine, halofantrine, pyrimethamine (sometimes tested with RPMI containing reduced levels of *p*-aminobenzoic acid and folic acid), and amodiaquine. Variations of this test, such as systems in which drug inhibition of [^{3}H]hypoxanthine uptake or bromodeoxyuridine incorporation into the DNA of cultured parasites is measured, have also been described (2, 7). The cumulative data from studies employing these tests suggest that drugs that show no efficacy in vitro are ineffective in treating patients with *P. falciparum* infection. In general, these in vitro tests of *P. falciparum* drug resistance have proven highly useful for epidemiologic studies, but they are rarely performed to guide the treatment of individual patients with malaria.

T. VAGINALIS

T. vaginalis is a falcultatively anaerobic protozoan that is a major cause of human vaginitis worldwide. *T. vaginalis* infections are found in both developing and industrialized countries, with an estimated 3 million cases in American women yearly (5). The only drug approved for the treatment of *T. vaginalis* infection in the United States is metronidazole [1-(2-hydroxyethyl)-2-methyl-5-nitroimidazole], a compound that selectively kills anaerobic bacteria and protozoans. For activity, metronidazole requires reduction of its nitro group to a hydroxylamine through receipt of electrons. Anaerobic organisms possess electron transport proteins of sufficiently low redox potential to allow drug activation. Metronidazole-resistant *T. vaginalis* was first isolated more than a decade ago from patients in whom metronidazole therapy repeatedly failed (23). Because no effective alternatives to metronidazole in the treatment of trichomonas infection are available, resistant *T. vaginalis* represents a growing health problem.

Two forms of metronidazole resistance in trichomonads have been described: aerobic, which is present only when trichomonads are grown aerobically and is the form of resistance seen in isolates from patients in whom metronidazole therapy has failed, and anaerobic, which can be seen under anaerobic conditions and to date has been generated only in the laboratory (19). The basis for aerobic resistance may lie in changes in intracellular levels of parasite oxidases. In studies of metronidazole-resistant and -susceptible

isolates, intracellular levels of ferredoxin were decreased by more than 50% in resistant strains (34). Lowered levels of ferredoxin could result in a higher local concentration of O$_2$; this increased O$_2$ may serve to protect cells from metronidazole through competition between O$_2$ and the drug for free electrons or through the reoxidation of NO$_2$ free radicals (19). The decreased intracellular levels of ferredoxin seen in *T. vaginalis* appear to be secondary to a reduction in ferredoxin gene transcription in resistant strains (34). The finding of lowered levels but not a complete absence of ferredoxin in resistant strains is consistent with the clinical finding that higher doses of metronidazole and more prolonged courses of therapy may cure some patients with resistant *T. vaginalis* (21). *T. vaginalis* strains exhibiting laboratory-generated anaerobic resistance appear to have decreased levels of or to completely lack the enzymes pyruvate:ferredoxin:oxidoreductase and hydrogenase, each of which may be important in metronidazole activation (19).

T. vaginalis can be cultivated axenically in a number of liquid or semisolid media and in tissue culture (8). Susceptibility testing is often performed by the methods outlined by Meingassner and Thurner (23). In vitro assays are done at 37°C in stoppered tubes or in multiwell plates incubated in a normal atmosphere or under anaerobic conditions. A known number of *T. vaginalis* are added to individual tubes or microwells that contain medium and serial dilutions of the compound to be tested (usually metronidazole or another member of the nitroimidazole family). After a 48-h incubation, the tubes and microwells are examined with an inverted microscope, and the highest dilution of the test compound in which no motile *T. vaginalis* is seen is recorded as the minimum lethal concentration. The minimal lethal concentrations of metronidazole for *T. vaginalis* isolates obtained from patients in whom metronidazole therapy failed are 6 to 131 times higher than those for metronidazole-susceptible *T. vaginalis* (21, 23, 25, 26).

LEISHMANIA SPP.

Approximately 10 to 15 million people are infected with *Leishmania* spp., the protozoan parasites that cause cutaneous leishmaniasis, mucocutaneous leishmaniasis, and visceral leishmaniasis (45). The recognition of *Leishmania* spp. as important pathogens in human immunodeficiency virus-infected people living in areas where *Leishmania* spp. are endemic has led to an increased interest in *Leishmania* infection. Treatment of leishmaniasis is problematic. Pentavalent antimonial compounds (e.g., sodium stibogluconate, N-methylglucamine) are the most widely used and probably the most effective drugs for therapy, but they are relatively toxic, difficult to administer, and expensive. This unfortunate situation has been exacerbated by the development of resistance to pentavalent antimonial compounds in *Leishmania* isolates.

The mechanism underlying antimonial compound resistance in *Leishmania* spp. is still being defined. Drug-resistant promastigotes appear to accumulate less radiolabeled drug than susceptible promastigotes (14), and it has been reported that transfection of a P glycoprotein-encoding gene into susceptible *Leishmania* spp. can confer low-level resistance to heavy metals (4). Resistance to pentavalent antimony in P glycoprotein-transfected *Leishmania* spp. has also been described (29). These findings implicate a P glycoprotein-mediated increase in drug efflux as one possible mechanism for *Leishmania* resistance to pentavalent antimonial compounds.

The promastigote form of *Leishmania* spp. can be grown in vitro at 25°C in media such as Schneider's *Drosophila* medium supplemented with 20% heat-inactivated fetal calf serum and 100 μg of gentamicin per ml without accompanying cells (16). The amastigote form (which is the stage that causes disease in humans) is an intracellular parasite and must be grown in tissue culture within host cells. Clinical isolates of *Leishmania* spp. have been examined for drug resistance by a radiorespirometric microtechnique in which promastigotes are incubated with serum-free medium either with or without the drug to be tested and then, under maximal growth conditions, all nutrients are removed and replaced by ^{14}C-labeled substrates (18). The amounts of $^{14}CO_2$ released by isolates incubated with drug and organisms incubated in drug-free medium are then compared. Clinical isolates that were resistant to more than 100 times the maximum pentavalent antimonial compound concentration achievable in human serum were found in this study (18). A second approach has been a semiautomated microdilution technique in which parasites are grown in medium containing serial dilutions of the drugs to be tested and drug effect is measured by an inhibition of parasite uptake of radiolabeled thymidine (15). Among patients treated with pentavalent antimonial compounds, a clear correlation between treatment outcome and the presence or absence of antimonial compound resistance in their isolates could be established.

The development of resistance to previously effective therapies is not limited to the three parasites discussed above. Isolates of the protozoan parasite *Giardia lamblia*, a major cause of diarrheal illness worldwide, that are resistant to metronidazole or furazolidone have recently been described (22, 41). The abilities to grow *G. lamblia* axenically (24) and to quantitate growth (by counting parasites or by measuring [^{3}H]thymidine incorporation) in the presence or absence of drug have made susceptibility studies feasible (13). Emetine resistance has been induced in axenically cultured *Entamoeba histolytica* and is mediated by a P glycoprotein homolog (36). Susceptibility testing of the highly drug resistant organism *Cryptosporidium parvum*, a coccidian parasite that can cause severe diarrhea in patients with AIDS, has been performed by measuring the abilities of agents to inhibit parasite infectivity of the human enterocyte cell line HT 29.74 (10).

Drug resistance in helminths has been studied by using animal models and short-term culture techniques. *Schistosoma mansoni* resistant to the agent hycanthone (tested with murine models of infection) has been isolated from patients failing hycanthone therapy; resistant worms appear to lack an enzyme required for activation of the drug (6, 33). Susceptibility testing of the filarial worm *Onchocerca volvulus*, a major cause of blindness worldwide, has been performed by culturing microfilariae in NCTC 135-Iscove modified Dulbecco medium buffered with HEPES and supplemented with 20% fetal calf serum and antibiotics and then examining the effect of compounds on larval motility (39).

CONCLUSIONS

Drug resistance in parasites represents a growing threat to health around the world. New techniques for susceptibility testing have been developed and may prove valuable in epidemiologic studies designed to identify populations at risk for drug-resistant organisms. However, in contrast to antimicrobial susceptibility testing in bacteria and viruses, susceptibility testing of parasites has yet to have any major impact on the treatment of individual patients. This failure reflects the lack of standardization of many of these tests and, more important, the fact that the areas in which susceptibility testing is most needed lack the resources to test individual isolates. Research laboratories must continue the search for reproducible, inexpensive, accurate tests to identify drug-resistant parasites.

REFERENCES

1. **Barnes, D. A., S. J. Foote, D. Galatis, D. J. Kemp, and A. F. Cowman.** 1992. Selection for high-level chloroquine resistance results in deamplification of the *pfmdr1* gene and increased sensitivity to mefloquine in *Plasmodium falciparum*. *EMBO J.* **11:**3067–3075.
2. **Brasseur, P., P. Druilhe, J. Kouamouo, O. Brandicort, M. Danis, and S. R. Moyou.** 1986. High level of sensitivity to chloroquine of 72 *Plasmodium falciparum* isolates from southern Cameroon in January 1985. *Am. J. Trop. Med. Hyg.* **35:**711–716.
3. **Bray, P. G., R. E. Howells, G. Y. Ritchie, and S. A. Ward.** 1992. Rapid chloroquine efflux phenotype in both chloroquine-sensitive and chloroquine-resistant *Plasmodium falciparum*. *Biochem. Pharmacol.* **44:**1317–1324.
4. **Callahan, H. L., and S. M. Beverley.** 1991. Heavy metal resistance: a new role for P-glycoproteins in *Leishmania*. *J. Biol. Chem.* **266:**18427–18430.
5. **Center for Disease Control.** 1979. Nonreported sexually transmitted diseases. *Morbid. Mortal. Weekly Rep.* **28:**61–66.
6. **Cioli, D., L. Pica-Mattoccia, and S. Archer.** 1993. Drug resistance in schistosomes. *Parasitol. Today* **9:**162–167.
7. **Desjardins, R. E., C. J. Canfield, J. D. Haynes, and J. D. Chulay.** 1979. Quantitative assessment of antimalarial activity in vitro by semiautomated microdilution technique. *Antimicrob. Agents Chemother.* **16:**710–718.
8. **Diamond, L. S.** 1957. The establishment of various trichomonads of animals and man in axenic culture. *J. Parasitol.* **43:**488–490.
9. **Fitch, C. D., N. G. Yunis, R. Chevli, and Y. Gonzalez.** 1974. High-affinity accumulation of chloroquine by mouse erythrocytes infected with *Plasmodium berghei*. *J. Clin. Invest.* **54:**24–33.
10. **Flanigan, T., R. Marshall, D. Redman, C. Kaetzel, and B. Unger.** 1991. In vitro screening of therapeutic agents against *Cryptosporidium parvum*: hyperimmune cow colostrum is highly inhibitory. *J. Protozool.* **38:**225S–227S.
11. **Foote, S. J., J. K. Thompson, A. F. Cowman, and D. J. Kemp.** 1989. Amplification of the multidrug resistance gene in some chloroquine-resistant isolates of *Plasmodium falciparum*. *Cell* **57:**921–930.
12. **Fritsche, T. R.** 1989. Pathogenic protozoa: an overview of in vitro cultivation and susceptibility to chemotherapeutic agents. *Clin. Lab. Med.* **9:**287–317.
13. **Gillin, F. D., and L. S. Diamond.** 1981. Inhibition of clonal growth of *Giardia lamblia* and *Entamoeba histolytica* by metronidazole, quinacrine, and other antimicrobial agents. *J. Antimicrob. Chemother.* **8:**305–316.
14. **Grogl, M., R. K. Martin, A. M. J. Oduola, W. K. Milhous, and D. E. Kyle.** 1991. Characteristics of multidrug resistance in *Plasmodium* and *Leishmania*: detection of p-glycoprotein-like components. *Am. J. Trop. Med. Hyg.* **45:**98–111.
15. **Grogl, M., T. N. Thomason, and E. D. Franke.** 1992. Drug resistance in leishmaniasis: its implication in systemic chemotherapy of cutaneous and mucocutaneous disease. *Am. J. Trop. Med. Hyg.* **47:**117–126.
16. **Hendricks, L. D., D. E. Wood, and M. E. Hajduk.** 1978. Haemoflagellates: commercially available liquid media for rapid cultivation. *Parasitology* **76:**309–316.
17. **Hitchings, G. H.** 1971. Folate antagonists as antibacterial and

antiprotozoal agents. VI. Present status and future prospects for chemotherapy with folate antagonists. *Ann. N.Y. Acad. Sci.* **186:**444–451.

18. **Jackson, J. E., J. D. Tally, W. Y. Ellis, Y. B. Mebrahtu, P. G. Lawyer, J. B. Were, S. G. Reed, D. M. Panisko, and B. L. Limmer.** 1990. Quantitative in vitro drug potency and drug susceptibility evaluation of *Leishmania* ssp. from patients unresponsive to pentavalent antimony therapy. *Am. J. Trop. Med. Hyg.* **43:**464–480.

19. **Johnson, P. J.** 1993. Metronidazole and drug resistance. *Parasitol. Today* **9:**183–186.

20. **Krogstad, D. J., I. Y. Gluzman, D. E. Kyle, A. M. J. Oduola, S. K. Martin, W. K. Milhous, and P. H. Schlesinger.** 1987. Efflux of chloroquine from *Plasmodium falciparum*: mechanism of chloroquine resistance. *Science* **238:**1283–1285.

21. **Lossick, J. G., M. Muller, and T. E. Gorrell.** 1986. In vitro drug susceptibility and doses of metronidazole required for cure in cases of refractory vaginal trichomoniasis. *J. Infect. Dis.* **153:**948–955.

22. **McIntyre, P., P. F. L. Boreham, R. E. Phillips, and R. W. Shepherd.** 1986. Chemotherapy in giardiasis: clinical responses and in vitro sensitivity of human isolates in axenic culture. *J. Pediatr.* **108:**1005–1010.

23. **Meingassner, J. G., and J. Thurner.** 1979. Strain of *Trichomonas vaginalis* resistant to metronidazole and other 5-nitroimidazoles. *Antimicrob. Agents Chemother.* **15:**254–257.

24. **Meyer, E. A.** 1976. *Giardia lamblia*: isolation and axenic cultivation. *Exp. Parasitol.* **39:**101–105.

25. **Muller, M., and T. E. Gorrell.** 1983. Metabolism and metronidazole uptake in *Trichomonas vaginalis* isolates with different metronidazole susceptibilities. *Antimicrob. Agents Chemother.* **24:**667–673.

26. **Muller, M., J. G. Lossick, and T. E. Gorrell.** 1987. In vitro susceptibility of *Trichomonas vaginalis* to metronidazole and treatment outcome in vaginal trichomoniasis. *Sex. Transm. Dis.* **15:**17–24.

27. **Nguyen-Dinh, P., R. Magloire, and W. Chin.** 1983. A simple field kit for the determination of drug susceptibility in *Plasmodium falciparum*. *Am. J. Trop. Med. Hyg.* **32:**452–455.

28. **Nguyen-Dinh, P., and W. Trager.** 1980. *Plasmodium falciparum* in vitro: determination of chloroquine sensitivity of three new strains by a modified 48-hour test. *Am. J. Trop. Med. Hyg.* **29:**339–342.

29. **Ouellette, M., and B. Papadopoulou.** 1993. Mechanisms of drug resistance in *Leishmania*. *Parasitol. Today* **9:**150–153.

30. **Payne, D., and W. H. Wernsdorfer.** 1989. Development of blood culture medium and a standard in vitro microtest for field-testing the response of *Plasmodium falciparum* to antifolate antimalarials. *Bull. W.H.O.* **67:**59–64.

31. **Peterson, D. S., W. K. Milhous, and T. E. Wellems.** 1990. Molecular basis of differential resistance to cycloguanil and pyrimethamine in *Plasmodium falciparum* malaria. *Proc. Natl. Acad. Sci. USA* **87:**3018–3022.

32. **Peterson, D. S., D. Walliker, and T. E. Wellems.** 1988. Evidence that a point mutation in dihydrofolate reductase-thymidylate synthase confers resistance to pyrimethamine in falciparum malaria. *Proc. Natl. Acad. Sci. USA* **85:**9114–9118.

33. **Pica-Mattoccia, L., S. Archer, and D. Cioli.** 1992. Hycanthone resistance in schistosomes correlates with the lack of an enzymatic activity which produces the covalent binding of hycanthone to parasite macromolecules. *Mol. Biochem. Parasitol.* **55:**167–176.

34. **Quon, D. V. K., C. E. D'Oliveira, and P. J. Johnson.** 1992. Reduced transcription of the ferredoxin gene in metronidazole-resistant *Trichomonas vaginalis*. *Proc. Natl. Acad. Sci. USA* **89:**4402–4406.

35. **Rieckmann, K. H., L. J. Sax, G. H. Campbell, and J. E. Mrema.** 1978. Drug sensitivity of *Plasmodium falciparum*. An in vitro microtechnique. *Lancet* **i:**22–23.

36. **Samuelson, J., P. Ayala, E. Orozco, and D. Wirth.** 1990. Emetine resistant mutants of *Entamoeba histolytica* overexpress mRNAs for multidrug resistance. *Mol. Biochem. Parasitol.* **38:**281–290.

37. **Schwartz, I. K., E. M. Lackritz, and L. C. Patchen.** 1991. Chloroquine-resistant *Plasmodium vivax* from Indonesia. *N. Engl. J. Med.* **324:**927.

38. **Slater, A. F. G., and A. Cerami.** 1992. Inhibition by chloroquine of a novel haem polymerase enzyme activity in malaria trophozoites. *Nature* (London) **355:**167–169.

39. **Strote, G.** 1987. In vitro study on the effect of the new chemotherapeutic agents CGP 6140 and CGP 20376 on the microfilariae of *Onchocerca volvulus*. *Trop. Med. Parasitol.* **38:**211–213.

40. **Trager, W., and J. B. Jensen.** 1976. Human malaria parasites in continuous culture. *Science* **193:**673–675.

41. **Upcroft, J. A., and P. Upcroft.** 1993. Drug resistance and *Giardia*. *Parasitol. Today* **9:**187–190.

42. **Wellems, T. E.** 1992. Malaria: how chloroquine works. *Nature* (London) **355:**108–109.

43. **Wellems, T. E., A. Walker-Jonah, and L. J. Panton.** 1991. Genetic mapping of the chloroquine-resistance locus on *Plasmodium falciparum* chromosome 7. *Proc. Natl. Acad. Sci. USA* **88:**3382–3386.

44. **WHO Expert Committee.** 1987. Prevention and control of intestinal parasitic infections. *W.H.O. Tech. Rep. Ser.* **749:**1–86.

45. **WHO Expert Committee.** 1990. Control of the leishmaniasis. *W.H.O. Tech. Rep. Ser.* **793:**1–155.

46. **Wilson, C. M., A. E. Serrano, A. Wasley, M. P. Bogenschutz, A. H. Shankar, and D. F. Wirth.** 1989. Amplification of a gene related to mammalian mdr genes in drug-resistant *Plasmodium falciparum*. *Science* **244:**1184–1186.

47. **Wilson, C. M., S. K. Volkman, S. Thaithong, R. K. Martin, D. E. Kyle, W. K. Milhous, and D. F. Wirth.** 1993. Amplification of pfmdr1 associated with mefloquine and halofantrine resistance in *Plasmodium falciparum* from Thailand. *Mol. Biochem. Parasitol.* **57:**151–160.

48. **World Health Organization.** 1992. World malaria situation in 1990. I. *Weekly Epidemiol. Rec.* **67:**161–167.

Assays for Antimicrobial Agents in Body Fluids

BETH E. OSTERGAARD, DAVID LAKATUA, AND JOHN C. ROTSCHAFER

123

The primary goals of monitoring concentrations of antimicrobial agents in body fluids are to optimize therapy and minimize drug toxicity. During the late 1970s and early 1980s, several investigators began to establish a relationship between antibiotic concentration, therapeutic effect, and antibiotic-induced adverse reactions. Therapeutic ranges correlating serum antibiotic concentrations with outcome and toxicity have been suggested for a number of antimicrobial agents. Antimicrobial agents such as aminoglycosides have narrow therapeutic ranges and are commonly monitored in order to minimize possible concentration-associated toxicities, including nephrotoxicity and ototoxicity.

In life-threatening illnesses, antimicrobial levels may be monitored to avoid underdosage and possible poor outcome associated with inadequate antimicrobial concentrations. Cystic fibrosis, burn, and dialysis patients as well as patients with changing renal function may benefit from monitoring of antimicrobial concentrations because of their significantly different and unpredictable pharmacokinetic parameters, which complicate the proper determination of dose. Antibiotic concentrations are useful monitoring parameters for patients with infections at sites where antibiotic penetration may be variable, as in meningitis, or when adequate absorption of oral agents is crucial to successful therapy, as in chronic osteomyelitis. Patients with AIDS frequently have malabsorption problems and may need concentration monitoring to guarantee satisfactory drug absorption. Monitoring antimycobacterial levels in tuberculosis patients is important to ensure compliance as well as antimicrobial absorption and proper dosing. Assays for antimicrobial agents are indicated for these clinical situations as well as for research purposes.

As a result of the observations correlating antibiotic concentration, therapeutic effect, and antibiotic-induced adverse reactions, there has been interest in developing rapid commercial methods for the accurate quantitation of antibiotic concentrations in sera and other biological fluids. Substantial technologic advances have provided a variety of methods for the sensitive, precise, accurate, specific, timely quantitation of antibiotic concentrations (Table 1). Over the years, these new technologies have essentially replaced the tedious and time-consuming methods of bioassay, radioenzymatic assay, and radioimmunoassay (RIA). While the laboratory's responsibilities have primarily focused on the proper collection and storage of specimens and the actual quantitation of antibiotic concentrations, there are additional tasks that must be addressed by the laboratory, especially now that current technologies are automated, accurate, and relatively simple to perform. Perhaps the most crucial steps in the current process of clinically quantitating antibiotic concentrations revolve around aspects not totally controlled by the laboratory. These include the administration of antimicrobial agents at the proper dose and time. In addition, the accurate documentation of the dose in milligrams, the administration times, and the serum sampling times are crucial for a meaningful interpretation of the data and for appropriate dosage adjustment. Without a precise record of these events, accurate quantitation of serum antibiotic concentrations is in most cases a meaningless exercise. The laboratory needs to ensure through the cooperation of physicians, nurses, and pharmacists that timely and appropriate interpretation of antimicrobial concentrations is actually taking place within their institutions.

CLINICAL SPECIMENS

Body Fluids

Various body fluids, including serum, plasma, urine, and cerebrospinal fluid, can be assayed to measure concentrations of antimicrobial agents. Of these, serum is the body fluid most frequently assayed. Plasma is also used, although heparinized plasma may interfere with gentamicin levels measured by certain bioassays and enzyme immunoassays. However, it does not interfere significantly with the fluorescence polarization immunoassay (FPI). The interference is unlikely to be clinically significant except in samples drawn from heparin locks and catheters, where heparin concentrations may be high (43). Assays of body fluids other than serum or plasma have not been well studied. When necessary, these assays may be performed by running standard antibiotic dilution curves in the same body fluid as the sample (15).

Collection

Proper specimen collection and proper documentation of times of administration and sampling are vital to the interpretation of test results and the pharmacokinetic applica-

TABLE 1 Comparison of antimicrobial assays

Assay	Specificity	Sensitivity (μg/ml)	Coefficient of variation (%)	Speed	Sample size	Advantages	Disadvantages
Bioassay	Other antibiotics may interfere	0.5–1	3–17	4–48 h	1–5 ml	Simple, versatile, inexpensive	Poor sensitivity, specificity, precision, and speed
RIA	May measure metabolites	0.1	8	2–3 h	50–200 μl	Sensitive, specific	Radioactive waste, expensive reagents and equipment
EMIT	Highly specific	1–2	5	10 min	<100 μl	Rapid, specific, minimal skill required	Limited antibiotic assays, poor accuracy at <2 μg/ml, expensive reagents
FPI	Highly specific	0.3–2	5	15–30 min	50 μl	Automated, specific, sensitive, precise	Expensive reagents, need to batch work, equipment cost
HPLC	Highly specific	0.1	5	30–60 min	<1 ml	Specific, rapid, sensitive, precise, versatile	Low capacity, preparatory work required, derivatization needed, expensive equipment
LA	Highly specific	0.1	5	5 min	40 μl	Automated, specific, precise, sensitive, inexpensive reagent	Very expensive equipment

tion of these results. If these procedures are not performed correctly, repeat studies may be required, which is time- and resource-consuming as well as discomforting to the patient. Generally, concentrations in serum are not measured until the antibiotic has reached steady state, the point at which the rate of drug accumulation equals the rate of drug elimination. However, when aggressive management is desired, pharmacokinetic studies can be performed on the first dose or during the period of non-steady-state conditions by employing a Bayesian or alternate sampling strategies such as series pharmacokinetics (17).

Several factors must be considered in determining whether samples are being drawn at the proper times. First, the patient must be receiving the same dose of antibiotic on a fixed schedule, with all doses during this period being administered on time. Second, the patient must be reasonably stable in regard to hydration status, renal function, and/or hepatic function. For water-soluble antibiotics, overhydration or dehydration of the patient dilutes or concentrates, respectively, serum antibiotic concentrations.

The next consideration is whether to draw a peak or a trough level or both. The peak concentration is defined as the concentration that occurs immediately at the end of an intravenous dose or at a specific time following oral or intramuscular (i.m.) antibiotic administration. The peak concentration is the highest point or largest concentration measured during the dosage interval. Practically, most laboratories do not actually attempt to measure true peak concentrations but rather draw "peak" or postinfusion serum samples sometime after antibiotic administration. The

appropriate postdose time to draw levels depends on many factors, including route of drug administration, patterns of absorption and distribution from the site of administration, half-life, bioavailability, drug formulation, and metabolism. Peak measurements of concentration in serum may be useful for antimicrobial agents that are concentration-dependent killers, such as aminoglycosides and fluoroquinolones. Peak concentrations are also monitored for flucytosine and chloramphenicol, for which drug-induced adverse reactions have been associated with high concentrations in serum (4). The monitoring of peak levels for parenterally administered chloramphenicol is somewhat unusual in that the antibiotic is administered as an ester, chloramphenicol succinate. For the monitoring of chloramphenicol levels, the ester must be hydrolyzed. This drug can take up to 2 h after the start of the infusion to reach peak concentration in serum. The time is dependent on the rate of hydrolysis, the rate of administration, and the site of chloramphenicol infusion. For these reasons, peak chloramphenicol levels are usually obtained 1 to 1.5 h following the 30- to 60-min parenteral administration of chloramphenicol. Agents given i.m. and orally are absorbed slowly, and in general, peak samples should be drawn 1 h following i.m. injection and 1 to 4 h after oral administration, depending on the drug. In diabetic patients, patients with peripheral vascular disease, or patients with limited muscle mass, antibiotic absorption following i.m. administration can vary substantially. Similarly, a variety of medical conditions, drug interactions, and food interactions can alter the absorption of different antibiotics from the gastrointestinal tract.

Trough level samples are drawn when the antibiotic concentration reaches the lowest point during the dosage interval. Almost by definition, this point occurs just prior to the next dose. Such data may be helpful in determining whether the patient is accumulating the drug or whether the antibiotic concentration remains above the MIC for the infecting organism. Trough concentrations can be artifactually influenced by early or late administration of the previous dose or by an error in dose preparation. Both peak and trough concentrations in serum are commonly determined for patients receiving aminoglycosides and vancomycin, as patient-specific dosing parameters can be determined by using these data (7, 26, 47).

Sampling strategies other than sampling to determine peak and trough levels are commonly performed in the clinical setting. Bayesian or series pharmacokinetics is used for patients requiring aggressive antibiotic concentration monitoring who are not yet at steady state. Most commonly, series kinetics is employed following the first dose of antibiotics in extremely ill patients. At least two and sometimes three postinfusion concentrations are determined with series pharmacokinetic studies (17).

Presently, the concept of single daily dosing or once-a-day dosing of aminoglycosides is gaining popularity in the clinical setting. Administering the entire daily dose of aminoglycoside at once may limit tissue accumulation and thus toxicity, as transport into the inner ear and kidney is thought to be saturable (19). Theoretically, dosing aminoglycosides once daily also allows for an aminoglycoside-free period during which concentrations in serum are essentially zero. This antibiotic-free period, the optimal duration of which is yet undefined, is thought to be important in overcoming bacterial adaptive resistance and limiting tissue accumulation, thus improving antibiotic performance (3, 11, 19). Previous studies by Moore and colleagues, who used conventional methods of aminoglycoside dosing, suggest that optimal aminoglycoside activity is present when the peak/MIC ratio reaches a magnitude of 8 to 12 (27). Most studies performed to date have not attempted to correlate a desired peak concentration in serum with clinical outcome or toxicity. Thus, while the simplicity of single daily dosing of aminoglycosides is an alluring feature, the lack of data demonstrating superiority of this method over conventional dosing methods, despite multiple comparative trials, leaves many questions surrounding this concept (19). For these reasons or until monitoring practices become more clearly defined and established, we advocate the use of traditional aminoglycoside-concentration-monitoring practices to screen appropriate patient candidates, to optimize serum peak/MIC ratios, and to limit cost and potential overexposure to aminoglycoside therapy (37).

Recently, there has also been interest in administering beta-lactam antibiotics with short half-lives in serum by continuous intravenous infusion to maintain the free or unbound serum antibiotic concentration above the bacterial MIC at all times. In this case, serum beta-lactam concentrations may generally be determined at any time after the patient has been on the infusion for 4 to 6 h, as most beta-lactam antibiotics have relatively short half-lives in serum (<1 to 2 h) and rapidly achieve steady-state conditions.

Information Needed by the Laboratory

In order for the laboratory to properly perform assays for various antimicrobial agents, additional information may be required. If bioassays are used, the laboratory needs to be aware of any additional antimicrobial agents or chemotherapeutic agents the patient has been receiving that could interfere with the assay and result in an erroneous value being reported. In order to interpret all assay results correctly, the laboratory needs accurate written documentation of collection times for sera, antibiotic administration times, antibiotic dose, and dosage interval. These data should also become part of the patient's permanent medical record.

Processing and Storage

Antimicrobial concentrations may also be affected by the timing of separation of the serum and plasma, the temperature at which the samples are stored, and the duration of storage at various temperatures. Generally, serum and plasma samples are centrifuged and separated immediately after they are collected. If assays are to be done that same day, storage in a refrigerator is usually sufficient. If assays are to be delayed, specimens need to be frozen. Cephalosporins in serum and plasma are labile in weak bases and strong acids and can be further stabilized for storage by the addition of buffers at a pH of 5 to 6.5 (38).

Aminoglycosides and beta-lactam antibiotics (primarily gentamicin, tobramycin, carbenicillin, and ticarcillin) have been extensively studied with regard to inactivation of the aminoglycoside component when these agents are used in combination (14, 15, 29, 33, 42). This interaction appears to be time, temperature, and concentration dependent. With commonly used doses of carbenicillin or ticarcillin, the interaction is unlikely to occur in vivo except in patients with severe renal failure (33). The inactivation of the aminoglycoside by ticarcillin or carbenicillin can also occur in vitro in the blood collection tube if the sample is not processed properly. This inactivation of the aminoglycoside results in aminoglycoside levels falsely low by as much as 40%, which if not recognized can translate into the patient receiving larger aminoglycoside doses that may increase the risk of toxicity (42). To avoid this interaction, samples containing aminoglycosides plus ticarcillin or carbenicillin should be placed on ice immediately after collection (14, 15). If the sample cannot be processed within 1 to 2 h, it should be refrigerated at 0 to 4°C until processing later that day. If the specimen is to be analyzed within 3 to 4 days, the sample should be stored at −20°C. If this process will take >4 days, the sample should be stored at −70°C (15, 29).

Concern that different specimen collection tubes might affect the measurement of antimicrobial concentrations has been voiced (2, 24, 32). However, studies that have examined measurements of aminoglycosides and chloramphenicol in specimens obtained in different types of collection tubes have failed to show any detrimental effects of different tubes (24, 32).

METHODS

There are many considerations in selecting an assay method for a specific antibiotic as well as for a specific institution. These considerations include specificity and sensitivity of the assay, precision, and speed. Required laboratory expertise, cost of equipment, cost of reagents, expiration dating, required assay frequency, and the broad-spectrum capability to measure other agents are all likely to affect the selection of a particular instrument or method. The methods used vary greatly depending on whether the institution is a large hospital that does frequent antimicrobial assays or a small community hospital that rarely performs such assay procedures. Features of the commonly used assay methods as well as newer methods that may become more prominent in the future are compared in Table 1.

Bioassay

Twenty years ago, most antibiotic assays were microbiologic assays. Techniques commonly used included agar or broth dilution and agar diffusion. Bioassays determine antibiotic concentrations by measuring graded responses of an organism and comparing unknown concentrations in serum of the specific antibiotic being measured against a standard curve. If additional antibiotics are present, use of an organism resistant to the interfering antibiotic and susceptible to the antibiotic being assayed or alteration of medium pH can eliminate interfering antibiotic activity. An alternative method involves the addition of an enzyme causing inactivation of the interfering antimicrobial agent (9, 15, 38). Presently, bioassays are rarely used to assay aminoglycosides or vancomycin, the most commonly assayed antimicrobial agents. The lack of biospecificity, sensitivity, precision, and speed and the subsequent development of newer assays make this method less desirable for frequent monitoring of antimicrobial agents.

The bioassay method is, however, still used for less commonly quantified antimicrobial agents such as those shown in Table 2 (15).

RIA

RIA consists of a competitive binding reaction between a radionuclide-labeled antibiotic and the unlabeled antibiotic in the sample for antibody-binding sites. A second antibody that acts specifically against the first antibody is added. The second antibody serves to separate the bound, labeled antibiotic from any free antibiotic by forming a precipitate that is separated out by centrifugation. The bound radionuclide is then measured either by liquid scintillation or with a gamma counter (9, 15, 21). This assay method is quantitative because of its specific immunoreactivity and has much greater sensitivity, specificity, and speed than bioassay methods. RIAs are more specific, although their antibodies may cross-react with metabolites as well as the parent compound. With growing environmental concerns about the radioactive waste produced and the cost of equipment, radioactivity labeling assays are being less frequently used for the quantification of a large number of antimicrobial and other pharmacologic agents.

EMIT

The enzyme-multiplied immunoassay technique (EMIT) involves a similar competitive binding assay with an enzyme-antibiotic conjugate and unlabeled antibiotic competing for antibody-binding sites. When binding occurs, the enzyme is inactivated, and the activity of the unbound enzyme is proportional to the concentrations of free and bound enzyme. This enzyme activity is quantified with a spectrophotometer. The advantages of this method include inexpensive instrumentation, rapid turnaround time, specificity, and simplicity. Problems such as expensive reagents, inability of the spectrophotometer to determine more than one concentration at a time, and lack of sensitivity for drug concentrations (<1 to 2 mg/dl) have limited the applicability of this assay method in larger hospitals (9, 15).

FPI

The FPI method involves a competitive binding reaction between unlabeled and fluorescein-labeled antibiotic for binding sites on antibodies to the drug being assayed. Binding of the drug-fluorescein conjugate results in a change in polarization of the fluorescence relative to that of the unbound drug. The polarization of fluorescence emitted is inversely proportional to the amount of unlabeled antibiotic present (15, 22, 34). FPI is an automated technique commonly used to assay aminoglycosides and vancomycin. The cost per sample is substantially decreased because of the stability of an established standard curve (stable for up to 1 month). Thus, clinical samples obtained on successive days can be analyzed against the previously generated standard curve. Other advantages include the speed involved with automation and a high sensitivity, which allows for dilutions to minimize interference (22).

For patients with renal failure, in whom vancomycin degradation products may have accumulated, FPI unfortunately measures both vancomycin (factor B) and vancomycin degradation products (CDP-1), thus indicating a falsely elevated concentration of vancomycin. This reaction occurs primarily in dialysis patients, who are often treated with vancomycin (1, 44). Hospitals monitoring vancomycin levels in these patients by FPI or RIA actually measure both vancomycin (factor B) and vancomycin degradation products (CDP-1). Assay results may be erroneously high by up to 39 and 67% for FPI and RIA, respectively (1, 44).

HPLC

High-performance liquid chromatography (HPLC) procedures require three main steps: extraction of the drug from the sample fluid, separation of the drug into a stationary phase from a mobile phase by column chromatography, and quantification by fluorometry, electron capture, or spectrometry (9, 45). HPLC is specific and sensitive because it utilizes chromatographic separation of drugs and metabolites in the determination of antimicrobial agent levels. A primary advantage of HPLC is its ability to separate metabolites and other closely related compounds (38). HPLC is not used routinely for frequent determination of antimicrobial agents levels, because the method is labor intensive, but once the samples are prepared, they can readily be assayed by automated methods. This method is commonly used in large clinical and research laboratories and performs with better sensitivity and accuracy than other assay methods (15). The reader is directed elsewhere for a complete review of HPLC principles and methods for specific antimicrobial agents (38, 45).

LA

Various assays using latex agglutination (LA) have been used for a number of years (10). The Immuno 1 system developed by Miles Technicon is an automated, homogeneous LA method that currently includes U.S. Food and Drug Administration-approved assay methods for gentamicin and tobramycin. This method involves a competitive binding reaction between the antibiotic and an antibiotic-conjugate for antibody-binding sites on latex particles. The reaction produces a turbid reaction in which the reaction rate is inversely proportional to the antibiotic level: the more antibiotic present, the lower the absorbance. This method has the same specificity as other immunoassays, is as sensitive and accurate as other methods, and is able to process up to 120 samples per h. The equipment required is

TABLE 2 Interpretation of antimicrobial agent levels in body fluids

Antimicrobial agent[a]	Therapeutic range of drug in serum (µg/ml)		Proper collection time(s)[b]	Most commonly used method(s)	Interfering substances[c]	Reference(s)
	Postdose/ peak	Predose/ trough				
Aminoglycosides			Depends on method used to interpret concentrations; peak, 0.25–1 h post i.v. dose; trough, 15–30 min prior to next dose	FPI	Penicillins[d]; N cross-reacts with G by LA; monoclonal antibodies and rheumatoid factor interfere by LA	12, 25, 28, 42, 47
G, T, N	4–8	0.5–2				
A, K	20–25	5–10				
Vancomycin	20–40	5–10	Depends on method used to interpret concentrations; peak, 0.25–1 h post i.v. dose; trough, 15–30 min prior to next dose	FPI	Vancomycin degradation products increase levels in FPI and RIA	1, 34, 44
Chloramphenicol	10–20	5–10	1 h post p.o. dose of base; 2 h post p.o. dose of palmitate; 0.5–1.5 h post i.v. dose	HPLC	Cefoxitin, aztreonam, sulfa drugs, imipenem, RIF, tetracycline by bioassay	4, 23, 41
Trimethoprim	<5		2–4 h post oral dose	HPLC		13
Sulfamethoxazole	<100		2 h post oral dose	HPLC		13
Flucytosine	80–100	25–50	2 h postdose	HPLC	Amphotericin B and ketoconazole by bioassay	45
Amphotericin B	1–2	0.25–1	1 h postinfusion	HPLC, RIA	Flucytosine, ketoconazole, vancomycin by bioassay	40
Beta-lactams	>MIC for organism being treated		Check random level at steady state or 3–5 times drug's half-life	Bioassay, HPLC	Vancomycin, aztreonam, imipenem, other penicillins, clindamycin, RIF, chloramphenicol, sulfa drugs by bioassay	15, 45
Antitubercular agents						16, 30, 39, 45, 46
Isoniazid	3–5		1–2 h postdose	HPLC	All antitubercular agents; *BA	
Rifampin	8–20		2–4 h postdose	HPLC	CS, C, K, SM; *BA	
Pyrazinamide	20–60		1–4 h postdose	HPLC	CS (by chemical assay)	
Ethambutol	3–5		2–4 h postdose	HPLC	All antitubercular agents; *BA	
Streptomycin	35–45		1–2 h postdose	HPLC	CS, C, K, RIF; *BA	
Amikacin	35–45		1–2 h postdose	FPI, EIA		
Kanamycin	35–45		1–2 h postdose	FPI, EIA	CS, C, K, RIF; *BA	
Capreomycin	20–50		1–2 h postdose	HPLC, EIA	CS, C, K, RIF; *BA	
Ofloxacin	8–10		1 h postdose	HPLC		
Ciprofloxacin	3–5		1 h postdose	HPLC		
Ethionamide	1–5		3 h postdose	HPLC	All antitubercular agents; *BA	
PAS	40–70		1.5–2 h postdose	HPLC	PZA, CS (by chemical assay)	
Cycloserine	20–35		2 h postdose	HPLC		

[a] G, gentamicin; T, tobramycin; N, netilmicin; A, amikacin; K, kanamycin; PAS, p-aminosalicylic acid.
[b] i.v., intravenous; p.o., peroral.
[c] SM, streptomycin; C, capreomycin; RIF, rifampin; PZA, pyrazinamide; CS, cycloserine; *BA, interference when bioassay method is used.
[d] Carbenicillin and ticarcillin have the greatest effect.

expensive, and the method is most practical for use in larger hospitals that work with a large volume of drug samples (25).

Capillary Electrophoresis

Capillary electrophoresis is a newer method of analysis under development for the detection of antimicrobial agents. The basic principle of this method involves the separation of molecules according to their hydrophobic properties, charges, sizes, and shapes. The use of capillaries as the electrophoretic migration channel improves the accuracy, speed, precision, and automation of this method compared with those of other electrophoretic methods. A more detailed discussion of this method can be found elsewhere (8, 18, 20).

INTERPRETATION OF RESULTS

Table 2 is a compilation of antimicrobial agents, their associated desirable concentrations in plasma, their proper collection times, the methods most commonly used to assay for them, and the substances that may interfere with detection of them depending on the assay used. This table is intended as a guide to interpretation of antimicrobial agent levels and offers some insight as to why antibiotic concentrations are not always as expected. Additionally, factors such as non-steady-state levels and missed doses may influence the validity of the reported drug levels. Patient-specific parameters such as weight, fluid status, protein binding, elimination rate, disease state, and drug interactions are additional factors to consider when interpreting assay results.

Consideration of the assay method used when interpreting antimicrobial agent levels is extremely important. When new assay methods are first studied, they are compared to traditional or accepted standard assay methods. The new assay method is validated if regression analysis yields a correlation of 1, a regression line slope of 1, and a y intercept of 0. Despite traditional validation using correlation coefficients, slopes, and y intercepts, significant differences have been reported in patient pharmacokinetic parameters and subsequently calculated dose and dosage intervals for gentamicin and amikacin determined by a variety of assay methods (5, 35, 36). Rotschafer et al. (35, 36) compared serum concentration-time data and the resultant pharmacokinetic parameters for gentamicin assayed by FPI and RIA and by RIA and EMIT. The pharmacokinetic parameters in different assays varied by up to 25% for volume of distribution, 11% for elimination rate constant, and 15% for total body clearance, even though the correlation between assays was good. These differences resulted in significant differences in dosage recommendations (35, 36). These studies demonstrate the need for the laboratory to evaluate the impact of new assay technology on therapeutic range and dosage regimens when switching to new assay methods. Another implication of variability between assay methods is the potential for reporting significantly different results if the serum drug concentration assays used during regular work hours are different from those used for statim or after-hour determinations.

Commercial pharmacokinetic software programs are provided as a value-added service by many assay manufacturers (6, 31). Such software usually provides an appropriate interpretation for most commonly monitored antibiotics, provided that such data are collected appropriately and that program users understand the underlying method used for that program's pharmacokinetic interpretation.

CONCLUSION

Assays for the determination of antimicrobial agent concentrations in various body fluids are commonly performed in both the clinical and the research settings. The choice of assay method is unique to each setting, and determining which assay method is appropriate for each setting should involve consideration of factors such as number of daily samples, speed, accuracy, precision, expense, number of drugs that can be assayed for, extent of automation, and extent of laboratory expertise required. In addition to the assay method itself, many extraneous factors affect the results of concentration measurements of antimicrobial agents. Accurate interpretation of antibiotic quantitative results requires an assessment of specimen type and possible interfering substances and appropriate collection and documentation of events surrounding sampling, processing, and storage procedures.

REFERENCES

1. **Anne, L., M. Hu, K. Chan, L. Colin, and K. Gottwald.** 1989. Potential problem with fluorescence polarization immunoassay cross-reactivity to vancomycin degradation product CDP-1: its detection in sera of renally impaired patients. *Ther. Drug Monit.* 11:585–591.
2. **Bailey, D. N., J. J. Coffee, and J. R. Briggs.** 1988. Stability of drug concentrations in plasma stored in serum separator blood collection tubes. *Ther. Drug Monit.* 10:352–354.
3. **Barclay, M. L., E. J. Begg, and S. T. Chambers.** 1992. Adaptive resistance following single doses of gentamicin in a dynamic in vitro model. *Antimicrob. Agents Chemother.* 36:1951–1957.
4. **Barriere, S. L.** 1988. Chloramphenicol, p. 340–341. *In* J. E. Knoben and P. O. Anderson (ed.), *Handbook of Clinical Drug Data*, 6th ed. Hamilton Press, Hamilton, Ill.
5. **Bleske, B. E., T. A. Larson, and J. C. Rotschafer.** 1987. Observed differences in amikacin pharmacokinetic parameters and dosage recommendations determined by enzyme immunoassay and fluorescence polarization immunoassay. *Ther. Drug Monit.* 9:48–52.
6. **Buffington, D. E., V. Lampasona, and M. H. H. Chandler.** 1993. Computers in pharmacokinetics: choosing software for clinical decision making. *Clin. Pharmacokinet.* 25:205–216.
7. **Cantu, T. G., N. A. Yamanaka-Yuen, and P. S. Lietman.** 1994. Serum vancomycin concentrations: reappraisal of their clinical value. *Clin. Infect. Dis.* 18:533–543.
8. **Chen, F.-T. A., C.-M. Liu, Y.-Z. Hsieh, and J. C. Sternberg.** 1991. Capillary electrophoresis—a new clinical tool. *Clin. Chem.* 37:14–19.
9. **Cipolle, R. J.** 1984. Antibiotics: methods for analysis, p. 197–203. *In* T. P. Moyer and R. L. Boeckx (ed.), *Applied Therapeutic Drug Monitoring*, vol. 2. *Review and Case Studies*, 2nd ed. The American Association for Clinical Chemistry, Washington, D.C.
10. **Conway, T. A., J. Landon, D. S. Smith, and E. J. Shaw.** 1983. Assessment of a latex-agglutination-inhibition card test for serum gentamicin, with a study of the effects of potential interfering factors. *Ther. Drug Monit.* 5:347–353.
11. **Daikos, G. L., V. T. Loians, and G. G. Jackson.** 1991. First-exposure adaptive resistance to aminoglycoside antibiotics in vivo with meaning for optimal clinical use. *Antimicrob. Agents Chemother.* 35:117–123.
12. **Davies, M., J. R. Morgan, and C. Anand.** 1975. Interactions of carbenicillin and ticarcillin with gentamicin. *Antimicrob. Agents Chemother.* 7:431–434.
13. **DeAngelis, D. V., J. L. Woolley, and C. W. Sigel.** 1990. High-performance liquid chromatographic assay for the simul-

taneous measurement of trimethoprim and sulfamethoxazole in plasma or urine. *Ther. Drug Monit.* **12:**387–392.

14. **Ebert, S. C., M. Leroy, and B. Darcey.** 1989. Comparison of aminoglycoside concentrations measured in plasma versus serum. *Ther. Drug Monit.* **11:**44–46.

15. **Edberg, S. C.** 1986. The measurement of antibiotics in human body fluids: techniques and significance, p. 381–476. *In* V. Lorian (ed.), *Antibiotics in Laboratory Medicine,* 2nd ed. The Williams & Wilkins Co., Baltimore.

16. **El-Yazigi, A., and S. Al-Rawithy.** 1990. A direct liquid chromatographic quantitation of ciprofloxacin in microsamples of plasma with fluorometric detection. *Ther. Drug Monit.* **12:**378–381.

17. **Erdman, S. M., K. A. Rodvold, and R. D. Pryka.** 1991. An updated comparison of drug dosing methods. III. Aminoglycoside antibiotics. *Clin. Pharmacokinet.* **20:**374–388.

18. **Evenson, M. A., and J. E. Wiktorowicz.** 1992. Automated capillary electrophoresis applied to therapeutic drug monitoring. *Clin. Chem.* **38:**1847–1852.

19. **Gilbert, D. N.** 1991. Once-daily aminoglycoside therapy. *Antimicrob. Agents Chemother.* **35:**399–405.

20. **Gordon, M. J., X. Huang, S. L. Pentoney, Jr., and R. N. Zare.** 1988. Capillary electrophoresis. *Science* **242:**224–228.

21. **Gosling, J. P.** 1990. A decade of development in immunoassay methodology. *Clin. Chem.* **36:**1408–1427.

22. **Jolley, M. E., S. D. Stroupe, C.-H. J. Wang, H. N. Panas, C. L. Keegan, R. L. Schmidt, and K. S. Schwenzer.** 1981. Fluorescence polarization immunoassay. I. Monitoring aminoglycoside antibiotics in serum and plasma. *Clin. Chem.* **27:**1190–1197.

23. **Koup, J., R., A. H. Lau, B. Brodsky, and R. L. Slaughter.** 1979. Chloramphenicol pharmacokinetics in hospitalized patients. *Antimicrob. Agents Chemother.* **15:**651–657.

24. **Landt, M., C. H. Smith, and G. L. Hortin.** 1993. Evaluation of evacuated blood-collection tubes: effects of three types of polymeric separators on therapeutic drug-monitoring specimens. *Clin. Chem.* **39:**1712–1717.

25. **Miles Technicon.** 1993. *Immuno 1 System Reference Manual,* unit 4, p. 2–4. Miles Technicon, Tarrytown, N.Y.

26. **Moellering, R. C., Jr.** 1994. Monitoring serum vancomycin levels: climbing the mountain because it is there? *Clin. Infect. Dis.* **18:**544–546. (Editorial.)

27. **Moore, R. D., P. S. Lietman, and C. R. Smith.** 1987. Clinical response to aminoglycoside therapy: importance of the peak concentration to minimum inhibitory concentration. *J. Infect. Dis.* **155:**93–98.

28. **Ngui-Yen, J. H., P. W. Doyle, and J. A. Smith.** 1984. Comparative analysis of two rapid, automated methods for determining aminoglycoside levels. *J. Clin. Microbiol.* **20:**962–965.

29. **Pickering, L. K., and P. Gearhart.** 1979. Effect of time and concentration upon interaction between gentamicin, tobramycin, netilmicin, or amikacin and carbenicillin or ticarcillin. *Antimicrob. Agents Chemother.* **15:**592–596.

30. **Pillay, S., and B. Maharaj.** 1992. The use of drug level measurements in adult patients receiving theophylline, antiepileptic drugs and amikacin. *Cent. Afr. J. Med.* **38:**331–332.

31. **Poirier, T. I., and R. A. Giudici.** 1992. Survey of clinical pharmacokinetic software for microcomputers. *Hosp. Pharm.* **27:**971–977.

32. **Quattrocchi, F., T. Karnes, J. D. Robinson, and L. Hen-**

deles. 1983. Effect of serum separator blood collection tubes on drug collection. *Ther. Drug Monit.* **5:**359–362.

33. **Riff, L. J., and G. G. Jackson.** 1972. Laboratory and clinical conditions for gentamicin inactivation by carbenicillin. *Arch. Intern. Med.* **130:**887–908.

34. **Ristuccia, P. A., A. M. Ristuccia, J. H. Bidanset, and B. A. Cunha.** 1984. Comparison of bioassay, high-performance liquid chromatography, and fluorescence polarization immunoassay for quantitative determination of vancomycin in serum. *Ther. Drug Monit.* **6:**238–242.

35. **Rotschafer, J. C., H. G. Berg, R. B. Nelson, L. Strand, and D. J. Lakatua.** 1983. Observed differences in gentamicin pharmacokinetic parameters and dosage recommendations determined by fluorescent polarization immunoassay and radioimmunoassay methods. *Ther. Drug Monit.* **5:**443–447.

36. **Rotschafer, J. C., C. Morlock, L. Strand, and K. Crossley.** 1982. Comparison of radioimmunoassay and enzyme immunoassay methods in determining gentamicin pharmacokinetic parameters and dosages. *Antimicrob. Agents Chemother.* **22:**648–651.

37. **Rotschafer, J. C., and M. J. Rybak.** 1994. Single daily dosing of aminoglycosides: a commentary. *Ann. Pharmacother.* **28:**797–801.

38. **Rouan, M. C.** 1985. Antibiotic monitoring in body fluids. *J. Chromatogr.* **340:**361–400.

39. **Seifart, H. I., P. B. Kruger, D. P. Parkin, P. P. Van Jaarsvels, and P. R. Donald.** 1993. Therapeutic monitoring of antituberculosis drugs by direct in-line extraction on a high-performance liquid chromatography system. *J. Chromatogr.* **619:**285–290.

40. **Shihabi, Z. K., B. L. Wasilauskas, and J. E. Peacock, Jr.** 1988. Serum amphotericin-B assay by scanning spectrophotometer. *Ther. Drug Monit.* **10:**486–489.

41. **Sood, S. P., V. I. Green, and C. L. Bailey.** 1987. Routine methods in toxicology and therapeutic drug monitoring by high performance liquid chromatography. II. A rapid microscale method for determination of chloramphenicol in blood and cerebrospinal fluid. *Ther. Drug Monit.* **9:**347–352.

42. **Tindula, R. J., P. J. Ambrose, and A. F. Harralson.** 1983. Aminoglycoside inactivation by penicillins and cephalosporins and its impact on drug-level monitoring. *Drug Intell. Clin. Pharm.* **17:**906–908.

43. **Walters, M. I., and W. H. Roberts.** 1984. Gentamicin/heparin interactions: effects on two immunoassays and on protein binding. *Ther. Drug Monit.* **6:**199–202.

44. **White, L. O.** 1988. The in-vitro degradation at 37°C of vancomycin in serum, CAPD fluid and phosphate-buffered saline. *J. Antimicrob. Chemother.* **22:**739–745.

45. **Yoshikawa, T. T., S. K. Maitra, M. C. Schotz, and L. B. Guze.** 1980. High-pressure liquid chromatography for quantitation of antimicrobial agents. *Rev. Infect. Dis.* **2:**169–181.

46. **Zaninotto, M., S. Secchiero, C. D. Paleari, and A. Burlina.** 1992. Performance of a fluorescence polarization immunoassay system evaluated by therapeutic monitoring of four drugs. *Ther. Drug Monit.* **14:**301–305.

47. **Zaske, D. E.** 1992. Aminoglycosides, p. 14-3–14-46. *In* W. E. Evans, J. J. Schentag, and W. J. Jusko (ed.), *Applied Pharmacokinetics—Principles of Therapeutic Drug Monitoring,* 3rd ed. Applied Therapeutics, Inc., Vancouver, British Columbia, Canada.

Author Index

Subject Index

Chlamydia, 671
human papillomavirus, 1084
in situ hybridization, 132, 134
interpretation, 49
Mycobacterium, 408
processing for bacteria, 277–278
processing for fungi, 710
processing for parasites, 1153–1154
processing for viruses, 871–873
specimen collection guidelines, 29
varicella-zoster virus, 896
TLC, *see* Thin-layer chromatography
Tobramycin, 1278, 1287–1288, 1302, 1314, 1328, 1333
Togaviridae, 860, 862, 864, 968, 987, 992
Toluidine blue O stain, 40, 46–47, 712, 745
Tongue worm, 1266
Tonometer, disinfection, 234–235
Tonsillitis, 947–948
Toroviridae, 860
Torovirus, 1018–1020
Torulopsis, 726, 734
Torulopsis candida, 725
Torulopsis glabrata, 11, 725, 728, 731, 734, 760, 1406, 1408, 1412
Torulopsis pintolopesii, 725
Toxic shock syndrome, 274, 284
Toxin
  Bacillus, 355
  C. botulinum, 578
  C. difficile, 579
  C. perfringens, 576–578
  cell culture assay, 164
  cholera, 471
  Clostridium, 576
  detection in cell culture, 164
  food-borne, 216
Toxocara, 15, 1163–1164
Toxocara canis, 1164, 1251
Toxocara cati, 1164, 1251
Toxocariasis, 1160, 1163–1164
Toxoplasma, 1142, 1164, 1166
Toxoplasma gondii, 15, 116, 219, 221, 756, 1141, 1143, 1154–1155, 1159, 1191–1192, 1292
Toxoplasmosis, 1154, 1160, 1191–1192
  antibody detection, 1164
  clinical disease, 1191–1192
  diagnosis, 1192
  epidemiology, 1192
  life cycle and morphology of parasite, 1191
  prevention, 1192
  treatment, 1192
Trabulsiella, 457
Trabulsiella guamensis, 438, 440, 459, 462
Tracheal aspirate
  identification of bacteria, 270
  processing for bacteria, 275
  specimen collection guidelines, 27
Tracheitis, 364
Trachoma, 670
Transbronchial biopsy, *P. carinii*, 741–742
Transcription-based amplification system, 138
Transfusion reaction, 511
Transmissible mink encephalopathy, 1121
Transposon, 195–196
Transtracheal aspirate
  identification of bacteria, 270
  processing for bacteria, 275
Trappers' ailment, *see* Tularemia
Traveler's diarrhea, 164
Trematoda, 1143
Trematode, 1235–1240, 1252–1253
Trench fever, 691, 1259
Treponema, 7, 13, 263, 626, 636–651

clinical significance, 636–638
description of genus, 636
diagnostic parameters and interpretation of tests, 638–640
genital region, 648
natural habitat and distribution, 636
oral, 647–648
taxonomy, 636
Treponema carateum, 636–637, 642
Treponema denticola, 636, 647
Treponema minutum, 647–648
Treponema pallidum, 11, 15, 119, 255, 274, 636–651, 1290, *see also* Syphilis
  antimicrobial susceptibility, 646–647
  dark-field microscopy, 641
  detection and identification, 641
  DFA-TP test, 641–642
  DFAT-TP test, 642
  FTA-ABS tests, 644–645
  isolation, 642
  MHA-TP test, 645
  nontreponemal tests, 643–644
  PCR techniques, 642
  serologic diagnosis, 642–644
  specimen collection, 641–42
  treponemal tests, 644
Treponema pectinovorum, 647
Treponema phagedenis, 636, 647
Treponema refringens, 636, 647–648
Treponema skoliodontum, 647
Treponema socranskii, 647
Treponema vincentii, 647
Treponemal test, *T. pallidum*, 644
Triatoma, 1259
Tributyrin hydrolysis test, *Branhamella*, 335
Trichinella, 1142
Trichinella spiralis, 10, 219, 1143, 1153–1154, 1164, 1247–1248
Trichinellosis, 224, 1160, 1164
Trichinosis, 1154, 1248
Trichoderma viride, 767
Trichomonas, 1166
Trichomonas hominis, 10, 1143, 1205, 1213, 1220, 1222–1223
Trichomonas tenax, 7, 1153, 1222–1223
Trichomonas vaginalis, 11, 131, 254, 273, 1143, 1153, 1155, 1159, 1165, 1204–1205, 1211, 1213–1217, 1224
  antigen detection, 1166
  drug resistance and susceptibility testing, 1297, 1425
Trichomoniasis, 1160, 1213
Trichophyton, 13, 700, 704, 707, 796
Trichophyton ajelloi, 792, 796, 800
Trichophyton concentricum, 791–792, 794, 800
Trichophyton equinum, 792, 800
Trichophyton fischeri, 799–800
Trichophyton flavescens, 792
Trichophyton georgiae, 792
Trichophyton gloriae, 792
Trichophyton gourvilii, 792–794, 800
Trichophyton kanei, 792, 796, 800
Trichophyton longifusum, 796
Trichophyton megninii, 792–793, 795, 800
Trichophyton mentagrophytes, 706, 791–793, 795–796, 800
Trichophyton raubitschekii, 792, 795–797, 800
Trichophyton rubrum, 706, 791–792, 795, 797, 800–801
Trichophyton schoenleinii, 791–794, 796, 801
Trichophyton simii, 792, 801
Trichophyton soudanense, 792–793, 796, 801
Trichophyton terrestre, 792, 795, 797, 801
Trichophyton tonsurans, 706, 791–794, 796–797, 801
Trichophyton vanbreuseghemii, 792
Trichophyton verrucosum, 791–794, 796–797, 801
Trichophyton violaceum, 791–794, 802
Trichophyton yaoundei, 792–793, 802
Trichosporon, 366, 700, 714, 726, 728, 730, 734–735, 806
Trichosporon asahii, 806

preparation for molecular diagnosis, 150
processing for bacteria, 268, 278–279
processing for fungi, 710
processing for viruses, 869, 871, 873
specimen collection guidelines, 29
Urogenital specimen, processing for parasites, 1153
Ustilago, 730, 786
Uta, 1185
Uukuvirus, 865, 987
UV light
DNA inactivation, 143
radiation sterilization, 87

Vaccinia virus, 1131–1134
antiviral susceptibility, 1415–1416
Vagina, microorganisms, 12
Vaginal specimen
processing for bacteria, 268, 273–274
processing for fungi, 710
specimen collection guidelines, 26
Vaginosis, 12
Vagococcus, 258, 308–309
Vancomycin, 1278–1279, 1302, 1328, 1333
body fluids, 1432
resistance, 173, 1319–1320
resistance genes, 1370–1371
resistance in enterococci, 1358
Vancomycin susceptibility test, gram-positive cocci, 318
Varicella, 895
Varicella-zoster virus (VZV), 7, 13, 15–16, 119, 175, 869, 895–904
antisera, 899
antiviral susceptibility, 1415–1420
cell culture, 898–899
clinical background, 895
description, 895
evaluation and interpretation of test results, 902–903
identification, 899
immunofluorescence staining, 899–900
isolation, 898–899
serologic diagnosis, 900–902
shell vial assay, 900
specimen handling, 896–898
Varicellovirus, 865
Vector-borne transmission, 185
Veillonella, 7, 10, 13, 15–16, 262, 589, 603–604, 608, 613, 615, 617–618
Veillonella atypica, 608
Veillonella dispar, 608
Veillonella parvula, 608
Venipuncture, 76, 81
Ventilation hood, 72
Verocytotoxin, 164
Vesication, 1258
Vesicle fluid, processing for viruses, 870–871
Vesicular lesion
processing for viruses, 869, 874
varicella-zoster virus, 896–897
Vesiculovirus, 864, 987
V-factor requirement, *Haemophilus*, 560
Vibrio, 7, 10, 13, 16, 259, 261, 272, 445, 465–478, 499, 503
antimicrobial susceptibility, 473, 1290
clinical significance, 465–476
culture, 468
description of genus, 465
direct examination of specimen, 468
identification, 468–473
isolation, 468
natural habitats, 465
specimen handling, 468
taxonomy, 465

Vibrio alginolyticus, 466–473
Vibrio carchariae, 259, 466–473
Vibrio cholerae, 164, 216, 220–223, 272, 465–476, 480, 551, 1290
non-O1, 466
O1, 466
O139, 466–467
Vibrio cincinnatiensis, 466–471, 473
Vibrio damsela, 466–473
Vibrio fluvialis, 466–473, 479
Vibrio furnissii, 466–473
Vibrio hollisae, 466–471, 473
Vibrio metschnikovii, 261, 465–471, 473
Vibrio mimicus, 466–472
Vibrio parahaemolyticus, 216, 220, 223, 466–473
Vibrio vulnificus, 221, 223, 443, 466–473
Vibrionaceae, 465, 477
Vibriostatic test, 471–472
Vidarabine, 1415–1421
Vincent's angina, 6, 276
Vincent's stomatitis, 23
Viomycin, 1388
Virogen Rotatest, 1013
Virus
direct staining of virus and viral antigens, 873–874
DNA, 863
enhancement of infectivity, 873
factories, 1132
food-borne illness, 217
immunostaining, 163
isolation, 872–873
isolation and detection in cell culture, 162, 870
nosocomial infections, 175
nucleic acid probes, 163
RNA, 862
shell vial isolation, 163–164
specimen for serologic tests, 872
specimen processing, 868–875
taxonomy, 859–867
unconventional, 1121–1130
VISUWELL reagin test, 645
Vitek ANI system, 599
Vitek Jr. system, 107
Vitek system, 107, 369, 1380–1381
Vitek UniScept, 1379
Voges-Proskauer test, *Streptococcus*, 302, 304
Volatile acid detection, 103
Volatile fatty acids, 254
obligate anaerobes, 125–126
Volutella cinerescens, 767
Vulvovaginitis, 273
VZV, *see* Varicella-zoster virus

"Waived" tests, 69
Walk-Away Rapid Neg Combo 1 Panel, 515
WalkAway System, 107, 1380–1381
Wangiella, 700, 831, 843
Wangiella dermatitidis, 13, 706–707, 714, 749, 836, 838, 843
Wasp, 1257, 1259, 1265
Water, sampling, 212
Water supply, laboratory, 159
Waterhouse-Friderichsen syndrome, 326
Wayson stain, *Haemophilus*, 558
Weeksella, 259, 528–530
Weeksella virosa, 529–530
Weeksella zoohelcum, 529–530
Weil-Felix test, 682
West Nile fever, 980
Western blot, *see also* Immunoblot
arenavirus, 1078–1079
filovirus, 1078–1079